GARDNER'S
CHEMICAL SYNONYMS AND
TRADE NAMES

Gardner's Chemical Synonyms and Trade Names

Tenth Edition
Revised and Enlarged

EXECUTIVE EDITORS Michael and Irene Ash

Gower

Published by
Gower Publishing Limited
Gower House
Croft Road
Aldershot
Hants GU11 3HR
England

Gower
Old Post Road
Brookfield
Vermont 05036
U.S.A.

British Library Of Cataloging in Publication Data
Gardner, William
   Gardner's Chemical Synonyms and Trade Names. — 10Rev.ed.
   I. Title   II. Ash, Michael   III. Ash, Irene
   661.0321

Library of Congress Cataloging-in-Publication Data
Gardner's chemical synonyms and trade names. — 10th ed., rev. and enl. / executive editors, Michael
   and Irene Ash.
      1310 p. 23.4 cm
      Includes index.
      1. Chemicals—Dictionaries. 2. Chemicals—Trademarks. I. Ash, Michael. II. Ash, Irene.
   TP9.G286 1994   93-44594
   660'.03—dc20   CIP

ISBN 0-566-07491-5

Typeset in Arial Narrow by Synapse Information Resources, Inc.

Printed and Bound in Great Britain by
Hartnolls Limited, Bodmin, Cornwall.

TP
9
G228c
1994

Contents

Preface

For more than 65 years and through nine previous editions, *Gardner's Chemical Synonyms and Trade Names* has distinguished itself as the best known and most widely used source of information in the field of chemicals for commerce. In the Tenth Edition, the publishers have supported the extensive revisions and research necessary for this important reference work to maintain and enhance a continuing tradition of excellence.

Many of the entries from previous editions have been revised, expanded, or deleted. This edition includes more than 40,000 trade names and chemicals of which approximately 18,000 entries are completely new; one third of the entries now contain identifying CAS numbers and almost 2000 chemicals are identified by EINECS (European Inventory of Existing Commercial Chemical Substances) numbers.

Entries, whenever possible, contain detailed information on definitions, classification, chemical formulas/descriptions, functions/applications, and manufacturers. The trade entries have been obtained directly from chemical manufacturers world-wide and supplemented by a research program into other secondary sources, including manufacturers' advertisements in trade publications.

Gardner's is now divided into three parts. Part I is an alphabetical listing of synonyms and trade names with extensive cross referencing to the main entries. Part II is the Manufacturers Directory and represents a considerable expansion over the Ninth Edition (more than twice the number of manufacturers). It includes telephone, telefax, and telex numbers wherever possible as well as world-wide subsidiaries for the manufacturers. The third section, the Appendices, contain valuable cross references for CAS numbers to trade name chemicals, CAS numbers to chemicals, and EINECS numbers to chemicals. The Manufacturer Successor list is useful in tracing chemical manufacturer acquisitions. Part II and each section of the Appendices are edge-tinted for ease of reference.

Special notations are used in Part I to give additional information about trade name chemicals:

1. Trade names associated with active chemical manufacturers who have not verified the current availability of the entry *are marked with* *.

2. Trade names that are verified as obsolete by their manufacturer *are marked with* †.

3. Trade names with no verification for manufacturer or product, i.e. returned correspondence or addresses unavailable, *are marked with* ‡.

4. Trade names for which a manufacturer is not identified *are marked with* §.

There is *no special notation* following trade name entries that are verified as available products. Two-thirds of the trade name entries in this Tenth Edition are in this category.

Proprietary considerations

Every attempt has been made to ensure the accuracy of the information given in this new edition of *Gardner's*. However, the publishers cannot be held responsible for the accuracy of the information, and the users are expected to bear in mind the following considerations:

The reporting of a name in *Gardner's* cannot imply definitive legality in establishing proprietary usage. Questions of legal ownership of a particular name can be resolved by due legal process.

A manufacturer in some countries may market its product under alternative names to those cited in *Gardner's*. Similarly, manufacture or marketing of a product may be licensed to a separate company in another country either under the same or a different name.

We are confident that users will find that this new edition contains a wealth of information which is difficult to obtain from any other source. This book is the culmination of more than a year of research. We are especially grateful to Roberta Dakan for her skills in chemical information database management. Her tireless efforts have been instrumental in the production of this reference.

It is the intention of the publishers to produce regularly updated editions at suitable intervals. Companies wishing to submit material for inclusion in future editions should write to: Irene Ash, Gower Publishing Limited, Croft Road, Aldershot, Hampshire GU11 3HR, UK or Synapse Information Resources, Inc., 1247 Taft Ave., Endicott, NY, USA.

Abbreviations

ABS	acrylonitrile-butadiene-styrene
absorp.	absorption
ACE	acetylcholinesterase
ACN	acrylonitrile
agric.	agricultural
alc.	alcohol
AMP	2-amino -2-methyl-1- propanol
anhyd.	anhydrous
applic(s).	application(s)
aq.	aqueous
ASA	acrylic-styrene-acrylonitrile
aux.	auxiliary
avail.	available
BHA	butylated hydroxyanisole
BHT	butylated hydroxytoluene
biodeg.	biodegradable
blk.	black
BMC	bulk molding compound
b.p.	boiling point
BP	British Pharmacopeia
BR	butadiene rubbers, polybutadienes
B/S	butadiene/styrene
C	degrees Centigrade
CAB	cellulose acetate butyrate
CAS	Chemical Abstracts Service
cc	cubic centimeter(s)
CDA	completely denatured alcohol
chem.	chemical
cm	centimeter(s)
cm^3	cubic centimeter(s)
CMC	carboxymethylcellulose, critical Micelle concentration
CNS	central nervous system
Co.	Company
coeff.	coefficient
compat.	compatible
compd.	compound
compr.	compression
conc.	concentrated, concentration
cosolv.	cosolvent
CPE	chlorinated polyethylene

CPVC	chlorinated polyvinyl chloride
CR	chloroprene rubber, polychloroprene
cryst.	crystalline, crystallization
cs or cSt	centistoke(s)
CTFA	Cosmetic, Toiletry and Fragrance Association
DAP	diallyl phthalate, diammonium phosphate
DB	dichlorophenoxy butyric acid
DEA	diethanolamide, diethanolamine
DEDM	diethylol dimethyl
dens.	density
deriv.	derivative(s)
DIBA	diisobutyl adipate
DIDA	diisodecyl adipate
dielec.	dielectric
disp.	dispersible, dispersion
dist.	distilled
distort.	distortion
DMC	4,4´-dichloro(methylbenzhydrol)
DMDM	dimethylol dimethyl
DMF	dimethyl formamide
DMSO	dimethyl sulfoxide
DNPT	dinitrosopentamethylenetetramine
DOP	dioctyl phthalate
DOT	Department of Transportation
DP acid	diphenolic acid
DPG	diphenyl guanidine
DTPA	diethylene triamine pentaacetic acid
ECTFE	ethylene/chlorotrifluoroethylene copolymer
EDTA	ethylene diamine tetraacetic acid
EINECS	European Inventory of Existing Commercial Chemical Substances
elec.	electrical
elong.	elongation
EMC	electromagnetic conductive
EMI	electromagnetic interference
EO	ethylene oxide
EP	extreme pressure
EPDM	ethylene-propylene-diene rubbers
EPM	ethylene-propylene rubbers
EPR	ethylene-propylene rubber
equip.	equipment
ESCR	environmental stress crack resistance
esp.	especially
ETFE	ethylene tetrafluoroethylene
ETU	ethylene thiourea
EVA	ethylene vinyl acetate
exc.	excellent

F	degrees Fahrenheit
FA	fatty acid
FDA	Food and Drug Administration
FEP	fluorinated ethylene propylene
FG	food grade
flamm.	flammable, flammability
flex.	flexural
f.p.	freezing point
FRP	fiberglass-reinforced plastics
ft	foot, feet
g	gram(s)
GFRP	glass fiber-reinforced plastic
gran.	granules, granular
GRAS	generally recognized as safe
grn.	green
GRP	glass-reinforced plastics, glass-reinforced polyester
HC	hydrocarbon
HCl	hydrochloric acid
HDI	hexamethylene diisocyanate
HDL	high density lipids
HDPE	high-density polyethylene
Hg	mercury
HIPS	high-impact polystyrene
HLB	hydrophilic lipophilic balance
HPLC	high performance liquid chromatography
hydrog.	hydrogenated
IC	integrated circuit
IIR	isobutylene-isoprene rubber
in.	inch(es)
incl.	including
incompat.	incompatible
ingred.	ingredient(s)
inj.	injection
inorg.	inorganic
insol.	insoluble
int'l	international
IPA	isopropyl alcohol, isopropanol
IPM	isopropyl myristate
IPP	isopropyl palmitate
IR	isoprene rubber (synthetic)
IU	international units
IV	intravenous
J	joule
kg	kilogram(s)
KTPP	potassium tripolyphosphate
l	liter(s)

lb	pound(s)
LDL	low density lipids
LDPE	low-density polyethylene
LED	light-emitting diode
liq.	liquid
LLDPE	linear low-density polyethylene
lt.	light
Ltd.	Limited
m	milli or meter(s)
m-	meta-
MA	methacrylic acid
max.	maximum
MBCA	4,4´-methylene bis(orthochloroaniline)
MBT	mercaptobenzothiazole
MBTS	2-mercaptobenzothiazole disulfide
MCPA	(4-chloro-2-methylphenoxy) acetic acid
MDI	methylene diphenylene isocyanate
MDM	monomethylol dimethyl
MDPE	medium density polyethylene
MEA	monoethanolamine, monoethanolamide
mech.	mechanial
med.	medium
MEK	methyl ethyl ketone
mfg.	manufacture
mg	milligram(s)
MIBK	methyl isobutyl ketone
min.	minute(s), mineral, minimum
MIPA	monoisopropanolamine, monoisopropanolamide
misc.	miscible
mixt.	mixture(s)
MKP	monopotassium phosphate
ml	milliliter(s)
mm	millimeter(s)
MMW-HDPE	medium molecular weight high density polyethylene
MOCA	methylene bis-orthochloroaniline
mod.	modulus, moderately
m.p.	melting point
MPK	methyl propyl ketone
MVTR	moisture vapor transmission rate
m.w.	molecular weight
N	normal
nat.	natural
nat'l.	national
NBR	nitrile-butadiene rubber
NC	nitrocellulose
NCR	nitrile-chloroprene rubber

NEMA	National Electrical Manufacturers Association
NF	National Formulary
N/F	nonflammatory
no.	number
nonflamm.	nonflammable
nonyel.	nonyellowing
NR	isoprene rubber (natural)
NSF	National Science Foundation
NTA	nitrilotriacetic acid
o-	ortho-
OEM	original equipment manufacturer
OPP	oriented polypropylene
org.	organic
OTC	over-the-counter
o/w	oil-in-water
p-	para-
Pa	Pascal
PAN	polyacrylonitrile
PBT	polybutylene terephthalate
pbw	parts by weight
PC	polycarbonate
PCA	2-pyrrolidone-5-carboxylic acid
PCP	pentachlorophenol
PCTFE	polychlorotrifluoroethylene
PE	polyethylene
PEEK	polyetheretherketone
PEG	polyethylene glycol
PEI	polyetherimide
PEK	polyetherketone
PES	polyether sulfone
PET	polyethylene terephthalate
petrol.	petroleum
PFA	perfluoroalkoxy
PG	polypropylene glycol
pH	hydrogen-ion concentration
phr	parts per hundred of rubber or resin
PIB	polyisobutylene
pkg.	packaging
PMA	phosphomolybdic acid
PMMA	polymethyl methacrylate
PO	propylene oxide
POE	polyoxyethylene, polyoxyethylated
POM	polyoxymethylene
POP	polyoxypropylene, polyoxypropylated
powd.	powder
PP	polypropylene

PPE	polyphenylene ether
PPG	polypropylene glycol
ppm	parts per million
PPO	polyphenylene oxide
PPS	polyphenylene sulfide
pract.	practically
prep.	preparation(s)
prod.	product(s), production
props.	properties
PS	polystyrene
pt.	point
PTFE	polytetrafluoroethylene
PTMEG	polytetramethylene ether glycol
PU	polyurethane
PUR	polyurethane
PVA	polyvinyl alcohol
PVAc	polyvinyl acetate
PVAL	polyvinyl alcohol
PVB	polyvinyl butyral
PVC	polyvinyl chloride
PVDC	polyvinylidene chloride
PVDF	polyvinylidene fluoride
PVE	polyvinyl ethyl ether
PVF	polyvinyl fluoride
PVM	polyvinyl methyl ether
PVP	polyvinylpyrrolidone
quat.	quaternary
qv	*quod vide* (which see)
R&D	research and development
resist.	resistance, resistant, resistivity
RFI	radio frequency interference
RIM	reaction injection molded/molding
RT	room temperature
RTM	resin transfer molding
RTV	room temperature vulcanizing
RV	recreational vehicle
SAN	styrene-acrylonitrile
sat.	saturated
S/B	styrene/butadiene
SBR	styrene/butadiene rubber
SBS	styrene-butadiene-styrene
SDA	specially denatured alcohol
SE	self-emulsifying
sec.	secondary
sl.	slightly
SMA	styrene maleic anhydride

SMC	sheet molding compound
sol.	soluble, solubility
sol'n.	solution
solv(s).	solvent(s)
sp.	specific
spec.	specification
SPF	sun protection factor
SR	styrene rubber
SRF	semireinforced furnace
std.	standard
str.	strength
surf.	surface
syn.	synthetic
t	tertiary
TBHQ	tert-butyl hydroquinone
TDI	toluene diisocyanate
TEA	triethanolamine, triethanolamide
tech.	technical
temp.	temperature
tens.	tensile or tension
tert.	tertiary
TFE	tetrafluoroethylene
THF	tetrahydrofuran
TIPA	triisopropanolamine
TMC	thick molding compound
TMPTA	trimethylolpropane triacrylate
TPGDA	triporpylene glycol diacrylate
TPO	thermoplastic polyolefin
UF	urea-formaldehyde
UHF	ultra high frequency
UHMW	ultra high molecular weight
UHMWPE	ultra high molecular weight polyethylene
UL	Underwriter's Laboratory
unsat.	unsaturated
UPVC	unplasticized polyvinyl chloride
USDA	United States Department of Agriculture
USP	Unites States Pharmacopeia
uv	ultraviolet
VA	vinyl acetate
VAE	vinyl acetate ethylene
VC	vinyl chloride
VdC, VDC	vinylidene chloride
veg.	vegetable
VHF	very high frequency
visc.	viscous, viscosity
VOC	volatile organic compounds

v/v	volume by volume
wh.	white
w/o	water-in-oil
wt.	weight
w/w	weight by weight
XLPE	crosslinked polyethylene
yel.	yellow
ylsh.	yellowish
#	number
%	percent
<	less than
>	greater than
@	at
≈	approximately
α	alpha
β	beta
ε	epsilon
γ	gamma
ω	omega

Part I

Chemical Synonyms and Trade Names

A

A-1. CAS 102-08-9; N,N´-Diphenylthiourea (thiocarbanilide); primary accelerator for latex and repair stocks, CR, CR latex, and ethylene-propylene-diene rubbers sponge compounds. [Monsanto Co]

A-1 Thlocarbanilide. 1,3-Diphenyl-2-thiourea; N,N-diphenyl-thiourea; vulcanization acelerator for fast-curing repair stocks; neoprene latex, natural rubber latex and cements; an activator for thiazole accelerators. [Monsanto Co] *

A-2. CAS 1344-28-1; Activated alumina; used in scavenging, dehydration and catalytic applications; in hydrogen peroxide, polyethylene and drying industries, and for scavenging fluoride, chloride, and trace heavy metals. [La Roche Chem.]

A-17. CAS 106-97-8; Butane; hydrocarbon propellant. [Phillips] *

A-0020. Styrene/butadiene rubber (100 parts), N-300 carbon black (82.5 parts), high aromatic oil (62.5 parts) black masterbatch. [Goldsmith & Eggleton]

A-31. CAS 75-28-5; Isobutane; hydrocarbon propellant. [Phillips] *

A-108. CAS 74-98-6; Propane; hydrocarbon propellant. [Phillips] *

A-625/641. CAS 50-70-4; Sorbitol. [ICI Am.]

AA. CAS 118-92-3; Anthranilic acid; antioxidant for fats, greases, lube oils, and polyamides; sludge preventative in furnace and lube oils; chelating agent and sequestrant; corrosion inhibitor; stabilizer of can lacquers, oils, and lubricants. [PMC Specialties]

AA#2 Lime Additive. Surfactants/antifoams blend; for lime or borax coating operations. [Crown Tech.]

A.A.A. Ointment. A proprietary mercurial ointment containing salicylic acid, boric acid and zinc oxide; used to treat skin irritations. [Jenkins Laboratories Inc] ‡

A.A.A. Spray. A proprietary preparation of benzocaine and cetalkonium chloride in an aerosol; a throat spray. [Armour Pharmaceutical Co] *

A-acid. 2-Aminonaphthol,5-sulfonic acid; a dye intermediate.

AAprotect. CAS 137-30-4; Ziram; bird and animal repellent. [Universal Crop Protection Ltd]

AA Standard. Castor oil; emollient for industrial applications where light color, high purity, and low acidity are desirable; plasticizer, wetting agent, lubricant for rubber compounding. [CasChem]

Aaterra WP. CAS 2593-15-9; Etridiazole; protective thiazole fungicide which is incorporated into soil or compost. [ICI Agrochemicals]

Aatex. Natural rubber based adhesive; for leather, paper and general industrial uses. [Anglo Speciality Adhesives]

AA USP. Castor oil; emollient for cosmetic and pharmaceutical purposes; lubricant for food processing release aid, protective coatings for vitamins, tableting. [CasChem]

AB®. CAS 96-20-8; 2-Amino-1-butanol; pigment dispersant, neutralizing/emulsifying amine, corrosion inhibitor, acid-salt catalyst, pH buffer, chemical and pharmaceutical intermediate, solubilizer. [Angus]

AB1000-F. Thermoset composite molding compound extremely high heat resist. electrical grade for compression, transfer and injection molding. [Cuyahoga Plastics]

AB3500. Polyester, thermoset, mineral-filled, glass-reinforced; thermoset for compression, transfer and injection molding. [Cuyahoga Plastics]

AB3500-AR. Polyester, thermoset, glass-reinforced; abrasion-resistant thermoset for compression, transfer, and injection molding. [Cuyahoga Plastics]

abaca. Manila hemp, the inner fiber of *Musa textilis*.

Abacid. An antacid preparation containing aluminum hydroxide gel and magnesium trisilicate. [Unichem] *

Abacid Plus. An antacid preparation containing aluminum hydroxide, magnesium trisilicate, dimethicone and dicyclomine hydrochloride. [Unichem] *

Abalon. Nonmetallic floor tiles for use in building. [Courtaulds plc]

Abalyn®. CAS 68186-14-1; Methyl rosinate; resin with compatibility, surface-wetting properties, visc., and tack used in lacquers, inks, paper coatings, varnishes, adhesives, sealing compounds, plastics, wood preservatives, and perfumes; with Gardner color 6, Gardner Holt viscosity 21 and acid number 6. [Hercules]

aba-odo. An African term for a mixture of rubber latexes, probably those from *Funtumia elastica* and *Ficus Vogelii*.

Abasin. A proprietary preparation of acetyl carbromal; a sedative. [Bayer AG] †

Abate® 1-SG, 2-CG, 4-E, 5CG. CAS 3383-96-8; Temephos; gran. and emulsifiable conc. herbicide for control of mosquito larvae in standing water, ponds, etc. [Am. Cyanamid/Ag] *

Abatia. The leaves of *Abatia rugosa;* used as a black dye.

Abavit B. Organo-mercury seed dressing. [Murphy Chemical Company Ltd] †

Abavit S. Organo-mercurial dip. [Murphy Chemical Company Ltd]†

Abbaflox. Flucloxacillin; antibiotic. [Abbott Laboratories] *

Abbalgesic. Paracetemol with dextropropoxyphene; analgesic. [Abbott Laboratories] *

Abbalide®. 1,3,4,6,7,8-Hexahydro-4,6,6,7,8,8-hexamethylcyclopenta[s]-2-benzopyran; musk odor for use in fragrances at all price levels. [Bush Boake Allen Ltd]

Abbarome. Perfumery base. [Bush Boake Allen Ltd]

Abbavert. 2-Ethyl hexanal 1,2-ethane diol cyclic acetal. [Bush Boake Allen Ltd]

Abbcite No. 2. An explosive for coal mines containing ammonium nitrate, nitroglycerine and dinitrotoluene. §

Abbenclamide. A proprietary preparation of glibenclamide; oral treatment of hypoglycemia. [Abbott Laboratories] *

Abbloraz. A proprietary preparation of lorazepam; anxiolytic. [Abbott Laboratories] *

Abbocillin-DC. Penicillin G Procaine; antibacterial. [Abbott Laboratories] *

Abbokinase. Urokinase; plasminogen activator. [Abbott Laboratories] *

Abbolactone. A proprietary preparation of spironolactone; used for edema in cirrhosis of the liver, nephrotic syndrome, congestive heart failure, potentiation of thiazide and loop diuretics, hypertension, Conn's syndrome. [Abbott Laboratories] *

Abbopramide. A proprietary preparation of metoclopramide; to promote gastric motility and emptying. [Abbott Laboratories] *

Abbopurin. A proprietary preparation of allopurinol; for gout prophylaxis, hyperuricemia. [Abbott Laboratories] *

Abboxapam. A proprietary preparation of oxazepam; anxiolytic. [Abbott Laboratories] *

Abboxide. A proprietary preparation of chlordiazepoxide; anxiolytic. [Abbott Laboratories] *

ABC Liniment. Compound liniment of aconite (*Linimentum aconiti compositum BPC*). §

ABC-Trieb®. CAS 1066-33-7; Ammonium bicarbonate; baking raising agent. [BASF AG] *

Abcure S-40-25 CAS 94-36-0; Benzoyl peroxide disp. in an inorganic medium; catalyst; alternative to MEK peroxide, benzoyl peroxide pastes or granules for catalyzation of unsat. polyester resins; well suited for spray applic. [Abco Industries] *

Abelite. Blasting explosives containing ammonium nitrate and trinitrotoluene. §

Abel's reagent. A 10% solution of normal chromic acid; used in the micro-analysis of carbon steels, for etching.

Abequito. Nimidane; acaricide. [Am. Cyanamid] *

Aberel. Retinoic acid; a proprietary preparation for treatment of acne. [McNeil Pharmaceuticals] *

Aberoid. A proprietary casein plastic material. §

Abex® 12S. Surfactant; emulsifier for vinyl acrylics and acrylic polymerization. [Rhone-Poulenc Surf.]

Abex® 23S. CAS 9004-82-4; Sodium laureth sulfate; emulsifier for emulsion polymerization. [Rhone-Poulenc Surf.]

Abex® EP-110. CAS 9051-57-4; Ammonium nonoxynol-9 sulfate; primary emulsifier and stabilizing agent for the preparation of vinyl acetate, vinyl acetate/acrylic, all acrylic, styrene/acrylic, and S/B emulsion copolymers; wetting agent, dispersant for agricultural formulations. [Rhone-Poulenc Surf.; Rhone-Poulenc France]

Abex® JKB. Proprietary surfactant; emulsifier for high-acid polymerization systems. [Rhone-Poulenc Surf.; Rhone-Poulenc France]

Abex® LIV/30. Ammonium alkylaryl ether sulfate; emulsifier for emulsion polymerization of acrylic, styrene-acrylic, vinyl acetate; detergent, emulsifier, foam stabilizer, wetting agent in household and industrial detergents, shampoos, bubble baths. [Rhone-Poulenc Geronazzo]

Abex® VA 50. Octoxynol-33, sodium laureth sulfate; emulsifier for high solids vinyl acetate emulsions. [Rhone-Poulenc Surf.; Rhone-Poulenc France]

Abicol. A proprietary preparation of reserpine and bendrofluazide; used for control of hypertension. [The Boots Co plc] †

Abidec. A vitamin preparation containing vitamin A, vitamin D, thiamine, riboflavine, nicotinamide and ascorbic acid. [Parke-Davis] *

abietic anhydride. Gum rosin; used in manufacturing of soaps, varnish, paint driers, wax comps., printing inks, artificial amber, and medicine.

Abil® 10-10000. Dimethicone; conditioner for skin and hair care, sunscreens, tanning creams, or lotions; aftershave, aerosol preparations. [Goldschmidt AG]

Abil® 281. Polysiloxane polyether copolymer; silicone surfactant used in personal care products. [Goldschmidt] *

Abil® AV 20-1000. CAS 2116-84-9; Phenyl trimethicone; conditioner for skin and hair care, sunscreen, tanning creams or lotions; aftershave, aerosol preparations. [Goldschmidt; Goldschmidt AG]

Abil®AV 8853. CAS 2116-84-9; Phenyl trimethicone; emollient providing skin protection, barrier against aq. media; perfume ingred. and fixative; provides improved rub and spreadability, faster penetration; nonsticky; prevents aerosol clogging. [Goldschmidt AG]

Abil® B 8839. CAS 69430-24-6; Cyclomethicone; conditioner for hair care products, aerosols, sticks, shaving preparations, deodorants, antiperspirants. [Goldschmidt AG]

Abil® B 8851. CAS 68937-55-3; Dimethicone copolyol; surfactant, conditioner for personal care products; emollient for hair and skin care products, aerosol shaving lather, deodorants, antiperspirants, creams and lotions, perfumes and colognes. [Goldschmidt AG]

Abil® B 8852. CAS 68937-55-3; Dimethicone copolyol; surfactant, conditioner for personal care products; emollient for hair and skin care products, aerosol shaving lather, deodorants, antiperspirants, creams and lotions, perfumes and colognes. [Goldschmidt AG]

Abil® B 9950. CAS 102523-96-6; Dimethicone propyl PG-betaine; silicone surfactant, conditioner; used in hair and skin care products. [Goldschmidt; Goldschmidt AG]

Abil® B 88183. CAS 68937-55-3; Dimethicone copolyol; surfactant used as foam formers and providing lubricating and gloss properties; refatting agent for skin products; increases slip in shaving creams; also for shampoos, shower gels, hand cleaners, aerosols, antiperspirants. [Goldschmidt; Goldschmidt AG]

Abil® B 88184. CAS 68937-55-3; Dimethicone copolyol; surfactant for hair and skin care products; anticracking agent for soap bars. [Goldschmidt; Goldschmidt AG]

Abil®EM-90. Cetyl dimethicone copolyol; conditioner, emulsifier for w/o type creams and lotions, roll-ons, suncare products, shampoos and conditioners. [Goldschmidt; Goldschmidt AG]

Abil® K 4. CAS 69430-24-6; Cyclomethicone; emollient, conditioner used in hair care products, aerosols, shaving preparations, deodorants, antiperspirants. [Goldschmidt AG]

Abil® OSW 12, OSW 13. Cyclomethicone, dimethiconol, dimethicone; glossing and conditioning agent for hair care products [Goldschmidt]

Abil®S201. Sodium poly PG-propyl dimethicone thiosulfate; hair conditioner and setting lotion; improves gloss and sheen of shampoos, conditioners, mousses, gels, styling aids. [Goldschmidt]

Abil® S255. Oleoyl/palmitoyl/palmitoleamidopropyl/ silkhydroxypropyldimonium chloride; conditioner and setting lotion; improves gloss and sheen of shampoos and conditioners. [Goldschmidt] *

Abil® WE 09. Polyglyceryl-4 isostearate, cetyl dimethicone copolyol, hexyl laurate; emulsifier for highly stable w/o creams and lotions; improves uv protection in sunscreens. [Goldschmidt AG]

Abil® WS 08. Cetyl dimethicone copolyol, cetyl dimethicone, polyglyceryl-3 oleate, hexyl laurate; emulsifier for w/o creams and lotions. [Goldschmidt AG]

Abil® ZP 2434. Polysiloxane polyalkyl copolymer; emollient used in cosmetic creams and lotions, pigmented products and nonaq. systems. [Goldschmidt] *

Abil® Quat 3270, 3272. Quaternium-80; conditioner, antistat for shampoos and hair rinses; also refatting agent for skin cleansers. [Goldschmidt, Goldschmidt AG]

Abil® Wax 2434. CAS 68554-53-0; Stearoxy dimethicone; wax improving applic. and skin care properties of emulsions; spreading and emollient properties for protection against aq. media; water barrier for creams and lotions. [Goldschmidt; Goldschmidt AG]

Abil® Wax 2440. Behenoxy dimethicone; wax improving applic. and skin care properties of emulsions; spreading and emollient properties for protection against aq. media; reduces whitening during applic. of creams and lotions; pigment solubilizer. [Goldschmidt; Goldschmidt AG]

Abil® Wax 9800. Stearyl dimethicone; wax improving color, luster, and spreadability of pigmented products; spreading, penetrating, and emollient properties for skin care products. [Goldschmidt; Goldschmidt AG]

Abil® Wax 9801. Cetyl dimethicone; wax providing emolliency and applic. benefits for antiperspirants; pigment solubilizer; also for skin care products. [Goldschmidt; Goldschmidt AG]

Abil® Wax 9809. CAS 68607-75-0; Stearyl methicone; wax providing water barrier for night creams and protective lotions. [Goldschmidt] *

Abil® Wax 9814. Cetyl dimethicone; emollient and spreading agent for cosmetic esters and oils; pigment grinding aid and dispersant especially for titanium dioxide (prevents reagglomeration); used in sunscreens and pigmented products such as pressed powds. [Goldschmidt]

Abiol. CAS 39236-46-9; Imidazolidinyl urea NF; preservative. [3V-Sigma] *

Abisol. A 40% solution of sodium hydrogen sulfite, $NaHSO_3$; a disinfectant and preservative. §

Abitol®. CAS 26266-77-3; Dihydroabietyl alcohol, technical grade; resinous plasticizer and tackifier in plastics, lacquers, inks, and adhesives; chemical intermediate. [Hercules]

Ablefilm® 550. Glass supported, toughened epoxy adhesive film; moisture-resist. adhesive film designed for substrate attach and sealing microelectronic pkgs. [Ablestik]

Ablufoam HT. Organic compound; antifoamer for high temp. jet dyeing. [Taiwan Surf.]

Ablufoam SAE. Silicone emulsion; textile antifoamer effective over wide pH range. [Taiwan Surf.]

Abluhide DS. Blend; soaking, degreasing and rewetting agent for leather manufacturing. [Taiwan Surf.]

Abluhide F Series. Surfactant; fatliquoring agent for leather manufacturing. [Taiwan Surf.]

Ablumide CDE. Cocamide DEA (1:1); foam stabilizer, thickener for shampoos, bubble baths, liquid detergents, toiletries. [Taiwan Surf.]

Ablumide CME. CAS 68140-00-1; Cocamide MEA (1:1); foam stabilizer, thickener for shampoos, bubble baths, liquid detergents, toiletries. [Taiwan Surf.]

Ablumide LDE. CAS 120-40-1; EINECS 204-393-1; Lauramide DEA (1:1); foam stabilizer, thickener for shampoos, bubble baths, liquid detergents, toiletries. [Taiwan Surf.]

Ablumide LME. CAS 142-78-9; EINECS 205-560-1; Lauramide MEA (1:1); foam stabilizer, thickener for shampoos, bubble baths, liquid detergents, toiletries. [Taiwan Surf.]

Ablumide SDE. CAS 93-82-3; EINECS 202-280-1; Stearamide DEA (1:1); thickener, emulsifier for min. and vegetable oils, microcrystalline wax. [Taiwan Surf.]

Ablumide SME. CAS 111-57-9; EINECS 203-883-2; Stearamide MEA (1:1); opacifier, thickener for shampoo, cream rinse, bubble bath. [Taiwan Surf.]

Ablumine 08. Alkyl (99% C8) benzyl dimethyl quaternary; leveling agent for acrylic fiber. [Taiwan Surf.]

Ablumine 230. N-alkyl dimethyl ammonium chloride; algicide for industrial cooling towers and swimming pools. [Taiwan Surf.] *

Ablumine 280. CAS 122-19-0; EINECS 204-527-9; Stearyl dimethyl benzyl ammonium chloride; antistat and hair conditioner. [Taiwan Surf.] *

Ablumine D10. Didecyl dimethyl ammonium methosulfate; disinfectant, sanitizer, germicide. [Taiwan Surf.]

Ablumine DHT75. Dihydrogenated tallow dimethyl ammonium methosulfate; antistat, fabric softener suitable for dryer sheets. [Taiwan Surf.]

Ablumine DT. Ditallow dimethyl ammonium methosulfate; household softener with wide dispersion stability. [Taiwan Surf.]

Ablumox C-7. CAS 61791-14-8; PEG-7 cocamine; corrosion inhibitor for acid cleaners. [Taiwan Surf.]

Ablumox CAPO. CAS 68155-09-9; Cocamidopropyl amine oxide; foamer, wetting agent, foam stabilizer, antistat, detergent, emollient. [Taiwan Surf.]

Ablumox LO. Lauramine oxide; foamer, wetting agent, foam stabilizer, antistat, detergent, emollient. [Taiwan Surf.]

Ablumox T-15. CAS 61791-44-4; PEG-15 tallow amine; leveling agent for dyeing of nylon with anionic dyes; controls rate of strike and dye migration. [Taiwan Surf.]

Ablumul AG-306. Blend; emulsifier for Acephate (agricultural). [Taiwan Surf.]

Ablumul EP. Surfactant; emulsifier, thickener for textile printing. [Taiwan Surf.]

Ablunol 200ML. CAS 9004-81-3; PEG 200 laurate; emulsifier, lubricant, dispersing and leveling agent, defoamer used in cosmetic, textile, paint, dyestuffs, and other industrial uses. [Taiwan Surf.]

Ablunol 200MO. CAS 9004-96-0; PEG 200 oleate; emulsifier, lubricant, dispersing and leveling agents used in cosmetic, textile, leather, paint and other industrial uses. [Taiwan Surf.]

Ablunol 200MS. CAS 9004-99-3; PEG 200 stearate; emulsifier, thickener, lubricant, softener, defoamer, dispersing and leveling agent used in cosmetic, textile, paint and other industrial uses. [Taiwan Surf.]

---

‡ Trade name and manufacturer not verified   § Trade name without identified manufacturer

Ablunol CO 10. CAS 61791-12-6; PEG-10 castor oil; emulsifier for oils, solvs. and waxes, lubricant, antistat. [Taiwan Surf.]

Ablunol DEGMS. CAS 9004-99-3; Diethylene glycol stearate; opacifier, pearlescent for cosmetics, detergents. [Taiwan Surf.]

Ablunol EGMS. Glycol stearate; opacifier, pearlescent for cosmetics, detergents. [Taiwan Surf.]

Ablunol GML. CAS 142-18-7; EINECS 205-526-6; Glyceryl laurate; component in mold release agents. [Taiwan Surf.]

Ablunol GMO. CAS 111-03-5; Glyceryl oleate; internal lubricant, antistat, antifogging agent for PVC film; mold release agent and rust prevention for compounded oils. [Taiwan Surf.]

Ablunol GMS. Glyceryl stearate; emulsifier for hand creams, lotions, cosmetics; textile lubricant-softener. [Taiwan Surf.]

Ablunol LA-3. Laureth-3; emulsifier, dispersant, detergent, wetting agent used in textile processing, cosmetics, metalworking compounds, agriculture, industrial cleaners. [Taiwan Surf.]

Ablunol LMO. Surfactant; leveling agent for dyeing polyester fibers with disperse dyes. [Taiwan Surf.]

Ablunol NP4. CAS 9016-45-9; Nonoxynol-4; detergent and dispersant for use in petrol. oils; intermediate for manufacturing of surfactants and antistatic agents; co-emulsifier for fats, oils and waxes. [Taiwan Surf.]

Ablunol OA-6. CAS 9004-98-2; Oleth-6; emulsifier for mineral oil and cosmetics. [Taiwan Surf.]

Ablunol S-20. CAS 1338-39-2; EINECS 215-663-3; Sorbitan laurate; emulsifier, emulsion stabilizer, thickener for cosmetic, pharmaceutical, food applications; textile fiber lubricant, softener; antifog agent. [Taiwan Surf.]

Ablunol S-40. CAS 26266-57-9; EINECS 247-568-8; Sorbitan palmitate; emulsifier, emulsion stabilizer, thickener for cosmetic, pharmaceutical, food applications; textile fiber lubricant, softener; antifog agent. [Taiwan Surf.]

Ablunol S-60. CAS 1338-41-6; EINECS 215-664-9; Sorbitan stearate; emulsifier, emulsion stabilizer, thickener for cosmetic, pharmaceutical, food applications; textile fiber lubricant, softener; antifog agent; silicone defoamer emulsions. [Taiwan Surf.]

Ablunol S-80. CAS 1338-43-8; EINECS 215-665-4; Sorbitan oleate; emulsifier, emulsion stabilizer, thickener for cosmetic, pharmaceutical, food applications; textile fiber lubricant, softener; antifog agent; wet processing of synthetic PU leather. [Taiwan Surf.]

Ablunol S-85. CAS 26266-58-0; EINECS 247-569-3; Sorbitan trioleate; emulsifier, emulsion stabilizer, thickener for cosmetic, pharmaceutical, food applications; textile fiber lubricant, softener; antifog agent. [Taiwan Surf.]

Ablunol SA-7. CAS 9005-00-9; Steareth-7; emulsifier for wax and cosmetics. [Taiwan Surf.]

Ablunol T-20. POE sorbitan laurate; o/w emulsifier for cosmetic, pharmaceutical and food applications. [Taiwan Surf.]

Abluphat AP Series. Complex org. phosphate ester, free acid; antistat, penetrant, wetting agent, solubilizer, detergent; for cotton and synthetics processing; high alkaline and acid tolerance. [Taiwan Surf.]

Abluphat LP Series. Phosphate ester sodium salt; antistat for synthetic fibers; penetrant, wetting agent, solubilizer, detergent. [Taiwan Surf.]

Ablupol AF. Mixture; antifoaming agent for coating color formulation of paper. [Taiwan Surf.]

Ablusoft A. Polyamide type surfactant; softener for acrylic fibers, cotton and synthetics. [Taiwan Surf.]

Ablusoft ES. Epoxy-modified silicone; finishing agent imparting durable softness, wrinkle resistance to cotton and synthetic fiber blends. [Taiwan Surf.]

Ablusoft PE. Polyethylene emulsion; finishing agent improving tear strength, abrasion resistance and handle of fabrics. [Taiwan Surf.]

Ablusol C-70. Dioctyl sulfosuccinate; wetting agent for industrial applications; penetrant/wetting agent for bleaching of cotton, agricutural applications. [Taiwan Surf.]

Ablusol DA. Ethoxylated decyl alcohol sulfosuccinate monoester; emulsifier for polyacrylate emulsion polymerization. [Taiwan Surf.]

Ablusol DBC. Calcium dodecylbenzene sulfonate; emulsifier for agricultural chem., oils, solvs. [Taiwan Surf.]

Ablusol DBD. CAS 26545-53-9; EINECS 247-784-2; DEA dodecylbenzene sulfonate; surfactant, emulsifier, wetting agent for bubble bath, shampoos, detergents. [Taiwan Surf.]

Ablusol DBM. CAS 1331-61-9; EINECS 215-559-8; Ammonium dodecylbenzene sulfonate; surfactant, emulsifier, wetting agent for bubble bath, shampoos, detergents. [Taiwan Surf.]

Ablusol DBT. TEA dodecylbenzene sulfonate; surfactant, emulsifier, wetting agent for bubble bath, shampoos, detergents. [Taiwan Surf.]

Ablusol LDE. Lauramide DEA sulfosuccinate monoester; detergent, foam booster/stabilizer for low irritation shampoos, bubble baths, liquid detergents. [Taiwan Surf.]

Ablusol LME. Lauramide MEA sulfosuccinate monoester; detergent producing copious lather and extra dry residue for high foaming rug shampoos. [Taiwan Surf.]

Ablusol ML. CAS 9084-06-4; Sodium naphthalene sulfonate formaldehyde condensate; plasticizer and water reducing agent used in pourable and high strength concrete. [Taiwan Surf.]

Ablusol PM. Alkyl sulfate; penetrant, scouring agent for mercerizing of cotton. [Taiwan Surf.]

Abluter BE. Cocamidopropyl betaine; detergent for mild cleansing products; antistat, softener, germicide, spreading/wetting agent. [Taiwan Surf.]

Abluter GL Series. Glycine derivs.; antistat, softener, germicide, spreading/wetting agent. [Taiwan Surf.]

Abluton A. Surfactant blend; penetrant, scouring agent for cotton and synthetics. [Taiwan Surf.]

Abluton CAT. Catalyst; durable water repellent for polyamide, cotton and synthetic blends. [Taiwan Surf.]

Abluton CMN. Methyl naphthalene; dye carrier for polyester fiber and its blends. [Taiwan Surf.]

Abluton CTP. Trichlorobenzene; carrier for bleaching polyester fiber and its blends; emulsifiable. [Taiwan Surf.]

Abluton MK9. Wax emulsion with zirconium salt; water repellent agent for natural and synthetic fibers. [Taiwan Surf.]

Abluton SR. Reactive silicone emulsion; durable water repellent for polyamide, cotton and synthetic blends. [Taiwan Surf.]

Abluton T30. CAS 108-90-7; Chlorobenzene; carrier for textile dyeing with low stain to wool. [Taiwan Surf.]

Abluwax EBS. CAS 110-30-5; EINECS 203-755-6; Ethyl-

---

* Trade name not verified as available   † Trade name verified as obsolete

ene bisstearamide; lubricant for ABS, PS, and PVC; defoamer and mold-releasing agent. [Taiwan Surf.]

ABM 5C Chlormequat Plus. Mixture of chlormequat and choline chloride; plant growth regulator for use in cereals and ornamentals. [ABM Chemicals Ltd]

ABM Chlormequat 40. CAS 999-81-5; Soluble concentrate containing 400 g/l chlormequat; plant growth regulator. [ABM Chemicals Ltd]

ABM Chlormequat 72.5. CAS 999-81-5; Soluble concentrate containing 725 g/l chlormequat; plant growth regulator. [ABM Chemicals Ltd]

Abocast. Adhesives, casting epoxies, solventless coatings. [Abatron Inc] *

Abocrete. Concrete and masonry patching and resurfacing compound. [Abatron Inc] *

Abocure. Catalysts, curing and hardening agents. [Abatron Inc] *

Abol. Garden aphicide. [ICI Chem. & Polymers Ltd]

Abopon. A proprietary liquid inorganic resinous product which forms films in a few minutes on drying in air; recommended as an adhesive, a suspended medium for pigments and abrasives, and for sealing of surfaces to be lacquered or painted; stated to be a boro-phosphate. §

Aboseal. Sealants, caulks. [Abatron Inc] *

Abracol®. A registered trademark for toluene sulfonic acid esters; used as plasticizers, emulsifying agents. [Bush Boake Allen Ltd]

Abradux. Extra tough and dense types of aluminum oxide; used for abrasive industries, super refractories, sandblasting and for safes. [Lonza AG]

Abramant. Semi-friable aluminum oxide; used for abrasive industries, super refractories, sandblasting and for safes. [Lonza AG]

Abramax. Pure white aluminum oxide; used for abrasive industries, super refractories, sandblasting and for safes. [Lonza AG]

Abrarex. Special grades of pure aluminum oxide; used for abrasive industries, super refractories, sandblasting and for safes. [Lonza AG]

Abrasit. Regular aluminum oxide, finely crystallized; used for abrasive industries, super refractories, sandblasting and for safes. [Lonza AG]

abrastol. (asaprol, asaprol-etrasol). Calcium-β- naphthol-γ-sulfonate; used as a clarifier for wines.

Abraum salts. (stripping salt, Stassfurt salts, potash salts). The names applied to the upper layers of mixed chlorides of magnesium, potassium, and sodium, overlying the beds of rock-salt at Stassfurt.

Abril. Synthetic waxes. [Abril Industrial Waxes]

Abrodil. A proprietary preparation of sodium iodomethanesulfonate. §

Abros. A heat-resisting alloy containing 88% nickel, 10% chromium, and 2% manganese. §

ABS resin. Abbreviation for acrylonitrile butadiene styrene resin.

A.B.S. 87%. Alkylbenzene sulfonic acid; detergent intermediate. [Triantaphyllou]

ABS 124ESG. CAS 9003-56-9; ABS copolymer; general purpose molding grade for spools and molded parts requiring high tensile modulus. [Mobil/Polystyrene] *

ABS 236F. CAS 9003-56-9; ABS copolymer; general purpose automotive grade for injection molding of interior parts. [Mobil/Polystyrene] *

ABS 236MA. CAS 9003-56-9; ABS copolymer; med. impact

copolymer for profile extrusions for automotive and appliance applications; natural solid; sp.gr. 1.05; soften. pt. (Vicat) 220 F; tensile strength 5700 psi; tensile elong. 35%; flex. strength 9500 psi; impact strength (Izod) 4.5 ft lb/in. notched; hardness (Rockwell) 106 R; distort. temp. 208 F (264 psi). [Mobil/Polystyrene] *

ABS 301K. CAS 9003-56-9; ABS copolymer; injection molding, pipe fitting grade for fittings for DWV, sewer pipe, and conduit. [Mobil/Polystyrene] *

ABS 500FR-1. CAS 9003-56-9; ABS copolymer; flame-retardant, high impact grade for aircraft, automotive, appliance parts, furniture, smoke detector housings. [Mobil/Polystyrene] *

Absaglas®. Flame retardant acrylonitrile- butadiene styrene. §

absolute alcohol. *See* ethyl alcohol

Abson A.B.S. 213. A proprietary general-purpose grade of ABS; used in injection molding and extrusion. [BFGoodrich Canada] *

Abson A.B.S. 230. A proprietary general-purpose grade of ABS possessing higher toughness; used in applications requiring a higher gloss. [BFGoodrich Canada] *

Abson A.B.S. 300. A grade of ABS having medium impact resistance; its good gloss makes it suitable for refrigeration applications. [BFGoodrich Canada] *

Abson A.B.S. 500. A grade of ABS having high-to-medium impact resistance and good strength over a wide temperature range. [BFGoodrich Canada] *

ABT-2500®. CAS 14807-96-6; Talc; antiblocking agent for polyolefin films. [Pfizer]

Abyssinian gold. Gold (Talmi gold, Cuivre poli). A yellow alloy of copper and zinc. It usually consists of about 91% copper and 8% zinc, but sometimes contains 86% copper, 12% zinc, and 1% tin; employed in the manufacture of cheap jewelry.

A-C® 6. CAS 9002-88-4; Polyethylene homopolymer; for adhesives, ink, floor finsihes, paper coatings, personal care, plastics, rubber, textiles, wax blends. [Allied-Signal]

A-C® 316. Oxidized HDPE homopolymer; for ink, floor finsihes, personal care, plastics, textiles, wax blends. [Allied-Signal]

A-C® 629. Oxidized polyethylene homopolymer; for adhesives, ink, floor finsihes, paper coatings, personal care, plastics, textiles, wax blends. [Allied-Signal]

ACA®. For medical industry. [DuPont UK]

acacia gum. *See* gum arabic

Acagine. A mixture of lead chromate and bleaching powder; used to purify acetylene. §

acajou balsam. (Cardol). A material obtained from the fruits of *Anacardium occidentale* (mahogany nuts, elephant nuts) or *A. orientale* by the extraction of the powdered nuts. employed in the preparation of indelible inks and colors for die-sink.

Acaprin®. Chemotherapeutic against piroplasmosis (babesiasis); veterinary medicine. [Bayer AG]

Acardite 2. N-Methyl-N´,N´-diphenylurea; stabilizer improving storage stability of powders and propellants; plasticizer for celluloid. [Lowi] *

Acarin. CAS 115-32-2; Active ingredient: dicofol; miticide. [Makhteshim Chemical Works Ltd]

Accelemal. Possibly thiocarbamide; a proprietary rubber vulcanization accelerator. §

Accelerase. Pancrelipase. A concentrate of pancreatic

---

‡ Trade name and manufacturer not verified   § Trade name without identified manufacturer

enzymes standardized for lipase content; enzyme. [Organon Inc] *

Accelerator. Range of amine and cobalt-based accelerators for use in unsaturated polyester resins. [Akzo Chemie UK Ltd]

Accelerator 2P. CAS 98-77-1; Piperidinium pentamethylene dithiocarbamate; a proprietary rubber vulcanizing accelerator. [Pacific Anchor Chemical Corp] *

Accelerator 4P. Dipentamethylene thiuram disulfide; a proprietary rubber vulcanizing accelerator. [Pacific Anchor Chemical Corp] *

Accelerator 100. Aldehyde derivative of a Schiff's base, made from both butyraldehyde and acetaldehyde. §

Accelerator 108. Tetramethyl thiuram disulfide 2/3, 2-merceapto-benzthiazole 1/3; a proprietary rubber vulcanizing accelerator. [Naugatuck (US Rubber)] *

Accelerator 399. Epoxy curing promoter for use with amine hardeners. [Texaco]

Accelerator 808. Butyraldehyde aniline; a proprietary rubber vulcanizing accelerator. §

Accelerator 55028. Cobalt octoate mixture, hydrocarbon solv; accelerator developed for filled systems to decrease cobalt absorption and geltime drift. [Akzo] *

Accelerator A1. Thiocarbanilide; a proprietary rubber vulcanizing accelerator. §

Accelerator A5-10. Formaldehyde-anilines; a proprietary rubber vulcanizing accelerator. §

Accelerator A7. Made from 2 molecules of ethylidene aniline condensed with 1 molecule of acetaldehyde; a proprietary rubber vulcanizing accelerator. §

Accelerator A11, 16E. Aldehyde Schiff's bases; a proprietary rubber vulcanizing accelerator. §

Accelerator A17. Methylene-p-toluidine; a proprietary rubber vulcanizing accelerator. §

Accelerator A19. Modified ethylidene-aniline; a proprietary rubber vulcanizing accelerator. §

Accelerator A20. A proprietary rubber vulcanizing accelerator. §

Accelerator A22. Di-o-tolyl thiourea; a proprietary rubber vulcanizing accelerator. §

Accelerator A32. Condensation products of aldehydes and Schiff's base, e.g., butyraldehyde and butylidine aniline; proprietary rubber vulcanizing accelerator. §

Accelerator A50. Adehyde-amine; a proprietary rubber vulcanizing accelerator. §

Accelerator A1010. Formaldehyde-aniline; a proprietary rubber vulcanizing accelerator. §

Accelerator BB. Butyraldehyde p-aminodimethylaniline; a proprietary rubber vulcanizing accelerator. §

Accelerator BZ Powder. CAS 136-23-2; Zinc dibutyl dithiocarbamate; nondiscoloring, nonstaining accelerator for ethylene-propylene-diene rubbers, natural rubber and SR latexes; stabilizer and antioxidant in uncured rubber. [Akrochem]

Accelerator D. Cyclo amine blended with organo metallic salt; accelerator recommended for use in light-colored compounds, especially shoe soling industry. [Anchor] *

Accelerator DBA. CAS 103-49-1; Dibenzylamine; a proprietary rubber vulcanizing accelerator. §

Accelerator DT. CAS 97-39-2; Di-o-tolylguanidine; a French proprietary rubber vulcanizing accelerator. §

Accelerator E. 0.4% active cobalt sol'n. in styrene; accelerator. [Scott Bader]

Accelerator E-A. Ethylidene-aniline; a proprietary rubber vulcanizing accelerator. §

Accelerator EZ Powder. CAS 14323-55-1; Zinc diethyl dithiocarbamate; nondiscoloring, nonstaining accelerator for natural rubber and SR latexes. [Akrochem]

Accelerator F-A. The high melting one is formaldehyde-aniline, and the low melting one consists of methylene-dianilide; a proprietary rubber vulcanizing accelerator. §

Accelerator G.M.F. Quinone dioxime; a proprietary rubber vulcanizing accelerator. §

Accelerator L. Thiocarbanilide; a French proprietary rubber vulcanizing accelerator. §

Accelerator Mercapto. CAS 149-30-4; Mercapto-benzthiazole; a German proprietary rubber vulcanizing accelerator. §

Accelerator MF. CAS 102-77-2; 2-Benzothiazyl-N-morpholine disulfide; accelerator in natural and synthetic rubbers, e.g., tire and mech. goods. [Akrochem] *

Accelerator MZ Powder. CAS 137-30-4; Zinc dimethyl dithiocarbamate; nondiscoloring, nonstaining accelerator for natural rubber, isoprene rubber, BR, styrene/butadiene rubber, isobutylene-isoprene rubber, and ethylene-propylene-diene rubbers rubbers and natural rubber latex. [Akrochem]

Accelerator NL-6. Cobalt octoate, hydrocarbon solv.; accelerator mainly used in combination with ketone peroxides for cure of unsat. polyester resins at room temperature and elevated temps. [Akzo] *

Accelerator NL-12. Cobalt octoate, hydrocarbon solv.; accelerator mainly used in combination with ketone peroxides for cure of unsat. polyester resins at room temperature and elevated temps. [Akzo] *

Accelerator NL49P. Cobalt octoate, phthalate solv.; accelerator mainly used in combination with ketone peroxides for cure of unsat. polyester resins at room temperature and elevated temps. [Akzo] *

Accelerator NL51P. Cobalt octoate, phthalate solv.; accelerator mainly used in combination with ketone peroxides for cure of unsat. polyester resins at room temperature and elevated temps. [Akzo] *

Accelerator NL53. Cobalt octoate, hydrocarbon solv.; accelerator mainly used in combination with ketone peroxides for cure of unsat. polyester resins at room temperature and elevated temps. [Akzo] *

Accelerator PTX. Phenyl-tolyl-xylyl-guanidine; a proprietary rubber vulcanizing accelerator. §

Accelerator R2. Condensation product from methlylene-dipiperidine and carbon disulfide; a proprietary rubber vulcanizing accelerator. §

Accelerator R3. Zinc salt of a dithio-carbaminic acid; a proprietary rubber vulcanizing accelerator. §

Accelerator R5. Dithiocarbamate; a proprietary rubber vulcanizing accelerator. §

Accelerator VN-2. Vanadium salt; accelerator for ketone peroxides, hydroperoxides, and peroxy esters; short gel times and a very high speed of cure can be achieved. [Akzo] *

Accelerator W29. Compound of diphenyl guanidine with dibenzyl dithiocarbaminic acid; a proprietary rubber vulcanizing accelerator. §

Accelerator W80. Diphenyl-guanidine salt of mercapto-benzthiazole; a proprietary rubber vulcanizing accelerator. §

Accelerator X28. Stated to be impure tarry diphenylguanidine; a proprietary rubber vulcanizing

---

accelerator. §

Accelerator Z88. Ammonia salt of mercapto-benzthiazole mixed with a softener; a proprietary rubber vulcanizing accelerator. §

Accelerator ZBX. Zinc butyl xanthate; a proprietary rubber vulcanizing accelerator. §

Accelerator ZPD. Zinc pentamethylene dithiocarbamate; a proprietary rubber vulcanizing accelerator. §

Accelerene V 1. Consists of equi-molecular proportions of p-nitroso-dimethylaniline and β-naphthol; a proprietary rubber vulcanizing accelerator. §

Accelguard 80. Chloride-free set accelerator for concrete and mortar; liquid form. [Feb Ltd]

Accobetaine CL. CAS 68424-94-2; EINECS 270-329-4; complex coco betaine; detergent, wetting agent, emulsifier, high foaming agent, solubilizer, household and cosmetic uses. [Karlshamns] *

Accobond. Melamine formaldehyde resins; film resins for process industries. [Cyanamid BV]

Accomeen C2. CAS 61791-14-8; PEG-2 cocamine; emulsifier, antistat, surfactant. [Karlshamns]

Accomeen C5. CAS 61791-14-8; PEG-5 cocamine; emulsifier, antistat, surfactant. [Karlshamns]

Accomeen C10. CAS 61791-14-8; PEG-10 cocamine; emulsifier, antistat, surfactant. [Karlshamns]

Accomeen C15. CAS 61791-14-8; PEG-15 cocamine; emulsifier, antistat, surfactant. [Karlshamns]

Accomeen S2. CAS 61791-24-0; PEG-2 soyamine; emulsifier, antistat, surfactant. [Karlshamns] *

Accomeen S10. CAS 61791-24-0; PEG-10 soyamine; emulsifier, antistat, surfactant. [Karlshamns]

Accomeen S15. CAS 61791-24-0; PEG-15 soyamine; surfactant. [Karlshamns] *

Accomeen T2. CAS 61791-44-4; PEG-2 tallow amine; emulsifier, antistat, surfactant, dispersant. [Karlshamns]

Accomeen T5. CAS 61791-44-4; PEG-5 tallow amine; emulsifier, antistat, surfactant, dispersant. [Karlshamns]

Accomeen T15. CAS 61791-44-4; PEG-15 tallow amine; emulsifier, antistat, surfactant, dispersant. [Karlshamns]

Accomet. Paint pretreatment; used for steel strip and coil. [Albright & Wilson Ltd]

Accomet C. Metal precleaning process. [Albright & Wilson Ltd, Phosphates & Speciality Business]

Accomid 50. Palm kernelamide DEA. [Karlshamns] *

Accomid C. Cocamide DEA; detergent, stabilizer, visc. improver, foam booster for shampoos and dishwashes; biodeg. [Karlshamns]

Accomid PK. Palm kernelamide DEA (1:1); visc. builder, foam booster/stabilizer, emulsifier for shampoos, liquid soaps, dish detergents, bubble bath products. [Karlshamns]

Acconon 200-DL. CAS 9005-02-1; PEG-4 dilaurate; surfactant used as emulsifier, dispersant, solubilizer, visc. control agent for cosmetics, pharmaceuticals, and industrial applications. [Karlshamns] *

Acconon 200-MS. CAS 9004-99-3; PEG-4 stearate; surfactant used as emulsifier, dispersant, solubilizer, visc. control agent for cosmetics, pharmaceuticals, and industrial applications. [Karlshamns]

Acconon 400-MO. CAS 9004-96-0; PEG-8 oleate; emulsifier, dispersant, lubricant, chemical intermediate, solubilizer, visc. control agent; for cosmetics, pharmaceuticals, food, agricuture, plastics. [Karlshamns]

Acconon 400-MS. CAS 9004-99-3; PEG-8 stearate; surfactant used as emulsifier, dispersant, solubilizer, visc. control agent for cosmetics, pharmaceuticals, and industrial applications. [Karlshamns]

Acconon 1300. PPG-3-laureth-9; surfactant used as emulsifier, dispersant, solubilizer, visc. control agent for cosmetics, pharmaceuticals, and industrial applications. [Karlshamns]

Acconon CA-5. CAS 61791-12-6; PEG-5 castor oil; surfactant used as emulsifier, dispersant, solubilizer, visc. control agent for cosmetics, pharmaceuticals, and industrial applications. [Karlshamns]

Acconon CA-9. CAS 61791-12-6; PEG-9 castor oil; surfactant used as emulsifier, lubricant, dispersant, solubilizer, visc. control agent for cosmetics, pharmaceuticals, and industrial applications. [Karlshamns]

Acconon CA-15. CAS 61791-12-6; PEG-15 castor oil; surfactant used as emulsifier, lubricant, dispersant, solubilizer, visc. control agent for cosmetics, pharmaceuticals, and industrial applications. [Karlshamns]

Acconon CON. PEG-10 propylene glycol glyceryl laurate; surfactant used as emulsifier, dispersant, solubilizer, visc. control agent for cosmetics, pharmaceuticals, and industrial applications. [Karlshamns]

Acconon E. CAS 25231-21-4; PPG-15 stearyl ether; surfactant used as emulsifier, dispersant, solubilizer, visc. control agent for cosmetics, pharmaceuticals, and industrial applications. [Karlshamns]

Acconon ETG. CAS 31694-55-0; Glycereth-26; humectant; lubricant for skin care products, creams, lotions, industrial applications. [Karlshamns]

Acconon TGH. PEG-20-PPG-10 glyceryl stearate; surfactant used as emulsifier, dispersant, solubilizer, visc. control agent, wetting and foaming agent for cosmetics, pharmaceuticals, and industrial applications; biodeg. [Karlshamns]

Acconon W230. CAS 68439-49-6; Ceteareth-20; surfactant used as emulsifier, dispersant, solubilizer, visc. control agent for cosmetics, pharmaceuticals, and industrial applications. [Karlshamns]

A-C®Copolymer 400A. Low molecular weight EVA copolymer; pigment dispersant; PS color concs. [Allied-Signal/A-C®Perf. Addit.]

A-C®Copolymer 540, 540A. CAS 9010-77-9; low molecular weight ethylene-acrylic acid copolymer; plastics lubricant and processing aid, pigment dispersant; internal lubricant PVC, nylon 6, nylon color concs.; for adhesives, floor finishes, personal care, plastics, wax blends. [Allied-Signal/A-C®Perf. Addit.]

A-C® Copolymer 580. CAS 9010-77-9; Ethylene/acrylic acid copolymer; alkali-dispersible additive for recyclable hot-melt and aq. adhesives and coatings. [Allied-Signal/A-C®Perf. Addit.]

A-C® Copolymer 5120. CAS 9010-77-9; Low molecular weight ethylene-acrylic acid copolymer; alkali-dispersible additive for recyclable hot-melt and aq. adhesives and coatings. [Allied-Signal/A-C®Perf. Addit.]

A-C® Copolymer 5180. CAS 9010-77-9; Low molecular weight ethylene-acrylic acid copolymer; alkali-dispersible additive for recyclable hot-melt and aq. adhesives and coatings. [Allied-Signal/A-C®Perf. Addit.]

Accoquat 2C-75. CAS 61789-77-3; EINECS 263-087-6; Dicocodimonium chloride; emulsifier, coupling agent; used for car spray waxes, dust control oil, spot removal. [Karlshamns] *

---

‡ Trade name and manufacturer not verified   § Trade name without identified manufacturer

Accoquat 2C-75H. CAS 61789-77-3; EINECS 263-087-6; Dicocodimonium chloride; emulsifier, coupling agent. [Karlshamns] *

Accosize. Modified alkenylsuccinic anhydride; used in the paper industry. [Cyanamid BV]

Accosoft 440-75. Methyl bis (hydrogenated tallow-amidoethyl) 2-hydroxyethyl ammonium methyl sulfate; fabric softener quaternary for textile industry, household and commercial use; good lubricity and scorch resistance; nonyellowing. [Stepan; Stepan Canada]

Accosoft 540 HC. Methyl bis (modified tallowamidoethyl) 2-hydroxyethyl ammonium methyl sulfate; fabric softener quaternary for household products and textile processing. [Stepan; Stepan Canada]

Accosoft 550-90 HHV. Methyl bis (tallowamidoethyl) 2-hydroxyethyl ammonium methyl sulfate; fabric softener and antistat with good rewet props. for laundry products, industrial textile processing. [Stepan; Stepan Canada]

Accosoft 550L-90. Methyl bis (tallowamidoethyl) 2-hydroxyethyl ammonium methyl sulfate; fabric softener quaternary for household laundry products, textile processing. [Stepan; Stepan Canada]

Accosoft 620-90. Methyl bis (tallowamidoethyl) 2-hydroxypropyl ammonium methyl sulfate; fabric softener and antistat for industrial and household products; excellent rewet props.; nonyellowing; U.S. patent #3,933,871. [Stepan; Stepan Canada]

Accosoft 750. Methyl bis (oleylamidoethyl) 2-hydroxyethyl ammonium methyl sulfate; fabric softener quaternary for heavy duty liquid laundry detergents. [Stepan; Stepan Canada]

Accosoft 808HT. Methyl-1-hydrogenated tallowamido-ethyl-2-hydrogenated tallow imidazolinium-methyl sulfate; fabric softener and antistat for household and industrial applications, dryer products. [Stepan; Stepan Canada]

Accosperse 20. CAS 9005-64-5; PEG-20 sorbitan laurate; emulsifier, solubilizer; biodeg. [Karlshamns] *

Accosperse 60. CAS 9005-67-8; PEG-20 sorbitan stearate; emulsifier, solubilizer; biodeg. [Karlshamns]

Accosperse 80. CAS 9005-65-6; PEG-20 sorbitan oleate; emulsifier, solubilizer; biodeg. [Karlshamns]

Accostrength. Anionic/cationic polyacrylamide; used in the paper industry. [Cyanamid BV]

Accostrength 72. Acrylamide copolymers (anionic); used in the paper industry. [Cyanamid BV]

Accostrength 711. Cationic polyacrylamide; for paper industry. [Cyanamid BV]

Accrolon® 9039. Self-lubricated bearing engineering thermoplastic; bearing material featuring high continuous service temp., excellent wear, abrasion and fatigue resist., outstanding mech. and excellent chem. resist. [Accro-Seal]

Accrolube. Grease with Teflon®; high-efficiency lubricant which creates a boundary lubrication film that reduces wear bewteen metal surfaces and protects against corrosion; for pneumatic cylinders, valves, hydarulic equip., conveyors, marine equip., bearas, bearings, etc. [Accro-Seal]

Accrolube-FG. Grease with Teflon®; biodeg. food-grade lubricant featuring waterproof props., chem. resist.; for conveyor chains, seamers, filling equip., sterilziers, pumps, valves, cylinders, mixers, bearings, pkg. equip. [Accro-Seal]

Accrotan. Self-basifying chrome tanning material. [British Chrome & Chemicals Ltd]

Accrox. CAS 1308-38-9; Special refractory grades of chromic oxide. [British Chrome & Chemicals Ltd]

Acctuf 3045. CAS 9003-07-0; PP impact copolymer; high impact general-purpose grade for extrusion, compression molding. [Amoco Chemical Co]

Accuglass. Spin-on Glass; solutions of inorganic polymers (siloxanes or silicates) spun applied to silicon wafers to form insulating or passivating films used in the manufacture of integrated circuits. [Allied-Signal, Planarization & Diffusion Products Division] *

Acculog. Standard volumetric concentrated solutions; quantitative analytical chemistry. [Schweizerhall]

Accurac. Polymers based on acrylamide; can be described as nonionic polyacrylamide, anionic polyacrylamide and cationic polyacrylamide; used in the paper industry. [Cyanamid BV]

Accurac 33/35/41. Polyamine condensation products; used in the paper industry. [Cyanamid BV]

Accurac 39-S. Cationic polyacrylamide solution; used in the paper industry. [Cyanamid BV]

Accuspin. Spin-on dopant; solutions of impurity atoms (boron, phosphorus, arsenic or antimony) spun applied to silicon wafers used to dope silicon to form transistors and integrated circuits. [Allied-Signal, Planarization & Diffusion Products Division] *

Accutane. Isotretinoin; keratolytic. [Hoffmann-LaRoche Inc] *

Accuthane® UR-1100. One-component urethane adhesive; patented adhesive for difficult-to-bond substrates such as nylon, SMC/FRP, galvanized and stainless steel; good heat resist. [H.B. Fuller] *

Accuvette. Plastic sample vial; used to hold blood cell dilution for counting in semiautomatic analyzers. [Coulter Electronics Ltd]

Ac-Di-Sol. Croscarmellose sodium; carboxymethylcellulose sodium that has been internally cross-linked; pharmaceutic aid. [FMC UK] *

Ace-ite. A proprietary bituminous or asphalt composition. §

Acelan A. Cetyl acetate, acrylated lanolin alcohol. [Fabriquimica] *

Acelan L. CAS 61788-48-5; EINECS 262-979-2; Acetylated lanolin. [Fabriquimica] *

Aceloid. A proprietary cellulose acetate material; used as a molding composition. §

Acelon. Cellulose acetate coated fabric. [May & Baker Ltd]*

Acelose. A proprietary cellulose acetate material. §

Aceplus. A proprietary cellulose acetate material. §

Acerado. (Fierroso). Names used for mercurial earths. §

Acerdol. CAS 10118-76-0; Calcium permanganate, $CaMn_2O_{[8}$; used in gastro-enteritis and diarrhea. §

Ace-Sil. A proprietary trade name for microporous rubber; used for battery separators and filters. §

Aceta. A proprietary brand of cellulose rayon; the name is also applied to a French nitrocellulose lacquer. §

Acetadeps. CAS 61788-48-5; EINECS 262-979-2; Acetylated lanolin; emollient, moisturizer for antiperspirants, baby oils, cleansers, shampoos, hair conditioners, sunscreen; binder for pressed powds. [Westbrook Lanolin]

acetaldehyde. (acetic aldehyde; ethyl aldehyde; ethanal) CH_3CHO; CAS 75-07-0; EINECS 200-836-8; Manufacture of acetic acid, acetic anhydride, n-butanol, peracetic acid, pentaerythritol, pyridines, 1,3-butylene glycol, tri-

---

* Trade name not verified as available   † Trade name verified as obsolete

methylolpropane; synthetic flavors. [BP Chem. Ltd.; Eastman; Hoechst-Celanese; Hüls UK; Mitsui Petrochem. Ind.]

Acetaloid. A proprietary cellulose acetate material in the form of rods, sheet, and molding composition. §

acetamide MEA. (N-acetyl ethanolamine; N-(2-hydroxyethyl) acetamide) Aliphatic amide; $C_4H_9NO_2$; CAS 142-26-7; EINECS 205-530-8; solvent, humectant, skin and hair conditioner, intermediate, coupling agent, pigment dispersant, clarifying agent for shampoos, moisturizer. [Heterene; McIntyre; Scher]

Acetamide MEA. CAS 142-26-7; EINECS 205-530-8; Acetamide MEA; solvent, humectant for cosmetics. [Witco/Organics]

Acetamin 24. CAS 2016-56-0; Cocamine acetate; surface coating agent for pigments, anticaking agent for fertilizer; emulsifier, dispersant, and softening agent for textiles; min. flotation reagent. [Kao Corp. SA] *

Acetamin 86. CAS 2190-04-7; Stearamine acetate; surface coating agent for pigments, anticaking agent for fertilizer; emulsifier, dispersant, and softening agent for textiles; min. flotation reagent. [Kao Corp. SA] *

Acetamin C. CAS 61790-57-6; EINECS 263-147-1; n-Cocamine acetate; flotation of mins., anticaking agents, emulsifier bactericide. [Kao Corp. SA]

Acetamin HT. CAS 61790-59-8; n-Hyd. tallow amine acetate; flotation of mins., anticaking agents, emulsifier bactericide. [Kao Corp. SA]

Acetamin T. CAS 2190-04-7; n-Tallow amine acetate; flotation of mins., anticaking agents, emulsifier bactericide. [Kao Corp. SA]

acetannin. (diacetyl tannin). Acetyltannic acid.

Acetargol. A mixture of formic and acetic acids.

Acetazolamide. 5-Acetamido-1,3,4-thiadiazole-2-sulfonamide. diamox. [Lederle Laboratories] *

Acetest. A mixture of sodium nitroprusside, aminoacetic acid, disodium phosphate, and lactose, in tablet form; used to test for the presence of ketones in blood and urine. [B C Ames] *

Acetex. A proprietary safety-glass. §

acetic acid. (ethanoic acid; vinegar acid; methane-carboxylic acid) CH_3COOH; CAS 64-19-7; EINECS 200-580-7; Manufacture of acetic anhydride, cellulose acetate, vinyl acetate monomer; acetic esters; production of plastics, pharmaceuticals, dyes, insecticides, photographic chemicals, food additives; solvent reagent. [Air Prods.; BASF; BP Chem.; General Chem.; Hoechst Celanese; Janssen Chimica; Quantum/USI]

acetic acid, sec-butyl ester. See s-butyl acetate

acetic acid t-butyl ester. See t-butyl acetate

acetic acid 1,1-dimethylethyl ester. See t-butyl acetate

acetic acid, ethenyl ester. See vinyl acetate

acetic acid, ethenyl ester, homopolymer. See polyvinyl acetate (homopolymer)

acetic acid, ethyl ester. See ethyl acetate

acetic acid, lead salight See lead acetate

acetic acid manganese (2+) salt. See manganese acetate

acetic acid, methyl ester. See methyl acetate

acetic acid 1-methylpropyl ester. See s-butyl acetate

acetic acid silver salt. See silver acetate

acetic acid sodium salt anhydrous. See sodium acetate

acetic acid, vinyl ester. See vinyl acetate

acetic acid vinyl ester polymers. See polyvinyl acetate (homopolymer)

acetic anhydride. (acetyl oxide; acetic oxide) $C_4H_6O_3$; CAS 108-24-7; EINECS 203-564-8; Cellulose acetate fibers and plastics; vinyl acetate; dehydrating and acetylating agent for pharmaceuticals, dyes, perfumes, explosives; aspirin; esterifying agent for food starch. [Ashland; BP Chem. Ltd.; Chisso; CPS; Eastman; Hoechst-Celanese; Schweizerhall; Union Carbide]

acetic ether. See ethyl acetate

acetin blue. Solutions of indulines in acetins (acetic esters of glycerol). See Induline, spirit soluble.

Acetodin. See aspirin. §

Acetohexamide. Dimelor; an oral hypoglycemic agent. §

Acetol. A proprietary cellulose acetate material in the form of flake. §

Acetol® 1706. Cetyl acetate, acetylated lanolin alcohol; water repellent; strongly hydrophobic emollient, penetrant, lubricant, and cosolv. used in hair products, creams, lotions, suntan preparations and baby products. [Henkel/Cospha; Henkel Canada]

acetone. (dimethylketone; 2-propanone) CH_3COCH_3; CAS 67-64-1; EINECS 200-662-2; Solvent for paints, varnishes and lacquers; for cleaning and drying precision equip.; delustrant for cellulose acetate fibers. [Allied-Signal; Ashland; BASF; BP Chem. Ltd; Dow; Eastman; Exxon; Mitsui Petrochem.; Montedipe SpA; Shell; Texaco; Union Carbide]

acetone bromoform. See Brometone.

acetone chloroform. See chlorobutanol

Acetonedicarboxylic acid. See β-ketoglutaric acid

acetone dimethyl acetal. See dimethoxypropane

acetonitrile. (methyl cyanide) CH_3CN; CAS 75-05-8; EINECS 200-835-2; solvent for hydrocarbon extraction processes, especially for butadiene; intermediate; catalyst; for separation of fatty acids from vegetable ols; manufacture of synthetic pharmaceuticals. [BP Chem. Ltd; Du Pont; R. W. Greeff; ICC Ind.; Penta Mfg.]

Acetonyl. Sodium salt of aspirin with potassium and sodium tartrates; a proprietary analgesic. [Upjohn Ltd] *

Acetophen. A proprietary preparation of aspirin, phenacetin and caffeine in capsule form. [Jenkins Laboratories Inc]‡

acetophenetidin. CAS 62-44-2; (phenedidin; ethoxyacetanilide); used as an antipyretic and analgesic.

acetophenone. (ketone methyl phenyl; acetylbenzene; hypnone) $C_6H_5COCH_3$; CAS 98-86-2; EINECS 202-708-7; Perfumery, solvent, intermediate for pharmaceuticals, resins, etc.; flavoring; polymerization catalyst, organic synthesis. [BP Chem. Ltd.; Enichem Am.; Janssen Chimica; Mitsui Petrochem. Ind.; Mitsui Toatsu Chem.; Montedipe SpA; Penta Mfg.]

Acetoquat CPB. CAS 140-72-7; Cetyl pyridinium bromide; germicide, sanitizing agent. [Aceto] *

Acetoquat CPC. CAS 123-03-5; EINECS 204-593-9; Cetyl pyridinium chloride; germicide, sanitizing agent. [Aceto]*

Acetoquat CTAB. CAS 57-09-0; EINECS 200-311-3; Cetrimonium bromide; germicide, sanitizing agent. [Aceto]*

Acetosol. CAS 79-34-5; (Westrol). A trade name for tetrachloroethane. §

Acetosulfone. Promacetin. [Parke-Davis] *

AcetOxyl 2.5 and 5. Contains benzoyl peroxide in two strengths (2.5% or 5%) in an aqueous gel base; as an aid in the treatment of acne vulgaris. [Stiefel Laboratories (UK) Ltd] *

acetoxyphenylmercury. See phenylmercuric acetate

---

‡ Trade name and manufacturer not verified   § Trade name without identified manufacturer

Acetron®. Material for bearing and wear applications with water/moisture present. [Polymer Corp]

Acetron®GP. CAS 105-57-7; Acetal produced from copolymer and homopolymer resins; low moisture absorp., machinability, high stiffness, and modulus of elasticity; used in bearings, gears, antifriction parts, elec. components, etc.; strength at above-normal temp. or moisture levels; avail. in rod, plate, bar, film, etc. [Polymer Corp.]

Acetron® NS. CAS 105-57-7; Acetal compound with solid lubricants; for bearing and wear applications, e.g., bearings, bushings, valve seats, seals, wear surfaces, rollers, gears, cams, liners, tooling fixtures, forming dies. [Polymer Corp.]

Acetryptin . 5-Acetyltryptamine; a proprietary antihypersensitive agent. [Warner-Chilcott] *

Acetulan®. Cetyl acetate, acetylated lanolin alcohol; binder for pressed powds.; emollient, plasticizer, cosolv., NV and sebum solv. for personal care products; lubricant for clay, talc, and starch; stabilizer for lanolin; solubilizer in aerosols; penetrant and spreading agent. [Amerchol Corp]

Acetyl. Anionic dyestuffs (level dyeing); used for wool and wool blends. [Holliday Dyes & Chemicals Ltd] †

acetylacetone. (2,4-pentanedione) $CH_3COCH_2COCH_3$; CAS 123-54-6; EINECS 204-634-0; solvent for cellulose acetate; intermediate; chelating agent for metals, paint driers, lubricant additives, pesticides. [Aldrich; Penta Mfg.; Union Carbide; Wacker Chem.]

Acetylarsan. Dimethylamine acetarsol. [May & Baker Ltd]*

acetylcellulose. See cellulose acetate

acetyl chloride. (ethanoyl chloride) CH_3COCl; CAS 75-36-5; EINECS 200-865-6; acetylating agent for organic preparations; dyestuffs; pharmaceuticals. [Hoechst-Celanese; Penta Mfg.; Saurefabrik Schweizerhall]

acetyl-choline hydrochloride. (acecholine); Used for lowering blood pressure and increasing intestinal movement.

acetyl-dinitro-butyl-xylene. See ketone musk.

acetylene. (ethyne) An asphyxiant gas; HC≡CH; CAS 74-86-2; intermediate for manufacture of vinyl chloride, vinylidene chloride, vinyl acetate, acrylates, acrylonitrile, acetaldehyde, perchloroethylene, trichloroethylene, 1,4-butanediol, carbon black, welding and cutting metals. [Air Prods.; BASF; DMS NV; Union Carbide]

acetylene black. (Shawinigan black). CAS 1333-86-4; A carbon black made by incomplete combustion of acetylene.

acetylformic acid. See pyruvic acid

acetyl methyl carbinol. (acetoin; 3-hydroxy-2-butanone) $CH_3COCH(OH)CH_3$; CAS 513-86-0; EINECS 208-174-1; Aroma carrier; preparation of flavors and essences. [BASF; Penta Mfg.]

acetyl oxide. See acetic anhydride

acetylpropionic acid. See levulinic acid

2-acetyl pyridine. (methyl 2-pyridyl ketone) C_7H_7NO; CAS 1122-62-9; EINECS 214-355-6. [Aldrich; Penta Mfg; Raschig; Reilly Ind.; Schweizerhall]

acetyl tributyl citrate. (2-(acetyloxy)-1,2,3-propane tricarboxylic acid, tributyl ester; 1,2,3-propanetricarboxylic acid, 2-acetyloxy)-, tributyl ester) Aliphatic ester; $CH_3COOC_3H_4(COOC_4H_9)_3$; CAS 77-90-7; plasticizer for vinyl resins. [Morflex; Pfizer Spec.; Unitex]

Acetysal. See aspirin. §

Acheson's Deflocculated Graphite. A lubricant obtained by macerating graphite with a solution of tannin for several weeks, forming a permanent emulsion; the graphite with water is called Aquadag; Oildag is prepared by pouring oil over the filtered dag, then freeing the material from moisture. [Acheson Colloids] *

Achilles Dipentene. CAS 138-86-3; Dipentene, commercial grade; solvent for resins, waxes and oils, perfumery, wetting agent and antiskinning for paint. [Langley Smith & Co Ltd]

Achilles Pine Oil. Pine oil (mixed isomers of terpene alcohol); disinfectants/cleaners, plasticizer for epoxy resin, solvent and leveller in paint formulations, wetting agent for pigments. [Langley Smith & Co Ltd]

Achilles Tall Oil Fatty Acid. Oleic acid/linoleic acid mixture with a rosin acid content; used in alkyd resins, detergents, disinfectants, soaps and core oils. [Langley Smith & Co Ltd]

Achromycin. A proprietary preparation containing tetracycline hydrochloride; an antibiotic. [Lederle Laboratories]

Aciculite. CAS 12141-46-7; Acicular aluminum silicate; high length-to-thickness ratio in naturally occurring mineral; reinforcement for plastic systems, coatings, caulks, sealants, and mastics. [Kaopolite]

acid, acetosalic. See aspirin.

Acid Aid 5LXS-IH. Lowers surface tension between steel and acid yielding faster pickling rates; protects base metal; excellent detergency props.; reduces HCl acid consumption. [Crown Tech.]

Acid Aid X. Accelerator and extender for hydrochloric and sulfuric acids; detergency props.; for metal cleaning, pickling. [Crown Tech.]

Acidan. CAS 97593-31-2; Monoglyceride citric acid ester; emulsifier, surfactant; used in food industry. [Grindsted Prods.; Grindsted Prods. Denmark]

Acidan N 12. CAS 68990-59-0; EINECS 273-613-6; Hydrogenated tallow glyceride citrate. [Grindsted Prods.] *

Acidax. A stearic acid for use in the rubber industry. §

acid bronze. Alloys containing from 82-88% copper, 8-10% tin, 2-8% lead, and 0-2% zinc. A metal containing 90% copper and 10% aluminum is also known as acid bronze.

acid calcium phoshate. See calcium phosphate (monobasic)

Aciderm®. Dyestuffs; acid dyes for all kinds of leather. [Bayer AG; Bayer plc]

Acid Felt Scour. Specially formulated product; pulp and paper felt cleaner for batch and continuous processing. [Hart Chem. Ltd.]

Acid Foamer. Quaternary ammonium chloride; surfactant; foaming agent for strong acids; used in aluminum trailer cleaner, brightener, acid inhibitors and cleaners, chrome plating baths. [Exxon/Tomah]

acid fuchsine. See acid magenta

acid magnesium citrate. See magnesium citrate, dibasic

acid of amber. Succinic acid, CH_2CH_2COOH.

acid of lemons. Citric acid.

acid of sugar. Oxalic acid.

Acidol®. Acid dyes for dyeing and printing wool, polyamide, silk. [BASF]

Acidol®M. Sulfo-group containing 1:2 metal complex dyes for dyeing and printing wool, polyamide, silk. [BASF]

Acidol-Pepsin. A proprietary preparation of betaine hydrochloride, and pepsin; used in the treatment of achlorhydria gastritis. [Bayer AG] †

acid tar. The waste acid from the washing of crude light oils

---

* Trade name not verified as available   † Trade name verified as obsolete

of coal tar.

Acid Thickener. Surfactant; visc. builder, wetting agent, corrosion inhibitor for acid-based cleaners, e.g., acid bowl cleaners, truck cleaners, building restoration cleaners; perfume solubilizer. [Exxon/Tomah]

Acihib A9. Amine deriv.; controls corrosion in HCl, sulfuric, phosphoric, and other acids; acid restrainer. [ICI Australia] *

Acilan. Acid wool dyestuffs with good leveling power. [Bayer AG]

Acilef. *See* Zithromax. [Pfizer International]

Acillin. CAS 69-53-4; Ampicillin; antibacterial. [ICN Nutritional Biochemicals Corp] *

acinitrazole. (aminitrazole. trichorad. tritheon) 2-Acetamido-5-nitro-thiazole.

Acintene®. Terpene products including α–pinene, β-pinene middle boilers containing dipentene, technical grade anctholc. α pinene: solvent, terpene resin monomer, intermediate for flavour and fragrance chemicals, intermediate for metal lubricant additive manufacture, camphor and camphene manufacture, petroleum extraction chemical. β-pinene: terpene resin monomer, intermediate for flavor and fragrance chemicals, solvent; middle-boilers containing dipentene: pine oil extender, high kauri butanol value solvent; technical grade anethole: anise flavor intermediate. [Arizona]

Acintol® **208.** Tall oil acid isooctyl ester; plasticizer for chloroprene, nitrile and Hypalon® elastomer systems. [Arizona]

Acintol®**736.** CAS 61790-12-3; EINECS 263-107-3; Tall oil acid; emulsifier for styrene/butadiene rubber polymerization. [Arizona]

Acintol®**2122.** CAS 61790-12-3; EINECS 263-107-3; Tall oil heads; surfactant. [Arizona]

Acintol®**D25LR.** CAS 61790-12-3; EINECS 263-107-3; Tall oil acid, distilled; surfactant for asphalt emulsifiers, concrete form release and air entraining agents, metalworking fluids, varnishes, printing inks, soaps, cleaners, degreasers. [Arizona]

Acintol®**D30E.** CAS 61790-12-3; EINECS 263-107-3; Dist. tall oil fatty acid; surfactant for asphalt emulsifiers, concrete form release and air entraining agents, metalworking fluids, varnishes, printing inks, soaps, cleaners, degreasers. [Arizona]

Acintol®**DFA.** CAS 61790-12-3; EINECS 263-107-3; Tall oil acid; for metalworking fluids, lubricant additives, oilfield chems., asphalt emulsifiers, alkyd resins, industrial/household cleaners, plasticizers, textile drawing lubricants, surface coatings, rubber products. [Arizona]

Acintol®**EPG.** CAS 61790-12-3; EINECS 263-107-3; Tall oil acid; surfactant for metalworking fluids, lubricant additives, oilfield chems., asphalt emulsifiers, alkyd resins, industrial/household cleaners, plasticizers, textile drawing lubricants, surface coatings, rubber products; epoxy grade. [Arizona]

Acintol®**FA-1.** CAS 61790-12-3; EINECS 263-107-3; Tall oil acid; surfactant for metalworking fluids, lubricant additives, oilfield chems., asphalt emulsifiers, alkyd resins, industrial/household cleaners, plasticizers, textile drawing lubricants, surface coatings, rubber products. [Arizona]

Acintol®**FA-2.** CAS 61790-12-3; EINECS 263-107-3; Tall oil acid; surfactant for metalworking fluids, lubricant additives, oilfield chems., asphalt emulsifiers, alkyd res-

ins, industrial/household cleaners, plasticizers, textile drawing lubricants, surface coatings, rubber products. [Arizona]

Acintol®**Liquaros.** Tall oil rosin; pigment dispersing agent, fortified paper size, compounding cutting oils and buffing materials, surface coatings, printing ink vehicles and base stock for surfactants. [Arizona]

Acintol®**R Type 3A.** Tall oil rosin; printing ink binder as resin or salt, paper sizing agent, emulsifier for styrene/butadiene rubber polymerization as soap, tackifier resin in adhesives, imidazoline modifier in corrosion inhibitors, elastomer modifier in emulsion polymerization, dust control additive; film former/plasticizer in lacquers and varnishes. [Arizona]

Acintol®**R Type LO-3A.** Tall oil rosin. [Arizona]

Acintol®**R Type S.** Tall oil rosin; printing ink binder as resin or salt, paper sizing agent, emulsifier for styrene/butadiene rubber polymerization as soap, tackifier resin in adhesives, imidazoline modifier in corrosion inhibitors, elastomer modifier in emulsion polymerization, dust control additive; film former/plasticizer in lacquers and varnishes. [Arizona]

Acintol®**R Type SB.** Tall oil rosin; printing ink binder as resin or salt, paper sizing agent, emulsifier for styrene/butadiene rubber polymerization as soap, tackifier resin in adhesives, imidazoline modifier in corrosion inhibitors, elastomer modifier in emulsion polymerization, dust control additive; film former/plasticizer in lacquers and varnishes. [Arizona]

Acintol®**R Type SFS.** Tall oil rosin, formaldehyde-treated. [Arizona]

Acintol®**R Type SM4.** Tall oil rosin, maleated. [Arizona]

Aciplex®. Ion exchange membrane. for food industry. [Asahi Chem. Industry]

Aciquel. Potassium glucaldrate; antacid. [McNeil Pharmaceuticals] *

Acithrocin. *See* Zithromax. [Pfizer International]

ACL. Chlorinated s-triazine triones; machine dishwashing compounds, disinfection of public and private swimming pools, household and industrial cleaners, bleaching agents for both domestic and industrial washing, formulation of industrial bacteriocides. [Monsanto Co]

ACL 56. CAS 2244-21-5; Sodium dichloroisocyanurate dihydrate; bleaching compound, sanitizer, disinfectant. [Monsanto Co]

ACL 59. CAS 2244-21-5; Potassium dichloroisocyanurate; used for swimming pool disinfection and sanitizing; used as oxidizer in mechanical dishwashing compounds and other detergent applications. [Monsanto Co]

ACL 60. CAS 2244-21-5; Sodium dichloroisocyanurate; bleaching compound, sanitizer, disinfectant, oxidizer in dishwashing compounds. [Monsanto Co]

ACL 66. (Monotrichloro) tetra (monopotassium dichloro) pentaisocyanurate; bleaching compound, sanitizer, detergent. [Monsanto Co]

ACL 85. CAS 87-90-1; Trichloroisocyanuric acid; bleaching compound, sanitizer, disinfectant, detergent. [Monsanto Co]

ACL 90 Plus. CAS 87-90-1; Trichloroisocyanuric acid, 90% chlorine product; used for pool disinfection in compacted forms, commercial bleaches and scouring powders. [Monsanto Co]

Aclarat 8678 Granules, Liquid. Fluorescent whitener for commerical/industrial laundry detergents, rug/uphol-

---

‡ Trade name and manufacturer not verified   § Trade name without identified manufacturer

stery cleaners, fabric softeners, laundry bleach, whitening soap, brightening polymers and plastics; especially for synthetics and wool. [Sandoz]

Aclar Films. Flurocarbon thermoplastic; high moisture barrier flexible film for pkg. and container liners for pharmaceutical and cosmetic pkg., clean room, DOD, electronic applications; extruded films in varying widths and thicknesses; optically clear. [Allied-Signal] *

Aclon PCTFE. Homopolymers and copolymers of chlorotrifluoroethylene. [Allied-Signal Inc]

AClyn® 201A. Ethylene/calcium acrylate copolymer; low molecular weight ionomers used as processing and performance additives; improves dispersion of additives in plastics; adhesion to variety of substrates. [Allied-Signal]

AClyn® 246A. Ethylene/magnesium acrylate copolymer; low molecular weight ionomers used as processing and performance additives; improves dispersion of additives in plastics; adhesion to variety of substrates. [Allied-Signal]

AClyn® 250. CAS 9010-77-9; Low molecular weight ethylene-acrylic acid copolymer, Mg ionomer; alkali-dispersible additive for recyclable hot-melt and aq. adhesives and coatings. [Allied-Signal/A-C® Perf. Addit.]

AClyn® 262. CAS 9010-77-9; Low molecular weight ethylene-acrylic acid copolymer, Na ionomer; water-dispersible additive for recyclable hot-melt and aq. adhesives and coatings. [Allied-Signal/A-C® Perf. Addit.]

AClyn® 296. CAS 9010-77-9; Low molecular weight ethylene-acrylic acid copolymer, Zn ionomer; alkali-dispersible additive for recyclable hot-melt and aq. adhesives and coatings. [Allied-Signal/A-C® Perf. Addit.]

Acne-Aid Detergent Soap. A blend of high molecular weight fatty acids and selected detergents and contains sulfated surfactant blend; as an aid in the management of acne and any condition where greasy skin predominates. [Stiefel Laboratories (UK) Ltd] *

Acnidazil. A proprietary preparation of miconazole nitrate and benzoyl peroxide; for acne vulgaris. [Janssen Pharmaceutical Ltd] *

Acnil. A proprietary preparation of resorcinol, precipitated sulfur, and cetrimide; an acne treatment. [Fisons plc, Pharmaceuticals Div] *

Acofor. CAS 61790-12-3; EINECS 263-107-3; Dist. tall oil fatty acids; latex stabilizer, dispersant (as soap) for pigments and fillers. [Reichhold] *

Aconol X6. CAS 61791-00-2; PEG-6 tallate; emulsifier for mineral oil. [Hart Chem. Ltd.]

Acopyrin. Antipyrine acetylsalicylate; an antipyretic. §

Acorga. Range of chemicals; used in the extraction and treatment of metals and metalloids. [Acorga Ltd]

acorn sugar. Quercitol, $C_6H_7(OH)_5$, found in acorns.

ACP 99. Acetphenone; solvent. [Arco Chemical Co]

Acpol. Polyester resin. [Freeman Chemical Corp] *

A-C®Polyethylene 6, 6A, 7, 7A, 8, 8A, 9, 9A. CAS 9002-88-4; Polyethylene wax; processing lubricant, melt index modifier, pigment dispersant, mold release aid; external lubricant PVC, color concs., polyolefin flow modifiers; thickener for cosmetic and pharmaceutical gels. [Allied-Signal] *

A-C® Polyethylene 316, 316A, 325, 330, 392, 395. Oxidized polyethylene; wax for polishes, finishes, and emulsions; heel-mark resistance and compatibility. [Allied-Signal] *

A-C®Polyethylene 617, 617A. CAS 9002-88-4; Polyethylene; see A-C Polyethylene 6. [Allied-Signal] *

A-C® Polyethylene 629, 680. Oxidized polyethylene; wax for polishes, finishes, and emulsions; slip resistance; emulsifiable, atmospheric. [Allied-Signal] *

A-C® Polyethylene 629A. Oxidized low molecular weight polyethylene; processing lubricant, mold release aid; PVC lubricant. [Allied-Signal] *

A-C® Polyethylene 6702. Oxidized polyethylene. [Allied-Signal] *

acquerite. A native alloy of silver and mercury, found in Chile, approximating to the formula $Ag_{12}Hg$.

Acquit®. For agriculture industry. [DuPont UK]

Acra-500. Fatty amine complex; asphalt wetting and antistripping agent. [Exxon/Tomah]

Acraconc®. A range of printing concentrates; dispersion thickeners for printing systems with a low white spirit content or without white spirit. [Bayer AG]

Acraconz®. Printing concentrates; dispersion thickeners for printing systems containing little or no white spirit. [Bayer AG]

Acrafil®. Flame retardant styrene-acrylonitrile. §

Acrafloc®. Range of binders; for flocking of textiles. [Bayer AG; Bayer plc]

Acralen® A. Aqueous dispersions of polyacrylic resins; suitable for impregnated or coated substrates that must be fast to light; especially suitable as binders for nonwoven fabrics and padding materials. [Bayer AG]

Acralen® AFR. Acrylic acid ester, acrylonitrile; self-crosslinking binder for nonwoven fabrics, textile finishing, lamination. [Bayer AG]

Acralen® ATR. Acrylic acid ester, styrene latex; self-crosslinking binder for nonwoven fabrics, lamination. [Bayer AG] *

Acralen® BS. Butadiene, styrene, acrylonitrile latex; self-crosslinking binder for nonwoven fabrics, tech. fabrics. [Bayer AG] *

Acralux®. Auxiliary to increase the lightfastness of polyamide dyeings. [Bayer AG]

Acramin® Binders. Auxiliary for pigment printing. [Bayer AG]

Acramin®Dyestuffs. Pigment dyestuffs; for the pad dyeing of fabrics made of cellulosics, synthetic fibers and their blends. [Bayer AG; Bayer plc]

Acrawax®. A proprietary synthetic wax for lubrication and blending with other waxes to increase their melting point. [Lonza Inc]

Acrawax® B. Amide wax; synthetic wax and plastic lubricants. [Lonza Inc] *

Acrawax® C. CAS 110-30-5; EINECS 203-755-6; N,N´-Ethylene bisstearamide; internal and surface lubricant in resins and plastics; processing aid; plasticizer for resin; flow improver; pigment dispersant; used in hot-melt adhesives and coatings; powd. grade used as lubricant, processing aid, detackifier, mold release, and antiblocking agent. [Lonza Inc]

Acrex. CAS 973-21-7; Dinobuton. [Murphy Chemical Co Ltd] †

Acriflex. CAS 18472-51-0; A proprietary preparation containing chlorhexidine gluconate; an antiseptic cream. [The Boots Co plc]

Acrilan. Acrylic fiber; for sweaters, handcraft yarns and carpets. [Monsanto Co] *

Acrilester. Two part polyester coating resin which is water-

---

based and catalysed by the addition of equal parts of the two components; high gloss overcoat for a variety of surfaces which exhibits extremely high heat resistance, scuff resistance and smooth durable surface. [ADM Tronics Unlimited Inc]

Acrilev ADK Special. Potassium salt of phosphate ester; caustic stable wetting agent, leveling agent, scouring agent. [Finetex]

Acrilev AM, AM-Special. Phosphate ester, potassium salt; detergent, wetter, dye leveler used in textiles. [Finetex]

Acrilev OJP-25N. Alkylalkoxylated phosphate ester, sodium salt; emulsifier, detergent, wetting agent, dispersant. [Finetex]

Acrilpact. Acrylic sealant; sealant in building. [Siliconas Hispania SA]

Acrisint 400, 410, 430. Carbomer. [3V-Sigma] *

Acritamer 934, 940, 941. Carbomer; suspending and visc. agent. [RITA]

Acrodel. An insecticide preparation. [ICI Chem. & Polymers Ltd]

acroleic acid. See acrylic acid

Acrolex. Acrylic resin. [Ferro/Engineering Thermoplastics]

Acronal®. Acrylate homo- and copolymers; binders for paper and board coating; binders and coating agents for production of materials based on leather fibers. [BASF AG; BASF plc]

Acronal® 14D. A plasticizer-free acrylic copolymer in the form of a 55% dispersion; adhesive. [BASF plc]

Acronal® 21D, 27D, 30D. Plasticizer free dispersions of thermosetting copolymers of various acrylic esters; adhesive. [BASF plc]

Acronal® 160D. A plasticizer-free acrylic copolymer in the form of a 40% dispersion; adhesive. [BASF plc]

Acronal® 350D. A plasticizer free acrylic copolymer in the form of a 45% dispersion; adhesive. [BASF plc]

Acrosol®. Acrylic acid ester copolymers; cobinders for paper coating. [BASF AG]

Acrosyl. A saponified cresol disinfectant. §

Acry-Ace. Polymethyl methacrylate; a proprietary molding powder. [Fudow] *

Acrycal® MP CP-41. Acrylic resin; injection molding resin with lowest heat resist. and max. flow. [Continental Polymers]

Acrycal® MP CP-924. Acrylic resin, impact-modified; injection molding grade with good impact strength, excellent light transmission; for medical, appliance, industrial, personal accessory, and RV exterior applications. [Continental Polymers]

Acrycal® MP CP-1000-E. Acrylic resin, impact-modified; extrusion grade with high impact strength; excellent blend stock material; for medical, appliance, industrial, personal accessory, and RV exterior applications. [Continental Polymers]

Acrydur. Reactive acrylic resin systems; for seamless, industrial and commercial floorings. [Ulfcar International A/S] *

Acryl. Dyestuffs for acrylic fibers. [BASF plc] *

Acrylafil G-40/20/FR and G-40/30/FR. Proprietary flame-retardant grades of styrene-acrylonitrile, reinforced with glass fiber. [Dart Industries Inc] *

Acrylamide. Acrylamide (2-propenamide); available as solid, 50% or 30%; used in the process industry. [Cyanamid BV]

Acryl/bis. Solution or powder of acrylamide/bis-acrylamide.

[Am. Research Prods.] ‡

acrylic acid. (acroleic acid; propenoic acid; ethylene carboxylic acid) $H_2C:CHCOOH$; CAS 79-10-7; EINECS 201-177-9; Monomer for polyacrylic and polymethacrylic acids, other acrylic acids, acrylic polymers. [BASF; Hoechst-Celanese; Hüls UK; Penta Mfg.; Rohm & Haas; Union Carbide]

acrylic acid 2,3-epoxypropyl ester. See glycidyl acrylate

acrylic acid, 2-ethylhexyl ester. See octyl acrylate

acrylic acid methyl ester. See methyl acrylate (monomer)

acrylic resin. See polyacrylate

Acrylic Resin AS. Acrylic copolymer aq. disp.; finishing agent giving full supple hand on all fibers. [ICI Surf. UK]

acrylite. See methyl methacrylate polymer

Acrylite®. Acrylic sheet, and molding and extrusion compounds. [Cyro Industries] *

Acrylite® FF. Acrylic sheet; clear, lightweight, rigid, dimensionally stable, and weather-resistant thermoplastic for use in skylights, recreational vehicles, boat and motorcycle fairings, signs, displays, boutique items; avail. in thicknesses from 1.5-12.7 mm. [Cyro Industries]

Acrylite® GP. Cast acrylic sheet; used in industrial plants and building products, e.g., window glazing, skylights, tub enclosures. [Cyro Industries]

Acrylite® H. Proprietary acrylic pellets; used for injection molding or extrusion. [Cyro Industries] *

Acrylite® M. Proprietary acrylic pellets providing medium heat resistance; used in injection molding and extrusion where greater flow is required. [Cyro Industries] *

Acrylo-40. A 40% solution of acrylamide. [Am. Research Prods.] ‡

Acryloft, Acryloft Conc. Modified quaternary ammonium compound; textile softener. [Rhone-Poulenc] *

Acryloid®. Acrylic ester resin; used for surface coatings. [Rohm & Haas]

Acryloid® 150. Alkyl methacrylate copolymer in solv. refined neutral oil; pour-pt. depressant for use in motor oils, gear lubricants, hydraulic fluids, other lubricant applications; dewaxing aid in solv. dewaxing processes for manufacturing of lubricating oils. [Rohm & Haas]

Acryloid® 702. Methacrylate copolymer; visc. index improver and pour pt. depressant for gasoline and diesel engine oils, hydraulic fluids, industrial lubricants. [Rohm & Haas]

Acryloid® A-21. Methyl methacrylate polymer in toluene/butanol (90/10); thermoplastic resin for automotive finishing, vinyl clear coatings, plastisol primers, specification lacquers, rail car and aircraft finishes. [Rohm & Haas]

Acryloid® A-21LV. Methyl methacrylate polymer in toluene/MEK/butanol (50/40/10); thermoplastic resin used in bake-sand-bake finishes, auto refinishes, aircraft and rail care finishes, vinyl clear coatings, plastisol primers. [Rohm & Haas]

Acryloid® A-30. CAS 9011-14-7; Methyl methacrylate polymer; thermoplastic resin used in vinyl topcoating with vinyl copolymers and homopolymers. [Rohm & Haas]

Acryloid® A-101. 40% Solids methyl methacrylate polymer in MEK; thermoplastic resin used for vinyl topcoating and printing inks. [Rohm & Haas]

Acryloid® AT-51. Hydroxyl-type acrylic polymer, thermosetting; thermosetting resin for applications requiring hardness and chemical, stain, and solv. resistance, e.g.,

---

‡ Trade name and manufacturer not verified   § Trade name without identified manufacturer

product finishing, light appliance finishing. [Rohm & Haas]

Acryloid® AT-63. Hydroxyl-type acrylic polymer; thermosetting resin with fast cure; for general product finishing, exterior coil coating. [Rohm & Haas]

Acryloid® B-44. CAS 9011-14-7; Methyl methacrylate copolymer; thermoplastic resin for finishes for vacuum-metallized plastics, aluminum sash and chrome plate, luminescent finishes, metal and masonry clear coatings, farm and factory machinery, product finishing. [Rohm & Haas]

Acryloid® B-67. Isobutyl methacrylate polymer in VM&P naphtha; thermoplastic resin which improves properties of alkyd and oleo-resinous varnishes; used in enamels, aerosols, clear coatings, farm and factory machinery, fluorescent pigment disp., plastic top coats, marine finishes. [Rohm & Haas]

Acryloid® B-99. Methyl methacrylate copolymer, xylene, toluene (70/30); thermoplastic resin used in pigment grind to improve gloss; used for refinish lacquer maintenance parts, aerosols, farm and factory machinery, pigment disp., printing inks, product finishing, auto refinishes, marine finishes. [Rohm & Haas]

Acryloid® F-10. Acrylates copolymer (butyl methacrylate polymer), min. thinner/Aromatic 100 (90/10); thermoplastic resin used in clear and pigmented coatings on metal, luminescent pigment vehicles, aerosols, printing inks. [Rohm & Haas]

Acryloid® WR-97. Hydroxyl-type acrylic polymer; thermosetting sol'n. resin reducible with water or solv.; crosslinkable with urea and melamine resins; pigment dispersant; also for exterior coil coating, product and appliance finishing. [Rohm & Haas]

acrylonitrile. (propenenenitrile; vinyl cyanide; cyanoethylene) $H_2C:CHCN$; CAS 107-13-1; EINECS 203-466-5; Monomer for acrylic and modacrylid fibers; in production of ABS and acrylonitrile styrene copolymers. [Aldrich; Am. Cyanamid; Asahi Chem Industry Co Ltd; BP Chem. Ltd; DSM NV; Du Pont; Mitsui Toatsu Chem.]

acrylonitrile-butadiene rubber. (NBR; acrylonitrile rubber; acrylonitrile-butadiene copolymer; nitrile rubber) Synthetic rubber. $[CH_2CH=CHCH_2CH_2CH(CN)]_x$; used for oil well parts, general-purpose oil-resistant applications, gaskets, grommets, o-rings.

acrylonitrile-butadiene-styrene. (ABS; acrylonitrile polymer with 1,3-butadiene) Thermoplastic resin grafted from the 3 monomers from which its name is derived; CAS 9003-56-9; used for automotive body parts, telephones, pkg., shower stalls, etc. [Aiscondel SA; Bamberger Polymers; BASF; LNP; Mitsui Toatsu Chem.; Monsanto; Reichhold/Emulsion Polymers]

Acrylweld. Two-part acrylic adhesive. [Hardman]

Acrymul AM 123R. Acrylates copolymer; a proprietary trade name for a self cross-linking acrylic emulsion. [Protex] *

Acrysol®. Acrylic and urethane thickeners; used for coatings. [Rohm & Haas; Rohm & Haas UK]

Acrysol®ASE-60. Crosslinked acrylic emulsion copolymer; used to suspend pigments and fillers in water-based paints, inks, or other coatings, and the abrasive particles in waxes or polishes; visc. modifier for emulsion and latex compounds; binder; thickener. [Rohm & Haas]

Acrysol®ASE-75. CAS 9003-01-4; Polyacrylic acid; thickener for latexes and emulsions for paints, flocking adhe-

sives. [Rohm & Haas]

Acrysol®ASE-95. Acrylic copolymer emulsion; thickener for fabric laminants, pigment disp. polar solvs., surfactant sol'ns., paints, inks, waxes, and polishes. [Rohm & Haas]

Acrysol®G-110. CAS 9003-03-6; Ammonium polyacrylate sol'n.; thickening and stabilizing agent for synthetic latices; used in coatings, adhesives, dipped, cast, and molded goods, cements for rug backing, spraying, spreading, brushing, and extruding compounds. [Rohm & Haas]

Acrysol®GS. CAS 9003-04-7; Sodium polyacrylate; thickener for natural and synthetic latexes for paints, films, coatings, and adhesives. [Rohm & Haas]

Acrysol®HV-1. CAS 9003-04-7; Sodium polyacrylate; all-purpose thickener for rubber rug backing, unholstery backing. [Rohm & Haas]

Acrysol®LMW. Polyacrylic acid homo/copolymer; detergent, water treatment and textiles. [Rohm & Haas]

Acrysol®TT-615. Acrylic copolymer emulsion; thickener for coatings, textile printing pastes, and adhesives. [Rohm & Haas]

Acrysol®WS-24. Acrylic copolymer. [Rohm & Haas]

Acrythane. Paint product available in all colors for automotive/commercial vehicle use; vehicle coating for original and refinishing; also for high quality chemically resistant coating of machinery. [H Marcel Guest Ltd] *

ACS. See styrene-acrylonitrile copolymer

ACS 60. CAS 37475-88-0; EINECS 253-519-1; Ammonium cumenesulfonate. [Witco SA] *

Acsil. Sodium orthosilicate, sodium sesquisilicate. [Crosfield Chemicals Ltd]

Acsium®. For polymer industry. [DuPont UK]

Acta. Throat drops. [Richardson-Vicks Inc] *

Actafoam®F-2. Activator-stabilizer for vinyl foams containing Kempore blowing agents; gas release accelerator. [Uniroyal]

Actafoam®R-3. Activator-stabilizer for vinyl foams containing Kempore blowing agents; gas release accelerator. [Uniroyal]

Actal. A proprietary preparation containing sodium polyhydroxy aluminum monocarbonate hexitol complex; an antacid; activated aluminas. [Bayer AG] †

Actal. Acetal; activated aluminas. [Laporte Absorbants]

Actan SP. Tannic acid conc.; corrosion inhibitor used in surface pretreatment for field applic.; base for applic. of coatings systems. [Troy] *

Actellic. CAS 29232-93-7; Pirimiphos-methyl; contact fumigant and organophosphorus insecticide which is available as a dust, liquid or smoke. [ICI Agrochemicals; ICI Chem. & Polymers Ltd]

Actellifog. CAS 29232-93-7; Pirimiphos-methyl; fumigant and insecticide for glasshouse crops. [ICI Agrochemicals; ICI Chem. & Polymers Ltd]

ACter 1450, 1450A. Ethylene/acrylic acid/vinyl acetate copolymer. [Allied-Signal] *

Acterol. An irradiated ergosterol, a vitamin D concentrate. §

ACTH 40. Corticotropin; hormone; glucocorticoid; diagnostic aid. [O'Neal, Jones & Feldman Pharmaceuticals] *

Acthar. Corticotropin; hormone; glucocorticoid; diagnostic aid. [Armour Pharmaceutical Co] *

Acthar Gel. A proprietary preparation of corticotrophin gelatin for injection. [Armour Pharmaceutical Co] *

Acticarbone. Powdered and granulated activated carbon;

---

* Trade name not verified as available
† Trade name verified as obsolete

used for purification, decolorization, deodorization, separation and recovery in liquid or gas phase, in the chemical, petrochemical, pharmaceutical and food industries (glucose factories, sugar refiners, oil refining, wine treatment); for the treatment of drinking and industrial water, etc.; catalyst supports. [Atochem UK/Ceca]

Acticide 50. Potassium orthophenyl phenate solution; sterilizing wash concentrate for surfaces prior to re-coating, wet-state protection of water-based compositions, prevention of fungal growth on leather in storage. [Thor Chemicals (UK) Ltd]

Acticide 50X. Blend of benzalkonium chloride and phenyl phenoxide; fungicide/algicide wash concentrate for sterilization of surfaces prior to re-coating and for industrial plant cleaning. [Thor Chemicals (UK) Ltd]

Acticide APA. Fluorinated sulfonamide-based powder available in standard and superfine forms; insoluble bactericide/fungicide for emulsion paints, textured coatings, renovating plasters, plasterboard jointing adhesives, etc. [Thor Chemicals (UK) Ltd]

Acticide AZ. Synergistic blend of aliphatic nitrogen and heterocyclic sulfur containing compounds; liquid; wet-state bactericide/fungicide for emulsion paints, adhesives, cellulose solutions, polymer emusions, fillers, pottery glazes, inks, etc. [Thor Chemicals (UK) Ltd]

Acticide CPC. Synergistic mixture of FDA approved chlorinated phenols; bactericide/fungicide for water-based products in the wet state, e.g., adhesives, carpet backing compounds. [Thor Chemicals (UK) Ltd]

Acticide DDM. CAS 97-23-4; Dichlorophen; bactericide/fungicide for textiles, cellulose solutions, proteins, adhesives and soaps; bactreicide/algicide for water treatment. [Thor Chemicals (UK) Ltd]

Acticide DDM/S. Dichlorophen sodium salt; bactericide/fungicide for textiles, cellulose solutions, proteins, adhesives and soaps; bactreicide/algicide for water treatment. [Thor Chemicals (UK) Ltd]

Acticide LG. Blend of chlorinated and nonchlorinated methyl isothiazolones with nonmetal salts stabilizing system; latex grade biocide for wet state protection of polymer emulsions and other aqueous based formulations. [Thor Chemicals (UK) Ltd]

Acticide MPM. A 10% mercury metal glycol solution of phenyl mercury nonane-2-ol; bactericide/fungicide for wet-state and dry-film protection of interior and exterior aqueous and solvent-based paints, woodstains, textured coatings, adhesive fillers, sealants, etc. [Thor Chemicals (UK) Ltd]

Acticide PMA 100. CAS 62-38-4; Phenyl mercury acetate powder; wet-state bactericide/fungicide for emulsion paints, plasters, wood-pulp, etc. [Thor Chemicals (UK) Ltd]

Acticide PMDDS. A 10% phenyl mercury dodecenyl succinate solvent solution; bactericide/fungicide for wet-state and dry-film protection of interior and exterior aqueous and solvent-based paints, woodstains, textured coatings, adhesives, fillers, sealants, etc. [Thor Chemicals (UK) Ltd]

Acticrom. Dry powder triazynil-type reactive dyes; used in the textile industry, mainly for dyeing cotton. [Multicrom SA]

Acticryl CL 959. Acrylic carbamate; radiation-curable monomer. [SNPE Chimie]

Acticulum. Conjugated glycopolypeptides; cosmetic products to help stimulate cell metabolism; strengthens skin against inflammation. [Active Organics] *

Actidil. CAS 6138-79-0; Proprietary formulations of triprolidine hydrochloride; tablets and syrup; for temporary relief of symptoms associated with allergic rhinitis and allergic skin conditions. [The Wellcome Foundation Ltd]

Acti-Dione. Cycloheximide; antipsoriatic. [Upjohn] *

Actif8®. Sodium aluminum phosphate, acidic with monocalcium phosphate, anhydrous; leavening agent for baking, cereals. [Rhone-Poulenc Food Ingreds.]

Actifed. Proprietary formulations of pseudoephedrine and triprolidine hydrochlorides; tablets, syrup; for temporary relief of upper respiratory symptoms, including nasal congestions, associated with allergy or common cold. [The Wellcome Foundation Ltd]

Actifed Compound Linctus/Oral Solution. A proprietary preparation containing pseudoephedrine hydrochloride, triprolodine hydrochloride and codeine phosphate; for temporary relief of coughs and upper respiratory symptoms including nasal congestion, associated with allergy or the common cold accompanied by a nonproductive cough. [The Wellcome Foundation Ltd]

Actifed DM Cough Elixir. A proprietary formulation of pseudoephedrine hydrochloride, triprolidine hydrochloride and dextromethorphan hydrobromide; for temporary relief of upper respiratory tract symptoms where nasal congestion is associated with a nonproductive cough. [The Wellcome Foundation Ltd]

Actifed Expectorant. A proprietary formulation of pseudoephedrine hydrochloride, triprolidine hydrochloride and guaifenesin; for temporary relief of upper respiratory tract symptoms where nasal congestion is associated with a productive cough. [The Wellcome Foundation Ltd]

Actifed Junior Cough Relief. A proprietary formulation of triprolidine hydrochloride and dextromethorphan hydrobromide; for temporary relief of nonproductive coughs in children. [The Wellcome Foundation Ltd]

Actiflo®68, 70. CAS 8002-43-5; EINECS 232-307-2; Natural lecithin; emulsifier, wetting agent, dispersant; used in the food industry. [Central Soya]

Actigen. Live cell animal extraction; used for cosmetic products (skin, body preparations and hair care products). [Active Organics] *

Actigen C. CAS 9007-34-5; Soluble animal collagen. [Active Organics] *

Actigen E. Hydrolyzed animal elastin. [Active Organics] *

Actigesic. A proprietary formulation of; for temporary relief of upper respiratory symptoms including nasal congestion associated with pyrexia and/or pain, e.g., the common cold, influenza, sinusitis and naso-pharyngitis. [The Wellcome Foundation Ltd]

Actiglow. Hydrolyzed mucopolysaccharides. [Active Organics] *

Actilex. Ion exchange materials. [Courtaulds plc]

Actilyse. A proprietary preparation of recombinant tissue-type plasminogen activator; a thrombolytic agent. [Boehringer Ingelheim Ltd]

Actimer FR-803. CAS 61368-34-1; Tribromostyrene; reactive monomer which can be copolymerized with styrene, acrylonitrile and maleic anhydride to impart flame retardancy to HIPS, ABS, SAN, SMA; crosslinking agent providing flame retardancy to thermoset systems.

---

‡ Trade name and manufacturer not verified§ Trade name without identified manufacturer

[AmeriHaas; Dead Sea Bromine]

Actimer FR-1025M. CAS 594477-55-11; Pentabromo-benzyl acrylate; flame retardant for engineering thermo-plastics (PET, PBT, PC, nylon 6 and 6/6); processing aid; maintains transparency of PC and HIPS resins. [AmeriHaas; Dead Sea Bromine]

Actimer FR-1033. CAS 59789-51-4; Tribromophenyl maleimide; reactive flame retardant for polymeric sys-tems which undergo crosslinking (XLPE, ethylene-pro-pylene-diene rubbers); can be grafted onto unsat. sites in ABS or copolymerized with styrenics to yield flame retardant systems with improved thermal props. [AmeriHaas; Dead Sea Bromine]

ActiMoist. CAS 9067-32-7; Sodium hyaluronate (hyaluronic acid); cosmetic products for high moisture retention/absorbtion. [Active Organics]

Actiphyte. A plant extract in propylene glycol water; for cosmetic products (skin, body preparations and hair care products). [Active Organics] *

Actiplex. Actiblend, a combination of actiphytes; for cos-metic products (skin, body preparations and hair care products). [Active Organics] *

Actipol® E6. CAS 9003-28-5; Activated polybutene; through grafting onto a base polymer, impact resistance, low temp. flexibility, and water resistance are enhanced in adhesives, sealants, coatings, unsat. polyesters, elec. compounds, foams, and other applications. [Amoco Chemical Co]

Actipron. Adjuvant containing 97% refined mineral oil; wetting agent for herbicides and fungicides. [Bayer plc]

Actipron. Adjuvant containing 97% refined mineral oil; wetting agent for herbicides and fungicides. [BP Oil Ltd]

Actiron NX 3. 2,4,6-Tri (dimethylaminomethyl) phenol; accelerator and hardener for epoxy resins. [Protex]

Actisize. Modified starch; used for adhesives and paper. [Roquette (UK) Ltd]

Activ-8®, Activ-8 in Hexylene Glycol. CAS 66-71-7; Solution forms containing 1,10-phenanthroline; drier ac-celerator and stabilizer used in combination with manga-nese and/or cobalt in coating systems that cure by oxidative polymerization. [R.T. Vanderbilt]

activated alumina. Partially dehydrated aluminum trihydrate in the form of hard porous lumps. has a strong affinity for water and will remove it from air.

activated carbon. See carbon, activated

activated charcoal. See carbon, activated

activated sludge. A material obtained by allowing the growth of microorganisms in the sludge deposited by sewage. it is used in the treatment of sewage.

Activator 736. Surface treated urea for easy dispersion in elastomers; activator for thiazole, thiuram and dithiocar-bamate accelerators; odor reducer when used with nitrosoamine type blowing agents. [Uniroyal] *

Activator 1102. Dibutyl ammonium oleate; accelerator/activator for natural and synthetic rubbers; lubricant. [Anchor] *

Activator STAG. Complex sec. amine, surface treated; relatively nonstaining, nondiscoloring activator for thiaz-ole-type accelerators; primary accelerator for natural rubber; strong sec. accelerator in styrene/butadiene rubber. [Akrochem]

Activax. A proprietary vaccine used in the treatment of fowl pox. [Coopers Animal Health Ltd] *

Active #2. Cocamide DEA. [Blew Chem.] *

Active #4. Cocamide DEA and DEA-dodecylbenzenesulfo-nate. [Blew Chem.] *

Active #18. CAS 93-83-4; EINECS 202-281-7; Oleamide DEA. [Blew Chem.] *

Activex. Fungicides for garden use. [ICI Garden Products]†

Activit. Thiocarbanilide; a proprietary rubber vulcanization accelerator; §

Activol. CAS 77-06-5; Gibberellic acid; a hormone growth regulator. [ICI Chem. & Polymers Ltd]

Activox. CAS 1314-13-2; Zinc oxides. [Durham Chemicals Ltd] *

Activox B. CAS 1314-13-2; Colloidal zinc oxide. [Harcros UK]

Acto 450, 500, 630, 632, 636, 639. Alkylaryl sodium sul-fonate; detergent, wetting agent, emulsifier, rust preven-tative. [Exxon] *

Actomol. A monoamine oxidase inhibitor containing mebanazine; antidepressant. [ICI Chem. & Polymers Ltd] †

ACtone® 1. Low molecular weight proprietary ionomer; pigment wetting agent, dispersion aid to increase color strength; used in thermoplastic and thermoset resins. [Allied-Signal/A-C® Perf. Addit.]

ACtone® 2010, 2010P. Low molecular weight proprietary ionomer resins; pigment wetting agent, dispersion aid for color concs. and masterbatches for polyester and styrenics. [Allied-Signal/A-C® Perf. Addit.]

ACtone® N. Low molecular weight proprietary ionomer; pigment wetting agent, dispersion aid, color enhancer for color concs. for nylon and polyester. [Allied-Signal/A-C® Perf. Addit.]

ACtone® P. CAS 9002-88-4; Low molecular weight branched polyethylene; vehicle for pigment pastes and presscakes; color enhancer. [Allied-Signal/A-C® Perf. Addit.]

Actos 50. A proprietary preparation of prolonged action adrenocorticotrophic hormone. [Consolidated Chemi-cals Ltd] *

Actrabase 31-A. Soap sulfonate; primary emulsifier for oil systems, metalworking fluids; base for naphthenic oils. [Climax Fluids Additives]

Actrabase 215. Emulsifier for metalworking fluids; low use level base for paraffinic oils. [Climax Fluids Additives]

Actrabase 264. Emulsifier for metalworking fluids; low use level base for naphthenic oils. [Climax Fluids Additives]

Actrabase PS-470. Med. molecular weight petrol. sulfonate, sodium salt; emulsifier and rust inhibitor for cutting and lube oils; dispersant for sol. oil and semi-syns. for metal-working fluids. [Climax Fluids Additives]

Actrabase SS-503. Semisynthetic conc. base, emulsifier for metalworking fluids. [Climax Fluids Additives]

Actracor 129, 856. Carboxylic acid amine salt; corrosion inhibitor for synthetic and semi-synthetic metal working fluids. [Climax Fluids Additives]

Actracor 401. Corrosion inhibitor for water-glycol hydraulic fluids. [Climax Fluids Additives]

Actracor 800. Hydrocarbon; corrosion inhibitor; metal work-ing additive. [Climax Fluids Additives]

Actracor 1987. Amine carboxylate; corrosion inhibitor for metalworking. [Climax Fluids Additives]

Actracor M. Ethanolamine-borate ester; corrosion inhibitor in cutting oils. [Climax Fluids Additives]

Actracor T. Triethanolamine borate ester; corrosion inhibi-tor in cutting oils. [Climax Fluids Additives]

---

* Trade name not verified as available   † Trade name verified as obsolete

Actrafoam A, B, C, S. Blend of glycols, fatty acids, and nonionic surfactants in a hydrocarbon base; general purpose defoamer (A, B); defoamer for water sewage applications (C, S). [Climax Fluids Additives]

Actrafos 104, 109. Phosphate ester; coupler for sulfated oils in cleaner formulations; lubricant for syn, semi-synthetic and water-based cutting, grinding, and drawing fluids. [Climax Fluids Additives] *

Actrafos 110, 110A. Complex aliphatic hydroxyl compound phosphate ester; pressure additive for cutting and rolling oils; hydrotrope for cleaning compounds; lubricant emulsifier and rust inhibitor; excellent for aluminum; 110A has higher melting point. [Climax Fluids Additives]

Actrafos 152A. Org. phosphate ester; extreme pressure lubricant for cutting oils; high phosphorus content. [Climax Fluids Additives]

Actrafos SA-216. Phosphate ester; lubricant and emulsifier for cutting oils. [Climax Fluids Additives]

Actrafos SN-315. Phosphate ester; lubricant, emulsifier for metal working; aluminum corrosion inhibitor. [Climax Fluids Additives]

Actrafos SP-407. Phosphate ester; low foaming lubricant and rust inhibitor for synthetic metalworking fluids. [Climax Fluids Additives]

Actrafos T. Tridecyl alcohol phosphate ester; extreme pressure lubricant and release agent for cutting oils. [Climax Fluids Additives]

Actrafos TDA. Org. phosphate ester; oil-sol. lubricant and release agent. [Climax Fluids Additives]

Actralube 21. Synthetic lubricant with moderate foaming props. [Climax Fluids Additives]

Actralube 100. Synthetic ester; lubricant for water-sol. cutting fluids. [Climax Fluids Additives] *

Actralube 310. Fatty ester; lubricity additive for cutting oil. [Climax Fluids Additives]

Actralube 1200. Modified triglyceride; metal lubricity additive. [Climax Fluids Additives]

Actralube 7142. CAS 8002-13-9; Blown rapeseed oil; metal lubricant additive. [Climax Fluids Additives]

Actralube SOS. Ester; emulsifier, lubricant, metalworking additive; substitute for sperm oil. [Climax Fluids Additives] *

Actralube Syn-147. Complex diester; lubricity additive for synthetic cutting fluids. [Climax Fluids Additives]

Actralube Syn-153. Soap; light duty synthetic lubricant and rust inhibitor for synthetic cutting fluids. [Climax Fluids Additives]

Actramide 176. Alkanolamide; cutting fluids. [Climax Fluids Additives] *

Actramide 202. 2:1 Tall oil fatty acid alkanolamide; emulsifier for sol. oils, metalworking fluids and emulsion cleaners; corrosion inhibitor. [Climax Fluids Additives]

Actramide 410. Alkanolamide; sec. emulsifier with lubricating properties. [Climax Fluids Additives]

Actramide 5264. Modified 2:1 tall oil fatty acid alkanolamide; emulsifier, lubricant, rust inhibitor. [Climax Fluids Additives]

Actran Extra. Lard oil; for industrial use. [Climax Fluids Additives]

Actrapid. Insulin; antidiabetic. [Bristol-Myers Squibb Co Inc] *

Actrapid Human. Human insulin, a protein that has the normal structure of the natural antidiabetic principle produced by the human pancreas; antidiabetic. [Bristol-

Myers Squibb Co Inc] *

Actrasol 167A. Sulfated castor oil, sodium neutralized; acid etching additive. [Climax Fluids Additives] *

Actrasol 6092. CAS 617788-68-9; Sulfated rapeseed oil; lubricant, emulsifier in pigment flushing, cleaners, textiles, paper processing. [Climax Fluids Additives]

Actrasol C-50, C-75, C-85. CAS 8002-33-3; EINECS 232-306-7; Sulfated castor oil; pigment wetting and disp.; lubricant and emulsifier for metalworking fluids; for pigment flushing, cleaners, textile, and paper processing. [Climax Fluids Additives]

Actrasol CS-75. Sulfonated soyabean oil, sodium neutralized; lubricant, emulsifier in pigment flushing, cleaners, textiles, paper processing. [Climax Fluids Additives]

Actrasol EO. Sulfated glyceryl trioleate, sodium neutralized; surfactant for shampoos, metalworking. [Climax Fluids Additives]

Actrasol KAP. Sulfated blend of oils, sodium neutralized; sperm oil substitute; fat liquor for leather, metalworking. [Climax Fluids Additives]

Actrasol MY-75. Sulfated methyl ester of soya fatty acid, sodium neutralized; lubricant and emulsifier in metalworking fluids, water-based drilling muds; oil field defoamer; biodeg. [Climax Fluids Additives]

Actrasol OY-75. Sulfated soyabean oil, sodium neutralized; lubricant, emulsifier in pigment flushing, cleaners, textiles, paper processing; biodeg. [Climax Fluids Additives]

Actrasol PSR. Sulfated ricinoleic acid, potassium neutralized; pigment wetting and disp.; aluminum lubricant; lubricant, emulsifier for pigment flushing, cleaners, textiles, paper processing. [Climax Fluids Additives]

Actrasol SBO. Sulfated butyl oleate, sodium neutralized; lubricant, emulsifier in pigment flushing, cleaners, textiles, paper processing. [Climax Fluids Additives]

Actrasol SP. Sulfonated tall oil fatty acid, sodium neutralized; wet process phosphoric acid defoamer; lubricant, emulsifier for pigment flushing, cleaners, textiles, paper processing. [Climax Fluids Additives]

Actrasol SP 175K. Sulfated tall oil, potassium neutralized; lubricant, emulsifier for pigment flushing, cleaners, textiles, paper processing; biodeg. [Climax Fluids Additives]

Actrasol SR 75. Sulfated oleic acid, ammonium neutralized; mold release agent; lubricant, emulsifier for pigment flushing, cleaners, textiles, paper processing; biodeg. [Climax Fluids Additives]

Actrasol SRK 75. Sulfated oleic acid, potassium neutralized; mold release agent; lubricant, emulsifier for pigment flushing, cleaners, textiles, paper processing; biodeg. [Climax Fluids Additives]

Actrasol SS. CAS 61790-35-0; Sulfated tall oil; rust preventative, lubricant, metal polish. [Climax Fluids Additives]

Actrel. Synthetic high boiling aromatic fluid. [Exxon Int'l.]

Actril. Selective weed killer. [May & Baker Ltd]

Actril S. Bromoxynil + dichloroprop + ioxyril + MCPA; broad-spectrum, post emergence contact and translocated herbicide. [Rhone-Poulenc Crop Protection Ltd]

Actrilawn. Selective weedkiller. [May & Baker Ltd]

Actrilawn 10. CAS 1689-83-4; Ioxynil; contact herbicide for use in turf. [Rhone-Poulenc Environmental Prods. Ltd]

Actrol 4DP. PEG diester; emulsifier and lubricity additive for metalworking fluids. [Climax Fluids Additives] *

Actrol 4MP, 628. PEG ester; emulsifier and lubricity additive for metalworking fluids. [Climax Fluids Additives] *

Actrol 6M25P. CAS 61791-00-2; PEG 600 tallate; low foaming surfactant with good lubricating chars.; adds emulsification, cleaning, hard water stability to sol. oils, cutting and grinding fluids, emulsion cleaners. [Climax Fluids Additives]

Acturin. See Diurexan. [Degussa AG] *

Actylon. Yarns and threads. [Courtaulds plc]

ACuflow AF-1. CAS 9010-77-9, CAS 300-92-5; Ethylene copolymer (CAS #9010-77-9) and aluminum stearate (CAS #300-92-5) mixture processing additive. [Allied-Signal]

Aculyn 22. Acrylates/steareth-20 methacrylate copolymer; thickener for cosmetics and toiletries (hair care products, hand creams, lotions, waterless hand cleaners). [Rohm & Haas]

Acumen. Bentazone + MCPA + MCPB; post-emergence contact and translocated herbicide for undersown cereals. [BASF plc]

Acumer. Water treatment polymers. [Rohm & Haas]

Acumer 1000. CAS 9003-01-4; Low molecular weight polyacrylic acid; scale inhibitor in industrial water treatment and oil production. [Rohm & Haas]

Acumer 5000. Polymer; scale inhibitor, dispersant for silica and magnesium silicate scale control; for water treatment, geothermal wells, reverse osmosis. [Rohm & Haas]

Acumer A-10. Acidic acrylic emulsion; flocculant, oil/water separator. [Rohm & Haas] *

Acumer C-3. Polyamine; flocculant. [Rohm & Haas] *

Acumer QR-1010. Acrylates copolymer; kaolin clay dispersant. [Rohm & Haas]

ACumist A-12, A-18. Oxidized polyethylene; for adhesives, inks, personal care, rubber applications. [Allied-Signal]

ACumist B-6, B-12, B-18, C-5, C-12, C-18. CAS 9002-88-4; Polyethylene; for adhesives, inks, personal care, rubber applications. [Allied-Signal]

Acupan. CAS 23327-57-3; Nefopam hydrochloride; analgesic. [3M Pharmaceuticals] †

Acusol®410N. CAS 9003-04-7; Sodium polyacrylate; detergent polymer, dispersant for cleaners, water treatment, mineral processing. [Rohm & Haas]

Acusol® 445. CAS 9003-01-4; Polyacrylic acid; detergent polymer for detergents and cleaners, water treatment, mineral processing, other industrial markets. [Rohm & Haas]

Acusol®445N. CAS 9003-04-7; Sodium polyacrylate; detergent polymer for detergents and cleaners, water treatment, mineral processing, other industrial markets. [Rohm & Haas]

Acusol® 445ND. CAS 9003-04-7; Sodium polyacrylate; detergent polymer for detergents and cleaners, water treatment, mineral processing, other industrial markets. [Rohm & Haas]

Acusol®460ND. Maleic acid/olefin copolymer, sodium salt; detergent polymer for detergents and cleaners, water treatment, mineral processing, other industrial markets. [Rohm & Haas]

Acusol®479N. Sodium acrylic acid/maleic acid copolymer; detergent polymer for detergents and cleaners, water treatment, mineral processing, other industrial markets. [Rohm & Haas]

Acusol® 479ND. Sodium acrylic acid/maleic acid copolymer; detergent polymer for detergents and cleaners,

water treatment, mineral processing, other industrial markets. [Rohm & Haas]

Acusol® 480N. CAS 9003-04-7; modified polyacrylic acid, sodium salt; detergent polymer for detergents and cleaners, water treatment, mineral processing, other industrial markets. [Rohm & Haas]

Acusol® 810. Acrylic crosslinked copolymer; detergent polymer, processing aid, thickener for detergents and cleaners, water treatment, mineral processing, other industrial markets. [Rohm & Haas]

Acusol® 820. Acrylic copolymer; detergent polymer, processing aid, thickener for detergents and cleaners, water treatment, mineral processing, other industrial markets. [Rohm & Haas]

Acusol® 830. Acrylates copolymer; detergent polymer, thickener for detergents, water treatment, mineral processing, other industrial markets. [Rohm & Haas]

Acusol® 840. Acrylic copolymer; detergent polymer, processing aid, thickener for detergents and cleaners, water treatment, mineral processing, other industrial markets. [Rohm & Haas]

Acusol®860N. CAS 9003-04-7; Sodium polyacrylate; detergent polymer, processing aid for detergents, water treatment, mineral processing, other industrial markets. [Rohm & Haas]

Acylan. CAS 61788-48-5; EINECS 262-979-2; Acetylated lanolin; lipid emollient for personal care and pharmaceutical products; forms water-repellent films. [Croda Chem. Ltd; Croda Inc]

Acylanid. Acetyl digitoxin; α–digitoxin monoacetate. [Sandoz] *

Acylglutamate AS-12. Sodium oleoyl glutamate, sodium cocoyl glutamate; good detergency for liquid dishwashing detergents, textile detergents, etc.; biodeg. [Ajinomoto] *

Acylglutamate CS-11. CAS 68187-32-6; EINECS:269-087-2; Sodium cocoyl glutamate; detergent, emollient for personal care products; bacteriostatic effect; biodeg. [Ajinomoto] *

Acylglutamate CS-21. Disodium cocoyl glutamate; detergent, emollient for personal care products; bacteriostatic effect; corrosion inhibitor; biodeg. [Ajinomoto] *

Acylglutamate CT-12. CAS 68187-29-1; EINECS 269-084-6; TEA N-cocoyl-L-glutamate; detergent, emollient for personal care products; bacteriostatic effect; biodeg. [Ajinomoto] *

Acylglutamate DL-12. Monolithium salt of N-distilled cocoyl-L-glutamic acid; foamer for carpet cleansers; easily becomes a powd. after drying. [Ajinomoto] *

Acylglutamate GS-11. Sodium hydrogenated tallow glutamate, sodium cocoyl glutamate; detergent, emollient for personal care products; bacteriostatic effect; biodeg. [Ajinomoto] *

Acylglutamate GS-21. Disodium cocoyl/tallowyl glutamate; basic material for heavy-duty detergents; reduces adverse reactions on human skin in toiletries; capturing agent of heavy metal ions; biodeg. [Ajinomoto] *

Acylglutamate HS-11. CAS 38517-23-6; Sodium hydrogenated tallow glutamate; detergent, emollient for personal care products; bacteriostatic effect; biodeg. [Ajinomoto] *

Acylglutamate HS-21. Disodium stearoyl glutamate; basic material for heavy-duty detergents; reduces adverse reactions on human skin in toiletries; capturing agent of heavy metal ions; biodeg. [Ajinomoto] *

Acylglutamate LS-11. Sodium lauroyl glutamate; detergent, emollient for personal care products; bacteriostatic effect; biodeg. [Ajinomoto] *

Acylglutamate LT-12. TEA lauroyl glutamate; detergent, emollient for personal care products; bacteriostatic effect; biodeg. [Ajinomoto] *

Acylglutamate MS-11. Sodium myristoyl glutamate; detergent, emollient for personal care products; bacteriostatic effect; biodeg. [Ajinomoto] *

AD-700. Surfactant/coupler/alkaline blend; biodeg. steam cleaner. [Anedco]

AD-709. Alkyl sulfonate; paraffin dispersant. [Anedco]

AD-710. Terpenes/amides/surfactants/solvent blend; paraffin dispersant to prevent or remove paraffin. [Anedco]

AD-713C. Diamines/amides/surfactant blend; conc. paraffin dispersant to remove or prevent paraffin. [Anedco]

AD-742C. Surfactant/coupler blend; biodeg. heavy-duty degreaser. [Anedco]

AD-749. Ethoxylated nonyl phenol and surfactant; drilling detergent. [Anedco]

AD-763. Pine oil, surfactants, couplers; for pine oil cleaners. [Anedco]

ADA. See β-ketoglutaric acid

Ad. A.M. A proprietary preparation of ephedrine hydrochloride and butethamate citrate; a bronchial antispasmodic. [Rybar Laboratories Ltd] *

Adabee. A proprietary preparation containing vitamins A and C, thiamine, riboflavine, pyridoxine, and nicotinamide; vitamin supplement. [Wyeth Laboratories] *

Adal. Alloying elements into aluminum alloys. [Foseco (F.S.) Ltd]

Adalat®. Dimethyl 1,4-dihydro-2,6-dimethyl-4-(2-nitrophenyl) pyridine 3,5 dicarboxylate; a cardioprotective coronary therapeutic and antihypertensive agent. [Bayer AG; Miles Pharmaceuticals]

Adalin. Polyethylene emulsion nonionic; softener for resin finishing; improves sewability, abrasion resistance, crease angles and softness. [Henkel Chemicals Ltd] *

Adalin. Bromodiethylacetylurea; soporific. [Bayer AG]

Adalox. A proprietary trade name for coated abrasives for sanding metal or plastics. §

Ad-aluminum. A proprietary trade name for an alloy of 82% copper, 15% zinc, 2% aluminum, and 1% tin. §

Adamac. Specially prepared tar for roads. [Thomas Ness Ltd] *

Adamant. Diamond or corundum. §

Adame. CAS 2439-35-2; Dimethylaminoethyl acrylate. [Rhone-Poulenc Surf.]

Adamite. A proprietary high-carbon nickel-chromium iron alloy; used for dyes. §

Adamon®. An antispasmodic. [AG Chemische Fabrik] ‡

Adamsite. 10-Chloro-5 : 10-dihydrophenarsazine; a poison gas. §

Adanon. A proprietary trade name for methadone. §

Adansonia Fiber. A fiber obtained from the bark of *Adansonia digitata*; used for making rope and sacking, and also for special paper. §

Adaphax 758. A proprietary factice containing neither sulfur nor chlorine used as a processing aid in the manufacture of polyurethanes and PVC. enables good electrical properties to be maintained in PVC and tackiness to be reduced; also used in white nitrile mixes. [Hubron Rubber Chemicals] *

Adapin. CAS 1229-29-4; Doxepin hydrochloride; antidepressant. [Pennwalt Corp] *

Adaprin. A proprietary preparation of nicotinamide and acetomenaphthone; used to treat chilblains. [Octel Chemicals Ltd] †

Adaptinol. Xanthophyl dipalmitate; an antidazzle preparation. [Bayer AG]

Adarola. A proprietary casein plastic. §

ADC. Epoxy formulations used in tooling, adhesives, electronic encapsulants, coatings, sealants and ablatives; ultra high temperature laminating resins, fast setting high temperature adhesives, pinhole-free RT laminating resin (requires no vacuum bag), chemical resistant coatings; thermally conductive and resistant coatings and encapsulants, construction sealants, binders and adhesives, soft tooling compounds, specialty coating for food industry and chemical industry. [ADC Resins] *

Adcora A3. A hard coal tar epoxy coating for tanks and pipes for effluent and for resistance to fumes and weathering.§

Adcora H2. A Hypalon (*qv*) coating for tanks etc.; possesses good resistance and good flexibility. §

Adcora P6. A proprietary trade name for a neoprene coating with good chemical and abrasion resistance and good flexibility. §

Adcora SP. A solventless epoxide-based coating giving a hard finish resembling enamel. §

Adcora V. A Viton (fluoroelastomer) coating with very good heat stability resistant to temperatures up to 260 C. §

Adcortyl-A. CAS 76-25-5; A proprietary name for the acetonide of triamcinolone. §

Adcortyl Cream. CAS 76-25-5; Triamcinolone acetonide in cream base. [Bristol-Myers Squibb Pharmaceuticals Ltd]

Adcortyl Injection. CAS 76-25-5; Triamcinolone acetonide in aqueous vehicle. [Bristol-Myers Squibb Pharmaceuticals Ltd]

Adcortyl in Orabase. CAS 76-25-5; Triamcinolone acetonide in ointment base. [E R Squibb & Sons Ltd]

Adcortyl Ointment. CAS 76-25-5; Triamcinolone acetonide in ointment base. [Bristol-Myers Squibb Pharmaceuticals Ltd]

Adcortyl Spray. CAS 76-25-5; Triamcinolone acetonide in aerosol spray. [Bristol-Myers Squibb Pharmaceuticals Ltd] †

Adcortyl with Graneodin Cream. Triamcinolone acetonide, neomycin sulfate and gramicidin in cream base. [Bristol-Myers Squibb Pharmaceuticals Ltd]

Adcortyl with Graneodin Ointment. Triamcinolone acetonide, neomycin sulfate and gramicidin in ointment base. [Bristol-Myers Squibb Pharmaceuticals Ltd]

Adcote. Adhesives and primers for flexible packaging. [Morton Int'l. Ltd]

Add F. CAS 64-18-6; Formic acid technical; liquid silage additive. [BP Chemicals Ltd]

Add-H. Liquid hay additive. [BP Chemicals Ltd]

Add-M. Liquid antimolding compound. [BP Chemicals Ltd]†

Addabond. Substance for bonding new concrete to old; particularly useful where it is impossible to dry out the substrate. [Addagrip Surface Treatments UK Ltd] *

Addacoat. Two part pack; colored material, giving a tile-like finish; applied to walls where cleaning is of paramount importance; used in damp proofing of cellars and basements. [Addagrip Surface Treatments UK Ltd] *

Addacol. Organic and inorganic pigments; used for coloring

---

‡ Trade name and manufacturer not verified   § Trade name without identified manufacturer

cement, concrete, bricks, etc. [Calder Colours (Ashby) Ltd]

Add-add. The leaves of *Celastraceae serratus*, of Abysinnia; used as an antiperiodic.

Addaflex. Flexible epoxy compound; for filling cracks and expansion joints in concrete floors. [Addagrip Surface Treatments UK Ltd] *

Addaflor. Solvent free colored epoxy top coat; oil, acid and chemical resistant; strong enough to hold antislip chippings; easily cleaned; applied to concrete, brick, wood and steel. [Addagrip Surface Treatments UK Ltd]*

Addagrout. Three-part pack epoxy-based grouting material; for bedding machinery, filling anchor points and repairing cracks and depressions in floors. [Addagrip Surface Treatments UK Ltd] *

Addalevel. Leveling compound for laying between 3 and 6 mm thick; applied to concrete and brick. [Addagrip Surface Treatments UK Ltd] *

Addamortar. Epoxy mortar (three part pack: fillers, hardener and resin); for filling cracks, holes and undulations, and for screeds; can be power floated. [Addagrip Surface Treatments UK Ltd] *

Addapitch. Pitch epoxy material; suitable for waterproofing and also for holding antislip chippings on ramps and loading bays etc. [Addagrip Surface Treatments UK Ltd]*

Addaprime. Two part pack: resin and hardener; solvent-free deep-penetrating epoxy primer.; can also be used as a coating; strong enough to hold antislip chippings; applied to concrete, brick, wood and steel. [Addagrip Surface Treatments UK Ltd] *

Addaseal. Polyurethane material of high quality (50% solids); clear or colored; for dust sealing concrete floors; strong enough to hold small particles of antislip aggregate. [Addagrip Surface Treatments UK Ltd] *

Addasure. Solvent based epoxy coating; two part pack; used for coating walls and floors. [Addagrip Surface Treatments UK Ltd] *

Adder. An asbestos fabric grade of Tufnol industrial laminates. [Tufnol Ltd]

Adder. Adjuvant containing 97% refined mineral oil; wetting agent for herbicides. [Embetec Crop Protection Ltd]

Addipast. CAS 1333-86-4; Carbon black pigment pastes. [Brockhues AG]

Additin. Antioxidants for the mineral oil industries; gum inhibitor. [Bayer AG]

Additin 30. CAS 90-30-3; Phenyl-α-naphthylamine; staining antioxidant for rubber tech. goods and heavily stressed goods; antiflexcracking agent for natural rubber and isoprene rubber; storage stabilizer for petrol. products. [Miles/Polysar Rubber]

Additive-A. CAS 8061-52-7; Modified calcium lignosulfonates; clay conditioners for production of bricks and tiles; act as plasticizers, lubricants, binders, and antiscumming agents. [Borregaard LignoTech]

Additol. Range of additives to improve properties such as flow and eliminate film defects etc; used in paints and printing inks. [Hoechst UK]

Additol. Range of additives to improve properties such as flow and eliminate film defects, etc.; used for paints and printing inks. [Resinous Chemicals Ltd]

Add It To Oil. Alkylaryl poly (ethylene oxy); natural oil emulsifier for use as an additive for conditioning oil for crop spray use. [Doyle Specialties] *

Adeka CR-5. Chlorinated rubber; for printing ink, overprint varnish, paint applications. [Asahi Denka Kogyo]

Adeka ED-505. TGE of trimethylol propane; epoxy resin diluent. [Asahi Denka Kogyo]

Adeka EP-4100. Liquid epoxy resin for general applications. [Asahi Denka Kogyo]

Adeka GH-200. CAS 51258-15-2; PPG-24-glycereth-24. [Asahi Denka Kogyo] *

Adeka P-400. Polyoxypropylene glycol; diol for polyurethane production. [Asahi Denka Kogyo]

Adeka PR-3007. Water-soluble polyether polyol; for water-absorbing polyurethane flexible foams, adhesives. [Asahi Denka Kogyo]

Adeka Catioace DM, PD Series. Tetraamonium salt type polymer; surfactant, antistat for electrification, low foaming, coating applications. [Asahi Denka Kogyo]

Adekacol CS, PS, TS. Phosphate ester type surfactants. [Asahi Denka Kogyo]

Adekacol EC Series. Dialkyl sulfosuccinate ester; emulsifier, detergent. [Asahi Denka Kogyo]

Adeka Estol. Fatty ester surfactants; for emulsification, solubilization and dispersing. [Asahi Denka Kogyo]

Adeka Hypote. CAS 7681-52-9; Sodium hypochlorite; for bleaching fiber, pulp and paper, bleaching chemicals and food, water treatment. [Asahi Denka Kogyo]

Adeka Kiku-Lube. Pour point depressants and viscosity index improvers. [Asahi Denka Kogyo]

Adeka Lub E-500. Chlorinated N-paraffin; extreme pressure agent for metalworking, especially formulation of cutting oil, grinding oil; plasticizer for PVC. [Adeka Fine Chem.]

Adekamine E Series. Mono- or di-alkyl tetraamonium salts; cationic surfactant, softener, antistat, emulsifier; mildly irritating. [Asahi Denka Kogyo]

Adekanol. EO-PO block copolymer; for detergency, emulsification, dispersing, wetting. [Asahi Denka Kogyo]

Adeka Optomer KR Series. One-component epoxy resin, uv curing; primers for materials such as metal/plastic/paper, finishing varnishes, protective varnishes, inks, adhesives, insulating coatings, conductive coatings. [Asahi Denka Kogyo]

Adeka Sakura-Lube 100. Oil-sol. organo molybdenum additive; for gasoline engine oil, diesel engine oil, gear oil, industrial lubricating oil, greases. [Asahi Denka Kogyo]

Adeka Sole CO. Cocamide DEA (1:2); surfactant providing detergency, stable foaming, anti-rust, dispersibility and thickening properties with only mild irritation. [Asahi Denka Kogyo]

Adeka Sole YA. CAS 68140-00-1; Cocamide MEA; surfactant providing detergency, stable foaming, anti-rust, dispersibility and thickening properties with only mild irritation. [Asahi Denka Kogyo]

Adekatol DES, DS, HAN, LS, TR, SAN, YES. Sulfate type surfactants; anionic surfactants. [Asahi Denka Kogyo]

Adekatol LA, LO, NP, OA, PC, SO Series. Ethoxylated nonionic surfactants. [Asahi Denka Kogyo]

Adelite. Colloidal silica; antislipping agent for paper and fiber binder for metal casting and refractory materials, catalyst carrier, reinforcing material for adhesives and paints. [Asahi Denka Kogyo]

Adelphane. A proprietary preparation of reserpine, and dihydrallazine sulfate; an antihypertensive. [Ciba plc] *

adenine. (vitamin B$_4$; 6-aminopurine) C$_5$H$_5$N$_5$; CAS 73-24-5; EINECS 200-796-1; Medicine and biochemical re-

---

search. [Aldrich; Lonza AG; Penta Mfg.; Schweizerhall; U.S. Biochemical]

Adeno. Adenosine phosphate; nutrient. [O'Neal, Jones & Feldman Pharmaceuticals] *

adenosine triphosphate. (ATP; adenosine, 5´-(tetrahydrogen triphosphate)) Organic compound; $C_{10}H_{16}N_5OP_3$; CAS 56-65-5; biochemical research. [Asahi Chem Industry Co Ltd; R. W. Greeff; Penta Mfg.]

Adenotriphos. A proprietary preparation of adenosine triphosphoric acid sodium. [Rona Laboratories] ‡

Adenyl. A proprietary preparation of adenosine monophosphoric acid; used in the treatment of rheumatic diseases. [Rona Laboratories] ‡

adepsine oil. *See* paraffin, liquid.

Adequan. Polysulfated glycosaminoglycan for intra-articular injection; for treatment of lameness in horses. [Luitpold-Werk] *

Adexolin. Water miscible pale yellow, sugar free concentrate containing vitamins A, C and D; for supplementation and prevention of deficiency of vitamins A, C and D which may occur during infancy and early childhood and in expectant and nursing mothers. [Evans Medical] *

ADF-600. Alcohol-based; general purpose defoamer for water-based drilling muds. [Anedco]

ADF-610. Silicone-based; defoamer. [Anedco]

Adflex. Compounded polymeric emulsion; flexing agent for cement based adhesives. [Howlett Adhesives Ltd] *

Adheso. A proprietary trade name for a synthetic wax consisting of a modified polymerized terpene. §

Adigan. A trademark for a digitalis preparation containing all the active principles except digitonin, this having been removed. §

Adimoll®BO. Benzoyl octyl adipate; monomeric plasticizer. [Bayer AG]

Adimoll® DB. CAS 105-99-7; Dibutyl adipate; monomeric plasticizer. [Bayer AG]

Adimoll® DH. Di-N-hexyl adipate. monomeric plasticizer. [Bayer AG]

Adimoll®DN. CAS 33703-08-1; Diisononyl adipate; monomeric plasticizer. [Bayer AG]

Adimoll®DO. CAS 103-23-1; Dioctyl adipate; plasticizer for PVC, applications requiring good low-temperature resistance; rubber plasticizer imparting high elasticity and low-temperature flexibility in conjunction with minimum volatility. [Bayer AG]

Adinol. A range of taurates; used as anionic surfactants, foaming agents in the cosmetic industry. [Croda Chem. Ltd]

Adinol. Textile auxiliaries based on triethyl citrate. [Fine Dyestuffs & Chemicals Ltd] *

Adinol CT. Sodium N-methyl N-cocoyl taurate in white powder form; anionic surfactant with high foaming and cleansing capacity, chemically stable, lime soap dispersant; used in cosmetics and toiletries; pharmaceutical preparations. [Croda Chem. Ltd]

Adinol OT. CAS 137-20-2; EINECS 205-285-7; Sodium N-methyl N-oleyl taurate in powder, paste, gel or liquid form; biodegradable anionic surfactant with detergent, wetting, emulsifying, dispersing and foaming properties; used in the textile and dyeing industries; leather; paper. [Croda Chem. Ltd]

adipic acid. (dicarboxylic acid C_6) HOOC(CH$_2$)$_4$COOH; CAS 124-04-9; EINECS 204-673-3; Manufacture of nylon and polyurethane foams; preparation of esters for

use as plasticizers and lubricants; food additive (acidulant); baking powders; adhesives. [Allied-Signal; Asahi Chem Industry Co Ltd; Du Pont; Monsanto; Penta Mfg.; Rhone-Poulenc; UCB SA]

adipocere. A wax-like mass left when animal bodies decompose in the earth. It consists of the fatty acid salts of calcium and potassium.

Adipon. Fatty alcohol sulfate, anionic; detergent for scouring and milling of worsted fabrics; softening effect. [Henkel Chemicals Ltd] *

Adiprene®. A range of polyether-based prepolymers offering high abrasion resistance, chemical resistance and electrical properties. [Uniroyal]

Adiprene® L-42. Urethane rubber; for products with high flexibility at low temps. [Uniroyal]

Adiprene® L-100. Urethane rubber; for products of high hardness, load-bearing capacity, and resistance to abrasion, oils, oxidation, ozone. [Uniroyal]

Adiprene® LW-570. Urethane rubber; for products of extreme hardness and high impact resistance, load-bearing capacity, hydrolytic stability; curable with MDA. [Uniroyal]

Adiprene® M-400. Urethane rubber; non-MBCA; offers improved processing. [Uniroyal]

Adipure®. For chemical industry. [DuPont UK]

Adiro. Acetylsalicylic acid; a platelet aggregation inhibitor, antithrombotic and anti-inflammatory drug. [Bayer AG]

Adirondackite. A rubber substitute made from sulfurized oils; used in the proofing of cloth, and as an insulator. §

Adju-Fluax. Influenza virus vaccine, bivalent types A and B with Adjuvant 65; for primary immunization and seasonal booster effect against influenza. [Merck & Co Inc] †

Adjunct B. Disodium phosphate; deposit inhibitor; precipitates sol. calcium hardness from boiler water to prevent scale deposits. [Drew Ind. Div.]

Adjust 4. Thixotrope that prevents settling by pigment wetting and controlled flocculation without appreciably increasing apparent visc.; also improves sag resist. as post-additive. [United Catalysts]

ADK CIZER O-130P. CAS 8013-07-8; EINECS 232-391-0; Epoxidized soybean oil; stabilizer and plasticizer for PVC; provides excellent heat and weathering stability and applicable to all PVC formulations; especially for improving heat stability in Ba-Zn or Ca-Zn formulation. [Asahi Denka Kogyo]

ADK CIZER O-180A. Epoxidized linseed oil; stabilizer providing higher heat weathering stability than O-130P; suitable for rigid calendering sheet and rigid bottle formulation. [Asahi Denka Kogyo]

ADK STAB 144. CAS 7440-66-6; Zn complex; PVC stabilizer for homogeneous tile flooring; provides excellent heat stability, improves initial color; effective blended with phosphite. [Asahi Denka Kogyo]

ADK STAB 465. CAS 58229-88-2; Di-n-octyltin mercaptide; PVC stabilizer providing excellent heat stability, color and transparency; suitable for rigid calendering sheet and blow formulation. [Asahi Denka Kogyo]

ADK STAB 466. Octyltin mercaptide; PVC stabilizer providing excellent heat stability, color and transparency; suitable for rigid calendering sheet formulation. [Asahi Denka Kogyo]

ADK STAB 1292. Dibutyltin mercaptide; PVC stabilizer for rigid transparent PVC with highest heat stability; applicable to rigid water pipe requiring NSF approval for high

---

‡ Trade name and manufacturer not verified   § Trade name without identified manufacturer

resist. to water extraction. [Asahi Denka Kogyo]

ADK STAB 1413. Benzophenone; uv absorber for polyolefins, PVC, etc.; good compatible with polymers. [Asahi Denka Kogyo]

ADK STAB 1500. Special phosphite; provides excellent heat, color and weathering stability; for rigid calendering sheet and blow bottle. [Asahi Denka Kogyo]

ADK STAB 2335. CAS 12513-27-8; Zinc borate; flame retardant, smoke preventer for flooring and cable formulations. [Asahi Denka Kogyo]

ADK STAB AC-122. Ba-Zn; general purpose PVC stabilizer providing good heat stability and color resist. [Asahi Denka Kogyo]

ADK STAB AC-133. Ba-Cd-Zn; PVC stabilizer providing excellent initial color, transparency and heat stability; for calendering sheet, leather and plasticized PVC compound by extrusion and injection molding processes. [Asahi Denka Kogyo]

ADK STAB AC-169. Ba-Zn; one-pkg. PVC stabilizer providing excellent heat stability and transparency; for calender sheets of plasticized PVC. [Asahi Denka Kogyo]

ADK STAB AP-536. Ba-Zn; general purpose PVC stabilizer providing good heat stability, color retention. [Asahi Denka Kogyo]

ADK STAB BT-11. CAS 77-58-7; Dibutyltin dilaurate; PVC stabilizer providing good heat and weathering resist.; improves processability in rigid transparent formulation. [Asahi Denka Kogyo]

ADK STAB BT-31. CAS 15535-69-0; Dibutyltin maleate; PVC stabilizer providing excellent initial color transparency; for whole rigid PVC products, extrusion, press and blow molding. [Asahi Denka Kogyo]

ADK STAB BT-83. Dibutyltin mercaptide; PVC stabilizer with excellent heat resist. and color stability; for injection molding and rigid PVC extrusion (T-die, contour). [Asahi Denka Kogyo]

ADK STAB EC-14. Zn-phosphite; improves initial color with Ba-Zn stabilizers in formulations containing TiO_2 and $CaCO_3$; suitable for plasticized and rigid calendering formulation and paste formulation. [Asahi Denka Kogyo]

ADK STAB FL-21. Na-Zn; stabilizer for foam PVC at low expansion. [Asahi Denka Kogyo]

ADK STAB GR-16. Ca-Zn; PVC stabilizer providing excellent long term heat stability; for extrusion and injection molding by filler loading formulation; effective with ADK STAB 1500. [Asahi Denka Kogyo]

ADK STAB LA-32. Benzotriazole; uv absorber for PVC, ABS, PS, PU, NMA, etc. [Asahi Denka Kogyo]

ADK STAB LA-57. Hindered amine light stabilizer; light stabilizer for polyolefins, PVC, ABS, etc.; superior light stability. [Asahi Denka Kogyo]

ADK STAB LS-2. CAS 15535-69-0; Dibutyltin maleate; lubricant and stabilizer for PVC; provides excellent transparency and gives good props. with tin maleate stabilizer; for extrusion and blow molding. [Asahi Denka Kogyo]

ADK STAB LS-8. CAS 123-95-5; EINECS 204-666-5; Butyl stearate; plastics lubricant. [Asahi Denka Kogyo]

ADK STAB NA-11. CAS 85209-91-2; EINECS 286-344-4; Sodium 2,2´-methylenebis-(4,6-di-t-butylphenyl) phosphate; nucleating agent upgrading heat deflection temp., flex. modulus, impact strength of PP, PET, PBT, polyamide; gives high transparency at low concs.; raises crystallization temp. [Asahi Denka Kogyo]

ADK STAB OF-14. Ba-Zn; general purpose foam PVC stabilizer for expanded leathercloth and wall paper. [Asahi Denka Kogyo]

ADK STAB OT-1. Di-n-octyltin dilaurate; PVC stabilizer providing excellent external lubricity, heat and process stability with Ca-Zn stabilizers; suitable for rigid sheet and blow formulation. [Asahi Denka Kogyo]

ADK STAB OT-9. Di-n-octyltin maleate; PVC stabilizer providing excellent heat stability and weatherability; suitable for rigid extrusion and calendering formulation. [Asahi Denka Kogyo]

ADK STAB RUP-9. Ba-Zn; PVC stabilizer for nonigrating elec. wire using polyester plasticizer; provides excellent heat stability and insulating props. [Asahi Denka Kogyo]

ADM. Ammonium dimolybdate; corrosion inhibitor for vapor phase inhibitor programs. [Climax Performance] *

ADM-407. Alkyl benzyl sulfonic acid; demulsifier. [Anedco]

ADM-407C. Alkyl benzyl sulfonic acid; demulsifier. [Anedco]

Adma®. Alkyl dimethyl amines. [Ethyl Corp]

Adma® 8, 10, 12. CAS 7378-99-6; Octyldimethylamine; intermediate for quaternary ammonium compounds, amine oxides, betaines. [Ethyl Corp]

Adma® 14. CAS 112-75-4; Tetradecyl dimethylamine; intermediate for quaternary ammonium compounds, amine oxides, betaines. [Ethyl Corp]

Adma®16. CAS 112-69-6; Hexadecyl dimethylamine; intermediate for quaternary ammonium compounds, amine oxides, betaines. [Ethyl Corp]

Adma® 18. CAS 124-28-7; EINECS 204-694-8; Octadecyl dimethylamine; intermediate for manufacturing of quaternary ammonium compounds for biocides, textile chems., oilfield chems., amine oxides, betaines, polyurethane foam catalysts, epoxy curing agents. [Ethyl Corp]

Adma®246-451. Dodecyl dimethylamine (40%), tetradecyl dimethylamine (50%), hexadecyl dimethylamine (10%); intermediate for quaternary ammonium compounds, amine oxides, betaines. [Ethyl Corp]

Adma® 1214. CAS 112-18-5, 112-75-4; Dodecyl dimethylamine (65%), tetradecyl dimethylamine (35%); intermediate for quaternary ammonium compounds, amine oxides. [Ethyl Corp]

Adma® WC. Octyl dimethylamine (7%), decyl dimethylamine (6%), dodecyl dimethylamine (53%), tetradecyl dimethylamine (19%), hexadecyl dimethylamine (9%), octadecyl dimethylamine (6%); intermediate for quaternary ammonium compounds, amine oxides, betaines. [Ethyl Corp]

Admerol®. Six common metals and their alloys; metal building materials; transportable buildings; railway tracks; nonelectric cables and wires; small hardware items; pipes and tubes; safes. [Reichhold]

Admerol®75-M-70. Varnish, modified oil. [Reichhold]

Admex®. Polymeric plasticizers; for plasticizing PVC and other polymers. [Hüls Am.]

Admiral®FPS Type 3089 FS. High solids fluidized polymer suspension; viscosifier for paper coatings. [Aqualon]

Admiralty brass. Cu-Ca-Zn-Sn alloy.

Admiralty gun metal. Alloys. Some contain from 87-90% copper and from 10-13% tin, while others consist of from 86-88% copper, 6-10% tin, and 2-6% zinc.

Admiralty nickel. See Adnic.

Admiralty white metal. A bearing alloy containing 86% tin, 8.5% antimony, and 5.5% copper.

Admire. CAS 105827-78-9; Imidacloprid; insecticidal seed

---

* Trade name not verified as available   † Trade name verified as obsolete

dressing with excellent root systemic properties for control of sucking pests (virus vectors), e.g., aphids, thrips, leafhoppers, white flies and some biting insects such as rice water weevil, Colorado potato beetle, wireworms, *Diabrotica* species, fruit fly, beet fly and onion fly on various crops, particularly on rice, cereals, maize, cotton, potatoes, beets and vegetables. [Bayer AG]

Admos Alloys. Brass alloys of varying composition containing small amounts of tin, nickel, lead, and iron.

Admox®. Alkyl dimethyl amine oxides. [Ethyl Corp]

Admox®14-85. CAS 3332-27-2; EINECS 222-059-3; Myristamine oxide; for soap bars, shaving creams, fabric softeners, hard surface cleaners, laundry detergents, oxygen bleach powds., toothpaste, agricuture, automatic dishwash, cellulose extraction, gasoline additives, bubble baths. [Ethyl Corp]

Admox® 18-85. CAS 2571-88-2; EINECS 219-919-5; Stearamine oxide; for soap bars, shaving creams, fabric softeners, hard surface cleaners, laundry detergents, oxygen bleach powds., toothpaste, agricuture, automatic dishwash, cellulose extraction, gasoline additives, bubble baths. [Ethyl Corp]

Admul. Mono and diglycerides and their acid derivatives. [Quest Int'l.]

Admune. A proprietary preparation of inactivated influenza virus used to confer immunity to the disease. [Duncan Flockhard Ltd] *

Adnic. (Admiralty nickel). An alloy containing 70% copper, 29% nickel, and 1% tin; it is resistant to corrosion and heat.

Adogen®66. Alkoxylated quaternary; nylon retarding agent; antistat for fabric, fiber, and yarn finish formulations. [Sherex/Div. of Witco]

Adogen®137. Dihydrogenated tallow dimethyl ammonium methyl sulfate; very high melting textile softener, antistat; nonyellowing; minimizes metal corrosion. [Sherex/Div. of Witco]

Adogen® 170. CAS 61790-33-8; EINECS 263-125-1; Tallowamine; reacted with maleic anhydride to make tallow succinamate, a high-foaming surfactant used to froth latex carpet backing. [Sherex/Div. of Witco]

Adogen® 185. C12-15 ether amine; reacted with maleic anhydride to make tallow succinamate, a high-foaming surfactant used to froth latex carpet backing. [Sherex/Div. of Witco]

Adogen® 412. CAS 112-00-5; EINECS 203-927-0; Lauryl trimethyl ammonium chloride; softener for textile, laundry, paper, etc. [Sherex/Div. of Witco]

Adogen®417. CAS 61790-41-8; EINECS 263-134-0; Soya trimethyl ammonium chloride; retardant and leveling agent for textiles. [Sherex/Div. of Witco]

Adogen®442. CAS 61789-80-8; EINECS 263-090-2; Quaternium-18; fabric softener conc. for home and commercial laundries; textile processing. [Sherex/Div. of Witco]

Adogen®442-P100. CAS 61789-80-8; EINECS 263-090-2; Dihydrogenated tallow dimethyl ammonium chloride; textile softener producing very slick cationic hand, max. softness. [Sherex/Div. of Witco]

Adogen®461. CAS 61789-18-2; EINECS 263-038-9; Coco trimethyl ammonium chloride; emulsifier, dispersant; used in corrosion inhibitor formulations for oil-field brines and HCl acidizing systems; textile antistat. [Sherex/Div. of Witco]

Adogen®462. CAS 61789-77-3; EINECS 263-087-6; Dicodimonium chloride, IPA/water; antistat, emulsifier, flocculating agent, dispersant used in corrosion inhibitor formulations for oil-field chemicals. [Sherex/Div. of Witco]

Adogen®470. Ditallowdimonium chloride; specialty quaternary for nonionic laundry detergent-softeners. [Sherex/Div. of Witco]

Adogen® 471. Tallow trimonium chloride, IPA; dispersant, antistat, emulsifier; used in corrosion inhibitor formulations for oilfield brines and HCl acidizing systems; textile antistat. [Sherex/Div. of Witco]

Adogen® 477. Tallow diamine diammonium dichloride; emulsifier, dispersant; retardant, dyeing assistant. [Sherex/Div. of Witco]

Adogen®MA-108 SF. CAS 124-28-7; EINECS 204-694-8; Dimethyl stearamine; neutralizer, conditioner, coemulsifier for personal care products. [Sherex/Div. of Witco]

Adogen®MA-112 SF. Dimethyl behenamine; neutralizer, conditioner, coemulsifier for personal care products. [Sherex/Div. of Witco]

Adogen® S-18 V. CAS 7651-02-7; EINECS 231-609-1; Stearamidopropyl dimethylamine; conditioner, antistat, coemulsifier, plasticizer, neutralizer for personal care products. [Sherex/Div. of Witco]

Adogen® TA-100. CAS 107-64-2; EINECS 203-508-2; Distearyl dimonium chloride; textile softener producing very slick cationic hand, max. softness; good nonyellowing, absorbency; can replace wax in sizing formulations. [Sherex/Div. of Witco]

Adogen® TA-101. CAS 107-64-2; EINECS 203-508-2; Distearyl dimonium chloride; textile softener producing very slick cationic hand, max. softness; good nonyellowing, absorbency; can replace wax in sizing formulations; disperses at lower temp. than TA-100. [Sherex/Div. of Witco]

Adol® 52 NF. CAS 36653-82-4; EINECS 253-149-0; Cetyl alcohol; coemulsifier, lubricant, foam control agent, cosolvent, plasticizer, stabilizer, emollient, intermediate; for metal lubricants, inks, textiles, emulsions, paper, cosmetics, mineral processing, oil field chemicals, fabric softeners. [Sherex/Div. of Witco]

Adol® 62 NF. CAS 112-92-5; EINECS 204-017-6; Stearyl alcohol; emollient, glass frit binders, waxes, emulsion stabilizers, esters, tertiary amines, surfactants, polymers, chemical intermediate; cosmetic formulations; see also Adol 42. [Sherex/Div. of Witco]

Adol® 63. Cetearyl alcohol; emulsifiers; prime base for detergents; used in plasticizers, tert. amines, lube oil additives, textile auxiliaries, mold lubricants, polymers, org. synthesis, chemical intermediates; see also Adol 42. [Sherex/Div. of Witco]

Adol® 64. Cetearyl alcohol; emollient, emulsion stabilizer, visc. modifier for skin care products; conditioner imparting velvety feel; opacifier for creams and lotions. [Sherex/Div. of Witco]

Adol®66. EINECS 248-470-8; Isostearyl alcohol; coemulsifier, lubricant, foam control agent, cosolvent, plasticizer, stabilizer, emollient, intermediate; for metal lubricants, inks, textiles, emulsions, paper, cosmetics, mineral processing, oil field chemicals, fabric softeners. [Sherex/Div. of Witco]

Adol®85. CAS 143-28-2; EINECS 205-597-3; Oleyl alcohol; emulsifier, lubricant, foam control agent, cosolv., plasticizer, emollient. [Sherex/Div. of Witco]

---

‡ Trade name and manufacturer not verified   § Trade name without identified manufacturer

Adox 3125. CAS 7758-19-2; Sodium chlorite; antimicrobial for water and waste water treatment. [Int'l. Dioxcide]

Adpro AP 2112-GP. CAS 9003-07-0; Polypropylene homopolymer; features high flow rate, efficient cycle times, UL and FDA approval; for injection molded consumer and pkg. items, medical components. [Genesis Polymers]

Adpro AP 8210-HS. CAS 9003-07-0; Polypropylene; ultra high impact resist., high uniform flow rate; for automotive and extreme impact lens; as impact modifier. [Genesis Polymers]

Ad-Pro-MTS. Adhesion promoter for use in adhesives, inks, coatings, and lacquers; improves adhesion especially for difficult substrates, aids gloss. [Rit-Chem]

Adrenalin. CAS 51-43-4; Epinephrine; adrenergic. [Parke-Davis] *

Adrenocorticotrophic Hormone. Corticotrophin. §

Adrenocortrophin ACTH. Corticotrophin. §

Adrenoxyl. A proprietary preparation of adrenochrome monosemicarbazone dihydrate. [Horlicks] *

Adriamycin. A freeze-dried powder containing lactose and doxorubicin hydrochloride for injection; a prescription drug. [Hercules] *

Adriblastina. CAS 23214-92-8; Doxorubicin; antineoplastic. [Farmitalia (Farmaceutici Italia)] *

Adrin. A proprietary brand of adrenalin (qv). §

Adroit. Cutting oils. [S & D Chemicals Ltd] *

adronal. Cyclohexanol.

adronal acetate. (hexalin acetate). Cyclohexanol acetate, a resin solvent.

Adroyd. CAS 434-07-1; A proprietary preparation of oxymetholone; an anabolic agent. [Parke-Davis] *

Adrucil. CAS 51-21-8; Injectable fluorouracil; a prescription drug. [Hercules] *

Adsee® 775. POE ethers and special resins; spreader sticker, wetting agent, and penetrant for agricutural spray; surfactant for monosodium methane arsonate formulations. [Witco/Organics]

Adsee® 799. Alkyl POE ether; agricture surfactant; soil penetrant. [Witco/Organics]

Adsorbocarpine. CAS 54-71-7; Pilocarpine hydrochloride; cholinergic. [Alcon Laboratories Inc]

Adtac. Range of low molecular weight aliphatic hydrocarbon resins. contributes a balance of tack and adhesive properties to elastomer systems; typical softening point grades range from 10-25C; used in adhesives, in coatings and as a waterproofing agent. [Hercules] *

Adurol. Monocloro and monobromo hydroquinones, $C_6H_3Cl(OH)_2$ and $C_6H_3Br(OH)_2$; photographic developers. §

Aduvex®. A range of benzophenone derivatives; ultra violet absorbers for the protection of polymers. [Octel Chemicals Ltd]

Advagum. A terpene resin; a proprietary plasticizer. §

Advance. Broad-spectrum herbicide. (Sold in UK for Dow Elanco). [ICI Chem. & Polymers Ltd]

Advance. Bromoxynil + fluroypyr + ioxynil; post-emergence contact herbicide for cereals. [ICI Agrochemicals]

Advantage; Trade name for nonpigment, polymeric colorants used for molded opaque polyolefin parts. [Milliken]

Advantage 5 Defoamer. Defoamer for acid and alkaline papermaking systems, deinking systems. [Hercules]

Advantage 52-B. Oil-based defoamer for pulp mill brownstock washing operations or other high-temp. surfactant-stabilized foam systems; improves drainage. [Hercules]

Advantage 70DYX. Silicone-based drainage aid for improved drainage in kraft pulpmill brownstock washing operations by removing entrained air and surface foam; also for cold-stock systems, pulpmill bleaching and screening. [Hercules]

Advantage® 101M. Contains Mekor® volatile oxygen scavenger/metal passivator, DPB-42 antiscalant, sequestrants, organics, and antifoam; deposit and corrosion inhibitor for treating steam generating systems. [Drew Ind. Div.]

Advantage® 124. DPB-42 antiscalant, sequesterants; polymeric deposit and corrosion inhibitor for steam generating systems. [Drew Ind. Div.]

Advantage 136 Defoamer. Hydrocarbon oil-based defoamer for use as drainage aid and foam killer in kraft pulpmill brownstock washing operations. [Hercules]

Advantage 1007B Defoamer. Water-based all-purpose defoamer for solid and surfactant-stabilized aq. foaming systems, e.g., acid and alkaline paper machines, pulpmill screening, deinking operations, size press and calender sol'ns., mill effluent systems. [Hercules]

Advantage CP. Vinyl acetate/butyl maleate/isobornyl acrylate copolymer, ethanol SDA-40B. [ISP]

Advantage DF 110. Hydrocarbon oil-based defoamer for papermaking systems. [Hercules]

Advantage M104 Defoamer. Drainage aid and foam killer for kraft pulpmill brownstock washing systems. [Hercules]

Advantage M1251 Production Aid. Production aid removing entrained air and surface foam in papermaking applications, and effluent treatment. [Hercules]

Advapak® ML-1325. Ester; lubricant pkg. for injection molding formulations; NSF-accepted for use @ 1.3-2.5 phr in PVC for potable water contact. [Morton Int'l./Specialty Chem.]

Advastab® LS-203. Organotin compound; lubricating stabilizer for high-output PVC pipe production on multiple screw extruders. [Morton Int'l./Specialty Chem.]

Advastab® TM-181. Methyltin mercaptide; stabilizer for rigid PVC processes (extrusion, calendering, injection and blow molding) and PVC-PVA formulations with prolonged/severe processing temps., for chlorinated PVC compds; tin catalyst in PU polymerizations. [Morton Int'l./Specialty Chem.]

Advastab® TM-692. Methyltin mercaptide; stabilizer for most rigid PVC applications for potable water, sewer, irrigation pipe, conduit, and duct, and extrusion applications, profile, foamed profile. [Morton Int'l./Specialty Chem.]

Advastab® TM-790 Series. Heat stabilizer/external lubricant for single and multiscrew extrusion of PVC potable water, sewer and irrigation pipe, conduit, telephone duct. [Morton Int'l./Specialty Chem.]

Advastab® WS-499. Organotin; heat stabilizer for extrusion of weatherable, rigid PVC articles. [Morton Int'l./Specialty Chem.] *

Advawax® 240. CAS 110-31-6; EINECS 203-756-1; N,N´-Ethylene bisoleamide; synthetic wax used as plastics processing lubricant and release agent, antistat, melting point modifier for waxes, industrial asphalts and tar, pigment dispersing agent for resin systems; polyamide-

---

* Trade name not verified as available

† Trade name verified as obsolete

paraffin coupling agent; used in adhesive tapes, coatings, food pkg. materials. [Morton Int'l./Specialty Chem.]

Advawax®290. CAS 110-30-5; EINECS 203-755-6; N,N´-Ethylene bisstearamide; synthetic wax used as plastics processing lubricant and release agent; melting point modifier for waxes and resin blends and industrial asphalt and tar; pigment dispersing agent for resin systems; paper-making defoamer; used in adhesive tapes, coatings, PP; also for food pkg. materials. [Morton Int'l./Specialty Chem.]

Advex 91025. CAS 9002-86-2; PVC compound; interior grade, high modulus, low expansion rigid cube profile extrusion compound. [BFGoodrich/Geon Vinyl]

Advil. CAS 15687-27-1; Ibuprofen; anti-inflammatory. [Whitehall Laboratories]

Advitacon. Oil soluble vitamins. [Quest Int'l.] *

Advitagel. Flour and confectionery emulsifier. [Quest Int'l.]

Advitamix. Animal food supplements. [Quest Int'l.] *

Advitaroma. Butter and meat flavoring. [Quest Int'l.] *

Advitrol. Antisettling and thickening agents for paints, varnishes, lubricants, adhesives, coatings, putties, and cosmetics. [Süd-Chemie AG] *

Advizor. Herbicide containing lenacil and chloridazon. [ICI Chem. & Polymers Ltd]

Advizor. Chloridazon + lenacil; pre-emergence herbicide for use in sugar beet. [Farm Protection Ltd]

AE-1. Ethoxylated lauryl alcohol; intermediate in manufacturing of surfactants. [Procter & Gamble]

Aegerite. A trade name for Elaterite. §

AEI Compound 403/401. Silane crosslinkable polyethylene compound; flame retardant compound for cable insulation. [AEI Compounds] *

AEI Compound 407/424. Chemically crosslinkable low-smoke, low-toxicity, halogen-free flame-retardant compound; for insulation of LV power cables and sheathing of all types of cables. [AEI Compounds] *

AEI Compound 505/401. Crosslinkable ethylene-propylene-diene rubbers; for low and medium voltage cable insulation. [AEI Compounds] *

Aeonite. A nickel silver containing 20% nickel; a trade name for Elaterite. §

AEPD®. CAS 115-70-8; 2-Amino-2-ethyl-1,3-propanediol; pigment dispersant, neutralizing amine, corrosion inhibitor, acid-salt catalyst, pH buffer, chemical and pharmaceutical intermediate, solubilizer. [Angus]

AEPD®-85. CAS 115-70-8; 2-Amino-2-ethyl-1,3-propanediol; chemical intermediate, formaldehyde scavenger, acid-salt catalyst for permanent-press resins, corrosion inhibitor. [Angus]

Aerelle®. For fibers industry. [DuPont UK]

Aerex®. De-icing and anti-icing fluid for airplanes. [BASF AG]

aerial cement. A term applied to cements which set in air, the setting being due to desiccation and carbonation.

Aerialite. A proprietary synthetic resin. §

Aero. A proprietary trade name for rosin-glycerol varnish and lacquer resins. §

Aero 301 Xanthate. Sodium sec. butyl xanthate; used in the mining industry. [Cyanamid BV]

Aero 303 Xanthate. Potassium ethyl xanthate; used in the mining industry. [Cyanamid BV]

Aero 317 Xanthate. Sodium isobutyl xanthate; used in the mining industry. [Cyanamid BV]

Aero 343 Xanthate. CAS 140-93-2; Sodium isopropyl xanthate; used in the mining industry. [Cyanamid BV]

Aero 350 Xanthate. Potassium amyl xanthate; used in the mining industry. [Cyanamid BV]

Aero 3477 Promoter. Sodium diisobutyl dithiophosphate; used in the mining industry. [Cyanamid BV]

Aero 3501 Promoter. Sodium diisoamyldithiophosphate; used in the mining industry. [Cyanamid BV]

Aero X. A proprietary rubber vulcanization accelerator. §

Aerocol. Polyvinyl acetate adhesives. [Ciba plc]

Aerodri 100 104. Modified dioctylsulfosuccinate; used in the mining industry. [Cyanamid BV]

Aerodri 200. Mixture of surfactants; used in the mining industry. [Cyanamid BV]

Aerodux. Resorcinol/formaldehyde resins; for wood adhesives, glass reinforced plastics, abrasives, and foundry resins. [Ciba plc]

Aerodux. Resorcinol formaldehyde resins. [Dynochem UK Ltd]

Aerofloat 208 Promoter. Sodium diethyl and sodium di-sec. butyl dithiophosphate mixture; used in the mining industry. [Cyanamid BV]

Aerofloat 211 Promoter. Sodium diisopropyl dithiophosphate; used in the mining industry. [Cyanamid BV]

Aerofloat 238 Promoter. Sodium di-s-butyl dithiophosphate; used in the mining industry. [Cyanamid BV]

Aerofonic. Compressed polyether acoustical foam. [ScotFoam Corp] *

Aerofroth 65. Polypropylene glycol; used in the mining industry. [Cyanamid BV]

Aerofroth 76. Mixture of higher alcohols; used in the mining industry. [Cyanamid BV]

Aerofroth 88. CAS 104-76-7; EINECS 203-234-3; 2-Ethylhexanol; used in the mining industry. [Cyanamid BV]

Aerofroth 99. 2-Ethylhexanol tails; used in the mining industry. [Cyanamid BV]

Aerolite. Urea/formaldehyde resins. [Ciba plc]

Aerolite. Phenol formaldehyde resins. [Dynochem UK Ltd]

Aeromatt. CAS 471-34-1; Precipitated calcium carbonate for cosmetics. [Rhone-Poulenc Sturge Lifford] †

Aero metal. An aluminum alloy, consisting mainly of aluminum with from 2.1-2.9% magnesium, 0.3-1.3% iron, and 0.2-0.6% copper.

Aeromin. An alloy of 91.6% aluminum and 8.4% magnesium.

Aeron. Alloys of 95% aluminum, 4% copper, and 1% silicon.

Aerophen. Phenol/formaldehyde resins. [Ciba plc]

Aerophen. Phenol/formaldehyde resins. [Dynochem UK Ltd]

Aerophine 3418A. Sodium diisobutyldithiophosphinate; used in the mining industry. [Cyanamid BV]

Aeroplex. A proprietary safety-glass. §

Aeroseb-Dex®. CAS 50-02-2; Dexamethasone; cortocosteroid. [Allergan, Inc]

Aeroseb-HC®. CAS 50-23-7; Hydrocortisone; corticosteroid. [Allergan, Inc]

Aerosil®. Highly dispersed pyrogenic silica; highly active filler for natural and synthetic rubber, especially for silicone rubber; as thickening agent for ointments, creams, toothpastes etc.; tableting and dragée production auxiliary; thixotropizing agent; for polyester resins, lacquers and printing inks; for maintenance of free-flowing characteristics of substances which tend to cake (free-flow agent); for the production of highest purity

---

‡ Trade name and manufacturer not verified   § Trade name without identified manufacturer

silicates; for blueprint papers; thickening agent for oils. [Degussa; Degussa Ltd]

Aerosil®130, 150. Silica; highly active filler for natural and synthetic rubber; thickening agent; tabletting and dragée production auxiliary; thixotropizing agent for polyester resins; antisetting agent. [Degussa; Degussa AG]

Aerosil®200. Fumed silica; anticaking and free-flow agent with high absorp. capacity; for adhesive, food, cosmetics, paint, paper, film, pesticides, pharmaceuticals, plastics, silicone rubber, inks, sealants industries; thixotrope for greases and mineral oils. [Degussa; Degussa AG]

Aerosil COK 84. A trade name for a mixture of Aerosil and alumina in 5:1 ratio. it is suited particularly for thickening aqueous and other polar systems. [Bush Beach Ltd] *

Aerosil Composition. A trade name for a mixture of Aerosil with 15% starch; specially designed for tableting. [Bush Beach Ltd] *

Aerosil®OX50. Fumed silica; for antiblocking PET, PP and PE films and tapes; for coating of elec. light bulbs. [Degussa]

Aerosil®R972. Fumed silica; anticaking and free-flow agent for adhesive, elec., cosmetics, paint, pesticides, pharmaceuticals, plastics, inks industries; improves water resist. of greases. [Degussa; Degussa AG]

Aerosil® R972V. Fumed silica; reinforces and improves storage stability of RTV compounds; yields softer silicone rubbers. [Degussa]

Aerosol® 18. CAS 14481-60-8; EINECS 238-479-5; Disodium stearyl sulfosuccinamate; emulsifier, dispersant, foamer, detergent, solubilizer for soaps and surfactants; alkaline cleaner formulations, brick and tile cleaners, emulsion polymerization of vinyl chloride and styrene/butadiene rubbers; emulsifying oils and waxes, household detergents, cleaning paper mill felts; foamer for foamed latexes and plastics; biodeg. [Am. Cyanamid; Cyanamid BV]

Aerosol®22. CAS 38916-42-6; Tetrasodium dicarboxyethyl stearyl sulfosuccinamate; emulsifier, dispersant, solubilizer, surfactant; emulsion polymerization of vinyl monomers; polishing waxes; surface tension depressant for writing and drawing inks; demulsifier for w/o emulsions; cleaning of paper mill felts; industrial, household, and metal cleaners; biodegradable. [Am. Cyanamid; Cyanamid BV]

Aerosol 200. Disodium alkyl amidopolyethoxy sulfosuccinate. [Cyanamid BV] †

Aerosol® 501. Disodium alkyl sulfosuccinate; dispersant, emulsifier, wetting agent, dispersant, foaming agent; used for acrylic and vinyl acetate emulsions; self-crosslinking latexes; textile wetting and foaming applications; biodeg. [Am. Cyanamid; Cyanamid BV]

Aerosol® A-102. CAS 39354-45-5; Disodium deceth-6 sulfosuccinate; emulsifier, solubilizer, foamer, dispersant, surfactant, wetting agent; used in emulsion polymerization of PVAc/acrylics, textiles, cosmetics, shampoos, wallboard, adhesives; biodeg.; stable to acid media. [Am. Cyanamid; Cyanamid BV]

Aerosol®A-103. CAS 9040-38-4; Disodium nonoxynol-10 sulfosuccinate; emulsifier, solubilizer, wetting agent, surfactant, surface tensile depressant; used in PVAc acrylic emulsions; textile emulsions, pad-bath additive; textile wetting; cosmetics, shampoos, wallboard, adhesives. [Am. Cyanamid; Cyanamid BV; Cyanamid of Great Britain Ltd]

Aerosol®A-196-85. CAS 23386-52-9; EINECS 245-629-3; Dicyclohexyl sodium sulfosuccinate; dispersant, surfactant; sole emulsifier for modified S/B; post additive to stabilize latex and promote adhesion; biodeg. [Am. Cyanamid; Cyanamid BV]

Aerosol®A-268. CAS 37294-49-8; Disodium isodecyl sulfosuccinate; surfactant, sole emulsifier for PVC latexes-vinyl, vinylidene chloride, acrylics, surface tensile depressant, solubilizer. [Am. Cyanamid; Cyanamid BV]

Aerosol® AY-65. CAS 922-80-5; EINECS 213-085-6; Diamyl sodium sulfosuccinate; wetting agent, dispersant, surfactant; used in agriculture, emulsion polymerization, electroplating, ore leaching, cleaning of porcelain, tile, brick, cement; biodeg. [Am. Cyanamid; Cyanamid BV]

Aerosol® C-61. Ethoxylated alkyl guanidine-amine complex; antistat, pigment dispersant, flushing agent, wetting agent, settling agent; alkaline, cement, brick, and tile cleaner formulations for crystal growth control, emulsion breaking, alkaline metal and paint brush cleaners; paint removers; textile softener; demulsifying agent; foaming agent; for plastics, paper, textiles, adhesive industries. [Am. Cyanamid; Cyanamid BV]

Aerosol®DPOS-45. CAS 25167-32-2; Disodium mono- and didodecyl diphenyl oxide disulfonate; emulsifier, dispersant, solubilizer, primary surfactant for emulsion polymerization systems; coupling agent; high electrolyte tolerance; stable in highly acidic and alkaline sol'ns. and at elevated temps. [Am. Cyanamid; Cyanamid BV]

Aerosol®GPG. CAS 577-11-7; EINECS 209-406-4; Sodium dioctyl sulfosuccinate; wetting agent, surface tensile depressant, emulsifier, dispersant; for dust control, industrial cleaners, emulsifying waxes; biodeg. [Am. Cyanamid; Cyanamid BV; Cyanamid of Great Britain Ltd]

Aerosol®IB-45. CAS 127-39-9; Sodium diisobutyl sulfosuccinate; emulsifier, wetting agent; emulsion polymerization of styrene, butadiene and copolymers; dye and pigment dispersant; for leaching, electroplating; biodeg. [Am. Cyanamid; Cyanamid BV]

Aerosol® MA-80. Dihexyl sodium sulfosuccinate; dispersant, textile wetting agent, emulsifier, solubilizer, penetrant; used for emulsion polymerization, battery separators, electroplating, ore leaching; germicidal act.; not as rapidly biodeg. as Aerosol 18 and 22. [Am. Cyanamid; Cyanamid BV]

Aerosol® NPES 458. CAS 9051-57-4; Ammonium salt of sulfated nonylphenoxy POE ethanol; high foaming surfactant for emulsion polymerization of acrylic, styrene and vinyl acetate systems, dishwashing detergents, germicides, pesticides, general purpose cleaners, cosmetics, and textile wet processing applications. [Am. Cyanamid]

Aerosol® NS. CAS 9084-06-4; Sodium naphthalene sulfonate; dispersant for pigments, extenders, and fillers in aq. media over broad pH range. [Am. Cyanamid]

Aerosol® OS. CAS 1322-93-6; Sodium diisopropyl-naphthalene sulfonate; emulsifier, dispersant and wetting agent; used in alkaline cleaning formulations; antigelling agents, automotive radiator cleaners, metal, cement, brick, and tile cleaners for crystal growth control; electroplating, filtration, glass cleaning, household detergents, leaching ores and slags, pigment disps.; soap additive; adjuvant in agricultural chem. products; slowly biodeg. [Am. Cyanamid; Cyanamid BV]

---

* Trade name not verified as available

† Trade name verified as obsolete

Aerosol® OT-70 PG. CAS 577-11-7; Sodium dioctyl sulfo-succinate, propylene glycol/water; wetting agent, surface tensile depressant, emulsifier, surfactant; for use where high flash required; biodeg. [Am. Cyanamid; Cyanamid BV]

Aerosol® OT-75%. CAS 577-11-7; EINECS 209-406-4; Dioctyl sodium sulfosuccinate; wetting agent and surface tensile depressant used in textile, rubber, petrol., paper, metal, paint, plastic, and agricultural industries; antistat for cosmetics, dry cleaning detergents, emulsion, plastic, pipelines, and suspension polymerization; emulsifier wax for polish, firefighting, germicide, metal cleaner, mold release agent; dispersant in paints and inks, paper, photography, process aid, rust preventative, soldering flux, and wallpaper removal. [Am. Cyanamid; Cyanamid BV]

Aerosol® OT-MSO. CAS 577-11-7; Dioctyl sodium sulfo-succinate in min. seal oil; wetting agent, lubricant, detergent for dry cleaning, corrosion resist. lubricants, agricultural emulsions, org. solvent systems; used when a higher flash is required. [Am. Cyanamid]

Aerosol® TR-70. CAS 2673-22-5; EINECS 220-219-7; Ditridecyl sodium sulfosuccinate; emulsifier, surfactant; used in emulsion polymerization of vinyl chloride and vinyl acetate, suspension polymerization of vinyl chloride; dispersant for resins, pigments, polymers, and dyes in org. systems; pigment dispersant in printing inks; rust preventative; biodeg. [Am. Cyanamid; Cyanamid BV]

Aerosporin Sterile Powder. CAS 1405-20-5; A proprietary formulation of polymyxin B sulfate; for treatment of acute infections caused by susceptible strains of bacteria including *Pseudomonas aeruginosa*. [The Wellcome Foundation Ltd]

Aerothene. CAS 75-09-2; Methylene chloride; a vapor-pressure depressants and carrier solvent. [Dow UK]

Aerotrol. A proprietary trade name for the hydrochloride of isoprenaline; an aerosol bronchodilator. [Abbott Laboratories] *

Aerotru 23. Modified melamine formaldehyde resin; used in the paper industry. [Cyanamid BV]

Aerozine A-50. Rocket propellant. [Olin]

Aeternol. A proprietary synthetic resin. §

Aethoxal. Fattening agent, oil components and dispersing agent; shampoos; foam baths; nonfoaming oil baths. [Henkel Chemicals Ltd] *

Aethoxal B. PPG-5-laureth-5; superfatting agent, emollient for bath oils, skin and personal care products, pharmaceuticals; biodeg. [Henkel/Cospha; Henkel KGaA]

Aethrol. A plastic of the pyroxylin-cellulose acetate type. §

AF 10 FG. Silicone antifoam; antifoam agent used for general food, poultry, and meat processing applications. [Harcros]

AF 10 IND. Silicone antifoam; antifoam for agriculture, cutting oils, drilling muds, effluent, inks, chemicals, detergents and textiles. [Harcros]

AF 60. Dimethyl polysiloxane aq. emulsion; defoaming agent used in adhesive, ink, latex, soap, starch, and paint manufacturing and other aq. industrial systems. [GE Silicones]

AF 70. 100% silicone compound; defoamer in petrol. refining, cutting oils, chem. processing, antifoam formulating; food additive for food processing. [GE Silicones]

AF 72. PEG-40 stearate, sorbitan stearate, and silica; food-grade antifoam agent, surfactant; also for industrial ap-

plications such as textile dyeing and finishing, leather finishing, latex processing, soap and detergent manufacturing, adhesive manufacturing, and as a boiler feed water defoamer. [GE Silicones]

AF 1025. Silicone antifoam; antifoam for delayed coker units. [Harcros]

AF 8820. Silicone antifoam; antifoam for effluent, agricuture, antifreeze, detergent applications; dilutable. [Harcros]

AF 9020. Dimethicone aq. emulsion; defoamer for industrial and food-processing systems including chemical processing (adhesive manufacturing, water-based ink manufacturing, latex processing, soap manufacturing, starch processing, paint additive, alcohol fermentation), waste treatment; petrochemical (resin polymerization, glycol dehydrators, ethylene oxide production, urea production). [GE Silicones]

AF GN-11-P. Nonsilicone antifoam, used for fermentation, drilling muds, effluent, adhesives, gas treating. [Harcros]

AF HL-36. Nonsilicone antifoam; for fermentation, processing beet sugar and yeast, distillation; Kosher. [Harcros]

AF HL-52. Nonsilicone antifoam; for solvs., latex paints, inks, chemical processing, adhesives, paper, paper coatings. [Harcros]

Afalon. CAS 330-55-3; Linuron; a residual urea herbicide for the control of weeds in field crops including potatoes and carrots. [Hoechst UK]

Afax. Continuous casting mold flux for all steel grades. [Foseco (F.S.) Ltd] †

A-Fax®. A range of amorphous polypropylenes; used in adhesives and sealants, used for asphalt modifications in construction and building industry, in carpet backing, in polyolefin modification, as sound deadening and rubber processing agents. [Hercules]

Afco-Chem B. CAS 822-16-2; Sodium stearate; lubricant for metallic sintering; lubricant and stabilizer for resins; pigment dispersant; mold release; waterproofing agent; lubricant additive. [Adeka Fine Chem.]

Afco-Chem CS. CAS 1592-23-0; EINECS 216-472-8; Calcium stearate; lubricant for metallic sintering; lubricant and stabilizer for resins; pigment dispersant; mold release; wateproofing agent; lubricant additive. [Adeka Fine Chem.]

Afco-Chem LIS. CAS 4485-12-5; Lithium stearate; lubricant for metallic sintering; lubricant and stabilizer for resins; pigment dispersant; mold release; wateproofing agent; lubricant additive. [Adeka Fine Chem.]

Afco-Chem MGS. CAS 557-04-0; Magnesium stearate; lubricant for metallic sintering; lubricant and stabilizer for resins; pigment dispersant; mold release; wateproofing agent; lubricant additive. [Adeka Fine Chem.]

Afco-Chem ZNS. CAS 557-05-1; Zinc stearate; lubricant for metallic sintering; lubricant and stabilizer for resins; pigment dispersant; mold release; wateproofing agent; lubricant additive. [Adeka Fine Chem.]

Afco-Coat. Inorganic powder; water-sol. precoating agent for stainless steel. [Adeka Fine Chem.]

Afcolene. A proprietary polystyrene. [Atochimie] ‡

Afco-Lube Series. Lubricants for wet wire drawing. [Adeka Fine Chem.]

Afco-Met Series. Lubricants for dry wire drawing. [Adeka Fine Chem.]

Afenil. Calcium chloride urea. §

Afflair® Lustre Pigments. Mica platelets coated with titanium dioxide and/or iron oxide; luster pigments for coat-

---

ings, inks, and plastics. [EM Industries]

Afilan EHS. CAS 22047-49-0; EINECS 244-754-0; 2-Ethyl-hexyl stearate; surfactant for textile processing. [Hoechst Celanese/Colorants & Surf.]

Afilan ICS. CAS 25339-09-7; EINECS 246-868-6; Isocetyl stearate; surfactant for textile processing. [Hoechst Celanese/Colorants & Surf.]

Afilan PP. CAS 14450-05-6; EINECS 238-430-8; Penta-erythrityl tetrapelargonate; surfactant for textile processing. [Hoechst Celanese/Colorants & Surf.]

Afilan TDA. Ditridecyl adipate; surfactant for textile processing. [Hoechst Celanese/Colorants & Surf.]

Afilan TDS. CAS 31556-45-3; EINECS 250-696-7; Tridecyl stearate; surfactant for textile processing. [Hoechst Celanese/Colorants & Surf.]

Afilan TMPP. Trimethylolpropane tripelargonate; surfactant for textile processing. [Hoechst Celanese/Colorants & Surf.]

Aflaban. Feed preservative based on sorbic acid; growth inhibitor for molds, yeasts and bacteria in animal feeds. [Monsanto Co] *

Aflamman CN. Halogen compound with metal oxide and binder; weakly anionic; environmentally nonhazardous; water-resistant flameproofing agent for treatment of technical fabrics from polyester; also suitable for backcoating. [Thor Chemicals (UK) Ltd]

Aflammit P. Organic phosphorus/nitrogen compound; wash and dry-cleaning resistant flameproofing agent for cotton. [Thor Chemicals (UK) Ltd]

Aflammit TI. Potassium hexafluoro titanate; flameproofing agent for processing of wool. [Thor Chemicals (UK) Ltd]

Aflammit ZAL. Zirconium acetate solution; flameproofing agent for processing of wool. [Thor Chemicals (UK) Ltd]

Aflammit ZR. Potassium hexafluoro zirconate; flame-proofing agent for processing of wool. [Thor Chemicals (UK) Ltd]

Aflunox. Perfluoropolyether fluids. [PCR]

Aflux. A range of fatty acid derivatives partly bound to highly active silica used as dispersing agents and internal lubricants in the rubber industry; used for molded and extruded technical articles. [Bayer plc]

Afonic. Embossed sound absorbing foam. [ScotFoam Corp] *

AFP®. Solid photopolymer. [Asahi Chem. Industry]

AFP 2000. Fungal protease; enzyme for hydrolysis of proteins under acid conditions; prevents haze in fruit juice. [Solvay Enzymes]

Afranil®. Alcohol and fatty acid derivs.; grease and foam inhibitor, pulp deaerator for papermaking. [BASF AG; BASF plc]

African phosphates. Mineral phosphates found in Tunis and Algeria. They contain from 55-65% calcium phosphate. Others found at Safaga and Kosseir contain 60-70% calcium phosphate; fertilizers.

Afrin® 12 Hour Nasal Spray. CAS 2315-02-8; Oxy-metazoline hydrochloride 0.05%; adrenergic. [Schering Corp]

Afrinol. Pseudoephedrine sulfate; adrenergic. [Schering Corp] *

Afrisect. Insecticidal formulation. [Mitchell Cotts Chemicals Ltd]

Afrol. Timber insecticide. [ICI Chem. & Polymers Ltd]

Afror Tyne Powder. A low-freezing explosive containing a nitrated mixture of glycerine and ethylene glycol and ammonium nitrate. §

Aftate. CAS 2398-96-1; Tolnaftate; antifungal. [Schering Corp]

Afugan. CAS 13457-18-6; Pyrazophos; systemic organo-phosphorus fungicide. [Hoechst UK]

Agalite. (mineral pulp, asbestine pulp). A variety of talc (hydrated magnesium silicate); used by papermakers. §

Agallol; Dressing for seed potatoes, flower bulbs and sugar cane cuttings. [Bayer AG]

Agalyn. A proprietary material used for dentures. §

agar. (agar-agar) A phycolloid derived from red algae; polysaccharide mixture of agarose and agaropectin; culture medium in microbiology and bacteriology; antistaling agent in baking, confections, meats, poultry; gelation agent in desserts and beverages; protective colloid in foods, pharmaceuticals, laboratory reagents, photographic emulsions. [Meer; Schweizerhall; Spice King; U.S. Biochemical]

agar-agar. *See* Agar.

agaricin. (agaric acid) A resin acid, obtained by extraction with alcohol of the fruit bodies of *Polyporus officinalis* and *Agaricus albus.* A febrifuge.

Agarol. A proprietary preparation containing liquid paraffin, phenolphthalein and agar; a laxative. [Warner] *

agate ware. (enamelled iron-granite). Enamelled iron.

Agathine. Salicyl-α–methyl-phenyl-hydrazide, $C_6H_5(CH_3)$; N · NHCO(C_6H_4OH); used in the treatment of neuralgia and rheumatism. §

Agatine. A proprietary phenol-formaldehyde resin in the form of sheet, rods, tubes, etc. §

Agavin. Thiosolucin-dihydrostreptomycin preparation for the veterinary field. [May & Baker Ltd] *

age. (axin). The fat of *Coccus Axin,* growing in Mexico. It consists of the glycerides of lauric and axinic acids.

Ageflex. Acrylate and methacrylate monomers. [CPS Chemical Co Inc]

Ageflex AGE. CAS 106-92-3; Allyl glycidyl ether; modifier for elastomer, epoxies, adhesives, fibers; reactive interme-diate for coatings, sizing/finishing agent for fiberglass; silane intermediate in elec. coatings. [CPS Chemical Co Inc]

Ageflex AMA. CAS 96-05-9; Allyl methacrylate; silane monomer intermediate; crosslinker offering two-stage polymerization, abrasion and solv. resist.; polymer modi-fier for high impact plastics, adhesives, acrylic elas-tomers, photoresists, optical polymers. [CPS Chemical Co Inc]

Ageflex 1,3 BGDMA. CAS 1189-098-8; 1,3-Butylene glycol dimethacrylate; crosslinker for plastisols, hard rubber rolls, cast acrylic sheet/rods, coagent for rubber com-pounding, impregnant for metal and wood composites, adhesives, glass-reinforced plastics. [CPS Chemical Co Inc]

Ageflex BGE. CAS 2426-08-6; Butyl glycidyl ether; reactive diluent in epoxy resins, laminating, flooring, elec. casting and encapsulants. [CPS Chemical Co Inc]

Ageflex CHMA. CAS 101-43-9; Cyclohexyl methacrylate; polymer modifier for optical lens coatings, adhesives, floor polishes, vinyl polymerization, anaerobic adhe-sives. [CPS Chemical Co Inc]

Ageflex DEGDMA. CAS 2358-84-1; Diethylene glycol dimethacrylate; crosslinking for rubber vulcanization, moisture barrier films and coatings, photopolymer print-ing plates and letterpress inks, conversion coatings and

---

* Trade name not verified as available   † Trade name verified as obsolete

adhesives. [CPS Chemical Co Inc]

Ageflex EGDMA. CAS 97-90-5; Ethylene glycol dimethacrylate; crosslinker and modifier of ABS, acrylic and PVC, ion exchange resins, encapsulation of smokeless powd., glaze coatings, dental polymers, paper processing aids, rubber modifier, adhesives, optical polymers, leather finishing, moisture barrier films; fiberglass-reinforced polyesters, emulsion polymerization. [CPS Chemical Co Inc]

Ageflex EOTMPTA. CAS 28961-43-5; Ethoxylated trimethylolpropane triacrylate; monomer having flexibility and fast cure response with other acrylated resins; radiation cured systems offer chem. and abrasion resist. and high gloss to inks, coatings, overprint varnishes. [CPS Chemical Co Inc]

Ageflex FA-1. CAS 2439-35-2; Dimethylaminoethyl acrylate; adhesion promoter in uv and eb cured coatings for metal, plastic, paper, and wood surfs.; catalyst for epoxy molding and extrusion resins; intermediate for water treatment chems., quaternary monomers, silane coupling agents, conductive paper coatings. [CPS Chemical Co Inc]

Ageflex FA-1Q75MC. CAS 44992-01-0; Dimethylaminoethyl acrylate methyl chloride quaternary antistatic finish for polyester fibers, flocculant and coagulant for industrial process water treatment, flocculant for mineral recovery, ion exchange resins, adhesives, acid dye receptivity, electrostatic coatings on wood, retention aids for paper. [CPS Chemical Co Inc]

Ageflex FA-1Q80DMS. CAS 13106-44-0; Dimethylaminoethyl acrylate dimethyl sulfate quaternary; antistatic finish for polyester fibers, flocculant and coagulant for industrial process water treatment, flocculant for mineral recovery, ion exchange resins, adhesives, acid dye receptivity, electrostatic coatings on wood, retention aids for paper. [CPS Chemical Co Inc]

Ageflex FA-2. CAS 2426-54-2; Diethylaminoethyl acrylate; industrial and automotive coatings, electronic photo resists, dye additives, lube oil additives; intermediate for water treatment chems., silane coupling agents, conductive paper coatings; retention aids for paper manufacturing; flocculant and coagulant. [CPS Chemical Co Inc]

Ageflex FA-2Q50DMS. CAS 21810-39-9; Diethylaminoethyl acrylate dimethyl sulfate quaternary antistatic finish for polyester fibers, flocculant and coagulant for industrial process water treatment, flocculant for mineral recovery, ion exchange resins, adhesives, acid dye receptivity, electrostatic coatings on wood, retention aids for paper. [CPS Chemical Co Inc]

Ageflex FA-6. CAS 2499-95-8; n-Hexyl acrylate; uv-cured inks/coatings, glass coating, visc. index improver for functional oils, polymer cements and sealants. [CPS Chemical Co Inc]

Ageflex FA-8. CAS 29590-42-9; Isooctyl acrylate; pressure-sensitive adhesives, coatings, caulks and sealants. [CPS Chemical Co Inc]

Ageflex FA-10. CAS 1330-61-6; Isodecyl acrylate; adhesives, coatings, uv-curable reactive diluent in inks and coatings, visc. index improver. [CPS Chemical Co Inc]

Ageflex FA-12. CAS 2156-97-0; Lauryl acrylate; uv-curable reactive diluent in inks and coatings, adhesives, visc. index improver, finishing aid for leather. [CPS Chemical Co Inc]

Ageflex FM-1. CAS 2867-47-2; Dimethylaminoethyl meth-

acrylate; detergent and sludge dispersant in lubricants; visc. index improver; flocculant for waste water treatment; retention aid for paper manufacturing; acid scavenger in PU foams; corrosion inhibitor; resin and rubber modifier; used in acrylic polishes and paints, hair prep. copolymers, sugar clarification, adhesives, water clarification. [CPS Chemical Co Inc]

Ageflex FM-1Q80DMS. CAS 6891-44-7; Dimethylaminoethyl methacrylate dimethyl sulfate quaternary; antistatic finish for polyester fibers, flocculant and coagulant for industrial process water treatment, flocculant for mineral recovery, ion exchange resins, adhesives, acid dye receptivity, electrostatic coatings on wood, retention aids for paper. [CPS Chemical Co Inc]

Ageflex FM-1Q80MC. CAS 5039-78-1; Dimethylaminoethyl methacrylate methyl chloride quaternary antistatic finish for polyester fibers, flocculant and coagulant for industrial process water treatment, flocculant for mineral recovery, ion exchange resins, adhesives, acid dye receptivity, electrostatic coatings on wood, retention aids for paper. [CPS Chemical Co Inc]

Ageflex FM-2. CAS 105-16-8; Diethylaminoethyl methacrylate; industrial and automotive clear coatings, dye additives, intermediate for water treatment and oilfield chems., stabilizer for fuel oils, sweetening agent for various hydrocarbon oils. [CPS Chemical Co Inc]

Ageflex FM-4. CAS 3775-90-4; t-Butylaminoethyl methacrylate; automotive dip tanks, coatings, industrial/consumer adhesives and coatings, dye and lube oil additives, intermediate for water treatment chems., o/w separations. [CPS Chemical Co Inc]

Ageflex FM-10. CAS 29964-84-9; Isodecyl methacrylate; pressure-sensitive adhesives, coatings for leather, textiles, paper, nonwovens, polymer modifier/stabilizer, visc. index improver, dispersion for plastics and rubber, floor waxes, potting compounds, sealants, adhesives. [CPS Chemical Co Inc]

Ageflex FM-12. CAS 142-90-5; Lauryl methacrylate (synthetic); lube oil additives, coatings for nonwoven fiber, floor waxes, paints, adhesives, varnishes, sealants, caulks, stabilizer for nonaq. dispersions and inks. [CPS Chemical Co Inc]

Ageflex FM-25. Undecyl methacrylate, lauryl methacrylate, tridecyl methacrylate, tetradecyl methacrylate, pentadecyl methacrylate (C12-15 methacrylate). [CPS Chemical Co Inc] *

Ageflex FM-68. CAS 32360-05-7; Stearyl methacrylate (natural, C16-18 methacrylates); lube oil additive, pour pt. depressant, paper coatings, textile finishes, paints, varnishes, pressure-sensitive adhesives. [CPS Chemical Co Inc]

Ageflex FM-1620. Hexadecyl methacrylate, stearyl methacrylate, eicosyl methacrylate (C16-20 methacrylate). [CPS Chemical Co Inc] *

Ageflex HDDA. CAS 13048-33-4; 1,6-Hexanediol diacrylate; fast curing monomer providing adhesion to metal and glass, flexibility in inks and coatings, water resist., good weatherability; reactive diluent for radiation-curable oligomers. [CPS Chemical Co Inc]

Ageflex IBOA. CAS 5888-33-5; Isobornyl acrylate; monomers when cured providing hardness, low shrinkage, abrasion resist., heat and water resist., good weatherability in automotive coatings, electronics, adhesives, and other acrylic polymers. [CPS Chemical Co Inc]

---

‡ Trade name and manufacturer not verified   § Trade name without identified manufacturer

Ageflex IBOMA. CAS 7534-94-3; Isobornyl methacrylate; monomers when cured providing hardness, low shrinkage, abrasion resist., heat and water resist., good weatherability in automotive coatings, electronics, adhesives, and other acrylic polymers. [CPS Chemical Co Inc]

Ageflex mDMDAC. CAS 7398-69-8; Dimethyl diallyl ammonium chloride; monomer for synthesis of homo and copolymers used as coagulant and flocculants for water treatment, mineral processing, demulsifier for petrol. recovery, elec. conductive paper and coatings, wet and dry strength resins, antistatic additives and coatings; cosmetic additives in hair conditioners, biocides, detergent additives, water-sol. polymers, electrographic paper and film. [CPS Chemical Co Inc]

Ageflex MEA. CAS 3121-61-7; Methoxyethyl acrylate; solv.-resist. elastomer, polyacrylate rubber, uv-curable reactive diluent, soft contact lenses, PVC impact modifier, fabric coatings, barrier coatings for polyethylene, textile coatings. [CPS Chemical Co Inc]

Ageflex NB-50. CAS 7398-69-8; Dimethyl diallyl ammonium chloride monomer, n-butanol; antistatic finish for polyester fibers, flocculant and coagulant for industrial process water treatment, flocculant for mineral recovery, ion exchange resins, adhesives, acid dye receptivity, electrostatic coatings on wood, retention aids for paper. [CPS Chemical Co Inc]

Ageflex n-HA. CAS 2499-95-8; n-Hexyl acrylate. [CPS Chemical Co Inc] *

Ageflex PEA. CAS 48145-04-6; Phenoxyethyl acrylate; uv-curable reactive diluent in inks and coatings, adhesives, visc. index improver, tile coating. [CPS Chemical Co Inc]

Ageflex T4EGDA. CAS 17831-71-9; Tetraethyleneglycol diacrylate; fast curing monomer providing good adhesion and flexibility, low shrinkage, and good impact strength in inks, coatings, adhesives, photo resists, and rubber products. [CPS Chemical Co Inc]

Ageflex TBGE. CAS 7665-72-7; t-Butyl glycidyl ether; reactive diluent in epoxy resins, corrosion inhibitor in some solvs., modifier for amines, acids and thiols. [CPS Chemical Co Inc]

Ageflex THFMA. CAS 2455-24-5; Tetrahydrofurfuryl methacrylate. [CPS Chemical Co Inc] *

Ageflex TM 402, 403, 404, 410, 421, 423, 451, 461, 462. CAS 3290-92-4; Trimethylolpropane trimethacrylate blends; processing aid for extrusion and molding of plastisols and rubber compounds (improves abrasion resist., adhesion in PVC plastisols, scorch and chem. resist., elevated temp. stability). [CPS Chemical Co Inc]

Ageflex TMPTA. CAS 15625-89-5; Trimethylolpropane triacrylate; uv-cured adhesives, wood fillers, inks, coatings, dry film photo polymer resists, flexographic, offset and screen printing inks, vinyl acrylic latex paint, exterior coatings, highly crosslinked polybutadiene rubber. [CPS Chemical Co Inc]

Ageflex TMPTMA. CAS 3290-92-4; Trimethylolpropane trimethacrylate; coagents for wire and cable, hard rubber rolls, polybutadiene and polyethylene, moisture barrier films and coatings, plastisols and vinyl acetate latexes, adhesives, molding compounds, textile products. [CPS Chemical Co Inc]

Ageflex TPGDA. CAS 68901-05-3; Tripropylene glycol diacrylate; uv-curable inks/coatings, floor tiles, wood coatings and fillers, adhesives, textile finishes, and rubber compounds. [CPS Chemical Co Inc]

Ageflex ZDA. CAS 14643-87-9; Zinc diacrylate. [CPS Chemical Co Inc] *

Agefloc A-50. Poly (2, hydroxypropyl-N,N-dimethyl ammonium chloride); flocculant and coagulant, dewatering aids in centrifugation, filtration, and flotation of both industrial and municipal waste sludges, potable water treatment. [CPS Chemical Co Inc]

Agefloc B-50LV. Dimethylamine/epichlorohydrin copolymer; for color removal, low turbidity water clarification, sugar cane/sugar beet processing, latex coagulation, latex waste clarification, liquid/solid separation, sludge dewatering, wire and felt cleaning compounds for paper manufacturing, stabilizer, retention aid. [CPS Chemical Co Inc]

Agefloc CF50. Inorganic-aluminum complex; for meat and poultry plant effluents, emulsion breaking, fat and grease separation, oil/water separation, raw water clarification, mining recycle water, replacement for alum or ferric chloride; reduces phosphorus and heavy metal levels. [CPS Chemical Co Inc]

Agefloc PC20HV. CAS 7398-69-8; Dimethyl diallyl ammonium chloride; paper industry retention aid, pigment dispersion, drainage aid, fiber dewatering, stabilizer for sizes, electroconductive polymer, recycling operatins, raw and waste water clarification. [CPS Chemical Co Inc]

Agefloc WT-20. CAS 26062-79-3; Poly(dimethyl diallyl ammonium chloride). [CPS Chemical Co Inc]

Agefloc WT-40. CAS 26062-79-3; Poly(dimethyl diallyl ammonium chloride); coagulant for water clarification, potable water treatment, waste water treatment, oil field, flotation enhancement, mining filtration aid. [CPS Chemical Co Inc]

Agenap HMW-H. CAS 1338-24-5; Naphthenic acid. [CPS Chemical Co Inc] *

Agent AT-804. Household blend; toilet bowl block. [Rhone-Poulenc Surf.]

Agequat 400. CAS 26062-79-3; Polyquaternium-6; for personal care formulations including hair sprays, shampoos, conditioners, mousses, and rinses. [CPS Chemical Co Inc]

Agequat 500, -5008. CAS 26590-05-6; Polyquaternium-7. for personal care formulations including hair sprays, shampoos, conditioners, mousses, and rinses. [CPS Chemical Co Inc]

Agequat C505. CAS 7398-69-8; Dimethyl diallyl ammonium chloride; drainage and retention aid, sludge dewatering. [CPS Chemical Co Inc]

Agerite. A full line of phenol and amine rubber antioxidants both primary and secondary; used in all forms of rubber. [BFGoodrich] *

Agerite® DPPD. CAS 74-31-7; Diphenyl-p-phenylene diamine; antioxidant used in rubber and mech. products. [R.T. Vanderbilt]

Agerite® HP-S. Dioctylated diphenylamine, diphenyl-p-phenylenediamine, 65:35 ratio; antioxidant; also used in CR compounds for outdoor service. [R.T. Vanderbilt]

Agerite® MA. CAS 26780-96-1; Polymerized 1,2-dihydro-2,2,4-trimethyl quinoline; antioxidant; aging protection to XLPE. [R.T. Vanderbilt]

Agerite® Hipar. A proprietary trade name for a mixture of isopropoxy diphenylamine, diphenyl-p-phenylamine and phenyl-b-naphthylamine; an antioxidant for rubber. [R.T. Vanderbilt]

---

* Trade name not verified as available   † Trade name verified as obsolete

Agerite® Resin D. CAS 26780-96-1; Polymerized 1,2-dihydro-2,2,4-trimethylquinoline; antioxidant for natural rubber and synthetic rubbers. [R.T. Vanderbilt]

Agerite®Spar. Styrenated phenol; a proprietary antioxidant. [R.T. Vanderbilt] *

Agerite® Stalite. A proprietary trade name for polymerized trimethyldihydroquinoline; an antioxidant. [R.T. Vanderbilt]

Agerite®Stalite S. Octylated diphenylamines; a proprietary antioxidant. [R.T. Vanderbilt]

Agerite® Superflex, Superflex Solid G. CAS 9003-79-6; Diphenylamine-acetone; antioxidant used in rubber and mech. products. [R.T. Vanderbilt]

Agerite®White. CAS 93-46-9; sym-Di-β-naphthyl-p-phenylenediamine; antioxidant. [R.T. Vanderbilt]

Agesperse 71. Polymeric carboxylic acid, sodium salt; low foaming dispersant, emulsifier, stabilizer for paper, paints, carpet backcoating, rubber, mining, textiles, ceramic slip, detergents, boiler water compounds, cooling water compounds, adhesives. [CPS Chemical Co Inc]

Agesperse 80. Polymeric carboxylic acid, sodium salt; dispersant, emulsifier, stabilizer for paper, rubber, mining, textiles, ceramic slip, detergents, boiler water compounds, cooling water compounds, adhesives. [CPS Chemical Co Inc]

Agestan®68. Silver amalgam in tablet form. [Bayer AG]

Agestat 41. CAS 7398-69-8; Dimethyl diallyl ammonium chloride; paper industry retention aid, pigment dispersion, drainage aid, fiber dewatering, stabilizer for sizes, electroconductive polymer, recycling operatins, raw and waste water clarification. [CPS Chemical Co Inc]

Agfa-Gevaert. Imaging systems; for graphic and reprographic systems, x-ray, cinematography, office systems, photography. [Agfa-Gevaert NV]

Agidex. Glucoamylase for conversion of starch into dextrose. [Glaxo Laboratories] *

Agitan 217. Silicone defoamer; for emulsion paints, emulsion polymers, adhesives. [Münzing Chemie GmbH]

Agitan 281. Silicone-free defoamer for emulsion paints, emulsion polymers, adhesives, aq. systems, silicate paints. [Münzing Chemie GmbH]

Agitan 301. Silicone defoamer; biodeg. defoamer for emulsion paints, emulsion polymers, synthetic renderings, adhesives. [Münzing Chemie GmbH]

Agitan E 255. Silicone emulsion; defoamer for emulsion paints, gloss emulsion paints, synthetic renderings, adhesives, aq. systems, glazes, aq. printing inks, polymerization processes. [Münzing Chemie GmbH]

Agitan P 800. Defoamer for powder systems, powder coatings, synthetic renderings, plasters, fillers, mortars, cements. [Münzing Chemie GmbH]

Agitan VP 725. Silicone compound; defoamer for lacquers, solv.-free systems, printing inks. [Münzing Chemie GmbH]

Agma. Calcinated magnesite. [ICI Chem. & Polymers Ltd]

agnin. (agnolin). Purified wool fat.

Agnowax. Wool wax alcohols. [Croda Chem. Ltd] †

Agomet. Methylmethacrylate adhesives. [Degussa Ltd]

Agral. Nonionic spreader containing 900 g/l alkylphenol ethoxylate; wetting, spreading and emulsifying agent for agricultural and horticultural pest control products. [ICI Chem. & Polymers Ltd]

Agramm. Nitrogenous fertilizers. [ICI Chem. & Polymers Ltd] †

Agriben. Manure composter for processing liquid and solid manure for agriculture. [Süd-Chemie AG] *

Agricastrol. A range of lubricants and hydraulic fluids for use with farm machinery; for engine oils and hydraulic fluids for tractors of all kinds. [Burmah-Castrol Ltd] *

Agrichem. CAS 1702-17-6; 2,4-D; translocated herbicide for cereals and established grassland. [Agrichem (International) Ltd]

Agrichem DB Plus. 2,4-DB + MCPA; translocated herbicide for cereal crops. [Agrichem (International) Ltd]

Agrichem Flowable Thiram. CAS 137-26-8; Thiram; fungicide with animal repellent properties. [Agrichem (International) Ltd]

Agrichem MCPA-25, 50. CAS 94-74-6; MCPA; herbicide for cereals and grassland. [Agrichem (International) Ltd]

Agricol. A proprietary range of alginates used for root dipping. [Alginate Industries Ltd] *

Agricorn 500. CAS 94-74-6; MCPA; herbicide for cereals and grassland. [Farmers Crop Chemicals Ltd]

Agricorn D. CAS 1702-17-6; 2,4-D; translocated herbicide for cereals and established grassland. [Farmers Crop Chemicals Ltd]

agricultural limestone. See limestone

Agricur®. Systemic nematicide and insecticide with good residual activity against free-living, cyt and root-knot nematodes, soil larvae, banana weevil borers (*Cosmpolites sordidus*) and other sucking and biting pests on the foliage of plants. [Bayer AG]

Agridin 60. An emulsifiable insecticide; contains active component diazinone $C_{12}H_{21}O_3N_2SP$, different emulgators and organic solvents; a light to dark brown liquid liquid, 60% active component; applied as an active insecticide against worms. [Chemical Combine] *

Agrilan®AEC123. Alkoxylate; coemulsifier, emulsion stabilizer for agricutural toxicant emulsifiable concs. [Harcros UK]

Agrilan® AEC178. EO/PO copolymer complex; coemulsifier, emulsion stabilizer for agricultural toxicants. [Harcros UK]

Agrilan®AEC266. CAS 577-11-7; Sodium dioctyl sulfosuccinate, ethanol; emulsifier for production of insecticide emulsifiable concs. and microemulsion formulations. [Harcros UK]

Agrilan®DG102. Alkylaryl ether sulfate; wetting agents for production of water-disp. gran. agrochemical toxicants. [Harcros UK]

Agrilan®EA14. Surfactant, activity optimizer for agrochem. toxicant formulations. [Harcros UK]

Agrilan® F513. Polyaromatic ethoxylate phosphate ester; wetting and dispersing agent for agricultural toxicants. [Harcros UK]

Agrilan®F546. Complex phosphate ester, free acid; emulsifier for agrochem. flowable/emulsifiable conc. blends. [Harcros UK]

Agrilan®FS101. Epoxidized vegetable oils; stabilizers for emulsifiable concs. of agricultural toxicants. [Harcros UK]

Agrilan®TKA103. Quaternary ammonium compound; wetting and compatibility agent for quaternary herbicides. [Harcros UK]

Agrilan® WP101. Aromatic sulfonate; wetting agent for production of water-disp. granules and wettable powds. of agrochem. toxicants. [Harcros UK]

Agrilan®X98. Calcium dodecylbenzene sulfonate; emulsi-

---

‡ Trade name and manufacturer not verified   § Trade name without identified manufacturer

fier for herbicide and pesticide formulations. [Harcros UK]

Agrilite alloy. A copper-lead alloy containing small amounts of tin.

Agrimer 15L. CAS 9003-39-8; Polyvinylpyrrolidone, 0.15% sorbic acid (preservative). [ISP]

Agrimer AL-22. CAS 9003-39-8; Alkylated polyvinyl-pyrrolidone. [ISP]

Agrimer VA 6. CAS 25086-89-9; Vinyl pyrrolidone/vinyl acetate copolymer. [ISP]

Agrimer VEMA-H-240. CAS 9011-16-9; Methyl vinyl ether/maleic anhydride copolymer, free acid. [ISP]

Agrimul. Anionic/nonionic emulsifiers; agricultural toxicant dispersersants. [Harcros UK] *

Agriphlan 24. Trade name for an orange-red liquid containing 240 g/liter trifluoraline; herbicide against cottonweeds, bean, tomato, pear, garlic and sunflower weeds; very efficient against amaranth, bristle-grass, knapweed etc. [Chemical Combine] *

Agrisol PX401. Aromatic alkoxylates; cosolv., penetrant, flow promotor for agrochemical toxicants. [Harcros UK]

Agrisorb. CAS 61791-44-4; A spreader containing 850 g/l tallow amine ethoxylate; wetting agent for glyphosate-based herbicides. [ABM Chemicals Ltd]

Agrispon. Mineral and plant extracts in a water base containing cytokinin, B-vitamin, morphogenic and porphyrin activity to aid in increased plant metabolism and soil nutrient availability; used for all agricultural, horticultural and forestry products. [SN Corp/Appropriate Technology Ltd]

Agrisynth BLO. CAS 96-48-0; γ-Butyrolactone; solv. for PAN, PS, fluorinated hydrocarbons, cellulose triacetate, shellac; used in paint removers, petrol. processing, hectograph process, specialty inks; intermediate for aliphatic and cyclic compounds; reaction and diluent solv. for pesticides; used in dyeing of acetate; wetting agent for cellulose acetate films, fibers, solv. welding of plastic films in adhesive applications. [ISP]

Agritol. A proprietary ammonium dynamite explosive. §

Agritox. Selective weedkiller. [May & Baker Ltd]

Agritox 50. CAS 94-74-6; MCPA; herbicide for cereals and grassland. [Rhone-Poulenc Crop Protection Ltd]

Agriwet. Nonionic spreader containing 25% alkylphenol ethoxylate; wetting agent for insecticides, fungicides and foliar feeds. [ABM Chemicals Ltd]

Agrocide. Insecticide. [ICI Chem. & Polymers Ltd]

Agropen. Adjuvant containing 95% emulsifiable vegetable oil; wetting agent for insecticides. [Ideal Manufacturing Ltd]

Agrosan. Organo mercury powder seed dressing. [ICI Chem. & Polymers Ltd]

Agrosan D. CAS 62-38-4; Phenylmercury acetate; organomercury fungicide seed dressing for cereals and fodder beet. [ICI Agrochemicals]

Agrosil® LR. Colloidal silicate; encourages intensive root development, improves irrigation efficiency, improves soils. [BASF AG] *

Agrosol. Liquid mercury seed dressing. [Plant Protection]*

Agrothion. Insecticide. [ICI Chem. & Polymers Ltd]

Agro-Vita. Mineral and plant extracts in a water base containing cytokinin, B-vitamin, morphogenic and porphyrin activity to aid in increased plant metabolism and yield; for all agricultural, horticultural and forestry products. [SN Corp/Appropriate Technology Ltd] †

Agroxone 50. CAS 94-74-6; MCPA; herbicide for cereals and grassland. [ICI Agrochemicals]

Agsol Ex 1. CAS 2555-05-7; N-Methyl-2-pyrrolidone. [ISP]

Agsol Ex 2. N-Ethyl-2-pyrrolidone. [ISP]

Agsol Ex 6C. N-2-Cyclohexyl-2-pyrrolidone. [ISP]

Agsol Ex 8. N-Octyl-2-pyrrolidone. [ISP]

Agsol Ex 12. N-Dodecyl-2-pyrrolidone. [ISP]

Agsol Ex BLO. CAS 96-48-0; γ-Butyrolactone; solv. for PAN, PS, fluorinated hydrocarbons, cellulose triacetate, shellac; used in paint removers, petrol. processing, hectograph process, specialty inks; intermediate for aliphatic and cyclic compounds; reaction and diluent solv. for pesticides; used in dyeing of acetate; wetting agent for cellulose acetate films, fibers, solv. welding of plastic films in adhesive applications. [ISP]

Agulin. Sheep vaccine. [Glaxo Laboratories] *

agurin. Theobromine-sodium-acetate; a diuretic.

AH Salt. Hexamethylenediamine adipate; monomer for production of polyamide 6/6. [BASF AG]

A-hydroCort. CAS 125-04-2; Hydrocortisone sodium succinate; glucocorticoid. [Abbott Laboratories]

Aiathesin. o-Hydroxy-benzoyl-alcohol (salicyl alcohol); antirheumatic and antiseptic. §

Aicello. Photo-sensitive diazo film in screen printing technology; stencil film for making photo-stencils in screen printing. [Aicello Chemical Co Ltd] *

Aich metal. (Gedge's metal, Sterro metal). An alloy similar to Delta metal, except that it contains iron. It usually consists of 60% copper, 38% zinc, and 1.5-2% iron, and is used for sheathing ships.

AID. Mixture of polyols and salts; used as a detergent; for stabilization of motor fuels carburettor cleaning. [UOP Inc] *

Ai Hao. A Chinese drug prepared from the leaves of *Artemisia vulgaris*. A remedy for hemorrhage and diarrhea.

Airbron. CAS 616-91-1; A proprietary preparation of acetylcysteine; a broncial inhalant. [British Drug Houses] *

Airedale. A range of dyes of various classes; for dyeing of leather. [Yorkshire Chemicals plc] *

Airets. Throat drops. [Richardson-Vicks Inc] *

Airex. PVC foam (soft and rigid), with a density from 50-400 kg/m³; used as protective padding and life jackets, as core material in sandwich construction used in boat building, automotive and aviation industries as well as in off-shore oil platforms, used in gymnastic mats, sealings and insulation. [Lonza AG]

Airflex®323. Self-crosslinking vinyl acetate-ethylene emulsion polymer; base for high-speed paper pkg. adhesives; nonwoven binder for wipes, towels, cover stock; formaldehyde-free. [Air Prods./Polymers]

Airflex® 426. Vinyl acetate-ethylene emulsion, carboxylated; base for adhesives; cellulosic protection; excellent adhesion to metal, polyester, other plastic films; sprayable, excellent water resist., heat sealable. [Air Prods./Polymers]

Airflex® 4500. Ethylene-vinyl chloride copolymer latex; crosslinkable flexible binder and saturant for paper and paperboard applications; excellent water, alcohol, and grease resist.; inherently flame retardant; nonwoven binder for fiberfill and high loft stock requiring flame retardancy; imparts flexibility and water resist. to caulks, mastics, barrier coats in building applications. [Air

---

* Trade name not verified as available   † Trade name verified as obsolete

Prods./Polymers]

Airflex® 4514. Ethylene-vinyl chloride copolymer latex; coating binder and saturant for paper and paperboard applications; inherently flame retardant; binder for flame retardant fabrics, heat sealable nonwovens; imparts flexibility and water resist. to caulks, mastics, barrier coats in building applications. [Air Prods./Polymers]

Airflex® 4530. Ethylene-vinyl chloride copolymer latex; coating binder and saturant for paper and paperboard applications; highest stiffness, tensile strength, and flame retardancy in series; excellent water, alcohol, and grease resist.; binder for filters, stiff flame retardant nonwovens; imparts flexibility and water resist. to caulks, mastics, barrier coats in building applications. [Air Prods./Polymers]

Airflex® 4814. Ethylene-vinyl chloride copolymer latex; carboxylated version of Airflex 4514; coating binder and saturant for paper and paperboard applications; improved heat and uv stability; also for nonwovens and textile applications. [Air Prods./Polymers]

Airflex® CA-50. Vinyl acetate-ethylene latex; proprietary stabilization; for carpetbacking applications; moderately stiff hand; high filler loads; formaldehyde-free. [Air Prods./Polymers]

Airflex® RB-8. Vinyl acetate-ethylene emulsion; PVAL stabilization; for carpetbacking applications; good fire retardant chars. [Air Prods./Polymers]

Airflex® RB-11. Vinyl acetate-ethylene latex emulsion; PVAL stabilization; for carpetbacking applications; for systems where firmer hand is desired. [Air Prods./Polymers]

Airflex® RB-35. Ethylene-vinyl chloride emulsion; surfactant stabilization; for carpetbacking applications; max. fire retardant chars. [Air Prods./Polymers]

Airflex® RB-40. Ethylene-vinyl chloride emulsion; surfactant stabilization; for carpetbacking applications; moderately stiff hand, excellent fire retardancy. [Air Prods./Polymers]

Airflex® TL-30. Vinyl acetate-ethylene latex; PVAL stabilization; for carpetbacking applications. [Air Prods./Polymers]

Airglow. Bright nickel plating process. [Hanshaw Chemicals] ‡

air-hardening steel. A trade term applied to a manganese tool steel containing some tungsten, which hardens when cooled in air.

Airlift® WB-1210, 1222, 1220, 1270, 1282, 1290. Water-based release agents; benefits including elimination of solv. emissions, reduction or elimination of wax build-up on molds, release and surf. quality equiv. to solv.-based products. [Air Prods./Polyurethanes]

Airol Roche. A proprietary preparation of tretinoin for treatment of acne. [Roche Products Ltd] *

Air Saltpeter. (Norwegian saltpeter). A mixture of calcium nitrite and nitrate. it is produced by passing air over a series of high intensity alternating arcs, absorbing the gases produced by lime water, and evaporating.

Airstrip. Waterproof nonocclusive plaster. [Smith & Nephew Pharmaceuticals Ltd] *

Airthane® PET-60D; PET-70D; PET-75D; PET-80A; PET-90A; PET-93A; PET-95A. TDI-PTMEG polyurethane prepolymer. [Air Prods./Polyurethanes]

Airthane® PPT-80A; PPT-95A;. TDI-PPG; polyurethane prepolymer. [Air Prods./Polyurethanes]

Airthane® PST-60D. TDI-polyester polyurethane prepolymer. [Air Prods./Polyurethanes]

Airthane® PST-80A; PST-90A. TDI-ester; polyurethane prepolymer. [Air Prods./Polyurethanes]

Airvol®. Polyvinyl alcohol. [Alfa Chemicals Ltd, distributor]

Airvol® 103, 107. CAS 9002-89-5; Polyvinyl alcohol, fully hydrolyzed; CAS 9002-89-5 super and fully hydrolyzed; offers high tensile strength and ease of film formation, excellent adhesive chars; super hydrolyzed grades for max. water and humidity resist. [Air Prods./Polymers]

Airvol® 125. CAS 9002-89-5; Polyvinyl alcohol, super hydrolyzed; see Airvol 103 for uses. [Air Prods./Polymers]

Airvol® 165. CAS 9002-89-5; Polyvinyl alcohol, super hydrolyzed; see Airvol 103 for uses. [Air Prods./Polymers]

Airvol® 203, 205. CAS 25213-24-5; Polyvinyl alcohol, partially hydrolyzed; CAS 25213-24-5; see Airvol 103 for uses. [Air Prods./Polymers]

Airvol® 205S, 523S, 540S. CAS 25213-24-5; Polyvinyl alcohol, partially hydrolyzed; see Airvol 103 for uses. [Air Prods./Polymers]

Airvol® 321, 325, 350. CAS 9002-89-5; Polyvinyl alcohol, fully hydrolyzed; see Airvol 103 for uses. [Air Prods./Polymers]

Airvol® 425, WS 42. CAS 25213-24-5; Polyvinyl alcohol, intermediate hydrolyzed; see Airvol 103. [Air Prods./Polymers]

Airvol® 523, 540. CAS 25213-24-5; Polyvinyl alcohol, partially hydrolyzed; see Airvol 103 for uses. [Air Prods./Polymers]

Airvol® 705, 723, 740. CAS 25213-24-5; Polyvinyl alcohol, partially hydrolyzed; low foam grades; see Airvol 103. [Air Prods./Polymers]

Airvol® MH-82, MM-14, MM-51, MM-81. CAS 9002-89-5; Polyvinyl alcohol, tackified; derived from fully hydrolyzed grades; tackified grades yield visc. aq. sol'ns. which possess tack when applied onto surfaces such as paper, reducing penetration. [Air Prods./Polymers]

Airvol® SH-72, SM-73. CAS 9002-89-5; Polyvinyl alcohol, tackified; derived from super hydrolyzed grades. [Air Prods./Polymers]

AIT. Pigment dispersions; for in-plant tinting of aqueous coatings. [Pacific Dispersions Inc] *

Aither's Lawn Sand Plus. Dichlorophen + ferrous sulfate; a moss killer/fertilizer mixture for turf. [R Aitken]

Aithesin. A proprietary preparation of alphaxalone and alphadolone acetate used in the induction of anesthesia. [Glaxo Laboratories] *

Aix Oil. Commercial varieties of edible olive oil. §

Ajax powder. An explosive consisting of potassium perchlorate, nitroglycerol, ammonium oxalate, wood meal, and small quantities of collodion cotton, and nitrotoluenes.

Ajicure® MY-24; Proprietary; accelerator for latent epoxy resin systems providing high storage stability, longer pot life; curable at lower temps. [Ajinomoto]

Ajicure® PN-23; Proprietary; accelerator for latent epoxy resin systems providing high storage stability, longer pot life; curable at lower temps. [Ajinomoto]

Ajidew A-100. CAS 98-79-3; PCA; nat. humectant used in cosmetics, soaps, dentifrices, medicinal supplies, tobacco, cellulose film, paper products, fiber products, paints; additive to dyeing agent, softening agent, finishing agent, and antistatic agent; intermediate for synthe-

---

‡ Trade name and manufacturer not verified   § Trade name without identified manufacturer

sis. [Ajinomoto] *

Ajidew N-50. CAS 28874-51-3; Sodium PCA; see Ajidew A-100. [Ajinomoto] *

ajkaite. (ajkite). A fossil resin found in Hungary.

Akarittom fat. A solid fat from *Parinarium laurinum*. It melts at 49-50 C and has an iodine value of 214.

Akaustan® A. Flameproofing agent for vegetable and animal fibers. [BASF AG; BASF plc]

Akbar. A condensation product of formaldehyde and p-toluidine; a rubber vulcanization accelerator; §

Akco Resins. A proprietary synthetic phenolic resin for varnish manufacture. §

Akfen Tablets. CAS 29110-48-3; Guanfaceine hydrochloride; used for the treatment of mild to moderate hypertension. (Available only in Ireland). [Wyeth Laboratories]

Akineton. CAS 514-65-8; Biperiden; anticholinergic; antiparkinsonian. [Knoll Pharmaceutical Co]

Akinetone. A trademark for a benzyl compound, $C_6H_5CH_2NH_2 \cdot CO \cdot C_6H_4 \cdot CO_2H$; an antispasmodic. [Knoll AG] *

Ak Mudar. (Akanda, Akra Rui, Erukku Erukkam). The bark of *Calotropis gigantea* and *C. procera;* an important Indian drug.

Akorex. CAS 8016-70-4; EINECS 232-410-2; Partially hydrogenated soybean oil. [Karlshamns]

Akorol®. Brake fluid components. [BASF AG]

Akrinol. CAS 7527-91-5; Acrisorcin; antifungal. [Schering Corp] *

Akrochem®9930 Zinc Oxide Transparent. CAS 3486-35-9; Precipitated basic zinc carbonate; accelerator-activator for transparent nat. and synthetic rubber goods, adhesives. [Akrochem]

Akrochem® Accelerator CZ-1. N,N dimethyl cyclohexyl ammonium dibutyl dithiocarbamate; accelerator for natural rubber, styrene/butadiene rubber, or latexes. [Akrochem]

Akrochem® Accelerator R. CAS 103-34-4; 4,4´-Dithio dimorpholine; accelerator for uses where a nonblooming or nonstaining sulfur donor is required; used for EV or semi-EV compounds, in synthetic and nat. rubbers. [Akrochem]

Akrochem® Accelerator VS. Zinc salt of dibutyl dithiophosphoric acid on silica carrier; nonblooming and nonstaining accelerator for ethylene-propylene-diene rubbers compounds and other sulfur-curable elastomers, especially hose and belt applications. [Akrochem]

Akrochem®Accelerator ZIPPAC. Zinc-amine dithiophosphate complex coated with mineral oil; accelerator; used with thiazoles and thiurams for fast cure rates in ENB-type ethylene-propylene-diene rubberss; nondiscoloring, nonstaining. [Akrochem]

Akrochem®Antioxidant 12. Butylated reaction production of p-cresol and dicyclopentadiene; nonstaining antioxidant, stabilizer, and antiozonant in polymers, including nat. and synthetic polyisoprene, neoprene, nitrile, styrene/butadiene rubber rubber, and latex. [Akrochem]

Akrochem®Antioxidant 16. Alkylated phenol; nonstaining antioxidant used in styrene/butadiene rubber latex compounding urethanes, ABS, and ethylene-propylene-diene rubbers. [Akrochem]

Akrochem®Antioxidant 33. Phenolic; nonstaining antioxidant for elastomers and latex. [Akrochem]

Akrochem®Antioxidant 58. 2-Mercapto-4(5)-methyl ben-zimidazole, zinc salt; nonstaining antioxidant for use in rubber compounds to improve heat resistance. [Akrochem]

Akrochem® Antioxidant DQ. 2,2,4-Trimethyl-1,2-dihydroquinoline, polymerized; antioxidant in rubber goods requiring resistance to high temps.; copper and manganese inhibitor; antiozonant. [Akrochem]

Akrochem®Antioxidant PANA. CAS 90-30-3; Phenyl-α–naphthylamine; antioxidant for rubber products; antiflex-cracking under dynamic stress. [Akrochem]

Akrochem® Antioxidant S. Octylated diphenylamine; antioxidant protecting dynamically stressed rubber products; antiozonant; antiflexcracking. [Akrochem]

Akrochem® Antiozonant MPD-100. Mixed diaryl p-phenylene diamine; antiozonant, antidegradant, antioxidant, and antiflexcracking agent for most diene polymers. [Akrochem]

Akrochem®Antiozonant PD-2. N-(1,3-dimethyl butyl)-N´-phenyl-p-phenylene diamine; antiozonant protecting rubber polymers against heat, oxidation, and flex cracking; copper, manganese inhibitor and styrene/butadiene rubber stabilizer. [Akrochem]

Akrochem® Cu.D.D. CAS 137-29-1; Copper dimethyl dithiocarbamate; accelerator for butyl rubber, ethylene-propylene-diene rubbers rubbers. [Akrochem]

Akrochem® DCBS Granules. Benzothiazyl 1-2-dicyclohexyl sulfenamide; accelerator for rubber industry. [Akrochem]

Akrochem®DOTG. CAS 97-39-2; Diorthotolyl guanidine; accelerator; provides activation for MBTS, MBT, and other thiazole accelerators in nat. rubber, styrene/butadiene rubber, NBR, and CR. [Akrochem]

Akrochem® DPG. CAS 102-06-7; Diphenyl guanidine; accelerator-activator for nat. rubber, styrene/butadiene rubber, and NBR. [Akrochem]

Akrochem® PEG 3350. Polyethylene glycol; activator for compounding with silica fillers; process aid, lubricant for rubber compounds (natural and synthetic); mold release for foam and mech. goods. [Akrochem]

Akrochem® Peptizer PTP. CAS 133-49-3; Penta-chlorothiophenol with activating and dispersing agents; peptizer for natural rubber, polyisoprene, styrene/butadiene rubber, polybutadiene, NBR, butyl, chloroprene and blends. [Akrochem]

Akrochem®Plasticizer LN. Naphthenic rubber process oil; low-visc. plasticizer. [Akrochem]

Akrochem® Powder Colors. Organic and inorganic pigments; colorants for rubber and certain thermoplastic polymers. [Akrochem]

Akrochem® P.P.D. CAS 98-77-1; Piperidinium penta-methylene dithiocarbamate; accelerator for latex; in rubber cements, compounds containing white factice and for tank linings; peptizing agent; nondiscoloring and nonstaining. [Akrochem]

Akrochem® Proaid FILL. Mixture of partially oxidized hydrocarbons; processing aid for injection molding of synthetic and natural rubber. [Akrochem]

Akrochem®Proaid FLOW. Mixture of fatty acids; processing aid for extrusion of synthetic and natural rubber. [Akrochem]

Akrochem®Proaid PEP. Zinc salts of high molecular weight fatty acids; processing aid for natural and synthetic rubber. [Akrochem]

Akrochem® Rubbersil RS-150/RS-200. Precipitated

---

* Trade name not verified as available   † Trade name verified as obsolete

amorphous silica; highly reinforcing filler for use in synthetic and natural rubber compounding, for mech. rubber goods, tires, adhesives, footwear, soling. [Akrochem]

Akrochem® TBUT. CAS 1634-02-2; Tetrabutyl thiuram disulfide; accelerator for rubber industry. [Akrochem]

Akrochem®TDEC. Tellurium diethyl dithiocarbamate; fast-curing primary or sec. accelerator for use in natural rubber, styrene/butadiene rubber, NBR, ethylene-propylene-diene rubbers, and butyl. [Akrochem]

Akrochem® TETD. CAS 97-77-8; Tetraethyl thiuram disulfide; accelerator for rubber industry. [Akrochem]

Akrochem® Thio No. 1. CAS 102-08-9; N,N´ diphenyl thiourea; accelerator for CR latex, natural rubber latex, and cements, ethylene-propylene-diene rubbers sponge compounds; activates thiazole accelerators; essentially nondiscoloring. [Akrochem]

Akrochem® TMTD. CAS 137-26-8; Tetramethylthiuram disulfide; very act., sulfur-bearing, nondiscoloring org. accelerator and activator; for curing systems requiring very low or no sulfur and for butyl and ethylene-propylene-diene rubbers compounds. [Akrochem]

Akrochem®Z.B.E.D. Zinc dibenzyl dithiocarbamate; accelerator; nondiscoloring and nonstaining. [Akrochem]

Akrochem®ZDA Powd. CAS 14643-87-9; Zinc diacrylate; activator for rubber compounding. [Akrochem]

Akrochem®Z.P.D. Zinc pentamethylene dithiocarbamate; ultra-accelerator for dry, natural rubber, especially footwear; nonpigmenting and nonstaining. [Akrochem] *

Akrochlor. Chlorinated paraffins; fire retardants. [Akrochem]

Akrodye. Dyes. [Akrochem]

Akrofax 900C. Vulcanized vegetable oil; rubber processing aid; absorbent for mineral oils and other liquid plasticizers on mill and Banbury equip.; speeds incorporation of fillers; flow promoter; provides unique surface finish to vulcanized rubber goods; improves ozone resistance. [Akrochem]

Akrofax A. Vulcanized vegetable oil; rubber processing aid; absorbent for mineral oils and other liquid plasticizers on mill and Banbury equip.; speeds incorporation of fillers; flow promoter; provides unique surface finish to vulcanized rubber goods; improves ozone resistance. [Akrochem]

Akrofax B Light. Sulfur-type vulcanized vegetable oil; extender, processing aid and softener for colored rubber products, especially for dimensional stability in tubing and calendered goods. [Akrochem]

Akroflex DAZ. A proprietary amine blend used as a combined antioxidant and antiozonant. [DuPont UK] *

Akroform® ETU-22 PM. Ethylene thiourea on compatible polymeric binder carrier; accelerator imparting a high state of cure to neoprene compounds; nonstaining and nondiscoloring. [Akrochem] *

Akrol. A proprietary synthetic resin. §

Akrolease®E-9410. Silicone polymer; release agent for PU, polyester, or epoxy materials from plastic or metal molds; corrosion inhibitor for metal. [Akrochem]

Akroplast®. Thermoplastic pigment dispersions; colorants for vinyls, acrylics, polyethylene, cellulose acetate butyrate, rubbers, urethanes, and phthalate pastes. [Akrochem]

Akrosperse® Color Masterbatches. Dispersions of Akrochem®Powder Colors in various elastomers, most commonly styrene/butadiene rubber and EPR. [Akrochem]

Akrosperse®Plasticizer Paste Colors. Org. and inorganic pigments; plasticizer pigment dispersions. [Akrochem]

Akrosperse® Water Paste Colors. Aq. pigment dispersions; colorants for latex products and adhesives. [Akrochem]

Akrotak 100. Pentaerythritol ester; tackifier for synthetic and natural rubber, adhesives and hot melts; provides wetting for filler/rubber; speeds up incorporation of min. fillers. [Akrochem]

Akrotherm. Histamine, acetylcholine and cholesterol; used for chilblains. [Napp Laboratories Ltd] *

Akrowax PE. CAS 9002-88-4; Low molecular weight polyethylene; process aid in natural and synthetic rubbers. [Akrochem]

Aktiplast. A range of zinc salts of unsaturated acids used as peptizing agents and dispersing agents and aromatic disulfides used as reclaiming agents in the rubber industry; used for molded and extruded articles, production of reclaims. [Bayer plc]

Aktisil AM. CAS 919-30-2; EINECS 213-048-4; γ-Aminopropyltriethoxysilane; filler for thermosets, thermoplastics. [Hoffmann Min.]

Aktisil EM. CAS 2530-83-8; EINECS 219-784-2; γ-Glycidoxypropyltrimethoxysilane; filler for thermosets. [Hoffmann Min.]

Aktisil MM. CAS 4420-74-0; EINECS 224-588-5; γ-Mercaptopropyltrimethoxysilane; filler for sulfur and metal oxide-cured systems. [Hoffmann Min.]

Aktisil PF 216. CAS 40372-72-3; EINECS 254-896-5; Bis-(3-(triethoxysilyl)propyl) tetrasulfane; filler for sulfur-cured systems. [Hoffmann Min.]

Aktisil VM. CAS 1067-53-4; EINECS 213-934-0; Vinyltris(β-methoxyethoxy)silane; filler for peroxide-cured systems. [Hoffmann Min.]

Aktren®. An analgesic, antipyretic drug. [Bayer AG]

Akulon®. Nylon 6 and 66. [AKU Holland] ‡

Akulon K and M. Proprietary grades of Nylon 6. [Algemene Industriele] ‡

Akulon R2. A grade of Nylon 66. [Algemene Industriele] ‡

Akuloy®J-75/30. Nylon 6 alloy with functionalized polyolefins, glass fiber-reinforced; engineering thermoplastic with improved dimensional stability for automotive, consumer products (power tool housings, furniture components), pump housings, impellers, fans, bearing retainers, gears, fasteners. [DSM]

Akuloy®J-75/30/HI. Nylon 6 alloy with functionalized polyolefins, glass fiber-reinforced; engineering thermoplastic with improved dimensional stability for automotive, consumer products (power tool housings, furniture components), pump housings, impellers, fans, bearing retainers, gears, fasteners. [DSM]

Akuloy®NY-75. Nylon 6 alloy with functionalized polyolefins; engineering thermoplastic with improved dimensional stability for automotive, consumer products (power tool housings, furniture components), pump housings, impellers, fans, bearing retainers, gears, fasteners. [DSM]

Akund. A vegetable down of the kapok class, from *Asclepias* species of South Africa.

Akwilox 133. CAS 68952-98-7; Brominated soybean oil; food additive in soft drinks for visc. adjustment. [Am. Chem. Services]

Akypo 1690 S. Laureth-5 carboxylic acid; additive for liquid heavy-duty detergent formulations; thickener for NaOH.

---

‡ Trade name and manufacturer not verified   § Trade name without identified manufacturer

[Chem-Y GmbH]

Akypo AD 100 SPC. Sodium PEG-3 lauramide carboxylate; surfactant for mild cosmetic products. [Chem-Y GmbH]

Akypo ITD 30 N. CAS 68891-17-8; Sodium trideceth-3 carboxylic acid; emulsifier for silicone oil; biodeg. [Chem-Y GmbH]

Akypo LF 1. Capryleth-6 carboxylic acid; low foaming surfactant for industrial, institutional and household cleaning; alkaline and acid stable; biodeg. [Chem-Y GmbH]

Akypo LF 2. CAS 107600-33-9; Capryleth-9 carboxylic acid; low foaming surfactant for industrial, institutional and household cleaning, cooling tower cleaners, disinfectant cleaners, high-pressure cleaners, metalworking fluids, electroplating, PU foam for orthopedic uses; alkaline and acid stable; biodeg. [Chem-Y GmbH]

Akypo LF 3. CAS 105391-15-9; Hexeth-4 carboxylic acid; low foaming surfactant for industrial, institutional and household cleaning; alkaline and acid stable; biodeg. [Chem-Y GmbH]

Akypo LF 4. Capryleth-9 carboxylic acid, hexeth-4 carboxylic acid; low foaming surfactant for industrial, institutional and household cleaning, cutting and drilling oils, drawing and rolling oils, water treatment products, high-pressure cleaners, cooling tower cleaners, electroplating, film developing; solubilizer; alkaline and acid stable; biodeg. [Chem-Y GmbH]

Akypo LF 4N. Sodium capryleth-9 carboxylate, sodium hexeth-4 carboxylate; low foaming surfactant for industrial cleaning; alkaline and acid stable; biodeg. [Chem-Y GmbH]

Akypo LF 5. CAS 105391-15-9; Buteth-2 carboxylic acid; low foaming surfactant for industrial, institutional and household cleaning; alkaline and acid stable; biodeg. [Chem-Y GmbH]

Akypo LF 6. Buteth-2 carboxylic acid, capryleth-9 carboxylic acid; low foaming surfactant for industrial, institutional and household cleaning, metalworking fluids, disinfectant cleaners, high-pressure cleaners, engine cleaners, automatic dishwash, cooling water systems, electroplating, textile pretreatment; alkaline and acid stable; biodeg. [Chem-Y GmbH]

Akypo MB 1614/1. Capryleth-4 carboxylic acid; surfactant for industrial cleaning. [Chem-Y GmbH] *

Akypo MB 2528S. CAS 105391-15-9; Hexeth-4 carboxylic acid; surfactant for industrial cleaning. [Chem-Y GmbH]*

Akypo NP 70. CAS 3115-49-9; Nonoxynol-8 carboxylic acid; emulsifier. [Chem-Y GmbH]

Akypo NTS. CAS 68987-89-3; Sodium laureth-6 carboxylate; detergent for carpet and upholstery cleaners especially aerosols; leak detector spray. [Chem-Y GmbH]

Akypo OCD 10 NV. Sodium deceth-2 carboxylate, sodium capryleth-2 carboxylate; defoamer for phosphoric acid industry (suitable for nitrate process). [Chem-Y GmbH]

Akypo OP 80. CAS 107628-08-0; 72160-13-5; Octoxynol-9 carboxylic acid; detergent, emulsifier; used in aq. solns. metalworking fluids, emulsion polymerization, film developing baths; lime soap dispersant; moderate foam. [Chem-Y GmbH]

Akypo OP 190. CAS 107628-08-0; 72160-13-5; Octoxynol-20 carboxylic acid; detergent, emulsifier; used in aq. sol'ns., metalworking fluids, electroplating; lime soap diseprsant; moderate foam. [Chem-Y GmbH]

Akypo RCS 60. CAS 68954-89-2; Ceteareth-7 carboxylic

acid; surfactant. [Chem-Y GmbH]

Akypo RLM 25. CAS 68954-89-2; Laureth-4 carboxylic acid; emulsifier, dispersant for emulsion and dispersion use; metalworking fluids; lime soap dispersant; moderate foam. [Chem-Y GmbH]

Akypo RLM 38. Laureth-5 carboxylic acid; surfactant. [Chem-Y GmbH]

Akypo RLM 100. Laureth-11 carboxylic acid; emulsifier for cosmetics applications; foam booster for cleaners, heavy-duty detergent formulations. [Chem-Y GmbH]

Akypo RLM 100 NV. CAS 33939-64-9; Sodium laureth-11 carboxylate; surfactant and additive for personal care products. [Chem-Y GmbH]

Akypo RLM 130. CAS 68954-89-2; Laureth-14 carboxylic acid; emulsifier. [Chem-Y GmbH]

Akypo RLM 160. CAS 27306-90-7; Laureth-17 carboxylic acid; emulsifier. [Chem-Y GmbH]

Akypo RLMQ 38. CAS 68954-89-2; Laureth-5 carboxylic acid; emulsifier, dispersant, additive for personal care products, household and industrial formulas; primary emulsifier for synthetic latex. [Chem-Y GmbH]

Akypo RO 20. Oleth-3 carboxylic acid; emulsifier for metalworking fluids; lime soap dispersant; very slow foaming. [Chem-Y GmbH]

Akypo RO 50. CAS 57635-48-0; Oleth-6 carboxylic acid; emulsifier for cleaning agents and metal cooling liqs.; chain lubricant; lime soap dispersant; slow foaming. [Chem-Y GmbH]

Akypo RO 90. CAS 57635-48-0; Oleth-10 carboxylic acid; emulsifier for metalworking fluids; lime soap dispersant; slow foaming; stable to hard water; biodeg. [Chem-Y GmbH]

Akypo RS 60. CAS 68954-89-2; 59559-30-7; Steareth-7 carboxylic acid; surfactant. [Chem-Y GmbH]

Akypo RS 100. CAS 68954-89-2; Steareth-11 carboxylic acid; surfactant. [Chem-Y GmbH]

Akypo RT 60. CAS 68954-89-2; Talloweth-7 carboxylic acid; surfactant. [Chem-Y GmbH]

Akypo TBP 40. Butoxynol-5 carboxylic acid; surfactant for metalworking fluids. [Chem-Y GmbH]

Akypo TBP 180. CAS 104909-82-2; Butoxynol-19 carboxylic acid; surfactant for zinc galvanization processes. [Chem-Y GmbH]

Akypo TFC-S. Laureth-5 carboxylic acid and sodium octyl sulfate; biodeg. foaming detergent, thickener, disinfectant for dairy, brewery, sanitary cleaning. [Chem-Y GmbH]

Akypo TPR. Sodium hexeth-4 carboxylate and trideceth-2; foam-suppressant surfactant for powd. cleaners for dishwash, steam carpet cleaners, metalworking industry. [Chem-Y GmbH]

Akypogene FP 35 T. CAS 61790-64-5; EINECS 263-155-5; TEA cocoate; cosmetics surfactant for shower, shampoo and bath formulations, liquid hand cleaner. [Chem-Y GmbH]

Akypogene HM 8. MEA lauryl sulfate, sodium PEG-6 cocamide carboxylate, disodium laureth sulfosuccinate; economical surfactant for manufacturing of mild shampoos, foam baths, shower baths, liquid soaps not irritating to optic mucosa. [Chem-Y GmbH]

Akypogene HM 12. Sodium C12-13 pareth sulfate, sodium PEG-6 cocamide carboxylate, disodium laureth sulfosuccinate, trideceth-2 carboxamide MEA; economical surfactant for prep. of mild shampoos, foam baths,

---

* Trade name not verified as available   † Trade name verified as obsolete

shower baths, and liquid soaps not irritating to optic mucosa. [Chem-Y GmbH]

Akypogene Jod F. Sodium laureth-11 carboxylate, iodine; antibacterial surfactant for food, dairy, beverage industries, agriculture., and hospitals. [Chem-Y GmbH]

Akypogene KTS. Ammonium polyacrylate and sodium laureth-6 carboxylate; base for rug shampoo and upholstery cleaner with antistatic and anticorrosive properties; especially for aerosols. [Chem-Y GmbH]

Akypogene SO. MEA-PPG-6-laureth-7 carboxylate, nonoxynol-2; surfactant for metalworking cooling lubricants; emulsifier, lime soap dispersant. [Chem-Y GmbH]

Akypogene VSM-N. Trideceth-2, sodium dodecylbenzene sulfonate; for manufacturing of liquid scour creams with excellent stability. [Chem-Y GmbH]

Akypogene WSW-W. Sodium laureth sulfate, sodium dodecylbenzene sulfonate, MEA laureth-6 carboxylate, cocamide DEA; surfactant for mild wool detergent formulations. [Chem-Y GmbH]

Akypogene ZA 97 SP. Potassium xylene sulfonate, potassium tallate, potassium cocoate; surfactant blend for liquid soap. [Chem-Y GmbH]

Akypomine®BC 50. CAS 68130-43-8; Fatty alcohol ether sulfate; surfactant for mineral industry; flotation agent for barite (selective); collector for flotation of typical salt minerals such as fluorspar, magnesite or scheelite. [Chem-Y GmbH]

Akypomine®BC/S. Fatty alcohol sulfate; flotation agent for barite. [Chem-Y GmbH]

Akypomine® MW 05. CAS 39464-66-9; Laureth-7 phosphate; surfactant for mining industry; fluorspar collector selective for barite. [Chem-Y GmbH]

Akypomine®P 191. Oleamine hydroxypropyl bistrimonium chloride, polyacrylamide; filtration auxiliary with flocculant for mining industry. [Chem-Y GmbH]

Akypo®-Muls 400. CAS 90453-59-1; PEG-9 stearamide carboxylic acid; nontoxic, biodeg. emulsifier for cosmetics, o/w emulsions. [Chem-Y GmbH]

Akypopress DB. CAS 9002-97-5; Synthetic polymer; depressant for metal ions. [Chem-Y GmbH]

Akypoquat 40. Oleoyl PG-trimonium chloride, stearoyl PG-trimonium chloride, behenoyl PG-trimonium chloride, palmitoyl PG-trimonium chloride, trideceth-2; environmentally safe laundry softener conc., textile softener; antistat, rewetting agent; fully biodeg. [Chem-Y GmbH]

Akypoquat 129. Isostearoyl PG-trimonium chloride and behenoyl PG-trimonium chloride; raw material for manufacturing of laundry softeners; molecule is fully biodeg. [Chem-Y GmbH] *

Akypoquat 131. CAS 69537-38-8; Behenoyl PG-trimonium chloride; raw material for cosmetic hair products, cream rinses; antistat; good wet and dry combing props.; molecule is fully biodeg. [Chem-Y GmbH]

Akypoquat 132. Lauroyl PG-trimonium chloride, hexylene glycol; raw material for cosmetic hair and skin products; molecule is fully biodeg. [Chem-Y GmbH]

Akypo CO 400. CAS 61788-85-0; PEG-40 hydrogenated castor oil; perfume solubilizer, o/w emulsifier for cosmetic products; eliminates oil bath turbidity. [Chem-Y GmbH]

Akyporox NP 15. CAS 27986-36-3; Nonoxynol-1; emulsifier for emulsion polymerization. [Chem-Y GmbH]

Akyporox NP 30. CAS 9016-45-9; Nonoxynol-3; for manufacturing of hair dye formulations; emulsifier for emulsion

polymerization. [Chem-Y GmbH]

Akyporox NP 40. CAS 9016-45-9; Nonoxynol-4; emulsifier for film developing baths. [Chem-Y GmbH]

Akyporox NP 90. CAS 9016-45-9; Nonoxynol-9; emulsifier. [Chem-Y GmbH]

Akyporox NP 95. CAS 9016-45-9; Nonoxynol-10; emulsifier for calcium stearate. [Chem-Y GmbH]

Akyporox NP 105. CAS 9016-45-9; Nonoxynol-10; emulsifier. [Chem-Y GmbH]

Akyporox NP 150. CAS 9016-45-9; 26027-38-3; Nonoxynol-15; emulsifier. [Chem-Y GmbH]

Akyporox NP 200. CAS 9016-45-9; Nonoxynol-20; emulsifier, wetting agent used in textile products, emulsion polymerization, degreasing baths, electroplating industry. [Chem-Y GmbH]

Akyporox NP 300V. CAS 9016-45-9; Nonoxynol-30; emulsifier for calcium stearate, emulsion polymerization. [Chem-Y GmbH]

Akyporox NP 1200V. CAS 9016-45-9; Nonoxynol-120; emulsifier for emulsion polymerization. [Chem-Y GmbH]

Akyporox OP 100. CAS 9002-93-1; Octoxynol-10; emulsifier; dust suppressant for coal mining industry. [Chem-Y GmbH]

Akyporox OP 115 SPC. CAS 9002-93-1; Octoxynol-12; emulsifier, wetting agent. [Chem-Y GmbH]

Akyporox OP 200. CAS 9002-93-1; Octoxynol-20. emulsifier. [Chem-Y GmbH] *

Akyporox OP 250 V. CAS 9002-93-1; Octoxynol-25; emulsifier for emulsion polymerization; perfume solubilizer. [Chem-Y GmbH]

Akyporox OP 400V. CAS 9002-93-1; Octoxynol-40; emulsifier for emulsion polymerization. [Chem-Y GmbH]

Akyporox RC 200. CAS 9004-95-9; Ceteth-20; solubilizer for cosmetic products. [Chem-Y GmbH]

Akyporox RLM 22. CAS 3055-93-4; Laureth-2; emulsifier for manufacturing of hair dye formulations; additive for manufacturing of snow from spray cans. [Chem-Y GmbH]

Akyporox RLM 40. CAS 5274-68-0; Laureth-4; emulsifier for manufacturing of hair dye formulations, cosmetic aerosols, oil bath formulations, window cleaners, hand cleaners, heavy-duty detergents. [Chem-Y GmbH]

Akyporox RLM 80V. Laureth-8; emulsifier for manufacturing of cosmetic aerosols, heavy-duty detergents, all-purpose cleaners. [Chem-Y GmbH]

Akyporox RLM 160. CAS 9002-92-0; Laureth-16; emulsifier for emulsion polymerization. [Chem-Y GmbH]

Akyporox RO 90. CAS 9004-98-2; Oleth-9; emulsifier, wetting agent for textile products, metalworking fluids, all-purpose cleaners, hand cleaners, creams and lotions. [Chem-Y GmbH]

Akyporox RTO 70. CAS 9004-98-2; Oleth-7; emulsifier for emulsion polymerization. [Chem-Y GmbH]

Akyporox RZO 30. CAS 61791-12-6; PEG-3 castor oil; surfactant for metalworking cooling lubricants. [Chem-Y GmbH]

Akyporox SAL SAS. CAS 142-87-0; Sodium lauryl sulfate; detergent, shampoo base, emulsifier for emulsion polymerization. [Chem-Y GmbH]

Akyposal 23 ST 70. Sodium C12-13 pareth sulfate; detergent; shampoo base. [Chem-Y GmbH]

Akyposal 100 DAL. CAS 107600-36-2; TIPA-laureth sulfate; emulsifier for bath oils; suited for anhydrous formulations; biodeg. [Chem-Y GmbH]

---

‡ Trade name and manufacturer not verified   § Trade name without identified manufacturer

Akyposal 2010 S. Sodium laureth sulfate, cocamide DEA, glycol distearate; surfactant, pearling agent, foam stabilizer for cosmetics; biodeg. [Chem-Y GmbH]

Akyposal 2010 SD. Sodium PEG-6 cocamide carboxylate, glycol distearate; surfactant, pearling agent for cosmetics; biodeg. [Chem-Y GmbH]

Akyposal 9278 R. CAS 9004-82-4; Sodium laureth sulfate; emulsifier for emulsion polymerization. [Chem-Y GmbH]

Akyposal ALS 33. CAS 2235-54-3; EINECS 218-793-9; Ammonium lauryl sulfate; detergent, emulsifier; shampoo base; used in emulsion polymerization. [Chem-Y GmbH]

Akyposal BA 28. Sodium trideceth sulfate; detergent for personal care, dishwashing, and textile products. [Chem-Y GmbH] *

Akyposal BD. CAS 69011-84-3; Sodium octoxynol-6 sulfate; emulsifier for emulsion polymerization. [Chem-Y GmbH]

Akyposal DS 28. CAS 68957-18-6; Sodium laureth sulfate; detergent; shampoo base. [Chem-Y GmbH]

Akyposal DS 56. CAS 68957-18-6; Sodium C12-13 pareth sulfate; detergent; shampoo base. [Chem-Y GmbH]

Akyposal EO 20 MW. CAS 9004-82-4; Sodium laureth sulfate; detergent, base for shampoos and bubble baths, emulsifier for emulsion polymerization. [Chem-Y GmbH]

Akyposal HF 28. Sodium laureth sulfate, magnesium laureth-16 sulfate; base for mild shampoos and shower products. [Chem-Y GmbH]

Akyposal MGLS. CAS 3097-08-3; Magnesium lauryl sulfate; detergent for shampoos, foam bath. [Chem-Y GmbH]

Akyposal MLES 35. CAS 68184-04-3; MEA-laureth sulfate; base, foamer for shampoo and bubble baths. [Chem-Y GmbH]

Akyposal MLS 30. CAS 4722-98-9; MEA-lauryl sulfate; detergent, shampoo base, shower baths. [Chem-Y GmbH]

Akyposal MS SPC. CAS 9004-82-4; Sodium laureth sulfate; detergent. [Chem-Y GmbH]

Akyposal NAF. Sodium dodecylbenzene sulfonate; emulsifier for emulsion polymerization. [Chem-Y GmbH]

Akyposal NLS. CAS 151-21-3; Sodium lauryl sulfate; detergent. [Chem-Y GmbH]

Akyposal NPS 60. CAS 9014-90-8; Sodium nonoxynol-6 sulfate; emulsifier for emulsion polymerization. [Chem-Y GmbH]

Akyposal NPS 100. CAS 9014-90-8; Sodium nonoxynol-10 sulfate; emulsifier for polymerization. [Chem-Y GmbH]

Akyposal OP 80. Octoxynol-9 carboxylic acid; emulsifier for emulsion polymerization. [Chem-Y GmbH]

Akyposal OPS 85. Sodium octoxynol-9 sulfate. [Chem-Y GmbH] *

Akyposal RLM 56 S. CAS 3088-31-1; Fatty alcohol ether sulfate; detergent for personal care products, liquid soaps, dishwashing products. [Chem-Y GmbH]

Akyposal RLM 70. CAS 9004-82-4; Fatty alcohol ether sulfate; detergent; shampoo base. [Chem-Y GmbH]

Akyposal TIPA 45. CAS 661-61-6; TIPA lauryl sulfate; surfactant for cosmetics, shampoos, shower and foam baths. [Chemsal]

Akyposal TLS 42. CAS 139-96-8; EINECS 205-388-7; TEA-lauryl sulfate; detergent; base for personal care products and car shampoos; foaming agent for agrochemicals, fire extinguishers. [Chem-Y GmbH]

Akyposept B. Benzylhemiformal. [Chem-Y GmbH]

Akypo®Soft 45 NV. CAS 53610-02-9; Sodium laureth-6 carboxylate; cosmetics surfactant for baby care products, shampoos, foam baths, medicinal liquid soaps; biodeg. [Chem-Y GmbH]

Akypo®Soft 100 MgV. CAS 99330-44-6; Magnesium laureth-11 carboxylate; detergent, emulsifier, wetting agent for shampoos, liquid soaps, foam baths, low irritation formulas; biodeg. [Chem-Y GmbH]

Akypo®Soft 100 NV. CAS 68987-89-3; Sodium laureth-11 carboxylate; detergent, emulsifier, wetting agent for shampoos, liquid soaps, foam baths, feminine hygiene products, low irritation formulas; PU foam for orthopedic use; biodeg. [Chem-Y GmbH]

Akypo®Soft 160 NV. CAS 33939-64-9; Sodium laureth-17 carboxylate; surfactant for contact lens cleaning fluids; very low eye irritation. [Chem-Y GmbH]

Akypo®Soft KA 250 BV. CAS 107628-03-5; Sodium PEG-6 cocamide carboxylate; economical surfactant for formulation of mild shampoos, foam baths, shower baths, liquid soaps; mild to skin and eyes; biodeg. [Chem-Y GmbH]

Akypostat MA 35. CAS 38720-61-5; Myreth-5 carboxylic acid; antistat and antifogging agent for polyester film; anticondensation agent for foil; biodeg. [Chem-Y GmbH]

alabaster. A form of gypsum, $CaSO_4 \cdot 2H_2O$ used for ornamental carvings. *Also see* onyx of Tecali.

alabaster, Oriental. A compact form of marble, $CaCO_3$.

Alacsan T. Quaternary sulfate. [Rhone-Poulenc Surf. Canada]

Alagan. CAS 15972-60-8; Alachlor; herbicide. [Agan Chemical Manufacturers Ltd]

Alamask. Industrial deodorants. [May & Baker Ltd] *

Alanex. CAS 15972-60-8; Active ingredient: alachlor; 2-chlor-2′,6′-diethyl-N-(methoxymethyl)-acetanilide; pre-emergence and pre-plant incorporated herbicide for the control of most annual grasses and certain broadleaf weeds. [Agan Chemical Manufacturers Ltd]

Alan-gilan. Cananga oil, a neutral oil from *Cananga odorata*.

α-alanine. (2-aminopropionic acid; 2-aminopropanoic acid) Amino acid; CH_3CHNH_2COOH; CAS 56-41-7 (L-form); EINECS 200-273-8; microbiological research, biochemical research, dietary supplement. [Penta Mfg.; U.S. Biochemical]

Alar. Growth regulator. [Murphy Chemical Ltd] †

Alargan. An alloy of aluminum and silver, the surface having been dusted with platinum black and hammered or subjected to pressure; a platinum substitute. §

Alasil. A proprietary preparation of calcium acetyl-salicylate and colloidal aluminum hydroxide. §

Alathon® H5234. CAS 9002-88-4; HDPE copolymer; high flow injection molding resin for frozen food containers, drink cups, housewares. [OxyChem/Alathon]

Alathon® L5440. CAS 9002-88-4; HDPE resin; blow molding resin. [OxyChem/Alathon]

Alathon® M5560. CAS 9002-88-4; HDPE copolymer resin; film-grade resin for water bath or chill roll film; food contact applications. [OxyChem/Alathon]

Alathon® M6062. CAS 9002-88-4; HDPE homopolymer resin; injection molding resin for crates, cases. [OxyChem/Alathon]

Alazine. CAS 15972-60-8, 1912-24-9; Active ingredients: alanex plus atranex; ready formulated mixture of alachlor plus atrazine for use as a selective pre-emergence

---

* Trade name not verified as available   † Trade name verified as obsolete

herbicide. [Agan Chemical Manufacturers Ltd]

Alba. Petrolatum USP. [Witco]

Albacar. Highly refined calcite. [Pfizer International]

Albacer. A proprietary synthetic wax for increasing melting point of waxes. §

Albalan. CAS 68201-49-0; EINECS 269-220-4; Lanolin wax; emollient, emulsifier; forms stable w/o emulsions. [Westbrook Lanolin]

Albalith. A white, light-resisting lithopone; used in the paint and rubber industries. §

Albalon-A Liquifilm®. Naphazoline hydrochloride, antazoline phosphate; decongestant, anti-inflammatory. [Allergan, Inc]

Albalon Liquifilm®. CAS 550-29-2; Naphazoline hydrochloride ophthalmic solution; vasoconstrictor. [Allergan, Inc]

Albamycin. A proprietary trade name for the calcium salt of Novobiocin. [Upjohn] *

Albamycin Capsules. CAS 1476-53-5; Novobiocin sodium; antibacterial. [Upjohn]

Albamycin G.U. A proprietary preparation of novobiocin calcium and sulfamethizole; urinary antiseptic. [Upjohn Ltd] *

Albamycin T. A proprietary preparation containing novobiocin and tetracycline; an antibiotic. [Upjohn Ltd] *

Albanite. A bituminous material found in Albania.

Albanose. A leucite rock.

Albaphos Dental Na 211. A proprietary trade name for sodium monofluorophosphate, a fluorine component for toothpastes, the toxic effects of which are only 1/3 of those of sodium fluoride. [Hoechst UK] *

Albar-40. CAS 1702-17-6; 2,4-D; herbicide. [Makhteshim Chemical Works Ltd]

Albar-M. CAS 94-74-6; MCPA; herbicide. [Makhteshim Chemical Works Ltd]

albarium. A lime obtained by burning marble; used for stucco.

Albar-Super. CAS 1702-17-6; 2,4-D; herbicide. [Makhteshim Chemical Works Ltd]

Albaryt. Finely divided barytes. [Sachtleben Chemie GmbH]

Albata. A nickel-brass or low nickel-silver containing about 8% nickel.

Albatex. Dyeing and printing assistant. [Ciba plc] *

Albatex OR. A proprietary trade name for a nonfoaming polyvalent amide; used as a leveling agent for vat dyes. [Ciba plc] *

Albatra metal. A nickel silver; it contains 57.5% copper, 22.5% zinc, l8.75% nickel, and 1.25% lead.

Albegal. Dyeing and printing assistant. [Ciba plc] *

Albegal CL. A proprietary trade name for an ester of sulfonated fat; used as a leveling agent in wool dyeing. [Ciba plc] *

Alberene. A blue-grey soapstone mined in Virginia.

Alberger® Natural Flake. CAS 7647-14-5; Sodium chloride; crystalline salt refined by Alberger process. [Akzo Salt]

Alberit MF. A proprietary melamine formaldehyde thermosetting molding compound; used in the manufacture of tracking-resistant moldings for the electrical industry. [Canadian Hoechst] *

Alberit MP. A propretary melamine/phenol-formaldehyde thermosetting molding compound. [Canadian Hoechst]*

Alberit PF. A proprietary phenol-formaldehyde resin thermosetting molding compound. [Canadian Hoechst] *

Alberit VP. A proprietary unsaturated polyester thermosetting molding compound used in the production of impact-resistant moldings. [Canadian Hoechst] *

Albert. Basic slag for fertilizing purposes. [Fisons plc, Horticulture Div] *

Albertat. A proprietary range of chemical fillers, extenders and additives for products containing synthetic resins, such as additives for thickening and preventing setting in paints and varnish. [Chemische Werke, Albert] *

Albertol. Rosin modified phenolic resins; used in printing inks, paints and varnishes. [Hoechst UK]

Albertol 142-R. Butyl phenol formaldehyde. [Chemische Werke, Albert] *

Albertol 175-A. The aluminum salt of unesterified Albertol IIlL (qv). [Chemische Werke, Albert] *

Albertol 237-R. Di-iso butylphenol-formaldehyde. [Chemische Werke, Albert] *

Albertol 326-R (387L). A rosin modified phenolic resin used in aircraft primers made from 1 part diane (qv), 1 part rosin and 0.1 part paraformaldehyde. [Chemische Werke, Albert] *

Albertol 347Q. An ester gum-phenolic combination made from xylenol-formaldehyde rosin, pentaerythritol, and glycerogen (qv). [Chemische Werke, Albert] *

Albertol 369-Q (209-L). An ester gum-phenolic combination made from phenol-formaldehyde, rosin, pentaerythritol, and glycerogen (qv). [Chemische Werke, Albert] *

Albertol IIlL. A phenol-resin condensation product melting at 106-133 C; saponification value 15.8; insoluble in alcohol, but soluble in linseed oil; stated to be a good substitute for kauri gum in the manufacture of oil varnishes; rosin glycerine-diane-formaldehyde condensate in the presence of alkali. [Chemische Werke, Albert] *

Albigen®A. Textile stripping agent, leveling agent, washing-off agent; prevents re-exhaustion of dyes in washing of prints. [BASF; BASF AG]

Albiogen. Tetramethylammonium oxalate. §

albion metal. A sheet of metal containing tin and lead; it is formed by pressing together sheets of these metals.

Aloferrin. (Aboferine) A phospho-albumen preparation of iron, a tonic. §

Aboleum. Oil insecticide. [Plant Protection] *

Abolineum. Oil insecticide. [ICI Chem. & Polymers Ltd]

Abolit. A proprietary phenol-formaldehyde synthetic resin.§

Abondur. A Bondur alloy (See Bondur) coated on each side with pure aluminum to improve corrosion resistance. §

Albor Die Steel. A proprietary steel containing small amounts of chromium, molybdenum, and carbon. §

Aboresin. A proprietary urea-formaldehyde synthetic resin; molding composition. §

Abral. Flux for use with aluminum bronzes, silicon bronzes and high tensile brasses. [Foseco (F.S.) Ltd]

Abras Propachlor. A pre-emergence herbicide for various horticultural crops. [ICI Chem. & Polymers Ltd]

Abrass. CAS 1918-16-7; Propachlor; pre-emergence herbicide for various horticultural crops. [ICI Agrochemicals]

Albrichrome. Textile dyestuffs. [Albright & Wilson Ltd, Phosphates & Speciality Business] *

Albicide. Fungicide. [Albright & Wilson Ltd, Phosphates & Speciality Business]

Albrifloc. Flocculating agent. [Albright & Wilson Ltd, Phosphates & Speciality Business]

---

‡ Trade name and manufacturer not verified

§ Trade name without identified manufacturer

Albrightex. Textile optical brighteners. [Albright & Wilson Ltd, Phosphates & Speciality Business]

Albrilan. Textile dyestuffs. [Albright & Wilson Ltd, Phosphates & Speciality Business] *

Albrilene. Textile dyestuffs. [Albright & Wilson Ltd, Phosphates & Speciality Business] *

Albrilon. Textile dyestuffs. [Albright & Wilson Ltd, Phosphates & Speciality Business] *

Albrilube. Textile lubricant. [Albright & Wilson Ltd, Phosphates & Speciality Business] *

Albrinol. Textile dyestuffs. [Albright & Wilson Ltd, Phosphates & Speciality Business] *

Albrinyl. Textile dyestuffs. [Albright & Wilson Ltd, Phosphates & Speciality Business] *

Albriquest. Sequestering agent. [Albright & Wilson Ltd, Phosphates & Speciality Business] *

Albriscour. Textile scouring agent. [Albright & Wilson Ltd, Phosphates & Speciality Business] *

Albrisolve. Textile dyeing agent. [Albright & Wilson Ltd, Phosphates & Speciality Business] *

Albrisperse. Textile dispersing agent. [Albright & Wilson Ltd, Phosphates & Speciality Business] *

Albritone. Textile leveling agent. [Albright & Wilson Ltd, Phosphates & Speciality Business] *

Albrivap. Boiler scale inhibitor. [Albright & Wilson Ltd, Phosphates & Speciality Business] *

Albumaid Preparations. A group of proprietary preparations of beef protein hydrolysate, amino acids, carbohydrates, vitamins and minerals; used in the treatment of aminoacidaeraias and mal-absorption syndromes. [Scientific Hospital Supplies] *

albumen. (dried egg white; egg albumin; albumin) CAS 9006-50-2; Protective colloid and emulsifier in baking; textile dye mordant; adhesives and veneers. [Penta Mfg.; Schweizerhall; Spice King; U.S. Biochemical]

Albuminar. Albumin human; blood volume supporter. [Armour Pharmaceutical Co]

Albumotope I-131. Albumin, iodinated I 131 serum; diagnostic aid; radioactive agent. [Bristol-Myers Squibb Co Inc] *

Alburex. Vegetable proteins; used in animal feedstuff. [Roquette (UK) Ltd]

Albustix. A prepared test strip of tetra-bromphenol blue with a citrate buffer; used to detect protein in urine. [B C Ames] *

Albutannin. (Protan). Albumen tannate.

Alcaine. CAS 5875-06-9; Proparacaine hydrochloride; anesthetic. [Alcon Laboratories Inc]

Alcalase®0.6 L. Proteolytic enzyme prepared from *Bacillus licheniformis*; major component: Subtilisin Carlsberg; food-grade enzyme for hydrolysis of proteins; used for improvement of functionality of vegetable and animal proteins; endopeptidase action on serine sites. [Novo Nordisk] *

Alcalase® 2.0 T. Proteinase; enzyme for laundry powd. detergent. [Novo Nordisk] *

Alcamine®. A versatile range of cationic softening agents for all textile fibers; includes dyebath stable and durable types. [Allied Colloids Ltd]

Alcamizer 1. CAS 12539-23-0, 11097-59-9; EINECS 234-319-3; Magnesium aluminum carbonate; heat stabilizer for PVC that reacts in a unique way with HCl in PVC; offers high heat stability, nontoxicity, and high transparency. [Kyowa Chem. Industry]

Alcamizer 2. CAS 96492-31-8, 11097-59-9; EINECS 234-319-3; Magnesium aluminum carbonate; heat stabilizer for PVC. [Kyowa Chem. Industry]

Alcan AA-100. CAS 1344-28-1; Activated alumina; in selective absorption processes; as starting material for catalyst. [Alcan]

Alcan C-70, C-71, C-72, C-73, C-75. CAS 1344-28-1; Calcined alumina; ceramic grade alumina for refractory bricks, whitewares. [Alcan]

Alcan FRF 5, 10, 20, 30, 40, 60, 80, 85. Alumina hydrate; flame retardant and smoke suppressant props. for plastics and rubber industries. [Alcan]

Alcan FRF LV2, LV4, LV5, LV6, LV7, LV8, LV9. Alumina hydrate; flame retardant and smoke suppressant. [Alcan]

Alcan GB-1S. Wrought, non heat-treatable 99.5% aluminum alloy. [Hoechst UK] *

Alcan GB-2S. Wrought, non heat-treatable commercially pure aluminum alloy. [Hoechst UK] *

Alcan GB-3S. Wrought, non heat-treatable aluminum alloy; stronger and harder than 2S. [Hoechst UK] *

Alcan GB-50S. Wrought, heat-treatable aluminum alloy; forms well in the W (solution heat-treated) condition. [Hoechst UK] *

Alcan GB-99.8%. Wrought, non heat-treatable high purity aluminum alloy. [Hoechst UK] *

Alcan GB-100. Cast, non heat-treatable commercially pure aluminum alloy. [Hoechst UK] *

Alcan GB-160. Cast, non heat-treatable aluminum alloy. [Hoechst UK] *

Alcan GB-350. Cast, heat-treatable aluminum alloy specially impact resisant; good resistance to marine conditions. [Hoechst UK] *

Alcan GB-B51S, 65S. Wrought, heat-treatable medium strength aluminum alloys. [Hoechst UK] *

Alcan GB-B53S, 54S, D54S, A56S, M57S. Wrought, non heat-treatable aluminum alloys in which magnesium is the main additive. [Hoechst UK] *

Alcan GB-B116. Cast, heat-treatable aluminum alloy available in four conditions: M (as cast), P (precipitation treated), W (solution treated) and WP (fully heat treated). [Hoechst UK] *

Alcan GB-B320. Cast, non heat-treatable medium strength aluminum alloy. [Hoechst UK] *

Alcan H-10. Alumina hydrate; for production of low-iron aluminum sulfate, alumina-based catalysts, sodium aluminate and other aluminum salts, coated titanium dioxide pigments, ceramic and glass products, fire-retardant carpet backing. [Alcan]

Alcan H-10-08; Alumina trihydrate; filler and extender in plastics, resins, rubber, latex foams, especially where flame retardance is important. [Alcan]

Alcan Aluminum Fluoride. Aluminum fluoride, aluminum oxide, silica; as electrolyte in reduction of alumina to aluminum metal; as flux in remelting and refining of aluminum and its alloys; opacifier aid in production of ceramic enamels, glass, and glazes. [Alcan]

Alcan Aluminum Sulfate Liquid. CAS 10043-01-3; Aluminum sulfate; in pulp and paper mills, water purification plants. [Alcan]

Alcan Recovered Cryolite. Sodium fluoroaluminate (87-90%), aluminum oxide (2%), sodium sulfate (4%), sodium carbonate (1-1.5%); source of fluorine; as ceramic flux and opacifier aid in the production of vitreous enam-

---

* Trade name not verified as available   † Trade name verified as obsolete

els, glass, glazes, abrasives; metallurgical flux in refining of aluminum and its alloys. [Alcan]

Alcan Superfine 4, 7, 11. Alumina hydrate; filler for fire retardants/smoke suppressants for plastics, rubber, paints, adhesives, adhesive tapes and in toothpaste, cosmetics, polishes and waxes; in paper coatings. [Alcan]

Alcaphos 24. Strongly alkaline silicated solid; formulated for heavy duty soak cleaning in steel fabricated metals. [Invequimica & CIA SCA] *

Alcapsol®. A range of synthetic microencapsulation products for the replacement of gelatine and natural compounds in paper manufacture. [Allied Colloids Ltd]

Alcement. (Alcacement). Fused cement prepared in the electric furnace from bauxite and lime. It contains approximately 40% CaO, 40% Al_2O_3, 10% SiO_2, and 10% Fe_2O_3. §

Alchemie. Epoxy resins, polyurethane resins and RTV silicone rubber systems; used in casting, moldmaking and laminated structures, engineering patterns and tool-making, electrical potting and encapsulations. [Alchemie Ltd]

Alchemix. Epoxy and polyurethane resins and RTV silicone rubber systems; used in casting, moldmaking and laminated structures, engineering patterns and toolmaking, electrical potting and encapsulations. [Alchemie Ltd]

Alchemy. Metal carboxylates; driers for paint or printing ink. [Manchem Ltd] *

Alcian. Dyestuffs, blues, greens and yellow dyes formed by introducing chloromethyl groups into phthalocyanin and its derivatives by means of dichlorodimethyl either in pyridine containing aluminum chloride. [ICI Chem. & Polymers Ltd] †

Alcin. A proprietary preparation containing sodium magnesium aluminum silicate and basic magnesium aluminate; an antacid. [Reckitts Colours Ltd] *

Alclar®. A range of synthetic microencapsulation products for the replacement of gelatine and natural compounds in paper manufacture. [Allied Colloids Ltd]

Alcoa 2-S. A proprietary trade name for a commercially pure aluminum. §

Alcoa 3-S. A proprietary alloy of aluminum, containing small amounts of copper, iron, silicon, and zinc. §

Alcoa 24-S. A proprietary wrought aluminum alloy containing 93.7% aluminum, 1.5% magnesium, 4.2% copper, 0.6% manganese. §

Alcoa 32-S. A proprietary alloy of aluminum with 12% silicon, 0.8% nickel, 1% magnesium, and 0.8% copper.§

Alcoa 43. A proprietary alloy of aluminum and silicon; contains 5% silicon. §

Alcoa 47. A proprietary alloy of aluminum with 12.5% silicon.§

Alcoa 108. A proprietary aluminum alloy with 3% silicon and 4% copper. §

Alcoa 112. A proprietary aluminum alloy containing 7-8.5% copper, 1-2% zinc, and up to 1.7% of other metals, mostly iron. §

Alcoa 122. A proprietary alloy of aluminum with 10% copper, 0.2% magnesium, and 1.2% iron. §

Alcoa 145. A proprietary aluminum alloy containing 10% zinc, 2.5% copper, and 1.2% iron. §

Alcoa 220-TA. A proprietary aluminum alloy containing aluminum with 10% magnesium. §

Alcoa 356. A proprietary alloy of aluminum containing 4%

silicon, 0.3% magnesium. §

Alcoa 515. A proprietary alloy containing aluminum with magnesium, silicon, and iron. §

Alcoa 535. A proprietary alloy of aluminum with 0.25% chromium, 1.25% magnesium, and 0.7% silicon. §

Alcobez. A comprehensive range of compounds to control corrosion, scale build-up and biofouling in boiler and cooling water systems. [Allied Colloids Ltd]

Alcobon. CAS 2022-85-7; A proprietary preparation of flucytosine; a systemic antifungal agent. [Roche Products Ltd] *

Alcocare® 1000. Surfactant/humectant blend; provides foam, feel, visc. and mildness to health care formulations. [Rhone-Poulenc Surf.]

Alcocare® 2011. Chloroxylenol PCMX concs.; high foaming antimicrobial conc.; alkali stable; for handwash, surgical scrubs, health care. [Rhone-Poulenc Surf.]

Alcocare® 3020. Iodine conc.; low foaming antimicrobial conc.; acid stable; for surgical scrubs, health care. [Rhone-Poulenc Surf.]

Alcodet® 218. CAS 9004-83-5; PEG-10 isolauryl thioether; emulsifier, wetting agent, detergent, carbon soil and grease cleaners, metal cleaning specialties, steel processing, textile scouring, insecticide emulsions, cosmetics, wood pulp and paper industries. [Rhone-Poulenc Surf.; Rhone-Poulenc France]

Alcodrill HPD-D. Dry, free-flowing, powdered carboxylate copolymers; deflocculant in water based drilling. [Alco Chemical Corp]

Alcodrill HPD-L. A polycarboxylate copolymer solution; deflocculant for water based drilling. [Alco Chemical Corp]

Alcofix®. Afterfixing agents for improving the wet-fastness properties of dyes and prints and as a pitch fixative in paper manufacture. [Allied Colloids Ltd]

Alcoflood®. A complete range of anionic, nonionic, and cationic polymers for enhanced oil recovery in a choice of molecular weights to suit both tight and open, oil or water wet formations; suitable for long- range mobility control or the various short range treatments used for water shut-off. [Allied Colloids Ltd]

Alcoform. A solution of formaldehyde in one of a variety of alcohols; used for production of butylated resins and methylated resins; methyl alcoforms are used for the production of ion-exchange resins. [Synthite Ltd]

Alcogas. Fuels containing alcohol and liquid fuels containing alcohol and a hydrocarbon; for chemicals, medicines and pharmaceutical preparations. [Quantum Chemical Corp]

Alcogum 296-W. A high viscosity sodium polyacrylate thickener; used in adhesives, paint, cement additive, protective colloid. [Alco Chemical Corp]

Alcogum 310. A high molecular weight inverse-emulsion copolymer (organic phase medical white oil); for adhesives, coatings. [Alco Chemical Corp] †

Alcogum 9639. CAS 9003-03-6; Ammonium polyacrylate; thickener for natural and synthetic latexes; used in dipped molded or cast goods; water solution [Alco Chemical Corp]

Alcogum 9710. A sodium polyacrylate thickener; used for coatings, packaging, adhesives, latex thickening. [Alco Chemical Corp]

Alcogum AN 10. High viscosity sodium polyacrylate thickener for latex systems; used in carpet backing, latex

---

‡ Trade name and manufacturer not verified   § Trade name without identified manufacturer

foam, adhesives, dispersants. [Alco Chemical Corp]

Alcogum L-11. A high efficiency alkali-swellable acrylic emulsion thickener; used for adhesives, latex thickening, paint thickening. [Alco Chemical Corp]

Alcogum L-15, L-26, L-28, L-31, L-35, L-36. CAS 9003-01-4; Polyacrylic acid; thickener for adhesives, paints, paper coatings, natural and synthetic latexes, alkali reactive emulsion; FDA approved. [Alco Chemical Corp]

Alcogum L-27. A high efficiency alkali-swellable acrylic emulsion thickener; used for adhesives, latex thickening, dispersion thickening. [Alco Chemical Corp]

Alcogum L-28, L-29, L-35. A reactive alkali activated acrylic emulsion polymer; used for paper coating, latex compounding. [Alco Chemical Corp]

Alcogum L-52. A self-crosslinking alkali activated emulsion thickener; used for adhesives, latex thickening. [Alco Chemical Corp]

Alcogum PT-33. An alkali activated 'associative' emulsion thickener; used for paint, adhesives, cleaners, wall joint compounds. [Alco Chemical Corp]

Alcogum TSB. A high viscosity sodium polyacrylate thickener; used for textile coatings for upholstery. [Alco Chemical Corp] *

Alcogum VEP-II. A high viscosity sodium polyacrylate thickener; latex adhesive in the tufted carpet industry. [Alco Chemical Corp]

alcohol (CTFA). See ethyl alcohol

alcohol C₇. See heptylalcohol

alcohol C₈. See capryl alcohol

alcohol C₈. See 2-ethylhexanol

alcohol C₈. See isooctyl alcohol

alcohol C₁₀. See n-decyl alcohol

alcohol C₂₂. See behenyl alcohol

alcohols, lanolin. See lanolin alcohol

Alcojet®. Sodium metasilicate, sodium carbonate, POE ester of mixed fatty and resin acids; detergent used in mechanical washers for stain removal; for cleaning healthcare instruments, laboratory ware, electronic components, pharmaceutical apparatus, industrial parts, etc. [Alconox]

Alcolec®439-C. CAS 8002-43-5; EINECS 232-307-2; Lecithin; wetting agent, emulsifier, release agent; for water-based paints, coatings, textiles. [Am. Lecithin]

Alcolec® 440-WD. CAS 8002-43-5; EINECS 232-307-2; Lecithin; emulsifier, stabilizer, visc. control agent, wetting agent, pigment grinding aid and dispersant for water-based paints, coatings. [Am. Lecithin]

Alcolec® 495. CAS 8002-43-5; EINECS 232-307-2; Lecithin; o/w emulsifier, wetting agent for aq. and oil-base systems; approved for food use. [Am. Lecithin]

Alcolec 532. A proprietary preparation of vinyl-based resin in bead form; a carboxylated vinyl copolymer; used in the formulation of flexographic printing inks and paper lacquers. [Allied Colloids Ltd] *

Alcolec®BS. CAS 8002-43-5; EINECS 232-307-2; Single bleached lecithin; commercial lecithin emulsifier, wetting and dispersing agent, stabilizer, release and lubricating agent, foam suppressant, solubilizer for food and industrial applications; choline source. [Am. Lecithin]

Alcolec®F-100. CAS 8002-43-5; EINECS 232-307-2; De-oiled lecithin; emulsifier for industrial and food applications; instantizing for milk powd., cake mixes, etc.; choline source. [Am. Lecithin]

Alcolec® S. CAS 8002-43-5; EINECS 232-307-2; Un-bleached lecithin; commercial lecithin emulsifier, wetting and dispersing agent, stabilizer, release and lubricating agent, foam suppressant, solubilizer for food and industrial applications; choline source. [Am. Lecithin]

Alcolec® Z-3. CAS 8029-76-3; EINECS 232-440-6; Hydroxylated lecithin; wetting agent, emulsifier for personal care products, pharmaceuticals, food use; improves dispersability of colors and flavors in aq. systems, improves wetting of fatty powds. [Am. Lecithin]

Alcolite. A proprietary product used for denture purposes. §

Alcolube®. Textile lubricants and raising assistants. [Allied Colloids Ltd]

Alcomer®. Synthetic water-soluble polymers designed to offer a wide range of fluid properties for specialized systems for drilling, completion and workover fluids; used for viscosity shale encapsulation, flocculation thinning, fluid loss, deflocculation, friction reduction, bentonite extension, and gellation in water, brine, acid and cement systems. [Allied Colloids Ltd]

Alconate L-80. Concentrated form of Petronate L, an anionic surfactant of the petroleum sulfonate type; emulsifier and wetting agent used when low oil content is required. [Witco Chemical Ltd] *

Alcon-Efrin. CAS 61-76-7; Phenylephrine hydrochloride; adrenergic. [Alcon Laboratories Inc] *

Alconox®. Blend of alkylaryl sulfonates, lauryl alcohol sulfates, phosphates, carbonates; detergent with wetting, sequestering and synergistic agents; for manual cleaning of laboratory and hospital glassware and instruments. [Alconox]

Alcopal FA. A foaming agent for aqueous systems used for carpet backing. §

Alcopar. CAS 3818-50-6; A proprietary preparation containing bephenium hydroxynaphthoate; used as an anthelmintic. [The Wellcome Foundation Ltd]

Alcophor AC. Tannin compounds, modified; corrosion inhibitor used in petrol.-based paint systems. [Henkel KgaA]

Alcopol. A wide range of nonionic and anionic wetting agents and surfactants. [Allied Colloids Ltd]

Alcopol AH New. Amine salt of sulfated higher fatty acid ester in the form of a brown fluid oil; powerful wetting, penetrating and emulsifying agent used in pigments; paint; leather; emulsion polymerization; dry cleaning; cutting oils; agricultural chemicals. [Allied Colloids Ltd]

Alcopol FA. Anionic surfactant in which the anion is a long chain sulfosuccinamate, in the form of a fluid dispersion; powerful surface active agent used in the production of low density latex foams with good wet stability. [Allied Colloids Ltd]

Alcopol O. Range of anionic surfactants in which the anion is dioctyl sulfosuccinate; emulsifiers and powerful wetting agents with applications in industries such as paper, textiles, asbestos, plastics and photographic film, metals, pest control, detergents and degreasing, dust control, glass cleaning oils, lubricants, paints, pigments; printing inks and a wide range of proprietary products, e.g., hand cleansers, cosmetics. [Allied Colloids Ltd]

Alcopol OB. Sodium diisobutyl sulfosuccinate as a water/alcohol solution; powerful wetting agent in the presence of electrolytes. [Allied Colloids Ltd]

Alcopol OD. Sodium ditridecyl sulfosuccinate in water/alcohol solution; emulsifier for oils, solvents, waxes and polymers. [Allied Colloids Ltd]

---

* Trade name not verified as available   † Trade name verified as obsolete

Alcopol OS. Sodium dihexyl sulfosuccinate in water/alcohol solution; emulsifier and powerful wetting agent in the presence of electrolytes; used for oils, solvents, waxes, polymers and windscreen wash concentrate. [Allied Colloids Ltd]

Alcopol T. Sodium salt of a sulfated higher fatty acid ester as a low viscosity pale yellow liquid; powerful wetting, penetrating and emulsifying agent used in pigments; paints; leather; emulsion polymerization; dry cleaning; cutting oils, agricultural chemicals. [Allied Colloids Ltd]

Alcoprint®. A comprehensive range of novel polyacrylic thickening agents for pigment, disperse and reactive printing on textiles; can be used in emulsion systems but specially developed for completely aqueous systems. [Allied Colloids Ltd]

Alcoproof®. Synthetic sizing agents used in papermaking. [Allied Colloids Ltd]

Alcor 7. A clad combination of stainless steel and aluminum-containing magnetic stainless interlayer; used in the production of high quality cookware and magnetic induction heating stoves. [Pfizer International] †

Alcoseal®. Tail seal adhesive for the tissue industry. [Allied Colloids Ltd]

Alcoset®. Crosslinking acrylic resins; used as a wash-resistant finish to improve weave locking and fabric stability; fixed in a relatively short time above 150 C, no catalyst needed. [Allied Colloids Ltd]

Alcosist®. Dyeing assistants for wool, acrylic and polyamide fibers. [Allied Colloids Ltd]

Alcosize®. Acrylic-based blended products for sizing staple yarns, spun synthetics, cotton yarns; excellent performance at low application levels. [Allied Colloids Ltd]

Alcosol 5. An industrial methylated spirit; solvent. [Sasolchem]

Alcosol EM. Ethanol/MEK blend; solvent for the paint industry. [Sasolchem]

Alcosol PI. Ethanol, ethyl acetate blend; solvent for the printing industry. [Sasolchem]

Alcosperse®. Levelling and dispersing agents for disperse dyeing systems for textiles. [Allied Colloids Ltd]

Alcosperse 104. A polycarboxylate solution polymer; anti-redeposition agent, slurry stabilization. [Alco Chemical Corp]

Alcosperse 107, 124, 149, 157. CAS 9003-04-7; Sodium polyacrylate; dispersant for pigments, high solids slurries, paper coating, paint, textile, mining, and ceramic applications. [Alco Chemical Corp]

Alcosperse 107-D. A dry powdered polycarboxylate polymer; dispersant in paper, board, coatings and paint. [Alco Chemical Corp]

Alcosperse 144. An acrylic solution polymer; used for slurry preparation. [Alco Chemical Corp] †

Alcosperse 149-C. An acrylate solution polymer; used for paper coatings, pigment slurries. [Alco Chemical Corp]

Alcosperse 169. A sodium polyacrylate solution polymer; dispersant in adhesives, paint, kaolin and calcium carbonate. [Alco Chemical Corp]

Alcosperse 175. A low molecular weight carboxylate co-polymer solution; used in laundry detergent, incrustation inhibitor; phosphate replacement. [Alco Chemical Corp]

Alcosperse 249. Ammonium polyacrylate solution polymer; dispersant in paint, adhesives. [Alco Chemical Corp]

Alcosperse 602. Polycarboxylate solution polymer; antiredesposition agent; detergents; phosphate replace-

ment. [Alco Chemical Corp]

Alcostat®. Conductive resins and antistatic agents for general uses including plastics, surface coatings and latex compounds. [Allied Colloids Ltd]

Alcotabs®. Blend of alkylaryl sulfonates, lauryl alcohol sulfates, phosphates, carbonates; detergent with wetting, sequestering and synergistic agents; for cleaning pipettes and tubes in hospital, clinical, education, R&D, and industrial laboratories. [Alconox]

Alcotac. A range of organic binders for all mineral agglomeration, including pelletization and briquetting applications. [Allied Colloids Ltd]

Alcotex®. Polyvinyl alcohol (partially hydrolyzed polyvinyl acetate); used for suspension PVC polymerization, reprographic film/paper, and adhesives. [Harlow Chemical Co Ltd]

Alcotreat 182. A high molecular weight inverse-emulsion polymer; used for drilling fluids, mineral extraction. [Alco Chemical Corp] †

Alcotreat PC 95. A high molecular weight cationic polymer solution; used for water/oil clarification in oil field applications. [Alco Chemical Corp] †

Alcovar. Fast dyes for spirit and cellulose varnishes. [Morton Int'l. Ltd]

Alcowipe®. CAS 67-63-0; EINECS 200-661-7; 70% W/v isopropyl alcohol wipe; hard surface cleanser. [Seton Healthcare Group plc]

Alcryn®. Halogenated ethylene interpolymer alloy; thermoplastic elastomer; melt processable synthetic rubber which does not require compounding or vulcanization and can be formed on most plastics equip.; offers excellent weather, ozone, heat, and oil resist. [DuPont; DuPont UK]

Alcumite. A proprietary corrosion-resisting alloy containing 87.5% copper, 7.5% aluminum, 3.5% iron, and 1.5% nickel. §

alcuronium chloride. Diallyldinortoxiferin dichloride. Alloferin.

Aldactide. A proprietary preparation of spironolactone and hydroflumethiazide; a diuretic. [G D Searle & Co Ltd] *

Aldactone. CAS 52-01-7; A proprietary preparation of spironolactone; a diuretic. [G D Searle & Co Ltd]

Aldamine. A proprietary trade name for acetaldehyde-ammonia $CH_3COCH \cdot (OH) \cdot NH_2$; used in the manufacture of plastics and as a pickling inhibitor for steel, and is a rubber vulcanizing accelerator. §

Aldehol A. An oxidized kerosene to be used in U.S.A. for denaturing methylated spirit. §

aldehyde. Acetaldehyde, CH_3CHO.

aldehyde C1. See formaldehyde

aldehyde C14. Undecalactone; a proprietary flaming material; §

Alder bark. The bark of *Alnus glutinosa*; used for fixing yellow dyes and as a tanning material.

Alderton's solution. A solution of ammonium ichthosulfonate, in glycerol.

Aldo® HMS KFG. Glyceryl stearate, high mono; kosher food grade emulsifier, softener. [Lonza Inc]

Aldo® MLD. CAS 142-18-7; EINECS 205-526-6; Glyceryl laurate, dispersible; emulsifier for cosmetic, pharmaceutical and industrial use. [Lonza Inc]

Aldo® MO. CAS 111-03-5; Glyceryl oleate; emulsifier, defoamer. [Lonza Inc]

Aldo® MR. CAS 141-08-2; EINECS 205-455-0; Glyceryl

---

‡ Trade name and manufacturer not verified   § Trade name without identified manufacturer

ricinoleate; emulsifier, solubilizer for cosmetic, pharmaceutical and industrial applications. [Lonza Inc]

Aldo® MS FG. Glyceryl stearate; emulsifier for general use in foods. [Lonza Inc]

Aldo® PGHMS KFG. CAS 1323-39-3; EINECS 215-354-3; Propylene glycol stearate, high mono; kosher food grade emulsifier, whipping agent. [Lonza Inc]

Aldobond. Blend of synthetic or natural elastomers and resins in a solvent or aqueous medium; adhesive compositions for industrial fabricating operations. [Aldo Products Co Inc]

Aldocoat. Blend of synthetic or natural elastomers and resins in a solvent of aqueous medium; used for industrial specialty coatings. [Aldo Products Co Inc]

Aldocorten. CAS 52-39-1; A proprietary preparation of aldosterone. [Ciba plc] *

Aldogen. A mixture of trioxymethylene and bleaching powder; an antiseptic. §

Aldol. 3-Hydroxybutanal, $H_3CCH(OH)CH_2CHO$; formerly used as a hypnotic. §

Aldomax GA-100. CAS 9032-08-0; Immobilized amyloglucosidase enzyme; enzyme; for saccharification of low visc., higher DE (50-95) starch containing streams. [UOP]

Aldomet. CAS 555-30-6; Methyldopa; antihypertensive. [Merck & Co Inc]

Aldomet Ester Hydrochloride. Methyldopate hydrochloride; antihypertensive. [Merck & Co Inc]

Aldones. Flavor bases. [Bush Boake Allen Ltd]

Aldoretic. Methyldopa, amiloride hydrochlorothiazide; antihypertensive. [Merck & Co Inc]

Aldoril. Methyldopa, hydrochlorothiazide; antihypertensive. [Merck & Co Inc]

Aldosperse. Emulsifiers; for food. [Jan Dekker BV] *

Aldosperse® 40/60 FG. 40% PEG-20 glyceryl stearate, 60% glyceryl stearate (high mono); bakery and food emulsifier; dough strengthener, softener. [Lonza Inc]

Aldosperse® ML 23. CAS 59070-56-3; PEG-23 glyceryl laurate; emulsifier, solubilizer, suspending and dispersing agent used in personal care products, textiles. [Lonza Inc]

Aldosperse® MO-50. Glycol monooleate and polysorbate 80; emulsifier for frozen desserts; antifog for PVC. [Lonza Inc]

Aldosperse® MS-20 FG. CAS 51158-08-8; PEG-20 glyceryl stearate; bakery and food emulsifier. [Lonza Inc]

Aldosperse® O-20 KFG. 80% Glyceryl stearate, 20% Polysorbate 80; kosher food grade emulsifier; for ice cream. [Lonza Inc]

Aldosperse® TS-40 KFG. 60% Glyceryl stearate, 40% Polysorbate 65; kosher grade food emulsifier; for frozen desserts. [Lonza Inc]

Aldosterone. 11β,21-Dihydroxy-3,20-dioxopregn-4-en-18-al; aldocorten; electrocortin. §

Aldrey. An aluminum alloy of Swiss origin used for electrical conductors; contains 98.7% aluminum, 0.6% silicon, 0.4% magnesium, and 0.3% iron. §

Aldrin Dust. Insecticide. [Murphy Chemical Ltd] †

Aldur. A urea-formaldehyde resin. §

Al-dur-ba. A patented alloy of zinc, copper, and aluminum. §

Aldydale. A proprietary phenol-formaldehyde synthetic resin. §

Aldyl®. For polymer industry. [DuPont UK]

Alecra. Chromium plating processes. [Albright & Wilson Ltd, Phosphates & Speciality Business]

Alepol. A proprietary preparation of selected sodium salts of hydrocarpus oil acids. §

Aletodin. See aspirin. §

Aleudrin. A proprietary trade name for the sulfate of isoprenaline; an analgesic and sedative, a bronchodilator. [Lewis Laboratories] ‡

Alevaire. CAS 25301-02-4; A proprietary preparation containing tyloxapol; a mucolytic. [Bayer AG] †

Alexan. CAS 147-94-4; A proprietary preparation of cytarabine; an antineoplastic agent. [Pfizer International]

Alexandrianl laurel oil. See laurel nut oil.

Alexis. Aluminum etchant. [Albright & Wilson Ltd, Phosphates & Speciality Business]

Alexis Antibloom. Proprietary additive for metal finishing for hot water sealing of anodized aluminum to prevent sealing bloom. [Albright & Wilson Am.] *

Alexite. CAS 1344-28-1; A proprietary trade name for an aluminum oxide abrasive. §

Alfa®. Proprietary name for extensive range of research chemicals and materials. [Johnson Matthey plc]

alfa. A variety of esparto grass used in the manufacture of paper. It is also the term for a synthetic tannin, a redbrown liquid containing 23% tanning substance, 11% non-tannins, 66% water, and a trace of sulfuric acid.

Alfacron. Poultry house insecticide. [Ciba plc]

Alfacron 10WP. CAS 35575-96-3; Azamethiphos; used for fly control. [Ciba-Geigy Agrochemicals]

Alfadex. Pyrethrin; contact insecticide. [Ciba-Geigy Agrochemicals]

Alfadil. See Cardura. [Pfizer International]

Alfavet. CAS 74176-31-1; Alfaprostol; prostaglandin. [Hoffmann-LaRoche Inc] *

Alferium. A proprietary alloy of aluminum with 2.5% copper, 0.62% magnesium, 0.5% manganese, and 0.3% silicon. §

Alferon. Alpha interferon, natural (injectable form); used for the treatment of genital warts. [Interferon Sciences Inc]*

Alferric. CAS 10043-01-3; Aluminum sulfate. [Laporte Industries Ltd] *

Alflorone. CAS 514-36-3; A proprietary trade name for the 21-acetate of fludrocortisone; §

Alfodex. Fungal-α-amylase. [Glaxo Laboratories] *

Alfol. A proprietary trade name for an aluminum foil in a crumpled condition used for heat insulation. §

Alforder. A proprietary synthetic resin. §

Alformin. A 16% solution of basic aluminum formate, $Al_2(OH)_2 (HCO_2)_4$; an astringent and antiseptic. §

Alfralat. A proprietary name for glyceryl phthalate resins. [Chemische Werke, Albert] *

Alfrax B301. CAS 1344-28-1; A commercial grade of bubble aluminum oxide. [Carborundum] *

Alftalat. Alkyd resins (oil modified polyesters); used for air drying decorative paints, air drying and stoving industrial finishes. [Resinous Chemicals Ltd]

Algae Treat. 30% Quatenary ammonia compound; used for algae control in cooling water systems. [Delaware Chemical Corp] *

Algafen. CAS 97-23-4; Dichlorophen; a fungicide, bactericide, and algicide used as a moss-killer. [Geeco]

Algalex 104. A solution of the sodium salt of a chlorinated phenyl derivative of methane containing surface active and antifoam agents; a nontoxic, noncorrosive bacteri-

---

* Trade name not verified as available   † Trade name verified as obsolete

cide for water systems. [Kinnis and Brown] *

Algalith. A proprietary algine plastic. §

algaroba. See locust bean gum

Algarobilla. A vegetable tanning material; it consists of the pods of *Caesulpinias brevifolia* of Chile, and contains about 60% tannin.

Algarobillin. A dye product obtained from the carob tree, *Ceratonia siliqua*, found in the Argentina; it is employed for dyeing cloth, khaki.

Algarovilla. A Columbian name for a copal resin obtained there.

Alger metal. An alloy of 90% tin and 10% antimony; a silvery-white alloy used in making jewelry.

Algestone acetonide. A progestational steroid; 16α, 17α-isopropylidenedi-oxypregn-4-ene-3,20-dione. §

Algier's metal. A jeweler's alloy. (a) Consists of 90% tin and 10% antimony; used for the manufacture of forks and spoons; (b) contains 94.5% tin, 5% copper, and 0.5% antimony; used for making hand bells.

algin. (sodium alginate; sodium polymannuronate) CAS 9005-38-3; Hydrophilic polysaccharide; stabilizer in manufacture of ice cream; emulsifying agent in foods and paints. [Aldrich; Kelco, Div of Merck]

Alginade MR, MRE. Alginate blend; ice cream stabilizer. [Kelco Int'l. Ltd]

alginic acid. (norgine) Polysaccharide composed of β-d-mannuronic acid residues; $(C_6H_8O_6)_n$; CAS 9005-32-7; EINECS 232-680-1; suspending, thickening, emulsifying, and stabilizing agent. [Kelco; Mendell; Penta Mfg.; Protan Ltd.]

alginic acid, potassium salt. See potassium alginate

Alginoplast®. Alginate impression material for dentistry. [Bayer AG]

Algipan Balm. A proprietary preparation of methyl-nicotinate, glycol salicylate, histamine, and capsicin; an embrocation. [Wyeth Laboratories] *

Algiron. (alginoid iron). An iron compound of alginic acid (from seaweed); it contains 11% iron.

Algisium-C. Methylsilanol mannuronate; provides cutaneous hydration, lipolytic action, skin regeneration and maintenance, for cosmetic and health products, milks, emulsions, creams, lotions, anti-aging formulations. [Exsymol]

Algistat. Preparations for water treatment. [BDH Chemicals Ltd]

Algitox®. Algicide for swimming pools. [Fargro Ltd]

Algodon. Cotton wool.

Algodon de Seda. The fiber of *Calotropis gigantea* is known in Venezuela by this name.

Algofen. CAS 97-23-4; Dichlorophen; fungicide, bactericide and algicide used as a moss-killer. [Geeco]

Algoflon®. PTFE resins. [Fluorocarbon Co. Ltd.; Ausimont; Montedison UK Ltd]

Algol. Vat dyestuffs; used for textile dyeing and printing. [Hoechst UK]

Algol®. Vat dyestuffs; used for textile dyeing and printing. [Cassella AG]

Algon 100. Synergistic blend of a substituted urea and an acid amide; powder; bactericide/fungicide/algicide for wet-state and dry-film protection of interior and exterior water and solvent-based coatings, woodstains, adhesives, fillers, sealants, etc.; recommended for Pliolite®-based coatings. [Thor Chemicals (UK) Ltd]

Algulose. A very pure cellulose used in papermaking;

obtained from kelp. §

alibated iron. Iron coated with alumininum to form a protective covering.

Alibi. Herbicide containing bifenox and linuron. [ICI Chem. & Polymers Ltd]

Alicep®. Chloridazon, chlorbufam; for pre- and post-emergence weed control in onions, leeks, chives, and flower bulbs. [BASF AG; BASF plc] *

Ali-Clean. Acid based aluminum cleaner; for cleaning aluminum and magnesium alloy products especially car wheels; removes dirt, brake dust, and oxidation just by brushing on and washing off. [Hermetite Products Ltd] *

Alidine. CAS 126-12-5; Dihydrochloride of anileridine. §

Aliette. Fungicide. [May & Baker Ltd]

Aliette. CAS 39148-24-8; Fasetyl-aluminum; a systemic phosphonate fungicide for horticultural crops. [Embetec Crop Protection Ltd]

Aliette Extra. Captan + fosetyl-aluminum + thiabendazole; fungicide seed dressing for peas. [Embetec Crop Protection Ltd]

Alimemazine. CAS 84-96-8; Trimeprazine. §

Alimet. Methionine hydroxy analog feed supplement; liquid source of amino acid (methionine) activity for poultry and other animal feeds. [Monsanto Co] *

Alipal®. Redesignated Abex or Rhodapex®. [Rhone-Poulenc Surf.]

Aliso. CAS 555-31-7; Aluminum isopropoxide; used for cosmetics, pharmaceuticals. [Rhone-Poulenc UK]

Alisol®. For agriculture industry. [DuPont UK]

Alistell. Herbicide containing linuron, 2,4-DB and MCPA. [ICI Chem. & Polymers Ltd]

Alistell. Emulsifiable concentrate containing 220 g 2,4-DB, 30 g linuron and 30 g MCPA per liter; used to control weeds in undersown cereals and seedling grassland. [Farm Protection Ltd]

alival. 3-Iodo-1,2-propanediol, $CH_2I \cdot CHOH \cdot CH_2OH$; used in medicine.

Alizarin. Acid wool dyes. [Bayer AG; Bayer plc]

alizarin oil. See Turkey red oils.

Alizarine. Anionic dyestuffs (level dyeing); used for wool and wool blends. [Holliday Dyes & Chemicals Ltd]

Alka. Range of alkali detergents for the food industry; for bottlewashing and tank cleaning. [Harshaw Chemicals Ltd] *

Alka-Donna. A proprietary preparation containing magnesium trisilicate, aluminum hydroxide and belladonna extract; an antacid. [Carlton Laboratories (UK) Ltd] *

Alka-Donna P. A proprietary preparation of Alka-Donna with phenobarbitone; an antacid and sedative. [Carlton Laboratories (UK) Ltd] *

Alkaflo®. Patented phosphate salt mixture; alkalizer for reactive dyes in textile industry. [Sybron]

Alkafoam D. Alkanolamide/ethoxylated alcohol blend; foaming agent for foam dyed carpets. [Rhone-Poulenc Surf.]

Alkagel. A trade name for alginates made for waterproofing.§

alkali cellulose. (hydrated cellulose). The product of the reaction between cotton and caustic soda. when hydrolyzed by water it gives hydrated cellulose.

Alkali Surfactant NM. Surfactant, wetting and coupling agent used in alkaline formulations for hard surface cleaning, floor strippers, heavy-duty degreasers, steam, soak tank, and household/institutional cleaners; solubi-

---

‡ Trade name and manufacturer not verified   § Trade name without identified manufacturer

lizer for nonionics into high electrolyte cleaners. [Exxon/ Tomah]

Alkalit. A proprietary synthetic resin obtained by heating the sodium salt of phenolphthalein with toluoyl chloride. §

Alkalsite. An explosive containing 25-32% potassium perchlorate, ammonium nitrate, trinitrotoluene, and other constituents. §

Alkamide® 101 CG. CAS 68603-42-9; Cocamide DEA, DEA; thickener, detergent, emulsifier for lower boiling aliphatic hydrocarbons; for cosmetic and industrial applications; produces emulsions stable in presence of alcohols, glycols, and phenols. [Rhone-Poulenc Surf.]

Alkamide®200 CGN. Cocamide DEA; thickener, detergent, emulsifier, foam stabilizer; for hard surface cleaners, floor cleaners, rinsable degreasers, metal cleaners, metalworking compounds; corrosion inhibition characteristics. [Rhone-Poulenc Surf.]

Alkamide® 327. CAS 120-40-1; EINECS 204-393-1; Lauramide DEA; foam and visc. modifier for personal care products. [Rhone-Poulenc Surf. Canada]

Alkamide®2104. 2:1 Cocamide DEA; detergent, emulsifier, foam booster; base for floor and general purpose cleaners; lubricant for synthetic grinding and cutting fluids. [Rhone-Poulenc Surf. Canada]

Alkamide®C-5. CAS 61791-08-0; PEG-5 cocamide; thickener, foam stabilizer, and emulsifier for formulated detergents and cosmetics. [Rhone-Poulenc Surf. Canada]

Alkamide®C-212. CAS 68140-00-1; Cocamide MEA (1:1); thickener, foam builder and stabilizer for soap or synthetic based washing powds. [Rhone-Poulenc Surf.]

Alkamide® CDE. CAS 68603-42-9; Cocamide DEA; detergent, emulsifier, stabilizer, thickener, foam stabilizer for personal care and detergent products. [Rhone-Poulenc Surf. Canada]

Alkamide® CP-1255. CAS 27883-12-1; Linolamide DEA, DEA. [Rhone-Poulenc Surf.]

Alkamide® DC-212/MP. Coconut poly-diethanolamide; used where alkaline builders are incorporated into aq. systems; visc. builder for liquid potash soaps; conveyor belt lubricant; corrosion inhibitor props. [Rhone-Poulenc Surf.]

Alkamide®DIN-295/S. CAS 68425-47-8; Linoleamide DEA (1:1); foam booster, emulsifier, visc. builder, thickener for shampoos, industrial cleaners; conditioning to hair. [Rhone-Poulenc Surf.; Rhone-Poulenc France]

Alkamide®DO-280. CAS 93-83-4; 2:1 Oleamide DEA and diethanolamine; emulsifier for sol. oils; corrosive inhibitor. [Rhone-Poulenc Surf.]

Alkamide® DS-280/S. CAS 93-82-3; EINECS 202-280-1; Stearamide DEA (1:1); visc. builder, thickener, foam booster, dispersant for nonionic and cationic systems, shampoos, bath preps., industrial cleaners; emulsifier, corrosion inhibitor, lubricant for metalworking fluids; conditioner for hair and skin care products. [Rhone-Poulenc Surf.]

Alkamide® HTDE. CAS 93-82-3; EINECS 202-280-1; Stearamide DEA; detergent, thickener, visc. builder, emulsifier for kerosene, vegetable and mineral oil, microcryst. wax. [Rhone-Poulenc Surf. Canada]

Alkamide® L7DE. Lauric-myristic DEA (1:1); detergent, foam booster/stabilizer, superfatting and thickening agent for toiletry and cleaning formulations; fortifier for perfumes in soaps. [Rhone-Poulenc Surf. Canada]

Alkamide® L-203. CAS 142-78-9; EINECS 205-560-1;

Lauramide MEA (1:1); foam builder/stabilizer for soap and synthetic washing powds.; visc. builder. [Rhone-Poulenc Surf.; Rhone-Poulenc France]

Alkamide® LE. CAS 120-40-1; EINECS 204-393-1; 1:1 Lauramide DEA; visc. booster, foam stabilizer in shampoo. [Rhone-Poulenc Surf.]

Alkamide® LIPA/C. CAS 142-54-1; EINECS 205-541-8; Lauramide MIPA; high foaming lubricant for metalworking fluids. [Rhone-Poulenc Surf.]

Alkamide® OIP. CAS 111-05-7; EINECS 203-828-2; Oleamide MIPA; foam modifier for high-temp. cleaners, especially liquid and powd. laundry detergents; lubricant for metalworking fluids. [Rhone-Poulenc Surf. Canada]

Alkamide® R-280. CAS 106-16-1; EINECS 203-368-2; Ricinoleamide MEA. [Rhone-Poulenc Surf.]

Alkamide® S-280. CAS 111-57-9; EINECS 203-883-2; Stearamide MEA (1:1); viscosifier for industrial cleaners; skin protectant in toilet bars, creams, lotions, pastes. [Rhone-Poulenc Surf.]

Alkamide® SDO. CAS 68425-47-8; EINECS 270-355-6; Soyamide DEA; foam stabilizer, visc. builder, superfatting agent for toiletries, cutting and sol. oils, textiles, household and industrial cleaners, corrosion inhibitor. [Rhone-Poulenc Surf. Canada]

Alkamide® STEDA. CAS 110-30-5; EINECS 203-755-6; Ethylene bis-stearamide; additive in pulp and paper defoamer formulations; lubricant, plasticizer, antistat, pigment dispersant for resins and plastics. [Rhone-Poulenc Surf. Canada]

Alkamide® WRS 1-66. CAS 93-83-4; EINECS 202-281-7; Oleamide DEA; emulsifier for highly nonpolar aliphatic hydrocarbons and chlorinated aliphatic hydrocarbons; rust inhibitor. [Rhone-Poulenc Surf.]

Alkaminox®C-2. CAS 61791-14-8; PEG-2 cocamine; textile scouring, dyeing assistant, softener, antistatic agent; used as corrosion inhibitor in steam generating and circulating systems; coemulsifier. [Rhone-Poulenc Surf.]

Alkamuls® 14/R. CAS 61791-12-6; PEG-60 castor oil; surfactant. [Rhone-Poulenc France]

Alkamuls®400-DO. CAS 9005-07-6; PEG-8 dioleate; emulsifier, solubilizer, lubricant, wetting agent for cosmetic, textile, metalworking, and agricultural uses. [Rhone-Poulenc Surf.]

Alkamuls®400-MO. CAS 9004-96-0; PEG-9 oleate; emulsifier for fats, wetting agent, dispersant, lubricant used in dairy industry, cosmetic, metalworking, and industrial applications. [Rhone-Poulenc Surf.]

Alkamuls® 600-DO. CAS 9005-07-6; PEG-12 dioleate; dispersant, emulsifier for o/w emulsions; for cosmetic, metalworking, and industrial use. [Rhone-Poulenc Surf.]

Alkamuls® A. CAS 9004-96-0; PEG-6 oleate; emulsifier, lubricant for sol. oils, most aliphatic solvs.; for lubricating and cutting oils, agricutural formulations. [Rhone-Poulenc Surf.; Rhone-Poulenc France]

Alkamuls® AG-900. CAS 9016-45-9; Ethoxylate blend; spreading and wetting agents for aq. pesticide systems. [Rhone-Poulenc Surf. Canada]

Alkamuls®B. CAS 61791-12-6; PEG-33 castor oil; emulsifier, dispersant for textiles, metallurgy, metal degreasing, personal care products; dye leveler, fabric softener. [Rhone-Poulenc France]

Alkamuls®COH-5. CAS 61788-85-0; PEG-5 hydrogenated castor oil; lubricant, softener, antistat, emulsifier, deter-

---

* Trade name not verified as available   † Trade name verified as obsolete

gent. [Rhone-Poulenc Surf.]

Alkamuls® EGDS. CAS 627-83-8; Glycol distearate pure; thickener, opacifier, pearlizing agent used in shampoos and cosmetic lotions. [Rhone-Poulenc Surf.]

Alkamuls®EGMS/C. CAS 111-60-4; Ethylene glycol monostearate; visc. booster, opacifying and pearlescing agent for liquid cosmetic and detergent compounds. [Rhone-Poulenc Surf.]

Alkamuls® EL-620. CAS 61791-12-6; PEG-30 castor oil; emulsifier, wetting agent, pigment dispersant, antistat, lubricant, solubilizer for industrial/household cleaners, cosmetics, pharmaceuticals, metalworking fluids, leather, pesticides, herbicides, paper industries. [Rhone-Poulenc Surf.; Rhone-Poulenc France]

Alkamuls®GMR-55LG. Glyceryl mono/dioleate; coemulsifier, lubricant, softener, emollient, rust preventative additive for mold release agents, synthetic fiber spln finishes, compounded oils; antistat, antifog for PVC film processing. [Rhone-Poulenc Surf. Canada]

Alkamuls® GMS/C. CAS 31566-31-1; Glyceryl stearate; emulsifier, wetting agent for cosmetic, agricuture, textile industries; coupler used to bind waxes together; emollient and thickener in cosmetic creams. [Rhone-Poulenc Surf.]

Alkamuls®L-9. CAS 9004-81-3; PEG-9 laurate; emulsifier, coemulsifier for cosmetic and toiletry preps.; defoamer, leveling agent for latex paints; dispersant for dyes and pigments. [Rhone-Poulenc Surf.]

Alkamuls® MM/M. CAS 3234-85-3; EINECS 221-787-9; Myristyl myristate; emollient, moisturizer, lubricant, and conditioner for hair and skin care products; visc. builders and gelling/stiffening agents for makeup and deodorant applications. [Rhone-Poulenc Surf.]

Alkamuls® PSML-20. CAS 9005-64-5; Polysorbate 20; emulsifier, solubilizer, antistat, visc. modifier, lubricant for textiles, cosmetics, pharmaceuticals. [Rhone-Poulenc Surf.]

Alkamuls® PSMO-5. CAS 9005-65-6; Polysorbate 81; emulsifier, solubilizer, antistat, lubricant for paint, food, cosmetic, insecticides, herbicides, fungicides, textiles, cutting oils. [Rhone-Poulenc Surf.]

Alkamuls®PSMO-20. CAS 9005-65-6; Polysorbate 80; emulsifier, wetting agent for cosmetic, food, agricutural applications; coemulsifier for aliphatic alcohols, petrol. oils, fats, solvs., waxes. [Rhone-Poulenc Surf.]

Alkamuls® PSMS-20. CAS 9005-67-8; Polysorbate 60; wetting agent, emulsifier for cosmetic and food applications, textiles, paper coatings; fiber-to-metal lubricant for fibers and yarns. [Rhone-Poulenc Surf.]

Alkamuls® PSTO-20. CAS 9005-70-3; Polysorbate 85; emulsifier for cosmetic and food applications; textile and leather lubricant. [Rhone-Poulenc Surf.]

Alkamuls® S-20. CAS 1338-39-2; EINECS 215-663-3; Sorbitan laurate; w/o emulsifier, lubricant and softener for the textile industry; secondary suspending agent, porosity modifier in PVC suspensions. [Rhone-Poulenc Geronazzo]

Alkamuls® S-60. CAS 1338-41-6; EINECS 215-664-9; Sorbitan stearate; w/o emulsifier, lubricant and softener for the textile industry; secondary suspending agent, porosity modifier in PVC suspensions. [Rhone-Poulenc Geronazzo]

Alkamuls® S-65. CAS 26658-19-5; EINECS 247-891-4; Sorbitan tristearate; w/o emulsifier, lubricant and soft-

ener for the textile industry; sec. suspending agent, porosity modifier in PVC suspensions. [Rhone-Poulenc Geronazzo]

Alkamuls®S-65-8. CAS 9004-99-3; PEG-8 stearate; emulsifier, self-emulsifying lubricant and softener for synthetic fibers. [Rhone-Poulenc Surf.]

Alkamuls® S-65-40. CAS 9004-99-3; PEG-40 stearate; emulsifier, self-emulsifying lubricant and softener for synthetic fibers. [Rhone-Poulenc Surf.]

Alkamuls® S-80. CAS 1338-43-8; EINECS 215-665-4; Sorbitan oleate; w/o emulsifier for mineral and vegetable oils, in metalworking; oil spill dispersant. [Rhone-Poulenc Geronazzo]

Alkamuls® S-85. CAS 26266-58-0; EINECS 247-569-3; Sorbitan trioleate; w/o emulsifier, lubricant and softener for the textile industry; sec. suspending agent, porosity modifier in PVC suspensions. [Rhone-Poulenc Geronazzo]

Alkamuls®SDG. CAS 106-11-6; PEG-2 stearate; emollient, moisturizer, lubricant for skin and hair care systems. [Rhone-Poulenc Surf.]

Alkamuls® SEG. CAS 111-60-4; Ethylene glycol monostearate; opacifier and pearling agent for shampoos, creams, liquid hand soaps, liquid detergents; emulsion stabilizer, visc. builder. [Rhone-Poulenc Surf.]

Alkamuls® SML. CAS 1338-39-2; EINECS 215-663-3; Sorbitan laurate; emulsifier for oils and fats in cosmetic, metalworking and industrial oil products; corrosion inhibitor; antistat for PVC. [Rhone-Poulenc Surf.]

Alkamuls® SMO. CAS 1338-43-8; EINECS 215-665-4; Sorbitan oleate; emulsifier, coupling agent, wetting agent for medicants, petrol. oils, fats, and waxes in the industrial, textile, metalworking, and cosmetic industries; textile and leather lubricant and softener; corrosion inhibitor. [Rhone-Poulenc Surf.]

Alkamuls® SMS. CAS 1338-41-6; EINECS 215-664-9; Sorbitan stearate; emulsifier and coupling agent; used to prepare silicone defoamer emulsions for industrial applications, paraffin wax emulsions for processing paper coatings; textile process lubricant; internal PVC film lubricant; cosmetics; foods. [Rhone-Poulenc Surf.]

Alkamuls® SS. CAS 2778-96-3; EINECS 220-476-5; Stearyl stearate; emollient, moisturizer, lubricant, and conditioner for hair and skin care products; visc. builders and gelling/stiffening agents for makeup and deodorant applications. [Rhone-Poulenc Surf.]

Alkamuls® STO. CAS 26266-58-0; EINECS 247-569-3; Sorbitan trioleate; emulsifier and coupling agent; used to compound textile and leather softener finishes; in metalworking fluids. [Rhone-Poulenc Surf.]

Alkamuls® STS. CAS 26658-19-5; EINECS 247-891-4; Sorbitan tristearate; hydrophobic emulsifier for use as a fiber-to-metal lubricant for synthetic and cotton fibers; cosmetics, foods. [Rhone-Poulenc Surf.]

Alkamuls® T-20. CAS 9005-64-5; Polysorbate 20; emulsifier, solubilizer, antistat and lubricant for textile industry; solubilizer for essential oils; raw material for no-tears shampoo. [Rhone-Poulenc Geronazzo]

Alkamuls® T-60. CAS 9005-67-8; PEG-20 sorbitan stearate; emulsifier, solubilizer, antistat and lubricant for textile industry; solubilizer for essential oils; raw material for no-tears shampoo. [Rhone-Poulenc Geronazzo]

Alkamuls®T-80. CAS 9005-65-6; Polysorbate 80; o/w emulsifier, solubilizer, textile fiber antistat/lubricant; used in

---

‡ Trade name and manufacturer not verified   § Trade name without identified manufacturer

hot and cold rolling formulations. [Rhone-Poulenc Geronazzo]

Alkamuls® T-85. CAS 9005-70-3; PEG-20 sorbitan trioleate; emulsifier, solubilizer, antistat and lubricant for textile industry; solubilizer for essential oils; raw material for no-tears shampoo. [Rhone-Poulenc Geronazzo]

alkane C4. See butane

alkane C5. See n-pentane

alkane C6. See hexane

alkane C7. See heptane

Alkanet. (Alkanna, Anchusin). Terms applied to two different plants, *Lawsonia inermis* and *Anchusa tintoria,* whose roots are the source of a red dye, anchusine (alkannin), $C_{15}H_{14}O_4$; the name is applied to the dye as well as to the plant.

Alkanol®. Surfactants. [DuPont UK]

Alkanol® 189-S. Sodium alkyl sulfonate; wetting agent, detergent, penetrant, foamer for textiles, elastomers, plastics, film, metal cleaning and pickling, hard surface cleaning, and chemical manufacturing; effective in acid and alkali media. [DuPont]

Alkanol®ND. Sodium alkyl diaryl sulfonate; foaming agent, dyeing assistant, surfactant; for textiles, chemical manufacturing; leveling agent for acid dyes on nylon. [DuPont]

Alkanol®WXN. Sodium alkylbenzene sulfonate; wetting, rewetting agent, foaming agent, emulsifier, dyeing assistant; for textiles, paper, chemical manufacturing, alkaline and acid cleaners; leveling agent for acid dyes; stable to acid or alkaline media. [DuPont]

Alkanol®XC. Sodium alkylnaphthalene sulfonate; wetting agent, dispersant, penetrant, low foaming; used in bleaching and dyeing of textiles, leather, paper, chemical manufacturing, photography; reduces shrinkage in ceramics manufacturing; dry colors manufacturing. [DuPont]

Alkanolamine. Dimethyl ethanolamine; solubilizer of synthetic resins for water soluble paints, raw material for ion exchange resins and coagulants. [Yokkaichi Chemical Co Ltd] *

Alkanox® 24-44. CAS 38613-77-3; Tetrakis (2,4-di-t-butylphenyl) 4,4´-biphenylylene diphosphonite; processing antioxidant for PE, PP, PC, PS, polyesters; protects against thermo-oxidative degradation during long term aging. [Enichem Synthesis SpA]

Alkanox®240. CAS 31570-04-4; Tris (2,4-di-t-butylphenyl) phosphite; high performance antioxidant for stabilization of polymers including PP, HDPE, LDPE, LLDPE, PC, ABS, and polyesters. [Enichem Synthesis SpA]

Alkanox® 240-3T. Tris (2,4-di-t-butylphenyl) phosphite, distearylthiodipropionate; antioxidant for stabilization of PP, PE. [Enichem Synthesis SpA]

Alkanox®P-24. CAS 26741-53-7; Bis (2,4-di-t-butylphenyl) pentaerythritol diphosphite; antioxidant for stabilization of ABS, PVC, PC; color stabilizer. [Enichem Synthesis SpA]

Alkapol PEG 300. CAS 25322-68-3; PEG-6; intermediate for surfactants; binder/lubricant in pharmaceuticals; plasticizer; paper softener; humectant; solvent; antistat; for cosmetics, textile, plastics processing, dyes and inks. [Rhone-Poulenc Surf.] *

Alkaquat® DMB-451-50, DMB-451-80. CAS 61789-71-7; Benzalkonium chloride; wetting agent, emulsifier, biocide, disinfectant for use in beverage industry, dairy industry, food processing, water treatment, paper indus-

try, pest control, preservatives, antidandruff rinses; general disinfection and sanitization for hospitals, and laundries. [Rhone-Poulenc Surf. Canada]

Alka-Seltzer®. Relief from pain, e.g., headache; also for short-term use in headache with hyperacidity of the stomach. [Bayer AG]

Alkasil® HNM 1223-15 (70%). Organo-modified polydimethylsiloxane in aromatic hydrocarbon solvents; for surface modification, waterproofing, release props. on wood, masonry, silica, mineral granules, paper, etc. [Rhone-Poulenc Surf. Canada]

Alkasil®NE 58-50. CAS 63148-55-0; Dimethicone copolyol. [Rhone-Poulenc Surf. Canada]

Alkasit. A proprietary cellulose adhesive. §

Alkasperse 25. A proprietary series of pigment dispersions based on a short oil xylol-thinned alkyd used in the coloring of medium to fast air-drying surface coatings. [Collinda Ltd]

Alkasperse®A-20. CAS 9003-04-7; Sodium polyacrylate; dispersant for boiler water compounds. [Rhone-Poulenc Surf. Canada]

Alkasperse®M-10. CAS 25086-62-8; Sodium polymethacrylate; polymeric dispersant for clay slurries and drilling muds; antiredeposition agent in phosphate-free detergents; conductivity aid for Electrofax paper. [Rhone-Poulenc Surf.] *

Alkastar 83. Organic brightener system; used for alkaline noncyanide zinc electroplating. [Harshaw Chemicals Ltd]

Alkasurf®NP-4. CAS 9016-45-9; Nonoxynol-4; emulsifier, detergent, dispersant, intermediate, stabilizer; plasticizer, antistat for plastics, surfactants, household, industrial, and cosmetic use, fat liquoring, cutting and sol. oils. [Rhone-Poulenc Surf. Canada]

Alkaterge®C. Oxazolidine; surfactant, emulsifier, emulsion stabilizer, wetting agent, acid acceptor; pigment grinding and dispersion aid; penetrant for textile and paper industries, metal cleaners; coatings; antifoam for antibiotic fermentation; antioxidant. [Angus]

Alkaterge®E. CAS 68140-98-7; Ethyl hydroxymethyl oleyl oxazoline; detergent, emulsifier, wetting agent, antifoamer, antioxidant; used in salt, soap, paper, textiles, and metal cleaners; emulsion stabilizer; acid acceptor; pigment grinding and disp. [Angus]

Alkaterge®T. CAS 28984-69-2; Oxazoline-type compound; detergent, invert emulsifier, pigment dispersant, corrosion inhibitor; wetting agent; emulsion stabilizer; grinding aid; acid acceptor; antifoam agent; antioxidant; lubricant. [Angus]

Alkaterge®T-IV. CAS 95706-86-8; Oxazoline deriv., ethoxylated; acid scavenger, offers filming protection to metal surfs.; corrosion inhibitor; o/w emulsifier; dispersant in aq. and nonaq. systems; wetting agent. [Angus]

Alkateric®A2P-OS. Octyl propionate; low-foaming surfactant for acid and alkaline cleaners; pH stable. [Rhone-Poulenc Surf.] *

Alkateric®AP-C. Cocopropionate; foaming agent for alkaline cleaning compound formulations. [Rhone-Poulenc Surf.] *

Alkateric®PB. CAS 693-33-4; EINECS 211-748-4; Cetyl betaine; mild substantive surfactant for personal care products; conditioner, antistat, emollient; as solubilizer, visc. builder, foam booster with lauryl sulfates; stable to acid and alkali media. [Rhone-Poulenc Surf.] *

---

* Trade name not verified as available

† Trade name verified as obsolete

Alkathene. Solid polymers of ethylene prepared by subjecting ethylene to extremely high pressures under carefully controlled conditions of temperature. [ICI Chem. & Polymers Ltd]

Alkawet®. Amphoteric surfactant; used as a wetting agent. [Lonza AG]

Alkawet® AA-60. Alkoxylated alcohol; low foam wetting agent for continuous dyeing of carpet. [Rhone-Poulenc Surf.]

Alkawet® CF. Proprietary; wetting agent, detergent for industrial use; electrolyte-tolerant, controlled foaming. [Lonza Inc]

Alkazid®Lye DIK, M. For removal of hydrogen sulfide and carbon dioxide from synthesis gas and cracked gas for production of ethylene. [BASF AG]

Al-kenna. The powdered roots and leaves of *Lawsonia inermis;* used in the East for dyeing the nails, teeth, and hair.

Alkeran Injection. CAS 148-82-3; An intravenous formulation of melphalan; used for the treatment of localized malignant melanoma of the extremities and localized soft tissue sarcoma of the extremities by regional arterial perfusion. [The Wellcome Foundation Ltd]

Alkeran Tablets. CAS 148-82-3; A proprietary formulation of melphalan; for the palliative treatment of multiple myeloma and advanced ovarian adrenocarcinoma. [The Wellcome Foundation Ltd]

Alkolite. A proprietary phenophthalein resin. §

Alkydal®. Phthalate resins modified with oil or fatty acids; for use in the formulation of paints and varnishes. [Bayer AG; Bayer plc]

alkyd resin. Thermosetting coating polymer; vehicles in exterior house paints, marine paints, and baking enamels; elec. components, encapsulation. [Bayer SA; Croda Resins Ltd; DSM GmbH; PPG Industries SA; Reichhold Chemie AG; Scott Bader]

Alkylate 215. CAS 68648-87-3; Linear dodecylbenzene; detergent intermediate used in light-duty detergents, dishwash, laundry, industrial cleaners. [Monsanto/Detergents & Phosphates]

Alkylate 230. Linear tridecylbenzene; detergent intermediate used in heavy duty detergents. [Monsanto/Detergents & Phosphates]

alkyl dimethyl benzyl ammonium chloride. *See* benzalkonium chloride

Alkynol®. Saturated polyesters free from oil and fatty acids; for use in the formulation of coil coatings and high-grade stoving finishes. [Bayer AG; Bayer plc]

Allabond Twenty/twenty Adhesive. CAS 25928-94-3; Epoxy adhesive; for production applications; for use with Activator BA-66B; also avail. in clear and conductive grades. [Bacon] *

Allabond Twenty/twenty NM. CAS 25928-94-3; Epoxy adhesive; nonmagnetic version of Allabond Twenty/twenty Adhesive. [Bacon] *

Allactol. Aluminum lactotartrate. §

allantoin. (2,5-dioxo-4-imidazolidinyl urea; glyoxyldiureide; 5-ureidohydrantoin) Heterocyclic organic compound; production of animal metabolism, excreted in urine; $C_4H_6N_4O_3$; CAS 97-59-6; EINECS 202-592-8; biochemical research, medicine; soothing agent, skin protectant; stimulates growth of healthy tissue. [3-V; Atomergic Chemetals; EM Ind.; R.W. Greeff; Hommel GmbH; ICI Am.; Penta Mfg.; Schweizerhall; Sutton Lab;

Tri-K Ind.]

Allbee with C. A proprietary preparation of Vitamins B and C. [Wyeth Laboratories] *

Allegheny 33, 44, 55, 66. Corrosion resisting alloys (formerly Ascoloy 33, 44, 55, 66); they contain iron and chromium : 33 contains 12-16% chromium, and 55 contains 26-30% chromium. §

Allegheny Metal. A proprietary corrosion-resisting alloycontaining iron with 17-20% chromium and 7-10% nickel. §

Allegron. CAS 894-71-3; A proprietary preparation of nortriptyline hydrochloride; an antidepressant. [Dista Products Ltd]

Allenite. Synonym for pentahydrite. §

Allenoy. A proprietary molybdenum steel. §

Allen's metal. An alloy of 55.3% copper, 44 6% lead, and 0.1% tin.

Allergospasmin. Reproterolhydrochloride and DNCG; aerosol; for prophylaxis and therapy of asthma bronchiale. [Degussa AG] *

allethrin. A synthetic insecticide structurally similar to pyrethrin.

Alletorphine. A semisynthetic morphine analgesic. §

Allguard. Water repellent coating for concrete, stone and masonry building walls. [Dow Corning]

Alligator wood. The wood of *Guarea grandifolia,* of West India.

Allisan. Horticultural fungicide. [The Boots Co plc] †

All-O. A proprietary liquid soap for use as a rubber lubricant.§

Alloferin. CAS 15180-03-7; A proprietary preparation of alcuronium chloride; a muscle relaxant. [Roche Products Ltd] *

Alloprene. Chlorinated rubber. [ICI Chem. & Polymers Ltd]

Allopurinol. 1H-Pyrazolo[3,4-d]-pyrimidin-4-ol; an uricosuric drug. §

Alloxan. Mesoxalylurea, $C_4H_4N_2O_5$. §

Alloy 2L5. An aluminum alloy with 12.5-14.5% zinc and 2.5-3% copper. §

Alloy 2L8. An aluminum alloy containing 11-13% copper. §

Alloy 3L11. An aluminum alloy containing 6-8% copper and tin may be added up to 1%. §

Alloy 39. An alloy containing aluminum with 3.75-4.25% copper, 1.2-1.7% magnesium, and 1.8-2.3% nickel. §

Alloy 109. An alloy containing 88% aluminum and 12% copper. §

Alloy 122. An alloy containing 88.6% aluminum, 10% copper, 1.2% iron, and 0.25% magnesium. §

Alloy 142. An alloy containing 92.5% aluminum, 4% copper, 2% nickel, and 1.5% magnesium. §

Alloy 145. An aluminum alloy containing 10-11% zinc, 2-3% copper, and 1-1.5% iron. §

Alloy 195. An aluminum alloy containing 4-5% copper, and not more than 1.2% silicon, 1.2% iron, 0.35% magnesium, and 0.35% zinc. §

Alloy 2129. A special nickel-iron alloy having fairly high permeability and excellent mechanical properties. §

Alloy AM4-4. An aluminum-magnesium alloy; it contains magnesium with 4% aluminum, 0-4% manganese, and 0-15% silicon. §

Alloy AM7-4. An alloy containing magnesium with 7% aluminum, 0-4% manganese, and 0-15% silicon. §

Alloy AMF. An alloy containing 50-60% nickel for low temperature use. §

Alloy AP33. An aluminum alloy containing 4.5% copper and

0-4% titanium. §

Alloy JL. An aluminum alloy containing 4-5% copper, 0-41% iron, and 0-35% silicon. §

Alloy L5. An aluminum alloy containing, in addition to aluminum, 13% zinc and 2.8% copper. §

Alloy L7. An alloy containing aluminum with 14% copper, and 1% manganese. §

Alloy L8. An alloy containing aluminum with 12% copper. §

Alloy L10. An alloy containing aluminum with 10% copper and 1% tin. §

Alloy L11. An alloy containing aluminum with 7% copper and 1% tin. §

Alloy L24. See Alloy Y. §

Alloy MG7. An alloy consisting mainly of aluminum with magnesium and manganese; it has mechanical properties similar to Duralmin, and is stated to be highly resistant to corrosion. §

Alloy N. An alloy of 91% aluminum, 6% copper, and 3% manganese. §

Alloy NCT3. An alloy containing 44.5% iron, 37.5% chromium, 17-5% nickel, and 0.5% manganese. §

Alloy Steel. The term applied to a steel containing one or more elements in addition to carbon; also see Steels V and W. §

Alloy T. An alloy containing aluminum with 3.8% magnesium, 0.5% iron, 0.5% silicon, and 0.1% copper. §

Alloy W.7. 1% Silicon, 4.5% manganese, balance nickel with controlled zirconium addition. [Wiggin Alloys Ltd] ‡

Alloy W.9. 1% Silicon, 4-5% manganese, balance nickel with controlled zirconium addition. [Wiggin Alloys Ltd] ‡

Alloy Y. (Alloy 24). An alloy of aluminum with 4% copper, 2% nickel, and 1.5% magnesium. §

Alloys RR. A series of aluminum alloys containing aluminum with 0.5-5% copper, 0.2-2.5% nickel, 0.05-5% magnesium, 0.6-1.5% iron, 0.05-0.5% titanium, and 0.2-5% silicon. §

Alloys Wm. White-bearing metals; Wm5 contains 78.5% lead, 15% antimony, 5% tin, and 1.5% copper, with a specific gravity of 10.1; Wm10 consists of 73.5% lead, 15% antimony, 10% tin, and 1.5% copper; Wm42 contains 42% tin, 41% led, 14% antimony and 3% copper; Wm80 consists of 80% tin, 10% antimony, 10% copper.§

Allpyral®. Allergy relief treatment. [Bayer AG]

Alluman. An alloy of aluminum with 10-20% tin and 4-6% copper. §

Ally®. CAS 74223-64-6; Metsulfuron-methyl; used for control of annual dicotyledons in cereals. [DuPont UK]

allyl alcohol dibromide. See 2,3-dibromo-1-propanol

4-allyl-1,2-dimethoxybenzene. See methyl eugenol

allyl methacrylate. $CH_2:C(CH_3)COOCH_2CH:CH_2$; CAS 96-05-9; EINECS 202-473-0; Silane monomer intermediate, crosslinker, polymer modifier for high impact plastics, adhesives, acrylic elastomers, optical polymers. [Aldrich; CPS; Monomer-Polymer & Dajac]

4-allyl-2-methoxyphenol. See eugenol

allyloestrenol. 17α-Allyloestr-4-en-17β-ol; gestanin.

allylprodine. 3-Allyl-1-methyl-4-phenyl-4-propionyloxy-piperidine.

4-allyl veratrole. See methyl eugenol

Almacarb. CAS 21645-51-2, CAS 546-93-0; Tablets of aluminum hydroxide-magnesium carbonate co-dried gel; antacid. [British Drug Houses] *

Almag. A proprietary aluminum alloy similar in composition to Alferium (qv). §

Almagel. CAS 21645-51-2, CAS 1309-42-8; A proprietary preparation of aluminum hydroxide gel and magnesium hydroxide; an antacid. [Ayrton Saunders plc] ‡

Almasilium. An aluminum alloy containing 1% magnesium and 2% silicon. §

Almazine. CAS 846-49-1; Lorazepam; pharmaceutical preparation for the treatment of anxiety. [M A Steinhard Ltd] *

Almelec. A proprietary alloy of aluminum containing 0.7% magnesium, 0.5% silicon, and 0.3% iron. §

Almen's reagent. A solution containing 5 g tannic acid in 240 cc of 50% alcohol, to which has been added 10 cc of a 25% solution of acetic acid; a precipitate is given with nucleoproteins.

Almevax. A proprietary preparation of live rubella vaccine to immunize against rubella. [The Wellcome Foundation Ltd]

Alminate. CAS 13682-92-3; Dihydroxyaluminum aminoacetate; antacid. [Bristol-Myers Squibb Co Inc]

Almocarpine. CAS 54-71-7; Pilocarpine hydrochloride; cholinergic. [Wyeth Laboratories] *

Almora. Magnesium gluconate; replenisher. [O'Neal, Jones & Feldman Pharmaceuticals] *

Almo Steel. Proprietary chrome-molybdenum steels. §

Almstab. Stabilizers for use in PVC compositions. [Associated Lead Manufacturers Ltd] *

Alnovin. CAS 53597-27-6; Fendosal; anti-inflammatory. [Hoechst-Roussel Pharmaceuticals Inc] *

Alnovol; Nonheat hardening phenol/formaldehyde resins; used for printing inks and rubber reinforcement. [Resinous Chemicals Ltd]

Alnovol. Nonheat hardening phenol/formaldehyde resins, novolak types; used in printing inks and rubber reinforcement. [Hoechst UK]

Alocrom. Chromating pretreatment for aluminum. [ICI Chem. & Polymers Ltd]

Aloe Vera Powd. 200XXX Extract-Microfine. Aloe vera gel; rapid dissolving ingred. for cosmetic, health and pharmaceutical industries. [Tri-K Industries]

Alomite. A trade name applied to a variety of sodalite used as an ornamental stone. §

Alon®. Fumed alumina. §

Alophen. A proprietary preparation of aloin, phenolphthalein, ipecacuanha, strychnine, and belladonna; a laxative. [Parke-Davis] *

Aloral. CAS 315-30-0; Tablets containing 100 mg and 300 mg allopurinol BP; used for the treatment of gout; primary hyperuricemia; secondary hyperuricemia; prophylaxis of uric acid and calcium oxalate stones. [Lagap Pharmaceuticals Ltd]

Alox. A proprietary trade name for a series of methyl esters of higher alcohols. §

Alox® 111. Mold release or parting compound for cast concrete products; rust preventive protecting metal forms. [Alox] *

Alox® 152. Lubricity agent in mineral oil sol'ns., cutting and metalworking formulations, quench oil additives, for automotive and industrial lubricants. [Alox] *

Alox® 318F. Corrosion inhibitor for use on ferrous and nonferrous metals, especially for bright or highly polished steel surfaces. [Alox] *

Alox® 350. Oxidized petrol. fraction esters; corrosion inhibitor and lubricant for making cutting and sol. oils. [Alox] *

---

* Trade name not verified as available

† Trade name verified as obsolete

Alox®436A. Mold release, lubricant, corrosion inhibitor for concrete molds. [Alox] *

Alox®488. Oxidized petrol. fractions; upper cylinder lubricant for internal combustion engines; cleans and prevents formation of carbon and lacquer deposits in engines. [Alox] *

Alox® 575. Oxygenated hydrocarbon with barium and a sodium petrol. sulfonate; emulsifier, corrosion inhibitor for cutting and sol. oils. [Alox] *

Alox® 606, 606-55, 606-70. Heavy-duty rust preventive, protective coating for metal surfaces., especially for automotive rustproofing; 606-55 is a total solv. cutback grade with 55% Alox 606 and 45% min. spirits; 606-70 is a partial solv. cutback grade with 70% Alox 606 and 30% min. [Alox] *

Alox®904. Base for formulation of fingerprint removers or rifle bore cleaners; corrosion preventive. [Alox] *

Alox® 1680. Oxygenated hydrocarbon containing minor amt. of phosphatide; corrosion inhibitor; penetrating oil additive. [Alox] *

Alox® 2000. Corrosion inhibitor and penetrant for metal conditioning compounds. [Alox] *

Alox® 2028L. Oxygenated hydrocarbon, calcium soap; nonemulsifiable, nonstaining, water-displacing rust preventive, lubricant. [Alox] *

Alox® 2211Y. Thixotropic rust preventive which deposits film providing long-term protection against humidity and salt fog; can be cutback in solv. or blended 50/50 with naphthenic oil. [Alox] *

Alox®2301. Oxidized microcrystalline wax; for prep. of wax emulsions for textile sizing, finishing and waterproofing. [Alox] *

Aloxicoll. Aluminum chlorohydrate. [Giulini Corp]

aloxidone. 3-Allyl-5-methyl oxazolidine-2,4-dione; allomethadione; malidone.

Aloxite. A trademark for abrasive and refractory materials consisting essentially of alumina. §

Alpacca. An alloy of 64% copper, 19% zinc, 14.5% nickel, 2% silver, 0.4% iron, and 0.12% tin; it is a nickel silver. §

Alpakka. See Alpacca. §

Alperox C. Lauroyl peroxide; polymerization catalyst. [Atochem North America Inc/Organic Peroxides Div]

Alperox-F. CAS 105-74-8; Lauroyl peroxide; initiator for bulk, solution, and suspension polymerization, high-temp. curing of polyester resins, and cure of acrylic syrup. [Atochem]

Alpex. Cyclized rubber resins; used for chemical resistant coatings. [Resinous Chemicals Ltd]

Alpex. Cyclized rubber resins; used for chemical resistant coatings. [Hoechst UK]

Alpfa. Proprietary name for extensive range of research chemicals and materials. [Johnson Matthey plc]

Alphachloralose. Rodenticide. [Rentokil Ltd]

Alphachroic. Chrome dyestuffs. [J C Bottomley]*

Alpha Chymar. Chymotrypsin; enzyme. [Barnes-Hind Inc]*

Alpha-Cillin. CAS 26309-95-5; Pivampicillin hydrochloride; oral antibiotic with broad-spectrum activity against Gram-positive and Gram-negative bacteria. [Merck & Co Inc] †

Alpha Daphnone. Pure α-isomethylionone. [Bush Boake Allen Ltd]

Alphadim® 90AB. CAS 67701-33-1; High-purity, molecularly dist. monoglyceride prepared from fully hardened edible fats and glycerin; food additive providing functional improvements in processing and storage stability; stabilizes and disperses fat particles in coffee whiteners; as starch complexing and softening aid in breads. [Am. Ingredients/Patco]

Alphadim® 90LC. CAS 61789-10-4; High-purity, molecularly distilled monoglyceride prepared from edible fats and glycerin; food additive providing a stable emulsion of finely dispersed water droplets for margarine and coffee whiteners. [Am. Ingredients/Patco]

Alphadim®90NLK. CAS 67701-32-0; High-purity, molecularly dist. monoglyceride prepared from refined sunflower oil and glycerin with BHA and citric acid; Kosher; food additive providing a stable, finely dispersed emulsion in diet margarines. [Am. Ingredients/Patco]

Alphadim®90SBK. High-purity, molecularly distilled monoglyceride prepared from fully hardened soybean oil and glycerin; Kosher; food additive providing a stable, finely dispersed emulsion in margarines; stabilizes and disperses fat particles in coffee whiteners; starch complexing agent. [Am. Ingredients/Patco]

Alphadolone. 3α,21-Dihydroxy-5α-pregnane-11,20-dione; an anasthetic component present in althesin. §

Alphadrol. CAS 53-34-9; Fluprednisolone; glucocorticoid. [Upjohn] *

Alphalin. Vitamin A; antixerophthalmic. [Eli Lilly & Co] †

Alphamine. CAS 30902-17-9; Midodrine hydrochloride; antihypertensive; vasoconstrictor. [Centerchem Products Inc] ‡

Alphamint. Peppermint blend. [Bush Boake Allen Ltd]

Alphanol. Acid dyestuffs; used for wool dyeing. [Hoechst UK]

Alphanol. Medium chain length alcohols, forming a plasticizer. [ICI Chem. & Polymers Ltd]

Alphanol®. Acid dyestuffs; used for wool dyeing. [Cassella AG]

AlphaRedisol. CAS 13422-51-0; Hydroxocobalamin; vitamin. [Merck & Co Inc] *

Alpha-Ruvite. CAS 13422-51-0; Hydroxocobalamin; vitamin. [Savage Laboratories] †

Alphasol OT. A proprietary trade name for the sodium salt of an alkyl ester of sulfosuccinic acid. §

Alpha-Step® MC-48. Sodium α sulfomethyl cocoate; surfactant, foam booster/stabilizer for dishwashing liqs. [Stepan; Stepan Canada]

Alpha-Step® ML-40. Sodium methyl-2 sulfolaurate and sodium ethyl-2 sulfolaurate; biodeg. surfactant, foaming agent, hydrotrope for dishwashing liqs., fine fabric washes, hard surface cleaners and bubble baths; scouring, leveling, coupling and foaming agent for textiles; metalworking formulations. [Stepan; Stepan Canada; Stepan Europe]

Alphatex®. CAS 12141-46-7; Metakaolinitic aluminum silicate produced by calcining kaolin clay; paper coating pigment, paper filler. [ECC International Ltd]

Alphatrex. CAS 5593-20-4; Betamethasone dipropionate USP; indicated for relief of the inflammatory and pruritic manifestations of corticosteroid responsive dermatoses. [Altana Inc] ‡

Alphenate. A series of phenolic-alkyd plasticized resins. [Chemische Werke, Albert] *

Alphide. A proprietary trade name for a cold molded refractory ceramic. §

Alphogen. (Alphozone, Succinoxate). Succinyl peroxide, $(COOH \cdot CH_2 \cdot CH_2CO)_2O_2$; an antiseptic. §

---

‡ Trade name and manufacturer not verified   § Trade name without identified manufacturer

Alphol. α-Naphthol salicylate, $C_6H_4(OH)$ $(COOC_{10}H_7)$; an antiseptic and antirheumatic. §

Alphosyl. A range of proprietary preparations containing allantoin and coal tar extract; antipsoriatic for dermatological use. [Stafford-Miller] ‡

Alphosyl HC. Proprietary preparation containing allantoin, coal tar extract and hydrocortisone; antipsoriatic for dermatological use. [Stafford-Miller] ‡

Alphoxat O 105. CAS 9004-96-0; PEG-5 oleate; basic material for textile industry. [Zschimmer & Schwarz]

Alphoxat S 110. CAS 9004-99-3; PEG-10 stearate; emulsifier and dispersant, greasing agent for textile and chemical technical industries. [Zschimmer & Schwarz]

Alphozone. See Alphogen. §

Alpivicin. CAS 26309-95-5; Pivampicillin hydrochloride; oral antibiotic with broad-spectrum activity against Gram-positive and Gram-negative bacteria. [Merck & Co Inc] †

Alplate. A proprietary aluminum coated steel. §

Alpolit. Unsaturated polyester resin in styrene; used for glass fiber reinforced laminate, casting and potting. [Resinous Chemicals Ltd]

Alprenolol. 1-(2-Allylphenoxy)-3-isopropylaminopropan-2-ol; a beta adrenergic receptor blocking agent; §

Alprokyds. Proprietary drying oil and non-drying oil modified alkyd resins. §

alquifon. (Black Lead Ore, Potter's Ore). A mineral. It consists of zinc sulfide; used in pottery to give a green glaze.

Alreco. Aluminum and aluminum alloy ingot; for aluminum die cast Industry. [Reynolds Metal Co]

Alresat. Maleinized rosin esters; used in nitrocellulose lacquers, printing inks and to improve gloss in air drying paints. [Hoechst AG; Hoechst UK]

Alresen. Alkyl phenol formaldehyde resins, terpene phenolic resins; used for oil varnishes and adhesives. [Hoechst UK]

Alrheumun®. 2-(3-Benzoylphenyl)propionic acid; antiflammatory, antirheumatic and analgesic. [Bayer AG]

Alromin Ru 1000. Antistatic agent. [Ciba plc] *

Alrosperse® 100. Surfactant blend; surfactant, interfacial tensile depressant, dispersant, deflocculant, solubilizer, emulsifier, corrosion inhibitor, antistat for metal processing, petrol. products, drycleaning, spotting compounds, leather and upholstery cleaners, emulsions, paints, inks. [M.S. Paisner]

Alscoap AF Series. Surfactants; foaming agent for air drilling. [Toho Chem. Industry]

Alscoap LN-40, LN-90. CAS 151-21-3; Sodium lauryl sulfate; detergent, shampoo base, toothpaste; polymerization emulsifier for synthetic resins and latex. [Toho Chem. Industry]

Alsi. A pigment consisting of finely ground aluminum-silicon alloy used to give durable and rust-preventative paints. §

Alsibronz. CAS 12001-26-2; Wet ground muscovite mica; used for paint, rubber, plastics, pearlescent pigments. [Franklin Mineral Products Co]

Alsica Alloys. Aluminum-silicon-copper alloys. §

Alsifer. A proprietary alloy of 40% silicon, 40% iron, and 20% aluminum; a hardener alloy for adding silicon to aluminum alloys. §

Alsifilm. A material made from bentonite; used in place of mica. §

Alsimag. Ceramic materials; used for insulation and for the dielectric of condensers. [3M] *

Alsimag 754. A beryllia ceramic. [3M] *

Alsimag 779. Leachable ceramic cores for precision metal castings. [3M] *

Alsimin. A similar alloy to Alsifer. §

Alsol. Aluminum acetotartrate; a germicide. §

Alstat. Very powerful antistatic agent; for all stages of textile processing. [Altex Chemical Co Ltd] *

Alstromed A 18 LV. N-Octadecyl sodium sulfosuccinamate solution; foaming agent, frothing agent and emulsifying agent. [Alco Chemical Corp] †

Alsynates. Metal carboxylates based on C_8-C_{10} branched chain synthetic aliphatic carboxylic acids; driers for paint or printing ink catalyst for unsaturated polyester. [Manchem Ltd] *

Alsynol RS-47. Cyclized rubber resins for coatings. [Daniel Prods.]

Alsystin. CAS 64628-44-0; Triflumuron; insect growth regulator with stomach action against biting insects (particularly caterpillars) on coffee, cotton, forest plants, maize, groundnuts, pome fruit, citrus and vegetables; also against migratory locusts and grasshoppers in the breeding areas. [Bayer AG]

Altacaps. Hydrotalcite plus dimethicone; used for indigestion, peptic ulceration. [Roussel Laboratories Ltd] †

Altacite Plus. Proprietary product of hydrotalcite, dimethicone; antacid, antiflatulent. [Roussel Laboratories Ltd]

Altal. A proprietary trade name for triphenylphosphate. §

Altalc 200 USP. CAS 14807-96-6; Talc; excellent color and purity; for pharmaceutical and cosmetic applications including baby powds., medicated foot powds.; glidant, lubricant, pigment carrier. [Cyprus Industrial Min.]

Altan. Veterinary laxative. [May & Baker Ltd] *

Altax®. CAS 120-78-5, CAS 557-05-1; Benzothiazyl disulfide, zinc stearate, petrol. process oil; accelerator for nat. and synthetic rubbers; primary accelerator and scorch modifying sec. accelerator in natural rubber and styrene/butadiene rubber copolymers; retarder-plasticizer in neoprene (G types); cure modifier in W types. [R.T. Vanderbilt]

Altene DG. CAS 79-01-6; Trichloroethylene; degreasing solvent. [Atochem Inc] *

AL terna GEL. CAS 21645-51-2; Aluminum hydroxide, dried; antacid. [Stuart Pharmaceuticals] †

Altolube. Excellent scrooping agent; for processing of warp knit nylon. [Altex Chemical Co Ltd] *

Altowhite LL. CAS 12141-46-7; Calcined aluminum silicate. extender pigment; exhibits improved optical properties in paint systems. [Dry Branch Kaolin]

Alubarb. A proprietary preparation of aluminum hydroxide, phenobarbitone and belladonna; a gastrointestinal sedative. [H N Norton & Co Ltd] ‡

Alubrasoft® 12-N. Softener substantive to cotton, wool, acrylics, nylon, and other synthetics; imparts softness, lubricity, antistatic props. to yarns and fabrics. [PPG/Specialty Chem.]

Alubrasoft®Super 100. Fatty polyamide; softener, antistat, conditioner, lubricant; imparts soft hand to synthetic and natural fibers. [PPG/Specialty Chem.]

Alu-Cap™. CAS 21645-51-2; Each capsule contains dried aluminum hydroxide gel equivalent to 400 mg of aluminum hyroxide; OTC antacid to alleviate heartburn, sour

---

* Trade name not verified as available   † Trade name verified as obsolete

stomach and/or acid indigestion. [3M Pharmaceuticals]

Aludone®. CAS 59792-81-3; Aluminum PCA; astringent, antiseptic; peripheral antiperspirant; for spray or stick deodorants, shower gel, hair comb-out balm. [UCIB]

Aludrox. CAS 21645-51-2; A proprietary preparation containing aluminum hydroxide gel; an antacid. [Wyeth Laboratories]

Aludrox CO. A proprietary preparation containing alumina-sucrose powder and magnesium hydroxide; an antacid. [Wyeth Laboratories] *

Aludrox SA. A proprietary antacid preparation containing aluminum hydroxide gel, magnesium hydroxide, gel, magnesium hydroxide, secondary butobarbitone and ambutonium bromide. [Wyeth Laboratories] *

Aludur. An alloy of aluminum and silicon, containing from 5-20% silicon. §

Alufrit. CAS 1344-28-1; Alumina microgrits. [Atomergic Chemetals Corp]

Alugan. A proprietary preparation of bromocyclen; a veterinary pesticide. §

Alugel. CAS 21645-51-2; Aluminum hydroxide. [Giulini Corp]

Aluline. CAS 315-30-0; Allopurinol; pharmaceutical preparation for the treatment of gout. [M A Steinhard Ltd] *

alum. Potassium aluminum sulfate, $K_2SO_4Al_2(SO_4)_3$ $\cdot 24H_2O$, ammonium aluminum sulfate $AlNH_4(SO_4)_2 \cdot 12H_2O$, and aluminum sulfate, $Al_2(SO_4)_3 \cdot 9H_2O$; are all known by this name, but it is usually appled to the potassium salt. *Also see* aluminum sulfate

Alumail®. A range of inorganic smelted products for use in the surface coating of metals; used for special purpose enamels for various uses as well as for electrophorectic or powder electrostatic application. [Bayer AG]

Aluman. An alloy of 88% aluminum, 10% zinc, and 2% copper.

Alumantine. A proprietary refractory containing 60-65% Al_2O_3. §

Alumbro. A trademark for a 2% aluminum-brass alloy having good resistance to corrosion by sea water and marine atmospheres. [ICI plc] *

Alumedia. Complex of alkyd resin and aluminum alkoxide; for preparation of high solids paint. [Manchem Ltd] *

Alumel. An electrical resistance alloy containing 94% nickel, 1% silicon, 2% aluminum, 2.5% manganese, and 0-5% iron. §

Alumilite. A proprietary trade name for chemical coatings applied to aluminum electrically. §

alumina. (aluminum oxide; calcined alumina; alumite) Inorganic compound; Al_2O_3; CAS 1344-28-1; EINECS 215-691-6; production of aluminum, abrasives, refractories, ceramics, elec. insulators, catalysts and catalyst supports, paper, spark plugs, crucibles and lab ware, adsorbent for gases/water vapors, chromatographic analysis, heat-resist. fibers, food additives. [Air Prods.; Alcan; Alcoa; Aldrich; Atomergic Chemetals; Lonza Sarl; Nissan Chem. Ind.; Rhone-Poulenc; Vista]

alumina, hydrate. (alumina trihydrate; aluminum hydroxide; aluminum trihydroxide) Inorganic compound; $Al_2O_3 \cdot 3H_2O$; CAS 1333-84-2; 21645-51-2; production of aluminum, abrasives, refractories, ceramics, elec. insulators, catalysts and catalyst supports, paper, spark plugs, crucibles and lab ware, adsorbent for gases/water vapors, chromatographic analysis, heat-resist. fibers, food additives. [Alcan; Alcoa; Atomergic Chemetals; Climax

Performance; Croxton & Garry Ltd; Nyco Minerals; Reheis; Solem]

Aluminac. A similar alloy to Alpax (*qv*), for making die castings. §

Aluminum Cloflbrate. Di-[2-(4-chlorophenoxy)-2-methylpropionato] hydroxyaluminum; a pharmaceutical used in the treatment of arteriosclerosis; §

Aluminoferric. CAS 10043-01-3; Consists of crude aluminum sulfate, and contains some iron sulfate; used as a precipitating agent in sewage and refuse liquids treatment, and also for removing suspended matter from boiler feed water. [Laporte Industries Ltd]

Aluminoid. A trademark for goods of the abrasive and refractory type, the essential constituent being crystalline alumina. §

aluminosilicic acid, magnesium salt. *See* magnesium aluminum silicato

alumino-vanadium. An alloy of aluminum and vanadium, obtained by adding a mixture of vanadium pentoxide and powdered aluminum to liquid aluminum; used as a deoxidizing agent.

Aluminox. A trademark for articles as abrasives and refactories; the essential constituent is crystalline alumina. §

aluminum. Al; Metallic element; CAS 7429-90-5; EINECS 231-072-3; building and construction, corrosion-resistant chemical equip. (desalination plants), die-cast auto parts, elec. industry (power transmission lines), photoengraving plates, permanent magnets, cryogenic technology, machinery, tubes for ointments. [Alcan GmbH; Norsk Hydro AS]

aluminum ammonium sulfate. *See* ammonium alum.

aluminum brass. Alloys of from 59-70% copper, 26-40% zinc, 0.3-5.2% aluminum, and sometimes a little iron.

aluminum bronze. There are various alloys under this name. Those containing a high percentage of aluminum are termed light, and have from 83-89% aluminum and 11-17% copper. The other type is called heavy, and contains from 85-95% copper and.

aluminum chloride, anhydrous. $AlCl_3$; CAS 7446-70-0; EINECS 231-208-1; Ethylbenzene catalyst, dyestuff intermediate, detergent alkylate, ethyl chloride, pharmaceuticals and organics, butyl rubber, petroleum refining, hydrocarbon resins, nucleating agent for titanium dioxide pigments. [Aldrich; Asada Chem Industry Co Ltd; Atochem N. Am.; Fluka; Harcros Durham; Witco/Argus]

aluminum chlorohydrate. (aluminum chloride hydroxide; aluminum chlorohydrol; aluminum chlorohydroxide) Inorganic salt. $[Al_2(OH)_5Cl]_x$; CAS 1327-41-9; 12042-91-0; commercial antiperspirant and deodorant; water purification; treatment of sewage and plant effluent. [Catomance Ltd; Reheis]

Aluminum Grade Bone Ash, BCP 600. Calcium hydroxyapatite; used to coat and protect all surfaces contacted by molten nonferrous metals; used extensively in aluminum industries. [Murlin Chemical Inc]

aluminum hydroxide. (aluminum oxide trihydrate; aluminum trihydrate) $Al(OH)_3$; CAS 21645-51-2; EINECS 244-492-7; Dyes, paints, textile finishing. [Alcan; Alcoa; Atomergic Chemetals; BA Chem Ltd; Nyco Minerals; Reheis; Rhone-Poulenc; Seimi Chem.; Solem; Vista; Whittaker, Clark & Daniels]

aluminum iron. (ferro-aluminum). An alloy of iron and aluminum; used for refining iron, also as a permanent

---

‡ Trade name and manufacturer not verified   § Trade name without identified manufacturer

ingredient for increasing the strength. A 15% alloy has been used for crucibles exposed to high temperatures.

aluminum iron brass. An alloy containing 61.1% copper, 35.3% zinc, 1.1% iron, and 2.3% aluminum.

aluminum iron bronze. An alloy containing 85-89% copper, 6-9% aluminum, and 3-7% iron.

aluminum magnesium bronze. An alloy of 89-94% copper, 5-10% aluminum, and 0.5% magnesium.

aluminum magnesium silicate. See magnesium aluminum silicate

aluminum manganese. An alloy of aluminum with 2-3% manganese.

aluminum manganese brass. An alloy consisting of 56-3% copper,40% zinc, 2.7% manganese, and 1% aluminum.

aluminum manganese bronze. An alloy of 89% copper, 9.6% aluminum, and 1-2% manganese.

aluminum nickel. An alloy containing varying amounts of nickel with aluminum. One alloy consists of 76.4% nickel, and 23.6% aluminum.

aluminum nickel bronze. An alloy containing 85% copper, with from 5-10% aluminum and 5-10% nickel.

aluminum-nickel-titanium. An alloy of 97.6% aluminum, 2% nickel, and 0.4% titanium.

aluminum zinc. An alloy consisting of 85% aluminum, 10% nickel, and 5% zinc.

aluminum nitrate. $Al(NO_3)_3 \cdot 9H_2O$; mordant for textiles, leather tanning, manufacture of incandescent filaments, catalyst in petroleum refining, nucleonics, anticorrosion agent, antiperspirant. [Aldrich; EM Ind.; Hoechst-Celanese; Sherman Chem. Ltd.; Spectrum Chem. Manufacture]

aluminum nitride. AlN; CAS 24304-00-5; as semiconductor in electronics, nitriding of steel. [Aldrich; Atomergic Chemetals; Carborundum; Dow; Mandoval Ltd]

aluminum oxide. Al_2O_3; CAS 1344-28-1; EINECS 215-691-6; Manufacture of aluminum, abrasives, refractories, ceramics, electrical insulators, catalyst and catalyst supports, paper, spark plugs, laboratoryware, adsorbent for gases, chromatographic analysis, fluxes, fibers, food additive (dispersant). [Alcan; Aldrich; Atomergic Chemetals; BA Chem. Ltd; Degussa; Ferro/Transelco; Hüls Am.; Lonza; Norton Chem. Process Prods.; Rhone-Poulenc; Vista]

Aluminum Oxide C. CAS 1344-28-1; Alumina; free-flow and anticaking agent; aids in reducing electrostatic charges of powder substances; for elec. industry. [Degussa]

aluminum phosphate. (MALP; monoaluminum phosphate, acidic solution) $Al(H_2PO_4)_3$; CAS 13530-50-2; Refractory bonding agent, metal processing. [Albright & Wilson; Rasa Ind.; Rhone-Poulenc Basic; Superfos Biosector A/S]

Aluminum Silicate P820. CAS 12141-46-7; Precipitated aluminum silicate; titanium dioxide extender in powd. coatings, decorative paints. [Degussa]

Aluminum Silicon Alloy C. A British Chemical Standard alloy; it contains 12.74% silicon, 0.34% iron, 0.005% manganese, 0.020% zinc, 0.006% titanium, and 0.010% copper.

aluminum silver. (silver metal). An alloy consisting of 57% copper, 20% nickel, 20% zinc, and 3% aluminum. The name is also applied to an alloy of 95% aluminum, and 5% silver.

aluminum stearate. (aluminum, dihydroxy (octadecanoato-

o-)). Aluminum salt of stearic acid; $CH_3(CH_2)_{16}COO-Al(OH)_2$; CAS 7047-84-9; 637-12-7; paint, varnish drier, greases, waterproofing agent, cement additive, lubricants, cutting compounds, flatting agent, cosmetics, pharmaceuticals, and defoaming agent. [Atochem/Wire Mill; Ferro/Grant; Magnesia GmbH; Norac; Synthetic Prods.; Witco]

aluminum sulfate. (alum; aluminum trisulfate; cake alum) Inorganic salt; $Al_2(SO_4)_3 \cdot 14H_2O$; CAS 10043-01-3; in pulp and paper mills, water purification plants, leather, textile, gypsum treatment, in fire retardants; deodorizer, decolorizer, food additive. [Alcan; Aldrich; Am. Cyanamid; Asada Chem Industry Co Ltd; Ashland; BA Chem. Ltd.; Ethyl; General Chem.; Rasa Ind.; Rhone-Poulenc Basic]

aluminum tin bronze. An alloy of 85% copper, 10% tin, 2-5% aluminum, and 2% zinc.

Alumite. CAS 1344-28-1; Alumina microgrits. [Atomergic Chemetals Corp]

Alundum®. A registered trademark for various types of goods, such as grinding wheels, abrasive and refractory grain, refractory articles and cement, porous plates, crucibles, and other articles made from crystalline alumina; or alumina which has been electrically fused and crystallized. §

Alunex. CAS 2438-32-6; Chlorpheniramine maleate; pharmaceutical preparation for the treatment of allergy. [M A Steinhard Ltd] *

Aluni. An aluminum-nickel alloy used as an anode for deposition of the alloy coating. §

Alupent. A proprietary preparation of orciprenaline sulfate; a beta$_2$-agonist bronchodilator. [Boehringer Ingelheim Ltd]

Alupent Expectorant. A proprietary preparation of orciprenaline sulfate and bisolvon; a bronchial antispasmodic. [Boehringer Ingelheim GmbH] *

Alupent Obstetric. A proprietary preparation of orciprenaline; used in the management of premature labor. [Boehringer Ingelheim GmbH] *

Alupent-Sed. A proprietary preparation of amylobarbitone and orciprenaline sulfate; a bronchial antispasmodic. [Boehringer Ingelheim Pharmaceuticals Inc]

Aluphos. CAS 13530-50-2; A proprietary preparation containing aluminum phosphate; an antacid. [Fisons plc, Pharmaceuticals Div] *

Alupram. CAS 439-14-5; Diazepam; pharmaceutical preparation for the treatment of depression. [M A Steinhard Ltd] *

Alusec. Aluminum organic complexes; rheology modifiers for high solids, air drying paint systems. [Manchem Ltd]*

Alusil. CAS 12141-46-7; Aluminum silicate. [Crosfield Chemicals Ltd]

Alusil ET. CAS 12141-46-7; Synthetic aluminum silicate of controlled particle size; used an extender in emulsion paint. [Crosfield Chemicals Ltd]

Alu-Tab™. CAS 21645-51-2; Each tablet contains dried aluminum hydroxide gel equivalent to 500 mg of aluminum hydroxide; antacid to alleviate heartburn, sour stomach and/or acid indigestion. [3M Pharmaceuticals]

Aluzyme. A proprietary preparation of aneurine hydrochloride, riboflavin, pyridoxine, niacin, pantothenic acid and folic acid; vitamin supplement. [Phillips Yeast Products Ltd] *

Alveograf. CAS 1306-06-5; Durapatite; prosthetic aid.

* Trade name not verified as available   † Trade name verified as obsolete

[Sterling Drug Inc] *

Alvex. Highly alkaline detergent. [Crosfield Chemicals Ltd]†

Alysine. CAS 54-21-7; Sodium salicylate; analgesic. [Merrell Dow Pharmaceuticals Inc] *

Alytol. Bitumen mastic; for roof repair and maintenance. [Vedag GmbH] *

Alzen. An alloy of 66% aluminum and 33% zinc.

amalgam. A name applied to alloys of metals with mercury; it is also used as a term for a native alloy of mercury and silver with a formula varying between AgHg and Ag_2Hg_3.

Amandol. A proprietary grade of benzaldehyde. [May & Baker Ltd] *

Amargosite. A trade name for a clay of the bentonite type.§

Amarin. Triphenyldihydroglyoxaline, $(C_6H_5CNH)_2$:CH C_6H_5. §

Amasil®. CAS 64-18-6; Formic acid; for ensiling and feed preservation. [BASF AG] *

Amasil®-P. Calcium propionate, calcium formate; ensiling agent for preservation of feeds. [BASF AG]

Amatols. Mixtures of trinitrotoluene and ammonium nitrate; an 80/20 amatol contains 80 parts ammonium nitrate and 20 parts trinitrotoluene. high explosives. §

Amax®, Amax No 1. CAS 102-77-2; N-Oxydiethylene 2-benzothiazole-sulfenamide; primary and secondary accelerators for rubber; safe at processing temperature and active over a wide curing range; particularly advantageous in styrene/butadiene rubber tires compounded with fine particle furnace blacks. [R T Vanderbilt Co Inc]

Amax XLP. Low phosphorus copper; oxygen-free copper plus 0.001-0.005% phosphorus; conductivity 98% IACS; copper + silver 99.95%, phosphorus-0.001-0.005%; ideal for applications where a low phosphorus content is beneficial and good conductivity with resistance to embrittlement must be ensured. [Amax Inc] *

Amaze®. CAS 25311-71-1; Isofenphos; soil-applied insecticide used for control of insects in rice crops and pear sucker. [Bayer AG]

Amazin®. Maneb with zinc; fungicide. [Aceto]

Ambazyme. Amyloglucosidase. [Hoechst UK]

amber. A fossil resin formed in certain beds of clay and sand, stated to be derived from *Pinites succinifer*. The following are varieties of amber: Succinite (m.p. 250-300 C), Gedanite (m.p. 150-180 C), Glessite (m.p. 250-300 C), Beck.

amber acid. *See* succinic acid

Amberdeen. *See* Bakelite.

Amberglow. A proprietary phenol formaldehyde synthetic resin. §

ambergris. A grey, wax-like product found in the sea. It occurs in certain conditions of the intestines of the sperm whale. The chief constituent is ambrein, $C_{23}H_{40}O$; used in perfumery.

amber-guaiacum resin. A variety of guaiacum resin; it is not an amber.

Ambergum®721. Cellulose deriv.; replaces gum arabic in lithographic printing processes. [Aqualon]

Amberite. A smokeless powder consisting of 71% nitro-cotton, 18.6% barium nitrate, 1.3% potassium nitrate, 1.4% wood meal, and 5.8% petroleum jelly.

Amberlac®. Synthetic resinous material in solid form or solutions; for use in industrial arts and coatings. [Reichhold]

Amberlac® 13-801. Acrylic monomer-modified alkyd. [Reichhold]

Amberlite®. Ion-exchange resins. [Rohm & Haas UK]

Amberlite® IRA-68. Acrylic; weakly basic anion exchange resin for industrial water treatment, pharmaceutical, chem. and food processing industries. [Rohm & Haas]

Amberlite® IRP-64. Polacrilin; a synthetic ion-exchange resin, supplied in the hydrogen or free acid form; pharmaceutic aid. [Rohm & Haas] *

Amberlite®IRP-88. Polacrilin potassium; pharmaceutic aid. [Rohm & Haas] *

Amberol. A phenol-formaldehyde resin combined with rosin or other resin; used in the varnish industry. §

Ambersil. Silicone emulsion for shell molding and hot box processes. [Foseco (F.S.) Ltd]

Ambiflo. Synthetic lubricants which are colorless, viscous liquids made by combining or polymerizing propylene oxide or propylene oxide and ethylene oxide; used in equipment for metal working, heat transfer and as automotive brake fluids, internal combustion engines, gears and bearings. [Dow UK]

Ambilhar. CAS 61-57-4; A proprietary preparation of niridazole; an antibilharzal agent. [Ciba plc] *

Ambiteric. Ampholytic surfactant. [Hoechst UK]

Ambiteric D. High molecular weight substituted betaine as a creamy unctuous mass; good alkali stability, wetting, foaming, detergency and solubilizing properties; used in industrial cleaner formulations; perfume solubilization; antistat. [Hoechst UK]

Ambitrol. A series of formulated engine coolants made from glycols, deionized water and suitable inhibitors; stationary engines operating for transmission of natural gas and petroleum products, electrical power generated systems, irrigation systems and drilling operations. [Dow UK]

Ambodryl. CAS 1808-12-4; A proprietary preparation of bromodiphenhydramine hydrochloride. [Parke-Davis] *

Amborate. Formoxy methyl isolongifolene. [Bush Boake Allen Ltd]

Amborol. Hydroxy methyl isolongifolene. [Bush Boake Allen Ltd]

Amboryl Acetate. Acetoxymethyl isolongifolene. [Bush Boake Allen Ltd]

Ambra. A phenol-formaldehyde synthetic resin. §

Ambrac Metal. A corrosion resistant nickel silver containing copper, nickel, zinc, and manganese. §

Ambraloys. A proprietary trade name for alloys of copper and aluminum with zincorion. §

Ambramycin. A proprietary trade name for tetracycline. §

Ambrasite. *See* Bakelite. §

Ambrene. Dinitro-t-butyl-m-cresol-methyl ether; an artificial musk perfume; §

ambrite. A resin found in the lignite of Auckland, New Zealand.

ambroid. (pressed amber). A product consisting of small fragments of amber heated under pressure; adhesive.

Ambrol. A phenol-formaldehyde synthetic resin. §

Ambroxan. 8-α-12-oxido-13,14,15,16 tetra-norlabdane; fragrance raw material; ambergris type. [Henkel; Cospha; Henkel Canada]

Ambush. CAS 52645-53-1; Insecticide containing permethrin. [ICI Chem. & Polymers Ltd]

Ambush C. CAS 66841-24-5; An emulsifiable concentrate containing 100 g cypermethrin per liter; a pyrethroid insecticide. [ICI Agrochemicals]

Amcar CL. Alkyl benzoate ester; carrier for disperse and

---

‡ Trade name and manufacturer not verified   § Trade name without identified manufacturer

cationic dyeing of polyester. [Am. Emulsions]

Amcar OCP. Chlorinated benzene, anionically emulsified; carrier for disperse dyeing of polyester. [Am. Emulsions]

Amcide. CAS 7773-06-0; EINECS 231-871-7; Ammonium sulfamate; an inorganic herbicide to control weeds and grasses in vegetables and ornamentals prior to planting and as a tree-killer. [Battle, Hayward & Bower Ltd]

Amciderm®. Amcinonid. [E Merck]

Amcill. CAS 69-53-4; Ampicillin; antibacterial. [Parke-Davis] *

Amcron. Oxygen-free copper plus 0.7 to 1.2% chromium; conductivity 82% IACS; chromium 0.7 to 1.2%, copper + silver + chromium 99.95%; principal uses based on its good compressive yield strength, creep resistance and thermal fatigue properties at moderately elevated temperatures. [Amax Inc] *

Amdye PH-12. Inorganic salts; replacement for trisodium phosphate in liquid form. [Am. Emulsions]

AME 4000®. Modified epoxy resin with superior strength/weight characteristics; high performance marine resin for fiberglass power boats and sailboats. [Ashland]

Ameen. Long chain aliphatic amines. [Akzo Chemie UK Ltd.]

Ameenex 70 WS. Amine salt; conc. for oilfield down-hole corrosion inhibition. [Chemron]

Ameenex C-18. Tall oil amido-amine; film-forming corrosion inhibitor; wetting, emulsifying and antistripping agent with asphalt compounds, coal tar pitches; drilling fluid additive; useful in non-metallic min. flotation. [Chemron]

Ameenex Polymer. Complex resinous polyamine; corrosion inhibitor intermediate; excellent film persistency; low emulsification tendency; high temp. stability; for oilfield applications. [Chemron]

Amenide. CAS 62265-68-3; Quinfamide; antiamebic. [Sterling Drug Inc] *

Amercell Polymer HM-1500. Nonoxynol hydroxyethylcellulose; nontacky thickener for aq. sol'ns., surfactant, emulsion stabilizer, film-former; substantive to skin and hair; for body lotions, moisturizers, hydroalcoholic products, sun lotions, hair conditioners, mousses, liquid makeup. [Amerchol Corp]

Amerchol® 400. Petrolatum, lanolin alcohol, cetyl alcohol, lanolin, stearone; auxiliary emulsifier, emulsion stabilizer for o/w and w/o systems, including makeup, pharmaceuticals; emollient, lubricant. [Amerchol; Amerchol Europe]

Amerchol®BL. Multisterol absorption base of lanolin sterol esters and higher alcohols; nonionic w/o emulsifier. [Amerchol Corp]

Amerchol®C. Petrolatum, lanolin, lanolin alcohol; absorp. base, auxiliary emulsifier for o/w systems, conditioner, emollient, moisturizer, stabilizer for cosmetics and pharmaceuticals, textile finishes. [Amerchol; Amerchol Europe]

Amerchol® CAB. Solid emollient multisterol extract of lanolin alcohols in petrolatum; w/o emulsifier activity, ideal for pharmaceutical vehicles well tolerated on dry and injured skin. [Amerchol Corp]

Amerchol® H-9. Absorption base containing cholesterol esters and free sterols; natural emollient and nonionic w/o emulsifier. [Amerchol Corp]

Amerchol® L-99. Mineral oil, lanolin alcohol; emulsifier, stabilizer, conditioner, emollient, moisturizer for creams and lotions, hair and skin preparations, dermatological specialties. [Amerchol; Amerchol Europe]

Amerchol® L-101. Mineral oil, lanolin alcohol; emollient, penetrant, emulsifier, moisturizer, softener, stabilizer for cosmetics, creams, makeup, hair dressing, pharmaceuticals, aerosols, baby products, textile finishes; plasticizer for hair sprays. [Amerchol Corp]

Amerchol® L-500. Mineral oil, lanolin alcohol, octyl-dodecanol; emulsifier, stabilizer, emollient, moisturizer, conditioner for hair and skin products, creams, lotions, makeup, aerosols, pharmaceutical vehicles, baby toiletries. [Amerchol Corp]

Amerchol Polysorbate. Emulsifying and solubilizing agent in foods; defoamer; used for vegetable oils, vitamins, beet sugar, yeast, cottage cheese. [D F Anstead Ltd] *

Amerchol® RC. Concentrated lipophilic lanolin alcohol fraction with lubricating, nontacky, barrier properties, particularly suited for makeup systems. [Amerchol Corp]

Amercor®8730. Blend of volatile amines; corrosion inhibitor providing protection in low, med. and high pressure sections of a steam/condensate system. [Drew Ind. Div.]

Amerfloc® 2. CAS 1302-42-7; Stabilized sol'n. of sodium aluminate; coagulant for treatment of potable water, water clarification, hot or cold lime softening applications. [Drew Ind. Div.]

Amerfloc® 275. High molecular weight polymer; flocculant and coagulant aid for potable water clarification, water plant sludge dewatering applications. [Drew Ind. Div.]

Amerfloc Plus® 5270. High molecular weight, highly anionic fluid polymer; flocculant and coagulant aid for sludge conditioning; for dewatering industrial slurries and water clarification applications in paper industry, food processing, nonpotable water clarification, oily waste treatment. [Drew Ind. Div.]

Amergel®100. Organics, surfactants, water blend; antifoam for aq. process systems where oils, solvs., waxes, silicas are undesirable, e.g., pulp and paper, nonwovens, effluent systems. [Drew Ind. Div.]

Amergize. Deposit modifier/combustion improver; a unique blend of oil soluble organometallic compounds. [Ashland Chemical Company] *

Amergy® 5400. Organometallic; fuel oil treatment; slag modifier preventing deposits and corrosion on boiler firesides. [Drew Ind. Div.]

Ameribond. CAS 8061-52-7; Modified calcium lignosulfonates; pelleting aids for animal feeds. [Borregaard LignoTech]

Ameribond 2000. Lignosulfonate; animal feed binder. [Borregaard LignoTech]

Americaine. CAS 94-09-7; Benzocaine; anesthetic. [Baxter Health Care] *

Ameripol Synpol 1009. Hot polymerized pre-crosslinked styrene/butadiene rubber; produces a smooth, nonstringy adhesive that breaks clean when gunned or troweled in place. [Ameripol Synpol]

Ameripol Synpol 1013/8000. Hot polymerized non-crosslinked styrene/butadiene rubber; used for paper saturation, as barrier coat prior to applic. of pressure-sensitive adhesive; useful in blends with other elastomer to increase cohesive strength, green strength. [Ameripol Synpol]

Amerite. A proprietary trade name for rubber derivatives and rubber-like resins in aqueous dispersion. §

Amerlate® LFA. CAS 68424-43-1; EINECS 270-302-7; Lanolin acid; emulsifier, stabilizer, emollient for fatty acid systems, aerosol shave creams, cream shampoos, wax

---

* Trade name not verified as available   † Trade name verified as obsolete

systems, household products; pigment dispersant; increases tack and plasticity of wax films. [Amerchol; Amerchol Europe]

Amerlate®P. CAS 63393-93-1; EINECS 264-119-1; Isopropyl lanolate; conditioner, penetrant, lubricant, moisturizer, emollient, w/o emulsifier, stabilizer, opacifier for cosmetics and pharmaceuticals; pigment dispersant; wetting agent and dispersant for solids; plasticizer for wax and pigment systems. [Amerchol; Amerchol Europe]

Amerlate® WFA. CAS 68424-43-1; EINECS 270-302-7; Lanolin acid; emulsifier, stabilizer for emulsions, aerosols, shampoos; stabilizer for conventional soap emulsions; wets and disperses pigments in makeups. [Amerchol; Amerchol Europe]

Amerol. The methyl ester of saccharin. §

Amerone. Perfumery base. [PPF International Ltd] *

Ameroxol® OE-2. CAS 9004-98-2; Oleth-2; solubilizer, emulsifier, dispersant, stabilizer, lipophilic cosolv. for creams and lotions, shampoos, and detergents, fluid and gelled transparent emulsions, fragrance products, and aerosols. [Amerchol; Amerchol Europe]

Ameroxol® OE-5. CAS 9004-98-2; Oleth-5. [Amerchol Corp]

Ameroyal. A concentrated liquid blend of polyelectrolyte scale inhibitors and antifoam agents; an evaporator treatment used to prevent scale deposition and foaming in conventional marine evaporators thereby minimizing the need for acid cleaning. [Ashland Chemical Company] *

Amerplex®605. Amerzine®corrosion inhibitor, Isoquest HT scale inhibitor, modified natural org. components, and antifoam; deposit and corrosion inhibitor for steam generating systems. [Drew Ind. Div.]

Amerscan MDP Kit. Medronate disodium; pharmaceutic aid. [Amersham Corp] *

Amerscent 86. Neutralizing/masking agent for odor control in waste water holding areas. [Drew Ind. Div.]

Amerscreen. Uv absorber; sunscreen. [D F Anstead Ltd] *

Amersep® MP-3. Sodium dimethyldithiocarbamates; metals precipitant for use in plating and metal finishing operations. [Drew Ind. Div.]

Amersil® DMC-287, DMC-357. Dimethicone copolyol. [Amerchol Corp]

Amersil® L-45 Grades. Dimethicone. [Amerchol Corp]

Amersil® ME-358. Cyclomethicone and dimethicone copolyol; emulsifier for prep. of water-in-silicone oil for personal care products. [Amerchol Corp]

Amersil® Simethicone. CAS 8050-81-5; Simethicone. [Amerchol Corp]

Amersil® VS-7207. CAS 69430-24-6; Cyclomethicone. [Amerchol Corp]

Amersite®2. CAS 7631-90-5; Sodium bisulfite and selected catalytic agents; corrosion inhibitor for steam generating systems. [Drew Ind. Div.]

Amersperse 1200. Scale inhibitor for evaporators, vacuum pans, and juice heating equip. [Drew Ind. Div.]

Amerstat®. Liquid and solid microbial control agents used for control of slime in the papermaking process for bacterial control in the sugar process and for a preservative in aqueous systems; used in paper mills, sugar mills, preservation of paints, coatings, adhesives, mineral slurries, drilling muds, animal glues, latent metal working fluids and paper coatings. [Ashland Chemical Company]

Amerstat®233. CAS 533-74-4; 3,5-Dimethyl tetrahydro-2-H,1,3,5-thiadiazone-2-thione; antimicrobial in industrial water systems, preservative in aq. systems. [Drew Ind. Div.]

Amerstat®250. 2-Methyl-4-isothiazolin 3-one and 5-chloro-2 methyl-4-isothiazolin-3-one; paper mill slimicide. [Drew Ind. Div.]

Amerstat® 251. 5-Chloro-2-methyl-4-isothiazolin-3-one and 2-methyl-4-isothiazolin-3-one; antimicrobial, preservative in latex. [Drew Ind. Div.]

Amerstat® 272. Sodium dimethyl dithiocarbamate and disodium ethylene bisdithiocarbamate; paper and sugar mill slimicide. [Drew Ind. Div.]

Amerstat®282. CAS 6317-18-6; Methylene bis (thiocyanate); antimicrobial in industrial water systems; preservative in water-containing systems. [Drew Ind. Div.]

Amerstat® 294. CAS 3064-70-8; Bis (trichloromethyl) sulfone and dispersants; paper mill slimicide. [Drew Ind. Div.]

Amerstat®300. 2,2, Dibromo-3-nitrilo propionamide; paper mill slimicide, antimicrobial agent for enhanced oil recovery systems; preservative for metal working fluids containing water. [Drew Ind. Div.]

Amertrol. Deposit inhibitors; deposit control agents for boiler deposit control. [Ashland Chemical Company]

Amerzine®. CAS 302-01-2; Organically catalyzed hydrazine; corrosion inhibitors for preboiler and afterboiler corrosion control, oil field line corrosion control and chromate reduction. [Ashland Chemical Company]

Amesec. A proprietary preparation of aminophylline, ephedrine hydrochloride and amylobarbitone; a bronchial antispasmodic. [Eli Lilly & Co] †

amesite. A chloritic mineral.

A Metal. A nickel-iron-copper alloy containing 6-8% copper; used for making audio-frequency transformers. *

A-methaPred. CAS 2375-03-3; Methylprednisolone sodium succinate; glucocorticoid. [Abbott Laboratories] *

Amethopterin. A proprietary trade name for methotrexate.§

Ametox. CAS 7772-98-7; Specially purified and sterilized sodium thiosulfate for use in metallic poisoning. [May & Baker Ltd] *

Ametrex. CAS 834-12-8; Active ingredient: ametryne; 2-ethylamino-4-isopropylamino-6-methylthio-1,3,5-triazine; selective pre-and post-emergence herbicide, also used as an aquatic herbicide and vine desiccant. [Agan Chemical Manufacturers Ltd]

Amfac. A proprietary preparation of antimenorrhagic factor; used in the treatment of functional uterine hemorrhage. [Armour Pharmaceutical Co] *

Amfaid. Surface active agents and detergents. [ABM Chemicals Ltd] *

amfecloral. α-Methyl-N-(2,2,2-trichloroethylidene) phenethylamine; an appetite suppressant.

Amfipen. CAS 69-53-4; A proprietary preparation of ampicillin; an antibiotic. [Brocades Pharma]

Amfix. High-speed fixer for photographic processing. [May & Baker Ltd] *

Amfix FRL. Formaldehyde-free fixing agent for improving wetfastness props. of cellulosics dyed or printed with direct or reactive dyes. [Am. Emulsions]

amfonelic acid. 7-Benzyl-1-ethyl-4-oxo-1,8-naphthyridine-3-carboxylic acid; a stimulant for the central nervous system.

Amgard® CPC 452. Red phosphorous-based additive;

---

‡ Trade name and manufacturer not verified   § Trade name without identified manufacturer

flame retardant additive for nylon resins; for injection molding, glass-filled, and extrusion resins and elec. applications. [Albright & Wilson Am.]

Amgard TBEP. CAS 78-51-3; Tri-(2-butoxyethyl) phosphate. [Surfachem Ltd]

amianth. See Amianthus.

amianthus. (amianth; mountain flax). CAS 1322-31-4; A white and satiny variety of asbestos.

Amical® 48. Diiodomethyl p-tolyl sulfone; mildewcide, fungicide for latex paints, emulsions, caulks, adhesives and sealants, and in lumber, construction, home improvement, textile, and automotive industries. [Angus]

Amical® 85. CAS 471-34-1; Calcium carbonate; filler designed for max. loading in resin systems. [Franklin Industrial Minerals]

Amical® 101. CAS 471-34-1; Calcium carbonate; filler designed for max. loading in resin systems. [Franklin Industrial Minerals]

Amical® Flowable. Diiodomethyl-p-tolyl sulfone aq. suspension; preservative, mildewcide, algicide for polymeric systems, especially latex paints, caulks, adhesives, leather. [Angus]

Amical® SC. CAS 471-34-1; Stearate surface-coated calcium carbonate; enhanced porcessability; surface coating is compatible with PVC, polyolefins, silicones, and engineering plastics. [Franklin Industrial Minerals]

Amicar. Amino caproic acid; hemostatic. [Lederle Laboratories USA]

amicarbalide. . 3,3´-Diamidinocarbanilide; Diampron is the isethionate; an antiprotozoan for veterinary use.

Amichrome. Premetallized dyes. [ICI Chem. & Polymers Ltd]

Amicon® C-860-4. CAS 25928-94-3; Conductive epoxy; high Tg die attach adhesive for IC assembly. [Emerson & Cuming] *

Amicon® C-940-4. Conductive polyimide; two-step cure die attach adhesive for IC assembly. [Emerson & Cuming]*

Amicon® CT-4042-5. CAS 25928-94-3; Two-component conductive epoxy; die attach adhesive for IC assembly. [Emerson & Cuming] *

Amicon® ECT-86. CAS 25928-94-3; Epoxy tape adhesive; elec. conductive version of TG-86; meets MIL-Std. 883C, Method 5011. [Emerson & Cuming] *

Amicon® ME-868. CAS 25928-94-3; Nonconductive epoxy; oxide-filled version of C-868-1; meets MIL-Std. 883C, Method 5011. [Emerson & Cuming] *

Amicon® SC-220. High purity silicone; blob top for high-rel circuits. [Emerson & Cuming] *

Amicon® SC-2634A/B. Two-component silicone; die coating. [Emerson & Cuming] *

Amicon® SC-3613. One-component silicone; electronic grade silicone; die attach adhesive for IC assembly. [Emerson & Cuming] *

Amicon® TG-86. CAS 25928-94-3; Epoxy tape adhesive; meets MIL-Std. 883C, Method 5011. [Emerson & Cuming] *

Amicure® 33-LV. CAS 280-57-9; Triethylene diamine in dipropylene glycol; catalyst for coatings, adhesives, sealants. [Air Prods. & Chems. Inc] *

Amicure® 352. Cycloaliphatic amines; curing agents. [Air Prods. & Chems. Inc]

Amicure® AEP. Cycloaliphatic amines; curing agents. [Air Prods. & Chems. Inc]

Amicure® CL-485. Aliphatic amine tetrol; crosslinker and

reactivity enhancer for PU coatings, adhesives, and sealants. [Air Prods. & Chems. Inc]

Amicure® DBU. Tertiary amine accelerator. [Air Prods. & Chems. Inc]

Amicure® PACM. Cycloaliphatic amines; curing agents. [Air Prods. & Chems. Inc]

Amicure® SA. Tertiary amine salts. [Air Prods. & Chems. Inc]

Amicure® TEDA. Tertiary amine accelerator. [Air Prods. & Chems. Inc]

Amicure® TMR 30. Tertiary amine accelerator. [Air Prods. & Chems. Inc]

Amidan. Dist. monoglycerides; emulsifier, dough conditioner, starch complexing agent in bread rolls, bread improvers. [Grindsted Prods.; Grindsted Prods. Denmark]

Amidate. CAS 33125-97-2; Etomidate; hypnotic. [Abbott Laboratories]

amide C$_1$. See formamide

amide C$_{18}$. See stearamide

Amide CMA-2. CAS 61791-08-0; PEG-2 coco MEA; thickener producing stable foam for personal care formulations. [Berol Nobel AB]

Amide RMA-2. PEG-2 rapeseedamide; thickener, foaming agent for personal care formulations. [Berol Nobel AB]

amidephrine. A vasoconstrictor and nasal decongestant; 3-(1-Hydroxy-2-methylaminoethyl) methanesulfonanilide. Dricol is the mesylate.

Amidex 1285. Modified coco diethanolamide (2:1); phosphate-compatible detergent, wetting agent, emulsifier for high-alkaline industrial and specialty cleaning compounds, e.g., degreasers, floor strippers; compatible with high conc. of inorg. in aq. systems without need for hydrotrope. [Chemron]

Amidex AME. CAS 142-26-7; EINECS 205-530-8; Acetamide MEA; antistat, humectant, conditioner for skin and hair products. [Chemron]

Amidex C. Modified coco diethanolamide; detergent, thickener, emulsifier, wetting agent, foam stabilizer, visc. builder; for industrial and household cleaners. [Chemron]

Amidex CE. Cocamide DEA (1:1); detergent, thickener, visc. builder, foam stabilizer for shampoos, cleaners, bubble baths, industrial cleaners, car shampoos, dishwashes, drycleaning detergents, waterless cleaners, solv. cleaners. [Chemron]

Amidex CIPA. Cocamide MIPA; antidefatting surfactant; for shampoos, skin cleansers, bubble baths. [Chemron]

Amidex CME. CAS 68140-00-1; Cocamide MEA; visc. builder, foam enhancer for personal care products, soap systems, synthetic powd. detergents, liquid dishwashing formulations. [Chemron]

Amidex CP. CAS 136-26-5; EINECS 205-234-9; Capramide DEA; flash foaming detergent, wetting agent for use in pigmented personal care systems. [Chemron]

Amidex KD. Cocamide DEA; surfactant for ethoxy sulfate systems; yields high stable viscosities at low conc.; flash foamer, foam stabilizer; for gelled shampoos, bath gels, liquid soaps, facial cleansers. [Chemron]

Amidex KME. CAS 68140-00-1; Cocamide MEA; foam builder, visc. booster, stabilizer for personal care products, synthetic powd. and liquid detergent systems. [Chemron]

Amidex L-9. CAS 120-40-1; EINECS 204-393-1; Lauramide

---

* Trade name not verified as available　　　† Trade name verified as obsolete

DEA; thickener, flash foamer, visc. enhancer, foam stabilizer/builder; for liquid detergents, household, institutional and industrial cleaning compounds. [Chemron]

Amidex LD. CAS 120-40-1; EINECS 204-393-1; Lauramide DEA; thickener, visc. builder, foam booster/stabilizer, detergent, emulsifier; for household, institutional and industrial cleaners, personal care products. [Chemron]

Amidex LIPA. CAS 142-54-1; EINECS 205-541-8; Lauramide MIPA; mild, low melting, fully active foam booster/stabilizer; for shampoo and detergent systems. [Chemron]

Amidex LMMEA. CAS 142-78-9; EINECS 205-560-1; Lauramide MEA; visc. builder, foam booster; for bath products, shampoos, skin cleansers. [Chemron]

Amidex LN. Linoleamide DEA; thickener, foam builder, emulsifier, conditioner; substantive to hair; for bath and skin care products, shampoos, conditioners. [Chemron]

Amidex O. CAS 93-83-4; EINECS 202-281-7; Oleamide DEA; thickener, emulsifier, lubricant, conditioner; for shampoos, mineral oil emulsions; compatible with hair dye systems. [Chemron]

Amidex PK. Palm kernelamide DEA; visc. builder and foamer for conditioning shampoos, mousses, styling gels. [Chemron]

Amidex RC. CAS 40716-42-5; EINECS 255-051-3; Ricinoleamide DEA; low foaming surfactant with wetting and softening props.; emulsifier with lubricity; for hair conditioners, shampoos, skin creams and lotions. [Chemron]

Amidex S. CAS 68425-47-8; EINECS 270-355-6; Soyamide DEA; foamer, visc. builder, emulsifier with skin feel props.; for shower and facial cleansers, liquid soaps, bath gels. [Chemron]

Amidex SE. CAS 93-82-3; EINECS 202-280-1; Stearamide DEA; thickener, emulsifier for personal care products including cold wave neutralizers, vegetable oil emulsions, conditioning shampoos and mousses. [Chemron]

Amidex SME. CAS 111-57-9; EINECS 203-883-2; Stearamide MEA; thickener, emulsifier for mineral oil and vegetable oil systems; for toiletry bars, creams, lotions. [Chemron]

Amidex TD. CAS 68140-08-9; EINECS 268-772-3; Tallowamide DEA; detergent for dry laundry compounds and specialty cleaners. [Chemron]

Amido-G-Acid. (Amido-G-Salt). β-Naphthylamine-6,8-disulfonic acid; a dyestuff intermediate. §

Amido-G-Salt. See Amido-G-Acid. §

Amido-R-Acid. (Amido-R-Salt). 2-Naphthylamine-3:6-disulfonic acid. §

Amido-R-Salt. See Amido-R-Acid. §

Amido Betaine C. Coconut amido alkyl betaine; component of personal care products; industrial foamer. [Zohar Detergent Factory]

Amidocid®. CAS 25311-71-1; Isofenphos; soil-applied insecticide used for control of insects in rice crops and pear sucker. [Bayer AG]

Amidogene. An explosive consisting of potassium nitrate, magnesium sulfate, wood charcoal, bran, and sulfur. §

amidol. 2:4-Diamino-phenol; used as a photographic developer.

Amidone. A proprietary trade name for methadone. §

amidosulfonic acid. See sulfamic acid

Amidox®. Ethoxylated alkylolamides; emulsifiers, detergents, wetting agents. [Stepan] *

Amidox® C-2. CAS 61791-08-0; PEG-3 cocamide; emulsi-

fier, detergent, wetting agent for dishwashing detergents, shampoos, emulsions; textile wetting and leveling agent. [Stepan; Stepan Canada]

Amidox® L-2. CAS 26635-75-6; PEG-3 lauramide; emulsifier, detergent, wetting agent for dishwashing detergents, shampoos, emulsions. [Stepan; Stepan Canada]

Amidozid. Soil applied insecticide; used for control of insects in rice crops and pear sucker. [Bayer AG]

Amiema MA-OD. Octyldodecyl N-myristoyl-N-methyl alanate; oil-phase cosmetics ingred. [Nihon Emulsion]

Amiema MA-OL. Oleyl N-myristoyl-N-methyl alanate; oil-phase cosmetics ingred. [Nihon Emulsion]

Amiesite. A proprietary asphalt-rubber product used in road surfacing. §

Amietol. Diethylethanolamine and a range of aminoethanols. [Imperial Chemical Industries plc]

Amigan. CAS 834-12-8, CAS 886-50-0; Ametryn + terbutryn; herbicide. [Agan Chemical Manufacturers Ltd]

Amigel. Sclerotium gum; natural gellifying agent; stable in acid pH range. [Alban Muller]

Amigen. Protein hydrolysate; replenisher. [Travenol Laboratories Inc] *

Amihope LL-11. CAS 52315-75-0; Lauroyl lysine; surface modifier, coemulsifier, codispersant; in cosmetics, medical, painting and other fields; filler for ink and paint; chelating agent. [Ajinomoto] *

Amikapron. A proprietary trade name for tranexamic acid. §

Amikin. CAS 39831-55-5; Amikacin sulfate; antibacterial. [Bristol Laboratories]

amilee. See paraffin, liquid.

Amilorin. CAS 17440-83-4; Amiloride hydrochloride; potassium conserving agent, diuretic. [Merck & Co Inc] †

Amine. Large range of cationic surfactants composed of primary amines, secondary amines or tertiary amines in liquid, solid or paste form; used in industry and in household and personal care formulations, though mostly in the form of derived quaternaries and various salts. [Keno Gard (UK) Ltd] ‡

amine C4. See n-butylamine

Amine 0. Dewatering agent and corrosion inhibitor. [Ciba plc] *

Amine 2HBG. CAS 61789-79-5; EINECS 263-089-7; N,N-Di(hydrogenated tallow) amine; surfactant intermediate; for pour pt. depressant formulations for diesel fuel, paper chem. auxiliary, personal care products. [Berol Nobel AB]

Amine 2M12D. CAS 67700-98-5; Dimethyl dodecyl amine; surfactant intermediate; quaternized end products used as bactericides, textile auxiliaries. [Berol Nobel AB]

Amine 2M14-50D. N,N-Dimethyl tetradecylamine; chemical intermediate; quaternized end products used as bactericides. [Berol Nobel AB]

Amine 2M14D. CAS 68439-70-3; Dimethyl tetradecylamine; surfactant intermediate; end products including quaternaries and amine oxides for detergent, disinfectant and cosmetic formulations. [Berol Nobel AB]

Amine 2M16D. CAS 68037-93-4; Dimethyl hexadecyl amine; surfactant intermediate; quaternized end products used in cosmetic formulations. [Berol Nobel AB]

Amine 2M18D. CAS 124-28-7; EINECS 204-694-8; Dimethyl octadecyl amine; surfactant intermediate; quaternized end products used in cosmetic formulations. [Berol Nobel AB]

Amine 2M 810 D. Methyloctyldecylamine; chemical intermediate for quaternaries used as bactericides. [Berol Nobel AB]

Amine 2M1214D. N,N-Dimethyl dodecyl tetradecylamine; chemical intermediate; end products including quaternaries and amine oxides used in disinfectant and cosmetic applications. [Berol Nobel AB]

Amine 2M1218D. CAS 61788-93-0; EINECS 263-020-0; N,N-Dimethyl cocamine; chemical intermediate; end products including quaternaries and amine oxides used in detergent and cosmetic applications. [Berol Nobel AB]

Amine 2MBGD-M. CAS 68814-69-7; N,N-Dimethyl tallowmine; chemical intermediate; quaternized end products used in detergent applications. [Berol Nobel AB]

Amine 2MHBGD. CAS 61788-95-2; Dimethyl hydrogenated tallowamine; surfactant intermediate; quaternized end products used in detergent applications. [Berol Nobel AB]

Amine 2MKKD. CAS 61788-93-0; EINECS 263-020-0; Dimethyl coco amine; surfactant intermediate; quaternized end products used as textile auxiliaries; betaines for cosmetic applications. [Berol Nobel AB]

Amine 2MOLD. CAS 68814-69-7; N,N-Dimethyl tallowamine; chemical intermediate; quaternized end products used in detergent applications. [Berol Nobel AB]

Amine 2 VT. CAS 61789-79-5; EINECS 263-089-7; Dihydrogenated tallowamine; chemical intermediate; end products including pour pt. depressant formulations for diesel fuel, paper chem. auxiliaries. [Berol Nobel AB]

Amine 8 D. CAS 111-86-4; 1-Octanamine; chemical intermediate. [Berol Nobel AB]

Amine 12. CAS 2016-57-1; EINECS 204-690-6; N-Dodecylamine; chemical intermediate; ethoxylated end-products used in detergent, cosmetic, and agricutural applications. [Berol Nobel AB]

Amine 12-98D. CAS 2016-57-1; EINECS 204-690-6; n-Dodecylamine; emulsifier; chemical intermediate; ethoxylated or guanidated end-products used in detergent, cosmetic, and agricutural formulations. [Berol Nobel AB]

Amine 14D. CAS 2016-42-4; n-Tetradecylamine, dist.; emulsifier; chemical intermediate; quaternized end-products used as bactericides. [Berol Nobel AB]

Amine 16D. CAS 143-27-1; EINECS 205-596-8; n-Hexadecylamine; emulsifier; chemical intermediate; end products such as quaternary ammonium compounds used as bactericides and in shampoo formulations. [Berol Nobel AB]

Amine 18-90. CAS 124-30-1; EINECS 204-695-3; N-Octadeylamine; chemical intermediate; ethoxylated end-products used in detergent, cosmetic, and agricutural applications. [Berol Nobel AB]

Amine 660. CAS 61791-57-9; EINECS 263-191-1; Tallow dipropylene triamine; chemical intermediate; end-products such as acetates used for emulsifiers and dispersants. [Berol Nobel AB]

Amine 740. CAS 68911-79-5; EINECS 272-787-0; N-Oleoalkyltripropylene tetraamine; chemical intermediate for production of surface active agents. [Berol Nobel AB]

Amine 760. Tallow tripropylene tetramine; chemical intermediate; end-products such as acetates used for emulsifiers and dispersants. [Berol Nobel AB]

Amine 780. CAS 97808-04-3; EINECS 307-919-9; N-Cocoalkyltripropylenetetraamine; chemical intermediate for production of surface active agents. [Berol Nobel AB]

Amine Acetate HBG. CAS 61790-59-8; Hydrogenated tallowamine acetate; reagent for pigment flushing, flocculation. [Berol Nobel AB]

Amine Acetate KK. CAS 61790-57-6; EINECS 263-147-1; Cocamine acetate; reagent for pigment flushing, flocculation. [Berol Nobel AB]

Amine B11. CAS 68037-92-3; EINECS 268-215-4; Eicosyl docosylamine; chemical intermediate; end-products such as ethoxylates used for detergent applications, acetates for emulsifiers and dispersants. [Berol Nobel AB]

Amine BG. CAS 61790-33-8; EINECS 263-125-1; Tallow amine; emulsifier, corrosion inhibitor; chemical intermediate producing sulfosuccinimides and textile auxiliaries. [Berol Nobel AB]

Amine CS-1135®. CAS 51200-87-4; Oxazolidine; emulsifying amine, corrosion inhibitor, alkaline pH stabilizer; for metalworking fluids and aq. systems. [Angus]

Amine CS-1246. CAS 7747-35-5; Oxazolidine; catalyst, resin reactant, formaldehyde substitute, crosslinking agent, corrosion inhibitor; raw material for synthesis. [Angus]

Amine D. CAS 1446-61-3; Dehydroabietylamine; used as asphalt additive, as cationic collectors for calcite, sylrite, mica, feldspar, vermicilulite and phosphate rock concentration operations. [Hercules] *

Amine HBG. CAS 61788-45-2; EINECS 262-976-6; Hydrogenated tallow amine; emulsifier, corrosion inhibitor; chemical intermediate producing ethoxylates for textile auxiliaries and acetates for emulsifiers and dispersants. [Berol Nobel AB]

Amine HBGD. CAS 61788-45-2; EINECS 262-976-6; Distilled hydrogenated tallow amine; emulsifier, corrosion inhibitor; chemical intermediate producing ethoxylated for textile auxiliaries and acetates for emulsifiers and dispersants. [Berol Nobel AB]

Amine KK. CAS 61788-46-3; EINECS 262-977-1; Cocamine; emulsifier, corrosion inhibitor; chemical intermediate producing ethoxylates used as detergents or textile auxiliaries. [Berol Nobel AB]

Amine M2HBG. CAS 61788-63-4; EINECS 262-991-8; Methyl di-(hydrogenated tallow) amine; surfactant intermediate. [Berol Nobel AB]

Amine M210D. CAS 7396-58-9; EINECS 230-990-1; Methyldidecylamine; chemical intermediate; quaternized end-products used as bactericides. [Berol Nobel AB]

Amine M218. CAS 4088-22-6; EINECS 223-819-7; Methyl dioctadecyl amine; surfactant intermediate. [Berol Nobel AB]

Amine OL. CAS 112-90-3; EINECS 204-015-5; Oleamine; emulsifier, corrosion inhibitor; chemical intermediate producing sulfosuccinimides for carpetback binding, engine-oil additives, ethoxylates for detergent applications. [Berol Nobel AB]

Aminess. Parenteral solution, tablets; essential amino acid for the treatment of ureamia. [KabiVitrum AB] *

aminic acid. Formic acid, HCOOH.

Amino-PF. A proprietary preparation of amino acids; an amino-acid supplement. [Pfizer International] †

aminoacetic acid. See glycine

Amino Acid Gelatinization Agent. N-Acyl glutamic acid diamide; gelatinization agent for oil for solidifying almost

---

all oils ranging from petrol. to vegetable oils. [Ajinomoto]*

3-aminoaniline. *See* m-phenylenediamine

4-aminobenzenesulfonamide. *See* sulfanilamide

p-aminobenzenesulfonic acid. *See* sulfanilic acid

p-aminobenzoic acid. (PABA) $NH_2C_6H_4CO_2H$; CAS 150-13-0; Dye intermediate, uv absorber in suntan lotions, pharmaceuticals, nutrition. [Am. Biorganics; R.W. Greeff; Natl. Starch & Chem.; Schweizerhall]

2-aminobenzoic acid, methyl ester. *See* methyl anthranilate

1-aminobutane. *See* n-butylamine

o-aminochlorobenzene. *See* o-chloroaniline

p-aminochlorobenzene. *See* p-chloroaniline

Amino-Collagen-25, -40. CAS 9105-54-7, 7732-18-5; Collagen amino acids; substantivity agent, penetrant, moisturizer for skin and hair care products, especially conditioners, shampoos, styling and setting products, nutritive skin products. [Maybrook]

aminocyclohexane. *See* cyclohexylamine

Aminodermin CLR. Sulfur rich amino acid conc.; conditioner for structurally damaged hair, oily skin care products. [Dr. Kurt Richter; Henkel/Cospha]

2-aminoethanesulfonic acid. *See* taurine

2-aminoethanol. *See* ethanolamine

2-aminoethyl alcohol. *See* ethanolamine

aminoethylethandiamine. *See* diethylenetriamine

aminoethylethanolamine. (hydroxyethylethylenediamine) $NH_2CH_2CH_2NHCH_2CH_2OH$; CAS 111-41-1; EINECS 203-867-5; Textile finishing compounds (antifumng agents, dyestuffs, cationic surfactants), resins, rubber, insecticides, medicinals. [BASF; Dow; Nippon Nyukazai; Schweizerhall; Union Carbide]

aminoethylpiperazine. (AEP; 1-(2-aminoethyl) piperazine; 2-piperazinoethylamione) $H_2NC_2H_4NCH_2CH_2NH-CH_2CH_2$; CAS 140-31-8; EINECS 205-411-0; Epoxy curing agent, intermediate for pharmaceuticals, anthelmintics, surface-active agents, synthetic fibers. [Dow; Texaco; Tosoh; Union Carbide]

Aminofoam C. TEA-lauroyl animal collagen amino acids; mild protein surfactant, detergent, conditioner for skin and hair care cleansing systems, shaving creams. [Croda Inc.; Croda Chem. Ltd.]

Aminofoam K. TEA-lauroyl animal keratin amino acids; mild protein surfactant, foaming agent for shampoos, conditioners, facial cleansers. [Croda Inc.]

aminoform. *See* hexamine

aminoformamidine hydrochloride. *See* guanidine hydrochloride

Aminogen I. α-Naphthylamine, $C_{10}H_7N$; a rubber vulcanization accelerator. §

Aminogen II. p-Phenylenediamine, $C_6H_4(NH_2)_2$; a rubber vulcanization accelerator. §

L-2-aminoglutaric acid. *See* glutamic acid

Amino Gluten MG. Maize gluten amino acids, sodium chloride; conditioner for skin creams and lotions and hair conditioners; humectant for cosmetics and pharmaceuticals. [Croda Inc.]

Aminogran. A proprietary phenylalanine-free food used in the treatment of phenylketonuria. [Allen & Hanburys Ltd]*

α-amino-β-hydroxybutyric acid. *See* L-threonine

2-amino-6-hydroxypurine. *See* guanine

2-aminohypoxanthine. *See* guanine

α-amino-β-imidazolepropionic acid. *See* histidine

1-α-amino-3-indolepropionic acid. *See* DL-tryptophan

α-aminoisocaproic acid. *See* L-leucine

α-aminoisovaleric acid. *See* L-valine

Aminol A-15. Trideceth-2 carboxamide MEA; biodeg. cosmetics surfactant, thickener, foam stabilizer; excellent dermatological props. [Chem-Y GmbH]

Aminol KDE. CAS 68603-42-9; Cocamide DEA; foam booster/stabilizer, superfatting agent for personal care products; solubilizer for perfumes, vegetable oils. [Chem-Y GmbH]

Aminol N. CAS 85536-23-8; PEG-4 rapeseedamide; biodeg. cosmetics surfactant, thickener; for shower bath, foam bath, shampoo, soap gel, and other surfactant formulations. [Chem-Y GmbH]

Aminol TEC N. CAS 85536-23-8; PEG-4 rapeseedamide; biodeg. emulsifier for metalworking fluids and lubricants, conveyor chain lubricant, in anticorrosive formulations. [Chem-Y GmbH]

2-amino-3-methylpentanoic acid. *See* L-isoleucine

α-amino-γ-methylvaleric acid. *See* L-leucine

2-aminopentanedioic acid. *See* glutamic acid

Aminophyllin. CAS 317-34-0; Aminophylline; relaxant. [G D Searle & Co] *

Aminoplex. A proprietary preparation of amino-acids, sorbitol, ethanol, vitamins and electrolytes used for parenteral nutrition. [Geistlich Sohne AG] *

1-amino-2-propanol. *See* isopropanolamine

2-aminopropionic acid. *See* α-alanine

Aminorex. 2-Amino-5-phenyl-2-oxazoline; an appetite suppressant; Apiquel is the fumarate. §

Amino-Silk SF. CAS 977077-71-6; Silk amino acids; substantive protein for elegant skin and hair preps.; penetrant, moisturizer. [Maybrook]

Aminosol. Protein hydrolysate; replenisher. [Abbott Laboratories] *

aminosuccinic acid. *See* aspartic acid

α-amino-β-thiolpropionic acid. *See* L-cysteine

Aminotriazole Bayer. CAS 61-82-5; Amitrole; fast-acting herbicide, for control of hard-to-kill grass and broadleaved weeds; mainly used in mixtures with other compounds (Ustinex products). [Bayer AG]

Aminox. Reaction product of di-phenylamine and acetone; gives protection against oxygen and heat deterioration; used in tire carcass, heels, soles, mechanicals, proofing sundries and wire insulation; effective in natural and nitrile rubbers and nylon. [Uniroyal] *

Aminox® Flake, Powd. CAS 9003-79-6; Diphenylamine-acetone reaction production; antioxidant. [Uniroyal]

Aminox®Naugard A. Antioxidant protecting polymers from loss of physical props. on extended exposure to heat; for EVA and polyamide hot-melt adhesives, nylon 6. [Uniroyal]

Aminoxid. Foam stabilizer; wetting; auxiliary; polymerization accelerator; used for shampoos, detergents; aqueous dispersions; photography; electroplating; vulcanizing. [Goldschmidt Ltd] *

Aminoxid WS 35. CAS 68155-09-9; Cocamidopropylamine oxide; detergent, emulsifier, wetting agent, softener, foam stabilizer for detergent preparations, cosmetic and pharmaceutical emulsions. [Goldschmidt; Goldschmidt AG]

1,2-aminozophenylene. *See* 1H-benzotriazole

Aminutrin. A proprietary preparation of aminoacids for oral nutrition. [Geistlich Sohne AG] *

---

‡ Trade name and manufacturer not verified

§ Trade name without identified manufacturer

Amipaque. CAS 31112-62-6; Metrizamide; diagnostic aid. [Sterling Drug Inc] *

amiphenazole. 2,4-Diamino-5-phenylthiazole; daptazole.

Amisynthetic A proprietary preparation of aceto-menaphthone and nicotinamide; used in the treatment of chilblains. [Armour Pharmaceutical Co] *

Amiter LGOD. CAS 82204-94-2; Dioctyldodecyl lauroyl glutamate; oily surfactant used in personal care products. [Ajinomoto; Ajinomoto USA; Nihon Emulsion]

Amiter LGOD-2. Dioctyldodeceth-2 lauroyl glutamate; emulsifer used in cosmetics; oil ingredient. [Ajinomoto; Ajinomoto USA; Nihon Emulsion]

Amiter LGS-2. Disteareth-2 lauroyl glutamate; emulsifier used in cosmetics. [Ajinomoto; Ajinomoto USA; Nihon Emulsion]

Amiter SG-OD. Di(2-octyldodecyl) N-stearoyl-L-glutamate; surfactant for cosmetic goods with high affinity to skin or hair. [Nihon Emulsion]

Amitraz. N-Methylbis-(2,4-xylyliminomethy1)-amine; Taktic; Triatrix; Triatox; a veterinary ascaricide. §

Amitril. CAS 549-18-8; Amitriptyline hydrochloride; antidepressant. [Parke-Davis] *

Amjet A-4. Leveling carrier for pressure dyeing of stock yarn and piece goods. [Am. Emulsions]

Amlev ACY-Super. Complex inorganic salt of org. amine; nonretarding leveling agent for polyacrylic dyeing. [Am. Emulsions]

Amlev CH641. Surfactant blend; leveling agent for dyeing nylon fiber, yarns, fabric, carpet. [Am. Emulsions]

Amlev DAS. Quaternary ammonium compound; leveling and retarding agent for Dacro 62 with cationic dyes; also suitable for acrylics. [Am. Emulsions]

Amlev HBL. Proprietary; low foaming penetrant, dispersant for jet machine dyeing of polyester, polyester/nylon carpet. [Am. Emulsions]

Amlev MRC. Sulfonated surfactant; acid dye leveler for continuous and batch dye applications [Am. Emulsions]

Amlight M-2. Mixture of inorganic compounds; peroxide bleaching stabilizer for cellulosic fibers and their blends. [Am. Emulsions]

Amlocor. See Norvasc. [Pfizer International]

Amlogard. See Norvasc. [Pfizer International]

Amlor. See Norvasc. [Pfizer International]

Amlovar. See Norvasc. [Pfizer International]

Amlube AEC. Blend; lubricant to prevent crack marks on polyester or nylons; improves backwinding of space-dyed yarns. [Am. Emulsions]

AAmlube V. Inorganic salt; all-purpose sequestering agent for beck dyeing; improves solubility and dispersibility of all dyes. [Am. Emulsions]

Ammonal. An explosive consisting of 30% trinitrotoluene, 47% ammonium nitrate, 22% aluminum powder, and 1% charcoal. §

ammonaldehyde. See hexamine.

Ammon-Carbonite. An explosive containing ammonium nitrate, flour, nitroglycerin, and collodion wool. §

Ammon-Dynamite. An explosive containing 40% nitroglycerin, 10% wood meal, 10% sodium nitrate, and 40% ammonium nitrate. §

Ammondyne. A coal-mine explosive containing 9-11% of nitroglycerin, 45-51% of ammonium nitrate, 8-10% of sodium nitrate, 17-19% of ammonium oxalate, and 11-13% of wood meal. §

Ammon-Foerdite I. An explosive containing ammonium nitrate, flour, nitroglycerin, collodion wool, glycerin, di-phenylamine, and potassium chloride. §

Ammon-Gelatin-Dynamite. A blasting explosive consisting of 50% nitroglycerin, 2.5% collodion cotton, 45% ammonium nitrate, and 2.5% rye meal. §

Ammon-Halalit. An explosive containing nitroglycerin, ammonium nitrate, vegetable meal, nitro-compounds, and potassium perchlorate. §

ammonia. (ammonia gas; ammonia anhydrous; spirit of Hartshorn) NH_3; CAS 7664-41-7; EINECS 231-635-3; Fertilizers, refrigerant, nitriding of steel, condensation catalyst, neutralizing agent, petroleum industry, latex preservative, explosives. [Air Prods.; Allied-Signal; Am. Cyanamid; Asahi Chem Industry Co Ltd; Chevron; General Chem.; La Roche Ind.; Mitsui Toatsu Chem.; Monsanto; Nissan Chem. Ind.; Norsk Hydro A/S; OxyChem; PPG Ind.; Unocal]

ammoniacal Turpethum. Hydrated dimercuri-ammonium sulfate, $(NHg_2)_2SO_4 \cdot 2H_2O$.

ammonia dynamites. Explosives usually containing nitroglycerin, wood pulp, ammonium nitrate, sodium nitrate, and calcium or magnesium carbonate. *

Ammonia Gelignite. An explosive containing 29.3% nitroglycerin, 0.7% nitro-cotton, and 70% ammonium nitrate.§

Ammonia-Olein. The trade name for a form of sulfonated castor oil. §

Ammonia-superphosphate. See nitrophosphate. §

Ammonioformaldehyde. See hexamine.

Ammonit C (Anfo-explosives). Primed from the bottom of the borehole by Gelatine Donarit 1 and detonating fuse. [Dynamit Nobel Wien GmbH] *

Ammonite. An explosive; contains ammonium nitrate, trinitrotoluene, and sodium chloride.

ammonium acetate. CH_3COONH_4; CAS 631-61-8; EINECS 211-162-9; Reagent in analytical chemistry, drugs, textile dyeing, preserving meats, foam rubbers, vinyl plastics, and explosives. [Aldrich; General Chem.; Schaefer Salt & Chem.; Verdugt BV]

ammonium alginate. (ammonium polymannuronate; alginic acid, ammonium salt) Ammonium salt of alginic acid; $C_6H_7O_6 \cdot NH_4$; CAS 9005-34-9; thickening agent and stabilizer in food products [Kelco Int'l.]

ammonium alum. (aluminum ammonium sulfate; ammonium aluminum sulfate; sulfuric acid, aluminum ammonium salt (2:1:1), dodecahydrate) Inorganic salt; $AlNH_4(SO_4)_2 \cdot 12H_2O$; CAS 7784-25-0; mordant in dyeing, water and sewage purification, sizing paper, retanning leather, clarifying agent, food additive, manufacture of lakes and pigments, fur treatment.

ammonium benzoate. $C_6H_5COONH_4$; Medicine, latex preservative. [Hart Prod. Corp; Verdugt BV]

ammonium bicarbonate. (ammonium hydrogen carbonate; acid ammonium carbonate; carbonic acid, monoammonium salt) Inorganic salt; NH_4HCO_3; CAS 1066-33-7; production of ammonium salts, dyes; leavening agent for cookies, crackers; fire-extinguishing compounds; pharmaceuticals, degreasing textiles, blowing agent for foam rubber, boiler scale removal, compost treatment. [BASF; General Chem.; Nissan Chem. Ind.; Norsk Hydro A/S; Rhone-Poulenc Basic]

ammonium bifluoride. (ammonium acid fluoride; ammonium hydrogen fluoride) NH_4HF_2; CAS 1341-49-7; Ceramics, chemical reagent, etching glass, sterilizer for

---

brewery, dairy, etc.; electroplating processing beryllium; laundry sour. [Bayer UK; Hoechst-Celanese; Miles; Solvay GmbH]

ammonium bisulfite. H_5NO_3S; Preservative. [Brotherton Ltd; General Chem.; Heico Chem.]

ammonium bromide. NH_4Br; CAS 12124-97-9; EINECS 235-183-8; Flame retardant for textiles, wood, chipboard, plywood. [Aldrich; Great Lakes Chem.; Johnson Matthey SA]

ammonium chloride. (sal ammoniac; salmiac; ammonium muriate) NH_4Cl; CAS 12125-02-9; EINECS 235-186-4; In dry batteries; mordant (dyeing and printing); safety explosives; flux for coating sheet and iron with zinc; manufacture of various ammonia compounds, fertilizer, pickling agent; in washing powders. [Aldrich; BASF; EM Ind.; General Chem.; Heico; Hüls Am.; Montefluos SpA]

ammonium dichromate. (ammonium bichromate) $(NH_4)_2Cr_2O_7$; CAS 7789-09-5; EINECS 232-143-1; Mordant for dyeing, pigments, manufacture of alizarin, chrome alum, catalysts, oil purification, pickling, leather tanning, synthetic perfumes, photography, lithography, pyrotechnics. [British Chrome & Chem.; EM Ind.]

ammonium fluoride. NH_4F; CAS 12125-01-8; EINECS 235-185-9; Manufacture of fluorides, analytical chemistry, antiseptic in brewing, etching glass, textile mordant, wood preservative, mothproofing agent. [Aldrich; Flexchemie BV; General Chem.; GE; Hoechst-Celanese; Olin Hunt]

ammonium hydrogen fluoride. *See* ammonium bifluoride

ammonium iodide. NH_4I; CAS 12027-06-4; EINECS 234-717-7; Iodides, medicine (expectorant), photography. [Rhone-Poulenc]

ammonium lauryl sulfate. (sulfuric acid, monododecyl ester, ammonium salt) Ammonium salt of lauryl sulfate; $C_{12}H_{26}O_4S \cdot H_3N$; CAS 2235-54-3; 68081-96-9; EINECS 218-793-9; detergent, emulsifier, foaming agent, dispersant, wetting agent; for personal care products, carpet shampoos, firefighting, dry wall manufacture, dyes, chemical specialties. [Lonza; Sandoz; Stepan; Witco]

ammonium molybdate. $(NH_4)_6Mo_7O_{24} \cdot 4H_2O$; In soil additives, enamel bonding agents, protective and decorative metal coatings, iron and steel alloys, lubricants, petrol. refining catalysts, pigments, corrosion inhibitors, smoke suppressants, production of molybdenum metal;. [AAA Molybdenum Prods.; Climax Molybdenum BV; Climax Performance]

ammonium nitrate. NH_4NO_3; CAS 6484-52-2; EINECS 229-347-8; Fertilizer, explosives, pyrotechnics, herbicides/insecticides, manufacture of nitrous oxide, absorbent for nitrogen oxides, ingredient of freezing mixtures, oxidizer in solid rocket propellants, nutrient for antibiotics and yeast, catalyst. [Air Prods.; Chevron; La Roche Ind.; Norsk Hydro AS; Unocal]

ammonium oxalate. $(NH_4)_2CO_4 \cdot H_2O$; Analytical chemistry, safety explosives, manufacture of oxalates, rust and scale removal. [Brotherton Ltd; General Chem.; Heico; Rhone-Poulenc]

ammonium persulfate. $(NH_4)_2S_2O_8$; Oxidizer, bleaching agent; photography; etchant for printed circuit boards, copper; electroplating; deodorizing oils; aniline dyes; food preservative; depolarizer in batteries; washing infected yeast; manufacture of other persulfates. [Aldrich; Degussa; EM Ind.; FMC; Interox Chem. Ltd]

ammonium phosphate. (MAP; monoammonium phosphate; ammonium phosphate, monobasic; ammonium dihydrogen phosphate) $NH_4H_2PO_4$; CAS 7722-76-1; Food products, fertilizer, flame retardants, plant nutrient sol'ns., manufacture of yeast, vinegar, yeast foods and bread improvers, food additive, analytical chemistry. [Albright & Wilson; Aldrich; Chisso; EniChem SpA; Heico; IMC Fertilizer; Monsanto; OxyChem; Rhone-Poulenc Basic; Showa Denko]

ammonium phosphate, dibasic. (DAP; diammonium phosphate; diammonium monohydrogen phosphate) . $(NH_4)_2HPO_4$; CAS 7783-28-0; Flame retardant for wood, paper, textiles fertilizer, plant nutrient sol'ns., feed additive; flux for soldering, purifying sugar; in ammoniacal dentifrices; manufacture of yeast, vinegar, bread improvers; foods, pharmaceuticals. [Albright & Wilson; Aldrich; Chisso; Heico, IMC Fertilizer; La Roche Ind.; Monsanto; OxyChem; Rhone-Poulenc Basic]

ammonium polyacrylate. (poly(acrylic acid), ammonium salt; 2-propenoic acid, homopolymer, ammonium salt) Ammonium salt of polyacrylic acid. $(C_3H_4O_2)_x \cdot xH_3N$; CAS 9003-03-6; dispersant for paints and coatings; thickening and stabilizing agent for synthetic latices; used in coatings, adhesives, dipped, cast, and molded goods, cements for rug backing, spraying, spreading, brushing, and extruding compounds

ammonium stearate. (octadecanoic acid, ammonium salt) Ammonium salt of stearic acid; $C_{17}H_{35}COONH_4$; CAS 1002-89-7; EINECS 213-695-2; vanishing creams, brushless shaving creams, other cosmetic products, waterproofing of cements, concrete, stucco, paper, textiles. [Magnesia GmbH; Original Bradford Soap Works]

Ammonium Stearate 33% Liquid CAS 1002-89-7; EINECS 213-695-2; Ammonium stearate; foam stabilizer, hand modifier for frothed latex systems. [Hart Chem. Ltd.]

ammonium sulfamate. $NH_2SO_3NH_4$; CAS 7773-06-0; EINECS 231-871-7; Flameproofing agent for textiles and paper; weed and brush killer; electroplating; generation of nitrous oxide. [Heico; Nissan Chem. Ind.; Spartan Flame Retardants]

ammonium sulfate. $(NH_4)_2SO_4$; CAS 7783-20-2; EINECS 231-984-1; Fertilizers, water treatment, fermentation, fireproofing compositions, viscose rayon, tanning, food additive. [Accurate Chem. & Scientific; Aldrich; Allied-Signal; BASF; DSM NV; General Chem.; Heico; Nissan Chem. Ind.; Schaefer Salt & Chem.; Showa Denko]

ammonium thiocyanate. (ammonium rhodanide) NH_4SCN; CAS 1762-95-4; EINECS 217-175-6; Analytical chemistry; thiourea; fertilizers; photography; in liquid rocket propellants; fabric dyeing; zinc coating; weed killer, defoliant; adhesives; curing resins; pickling iron and steel; electroplating; polymerization catalyst; metals separation. [Carbo-Tech GmbH; Degussa; Witco/Argus]

ammonium thiosulfate. $(NH_4)_2S_2O_3$; CAS 7783-18-8; EINECS 231-982-0; Photographic fixing agent; analytical reagent; fungicide; reducing agent; brightener in silver plating baths; cleaning compounds for zinc-base die-cast metals; hair waving preparations; fog screens. [Blythe, William Ltd; Du Pont; General Chem.]

ammonium tungstate. (ammonium wolframate; ammonium paratungstate). $(NH_4)_{10}H_2(W_2O_7)_6$; CAS 11120-25-5; EINECS 234-364-9; Preparation of ammonium phosphotungstate and tungsten alloys. [Aldrich; Climax Molybdenum]

---

‡ Trade name and manufacturer not verified   § Trade name without identified manufacturer

ammonocarbonous acid. Hydrocyanic acid, HCN.

Ammonyx®4, 4B, 485, 4002. CAS 122-19-0; EINECS 204-527-9; Stearalkonium chloride; emulsifier, conditioner, softener, emollient for cosmetics. [Stepan; Stepan Canada]

Ammonyx® CETAC, CETAC-30. CAS 112-02-7; EINECS 203-928-6; Cetrimonium chloride; emulsifier, conditioner, softener, emollient for cosmetics. [Stepan; Stepan Canada]

Ammonyx® CO. CAS 7128-91-8; EINECS 230-429-0; Palmitamine oxide; conditioner, detergent, foam stabilizer, visc. builder used in cosmetic, household, and janitorial products; wetting agent in conc. electrolyte sol'ns.; textile lubricant, emulsifier, wetter, dye dispersant. [Stepan; Stepan Canada]

Ammonyx® KP. CAS 37139-99-4; Olealkonium chloride; conditioner, antistat in clear hair rinses. [Stepan; Stepan Canada]

Ammonyx® LO. Lauramine oxide; foamer/foam stabilizer, wetting agent, visc. builder, grease emulsifier for shampoos, bath products, fine fabric cleaners, hard surface cleaners containing acids or bleach, dishwash, shaving creams, lotions; textile lubricant, emulsifier, dye dispersant. [Stepan; Stepan Canada]

Ammonyx® OAO. CAS 14351-50-9; EINECS 238-311-0; Oleamine oxide; wetting agent, foam booster/stabilizer, conditioner, visc. builder for shampoos, bubble baths, hand soaps, conditioners. [Stepan; Stepan Canada]

Amo® Balanced Salt Solution. Balanced salt solution; sterile intraocular surgical irrigation. [Allergan, Inc]

Amo®Endosol Extra®. Balanced salt solution enriched with bicarbonate, dextrose and glutathione; sterile intraocular surgical irrigation solution. [Allergan, Inc]

Amoco® 1012. CAS 9003-07-0; PP homopolymer resin; extrusion grade resin. [Amoco Chemical Co]

Amoco® 1016. CAS 9003-07-0; PP homopolymer resin; general purpose injection molding grade; FDA compliance. [Amoco Chemical Co]

Amoco® 1246. CAS 9003-07-0; PP homopolymer resin; LTHA injection molding grade; FDA compliance. [Amoco Chemical Co]

Amoco® 4018. CAS 9003-07-0; PP homopolymer resin; general-purpose injection molding/extrusion grade. [Amoco Chemical Co]

Amoco® 5016. CAS 9003-07-0; PP homopolymer fiber resin; for fiber/film applications; FDA compliance. [Amoco Chemical Co]

Amoco®6114. CAS 9003-07-0; PP homopolymer film resin; for oriented film applications; FDA compliance. [Amoco Chemical Co]

Amoco®6400P. CAS 9003-07-0; PP homopolymer resin; powd. grade resin for general purpose applications. [Amoco Chemical Co]

Amoco® 7234. CAS 9003-07-0; PP homopolymer resin, nucleated, antistat; injection molding grade; FDA compliance. [Amoco Chemical Co]

Amoco® 7239. CAS 9003-07-0; PP homopolymer; nucleated, antistat; injection molding grade with improved clarity, high rigidity and tensile strength properties for pkg., disposable medical, houseware products; FDA compliance. [Amoco Chemical Co]

Amoco® 7728. CAS 9003-07-0; PP homopolymer resin; general-purpose radiation-stable grade. [Amoco Chemical Co]

Amoco® 8244. CAS 9010-79-1; Ethylene/propylene copolymer; general purpose grade suitable for extrusion blow molding applications; improved toughness over Amoco 8217; FDA compliance. [Amoco Chemical Co]

Amoco® 8410. CAS 9010-79-1; Ethylene/propylene copolymer; slip and antiblock; for cast film applications; FDA compliance. [Amoco Chemical Co]

Amoco®9119. CAS 9003-07-0; PP; enhanced grade offering high stiffness, higher heat deflection temps., good gloss on finished parts, improved processability; for sheet, extrusion and thermoforming. [Amoco Chemical Co] *

Amoco® BR-310. CAS 9003-28-5; Polybutene/polyolefin blend; acts as a moisture barrier and corrosion inhibitor when used to flood the area between the layers of composite metal-plastic sheath in elec. cables; its adhesive quality prevents slippage between the layers. [Amoco Chemical Co]

Amoco® CI-500. CAS 9003-28-5; Polybutene/polyolefin blend; cable filling compound acting as a moisture barrier and corrosion inhibitor in cable core containing paired wires; also to fill the interstitial space in the core. [Amoco Chemical Co]

Amoco®H2R. CAS 9003-53-6; HIPS; impact-grade resin for use in food containers, plates, trays, thermoformed panels. [Amoco Chemical Co]

Amoco®H3E. CAS 9003-53-6; HIPS; impact-grade resin for use in thin-wall containers for food and dairy. [Amoco Chemical Co]

Amoco® H-15. CAS 9003-28-5; Polybutene (isobutylene butene copolymer); used as tackifier, strengthener, and extender in adhesives, as plasticizer for rubber, as vehicle and fugitive binder for coatings, as cling additive for LLDPE stretch wrap films, as reactive intermediate for specialty chemicals; as leather impregnant, as vehicle or modifier for caulks, sealants, and glazing compounds, and in lubricants, paper treatments, elec. compounds; FDA compliance. [Amoco Chemical Co]

Amoco® L-14. CAS 9003-28-5; Polybutene; used as tackifier, strengthener, and extender in adhesives, as plasticizer for rubber, as vehicle and fugitive binder for coatings, as cling additive for LLDPE stretch wrap films, as reactive intermediate for specialty chemicals; as leather impregnant, as vehicle or modifier for caulks, sealants, and glazing compounds, and in lubricants, paper treatments, elec. compounds; FDA compliance. [Amoco Chemical Co]

Amoco®PIA. Purified isophthalic acid (1,3-benzene dicarboxylic acid); high purity product reacted to form esters, amides, salts, acid chlorides and other org. intermediates; in prep. of low color sat. and unsat. polyesters, engineering resins, PET copolymers; derivs. used in thermoset composites, paints, coatings, bottle and fiber resins, adhesives. [Amoco Chemical Co]

Amoco®R1. CAS 9003-53-6; Crystal PS resin; for use in foam sheet, food containers, meat and produce trays, jackets for thin-wall glass bottles. [Amoco Chemical Co]

Amoco®R5. CAS 9003-53-6; Crystal PS resin; for use in injection blow-molded containers. [Amoco Chemical Co]

Amodel®A-1115HS, A-1145HS. Polyphthalamide, glass-reinforced; semicryst. thermoplastic resin with outstanding dimensional stability and processing chars., high strength and stiffness, high thermal props., excellent chem. resist. for automotive under-the-hood parts, gears, bearings, and industrial parts. [Amoco Chemical Co]

Amodel® A-1340HS. Polyphthalamide, min./glass-reinforced; semicryst. thermoplastic resin with lower warp than glass-reinforced grades, higher strength and stiffness than min.-reinforced grades; used for small engine components, power tools, ignition components. [Amoco Chemical Co]

Amodel® AF-1115VO, AF-1133VO, AF-1145VO. Polyphthalamide, glass-reinforced; flame-retarded; semicrystalline thermoplastic resin with high strength and stiffness; used for elec. components, e.g., connectors, switches, sockets, circuit breakers. [Amoco Chemical Co]

Amodel® ET-1000. Polyphthalamide, impact-modified; semicryst. thermoplastic resin with high toughness and impact, high strength and stiffness, low moisture sensitivity; used for power tools, recreational equip., clips and fasteners, caster wheels. [Amoco Chemical Co]

Amollan®A. Oxyethylated fatty amine; wetting and emulsifying agent for bating and degreasing leather. [BASF AG]

Amollan®L. Org. esters and fatty acids; leveling agent for applic. of pigment finishes. [BASF AG]

Amoloid HV. CAS 9005-34-9; Ammonium alginates; used for textile printing, ceramic binding, can sealant. [Kelco]

Amoloid LV. CAS 9005-34-9; Ammonium alginates; used for textile printing, ceramic binding, can sealant. [Kelco]

Amonyl 380 BA. Cocamidopropyl betaine; detergent for shampoos. [Seppic]

Amonyl 675 SB. CAS 68139-30-0; Cocamidopropylhydroxysultaine; surfactant for shampoos. [Seppic]

Amonyl DM. Quaternium-82; cosmetic ingred. [Seppic]

Amo Vitrax®. CAS 9067-32-7; Sodium hyaluronate; surgical aid. [Allergan, Inc]

Amoxicap. CAS 26787-78-0; A proprietary preparation of amoxicillin; an antibiotic. [Pfizer International]

Amoxil. CAS 26787-78-0; Amoxicillin; antibacterial. [SmithKline Beecham]

Amoxisyrup. CAS 26787-78-0; A proprietary preparation of amoxicillin; an antibiotic. [Pfizer International]

AMP. CAS 124-68-5; 2-Amino-2-methyl-1-propanol; emulsifier, catalyst; dispersant for pigments and latex paints; corrosion inhibitor; stabilizer; resin solubilizer. [Angus]

AMP. Adenosine phosphate.

AMP-95. CAS 124-68-5; 2-Amino-2-methyl-1-propanol; emulsifier, catalyst; dispersant for pigments and latex paints; corrosion inhibitor; stabilizer; resin solubilizer. [Angus]

Ampco. An aluminum bronze containing from 86-92% copper, 7-11% aluminum, and 1.3% iron. §

AMPD. CAS 115-69-5; 2-Amino-2-methyl-1,3-propanediol; pigment dispersant, neutralizing amine, corrosion inhibitor, acid-salt catalyst, pH buffer, chemical and pharmaceutical intermediate; solubilizer or emulsifier system component in personal care products. [Angus]

Ampec. See Norvasc. [Pfizer International]

Ampen. CAS 69-53-4; A proprietary preparation of ampicillin; an antibiotic. [Pfizer International]

Amphenol. A proprietary trade name for polystyrene products. §

Amphionic. Ampholytic surfactant. [Rhone-Poulenc UK]

Amphionic 25B. High molecular weight amino-acid derivative, supplied as a golden liquid; alkaline cleaning and sanitizing formulations and biocidal soaps; an efficient biocide with a broad spectrum of kill; good stability in the presence of electrolytes; compatability with other types of surface agent; dispersant. [Rhone-Poulenc UK]

Amphisol. CAS 69331-39-1; DEA-cetyl phosphate; acid pH emulsifier. [Bernel]

Amphisol K. Potassium cetyl phosphate; emulsifier; stable over wide pH range. [Bernel]

Amphobac. Bactericidal amphoterics; used in shampoos and industrial cleaners. [Lonza AG]

Amphocerin. Mixture of higher molecular fatty alcohol and wax esters; water-in-oil type ointments and creams with good spreading properties. [Henkel Chemicals Ltd] *

Amphocerin K. Cetearyl alcohol, lanolin, hydrogenated peanut oil, vegetable oil, mineral oil, petrolatum; cream base for manufacturing of light and smooth creams and ointments of the w/o type. [Henkel/Cospha; Henkel KGaA]

Amphojel. CAS 21645-51-2; Aluminum hydroxide; antacid. [Wyeth Laboratories]

Ampholak 7TX. CAS 97659-53-5; EINECS 307-458-3; Tallowamphopolycarboxyglycinate; med. foaming detergent used in detergent applications, nonirritating toiletries, cosmetics, shampoos, liquid soaps; reduces irritation of anionics; softening agent. [Berol Nobel AB]

Ampholak 7TX/C. CAS 97659-53-5; EINECS 307-458-3; Stearylamphopolycarboxyglycinate; used in detergents, shampoos, liquid soaps; conditioner in shampoos; softener; reduces irritation of anionics. [Berol Nobel AB]

Ampholak 7TX-SD 55. CAS 97659-53-5; EINECS 307-458-3; Tallowamphopolycarboxyglycinate; detergent, softener for laundry and hard surface cleaners; also for cosmetics. [Berol Nobel AB]

Ampholak 7TX-T. CAS 97659-53-5; EINECS 307-458-3; Tallowamphopolycarboxyglycinate; detergent, softener for detergent applications, cosmetics, shampoos, liquid soaps; reduces irritation of anionics; conditioner in shampoos. [Berol Nobel AB]

Ampholak 7TY. CAS 97488-62-5; EINECS 306-998-7; Tallowamphopolycarboxypropionic acid; low foaming surfactant for alkaline cleaners, toiletries. [Berol Nobel AB]

Ampholak CCA. Complex N-alkylamino propionic acid and alkyldimethylbenzyl ammonium chloride; corrosion inhibitor at all pH values against corrosion by H_2 and CO_2 in presence of salt water; algicide, bactericide. [Berol Nobel AB]

Ampholak MDX-1. Blend of amphoteric surfactants; designed for mild washing up liqs.; reduces irritation to skin from anionics; bacteriostatic props., preservative for formulations. [Berol Nobel AB]

Ampholak XCE. CAS 97659-51-3; EINECS 307-456-2; Cocoiminodiglycinate; med. foaming surfactant, hydrotrope, detergent for industrial applications in strong alkaline sol'ns. [Berol Nobel AB]

Ampholak XCO-30. CAS 68608-65-1; EINECS 271-793-0; Sodium cocoamphoacetate; med. foaming surfactant for toiletries, nonirritating shampoos, acid hard surface cleaners. [Berol Nobel AB]

Ampholak XJO. CAS 68608-64-0; EINECS 371-792-5; Disodium capryloamphodiacetate; low foaming wetting agent, hydrotrope for high alkaline industrial hard surface cleaners. [Berol Nobel AB]

Ampholak XO7. CAS 97659-53-5; EINECS 307-458-3;

---

‡ Trade name and manufacturer not verified § Trade name without identified manufacturer

Oleoamphocarboxyglycinate; med. foaming, multipurpose cleaner component; for nonirritating toiletries, conditioners, liquid soap, as softener. [Berol Nobel AB]

Ampholak XOO-30P. CAS 70024-77-0; EINECS 274-267-9; Oleoamphocarboxyglycinate; surfactant for laundry detergents, hard surface cleaners where high visc. is desirable. [Berol Nobel AB]

Ampholak XTP. N-Tallowamido-polyamino-polygincate; for detergent applications, cosmetics, shampoos, liquid soaps; conditioner in shampoos; softergent in detergent formulations; anti-irritant for anionics. [Berol Nobel AB]

Ampholak YCA/P. CAS 91995-05-0; EINECS 295-264-9; Cocoiminodipropionate half sodium salt; med. to high foaming surfactant for alkaline cleaners; high stability to alkali. [Berol Nobel AB]

Ampholak YCE. CAS 97659-50-2; EINECS 307-455-7; Cocoiminodipropionate; med. foaming surfactant for industrial alkaline cleaners, cosmetic preps.; hydrotrope. [Berol Nobel AB]

Ampholak YCO-40. CAS 68919-40-4; EINECS 272-897-9; Cocoamphocarboxypropionate; mild, stable surfactant for highly alkaline systems, hard surface cleaners, industrial laundry detergents. [Berol Nobel AB]

Ampholak YJH-40. CAS 94441-92-6; EINECS 305-318-6; Sodium alkyliminodipropionate; surfactant for strong alkali cleaners where low foam is required; high caustic stability. [Berol Nobel AB]

Ampholan®. Complex amphoteric surfactants in high foam formulations; used for froth flotation, fire fighting, toiletries. [Harcros UK]

Ampholan B171. Cocoamido propyl betaine in liquid form; foaming and wetting agent, for toiletries, industrial cleaners, cement, gypsum & latex. [Henkel Europe] *

Ampholyt JA 140. Sodium lauroamphoacetate; mild surfactant for cosmetics, shampoos, detergents, baby care products. [Hüls Am.; Hüls AG]

Ampholyt JB 130. Cocamidopropyl betaine; surfactant for cosmetics, shampoos, detergents, hair shampoos, foam baths, shower foams, liquid soaps. [Hüls Am.; Hüls AG]

Ampholyte KKDP-60. CAS 84812-94-2; EINECS 284-219-9; N-Cocoalkylaminopropionic acid; emulsifier, dispersant, corrosion inhibitor. [Berol Nobel AB]

Ampholyte KKE-70. CAS 84812-94-2; EINECS 284-219-9; Coco alkyl aminopropionic acid; surfactant for detergents, toiletries; emulsifier, dispersant, corrosion inhibitor. [Berol Nobel AB]

Ampholyte SKKP 70. Amphoteric surfactant with dispersing and corrosion inhibiting properties; for emulsion paints, pigment grinding. [Keno Gard (UK) Ltd] ‡

Amphomer® LV-71. Octyl acrylamide/acrylates/butylaminoethyl methacrylate copolymer; hair fixative resin enhancing stiffness, holding and moisture resist. [Natl. Starch]

Amphoram. Alkyl amino acids based on coco and tallow alkyl chains; amphoteric surface active agent used in cosmetics, detergents, paint and pigment industries. [Atochem UK/Ceca]

Amphosol®CA. Cocamidopropyl betaine; mild conditioner, detergent, wetting agent, visc. builder, foam enhancer, base for cosmetics and household and industrial liquid detergents. [Stepan; Stepan Canada; Stepan Europe]

Amphosol CA. Derivative of alkyl amido propyl N-dimethyl amino acetic acid, as a clear yellow liquid; for shampoos and bubble baths. [KWR Chemicals Ltd] *

Amphosol® CB3. C8-18 alkylamido betaine; detergent, foaming and wetting agent for household and industrial cleaners. [Stepan Europe]

Amphosol DM and DMA. Acetyldimethyl alkylammonium chloride, sodium salt or in acid form; clear yellow liquid; for bacterial detergent preparations. [KWR Chemicals Ltd] *

Amphoteen 24. CAS 66455-29-6; EINECS 266-368-1; C12-14 alkyldimethyl betaine; surfactant for low-irritation shampoos, washing-up liqs., hard surface cleaners, vehicle cleaners. [Berol Nobel AB]

Amphoteen BCA-30. CAS 70851-07-9; EINECS 274-923-4; Cocoamidopropyl betaine; foam enhancer, visc. builder, thickener, mild surfactant for liquid soaps and washing-up liqs. [Berol Nobel AB]

Amphoteen BCM-30. CAS 68424-94-2; EINECS 270-329-4; Cocoalkyl dimethyl betaine; surfactant for low irritation shampoos and dishwashing liqs. [Berol Nobel AB]

Amphoteen BTH-35. CAS 70750-46-8; EINECS 274-845-0; Tallow bis(hydroxyethyl) betaine; surfactant for low irritation shampoos and dishwashing liqs.; thickener for household acidic cleaners; stable over wide pH range. [Berol Nobel AB]

Amphotensid 9M. Disodium cocoamphodiacetate and sodium laureth sulfate; detergent for personal care products. [Zschimmer & Schwarz]

Amphotensid B4. Cocamidopropyl betaine; surfactant for cosmetics, shampoos, detergents. [Zschimmer & Schwarz]

Amphoterge®. Substituted imidazoline amphoterics; used in shampoos and industrial cleaners. [Lonza AG]

Amphoterge® J-2. Disodium capryloamphodiacetate; wetting agent and detergent for personal care and industrial applications. [Lonza Inc]

Amphoterge®K. CAS 68919-41-5; Sodium cocoamphopropionate; detergents used in shampoos, skin cleansers, dishwashing; salt-free. [Lonza Inc]

Amphoterge® K-2. Disodium cocoamphodipropionate; detergents used in shampoos, skin cleansers, dishwashing, heavy duty liquid cleaners. [Lonza Inc]

Amphoterge® KJ-2. CAS 68815-55-4; Disodium caprylomphodipropionate; salt-free version of Amphoterge J-2; wetting agent, detergent for personal care and industrial applications. [Lonza Inc]

Amphoterge® L Special. Disodium lauroamphodiacetate; mild shampoo conc. [Lonza Inc]

Amphoterge® NX. Coco imidazoline dicarboxylate; industrial detergent. [Lonza Inc]

Amphoterge® W. Sodium cocoamphoacetate; surfactant for mild shampoos, skin cleansers, heavy duty cleaners, dishwashing preps. [Lonza Inc]

Amphoterge®W-2. Disodium cocoamphodiacetate; surfactant for nonirritating shampoos and skin cleansers, heavy duty liquid cleaners. [Lonza Inc]

Amphoteric 300. Sodium eicosyloxypropyliminodipropionate; general surfactant. [Exxon/Tomah]

Amphoteric 400. Iminopropionate, partial sodium salt; low foam detergent, coupler for hard surface alkaline or acid detergents, laundry, metal, acid bowl cleaners; defoamer in latex paints; corrosion inhibitor in metalworking lubricants; leather lubricant; stable in acid, alkali and conc. electrolytes. [Exxon/Tomah]

Amphoteric L. Coco derivative; detergent, foam stabilizer/booster, wetting agent, mild surfactant for liquid deter-

---

* Trade name not verified as available   † Trade name verified as obsolete

gents, shampoos, hand soaps, mech. foaming systems, dishwash; stable in mildly acid and alkaline media. [Exxon/Tomah]

Amphoteric N. Sodium C12-15 alkoxypropyl iminodi-propionate; high foam wetting agent, coupler for shampoos, detergents; corrosion inhibitor in metalworking lubricants; visc. builder; fire fighting foams. [Exxon/Tomah]

amphotropin. Hexamethylene-tetramine-camphorate, $(C_6H_{12}N_4)_2 \cdot C_8H_{14}$ (COOH)$_2$; a urinary antiseptic.

Ampiclox. CAS 69-53-4, CAS 61-72-3; A proprietary preparation of ampicillin and cloxacillin; an antibiotic. [SmithKline Beecham] *

Ampilar®. CAS 69-53-4; Capsules containing 250 mg and 500 mg ampicillin as ampicillin trihydrate; syrup: powder for reconstitution to contain 125 mg/5ml and 250 mg/5ml ampicillin as ampicillin trihydrate; used for treatment of infections of respiratory, urinary and digestive tract where infection may be due to more than one pathogen; antitiobic active against shigella, *E coli*, *proteus mirabilis*, *hemophilus influenza*, etc. [Lagap Pharmaceuticals Ltd]

Amplex. CAS 1406-65-1; A proprietary preparation of chlorophyll; a deodorant. [Ashe Chemicals] *

AmpliWax PCR Gems. Specially formulated wax beads to enhance PCR process; replaces mineral oil as vapor barrier in PCR amplifications. [Perkin-Elmer]

Amprol. CAS 121-25-5; Amprolium; coccidiostat. [Merck & Co Inc] *

Amron. A proprietary vinylite base plastic for coatings. §

Amsco Steel. A proprietary high manganese steel containing 12-13% manganese and 1.2% carbon. §

Amsidyl. CAS 51264-14-3; Amsacrine; antineoplastic. [Parke-Davis] *

Amsil. Oxygen free copper plus 8 oz to 30 oz per ton silver; conductivity 100% IACS. principal uses are based on its good creep strength at elevated temperatures and its high softening point. [Amax Inc] *

Amsoft FA. Fatty ester; nonyellowing softener for cotton, knits, yarn and fabric lubricant. [Am. Emulsions]

Amsoft MDH-20. CAS 9002-88-4; Polyethylene emulsion; general-purpose nonyellowing hand modifier for improved abrasion and sewability. [Am. Emulsions]

Amsol GMS. Surfactant blend; antigelling agent for conc. dye sol'ns. [Am. Emulsions]

Amsperse 109. Dispersing assistant for disperse dyes; prevents agglomeration; compatibilizes dye systems. [Am. Emulsions]

Amstat. Tranexamic acid; hemostatic. [Lederle Laboratories USA] *

Amsulf. Sulfur copper alloy containing oxygen-free copper and 0.3% sulfur; conductivity 96% IACS. [Amax Inc]*

Amtel. Tellurium-copper alloy containing oxygen-free copper and 0.5% tellurium; conductivity 93% IACS. [Amax Inc] *

Amterge TC. Blend of org. solvs., detergents, emulsifiers; scouring and cleaning aid. [Am. Emulsions]

Amuno. CAS 53-86-1; Indomethacin; analgesic and anti-inflammatory agent for the prompt and sustained relief of certain rheumatic conditions. (Germany). [Merck & Co Inc]

Amvis. An explosive containing ammonium nitrate. §

Amwet DAD. Ethoxylated alcohol; detergent, nonrewetting wetting agent, penetrant. [Am. Emulsions]

Amwet DOSS. Dioctyl sulfosuccinate; fast wetting agent for synthetics, cotton and blends; excellent for space dyeing. [Am. Emulsions]

Amwet MS-100. Solvent-based penetrant, scour, wetting agent for bleaching, alkaline scouring and dyeing operations. [Am. Emulsions]

amygdalic acid. See mandelic acid

amyl acetate. (amylacetic ester; banana oil; pear oil) $CH_3COOC_5H_{11}$; CAS 628-63-7; Solvent for lacquers and paints, extraction of penicilin, photographic film, leather and nail polishes, warning odor, flavoring agent, printing and finishing fabrics, solvent for phosphors in fluorescent lamps. [Aldrich; BP Chem. Ltd; Penta Mfg; Union Carbide]

amyl alcohol. (amyl hydrate; 2-methyl-1-butanol; 2-pentanol) $C_5H_{11}OH$; CAS 71-41-0 (n-); 75-85-4 (t-) Eight isomers possible; raw material for pharmaceutical preparations, organic synthesis solvent; t-: solvent, flotation agent, organic synthesis, medicine (sedative). [Ashland; Hoechst-Celanese; MTM Spec. Ltd; Union Carbide; Vista]

amylase. Diastase, an enzyme which renders starch soluble, by converting it into maltose.

amylcarbinol. See hexyl alcohol

amyl hydrate. See amyl alcohol

amyl hydride. See n-pentane

Amylit. A diamalt compound, which is an enzymic product; used for desizing in the textile industry. §

Amyl Ledate®. CAS 19010-66-3; Lead dimethyldithiocarbamate 50% in oil; liquid; rubber accelerator recommended for improved dynamic properties in natural and polyisoprene rubbers. [R T Vanderbilt Co Inc]

amyl mustard oils. Amyl thiocarbimides, $C_5H_{11}N \cdot CS$. *

amylodextrin. See soluble starch.

Amylogen. A soluble starch. §

amyloid. Concentrated sulfuric acid dissolves cellulose, gradually converting it into dextrin, and ultimately into dextrose; if the solution, as soon as it is made, is diluted with water, a gelatinous hydrate is produced; this substance is known as amyloid.

Amylopsin. Pancreatic diastase. §

Amylozine Spansule. A proprietary preparation of trifluoperazine and amylobarbitone; a sedative. [SmithKline Beecham] *

Amylozyme. Amylase starch converting enzyme. [Rhone-Poulenc UK]

Amylum. Starch $(C_{12}H_{20}O_{10})_n$. §

Amysal. Aryl salicylate, C_6H_4(OH) COOC$_5$H$_{11}$. §

Amytal. Isoamyl-ethyl-barbituric acid, a proprietary preparation of amylobarbitone; anesthetic applied intravenously; a hypnotic. [Eli Lilly & Co]

Amytal Sodium. CAS 64-43-7; Amobarbital sodium; hypnotic; sedative. [Eli Lilly & Co]

Amyx A-25-S 0040. CAS 122-19-0; EINECS 204-527-9; Stearyl dimethyl benzyl ammonium chloride; conditioner, softener, and emollient for hair rinses, skin creams, and lotions; emulsifier. [Clough]

Amyx CDO 3599. CAS 68155-09-9; Cocamidopropylamine oxide; mild high foaming surfactant, foam booster/stabilizer, wetting agent, hair conditioner for personal care, household, and janitorial products. [Clough]

Amyx CO 3764. CAS 7128-91-8; EINECS 230-429-0; Cetamine oxide; mild high foaming surfactant, foam booster/stabilizer, conditioner for personal care, household, and janitorial products. [Clough]

---

‡ Trade name and manufacturer not verified   § Trade name without identified manufacturer

Amyx LO 3594. Lauramine oxide; mild high foaming surfactant, foam booster/stabilizer, wetting agent, grease emulsifier for personal care, household and janitorial products. [Clough]

Amyx SO 3734. CAS 2571-88-2; EINECS 219-919-5; Stearamine oxide; mild high foaming surfactant, foam booster/stabilizer, conditioner, emulsifier for personal care, household, and janitorial products. [Clough]

Amyx ST 3837. Cetearyl alcohol, PEG-40 hydrogenated castor oil, stearalkonium chloride; conc. for prep. of hair conditioners. [Clough]

Amzirc. Zirconium copper alloys containing oxygen-free copper and 0.13-0.20% zirconium. [Amax Inc] *

AN-30GF. Amorphous nylon (modified nylon 12), 30% glass fiber-reinforced; exhibits low moisture absorp. and good dimensional stability, improved mech., thermal, and chem. props. [Compounding Tech.]

Anabalm. A proprietary preparation of oleoresin capsicum, menthol, histamine hydrochloride, beta-chloroethyl salicylate and squalene; an embrocation. [Crookes Laboratories] *

Anabolex. CAS 521-18-6; A proprietary preparation of stanolone; anabolic agent. [Lloyd, Hamol] *

Anacal. A proprietary preparation containing mucopolysaccharide, polysulfuric acid ester, prednisolone lauromacrogol-400 and hexachlorophene; used in the treatment of hemorrhoids. [Luitpold-Werk] *

anacardic acid. Obtained from *Anacardium occidenale* (cashew nut); the ammonium salt is used as a vermifuge.

Anacobin. CAS 68-19-9; A proprietary trade name for cyanocobalamin. §

Anacotine. Narcotine.

Anadoucissant 88210 T. Softener for polyamide and cellulose fibers; compatible with electrostatic processes. [Ceca SA]

Anadrol. CAS 434-07-1; Oxymetholone; androgen. [Syntex Laboratories Inc] *

Anaesthyl. *See* Anestile. §

Anaflex. A proprietary preparation of polynoxylin. [Geistlich Sohne AG] *

Anafranil. A proprietary preparation of chlomipramine hydrochloride; an antidepressant drug. [Ciba plc] *

Anahaemin. A proprietary preparation of a solution of an active erythropoietic fraction of liver; a hematinic. [British Drug Houses] *

Analar. Laboratory reagents and chemicals. [British Drug Houses]

Analgesic Balm (AS). A proprietary preparation of methyl salicylate, capsicum oleoresin, methyl nicotinate, menthol and camphor; a rubifacient. [Ayrton Saunders plc] ‡

Analgin. A proprietary preparation of aspirin, phenacetin, codeine, caffeine and phenolphthalein; an analgesic. [H N Norton & Co Ltd] ‡

Analoam. Soil testing reagent. [Murphy Chemical Co Ltd] †

Analock. A proprietary preparation of mepirizole; for the relief of pain and inflammation. [Pfizer International]

Ananase. A proprietary preparation of bromelains; used as an anti-inflammatory agent. [Rorer] †

Anapolon. CAS 434-07-1; A proprietary preparation of oxymetholone; an anabolic agent. [Syntex Laboratories Inc] *

Anaprox. CAS 26159-34-2; Naproxen sodium; anti-inflammatory; analgesic; antipyretic. [Syntex Laboratories Inc]

Anaroids. A proprietary preparation containing resorcinol, powdered gall, bismuth subgallate, titanium dioxide, zinc oxide, boric acid, balsam of peru and kaolin; an antipruritic. [Rybar Laboratories Ltd] *

Anasite. An explosive consisting of ammonium perchlorate and myrabolans, usually with some sodium or potassium nitrate, and a small quantity of agar-agar. §

Anasthol. A mixture of ethyl and methyl chlorides; a local anasthetic. §

Anatola. Vitamin A; antixerophthalmic. [Parke-Davis] *

anatomical alloy. An alloy of 53.5% bismuth, 19% tin, 17% lead, and 10.5% mercury.

anatto. *See* annatto.

Anavar. CAS 53-39-4; Oxandrolone; androgen. [G D Searle & Co] *

Anazol. Ethyl phthalate. §

Ancaflex. A proprietary trade name for a series of boron trifluoride-based polymers; used as curing agents for epoxy resins to give flexible products. [Pacific Anchor Chemical Corp] *

Ancaflex 70 and 150. A proprietary range of polymeric hardeners based on boron trifluoride. [Pacific Anchor Chemical Corp] *

Ancamide 280, 400. A proprietary trade name for a fluid complex polyamide; used as a curing agent for epoxy resins. [Anchor Chemical (UK) Ltd]

Ancamine LO. An epoxy hardener for use at low temperatures; free from phenolic odor and processing a low irritation index. [Anchor Chemical (UK) Ltd]

Ancamine LT. A proprietary trade name for an activated aromatic amine; curing agent for epoxy resins. [Anchor Chemical (UK) Ltd]

Ancamine MCA. Modified cyclo-aliphatic polyamine; used in floorings and coatings. [Anchor Chemical (UK) Ltd]

Ancaris. Thenium compound. [The Wellcome Foundation Ltd] †

Ancatax. Dibenzthiazyl disulfide; a proprietary rubber accelerator; §

Ancazate BU. A self-dispersible zinc butyl dithiocarbamate; used as an accelerator for vulcanization and an antioxidant for rubbers. [Anchor Chemical Group plc]

Ancazate EPH. Zinc ethyl phenyl dithiocarbamate. [Anchor Chemical Group plc]

Ancazate ET. CAS 14323-55-1; Zinc diethyl dithiocarbamate. [Anchor Chemical Group plc]

Ancazate ME. CAS 137-30-4; Zinc dimethyl dithiocarbamate; used as an accelerator. [Anchor Chemical Group plc]

Ancazate Q. A proprietary complex of zinc dithiocarbamate used as an accelerator. [Anchor Chemical Group plc]

Ancazate XX. Butyl dithiocarbamate; used as an accelerator. [Anchor Chemical Group plc]

Ancazide ET. CAS 97-77-8; Tetraethyl thiuram disulfide; a proprietary rubber accelerator; §

Ancazide IS. CAS 97-74-5; Tetramethyl thiuram monosulfide; a proprietary rubber accelerator. §

Ancazide ME. CAS 137-26-8; Tetramethyl thiuram disulfide; a proprietary rubber accelerator. §

Ancef. CAS 27164-46-1; Cefazolin sodium; antibacterial. [SmithKline Beecham]

anchoic acid. (lepargylic acid; azelaic acid) $C_9H_{16}O_4$.

Anchor. A proprietary trade name for a vanadium tool steel.§

Anchor 1040, 1115, 1170, 1171 and 1222. A proprietary group of boron trifluoride epoxy hardener curing agents;

---

* Trade name not verified as available   † Trade name verified as obsolete

they consist of modified amine complexes of boron trifluoride. [Pacific Anchor Chemical Corp] *

Anchor DLCT, PLCT. Liquid tackifying resins; adhesives. [Anchor Chemical (UK) Ltd]

Anchoracel. Thiocarbanilide; a proprietary rubber vulcanization accelerator. §

Anchor-bac®. . [DuPont UK]

Anchorite. A light rubber product used in rubber mixings. §

Anchorlube G-771. Animal/vegetable oil in water emulsion; metal cutting and tapping compound for stainless and other hard to work metals or operations on sensitive appliances. [Anchor Chemical Co]

Anchred. A red oxide used in rubber mixings. §

Ancobon. CAS 2022-85-7; Flucytosine; antifungal. [Hoffmann-LaRoche Inc] *

Ancofen. Tablets of meclozine hydrochloride, ergotamine tartrate and caffeine for relief of migraine. [British Drug Houses] *

Ancolan. CAS 31884-77-2; A proprietary trade name for the dihydrochloride of meclozine; antihistamine and anti-emetic. [British Drug Houses] *

Ancoloxin. Tablets of meclozine hydrochloride and pyridoxine hydrochloride; taken to combat nausea and vomiting of pregnancy. [British Drug Houses] *

Ancor® CR-538, CR-539. Proprietary acetylenic alcohol-based; corrosion inhibitor for metals, acid pickling, industrial cleaning, coatings and inks. [Air Prods. & Chems. Inc]

Ancor®LB-503, LB-504. Proprietary blends of amine salts; contains no nitrites or nitrates; corrosion inhibitor for metalworking fluids, water-based coatings and inks. [Air Prods. & Chems. Inc]

Ancor® OW-1. Secondary acetylenic alcohol; corrosion inhibitor for min. acid systems, oil well acidizing, steel pickling. [Air Prods. & Chems. Inc]

Ancor®OW-9. Proprietary blend of acetylenic alcohols and diols; corrosion inhibitor for oil well acidizing, steel pickling, industrial cleaning and inks. [Air Prods. & Chems. Inc]

Ancotil. *See* Alcobon. [Roche Products Ltd] *

Ancovert. Tablets of meclozine hydrochloride with nicotinic acid; for treatment of vertigo and Meniere's disease. [British Drug Houses] *

Ancrack. Naptalam plus dinitro; pre-emergent herbicide for use on peanuts and soybeans. [Draxel Chemical Company] ‡

Ancrod. An anticoagulant whose active principle is obtained from the venom of the Malayan pitviper A*gkistrodon rhodostoma,* acting specifically on fibrinogen; §

Ancyte. CAS 2608-24-4; Piposulfan; antineoplastic. [Abbott Laboratories] *

andalusite. A mineral; it is a silicate of aluminum, Al_2SiO_5.

Andaria®. Novelty rayon filament. [Asahi Chem. Industry]

Anderol®. Synthetic lubricants; for lubrication of compressors, crankcases and other industrial uses. [Hüls Am.]

Anderol® Premium Plus. Synthetic lubricant and coolant; for lubrication and cooling of rotary compressors. [Hüls Am.]

Andersil. High flash point liquid polysilicate; stable liquid binders for coatings and precision casting molds. [Anderson Development Company] *

andersonite. A mineral, $6[Na_2CaVO_2(CO_3)_2 \cdot 6H_2O]$.

Andozac®. . [DuPont Merck Pharmaceuticals]

Andrez 8000. 85% bound styrene/butadiene; used to im-

prove the processing properties and increase the hardness and modulus of many rubber compounds. [Anderson Development Company] *

Andro 100. CAS 58-22-0; Testosterone; androgen. [O'Neal, Jones & Feldman Pharmaceuticals] *

Andro LA 200. CAS 315-37-7; Testosterone enanthate; androgen. [O'Neal, Jones & Feldman Pharmaceuticals]*

Androcur. CAS 427-51-0; A proprietary preparation of cyproterone acetate; used in the control of hypersexual and deviant behavior in males. [Schering Health Care Ltd]

Androgyn LA. CAS 315-37-7; Testosterone enanthate; androgen. [O'Neal, Jones & Feldman Pharmaceuticals]*

Androx 3961. Shield; a water displacing fluid meeting DTD 900/4942. §

Andur. Urethane based prepolymers, water emulsion thermoplastics and coatings; for use in the manufacturing of urethane elastomers, varnishes and paints, adhesives, etc. [Anderson Development Company] *

Andursil. A proprietary preparation of aluminum hydroxide gel, magnesium hydroxide and carbonate, and simethicone; an antacid. [Ciba plc] *

Anectine Injection. A proprietary formulation of suxemethonium chloride; a muscle relaxant used in general anesthesia to facilitate endotracheal intubation and to provide skeletal muscle relaxation during surgery. [The Wellcome Foundation Ltd]

Anedco ADM-407. Long-chain benzyl sulfonate; surfactant, demulsifier for bad tank bottoms and slop oil in the refinery. [Anedco]

Anedco ADM-407C. Alkylbenzyl sulfonic acid; surfactant, demulsifier for bad tank bottoms. [Anedco]

Anedco AF-800. Ammonium salt of an alcohol ether sulfate; fresh-water foamer for air/gas drilling and well clean-out operations. [Anedco]

Anedco AF-801. Alkyl sulfate salt; brine water foamer for air/gas drilling and well clean-out operations. [Anedco]

Anedco AW-395. Coco alkyl dimethyl benzyl quaternary amine; quaternary surfactant; anti-clay swelling agent; corrosion inhibitor for waterfloods, oil and gas wells, pipelines; foamer additive for surfactants, cleaners, water treating. [Anedco]

Anedco AW-396. Quaternary amine/nonyl phenol ethoxylate blend; surfactant flush aid; anti-clay swelling agent; corrosion inhibitor for waterfloods, oil and gas wells, pipelines; foamer for wetting agents, surfactants, cleaners, water treating. [Anedco]

Anedco AW-397. Nonyl phenol ethoxylate; surfactant for manufacturing of corrosion inhibitors; emulsifier, detergent, wetting agent, penetrant, antistat, coupling agent. [Anedco]

Anedco DF-6002. Nonsilicone, alcohol-based; defoamer for drilling muds. [Anedco]

Anedco DF-6031. Silicone emulsion; antifoamer/defoamer for aq. systems; effective in acid and alkaline media; for cooling towers, amine scrubbers, glycol dehydrators, water-based drilling muds, cleaning compounds, effluents, cutting oils, abrasive slurries. [Anedco]

Anedco DF-6130. Dimethyl silicone fluid; antifoamer for gas-oil separators. [Anedco]

Anergan 25. CAS 58-33-3; Promethazine hydrochloride; antiemetic; antihistaminic. [O'Neal, Jones & Feldman Pharmaceuticals] *

Anergan 50. Promethazine hydrochloride; antiemetic;

---

‡ Trade name and manufacturer not verified   § Trade name without identified manufacturer

antihistaminic. [O'Neal, Jones & Feldman Pharmaceuticals] *

Anestacon. CAS 137-58-6; Lidocaine; anesthetic. [Alcon Laboratories Inc] *

Anestan. CAS 136-47-0; A proprietary preparation containing tetracaine hydrochloride; an antipruritic; bronchial asthma preparation. [The Boots Co plc]

anesthesol. See benzocaine

Anesthol. See Anestile.

Anesthyl. See Anestile.

Anestile. (Anaesthyl, Anesthyl, Anesthol). A mixture of ethyl and methyl chlorides; an anasthetic. §

Anethaine. CAS 136-47-0; An emollient, white, vanishing cream containing tetracaine hydrochloride BP 1% w/w; quickly and effectively relieves itching and stinging to check scratching; gives prompt temporary relief from insect bites, stings, nettle rash, heat rash, skin itches, chapping and chaffing. [Evans Medical] *

aneurine hydrochloride. See thiamine hydrochloride

Angarite. A Russian cast basalt (qv) used as an electrical insulator.

Angio-Conray. CAS 1225-20-3; Iothalamate sodium; diagnostic aid. [Mallinckrodt Inc] *

angiotensin amide. Val⁵-hypertensin II-Asp-β-amide. Asn-Arg-Val-Tyr-Val-His-Pro-Phe; a hypertensive.

Angiovist 282. Diatrizoate meglumine; diagnostic aid. [Berlex Laboratories Inc] *

Angioxyl. A proprietary preparation of pancreatic extract insulin free. §

Angised. A proprietary preparation of glyceryl trinitrate in a stabilized base; a vasodilator used in the treatment of angina pectoris. [The Wellcome Foundation Ltd]

Anhydrite. CAS 7778-18-9; Calcium sulfate; concrete substitute and soil conditioner. [Pacific Chemical Industries Pty Ltd] *

anhydrite (natural form). See calcium sulfate (anhydrous)

Anhydrol. CAS 1332-58-7; Specially processed calcined kaolin (aluminum silicate); used as a molecular sieve support. [Engelhard]

Anhydrol Forte. CAS 7446-70-0; Colorless evaporative solution containing 20% w/v aluminum chloride hexahydrate; used for the topical treatment of hyperhidrosis specifically involving axillae, hands or feet. [Dermal Laboratories Ltd]

Anhydron. CAS 2259-96-3; Cyclothiazide; diuretic; antihypertensive. [Eli Lilly & Co]

Anhydrone. A proprietary name for a perchlorate of magnesium, a drying agent for gases. §

anhydrosorbitol tristearate. See sorbitan tristearate

anhydrous gypsum. See calcium sulfate (anhydrous)

anhydrous lanolin. See lanolin

Anhydrous Lanolin HP-2050. CAS 8006-54-0; Lanolin; high purity lanolin for cosmetics applications; emulsifier, emollient, conditioner, lubricant for creams, lip products, hair grooms, makeup, suncare products. [Henkel/Cospha; Henkel Canada] *

Anhydrous Lanolin P9SRA. CAS 8006-54-0; Anhydrous lanolin; emollient, moisturizer, emulsifier for pharmaceuticals and cosmetics (baby creams, cleansers, eye preps., foundation, hypo-allergenic cosmetics, lipstick, sunscreen preps.). [Westbrook Lanolin]

Anhydrous Lanolin P80. CAS 8006-54-0; Anhydrous lanolin; emulsifier, emollient, moisturizer for pharmaceuticals and cosmetics (baby creams, cleansers, eye preps.,

foundation, lipstick, sunscreen preps.). [Westbrook Lanolin]

Anidrisorb. Anhydro-sorbitols; used for pharmaceutical encapsulation. [Roquette (UK) Ltd]

aniline black. (aniline black in paste, fine black, oxidation black). A black dyestuff produced on the fiber by the oxidation of an aniline salt.

p-anilinesulfonic acid. See sulfanilic acid

anilotic acid. Nitrosalicylic acid.

animal charcoal. (bone charcoal). This term is used for all charcoal produced by the ignition of animal substances with exclusion of air, but more particularly to that obtained from bones. This material contains approximately 10% carbon, and 90% mineral matter, mainly calcium phosphate and is used for absorbing dyes.

animal glycerin. Neatsfoot oil.

animal oil. See bone oiL.

animal starch. Glycogen, found in the blood and liver of mammals.

anime resin. (gum anime). A fossil copal resin from South America.

Anionyx®12S. CAS 56388-43-3; EINECS 260-143-1; Disodium oleamido PEG-2 sulfosuccinate; detergent for personal care products, bubble baths, dishwashing liqs. [Stepan; Stepan Canada]

Aniscol. Antiscale paints to reduce oxidation losses and surface decarburization. [Foseco (F.S.) Ltd]

m-, o-, p-anisidine. C₇H₉NO; CAS 536-90-3, 90-04-0, 100-09-4 respectively; EINECS 208-651-4, 201-963-1, 202-818-4 respectively; Intermediate for azo dyestuffs, guaiacol. [Aldrich; Penta Mfg.; Rhone-Poulenc]

Anisoline. See Rhodamine 3B. §

p-anisoyl chloride. (4-methoxybenzoyl chloride) CH₃OC₆H₄COCl; CAS 100-07-2; EINECS 202-816-4; Intermediate for dyes and medicines. [Aldrich; Eastman; Schweizerhall]

Anka steel. (V2A Steel). Nickel-chromium steels containing from 15-16% chromium and from 7-10% nickel.

annaline. See gypsum.

annatto. (anatto, anotto, arnatto, arnotto, Orleans, rocou). Vegetable dye containing ethyl bixin, obtained from the fleshy covering of the seeds of the ruccu tree, Bixa orellana; C₂₇H₃₄O₄; as extract for coloring foods, food-marking inks. [Haarman & Reimer/Food Ingred.; Meer; Penta Mfg.; Pfizer; Warner-Jenkinson]

annotto. See annatto.

annulene. See benzene

Anobolex. CAS 521-18-6; A proprietary trade name for stanolone. §

anode mud. (anode slime). The material which falls to the bottom of the electrolysing vessel during the electrolytic refining of copper; it contains copper (10-25%) gold (0.7-2%) silver (5-40%) tellurium, antimony, and arsenic.

Anodesynthetic A proprietary preparation containing ephedrine hydrochloride, lignocaine hydrochloride and allantoin; a preparation of hemorrhoids. [The Boots Co plc]

Anodyne. See Antipyrine. §

anogon. The mercury salt of 2,6-diiodophenol-4-sulfonic acid; used in medicine.

anol. p-Propenyl alcohol.

Anonaid. Anionic surfactants and detergents. [Rhone-Poulenc UK]

Anonaid TH. CAS 577-11-7; EINECS 209-406-4; Sodium

---

dioctyl sulfosuccinate in liquid form; wetting and emulsifying agent for the textile, leather and paper industries and for emulsion polymerization. [Rhone-Poulenc UK]

anone. Cyclohexanol.

Anorvit. Tablets of ferrous sulfate, ascorbic acid, and acetomenaphthone; for hypochromic anemia. [British Drug Houses] *

Anotex. Dyes for anodized metal. [Pointing Ltd] *

Anovlar. A proprietary preparation of norethisterone acetate and ethinyl estradiol; oral contraceptive. [Schering Chemicals Ltd] *

Anox® 20. CAS 6683-19-8; Tetrakis methylene (3,5-di-t-butyl-4-hydroxyhydrocinnamate) methane; antioxidant, processing stabilizer for thermoplastic polymers; especially effective against polymer thermo oxidative degradation during long term aging. [Enichem Synthesis SpA]

Anox® 70. CAS 41484-35-9; 2,2´-Thiodiethyl bis-(3,5-di-t-butyl-4-hydroxyphenyl)propionate; antioxidant providing processing and end-use stability for LDPE copper wire insulation and cable jacketing, carbon black loaded polyolefins, chem. crosslinked PE, HIPS, ABS, ethylene-propylene-diene rubbers, styrene/butadiene rubber, neoprene, natural and synthetic elastomers. [Enichem Synthesis SpA]

Anox® IC-14. CAS 27676-62-6; Tris (3,5-di-t-butyl-4-hydroxybenzyl) isocyanurate; nonvolatile, nondiscoloring antioxidant protecting polymers against high temp. degradation during processing and thermal stability during service life. [Enichem Synthesis SpA]

Anox®PP 18. CAS 2082-79-3; Octadecyl-3-(3´,5´-di-t-butyl-4´-hydroxyphenyl) propionate; antioxidant retarding oxidative degradation during polymerization, processing and in end-use applications; stabilizer for polyolefins, impact styrenics, blocked copolymers, elastomers, adhesives, PVC, PU. [Enichem Synthesis SpA]

Anquil. CAS 2062-84-2; A proprietary preparation of benperidol used in the control of deviant sexual behavior. [Janssen Pharmaceutical Ltd] *

Ansa. Alkyl naphthaline sodium sulfonate. [Albright & Wilson Ltd] *

Ansaid. CAS 5104-49-4; Flurbiprofen; analgesic; anti-inflammatory. [Upjohn]

Ansax. CAS 6484-52-2; Fertilizer prilled ammonium nitrate. [L & K Fertilisers Ltd] *

Anscor®. Lightly calcined sea water magnesia. [Steetley Magnesia Products Ltd]

anserine. A natural peptide from muscle. It is β-alanyl methyl histidine.

Ansol. Solvents and solvent mixtures containing anhydrous alcohol; especially those adpated for use a general organic solvents, solvents for nitrocellulose resins, wood and metal lacquers, paint and varnish removers, cleaners, shellac and floor finishes and alcohol substitutes. [Quantum Chemical Corp]

Ansol A. A proprietary blend of anhydrous denatured ethyl alcohol with small percentages of esters, other alcohols, and hydrocarbons; a nitrocellulose and resin solvent. §

Ansol B. A preparation similar to A, except that B contains a small amount of normal butyl alcohol in the place of amyl alcohol. §

Ansol E-121. A trademark for ethylene glycol dimethyl ether. [Ansul Co] *

Ansol E-161. A trademark for triethylene glycol dimethyl ether. [Ansul Co] *

Ansol E-181. A trademark for tetraethylene glycol dimethyl ether (tetraglyme). [Ansul Co] *

Ansolysen. CAS 52-62-0; Pentolinium tartrate. [May & Baker Ltd] *

Anspor. CAS 38821-53-3; Cephradine; antibacterial. [SmithKline Beecham] *

Anstex AK-25. Special phosphate; antistat for synthetic fibers, plastics. [Toho Chem. Industry]

Antabuse. CAS 99-77-8; Disulfiram; alcohol deterrent. [Wyeth Laboratories]

Antak. Alcohol; contact tobacco sucker control material. [Draxel Chemical Company] ‡

Antara®. Redesignated Lubrhophos®. [Rhone-Poulenc Surf.]

Antaron ET-201. PVP/decene copolymer. [ISP]

Antarox® 17-R-2. CAS 9003-11-6; Meroxapol 172; defoamer, dispersant, wetting agent, emulsifier, demulsifier, leveling agent, detergent for industrial/household cleaners, fermentation, paper processing, rinse aids, automatic dishwashing, metal cleaning. [Rhone-Poulenc Surf.]

Antarox®461/P. EO/PO alkylphenol block polymer; emulsifier, dispersant used in agricultural industry for prep. of emulsifiable concs. and toxicant flowable systems. [Rhone-Poulenc Surf.; Rhone-Poulenc Geronazzo]

Antarox®497/P. EO/PO alkylphenol block polymer; emulsifier, dispersant used in agricultural industry for prep. of emulsifiable concs. and toxicant flowable systems. [Rhone-Poulenc Surf.; Rhone-Poulenc Geronazzo]

Antarox® B-10. Low molecular weight block copolymer; surfactant. [Rhone-Poulenc France]

Antarox® BL-214. CAS 68603-25-8; Octyl/decyl alcohol, ethoxylated and propoxylated; wetting, rewetting agent, detergent for metal cleaning, textile finishing, industrial/household cleaners. [Rhone-Poulenc Surf.; Rhone-Poulenc France]

Antarox® E-100. CAS 9003-11-6; Poloxamer 401; surfactant. [Rhone-Poulenc France]

Antarox®L-61. CAS 9003-11-6; Poloxamer 181; defoamer, dispersant, wetting agent, emulsifier, demulsifier, leveling agent, detergent, lubricant for household/industrial cleaners, metalworking fluids, agricultural formulations, rinse aids, automatic dishwashing, water treatment. [Rhone-Poulenc Surf.; Rhone-Poulenc France]

Antarox® LA-EP 15. Modified oxyethylated straight chain alcohol; detergent, dispersant, wetting agent, emulsifier; for controlled foam applications, machine dishwashing, rinse aid compositions. [Rhone-Poulenc Surf.]

Antarox® PGP 23-7. CAS 9003-11-6; Poloxamer 237; coemulsifier for cosmetics, toiletries, pulp and paper defoamers; dispersant, visc. control agent. [Rhone-Poulenc Surf. Canada]

Antas. A proprietary preparation containing pepsin, aluminum hydroxide, magnesium trisilicate, belladonna, magnesium hydroxide; an antacid. [Consolidated Chemicals Ltd] *

Antasil. An antacid/deflatulent. [ICI Chem. & Polymers Ltd]†

Antec Farm Fluid S. A mixture of organic acids, high molecular weight phenols, low molecular phenols and surfactant; used for the disinfection of all types of livestock buildings. [Antec International Ltd]

Antec Longlife 250 S. A mixture of organic acids, high molecular weight phenols, synthetic boicide and surfactant; used for the disinfection of all types of livestock

---

‡ Trade name and manufacturer not verified   § Trade name without identified manufacturer

buildings. [Antec International Ltd]

Antec OO-Cide. Two part ammonia release system incorporating a biocide; coccidiacide; for the treatment of all livestock buildings. [Antec International Ltd]

Antec Virkon S. Oxidizing detergent/disinfectant system with particularly broad spectrum activity against viruses and safety in use; for disinfection of hard surfaces, water, environment and instruments in agriculture, industrial and medical situations. [Antec International Ltd]

Antelope. Acid sodium pyrophosphate. [Albright & Wilson Ltd] *

Antepan. Piperazine anthelmintic. [Coopers Animal Health Ltd] *

Antepar. CAS 110-85-0; Piperazine citrate; anthelmintic. [The Wellcome Foundation Ltd]

Antepsin. CAS 54182-58-0; Sucralfate; tablets and suspension; for treatment of duodenal ulcer, gastric ulcer, chronic gastritis and the prophylaxis of gastrointestinal hemorrhage from stress ulceration in seriously ill patients. [Wyeth Laboratories]

Ant Flip. CAS 10043-35-3; A gel bait with boric acid as the active ingredient; for household, hospital, restaurant use for controlling pharoah and other sweet eating ants; workers carry back to nest to kill the colony. [Colonial Products Inc]

Ant Gun! Contains diazinon and pyrethrins; ready-for-use spray for control of ants and household insects. [ICI Garden Products]

Anthelvet. CAS 522-48-5; Tetrahydrozoline hydrochloride; adrenergic. [McNeil Pharmaceuticals] *

Anthical. A proprietary preparation of mepyramine maleate and zinc oxide; an antihistamine skin cream. [May & Baker Ltd] *

Anthiomaline. Hexa lithium antimony tri(mercapto succinate). [RMB Animal Health Ltd]

Anthion. CAS 7727-21-1; Potassium persulfate; used as a hypo eliminator in photography. §

Anthiphen. CAS 97-23-4; Dichlorophen. [May & Baker Ltd]*

Anthisan. Mepyramine maleate. [Rhone-Poulenc Rorer Ltd]

Anthium Dioxcide. CAS 10049-04-4; Chlorine dioxide, stabilized; broad spectrum biocide, preservative; deodorant designed to remove odors caused by residual monomers in resins; works by oxidation. [Int'l. Dioxcide]

anthocyanins. The red, blue, and violet coloring matters of plants.

Anthosin®. Based on acid dyes; for paper coloring in papermaking. [BASF AG]

Anthoxan. 4-Isopropyl-5, 5-dimethyl-1, 3-dioxane; raw material for herbal fragrances. [Henkel]

Anthra-Derm. CAS 1143-38-0; Anthralin; antipsoriatic. [Dermik Laboratories Inc] *

anthracite. A hard coal, containing 85-95% carbon. It burns with little smoke.

anthraflavic acid. 2,6-Dihydroxyanthra-quinone.

anthraflavone. A dyestuff which is prepared by treating 2-methylanthraquinone with condensing agents; used to dye cotton yellow shades.

anthragalanthranol. Trioxyanthranol, $C_{14}H_{10}O_4$.

anthranilic acid. o-Aminobenzoic acid, $C_6H_4(NH_2) \cdot COOH$; used as a dye intermediate.

Anthranol. A smooth soft ointment containing dithranol (in 0.4, 1.0 and 2.0 strengths) w/w in a base containing cetyl alcohol, liquid paraffin, soft white paraffin and sodium

sulfate with salicylic acid; for the topical treatment of subacute and chronic psoriasis including psoriasis of the scalp. [Stiefel Laboratories (UK) Ltd] *

anthrapurpurin. (Isopurpurin, Alizarin GD, Alizarin RF, Alizarin RT, Alizarin RX, Alizarin SC, Alizarin SSA, Alizarin SX, Alizarin SX Extra, Alizarin WG). Trioxyanthraquinone, $C_{14}H_5O_2(OH)_3$, and dyes red shades on alumina; a mordant dyestuff.

anthraquinone. $C_{14}H_8O_2$; CAS 84-65-1; EINECS 201-549-0; Intermediate for dyes and organics, organic inhibitor, bird repellent for seeds. [Buckton Scott Ltd; ICI Am.]

Anthraquinone. Discharging auxiliary for textiles. [BASF plc] *

anthrarobin. Dihydroxyanthranol, $C_{14}H_{10}O_2$; used in the treatment of skin diseases.

anthrarufin. 1,5-Dihydroxy-anthraquinone.

Anthraxolite. A variety of anthracite. §

anthropic acid. A mixture of palmitic and stearic acids.

antiar. The milky juice of the upas tree; used as an arrow poison.

Antibacterin. Chloramine T.

Antiblu/Antiboror. Fungicide/insecticide; used for freshly sawn timber. [Hickson & Welch Ltd]

Anticor 70. Anticorrosion pigments containing zinc ferrite as the active antirust pigment for the formulation of chromate-free primers. [Bayer AG]

Anticorodal. An aluminum alloy containing 1% silicon, 0.06% magnesium, and 0.06% manganese.

Antidol. A proprietary preparation of ethosalamide, paracetamol, and caffeine; an analgesic. [Lewis Laboratories] ‡

Antidust 2. CAS 557-05-1; A proprietary preparation similar to Antidust F, but containing a particularly fine zinc stearate in aqueous dispersion. [Rhein-Chemie Rheinau] *

Antidust F. A proprietary group of surfactants in combination with polyvalent alcohols; used in a concentrated aqueous solution to prevent undesirable surface tackiness in sheets and extrudates of plastics and rubber materials. [Rhein-Chemie Rheinau] *

Antifoam. Silicone and nonsilicone antifoams for industrial applications. [Bayer plc]

Antifoam 20WB. Water-based antifoam for screen room, paper machine, bleaching applications, waste water treatment, chem. processing industry, latex, coatings, metal treating, and electroplating. [Stockhausen]

AntiFoam 55. Silicone defoamer; free-rinsing defoamer for garment dyeing. [Yorkshire Pat-Chem]

Antifoam 6031. Sulfocarboxylic acid/complex fatty blend; defoamer for manufacturing of phos-acid and fertilizer. [Stockhausen]

Antifoam 7800 New. Higher hydrocarbons and their sulfonic acid derivs.; antifoam for aq. systems in sugar, fertilizer, phosphoric acid, dyestuffs, paper, leather, plastics, and chemical industries; resistant to acid and weak alkalis. [Miles/Organic Prods.]

Antifoam ET. An emulsified form of Antifoam T. [Bayer AG]†

Antifoam FRS. Blended silicones and emulsifiers; highly effective defoamer and antifoam for ambient and high temp. operations; free rinsing on carpets. [Am. Emulsions]

Antifoam GEB. Silicone defoamer; free-rinsing defoamer for continuous and beck dyeing of carpets. [Am. Emul-

---

sions]

Antifoam T. CAS 126-73-8; Tri-n-butyl phosphate; an antifoaming agent used in the manufacture of paper coatings. [Bayer AG]

Anti-foam TP. Antifoam for pigment printing of textiles. [BASF AG]

Antifoam VOL. Polysiloxane; defoamer for use in dyebaths, finish mixes where stability is essential. [Am. Emulsions]

Antifungin. The trade name for magnesium borate, a fungicide. §

Antigrison. An explosive; it consists of 27% nitro-glycerin, 1% nitro-cotton, and 72% ammonium nitrate.

Antihypo. Potassium percarbonate, $K_2C_2O_6$; used as a bleaching agent and hypo eliminator in photography. §

Antikamnia. See Ammonal. §

Antil® 141 Liquid Propylene glycol, PEG-55 propylene glycol oleate; thickener for aq. sol'ns. of surfactants, e.g., shampoos, foam baths, shower preps., liquid soaps; solubilizes essential oils into aq. surfactant systems; liquid version developed for cold process systems. [Goldschmidt; Goldschmidt AG]

Antil®141 Solid. PEG-55 propylene glycol oleate; thickener for surfactant systems, e.g., shampoos, shower gels; solubilizer for essential oils, fragrances, and perfumes. [Goldschmidt]

Antil® 208. Carbomer 208; highly effective thickener for detergent/soap-based products; refatting agent in hair or skin cleansing products. [Goldschmidt]

Antilirium. CAS 57-64-7; Physostigmine salicylate; cholinergic. [O'Neal, Jones & Feldman Pharmaceuticals] *

Antiluetin. Potassium-ammonium-antimonyl-tartrate, [SbO(C₄H₄O₆)₂DFKNH₄] · H₂O; a trypanocide. §

Antilux. Blends of selected paraffins and micro waxes, which protect rubber articles from damage by the sun, ozone and weathering; used for technical molded and extruded articles, articles subjected to dynamic stress, tires, conveyor belts, cable coverings, articles fit for foodstuffs quality. [Bayer plc]

Antilux AOL. A proprietary blend of paraffinic hydrocarbons used as an antiweathering agent in natural and synthetic rubbers. [Rhein-Chemie Rheinau] *

Antimigrant 157. Antimigrant for continuous dyeing of textiles. [Catawba-Charlab]

Antimigrant C-45. Sodium alginate; antimigrant for dyeing. [Yorkshire Pat-Chem]

Anti-Migrant CAS; Proprietary blend of high molecular weight polymers and sequestrants; antimigrant with excellent lubricity for applications such as pigment padding and continous disperse dyeing. [Arol Chem. Prods.]

Antiminth. CAS 22204-24-6; A proprietary preparation of pyrantel pamoate; an anthelmintic. [Pfizer International]

antimonial lead. An alloy of 87% lead and 13% antimony.

antimony. Sb; CAS 7440-36-0; EINECS 231-146-5; hardening alloy for lead, bearing metal, type metal, solder, collapsible tubes and foil, sheet and pipe, semiconductor technology, pyrotechnics. [Aldrich; Amspec Chem.; Atomergic Chemetals]

antimony oxide. See antimony trioxide

antimony pentachloride. (antimony (V) chloride) CAS 7647-18-9; Cl₅Sb; EINECS 231-601-8; For analytical testing of alkaloids and cesium; dyeing intermediates; as chlorine carrier in organic chlorinations. [Aldrich; Atomergic Chemetals; Hoechst-Celanese]

antimony pentafluoride. (antimony (V) fluoride) F₅Sb;

CAS 7783-70-2; EINECS 232-021-8; Catalyst and/or source of fluorine in fluorination reactions. [Allied-Signal; Atochem N. Am.; Atomergic Chemetals]

antimony pentasulfide. (antimony (V) sulfide) Sb₂S₅; CAS 1315-04-4; EINECS 215-255-5; Red pigment, rubber accelerator. [Atomergic Chemetals]

antimony potassium tartrate. (tartar emetic; potassium antimonyl tartrate; tartrated antimony) K(SbO)-C₄H=O₆·¹/₂H₂O; Textile and leather mordant; medicine; insecticide. [Aldrich]

antimony salt. Double salts of antimony fluoride, SbF₂, with alkali sulfates, or with alkali fluorides; used as mordants in dyeing.

antimony trichloride. SbCl₃; CAS 10025-91-9; EINECS 233-047-2; Antimony salts, bronzing iron, mordant, manufacture of lakes, chlorinating agent in organic synthesis, pharmaceuticals, fireproofing textiles, analytical reagent. [Akzo; Aldrich; Hoechst-Celanese; Nihon Kagaku Sangyo]

antimony trioxide. (antimony white; antimony (III) oxide; antimony peroxide; antimony sesquioxide) Sb₂O₃; CAS 1309-64-4; EINECS 215-175-0; Flameproofing of textiles, paper, plastics; as paint pigment; ceramic opacifier; catalyst; intermediate staining iron and copper; phosphorus; mordant; glass decolorizer. [Aldrich; Aspec Chem.; Asarco; Atochem N. Am.; Atomergic Chemetals; Chemisphere Ltd; Hoechst-Celanese; Laurel Ind.; Nihon Kagaku Sangyo]

antimony trisulfide. (antimony (III) sulfide) Sb₂S₃; CAS 1345-44-6; EINECS 215-713-4; Vermilion or yellow pigment, antimony salts, pyrotechnics, matches, percussion caps, camouflage paints, ruby glass. [Atomergic Chemetals; BASF]

antimony white. See antimony trioxide

Anti-oxidant 425. CAS 88-24-4; A proprietary preparation of 2,2-methylene-bis(4-ethyl-6-t-butyl) phenol. [Ciba-Geigy Europe] *

Antioxidant 431. Nondiscoloring and nonstaining antioxidant for dry rubber and latex. [Uniroyal] *

Antioxidant 449. A phenolic phosphite antioxidant; a nondiscoloring stabilizer for EDPM polymers; also economical replacement for BHT in compound work. [Uniroyal] *

Antioxidant 451. Alkylated hydroquinone; for synthetic rubbers and plastics; used as a stabilizer for synthetic rubbers, such as polybutadiene and as an antioxidant for uncured adhesives; it functions as an antioxidant in both black and nonblack cured compounds and latex compounds. [Uniroyal] *

Anti-oxidant 2246. CAS 119-47-1; A proprietary preparation of 2,3-methylenebis (4-methyl-6-t-butyl) phenol. [Ciba-Geigy Europe] *

Anti-oxidant 4010. A proprietary preparation of N-phenyl-N´-cyclohexyl-p-phenylenediamine. §

Anti-Oxidant AH. A proprietary aldol-α-naphthylamine resin. [Bayer AG] †

Anti-Oxidant AP. A proprietary aldol α-naphthylamine powder. [Bayer AG] †

Antioxidant BA. Aldol-α-naphthylamine powder; a proprietary antioxidant; §

Anti-Oxydant Bayer. CAS 128-37-0; BHT; for the protection of animal fats, feedstuffs, concentrates and mixed feed from oxidative decomposition and loss of essential nutrients. [Bayer AG]

Anti-Oxidant DDA. A proprietary derivative of diphenylamine. [Bayer AG] †

Anti-Oxidant DNP. Di-β-naphthyl-p-phenylenediamine. [Bayer AG] †

Anti-Oxidant DOD. 4, 4´-Dioxydiphenyl. [Bayer AG] †

Anti-Oxidant EM. A 30% aqueous emulsion of a diphenylamine derivative. [Bayer AG] †

Anti-Oxidant MB. Phenylenethiourea; a proprietary antioxidant; 2-mercaptolbenzimidazole. [Bayer AG] †

Anti-Oxidant PAN. Phenyl-α-naphthylamine. [Bayer AG] †

Antioxidant PBN. Phenyl-β-naphthylamine; a proprietary antioxidant. §

Antioxidant RES. Aldol-α-naphthylamine resin; a proprietary antioxidant. §

Anti-Oxidant RR 10 N. A proprietary range of alkylated phenols. [Bayer AG]

Anti-Oxidant SP. A proprietary styrenated phenol. [Bayer AG]

Antioxydant NV 3. Alkylphenol; used in tire industry for tech. goods. [BASF AG]

Antioxygene A. A proprietary trade name for phenyl-a-naphthylamine; an antioxidant. [Allied Colloids Ltd] *

Antioxygene AFL. A proprietary trade name for a ketone-amine reaction product. [Allied Colloids Ltd] *

Antioxygene AN. A proprietary trade name for aldol-α-naphthylamine paste; an antioxidant. [Allied Colloids Ltd] *

Antioxygene BN. A proprietary trade name for β-naphthol; an antioxidant. [Allied Colloids Ltd] *

Antioxygene CAS. A proprietary trade name for a mixture of phenyl-α-naphthylarnine and meta-toluylene-diamine. [Allied Colloids Ltd] *

Antioxygene INC. A proprietary trade name for aldol-α-naphthylamine in powder form; an antioxidant. [Allied Colloids Ltd] *

Antioxygene MC. A proprietary trade name for phenyl-β-naphthylamine; an antioxidant. [Allied Colloids Ltd] *

Antioxygene RA. A proprietary trade name for aldol naphthylamines; antioxidants. §

Antioxygene RES. A proprietary trade name for an aldol-α-naphthylamine resin; an antioxidant. [Allied Colloids Ltd]*

Antioxygene RM. A proprietary trade name for aldol naphthylamine; an antioxidant. [Allied Colloids Ltd] *

Antioxygene RO. A proprietary trade name for aldol naphthylamine; an antioxidant. [Allied Colloids Ltd] *

Antioxygene STN. A proprietary trade name for phenyl-α-naphthylamine mixed with m-toluylene-diamine and stearic acid; an antioxidant. [Allied Colloids Ltd] *

Antioxygene WBC. A proprietary trade name for an antioxidant. [Allied Colloids Ltd] *

Antiozonant AFD. Nonstaining antiozonant. [Bayer plc]

Antiprex®. A range of water-soluble polymers developed for the control of scale and deposition formation in aqueous systems; widely used in detergent formulation as builders and in cooling water and evaporation systems. [Allied Colloids Ltd]

antipyoninum. Neutral sodium tetraborate, prepared by fusing together borax and boric acid.

Antipyrine. 1-Phenyl-2,3-dimethyl-pyrazolone; a febrifuge and analgesic. §

Antiquax. Wax polishes; for furniture. [James Briggs & Sons Ltd] *

Antisepsin. (Bromanilide). See Ammonal. §

Antiseptin. A mixture of boric acid zinc iodide, and thymol; an antiseptic dusting powder.

Antisettle. Thixotrope. [Cray Valley Ltd]

Antispumin ZU. Ethoxylated/propoxylated fatty alcohols; defoamer for beet sugar industry for flume water, diffusion, liming, and carbonation; also for paper mill effluent. [Stockhausen]

Antistat 100. Complex blend; nondurable antistat for synthetics; nonyellowing and low foaming. [Am. Emulsions]

Antistat 7220. Oxyalkylated polyester; antistat. [BASF]

Antistat BT. Static depressant for natural and synthetic fibers and fabrics. [Yorkshire Pat-Chem]

Antistat RD2-2351. Quaternized fatty amine; antistat for all textile fibers. [Marlowe-Van Loan]

Antistatic 812 and 813. 100% Active phenolic ethoxylates; antistatic compounds for plastics used in proportions of 5-7%. [Farbwerke Hoechst] *

Antistatic 816. A proprietary 95% active lauric imide; antistatic compound for plastics used in polyethylene, PVC and polystyrene. [Farbwerke Hoechst] *

Antistatin. Antistatic finishing agent for textiles. [BASF plc]*

Antistin. CAS 2508-72-7; A proprietary preparation of antazoline hydrochloride. [Ciba plc] *

Antistitin-Privine. A proprietary preparation of antazoline sulfate and naphazoline nitrate; used in the treatment of allergic rhinitis. [Ciba plc] *

Antitan. A tannin remover. [S & D Chemicals Ltd] *

Anti-Terra®-202. Alkylammonium salt of a higher molecular weight polycarboxylic acid sol'n.; wetting and dispersing additive to prevent settling and flooding of pigments; gellant for organophilic bentonites; for coating systems. [Byk-Chemie USA]

Anti-Terra®-P. Phosphoric acid salt of long chain carboxylic acid polyamine amides; wetting and dispersing additive to prevent settling and flooding of pigments in alkyds, alkyd-melamine, chlorinated rubber systems, PVC copolymers, acid catalyzed paints. [Byk-Chemie USA]

Antithrombin. Infusion substance; cofactor of heparin and major guardian of the hemostatic balance. [KabiVitrum AB] *

Antitoxine. See Ammonal. §

Antitrem. CAS 52-49-3; Trihexylphenidyl hydrochloride; anticholinergic; antiParkinsonian. [J B Roerig] *

Antivert. Meclizine hydrochloride; anti-emetic. [J B Roerig]*

anti-white lead. See oil white.

Antiwick OP. Thickener for pigment print pastes to control wicking and flushing. [Yorkshire Pat-Chem]

Ant Killer. Insecticide. [Fisons plc, Horticulture Div] *

Ant Killer Dust. Insecticide. [ICI Agrochemicals]

Antlak. CAS 333-41-5; 1.74% Diazinon ant and crawling insect lacquer (aerosol); used for household crawling insect control on hard surfaces. [Doff Portland Ltd]

Antoban. A veterinary anthelmintic. [Wellcome Foundation Ltd]

ant oil. See oil of ants.

Antoin. A proprietary preparation of aspirin, calcium carbonate, citric acid, phenacetin, codeine phosphate and caffeine citrate; an analgesic. [Cox Pharmaceuticals]

Anton N. Octylated diphenylamine; a trademark for a general-purpose antioxidant used in many elastomers. [DuPont UK] *

Antox® N. Octylated diphenylamine; a registered trade name for a general purpose antioxidant with only mild discoloring and staining characteristics. [DuPont UK] *

---

*   Trade name not verified as available

†   Trade name verified as obsolete

Antozite®. Phenylenediamines; rubber antiozonants used in natural rubber, isoprene rubber, BR, styrene/butadiene rubber and CR rubbers. [R T Vanderbilt Co Inc]

Antracol®. CAS 12071-83-9; Propineb; fungicide for protective control of potato blight, hop downy mildew, apple scab, leafspot on celery, blackcurrants and gooseberries, downy mildew on grapes and suppression of yellow rust on winter wheat. [Bayer AG]

Antraderm. A proprietary preparation of dithranol; a dermatological product. [Brocades Pharma] †

Antrenyl Duplex. CAS 50-10-2; A proprietary preparation of oxyphenonium bromide. gastrointestinal sedative. [Ciba plc] *

Antromid-S. CAS 637-07-0; Clofibrate; antihyperlipoproteinemic. [Wyeth Laboratories]

Antron®. Fibers. [DuPont UK]

Antron®Stainmaster®. Fibers. [DuPont UK]

Anturan. CAS 57-96-5; A proprietary preparation of sulfinpyrazone; an antirheumatic. [Ciba plc] *

Anturane. CAS 57-96-5; Sulfinpyrazone; uricosuric. [Ciba plc] *

Anugesic. A proprietary preparation of boric acid, balsam of Peru, bismuth oxide, bismuth subgallate, resorcinol, zinc oxide and pramoxine; analgesic suppositories. [Warner]*

Anugesic-HC. A proprietary preparation of pramoxine hydrochloride, hydrocortisone acetate, zinc oxide, Peru balsam, benzyl benzoate, bismuth oxide and resorcinol; used in the treatment of hemorrhoids. [Warner] *

Anusol. A proprietary preparation containing boric acid, zinc oxide, bismuth subgallate, bismuth oxide, Peru balsam and resorcinol; an antipruritic. [Warner] *

Anusol-HC. A proprietary preparation of benzyl benzoate, bismuth subgallate, bismuth oxide, resorcinol, Peru balsam, zinc oxide and hydrocortisone acetate; used in the treatment of hemorrhoids. [Warner] *

Anxine. Tablets of dexamphetamine sulfate, cyclobarbitone, and mephenesin. [Allen & Hanburys Ltd] *

Anzon. Flame retardant compositions. [Anzon Ltd] *

Aolept®. 10-[3-(4-Hydroxypiperidino)propyl]-phenothiazine-2-carbonitrile; a neuroleptic. [Bayer AG]

AOR/GR. A proprietary name for technically-comminuted rubber. [Societ Indochine de Plantations d'Hvas (SIPH), Ivory Coast] ‡

Aosoft. Tetrafilcon A; contact lens material. [Am. Optical] *

Apagallin. Tetraiodophenol-phthalein; an antiseptic and indicator. §

Apatef. CAS 69712-56-7; Injectable antibiotic, containing cefotetan. [ICI Chem. & Polymers Ltd]

Apec®. Polyarylate (aromatic polyester carbonate) and copolycarbonate based on bisphenol A and bisphenol TMC (trimethylcyclohexanon); high temperature resistant engineering thermoplastic for the manufacture of injection moldings; used for components exposed to high temperatures in electrical systems for automobiles, lighting engineering; the electrical and electronic industries, medical engineering and domestic appliances. [Bayer plc]

Apec®DP9-9330. PC; high-heat amorphous thermoplastic with high strength and optical props. for extrusion and injection molding applications including demanding transparent and opaque applications, lenses in auto and industrial lighting, fuse housings, microwave oven doors. [Miles]

Apec® HT DP9-9350. PC; high-heat amorphous thermoplastic resin suitable for demanding applications such as lenses and reflectors used in high heat-output lighting devices and automotive fuses; injection molding and extrusion grade. [Bayer; Miles]

Apec® HT KU 1-9350. Amorphous aromatic PC; thermoplastic offering high heat resist. and improved flowability, transparency, and uv-stability for automotive elec. components, lamp housings, lamp reflectors, domestic appliance components. [Bayer AG] *

Apec®KL 1-9306. Polyester carbonate; amorphous thermoplastic providing heat resist. between 150 and 184 C for lighting engineering, electronics, elec. engineering applications. [Bayer AG] *

Apec® KU 1-9309. Polyester carbonate; flame-retarded grade; see Apec KL 1-9306. [Bayer AG] *

Apernyl®. Socket pellets for prophylactic and therapeutic care of extraction wounds; used in dentistry. [Bayer AG]

Apex 400. A proprietary alloy of aluminum with silicon. §

Apexior. Heat-resistant organic coating for water-side corrosion protection of steam generating equipment and auxiliaries; for steam generating equipment, feed water heaters, evaporators, steam turbines, diesel cylinder liners, condensate tanks. [Dampney Company Inc] *

APG® 225 Glycoside. CAS 68515-73-1; C8-10 alkyl polysaccharide ether; caustic-stable wetting agent and coupler for agricutural formulations; biodeg. [Henkel/ Emery]

APG® 300 CS. Decyl polyglucose; cosurfactant, auxiliary foaming agent for mild shampoos and other personal care cleansers. [Henkel/Cospha; Henkel Canada]

APG® 300 Glycoside. CAS 113976-90-2; C9-11 alkyl polysaccharide ether; caustic-stable degreaser, emulsifier, dispersing and wetting agent; for general purpose and hard surface cleaners, agricutural formulations; biodeg. [Henkel/Emery]

APG®600 CS. Lauryl polyglucose; cosurfactant, visc. modifier and thickener for mild shampoos, other personal care cleansers. [Henkel/Cospha; Henkel Canada]

APG® 600 Glycoside. CAS 110615-47-9; C12-14 alkyl polysaccharide ether; caustic-stable detergent active; visc. modifier for crutcher spray-dryer slurries; biodeg. [Henkel/Emery]

Aphox. CAS 23103-98-2; Granules containing 50% w/w pirimicarb; used for control of aphids. [ICI Chem. & Polymers Ltd]

Aphrogene. Premetallized dyes. [ICI Chem. & Polymers Ltd] †

Aphthite. An alloy of 800 parts copper, 25 parts platinum, 10 parts tungsten, and 170 parts gold. §

Aphtite. Zinc and cadmium-containing nickel bronzes; used for high-grade imitation silver products.

Apiezon. A range of high quality oils, waxes and greases prepared by molecular distillation of low volatility hydrocarbon feedstocks. originally developed for high vacuum applications but suitable in several other sectors of industry. [Shell] *

Apifac. Polyglyceryl-2 isostearate; self-emulsifying base for w/o cosmetic creams. [Gattefosse; Gattefosse SA]

Apifil. PEG-8 beeswax; self-emulsifying base for o/w emulsions in cosmetics and pharmaceuticals. [Gattefosse; Gattefosse SA]

apigenine. A yellow dyestuff obtained by decomposing the glucoside apiine found in parsley.

---

‡ Trade name and manufacturer not verified   § Trade name without identified manufacturer

apiol. (apiol, white). The crystalline constituent of parsley oil; liquid apiol is essential oil derived from an apiol-bearing variety of parsley; sometimes used as a diuretic.

Apisate. Tablets containing diethylpropion with vitamin B complex in a delayed-release base; appetite suppressant used in the short-term treatment of obesity. [Wyeth Laboratories]

APL. Gonadotropin, chorionic; gonad-stimulating principle. [Wyeth Laboratories] *

Aplisol. Tuberculin; diagnostic aid (dermal reactivity indicator). [Parke-Davis] *

Aplitest. Tuberculin; diagnostic aid (dermal reactivity indicator). [Parke-Davis] *

Apollo 50C. CAS 74115-24-5; Suspension concentrate containing 500 g clofentezine per liter; an acaricide for use on top fruit. [Schering Agrochemicals Ltd]

Apolloy. A proprietary copper-iron alloy containing 0.25% copper and 0.08% carbon. §

Apolomine. A proprietary preparation of hyoscine hydrobromide, benzocaine, riboflavine, pyridoxine hydrochloride, and nicotinamide; an antinauseant. §

aposafranine. A dyestuff. It is diazotized safranine boiled with alcohol.

Aposet 707. A proprietary trade name for a ketone peroxide catalyst for polyesters. §

Appeel®. Polymer. [DuPont UK]

Apperitive Saffron of Iron. Ferric subcarbonate. §

apple acid. See N-hydroxysuccinic acid

Appretan. Washfast finishing agents. [Hoechst UK]

Appretan Ant. An acrylate-based copolymer; a proprietary dispersant surfactant for finishing woven fabrics. [Alma Paint & Varnish Co] ‡

Appretan CPF. A proprietary polyvinyl acetate dispersion surfactant used for finishing woven, nonwoven and knitted fabrics. [Alma Paint & Varnish Co] *

Appretan GM. A proprietary polyvinyl acetate-based dispersion surfactant used in the finishing of especially lightweight fabrics, knitted fabrics and nonwovens. [Alma Paint & Varnish Co] *

Appretan TN. A vinyl acetate copolymer; a proprietary dispersion surfactant for finishing woven, nonwoven and knitted fabrics. [Alma Paint & Varnish Co] *

Appreteen. Water soluble size. [S & D Chemicals Ltd] *

A.P.P. Stomach Powder. A proprietary preparation containing calcium carbonate, magnesium carbonate, magnesium trisilicate, bismuth carbonate, aluminum hydroxide gel, homatropine methylbroroide, papaverine, and phenobarbitone; an antacid. [Consolidated Chemicals Ltd] *

APR®. Liquid photopolymer, printing plate making system. [Asahi Chem. Industry]

Apresoline. CAS 304-20-1; A proprietary preparation of hydralazine hydrochloride; antihypertensive. [Ciba plc]*

Apresoline Hydrochloride. CAS 304-20-1; Hydralazine hydrochloride; antihypertensive. [Ciba-Geigy Corp]

Aprinox. A proprietary preparation of bendrofluazide; a diuretic. [The Boots Co plc]

Apron Combi. Mixture of metalaxyl, thiram and thiabendazole; protectant fungicide for pea and bean seeds. [Ciba-Geigy Agrochemicals]

Apron T. Mixture of metalaxyl and thiabendazole; protectant fungicide for grass seed. [Ciba-Geigy Agrochemicals]

Aptine. CAS 13707-88-5; Alprenolol hydrochloride; antiadrenergic. [Astra Pharmaceutical Products Inc] *

Apyrogen. A proprietary preparation of pyrogen-free water for injection. [G D Searle & Co Ltd] *

Aquabase. Fatty alcohols/PEG ester blend; base for o/w emulsions; emulsifier for baby creams, cleansers, day creams/lotions, foundations, night creams, sunscreen preps. [Westbrook Lanolin]

Aquabase. Waterborne automotive paints. [ICI Chem. & Polymers Ltd]

Aquabloc. Water-based acrylic resins and coatings; for use as a replacement to polyethylene or wax coated paper and board due to its high water barrier and high MVTR barrier properties; used in food packaging, bakery applications, frozen foods; has water, grease, oil and other product resistance properties for paper and boards. [ADM Tronics Unlimited Inc]

Aquabrome. Swimming pool water disinfectant. [Great Lakes Europe]

Aqua-Chem®. Colorant dispersions; for coloring of emulsion and other water-borne coating compositions. [Hüls Am.]

Aquacillin. CAS 6130-64-9; Penicillin G Procaine; antibacterial. [Schering Corp] *

Aquaclene. Tablets containing chloramine. [British Drug Houses] *

Aquacoat. Water-based overlacquers. [The Scottish Adhesives Co Ltd]

Aquadag. CAS 7782-42-5; Colloidal graphite in water; used as a lubricant for drawing tungsten and molybdenum filament wires, for metal forming operations such as extrusion, as an aid to cutting and for forming opaque coatings for face plates of cathode ray tubes. [Acheson Colloids]

Aquafilm. Fluorocarbon; water and grease proofing for specialty papers and nonwovens. [CNC Int'l. L.P.]

Aquaflex. Tetrafilcon A; contact lens material. [UCO Optics Inc] *

Aquafloc. Flocculants. [Dearborn Chemicals Ltd]

Aquaforte. Water-based primer/adhesive with low solids and containing no organic solvents; extrusion primer for film to film, paper and foil, adhesion promoter of inks to foil. [ADM Tronics Unlimited Inc]

aqua fortis. See nitric acid

Aquagel. A proprietary trade name for a colloidal bentonite; also a proprietary hydrated silicate of alumina for waterproofing cement. §

Aqua Gro G Granular. Blended nonionic soil wetting agent; 100% active Aqua-Gro 40% wt. (polyoxyethylene ester of cyclic acids—47%, polyoxyethylene ether of alkylated phenols—47%, silicone antifoam emulsion—6%), vermiculite—60% wt; granular wetting agent to aid water penetration and drainage; used by greenhouses, nurseries and interior plantscapers in the manufacture of horticulture growing and potting media. [Aquatrols Corp of Am.] *

Aqua-Gro L Liquid. Blended nonionic soil wetting agent; polyoxyethylene esters of cyclic acids—47%, polyoxyethylene esters of alkylated phenols—47%, silicone antifoam emulsion—6%; provides increased water penetration into and out of soils and horticultural media; used in manufacture of horticultural growing and potting media; on golf courses, sports turf, lawns and exterior landscapes; drainage of compacted soils and puddles; poa annua seedhead inhibition; spreader-activator; adjuvant; hydroseeding; dew and frost control. [Aquatrols

Corp of Am.] *

Aqua-Gro S Spreadable. Blended nonionic soil wetting agent; 100% active Aqua-Gro 15% wt; (polyoxyethylene ester of cyclic acids—47%, polyoxyethylene ether of alkylated phenols—47%, silicone antifoam emulsion 6%) ground corn cobs—85%; granular soil wetting agent to aid water penetration and drainge; used on golf courses, sprots turf, lawns and exterior landscapes. [Aquatrols Corp of Am.] *

Aqua Ivy, AP. Poison ivy extract, alum precipitated; ivy poisoning counteractant. [Miles Pharmaceuticals] *

Aqualac. A proprietary shellac. §

Aqualease 2802. High molecular weight silicone polymer; water-based heavy-duty release agent for cast urethane elastomer and integral skin urethane foams. [George Mann]

Aqualease 6102. Silicone emulsion; release agent with better corrosion protection and excellent film formation; releases synthetic microcellular and solid elastomers, epoxies and polyester systems. [George Mann]

Aqualipid 95. CAS 8002-43-5; EINECS 232-307-2; Deoiled lecithin; marine animal feed additive. [Central Soya]

Aqualite. A proprietary phenol-formaldehyde synthetic resin laminated product and bearing material requiring only water as lubricant. §

Aqualon®Cellulose Gum. CAS 9004-32-4; Sodium CMC; suspending agent for abrasive and polishing agents, and prevents syneresis in toothpaste; rheology control agent in creams and lotions; adhesive and cohesive agent used in denture adhesive and ostomy adhesive products. [Aqualon]

Aqualon® CMC-T. CAS 9004-32-4; Sodium CMC, tech. grades; binder, thickener, stabilizer, suspending agent, film-former, rheology control aid, water-retention aid used for adhesives, aerial-drop fluids, ceramics, coatings, detergents, lithography, paper, textiles, and tobacco. [Aqualon]

Aqualon® CMHEC-37L. Carboxymethyl hydroxyethyl cellulose; hydrophilic colloid with good flocculating action on suspended solids and water-binding capacity; possibility for complexing and crosslinking reactions. [Aqualon]

Aqualon®EHEC. CAS 9004-58-4; Ethyl hydroxyethylcellulose; film-forming polymer for inks and coatings including printing inks, lacquers, varnishes, specialty finishes; avail. in four visc. types. [Aqualon]

Aqualose. Alkoxylated lanolin and lanolin derivatives; water-soluble emollients, o/w emulsifiers and wetting agents for cosmetics. [Westbrook Lanolin] *

Aqualose L30. CAS 61790-81-6; PEG-30 lanolin; emollient, emulsifier, plasticizer, solubilizer. [Westbrook Lanolin]

Aqualose LL100. PPG-40-PEG-60 lanolin oil; emollient, emulsifier, plasticizer, solubilizer used in aq./alcoholic preparations of alcohol content. [Westbrook Lanolin]

Aqualose W20. CAS 61791-20-6; Laneth-20; plasticizer and solubilizer for hydrophobic substances; emollient, emulsifier, solubilizer for aftershaves, antiperspirants, cleansers, foam baths, shampoos, nailcare; carrier for foam baths, shampoos, nailcare. [Westbrook Lanolin]

Aqualox® 225-100, 225A-100. Amine salt; surfactant, corrosion inhibitor, lubricant, antiwear additive effective in inhibiting the attack of ferrous metals by aq. sol'ns.; for metalworking formulations; use 225-100 for EP/antiwear

use; 225A-100 grade contains no phosphorous. [Alox] *

Aqualox® 232. Amine salts of organic acids; corrosion inhibitor, low foaming surfactant for synthetic metalworking formulations, especially in aqueous solution. [Alox*

Aqualox®2268. Low foaming corrosion inhibitor, lubricity and antiwear agent protecting ferrous metals from aqueous sol'ns. and steel against high humidity conditions during extended indoor storage; emulsions deposit very thin waxy films. [Alox] *

Aqualube. A proprietary plasticizer for water soluble materials. §

Aqua Magic. Silicones; stain preventive for fabrics. [Adasco Inc]

Aquamephyton. Phytonadione; an aqueous colloidal solution of vitamin K_1 for parenteral administration. [Merck & Co Inc]

Aqua Mer. A totally aqueous dry film photoresist, which comprises a three-layer sandwich construction of polyolefin/photopolymer/polyester. [Hercules] *

Aquamet M. A dithiocarbamate, liquid formulation suitable for precipitating metal ions from solution; used in waste water treatment. [Alco Chemical Corp]

Aquamollin®. Water softening agents. [Cassella AG]

Aquanol. Liquid blend of polymeric resins and modified siloxane in a petroleum solvent base; used as an internal sealer and an external water repellant for concrete, brick, treated concrete block, stucco, wood, and stone. [Secure Inc] †

Aqua Paste™. Inhibited aluminum pigments; aluminum pigments for aqueous paints and coatings used for decorative metallic effects as well asd a wide range of protective coatings applications. [Silberline Mfg Co Inc]

Aquapel®. Fatty acid ketene dimer emulsions; sizing agent for paper and paperboard under neutral conditions. [Hercules; Hercules Ltd]

Aquapel®360XC. Reactive sizing emulsion for use against a wide variety of penetrants; for papermaking industry. [Hercules]

Aquaperle. Textile auxiliary chemicals. [ICI Chem. & Polymers Ltd]

Aquaphil K. Lanolin and lanolin alcohol; emollient, emulsifier with enhanced w/o emulsion stability. [Westbrook Lanolin]

Aquaphoril. See Diurexan. [Degussa AG] *

Aquaplex. A proprietary trade name for alkyd resin dispersed in an aqueous medium; an emulsion varnish and lacquer resin vehicle for stucco, etc. §

Aquapol. Hydrophilic urethane prepolymer. [Freeman Chemical Corp] *

Aquaprint. Moldable impression material, synthetic resins. [BP Chemicals Ltd]

aqua regia. (chlorazotic acid; chloronitrous acid; nitrohydrochloric acid; nitromuriatic acid) CAS 8007-56-5; A mixture of one volume of nitric acid and three volumes of hydrochloric acid; used in metallurgy, testing metals, in dissolving metals such as platinum and gold.

Aquaresin. Glyceryl boriborate; a plasticizer. §

Aquarite. Water treatment chemicals. [Albright & Wilson Ltd, Phosphates & Speciality Business]

Aquaseal. Bitumen-based emulsions for waterproofing roofs, asphalt, asbestos-cement, concrete. [Feb Ltd]

Aquaseal Aquaflex Liquid Felt. Polychloroprene bitumen emulsion with excellent elastomeric properties; for coating asphalt, asbestos cement, roofing felt, concrete,

---

‡ Trade name and manufacturer not verified   § Trade name without identified manufacturer

corrugated iron, wood, etc. [Feb Ltd]

Aquaseal Firmafix. A cold-applied bituminous felt adhesive for bonding roofing felts to metal, timber, concrete screeds, asphalt and built-up felt. [Feb Ltd]

Aquaseal Reflect. White solar reflective roof coating that prevents heat build-up on roofs and protects roofing membranes from uv attack and environmental pollution. [Feb Ltd]

Aquaseal Weatherwise Standard. All-weather roofing treatment providing waterproofing to asphalt, asbestos-cement, concrete, corrugated iron, roofing felt, etc. [Feb Ltd]

Aquasil. Window desiccants. [Laporte Industries Ltd] *

Aquasil®. Inhibited aluminum pigments; used for aqueous paints and coatings used for decorative metallic effects as well as a wide range of protective coatings applications. [Silberline Mfg Co Inc]

Aquasoft®. Water-based textile ink; direct silk screen for T-shirts, athletic garments, aprons, tote bags, draperies, tablecloths, caps, wallhangings; excellent adhesion to cotton, blends and most synthetic fabrics. [Int'l. Coatings Co Inc]

Aquasol. Aqueous dispersants; used for pigment dispersions. [Tennant-KVK Ltd]

Aquasol A. Vitamin A; antixerophthalmic. [Armour Pharmaceutical Co] *

Aquasol E. Vitamin E supplement. [Armour Pharmaceutical Co] *

Aquasorb® A250. CAS 9004-32-4; Carboxymethylcellulose; absorbent for urine, blood, and other body fluids; used in feminine hygiene products, medical disposables, disposable diapers. [Aqualon]

Aquasperse. Colorant dispersions; for coloring of emulsion and other water-borne coating compositions. [Hüls Am.]

Aquastab PA 48. Dilauryl thiodipropionate; stabilizer/antioxidant for polymers. [Eastman]

Aquastore. Crosslinked polyacrylamide. [Cyanamid BV]

Aquasun. Sun protection products, to protect skin while promoting a tan. [Richardson-Vicks Inc] *

Aquatac® 5527. Water-based resin; freeze/thaw-stable disp. for removable and permanent label adhesive applications. [Arizona]

Aquatac® 6085. Glyceryl rosinate aq. dispersion; tackifier for pressure-sensitive adhesives; produces formulations with aggressive tack and peel, excellent shear props. after aging; for waterborne labels, decals, shelf liners, construction adhesives, tapes. [Arizona]

Aquatac® 8005. Water-dispersible resin; designed for freezer-grade label applications and as low softening point modifier for water-based adhesives. [Arizona]

Aquatec. A paraffin wax emulsion sometimes with aluminum acetate; proprietary trade name for a waterproofing material. §

Aqua-Tein C. Collagen amino acids, acetamide MEA, propylene glycol; substantive moisturizer, emollient for hair and skin care products (shampoos, conditioners, ethnic products, nutritive eye creams, face creams and lotions; anti-irritant for anionic shampoos. [Maybrook]

Aquathane Series. Fluorochemical polyurethane aq. copolymer; finishes for natural, synthetic and blended constructions; forms tough films with resistance to degradation. [CNC Int'l. L.P.]

Aquathene AQ 120-000. Ethylene vinylsilane copolymer; wire and cable resin for use in low voltage power cable applications. [Quantum/USI]

Aquathene MP. Structural waterproofing based on 125 micron thick polyethylene film; waterproofs solid floors where mechanical damage is unlikely. [Feb Ltd]

Aquathene MV. Structural waterproofing based on 300 micron black polyvinyl chloride film; provides protection for underground structures and roofing. [Feb Ltd]

Aquatherm. High-density insulation board specifically designed for use as a roof overlay board. [Feb Ltd]

Aqua Thix. Modified polysaccharide thixotrope; thickening agent and protective colloid for water-based coating compositions. [Hüls Am.]

Aquatreat AR-7-H. A high molecular weight acrylic acid polymer; used for adhesives, lithographics, latex stabilization. [Alco Chemical Corp]

Aquatreat AR-225-D. A free flowing low molecular weight sodium polymethacrylate; dispersant and desludger in water systems, cooling towers, boilers and heat exchangers. [Alco Chemical Corp]

Aquatreat AR-232. A low molecular weight sodium polymethacrylate solution; dispersant and desludger in water systems, cooling towers, boilers and heat exchangers. [Alco Chemical Corp]

Aquatreat AR-626. A low molecular weight acrylate copolymer solution; used for scale prevention in cooling towers, boilers, heat exchangers, oil field applications. [Alco Chemical Corp]

Aquatreat AR-648. A polycarboxylate copolymer solution; used for scale prevention in cooling towers, boilers and heat exchangers. [Alco Chemical Corp]

Aquatreat AR-900. A low molecular weight sodium polyacrylate; used for scale prevention in cooling towers and heat exchangers. [Alco Chemical Corp]

Aquatreat DNM-9, DNM-25, DMN-360. Dithiocarbamate salts; short-stop in emulsion polymerization of rubber; biocide, fungicide, and algicide used in water treatment, paper, sugar, and petrol. applications. [Alco Chemical Corp]

Aquatreat DNM-30; Sodium dimethyldithiocarbamate (15-16.6%), nabam (15-16.6%), inert; short-stop in emulsion polymerization of rubber; fungicide and bactericide for use in pulp/paper mills, sugar mills, drilling fluids, petrol. recovery; algicide for use in industrial recirculating water cooling towers, etc. [Alco Chemical Corp]

Aquatreat KM. CAS 128-03-0; Potassium dimethyldithiocarbamate; polymerization short stops in the copolymerization of styrene and butadiene. [Alco Chemical Corp]

Aquatreat SDM. Sodium dimethyldithiocarbamate; see Aquatreat DNM-9. [Alco Chemical Corp]

Aquatrend®. Colorant dispersions; for in-plant coloring of latex emulsion and other water-based coating compositions. [Hüls Am.]

Aqua-Trete®. Alkylalkoxysilane; weatherproofing agents for concrete terra cotta, brick, stucco and masonry surfaces. [Hüls Am.]

Aquazym. α-Amylase produced by submerged fermentation of a selected strain of *Bacillus subtilis*; intended for use in the desizing of textiles. [Novo Nordisk]

arabic gum. See gum arabic

D-arabinose. $C_5H_{10}O_5$; CAS 10323-20-3; EINECS 233-708-5; Culture medium. [Lonza; U.S. Biochemical]

L-arabinose. $C_5H_{10}O_5$; CAS 5328-37-0; EINECS 226-214-6; Culture medium. [Penta Mfg.; Pfanstiehl Labs; Schweizerhall; U.S. Biochemical]

Aracast. Heterocyclic epoxide resins. [Ciba plc] *
arachis oil. See peanut oil
Arakote®3000. Polyester resin; excellent flow grade. [Ciba-Geigy/Plastics] *
Araldite®. Epoxy resin systems; used for casting, encapsulating, laminating, surface coating and as an adhesive. [Ciba plc]
Araldite®2001. CAS 25928-94-3; Two-component epoxy adhesive; used for light engineering structures, boat and vehicle parts, sporting goods; bonds well to glass fiber laminates. [Ciba-Geigy GmbH] *
Araldite® CY 225. CAS 25928-94-3; Liquid epoxy resin; used for casting systems, elec. insulated components, high strength structural applications. [Ciba-Geigy/Plastics] *
Araldite®ECN 1235. CAS 25928-94-3; Solid epoxy cresol novolac; for high temp. adhesives, coatings, elec. and laminating applications. [Ciba-Geigy/Plastics] *
Araldite® GT 6060. CAS 25928-94-3; Solid bisphenol A epoxy; for castings, elec. encapsulating, laminating and adhesive applications. [Ciba-Geigy/Plastics] *
Araldite® GZ 540 X-90. CAS 25928-94-3; Epoxy sol'n. in xylene; two-pkg. epoxy sol'n. for maintenance and architectural coatings. [Ciba-Geigy/Plastics] *
Araldite®LT 8052. Solid brominated epoxy; flame retardant epoxy for impregnating, casting applications. [Ciba-Geigy/Plastics] *
Araldite® LY 8047. Brominated epoxy liquid; for prepreg laminating applications. [Ciba-Geigy/Plastics] *
Araldite®PT 810. CAS 25928-94-3; Epoxy resin; unmodified epoxy with color stability at high temps. and good weathering; good thermal, adhesive and chem. resist. [Ciba-Geigy Plastics UK] *
Araldite®PY 306. CAS 25928-94-3; Bisphenol F epoxy liquid modifier of other resins for lower visc., higher solids coatings. [Ciba-Geigy/Plastics, Plastics UK] *
Araldite®XD 897. CAS 25928-94-3; Epoxy resin; stabilizer for chlorinated vinyl resins. [Ciba-Geigy Plastics UK] *
Araldite® XD 4955. CAS 25928-94-3; Bisphenol F epoxy liquid for civil engineering and coatings requiring higher solids and performance. [Ciba-Geigy/Plastics, Plastics UK] *
Araldite® XU GY 358. CAS 25928-94-3; Epoxy resin; weatherable epoxy for maintenance and marine coatings, automotive refinishing. [Ciba-Geigy/Plastics] *
Aralen Hydrochloride. Chloroquine hydrochloride; antiamebic; antimalarial. [Sterling Drug Inc] *
Aralen Phosphate. CAS 50-63-5; Chloroquine phosphate; antimalarial; anti-amebic; suppressant. [Sterling Drug Inc] *
Aramine. CAS 54-49-9; Metaraminol; vasopressor agent for the treatment of acute shock. [Merck & Co Inc]
Aranox®. p-(p-toluenesulfonyl amido) Diphenylamine; antioxidant protecting EVA and polyamide hot-melt adhesives, PP, LDPE, LLDPE, HDPE against thermal degradation. [Uniroyal]
Arassist APH. Organic acid; buffering agent and bath stabilizer in acid dyebaths. [Arol Chem. Prods.]
Arassist HKM. Peroxide stabilizer and sequestrant for use in continuous bleaching of cotton or poly/cotton blends. [Arol Chem. Prods.]
Aratex. CAS 68334-28-1; Partially hydrogenated vegetable oil (cottonseed, soybean); icing stabilizer; syrups; donut glazes; bakery dry mixes. [Van Den Bergh Foods]

Aratronic® 5001. CAS 25928-94-3; Bisphenol A epoxy liquid; electronic grade. [Ciba-Geigy/Plastics] *
Aratronic® 5040. CAS 25928-94-3; Bisphenol F epoxy liquid; electronic grade. [Ciba-Geigy/Plastics] *
Aravite® 3001. Cyanoacrylates; structural adhesive for manufacturing of loudspeakers, optical instruments, jewelry, toys, rubber seals, domestic appliances, computers. [Ciba-Geigy Plastics UK] *
Arazate®. Zinc dibenzyldithiocarbamate; activator. [Uniroyal]
Arbacet. CAS 55028-70-1; Arbaprostil; antisecretory. [Upjohn] *
Arbeflex. Plasticizers. [Robinson Brothers Ltd] †
Arbestab. Antioxidants and uv-stabilizers for polymers; useful in a number of plastics materials but particularly effective in polyolefins. [Robinson Brothers Ltd]
Arbid®. Relief from rhinitis. [Bayer AG]
Arbo. Generic name for a range of putties, mastics, and sealants. [Adshead Ratcliffe & Co Ltd]
Arbocaulk. An acrylic emulsion-based sealant for gun application; principally used for internal pointing applications. [Adshead Ratcliffe & Co Ltd]
Arbocel. Wood cellulose; for use in bitumen products, adhesives, plastics, and sealants. [ICI Chem. & Polymers Ltd]
Arbocrylic. An acrylic solvent-based sealant for gun application; principally used for sealing external joints in building structures. [Adshead Ratcliffe & Co Ltd]
Arboflex. A glazing compound based on a blend of vegetable oils, plasticizers and butyl rubber; used for bead glazing aluminum and sealed timber window frames. [Adshead Ratcliffe & Co Ltd]
Arbofoam. Polyurethane foam packed in an aerosol dispenser; used to seal and insulate gaps around pipes and duct work and as a fixing, gap filling adhesive for doors and windows. [Adshead Ratcliffe & Co Ltd]
Arbogard. Herbicides. [ICI Chem. & Polymers Ltd]
Arbokol. A range of single and two component polysulfide and epoxy/polysulfide sealants; used for sealants in building joints, floor joints and double glaze unit construction. [Adshead Ratcliffe & Co Ltd]
Arbolite. A putty composition based on a blend of vegetable oils; used for face glazing steel window frames. [Adshead Ratcliffe & Co Ltd]
Arbomast. A range of gun-applied sealants based on vegetable oils or butyl rubber; low-cost, general-purpose sealants for a range of applications. [Adshead Ratcliffe & Co Ltd]
Arborsan®. Substitute for creosote; wood preservative. [Lanstar Ltd]
Arboseal. A range of preformed mastics strips based on butyl rubber and polybutenes; used for making watertight and dustproof seals between components, where the joint is under compression. [Adshead Ratcliffe & Co Ltd]
Arbosil. A range of RTV silicone-based single component sealants; used for sealing a wide range of industrial and building applications between a variety of substrates. [Adshead Ratcliffe & Co Ltd]
Arbostrip. Self-adhesive foam strips based on plasticized PVC; compression sealants for draft-proofing and similar applications. [Adshead Ratcliffe & Co Ltd]
Arbyl. Dispersing and leveling agent; used for dyeing, a wetting agent and detergent for pretreatment, designing

---

‡ Trade name and manufacturer not verified   § Trade name without identified manufacturer

and dyeing in the textile industry. [Degussa AG] *

Arbylen. A wetting agent and detergent; used for pretreatment, desizing and dyeing in the textile industry. [Degussa AG] *

ARcare® On/OFF 7810. Medical grade tape consisting of on/off Ma-48 medical grade adhesive coated on one side of flexible nonwoven fabric; intended for mounting medical devices to the skin when good long-term adhesion and/or ease of removal after soaking with water are desired. [Adhesives Research]

Arcel Moldable Polyethylene Copolymers. CAS 9002-88-4; Polyethylene copolymers; resins for molded resilient foam pkg. (for computers, military instruments, home entertainment electronics, medical devices), automotive, marine, and fiberglass laminate applications. [Arco Chemical Co]

archil. (orchil, orseille, Persio, cudbear, orchellin). A natural coloring matter obtained from *Roccela tinctoria* and other lichens. The coloring principle is orcin, $C_6H_3(CH_3)(OH)_2$, which, in the presence of air and ammonia, oxidizes to a violet dye.]

Arcolloy. A nonmagnetic alloy of iron with 12-16% chromium, and less than 0.12% carbon, 0.5% manganese, 0.025% phosphorus, 0.025% sulfur, and 0.5% silicon. §

Arcoloy. A proprietary copper-silicon casting alloy containing 97.25% copper, 2.63% silicon, 0.12% iron, and 0.01% phosphorus. §

Arconate® Propylene Carbonate. CAS 108-32-7; Propylene carbonate; solv. with high boiling pt., low toxicity, broad range of applications; reactive diluent for woodbinders, urethane foams and coatings, foundry sand binders, in textile and synthetic fiber industry, natural gas treating; lubricant in cosmetics. [Arco Chemical Co]

Arcosolv® DPM. CAS 34590-94-8; Dipropylene glycol monomethyl ether; solv. for coatings, cleaners, inks, agricultural products, cosmetics, chemical intermediate applications. [Arco Chemical Co]

Arcosolv® DPMA. CAS 88917-22-0; Dipropylene glycol methyl ether acetate; solv. where a slow evaporating nonhydroxylic solv. is required; effective in coatings; as a coalescent in waterborne emulsion systems. [Arco Chemical Co]

Arcosolv® PM. CAS 107-98-2; Propylene glycol methyl ether; solv. for coatings, cleaners, inks, agricultural products, cosmetics, chemical intermediate applications. [Arco Chemical Co]

Arcosolv® PMA. CAS 108-65-6; Propylene glycol methyl ether acetate; slow-evaporating solv. with good solvency for many commonly used coating resins, e.g., acrylics, NC, and urethanes; used in lacquers, water-based paints. [Arco Chemical Co]

Arcosolv® PTB. CAS 57018-52-7; Propylene glycol mono-t-butyl ether; solv. offering a blend of hydrophobicity and hydrophilicity for coatings, cleaners, electronic, and ink applications; strong coupling ability; used in light-duty and hard-surface cleaners, water-reducible polyester and alkyd resin production; chemical intermediate in the synthesis of monomeric and polymeric products; cosolv. [Arco Chemical Co]

Arcosolv® TPM. CAS 20324-33-8; PPG-3 methyl ether; solv. for coatings, cleaners, inks, agricultural products, cosmetics, chemical intermediate applications; slow evaporating. [Arco Chemical Co]

Arctite Injection Mortar. Expanding cement; used for brickwork; prevents rising damp (DPC); nontoxic. [Arcmann-Denmark A/S]

Arctite Quickbinder. A liquid for treating concrete for severe leaks. [Arcmann-Denmark A/S]

Arctite Slurry 200 B. Waterproofing cement; used for concrete, high water pressure applications; nontoxic. [Arcmann-Denmark A/S]

Arctite Tanking Mortar 500. Cement based waterproofing compounds, nontoxic; used for concrete and brick structures. [Arcmann-Denmark A/S]

Arcton. A range of fluorinated hydrocarbon refrigerants and aerosol propellants (fluorocarbons) (*qv*). [ICI Chem. & Polymers Ltd]

Ardeer Powder. An explosive containing 31-34% nitroglycerin, 11-14% kieselguhr, 47-51% magnesium sulfate, 4-6% potassium nitrate, and 0.5% ammonium or calcium carbonate. §

Ardel® D-100. Polyarylate resin; engineering plastic for automotive, elec./electronic, glazing/solar energy, safety equip., and plumbing applications; features excellent uv resist., toughness, transparency, good electricals. [Amoco Chemical Co]

Ardenite. A proprietary synthetic resin; a molding composition. §

Ardent. Suspension concentrate containing 40 g diflufenican and 400 g trifluralin per liter; used for control of weeds in winter cereals. [Embetec Crop Protection Ltd]

Ardinex. A proprietary preparation of guaifenesin, methaqualone hydrochloride, and ephedrine hydrochloride; a bronchial antispasmodic. [The Boots Co plc]†

Ardmorite. A variety of bentonite (*qv*), found in the Pierre shales at Ardmore, South Dakota.

Ardux. Modified urea/formaldehyde resin. [Ciba plc] *

arecaidine. N-methyl-Δ−3-tetrahydro-pyridine-3-carboxylic acid.

arecoline. The methyl ester of N-methyl-Δ-3-tetrahydropyridine-3-carboxylic acid.

Arelon. CAS 34123-59-6; Suspension concentrate containing 553 g isoproturon per liter; used for annual weed control in cereals. [Hoechst UK]

Aremco-Bond. High temperature organic adhesive. [Merlec Co] ‡

Aremco-Cast. High temperature casting material. [Merlec Co] ‡

Aremco-Coat. High temperature coating material. [Merlec Co] ‡

Aremsol A. Cocamidopropyl betaine; foam booster for conventional shampoos, conditioning shampoos and conditioning rinses. [Ronsheim & Moore] *

Aremsol MA. MEA lauryl sulfate; base for preparation of liquid shampoos; biodeg. [Ronsheim & Moore]

Arenka. A proprietary high-strength yarn manufactured from aramides (aromatic polyamides). [Enka Glanzstoff AG]*

Arenolite. An artificial siliceous-argillaceous-calcareous stone.

Aresenid. CAS 7778-39-4; Arsenic acid solution; wood preservative. [Mechema Chemicals Ltd] *

Aresin. Herbicide for potatoes and leeks. [Hoechst UK] *

Areskap. A proprietary trade name for a butyl-phenyl-phenol sodium sulfonate. §

---

* Trade name not verified as available   † Trade name verified as obsolete

Aresket. A proprietary trade name for a wetting agent; stated to be a butyl-diphenyl sodium sulfonate. §

Aresklene. A proprietary trade name for a dibutyl-phenyl-phenol sulfonate; a mold lubricant and emulsifier. §

Aretan. A mercurial fungicide. [ICI Chem. & Polymers Ltd]

Aretone 270. A proprietary trade name for a fine mica used as a pigment in paints. §

Arfonad. A proprietary preparation of the (+)-camsylate of trimetaphan; a vasodilator. [Roche Laboratories; Roche Products Ltd]

Argal. See argol.

Argentai. An alloy of 85% copper, 10% tin, and 5% cobalt.

Argentalium. An aluminum alloy containing antimony.

argentan. A nickel silver. It consists of 56% copper, 26% nickel, 18% zinc, and 1% iron, and is used as an electrical resistance alloy.

argentous oxide. See silver oxide (ous)

Argidone®. CAS 56265-06-6; EINECS 260-081-5; Arginine PCA; moisturizing adjuvant for nutritive or generative creams or lotions; activates cell metabolism. [UCIB]

argilla. See China Clay.

arginine. Amino acid; $H_2NCNHNHCH_2CH_2CH_2CH-NH_2COOH$; CAS 74-79-3 (L-form); EINECS 200-811-1; Biochemical research, medicine, pharmaceuticals, dietary supplement. [Degussa; Penta; U.S. Biochemical]

argipressin. An antidiuretic hormone.

Argobase. W/o absorption bases containing lanolin and/or lanolin alcohols; emulsifiers for cosmetics and ointments. [Westbrook Lanolin] *

Argobase 125. CAS 8027-33-6; EINECS 232-430-1; Lanolin alcohols extract; emollient, w/o emulsifier used in personal care products. [Westbrook Lanolin]

Argobase EU. Sterols and sterol esters lanolin extracts; emollient, emulsifier, stabilizer; absorp. base. [Westbrook Lanolin]

argol. (argal) CAS 868-14-4; A crystalline crust deposited on the sides of the vat in which grape juice has been fermented; it contains 40-70% tartaric acid, principally as potassium hydrogen tartrate.

Argonol. Liquid lanolin derivatives; fluid emollients and moisturizers for cosmetics. [Westbrook Lanolin] *

Argonol 1SO. Emollient, moisturizer for aftershaves, baby oils, foam baths, nailcare, sunscreen preps.; plasticizer for hair sprays. [Westbrook Lanolin]

Argonol 40. CAS 85005-47-6; Isobutylated lanolin oil; w/o emulsifier, emollient used in personal care products. [Westbrook Lanolin]

Argonol 50 Pharmaceutical. Lanolin oil; emollient, lubricant. [Westbrook Lanolin]

Argonol ACE5. Cetyl acetate, acetylated lanolin alcohol; emollient, moisturizer for baby oils, eye preps., foam baths, sunscreen preps.; lubricant/glossing aid for hairsprays, lipsticks; binder for lipsticks, pressed powds. [Westbrook Lanolin]

Argotone. A proprietary preparation of ephedrine hydrochloride, and silver vitellin. nasal drops. [Rona Laboratories] ‡

Argowax. Refined lanolin alcohols; w/o emulsifier in cosmetics and pharmaceuticals. [Westbrook Lanolin] *

Argowax Standard. CAS 8027-33-6; EINECS 232-430-1; Lanolin alcohol BP; gelling agent, w/o emulsifier, emollient, moisturizer for baby creams, day creams/lotions, face masks, foundation, sunscreen preps.; plasticizer for hairsprays. [Westbrook Lanolin]

Argozie. (Arguzoid, Argozoil). An alloy of from 54-56% copper, 23-38% zinc, 2-4% tin, 2-3.5% lead, and 13.5-14% nickel. §

Argus DLTDP. CAS 123-28-4; Dilauryl thiodipropionate; antioxidant used for polyoelfins, thermoplastic elastomers, synthetic fubber; antioxidant for cosmetics and pharmaceuticals. [Witco/Argus]

Argus DMTDP. CAS 16545-54-3; Dimyristyl thiodipropionate; antioxidant for polyolefins and other polymeric systems. [Witco/Argus]

Argus DSTDP. CAS 693-36-7; Distearyl thiodipropionate; antioxidant for polyolefins and other polymeric systems where long-term heat stability is required; also for pharmaceutical and cosmetic products, oils, greases, and lubricants. [Witco/Argus]

Argus DTDTDP. CAS 10595-72-9; Ditridecyl thiodipropionate; antioxidant for polyolefins and thermoplastic elastomers, especially latexes where a liquid antioxidant/stabilizer is dispersed more effectively. [Witco/Argus]

Arguzoid. An alloy of 56% copper, 23% zinc, 4% tin, 3.5% lead, and 13.5% nickel. §

Arguzoil. See Argozie. §

Argylene. Powder containing 8% w/w sodium silver thiosulfate; used to prolong flower life in pot plants. [Fargro Ltd]

Argyrolith. (China silver, electroplate). Names given to alloys containing 50-70% copper, 10-20% nickel, and 5-30% zinc. Alfenide and Argentan are similar alloys; they are nickel silvers (German silver). §

Argyrol®S.S.10% (stabilized solution). Mild silver protein; indicated in the treatment of eye infections and also preoperative in eye surgery. [Iolab Corp]

Ariabel. Inorganic and organic cosmetic pigments; for coloring of cosmetic products. [Morton Int'l. Ltd] †

Ariagran. High strength water soluble, granular food colors; used for coloring of foodstuffs and pharmaceuticals. [Morton Int'l. Ltd] †

Arianor. Semipermanent, water soluble hair colors; incorporated in hair products intended to color hair. [Morton Int'l. Ltd] †

Ariavit. High strength water soluble powder food colors; for coloring of foodstuffs and pharmaceuticals. [Morton Int'l. Ltd] †

Aricel. Melamine-formaldehyde resins, usually 80% solids; cross-linking polymer systems, textile finish, filter papers, nonwoven binder. [Astro Industries Inc] *

Aricyl. Veterinary product; injectable deodorant and tonic. [Bayer AG]

Aridex. A retexturing and reproofing aid. [Laporte Industries Ltd] *

Aridry B. Methylol stearamide; water repellent durable to washing and dry cleaning; extender for use with fluorocarbons. [CNC Int'l. L.P.]

Arigal PMP. A proprietary trade name for a solution of an organic mercuric compound which, when used together with Arigal C, imparts a mildew resistant and rot-proof finish on cellulosic fibers; fast to water, washing and dry cleaning. [Ciba plc] *

Arigran. Granular food colors of guaranteed purity. [Morton Int'l. Ltd] †

Arikrome S. Chrome complex sol'n.; water repellent for specialty papers and nonwovens; may be used with dyes or pigmented papers to improve resist. to leaching or crocking. [CNC Int'l. L.P.]

Arilvax. A proprietary vaccine against yellow fever. [The

---

‡ Trade name and manufacturer not verified   § Trade name without identified manufacturer

Wellcome Foundation Ltd]

Arimid®. Fiber. [DuPont UK]

Aripol. Polyelectrolyte flocculating agent. [Steetley Chemicals Ltd] *

Aristar. Ultrapure reagents and solvents. [BDH Chemicals Ltd]

Aristocort. CAS 124-94-7; Triamcinolone; glucocorticoid. [Lederle Laboratories USA] *

Aristocort Acetonide. CAS 76-25-5; Triamcinolone acetonide; glucocorticoid. [Lederle Laboratories USA] *

Aristocort Forte Parenteral. Triamcinolone diacetate; glucocorticoid. [Lederle Laboratories USA] *

Aristocort Syrup. Triamcinolone diacetate; glucocorticoid. [Lederle Laboratories USA] *

Aristoflex. Hair lacquer resins. [Hoechst UK]

Aristol A. 60% mono substituted C20-24 benzene; lubricant; lube oil additive; chemical feedstock for sulfonation to produce emulsifiers and corrosion preventatives. [Pilot]

Aristonate H. Sodium petroleum sulfonate (CAS 78330-12-8); surfactant for formulating drycleaning soaps, cutting oils, textile oils, leather oils, rust preventative and fuel oil compositions; ore floation collectors; emulsifiers for agricultural sprays. [Pilot]

Aristospan. CAS 5611-51-8; Triamcinolone hexacetonide; glucocorticoid. [Lederle Laboratories USA] *

Arizole® Anethole Extra. CAS 104-46-1; Anethole; anise flavoring material. [Arizona]

Arizole® Pine Oil. Pine oil; disinfectants and cleaners, odorant, frothing agent in mineral flotation, solvent. [Arizona]

Arizona 208. Tall oil fatty acid ester; for plasticizers, extenders, surfactants in grinding and cutting oils, specialty lubricant additives, corrosion inhibitors, specialty solvs. for printing inks, metalworking, and oil well servicing. [Arizona]

Arizona DR-22. Disproportionated rosin; emulsifier, detergent, wetting agent; used to prepare emulsifiers for styrene/butadiene rubber polymerization, as shortstop for solv. polymerizations of rubber, as plasticizer/tackifier; used to make Arizona disproportionated tall oil rosin soaps (DRS-40, 42, 43,44). [Arizona]

Arizona DR-24. Disproportionated rosin; emulsifier for styrene/butadiene rubber production; intermediate for production of disproportionated rosin soaps. [Arizona]

Arizona DRS-40. CAS 61790-50-9; Potassium soap of disproportionated tall oil rosin; emulsifier, detergent, wetting agent; for ABS, styrene/butadiene rubber, other synthetic elastomers. [Arizona]

Arizona DRS-43. CAS 61790-51-0; Sodium salt of disproportionated tall oil rosin; emulsifier, detergent, wetting agent; emulsifier for ABS, styrene/butadiene rubber, synthetic elastomers. [Arizona]

Arizona DRS-50. CAS 61790-50-9; Potassium soap of disproportioned tall oil rosin; polymerization emulsifier for the styrene/butadiene rubber industry. [Arizona]

Arizona FA-7001. Tall oil dimer acid; for manufacturing of many dimer acid-based derivatives. [Arizona]

Arklone. CAS 76-13-1; Trichlorotrifluoroethane; a cleaning solvent used in aerosols. [ICI Chem. & Polymers Ltd]

Arko metal. An alloy of 80% copper, and 20% zinc.

Arkopal. Alkylphenol polyglycol detergent bases. [Hoechst UK]

Arkopal N. Range of nonionic surfactants of the nonylphenol

ethoxylate type in liquid, paste or wax form; wetting, dispersing, foaming, emulsification, detergent and cleaning agent used in domestic and industrial cleaners, disinfectant cleaners, auxiliaries for textile, leather and fur dressing, metal working, rubber, electroplating, pesticides; plant protection, building industry, antidusting and many other uses. [Hoechst UK]

Arkopon T. CAS 137-20-2; Sodium oleoyl methyl tauride in powder form; dispersing and wetting agent used in wettable powders, plant protection and pest control. [Hoechst UK] *

Arlacel® 20. CAS 1338-39-2; EINECS 215-663-3; Sorbitan laurate; emulsifier for cosmetics, pharmaceuticals. [ICI Spec. Chem.; ICI Surf. Belgium]

Arlacel® 40. CAS 26266-57-9; EINECS 247-568-8; Sorbitan palmitate; emulsifier for cosmetics, pharmaceuticals. [ICI Spec. Chem.; ICI Surf. Belgium]

Arlacel® 60. CAS 1338-41-6; EINECS 215-664-9; Sorbitan stearate; emulsifier for cosmetics, pharmaceuticals. [ICI Spec. Chem.; ICI Surf. Belgium]

Arlacel® 80. CAS 1338-43-8; EINECS 215-665-4; Sorbitan oleate; emulsifier for cosmetics, pharmaceuticals. [ICI Spec. Chem.; ICI Surf. Belgium]

Arlacel® 83. CAS 8007-43-0; EINECS 232-360-1; Sorbitan sesquioleate; emulsifier; cosmetic and pharmaceutical grade of Arlacel®C. [ICI Spec. Chem.; ICI Surf. Belgium]

Arlacel® 85. CAS 26266-58-0; EINECS 247-569-3; Sorbitan trioleate; surfactant for cosmetics and pharmaceuticals. [ICI Spec. Chem.; ICI Surf. Belgium]

Arlacel® 165. Glyceryl stearate, PEG-100 stearate; surfactant, emulsifier, thickener, opacifier for cosmetics and allied fields; acid-stable; self-emulsifying. [ICI Spec. Chem.; ICI Surf. Belgium]

Arlacel® 186. Glyceryl oleate, propylene glycol, 0.02% BHA and 0.01% citric acid as preservatives; surfactant, emulsifier, thickener for personal care products; defoamer for oral pharmaceutical products. [ICI Spec. Chem.; ICI Surf. Belgium]

Arlacide A. Chlorhexidine acetate; preservative/bactericide for liquid and powd. preparations. [ICI Spec. Chem.] *

Arlagard. Bacteriostat. [ICI Am.]

Arlamol. Blends of specific fatty acid esters; nonionic surfactants. [ICI Am.]

Arlamol® E. CAS 25231-21-4; PPG-15 stearyl ether with preservatives; emollient, solv. for personal care products. [ICI Spec. Chem.; ICI Surf. Belgium]

Arlamol® ISML. Isosorbide laurate; surfactant. [ICI Australia]

Arlasolve® 200. Isoceteth-20; surfactant, emulsifier, solubilizer for cosmetics. [ICI Spec. Chem.]

Arlasolve® DMI. CAS 5306-85-4; EINECS 226-159-8; Dimethyl isosorbide; surfactant, emollient. [ICI Spec. Chem.]

Arlatone® B. Polysorbate 85 and dinonyl phenol; emulsifier for cosmetic cleansers, waterless hand cleaners, blooming bath oils. [ICI Spec. Chem.]

Arlatone® G. CAS 61788-85-0; PEG-25 hydrogenated castor oil; surfactant, solubilizer, emollient; formulates clear gels. [ICI Spec. Chem.; ICI Surf. Belgium]

Arlatone® T. PEG-40 sorbitan peroleate; emulsifier, solubilizer, antistat, lubricant, spreading agent; used for bath oils, and in the textile industry. [ICI Spec. Chem.; ICI Surf. Belgium]

Arlef-100. CAS 530-78-9; A proprietary preparation of

---

* Trade name not verified as available   † Trade name verified as obsolete

flufenamic acid; an antirheumatic. [Parke-Davis] *

Arlidin. CAS 849-55-8; Nylidrin hydrochloride; vasodilator. [USV Pharmaceutical Corp] ‡

Arlin. A proprietary name for polyethylene film. [Poly-Version Inc] ‡

Arlinflex. Rigid vinyl. [Arlington Mills Inc]

Arlix. CAS 55837-27-9; Piretanide; diuretic. [Hoechst-Roussel Pharmaceuticals Inc] *

Arlon®. Polyetheretherketone; used for structures requiring toughness and chemical resistance, e.g., valve seats, compressor plates. [Greene, Tweed & Co]

Arloy®. Blends of Dylark SMA and polycarbonate; engineering resins providing heat resist., high impact strength and moldability; for automotive instrument panels, seat belt retractors, speaker grills, surgical appliances, institutional feeding trays, camera components, power tool housings. [Arco Chemical Co]

Armac®. Series of cationic surfactants which consist of acetate salts of primary amines, and which are soluble; used for pigment flushing, froth flotation and flocculation, particularly in mineral flotation; petroleum processing; leather; paper; pigments and surface coatings; ceramics. [Akzo Chemie UK Ltd]

Armco. A trademark for ingot iron and stainless steel. §

Armco Ingot Iron. A trademark for a very pure iron, 99.84% pure. §

Armeen®. Range of cationic surfactants composed mainly of primary amines, in liquid, paste or solid form; degree of surface activity varies with composition; they have the ability to change mineral surfaces from hydrophobic to hydrophilic; used in petroleum; road emulsions; plastics; rubber; textiles; leather; herbicides; fungicides; rodent repellents; mineral flotation; paper; pigments and surface coatings; water and sewage treatment; wax; sealant formulations; cement curing. [Akzo Chemie UK Ltd]

Armeen. A range of coco, oleyl and stearyl amines; used as chemical intermediates, anticaking agents and in secondary oil recovery. [Harcros Australia]

Armeen® 2-10. CAS 1120-49-6; Didecylamine; industrial surfactant. [Akzo]

Armeen®2-18; CAS 112-99-2; Dioctadecylamine; industrial surfactant. [Akzo]

Armeen® 2C. CAS 61789-76-2; Dicocamine (sec. amine); emulsifier, flotation agent, corrosion inhibitor. [Akzo]

Armeen® 2HT. CAS 61789-79-5; EINECS 263-089-7; Di(hydrogenated tallow) amine (sec. amine); emulsifier, flotation agent, corrosion inhibitor. [Akzo]

Armeen® 2T; CAS 68783-24-4; Ditallowamine; emulsifier, flotation agent, corrosion inhibitor. [Akzo]

Armeen® 3-12; CAS 102-87-4; EINECS 203-063-4; Tridodecylamine; chemical intermediate for manufacturing of sol. betaines and quaternary ammonium salts; carrier for manufacturing of citric acid and oil. [Akzo]

Armeen® 3-16; CAS 67701-00-2; Trihexadecylamine; chemical intermediate for manufacturing of oil-sol. betaines and quaternary ammonium salts. [Akzo]

Armeen® 12. CAS 124-22-1; Lauramine; industrial surfactant. [Akzo]

Armeen®12D. CAS 124-22-1; Lauramine (primary amine); emulsifier, flotation agent, corrosion inhibitor; lubricant for metal treatment. [Akzo; Akzo Chem. BV]

Armeen® 16. CAS 143-27-1; EINECS 205-596-8; Palmitamine; industrial surfactant. [Akzo]

Armeen® 16D. CAS 143-27-1; EINECS 205-596-8;

Palmitamine (primary amine); emulsifier, flotation agent, corrosion inhibitor. [Akzo; Akzo Chem. BV]

Armeen® 18. CAS 124-30-1; EINECS 204-695-3; Stearamine (primary amine); emulsifier, flotation agent, corrosion inhibitor, anticaking agent; hard rubber mold release agent. [Akzo; Akzo Chem. BV]

Armeen® 18D. CAS 124-30-1; EINECS 204-695-3; Stearamine, dist.; emulsifier, flotation agent, corrosion inhibitor, anticaking agent; rubber processing auxiliary; mold release agent for plastics and rubber. [Akzo]

Armeen® C. CAS 61788-46-3; EINECS 262-977-1; Cocamine (primary amine); emulsifier, flotation agent, corrosion inhibitor, stripping agent for paints. [Akzo; Akzo Chem. BV]

Armeen® CD. CAS 61788-46-3; EINECS 262-977-1; Cocamine (primary amine); emulsifier, flotation agent, corrosion inhibitor, stripping agent for paints. [Akzo]

Armeen® DM12D. CAS 112-18-5; Dimethyl lauramine; surfactant intermediate. [Akzo]

Armeen® DM16D. CAS 112-69-6; Dimethyl palmitamine; chemical intermediate, raw material for surfactants. [Akzo]

Armeen® DM18D. CAS 124-28-7; EINECS 204-694-8; Dimethyl stearamine; chemical intermediate, raw material for surfactants. [Akzo]

Armeen® DMCD. CAS 61788-93-0; EINECS 263-020-0; Dimethyl cocamine; chemical intermediate, raw material for surfactants. [Akzo; Akzo Chem. BV]

Armeen® DMHTD. CAS 61788-95-2; Dimethyl hydrogenated tallow amine; chemical intermediate, raw material for surfactants. [Akzo; Akzo Chem. BV]

Armeen® DMOD. CAS 28061-69-0; Oleyl dimethylamine; surfactant intermediate. [Akzo]

Armeen® DMSD. CAS 61788-91-8; Soyaalkyl dimethylamine; surfactant intermediate. [Akzo]

Armeen® DMTD. CAS 68814-69-7; Tallowalkyl dimethylamine; surfactant intermediate. [Akzo]

Armeen® HT. CAS 61788-45-2; EINECS 262-976-6; (Hydrogenated tallow) amine (primary amine); emulsifier, flotation agent, corrosion inhibitor, chemical intermediate, anticaking agent. [Akzo; Akzo Chem. BV]

Armeen® HTD. CAS 61788-45-2; EINECS 262-976-6; (Hydrogenated tallow) amine (primary amine); emulsifier, flotation agent, corrosion inhibitor, chemical intermediate, anticaking agent. [Akzo; Akzo Chem. BV]

Armeen® L8D. CAS 104-75-6; 2-Ethylhexylamine, distilled; chemical intermediate for vapor phase corrosion inhibitors. [Akzo]

Armeen® M2-10D. CAS 7396-58-9; Didecyl methylamine; chemical intermediate for water-sol. betaines; catalyst for urethane resins. [Akzo]

Armeen® M2C. CAS 61788-62-3; Dicoco methylamine; chemical intermediate; surfactant; for manufacturing of oil-sol. betaines and quaternary ammonium salts. [Akzo; Akzo Chem. BV]

Armeen® M2HT; CAS 61788-63-4; Dihydrogenated tallow methylamine; chemical intermediate; for manufacturing of oil-sol. betaines and quaternary ammonium salts. [Akzo; Akzo Chem. BV]

Armeen® OL. CAS 112-90-3; EINECS 204-015-5; Oleamine; emulsifier, flotation agent, corrosion inhibitor. [Akzo]

Armeen® OLD. CAS 112-90-3; EINECS 204-015-5; Oleamine, dist.; emulsifier, flotation reagent, corrosion

---

‡ Trade name and manufacturer not verified   § Trade name without identified manufacturer

inhibitor. [Akzo]

Armeen® S. CAS 61790-18-9; Soyamine; emulsifier, flotation agent, corrosion inhibitor. [Akzo]

Armeen® SD. CAS 61970-18-9; Soyamine, dist.; emulsifier, flotation agent, corrosion inhibitor. [Akzo]

Armeen® T. CAS 61790-33-8; EINECS 263-125-1; Tallowamine (primary amine); emulsifier, flotation reagent, corrosion inhibitor, dispersant, anticaking agent, chemical intermediate, cosmetics ingredient. [Akzo; Akzo Chem. BV]

Armeen® TD. CAS 61790-33-8; EINECS 263-125-1; Tallowamine, dist.; emulsifier, flotation agent, corrosion inhibitor. [Akzo]

Armenian cement. A jeweler's cement containing gum mastic, isinglass, gum ammoniac, alcohol, and water; it is made by soaking the isinglass in water and mixing it with the spirit containing the gums.

Armid® E. CAS 112-84-5; EINECS 204-009-2; Erucamide; mold release agent for rubber and plastics; auxiliary for processing rubber. [Akzo]

Armid® HT. CAS 61790-31-6; EINECS 263-123-0; Hydrogenated tallow amide; see Armid 18; also antifoam in steam generator systems, lubricant additive; auxiliary for rubber processing. [Akzo]

Armid® HTD. CAS 124-30-1; EINECS 204-695-3; Stearylamine; processing aid for high visc. rubber compounds; facilitating flow behavior and improving mold release. [Akzo] *

Armid® O. CAS 301-02-0; EINECS 206-103-9; Oleamide; see Armid 18; also release agent in cosmetics, penetrant in paper manufacture. [Akzo]

Armide. See Camite. §

Armillatox®. Emulsion of polyhydric phenols in soap; home garden lawn treatment and fungicide; controls moss in lawns; reduces the severity of club root; hinders the spread of honey funfus. [Armillatox Ltd]

Armite. Vulcanized fiber. [Spaulding Fibre Co] *

Armix 146. Formulated product; specialty product designed to enhance the effectiveness of MSMA formulations. [Witco/Organics]

Armix 176. Formulated product; tank mix which provides a combination of wetting, sticking, spreading and penetration. [Witco/Organics]

Armofilm. Long chain filming amine emulsion. [Akzo Chemie UK Ltd]

Armoflo®. Conditioner; hygroscopic salts and fertilizers. [Akzo Chemie UK Ltd]

Armofog. Anticondensing agent for polyolefins. [Akzo Chemie UK Ltd]

Armogard. Fuel oil additives. [Akzo Chemie UK Ltd]

Armogel. Thickening agent. [Akzo Chemie UK Ltd] †

Armogloss. Cationic car wash additive. [Akzo Chemie UK Ltd] †

Armohib®. Acid inhibitors. [Akzo Chemie UK Ltd]

Armohib® 18, 28. Formulated products containing cationic surfactants consisting of aliphatic nitrogen derived materials; inhibitors used in acid pickling; plant cleaning; oil well acidizing; 18 developed for use with sulfuric, phosphoric, citric and sulfamic acids; 28 for use with hydrochloric acid. [Akzo Chemie UK Ltd]

Armor-Kote. Emulsified coal tar pitch. [Crowley Chem.]

Armor-ply. A proprietary trade name for metal bonded plywood. §

Armoslip®. Slip and antiblocking agents for polyolefins.

[Akzo Chemie UK Ltd]

Armostat®. Antistatic agent for polyolefins. [Akzo Chemie UK Ltd]

Armoteric LB. Amphoteric surfactant supplied as a yellowish liquid; for baby shampoos; bubble baths; strong acid, and alkaline cleaning detergents. [Akzo Chemie Nederland BV]

Armoteric SB. Amphoteric surfactant supplied as a yellowish paste; for baby shampoos; bubble baths; strong acid and alkaline cleaning detergents. [Akzo Chemie Nederland BV] *

Armourcote. Unreinforced fluorocarbon coatings and coatings reinforced with stainless steel, molybdenum and ceramic; for low friction and nonstick surfaces in baking and food processing, and general industrial applications. [Fothergill Tygaflor Ltd] *

Armowax. Synthetic waxes. [Armour Hess Chemicals] ‡

Armowax. Processing aid for highly filled polyolefins. [Akzo Chemie UK Ltd]

Armowax EBS. CAS 110-30-5; EINECS 203-755-6; A proprietary trade name for N,N´-ethylene bisstearamide. [Armour Hess Chemicals] ‡

Armul 17. Formulated product; emulsifier for paraffinic hydrocarbon crop oils. [Witco/Organics]

Armul 22, 44, 66, 88. Blended emulsifiers; emulsifiers which, when used in combinations, can be used to formulate a wide variety of pesticide products. [Witco/Organics]

Armyl. A proprietary trade name for Lymecycline. §

arnatto. See annatto.

Arneel DN. A proprietary trade name for the dimerized product of octadecene and octadecadiene nitriles; a vinyl plasticizer. [Armour Pharmaceutical Co] *

Arneel HF. A proprietary trade name for the 18, 20 and 24 carbon atom fatty acid nitriles; vinyl plasticizers. [Armour Pharmaceutical Co] *

Arneel® OD. Octadecene nitrile; detergent, wetting agent, rust inhibitor. [Akzo]

Arneel S. A proprietary trade name for a derivative of octadecene and octadecadiene nitriles; a vinyl plasticizer. [Armour Pharmaceutical Co] *

Arneel TOD. A proprietary trade name for a derivative of octadecene and octadecadiene nitriles. A vinyl plasticizer. [Armour Pharmaceutical Co] *

Arnica Oil CLR. Arnica extract, soybean oil, tocopherol; emollient, conditioner; protective skin and hair care products. [Dr. Kurt Richter; Henkel/Cospha]

Arnica Yellow. An azo dyestuff prepared by condensing p-nitrotoluenesulfonic acid with p-aminophenol, in the presence of aqueous caustic soda; dyes cotton golden-yellow from a salt bath.

Arnite A.K.U. A trade name for a polyethylene glycol terephthalate; injection molding material. [Algemene Industriele] ‡

Arnite G. Thermoplastic polyester grade; for injection molding and extrusion. [Algemene Industriele] ‡

arnotto. See annatto.

ARO. Asbestos and nonasbestos friction material; friction material for brakes and clutches. [Caramba Chemie GmbH] *

Arobleach HW. Blend of oxygenated inorganics, bleach, and surfactants; for single-bath scouring, bleaching, and dyeing of polyester/cotton blends. [Arol Chem. Prods.]

Arobleach MX. Peroxygen compound; one-bath bleaching and dyeing agent for knit or woven fabrics of cotton or

---

* Trade name not verified as available   † Trade name verified as obsolete

cotton/synthetic blends. [Arol Chem. Prods.]

Arochem. (Aroplax). A proprietary trade name for soft oil modified alkyds. §

Aroclean MC-4. Heavy-duty industrial cleaner for metal surfaces, plastic, concrete; solubilizes trimer build-up in textile processing equip. [Arol Chem. Prods.]

Aroclear. Proprietary blend with alkaline builders; one-step clearing agent for reduction clearing of disperse dyes on polyester; biodeg.; removes excess dye concs. on fabric, anti-redeposition agent on fabric and equip.; stripping agent for overdyed fabric lots. [Arol Chem. Prods.]

Arodet AA-350. Alkylaryl sulfonate with glycol coupling agents, detergent, wetting agent, dyeing assistant, scouring agent, leveler, retarder, dye dispersant, finishing agent, emulsifier. [Arol Chem. Prods.]

Arodet AN-100. Modified nonionic derivative; detergent, wetting agent, scouring agent, dye dispersant. [Arol Chem. Prods.]

Arodet AN-160. Detergent, wetting agent for natural and synthetic fiber processing, textile scouring, cotton desizing, kier boiling; dye dispersant, leveling and penetrating agent; stable to acid and alkalies, hard water, bleaching agents. [Arol Chem. Prods.]

Arodet BLN Special. Blend of long-chain ethoxylates; detergent, wetting agent, penetrant for natural and synthetic fibers; aids dyestuff dispersion; fulling agent for wool; solv. emulsifier. [Arol Chem. Prods.]

Arodet BN-100. Ethoxylated alcohol; detergent, wetting agent, emulsifier, dyeing assistant, dispersant for dyeing, finishing, textiles, pigments, resins. [Arol Chem. Prods.]

Arodet E-15. Blend; one-bath scouring agent for dyebaths for woolen fabrics; imparts wetting, scouring and leveling without interfering with subsequent dye procedures. [Arol Chem. Prods.]

Arodet HCS. Ethoxylate; general-purpose detergent, wetting agent, scouring agent for synthetic and natural fibers; high cloud pt., relatively low foaming, high temp. operating stability. [Arol Chem. Prods.]

Arodet MKD. Blend of nonionic and neutralized phosphate ester surfactants; low temp. textile scouring agent for removal of sizes, waxes, and natural or synthetic oils from fabrics. [Arol Chem. Prods.]

Arodet N-100. Nonylphenol PEG ester; scouring and soaping off agent for natural and synthetic fibers. [Arol Chem. Prods.]

Arodet TA-8. Phosphate ester; multipurpose surfactant for textile processing; for desizing, kier boiling, bleaching, wetting and dispersion in jig or beck, after-scouring; emulsifier for polar and nonpolar solvs. used as carriers (trichlorobenzene, butyl benzoate, etc.). [Arol Chem. Prods.]

Arofene. Formaldehyde polymers with phenol and substituted phenols usually supplied in solvent other than water; used for paper impregnation: air, oil and fuel filters; fiber bonding: nonwovens of all types; laminates: rolled and flat stock; adhesives: high performance. [Ashland Chemical Company] *

Arofix F-6. Fixative for acid dyes on nylon when applied as an after-treatment. [Arol Chem. Prods.]

Arofix SRN. Modified formaldehyde resin; fixative for direct dyes where complete removal of inorganic salts from the dyed goods is impractical, e.g., in package, beam, or skein work. [Arol Chem. Prods.]

Aroflat®. Alkyd synthetic resin; for manufacture of flat wall paints. [Reichhold]

Aroflat® 3113-P-30. Flat alkyd. [Reichhold]

Aroflint®. Resin solutions. [Reichhold]

Aroflint® 202-A6X-60. Polyester resin. [Reichhold]

Aroflint® 303-X-90. CAS 25928-94-3; Epoxy resin. [Reichhold]

Arofoam. Unsaturated polyester, two component system; used for structural applications such as acrylic tubes and showers with fiberglass reinforcement. [Ashland Chemical Company] *

Arofoam SNI. Ethoxylates blend; micro-foam surfactant for foam dyeing procedures on synthetic fibers; dispersant for dyestuffs; dye leveling agent; antiprecipitant. [Arol Chem. Prods.]

Arofos 200 Conc. Phosphate ester blend; detergent, wetting agent, emulsifier, penetrant, dye leveling agent, dispersant. [Arol Chem. Prods.]

Arofos 326. Polyphosphorylated surfactant; detergent, emulsifier, wetting agent. [Arol Chem. Prods.]

Aroful BV-50. Amine condensate/ethoxylates blend; fulling detergent for carbonized woolen fabrics under acidic conditions; wetting agent, penetrant. [Arol Chem. Prods.]

Arogrip. Adhesive for marine market. [Ashland] *

Arol Biodet. Biodeg. scouring agent, rapid wetting agent for textile processing including degreasing, desizing, bleaching, dyeing, and finishing operations. [Arol Chem. Prods.]

Arol Defoamer NA2X. Silicone-stearate blend; defoamer for atmospheric dyeing operations in textile industry. [Arol Chem. Prods.]

Arolev ADL-30. Long-chain derivative; leveling and retarding agent for wool and acrylic fibers; stable to dilute acids and alkalies, hard water, salts. [Arol Chem. Prods.]

Arolev CDD. Sulfated fatty ester blend; surfactant, fast wetting/rewetting agent, emulsifier, leveling agent for cellulosics dyed with direct dyes, various synthetics and blends; for scouring, solv. scouring, sizing, kier bleaching, etc.; biodeg. [Arol Chem. Prods.]

Arolev MTR-7. Low foam leveling agent, dye dispersant, lubricant; synergistic with polyester dye carriers; for piece dyeing, yarn dyeing, atmospheric or pressure equip. [Arol Chem. Prods.]

Arolon®. Liquid copolymers for paints. [Reichhold]

Arolon® 580-W-42. Waterborne dispersion alkyd. [Reichhold]

Arolterge 100M. Blend of fatty acid alkanolamides and phosphate esters; detergent, emulsifier, wetting agent. [Arol Chem. Prods.]

Arolube MIT-1. Ethoxylates blend; softener, lubricant, crack mark inhibitor for all fibers; for dye bath, bleach bath, scouring bath; promotes leveling; stable under high temp. and pressure. [Arol Chem. Prods.]

Arol Woolbrite. Specialty reducing agent for wool treatment. [Arol Chem. Prods.]

Aromabator PC-80. A tamed and stabilized nontoxic chlorine dioxide complex concentrate formulated for use as an additive deodorant without free chlorine release. formulated for use as an airborne spray for industrial applications; effectively arrests malodors caused by viruses, fungi, bacteria, and coliform densities; added to coolants, cutting oils, industrial sumps, sludge pits, cooling towers, waste water, marine holding stations and

---

‡ Trade name and manufacturer not verified   § Trade name without identified manufacturer

ships' bilging areas, animal housing and restroom surfaces; for spray application. [Punati Chemical Corp] ‡

Aromabator PC-88. A tamed and stabilized nontoxic chlorine dioxide complex concentrate formulated for use as an additive deodorant without free chlorine release; as an airborne spray in home and farm applications; for use in kitchens, toilets, outhouses, pet and animal housing; can be safely sprayed on fabric and nonfabric surfaces, on air conditioning and humidifier filters. [Punati Chemical Corp] ‡

Aromaplas. Range of perfumes for plastics. [PPF International Ltd] *

Aromasol. Solvents. [ICI Chem. & Polymers Ltd]

Aromasol 17. A narrow distillation range white spirit substitute with 17% aromatic content. [Sasolchem]

Aromatic Oil 745. Aromatic plasticizer; used in adhesives, rubber (cements, mech. and molded goods, tires), caulking compounds. [Neville]

Aromatic Solvent 150. CAS 8030-30-6; 150° flash aromatic naphtha with narrow distillation range; high solvency, high flash pt. for paint and protective coatings, herbicide and pesticide carrier, synthetic resin manufacturing, degreasing applications. [Texaco]

Aromex. Powdered perfumery compounds. [Bush Boake Allen Ltd]

Aromix. Solvent emulsifier concentrates for pesticide formulations. [Plant Protection] *

Aromix. Mixture of heavy aromatics and aliphatics; solvent. [Sasolchem]

Aromox®. Long chain aliphatic amine oxides. [Akzo Chemie UK Ltd]

Aromox® C/12-W. CAS 61791-47-7; EINECS 263-180-1; Dihydroxyethyl cocamine oxide; wetting agent, emulsifier, stabilizer, antistat, foaming agent for detergents, shampoos, cosmetics, textiles, metal plating, petrol. additives, paper, plastics, rubber; gel sensitizer for latex foam; biodeg. [Akzo; Akzo Chem. BV]

Aron Alpha. A proprietary cyano-acrylate adhesive. [Toagosie Chemical Co] *

Aroplax. See Arochem. §

Aroplaz®. Synthetic resins for use in compounding protective and decorative coatings, printing, textile inks, and for general industrial use. [Reichhold]

Aroplaz® 3667-Z-80. High solids alkyd. [Reichhold]

Aroplaz® 6820-100. High solids polyester. [Reichhold]

Aropol. Unsaturated polyester resins, including orthophthalic, isophthalic, and other specialty polymer types; used for fiberglass reinforced polyester applications in construction, transportation, gel coats, marine, consumer, electrical and corrosion resistant markets. [Ashland] *

Aropol 2036. Isophthalic polyester resin; resin for molding and pultrusion applications. [Ashland]

Aropol 7020. High reactivity isophthalic polyester resin; can be chemically thickened for use in low-shrink and controlled-shrink SMC and BMC applications. [Ashland]

Aropol 7240T-15. Isophthalic polyester resin; thixotropic, prepromoted, chemical-resistant resin for use in hand lay-up and spray lay-up, fume hoods and ducts, tanks, pipes. [Ashland]

Aropol 7710. Isophthalic polyester resin; flexible resin with outstanding toughness and tensile strength; for blending with rigid resis; used in potting compounds for electronic components, polyester gaskets, caulking and sealing compounds. [Ashland]

Aropol 8321. Polyester resin; resilient resin for general purpose matched-die molding, preform molded chairs, tote bins, pultrusion applications. [Ashland]

Aropol 8420. Polyester resin; nonpromoted, resilient resin for use in matched-die molding, pultrusion, architectural sheeting, gel coats, casting. [Ashland]

Aropol Phase Alpha. Low profile unsaturated polyester resins; sheet molding compound for compression molding into Class A exterior automotive panels (hoods, roofs, and deck lids) and truck panels (hoods, tilt cabs). [Ashland] *

Aropol Phase II. Low profile unsaturated polyester resins; sheet molding compound for compression molding into Class A exterior automotive panels (hoods, roofs, deck lids) and truck panels (hoods, tilt cabs). [Ashland] *

Aropol WEP. Unsaturated polyester resins which can form water emulsions; for casting of decorative art and other applications. [Ashland] *

Aroquest 75. CAS 60-00-4; Ethylene diamine tetraacetic acid; chelating agent for sequestering metal ions over wide pH range, e.g., in peroxide bleaching baths. [Arol Chem. Prods.]

Aroquest 100. CAS 64-02-8; EINECS 200-573-9; Tetrasodium EDTA; sequestering agent for textile industry, boiler water treatment (water softening, scale removal and prevention). [Arol Chem. Prods.]

Aroquest 120. Sequestering agent for calcium and iron at neutral and mildly alkaline conditions; for textile scouring, dyeing assistant, industrial cleaning compounds, textile bleaching. [Arol Chem. Prods.]

Aroquest MLC. Chelating agent for peroxide bleach baths. [Arol Chem. Prods.]

Aroquest M Special. CAS 67-43-6; Diethylene triamine pentaacetic acid adduct; sequestrant for use in peroxide bleach baths. [Arol Chem. Prods.]

Aroset®. Acrylic sol'ns. and emulsions; crosslinking and thermoplastic resins with outstanding clarity and uv and oxidation resist.; for adhesives. [Ashland]

Arosoft Base LCS-2. Blend of fatty amides and nonionic softeners; all-purpose softener base for applic. on nylon, polyester and other synthetics and blends. [Arol Chem. Prods.]

Arosoft GSE-D. Ethoxylated glyceride plus esters; softener and lubricant for cotton and other textile fibers, yarns, and fabrics; imparts full, soft hand with excellent drape. [Arol Chem. Prods.]

Arosoft LC-15. Fatty amide/nonionic softener blend; non-yel. softener for nylon and other synthetic fibers. [Arol Chem. Prods.]

Arosolve 9D5R. Trichlorobenzene, butyl benzoate; high exhaustion carrier for use on polyester and blends where extremely effective swelling, penetration and leveling are required. [Arol Chem. Prods.]

Arosolve 570-HF. Aromatic solvs./detergent blend; non-red label solv. scouring agent, detergent for textiles; roller cleaner, tar remover. [Arol Chem. Prods.]

Arosolve B-950. Chlorinated solvs./emulsifiers blend; solv. scour for removal of grease, graphite and oil stains on polyester, nylon and other synthetic and natural fibers; degreaser and tar remover; stable to most acids, alkalies, metal salts. [Arol Chem. Prods.]

Arosolve MN-LF. CAS 1321-94-4; Emulsified aromatic naphthas; low foaming polyester dye carrier for use in

---

* Trade name not verified as available   † Trade name verified as obsolete

jets and other high-pressure dyeing equip.; produces level and bright shades on polyester. [Arol Chem. Prods.]

Arosolve RCB. Low-foaming emfulsifiers/aromatic petrol. distillate/biphenyl blend; solv. carrier for use on synthetic fabrics and their blends; low foaming for high shear jets. [Arol Chem. Prods.]

Arosolve XNF-1. Solvs./low foaming detergent blend; low foaming pressure jet solv. scour for difficult grease, graphite and oil stains on polyester, nylon and other synthetic and natural fibers; general degreaser and tar remover; stable to most acids and alkalies. [Arol Chem. Prods.]

Arosulf SBO-65. Sulfated fatty acid ester; wetting, rewetting agent, lubricant for textile dyeing operations; dye leveling agent; emulsifier for solv. systems. [Arol Chem. Prods.]

Arosulf SCO-75%. Castor oil deriv.; textile processing auxiliary, bleaching, leveling and dyeing assistant; emulsifier, finishing agent; also for industrial waxes, polishes, paints. [Arol Chem. Prods.]

Arosurf®66-E2. CAS 52292-17-8; Isosteareth-2; emulsifier, emollient for personal care products, cutting oils; o/w and w/o systems; coupling agent, emulsion stabilizer, perfume stabilizer. [Sherex/Div. of Witco]

Arosurf® 66-PE12. PPG-3-isosteareth-9; low cloud pt. emulsifier, emollient, dispersant, bath oil spreading agent, perfume solubilizer. [Sherex/Div. of Witco]

Arosurf® AA 23. Diamine; asphalt emulsifier for rapid set and mixing grade emulsions. [Sherex/Div. of Witco]

Arosurf®MG-70. Primary ether amine; flotation reagent for the iron mining industry. [Sherex/Div. of Witco]

Arosurf® TA-100. CAS 107-64-2; EINECS 203-508-2; Distearyl dimonium chloride; fabric softener conc., conditioner for home and commerical laundry and textile processing. [Sherex/Div. of Witco]

Arotap®. Acrylic and alkyd copolymers; for photoconductive applications in paper industry. [Ashland]

Arotech. Acrylamate polymer; for fiberglass reinforced parts requiring superior strength properties for automotive and other applications. [Ashland Chemical Company] *

Arotex. Growth regulator containing 644 g chlormequat and 32.2 g choline chloride per liter; for use on wheat, oats or rye. [ICI Agrochemicals]

Arothix. Sixteen thixotropic vehicles for wall paints. [Reichhold]

Arothix 4000-P-40. Flat alkyd. [Reichhold]

Arotran 50437-8. Polyester resin; for resin transfer molding; used for premium Class A automotive body panels, inner structures. [Ashland]

Arova 16. 1,4-Dioxacyclohexadecane-5,16-dione; musk perfume. [Hüls AG]

Arovit. See Ro-A-Vit. [Roche Products Ltd] *

Arowet 70 E. Sulfonated ester; wetting/rewetting agent, penetrant for textile processing, desizing, scouring, bleaching, level dyeing and printing, finishing operations. [Arol Chem. Prods.]

Arowet ODA. Biodeg. wetting agent for batch and continous operations on natural and synthetic fibers and blends; relatively low foaming. [Arol Chem. Prods.]

Arowet SC-75. CAS 577-11-7; EINECS 209-406-4; Dioctyl sodium sulfosuccinate; fast wetting agent, penetrant, and dyeing assistant for mild acidic or alkaline textile processing; dye leveling agent. [Arol Chem. Prods.]

Arozyme TD. CAS 9000-92-4; α Amylase; thermo-stable enzyme, textile desizing agent. [Arol Chem. Prods.]

ARP®. Addition agents, wetting agents, other specialty chemicals designed to solve problems of adhesion, leveling, pitting and other conditions affecting quality of surface treatment in metal finishing industry. [Witco/Allied-Kelite]

Arpak 4322. CAS 9002-88-4; Expanded polyethylene beads; produces very flexible closed-cell foam offering high resiliency; foams maintain dimensions and shock absorbence after repeated deformations; for dynamic cushioning applications. [Arco Chemical Co]

Arpal Non Selex. CAS 7775-09-9; Powder containing 58.2% w/w sodium chlorate; used for total weed control for paths, drives and noncrop areas. [R P Adams Ltd]

Arpocox. CAS 55779-18-5; Arprinocid; coccidiostat. [Merck & Co Inc] *

Arpro 3313. CAS 9003-07-0; Expanded PP beads; produces low-dens., closed-cell foam with excellent energy absorp. chars.; excellent recoverability from repeated shocks; foams maintain dimensional stability when exposed to temp. extremes. [Arco Chemical Co]

Arpylene. Propylene compounds. [Norsk Hydro Polymers Ltd] ‡

Arquad®. Range of cationic surfactants composed of alkyl quaternary ammonium chlorides in mainly liquid form; effective in killing micro-organisms at low concentrations; used in sanitizing foodstuffs, catering, blanket sterilization; algal control; mold inhibition; air conditioning; softening agents for textiles; laundry, dry cleaning; paper; corrosion inhibition; petroleum, e.g., in drilling; emulsification, e.g., in road making, metal cleaning, insecticides; antistats, e.g., plastics, rubber; latex foam rubber; cosmetics; leather; flocculants; conditioning agents for toiletries. [Akzo Chemie UK Ltd]

Arquad. Quaternary ammonium salts. [A & E Connock (Perfumery & Cosmetics) Ltd]

Arquad® 2C-70 Nitrite. CAS 71487-01-9; Dicoco nitrite, methanol/isopropanol; biodeg. surfactant, dispersant for protective coatings, pigments, inks, textiles, agricuture, acid pickling baths, marine applics, metalworking, electroplating, fuel treatment, emulsion/plastic manufacturing, waste water treatment, mineral processing, paper. [Akzo]

Arquad® 2C-75. CAS 61789-77-3; EINECS 263-087-6; Dicocodimonium chloride, aq. IPA; biodeg. emulsifier, foaming, wetting, dispersing agents, corrosion inhibitor, softener, dyeing aid, antistat for textiles, paper, cosmetics; industrial, agriculture, plastics, petrol. industry, acid pickling baths; bactericide, algicide. [Akzo; Akzo Chem. BV]

Arquad® 2HT-75. CAS 61789-80-8; Quaternium-18, aq. IPA; biodeg. emulsifier, foaming, wetting, dispersing agents, corrosion inhibitor, antistat, bacteriostat for paper softening, household laundry, hair conditioning. [Akzo]

Arquad® 2T-75. CAS 68783-78-8; Ditallow dimonium chloride, aq. ethanol; biodeg. surfactant, dispersant for protective coatings, pigments, inks, textiles, agricuture, acid pickling baths, marine applics, metalworking, electroplating, fuel treatment, emulsion/plastic manufacturing, waste water treatment, mineral processing, paper. [Akzo]

Arquad®12-37W. CAS 112-00-5; EINECS 203-927-0; Laurtrimonium chloride; emulsifier, corrosion inhibitor, textile

---

‡ Trade name and manufacturer not verified   § Trade name without identified manufacturer

softener, antistat, hair conditioner and combing aid emulsifier; biodeg. [Akzo]

Arquad® 12-50. CAS 112-00-5; EINECS 203-927-0; Laurtrimonium chloride, IPA; biodeg. emulsifier, foaming, wetting, dispersing agents, corrosion inhibitor, softener, dyeing aid, antistat for textiles, paper, cosmetics; industrial, agriculture, plastics, petrol. industry, acid pickling baths; bactericide, algicide; gel sensitizer for latex foam. [Akzo]

Arquad® 16-29. CAS 112-02-7; EINECS 203-928-6; Cetrimonium chloride; emulsifier, foaming, wetting, dispersion agents, corrosion inhibitor, antistat for textiles, cosmetics, industrial, agricuture; bactericide, algicide. [Akzo]

Arquad® 16-50. CAS 112-02-7; EINECS 203-928-6; Cetrimonium chloride, IPA; emulsifier, foaming, wetting, dispersing agents, corrosion inhibitor, softener, dyeing aid, antistat for textiles, paper, cosmetics; industrial, agriculture, plastics, petrol. industry, acid pickling baths; bactericide, algicide; rubber to textile bonding agent; biodeg. [Akzo; Akzo Chem. BV]

Arquad® 18-50. CAS 112-03-8; EINECS 203-929-1; Steartrimonium chloride, IPA; emulsifier, foaming, wetting, dispersing agents, corrosion inhibitor, softener, dyeing aid, antistat for textiles, paper, cosmetics; industrial, agriculture, plastics, petrol. industry, acid pickling baths; bactericide, algicide; dye leveling agent, visc. stabilizer, in lubricant compounding; biodeg. [Akzo; Akzo Chem. BV]

Arquad® 210-50. CAS 7173-51-5; Didecyl dimonium chloride, aq. ethanol; biodeg. surfactant, dispersant for protective coatings, pigments, inks, textiles, agricuture, acid pickling baths, marine applics, metalworking, electroplating, fuel treatment, emulsion/plastic manufacturing, waste water treatment, mineral processing, paper. [Akzo; Akzo Chem. BV]

Arquad® 218-75; CAS 107-64-2; EINECS 203-508-2; Dioctadecyl dimethyl ammonium chloride, aq. IPA; biodeg. surfactant, dispersant for protective coatings, pigments, inks, textiles, agricuture, acid pickling baths, marine applics, metalworking, electroplating, fuel treatment, emulsion/plastic manufacturing, waste water treatment, mineral processing, paper. [Akzo]

Arquad® 218-100; CAS 107-64-2; EINECS 203-508-2; Dioctadecyl dimethyl ammonium chloride; biodeg. surfactant, dispersant for protective coatings, pigments, inks, textiles, agricuture, acid pickling baths, marine applics, metalworking, electroplating, fuel treatment, emulsion/plastic manufacturing, waste water treatment, mineral processing, paper. [Akzo]

Arquad® 316(W). CAS 71060-72-5; Trihexadecylmethyl ammonium chloride, water; industrial surfactant for pigment dispersing, coatings, inks, paper processing. [Akzo]

Arquad® B-100. CAS 68391-01-5; Benzalkonium chloride, aq. IPA; antimicrobial for industrial applications, sec. oil recovery, textiles, cosmetics, pharmaceuticals, sanitizers. [Akzo]

Arquad® C-33W. CAS 61789-18-2; EINECS 263-038-9; Cocotrimonium chloride; emulsifier, corrosion inhibitor, textile softener, antistat; hair conditioning and combing aid emulsifier; emulsion-break retardant in cosmetics; biodeg. [Akzo]

Arquad® C-50. CAS 61789-18-2; EINECS 263-038-9; Co-

cotrimonium chloride, IPA; biodeg. emulsifier, foaming, wetting, dispersing agents, corrosion inhibitor, softener, dyeing aid, antistat for textiles, paper, cosmetics; industrial, agriculture, plastics, petrol. industry, acid pickling baths; bactericide, algicide; gel sensitizer for latex foam. [Akzo; Akzo Chem. BV]

Arquad® DMCB-80. CAS 61789-71-7; EINECS 263-080-8; Cocoalkyl dimethyl benzyl ammonium chloride, aq. IPA; microbicide for disinfectants, sanitizers, algicides for use in swimming pools, air conditioning cooling towers, bathroom cleaners, petrol. recovery. [Akzo]

Arquad® DMHTB-75. CAS 61789-72-8; EINECS 263-081-3; Hydrogenated tallow dimethylbenzyl ammonium chloride, aq. IPA; bactericide, disinfectant, softening agent for textiles. [Akzo]

Arquad® HTL8(W) MS-85. 2-Ethylhexyl hydrogenated tallowalkyl methosulfate; biodeg. surfactant, dispersant for protective coatings, pigments, inks, textiles, agriculture, acid pickling baths, marine applics, metalworking, electroplating, fuel treatment, emulsion/plastic manufacturing, waste water treatment, mineral processing, paper. [Akzo]

Arquad® M2HTB-80; CAS 61789-73-9; Di(hydrogenated tallow)benzyl methyl ammonium chloride, aq. IPA; industrial surfactant for prep. of organophilic clays. [Akzo]

Arquad® S-50. CAS 61790-41-8; EINECS 263-134-0; Soytrimonium chloride, IPA; emulsifier, corrosion inhibitor, textile softener, antistat; hair conditioning and combing aid emulsifier; bitumen emulsions; slime control agent in water systems; biodeg. [Akzo; Akzo Chem. BV]

Arquad® T-27W. CAS 8030-78-2; Tallow trimonium chloride; biodeg. emulsifier, foaming, wetting, dispersing agents, corrosion inhibitor, softener, dyeing aid, antistat for textiles, paper, cosmetics; industrial, agriculture, plastics, petrol. industry, acid pickling baths; bactericide, algicide. [Akzo]

Arquad® T-50. CAS 8030-78-2; Tallowtrimonium chloride, aq. IPA; emulsifier, corrosion inhibitor, textile softener, antistat; hair conditioning and combing aid emulsifier; also used in manufacturing of antibiotics; gel sensitizer for latex foam; biodeg. [Akzo; Akzo Chem. BV]

Arquel. CAS 644-62-2; Meclofenamic acid; anti-inflammatory. [Parke-Davis] *

Arrconox AHT, DNL and DNP. A proprietary range of antioxidants used in the manufacture or processing of rubber. [Rubber Regenerating Co] ‡

Arrconox S.P. A nonstaining antioxidant. [Rubber Regenerating Co] ‡

Arrcorez 16. A butyl rubber curing resin. [Rubber Regenerating Co] ‡

Arrcorez 17. A tackifying resin. [Rubber Regenerating Co]‡

Arresin. CAS 330-55-3; Emulsifiable concentrate containing 200 g/l monolinuron; used for control of annual dicotyledons in potatoes, french beans and leeks. [Hoechst UK]

Arret. CAS 53179-11-6; A proprietary preparation of loperamide; an antidiarrhoeal. [Janssen Pharmaceutical Ltd] *

arrhenal. (arsinyl, new cacodyl). Sodium methylarsinate; $AsO(CH_3)(ONa)_2 \cdot 6H_2O$.

Arrow Tool Steels. Proprietary steels containing 0.9-1.02% chromium, 0.16-0.20% vanadium, 0.5-0.6% manganese, and 0.20-0.30% carbon. §

Arsan 600. Moldable resilient resin; produces resilient, low-

---

dens., closed-cell foam offering low dynamic set; foams maintain shock absorbence after repeated impacts. [Arco Chemical Co]

Arsenal. Soluble concentrate containing imazapyr; used for bracken control in noncrop areas. [Chipman Ltd]

Arsenal®. 2-[4,5-Dihydro-4-methyl-4-(1-methylethyl)-5-oxo-1H-imidazol-2-yl]-3-pyridinecarboxylic acid with 2-propanamine (1:1) salt; used for bracken control in noncrop areas. [Am. Cyanamid/Ag; Cyanamid of Great Britain Ltd]

Arsenal XL. A soluble concentrate containing 300 g atrazine and 12.5 g imazapyr per liter; used for total weed control in non crop areas. [Chipman Ltd]

Arsenal®XL. Soluble concentrate containing 300 g atrazine and 12.5 g imazapyr per liter; for total weed control in noncrop areas. [Cyanamid of Great Britain Ltd]

arsenic. As; CAS 7440-38-2; EINECS 231-148-6; Metallic form; alloying additive for metals, especially lead and copper as shot, battery grids, cable sheaths; high-purity semiconductor grade: manufacture of gallium arsenide for dipoles and other electronic devices; doping agent; soliders; medicine. [Aldrich; Atomergic Chemetals; Whiting, Peter Ltd]

arsenic acid. See arsenic pentoxide

arsenic bronze. An alloy of 80% copper, 10% tin, 9.2% lead, and 0.8% arsenic.

arsenic hydride. See arsine

arsenic pentasulfide. As_2S_5; Paint pigments, light filters, other arsenic compounds. [Atomergic Chemetals]

arsenic pentoxide. (arsenic oxide; arsenic anhydride; arsenic acid) As_2O_5; CAS 1303-28-2; Arsenates, insecticides, dyeing and printing, weed killer, colored glass, metal adhesives. [Atomergic Chemetals; Spectrum Chem. Manufacture]

arsenic trichloride. (arsenic chloride; arsenious chloride; arsenous chloride; fuming liquid arsenic) $AsCl_3$; CAS 7784-34-1; Intermediate for organic arsenicals (pharmaceuticals, insecticides), ceramics. [Atomergic Chemetals; Noah Chem.]

arsenic trifluoride. (arsenious fluoride) AsF_3; CAS 7784-35-2; Fluorinating reagent, catalyst, ion implantation source, dopant. [Atomergic Chemetals; Elf Atochem N. Am.; Noah Chem.]

arsenic trioxide. (crude arsenic; white arsenic; arsenious acid; arsenious oxide; arsenous anhydride) As_2O_3; CAS 1327-53-3; EINECS 215-481-4; Pigments, ceramic enamels, aniline colors, decolorizing agent in glass, insecticide, rodenticide, herbicide, sheep and cattle dip, hide preservative, wood preservative, preparation of other arsenic compounds. [Atomergic Chemetals; Noah Chem.; Outokumpu Oy; Transene]

arsenious acid. See arsenic trioxide

arsine. (arsenic hydride) AsH_3; CAS 7784-42-1; Organic synthesis, military poison, doping agent for solid-state electronic components. [Air Prods.; Atomergic Chemetals]

Arsinette. Arsenate insecticides. [Plant Protection] *

Artane. CAS 52-49-3; Trihexylphenidyl hydrochloride; anticholinergic; antiparkinsonian. [Lederle Laboratories USA]

Artane Sustets. A proprietary preparation of benzhexol hydrochloride, in a sustained release form; used in the treatment of Parkinson's disease. [Lederle Laboratories]

Art Bronze. An alloy of 80-90% copper and 5-8% tin.

Arthripax Cream. A proprietary preparation of benzyl salicylate, glycol salicylate, terebene, menthol, ephedrine hydrochloride and capsicin; an embrocation. [Nicholas Laboratories Ltd] *

Arthrobid. CAS 38194-50-2; Sulindac; antirheumatic. [Merck & Co Inc] †

Artic. A proprietary trade name for methyl chloride used in refrigeration. §

Artic Mist. CAS 14807-96-6; Talc. [Steetley Minerals Ltd]

artificial cinnabar. See mercury sulfide (ic), red and black

artificial oil of ants. See furfural

artificial rubber. See Thinoline.

Artisil®. Pigment for coloring oils, waxes, and solvs. [Sandoz]

Artodan SP 55 Kosher. CAS 25383-99-7; EINECS 246-929-7; Sodium stearyl-2-lactylate; food emulsifier, dough conditioner, starch complexing agent, bread improver, freeze/thaw emulsions. [Grindsted Prods.; Grindsted Prods. Denmark]

Arton F. Transparent polymer with improved flowability for optical discs, optical fibers, lens applications. [Japan Synthetic Rubber]

Arton G. Transparent polymer with high heat resist. for optical discs, optical fibers, lens applications. [Japan Synthetic Rubber]

Artribid. CAS 38194-50-2; Sulindac; antirheumatic. (Portugal). [Merck & Co Inc]

Arubren®. Chlorinated paraffin plasticizer. [Bayer AG; Bayer plc]

Arvetane. A proprietary adhesive containing polyurethane. [Arveta SA] ‡

Arvin. CAS 9046-56-4; A proprietary preparation of ancrod; an anticoagulant. [Berk Pharmaceuticals Ltd] *

Arylan®. Alkaryl sulfonic acids, salts and blends with nonionics; for anionic detergency, emulsification, emulsion polymerization. [Harcros UK]

Arylan®CA. Calcium dodecylbenzene sulfonate; emulsifier for degreasers, herbicides, pesticides, waxes, hydrocarbon solvs. [Harcros; Harcros UK]

Arylan® PWS. CAS 26264-05-1; Isopropylamine dodecylbenzene sulfonate; surfactant; emulsifier for mineral oils, kerosene, waxes, and chlorinated solvs., herbicides and insecticides; for manufacturing of emulsion degreasers and kerosene-based hand cleaning gels. [Harcros; Harcros UK]

Arylan S. Anionic surfactant as pale cream flakes; primary emulsifier and wetting agent for emulsion polymerization and wettable powders. [Harcros UK] *

Arylan® SBC Acid. Straighter chain dodecylbenzene sulfonic acid; surfactant, detergent, foaming agent, emulsifier for phenolic materials; intermediate for liquid detergents, dishwash, emulsifiers; biodeg. [Harcros; Harcros UK]

Arylan®SC15. Sodium dodecylbenzene sulfonate; biodeg. wetting agent, detergent base, emulsifier for emulsion polymerization. [Harcros; Harcros UK]

Arylan®SNS. Anionic surfactant supplied as a buff powder; dispersing agent. [Harcros UK] *

Arylan® SP. Anionic surfactant in acid form, supplied as a brown viscous liquid; low free oil and inorganic content; biodegradable intermediate for detergents, especially liquids. [Harcros UK] *

Arylan® SX. Anionic surfactant in flake form; base for detergents, wetting agent for detergent powders.

---

‡ Trade name and manufacturer not verified   § Trade name without identified manufacturer

[Harcros UK] *

Arylan®TE/C. Anionic surfactant in liquid form; emulsifier for specialty waxes, chlorinated solvents. [Harcros UK]*

Arylene M40. CAS 577-11-7; EINECS 209-406-4; Dioctyl sodium sulfosuccinate; wetting agent, rewetting, dewatering surfactant, filtration aids. [Hart Chem. Ltd.]

Arylmate®. CAS 29973-13-5; Ethiofencarb; insecticide used for control of aphids. [Bayer AG]

Arylon® LP 401 NC10. Polyarylate resin, unreinforced. [DuPont; DuPont UK]

AS. CAS 9004-70-0; Nitrocellulose with nitrogen content of 11.3 to 11.7%; for coatings on cellophane and in converting operations for paper coatings. [Hercules] *

AS-10GF. CAS 9003-56-9; ABS resin, 10% glass fiber-reinforced; thermoplastic with dimensional stability and toughness; for business machine housings, cabinetry, tool housings, appliances, automotive, and construction materials. [Compounding Tech.]

AS-15CF/000. CAS 9003-56-9; ABS, 15% carbon fiber-reinforced; thermoplastic resin. [Compounding Tech.]

ASA. Alkenyl succinic anhydride; intermediate for defoamers, demulsifiers, emulsifiers, foam boosters, wetting agents, detergents, dispersants; sizing agent for paper. [Ethyl Corp] *

ASA. CAS 50-78-2; Aspirin; analgesic; antipyretic; antirheumatic. [Eli Lilly & Co] †

Asadene®. BR, styrene/butadiene rubber thermoplastic elastomers. [Asahi Chem. Industry]

Asa Dulcis. Benzoin.

Asaflex®. Transparent styrenic resin. [Asahi Chem. Industry]

asafoetida. A gum-resin; it is the dried juice of the roots of *Ferula narthex* and *F. Foetida;* a nerve stimulant and antispasmodic.

Asagran. Acetylsalicylic acid BPC granules. [Rhone-Poulenc UK]

Asahi Aji®. CAS 527-07-1; Monosodiumglutamate; for food industry. [Asahi Chem. Industry]

Asaprene®. BR, styrene/butadiene rubber thermoplastic elastomers. [Asahi Chem. Industry]

asarol. (asaronic camphor, asarum camphor). Propenyl-trimethoxybenzene, $C_6H_2(CH:CHCH_3)(OCH_3)_3$; an emetic and cathartic.

Asbesto-Wet. Blend of polyoxyethylene esters of mixed organic acids (47%) and polyoxyethylene ether of alkylated phenols (47%) containing a silicone defoamer (6%) for ease of handling; for dust control and wet removal of asbestos. [Aquatrols Corp of Am.] *

Ascabiol. CAS 120-51-4; A proprietary preparation of benzyl benzoate; used in the treatment of scabies and pediculosis. [May & Baker Ltd]

Ascinin®P, R, Special. Antiskinning and stabilizing agent used in oil based paints and varnishes. [Bayer AG; Bayer plc]

Ascon. A proprietary preparation containing dried aluminum hydroxide gel, magnesium trisilicate, and hyoscyamine hydrobromide; an antacid. [Cox Pharmaceuticals]

ascorbic acid. (1-ascorbic acid; vitamin C) $C_6H_8O_6$; CAS 50-81-7; Nutrition, color fixing, flavoring and preservative, oxidant, abscission of citrus fruit in harvesting, reducing agent. [BASF; Gist-Brocades Food Ingreds.; R.W. Greeff; Hoffmann-LaRoche; Penta Mfg.; Schweizerhall; Takeda USA]

L(+)-ascorbic acid sodium salt. *See* sodium ascorbate

Ascorbin. CAS 134-03-2; Sodium ascorbate; vitamin. [Merrell Dow Pharmaceuticals Inc] *

Ascorbosilane C. Ascorbyl methylsilanol pectinate; cosmetic ingred. for anti-aging formulations, after sun bath treatments, superficial burn treatments. [Exsymol]

Ascot. CAS 148-79-8, 137-26-8; Mixture of thiabendazole and thiram; fungicide seed dressing. [Ciba-Geigy Agrochemicals]

Asellacrin. CAS 9002-72-6; Somatropin; hormone. [Calbiochem-Behring Corp] *

Aseptisil. An alkaline bottle washing detergent. [Staveley Chemicals Ltd] *

Aseptoforms. p-Hydroxybenzoates. [R W Greeff & Co Inc]

Aserbine. A proprietary preparation of malic acid ester of propylene glycol, malic acid, benzoic acid and salicylic acid used as a desloughing agent. [Horlicks] *

Ashberry metal. (Ashbury Metal). An alloy of 80% tin, 14% antimony, 2% copper, and 1% zinc.

Ashbury Metal. *See* Ashberry Metal.

Ashlade 4% At Gran. CAS 1912-24-9; Atrazine; a residual herbicide. [Ashlade Formulations Ltd]

Ashlade 4-60 CCC. CAS 999-81-5; Soluble concentrate containing 460 g/l chlormequat; plant growth regulator. [ABM Chemicals Ltd]

Ashlade 5C. Mixture of chlormequat and choline chloride; plant growth regulator. [Ashlade Formulations Ltd]

Ashlade 700 CCC. CAS 999-81-5; Soluble concentrate containing 700 g/l chlormequat; plant growth regulator. [ABM Chemicals Ltd]

Ashlade Adjuvant Oil. Adjuvant containing 99% refined mineral oil; herbicide wetting agent. [Ashlade Formulations Ltd]

Ashlade Atrazine 50 FL. CAS 1912-24-9; Atrazine; a residual herbicide. [Ashlade Formulations Ltd]

Ashlade Blight Fungicide. Mixture of cymoxanil and mancozeb; used to control potato blight. [Ashlade Formulations Ltd]

Ashlade Cosmic FL. A suspension concentrate containing 40 g carbendazim, 320 g maneb and 90 g tridemorph per liter. systemic fungicide for cereals. [Ashlade Formulations Ltd]

Ashlade CP. Suspension concentrate containing 86 g chloridazon and 400 g propachlor per liter; a residual herbicide for beet crops. [Ashlade Formulations Ltd]

Ashlade D-Moss. Mixture of chloroxuron and ferrous sulfate; a lawn sand herbicide to control mosses in turf. [Ashlade Formulations Ltd]

Ashlade Flotin. CAS 76-87-9; Suspension concentrate containing 625 g fentin hydroxide per liter; used for control of potato blight. [Ashlade Formulations Ltd]

Ashlade Halt. CAS 66841-24-5; Emulsifiable concentrate containing 100 g cypermethrin per liter; a pyrethroid insecticide. [Ashlade Formulations Ltd]

Ashlade Linuron. CAS 330-55-3; Linuron; a residual urea herbicide for the control of weeds in field crops. [Ashlade Formulations Ltd]

Ashlade M. CAS 83601-81-4, 12427-38-2; Carbendazim + maneb; systemic fungicide for cereals. [Ashlade Formulations Ltd]

Ashlade Mancarb FL. CAS 83601-81-4, 12427-38-2; A suspension concentrate containing 50 g carbendazim and 320 g maneb per liter; systemic fungicide for cereals. [Ashlade Formulations Ltd]

Ashlade SMC. Copper oxychloride + maneb + sulfur; a

---

* Trade name not verified as available　　　　† Trade name verified as obsolete

fungicide for wheat and barley which also stimulates yields. [Ashlade Formulations Ltd]

Ashlade TCNB. CAS 117-18-0; Granules or dustable powder containing tecnazene; protectant fungicide and potato sprout suppressant. [Ashlade Formulations Ltd]

Ashland Hi-Sol 10. An aromatic hydrocarbon solvent; with high solvency for paints, varnishes, resins and insecticides. [Ashland Chemical Company] *

Ashland Hi-Sol 15. An aromatic hydrocarbon solvent; used in baked enamels and in chlorinated rubber finishes. [Ashland Chemical Company] *

Ashland Kwik-Dri. An aliphatic hydrocarbon solvent; used in cleaning compounds, waxes, polishes and as a resin solvent. [Ashland Chemical Company] *

Ashland Lacolene. An aliphatic hydrocarbon solvent; used as a diluent for lacquer. [Ashland Chemical Company] *

Ashlene®. Thermoplastic engineering resins including polyamide (nylon), polycarbonate, polyester PBT, polyphenylene oxide; used for injection molding and extrusion. [Ashley Polymers]

Ashlene® 61-2M. CAS 32131-17-2; Nylon 66 resin, 40% mineral-reinforced; offers improved dimensional stability, high stiffness, and high heat resist, improved impact; chrome plateable; for lamp and instrument housings, automotive and marine hardware, range knobs. [Ashley Polymers]

Ashlene® 520. CAS 32131-17-2; Nylon 66, slightly lubricated; general purpose economy injection molding resin. [Ashley Polymers]

Ashlene® 520-13G. CAS 32131-17-2; Nylon 66, 13% glass fiber-reinforced, lubricated; general purpose economy injection molding resin designed for parts requiring excellent thermal and dimensional stability and higher strength than conventional resins; lubricated for improved machine feed and mold release. [Ashley Polymers]

Ashlene® 520BU. CAS 32131-17-2; Nylon 66; utility grade based on reprocessed resin; injection molding. [Ashley Polymers]

Ashlene® 520MS. CAS 32131-17-2; Nylon 66, molybdenum disulfide-modified; self-lubricating economy injection molding resin with excellent wear resist. [Ashley Polymers]

Ashlene® 525-13G. CAS 32131-17-2; Nylon 66, 13% glass fiber-reinforced, lubricated; high impact. [Ashley Polymers] *

Ashlene® 527. CAS 32131-17-2; Nylon 66; economy injection molding resin; very high impact resist. [Ashley Polymers]

Ashlene® 527LD-13G; CAS 32131-17-2; Nylon 66, 13% glass fiber-reinforced, lubricated; very high impact. [Ashley Polymers]

Ashlene® 528BR-WO. CAS 32131-17-2; Nylon 66, flame retardant. [Ashley Polymers]

Ashlene® 528L-13G. CAS 32131-17-2; Nylon 66, 13% glass fiber-reinforced, lubricated; general purpose reinforced resin for parts requiring excellent thermal and dimensional stability and higher strength than conventional resins; lubricated for improved machine feed and mold release. [Ashley Polymers]

Ashlene® 541. CAS 32131-17-2; Nylon 66; extrusion grade; very high visc. for thick slab material, large rod stock and large complex profiles, pipes, and slabs. [Ashley Polymers]

Ashlene® 541S. CAS 32131-17-2; Nylon 66, heat-stabilized; extrusion grade; high visc. for sheet, film, and blown film extrusion. [Ashley Polymers]

Ashlene® 630-33G. CAS 25038-54-4; Nylon 6, 30% glass fiber-reinforced, lubricated; general purpose economy injection molding resin. [Ashley Polymers]

Ashlene® 735. CAS 25038-54-4; Nylon 6 plasticized copolymer; economy injection molding resin; improved impact resist. [Ashley Polymers]

Ashlene® 830L. CAS 25038-54-4; Nylon 6, lubricated; for improved machine feed and mold release. [Ashley Polymers]

Ashlene® 840. CAS 25038-54-4; Nylon 6; extrusion grade; high visc. general purpose for film, rigid pipe, large profile, and thick slab extrusions. [Ashley Polymers]

Ashlene® 858. CAS 25038-54-4; Nylon 6; resin with good surface finish and high strength, rigidity, toughness, and heat, abrasion and chem. resist.; for rotomolding. [Ashley Polymers]

Ashlene® 870. Amorphous nylon; general-purpose transparent nylon with good dimensional stability and chem. resist. [Ashley Polymers]

Ashlene® 980L. Nylon 6/12; general-purpose resin with low moisture absorp., excellent dimensional stability. [Ashley Polymers]

Ashlene® 980LS-40G. Nylon 6/12, short glass fiber-reinforced; general-purpose, heat-stabilized resin with low moisture absorp. and good dimensional stability. [Ashley Polymers]

Ashlene® 981S. Nylon 6/12; heat-stabilized, flexible extrusion grade for cable jacketing; low moisture absorp. and dimensional stability. [Ashley Polymers]

Asilone. Proprietary preparations containing aluminum hydroxide, magnesium oxide and activated dimethicone; antiflatulent and antacids. [The Boots Co plc]

Asilone Paediatric. A proprietary preparation containing polymethylsiloxane and carob flower; an antacid. [Berk Pharmaceuticals Ltd] *

Askure. Acid catalyst used in conjunction with furan resin binders for the cold set bonding of sand molds and cores. [Foseco (F.S.) Ltd]

Asmal. A proprietary preparation of theophylline, ephedrine; and phenobarbitone; a bronchial antispasmodic. [H N Norton & Co Ltd] ‡

Asmapax. A proprietary preparation of ephedrine resinate, theophylline and bromvaletone; a bronchial antispasmodic. [Nicholas Laboratories Ltd] *

Asmatane Mist. CAS 51-42-3; Epinephrine bitartrate; adrenergic. [3M Pharmaceuticals] †

Asma-Vydrin. A proprietary preparation of adrenaline, atropine, papaverine, pituitary extract and chlorbutol; a bronchial inhalation. [Lewis Laboratories] ‡

Asmer. Shape-memory resin. [Asahi Chem. Industry]

A-Sol. CAS 68-26-8; Vitamin A; antixerophthalmic. [The Purdue Frederick Co] *

Asp. An asbestos paper grade of Tufnol industrial laminates. [Tufnol Ltd]

ASP®. CAS 1332-58-7; Very fine to coarse particle size hydrous kaolin (aluminum silicate); used as an extender pigment in coatings, adhesives, caulks, inks, polishes, molecular sieve support, specialty ceramics. [Engelhard]

Aspac®. Polystyrene foam loose fill. [Asahi Chem. Industry]

asparagine , anhydrous D- and L-. $NH_2COCH_2CH(NH_2)$-

---

‡ Trade name and manufacturer not verified   § Trade name without identified manufacturer

COOH; CAS 2058-58-4 and 70-47-3, respectively; EINECS 218-163-3, 200-735-9, respectively; Biochemical research, preparation of culture media, medicine. [Aldrich; Degussa; Penta Mfg.; Spectrum Chem. Mfg.; Tanabe USA; U.S. Biochemical]

asparaginic acid. See aspartic acid

Aspartame. (3-amino-N-(α-methoxycarbonylphenethyl) succinamic acid; L-aspartyl-L-phenylalanine methyl ester) A sweetening agent; §

aspartic acid, D- and L-. (asparaginic acid; asparagic acid; aminosuccinic acid) COOHCH$_2$CH(NH$_2$)COOH; CAS 1783-96-6 and 56-84-8, respectively; EINECS 217-234-6 and 200-291-6, respectively; Biological and clinical studies, preparation of culture media, organic intermediate, ingredient of aspartame, detergents, fungicides, germicides, mtal complexation. [Atomergic Chemetals; Penta Mfg.; Schweizerhall; Tanabe USA; U.S. Biochemical]

Aspect® TPPE. CAS 9009-54-5; Thermoplastic polyester compounds; natural and flame-retarded grades avail.; offers easy processing, excellent toughness, outstanding mech. props. [Phillips] *

Aspellin. A proprietary preparation containing menthol, camphor, aspirin, methyl salicylate, glycerin ammonia, citronella oil and methylated spirit; a liniment. [Radiol Chemicals Ltd] *

Aspergurn. A proprietary preparation of aspirin, phenacetin, codeine phosphate, and caffeine; an analgesic. [H N Norton & Co Ltd] ‡

asphaltenes. Constituents of bitumen insoluble in hexane, but soluble in carbon tetrachloride.

Aspiquinol. An antirheumatic. [Bayer AG]

Aspirin®. Trade name for acetylsalicylic acid; an analgesic for slight to moderate pain, e.g., headache, dental and period pain, inflammation and fever, including colds; also a platelet aggregation inhibitor. [Bayer AG]

Aspisol®. An injectable analgesic. [Bayer AG]

Asplit. A range of acid and chemical resisting cements. [Prodorite Ltd]

Asplosal. See Aspirin. §

Aspon. Acid sodium orthophosphate for laundry use. [Albright & Wilson Ltd] *

Aspro. See Aspirin. §

Aspulum. A mercury derivative of chlorophenol; a seed preservative. §

Aspumit AP. Silicone emulsion; nonionic; defoamer for all textile finishing processes. [Thor Chemicals (UK) Ltd]

Aspumit SDM. Synergistic blend of deaerating products and surfactants; weakly anionic; deaerating agent and penetration accelerator for all fibers. [Thor Chemicals (UK) Ltd]

Assaf. Silicone foam control agent. [Rhone-Poulenc UK]

Asset. An emulsifiable concentrate containing 50 g benazolin, 125 g bromoxynil and 62.5 g ioxynil per liter; a post-emergence herbicide for cereal crops and grass. [Schering Agrochemicals Ltd]

Astacin® Finish PUD. Polyester-polyurethane dispersions; bottoming and top wating agent for leather and fur industry. [BASF AG]

Asterite. Filled methyl methacrylate dispersion. [ICI Chem. & Polymers Ltd]

Asterol. Proprietary antifungal preparations containing 2-dimethylamino-6-(β-diethylamino-ethoxy)-benzothiazole and its salts; a skin fungicide. [Roche Products Ltd]*

Asthmatussin. A proprietary preparation of guaifenesin, ephedrine sulfate, and phenobarbitone; a bronchial antispasmodic. [Wyeth Laboratories] *

Asthmatussin-T. A proprietary preparation of guaifenesin, ephedrine sulfate, phenobarbitone, and theophylline; a bronchial antispasolodic. [Wyeth Laboratories] *

Astick. Adhesive promoter for asphaltic shingle; shingle tab adhesive additive. [Chemseco] *

Astingol. A proprietary preparation of dimethyl phthalate and diethyl toluamide; an insect repellant. [Ayrton Saunders plc] ‡

Astix. CAS 93-65-2; Soluble concentrate containing 600 g/l mecoprop-P; used for control of weeds in undersown cereals and grassland. [Rhone-Poulenc Crop Protection Ltd]

Aston 123. Thermosetting polyamine; durable textile antistat with high resistance to laundering and dry cleaning on all substrates. [Rhone-Poulenc Surf.]

Aston RC. Special cationic surfactant in paste or liquid form; antistat for rugs; reduces resoiling after shampooing. [Millmaster-Onyx UK] *

Astra®. Dyestuff; for the paper, printing ink, surface coatings and office supplies industries. [Bayer AG; Bayer plc]

Astradur® A and T. A registered trademark for high impact PVC. §

Astraflex®. Dyestuffs; for the printing ink industry. [Bayer AG]

Astragal®. Retarders for dyeing polyacrylonitrile fibers. [Bayer AG; Bayer plc]

Astralex. A proprietary range of chemicals used in the bright plating of nickel. [Albright & Wilson Ltd] *

Astralon®. A registered trademark for PVC polymers in sheet form. §

Astrazon®. Cationic dyestuffs; for dyeing and printing polyacrylonitrile fibers. [Bayer AG; Bayer plc]

Astro Floctite. Acrylic and other polymer blends; adhesive for flocking auto parts, carpets, mats, wall plaques, assorted items. [Astro Industries Inc] *

Astrol. Bromoxynil + ioxynil + isoproturon; a contact herbicide for cereal crops. [Embetec Crop Protection Ltd]

Astrolith. A proprietary trade name for a special lithopone; a pigment. §

Astrolok. Sprayable, moisture curing adhesive; adhesive for laminates. [Apollo Chemicals Ltd]

Astro Mel. Melamine-formaldehyde resins, usually 80% solids; water resistant corrugated boxes, abrasive nonwoven pads, textile finish. [Astro Industries Inc] *

Astroplax. Sodium p- glycollylarsanilate; a hydraulic gypsum cement. [May & Baker Ltd] *

Astroturf. Polyethylene; for doormats. [Monsanto Co] *

Astrowet O-70-PG. CAS 577-11-7; Sodium dioctyl sulfosuccinate, propylene glycol; wetting, emulsifying agent; for high flash point applications. [Alco Chemical Corp]

Astrowet 0-75. CAS 577-11-7; EINECS 209-406-4; Dioctyl sodium sulfosuccinate solution; wetting agent and emulsifier. [Alco Chemical Corp]

Astrowet O-75. CAS 577-11-7; EINECS 209-406-4; Sodium dioctyl sulfosuccinate; wetting, emulsifying agent. [Alco Chemical Corp]

Astryl. Sodium p-glycollylarsanilate. [Rhone-Poulenc UK]

Astryn® 63A6-2. CAS 9003-07-0; PP, 20% mineral-reinforced; avail. in homopolymer and copolymer grades; offers high gloss, best surface quality, excellent impact resist., outstanding heat-aging resist., high solv./chem.

---

resist., excellent stress crack resist., low shrinkage; used for ABS replacement, lawn and garden tools/housings, lawn mower decks, appliance housings. [Himont] *

Astryn® 63F4-2. CAS 9003-07-0; PP homopolymer, 20% talc-reinforced; aesthetic polymer offering max. stiffness, excellent heat-aging resist., high solv./chem. resist., excellent stress crack resist., good dimensional stability, low shrinkage; injection molding grade used for automotive parts, small and large appliances, elec. parts, housewares, utility products. [Himont] *

Astryn® 65F4-4. CAS 9003-07-0; PP, 40% talc-filled; max. stiffness and high temp. performance; UL (115 C continuous use); in automotive, appliances, industrial components. [Himont] *

Astryn® 65F5-4. CAS 9003-07-0; PP, 40% calcium carbonate-filled; best high flex modulus/impact balance, good colorability, surface finish; in housewares, small appliances. [Himont] *

Astryn® 73F4-2. CAS 9003-07-0; PP copolymer, 20% talc-reinforced; injection molding resin offering outstanding heat-aging resist., high solv./chem. resist., excellent stress crack resist., low shrinkage, optimum balance of stiffness and impact resist.; used for automotive parts, small and large appliances, elec. parts, housewares, furniture, utility products. [Himont] *

Astryn® 73F5-2. CAS 9003-07-0; PP copolymer, 20% calcium carbonate-reinforced; high-impact injection molding resin offering good balance of stiffness and impact resist., outstanding heat-aging resist., high solv./chem. resist., excellent stress crack resist., low shrinkage; used for automotive parts, small and large appliances, elec. parts, housewares, furniture, utility products. [Himont] *

Astryn® 78F4-2. CAS 9003-07-0; PP copolymer, 20% reinforced; high-impact extrusion grade resin with good surface quality, outstanding heat-aging resist., high solv./chem. resist., excellent stress crack resist.; used for extruded profiles, blow-molded ducts, thermoformed trays. [Himont] *

Astryn® BA16G. CAS 9003-07-0; PP copolymer, uv-stabilized; extrusion grade offering impact resist., good dimensional stability, outstanding heat-aging resist., high solvent/chemical resist., excellent stress crack resist.; used for weatherstripping, extruded profiles. [Himont] *

Astryn® SD068-4. CAS 9003-07-0; PP, 40% calcium carbonate-reinforced; blow molding resin with high melt strength, high impact resist., high stiffness, good long-term heat aging, high solv./chem. resist., excellent stress crack resist.; used for vacuum-formed wiring channels, extruded profiles, blow-molded ducts. [Himont] *

Asulox. Selective weedkiller. [May & Baker Ltd]

Asulox. CAS 3337-71-1; A soluble concentrate containing 400 g asulam per liter; herbicide for control of docks and bracken. [Embetec Crop Protection Ltd; Rhone-Poulenc Environmental Prods. Ltd]

Asuntol®. Preparation for the control of ectoparasites, mange mites included, on all domestic animals; veterinary medicine. [Bayer AG]

A.T. 10. A proprietary preparation of dihydrotachysterol and ergocalciferolum; used in vitamin D deficiency. [Bayer AG]

AT-20GF. CAS 105-57-7; Acetal resin, 20% glass fiber-reinforced; offers lubricity and chemical and hot water resistance for automotive, hardware, plumbing applications. [Compounding Tech.]

AT 1806M; AT 4030M. CAS 24937-78-8; EVA copolymer; for hot-melt adhesives, coatings, and sealants. [AT Plastics]

Atabrine Hydrochloride. Quinacrine hydrochloride; anthelmintic; antimalarial. [Sterling Drug Inc] *

Atarax. CAS 68-88-2; A proprietary preparation of hydroxyzine; an ataractic. [Pfizer International]

Atar Phenol. A natural phenol derived by fractionation from crude tar acids; a colorless, crystalline solid at ambient temperatures, with a distinct cresylic odor; used in the manufacture of phenol-formaldehyde resins and novolacs, in disinfectants, in selective weedkillers, as a preservative, and in epoxies, e.g., Bisphenol A. [Sasolchem] *

Atasorb. CAS 1337-76-4; A proprietary preparation of activated attapulgite; an antidiarrhoeal. [Eli Lilly & Co] †

Atasorb-N. A proprietary preparation of activated attapulgite, neomycin sulfate, and pectin; an antidiarrhoeal. [Eli Lilly & Co] †

Atensine. CAS 439-14-5; A proprietary preparation of diazepam; a tranquilizer. [Berk Pharmaceuticals Ltd] *

Aterite. A nickel silver;. usually contains from 47-68% copper, 17-38% zinc, 10-14% nickel, 1.5-1.9% iron, and 0.16-2.2% manganese.

Atgard. CAS 62-73-7; A proprietary preparation of dichlorvos; an insecticide. §

Athrombin-K. Warfarin potassium; anticoagulant. [The Purdue Frederick Co] *

Atinosol. A proprietary thallium acetate solution. §

Atiran. A potato fungicide. [Plant Protection] *

Ativan Injection. CAS 846-49-1; Clear colorless solution containing 4 mg lorazepam per ml; preoperative medication before surgery or prolonged or uncomfortable procedures, e.g., bronchoscopy, endoscopy; also for acute anxiety states and for status epilepticus. [Wyeth Laboratories]

Ativan Tablets. CAS 846-49-1; Lorazepam; benzodiazepine tranquilizer used in the short-term treatment of moderate to severe anxiety. [Wyeth Laboratories]

Atlac®. Synthetic polyester resins. [Reichhold]

Atlac® 382-05A. Bisphenol A fumarate polyester resin; resin with superior resist. to hydrolysis and chem. attack, resist. to deformation in high temperatures; used in fiberglass reinforced structures, coatings, mortars in pulp/paper, metal treatment, etc. [Reichhold/Reactive Polymers] *

Atlac® 797CT. Neopentyl-chlorendic polyester resin; high thermo-oxidative resist.; provides fire protection in structures at high operating temps. and in presence of oxidizing materials; for large fabrications, e.g., stack linings and scrubber systems in power and chlor alkali industries. [Reichhold/Reactive Polymers] *

Atlacide. CAS 7775-09-9; Powder containing 58.2% w/w sodium chlorate; for total weed control for paths, drives and noncrop areas. [Chipman Ltd]

Atlacide Extra. Atrazine + sodium chlorate; used for total weed control in non crop areas. [Chipman Ltd]

Atladox HI. Soluble concentrate containing 240 g 2,4-D and 65 g picloram per liter; used to control weeds in non crop grass and grass verges. [Chipman Ltd]

Atlas®. Man-made fibers, monofilament (wire); used for ropes and other industrial uses. [Bayer AG]

---

‡ Trade name and manufacturer not verified   § Trade name without identified manufacturer

Atlas 5C Chlormequat. Soluble concentrate containing 460 g chlormequat and 320 g choline chloride per liter; plant growth regulator. [Atlas Interlates Ltd]

Atlas 10 Bronze. A proprietary trade name for an aluminum bronze containing 9.0% aluminum, 7.0% lead with copper. §

Atlas 89. A proprietary trade name for an alloy of copper with 9.0% aluminum and 1.0% iron. §

Atlas 90. A proprietary trade name for an aluminum bronze containing 90.0% copper with 10.0% aluminum. §

Atlas CIPC 40. CAS 101-21-3; Emulsifiable concentrate containing 400 g chlorpropham per liter; a carbamate herbicide. [Atlas Interlates Ltd]

Atlas D. CAS 1702-17-6; 2,4-D; Translocated herbicide for cereals and established grassland. [Atlas Interlates Ltd]

Atlas EM-2. Glycol ester; fiber lubricant, emulsifier. [Atlas Refinery]

Atlas EMJ-2. Nonylphenoxyl polyethoxy ethanol; detergent, dispersant, emulsifier, wetting agent, penetrant; grease dispersant; for leather, textile sizing, bleaching operations, paper industry. [Atlas Refinery]

Atlas JG #1. Raw, refined, and standardized fish oil; leather additive; aids fiber lubrication; masks undesirable finish odors. [Atlas Refinery]

Atlas M 130. Methacrylate-based resin filled with 60 or 85% aluminum; for thermoforming large molds, prototype injection and blow molds, soft foam molds, open mold laminating, molds for RIM and other urethane processes, fiber-reinforced processes. [Degussa]

Atlas MCPA. CAS 94-74-6; MCPA; herbicide for cereals and grassland. [Atlas Interlates Ltd]

Atlas Adherbe®. Crop chemical enhancer/additive. [Allied Colloids Ltd]

Atlas Adherbe. Adjuvant containing 83% refined mineral oil; wetting agent for herbicides. [Atlas Interlates Ltd]

Atlas Adjuvant Oil. Adjuvant containing 95% refined mineral oil; wetting agent for herbicides. [Atlas Interlates Ltd]

Atlasbeam #1. Odorless dehairing assistant for leathers. [Atlas Refinery]

Atlas Brown. Pesticides for vegetables. [Allied Colloids Ltd]

Atlas Brown. Emulsifiable concentrate containing 150 g chlorpropham and 100 g pentanochlor per liter; a residual herbicide. [Atlas Interlates Ltd]

Atlas Chlormequat 46, 700. CAS 999-81-5; Chlormequat; plant growth regulator. [Atlas Interlates Ltd]

Atlas Defoamer AFC. Hydrophobized silicone; defoamer for textile and leather industries. [Atlas Refinery]

Atlas Electrum. Pesticides for sugar beet. [Allied Colloids Ltd]

Atlas Electrum. Suspension concentrate containing 200 g chloridazon, 30 g chlorpropham, 20 g fenuron and 120 g propham per liter; a residual herbicide for beet crops. [Atlas Interlates Ltd]

Atlas Gold. Pesticides for sugar beet. [Allied Colloids Ltd]

Atlas Gold. Suspension concentrate containing 37.5 g chlorpropham 25 g fenuron and 150 g propham per liter; an herbicide for use on beet crops. [Atlas Interlates Ltd]

Atlas Indigo. Pesticides for vegetables. [Allied Colloids Ltd]

Atlas Indigo. Mixture of chlorpropam and propham; plant growth regulator to suppress sprout growth in stored potatoes. [Atlas Interlates Ltd]

Atlas Leather Odor. Compounded. masking agent exhibiting a traditional leather aroma. [Atlas Refinery]

Atlas Libsorb. Nonionic spreader containing 900 g/l alkyl alcohol ethoxylate; wetting agent for herbicides. [Atlas Interlates Ltd]

Atlas Lignum. Pesticides for forestry and amenity products. [Allied Colloids Ltd]

Atlas Lignum Granules. Atrazine + dalapon; granular soil acting herbicide for use in forestry plantations. [Atlas Interlates Ltd]

Atlas Linuron. CAS 330-55-3; Linuron; a residual urea herbicide for the control of weeds in field crops. [Atlas Interlates Ltd]

Atlasol 103. Sodium decyl sulfate; emulsifier, wetting agent, dispersant, fiber lubricant, synthetic fatliquor; for textile, leather, and general industrial applications. [Atlas Refinery]

Atlasol 118-U. Sulfated neatsfoot oil, hydrocarbon, relatively high boiling solv.; fatliquor for leather, especially white leather. [Atlas Refinery]

Atlasol 170. Emulsified neatsfoot oil, refined coconut oil, and synthetic oils; fatliquor for snow white alum tanned leather. [Atlas Refinery]

Atlasol 177. Sulfonated oil, fatty alcohols, and synthetic lubricants; fatliquor for light, fluffy leathers. [Atlas Refinery]

Atlasol 178. Sulfated neatsfoot oil, sperm oil; lightfast fatliquor for upper leathers. [Atlas Refinery] *

Atlasol BSC. Sulfonated sperm oil and chlorinated ester; fatliquor for leather including hair-on skins; chrome and alum stable. [Atlas Refinery]

Atlasol CSN. Modified chlorosulfonated hydrocarbon; fatliquor for leathers, especially for white or pale colored leather where lightfastness and prolonged heat stability are important. [Atlas Refinery]

Atlasol KAD. CAS 95-19-2; EINECS 202-397-8; Stearic imidazoline; lightfast fiber lubricant, fatliquor, and softener for textile and leather applications. [Atlas Refinery]

Atlasol KMM. Ricinoleic acid, triethanolamine salt; leather tanning surfactant, lubricant. [Atlas Refinery]

Atlas Orange. CAS 1918-16-7; Propachlor; a pre-emergence herbicide for various horticultural crops. [Atlas Interlates Ltd]

Atlas Pink C. Pesticides for vegetables. [Allied Colloids Ltd]

Atlas Pink C. Suspension concentrate containing 25 g chlorpropham, 6 g diuron and 100 g propham per liter; an herbicide for use on outdoor lettuce. [Atlas Interlates Ltd]

Atlas Protrum®K. Pesticides for sugar beet. [Allied Colloids Ltd]

Atlas Red. Pesticides for vegetables. [Allied Colloids Ltd]

Atlas Red. Suspension concentrate containing 200 g chlorpropham, cresylic acid and 50 g fenuron per liter; an herbicide for use on ornamentals and vegetables. [Atlas Interlates Ltd]

Atlas Sheriff®. Pesticides for vegetables. [Allied Colloids Ltd]

Atlas Sheriff. CAS 2921-88-2, CAS 60-51-5; Mixture of chlorpyrifos and dimethoate; systemic and fumigant insecticide for brassica crops. [Atlas Interlates Ltd]

Atlas Silver. CAS 1698-60-8; Suspension concentrate containing 430 g chloridazon per liter; a pyridazinone herbicide for beet crops. [Atlas Interlates Ltd]

Atlas Solan. Pesticides for vegetables. [Allied Colloids Ltd]

Atlas Steel. A proprietary hot die steel containing 9-11% tungsten, 3.25-3.5% chromium, and a little vanadium. §

Atlas Steward®. Herbicide for cereals. [Allied Colloids Ltd]

Atlas Steward. Suspension concentrate containing 560 g γ-

---

* Trade name not verified as available

† Trade name verified as obsolete

HCH per liter; an organochlorine insecticide. [Atlas Interlates Ltd]

Atlastan AR. Low molecular weight acrylic polymer; tanning agent. [Atlas Refinery]

Atlastan LC. Aromatic carboxylic acid deriv.; auxiliary synthetic tanning agent for chrome leathers. [Atlas Refinery]

Atlas Tanked Cod Oil. CAS 8001-69-2; Cod oil; leather additive providing fullness and mellowness to vegetable tanned sole leather, shoe linings, shoe uppers; ingred. in stuffing compounds and for chamois leather processing. [Atlas Refinery]

Atlas Tecgran. Pesticides for vegetables. [Allied Colloids Ltd]

Atlas Total A, Total S. Pesticides for forestry and amenity products. [Allied Colloids Ltd]

Atlavar. Atrazine + 2,4-D + sodium chlorate; used for total weed control in non crop areas. [Chipman Ltd]

Atlazin. CAS 1912-24-9, 61-82-5; A liquid formulation containing 250 g atrazine and 218 g aminotriazole per liter as a suspension concentrate; used for total weed control on industrial sites, paths, kerbs and channels, drives and hard tennis courts, hardstanding and storage areas. [Chipman Ltd]

Atlox. Blends of anionic and ionic surfactants. [ICI Am.]

Atmer®. Surfactants. [Imperial Chemical Industries plc]

Atmer® 100. CAS 1338-39-2; EINECS 215-663-3; Sorbitan laurate; antifog agent for PE and EVA food-wrapping films; antistat; cling additive. [ICI Polymer Additives]

Atmer® 103. Sorbitan ester; antifog agent for long-lasting properties in LDPE, EVA, and PVC agricultural film; wetting agent for PP and PE films. [ICI Polymer Additives]

Atmer® 105. CAS 1338-43-8; EINECS 215-665-4; Sorbitan oleate; antifog, antistat, cling additive for low-temp. LDPE film applications. [ICI Polymer Additives]

Atmer® 106. CAS 26266-58-0; EINECS 247-569-3; Sorbitan trioleate; antifog, cling additive for LDPE film. [ICI Polymer Additives]

Atmer® 121. Glyceryl oleate; antifog agent and cling additive for PVC food-wrapping film. [ICI Polymer Additives]

Atmer® 122. Glyceryl stearate; antistat for LDPE, PP, flex, PVC; antifog. [ICI Polymer Additives]

Atmer® 1007. Glyceryl oleate; antifog, cling agent for LDPE and PVC films. [ICI Polymer Additives]

Atmer® 7001. 50% Atmer 129/163 (2:1) in PP; antistat for injection molding PP. [ICI Polymer Additives]

Atmer® 8112. 20% in PE; antifog for film. [ICI Polymer Additives]

Atmido. A siliceous earth; used as a filtering medium, also as a rubber filler. §

Atmos. Glycerine fatty acid esters. [ICI Am.]

Atmos 150. Glyceryl stearate; food emulsifier; antistat for plastics (PP, PS) useful in food pkg. [Witco/Humko]

Atmos 300. Glyceryl oleate; food emulsifier. [Witco/Humko]

Atmos 659 K. Blend of propylene glycol mixed esters, mono/diglycerides and lecithin; food emulsifier for cakes. [Witco/Humko]

Atmul 80. Mono- and diglycerides; food emulsifier. [Witco/Humko]

Atmul 124. Glyceryl monoester; antistat for PP and PS useful in food pkg. [Witco/Humko]

Atmul 2622 K. Glyceryl-lactostearate; lipophilic food emulsifier for aerating applications. [Witco/Humko]

Atolex ASL/C. Cationic surfactant in liquid form; lubricating and antistatic agent. [Standard Chemical Company]

Atolex ASL/C100. Cationic surfactant in the form of a thick liquid; antistatic lubricant. [Standard Chemical Company]

Atolex DA/25. Naphthalene sulfonate in liquid form; anionic dispersing liquid and leveling assistant particularly for disperse or acid dyes on acrylics or polyester. [Standard Chemical Company]

Atolex Polythene Emulsions. Full range of cationic surfactants in liquid form; softening and lubricating agents used in additives to resin finishes. [Standard Chemical Company]

Atolex QE. Cationic surfactant in liquid form; retarding agent. [Standard Chemical Company]

Atomite. CAS 471-34-1; Fine ground calcium carbonate. [Thompson, Weinman & Co] *

Atomite®. CAS 1317-65-3; Calcium carbonate; high brightness, controlled particle size, easy dispersing grade for water and solv.-based coatings, rubber, plastics, caulks, sealants, adhesives, etc. [ECC]

Atomol. Nasal decongestant with prolonged action. [Allen & Hanburys Ltd] *

Atonin. See aspirin. §

Atophan. 2-Phenylquinoline-4-carboxylic acid; used in the treatment of gout and sciatica. §

atoquinol. Allylphenylcinchoninic ester. A powerful uric acid solvent and eliminator.

Atosil®. CAS 58-33-3; Promethazine hydrochloride; a phenothiazine derivative with sedative effect on autonomic nervous system and antihistaminic properties. [Bayer AG]

Atpet. Range of surfactant properties including emulsifying, wetting, dispersing, antifoam, antirust. [ICI Am.]

Atpet 300. CAS 25322-68-3; Polyethylene glycol; pharmaceutical aid; ointment base; suppository base; solvent; tablet excipient; tablet and/or capsule lubricant. [ICI Am.]*

Atpet 400. CAS 25322-68-3; Polyethylene glycol; ointment base; suppository base; solvent; tablet excipient; tablet and/or capsule lubricant. [ICI Am.]

Atpet 600. CAS 25322-68-3; Polyethylene glycol; Pharmaceutical aid; ointment base; suppository base; solvent; tablet excipient; tablet and/or capsule lubricant. [ICI Am.]*

ATP Nucleotides. Propylene glycol, collagen amino acids, adenosine triphosphate; moisturizer. [Croda Inc.]

Atprime®. Bonding agent for reinforced plastic. [Reichhold]

Atraflow. CAS 1912-24-9; A soluble concentrate containing 500 g atrazine per liter; a residual herbicide. [Rhone-Poulenc Environmental Prods. Ltd]

Atraflow Plus. CAS 1912-24-9, 61-82-5; A liquid formulation containing 264g atrazine and 214g aminotriazole per liter as a suspension concentrate; for used where total weed control is required including industrial sites, paths, curbs, and channels, drives and hard tennis courts, hardstanding and storage areas. [Burts & Harvey]

Atraflow Plus. CAS 1912-24-9, 61-82-5; A liquid formulation containing 270 g atrazine and 160 g aminotriazole per liter as a suspension concentrate; for use where total weed control is required including industrial sites, paths, curbs, and channels, drives and hard tennis courts, hardstanding and storage areas. [Rhone-Poulenc Environmental Prods. Ltd]

---

‡ Trade name and manufacturer not verified

§ Trade name without identified manufacturer

Atragan. Herbicide. [Agan Chemical Manufacturers Ltd]

Atramentum Stone. A mixture of ferric and ferrous sulfates with ferric oxide; used in the manufacture of inks.

Atramet Combi. Active ingredients: atranex plus ametrex; ready formulated mixture of atrazine plus ametryne for use as a selective pre- and post-emergence herbicide. [Agan Chemical Manufacturers Ltd]

Atranex. CAS 1912-24-9; Active ingredient: atrazine; 2-chloro-4-ethylamino-6-isopropylamino-1,3,5-triazine; pre- and post-emergence herbicide. [Agan Chemical Manufacturers Ltd]

Atrinal. CAS 52508-35-7; Soluble concentrate of 200 g dikegulac per liter; plant growth regulator for hedges. [Rhone-Poulenc Environmental Prods. Ltd]

Atrixo. A silicone hand cream. [Smith and Nephew Pharmaceuticals Ltd] ‡

Atromid-S. CAS 637-07-0; A proprietary preparation of clofibrate; used to reduce blood cholesterol levels. [ICI Chem. & Polymers Ltd]

Atropisol. Atropine sulfate; anticholinergic. [CooperVision Inc] ‡

Atropisol® Ophthalmic Solution. Atropine sulfate; indicated for the treatment of inflammatory conditions of the iris or uveal tract and also as a cycloplegic or mydriatic for refraction. [Iolab Corp]

Atroscine. Optically inactive scopolamine (dl-hyoscine). §

Atrovent. CAS 22254-24-6; Ipratropium bromide; an anticholinergic bronchodilator. [Boehringer Ingelheim Ltd]

atroxindol. The anhydride of o-amino-α–phenylpropionic acid, $C_6H_4(NH)(CH\ CH_3)CO$.

Atrust. Rust converter. [Imperial Chemical Industries plc]

A/T/S. Erythromycin; antibacterial. [Hoechst-Roussel Pharmaceuticals Inc] *

Attaclay. CAS 1337-76-4; Fine particle size sorbent attapulgite (magnesium aluminum silicate); used as a chemical conditioning agent (free-flow agent); for prilled ammonium nitrate and urea fertilizers and other bulk granular chemicals. [Engelhard]

Attacote. CAS 1337-76-4; Fine particle size sorbent attapulgite (magnesium aluminum silicate); used as a chemical conditioning agent (free-flow agent); for fire extinguishing chemicals. [Engelhard]

Attaflow. CAS 1337-76-4; Liquid (slurry) attapulgite (hydrous magnesium aluminum silicate); used as a suspending agent for liquid suspension fertilizers, flowable pesticides. [Engelhard]

Attagel. CAS 1337-76-4; Colloidal attapulgite (hydrous magnesium aluminum silicate); used as a thixotropic agent, gellant and suspending aid for paints, coatings, adhesives, inks, caulks and liquid suspension fertilizers. [Engelhard]

Attane. Ultra-low density polyethylene copolymer. [Dow UK]

Attane 4601, 4802. CAS 9002-88-4; Ultra low dens. LLDPE; blown film extrusion resin. [Dow Plastics]

attapulgite. (Fullers earth; palygorskite; dioctrahedral smectite) CAS 1337-76-4; A hydrated aluminum-magnesium silicate, chief ingredient in Fullers earth; drilling fluids, decolorizing oils, filter medium, absorbent, adsorbent. [Bromhead & Denison Ltd]

Attapulgus. CAS 1337-76-4; Colloidal attapulgite (hydrous magnesium aluminum silicate); used as suspending agent for oil well drilling clay, particularly salt water formations. [Engelhard]

Attasorb. CAS 1337-76-4; Fine particle size sorbent attapulgite (magnesium aluminum silicate); used as a chemical conditioning agent (free-flow agent) for powdered detergents, agricultural chemicals. [Engelhard]

Attenuvax. Attenuated line of measles virus derived from Enders' attenuated Edmonston strain; measles vaccine. [Merck & Co Inc]

Aturbane. A proprietary preparation of phenglutarimide hydrochloride; used in Parkinsonism. [Ciba plc] *

aubepine. Anisaldehyde; used in perfumery.

Audax. A proprietary preparation of choline salicylate, ethylene oxide polyoxy propylene condensate; analgesic ear drops. [Napp Laboratories Ltd] *

Audicort. A proprietary preparation of triamcinolone acetonide, neomycin undecylenate and benzocaine; used as ear-drops in the treatment of otitis externa. [Lederle Laboratories] *

Audrey. Automatic dielectric analyzer. [Tetrahedron Association Inc] *

Auel solder. An alloy of 63% tin, 35% zinc, 1.7% copper, and 0.3% aluminum.

Augmentin. A proprietary preparation of amoxycillin with potassium clavulanate; an antibiotic. [SmithKline Beecham]

Augsburg metal. A brass containing 72% copper, and 28% zinc.

Aunativ. Solution for injection; human immunoglobulin antihepatitis B. [KabiVitrum AB] *

Auracet. A proprietary preparation of aluminum acetotartrate and lead subacetate; ear drops. [Octel Chemicals Ltd] †

Auracryl. Aqueous color dispersions for use in water-borne industrial coatings. [Engelhard]

Auralgan. Glycerol containing phenazone and benzocaine; ear drops to relieve pain and inflammation of acute otitis externa and facilitate the removal of ear wax. [Wyeth Laboratories] †

Auralgicin. A proprietary preparation of ephedrine, benzocaine, chlorbutol, potassium hydroxyquinoline sulfate, phenazone and glycerine. ear drops. [Fisons plc, Pharmaceuticals Div] *

Aurantine. The trade name for osage orange extract (from the bark of a shrub); used in the textile industry for tanning. §

Aurantine. A residue containing terpenes left behind in the refining of orange oil; used as a perfume for soaps. §

Aurantiol®. Schiff base of hydroxycitronellal and methyl anthranilate; fragrance; sweetly floral. [BASF AG]

Aurasperse. Aqueous color dispersions for use in water based coatings. [Engelhard]

Aureocort. A proprietary preparation of triamcinolone acetomide and chlortetracycline hydrochloride; used in the treatment of skin disorders. [Lederle Laboratories] *

Aureomycin. CAS 64-72-2; Chlortetracycline hydrochloride; antibacterial; antiprotozoal. [Lederle Laboratories USA] *

auric chloride. See gold chloride

Aurocyanase. A colloidal gold and potassium double cyanide.

Auromet 55. A proprietary trade name for an alloy containing 76-80% copper, 10-12% aluminum, 4-6% iron, and 4-6% nickel. §

aurum. (gold) Au.

Aurum Series. Thermoplastic polyimide resin; super heat-resistant resin for injection and extrusion molding; radia-

---

* Trade name not verified as available   † Trade name verified as obsolete

tion resistant, good chem. and oil resist., elec. props. and weatherability; for mech. and precision parts, elec./electronic parts, automotive parts, wire extrusion coating, film, fiber. [Advanced Web Prods.]

Aurum 400, 450, 500. Thermoplastic polyimide resin; high med. and low flow non-filled grades. [Advanced Web Prods.]

Aurum JAF 3040. Thermoplastic polyimide resin, aramid-fiber filled; low wear, high PV value grade. [Advanced Web Prods.]

Aurum JGN 3030. Thermoplastic polyimide resin, carbon fiber filled; high modulus, high strength, low wear grade. [Advanced Web Prods.]

Aurum JCN 6030. Thermoplastic polyimide resin, carbon fiber filled; high flow, high modulus, high strength reinforced grade. [Advanced Web Prods.]

Aurum JGN 3030. Thermoplastic polyimide resin, glass fiber filled; electricaly insulated, high modulus reinforced grade. [Advanced Web Prods.]

Aurum JNF 3010. Thermoplastic polyimide resin, fluoropolymer filled; low friction reinforced grade. [Advanced Web Prods.]

Aurum JNF 3020. Thermoplastic polyimide resin, fluoropolymer filled; low friction, high PV value grade. [Advanced Web Prods.]

Aurum JQF 3025. Thermoplastic polyimide resin, fluoropolymer filled; low friction, med. modulus grade. [Advanced Web Prods.]

Aurum JRF 3025. Thermoplastic polyimide resin, graphite filled; low friction, low wear grade. [Advanced Web Prods.]

Aurum JRN 3015. Thermoplastic polyimide resin, graphite filled; low wear grade. [Advanced Web Prods.]

austenite. A characteristic constituent of very highly carbonized steel, containing more than 1.1% carbon.

Australian gold. A gold-silver alloy containing 8.33% silver; used for coinage.

Austrapol. Styrene/butadiene and polybutadiene polymers; used for tires, retread, general rubber goods and plastics modification. [Australian Synthetic Rubber Co Ltd]

Austratex. Styrene-butadiene high solids latex; used for carpet underlay foams, bitumen modification. [Australian Synthetic Rubber Co Ltd]

Austrostab. A full range of stabilizer systems for PVC, containing stabilizers (Pb, Pb/Ba, Cd, Ba/Cd, Ca/Zn), lubricants, pigments, fillers, and modifiers; completely ready-for-use formulations for applications such as PVC pipe, cable, profile. [Chemtech (Crop Protection) Ltd] *

Austrox. Melted lead oxide granules; used as raw material for specialty glasses such as crystal glass and glass for electronic tubes. [Chemson Polymer Additiv GmbH] *

AuSub. Gold substitute inorganic die attach pastes; for electronic applications in computer industry and for military and aerospace uses. [Johnson Matthey plc]

Autan®. Consumer insect repellent against mosquitoes, gnats and other flying insects. [Bayer AG]

Auto Command®. For automotive industry. [DuPont UK]

Autofroth. CAS 9009-54-5; Polyurethane foam; rigid foam systems (forth-in-place) for insulation, flotation, molding. [Olin]

Autogal. A trademark for a flux used for soldering and welding aluminum; a mixture of the halogen salts of the alkali metals. §

Automate. Liquid dyestuffs for petroleum. [Morton Int'l. Ltd]

automolite. A mineral, ZnO Al$_2$O$_3$.

Autopak. CAS 9009-54-5; Polyurethane foam; rigid/flexible foam system (pour-in-place) for pre-mold pkg. [Olin]

Autopoon Nl. Cationic surfactants, solvs., and solubilizers; water-repellent conc. for preps. for lacquered surfaces, car finishing. [Zschimmer & Schwarz]

Autopour. CAS 9009-54-5; Polyurethane foam; pressurized rigid foam system (pour-in-place) for insulation, molding applications. [Olin]

Autopur WK 4121. Surfactant blend; basic material for car shampoo formulations with water-repellent and gloss effects. [Zschimmer & Schwarz]

Autovisuel®. For automotive industry. [DuPont UK]

Autumn Kite. CAS 34123-59-6, 1582-09-8; Emulsifiable concentrate of 300 g isoproturon and 200 g trifluralin per liter; used for control of annual grasses in winter wheat and barley. [Schering Agrochemicals Ltd]

Autumn Lawn Food. Lawn fertilizer. [Fisons plc, Horticulture Div] *

Auxiliary PR-10BT. Antiwicking agent and thickener for textile pigment printing. [Catawba-Charlab]

Avabond. Adhesives for packaging, woodworking, textiles etc. [Avalon Chemical Co Ltd]

Avadex. 2,3-Dichloroallyl diisopropyl thiocarbamate; herbicide for wild oats. [Monsanto Co; Monsanto plc]

Avadex® BW. Herbicide with tri-allate as active ingred.; for control of wild oat, slender foxtail and bent grass in sugar beet and feed turnips, summer and winter barley, winter rye. [BASF AG]

Avadyne AV1200/CA100. Two-part urethane prepolymer emulsion adhesive; for lamination of film-to-film and film-to-metallized film structures. [Pierce & Stevens]

Avalon. Thermoplastic polyurethanes for injection molding and as adhesives and coatings. [Avalon Chemical Co Ltd]

Avamid. Polyimide prepregs. [DuPont UK]

Avamid 150. Avocadamide DEA, avocado oil; biodeg. SE foam stabilizer, visc. builder, lubricant for conditioning shampoos, hair rinses, creams and lotions; imparts smooth, silky feel to skin and hair. [Mona Industries]

Avanel® S-30. Sodium C12-15 pareth-3 sulfonate; biodeg. mild surfactant, emulsifier for personal care, household, institutional and industrial products; stable in presence of hypochlorite and over entire pH range. [PPG/Specialty Chem.]

Avantine. CAS 67-63-0; EINECS 200-661-7; A brand of isopropyl alcohol; an anesthetic. [Laporte Industries Ltd]*

Avatec (as sodium). Lasalocid; coccidiostat. [Hoffmann-LaRoche Inc] *

Avaunt®. For agriculture industry. [DuPont UK]

Avazyme. Chymotrypsin; enzyme. [Wallace Laboratories]*

Avecolite. A proprietary phenol-formaldehyde synthetic resin. §

Avenge 2. CAS 43222-48-6; Soluble concentrate of 150 g difenzoquat per liter; used for control of wild oats in cereals. [Cyanamid of Great Britain Ltd; Schering Agrochemicals Ltd]

Aventox SC. Simazine + trietazine; herbicide for peas and beans. [DowElanco Ltd]

Aventyl. A proprietary preparation of nortriptyline hydrochloride; an antidepressant. [Eli Lilly & Co] †

Aventyl Hydrochloride. Nortriptyline hydrochloride; antidepressant. [Eli Lilly & Co]

---

‡ Trade name and manufacturer not verified   § Trade name without identified manufacturer

Aveonal. An alloy of aluminum with 4% copper, 0.05% magnesium, 0.05% manganese, and 0.05% silicon.

Aversin. Paraffin wax emulsion with zirconium salts; non-permanent water-repellent finish compatible with resins, suitable for all kinds of fibers. [Henkel Chemicals Ltd] *

Avertin. A proprietary preparation of bromethol; used to control eclampsia in toxemia of pregnancy. [Bayer AG]†

Avgard. CAS 7601-54-9; Trisodium phosphate; food additive for meat, poultry, and seafood industries. [Rhone-Poulenc Food Ingreds.]

Avgard. Antimisting kerosene. [ICI Chem. & Polymers Ltd]†

Avialite. A proprietary trade name for an alloy of copper with about 9.0% aluminum and 1.0% iron. §

Aviamide-6. Policapram; pharmaceutic aid. [Avicon Inc] *

Aviashine. A blend of solvents, carriers, abrasives and waxes; aircraft maintenance chemical; provides effective cleaning for paintwork, chrome and other metal surfaces; can be polished to give a durable bright and protective finish. [The Kent Chemical Company Ltd]

Aviawash. A blend of detergents and surfactants with inhibitors; aircraft maintenance chemical for aircraft exterior cleaning; for cleaning painted and unpainted external surfaces of aircraft and ground equipment. [The Kent Chemical Company Ltd]

Avicel. Microcrystalline cellulose; pharmaceutic aid. [FMC; FMC UK] *

Avicell-RC. A proprietary trade name for a chemically pure colloidal cellulose; forms thixotropic dispersions both mechanically and thermally stable; edible and metabolically alert; used as a thickening agent. [BP Chemicals]*

Aviester. Pegoterate; pharmaceutic aid. [Avicon Inc] *

Avilon. Metal complex dyes. [Ciba plc] *

Avional D. An alloy of aluminum with 3.9% copper, 0.5% nickel, 0.5% magnesium, 0.55% silicon, and 0.3% iron.§

Avirol. Textile and leather auxiliary. [Hickson & Welch Ltd]

Avirol® 125 E. Sodium alkyl ether sulfate; emulsifier for vinyl acetate copolymers, S/B latexes, vinyl chloride copolymers, acrylate homo- and copolymers. [Henkel/Functional Prods.]

Avirol® A. CAS 2235-54-3; EINECS 218-793-9; Ammonium lauryl sulfate; emulsifier for emulsion polymerization; additive for mech. latex foaming; foaming agent for acrylate disps., carpet and upholstery cleaners; air entraining agent for mortars. [Henkel/Functional Prods.]

Avirol® AE 3003. CAS 67762-19-0; Ammonium laureth sulfate; emulsifier for vinyl acetate copolymers, S/B latexes, vinyl chloride copolymers, acrylate homo- and copolymers. [Henkel/Functional Prods.]

Avirol® FES 996. CAS 9004-82-4; Sodium laureth sulfate; emulsifier for vinyl acetate copolymers, S/B latexes, vinyl chloride copolymers, acrylate homo- and copolymers. [Henkel/Functional Prods.]

Avirol® SA 4106. Sodium 2-ethylhexyl sulfate; wetting agent, stabilizer for plastics, rubber, adhesives, food contact paper; coemulsifier for vinyl chloride, acrylics, vinyl acetate copolymers; biodeg. [Henkel/Functional Prods.]

Avirol® SA 4108. Sodium n-octyl sulfate; emulsifier, low foaming surfactant, wetting agent. [Henkel/Functional Prods.]

Avirol® SA 4110. Sodium n-decyl sulfate; wetting and emulsifying agent for plastics. [Henkel/Functional Prods.]

Avirol® SA 4113. CAS 3026-63-9; Sodium tridecyl sulfate;

emulsifier for S/B and vinyl chloride copolymers. [Henkel/Functional Prods.]

Avirol® SE 3002. CAS 9004-82-4; 1335-72-4; Sodium laureth sulfate; emulsifier for vinyl acetate copolymers, S/B latexes, vinyl chloride copolymers, acrylate homo-and copolymers. [Henkel/Functional Prods.]

Avirol® T 40. CAS 139-96-8; EINECS 205-388-7; TEA lauryl sulfate; emulsifier for emulsion polymerization; additive for mech. latex foaming; foaming agent for acrylate dispersions, carpet and upholstery cleaners; air entraining agent for mortars. [Henkel/Functional Prods.]

Aviso® G. Metiram, cymoxanil; fungicide for potatoes, vines, and other crops. [BASF AG] *

Avisol. Neutral soluble sulfathiazole for poultry. [May & Baker Ltd] *

Avistin®. A range of basic fatty acid condensation products; used for the manufacture of cationic preparing and finishing agents for textiles. [Hüls AG]

Avistin® PN. CAS 141-21-9; EINECS 205-469-7; Stearamidoethyl ethanolamine; for production of cationic textile auxiliary agents. [Hüls Am.; Hüls AG]

A-Vitan. Vitamin A; antixerophthalmic. [Janssen Pharmaceutica Inc] *

Avitex. Surface active agents. [DuPont UK]

Avitige®. Fiber. [DuPont UK]

Avitone®. Dyeing assistants. [DuPont UK]

Avitone® A. Sodium alkyl sulfonate; finishing agent, softener, lubricant for improving texture and hand of textiles, leather, and paper; also for elastomers; highly stable to chemicals and oxidation. [DuPont]

Avitrol. CAS 504-24-5; 4-Aminopyridine; treated grain baits for pest bird control; classified as a 'restricted use' pesticide. [Avitrol Corporation]

Avivage. Combination of sulfated fats with special additives; softener, stabilizer in bleaching liquors. [Henkel Chemicals Ltd] *

Avivan. Finishing agent. [Ciba plc] *

Avloclor. CAS 50-63-5; A proprietary preparation of chloroquine phosphate; antimalarial. [ICI Chem. & Polymers Ltd]

Avlosulfon. A proprietary preparation of dapsone; used in the treatment of leprosy. [ICI Chem. & Polymers Ltd]

Avoca. CAS 7761-88-8; Toughened silver nitrate; used for removal of warts and granulation tissue, for cautery, as a caustic when applied to mucous membrane; to be used only under medical supervision. [Bray Health & Leisure Ltd]

Avocado Oil CLR. CAS 8024-32-6; Avocado fatty oil; emollient; conditioner for skin and hair care preparations. [Dr. Kurt Richter; Henkel/Cospha]

Avoilefin. Polypropene 25; pharmaceutic aid. [Avicon Inc]*

Avolan®. Leveling agents for wool; brightening agents to correct faulty dyeings on wool; dispersant for dyeing polyester fibers. [Bayer AG; Bayer plc]

Avomine. CAS 17693-51-5; A proprietary preparation of promethazine theoclate (promethazine 8-chlorotheophyllinate); an anti-emetic. [May & Baker Ltd]

avoparcin. A growth promoter; a glycopeptide antibiotic obtained from cultures of Streptomyces candilus.

Avotan. CAS 37332-99-3; Avoparcin; antibacterial. [Lederle Laboratories USA] *

Avron. Acrylic dispersion. [ICI Chem. & Polymers Ltd]

AVT-75. CAS 124-68-5; 2-Amino-2-methyl-1-propanol; boiler water treatment chemical, corrosion inhibitor, car-

---

* Trade name not verified as available   † Trade name verified as obsolete

bon dioxide absorber. [Angus]

Axall. Herbicide. [May & Baker Ltd] *

Axarel®. For electrical industry. [DuPont UK]

Axelcon. Thermoplastic lubricant concentrates available in various carrier resins for specific applications; high lubricant concentrations based on the Mold Wiz internal lubricants with a complex polymeric base. [Axel Plastics Research Laboratories Inc]

Axelglo. Proprietary, polymeric based polishes to maintain and renew fiberglass and metal (painted and unpainted) surface luster; used for molded fiberglass products (yachts, camper tops etc) and automotive. [Axel Plastics Research Laboratories Inc]

axerophthol. See vitamin A

Axetin. A proprietary preparation of hydroxizine hydrochloride, ephedrine sulfate and theophylline; an antiasthmatic. [Pfizer International] †

Axid. CAS 76963-41-2; Nizatidine; anti-ulcer treatment. [Eli Lilly & Co]

Axiquel. CAS 4171-13-5; Valnoctamide; tranquilizer. [McNeil Pharmaceuticals] *

Axite. An explosive; smokeless powder which contains guncotton, nitro-glycerine, petroleum jelly, and a little potassium nitrate. §

Axol® C 62. CAS 68990-05-6; Citric acid ester of glycerol mono/distearates; food emulsifier. [Goldschmidt AG]

Axol® E 61. CAS 8029-91-2; Acetylated hydrogenated lard glyceride; food emulsifier, lubricant, solv., plasticizer and coating material for foodstuffs and cosmetics. [Goldschmidt AG]

Axol® L 61, L62. CAS 689990-06-7; Lactic acid fatty acid glyceride; food emulsifier. [Goldschmidt AG]

Axoridin. Latamoxef; semi-synthetic broad-spectrum beta-lactam antibiotic for parenteral administration. [Merck & Co Inc] *

Aygestin. CAS 51-98-9; Norethindrone acetate; progestin. [Wyeth Laboratories]

Ayrtol. CAS 88-04-0; A proprietary preparation of chloroxylenol; a disinfectant. [Ayrton Saunders plc] ‡

AZ. Toothpaste to help fight plaque and cavities. [Richardson-Vicks Inc] *

Azactam. CAS 78110-38-0; A proprietary preparation of aztreonam; an antibiotic. [Bristol-Myers Squibb Pharmaceuticals Ltd] *

azacyclonal. Diphenyl-4-piperidylmethanol.

azacyclotridecane-2-one polyamide. See nylon 12

azalomycin. A mixture of related antibiotics produced by Streptomyces hygroscopicus, var. asalomyceticus.

Azapen. CAS 132-92-3; Methicillin sodium; antibacterial. [Pfizer Inc] †

azapetine. 1-Allyl-2,7-dihydro-3,4:5,6-dibenzazepine.

azapropazone. (5-Dimethylamino-9-methyl-2-propyl-1H-pyrazolo-[1,2-a][1,2,4]-benzotriazine-1,3(2H)-dione) An analgesic and anti-inflammatory.

azaribine. (-β-D-ribofuranosyl-1,2,4-triazine-3,5-(2H, 4H)dione 2′,3′,5′-triacetate; 6-azauridine 2′,3′,5′,- triacetate) A preparation used in the treatment of psoriasis.

azatadine. (5,6-dihydro-11-(1-methyl-4-piperidylidene) benzo(h)cyclohepta(b)-pyridine) An antihistamine.

Azilex. CAS 275-51-4; A proprietary preparation of azulene; an antipruritic. [Ingasetter Ltd] †

azimidobenzene. See 1H-benzotriazole

Aziplex. Blended metal chelate. [ABM Chemicals Ltd] *

Azitrocin. See Zithromax. [Pfizer International]

Azitrocina. See Zithromax. [Pfizer International]

Azitromax. See Zithromax. [Pfizer International]

Azlin. CAS 37091-66-0; Azlocillin; antibacterial. [Miles Pharmaceuticals] *

2,2′-azobis (2,4-dimethylvaleronitrile). Initiator for suspension polymerization of vinyl chlorides, solution polymerization of various monomers. [Spectrum Chem. Manufacture]

1,1′-azobisformamide. See azodicarbonamide

Azocoll®. Substrate. [Calbiochem Corp]

azodicarbonamide. (1,1′-azobisformamide; diazenedicarboxamide; azodicarboxamide) $H_2NCON=NCONH_2$; CAS 123-77-3; EINECS 204-650-8; Blowing and foaming agent for plastics; bleaching agent in cereal flour. [Atochem SA; Gist-Brocades Food Ingreds.; Olin; Schering Berlin Polymers; Uniroyal]

Azoene. Fast red salt (ponceau fast L salt); a pinkish cream powder for use in automatic SGO-T assays. [British Drug Houses] *

Azofix. Azoic colors. [Fine Dyestuffs & Chemicals Ltd] *

Azoground. Azoic colors. [Fine Dyestuffs & Chemicals Ltd]*

Azoguard. A stabilizer for diazo compounds. [Imperial Chemical Industries plc] *

Azol. 4-Aminophenol hydrochloride in solution. [Johnsons of Hendon] *

Azolan. CAS 61-82-5; 1,2,4-Triazol-3-ylamine; active ingredient: aminotriazole; weedkiller with good translocation characteristics for the control of perennial and annual weeds. [Agan Chemical Manufacturers Ltd]

Azolid. CAS 50-33-9; Phenylbutazone; antirheumatic. [USV Pharmaceutical Corp] ‡

Azolith. A proprietary pigment containing 71.0% $BaSO_4$ and 29.0% ZnS. §

Azomagenta G. A dyestuff obtained by diazotizing sulfanilic acid, and treating the product with S acid. §

Azoman. A proprietary trade name for hexazole. §

Azone. CAS 59227-89-3; Laurocapram; pharmaceutic aid. [Nelson Research] *

azophor red. See paranitraniline red.

Azoprint. Azoic printing colors. [Fine Dyestuffs & Chemicals Ltd] *

Azorapid. Azoic printing colors. [Fine Dyestuffs & Chemicals Ltd] *

azoresorcin. See diazoresorcin.

Azosan. A fungicide. [May & Baker Ltd] *

Azo-Standard. CAS 136-40-3; Phenazopyridine hydrochloride; an analgesic. [Alcon Laboratories Inc] *

Azostix. A prepared test strip of urease, bromothymol blue and buffers, for the semi-quantitative determination of blood urea levels. [B C Ames] *

azotic acid. See nitric acid

Azotox. Insecticides. [ICI Chem. & Polymers Ltd] †

Azotoz 580. CAS 919-86-8; Emulsifiable concentrate containing 580 g demeton-S-methyl per liter; a systemic organophosphorus insecticide. [BritAg Industries Ltd]

Azromax. See Zithromax. [Pfizer International]

Aztec®t-Amyl Peroxypivalate-75 OMS. CAS 29240-17-3; t-Amyl peroxypivalate, odorless min. spirits. [Catalyst Resources] *

Aztec® Benzoyl Peroxide-70 -77. CAS 94-36-0; Benzoyl peroxide, water. [Catalyst Resources] *

Aztec® Benzoyl Peroxide-Dry. CAS 94-36-0; Benzoyl peroxide. [Catalyst Resources] *

Aztec® 1,1-Bis(t-Butylperoxy)Cyclohexane-80 BBP.

---

‡ Trade name and manufacturer not verified   § Trade name without identified manufacturer

CAS 3006-86-8; 1,1-Bis(t-butylperoxy)cyclohexane, butylbenzyl phthalate. [Catalyst Resources] *

Aztec® 1,1-Bis(t-Butylperoxy)-3,3,5-Trimethyl Cyclohexane. CAS 6731-36-8; 1,1-Bis(t-butylperoxy)3,3,5-trimethyl cyclohexane. [Catalyst Resources] *

Aztec® 1,1-Bis(t-Butylperoxy)-3,3,5 Trimethyl Cyclohexane-75 DBP. CAS 6731-36-8; 1,1-Bis(t-butylperoxy)3,3,5-trimethyl cyclohexane in dibutyl phthalate. [Catalyst Resources] *

Aztec® t-Butyl Hydroperoxide-70, Aq. CAS 75-91-2; t-Butyl hydroperoxide, water. [Catalyst Resources] *

Aztec® t-Butyl Peracetate-50 OMS, 60 OMS, 75 OMS. CAS 107-71-1; t-Butyl peroxyacetate in odorless min. spirits. [Catalyst Resources] *

Aztec® t-Butyl Perbenzoate. CAS 614-45-9; t-Butyl perbenzoate. [Catalyst Resources] *

Aztec® t-Butyl Peroctoate. CAS 3006-82-4; t-Butyl peroctoate. [Catalyst Resources] *

Aztec® t-Butyl Peroctoate-50 DOP. t-Butyl peroctoate in dioctyl phthalate. [Catalyst Resources] *

Aztec® t-Butyl Peroctoate-50 OMS. CAS 3006-82-4; t-Butyl peroctoate in odorless min. spirits. [Catalyst Resources] *

Aztec® t-Butyl Peroxyisobutyrate-75 OMS. CAS 109-13-7; t-Butyl peroxyisobutyrate, odorless min. spirits. [Catalyst Resources] *

Aztec® t-Butyl Peroxyneodecanoate-50 OMS, 75 OMS. CAS 26748-41-4; t-Butyl peroxyneodecanoate, odorless min. spirits. [Catalyst Resources] *

Aztec® t-Butyl Peroxypivalate-75 OMS. CAS 927-07-1; t-Butyl peroxypivalate, odorless min. spirits. [Catalyst Resources]*

Aztec® CHP-80. CAS 80-15-9; Cumene hydroperoxide. [Catalyst Resources] *

Aztec® DCP-R. CAS 80-43-3; Dicumyl peroxide. [Catalyst Resources] *

Aztec® 2,5-Di. CAS 78-63-7; 2,5-Dimethyl-2,5-di(t-butylperoxy)hexane. [Catalyst Resources] *

Aztec® Di-t-Butyl Peroxide. CAS 110-05-4; Di-t-butyl peroxide. [Catalyst Resources] *

Aztec® TKB. Mixed peroxide sol'n. in butyl benzyl phthalate. [Catalyst Resources] *

Aztec® 2,5-Tri. CAS 1068-27-5; 2,5-Dimethyl-2,5-di(t-butylperoxy)hexyne-3. [Catalyst Resources] *

Azurico. Blue on glaze decoration. [Degussa Ltd]

---

* Trade name not verified as available   † Trade name verified as obsolete

B

B-2. CAS 63394-02-5; Silicone elastomer; general purpose material with good filler acceptance, heat resist., compression set; for extrusion, molding. [Wacker Silicones]

B-147. Reclaimed rubber from mechanically defibered passenger tires; used in carcass, sidewall, and undertread of passenger, light truck, and off-road tires, and general purpose mechanical goods. [Midwest Rubber Reclaiming] *

B-182. CAS 63394-02-5; Silicone elastomer; economical grade with good filler acceptance, low compression set; for extrusion, molding applications. [Wacker Silicones]

B-618. Reclaimed rubber from modified whole tire, with whiting, clay, and min. rubber added; used in automotive floor mats, semipneumatic tires. [Midwest Rubber Reclaiming]

B-8880-50%. High molecular weight polyoxyalkylene polymer; dyeing assistant for acrylics; stabilizer for aq. emulsions. [Ethox]

BA 27. A non-heat-treatable alloy of aluminum containing 3.5% magnesium and 0.4% manganese.

Bacacil. CAS 37661-08-8; A proprietary preparation of bacampicillin hydrochloride; an antibiotic. [Pfizer International]

Bacdip®. Preparation for the control of ectoparasites, especially of multi-host ticks on all domestic animals; veterinary medicine. [Bayer AG]

Bachite. A specially prepared carbon made from anthracite.

Baciguent. CAS 1405-87-4; Bacitracin; antibacterial. [Upjohn] *

Baclad®. Metal clad products. [Asahi Chem. Industry]

Bacnelo. A proprietary mixture of 3% sulfur, 2% coal tar solution, zinc oxide, and calcium hydroxide; used in the treatment of skin disorders. [B F Ascher & Co Inc] †

Bac-N-Fos®. Sodium hexametaphosphate and sodium bicarbonate; food additive for meat, poultry, and seafood industries. [Rhone-Poulenc Food Ingreds.]

Baco AF. Alumina trihydrate; high purity grade for use in toothpaste, pharmaceuticals, cosmetics; excellent cleaning and polishing props.; stabilizes sol. of fluorine compounds in toothpaste. [BA Chem. Ltd.]

Bacote®. Paper coating insolubilizers. [Magnesium Elektron Ltd]

Bacterase. Bacterial enzyme or diastase. [Rhone-Poulenc UK]

Bactesyn. See Unasyn. [Pfizer International]

Bactipront. A proprietary preparation of cotrimoxazole; a urinary tract antibacterial. [Pfizer International]

Bactiram. Specialty biocides for use in crude oil production, water treatment. [Atochem UK/Ceca]

Bactistep® MH 80. Dialkyl dimethyl ammonium methoxy sulfate; sanitizer, germicide, algicide, fungicide, deodorizing agent, antistat for water treatment. [Stepan Europe]

Bactocill. CAS 7240-38-2; Oxacillin sodium; antibacterial. [SmithKline Beecham]

Bactospeine. Bacillus thuringiensis; a bacterial insecticide for control of caterpillars. [Fargro Ltd]

Bactospeine WP. Bacillus thuringiensis; a bacterial insecticide for control of caterpillars. [Koppert (UK) Ltd]

Bactratycin. CAS 1404-88-2; Tyrothricin; antibacterial. [Wallace Laboratories] *

Bactria. Proprietary range of industrial biocides based on isothiazolinones, formal donors, etc.; used for latex and emulsions, metal working fluids, paints, adhesives, graphic arts, etc. [Reichhold]

Bactrim. A proprietary preparation of trimethoprim and sulfamethoxazole; an antibiotic. [Roche Laboratories; Roche Products Ltd]

Bactrim Roche. Bactericide. [Roche Products Ltd] *

Badin metal. A name used for an alloy of iron with 8-10% aluminum, 18-20% silicon, and 4-6% titanium; used for adding silicon to steel.

Bafixan®Dyes. Disperse dyes for transfer paper printing on textile printing machines. [BASF]

Bahn metal. A lead base bearing alloy containing copper, 0.7% calcium, 0.6% sodium, and 0.04% lithium; also known as Railway metal.

Baiculein. 5:6:7-Trihydroxyflavone.

Bailey solder. An alloy containing 70% tin, 16% zinc, 10% aluminum, and 4% phosphor tin.

Bakelite. Phenolic molding powders. [Bakelite Polymers (UK) Ltd] *

Bakelite. Phenolic molding materials. [Rutgerswerke]

Bakelite A and B. Phenol-formaldehyde resins, soluble in certain solvents. [Bakelite Corporation] *

Bakelite C-9. A proprietary trade name for soft oil modified alkyds. [Bakelite Corporation] *

Bakelite Dilecto. A laminated product consisting of paper or fiber cemented together with a phenol-formaldehyde resin. [Bakelite Corporation] *

Bakelite DQD-3269. A proprietary trade name for an ethylene-vinyl acetate copolymer; a plastics material which retains its flexibility and toughness at low temperatures. [Bakelite Corporation]

Bakelite Micarta. A product similar to Bakelite Dilecto. [Bakelite Corporation] *

Baker P-2C. A proprietary trade name for cellosolve ricinoleate; a vinyl plasticizer. §

Baker P-8. A proprietary trade name for glyceryl triacetyl ricinoleate; a plasticizer for vinyl polymers and GR/S, neoprene GN, ethyl cellulose, and perbunan. §

---

‡ Trade name and manufacturer not verified

§ Trade name without identified manufacturer

Baker's Anaesthetic Ether. Diethyl ether or anesthetic ether. [May & Baker Ltd] *

Baker's P and S Liquid. A proprietary mixture of 1% phenol, sodium chloride, liquid paraffin, and water. [Baker Laboratories] *

baker's salt. Ammonium carbonate, $(NH_4)_2CO_3$.

Bakfil. Granular infill materials for use behind Garnex boards in continuous casting tundishes. [Foseco (F.S.) Ltd]

baking soda. See sodium bicarbonate

Bakurol. See paraffin, liquid.

B.A.L. CAS 59-52-9; A proprietary preparation of dimercaprol; used to treat heavy metal intoxication. [The Boots Co plc]

BAL. Adhesive; used for fixing ceramic tiles. [Building Adhesives Ltd]

Balance®. For agriculture industry. [DuPont UK]

Balanced Salt Solution. A proprietary solution of 0.49% sodium chloride, 0.075% potassium chloride, and 0.036% calcium chloride. [Alcon Universal] *

Balata. The coagulated milky juice of Mimusops globosa (the balata or bullet tree) of South America; used for electrical insulation, for transmission belts, and as soles for shoes.

Baldwin's phosphorus. Calcium nitrate, $Ca(NO_3)_2$.

BAL in Oil. CAS 59-52-9; Dimercaprol; antidote to arsenic and gold and mercury poisoning. [Hynson Westcott & Dunning] *

Ballistite. A trademark for a smokeless powder containing nitroglycerin and collodion cotton. §

Balmex Medicated Lotion. A proprietary protective lotion of lanolin, hexachlorophene, allantoin, balsam Peru and silicone oil in a nonmineral oil base. [Macsil Inc] *

Balmosa®. Menthol, camphor, methyl salicylate; rubifacient cream. [Pharmax Ltd]

Bal-nela. An alloy of 28% nickel, 66% copper, with 6% manganese, silicon, iron, and zinc.

Balneol. Mineral oil; laxative; pharmaceutic aid. [Rowell Laboratories Inc] *

Balnetar Liquid. A mixture of 2.5% crude coal tar, a lanolin derivative, mineral oil and a nonionic emulsifier; a proprietary preparation used to treat skin disorders. [Westwood Pharmaceuticals Inc] *

Balsamarome®. Resinoids; used for perfumery. [Laserson & Sabetay]

balsam of Peru. An oleoresin obtained from the bark of Myroxylon pereinaeg. It contains esters of cinnamic and benzoic acids. It is antiseptic, and is used in medicine and perfumery.

balsam of Tolu. The product of the tree Myroxylon toluifera; used in medicine and in perfumery.

balsams. Resins containing benzoic or cinnamic acids.

Balsan. Proprietary name for a specially purified preparation of balsam Peru. [Macsil Inc] *

Baltane CF. CAS 71-55-6; 1,1,1-Trichloroethane; degreasing solvent. [Nickerson Chemicals Ltd]

Baltane D. CAS 71-55-6; 1,1,1-Trichloroethane. [Nickerson Chemicals Ltd]

BAM. Hot melt adhesive. [Beardow & Adams (Adhesives) Ltd]

Bamate. CAS 57-53-4; A proprietary name for meprobamate; a tranquilizer. [Century Pharmaceuticals Inc] *

Bamo. CAS 57-53-4; A proprietary name for meprobamate; a tranquilizer. [Misemer Pharmaceuticals Inc] *

BAN 120 L. CAS 9000-92-4; α-amylase derived from

Bacillus subtilis; enzyme used for partial breakdown of gelatinized starch into dextrins; used in starch industry for batchwise liquefaction of starch in prod. of dextrins, dextrose, and glucose syrups; used in alcohol industry to break down the starch contained in mash; paper industry for production of starch pastes of various viscs. for sizing and coating preparations. [Novo Nordisk] *

Banamine. CAS 42461-84-7; Flunixin meglumine; anti-inflammatory; analgesic. [Schering Corp] *

bandine. See paraffin, liquid.

Band-Lock. Hot-melt adhesives. [Reichhold]

Bandrift. A spray drift retardant suitable for ground and aerial spraying of pesticides and other agrochemicals. [Allied Colloids Ltd]

Banflex. CAS 4682-36-4; Orphenadrine citrate; relaxant; antihistaminic. [O'Neal, Jones & Feldman Pharmaceuticals] *

Banistyl. Dimethothiazine. [May & Baker Ltd] *

Banlene Plus. Soluble concentrate of 18 g dicamba, 252 g MCPA and 84 g mecoprop per liter; used for weed control in cereals and grassland. [Schering Agrochemicals Ltd]

Banminth. CAS 33401-94-4; Pyrantel tartrate; anthelmintic. [Pfizer Inc]

Banocide. CAS 1642-54-2; A proprietary preparation of diethylcarbamazine citrate; used in the treatment of filariasis. [The Wellcome Foundation Ltd]

Banox. Additive to de-icing salts. [Albright & Wilson Ltd, Phosphates & Speciality Business] *

Banox. Food grade antioxidants. [UOP Speciality Products]

Banox ES. BHT, lecithin, soybean oil, min. oil; antioxidant blend for animal feeds. [UOP]

Banthine. CAS 53-46-3; Methantheline bromide; anticholinergic. [G D Searle & Co] *

Banwee. Residual herbicide. [Fisons plc, Horticulture Div]*

Banweed. CAS 15299-99-7; A suspension concentrate containing napropamide; soil-applied residual for control of annual grasses and annual broad-leaved weeds in field and container grown nursery stock. [Fisons plc, Horticulture Div] *

Banweed-S. A suspension concentrate containing napropamide and simazine; soil-applied residual for control of annual grasses and annual broad-leaved weeds in field and container grown nursery stock. [Fisons plc, Horticulture Div] *

Baquacil. CAS 56-03-1; Polymeric biguanide swimming pool sanitizer. [ICI Chem. & Polymers Ltd]

Baquagold. Swimming pool algicide. [Imperial Chemical Industries plc]

Baquatop. CAS 7722-84-1; Hydrogen peroxide for the treatment of swimming pools. [ICI Chem. & Polymers Ltd]

B.A.R. Butyl acetyl ricinoleate; a vinyl plasticizer. [Harcros UK] *

Barabar. Oxidation and corrosion inhibitors. [Exxon Int'l.] *

Baragel. Rheological additives; used for grease, underbody coatings. [Rheox Inc]

Baragel. Thickening and thixotropic additives; used for grease, underbody coatings. [Steetley Minerals Ltd]

Baratol Tablets. CAS 26844-12-2; Indoramin; alphablocker used in the treatment of hypertension. [Wyeth Laboratories; Monmouth Pharmaceuticals Ltd]

Barazan. CAS 70458-96-7; Norfloxacin; urinary antiseptic. (Germany) [Merck & Co Inc]

Barberite. A corrosion-resisting alloy containing 88.5%

---

copper, 5% nickel, 5% tin, and 1.5% silicon; has a high tensile strength in addition to good corrosion-resisting properties.

Barbidex. A proprietary preparation of dexamphetamine resinate, and phenobarbitone; antiobesity agent. [Nicholas Laboratories Ltd] *

Barbivis. CAS 50-06-6; Phenobarbital; anticonvulsant; hypnotic; sedative. [Alcon Laboratories Inc] *

Barbouze's Alloy. An alloy of aluminum with 10% tin.

Barchlor. Alkyl chlorides. [Lonza Ltd]

Bardac. Twin chain quaternary ammonium compounds; germicidal applications in hospitals, institutions, and industrial water treatment. [Lonza AG]

Bardase. A proprietary preparation containing phenobarbitone, hyoscyamine sulfate, hyoscine hydrobromide, atropine sulfate, and aspergillus oryzae enzymes; gastric sedative. [Parke-Davis] *

Bardew. Fungicide. [Schering Agrochemicals Ltd] *

Bardyne. Isdopher; a biocide with useful hospital and animal health applications. [Lonza AG]

Barex® 210. Rubber-modified copolymer containing 75% acrylonitrile and 25% methacrylate; high-barrier resins in injection molding and extrusion grades for film, sheet, blow molding and injection molding applications; used for packaging for food, pharmaceuticals, household and industrial chemicals; (lighter fluid, polishes, insecticides, detergents), cleaning solvs., waxes, cosmetics, and other toiletries. [BP Chem. Inc.]

Barfoed's Reagent. Copper acetate, 1 part, is dissolved in water, 15 parts. To 200 cc of this solution, 50 cc of 68% acetic acid are added; used to distinguish glucose and other monosaccharides from disaccharides, such as lactose and maltose. With glucose red c

Bariform. CAS 7727-43-7; Barium sulfate for x-ray diagnosis. [Laporte Industries Ltd] *

Bario metal. An alloy containing 90% nickel, 1.22% tungsten, 0.29% silicon, and 4.25% chromium; acid and heat-resistant.

Barisol Super BRM. Complex multicarbon alcohol phosphate potassium salt; textile wetting agent for dye leveling, pectin removal from cottons, scouring of cotton and synthetic blends; dispersant. [Dexter]

barium. Alkaline-earth element; Ba; CAS 7440-39-3; EINECS 231-149-1; Alloys in vacuum tubes, deoxidizer for copper, lubricant for anode rotors in x-ray tubes, spark-plug alloys. [Aldrich; Atomergic Chemetals; Degussa; Noah Chem.]

barium acetate. $(CH_3COO)_2Ba$; CAS 543-80-6; EINECS 208-849-0; Chemical reagent, acetates, textile mordant, catalyst manufacture, paint and varnish driers. [Aldrich; Bernardy Chimie SA; Hoechst Celanese; Mallinckrodt]

barium binoxide. See barium peroxide

barium bromide. BaBr · $2H_2O$; manufacture of bromides, photographic compounds, phosphors. [Atomergic Chemetals]

barium carbonate. $BaCO_3$; CAS 513-77-9; EINECS 208-167-3; Treatment of brines in chlorine-alkali cells to remove sulfates, rodenticide, production of barium salts, ceramic flux, optical glass, case-hardening baths, ferrites, in radiation-resistant glass for color television tubes. [Bernardy Chimie SA; Mallinckrodt; Solvay GmbH]

barium chloride. $BaCl_2$ · $2H_2O$; CAS 10361-37-2; Production of artificial barium sulfate, other barium salts; re-

agents; lubrication oil additives; boiler compounds; textile dyeing; pigments; manufacture of white leather. [Aldrich; EM Industries; Hoechst Celanese; Mallinckrodt; Sachtleben Chemie GmbH; Solvay GmbH]

barium chromate. (lemon chrome; ultramarine yellow; baryta yellow) $BaCrO_4$; CAS 10294-40-3; EINECS 233-660-5; Safety matches, corrosion inhibitor in metal-joining compounds, pigment for paints, ceramics, fuses, pyrotechnics, metal primers, ignition control devices. [Atomergic Chemetals; BASF AG; Noah Chem.]

barium dioxide. See barium peroxide

barium monoxide. See barium oxide

barium nitrate. $Ba(NO_3)_2$; CAS 10022-31-8; Pyrotechnics, incendiaries, chemicals (barium peroxide), ceramic glazes, rodenticide, electronics. [Aldrich; Berk Chem. Ltd; Hoechst Celanese; Noah Chem ; San Yuan Chem. Co. Ltd.]

barium oxide. (barium monoxide; barium protoxide; calcined baryta) BaO; CAS 1304-28-5; EINECS 215-127-9; Dehydrating agent for solvents, detergent for lubricating oils. [Atomergic Chemetals; Hoechst Celanese; Noah Chem.; Spectrum Chem. Mfg.]

barium peroxide. (barium binoxide; barium dioxide; barium superoxide) BaO_2; CAS 1304-29-6; EINECS 215-128-4; Bleaching, decoloring glass, thermal welding of aluminum, manufacture of hydrogen peroxide, oxidizing agent, dyeing textiles. [Bernardy Chimie SA; Hocehst Celanese; Spectrum Chem. Mfg.]

barium protoxide. See barium oxide

barium stearate. $Ba(C_{18}H_{35}O_2)_2$; CAS 6865-35-6; Waterproofing agent, lubricant in metalworking, plastics, and rubber; wax compounding; prep. of greases; heat and light stabilizer in plastics. [Adeka Fine Chem.; Reagens SpA; Syn. Prods.; Witco]

barium sulfate. (barytes, natural; blanc fixe, artificial; precipitated; basofor) Inorganic salt; $BaSO_4$; CAS 7727-43-7; EINECS 231-784-4; Weighting mud in oil drilling, paper coating, paints; filler and delustrant for textiles, rubber, plastics, and lithographic inks; radiation shield. [Cyprus Industrial Minerals; J.M. Huber; Mallinckrodt; Sachtlebe Chemie GmbH]

barium sulfonate. Rust preventive. [Atomergic Chemetals; Lubrizol; Witco/Sonneborn]

barium superoxide. See barium peroxide

barium titanate. $BaTiO_3$; CAS 12047-27-7; EINECS 234-975-0; Ferroelectric ceramics used in storage devices, dielectric amplifiers, digital calculators. [Aldrich; Atomergic Chemetals; Ferro/Transelco; Noah Chem.]

Barkite B. Di(dimethylcyclohexyl)oxalate; a plasticizer for cellulose lacquers. [Laporte Industries Ltd] *

Barlene. Tertiary amines. [Lonza Ltd]

Barlene® 12S. Lauryl dimethyl amine; chemical intermediate; personal care additive. [Lonza Inc]

Barlene® 14S. Myristyl dimethyl amine; chemical intermediate; personal care additive. [Lonza Inc]

Barlene® 16S. Cetyl dimethyl amine; chemical intermediate; personal care additive. [Lonza Inc]

Barlene® 18S. CAS 124-28-7; EINECS 204-694-8; Stearyl dimethyl amine; chemical intermediate; personal care additive. [Lonza Inc]

Barlox. A range of alkyl dimethyl amine oxides; used for detergents, shampoos, and textile processing applications. [Lonza AG]

Barlox® 12. CAS 61788-90-7; EINECS 263-016-9;

---

‡ Trade name and manufacturer not verified   § Trade name without identified manufacturer

Cocamine oxide; detergent; visc. builder, emollient. [Lonza Inc]

Barlox® 14. CAS 3332-27-2; EINECS 222-059-3; Myristamine oxide; detergent; visc. builder, emollient. [Lonza Inc]

Barlox® 16S. CAS 7128-91-8; EINECS 230-429-0; Cetamine oxide; biodeg. foam stabilizer, visc. builder, emulsifier, conditioner for personal care and industrial products. [Lonza Inc]

Barlox® 18S. CAS 2571-88-2; EINECS 219-919-5; Stearyl dimethyl amine oxide; biodeg. foam stabilizer, visc. builder, emulsifier, conditioner for personal care and industrial products. [Lonza Inc]

Barlox® C. CAS 68155-09-9; Cocamidopropylamine oxide; foam stabilizer and visc. builder for shampoos, industrial products. [Lonza Inc]

barm. See yeast

Bäropan MC 8046 SP. Ca/Zn; Stabilizer for shoe injection molding of plasticized PVC. [Bärlocher GmbH]

Bäropan TX 296 KA. Stabilizer for cable compounds. [Bärlocher GmbH]

Bäropol. Additives for polyolefins and polystyrenes (antioxidants, lubricants, release agents, uv stabilizers, antiblocking agents, antistats, flame retardants, fillers and pigments). [Bärlocher GmbH]

Baros. A heat-resisting alloy containing 90% nickel and 10% chromium.

Barosperse. CAS 7727-43-7; Barium sulfate; diagnostic aid. [Mallinckrodt Inc] *

Bärostab® CT 901. Ca/Zn product; stabilizer offering very good color props. and excellent transparency for injection molding of plasticized PVC; for extrusion and shoe injection molding. [Bärlocher GmbH]

Bärostab® KK 47 S. K/Zn; fast-kicking stabilizer for foam processing in production of wallpaper, artificial leather; highest degree of whiteness. [Bärlocher GmbH]

Bärostab® L 230. Zinc octoate; fast-kicking stabilizer for foam processing in production of sealings for caps; good organoleptic props., excellent color; approved for contact with foodstuffs. [Bärlocher GmbH]

Bärostab® NT 1005. Mg/Al/Zn complex; stabilizer for plastisol processing; especially for artificial leather, wallpaper, conveyor belts, nontoxic applications. [Bärlocher GmbH]

Bärostab® UBZ 76 BX. Ba/Zn complex with self-lubrication; stabilizer for transparent cable sheathing compounds; hydrolysis-resist.; imparts high transparency and excellent light stability. [Bärlocher GmbH]

Bärostab® UBZ 632. Barium zinc complex; stabilizer for PVC plastisols, especially topcoats for light pigmented cushion vinyl flooring; features good color; especially suitable in presence of inhibitors such as TMA. [Bärlocher; Vanderbilt]

Bärostab® UBZ 820 KA. Ba/Zn complex with self-lubrication; stabilizer for transparent cable compounds; imparts excellent heat stability. [Bärlocher GmbH]

Barotrast. CAS 7727-43-7; Barium sulfate; diagnostic aid. [Armour Pharmaceutical Co] *

Barquat®. Alkyl dimethyl benzyl quaternary ammonium compounds; for germicidal applications in hospitals, institutions and industrial water treatment, also used as antistatic agents. [Lonza Inc]

Barquat® CME-35. CAS 78-21-7; EINECS 201-094-8; Cetethyl morpholinium ethosulfate; antistat, combing aid

and detangling agent, textile lubricant, odor counteractant. [Lonza Inc]

Barquat® CT-29. CAS 112-02-7; EINECS 203-928-6; Cetrimonium chloride; coagulating agent in manufacturing of antibiotics. [Lonza Inc]

Barquinol HC. A proprietary preparation of hydrocortisone and clioquinol; an antibacterial dermatological agent. [Fisons plc, Pharmaceuticals Div] *

Barras. A pine resin.

Barrialon® CX. Polyvinylidene chloride coextruded multilayer film. [Asahi Chem. Industry]

Barrialon® S. Biaxially oriented multilayer film. [Asahi Chem. Industry]

Barrialon® SF. Polyvinylidene chloride film. [Asahi Chem. Industry]

Barrialon® UB. Vinylidene chloride copolymer; ultra-high barrier film for lamination. [Asahi Chem. Industry]

Barrier. Suspension concentrate containing 300 g chloridazon, 30 g fenuron and 170 g propham per liter; a residual herbicide for beet crops. [Truchem Ltd]

Barrier-Guard®. Polymer preparations providing moistureproof film to textile fabrics. [Reichhold]

Barseb. CAS 50-03-3; Hydrocortisone acetate; glucocorticoid. [Barnes-Hind Inc] *

Barseb HC. CAS 50-23-7; Hydrocortisone; glucocorticoid. [Barnes-Hind Inc] *

Barsilowsky's base. Aminoditolyl p-toluquinonediimine.

Bartex® 80. CAS 7727-43-7; Barium sulfate; extender pigment for gloss enamels, powd. coatings, semigloss latexes, house paints, primers; filler in many plastics applications, in rubber goods, and in ceramics. [Hitox]

barwood. See redwoods.

baryta. BaO; Barium monoxide.

baryta yellow. See barium chromate

barytes, natural. See barium sulfate

BAS 438. Suspension concentrate containing 250 g chlorothalonil and 187 g fenpropimorph per liter; a systemic fungicide for winter wheat. [BASF plc]

BAS 46402F. Systemic fungicide for use in winter wheat and barley. [BASF plc]

Basacid® Dyes. Anionic dyes; for production of inks, coloring of cleaners, wood preservatives in paint and varnish industry. [BASF AG]

Basacryl/Bafixan. Dyestuffs for textile transfer printing. [BASF plc] *

Basacryl® Dyes. Cationic dyes for dyeing and printing acrylic fibers and cationic dyeable polyester fibers. [BASF]

Basacryl Salt. Levelling agent for textile dyeing. [BASF plc]

Basacryl® Salt NB-KU. Dye retarder for dyeing of acrylic fibers. [BASF]

Basacryl® Salt TX-412. Resist agent for printing nylon with disperse/acid dyes. [BASF]

Basagran®. CAS 25057-89-0; Bentazone; a contact herbicide. [BASF plc]

Basagran®. CAS 25057-89-0; Bentazon; for post-emergence control of broadleaf weeds in rice, soybeans, legumes, etc. [BASF AG] *

Basagran® M 60. Bentazon, propanil; for post-emergence control of broadleaf weeds in cereals [BASF AG] *

Basamid®. Granules containing 98-99% dazomet; soil fumigant. [BASF plc]

Basamid® Granular. Dazomet (DMTT); soil fumigant for control of nematodes, soil-borne fungi, soil insects and

---

* Trade name not verified as available

† Trade name verified as obsolete

germinating weeds. [BASF AG]

Basammon® Extra 25. Ammonium sulfate nitrate with dicyandiamide as stabilizer; for improved nitrogen utilization and sustained action on agricultural crops. [BASF AG]

Basantol® Dyes. Anionic dyes for paint and varnish industry. [BASF AG]

Basazol®. Cationic liquid dye. based on basic dyes; coloring for papermaking. [BASF AG; BASF plc]

Bascal®. Acid donor for textile dyeing. [BASF plc] *

Bascal® S. Mixture of aliphatic dicarboxylic acids; deliming agent and pickling acid; for fur industry. [BASF AG]

Base 10-L. Sulfurized lard compounds; EP agent for drawing compounds, sol. and cutting oils. [Ferro/Keil]

Base 36. Sulfurized oleic acid; can be neutralized with bases to form water-sol. or -disp. soaps; also used in lubricating oils, cutting and sol. oils. [Ferro/Keil]

Base 75. Sodium sulfonate-soap deriv.; o/w emulsifier for conventional and EP sol. oils and pastes for metalworking industry; imparts stability and wetting props. [Ferro/Keil]

Base 104. 2-Ethyl hexanol phosphate, free acid; surfactant base for making low foam wetters, penetrants, antistats, lubricants, rust preventatives. [Clark]

Base 500-A. Used for prep. of catiaonic and anionic textile softeners. [CNC Int'l. L.P.]

Base 865. Sulfur chlorinated base; extreme pressure additive, detergent, lubricant for high performance fluids, machining and grinding operations. [Ferro/Keil]

Base 7800. Sulfonate soap; o/w emulsifier for sol. oils and pastes for metalworking industry, emulsion cleaners, synthetic and semi-synthetic coolants; hard water tolerance; aids rust protection. [Ferro/Keil]

Base HS. Sulfur base; high in active sulfur; for use in difficult metalworking applications, cutting oils. [Ferro/Keil]

Base ML. Methyl lardate; wetting/oiliness/lubricity agent for metalworking, lubricating, motor, and rolling oils; antiwear additive, process aid, release additive. [Ferro/Keil]

Base MT. Methyl tallowate; wetting/oiliness/lubricity agent for metalworking, lubricating, motor, and rolling oils; antiwear additive, process aid, release additive. [Ferro/Keil]

Base Nacrante 2078. Cocamidopropyl betaine, cocamide DEA, glycol stearate; foaming base for cosmetic pearlescent preps. [Seppic]

Base Nacrante 6030 CP. Sodium coceth-2 sulfate, triethylene glycol distearate; foaming base for cosmetic pearlescent preps. [Seppic]

Base Nacrante 9578. Sodium laureth sulfate, cocamide DEA, glycol stearate; foaming base for cosmetic pearlescent preps. [Seppic]

Basensol®. Functional block polymers based on propylene oxide/ethylene oxide or polyalkylene glycol ether; sensitizing agents for polymer dispersions. [BASF AG]

base oil. See blown oils.

Base Wax 36-AG. Substituted amide wax concentrate; substantive softener wax for synthetic fibers and fabrics; may be used to prepare stock sol'n. of softeners or softeners themselves. [Eastern Color & Chem.]

Basex. A proprietary base-exchange material for water softening; approximate composition $Na_2Al_2O_3 \cdot 14SiO_2$; stated to have high softening capacity. §

BASF Dimethoate 40. Systemic/contact insecticide. [BASF plc]

BASF MCPA Amine 50. CAS 94-74-6; MCPA; herbicide for cereals and grassland. [BASF plc]

BASF Reactive Resist Liq. Auxiliary for resist prints with Basilen P dyes under Primazin dyes. [BASF]

Basfapon®. Dalapon-sodium; post-emergence systemic herbicide for control of grasses in annual and perennial crops, on nonagricultural land, in ditches, and pastures [BASF AG]

Basfoliar® 6-12-6. Liq. foliar fertilizer with phosphate, zinc copper, and other micronutrients; for Indian corn and other crops with high phosphate requirements. [BASF AG]

Basfoliar® 34. Liq. nitrogenous foliar fertilizer with magnesium and micronutrients; for agricultural crops, vines, fruit, hops, and field vegetables. [BASF AG]

basic bismuth chloride. See bismuth oxychloride

Basicop. CAS 7758-98-7; Basic copper sulfate; wettable powd. agricultural fungicide for wide range of crops. [Griffin]

Basilen®. Reactive dyes for dyeing cellulosic fibers. [BASF; BASF plc]

Basilex. A wettable powder containing tolclofos-methyl; protective fungicide for use on all ornamentals and some edible crops against *Rhizoctonia*. [Fisons plc, Horticulture Div] *

Basocoll® CM. CAS 9003-08-1; Melamine-formaldehyde condensate; sizing assistant for papermaking. [BASF AG]

basofor. See barium sulfate

Basofor. CAS 7727-43-7; A proprietary trade name for a specially precipitated barium sulfate. §

Basoform®. Nonvolatile formaldehyde bonding agent; formaldehyde catcher for cured UF foams. [BASF AG]

Basogal® C. Nonfoaming leveling agent for vat dyes. [BASF]

Basojet® PEL 200%. EO condensate/anionic blend; nonfoaming dispersing and leveling agent for dyeing of polyester fibers with disperse dyes under high temp. conditions. [BASF]

Basokol® NB-S. Polymeric organic compound; sequestering agent for dyeing cotton and polyester/cotton; protective colloid; dispersant for dyeing, bleaching, afterscouring; stable to acids, alkalies, and electrolytes. [BASF]

Basol® WS. Aromatic sulfonic acid condensate; dispersant, protective colloid, dyeing assistant especially for exhaust dyeing of polyester; stabilizes dye dispersions; stable to acids, alkalies, hard water, electrolytes. [BASF]

Basolan®. Wool finishing agents. [BASF plc]

Basolan® F. Auxiliary for improving wet fastness of acid dyes on wool, silk, and polyamide fibers. [BASF]

Basomol®. Synthetic wetting agent containing phosphoric acid for manufacturing of synthetic resin foam in conjunction with Basopor®. [BASF AG]

Basonat®. Isocyanate-based; for adhesive and sealant industry; hardener components for polyols, prep. of prepolymers containing NCO groups; polyurethane adhesives and binders for granular materials. [BASF AG]

Basonyl® Dyes. Cationic dyes for production of inks, carbon paper coatings, for dyeing natural fibers such as jute, sisal, wool; paint and varnish industry for production of daylight fluorescent pigments and gloss paints. [BASF AG]

---

‡ Trade name and manufacturer not verified   § Trade name without identified manufacturer

Basopal®. Detergent for washing and cleaning processes in textile industry. [BASF AG; BASF plc]

Basophen®M. Alkyl ester of phosphoric acid with anionic/nonionic emulsifiers; nonfoaming wetting agent, foam suppressant for textile applications, package dyeing of cotton, cotton/polyester, wool, desizing, latex penetration in carpet backing; stable in hard water and liquors containing alkali, acetic acid, or sulfuric acid. [BASF/Fibers]

Basophen®RA. Sulfosuccinate; wetting agent for desizing, pretreatment, and bleaching of cellulosic fibers; resistant to chlorine, hard water, weak acids. [BASF/Fibers]

Basophob®. Aq. paraffin or polyethylene wax dispersions, solns. of fatty acid derivs.; hydrophobic agents for internal or surface treatment of paints, mortar, concrete and paper. [BASF AG]

Basoplast®. Alkyl ketene diamide copolymers; cationic internal and surface sizes for papermaking. [BASF AG; BASF plc]

Basopon®LN. Alkylphenol ethoxylate; wetting agent, detergent with dispersing effects for wool and synthetic fibers, desizing and scouring applications; stain removing agent; stable to hard water, alkalis, acids, reducing and oxidizing agents. [BASF/Fibers]

Basopon®TX-110. Proprietary; nonfoaming scouring agent with emulsifying properties for polyester knit goods. [BASF]

Basopor®. Urea-formaldehyde resin precondensate; for production of syntheticresin foam (in-situ UF foam), for use in building (as thermal insulation) and in mining (e.g., for protection against explosions). [BASF AG]

Basoset® 162. CAS 25928-94-3; Epoxy resin based on epichlorohydrin and an aliphatic polyol; as casting resin, optionally in combination with glass fibers for production of molded articles; for potting/encapsulation in elec. industry; for bonding glass fiber reinforced polyester resin parts to each other; for binding expanded PS. [BASF AG]

Basosoft®. Brighteners and softeners for textile industry. [BASF AG; BASF plc]

Basotect®. Open-cell, resilient foam plastic based on melamine resin; for soundproofing and thermal insulation where good fireproofing, high heat stability and low weight are required. [BASF AG]

Basotol®. Oxidizer for vat dyes; auxiliary for protecting dyes against reduction. [BASF]

Basotrope® W. Auxiliary for producing wh. discharges on dyed grounds that are difficult to discharge and for stripping dyed shades. [BASF]

Basovit®Dyes. Anionic dyes for foodstuffs and cosmetics industry for fiber pen inks, detergents, and seeds. [BASF AG]

Bassorin. (Adraganthin). Tragacanthin, the mucilage from gum tragacanth.

Bastamol®. Dyeing auxiliaries for fixing and deepening anionic dyeings and for leveling cationic dyeings in fur and leather industries. [BASF AG]

Bastanet. A decolorizing carbon.

bastose. The cellulose of jute.

Basudin. CAS 333-41-5; Diazinon; an organophosphorus insecticide. [Ciba-Geigy Agrochemicals]

Basyntyn®. Syn. replacement tanning agents and auxiliary tanning agents. [BASF AG; BASF plc]

Batchite. An activated carbon produced from coco-nut shells.

Bath metal. A brass containing 83% copper and 17% zinc. Another alloy consists of 55% copper and 45% zinc.

Battal. CAS 83601-81-4; Fungicide containing carbendazim. [ICI Chem. & Polymers Ltd]

Battal FL. CAS 83601-81-4; A liquid formulation containing 500 g carbendazim per liter as a suspension concentrate; systemic fungicide. [Farm Protection Ltd]

battery acid. See sulfuric acid

battery copper. A brass containing 94% copper and 6% zinc.

Baudoin's metal. Complex nickel silvers. One contains 72% copper, 16.6% nickel, 1.8% cobalt, 2.25% tin, and 7.1% zinc, and another 75% copper, 16% nickel, 2.25% zinc, 2.75% tin, 2% cobalt

Baum's Acid. See Schaeffer's acid.

Bavistin®. CAS 83601-81-4; Carbendazim; systemic fungicide for cane fruit. [BASF plc]

Bavistin® DF. CAS 83601-81-4; Carbendazim; systemic fungicide for control of fungus in vines, fruit, vegetables, ornamentals [BASF AG]

Baxton®. For chemical industry. [DuPont UK]

Bayblend®. Thermoplastic polycarbonate/ABS or polycarbonate/ASA blends; engineering thermoplastics for the manufacture of injection moldings for use in the electrical and automotive industries and for domestic appliances. [Bayer AG; Bayer plc; Miles]

Bayblend® DP2-1448. PC/ABS blend; bromine/chorine-free flame-retardant thermoplastic with easy processing for business machine housings, portable computer housings, consumer products, personal care items and appliances. [Miles] *

Bayblend® FR 1439. PC/ABS blend, flame retardant; thermoplastic resin offering good impact resist. even at low temp., rigidity and dimensional stability, easy processing; ideal for business machine and automotive markets. [Miles]

Bayblend®T 44. PC/ABS blend; high-productivity general purpose thermoplastic resin offering good impact resist. even at low temp., rigidity and dimensional stability, easy processing; ideal for business machine and automotive markets. [Bayer; Miles]

Bayblend®T 45 MN. PC/ABS blend; thermoplastic blend for injection molding, extrusion, blow molding, thermoforming, machining, welding, bonding, screwing, coating, printing, vacuum metallizing, and electroplating; for automotive, elec., domestic appliances, camera bodies, electroplated components and other applications. [Bayer; Miles]

Bayblend®T 88-2N. PC/ABS blend, 10% glass-reinforced; thermoplastic resin offering good impact resist. even at low temp., rigidity and dimensional stability, easy processing; ideal for business machine and automotive markets. [Bayer; Miles]

Baybond®. Hydrous polyurethane dispersion; for glass fiber sizes. [Bayer AG; Bayer plc]

Bayboran®. Selective reducing agent for aldehydes, ketones, acid chlorides, peroxides, Schiff bases, azides, etc.; used for electroless metallizing, purification and stabilization of process streams (odor removal, decolorization). [Bayer AG]

Baycarb®. Fenobucarb; insecticide used for control of hoppers, bugs and leaf beetles in rice. [Bayer AG]

Baycaron®. 4-Chloro-N-methyl-N-(tetrahydro-2-methylfurfuryl)benzene-1,3-disulfonamide; diuretic; saluretic, and antihypertensive drug. [Bayer AG]

Baycast. Biomedical preparation; polyurethane supportive dressing. [Bayer AG]

Baychrom®. Different types of mineral tanning agents containing chrome oxide, some of them self-basifying [Bayer AG]

Baycid®. CAS 55-38-9; Fenthion; insecticide used for control of biting and sucking pests. [Bayer AG]

Baycidal. Insect growth regulator for the control of fly larvae. [Bayer AG]

Baycillin®. α-Phenoxypropylpenicillin; a semisynthetic oral penicillin preparation. [Bayer AG]

Bayclin. Perfumed all-purpose cleaner with disinfecting properties for dissolution or concentrated application. [Bayer AG]

Bayco®. Monofilament; used for wine and fruit growing. [Bayer AG]

Baycoll®. Hydroxyl polyethers and hydroxyl polyesters; used in conjunction with Desmodur in the production of reaction adhesives. [Bayer AG]

Baycoll® 17. Linear saturated hydroxyl polyester; polyol used in two-component reaction adhesives in combination with Desmodur polyisocyanates; suitable for bonding plastic, foam, fabric, and wood substrates; modifier in solv.-based Desmocoll PU adhesives. [Bayer; Miles] *

Baycoll® MD 3040. Linear polyester polyol; used for two-part PU adhesives; as modifiers and building blocks for PU. [Miles; Polysar]

Baycor®. CAS 55179-31-2; Bitertanol; broad spectrum fungicide. [Bayer AG]

Baycox. Treatment of coccidiosis in poultry. [Bayer AG]

Baycryl®. A range of acrylic resins. [Bayer AG]

Baycuten®, Baycuten N. An anti-eczematous preparation. [Bayer AG]

Bayderm® A. Solutions of anionic dyestuffs; for spraying, curtain coating and printing as well as for shading the finish. [Bayer AG]

Bayderm® KF. Solutions of cationic dyestuffs for brilliant spray dyeings and for shading finishing liquors; used in leather industry. [Bayer AG]

Bayderm® Colours B-TO. Transparent pigment dispersions with low binder content; used for organic and aqueous finishes and for coating textiles. [Bayer AG]

Bayderm® Colours C-TO. Covering pigment dispersions with low binder content; used for organic and aqueous finishes and for coating textiles. [Bayer AG]

Bayderm® Lacquers Auxiliaries. Polyurethane-based products for leathers. [Bayer plc]

Baydur®. Rigid integral skin polyurethane foams; for covers for machinery, housings for computers, sporting goods, furniture, engineering components, automotive interior trim. [Bayer AG]

Baydur® STR. Polyurethane resin systems. structural composites. [Miles]

Bayer 5072. CAS 140-56-7; p-Dimethylamino-benzene-diazo sodium sulfonate; fungicide for the prevention of crop damage by soil fungi. [Bayer AG]

Bayer CM. Chlorinated polyethylene; used for the rubber and cable industries; very good resistance to ageing, ozone, weathering and chemicals. [Bayer AG]

Bayer SBR Latex 200 C. CAS 9003-55-8; S/B latex; nonstaining hard component for blended latex goods. [Bayer AG]

Bayer UV Absorber 325, 340. UV absorbers for protection of rigid and plasticized PVC, cellulose derivatives, polystyrene, polyacrylate; used in clear finishes to protect substrate against light. [Bayer AG]

Bayer Base Plates Glass-Clear. Plastic plates; used for the preparation of individual trays, for taking bites and trial fittings and for protective plates for surgical wounds. [Bayer AG] †

Bayer Perlon®. Man-made staple fiber and monofilament. [Bayer AG]

Bayer's Tonic. A tonic and restorative. [Bayer AG] †

Bayertitan. CAS 13463-67-7; EINECS 236-675-5; Titanium dioxide pigments. [Bayer plc]

Bayferon®. Interferon inducer for cattle for prophylaxis and treatment of infections due to interferon-sensitive infectious agents and for paramunisation; veterinary medicine. [Bayer AG]

Bayferrox®. A range of iron oxides; used as pigments in plastics and linoleum and in the rubber industry; used in electroceramics and in the manufacture of magnetic tapes. [Bayer AG; Bayer plc]

Bayfidan®. CAS 55219-65-3; An emulsifiable concentrate containing 250 g/l triadimenol; fungicide to control powdery mildew, rusts and rhychosporium in winter and spring crops of wheat, barley, oats and rye. [Bayer AG; Bayer plc]

Bayfill®. Semirigid polyurethane filling foams; for automotive engineering, instrument panels, consoles, arm rests, bumpers. [Bayer AG; Bayer plc]

Bayfit®. Flexible polyurethane foams; flexible foam for cushioning applications, upholstered furniture, mattresses, automotive engineering, textiles, packaging. [Bayer AG; Bayer plc]

Bayflex®. Elastic integral skin PU foams, microcellular RIM elastomers; for interior and exterior automotive applications, shoe soles, packaging, engineering components, domestic appliances. [Bayer AG; Bayer plc]

Bayfol®. Film made from blends based on polycarbonates, other engineering thermoplastics or composites; used for special overlay and decoration film for membrane keyboards, for front facings with integrated membrane switches and for all applications requiring a high level of chemical resistance. [Bayer AG; Miles]

Bayfol® CR 6-2. PC blend film; tech. film with high dynamic strength (especially for membrane switch overlays), increased chem. resist., scratch resist.. [Bayer AG]

Bayfolan®. Foliar feed containing macro and micro nutrients; for all agricultural and horticultural crops to help recover from the effects of adverse conditions such as drought, low temperatures or waterlogging. [Bayer AG; Bayer plc]

Bayfresh®. Perfumed air freshener as aerosol, liquid or on solid base. [Bayer AG]

Baygal®. A polyurethane based casting resin; used in the electrical industry and for rock consolidation. [Bayer AG]

Baygal® K30. Branched polyether polyol; used for producing cast PU resin compounds with high heat resist. for electronics and elec. engineering. [Bayer AG] *

Baygal® K115. Polyether-polyester polyol; used for producing PU casting compounds for electronics and elec. engineering; improved compatibility with Baymidur isocyanates, reduced sensitivity to moisture. [Bayer AG] *

Baygal® K190. Linear, hydroxyl-bearing polyether polyol; used as flexibilizer in combination with other polyols for

---

‡ Trade name and manufacturer not verified   § Trade name without identified manufacturer

PU industry. [Bayer AG] *

Baygal® K390. Branched polyether polyol; used in combination with other polyols as flexibilizer. [Bayer AG] *

Baygard®. Textile finishing product; used for floor coverings; soil, oil and water repellent; pile stability. [Bayer AG]

Baygen®. Lacquers and auxiliaries based on polyurethane reactive lacquers; for cold lacquer finishes. [Bayer AG; Bayer plc]

Baygenal®. Dyestuffs; for high-grade chrome upper leathers. [Bayer AG]

Bayglaze. Ready-to-use glazes. [Bayer AG]

Baygon®. CAS 114-26-1; Propoxur; broad spectrum insecticide for the control of household and hygiene pests. [Bayer AG] *

Baygon®. Household insecticide; for control of stored product pests, vector- and nuisance insects, cockroaches, ants, beetles, midges, mites, flies, and other crawling and flying pests. [Bayer AG]

Baygon® MEB Spray. Household insecticide; used for control of flies, moths, wasps and other flying pests. [Bayer AG]

Bayguard. Waterproofing agents for textiles. [Bayer plc]

Bayhibit® AM. CAS 37971-36-1; 2-Phosphonobutane-1,2,4-tricarboxylic acid; corrosion and scale inhibitor in cooling systems; sea water evaporation units, water used for flooding in oil drilling; deflocculation of ceramic slips and oil drilling sludges, stabilization of pigment suspensions; formulation of cleaning agents; pickling and cleaning agent for oxidized metal surfaces; sequestrant; dispersant. [Bayer AG; Miles]

Bayhydrol. Water-thinnable binders for industrial coatings. [Bayer plc]

Bayhydrol 140 AQ. Aliphatic polyester PU disp. in water/toluene; adhesion promoter in metal/plastic composite structures, textile and leather coatings, primers for rigid surface coatings. [Miles]

Baykanol® AK, HLX, SL. Dyeing auxilliaries; high quality, almost neutral syntans for leveling aniline dyeing; the products facilitate the neutralization of chrome leathers and improve the fullness. [Bayer AG]

Baykanol® Liquor TN. Light fast special liquor for suede and white leathers; used in leather industry. [Bayer AG]

Baykisol®. Silica sol; used for the fining of apple juice, lemon juice, other juices, wines. [Bayer AG]

Baylan®. Low temperature dyeing auxiliaries for wool. [Bayer AG; Bayer plc]

Baylectrol®. Chlorine-free electrical insulating fluid for power capacitors. [Bayer AG]

Bayleton®. CAS 43121-43-3; A wettable powder containing 25% w/w triadimefon; systemic fungicide applied to winter and spring crops of wheat, barley, oats or rye for the control of powdery mildew, rhynchosporium, yellow rust, brown rust, crown rust (oats) and snow rot (winter barley). [Bayer AG; Bayer plc]

Bayleton® 5. CAS 43121-43-3; Wettable systemic fungicide powder containing 5% w/w triadimefon; used to control powdery mildew on apples, hops, raspberries, strawberries and other cane fruits plus American gooseberry mildew on all varieties of blackcurrants and gooseberries. [Bayer plc]

Bayleton® BM. A wettable powder systemic fungicide containing 12.5% w/w triadimenol and 25% w/w carbendazim; used to control eyespot, mildew and early

attacks of yellow and brown rust on winter wheat and winter barley, rhynchosporium on winter barley and eyespot and mildew on winter rye. [Bayer AG]

Bayleton® CF. Wettable powder fungicide with contact and systemic properties containing 6.25% w/w triadimefon and 65% w/w captafol; used to control powdery mildew, yellow and brown rust, leaf spot and glume blotch and to reduce the late-season ear disease complex on spring and winter wheat. [Bayer AG]

Baylith®. Techical grade oxides and zeolites; uses include intensive driers for use in polyurethane systems, paints and varnishes, plastics and solvents. [Bayer AG]

Baylon. Low density polyethylene homopolymers and ethylene/vinyl acetate copolymers; used for film, extrusion coating and cables. [Bayer AG] †

Baylube®. Polyether bases for synthetic lubricants. [Bayer AG; Bayer plc]

Bayluscid. CAS 1420-04-8; Molluscicide for the control of freshwater, disease-carrying snails. [Bayer AG]

Bayluscide®. Clonitralid; molluscicide for the control of water snails. [Bayer AG]

Baymat®. Fungicide with good penetrant, protective, curative and eradicative activity for control of scab and blossom wilt on pome and stone fruit, rust, leafspot diseases and mildews on pome and stone fruit, bananas, vegetables, sugar beet, ornamentals. [Bayer AG]

Baymer®. Rigid polyurethane foams; rigid foams for heating and refrigeration engineering, building industry, insulation, sporting goods, automotive engineering. [Bayer AG; Bayer plc]

Baymetex®. Metallized textile fabrics, primarily fabrics woven from raw materials such as polyamide, aramid, glass fiber, carbon fiber and polyester/cotton for EMI shielding, reflection of electromagnetic waves and lightning protection of sandwich elements. [Bayer AG]

Baymicin. 4-O-[(2S,3S)-3-Amino-6-aminomethyl-3,4-dihydro-2H-pyran-2-yl]-2-deoxy-6-O-(3-deoxy-4-C-methyl-3-methylamino-β-L-arabinopyranosyl)-D-steptamine sulfate; a bactericidal broad spectrum antibiotic with specific action against gram-negative microorganisms. [Bayer AG]

Baymicron®. Aqueous microcapsule dispersions for the manufacture of carbonless copying papers giving clear blue and black copies. [Bayer AG]

Baymid. PCD foam; polycarbodiimide foam. [Bayer plc]

Baymidur®. Polyurethane based casting resin; for use in the electrical industry, for rock consolidation and for the formulation of core sand binders. [Bayer AG]

Baymidur® K88. Aromatic diisocyanate; for production of PU cast compounds for electronics and elec. engineering. [Bayer AG] *

Baymidur® KL 3-5001. CAS 4098-71-9; Isophorone diisocyanate; nonaromatic resin used with Baygal KL 3-5004 to produce weather-resist. compounds with high insulating strength especially under outdoor conditions. [Bayer AG] *

Baymin®. Flotation chemicals. [Bayer AG; Bayer plc]

Baymix®. Preparation for the control of gastrointestinal worms in cattle; veterinary medicine. [Bayer AG]

Baymod®. Ethylene-vinyl acetate copolymer. [Bayer AG]

Baymod® A. Acrylo-butadiene-styrene polymer, PVC modifier. [Bayer AG]

Baymod® A95. CAS 9003-54-7; SAN copolymer; modifier improving the thermoforming props. and heat resist. of

---

semirigid and rigid PVC; suitable for weatherable applications. [Miles; Polysar]

Baymod® A KU3-2086. CAS 9003-56-9; ABS; high efficiency modifier improving the thermal stability and impact resist. of nonweatherable rigid PVC. [Miles; Polysar]

Baymod® L450 N. CAS 24937-78-8; EVA copolymer; impact modifier for rigid weatherable PVC. [Miles; Polysar]

Baymod® PU. Powdered plasticizing polymers for PVC. [Bayer AG]

Baymoflex A. Acrylo-styrene-acrylate polymers blend. [Bayer AG]

Baymoflex A KU3-2069.A. ASA; modifier for semirigid halogen-free sheeting, especially for automotive instrument panels. [Miles; Polysar]

Baymol® A, D. Range of nonionic tanning auxilliaries with grease emulsifying and degreasing effect. [Bayer AG; Bayer plc]

Baymosthrin. Pyrethorid with exceptionally fast knockdown action for control of insects in public health sector. [Bayer AG]

Baymycard. CAS 63675-72-9; Pharmaceutical preparation containing nisoldipine; also known as Syscor; coronary selective calcium antagonist, especially for the treatment of angina pectoris. [Bayer AG]

Baynat®. Rigid polyurethane foams; rigid foams for heating and refrigeration engineering, building industry, insulation, sporting goods, automotive engineering. [Bayer AG]

Bayofly®. Specific for the control of flies in cattle. [Bayer AG]

Bayolin®. Ethylene glycol monosalicylate; a percutaneous antirheumatic and anti-inflammatory agent. [Bayer AG]

Bayo-n-ox®. Growth promoter for pigs with antibiotic activity; increases weight gain, improves feed conversion and reduces rearing losses. [Bayer AG]

Bayotensin®. CAS 39562-70-4; Pharmaceutical preparation containing nitrendipine; also known as Baypress; calcium antagonist specific for the treatment of hypertension. [Bayer AG]

Bayothrin. Pyrethorid with exceptionally fast knockdown action for control of insects in public health sector. [Bayer AG]

Bayovac. Various vaccines for animals. [Bayer AG]

Baypen®. Sodium (6R)-6-[D-2-(3-mesyl-2-oxoimidazolidine-1- carboxamido)-2-phenylacetamido] penicillanate monohydrate; broad spectrum penicillin. [Bayer AG]

Bayplast®. Organic pigments; for the plastics industry. [Bayer AG]

Baypreg®. Polyols for the Baypreg system (filler and glass fibers containing resin compounds for SMC technology). [Bayer AG]

Baypren®. Polychloroprene rubber; raw material in the adhesive and rubber industries; used for rubber goods with excellent resistance to weathering and aging, good flame-retardant behavior and insensitivity to many chemicals. [Bayer AG; Bayer plc]

Baypren® 110. CAS 126-99-8; CR; synthetic polymer used for moldings and extrudates, reinforced hoses, roll covers, belting, cable sheathings and insulation, sponge rubber, sheeting, fabric proofings, footwear, food-contact goods, adhesives for footwear, furniture, building industry. [Bayer; Miles]

Baypren® 110 VSC. CAS 126-99-8; CR; modified grade providing reduced tendency to cryst. and harden at low temps.; offers good resist. to aging, weathering, ozone. [Bayer; Miles] *

Baypren® 216. CAS 126-99-8; CR; modified grade of Baypren 210 offering superior tensile strength and tear resist.; recommended where improved physical props. or cost reductions are necessary; applications including cable jackets, conveyor belts, hose covers, seals, bellows, and mechanical goods. [Bayer; Miles]

Baypren® 310. CAS 126-99-8; CR elastomer; for adhesive compding.; used for contact adhesives for bonding high-pressure laminates to wood in the manufacturing of furniture, sole attaching in the footwear industry, cements for building and automotive trades, and other operations involving fibrous substrates, plastics, foam materials, and metals. [Bayer; Miles]

Baypren® AT-H, AT-M, AT-S. CAS 126-99-8; CR elastomer, thiuram modified; for formulating contact adhesives used in boding of high-pressure laminates to wood in manufacturing of furniture, sole attaching in footwear industry, cements for building and automotive trades, and other operations involving fibrous substrates, plastics, foam materials, and metals. [Bayer; Miles]

Baypren® EM1. CAS 126-99-8; Precrosslinked CR; elastomer offering favorable processing chars. in mixing, calendering, extruding, and, to some extent, in injection molding; vulcanizates have good resist. to aging, weathering, and ozone attack, but do not exhibit outstanding low temp. props.; applications including extruded profiles, hose, cable, and calendered sheeting. [Bayer; Miles]

Baypren® M1. CAS 126-99-8; CR, mercaptan-modified; for rubber goods of moderate hardness and med. filler loadings, especially for applications which do not have to meet stringent low temp. requirements; vulcanizates have good resist. to aging, weathering, and ozone attack, and moderate oil resist.; applications including molded and extruded goods, hose, cable jackets, closed cell sponge and elastomeric linings. [Bayer; Miles]

Baypren® Latex KA 8348. CAS 126-99-8; Polychloroprene aq. colloidal disp.; rosin acid soap emulsifier system; for manufacturing of molded foam; vulcanizates have slight to med. cryst. tendency and show little discoloration when exposed to light. [Bayer; Miles] *

Baypren® Latex L 200A. CAS 126-99-8; Polychloroprene aq. colloidal disp.; used in dipped goods, paper and fabric saturants, and coatings. [Bayer; Miles]

Bayprena. Sulfamethoxydiazine; long acting sulfonamide. [Bayer AG]

Baypress®. CAS 39562-70-4; Pharmaceutical preparation containing nitrendipine; also known as Bayotensin; calcium antagonist specific for treatment of hypertension. [Bayer AG; Miles Pharmaceuticals] *

Bayprint®. Screen printing pastes for the manufacture of printed circuits. [Bayer AG]

Bayrena®. Sulfonamide; preparation for the treatment of bacterial infections; veterinary medicine. [Bayer AG]

Bayrotren®. Slow-release antihypotensive agent. [Bayer AG]

BayroVas®. Pain-relieving, anti-inflammatory therapeutic agent for veins. [Bayer AG]

Bayrusil®. Diethchinalphion; insecticide for the control of biting and sucking pests. [Bayer AG]

---

‡ Trade name and manufacturer not verified   § Trade name without identified manufacturer

Baysan®. Range of household cleaning products with disinfecting properties. [Bayer AG]

Bayscript®. Dyes for ink-jet printing. [Bayer AG; Bayer plc]

Baysical®. Precipitated silicates; inorganic fillers improving the whiteness and opacity of paper; preserves the free-flowing capacity of table salt; as a grinding aid in powders with the tendency to cake and as an extender in emulsion paints. [Bayer plc]

Baysilex®. Addition curing elastomeric high precision impression for the single phase technique two paste system material; used in dentistry. [Bayer AG]

Baysilone®. A range of silicone fluids; used in the production of polishes and water repellants, transfer media, dampening fluids, hydraulic fluids, dielectrics, lubricants in the production of plastics and man-made fibers and in the metal industry. [Bayer AG]

Baysin®. Finish for full grain/corrected grain side leather and splits. [Bayer AG; Bayer plc]

Baysolvex®. Solvent extractants for nonferrous metals, precious metals and rare earths. [Bayer AG]

Baysport®. PU elastomer; for building, furniture, shoes, automotive and mechanical engineering, sports surfacing, sporting goods, textiles, elec. industry, domestic appliances. [Bayer AG; Bayer plc]

Baystal®. Butadiene-styrene copolymers with self-cross-linking groups and possible additional comonomers; used for special applications in a variety of industries e.g. the textile, paper and rubber industries. [Bayer AG; Bayer plc]

Baysynthol®. Synthetic sizing agents; used in the paper industry. [Bayer AG; Bayer plc]

Baytan®. Dry powder containing 25% w/w triadimenol and 3% w/w fuberidazole; a seed treatment for barley, wheat, oats and rye; controls the important seed and certain soil borne diseases including loose smut, covered smut, foot rot, leaf stripe, bunt and early attacks of mildew and rhynchosporium. [Bayer AG; Bayer plc]

Baytan. Powder mixture of fuberidazole and triadimenol; seed dressing for cereals. [ICI Agrochemicals]

Baytan® IM. Dry powder mixture of fuberidazole, imazalil and triadimenol; seed dressing for barley. [Bayer plc]

Baytec®. PU elastomer; for building, furniture, shoes, automotive and mechanical engineering, sports surfacing, sporting goods, textiles, elec. industry, domestic appliances. [Bayer AG; Bayer plc]

Bayteroid®. Fast-acting synthetic pyrethroid with a long residual effect and broad spectrum of activity; for use in controlling caterpillars, beetles and their larvae, cutworms, sucking pests on vegetables, pome and stone fruit, grapes, maize, soya, tobacco, cotton and other crops. [Bayer AG]

Baytex®. CAS 55-38-9; Fenthion; insecticide for control of hygiene pests. [Bayer AG]

Baytherm®. Rigid polyurethane foams; rigid foams for heating and refrigeration engineering, building industry, insulation, sporting goods, automotive engineering. [Bayer AG; Bayer plc]

Baythion®. CAS 14816-18-3; Phoxim; insecticide for the control of stored-product pests. [Bayer AG]

Baythroid®. CAS 68359-37-5; Cyfluthrin; a pyrethroid insecticide with mainly contact action particularly against biting insects on cotton, vegetables, pome and stone fruits, citrus, maize, soybeans, and other crops; also against migratory locusts and grasshoppers. [Bayer AG;

Bayer plc]

Bayticol®. Preparation specifically for control of ticks and other ectoparasites on domestic animals; veterinary medicine. [Bayer AG]

Baytigan® AR. Softening, lightfast, anionic polymer retanning material for chrome leather. [Bayer AG]

Baytril®. Anti-infective agent for the therapy and prophylaxis of infectious diseases in animals. [Bayer AG]

Baytroid®. Fast-acting synthetic pyrethroid with a long residual effect and broad spectrum of activity; for use in controlling caterpillars, beetles and their larvae, cutworms, sucking pests on vegetables, pome and stone fruit, grapes, maize, soya, tobacco, cotton and other crops. [Bayer AG]

Bayvap®. Electric vaporizer tablets to combat gnats and mosquitos in rooms. [Bayer AG]

Bayvarol. For the control of ectoparasites in bees (varroa). [Bayer AG]

Bazak. Photographic chemicals. [Makhteshim Chemical Works Ltd]

BB10GF/15T. Bayblend, 10% glass fiber, 15% Teflon. [Compounding Tech.]

BB Accelerator. Derivative of dimethyl-p-phenylenediamine; a rubber vulcanization accelerator. [BFGoodrich]*

BB Chlorothalonil; CAS 1897-45-6; Chlorothalonil; a fungicide for wide range of agricultural crops. [Brown Butlin Ltd]

B.B.D.C. Standard Alloy. An alloy of 88.5% copper, 10% tin, 1% nickel, 0.25% lead, and 0.25% phosphorus.

BBP. See butyl benzyl phthalate

BBS. CAS 68334-28-1; Partially hydrogenated vegetable oil. [Karlshamns]

BBTS. CAS 95-31-8; N-t-Butyl-2-benzothioazole sulfenamide; delayed-action accelerator for natural and synthetic rubbers. [Akrochem]

B.C. 500. A proprietary vitamin supplement containing thiamine, riboflavine, pyridoxine, calcium partothenate, cyanocobalamin and ascorbic acid. [Whitehall Laboratories Ltd]

B.C. 500 With Iron. Vitamin supplement containing ferrous fumarate, thiamine, riboflavine, nicotinamide, pyridoxine, calcium pantothenate, ascorbic acid; for treatment of iron deficiency when therapeutic doses of water-soluble vitamins are also required. [Whitehall Laboratories Ltd]

BCF. Bromochlorodifluoromethane/Halon 1211 fire extinguishant. [ICI Chem. & Polymers Ltd]

BDO. See 1,4-butanediol

Beanfeast. Proteinaceous substances used as food or as ingredients for food. [Courtaulds plc]

Bear. A cotton fabric grade of Tufnol industrial laminates. [Tufnol Ltd]

Bearflex® LAO. Extender oil offering light color and low aniline pts.; for elastomer compounding, especially nitrile, neoprene, SBR, and natural rubber. [Witco/Golden Bear]

Bearium. Proprietary alloys of copper with 17.5-28% lead and 10% tin. Bearing metal. §

Beatin. A trademark used in Germany for the sale of famel syrup (qv). [The Boots Co plc]

Beaudouin's reagent. 1% furfural in alcohol.

Beaver steel. A proprietary nickel chromium steel containing 1.5% nickel, 0.75% chromium, 0.6% manganese, and 0.55% carbon. §

---

* Trade name not verified as available   † Trade name verified as obsolete

Beaverwhite 200. CAS 14807-96-6; Talc; high purity platy talc pigment for paints, rubber dusting, building products, adhesives, caulks, and sealants. [Luzenac Am.]

Bebate. CAS 22298-29-9; A proprietary preparation containing betamethasone 17-benzoate; a steroid skin preparation. [Warner] *

BE Buffer. Tris-borate-EDTA solution or powder. [Am. Research Prods.] ‡

Becalin. See Unasyn [Pfizer International]

Becantyl. A proprietary preparation containing sodium dibunate; a cough syrup. [Horlicks] *

Beckacite® 110, 115. Phenolic modified rosin ester resin; resins for paste ink applications. [Arizona]

Beckacite®425. Maleic modified rosin ester resin; resins for paste ink applications. [Arizona]

Beckacite®4900. Modified tall oil rosin; for manufacturing of paper size, intermediate in rosin derivative production, printing ink binders as resins, tackifier resin in sealants and mastics, as starting point rosin for resin esters. [Arizona]

Beckamine®. Synthetic resins for use in paints, varnishes, enamels, etc. and in industrial arts. [Reichhold]

Beckamine® 21-500. Urea-formaldehyde resin. [Reichhold]

Beckol. A proprietary trade name for synthetic alkyd resin.§

Beckolin®. Synthetic oils for use in industrial arts, paints, varnishes, lacquers, linoleum, oil cloth, inks, etc. [Reichhold]

Beckolloid. A proprietary synthetic resin plastic. §

Beckopol®. Synthetic resins for use in the industrial arts, paints, varnish, enamel, lacquer, etc. [Reichhold]

Beckosol. Plastic synthetic resins for paints, varnishes, lacquers, paper/cardboard goods. §

Beckosol 13-400. Waterborne med. oil alkyd. [Reichhold]

Becksol. A proprietary trade name for alkyd synthetic resins.§

Beckton white. See lithopone.

Beclovent Inhaler. CAS 5534-09-8; Beclomethasone dipropionate; glucocorticoid. [Glaxo Inc] *

Becomel. A proprietary preparation containing nicotinamide, thiamine, pyridoxine and riboflavine; vitamin supplement. [Crookes Laboratories] *

Beconase. CAS 5534-09-8; A proprietary preparation of beclomethasone dipropionate as an aerosol; used in the treatment of allergic rhinitis. [Allen & Hanburys Ltd]

Beconase Nasal Inhaler. CAS 5534-09-8; Beclomethasone dipropionate; glucocorticoid. [Glaxo Inc] *

Becosal. A range of fire extinguishing powders which are also compatible with foam. [Degussa AG] *

Becosed. A proprietary preparation of phenobarbitone, aneurine, riboflavin and nicotinamide. [H N Norton & Co Ltd] ‡

Becosym. A proprietary preparation containing thiamine, riboflavine, nicotinamide and pyridoxine; a vitamin B complex. [Roche Products Ltd] *

Becotide. CAS 5534-09-8; A proprietary preparation of beclomethasone dipropionate as an aerosol; used in asthma therapy. [Allen & Hanburys Ltd]

Becxopox. Epoxide resins; for paints, adhesives, sealants and encapsulants. [Resinous Chemicals Ltd] †

Bedesol. Synthetic resins. §

Bedranol®. CAS 318-98-9; Pink film coated tablets containing 10mg, 40mg, 80mg and 160mg propranolol hydrochloride BP; used in the treatment of hypertension,

angina pectoris, cardiac dysrhythmias, tachycardia, anxiety, essential tremor, prophylaxis of migraine, adjunctive therapy in thyrotoxicosis. [Lagap Pharmaceuticals Ltd]

beef tallow. See tallow.

Beehive Balsam. A proprietary preparation of glycerin, honey, lemon and ipecacuanha; a cough linctus. [Ayrton Saunders plc] ‡

Beesix. CAS 58-56-0; Pyridoxine hydrochloride; vitamin. [O'Neal, Jones & Feldman Pharmaceuticals] *

beeswax. (cera alba) CAS 8006-40-4 (white), 8012-89-3 (yellow); Purified wax from the honeycomb; food additive, furniture and floor waxes, shoe polishes, leather dressings, anatomical specimens, artificial fruit, textile sizes and finishes, church candles, cosmetic creams, adhesive compounds [British Wax Refining; ICI Spec.; Koster Keunen; Maruzen Fine Chem.; Strahl & Pitsch]

Beetafil. Foam resins for thermal insulation. [BIP Chemicals Ltd]

Beetafin. Aqueous polyurethane resins; used in leather and textile industries. [BIP Chemicals Ltd]

Beetle. Thermosetting molding powders (urea-formaldehyde, polyester), thermoplastic molding materials (nylon 6 and 66, PET and PBT polyesters, polycarbonate, acetal), resins (unsaturated polyester, urea-formaldehyde, melamine-formaldehyde and alkyds for paints, textiles, adhesives. [BIP Chemicals Ltd]

Beetle Resin BT 333. A cyclic reactant; recommended for soft mechanical finishes on cellulosic fabrics, also soft crease resistant and shrink resistant finishes on cellulose/synthetic fiber blends. [BIP Chemicals Ltd]

Beetle Resin BT 334. A cyclic reactant-modified melamine cross linking agent; recommended for chlorine resistant finishes on cotton fabrics and easy care finishes on cellulose/synthetic fiber blends. [BIP Chemicals Ltd]

Beetle Resin W69. A modified urea-formaldehyde resin; recommended for glueing "Vac-Vac" treated timbers where difficulties may be experienced with standard adhesives. [BIP Chemicals Ltd]

beet molasses. See molasses.

Beetomax. CAS 13684-63-4; Emulsifiable concentrate containing 114 g/l phenmedipham; used for weed control for beet crops. [Fine Agrochemicals Ltd]

beet sugar. See sucrose.

Beetup. CAS 13684-63-4; Emulsifiable concentrate containing 114 g/l phenmedipham; weed control for beet crops. [MTM Agrochemicals Ltd]

Beflavine Roche. CAS 83-88-5; A proprietary preparation of riboflavin (vitamin B_2); vitamin supplement. [Roche Products Ltd] *

Beflavit. CAS 83-88-5; vitamin B_2 (riboflavin) preparations. [Roche Products Ltd] *

behenyl alcohol. (1-docosanol; alcohol C_{22}) Mixture of fatty alcohols chiefly of n-docosanol; $CH_3(CH_2)_{20}CH_2OH$; CAS 661-19-8; EINECS 211-546-6; synthetic fibers, lubricants, evaporation retardant on water surfaces. [M. Michel; Schweizerhall; Sherex; Vista]

Belclene. Water treatment chemicals. [Ciba plc] *

Belco. Cellulose car finish paints and resins. [ICI Chem. & Polymers Ltd]

Beldex. ASS modifiers for PVC. [GE Plastics ABS Ltd] *

Belfasin 320 Crushed. Softener concentrate for use on package dyed yarns. [Henkel/Textiles]

Belgard. Water treatment chemicals. [Ciba plc] *

---

‡ Trade name and manufacturer not verified   § Trade name without identified manufacturer

Beligno seeds. *See* olives of Java.

Belite. Antifoaming agents. [Ciba plc] *

Belladenal. A proprietary preparation of phenobarbitone and belladonna leaf; a sedative. [Sandoz] *

Bellasol. Foam preventives. [Ciba plc] *

Bellauxine. Water treatment chemicals. [Ciba plc] *

Bellclo. Soluble concentrate containing 250 g 2,4-DB and 53 g mecoprop per liter; a translocated herbicide. [MTM Agrochemicals Ltd]

Bellergal. A proprietary preparation of belladonna, ergotamine tartrate and phenobarbitone; a sedative. [Sandoz]*

Bellite. Explosives for coal mine consisting of ammonium nitrate and dinitrobenzene with or without sodium chloride.

Belloid. Dispersing agents. [Ciba plc] *

bell pepper. The fruit of *Capsicum grossum.*

Belmac Plus. Soluble concentrate containing 38 g MCPA and 262 g MCPB per liter; for control of weeds in undersown cereals and grassland. [MTM Agrochemicals Ltd]

Belmac Straight. CAS 94-81-5; Soluble concentrate containing 400 g/l MCPB; for control of weeds in undersown cereals and grassland. [MTM Agrochemicals Ltd]

Beloc. CAS 56392-17-7; A proprietary preparation of metoprolol tartrate; used for hypertension, cardiac arrythmias. [Pfizer International]

Belpro. A water-based acrylic copolymer temporary coating and remover (also known as Tempro). [ICI Chem. & Polymers Ltd]

Belro. Wood rosin derivative insoluble in water, soluble in organic solvents, fats and oils; acid number 114; drop softening point of 90 C; construction adhesive. [Hercules] *

Belsil ADM 6041 E. Amodimethicone emulsion; substantive conditioner for hair conditioners and shampoos. [Wacker-Chemie]

Belsil CM 020. CAS 69430-24-6; Cyclomethicone; used in deodorants to replace alcohol and IPM, prevent stickiness of aerosol nozzles and roll-on balls; also for hair and skin care products and decorative cosmetics. [Wacker-Chemie]

Belsil CM 1000. Dimethyl polysiloxane; used in skin and hair care formulas; gives skin a pleasant, supple feel. [Wacker-Chemie]

Belsil DM 0.65. Dimethicone; used in skin and hair care products and decorative cosmetics; enhances suppleness and gives soft, velvety feel to skin; prevents stickiness and increases water resist. of cosmetics. [Wacker-Chemie]

Belsil DMC 6031. Dimethicone copolyol; wetting aid improving surface slip, emollient, moisturizer, fatting agent for cosmetics; stabilizer for foams and emulsions; plasticizer for hair spray resins. [Wacker-Chemie]

Belsil PDM 20. Phenyl dimethicone; used in skin and hair care products and decorative cosmetics; give excellent feel to skin; high penetrating and water repellent props.; impart suppleness and depth of color to hair. [Wacker-Chemie]

Belsil SDM 6021. CAS 68554-53-0; Stearoxydimethicone; give nongreasy, soft, velvety feel to skin; enhance gloss and color brightness in decorative cosmetics; good spreading props., protection against aq. media. [Wacker-Chemie]

Belsoft. Blend of acid amides and fatty alcohol sulfates, anionic; softener for cellulosic and synthetic fibers, especially suitable for terry cloths and textured polyester. [Henkel Chemicals Ltd] *

Belsol. Wetting and penetrating agents. [J C Bottomley] *

Beltherm. Fuel oil additive. [Ciba plc] *

Belzak AC. α-Sodium glucoheptonate dihydrate crystals; sequestering agent for metal cleaning and processing, various industrial cleaning compounds, bottle wash, set retarder in concrete and trace metals for agriculture. [Belzak Corporation]

Belzak BL-50. Beta sodium glucoheptonate 50% liquid; used for metal cleaning and processing, various industrial cleaning compounds, bottle wash, set retarder in concrete and trace metals for agriculture. [Belzak Corporation]

Bemal. An alloy of 70% copper, 29% zinc, and traces of lead and iron.

Bemberg®. Cuprammonium rayon; viscose rayon for textile industry. [Asahi Chem. Industry]

Bemiliese®. Nonwoven fabric of cuprammonium rayon. [Asahi Chem. Industry]

Benacine. A proprietary preparation of diphenhydramine and hyoscine hydrobromide; antinauseant. [Parke-Davis] *

Benadon. A proprietary preparation of pyridoxine; an antinauseant. [Roche Products Ltd] *

Benadryl. CAS 147-24-0; Diphenhydramine hydrochloride; antihistaminic. [Parke-Davis]

Benafed. A proprietary preparation of diphenhydramine hydrochloride, dextramethorphan hydrobromide, pseudoephedrine hydrochloride, ammonium chloride, sodium citrate, chloroform and menthol in a syrup; a cough suppressant. [Parke-Davis] *

Ben-A-Gel®. CAS 1302-78-9; Beneficiated bentonite clay; for use as thickening, gelling and emulsifying agent for water systems. [Rheox Inc]

Benalite. A proprietary trade name for a lignin plastic; cured lignin sheets. §

Benaloid. A proprietary trade name for uncured lignin sheets. §

Benapen. Benethamine penicillin. [Glaxo Laboratories] *

Benaqua. Rheological additives for aqueous systems. [Rheox Inc]

Benathix. Rheological additives; for unsaturated polyester laminating resins. [Rheox Inc]

Benathix. Thixotropic and thickening additives; used in unsaturated polyester laminating resins. [Steetley Minerals Ltd]

Benatol. CAS 100-51-6; Pure benzyl alcohol. [Bush Boake Allen Ltd] †

Benazalox. CAS 3813-05-6, 57754-85-5; Benazolin + clopyralid; an herbicide mixture for use in oilseed rape. [Schering Agrochemicals Ltd]

Bendalite. A proprietary trade name for a cast styrene synthetic resin. §

Bendigon®. An antihypertensive with a depot effect. [Bayer AG]

Bendopa. CAS 59-92-7; Levodopa; antiParkinsonian. [ICN Pharmaceuticals Inc] *

Benecel®Hydroxypropylmethylcellulose. CAS 9004-65-3; Methylhydroxypropylcellulose; *see* Benecel CM for applications and props. [Aqualon]

Benecel® Methylcellulose. Carboxymethylmethylcellu-

---

lose; thickener, stabilizer, rheology control agent, film-former, suspending agent, water-retention aid, binder for food, pharmaceutical, and cosmetic industries. [Aqualon]

Benedict Plate. A nickel silver containing 57% copper, 28% zinc, and 15% nickel.

Benefex. CAS 1861-40-1; Active ingredient: benfluralin; N-butyl-N-ethyl-2,6-dinitro-4-trifluoromethylaniline; pre-emergence herbicide with a wide range of weed control both of annual grass weeds and broad-leaved weeds. [Agan Chemical Manufacturers Ltd]

Benemid. CAS 57-66-9; Probenecid; uricosuric agent for the treatment of chronic gout. [Merck & Co Inc]

Benerva. CAS 67-03-8; A proprietary preparation of thiamine hydrochloride (vitamin B_1). [Roche Products Ltd]*

Benerva Compound. Vitamin B_1, B_2, and nicotinamide. [Roche Products Ltd] *

Benexol Roche. A proprietary preparation of vitamins B_1 and B_6. [Roche Products Ltd] *

Bengal catechu. See cutch.

Bengal isinglass. See agar-agar.

Bengue's balsam. 20% menthol, 20% methyl salicylate in a lanolin base; rubefacient - for the symptomatic relief of muscular pain and stiffness (including rheumatic pains), also as a decongestant by application to the chest or by inhalation. [Bengue & Co Ltd] *

Benical. A proprietary preparation of dextromethorphan, pseudoephedrine and chlorphenamine. cough and cold preparation. [Roche Products Ltd] *

beni oil. See gingelly oil.

beniseed oil. See gingelly oil.

Benlate. CAS 17804-35-2; Benomyl; fungicide for garden use. (Sold in UK on behalf of Du Pont). [ICI Chem. & Polymers Ltd]

Benlate®. CAS 17804-35-2; Benomyl; a systemic fungicide. [DuPont UK]

Bennatate. A proprietary trade name for a cellulose acetate plastic. §

benné oil. See gingelly oil.

Benol. White mineral oil. [Rheox Inc]

Benoquin. Monobenzone; depigmentor. [ICN Pharmaceuticals Inc]

Benoral. CAS 5003-48-5; A proprietary preparation of benorylate; an antirheumatic drug. [Sterling Research Laboratories] *

Benox® L-40LV. CAS 94-36-0; Benzoyl peroxide dispersion; catalyst for unsaturated polyester and vinyl ester resins; faster gel to peak and a lower exotherm than methyl ethyl ketone peroxide systems; polymerization initiator for vinyl monomers, e.g., styrene, acrylate. [Norac]

Benoxyl. CAS 94-36-0; A proprietary preparation of benzoyl peroxide; used in dermatology as an antibacterial agent. [Stiefel Laboratories Inc; Stiefel Laboratories (UK) Ltd]

Benpon®. Psychotropic drug. [Bayer AG]

Bensapol. A proprietary trade name for a wetting agent consisting of sulfonated oils and a solvent. §

Bensuccin. A proprietary preparation of benzyl succinate. §

Bentalol. CAS 100-51-6; Specially pure benzyl alcohol. [Bush Boake Allen Ltd] †

Benteine. CAS 140-11-4; Specially pure benzyl acetate. [Bush Boake Allen Ltd]

Bentene. Butyraldehyde-aniline; a proprietary rubber accelerator. [Naugatuck (US Rubber)] *

Bentobrite® 770. Micronized white powd. used as general purpose suspension agent and gellant for household and industrial products. [Am. Colloid]

Bentokol. Foundry coal dust replacement additive. [Foseco (F.S.) Ltd]

Bentone. Rheological additives; used for paint, ink, grease, caulks, sealants, cosmetics and adhesives. [Rheox Inc]

Bentone. Thixotropic and thickening additives; used in paint, ink, grease, caulks, sealants, cosmetics and adhesives. [Steetley Minerals Ltd]

Bentone-34. Rheological additives. [Rheox Inc]

Bentone Gel. Rheological additives; pregelled bentone additive for cosmetics. [Rheox Inc]

Bentone Gel. Thixotropic and thickening additives. pregelled bentone additive for cosmetics. [Steetley Minerals Ltd]

Bentone SD. Super dispersing rheological additives; used for paint, ink, caulks, sealants and adhesives. [Rheox Inc]

bentonite. (colloidal clay; soap clay; wilkinite; tonsil L80) CAS 1302-78-9; Native hydrated colloidal aluminum silicate clay; oil-well drilling fluids, cement slurries for oil-well casings, thickener, fireproofing, cosmetics, decolorizing agent, filler in ceramics, emulsifier for oils, suspending agent in pharmaceuticals, base for plasters. [Am. Colloid; Dry Branch Kaolin; Norsk Hydro AS; L. A. Salomon; Southern Clay; R.T. Vanderbilt]

BentoPharm. CAS 1302-78-9; Bentonite, pharmaceutical quality; suspension and thickening agent. [Bromhead & Denison Ltd]

Bentyl. CAS 67-92-5; Dicyclomine hydrochloride; anticholinergic. [Merrell Dow Pharmaceuticals Inc] *

Benvic. Granulated PVC compound. [Laporte Industries Ltd]

Benylate. CAS 120-51-4; Benzyl benzoate; pharmaceutic necessity for dimercaprol. [Sterling Drug Inc] *

Benylin DM. Dextromethorphan hydrobromide; antitussive. [Parke-Davis] *

Benylin Expectorant. A proprietary preparation containing diphenhydramine hydrochloride, ammonium chloride, sodium citrate, chloroform and menthol; a cough linctus. [Parke-Davis] *

Benzac. CAS 94-36-0; Benzoyl peroxide; keratolytic. [Galderma Laboratories]

benzalacetone. See benzylidene acetone

benzaldehyde. (benzoic aldehyde; synthetic oil of bitter almond) C_6H_5CHO; CAS 100-52-7; EINECS 202-860-4; Chemical intermediate for dyes, flavors, perfumes, aromatic alcohols; solvent for oils, resins, cellulose acetate and nitrate; manufacture of cinnamic acid, benzoic acid; pharmaceuticals; photographic chemicals. [Aceto; Aldrich; DSM BV; R.W. Greeff; Haarmann & Reimer Janssen Chimica; Penta Mfg.; Snia (UK); Spectrum Chem. Mfg.]

benzalkonium chloride. (alkyl dimethyl benzyl ammonium chloride) Quaternary ammonium salt; mixture of alkylbenzyldimethylammonium chlorides; $C_6H_5CH_2N$-$(CH_3)_2RCl$, $R = C_8H_{17}$ to $C_{18}H_{37}$; CAS 8001-54-5; 61789-71-7; 68391-01-5; 68424-85-1; 68989-00-4; 85409-22-9; EINECS 263-080-8; 269-919-4; 270-325-2; 287-089-1; Cationic detergent; surface antiseptic, fungicide, preservative in pharmaceuticals. [Akzo; Chemron; EM Industries; Lonza AG; Sherex; Stepan; Witco/Humko]

Benzamin®. Diazotizing dyes for cotton and other vegetable

---

fibers, textiles from regenerated cellulose. [Bayer AG]

Benzanil. A range of direct dyes; for dyeing of cotton and viscose fibers. [Yorkshire Chemicals plc] *

Benzedrex. CAS 3595-11-7; Propylhexedrine; adrenergic. [SmithKline Beecham] *

benzene. (benzol; cyclohexatriene; annulene) C_6H_6; CAS 71-43-2; EINECS 200-753-7; Manufacture of ethylbenzene, dodecylbenzene, cyclohexane, phenol, nitrobenzene, chlorobenzene; dyes, medicinals, artificial leather, airplane dopes, lacquers; solv. for waxes, resins, oils. [BP Chem. Ltd; Exxon; Janssen Chimica; Mitsubishi Petrochem.; Mitsui Petrochem. Ind.; Mobil; Oxychem; Shell]

benzene carboxylic acid. See benzoic acid.

benzene chloride. See chlorobenzene.

1,3-benzenediamine. See m-phenylenediamine.

1,4-benzenediamine. See p-phenylenediamine.

1,2-benzenedicarboxylic acid dibutyl ester. See dibutyl phthalate.

1,2-benzenedicarboxylic acid dimethyl ester. See dimethyl phthalate.

1,2-benzenedicarboxylic acid dioctyl ester. See dioctyl phthalate.

1,2-benzenediol. See pyrocatechol.

1,3-benzenediol. See resorcinol.

1,4-benzenediol. See hydroquinone.

1,3-benzenediol, 4-chlor-. See 4-chlororesorcinol.

1,4-benzenediol, 2-(1,1-dimethylethyl)-. See t-butyl hydroquinone.

benzenesulfonamide. $C_6H_5SO_2NH_2$; CAS 98-10-2; EINECS 202-637-1. [Unitex]

1,2,3-benzenetriol. See pyrogallol.

Benzets. A proprietary preparation of benzalkonium chloride and benzocaine; throat lozenges. [H N Norton & Co Ltd] ‡

Benzex. Benzyl cellulose. §

benzhydrol. Diphenyl methanol, $(C_6H_5)_2$.CH.OH.

Benzilan. CAS 510-15-6; Active ingredient: chlorobenzilate; agricultural acaricide. [Makhteshim Chemical Works Ltd]

benzimidazole. (1,3-benzodiazole) $C_7H_6N_2$; CAS 51-17-2; EINECS 200-081-4. [Atomergic Chemetals; Janssen Chimica; Penta Mfg.; Schweizerhall]

2-benzimidazolethiol. See 2-mercaptobenzimidazole.

benzin. See naphtha.

benzisotriazole. See 1H-benzotriazole.

Benzo®. Direct dyes; suitable for cotton and rayon goods where no special demands are made on fastness properties. [Bayer AG; Bayer plc]

benzocaine. (ethyl-p-aminobenzoate hydrochloride; procaine hydrochloride; anesthesol) CAS 51-05-8; Local anesthetic in medicine, suntan preparations. [R.W. Greeff; Natl. Starch & Chem.; Nipa Labs; Roussel Labs Ltd; Schweizerhall]

Benzo® Cuprol. Direct dyestuffs whose wet fastness and light fastness are considerably improved by an aftertreatment with copper salts. [Bayer AG]

Benzodent. Analgesic denture ointment, for denture discomfort. [Richardson-Vicks Inc] *

1,3-benzodiazole. See benzimidazole.

Benzoflex. Plasticizer. [Velsicol] *

Benzoflex 2-45. CAS 120-55-8; A proprietary trade name for diethylene glycol dibenzoate. [Tennessee Corp] *

Benzofloc. Organic and inorganic specialty flocculant and coagulant polymers; used for solids flocculation in industrial and municipal solids, dissolved or collodial form, color, turbidity, algae removal; specialty application. [Benzsay & Harrison Inc]

benzoglycolic acid. See mandelic acid.

benzoic acid. (benzenecarboxylic acid; phenylformic acid; dracylic acid) $C_7H_6O_2$; CAS 65-85-0; EINECS 200-618-2; Sodium and butyl benzoates, plasticizers, benzoyl chlorides, food preservatives, flavors, perfumes, antifungal agent. [Atochem SA; R.W. Greeff; Mallinckrodt; Penta Mfg.; Schaefer Salt; Velsicol]

Benzoic Acid K. Gloss and flow promoter for use in oil and synthetic resin-based topcoats; hardener for rubber soles, heels, floor coverings. [Bayer AG]

benzoic acid, butyl ester. See butyl benzoate.

benzoic acid, 2-hydroxy-. See salicylic acid.

benzoic acid methyl ester. See methyl benzoate.

benzoic acid potassium salt. See potassium benzoate.

benzoic acid, sodium salt. See sodium benzoate.

benzoic aldehyde. See benzaldehyde.

benzoin. (α-hydroxybenzyl phenyl ketone; α-hydroxy-α-phenylacetophenone) $C_6H_5CH(OH)COC_6H_5$; CAS 119-53-9; EINECS 204-331-3; Organic synthesis; intermediate; photopolymerization catalyst. [Aldrich; Janssen Chimica; Dr. Madis Labs; Snia (UK); Spectrum Chem. Mfg.]

benzol. See benzene

benzonitrile. (phenyl cyanide) C_6H_5CN; CAS 100-47-0; EINECS 202-855-7; Manufacture of benzoquanamine; intermediate for rubber chemicals; solvent for nitrile rubber, specialty lacquers, resins and polymers, anhydrous metallic salts. [Penta Mfg.; PMC Specialties; Spectrum Chem. Mfg.]

benzoperoxide. See benzoyl peroxide

benzophenone-1. (2,4-dihydroxybenzophenone; benzoresorcinol; 4-benzoyl resorcinol) Organic benzophenone derivative.; $C_6H_5COC_6H_3(OH)_2$; CAS 131-56-6; EINECS 205-029-4; uv absorber in polymers. [EM Industries; Ferro/Bedford; R.W. Greeff; Haarmann & Reimer; Hoechst Celanese; Quest Int'l.; Sartomer]

benzophenone-2. (2,2´,4,4´-tetrahydroxy benzophenone) Organic benzophenone derivative; $C_{13}H_{10}O_5$; CAS 131-55-5; EINECS 205-028-9; used as commercial uv absorber. [EM Industries; Ferro/Bedford; R.W. Greeff; Haarmann & Reimer; Hoechst Celanese; Quest Int'l.; Sartomer]

benzophenone-3. (2-hydroxy-4-methoxybenzophenone; 2-hydroxy-4-methoxyphenyl; phenylmethanone; oxybenzone) Organic benzophenone derivative; $C_{14}H_{12}O_3$; CAS 131-57-7; EINECS 205-031-5; used as a sunscreen. [EM Industries; Ferro/Bedford; R.W. Greeff; Haarmann & Reimer; Hoechst Celanese; Quest Int'l.; Sartomer]

benzophenone-4. (2-hydroxy-4-methoxybenzophenone-5-sulfonic acid; sulisobenzone) Organic benzophenone derivative; $C_{14}H_{12}O_6S$; CAS 4065-45-6; EINECS 223-772-2; uv absorber in sunscreen products and in hair sprays and shampoos for dyed and tinted hair; for leather and textile fibers. [EM Industries; Ferro/Bedford; R.W. Greeff; Haarmann & Reimer; Hoechst Celanese; Quest Int'l.; Sartomer]

benzophenone-6. (2,2´-dihydroxy-4,4´-dimethoxy-benzophenone; bis(2-hydroxy-4-methoxyphenyl) methanone) Organic benzophenone derivative;

$C_{15}H_{14}O_5$; CAS 131-54-4; EINECS 205-027-3; uv light absorber, especially in paints, plastics. [EM Industries; Ferro/Bedford; R.W. Greeff; Haarmann & Reimer; Hoechst Celanese; Quest Int'l.; Sartomer]

benzophenone-8. (2,2'-dihydroxy-4-methoxybenzophenone; dioxybenzone) Organic benzophenone derivative; $C_{14}H_{12}O_4$; CAS 131-53-3; EINECS 205-026-8; uv absorber, light stabilizer. [EM Industries; Ferro/Bedford; R.W. Greeff; Haarmann & Reimer; Hoechst Celanese; Quest Int'l.; Sartomer]

benzophenone-9. (disodium 2,2'-dihydroxy-4,4'-dimethoxy-5,5'-disulfobenzophenone) Organic benzophenone derivative.; $C_{15}H_{14}O_8S \cdot Na$; CAS 3121-60-6; EINECS 221-498-8; uv absorber in cosmetic formulations, textiles and water-based paints. [EM Industries; Ferro/Bedford; R.W. Greeff; Haarmann & Reimer; Hoechst Celanese; Quest Int'l.; Sartomer]

benzophenone-11. Organic benzophenone derivative.; mixture of benzophenone-6 and -2 and other tetra-substituted benzophenone materials; CAS 1341-54-4; UV absorber used in NC lacquer, fluorescent paint, inks, and for protecting furniture woods, colored liq. toiletries and cleaning agents, isocyanate systems, and butyrate metal lacquers. [EM Industries; Ferro/Bedford; R.W. Greeff; Haarmann & Reimer; Hoechst Celanese; Quest Int'l.; Sartomer]

benzophenone-12. (2-hydroxy-4-(octyloxy) benzophenone. [2-hydroxy-4-(octyloxy)phenyl]phenylmethanone; octabenzone) Organic benzophenone derivative.; $C_{21}H_{26}O_3$; CAS 1843-05-6; EINECS 217-421-2; uv stabilizer for polyethylene. [EM Industries; Ferro/Bedford; R.W. Greeff; Haarmann & Reimer; Hoechst Celanese; Quest Int'l.; Sartomer]

benzoquinone. Quinone, $C_6H_4O_2$.

benzoresorcin. Dihydroxy-benzophenone.

benzoresorcinol. See benzophenone-1.

2-benzothiazolethiol. See 2-mercaptobenzothiazole.

2(3H)-benzothiazolethione. See 2-mercaptobenzothiazole.

1H-benzotriazole. (benzisotriazole; 1,2-aminozophenylene; azimidobenzene) $C_6H_5N_3$; CAS 95-14-7; EINECS 202-394-1; Chelating agent and sesquestrant for copper ions; corrosion inhibitor for copper, brass, bronze; used in antifreeze, cleaners, coatings, detergents, functional fluids, metalworking fluids, packaging materials, polishes. [Atomergic Chemetals; Miles; PMC Specialties; Sandoz]

benzoyl peroxide. (dibenzoyl peroxide; benzoperoxide; benzoyl superoxide) Organic peroxide.; $[C_6H_5C(O)]_2O_2$; CAS 94-36-0; EINECS 202-327-6; oxidizing agent in bleaching oils, flours; catalyst for plastics; initiator in polymerization. [Akzo Chemie SA; Atochem N. Am.; Norac]

4-benzoyl resorcinol. See benzophenone-1.

benzoyl superoxide. See benzoyl peroxide.

benztrimonium chloride. See benzyl trimethyl ammonium chloride.

Benztrone. Oestradiol benzoate BP in ethyl oleate BP for injection; may be used for estrogen replacement therapy associated with the menopause, female hypogonadiom and dysmenorrhoea; also used for the treatment of malignant neoplasm of the breast of post menopausal women and of the prostate in men. [Paines & Byrne Ltd]*

Benzyl. Acid wool dyestuffs. [Ciba-Geigy] *

benzyl acetate. (phenylmethyl acetate) CH_3COO-$CH_2C_6H_5$; CAS 140-11-4; EINECS 205-399-7; Artificial jasmine and other perfumes; soap perfume; flavors; solvent and high boiler for cellulose acetate and nitrate, natural and synthetic resins; oils; lacquers; polishes; printing inks; varnish removers. [Haarmann & Reimer; Janssen Chimica; MTM Spec. Ltd; Penta Mfg.; Quest Int'l.]

benzyl alcohol. (hydroxytoluene; phenylmethanol; phenylcarbinol) $C_6H_5CH_2OH$; CAS 100-51-6; EINECS 202-859-9; Perfumes; flavors; photographic developer; dyeing nylon, textiles, sheet plastics; solvent for dyestuffs, cellulose esters, casein, waxes; heat-sealing polyethylene films; intermediate for benzyl esters and ethers; bacteriostat; cosmetics; inks. [Givaudan Iberica SA; R.W. Greeff; Janssen Chimica; Penta Mfg.; Quest Int'l.]

benzyl benzoate. $C_6H_5COOCH_2C_6H_5$; CAS 120-51-4; EINECS 204-402-9; Fixative and solvent for musk in perfumes and flavors; external medicine; plasticizer for nitrocellulose and cellulose acetate; miticide. [Haarmann & Reimer; Janssen Chimica; Penta Mfg.; Schweizerhall]

benzyl butyl phthalate. See butyl benzyl phthalate

benzyl chlorocarbonate. See benzyl chloroformate

benzyl chloroformate. (benzyl chlorocarbonate) $ClCOOCH_2C_6H_5$; CAS 501-53-1; EINECS 207-925-0; Peptide synthesis. [Janssen Chimica; PPG Industries]

N-benzyldimethylamine. (DMBA; N-(phenylmethyl) dimethylamine; N,N-dimethylbenzylamine) $C_6H_5CH_2N$-$(CH_3)_2$; CAS 103-83-3; EINECS 203-149-1; Intermediate, dehydrohalogenating catalyst, corrosion inhibitor, acid neutralizer, potting compounds, adhesives, cellulose modifier. [Aceto; R.W. Greeff; Penta Mfg.; PMC Specialties; Zeeland]

benzylidene acetone. (benzalacetone; methyl styryl ketone; trans-4-phenyl-3-buten-2-one) $C_6H_5CH:CH$-$COCH_3$; CAS 122-57-6; EINECS 204-555-1; Organic synthesis; perfumery; fixative; flavors. [Penta Mfg.; Raschig; Schweizerhall]

benzyltriethyl ammonium chloride. (N,N,N-triethylbenzenemethanaminium chloride) Quaternary ammonium salt; $C_6H_5CH_2N(Cl)(C_2H_5)_3$; CAS 56-37-1; EINECS 200-270-1; solvent for cellulose; gelling inhibitor in polyester resins; intermediate. [Janssen Chimica; Zeeland]

benzyl trimethyl ammonium chloride. (TMBAC; trimethylbenzylammonium chloride; benzyl trimethyl ammonium chloride; benztrimonium chloride) Quaternary ammonium salt; $C_6H_5CH_2N(CH_3)_3Cl$; CAS 56-93-9; EINECS 200-300-3; dispersant, dye leveler and retarder, emulsifier used in textile industry; solvent for cellulose; gelling inhibitor in polyester resins; intermediate. [Janssen Chimica; Sherex; Sybron]

Benzyl Tuex®. CAS 10591-85-2; N,N,N',N'-Tetrabenzylthiuram disulfide; primary or sec. accelerator for natural and synthetic rubbers; for tire compounds, wire insulation, mechanical goods, sponge, footwear, sheeting, hose; activator for sulfenamide and thiazole accelerators; sulfur donor minimizing reversion in natural rubber. [Uniroyal]

Beogex. A proprietary preparation containing sodium acid phosphate and sodium bicarbonate; a laxative. [Pharmax Ltd] *

---

‡ Trade name and manufacturer not verified   § Trade name without identified manufacturer

Bepadin. CAS 74764-40-2; Bepridil hydrochloride; vasodilator. [Wallace Laboratories] *

Bepanthen. A proprietary preparation of pantothenic alcohol or dexpanthenol; used in the treatment of paralytic ileus. [Roche Products Ltd] *

Beplete. A proprietary preparation of phenobarbitone, thiamine hydrochloride, riboflavine, pyridoxine hydrochloride, nicotinamide, and alcohol. [Wyeth Laboratories] *

Beplex. A proprietary preparation of thiamine, riboflavine, and nicotinamide. [Wyeth Laboratories] *

Bercotox. Cattle dips and sprays. [The Wellcome Foundation Ltd] *

Berelex. CAS 77-06-5; Gibberellic acid; a plant growth regulator to increase cropping in apples and pears. [ICI Chem. & Polymers Ltd]

Bergaminol. (Bergamiol). Artificial oil of bergamot. Linalyl acetate; used in perfumery.

Bergauf. Skin protective soap with 0-48-G skin protective agent;. medicated foot-spray with bactericidal and fungicidal properties; eye cleansing cream for intensive and soothing cleansing of the area around eyes; for protection and care of the skin under environmental stress. [Dynamit Nobel Wien GmbH] *

Berger Colorizer - Full Gloss/Vinyl Matte/Vinyl Silk. Gloss: alkyd resin base thinned with white spirit. Matte: V.A. copolymer emulsion, water based. Silk: V.A. copolymer emulsion, water based; 450 colors available, high gloss external and internal application, matt and silk interior application. [Berger Jenson & Nicholson Ltd] *

Berger Cuprinol Woodpaints and Woodstains. Primer, Matte and Gloss: acrylic copolymer emulsion, water thinned; sheen finishes: alkyd based, white spirit thinned; exterior microporous wood paints and stains giving protection and decoration. [Berger Jenson & Nicholson Ltd] *

Berger mixture. A mixture of zinc, potassium chlorate, a chlorinating agent, such as carbon tetrachloride, and a filler, such as kieselguhr; a smoke screen material.

Berkatekt. A range of products designed to protect titanium, steel, and super alloys during heating and during the forging process. [Acheson Colloids]

Berkatens. CAS 52-53-9; Verapamil; for the treatment of angina pectoris, supraventricular tachycardia, atrial fibrillation and flutter; and hypertension. [Berk Pharmaceuticals Ltd] *

Berkazide. Bendrofluazide. [Berk Pharmaceuticals Ltd] *

Berkmycen. CAS 79-57-2; A proprietary preparation of oxytetracycline; an antibiotic. [Berk Pharmaceuticals Ltd] *

Berkolol. CAS 525-66-6; Propranolol; for the treatment of hypertension, angina pectoris, cardiac dysrhythmias and tachycardia. [Berk Pharmaceuticals Ltd] *

Berkozide. A proprietary preparation of bendrofluazide; a diuretic. [Berk Pharmaceuticals Ltd] *

Berkstop. A range of coatings to assist in the heat treatment (carburizing and nitriding) of steel. [Acheson Colloids]

Berlin Brown. A pigment produced by charring Prussian blue; a mixture of ferroso-ferric oxide and charcoal, and is used as an artist's color.

Bermacoll. CAS 9004-58-4; Ethyl hydroxyethyl cellulose; for adhesives. [Berol Nobel Ltd]

Bernel® Ester 168. Isocetyl octanoate; emollient with dry, silky feel. [Bernel]

Bernel®Ester 2014. Octyldodecyl myristate; rich emollient,

pigment disperser, lipstick component. [Bernel]

Bernel®Ester CO. CAS 59130-69-7; EINECS 261-619-1; Cetyl octanoate; noncomedogenic emollient; low visc. ester. [Bernel]

Bernel® Ester DID. CAS 103213-20-3; Diisopropyl dimer dilinoleate; emollient, film-former. [Bernel]

Bernel® Ester DISM. Diisostearyl malate; high visc. emollient for anhyd. systems. [Bernel]

Bernel® Ester DOM. Dioctyl maleate; emollient, oxybenzone solubilizer, cleanser and wax solv.; noncomedogenic; imparts shine in hair conditioners. [Bernel]

Bernel®Ester EHP. CAS 29806-73-3; EINECS 249-862-1; Octyl palmitate; cost-effective emollient. [Bernel]

Bernel®Ester NPDC. Neopentyl dicaprate; light, dry emollient; pigment wetter and binder. [Bernel]

Bernel®Ester TOC. Trioctyl citrate; noncomedogenic emollient, pigment wetter. [Bernel]

Bernel®OPG. CAS 59587-44-9; EINECS 261-819-9; Octyl pelargonate; dry, nonoily emollient. [Bernel]

Bernit. A proprietary trade name for a cellulose acetate plastic. §

Berocca. Proprietary preparation of vitamins $B_{1, 2, 5, 6}$, and B_{12}, vitamin C, biotin and nicotinamide. [Roche Laboratories; Roche Products Ltd]

Berol. Block polymers; machine dishwashing and rinse aids; emulsion polymerization. [Berol Kemi (UK) Ltd] *

Berol 02. CAS 68412-54-4; Nonylphenol ethoxylate; emulsifier; solv. cleaner; emulsion polymerization; biodeg. [Berol Nobel AB]

Berol 09. CAS 68412-54-4; Nonyl phenol ethoxylate; detergent, wetting agent, emulsifier, dispersant; biodeg. [Berol Nobel AB]

Berol 26. Nonionic surfactant of the nonylphenol ethoxylate type, in liquid form; emulsifiers; solvent cleaners. [Berol Kemi (UK) Ltd] *

Berol 28. CAS 26635-93-8; PEG-7 oleamine; emulsifier, dispersant, wetting agent, antistat, anticorrosive for agriculture, leather, textiles, metalworking and plastic industries. [Berol Nobel AB]

Berol 048. CAS 9043-30-5; Ethoxylated tridecyl fatty alcohol; wetting agent, detergent; for alkaline and acid industrial cleaners; gives high brittle foam. [Berol Nobel AB]

Berol 108. CAS 61791-12-6; PEG-40 castor oil; surfactant, emulsifier for chemical industry; as softener, rewetting agent, pigment dispersant, dye assistant, leveling agent for paints, textiles, leather; lubricant additive and emulsifier in lubricants for plastics, metals, textiles. [Berol Nobel AB]

Berol 191. CAS 61791-12-6; PEG-200 castor oil; surfactant, emulsifier for chemical industry; as softener, rewetting agent, pigment dipsersant, dye assistant, leveling agent for paint, textile, leather; lubricant additive/emulsifier in lubricants for plastics, metals, textiles. [Berol Nobel AB]

Berol 195. CAS 61791-12-6; PEG-32 castor oil; surfactant, emulsifier for chemical industry. [Berol Nobel AB]

Berol 198. CAS 61791-12-6; PEG-160 castor oil; surfactant, emulsifier for chemical industry; as softener, rewetting agent, pigment dispersant, dye assistant, leveling agent for paint, textile, leather; lubricant additive/emulsifier in lubricants for plastics, metals, and textiles. [Berol Nobel AB]

Berol 199. CAS 61791-12-6; PEG-32 castor oil; surfactant, emulsifier for chemical industry; as softener, rewetting

---

* Trade name not verified as available   † Trade name verified as obsolete

agent, pigment dipsersant, dye assistant, leveling agent for paint, textile, leather; lubricant additive/emulsifier in lubricants for plastics, metals, textiles. [Berol Nobel AB]

Berol 259. Nonionic surfactant consisting of nonylphenol ethoxylate in liquid form; foam depressor. [Berol Kemi (UK) Ltd] *

Berol 260. CAS 68439-46-3; C9-11 pareth-4; surfactant for alkaline systems, hard surf. cleaners, industrial and institutional cleaners, vehicle cleaners. [Berol Nobel AB]

Berol 267. Nonionic surfactant of the nonylphenol ethoxylate type, in liquid form; liquid cleaners; wetting agents; emulsifiers. [Berol Kemi (UK) Ltd] *

Berol 269. CAS 68891-21-4; Dinonylphenol ethoxylate; nonionic emulsifier for polymerization; detergent; biodeg. [Berol Nobel AB; Berol Kemi (UK) Ltd]

Berol 272 and 716. Dinonylphenol ethoxylate in liquid or paste form; nonionic low foaming detergents. [Berol Kemi (UK) Ltd] *

Berol 278, 281, 282, 291 and 292. Nonionic surfactants of the nonylphenol ethoxylate type in liquid or wax form; used for emulsion polymerization. [Berol Kemi (UK) Ltd]*

Berol 302. CAS 13127-82-7; PEG-2 oleamine; emulsifier, dispersant, wetting agent, antistat, anticorrosive for agriculture, leather, textiles, metalworking and plastic industries. [Berol Nobel AB]

Berol 303. CAS 26635-93-8; PEG-12 oleamine; emulsifier, wetting agent, antistat, anticorrosive for agriculture, leather, textiles, metalworking and plastic industries. [Berol Nobel AB]

Berol 307. CAS 61791-14-8; 61791-31-9; PEG-2 cocamine; emulsifier, wetting agent, antistat, anticorrosive for agriculture, leather, textiles, metalworking and plastic industries. [Berol Nobel AB]

Berol 374. CAS 9003-11-6; EO/PO block polymer; low-foaming surfactant, emulsifier for emulsion polymerization especially for latex paints; foam depressant/detergent for foodstuffs industry; emollient. [Berol Nobel AB]

Berol 381. CAS 61791-26-2; PEG-15 tallowamine; emulsifier, wetting agent, antistat, anticorrosive for agriculture, leather, textiles, metalworking and plastic industries. [Berol Nobel AB]

Berol 386. CAS 61791-26-2; PEG-20 tallowamine; emulsifier, wetting agent, antistat, anticorrosive for agriculture, leather, textiles, metalworking and plastics industries. [Berol Nobel AB]

Berol 391. CAS 61791-26-2; PEG-5 tallowamine; emulsifier, wetting agent, antistat, anticorrosive for agriculture, leather, textiles, metalworking and plastics industries. [Berol Nobel AB]

Berol 397. CAS 61791-14-8; PEG-15 cocamine; emulsifier, wetting agent, antistat, anticorrosive for agriculture, leather, textiles, metalworking and plastic industries. [Berol Nobel AB]

Berol 452. CAS 68891-38-3; Sodium laureth sulfate; wetting and foaming agent; emulsifier for shampoos, bath preps.; biodeg. [Berol Nobel AB]

Berol 455. PEG-3 tallow diamine; emulsifier, wetting agent, antistat, anticorrosive for agriculture, leather, textiles, metalworking and plastic industries. [Berol Nobel AB]

Berol 456. CAS 61791-44-4; PEG-2 tallowamine; emulsifier, wetting agent, antistat, anticorrosive for agriculture, leather, textiles, metalworking and plastic industries. [Berol Nobel AB]

Berol 457. CAS 61791-26-2; PEG-5 tallowamine; emulsifier,

wetting agent, antistat, anticorrosive for agriculture, leather, textiles, metalworking and plastic industries. [Berol Nobel AB]

Berol 472. CAS 1120-01-0; EINECS 214-292-4; Sodium cetyl sulfate in liquid form; anionic surfactant used in emulsion polymerization. [Berol Kemi (UK) Ltd] *

Berol 474. Sodium lauryl sulfate in paste form; anionic surfactant used in emulsion polymerization. [Berol Kemi (UK) Ltd] *

Berol 475. Sodium alkyl ether sulfate in liquid or paste form; anionic surfactants used in shampoos and bath preparations and the textile industry. [Berol Kemi (UK) Ltd] *

Berol 480. CAS 139-96-8; EINECS 205-388-7; TEA lauryl sulfate; detergent, foaming agent for hair shampoos, foam baths, hand cleaners; biodeg. [Berol Nobel AB; Berol Kemi (UK) Ltd]

Berol 484. Monoethanolamine lauryl sulfate in liquid form; anionic surfactant used in shampoos and bath preparations. [Berol Kemi (UK) Ltd] *

Berol 490. Anionic surfactant in liquid form; for emulsion polymerization. [Berol Kemi (UK) Ltd] *

Berol 496. Anionic surfactant in the form of a paste; for powder and liquid detergents. [Berol Kemi (UK) Ltd] *

Berol 513 and 525. Anionic surfactant in acid form, in which the anion is a phosphate ester, supplied as a paste; detergent auxiliary. [Berol Kemi (UK) Ltd] *

Berol 518. Anionic surfactant in acid form in which the anion is a phosphate ester, supplied as a wax; foam regulator. [Berol Kemi (UK) Ltd] *

Berol 521. Potassium alkyl phosphate ester; corrosion inhibitor, solubilizer of nonionic surfactants in presence of high electrolyte concentrate; for liq. alkaline hard surf. cleaners; biodeg. [Berol Nobel AB]

Berol 563. Alkyl polyglycol ether ammonium methyl sulfate; detergent, alkaline degreasing and cleaning agent; hydrotrope for aq. alkaline cleaners for hard surf. cleaning, acid cleaners. [Berol Nobel AB; Berol Kemi (UK) Ltd]

Berol 594. Hydroxy-ethyl 2 alkylimidazoline in liquid form; cationic surfactant used in water displacing acid cleaners in industrial cleaning; corrosion inhibition. [Berol Kemi (UK) Ltd] *

Berol 733. Potassium phosphate ester in liquid form; hydrotrope for nonionics. [Berol Kemi (UK) Ltd] *

Berol 784. Alkylaryl sulfonate/fatty alcohol ethoxylate blend; high foaming surfactant blend for neutral cleaning products. [Berol Nobel AB]

Berol 822. Anionic surfactant in liquid form. [Berol Kemi (UK) Ltd] *

Berol 829. CAS 61791-12-6; PEG-20 castor oil; surfactant, emulsifier for chemical industry. [Berol Nobel AB]

Berol WASC. CAS 68412-54-4; Nonylphenol ethoxylate; biodeg. household and industrial detergent, textile washing agent, dispersant for paints, varnishes. [Berol Nobel AB]

Berotec. Fenoterol hydrobromide; a beta$_2$ bronchodilator. [Boehringer Ingelheim Ltd]

Berpak. A two-component cold curing polyurethane encapsulant; sold in kits containing molds and electrical accessories; used for jointing of underground mains electrical cables and telecommunications cables. [Berger Elastomers]

Berries. Various vitamins, minerals and nutritional supplements; chewable dietary supplements. [Seven Seas Ltd]

Bersch bearing metal. An alloy of 93% aluminum and 7%

---

‡ Trade name and manufacturer not verified   § Trade name without identified manufacturer

nickel.

Berthier's alloy. A copper-nickel alloy containing approximately 32% nickel.

Berubigen. CAS 68-19-9; Cyanocobalamin; vitamin. [Upjohn] *

berylla. Beryllium oxide.

Beryllium. Metallic element; Be; CAS 7440-41-7; EINECS 231-150-7; Structural material in space technology; moderator in nuclear reactors; source of neutrons; windows for x-ray tubes; in gyroscopes, computer parts, inertial guidance systems; additive in solid-propellant rocket fuels; beryllium-copper alloys. [Atomergic Chemetals; Degussa; Noah Chem.]

beryllium bronze. An alloy of 97.5% copper with 2.5% benyllium.

Besconus. Textile processing aid. [Crosfield Chemicals Ltd] †

Besiege®. For agriculture industry. [DuPont UK]

Be Square® 185. CAS 63231-60-7; Hard microcryst. wax consisting of n-paraffinic, branched paraffinic, and naphthenic hydrocarbons; wax used in hot-melt coatings and adhesives, cup and paper coatings, printing inks, plastic modification (as lubricant and processing aid), lacquers, paints, and varnishes, as binder in ceramics, for potting in elec./electronic; components, in investment casting, rubber and elastomers (plasticizer, antisunchecking, antiozonant), as emulsion wax size in papermaking, as fabric softener ingred., in cosmetic hand creams and lipsticks. [Petrolite]

Beta-Air. CAS 59828-07-8; Procaterol hydrochloride; bronchodilator. [Parke-Davis] *

Beta-Cardone. CAS 959-24-0; A proprietary preparation of sotalol hydrochloride; used in the treatment of cardiac dysrhythmias. [Duncan Flockhard Ltd]

Betacortril. CAS 378-44-9; A proprietary preparation of betamethasone; a corticosteroid. [Pfizer International]

Betacortril Forte. CAS 378-44-9; A proprietary preparation of betamethasone; a corticosteroid. [Pfizer International]

Betadine. CAS 25655-41-8; Povidone-iodine; used for infection as a topical antiseptic. [Napp Laboratories Ltd]*

Betagan Liquifilm®. Levobunolol hydrochloride; antiglaucoma agent. [Allergan, Inc]

Betalin 12 Crystalline. CAS 68-19-9; Cyanocobalamin; vitamin. [Eli Lilly & Co] †

Betalion. CAS 13684-63-4; Emulsifiable concentrate containing 114 g/l phenmedipham; used for weed control for beet crops. [Portman Agrochemicals Ltd]

Beta Lite® 3503. Hydrocarbon resin; resin for paste ink applications. [Arizona]

Betaloc. A proprietary preparation of metaprobol tartrate; used in the treatment of angina pectoris. [Astra Chemicals Ltd] ‡

Betamas. See Betamaze. [Pfizer International]

Betamase. See Betamaze. [Pfizer International]

Betamaz. See Betamaze. [Pfizer International]

Betamaze® IM/IV. CAS 68373-14-8; Sulbactam; antibiotic. Also known as Betamas (Sweden), Betamase (Switzerland), Betamaz (Germany, Guatemala, Honduras), Combactam (Austria). [Pfizer International]

Betanal E. CAS 13684-63-4; Phenmedipham; selective herbicide for weed control for beet crops and strawberries. [Schering Agrochemicals Ltd]

Betanal E. CAS 13684-63-4; Emulsifiable concentrate containing 114 g/l phenmedipham; used for weed control for beet crops. [ICI Agrochemicals]

Betanal Tandem. CAS 26225-79-6, 13684-63-4; Emulsifiable concentrate of 100 g ethofumesate and 80 g phenmedipham per liter; used for weed control in beet crops. [Schering Agrochemicals Ltd]

Betapal Concentrate. Soluble concentrate containing 16 g/l (2-naphthyloxy) acetic acid; plant growth regulator. [Synchemicals Ltd]

Betapen-VK. CAS 132-98-9; Penicillin V Potassium; antibacterial. [Bristol Laboratories]

Beta Plus. Partially hydrogenated soybean oil with mono and diglycerides, sodium stearoyl lactylate, ethoxylated mono and diglycerides, TBHQ; kosher; high performance fat for continuous and conventional breads and other yeast-raised goods. [Van Den Bergh Foods]

Betaprene® 253. Hydrocarbon resin; resin for paste ink applications; tackifier for adhesives. [Arizona]

Betaprone. CAS 57-57-8; Propiolactone; disinfectant. [O'Neal, Jones & Feldman Pharmaceuticals] *

Betaseal® 43518. Solv.-based silane blend type primer; conditions the surface and promotes adhesion to glass, e.g., windshields of automobiles, trucks, buses, off-road vehicles. [Essex Specialty Prods.]

Betaseal® 43520A. Polyurethane-based solv. release type primer; screens out uv rays and promotes adhesion between polyurethane adhesive and glass; for windshield installation and backlite bonding in trucks, buses, off-road vehicles. [Essex Specialty Prods.]

Betaseal® 43555. Primer promoting adhesion between Betaseal® adhesives and PVC substrates; for bonding automotive vinyl trim to glass and various painted substrates. [Essex Specialty Prods.]

Betaseal® 58702. One-component polyurethane adhesive; fast moisture curing adhesive for stationary glass bonding when used with appropriate glass primers. [Essex Specialty Prods.]

Betaserc. CAS 5638-76-6; Betahistine; counteracts disorders of the inner ear, Menire's disease. [Duphar BV] *

Betasol Ot-A. A sulfonated ester; a proprietary trade name for a wetting agent §

Betathane. Solid polyurethane elastomers. [Hallam Polymer Engineering Ltd] *

Betatrex. CAS 2152-44-5; Betamethasone valerate USP; indicated for the relief of the inflammatory and pruritic manifestations of corticosteroid responsive dermatoses. [Altana Inc] ‡

Betnelan. CAS 378-44-9; A proprietary preparation of betamethasone. [Glaxo Laboratories]

Betnesol. A proprietary preparation of betamethasone disodium phosphate. [Glaxo Laboratories]

Betnesol-N. A proprietary preparation of betamethasone sodium phosphate and neomycin sulfate. [Glaxo Laboratories]

Betnovate. CAS 2152-44-5; A proprietary preparation of betamethasone 17-valerate; a dermatological corticosteroid. [Glaxo Laboratories]

Betnovate-C. A proprietary preparation of betamethasone valerate and clioquinol used in dermatology. [Glaxo Laboratories]

Betnovate-N. A proprietary preparation of betamethasone valerate and neomycin sulfate used in dermatology. [Glaxo Laboratories]

betol. (naphthosalol, naphtholsalol, salinaphthol). β-Naphthyl salicylate; used in medicine.

Betricing. Partially hydrogenated vegetable oil (soybean, cottonseed), mono and diglcyerides, < 0.9% polysorbate 60; kosher; multipurpose shortening for icings, fillings, yeast-raised goods. [Van Den Bergh Foods]

Betrkake. Partially hydrogenated vegetable oil (soybean, cottonseed), mono and diglcyerides; kosher; high quality emulsified shortening for cakes, icings, and sweet doughs. [Van Den Bergh Foods]

Betrox. CAS 7647-14-5; Sodium chloride fertilizer. [ICI Chem. & Polymers Ltd]

Betsolan. Betamethasone 21 phosphate veterinary preparations. [Glaxo Laboratories] *

Better Flowable. CAS 1698-60-8; Suspension concentrate containing 430 g chloridazon per liter; a pyridazinone herbicide for beet crops. [Sipcam UK Ltd]

Beutene. Butyraldehyde-aniline condensation product; fast-curing accelerator most active above 250 F (121 C); moderately nonscorchy at processing temperature; compatible with channel and furnace blacks, reclaimed rubber and acidic softeners; hard clays retard its activity; fast cures in hard rubber, wire insulation and mechanical goods. [Uniroyal] *

Bevacid. Tall oil fatty acids and distilled tall oil; used in surface coatings, soaps, oils and synthetic resins. [Bergvik Sales Ltd]

Bevaloid. Emulsifier for natural oils; dispersant; for leather industry, nonaqueous systems. [Bevaloid Ltd]

Bevaloid. Emulsifier for natural oils; dispersant; foam control agent; used in leather industry; nonaqueous systems. [Rhone-Poulenc UK]

Bevaloid 35 and 36. Anionic surfactant in powder form; Bevaloid 36 is a higher molecular weight version of Bevaloid 35; dispersant for dyestuffs and pigments, and in concrete and resin technology. [Bevaloid Ltd] *

Bevaloid 111. Sodium polycarboxylate in liquid form; long term stability dispersant for paint. [Bevaloid Ltd] *

Bevaloid 211. Sodium polycarboxylate of low molecular weight, in liquid form; dispersant for paper coating and paint, sometimes used in conjunction with polyphosphates. [Bevaloid Ltd] *

Bevaloid 1299. CAS 577-11-7; EINECS 209-406-4; Sodium dioctyl sulfosuccinate in liquid form; powerful wetting agent used in textiles and detergent dry cleaning. [Bevaloid Ltd] *

Bevaloid 6423. CAS 127-39-9; Sodium di-isobutyl sulfosuccinate in flake form; anionic surfactant having stability to electrolytes; used in emulsion polymerization and electroplating. [Bevaloid Ltd] *

Bevaloid 6522. Cationic surfactant in liquid form; softener for acrylic fibers. [Bevaloid Ltd] *

Bevaloid 6703. Low molecular weight polycarboxylate in liquid form; dispersant for calcium carbonate, china clay. [Bevaloid Ltd] *

Bevaloid 6744. Sodium polycarboxylate of low molecular weight, in liquid form; dispersant for drilling muds. [Bevaloid Ltd] *

Bevaloid DA 6805. Cationic surfactant in liquid form; leveling agent used in dyeing polyamide with acid dyes. [Bevaloid Ltd] *

Beviros. Tall oil resins; used for surface coatings, synthetic resins and binders. [Bergvik Sales Ltd]

Bevitack Resins. Esters and derivatives of distilled tall oil resin; components of adhesives and coatings. [Bergvik Sales Ltd]

Bewoid. Modified rosin emulsion. [Tenneco Malros Ltd] *

Bewopac. Modified rosin emulsion. [Tenneco Malros Ltd] *

Bex. Bakery compound. [Albright & Wilson Ltd, Phosphates & Speciality Business]

Bexfilm. Acetate films, polyester films. [ICI Chem. & Polymers Ltd] †

Bexloy®. Automotive engineering resins. [DuPont UK]

Bexphane. Polypropylene film. [Hercules Ltd] †

Bextasol. CAS 2152-44-5; A proprietary preparation of betamethasone valerate as an aerosol; used in asthma therapy. [Glaxo Laboratories] *

Bexton. CAS 1918-16-7; Propachlor; herbicide is used for pre-emergence grass and broadleaf weed control in corn and grain sorghum. [Dow UK] *

Beycopon. Specialty blends of anionic surfactants; emulsifiers, industrial detergents, textiles. [Atochem UK/Ceca]

Beycostat. Fatty alcohol phosphate esters; used for detergents, antifoams, emulsifiers, wetting agents for agrochemicals. [Atochem UK/Ceca]

Beycostat 148 K. CAS 7778-77-0; EINECS 231-913-4; Potassium phosphate; surfactant. [Ceca SA]

Beycostat 231. Phosphate ester, potassium salt; antistat, wetting agent. [Ceca SA]

Beycostat 256A. C8 fatty alcohol phosphate ester; surfactant, release agent, corrosion inhibitor, antifoam, intermediate. [Ceca SA]

Beycostat 273 P. CAS 7778-77-0; EINECS 231-913-4; Potassium phosphate; surfactant, wetting agent, detergent. [Ceca SA]

Beycostat 319 A. PEG-6 C13 fatty alcohol phosphate ester; antistat, degreaser. [Ceca SA]

Beycostat 656 A. PEG-8 alkylphenol phosphate ester; emulsifier for emulsion polymerization. [Ceca SA]

Beycostat B 070 A. Phosphate ester; for acid detergent formulations. [Ceca SA]

Beycostat B 706 A. C13 fatty alcohol phosphate ester; corrosion inhibitor, EP additive; intermediate. [Ceca SA]

Beycostat B 706 E. Phosphate ester, TEA salt; antistat for textile lubricants. [Ceca SA]

Beycostat LP 4 A. PEG-4 C12 fatty alcohol phosphate ester; antistat, degreaser; additive for acid or basic detergents. [Ceca SA]

Beycostat NE. Alkyl ether sulfate and solv.; antistatic for PVC, wetting agent. [Ceca SA]

Beycostat QA. PEG-4 alkylphenol phosphate ester; emulsifier, detergent base. [Ceca SA]

Bezalip. CAS 41859-67-0; Bezafibrate; antihyperlipoproteinemic. [Norwich Eaton Pharmaceuticals Inc] *

BFP 64K. Mono/diglyceride from hydrogenated soybean oil and glycerin, TBHQ, citric acid; food ingred., emulsifier, crumb softener; for bread, sweet goods, bakery mixes, shortening, margarine. [Am. Ingredients/Patco]

BFP 65. Lard glycerides, TBHQ, citric acid; food additive, emulsifier for baked products, mixes, shortenings, icings. [Am. Ingredients/Patco]

BFP 74. Lard glycerides, citric acid; food additive, emulsifier for coffee whiteners, whipped toppings, snack food, chewing gum, margarine. [Am. Ingredients/Patco]

BFP 75. Lard glycerides, citric acid; food additive, emulsifier for coffee whiteners, whipped toppings, snack food, chewing gum, margarine. [Am. Ingredients/Patco]

BH 2,4-D Ester 40. CAS 1702-17-6; 2,4-D; Translocated herbicide for cereals and established grassland. [Rhone-Poulenc Environmental Prods. Ltd]

---

‡ Trade name and manufacturer not verified   § Trade name without identified manufacturer

BHA. (butylated hydroxyanisole. (1,1-dimethylethyl)-4-methoxyphenol; 3-tert-butyl-4-hydroxyanisole) Mixture of isomers of tertiary butyl-substituted 4-methoxyphenols; $C_{11}H_{16}O_2$; CAS 25013-16-5; EINECS 204-442-7; antioxidant for foods, etc. [Eastman; Penta Mfg.; UOP]

BH CMPP/2,4-D. Soluble concentrate containing 116 g 2,4-D and 250 g mecoprop per liter; used to control weeds in grassland. [Rhone-Poulenc/Agri]

BH Dalapon. CAS 75-99-0; Water soluble powder containing 85% dalapon; a translocated herbicide. [Rhone-Poulenc Environmental Prods. Ltd]

BH MCPA 75. CAS 94-74-6; MCPA; herbicide for cereals and grassland. [Rhone-Poulenc Environmental Prods. Ltd]

BH Dockmaster. Soluble concentrate of 125 g dicamba and 125 g maleic hydrazide per liter; used to control docks in road verges and noncrop grass. [Rhone-Poulenc Environmental Prods. Ltd]

BH Prefix D. CAS 1194-65-6; granular herbicide containing dichlobenil; used to control weeds in woody crops and noncrop areas. [Rhone-Poulenc Environmental Prods. Ltd]

BHT. (DBPC; butylated hydroxytoluene; 2,6-di-t-butyl-p-cresol; 2,6-bis (1,1-dimethylethyl)-4-methylphenol) Substituted toluene; $C_{15}H_{24}O$; CAS 128-37-0; EINECS 204-881-4; antioxidant for foods, animal feed, petrol. products, synthetic rubbers, plastics, soaps; antiskinning agent in paints and inks. [Penta Mfg.; PMC Specialties; Raschig; Uniroyal]

biacetyl. See diacetyl.

Biactol. Antibacterial face wash to help clear acne blemishes. [Richardson-Vicks Inc] *

Biakmetals. A group of alloys some of which are zinc-copper alloys with small amounts of nickel or manganese or both, and others are aluminum-zinc-copper alloys, or zinc-aluminum alloys.

BI Ammonium Phosphate. Used for other chemicals, water treatment, fire fighting, cosmetic products, food industry, pharmaceutical products. [Rhone-Poulenc NV/CdF Chimie AZF] *

Biarsan. CAS 22760-18-5; Proquazone; anti-inflammatory. [Sandoz Pharmaceuticals] *

Biasbeston. A synthetic varnish.

Biavax II. Rubella and mumps vaccine, live attenuated; for immunization against German measles and mumps. [Merck & Co Inc]

bibenzene. See diphenyl

bicarbonate of soda. See sodium bicarbonate

Bicillin. A proprietary preparation of penicillin and procaine penicillin; an antibiotic. [Brocades Pharma]

Bicillin L-A. CAS 1538-09-6; Penicillin G Benzathine; antibacterial. [Wyeth Laboratories]

BiCNU. CAS 154-93-8; Carmustine; antineoplastic. [Bristol Laboratories]

Bicor®. Oriented and nonoriented polypropylene films; for packaging of food and nonfood products and special industrial applications. [Mobil Plastics Europe] *

Bicor®70 PXS. Oriented PP film, 1 side PVDC coated; used as the inside sealant web in laminations; superior seal strength, wide sealing range, excellent hot tack; lap seals to acrylic coated OPP or to sealable PVDC coated films; aroma barrier and moderate oxygen barrier. [Mobil] *

Bicor®220 AB, 250 AB, 310 AB, 380 AB, 420 HS. Oriented PP film, 2 side acrylic coated; excellent gloss, aroma barrier, and high speed machinability; designed for printing with solv. or water-based inks; excellent film for replacing cellophane where oxygen barrier is not required. [Mobil] *

Bicor® 240 B, 306 B, 420 B, 470 B. CAS 9003-07-0; PP homopolymer film, unmodified; base sheet designed for lamination where slip is not required; one side treated; nonsealable. [Mobil] *

Bicor®318 ASB, 252 ASB. Oriented PP film, 1 side PVDC coated, 1 side acrylic coated; film for horizontal, vertical, and overwrap applications; ideal for replacing light gauge cellophane; excellent machinability, gloss, aroma barrier, printability, hot tack, moderate oxygen barrier. [Mobil] *

Bidisin. Control of wild oats in cereals, maize and sugar beet. [Bayer AG]

Bidormal. A proprietary preparation of pentobarbitone sodium and butobarbitone; a hypnotic. [Allen & Hanburys Ltd] *

Bielzite. A resin-asphalt containing sulfur and nitrogen.

Bife. A proprietary mixture of thiamine hydrochloride and ferrous sulfate; used in the treatment of iron deficiency. [Jenkins Laboratories Inc] ‡

Bifiteral. CAS 4618-18-2; Lactulose; for treatment of obstipation and (pre-) hepatic coma. [Duphar BV] *

Biju®. CAS 7787-59-9; Bismuth oxychloride; colorant and pearlescent for frosted cosmetics; pearl nail enamel because of brilliance and smoothness. [Mearl]

BIK. Surface treated urea for easy dispersion in elastomers; activator for thiazole, thiuram, and dithiocarbamate accelerators; odor reducer when used with nitrosoamine type blowing agents. [Uniroyal] *

Bikini Cream. A proprietary suntan cream. [Ayrton Saunders plc] ‡

Bikorit. White special fused corundum (crystalline aluminum oxide); used for production of ramming mixes, shape bricks and crucibles for the lining of high temperature furnaces; molding material for precision casting molds and the casting of aggressive special steels; raw material for the electroslag remelting process, separating agent for the annealing processes. [Hüls UK Ltd]

Bilarcil®. Dimethyl 2,2,2-trichloro-1-hydroxyethyl phosphate; an antischistosomal preparation. [Bayer AG]

Bilevon®-Solution. Preparation against liver flukes (Fasciola hepactica and Fasciola gigantica), for subcutaneous injection in cattle; veterinary medicine. [Bayer AG]

Bilgen bronze. An alloy of 97% copper, 1.9% tin, 0.52% iron, and 0.24% lead.

Biligrafin. A proprietary trade name for the bis-meglumine salt of Iodipamide. [Schering Chemicals Ltd] *

Bilimiro. CAS 41473-08-9; Iopronic acid; diagnostic aid. [Bracco Industria Chimica SpA] *

Bilimiron. CAS 41473-08-9; Iopronic acid; diagnostic aid. [Bracco Industria Chimica SpA] *

Biliscopin for infusion. Sterile solution of meglumine iotroxinate containing 50 mg iodine/ml; x-ray contrast media. [Schering Health Care Ltd]

Bi-Lite®. Bismuth oxychloride/mica; pearlescent pigments for cosmetic eye, face, lip, and body make-up. [Van Dyk]

Bilivist. CAS 1221-56-3; Ipodate sodium; diagnostic aid. [Berlex Laboratories Inc] *

Bilopaque. CAS 7246-21-1; Tyropanoate sodium; diagnostic aid, radiopaque medium, cholecystographic. [Sterling Drug Inc] *

---

* Trade name not verified as available   † Trade name verified as obsolete

Biloptin. A proprietary trade name for sodium ipodate. [Schering Health Care Ltd]

Bilostat. A proprietary trade name for dehydrochloric acid.§

Bilston. A basic slag for fertilizing purposes. [Fisons plc, Horticulture Div] *

Bilt-Cote®. CAS 1332-58-7; Kaolin clay; used for agricultural prill conditioning. [R T Vanderbilt Co Inc]

Bilt-Plates®. CAS 1332-58-7; Kaolin clay; used as filler when high brightness is not a prerequisite; for paints, cosmetics and pharmaceuticals, pitch control for paper processing. [R T Vanderbilt Co Inc]

Bilt-Rex®. Carboxylated polyethylene resin; used for coatings for paper or paperboard. [R T Vanderbilt Co Inc]

Biltrlclde® . 2-Cyclohexylcarbonyl-1,2,3,6,7-11b-hexahydropyrazino[2,1-a]isoquinolin-4-one; an antischistosomal preparation. [Bayer AG; Miles Pharmaceuticals]

Bima's redwood. See redwoods.

Binab T. Trichoderma viride; biological fungicide to control silver leaf in fruit trees. [Henry Doubleday Research Association]

Binarite. See marcasite.

Bind. Isocyanates binders. [Dow UK]

Bi-nell®. Fibers. [DuPont UK]

Binotal®. A broad-spectrum penicillin preparation. [Bayer AG]

Bioacid. A range of alkylaryl sulfonic acids; detergent for use in the manufacture of liquid and powdered cleaning compounds. [Harcros Australia] *

Bio-add. Fish/meat silage additive. [BP Chemicals Ltd]

Bioban®BNPD-40. CAS 52-51-7; 2-Bromo-2-nitropropane-1,3-diol; broad spectrum antimicrobial, preservative for metalworking fluids. [Angus]

Bioban® CS-1135. 4,4-Dimethyloxazoline (74.7%), 3,4,4-trimethyloxazolidine (2.5%); preservative, antibacterial agent for water-based paints, latexes, emulsions, metalworking fluids; for oilfield water-flooding operations; aids corrosion protection. [Angus]

Bioban®GK. Hexahydro-1,3,5-tris(2-hydroxyethyl)-s-triazine; biocide for metalworking fluids, latex paints, emulsions, caulks, and adhesives. [Angus]

Bioban® N-95. 5-Hydroxymethoxymethyl-1-aza-3,7-dioxabicylco(3.3.0)octane (24.5%), 5-hydroxymethyl-1-aza-3,7-dioxabicylco(3.3.0)octane (17.7%), 5-hydroxypoly[methyleneoxy]methyl-1-aza-3,7-dioxabicyclo (3.3.0)octane (7.8%); preservative for aq. metalworking fluids. [Angus]

Bioban® P-1487® . CAS 2224-44-4; 1854-23-5; 4-(2-Nitrobutyl) morpholine and 4,4´-(2-ethyl-2-nitrotrimethylene) dimorpholine; preservative, antibacterial and antifungal agent used in metalworking fluids, lubricants and cutting oils. [Angus]

Bioban® TS. 4-(2-Nitrobutyl) morpholine, 4,4´-(2-ethyl-2-nitrotrimethylene)dimorpholine, tripropylene glycol monomethyl ether; Preservative for metalworking fluids. [Angus]

Biobor. Diesel fuel fungicide. [U.S. Borax & Chem.]

Biobrom C-103L. 2,2-Dibromo-3-nitrilopropionamide; biocide for industrial water systems, including cooling towers, pulp and papermill effluents, oil-recovery systems, metal-cutting coolants, air conditioning systems. [Dead Sea Bromine]

BioCare® Polymer HA-24. Polyquaternium-24, hyaluronic acid; emollient, humectant, conditioner, softener, moisturizer, lubricant for hair and skin; substantive to protein substrates. [Amerchol Corp]

BioCare® SA. Albumen, hyaluronic acid, and dextran sulfate; polymer providing hydration and revitalization to surface skin; lifts wrinkles; substantive to protein substrates; for eye gels, facial treatments, skin toners, makeup, moisturizers, after-sun products, cleansing lotions. [Amerchol Corp]

Biocide. Animal feed preservative. [BP Chemicals Ltd]

Biocyde. Industrial biocides. [Imperial Chemical Industries plc]

Biodynes® TRF Ultra-5. Live yeast cell deriv.; moisturizer for skin; minimizes dryness and pain of sunburn and wind chapped skin. [Brooks Industries]

Bio-Feed. A range of nonstarch polysaccharide degrading and nutrient hydrolyzing enzyme preparations produced by submerged fermentation; added to animal feeds in order to boost feed digestability and improve feed utilisation. [Novo Nordisk]

Biogastrone. CAS 7421-40-1; A proprietary preparation of carbenoxolone sodium; a gastro-intestinal sedative. [Berk Pharmaceuticals Ltd] *

Biogen. (magnesium perhydrol). A mixture of magnesium oxide and magnesium peroxide; contains from 15-25% magnesium peroxide, MgO_2, and is used as a bleaching agent and antiseptic.

Biomart. Diesel fuel bactericide. [Crowley Chem.]

Biomate. Microbiological growth control agents; biocide for cooling water. [Grace Dearborn Ltd]

Biomatrix®. CAS 9004-61-9; Hyaluronic acid. [Amerchol Corp]

bioMeT 14. 10% Diphenylstibine 2-ethyl hexoate sol'n. in dioctylphthalate; antimicrobial compound for protection of PVC systems. [Atochem]

bioMeT 204. Triphenyltin fluoride; marine antifoulant additive for commercial ship bottom paints. [Atochem]

bioMeT 430/45. Tri-n-butyltinfluoride disp. in org. solv.; antifoulant for marine paints. [Atochem]

bioMeT TBTF. Tributyltin fluoride; antifoulant for marine paints. [Atochem]

bioMeT TBTO. CAS 56-35-9; Bis (tributyltin) oxide; antimicrobial for paper mill slime control, industrial cooling water, sec. oil recovery, hospital use, textiles, plastics, urethane foam, paper preservation; antifoulant for shipbottom paints; wood adhesives. [Atochem]

Biomin®Cinque. Silicon, zinc, copper, iron and magnesium glyconucleopeptides; five essential minerals bound in a complex matrix of a low molecular weight peptide/mineral/nucleotide/carbohydrate. [Brooks Industries]

Biomin® Marine. Sea minerals yeast deriv.; marine elements in substantive moisturizing form; for skin care cosmetics. [Brooks Industries]

Biomydrin. CAS 61-76-7; Phenylephrine hydrochloride; adrenergic. [Parke-Davis] *

Biopal® NR-20. CAS 11096-42-7; Nonoxynol-12 iodine; iodophor used in formulating no rinse sanitizing solns. [Rhone-Poulenc Surf.]

Biopal®NR-20 W. Nonylphenoxypoly(ethyleneoxy) ethanol iodine complex; concentrated iodophor suitable for formulating "no rinse" sanitizing solutions. [Rhone-Poulenc Surf.] *

Biopal® VRO-20. Nonylphenoxypoly(ethyleneoxy) ethanol iodine complex; concentrated iodophor used for cleaning, sanitizing and disinfecting hospital, biological labo-

---

‡ Trade name and manufacturer not verified   § Trade name without identified manufacturer

ratory and dairy equipment, breweries, multiple-use eating and drinking utensils in bars, restaurants, etc. [Rhone-Poulenc Surf.] *

Biopar Forte. A proprietary preparation of Vitamin B$_{12}$ and intrinsic factor. hematinic. [Armour Pharmaceutical Co]*

Biophos 35. Phosphoglycoproteins, adenosine triphosphate, magnesium and potassium glycoprotein; cosmetic ingred. for moisturizing. [Brooks Industries]

Bioplex RNA. Propylene glycol, hydrolyzed RNA, hydrolyzed DNA; cosmetic ingred. for skin and hair care products. [Brooks Industries]

Biopol. Poly B-hydroxybutyrate (PHB) biodegradable thermoplastic. [ICI Chem. & Polymers Ltd]

Bio-Pol® **OE.** Sodium C8-16 isoalkyl succinyl lactoglobulin sulfonate; oil-absorbing polymer designed to entrap surface oil of the skin; film-former enhancing skin feel; excellent pigment dispersant and color enhancer. [Brooks Industries]

Biopol® **TE.** Dermal tissular extract; cosmetic ingred. for moisturizing applications. [Brooks Industries]

Bio-Pruf®. Antimicrobials for plastic products for healthcare, sanitary maintenance, construction, transportation and agriculture markets; protects against cosmetic/hygienic and structural degradation. [Morton Int'l./Plastics Additives]

Bioques. A series of enzyme producing biological cultures used to degrade industrial and municipal organically fouled waste water; both dry and liquid compositions available. [Ques Industries]

Bioques Q. A liquid chemical formulation which can liquify and digest complex fats, oils, grease, cellulose, proteins and starch; for use in toilets, drains and grease traps. [Ques Industries]

Bioques Z. A blend of highly active, broad spectrum bioactive cultures that have the ability to digest and liquify organic wastes; designed for use in all sewage systems, lagoons, sink drains and traps, and grease traps. [Ques Industries]

Bioral. CAS 7421-40-1; A proprietary preparation of carbenoxolone sodium; used to treat peptic ulceration. [Berk Pharmaceuticals Ltd] *

Biorion 450 Super. Iron supplement; for food industry. [Asahi Chem. Industry]

Biosil. Cobalt-chrome dental alloy. [Degussa Ltd]

Biosint Supra. Dental chrome investment. [Degussa Ltd]

Bio-Soft®. Alkylbenzene sulfonic acid and alkylbenzene sulfonate; detergents. [Stepan] *

Bio-Soft® **9283.** Surfactant blend; all-purpose detergent base for industrial/household cleaners. [Stepan Europe]

Biosoft C100. Anionic surfactant as an ivory paste; for detergent paste or liquids. [KWR Chemicals Ltd] *

Biosoft D. Anionic surfactant in slurry form; for liquid detergents. [KWR Chemicals Ltd] *

Bio-Soft®**D-40.** Sodium dodecylbenzene sulfonate; biodeg. detergent with good foaming, wetting, emulsifying properties; detergent base; for household, industrial and institutional cleaners; emulsifier, penetrant, and dye dispersant for textile applications. [Stepan; Stepan Canada]

Bio-Soft® **E-200.** Laureth-1; emulsifier, detergent, wetting and foam stabilizing. [Stepan Canada]

Bio-Soft®**E-400.** CAS 68131-39-5; C12-15 pareth-3; emulsifier, detergent, wetting and foam stabilizing. [Stepan Canada]

Bio-Soft®**EA-4.** Linear primary alkoxylate; surfactant, emulsifier, detergent. [Stepan; Stepan Canada]

Bio-Soft® **FF 400.** CAS 26183-52-8; Deceth-2; surfactant. [Stepan Canada]

Bio-Soft® **LD-95.** Sodium dodecylbenzene sulfonate, sodium laureth sulfate, lauramide DEA, and urea; base for dishwashing, carwashing, and other light duty hard surface detergents; biodeg. [Stepan; Stepan Canada]

Bio-Soft® **MT 40.** CAS 68952-16-9; TEA-coco hydrolyzed animal protein; very mild surfactant, conditioner, foaming agent for medicated and conditioning shampoos, creams, baby products. [Stepan Europe]

Bio-Soft® **N-300.** TEA-dodecylbenzene sulfonate; detergent, wetting and foaming agent for dishwash, carwash detergents, oil hair shampoos. [Stepan; Stepan Canada]

Biosoft N-300. Anionic surfactant, in liquid form; for liquid detergents. [KWR Chemicals Ltd] *

Biosoft S and D-35X. Anionic surfactant in liquid form; for liquid detergents. [KWR Chemicals Ltd] *

Bio-Soft® **S-100.** Linear dodecylbenzene sulfonic acid; emulsifier, detergent intermediate for formulation of built detergents, dishwash, all-purpose cleaners, acid cleaners, degreasers, industrial cleaners; textile scouring, wetting, bleaching, dyeing assistant; high foamer when neutralized; biodeg. [Stepan; Stepan Canada; Stepan Europe]

Biosoft S-100 and JN. Anionic surfactant in acid form, supplied as a viscous liquid; intermediate in detergent preparations. [KWR Chemicals Ltd] *

Bio-Soft®**TD 400.** CAS 24938-91-8; Trideceth-3; emulsifier. [Stepan Canada]

Biosol. Water soluble coolant for use in glass manufacturing. [Specialty Products Co] *

Biosol. CAS 1405-10-3; Neomycin sulfate; antibacterial. [Upjohn] *

Biosone. CAS 471-53-4; A proprietary preparation of enoxolone; antipruritic skin ointment. §

Biosperse®. Microbiocidal agents used to control microbiological growth; used for cooling towers, air washers, pasteurizers, cooling water systems, oil field water systems and metal working fluids. [Drew Ind. Div.]

Biosperse® **240.** 2,2-Dibromo-3-nitrilopropionamide and solubilizing agents; antimicrobial for control of bacteria and algae in recirculating cooling water systems, air washer systems, evaporative condensers. [Drew Ind. Div.]

Biosperse® **250.** 5-Chloro-2-methyl-4-isothiazolin-3-one and 2-methyl-4-isothiazolin-3-one; broad-spectrum antimicrobial for control of bacteria, fungi, and algae in industrial recirculating cooling water systems, oil field aq. systems; preservative in aq. metalworking fluids. [Drew Ind. Div.]

Biostat A.1. CAS 79-57-2; Oxytetracycline for fish preparation. [Pfizer International] †

Biosulphur Powder. CAS 7704-34-9; Micro grained act. sulfur (96.5% S) with protective colloid; products for impure skin, oily hair and dandruff. [Dr. Kurt Richter; Henkel/Cospha]

Bio-Surf I-20. Nonylphenoxypoly (ethyleneoxy) ethanoliodine complex, iodophor concentrate; antimicrobial, germicide, disinfectant, sanitizer for cleaning, sanitizing, disinfecting in hospital, food and beverage plants, breweries, restaurants. [Lonza Inc]

Bio Terge®. Alpha olefin sulfonate; detergent, shampoo,

---

* Trade name not verified as available   † Trade name verified as obsolete

bubble bath. [Stepan] *

Bio-Terge® AS-40. CAS 68439-57-6; EINECS 270-407-8; Sodium C_{14}-C_{16} olefin sulfonate; detergent, foaming agent for personal care, commercial and industrial formulations. [Stepan; Stepan Canada]

Bio-Terge® AS-40 and AS-90F. Sodium alpha-olefin sulfonate in liquid or flake form; liquid form used for dishwashing detergents and car washing; flake form used in powdered detergents. [KWR Chemicals Ltd] *

Bio-Terge® PAS-8S. Sodium 1-octane sulfonate; hydrotrope and detergent used in acid, alkaline, high electrolyte or bleach containing cleaners for industrial, institutional and household markets, e.g., acid cleaners, carpet steam cleaners, automatic dishwash; textile penetrant, dye dispersant; metalworking formulations. [Stepan; Stepan Canada; Stepan Europe]

Biotexin P. Novobiocin and penicillin veterinary preparation. [Glaxo Laboratories] *

Biothane System 228. Polyurethane system; for biomedical applications, e.g., potting/encapsulating compounds, adhesives, coatings, and sealants for artificial kidneys, blood oxygenators, blood filters, catheters, industrial filters. [CasChem] *

Biotrase. A proprietary preparation of trypsin and bithional; for treatment of ulcers and wounds. [Lloyd, Hamol] *

Biotren. A proprietary preparation of glycine, zinc bacitracin, neomycin sulfate, l- cysteine and di-threonine; an antibacterial powder. [Carlton Laboratories (UK) Ltd] *

biphenyl. (PHPH; diphenyl; xenene; phenyl benzene) $C_6H_5C_6H_5$; CAS 92-52-4; EINECS 202-163-5; Organic synthesis; heat transfer agent; fungistat in packaging citrus fruit; plant disease control. [Aldrich; Monsanto; Sybron]

1,1´-biphenyl. See diphenyl.

Bi-play®. For agriculture industry. [DuPont UK]

Birgin®. Sprout suppressant for potatoes in storage. [Bayer AG]

Birlane. CAS 470-90-6; Liquid seed dressing containing chlorfenvinphos for winter wheat. [ICI Chem. & Polymers Ltd] †

Birlane. CAS 470-90-6; Liquid seed dressing containing 240 g chlorfenvinphos per liter; used for winter wheat. [Shell UK]

Birmabright. A corrosion-resisting alloy containing 96% aluminum, 3.5% magnesium, and 0.5% manganese.

Birmasil alloy. A special alloy which is a nickel-aluminum-silicon alloy containing up to 3.5% nickel and from 8.13% silicon. It has high tensile strength.

Birmidium. A proprietary trade name for an alloy of aluminum with smaller amounts of copper, nickel, and magnesium; similar to Y-alloy. §

Birmingham platina. A brass containing 47% copper, 53% zinc, and 0.25% iron.

Birox®. Resistor compositions. [DuPont UK]

Bis-2. 2-Bis-acrylamide solution. [Am. Research Prods.] ‡

bis(2-aminoethyl) amine. See diethylenetriamine.

bis(4-aminophenyl) sulfone. See 4,4´-diaminodiphenyl sulfone.

2,6-bis (1,1-dimethylethyl)-4-methylphenol. See BHT.

bis (1,1-dimethylethyl) peroxide. See di-t-butyl peroxide.

bis(2-ethylhexyl) hexanedioate. See dioctyl adipate.

bis(D-gluconato) copper. See copper gluconate.

bis(2-hydroxyethyl) sulfide. See thiodiglycol.

bis(2-hydroxy-4-methoxyphenyl) methanone. See benzophenone-6.

2,2-bis (4-hydroxyphenol) propane. See bisphenol A.

Bisiumina Suspension. CAS 12284-76-3; A proprietary preparation containing bismuth aluminate; an antacid. [MCP Pharmaceuticals] *

Bislumina. CAS 16230-35-6, CAS CAS 1309-4; A proprietary preparation containing bismuth aluminate and magnesium oxide; an antacid. [MCP Pharmaceuticals]*

Bismate®. CAS 21260-46-8; Bismuth dimethyldithiocarbamate; ultra accelerator for NR, IR, BR, and SBR; for high temp., high speed vulcanization. [R.T. Vanderbilt]

Bismet. CAS 21260-46-8; Bismuth dimethyldithiocarbamate; accelerator for SBR, NR, IR, and BR compounds that are high-temp. cured; nonstaining. [Akrochem]

bis(2-methylpropyl) hexanedioate. See diisobutyl adipate

Bismica 46. CAS 12001-26-2, 7787-59-9; Mica, bismuth oxychloride; pearlescent pigment. [Presperse]

Bismucyn. CAS 31586-77-3; A proprietary preparation of bismuth sodium tartrate; throat lozenges. [Rybar Laboratories Ltd] *

bismuth. Bi; CAS 7440-69-9; EINECS 231-177-4; Pharmaceuticals and medicinals, cosmetics, alloys, catalyst in making acrylonitrile, additive, coating selenium. [Asarco; Atomergic Chemetals; Frys Metals Ltd; Noah Chem.]

bismuth bronze. An alloy. a) Consists of 1 part bismuth with 16 parts tin; b) contains 1 part bismuth, 63 parts copper, 21 parts spelter, and 9 parts nickel; resists sea-water.

bismuth carbonate basic. See bismuth subcarbonate.

bismuth chloride oxide. See bismuth oxychloride.

bismuth citrate. $BiC_6H_5O_7$; used in medicine. [Atomergic Chemetals; Celtic Chem. Ltd.; Spectrum Chem. Mfg.]

bismuth nitrate. (bismuthoxy nitrate; bismuthyl nitrate; bismuth trinitrate) $Bi(NO_3)_3$; CAS 10361-46-3; EINECS 233-792-3; Preparation of other bismuth salts, bismuth luster on tin, luminous paints and enamels, precipitation of alkaloids. [Atomergic Chemetals; Mallinckrodt; Nihon Kagaku Sangyo; Noah Chem.]

bismuth nitrate, basic. See bismuth subnitrate.

bismuth oxycarbonate. See bismuth subcarbonate.

bismuth oxychloride. (bismuth chloride oxide; basic bismuth chloride; bismuth subchloride) Inorganic pigment; BiOCl; CAS 7787-59-9; cosmetics, pigment, dry cell cathodes, artificial pearls. [Atomergic Chemetals; Mallinckrodt; Mearl; Van Dyk]

bismuth subcarbonate. (bismuth oxycarbonate; bismuth carbonate basic) $(BiO)_2CO_3$; Bismuth compounds, cosmetics, opacifier in x-ray diagnosis, enamel fluxes, ceramic glazes. [Atomergic Chemetals; Mallinckrodt; Spectrum Chem. Mfg.]

bismuth subchloride. See bismuth oxychloride.

bismuth subnitrate. (bismuth nitrate, basic; bismuth oxynitrate) Inorganic salt; $Bi_5(OH)_9(NO_3)_4O$; CAS 1304-85-4; Cosmetics, ceramic glazes, enamel fluxes. [Atomergic Chemetals; Celtic Chem. Ltd; R.W. Greeff; Mallinckrodt]

bismuth subsalicylate. (bismuth salicylate basic) $Bi(C_7H_5O_3)_3Bi_2O_3$; Surface-coating plastics and copying paper. [Atomergic Chemetals; R.W. Greeff; Mallinckrodt; Spectrum Chem. Mfg.]

bismuth trinitrate. See bismuth nitrate.

bismuth trioxide. (bismuth oxide; bismuth yellow; bismite) Bi_2O_3; CAS 1304-76-3; EINECS 215-134-7; Enameling cast iron, ceramic, porcelain colors. [Atomergic Chemetals; Ferro/Transelco; Mallinckrodt; Nihon

Kagaku Sangyo]

Bisoflex. Plasticizers. [BP Chemicals Ltd]

Bisoflex 8N. Condensation product of 2-ethyl hexyl urethane and formaldehyde; a vinyl plasticizer. [BP Chemicals Ltd]

Bisoflex 79A. The adipate of mixed C_7-C_9 alcohols; a vinyl plasticizer. [BP Chemicals Ltd]

Bisoflex 81. CAS 117-81-7; Dioctyl phthalate; a vinyl plasticizer. [BP Chemicals Ltd]

Bisoflex 82. CAS 117-81-7; Dioctyl phthalate; a vinyl plasticizer. [BP Chemicals Ltd]

Bisoflex 88. The phthalate of a C_8 alcohol; a vinyl plasticizer. [BP Chemicals Ltd]

Bisoflex 100. CAS 68515-49-1; Diisodecyl phthalate; a plasticizer. [BP Chemicals Ltd] *

Bisoflex 102a and DOA. Dioctyl adipate; a vinyl plasticizer. [BP Chemicals Ltd] *

Bisoflex 104. A trademark for an adipic ester plasticizer. [BP Chemicals Ltd] *

Bisoflex 106. A primary low temperature plasticizer for PVC and synthetic rubber. [BP Chemicals Ltd]

Bisoflex 130. A trademark for ditridecyl phthalate; a plasticizer. [BP Chemicals Ltd] *

Bisoflex 610. C_6-C_{10} aliphatic alcohol phthalate; a plasticizer. [BP Chemicals Ltd] *

Bisoflex 619. C_6-C_{10} trimellitates; plasticizer. [BP Chemicals Ltd] *

Bisoflex 791. Dialkyl (C_7-C_9) phthalate; plasticizer. [BP Chemicals Ltd]

Bisoflex 799. A trimellitate plasticizer for higher temperature resistant PVC for cable use. [BP Chemicals Ltd]

Bisoflex 810. C_8-C_{10} aliphatic phthalate; a plasticizer. [BP Chemicals Ltd] *

Bisoflex 819. C_8-C_{10} trimellitates; plasticizer. [BP Chemicals Ltd] *

Bisoflex 1002. A trademark for a polymeric plasticizer. [BP Chemicals Ltd] *

Bisoflex 1007. A trademark for a polymeric plasticizer. [BP Chemicals Ltd] *

Bisoflex DNA. Dinonyl adipate; a vinyl plasticizer. [BP Chemicals Ltd]

Bisoflex L79. A higher straight chain phthalate plasticizer. [BP Chemicals Ltd]

Bisoflex L911. A higher straight chain phthalate plasticizer. [BP Chemicals Ltd]

Bisoflex ODN. A trade mark for an adipic ester plasticizer. [BP Chemicals Ltd] *

Bisoivomycin. A proprietary preparation of Bisolvon and oxytetracycline hydrochloride; an antibiotic. [Boehringer Ingelheim GmbH] *

Bisol. Solvents, plasticizers, intermediates. [BP Chemicals Ltd]

Bisolene. Proprietary liquid fuels. [BP Chemicals Ltd]

Bisolite. Solid fuels, e.g., metaldehyde. [BP Chemicals Ltd]

Bisolube. Oil additives. [BP Chemicals Ltd]

Bisolvon. CAS 611-75-6; A proprietary preparation of bromhexine hydrochloride; a mucolytic agent. [Boehringer Ingelheim GmbH] *

Bisomer. Specialty chemicals for use as intermediates in the production of various products in industry, particularly surface coatings, adhesives and sealants. [BP Chemicals Ltd]

Bisomer 2HEA. 2-Hydroxyethylacrylate; a monomer which permits the production of polymers with side chain hydroxyl groups suitable for further reaction (crosslinking) and the production of thermosetting acrylic adhesives. [BP Chemicals Ltd]

Bisomer 2HEMA. CAS 868-77-9; 2-Hydroxyethyl methacrylate; a monomer which permits the production of polymers with side chain hydroxyl groups suitable for crosslinking and the production of thermosetting acrylic surface coating adhesives. [BP Chemicals Ltd]

Bisomer 2HPMA. CAS 27813-02-1; 2-Hydroxypropyl methacrylate. [BP Chemicals Ltd]

Bisomer D10M. CAS 1330-76-3; Diisooctyl maleate plasticizer. [BP Chemicals Ltd]

Bisomer DALP. Diallylphthalate plasticizer. [BP Chemicals Ltd]

Bisomer DAM. Dialkyl maleate (C_7-C_9 alcohols); plasticizer. [BP Chemicals Ltd]

Bisomer DBF. CAS 105-76-9; Dibutyl fumarate plasticizer. [BP Chemicals Ltd]

Bisomer DBM. CAS 105-76-0; EINECS 203-328-4; Dibutyl maleate plasticizer. [BP Chemicals Ltd]

Bisomer DNM. Dinonyl maleate plasticizer. [BP Chemicals Ltd]

Bisomer DOM. Diethyl hexyl maleate plasticizer. [BP Chemicals Ltd]

Bisoprufe. Preparations for waterproofing cements. [BP Chemicals Ltd] †

Bisoxyl. CAS 7787-59-9; Bismuth oxychloride in liquid suspension; an antisyphilitic. [British Drug Houses] *

bisphenol A. (4,4´-Isopropylidenediphenol (CTFA); 2,2-bis (4-hydroxyphenol) propane; 4,4´-(1-Methylethylidene) bisphenol) $(CH_3)_2C(C_6H_4OH)_2$; CAS 80-05-7; EINECS 201-245-8; Intermediate in manufacture of epoxy, polycarbonate, phenoxy, polysulfone, polyester resins, flame retardant, rubber chemicals, fungicide. [Aristech; Mitsui Petrochem. Ind.; Mitsui Toatsu Chem.; Shell]

Bistabillin. Penicillin preparation. [The Boots Co plc] †

bis(tributyltin) oxide. See tributyltin oxide

bis (trimethylsilyl) amine. See hexamethyldisilazane

Bi-Tarco. Tar compounds for roads. [Thomas Ness, North Thames Gas Board] *

Bitran. Cationic surfactant range, composed of high molecular weight imidazoline compounds and their salts; viscous liquids/brown aqueous solutions; used as adhesion aids, anticorrosives, dispersants, flocculants, dewatering agents; used in road maintenance; metalworking; paints; emulsion cracking; effluent treatment; pigment dispersion. [Rhone-Poulenc UK]

Bitran H. A proprietary long-chain cyclic polyamine. [Glover (Chemicals) Ltd] ‡

Bitrex®. CAS 3734-33-6; Denatonium benzoate; available as powder and in aqueous ethanol, methanol and N-propanol; a denaturant for use in alcohols; aversive agent in hazardous household, garden and automotive products as an aid to preventing accidental poisoning by ingestion; animal repellent. [Macfarlan Smith Ltd]

Bitrexene®. CAS 3734-33-6; Solution of denatonium benzoate in a special blend of surface active agents; designed to be diluted in white spirit to give a concentration of 40 ppm active ingredient in the final product, a stable solution; aversive agent for incorporation into most hydrocarbon solvents as an aid to help prevent accidental poisoning by ingestion. [3M Pharmaceuticals]

bittern. The mother liquor remaining after the Crystallization of sodium chloride from sea-water. It is a source of

---

* Trade name not verified as available   † Trade name verified as obsolete

magnesium, and also contains bromides and iodides.

Bitumastic. A proprietary trade name for a spirit paint made from refined coal-tar pitch, etc. §

bitumen. (mineral pitch, natural asphalt, compact bitumen, naphthine) CAS 8052-42-4; A hard pitchy material found at the surface of the Dead Sea, and in the pitch lake at Trinidad; used in hot-melt adhesive, coatings, paints, sealants, roofing and road coatings.

Bitumuls. Asphalt products. [Chevron] *

Bitusize. Asphalt products. [Chevron] *

Bituvar. Anticorrosive paint. [J C Bottomley] *

Bivert. Amine salts of organic acids, aromatic acid, aromatic and aliphatic petroleum distillate; for use with all pesticides (herbicides, insecticides and fungicides) to control evaporation, increase plant coverage and control drift. [Stull Chemical Company] *

BL 3. Semipermanent release agent for application to flexible molds and metal molds; effective for halogenated or peroxide-cured elastomers. [Releasomers]

BL-60®. Sodium aluminum phosphate, acidic with aluminum sulfate, anhydrous; leavening agent for baking, cereals. [Rhone-Poulenc Food Ingreds.]

B & L 70. Lidofilcon A and containing 70% water; contact lens material. [Bausch & Lomb, Professional Products Div] *

Black 103. CAS 1309-37-1, 7787-59-9; Iron oxides, bismuth oxychloride; inorg. colorant. [Presperse]

Black and White Bleaching Cream. CAS 123-31-9; Hydroquinone; depigmentor. [Schering Corp] †

Black catechu. See cutch.

Black Grip. Hard polyurethane elastomer resin, MBOCA, silica; for solid fork truck tires, belting (conveyor), friction drive wheels, drum rotators. [Royale Polymers Ltd] *

black iron. See pyrolignite of iron.

black iron liquor. See pyrolignite of iron.

black lead. See graphite.

black liquor. See pyrolignite of iron.

black mordant. See pyrolignite of iron.

Black-Out®. Modified elastomeric base material in solvents; finishing agent for decorative and protective flexible finishes on rubber products. [R T Vanderbilt Co Inc]

Black Out® Black. CAS 108-88-3; Toluene (77.5%), 2-butanone (10.9%), carbon black (2.2%); decorative and protective finishing agent for rubber; deposited films are resistant to ozone, acids, alkalies, and paraffinic hydrocarbons. [R.T. Vanderbilt]

Blackox. Fe_3O_4; CAS 1309-37-1; Foundry grade black iron oxide for core and mold use with particular emphasis on elimination of sub-surface porosity and carbon streaking when phenolic urethane sand binder systems are in use. [DCS Color & Supply Co Inc] *

black oxide of iron. See magnetic iron ore.

Black Pearls® 1100. CAS 1333-86-4; Carbon black; for coloring plastics. [Cabot]

Bladafum®. CAS 3689-24-5; Sulfotep; insecticidal fumigant used for control of greenhouse pests, e.g., aphids, whiteflies, thrips, spider mites, mealybugs, mobile stages of scale insects on vegetables and ornamentals. [Bayer AG]

Bladan®. CAS 56-38-2; Parathion; spray and dust formulations for control of biting and sucking pests. [Bayer AG]

Blade. CAS 2135-22-0; Granules containing 10% w/w oxamyl; systemic insecticide and nematicide. [Kommer-Brookwick Ltd]

Blagdenite. Unsaturated polyester resins; used for rein-

forced plastics, marine, transport and industrial. [Blagden Chemicals Ltd] *

Blagden Resins. Synthetic resins for the coatings industry; used for industrial and decorative paints and printing inks. [Blagden Chemicals Ltd] *

blanc fixe, artificial, precipitated. See barium sulfate

Blanc Fixe Micro. CAS 7727-43-7; Barium sulfate, $BaSO_4$; filler for paints and coatings. [Sachtleben Chemie GmbH]

blanchite. See hydrosulfite.

Blancol. Optical brightening agents; for whitening of textiles. [Holliday Dyes & Chemicals Ltd] †

Blancol N. The sodium salt of sulfonated naphthalene-formaldehyde condensate; in papermaking disperses pigments, clays and other solids, prevents pitch coagulation, reduces two sideness, improves sizing; bleaching, dispersing, leveling and neutralizing agent for leather. [ISP] *

Blancorol®. Different type products for retanning; the majority contain chrome oxide. [Bayer AG]

Blandlax. A laxative. [The Boots Co plc]

Blandofen CAZ. Complex polyalkyl amido imidazolinium sulfate as a white-yellow viscous liquid; cold water dispersible cationic surfactant, stable to freeze thaw cycles; used in softeners for fabrics and paper. [ISP] *

Blandofen CT. CAS 107-64-2; EINECS 203-508-2; Distearyl dimethyl ammonium chloride concentrate in isopropanol/water; supplied as a white-yellow soft paste; not cold water dispersible; higher viscosity than other cationics in the range; cationic surfactant used for fabric softeners. [GAF Great Britain] *

Blandofen FA. Complex polyalkyl amido imidazolinium sulfate as a yellow viscous liquid; cationic surfactant, thixotropic at low temperatures, specially developed for bulk deliveries; used in fabric softeners. [ISP] *

Blandol®. White mineral oil. [Witco/Sonneborn]

Blandthax. A blackquarter/anthrax vaccine. [The Boots Co plc]

Blanket Adhesive H-98. Blanket adhesive for flat bed and rotary screen print machines. [Catawba-Charlab]

blankit. See hydrosulfite.

Blankit®. CAS 7775-14-6; Sodium dithionite; stabilized bleaches for textile, leather and fur, pulp and paper industries. [BASF AG; BASF plc]

Blanko-Blech. An alloy of 80% copper and 20% nickel.

Blankophor®. Fluorescent whitening agents for textiles and detergents; textile finishing agent. [Bayer AG; Bayer plc]

Blanose. CAS 9004-32-4; Sodium salt of carboxymethyl cellulose; stabilizer and thickener in aqueous systems for food and nonfood uses. [Hercules] *

Blanthax. Blackquarter-anthrax vaccine. [The Wellcome Foundation Ltd] *

blasting gelatin. A blasting explosive; a jelly-like mass is obtained when nitro-cotton is added to nitro-glycerin, and contains from 90-93% nitro-glycerin, and 7-10% dry nitro-cotton.

Blattanex®. CAS 114-26-1; Emulsifiable concentrate containing 216 g/l propoxur; a carbamate insecticide. [Bayer AG; Bayer plc]

Blattanex® 20. CAS 114-26-1; Emulsifiable concentrate containing 20% propoxur; used for quick knock down and lasting residual control of crawling and flying pests. [Bayer plc]

Blattanex® Residual Spray. An aerosol containing 0.5%

---

‡ Trade name and manufacturer not verified   § Trade name without identified manufacturer

dichlorvos and 2% propoxur; used for quick knock down and lasting residual control of crawling and flying pests. [Bayer plc]

Blatt gold. A brass containing 77% copper and 23% zinc.

Blatt silver. An alloy of 91.1% tin, 8.25% zinc, 0.35% lead, and 0.23% iron.

Blaxon LT. Acid phosphating concentrate for coating of metals with a corrosion-resist. finish. [Eastern Color & Chem.]

Blazer®. CAS 62476-59-9; Acifluorfen; for post-emergence control of broad-leaved weeds and suppression of some annual grasses in soybeans. [BASF AG]

Blazon. Water soluble, biodegradable colorants; for spray pattern indicators for application of herbicides, pesticides in the lawn and turf, forestry and industrial weed control areas. [Milliken] *

BLE. Reaction product of diphenylamine and acetone; used in natural, IR, SBR, BR, neoprene, and nitrile rubbers; particularly recommended in tire treads and carcass to combat the effects of heat and mechanical flexing; a general purpose antioxidant where discoloration and staining are not factors; also available as a 75% active powder on silica. [Uniroyal] *

Bleachit® 1A. Auxiliary for bleaching and optical whitening wool and polyamide. [BASF]

bleach liquor. (bleaching liquid, liquid chloride of lime). Prepared by passing chlorine through milk of lime; a solution of chlorinated lime.

blenda. See oil white.

Blenderm. A proprietary preparation of polythene adhesive tape for surgical use. [3M] *

Blendex® 101. CAS 9003-56-9; ABS; modifier resin providing good impact strength and improved processing to PVC compounds; used in calendered films, semiflexible PVC film. [GE Specialty]

Blendex® 310. CAS 9003-56-9; ABS; impact modifier for rigid PVC, epoxies, PU, polyesters. [GE Specialty]

Blendex® 340. CAS 26741-53-7; High efficiency impact modifier for PVC siding and profile substrate applications. [GE Specialty]

Blendex® 586. Poly (α-methylstyrene-styrene acrylonitrile); high heat modifier resin used to upgrade PVC compounds. [GE Specialty]

Blendex® 975. ASA copolymer; weatherable modifier for use in rigid polymer applications; can be alloyed with PVC and other miscible thermoplastics in molded, extruded or calendered applications; provides impact and higher heat distort. props. [GE Specialty]

Blendex® HPP 801. Polycarbonate modifier resin; modifier for other polymers; base for color concentrates; reinforcing modifier for recycled polymers and alloys. [GE Specialty]

Blendmax 322. CAS 8002-43-5; EINECS 232-307-2; Enzyme-modified lecithin; emulsifier with enhanced water dispersibility; for instantizing whole milk powds., emulsifying vegetable and animal fats in milk replacer products; dough conditioner, antistaling agent in baking. [Central Soya]

Blendur®. High-quality raw materials based on epoxy, acrylic and polyurethane resins; for thermoset applications in electrical/electronics industry, automotive engineering. [Bayer AG]

Blendur® KU 3-4513. Polyether polyol; low-visc. polyol used in combination with isocyanates to produce soft,

flexible compounds with improved thermal endurance. [Bayer AG] *

Blenoxane. CAS 9041-93-4; Bleomycin sulfate; antineoplastic. [Bristol Laboratories]

bleomycin. An antineoplastic antibiotic produced by *Streptomyces verticillus*.

Bleph-10®. Sulfacetamide sodium; antibacterial. [Allergan, Inc]

Bleph-10 Liquifilm®. Sulfacetamide sodium; antibacterial. [Allergan, Inc]

Blephamide Liquifilm®. Sulfacetamide sodium, prednisolone acetate; antibacterial. [Allergan, Inc]

Blephamide S.O.P.®. Sulfacetamide sodium, prednisolone acetate; antibacterial. [Allergan, Inc]

BLE® 25. Acetone/diphenylamine reaction product; superaging and flex resist. antioxidant for natural and synthetic rubber. [Uniroyal] *

Blex. CAS 29232-93-7; Pirimiphos-methyl; insecticide and contact fumigant. [ICI Agrochemicals; ICI Chem. & Polymers Ltd]

BLO®. CAS 96-48-0; γ-Butyrolactone; solv. for PAN, PS, fluorinated hydrocarbons, cellulose triacetate, shellac; used in paint removers, petrol. processing, hectograph process, speciality inks; intermediate for aliphatic and cyclic compounds; reaction and diluent solv. for pesticides; used in dyeing of acetate; wetting agent for cellulose acetate films, fibers, solv. welding of plastic films in adhesive applications. [ISP]

Blocadren. CAS 26921-17-5; Timolol maleate; beta-blocker for the relief of hypertension, angina pectoris and migraine. [Merck & Co Inc]

Blocazide. Timolol maleate and hydrochlorothiazide; for the relief of hypertension. [Merck & Co Inc] †

Blockain Hydrochloride. CAS 550-83-4; Propoxycaine hydrochloride; anesthetic. [Sterling Drug Inc] *

Block-Out A-SF. Smoke-free auxiliary in piment printing to achieve a discharge effect on dyed grounds. [Catawba-Charlab]

Blocuretic. Timolol maleate combined with hydrochlorothiazide and amiloride hydrochloride; for the relief of hypertension. [Merck & Co Inc] †

Bloodit. A synthetic gum for process engraving. [Johnsons of Hendon] *

blood meal. A nitrogenous fertilizer prepared by coagulating blood, drying and grinding the product; contains on an average, from 11-14% nitrogen, and 0.75% phosphorus.

blown oils. (oxidized oils, polymerized oils, soluble castor oils, base oil). When semi-drying vegetable oils, marine animal oils, and liquid waxes are warmed at from 70-120 C., and a current of air blown through them, the oils oxidize to thickened fluids.

Bluboro®. Aluminum sulfate, calcium acetate, boric acid; astringent. [Allergan, Inc]

Blue Basic Lead Sulphate. A basic lead sulfate containing lead oxide, sulfite, sulfide, zinc oxide, and carbon, produced from lead ore by volatilization; used in rubber mixing and in priming paint.

blue copperas. See copper sulfate (ic)

Blue Dot. A double-base-type formulation which minimizes charge weight and moisture absorption; graphite glaze enables smooth granule flow; used for ammunition; designed for use in magnum shotshell loads; also for use in magnum handgun loads. [Hercules] *

blue gold. A jeweler's alloy, consisting of 75% gold and

---

* Trade name not verified as available   † Trade name verified as obsolete

25% iron.

Blueminster Resin Emulsions. Dispersions of rosin and rosin derivatives in water; tackifier for water-based adhesives. [Blueminster Ltd]

Blue Mold. *Penicillium Glaucum,* a fungus.

Blue Powder. (zinc dust, zinc fume). A by-product in the smelting of zinc; consists of a mixture of finely divided zinc and zinc oxide. It has the power of absorbing hydrogen; used in the chemical industries; employed to discharge locally the color of dyed cotton goods; used for the recovery of gold from the cyanide solution of the metal.

blue stone. *See* copper sulfate (ic).

blue vitriol. *See* copper sulfate (ic).

bluish eosin. *See* erythrosin.

BMC 100, 102. Thermoset BMC; low-cost, general-purpose compound with excellent heat distort. resist., dimensional stability, flow, controlled reactivity; for injection molding of very intricate parts; applications including elec. brush holders, high temp. coil bobbins; 102 complies with Ford ESFM3D-56A. [BMC] *

BMC 200. Thermoset BMC reinforced with high strand integrity fiberglass; flame-retardant, med.-strength elec. grade formulated to provide proper filling of intricate mold cavities; applications including internal TV parts, elec. connectors, coil bobbins. [BMC] *

BMC 800. Thermoset polyester BMC, 10-20% glass-reinforced; corrosion- and weather-resist. compound for marine, outdoor elec., corrosive pump parts, and chemical barriers. [BMC]

BMC 1000. Thermoset polyester BMC, glass-reinforced; food grade compound for molding of dishware for microwave oven use and conventional ovens to 410 F; also for other houseware items. [BMC] *

B.M. Mixture. A mixture for producing smoke screens; contains 35% zinc, 42% carbon tetrachloride, 9% sodium chlorate, 5% ammonium chloride, and 8% magnesium carbonate.

B-Nine. CAS 1596-84-5; Water soluble powder containing 85% daminozide; a plant growth regulator. [Fargro Ltd]

Bobierre metal. A brass containing 58-66% copper and 34-42% zinc. Ships' sheathing metal.

BOC anhydride. *See* di-t-butyl dicarbonate.

Bodenstein. Synonym for amber.

Bodryl. A proprietary preparation of bromodiphenhydramine hydrochloride, aspirin, phenacetin, phenylephrine, caffeine and aluminum hydroxide dried gel; cold remedy. [Parke-Davis] *

Bohemian earth. (Veronese earth, Tyrolean earth, Seladon green, terre verte, stone green). Green earths, which are products of the disintegration of minerals, chiefly of the hornblende type. Stone green is a mixture of ground green earth and white clay; used in the manufacture of waterproofing paints

Bohemian topaz. Synonym for citrine.

Bohnalite B. A proprietary trade name for an aluminum alloy containing 4.5% copper and 0.3% magnesium. §

Bohnalite J. A proprietary trade name for an alloy of aluminum with 10% copper. §

Bohnalite S43. A proprietary trade name for an alloy of aluminum with about 5% silicon. §

Bohnalite S51. A proprietary trade name for an alloy of aluminum with small amounts of magnesium, silicon, and iron. §

Bohnalite U. A proprietary trade name for an alloy of aluminum with 13% silicon. §

Bohnalite Y. A proprietary aluminum alloy containing small quantities of copper, nickel, and magnesium. §

Bohrmittel Hoechst. Alkylsulfamido carboxylic acid, sodium salt; corrosion inhibitor, lubricant, wetting agent, and emulsifier for metalworking fluids; extreme pressure props. [Hoechst Celanese/Colorants & Surf.; Hoechst AG]

boiled oil. Linseed oil, which has been boiled with litharge to render the oil more drying. The term is also applied to linseed oil which has been heated for some time. The tendency to oxidize is thereby increased.

Boiler-Aid. Various liquid or powder blends of boiler water deposit and corrosion inhibitors; used for steam and hot water boiler treatment. [Schaefer Technologies Inc]

boiler plug alloy. An alloy of 8 parts bismuth, 5-30 parts lead, and 3-24 parts tin.

Boisambrene Forte. Formaldehyde ethyl cyclododecylacetal; see Boisambrene. [Henkel/Cospha]

Bolda. Fungicide containing carbendazim, maneb, and sulfur. [ICI Chem. & Polymers Ltd]

Bolda FL. A suspension concentrate containing 50 g carbendazim, 320 g maneb and 100 g sulfur per liter; systemic fungicide for cereals. [Farm Protection Ltd]

Bolfo®. Preparation for external use on dogs and cats against ecto parasite infestation; veterinary medicine. [Bayer AG]

Bolstar. CAS 35400-43-2; Sulprofos; insecticide with stomach and contact action mainly against caterpillars of *Heliothis spp.* and *Spodoptera spp.* on cotton and other crops. [Bayer AG]

bolster silver. A nickel silver. It is an alloy of 65.5% copper, 16% zinc, 18% nickel, and 0.5% lead.

Bolus alba. *See* kaolin.

Bombardier. CAS 1897-45-6; Suspension concentrate formulation containing 50% w/v chlorothalonil; for use as an agricultural/horticultural fungicide. [Universal Crop Protection Ltd]

Bombardier. CAS 1897-45-6; Chlorothalonil; a fungicide for a wide range of agricultural crops. [Farm Protection Ltd]

Bombay catechu. *See* cutch.

Bond-A-Tint. Pigment dispersions; for coloration of bonded polyurethane carpet underlay. [Pacific Dispersions Inc]*

Bonder/Gardobond. Phosphate systems for iron, steel, zinc, aluminum and its alloys; for corrosion protection, pretreatment prior to painting, assisting cold forming and improving sliding friction properties and running-in characteristics. [Sachtleben Chemie GmbH]

Bonder/Gardoclean. Alkaline, neutral, acid-passivating cleaners for spray and dip application. [Sachtleben Chemie GmbH]

Bonderite. Phosphating process for iron, steel, aluminum, and zinc which converts the metal surface into a zinc phosphate coating; corrosion inhibiting and paint bonding treatments. [Brent Chemicals International plc]

Bonderlube/Gardolube. Water-based systems oils, fatty emulsions; for cold forming solid and hollow ware; drawing, cold extrusion, wall ironing; for steel, stainless steel, aluminum. [Sachtleben Chemie GmbH]

Bonding Agent 2001. Polymethylene polyphenyl isocyanate sol'n. in dibutyl phthalate; bonding agent for PVC coatings. [Miles; Polysar]

Bonding Agent M 3. A methylene donor; used in combina-

---

‡ Trade name and manufacturer not verified   § Trade name without identified manufacturer

tion with Bonding Agents R6 in natural, IR, BR, SBR, neoprene, or nitrile rubbers to give improved adhesive bonds to fabrics such as cotton, nylon, polyester, rayon and glass, as well as to metals. [Uniroyal] *

Bonding Agent P 1. 4,4′-Methylene-bis-(phenyl carbanilate). For pre-dip solutioning of polyester tire and industrial cord prior to a secondary treatment with an RFL system; the two dip treatment provides superior adhesion to other known treatment systems. [Uniroyal] *

Bonding Agent R 6. A resorcinol donor; used in combination with Bonding Agent M3 in natural rubber, IR, BR, SBR, neoprene or nitrile rubber to give improved adhesive bonds to fabrics such as cotton, rayon, nylon, polyester and glass, as well as to metals. [Uniroyal] *

Bonding Agent TN. 70% solution of a polyester containing hydroxyl groups; used to improve adhesion of PVC coatings on polyester and polyamide coatings. [Bayer AG]

Bondogen. A mixture of an oil soluble sulfonic acid of high molecular weight with a high boiling alcohol and a paraffin oil; used as a peptizing agent and strong plasticizer for all elastomers; functions as a scorch retarder. [King Industries]

Bond-Plus. Adhesives, industrial coatings for packaging, laminating, paper converting, wood-working, labelling, foil laminating, paper cups, FDA approved adhesives and coatings; used for packaging: case sealing, labelling, lap-gluing; paper converting: foil laminating, paper to paper, paper to film laminations, paper cups, coatings; woodworking: water-resistant adhesives. [Industrial Adhesives Company] *

Bond-Plus HM. Adhesives, industrial coatings for packaging, laminating, paper converting, wood-working, labeling, foil laminating, paper cups, FDA approved adhesives and coatings; used for packaging: case sealing, labeling, lap-gluing; paper converting: foil laminating, paper to paper, paper to film laminates, paper cups, coatings; woodworking: water-resistant adhesives. [Industrial Adhesives Company] *

BondTint. Family of polymeric, liquid, reactive colorants used to color polyurethane-based adhesives for rebonded carpet underlay. [Milliken]

Bondur. An aluminum alloy containing 4.2% copper, 0.3-0.6% manganese, and 0.5-0.9% magnesium; a corrosion-resisting alloy. §

Bonefos. A proprietary preparation of sodium clodronate; used in the management of hypercalcemia of malignancy. [Boehringer Ingelheim Ltd]

bone oil. CAS 8001-85-2; (animal oil, Dippel's oil, oil of Hartshorn, bone tar). A dark brown oil, rich in pyridine bases, obtained by the distillation of bones; used for denaturing spirits.

bone tar. See bone oil.

Bonine. Meclizine hydrochloride; antiemetic. [Pfizer Inc]

Bonjela. A proprietary preparation containing choline salicylate and cetyl dimethyl benzyl ammonium chloride. [Lloyd's Pharmaceuticals Ltd] *

Bonner L-894. A proprietary trade name for a polyester resin. §

Bonomycin. CAS 808-26-4; Sancycline; antibacterial. [Pfizer Inc] †

Bonosol. Acrylic resins. [Ernst Jager GmbH]

Bonotex. Acrylic latexes. [Berol Nobel Ltd]

Bontex. Range of synthetic blended detergents. [Unilever]*

Bonzi. CAS 76738-62-0; Suspension concentrate containing 4 g/l paclobutrazol; plant growth regulator for use on ornamental plants. [ICI Chem. & Polymers Ltd]

Boost. Fish oil concentrates; source of omega 3 fatty acids in fish farm feeds. [Seven Seas Ltd]

boracic acid. See boric acid.

Boral. Aluminum-boro-tartrate; an antiseptic and astringent prescribed in skin diseases, and in inflammation of the ear.

Borascu. General nonselective weedkillers. [Borax Consolidated Ltd]

Borateem. Borates. [Borax Consolidated Ltd]

borax. See sodium borate

Borax. CAS 1303-96-4; Sodium tetraborate decahydrate; dispersant, wetting agent for NR, SR latexes; mold lubricant for general dry rubber molding. [U.S. Borax & Chem.]

Boraxo. Industrial hand cleaners. [Borax Consolidated Ltd]

boraxusta. Calcined borax.

Borcher's metal. Noncorrosive alloys. One alloy contains 64.6 5 nickel, 32.3% chromium, 1.8% molybdenum, and 0.5% silver; others are stated to contain 30% chromium, 34-35% cobalt, and 34-35% nickel.

Bordeaux Mixture. Wettable powder copper fungicide; garden fungicide for crop and noncrop plantings. [Vitax Ltd]

Borden. Foundry sand binders. [Borden (UK) Ltd] *

Borderland Black. Active ingredient: Mesurol-3,5-dimethyl-y-(methylthio) phenol methylcarbamate; seed protectant to prevent sprout pulling by birds in newly planted corn. [Borderland Products Inc] †

Borester. Organic boron compounds. [Borax Consolidated Ltd]

Borester. Organic borates. [Rhone-Poulenc UK]

Borester 7. Trihexylene glycol diborate; used for timber preservation. [Manchem Ltd] *

Boresters. A series of esters of boric acid. They are used as a convenient means of introducing boron into organic media such as paints and plastics. §

borethyl. Triethylboron, $(C_2H_5)_3B$.

Borfax. Shaped lightweight sideliner lines for steel ingot heading. [Foseco (F.S.) Ltd] †

boric acid. (boracic acid; orthoboric acid) H_3BO_3; CAS 10043-35-3; EINECS 233-139-2; Heat-resistant glass, porcelain enamels, boron chemicals, metallurgy, flame retardant in cellulosic insulation, mattress batting and cotton textiles, fungus control on citrus fruits, ointment and eyewash, nickel electroplating baths. [Janssen Chimica; OxyChem; U.S. Borax & Chem.]

boric acid, zinc salt. See zinc borate

boric anhydride. See boron oxide

boric oxide. See boron oxide

Borite. Consists essentially of crystalline alumina; a trademark for goods used as abrasives and refractories §

borium. A fused tungsten carbide diamond substitute formed by exposing tungsten at high temperatures to carbon monoxide or hydrocarbon gases.

borneol camphol. (Baros camphor; Borneol; Bhimsiam camphor; Borneo camphor; Dryobalanops camphor; Malay camphor; Sumatra camphor) $C_{10}H_{17}OH$; A terpene from Dryobalanops camphora; used in perfumery, in celluloid manufacture, and in medicine as an antiseptic; used in perfumery, in celluloid manufacture, and in medicine as an antiseptic.

---

* Trade name not verified as available † Trade name verified as obsolete

Bor-Nitrophoska® 13-13-21 + 0.1 B. Complex fertilizer with 13% nitrogen, 13% phosphate, 21% potash, and 0.1% boron; for all agricultural crops requiring boron and horticultural crops which are not sensitive to chloride. [BASF AG]

Borocil. Nonselective weedkiller. [Borax Consolidated Ltd]

Borocil A. CAS 1912-24-9; Atrazine; a residual herbicide. [ABM Chemicals Ltd]

Borocil Extra. Atrazine + bromacil + diuron; used for total weed control in non crop areas. [ABM Chemicals Ltd]

Borocil K. Bromacil + diuron; used for total weed control in non crop areas. [ABM Chemicals Ltd]

Borofax Ointment. A proprietary formulation of boric acid and lanolin; an externally applied emollient for applications to burns, abrasions etc. [The Wellcome Foundation Ltd]

Boroflow. CAS 122-34-9; Suspension concentrate containing 500 g/l simazine; a triazine herbicide to control weeds and grasses in cane fruit, roses and some vegetables. [ABM Chemicals Ltd]

Boroflow A. CAS 1912-24-9; A liquid formulation containing 500 g atrazine per liter as a suspension concentrate; a residual herbicide. [ABM Chemicals Ltd]

Boroflow A/ATA. CAS 1912-24-9; 61-82-5; A liquid formulation containing 270 g atrazine and 160 g aminotriazole per liter as a suspension concentrate; used for total weed control on industrial sites, paths, kerbs and channels, drives and hard tennis courts, hardstanding and storage areas. [ABM Chemicals Ltd]

Boroflow S/ATA. CAS 61-82-5; 122-34-9; A suspension concentrate containing 160 g aminotriazole and 270 g simazine per liter; used for total weed control in non crop areas and fruit orchards. [ABM Chemicals Ltd]

Boroflux. A mixture of boron suboxide with boric anhydride and magnesia; used to the extent of about 1% to deoxidize copper during purification.

Borogard® ZB. CAS 12513-27-8; Zinc borate; corrosion inhibitor, biocide, fungicide, in-can preservative for paints and coatings. [U.S. Borax & Chem.]

boroglyceride. (boroglycerine). Glyceryl borate; $C_3H_5BO_3$; used as a preservative for wines and fruits.

Borolon. Consists essentially of crystalline alumina; a trademark for articles made for abrasive and refractory purposes. §

boron. Nonmetallic element; B; CAS 7440-42-8; EINECS 231-151-2; special-purpose alloys; cementation of iron; neutron absorber in reactor controls; oxygen scavenger for copper and other metals; fibers and filaments in composites; semiconductors; rocket propellant mixtures. [Atomergic Chemetals; Noah Chem.]

boron bromide. See boron tribromide

boron chloride. See boron trichloride

boron fluoride. See boron trifluoride

boron nitride. BN; CAS 10043-11-5; Manufacture of alloys, in semiconductors, nuclear reactors, lubricants. [Atomergic Chemetals; Carborundum Co.; New Metals & Chems. Ltd; Noah Chem.]

boron oxide. (boron trioxide; boric anhydride; boric oxide) B_2O_3; CAS 1303-86-2; EINECS 215-125-8; Production of boron, heat-resistant glassware, fire-resistant additive for paints, electronics, liquid encapsulation techniques, herbicide. [Atomergic Chemetals; Noah Chem.; Spectrum Chem. Mfg.; U.S. Borax & Chem.]

boron tribromide. (boron bromide) BBr_3; CAS 10294-33-

4; EINECS 233-657-9; Catalyst in organic synthesis manufacture of diborane. [Aldrich; Atomergic Chemetals; Janssen Chimica; Kerr-McGee]

boron trichloride. (boron chloride) BCl_3; CAS 10294-34-5; EINECS 233-658-4; Catalyst in organic synthesis; source of boron compounds; refining of alloys; soldering flux; electrical resistors; extinguishing magnesium fires in heat-treating furnaces; manufacture of diborane. [Air Prods.; Aldrich; Atomergic Chemetals; Kerr-McGee]

boron trifluoride. (boron fluoride) BF_3; CAS 7637-07-2; Catalyst in organic synthesis; production of diborane; instruments for measuring neutron intensity; soldering fluxes; gas brazing. [Air Prods.; Akzo; Aldrich; Allied-Signal; Atomergic Chemetals]

boron trioxide. See boron oxide

borosalicylic acid. A solution containing 4% each boric and salicylic acids. Has been used as an antiseptic.

Boroxo. A heavy-duty soap powder hand-cleanser, based on borax. [Borax Consolidated Ltd] *

Borrebond. CAS 8061-52-7; Calcium lignosulfonate; binder, low-cost dispersant. [Borregaard LignoTech]

Borrechel. Lignosulfonate containing chelated trace elements; used as micronutrients. [Borregaard LignoTech]

Borresperse CA/CAF. CAS 8061-52-7; Calcium lignosulfonate, desugared; dispersant for pesticide formulations, concrete admixtures. [Borregaard LignoTech]

Borrewell C. Chrome lignosulfonate; conditioner in water-based oil well drilling mud systems. [Borregaard LignoTech]

Borrewell FC. Ferrochrome lignosulfonate; conditioner in water-based oil well drilling mud systems. [Borregaard LignoTech]

Borrewell FE. Iron lignosulfonate; conditioner in water-based oil well drilling mud systems. [Borregaard LignoTech]

Bortin 45. Brucella abortus vaccine for veterinary purposes. [Glaxo Laboratories] *

Borvicote. A proprietary range of emulsions of vinyl homopolymers and copolymers. [Borregaard LignoTech] *

Bos MH. CAS 123-33-1; Maleic hydrazide; a plant growth regulator for grass and to reduce bud growth in trees, hedges and vegetables. [Bos Chemicals Ltd]

Boselon. Packaging materials; plastic film made of low density polyethylene or other thermoplastic, containing a volatile corrosion inhibitor effective for rust protection of ferrous metals; used for packaging for metal parts for the purpose of corrosion protection. [Aicello Chemical Co Ltd] *

Bostik. Wide variety of formulations based on natural and synthetic materials (rubbers, resins, polymers, etc.); used for adhesives, sealants and coating compounds for a wide variety of bonding, sealing and productive coating applications in all types. [Bostik Ltd] *

Bostik 1100FS. One-part urethane sealant; cures by reaction with atmospheric moisture to form tough, flexible seal; for sealing and bonding applications in general industry, HVAC, marine industry, truck trailer, container construction, log home, RV. [Bostik]

Boszamet. Granules containing 98-99% dazomet; soil fumigant. [Bos Chemicals Ltd]

Botox®. Botulinum toxin; neuromuscular blocker. [Allergan, Inc]

Botrilex. CAS 82-68-8; A horticultural fungicide containing quintozene. [ICI Chem. & Polymers Ltd]

---

‡ Trade name and manufacturer not verified   § Trade name without identified manufacturer

Bouchardt's reagent. A solution of 1 part iodine and 2 parts potassium iodide in 20 parts water. An alkaloid reagent giving a brown precipitate; used in chemical analysis.

Bourbonne's metal. An alloy of 50.48% tin, 48.8% aluminum, 0.25% copper, and 0.33 percent iron.

Bourbouze aluminum solder. An alloy of 47% zinc, 37% tin, 10% aluminum, and 5% copper.

Bourbouze solder. An alloy of 83% tin and 17% aluminum.

Bovatec (as sodium). Lasalocid; coccidiostat. [Hoffmann-LaRoche Inc] *

Bovinal-20. CAS 9048-46-8; Serum albumin; protein for use in skin and hair care preps. [RITA]

Bovinox. Cattle dip and spray. [The Wellcome Foundation Ltd] *

Bowhill's stain. A microscopic stain; contains 15 cc of a saturated alcoholic solution of orcin, 10 cc of 20% tannic acid solution, and 30 cc water.

Bow-wire brass. See brass.

Boxite. Consists essentially of crystalline alumina; a trademark for articles of the abrasive and refractory class. §

Boxolon. Herbicide containing clopyralid, bromoxynil and mecoprop. [ICI Chem. & Polymers Ltd] †

Bozefloc. Polyacrylic flocculants. [Hoechst UK]

Bozetol. A sulfonated castor oil product; a proprietary trade name for a wetting agent. §

Bozzle. Container for agrochemicals; used with the 'Electrodyn' sprayer. [ICI Chem. & Polymers Ltd]

BP LDPE. Low density polyethylene. [BP Chemicals Ltd]

BP Mycocide. Liquid preservative based on propionic acid. [BP Chemicals Ltd] †

BP polystirene. Polystyrene. [BP Chemicals Ltd]

B.P. Pyro®. CAS 7758-16-9; Sodium acid pyrophosphate; leavening agent for baking, cereals. [Rhone-Poulenc Food Ingreds.]

BR. See polybutadiene

Brabant PCNB. CAS 82-68-8; Dustable powder containing 20% w/w quintozene; fungicide for various agricultural crops. [Bos Chemicals Ltd]

Bradophen. Quaternary ammonium bacericide. [Ciba plc]*

Bradosol. CAS 538-71-6; A proprietary preparation of domiphen bromide. throat lozenges [Ciba plc] *

Bradsil. Silicone-based softeners and lubricants; for use as textile softeners and lubricants. [Hickson & Welch Ltd]

Bradsyn. Polyethylene softeners for textile finishing. [Hickson & Welch Ltd]

Braemer's reagent. A tannin reagent. It consists of 1 gram sodium tungstate and 2 g sodium acetate in 10 cc of water.

Brakol. A water treatment alkali. [Laporte Industries Ltd] *

Brasivol. CAS 1344-28-1; An abrasive cleaning paste in three grades, each containing synthetic aluminum oxide in a nonirritant soap-detergent base; for the treatment of acne vulgaris. [Stiefel Laboratories Inc; Stiefel Laboratories (UK) Ltd]

Brasoran 50 WP. CAS 4658-28-0; Aziprotryne; a selective herbicide. [Ciba-Geigy Agrochemicals]

brass. A copper-zinc alloy of varying proportions. Usually it contains more than 18% zinc, and lead is sometimes added to the extent of 1-2%. Ordinary brass contains 67% copper and 33% zinc; used in condenser tube plates, piping, hose nozzles, couplings, oil gauges, flow indicators, air and drain cocks, marine equipment, jewelry, etc.

Brassica campestris oil. See rapeseed oil

Bras-sicol. CAS 82-68-8; A contact fungicide containing 50% w/w quintozene; used for the control of diseases in turf. [Burts & Harvey] *

Brassicol. Fungicide for soil and turf treatment. [Hoechst UK] *

brass, iron. A brass containing from 1-9% iron.

brass, iserlohn. An alloy of 64% copper, 33.5% zinc, and 2.5% tin.

brass, leaded. Alloys of from 71-79% copper, 4.5-9.5% lead, 8.5-23% zinc, and traces to 3% tin.

Bravit. A proprietary preparation of thiamine hydrochloride, riboflavine, pyridoxine hydrochloride, nicotinamide and ascorbic acid; a vitamin supplement. [Galen Ltd] *

Bravo 500. CAS 1897-45-6; Chlorothalonil; a fungicide for a wide range of agricultural crops. [BASF plc]

Bravocarb. A liquid formulation containing 100 g carbendazim and 450 g chlorothalonil per liter as a suspension concentrate; systemic fungicide. [Fermenta ASC Europe Ltd]

Braxo. Industrial hand cleanser. [Borax Consolidated Ltd]

Braze®. Xylene; dimethylbenzene; xylol. [R.T. Vanderbilt]

Brazilian Elemi. Elemi from Icica icariba.

Brazil wax. See carnauba

Brazil wood. See redwoods.

brazing solder. Alloys of from 35-45% copper, 45-57% zinc, and 8-10% nickel.

BR Destral. Heavy-duty nonselective herbicide; used for weedkilling. [Borax Consolidated Ltd]

BR Destral. Aminotriazole + bromacil + diuron; used for total weed control in non crop areas and on railway tracks. [ABM Chemicals Ltd]

Breaker F. Enzyme for hydrolyzing water-sol. polysaccharides such as derivatized guar, guar, cellulose ethers. [Rhone-Poulenc/Water Soluble Polymers]

Breaxit. Performance chemicals for oil field. [Exxon Int'l.] *

Brebent. CAS 1302-78-9; Bentonite clay for binding, suspending and emulsifying sands. [Laporte Industries Ltd]

Brebond. Bonding clay for foundry sands. [Laporte Industries Ltd]

Bredinin®. Immunosuppressant. [Asahi Chem. Industry]

Breecht's double salt. Potassium dimagnesium sulfate.

Breedervac I Plus, II Plus, III Plus, IV Plus. For immunization of poultry. [Intervet Inc]

Breedervac-Reo-Plus, RN Plus. For immunization of poultry. [Intervet Inc]

Bregel. CAS 1302-78-9; Drilling bentonite. [Laporte Industries Ltd]

Brentan. CAS 22832-87-7; A proprietary preparation containing miconazole nitrate; antifungal. [Janssen Pharmaceutical Ltd] *

Breokinase. Urokinase; plasminogen activator. [Sterling Drug Inc] *

Breon. Vinyl materials; nitrile and acrylic rubbers. [British Geon] *

Breon GA 301A. A proprietary high-quality insulation grade of polyvinyl chloride suitable for use at temperatures up to 105 C. [British Geon] *

Breon GA 302A. A hard insulation grade of polyvinyl chloride suitable for use at 85 C. [British Geon] *

Breon GA 304A. A soft insulation and sheathing grade of heat-resisting polyvinyl chloride; suitable for use at 85 C. [British Geon] *

Breon GA 314A. A high-quality soft insulation and sheathing

---

* Trade name not verified as available   † Trade name verified as obsolete

grade of polyvinyl chloride able to withstand high temperatures. typically used to sheathe the insulated wiring in electric blankets. [British Geon] *

Breox. Polyalkylene glycols and polyethylene glycols; brake fluids, lubricants. [BP Chemicals Ltd]

Bresille wood. See redwoods.

Bresin 2, 2E. Thermoplastic resins derived from natural materials extracted from pinewood; used in construction adhesives and mastics. [Hercules] *

Brestan. Mixture of fentin acetate and maneb; used for control of potato blight. [Hoechst UK] *

Bretol®. CAS 124-03-8; Cetyldimethylethylammonium bromide. [Zeeland]

bretonite. Iodoacetone; $CH_3CO \cdot CH_2I$.

Bretylate. CAS 61-75-6; A proprietary preparation of bretylium tosylate; used in the treatment of cardiac arrhythmias. [The Wellcome Foundation Ltd]

Bretylol. CAS 61-75-6; Bretylium tosylate; anti-adrenergic, cardiac depressant. [Baxter Health Care] *

Brevibloc®. CAS 81161-17-3; Esmolol hydrochloride, anti-adrenergic. [DuPont Merck Pharmaceuticals]

Brevidil. A proprietary preparation of suxemethonium bromide; a muscle relaxant. [May & Baker Ltd] *

Brevidil E. Short acting muscle relaxant. [May & Baker Ltd]*

Brevidil M. Suxamethoniom salts. [May & Baker Ltd] *

Breviol. Dyeing auxiliary; dispersing and leveling agent for dyeing blends of polyester/wool and polyester/acrylic; dispersing agent for cationic dyestuffs, dyeing in general. [Henkel Chemicals Ltd] *

Brevital. CAS 22151-68-4; A proprietary preparation of methohexitone sodium; an intravenous barbiturate anesthetic. [Eli Lilly & Co]

Brevital Sodium. CAS 22151-68-4; Methohexital sodium; anesthetic. [Eli Lilly & Co] *

Brexin. Chlorpheniramine maleate USP; pseudoephedrine hydrochloride USP; indicated for the temporary relief of symptons of the common cold; allergic rhinitis and sinusitis. [Altana Inc] ‡

Bricanyl. CAS 23031-32-5; A proprietary preparation of terbutaline sulfate used in asthma therapy. [Astra Chemicals Ltd] ‡

Bricanyl Expectorant. A proprietary preparation of terbutaline sulfate and guaifenesin; an expectorant cough medicine. [Astra Chemicals Ltd] ‡

Brightray Alloy B. A trademark for an electrical resistance alloy of 59% nickel, 16% chromium, 0.3% silicon and the balance iron. [Wiggin Alloys Ltd] ‡

Brightray Alloy C. A trademark for an electrical resistance alloy of 19.5% chromium, 1.5% silicon, 0.04% rare earth elements and the balance nickel. [Wiggin Alloys Ltd] ‡

Brightray Alloy F. A trademark for an electrical resistance alloy of 37% nickel, 18% chromium, 2.2% silicon, and the balance iron. [Wiggin Alloys Ltd] ‡

Brightray Alloy H. A trademark for an electrical resistance alloy of 19.5% chromium, 3.6% aluminum, and the balance nickel. [Wiggin Alloys Ltd] ‡

Brightray Alloy S. A trademark for an electrical resistance alloy of 20% chromium, and the balance nickel. [Wiggin Alloys Ltd] ‡

Brij®. Polyoxyethylene alkyl ethers. emulsifiers. [ICI Am.]

Brij®30. Laureth-4; o/w emulsifier for topical cosmetics. [ICI Spec. Chem.; ICI Surf. Belgium]

Brij® 30SP. Laureth-4; preservatives; o/w emulsifier for topical cosmetics. [ICI Spec. Chem.]

Brij® 52. CAS 9004-95-9; Ceteth-2 (antioxidants added); surfactant, emulsifier for topical cosmetics. [ICI Spec. Chem.; ICI Surf. Belgium]

Brij® 72. CAS 9005-00-9; Steareth-2 with preservatives; surfactant, emulsifier especially for topical cosmetic applications. [ICI Spec. Chem.; ICI Surf. Belgium]

Brij®93. CAS 9004-98-2; Oleth-2 with preservatives; surfactant with low color and odor; disperses emollients, perfume oils, and surfactants; especially for blooming bath oils. [ICI Spec. Chem.]

Brij®96. Polyoxyl 10 oleyl ether; pharmaceutic aid. [ICI Am.]

Brij®97. Polyoxyl 10 oleyl ether; pharmaceutic aid. [ICI Am.]

Brij® 700 S. CAS 9005-00-9; Steareth-100; emulsifier, solubilizer. [ICI Spec. Chem.]

Brij®721. CAS 9005-00-9; Steareth-21; cosmetic emulsifier, solubilizer for fragrances. [ICI Spec. Chem.; ICI Surf. Belgium]

Briklens. Laundry detergents. [Laporte Industries Ltd] *

Briline. Diazo compounds and coupling agents. [Bridge Pharmaceuticals] *

Brilliant Indigo. Dyestuffs for dyeing and printing of cellulosic fibers. [BASF plc] *

brimstone. See sulfur

Brimstone Plus. Mixture of potassium sorbate, sodium metabisulfate and sodium propionate; a broad-spectrum fungicide for field crops. [Mandops (UK) Ltd]

Bri-Nylon. Nylon yarns. [ICI Chem. & Polymers Ltd]

Briosil. A nonprecious metal alloy; for dentistry and dental engineering. [Degussa AG] *

Briotril. Bromoxynil octanoate + ioxynil octanoate; herbicide used for selective post-emergence weed control. [Agan Chemical Manufacturers Ltd]

Briotril Plus. An emulsifiable concentrate containing 200 g bromoxynil and 200 g ioxynil per liter; a post-emergence contact herbicide for cereal crops. [Pan Britannica Industries Ltd]

Briphos. Range of anionic surfactants composed of aliphatic or nonylphenol based ethoxylated phosphate esters (mixture of mono- and di-esters) in acid form; viscous liquids or paste; used for agricultural chemicals, cleaners, cosmetics, toiletries, dry cleaning, inks, lubricants, surface coatings. [Albright & Wilson Ltd, Phosphates & Speciality Business]

Briquest®. Phosphonate derivaties. [Albright & Wilson Ltd, Phosphates & Speciality Business]

Briquest® 301-30SH. Sodium nitrilotris (methylene phosphate); sequestrant for scale inhibition and corrosion control. [Albright & Wilson UK]

Briquest®301-50A. Nitrilotris (methylene phosphonic acid); sequestrant for water treatment, oil-drilling muds, powd. detergents, photographic applications. [Albright & Wilson UK]

Briquest® 462-23K. Potassium hexamethylene diamine tetrakis (methylene phosphate); sequestrant for water treatment. [Albright & Wilson UK]

Briquest® 543-45AS. CAS 22042-96-2; Diethylene-triamine-pentakis (methylene phosphonic acid); sequestrant for peroxide stabilization in pulp bleaching and de-inking, in liq. detergents and oil-field chemicals. [Albright & Wilson UK]

Briquest® ADPA-60A. CAS 2809-21-4; 1-Hydroxy-ethylidene-1; 1-diphosphonic acid; used for water treatment and oil-drilling muds, in powd. detergents and photographic applications, sequestering agent for cal-

---

‡ Trade name and manufacturer not verified   § Trade name without identified manufacturer

cium carbonate. [Albright & Wilson Am.]

Brisgo II. A thermoplastic resin derived from rosin acids; hog carcass dehairing composition. [Hercules] *

Bristagen. CAS 1405-41-0; Gentamicin sulfate; a complex antibiotic substance, produced by *Micromonospora purpurea nsp*; three components: sulfates of gentamicin C_1, gentamicin C_2, and gentamicin C_{1A}; antibacterial. [Bristol Laboratories] *

Bristamycin. CAS 643-22-1; Erythromycin stearate; antibacterial. [Bristol Laboratories] *

Bristocycline. CAS 60-54-8; A proprietary preparation of tetracycline; an antibiotic. [Bristol-Myers Squibb Co Inc]

Bristol brass. See brass.

Britannia metal. An alloy of from 74-91% tin, 5-24% antimony, and 0.15-3.68% copper, sometimes with small quantities of zinc, lead, and bismuth; a Britannia metal containing 90% tin and 10% antimony has a specific gravity of 7.

Britesil. Hydrous sodium polysilicate; used as an alkaline builder in powdered or granular laundry detergents, dishwashing detergents, household cleaners. [PQ Corp]*

Britesorb. Hydrous silica; clarifying and chill proofing agent for beer, clarifying or fining agent for wines, fruit juices etc.; selective absorbent of proteins and metals. [PQ Corp] *

British gum. (starch gum; vegetable gum; gommeline; artificial gum). Dextrin; $n(C_6H_{10}O_5)$; obtained by the action of diastase on starch paste, or by heating starch with a trace of acid; used as an adhesive.

Britomya. Ground chalk; whitening fillers. [Croxton & Garry Ltd]

Britonite. An explosive containing nitro-glycerin, potassium nitrate, wood meal, and ammonium oxalate.

Brittox. Herbicide. [May & Baker Ltd] *

Brix. Cupola flux. [Foseco (F.S.) Ltd] †

Brixil. Cupola flux and silicon additive. [Foseco (F.S.) Ltd]†

Brix metal. Alloys of from 60-75% nickel, 15-20% chromium, 5% copper, 1-4% tungsten, 4% silicon, 3% titanium, 2% aluminum, and 1% bismuth; noncorrosive.

Briz. A scouring powder. [Unilever] *

broad salt. Ground rock salt.

Broadshot. Emulsifiable concentrate containing 200 g 2,4-D, 85 g dicamba and 65 g triclopyr per liter; an herbicide to control perennial and woody weeds. [Shell UK]

Brocadopa. CAS 59-92-7; A proprietary preparation of levodopa used in the treatment of Parkinson's disease. [Brocades Pharma]

Broenner's acid. 2-Naphthylamine-6-sulfonic acid.

Brolac Dualcote Acrylic Primer/Undercoat. Acrylic copolymer emulsion, water thinned, interior primer for soft and hardwoods and an interior and exterior undercoat for Brolac Full Gloss. [Berger Jenson & Nicholson Ltd] *

Brolac Eggshell Low Odour. Alkyd resin based, white spirit thinned; for interior application; can be used in kitchens and bathrooms giving satin sheen finish. [Berger Jenson & Nicholson Ltd] *

Brolac Full Gloss. Alkyd resin based, white spirit thinned; high gloss protective finish for interior and exterior use by the professional decorator and specifier. [Berger Jenson & Nicholson Ltd] *

Brolac PEP Vinyl Matte & Vinyl Silk Emulsions. VA copolymer emulsion based, water thinned; Vinyl Matte is for both exterior and interior application giving a matt finish; Vinyl Silk is for interior application giving a silk sheen. [Berger Jenson & Nicholson Ltd] *

Brolac Primers, Sealers and Surface Preparation Products. Various alkyd/oleo resinous, white spirit thinned; protective primers/sealers for painting surfaces. [Berger Jenson & Nicholson Ltd] *

Brolac Specialist Coatings. Various specialty resins; for maintenance and protection of steel, floors, concrete and other surfaces requiring extra resistance and performance. [Berger Jenson & Nicholson Ltd] *

Brolac Superflat Emulsion. VA copolymer emulsion, water thinned; for interior application, especially suited to new plasterwork. [Berger Jenson & Nicholson Ltd] *

Brolac Tartaruga. VA copolymer emulsion, water thinned; for high build textured wall finish for interior and exterior use. [Berger Jenson & Nicholson Ltd] *

Brolac Undercoat. Alkyd resin based, white spirit thinned; used in preparation for all types of Brolac finishes. [Berger Jenson & Nicholson Ltd] *

Brolac Varnishes. Alkyd resin base (some P.U. modified), thinned with white spirit; provides surface sheen to soft and hardwoods. [Berger Jenson & Nicholson Ltd] *

Brolac Weathercoat No. 1 - finely textured. VA copolymer emulsion, water thinned; for masonry exterior coatings. [Berger Jenson & Nicholson Ltd] *

Brolac Weathercoat No. 2 - smooth. VA copolymer emulsion, water thinned; for exterior masonry, ideal for airless spray equipment; suitable for buildings subjected to atmospheric pollution. [Berger Jenson & Nicholson Ltd] *

Brolac Weathercoat No. 3. Pliolite resin based, white spirit thinned; provides smooth finish to masonry; can be applied in cold, wet and damp conditions with no separate sealer requirement. [Berger Jenson & Nicholson Ltd] *

Brolene. CAS 140-63-6; A proprietary preparation of propamidine isethionate; an ocular antiseptic. [May & Baker Ltd]

bromal. Tribromoacetaldehyde, CBr_3CHO; formerly used in medicine.

Bromat®. CAS 57-09-0; EINECS 200-311-3; Cetyl trimethyl ammonium bromide; surfactant, emulsifier, germicide. [Zeeland]

brombutol. See brometone.

bromcresol purple. Dibromo-o-cresolsulfonphthalein. An indicator.

Bromeikon. Sodium tetrabromophenolphthalein.

Bromelain Conc. Mixture of proteases; enzyme for hydrolysis of plant and animal proteins to peptides and amino acids; for tenderizer formulations for meat; in baking; pharmaceuticals (wound debriding agent, blood typing studies); fish processing; eliminates protein haze in foods. [Solvay Enzymes]

Bromelains. A concentrate of proteolytic enzymes derived from Ananas comosus Merr. Ananase; a pale brown powder; recommended as a substitute for pancreatin; prepared from pineapple juice. [British Drug Houses] *

brometone. (acetone-bromoform, brombutol). Tribromo-tert-butyl alcohol, $CBr_3C(CH_3)_2OH$). Formerly used in medicine as a sedative.

Bromex. CAS 300-76-5; Active ingredient: naled; fast-acting agricultural insecticide of short to moderate residual action. [Makhteshim Chemical Works Ltd]

bromic acid, sodium salt. See sodium bromate.

Bromicide. Cooling water biocide. [Great Lakes Europe]

bromidine. A dry mixture of sodium bisulfate with sodium or potassium bromide and bromate; a disinfectant.

bromindione. 2-(4-Bromophenyl)indane-1,3-dione; Circladin) An anticoagulant.

bromine. Halogen; Br_2; CAS 7726-95-6; EINECS 231-778-1; manufacture of ethylene dibromide, organic synthesis; in water disinfection; bleaching fibers; medicinals; dyestuffs. [Bromine & Chems. Ltd; Ethyl; Great Lakes Chem.; Janssen Chimica; Spectrum Chem. Mfg.]

bromine cyanide. See cyanogen bromide

bromine salt. A mixture made by saturating caustic soda with bromine. draining off the mother liquor, and adding sodium bromate; used in the extraction of gold ores.

1-bromobutane. See n-butyl bromide.

Bromocol®. A range of pharmaceutical products. [Cassella AG] *

bromocresol green. Tetrabromo-m-cresolsulfonphthalein; an indicator.

Bromodan. A brominating agent. [Fisons plc, Horticulture Div] *

1-bromodecane. See decyl bromide

Bromodeine. A proprietary preparation containing bromoform, codeine hydrochloride, liquid extract of krameria, wild cherry and senega; a cough linctus. [Crookes Laboratories] *

Bromoform. CAS 75-25-2; Used for gem and mineral testing. [Geoliquids] *

bromoform. Tribromomethane, $CHBr_3$; used in medicine

Bromo-Gas. Methyl bromide with chloropicrin [Great Lakes Europe]

Bromoklor 50. Halogenated aliphatic liqs. containing bromine and chlorine; flame retardant used in plasticized PVC; uses including disp. molded or coated automotive parts, interior trim, package closures, boots, hand grips, flooring, upholstery, carpet backing, furniture and wall covering, and packaging film; plasticizing properties in PVC (not primary). [Ferro]

bromol. Tribromophenol, $C_6H_2(OH)Br_3$; used as a caustic and disinfectant.

bromolaurionite. A mono-hydrated lead oxydibromide, $PbO \cdot PbBr_2 \cdot H_2O$, obtained by heating a solution of lead acetate and sodium bromide for 12 hours.

bromomethane. See methyl bromide.

bromonitroform. Bromotrinitromethane, $CBr(NO_2)_3$.

Bromopicrin. Tribromonitromethane, $CBr_3(NO_2)$. organic synthesis; military poison.

bromo-purpurin. Bromotrihydroxyanthraquinone.

bromopyrin. Monobromoantipyrine, $C_{11}H_{11}BrN_2O$.

Bromotril. Active ingredients: bromoxynil octanoate; 2,6-dibromo-4-cyanophenyl octanoate; selective post-emergence control of a wide range of annual broadleaf weeds in winter and spring cereals and in corn. [Agan Chemical Manufacturers Ltd]

bromowagnerite. A compound of calcium bromide and phosphate, $Ca_3(PO_4)_3 \cdot CaBr_2$.

Bromox. CAS 7758-01-2; Potassium bromate dispersion in food grade filler; flour additive. [Diaflex Ltd] *

bromoxynil. (1,4-dibromo-3-cyano phenol) A specific herbicide for use in cereal crops.

Bromsulphalein. CAS 71-67-0; Sulphobromophthalein sodium; diagnostic aid. [Hynson Westcott & Dunning] *

Bronchilator. A proprietary preparation of isoetharine methanesulfonate, phenylephrine hydrochloride and thenyldiamine hydrochloride; a bronchial antispasmodic. [Bayer AG]

Broncho-Binotal®. A preparation for bacterial bronchopumlonary infections. [Bayer AG]

Bronchodil. See Bronchospasmin. [Degussa AG] *

Bronchopront. A proprietary preparation of ambroxol hydrochloride; a mucolytic. [Pfizer International]

Bronchospasmin. Reproterol; B_2-Mimetikum; tablets, ampoules and aerosol. [Degussa AG] *

Bronco. Active ingredients are 2.6 lb of 2-chloro-2'6'-diethyl-N-(methoxymethyl) acetanilide and 1.4 lb of the isopropylamine salt of glyphosate; herbicide for no-till farming. [Monsanto Co] *

Bronidox L. 5-Bromo-5-nitro-1,3-dioxane dissolved in 1,2 propylene glycol; preservative for shampoos, foam bath and other surfactant preparations. [Henkel/Cospha; Henkel Canada]

Bronkaid Mist. CAS 51-43-4; Epinephrine; adrenergic. [Sterling Drug Inc] *

Bronkephrine. CAS 3198-07-0; Ethylnorepinephrine hydrochloride; bronchodilator. [Sterling Drug Inc] *

Bronkometer. Isoetharine mesylate; bronchodilator. [Sterling Drug Inc] *

Bronkosol. CAS 2576-92-3; Isoetharine hydrochloride; bronchodilator. [Sterling Drug Inc] *

Bronner's acid. 2-Naphthylamine-6-sulfonic acid.

Bron-Newcavac-M. Bronchitis-Newcastle disease, inactivated vaccine; for immunization of poultry. [Intervet Inc]

Bron-Newcavac-MG. For immunization of poultry. [Intervet Inc]

Bronocot. Cotton seed dressings. [ICI Chem. & Polymers Ltd]

Bronopol. CAS 52-51-7; 2-Bromo-2-nitropropane-1,3-diol; broad spectrum antimicrobial agent; preservative for topical cosmetics, pharmaceuticals, toiletries. [Angus; Inolex]

Bronopol-Boots. Pharmaceutical and cosmetic preservative. [The Boots Co plc]

Bronotabs. Milk testing preservative. [The Boots Co plc] *

Bronox. Wettable powder containing linuron and tritrazine; used for control of annual dicotyledons and annual grasses in potatoes and nursery stock. [Schering Agrochemicals Ltd]

Brontyl. CAS 603-00-9; A proprietary preparation of proxyphylline; bronchial antispasmodic. [Lloyd, Hamol]*

bronze. Alloys usually consisting of copper and tin in varying proportions, often with zinc, and occasionally with lead. The copper varies from about 74-95%, the tin from 1-20%, zinc from 0-17%, and the lead from 0-18%; spark-resistant tools, vacuum dryers, water gauges, flow indicators, valve, drain cocks, fine arts.

Bronze A. A British chemical standard. It contains 85.5% copper, 9.96% tin, 1.86% zinc, 1.83% lead, 0.25% phosphorus, 0.24% antimony, 0.07% iron, 0.04% nickel, and 0.06% arsenic.

bronze acetate. A calcium acetate prepared from crude pyroligneous acid and lime. It contains from 60-70% acetate. The name is also applied to an impure variety of lead acetate prepared from the same acid.

bronze bearing metals. Very variable alloys; One type contains from 70-91% tin, 7-26% antimony, and 2-22% copper; while another class contains from 70-86% copper, 4-20% tin, 0-30% zinc, and 0-15% lead.

bronze, Durbar. See Durbar bronze. §

---

‡ Trade name and manufacturer not verified   § Trade name without identified manufacturer

bronze, Eclipse. See Eclipse bronze. §
bronze, Olympic. See Olympic bronze. §
bronze, Phono. See Phono bronze. §
bronze powder. See copper.
bronze, Vulcan. See Vulcan bronze. §
bronze wire. An alloy of 98.75% copper, 1.2% tin, and 0.05% phosphorus.
bronzing liquids. Consist of volatile liquids which will hold up the metal and some material which will keep the metallic powder from rubbing off after it has been applied. The best one contains pyroxylin dissolved in amyl acetate to which the metallic powder is added
bronzing solder. An alloy of 50% zinc and 50% copper.
bronzite. A pyroxene mineral; $16[(Mg,Fe)SiO_3]$.
Brookosome®EFA. Phospholipids, omega 6 linoleic acid, omega 3 linolenic acid; moisturizer for skin cosmetics. [Brooks Industries]
Brookosome® EPO. Phospholipids and evening primrose oil; moisturizer for skin cosmetics. [Brooks Industries]
Brookosome® Fucus. Phospholipids, fucus extract; skin moisturizer for use in slimming creams; derived from succulent giant kelp (Fucus vesiculosus) [Brooks Industries]
Brookosome®TRF. Live yeast cell deriv. and phospholipids; tissue respiratory factors promoting wound healing and anti-inflammatory effects on skin. [Brooks Industries]
Brookswax D. Cetearyl alcohol, ceteareth-20; emulsifying wax substantive to hair. [Brooks Industries]
Brookswax P. Emulsifying wax NF; emulsifying wax substantive to hair. [Brooks Industries]
Brophos 5C10. CAS 50643-20-4; PPG-5 ceteth-10 phosphate; surfactant, emulsifier substantive to hair. [Brooks Industries]
Brophos OL-3. CAS 39464-69-2; Oleth-3 phosphate; surfactant, emulsifier substantive to hair. [Brooks Industries]
Brophos OL-3N. CAS 58855-63-3; DEA oleth-3 phosphate; surfactant, emulsifier substantive to hair. [Brooks Industries]
Brotopon. CAS 52-86-8; A proprietary preparation of haloperidol used for psychotic disorders. [Pfizer International]
Brovon. Atropine methonitrate, adrenaline and papaverine; for bronchospasm. [Napp Laboratories Ltd] *
Brown 208. CAS 1309-37-1, 7787-59-9; Iron oxides, bismuthoxychloride; inorg. colorant. [Presperse]
brown barberry gum. See Morocco gum.
Brown Copp. CAS 1317-39-1; Brown cuprous oxide. [Am. Chemet]
brown lead ore. (Linnets). A mixture of lead phosphate and lead chloride, $3[Pb_3(PO_4)_2]+PbCl_2$.
brown lead oxide. Lead dioxide, PbO_2.
brown ore. A variable mixture of hydrated oxides of iron, usually $2Fe_2O_{33} \cdot H_2O$.
brown oxide of tungsten. Tungsten dioxide, WO_2.
brown precipitate. Iodine dissolved in potassium iodide.
Broxli. CAS 132-93-4; A proprietary preparation containing phenethicillin potassium; an antibiotic. [SmithKline Beecham] *
Brozgerite. See Clevite. §
Brufen. CAS 15687-27-1; A proprietary preparation containing ibuprofen; an antirheumatic drug. [The Boots Co plc]
Brugre powder. An explosive; a priming composition,

containing 54% ammonium picrate, and 46% potassium nitrate.
Brulidine. CAS 614-87-9; A proprietary preparation of dibromopropamidine isethionate; antiseptic skin cream. [Rhone-Poulenc Rorer Ltd]
Brunner's salt. $HgSK_2S \cdot 5H_{2O}$; Obtained by dissolving vermilion in potassium mono-sulfide.
Brunol. A proprietary preparation of n-butyl salicylate. [Catomance Group]
Brunswick black. Asphalt or pitch mixed with turpentine and linseed oil, and heated.
Brunswick blue. (Celestial blue). A pigment produced by mixing 50-90% barytes with Prussian blue.
Brush-B-Gon. Brush killer. [Chevron] *
Brush Buster. 2,4-D and 2,4-DP; A post-emergence herbicide used to control woody species such as poison oak/ivy and brambles; applied as a foliar spray or straight from the bottle to the cut stump of woody species. [Lawn & Garden Products Inc]
Brushcrete. Two-component cementitious coating providing a hard-wearing, seamless, waterproof membrane which protects and resurfaces concrete masonry and other construction materials. [Feb Ltd]
brush wire. A brass wire containing 64.25% copper, 35% zinc, and 0.75% tin.
Brussels System. A systemic insecticide. [Murphy Chemical Co Ltd] †
B.R.V. A coal-tar distillate consisting chiefly of high boilng constituents; used as a rubber softener.
Bryrel. CAS 144-29-6; Piperazine citrate; anthelmintic. [Sterling Drug Inc] *
B/S. See butadiene/styrene copolymer
B.S. Sea Water Alloy. An aluminum alloy containing 7.5-9.5% magnesium, 0.2% silicon, and 0.2-0.6% manganese; tensile strength is 45-55 kg/mm^2.
BSWL 202. CAS 10099-76-0; Basic silicate wh. lead; wh. pigment acting as heat stabilizer for chlorinated polyethylene, chlorosulfonated polyethylene, PVC, and polyepichlorohydrin; rust-inhibitive pigment in the automobile industry; used in industrial or maintenance paints. [Eagle-Picher] *
BTC®50 USP. Benzalkonium chloride; antimicrobial for hard surf. disinfection, sanitization, deodorization. [Stepan; Stepan Canada]
BTC® 99. CAS 7173-51-5; Didecyl dimonium chloride; low foaming algicide and slimicide for swimming pool and industrial water treatment. [Stepan; Stepan Canada]
BTC® 818. CAS 32426-11-2; Quaternium-24; disinfectant, sanitizer, and fungicide for hard surfaces; excellent sanitizer in hard water to 800 ppm as $CaCO_3$. [Stepan; Stepan Canada]
BTC®824. CAS 139-08-2; EINECS 205-352-0; Myristalkonium chloride; antimicrobial for hard surf. disinfection and sanitization; algicide for swimming pools and industrial water treatment. [Stepan; Stepan Canada]
BTC® 885 P40. Quaternium-24, benzalkonium chloride; germicide for formulation of disinfectant, sanitizer and fungicidal products for hospitals, nursing homes, public institutions. [Stepan; Stepan Canada]
BTC® 1010-80. Quaternium-12; fungicide for hard-surf. disinfection and sanitization; algicide in swimming pool and industrial water treatment; deodorizer. [Stepan; Stepan Canada]
BTC®2565. CAS 139-08-2; EINECS 205-352-0; Myristalko-

---

nium chloride; algicide and slimicide for swimming pool and industrial water treatment. [Stepan; Stepan Canada]

BTG alloy. A heat-resisting alloy containing 60% nickel, 12% chromium, 1-4% tungsten, and balance iron.

Bubber Shet. CAS 57-13-6; EINECS 200-315-5; Prilled urea containing minimum of 46% available nitrogen; used as a fertilizer to supply nitrogen to the crops for better yield per acre. [Dawood Hercules Chemicals Ltd]*

Bubble Breaker® 748. Silicone-free blend; defoamer for water-based systems, paints/coatings. [Witco/Organics]

Bubble Breaker®3056A. Disp. of reacted silica in hydrocarbon solv.; defoamer used in latex manufacturing operations, formulation of water-based paints and adhesives, effluent water, asphalt emulsions, PVC monomer stripping. [Witco/Organics]

Bubble Breaker®DMD-1. Complex surfactant; oilfield surfactant, defoamer. [Witco/Organics]

Bubblefil. A proprietary trade name for regenerated cellulose. §

Buca. CAS 1332-58-7; Very fine particle size ultra pulverized hydrous kaolin (aluminum silicate); used as an extender pigment in coatings, inks, caulks, rubber. [Engelhard]

Bucarpolate. A pyrethrum synergist. [Bush Boake Allen Ltd]

buchu camphor. Diosphenol, $C_{10}H_{15}O_2$, the chief constituent of the essential oil obtained from Buchu leaves from *Barosma beturima;* used as an antiseptic.

Buckland's cement. A label cement consisting of 50% gum arabic, 37.5% starch, and 12.5% white sugar. It is mixed with a little water for use. §

Buckroid. A very tough form of pure vulcanized rubber; used for making mats, and for other purposes.

Bucrol. Carbutamide.

Buctril. Selective weedkiller. [May & Baker Ltd] *

Budale. A proprietary preparation of paracetamol, codeine phosphate and butobarbitone; a sedative for veterinary use. [Dales Pharmaceuticals Ltd] *

Budene®1207. CAS 9003-17-2; Polybutadiene; nonstaining, sol'n. polymerized; used in tire tread and carcass compounds, v-belts, conveyor belt covering, hose covers and tubes, solid golf balls, footwear, sponge rubber, mechanical goods; imparts abrasion resist., low temp. props., resilience, durabilit [Goodyear] *

Budene®1254. CAS 9003-17-2; Polybutadiene CAS 9003-17-2, extended with 18.5-21.5% petrol. process oil CAS 64742-04-7; used for treads, belts, retread stocks, mechanical goods; good dynamic props., enhanced compound processing. [Goodyear] *

Bufa. CAS 84-74-2; A proprietary trade name for dibutyl phthalate. [Koninklijke Maastrichtsche] *

Bufferight. CAS 144-55-8; Sodium bicarbonate. [Kerr-McGee Chemical Corp]

Bufferin. A proprietary preparation of aspirin, magnesium carbonate and aluminum glycinate; an analgesic. [Bristol-Myers Squibb Co Inc]

Bu-Gas. CAS 106-97-8; Butane. [Chevron] *

Bug Check®BF. Bacterial and fungal contamination indicator. [Angus]

Bug Check® SRB. Simplified sulfate-reducing bacteria viable counts. [Angus]

Bug-Geta. Slug and snail pellets. [Chevron] *

Bug Gun! Ready-for-use insecticide spray. [ICI Garden Products]

Bulldock. Beta-cyfluthrin; insecticide with mainly contact action against biting insects on maize, cotton, deciduous fruit, groundnuts, potatoes, vegetables and other crops; also against migratory locusts and grasshoppers. [Bayer AG]

Bullet Brass. See brass.

Bullion®. For agriculture industry. [DuPont UK]

Bull metal. An alloy similar to Delta metal in composition.

Bullseye. A double-base-type formulation minimizing charge weight and moisture absorption; graphite glaze enables smooth granule flow; a smokeless powder used for ammunition. [Hercules] *

Bullseye. Water soluble, biodegradable colorants; for spray pattern indicators for application of herbicides, pesticides in the lawn and turf, forestry and industrial weed control areas. [Milliken] *

Bullseye CDA. Amitrole + atrazine + diuron; a liquid mixture of herbicides for weed control. [ICI Agrochemicals]

Bumal. Modified rosin emulsion. [Tenneco Malros Ltd] *

Bumex. CAS 28395-03-1; Bumetanide; diuretic. [Hoffmann-LaRoche Inc] *

Buminate. Albumin human; blood volume supporter. [Hyland Therapeutics] *

Bumper. CAS 60207-90-1; Propiconazole; fungicide. [Makhteshim Chemical Works Ltd]

Bumyr. CAS 110-36-1; Butyl myristate; emollient; cosmetic ingred.. [Amerchol Corp]

Buna®CB. Cis-1,4-polybutadiene polymers; used for tires, conveyor belting, mountings and roll covers and plastics modification. [Bayer AG]

Buna® CB 11. Butadiene rubber, organometallic titanium catalyst; used in tires, conveyor belting, caterpiller tread blocks, footwear soles, transmission belting; for blending with NR in buffers, roll covers, seals, and injection molded goods. [Bayer; Miles; Polysar]

Buna®CB 22. Butadiene rubber, neodymium catalyst type, nonstaining stabilizer; used in blends with other general purpose rubbers (NR and IR); imparts excellent abrasion resist., rebound resilience, dynamic fatigue resist.; in NBR and CR, improves the low temp. brittleness pt. and helps to reduce mill sticking respecially; applications including tires, conveyor belts, power transmission belting, roll covers, shoe soling, hose, and extruded and molded goods. [Bayer; Miles; Polysar]

Buna Hüls AP. A range of ethylene/propylene rubbers; used as modifiers for polyolefins, molded articles and hoses, blend component for improving flowability and green strength. [Hüls AG]

Buna Hüls EM. A range of styrene/butadiene rubbers; emulsion polymerization; used for tires, injection moldings and industrial rubber goods. [Hüls AG]

Buna S. A butadiene-styrene copolymer; a proprietary trade name for a vulcanizable synthetic rubber. §

Buna SL. A range of styrene/butadiene rubbers; solution polymerization; used for tire treads, industrial rubber goods, shoe soles and heels. [Bayer AG] *

Buna SS. An important series of synthetic products is made by copolymerizing butadiene with 10-30% of another polymerizable substance such as styrene or acetonitrile, e.g., Buna N, etc. Contains a larger percentage of styrene. §

Bunatex. A range of styrene/butadiene rubber latex; used as raw materials for foam backing for carpets and textiles, mattresses, upholstery materials for the furniture indus-

---

‡ Trade name and manufacturer not verified   § Trade name without identified manufacturer

try, shoe components, rubberized hair bitumen modifiers and friction li [Hüls AG]

Buna Vi. A range of vinyl/butadiene rubbers; used for industrial goods and tires. [Bayer AG] *

Bunte's salt. Sodium-ethyl thiosulfate, (qv).

Buprenex. CAS 53152-21-9; Buprenorphine hydrochloride; analgesic. [Norwich Eaton Pharmaceuticals Inc] *

Bur-A-Loy®3873. NBR/PVC (60:40), 70 parts DOP plasticizer; highly plasticized; soft roll and printing blanket stocks; excellent PVC disp. and end-processing. [Mach-1 Compounding]

Bur-A-Loy®3874. NBR/PVC (60:40), 85 parts DOP; used for low durometer rolls and molded goods; good oil, abrasion, and ozone resist. [Mach-1 Compounding]

Bur-A-Loy®5915. NBR/PVC (50:50), 15 parts DOP plasticizer; thermoset or thermoplastic elastomer; for heels, soles, mats, and extruded goods where oil, abrasion, and ozone resist. is required. [Mach-1 Compounding]

Bur-A-Loy®7130. NBR/PVC (70:30), 30 parts DOP plasticizer; for hose, oil seals, and general molded and extruded goods; exhibits excellent oil and fuel resist. [Mach-1 Compounding]

Burco CS-LF. Phosphate ester of EO/PO block polymer, halogen-capped ethoxylated polyether; biodeg. low foaming synergistic surfactant blend, scouring agent for textile processing, dishwash rinse aid, hard surf. cleaners, liq. laundry products, steam cleaners; stable to 10% caustic. [Burlington Chem.]

Burco DFE-45. Polyol ester; emulsifier, defoamer for paper, textiles, water treatment, coatings, metalworking applications and low-foaming emulsions. [Burlington Chem.]

Burco LAF-6. Aliphatic alcohol alkoxylate; biodeg. low foam detergent, wetting agent for rinse aid formulations, automatic dishwash, metal cleaners, textile scouring agents; stable to acids and alkalies. [Burlington Chem.]

Burco NCS-80. Surfactant blend; moderately foaming detergent intermediate for formulating mildly alkaline cleaners for use as vehicle cleaners, floor cleaners, car wash. [Burlington Chem.]

Burco NPS-225. CAS 68515-73-1; Alkyl polyglycoside; dispersant, wetting agent, coupling agent in high electrolyte solns.; for hard surf. cleaners (acid or alkali), all-purpose cleaners, metal cleaners; hydrotrope in highly alkaline formulations; biodeg.; stable to acids or alkalies. [Burlington Chem.]

Burco TME. Ethoxylated dodecylmercaptan; detergent for aq. cleaning systems, wool scouring, hard surf. cleaners; oil splitter; replaces chlorinated hydrocarbons in metal degreasing. [Burlington Chem.]

Burco Anionic APS. Ethoxylated sulfonate; hypochlorite-stable surfactant, emulsifier for acid and alkaline cleaners, disinfectants, personal care products, household cleaners, tub and tile cleaners, mildew removers, textile scours, dairy cleaners; stable over entire pH range. [Burlington Chem.]

Burcofac 1060. Organic phosphate ester; wetting agent, detergent, emulsifier, hydrotrope for other surfactants, glycol ethers; for mildly alkaline cleaner formulations, all-purpose cleaners, floor cleaners/wax strippers, carwash, textile scouring; alkaline stable. [Burlington Chem.]

Burcol BP-181. Low-foaming block polymer; defoamer for textiles, paper, metalworking, antifreeze, and other applications; lubricant base; rinse aid formulations. [Bur-

lington Chem.]

Burcop. Soda/Bordeaux fungicide. [McKechnie Chemicals Ltd] *

Burcosperse AP Liq. CAS 9003-04-7; Low molecular weight sodium polyacrylate sol'n.; chelating agent, anti-redeposition agent. [Burlington Chem.]

Burcoterge DG-40. Linear alcohol derivs.; wetting agent, detergent, emulsifier for laundry and hard surf. cleaners, microemulsions with solvs., degreasing formulations; excellent performance in hot or cold systems. [Burlington Chem.]

Burcotreat 900-A. CAS 9003-01-4; Polyacrylic acid; chelating agent, antiredeposition agent; free acid version of Burcosperse AP Liq. [Burlington Chem.]

Burcowet TM-LF. Alkoxylated linear alcohol; defoamer at high temps.; adds wetting, detergency. [Burlington Chem.]

Burez. Disproportionated rosin and rosin derivatives. [Albright & Wilson Ltd; Tenneco Malros Ltd]

Burgess 30-P. CAS 12141-46-7; Aluminum silicate, thermo-optic; used in PVC compounds; combines ease of incorpiation, uniformity of compound color, excellent elec. props., low sp.gr. [Burgess Pigment]

Burgess 2211. CAS 12141-46-7; Aluminum silicate, anhyd., surf. treated; used in min.-filled nylon applications and polyterephthalate, urethane, PVC, thermoplastic polyester; features low warpage and high impact strength. [Burgess Pigment]

Burgess KE. CAS 12141-46-7; Aluminum silicate, anhyd., surf. treated; very pure, high brightness clay for use in EPR, EPT, crosslinked polyethylene and polyester systems; excellent wet and dry, initial and long term elec. chars.; increases tensile strength and compression set. [Burgess Pigment]

Burgess-Hambuechen solution. A solution containing 275 g ferrous ammonium sulfate and 1000 cc water; used for the electrodeposition of iron, using a current density of 6-10 amps/ft^2 at 30 C.

Burgess solder. An alloy of 76% tin, 21% zinc, and 3% aluminum.

Burgundy Lake. A proprietary trade name for a red lake containing organic colors, aluminum hydroxide, and Blanc fixe[ac]. §

Burgundy pitch. The resinous exudation of the European silver fir; artificial substitutes have been made; used in medicine to make plasters.

Burinex K. A proprietary preparation of bumetanide with potassium chloride in slow-release form; a diuretic. [Leo Laboratories] *

Burkeite. (Gauslinite). A double salt of sodium carbonate and sulfate.

Burmite. (Birmite). Burmese amber of a reddish-brown color.

Burmol®. For stripping of dyeings and for removing discoloration and stains from textiles; for commercial laundries. [BASF AG]

Burnol Acriflavine Cream. CAS 8048-52-0; An emulsified cream containing neutral acriflavine for use in cases of burns, wounds, etc. [The Boots Co plc]

burnt carmine. A pigment obtained by calcining carmine.

burnt hypo. (black hypo, Eureka compound). A mixture of lead thiosulfate and sulfide, and sulfur; used in the vulcanization of rubber.

burnt iron. Iron which has been heated to a high tempera-

---

ture for a long time. It is brittle.

burnt magnesia. CAS 1309-48-4; Magnesium oxide; used in refractories, especially for steel furnace linings, polycrystalline ceramic for aircraft windshields; removal of sulfur dioxide from stack gases.

burnt nickel. A term used for a grey pulverulent nickel precipitated by too strong a current during its electrodeposition.

burnt pyrites. Pyrites which have been burnt until 70% of ash is left; consists of iron oxide, Fe_2O_3.

burnt topaz. When Brazilian topaz is heated, it changes from a cherry-yellow to a rose-pink, being then known as burnt topaz.

burnt umber. Umber which has been heated, whereby its color is somewhat reddened. Raw umber is a brown earthy variety of ocher, colored by oxides of iron.

Buro-sol Concentrate. CAS 8006-13-1; Aluminum acetate; astringent. [Doak Pharmacal Co] ‡

Burow's solution. A 7.5-8% solution of aluminum acetate. (*Liquor alumini acetatis, BPC*); an astringent and antiseptic.

Burr Brass. See brass.

Bursoline. Sulfonated oil for tanners. [Clayton Aniline Co Ltd] *

Burtolin. CAS 123-33-1; A tree growth inhibitor containing 185 g/liter maleic hydrazide (as the potassium salt); used to control shoots on the trunk and suckers around the base of street trees; it also inhibits the development of buds on the trunk which remain dormant following treatment.   [Rhone-Poulenc Environmental Prods. Ltd]

Buscopan. A proprietary preparation containing hyoscine-N-butylbromide; an antispasmodic. [Boehringer Ingelheim Ltd]

Bush metal. An alloy of 72% copper, 14% tin, and 14% yellow brass.

Bushwacker. CAS 34014-18-1; Wettable powder or granules containing tebuthiuron; used for total weed control in noncrop areas.   [Rhone-Poulenc Environmental Prods. Ltd]

Buspar. CAS 33386-08-2; Buspirone hydrochloride; tranquilizer. [Mead Johnson & Co]

Butac®. Resin acids-amine resin soaps blend; tackifier for IIR, NR, SBR, molding aid for NBR, SBR; improves pigment disps.; activates cure slightly. [Whitney & Oettler] *

Butacite®. Polyvinyl butyral. [DuPont UK]

Butaclor. Polychloroprene rubber. [BP Chemicals Ltd]

Butacote. CAS 50-33-9; A proprietary preparation of phenyl butazone; an antirheumatic drug. [Ciba plc] *

butadiene rubber. See polybutadiene

butadiene/styrene copolymer. (B/S) Mechanical rubber goods, flooring tile, tire tread, rug backing, adhesives, asphalt modification, injection molded items, medical devices, containers, toys, food containers; blendable modifier. [BASF; Firestone Syn. Rubber & Latex; Goodyear Tire & Rubber; Reichhold/Emulsion Polymers; Shell]

Butakon A2554. A proprietary butadiene copolymer rubber. [Revertex Ltd] †

Butakon ML 577/1. A proprietary butadiene/methyl methacrylate latex. [Revertex Ltd] †

butane. (n-butane; alkane C_4) $CH_3CH_2CH_2CH_3$; CAS 106-97-8; EINECS 203-448-7; hydrocarbon propellant. [Air

Prods.; Electrochem Ltd; Phillips 66 Co.]

butanedioic acid. See succinic acid.

1,4-butanediol. (BDO; 1,4-butylene glycol; tetramethylene glycol) $HO(CH_2)_4OH$; CAS 110-63-4; EINECS 203-786-5; intermediate; used in polyurethane formulation in the hard segment as a curative. [Arco; BASF; DuPont; Hüls UK; ISP]

2,3-butanedione. See diacetyl.

butane dioxime. See dimethyl glyoxime.

butanenitrile. See butyronitrile.

1-butanethiol. See n-butyl mercaptan.

Butanex. CAS 23184-66-9; Active ingredient: butachlor; N-(butoxymethyl)-2-chloro-2′,6′-diethylacetanilide; selective pre-emergence and early post-emergence weed control in transplanted, direct seeded and upland rice. [Agan Chemical Manufacturers Ltd]

1-butanol. See butyl alcohol.

2-butanone. See methyl ethyl ketone.

2-butanone peroxide. See methyl ethyl ketone peroxide.

Butanox. CAS 1338-23-4; Methylethylketone peroxide; used for the ambient temperature curing of unsaturated polyester resins [Akzo Chemie UK Ltd]

Butarez. Liquid polybutadienes. [Phillips]

Butasan Vulcanization Accelerator. CAS 136-23-2; Zinc dibutyldithiocarbamate; nondiscoloring stabilizer in noncuring applications and in butyl rubber. [Monsanto Co] *

Butazate. CAS 136-23-2; Zinc dibutyldithiocarbamate; an ultra accelerator, fast curing from 212 F (100 C) and up; low critical temperature, medium curing range below 250 F (121 C); nonstaining and nondiscoloring; for latex compounding; has strong tendency to precure. [Uniroyal] *

Butazate 50D. CAS 136-23-2; A 50% active water dispersion of Butazate; ready-to-use form for latex compounding; used in combination with Ethazate 50D and Oxaf for greater economy and improved physical properties. [Uniroyal] *

Butazolidin. CAS 50-33-9; Phenylbutazone; antirheumatic. [Ciba-Geigy Corp] *

Butazolidin Alka. A proprietary preparation of aluminum hydroxide gel, magnesium trisilicate and Butazolidine. [Ciba plc] *

Butazolidin with Xylocaine. A proprietary preparation of phenylbutazone and lignocaine. [Ciba plc] *

2-butenedioic acid, dibutyl ester. See dibutyl maleate.

1-butene, homopolymer. See polybutene.

2-butenoic acid. See crotonic acid.

Butesin Picrate. CAS 577-48-0; Butamben picrate; anesthetic. [Abbott Laboratories]

Butex. CAS 9010-85-9; Reclaimed rubber from butyl inner tubes; used in tire inner liners, inner tubes, and in butyl tapes and sealants. [Midwest Rubber Reclaiming]

Butex. Esters of 4-hydroxybenzoic acid. [Bush Boake Allen Ltd]

Butex. A proprietary trade name for a synthetic rubber. §

Butisan. Suspension concentrate containing 500 g/l metazachlor; used for weed control in brassicas and ornamental crops. [Bayer plc]

Butisan®S. Metazachlor; systemic herbicide against annual grasses, broadleaf weeds [BASF AG]

Butofan®. Butadiene polymer dispersions; paper impregnation and coating; binders for paper, adhesives, and textile coatings; binders for production of materials

---

‡ Trade name and manufacturer not verified   § Trade name without identified manufacturer

based on leather fibers. [BASF AG]

Butofan® D. Polybutadiene dispersion. [BASF plc]

Butonal®. Butadiene/styrene polymer dispersions; binders for production of adhesives; for treatment of asphalt. [BASF AG]

Butox. A proprietary preparation of polyisobutylene-isoprene. [Hardman]

Butoxone. Herbicide. [ICI Chem. & Polymers Ltd]

butoxyethanol. (2-butoxyethanol; ethylene glycol monobutyl ether; ethylene glycol butyl ether) Ether alcohol; $HOCH_2CH_2OC_4H_9$; CAS 111-76-2; EINECS 203-905-0; solvent for nitrocellulose resins, spray lacquers, cosolvent, gas chromatography. [Arco]

butoxyethanol acetate. (ethylene glycol butyl ether acetate) $CH_3CH_2CH_2CH_2OCH_2CH_2OCOCH_3$; CAS 112-07-2; Solvent. [Arco; Eastman; OxyChem]

butoxyl. 1-Methoxybutyl acetate. solvent for lacquers.

Butoxyne 497. A mixture of hydroxyethyl ethers of butynediol; nickel brightener in electroplating; pickling inhibitor used prior to plating copper; corrosion inhibitor for specialty application such as in aerosol cans. [ISP]

butter color. See annatto.

butter of paraffin. Soft paraffin.

butter of sulfur. Precipitated sulfur.

butter yellow. (oil yellow, butyro flavine). An azo dyestuff; dimethylaminoazobenzene, $C_6H_5N_2 \cdot C_6H_4N(CH_3)_2$. Formerly used for coloring butter and oils.

button metal. An alloy of 57% zinc and 43% copper.

button solder. (white solder). Usually contains 50% tin, 30% copper, and 20% brass; or 33% copper, 27% brass, and 40% zinc.

Butvar. Polyvinyl butyral resins; white, free-flowing powders supplied in seven resin types or grade forms with various hydroxyl content whose solutions provide a wide range of viscosities; recommended for upgrading the coating performance of thermosetting phenolics, ureas, melamines, epoxies, alkyds, polyurethanes and polyesters. [Monsanto Co]

n-butyl acetate. (acetic acid, butyl ester) Ester of butyl alcohol and acetic acid; $CH_3COOC_4H_9$; CAS 123-86-4; EINECS 204-658-1; manufacture of lacquers, artificial leather, photographic films, plastics, safety glass. [BASF; BP Chem Ltd; Eastman Chem.; General Chem.; Janssen Chimica; Penta Mfg.; Union Carbide]

s-butyl acetate. (acetic acid, sec-butyl ester; acetic acid 1-methylpropyl ester) Ester of butyl alcohol and acetic acid; $CH_3COOCH(CH_3)(C_2H_5)$; CAS 105-46-4 (s-); solv. for nitrocellulose lacquers, thinners, nail enamels, leather finishes. [Schweizerhall]

t-butyl acetate. (acetic acid 1,1-dimethylethyl ester; acetic acid t-butyl ester) Ester of t-butyl alcohol and acetic acid; $CH_3COOC(CH_3)_3$; CAS 540-88-5; EINECS 208-760-7; a gasoline additive. [Aldrich; Schweizerhall]

butyl acrylate. (butyl-2-propenoate) $CH_2{:}CHCOOC_4H_9$; CAS 141-32-2; EINECS 205-480-7; Intermediate in organic synthesis, polymers and copolymers for solvent coatings, adhesives, paints, binders, emulsifiers. [BASF]

butyl alcohol. (n-butyl alcohol; 1-butanol; propyl carbinol) Aliphatic alcohol; $CH_3(CH_2)_2CH_2OH$; CAS 71-36-3; EINECS 200-751-6; preparation of esters, solv. for resins, plasticizers. [Ashland; BASF; BP Chem Ltd; Eastman; Hoechst Celanese; Shell; Union Carbide; Vista]

n-butylamine. (amine C_4; 1-aminobutane) $CH_3(CH_2)_3NH_2$; CAS 109-73-9; EINECS 203-699-2; Intermediate for

emulsifying agents, pharmaceuticals, insecticides, rubber chemicals, dyes, tanning agents. [Air Prods.; BASF; Atochem N. Am.]

butylated hydroxytoluene. See BHT.

butyl benzoate. (benzoic acid, butyl ester; n-butyl benzoate) $C_6H_5COOC_4H_9$; CAS 136-60-7; EINECS 205-252-7; Solv. for celulose ether, plasticizer, perfume ingredient, dyeing of textiles. [Novachem; Pentagon Chem. Ltd; Penta Mfg.; Stockhausen; Sybron; Unitex]

butyl benzyl phthalate. (BBP; 1,2-benzenedicarboxylic acid, butyl phenylmethyl ester; benzyl butyl phthalate) Aromatic ester; $C_4H_9OOCC_6H_4COOC_7H_7$; CAS 85-68-7; plasticizer for polyvinyl and cellulosic resins, organic intermediate. [Ashland; Monsanto]

n-butyl bromide. (1-bromobutane) C_4H_9Br; CAS 109-65-9; Alkylating agent. [Great Lakes; Schweizerhall]

Butyl Carbitol. Diethylene glycol monobutyl ether. A lacquer solvent boiling at 240 C.

Butyl Cellosolve. Ethylene glycol monobutyl ether; a nitrocellulose solvent, and is used in the manufacture of brushing lacquers.

Butyl Di-Icinol. CAS 112-34-5; EINECS 203-961-6; 2-(2-Butoxyethoxy)-ethanol; solv. for use in protective coatings, inks, cleaning products, agricultural chems.; aids wetting, penetration, and soil removal; coupling solv. [ICI Australia]

Butyl Eight®. Dithiocarbamate compound in a solvent; ultra accelerator for vulcanization. [R T Vanderbilt Co Inc]

1, 4-butylene glycol. See 1,4-butanediol.

butylethylacetic acid. See 2-ethylhexoic acid.

butyl ethylene. See 1-hexene.

t-butyl hydroperoxide. (TBHP; 1,1-dimethylethyl-hydroperoxide) Organic peroxide; $(CH_3)_3COOH$; CAS 75-91-2; EINECS 200-915-7; catalyst in polymerization reactions; oxidative membrane in RBC suspensions; introduces peroxy group into organic molecules. [Akzo; Arco; Atocem N. Am.; Witco/Argus]

t-butyl hydroquinone. (TBHQ; mono-tert-butyl hydroquinone; 1,4-benzenediol, 2-(1,1-dimethylethyl)-) Aromatic organic compound.; $C_{10}H_{14}O_2$; CAS 1948-33-0; EINECS 217-752-2; polymerization inhibitor, industrial antioxidant. [Eastman; Penta Mfg.; Schweizerhall; Showa Denko; UOP]

3-tert-butyl-4-hydroxyanisole. See BHA.

butyl p-hydroxybenzoate. See butylparaben.

Butyl Icinol. CAS 111-76-2; EINECS 203-905-0; 2-Butoxy ethanol; solv. for lacquers, enamels, water-borne systems, inks, industrial cleaners, waterless hand cleaners. [ICI Australia]

Butylite. A proprietary butyl rubber sealant. [Polymeric Systems Inc] *

Butyl Kamate®. Potassium dibutyldithiocarbamate; accelerator for vulcanization of rubber. [R T Vanderbilt Co Inc]

butyl lactate. (n-butyl lactate) Ester; $CH_3CHOHCOOC_4H_9$; CAS 138-22-7; dipped latex products, coatings, specialty finishes, inks, diluents, adhesives, intermediates, photoresist solvs., screen printing of electronic parts, flavors and fragrances, solvs. for nitrocellulose, antiskinning agent, dry cleaning fluids. [CPS; Penta Mfg.; Purac Biochem BV]

n-butyl mercaptan. (n-butanethiol; 1-butanethiol) n-C_4H_9SH; CAS 109-79-5; EINECS 203-705-3; intermediate, solvent. [Aldrich; Atochem N. Am.; Phillips 66]

butyl methacrylate. (butyl 2-methyl-2-propenoate) Ester of

---

* Trade name not verified as available   † Trade name verified as obsolete

n-butyl alcohol and methacrylic acid; $CH2:C(CH_3)$-$COO(CH_2)_3CH_3$; CAS 97-88-1; EINECS 202-615-1; monomer for resins, solvent coatings, adhesives, oil additives; emulsions for textiles, leather, paper finishing. [Degussa; IOI Spec.; Janssen Chimica; Mitsubishi Gas Chem.; Rohm & Haas]

butyl 2-methyl-2-propenoate. *See* butyl methacrylate.

Butyl Namate®. CAS 136-30-1; Sodium di-n-butyl-dithiocarbamate aq. sol'n.; accelerator for latexes of all elastomers. [R.T. Vanderbilt]

butyl octadecanoate. *See* butyl stearate.

butyl 9-octadecenoate. *See* butyl oleate.

butyl octyl phthalate. Used as a plasticizer. [Aristech; BASF]

Butylol. Mixture of normal and isobutyl alcohol; solvent. [Sasolchem]

butyl oleate. (butyl 9-octadecenoate) Ester of butyl alcohol and oleic acid; $C_{22}H_{42}O_2$; CAS 142-77-8; EINECS 205-559-6; plasticizer. [Ferro/Keil; Inolex; Sybron NV; Witco/Humko]

Butyl Oleate C-914. CAS 142-77-8; EINECS 205-559-6; Butyl oleate; plasticizer. [C.P. Hall]

butylparaben. (butyl p-hydroxybenzoate; 4-hydroxy-benzoic acid butyl ester; n-butyl p-hydroxybenzoate) Ester of butyl alcohol and p-hydroxybenzoic acid; $C_{11}H_{14}O_3$; CAS 94-26-8; antifungal for pharmaceuticals; preservative in foods. [Inolex; Nipa Labs; Penta Mfg.]

t-butyl peracetate. (t-butyl peroxyacetate) Organic perox-ide; $(CH_3)_3COOCOCH_3$; CAS 107-71-1; initiator. [Akzo; Atochem N. Am.; Witco/Argus]

t-butyl perbenzoate. (t-butyl peroxybenzoate) Organic peroxide; $(CH_3)_3COOCOC_6H_5$; CAS 614-45-9; initiator for polymerization and/or crosslinking of monomers and unsaturated polymers. [Akzo; Atochem N. Am.; Norac; Witco/Argus]

t-butyl peroxypivalate. Organic peroxide; $(CH_3)_3COO$-$COC(CH_3)_3$; CAS 927-07-1; low-temp. initiator for radi-cal catalyzed polymerization of vinyl monomers. [Akzo; Atochem N. Am.; Witco/Argus]

4-t-butyl phenol. $(CH_3)_3CC_6H_4OH$; CAS 98-54-4; EINECS 202-679-0; Antioxidant; chemical intermediate for syn-thetic resins, plasticizers, surface active agents. [Janssen Chimica; PMC Specialties; Schenectady]

butyl phthalate. *See* dibutyl phthalate

butyl-2-propenoate. *See* butyl acrylate

n-butyl propionate. $C_2H_5CO_2C_4H_9$; CAS 590-01-2; Solvent for nitrocellulose, retarder in lacquer thinner, ingredient of perfumes, flavors. [Penta Mfg.; Union Carbide]

butyl rubber. A proprietary trade name for a copolymer of isobutylene with a small percentage of diene such as butadiene; an unsaturated synthetic rubber possessing the minimum unsaturation required for vulcanization. *

butyl stearate. (butyl octadecanoate; n-butyl octa-decanoate; octadecanoic acid butyl ester) Ester of butyl alcohol and stearic acid; $CH_3(CH_2)_{16}COO(CH_2)_3CH_3$; CAS 123-95-5; EINECS 204-666-5; solv., spreading and softening agent in plastics, textiles, cosmetics, rubbers. [Amerchol; Henkel/Organic Prods.; Inolex; Mosselman NV; Penta Mfg.; Stepan; Witco/Humko]

Butyl Stearate C-895. CAS 123-95-5; EINECS 204-666-5; Butyl stearate; plasticizer. [C.P. Hall]

Butyl Zimate®. CAS 136-23-2; Zinc di-n-butyldithio-carbamate; ultra accelerator for EPDM and natural and synthetic latexes; provides fast, flat cures in SBR, nitrile,

and neoprene latexes; functions as nondiscoloring anti-oxidant in noncuring applications and stabilizer in IIR; antioxidant in thermoplastic rubbers and hot melts. [R.T. Vanderbilt]

but-2-yne-1,4-diol. (1,4-butynediol) $HOCH_2=CCH_2OH$; CAS 110-65-6; EINECS 203-788-6; Corrosion inhibitor in acid pickles and cleaners. [BASF; ISP]

n-butyraldehyde. (n-butylaldehyde; butyric aldehyde; n-butanal) $CH_3CH_2CH_2CHO$; CAS 123-72-8; EINECS 204-646-6; Plasticizers, rubber accelerators, solvents, and high polymers. [Eastman; Hoechst Celanese; Neste UK; Penta Mfg.; Union Carbide]

butyric acid sodium salt. *See* sodium butyrate.

butyric aldehyde. *See* n-butyraldehyde.

butyroin. $C_3H_7 \cdot CO \cdot CH(OH)C_3H_7$; 5-Hydroxy-4-octanone

butyrolactam. *See* 2-pyrrolidone.

butyrolactone. (dihydro-2(3H)-furanone; -butyrolactone; 4-hydroxybutyric acid, -lactone) Lactone; $C_4H_6O_2$; CAS 96-48-0; EINECS 202-509-5; intermediate for synthesis of polyvinylpyrrolidone, piperidine, phenylbutyric acid, etc.; solv. for polyacrylonitrile, cellulose acetate, methyl methacrylate polymers, polystyrene; constituent of paint removers, textile aids, drilling oils. [Aldrich; BASF; ISP; Janssen Chimica; Spectrum Chem. Mfg.; UCB SA]

butyrone. Dipropyl ketone, $CO(CH_2 CH_2 CH_3)_2$.

butyronitrile. (propyl cyanide; nitrile C_4; butanenitrile) $CH_3CH_2CH_2CN$; CAS 109-74-0; EINECS 203-700-6; Basic material in industrial, chemical and pharmaceuti-cal intermediates and products, poultry medicines. [Air Prods.; Eastman; Janssen Chimica; Lonza]

Bu-White. Disproportionated resin derivatives. [Tenneco Malros Ltd] *

B₁Vac. B_1 type, B_1 strain, Newcastle vaccine; for immuniza-tion of poultry. [Intervet Inc]

BWF. Extruded acrylic and polycarbonate profiles; for shopfitting, sign work. [Cornelius Chemical Co Ltd] *

BX 310. A range of polypropylene films of different gauge thicknesses and widths; for packaging applications. [Hercules] *

B-X-A. A reaction product of diarylamine-ketone-aldehyde; a proprietary antioxidant. [Rubber Regenerating Co] ‡

BXA. Diarylamine-ketone-aldehyde reaction product; for CR, NBR, SBR; nonblooming, easily disperses; protects against heat and oxygen; brown discoloration in stocks exposed to light, may stain light colored materials in contact with cured stocks; used in tire treads, carcass, inner tubes, insulated wire, soles, heels and mechani-cals. [Uniroyal] *

BXA Flake. CAS 9003-79-6; Diphenylamine-acetone reac-tion product; antioxidant. [Uniroyal]

BXT. Polypropylene film; for wrapping tobacco products. [Hercules] *

BY-59-18. Hydrocarbon resin, modified; used as replace-ment for rosin derivs., and in inks. [Neville]

Byacin. Iodophor in liquid solution; for controlling storage rots in potatoes. [Wheatley Chemical Co Ltd] *

Byakisol 30. Silica sol'n; used for binders, surface treat-ment, wine and fruit juice fining, and the textile industry. [Bayer AG]

Byatran. Iodophor and TBZ as dry granules; for controlling soil borne diseases in growing potato crops. [Wheatley Chemical Co Ltd] *

Byatran. Granules containing nonylphenoxypoly (ethylene-oxy) ethanol-iodine complex and tecnazene; used to

---

‡ Trade name and manufacturer not verified   § Trade name without identified manufacturer

control soilborne diseases in growing potato crops. [Dean Agrochemicals Ltd]

Byco A, C, O. CAS 9000-70-8; Gelatin NF; binders in pharmaceutical tableting; excipient, film former, coating agent; emulsion stabilizer; adjuvant protein in nutritional supplement. [Croda Inc.]

Bygran F. Tecnazene and iodophor as dry granules; for controlling sprouting, dry rot and other storage diseases in potatoes. [Wheatley Chemical Co Ltd] *

Bygran F. Granules containing nonylphenoxypoly (ethyleneoxy) ethanol-iodine complex and thiabendazole; used for controlling sprouting, dry rot and other storage diseases in potatoes. [Dean Agrochemicals Ltd]

Bygran S. CAS 117-18-0; Tecnazene as dry granules; for controlling sprouting and dry rot in stored potatoes. [Wheatley Chemical Co Ltd] *

Bygran S. CAS 117-18-0; Granules containing 10% w/w tecnazene; used for controlling sprouting and dry rot in stored potatoes. [Dean Agrochemicals Ltd]

BYK®-020. Sol'n. of a modified polysiloxane copolymer; defoamer for water-reducible coating systems, e.g., alkyds, polyester, epoxy esters, acrylics. [Byk-Chemie USA]

BYK®-024. Polysiloxanes/polymer mixture; defoamer for water-based systems including polyurethane, acrylate/polyurethane paints, wood varnishes, furniture paints, pigmented dispersion paints, plastic coatings. [Byk-Chemie USA]

BYK®-045. Emulsion of hydrophobic solids, emulsifiers, foam-destroying polysiloxanes; defoamer for emulsion paints, paper coatings, foil coatings, stains. [Byk-Chemie USA]

BYK®-151. Sol'n. of an alkylammonium salt of a polyfunctional polymer; dispersant, wetting agent for inorg. and org. pigments, fillers in aq. systems; improves color strength development and rheological props. of pigment pastes. [Byk-Chemie USA]

BYK®-156. CAS 9003-03-6; Sol'n. of an ammonium salt of an acrylic acid copolymer; dispersant for gloss and semigloss latex systems; improves gloss and stability on heat aging. [Byk-Chemie USA]

BYK®-307. Polyether modified dimethyl polysiloxane copolymer; additive to increase surface slip, substrate wetting, scratch and mar resistance for paints, printing inks. [Byk-Chemie USA]

BYK®-A500. Silicone-free polymeric sol'n.; defoamer, air release agent for laminating, spray-up, hand lay-up molding, gel coats, and solv.-free epoxy flooring systems; prevents air entrapment and porosity. [Byk-Chemie USA]

BYK®-ES 80. Sol'n. of an alkylammonium salt of an unsaturated acidic carboxylic acid ester; additive for increasing the conductivity of electrostatically sprayed coatings; used for solv.-based air dry and baking enamels. [Byk-Chemie USA]

BYK®-P104. Sol'n. of higher molecular weight unsaturated polycarboxylic acid; wetting, dispersing, antiflooding and antisettling additive; for coating systems. [Byk-Chemie USA]

BYK®-Catalyst 450. Sol'n. of an amine salt of p-toluene sulfonic acid; additive to improve curing in acid catalyzable org. systems. [Byk-Chemie USA]

Bykanol®-N. Sol'n. of an alkylolammonium salt of acidic phosphoric acid esters and ketoxime; antigelling agent and visc. stabilizer for solv.-based coatings which have a tendency to gel on aging. [Byk-Chemie USA]

Byketol®-OK. Mixture of high boiling aromatics, ketones and esters; additive to counteract surface defects and improve leveling for solv-based coatings, chlorinated rubber systems, silk screen inks. [Byk-Chemie USA]

Bykumen®. Sol'n. of a higher molecular weight unsaturated acidic polycarboxylic acid ester; wetting, dispersing additive to improve pigment wetting and prevent settling of pigments; for solv. and solv.-free systems, alkyd trade sales systems, acrylics, polyesters. [Byk-Chemie USA]

Bynel®. Coextrudable adhesive resins. [DuPont UK]

Bynin Amara. A proprietary preparation of iron phosphorus and nux vomica in Bynin liquid malt; tonic. [Allen & Hanburys Ltd] *

C

C-1. CAS 1344-28-1; Calcined alumina; for manufacturing of abrasives, fused alumina, high-alumina refractories, technical ceramics, kiln furniture, whitewares, fiberglass, ceramic fiber, elec. insulators and supports. [La Roche Chem.]

C-2. CAS 7758-19-2; Sodium chlorite. [Olin]

C_6 linear alpha olefin. See 1-hexene

C_{12} alpha olefin. See dodecene-1

C_{12} linear primary alcohol. See lauryl alcohol

C_{13} linear primary alcohol. See tridecyl alcohol

C_{14} linear primary alcohol. See myristyl alcohol

C_{16} linear primary alcohol. See cetyl alcohol

C_{18} linear alcohol. See stearyl alcohol

C-84. CAS 25928-94-3; Epoxy compound; general purpose potting compound for potting and casting applications; inherently high degree of flame resistance; used with 7.3 phr Activator BA-63. [Bacon]

C-715u. CAS 63394-02-5; Silicone elastomer; features high green strength, high resilience; for extrusion, molding, calendering. [Wacker Silicones]

C-920u. CAS 63394-02-5; Silicone elastomer; high temp. grade remaining flexible after 24 h @ 371 C. [Wacker Silicones]

CA. See cellulose acetate

CA-25. Calcium aluminate; refractory bonding agent fur use in refractories designed for service above 3000 F. [Alcoa]

CA-394-60S. CAS 9004-35-7; Cellulose acetate; used where high strength and good resistance to heat, uv light, oils, and greases are required. [Eastman]

CA0397. CAS 4369-14-6; 3-Acryloxypropyltrimethoxysilane; coupling agent, chem. intermediate, blocking agent, release agent, lubricant, primer, reducing agent. [Hüls Am.]

CA0567. CAS 2551-83-9; Allyltrimethoxy silane; coupling agent, chem. intermediate, blocking agent, release agent, lubricant, primer, reducing agent. [Hüls Am.]

CA0570. CAS 762-72-1; Allyltrimethyl silane; coupling agent, chem. intermediate, blocking agent, release agent, lubricant, primer, reducing agent. [Hüls Am.]

CA0699. CAS 3069-29-2; N-(2-Aminoethyl)-3-aminopropylmethyldimethoxy silane; coupling agent, chem. intermediate, blocking agent, release agent, lubricant, primer, reducing agent. [Hüls Am.]

CA0700. CAS 1760-24-3; EINECS 212-164-2; N-2-Aminoethyl-3-aminopropyltrimethoxysilane; coupling agent, chem. intermediate, blocking agent, release agent, lubricant, primer, reducing agent. [Hüls Am.]

CA0742. CAS 3179-76-8; 3-Aminopropylmethyldiethoxy silane; coupling agent, chem. intermediate, blocking agent, release agent, lubricant, primer, reducing agent. [Hüls Am.]

CA0750. CAS 919-30-2; EINECS 213-048-4; 3-Aminopropyltriethoxysilane; coupling agent, chem. intermediate, blocking agent, release agent, lubricant, primer, reducing agent. [Hüls Am.]

CA0880. CAS 13822-56-5; EINECS 237-511-5; 3-Aminopropyltrimethoxysilane; coupling agent, chem. intermediate, blocking agent, release agent, lubricant, primer, reducing agent. [Hüls Am.]

CA0900. CAS 107-72-2; Amyltrichlorosilane; coupling agent, chem. intermediate, blocking agent, release agent, lubricant, primer, reducing agent. [Hüls Am.]

CAB. See cellulose acetate butyrate

CAB-171-15S. CAS 9004-36-8; Cellulose acetate butyrate; used in cloth coatings for airplanes, wire, leather, and plastics. [Eastman]

CAB-381-0.1, 381-0.5, 381-2, 381-20. CAS 9004-36-8; Cellulose acetate butyrate; used in lacquers. [Eastman]

CAB-500-5. CAS 9004-36-8; Cellulose acetate butyrate; used for hot melts and additives for polyurethanes. [Eastman]

Cabelec®. Compounds capable of conducting electricity; used in the manufacture of electrically conductive and antistatic articles. [Cabot Plastics Ltd]

Cabelec® 1015. CAS 25038-54-4; Nylon 6/carbon blk. compound; conductive compound for injection molding applications; gives rigidity with permanent elec. conductivity; suggested for packaging and electronic production handling applications, e.g., for handling explosives, electronic components, pigments. [Cabot Plastics Ltd.]

Cabelec® 1017. CAS 9002-88-4; Modified polyethylene/carbon blk. compound; conductive compound for thin film extrusion; for packaging and production handling where freedom from hazard of electrostatic discharge is necessary, e.g., for handling explosive powds., pigments, electronic components. [Cabot Plastics Ltd.]

Cabelec®3004. CAS 9003-07-0; PP copolymer/carbon blk. compound; permanently static dissipative compound for injection molding; for applications where slower discharge of static elec. is required such as unearthed containers and specialized moldings; suggested for ammunition works, hospitals, mines, petrol. plants, electronics, etc. [Cabot Plastics Ltd.]

Cabelec® 3172. CAS 9002-88-4; HDPE/carbon blk. compound; conductive compound for molding and sheet extrusion; for production handling applications where freedom from electrostatic discharge hazards is necessary, e.g., in handling of explosive powds., pigments, electronic components. [Cabot Plastics Ltd.]

---

‡ Trade name and manufacturer not verified   § Trade name without identified manufacturer

Cabelec®3464. CAS 9003-07-0; PP copolymer/carbon blk. compound; conductive compound for extrusion applications incl. extruded tapes for sacking, fiber/twine for fabric or interweaving into fabric for chair coverings, for flexible pipes and tubing, and as corrugated sheet for packaging electronics. [Cabot Plastics Ltd.]

Cabflex DIOZ. Diisooctyl azelate. §

Cablinol. A speciality perfumery chemical. [PPF International Ltd] *

Cabol 100. A hydrocarbon oil type vinyl plasticizer. §

Cab-O-Sil®L-90. CAS 112945-52-5; Fumed silica; dispersant, anticaking agent for foods, agric. products, and powds. for cosmetics and coatings industries. [Cabot]

Cab-O-Sil® TS-530. CAS 68909-20-6; Fumed silica, hexamethyldisilazane-surface treated; reinforcing filler for elastomers; free flow agent for toners and powd. coatings; antisettling agent in coatings; dry powd. carrier for perfumes, pesticides, veterinary products, etc. [Cabot]

Cab-O-Sperse® A 105. Fumed silica aq. disp., ammonia-stabilized; thickener; rheological control in water-based systems. [Cabot] *

Cabot® PE 6008. 30% trimethyl quinoline antioxidant in LDPE carrier; antioxidant masterbatch for polymeric cable applications incl. cable sheath insulation layers and semiconductive screens; compatible with LDPE, EVA, EBA, LLDPE, HDPE, EPDM, ACN. [Cabot Plastics Ltd.]

Cabot®PE 9006. 75% calcium carbonate plus antioxidant in LDPE carrier; antifibrilation masterbatch for use in PP weaving tapes, PP strapping bands, PP string and twine, LDPE shrink film; effective antiblock; compatible with LPDE, LLDPE, HDPE, PP. [Cabot Plastics Ltd.]

Cabot®PE 9007. CAS 9002-88-4; LDPE polymer with mildly abrasive fine natural silica; antiblock masterbatch for blown LDPE and LLDPE film extrusion; also as cleaning compound for extruders; compatible with LDPE, LLDPE, HDPE, PP, EVA. [Cabot Plastics Ltd.]

Cabot® PE 9138. Polymer masterbatch with min. additive; infra-red absorber masterbatch; used to improve thermal barrier properties of polyethylene film in greenhouse and other agric. applications; optimum addition level is 11-12% in polyethylene. [Cabot Plastics Ltd.]

Cabot®PE 9166. 5% erucamide slip agent in LPDE carrier; slip additive masterbatch providing excellent slip and good antiblock effect to LDPE and LLDPE films; slower migration of slip agent is advantageous when printing, sealing, or laminating processes are carried out immediately after film production; suggested addition level 2-3%. [Cabot Plastics Ltd.]

Cabuflx. An adhesive for cellulose acetate butyrate. [May & Baker Ltd]

Cabulite. CAS 9004-36-8; Cellulose acetate butyrate film and sheet. [May & Baker Ltd] *

Cachalot® C-50. CAS 36653-82-4; EINECS 253-149-0; Cetyl alcohol NF; emollient used in cosmetics. [M. Michel]

Cachalot® M-43. CAS 112-72-1; EINECS 204-000-3; Myristyl alcohol; emollient used in cosmetics. [M. Michel]

Cachalot® O-15. CAS 143-28-2; EINECS 205-597-3; Oleyl alcohol; conditioner, lubricant for cosmetics; corrosion inhibitor additive to lube oils. [M. Michel]

Cachalot® S-56. CAS 112-92-5; EINECS 204-017-6; Stearyl alcohol USP; emollient used in cosmetics. [M. Michel]

cachou. See cutch.

C acid. 2-Naphthylamine-4,8-disulfonic acid.

cacodyl. (Alkarsin). Tetramethyldiarsine, $As_2(CH_3)_4$.

cacodylic acid. Dimethylarsinic acid, $As(CH_3)_2O(OH)$.

cacodylic acid sodium salt trihydrate. See sodium cacodylate

cadaverine. (Pentamethylenediamine) $NH_2(CH_2)_5NH_2$; A base found in ergot and formed by bacterial decomposition; used in the preparation of high polymers; in biomedical research.

Cadco®Acetal. CAS 105-57-7; Acetal resin; high resist. to moisture, gasoline, org. solvs.; for antifriction parts, bearings, bushings, rollers, cams, dimensionally stable parts for business machines, elec. components, gears, plumbing parts; for automotive, appliance, plumbing, construction, hardware, electronics, and consumer goods. [Cadillac Plastic & Chem.]

Cadco®Cast Acrylic. Cast acrylic rods, tubes, and blocks; weather/chem. resist. strong material with optical clarity; for bathroom fixtures, chem. retorts, display items, elec. components, food canisters, furniture components, knobs, lamp shades, lighting fixtures, medical supplies, surgical instruments. [Cadillac Plastic & Chem.]

Cadco®Cast Nylon. CAS 25038-54-4; Cast nylon 6; strong, lightweight, self-lubricating thermoplastic resist. to wear, abrasion and chems.; for bearings, bushings, cams, gears, hydraulic seals, insulators, rollers, seals, slide bearings, tooling fixtures, valves, wheels, etc. [Cadillac Plastic & Chem.]

Cadco®Nylon. Nylon; good abrasion and wear resist., elec. insulating chars., noise dampening props. for unlubricated gears, bearings, and antifriction parts, mech. parts, auto body parts, elec. parts, high impact parts, food processing parts. [Cadillac Plastic & Chem.]

Cadco®Teflon. PTFE and FEP fluorocarbons; high melting thermoplastic for valve and pump components, gaskets, mech. components, food processing, chem. equip., tape-wrapped wire, transformers, coils, electronics equip. [Cadillac Plastic & Chem.]

Cadco® UHMW. CAS 9002-88-4; Ultra-high molecular weight polyethylene; high strength, abrasion and impact props.; for various applications from suction box covers for high-speed paper machines to surfaces on snow skis. [Cadillac Plastic & Chem.]

Caddy. Cadmium turf fungicide. [W A Cleary]

Cadet®BPO-70W. CAS 94-36-0; Benzoyl peroxide; initiator for elevated-temp. curing of unsat. polyester resins in applications such as matched metal die molding, pultrusion, vacuum bag molding, continuous laminating, hot-cure casting, injection molding; also for PS, specialized PVC resin, acrylic polymers. [Akzo] *

CADG. See cocamidopropyl betaine.

Cadmate®. CAS 14239-68-0; Activated cadmium diethyldithiocarbamate; accelerator for vulcanization of synthetic rubber. [R T Vanderbilt Co Inc]

cadmium. Cd; CAS 7440-43-9; EINECS 231-152-8; In alloys; electrodeposited and dipped coatings on metals. [Aldrich; Asarco; Atomergic Chemetals; Pasminco Europe; Zinc Corp. of Am.]

cadmium bronze. A copper alloy containing 0.5-1.2% cadmium; used for telephone and trolley wire.

cadmium chloride. $CdCl_2$; CAS 10108-64-2; Preparation of cadmium sulfide, analytical chemistry, photography,

---

* Trade name not verified as available   † Trade name verified as obsolete

dyeing and calco printing, in electroplating baths and tinning solutions, manufacture of special mirrors, cadmium yellow. [Atomergic Chemetals; Mallinckrodt; Nihon Kagaku Sangyo; Noah Chem.]

cadmium fluoride. CdF_2; CAS 7790-79-6; EINECS 232-222-0; Electronic and optical applications, high-temperature dry-film lubricants, starting material for crystals for lasers, phosphors. [Atomergic Chemetals; Cerac; Noah Chem.]

cadmium hydroxide. (cadmium hydrate) $Cd(OH)_2$; Cadmium salts, cadmium plating, storage-battery electrodes. [Noah Chem.]

cadmium iodide. CdI_2; CAS 7790-80-9; EINECS 232-223-6; Photography, process engraving and lithography, analytical chemistry, electroplating, lubricants, phosphors, nematocide. [Atomergic Chemetals; Cerac; Spectrum Chem. Mfg.]

cadmiumized zinc. Zinc metal placed in a 2% cadmium sulfate solution for five minutes, then well washed; used as a reducing agent.

cadmium lithopone. (cadmopone). A pigment analogous to lithopone, in which cadmium replaces zinc. It is made by the precipitation of cadmium sulfate solution with barium sulfide, and contains 38% cadmium sulfide.

cadmium oxide. (cadmium oxide brown) CdO; CAS1306-19-0; EINECS 215-146-2; Cadmium plating baths, electrodes for storage batteries, cadmium salts, catalyst, ceramic glazes, phosphors, nematocide. [Amax Inc; Asarco; Atomergic Chemetals; Chemisphere Ltd; Mallinckrodt; Nihon Kagaku Sangyo]

Cadmopone®. See cadmium lithopone. §

Cadmopur. A range of cadmium pigments; especially for use in plastics such as polyethylene, polystyrene, polyamides, PVC, etc., as well as in rubber. [Bayer AG]

Cadon. Engineering thermoplastics, impact-modified styrene-maleic anhydride terpolymers; used in automotive interior/exterior parts, business machines and appliance housings, electrical equipment, electronic parts, plumbing and industrial parts. [Monsanto Co] *

Cadoussant. Blends of cationic surfactants; softening and antistatic agent for synthetic fibers. [Atochem UK/Ceca]

Cadox® 40E. CAS 94-36-0; Benzoyl peroxide with plasticizer; initiator for ambient-temp. polyester cures. [Akzo] *

Cadox® BFF-50. Benzoyl peroxide with dicyclohexyl phthalate; initiator for elevated-temp. curing of unsat. polyester resins in applications such as matched metal die molding, pultrusion, vacuum bag molding, continuous laminating, hot-cure casting, injection molding [Akzo] *

Cadox® BPO-W40. CAS 94-36-0; Benzoyl peroxide. [Akzo]*

Cadox® BS. CAS 94-36-0; Benzoyl peroxide; cross-linking agent used for curing silicone rubbers. [Akzo] *

Cadox® BTA. Benzoyl peroxide, calcium sulfate; initiator for ambient-temp. polyester cures. [Akzo] *

Cadox® BTW-50. CAS 94-36-0; Benzoyl peroxide; initiator for cure of unsaturated polyester resins [Akzo] *

Cadox® F-85. Ketone peroxide; initiator for ambient-temp. polyester cures. [Akzo] *

Cadox® HBO-50. CAS 1338-23-4; MEK peroxide; initiator for cure of unsaturated polyester resins [Akzo] *

Cadox® L-30. CAS 1338-23-4; MEK peroxide; initiator for cure of unsaturated polyester resins [Akzo] *

Cadox® M-30. MEK peroxide in DMP/DAP; initiator for ambient-temp. polyester cures; used for spray-up applications. [Akzo] *

Cadox® MDA-30. MEK peroxide in DAP; initiator for ambient-temp. polyester cures; low visc.; used for airless spray-gun application. [Akzo] *

Cadox® TDP. CAS 94-17-7; 2,4-Dichlorobenzoyl peroxide in plasticizer; highly reactive initiator used with amine-accelerated polyester resins. [Akzo] *

Cadox® TS-50S. CAS 94-17-7; 2,4-Dichlorobenzoyl peroxide, phthalate-free; cross-linking agent for polymers; co-vulcanizing agent [Akzo] *

CAE. Ethyl N-cocoyl-L-arginate PCA salt; surfactant, foamer, and antistat; preservative; antiseptic; germicide; disinfectant in cosmetics, detergents, dentifrices, medical supplies. [Ajinomoto; Ajinomoto USA] *

caesium. See cesium

Cafadol. A proprietary preparation of paracetamol and caffeine citrate; an analgesic. [Typharm] *

Cafaspin®. A restorative and analgesic. [Bayer AG]

Cafergot. A proprietary preparation of ergotamine tartrate and caffeine; used for migraine. [Sandoz] *

caffeine. (theine; methyltheobromine; 1,3,7-trimethyl-xanthine) $C_8H_{10}N_4O_2$; CAS 5808-2; EINECS 200-362-1; Beverages, medicine (CNS stimulant). [Am. Bio-Synthetics; Bell Flavors & Fragrances; R.W. Greeff; Janssen Chimica; Knoll AG; Penta Mfg.]

Cafiaspirina. An analgesic, anti-inflammatory drug. [Bayer AG]

Caflon. Foam stabilizer; for liquid detergents; shampoos; bubble baths; hand cleaners. [Cargo Fleet Chemical Co Ltd] *

Caflon MIS. Anionic surfactant as a pale amber clear liquid; emulsifier for hand cleaning gels and degreasers. [Cargo Fleet Chemical Co Ltd] *

Caflon MS33. Monoethanolamine alkyl sulfate in liquid form; for liquid detergents and shampoos. [Cargo Fleet Chemical Co Ltd] *

Caflon NAS 25. Anionic surfactant as a pale yellow liquid; high foaming wetting agent for liquid detergents. [Cargo Fleet Chemical Co Ltd] *

Caflon SA and SNA. Anionic surfactant in acid form, supplied as a dark brown liquid; detergent intermediate. [Cargo Fleet Chemical Co Ltd] *

Caflon SS28. Sodium alkyl sulfate in liquid form; for liquid detergents and shampoos. [Cargo Fleet Chemical Co Ltd] *

cajeputene. See dipentene.

cajeputol. See eucalyptol.

cake lac. See lac.

Cake Mix 96. Partially hydrogenated soybean oil, propylene glycol mono and diesters of fats and fatty acids, mono and diglycerides, optional lecithin; kosher; high performance fat for single-stage household cake mixes. [Van Den Bergh Foods]

calaba oil. See laurel nut oil.

Calaband. A proprietary zinc paste and calamine bandage; indicated for fairly acute and subacute eczema, erythema and dermatitis after plaster removal. [Seton Healthcare Group plc]

Calac. CAS 62-54-4; Calcium acetate. [Mechema Chemicals Ltd] *

Caladryl. A proprietary preparation of diphenhydramine, camphor and calamine in an aerosol spray, for dermatological use. [Parke-Davis] *

Calamide C. Cocamide DEA superamide; foam stabilizer,

---

‡ Trade name and manufacturer not verified   § Trade name without identified manufacturer

emulsifier for liq. dishwash, bubble baths, shampoos, all-purpose cleaners; thickener; imparts mildness. [Pilot]

Calamide O. Modified coco-oleic DEA superamide; surfactant for janitorial cleaners, rust preventives, lubricants, wetting, foaming, cleaning sol'ns. [Pilot]

calamine. (zinc spar, spathic zinc ore, smithsonite). CAS 1314-13-2; A term applied to both the silicate and carbonate of zinc, usually the carbonate, $ZnCO_3$, found native. In mineralogy, the term is used for the silicate, smithsonite is used for the carbonate.

Calaton. A textile finishing agent. [ICI Chem. & Polymers Ltd]

Calavite. A proprietary preparation containing vitamin A, ascorbic acid, cyanocobalamin, nicotinamide, thiamine and calciferol; a vitamn supplement. [Carlton Laboratories (UK) Ltd] *

Calbiochem®. Biochemicals. [Calbiochem Corp]

Calbiosorb™. Absorbent. [Calbiochem Corp]

Calbor. Borated animal bone and borate containing flux. [Borax Consolidated Ltd] *

Calbrite. CAS 7757-93-9; Dentifrice grade of dicalcium phosphate. [Albright & Wilson Ltd, Phosphates & Speciality Business]

Calbux. CAS 1317-65-3; A proprietary ground limestone. [ICI Chem. & Polymers Ltd] †

Calcars. CAS 7778-44-1; Calcium arsenate. [Mechema Chemicals Ltd] *

Calcene. A trade name applied to a precipitated calcium carbonate with 2% organic material; prepared for use as a rubber filler. §

Calcet. Contains calcium gluconate and calcium lactate; replenisher. [Mission Pharmacal Co] *

calcetal. Calcium acetylsalicylate.

Calcibind. Cellulose sodium phosphate, an insoluble, non-absorbable ion- exchange resin made by phosphorylation of cellulose; contains approximately 34% of inorganic phosphate and approximately 11% of sodium; antiurolithic. [Mission Pharmacal Co]

Calcibronat. CAS 33659-28-8; A proprietary preparation of calcium bromidolactobionate; a sedative. [Sandoz] *

Calcichrome. Cyclo tris-7-(1-azo-8-hydroxynaphthalene 3:6 disulfonic acid); a mauve crystalline powder; a sensitive and specific reagent for calcium giving a red color with Ca 2^+ ions in alkaline solution. [British Drug Houses] *

calcidine. Calcium iodide, CaI_2.

calciferol. Synthetic vitamin D; vitamin D_2.

Calcimar. Calcitonin: a polypeptide hormone that lowers the calcium concentration in plasma; regulator. [USV Pharmaceutical Corp] ‡

calcined baryta. See barium oxide

calcined magnesia. See magnesium oxide

Calcinite. See Carbora. §

calcinol. Calcium iodate, $Ca(IO_3)_2$.

calciofon. See calcium gluconate

Calciparine. Heparin calcium; anticoagulant. [Baxter Health Care] *

Calcisorb. CAS 9038-41-9; Each sachet contains sodium cellulose phosphate as a white to beige fibrous powder. [3M Pharmaceuticals] †

Calcitare. A proprietary preparation of porcine calcitonin; used in the treatment of Paget's disease. [Armour Pharmaceutical Co] *

calcitonin. A polypeptide hormone of ultimobranchial origin,

extractable from the thyroid gland of mammalian species or the ultimobranchial gland of non-mammals. It lowers the calcium concentration in the plasma of mammals.

calcium. Ca; CAS 7440-70-2; EINECS 231-179-5; Alloying agent for aluminum, copper, lead; reducing agent for Be. [Atomergic Chemetals; Cerac; Leverton-Clarke Ltd; Noah Chem.; Pfizer]

calcium acetate. (vinegar salts; gray acetate; lime acetate) $(CH_3COO)_2Ca \cdot H_2O$; CAS 62-54-4; EINECS 200-540-9; Manufacture of acetone, acetic acid, acetates, mordant in dyeing and printing of textiles, stabilizer in resins, additive to calcium soap lubricants, food additive, corrosion inhibitor. [General Chem.; Mallinckrodt; Niacet; Schaefer Salt; Verdugt BV]

calcium bromide. $CaBr_2 \cdot 2H_2O$; CAS 7789-41-5; EINECS 232-164-6; Photography, medicine, dehydrating agent, food preservative, road treatment, freezing mixtures, sizing compounds, wood preservative, fire retardant. [Atomergic Chemetals; Broomchemie BV; Cerac; Ethyl; Great Lakes; Spectrum Chem. Mfg.]

calcium carbide. CaC_2; CAS 75-20-7; Generation of acetylene gas for welding, chloroethylenes, vinyl acetate monomer, acetylene chemicals, reducing agent. [Pechiney Electrometallurgie; Spectrum Chem. Mfg.]

calcium carbonate. Also see limestone. (carbonic acid, calcium salt; precipitated calcium carbonate, commercial form; prepared calcium carbonate, native purified form) Inorganic salt; $CaCO_3$; CAS 471-34-1; 1317-65-3; EINECS 207-439-9; source of lime; neutralizing agent; opacifying agent in paper; fortification of bread; putty; tooth powds.; antacid; whitewash; portland cement; paint; rubber; plastics; insecticides; in chemical analysis. [BASF; ECC Int'l.; EM Industries; Genstar Stone Prods.; Georgia Marble; J.M. Huber; Mallinckrodt; Nichia Kagaku Kogyo; Pfizer; Whittaker, Clark & Daniels]

calcium carbonate, natural. See limestone.

Calcium Chel 330. Pentetate calcium trisodium; chelating agent. [Ciba-Geigy Corp] *

calcium chloride. (calcium chloride, dihydrate) Inorganic salt; $CaCl_2.2H_2O$; CAS 10043-52-4; EINECS 233-140-8; deicing and dust control of roads, drilling muds, dustproofing, freezeproofing, and thawing coal, coke, stone, sand, ore, concrete conditioning; drying and desiccating agent; sequestrant in foods. [Akzo Salt; Allied-Signal; EM Industries; Gist-Brocades Food Ingreds.; Kemira Kemi UK; Mallinckrodt; Nichia Kagaku Kogyo; OxyChem; Schaefer Salt]

calcium chloride, dihydrate. See calcium chloride.

calcium citrate. (lime citrate; tricalcium citrate) $Ca_3(C_6H_5O_7)_2 \cdot 4H_2O$; Dietary supplement, sequestrant, buffer, and firming agent in foods. [EM Industries; Rit-Chem; Rottapharm SpA]

calcium dioxide. See calcium peroxide.

Calcium Disodium Versenate. CAS 62-33-9; Edetate calcium disodium injection USP; lead chelating agent. [3M Pharmaceuticals]

calcium diuretin. Theobromine and calcium salicylate.

calcium fluoride. CaF_2; CAS 7789-75-5; EINECS 232-188-7; Source of fluorine, flux, ceramics, phosphors, paint pigment, catalyst in wood preservative, spectroscopy, electronics, lasers, high-temperature dry-film lubricants. [Cerac; GE; Noah Chem.; Solvay GmbH]

calcium gluconate. (D-gluconic acid calcium salt; calciofon; glucal) $Ca[HOCH_2(CHOH)_4COO]_2$; CAS 299-28-5;

EINECS 206-075-8; Mineral source for pharmaceutical and food products; anticaking agent in coffee powds.; food additive, buffer, sequestering agent; vitamin tablets; in sewage purification. [Akzo; EM Industries; Lipo; Mallinckrodt; Pfizer]

calcium hydrate. See calcium hydroxide.

calcium hydrochlorphosphate. A mixture of calcium chloride and phosphate.

calcium hydrogen orthophosphate. See calcium phosphate, dibasic.

calcium hydroxide. (calcium hydrate; hydrated lime; slaked lime) Inorganic base; $Ca(OH)_2$; CAS 1305-62-0; EINECS 215-137-3; mortar, plasters, cements, calcium salts, disinfectant, food additives, lubricants, pesticides, manufacture of paper pulp, in SBR vulcanization, in water treatment. [EM Industries; Janssen Chimica; Mallinckrodt, Pfizer; U.S. Gypsum]

calcium hypochlorite. (calcium oxychloride) $Ca(OCl)_2$; CAS 7778-54-3; EINECS 231-908-7; Algicide, bactericide, deodorant, potable water purification, disinfectant for pools, bleaching agent, oxidizing agent; commercial grade usu. 50% or more $Ca(OCl)_2$. [Olin; PPG Industries; Surex Int'l. Ltd]

calcium iodate. $Ca(IO_3)_2$; Deodorant, mouthwashes, food additive, dough conditioner. [Atomergic Chemetals; Blythe, William Ltd; R.W. Greeff; Mitsui Toatsu; Spectrum Chem. Mfg.]

calcium lactophosphate. A mixture of calcium lactate, calcium acid lactate, and calcium acid phosphate. It contains about 2% P_2O_5.

calcium molybdate. $CaMoO_4$; CAS 7789-82-4; Molbydic acid, alloying agent in production of iron and steel, crystals in optical and electronic applications, phosphors. [AAA Molybdenum Prods.; Atomergic Chemetals; Cerac; Noah Chem.]

calcium octadecanoate. See calcium stearate

calcium oxide (lime; quicklime; calx) Inorganic oxide; CaO; CAS 1305-78-8; EINECS 215-138-9; refractory, sewage treatment, insecticides, fungicides, manufacture of steel and aluminum; flotation of nonferrous ores; manufacture of glass, paper, Ca salts; in drilling fluids, lubricants; laboratory. [Cerac; GE; Hüls Am.; Mallinckrodt; Pfizer; U.S. Gypsum]

calcium oxychloride. See calcium hypochlorite

calcium pantothenate. (calcium N-(2,4-dihydroxy-3,3-dimethyl-1-oxobutyl-β-alanine; vitamin B_5, calcium salt; calcium D-pantothenate) Calcium salt of pantothenic acid; $C_9H_{17}NO_5 \cdot 1/2$ Ca; CAS 137-08-6 (D-form); medicine, animal feeds, dietary supplement. [BASF; Hoffmann-La Roche; Schweizerhall; Takeda USA; Tanabe USA]

calcium peroxide. (calcium superoxide; calcium dioxide) CAS 1305-79-9; Seed disinfectant, dentifrices, dough conditioners, antiseptic, bleaching of oils, modification of starches, high-temperature oxidation. [Chemoxal SA; Eagle-Picher; FMC; Henley; Interox Am.]

Calcium Petronate 25H, 25C and 300. Calcium petroleum sulfonate; detergents in lube oil additives, rust inhibitors, emulsifiers. [Witco Chemical Ltd] *

calcium phosphate (monobasic). (MCP; monocalcium phosphate, monohydrate; acid calcium phoshate) Phosphoric acid, calcium salt (2:1); $CaH_4(PO_4)_2$; CAS 7758-23-8; EINECS 231-837-1; leavening acid in food products, mineral supplement, fertilizer, stabilizer for plastics,

to control pH in malt, glass manufacture, firming agent. [Albright & Wilson; FMC; Kemira Kemi UK; Mallinckrodt; Monsanto; OxyChem; Nichia Kagaku Kogyo; Rhone-Poulenc Basic]

calcium phosphate, dibasic. (DCP-0; dicalcium phosphate, anhydr.; calcium hydrogen orthophosphate) $CaHPO_4$ or $CaHPO_4 \cdot 2H_2O$; CAS 7757-93-9; Foods, pharmaceuticals, dentifrice, medicine, glass, fertilizer, stabilizer for plastics, dough conditioner, yeast food. [Albright & Wilson; EM Industries; FMC; GE; Janssen Chimica; Mallinckrodt; OxyChem; Rhone-Poulenc Basic]

calcium phosphate, tribasic. (TCP; tricalcium phosphate; tricalcium orthophosphate) $3Ca_3(PO_4)_2 \cdot Ca(OH)_2$; CAS 7758-87-4; Foods, pharmaceuticals, polystyrene, ceramics, mordant, fertilizers, dentifrices, stabilizer for plastics, in meat tenderizer, anticaking agent, nutrient supplement. [Albright & Wilson; FMC; Mallinckrodt; Monsanto; Rhone-Poulenc Basic]

calcium propionate. (propionic acid, calcium salt) Calcium salt of propionic acid; $Ca(OOCCH_2CH_3)_2$; CAS 4075-81-4; EINECS 223-795-8; mold-inhibiting; food additive; antifungal agent; improves scorch resistance and processability of butyl rubber. [Gis-Brocades Food Ingreds.; Niacet; Verdugt BV]

Calcium Resonium. A proprietary preparation of calcium polystyrene sulfonate; used in the treatment of hyperkalemia. [Winthrop Laboratories] *

calcium silicate. (silicic acid, calcium salt) Hydrous or anhydrous silicate with varying proportions of calcium oxide and silica; Common forms: $CaSiO_3$, Ca_2SiO_4, Ca_3SiO_5; CAS 10101-39-0; constituent of lime glass, portland cement; reinforcing filler in elastomers and plastics; absorbent for liqs., gases, vapors; anticaking agent; suspending agent; pigment and pigment extender; binder for refractories; in chromatography, road building. [Celite; Crosfield; Degussa; R.T. Vanderbilt]

calcium stearate. (calcium octadecanoate; stearic acid, calcium salt) Calcium salt of stearic acid; $Ca(C_{18}H_{35}O_2)_2$; CAS 1592-23-0; EINECS 216-472-8; water repellent; flatting agent in paints, emulsions; release agent for plastic molding powds.; stabilizer for PVC resins; lubricant; in pencils and crayons; food grade as conditioning agent in foods and pharmaceuticals. [Adeka Fine Chem.; Atochem N. Am./Wire Mill; Eka Nobel Ltd; Ferro/Grant; Henkel/Organic Prods.; Mallinckrodt; PPG Industries; R.T. Vanderbilt; Witco]

calcium sulfate (anhydrous). (anhydrite (natural form); calcium sulfonate; anhydrous gypsum) Inorganic salt; $Ca \cdot H_2O_4S$; CAS 7778-18-9; EINECS 231-900-3; insol. anhydride: in cement formulations, as paper filler; sol. anhydride: drying agent, desiccant. [Kemira Kemi AB; U.S. Gypsum]

calcium sulfonate. CAS 61789-86-4; Rust preventive, detergent for diesels; pigment dispersant for thermoplastic and epoxy color concs.; visc. stabilizer. [Lubrizol; Stepan; Witco/Sonneborn]

calcium superoxide. See calcium peroxide

calcium titanate. (perovskite) Inorganic compound; CaO_3Ti; CAS 12049-50-2; electronics. [Atomergic Chemetals; Cerac; Ferro/Transelco; TAM Ceramics]

calcium zirconate. $CaZrO_3$. [Atomergic Chemetals; Cerac; Ferro/Transelco; TAM Ceramics]

calcreose. Calcium creosotate.

---

‡ Trade name and manufacturer not verified   § Trade name without identified manufacturer

Cal-C-Vita. Proprietary preparation of vitamin B_6, vitamin C, vitamin D, and calcium. [Roche Products Ltd] *

Calcydic. Calcium monohydrogen phosphate and Vitamin D. [Allen & Hanburys Ltd] *

Calderol. CAS 19356-17-3; Calcifediol; regulator. [Upjohn]*

Caldiox. Calcium plumbate. [Associated Lead Manufacturers Ltd] *

Caldo Bordeles Valles. Bordeaux mixture plus adjuvants; wettable powder used as protective fungicide for foliage application to ornamental and crop plants. [Industrias Quimicas Del Valles SA] *

Caldura. A high temperature (240 C) resistant resin, containing aromatic hydrocarbon groups linked by oxygen and methylene bridges. [Associated Electrical Industries (GEC)] ‡

Caledon. Vat dyes. [ICI Chem. & Polymers Ltd]

Calendula Oil CLR. Soybean oil, calendula extract, tocopherol; emollient, conditioner; emulsified and oily preparation for skin care. [Dr. Kurt Richter; Henkel/Cospha]

Calester. Alpha sulfo methyl laurate; surfactant for high-quality toilet soaps, laundry detergents, automobile cleaners, spray cleaners, foamers, emulsifiers [Pilot]

Calfax 10L-45. CAS 36445-71-3; Sodium linear decyl diphenyl oxide disulfonate; biodeg. surfactant, solubilizer, dispersant for dye bath leveling, pigment dispersion, heavy-duty cleaners, latex emulsification, agric. chemicals, phenolic germicides, bottle washing. [Pilot]

Calfax DB-45. CAS 28519-02-0; Sodium alkyl diphenyl oxide disulfonate; detergent, solubilizer for dye bath and other strongly polar applics, e.g., dip tank cleaners, electroplating baths, heavy-duty cleaners, latex emulsifiers, agric. chemicals; tolerant of high alkalinity, high acidity, high levels of electrolyte. [Pilot]

Calfoam ES-30. CAS 9004-82-4; Sodium laureth sulfate; detergent, foam stabilizer, flash foamer, wetter for detergent systems, personal care products, wool washing; emulsion polymerization. [Pilot]

Calfoam NEL-60. CAS 67762-19-0; Ammonium lauryl ether sulfate; flash foamer, foam stabilizer, detergent, emulsifier, wetter for liq. detergents, bubble baths, shampoos, car washing; lime soap dispersant. [Pilot]

Calfoam SLS-30. CAS 151-21-3; Sodium lauryl sulfate; mild detergent, foamer for personal care products; rug/upholstery shampoos; emulsifier for cosmetics, emulsion polymerization of latex, SBR rubber, polyacrylates, elastomers; foaming agent for foamed rubber. [Pilot]

Calgel. A proprietary formulation of lignocaine hydrochloride and cetylpyridium chloride; indicated for use in teething. [The Wellcome Foundation Ltd]

Calgon. Water softening compound. [Albright & Wilson Ltd, Phosphates & Speciality Business]

Calgon® GW 12x40. Activated carbon; filter medium for dechlorination and removal of trace dissolved organics from water supplies. [Calgon Carbon] *

Calgon® Type BPL 4x10, 6x16, 12x30. Activated carbon; for vapor phase applications. [Calgon Carbon] *

Calgon®Type HGR. Activated carbon, sulfur-impregnated; for mercury removal. [Calgon Carbon] *

Calgon® Type RB. Activated carbon; pulverized form for purification applications in chemical, food and pharmaceutical industries. [Calgon Carbon] *

Calgon®Type SGL. CAS 7440-44-0; Carbon; for purification and decolorization of aq. and org. liqs. [Calgon Carbon] *

Calgonite. Dishwashing machine detergents. [Albright & Wilson Ltd, Phosphates & Speciality Business] *

Cal-Grid. A calcium-aluminum alloy used in the production of calcium-containing long-life lead battery plates. [Pfizer International] †

Calibre. A family of polycarbonate resins used widely in automotive, electronics, business machines, medical and housewares applications. [Dow UK]

Calibre®. For agriculture industry. [DuPont UK]

Calibre 200-4. PC resin; FDA compliance; engineering thermoplastic for food contact and medical applications. [Dow Plastics]

Calibre 302-E. PC resin; excellent clarity and uv/weatherability props. for profile extrusion, sheet and lighting applications. [Dow Plastics]

Calibre 400-10. PC resin; impact modified grade. [Dow Plastics]

Calibre 510, 550. PC resin, glass-reinforced; applications are in transportation, electronics, and service parts. [Dow Plastics] *

Calibre 700-4. PC resin; ignition-resistant and clarity grade; for electronics housings, optical parts, transportation, and appliances. [Dow Plastics]

Calibre 1001CD. PC resin; for the compact disc market; offers proven purity, increased clarity, processing ease, excellent physical properties and dimensional stability. [Dow Plastics]

Calibre 2060-4. PC resin; formulated to meet sterilization needs of the health care industry, providing exceptional clarity, heat resist., impact. strength, and processability; 2060 resins are for applications using steam or ethylene oxide sterilization; 2080 resins are for radiation sterilization. [Dow Plastics]

Calibre LG2010. PC resin; opthalmic grade. [Dow Plastics]

Calibrite. A trademark for series of aluminum pigments for plastics. §

Caliche. Crude nitrate of soda (Chile saltpeter). The soluble salts which cement together the sand, clay, and stones in the mixture known as caliche are mainly sodium nitrate and sodium chloride.

Calido. An alloy of 64% nickel with from 15-25% iron, 8-12% chromium, and 3-8% manganese.

Calido brass. An alloy of 70% copper, 30% zinc, and traces of iron.

Calido-elalco. A heat-resisting alloy containing 60% nickel, 24% iron, and 16% chromium.

Califlux® 90. Aromatic oil; extender oil offering low aniline pts.; for elastomer compounding, esp. nitrile, neoprene, SBR, and natural rubber. [Witco/Golden Bear]

Caligesic. Benzocaine, calamine and metacresol; topical relief of inflammatory and pruritic conditions of the skin. [Merck & Co Inc] †

Calight RPO. Naphthenic process oil; for rubber industry, resin extending, PVC, textiles, caulking compounds, etc. [Calumet Lubricants]

Caliment. CAS 7757-93-9; Food grade of dicalcium phosphates. [Albright & Wilson Ltd, Phosphates & Speciality Business] *

Calimmunochem™. Immunochemicals. [Calbiochem Corp]

Calimulse PRS. Isopropylamine dodecylbenzene sulfonate; biodeg. emulsifier, solubilizer; dry cleaning; degreasers; latex emulsifier; pigment dispersant; agric. sprays, oil slick emulsifiers. [Pilot]

---

* Trade name not verified as available

† Trade name verified as obsolete

Calipharm. CAS 7757-93-9; Pharmaceutical grade of dicalcium phosphate. [Albright & Wilson Ltd, Phosphates & Speciality Business] *

Calirus. CAS 15310-01-7; Benodanil; a systemic fungicide. [BASF plc]

Calite. An alloy of 50% iron, 35% nickel, 10% aluminum, and 5% chromium. It is very resistant to heat, and melts at 2777 F.

Calixin®. CAS 24602-86-6; Tridemorph; systemic fungicide for control of powdery mildew in cereals, vegetables, etc. [BASF AG; BASF plc]

Calixin® M. Tridemorph, maneb; for control of fungus diseases in various crops [BASF AG]

Calkleen. Industrial cleaning compound. [The Wellcome Foundation Ltd] *

Callaica. Synonym for turquoise. §

Callaway 4000 Series. Series of formaldehyde-free cationic polymers; dye fixatives for direct, reactive, sulfur and pigment dyes; minimal effect of lightfastness; for textile use. [Callaway]

Calloseal. Rubber sheet or strip containing expandable graphite for sealing fire doors, pipe joints etc. [Foseco (F.S.) Ltd]

Callotek. Expandable, acid treated graphite. [Foseco (F.S.) Ltd]

Callusolve. Amber colored paint containing 25% benzalkonium chloride bromine; used for topical use in the treatment of warts, especially multiple or mosaic warts. [Dermal Laboratories Ltd]

Calmitol. A proprietary preparation of chloral hydrate, camphor, menthol, iodine, hyoscyamine, and zinc oxide in an ointment base; antipruritic. [Horlicks] *

Calofil. CAS 471-34-1; Calcium carbonate. [Rhone-Poulenc UK]

Calofort®. CAS 471-34-1; Precipitated calcium carbonate, $CaCO_3$; used for sealants, plastics, rubber. [Rhone-Poulenc Sturge Lifford]

Calomel. CAS 10112-91-1; Mercurous chloride, Hg_2Cl_2; used for control of clubroot in brassicas and white rot in onions. [Hortichem Ltd]

calomic. A nickel-iron-chromium resistance alloy; used for electric heater elements operating up to 900 C.

Calonutrin. A proprietary preparation of mono-, poly- and disaccharides used in the tube feeding of invalids. [Geistlich Sohne AG] *

Calopake®. CAS 471-34-1; Precipitated calcium carbonate, $CaCO_3$; used for paper, paint. [Rhone-Poulenc Sturge Lifford]

Caloreen. Glucose polymer; nutritional supplement. [Roussel Laboratories Ltd] †

Caloride. A proprietary trade name for cakes of calcium chloride for drying purposes. §

calorite. Alloys. One contains 65% nickel, 15% iron, 12% chromium, and 8% manganese ; and another 65% nickel, 23% iron, and 12% chromium.

Caloxal CLP 45. A dispersion of high grade calcium oxide in chlorinated paraffin used as a desiccant for rubber and plastics. [Rhone-Poulenc Sturge Lifford]

Caloxol. A range of dispersions of calcium oxide in oil wax or plasticizer; dessicant for use in rubber and plastics industries. [Rhone-Poulenc UK]

Caloxol CP2. CAS 1305-78-8; A proprietary calcium oxide desiccant for rubber. [Rhone-Poulenc Sturge Lifford] *

Calphon. Calcium salt solution. [Bayer AG]

Cal Plus. CAS 10043-52-4; Calcium chloride dihydrate, $CaCl_2 \cdot 2H_2O$; replenisher. [Mallinckrodt Inc] *

Calpol. CAS 103-90-2; Paracetamol compounds; for treatment of mild to moderate pain and pyrexia. [The Wellcome Foundation Ltd]

Calprona K. Mold inhibitor. [BP Chemicals Ltd]

Calquat. Industrial cleaning compound. [The Wellcome Foundation Ltd] *

Calsan. Sanitary disinfectant. [The Wellcome Foundation Ltd] *

Calsept. A proprietary preparation of calamine, menthol, cetrimide, camphor, and zinc oxide; antiseptic skin cream. [H N Norton & Co Ltd] ‡

Calsil. Precipitated silica with a proportion of calcium oxide; medium-active reinforcing filler with outstanding extrusion characteristics, good processibility for technical rubber articles. [Degussa AG] *

Calsoft AOS-40. Sodium alpha olefin sulfonate; surfactant for hand soaps, shampoos, hard surface cleaners, household and industrial cleaners. [Pilot]

Calsoft F-90. Sodium dodecylbenzene sulfonate; detergent, emulsifier, wetter for all-purpose and hard surface cleaners, bubble baths, degreasers, laundry powds., textile scouring aids, emulsion polymers, sanitation, emulsion paints, wettable powds., ore flotation, metal pickling. [Pilot]

Calsoft LAS-99. Dodecylbenzene sulfonic acid, linear; biodeg. detergent, emulsifier, intermediate for liq. and dry detergents, hard surface cleaners, stripping, wetting, foaming. [Pilot]

Calsoft T-60. TEA-dodecylbenzene sulfonate; biodeg. detergent, wetting agent, flash foamer; liq. detergents, wool wash compounds, cosmetics and shampoos, agric. emulsifiers, industrial cleaners, textile scouring, and car wash compounds [Pilot]

Calsol 510. Naphthenic process oil; provides excellent initial color and color stability for thermoplastics, radial and styrene block elastomers. [Calumet Lubricants]

Calsol 804. Naphthenic process oil; provides excellent color stability in resin extending, PVC, textiles, caulking compounds and other applications. [Calumet Lubricants]

Calsolene. General industrial wetting agent, emulsifier and emulsion breaker; biodegradable; used in oil and petroleum industry; gas manufacture; control of foam generation. [ICI Chem. & Polymers Ltd]

Calstrip. Industrial cleaning compound. [The Wellcome Foundation Ltd] *

Calsuds 81. Formulated product; concentrate base for liq. detergents, all-purpose and hard surface cleaning, dishwash, car wash, foaming and wetting sol'ns. [Pilot]

Calsun Bronze. A proprietary trade name for an aluminum bronze containing copper with 2.5% aluminum and 2% tin. §

Calsynar. A proprietary preparation of synthetic salmon calcitonin; used in the treatment of Paget's disease. [Armour Pharmaceutical Co] *

Calthane. Two-part solvent-free urethane elastomer systems; for molds for making plastic, concrete and plaster parts. [Cal Polymers] *

Calthane ND 1100. Two-part urethane elastomer system; used in flexible and rigid castings, prosthetics, underwater sports equipment, optical lenses, art objects, mech. devices, flexible windshields, industrial tubing or sheeting, toys, adhesives; mix ratio: 1A:1B. [Cal Polymers] *

---

‡ Trade name and manufacturer not verified   § Trade name without identified manufacturer

Calthane NF 0710. Two-part nonfilled urethane elastomer (MDI-based prepolymerPart A; polyester resin Part B); tougher than filled Calthane systems; low mixed visc. suggests use for machine-dispensed urethane sheets, skate board wheels, mech. parts; mix ratio 7A/10B. [Cal Polymers] *

Calthane NF 1900. Two-part urethane elastomer (A is an MDI-based prepolymer, B is a wh. resin); ideal for making tough, flexible molds; produces solid castings without degassing; other applications incl. casting industrial and print rollers, bumper pads, diaphragms, dust boots, engine mounts, gaskets, liners, wear pads; cures at R.T.; mix ratio: 1A/9B; green (A), wh. (B), mixed wh. (B may be pigmented to obtain any color). [Cal Polymers] *

Calthor Suspension. Ciclacillin; orally active broad spectrum amino-penicillin. [Wyeth Laboratories] †

Calthor Tablets. Ciclacillin; orally active broad spectrum antibiotic (amino-penicillin). [Wyeth Laboratories] †

Cal-Tint. Colourant dispersions; for coloring of aqueous and nonaqueous coating compositions. [Hüls Am.]

Calurea. A nitrogenous fertilizer. It is a mixture of urea and calcium nitrate containing 34% nitrogen.

calx. *See* calcium oxide

Calyptol. A proprietary preparation of eucalyptol, terpineol, oil of pine needles, thyme, and rosemary; a bronchial inhalant. [Smith & Nephew Pharmaceuticals Ltd] *

C.A.M. A proprietary preparation of ephedrine hydrochloride and butethamate citrate; a bronchial antispasmodic. [Rybar Laboratories Ltd] *

Camalox. Contains aluminum hydroxide gel, precipitated calcium carbonate, and magnesium hydroxide; an antacid and laxative. [William H Rorer Inc] *

Cambe wood. *See* redwoods.

Cambilene. A selective weedkiller. [Fisons plc, Horticulture Div] *

Cambison. A proprietary preparation of prednisolone, neomycin and quinolyl denrivatives; used in dermatology as an antibacterial agent. [Hoechst UK] *

Cambrelle. Melded fabrics. [Camtex Fabrics Ltd]

Cambrite. A smokeless powder, containing 22-24% nitroglycerin, 3-4.5% barium nitrate, 26-29% potassium nitrate, 32-35% dried wood meal, 1% calcium carbonate, and 7-9% calcium chloride.

Camel-CAL®. CAS 471-34-1; Calcium carbonate; filler for water-based coatings and inks, paper, and PVC pipe. [Genstar Stone Prods.]

Camel-CARB®. CAS 471-34-1; Calcium carbonate; filler/extender used in interior flat paint and exterior house paints, rubber compounds, putty and caulk, ceramics, adhesives, linoleum, floor tile, and textile coatings. [Genstar Stone Prods.]

Camel-FIL. CAS 471-34-1; Calcium carbonate; filler designed for high filler loading in glass-reinforced polyester; also for PVC (rigid and flexible), PP, rubber automotive goods and floor tiles, caulks, sealants, and adhesives. [Genstar Stone Prods.]

Camel-FINE. CAS 471-34-1; Calcium carbonate. [Genstar Stone Prods.]

Camel-TEX®. CAS 471-34-1; Calcium carbonate; fine-ground, general-purpose filler used in interior flat paints, primers, and sealers, polyester fiberglass premixes, preforms, and hand lay-up gel coats, rubber automotive products, household products, tubing, medical products, and closures, putty, caulk, bath tub sealers, body dead-eners, and adhesives. [Genstar Stone Prods.]

Camel-WITE®. CAS 471-34-1; Calcium carbonate; filler for paint, paper, paper coating, PVC, rubber (automotive goods, footwear, medical supplies), thermoplastics, thermosets, and in caulks, glazing compounds, ceramics, adhesives, food processing. [Genstar Stone Prods.]

Camelia metal. An alloy of 70.2% copper, 14.7% lead, 10.2% zinc, 4.2% tin, and 0-5% iron.

camenthol. A mixture of camphor and menthol for inhalation.

Cameo®. For agricultural industry. [DuPont UK]

Camflex. Plastics in the form of sheets. [Courtaulds plc]

Camie 300. General-purpose spray adhesive with quick adhesion to metal, plastic, wood, paper, cardboard, fabric, leather; for garment, furniture, upholstery, and shipping applications. [Camie-Campbell]

Camilol. α-Bisabolol; cosmetic ingred. [Maybrook]

Camite. Proprietary trade name for tungsten carbide materials. §

Camoquin. A proprietary preparation containing amodiaquine and hydrochloride; an antimalarial drug. [Parke-Davis] *

Campaign. Post-emergence herbicide. [Murphy Chemical Ltd] †

Campbell's CIPC 40%. CAS 101-21-3; Emulsifiable concentrate containing 400 g chlorpropham per liter; a carbamate herbicide. [MTM Agrochemicals Ltd]

Campbell's DB Straight. CAS 94-82-6; Soluble concentrate containing 300 g 2,4-DB per liter; used to control weeds in lucerne. [MTM Agrochemicals Ltd]

Campbell's Destox. CAS 1702-17-6; 2,4-D; Translocated herbicide for cereals and established grassland. [MTM Agrochemicals Ltd]

Campbell's Dioweed 50. CAS 1702-17-6; 2,4-D; Translocated herbicide for cereals and established grassland. [MTM Agrochemicals Ltd]

Campbell's DSM. CAS 8022-00-2; Emulsifiable concentrate containing 580 g demeton-S-methyl per liter; a systemic organophosphorus insecticide. [MTM Agrochemicals Ltd]

Campbell's Field Marshal. Mixture of dicamba, MCPA and mecoprop; used for weed control in cereals and grassland. [MTM Agrochemicals Ltd]

Campbell's Grassland Herbicide. Mixture of dicamba, MCPA and mecoprop; used for weed control in cereals and grassland. [MTM Agrochemicals Ltd]

Campbells Linuron 45%. CAS 330-55-3; Linuron; a residual urea herbicide for the control of weeds in field crops. [MTM Agrochemicals Ltd]

Campbell's MC Flowable. CAS 83601-81-4, 12427-38-2; A suspension concentrate containing 62 g carbendazim and 400 g maneb per liter. systemic fungicide for cereals. [MTM Agrochemicals Ltd]

Campbell's MCPA 25, 50. CAS 94-74-6; MCPA; herbicide for cereals and grassland. [MTM Agrochemicals Ltd]

Campbells Nabam Soil Fungicide. CAS 142-59-6; Soluble concentrate containing 320 g/l nabam; used for control of root rot in tomatoes and chrysanthemums. [MTM Agrochemicals Ltd]

Campbell's New Camppex. Soluble concentrate containing 34 g 2,4-D, 133 g dichlorprop, 53 g MCPA, and 164 g mecoprop per liter; an herbicide for use in cereals. [MTM Agrochemicals Ltd]

Campbells Nico-Soap. CAS 54-11-5; Nicotine; alkaloid

---

* Trade name not verified as available   † Trade name verified as obsolete

insecticide. [MTM Agrochemicals Ltd]

Campbells Rapier. CAS 23950-58-5; Propyzamide; a residual herbicide for agricultural crops. [MTM Agrochemicals Ltd]

Campbell's Redipon. CAS 120-36-5; Soluble concentrate of 500 g dichlorprop per liter; used for control of weeds in barley, wheat and oats. [MTM Agrochemicals Ltd]

Campbell's Redipon Extra. Soluble concentrate of 350 g dichlorprop and 150 g MCPA per liter; used for control of weeds in cereals and grass. [MTM Agrochemicals Ltd]

Campbell's Redlegor. CAS 94-82-6, 94-74-6; 2,4-DB + MCPA; translocated herbicide for cereal crops. [MTM Agrochemicals Ltd]

Campbell's Sugar Beet Herbicide. Emulsifiable concentrate containing 27.5 g chlorpropham, 28.5 g fenuron, and 147 g propham per liter; an herbicide. [MTM Agrochemicals Ltd]

Campbell's Trifluron. CAS 330-55-2, 1582-09-8; Mixture of linuron and trifluralin; herbicide for winter cereals. [MTM Agrochemicals Ltd]

Campbell's X-Spor. CAS 12427-38-2; Suspension concentrate containing 480 g maneb per liter; a dithiocarbamate fungicide to control blight, rusts and mildew. [MTM Agrochemicals Ltd]

Campeachy wood. See logwood.

2-camphanone. See camphor.

camphor. (1,7,7-trimethylbicyclo[2.2.1] heptan-2-one; gum camphor; 2-camphanone) $C_{10}H_{16}O$; CAS 76-22-2, 464-49-3; EINECS 207-355-2; Medicine, plasticizer for cellulose nitrate, other explosives and lacquers, insecticides, moth and mildew proofing, tooth powders, flavoring, embalming, pyrotechnics, intermediate. [Buckton Scott Ltd; R.W. Greeff; Lonza; Penta Mfg.; Quest Int'l.; Schweizerhall]

camphorated oil. (*Linimentum camphoroe B.P.*) It consists of 200 parts camphor, and 800 parts olive oil.

Campolon. A proprietary preparation of liver extract with Vitamin B_{12}; a hematinic. [Bayer AG] †

Campovit. An injectable vitamin B complex with liver extract. [Bayer AG] †

Camtex. Melded fabrics. [Camtex Fabrics Ltd]

Camwood. (Cambe wood). See redwoods.

Camyna. A proprietary preparation of thioxolone. [Boehringer Ingelheim GmbH] *

Canacert. Foodstuffs coloring matter meeting Canadian regulations. [Pointing Ltd] *

Canadian Certicol. Food colors meeting Canadian specifications. [Morton Int'l. Ltd] †

Canadium. An alloy of 1 part palladium, 2 parts platinim, and 6 parts nickel; used as a substitute for platinum.

Canagel 75. Semigelatin dynamite with good water resistance, high weight and volume energy and relatively high detonation velocity; used for construction and building industry, mining, explosives. [Hercules] *

Canarin. (persulfocyanogen yellow). A yellow coloring matter obtained by the action of bromine upon potassium or ammonium thiocyanate; used in calico printing.

candelilla wax. Wax from various *Euphorbiaceae* species; CAS 8006-44-8; manufacture of cosmetics, rubber substitutes, polishes, candles, sealing waxes, varnishes, leather, creams; for waterproofing; elec. insulation; inks; molding compositions; sizing paper; protective coating for citrus; hardening other waxes. [Penta Mfg.; Sevenson Bros.; Strahl & Pitsch]

Candeptin. A proprietary preparation of candicidin in a petrolatum base; antimonial vaginal ointment. [LRC Products Ltd] *

Candeptin-N. A proprietary preparation of candicidin and neomycin, in a petrolatum base; antibiotic skin ointment. [LRC Products Ltd] *

Candex®. CAS 50-99-7; Dextrose with small amounts of higher glucose saccharides; offers sweet, nongritty taste and is easily blended with flavors, lubricants, and other dry additives; excellent flow and compaction props.; for use in chewable tablets, esp. those made by direct compression. [Mendell]

Candigyn. See Diflucan. [Pfizer International]

Canesten®. CAS 23593-75-1; Clotrimazole; broad spectrum antifungal and tuchomonicidal drug. [Bayer AG]

Canfelzo. CAS 1314-13-2; Zinc oxide. [Pigment & Chemical Inc]

Canguard® 327. CAS 51200-87-4; 4,4-Dimethyl-oxazolidine; in-can preservative for water-containing systems, e.g., latex paint, resin emulsions, caulks, adhesives; broad spectrum antimicrobial activity; emulsifying capability when used with fatty acids; alkaline buffering capability. [Angus]

Canguard® 409. CAS 52-51-7; 2-Bromo-2-nitropropane-1,3-diol; preservative for adhesives, coatings, paints, starch, pigment and extender slurries, latex, antifoam emulsions, and inks. [Angus]

Canguard® 454. Hexahydro-1,3,5-tris(2-hydroxyethyl)-s-triazine; in-can preservative for water-containing systems, e.g., latex paint, resin emulsions, adhesives, inks; broad spectrum antimicrobial activity; nonfoaming. [Angus]

Canilep. Vaccines for dogs. [Glaxo Laboratories] *

Canilep D.D. A combined double-dose vaccine for dogs. [Glaxo Laboratories] *

Canopar. CAS 4304-40-9; Thenium closylate; anthelmintic. [Burroughs Wellcome Co] *

Cantabiline. CAS 90-33-5; Hymecromone; choleretic. [Lipha SA] *

Cantamega 1000. A fiber based vitamin and mineral supplement; dietary supplement. [Larkhall Laboratories plc] *

Cantamega 2000. A comprehensive vitamin and mineral dietary supplement. [Larkhall Laboratories plc] *

Cantil. CAS 76-90-4; Mepenzolate bromide; anticholinergic. [Merrell Dow Pharmaceuticals Inc] *

Cantreece®. Fibers. [DuPont UK]

Canzler Wire. An alloy of 98.8% copper, 1% silver, and 0.2% phosphorus.

Cao. CAS 2409-55-4; Butylated paracresol; antioxidants for food, plastic, rubbers, and other general purpose requirements. [PMC Specialities Group Inc]

caoutchouc. Coagulated latex of various rubber trees and shrubs; used in the manufacture of electrical insulation, toys, and beltings.

CAP. See cellulose acetate propionate.

C-A-P. Cellulose acetate phthalate; pharmaceutic aid. [Eastman]

CA-P20. Clarifying agent concentrate in PP carrier; nucleating agent in PP homopolymer and random copolymers; provides improved clarity, better gloss, improved tensile and flex. strength and dimensional stability. [Polyvel]

CAP-482-0.5. CAS 9004-39-1; Cellulose acetate propionate ester; used in printing inks, paper coatings, cloth coatings; excellent grease barrier properties. [Eastman]

---

‡ Trade name and manufacturer not verified   § Trade name without identified manufacturer

CAP-482-20. CAS 9004-39-1; Cellulose acetate propionate ester; for blending with CAP-482-0.5. [Eastman]

Capa. Caprolactone and polymers based on it. [Interox Chemicals Ltd]

Capasal. Brown colored foaming shampoo containing 0.05% w/w salicylic acid BP, 1.0% w/w coconut oil BP, 1.0% w/w distilled coal tar; used for the treatment of dry scaly scalp conditions. [Dermal Laboratories Ltd]

Capastat. CAS 1405-37-4; Capreomycin sulfate; antituberculant agent. [Dista Products Ltd]

Capastat Sulfate. CAS 1405-37-4; Capreomycin sulfate, an antibiotic produced by streptomyces capreolus; antibacterial. [Eli Lilly & Co]

Cap copper. An alloy of 95-97% copper and 3-5% zinc; used for deep drawing and stamping.

Capcure Emulsifier 37S. Ethoxylate; epoxy resin emulsifier. [Henkel/Functional Prods.]

Capella®. For agriculture industry. [DuPont UK]

C-A-P Enteric Coating Polymer. Cellulose acetate phthalate. [Eastman] *

Capexco. A coal mine explosive consisting of 32-34% nitroglycerin, 0.5-1.5% nitro-cotton, 24-25% sodium nitrate, 30-32% ammonium oxalate, and 8-10% wood meal.

Capital 170. Coconut fatty acids. [Karlshamns]

Capitol. Clear aqueous gel containing 0.5% benzalkonium chloride; used for the topical treatment of pityriasis capitis and other seborrheic scalp conditions where there is scaling and dandruff. [Dermal Laboratories Ltd]

Capla. CAS 64-55-1; A proprietary preparation of mebutamate; an antihypertensive. [Horlicks] *

Caplenal. CAS 315-30-0; Allopurinol; for the treatment of gout, primary hyperuricemia, secondary hyperuricemia, prophylaxis of uric acid and calcium oxalate stones. [Berk Pharmaceuticals Ltd] *

Capmul®EMG. CAS 51158-08-8; PEG-20 glyceryl stearate; dough conditioner for yeast-raised baked goods. [Karlshamns]

Capmul®GDL. CAS 27638-00-2; EINECS 248-586-9; Glyceryl dilaurate; emulsifier used in fats and oils; for cosmetics, pharmaceuticals. [Karlshamns]

Capmul®GMO. CAS 111-03-5; Glyceryl oleate; food emulsifier, wetting control agent; dispersant for pigments, solids; defoamer. [Karlshamns]

Capmul® GMS. Glyceryl stearate; stabilizer; internal lubricant for cosmetics; food emulsifier used in margarine, yeast-raised baked goods. [Karlshamns]

Capmul® MCM. Glyceryl caprate/caprylate; co-solv. and coupler for org. compounds; w/o emulsifier. [Karlshamns]

Capmul®O. CAS 1338-43-8; EINECS 215-665-4; Sorbitan oleate; food emulsifier and dispersant. [Karlshamns]

Capmul® POE-L. CAS 9005-64-5; Polysorbate 20; food emulsifier and solubilizer for flavors. [Karlshamns]

Capmul® POE-O. CAS 9005-65-6; Polysorbate 80; food emulsifier for frozen desserts; solubilizer for oils into water systems; used for flavors, fragrances, vitamins, pharmaceuticals, solv. [Karlshamns]

Capmul® POE-S. CAS 9005-67-8; Polysorbate 60; food emulsifier; solubilizer for oils into water systems. [Karlshamns]

Capmul®S. CAS 1338-41-6; EINECS 215-664-9; Sorbitan stearate; food emulsifier for chocolate and confectionary coatings, shortenings. [Karlshamns]

Caposil. A trademark for calcium silicate and asbestos based insulating materials. Caposil 1400 withstands 1400 F. (760 C.) and Caposil HT withstands 1850 F. (1000 C.). [Cape Insulation Cape Asbestos Co] *

Capoten. CAS 62571-86-2; A proprietary preparation of captopril; an ACE inhibitor used in the treatment of cardiovascular disorders. [Bristol-Myers Squibb Pharmaceuticals Ltd]

Capozide. Captopril and hydroflumethiazide; antihypertensive; enzyme inhibitor; diuretic. [Bristol-Myers Squibb Co Inc]

Cappagh brown. (euchrome, mineral brown). A natural pigment used in oil-painting, containing hydrated oxide of iron and 27% manganese dioxide.

Capramin. Decongestant and antipyretic tablets. [Allen & Hanburys Ltd] *

Capran®. Polyamide 6 film useful for food packaging, composites manufacturing and other industrial applications. [Allied-Signal Inc]

Capran® 77C. CAS 25038-54-4; Nylon 6 film, unoriented; thermoplastic film for general purpose packaging, sterilization packaging; clear extruded films in varous widths and thicknesses. [Allied-Signal Engineered Plastics]

Capran®Emblem Film. Biaxially oriented nylon film for food packaging and industrial applications. [Allied-Signal Inc]

Capran®Unidraw®. CAS 25038-54-4; Nylon 6, monaxially oriented; specifically engineered as a carrier web for molding fiberglass-reinforced panels; exceptional tear and tensile strength. [Allied-Signal Engineered Plastics]

Capreomycin. A proprietary antibiotic produced by *Streptomyces capreolus* present as the sulfate in capastat. [Dista Products Ltd] *

Capri blue gon. A dyestuff. It is the zinc double chloride of dimethyldiethyldiaminotoluphenazonium chloride, $C_{17}H_{20}N_3OCl$. Dyes cotton greenish-blue.

capric acid. (n-decanoic acid; n-capric acid; carboxylic acid C_{10}) Fatty acid; $CH_3(CH_2)_8COOH$; CAS 334-48-5; EINECS 206-376-4; esters for perfumes and flavors; base for wetting agents; intermediate for chemical synthesis. [Akzo; Aldrich; Henkel/Emery; Mirachem Srl; Procter & Gamble; Witco/Humko]

capric acid chloride. See decanoyl chloride.

Caprinol®. An antihypertensive preparation. [Bayer AG]

Capriton. A proprietary preparation of chlorpheniramine maleate, phenylephine hydrochloride, aspirin, caffeine; decongestant and antipyretic tablets. [Allen & Hanburys Ltd] *

caproic acid. (carboxylic acid C_6; hexanoic acid) $CH_3(CH_2)_4COOH$; CAS 142-62-1; EINECS 205-550-7; Analytical chemistry, flavors, manufacture of rubber chemicals, varnish driers, resins, pharmaceuticals. [Aldrich; Chisso Am.; Jansse Chimica; Penta Mfg.; Schweizerhall]

Caprol® 2G4S. CAS 72347-89-8; Polyglyceryl-2 tetrastearate; food emulsifier; opacifier; wax modifier; thickener. [Karlshamns]

Caprol®3GO. CAS 9007-48-1; Polyglyceryl-3 oleate; food emulsifier for frozen desserts, vegetable dairy products, diet spreads; wetting agent for dyes and pigments in cosmetics; defoamer. [Karlshamns]

Caprol® 3GS. CAS 37349-34-1; Polyglyceryl-3 stearate; food emulsifier, stabilizer and whipping agent used in frozen desserts and fat reduction. [Karlshamns]

Caprol®6G2O. Polyglyceryl-6 dioleate; food emulsifier for frozen desserts. [Karlshamns]

---

* Trade name not verified as available   † Trade name verified as obsolete

Caprol®6G2S. Polyglyceryl-6 distearate; food emulsifier for whipped toppings, frozen desserts, coffee whiteners. [Karlshamns]

Caprol® 10G2O. Polyglyceryl-10 dioleate; o/w emulsifier, humectant, lubricant; for frozen desserts. [Karlshamns]

Caprol® 10G4O. CAS 34424-98-1; Polyglyceryl-10 tetraoleate; food emulsifier, visc. control, stabilizer. [Karlshamns]

Caprol®10G10O. CAS 11094-60-3; Polyglyceryl-10 decaoleate; food emulsifier, solubilizer, lubricant, and dispersant. [Karlshamns]

Caprol® 10G10S. CAS 39529-26-5; EINECS 254-495-5; Polyglyceryl-10 decastearate; lubricant for thread finishes, wax additive, crystal modifier; thickener; opacifier. [Karlshamns]

Caprol®ET. Polyglyceryl mixed esters; food emulsifier and crystal inhibitor in oils. [Karlshamns]

ε-caprolactone monomer. (6-hexanalactone, 2-oxepanone; 6-caprolactone monomer) $C_6H_{10}O_2$; CAS 502-44-3; EINECS 207-938-1; Intermediate in adhesives, urethane coatings, elastomers; solvent; diluent for epoxy resins; synthetic fibers; organic synthesis. [Union Carbide]

Capron. Range of type 6 nylon polymers including reinforced grades of nylon 6 and either glass fibers or a blend of glass fibers and selected minerals; used for power tool housings, chain saw housings and other components, housings for lawn and garden equipment, automotive cooling fans and other engine compartment applications and exterior automotive body parts. [Allied-Signal Inc]

Capron 8253, 8350, 8351, 8352. A proprietary range of copolymer grades of nylon possessing high impact strength; for extrusion and injection molding. [Allied-Signal Inc]

Capron 8270 HS. A proprietary modified nylon 6 compound used in blow molding. [Allied-Signal Inc]

Capron 8331G, 8332G, 8333G, 8334G. A family of high impact, glass reinforced materials exhibiting improved dry-as-moled toughness. [Allied-Signal Inc]

Capron 8200. A proprietary nylon used for injection molding applications requiring improved toughness over standard nylons. [Allied-Signal Inc]

Capron®8202. CAS 25038-54-4; Nylon 6 homopolymer; for consumer and industrial parts, e.g., elec. hair brush, chain saw housings, elec. parts; features rigidity, heat resist., excellent surface appearance. [Allied-Signal Engineered Plastics]

Capron 8202C, 8203C, 8202CQ. A proprietary range of crystalline type nylon molding compounds; for improved cycle time and higher stiffness. [Allied-Signal Inc]

Capron®8203C HS. CAS 25038-54-4; Nylon 6 homopolymer, nucleated; heat-stabilized material for automotive push-pull cable assembly, inner tube; featuring stiffness, wear resist. [Allied-Signal Engineered Plastics]

Capron 8206 S. A proprietary nylon of good flexibility used in extrusion and molding operations. [Allied-Signal Inc]†

Capron 8230G. Nylon with 6% glass fiber; improved stiffness and dimensional stability. [Allied-Signal Inc]

Capron 8231G, 8323G, 8233G. A range of glass-filled nylons (14%, 25%, and 33% respectively). [Allied-Signal Inc]

Capron®8232G HS FR. CAS 25038-54-4; Nylon 6, 25% glass-reinforced, flame-retarded; flame-retarded grade for tool housings; features excellent surface appearance, strength, stiffness, impact resist., creep resist., heat and chem. resist. [Allied-Signal Engineered Plastics]

Capron®8233G HS. CAS 25038-54-4; Nylon 6, 33% glass-reinforced; for automotive end fittings; features rigidity, chem. resist. [Allied-Signal Engineered Plastics]

Capron® 8253. CAS 25038-54-4; Nylon 6 copolymer; for automotive fasteners and clips; features flexibility, toughness, low temp. impact resist. [Allied-Signal Engineered Plastics]

Capron® 8259. CAS 25038-54-4; Nylon 6 copolymer; for consumer/industrial parts, e.g., fasteners; features flexibility, low temp. impact resist. [Allied-Signal Engineered Plastics]

Capron® 8266G HS. CAS 25038-54-4; Nylon 6, 40% mineral/glass-reinforced; for automotive grills. [Allied-Signal Engineered Plastics]

Capron 8267 G HS. Mineral and glass reinforced nylon for improved mechanical properties and low warpage. [Allied-Signal Inc]

Capron®8280. CAS 25038-54-4; Nylon 6 homopolymer; for consumer/industrial parts, e.g., roto-molded reservoirs, tanks, shapes; processability for large parts, chem. resist., impact strength, rigidity. [Allied-Signal Engineered Plastics]

Caprosem. A proprietary preparation containing Testosorone, 17-chloral hemi-acetal acetate; male sex hormone. [Leo Laboratories] *

capryl alcohol. (1-octanol; alcohol C_8; octyl alcohol) $CH_3(CH_2)_7OH$; CAS 111-87-5; EINECS 203-917-6. [Ethyl; M. Michel; Penta Mfg.; Schweizerhall; Vista]

caprylic acid. (n-octanoic acid; octoic acid) Fatty acid; $CH_2(CH_2)_6COOH$; CAS 124-07-2; EINECS 204-677-5; Manufacture of dyes, drugs, perfumes. [Akzo; Aldrich; Henkel/Emery; Procter & Gamble; Unichema]

caprylic acid sodium salt. See sodium caprylate.

Capsolin. A proprietary preparation of oleoresin capsicum, camphor, oil of turpentine and eucalyptus; a rubefacient. [Parke-Davis] *

capsule metal. An alloy of 92% lead and 8% tin.

captan. (N-trichloromethylthiotetrahydrophthalimide; N-trichloromethylthio-4-cyclohexene-1,2-dicarboximide) Organic compound; $C_9H_8Cl_3NO_2S$; CAS 133-06-2; seed treatment, fungicide, and bacteriostat in plants, plastics, leather, fabrics, and fruit preservation; gas odorant. [Industrias Quimicas del Valles SA; R.T. Vanderbilt Co Inc]

Captan. A foliage fungicide. [ICI Chem. & Polymers Ltd] †

Captan-50. A foliage fungicide. [Plant Protection] *

Captan 83 WP. Fungicide for apple and pear scab. [ICI Chem. & Polymers Ltd]

Captan-Col. CAS 133-06-2; A captan fungicide. [Plant Protection] *

Captan Granular. CAS 133-06-2; Captan as dry granules; used for controlling soil borne fungal diseases in tomato, lettuce and strawberry. [Wheatley Chemical Co Ltd]

Captax. CAS 149-30-4; 2-Mercaptobenzothiazole; primary accelerator for both natural and synthetic rubbers. [R T Vanderbilt Co Inc]

Captax-disulfide. See Altax. §

Captex® 200. Propylene glycol dicaprylate/dicaprate; carrier, coupler, solv. for flavors, fragrance oil, sol. colorants, vitamins, medicinals, cosmetics; emollient for creams, lotions, makeup. [Karlshamns]

---

‡ Trade name and manufacturer not verified   § Trade name without identified manufacturer

Captex®300. CAS 65381-09-1; Caprylic/capric triglyceride; solv. for colors and perfumes; emollient, moisturizer in cosmetics, toiletries, pharmaceuticals; plasticizer. [Karlshamns]

Captex® 800. CAS 56519-71-2; Propylene glycol dioctanoate; nonoily lubricant imparting rich feel to skin in cosmetics and pharmaceuticals; carrier for essential oils, flavors; vehicle for vitamins, medicinals, nutritional products [Karlshamns]

Captex® 810B. Caprylic/capric/linoleic triglyceride; emollient, solv., carrier, fixative, and extender in pharmaceuticals, nutritional, and cosmetic applications. [Karlshamns]

Captex®8000. CAS 538-23-8; Caprylic triglyceride; nonoily lubricant imparting rich feel to the skin; for cosmetics and pharmaceuticals; carrier for essential oils, flavors; vehicle for vitamins, medicinals, nutritional products [Karlshamns]

Captor. Surfactants for use in oil recovery. [Imperial Chemical Industries plc]

Carace®. CAS 83915-83-7; Lisinopril; for the treatment of hypertension. [Merck & Co Inc]

Carace®Plus. Lisinopril, hydrochlorothiazide; for the treatment of hypertension in patients for whom combination therapy is appropriate. [Merck & Co Inc]

Caradate. A range of polyether polyols and isocyanates used for the production of flexible and rigid polyurethane foams; may be used in furniture and automotive sealing, in pipes and tanks; effective insulators; also find use in domestic freezers and refrigerators. [Shell UK]

Caradol. A range of polyether polyols and isocyanates used for the production of flexible and rigid polyurethane foams; used in furniture and automotive sealing, in pipes and tanks (effective insulators.), in domestic freezers and refrigerators. [Shell] *

Carafate. CAS 54182-58-0; Sucralfate; anti-ulcerative. [Marion Merrell Dow Inc]

Caramba. Rust solvents, penetrating oils, anticorrosives, preservative preparations, anticorrosion lacquers, underbody coatings and cavity preservation compounds for vehicles; cleaning, polishing, scouring and abrasive preparations for metals, motors and machine parts, for window panes of motor cars, antifreeze and deicer. [Caramba Chemie GmbH] *

Caramba Felgenglanz. Detergent for cleansing wheel-rims. [Caramba Chemie GmbH] *

Caramba Felgenneu. Detergent for cleansing wheel-rims. [Caramba Chemie GmbH] *

Caramba Lackkrone. Cleans, polishes and seals in one operation paintwork and motor cars. [Caramba Chemie GmbH] *

Caramba Perlglanz. Highly effective and concentrated product for lacquer care; sealant for motor cars. [Caramba Chemie GmbH] *

Carbagel. An activated carbon containing calcium chloride; a proprietary trade name for a drying agent. §

Carbalax®. A laxative (suppository). [Pharmax Ltd]

carbamide. See urea.

carbanilide. CAS 102-07-8; Diphenylurea, $C_6N_5NH \cdot CO \cdot NHC_6H_5$; organic synthesis.

carbanillic ether. CAS 101-99-5; Phenylurethane.

carbarsone. CAS 121-59-5; 4-Carbaminophenylarsonic acid; used in medicine.

Carbaryl. CAS 63-25-2; 1-Naphthyl methylcarbamate; used as an insecticide.

Carbate Flowable. CAS 83601-81-4; A liquid formulation containing 500 g carbendazim per liter as a suspension concentrate; systemic fungicide. [Pan Britannica Industries Ltd]

Carbathene. A proprietary trade name for an ethylene N-vinyl carbazole copolymer; a molding compound form stable over 240 C. [Standard Telecommunications Laboratories] ‡

Carbazole Blue. A dyestuff, obtained by fusing carbazole with oxalic acid.

Carbazole Violet. A dyestuff obtained by fusing 9-ethylcarbazole with oxalic acid.

Carbergan. Carbon tetrachloride and promethazine in oil. [May & Baker Ltd] *

Carbetamex. Herbicide. [Rhone-Poulenc UK]

Carbetamex. CAS 16118-49-3; Carbetamide; a residual herbicide. [Embetec Crop Protection Ltd]

Carbex. See Carbora. §

Carbicon. See Carbora. §

Carbilys. Used to render starch insoluble in the manufacture of paper and board, and also in the textile and spinning process industries. [Roquette (UK) Ltd]

carbinol. See methyl alcohol.

Carbital®. CAS 471-34-1; Calcium carbonate; finely ground natural marble sold as a slurry in water; paper coating pigments; fillers for paper; fillers and extenders for paints. [ECC International Ltd]

Carbital®35. CAS 1317-65-3; Calcium carbonate; med.-fine wet-ground filler pigment; also for matte or dull coatings. [ECC]

Carbital®50. CAS 1317-65-3; Calcium carbonate; fine, wet-ground filling pigment offering excellent retention in the sheet, excellent optical props.; also for matte or dull coatings. [ECC]

Carbitol. Diethylene glycol monoethyl ether, $C_2H_5O \cdot CH_2 \cdot CH_2O \cdot CH_2 \cdot CH_2OH$; a solvent for cellulose nitrate, shellac, copal, rosin, etc.; used in dopes for artificial leather and is added to brushing lacquers. [Union Carbide]

Carbo Alumina. Consists mainly of crystalline alumina; a trademark for goods of the abrasive and refractory class§

Carbobrant. See Carbora. §

Carbocaine. CAS 1722-62-9; Mepivacaine hydrochloride; anesthetic. [Sterling Drug Inc] *

Carbo-corundum. A trademark for articles made from crystalline alumina; refractories and abrasives. §

Carbodan. CAS 1563-66-2; Carbofuran; insecticide. [Makhteshim Chemical Works Ltd]

Carbofin. Lamp black. [Brockhues AG]

Carboform. Plastic molding material made from carbon fibers impregnated with resin, in the form of sheets, tape, matting or fabric, and tow for packing and jointing. [Cyanamid Fothergill Ltd]

Carbofrax. A carborundum refractory containing more than 90% silicon carbide.

carbo-gel. Silica gel impregnated with carbon, for use to absorb organic vapours.

Carbogran. CAS 409-21-2; Silicon carbide; used in abrasive industries. [Lonza AG]

Carbogran E. CAS 409-21-2; Silicon carbide; used in the electrical indsutry. [Lonza AG]

Carbogran UF. Silicon carbide; for use in sintered ceramics and for abrasion resistant surfaces. [Lonza AG]

---

* Trade name not verified as available   † Trade name verified as obsolete

Carbokaylene. A proprietary preparation of vegetable charcoal with colloidal aluminum silicate. §

Carbolan. Super-milling acid dyes. [ICI Chem. & Polymers Ltd]

Carbolfuchsine. (Ziehl's stain). A microscopic stain for bacteria. It contains 5 parts fuchsine, 25 parts phenol, 50 parts alcohol, and 500 parts water.

Carbolic Acid. CAS 108-95-2; Phenol, C_6H_5OH. In trade, the term is used for pure phenol, the cresols and their mixtures with phenol, and also for crude tar oils.

carbolic acid crystals. Pure phenol.

Carbolon. A trademark for abrasive articles consisting essentially of silicon carbide. §

Carboloy. A trademark for hard metal compounds consisting of tungsten carbide and cobalt for cutting glass and for high-speed tools. §

carbol xylene. A microscopic clearing solution containing 3 parts xylene and 1 part phenol.

Carbomang. A proprietary trade name for a steel containing 1-1.25% manganese, 0.45% chromium, 0.5% tungsten, and 1% carbon. §

Carbomant. CAS 409-21-2; Silicon carbide; used in abrasive industries and wire sawing. [Lonza AG]

Carbomer. A suspension agent. A polymer of acrylic acid crosslinked with allyl sucrose.

Carbomix® 1605. Cold SBR blk. masterbatch with 50 phr N550 black; nonstaining masterbatch for high-quality extruded and molded mech. goods. [Copolymer Rubber]

Carbomix® 1609. Cold SBR blk. masterbatch with 40 phr N110 black, 5 phr highly aromatic oil; for applications requiring high tens. strength and low heat buildup; for high-quality extruded and molded goods. [Copolymer Rubber]

Carbomix® 3651. Cold SBR blk. masterbatch with 52 phr N234 black, 10 phr highly aromatic oil; for high-quality tread compounds, extruded and molded goods. [Copolymer Rubber]

Carbomucil. A proprietary preparation of charcoal, magnesium carbonate, and sterculia; an antidiarrhoeal. [Norgine Ltd] *

Carbon 4E. A liquid formulation containing 48% trichloropyrester; controls woody weeds and perennial broad-leaved weeds in forestry and noncropped areas. [Burts & Harvey] *

carbon, activated. (active carbon; activated charcoal) Clarifying, deodorizing, decolorizing, filtering; activated charcoal in medicine as antidote, adsorptive. [Allied-Signal; Am. Norit; Atochem N. Am.; Calgon Carbon; Ceca SA; United Catalysts; Westvaco]

Carbonado. (Black Diamond). A black, compact variety of diamond; used in the steel crowns of rock-drills.

carbon bisulfide. See carbon disulfide.

carbon black. (thermal black; channel black; furnace black) Finely divided particles of elemental carbon obtained by incomplete combustion of hydrocarbons (channel or impingement process); C; CAS 1333-86-4; tire treads, belt covers, other abrasion-resistant rubber products; uv light absorber; colorant for printing inks. [Akrochem; Akzo; Cabot; Degussa; Exxon; R.T. Vanderbilt; Witco/Cancarb]

Carbon Bronze. An alloy containing 75.5% copper, 9.75% tin, and 14.5% lead; a white metal used for bearings.

Carbondale silver. An alloy of 66% copper, 18% nickel, and 16% zinc.

carbon dioxide. CO_2; CAS 124-38-9; EINECS 204-696-9; Refrigeration, carbonated beverages, aerosol propellant, chemical intermediate, fire extinguishing, inert atmospheres, municipal water treatment, medicine, oil wells, mining, blowing agent. [Air Prods.; Nissan Chem. Ind.; Norsk Hydro AS; Showa Denko]

carbon disulfide. (carbon bisulfide) CS_2; CAS 75-15-0; EINECS 200-843-6; Viscose rayon, cellophane, manufacture of carbon tetrachloride and flotation agents, solvent. [Akzo; PPG Industries; Rhone-Poulenc Chemie NV]

Carbonet. Nonadherent gauze dressing. [Smith & Nephew Pharmaceuticals Ltd] *

carbonic acid, dipotassium salt. See potassium carbonate.

carbonic acid, disodium salt. See sodium carbonate.

carbonic acid, magnesium salt (1:1). See magnesium carbonate.

carbonic acid, magnesium salt (2:1). See magnesium carbonate.

carbonic acid, monosodium salt. See sodium bicarbonate.

carbonic acid, sodium salt (2:3). See sodium sesquicarbonate.

carbonic acid, zinc salt (1:1). See zinc carbonate.

Carbonin. Granular carbonaceous additive for molten iron and steel. [Foseco (F.S.) Ltd] †

carbon tetrachloride. (tetrachloromethane) Chlorinated hydrocarbon; CCl_4; CAS 56-23-5; EINECS 200-262-8; refrigerants, metal degreasing, agric. fumigant, chlorinating organic compounds, production of semiconductors, solvents. [Ashland; Janssen Chimica; Mitsui Toatsu Chem.; Montefluos SpA; Solvay SA]

carbonyldiamide. See urea.

carbonyl-iron powder. See iron.

Carboplastic. A proprietary fireproof plastic cement, the main constituent of which is carborundum; suitable for repairing furnaces. §

Carbopol® 613, 614. CAS 9003-01-4; Polyacrylic acid; emulsifier for solv. cleaners, emulsion stabilizer, suspending agent, thickener for detergent formulations. [BFGoodrich]

Carbopol® 910. Carbomer 910, a polymer of acrylic acid, crosslinked with a polyfunctional agent; the viscosity of a neutralized preparation containing 2.5 g of Carbomer 910 in 500 ml water is not less than 2,500 centipoises and not more than 7000 centipoises; pharmaceutic aid. [BFGoodrich] *

Carbopol® 934. Carbomer 934, a polymer of acrylic acid, crosslinked with a polyfunctional agent; the viscosity of a neutralized preparation containing 2.5 g of carbomer 934 in 500 ml water is not less than 3,000 centipoises and not more than 4,000 centipoises; pharmaceutic aid. [BFGoodrich]

Carbopol® 934P. Carbomer 934P, a polymer of acrylic acid, crosslinked with a polyfunctional agent; pharmaceutic aid. [BFGoodrich]

Carbopol® 940. Carbomer 940, a polymer of acrylic acid, crosslinked with a polyfunctional agent; pharmaceutic aid. [BFGoodrich] *

Carbopol® 941. Carbomer 941, a polymer of acrylic acid, crosslinked with a polyfunctional agent; a neutralized preparation containing 2.5 g of carbomer 941 in 500 ml water is not less than 4,000 centipoises and not more than 11,000 centipoises; pharmaceutic aid. [BF

---

‡ Trade name and manufacturer not verified   § Trade name without identified manufacturer

Goodrich] *

Carbo-Pulbit®. Oral antidiarrhoeic; veterinary medicine. [Bayer AG]

Carbora. Proprietary name for silicon carbide materials; abrasive. §

Carboraffin. CAS 7440-44-0; Powdered activated carbon; used for decolorizing and improving the odor and taste of liquids mainly in the chemical and food industries. [Bayer AG] †

Carborex. CAS 409-21-2; Silicon carbide, SiC; for abrasive, refractory, metallurgical and other usages. [Orkla Exolon A/S & Co] *

Carborite. See Carbora. §

Carborundum. The name applied to abrasives, the main constituent of which is silicon carbide, SiC. [Harbison-Carborundum Corp.] ‡

Carbosal. A range of multipurpose fire exinguishing powders for extinguishing fires of incandescent solid, liquid or gaseous materials. [Degussa AG] *

Carb-O-Sep. CAS 121-59-5; Carbarsone; N-carbamoyl-arsanilic acid; antiamebic. [Whitmoyer Laboratories Inc]‡

Carboset 514A. Acrylic resin, IPA; thermoplastic film-forming resin used in protective metal coatings, paints, ceramics, adhesives, textiles, paper, leather, cosmetics, floor polishes, chemical specialties; excellent dispersant, leveling, and binding char. [BFGoodrich] *

Carboset 531. Acrylic resin, ammonia water; thermoset film-forming resin used in protective metal coatings, paints, ceramics, adhesives, textiles, paper, leather, cosmetics, floor polishes, chemical specialties; excellent dispersant, leveling, and binding char. [BFGoodrich] *

Carboset Resins. Acrylic copolymers: alkaline water dispersions, dry, granular solid, polymer solutions in organic solvents and high viscosity dry liquid polymer; used for paper coatings, electronics, industrial coatings, printing inks, adhesives and cosmetics. [BFGoodrich UK] *

Carbosorb. Soda synthetic silicate carbon dioxide absorbent. [BDH Chemicals Ltd]

Carbostat 2203. Quaternary amine; surfactant for textile processing. [Hoechst Celanese/Colorants & Surf.]

carbostyril. (2-quinoline; o-amino cinnamic acid lactam) 2-Hydroxyquinoline.

Carbotex. A brand of natural rubber and carbon black; used in rubber mixings. §

Carbowax®. Polyethylene glycols. [Union Carbide (UK) Ltd] *

Carbowax® Compound 20M. CAS 25322-68-3; PEG-350; binder, lubricant for ceramics, powd. metallurgy, toilet bowl cleaners. [Union Carbide]

Carbowax® MPEG 350. CAS 9004-74-4; PEG-6 methyl ether; surfactant intermediate, lubricant for adhesives, inks, mining, soaps and detergents. [Union Carbide]

Carbowax® PEG 200. CAS 25322-68-3; PEG-4; intermediate for surfactants, lubricants, urethanes; antistat, humectant, mold release agent, plasticizer for adhesives, inks, lubricants. [Union Carbide]

Carbowax® PEG 540 Blend. CAS 25322-68-3; PEG-6 and PEG-32 (41:59); base for ointments and suppositories; also for adhesives, creams and lotions, deodorant sticks. [Union Carbide]

Carbowax® PEG 8000. CAS 25322-68-3; PEG-150; antistat, ceramic binder, surfactant intermediate, dye carrier, lubricant, release agent, tablet binder for adhesives,

creams and lotions, mining, powd. metallurgy, soaps and detergents, tablet coating, toilet bowl cleaners. [Union Carbide]

Carbowax® Sentry. CAS 25322-68-3; Polyethylene glycol; pharmaceutic aid: ointment base; suppository base; solvent; tablet excipient; tablet and/or capsule lubricant. [Union Carbide] *

carbox metal. An alloy of 84% lead, 14% antimony, 1% iron, and 1% zinc.

Carboxide. A proprietary fumigant containing 1 part ethylene oxide and 8 parts carbon dioxide. §

carboxylic acid C5. See n-valeric acid.

carboxylic acid C6. See caproic acid.

carboxylic acid C7. See heptanoic acid.

carboxylic acid C10. See capric acid.

carboxylic acid C18. See stearic acid.

carboxymethylcellulose sodium. (CMC; cellulose gum (CTFA); sodium carboxymethylcellulose; sodium CMC) Sodium salt of the carboxylic acid $R\text{-}O\text{-}CH_2COONa$; $[(C_6H_7O_2(OH)_2OCH_2COOH]_n$; CAS 9004-32-4; in drilling muds; in detergents as soil-suspending agent; in emulsion paints, adhesives, inks, textile sizes; as protective colloid; food stabilizer, binder, thickener; cosmetics; in pharmaceuticals as suspending agent, excipient, viscosity modifier. [Aqualon; Courtaulds Water Soluble Polymers; J.W.S. Delavau; FMC; Hercules]

Carbrital. A proprietary preparation of carbromal and pentobarbitone sodium; a hypnotic. [Parke-Davis] *

carburet of iron. See graphite.

carburetted hydrogen, heavy. See olefiant gas.

Carcanol G23. A proprietary trade name for a fraction of cashew nut shell liquid. §

Cardamist. A proprietary preparation of glyceryl trinitrate in aerosol form; used in angina pectoris. [Nicholas Laboratories Ltd] *

Cardenalin. See Cardura. [Pfizer International]

Cardiacap. CAS 78-11-5; A proprietary preparation of pentaerythritol tetranitrate; a vasodilator used for angina pectoris. [Consolidated Chemicals Ltd] *

Cardiacap A. A proprietary preparation of pentaerythritol tetranitrate and amylobarbitone in a sustained release form; used in angina pectoris. [Consolidated Chemicals Ltd] *

Cardice. CAS 124-38-9; Solid carbon dioxide. [Distillers MG Ltd]

Cardilan. CAS 395-28-8; Isoxsuprine; promotes blood flow rate. [Duphar BV] *

Cardilate. CAS 7297-25-8; Oral/sublingual/buccal tablets; a proprietary formulation of erythrityl tetranitrate; for treatment of angina pectoris. [The Wellcome Foundation Ltd]

Cardinal. A dyestuff; a mixture of chrysoidine and fuchsine; used for dyeing cotton red.

Cardinal Red J. A dyestuff; a British brand of Fast red. §

Cardio-Green. Indocyanine green; diagnostic aid. [Hynson Westcott & Dunning] *

Cardiomax®. Garlic oil; for maintenance of a healthy cardiovascular system. [Seven Seas Ltd]

Cardipol® LP. Oxidized polyethylene wax; for formulating hot-melt adhesives. [Petrolite]

Cardis® 36. Oxidized microcryst. wax; used in the formulation of polishes. [Petrolite]

Cardolite. A proprietary cashew nut derivative of the phenol aldehyde polymer class; also proprietary trade name for plasticisers, resins, rubber-like polymers, and solvents. §

---

* Trade name not verified as available   † Trade name verified as obsolete

Cardolite® NC-507. CAS 501-24-6, 3158-56-3; 3-(n-Pentadecyl) phenol; starting raw material for surfactants, antioxidants, anticorrosives; lubricant additive; cosolv. for insecticides, germicides; resin modifier. [Cardolite] *

Cardolite® NC-511. CAS 8007-24-7; 3-(n-Penta-8´-decenyl) phenol; brn. to amber oily liq.; m.w. 302; sol. in org. solvs. incl. aliphatic hydrocarbons; sp.gr. 0.930; b.p. 223-227 C (10 mm Hg); ref. index 1.5112. [Cardolite] *

Cardolite® NC-1307. Terpene-based extender/flexibilizer/diluent for epoxy manufacturing; also for concrete coatings. [Cardolite] *

Cardophylin. Aminophylin tablets, ampoules and suppositories. [Fisons plc, Pharmaceuticals Div] *

Cardox. An explosive utilizing liquid carbon dioxide.

Cardular. See Cardura. [Pfizer International]

Cardura. A proprietary trade name for nondrying alkyd resins with excellent weathering properties used for surface coatings. [Shell]

Cardura®. CAS 74191-85-8; Doxazosin; antihypertensive. Also known as Alfadil (Sweden), Cardenalin (Japan), Cardular (Germany), Carduran (Denmark, France, Norway, New Zealand), Daxiren (Greece), Doxaben (Taiwan), Supressin (Austria). [Pfizer International]

Cardura E. Glycidyl ester of the synthetic fatty acid, Versatic 10; versatile intermediate for the production of stoving enamels, nitrocellulose lacquers and urethane paints. [Shell UK]

Carduran. See Cardura. [Pfizer International]

Cariflex. A range of synthetic rubbers (styrene butadiene, cis-polyisoprene and polybutadiene). [Shell UK]

Cariflex Butadiene Rubber (BR). Polybutadienes with resiliency and very good abrasion resistance; available in straight and oil-extended versions; used in car tires and in the manufacture of high-impact polystyrenes as well as many industrial applications. [Shell UK]

Cariflex Isoprene Rubbers (IR). Chemically very similar to natural rubber; available in straight and oil-extended versions; used in tire carcasses, footwear and belting. [Shell UK]

Cariflex S. A proprietary styrene-butadiene rubber. [Shell]*

Cariflex Styrene-Butadiene Rubbers (SBR). A family of general purpose synthetic rubbers available as straight, oil-extended, carbon-black masterbatch or resin/rubber masterbatch; widely used in car tires, footwear, adhesives and a range of industrial and domestic products. [Shell UK]

Cariflex Thermoplastic Rubbers (TR). Materials possessing the inherent elasticity of rubbers yet processable as thermoplastics; used in the manufacture of adhesives or blended with thermoplastics for improved impact properties; when compounded, used as carpet-backing or injection-molded for footwear and a range of industrial and domestic products. [Shell UK]

Carin. See hexamine.

Carina. A proprietary trade name for polyvinyl chloride (qv). [Shell] *

Carindaden. A proprietary preparation of carindacillin sodium; an antibiotic used in the treatment of genito-urinary tract infections. [Pfizer International] †

Carinex. A proprietary trade name for a series of polystyrenes and toughened polystyrenes. [Shell] *

Carinex SB41. A proprietary trade name for an easy processing polystyrene. [Shell] *

Carinex SI 73. A proprietary high-impact grade of polystyrene possessing good flow properties in molding. [Shell Chemicals Ireland Ltd] *

Cariod. Papain; enzyme. [Sterling Drug Inc] *

Carisoma®. CAS 78-44-4; Carisoprodol; a proprietary preparation for use as a muscle relaxant. [Pharmax Ltd]

Caritrol. CAS 1642-54-2; Diethylcarbamazine citrate. [Rhone-Poulenc Rorer Ltd]

Carletti's indicator. Phenolphthalein which has been reduced by caustic soda and zinc dust.

Carlona 55-004. A proprietary high density polyethylene possessing high resistance to stress cracking; used in the blow-molding of containers. [Shell] *

Carlona 60-010. A proprietary high density polyethylene used in the extrusion blow-molding of thin-walled bottles. [Shell] *

Carlona 60-060. A proprietary high density polyethylene used in the injection-molding of heavy-duty containers. [Shell] *

Carlona 60-120. A proprietary high density polyethylene used in the injection-molding of thin sections. [Shell] *

Carlona 460. A proprietary polyethylene used in the production of film; has high impact strength but contains a slip additive. [Shell] *

Carlona 462. A proprietary polyethylene containing slip additives, for optical and mechanical use. [Shell] *

Carlona 463. A proprietary polyethylene containing slip additives for optical and mechanical film. [Shell] *

Carlona LB 157. A proprietary low-density polyethylene used in the production of film. [Shell] *

Carlona LF 456. A proprietary polyethylene having high impact strength; used in the production of film. [Shell] *

Carlona LF 459. A proprietary low-density polyethylene possessing good optical and mechanical properties for the production of film. [Shell] *

Carlona P PLZ 532. A proprietary talc-filled, heat-stabilized polyethylene used in the production of rigid moldings. [Shell] *

Carlona P PY 61. A proprietary polypropylene homopolymer used in the production of film, especially stretched tapes for use as plastic string. [Shell] *

carloon bronze. An alloy of 75.5% copper, 14.5% lead, and 10% tin.

Carlton Suspension N.K. A proprietary preparation containing neomycin sulfate and light kaolin; an antidiarrheal. [Carlton Laboratories (UK) Ltd] *

carmalum. A microscopic stain; consists of carminic acid in aqueous alum.

Carmargo® White. Calcium bentonite; filler and binder for technical applications. [Am. Colloid]

carmine. Aluminum lake of the coloring agent, cochineal; cochineal is a natural pigment derived from the dried female insect Coccus cacti; CAS 1390-65-4; EINECS 215-724-6; dyes, inks, indicator in chemical analysis, coloring food, medicine. [Aceto; R.W. Greeff; Penta Mfg.; Warner-Jenkinson]

Carmine 40. CAS 1390-65-4; Carmine; cosmetic pigment. [RITA]

Carmine 224. CAS 1390-65-4, 7787-59-9; Carmine, bismuth oxychloride; natural colorant. [Presperse]

carmine lake. (lac lake, Indian lake). Cochineal carmine prepared by precipitating a decoction of cochineal with alum or stannic chloride, with addition of acid oxalate or tartrate of potassium. See cochineaL.

carmine red. Obtained by boiling a dilute aqueous solution

---

‡ Trade name and manufacturer not verified   § Trade name without identified manufacturer

of carminic acid with a few drops of a mineral acid. It gives colored lakes.

carnallite. A double chloride of potassium and magnesium, $MgCl_2 \cdot KCl \cdot 6H_2O$, found in the Stassfurt deposits; a source of potassium chloride and potash manures.

Carnation. White mineral oil. [Witco/Sonneborn]

carnauba. (Brazil wax) CAS 8015-86-9; Exudate from leaves of Brazilian wax palm tree *Copernicia prunifera*; greyish, yellowish, or greenish in color, and when freshly purified melts at 85-86 C; shoe polishes, leather finishes, varnishes, waterproofing, furniture and floor polishes; hardening candles; substitue for beeswax; plasticizer for dental compounds; purified form in cosmetics, pharmaceuticals. [Industrial Waxes Ltd; Penta Mfg.; Stevenson Bros.; Strahl & Pitsch]

Carnauba Spray 200. Quaternary blend; car rinse with carnauba wax. [Sherex/Div. of Witco]

carnauba wax. See carnauba.

Carnoid. A proprietary trade name for a casein plastic. §

Carnot's reagent. Basic bismuth nitrate (100 g) is dissolved in hot, concentrated hydrochloric acid, and diluted to a liter with 92% alcohol.

Carnoy's fluid. A mixture of absolute alcohol and glacial acetic acid; used for fixing animal tissue.

Caroat. A peroxomonosulfate compound with about 4.5% active oxygen; oxidizing agent used for production of cleansers of all types, e.g., denture cleansers, household and toilet cleaners, for detoxicating cyanidic waste water. [Degussa Ltd]

carob flour. See locust bean gum.

Caro bronze. An alloy of 92% copper, 8% tin, and 0.25% phosphorus. A bearing metal.

carob seed gum. See locust bean gum.

Carolid® MN-1. Methyl naphthalene carrier; carrier for polyester and blends; low odor. [Sybron]

Caromax. A range of close cut aromatic hydrocarbon solvents. [Carless Refining & Marketing Ltd]

Carophyll. Pigments for animal feedstuffs. [Roche Products Ltd] *

Caro's acid. Persulfuric acid, H_2SO_2; used in dye manufacture, oxidizing agent, bleaching.

Caro's reagent. Obtained by dissolving ammonium or potassium persulfate in concentrated sulfuric acid; a pasty oxidizing agent; used for testing alkaloids.

carotene. (β-Carotene; provitamin A) $C_{40}H_{56}$; CAS 7235-40-7; EINECS 230-636-6; Color additive for foods; vitamin A precursor; uv screen. [BASF; Hoffmann-La Roche; Penta Mfg.; Schweizerhall; Warner-Jenkinson]

Caroubier. An acid dyestuff, giving crimson shades on wool.

Carovax. A proprietary pasteurella vaccine used in veterinary work. [Coopers Animal Health Ltd] *

Carp. A cotton fabric grade of Tufnol industrial laminates. [Tufnol Ltd]

Carp Brand. Salt. [ICI Chem. & Polymers Ltd] †

Carpenter 22Cr-13Ni-5Mn. Nitrogen-strengthened austenitic stainless steel; alloy providing good corrosion resist. and high strength; for valve shafts, pumps and fittings for chem. equip., fasteners, cables, chains, screens, wire cloth, marine hardware, heat exchanger parts, photographic process equip. [Carpenter Tech.]

Carpenters Wood Glue. PVA liquid; fast grab wood glue. [Wessex Resins & Adhesives Ltd]

carrageenan. (chondrus; Irish moss extract; carrageen) Sulfated polysaccharide; CAS 9000-07-1; emulsifier ,

gelling agent in food products, toothpaste, cosmetics, pharmaceuticals; stabilizing aid in ice cream. [G Fiske & Co Ltd; FMC; Hercules]

Carriant Series. Oil mixt.; textile dyeing assistants; carriers for polyester fibers. [Toho Chem. Industry]

Carrisorb. CAS 1337-76-4; Attapulgite clay; an absorbent used in treatment and bleaching of oils (mainly mineral). [Bromhead & Denison Ltd]

Carrot Oil CLR. Soybean oil, carrot oil, carrot extract, β-carotene, tocopherol; emollient, conditioner, superfatting agent; emulsified and oily preparations for care of skin and hair. [Dr. Kurt Richter; Henkel/Cospha; Henkel Canada]

Carsamide®. Alkylamines. [Lonza Inc]

Carsamide® AMEA. CAS 142-26-7; EINECS 205-530-8; Acetamide MEA (1:1); hair conditioner and antistat. [Lonza Inc]

Carsamide®CA. Cocamide DEA (1:1); detergent, dispersant, emulsifier, wetting agent, foam booster, thickener, softener for industrial, cosmetic, and household cleaners; biodeg.. [Lonza Inc]

Carsamide® CMEA. CAS 68140-00-1; Cocamide MEA (1:1); foam booster, stabilizer, visc. builder for shampoos and detergents. [Lonza Inc]

Carsamide® SAL-7. CAS 120-40-1; EINECS 204-393-1; Lauramide DEA (1:1); detergent, emulsifier, foaming agent, foam stabilizer, thickener for shampoos, bath products, household, institutional and industrial detergents. [Lonza Inc]

Carset. Hardener for cold-set sand molds. [Foseco (F.S.) Ltd]

Carsil. Binder for the CO_2 process. [Foseco (F.S.) Ltd]

Carsilon. CAS 409-21-2; Silicon carbide; used in special refractories. [Lonza AG]

Carsofoam® 1618. Cetearyl alcohol; bodying agent, visc. modifier for cosmetic, personal care, and household products [Lonza Inc]

Carsofoam®BS-I. PEG-80 sorbitan laurate, sodium trideceth sulfate, PEG-150 distearate, disodium lauroamphodiacetate, cocamidopropylhydroxysultaine, sodium laureth-13 carboxylate; low irritation concentrate for baby shampoos. [Lonza Inc]

Carsofoam®MSP. TEA-lauryl sulfate, cocamide DEA, cocamidopropyl betaine, methyl paraben; low irritation concentrate for baby shampoo. [Lonza Inc]

Carsofoam®T-60-L. TEA dodecylbenzene sulfonate; high foaming detergent, wetting agent for cosmetic, industrial, and institutional usage. [Lonza Inc]

Carsonol®. Alcohol sulfates; used for shampoos and bubble baths where high foaming and mildness are desired. [Lonza AG]

Carsonol® ALS-R. CAS 2235-54-3; EINECS 218-793-9; Ammonium lauryl sulfate; detergent with high foam, good wetting and emulsifying properties used for cosmetics, chemical specialties. [Lonza Inc]

Carsonol® AOS. CAS 68439-57-6; EINECS 270-407-8; Sodium C14-C16 olefin sulfonate; surfactant for shampoos, liq. soaps, industrial cleaners. [Lonza Inc]

Carsonol® DLS. DEA lauryl sulfate; biodeg. detergent, foaming agent, wetting agent, emulsifier for personal care products [Lonza Inc]

Carsonol®MLS. Magnesium lauryl sulfate; detergent with high foam, good wetting and emulsifying properties used for cosmetics, chemical specialties, rug and upholstery

---

* Trade name not verified as available

† Trade name verified as obsolete

formulations. [Lonza Inc]

Carsonol® SES-A. CAS 67762-19-0; Ammonium laureth sulfate; surfactant with excellent foaming in hard and soft water, for cosmetic, household, and industrial uses, shampoos, bubble baths, liq. cleaners. [Lonza Inc]

Carsonol®SES-S. CAS 9004-82-4; Sodium laureth sulfate; surfactant with excellent foaming in hard and soft water, for cosmetic, household, and industrial uses, liq. carwash, laundry detergents. [Lonza Inc]

Carsonol® SHS. Sodium 2-ethylhexyl sulfate; low foaming detergent, wetting agent, penetrant, emulsifier used in caustic sol'ns. for peeling of fruits and vegetables; stable to high concs. of electrolytes. [Lonza Inc]

Carsonol®SLES-2. Sodium laureth sulfate; mild surfactant with excellent foaming in hard and soft water, for cosmetic, shampoo, bubble bath. [Lonza Inc]

Carsonol® SLS Paste B. CAS 151-21-3; Sodium lauryl sulfate; detergent with high foam, good wetting and emulsifying properties used for cosmetics, chemical specialties, shampoo bases, textile scouring. [Lonza Inc]

Carsonol® TLS. CAS 139-96-8; EINECS 205-388-7; TEA-lauryl sulfate; biodeg. detergent with high foam, good wetting and emulsifying properties used for cosmetics, mild shampoos, bubble baths, chemical specialties, emulsion polymerization. [Lonza Inc]

Carsonon®. Nonionic sulfates and wetting agents; used in industrial and household cleaning products where good detergency is required. [Lonza AG]

Carsonon®144-P. CAS 63793-60-2; PPG-3 myristyl ether; lubricant, emollient, solubilizer for cosmetics incl. silicon systems; aids low temp. stability, antistatic and conditioning effects; contributes to mildness and spreading behavior. [Lonza Inc]

Carsonon® 169-P. CAS 9035-85-2; PPG-10 cetyl ether; lubricant, emollient, solubilizer for cosmetics incl. silicon systems; aids low temp. stability, antistatic and conditioning effects; contributes to mildness and spreading behavior. [Lonza Inc]

Carsonon®N-4. CAS 9016-45-9; Nonoxynol-4; emulsifier, detergent, wetting agent, dispersant for household and industrial uses; intermediate; drycleaning detergent. [Lonza Inc]

Carsoquat®. Quaternary ammonium compounds; used as creme rinses in personal care applications. [Lonza AG]

Carsoquat® 816-C. Cetearyl alcohol, PEG-40 castor oil, and stearalkonium chloride; formulated base, cream rinse concentrate. [Lonza Inc]

Carsoquat® 868 E. Dicetyl dimonium chloride; hair conditioner, cream rinse, fabric softener. [Lonza Inc]

Carsoquat® CB. Cetyl alcohol, glyceryl stearate, dicetyl dimonium chloride, cetrimonium chloride, polysorbate 85, PEG-40 castor oil; cream rinse concentrate. [Lonza Inc]

Carsoquat® CT-429. CAS 112-02-7; EINECS 203-928-6; Cetyl trimethyl ammonium chloride; surfactant; conditioner in hair and skin care preparations; antistat. [Lonza Inc]

Carsoquat® SDQ-25. CAS 122-19-0; EINECS 204-527-9; Stearalkonium chloride; conditioner, softener, creme hair rinse and conditioner for cosmetics and natural fibers. [Lonza Inc]

Carsosoft®. Quaternary ammonium compounds; used as fabric softeners. [Lonza AG]

Carsosoft®CFI-90. Alkyl imidazolinium methosulfate; fabric softener for prep. of clear liq. concs.. [Lonza Inc]

Carsosoft®S-90. Quaternium-27; fabric softener and antistat. [Lonza Inc]

Carsosoft®T-90. CAS 130124-24-2; Quaternium-53; fabric softener, rewetting agent, antistat, and antilint for home and commercial laundries. [Lonza Inc]

Carspray #2. Dicoco quaternary; emulsifier plus glycol ether for car rinses. [Sherex/Div. of Witco]

Carspray CW. Quaternary/surfactant blend; wash and wax concentrate for autos. [Sherex/Div. of Witco]

Carstab® DLTDP. CAS 123-28-4; Dilauryl thiodipropionate; heat aging stabilizer in conjunction with primary antioxidants; for PP, HDPE. [Morton Int'l./Specialty Chem.]

Carstab® DSTDP. CAS 693-36-7; Distearyl thiodipropionate; long-term heat aging stabilizer in conjunction with primary antioxidants; for PP, HDPE. [Morton Int'l./Specialty Chem.]

Cartaretin F-4. CAS 61840-27-5; Adipic acid/dimethylaminohydroxypropyl diethylenetriamine copolymer; polymer substantive to hair; for shampoo systems; imparts lubricity. [Sandoz]

Carterite. A proprietary trade name for resin-pulp molding compound. §

Carthamus tinctorious oil. See safflower oil.

Cartolac. General-purpose carton lacquers. [The Scottish Adhesives Co Ltd]

Cartose. Dextrose; replenisher. [Sterling Drug Inc] *

Cartrol. CAS 51781-21-6; Carteolol hydrochloride; anti-adrenergic. [Abbott Laboratories]

carvacrol. CAS 499-75-2; 2-Methyl-5-isopropyl-phenol. $C_{10}H_{14}O$; used in perfumes; fungicides, disinfectant, flavoring, organic synthesis.

carvone. $C_{10}H_{14}O$; CAS 99-49-0, 2244-16-8; EINECS 218-827-2; Flavoring, liqueurs, perfumery, soaps. [Penta Mfg.; Schweizerhall]

Carylderm. CAS 63-25-2; Carbaryl; used for infestation by lice. [Napp Laboratories Ltd] *

caryophyllic acid. See eugenol.

Casabet. Coco amido betaine and coco amido sulfo betaine; betaine type surfactants characterized by low toxicity and irritancy and high foam production properties; used in the formulation of household cleaning products, in industrial and institutional detergent products and in personal care products. [Thomas Swan & Co Ltd]

Casabet 655. Coco amido sulfo betaine as a semi-viscous liquid; foam booster, irritancy depressant, stable over wide pH range used in shampoos, bubble baths and liquid soaps. [Thomas Swan & Co Ltd]

Casahib. Amine functional compounds; corrosion inhibitors for crude oil and gas production. [Thomas Swan & Co Ltd]

Casamer. Epoxy and urethane acrylates; electron beam and ultraviolet curing systems. [Thomas Swan & Co Ltd] †

Casamid. Nonreactive polyamides, dicydiamide, accelerated dicydiamide and sustituted dicydiamide; epoxy resin curing agents used in surface coatings, powder coatings, electrical encapsulation and potting, filament winding, flexographic and gravure ink binders. [Thomas Swan & Co Ltd]

Casamine. Aminoethyl alkyl imidazolines and hydroxyethyl alkyl imidazolines; antifungal agents, emulsifiers in agricultural sprays, ore flotation, and acid detergent formu-

---

‡ Trade name and manufacturer not verified   § Trade name without identified manufacturer

lations. [Thomas Swan & Co Ltd]

Casamox. Amine oxides; nonionic detergent of versatile properties for the formulation of a wide range of cosmetic, detergent and industrial products; lime soap dispersant, detergent thickener and foam booster. [Thomas Swan & Co Ltd]

casanthranol. A purified mixture of the anthranol glycosides derived from *Casara sagrada.*

Casaquat. Imidazoline derived quaternary ammonium compounds; widely used in shampoos, hair conditioners, skin preparations, bacteriacidal and algacidal application; especially for the formulation of fabric softeners, both domestic and commercial. [Thomas Swan & Co Ltd]

Casateric. Fatty acid imidazoline derived carboxylate amphoteric surfactants; used to reduce irritancy of other surfactants used in shampoo, personal care and domestic applications; also used in industrial cleaning, textile processing, ore benefaction, oil-gas production. [Thomas Swan & Co Ltd]

Casathane. Prepolymers based on polyether and polyester diols; used for casting systems for the engineering industry. [Thomas Swan & Co Ltd]

Cascade. CAS 55-63-0; A nitroglycerin dynamite; for construction and building, explosives, mining. [Hercules] *

Cascade. Photographic wetting agent. [May & Baker Ltd] *

Cascamite. Precatalyzed urea-formaldehyde powdered resin glue; used for general purpose wood glue; waterproof. [Wessex Resins & Adhesives Ltd]

cascara sagrada. The dried bark of *Rhamnus Purshiana;* used as a laxative.

Casco. Casein-based adhesives. [Borden (UK) Ltd] *

Cascogel TM. Animal glue based adhesives. [Borden (UK) Ltd]

Cascomelt. Hot melt adhesives. [Borden (UK) Ltd]

Cascophen. Phenolic resins. [Borden (UK) Ltd] *

Cascophen Resorcinol Resin RS 216/RXS-8. Liquid resorcinol-formaldehyde resin with powder catalyst; general purpose wood glue; weatherproof. [Wessex Resins & Adhesives Ltd]

Casco-resin. Urea-formaldehyde resins. [Borden (UK) Ltd]

Casco-Resin TM. Urea-formaldehyde resins and adhesives. [Borden (UK) Ltd]

Cascorez TM. Vinyl acetate-based adhesives. woodwork adhesive. [Borden (UK) Ltd]

Cascosel TM. Aluminum foil laminating adhesives. [Borden (UK) Ltd]

Cascotak TM. Pressure sensitive adhesives. [Borden (UK) Ltd]

casein. (milk protein) Mixture of phosphoproteins from cows milk; CAS 9005-46-3, 9000-71-9; EINECS 232-555-1; cheesemaking, plastic items, paper coatings, water-dispersed paints, adhesives, textile sizing, foods and feeds, textile fibers, dietetic preparations, binder in foundry sands. [Meggle Marketing GmbH; Nat'l. Casein; U.S. Biochemical; Worthington Biochemical]

casein glue. Made by stirring casein with 25% distilled water, and 1-4% sodium bicarbonate, adding another 25% distilled water, standing, and adding an antiseptic to prevent mold; can be applied cold.

casein magnesia. A preparation which consists of powdered casein, water, and magnesia; it fixes mineral pigments.

casein paints. Paints formed by the addition of a powder containing casein and alkali, to water; coloring matter and lime are added.

casein silk. Casein dissolved in alkali or zinc chloride solution, and spun into an acid bath.

casein, sodium salt. *See* sodium caseinate.

caseogum. A solution of casein in lime water; used as an adhesive, or for impregnating linen and cotton fabrics.

Cashmilon®. Acrylic staple fiber. [Asahi Chem. Industry]

Casilan. Calcium caseinate. [Glaxo Laboratories] *

Casolithe. *See* Gallatite.

Casoron. CAS 1194-65-6; Granular herbicide containing dichlobenil; used to control weeds in woody crops and noncrop areas. [Chipman Ltd, ICI Agrochemicals, Synchemicals Ltd]

Casoron G. CAS 1194-65-6; Dichlobenil; direct, selective weed killing action in orchards, vineyards, flower beds and parkland areas and along rail tracks, motorways and waterways. [Duphar BV] *

Casoron G4. CAS 1194-65-6; Granule containing dichlobenil; residual herbicide for use among established trees and shrubs, paths, hard surfaces and vacant ground. [Vitax Ltd]

Cassappret®. Textile finishing agents. [Cassella AG]

Cassastat®. Antistatic finishing agents for textiles. [Cassella AG]

Cassatan®. Tanning agents. [Cassella AG]

cassava. A food product. It consists of the starch obtained from the roots of the Manioc, *Manihot utilissima;* used to make tapioca, laundry starch; and adhesives.

cassava meal. Ground cassava root.

Cassella's acid. CAS 92-40-0; β-Naphtholsulfonic acid δ; used as an intermediate for azo dyes.

Casselmam's green. A pigment made by mixing boiling solutions of copper sulfate and an alkaline acetate; consists of $CuSO_4 \cdot 3Cu(OH)_2 \cdot 4H_2O$.

cassia oil. (Chinese cinnamon oil; cinnamon oil) Used in flavoring, perfumery, medicine, salt.

cassiterite. *See* tin oxide (ic).

Cassulfon®. Water-soluble sulfur dyestuffs; used for textile dyeing and printing. [Cassella AG]

Cassurit®. Melamine resin products; for textile finishing. [Cassella AG]

Castaldo. Natural rubber (CIS-1, 4-polyisoprene compound); for mold making for jewelry casting. [F E Knight Inc] *

Castaloy. A proprietary trade name for high carbon, high chromium steel. §

Castaway Plus. Suspension concentrate containing 60 g γ-HCH and 500 g thiophanate-methyl per liter; used for control of earthworms and leatherjackets in turf. [Rhone-Poulenc Environmental Prods. Ltd]

cast brass. Brass which is not required to be spun, rolled, drawn or hammered. It is made by melting together the copper and zinc. This usually contains 66% copper with zinc, and the lead varies from 1-3%.

Castellanos powder. A dynamite containing nitro-glycerin, nitrobenzene, fibrous material, and kieselguhr.

Castethane. Two-component systems used in the manufacture of cast polyurethane elastomers. [Dow UK]

casting copper. An American copper containing 98.5-99.75% copper; used for casting. §

cast iron D2. A British chemical standard; a grey phosphoric cast iron in the form of fine turnings. It contains 1.31% silicon, 1.07% phosphorus, and 1.64% manganese, also

---

about 2.5% graphitic carbon, 0.8% combined carbon, and 0.03% sulfur.

Casto-Magic. A liquid oil derivative applied to concrete forms to release forms from concrete and improve appearance of concrete. [Rostine Manufacturing & Supply Co] *

Castomer. Two-component high duty polyurethane elastomer systems. [Baxenden Chemicals Ltd]

castor oil. (ricinus oil; oil of Palma Christi; tangantangan oil) Fixed oil obtained from seeds of *Ricinus communis*; CAS 1323-38-2; 8001-79-4; EINECS 232-293-8; plasticizer in lacquers and nitrocellulose; polyurethane coatings; hydraulic fluids, industrial lubricants; used in manufacture of Turkey red oil, cosmetics, leather treatment. [Ashland; CasChem; Climax Performance; Fanning; Harcros; Lipo; Norman, Fox]

castor oil acid, methyl ester. *See* methyl ricinoleate.

castor oil, hydrogenated. *See* hydrogenated castor oil.

castor oil, soluble. *See* blown oils.

castorwax. *See* hydrogenated castor oil.

Castorwax® MP-70. Hydrogenated castor oil; wax for anhyd. products requiring a soft creamy texture; for cosmetics and pharmaceuticals. [CasChem]

Castorwax® MP-80. Hydrogenated castor oil; release agent; wax used for formulating antiperspirant sticks, suspending aid for aluminum chlorhydrate. [CasChem]

Castorwax® NF. Hydrogenated castor oil NF; wax for pharmaceutical and food applications. [CasChem]

Castrix Grains. Control of field and house mice. [Bayer AG]

Castrol GTX. High performance motor oil; for all types of motor car engines. [Burmah-Castrol Ltd] *

Castrol Turbomax. A heavy duty high performance mineral-based crankcase lubricant for commercial vehicles; used for service, fill and top-up turbocharged and naturally aspirated diesel engines (except Detroit Diesel) and virtually all petrol units. [Burmah-Castrol Ltd] *

Castung 103 G-H. Castor oil, dehydrated; used with phenolics to obtain fast dry coatings with max. alkali resist., e.g., for sanitary can linings, corrosion resist. coatings, traffic paints, varnishes, ink vehicles, marine finishes; provides effective internal plasticization. [CasChem]

Cast yellow brass. Usually an alloy of 67% copper, 31% zinc, and 2% lead. It melts at 895 C.

Casul® 70 HF. Calcium dodecylbenzene sulfonate; coemulsifier for o/w and w/o formulations, agric. emulsions. [Harcros]

Caswell Adhesives. A range of industrial adhesive products. [Caswell & Co Ltd]

Catabond. A proprietary trade name for a phenol-formaldehyde liquid resin used for plywood manufacture where a waterproof bond is required. §

Cata-Chek. For catalysts for polymerization and adhesion of urethanes and other synthetic organic polymers, in Int Class 1. [Ferro] *

Cataflot. Quaternary ammonium compounds; flotation collectors. [Atochem UK/Ceca]

Catafor. Electrostatic paint and PU foam additive. [Rhone-Poulenc UK]

Cataid. Surface active agents and detergents. [ABM Chemicals Ltd] *

Catalase L. CAS 9001-05-2; Bovine catalase; enzyme which catalyzes the decomposition of hydrogen peroxide to water and oxygen; for milk production [Solvay Enzymes]

Catalazuli. A proprietary trade name for a phenol-formaldehyde synthetic resin product. §

Catalex. A proprietary trade name for an expanded phenol formaldehyde plastic. §

Catalpo. CAS 1332-58-7; Fine particle size ultra pulverized hydrous kaolin (aluminum silicate); used as and extender and reinforcer in coatings, adhesives, rubber. [Engelhard]

Catalyst 9. Magnesium chloride-based; fast cure catalyst for textile resin finishes. [BASF]

Catalyst CC. Amine hydrochloride; catalyst for urea-formaldehyde resins. [CNC Int'l. L.P.]

Catalyst RD Liq. Buffered inorg. salt catalyst; fast-acting resin catalyst. [Eastern Color & Chem.]

Catalyst ZA. Fast-acting accelerator for silicone water repellents. [CNC Int'l. L.P.]

Catapol SR. A proprietary range of polyurethane elastomers ranging in hardness from 20-60 Durometer A. [Longfield Chemicals] *

Catapres. CAS 4205-91-8; A proprietary preparation of clonidine hydrochloride; an antihypertensive. [Boehringer Ingelheim Ltd]

Catapres-TTS. CAS 4205-90-7; Clonidine; antihypertensive. [Boehringer Ingelheim Pharmaceuticals Inc] *

Catarase®. 1:5,000 Ophthalmic solution when reconstituted; chymotrypsin; for enzymatic zonulysis prior to intracapsular lens extraction. [Iolab Corp]

Catavar. A proprietary phenol-formaldehyde surface coating or laminating varnish. §

Catavat Black N-JBB. Vat dyestuffs. [Catawba-Charlab]

CatCO 600. Catalyst for abatement of carbon monoxide and unburned hydrocarbons; for gas turbines, heaters and boilers. [Engelhard]

CatCO 610ST. Catalyst for abatement of carbon monoxide and unburned hydrocarbons in the presence of high levels of sulfur compounds; for any industrial fossil fuel-fired equip. [Engelhard]

Catechol. *See* pyrocatechin.

Catechu. *See* cutch.

C-A-T Enteric Coating Polymer. Cellulose acetate trimellitate. [Eastman] *

Catex. Imidazoline derived cationic and amphoteric surfactants; used for textile softeners, antistats, foaming agents and solubilizing detergents. [Thomas Swan & Co Ltd] †

Catigene® 50 USP. n-Alkyl (50% C12, 30% C14, 17% C16, 3% C18) dimethylbenzyl ammonium chloride; germicide, algicide, fungicide, deodorizing agent, antistat. [Stepan Europe]

Catigene® 818. Octyl decyl dimethyl ammonium chloride, dioctyl dimethyl ammonium chloride, didecyl dimethyl ammonium chloride; germicide, algicide, fungicide, deodorizing agent, antistat. [Stepan Europe]

Catigene® 1011. CAS 7173-51-5; Didecyl dimethyl ammonium chloride; germicide, algicide, fungicide, deodorizing agent, antistat; also for oilfield production [Stepan Europe]

Catigene 4513. Mixture of the chlorides of alkyl dimethyl benzyl ammonium and alkyl dimethyl ethyl benzyl ammonium as a clear yellow liquid; for algicides, bactericides, fungicides, deodorants. [KWR Chemicals Ltd] *

Catigene® B 50. n-Alkyl (40% C12, 50% C14, 10% C16) dimethylbenzyl ammonium chloride; germicide, algicide, fungicide, deodorizing agent, antistat. [Stepan Europe]

Catigene BR 80 B. CAS 7281-04-1; EINECS 230-698-4; Lauryl dimethyl benzyl ammonium bromide as a clear yellow liquid; for algicides, bactericides, fungicides, deodorants. [KWR Chemicals Ltd] *

Catigene® CA 56. Alkyl amidopropyl trimethyl ammonium methoxysulfate; antistat additive for carpet latex and bitumen. [Stepan Europe]

Catigene® CETAC 30. Alkyl trimethylammonium chloride; antistat, lubricant, emulsifier, softener, germicide, algicide, fungicide, deodorizing agent for household/industrial cleaners, oilfield production [Stepan Europe]

Catigene® DC 100. CAS 139-08-2; EINECS 205-352-0; n-Alkyl (3% C12, 95% C14, 2% C16) dimethylbenzyl ammonium chloride; germicide, algicide, fungicide, deodorizing agent, antistat. [Stepan Europe]

Catigene DC/100. CAS 139-08-2; EINECS 205-352-0; Myristyl dimethyl benzyl ammonium chloride as a white powder; bactericides for pharmaceuticals. [KWR Chemicals Ltd] *

Catigene Red-brown. A dyestuff obtained by the action of alkali sulfides and sulfur upon aminohydroxy-phenazines.

Catigene SR. Quaternary alkylamine acetate as a white fluid paste; cationic surfactant used in textile softeners. [KWR Chemicals Ltd] *

Catigene T80. Alkyl dimethyl benzyl ammonium chloride as a clear yellow liquid; for bactericides, fungicides, algicides. [KWR Chemicals Ltd] *

Catinal CB-50. CAS 139-07-1; EINECS 203-351-5; Lauralkonium chloride; disinfectant, germicide, antistat. [Toho Chem. Industry]

Catinal HTB-70. CAS 57-09-0; EINECS 200-311-3; Hexadecyl trimethyl ammonium bromide; base material for hair rinse, antistat, germicide. [Toho Chem. Industry]

Catinal OB-80E. CAS 122-19-0; EINECS 204-527-9; Stearyl dimethyl benzyl ammonium chloride; base material for hair rinse. [Toho Chem. Industry]

Catiomaster-C. CAS 3327-22-8; 3-Chloro-2-hydroxypropyl trimethylammonium chloride (50% aq solution); quaternary cationic agent for starch and cellulose. [Yokkaichi Chemical Co Ltd] *

Cationic Collagen Polypeptides. CAS 9007-34-5; Cationic collagen polypeptides; substantive film-former for hair and skin care products (shampoos, conditioners, mousses, shave preps.); protective colloid effect. [Maybrook]

Cationic Softener X Concentrate. Alkyl-imidazoline derivative in paste form; softener with good substantivity; used as glass fiber mordant and lubricant. [Millmaster-Onyx UK] *

Catisol AO 100. Oleylamine acetate; emulsifier, wetting agent, antistat, anticorrosive agent, lubricant for textile lubricants for mineral and synthetic fibers. [Stepan Europe]

Catisol AO 100. Oleylamine acetate as a yellow soft wax; cationic surfactant used as an antistatic agent for glass fibers. [KWR Chemicals Ltd] *

Catomer VA. CAS 9003-20-7; Polyvinyl acetate homopolymer emulsion; textile hand builder, stiffener, cationic exhaustible. [Sybron]

Catosal®. Organic phosphorus preparation; a stimulator of metabolism used in veterinary medicine. [Bayer AG]

cat's eye dammar. See dammar resin.

cat's-eye resin. An East Indian dammara resin obtained from *Pinus dammara* or *Dammara alba*.

cat's gold. (cat's silver). Very finely powdered mica is sometimes called by these names. The term cat's gold is also used for mosaic gold (a tin sulfide).

cat's silver. See cat's gold. Synonym for Mica.

Caucho Blanco. A rubber obtained from different species of the genus *Sapium* which belongs to the *Enphorboaceoe* family and found in the northern part of South America.

Caust X. Sodium phosphates, silicates and precipitating agents which are compounded into solid bricks; used as a descalant in the caustic removal sections of bottle washers. [Delaware Chemical Corp] *

caustic baryta. Ba(OH)$_2$; CAS 17194-00-2; Barium hydroxide.

caustic lime. (lime; calcium hydroxide).

caustic potash. See potassium hydroxide.

caustic soda. See sodium hydroxide.

Causul. A nickel-copper-chromium cast iron of marked acid and alkaline resistance; used in the manufacture of valves intended particularly for handling sulfuric acid and caustic soda.

Caved-S. A proprietary preparation containing glycyrrhizinic acid as powdered block liquorice, bismuth sub-nitrate, aluminum hydroxide gel, light magnesium carbonate, sodium bicarbonate and powdered frangula bark. [AKU Holland] ‡

Caytur 4. A proprietary partial complex of zinc chloride with benzothiazyl disulfide; used as a cross-linking agent in sulfur-curable urethane elastomers; formerly known as LD-55. [DuPont UK] *

Caytur 21 & 22. A trademark for a range of curing agents used in urethane elastomers. [DuPont UK] *

Cazin. An alloy of cadmium and zinc, containing 82.6% cadmium.

CB-4-34. Aromatic plasticizer; used in adhesives, rubber (cements, mech. and molded goods, tires), caulking compounds [Neville]

CB2100. CAS 3768-58-9; Bis (dimethylamino) dimethylsilane; coupling agent, chem. intermediate, blocking agent, release agent, lubricant, primer, reducing agent. [Hüls Am.]

CB2405. CAS 126-80-7; Bis (glycidoxypropyl) tetramethyldisiloxane; coupling agent, chem. intermediate, blocking agent, release agent, lubricant, primer, reducing agent. [Hüls Am.]

CB2408. Bis (hydroxyethyl) aminopropyltriethoxy silane, ethanol; coupling agent, release agent, lubricant, blocking agent, chemical intermediate. [Hüls Am.]

CB2409.5. CAS 16230-35-6; Bis-(N-methylbenzamide) ethoxymethyl silane; coupling agent, release agent, lubricant, blocking agent, chemical intermediate. [Hüls Am.]

CB2493. Bis (trimethoxysilylpropyl) ethylene diamine, methanol; coupling agent, release agent, lubricant, blocking agent, chemical intermediate. [Hüls Am.]

CB2494. CAS 40372-72-3; Bis [3-(triethoxysilyl) propyl] tetrasulfide; coupling agent, release agent, lubricant, blocking agent, chemical intermediate. [Hüls Am.]

CB2500. CAS 10416-59-8; Bis(trimethylsilyl)acetamide; coupling agent, chem. intermediate, blocking agent, release agent, lubricant, primer, reducing agent. [Hüls Am.]

CB2785. CAS 1000-50-6; n-Butyldimethylchlorosilane; coupling agent, chem. intermediate, blocking agent, release

---

agent, lubricant, primer, reducing agent. [Hüls Am.]

CB2790. CAS 18162-48-6; t-Butyldimethylchloro silane; coupling agent, chem. intermediate, blocking agent, release agent, lubricant, primer, reducing agent. [Hüls Am.]

CB2805. CAS 58479-61-1; t-Butyldiphenylchloro silane; CAS 58479-61-1; coupling agent, chem. intermediate, blocking agent, release agent, lubricant, primer, reducing agent. [Hüls Am.]

CBTS. CAS 95-33-0; N-Cyclohexyl-2-benzothiazole sulfenamide; delayed-action sulfenamide accelerator for natural, reclaim, and synthetic rubbers; nondiscoloring. [Akrochem]

CC-103. CAS 471-34-1, 1317-65-3; Calcium carbonate; coarse ground filler for polyolefins, carpet backing, caulks, sealants, putties, as mild abrasives in cleaners. [ECC]

CC3005. CAS 7787-85-1; 2-Chloroethylmethyldichloro silane; coupling agent, chem. intermediate, blocking agent, release agent, lubricant, primer, reducing agent. [Hüls Am.]

CC3270. CAS 1719-57-9; Chloromethyldimethylchloro silane; coupling agent, chem. intermediate, blocking agent, release agent, lubricant, primer, reducing agent. [Hüls Am.]

CC3275. CAS 1558-33-4; Chloromethylethyldichloro silane; coupling agent, chem. intermediate, blocking agent, release agent, lubricant, primer, reducing agent. [Hüls Am.]

CC3285. CAS 2344-80-1; Chloromethyltrimethyl silane; coupling agent, chem. intermediate, blocking agent, release agent, lubricant, primer, reducing agent. [Hüls Am.]

CC3290. CAS 18171-19-2; 3-Chloropropylmethyldimethoxy silane; coupling agent, chem. intermediate, blocking agent, release agent, lubricant, primer, reducing agent. [Hüls Am.]

CC3291. CAS 2550-06-3; Chloropropyltrichloro silane; coupling agent, chem. intermediate, blocking agent, release agent, lubricant, primer, reducing agent. [Hüls Am.]

CC3292. CAS 5089-70-3; 3-Chloropropyltriethoxysilane; coupling agent, chem. intermediate, blocking agent, release agent, lubricant, primer, reducing agent. [Hüls Am.]

CC3300. CAS 2530-87-2; 3-Chloropropyltrimethoxysilane; coupling agent, chem. intermediate, blocking agent, release agent, lubricant, primer, reducing agent. [Hüls Am.]

CC3433. CAS 17932-62-6; 2-Cyanoethyltriethoxy silane; coupling agent, chem. intermediate, blocking agent, release agent, lubricant, primer, reducing agent. [Hüls Am.]

CC3555. CAS 1071-27-8; 3-Cyanopropyltrichloro silane; coupling agent, chem. intermediate, blocking agent, release agent, lubricant, primer, reducing agent. [Hüls Am.]

CCA Type C Wood Preservative 50-60%. CAS 11125-95-4; Chromate copper arsenate in water; wood preservative. [CSI] *

CCC 700. CAS 999-81-5; Chlormequat; plant growth regulator. [Farmers Crop Chemicals Ltd]

CD480. CAS 41637-38-1; 10 mole ethoxylated bisphenol A dimethacrylate; difunctional monomer providing excellent hardness and good adhesion props. when cured; for use in electronic applications, e.g., dry film resists and solder masks. [Sartomer]

CD492. CAS 53879-54-2; Propoxylated trimethylolpropane triacrylate; monomer with excellent solvency, good flexibility, improved impct resist. to uv and electron beam curing formulations; for use in wood, metal and vinyl flooring coatings, printing inks, overprint varnishes, solder masks. [Sartomer]

CD3770. CAS 541-02-6; Decamethylcyclopentasiloxane; coupling agent, chem. intermediate, blocking agent, release agent, lubricant, primer, reducing agent. [Hüls Am.]

CD3780. CAS 141-62-8; Decamethyltetrasiloxane; CAS 141-62-8; coupling agent, chem. intermediate, blocking agent, release agent, lubricant, primer, reducing agent. [Hüls Am.]

CD4153. CAS 13170-23-5; Di-t-butoxydiacetoxysilane; coupling agent, chem. intermediate, blocking agent, release agent, lubricant, primer, reducing agent. [Hüls Am.]

CD4368. CAS 68304-37-6; 1,3-Dichlorotetraisopropyldisiloxane; coupling agent, chem. intermediate, blocking agent, release agent, lubricant, primer, reducing agent. [Hüls Am.]

CD4450. CAS 996-50-9; N,N-Diethylaminotrimethyl silane; coupling agent, chem. intermediate, blocking agent, release agent, lubricant, primer, reducing agent. [Hüls Am.]

CD5400. CAS 2083-91-2; Dimethylaminotrimethyl silane; CAS 2083-91-2; coupling agent, chem. intermediate, blocking agent, release agent, lubricant, primer, reducing agent. [Hüls Am.]

CD5430. 3,3-Dimethylbutyldimethylchloro silane; coupling agent, chem. intermediate, blocking agent, release agent, lubricant, primer, reducing agent. [Hüls Am.]

CD5470. CAS 1066-35-9; Dimethylchloro silane; coupling agent, chem. intermediate, blocking agent, release agent, lubricant, primer, reducing agent. [Hüls Am.]

CD5550. Dimethyldichloro silane; coupling agent, chem. intermediate, blocking agent, release agent, lubricant, primer, reducing agent. [Hüls Am.]

CD5600. CAS 78-62-6; Dimethyldiethoxy silane; CAS coupling agent, chem. intermediate, blocking agent, release agent, lubricant, primer, reducing agent. [Hüls Am.]

CD5605. CAS 1112-39-6; Dimethyldimethoxysilane; coupling agent, chem. intermediate, blocking agent, release agent, lubricant, primer, reducing agent. [Hüls Am.]

CD5610. CAS 71864-46-5; (2,3-Dimethylpropyl) dimethylchloro silane; coupling agent, chem. intermediate, blocking agent, release agent, lubricant, primer, reducing agent. [Hüls Am.]

CD5635. CAS 14857-34-2; Dimethylethoxysilane; coupling agent, chem. intermediate, blocking agent, release agent, lubricant, primer, reducing agent. [Hüls Am.]

CD5636. CAS 18643-08-8; Dimethyloctadecylchloro silane; coupling agent, chem. intermediate, blocking agent, release agent, lubricant, primer, reducing agent. [Hüls Am.]

CD5950. CAS 80-10-4; Diphenyldichloro silane; coupling agent, chem. intermediate, blocking agent, release agent, lubricant, primer, reducing agent. [Hüls Am.]

CD6000. CAS 2553-19-7; Diphenyldiethoxy silane; coupling agent, chem. intermediate, blocking agent, release

---

‡ Trade name and manufacturer not verified   § Trade name without identified manufacturer

agent, lubricant, primer, reducing agent. [Hüls Am.]

CD6010. CAS 6843-66-9; Diphenyldimethoxy silane; coupling agent, chem. intermediate, blocking agent, release agent, lubricant, primer, reducing agent. [Hüls Am.]

CD6150. CAS 947-42-2; Diphenylsilanediol; coupling agent, chem. intermediate, blocking agent, release agent, lubricant, primer, reducing agent. [Hüls Am.]

CD6210. CAS 2627-95-4; Divinyltetramethyldisiloxane; coupling agent, chem. intermediate, blocking agent, release agent, lubricant, primer, reducing agent. [Hüls Am.]

CD6220. CAS 4484-72-4; Dodecyltrichloro silane; coupling agent, chem. intermediate, blocking agent, release agent, lubricant, primer, reducing agent. [Hüls Am.]

CDA Dicotox Extra. CAS 1702-17-6; 2,4-D; translocated herbicide for cereals and established grassland. [Rhone-Poulenc Environmental Prods. Ltd]

CDA Mildothane. CAS 23564-05-8; Thiophanate-methyl; a systemic insecticide. [Rhone-Poulenc Environmental Prods. Ltd]

CDA Roval. CAS 36734-19-7; Iprodione; a fungicide with protectant activity for use in turf and amenity grasses. [Rhone-Poulenc Environmental Prods. Ltd]

CDA Simflow Plus. CAS 61-82-5, 122-34-9; A suspension concentrate containing 100 g aminotriazole and 300 g simazine per liter; used for total weed control in non crop areas and fruit orchards. [Rhone-Poulenc Environmental Prods. Ltd]

CDA Viper. Amitrole + atrazine + diuron; a liquid mixture of herbicides for weed control. [CDA Chemicals Ltd]

CDB 90. CAS 87-90-1; Trichloroisocyanuric acid; chlorinating agent for automatic dishwashing detergents, scouring cleaners, chlorinated detergents, sanitizers, dry bleaches, wool shrink, cooling tower, and toilet water. [Olin]

CDB Clearon. CAS 2244-21-5; Sodium dichloroisocyanurate dihydrate; chlorinating agent for automatic dishwashing detergents, scouring cleaners, chlorinated detergents, sanitizers, dry bleaches, wool shrink. [Olin]

CDS-1801. CAS 1150047-92-2; Behenyl polyethoxy ethylmethacrylate, unneutralized [Rhone-Poulenc Surf.]

CE6250. CAS 3388-04-3; EINECS 222-217-1; 2-(3,4-Epoxycyclohexyl) ethyltrimethoxysilane; coupling agent, chem. intermediate, blocking agent, release agent, lubricant, primer, reducing agent. [Hüls Am.]

CE6345. CAS 17689-77-9; Ethyltriacetoxysilane; coupling agent, chem. intermediate, blocking agent, release agent, lubricant, primer, reducing agent. [Hüls Am.]

CE6350. CAS 115-21-9; Ethyltrichloro silane; coupling agent, chem. intermediate, blocking agent, release agent, lubricant, primer, reducing agent. [Hüls Am.]

Ceanel. A proprietary preparation of phenylethyl alcohol, cetrimide and undecenoic acid; scalp antiseptic used as shampoo. [Quinoderm Ltd]

ceara wax. See carnauba wax.

Cebion®. A proprietary vitamin C preparation. [E Merck]

Cecagel. Silica gel; for gas drying; catalyst support. [Atochem UK/Ceca]

Cecaperl. Expanded perlite; used for industrial and cryogenic insulation (tanks and methane tankers). [Atochem UK/Ceca]

Cecarbon. CAS 7440-44-0; Granular activated carbon; used for purification, decolorization, deodorization, separation and recovery in liquid or gas phase, in the chemical, petrochemical, pharmaceutical and food industries (glucose factories, sugar refiners, oil refining, wine treatment); for the treatment of drinking and industrial water, etc.; catalyst supports. [Atochem UK/Ceca]

Cecasil. CAS 10101-39-0; Calcium silicate; industrial fillers. [Atochem UK/Ceca]

Ceclor. Cefaclor; antibacterial. [Eli Lilly & Co]

Ce-Cobalin. Syrup containing vitamins B_{12} and C; provides vitamin supplementation in deficiency states such as self imposed dietary restrictions as in strict vegetarianism or following therapeutic diets for gastro-intestinal ulceration. [Paines & Byrne Ltd] *

Cecolene 1. A proprietary trade name for trichlorethylene. §

Cecolene 2. A proprietary trade name for perchlorethylene.§

Cedarite. See Chemawinite.

Cedepal FA-406. Ammonium ether sulfate; foaming agent, emulsifier, detergent, wetting and foam stabilizing; for gypsum board production [Stepan; Stepan Canada]

Cedepal FS-406. Sodium deceth sulfate; surfactant for high electrolyte applications. [Stepan Canada]

Cedepal TD-403. Sodium trideceth sulfate; surfactant, wetting agent, foamer for shampoo, bath products, mild baby products [Stepan; Stepan Canada]

Cedepal®Range. Fat carrier compounds; fat components in premixes and cake ready mixes and in yeast raised baked goods. [Grünau]

Cedephos®FA600. Deceth-4 phosphate; detergent, emulsifier, wetting agent, hard surface detergent for industrial cleaners, metal cleaners, janitorial products, agric., textile wetting, emulsion polymerization, oil emulsification, lubricants, corrosion inhibitor, dedusting agent; coupling agent in highly alkaline industrial detergent systems. [Stepan; Stepan Canada]

Cederan® P 23. Phosphate single fertilizer (23% phosphate); for all crops and soil types. [BASF AG]

Cedilanid. A proprietary preparation of lanatoside C. [Sandoz] *

Cedilanid Ampoules. CAS 17598-65-1; A proprietary preparation of deslanoside. [Sandoz] *

Cedocard. A proprietary preparation of sorbide nitrate; used in the treatment of angina pectoris. [Tillots Laboratories]*

Ceduran. A proprietary preparation of nitrofurantoin with deglyceryr-rhizinated liquorice; an antibiotic. [Tillots Laboratories] *

CeeNU. CAS 13010-47-4; Lomustine; antineoplastic. [Bristol Laboratories]

Ceepree®C200. Blend of vitreous materials with very broad, almost continuous melting range; fire barrier/smoke suppressant additive for polymeric composites, esp. used in conjunction with fiber, glass reinforcement or in adhesives, coatings, and mastics. [Brunner Mond]

Cefadyl. CAS 24356-60-3; Cephapirin sodium; antibacterial. [Bristol Laboratories] *

Cefa-Lake. CAS 24356-60-3; Cephapirin sodium; antibacterial. [Bristol Laboratories] *

Cefizox. CAS 68401-82-1; Ceftizoxime sodium; antibacterial. [SmithKline Beecham] *

Cefkanat®. Mineral single feed for poultry. [BASF AG]

Cefkaphos®. CAS 7758-23-8; Calcium phosphate; feed phosphate for the mixed feed industry. [BASF AG]

Cefmax. CAS 75738-58-8; Cefmenoxime hydrochloride; antibacterial. [TAP Pharmaceuticals] *

Cefobid. CAS 62893-20-3; A proprietary preparation of cefoperazone sodium; an antibiotic. [Pfizer Interna-

---

tional]

Cefobine. CAS 62893-20-3; A proprietary preparation of cefoperazone sodium; an antibiotic. [Pfizer International] †

Cefobis. CAS 62893-20-3; A proprietary preparation of cefoperazone sodium; an antibiotic. [Pfizer International]

Cefomonil. CAS 52152-93-9; Cefsulodin sodium; antibacterial. [TAP Pharmaceuticals] *

Cegemett®Range. Mono and diglycerides, partially esterified with lactic and/or citric acid; prevents gelation and fat separation in manufacturing of sausages. [Grünau]

Cegepal®Range. Fat powds.; food additive for dessert and cake mixes, soups. [Grünau]

Cegeprot® Range. Protein concs.; food emulsifier, stabilizer, protein enrichment for meat and sausage industry. [Grünau]

Cegeskin®Range. Acetic acid esters of mono and diglycerides of edible fatty acids; coating for shelf life and reduction of weight loss for raw and cooked sausages. [Grünau]

Cegesol® Range. Medium chain triglycerides; dust preventer for spice mixtures and other powd. blends in meat/sausage industry. [Grünau]

Cegesterin® Range. Monoglyceride hydrate dispersions; antistaling effect, formation of starch complexes. [Grünau]

Ceilthane. One and two-component polyurethanes; coatings for industrial and specialty applications; also for elastomers for encapsulation, structural enhancement. [Flexible Prods.]

Cekas. A heat-resisting alloy containing 59.7% nickel, 11.2% chromium, 28% iron, and 2% manganese.

Cekol. CAS 9004-32-4; Sodium carboxy methyl cellulose; thickening agent. [Berol Nobel Ltd]

Celacol. Hydroxypropyl methyl/methyl/methylethyl cellulose; water-soluble polymers with film-forming and water-retaining properties; for plaster and cement products, emulsion paints, foods, pharmaceuticals and cosmetics. [Courtaulds Chemicals, Water Soluble Polymers]

Celafuse. A proprietary range of polyamide resins. Celafuse 100 is a terpolyamide with good melt flow properties (melting point 103-108 C); Celafuse T is a terpolyamide with a higher viscosity (melting point 115-125 C); Celafuse CP has a melting pont of 145-150 C with good hydrophobic properties; Celafuse SG is a plasticized copolyamide with a melting point of 110-120 C. [Hoechst Celanese] *

Celanese. Celanese is an English name for cellulose acetate silk; the same type of silk is known in America as Lustron. §

Celanese® Nylon 6/6. Polyamide (PA); used for coil bobbins, window winder mechanisms, lighter bodies and valve rocker covers. [Hoechst UK]

Celanese® Nylon 1000-1. CAS 32131-17-2; Nylon 6/6 resin; general purpose resin for injection molding and extrusion; used for mech. parts, gears, bearings, hardware. [Hoechst Celanese] *

Celanese® Nylon 1003-1. CAS 32131-17-2; Nylon 6/6 resin, heat-stabilized; for high-temp. resistance applications, e.g., under-the-hood automotive parts. [Hoechst Celanese] *

Celanese®Nylon 1500-1. CAS 32131-17-2; Nylon 6/6, 33% glass fiber-reinforced; injection molding resin for use in screw injection molding machines; avail. in surface-lubricated grade (Celanese Nylon 1500-2), heat-stabilized grade (Celanese Nylon 1503-1), surface-lubricated and heat-stabilized (Celanese Nylon 1503-2) with same props. [Hoechst Celanese] *

Celanese®Nylon 7420. CAS 32131-17-2; Nylon 6/6 resin, 13% glass fiber-reinforced; glass-reinforced version of Grade 7020. [Hoechst Celanese] *

Celanese®Nylon 7423. CAS 32131-17-2; Nylon 6/6 resin, heat stabilized; heat stabilized version of Grade 7420. [Hoechst Celanese] *

Celanese®Nylon N-186. CAS 32131-17-2; Nylon 6/6 resin, heat stabilized; high molecular weight extrusion grade. [Hoechst Celanese] *

Celanex®. Thermoplastic polyester (PBT) polybutylene terephthalate; used for domestic equipment housings, plugs and sockets, switches, telecommunications, keyswitches, switchboard components, automotive, distribution housings, exterior door handles and wiper components. [Hoechst UK]

Celanex®1300A. CAS 26062-94-2; Thermoplastic polyester resin (PBT); unmodified, super flow. [Hoechst Celanese] *

Celanex®1600A. CAS 26062-94-2; Thermoplastic polyester resin (PBT); unmodified, extrusion grade. [Hoechst Celanese] *

Celanex®2000K. CAS 26062-94-2; Thermoplastic polyester resin (PBT); unfilled, key cap grade. [Hoechst Celanese] *

Celanex®3310. CAS 26062-94-2; Thermoplastic polyester resin (PBT), 30% glass-reinforced; for aggressive end-use conditions. [Hoechst Celanese] *

Celanex®5300. CAS 26062-94-2; Thermoplastic polyester resin (PBT), 30% fiberglass-filled; improved surface finish. [Hoechst Celanese] *

Celanex®J600. CAS 26062-94-2; Thermoplastic polyester resin (PBT), 40% glass/mineral-filled; low warpage, improved impact. [Hoechst Celanese] *

Celastic. A proprietary trade name for a pyroxylin product.§

Celasyl. Fast dyestuffs for artificial fibers. [J C Bottomley] *

Celatene Colors. Proprietary colors which are aminoanthraquinone derivatives. §

Celatom. CAS 7631-86-9; Diatomaceous earth; filter powder. [Flexibulk Ltd]

Celbenin. CAS 132-92-3; A proprietary preparation containing methicillin sodium; an antibiotic. [SmithKline Beecham] *

Celcon®EC90+. Acetal copolymer; semiconductive grade for applications requiring rapid dissipation of static build-up. [Hoechst Celanese]

Celcon® GB25. Acetal copolymer; 25% glass bead-filled; for low shrinkage and warp resist. in large, flat, and thin-walled parts. [Hoechst Celanese]

Celcon®LW90. Acetal copolymer; low-wear grade for high-speed, low-load service against metals; used in bearings, slide plates, bushings, wear surfaces, and conveyor links or plates. [Hoechst Celanese]

Celcon® M25. Cryst. acetal copolymer; thermoplastic engineering resin used for extrusion and selected injection molding applications in easy-to-fill molds; end-uses incl. wire coatings, rod, tube, sheet, and slab, and injection molded items requiring extra toughness and elongation, e.g., chain links, plumbing fittings, and ski bindings. [Hoechst Celanese]

---

‡ Trade name and manufacturer not verified   § Trade name without identified manufacturer

Celcon® M270. Cryst. acetal copolymer; high-flow grade thermoplastic engineering resin designed for superior moldability in multi-cavity, intricate, or hard-to-fill mold applications, e.g., combs, marking pen bodies, and housings. [Hoechst Celanese]

Celcon® U10. Acetal copolymer; high melt strength for extrusion and blow molding applications; used for aerosol containers, industrial tanks and floats, rod, tube, slab, and profiles. [Hoechst Celanese]

Celcon® UV25Z. Acetal copolymer, uv-stabilized; Celcon® M25-based material stabilized for use where uv radiation degradation is a problem; used for automotive interiors and recreational items exposed to sunlight, e.g., knobs, buttons, toys, cams, and levers. [Hoechst Celanese]

Celeron. A proprietary trade name for a synthetic resin. §

Celescot. A proprietary trade name for a pyroxylin product.§

Celestol. A proprietary trade name for an alkyd synthetic resin. §

Celestols. A proprietary trade name for polybasic acid-polyhydric alcohol fatty acid type synthetic resins. §

Celestone. CAS 378-44-9; Betamethasone; glucocorticoid. [Schering Corp]

Celestron. A condensation product of phenol and formaldehyde with fillers; used as an insulator.

Celevac. A proprietary preparation containing oxyphenisatin diacetate; a laxative. [Parke-Davis] *

Celite®. A registered trademark for a material for separating impurities and other matters from fluids, and used as an aid in filtering, dehydrating, and demulsifying of liquids; also used as a filler in the plastics industry; the term is also applied to a constituent of Portland cement and clinker; it consists of a solution of dicalcium silicate in dicalcium aluminate. §

Celite® 110. CAS 7631-86-9; Diatomaceous silica, flux calcined; extender pigment and flatting agent for solv.- and water-thinned paints; increases toughness and durability; add "tooth" for adhesion of subsequent coats, improves sanding properties, control vapor permeability, reduces blistering and peeling; aids faster drying with more rapid solv. release, and in traffic paints, improves night reflectivity. [Celite]

Celite® 209. CAS 7631-86-9; Diatomaceous silica; functional filler used for agric. chemicals; grinding aid, and conditioner; low concentrate toxicants (up to 50%) where greater inertness is required; carrier for liq. seed inoculants; improves flow properties in fertilizers. [Celite]

Celite® 270. CAS 7631-86-9; Diatomaceous silica, calcined; functional filler for rubber industry; processing aid in rubber compounds; semireinforcing filler in mechanical rubber goods. [Celite]

Celite® HSC. CAS 7631-86-9; Diatomaceous silica, flux calcined; functional filler used in polishes and cleaners; suitable for fast-cutting buffing compounds [Celite]

Celite® R-625. CAS 7631-86-9; Diatomite; filler for use as inert, porous support/carrier for catalysts in industrial processes. [Celite]

Celite® Snow Floss. CAS 7631-86-9; Diatomaceous silica; functional filler used in polishes and cleaners; suitable for fast-cutting buffing compounds [Celite]

Celkate T-21. CAS 1343-88-0; Synthetic hydrous magnesium silicate; functional filler. [Celite]

Cellacephate. A partial mixed acetate and hydrogen phthalate ester of cellulose; used as an enteric coating.

Cellamine PAD. Semidurable gas-fading inhibitor for printed material [Eastern Color & Chem.]

Cellanite. A proprietary trade name for a synthetic resin paper product. §

Cellastine. A proprietary trade name for a cellulose-acetate product. §

Cellasto®. Cellular PU elastomers; for casting. [BASF AG]

Cellesta. A proprietary trade name for a pyroxylin product.§

Cellestren. Dyestuffs for cellulosic/polyester blends. [BASF plc] *

Cellidor. Cellulose acetate, acetate butyrate and propionate compounds; for the manufacture by injection molding and/or extrusion of spectacle frames, tool handles, seating furniture etc. and the production of fluidized bed coating powders. [Bayer AG] †

Cellit. Cellulose acetate butyrate and propionate; used for photographic films, electrical insulating films as well as block casting. [Bayer AG] †

Cellitazol® STN. Disperse dyes for dyeing and printing acetate and triacetate and synthetic fibers. [BASF AG]

Celliton® Dyes. Disperse dyes for dyeing and printing acetate, triacetate and synthetic fibers. [BASF AG; BASF plc]

Cellmore. Polyvinylidene chloride expandable beads. [Asahi Chem. Industry]

Cellobond. Phenolic products. [BP Chemicals Ltd]

Cellocaps. A proprietary viscose product. §

celloidin. (photoxylin) CAS 9004-70-0; . A substance which is obtained from collodion by precipitating it from its solution in alcohol and ether. It consists of pure nitrocellulose; used in microscopy and in surgery.

Cellokyd®. Alkyd resins for use in the manufacture of protective and decorative coatings. [Reichhold]

Cellokyd® 2708. High solids alkyd. [Reichhold]

Cellolyn. A range of dibasic-acid-modified rosin esters; tackifier for adhesives; used for type C gravure inks, nitrocellulose coatings.

Cellomold®. A cellulose acetate molding material and other plastics. §

Cellophane. Regenerated celulose film, plain and colored; P: uncoated films; M: nitrocellulose-coated films; X: copolymer-coated films; film for use on high speed automatic packaging machines and for incorporation into laminates. [UCB nv Film Sector] *

Cellosize. CAS 9004-62-0; Hydroxyethyl cellulose. thickening agent and adhesive. [Union Carbide]

Cellosize® HEC QP Grades, HEC WP Grades, Polymer PCG-10. CAS 9004-62-0; Hydroxyethylcellulose; water-sol. polymer. [Amerchol Corp]

Cellosolve. Ethanediol ethers and ether esters. [Union Carbide; Union Carbide (UK) Ltd]

Cellovar 8076-M. Varnish, modified oil. [Reichhold]

Celluclast. A cellulase preparation made by submerged fermentation of a selected strain of the fungus *Trichoderma reesei*; can be used in any case where the aim is break-down of cellulosic matter for production of fermentable sugar, reduction of viscosity or increase in yield of valuable products of plant origin. [Novo Nordisk]

Cellucraft. A proprietary trade name for nitro-cellulose spray coating. §

Celluflex M179. A proprietary trade name for alkyl-aryl plasticizer. §

Celluflow C-25. CAS 9004-34-6; Cellulose; cosmetic ingred. with excellent oil absorbancy, moisture retention, high lubricity for powds., emulsions and anhydrous sys-

---

* Trade name not verified as available   † Trade name verified as obsolete

tems. [Presperse]

Celluflow TA-25. CAS 9004-35-7; Cellulose triacetate; cosmetic ingred. with excellent oil absorbancy, moisture retention, high lubricity for powds., emulsions and anhyd. systems. [Presperse]

Cellufluor. Chemical equivalent for calcofluor white; fungal stain for cytopathology. [Polysciences Inc]

Cellufresh™. CAS 9004-32-4; Carboxymethylcellulose sodium; ocular lubricant. [Allergan, Inc]

cellulase. Enzyme complex; derived from *Aspergillus niger*; CAS 9012-54-8; EINECS 232-734-4; digestive aid in medicine and brewing industry; aids bacteria in the hydrolysis of cellulose. [Schweizerhall]

Cellulase 4000. CAS 9012-54-8; Fungal cellulase; enzyme for pharmaceuticals (aids digestion of cellulosics), animal feeds, brewing, fruit juices, essential oils and spices, paper and other waste treatment. [Solvay Enzymes]

Cellulase AC. CAS 9012-54-8; Fungal cellulase; enzyme for extraction of grains, vegetables, fruits, and other plant tissues. [Solvay Enzymes]

Cellulate. A proprietary trade name for a cellulose acetate plastic. §

cellulith. A material made by grinding cellulose in water until a jelly is produced and boiling until hard; used as a binding agent for carborundum wheels, also as a packing material.

Celluloid. Composed of a soluble nitrocellulose mixed with camphor, obtained by gelatinising nitrocellulose by means of a solution of camphor in ethyl alcohol; can be molded, and was formerly used extensively for making toys, combs, and other articles. §

celluloid-caoutchouc. A material prepared by dissolving rubber and celluloid in cyclohexanol and mixing the solutions. An elastic substance is produced.

cellulose. (wood pulp, bleached; cotton fiber) Natural polysaccharide derived from plant fibers; CAS 9004-34-6; EINECS 232-674-9; fibers: basic material for textile, paper; for manufacture of nitrocellulose, cellulose acetate, etc.; in chromatography; as ion exchange material; microcrystalline: as binder-disintegrants in tableting; colloidal: stabilizer, emulsifier, thickener in foods. [Degussa; Eastman; FMC Int'l.; Hercules Ltd; Edw. Mendell]

cellulose acetate. (CA; cellulose acetate ester; acetylcellulose) Cellulosic; CAS 9004-35-7; manufacture of rubber and celluloid substitutes, nonflamm. photographic and cinema films, airplane dopes, varnishes and lacquers; waterproofing and sizing for textiles; coating skins; insulating elec. wires. [Aldrich; Courtaulds Acetate; Eastman; FMC]

cellulose acetate butyrate. (CAB; cellulose acetobutyrate; cellulose, acetate butanoate) Butyric acid ester of a partially acetylated cellulose; CAS 9004-36-8; thermoplastic for automotive, tool, building, furniture industries, domestic appliances, lighting, elec., radio, and TV industries, optical, photographic, stationery, toy, toiletries, packaging applications; coatings, printing inks, lacquers. [Eastman; FMC; Whitfield Chem. Ltd]

cellulose acetate ester. See cellulose acetate.

cellulose acetate propionate. (CAP; cellulose propionate; cellulose acetate propionate ester; cellulose, acetate propanoate) Propionic acid ester of a partially acetylated cellulose; CAS 9004-39-1; thermoplastic for automotive, tool, building, furniture industries, domestic appliances,

lighting, elec., radio, and TV industries, optical, photographic, stationery, toy, toiletries, and packaging applications; printing inks, paper and cloth coatings. [Aldrich; Eastman]

cellulose acetobutyrate. See cellulose acetate butyrate.

cellulose gum (CTFA). See carboxymethylcellulose sodium.

cellulose methyl ether. See methylcellulose.

cellulose, nitrate. See nitrocellulose.

cellulose pitch. The residue obtained from the evaporation of the waste sulfite lye, from the treatment of wood in the sulfite process; used for making briquettes.

cellulose propionate. See cellulose acetate propionate.

cellulose triacetate. (triacetyl cellulose) Also see cellulose acetate. CAS 9012-09-3; protective coatings resistant to most solvents; textile fibers; base for magnetic tape. [Courtaulds Acetate; Eastman; FMC]

Cellulosine. A proprietary trade name for a bleached celluloid. §

Cellulysin®. Cellulase. [Calbiochem Corp]

Celluplastic. A proprietary trade name for a plasticized cellulose used for containers. §

Cellushi. A proprietary viscose product used for packing. §

Celluvarno. A proprietary trade name for cellulose nitrate surfacing material. §

Celluvisc®. CAS 9004-32-4; Carboxymethylcellulose sodium; ocular lubricant. [Allergan, Inc]

Celluzyme®. A cellulytic enzyme preparation produced by submerged fermentation of the fungus *Humicola insolens*; used as color brightening, softening, and removal of particulate soil. [Novo Nordisk]

Celluzyme®2400 T. CAS 9012-54-8; Cellulase; enzyme for laundry powd. detergents and laundry additives. [Novo Nordisk] *

Celmar®. A trademark for polypropylene/glass fiber reinforced structures. [Hoechst Celanese] *

Celmontite. A coal mine explosive containing 65.5-68.5% ammonium oxalate, 10.5-12.5% trinitrotoluene, and 19.5-21.5% sodium chloride.

Celoden. Plastics materials in sheet form. [Courtaulds plc]

Celogen®. A range of nonstaining and nondiscoloring, nontoxic and odorless nitrogen blowing agents for sponge rubber and expanded plastics. [Uniroyal] *

Celogen® AZ 120, 130, 150, 180, 199. CAS 123-77-3; Azodicarbonamide; chemical blowing agent for thermoset and thermoplastic polymers; for injection-molding structural foam, extrusion of profiles, sheet, pipe, and wire coatings, and vinyl plastisol, coating, and calendering. [Uniroyal]

Celogen®OT. CAS 80-51-3; p,p-Oxybis benzene sulfonyl hydrazide; blowing agent for sponge rubber and expanded plastics (LDPE wire/cable, structural foam injection moldings, and rotational casting, flexible PVC structural foam injection molding). [Uniroyal]

Celogen®RA. p-Toluene sulfonyl semicarbazide; chemical blowing agent for polymers processing at 216-260 C range; for injection molding structural foam (HIPS, ABS, PP), extrusion (rigid PVC), plasticized vinyl. [Uniroyal] *

Celogen® TSH. CAS 877-66-7; p-Toluene sulfonyl hydrazide; chemical blowing agent for thermoset polyester. [Uniroyal]

Celogen® XP-100. Sulfonyl hydrazide; chemical blowing agent. [Uniroyal]

Celontin. CAS 77-41-8; Methsuximide; anticonvulsant.

---

‡ Trade name and manufacturer not verified   § Trade name without identified manufacturer

[Parke-Davis]

Celontin Kapseals. CAS 77-41-8; A proprietary preparation of methsuximide; an anticonvulsant. [Parke-Davis]

Celquat® H-100. Polyquaternium-4; substantive polymer providing gloss and antistatic properties to setting lotions, cream rinses, mousses, shampoos, conditioning soaps, skin lotions and creams. [Nat'l. Starch]

Celquat®SC-240. Polyquaternium-10; substantive polymer providing gloss and antistat props. to setting lotions, cream rinses, mousses, shampoos, conditioning soaps, skin lotions and creams; more compatible with anionics. [Nat'l. Starch]

Celsit. An alloy similar in composition to Stellite.

Cel-Soft #2. Softener/lubricant for towel, tissue and specialty creped papers; excellent for dryer and press release. [CNC Int'l. L.P.]

Celstran®ACG40-01-4. Acetal copolymer, 40% glass long fiber-reinforced; injection molding. [Polymer Composites]

Celstran® N6G30-01-4. CAS 25038-54-4; Nylon 6, 30% glass long fiber-reinforced; injection molding. [Polymer Composites]

Celstran®N66G30-01-4. CAS 32131-17-2; Nylon 6/6, 30% glass long fiber-reinforced; injection molding. [Polymer Composites]

Celstran® PBTG30-01-4. CAS 26062-94-2; PBT, 30% glass long fiber-reinforced; injection molding. [Polymer Composites]

Celstran® PCG30-01-4. PC, 30% glass long fiber-reinforced; injection molding. [Polymer Composites]

Celstran® PETG30-01-4. CAS 25038-59-9; PET, 30% glass long fiber-reinforced; injection molding. [Polymer Composites]

Celstran® PPG30-01-4. CAS 9003-07-0; PP, 30% glass long fiber-reinforced; injection molding. [Polymer Composites]

Celstran®PPSG30-01-4. CAS 9016-75-5; PPS, 30% glass long fiber-reinforced; injection molding. [Polymer Composites]

Celstran® PUG30-01-4. PU, 30% glass long fiber-reinforced; injection molding. [Polymer Composites]

Celstran®SMAG30-01-4. CAS 9011-13-6; SMA, 30% glass long fiber-reinforced; injection molding. [Polymer Composites] *

Celta. See Luftseide.

Celtex. Extruded cellular ceramic filters for cast irons. [Foseco (F.S.) Ltd] †

Celtid. A proprietary trade name for a pyroxylin product. §

Celtite. An explosive. It consists of 56-59% nitroglycerin, 2-3.5% nitro-cotton, 17-21% potassium nitrate, 8-9% wood meal, and 11-13% ammonium oxalate.

Celulon. Series of cellular polyurethane coating materials suitable for application by brush or spray. [Unitex Ltd]

cement, adamantine. A mixture of powdered pumice and silver amalgam; used in dentistry. *

cement, American. A rubber cement made from 10 parts rubber, 6 parts chloroform, and 2 parts mastic.

cementation steel. Obtained by heating bars of good malleable iron, packed with nitrogenous matter, or wood charcoal.

cement copper. (cementation copper, copper precipitate). Copper produced from copper liquors and mine liquors, by means of iron (pig iron, scrap iron, or spongy iron). It is usually contaminated with arsenic, antimony, and iron.

cementite. Triferrous carbide, Fe_3C, the hardest component of steel.

cementite, independent. Cementite in rectilinear lamellae.

cement mortar. A mixture of natural slag or Portland cement, sand, and water. Lime is also added.

Cement Prodor. A range of acid resisting cements. [Prodorite Ltd]

Cemset. Accelerators for use in cement bonded sand molds and cores. [Foseco (F.S.) Ltd] †

Cendevax. A proprietary rubella vaccine. [SmithKline Beecham] *

Ceneg. A proprietary trade name for dinonyl phthalate; a vinyl plasticizer [Koninklijke Nederlandsche Gist-En Spiritusfabriek] ‡

Cenegen® 7. Alkylaryl sulfonate; retarding and leveling agent for dyeing nylon with acid and neutral premetallized dyes; good barre coverage. [Crompton & Knowles]

Cenegen®CJB. Alkyl-ethoxy condensate; leveling, migrating, and retarding agent for dyeing nylon with acid dyes. [Crompton & Knowles]

Cenekol®1141. Sulfonated phenolic condensate; fixing and reserving agent for nylon in cellulosic blends and acid dye fixatives; for automotive nylon/rayon blends, nylon/wool blends. [Crompton & Knowles]

Cenekol® FT Supra. Aromatic condensate; surfactant for the textile industry; acid dye fixative on nylon; reserving agent in dyebaths. [Crompton & Knowles]

Centa™. β-Lactamase substrate. [Calbiochem Corp]

Centari®. Acrylic enamels; used for automotive industry. [DuPont UK]

Centex. CAS 7775-09-9; Soluble concentrate containing 6.4% w/w sodium chlorate per liter; used for total weed control for paths, drives and noncrop areas. [Chemsearch (UK) Ltd]

Centrac. Profadol hydrochloride; analgesic. [Parke-Davis]*

Centralite. Dimethyldiphenylurea; used in explosive powders.

Centrax. CAS 2955-38-6; Prazepam; sedative. [Parke-Davis]

Centrex® 811. Weatherable polymer; general-purpose injection molding grade providing good balance of physical props., gloss, and processing, and excellent color stability. [Monsanto Plastics]

Centrex® 833. Weatherable polymer; extrusion grade; suited for coextrusion over ABS; offers gloss, toughness, property retention. [Monsanto Plastics]

CentriCell. Centrifugal ultrafilter for laboratory use. [Polysciences Inc]

Centrifugal Syrup. A selective weedkiller. [A H Marks & Co Ltd] *

Centrocap®162SS, 162US. CAS 8002-43-5; EINECS 232-307-2; Special grade lecithin; designed for encapsulation where clarity and brilliance are required. [Central Soya]

Centrol® 2FSB, 2FUB, 3FSB, 3FUB. CAS 8002-43-5; EINECS 232-307-2; Standard. grade lecithin; emulsifiers, dispersants. [Central Soya]

Centrol®CA. CAS 8002-43-5; EINECS 232-307-2; Special grade lecithin; o/w emulsifier. [Central Soya]

Centrolene® A, S. CAS 8029-76-3; EINECS 232-440-6; Hydroxylated lecithin; o/w emulsifiers, increased hydrophilic props. [Central Soya]

---

* Trade name not verified as available   † Trade name verified as obsolete

Centrolex®C. CAS 8002-43-5; EINECS 232-307-2; Special grade lecithin; emulsifier, stabilizer, suspending agent for foods. [Central Soya]

Centromix®CPS. CAS 8002-43-5; Lecithin, polysorbate 80; o/w emulsifier. [Central Soya]

Centrophase® C. CAS 8002-43-5; EINECS 232-307-2; Special grade lecithin; wetting agents. [Central Soya]

Centrophase® HR. CAS 8002-43-5; EINECS 232-307-2; Lecithin; heat resistant; multifunctional ingredient; food substance; lubricant and release agent for heated surfaces. [Central Soya]

Centrophase® HR2B, HR2U. CAS 8002-43-5; EINECS 232-307-2; Special grade lecithin; wetting agents. [Central Soya]

Centrophil® K. CAS 8002-43-5; EINECS 232-307-2; Special grade lecithin. [Central Soya]

Centyl. A proprietary preparation of bendrofluazide; a diuretic. [Leo Laboratories] *

Centyl K. A proprietary preparation of bendrofluazide and potassium chloride in a slow release core; a diuretic. [Leo Laboratories] *

Cenwax®G. Hydrogenated castor oil; lubricant, wax modifier; used in coatings. [Union Camp] *

Cenwax®ME. CAS 141-23-1; EINECS 205-471-8; Methyl hydroxystearate; lubricant. [Union Camp] *

Cephreine. Pure citronellyl acetate. [Bush Boake Allen Ltd]†

Cephrol. Pure dextro citronellol. [Bush Boake Allen Ltd]

Cephthalothin. 7-(2-Thienylacetamido)cephalosporanic acid.

Cephulac. CAS 4618-18-2; Lactulose; for treatment of obstipation and (pre-) hepatic coma. [Duphar BV] *

Ceplac. Erythrocine; an aid to the teaching of oral hygiene. [Berk Pharmaceuticals Ltd] *

Ceporex. CAS 15686-71-2; A proprietary preparation containing cephalexin; an antibiotic. [Glaxo Laboratories]

Ceporin. CAS 50-59-9; A proprietary preparation containing cephaloridine; an antibiotic. [Glaxo Laboratories] *

C.E. powders. Explosive powders, containing tetryl- or tetranitromethylaniline.

Cepton. A range of medicated skin care products. [ICI Chem. & Polymers Ltd] †

cera alba. *See* beeswax.

Cerabond 18. Alumina-based ceramic adhesive; high temp. resist. adhesive used to bond selected inorg. substrates incl. metals. [Master Bond]

Cerabrit. Synthetic waxes. [Abril Industrial Waxes]

Cerachem®. Alumina-silica-zirconia composite; refractory bulk fibers for solving heat problems in furnaces (metal, petrochem., kilns in ceramic industry, boilers in utility industry). [Thermal Ceramics]

Cerachrome®. Alumina-silica-chromia composite; refractory bulk fibers for solving heat problems in furnaces (metal, petrochem., kilns in ceramic industry, boilers in utility industry). [Thermal Ceramics]

Ceracolor. On glaze colors in tubes ready for use for painting on porcelain, bone china and earthenware. [Degussa Ltd]

Cerafiber®. Blend of alumina and silica; refractory bulk fibers for solving heat problems in furnaces (metal, petrochem., kilns in ceramic industry, boilers in utility industry). [Thermal Ceramics]

Ceralan®. CAS 8027-33-6; EINECS 232-430-1; Lanolin alcohols; emollient; w/o emulsifier. [Amerchol Corp]

Ceralumin C. A proprietary alloy of aluminum with 2.5% copper, 1.5% nickel, 1.2% iron, 1.2% silicon, 0.8% magnesium, and 0.15% cerium. §

Ceramabond. High temperature ceramic adhesive. [Meclec Ltd]

Ceramcel. Hollow ceramic spheres; for chem. catalyst supports, polymer and metal matrix composites, automotive energy absorbing structures, plastic fillers, lightweight concrete fillers, thermal insulation, filter bed media, sound attenuation, industrial explosives, encapsulation. [Microcel Tech.]

Ceramer®67. Hard isopropyl maleate adduct of polyolefin wax; used in the manufacturing of emulsion coatings and floor polishes. [Petrolite]

Ceramitalc. CAS 14807-96-6; Finely ground industrial talc; used in ceramic applications where higher fired strength is required; also used as an auxilliary flux. [R T Vanderbilt Co Inc]

Ceramite. A solution of flurosilicates; used as a disinfectant, as a preservative for wood, and for hardening cements.§

Ceramol. Refractory mold and core coatings. [Foseco (F.S.) Ltd]

Ceramtex. Range of colored thermoplastic masterbatches achieving ceramic speckle or fleck effect when molded. [Collinda Ltd]

Ceranine PN Base. Emulsifier/conditioner for skin and hair care products conferring softening, lubricating and antistatic props. [Sandoz]

Ceraphyl® 28. CAS 35274-05-6; Cetyl lactate; emollient binder for pressed powds., lipsticks, hair products [Van Dyk]

Ceraphyl® 31. CAS 6283-92-7; Lauryl lactate; emollient binder for pressed powds., lipsticks, hair products; also antitack agent in antiperspirants. [Van Dyk]

Ceraphyl®41. C12-15 alcohols lactate; emollient; for sheen on hair; antitack in antiperspirants. [Van Dyk]

Ceraphyl®45. CAS 56235-92-8; Dioctyl maleate; binder, emollient for cosmetic product applications. [Van Dyk]

Ceraphyl® 50. CAS 1323-03-1; Myristyl lactate; lubricant, emollient for skin products, alcoholic preps., shaving lotions, colognes, makeup, medicated products [Van Dyk]

Ceraphyl®55. CAS 106436-39-9; Tridecyl neopentanoate; emollient for creams and lotions, binder for pressed powds. [Van Dyk]

Ceraphyl® 60. CAS 51812-80-7; Quaternium-22; conditioner, emollient, moisturizer, humectant, antistat; highly substantive to skin and hair. [Van Dyk]

Ceraphyl®65. CAS 68953-64-0; Quaternium-26; emollient; hair conditioner; mild foaming auxiliary emulsifier with antistatic, antitangle properties for shampoos, rinses and other hair products [Van Dyk]

Ceraphyl®70. CAS 68921-83-5; Quaternium-70, propylene glycol; antitangle, antistatic ingred. used in all types of hair conditioners; auxiliary emulsifier and emollient for skin creams and lotions. [Van Dyk]

Ceraphyl®85. CAS 87616-36-2; Stearamidopropyl cetearyl dimonium tosylate, propylene glycol; emollient; hair conditioner; mild foaming auxiliary emulsifier with antistatic, antitangle properties for shampoos, rinses and other hair products [Van Dyk]

Ceraphyl®140. CAS 3687-46-5; EINECS 222-981-6; Decyl oleate; emollient; binder for pressed powds.; pigment

---

‡ Trade name and manufacturer not verified   § Trade name without identified manufacturer

dispersant; co-solv. [Van Dyk]

Ceraphyl® 140-A. CAS 59231-34-4; EINECS 261-673-6; Isodecyl oleate; emollient binder for pressed powd.; make-up solubilizer; wetting agent for iron oxides; cleansing agent for emulsions. [Van Dyk]

Ceraphyl® 230. CAS 6938-94-9; EINECS 248-299-9; Diisopropyl adipate; emollient; coupler; increased spread of bath oils; reduces oiliness of mineral oil. [Van Dyk]

Ceraphyl® 368. CAS 29806-73-3; EINECS 249-862-1; Octyl palmitate; emollient binder for pressed powds., blushers; gloss agent in lipsticks; antitack for antiperspirants. [Van Dyk]

Ceraphyl® 375. CAS 58958-60-4; EINECS 261-521-9; Isostearyl neopentanoate; emollient binder for pressed powds., blushers; gloss agent in lipsticks; antitack for antiperspirants. [Van Dyk]

Ceraphyl® 424. CAS 3234-85-3; EINECS 221-787-9; Myristyl myristate; emollient; increases visc. of creams and lotions at low concs. [Van Dyk]

Ceraphyl® 494. CAS 25339-09-7; EINECS 246-868-6; Isocetyl stearate; emollient for skin and hair products [Van Dyk]

Ceraphyl®791. CAS 97338-28-8; Isocetyl stearoyl stearate; pigment dispersant, emollient, lubricant, spreading agent for lipsticks, etc. [Van Dyk]

Ceraphyl® 847. CAS 9005-275-8; Octyldodecyl stearoyl stearate; emollient binder for pigmented sticks and emulsions with dispersing and mold release properties. [Van Dyk]

Ceraphyl® GA. CAS 68648-66-8; Maleated soybean oil; emollient for creams and lotions; hair and skin conditioner. [Van Dyk]

Ceraphyl® ICA. CAS 36311-34-9; EINECS 252-964-9; Isocetyl alcohol; emollient for creams and lotions, binder for pressed powds., hair and skin conditioner. [Van Dyk]

Ceraphyl®IPL. CAS 22882-95-7; Isopropyl linoleate; superfatting agent for skin, hair, and cleansing products; skin conditioner leaving luxurious afterfeel on skin; imparts luster and softness in hair care products; reduces dry afterfeel of liq. soap formulas. [Van Dyk]

Cerasin(e)®. A trademark currently awaiting reallocation by its proprietors. [Cassella AG] *

Cerasynt® 303. CAS 3179-81-5; Diethylaminoethyl stearate; visc. builder in hair dyes; pharmaceutical emulsifier; dispersant, wetting agent. [Van Dyk]

Cerasynt®840. CAS 9004-99-3; PEG-20 stearate; emulsifier, visc. builder, stabilizer for creams, lotions, ointments; superfatting agent in shampoos; vehicle for stick products melting at body temp. [Van Dyk]

Cerasynt® D. CAS 14351-40-7; EINECS 238-310-5; Stearamide MEA-stearate; opacifier, thickener for liq. cream shampoos; auxiliary emulsifier in hydrocarbon aerosol systems such as shave creams. [Van Dyk]

Cerasynt®GMS. Glyceryl stearate; sec. o/w emulsifier for creams and lotions; visc. builder for emulsions. [Van Dyk]

Cerasynt®IP. Glycol stearate and other ingreds.; emulsifier, opacifier and pearling agent for lotion shampoos and liq. soaps. [Van Dyk]

Cerasynt®M. Glycol stearate; opacifier, thickener, pearlescent for liq. and cream shampoos; sec. emulsifier for cosmetics and pharmaceuticals. [Van Dyk]

Cerasynt® PA. CAS 1323-39-3; EINECS 215-354-3; Pro-pylene glycol stearate; opacifier for liq. and cream shampoos; sec. emulsifier for cosmetics. [Van Dyk]

Cerasynt®Q. Glyceryl stearate SE; emulsifier for soap o/w emulsions, creams and lotions. [Van Dyk]

Cerasynt® WM. Glyceryl stearate, stearyl alcohol, and sodium lauryl sulfate; emulsifier for o/w creams, lotions, ointments, antiperspirants; electrolyte tolerance and low pH stability. [Van Dyk]

Ceratex. A trademark for wax and rubber containing impregnating compounds for textiles and papers. §

Ceratotect. CAS 148-79-8; Soluble concentrate containing 266 g/l thiabendazole; a systemic insecticide. [MSD Agvet]

Cerazole. CAS 72-14-0; Sulfathiazole; antibacterial. [SmithKline Beecham] *

Cercobin. CAS 23564-05-8; Suspension concentrate containing 500 g/l thiophanate-methyl; a systemic insecticide. [Rhone-Poulenc Crop Protection Ltd]

Cerebrose. Galactose, $C_6H_{12}O_2$.

Cereclor. Series of secondary plasticizers manufactured from chlorinated waxes; the percentage of chlorine is indicated by the number after the name, e.g., Cereclor 70. [ICI Chem. & Polymers Ltd]

Cereflo. A purified bacterial β-glucanase preparation produced by submerged fermentation of a selected strain of *Bacillus subtilis*; used as a supplementary glucanase preparation when masking malt or mixtures of malt and barley. [Novo Nordisk]

Cerelose. CAS 50-99-7; A commercial glucose.

Ceremix. Contains the following enzyme actives: α-amylase, β-glucanase, and proteinase; used in the brewing process when a proportion of the malt is replaced by barley and for the production of malt extract and barley syrups. [Novo Nordisk]

Cerere. Tricresylmercuroacetate. It is a mixture of mono- and diacetate derivatives of the three cresols, with about 75% of the mono- derivative, and containing about 57% mercury. It accelerates the germination of grain and affords protection against animal and vegetable parasites.

Ceres®. Fat-soluble dyestuffs; used for shoe and floor polishes, office supplies, waxes, oils, fats, fuels, plastics, surface coatings, and printing inks. [Bayer AG]

Ceresan. Seed dressing for control of fungal diseases on cereals, rice, cotton and vegetables. [Bayer AG]

ceresin. (white ozokerite wax; earth wax; mineral wax) CAS 8001-75-0; Waxy mixture of hydrocarbons obtained by purification of ozokerite; candles, sizing; bottles for hydrofluoric acid; shoe and leather polishes; antifouling paints; cosmetics; waterproofing textiles; substitute for beeswax; dental wax compounds [Astor Wax; Jonk BV; Stevenson Bros.; Strah & Pitsch]

Ceresit. Waterproofing compounds for cement, consisting mainly of calcium carbonate, alum, and calcium soap, sometimes with more or less free oil or fat; sold in the form of powder to be mixed with dry cement, or as a paste to be mixed with water. §

Ceresol. CAS 62-38-4; Phenylmercury acetate; organomercury fungicide seed dressing for cereals and fodder beet. [ICI Agrochemicals; ICI Chem. & Polymers Ltd]

Cerespan. Papaverine hydrochloride; relaxant. [USV Pharmaceutical Corp] ‡

Ceretan. Seed dressing for control of fungal diseases on cereals, rice, cotton, and vegetables. [Bayer AG]

---

* Trade name not verified as available   † Trade name verified as obsolete

Cerevase. A standardized solution of refined papain or purified papain concentrate; used in the brewing industry for the prevention of colloidal haze caused by repeated chilling and warming of beer. [Pfizer International]

Cerevax. A flowable concentrate containing 360 g carboxin and 20 g thiabendazole; fungicide seed dressing for rye and wheat. [ICI Agrochemicals; ICI Chem. & Polymers Ltd]

Cerevax Extra. A flowable concentrate containing 300 g carboxin, 20 g imazalil and 25 g thiabendazole; fungicide seed dressing for oats and barley. [ICI Agrochemicals]

Cerex. A copolymer containing carbon, hydrogen, and nitrogen, probably of the acrylonitrile type; a proprietary thermoplastic stated to be resistant to deformation at 100 C. §

Cerfluorite. A compound, $(Ca_3Ce_2)F_6$, prepared artificially.§

Cergem. CAS 64318-79-2; Gemeprost; prostaglandin. [G D Searle & Co] *

Ceridust. Micronized polyethylene wax. [Hoechst UK]

cerin. See ozokerite.

Cérite. A French synthetic resin of the phenol-formaldehyde type.

Ceritone. An aromatic flavoring chemical. [PPF International Ltd] *

cerium. A rare-earth element; Ce; CAS 7440-45-1; EINECS 231-154-9; cerium salts, cerium-iron alloys, ignition devices, military signaling, illuminant in photography, reducing scavenger, catalyst, alloys for jet engines, solid-state devices, rocket propellants, vacuum tubes, diluent in plutonium nuclear fuels. [Aldrich; Cerac; Ferro/Transelco; Rhone-Poulenc Basic]

cerium sulfate. (ceric sulfate tetrahydrate) $Ce(SO_4)_2 \cdot 4H_2O$; CAS 10294-42-5; EINECS 237-029-5; Dyeing and printing textiles, analytical reagent, waterproofing, mildewproofing. [Atomergic Chemetals; Noah Chem.; Rhone-Poulenc Basic]

Cerone. Growth regulator containing 480 g 2-chlorethylphosphonic acid per liter; used for winter barley. (Sold under license from Union Carbide) [ICI Chem. & Polymers Ltd]

Cerone. Growth regulator containing 480 g 2-chlorethylphosphonic acid per liter; used for winter barley. [Embetec Crop Protection Ltd]

cerosin. See ozokerite.

cerotin. Ceryl cerotate, $C_{26}H_{53}COOC_{27}H_{55}$; occurs in Chinese wax.

Ceroxin GL. CAS 106-14-9; EINECS 203-366-1; A proprietary trade name for 12-hydroxy stearic acid; a lubricant for plastics processing. [J H Little] ‡

Ceroxin GMO. A proprietary trade name for a partially esterified fatty acid made from naturally occurring saturated or unsaturated fatty acids and a polyhydric alcohol; a lubricant for plastics processing. [J H Little] ‡

Ceroxin GMR. A proprietary trade name for naturally occurring saturated or unsaturated fatty acids partially esterified with a polyhydric alcohol; a lubricant for PVC processing. [J H Little] ‡

Ceroxin GMSI. A proprietary trade name for a solid lubricant for PVC processing made from naturally occurring saturated or unsaturated fatty acids partially esterified with a polyhydric alcohol. [J H Little] ‡

Ceroxin TRI. A proprietary trade name for hydroxystearic acid glyceride; a lubricant for PVC which does not cause discoloration and which is particularly suitable for compounds for electrical purposes. [J H Little] ‡

Cerrobase. An alloy of lead and bismuth used as a pattern metal in foundry work.

Cerrobend. An alloy of lead, bismuth, tin, and cadmium used for tube and section bending in foundry work.

Cerromatrix. A proprietary alloy of bismuth, lead, tin, and antimony; expands on cooling. §

Certi-fired. Resistor compositions. [DuPont UK]

Certincoat. Organotin; patented coating system for application to glass bottles to reduce breakage. [Atochem]

Certistain. High quality microscopy stains. [BDH Chemicals Ltd]

Certite. Polyester resin based concrete repair and grouting materials. [SBD Construction Products Ltd]

Certolake. Water insoluble, aluminum lake food colors; used for coloring of foodstuffs and pharmaceuticals. [Morton Int'l. Ltd] †

Certrol. Selective weedkiller. [A H Marks & Co Ltd]

Cerubidin. Rubidomycin. [Rhone-Poulenc Rorer Ltd]

Cerubidine. Daunorubicin hydrochloride; antineoplastic. [Ives Laboratories Inc] *

Cerumol. A proprietary preparation of p-dichlorbenzene, chlorbutol and oil of terebinth; ear drops. [LAB Ltd] ‡

ceruse. Basic lead carbonate

Cervagem. CAS 64318-79-2; A proprietary preparation of gemeprost; a prostaglandin analog for softening and dilitation of the cervix uteri in first trimester abortion. [May & Baker Ltd] *

Cervantal. A broad-spectrum penicillin combination with an extended range of action; also known as Totocillin. [Bayer AG]

cesium. (caesium) An alkali-metal element; Cs; CAS 7440-46-2; EINECS 231-155-4; photoelectric cells, vacuum tubes, hydrogenation catalyst, ion propulsion systems, rocket propellant, heat transfer fluid, thermochemical reactions. [Aldrich; Atomergic Chemetals; Cabot; Cerac]

cesium bromide. CsBr; CAS 7787-69-1; EINECS 232-130-0; Crystals for infrared spectroscopy, scintillation counters, fluorescent screens. [Atomergic Chemetals; Cabot; Cerac; Noah Chem.]

cesium carbonate. $CsCO_3$; CAS 534-17-8; EINECS 208-591-9; Brewing, mineral waters, in specialty glasses, polymerization catalyst for ethylene oxide. [Aldrich; Atomergic Chemetals; Cabot; Cerac]

cesium chloride. CsCl; CAS 7647-17-8; EINECS 231-600-2; Brewing; preparation of cesium compounds; mineral waters; evacuation of radio tubes; in ultracentrifuge separations; fluorescent screens; contrast medium. [Accurate Chem. & Scientific; Atomergic Chemetals; Cabot; Cerac; Janssen Chimica]

cesium fluoride. CsF; CAS 13400-13-0; EINECS 236-487-3; Optics, catalysis, specialty glasses. [Atomergic Chemetals; Cabot; Cerac; Spectrum Chem. Mfg.]

cesium hydrate. See cesium hydroxide.

cesium hydroxide. (cesium hydrate) CsOH; CAS 21351-79-1; EINECS 244-344-1; Electrolyte in alkaline storage batteries (esp. at subzero temperatures), polymerization catalyst for siloxanes. [Aldrich; Atomergic Chemetals; Cabot; Cerac; Noah Chem.]

cesium iodide. CsI; CAS 7789-17-5; EINECS 232-145-2; Crystals for infrared spectroscopy, scintillation counters, fluorescent screens. [Atomergic Chemetals; Cabot; Cerac; Noah Chem.]

---

‡ Trade name and manufacturer not verified   § Trade name without identified manufacturer

cesium nitrate. $CsNO_3$; CAS 7789-18-6; EINECS 232-146-8; Cesium salts. [Cabot; Cerac; Noah Chem.]

cesium sulfate. Cs_2SO_4; CAS 10294-54-9; EINECS 233-662-6; Brewing, mineral waters, for density gradient in ultracentrifuge separation. [Aldrich; Atomergic Chemetals; Cabot; Cerac]

Cesol®. Chlormethylpyridine-β-carbonic acid methyl ester. [E Merck] *

Cestarsol. Arecoline acetarsol. [May & Baker Ltd] *

cetaceum. Spermaceti.

Cetacourt. CAS 50-23-7; Hydrocortisone; glucocorticoid. [Galderma Laboratories] *

Cetaffine®. CAS 36653-82-4; EINECS 253-149-0; Cetyl alcohol; raw material for cosmetics and pharmaceuticals. [Laserson & Sabetay]

Cetal. CAS 36653-82-4; EINECS 253-149-0; Cetyl alcohol NF; emollient used in emulsions, oils, and makeup; visc. control in emulsions; auxiliary emulsifier. [Amerchol Corp]

Cetal. An alloy of 87% aluminum and 13% silicon.

Cetamide. Sulfacetamide sodium; antibacterial. [Alcon Laboratories Inc]

Cetamoll®. Plasticizers for polyamide-based surface coating resins and for polyamides. [BASF AG]

Cetaped. Veterinary antiseptic preparation. [ICI Chem. & Polymers Ltd] †

Cetats®. Cetrimonium p-toluene sulfonate. [Zeeland]

Cetavlex. A preparation of cetrimide in a cream base; antiseptic skin cream. [ICI Chem. & Polymers Ltd] *

Cetavlon. A preparation of cetrimide; skin disinfectant and wound cleanser. [ICI Chem. & Polymers Ltd]

Cetec. A proprietary trade name for a cold molding bituminous compound (nonrefractory.) §

Ceteprin. CAS 3614-30-0; A proprietary preparation of emepronium bromide; used in the control of frequency of micturition. [Goodrich-Gulf Chemicals] *

cetin. See cetyl palmitate.

cetine. See cetyl palmitate; occurs in spermaceti.

Cetiol®. CAS 3687-45-4; EINECS 222-980-4; Oleyl oleate; emollient; oily component of strong greasy char., for cosmetic/pharmaceutical products; carrier for lipid sol. ingreds. [Henkel/Cospha; Henkel Canada]

Cetiol® 868. CAS 22047-49-0; EINECS 244-754-0; Octyl stearate; emollient; superfatting oil for o/w and w/o emulsions; for cosmetics/pharmaceuticals. [Henkel/Cospha; Henkel Canada]

Cetiol® 1414E. CAS 59686-68-9; Myreth-3-myristate; self-emulsifying emollient; cosmetic preparations such as creams, lotions, lipsticks, and blooming bath oils. [Henkel/Cospha; Henkel Canada]

Cetiol® A. CAS 34316-64-8; Hexyl laurate; vehicle for lipid-sol. topical act. ingreds. used in skin lubricants and personal care products; mild emollient. [Henkel/Cospha; Henkel Canada]

Cetiol® B. CAS 105-99-7; Dibutyl adipate; emollient; oily component for day creams, liq. emulsions. [Henkel/Cospha; Henkel Canada]

Cetiol® G16S. CAS 25339-09-7; EINECS 246-868-6; Isocetyl stearate; emollient, lubricant for cosmetic preparations; lipsticks. [Henkel/Cospha; Henkel Canada]

Cetiol® HE. CAS 68201-46-7; PEG-7 glyceryl cocoate; emollient oil, superfatting agent for aq. formulations in personal care products; dispersant for biologically act. ingreds. [Henkel/Cospha; Henkel KGaA]

Cetiol® J600. CAS 17673-56-2; Oleyl erucate; emollient; fatty component for cosmetic preparations, jojoba oil substitute. [Henkel/Cospha; Henkel Canada]

Cetiol® LC. Coco caprylate/caprate; penetrating emollient oil used in personal care products [Henkel/Cospha; Henkel Canada]

Cetiol® MM. CAS 3234-85-3; EINECS 221-787-9; Myristyl myristate; emollient; wax ester with superfatting properties; for skin care and stick preps. [Henkel/Cospha; Henkel Canada]

Cetiol® R. Trihydroxy methoxystearin; emollient; fatty oil for makeup preparations; castor oil substitute. [Henkel/Cospha; Henkel Canada]

Cetiol® S. Dioctylcyclohexane; emollient, superfatting agent; used in cosmetic and pharmaceutical creams and emulsions. [Henkel/Cospha; Henkel Canada]

Cetiol® SB45. CAS 68424-60-2; Shea butter; emollient, consistency giving agent for o/w and w/o creams and emulsions; native fatting agent for creams, lotions, anhyd. creams. [Henkel/Cospha; Henkel Canada]

Cetiol® SN. Cetearyl isononanoate; emollient for application in skin care, massage and sun protection preparations; oily component with expressed hydrophobic effect. [Henkel/Cospha; Henkel Canada]

Cetiol® V. CAS 3687-46-5; EINECS 222-981-6; Decyl oleate; penetrating emollient, carrier for lipid sol. substances used in personal care products and pharmaceutical topical applications. [Henkel/Cospha; Henkel Canada]

Cetiprin. CAS 3614-30-0; A proprietary preparation of emepronium bromide; used in the control of frequency of micturition. [KabiVitrum AB] *

Cetiprin Novum. Emepronium carragenate; tablets, solution for injection, syrup; for the treatment of micturition disorders. [KabiVitrum AB] *

Cetodan 50-00P Kosher. Acetylated monoglyceride from hydrogenated refined fats; food emulsifier for cake shortenings and fats; edible coating agent for meat products, candy, fruit and nuts. [Grindsted Prods.]

Cetol® . Cetyldimethylbenzyl ammonium chloride. [Zeeland]

Cetomacrogol 1000. CAS 9004-95-9; Polyethylene glycol 1000 monocetyl ether. Polyoxyethylene glycol 1000 monocetyl ether.

Cetomacrogol 1000 BP. CAS 68439-49-6; Ceteareth-20 BP; pharmaceutical o/w emulsifier, solubilizer; wetting agent for stick formulations; for depilatories, antiperspirants, conditioning rinses. [Croda Inc.; Croda Chem. Ltd.]

Cetosan. A mixture of the higher alcohols of spermaceti, mainly cetyl and octodecyl alcohols, with petroleum jelly.

Cetostearyl Alcohol BP, Alcohol NF. Cetearyl alcohol. [Croda Inc.]

Cetrimide BP. CAS 57-09-0; EINECS 200-311-3; Cetyl trimethyl ammonium bromide; antistat in hair conditioners; biocidal applications; phase transfer catalysis. [Aceto]

cetrimonium bromide. (cetyltrimethylammonium bromide; hexadecyltrimethylammonium bromide; N,N,N-trimethyl-1-hexadecanaminium bromide) Quaternary ammonium salt; $CH_3(CH_2)_{15}N(CH_3)_3Br$; CAS 57-09-0; EINECS 200-311-3; cationic surface active agent, detergent; antiseptic; laboratory reagent; germicide. [Aceto; Aldrich; Chemron; Sherex; Zeeland]

---

* Trade name not verified as available

† Trade name verified as obsolete

cetyl alcohol. (palmityl alcohol; C_{16} linear primary alcohol; 1-hexadecanol) Fatty alcohol; $CH_3(CH_2)_{14}CH_2OH$; CAS 36653-82-4; EINECS 253-149-0; perfumery; emulsifier, emollient, coupling agent; foam stabilizer in detergents; opacifier; thickener; chemical intermediate; cosmetics and pharmaceuticals. [Aarhus Oliefabrik A/S; Amerchol; Chemron; Croda; Ethyl; Lipo; Lonza; M. Michel; Norman, Fox; Procter & Gamble; Sherex; Stepan; Vista]

cetylic acid. See palmitic acid.

cetyl lactate. (n-hexadecyl lactate; 2-hydroxypropanoic acid hexadecyl ester; 1-hexadecanol lactate) Ester of cetyl alcohol and lactic acid; $CH_3CHOHCOO$-$(CH_2)_{15}CH_3$; CAS 35274-05-6; nonionic emollient; cosmetics and pharmaceuticals. [Am. Biorganics]

cetyl palmitate. (hexadecanoic acid, hexadecyl ester; palmitic acid, hexadecyl ester; cetin) Ester of cetyl alcohol and palmitic acid; $C_{15}H_{31}COOC_{16}H_{33}$; CAS 540-10-3; EINECS 208-736-6; base for ointments; manufacture of candles and soaps. [Croda; Sherex; Werner G. Smith; Stpean; Witco/Humko]

cetylpyridinium chloride. (1-hexadecylpyridinium chloride) Quaternary ammonium salt; $C_{21}H_{38}CIN$; CAS 123-03-5; 6004-24-6; EINECS 204-593-9; emulsifier; antibacterial, preservative in cough syrups and lozenges; topical anti-infective. [Schweizerhall; Spectrum Chem. Mfg.; Weiders Farmasoytiske A/S; Zeeland]

cetyl stearate. (n-hexadecyl stearate) Ester of cetyl alcohol and stearic acid; $C_{34}H_{66}O_2$; CAS 1190-63-2. [Koster Keunen; Sherex]

Cevalin. CAS 134-03-2; Sodium ascorbate; vitamin. [Eli Lilly & Co]

Ceylon Isinglass or Gelatin. See agar-agar. §

ceyssatite. A white earth consisting of almost pure silica; an absorbent powder.

CF 1500. CAS 65996-61-4; Cellulose fibers; reinforcing filler; contributes excellent mech. props. in V-belt formulations, etc. [Custom Fibers] *

CF1-3510. Fluorosilicone; oil/solv. resist. material for electronic/aerospace application. [McGhan NuSil]

CF 31,000C Coarse. CAS 9004-34-6; Cellulose fibers; for use as asbestos replacement in industrial applications; relatively coarse fiber with extremely high oil absorp.; for use when max. visc. increase, sag resist., and fiber reinforcement is required; suited for solv. and water systems; applications incl. asphalt plastic roof cement, epoxy adhesives, roof coatings. [Custom Fibers]

CF 42,500T Medium. CAS 9004-34-6; Cellulose fibers; asbestos replacement fibers providing increased visc. and sag resistance, dispersion, and fiber reinforcement for asphalt plastic roof cement, caulks, putties, aluminum roof coating, adhesives. [Custom Fibers]

CF 70,000WDK, Ex. Superfine. CAS 9004-34-6; Cellulosic fibers containing an anionic wetting agent; asbestos replacement fibers providing increased visc. and sag resistance, dispersion, and fiber reinforcement for textured coatings, adhesives, roof coatings, caulks and sealants, paints, and flocking material. [Custom Fibers]

C-Flakes. CAS 68334-00-9; Hydrogenated cottonseed oil. [Karlshamns]

CG-80. Dairy pipeline cleaner/sterilizer. [Ciba plc] *

CG6710. CAS 2897-60-1; (3-Glycldoxypropyl)-methyldiethoxy silane; coupling agent, chem. intermediate, blocking agent, release agent, lubricant, primer, reducing agent. [Hüls Am.]

CG6720. CAS 2530-83-8; EINECS 219-784-2; 3-Glycidoxypropyltrimethoxysilane; coupling agent, chem. intermediate, blocking agent, release agent, lubricant, primer, and reducing agent. [Hüls Am.]

CH7250. CAS 1009-93-4; 1,1,3,3,5,5-Hexamethylcyclotrisilazane; coupling agent, chem. intermediate, blocking agent, release agent, lubricant, primer, and reducing agent. [Hüls Am.]

CH7260. CAS 541-05-9; Hexamethylcyclotrisiloxane; coupling agent, chem. intermediate, blocking agent, release agent, lubricant, primer, and reducing agent. [Hüls Am.]

CH7280. CAS 1450-14-2; Hexamethyldisilane; coupling agent, chem. intermediate, blocking agent, release agent, lubricant, primer, and reducing agent. [Hüls Am.]

CH7300. Hexamethyldisilazane; coupling agent, chem. intermediate, blocking agent, release agent, lubricant, primer, and reducing agent. [Hüls Am.]

CH7310. CAS 107-46-0; Hexamethyldisiloxane; coupling agent, chem. intermediate, blocking agent, release agent, lubricant, primer, and reducing agent. [Hüls Am.]

CH7332. CAS 928-65-4; Hexyltrichloro silane; coupling agent, chem. intermediate, blocking agent, release agent, lubricant, primer, and reducing agent. [Hüls Am.]

Chafer 5C Chlormequat. Soluble concentrate containing 640 g chlormequat and 64 g choline chloride per liter; plant growth regulator for use in cereals and ornamentals. [BritAg Industries Ltd]

Chafer Certrol-E. Bromoxynil + dichlorprop + ioxynil; herbicide mixture for weed control in spring cereals. [DuPont UK]

Chafer CMPP Super. CAS 7085-19-0; Soluble concentrate containing 570 g/l mecoprop; for control of weeds in cereals and grassland. [BritAg Industries Ltd]

Chafer MCPA 675. CAS 94-74-6; MCPA; herbicide for cereals and grassland. [BritAg Industries Ltd]

chalcostibnite. $CuSbS_2$; A mineral.

Chaldegal. Foam stabilizers, solubilizers; for detergent formulations; shampoos. [Hoechst UK] *

Chalkone. CAS 94-41-7; 1,3-Diphenyl propenone, $C_6H_5 \cdot CH:CH \cdot CO \cdot C_6H_5$.

Chalk, Prepared. (calcium carbonate. prepared; drop chalk); used in medicine, tooth powders, polishing powders, silicate cement.

Challenger®. For the automotive industry. [DuPont UK]

Chamotan®. Fish oils; dressing oils for chamois leather. [Seven Seas Ltd]

Chandor. CAS 330-55-2, 1582-09-8; Emulsifiable concentrate containing 120 g linuron and 240 g trifluralin per liter; herbicide for winter cereals. [DowElanco Ltd]

channel black. See carbon black.

Channelyzer. Dedicated pulse height analyzer and sorter; used to obtain rapid, high resolution, particle size distributions from semi-automatic cell or particle analyzers. [Coulter Electronics Ltd]

Chardot 815 T. Leveling agent for dispersed dyes on polyamide, possessing softening and lubricating props. [Ceca SA]

Chargemaster®R530. Derived from corn. [Grain Processing]

Chargepac®5. Org. and inorg. cationic polymers; coagulant for water and waste water clarification. [Drew Ind. Div.]

Charguard 329. Intumescent flame retardant. [Great Lakes]

Charisma®. For the agriculture industry. [DuPont UK]

Chartoon Character®. Polymer. [DuPont UK]

---

‡ Trade name and manufacturer not verified   § Trade name without identified manufacturer

Chatelier solder. An alloy of 70% tin, 25% zinc, 2% aluminum, and 1.5% phosphorus.

Chatterton's compound. Mixtures of tar, rosin, and gutta-percha; used for cementing gutta-percha to wood and metals.

Chaubert's oil. Consists of 75% oil of turpentine, and 25% oil of hartshorn.

chaulmestrol. Consists of the ethyl esters of the fatty acids of chaulmoogra oil.

chaulphosphate. Sodium dichaulmoogryl-β-glycero-phosphate; used in leprosy.

chavicol. p-Allylphenol; occurs in many essential oils.

Chavosol. A dental antiseptic containing p-allylphenol. §

Chavosote. *See* Chavosol. §

Checkmate. CAS 74051-80-2; Emulsifiable concentrate containing 193 g/l sethoxydim; a cyclohexene-oxime herbicide for annual grasses in field crops. [Embetec Crop Protection Ltd]

Checkmate. CAS 7681-49-4; Sodium fluoride; dental caries prophylactic. [Oral-B Laboratories Inc] *

Cheelox® 80. CAS 140-01-2; DTPA, pentasodium salt; all purpose chelating agent used in pulp bleaching applications using hydrogen peroxide. [Rhone-Poulenc Surf.]

Cheelox® 100. CAS 64-02-8; EINECS 200-573-9; Tetrasodium EDTA; sequesterant for metalworking fluids. [Rhone-Poulenc Surf.]

Cheelox® 120. CAS 139-89-9; Trisodium HEDTA; chelating agent used for iron from pH 1.13; sequestrant for magnesium. [Rhone-Poulenc Surf.; Rhone-Poulenc France]

Cheelox® 354. Sodium glucoheptonate; chelates iron In caustic sol'ns. [Rhone-Poulenc]

Cheelox® CG; Substitute for sodium hexametaphosphate; treatment chemical for complexing metal ions in the dye bath without affecting the shade of metal complex dyes. [Rhone-Poulenc Surf.]

Cheelox® HE-24. CAS 139-89-9; Trisodium HEDTA. [Rhone-Poulenc Surf.]

Cheelox® NTA-Na3. CAS 5064-31-3; Nitriloacetate trisodium salt; *see* Cheelox NTA-14. [Rhone-Poulenc Surf.]

Cheetah R. CAS 66441-23-4; An emulsion containing 60 g fenoxaprop-ethyl per liter; used for grass weed control in wheat. [Hoechst UK]

Chel. Chelating agents. [Ciba plc] *

Chel DM-41. CAS 139-89-9; Trisodium HEDTA; chelating agent used in bar soaps, photographic developer baths, textiles, and min. separations. [Ciba-Geigy]

Chel DTPA. CAS 67-43-6; Pentetic acid; chelating agent used for stabilizing peroxides, biological preparations, cosmetics, textiles, scale removal, and rare earth separations. [Ciba-Geigy]

Chel DTPA-41. CAS 140-01-2; Pentasodium diethylenetriaminepentaacetic acid; chelating agent for metals. [Ciba-Geigy]

Chelatex. Water-based adhesives. [Caswell & Co Ltd]

Chelsea. A range of solvent-based adhesives, dubbins and shoe polishes. [Caswell & Co Ltd]

Chelsea Melt. Hot melt adhesives. [Caswell & Co Ltd]

Cheltenham salts. A mixture of 34 parts sodium sulfate, 23 parts magnesium sulfate, and 50 parts sodium chloride.

Chemal 2EH-2. PEG-2 2-ethylhexyl ether; detergent, wetting agent, emulsifier, dispersant, solubilizer, defoamer for textiles, metal cleaners, industrial and institutional cleaners, household cleaners, hand cleaners, specialties. [Chemax]

Chemal BP 261. Difunctional block polymers ending in primary hydroxyl groups; defoamer, emulsifier, demulsifier, dispersant, binder, stabilizer, wetting agent, chemical intermediate; for metalworking, cosmetic, paper, textiles, dishwashing detergents, rinse aids, as lubricant bases. [Chemax]

Chemal DA-4. CAS 26183-52-8; Deceth-4; detergent, wetting and penetrating agent for textile processing, clay soils, and fire fighting products; emulsifier for polyethylene emulsions; dispersant, solubilizer, defoamer; metal cleaners, industrial, institutional, household, and hand cleaners. [Chemax]

Chemal LA-4. Laureth-4; o/w emulsifier, lubricant, detergent, dispersant, solubilizer, defoamer for cosmetic, household, silicone polish, and mold release products [Chemax]

Chemal LF 14B, 25B, 40B. Alkoxylated linear alcohol; low foaming biodeg. surfactant, wetting agent, detergent, defoamer used as rinse aids in mechanical dishwashing, spray metal cleaning formulations and detergent products [Chemax]

Chemal LFL-10, -17, -19, -28, -38, -47. Alkoxylated linear alcohol; low foaming biodeg. surfactant, wetting agent, detergent, defoamer used in rinse aids and mechanical dishwashing detergents, spray metal cleaning formulations and detergent products [Chemax]

Chemal OA-4. CAS 9004-98-2; Oleth-4; dispersant, detergent; emulsifier and solubilizer for topical cosmetic applications; stabilizer and anticoagulant for natural and synthetic latexes; emulsifier for waxes used in coating citrus fruit. [Chemax]

Chemal OA-5. CAS 9004-98-2; Oleth-5; emulsifier, lubricant, and solubilizer. [Chemax]

Chemal OA-20/70CWS. CAS 9004-98-2; Oleth-20; emulsifier, lubricant, solubilizer. [Chemax]

Chemal TDA-3. CAS 24938-91-8; Trideceth-3; wetting agent, detergent, emulsifier, dispersant, foam stabilizer; solubilizer, penetrant for scouring and dye leveling in textiles, in cleaning and dishwashing compounds [Chemax]

Chemalog. Specialty chemical catalogue; chemicals sold through Chemalog to researchers for research and development applications. [Chemical Dynamics Corp]

Chematex. Synthetic latex cement for bedding and jointing tiles and bricks. [Prodorite Ltd]

Chemawinite. (Cedarite). A pale yellow Canadian amber.

Chemax AR-497. PEG-15 rosin acid; emulsifier, detergent for acid cleaners, esp. for aluminum. [Chemax]

Chemax CO-5. CAS 61791-12-6; PEG-5 castor oil; emulsifier, lubricant for textiles; pigment dispersant in latex paints, paper; essential oils solubilizer. [Chemax]

Chemax CO-16. CAS 61791-12-6; PEG-16 castor oil; emulsifier for fiber lubricants; cutting oils and hydraulic fluids; clay and pigment dispersant, rewetting agent, softener, dyeing assistant for paint, paper, textile, and leather industries. [Chemax]

Chemax DF-10, DF-10A. Silicone defoamer; defoamer/antifoam for pulp and paper, textiles, paints, effluent treatment, commercial cleaning processes, adhesives, metalworking. [Chemax]

Chemax DNP-8. CAS 9014-93-1; Nonyl nonoxynol-8; emulsifier for nonpolar solv. and oils; detergent for cellulosic and synthetic fibers; dispersant for hard surface cleaners and laundry compounds; solubilizer. [Chemax]

---

* Trade name not verified as available   † Trade name verified as obsolete

Chemax DNP-18. CAS 9014-93-1; Nonyl nonoxynol-18; emulsifier, detergent, solubilizer. [Chemax]

Chemax DOSS/70. CAS 577-11-7; EINECS 209-406-4; Sodium dioctyl sulfosuccinate; wetting agent for textile, agric., detergent formulations, emulsion polymerization; pigment dispersant in paints and inks; solubilizer for drycleaning solvs. [Chemax]

Chemax E-200 ML. CAS 9004-81-3; PEG-4 laurate; emulsifier for mineral and cutting oils; dispersant, detergent, lubricant; coemulsifier and defoamer in water-based coatings; cosmetics ingred. [Chemax]

Chemax E-200 MO. CAS 9004-96-0; PEG-5 oleate; emulsifier for mineral and fatty oils, solv.; degreaser, dispersant, detergent, lubricant; metal, textile, cosmetic, plastisol formulations. [Chemax]

Chemax E-200 MS. CAS 9004-99-3; PEG-5 stearate; emulsifier for mineral oils and fats used in polishes and metal buffing compounds; dye assistant, lubricant, softener, antistat; for metal lubricants, textiles, cosmetic, plastisol formulations. [Chemax]

Chemax E-400 ML. CAS 9004-81-3; PEG-8 laurate; emulsifier, dispersant, detergent, lubricant; visc. control agent in plastisol formulations; wetting agent and defoamer in latex paint; cosmetics ingred. [Chemax]

Chemax E-400 MO. CAS 9004-96-0; PEG-9 oleate; emulsifier and lubricant for solv. and oils in pesticides and metal cleaners; detergent, dispersant; textile, cosmetics, plastisol formulations. [Chemax]

Chemax E-400 MS. CAS 9004-99-3; PEG-9 stearate; lubricant and softener for synthetic fibers; dye assistant, antistat, emulsifier; for metal lubricants, textiles, cosmetics, plastisols. [Chemax]

Chemax E-600 ML. CAS 9004-81-3; PEG-14 laurate; emulsifier, dispersant, detergent, lubricant; metal, textile, cosmetics, plastisol formulations. [Chemax]

Chemax E-600 MO. CAS 9004-96-0; PEG-14 oleate; surfactant used as coemulsifier and lubricant in industrial formulations, cosmetics, metal lubricants, textiles, plastisols; dispersant, detergent. [Chemax]

Chemax E-600 MS. CAS 9004-99-3; PEG-14 stearate; dye assistant, lubricant, softener, antistat, emulsifier for cosmetic and textile formulations. [Chemax]

Chemax E-1000 MO. CAS 9004-96-0; PEG-20 oleate; emulsifier for mineral and fatty oils, solv.; degreaser, dispersant, detergent, lubricant. [Chemax]

Chemax E-1000 MS. CAS 9004-99-3; PEG-20 stearate; emulsifier for cosmetic and textile formulations; dye assistant, lubricant, softener, antistat. [Chemax]

Chemax HCO-5. CAS 61788-85-0; PEG-5 hydrogenated castor oil; emulsifier, lubricant, softener, dispersant; coemulsifier for synthetic esters; for plastics, metals, textiles, leather, paint, and paper indutries. [Chemax]

Chemax HCO-200/50. CAS 61788-85-0; PEG-200 hydrogenated castor oil; emulsifier and lubricant for plastics, metals, textiles, paint, paper, leather industries. [Chemax]

Chemax NP-1.5. CAS 9016-45-9; Nonoxynol-1.5; emulsifier, dispersant, detergent, wetting agent, solubilizer, coupler for textile, metalworking, household, industrial, agric., paper, paint, and other industries. [Chemax]

Chemax NP-4. CAS 9016-45-9; Nonoxynol-4; emulsifier, dispersant, detergent, wetting agent, solubilizer, coupler for textile, metalworking, household, industrial, agric., paper, paint, and other industries. [Chemax]

Chemax NP-6. CAS 9016-45-9; Nonoxynol-6; emulsifier, dispersant, detergent, wetting agent, solubilizer, coupler for textile, metalworking, household, industrial, agric., paper, paint, and other industries. [Chemax]

Chemax NP-9. CAS 9016-45-9; Nonoxynol-9; emulsifier, dispersant, detergent, wetting agent, solubilizer, coupler for textile, metalworking, household, industrial, agric., paper, paint, and other industries. [Chemax]

Chemax NP-10. CAS 9016-45-9; Nonoxynol-10; emulsifier, dispersant, detergent, wetting agent, solubilizer, coupler for textile, metalworking, household, industrial, agric., paper, paint, and other industries. [Chemax]

Chemax NP-15. CAS 9016-45-9; Nonoxynol-15; surfactant, detergent, wetting and rewetting agent, emulsifier in textile, leather, paper, paint, and metal processing. [Chemax]

Chemax NP-30. CAS 9016-45-9; Nonoxynol-30; surfactant, detergent, wetting and rewetting agent, emulsifier in textile, leather, paper, paint, and metal processing. [Chemax]

Chemax NP-40. CAS 9016-45-9; Nonoxynol-40; polymerization emulsifier for vinyl acetate and acrylic emulsions; stabilizer for synthetic latices; wetting agent in electrolyte sol'ns. [Chemax]

Chemax OP-3. CAS 9002-93-1; Octoxynol-3; emulsifier, detergent, stabilizer, dispersant, wetting agent; pesticides and floor finishes. [Chemax] *

Chemax OP-7. CAS 9002-93-1; Octoxynol-7; detergent comps.; industrial metal cleaning, acid and waterless hand cleaners, and floor finishes. [Chemax] *

Chemax OP-30/70. CAS 9002-93-1; Octoxynol-30; emulsifier, dispersant, detergent, wetting agent, solubilizer, coupler for textile, metalworking, household, industrial, agric., paper, paint, and other industries. [Chemax]

Chemax OP-40. CAS 9002-93-1; Octoxynol-40; emulsifier, detergent, stabilizer, wetting agent, dispersant. [Chemax]

Chemax OP-40/70. CAS 9002-93-1; Octoxynol-40; emulsifier for vinyl acetate and acrylate polymerization. [Chemax]

Chemax PEG 200 DO. CAS 9005-07-6; PEG-4 dioleate; surfactant; coemulsifier for oils as mold release agent. [Chemax] *

Chemax PEG 400 DO. CAS 9005-07-6; PEG-8 dioleate; emulsifier and solubilizer for solv., fats, and mineral oils; in lubricant, softener, and defoamer formulations for agric., cosmetic, household, leather, metalworking, and textile industries. [Chemax]

Chemax PEG 600 DO. CAS 9005-07-6; PEG-12 dioleate; emulsifier in lubricant, softener, and defoamer formulations for agric., cosmetic, household, leather, metalworking, and textile industries. [Chemax]

Chemax SBO. Sulfated butyl oleate; softener, emulsifier, wetting agent in textile and metal working industries. [Chemax] *

Chemax SCO. CAS 8002-33-3; EINECS 232-306-7; Sulfated castor oil; softener, emulsifier, wetting agent in textile and metal working industries. [Chemax] *

Chemax TO-8. CAS 61791-00-2; PEG-8 tallate; emulsifier, lubricant, dye assistant. [Chemax]

Chemax TO-10. CAS 61791-00-2; PEG-10 tallate; emulsifier, detergent in industrial lubricants and degreasers. [Chemax]

Chemax TO-16. CAS 61791-00-2; PEG-16 tallate; foam

---

‡ Trade name and manufacturer not verified   § Trade name without identified manufacturer

detergent and emulsifier; lubricant, dye assistant. [Chemax]

Chembetaine BW. CAS 68424-94-2; EINECS 270-329-4; Coco betaine; visc. builder, gelling agent, industrial surfactant; lime soap dispersant; mild to skin; tolerant to hard water; stable in high-electrolyte sol'ns.; cationic in acid and anionic in alkaline media. [Chemron]

Chembetaine C. Cocamidopropyl betaine; high foaming mild industrial and personal care surfactant; foam and visc. builder; lime soap dispersant; foaming agent for water and acid systems; stable over wide pH range. [Chemron]

Chembetaine CAS. CAS 68139-30-0; Cocamidopropylhydroxysultaine; anti-irritant for other surfactants; esp. for baby shampoos and baby bath products; detergent for heavy-duty industrial alkaline cleaners (steam cleaners, wax remover, hard surface cleaner); wetting agent in acid pickling of metals; lime soap dispersant; visc. builder. [Chemron]

Chembetaine CB. CAS 68424-94-2; EINECS 270-329-4; Cocobetaine; foaming surfactant effective in hard and soft water; mild to hair and skin; for shampoos, soaps, conditioners. [Chemron]

Chembetaine CGF. Cocamidopropyl betaine; high purity, low color surfactant with foam and visc. building props.; for mediated shampoos and conditioners, facial cleansers, bubble baths, bath gels. [Chemron]

Chembetaine L. Lauramidopropyl betaine; foam booster, visc. builder; mild surfactant for shower gels, liq. soaps, skin cleansers, shampoos. [Chemron]

Chembetaine OL-30. CAS 871-37-4; EINECS 212-806-1; Oleyl betaine; gentle, substantive visc. builder for shampoos, conditioners, mousses. [Chemron]

Chembetaine S. Soyamidopropyl betaine; conditioner, foamer, visc. builder for shampoos, conditioners, bath products [Chemron]

Chembetaine TG. Dihydroxyethyl tallow glycinate; conditioner providing nonoily feel and resistance to build-up for cream rinses, shampoos, conditioning mousses, comb-out sprays. [Chemron]

Chem-Calk® 500. Two-component polyurethane; nonsag, high performance elastomeric joint sealant. [Bostik]

Chemcaulk. Two-component fluoroelastomer base caulk; corrosion resist. sealant for expansion joint flanges, expansion joint grouting in chem.-resist. flooring, expansion joint repair, formed in-place gaskets; recommended for contact with strong acids, chlorinated hydrocarbons and solvs. [Advanced Polymer Coatings]

Chemcoat. Two-component fluoroelastomer (Viton®) base coating system; recommended for harsh chem. environments; applied to valves, pipes, fittings, tanks, electronic components, cables; cures at R.T.; excellent adhesion props. [Advanced Polymer Coatings]

Chemcogen AC. CAS 28519-02-0; Disodium alkyl diphenyl oxide disulfonate [Rhone-Poulenc Surf.]

Chemdur. Liquid elastomeric urethane membrane for tanking rooms and pits. [Prodorite Ltd]

Chemeen 18-2. CAS 10213-78-2; PEG-2 stearamine; emulsifier and antistat in textiles, metal buffing, and rubber compounds; lubricant for fiber glass. [Chemax]

Chemeen C-2. CAS 61791-14-8; PEG-2 cocamine; emulsifier, antistat, dye leveler, wetting agent, lubricant, dispersant; substantive to metals, fibers, and clays. [Chemax]

Chemeen DT-3. PEG-3 tallow diamine; emulsifier, textile dyeing assistant, corrosion inhibitor used in preparation of asphalt and agric. chemical emulsions. [Chemax]

Chemeen HT-5. CAS 61791-26-2; PEG-5 hydrogenated tallow amine; emulsifier and antistat in textiles, metal buffing and rubber compounds, lubricant for fiber glass. [Chemax]

Chemeen O-30, O-30/80. PEG-30 oleamine; emulsifier, antistat, lubricant, and textile dyeing assistant; antiprecipitant in cross dyeing. [Chemax]

Chemeen S-2. CAS 61791-24-0; PEG-2 soya amine; emulsifier, antistat, lubricant. [Chemax] *

Chemeen T-2. CAS 61791-44-4; PEG-2 tallow amine; antistat for carpet shampoos; emulsifier, lubricant, dispersant, softener, antiprecipitant, leveling and migrating agent in textile dyeing process. [Chemax]

Chemfac NC-0910. POE alkyl phenol phosphate; wetting agent, detergent, hydrotrope, emulsifier, rust inhibitor, EP additive for alkaline detergents, metal cleaners, hard surface cleaners, textile scours, metal and textile lubricants, drycleaning soaps, emulsion polymerization, agric. formulations. [Chemax]

Chemfac PA-080. Alcohol phosphate ester; detergent, emulsifier, wetting agent, lubricant, antistat for alkaline detergents, metal cleaners, hard surface cleaners, textile scours, metal and textile lubricants, drycleaning soaps, emulsion polymerization, agric. formulations. [Chemax]

Chemfac PB-082. Phosphate esters; detergent, wetting, and coupling agent, antistat, emulsifier for alkaline detergents, metal cleaners, hard surface cleaners, textile scours, metal and textile lubricants, drycleaning soaps, emulsion polymerization, agric. formulations. [Chemax]

Chemfac PB-184. CAS 39464-69-2; Oleth-4 phosphate; wetting agent, detergent, hydrotrope, emulsifier, rust inhibitor, EP additive for alkaline detergents, metal cleaners, hard surface cleaners, textile scours, metal and textile lubricants, drycleaning soaps, emulsion polymerization, agric. formulations. [Chemax]

Chemfac PC-006. CAS 39464-70-5; Phosphate ester; wetting agent, lubricant, antistat, detergent, emulsifier; foaming hydrotrope. [Chemax]

Chemfac PD-600. CAS 52019-36-0; POE alkyl ether phosphate; wetting agent, detergent, hydrotrope, emulsifier, rust inhibitor, EP additive for alkaline detergents, metal cleaners, hard surface cleaners, textile scours, metal and textile lubricants, drycleaning soaps, emulsion polymerization, agric. formulations. [Chemax]

Chemfac PN-322. Phosphate ester, neutralized; hydrotrope for solubilizing nonionic surfactants in high concs. of alkali or other electrolytes; wetting agent, detergent, emulsifier, rust inhibitor, EP additive for alkaline detergents, textile scours, emulsion polymerization, lubricants and agric. [Chemax]

Chemfac PX-322. Phosphate ester; hydrotrope for solubilizing surfactants in alkali or other electrolytes. [Chemax]*

Chemfac RD-1200. Proprietary; environmentally safe industrial defoamer for offshore drilling. [Chemron]

Chemfax 5AM-100. Hydrocarbon resin polymerized from olefins; used in adhesives, coatings, tackifiers, etc. [Chemfax]

Chemglaze®. High performance single- and two-pack urethane coatings, both moisture cure and catalyzed cure; used to preserve, protect and beautify all types of

---

* Trade name not verified as available   † Trade name verified as obsolete

substrates including rubber, plastic, concrete, metal and glass. [Lord Corporation (UK) Ltd]

Chemglaze®Z004. Moisture-cured polyurethane coating for rubber. [Lord]

Chemical 39 Base. CAS 141-21-9; EINECS 205-469-7; Stearamidoethyl ethanolamine; cosmetic and toiletry base; emulsifying agent for creams and lotions; conditioning additive for hair products; lubricant in skin products [Sandoz]

Chemical Base 6532. CAS 16889-14-8; EINECS 240-924-3; Stearamidoethyl diethylamine; emulsifier, emollient, conditioning agent for skin and hair, shaving creams, hydro-alcoholic lotions, skin creams, cream rinse shampoos; substantive to hair. [Sandoz]

Chemidex B. CAS 60270-33-9; EINECS 262-134-8; Behenamidopropyl dimethylamine; substantive surfactant for cream rinses, conditioners, shampoos, creams, and lotions. [Chemron]

Chemidex C. CAS 68140-01-2; EINECS 268-771-8; Cocamidopropyl dimethylamine; substantive surfactant for cream rinses, conditioners, shampoos, creams, and lotions. [Chemron]

Chemidex L. CAS 3179-80-4; EINECS 221-661-3; Lauramidopropyl dimethylamine; substantive surfactant for cream rinses, conditioners, shampoos, creams, and lotions. [Chemron]

Chemidex M. CAS 45267-19-4; EINECS 256-214-1; Myristamidopropyl dimethylamine; substantive surfactant for cream rinses, conditioners, shampoos, creams, and lotions. [Chemron]

Chemidex O. CAS 109-28-4; EINECS 203-661-5; Oleamidopropyl dimethylamine; substantive surfactant for cream rinses, conditioners, shampoos, creams, and lotions. [Chemron]

Chemidex P. CAS 39669-97-1; Palmitamidopropyl dimethylamine; substantive surfactant for cream rinses, conditioners, shampoos, creams, and lotions. [Chemron]

Chemidex R. Ricinoleamidopropyl dimethylamine; substantive surfactant for cream rinses, conditioners, shampoos, creams, and lotions. [Chemron]

Chemidex S. CAS 7651-02-7; EINECS 231-609-1; Stearamidopropyl dimethylamine; low irritation emulsifier with substantivity to protein and cellulosic substrates; for cream rinses, conditioners, shampoos, creams, and lotions. [Chemron]

Chemidex SE. Stearamidoethyl dimethylamine; substantive surfactant for cream rinses, conditioners, shampoos, creams, and lotions. [Chemron]

Chemidex SI. CAS 67799-04-6; EINECS 267-101-1; Isostearamidopropyl dimethylamine; substantive surfactant for cream rinses, conditioners, shampoos, creams, and lotions. [Chemron]

Chemidex SO. CAS 68188-30-7; Soyamidopropyl dimethylamine; substantive surfactant for cream rinses, conditioners, shampoos, creams, and lotions. [Chemron]

Chemidex T. CAS 68425-50-3; Tallowamidopropyl dimethylamine; substantive surfactant for cream rinses, conditioners, shampoos, creams, and lotions. [Chemron]

Chemidex WC. CAS 68140-01-2; EINECS 268-771-8; Cocamidopropyl dimethylamine; substantive surfactant for cream rinses, conditioners, shampoos, creams, and lotions. [Chemron]

Chemigum® HR662. Butadiene-acrylonitrile-N-(4-anilinophenyl) methacrylamide terpolymer; NBR, cold polymer; used in molded and extruded goods such as gaskets and oil seals, oil well cable jackets and hose that require high resist. to oil and heat aging. [Goodyear] *

Chemigum® Latex 260. Nitrile latex; used as binder for asbestos and cellulose; provides good water resist.; sulfur vulcanizable. [Goodyear] *

Chemigum®N5. NBR, hot polymer, slightly staining antioxidant; for use in cements, grouts, adhesives. [Goodyear]*

Chemigum®N7. NBR, hot polymer, slightly staining antioxidant; for compounds which must withstand rugged or high-temp. service, e.g., hose, tubes, and belts. [Goodyear] *

Chemigum® N318B. NBR, cold polymer, nonstaining antioxidant; for extrusions and compr. molding, hose tubes; exceptional grn. strength; noncorrosive. [Goodyear] *

Chemigum® N917. NBR, cold polymer, slightly staining antioxidant; for gaskets, seals, o-rings for the aircraft industry and other applications requiring best low temp. flexibility and resistance to oil and fuel; noncorrosive. [Goodyear] *

Chemigum® NX-775. Carboxylated NBR, cold polymer, nonstaining antioxidant; FDA compliance for applications where outstanding abrasion and oil resist. and high strength are required; used for footwear, cable jackets, oil field parts, rollers and roll covers, shaft seals, pump liners, industrial wheels, conveyor belts. [Goodyear] *

Chemigum® P7-D. Branched NBR powd. partitioned with 9% calcium carbonate; for rubber, adhesives (high gel strength), dry or semiwet process disc-pads. [Goodyear]*

Chemigum® P83. NBR powd. partitioned with 9% PVC, lightly precrosslinked, stabilized; designed for PVC modification; improves hot-melt stability, plasticizer permanence, and processability for calendered and extruded film, sheet, gaskets and seals, hose and tubing, technical extrusion profiles. [Goodyear] *

Chemigum® TPE 03050. Nitrile-based thermoplastic elastomer; for automotive weatherstripping, tubing, architectural seals, coextruded glazing gaskets, appliance profiles and moldings, hose and tubing, cable sheathing, conveyor belting, flooring, roofing; modifier for vinyl, TPU and polar TPEs. [Goodyear]

Chemlease 55. Silicone blend; mold release for urethane elastomers. [Chemlease] *

Chemlease 88. Fluoropolymer blend; mold release for elec. potting, filament winding, phenolic board, adhesive laminate. [Chemlease] *

Chemlease 158R. Nonsilicone; water-based release system for epoxy composites, compr. molding, RIM urethane molding, rubbers. [Chemlease] *

Chemlease 906E. Silicone water-based emulsion; mold release for rubber and plastics industries. [Chemlease]*

Chemlease SP 40. General-purpose semi-permanent release system for plastic and rubber, elec. potting, filament winding, phenolic board, roto molding. [Chemlease] *

Chemline. A range of anti-abrasive glass tiles and ceramic tiles for coal and coke bunkers and hoppers. [Prodorite Ltd]

Chemlock. Rubber to metal heat activated bonding agents and adhesives. [Durham Chemicals Ltd]

Chemlok®205. One-coat bonding agent for NBR and primer for cover coat adhesive. [Lord]

---

‡ Trade name and manufacturer not verified   § Trade name without identified manufacturer

Chemlok® 220. General purpose cover coat bonding agent. [Lord]

Chemlok® 459. Bonding agent; primer promoting adhesion to thermoplastic elastomers and polyolefins. [Lord]

Chemlok® 607. Silane-based; one-coat adhesive for bonding elastomers, esp. silicone, fluroelastomers, EPDM, nitrile, epichlorohydrin, polyacrylate, hydrogenated nitriles. [Lord]

Chemlube. A series of diester and polyolester synthetic based oils; for compressor oils, chain oils, gear oils, automotive engine oils, high temp oils, etc. [Ultrachem Inc] *

Chemol®. A trademark currently awaiting reallocation by its proprietors. [Cassella AG] *

Chemoxide CAW. CAS 68155-09-9; Cocamidopropylamine oxide; surfactant for mild, low irritation personal care and industrial applications, e.g., shampoos, facial cleansers, bath products; foam and visc. builder, emollient over broad pH range. [Chemron]

Chemoxide L. Lauramidopropyl betaine; foam booster, visc. builder; mild surfactant. [Chemron]

Chemoxide LM-30. Lauramine oxide; visc. builder, foam enhancer for household and industrial cleaners, personal care products; tolerant to electrolytes for improved hard water performance. [Chemron]

Chemoxide O. CAS 14351-50-9; EINECS 238-311-0; Oleamine oxide; thickener, visc. builder; wetting agent for pigments and dyes; used in hair colorants, gels, permanent waves. [Chemron]

Chemoxide O1. Oleyl dimethylamine oxide; foamer, thickener, industrial surfactant; tolerant to electrolytes. [Chemron]

Chemoxide SAO. CAS 25066-20-0; Stearamidopropylamine oxide; wetting and foaming agent, visc. builder, conditioner, softener for hair; for shampoos, conditioners, mousses; emulsifier in creams and lotions. [Chemron]

Chemoxide ST. CAS 2571-88-2; EINECS 219-919-5; Stearamine oxide; conditioner, softener, visc. and foam builder for conditioning shampoos, rinses improving comb-out and manageability. [Chemron]

Chemoxide T. CAS 61791-46-6; EINECS 263-179-6; Dihydroxyethyl tallowamine oxide; detergent, wetting agent for low-pH shampoos and conditioners, alkaline media; for use in gels, liqs., or emulsions. [Chemron]

Chemoxide TAO. CAS 68647-77-8; Tallowamidopropylamine oxide; visc. builder, foam booster; improves manageability and luster in hair; for use in conditioners, sprays, mousses. [Chemron]

Chemoxide WC. CAS 61788-90-7; EINECS 263-016-9; Cocamine oxide; visc. builder, foam booster, emollient for shampoos, cleansers, bath products [Chemron]

Chemphonate AMP. CAS 6419-19-8; EINECS 229-146-5; Amino tris(methylene phosphonic acid); scale inhibitor for water injection, water disposal and production systems. [Chemron]

Chemphonate AMP-S. Sodium amino tris(methylene phosphonate); scale inhibitor for water systems; dispersant for drilling muds. [Chemron]

Chemphonate HEDP. 1-Hydroxyethyl-1-diphosphonic acid; temp.-stable scale inhibitor for oilfield application; sequestering agent for calcium carbonate. [Chemron]

Chemphonate NP. Sodium TEA phosphoric acid ester; scale inhibitor for oilfield prod. [Chemron]

Chemphos TC-227. Aromatic phosphate ester; detergent, wetting agent, emulsifier, coupling agent, surface tension reducer; for alkaline cleaners, heavy-duty all-purpose metalworking detergents, steam cleaning, dairy cleaners, bottle washing compounds, floor strippers; lubricant and detergent for drilling fluids; oil treating chemicals; emulsion polymerization of vinyl acetate, acrylates, SBR. [Chemron]

Chemphos TC-231S. Sodium nonoxynol-9 phosphate; emulsifier, solubilizer, antistat, substantivity agent for hair care products, perms, straighteners, depilatories; resistant to hydrolysis. [Chemron]

Chemphos TC-337. Nonoxynol-20 phosphate; emulsion polymerization surfactant for vinyl acetate, acrylates, SBR; emulsifier, solubilizer, antistat, substantivity agent for hair care products, perms, straighteners, depilatories; resistant to hydrolysis. [Chemron]

Chemphos TC-341. CAS 39464-64-7; Nonyl nonoxynol-10 phosphate; emulsifier, solubilizer, antistat, substantivity agent for hair care products, perms, straighteners, depilatories; resistant to hydrolysis. [Chemron]

Chemphos TC-444. Aliphatic phosphate ester; coupling agent for nonionic surfactants with liq. alkali detergent systems; surface tension reducer; for oil treating chemicals; compatible with high concs. of sodium hydroxide and silicate builders. [Chemron]

Chemphos TDAP. Alkyl acid phosphate; oil treating surfactant, asphaltine dispersant. [Chemron]

Chemphos TR-414W. Phosphate ester; lubricant and detergent for water-based drilling fluids. [Chemron]

Chemphos TR-505. CAS 39464-69-2; Oleth-10 phosphate; emulsifier, solubilizer, antistat, substantivity agent for hair care products, perms, straighteners, depilatories; resistant to hydrolysis. [Chemron]

Chemphos TR-505D. CAS 58855-63-3; DEA-oleth-10 phosphate; emulsifier, solubilizer, antistat, substantivity agent for hair care products, perms, straighteners, depilatories; resistant to hydrolysis. [Chemron]

Chemphos TR-510. CAS 39464-66-9; Laureth-4 phosphate; emulsifier, solubilizer, antistat, substantivity agent for hair care products, perms, straighteners, depilatories; resistant to hydrolysis. [Chemron]

Chemphos TR-510S. CAS 42612-52-2; Sodium laureth-4 phosphate; emulsifier, solubilizer, antistat, substantivity agent for hair care products, perms, straighteners, depilatories; resistant to hydrolysis. [Chemron]

Chemphos TR-515. CAS 39464-69-2; Oleth-3 phosphate; emulsifier, solubilizer, antistat, substantivity agent for hair care products, perms, straighteners, depilatories; resistant to hydrolysis. [Chemron]

Chemphos TR-515D. CAS 58855-63-3; DEA-oleth-3 phosphate; emulsifier, solubilizer, antistat, substantivity agent for hair care products, perms, straighteners, depilatories; resistant to hydrolysis. [Chemron]

Chemphos TR-541. CAS 39464-69-2; Oleth-4 phosphate; emulsifier, solubilizer, antistat, substantivity agent for hair care products, perms, straighteners, depilatories; resistant to hydrolysis. [Chemron]

Chemprene 50, 75. Isoprenoidal polymer; excellent water, acid, and alkali resistant, excellent oxidative stability. [Chemfax]

Chemprene R-10. Thermoplastic isoprenoidal polymer; used in compding. of synthetic, natural, and reclaim rubber; softener, tackifier, and reinforcing agent; used in calendering and extruding, hard rubber, molded goods, in camelback, tubes, and tire stocks, in rubber cements, wire and cable insulation, floor tile, shoe compounds, caulks. [Chemfax]

Chemquat 12-50. CAS 112-00-5; EINECS 203-927-0; Laurtrimonium chloride; surfactant, corrosion inhibitor, antistat for plastics, textile dyeing aid, gel sensitizer in latex foam production; visc. depressant in paper and textile softener formulations. [Chemax]

Chemquat 16-50. CAS 112-02-7; EINECS 203-928-6; Cetrimonium chloride; surfactant, corrosion inhibitor, antistat for plastics, textile dyeing aid, gel sensitizer in latex foam production; visc. depressant in paper and textile softener formulations. [Chemax]

Chemquat C/33W. CAS 61789-18-2; EINECS 263-038-9; Coco trimethyl ammonium chloride; surfactant, corrosion inhibitor, antistat for plastics, textile dyeing aid, gel sensitizer in latex foam production; visc. depressant in paper and textile softener formulations. [Chemax]

Chemraz®. Perfluoroelastomer (PTFE plus proprietary agents); extremely chemically resistant seals for aggressive chemical environments. [Greene, Tweed & Co]

Chem-Rez. Phenolic and furfuryl alcohol-based resin systems cured through the application of acid catalyst or with heat; for production of foundry cores and molds. [Ashland Chemical Company] *

Chemsalan NLS 30. CAS 151-21-3; Sodium lauryl sulfate; surfactant. [Chemsal]

Chemsalan RLM 28. CAS 9004-82-4; Sodium laureth sulfate; biodeg. surfactant. [Chemsal]

Chemset. Artificial and synthetic resins all adapted for setting; used for adhesives, chemical products for potting, impregnating, flooring, casting, molding, laminating, building, sheathing, coating, enamelling, waterproofing and for the hardening of resins. [R F Bright Enterprises Ltd]

Chemstat®106G/90. CAS 58767-50-3; Bis (2-hydroxyethyl) octyl methylammonium p-toluene sulfonate; permanent internal antistat for polystyrene and other thermoplastics where high processing temps. required; also as external antistat. [Chemax]

Chemstat® 122. Ethoxylated coco amine; internal antistat for polyolefins. [Chemax]

Chemstat®172. CAS 13127-82-7; Ethoxylated oleyl amine; internal antistat for polyolefins. [Chemax]

Chemstat® 192/NCP. Ethoxylated stearyl amine; permanent internal antistat effective in eliminating electrostatic problems in extrusion, injection and blow molding of thermoplastics. [Chemax]

Chemstat® 273-C. CAS 61791-31-9; N,N-Bis (2-hydroxyethyl) coco amine; permanent internal antistat for plastic film and molded products; eliminates electrostatic problems in extrusion, injection and blow molding. [Chemax]

Chemstat® 273-E. CAS 10213-78-2; N,N-Bis (2-hydroxyethyl) stearyl amine; permanent internal antistat for plastic film and molded products; eliminates electrostatic problems in extrusion, injection and blow molding. [Chemax]

Chemstat®9820A. Bis (2-hydroxyethyl) octyl methyl ammonium p-toluene sulfonate in ABS carrier resin; permanent internal antistat for ABS; recommended use 1-3% of act. antistat. [Chemax] *

Chemstat® AF-906. Antifog agent for polyethylene film. [Chemax]

Chemstat®HTSA#1. CAS 16260-09-6; Oleyl palmitamide; slip/antiblock agent, antistat, mold release with high thermal stability; for PP film. [Chemax]

Chemstat®HTSA#3. CAS 10094-45-8; Stearyl erucamide; slip/antiblock agent, mold release with high thermal stability. [Chemax]

Chemstat® P-400. CAS 25322-68-3; Polyethylene glycol; permanent internal antistat for polyethylene. [Chemax]

Chemstat®PS-101. CAS 68037-49-0; Sodium C10-18 alkyl sulfonate; permanent internal antistat for thermoplastic film and molded products; temporary external antistat; biodeg. [Chemax]

Chemsulf S2EH-Na. Sodium 2-ethylhexyl sulfate; low foaming surfactant, wetting agent, emulsifier, detergent; electrolyte tolerant. [Chemax]

Chemsulf SBO/65. Sulfated butyl oleate; softener, wetting agent, lubricant additive, emulsifier, solubilizer for textile, metalworking industries. [Chemax]

Chemsulf SCO/75. CAS 8002-33-3; EINECS 232-306-7; Sulfated castor oil; softener, wetting agent, lubricant additive, emulsifier, solubilizer for textile, metalworking industries. [Chemax]

Chemtac 20, 35. Recommended for harsh chem. environments when additional surface tack is required for laminating and/or joining substrates. [Advanced Polymer Coatings]

Chemtech Cypermethrin. CAS 66841-24-5; Emulsifiable concentrate containing 100 g cypermethrin per liter; a pyrethroid insecticide. [Chemtech (Crop Protection) Ltd]

Chem-Trete®. Alkylalkoxysilane; weatherproofing agents for concrete and masonry surfaces. [Hüls Am.]

Chemzoline 1411. Aminoethyl tall oil imidazoline; corrosion inhibitor base for high-temp. oilfield applications. [Chemron]

Chemzoline C-22. CAS 61791-38-6; EINECS 263-170-7; Hydroxyethyl coco imidazoline; emulsifier for oils; corrosion inhibitor for oilfield applications. [Chemron]

Chemzoline T-11. Aminoethyl tall oil imidazoline; intermediate for production of film-forming corrosion inhibitors; in automatic car wash rinse aids, oilfield applications; pigment dispersant for paints. [Chemron]

Chemzoline T-33. Aminoethyl tall oil imidazoline; industrial surfactant; produces self-emulsifiable mineral oil compositions; dispersant for mineral spirits, aromatic solvs.; film-forming corrosion inhibitor; antistripping agent for asphalt, coal tar pitches; pigment dispersant in paints. [Chemron]

Chemzoline T-44. CAS 61791-39-7; EINECS 263-171-2; Hydroxyethyl tall oil imidazoline; emulsifier for mineral, vegetable, and animal oils; emulsifier for oil-based drilling muds; corrosion inhibitor for oilfield applications. [Chemron]

Chenzinsky-Plehn's solution. A microscopic stain. It contains 0.25 gram eosin, 50 g of 70% alcohol, 100 g of a saturated solution of methylene blue, and 100 cc distilled water.

Cheque. CAS 3704-09-4; Mibolerone; anabolic; androgen. [Upjohn] *

Ches®500. Nonfat drymilk, xanthan gum, propylene glycol, alginate, glyceryl stearate, sodium glyceryl oleate phosphate; food grade stabilizer; cold mix emulsifier for cosmetics and pharmaceuticals; cold hot emulsion system; unique ambient temp. emulsifier which yields stable, aesthetic o/w emulsions. [CasChem]

---

‡ Trade name and manufacturer not verified   § Trade name without identified manufacturer

chestnut brown. *See* umber.

chia oil. An oil obtained from the Mexican plant *Salvia hispanica*. The raw oil dries slowly, but the boiled oil is a good drying oil.

chian turpentine. (Alk, Chio turpentine, Chios turpentine, Cyprian turpentine, Scian turpentine). Names applied to the oleo-resin obtained from the bark of *Pistachi terebinthus*, a tree in the Mediterranean and Asia Minor.

Chicago acid. *See* 2S acid.

chicle. A gum obtained from *Achras* species and others of Mexico, Belize, and Venezuela; has been used in the manufacture of chewing gum, but has now been considerably superseded by other gums, e.g., Jelutong.

Chierite. A proprietary molding material of urea formaldehyde. [Butese] ‡

Chierol. A proprietary phenolic molding material. [Butese]‡

Chilcote. Dressings for metal chills. [Foseco (F.S.) Ltd]

Childion. Emulsifiable concentrate of 166 g dicofol and 58.7 g tetradifon per liter; a contact acaricide. [Hortichem Ltd]

Childion. Emulsifiable concentrate of 170 g dicofol and 62.5 g tetradifon per liter; a contact acaricide. [ICI Agrochemicals]

Chile saltpeter. *See* sodium nitrate.

Chili niter. *See* saltpeter.

Chili saltpeter. CAS 7631-99-4; (Chili niter, soda salt peter, Peru saltpeter, cubic saltpeter, soda niter, cubic niter, Nitratine). Sodium nitrate, $NaNO_3$, found as deposits in Chile and Peru; used as an oxidizing agent; in solid rocket propellants, fertilizer, flux, glass manufacture, refrigerant, matches, dynamite, dyes, pharmaceuticals, anaphrodisiac.

Chillsa FE®. CAS 57-55-6; EINECS 200-338-0; Inhibited propylene glycol. [Arco Chemical Co]

Chiltern Cropspray 11E. Adjuvant containing 99% refined mineral oil; wetting agent for herbicides. [Chiltern Farm Chemicals Ltd]

Chiltern Cyperkill 10. CAS 66841-24-5; An emulsifiable concentrate containing 100 g cypermethrin per liter; a pyrethroid insecticide. [Chiltern Farm Chemicals Ltd]

Chiltern Fazor. CAS 123-33-1; Maleic hydrazide; a plant growth regulator for grass and to reduce bud growth in trees, hedges and vegetables. [Chiltern Farm Chemicals Ltd]

Chiltern IPU. CAS 34123-59-6; Suspension concentrate containing 500 g isoproturon per liter; used for annual weed control in cereals. [Chiltern Farm Chemicals Ltd]

Chiltern Kocide 101. CAS 20427-59-2; Copper hydroxide; a protectant fungicide. [Chiltern Farm Chemicals Ltd]

Chiltern Ole. CAS 1897-45-6; Chlorothalonil; a fungicide for a wide range of agricultural crops. [Chiltern Farm Chemicals Ltd]

Chiltern Pyrazol. CAS 1698-60-8; Suspension concentrate containing 430 g chloridazon per liter; a pyridazinone herbicide for beet crops. [Chiltern Farm Chemicals Ltd]

Chimassorb® 119FL. CAS 106990-43-6; 1,3,5-Triazine-2,4,6-triamine, N,N''''-[1,2-ethanediylbis[[[4,6-bis [butyl(1,2,2,6,6-pentamethyl-4-piperidinyl)amino]-1,3,5-triazine-2-yl]imino]-3,1-propanediyl]]-bis[N',N''-dibutyl-N',N''-bis(1,2,2,6,6-pentamethyl-4-piperidinyl); light and thermal stabilizer for PP fiber applications in automotive, marine and residential carpets, agric. films, fertilizer bags, thick section pigmented applications, rotational molding applications. [Ciba-Geigy/Additives]

China clay. *See* kaolin.

chinaldine. *See* quinaldine.

Chinaspin®. A preparation for the relief of colds and flu. [Bayer AG]

Chinese bronze. Alloys of from 72.5-74% copper, 15-18.5% lead, 10-14% zinc, and 1-5% tin. One alloy, also known as Chinese bronze, contains 78% copper and 22% tin.

Chinese glue. Shellac dissolved in alcohol; used for joining wood, earthenware, or glass.

Chinese isinglass. CAS 9002-18-0; *See* agar-agar.

Chinese red. *See* cinnabar and chrome red.

Chinese scarlet. *See* chrome red.

Chinese silver. (Peru silver). A German silver, containing a little aluminum.

Chinese tallow. (vegetable tallow). A waxy substance obtained from the outer coating of the fruit of *Stillingia sebifera*, of China.

Chinese wax. (vegetable spermaceti, tree wax). Also wrongly called Japanese wax. Insect wax obtained from the insect *Coccus ceriferus*, or *C. pela*, which deposits wax on certain trees.Its chief ingredient is ceryl certate.

Chinese white. (zinc white) The name is sometimes applied to barium sulfate.

Chinese white copper. An alloy of 40% copper, 31% nickel, 25% zinc, and 2% iron.

Chinese wood oil. (wood oil). Tung oil obtained from the seeds of *Aleurites* species in China and Japan.

Chinex®. Polymer. [DuPont UK]

Chinine. Quinine.

chinoline. *See* quinoline.

Chio(s) turpentine. *See* Chian turpentine.

Chipman Path Weedkiller. CAS 1912-24-9, 61-82-5; A liquid formulation containing 166 g atrazine and 137 g aminotriazole per liter as a suspension concentrate; used for total weed control on industrial sites, paths, kerbs, drives and hard standing areas. [Chipman Ltd]

Chisso-rite. A proprietary synthetic resin made by condensing formaldehyde and acid oils from low temperature coal carbonization. §

chitin. (poly(N-acetyl-1,4-β-D-glucopyranosamine)) A glucosamine polysaccharide $(C_8H_{13}NO_5)_n$; CAS 1398-61-4; EINECS 215-744-3; biological research, source of chitosan. [Ajinomoto USA; Amerchol; Atomergic Chemetals; Tri-K Industries]

Chloracel®. Sodium aluminum chlorhydroxy lactate; deodorant active. [Reheis Inc]

Chloractil. CAS 69-09-0; A proprietary preparation of chlorpromazine hydrochloride; a tranquilizer. [DDSA Pharmaceuticals Ltd] *

Chloral. CAS 75-87-6; Trichloroacetaldehyde,$Cl_3 \cdot CCHO$; used as a hypnotic.

Chloralamide. Chloral formamide, $CCl_3 \cdot CH(OH) \cdot NH \cdot CHO$; a mild hypnotic and sedative.

chloral hydrate. CAS 302-17-0; The covalent hydrate of chloral, $Cl_3CCH(OH)_2$; used in medicine as a sedative, manufacture of DDT, and liniments.

chloral iodine. A solution of chloral hydrate (50 g in 20 cc water) saturated with iodine; used for the detection of starch grains.

Chloramine B. CAS 127-52-6; Sodium benzenesulfochloroamide, $C_6H_5 SO_2 NNaCl+2H_2O$; used like Chloramine T.

Chloramine T. CAS 127-65-1; (tochlorine, tolamine, chloramine-Heyden, Pyrgos, Mianin, Aktivin). Sodium p-

---

* Trade name not verified as available   † Trade name verified as obsolete

toluenesulfochloramide, $CH_3C_6H_4$ (SO_2NClNa) · $3H_2O$; an antiseptic used in medicine; solutions of Chloramine T are also used as detergents and bleaching age

Chloramine-Heyden. *See* Chloramine T.

chloranil. CAS 118-75-2; Tetrachlorquinone, $C_6Cl_4O_2$; used as an agricultural fungicide, dye intermediate,, electrodes for pH measurement, reagent.

Chlorantine. Dyestuffs fast to light. [Clayton Aniline Co Ltd]*

chlorarsine. Cacodyl, $As_2(CH_3)_4$.

Chlorasol®. CAS 7681-52-9; Sodium hypochlorite solution; used for cleansing and desloughing wounds. [Seton Healthcare Group plc]

chlorate of soda. *See* sodium chlorate.

Chlorazol®Dyes. A registered trademark applied to certain dyestuffs. §

Chlorazone. *See* Chloramine T.

Chlorcahücit. A German explosive.

Chlorcosane. CAS 63449-39-8; A liquid chlorinated paraffin wax; the chlorine content varies from 27- 35%.

chlor cresol green. Tetrachlor-m-cresolsulfonphthalein.

Chlordispel. Dishwashing detergent. [The Wellcome Foundation Ltd] *

Chlorea. CAS 1912-24-9; Atrazine; a residual herbicide. [Chipman Ltd]

Chloresium. Chlorophyllin copper complex obtained from chlorophyll by replacing the methyl and phytyl ester groups with alkali and the magnesium with copper; deodorant. [Rystan Company Inc] *

chloretone. CAS 57-15-8; (chlorbutanol, acetone chloroform, methaform; 1,1,1-trichloro-2-methyl 2-propanol) $(CH_3)_2C(OH)CCl_3$; used a an hypnotic, soporific, as an inhalation anesthetic, and as a remedy for sea-sickness.

Chlorez®. Chlorinated paraffin, solid; flame retardant in plastics, rubber, coatings etc. [Dover]

Chlorez® 700. Resinous chlorinated paraffin; flame retardant for LDPE, in coatings, inks, plastics, foams, adhesives, paper, and fabrics. [Dover]

Chlorez® 700-DD. Resinous chlorinated paraffin; flame retardant for use in wh. coatings; superior color stability. [Dover]

Chlorez® 700-DF. Resinous chlorinated paraffin; flame retardant for paints, printing inks, plastics, foams, adhesives, paper and fabric coatings. [Dover]

Chlorez® 760. Resinous chlorinated paraffin; flame retardant for LDPE, PP, olefins, styrenes, adhesives, wire and cable and other applications. [Dover]

chlorfenvinphos. CAS 470-90-6; . 2-Chloro-1-(2,4-dichlorophenyl)-vinyl diethyl phosphate; an insecticide.

Chlorguard. Industrial control gear for effecting automatic closure of valves. [ICI Chem. & Polymers Ltd]

chlorhexidine digluconate. (N,N′-bis (4-chlorophenyl)-3,12-diimino-2,4,11,13-tetraazatetradecane-diimidamide compound with D-gluconic acid; chlorhexidine gluconate) Salt of chlorhexidine and gluconic acid; $C_{22}H_{30}Cl_2N_{10}$ · $2C_6H_{12}O_7$; CAS 18472-51-0; 14007-07-9. [Degussa; Lonza AG]

Chlorhydrol®. Aluminum chlorohydrate powder and solution; antiperspirant active; also used in water treatment. [Reheis Inc]

Chlorhydrol®50% Sol'n. Aluminum chlorohydrate; antiperspirant active. [Reheis Inc]

chloride of lime; CAS 7778-54-3; Calcium hypochlorite

Chlorina. Chloramine.

chlorinated rubber. CAS 9006-03-5; Elastomer to which

65% chlorine has been added to give a solid, film-forming resin; used in swimming pool, traffic, marine, and masonry paints.

chlorine. Cl_2; CAS 7782-50-5; Manufacture of CCl_4, trichloroethylene, chlorinated hydrocarbons, neoprene, PVC, etc.; water purification; shrinkproofing wool; in flame-retardant compounds; food processing; bleaching. [Air Prods.; Asahi Chem Industry Co Ltd; Asahi Denka Kogyo; Atochem; BASF; Georgia-Pacific Resins; Olin; OxyChem; PPG Industries; Showa Denko]

chlorine dioxide. (chlorine peroxide) ClO_2; CAS 10049-04-4; Bleaching wood pulp, fats and oils; biocide; odor control; water purification; oxidizing agent; bactericide, antiseptic. [Drew Ind. Div.; Int'l. Dioxcide]

chlorine peroxide. *See* chlorine dioxide.

Chloritane. CAS 7758-19-2; Sodium chlorite. [Interox Chemicals Ltd] *

Chlormytol. A proprietary preparation of chloramphenicol and prednisolone; used in dermatology as an antibacterial agent. [Parke-Davis] *

chloroacetic acid. *See* monochloroacetic acid.

chloroacetophenone. (α-chloroacetophenone) $C_6H_5CO-CH_2Cl$; CAS 532-27-4; Pharmaceutical intermediate, riot-control gas. [Janssen Chimica; Penta Mfg.; Schweizerhall]

m-chloroaniline. (m-aminochlorobenzene) $ClC_6H_4NH_2$; CAS 108-42-9; Intermdiate for azo dyes and pigments, pharmaceuticals, insecticides, agricultural chemicals. [Du Pont; Janssen Chimica; Schweizerhall]

o-chloroaniline. (o-aminochlorobenzene) $ClC_6H_4NH_2$; CAS 95-51-2; Dye intermediate, standards for colorimetric apparatus, manufacture of petroleum solvents and fungicides. [Du Pont]

p-chloroaniline. (p-aminochlorobenzene) $ClC_6H_4NH_2$; CAS 106-47-8; Dye intermediate, pharmaceuticals, agricultural chemicals. [Du Pont; Janssen Chimica; Mitsui Toatsu]

Chloroben. CAS 95-50-1; o-Dichlorobenzene; used in sewage treatment.

chlorobenzaldehyde. C_6H_4CHOCl; Intermediate for triphenyl methane and related dyes, organic intermediate. [Hoechst Celanese; Janssen Chimica; Penta Mfg.; Rit-Chem]

chlorobenzene. (monochlorobenzene; phenyl chloride; benzene chloride) C_6H_5Cl; CAS 108-90-7; EINECS 203-628-5; Pesticide intermediate; manufacture of phenol, aniline, DDT; solvent carrier for methylene diisocyanate; solvent for paints; heat transfer medium. [Aldrich; Atochem SA; Janssen Chimica; Monsanto; PPG Industries]

4-chloro-1,3-benzenediol. *See* 4-chlororesorcinol.

chlorobromal. Chlorodibromoacetaldehyde, $CCl·Br_2·CHO$.

Chlorobromhydrin. α-Chloro-α-bromo isopropyl alcohol, $CH_2Br · CH (OH) · CH_2Cl$.

chlorobromoform. Chlorodibromo methane, $CHCl · Br_2$.

2-chlorobutadiene 1,3. *See* polychloroprene.

chlorobutanol. (trichloro-t-butyl alcohol; 1,1-trichloro-2-methyl-2-propanol; acetone chloroform) $Cl_3CC-(CH_3)_2OH$; CAS 57-15-8; Plasticizer for cellulose esters and ethers, preservative for biological fluids and solutions, antimicrobial agent, anesthetic in dentistry. [EM Industries; R.W. Greeff; Penta Mfg.; Spectrum Chem. Mfg.]

chlorocyclohexane. *See* cyclohexyl chloride.

chloroethene homopolymer. *See* polyvinyl chloride.

---

‡ Trade name and manufacturer not verified   § Trade name without identified manufacturer

chloroethylene polymer. See polyvinyl chloride.
Chlorofin 42. Chlorinated paraffin extender for vinyl plastics containing 40% chlorine. [Hercules] *
Chloroflo® 40, 42. Chlorinated paraffin; lubricant additive. [Dover]
chloroform. (trichloromethane) CHCl₃; CAS 67-66-3; EINECS 200-663-8; Fluorocarbon plastics, solvent, analytical chemistry, fumigant, insecticides. [Atochem N. Am.; Hüls AG; Mallinckrodt; Mitsui Toatsu Chem.; OxyChem]
chloromethane. See methyl chloride.
chloromethyltrimethylsilane. (trimethylsilylmethyl chloride) (CH₃)₃SiCH₂Cl; CAS 2344-80-1; EINECS 219-058-5; Reagent for the Peterson olefination and the homologation of ketones and aldehydes via alpha, beta-epoxysilanes. [Hüls Am.; Janssen Chimica]
Chloromycetin. CAS 56-75-7; Chloramphenicol; antibacterial; antirickettsial. [Parke-Davis]
Chloromycetin Intramuscular Kapseals, Palmitate Suspension, Pure, Suppositories. CAS 56-75-7; A proprietary preparation containing chloramphenicol; an antibiotic. [Parke-Davis] *
Chloromycetin Succinate. CAS 56-75-7; A proprietary preparation containing chloramphenicol, (as the sodium salt of the monosuccinic ester); an antibiotic. [Parke-Davis] *
Chloroneb 65W Fungicide. CAS 2675-77-6; 65% Chloroneb wettable powder; seed treatment to suppress seeding blights, soreshin and pre- and post-emergence damp-off caused by rhizoctonia solani, pythium spp and sclerotium rolfsii on cotton, beans, soybeans and sugar beets. [Kincaid Enterprises Inc] *
Chloroneb Systemic Flowable Fungicide. CAS 2675-77-6; 30% Chloroneb; seed treatment for control of rhizoctonia solani, pythium spp and sclerotium rolfsii on cotton, beans, soybeans and sugar beets. [Kincaid Enterprises Inc] *
chlorophyll. (leaf green; chromule) CAS 479-61-8, 1406-65-1; The green coloring matter of plants, leaves, and stalks; it is a magnesium compound; colorant for soaps, oils, fats,waxes, liquors, confectionery, preserves, cosmetics, perfumes; dentistry; source of phytol; sensitizer for color film; toothpaste additive; deodorant. [Atomergic Chemetals; Biochim Srl; Penta Mfg.; Spectrum Chem. Mfg.]
Chloropicrin. (nitrochloroform; trichloronitromethane) CCl₃(NO₂); CAS 76-06-2; Used in organic synthesis, dyestuffs, fumigants, insecticides, rat extermination, tear gas.
chloroprene. CAS 126-99-8; 1:3-chlor-2-butadiene; used as a protective coating and in synthetic rubber manufacture by polymerization. The polymerized product bears the name of Neoprene.
chloroprene rubber. See polychloroprene.
chloropropylene oxide elastomer. See epichlorohydrin elastomer.
3-chloropropyltriethoxysilane. C₉H₂₁ClO₃Si; CAS 5089-70-3; EINECS 225-805-6; Coupling agent, release agent, lubricant, blocking agent, chemical intermediate. [Hüls Am.; Union Carbide]
3-chloropropyltrimethoxysilane. C₆H₁₅ClO₃Si; CAS 2530-87-2; EINECS 219-787-9; Coupling agent for epoxies, nylons, urethanes. [Dow Corning; Hüls Am.; PCR; Union Carbide]

Chloropryl. See Isopral. §
Chloroptic®. CAS 56-75-7; Chloramphenicol; antibacterial; antirickettsial. [Allergan, Inc]
Chloroptic S.O.P.®; CAS 56-75-7; Chloramphenicol; antibacterial. [Allergan, Inc]
Chloropyramine. Halopyramine.
2-chloropyridine. C₅H₄ClN; 109-09-1; EINECS 203-646-3; Production of antihistamines, germicides, pesticides, agricultural chemicals. [Expansia SA; Olin; Penta Mfg.; Schweizerhall]
4-chlororesorcinol. (1,3-benzenediol, 4-chlor-; 4-chloro-1,3-benzenediol) Halogenated phenol; C₆H₅ClO₂; CAS 95-88-5. [Rit-Chem]
Chloros. CAS 7681-52-9; Disinfectant containing sodium hypochlorite. [ICI Chem. & Polymers Ltd]
Chlorosoda. A proprietary form of solidified sodium hypochlorite for use as a bleaching agent; a small proportion of a saturated fatty acid, such as lauric acid, is incorporated. §
n-chlorosuccinimide. C₄H₄ClNO₂; CAS 128-09-6; EINECS 204-878-8; Chlorinating agent, disinfectant for swimming pools, bactericide. [Janssen Chimica; Penta Mfg.; Schweizerhall]
Chlorotex. Reagent for estimating residual chlorine in water. [BDH Chemicals Ltd] *
Chlorothalonil. CAS 1897-45-6; Tetrachloroisophthalonitrile; raw material for fungicides. [SNIA (UK) Ltd]
Chlorothene. CAS 71-55-6; A line of inhibited 1,1,1-trichloroethane solvents; used in industry. [Dow UK] *
Chlorothene (VG). CAS 71-55-6; An inhibited grade of 1,1,1,-trichlorethane. degreasing solvent. [Dow UK] *
2-chlorothiophene. (2-thienyl chloride) C₄H₃ClS; CAS 96-43-5; EINECS 202-505-3. [Janssen Chimica]
chlorotrifluoromethane. CClF₃; CAS 75-72-9; Refrigerant, dielectric and aerospace chemical, hardening of metals, pharmaceutical processing. [Atochem N. Am.; PCR]
chlorous acid, sodium salt. See sodium chlorite.
Chlorovin®. Polyvinyl chloride. §
Chlorovis 150A. Chlorinated paraffin; lubricant additive. [Dover]
Chlorowax. Chlorinated paraffins. [Occidental Chemical Corp]
Chlorowax 40. Liquid chlorinated paraffin; a vinyl plasticizer. [U.S. Industrial Chem.] *
Chlorowax 50. Chlorinated paraffin (50% chlorine); a vinyl plasticizer. [U.S. Industrial Chem.] *
Chlorowax 70. Resinous chlorinated paraffin; a vinyl plasticizer. [U.S. Industrial Chem.] *
Chlorowax LV. A chlorinated paraffin; a vinyl plasticizer. [U.S. Industrial Chem.] *
chloroxethose. Hexachlorodivinyl ether (CCl₂:CCl)₂O.
Chlorozone. CAS 127-65-1; A bleaching liquor prepared by passing chlorine into caustic soda.
chlorquinaldol. 5,7-Dichloro-8-hydroxy-2-methylquinoline; Steroxin; used in medicine as a bactericide and fungicide.
Chlor-Tabs. Effervescent chlorine tablets. [PPF International Ltd] *
Chlor-Trimeton. CAS 2438-32-6; Chlorpheniramine maleate; antihistaminic. [Schering Corp]
Chloryl. CAS 75-00-3; Ethyl chloride, C₂H₅Cl; used as an anesthetic.
Chlorylen. Trichlorethylene.

---

* Trade name not verified as available   † Trade name verified as obsolete

Chlotride. CAS 58-94-6; Chlorothiazide; for the relief of edema, edema accompanying premenstrual tension and control of hypertension. [Merck & Co Inc]

Chlumin. An aluminum alloy resistant to sea water. It contains chromium, a few% of magnesium, and iron.

Chobile. A proprietary extract of ox bile with oxidized oxbile acids; used as a stimulant to bile secretion. [Mallinckrodt Inc] *

Cholan DH. CAS 81-23-2; Dehydrocholic acid; choleretic. [Pennwalt Corp] *

Cholebrine. CAS 16034-77-8; Iocetamic acid; diagnostic aid. [Mallinckrodt Inc] *

Choledyl. Oxtriphylline; bronchodilator. [Parke-Davis]

cholest-5-en-3β-ol. See cholesterol.

cholesterol. (cholest-5-en-3-β-ol) Mono-unsaturated sec. alcohol of the cyclopentenophenanthrene system; $C_{27}H_{46}O$; CAS 57-88-5; EINECS 200-353-2; emulsifying agent in cosmetics and pharmaceutical products; source of estradiol. [Croda; EM Industries; Schweizerhall; Solvay Duphar BV; U.S. Biochemical]

Cholesterol. CAS 57-88-5; Cholesterol. [Croda Chem. Ltd]

Cholestrophane. Dimethylparabanic acid, $C_5N_6N_2O_3$.

Cholestyramine. CAS 11041-12-6; A styryl-divinylbenzene copolymer (about 2% divinyl-benzene) containing quaternary ammonium groups; used to eliminate from the body toxic organochlorine compounds.

Choletec. CAS 78266-06-5; Mebrofenin; diagnostic aid. [Bristol-Myers Squibb Co Inc] *

choline bitartrate. $(C_5H_{14}NO)C_4H_5O_6$; Medicine, dietary supplement, nutrient. [Am. Biorganics; R.W. Greeff; Pennwalt Italia SpA; Penta Mfg.; Schweizerhall]

choline chloride. (choline hydrochloride; 2-hydroxyethyl)trimethylammonium chloride) $(CH_3)_3N(Cl)$-CH_2CH_2OH; CAS 67-48-1; EINECS 200-655-4; Animal feed additive. [Am. Biorganics; Mitsubishi Gas Chem.; Penta Mfg.; Tanabe USA; UCB SA]

Cholografin. CAS 606-17-7; Iodipamide; pharmaceutic necessity for iodipamide meglumine injection. [Bristol-Myers Squibb Co Inc] *

Cholografin Meglumine. CAS 3521-84-4; Iodipamide meglumine; diagnostic aid. [Bristol-Myers Squibb Co Inc] *

Choloxin. Dextrothyroxine sodium; antihyperlipoproteinemic. [Boots Pharmaceuticals Inc; The Boots Co plc]

Cholumbrin. A proprietary preparation of sodium tetraiodophenolphthalein. §

chondroitin sulfate. (chondroitin sulfuric acid) Mucopolysaccharide; major constituent of the cartilagenous tissue in the body; CAS 9007-28-7. [Croda; Kraeber GmbH]

chondroitin sulfuric acid. See chondroitin sulfate.

chondrus. See carrageenan.

Choron. Gonadotropin, chorionic; gonad-stimulating principle. [O'Neal, Jones & Feldman Pharmaceuticals] *

CHP. N-Cyclohexyl-s-pyrrolidone; solvent, reaction intermediate, textile auxiliary, cosmetic ingredient. [ISP]

CHP-5. CAS 80-15-9; Cumene hydroperoxide sol'n.; initiator for room temperature curing of vinyl ester of other unsat. polyester resins; esp. useful used with vinyl ester resins that are cobalt promoted. [Witco/Argus]

CHP-158. CAS 80-15-9; 80% sol'n. of cumene hydroperoxide; initiator for vinyl monomers and copolymers and the crosslinking of unsat. polyester resins. [Witco/Argus]

CHPTA 65%. CAS 3327-22-8; 3-Chloro-2-hydroxypropyltrimethyl ammonium chloride; cationizing reagent for natural and synthetic polymers; starch modifier for textiles. [Chem-Y GmbH]

Christolit. A proprietary plastic of the phenol-formaldehyde type. §

Christophite. Synonym for Marmatite.

Chrogo U42. An alloy of 40% gold, 45% copper, 14% nickel, 1% chromium, and traces of platinum. A dental alloy. §

Chroma-Cal®. Colorant dispersions; for coloring of coating compositions. [Hüls Am.]

Chroma-Chem. Colorant dispersions; for coloring of industrial coating compositions. [Hüls Am.]

Chromaflo®. Pourable dispersions for high pigment loading, controlled visc.; for pumping and metering. [Plasticolors]

Chromagan. Nickel-chromium steel for the manufacture of tableware.

Chromagel. Chemical products for use in chromatography and chromtographic systems. [Courtaulds plc]

Chromagen. Ferrous fumarate USP, ascorbic acid USP, cyanocobalamin USP, desicated stomach substance; phosphorous-free vitamin and mineral dietary supplement; indicated for the treatment of all anemias responsive to oral iron therapy; for use during pregnancy and lactation. [Altana Inc] ‡

Chromaguard. Plastics materials in the form of films, sheets, strips or labels incorporating a photochromic image or pattern. [Courtaulds plc]

Chromalay. Materials for thin-layer chromatography. [May & Baker Ltd] *

Chromalbin. Albumin, chromated Cr51 serum; radioactive agent. [Bristol-Myers Squibb Co Inc] *

chromaline. A chrome mordant made by reducing chromic acid with glycerin; used for printing chrome colors on wool.

Chroma-Lite®. Bonded combinations of colored pigments and bismuth oxychloride on mica; low-dusting powds. imparting a subdued satiny luster to pressed powd. products [Van Dyk]

chromaloy. Nickel-chromium-iron alloys; the specific gravity varies from 8.15-8.35, and the melting point 1360-1390 C.

chromaluminum. An alloy of aluminum, chromium, and other metals. It has a specific gravity of 2.9.

Chroman B. A Rohn alloy containing 64% nickel, 20% iron, 15% chromium, and 1% manganese.

Chroman Co. A Rohn alloy containing 79% nickel, 20%. chromium, and 1% manganese.

Chromargans. Nonoxidizing steels, contain chromium; non-magnetic; used for the manufacture of turbine blades, valves, cutlery, etc.

Chromaset DF-100. Cotton fixative for indigo and direct dyes. [Henkel/Textiles]

Chromasist 1487A. CAS 9084-06-4; Sodium naphthalene sulfonate; dispersing and leveling agent for disperse dyes; esp. for high-temp. dyeable polyester; high molecular weight [Henkel/Textiles]

Chromastral. Pigments for paints and plastics. [ICI Chem. & Polymers Ltd]

chromatized gelatin. (chrome cement, chrome glue). Made by adding 1 part potassium dichromate to 5 parts of a solution (5-10%) of gelatin; a cement for glass.

Chromatogram. Materials for thin layer chromatography.

---

‡ Trade name and manufacturer not verified   § Trade name without identified manufacturer

[Kodak Ltd] *

chromax. An electrical resistance alloy containing 75% nickel and 25% chromium.

chromax bronze. An alloy of 15% nickel, 67% copper, 12% zinc, 3% aluminum, and 3% chromium.

Chrombral. Flux for copper/chromium alloys. [Foseco (F.S.) Ltd]

chrome. (acid alizarin, acid anthracene, diamond salicine). Acid chrome colors for wool.

chrome alum. Potassium chromium alum, $Cr_2(SO_4)_3 \cdot K_2SO_4 \cdot 24H_2O$.

chrome amalgam. An alloy of chromium and mercury obtained by the electrolysis of chromic chloride, using a mercury cathode.

chrome black. Anhydrous copper chromate.

chrome bronze. Crystalline chromium oxide obtained by heating potassium bichromate with sodium chloride or in a stream of hydrogen. It has a metallic sheen, a specific gravity of 5.61, and cuts glass.

chrome brown. Manganese chromate.

chrome cinnabar. *See* chrome red.

Chromeduol. Concentrated chrome sulfate powders; used for tanning, mineral dyeing and electrolytic purposes. [Lancashire Chemical Works Ltd]

chrome emerald green. *See* chromium green.

Chrome Fast Cyanine B, BN. Dyestuffs; British brands of Palatime chrome blue. §

chrome garnet. *See* chrome red.

chrome green. CAS 7758-97-6; Tetramethyldiamino-triphenylcarbinol-m-carboxylic acid, $C_{24}H_{25}N_2O_3$, is known as chrome green. It dyes chromed wool green, also used in cotton printing. The name has also been used for various other pigments.

Chrome Green 106. CAS 1308-38-9, 7787-59-9; Chromium hydroxide, bismuthoxychloride; inorg. colorant. [Presperse]

chrome iron. (ferrochrome). An alloy of iron and chromium, usually containing from 62-68% chromium; used in the manufacture of chrome steel.

chromels. Nickel/chromium or nickel/chromium/iron alloys. They are used for heating elements, and as molds for glass.

Chrome-nickel steel, high. These alloys usually contain 70-75% iron, 17-20% chromium, and 8-10% nickel.

chrome orange. *See* chrome red.

chrome prune. A mordant dyestuff. It gives claret shades with chrome mordants.

chrome red. CAS 18454-12-1; (Austrian cinnabar, Chinese red, Persian red, Victoria red. Vienna red, derby red, chrome cinnabar, chrome orange, American vermilion, chrome garnet, chrome ruby, chrome carmine, Chinese scarlet). Pigments consisting of basic lead chromate, $PbCrO_4Pb(OH)_2$

chrome ruby. *See* chrome red.

chrome steels. These steels represent a range of alloys containing from 0.2-2% carbon and from 0.5-15% chromium. Ball-bearing steel contains 1% carbon and 1% chromium, chrome die steel contains 2% carbon and 12% chrom

chrome steels, high. These usually contain from 64-81% iron and 18-35% chromium.

Chrometan. Chrome tanning powders. [British Chrome & Chemicals Ltd]

Chrome-tin Pink. Obtained by calcining a mixture of stannic oxide and a small amount of chromic oxide.

Chrometrace. CAS 10025-73-7; Chromic chloride hexahydrate; supplement. [Armour Pharmaceutical Co] *

chrome yellow. A pigment. It is normal lead chromate, $PbCrO_4$. The compounds, $PbCrO_4 \cdot PbSO_4$ and $PbCrO_4 \cdot 2PbSO_4$, are also known as chrome yellow.

Chromglaserite. A double salt, $3K_2CrO_4 \; Na_2CrO_4$.

chromia. *See* chromium oxide (ic).

chromic acetate. *See* chromium acetate (ic).

chromic acid. (chromium trioxide; chromic anhydride) CrO_3; CAS 7738-94-5; Chemicals (chromates, oxidizing agents, catalysts), medicine, process engraving, anodizing, ceramic glazes, colored glass, metal cleaning, inks, tanning, paints, textile mordant, etchant for plastics. [Aldrich; Atochem N. Am.; Biritsh Chrome & Chem.; OxyChem; Rit-Chem; Spectrum Chem. Mfg.]

chromic anhydride. *See* chromic acid.

chromic chloride. *See* chromium chloride (ic).

chromic oxide. *See* chromium oxide (ic).

chromidium. A nickel-chromium cast iron.

Chromiform. A reagent used for the preservation of milk samples. It consists of pastilles containing 0.25 g potassium dichromate and 0.25 g trioxymethylene.

Chromitan®. Basic chromium salts; for tanning and retanning leathers and furs. [BASF AG]

Chromitope Sodium. CAS 10039-53-9; Sodium chromate Cr 51; diagnostic aid; radioactive agent. [Bristol-Myers Squibb Co Inc] *

chromium. Metallic element; Cr; CAS 7440-47-3; EINECS 231-157-5; alloying and plating element for corrosion resistance, stainless steels, protective coatings, nuclear and high-temperature research, constituent of inorganic pigments. [Aldrich; Atomergic Chemetals; Cerac; Noah Chem.]

chromium acetate (ic). (chromic acetate) $Cr(C_2H_3O_2)_3 \cdot H_2O$; CAS 1066-30-4; Textile mordant, tanning, polymerization and oxidation catalyst, emulsion hardener. [Atomergic Chemetals; Noah Chem.; Spectrum Chem. Mfg.]

chromium bronze. The term applied to copper-zinc or copper-tin alloys to which chromium has been added up to 5%.

chromium (II) chloride anhydrous. *See* chromium chloride (ous).

chromium chloride (III) anhydrous. *See* chromium chloride (ic).

chromium chloride (ic). (chromium chloride (III) anhydrous; chromium trichloride; chromic chloride; chromium sesquichloride) $CrCl_3$; CAS 10025-73-7; EINECS 233-038-3; Chromium salts, intermediates, textile mordant, chromium plating, preparation of sponge chromium, catalyst for polymerizing olefins, waterproofing. [Atomergic Chemetals; Cerac; Hoechst Celanese]

chromium chloride (ous). (chromium (II) chloride anhydrous; chromous chloride) $CrCl_2$; CAS 10049-05-5; EINECS 233-163-3; Reducing agent, catalyst, reagent, chromizing. [Atomergic Chemetals; Cerac; Noah Chem.]

chromium copper. An alloy of copper and chromium, containing 10% chromium; used in the manufacture of hard steels. Also added to increase elasticity.

chromium green. (Guignet's green, chrome emerald green, Mittler's green, permanent green, emerald green, Veridian, chrome green,, French Veronese green). Green pigments consisting of hydrated sesquioxide of

---

* Trade name not verified as available   † Trade name verified as obsolete

chromium, with phosphate or borate of chromium.

chromium manganese. An alloy containing 30% chromium and 70% manganese; used in the manufacture of hard steels. Also added to copper to increase elasticity.

chromium molybdenum. An alloy of 50% chromium and 50% molybdenum; used in the manufacture of hard steels.

chromium-molybdenum steel. Alloys containing from 0.06-1.2% carbon, traces to 15% molybdenum, and traces to 6% chromium.

chromium nickel. An alloy of 10% chromium and 90% nickel, also 50% chromium, and 50% nickel; used in the manufacture of hard steels.

chromium (III) oxide. See chromium oxide (ic).

chromium (VI) oxide. See chromium oxide (ous).

chromium oxide (ic). (chromic oxide; chromia; chromium (III) oxide; green cinnabar; chromium sesquioxide) Cr_2O_3; CAS 1308-38-0; EINECS 215-160-9; Metallurgy, green paint pigment, ceramics, catalyst in organic synthesis, green granules in asphalt roofing, component of refractory brick, abrasive. [Atomergic Chemetals; British Chrome & Chem.; Noah Chem.]

chromium oxide (ous). (chromium (VI) oxide) CrO_3; CAS 1333-82-0; EINECS 215-607-8.

Chromium Sandoz. A proprietary preparation of chromium sesquioxide. [Sandoz] *

chromium sesquichloride. See chromium chloride (ic).

chromium sesquioxide. See chromium oxide (ic).

chromium sulfate (ous). (chromous sulfate) $CrSO_4 \cdot 5H_2O$; Oxygen scavenger, reducing agent, analytical reagent.

chromium trichloride. See chromium chloride (ic).

chromium trioxide. See chromic acid.

chromium-vanadium-molybdenum steels. Alloys of iron containing 0.1-0.55% carbon, 0.22-1.45% molybdenum 0.8-1.5% chromium, and 0.15-0.45% vanadium.

chromium-vanadium steel. An alloy of this type contains 0.3-0.4% carbon, 1-1.5% chromium, and 0.15-0.25% vanadium.

Chromol. Metachrome dyestuffs. [James Robinson & Co Ltd] †

Chromolay. Materials for thin layer chromatography. [May & Baker Ltd] *

Chromosal®. Different types of tanning materials containing chrome oxide. [Bayer AG]

Chromospun. Acetate spun yarn. [Eastman]

chromotrope acid. (Chromogen I, Chromogen C, LL). Dihydroxynaphthalene-disulfonic acid. An acid dye for wool.

chromous chloride. See chromium chloride (ous).

Chromovan Steel. A proprietary trade name for a non-sparking tool steel containing 12.5% chromium, 0.8% molybdenum, 1% vanadium, and 1.6% carbon. §

chronin. An alloy of 84% nickel and 15% chromium.

chronite. Heat-resisting alloys containing 63-67% nickel, 13-16% chromium, 12-20% iron, 0-0.4% silicon, 0.1% manganese, and 0-0.8% aluminum.

Chronogyn. Danazol; anterior pituitary suppressant. [Sterling Drug Inc] *

Chronulac. CAS 4618-18-2; Lactulose; for treatment of obstipation and (pre-) hepatic coma. [Duphar BV] *

Chronulac. CAS 4618-18-2; Lactulose; laxative. [Merrell Dow Pharmaceuticals Inc] *

chrysazin. CAS 117-10-2; Dihydroxy anthraquinone; used in medicine.

chrysazol. Dihydroxyanthracene.

chrysin. 1, 3-Dihydroxyflavone. A pigment found in poplar buds.

Chrysine. See Ozokerine. §

Chrysocale. A jeweler's alloy, containing 9 parts copper, 8 parts zinc, and 2 parts lead.

Chrysochalk. (Gold-copper). An alloy containing 59-93% copper, 8-39% zinc, and 1.6-1.9% lead. A jeweler's alloy.

Chrysoform. Dibromodiiodohexamethylene-tetramine, $C_6H_8Br_2I_2N_4$; used as an antiseptic.

Chrysoidine crystal. A dyestuff. It consists of the hydrochloride of phenylazo-m-phenylenediamine, with some of the homologues from o- and p- toluidine. Dyes wool and silk orange.

Chrysorin. An alloy of 66% copper and 34% zinc.

chrysotile. (Chrysotile asbestos, Canadian asbestos). An asbestos mineral, the average composition of which is 40.5% silica, 41.5% magnesium oxide, 14% water, 2.6% iron oxides, and 1.3% aluminum oxide; yields the best type of asbestos.

Chryzoplus. CAS 133-32-4; 4-Indol-3-ylbutyric acid; a root growth promoter. [Fargro Ltd]

Chryzopon. CAS 133-32-4; 4-Indol-3-ylbutyric acid; a root growth promoter. [Fargro Ltd]

Chryzosan. CAS 133-32-4; 4-Indol-3-ylbutyric acid; a root growth promoter. [Fargro Ltd]

Chryzotek. CAS 133-32-4; 4-Indol-3-ylbutyric acid; a root growth promoter. [Fargro Ltd]

CHT Activator NB. Hydrogen peroxide activator for low pH bleaching. [Catawba-Charlab]

CHT Antifoam MI. Mineral oil-based defoaming agent. [Catawba-Charlab]

CHT Biavin 109. Concentrate gliding, anticreasing and leveling agent for cotton. [Catawba-Charlab]

CHT Carrier GR-A. Carrier for dyeing of polyester and PES blends. [Catawba-Charlab]

CHT Contavan ALR. Org. stabilizer for alkaline peroxide bleaching. [Catawba-Charlab]

CHT Cotoblanc HTD-N. Low-foam alkali-stable scouring auxiliary and chelate. [Catawba-Charlab]

CHT Defoamer SC. Deaerating agent for transfer printing inks. [Catawba-Charlab]

CHT Egasol SP. Leveling and penetrating agent for rapid dyeing of polyester. [Catawba-Charlab]

CHT Felosan TAK-NO. Low foaming stain remover and detergent. [Catawba-Charlab]

CHT Heptol NWS. Polyphosphate; for textile washing. [Catawba-Charlab]

CHT Intensol TH-B. Dyestuff solv. and machine cleaner. [Catawba-Charlab]

CHT Lavotan DS. Wetting, washing and cleaning surfactant. [Catawba-Charlab]

CHT Lustraffin BA. In-bath yarn lubricant for textile knitting and weaving. [Catawba-Charlab]

CHT Meropan BRE. Peroxide neutralizer. [Catawba-Charlab]

CHT Prisulon SNP-113S. Synthetic thickener for transfer paper printing. [Catawba-Charlab]

CHT Rapidoprint M-4. Leveling agent for transfer paper printing. [Catawba-Charlab]

CHT Retinol M. Stripping and leveling agent with dyestuff affinity. [Catawba-Charlab]

CHT Rewin MRT. Nonformaldehyde cationic dye fixative. [Catawba-Charlab]

---

‡ Trade name and manufacturer not verified   § Trade name without identified manufacturer

CHT Sarabid DLO Conc. Auxiliary with dyestuff affinity for pre- and aftertreatment. [Catawba-Charlab]

CHT Subitol HLF Conc. Low foaming alkali-stable surfactant for continuous bleaching. [Catawba-Charlab]

CHT Thickener 8300E. Extremely pure synthetic thickener for pigment prints with soft hand; suitable for very fine mesh screens. [Catawba-Charlab]

CHT Tubingal 220A. Textile softener. [Catawba-Charlab]

CHT Tubiprint PERL C. Pigment printing paste for pearlescent effects. [Catawba-Charlab]

CHT Viscavin DMS. Anticreasing agent with leveling and dispersing props. for polyester. [Catawba-Charlab]

Chymacort. A proprietary preparation of hydrocortisone acetate, neomycin palmitate and pancreatic enzymes; used in dermatology. [Armour Pharmaceutical Co] *

Chymar. A proprietary preparation of α-chymotrypsin; for treatment of wounds. [Armour Pharmaceutical Co] *

Chymar Ointment. A proprietary preparation of hydrocortamine hydrochloride, neomycin palmitate and pancreatic enzymes; used in dermatology. [Armour Pharmaceutical Co] *

Chymex. CAS 37106-97-1; Bentiromide; diagnostic aid. [Adria Laboratories Inc] *

Chymocyclar. A proprietary preparation of tetracycline, trypsin, and chymotrypsin; antibiotic. [Armour Pharmaceutical Co] *

Chymoral. A proprietary preparation of trypsin and chymotrypsin; digestive enzymes. [Armour Pharmaceutical Co]*

Chymosin. Rennin.

CI-2. CAS 25928-94-3; Two-part epoxy compound; coil impregnant for elec. components; also for casting and coating applications. [Bacon]

CI7810. CAS 18395-30-7; Isobutyltrimethoxysilane; coupling agent, chem. intermediate, blocking agent, release agent, lubricant, primer, and reducing agent. [Hüls Am.]

CI7840. CAS 24801-88-5; Isocyanatopropyltriethoxy silane; CAS 24801-88-5; coupling agent, chem. intermediate, blocking agent, release agent, lubricant, primer, and reducing agent. [Hüls Am.]

Cialit. The sodium salt of 2-ethyl mercury mercaptobenzoxazol-5-carboxylic acid; an antifouling pigment.

Ciatyl®. Neuroleptic. [Bayer AG]

Ciba. Vat dyes. [Ciba plc] *

Ciba 1906. A proprietary preparation containing thiambutosine. [Ciba plc] *

Cibacet. Disperse dyes. [Ciba plc] *

Cibacoll™. α-Amylase substrate. [Calbiochem Corp]

Cibacrolan. Reactive dyestuffs. [Clayton Aniline Co Ltd] *

Cibacron. Reactive dyes. [Ciba plc] *

Cibalan. Wool dyestuffs. [Ciba plc] *

Cibalith-S. CAS 919-16-4; Lithium citrate tetrahydrate; antimanic. [Ciba-Geigy Corp]

Cibamin. Modified melamine/formaldehyde resins. [Ciba plc] *

Cibanite. A proprietary trade name for an aniline-formaldehyde resin resistant to water, oil, alkalis, and organic solvents. §

Cibanoid. A proprietary trade name for urea-formaldehyde synthetic resins. §

Cibanone. Vat dyes. [Ciba plc] *

Cibaphasol 6042. A proprietary trade name for a fatty acid amide derivative giving improved quality and appearance to continuous dyeings; used as a dyeing auxiliary

for polyacrylonitrile fibers. [Ciba plc] *

Cibatex 248. A proprietary trade name for a phenol sulfonic acid derivative which gives improved wet fastness properties of dyeings; used also as a synthetic tanning agent for dyes on nylon 66. [Ciba plc] *

Cibatex PA. A proprietary trade name for a synthetic tanning agent of the phenol sulfonic acid type used as an improver of wet fastness in the dyeing of nylon. [Ciba plc]*

Cicatrin. A proprietary formulation containing neomycin sulfate and bacitracin zinc; for treatment and prophylaxis of superficial bacterial skin infections. [The Wellcome Foundation Ltd]

Cicuta. Conium.

cidrase. Cider yeast.

cignolin. CAS 58-93-5; 1,8-Dihydroxyanthranol.

CIL. Silica product; corrosion inhibitor for water systems incl. potable water systems, cannery rinse water, etc.; forms barrier-type protective film on internal metal surfaces handling treated water. [Drew Ind. Div.]

Cilbond. A range of rubber to metal and rubber to plastics adhesives. [Compounding Ingredients Ltd]

Cilcast. A two-component cold cure polyurethane elastomer. [Compounding Ingredients Ltd]

Cilloral. CAS 113-98-4; Penicillin G potassium; antibacterial. [Bristol-Myers Squibb Co Inc]

Cimet. Stainless alloys containing about 48% chromium with iron.

Cimflx 606. CAS 14087-96-6; Talc; reinforcement in PP homopolymer; offers unique color and brightness in thermoplastics, good long-term heat resist. in polyolefins. [Luzenac Am.]

Cimmol. Cinnamyl hydride.

Cimpact 699. CAS 14807-96-6; Talc; reinforcer giving superior impact strength in polyolefins, excellent stability in gel coats. [Luzenac Am.]

Cimpact 710. CAS 14807-96-6; Talc; for high impact, TPO, auto, and appliance applications. [Luzenac Am.]

Cindal. An aluminum alloy containing zinc and small amounts of magnesium and chromium.

Cindol. CAS 63323-46-6; Ciramadol hydrochloride; analgesic. [Wyeth Laboratories]

Cindumix. Coal tar/bitumen mixture; surface dressing material for roads. [Cindu Chemicals BV] *

cinene. See dipentene.

cineol. See eucalyptol.

cinereine. An induline dyestuff obtained from azoxyaniline, aniline hydrochloride, and p-phenylenediamine.

cinnabar. (Chinese red, mercury blend vermilion, liver ore, patent red, cinnabarite). It is mercuric sulfide, HgS; used as a pigment.

cinnabar green. See chrome green.

cinnabarite. See cinnabar.

Cinnaloid. CAS 24815-24-5; A proprietary preparation of rescinnamine; an antihypertensive. [Pfizer International]

cinnamaldehyde. See cinnamic aldehyde.

cinnamein. CAS 103-41-3; A term applied to benzyl cinnamate. It is, however, also used for the mixture of ester and alcohol in the Balsams of Tolu and Peru, which are not extractable by alkalies from an ethereal solution.

cinnamic acid. (β-phenylacrylic acid; 3-phenylpropenoic acid; cinnamylic acid) C_6H_5CH:CHCOOH; CAS 140-10-3; EINECS 205-398-1; Medicine (anthelmintic), perfumes, intermdiate. [Aceto; Hüls Am.; Penta Mfg.; Raschig]

---

* Trade name not verified as available   † Trade name verified as obsolete

cinnamic aldehyde. (cinnamaldehyde; trans-3-phenyl-2-propenal) $C_6H_5CH:CHCHO$; CAS 14371-10-9; EINECS 203-213-9; Flavors, perfumery. [Nipa Labs; Penta Mfg.; Quest Int'l.]

cinnamoyl chloride. (trans-3-phenylacryloyl propenoyl chloride; phenylacrylyl chloride) $C_6H_5CH:CHCOCl$; CAS 102-92-1; EINECS 203-065-5; Reagent for determination of water, chemical intermediate. [ICI Specialties; Janssen Chimica; Penta Mfg.; Raschig; Schweizerhall]

cinnamylic acid. See cinnamic acid.

Cinnar. CAS 298-57-7; A proprietary preparation containing cinnarizine; anti-motion sickness. [Janssen Pharmaceutical Ltd] *

Cinobac. CAS 28657-80-9; Cinoxacin; antibacterial. [Eli Lilly & Co]

Cinopal. CAS 36330-85-5; Fenbufen; anti-inflammatory. [Lederle Laboratories USA] *

Cinquasia. High performance organic pigments. [Ciba plc]

Cin-Quin. CAS 50-54-4; Quinidine sulfate; cardiac depressant. [Rowell Laboratories Inc] *

Cipralan. Cibenzoline; cardiac depressant. [Hoffmann-LaRoche Inc] *

Ciprobay®. CAS 85721-33-1; Pharmaceutical preparation containing ciprofloxacin; also known as Ciproxin; broad spectrum antibiotic. [Bayer AG]

Ciproxin. CAS 85721-33-1; Pharmaceutical preparation containing ciprofloxacin; also known as Ciprobay; broad spectrum antibiotic. [Bayer AG]

Cirami No. 1. Beeswax, candelilla wax, shea butter; natural thickener. [Alban Muller]

Circacid. Dairy hygiene compound. [The Wellcome Foundation Ltd] *

Circadet. Surface active detergent. [Ciba plc] *

Circaline MK 11. Hypochlorite pipe line cleaner/sterilizer. [Ciba plc]

Circanol. Ergoloid mesylates; cognition adjuvant. [3M Pharmaceuticals] †

Circosan. Dairy hygiene detergent sterilizer. [The Wellcome Foundation Ltd] *

Circupon®. Vasoactive preparation. [Bayer AG]

Cire De Lanol CTO. Ceteareth-33 and cetearyl alcohol; cosmetic emulsifier. [Seppic]

Cirin. CAS 50-78-2, CAS 50-81-7; A proprietary preparation of aspirin and ascorbic acid 6-1; an analgesic. [Zemmer Co Inc] ‡

Cirol. CAS 68334-28-1; Partially hydrogenated vegetable oil (soybean, cottonseed); kosher; domestic oil for nut roasting, margarines, frozen desserts, coffee whiteners. [Van Den Bergh Foods]

Cirrasol. A textile softening agent, fiber lubricant or antistatic agent. [ICI Chem. & Polymers Ltd]

cis-butenedioic acid. See maleic acid.

cis-butenedioic anhydride. See maleic anhydride.

Cisdene®1203. CAS 9003-17-2; Stereospecific polybutadiene, high cis, nonstaining antioxidant; used in mixts. with SBR or NR to enhance abrasion resist., reduce shrinkage, improve low temp. props.; vulcanized with sulfur and conventional accelerator systems. [Am. Syn. Rubber]

cis 13-docosenamide. See erucamide.

Cismollan® BH. Highly bactericidal and fungicidal soaking agent for dried and salted raw hides, outstanding wetting action. [Bayer AG]

cis-9-octadecenoic acid. See oleic acid.

cis-9-octadecen-1-ol. See oleyl alcohol.

Cisplatin. CAS 15663-27-1; Injection containing cisplatin; antineoplastic. [Lederle Laboratories] *

cis-1,4-polyisoprene rubber. See polyisoprene.

Citanest. CAS 1786-81-8; Prilocaine hydrochloride; anesthetic. [Astra Pharmaceutical Products Inc] *

Citarin®L. Anthelmintic against lungworms and gastrointestinal nematodes in cattle, sheep, goats and pigs; veterinary medicine. [Bayer AG]

Cithrol. Polyethylene glycol esters; used as nonionic surfactants. [Croda Surf. Ltd]

Cithrol GMS A/S. Acid-stable glyceryl monostearate; a nonionic surfactant used in the cosmetic and pharmaceutical industries. [Croda Chem. Ltd]

Citmol 316. Triisocetyl citrate; emollient for cosmetics, cleansing creams, lipsticks; pigment dispersant. [Bernel; Heterene]

Citmol 320. Trioctyldodecyl citrate; noncomedogenic, high visc. emollient, film-former, castor oil replacement for lipsticks. [Bernel]

Citobaryum. CAS 7727-43-7; A proprietary preparation of special barium sulfate. §

Citotray. Light curing plastic for the preparation of custom impression trays; used in dentistry. [Bayer AG]

Citowett. Nonionic spreader containing 99-100% alkylaryl polyglycol ether; wetting agent for herbicide, fungicide and insecticide sprays. [BASF plc]

Citraclean. CAS 77-92-9; Citric acid metal cleaning process. [John & E Sturge Ltd, Selby] *

Citra-Gran. A proprietary preparation containing tartaric acid sodium bicarbonate, citric acid, potassium bicarbonate, magnesium sulfate, lithium benzoate, anhydrous sodium phosphate and calcium lactophosphate; an antacid. [AKU Holland] ‡

citral (cis and trans). (3,7-dimethyl-2,6-octadienal) $(CH_3)_2C:CHCH_2CHC(CH_3)CHCHO$; CAS 5392-40-5; EINECS 226-394-6; Perfumes, flavoring agents, intermediate for other fragrances, vitamin A synthesis. [Lucta SA; Penta Mfg.; Schweizerhall; SCM Glidco Organics]

Citralka. A proprietary preparation containing sodium acid citrate. [Parke-Davis] *

Citranova. Synthetic citrus oils.

Citranox®. Blend of organic acids, anionic and nonionic surfactants, and alkanolamines; phosphate-free biodeg. acid detergent, wetting agent for cleaning dairy equip., laboratory ware, clean rooms, optical and electronic parts, pharmaceutical apparatus, industrial parts, etc. [Alconox]

Citrest. A citric acid ester. [Rhone-Poulenc Surf.] *

citric acid. (2-hydroxy-1,2,3-propanetricarboxylic acid; β-hydroxytricarballylic acid) Organic acid; $HOC(COOH)(CH_2COOH)_2$; CAS 77-92-9; EINECS 201-069-1; preparation of citrates, flavoring extracts, confections, soft drinks; antioxidant in foods; sequestering agent; detergent builder; metal cleaner. [Cargill; R.W. Greeff; Haarmann & Reimer; Hoffmann-La Roche; Penta Mfg.]

citric acid trisodium salt dihydrate. See sodium citrate.

citrin. Vitarain P.

Citroflex 2. CAS 77-93-0; Triethyl citrate; plasticizer for cellulosics. [Morflex]

Citroflex 2. CAS 77-93-0; Triethyl citrate; a vinyl plasticizer. [Pfizer International]

Citroflex 4. CAS 77-94-1; Tri-n-butyl citrate; plasticizer for

---

‡ Trade name and manufacturer not verified   § Trade name without identified manufacturer

cellulosics. [Morflex]

Citroflex 4. CAS 77-94-1; Tributyl citrate; a vinyl plasticizer. [Pfizer International]

Citroflex A-2. CAS 77-89-4; Acetyltriethyl citrate; plasticizer for cellulosics. [Morflex]

Citroflex A-2. CAS 77-89-4; Acetyl triethyl citrate; a vinyl plasticizer. [Pfizer International]

Citroflex A-4. CAS 77-90-7; Acetyl tri-n-butyl citrate; plasticizer for PVC, PVDC, esp. food films, medical articles. [Morflex]

Citroflex A-4. CAS 77-90-7; Acetyl tributyl citrate; a vinyl plasticizer. [Pfizer International]

Citroflex A-6. CAS 24817-92-3; Acetyl tri-n-hexyl citrate; vinyl plasticizer for medical applications and other toxicologically sensitive areas. [Morflex]

Citroflex A-8. Acetyl tri-2-ethylhexyl citrate; a vinyl plasticizer. [Pfizer International]

Citroflex B-6. CAS 82469-79-2; n-Butyryl tri-n-hexyl citrate; vinyl plasticizer for medical applications. [Morflex]

citronellal. (3,7-dimethyl-6-octenal) $C_{10}H_{20}O$; CAS 106-23-0; EINECS 203-376-6; Soap perfumery, manufacture of hydroxycitronellal, insect repellant. [Penta Mfg.; PMC Specialties; Schweizerhall]

citronellol. (3,7-dimethyl-6-octen-1-ol) $(CH_3)_2C:CH-CH_2CH_2CH(CH_3)CH_2CH_2OH$; CAS 106-22-9; EINECS 203-375-0; Perfumery (floral odors). [Int'l. Flavors & Fragrances; Penta Mfg.; SCM Glidco Organics]

Citrozone. Vanadium sodium citrochloride.

Cladex®. Metal clad products. [Asahi Chem. Industry]

Claforan. CAS 64485-93-4; Cefotaxime sodium; antibiotic. [Roussel Laboratories Ltd]

Clairolite L. Calcium bentonite; clay with high brightness and low visc.; plasticizer and binding agent for white firing ceramic; moisture absorbent, anticaking agent. [Kaopolite]

Clairsol. Highly dearomatized aliphatic hydrocarbon solvents. [Carless Refining & Marketing Ltd]

Clampdown. Silage treatment. [May & Baker Ltd] *

Claradex CH-540. CAS 9003-56-9; ABS resin; general purpose for injection molding and extrusion. [Shin-A Chem. Mfg.; Syn. Rubber Tech.]

Claradin. CAS 50-78-2; A proprietary preparation of aspirin in an effervescent base; an analgesic [Nicholas Laboratories Ltd] *

Clar-Apel. A proprietary viscose plastic used as a packing material. §

Clarase® 5,000, 40,000. CAS 9000-92-4; α-Amylase; enzyme for hydrolysis of starch and dextrin; for starches, syrups, brewing, fruit juices. [Solvay Enzymes]

Clarcel. CAS 7631-86-9; Diatomaceous earth filter aid; for liquids filtration in chemical, pharmaceutical and food industries. [Atochem UK/Ceca]

Clarcel Flo. Perlite-based filter aid; for liquids filtration in chemical, pharmaceutical and food industries. [Atochem UK/Ceca]

Clarex® L. CAS 9032-75-1; Pectinase; enzyme; depectinization of fruit juices, grape processing, jams and jellies, berry processing and wine production [Solvay Enzymes]

Clarfina. Synthetic zeolite; raw material for detergents. [SNIA (UK) Ltd]

Clarifex 800. Clarifier. [ICI Polymer Additives]

Clarifloc. Flocculants and coagulants; clarification of water and waste water, sludge treatment, thickening and de-watering. [Allied-Signal Inc/Water Treatment Div] *

Clarifloc. CAS 9000-07-1; Carrageenan. [Rhone-Poulenc UK]

Clarifoil. Films or pellicles made from cellulose acetate or other cellulose derivatives or cellulose; sold in the form of sheets, bands or strips; for wrapping or packing or electrical insulating purposes. [Courtaulds plc]

Clarin. A proprietary preparation of hydroxizine hydrochloride, ephedrine sulfate and theophylline; an anti-asthmatic. [Pfizer International] †

Clarion®. For the agriculture industry. [DuPont UK]

Clar+Ion®A410P. CAS 10043-01-3; Aluminum sulfate and flocculating agent; coagulant and flocculant for industrial and potable water treatment applications; esp. where rapid settling is desired; emulsion breaker. [General Chem.]

Clarit. A particularly pure calcium bentonite for the adsorptive stabilization of wines, fruit juices, vinegar etc. [Süd-Chemie AG] *

Clarital®. Polyvinyl acetate; antitranspirant. [Fargro Ltd]

Clarite. Pretreatment agent. [Ciba plc] *

Clark's patent alloy. An alloy of 75% copper, 7.2% zinc, 14.5% nickel, 1.9% tin, and 1.9% cobalt.

Clar-O-Cel. Cellulose fiber filter aids; for liquids filtration in chemical, pharmaceutical, and food industries. [Atochem UK/Ceca]

Clarosan 1FG. CAS 886-50-0; Granules containing 1% w/w terbutryn; for aquatic weed control. [Ciba-Geigy Agrochemicals]

Clarstabil. Clarification of wines, beers, etc. [Minas de Gádor SA] †

Clarus metal. A light aluminum alloy used for sheets and tubes.

Clarvin. Clarification of wines, beers, etc. [Minas de Gádor SA]

Claudilithe. See Galalith. §

Clay Breaker. Mineral gypsum; clay breakdown agent. [Vitax Ltd]

Claymaster. Polystyrene boards of low density colored pink; compressible fill beneath concrete foundations. [Vencel Resil Ltd]

Clay, Porcelain. See China clay.

Claysil. A silicate adhesive. [Crosfield Chemicals Ltd]

Clay, White. See China clay.

CLD 2. Croscarmellose sodium; carboxymethylcellulose sodium that has been internally cross-linked; pharmaceutic aid. [Buckeye Cellulose Corp] *

Cleanacres CMPP. CAS 7085-19-0; Soluble concentrate containing 600 g/l mecoprop; for control of weeds in cereals and grassland. [Cleanacres Ltd]

Cleanacres PDR 675. CAS 999-81-5; Soluble concentrate containing 675 g/l chlormequat; plant growth regulator. [Cleanacres Ltd]

Clean-Up. Contains concentrated tar acids in emulsion; general moss-killer and sterilizer for garden use. [ICI Garden Products]

Clean Wiz. Proprietary solvent and aqueous formulations to clean and strip build-up and residue of resin and mold release from metal and composite molds; for fiberglass, polyurethane foam and rubber. [Axel Plastics Research Laboratories Inc]

Cleapact TI-100. Styrene-based copolymer; thermoplastic with excellent balance of transparency, impact strength, processibility; injection molding grade for copy machine

---

trays, printer covers, packaging materials for elec. equip. or food. [Dainippon Ink & Chem.]

Cleapact TI-100S. Styrene-based copolymer; thermoplastic with excellent balance of transparency, impact strength, processibility; extrusion grade for copy machine trays, printer covers, packaging materials for elec. equip. or food. [Dainippon Ink & Chem.]

Clearam. A range of modified starches that require heating in aqueous solution to develop their viscosity; used in the food industry to develop viscosity in a wide number of applications e.g. sauces, gravies, custards etc. [Roquette (UK) Ltd]

Clearasil Adult Care. Medication for adult skin blemishes. [Richardson-Vicks Inc] *

Clearasil Super Strength. CAS 94-36-0; Acne treatments with 10% benzoyl peroxide (tinted or vanishing cream or lotion formula), to help clear acne blemishes and absorb excess skin oil. [Richardson-Vicks Inc] *

Clearbreak TEB. Polymeric amine; reverse emulsion breaker for oil treating; may be combined with metal salts or anionic polymers for enhanced performance. [Chemron]

Clear by Design. CAS 94-36-0; Benzoyl penoxide; keratoytic. [Herbert Laboratories] *

Clearcol. CAS 9007-34-5; Soluble animal collagen; protein for moisturizing and conditioning of facial systems where clarity is important. [Croda Inc.]

Clear Conc. 7174. Syn. thickener for aq. pigment printing systems. [Catawba-Charlab]

Clear Eyes. CAS 550-29-2; Naphazoline hydrochloride; adrenergic. [Abbott Laboratories] *

Cleargum®. Thin-boiling food modified starch; used for foods, confectionery and coatings. [Roquette (UK) Ltd]

Clearine. A proprietary preparation containing distilled witch hazel and naphazoline hydrochloride; for ophthalmic use. [The Boots Co plc]

Clearon. CAS 2244-21-5; Sodium dichloroisocyanurate dihydrate. [Schering Agrochemicals Ltd] *

Clearon®. Terpene tackifying resin. [Aceto]

Clearpol. Alginate for brewing. [Alginate Industries Ltd] *

Clearsite. A proprietary trade name for cellulose acetate used for transparent containers. §

Clearsol®. Blend of surfactants and xylenols; disinfectants used for hospitals. [Coventry Chemicals Ltd]

Clear-Stat AS401. Nonamine antistatic concentrate for use with polyolefins. [Polycolors] *

ClearTint®. Nonpigmented, polymeric colorants used to obtain clear color in polyolefins and other plastics, particularly in clarified polypropylene; colorants for housewares, protective packaging, medical devices, blow molded bottles, and sheet. [Milliken]

Cleartuf Series. CAS 25038-59-9; PET polyester resin; thermoplastic providing clarity, good processability, chem. and solv. resist., high strength, impact resist. for bottles and cans; FDA compliance. [Goodyear] *

Clearway. CAS 61-82-5, 122-34-9; A suspension concentrate containing 100 g aminotriazole and 300 g simazine per liter; used for total weed control in non crop areas and fruit orchards. [Rhone-Poulenc Environmental Prods. Ltd]

Clearys Waterless Hand Cleaner. concentrate heavy-duty but gentle cleaning agent for removal of pesticide residues, grease, oil, tar, ink, paint, carbon, adhesives from hands, work clothes, cars. [W A Cleary]

Cleaval. Suspension concentrate containing 60 g cyanazine and 400 g mecoprop per liter; a contact herbicide. [Shell UK]

Clebrium alloys. Heat-resisting alloys. One contains 76.5% iron, 13% chromium, 3.6% molybdenum, 2.6% carbon, 2% nickel, 1.5% silicon, and 0.75% manganese.

Cleland reagent. See dithiothreitol.

Clelands Reagent. CAS 3483-12-3; Dithiothreitol; a white powder, m.p. 37 C; a reagent for protecting -SH groups in certain biochemical systems. [British Drug Houses] *

Clenecorn. CAS 7085-19-0; Soluble concentrate containing 570 g/l mecoprop; for control of weeds in cereals and grassland. [Farmers Crop Chemicals Ltd]

Clenesco. A proprietary trade name for anhydrous sodium metasilicate. §

Cleocin. CAS 18323-44-9; Clindamycin; antibacterial. [Upjohn]

Clerici solution. A molecular mixture of thallium malonate and formate; used for floating mineral specimens to determine the specific gravity.

Clerit. A horticultural fungicide. [ICI Chem. & Polymers Ltd]

Clerite. A horticultural fungicide. [Plant Protection] *

Clermait®. CAS 41083-11-8; Azocyclotin; acaricide used for control of red spider mites. [Bayer AG]

Cleroxide. A stabilizer for polyvinyl chloride. [Akzo Chemie]*

Cleve's acid. β-Acid is α-naphthylamine-6-sulfonic acid, or γ-Acid is α-naphthylamine 3-sulfonic acid, ϖ-Acid or J-acid is α-naphthylamine 7-sulfonic acid and ϖ- or δ-acid is α-nitronaphthalene-7-sulfonic acid.

Clevite®. A registered trademark. [Clevite Corporation] ‡

Cliché metal. An alloy of 33% tin, 46% lead, and 21% cadmium. Another alloy called by this name contains 48% tin, 32.5% lead, 9% bismuth, and 10.5% antimony.

Clifton Chlormequat 46. CAS 999-81-5; Soluble concentrate containing 460 g/l chlormequaternary plant growth regulator. [Clifton Chemicals Ltd]

Clifton CMPP Amine 60. CAS 7085-19-0; Soluble concentrate containing 600 g/l mecoprop; for control of weeds in cereals and grassland. [Clifton Chemicals Ltd]

Clifton Glyphosate Additive. CAS 61791-44-4; Wetting and spreading agent containing 850 g/l tallow amine ethoxylate for use with phosphonoglycine herbicides. [Clifton Chemicals Ltd]

Clifton Wetter. Nonionic adjuvant containing 250 g/l alkyl phenol ethoxylate; wetting and spreading agent for use in herbicides, fungicides, insecticides and foliar feeds. [Clifton Chemicals Ltd]

Climacel. Protective moisturizer. [Richardson-Vicks Inc] *

Climatone. A proprietary preparation of ethinoloestrodiol and methyl testosterone; used in the treatment of premenstrual menopausal symptoms. [Pharmax Ltd] *

Climax. A magnetic alloy containing 30% nickel and 70% iron. Another alloy containing 73% iron, 24.4% nickel, and 2.6% manganese has also been called Climax.

Climax 193. An alloy of 68% iron, 28% nickel, 2% chromium, and 1% manganese.

Climeline. A sodium phosphate used as a water softener.

Clinafarm. CAS 73790-28-0; A proprietary preparation containing imazalil; used for fumigation. [Janssen Pharmaceutical Ltd] *

Clinifeed 400. Protein, carbohydrate, fat, vitamins, and minerals; liquid feed. [Roussel Laboratories Ltd] †

Clinifeed Favour. Protein, carbohydrate, fat, vitamins and

---

‡ Trade name and manufacturer not verified   § Trade name without identified manufacturer

minerals; liquid feed. [Roussel Laboratories Ltd] †

Clinifeed ISO. Proprietary preparation of protein, carbohydrate, fat, vitamins and minerals; liquid feed. [Roussel Laboratories Ltd] †

Clinifeed Protein Rich. Protein, carbohydrate, fat, vitamins and minerals; liquid feed. [Roussel Laboratories Ltd] †

Clinimycin. CAS 79-57-2; A proprietary preparation of oxytetracycline; an antibiotic. [Glaxo Laboratories] *

Clinistix. A prepared test strip of glucose oxidase, peroxidase and o-toluidine; used for the detection of glucose in urine. [B C Ames] *

Clinitest. A prepared test tablet containing copper sulfate, sodium hydroxide, citric acid, and sodium carbonate; used for the detection of reducing substances in urine. [B C Ames] *

Clinitetrin. CAS 64-75-5; A proprietary preparation of tetracycline hydrochloride; an antibiotic. [Glaxo Laboratories] *

Clinium. CAS 3416-26-0; A proprietary preparation of lidoflazine; used for angina pectoris. [Janssen Pharmaceutical Ltd] *

Clinoril®. CAS 38194-50-2; Sulindac; antirheumatic. [Merck & Co Inc]

Clipper. Tree growth regulator. [ICI Chem. & Polymers Ltd]

Clistin. CAS 3505-38-2; Carbinoxamine maleate; antihistaminic. [McNeil Consumer Products Co] *

Clistin-D. Contains acetaminophen, carbinoxamine maleate and phenylephrine hydrochloride; analgesic; antipyretic; antihistaminic; adrenergic. [McNeil Consumer Products Co] *

Cloderm. CAS 34097-16-0; Clocortolone pivalate; glucocorticoid. [Ortho Pharmaceutical Corp] *

Cloisonné®. Highly lustrous pigments with deep colors produced by combination of light interference and light absorption. [Mearl]

Clomid. CAS 50-41-9; Clomiphene citrate; gonad-stimulating principle. [Merrell Dow Pharmaceuticals Inc] *

Clonevac-30. B₁ type, cloned lasota strain, Newcastle vaccine; for immunization of poultry. [Intervet Inc]

Clonevac-30-Ma5. B₁ type, clone 30 strain Newcastle and Ma5 Mass. type vaccine; for immunization of poultry. [Intervet Inc]

Clonevac-30T. B₁ type, cloned lasota strain, turkeys, Newcastle vaccine; for immunization of poultry. [Intervet Inc]

Clonevac D-78. Gumboro D-78 strain; for immunization of poultry. [Intervet Inc]

Clonopin. CAS 1622-61-3; Clonazepam; anticonvulsant. [Hoffmann-LaRoche Inc] *

Clont®. CAS 69198-10-3; Metronidazole hydrochloride; trichomonicidal and anaerobicidal preparation. [Bayer AG]

Cloparin. Chlorinated paraffins; plasticizers for PVC, paints; additives for lubricating and cutting oils and flame resistant additives. [Collinda Ltd]

Cloparin. Range of chlorinated paraffins. [SNIA (UK) Ltd]

Cloparol. Chlorinated paraffins. [SNIA (UK) Ltd]

Cloparten. Chlorosulfonated paraffins; raw material for the preparation of emulsifiable synthetic greasing agents for hides and leather. [Caffaro SpA] *

Cloparten Z. Chlorosulfonated paraffins. [SNIA (UK) Ltd]

Clophen. Synthetic liquid insulants and coolants for the electrical industry; used as high grade, flame-retardant impregnants for capacitors and as liquid coolants for transformers and rectifiers. [Bayer AG] †

Clorafin. Chlorinated paraffin; nonflammable, chemically resistant; good plasticizing properties; plasticizer in inks, additive in cutting oils and drawing compounds, plasticizing resin in chlorinated rubber corrosion-resistant coatings, plasticizer for vinyl resins, waterproofer and flameproofer for textiles. [Hercules] *

Cloran. A trademark for a chlorinated anhydride thermal and chemical stablilizer for polymers. [UOP Inc] *

Clorpactin WCS-90. CAS 52906-84-0; Oxychlorosene sodium; anti-infective, topical. [Guardian Chemical]

Clorpactin XCB. CAS 8031-14-9; Oxychlorosene; the hypochlorous acid complex of a mixture of the phenyl sulfonate derivatives of aliphatic hydrocarbons; antiinfective, topical. [Guardian Chemical]

Clortex. Chlorinated rubber; resin for the preparation of paints and varnishes. [Caffaro SpA] *

Clortol. Chlorine releasing sterilant. [Rhone-Poulenc UK]

Closyl 30 2089. CAS 61791-59-1; EINECS 263-193-2; Sodium-N-cocoyl sarcosinate; wetting, foaming detergent used in personal care and household products; corrosion inhibitor for mild steel. [Clough]

Closyl LA 3584. CAS 137-16-6; EINECS 205-281-5; Sodium lauroyl sarcosinate; soap-like detergent providing wetting and foaming; for personal care and household products; corrosion inhibitor for mild steel. [Clough]

cloth oil. That fraction obtained from the residue of Russian petroleum by distillation.

cloustonite. A variety of asphalt.

Clout. Herbicide. [May & Baker Ltd]

Clout. CAS 55635-13-7; Alloxydim-sodium; herbicide for control of annual weeds. [Embetec Crop Protection Ltd]

Clovacorn Extra. An emulsifiable concentrate containing 220 g 2,4-DB, 30 g linuron and 30 g MCPA per liter; used to control weeds in undersown cereals and seedling grassland. [Farmers Crop Chemicals Ltd]

Clovean. A detergent for bakehouse usage. [Laporte Industries Ltd] *

Clovotox. CAS 7085-19-0; Soluble concentrate containing 300 g/l mecoprop; for control of weeds in cereals and grassland. [Rhone-Poulenc Environmental Prods. Ltd]

Clowes' solution. A solution containing 160 g potassium hydroxide in 200 cc water and 10 g pyrogallol; used in the Hempel pipette for the absorption of oxygen.

Cloxapen. CAS 7081-44-9; Cloxacillin sodium; antibacterial. [SmithKline Beecham]

Cloxicap. CAS 61-72-3; A proprietary preparation of cloxacillin; an antibiotic. [Pfizer International]

Cloxisyrup. CAS 61-72-3; A proprietary preparation of cloxacillin; an antibiotic. [Pfizer International]

Clozan. CAS 33671-46-4; A proprietary preparation of clotiazepam; an antianxiety agent. [Pfizer International]

Clozaril. CAS 5786-21-0; Clozapine; sedative. [Sandoz Pharmaceuticals] *

CLSP 499. Palm kernel oil, lecithin; kosher; coating fat for pastel and chocolate flavored confectionery coatings. [Van Den Bergh Foods]

Club. CAS 2032-65-7; Pellets containing 4% w/w methiocarb; snail and slug bait [ICI Agrochemicals]

Clysar®. Shrink film. [DuPont UK]

Clysodrast. CAS 1336-29-4; Bisacodyl tannex; water soluble complex of biscacodyl and tannic acid; laxative. [Armour Pharmaceutical Co] *

CM8450. CAS 31001-77-1; 3-Mercaptopropylmethyl-

---

dimethoxy silane; CAS 31001-77-1; coupling agent, chem. intermediate, blocking agent, release agent, lubricant, primer, and reducing agent. [Hüls Am.]

CM8500. CAS 4420-74-0; EINECS 224-588-5; 3-Mercaptopropyltrimethoxysilane; coupling agent, chem. intermediate, blocking agent, release agent, lubricant, primer, and reducing agent. [Hüls Am.]

CM8550. CAS 2530-85-0; EINECS 219-785-8; 3-Methacryloxypropyltrimethoxysilane; coupling agent, chem. intermediate, blocking agent, release agent, lubricant, primer, and reducing agent. [Hüls Am.]

CM8620. CAS 3069-25-8; N-Methylaminopropyltrimethoxysilane; coupling agent, chem. intermediate, blocking agent, release agent, lubricant, primer, and reducing agent. [Hüls Am.]

CM8645. CAS 5578-42-7; Methylcyclohexyldichlorosilane; coupling agent, chem. intermediate, blocking agent, release agent, lubricant, primer, and reducing agent. [Hüls Am.]

CM8650. CAS 17865-32-6; Methylcyclohexyldimethoxysilane; coupling agent, chem. intermediate, blocking agent, release agent, lubricant, primer, and reducing agent. [Hüls Am.]

CM8750. CAS 75-54-7; Methyldichloro silane; coupling agent, chem. intermediate, blocking agent, release agent, lubricant, primer, and reducing agent. [Hüls Am.]

CM8930. CAS 149-74-6; Methylphenyldichloro silane; coupling agent, chem. intermediate, blocking agent, release agent, lubricant, primer, and reducing agent. [Hüls Am.]

CM8980. CAS 4253-34-3; Methyltriacetoxy silane; coupling agent, chem. intermediate, blocking agent, release agent, lubricant, primer, and reducing agent. [Hüls Am.]

CM9000. CAS 75-79-6; Methyltrichloro silane; coupling agent, chem. intermediate, blocking agent, release agent, lubricant, primer, and reducing agent. [Hüls Am.]

CM9050. CAS 2031-67-6; EINECS 217-983-9; Methyltriethoxysilane; coupling agent, chem. intermediate, blocking agent, release agent, lubricant, primer, and reducing agent. [Hüls Am.]

CM9100. CAS 1185-55-3; EINECS 214-685-0; Methyltrimethoxysilane; coupling agent, chem. intermediate, blocking agent, release agent, lubricant, primer, and reducing agent. [Hüls Am.]

CM9160. CAS 24589-78-4; N-Methyl-N-trimethylsilyltrifluoroacetamide; coupling agent, chem. intermediate, blocking agent, release agent, lubricant, primer, and reducing agent. [Hüls Am.]

CM9220. CAS 22984-54-9; Methyltris (methylethylketoxime) silane; coupling agent, chem. intermediate, blocking agent, release agent, lubricant, primer, and reducing agent. [Hüls Am.]

CMC. See carboxymethylcellulose sodium.

CMD 834. Modified cycloaliphatic amine adduct; room temperature curing agent for epoxy resins; end uses incl. industrial floor grouting, high build glaze, tank linings, general purpose castings and encapsulations. [Shell]

CN 103. Epoxy acrylate; for paper clear coatings, wood top coatings, screen inks, litho inks, polyethylene coatings, metal decorative coatings, adhesive papers, wood fillers, solder masks and photoresists. [Sartomer]

CN 104 A80. Epoxy acrylate/TPGDA; features excellent flow and gloss; used for paper coatings, wood coatings, inks. [Sartomer]

CN 104 B80. Epoxy acrylate/HDDA; offers flexibility for paper coatings, wood coatings, metal coatings, inks. [Sartomer]

CN 104 C75. Epoxy acrylate/TMPTA; used for paper coatings, wood coatings, metal coatings, inks. [Sartomer]

CN 104 D80. Epoxy acrylate/GPTA; used for paper coatings, wood coatings, metal coatings, inks. [Sartomer]

CN 104 F50. Epoxy acrylate/DIDA; used for paper coatings, wood coatings, metal coatings, inks. [Sartomer] *

CN 111. Epoxidized soybean oil acrylate; offers flexibility, pigment wetting, high adhesion for paper coatings, wood coatings, metal coatings, inks. [Sartomer]

CN 112 C60. Epoxy novolak acrylate/TMPTA; offers high heat resist.; used for inks, solder resists. [Sartomer]

CN 300. Polybutadiene diacrylate; used for optical fibers, pottants and encapsulants, conversion coatings, pressure-sensitive adhesives. [Sartomer]

CN 953. Aliphatic urethane acrylate, soft; used for printing inks, PVC floor coatings, wood coatings, anaerobic adhesives, paper, flexible overprint varnishes. [Sartomer]

CN 960. Urethane acrylate resin; hard; nonyellowing; provides excellent heat, light, and chem. resist. to coating formulations; recommended for use in overprint varnishes, laminating adhesives, plastic film and foil coatings, screen inks, and rigid plastic and coil coatings; supplied as base resin or as resin/monomer blends which offer hardness and water-wh. clarity to uv/eb cured products [Sartomer]

CN 962. Aliphatic urethane acrylate resin; flexible, nonyellowing; provides excellent falling sand and high impact resist. to coating and adhesive formulations; supplied as base resin or in resin/monomer blends which offer flexibility, clarity, and excellent heat and light stability to uv/eb cured products; used in screen inks, foil coatings, laminating adhesives, pressure-sensitive adhesives, and metal coatings requiring extremely flexible props. [Sartomer]

CN 966. Aliphatic urethane acrylate resin; highly flexible, nonyellowing; provides excellent high impact resist. to coating and adhesive formulations; supplied as base resin or as resin/monomer blends; which offer exceptional flexibility, clarity, and excellent heat and light stability to uv/eb-cured products [Sartomer]

CN 970. Aromatic urethane acrylate resin; hard; produces hard, fast curing, solv.-resistant uv/eb-cured products comparable to those made from acrylated epoxy resins; supplied as base resin or as resin/monomer blends; for use in adhesives, and paper, wood, and metal coatings. [Sartomer]

CN 974. Aromatic urethane methacrylate; high flexibility; hydrophilic; used for adhesives, paper, wood, ink, metal coatings. [Sartomer]

CNC Antifoam 30-FG. Silicone emulsion; antifoam/defoam for paper, textile, water-phase paints and other industrial uses; avail. in food grade versions. [CNC Int'l. L.P.]

CNC Antifoam A-107. Filled silicone fluid; defoamer for aq. and nonaq. systems; effective over very broad pH and temp. range. [CNC Int'l. L.P.]

CNC Defoamer 12, 34, 407. Defoamer for paper applications for difficult, foamy high solids, or high resin content systems. [CNC Int'l. L.P.]

CNC Defoamer 69, 97. Nonsilicone; water-based defoamer for paper applications. [CNC Int'l. L.P.]

CNC Detergent E. Coconut fatty acid amine condensate;

---

‡ Trade name and manufacturer not verified   § Trade name without identified manufacturer

detergent, wetting agent, dyeing assistant; leveling agent for acid dyestuffs on wool; rewetting and softening agent for sanforizing; scouring agent for synthetics; also for laundry processing of garments. [CNC Int'l. L.P.]

CNC Dispersant WB Series. surfactants, leveling agents, dispersants, dyeing assistants for textiles. [CNC Int'l. L.P.]

CNC Dispersion PE. Complex phosphate ester of nonionic ethoxylates; forms self-emulsfiable concs. when mixed with many solvs.; base for textile wetting agents. [CNC Int'l. L.P.]

CNC Foam Assist AA, AA-100. High solids foaming agents for textile foam finishing. [CNC Int'l. L.P.]

CNC Gel Series. Detergents for textile scouring, wetting, dispersing. [CNC Int'l. L.P.]

CNC Inhibitor 30. Dyeing and bleaching assistant providing protection against atmospheric yellowing to Spandex fabrics. [CNC Int'l. L.P.]

CNC Leveler JH. Sulfated PEG ester; low-foaming surfactant, wetting agent, dispersant for textile applications; high compatible with high concs. of acids. [CNC Int'l. L.P.]

CNComerse IMP. Alkyl phospho-sulfate; nonfoaming biodeg. mercerizing agent, wetting and penetrating agent. [CNC Int'l. L.P.]

CNC PAL 100, 200, 300. Surfactants; anionic, nonionic, and cationic low-foaming biodeg. surfactants for rewetting towels, tissue or corrugating medium, deinking printed wastes, magazine stocks; wetting and penetrating agent, detergent. [CNC Int'l. L.P.]

CNC PAL V-8 Supra. Detergent, emulsifier for textile processing. [CNC Int'l. L.P.]

CNC Product PW. Print wash used in conjunction with CNC Fix Concentrate for nylon printed with acid dyes. [CNC Int'l. L.P.]

CNC Product ST. Surfactant; print wash for use where high concs. of residual color and print paste are to be removed from cotton, rayon, or synthetics; wetting and scouring agent. [CNC Int'l. L.P.]

CNC Soft C-1. Synthetic softener for use where a soft, luxurious hand is required; all-purpose softener/lubricant for textile processing. [CNC Int'l. L.P.]

CNC Sol 72-N Series. Wetting and rewetting agents for textiles; stable to dilute acids and alkalies; not for use with water repellent or fluorochemical finishes. [CNC Int'l. L.P.]

CNC Sol BD. Solv./emulsifier blend; for removal of waxes, greases, oils, and greige-mill dirt from cotton, rayon, and blends. [CNC Int'l. L.P.]

CNC Sol XNN #11. Solv./emulsifier blend; for removal of waxes, greases, oils, and greige-mill dirt from cotton, rayon, and blends. [CNC Int'l. L.P.]

CNC Solv 809. Solv./surfactant blend; for deinking and repulping resin coated mixed waste furnishes; prevents redeposition of agglomerated inks and binders. [CNC Int'l. L.P.]

CNC Wet CP Conc. Phosphated long chain alcohol; wetting and dispersing agent, leveler for pigments and dyes, in peroxide bleach baths; dyeing assistant; suitable for cotton and synthetics processing; acid and alkali stable. [CNC Int'l. L.P.]

CO9745. Octadecyldimethyl [3-(trimethoxysilyl) propyl] ammonium chloride, CAS 27668-52-6, methanol; coupling agent, chem. intermediate, blocking agent, release agent, lubricant, primer, reducing agent. [Hüls Am.]

CO9750. CAS 112-04-9; Octadecyltrichloro silane; CAS 112-04-9; coupling agent, chem. intermediate, blocking agent, release agent, lubricant, primer, reducing agent. [Hüls Am.]

CO9800. CAS 1020-84-4; Octamethylcyclotetrasilazane; CAS 1020-84-4; coupling agent, chem. intermediate, blocking agent, release agent, lubricant, primer, reducing agent. [Hüls Am.]

CO9810. Octamethylcyclotetrasiloxane; CAS 556-67-2; coupling agent, chem. intermediate, blocking agent, release agent, lubricant, primer, reducing agent. [Hüls Am.]

CO9816. CAS 107-51-7; Octamethyltrisiloxane; coupling agent, chemical intermediate, blocking agent, release agent, lubricant, primer, and reducing agent. [Hüls Am.]

CO9817. CAS 546-56-5; Octaphenylcyclotetrasiloxane; CAS 546-56-5; coupling agent, chem. intermediate, blocking agent, release agent, lubricant, primer, reducing agent. [Hüls Am.]

CO9819. CAS 18162-84-0; Octyldimethylchloro silane; CAS 18162-84-0; coupling agent, chem. intermediate, blocking agent, release agent, lubricant, primer, reducing agent. [Hüls Am.]

CO9830. CAS 5283-66-9; Octyltrichlorosilane; coupling agent, chem. intermediate, blocking agent, release agent, lubricant, primer, reducing agent. [Hüls Am.]

CO9835. CAS 2943-75-1; EINECS 220-941-2; Octyltriethoxysilane; coupling agent, chem. intermediate, blocking agent, release agent, lubricant, primer, reducing agent. [Hüls Am.]

Coactabs. CAS 32886-97-8; Amdinocillin pivoxil; antibacterial. [Hoffmann-LaRoche Inc] *

Coactin. CAS 32887-01-7; Amdinocillin; antibacterial. [Hoffmann-LaRoche Inc] *

Coagulant CHA. Cyclohexyl amine acetate; coagulant used in dipped goods from natural latex. [Bayer AG]

Coagulant WS. Functional polyorganosiloxane; heat sensitizer permitting coagulation pt. of 32-60 C in mixes based on natural or synthetic latex. [Bayer; Miles; Polysar]

Coalatex. Primarily consists of oxalic acid and oxalate; a good coagulating agent in 10% solution; a proprietary coagulating material for use in coagulating rubber latex.§

coal black. See mineral black.

Coaldet. An electric blasting cap comprising copper-bronze alloy shell; designed especially for initiating explosives in coal mining. [Hercules] *

Coalite N.T.P. CAS 25155-23-1; Nontoxic trixylenyl phosphate plasticizer. [Coalite Fuels & Chemicals Ltd] *

coal oil. See kerosene.

coal tar naphtha. See naphtha.

Coaltec. Wood preservative. [Coalite Fuels & Chemicals Ltd]

coarse metal. (Matte). An impure mixture of ferrous and cuprous sulfides produced in copper smelting. It usually contains from 20-75% copper, 12-45% iron, and 19-25% sulfur.

coarse solder. See plumber's solder.

Coasol. Diisobutyl adipate/glutarate/succinate; coalescing solvent for emulsion paints/adhesives. [Chemoxy International Ltd]

Coate®. Metal salts of organic acids; driers for coating compositions. [Hüls Am.]

Coatmaster® K580. Binder chemical derived from corn. [Grain Processing]

Cobac. Cobalt acetate. [Mechema Chemicals Ltd] *

Cobadex Ointment. CAS 50-23-7; A proprietary preparation of hydrocortisone in a silicone base for dermatological use. [Cox Pharmaceuticals] *

Cobalin-H. Injectable hydroxobalamin BP; main indications include Addisonian pernicious anemia prophylaxis and treatment of other macrocytic anemias due to B_{12} deficiency and tobacco amblyopia and Leber's atrophy. [Paines & Byrne Ltd] *

cobalt. (super cobalt) Co; CAS 7440-48-4; EINECS 231-158-0; Oxidizing agent, lamp filaments, in manufacture of cobalt steel; in porcelain, glass, pottery, enamels. [Aldrich; Atomergic Chemetals; Cerac; Noah Chem.]

Cobalt 254. A paste containing compounds of cobalt; loss-of-dry inhibitor for nonaqueous coating compositions, inhibitor of gel-time drift in unsaturated polyester resins. [Hüls Am.]

cobalt acetate (ous). (cobaltous actate) $Co(C_2H_3O_2)_2 \cdot 4H_2O$; Inks, paint and varnish driers, catalyst, anodizing, mineral supplement in feed additives, foam stabilizer. [Atomergic Chemetals; Celtic Chem. Ltd; Mallinckrodt; Nihon Kagaku Sangyo; Noah Chem.]

cobalt black. See cobalt oxide (ic).

cobalt brass. Acid-resisting alloys of 52% copper, 17-25% zinc, and 22-30% cobalt.

cobalt bronze. A phosphate of cobalt and ammonium.

cobalt carbonate (ous). (cobaltous carbonate basic; cobaltous hydroxide carbonate) $CoCO_3[Co(OH)_2]_n$; CAS 12602-23-2; EINECS 235-714-3; Manufacture of cobaltous oxide, cobalt pigments, cobalt salts; intermediate. [Atomergic Chemetals; Celtic Chem. Ltd; Nihon Kagaku Sangyo; Noah Chem.]

cobalt chloride (ous). (cobaltous chloride anhydrous) $CoCl_2$; CAS 7646-79-9; EINECS 231-589-4; Absorbent for ammonia, gas masks, electroplating, inks, hygrometers, manufacture of vitamin B_{12}, flux for magnesium refining, solid lubricant, dye mordant, catalyst, barometers, laboratory reagent, fertilizer additive. [Celtic Chem. Ltd; Mallinckrodt; Nihon Kagaku Sangyo; Spectrum Chem. Mfg.]

cobalt-chromium-molybdenumsSteels. Usually an alloy of iron with 3.05% molybdenum, 2.16% chromium, 1.33% cobalt and 0.65% carbon.

cobalt fluoride (ic). (cobalt trifluoride; cobaltic fluoride) CoF_3; CAS 10026-18-3; EINECS 233-062-4; Fluorination of hydrocarbons. [Atochem N. Am.; Atomergic Chemetals; Cerac]

cobaltic fluoride. See cobalt fluoride (ic).

cobaltic oxide monohydrate. See cobalt oxide (ic).

cobalt monoxide. See cobalt oxide (ous).

cobalt naphthenate (ous). (cobaltous naphthenate) Paint and varnish drier, bonding rubber to steel and other metals. [Akzo; Nuodex Espanola SA; Troy]

cobalt nitrate (ous). $Co(NO_3)_2 \cdot 6H_2O$; CAS 10141-05-6; Oxidizing agent, dangerous fire risk in contact with organic materials. [Atomergic Chemetals; Celtic Chem. Ltd; Mallinckrodt; Nihon Kagaku Sangyo; Noah Chem.]

cobaltous actate. See cobalt acetate (ous).

cobaltous carbonate basic. See cobalt carbonate (ous).

cobaltous chloride anhydrous. See cobalt chloride (ous).

cobaltous hydroxide carbonate. See cobalt carbonate (ous).

cobaltous naphthenate. See cobalt naphthenate (ous).

cobaltous oxide. See cobalt oxide (ous).

cobaltous sulfate. See cobalt sulfate (ous).

cobalt oxide (ic). (cobaltic oxide monohydrate; cobalt black) Co_2O_3; CAS 1308-04-9; Pigment, coloring enamels, glazing pottery. [Atomergic Chemetals; Cerac; Noah Chem.]

cobalt oxide (ous). (cobaltous oxide; cobalt monoxide) CoO; CAS 1307-96-6; Pigments for ceramics and paints; oxidation catalyst for drying oils; preparation of cobalt-metal catalysts, coloring glass;,feed additive. [Atomergic Chemetals; Cerac; Noah Chem.]

Cobaltron Steel Alloy. A proprietary alloy containing iron with about 11% chromium, 2.25% cobalt, 1.5% carbon, 1.25% molybdenum, and 0.25% tungsten. §

cobalt steel. Alloys of steel with cobalt; used for certain parts of electrical machinery requiring a high permeability. An alloy containing 34.5% cobalt, at high inductions, has a higher permeability than pure iron.

cobalt sulfate (ous). (cobaltous sulfate) $CoSO_4$; CAS 10124-43-3; Ceramics, pigments, glazes, in plating baths for cobalt, additive to soils, catalyst, paint and ink drier, storage batteries. [Atochem N. Am.; Atomergic Chemetals; Johnson Matthey plc; Mallinckrodt; Nihon Kagaku Sangyo]

cobalt trifluoride. See cobalt fluoride (ic).

Coban. Monensin, produced by *Streptomyces cinnamonensis*, and used as the sodium salt; antibacterial; antifungal. [Eli Lilly & Co]

Cobastab. CAS 68-19-9; A proprietary preparation of cyanocobalamin for injection. [The Boots Co plc]

Cobatope-60. CAS 7646-79-9; Cobaltous chloride Co 60; radioactive agent. [Bristol-Myers Squibb Co Inc] *

Cobex. Weedkiller containing dimitramine. [ICI Chem. & Polymers Ltd] †

Cobox. Cobalt oxide. [Mechema Chemicals Ltd] *

Cobox® L. Ammoniacal copper polyacrylate; contact fungicide for control of fungus and bacterial diseases in coffee, cotton, fruits [BASF AG]

Cobrol. A photographic developer. [May & Baker Ltd] *

cocamidopropyl betaine. (CADG; cocamidopropyl dimethyl glycine) Zwitterion (inner salt); $RCONH-CH_2)_3N^+(CH_3)_2CH_2COO^-$, RCO represents the coconut acid radical; CAS 61789-40-0; 70851-07-9; 83138-08-3; 86438-79-1; EINECS 263-058-8; 274-923-4; amphoteric surfactant for shampoos, bath products, personal hygiene preps. [Chemron; Goldschmidt; Henkel/Emery; Huntington Labs; Inolex; McIntyre; Mona; Scher; Sherex]

cocamidopropyl dimethyl glycine. See cocamidopropyl betaine.

cochineal. (carmine, carmine lake). The dried bodies of the female shield louse, *Coccus Cacti*. It is the source of carmine, a red coloring matter; used for dyeing scarlet on tin mordanted wool, but chiefly for the preparation of pigments.

cochranite. Titanium dicyanide.

cochrome. An alloy of 60% cobalt, 12% chromium, 24% iron, and 2% manganese. It has been used in the place of Nichrome for the elements of electrical heating apparatus.

Coclopet. Copper chloride (petroleum refining grade). [Mechema Chemicals Ltd] *

Coclor. CAS 7447-39-4; Industrial cupric chloride (35/36%

---

‡ Trade name and manufacturer not verified   § Trade name without identified manufacturer

Cu). [Mechema Chemicals Ltd] *

Cocloran. Anhydrous copper chloride (46-47% Cu). [Mechema Chemicals Ltd] *

cocoa butter. CAS 8002-31-1; (cacao butter; theobroma oil). The fat extracted from the seeds of *Theobroma cacao*. The seeds contain from 35-45% of the butter. It has an acid value of 0.6-1.3 (oleic acid), a saponification value of 192-193, and melts at 33-34 C; used in confectionaries, suppositories, pharmaceuticals, soaps, and cosmetics.

cocoaline. *See* vegetable butter.

Coco-Diazine. CAS 68-35-9; Sulfadiazine; antibacterial. [Eli Lilly & Co] †

Cocoloid®. Alginate; used for chocolate milk, soft-serve frozen desserts, ice cream, custard, variegated syrup and fudge topping. [Kelco]

coconut butter. CAS 8001-31-8; (coconut oil, coconut milk, copra). The sweet watery liquid contained in the coconut is called coconut milk. This disappears and gives place to a soft edible pulp, which hardens in the air, and is sold as copra. Copra contains from 60-70% oil; which is extracted as coconut oil or butter.

Coconut oil. *See* coconut butter.

Coconut Oils® 76, 92, 110. CAS 8001-31-8; EINECS 232-282-8; Coconut oils, refined bleached, deodorized; base for ice cream coatings, candies, icings; clouding agent for beverages; emollient for creams and lotions; flavor solubilizer; pharmaceutical vehicle. [Stepan/PVO]

Coco-Quinine. CAS 804-63-7; Quinine sulfate; antimalarial. [Eli Lilly & Co] †

Codacide Oil. Adjuvant containing 95% emulsifiable vegetable oil; extender and wetting agent for herbicides, fungicides and insecticides. [Microcide Ltd]

Code 321. CAS 8016-70-4; EINECS 232-410-2; Partially hydrogenated soybean oil; kosher; high performance fat used for filler, snack frying, prepared foods, spray oil. [Van Den Bergh Foods]

Codelcortone. CAS 50-24-8; Prednisolone; used for certain endocrine and nonendocrine disorders responsive to corticosteroid therapy. [Merck & Co Inc] †

Codelsol®. CAS 125-02-0; Prednisolone sodium phosphate; parenteral corticosteroid for certain endocrine and nonendocrine disorders. [Merck & Co Inc]

Codelspray. Prednisolone phosphate; topical spray for corticosteroid-responsive skin conditions. [Merck & Co Inc] †

Co-Deltra. CAS 53-03-2; Prednisone; used for certain endocrine and nonendocrine disorders responsive to corticosteroid therapy. [Merck & Co Inc] †

Codiazine. CAS 68-35-9; Sulfadiazine; antibacterial. [SmithKline Beecham] *

Codidoxal. A proprietary preparation of doxycycline hyclate and codeine; an antibiotic and antitussive. [Pfizer International]

Codinyl. A proprietary preparation of ephedrine hydrochloride, menthol, codeine phosphate, syrup of prunes and kola; a cough linctus. [A C Hatrick Chemicals] *

Codiphen. CAS 50-78-2, CAS 52-28-8; A proprietary preparation of aspirin and codeine phosphate; an analgesic. [Pfizer International]

Codis. CAS 50-78-2, CAS 52-28-8; A proprietary preparation of aspirin and codeine phosphate; an analgesic. [Reckitts Colours Ltd] *

Codite. A proprietary trade name for a vulcanized fiber and pure cotton cellulose plastic tubing. §

cod liver oil. (oils, cod liver; morrhua oil) Fixed oil expressed from fresh livers of *Gadus morrhua* and other species of codfish; CAS 8001-69-2; medicine (vitamin A and D content), chamois-leather tanning. [Arista; R.W. Greeff; Penta Mfg.]

Codroxomin. CAS 13422-51-0; Hydroxocobalamin; vitamin. [O'Neal, Jones & Feldman Pharmaceuticals] *

Codur. A proprietary trade name for a synthetic clear or colored baking enamel. §

Co-Elorine Pulvules. A proprietary preparation of amylobarbitone and tricyclamol chloride. [Eli Lilly & Co] †

Co-Ferol. A proprietary preparation of ferrous fumarate and folic acid; a hematinic. [Cox Pharmaceuticals] *

Cofill. A finely ground mixture (50:50) of highly dispersed silica-resorcin; adhesion powder for bonding rubber mixtures to treated and untreated synthetic textiles and metal fabrics. [Degussa AG] *

Co-Gell® A2/B270. Mineral oil, aluminum isostearates/laurates/palmitates, isopropyl palmitate; stabilizer for continuous phase of w/o emulsion systems and gelling agent for oils and nonaq. fluids; waterproofing, lubricity props.; improved pigment adhesion and slip. [Rhone-Poulenc Surf.]

Cogentin. CAS 132-17-2; Benztropine mesylate; for the treatment of Parkinsonism and allied disorders of the nervous system. [Merck & Co Inc]

Cogesic. CAS 3734-16-5; Prodilidine hydrochloride; analgesic. [Mead Johnson & Co] *

Cogest. Pancreas substance, hog bile, papain-pepsin complex, diastase of malt; papain 48.6 mg and pepsin 48.6 mg specially granulated and stabilized; used as a digestive aid, for the relief of simple indigestion. [Pharmaceutical Basics Inc] *

Cohedur®. Bonding agents for rubber. [Bayer AG; Bayer plc]

Cohedur® RK. CAS 108-46-3; Resorcinol component with filler, 1:1 ratio; bonding agent in polychloroprene. [Bayer; Miles; Polysar]

Coherex®. Concentrate nonvolatile emulsion consisting of 60% natural petrol. resins and 40% wetting sol'n.; dust retardant providing dust control on ballparks, playgrounds, dirt roads, construction sites. [Witco/Golden Bear]

COHRlastic® 400. CAS 63394-02-5; Solid silicone rubber; general purpose molded sheet. [Furon/CHR] *

COHRlastic® 1010. Silicone coated on fiberglass; thin, tough, dimensionally stable, flexible fabric for applications incl. belting, vacuum blankets, thermal shielding, and diaphragms; 1000 series has superior elec. props. and good abrasion resist.; general purpose, elec. grade. [Furon/CHR] *

COHRlastic® 1867. CAS 63394-02-5; Conductive silicone rubber; thermally conductive coated fabrics providing thin, cost-effective heat transfer capability; avail. plain or with pressure-sensitive adhesive 1 side. [Furon/CHR] *

COHRlastic® 3320. CAS 63394-02-5; Solid silicone rubber, fiberglass-reinforced (20% for 1/16 in.; 14% 3/32 in., 10% 1/8 in.); dimensionally stable, durable material for press pads, belting, and gasketing; 3320 grade offers lubricating oil resist. and excellent compr. set resist. [Furon/CHR] *

COHRlastic® 8016. Aluminum mesh impregnated with

---

* Trade name not verified as available † Trade name verified as obsolete

neoprene rubber; electrically conductive RFI/EMI shield material; the mesh maintains elec. contact, while the elastomer provides an effective seal in gasket applications. [Furon/CHR] *

COHRlastic®9041. CAS 63394-02-5; Solid silicone rubber, phenyl-based; high strength, low temp. grade. [Furon/CHR] *

COHRlastic® F12. Silicone foam sheet; low dens., flame-retardant foam sheet for aviation, mass transit, automotive, electronics, construction, furnishings industries; applications incl. fireblocks, thermal barriers, noise and vibration dampeners, insulation, high performance gaskets or seals; nontoxic on burning. [Furon/CHR] *

COHRlastic® FR17. Silicone rubber foam bonded one side to a reinforcing silicone coated fiberglass fabric; lightweight, thermal insulating/fire blocking composite. [Furon/CHR]

COHRlastic® R10450. CAS 63394-02-5; Silicone closed cell sponge rubber, fiberglass reinforced; for high performance gasketing, thermal shielding, vibration mounts, press pads; offers. the compressbility of sponge and dimensional stability in the X-Y direction; for consistent size and shape of die cut parts; eliminates outward exrusion under pressure. [Furon/CHR] *

COHRlastic® R10490. Fluorosilicone sponge rubber sheet; for gasket, seal applications where it maintains integrity in presence of jet fuel, engine oil, etc.; military and commercial aircraft industry [Furon/CHR] *

COHRlastic® TC100. CAS 63394-02-5; Conductive silicone rubber; unsupported, thermally conductive grade providing thermal and mech. protection to electronic devices. [Furon/CHR] *

Co-Hydeltra. CAS 50-24-8; Prednisolone; indicated for certain endocrine and nonendocrine disorders responsive to corticosteroid therapy. [Merck & Co Inc] *

cohydrol. A colloidal graphite solution.

coinage bronze. An alloy of 95% copper, 4% tin, and 1% zinc.

Cojene. Analgesic and antipyretic tablets. [Fisons plc, Pharmaceuticals Div] *

coke. CAS 50-36-2; The carbonaceous residue from the distillation of coal.

Colac. Laxative tablets for relief of constipation. [Richardson-Vicks Inc] *

Colace. Dioctyl sulfosuccinate; a proprietary laxative. [Mead Johnson & Co]

Colacryl. Acrylic resin in solid beads or as a solution. [Bonar Polymers Ltd]

Colamine. CAS 141-43-5; Aminoethanol.

Colanyl. Pigments for resin dispersions; used in aqueous-based paints. [Hoechst UK]

Colasta. A proprietary trade name for a phenol-formaldehyde synthetic resin. §

Colastex. A patented mixture of Colas (cold asphalt) and rubber latex for improving and hardening roads. §

Colbenemid. Probenecid and colchicine; for the maintenance treatment of gout. [Merck & Co Inc]

Colcar®. A range of carriers for dyeing synthetic fibers and their blends. [Allied Colloids Ltd]

Colcolor. Carbon black/plastic concentrate; used for pigmenting of paper and cement for the plastics processing industry, pigmenting of PE, PP, EVA, PVC, PS, SAN, and ABS. [Degussa AG] *

cold varnishes. Varnishes obtained by heating linseed oil to 105 C. for four and a half hours, adding manganese borate, linoleate, or resinate, and stirring the mass with compressed air.

Coles solder. An alloy of 82% tin, 11% aluminum, 5% nickel, and 2% manganese.

Colestid. CAS 37296-80-3; Colestipol hydrochloride, a basic anion-exchange resin, highly cross-linked and insoluble; antihyperlipoproteinemic. [Upjohn]

Col-Evac. CAS 144-55-8; Sodium bicarbonate; replenisher; alkalizer. [O'Neal, Jones & Feldman Pharmaceuticals] *

Colex 900 BP. Acrylic bonding paste. [Colourex Ltd]

Colex 1000 FR. Fire retardent polyester gel coat; special gel coat resin for manufacture of imitation fuel effects for electric and gas. [Colourex Ltd]

Colex 2000. General purpose gel coat. [Colourex Ltd]

Colex 2200. General purpose white gel coat. [Colourex Ltd]

Colex 4300. Casting resin for cold casting. [Colourex Ltd]

Colfarit®. Acetylsalicylic acid; a platelet aggregation inhibitor, antithrombotic and anti-inflammatory drug. [Bayer AG]

Colfite. A proprietary trade name for a graphited laminate bearing material. §

Colicort. CAS 125-02-0; Prednisolone sodium phosphate; parenteral corticosteroid for certain endocrine and nonendocrine disorders. [Merck & Co Inc]

Colifoam. CAS 50-03-3; White odorless aerosol foam containing hydrocortisone acetate; topical treatment for ulcerative colitis. [Stafford-Miller] ‡

Collacral®. Thickeners for polymer dispersions and other aq. systems, paints, adhesives, sealants, for production of nonwovens. [BASF AG]

Collafix®. A range of acrylic copolymers for use as wallpaper adhesives; all are concentrated liquids; grades are available for both conventional and pre-pasting adhesives. [Allied Colloids Ltd]

collagen. (collagen fiber; ossein; soluble animal collagen) Fibrous protein derived from connective tissues in animals; clear to hazy, colorless; m.w. ≈ 100,000; CAS 9007-34-5; EINECS 232-697-4; as adhesive, in cosmetic industry for face creams, lotions, hair preps.; sausage casing; as fibers in sutures; in leather substitutes; as a gel in photographic emulsions. [Croda; Hormel; Inolex; Maybrook; RITA; U.S. Biochemical]

Collagen 15K. Hydrolyzed mixed glycosaminoglycans, hydrolyzed collagen, hydrolyzed dermal proteins; cosmetic ingred. for moisturizing applications. [Brooks Industries]

Collagen CLR. CAS 9007-34-5; Carrier of native sol. collagen in weakly acid hydrophilic medium; product for aging skin, wrinkle and after-sun treatment. [Dr. Kurt Richter; Henkel/Cospha; Henkel Canada]

Collagenase. Clostridiopeptidase A; 3.4.4.19; digests native collagen at about physiological pH. §

collagen fiber. See collagen.

Collagen Hydrolyzate Cosmetic 55. CAS 9015-54-7; Hydrolyzed collagen; protective colloid effect, anti-irritant props.; substantive film-former with dye leveling effects for skin and hair care products (shampoos, conditioners, color treatments, shave creams), liq. soaps, dish detergents. [Maybrook]

Collagen Native Extra 1%. CAS 9007-34-5; Soluble collagen; moisturizer, film-former for skin care products; protective barrier. [Maybrook]

Collamino 25. CAS 9015-54-7; Collagen amino acids; cosmetic ingred. [Brooks Industries]

---

‡ Trade name and manufacturer not verified   § Trade name without identified manufacturer

Colla-Moist CG. Collagen glycerides; moisturizing protein complexes for cosmetics. [Brooks Industries]

Colla-Moist WS. Propylene glycol, hydrolyzed collagen, PPG-12-PEG-65 lanolin oil; moisturizing protein complexes for cosmetics. [Brooks Industries]

Collargol. (Credes silver). Colloidal silver; has been used medicinally. §

collasan. Colloidal kaolin.

Collasol. CAS 9007-34-5; Sol. animal collagen; humectant; hygroscopic film former; conditioner; skin care products [Croda Inc; Croda Chemicals Ltd]

Collatex. CAS 9005-34-9; Ammonium alginate. [Kelco Int'l. Ltd]

collaurin. A form of colloidal gold.

collene. Colloidal silver.

Collet steel. A manganese steel with small amounts of chromium and 0.75% carbon.

Collex G. CAS 8061-52-7; Fermented calcium lignosulfonate; dispersant; low-cost concrete additive. [Borregaard LignoTech]

collidine. CAS 29611-84-5; Trimethyl- and/or methylethylpyridines; used as chemical intermediated, dehydrohalogenating agent.

Colliron. CAS 20344-49-4; A proprietary preparation of ferric hydroxide colloid; a hematinic. [British Drug Houses] *

Collocal-D. A proprietary preparation of calcium oleate and stearate and calciferol (vitamin D). [Crookes Laboratories] *

collodion. A solution of pyroxylin (nitrocellulose) in ether and alcohol; cements, coating wounds and abrasions, solvent for drugs, corn removers, process engraving, lithography, photography. [Spectrum Chem. Mfg.]

collodion cotton. CAS 9004-70-0; Pyroxyllin

collodion silk. An artificial silk made from collodion spun by precipitation in a liquid bath.

Colloid 106. Low molecular weight acrylic polymer aq. sol'n.; scale inhibitor used in water treatment and dishwashing compounds [Rhone-Poulenc]

Colloid 111. Sodium polycarboxylate; general purpose paint dispersant. [Rhone-Poulenc Surf.; Rhone-Poulenc/Water Soluble Polymers]

Colloid 111D. CAS 37199-81-8; Sodium salt of polycarboxylic acid; surfactant, dispersant for agric. formulations. [Rhone-Poulenc Surf.; Rhone-Poulenc/Water Soluble Polymers]

Colloid 202. CAS 9003-04-7; Med. molecular weight sodium polyacrylate sol'n.; builder additive for detergents; antiredeposition aid for laundry applications; dispersion aid for dishwash formulations; process and anticaking aid for dry products; sequestrant for liq. or dry formulations. [Rhone-Poulenc/Water Soluble Polymers]

colloidal selenium. See selenium.

Collokit. Technetium Tc 99m sulfur colloid; radioactive agent. [Abbott Laboratories] *

Collone. Emulsifying wax. [Rhone-Poulenc UK]

Colloresin D. A methyl cellulose soluble in cold water. §

Colloresin DK. A proprietary alkyl ether of cellulose soluble in cold water insoluble in hot water; for use as a thickening agent in textile printing. §

Collosol Argentum. CAS 7440-22-4; A proprietary preparation of colloidal silver; used as an ocular antiseptic. [Crookes Laboratories] *

Collosol Calamine. A proprietary preparation of colloidal calamine; a skin cream. [Crookes Laboratories] *

Collosol Manganese. A proprietary preparation of colloidal copper, chromium and manganese; skin antiseptic. [Crookes Laboratories] *

Collotone. A proprietary preparation of iron and ammonium citrate, iron and manganese citrate, potassium glycerophosphate, sodium glycerophosphate, thiamine hydrochloride, tincture of nux vomica and caffeine citrate; a tonic. §

Collozets. A proprietary preparation of tyrothricin and cetylethyldimethylammonium ethyl sulfate; used for sore throat. [Crookes Laboratories] *

collozine. Colloidal zinc hydroxide.

Collubarb. A proprietary preparation of aluminum hydroxide gel, phenobarbitone and atropine sulfate; an antacid. [British Drug Houses] *

Collys. Native and modified starches; used in the board industry. [Roquette (UK) Ltd]

Colmonoy. A proprietary trade name for a chromium boride; an abrasive. §

Colmonoy No. 6. A corrosion-resisting alloy with about 75% nickel base; an essential constituent is chromium boride.§

Colofac. CAS 3625-06-7; Mebeverine; spasmolytic, counteracts intestinal spasms. [Duphar BV] *

Cologel. A proprietary preparation containing methylcellulose; a laxative. [Eli Lilly & Co] †

Colomycin. CAS 1066-17-7; Colistin (various salts); an antibiotic for the treatment of Gram-negative infections. [Pharmax Ltd]

Colona Steel. A proprietary trade name for a nickel-chromium steel containing some manganese. §

colophony. See rosin.

Colorado silver. An alloy of 57% copper, 25% nickel, and 18% zinc. It is a nickel silver (German silver).

Colorlok. Inorg. salts; antimigrator for direct dyes on cellulosics. [Marlowe-Van Loan]

Colormatch®. High pigment loading dispersions for excellent color control and economy of use; for polyester, epoxy, vinyl, urethane, vinyl ester, gel coats. [Plasticolors]

Color-Max. Pigment/plasticizer dispersions; pourable colors for plastisols and other systems; offered custom matched to standard. [W J Ruscoe Co]

Colorol 70. Phosphoamino compounds; wetting and dispersing agent, binder for 2-part PU varnishes, epoxy-varnishes in systems based on PVC, PVA. [Lucas Meyer]

Colorol Rust Binder. Compound based on modified oils and carboxylic ester; rust binder, hardener, corrosion inhibitor; penetrates and wets rust and similar metal oxides allowing water-repellent film to coat and protect; contains xylene. [Lucas Meyer]

Colorol Standard. Phosphoamino compounds modified with film-formers; wetting and dispersing agent, binder for pigmented air-dry and stoving finishes, anticorrosive paints, alkyd and marine paints, zinc dust paints, bituminous paints, epoxy resins, putty, solventless floor sealers, two-part finishes, inks. [Lucas Meyer]

Color Seal. A two-step application of sealers; seals, hardens and dustproofs concrete, providing an attractive, colored, very high gloss, tough abrasion and chemical resistant floor. [Secure Inc] †

Colortrend. Colorant dispersions; for volumetric machine

---

* Trade name not verified as available   † Trade name verified as obsolete

coloring of coating compositions. [Hüls Am.]

Col-o-tex. A proprietary trade name for lacquer coated fabrics. §

Colour-Chem. Organic pigment powders; for paints, printing inks, plastics and rubber. [Colour-Chem Ltd] *

Col-o-vin. A proprietary trade name for polyvinyl synthetic resin coated fabrics. §

Colrex Compound. CAS 52-28-8; Codeine phosphate; antitussive; analgesic. [Rowell Laboratories Inc] *

Colrex Expectorant. CAS 93-14-1; Guaifenesin; expectorant. [Rowell Laboratories Inc] *

Colsol. Fatty amide; cotton and wool softener and lubricant for the textile industry; finishing agent. [Scher Chemicals Inc]

Colstar®. For the agriculture industry. [DuPont UK]

Coltapaste. Zinc paste and coal tar bandage. [Smith & Nephew Pharmaceuticals Ltd] *

Coltrock. A proprietary trade name for a phenolic plastic. §

Colturiet. A wide range of epoxy coatings, coal tar, solvent free and solventless coatings; also tank coatings. [Sigma Coatings] *

Coltwood. A proprietary trade name for a phenolic plastic.§

Colugel. CAS 21645-51-2; A proprietary aluminum hydroxide gel; used as an antacid. [Ulmer Pharmacal Co] *

Columbia Resin. A proprietary trade name for a transparent thermosetting synthetic resin. §

columbium. See niobium.

Colvinal®. Vinyl copolymer sizes for application to all continuous filament cellulosic fibers; used with starch or CMC for sizing synthetic staple yarns or blends. [Allied Colloids Ltd]

Coly-Mycin M Parenteral. CAS 8068-28-8; Colistimethate sodium; antibacterial. [Parke-Davis] *

Coly-Mycin S. CAS 1264-72-8; Colistin sulfate; antibacterial. [Parke-Davis] *

Colza oil. See rapeseed oil.

Comac. General agrochemicals. [McKechnie Chemicals Ltd]

Comac Bordeaux Plus. Copper sulfate/lime complex; a protectant fungicide for the control of blight and canker. [McKechnie Chemicals Ltd]

Comac Macuprax. Wettable powder containing 16.7% weight/weight copper sulfate and cufraneb (outside U.S.); a protectant fungicide against potato blight, canker in fruit trees and mildew in grapes. [McKechnie Chemicals Ltd]

Comac Parasol. CAS 20427-59-2; Copper hydroxide; a protectant fungicide. [McKechnie Chemicals Ltd]

Combactam. See Betamaze. [Pfizer International]

Combantrin. CAS 22204-24-6; A proprietary preparation of pyrantel pamoate; an anthelmintic. [Pfizer International]

Combelen®. Neuroplegic for use as tranquilizer in domestic animals; veterinary medicine. [Bayer AG]

Combined Seed Dressing. Fungicide/insecticide. [Murphy Chemical Ltd] †

Combismalt®. Special-purpose enamels for various uses (e.g., regenerative heat exchangers and electronic components); especially for electrophoretic or electrostatic powder application. [Bayer AG]

Comboloob 0609. Synthetic and hydrocarbon wax system; lubricant for processing under moderate to high temp. and shear conditions; for pipe, window profiles, injection molding. [Astor Wax] *

Combovac-30. B_1 type, lasota strain, Newcastle with Massachuttes and Connecticut types; for immunization of poultry. [Intervet Inc]

Comelian. See Cormelian. [Degussa AG] *

Comet. Liquid soaps. [Unilever] *

Comet metal. An alloy of 67% iron, 30% nickel, 2.2% chromium, with small quantities of manganese and copper.

Comforel®. Fibers. [DuPont UK]

Comforteam®. Fibers. [DuPont UK]

Comfort Eye Drops. CAS 550-29-2; Naphazoline hydrochloride; adrenergic. [Barnes-Hind Inc] *

Comital®, Comital L. Anticonvulsant preparations. [Bayer AG]

Commando. CAS 52756-22-6; Emulsifiable concentrate of 200 g flamprop-M-isopropyl per liter; used for control of wild oats in cereal crops. [Shell UK]

Common Degras. Anhydrous lanolin deriv. from crude wool grease; EP and slip aid for wire drawing compounds; in slushing oils, cutting oils and lubricants; long term rust preventive; in leather stuffing greases; waterproofing agent. [Fanning]

common salt. See sodium chloride.

Comodor. Emulsifiable concentrate containing 720 g/l tebutam; weed germination inhibitor. [Farm Protection Ltd]

Comodor 600. Emulsifiable concentrate containing 600 g/l tebutam; weed germination inhibitor. [ICI Agrochemicals]

Comoflastic®. Fibers. [DuPont UK]

Comox. Cobalt and molybdenum oxides on alumina. [Laporte Industries Ltd] *

compact bitumen. See bitumen.

Compactrol®. CAS 10101-41-4; Calcium sulfate, dihydrate NF; tablet and capsule filler for pharmaceutical tablets manufacturing by direct compression. [Mendell]

Compak. Composite kit of photographic processing chemicals. [May & Baker Ltd] *

Compalox. CAS 1344-28-1; Specialty aluminum oxide product; used for the purification of waste water. [Lonza AG; Lonza Inc]

Compass. Mixture of iprodione and thiophanate-methyl; systemic fungicide for field crops. [Rhone-Poulenc Crop Protection Ltd]

Compazine. CAS 58-38-8; Prochlorperazine; anti-emetic. [SmithKline Beecham]

Comperlan COD. Cocamide DEA; foam and visc. increasing agent with emulsifying properties; conditioner for shampoos and other cosmetic and pharmaceutical applications; solv. [Henkel KGaA; Pulcra SA]

Comperlan KD. Cocamide DEA; thickener, foam booster/stabilizer, superfatting agent for personal care products; hair conditioner. [Henkel; Henkel Canada; Henkel KGaA; Pulcra SA]

Comperlan LD. CAS 120-40-1; EINECS 204-393-1; Lauramide DEA superamide; foam booster, stabilizer, detergency and visc. builder, emulsifier for personal care products, household detergents. [Henkel; Henkel KGaA; Pulcra SA]

Comperlan P 100. CAS 68140-00-1; Cocamide MEA; foamer, thickener, visc. builder for powd. detergents, soaps; dispersant and blending agent for many cosmetic products [Pulcra SA]

Comperlan PKDA. Cocamide DEA (2:1); detergent base, foam booster/stabilizer, emulsifier, wetting agent for

---

household detergents. [Pulcra SA]

Compimide. High temperature bismalemide thermosetting resins. [The Boots Co plc] †

Compitox. A selective weedkiller. [May & Baker Ltd]

Compitox Extra. CAS 7085-19-0; Soluble concentrate containing 570 g/l mecoprop; for control of weeds in cereals and grassland. [Rhone-Poulenc Crop Protection Ltd]

Complamex. A proprietary preparation of xanthinol nicotinate; a peripheral vasodilator. [H N Norton & Co Ltd] ‡

Complamin. CAS 437-74-1; Xanthinol niacinate; vasodilator (peripheral). [3M Pharmaceuticals] †

Complemix. CAS 577-11-7; EINECS 209-406-4; Dioctyl sodium sulfosuccinate USP/BP; used in the process industry. [Cyanamid BV]

Complete. Denture cleanser toothpaste, to clean dentures. [Richardson-Vicks Inc] *

Comploment. Vitamin B_6 in a controlled release tablet; for Vitamin B_6 deficiency. [Napp Laboratories Ltd] *

Compoglas. A proprietary trade name for a thermoplastic composition containing glass fibers. §

Compolon Forte. An injectable liver vitamin B_{12} preparation. [Bayer AG] †

Compound 403/401. CAS 9002-88-4; Silane crosslinkable flame-retardant polyethylene; for cable insulation applications [AEI Compds.] *

Compralgyl. An all-purpose analgesic. [Bayer AG]

Comprena. Food color and flavor compounds. [PPF International Ltd] *

Compritol 888. Glyceryl mono/di/tribehenate; food emulsifier and additive for tablet manufacturing. [Gattefosse; Gattefosse SA]

Compron. Veterinary compressed products. [May & Baker Ltd] *

Comtek. Sheetfed offset printing inks; for commercial printing. [Allied-Signal Inc/Sinclair and Valentine Division] *

Com-Trol. Soluble concentrate containing 6.6% w/w 2,4-D and 250 g mecoprop; used to control weeds in grassland. [Certified Laboratories Ltd]

Conacure. Epoxy and polyurethane curing agents. [Conap]

Conadil. CAS 61-56-3; Sulthiame; anticonvulsant. [3M Pharmaceuticals] †

Conap® UC-21. Two-component modified urethane resin systems; non-MBOCA, non-TDI; nonmoisture-sensitive; produces tough, high impact-resist. elastoplastics; used as casting or laminating resins; applications incl. foundry patterns and linings, glass fiber-reinforced plastic tools, holding and checking fixtures, die faces, metal stamping pads, prototype parts; cures at R.T. or elevated temps.; mix ratio: 100/9.6. [Conap]

Conap® UC-28 (formerly Conap DPUC-11898). Two-component urethane system; surface coat with optimum abrasion resist.; for application to vertical surfaces with min. sag; suitable for fixtures, gauges, patterns, dies, and for repair and redesign. [Conap]

Conapoxy®. Epoxy resin systems; used for potting and encapsulating tooling applications. [Conap]

Conapoxy® FR-1270. CAS 25928-94-3; Two-component epoxy; potting and casting system with excellent thermal shock resist.; for encapsulation of strain sensitive devices, elec./electronic modules, transformers, coils. [Conap]

Conapoxy®TE-1257/Conacure®EA-08. CAS 25928-94-3;

Highly filled aluminum epoxy base; produces tough and dimensionally stable castings with excellent heat dissipation; for casting drill jigs, spotting blocks, chcuck jaws, dies, vacuum forming models, match plates, molds, core boxes; EA-08 curative for high operating temp., med. pot life, heat cure; mix ratio: 100/5. [Conap]

Conastic® AD-20. Two-part PU adhesive; flexible, high strength adhesive for bonding thermoplastics such as PC, Kapton, Hytrel, PVC, ABS, nylon; opaque amber; mix ratio 100/17.5. [Conap] *

Conastic® ST-115. Two-part PU adhesive; dielec. staking adhesive offering thixotropy, short pot life, good dielec. props.; ideal for holding PC board components in place, as PU patching compound and water barrier joint sealer; opaque; mix ratio 100/17.5. [Conap] *

Conathane®. Polyurethane resin systems and elastomers; used for tooling applications and potting and encapsulating applications. [Conap]

Conathane® CE-1155. Two-component PU dielec. conformal coating; insulating coating providing ultimate protection to aircraft avionics, instrumentation, missiles, spacecraft, fire and smoke detectors, electronic components, coils, and transformers. [Conap] *

Conathane® CE-1163. One-component PU dielec. conformal coating; insulating coating providing ultimate protection to aircraft avionics, instrumentation, missiles, spacecraft, fire and smoke detectors, electronic components, coils, and transformers. [Conap] *

Conathane® EN-4. Two-component polybutadiene-based liq. urethane casting and potting system; produces tough, stable elastomers with outstanding hydrolytic stability, elec. props., thermal shock resist., and low exotherm, shrinkage, and toxicity; applications incl. cable molding, cable connectors, underwater protection, strain sensitive potting, high voltage potting, cryogenic potting, vibration dampers, sealant bushings, transducer potting; mix ratio 100/17.5. [Conap] *

Conathane® EN-20. Two-part PU resin system; low visc., low toxicity, RT curing system for potting, casting, embedding, and encapsulation of electronic circuits, components, and power devices; non-TDI, non-MOCA. [Conap] *

Conathane® EN-2521. Two-component PU potting and encapsulating system; used for modules, strain-sensitive components, transformers, and coils; non-MBOCA, non-TDI; EN-2521 is UL recognized; mix ratio 20/100. [Conap] *

Conathane® RN-1501. TDI polyether elastomer; high-performance PTMG polyether prepolymer which produces flexible to semirigid urethanes; uses incl. rollers, tires, belts, gaskets, seals, wheels, molded parts, die pads, bushings, gears. [Conap] *

Conathane® RN-1558. TDI polyether elastomer; med.-performance polyether elastomer for rolls, seals, casters, bushings, gears, impellers. [Conap] *

Conathane® RN-1570. Aliphatic isocyanate polyether elastomer; high-performance prepolymer yielding tough elastomers with uv and hydrolysis resistance. [Conap] *

Conathane® TU-50A. Two-component liq. TDI/MBOCA-based casting systems; tooling resin system producing flexible elastomers of exceptional toughness, high elong., high tensile and tear strength, and excellent abrasion resist.; applications incl. cast-in-place linings for metal polishing and finishing equip., industrial

---

* Trade name not verified as available   † Trade name verified as obsolete

wheels, rollers; metal forming pads, caters, vibration, shock, and sound dampening pads, drop hammer faces, flexible molds, washers, gaskets, bushings, diaphragms, impellers, flexible couplings, belts, pulleys; mix ratio 100/94. [Conap] *

Conathane®TU-400. Two-component liq. casting systems; non-MBOCA, non-TDI; produces elastomers of exceptional toughness, high elong., high tensile and tear strength, and excellent abrasion resist.; cures at room temperature or elevated temps.; applications incl. long-lasting flexible molds, cast-in-place linings; for vibratory finishing and metal polishing equip., vibration, shock, and sound dampening pads, industrial wheels, belts, pulleys, casters, impellers; mix ratio: 100/87. [Conap]

Conathane® TU-4010. Two-part urethane elastomer system; filled, low-visc. liq. casting system for tooling applications. [Conap]

Conathane® UC-17. Two-component urethane resin systems; non-MBOCA, non-TDI; produces tough, high impact-resist., dimensionally stable castings; used for prototype parts and limited production runs; mix ratio: 25/100. [Conap]

Conathane® UC-34. Two-component urethane resin systems; casting system; non-TDI, non-BMOCA; 1:1 mix ratio and low visc. for ease of handling; fast set and cure chars.; ideal for casting duplicate masters, jigs, and patterns; mix ratio: 100/100. [Conap]

Concentrated Dariloid®. Alginate; used for puddings, pie fillings, custards, cheese dips, cheesecake, egg nog, milk shake, toppings, syrups, baked cream fillings. [Kelco]

Concentrated Dariloid® KB. CAS 9005-37-2; Propylene glycol alginate; used for sour cream, cottage cheese dressings, cream cheese, egg nog, ice cream fruit, fruit beverages, water ice, sherbet, lowfat novelties. [Kelco]

Concentrated Dariloid® XL. Alginate; used for egg nog, liquid dairy applications requiring a low viscosity stabilizer. [Kelco]

concentrated size. Powdered glue.

Conceptrol. CAS 9016-45-9; Nonoxynol 9; spermaticide. [Ortho Pharmaceutical Corp]

Concert®. For agriculture industry. [DuPont UK]

Concor®. CAS 66722-44-9; Bisoprolol. [E Merck]

Concordin®. CAS 1225-55-4; Protriptyline hydrochloride; antidepressant with anergic properties. [Merck & Co Inc]

Concurat® L. Anthelmintic against gastrointestinal nematodes and lungworms in cattle, sheep, goats and pigs; veterinary medicine. [Bayer AG]

Condens-Aid. Blends of amines; used for steam and return line corrosion inhibitors. [Schaefer Technologies Inc]

condenser foil. An alloy of 90% lead, 9.25% tin, and 0.75% antimony.

Condensite. See Bakelite.

Condensol®. Catalysts for textile high-grade finishing. [BASF AG; BASF plc]

Conditioner Base. Blend of conditioning and stabilizing agents. [Croda Chem. Ltd]

condor. See oil white.

Conductive Nylon 12. Conductive nylon 12; for static dissipation and high tensile strength applications [Americhem]

conductivity bronze. A bronze containing copper with 0.8% cadmium and 0.6% tin.

Conducto-Lube®. Pure silver powder and high grade petroleum oil; used in high-speed air blast breakers or virtually any application where a conductive lubricant is needed. [The Cool-Amp Conducto-Lube Co]

Conductomer ABS-22. CAS 9003-56-9; ABS with 22% conductive carbon black.; elec. conductive compound for fabricated parts capable of alleviating electrostatic discharge. [Syn. Rubber Tech.]

Conductomer HDC-22HLMI-M. CAS 9002-88-4; HDPE, 22% conductive carbon black; elec. conductive compound for fabricated parts capable of alleviating electrostatic discharge. [Syn. Rubber Tech.]

Conducto-Wrap. Copper foil tape with conductive pressure sensitive adhesive; for EMI shielding and electrical splicing. [Custom Coating & Laminating Corp] *

Condux. Electrically conductive elastomers. [Hardman]

Condyline. CAS 518-28-5; A proprietary preparation of podophyllotoxin; used in the treatment of condylomata. [Brocades Pharma]

Condy's fluid. A solution of aluminum permanganate with some aluminum sulfate. It is an oxidizing agent, and is used as a disinfectant and deodorizer.

Con-Fer. A proprietary preparation containing ethinylestradiol and norethisterone acetate with ferrous fumarate tablets; an oral contraceptive with an iron supplement. [Parke-Davis] *

Confidor. CAS 105827-78-9; Imidacloprid; systemic insecticide with contact and stomach action; applied as a foliar or soil treatment especially against sucking pests (virus vectors), e.g., aphids, thrips, whiteflies and leafhoppers on rice, potatoes, vegetables, cotton, tobacco, citrus, pome and stone fruit and other crops; also against some biting insects such as rice water weevil, Colorado potato beetle, wireworms, frit fly, beet fly, onion fly, citrus and apple leaf miners. [Bayer AG]

Conifer and Shrub Fertiliser. Powdered fertilizer containing NPK 10:7.5:10 plus 1.8% Mg and trace elements; all-purpose base and top dressing fertilizer. [Vitax Ltd]

Conlex. Specialized latices based on acrylic-styrene copolymers; used in floor polishes. [Morton Int'l. Ltd]

Conn's stain. A microscopic stain. It contains 1 gram rose bengal, 5 g phenol, and 100 cc distilled water.

Conoptic® UC-33. Two-part PU resin system; casting resin system formulated for applications requiring clarity; used for prototype parts, windows, display devices, lens materials. [Conap]

Conotrane. A proprietary preparation of benzalkonium chloride and dimethicone; an antiseptic skin barrier cream. [Boehringer Ingelheim Ltd]

Conovid. A proprietary preparation of norethynodrel and mestranol. oral contraceptive. [G D Searle & Co Ltd] *

Conovid-E. A proprietary preparation of norethynodrel and mestranol. oral contraceptive. [G D Searle & Co Ltd] *

Conray, 30, 43. CAS 13087-53-1; Iothalamate meglumine; diagnostic aid. [Mallinckrodt Inc] *

Conray 325, 400. CAS 1225-20-3; Iothalamate sodium; diagnostic aid. [Mallinckrodt Inc] *

Conray 420. A proprietary preparation of sodium and/or meglumine iothalamate; intravascular X-ray contrast medium. [May & Baker Ltd] *

Conrex. Acrylic leveling resin. [Morton Int'l. Ltd]

Consal. CAS 497-19-8; EINECS 207-838-8; Sodium carbonate hydrated. [Church & Dwight Co]

Constab. Specialized masterbatches for thermoplastics. [Cornelius Chemical Co Ltd] *

Constantan. An alloy of 60% copper and 40% nickel. It has a specific gravity of 8.9, and melts at 1290 C; used as an electrical resistance.

Constantin. Electrical resistance alloys. One contains 54% copper and 46% nickel, and another consists of 54% copper, 44% nickel, 1.3% manganese, and 0.4% iron.

constructal. An aluminum alloy containing 3% of alloying elements, chiefly zinc.

Construction 1200®. Silicone sealant; high strength acetoxy-cure sealant for structural glazing and premium general-purpose applications. [GE Silicones]

Contact 75. CAS 1897-45-6; Chlorothalonil; a fungicide for a wide range of agricultural crops. [Fermenta ASC Europe Ltd]

Contax. Weed killer. [Chevron] *

Contax. A proprietary preparation containing the diacetate of oxyphenisatin; a laxative. [Cox Pharmaceuticals] *

Contex. An antifluxing agent for gold- and silversmiths and in the jewelry industry. [Degussa AG] *

Continental® Clay. CAS 1332-58-7; Sec. kaolin clay; filler used in agric. application. [R.T. Vanderbilt]

Continex®LH-10. CAS 1333-86-4; Carbon black; improved abrasion resist. props. with lower hysteresis; for rubber, plastics, paper, and printing ink industries. [Witco/Concarb]

Continex®N351. CAS 1333-86-4; Carbon black; for rubber, plastics, paper, and printing ink industries. [Witco/Concarb]

contracid. A corrosion-resisting alloy stable to nitric acid, hydrochloric acid, sulfuric acid, and other reagents. It contains from 50-60% nickel, 15-20% chromium, 0-20% iron, and up to 10% molybdenum or tungsten.

Contractors 1000®. Silicone sealant; general-purpose glazing and sealing product [GE Silicones]

Contradet. A laundry contra-flow detergent. [Laporte Industries Ltd] *

Contrapar. Gloxazone; anaplasmodastat. [The Wellcome Foundation Ltd] *

Contraqua LF. Zirconium wax paraffin emulsion; cationic; hydrophobic agent for all types of fibers; water-repelling effect stable to gelling temperatures up to 190 C. [Thor Chemicals (UK) Ltd]

Contrast®. For agriculture industry. [DuPont UK]

Contrastol W. For contrast dyeings. [Bayer plc] †

Control I-100. CAS 9005-25-8; Corn starch. [Grain Processing]

Controx® KS. Tocopherol, citric esters; antioxidant for cosmetics. [Henkel/Cospha; Henkel Canada] *

Controx®Range. Antioxidants for animal and vegetable oils and fats in food industry. [Grünau]

Convol. Concentrated volumetric solutions. [BDH Chemicals Ltd]

Cook's alloys. One contains 68.5% antimony and 31.5% zinc, and another consists of 57% antimony and 43% zinc.

Cooksons. CAS 1327-33-9; White oxide of antimony. [Associated Lead Manufacturers Ltd] *

Cool-Amp. Sodium chloride, calcium chloride, silver chloride, potassium bitartrate; powder for silver plating electrical bus bars, connectors, etc. [The Cool-Amp Conducto-Lube Co]

Coolanol. Series of formulated silicate ester fluids used for a wide range of heat transfer, dielectric and hydraulic fluid applications; for an operating-temperature range of -90°F to 700°F; used primarily in aerospace and advanced electronic hardware as a coolant/dielectric. [Monsanto Co] *

Coolmax®. Fibers. [DuPont UK]

Cool-Treet. Various liquid blends of cooling water deposit and corrosion inhibitors; used for cooling water treatment. [Schaefer Technologies Inc]

Coomassie. Milling and half milling acid dyes. [ICI Chem. & Polymers Ltd]

Coopane. CAS 142-88-1; Piperazine adipate. [The Wellcome Foundation Ltd] *

Coopaphene. Hexachlorophane; anthelmintic. [The Wellcome Foundation Ltd] *

Cooper Coopex. CAS 52645-53-1; Permethrin; a pyrethroid insecticide. [The Wellcome Foundation Ltd]

Coopercote. Insecticidal varnish for paper, board etc. [The Wellcome Foundation Ltd] *

Cooper Graincote. CAS 5598-13-0; Chlorpyrifos-methyl; an organophosphate insecticide for the treatment of pests in stored grain and oilseed rape. [The Wellcome Foundation Ltd]

cooperite. An alloy of 80% nickel, 14% tungsten, and 6% zirconium; used for cutting tools. A modified alloy contains tantalum.

Coopermatic. Automated insecticide dispenser. [The Wellcome Foundation Ltd] *

Cooper's gold. An alloy of 19% platinum and 81% copper.

Cooper's pen metal. An alloy of 50% platinum, 37.5% silver, and 12.5% copper.

Cooper's speculum metal. An alloy of 58% copper, 27% tin, 10% platinum, 4% zinc, and 1% arsenic.

Copac. Metal salt complex; accelerator for unsaturated polyesters; drier for nonaqueous coating compositions. [Hüls Am.]

Copac®E. Ammoniacal copper sulfate; for control of bacterial diseases in pears, vegetables, ornamentals [BASF AG]

copal oils. Oils obtained by the dry distillation of copal; used for the preparation of oil varnishes.

Copaloy. A proprietary platinum-tin-antimony alloy with a small percentage of copper. A bearing metal. §

Copel alloy. An alloy of 55% copper and 45% nickel.

Copene. A proprietary trade name for a polyterpene copolymer resin used in lacquers, paints, and varnishes. §

Coperflex BR-45. 1,3-Butadiene polymer; light colored, nonstaining rubber. [Coperbo]

Coperflex SSBR-B18 4525. 1,3-Butadiene/styrene block polymer, sol'n. polymerized with an alkyl-lithium catalyst; light colored, nonstaining rubber. [Coperbo]

Copernick. An alloy similar to Hypernick; the permeability is constant over a wide range of flux densities. §

Copherol®1250. Tocopheryl acetate; antioxidant and moisturizer for sun protection and skin care products [Henkel/Cospha; Henkel Canada]

Copherol®F 1300. Tocopherol; antioxidant and moisturizer for sun protection and skin care products [Henkel/Cospha; Henkel Canada]

Copisil. Color developer for the production of carbonless copying papers, thermoreactive papers and chemical test papers. [Süd-Chemie AG] *

Copo®1500. Cold SBR, nonpigmented; general purpose, staining, cold SBR with good physicals and abrasion chars. [Copolymer Rubber]

Copo®1712. Cold SBR, nonpigmented with 37.5 phr oil;

---

* Trade name not verified as available   † Trade name verified as obsolete

general purpose, staining, oil-extended cold SBR. [Copolymer Rubber]

Copolymer 186. Castor oil, dehydrated; for prep. of quality coating vehicles by cold blending; compatible with phenolic, rosin esters, hydrocarbons and alkyd resins. [CasChem]

copper. (bronze powder; copper bronze; gold bronze) Cu; CAS 7440-50-8; EINECS 231-159-6; Electric wiring, switches, plumbing, heating, roofing; chemical and pharmaceutical machinery; alloys; coatings; cooking utensils. [Aldrich; Asarco; M&T Harshaw; Noah Chem.]

copper acetate (ic). (cupric acetate) $(CH_3CO_2)_2Cu$; CAS 142-71-2; EINECS 205-553-3; Pesticide, catalyst, fungicide, pigments, manufacture of Paris green. [Celtic Chem. Ltd; Mallinckrodt; Nihon Kagaku Sangyo]

Copper Antracol®. Combination product for prevention of fungal infections in fruit, grapes, vegetables and arable crops. [Bayer AG]

copperas. See iron sulfate (ous).

copper, bis(D-gluconato)-. See copper gluconate.

copper bromide (ic). (cupric bromide) $CuBr_2$; Photography (intensifier), organic synthesis (brominating agent), battery electrolyte, wood preservative. [Aldrich; Atomergic Chemetals; Cerac; Hoechst Celanese; Mallinckrodt]

copper bromide (ous). (cuprous bromide) CuBr; CAS 7787-70-4; EINECS 232-131-6; Catalyst in organic reactions. [Aldrich; Atomergic Chemetals; Cerac; Hoechst Celanese]

copper bronze. See copper.

copper carbonate (ic). (cupric carbonate basic) $CuCO_3 \cdot Cu(OH)_2$; CAS 12069-69-1; EINECS 235-113-6; Pigments, pyrotechnics, insecticides, copper salts, astringent in pomades, antidote for phosphorus poisoning, smut preventive, fungicide for seed treatment, feed additive. [Am. Chemet; Boliden Intertrade; Nihon Kagaku Sangyo]

copper chloride (ous). (cuprous chloride) CuCl; CAS 7758-89-6; EINECS 231-842-9; Catalyst, preservative and fungicide, desulfurizing and decolorizing agent for petroleum industry, absorbent for carbon monoxide. [Atomergic Chemetals; Cerac; Hoechst Celanese; Mallinckrodt; Nihon Kagaku Sangyo; Zinder SpA]

Copper-Count. Broad spectrum bactericide and fungicide. [Mineral Research & Development Corp]

Copper Euparen®. Combination product for prevention of fungal infections in fruit, grapes, vegetables and arable crops. [Bayer AG]

copper gluconate. (bis(D-gluconato) copper; copper, bis(D-gluconato)-) Copper salt of gluconic acid; $[HOCOO·CHOHCHCHOHCHOHCH_2OH]_2Cu^{++}$; CAS 527-09-3; mineral source for pharmaceutical and food products [Akzo; Glucona; Spectrum Chem. Mfg.]

copper green. A term applied to the mineral malachite.

copper hydrate. See copper hydroxide (ic).

copper hydroxide (ic). (cupric hydroxide; copper oxide hydrated; copper hydrate) $C_4(OH)_2$; CAS 20427-59-2; Copper salts, mordant, paper staining; pesticide, fungicide, catalyst. [Am. Chemet; Cuproquim; Faesy & Besthoff; Griffin]

copper iodide (ous). (cuprous iodide) CuI; CuI; CAS 7681-65-4; EINECS 231-674-6; Feed additive, in table salt as source of dietary iodine, catalyst, cloud seeding. [Aldrich; Atomergic Chemetals; Blythe, William Ltd; Cerac; R.W. Greeff; Mitsui Toatsu; Nihon Kagaku Sangyo]

Copper-Lonacol. Copper-zineb. combination spray for the control of fungal diseases. [Bayer AG]

copper naphthenate. Copper salt of petroleum naphthenic acids; CAS 1338-02-9; wood, canvas and rope preservative; antifouling in paints; insecticide, fungicide. [Akzo; KMZ Chem. Ltd; Troy]

copper-nickel. An alloy of 50% copper and 50% nickel; used in the manufacture of nickel copper alloys, is known by this name. The term is also applied to the mineral Niccolite, NiAs.

copper nitrate (ic). (cupric nitrate trihydrate) $Cu(NO_3)_2 \cdot 3H_2O$; CAS 10031-43-3; EINECS 221 838-5; Light-sensitive papers, analytical reagent, textile dyeing mordant, nitrating agent, insecticide, coloring copper black, electroplating, paints, varnishes, enamels, pharmaceuticals, catalyst. [Blythe, Williams Ltd; Mallinckrodt]

copper oxide hydrated. See copper hydroxide (ic).

copper oxide (ic). (cupric oxide) CuO; CAS 1317-38-0; EINECS 215-269-1; Ceramic colorant, reagent in analytical chemistry, insecticide, catalyst, purification of hydrogen, batteries and electrodes, electroplating, solvent, desulfurizing oils, rayon, metallurgical and welding fluxes, antifouling paints, phosphors. [Aldrich; Am. Chemet; Cerac; Chemisphere Ltd; Nihon Kagaku Sangyo; Noah Chem.]

copper silumin. An alloy of 85.6% aluminum, 13% silicon, 0.8% copper, and 0.6% iron.

copper soap. Copper resinate.

copper solder. An alloy of 2 parts lead and 5 parts tin.

copper steel. An alloy of steel with up to 1% copper, usually 0.5%. It resists corrosion.

copper sulfate (ic). (CSP; copper sulfate pentahydrate; cupric sulfate; blue vitriol; blue stone; blue copperas) $CuSO_4 \cdot 5H_2O$; CAS 7758-99-8; EINECS 231-847-6; Soil additive, pesticides, feed additive, germicides, textile mordant, leather, pigments, batteries, electroplated coatings, copper salts, analytical reagent, medicine, wood preservative, lithography, ore flotation, petroleum, rubber, steel manufacture [Allchem Industries; Farleyway Chem. Ltd]

copper sulfate pentahydrate. See copper sulfate (ic).

Coppertox. Livestock dips and sprays. [The Wellcome Foundation Ltd] *

Coppertrace. CAS 7447-39-4; Cupric chloride dihydrate; supplement, trace mineral. [Armour Pharmaceutical Co]*

copper water. Iron sulfate, $FeSO_4$.

Coppesan. A copper fungicide. [The Boots Co plc] †

Coprantex. Dyeing and printing assistant. [Ciba plc] *

Coprol. CAS 577-11-7; EINECS 209-406-4; A proprietary trade name for dioctyl sodium sulfosuccinate. §

Cops I. CAS 52556-42-0; Sodium 1-allyloxy-2-hydroxypropane sulfonate [Rhone-Poulenc Surf.]

Coptal. Textile auxiliary chemicals. [ICI Chem. & Polymers Ltd]

Co-Pyronil. A proprietary preparation containing pyrrobutamine, methapyriline and cyclopentamine hydrochloride. [Eli Lilly & Co]

Corafilm. Amine treatment; for steam lines. [Garvey Chemical Corp] *

Corangil. A proprietary preparation of glyceryl trinitrate, pentaerythritol tetra-nitrate, diprophylline and papaverine hydrochloride; used in the treatment of angina pectoris. [British Drug Houses] *

---

‡ Trade name and manufacturer not verified

§ Trade name without identified manufacturer

Corasole. Carbon black/plastic concentrate; used for pigmenting of paper and cement for the plastics processing industry, pigmenting of PE, PP, EVA, PVC, PS, SAN, and ABS. [Degussa AG] *

Coravol. Original amine process; for steam lines. [Garvey Chemical Corp] *

Corban. Corrosion inhibitors; used to reduce metal loss from oil and gas well equipment caused by hydrogen sulfide, carbon dioxide and organic acids; the inhibitor coats the metal surfaces with a thin protective film. [Dow UK] *

Corbel®. CAS 67306-03-0; Fenpropimorph; systemic fungicide [BASF AG; BASF plc]

Corbel® CL. Suspension concentrate containing 250 g chlorothalonil and 187 g fenpropimorph per liter; a systemic fungicide for winter wheat. [BASF plc]

Corbel®-Duo. Fenpropimorph, carbendazim; fungicide [BASF AG]

Corbel®-Star. Fenpropimorph, chlorothalonil; fungicide [BASF AG]

Corbrite. Disazo yellow and azo red pigments. [European Colour (Pigments) Ltd]

Corcert. Food colors meeting EEC specifications. [European Colour (Pigments) Ltd]

Cordetec 100. Phthalic anhydride catalyst made of vanadium pentoxide and titanium dioxide with some promoters supported on a uniform ceramic ring; the catalyst is used to produce phthalic anhydride from o-xylene reaching yields between 107-109 kg of phthalic anhydride from o-xyene (100%); the catalyst can be used in normal process and in low energy process. [Desarrollo Quimico Industrial Spa] *

Cordex. A proprietary preparation of prednisolone and aspirin; anti-inflammatory agent. [Upjohn Ltd] *

Cordilox. CAS 152-11-4; A proprietary preparation of verapamil hydrochloride; an antihypertensive. [Abbott Laboratories] *

Cordite. (Cordite MD; Maximite). An explosive used as powder and filaments. It is a mixture of 65% guncotton, Gelatinized by means of acetone, 30% nitroglycerine, and 5% petroleum jelly.

Cordran. CAS 1524-88-5; Flurandrenolide; glucocorticoid. [Eli Lilly & Co]

Cordura®. Fibers. [DuPont UK]

Corephen®10. Phenol formaldehyde; for acid-proof stoving finishes. [Bayer AG]

Coresize 630. Hydrocarbon emulsion; sizing agent imparting water resistance to core of gypsum board, esp. bathroom sheathing and other high-moisture-resist. boards. [Hercules]

Corex. See Carbora. §

Corexit. Oil slick dispersants. [Exxon UK] *

Corfast. Azo red and yellow pigments. [European Colour (Pigments) Ltd]

Corfix. A range of adhesives for all types of resin, silicate and oil-bonded cores. [Foseco (F.S.) Ltd]

Corgard. CAS 42200-33-9; Nadolol; used for the treatment of cardiovascular disorders. [Bristol-Myers Squibb Pharmaceuticals Ltd]

Corgaretic. A proprietary preparation of nadolol and bendrofluazide; used for use in the treatment of hypertension. [Bristol-Myers Squibb Pharmaceuticals Ltd]

Corgran. Granular organic pigments, toners, food colors. [European Colour (Pigments) Ltd]

Coriacide. Leather dyes. [ICI Chem. & Polymers Ltd]

Corial®. Special org. solv. for leather finishes. [BASF AG; BASF plc]

Corialbinder®. Polymer dispersions; binders for leather finishes. [BASF AG]

Corial Primer. Aqueous polymer dispersion for leather. [BASF plc]

Corian®. Polymer. [DuPont UK]

Coriban. Diamphenethide fasciolicide. [The Wellcome Foundation Ltd] *

Coric. CAS 83915-83-7; Lisinopril; for the treatment of hypertension. (Germany) [Merck & Co Inc]

corichrome. Titanium lactates; used as morrdants and strikers in the leather industry.

coridine. n-Propyllutidine.

Cori ester. Glucopyranose-1-monophosphate.

Corilene. An aqueous degreasing agent for leather. [ICI Chem. & Polymers Ltd]

Corinal. A trademark for a synthetic tannin prepared by condensing heavy tar oils with formaldehyde and then making the aluminum salt. [National Lead Co] *

Corindite. The trade name for an abrasive and refractory of the carborundum type; obtained from bauxite by heating it with anthracite, and contains 69%. Al_2O_3, with SiO_2 and Fe_2O_3. §

corioflavines. Red or reddish-brown dyestuffs; used for leather.

Coripact. Mineral fiber thermal-insulating boards and strips together with bitumen sheet; for non-ventilated roofs and for all roof coverings. [Vedag GmbH] *

coriphosphines. Dyestuffs. They are alkylated aminoacridines.

Corisol. A proprietary preparation of suprarenal gland hormone. §

Coriumine. Leather dyes. [ICI Chem. & Polymers Ltd]

cork. The outer bark of a tree, Quercus suber.

Corlake. Food color lakes on insoluble substrates. [European Colour (Pigments) Ltd]

Corlan. A proprietary preparation of hydrocortisone hemisuccinate sodium; for mouth ulcers. [Glaxo Laboratories]

Corlar®. Epoxy finish; used for automotive industry. [DuPont UK]

Cormelian. Dilazephydrochloride; tablet form; for treatment of ischemic heart disease. [Degussa AG] *

Cormix. Concrete admixtures for the construction industry. [Crosfield Chemicals Ltd]

Cormul. Range of cationic bitumen emulsifiers in liquid form; used in road surfacing: surface dressing tack coats; stone coating; grouting; slurry sealing. [Thomas Swan & Co Ltd] †

Cornalith. See Gallatite.

Cornish bronze. An alloy of 78% copper, 9.5% tin, and 12.5% lead; a white bearing metal.

Cornish clay. See China clay.

Cornite. A hard vulcanite.

corn oil. (maize oil; oils, corn; Zea mays oil) Refined fixed oil obtained from wet milling of corn, Zea mays; CAS 8001-30-7; EINECS 232-281-2; used in preparing foodstuffs, soap, lubricants, leather finishing, margarine, salad oil, hair dressing; solvent; paints. [Grain Processing; Karlshamns; Penta Mfg.; A.E. Staley Manufacture]

Cornox. A selective weedkiller. [The Boots Co plc] †

Cornox Plus. Selective herbicide. [Schering Agrochemicals Ltd] *

---

* Trade name not verified as available   † Trade name verified as obsolete

Cornoxynil. Selective herbicide. [Schering Agrochemicals Ltd] *

Corn-Pro 35. Hydrolyzed corn protein; cosmetic ingredient for skin and hair care products [Brooks Industries]

corn starch. (starch, corn) Obtained from grains of *Zea mays*; carbohydrate polymer consisting primarily of amylose and amylopectin; white powd.; CAS 9005-25-8; source of glucose; filler in baking powder; thickening agent in food products; adhesives; coatings; additive in plastics. [Am. Maize Prods.; Cerestar UK; Grain Processing; Nat'l. Starch & Chem.; A.E. Staley Manufacture]

corn sugar. *See* glucose.

corn sugar gum. *See* xanthan gum.

corn syrup. (glucose syrup) Mixture of D-glucose, maltose, and maltodextrins; CAS 8029-43-4; used in the food industry as sweetener, thickener, bodying agent in soft drinks. [Am. Maize Prods.; Gist-Brocades Food Ingreds.; A.E. Staley Manufacture]

Cornuite. A protein-like mineral found in the diatomaceous earth from Neu-Ohe. It is a gelatinous albuminous material with a 3% dry residue.

Cornutine. Impure ergotoxine.

Corolox. A trademark for abrasive and refractory materials; the essential constituent is crystalline alumina. §

Corona. Lanolin BP/EP; used for cosmetic and pharmaceutical preparations. [Croda Chem. Ltd]

Coronet. Refined lanolin; used as emulsifier, emollient, dispersing, solubilizing, and wetting agent. [Croda Chem. Ltd]

Coronium. An alloy containing 80% copper, 15% zinc, and 5% tin.

Coronium Bromide. Strontium bromide, SrBr₂. §

Corotox. A trademark for goods made for abrasive and refractory purposes; the essential constituent is crystalline alumina. §

Corowalt. A trademark for materials of the abrasive and refractory type, the essential constituent of which is crystalline alumina. §

Corox. CAS 1309-48-4; A proprietary insulating material having a great thermal conductivity and high electrical insulating power; consists essentially of magnesium oxide. §

Corozo. Vegetable ivory, the seeds of *Phytelephas macrocarpa*.

Correctol®. Yellow phenolphthalein, docusate sodium; combines a softening agent with a mild laxative for gentle, more natural relief. [Schering Corp]

Correctol®Extra Gentle. CAS 577-11-7; Docusate sodium; stool softener, stimulant-free laxative for sensitive systems. [Schering Corp]

Corro-Guard. Various liquid blends of deposit and corrosion inhibitors for nonpotable water systems; used for recirculating hot and chilled water systems. [Schaefer Technologies Inc]

Corrolite®. Vinyl esters. [Reichhold]

Corronel 220. A trademark for a nickel-molybdenum-vanadium alloy with good resistance to hydrochloric, sulfuric and phosphoric acids under reducing conditions. [Wiggin Alloys Ltd] ‡

Corronel Alloy 230. A trademark for an alloy of 35% chromium and 65% nickel; resistant to nitric and nitric/hydrochloric acid mixtures. [Wiggin Alloys Ltd] ‡

corronil. An alloy of 70% nickel, 26% copper, and 4% manganese. It is a corrosion resisting alloy.

Corrosalloy. A proprietary trade name for stainless steels.§

corrosiron. An iron-silicon alloy containing 12% silicon. It is stated to be very resistant to acids.

Corrosist. A proprietary trade name for nickel base corrosion resistant alloys; used for valves in chemical plant subject to contact with chlorine. §

corrosive sublimate. Mercuric chloride, HgCl₂.

Corry's Moss Remover. Mixture of copper sulfate and sodium tetraborate; used to control moss on paths and hard tennis courts. [Synchemicals Ltd]

Corseal. Refractory moldable products for repairing damaged cores or sealing core joints. [Foseco (F.S.) Ltd]

Corsodyl. A proprietary preparation of chlorhexidine gluconate used in the treatment of gingivitis and dental hygiene generally. [ICI Chem. & Polymers Ltd]

Cortacream. A proprietary preparation of hydrocortisone acetate and silicon fluid on a dressing; used for eczematous skin disorders. [Smith & Nephew Pharmaceuticals Ltd] *

Cortaid. CAS 50-03-3; Hydrocortisone acetate; glucocorticoid. [Upjohn]

Cort-Dome. CAS 50-23-7; Hydrocortisone; glucocorticoid. [Miles Pharmaceuticals]

Cortef. CAS 50-23-7; A proprietary preparation of hydrocortisone for local use on skin. [Upjohn; Upjohn Ltd]

Cortef Acetate. CAS 50-03-3; Hydrocortisone acetate; glucocorticoid. [Upjohn] *

Cortef Oral Suspension. Hydrocortisone cypionate; glucocorticoid. [Upjohn] *

Cortelan. CAS 50-04-4; A proprietary preparation of cortisone acetate. [Glaxo Laboratories] *

Corten. A proprietary trade name for a chromium steel containing 0.5% chromium, 0.10% carbon, 0.1% manganese, 0.5% silicon, and 0.3% copper. §

Cortenema. CAS 50-23-7; Aqueous suspension of hydrocortisone 100 mg per 60 ml; a steroid enema as adjunct in the treatment of idiopathic nonspecific ulcerative colitis. [Bengue & Co Ltd] *

Cortico-Gel. A proprietary preparation of corticotrophin gelatin. [Crookes Laboratories] *

Cortifoam. CAS 50-23-7; A proprietary preparation of hydrocortisone in aerosol form for dermatological use. [Pfizer International] †

Cortipix. A proprietary preparation of coal tar and hydrocortisone acetate; used in the treatment of eczema. [Fisons plc, Pharmaceuticals Div] *

Cortisporin. Proprietary formulations of polymyxin B sulfate, neomycin sulfate and hydrocortisone; topical ointment and cream, otic solution and suspension, ophthalmic suspension and ointment; topical preparations: for treatment of skin conditions in which bacterial infection is present, and for secondarily infected insect bites; otic preparations: for the treatment of otitis externa due to or complicated by bacterial infection; ophthalmic preparations: for treatment of susceptible bacterial infections of the eye and its adnexa; for pre- and postoperative use to prevent infection following surgical procedures. [The Wellcome Foundation Ltd]

Cortisporin Ointment. Proprietary formulations of polymyxin B sulfate, bacitracin zinc, neomycin sulfate and hydrocortisone; ophthalmic ointment for the treatment of corticosteroid-responsive dermatoses with secondary infection or steroid-responsive inflammatory ocular conditions where the risk of infections exists. [The Wellcome

---

‡ Trade name and manufacturer not verified   § Trade name without identified manufacturer

Foundation Ltd]

Cortistab. CAS 50-04-4; A proprietary preparation of cortisone acetate. [The Boots Co plc]

Cortistan. CAS 53-06-5; Cortisone in a solution of 25 mg/10 cc [Standex Laboratories] ‡

Cortitrane. A proprietary preparation of phenylmercuric dinaphthylmethane disulfonate and prednisolone; antibiotic steroid skin cream. [Octel Chemicals Ltd] †

Cortocaps. A proprietary preparation of hydrocortisone acetate and neomycin sulfate; used in eye infections. [Crookes Laboratories] *

Cortoderm. CAS 50-03-3; A proprietary preparation of hydrocortisone acetate for dermatological use. [Crookes Laboratories] *

Cortoderm-N. A proprietary preparation of hydrocortisone and neomycin used in dermatology as an antibacterial agent. [Crookes Laboratories] *

Corton A1. Rubber bitumen-based product designed for use where fuel resistance is important, e.g., airport runways. [Feb Ltd]

Cortone. Organic pigment toners. [European Colour (Pigments) Ltd]

Cortone Acetate. CAS 50-04-4; Cortisone acetate; glucocorticoid. [Merck & Co Inc]

Cortril. CAS 50-03-3; A proprietary preparation of hydrocortisone acetate; a corticosteriod. [Pfizer International]

Cortril Acetate-AS. CAS 50-03-3; Hydrocortisone acetate; glucocorticoid. [Pfizer Inc]

Cortrophin Gel ACTH. Corticotropin, repository; hormone; glucocorticoid; diagnostic aid. [Organon Inc] *

Cortrophin Zinc ACTH. Corticotropin zinc hydroxide; hormone; glucocorticoid; diagnostic aid. [Organon Inc] *

Cortrosyn. CAS 16960-16-0; Cosyntropin; hormone. [Organon Inc] *

Cortymol® LP. Phenol-free org. fungicide; prevents mold formation on leather. [BASF AG]

Cor-Tyzine. A proprietary preparation of tetrahydrozoline hydrochloride and prednisolone; a nasal decongestant and anti-inflammatory. [Pfizer International]

Corubin. An artificial corundum Al_2O_3. It is the alumina which constitutes the slag formed in the reaction between aluminum and metallic oxides (Thermit); used for polishing purposes, and in the manufacture of fireproof stones.

Corundite. A trademark used for abrasive and refractory materials the essential constituent of which is crystalline alumina. §

Corvasal. See Corvaton. (France, Ivory Coast, Tunisia) [Cassella AG]

Corvaton®. CAS 25717-80-0; Molsidomine; for coronary therapeutics. [Cassella AG]

Corvaton. CAS 25717-80-0; Molsidomine; anti-anginal; vasodilator. [Hoechst-Roussel Pharmaceuticals Inc] *

Corvic. PVC, emulsion polymers and copolymers. [ICI Chem. & Polymers Ltd] †

Cosbiol®. CAS 111-01-3; Perhydrosqualene (squalane); high quality oil for cosmetics and pharmaceuticals. [Laserson & Sabetay]

Coscopin. CAS 128-62-1; A proprietary preparation containing noscapine; a cough linctus. [British Drug Houses]*

Coscopin Paediatric. CAS 128-62-1; A proprietary preparation containing noscapine; a cough linctus. [British Drug Houses] *

Cosiderm Collagen Masks. Soluble collagen and collagen; cosmetic ingred. [Maybrook]

Coslettized steel. A steel whose surface has been rust-proofed by dipping in a solution of iron phosphate and phosphoric acid.

Cosmedia Guar C-261 N. CAS 65497-29-2; Guar hydroxypropyltrimonium chloride; visc. builder; personal care product formulating; substantivity provides hair conditioning; stabilizer and thickener for emulsions; provides slip to finished formulations. [Henkel/Cospha; Henkel Canada]

Cosmedia Polymer HSP 1180. Polyacrylamido-methylpropane sulfonic acid; smooth feel agent for creams, lotions, liq. antiperspirants, shaving creams, soaps, nail polish removers. [Henkel/Cospha; Henkel Canada]

Cosmegen. Dactinomycin; cytotoxic, antineoplastic antibiotic. [Merck & Co Inc]

Cosmegin Lyovac. A proprietary preparation of actinomycin D; a cytotoxic drug. [Montedison UK Ltd] *

Cosmetic Lanolin Anhydrous USP. Lanolin; superfatting emollient, emulsifier for creams, lotions, lipsticks, make-up, sunscreen products [Croda Inc.]

Cosmetol® X. Unrefined castor oil USP, with antioxidant, deodorized specially refined grade of castor oil containing a food grade antioxidant; emollient, pigment wetter, cosolv., lubricant for cosmetics, makeup, antiperspirant sticks. [CasChem]

Cosmic. Emulsifying agents. [Abril Industrial Waxes]

Cosmic®FL. Maneb, tridemorph, carbendazim; broad spectrum cereal fungicide [BASF AG; BASF plc]

Cosmica®. Pigment series where absorption colors are deposited directly on the mica, creating highly intense effects with minimal luster; suitable for most cosmetics. [Mearl]

Cosmin. See Agathin. §

Cosmocil. Preservatives for cosmetics. [ICI Chem. & Polymers Ltd]

Cosmos Alloy. A proprietary lead base alloy with small amounts of tin and antimony; a bearing metal. §

Cosmowax. Nonionic self emulsifying wax base. [Croda Chem. Ltd]

Cosmowax K. Stearyl alcohol and ceteareth-20; emulsifier, stabilizer with body, opacity and conditioning props. for personal care products and pharmaceuticals. [Croda Inc.]

Cosmowax P. Cetearyl alcohol, ceteareth-20; emulsifier, emulsion stabilizer for lotions, creams. [Croda Inc.]

Cosylan. A proprietary preparation containing tincture of cocillana, liquid extract of squill and senega, antimony potassium tartrate, cascarin, ethyl-morphine hydrochloride, menthol and syrup; a cough linctus. [Parke-Davis]*

Cotane. 2-Hydroxy biphenyl. [Coalite Chemicals Ltd]

Cotazym. Pancrelipase; a concentrate of pancreatic enzymes standardized for lipase content; enzyme. [Organon Inc] *

Coterpin. A proprietary preparation of codeine, terpin, menthol pine oil, and eucalyptus; a cough linctus. [Ayrton Saunders plc] ‡

Cothias metal. An alloy of 67% copper and 33% tin; used as a hardener for zinc alloys.

Cotinazin. CAS 54-85-3; Isoniazid; antibacterial. [Pfizer Inc]

Cotnion-Ethyl. CAS 2642-71-9; Active ingredient: azinphos-ethyl; persistent agricultural organo-

---

* Trade name not verified as available   † Trade name verified as obsolete

phosphorous insecticide. [Makhteshim Chemical Works Ltd]

Cotnion-Ethyl-Methyl. CAS 2642-71-9, 86-50-0; Active ingredients: cotnion-ethyl plus cotnion-methyl; persistent agricultural organophosphorous insecticide combination. [Makhteshim Chemical Works Ltd]

Cotnion-Methyl. CAS 86-50-0; Active ingredient: azinphosmethyl; persistent agricultural organophosphorous insecticide. [Makhteshim Chemical Works Ltd]

Cotolan Fast. Dyestuffs for the one-bath of union materials (wool/cellulosics). [Bayer plc] †

Cotonerol®. Azo dyestuffs; used for textile dyeing and printing. [Cassella AG]

Cotopa. A form of textile insulating material composed of acetylated cotton yarn.

Cottestren. Dyestuffs for cellulosic/polyester blends. [BASF plc]

Cottestren®C Dyes. Mixed dyes for one-bath tone-in-tone dyeing of polyester/cellulose fiber blends. [BASF AG]

Cottoclarin. Wetting agent; for scouring, bleaching, dyeing and wetting processes. [Henkel Chemicals Ltd] *

cotton black. A dyestuff obtained by the fusion of o-,p-dinitrodiphenylamine-sulfonic acid with sodium polysulfides. Dyes cotton brownish-black.

cotton blue. Many of the direct cotton blues, such as diamine, benzo, and Congo blues, are sold under this name.

Cottonex. CAS 2164-17-2; Active ingredient: fluometuron; 1,1-dimethyl-3-(α,α,α-trifluoro-m-tolyl)urea; residual herbicide effective against a wide range of both annual broafleaf weeds and grasses. [Agan Chemical Manufacturers Ltd]

cotton fiber. See cellulose.

Cotton-Pro®. CAS 7287-19-6; Prometryn suspension; flowable herbicide for selective weed control in cotton and celery crops. [Griffin]

cotton wax. The wax found in cotton fiber. It melts at 82-86 C., and appears similar to carnauba wax.

Couch and Grass Killer. CAS 75-99-0; Soluble powder containing dalapon; used for control of grasses in crop and noncrop areas. [Vitax Ltd]

Cougar. Water-soluble polymeric materials for use as thickening agents or as texturizers in chemical products. [Courtaulds plc]

Couloscope (Coulometric coating thickness gauge). Coulometric electrolytes; for measurement of metallic coating thickness of coatings deposited over metallic and non metallic substrates, or of metal foils. [Fischer Instrumentation (GB) Ltd] *

Coulter. Range of medical and scientific laboratory instruments and products; used for blood cell and particle counting and sizing, blood chemistry analyzis. [Coulter Electronics Ltd]

Coulter Clenz. Cleaning solution; for cleaning of automatic blood cell analyzers. [Coulter Electronics Ltd]

Coulter Clone. Range of reagents based on monoclonal antibodies; used to identify different cell types. [Coulter Electronics Ltd]

Coulter Counter. Range of cell and particle counting and sizing analyzers; used for particle and cell counting and size distribution analyzis. [Coulter Electronics Ltd]

Coumadin. CAS 129-06-6; Warfarin sodium; anticoagulant. [Endo Laboratories Inc] *

coumalic acid. α-Pyrone-3-carboxylic acid.

Coumalux®. 7-Diethylamino-4-methyl-2H-benzopyran-2-one; optical brightener. [Octel Chemicals Ltd]

coumarone-indene resin. Thermosetting resin; adhesives, printing inks, floor tile binder; friction tape; paints; varnishes, enamels; in chewing gum. [Allchem Industries; Natrochem; Neville; Spectar Ltd]

coumarone resin. See paracoumarone resin.

Countdown. CAS 41394-05-2; Metamitron; triazinone herbicide for weed control in beet crops. [Kommer-Brookwick Ltd]

Country Fresh Disinfectant. CAS 1330-85-4; Soluble concentrate of 150 g dodecylbenzyl trimethylammonium chloride per liter; used for algae control. [Dimex Ltd]

Coupler. Photosensitive coupler. [Rhone-Poulenc UK]

Coupler SC. Suspension concentrate containing 60 g clopyralid and 350 g cyanazine per liter; a contact herbicide. [Shell UK]

Coupsil VP 6109. Silica reacted with bis (3-triethoxysilylpropyl) tetrasulfane; reinforcing filler for rubber compounds incl. NR, IR, SBR, BR, NBR, EPDM, and EPM. [Degussa AG; Struktol]

Courcel®. Proprietary and blended gums and stabilizers for the food industry; used for pet foods, sauces/salad dressings, meat products, vegetarian products, batters, egg products, water ices. [Courtaulds Chemicals, Water Soluble Polymers]

Courgel®. Performance water-soluble polymers with gelling properties for industrial applications; used for ceramic tile adhesives, textured building finishes, ceramic glazes, horticultural products. [Courtaulds Chemicals, Water Soluble Polymers]

Courline. A proprietary trade name for polypropylene monofilament yarns. [Hoechst Celanese] *

Courlose®. CAS 9004-32-4; Sodium carboxymethyl cellulose; used for oil well drilling, detergents, wallpaper adhesives, ceramic glazes, emulsion paints, and paper. [Courtaulds Chemicals, Water Soluble Polymers]

Cournova. A proprietary trade name for polypropylene oriented slit film for the manufacture of strings, twines and ropes. [Hoechst Celanese] *

Courtaulds. CAS 9004-35-7; Cellulose acetate. [Courtaulds plc]

Courtek. Chemical products for use in industry; polymeric plastics. [Courtaulds plc]

Courthene. Plastics materials in the form of sheets (nontextile) for use in manufacture. [Courtaulds plc]

Courtochrome. Chemical products for use in photochromy; photosensitive paper, plates, films, microfilm and foils. [Courtaulds plc]

court plaster. Isinglass dissolved in water, alcohol, and glycerin, and painted on taffeta.

Covar. Aliphatic solvent containing emulsifier. [Exxon Int'l.]*

Coveral. Aluminum alloy cleansing flux. [Foseco (F.S.) Ltd]

Coverite. A wetting agent. [Murphy Chemical Co Ltd] *

Covermark. CAS 13463-67-7; EINECS 236-675-5; A proprietary preparation of titanium dioxide with different pigments in a cream base; a skin masking cream. [Stiefel Laboratories (UK) Ltd] *

Covexin. A proprietary combined sheep vaccine. [Coopers Animal Health Ltd] *

Covi-Ox T-50. Tocopherol; antioxidant and blocking agent for cosmetic formulations. [Henkel/Cospha; Henkel Canada]

Covipherol T-75. Mixed d-tocopherols; antioxidant in food

---

‡ Trade name and manufacturer not verified   § Trade name without identified manufacturer

and cosmetic products [Henkel/Cospha; Henkel Canada]

Covon. Glycol-free colorants; formulated for in-plant tinting of high-gloss latex, emulsion and water-based paints. [Hüls Am.]

Cow Gum. Adhesive for mounting of photographs and art work. [Cow Proofings Ltd] ‡

Cowles' aluminum bronze. An alloy containing 89-98.75% copper and 1.25-11% aluminum. One alloy contains small amounts of iron and silicon.

Coxistac. CAS 53003-10-4; Salinomycin; an ionophorous antibiotic used as an anticoccidial agent in poultry. [Pfizer International]

Coxpyrin. See aspirin.

Coyden. Coccidiostat is mixed with chicken feed to prevent coccidiosis (diarrhoea) in poultry. [RMB Animal Health Ltd]

Cozirc. Chemically combined cobalt-zirconium carboxylate; drier for lead and/or barium free paint or printing ink. [Rhone-Poulenc UK

CP0110. CAS 940-41-0; 2-Phenethyltrichloro silane; CAS 940-41-0; coupling agent, chem. intermediate, blocking agent, release agent, lubricant, primer, reducing agent. [Hüls Am.]

CP-153-2 (25% in xylene). Chlorinated polyolefin; used as primer for polyethylene; excellent adhesion char. [Eastman]

CP0156. N-Phenylaminopropyltrimethoxy silane; CAS 3068-76-6; coupling agent, chem. intermediate, blocking agent, release agent, lubricant, primer, reducing agent. [Hüls Am.]

CP0160. CAS 768-33-2; Phenyldimethylchloro silane; CAS 768-33-2; coupling agent, chem. intermediate, blocking agent, release agent, lubricant, primer, reducing agent. [Hüls Am.]

CP0280. CAS 98-13-5; Phenyltrichloro silane; CAS 98-13-5; coupling agent, chem. intermediate, blocking agent, release agent, lubricant, primer, reducing agent. [Hüls Am.]

CP0320. CAS 780-69-8; Phenyltriethoxy silane; CAS 780-69-8; coupling agent, chem. intermediate, blocking agent, release agent, lubricant, primer, reducing agent. [Hüls Am.]

CP0330. Phenyltrimethoxy silane; CAS 2996-92-1; coupling agent, chem. intermediate, blocking agent, release agent, lubricant, primer, reducing agent. [Hüls Am.]

CP-343-1 (100%). Chlorinated polyolefin; used as primer for PP, selected plastics, and metals where it promotes adhesion. [Eastman]

CP0800. CAS 141-57-1; n-Propyltrichloro silane; coupling agent, chem. intermediate, blocking agent, release agent, lubricant, primer, reducing agent. [Hüls Am.]

CP0810. CAS 1067-25-0; n-Propyltrimethoxy silane; coupling agent, chem. intermediate, blocking agent, release agent, lubricant, primer, reducing agent. [Hüls Am.]

CPB. Dibutylxanthogen disulfide; a nondiscoloring and nonstaining low temperature ultra-accelerator for natural, SBR, nitrile and neoprene rubbers. [Uniroyal] *

C.P.D. Cadmium pentamethylene dithiocarbamate. A rubber vulcanization accelerator.

CPE. (chlorinated polyethylene) resins. When added to other plastics, CPE can be calendered, injection molded or extruded into tough, chemical resistant, weather resistant products; typical end uses would be pond liners,

automotive hose and chemical transfer hose. [Dow UK]*

C-petroleum naphtha. (petroleum benzine, safety oil). That fraction of petroleum distilling at 80.100 C., of specific gravity 0.667-0.707.

CP Filler. CAS 471-34-1, 1317-65-3; Calcium carbonate; competitively priced filler for construction products where color is of secondary importance. [ECC]

CPH-31-N. Glyceryl oleate; antiblocking agent. [C.P. Hall]

CPH-43-N. CAS 9004-81-3; PEG-12 laurate; solubilizer and dispersant. [C.P. Hall]

CPH-52-SE. CAS 1323-39-3; EINECS 215-354-3; Propylene glycol stearate. [C.P. Hall]

CPH-205-NX. Glyceryl oleate; lube additive, rust preventive. [C.P. Hall]

CPH-211-N. CAS 9005-07-6; PEG-8 dioleate; water treatment, lubricant. [C.P. Hall]

CPH-250-SE. Glyceryl stearate; emulsifier. [C.P. Hall]

CPH-353-SE. Glyceryl oleate; corrosion inhibitor. [C.P. Hall]

C.P.R. Multi-component systems used in the manufacture of rigid foams. [Dow UK] *

CPS034. CAS 63148-62-9; Dimethicone; fluids for mech., heat transfer and dielec. applications. [Hüls Am.]

CPS 076. (N-Trimethoxysilylpropyl)-polyethylenimine; coupling agent esp. for mineral-filled and adhesive bonding applications of high molecular weight thermoplastic polyamides and polyesters. [Hüls Am.]

CPS120. CAS 63148-57-2; Polymethylhydrosiloxane; crosslinker, waterproofing agent. [Hüls Am.]

CPS130. CAS 68607-75-0; Polymethyloctadecylsiloxane. [Hüls Am.]

CPS140. CAS 68440-90-4; Polymethyloctylsiloxane. [Hüls Am.]

CPS340. CAS 67923-14-2; Dimethicone, silanol-terminated; reactive fluid, RTV intermediate. [Hüls Am.]

CPS925. CAS 68037-87-6; Polyvinylmethylsiloxane; coupling agent. [Hüls Am.]

CPS9120. Polydiethoxysiloxane; silicone dioxide source. [Hüls Am.]

C-Quens. A proprietary preparation of mestranol (14 yellow tablets) and mestranol and chlormadinone (7 pink tablets); oral contraceptive. [Eli Lilly & Co] †

CR. See polychloroprene.

CR-39. Allyl diglycol carbonate monomer. [PPG Industries]

Cradocap. Cetrimide; for treatment of cradle cap. [Napp Laboratories Ltd] *

Crafol AP-16. Ethoxylated C16 phosphate ester; surfactant, textile lubricant. [Pulcra SA]

Crafol AP-31. Potassium lauryl phosphate; manufacturing of man-made textile fibers; lubricant, antistat; spinning oil for carpets. [Pulcra SA]

Crafol AP-53. CAS 66197-78-2; Nonylphenol ether phosphate, free acid; lubricant, emulsifier, corrosion inhibitor for oil and water-based cutting fluids, hydraulic fluids, rolling oils. [Pulcra SA]

Crafol AP-64. Ethoxylated C12-18 phosphate (8 EO), sodium salt; wetting agent, detergent, humectant. [Pulcra SA]

Crafol AP-201. CAS 12751-23-4; Lauryl phosphate, free acid; lubricant, emulsifier for leather and textile auxiliaries. [Henkel KGaA; Pulcra SA]

Crafol AP-260. Sodium laureth (10) phosphate; emulsifier, wetting agent, antistat for leather, textile auxiliaries, hot baths containing cyanides. [Henkel KGaA; Pulcra SA]

Crafol AP-262. MEA laureth (6) phosphate; lubricant, anti-

---

stat for textile auxiliaries. [Henkel KGaA]

Craig gold. An alloy of 80% copper, 10% nickel, and 10% zinc. It is a nickel silver (German silver).

Cranco. Lacquers and enamels. [ICI Chem. & Polymers Ltd]†

Crasnitin®. Asparaginase; an enzyme with antileukemic effect and tumor-specific site of action. [Bayer AG]

Crastine®. Thermoplastic polyester molding compounds. [Ciba plc]

Crastine® S 600. CAS 26062-94-2; PBT polyester; for electronics, elec., automotive, mech. engineering, chem. and apparatus engineering, domestic appliances, medical appliances, sporting goods. [Ciba-Geigy GmbH] *

Crastine® SG 625. CAS 26062-94-2; PBT polyester, 30% long glass fiber-reinforced. [Ciba-Geigy GmbH] *

Crastine® SG 665 FR. CAS 26062-94-2; PBT polyester, 30% long glass fiber-reinforced [Ciba-Geigy GmbH] *

Crastine® SO 653. CAS 26062-94-2; PBT polyester, 20% glass bead-reinforced. [Ciba-Geigy GmbH] *

Crastine® XB 3035. CAS 26062-94-2; PBT polyester [Ciba-Geigy GmbH] *

Crastine® XMB 1068. CAS 25038-59-9; PET polyester, 30% glass fiber-reinforced. [Ciba-Geigy GmbH] *

Cravenette EFC. Durable water and oil repellent to be used with and without fluorocarbons; excellent resin/catalyst stability. [Yorkshire Pat-Chem]

Craymer. Dimer acids, dimer derivates. [Cray Valley Ltd]

Crayvallac. Thixotropes. [Cray Valley Ltd]

Cream E45. A proprietary preparation containing white, soft paraffin, light liquid paraffin and anhydrous lanolin; a dermatological cream. [The Boots Co plc]

Creamaffin. Liquid paraffin, BP, magnesium hydroxide BP laxative. [The Boots Co plc] *

cream of tartar. *See* potassium bitartrate.

Creamtex. CAS 68334-28-1; Partially hydrogenated vegetable oil (soybean, cottonseed); kosher; good quality bakery shortening for use where emulsifier is not required. [Van Den Bergh Foods]

Crelan®. Resins and curing agents for the formulation of electrostatic thermosetting powder coatings. [Bayer AG]

Cremalgex. A proprietary preparation of methyl nicotinate, capsicum oleoresin, glycol salicylate and histamine hydrochloride; a rubefacient. [H N Norton & Co Ltd] ‡

Cremalgin. A proprietary preparation of methyl nicotinate, capsici,n and glycol salicylate; an embrocation. [Berk Pharmaceuticals Ltd] *

Cremalys. Blends of gelatinized and nongelatinized starches; used in the food industry to develop creamy textures. [Roquette (UK) Ltd]

Cremba. Mineral oil, petrolatum, lanolin alcohol, and lanolin; emollient, moisturizer, and emulsifier for cosmetics and pharmaceuticals, detergent surgical scrubs. [Croda Inc; Croda Chemicals Ltd]

Cremerol HMG. Hydroxylated milk glycerides; emollient, auxiliary emulsifier for personal care products. [Amerchol Corp]

Cremodex. Sulfadiazine/sulfamethazine; bacteriostatic. [Merck & Co Inc] †

Cremomycin. Succinylsulfathiazole and neomycin; for the treatment of mild bacterial diarrheas. [Merck & Co Inc] †

Cremophor®. Emulsifiers and solubilizers. [BASF plc]

Cremophor®A 11. CAS 68439-49-6; Ceteareth-11; emulsifier for cosmetic and pharmaceutical preparations. [BASF AG]

Cremophor®A 25. CAS 68439-49-6; Ceteareth-25; emulsifier for cosmetic and pharmaceutical preparations. [BASF AG]

Cremophor® EL. CAS 61791-12-6; PEG-35 castor oil; solubilizer, emulsifier used for essential oils, pharmaceuticals, cosmetics, veterinary medicine. [BASF; BASF AG]

Cremophor®NP 10. CAS 9016-45-9; Nonoxynol-10; solubilizer for perfumes, essential oils, and flavors. [BASF AG]

Cremophor® NP 14. CAS 9016-45-9; Nonoxynol-14; *see* Cremophor NP 10. [BASF AG]

Cremophor® RH 40. CAS 61788-85-0; PEG-40 hydrogenated castor oil; solubilizer for essential oils and perfumery, emulsifier; for cosmetics and pharmaceuticals. [BASF; BASF AG]

Cremophor® RH 60. CAS 61788-85-0; PEG-60 hydrogenated castor oil; solubilizer, emulsifier for essential oils and perfumes; for cosmetics. [BASF; BASF AG] *

Cremophor®S 9. CAS 9004-99-3; PEG-9 stearate; emulsifier for o/w type, thickening agent; suspension stabilizer; lubricating and antitack effects; for cosmetics and pharmaceuticals. [BASF AG]

Cremophor® WO 7. CAS 61788-85-0; PEG-7 hydrogenated castor oil; emulsifier for cosmetic w/o preps. [BASF AG]

Cremosan. A proprietary preparation of zinc oxide, colophony cresot, formaldehyde and thymol; an antiseptic cream. [Ayrton Saunders plc] ‡

Cremostrep. Succinylsulfathiazole, streptomycin sulfate and kaolin; for relief of the mild bacterial diarrheas seen in general practice. [Merck & Co Inc] †

Cremosuxidine. Succinylsulfathiazole; for treatment of mild bacterial diarrhoeas. [Merck & Co Inc] †

crems white. *See* flake white.

Crenette. Nonwoven glass scrim fabric; for reinforcement of filler media, floor coverings, needle-felt, kraft paper and plastic sheeting. [Fothergill Tygaflor Ltd] *

Creolin. A coal tar black disinfectant containing 20% coal tar acids conforming to B.S.2462 BA; a disinfectant for public health, industrial and institutional application. [William Pearson Ltd] *

creosol. CAS 93-51-6; Homocatechol methyl ester.

Creosote. CAS 8001-58-9; A term used in reference to the mixed phenols obtained from wood tar, coal tar, and other sources; used for wood preservation.

Crepetrol® 190. Polymer; creping and adhesion aid for tissue machines. [Hercules]

Cresatin. m-Cresol acetate. External antiseptic and analgesic.

Cresavon. Hospital antiseptic soap. [Laporte Industries Ltd]*

Crescormon. CAS 9002-72-6; Somatropin; hormone. [KabiVitrum AB] *

Cresol. (Tricresol; cresylol)Crude cresol contains approximately 35% o-, 40% m-, and 25% p-cresol, C_6H_4-$(CH_3)OH$; used as an antiseptic.

m-cresol. (3-methylphenol; m-cresylic acid) $CH_3C_6H_4OH$; CAS 108-39-4; EINECS 203-577-9; In disinfectants, fumigants; in photographic developers, explosives; phenolic resins, ore flotation; textile scouring agent; manufacture of coumarin and salicylaldehyde; herbicides, surfactants. [Allchem Industries; Mitsui Petrochem. Ind.;

---

‡ Trade name and manufacturer not verified   § Trade name without identified manufacturer

Penta Mfg.; Schweizerhall; Spectrum Chem. Mfg.]

o-cresol. (2-methylphenol) C_7H_8O; CAS 95-48-7; EINECS 202-423-8; Disinfectant, phenolc resins, tricresyl phosphate, ore flotation, textile scouring, organic intermediate, manufacture of salicyladehyde, coumarin, and herbicides, surfactant. [Allchem Industries; Crowley Tar Prods.; Penta Mfg.; PMC Specialties]

p-cresol. (4-methylphenol; p-cresylic acid) $CH_3C_6H_4OH$; CAS 106-44-5; EINECS 203-398-6; Disinfectant, phenolic resins, ore flotation; textile scouring agent; manufacture of coumarin and salicylaldehyde; herbicides, surfactants, synthetic food flavors. [Allchem Industries; Am. Biorganics; Penta Mfg. PMC Specialties; Spectrum Chem. Mfg.]

Cresolox®. Blend of vegetable soaps, coal tar oils and tar acids; disinfectants. [Coventry Chemicals Ltd]

cresol purple. m-Cresolsulfonphthalein; used as an indicator.

cresol red. o-Cresolsulfonphthalein; used as an indicator.

cresotic acids. Cresol carboxylic acids; $CH_3C_6H_3(OH)$-COOH; ten possible isomers of these acids; dye intermediate; research on plant growth inhibition.

Cressylite. A mixture of picric acid and trinitrocresol; used in explosives.

Crestalan. Lanolin and isopropyl esters; used in cosmetic and household products. [Croda Chem. Ltd]

Crestavin. A proprietary trade name for vinyl acetate resins.§

Crester KZ. Polyglycerol esters; food emulsifier. [Croda Surf. Ltd]

Crestolan NF. Detergent for textile scouring. [Reilly-Whiteman]

Crestomer®. Unsaturated urethane acrylate resins; used for adhesives, binders; tough, impact resistant FRP applications. [Scott Bader]

Crestomer® 1066A. Tough resin (styrene monomer) with high filler tolerance and good adhesion; cured with benzoyl peroxide catalyst; for casting resin, binder for seamless flooring with good resilience and chem. resist. [Scott Bader]

Crestomer® 1080. Tough, flexible base resin in styrene sol'n.; for formulation of high performance, high impact adhesives and filled casting resins; impact modifier for polyester resins. [Scott Bader]

Crestopene 5X. Wetting agent for textile processing of cellulosics and blends. [Reilly-Whiteman]

Crestophen®. Phenolic resins, phenolic resin ancillaries; for highly fire resistant FRP applications. [Scott Bader]

Crestosolve 630. Detergent for textile scouring. [Reilly-Whiteman]

Cresylene. Dyestuffs. [Imperial Chemical Industries plc]

cresylic acid. Commercial mixtures of phenolic materials boiling above the cresol range; EINECS 215-293-2; phosphate esters, phenolic resins, wire enamel solvent, plasticizers, gasoline additives, laminates, coating for magnet wire, disinfectants, metal cleaning, flotation agents, surfactants, intermediates, oil additives, solvent refining, pesticides. [Allchem Industries; Crowley Tar Prods.; PMC Specialties; Spectar Ltd]

Creto. A cationic surface-active agent. [Croda Chem. Ltd]†

Crex. Mild laundry alkali. [ICI Chem. & Polymers Ltd]

Crexathix. Castor oil derivatives; thixotropic agent for paints and printing inks. [Blagden Chemicals Ltd] *

Cri-Line APC-718-75. Fluoroelastomer compound; elas-tomer for extruded profile, o-ring cord. [Cri-Tech]

Cri-Line FDA-612. Fluoroelastomer compound; elastomer for compr., transfer and injection molding, extrusion, calendering; for food contact applications. [Cri-Tech]

Cri-Line FDA-715. Fluoroelastomer compound; elastomer for compr., transfer and injection molding, extrusion, calendering; for food contact applications. [Cri-Tech]

Cri-Line GP-715. Fluoroelastomer copolymer; general-purpose elastomer for compr., transfer, injection molding, extrusion, calendering of valve stem seals, custom molded goods, extruded profiles, roll covers. [Cri-Tech]

Cri-Line HF-618-65. Fluoroelastomer terpolymer; elastomer for compr., transfer and injection molding, extrusion, calendering of fuel line and seals, diesel cylinder linders, diaphragms. [Cri-Tech]

Cri-Line IF-612. Hexafluoropropylene/vinylidene fluoride/tetrafluoroethylene terpolymer; elastomer for compr., transfer and injection molding, extrusion, calendering of fuel line and seals, expansion joints, shaft seals, valve stem seals. [Cri-Tech]

Cri-Line LC-508-55. Reprocessed and virgin fluoroelastomer compound; low-cost elastomer for compr., transfer, injection molding, extrusion and calendering of lathe cuts, custom molded goods. [Cri-Tech]

Cri-Line LC-612-THK. Reprocessed and virgin fluoroelastomer compound; low-cost elastomer for thick cross section curing; for compr., transfer, injection molding, extrusion, calendering of custom molded goods, rubber-to-metal bonded seals. [Cri-Tech]

Cri-Line LC-708. Reprocessed and virgin fluoroelastomer compound; low-cost elastomer for compr. molding and calendering; for calendered sheet packing. [Cri-Tech]

Cri-Line SP-508. Hexafluoroporpylene/vinylidene fluoride copolymer; elastomer for compr., transfer, injection molding, extrusion and calendering of o-rings, seals, rolls, covers, diaphragms. [Cri-Tech]

Cri-Line SP-815. Hexafluoroporpylene/vinylidene fluoride copolymer; elastomer for compr. and transfer molding of o-rings and seals. [Cri-Tech]

Criliprint 788FYN. Self-crosslinking acrylic copolymer; durable pigment and pad binder (textiles). [Sybron]

Crilitex H-50. Self-crosslinking acrylic copolymer; hand builder producing soft full hand, durable to washing and dry cleaning. [Sybron]

Crill. Sorbitan esters; used as nonionic surfactants and food emulsifiers. [Croda Surf. Ltd]

Crill 1. CAS 1338-39-2; EINECS 215-663-3; Sorbitan laurate; emulsifier, pigment dispersant, cosolv., wetting agent, antifoam, visc. reducer, mold release, antiblock agent, corrosion inhibitor, lubricant, antistat; used for cosmetics, food and food packaging, insecticides and herbicides, leather treatment, metalworking fluids, oil slick dispersing, paints and inks, pharmaceuticals, plastics, polishes, textiles. [Croda Inc.; Croda Surf. Ltd.]

Crill 2. CAS 26266-57-9; EINECS 247-568-8; Sorbitan palmitate; see Crill 1. [Croda Inc.; Croda Surf. Ltd.]

Crill 3. CAS 1338-41-6; EINECS 215-664-9; Sorbitan stearate; emulsifier, lubricant, antistat; o/w emulsions; cosmetic and pharmaceutical creams and lotions; polishes; insecticides, herbicides; metal cleaners; buffing compounds; textile lubricants; food applications. [Croda Inc.; Croda Surf. Ltd.]

Crill 4. CAS 1338-43-8; EINECS 215-665-4; Sorbitan oleate; w/o emulsifier, wetting agent, pigment dispersant,

---

* Trade name not verified as available   † Trade name verified as obsolete

coupler, antifoam; cosmetic and pharmaceutical application; aerosol polishes; insecticidal sprays; inks and surf. coatings; metal working lubricants; cutting oils; textile lubricants; dry cleaning operations; food process antifoam; oil slick dispersant. [Croda Inc.; Croda Surf. Ltd.]

Crill 6. CAS 71902-01-7; Sorbitan isostearate; w/o emulsifier, wetting agent, pigment dispersant for creams, lotions, aerosols. [Croda Inc.; Croda Surf. Ltd.]

Crill 35. CAS 26658-19-5; EINECS 247-891-4; Sorbitan tristearate; emulsifier, lubricant, antistat for cosmetic, pharmaceutical, food, and industrial applications. [Croda Inc.; Croda Surf. Ltd.]

Crill 43. CAS 8007-43-0; EINECS 232-360-1; Sorbitan sesquioleate; w/o emulsifier, wetting agent, pigment dispersant; for cosmetic, pharmaceutical, food, and industrial applications. [Croda Inc.; Croda Surf. Ltd.]

Crill 45. CAS 26266-58-0; EINECS 247-569-3; Sorbitan trioleate; w/o emulsifier, wetting agent, pigment dispersant; for cosmetic, pharmaceutical, food, and industrial applications. [Croda Inc.; Croda Surf. Ltd.]

Crillet. Polyethoxylated sorbitan esters; used as nonionic surfactants and food emulsifiers. [Croda Chem. Ltd]

Crillet 1. CAS 9005-64-5; Polysorbate 20; solubilizer, emulsifier, dispersant, wetting agent; often combined with a member of the Crill range in emulsification systems; used in cosmetics, food and food packaging, household products, insecticides, herbicides, metalworking fluids, paints and inks, pharmaceuticals, textiles. [Croda Inc.; Croda Chem. Ltd.]

Crillet 2. CAS 9005-66-7; PEG-20 sorbitan palmitate; emulsifier, solubilizer, wetting agent for cosmetic, pharmaceutical, food, and industrial applications. [Croda Chem. Ltd.]

Crillet 3. CAS 9005-67-8; Polysorbate 60 NF; o/w emulsifier for cosmetics and pharmaceuticals, dispersant for insecticides, herbicides, cattle dyes, penetrant, leveling agent, lubricant, antistat. [Croda Inc.; Croda Chem. Ltd.]

Crillet 4. CAS 9005-65-6; Polysorbate 80 NF; emulsifier, dispersant, solubilizer, lubricant, detergent, antistat, wetting agent for cosmetics, pharmaceuticals, polishes, insecticides, leather degreasing, veterinary products [Croda Inc.; Croda Chem. Ltd.]

Crillet 6. CAS 66794-58-9; PEG-20 sorbitan isostearate; o/w emulsifier, solubilizer for fragrances and perfumes; for creams, lotions, ointments; improved resistance to oxidation. [Croda Inc.; Croda Chem. Ltd.]

Crillet 11. CAS 9005-64-5; PEG-4 sorbitan laurate; emulsifier, solubilizer, wetting agent for cosmetic, pharmaceutical, food, and industrial applications. [Croda Chem. Ltd.]

Crillet 31. CAS 9005-67-8; PEG-4 sorbitan stearate; emulsifier, solubilizer, wetting agent for cosmetic, pharmaceutical, food, and industrial applications. [Croda Chem. Ltd.]

Crillet 35. CAS 9005-71-4; Polysorbate 65; emulsifier, solubilizer, wetting agent for cosmetic, pharmaceutical, food, and industrial applications. [Croda Chem. Ltd.]

Crillet 41. CAS 9005-65-6; PEG-5 sorbitan oleate; emulsifier, solubilizer, wetting agent for cosmetic, pharmaceutical, food, and industrial applications. [Croda Chem. Ltd.]

Crillet 45. CAS 9005-70-3; Polysorbate 85; emulsifier, solubilizer, wetting agent for cosmetic, pharmaceutical,

food, and industrial applications. [Croda Chem. Ltd.]

Crillon. Detergent and foam stabilizing properties; antistatic agents; anticorrosive properties; skin protecting agents; used in shampoos, bubble bath formulations; detergent cleaners; hand cleansers; cutting fluids and soluble cutting oil compositions. [Croda Surf. Ltd]

Crillon CDY. Cocamide DEA; sol. and cutting oils. [Croda Chem. Ltd.] *

Crillon LDE. CAS 120-40-1; EINECS 204-393-1; Lauramide DEA; detergent and foam stabilizer; o/w emulsifiers, antistat; anticorrosive. [Croda Surf. Ltd.]

Crillon LME. CAS 142-78-9; EINECS 205-560-1; Lauramide MEA; lubrication; skin protection agent. [Croda Chem. Ltd.] *

Crillon ODE. CAS 93-83-4; EINECS 202-281-7; Oleamide DEA; emulsifier, stabilizer, skin protectant, lubricant, anti-irritant used in personal care products; additive for cutting fluids and sol. cutting oils. [Croda Surf. Ltd.]

Crimidesa. CAS 7757-82-6; Anhydrous sodium sulfate; detergents, glass, dyestuffs, etc. [Bromhead & Denison Ltd]

Crimson Lake. A cochineal lake containing aluminum salts.§

Cristite. A proprietary trade name for a steel containing 10% chromium, 17% tungsten, 3.5% carbon, and 2.5% molybdenum. §

Criterion. Blend of pyridinium compounds, alcohols, and saccharin; nickel electroplating additive. [Taskem Inc] *

Croak. Fluometuron plus MSMA; herbicide for post-emergence control of broadleaf and grass weeds in cotton. [Draxel Chemical Company] ‡

Crocell. Hot-dip, thermoplastic strippable coating. [Croda Application Chemicals Ltd]

Crocidolite. (Blue asbestos; cape blue) CAS 61105-31-5; A fibrous mineral. It is a member of the group of minerals known as soda-amphiboles. Found in South Africa and Western Australia. It is a hydrated silicate of sodium and iron $(Fe.Na_2)_4Si_4O_{12}$. Pale blue asbestos, found in South Africa and Western Australia. This name is also incorrectly applied to a variety of quartz known as Tiger's eye.

crocus. Saffron, a coloring matter from the dried and powdered flowers of the saffron plant, Crocus sativus; used for coloring confectionery, also refers to ferric oxide having a bluish tint and used for polishing metals, both known as crocus.

Crocus Martius. A hydrated ferric oxide; used as a pigment for pottery and other purposes.

crocus Saturni. See red lead.

Croda Bath Oil Disperant. Synergistic blend of nonionic surfactants. [Croda Chem. Ltd]

Crodacel QL. Laurdimonium hydroxyethyl cellulose; conditioner improving foaming for skin and hair care products [Croda Inc.]

Crodacel QM. Cocodimonium hydroxyethyl cellulose; conditioner improving foaming and imparting body to skin and hair care products [Croda Inc.]

Crodacel QS. Steardimonium hydroxyethyl cellulose; conditioner improving foaming and imparting body to skin and hair care products [Croda Inc.]

Crodacid. Stearic, oleic, behenic, erucic and palmitic fatty acids. [Croda Universal Ltd]

Crodacid B. Behenic acid; gellant for stick formulations when neutralized. [Croda Inc.]

---

‡ Trade name and manufacturer not verified   § Trade name without identified manufacturer

Crodacol. A range of fatty alcohols used as nonionic surfactants; pigment dispersing agents for cosmetics, perfumes and toiletries. [Croda Chem. Ltd]

Crodacol C70. CAS 36653-82-4; EINECS 253-149-0; Cetyl alcohol; sec. emulsifier, thickener, opacifier, and structural agent in anhyd. stick systems. [Croda Inc.]

Crodacol C95NF. CAS 36653-82-4; EINECS 253-149-0; Cetyl alcohol NF; primary structural agent in antiperspirant sticks. [Croda Inc.]

Crodacol CS50. Cetearyl alcohol; emollient; hair and skin lubricant. [Croda Inc.]

Crodacol S70. CAS 112-92-5; EINECS 204-017-6; Stearyl alcohol; sec. emulsifier, thickener, opacifier, and structural agent in anhyd. stick systems. [Croda Inc.]

Crodacol S95NF. CAS 112-92-5; EINECS 204-017-6; Stearyl alcohol NF; primary structural agent in antiperspirant sticks. [Croda Inc.]

Crodacreme. Blend of emulsifiers incl. mono/diglycerides of fatty acids; ice cream emulsifier/stabilizer. [Croda Food Prods. Ltd.]

Croda Fluid. Temporary rust preventative. [Croda Application Chemicals Ltd]

Crodafos. Range of surfactants composed of alkyl ether phosphates in acid form; each is a mixture of mono- and di- esters and may be converted to salts by neutralization with e.g., alkanolamine or metal hydroxide; detergents and coupling agents (for nonionics) with wetting and emulsifying properties; used in heavy-duty clothes washing; hard surface cleaners; floor and industrial cleaners and steam cleaning systems; dry cleaning; dispersible solvent cleaners; cutting and rolling oils, grinding fluids, lubricants; as antistats, lubricants and softeners in synthetic fiber and wool processing; emulsion polymerization; pesticides, fertilizers; waxes, polishes, surface coatings; cosmetics, pharmaceuticals. [Croda Chem. Ltd]

Crodafos 25 D2 Acid. CAS 68071-35-2; PEG-2 C12-15 ether phosphate; hair conditioner for shampoo formulations. [Croda Chem. Ltd.]

Crodafos CAP. CAS 111019-03-5; PPG-10 cetyl ether phosphate; w/o emulsifier; antistat used in personal care products; modifies pH, thickening, emulsifying and suspending props.; enhances hair conditioning; useful for microemulsion systems. [Croda Inc.]

Crodafos CDP. CAS 90388-14-0; Cetyl diethanolamine phosphate; emulsifier and stabilizer for o/w emulsions. [Croda Chem. Ltd.]

Crodafos CS2 Acid. Ceteareth-2 phosphate; emulsifier and stabilizer for o/w emulsions. [Croda Chem. Ltd.]

Crodafos N3 Acid. CAS 39464-69-2; Oleth-3 phosphate; surfactant, conditioner, antistat, o/w emulsifier, gelling agent for surfactants, cosmetics, pharmaceuticals, and toiletries, microemulsion gels; corrosion inhibitor and anti-gelling agent in aerosol antiperspirant systems. [Croda Inc.; Croda Chem. Ltd.]

Crodafos N3 Neutral. CAS 58855-63-3; DEA-oleth-3 phosphate; o/w emulsifier, gelling agent for surfactants and for prep. of clear mineral oil gels and microemulsion gels; used in hair relaxers. [Croda Inc.]

Crodafos N5 Acid. CAS 39464-69-2; Oleth-5 phosphate; emulsifier and stabilizer for o/w emulsions. [Croda Chem. Ltd.]

Crodafos N10 Acid. CAS 39464-69-2; Oleth-10 phosphate; o/w emulsifier, gelling agent for surfactants, prep. of clear mineral oil gels, skin cleansers, clear microemulsion gels. [Croda Inc.]

Crodafos N10 Neutral. CAS 58855-63-3; DEA-oleth-10 phosphate; o/w emulsifier, gelling agent for surfactants, prep. of clear mineral oil gels, skin cleansers, clear microemulsion gels. [Croda Inc.]

Crodafos O2 Acid. CAS 39464-69-2; Oleth-2 phosphate; hair conditioners for shampoo formulations. [Croda Chem. Ltd.]

Crodafos O2 TEA. TEA-oleth-2 phosphate. [Croda Chem. Ltd.] *

Crodafos SG. CAS 50643-20-4; PPG-5 ceteth-10 phosphate; o/w emulsifier, conditioner, wet comb enhancer for shampoos and cream rinses; thickener, gellant, and pH adjuster for shampoos. [Croda Inc.; Croda Chem. Ltd.]

Crodafos T2 Acid. CAS 9046-01-9; Trideceth-2 phosphate; hair conditioners for shampoo formulations. [Croda Chem. Ltd.]

Crodakyd. Water-thinnable resins. [Croda Resins Ltd]

Crodalan. Surface active emollient; emulsifier and solubilizer; super-fatting, wetting and spreading agent; used in o/w cosmetics; perfume and essential oils for skin perfumes bath essences; all skin and hair care preparations; baby products; germicidal skin cleaners. [Croda Chem. Ltd]

Crodalan AWS. Polysorbate 80, cetyl acetate, acetylated lanolin alcohol; emollient, superfatting agent, conditioner, o/w emulsifier, dispersant, wetting agent, plasticizer, solubilizer used in cosmetics, pharmaceuticals, detergent systems. [Croda Inc.; Croda Chem. Ltd.]

Crodalan C24. Polyethoxylated cholesterol derivatives; used in the cosmetic industry. [Croda Chem. Ltd]

Crodalan IPL. Isopropyl lanolate; general purpose raw material for cosmetics. [Croda Chem. Ltd]

Crodalan LA. Cetyl acetate and cetylated lanolin alcohol; emollient, penetrant, wetting agent, conditioner, plasticizer used in cosmetics, pharmaceuticals. [Croda Inc; Croda Chemicals Ltd]

Crodamer. A proprietary trade name for a range of uv radiation curable resins for lacquers and inks. [Croda Resins Ltd]

Crodamet Series. Ethoxylated primary amines; emulsifier, antistat for plastics and fibers. [Croda Chem. Ltd.]

Crodamide. Fatty acid amides; primary, secondary, and bisamides based on stearic, oleic, erucic and behenic fatty acids; applications mainly in plastics industry as slip and antiblock agents. [Croda Universal Ltd]

Crodamide E, ER. CAS 112-84-5; EINECS 204-009-2; Erucamide; international slip and antiblock agent for polyolefins; internal release agent for moulded thermoplastic polymers. [Croda Chem. Ltd.] *

Crodamide O, OR. CAS 301-02-0; EINECS 206-103-9; Oleamide; international slip and antiblock agent for polyolefins; internal release agent for moulded thermoplastic polymers. [Croda Chem. Ltd.] *

Crodamide S, SR. CAS 124-26-5; EINECS 204-693-2; Stearamide; international slip and antiblock agent for polyolefins; internal release agent for moulded thermoplastic polymers. [Croda Chem. Ltd.] *

Crodamine. Tertiary amines. [Croda Universal Ltd] *

Crodamine 1. Series of cationic surfactants composed of primary amines RNH$_2$, where the fatty alkyl group, R, may be lauric, palmitic, stearic, coconut, soya, tallow, or

---

* Trade name not verified as available   † Trade name verified as obsolete

oleic in origin; solid, liquid or paste forms; D in the code number denotes distilled grade for superior color and heat stability; used for corrosion inhibition; emulsifiers for herbicides; mineral flotation; pigment dispersion; anticaking agents; auxiliaries for leather, textiles, rubber, plastics and metal industries. [Croda Universal Ltd]

Crodamine 1.16D. CAS 143-27-1; EINECS 205-596-8; Primary cetyl amine; emulsifier for herbicides, ore flotation, pigment dispersion; auxiliary for textiles, leather, rubber, plastics, and metal industries. [Croda Universal Ltd.]

Crodamine 1.18D. CAS 124-30-1; EINECS 204-695-3; Stearylamine; emulsifier for herbicides, ore flotation, pigment dispersion; auxiliary for textiles, leather, rubber, plastics, and metal industries. [Croda Universal Ltd.]

Crodamine 1.HT. CAS 61788-45-2; EINECS 262-976-6; Hydrogenated tallow amine; anticaking agent for fertilizers; emulsifier for herbicides, ore flotation, pigment dispersion; auxiliary for textiles, leather, rubber, plastics, and metal industries. [Croda Universal Ltd.] *

Crodamine 1.O, 1.OD. CAS 112-90-3; EINECS 204-015-5; Oleyl amine; emulsifier for herbicides, ore flotation, pigment dispersion; auxiliary for textiles, leather, rubber, plastics, and metal industries. [Croda Universal Ltd.]

Crodamine 1.T. CAS 61790-33-8; EINECS 263-125-1; Tallow amine; emulsifier for herbicides, ore flotation, pigment dispersion; auxiliary for textiles, leather, rubber, plastics, and metal industries. [Croda Universal Ltd.]

Crodamine 2.C, 2.S and 2.HT. Cationic surfactant consisting of secondary amines R_2NH, in solid form, where the fatty alkyl group, R, may be coconut, soya or hydrogenated tallow in origin respectively; used for corrosion inhibition; emulsifiers for herbicides; mineral flotation; pigment dispersion; anticaking agents; auxiliaries for leather, textiles, rubber, plastics and metal industries. [Croda Universal Ltd]

Crodamine 3.A16D. CAS 112-69-6; Palmityl dimethylamine; emulsifier for herbicides, ore flotation, pigment dispersion; auxiliary for textiles, leather, rubber, plastics, and metal industries. [Croda Universal Ltd]

Crodamine 3.A18D. CAS 124-28-7; EINECS 204-694-8; Stearyl dimethylamine, dist.; emulsifier for herbicides, ore flotation, pigment dispersion; auxiliary for textiles, leather, rubber, plastics, and metal industries. [Croda Universal Ltd.]

Crodamine 3.ABD. CAS 21542-96-1; Dimethyl behenylamine; emulsifier for herbicides, ore flotation, pigment dispersion; auxiliary for textiles, leather, rubber, plastics, and metal industries. [Croda Universal Ltd.]

Crodamine 3.AED. Dimethyl erucylamine; emulsifier for herbicides, ore flotation, pigment dispersion; auxiliary for textiles, leather, rubber, plastics, and metal industries; intermediate. [Croda Universal Ltd.]

Crodamine 3.AOD. CAS 28061-69-0; Dimethyl oleylamine; emulsifier for herbicides, ore flotation, pigment dispersion; auxiliary for textiles, leather, rubber, plastics, and metal industries. [Croda Universal Ltd.]

Crodamine 3A. Series of cationic surfactants of the tertiary amine type, where the fatty alkyl groups are lauric, palmitic, stearic and coconut in origin; supplied in liquid or paste form; used for corrosion inhibition; emulsifiers for herbicides and synthetic resins; mineral flotation; pigment dispersion; anticaking agents, e.g., for high analyzis NPK fertilizers; auxiliaries for leather, textiles,

rubber, plastics and metal industries; catalysts in the production of polyurethane foam. [Croda Universal Ltd]

Crodamine 3ABD. N,N-Dimethylbehenylamine; an intermediate for the preparation of quaternary ammonium compounds, betaines and amine oxides. Other uses include pharmaceuticals, solvents and anticorrosives. [Croda Universal Ltd]

Crodamine 3AED. N,N-Dimethylerucylamine; used as an intermediate, other uses include pharmaceuticals, solvents and anticorrosives. [Croda Universal Ltd]

Crodamine 3AHRD, 3ARD. N,N-Dimethyl C18-22 amines; used as intermediates, in pharmaceuticals, solvents and anticorrosives. [Croda Universal Ltd]

Crodamine 3AOD. N,N-Dimethyloleylamine; an intermediate for the preparation of quaternary ammonium compounds, betaines and amine oxides; also in pharmaceuticals, solvents and anticorrosives. [Croda Universal Ltd]

Crodamol. A range of straight and branched chain mono and dibasic esters; used as cosmetic emollients. [Croda Surf. Ltd]

Crodamol BE. CAS 18312-32-8; EINECS 242-201-8; Behenyl erucate; thickening and opacifying agent, emollient, stabilizer, modifier for lipsticks, powd. suspensions. [Croda Inc.]

Crodamol CAP. EINECS 261-619-1; Cetearyl octanoate; emollient simulating properties of preen gland oil; provides water repellency in skin care preps. [Croda Inc.]

Crodamol CP. CAS 540-10-3; EINECS 208-736-6; Cetyl palmitate; emollient for replacing spermaceti wax. [Croda Inc.]

Crodamol CSP. Cetearyl palmitate; improves feel and body of emulsions; replaces spermaceti wax and beeswax in personal care products [Croda Inc.]

Crodamol GTC/C. Glyceryl tricapryl caprylate; bakery lubricant, release agent, glazing agent, hypoallergenic baby food formulations, flavor carrier. [Croda Food Prods. Ltd.]

Crodamol IPM. CAS 110-27-0; EINECS 203-751-4; IPM, perfumery grade; spreading agent, emollient, cosolv. for cosmetic raw materials. [Croda Inc.]

Crodamol MM. CAS 3234-85-3; EINECS 221-787-9; Myristyl myristate; emollient, superfatting agent, visc. builder in emulsions; substitute for spermaceti wax and/or beeswax in cosmetic and pharmaceutical formulations. [Croda Inc.]

Crodamol PMP. PPG-2 myristyl ether propionate; emollient, lubricant with dry, lt. greaseless feel; coupling agent; solv. for sunscreen actives; for bath oils, creams, moisturizers, emulsions. [Croda Inc.]

Crodamol PTC. Pentaerythritol tetracaprylate/caprate; lubricant for creams and lotions; reduces tack in clear gel microemulsions. [Croda Inc.]

Crodamol PTIS. Pentaerythritol tetraisostearate; lubricant for creams and lotions; reduces tack in clear gel microemulsions. [Croda Inc.]

Crodamol SS. CAS 8002-23-1; Cetyl esters; synthetic spermaceti NF; emollient and visc. builder for cosmetic and pharmaceutical preparations. [Croda Inc.]

Crodamol W. CAS 66009-41-4; Stearyl heptanoate; nongreasy emollient and water repellent for cosmetics and toiletries, esp. stick formulations; melts rapidly on application to skin; synthetic preen gland wax. [Croda Inc.]

Crodapearl. Modified glycol ester used as a nonionic

---

‡ Trade name and manufacturer not verified

§ Trade name without identified manufacturer

surfactant; pearling agent for cosmetics, perfumes and toiletries. [Croda Surf. Ltd]

Crodapearl Liq. Sodium laureth sulfate, glycol MIPA stearate; pearling agent for shampoos, bubble baths, dishwashing liqs. [Croda Inc.]

Crodapearl NI Liquid. Hydroxyethyl stearamide-MIPA, PPG-5 ceteth-20; pearlescent for detergent systems, lotions, gels, clear rinses, bath products [Croda Inc.]

Crodaplast. A proprietary trade name for a range of hydroxy functional acrylic resins for high-quality automotive and kitchen appliance finishes. [Croda Resins Ltd]

Crodapol. A proprietary trade name for a range of saturated polyester resins for two pack and melamine cure. [Croda Resins Ltd]

Crodapur. Technical lanolin; plasticizer for tape adhesives, mastics, and putty. [Croda Chem. Ltd.] *

Crodarom Avocadin. Avocado oil unsaponifiables; botanical extract for personal care products [Croda Inc.]

Crodarom Calendula O. Soybean oil, calendula extract, tocopherol; botanical extract for personal care products [Croda Inc.]

Crodarom Carrot O. Peanut oil, carrot extract, isopropyl myristate, tocopherol; botanical extract for personal care products [Croda Inc.]

Crodarom Chamomile A. Propylene glycol and matricaria extract; botanical extract for personal care products [Croda Inc.]

Crodarom Chamomile EO. Caprylic/capric triglycerides and matricaria extract; botanical extract for personal care products [Croda Inc.]

Crodarom Chamomile O. Caprylic/capric triglycerides and matricaria extract; botanical extract for personal care products [Croda Inc.]

Crodarom Nut O. Peanut oil, mineral oil, walnut extract; botanical extract for personal care products [Croda Inc.]

Crodarom St. Johns Wort O. Olive oil, hypericum extract; botanical extract for personal care products [Croda Inc.]

Crodascoop. Mono/diglycerides of fatty acids; emulsifier, stabilizer for soft ice cream. [Croda Food Prods. Ltd.]

Crodasinic. N-Acyl sarcosines and their sodium salts; used as anionic surfactants; foaming agents for cosmetics, perfumes and toiletries. [Croda Chem. Ltd]

Crodasinic L. CAS 97-78-9; EINECS 202-608-3; Lauroyl sarcosine in acid form, as a white solid; properties include mild detergency, high foaming, bacteriostatic activity, enzyme inhibition, corrosion inhibition, hard water tolerance and stability in mildly acid formulations; anionic surfactant used in cosmetics and toiletries, dentifrices and shampoos, carpet shampoos, emulsion polymerization, metal treatment, food and food packaging, textiles, fine fabric detergents. [Croda Chem. Ltd]

Crodasinic LS30. CAS 137-16-6; EINECS 205-281-5; Sodium N-lauroyl sarcosinate; foaming, wetting agent and detergent for acidic conditions; corrosion inhibitor; bacteriastat and inhibitor; used in dental care preps., pharmaceuticals, personal care products, household and industrial applications. [Croda Chem. Ltd.]

Crodasinic LS35. CAS 137-16-6; EINECS 205-281-5; Sodium lauroyl sarcosinate in the form of a clear liquid; properties include mild detergency, high foaming, bacteriostatic activity, enzyme inhibition, corrosion inhibition, hard water tolerance and stability in mildly acid formulations; anionic surfactant used in cosmetics and toiletries, dentifrices and shampoos, carpet shampoos, emulsion

polymerization, metal treatment, food and food packaging, textiles, fine fabric detergents. [Croda Universal Ltd]

Crodasinic LT40. CAS 16693-53-1; TEA lauroyl sarcosinate; detergent, foaming agent, wetting agent, dispersant, emulsifier, anticorrosive, foam stabilizing synergist for carpet shampoos, textile and cosmetic detergent systems. [Croda Chem. Ltd.]

Crodasinic O. CAS 110-25-8; EINECS 203-749-3; N-Oleoyl sarcosine; corrosion inhibitor in oils, fuels, lubricants, greases, surface coatings; as antifog agent for food packaging polyolefin films. [Croda Chem. Ltd.]

Crodasinic OS35. Sodium N-oleoyl sarcosinate; foaming agent, wetting agent, detergent, lubricant, antistat, corrosion inhibitor, bacteriostat, penetrant for dental, pharmaceutical, shampoos, depilatories, shaving preps., household and industrial uses. [Croda Chem. Ltd.] *

Crodasone W. Hydrolyzed wheat protein polysiloxane copolymer; substantive film-forming copolymer with lubricity, gloss, conditioning props. for hair and skin care products [Croda Inc.]

Crodasub. Factice; a processing aid for rubber. [Croda Universal Ltd]

Crodatem. Diacetyl tartaric acid ester; antistaling agent, starch complexing agent in biscuits, cake mixes, coffee whitener, gravy mix, toffee emulsifier. [Croda Food Prods. Ltd.]

Crodateric. A range of imidazoline based amphoteric surfactants; properties include dense foaming and high foam stability; stability in acids, alkalies and strong electrolyte solutions; compatability with soaps, quaternary germicides; good detergency; wetting, emulsifying properties;sequestering properties; hard water solubility; germicidal and fungicidal properties; used for acid and alkali detergents; agricultural sprays; antistats; corrosion inhibitor; toiletries; dry cleaning; emulsions; polishes; polymers; textiles; metal treatment; paints and inks; petroleum; paper. [Croda Surf. Ltd]

Crodateric C. Derived from coconut fatty acid; surfactant. [Croda Universal Ltd.]

Crodateric Cy. CAS 63451-23-0; Derived from caprylic acid; surfactant. [Croda Universal Ltd.] *

Crodateric O, O.100. CAS 32456-28-6; Derived from oleic acid; surfactant. [Croda Universal Ltd.] *

Crodateric S. CAS 30342-62-2; Derived from stearic acid; surfactant. [Croda Universal Ltd.] *

Crodax. Rust preventatives. [Croda Chem. Ltd]

Crodax DP 50. Complex mixt. of org. acids, esters, and lactones; converted to metallic soaps and esters used in solv. or oil-based rust preventives and lubricating products [Croda Chem. Ltd.] *

Crodax DP 100. Calcium soap and oxidate; used in thick film rust preventives providing protection for steel in salt environments; used for underseal, wire rope dressings, waterproofing agents. [Croda Chem. Ltd.] *

Crodazoline Cy. CAS 37478-68-5; EINECS 253-521-2; Caprylic imidazoline. [Croda Universal Ltd.] *

Crodazoline O. CAS 95-38-5; Oleic imidazoline; surfactant, wetting agent, emulsifier, softener, for textiles, asphalt, tar emulsion breaker, paint additive, corrosion inhibitor, lubricating metal processing aids. [Croda Universal Ltd.]

Crodazoline S. CAS 95-19-2; EINECS 202-397-8; Stearic imidazoline; surfactant, wetting agent, emulsifier, softener, for textiles, asphalt, tar emulsion breaker, paint additive, corrosion inhibitor. [Croda Universal Ltd.]

---

* Trade name not verified as available

† Trade name verified as obsolete

Croderol G7000. CAS 56-81-5; Glycerin; conditioner, humectant, moisturizing agent in cosmetics and pharmaceuticals, diluent and plasticizer for many polar materials. [Croda Chem. Ltd.] *

Crodesta. A range of sucrose esters of fatty acids; used as nonionic surfactants; pigment dispersing agents for paint. [Croda Chem. Ltd] *

Crodesta DKS F10. CAS 27195-16-0; EINECS 248-317-5; Sucrose mono/di/tri palmitic/stearic acid; emulsifier, wetting agent, dispersant for use in cosmetics, pharmaceuticals. [Croda Surf. Ltd.]

Crodesta DKS F110. CAS 25168-73-4; EINECS 246-705-9; Sucrose mono/di/tri palmitic/stearic acid; emulsifier, wetting agent, dispersant for use in cosmetics, pharmaceuticals. [Croda Surf. Ltd.]

Crodesta F-10. CAS 27195-16-0; EINECS 248-317-5; Sucrose distearate; dispersant, emulsifier, wetting agent, solubilizer, emollient, detergent in cosmetics, toiletries, pharmaceuticals (suntan and baby lotions). [Croda Inc.]

Crodesta F-110. Sucrose distearate and sucrose stearate; dispersant, emulsifier, wetting agent, solubilizer, detergent in cosmetics, toiletries, pharmaceuticals; thickener and suspending agent. [Croda Inc.]

Crodesta F-160. CAS 25168-73-4; EINECS 246-705-9; Sucrose stearate; dispersant, emulsifier, wetting agent, solubilizer, detergent, foaming agent in cosmetics, toiletries, ingestible pharmaceuticals; thickener and suspending agent. [Croda Inc.]

Crodesta SL-40. Sucrose cocoate; dispersant, emulsifier, wetting agent, solubilizer, mild detergent, high foaming emollient in cosmetics, toiletries, pharmaceuticals. [Croda Inc.; Croda Surf. Ltd.]

Crodet. A range of polyethoxylated fatty acids; used as nonionic surfactants; used as fiberglass size and as lubricants in the textile industry. [Croda Chem. Ltd]

Crodet C10. CAS 61791-29-5; PEG-10 coconut fatty acid; surfactant for cosmetic and industrial applications. [Croda Chem. Ltd.]

Crodet L4. CAS 9004-81-3; PEG-4 laurate; o/w emulsifier for cosmetics and pharmaceutical creams, lotions and ointments, industrial applications, wetting agent, solubilizer for perfumes or aq. alcoholic preparations; dispersant; plasticizer for hair setting sprays. [Croda Chem. Ltd.]

Crodet O4. CAS 9004-96-0; PEG-4 oleate; surfactant for cosmetic and industrial applications. [Croda Chem. Ltd.]

Crodet S4. CAS 9004-99-3; PEG-4 stearate; o/w emulsifier for cosmetics and pharmaceutical creams, lotions and ointments, industrial applications; wetting agent, solubilizer for perfumes or aq. alcoholic preparations, dispersant. [Croda Chem. Ltd.]

Crodex. A range of emulsifying waxes; used as nonionic surfactants, emulsifiers, emollients, dispersing, solubilizing and wetting agents by the pharmaceutical and fine chemicals industries. [Croda Chem. Ltd]

Crodex A. Cetostearyl alcohol and sodium lauryl sulfate; emulsifying wax BP for pharmaceuticals and cosmetic uses. [Croda Chem. Ltd.]

Crodex C. Cetostearyl alcohol and Cetrimide BP; emulsifying wax BPC, bactericides, for pharmaceuticals and cosmetics, hair conditioning rinses. [Croda Chem. Ltd.]

Crodex N. Cetostearyl alcohol and ceteth-20; emulsifying wax BP, wetting agent, penetrant, emulsifier for most emollient materials in cosmetics and pharmaceuticals.

[Croda Chem. Ltd.]

Crodinhib. Amine borates; used as lubricants and plasticizers; corrosion inhibitors for paints and oils. [Croda Chem. Ltd]

Crodinhib RT70, RT70S. Borate derivs.; corrosion inhibitor used in metal working fluids and oils. [Croda Chem. Ltd.]*

Crodol. A barrier cream. [Croda Chem. Ltd] *

Crodolene. Stearic, oleic, behenic, erucic and palmitic fatty acids. [Croda Universal Ltd]

Crodon. The trade name for a type of chromium plate. §

Croduret 10. CAS 61788-85-0; PEG-10 hydrogenated castor oil; emulsifier, solubilizer, emollient, superfatting agent, detergent used for cosmetics, textiles, metalworking fluids, emulsion polymerization, insecticides, herbicides, household detergents. [Croda Chem. Ltd.]

Crodyne BY19. CAS 9000-70-8; Pharmaceutical gelatin NF; protective colloid, moisturizer, conditioner for skin and hair care products; humectant, thickener for pharmaceutical and food applications. [Croda Inc; Croda Chemicals Ltd]

Crolactil. Acyl lactylates; used as anionic surfactants, wetting agent for paint. [Croda Universal Ltd]

Crolactil CSL. CAS 5793-94-2; Calcium 2-stearoyl lactylate; w/o emulsifier, emollient for skin care and treatment products; ingredients in food products [Croda Surf. Ltd.]

Crolactil SISL. CAS 66988-04-3; EINECS 266-533-8; Sodium isostearoyl lactylate; o/w emulsifier, emollient for skin care and treatment products; ingredients in food products [Croda Surf. Ltd.]

Crolactil SSL. CAS 25383-99-7; EINECS 246-929-7; Sodium stearoyl lactylates; o/w emulsifier, emollient for skin care and treatment products; ingredients in food products [Croda Chem. Ltd.] *

Crolan. Lanolin fatty acid esters; used in the cosmetic industry. [Croda Chem. Ltd] *

Crolastin. Hydrolyzed animal elastin; moisturizer and conditioner for skin care products and cleansers. [Croda Inc; Croda Chemicals Ltd]

Crolastin 10 Powder. Partially hydrolyzed elastin; conditioner for skin care cosmetics. [Croda Inc.]

Crolax. A proprietary preparation containing dioctyl sodium sulfosuccinate and dihydroxyanthaquinine; a laxative. [Crookes Laboratories] *

Crolec 4135. CAS 8002-43-5; EINECS 232-307-2; Modified lecithin; conditioner, superfatting, emulsifier for cosmetics and pharmaceuticals. [Croda Chem. Ltd.]

Croloy. A proprietary trade name for high chromium steel containing molybdenum and vanadium. §

Crolyn®. For the electrical industry. [DuPont UK]

Cromal. An alloy of aluminum with 2-4% chromium and smaller amounts of nickel and manganese. It melts at 700 C., and is specially suitable for castings.

Cromalit. Concentrated chrome alums; tanning and drilling mud additive. [Lancashire Chemical Works Ltd]

Cromalit 150. Concentrated basic potash chrome alum powder. [Lancashire Chemical Works Ltd]

Cromaloy II. An alloy of 80% nickel, 15% chromium, and 5% iron.

Cromaloy III. An alloy of 85% nickel and 15% chromium.

Cromaloy IV. An alloy of 80% nickel and 20% chromium.

Cromax®. For the automotive industry. [DuPont UK]

Cromeen. Substituted alkylamine deriv. lanolin acids; mild

---

‡ Trade name and manufacturer not verified   § Trade name without identified manufacturer

multifunctional surfactant with high foaming, detergency, and emulsifying props.; for shampoos, detergents, and hand cleansing preps., aerosol skin and shaving foams. [Croda Chem. Ltd.]

Cromoist CS. Chondroitin sulfate, hydrolyzed animal protein; moisturizer, conditioner for face and body creams and lotions. [Croda Inc.]

Cromoist HYA. Hydrolyzed animal protein, hyaluronic acid; moisturizer, conditioner for skin care products, facial creams. [Croda Inc.]

Cromoist O25. Hydrolyzed whole oats; skin care ingred. improving feel props. of creams and lotions, imparting soft, cushiony feel on skin; moisturizer. [Croda Inc.]

Cromophtal. High performance organic pigments for plastics; a registered trade name. [Ciba plc]

Cromophtal C-20. Pigment preparations. [Ciba plc] *

Cromo Steel. A proprietary trade name for a chrome-molybdenum steel. §

Cromphytal M-20. Pigment preparations. [Ciba plc] *

Cromul 0685. Complex ethoxy (6) fatty alcohol ether; emulsifier and opacifier for cosmetic and pharmaceutical creams and lotions. [Croda Chem. Ltd.]

C-Ron. CAS 141-01-5; Ferrous fumarate; .hematinic. [Rowell Laboratories Inc] *

Cronectin. Hydrolyzed fibronectin; conditioner for skin and hair care products [Croda Chem. Ltd.] *

Croneton®. CAS 29973-13-5; Ethiofencarb; systemic insecticide used for control of aphids. [Bayer AG]

Cronex®. For the medical industry. [DuPont UK]

C-Ron Forte. CAS 141-01-5; Ferrous fumarate; hematinic. [Rowell Laboratories Inc] *

Cronite. An alloy of nickel and chromium. No. 1 contains 85% nickel, and 15% chromium.

Crop-Check®. For the agriculture industry. [DuPont UK]

Cropepsol. CAS 9015-54-7; Hydrolyzed collagen; conditioner for skin and hair conditioner. [Croda Chem. Ltd.]*

Cropeptide W. Hydrolyzed wheat protein, wheat oligosaccharides; film-forming conditioning protein for controlling moisture and strengthening hair. [Croda Inc.]

Cropeptone. CAS 9015-54-7; Hydrolyzed animal protein in aqueous solution; used in the cosmetic industry. [Croda Chem. Ltd]

Cropol 60. CAS 577-11-7; EINECS 209-406-4; Sodium dioctyl sulfosuccinate as an ethanol solution; emulsifier and powerful wetting agent used in a wide range of manufacturing industries, especially textiles. [Croda Chem. Ltd] *

Cropotex®. Contact acaricide for control of spider mites on pome and stone fruit, plums, damsons, citrus fruit and citrus rust mites. [Bayer AG]

Croptex Amber. CAS 1918-16-7; Propachlor; a pre-emergence herbicide for various horticultural crops. [Hortichem Ltd]

Croptex Bronze. CAS 2307-68-8; Emulsifiable concentrate containing 400 g/l pentanochlor; used to control weeds in horticultural crops. [Hortichem Ltd]

Croptex Chrome. Suspension concentrate containing 80 g chlorpropham and 15 g fenuron per liter; an herbicide for use on ornamentals and vegetables. [Hortichem Ltd]

Croptex Fungex. Cupric ammonium carbonate; a protectant fungicide. [Hortichem Ltd]

Croptex Pewter. A suspension concentrate containing 80 g cetrimide and 80 g chlorpropham per liter; soil acting herbicide for lettuce. [Hortichem Ltd]

Croptex Steel. Powder containing 95% w/w sodium monochloroacetate; for annual dicotyledons control in various horticultural crops. [Hortichem Ltd]

Croquat. Quaternized polypeptides. [Croda Chem. Ltd]

Croquat HH. CAS 68915-25-3; Cocodimonium hydroxypropyl hydrolyzed hair keratin; substantive conditioning protein for hair shampoos and conditioners. [Croda Inc.]

Croquat L. Laurdimonium hydrolyzed animal protein; substantive protein, conditioner for clear rinses, shampoos, conditioners. [Croda Inc.]

Croquat M. Cocodimonium hydrolyzed animal protein; substantive protein, conditioner for shampoos, perms, hair relaxers. [Croda Inc.]

Croquat S. Steardimonium hydrolyzed animal protein; substantive protein, conditioner for cream rinses. [Croda Inc.]

Croquat WKP. CAS 68915-25-3; Cocodimonium hydrolyzed animal keratin; permanent conditioning protein for cream rinses, shampoos, conditioners, perms, nail care products [Croda Inc.]

Crosanol. Fiber lubricants. [Croda Universal Ltd]

Croscolor. Dyeing auxiliaries. [Croda Universal Ltd]

Croscour. Scouring agents for textiles. [Croda Universal Ltd]

Crosdurn. Durable finishes. [Crosfield Chemicals Ltd] †

Crosfield. Chemicals. [Crosfield Chemicals Ltd]

Crosfield EP. Catalysts. [Crosfield Chemicals Ltd]

Crosfield HP. Matting agents. [Crosfield Chemicals Ltd]

Crosfield SP. Silicas. [Crosfield Chemicals Ltd]

Crosil. Textile processing system; for wool shrink proofing. [Croda Universal Ltd]

Crosil. Textile processing system. [Crosfield Chemicals Ltd]

Crosilk. CAS 977077-71-6; Silk amino acids. [Croda Chem. Ltd]

Crosilk 10,000. CAS 96690-41-4; Hydrolyzed silk; protein conditioner providing manageability, gloss, texture in hair care products, moisturizing and protection in skin care products [Croda Inc.]

Crosilk Liq. CAS 977077-71-6; Silk amino acids; conditioner, humectant for skin and hair care products [Croda Inc.]

Crosilk Powder. CAS 9009-99-8; Silk powder; protein for solid make-up, hair products where it absorbs oil, improves leveling, enhances spreading, gives elasticity and lubricity, modifies application properties and provides a silky, lustrous appearance; improves pigment binding and stability. [Croda Inc.]

Crosilkquat. Cocodimonium silk amino acids; substantive conditioner and moisturizer with excellent foaming props. for skin and hair conditioners, shampoos, styling mousses, perms and relaxers, night creams and lotions; effective over wide pH range. [Croda Inc.]

Croslube. Yarn lubricants. [Croda Universal Ltd, Crosfield Chemicals Ltd]

Crosoft. Softening agents for fiber softeners and finishes. [Croda Universal Ltd, Crosfield Chemicals Ltd]

Crossential EPO, Super Refined. Evening primrose oil; contains essential fatty acids for skin care. [Croda Inc.]

Crossfire. CAS 2921-88-2; Emulsifiable concentrate containing 228g chlorpyrifos per liter; an organophosphorus insecticide. [Rhone-Poulenc Crop Protection Ltd]

Crossil. Silicate chemicals. [Crosfield Chemicals Ltd]

Crostat. Antistatic agents. [Croda Universal Ltd]

Crostat. Antistatic agents. [Crosfield Chemicals Ltd]

---

* Trade name not verified as available   † Trade name verified as obsolete

Crosterene. Stearic, oleic, behenic, erucic and palmitic fatty acids. [Croda Universal Ltd; Croda Chem. Ltd]

Crosultaine C-50. CAS 68139-30-0; Cocamidopropyl hydroxysultaine; foam booster/stabilizer effective over wide range of pH and water hardness; for shampoos, baby shampoos; lime soap dispersant. [Croda Inc.]

Crosultaine E-30. Erucamidopropyl hydroxysultaine; visc. booster and conditioner giving silky feel, comb and static control at levels less than 2% active; improves creaminess and lubricity of lather; lime soap dispersant. [Croda Inc.]

Crosultaine T-30. Tallowamidopropyl hydroxysultaine; foam and visc. booster; improves wet combing and hair condition; lime soap dispersant. [Croda Inc.]

Crotein A,C, O. CAS 9015-54-7; Hydrolysed collagen. [Croda Chem. Ltd]

Crotein AD Anhyd. AMP isostearoyl hydrolyzed animal protein; conditioners for hair preparations, alcohol products, aerosol hair sprays, skin tonics, setting lotions, aftershaves. [Croda Inc.]

Crotein ADW. AMP isostearoyl hydrolyzed wheat protein; conditioning protein, film modifier for alcoholic and hydroalcoholic lotions, hair care products, skin tonics, aftershave, nail preps., quick-breaking foam aerosols; plasticizer for resins in hair products [Croda Inc.]

Crotein ASC. CAS 68951-89-3; Ethyl ester of hydrolyzed animal protein; protein conditioner and film modifier for alcoholic and hydroalcoholic lotions and hair care products [Croda Inc.]

Crotein ASK. CAS 69430-36-0; Hydrolyzed animal keratin; conditioner, film modifier for hair setting/conditioning systems. [Croda Inc.]

Crotein CAA. CAS 9015-54-7; Collagen amino acids. [Croda Chem. Ltd]

Crotein CAA/SF. CAS 9015-54-7; Animal collagen amino acids; moisturizer for skin creams and lotions. [Croda Inc.]

Crotein HKP. CAS 68238-35-7; Keratin amino acids. [Croda Chem. Ltd]

Crotein HKP Powd. Hair keratin amino acids, sodium chloride; substantive conditioner and moisturizer for shampoos, cream rinses. [Croda Inc.]

Crotein HWE. Hydrolysed whole eggs. [Croda Chem. Ltd]

Crotein IP. Isostearoyl hydrolyzed animal protein; conditioner for use in solvent-based nail polish and removers. [Croda Inc.]

Crotein K. CAS 69430-36-0; Hydrolyzed animal keratin; proteinic conditioner and moisturizer for hair and nail care products [Croda Inc.]

Crotein O. CAS 9015-54-7; Refined collagen hydrolysates; conditioning agent, foam booster and stabilizer, dye leveling agent in hair products, substantive to hair and skin. [Croda Inc.]

Crotein Q. CAS 11174-62-0; Steartrimonium hydrolyzed animal protein; substantive conditioner, body and gloss agent for hair care preparations. [Croda Inc; Croda Chemicals Ltd]

Crotein SPA. CAS 9015-54-7; Hydrolyzed animal protein; conditioner for hair care preparations; foam stabilizer/booster in shampoo; peptizing aid in shampoos; dye leveling aid in hair dyes and bleaches; also for skin treatment, in depilatories. [Croda Inc.]

Crotein WKP. CAS 69430-36-0; Hydrolyzed animal keratin; conditioner for hair products [Croda Inc.]

Crothix. Polyol alkoxy ester; mild thickener for aq. systems, shampoos, auxiliary emulsifier and bodying agent for creams and lotions. [Croda Inc.]

crotonic acid. (β-methacrylic acid; 2-butenoic acid) $CH_3CH:CHCOOH$; CAS 3724-65-0; Synthesis of resins, polymers, plasticizers, drugs. [Aldrich; Allchem Industries; Atomergic Chemetals; Chisso Am.; Eastman]

Crotorite. A cupro-manganese alloy. It contains 68% copper, 30% manganese, and 2% iron.

Cro-tung. A proprietary trade name for a chromium-tungsten steel. §

Crovol A40. PEG-20 almond glycerides; coemulsifier, superfatting agent, emollient, counter-irritant, wetting aid, solubilizer, dispersant for skin and hair products, soaps, bath oils, astringents, antiperspirants; plasticizer for styling mousses and aq. aerosols; fragrance solubilizer. [Croda Inc.; Croda Chem. Ltd.]

Crovol M40. PEG-20 corn glycerides; coemulsifier, superfatting agent, emollient, counter-irritant, wetting aid, solubilizer, dispersant for skin and hair care products, soaps, bath oils, astringents, antiperspirants; plasticizer for styling mousses and aq. aerosols; fragrance solubilizer. [Croda Inc.; Croda Chem. Ltd.]

Crovol PK40. PEG-12 palm kernel glycerides; coemulsifier, superfatting agent, emollient, counter-irritant, wetting aid, solubilizer, dispersant for skin and hair care products, soaps, bath oils, astringents, antiperspirants; plasticizer for styling mousses and aq. aerosols; fragrance solubilizer. [Croda Inc.; Croda Chem. Ltd.]

Crow. A cotton fabric grade of Tufnol industrial laminates. [Tufnol Ltd]

Crow Chex. Active ingredient: copper oxalate; seed protectant to prevent sprout pulling by birds in newly planted corn. [Borderland Products Inc] †

Crown. Tires. [Chevron] *

Crown L-60B Acid Inhibitor. Complex amine and thioureas; HCl acid inhibitor protecting metals through absorption to specific corrosion sites; for pickling operations, cleaning industrial boilers, heat exchangers and condensers, in copper cleaners and toilet bowl cleaners. [Crown Tech.]

Crown L-1011 Acid Inhibitor. Complex amine and thioureas; Inhibitor for sulfuric acid; used in metalworking industry in continuous and batch pickling operations. [Crown Tech.]

Crown Acid Aid X. Accelerator and extender for hydrochloric and sulfuric acid; increases pickling action of acid; excellent detergency props. [Crown Tech.]

Crown Anti-Foam. Silicone emulsion; antifoam for dehydrating, evaporating, fermentation processes, abrasive slurries, commercial cleaners, adhesives, latex emulsions, cutting oils, insecticides and pesticides; effective at concs. as low as 5 ppm; stable to high shear and high pH. [Crown Tech.]

Crown Foamer 20. Foamer, wetting agent producing a detergent-type foam for pickling sol'ns.; effective in sulfuric and hydrochloric acid baths. [Crown Tech.]

Crown solder. An alloy of 63% tin, 18% zinc, 13% aluminum, 1% lead, 3% copper, and 2% antimony.

CR Resins. A proprietary trade name for allyl resins. §

Cru Fax P9N. Antistat for paper manufacturing. [Crucible]

Cruisemaster. Tires, batteries and accessories. [Chevron]*

Crusader. Post-emergence herbicide. [Murphy Chemical Ltd] †

---

‡ Trade name and manufacturer not verified   § Trade name without identified manufacturer

Cruverlite. A proprietary trade name for a luminous plastic molding powder. §

Crylene. An acetaldehyde-aniline condensation product; a proprietary rubber vulcanization accelerator. §

CryOfine® Butyl. CAS 9010-85-9; Fine particle rubber produced by cryogenically grinding butyl scrap in a liq. nitrogen environment; blends easily into butyl and halobutyl compounds for improved processing, lower cost, reduced blisters, and increased air flow during curing; used in molded and mech. goods, sealants, membranes, tubes, dynamic parts, diaphragms, tire innerliners. [Midwest Elastomers]

CryOfine® EPDM. Fine particle rubber filler prepared by cryogenically grinding cured EPDM flash, trim, and other scrap; reduces cost of EPDM compounds; used in hose, molded and extruded goods, sponge, roofing, and thermoplastic elastomers. [Midwest Elastomers]

Cryoflex® 660. Dibutoxyethoxyethyl formal. [Sartomer]

Cryo-Heta®. For the medical industry. [DuPont UK]

Cryoseal. Modified natural rubber latex; water-based cold seal adhesives; for confectionery packaging, applied by gravure printing. [The Scottish Adhesives Co Ltd]

Cryptogil NA. Sodium pentachlorophenoxide; for control of fungus diseases in mushroom houses. [Darmycel UK]

Cryptolin. Gonadorelin acetate; luteinizing hormone-releasing factor acetate hydrate; gonad-stimulating principle. [Hoechst-Roussel Pharmaceuticals Inc] *

Cryptone. Proprietary trade names for white pigments containing lithopone and TiO_2. §

Cryptonol. CAS 134-31-6; 8-Hydroxyquinoline sulfate; used for control of soil borne diseases in ornamentals. [Fargro Ltd]

crysmalin. See paraffin, liquid.

Crysta. A bright zinc process. [Hanshaw Chemicals] ‡

Crystal. Sodium, potassium and lithium silicates. [Croda Universal Ltd]

Crystal. CAS 1344-09-8; Sodium silicates. [Crosfield Chemicals Ltd]

Crystal 1000. CAS 63148-62-9; Release blend in water/ethanol; semi-permanent mold release providing high slip for injection, compr. and transfer molding of urethane parts, natural, silicone and other synthetic elastomer parts. [TSE Industries]

Crystal® O. Refined castor oil USP; emollient, pigment wetter, cosolv., lubricant for cosmetics, makeup, antiperspirant sticks. [CasChem]

Crystal® Crown. Refined castor oil USP, with antioxidant; emollient, pigment wetter, cosolv., lubricant for cosmetics, makeup, antiperspirant sticks. [CasChem]

Crystalex. A proprietary trade name for an acrylic denture base. §

crystal glass. A glass composed of lead and potassium silicates.

Crystal Inhibitor #5. Blend; retards formation of crystals in 2,4-D amine formulation dilutions. [Harcros]

Crystalite. Organic brighteners; for cyanide zinc electroplating. [Engelhard Technologies Ltd]

Crystalite. A proprietary trade name for an acrylic molding powder. §

Crystalor BC-1. Polymethylpentene resin, 30% glass-filled; injection molding resin with outstanding impact props., high heat distort. temp., light weight, good chem. resist., excellent elec. props.; for automotive, appliances, elec./electronic applications. [Phillips] *

Crystalor DC-6. CAS 25068-26-2; Polymethylpentene resin; high impact props. and clarity, light weight, good elong.; for automotive, appliances, medical, housewares, elec./electronic and packaging applications. [Phillips] *

Crystal Polystyrene. Polymerization product of styrene monomer, with additives; general commodity plastic for extrusion and injection. [Lin-Pac International] *

Crystamet 1020. CAS 6834-92-0; EINECS 229-912-9; Sodium metasilicate, pentahydrate; alkaline silicate used for formulating specialty detergents and contributing buffering capacity and corrosion inhibition of soft metals and ceramic glazes. [Crosfield]

Crystamycin. A proprietary preparation containing benzylpenicillin sodium and streptomycin; an antibiotic. [Glaxo Laboratories] *

Crystamycin Forte. A proprietary preparation containing benzylpenicillin sodium and streptonlycin; an antibiotic. [Glaxo Laboratories] *

Crystapen. CAS 69-57-8; A proprietary preparation containing benzylpenicillin sodium; sodium penicillin G; an antibiotic. [Glaxo Laboratories] *

Crystapen G. A proprietary preparation containing benzylpenicillin (as the potassium salt); an antibiotic. [Glaxo Laboratories] *

Crystapen V. A proprietary preparation containing phenoxymethylpenicillin (as the calcium salt); an antibiotic. [Glaxo Laboratories] *

Crystex®. CAS 7704-34-9; Insoluble sulfur. [Akzo]

Crystex. A proprietary viscose product used for packing. §

Crystex® Regular. CAS 9035-99-8, 7704-34-9; Insoluble sulfur (polymeric sulfur and sulfur); prevents crystallization of sulfur on uncured rubber surfaces (sulfur bloom); detackifier; retards bin scorch; minimizes sulfur migration [Akzo] *

Crystic. A range of unsaturated polyester resins, gelcoats, pigment pastes, FRP ancillaries; used for the manufacture of reinforced and nonreinforced plastics articles and castings and coatings. [Scott Bader]

Crystic® 2-414PA. Orthophthalic polyester resin; for building, land transport, boat building applications; low exotherm; thixotropic. [Scott Bader]

Crystic® 39PA. Isophthalic/HET polyester resin; fire retardant gelcoat for building, land transport applications. [Scott Bader]

Crystic® 199. Isophthalic polyester resin; heat resist. resin for high temp. use and elec. applications, tanks and chem. plant. [Scott Bader]

Crystic® 471PALV. Orthophthalic polyester resin; preaccelerated, rapid-hardening resin with fast wetting of reinforcement and short mold release time, for contact molding applications and catalyst injection spray equip.; for automotive, marine, industrial, and general purpose applications. [Scott Bader]

Crystic® 581PA. Orthophthalic polyester resin; tough resin for formulation of body fillers in building and land transport industries; thixotropic; gel time 5 min @ 20 C. [Scott Bader]

Crystic® Fireguard. Unsaturated polyester gelcoat; fire retardant. [Scott Bader]

Crystic® Fireguard 75PA. DBNPG polyester resin; thixotropic, intumescent surface coating resin giving fire protection to GRP laminates; fire-retardant coating for timber and other slightly porous surfs. [Scott Bader]

Crysticillin. CAS 6130-64-9; Penicillin G procaine; antibacterial. [Bristol-Myers Squibb Co Inc] *

Crystic® Impel. Unsaturated polyester molding granules. [Scott Bader]

Crystic®Impreg. SMC; viscosity controlled by temperature. [Scott Bader]

Crystic® Prefil F. CAS 1309-64-4; Antimony trioxide and chlorinated org. compound dispersed in polyester resin; additive to facilitate the manufacturing of reduced fire hazard moldings. [Scott Bader]

Crystic® Pregel 17. Silica in a general purpose polyester resin disp.; resin additive to confer thixotropic properties to general purpose polyester resins for laminating or gelcoat applications. [Scott Bader]

Crystodigin. CAS 71-63-6; Digitoxin; cardiotonic. [Eli Lilly & Co]

Crystoids. CAS 136-77-6; Hexylresorcinol; antihelmintic preparation. [Merck & Co Inc] †

Crystolite. See Carbora. §

Crystolon®. A registered trade name for various types of goods such as grinding wheels, abrasives, and refractory grain, etc., made from silicon carbide. §

Crystran. Crystals for optical transmission. [BDH Chemicals Ltd]

CS1590. 3-(N-Styrylmethyl-2-aminoethylamino) propyltrimethoxy silane hydrochloride, methanol; coupling agent, chem. intermediate, blocking agent, release agent, lubricant, primer, reducing agent. [Hüls Am.]

CSC 2-Aminobutane. CAS 13952-84-6; 2-Aminobutane; fungicide for stored potatoes. [Chemical Spraying Co Ltd]

CSE-6000 Series. Cold set epoxies; for vessel linings and heavy-duty maintenance applications; high resist. to chems. [Heresite Protective Coatings]

CSI. Crofilcon A; contact lens material. [Syntex Ophthalmics Inc] *

CSP. See copper sulfate (ic).

CT-690, CT-700. Polyurethane elastomeric resins. [Unichem] *

CT1750. CAS 4766-57-8; Tetra-n-butoxysilane; coupling agent, chem. intermediate, blocking agent, release agent, lubricant, primer, reducing agent. [Hüls Am.]

CT1800. CAS 10026-04-7; Tetrachloro silane; coupling agent, chem. intermediate, blocking agent, release agent, lubricant, primer, reducing agent. [Hüls Am.]

CT2015. CAS 13528-93-3; 1,1,4,4-Tetramethyldichlorodisilethylene; coupling agent, chem. intermediate, blocking agent, release agent, lubricant, primer, reducing agent. [Hüls Am.]

CT2030. CAS 3277-26-7; Tetramethyldisiloxane; coupling agent, chem. intermediate, blocking agent, release agent, lubricant, primer, reducing agent. [Hüls Am.]

CT2050. CAS 75-76-3; Tetramethyl silane; coupling agent, chem. intermediate, blocking agent, release agent, lubricant, primer, reducing agent. [Hüls Am.]

CT2090. CAS 682-01-9; Tetrapropoxy silane; coupling agent, chem. intermediate, blocking agent, release agent, lubricant, primer, reducing agent. [Hüls Am.]

CT2500. CAS 998-30-1; Triethoxy silane; coupling agent, chem. intermediate, blocking agent, release agent, lubricant, primer, reducing agent. [Hüls Am.]

CT2507. CAS 23779-32-0; N-(Triethoxysilylpropyl) urea; coupling agent, chem. intermediate, blocking agent, release agent, lubricant, primer, reducing agent. [Hüls Am.]

CT2520. CAS 994-30-9; Triethylchloro silane; coupling agent, chem. intermediate, blocking agent, release agent, lubricant, primer, reducing agent. [Hüls Am.]

CT2523. CAS 617-86-7; Triethyl silane; coupling agent, chem. intermediate, blocking agent, release agent, lubricant, primer, reducing agent. [Hüls Am.]

CT2902. CAS 50975-76-3; 1-Trimethoxysilyl-2- (chloromethyl) phenylethane; coupling agent, chem. intermediate, blocking agent, release agent, lubricant, primer, reducing agent. [Hüls Am.]

CT2910. CAS 35141-30-1; Trimethoxysilylpropyl-diethylenetriamlne; coupling agent, chem. intermediate, blocking agent, release agent, lubricant, primer, reducing agent. [Hüls Am.]

CT2925. N-Trimethoxysilylpropyl-N,N,N-trimethyl ammonium chloride, methanol; coupling agent, chem. intermediate, blocking agent, release agent, lubricant, primer, reducing agent. [Hüls Am.]

CT2928. CAS 2857-97-8; Trimethylbromosilane; coupling agent, chem. intermediate, blocking agent, release agent, lubricant, primer, reducing agent. [Hüls Am.]

CT2950. CAS 75-77-4; Trimethylchloro silane; coupling agent, chem. intermediate, blocking agent, release agent, lubricant, primer, reducing agent. [Hüls Am.]

CT2970. CAS 1825-62-3; Trimethylethoxy silane; coupling agent, chem. intermediate, blocking agent, release agent, lubricant, primer, reducing agent. [Hüls Am.]

CT3250. CAS 13434-12-6; Trimethylsilyl acetamide; coupling agent, chem. intermediate, blocking agent, release agent, lubricant, primer, reducing agent. [Hüls Am.]

CT3254. CAS 2754-27-0; Trimethylsilylacetate; coupling agent, chem. intermediate, blocking agent, release agent, lubricant, primer, reducing agent. [Hüls Am.]

CT3600. CAS 18156-74-6; Trimethylsilyl imidazole; coupling agent, chem. intermediate, blocking agent, release agent, lubricant, primer, reducing agent. [Hüls Am.]

CT3610. CAS 16029-98-4; Trimethylsilyl iodide; coupling agent, chem. intermediate, blocking agent, release agent, lubricant, primer, reducing agent. [Hüls Am.]

CT3795. CAS 27607-77-8; Trimethylsilyl trifluoromethane sulfonate; coupling agent, chem. intermediate, blocking agent, release agent, lubricant, primer, reducing agent. [Hüls Am.]

CTC-3300. CAS 32131-17-2; Nylon 6/6, 33% glass fiber-reinforced. [Compounding Tech.] *

CTW. Epoxy resin based cements, screeds and coatings. [Prodorite Ltd]

CTX-308. 30% PBT/50% ABS, 20% glass fiber-reinforced. [Compounding Tech.] *

CTX-312. CAS 25038-54-4; Nylon 6, 10% glass fiber-reinforced, impact-modified. [Compounding Tech.] *

CTXC-020. Polyarylsulfone, carbon fiber-reinforced; for ESD applications. [Compounding Tech.] *

Cuba black. (Dianil Black). A dyestuff. Dyes cotton black from an alkaline bath.

cubanite. A mineral, $CuFe_2S_4$.

Cuba orange. A dyestuff prepared by the action of sodium sulfite upon diazo-naphthalene-sulfonic acid. Dyes wool orange.

cube ore. See Tyrolite.

cubical catechu. See cutch.

cubic niter. See sodium nitrate.

Cucar. Copper carbonate. [Mechema Chemicals Ltd] *

---

‡ Trade name and manufacturer not verified   § Trade name without identified manufacturer

Cuclat. Kerosene sweeting catalyst. [Mechema Chemicals Ltd] *

Cudbear. *See* Archil. §

Cudgel. CAS 944-22-9; A suspension of microcapsules containing fonofos; an insecticide for the control of vine weevil, sciarid fly, and cabbage root fly. [Fisons plc, Horticulture Div]

Cudgel. CAS 944-22-9; A suspension of microcapsules containing fonofos; an insecticide for the control of vine weevil, sciarid fly and cabbage root fly. [ICI Agrochemicals]

Cufenium. An alloy of 22% nickel, 72% copper, and 6% iron.

Cufor. Copper formate. [Mechema Chemicals Ltd] *

Cuite. Natural silk freed from the silk gum or sericin (*qv*). Raw silk consists of about 66% fibroin, forming the real silk substance, and silk gum or sericin. The silk gum is removed by treating with hot neutral soap solution.

cullet. Broken glass; used in the manufacture of glass.

Culminal® Hydroxypropylmethylcellulose. CAS 9004-65-3; Methylhydroxypropylcellulose [Aqualon]

Culminal® Methylcellulose. CAS 9004-67-5; Methylcellulose. [Aqualon]

Cultar. CAS 76738-62-0; Suspension concentrate containing 250 g/l paclobutrazol; plant growth regulator. [ICI Chem. & Polymers Ltd]

Cumal. CAS 122-03-2; EINECS 204-516-9; p-Isopropylbenzaldehyde; intermediate for pharmaceuticals, perfumes. [Mitsubishi Gas]

Cumar®. Coumarone-indene resins; for the production of varnishes, polishes, artificial leather and linoleum. [Neville Chemical Co]

cumar gum. *See* paracoumarone resin.

Cumar® P-10, R-1. Coumarone-indene resin; resin with excellent resist. to alkalies, dilute acids and moisture; for use in adhesives, aluminum paints, varnishes, rotogravure inks, rubber compounds [Neville]

Cumate®. CAS 137-29-1; Copper dimethyldithiocarbamate; accelerator for vulcanization of rubber. [R T Vanderbilt Co Inc]

cumene. (isopropylbenzene) $C_6H_5CH(CH_3)_2$; CAS 98-82-8; Production of phenol, acetone and α-methylstyrene; solvent. [Ashland; Chevron; Georgia Gulf; Hüls AG; Mitsubishi Petrochem.; Mitsui Petrochem. Ind.]

cumene peroxide. *See* dicumyl peroxide.

Cumitan®. For the medical industry. [DuPont UK]

cumyl phenol. (4-cumylphenol; 4-(2-phenylisopropyl) phenol) $C_6H_5C(CH_3)_2C_6H_4OH$; CAS 599-64-4; Intermediate for resins, insecticides, lubricants. [Hüls AG; ICI Specialties; PMC Specialties; Schenectady]

Cunilate 2174-NO. Antimicrobial for solv. systems, adhesives, latexes, FDA-approved preservatives. [Morton Int'l./Plastics Additives]

Cuniloy. An alloy of nickel, manganese, copper, and small quantities of lead.

Cuniphen 2778-1. Antimicrobial for in-can preservation; FDA approved. [Morton Int'l./Plastics Additives]

Cunitex. Bird and animal repellant. [May & Baker Ltd] *

Cupac. Copper acetate. [Mechema Chemicals Ltd] *

Cupalit. A softening agent and lubricant; used by the textile industry for the after treatment of all types of fibers. [Degussa AG] *

Cupar. Copper arsenate. [Mechema Chemicals Ltd] *

cuperatin. Copper albuminate.

Cupertine. Bordeaux mixture 84%, maneb 8%; wettable

powder used as protective fungicide for foliage application to ornamental and crop plants. [Industrias Quimicas Del Valles SA] *

Cupertine Folpet. Bordeaux mixture 80%, folpet 10%; wettable powder used as protective fungicide for foliage application to ornamental and crop plants. [Industrias Quimicas Del Valles SA] *

Cupertine Super. Bordeaux mixture 90%, cymoxanil 3%; wettable powder used as protective and curative fungicide for foliage application to ornamental and crop plants. [Industrias Quimicas Del Valles SA] *

cupferron. (copperone) CAS 135-20-6; Nitrosophenylhydroxylamine. The ammonium salt is used as a precipitating agent for copper, in the determination of copper; used as an analytical reagent, especially for the separation and precipitation of metals, e.g., copper, iron, vanadium.

Cupolloy. Ferro-molybdenum briquettes for cupola. [Foseco (F.S.) Ltd]

cuprammonium silk. (cuprate silk). Artificial silks made by dissolving cotton in a cuprammonium solution (copper hydrate dissolved in ammonia), and precipitating it in a fine thread.

cupranium. A name for certain brass and bronze alloys. The bronze contains tantalum and vanadium.

Cuprase. A form of colloidal copper; a proprietary trade name for a vulcanizing accelerator containing copper. §

Cupravit®. CAS 1332-40-7; Copper oxychloride; copper containing spray used for control of fungal diseases like downy mildews, late blight, early blight, apple scab and various leaf spot diseases on a wide range of crops. [Bayer AG]

Cuprex. Copper alloy general purpose flux. [Foseco (F.S.) Ltd]

cupric acetate. *See* copper acetate (ic).

cupric bromide. *See* copper bromide (ic).

cupric carbonate basic. *See* copper carbonate (ic).

cupric hydroxide. *See* copper hydroxide (ic).

cupric nitrate trihydrate. *See* copper nitrate (ic).

cupric oxide. *See* copper oxide (ic).

cupric sulfate. *See* copper sulfate (ic).

Cupridan. CAS 1317-39-1; Copper oxide; fungicide. [Makhteshim Chemical Works Ltd]

cupriferous calamine. *See* Tyrolite.

Cuprimine. CAS 52-67-5; Penicillamine; used for Wilson's disease, cystinuria. [Merck & Co Inc]

Cuprinol. A copper naphthenate or sodium pentachlorphenate preparation used as a preservative for timber and fabrics. §

Cuprit. Copper alloy cleansing flux. [Foseco (F.S.) Ltd]

cupro-aluminums. Alloys of aluminum and copper containing 1-20% aluminum. They are also wrongly called aluminum bronzes.

cuprocyan. A double cyanide of potassium and copper.

Cuprodine. A medium for coating steel with copper. [ICI Chem. & Polymers Ltd] †

Cuproid. CAS 7758-89-6; Cuprous chloride. [Mechema Chemicals Ltd] *

Cuprokylt. CAS 1332-40-7; Copper oxychloride 87% w/w wettable powder formulation; a protectant fungicide for agriculture and horticulture. [Universal Crop Protection Ltd]

Cupron. A proprietary trade name for a copper nickel alloy with a low temperature coefficient of resistance. §

---

* Trade name not verified as available   † Trade name verified as obsolete

Cuprophenyl. An after coppering direct dye. [Ciba plc] *
cupror. An aluminum bronze. It contains 94.2% copper and 5.8% aluminum.
Cuprosana. CAS 1332-40-7; A dustable powder containing 10% w/w copper oxychloride; a protectant fungicide for hops. [Universal Crop Protection Ltd]
cupro-steel. An alloy of steel with copper up to 4%. Occasionally used for printing rollers and projectiles.
cupro-titanium. Usually an alloy of copper with 10% titanium; used as a deoxidizer in making brass and bronze castings.
cuprous bromide. See copper bromide (ous).
cuprous chloride. See copper chloride (ous).
cuprous iodide. See copper iodide (ous).
cupro-vanadium. An alloy usually containing from 10-15% vanadium, 60-70% copper, 10-15% aluminum, and 2-3% nickel.
Cuprox. CAS 1317-39-1; Copper oxide; fungicide. [Makhteshim Chemical Works Ltd]
Curafos® Formula 11-2, 22-4. Sodium tripolyphosphate, sodium polyphosphate (glassy); food additive for meat, poultry, and seafood industries. [Rhone-Poulenc Food Ingreds.]
Curafos® STP. CAS 7758-29-4; Sodium tripolyphosphate; food additive for meat, poultry, and seafood industries, processed foods, confections. [Rhone-Poulenc Food Ingreds.]
Curaseal. Natural and synthetic rubber latex based adhesives and coatings; cohesive, pressure-sensitive and synthetic resin emulsions; used for numerous adhesive bonding applications. [Testworth Laboratories Inc]
Curasol. Soil erosion inhibitor. [Hoechst UK]
Curaterr®. CAS 1563-66-2; Carbofuran; granular soil-applied insecticide for use in sugar beet, tobacco, maize, rice, sugar cane and vegetables. [Bayer AG]
Curatin. CAS 1229-29-4; Doxepin hydrochloride; veterinary antipruritic. [Pfizer Inc] †
Curavis® 250/350. Sodium hexametaphosphate, sodium tripolyphosphate, sodium acid pyrophosphate; food additive for meat, poultry, and seafood industries. [Rhone-Poulenc Food Ingreds.]
Curb. CAS 7784-25-0; Aluminum ammonium sulfate; bird and animal repellent. [Sphere Laboratories (London) Ltd]
Curbetan®. For the agriculture industry. [DuPont UK]
Curbeton 0550. Desugared hardwood calcium/magnesium lignosulfonate; water-reducing and strength-increasing concrete additive. [Borregaard LignoTech]
Curene. Catalysts and light and heat stabilizers; catalysts and chain extenders for urethane polymers. [Anderson Development Company] *
Curenox-50. CAS 1332-40-7; Copper oxychloride formulated to 50% metallic copper contents; wettable powder used as protective fungicide for foliage application to ornamental and crop plants. [Industrias Quimicas Del Valles SA] *
Cure-Rite® 18. Thiocarbamyl sulfenamide; nonstaining primary accelerator for EPDM, SBR, nitrile, natural and butyl rubbers. [Akrochem]
Curgon. A proprietary trade name for naphthenate driers. §
Curithane. Reactive polyamines used as a hardener and curing agent for polyisocyanurate foams; rigid PU catalysts. [Dow UK]
Curithane 103. Polymethylene polyaniline; amine curing

agent for PU; intermediate in manufacturing of polyamides, polyimides, coatings, and plastics. [Dow Plastics]*
Curodex. Rubber odorants and deodorants. [Bush Boake Allen Ltd]
Curolac. Precatalysed and two-pack curing lacquers. [The Scottish Adhesives Co Ltd]
Curtexil 100S. Hardwood calcium/magnesium lignosulfonate; general binder and dispersant. [Borregaard LignoTech]
Curzate®. Fungicide. [DuPont UK]
Curzate® M. Mixture of cymoxanil and mancozeb; used to control potato blight. [DuPont UK]
Cusamon. Copper sulfate monohydrate. [Mechema Chemicals Ltd] *
Cusatrib. Tribasic copper sulfate. [Mechema Chemicals Ltd] *
Cusiloy A. A proprietary trade name for an alloy containing 95.5% copper, 3% silicon, 1% iron, and 0.5% tin. §
Custom Age 625 Plus® Alloy. Nickel-base alloy; precipitation-hardenable alloy with good corrosion resistance; used for deep sour gas wells, refineries, chem. process industry, high-temp. nuclear water. [Carpenter Tech.]
Cusyd. Copper thiocyanate. [Mechema Chemicals Ltd] *
Cut Aid. ASTM S-315 Oil, 1,1,1-trichlorethane, mask odor No 3; excellent cutting aid for general machine operations; exceptional for disc sanding, filing and punch press operations. [Doyle Specialties] *
Cutavit Richter. Complex of vitamins A and E and essential fatty acids in lipopholic medium; product for dry skin and hair. [Dr. Kurt Richter; Henkel/Cospha]
Cutch. (kutch, catechu, katechu, Gambier, Japan earth, cachou, cutt). Natural dyestuff. It consists of a brown or reddish-brown amorphous extract, obtained by boiling with water the wood of *Acacia spp., Mimosa spp.,* or *Uncaria spp.*; used in textile dying; tanning, and pharmaceuticals.
Cutina®. Self-emulsifying raw material; creams and emulsions. [Henkel Chemicals Ltd] *
Cutina® BW. Glyceryl hydroxystearate, cetyl palmitate, microcryst. wax, trihydroxystearin; wax for use as beeswax substitute; visc. agent for personal care products [Henkel/Cospha; Henkel Canada]
Cutina® CBS. Glyceryl stearate, cetearyl alcohol, cetyl palmitate, and cocoglycerides; cream base for manufacturing of creams and lotions of the o/w type; visc. agent and stabilizer. [Henkel/Cospha; Henkel KGaA]
Cutina® CP. CAS 540-10-3; EINECS 208-736-6; Cetyl palmitate; synthetic spermaceti; consistency factor for creams, ointments, liq. emulsions, fatty make-ups, and sticks. [Henkel/Cospha; Henkel Canada]
Cutina® E24. CAS 51158-08-8; PEG-20 glyceryl stearate; o/w emulsifier for mild creams and emulsions for baby and childrens preparations, sun preps. [Henkel Canada; Henkel/Emery/Cospha; Henkel KGaA]
Cutina® FS 25, FS 45. Fatty acid mixt.; consistency factor after saponification; o/w emulsifier for emulsions and ointments. [Henkel KGaA]
Cutina® GMS. Glyceryl stearate; nonself-emulsifying cream base for o/w and w/o emulsions; visc. agent for creams, ointments, sticks; beads. [Henkel/Cospha; Henkel Canada]
Cutina® HR. Hydrogenated castor oil; lubricant for tablets, high melting consistency giving factor, thickener for oils.

---

‡ Trade name and manufacturer not verified   § Trade name without identified manufacturer

[Henkel/Cospha; Henkel Canada]

Cutina® KD16. Glyceryl stearate SE; emulsifier and fatting co-agent for o/w ointments and cosmetic creams; base. [Henkel Canada; Henkel/Emery/Cospha; Henkel KGaA]

Cutina® MD-A. Glyceryl stearate; cream base, consistency factor for cosmetic and pharmaceutical creams and emulsions; good spreading props. [Henkel/Cospha; Henkel KGaA]

Cutlanego. An alloy of 50% bismuth and 50% tin; used for tempering steel tools.

Cutlass. CAS 18467-77-1; Contains dikegulac; growth regulator for hedges. [ICI Garden Products]

Cutonic. Micronutrient foliar sprays. [McKechnie Chemicals Ltd] *

Cutt. See cutch.

Cutterglobin®. Immunoglobulin. [Bayer AG]

Cuvan®. Proprietary formulations; corrosion inhibitors and metal deactivators for industrial and automotive oils and metalworking fluids. [R T Vanderbilt Co Inc]

CV-1142. One-part silicone adhesive/sealant; RTV non-corrosive low-temp. adhesive/sealant. [McGhan NuSil]

CV-2640. Two-part silicone adhesive/sealant; electrically conductive RTV RF/EMI shielding/grounding adhesive/sealant, potting compound [McGhan NuSil]

CV4720. CAS 1719-58-0; Vinyldimethylchlorosilane; coupling agent, chem. intermediate, blocking agent, release agent, lubricant, primer, reducing agent. [Hüls Am.]

CV4772. Vinylmethyldichloro silane; coupling agent, chem. intermediate, blocking agent, release agent, lubricant, primer, reducing agent. [Hüls Am.]

CV4800. CAS 4130-08-9; Vinyltriacetoxy silane; coupling agent, chem. intermediate, blocking agent, release agent, lubricant, primer, reducing agent. [Hüls Am.]

CV4900. CAS 75-94-5; Vinyltrichlorosilane; coupling agent, chem. intermediate, blocking agent, release agent, lubricant, primer, reducing agent. [Hüls Am.]

CV4910. CAS 78-08-0; EINECS 201-081-7; Vinyltriethoxysilane; coupling agent, chem. intermediate, blocking agent, release agent, lubricant, primer, reducing agent. [Hüls Am.]

CV4917. CAS 2768-02-7; EINECS 220-449-8; Vinyltrimethoxysilane; coupling agent, chem. intermediate, blocking agent, release agent, lubricant, primer, reducing agent. [Hüls Am.]

CV5000. CAS 1067-53-4; EINECS 213-934-0; Vinyltris(methoxyethoxy)silane; coupling agent, chem. intermediate, blocking agent, release agent, lubricant, primer, reducing agent. [Hüls Am.]

CV5050. CAS 2224-33-1; Vinyl tris (methylethylketoxime) silane; coupling agent, chem. intermediate, blocking agent, release agent, lubricant, primer, reducing agent. [Hüls Am.]

CV5100. CAS 5356-84-3; Vinyl tris (trimethylsiloxy) silane; coupling agent, chem. intermediate, blocking agent, release agent, lubricant, primer, reducing agent. [Hüls Am.]

CW 79. Lidofilcon B material containing 79% water; contact lens material. [Bausch & Lomb, Professional Products Div] *

CW-35. Chlorinated paraffin; offers good stability and resist. to thermal decomp.; used in metalworking applications providing corrosion inhibition; EP agent for sol. and cutting oils, some drawing compounds [Ferro/Keil]

CWT 102. Blend; corrosion inhibitor for open cooling water

systems. [Drew Ind. Div.]

CXL 400, 78T. General purpose adhesive for concrete or steel. [Colebrand Ltd]

Cyanacryl® 35. Acrylic elastomer. [Am. Cyanamid]

Cyanamer P-80. Sulfonated polycarboxylic acid polymer. [Cyanamid BV]

Cyanamer P35-P70. Acrylamide copolymers (anionic). [Cyanamid BV]

Cyanaprene® 1050. TDI-polyether urethane prepolymer; used for solid tires, coated materials, rolls, molded mech. goods; high abrasion resist., good resilience, hydrolytic stability, and load bearing capability, oil and fuel resist.; vulcanizable with diamines. [Am. Cyanamid; Air Prods.]

Cyanaprene® 1080. TDI-polyether urethane prepolymer; see Cyanaprene 1050; 1080 esp. suited for mining applications; vulcanizable with diamines. [Am. Cyanamid; Air Prods.]

Cyanaprene® 1090. TDI-polyether urethane prepolymer; see Cyanaprene 1050; vulcanizable with diamines. [Am. Cyanamid; Air Prods.]

Cyanaprene® 2070. TDI-polyether urethane prepolymer [Am. Cyanamid; Air Prods.]

Cyanaprene®, 2160, 2175, 2180. TDI-polyether urethane prepolymer [Am. Cyanamid; Air Prods.]

Cyanaprene® A-7QM. TDI-polyether urethane prepolymer; used in solid tires, coated materials, rolls, molded mech. goods; high abrasion resist., tear strength oil and fuel resist., good load bearing chars., hydrolytic stability; vulcanizable with diamines. [Am. Cyanamid; Air Prods.]

Cyanaprene® A-8. TDI-polyether urethane prepolymer; used in solid tires, coated materials, heels, soles, rubber-coved rolls, molded mech. goods, metal forming pads; high abrasion resist., tear strength oil and fuel resist.; vulcanizable with diamines. [Am. Cyanamid; Air Prods.]

Cyanaprene® A-8QM. TDI-polyether urethane prepolymer [Am. Cyanamid; Air Prods.]

Cyanaprene® A-9. Polyester-based PU; see Cyanaprene A-8; vulcanizable with diamines. [Am. Cyanamid; Air Prods.]

Cyanaprene® A-9QM. TDI-polyether urethane prepolymer [Am. Cyanamid; Air Prods.]

Cyanaprene® A-75QM. TDI-polyester urethane prepolymer; see Cyanaprene A-7QM. [Am. Cyanamid; Air Prods.]

Cyanaprene® A-85. TDI-polyester urethane prepolymer; castable urethane used in casters, lift truck and press-on tires; high abrasion resist., cutting and flex fatigue, solv. resist., exceptional thermal stability; vulcanizable with diamines. [Am. Cyanamid; Air Prods.]

Cyanaprene® D-5QM. TDI-polyester urethane prepolymer [Am. Cyanamid; Air Prods.]

Cyanaprene® D-6. TDI-polyester urethane prepolymer; castable urethane used in top lifts, pellet and caster wheels, molded mech. goods, rubber-covered rolls; high abrasion resist., tear strength, oil and fuel resist.; vulcanizable with diamines. [Am. Cyanamid; Air Prods.]

Cyanaprene® D-7. TDI-polyester urethane prepolymer; castable urethane used in rubber-covered rolls, bearing blocks, drilling jig fixtures, cutting pads; high abrasion resist., oil and fuel resist.; vulcanizable with diamines. [Am. Cyanamid; Air Prods.]

Cyanaprene® D-55. TDI-polyester urethane prepolymer [Am. Cyanamid; Air Prods.]

Cyanatrol. Modified polyacrylamide in water-in-oil emul-

---

* Trade name not verified as available   † Trade name verified as obsolete

sion. [Cyanamid BV]

Cyanex. Modified phosphine derivatives. [Cyanamid BV]

Cyanine. Anionic dyestuffs (level dyeing); for wool and wool blends. [Holliday Dyes & Chemicals Ltd]

cyanine. (Leitch's blue, quinoline blue). A blue pigment consisting of a mixture of cobalt blue and Prussian blue. A quinoline dyestuff,is also known as cyanine or quinoline blue; used as a panchromatic sensitizer, and also dyes silk.

Cyanine Fast. Acid dyes. [Holliday Dyes & Chemicals Ltd]†

Cyanine Moderns. Dyestuffs, which are condensation products of gallo-cyanines and allyldiamines.

cyanocobalamin. See vitamin B12.

cyanogen bromide. (bromine cyanide) CBrN; CAS 506-68-3; EINECS 208-051-2; Organic synthesis, parasiticide, fumigating compositions, rat extermination, cyaniding reagent in gold extraction processes. [Aldrich; Atomergic Chemetals; Eastman; Janssen Chimica]

cyanogran. See sodium cyanide.

Cyanogran®. For the chemical industry. [DuPont UK]

Cyanolime. CAS 592-01-8; Calcium cyanide. [Mechema Chemicals Ltd] *

Cyanolit-Hitemp. A proprietary range of high-temperature cyanoacrylate adhesives. [Unichem] *

Cyanosin A. (Cyanosin, Spirit Soluble). The alkali salt of tetrabromodichlorofluoresceine methyl ether; used in silk dyeing.

Cyanosin B. A dyestuff. It is the sodium salt of tetrabromotetrachlorofluoresceine ethyl ether. Dyes wool bluish-red.

Cyanox®425. CAS 88-24-4; 2,2´-Methylenebis (4-ethyl-6-tert-butylphenol); antioxidant for impact molding resins; stabilizes acrylics and ABS; end-uses incl. extruded and molded products [Am. Cyanamid]

Cyanox® 711. CAS 10595-72-9; Ditridecyl thiodipropionate; sec. antioxidant for stabilizing polyolefins, ABS, petrol. lubricants, SBR latex compositions. [Am. Cyanamid]

Cyanox®1212. Mixed lauryl-stearyl thiodipropionate; sec. antioxidant for stabilizing polyolefins; applications incl. pipe, hot-melt adhesives, and molded olefin products [Am. Cyanamid]

Cyanox®1790. 1,3,5-Tris (4-tert-butyl-3-hydroxy-2,6-dimethylbenzyl)-1,3,5-triazine-2,4,6-(1H,3H,5H)-trione; antioxidant for use in polyolefin pipe, film, household appliances, olefin and urethane fibers, styrenics and polyesters. [Am. Cyanamid]

Cyanox®2246. CAS 119-47-1; 2,2´-Methylenebis (4-methyl-6-tert-butylphenol); antioxidant preventing thermal oxidation of ABS, polyethylene, PP, and EVA; oxidation inhibitor for fats, oils, and paraffin wax; polymerization inhibitor in chemical processes; for ABS, hot-melt adhesives, latex carpet backing, and specialty olefin applications. [Am. Cyanamid]

Cyanox®2777. 1:2 blend of 1,3,5-tris (4-t-butyl-3-hydroxy-2,6-dimethylbenzyl)-1,3,5-triazine-2,4,6-(1H,3H,5H) trione and tris (2,4-di-t-butylphenyl) phosphite; antioxidant, stabilizer for polymers, esp. polyolefins in high-temp. processing. [Am. Cyanamid]

Cyanox®LTDP. CAS 123-28-4; Dilauryl thiodipropionate; sec. antioxidant in ABS, PP, and polyethylene; used in food packaging materials, automotive, appliance, battery casing, pipe; stabilization of oils, lubricants, sealants, and adhesives. [Am. Cyanamid]

Cyanox®MTDP. CAS 16545-54-3; Dimyristyl thiodipropionate; sec. antioxidant for protection of polyolefins. [Am. Cyanamid]

Cyanox®STDP. CAS 693-36-7; Distearyl thiodipropionate; sec. antioxidant used in polyolefins and other polymers; used in automotive, appliance, container film, sealant, and adhesive applications. [Am. Cyanamid]

cyanuric acid. (cyanuric acid, anhydride; 2,4,6-trihydroxy-1,3,5-triazine; tricarbimide; tricyanide) $C_3H_3N_3O_3$; CAS 108-80-5; EINECS 203-618-0; Intermediate for chlorinated bleaches, selective herbicide, whitening agents. [Allchem Industries; Monsanto; Orkem UK Ltd; 3-V]

cyanuric acid, anhydride. See cyanuric acid.

Cyasorb® UV 9. CAS 131-57-7; EINECS 205-031-5; 2-Hydroxy-4-methoxybenzophenone; light stabilizer and uv absorber for plastics and coatings; esp. for flexible and rigid PVC, unsat. polyesters, and acrylics; used in outdoor sheeting and glazing applications, molded products, adhesives. [Am. Cyanamid]

Cyasorb®UV 24. CAS 131-53-3; EINECS 205-026-8; 2,2´-Dihydroxy-4-methoxybenzophenone; light stabilizer and uv absorber for coatings and plastics, e.g., alkyds, phenolics, PU coatings; stabilizer for polyester film and PVC formulations. [Am. Cyanamid]

Cyasorb®UV 531. CAS 1843-05-6; EINECS 217-421-2; 2-Hydroxy-4-n-octoxybenzophenone; light stabilizer and uv absorber for plastics and coatings, e.g., polyethylene, PP, PVC, and EVA; uses incl. pipe, storage tanks, and auto, marine, garden products, auto refinish and industrial coatings, adhesives and sealants. [Am. Cyanamid]

Cyasorb® UV 1084. 2,2´-Thiobis (4-t-octylphenolato)-n-butylamine nickel II; light and heat stabilizer for polyolefins, e.g., PP fiber, LDPE agric. films and pool liners, and molded products [Am. Cyanamid]

Cyasorb® UV 2098. 2-Hydroxy-4-acryloyloxyethoxy benzophenone; light stabilizer, uv absorber which may be chemically bonded with monomers or polymers. [Am. Cyanamid]

Cyasorb®UV 2126. Polymer of 4-(2-acryloyloxyethoxy)-2-hydroxybenzophenone; light stabilizer and uv absorber for films and plastics in automotive, greenhouse, home siding, and solar applications. [Am. Cyanamid]

Cyasorb® UV 2908. 3,5-Di-t-butyl-4-hydroxybenzoic acid, n-hexadecyl ester; light stabilizer, free radical scavenger for polyolefins, esp. pigmented opaque formulations; antioxidant; for pipe, crates, drums, auto, marine, garden, recreational products [Am. Cyanamid]

Cyasorb®UV 3346. Oligomeric hindered amine; stabilizer for polymers alone or in combination with UV absorbers. [Am. Cyanamid]

Cyasorb® UV 5411. CAS 3147-75-9; 2-(2-Hydroxy-5-t-octylphenyl)-benzotriazole; light stabilizer and uv absorber for polymeric systems incl. polyester, PVC, styrenics, acrylics, PC, and polyvinyl butyral; end-uses incl. molding, sheet, and glazing materials for window, marine, and auto applications; also in coatings, photo products, sealants, and elastomeric materials. [Am. Cyanamid]

Cyastat® 609. N,N-Bis (2-hydroxyethyl)-N-(3-dodecyloxy-2-hydroxypropyl) methyl ammonium methosulfate, IPA; antistatic agent with good heat stability; for PVC phonograph records, specialty packaging. [Am. Cyanamid]

Cyastat® LS. (3-Lauramidopropyl) trimethylammonium methyl sulfate; antistatic agent for polymeric materials,

---

‡ Trade name and manufacturer not verified   § Trade name without identified manufacturer

i.e., coatings, PVC, PS, polyolefins, ABS. [Am. Cyanamid]

Cyastat® SN. Stearamidopropyl dimethyl-2-hydroxyethyl ammonium nitrate, IPA; antistatic agent for polymers; used for plastics, surface coatings, paper, glass, and other materials; dispersant in coatings. [Am. Cyanamid]

Cyastat® SP. Stearamidopropyl dimethyl-2-hydroxyethyl ammonium dihydrogen phosphate, IPA.; antistatic agent with surface-active properties; for plastics, waxes, textiles, and glass; emulsifier, settling, dispersing, and re-wetting agent. [Am. Cyanamid]

Cybis. CAS 389-08-2; Nalidixic acid; antibacterial. [Sterling Drug Inc] *

Cybond WD-4517 and WD-4521. Proprietary two-component polyurethane adhesives of the solvent type. [Unichem] *

Cyclaine. CAS 532-76-3; Hexylcaine hydrochloride; used for topical anesthesia of intact mucous membranes. [Merck & Co Inc] †

cyclamic acid. CAS 100-88-9; (-cyclohexylsulfamic acid; hexamic acid) $C_6H_{11}NHSO_3H$; used as a non-nutritive sweetener, acidulant.

Cyclamycin. A proprietary preparation of triaceto-oleandromycin; an antibiotic. [Wyeth Laboratories] *

Cyclanon®. Wetting agents for textiles. [BASF plc]

Cyclanon® R. After-treatment agent for polyester dyeings and prints. [BASF AG]

Cyclapen-W. CAS 3485-14-1; Cyclacillin; antibacterial. [Wyeth Laboratories] *

Cyclatex. A proprietary cyclized rubber. [Hubron Rubber Chemicals] *

Cyclimorph Injection. A proprietary preparation of cyclizine tartrate and morphine tartrate; an analgesic for treatment of severe pain. [The Wellcome Foundation Ltd]

Cyclite. A proprietary cyclized rubber. [Durham Raw Materials] *

Cyclo. Surfactant for cosmetics, toiletries, pharmaceutical, processing, agricultural and other industries. [Baxenden Chemicals Ltd]

cycloamylose. See cyclodextrin.

β-cycloamylose. See cyclodextrin.

Cyclochem. A fatty acid ester and emulsifying and wetting agent. [Rhone-Poulenc Surf.] *

Cyclocort. CAS 51022-69-6; Amcinonide; glucocorticoid. [Lederle Laboratories USA] *

cyclodextrin. (cycloamylose; β-cycloamylose; cyclomaltoheptaose) Cyclic polysaccharide comprised of six to eight glucopyranose units; CAS 7585-39-9; EINECS 231-493-2. [Am. Maize Prods.; Janssen Chimica; Pfanstiehl Labs; U.S. Biochemical]

Cyclo-Flo. Dispersant blend; fuel oil additive designed to remove and prevent formation of sludge and carbonaceous deposit when using heavy reisudal fuel oil. [Taiwan Surf.]

Cyclofor. A paint additive. [Rhone-Poulenc Surf.] *

Cyclogol. Low foaming detergent, rinse aid; for machine dishwashing, soluble bath oils. [Witco Chemical Ltd] *

Cyclogyl. CAS 5870-29-1; Cyclopentolate hydrochloride; anticholinergic. [Alcon Laboratories Inc]

cyclohexane. (hexahydrobenzene; hexamethylene; hexanaphthene) C_6H_{12}; CAS 110-82-7; EINECS 203-806-2; Manufacture of nylon; solvent for cellulose ethers, fats, oils, waxes; paint and varnish remover; glass substitutes; in analytic chemistry; chemical intermediate; in

fungicidal formulations. [Exxon BV; Phillips 66; Texaco]

1,2-cyclohexanediamine. See 1,2-diaminocyclohexane.

cyclohexanol. (hexahydrophenol; hexalin) $C_6H_{11}OH$; CAS 108-93-0; EINECS 203-630-6; Soap making to incorporate solvents and phenolic insecticides; source of adipic acid for nylon, textile finishing; solvent for alkyd and phenolic resins and rubber. [Allied-Signal; BASF; UCB SA]

cyclohexanone. (ketohexamethylene; pimelic ketone) $CO(CH_2)_4CH_2$; CAS 108-94-1; EINECS 203-631-1; Used as paint and varnish remover; solvent for cellulose acetate, nitrocellulose, natural resins, vinyl resins, rubber, waxes, fats; in production of adipic acid for nylon, cyclohexanone resins. [Aldrich; Allied-Signal; BASF; DSM BV; Union Carbide]

cyclohexatriene. See benzene.

cyclohexylamine. (aminocyclohexane) $C_6H_{13}N$; CAS 108-91-8; EINECS 203-629-0; Boiler water treatment, rubber accelerator, intermediate in organic synthesis. [Air Prods.; Atochem N. Am.; BASF; PMC Specialties]

cyclohexyl chloride. (chlorocyclohexane) $C_6H_{11}Cl$; CAS 542-18-7; EINECS 208-806-6. [Hüls AG; Janssen Chimica]

Cyclolac. ABS polymers; for injection molding, sheet extrusions. [GE Plastics ABS Ltd] *

Cyclolube® 62. Naphthenic oil distillate; process oils offering light color, low pour pts. for general-purpose compounding of elastomers, e.g., neoprene, SBR, isoprene, EDPM, butyl, and natural rubber. [Witco/Golden Bear]

Cyclolube® 85. Low aromatic naphthenic oil; process oil offering excellent color and heat stability for rubber compounding, esp. EPDM and butyl; also for SBR, isoprene, and natural rubber. [Witco/Golden Bear]

Cyclolube® NN-1. Naphthenic oil distillate; process oilsoffering light color, low pour pts. for general-purpose compounding of elastomers, e.g., neoprene, SBR, isoprene, EDPM, butyl, and natural rubber. [Witco/Golden Bear]

cyclomaltoheptaose. See cyclodextrin.

Cyclomatic. Blend of anionic detergents; used for degreasing of synthetic fibers under pressure. [Atochem UK/ Ceca]

Cyclomatic Dur. Wetting agent for use in hydrosulfite bleaching sol'ns. for textiles. [Ceca SA]

Cyclomide. A fatty acid alkanolamide. [Rhone-Poulenc Surf.] *

Cyclonal Sodium. A soluble hexabarbitone. [May & Baker Ltd] *

Cyclonette. Emulsifiable wax composition. [Rhone-Poulenc Surf.] *

Cyclonox. Cyclohexanone peroxide-based range of catalysts; for the ambient temperature curing of unsaturated polyester resins. [Akzo Chemie UK Ltd]

Cyclopar. CAS 64-75-5; Tetracycline hydrochloride; antiamebic; antibacterial; antirickettsial. [Parke-Davis] *

Cyclophos. A phosphate ester. [Rhone-Poulenc Surf.] *

Cyclopol. A detergent composition. [Rhone-Poulenc Surf.]*

Cyclo-Progynova 1 mg. 11 Tablets containing 1 mg estradiol valerate, 10 tablets containing 1 mg estradiol valerate and 250 mcg levonorgestrel. [Schering Health Care Ltd]

Cyclo-Prostin. Epoprostenol sodium; inhibitor. [Upjohn] *

Cyclops metal. A nickel-chromium-iron alloy, containing

18% nickel, 18% chromium, and the rest iron; resists corrosion.

Cyclorans. . They contain an alcohol of high boiling point emulsified with a potassium olein soap; wetting-out agents for textiles. §

Cyclorubbers. Thermoplastic products made by heating a mixture of rubber sheet with ~10% of its weight of an organic sulfonyl chloride or an organic sulfonic acid. These products resemble gutta-percha or can be made to resemble shellac according to treatment.

Cycloryl. A liquid surface active agent. [Rhone-Poulenc Surf.] *

Cycloryl 580 and 585N. Sodium lauryl sulfate, conforming to BP specification; powder or needle form; emulsification polymerization; emulsification. [Witco Chemical Ltd]*

Cyclosan. 4% Calomel dust. [May & Baker Ltd] *

Cycloserine Roche. Tuberculostatic drug. [Roche Products Ltd] *

Cyclospasmol. CAS 456-59-7; A proprietary preparation of cyclandelate; a vasodilator. [Brocades Pharma] †

Cycloteric. An amphoteric surface active agent. [Rhone-Poulenc Surf.] *

Cycloton®SCS. CAS 122-19-0; EINECS 204-527-9; Stearalkonium chloride and additives. [Rhone-Poulenc Surf.]

Cyclovertal. 3, 6-Dimethyl-3-cyclohexene-1-carbaldehyde; general fragrance raw material, green, fruity notes. [Henkel/Cospha]

Cycocel®. Clormequat chloride, choline chloride; improves resistance against lodging of oats, rye, and wheat [BASF AG]

Cycogan. CAS 999-81-5; Active ingredient: chlormequat; 2-chloroethyltrimethyl-ammonium chloride; versatile plant growth regulant widely used for the prevention of lodging in wheat. [Agan Chemical Manufacturers Ltd]

Cycolac®. ABS thermoplastic compounds. [GE Plastics Ltd]

Cycolac® CKM1. CAS 9003-56-9; ABS resin, flame-retardant; UL-recognized grade offering uv stability, high flow, excellent finished part aesthetics; for business machine, computer equip., and elec. applications. [GE Plastics]

Cycolac® DH. CAS 9003-56-9; ABS resin; med.-impact, high heat, modulus, and tensile grade for injection molding applications. [GE Plastics]

Cycolac® GPM4700. CAS 9003-56-9; ABS resin; general purpose grade with enhanced flow and more economical processing. [GE Plastics]

Cycolac®GPX2800. CAS 9003-56-9; ABS resin; extrusion grade resin for bathtub surrounds, recreational vehicle parts, high-abuse products (luggage, automotive components). [GE Plastics]

Cycolac®KCS. CAS 9003-56-9; ABS resin, flame-retardant; extrusion grade resin with superior thermal stability; ideal for thermoformed applications. [GE Plastics]

Cycolac® KJM. CAS 9003-56-9; ABS resin, flame-retardant; injection-molding polymer for many applications previously designed for flame retardant structural foam. [GE Plastics]

Cycolac® X-11. CAS 9003-56-9; ABS resin; high heat injection molding grade used in building products and interior automotive trim. [GE Plastics]

Cycolin® GCM1900, GCM2900. Chemically resistant G series; excellent processability, impact strength and heat resist. [GE Plastics]

Cycoloy®C1110. PC/ABS alloy; general purpose injection molding grade formulated to maintain impact and ductility below -30 C; provides high heat resist. and impact strength and still processes well into long flow lengths and complicated parts. [GE Plastics]

Cycoloy® C2800. PC/ABS alloy; formulated for flame retardance without chlorinated or brominated additives; superior flow for thin-wall sections. [GE Plastics]

Cycoloy® MC8100. PC/ABS alloy; designed for extrusion blow molding. [GE Plastics]

Cycom. Resin impregnated glass, carbon or aramid fabrics; advanced composite molded components for military and aerospace applications. [Fothergill Tygaflor Ltd] *

Cycom®MCG Fiber. Nickel-coated graphite fiber; for integrally heated tooling applications. [Am. Cyanamid]

Cydril. Cationic and anionic modified polyacrylamides. [Cyanamid BV]

Cy-Ex. Polyacrylate; used for oil field applications. [Cyanamid BV]

Cyfloc 6000. Modified polyamine; used for oil field applications. [Cyanamid BV]

Cyfol. A proprietary preparation of ferrous gluconate, folic acid and vitamin B_{12}; a hematinic. [Rybar Laboratories Ltd] *

Cyglas®. Thermoset polyester composites; molding compounds for industrial applications requiring elec. insulation, high strength, rigidity at elevated temps. and corrosion resist. [Am. Cyanamid] *

Cygna. Self-emulsifiable mineral oil based and water soluble synlube based products; fiber processing aids and lubricants. [Thomas Swan & Co Ltd] †

Cy-Guard. Modified polyacrylamide; used for oil field applications. [Cyanamid BV]

Cykelin. Drying oils for use in paints, varnishes, lacquers and enamels. [Reichhold]

Cyklokapron. Solution for injection, tablets, oral solution; antifibrinolytic agent. [KabiVitrum AB] *

Cylence. Parasiticide for the control of lice in cattle and sheep; also known as Synticol. [Bayer AG]

Cylert. CAS 2152-34-3; Pemoline; stimulant. [Abbott Laboratories]

Cylink ISOBU-M-AMD. N-(Isobutoxymethyl)acrylamide; used in the process industry. [Cyanamid BV]

Cylink M.B.A. N,N´-Methylenebisacrylamide; used in the process industry. [Cyanamid BV]

Cylink NBMA. N-(Butoxymethyl)acrylamide; used in the process industry. [Cyanamid BV]

Cylink NM-AMD. N-Methylolacrylmide; used in the process industry. [Cyanamid BV]

Cylok® GM. Cyanoacrylate adhesive; for high tensile strength bonds of metals and rubber to metal; bonds sl. dirty or oily metals. [Lord]

Cylok®R. Cyanoacrylate adhesive; splice adhesive for butt-and-mitre cut joints and o-ring repairs; adheres to natural and synthetic rubbers. [Lord]

Cymag. CAS 143-33-9; Poisonous gassing compound containing sodium cyanide; a rodent, rabbit and insect exterminator. [ICI Chem. & Polymers Ltd]

Cymbal brass. See brass.

Cymbal metal. A brass containing 78% copper and 22% zinc.

Cymbilide. Cyclamen aldehyde. [May & Baker Ltd] *

Cymbush. CAS 66841-24-5; Insecticide containing

---

‡ Trade name and manufacturer not verified   § Trade name without identified manufacturer

cypermethrin. [ICI Chem. & Polymers Ltd]

Cymogran. Phenylalanine-free casein hydrolysate. [Allen & Hanburys Ltd] *

Cymperator. CAS 66841-24-5; Insecticide containing cypermethrin. [ICI Chem. & Polymers Ltd]

Cymyl orange. An indicator. It is an azo dye obtained by combining Diazotized sulfonated amino-cymene with dimethyl-aniline.

Cynorex. A bright cyanide copper plating process. [Hanshaw Chemicals] ‡

Cyomin. CAS 68-19-9; Cyanocobalamin; vitamin [O'Neal, Jones & Feldman Pharmaceuticals] *

Cypan. Hydrolyzed polyacrylonitrile; used for oil field applications. [Cyanamid BV]

Cyperkill. CAS 66841-24-5; An emulsifiable concentrate containing 50 g cypermethrin per liter; a pyrethroid insecticide. [Mitchell Cotts Chemicals Ltd]

Cypersect. Insecticidal formulation. [Mitchell Cotts Chemicals Ltd]

Cypersect. CAS 66841-24-5; Emulsifiable concentrate containing 100 g cypermethrin per liter; a pyrethroid insecticide. [Barclay Chemicals UK]

Cypertox. CAS 66841-24-5; 100 g/l Cypermethrin; broadspectrum insecticide with many crop uses. [Farmers Crop Chemicals Ltd]

Cyprian turpentine. See Chian turpentine.

Cypromin. CAS 2829-19-8; Rolicyprine; antidepressant. [Schering Corp] *

Cypro Promoters. Polyamines; used in paper industry. [Cyanamid BV]

Cyprostat OP. CAS 427-51-0; 50 mg Cyproterone acetate [Schering Health Care Ltd]

Cyquest. Modified polyacrylamides; antiprecipitant for mining applications. [Cyanamid BV]

Cyracure. Ultra-violet curing agents. [Union Carbide; Union Carbide (UK) Ltd]

Cyrel®. For the chemical industry. [DuPont UK]

Cyren-A. Stilboestrol; a synthetic estrogen for implantation in carcinoma of the prostrate. [Bayer AG] †

cyrene. A resin used in adhesives, coatings, and molding compounds.

Cyrez 963. Hexamethoxymethylamine; adhesion promoter; used in rubber industry. [Cyanamid BV]

Cyrez 963/4 Powders. Hexamethoxymethylamine on silica/silicate carrier; adhesion promoter; used in rubber industry. [Cyanamid BV]

Cyrolite® G-20. Acrylic mutlipolymer; compound for extrusion, blow molding, vacuum forming, and injection molding applications where transparent material with good chem. resist. is required; ideal for food and medical packaging; FDA approved for food contact use. [Cyro Industries]

Cyrolite® G-20 Hiflo®. Acrylic multipolymer; molding/extrusion compound suited for hard-to-fill molds, small intricate moldings, as well as large-area moldings; FDA approved for food contact use. [Cyro Industries]

Cyrolon®UVP. PC sheet; weather-resist. sheet for glazing applications where long-term impact resist., high service temps., and optical clarity are required. [Cyro Industries]

Cyrolon®ZX. PC sheet; continuously manufacturing lightweight, rigid sheet for applications requiring excellent impact resist. or high service temps.; high optical quality, dimensional stability; used for safety glazing, machine guards, panels for vending machines, signs, RV windscreens and windows, displays. [Cyro Industries]

Cyrpon®. Tranquilizer. [Bayer AG]

cystamin. See hexamine.

L-cysteine. ((+)-2-amino-3-mercaptopropionic acid; α-amino-β-thiolpropionic acid) A nonessential amino acid; $HSCH_2CH(NH_2)COOH$; CAS 52-90-4; EINECS 200-158-2; biochemical and nutrition research, reducing agent in bread doughs. [Diamalt GmbH; Nippon Rikagakuyakuhin; Showa Denko]

cysteine hydrochloride. (L-cysteine hydrochloride anhydrous) $HSCH_2CH(NH_2)COOH.HCl$; CAS 52-89-1; EINECS 200-157-7. [Bretagne Chimie Fine SA; Degussa; EM Industries; R.W. Greeff; Nippon Rikagakuyakuhin; Penta Mfg.; Tanabe USA; U.S. Biochemical]

Cystemme. CAS 68-04-2; A proprietary preparation of sodium citrate as granules intended for dissolution in water before taking; to alleviate the symptons of cystitis. [Abbott Laboratories] *

cystine. (di(α-amino-β-thiolpropionic acid); β,β'-dithiobisalanine) $HOOCCH(NH_2)CH_2SSCH_2CH(NH_2)COOH$; CAS 56-89-3; EINECS 200-296-3; Biochemical and nutrition research, nutrient and dietary supplement. [Aldrich; Am. Biorganics; Bretagne Chimie Fine SA; Degussa; R.W. Greeff; Tanabe USA; U.S. Biochemical]

Cysto-Conray. CAS 13087-53-1; Iothalamate meglumine. diagnostic aid. [Mallinckrodt Inc] *

cystogen. See hexamine.

Cystografin. Diatrizoate meglumine; diagnostic aid. [Bristol-Myers Squibb Co Inc] *

Cystopurin. Urinary antiseptic tablets. [Fisons plc, Pharmaceuticals Div] *

Cystorelin. Gonadorelin acetate; luteinizing hormone-releasing factor acetate hydrate; gonad-stimulating principle. [Abbott Laboratories] *

Cystospaz. CAS 101-31-5; Hyoscyamine; anticholinergic. [Alcon Laboratories Inc] *

Cytacon. CAS 68-19-9; A proprietary preparation of cyanocobalamin; oral vitamin B_{12} preparations. [Duncan Flockhard Ltd]

Cytadren. CAS 125-84-8; Aminoglutethimide; adrenocortical suppressant; antineoplastic. [Ciba-Geigy Corp]

Cytamen. CAS 68-19-9; A proprietary preparation of cyanocobalamin Vitamin B_{12}. [Duncan Flockhard Ltd]

Cy-Temp. Modified polyacrylamide; used for oil field applications. [Cyanamid BV]

Cytomel. CAS 55-06-1; Liothyronine sodium; thyroid hormone. [SmithKline Beecham]

Cytop. Fluoropolymer; transparent, low ref. index, durable to uv light, good coating props.; as clad material for optics, electronics, chemical industry applications. [Asahi Glass] *

Cytosar. CAS 147-94-4; A proprietary preparation of cytarabine; a cytotoxic drug. [Upjohn Ltd] *

Cytosar-U. CAS 147-94-4; Cytarabine; antineoplastic; antiviral. [Upjohn]

Cytotec. CAS 59122-46-2; Misoprostol; anti-ulcerative. [G D Searle & Co]

Cytox 2050. Dodecylguanidine hydrochloride; quaternary ammonium biocide for water and oil field applications. [Cyanamid BV]

Cytoxan. See Endoxan. [Degussa AG] *

Cytrel. A proprietary tobacco substitute. [Hoechst-

Celanese] *

Cytro-Lane. CAS 950-10-7; Emulsifiable concentrate con-taining 250 g/l mephosfolan; used for control of damson-hop aphid in hops. [Cyanamid of Great Britain Ltd]

D

D1-SEA 210 Silicone. CAS 63394-02-5; Two-component silicone elastomeric adhesive; cures fast to a tough, durable, resilient silicone rubber at room temperature with primerless adhesion to many substrates; for deep sections, automotive headlamps, as gasketing material for clothes iron steam chambers. [GE Silicones]

D2T2. Dye diffusion thermal transfer process for printing. [Imperial Chemical Industries plc]

D12F. CAS 9005-25-8; Oxidized corn starch. [Grain Processing]

D-129. Propylene/hexene copolymer; adhesives, and sealants raw materials. [Eastman]

D-151. Propylene/ethylene copolymer; adhesives, and sealants raw materials. [Eastman]

D-201. CAS 1344-28-1; Activated alumina; abrasion-resist. desiccant and drying agent for use in pressure swing and other packaged dryers; also for drying org. liqs. and gases in industrial processes, and air in air conditioning systems; desiccant/adsorbent in reconditioning oils. [La Roche Chem.]

D-300 Conc. Electrolyte used to produce hard and durable architectural and engineering anodized surfaces. [Pilot]

D3770. CAS 541-02-6; Decamethylcyclopentasiloxane; as cleaning, polishing, and damping media; offers low toxicity, inertness. [Hüls Am.]

D3780. CAS 141-62-8; Decamethyltetrasiloxane; as cleaning, polishing and damping media; offers low toxicity, inertness. [Hüls Am.]

D6219.5. Dodecamethylpentasiloxane; as cleaning, polishing and damping media; offers low toxicity, inertness. [Hüls Am.]

D-7040. Silicone fluid; high purity diffusion pump fluid. [McGhan NuSil]

DAB. 3,3′-Diaminobenzidine; a proprietary intermediate for various high-temperature plastics, used to make polypyrones and polyquinoxalines. [Upjohn Ltd] *

Dabco®33-LV. CAS 280-57-9; 33% Triethylenediamine in dipropylene glycol; amine-based catalyst for PU high resiliency slabstock and molded foam, coating. [Air Prods./Polyurethanes]

Dabco® 120. Metal-based catalyst for PU flexible molded foam. [Air Prods./Polyurethanes]

Dabco® 1027, 1028. Catalysts for improved processing in microcellular polyurethane foams. [Air Prods./Polyurethanes]

Dabco®7928. Amine-based catalyst for flexible slabstock PU foam; optimized for conventional equip. [Air Prods./Polyurethanes]

Dabco®8154. Amine-based catalyst for improved processing of flexible slabstock PU foam; gives delayed action. [Air Prods./Polyurethanes]

Dabco®8264. Amine-based catalyst for improved processing of flexible slabstock PU foam. [Air Prods./Polyurethanes]

Dabco® B-16. Amine-based catalyst for flexible slabstock PU foam. [Air Prods./Polyurethanes]

Dabco® BDO. CAS 110-63-4; 1,4-Butanediol; curative, chain extender; provides reactive H-source in prepolymer production; used to provide hard segments in PU. [Air Prods./Polyurethanes]

Dabco®BL-11. Amine-based catalyst for flexible slabstock PU foam. [Air Prods./Polyurethanes]

Dabco®BL-17. Amine-based catalyst for improved processing of flexible slabstock PU foam; gives delayed action. [Air Prods./Polyurethanes]

Dabco®BL-22. Blowing catalyst for polyurethane; designed for use as co-catalyst with Dabco 33-LV, etc. [Air Prods./Polyurethanes]

Dabco® BLV. Amine-based catalyst for flexible slabstock PU foam; optimized preblend. [Air Prods./Polyurethanes]

Dabco® CS90. CAS 71-55-6; Trichloroethane; catalyst for chlorinated solv.-blown foams [Air Prods./Polyurethanes]

Dabco® Crystalline. CAS 280-57-9; Triethylenediamine; catalyst for PU coatings. [Air Prods./Polyurethanes]

Dabco® DC198. Silicone glycol copolymer; high-potency surfactant used in production of flexible slabstock polyurethane foam. [Air Prods./Polyurethanes]

Dabco® DC5043. Silicone glycol copolymer; surfactant for production of molded and slabstock high-resiliency polyurethane foam; nonhydrolyzable; broad processing latitude. [Air Prods./Polyurethanes]

Dabco®DC5125. Silicone glycol copolymer; wide-processing-latitude surfactant for production of conventional and flame-retarded flexible polyurethane slabstock foam systems. [Air Prods./Polyurethanes]

Dabco®DC5160. Silicone glycol copolymer; surfactant used in production of conventional and flame-retarded slabstock polyurethane foam. [Air Prods./Polyurethanes]

Dabco®DC5164. Silicone product; surfactant for production of difficult-to-stabilize high-resiliency molded PU foam formulations. [Air Prods./Polyurethanes]

Dabco®DC5169. Silicone product; surfactant for production of high-resiliency MDI-based PU molded foams. [Air Prods./Polyurethanes]

Dabco®DC5180. Surfactant for flexible slabstock PU foam. [Air Prods./Polyurethanes]

Dabco® DC5258. Surfactant; provides for max. airflow in

---

polyurethane foam without sacrificing surf. stability. [Air Prods./Polyurethanes]

Dabco®DC5365. Silicone surfactant; low-to-med. efficiency stabilizer for TDI/MDI flexible mold foams. [Air Prods./Polyurethanes]

Dabco® DC5425. Silicone surfactant; for flexible molded foams; med.-efficiency stabilizer for water-blown TDI-based molded foams. [Air Prods./Polyurethanes] *

Dabco® DC5450. Silicone surfactant; surfactant for water-blown MDI-based PU foams; provides enhanced bulk stability. [Air Prods./Polyurethanes]

Dabco® DC5885. Silicone surfactant; for flexible molded foams; med.-efficiency stabilizer for water-blown TDI-based molded foams. [Air Prods./Polyurethanes] *

Dabco®DC5890, DC5895. Surfactant for flexible slabstock PU foam processing. [Air Prods./Polyurethanes]

Dabco® DEOA-LF. Crosslinker for flexible slabstock PU foam processing. [Air Prods./Polyurethanes]

Dabco®DM9534, DM9793. Metal-based catalyst for flexible slabstock PU foam. [Air Prods./Polyurethanes]

Dabco®DMEA. Amine-based catalyst for flexible slabstock PU foam. [Air Prods./Polyurethanes]

Dabco® MC. catalyst for improved processing of flexible slabstock PU foam; for methylene chloride systems. [Air Prods./Polyurethanes]

Dabco®NCM, NEM, NMM. Amine-based catalyst for flexible slabstock PU foam. [Air Prods./Polyurethanes]

Dabco® T-9. Metal-based catalyst for PU flexible slabstock and molded foam. [Air Prods./Polyurethanes]

Dabco® T-10, T-11. Metal-based catalyst for flexible slabstock PU foam. [Air Prods./Polyurethanes]

Dabco® T-12. CAS 77-58-7; Dibutyltin dilaurate; metal-based catalyst for PU flexible molded foam, coatings. [Air Prods./Polyurethanes]

Dabco®T-95. Metal-based catalyst for flexible slabstock PU foam. [Air Prods./Polyurethanes]

Dabco® TL. Catalyst for improved processing of flexible slabstock PU foam; for methylene chloride systems. [Air Prods./Polyurethanes]

Dabco®X-542, X-543. Amine-based catalyst for PU flexible molded foam; provides fast cure. [Air Prods./Polyurethanes]

Dabco® X2-5357, X2-5367. Surfactants for reduced CFC and non-CFC blown rigid polyurethane foam systems. [Air Prods./Polyurethanes]

Dabco®XDM. Amine-based catalyst for PU flexible molded foam; for improved surface cure. [Air Prods./Polyurethanes]

Dabco® XF-C10-40. Amine-based catalyst for PU flexible molded foam; provides fast cure. [Air Prods./Polyurethanes]

Dabco®XF-F2002, XF-F2003. Amine catalyst; for all-water-blown MDI cold cure seating process in polyurethane production [Air Prods./Polyurethanes]

Dabinese. CAS 94-20-2; A proprietary preparation of chlorpropamide; an oral hypoglycemic agent. [Pfizer International] †

Daconil Turf. CAS 1897-45-6; Chlorothalonil; a fungicide for a wide range of agricultural crops. [ICI Agrochemicals Professional Products]

Dacospin 1735-A. PEG-30 castor oil glycerides; surfactant, lubricant for PP carpet backing giving excellent fiber-to-fiber lubricity and plasticizing properties. [Henkel/Textiles]

Dacospin LA-704. PEG-7/PPG-4 tridecyl alcohol; surfactant for textile use. [Henkel/Textiles]

Dacriose® Ophthalmic Irrigating Solution. An isotonic, buffered solution of purified water, sodium phosphate, potassium chloride, sodium hydroxide, edetate disodium, sodium chloride and benzalkonium chloride; for irrigating the eye to help relieve irritation by removing loose foreign material, air pollutants (smog or pollen) or chlorinated water. [Iolab Corp]

Dacron®. A polyethylene terephthalate fiber having high strength and low water absorption. [DuPont UK]

Dacthal. CAS 1861-32-1; Wettable powder containing 75% weight/weight chlorthal-dimethyl; an herbicide for use on ornamentals, turf, fruit and vegetables. [Fermenta ASC Europe Ltd]

Dactil. CAS 129-77-1; A proprietary preparation containing piperidolate hydrochloride; an antispasmodic. [MCP Pharmaceuticals] *

Dag 137. CAS 7782-42-5; Colloidal graphite in water; lubricant additive for dry gear, dry chain lubes, aerosols, machine oils, assembly lubes, thread lubes. [Acheson Colloids]

Dag 154. CAS 7782-42-5; Colloidal graphite in anhyd. IPA; lubricant additive for dry gear and chain lubes, aerosols, machine oils, assembly and thread lubes. [Acheson Colloids]

Dag 155. Colloidal graphite in trichloroethane; lubricant additive for dry gear and chain lubes, aerosols, machine oils, assembly and thread lubes. [Acheson Colloids]

Dag 197. CAS 7782-42-5; Colloidal graphite in polyglycol synthetic oil; lubricant additive for chain lubes. [Acheson Colloids]

Dag 243. Graphite/MoS$_2$ in 500 solv. refined paraffinic petrol. oil; lubricant additive for aerosols, thread lubes. [Acheson Colloids]

Dag 2412. Colloidal graphite in mineral spirits; lubricant additive for assembly and thread lubes. [Acheson Colloids]

Dagenite. A proprietary trade name for a bituminous asbestos-filled thermoplastic for accumulator cases. §

Dagger. Suspension concentrate containing 300 g imazamethabenz-methyl per liter; used for control of grass weeds in winter cereals. [Cyanamid of Great Britain Ltd]

Dahl's acids. Acid II, β-naphthylamine-4, 6-disulfonic acid. Acid III,β-naphthylamine-4,7-disulfonic acid.

Dahmenite A. An explosive containing ammonium nitrate, naphthalene, and potassium bichromate.

Daintex. A proprietary trade name for a wetting agent; contains miscible terpene alcohols. §

Dairene®. CAS 101-05-3; Anilazine; fungicide for the prevention of leaf and fruit diseases in vegetables, tobacco, coffee, cereals, berry fruit, and ornamental plants. [Bayer AG]

Dairin®. CAS 101-05-3; Anilazine; broad spectrum fungicide used for tobacco, potatoes, cereals and ornamentals. [Bayer AG]

Dairos. A dairy detergent. [Laporte Industries Ltd] *

Dairozon. A dairy sterilization agent. [Laporte Industries Ltd] *

Dairy Fly Spray. Synergized pyrethrins insecticide. [Ciba plc]

Dairy Flyspray. Mixture of γ-HCH, resmethrin and tetramethramethrin; controls flies in livestock houses.

---

‡ Trade name and manufacturer not verified　　　　　　§ Trade name without identified manufacturer

[Deosan Ltd]

Dakamballi starch. A starch prepared from the fruit of *Aldina insignis*, a tree of British Guiana.

Dakin's solution. CAS 7681-52-9; A mixture of hypochlorite and perborate of sodium, with small amounts of hypochlorous and boric acids; used as an antiseptic for wound treatments.

Daktacort. A proprietary preparation of miconazole nitrate and hydrocortisone; for inflamed and infected skin conditions. [Janssen Pharmaceutical Ltd] *

Daktarin. CAS 22832-87-7; A proprietary preparation of miconazole nitrate used as an antifungal agent. [Janssen Pharmaceutical Ltd] *

Dalacin C. CAS 18323-44-9; A proprietary preparation of clindamycin; an antibiotic. [Upjohn Ltd] *

Dalcaine. CAS 6108-05-0; Lidocaine hydrochloride; anesthetic. [O'Neal, Jones & Feldman Pharmaceuticals] *

Dalfratex. Flexible silica textiles. [The Chemical & Insulating Co Ltd]

Dalgan. CAS 53648-55-8; Dezocine; analgesic. [Wyeth Laboratories] *

Dalivit. Multivitamin preparations; indicated for the prevention and treatment of vitamin deficiency states. [Paines & Byrne Ltd] *

Dalmane. CAS 17617-23-1; A proprietary preparation of flurazepam; a benzodiazepine hypnotic. [Roche Laboratories; Roche Products Ltd]

Dalnate. CAS 27877-51-6; Tolindate; antifungal. [USV Pharmaceutical Corp] ‡

Dalpad. Coalescing agent assisting film formation of certain latexes; maintains that characteristic during formulation storage and resists tendency (of films) towards water sensitivity. [Dow UK]

Daltocel. Polyesters or polyethers for flexible foams. [ICI Chem. & Polymers Ltd]

Daltoflex. Polyurethane flexible foams. [ICI Chem. & Polymers Ltd]

Daltofoam. Rigid polyurethane foams. [Imperial Chemical Industries plc]

Daltogard. A polyurethane foam additive. [ICI Chem. & Polymers Ltd]

Daltolac. Polyurethane rigid foams. [ICI Chem. & Polymers Ltd]

Daltomold. Trademark for a range of plastic molding compounds. §

Daltoped. Polyurethane systems for shoe soling. [ICI Chem. & Polymers Ltd]

Daltorez. Polyester resins for adding to polyurethane foams and elastomers. [ICI Chem. & Polymers Ltd]

Daltorol. Polyester for printers rollers. [ICI Chem. & Polymers Ltd] †

Damar bronze. *See* Damascus bronze.

Damascenized steel. Steel made by repeatedly welding, drawing out, and doubling up a bar made of a mixture of steel and iron, the surface of which has been treated with an acid. The steel is left with a black coating of carbon, and the iron retains its metallic luster.

Damascus bronze. (Damar bronze). An alloy of 76% copper, 10.5% tin, and 12.5% lead; a white bearing metal.

Dama®810. Dioctyl/octyldecyl/didecyl methyl amines; intermediate for mfg. of quaternary ammonium compounds for biocides, textile chemicals, oil field chemicals, amine oxides, betaines, polyurethane foam catalysis, epoxy

curing agent; in fabric softeners, disinfectants, laundry detergents. [Ethyl Corp]

Dama®1010. CAS 7396-58-9; Didecyl methylamine; intermediate for mfg. of quaternary ammonium compounds for biocides, textile chemicals, oil field chemicals, amine oxides, betaines, polyurethane foam catalysis, epoxy curing agent; in fabric softeners, disinfectants, laundry detergents. [Ethyl Corp]

Damfin. CAS 62610-77-9; Emulsifiable concentrate containing 950 g/l methacrifos; for pest control in stored grain. [Ciba-Geigy Agrochemicals]

Damiana. The dried leaves of a Mexican plant, *Turnera diffusa*; a tonic.

Dammar Resin. CAS 9000-16-2; A resin obtained from *Hopea, Shorea,* and *Balanocarpus* species, mainly from Malaysia. The melting point usually varies from 90-200 C, acid value from 33-72, and the ash from 0.04-0.52%; used in varnishes, cellulosic lacquers, paper and textile coatings.

Damoil. Phytonomic oil; dormant and summer spray oil for use as a contact insecticide. [Draxel Chemical Company] ‡

Damox. Dialkylmethylamine oxides. [Ethyl Corp]

Danbar. A proprietary preparation of veratrum viride alkaloids; for scalp application. [Gerhardt Pharmaceuticals]*

Daneral. A proprietary preparation of pheniramine p-aminosalicylate. [Hoechst UK] *

Daneral-SA. CAS 132-20-7; A proprietary preparation of pheniramine maleate. [Hoechst UK] *

Danex. CAS 13463-41-7; Pyrithione zinc; antibacterial; antifungal; antiseborrheic. [Herbert Laboratories] *

Danex. CAS 52-68-6; Active ingredient: trichlorphon; organophosphate agricultural insecticide with a very broad range of activity. [Makhteshim Chemical Works Ltd]

Danocrine. Danazol; anterior pituitary suppressant. [Sterling Drug Inc] *

Danol. A proprietary preparation of danazol; used as a suppressant of gonadotrophins. [Winthrop Laboratories] *

Danol Diols. Propoxylated, ethoxylated esters. [Witco/Organics]

Dantafur. CAS 67-20-9; Nitrofurantoin; antibacterial. [Norwich Eaton Pharmaceuticals Inc] *

Danthron. CAS 117-10-2; 1,8-Dihydroxyanthraquinone.

Dantobrom®RW. Halogenated hydantoin; biocide for water treatment. [Lonza Inc]

Dantochlor®RW. Halogenated hydantoin; biocide for water treatment. [Lonza Inc]

Dantocol®DHE. CAS 26850-24-8; DEDM hydantoin; resin crosslinker in coatings and polymers; intermediate for epoxies, urethane resins, and antistatic lubricants for the textile and plastics industries. [Lonza Inc]

Dantoest®DHE DL. DEDM hydantoin dilaurate; fiber lubricant; intermediate for epoxies, urethane resins, and antistatic lubricants for the textiles industry. [Lonza Inc]

Dantogard®. MDM hydantoin and DMDM hydantoin; industrial preservative. [Lonza Inc]

Dantoin® BCDMH. CAS 126-06-7; 1-Bromo-3-chloro-5,5-dimethyl hydantoin; low temp. industrial bleach. [Lonza Inc]

Dantoin® DCDMH. CAS 118-52-5; 1,3-Dichloro-5,5-dimethyl hydantoin; intermediate for custom chemical synthesis, laundry bleach formulations, and automatic dishwashing compounds. [Lonza Inc]

---

* Trade name not verified as available   † Trade name verified as obsolete

Dantoin® GSD-550. CAS 126-06-7; 1-Bromo,3-chloro-5,5-dimethyl hydantoin; water treatment. [Lonza Inc]

Dantoin® MDMH. CAS 116-25-6; MDM hydantoin; intermediate for cosmetics and other applications; preservation and gelation agent. [Lonza Inc]

Dantosperse® DHE(15) DS. PEG-15 DEDM hydantoin distearate; fiber lubricant. [Lonza Inc]

Dantosperse® DHE (15) MO. PEG-15 DEDM hydantoin oleate; fiber lubricant; intermediate for epoxies, urethane resins, and antistatic lubricants for the textiles industry. [Lonza Inc]

Dantrium. Dantrolene sodium; relaxant. [Norwich Eaton Pharmaceuticals Inc]

Dantyl. A proprietary preparation of p-aminosalicylic acid phenylester, p-aminosalicylic acid and sucrose; antituberculous agent. [Leo Laboratories] *

Dantyl-Inah. A proprietary preparation of phenyl-p-aminosalicylic acid, p aminosalicylic acid, isoniazid and sucrose; antituberculous drug. [Leo Laboratories] *

Daonil. A proprietary preparation of glibenclamide; used in the treatment of late-onset diabetes. [Hoechst UK] *

Daotan. Polyurethane modified alkyd resins; used for air drying paints and varnishes. [Resinous Chemicals Ltd]

DAP. See diallyl phthalate.

daphnetin. 1,8-Dihydroxycoumarin. $C_{14}H_6O_2(OH)_2$; used in dyes and medicines.

Dapon 35. CAS 131-17-9; A trademark for diallyl phthalate; a molding material. [FMC UK] *

Dapon M. A trademark for diallyl isophthalate; a molding material. [FMC UK] *

Daponite Sheet. CAS 131-17-9; Diallyl phthalate resin; for decorative impregnated paper. [Sumitomo Bakelite Co Ltd] *

Dappol. Boiler and cooling water treatment. [Laporte Industries Ltd]

Daprisal. A proprietary preparation of dexamphetamine sulfate, amylobarbitone, aspirin and phenacetin. [SmithKline Beecham] *

Dapro-7. Visc. stabilizer, antigelling agent for solv.-thinned coatings. [Daniel Prods.]

Dapro 5005. Special drier, chelating catalyst replacing cobalt drier in modified polyurethanes and alkyds. [Daniel Prods.]

Dapro DF 880. Dispersion of metallic salt of fatty acid; foam suppressor for aq. coatings, inks, and adhesives. [Daniel Prods.]

Dapro DF 900. Dispersion of olefinic solids; foam suppressor for aq. coatings, inks and adhesives. [Daniel Prods.]

Dapro DF 1181. Silicone-modified dispersion of olefinic solids; foam suppressor for aq. coatings, inks, and adhesives. [Daniel Prods.]

Dapro S-65. Surfactant blend; interfacial tension modifier for minimizing crawling, fish-eyes and cratering in solv.-based coating systems. [Daniel Prods.]

Dapro U-99. Surfactant blend; interfacial tension modifier for minimizing crawling, fish-eyes and cratering in water- and solv.-based coating systems and inks. [Daniel Prods.]

Dapro W-77. Surfactant blend; interfacial tension modifier for minimizing crawling, fish-eyes and cratering in water-based coating systems. [Daniel Prods.]

Dapro Defoamer NA 1621. Blend of hydrocarbon liqs. and metallic soaps modified with silicone polymer; defoamer for nonaq. clear urethane wood finishes. [Daniel Prods.]

Dapsetyn. A chloramphenicol/dapsone veterinary preparation. [Allen & Hanburys Ltd] *

dapsone. CAS 80-08-0; Di-(4-aminophenyl)sulfone; diaphenylsulfone; Avlosulfon.

Dapsyvet. A chloramphenicol/dapsone veterinary preparation. [Allen & Hanburys Ltd] *

Daptazole. CAS 490-55-1; A proprietary preparation of amiphenazole; a central nervous stimulant. [Nicholas Laboratories Ltd] *

Daracide. Slimicide for use in the paper and pulp industry. [Grace Dearborn Ltd]

Daraclean. Cleaning materials. [Grace Dearborn Ltd]

Daraclor. A proprietary formulation of pyrimethamine and chloroquine; an antimalarial. [The Wellcome Foundation Ltd]

Daradefoam. Defoamers; used in paper and pulp industry. [Grace Dearborn Ltd]

Darafloc. Flocculants. [Grace Dearborn Ltd]

Daranide. CAS 120-97-8; Dichlorphenamide; potent carbonic anhydrase inhibitor for control of glaucoma. [Merck & Co Inc]

Daran® 229. PVDC emulsion; odor-free, high adhesion coating (paper and board); low heat seal temps., high barriers, non-blocking and excellent slip char. [W R Grace/Organics]

Daran® 8350. PVDC copolymer emulsion; for use as a water-borne laminating adhesive. [W R Grace/Organics]*

Daraprim Tablets. CAS 58-14-0; A proprietary formulation of pyrimethamine; for the chemoprophylaxis of malaria due to susceptible strains of plasmodia or conjoint with fast-acting schizonticides to initiate transmission control and suppressive cure; also for the treatment of toxoplasmosis used conjointly witha sulfonamide. [The Wellcome Foundation Ltd]

Darasperse. Dispersants; used in paper and pulp industry. [Grace Dearborn Ltd]

Daraspray. Dispersants; used in paper and pulp industry. [Grace Dearborn Ltd]

Daratak® 89L. Adhesive emulsion for formulation of remoistenable adhesives for envelopes, labels, stamps, tapes; also for textile warp sizing and adhesives for arts and crafts. [W R Grace/Organics]

Daratak® MX. Acrylic vinyl acetate terpolymer; adhesive polymer with fast grab and set; provides adhesion of vinyl films to wood as good as EVAs with tougher bonds. [W R Grace/Organics]

Daratak® RP2000. CAS 9003-20-7; Carboxylated polyvinyl acetate emulsion; adhesive base for paper/paper, paper/foil laminations, bottle labels, coated paperboard. [W R Grace/Organics]

Daratak® SP1011. CAS 9003-20-7; PVAc emulsion; highly water-resistant, borax-tolerant adhesive emulsion; can be formulated with borax to give fire-retardant adhesives and paints; also used in heat-sealing applications and in aluminum foil adhesives. [W R Grace/Organics]

Daratak® XB-3631. Vinylidene chloride-based adhesive emulsion; for use on difficult-to-adhere substrates incl. films, plastics, and vinyls. [W R Grace/Organics]

Darathane® WB-4000. Waterborne urethane prepolymer; dispersant; emulsifier; latex polymerization; yields cationic, cross linkable, emulsifier-free polymers on curing; waste water treatment. [W R Grace/Organics]

Darbid. Isopropamide iodide; anticholinergic. [SmithKline

---

‡ Trade name and manufacturer not verified   § Trade name without identified manufacturer

Beecham] *

D'Arcet's alloy. a) Consists of 50% bismuth, 25% lead, and 25% tin, melting-point 93 C; b) contains 50 parts bismuth, 25 parts lead, 25 parts tin, and 250 parts mercury.

Darcil. A proprietary preparation of potassium phenethicillin; an antibiotic. [Wyeth Laboratories] *

Darco. CAS 7440-44-0; Activated carbon, powdered and granular; for water treatment, glucose and sugar purification, chemical manufacture, gas and air treatment. [Norit NV] *

Darco. CAS 7440-44-0; A decolorizing and refining carbon; a substitute for bone char.

Darcoxin. See Norvasc. [Pfizer International]

Darex®110L. Nitrile latex; copolymer latex for paper and felt saturation; high delamination resist. and edge tear; used in tape and sandpaper bases; heat curable. [W R Grace/Organics]

Darex® 165L. CAS 9003-54-7; SAN latex; water-based latex for bonding and impregnation of fibers, webs, and films; forms stiff films by itself; produces flexible films by plasticizing with softer latices or compatible plasticizers e.g., the phthalyl glycollates. [W R Grace/Organics]

Darex®5281L. SB latex; low emulsifier latex providing clear, smooth, high gloss, water-resistant coatings with good resistance to blocking; dries rapidly, with little heat; gross water holdout after application (paper and paperboard substrates). [W R Grace/Organics] *

Darex®550L. Vinylidene chloride emulsion; for use in flame retardant adhesives; designed for adhesion to nonporous substrates (e.g., film-to-paper, foil-to-paper applications). [W R Grace/Organics]

Darex®636L. S/B latex; nonfilm forming latex used as box toe fabric saturant. [W R Grace/Organics]

Daricol. CAS 125-53-1; A proprietary preparation of oxyphencyclimine; an antispasmodic. [Pfizer International]

Daricon. CAS 125-53-1; A proprietary preparation of oxyphencyclimine; an antispasmodic. [Pfizer International]

Dariloid®. Alginates; dairy stabilizers and stabilizer/emulsifiers; thickener for food prep.; milk sol. at pasteurization temps. [Kelco]

Dariloid® 100. CAS 11138-66-2; Xanthan gum; used for sherbert, water ice, cottage cheese dressings and sour cream. [Kelco]

Dariloid®Q. Alginate; used for sour cream-based chip dip, cheese dip, cheese sauce, pie filling. [Kelco]

Dariloid®QH. Alginate; used for puddings, cheesecake mix, whipped toppings, bakery fillings, instant egg nog, milk shake, dessert dry mix. [Kelco]

Daritran. A proprietary preparation of oxyphencyclimine and meprobromate; an antispasmodic and tranquilizer. [Pfizer International]

Daritrax. A proprietary preparation of oxyphencyclimine and hydroxyzine; an antispasmodic and tranquilizer. [Pfizer International] †

Darmex. High technology lubricant; used for all types of industrial/automotive equipment. [Darmex Corp]

Darmex Plus. High technology lubricant; used for all types of industrial/automotive equipment. [Darmex Corp]

Darmycel Agarifume Smoke. CAS 52645-53-1; Permethrin; a pyrethroid insecticide. [Darmycel UK]

Dartalan. A proprietary preparation of thiopropazate dihydrochloride; a sedative. [G D Searle & Co Ltd] *

Dartex®. Fibers. [DuPont UK]

Darvan®. Nonionic dispersing agents and surfactants; used for dispersing materials in rubber, paint, ceramics, plastics, cosmetics, pharmaceuticals, agriculture and household products. [R T Vanderbilt Co Inc]

Darvan® No. 1. CAS 9084-06-4; Sodium polynaphthalene sulfonate; latex dispersant. [R.T. Vanderbilt]

Darvan® No. 2. CAS 8061-51-6; Sodium lignosulfonate; dispersing and emulsifying agent for rubber industry, esp. for zinc oxide, clays, and sulfur. [R.T. Vanderbilt]

Darvan® No. 404. CAS 8061-52-7; Calcium lignosulfonate; dispersant. [R.T. Vanderbilt] *

Darvan® SMO. Sodium salt of sulfated methyl oleate; dispersant; improves smoothness and gloss of dipped CR latex films. [R.T. Vanderbilt]

Darvisul. Sulfaquinoxaline/idveridine coccidiostat. [The Wellcome Foundation Ltd] *

Darvon. CAS 550-83-4; Propoxyphene hydrochloride; analgesic. [Eli Lilly & Co]

Darvon-N. CAS 26570-10-5; Propoxyphene napsylate; analgesic. [Eli Lilly & Co]

Dasag. See Surophosphate.

Dasanit®. Granular insecticide; used for treatment of biting insects and nematodes. [Bayer AG]

Dastar. CAS 57-88-5; Cholesterol. [Croda Chem. Ltd]

Datac. Industrial adhesives for packaging, paper-converting and woodworking. [Datac Adhesives Ltd]

Datagel. Activated datem gel. [Croda Chem. Ltd] *

Datamuls. Diacetyl tartaric acid esters; increases volume, fermenting stability and gas retention; improved crumb structure; prolonged shelf-life; improved dough compatability with machines; for bread products. [Thomas Goldschmidt Ltd]

Datem. Diacetyl tartaric esters of monoglycerides; edible emulsifiers for use in lipsticks and similar products. [Croda Chem. Ltd]

Datril. CAS 103-90-2; Acetaminophen; analgesic; antipyretic. [Bristol-Myers Squibb Co Inc]

Daubond DC-9200-A/DC-9200-B. Modified polyurethane latex system (polyurethane and polyisocyanate resp.); for vacuum forming operations; good adhesion to vinyl-backed foam and various substrates to mfg. dashboards, door panels for automotive industry. [Daubert]

Daubond DC-9300. CAS 126-99-8; Polychloroprene latex; adhesive for vacuum forming operations; good adhesion to shoddy and ABS substrates that are vacuum formed to foam-backed vinyl used on automotive dashboards, door panels, etc. [Daubert]

Daudelin solder. An alloy of 65.6% tin, 12.2% zinc, 1% aluminum, 17.4% lead, 3.1% copper, and 0.4% phosphorus.

Davenol. A proprietary preparation containing carbinoxamine maleate, ephedrine hydrochloride and pholcodine; a cough linctus. [Wyeth Laboratories] *

Davey's gray. A pigment prepared from siliceous earths; used principally in mixtures with other colors to reduce tones.

Davis metal. An alloy of 67% copper, 29% nickel, 2% iron, and 1.5% manganese.

Dawson bronze. An alloy of 83.9% copper, 15.9% tin, and traces of antimony, lead, iron, arsenic, and zinc.

Daxad®11. CAS 9084-06-4; Low molecular weight naphthalene sulfonate formaldehyde condensate, sodium salt; dispersant for pigments in aq. media; used in agricultural

---

* Trade name not verified as available			† Trade name verified as obsolete

chemicals, mastics, caulks, sealants, pigment slurries and disps. [W R Grace/Organics]

Daxad® 19L-40. High molecular weight naphthalene sulfonate formaldehyde condensate, sodium/potassium salt, in solution; cold weather-stable dispersant, water-reducing agent, and visc.-reducer for high-solids slurries like concrete, cement, gypsum, lime, coal. [W R Grace/Organics] *

Daxad® 23. Sodium salts of polymerized substituted benzoid alkyl sulfonic acids; dispersant for agricultural chemicals, concrete admixtures, dyes, high-strength cements, linoleum pastes, pigment slurries and disps., and water treatment chemicals. [W R Grace/Organics]

Daxad®30-30. Sodium polymethacrylate solution; dispersant esp. for pigments in aq. sol'ns.; used in paint formulations, emulsion polymerization, water treatment (as a scale control agent for boiler systems), in large particle suspensions. [W R Grace/Organics]

Daxad® 37LN10-35. CAS 9003-04-7; Sodium polyacrylate solution; dispersant for rapid dispersion of solid materials (esp. clays) in aq. systems; dispersant, deflocculant for pigment mfg., paper coating, latex paints, oil well drilling muds; antiredeposition agent in cleaning formulas; stable over wide pH range. [W R Grace/Organics]

Daxad®CP-2. Cationic polyelectrolyte; dispersant for fillers, pigments, other additives; precipitates wetting agents, detergent soaps, emulsifiers, and dispersants out of industrial waters; coagulates latexes; flocculates kaolin clay disps.; breaks o/w and w/o emulsions. [W R Grace/Organics] *

Daxid. CAS 37661-08-8; A proprietary preparation of bacampicillin hydrochloride; an antibiotic. [Pfizer International] †

Daxiren. See Cardura. [Pfizer International]

Dazide. CAS 1596-84-5; Water soluble powder containing 85% daminozide; a plant growth regulator. [Fine Agrochemicals Ltd]

Dazomet. Granules containing 98-99% dazomet; soil fumigant. [DowElanco Ltd]

DB-1 Defoamer. Organo-modified silicone polymer, silica, and waxes; defoamer for emulsifiable oils (metalworking lubricant sol. oils), paints, water treatment, coolants, agricultural sprays, rolling oils, coatings. [Genesee Polymers]

DB-19 Antifoam Compd. Silicone compound with water carrier; antifoam for aq. systems, hot or cold foaming systems, alkaline systems, industrial processing, chemical processing, cleaning products, paints, paper and latex processing, metalworking. [Genesee Polymers]

D.B.A. See Accelerator D.B.A. §

DBA Accelerator. Dibenzylamine and mono-benzylamine blend; a nonstaining and nondiscoloring activator for CPB, other xanthates or carbon disulfide; used in natural, synthetic and nitrile rubbers. [Uniroyal] *

DBM. See dibutyl maleate.

DB Oil. Castor oil; for applications requiring max. purity with min. acidity and moisture, e.g., sonar fluid, liq. dielec., urethane reactions. [CasChem]

DBP. CAS 84-74-2; Dibutyl phthalate; a plasticizer for vinyl and other plastics.

DBPC. See BHT.

DBPC. Di-t-butyl-para-cresol; a proprietary antioxidant. [Koppers Co Inc] †

DBU. See diazabicycloundecene.

DC 150. Chemical catalyst system; for decorative bright chromium plating (using low chromic acid concentrations). [Harshaw Chemicals Ltd] *

DC700. Chemical catalyst system; for decorative bright chromium plating. [Harshaw Chemicals Ltd] *

DCC. See dicyclohexyl carbodiimide.

DC Cristobalite. Cristobalite containing investment material for precious metals; dental preparation. [Bayer AG]

DCHP. See dicyclohexyl phthalate.

DCI-3. Corrosion inhibitor. [DuPont UK]

DCP. CAS 131-15-7; Dicapryl phthalate; a plasticizer for vinyl plastics and cellulosic resins.

DCP-0. See calcium phosphate, dibasic.

DDAVP. Desmopressin acetate; antidiuretic. [USV Pharmaceutical Corp] ‡

DDBS 100. Alkylbenzene sulfonic acid, branched; detergent intermediate. [Zohar Detergent Factory]

D.D.D. Dimethylamine dimethyldithiocarbamate; used as an accelerator for rubber vulcanization.

DDM. See n-dodecyl mercaptan.

4,4′-DDS. See 4,4′-diaminodiphenyl sulfone.

DDT. CAS 50-29-3; Dichlorodiphenyltrichlorethane; a powerful insecticide. *

DEA. See diethanolamine.

De-Acidite. An anion exchange material. [The Permutit Co] *

Dead Sea Bath Salts. Natural hygroscopic materials found in the Dead Sea; for therapeutic baths. [Int'l. Sourcing]

Deadline. Range of pesticides for public health use. [Rentokil Ltd]

dead silver. (frosted silver). Silver, whitened by heating in air, and immersed in dilute sulfuric acid.

Deanase. A proprietary preparation of desoxyribonuclease; for treatment of ulcers, bruises and abscess. [Consolidated Chemicals Ltd] *

Deanase D.C. A proprietary preparation of delta chymotrypsin; digestive enzyme. [Consolidated Chemicals Ltd]*

deanol. See dimethylethanolamine.

Deanox. CAS 1309-37-1; Iron oxides. [Deanshanger Oxides Ltd] *

Deapril-ST. Ergoloid mesylates; cognition adjuvant. [Mead Johnson & Co] *

Dearcide. Biocides used for cooling waters. [Grace Dearborn Ltd]

DEAS Base. Fatty alkanolamide; base for textile softeners; finished products yield a dry, hard, sl. waxy hand. [Clark]

Deasol. High boiling aromatic solvent used for fuel treatment. [Grace Dearborn Ltd]

Debenal®. Geriatric drug for dogs and cats; veterinary medicine. [Bayer AG]

Debendox. A proprietary preparation of dicyclomine hydrochloride, doxylamine succinate and pyridoxine hydrochloride; antiemetic. [Richardson-Vicks Inc] *

Debron 711. CAS 9016-75-5; A proprietary coating compound manufactured from polyphenylene sulfide. [de Beers Laboratories Inc] ‡

Debut®. For the agriculture industry. [DuPont UK]

Decaderm. CAS 50-02-2; Dexamethasone; glucocorticoid. [Merck & Co Inc]

Decadex. Water-based elastomeric weatherproofing compound applied by brush or spray; for roof refurbishment, anticarbonation/all application external coating in range of colors. [Liquid Plastics Ltd]

---

‡ Trade name and manufacturer not verified   § Trade name without identified manufacturer

Decadron. CAS 50-02-2; Dexamethasone; used for certain endocrine and nonendocrine disorders responsive to corticosteroid therapy. [Merck & Co Inc]

Decadron-LA. CAS 1177-87-3; Dexamethasone acetate; adrenocortical steroid. [Merck & Co Inc] *

Decadronal. CAS 1177-87-3; Dexamethasone acetate; parenteral therapy for sustained relief of certain endocrine and nonendocrine disorders responsive to corticosteroid therapy. [Merck & Co Inc]

Decadron Duofase. CAS 50-02-2; Dexamethasone; for prompt and sustained corticosteroid activity for the treatment of certain endocrine and nonendocrine disorders. [Merck & Co Inc] †

Decadron Shock-Pak®. CAS 2392-39-4; Dexamethasone sodium phosphate; for the adjunctive treatment of shock. [Merck & Co Inc]

Deca-Durabolin. CAS 360-70-3; Nandrolone decanoate; andogen. [Organon Inc]

Deca-Indocid. Dexamethasone and indomethacin; for the treatment of rheumatoid arthritis and certain nonarticular musculoskeletal disorders. [Merck & Co Inc] †

Decalex. A photographic developer. [May & Baker Ltd] *

Decalin. CAS 91-17-8; Decahydronaphthalene, $C_{10}H_{18}$, a paint and resin solvent; used as turpentine substitute. [DuPont] ‡

Decaltal®. Nonswelling org. acids and their salts; deliming and masking agents in chrome tanning. [BASF AG]

decamphorized oil of turpentine. (oxidized oil of turpentine). The residue from the manufacture of camphor; consists mainly of dipentene.

decanedioic acid. See sebacic acid.

1-decanol. See n-decyl alcohol.

Decanox-F. Decanoyl peroxide; initator for bulk, solution, and suspension polymerization, curing elastomers, and high-temp. cure of polyester resins. [Atochem]

decanoyl chloride. (capric acid chloride) $CH_3(CH_2)_8COCl$; CAS 112-13-0; EINECS 203-938-0; Intermediate, polymerization initiator. [Atochem N. Am.; Janssen Chimica]

Decap. CAS 67-68-5; DMSO-based; solv. for depotting, deflashing, decapsulation of epoxy castings, transfer moldings; used hot (150 C); nonselective. [Dynaloy]

Decapryn Succinate. CAS 562-10-7; Doxylamine succinate; antihistaminic. [Merrell Dow Pharmaceuticals Inc]*

Decaspray®. CAS 50-02-2; Dexamethasone; a topical spray for corticosteroid-responsive skin and eye conditions. [Merck & Co Inc]

Deccox. Coccidiostat for sheep. [RMB Animal Health Ltd]

Deccox. CAS 18507-89-6; Decoquinate. [May & Baker Ltd]

Decelox. CAS 1314-13-2; Zinc oxide, modified to give slow rubber cure. [Durham Chemicals Ltd] *

Dechlorane A-O. CAS 1327-33-9; Antimony oxide; contains halogens; a proprietary synergistic agent for fire-retardant plastics. [Kingsley & Keith Chemical Corp] *

Dechlorane®Plus 25. CAS 13560-89-9; Chlorine-containing cycloaliphatic compound; 1, 2, 3, 4, 7, 8, 9, 10, 13, 13, 14, 14-dodecachloro-1,4,4α,5,6,6α,7,10,10α, 11, 12, 12α - dodecahydro-1, 4:7,10-dimethanodibenzo (a,e) cyclooctene; flame retardant in polymer systems (thermoplastics, thermosets, and elastomers); usually combined with antimony oxide as a synergist. [OxyChem]

Decholin. CAS 81-23-2; Dehydrocholic acid; choleretic. [Miles Pharmaceuticals] *

Decimate. Suspension concentrate containing 225 g chlorthal-dimethyl and 216 g propachlor per liter; an herbicide for use on brassicas and onions. [Fermenta ASC Europe Ltd]

Decis. CAS 52820-00-5; Emulsifiable concentrate containing 25 g deltamethrin per liter; a pyrethroid insecticide. [Hoechst UK]

Decisquick. Emulsifiable concentrate containing 25 g deltamethrin and 400 g heptenophos per liter; a systemic aphicide. [Hoechst UK]

Deck Seal-PD. Polysiloxane and methylmethacrylate in an aromatic and aliphatic solvent vehicle system; sealer for parking decks, bridge decks, ramps, stadiums etc. [Nova Chemical Inc] *

Declar®. Thermoplastic sheet with Tedlar®PVF film surface finish; formulated to meet aircraft industry flamm. requirements for interior parts. [DuPont; Du Pont (*UK) Ltd; Sheffield Plastics] *

Declinax. A proprietary preparation of dibrisoquine sulfate; an antihypertensive. [Roche Products Ltd] *

Declomycin. CAS 74-73-3; Demeclocycline hydrochloride; antibacterial. [Lederle Laboratories USA]

Decoart. Phenolic resin; for decorative laminates. [Sumitomo Bakelite Co Ltd] *

Deco Board P. Polyester resin; for decorative laminates. [Sumitomo Bakelite Co Ltd] *

Decol. A range of alkyl aryl sulfonates; all purpose liquid detergent concentrates, detergent bases for powdered cleansers, foaming agents for plaster board production, concrete foaming agents, wetting agents in textile processing and raw wool scouring detergent. [Harcros Australia] *

Decola, F, FG, MA, MF. Melamine resin; for decorative laminates. [Sumitomo Bakelite Co Ltd] *

Decola PFC. Melamine resin; for postform. [Sumitomo Bakelite Co Ltd] *

Decola Back Sheet. Phenolic resin; for back sheet. [Sumitomo Bakelite Co Ltd] *

Decola Excel. Melamine resin; for decorative laminates. [Sumitomo Bakelite Co Ltd] *

Decolamide. A range of alkanolamides; used as foam boost additives in detergents and shampoos, superfatting agents, opacifiers, thickeners, demulsifiers and emulsifiers. [Harcros Australia] *

Decola New Marine. Melamine resin; for decorative laminates. [Sumitomo Bakelite Co Ltd] *

Deconyl. A proprietary trade name for a weatherproof nylon coating. [Plastic Coatings Ltd] *

Deco Poly. Polyester resin; for decorative laminates. [Sumitomo Bakelite Co Ltd] *

Decopress. Antithrombotic stocking. [Bayer AG]

Decorpa. CAS 9000-30-0; A proprietary preparation of guar gum granules; used as an antiobesity agent. [Norgine Ltd]

Decothane® SP. An elastomeric high-build single-pack polyurethane roofing compound; forms a seamless, durable, uv resistant, waterproof barrier that is immediately resistant to rain and ponded water; applied by brush roller or airless spray; for refurbishment of all roofs-flat or pitched. [Liquid Plastics Ltd]

Decrolin®. CAS 24887-06-7; Zinc formaldehyde sulfoxylate; reducing agent after catalytic bleaching of fur skins [BASF AG; BASF plc]

n-decyl alcohol. (alcohol C-10; noncarbinol; 1-decanol) Fatty alcohol; $CH_3(CH_2)_8CH_2OH$; CAS 112-30-1; 68526-85-2; EINECS 203-956-9; in manufacture of plas-

---

* Trade name not verified as available   † Trade name verified as obsolete

ticizers, lubricants, petrol. additives, herbicides, surfactants, solvents; moderate antifoaming capacity. [Ethyl; M. Michel; Penta Mfg.; Schweizerhall; Vista]

decyl bromide. (1-bromodecane) $CH_3(CH_2)_9Br$; CAS 112-29-8; EINECS 203-955-3. [Ethyl; Great Lakes; Humphrey]

De De Tane. CAS 50-29-3; DDT products. [Murphy Chemical Co Ltd] †

Dedevap®. CAS 62-73-7; DDVP; insecticide used for spray control of sucking, biting and mining insects in greenhouses. [Bayer AG]

Dedico 5981. Dehydrated castor oil fatty acid; for production of fast drying alkyd resins and epoxy esters incl. styrenated and maleinized resins; cured coatings show excellent mech. and chem. props.; for automotive lacquers, car repair lacquers, can coatings. [Unichema]

deeline. See paraffin, liquid.

Deenax. CAS 128-37-0; 2,6-Di-t-butyl 4-methylphenol; a proprietary antioxidant [Exxon] *

Deep Feed. Liquid fertilizer. [Fisons plc, Horticulture Div] *

DEET. See diethyl toluamide.

Def®. CAS 78-48-8; Tribufos; defoliant for use on cotton. [Bayer AG]

Defencin. A proprietary preparation of isoxuprine resinate; a peripheral vasodilator. [Bristol-Myers Squibb Co Inc]

Defirust. A proprietary trade name for a rustless iron containing 12-15% chromium and 0.1% carbon. §

Deflamene. CAS 2825-60-7; Formocortal; glucocorticoid. [Farmitalia (Farmaceutici Italia)] *

Deflavit® ZA. Stripping agent for wool and nylon dyeings. [BASF AG]

Defoamer 1713. Proprietary blend; paper machine defoamer for fine paper. [Hart Chem. Ltd.]

Defoamer A 50. Tall oil acid, octoxynol-200; defoamer for phosphate mineral processing for the sulfuric acid process in mfg. of artificial fertilizer; biodeg. [Chem-Y GmbH]

Defoamer B 90. Octeth-3 carboxylic acid; defoamer for phosphate mineral processing for the sulfuric acid process in mfg. of artificial fertilizer and for the regeneration of silver in electroplating baths. [Chem-Y GmbH]

Defoamer C5B. Nonsilicone mineral oil-based; defoamer for effluents in pulp and paper industry; effective over wide pH range. [Hart Chem. Ltd.]

Defoamer DF-160-L. Defoamer for synthetic latex coating systems, esp. acrylic emulsions and binders; also controls foaming of sodium lauryl sulfate sol'ns. [Henkel/Textiles]

Defoamer KCE/S. Proprietary blend; water-based defoamer for screen room, bleachery, and effluent applications in pulp and paper industry. [Hart Chem. Ltd.]

Defoamer NXZ. Defoamer for use with latex coating and latex printing; also in atmospheric dyeing, printing and textile finishing processes. [Henkel/Textiles]

Defoamer S. Silicone-based defoamer; multipurpose defoaming agent. [Hart Chem. Ltd.]

Defoamer SF. Perfluoro alkyl phosphinate/phosphonate; surfactant for agricultural formulations. [Hoechst Celanese/Colorants & Surf.]

Defoamer WB Series. Proprietary blend; water-based defoamers for screen room and linerboard applications in pulp/paper industry. [Hart Chem. Ltd.]

Defol. CAS 7775-09-9; Sodium chlorate; cotton defoliant, dessicant for corn, grain sorghum, sunflowers, safflowers, rice, soybeans, chili peppers and guar beans. [Draxel Chemical Company] ‡

Defolia. Defoliant for hops. [Murphy Chemical Co Ltd] †

Defomax. Nonionic surfactant/wax emulsions; antifoaming agents for rubbers and plastics. [Toho Chem. Industry]

Degadur. Methacrylic resins; self-leveling coatings, mortar, as floor and wall coverings, road construction, ship deck coatings. [Degussa Ltd]

Degalan. Polymethyl methacrylate compounds for injection molding and extrusion; used in production of light- and weather-resistant sheets, tubes and specially shaped profiles as well as finished parts such as car light assemblies, graduated dials, writing and drawing equipment, household utensils, watch crystals, optical lenses, lamp covers. [Degussa Ltd]

Degalex. Aqueous pure acrylic emulsions; used as binder for the production of high grade emulsion paints and synthetic resin plasters. [Degussa]

Degament. Cold-curing, low-viscosity methacrylate resins for precast elements (concrete/polymer/composite); used for coverings on buildings, walk on elements, pictures, sanitary components, stationary sport installations, machine mountings, stone bondings and marble agglomerates. [Degussa Ltd]

Degapas. Aqueous solutions of acrylic acid and methacrylic acid polymers (S series) and their sodium salts (N series); used for the dispersion of organic solid materials; as a builder in phosphate free or low phosphate cleaners; as thickening agents; for the complexing of multivalent metal cations; for the sizing of polyamide fibers. [Degussa Ltd]

Degaplast. Acrylic resins and foam resins; used in production of cast arm and leg prostheses, apparatus, support corsets, night splints, bed casts; as filler resin for last tip extensions and corrections. [Degussa AG]

Degaroute. Acrylic road marking. [Degussa Ltd]

Degaser. A degassing agent for aluminum alloys. [Foseco (F.S.) Ltd]

Degest-2. CAS 550-29-2; Naphazoline hydrochloride; adrenergic. [Barnes-Hind Inc] *

Deglas. Extruded acrylic sheet; used to cover lighting fixtures, for light domes, illuminated signs, wash basins and bathtubs, roof canopies, etc. [Degussa Ltd]

Degopol. Components for polyurethane shoe soling systems. [ICI Chem. & Polymers Ltd] †

Degopur. Polyurethane systems. [ICI Chem. & Polymers Ltd] †

Degras. Woolgrease [Croda Inc.]

Degras Special. CAS 8020-84-6; Lanolin from crude wool grease; EP and slip agent for wire drawing compounds, slushing oils, cutting oils, lubricants; long term rust preventive; in leather stuffing greases; waterproofing agent. [Fanning]

Degreez. Methyl oleate, methyl stearate, methyl palmitate, methyl laurate, methyl myristate; degreasing agent for removing oil stains from metal, textiles, etc.; biodeg. [Alzo]

Degressal® SD 20. Polypropoxylate; defoamer for cleaners and detergents. [BASF AG]

Degressal® SNC. Modified phosphoric acid monoester; defoamer for detergents and cleaners. [BASF AG]

Degubond. A precious metal alloy; for dentistry and dental engineering. [Degussa AG] *

Degucast. A precious metal alloy; used for dentistry and

---

‡ Trade name and manufacturer not verified   § Trade name without identified manufacturer

dental engineering. [Degussa Ltd]

Degudent. A range of precious metal alloys; used for dentistry and dental engineering. [Degussa Ltd]

Deguflex. Dental impression material. [Degussa Ltd]

Deguform. Dental silicone duplicating material. [Degussa Ltd]

Degulor. A range of precious metal alloys; used for dentistry and dental engineering. [Degussa Ltd]

Deguphos. Precious metal heterogeneous catalysts. [Degussa Ltd]

Degupress. Dental acrylic denture base material. [Degussa Ltd]

Degusorb. CAS 7440-44-0; A range of activated carbons; used for decolorizing, cleaning and deodorizing in the chemical, pharmaceutical, beverage and food industry. [Degussa AG]

Degutron. Dental casting machine. [Degussa Ltd]

Deguvest. Plaster free material for the dental casting process. [Degussa Ltd]

D.E.H. Epoxy curing agents (hardeners) specifically designed to provide a desirable range of properties and handling characteristics for the hardening of epoxy resins. [Dow UK]

D.E.H. 20. CAS 111-40-0; Diethylenetriamine; aliphatic polyamine curing agent for epoxy resins; used for civil engineering, adhesives, grouts, casting and elec. encapsulation; D.E.H. 20 is a general purpose curing agent. [Dow]

D.E.H. 40. Accelerated dicyandiamide; epoxy curing agent for powd. coating formulations. [Dow]

D.E.H. 52. CAS 111-40-0; Diethylenetriamine adduct; adducted aliphatic polyamine curing agent for epoxy resins. [Dow]

De Han salt. A double salt of antimony trifluoride and ammonium sulfate, $SbF_3(NH_4)_2SO_4$. A mordant.

Dehesive®. Coating, separating and impregnating media for paper, leather, cardboard, asbestos, synthetic plastics, metal and textiles; all containing synthetic resins and silicones. [Wacker-Chemie GmbH]

Dehesive®920. Silicone; solv.-free silicone system with high curing rate for high-quality release coatings. [Wacker Silicones]

Dehscofix 904. Substituted phenol ethoxylated phosphate ester, TEA salt; emulsifier, dispersant for agricultural chemicals, leather processing. [Albright & Wilson Am.]

Dehscofix 911. Naphthalene sulfonic acid, formaldehyde condensate; surfactant for leather processing. [Albright & Wilson Am.]

Dehscofix 912. CAS 9084-06-4; Sodium naphthalene-formaldehyde sulfonate; surfactant for leather processing. [Albright & Wilson Am.]

Dehscofix 916. CAS 1322-93-6; Sodium diisopropyl naphthalene sulfonate; wetting agents for agricultural chemicals, leather processing. [Albright & Wilson Am.]

Dehscofix 917. CAS 25417-20-3; Sodium di-n-butyl naphthalene sulfonate; wetting agents for agricultural chemicals, leather processing. [Albright & Wilson Am.]

Dehscofix 918. CAS 120-18-3; β-Naphthalene sulfonic acid; surfactant for leather processing. [Albright & Wilson Am.]

Dehscofix 920. CAS 9084-06-4; Sodium naphthalene-formaldehyde sulfonate; wetting agents for agricultural chemicals; pigment dispersant for leather processing; dispersant/emulsifier for emulsion polymerization, rub-

ber. [Albright & Wilson Am.]

Dehscofix 923. Sodium dimethyl naphthalene-formaldehyde sulfonate; wetting agents for agricultural chemicals, leather processing. [Albright & Wilson Am.]

Dehscofix 929. Ammonium naphthalene-formaldehyde sulfonate; surfactant for leather processing. [Albright & Wilson Am.]

Dehscofix CO Series. Castor oil ethoxylates; emulsifier for agricultural chemicals. [Albright & Wilson Am.]

Dehscotex. Formulated auxiliaries for textiles and leather. [Albright & Wilson Am.]

Dehscotex BA Series. Formulated surfactant blends; bleaching agent for textile processing. [Albright & Wilson Am.]

Dehscotex DT 809. Formulated surfactant; detergent for leather processing. [Albright & Wilson Am.]

Dehscotex DT Series. Formulated surfactant blends; washing/scouring detergent for textile processing. [Albright & Wilson Am.]

Dehscotex DY Series. Formulated surfactant blends; dye bath dispersant and leveling agent for textile processing. [Albright & Wilson Am.]

Dehscotex FW Series. Formulated surfactant blends; felting auxiliary for textile processing. [Albright & Wilson Am.]

Dehscotex MC Series. Formulated surfactant blends; wetting agent for textile mercerizing and causticizing. [Albright & Wilson Am.]

Dehscotex SN Series. Formulated surfactants; softener for textile processing. [Albright & Wilson Am.]

Dehscotex VP-PF. Nonsilicone surfactant blends; fiber antistat for textile processing. [Albright & Wilson Am.]

Dehscotex WA Series. Formulated surfactant blends; wetting agent for textile processing. [Albright & Wilson Am.]

Dehscoxid 700 Series. Synthetic alcohol ethoxylates; wetting agents for agricultural chemicals, textiles, leather. [Albright & Wilson Am.]

Dehscoxid 730/740 Series. C13 alcohol ethoxylates; surfactant. [Albright & Wilson UK]

Dehybor. Borax from which the water of crystallization has been removed by heat (anhydrous borax); it is widely used in glass, vitreous enamel, ceramic glaze and metallurgical industries where borax has to be melted. [Borax Consolidated Ltd]

Dehydag Wax 14. CAS 112-72-1; EINECS 204-000-3; Myristyl alcohol; consistency factor for cosmetic and pharmaceutical o/w and w/o creams, ointments, emulsions, liniments, and sticks. [Henkel] *

Dehydag Wax 16. CAS 36653-82-4; EINECS 253-149-0; Cetyl alcohol; consistency factor for cosmetic and pharmaceutical o/w and w/o creams, ointments, emulsions, liniments, and sticks. [Henkel] *

Dehydag Wax 18. CAS 112-92-5; EINECS 204-017-6; Stearyl alcohol; consistency factor for cosmetic and pharmaceutical o/w and w/o creams, ointments, emulsions, liniments, and sticks. [Henkel] *

Dehydag Wax 22 (Lanette). CAS 661-19-8; EINECS 211-546-6; Behenyl alcohol; consistency factor for cosmetic and pharmaceutical o/w and w/o creams, ointments, emulsions, liniments, and sticks. [Henkel] *

Dehydag Wax E. CAS 59186-41-3; Sodium cetearyl sulfate; o/w emulsifier used in personal care products and powd. cleaners. [Henkel] *

Dehydag Wax N. Cetearyl alcohol and sodium cetearyl

sulfate; mfg. of o/w creams and liq. emulsions; solid. [Henkel] *

Dehydag Wax O. Cetearyl alcohol; consistency modifier used in personal care products and pharmaceuticals. [Henkel] *

Dehydag Wax SX. Cetearyl alcohol, sodium lauryl sulfate (90:10 ratio); surfactant used in mfg. of o/w emulsions for personal care products; SE base. [Henkel] *

Dehydag Wax W. Cetearyl alcohol and sodium lauryl sulfate (90:10 ratio); base for mfg. of ointments, creams and liniments; gran. [Henkel] *

Dehydol LS 2. Laureth-2; emulsifier, solubilizer for solvs., oils, bases for production of sulfates; raw material for dishwashing, cleansing agent and cold cleaners; in bath oils, waterless hand cleaners. [Henkel Canada; Henkel KGaA; Pulcra SA]

Dehydol LT 3. CAS 61791-13-7; Coceth-27, emulsifier, solubilizer for solvs., oils; base for production of ether sulfates; raw material for low-foaming detergents, dishwashing, cleansing agents, cold cleaners; superfatting agent. [Henkel KGaA; Pulcra SA]

Dehydol PCS 6. CAS 68439-49-6; Cetareth-6; emulsifier; intermediate raw material for mfg. of ethoxysulfates for use in detergents and industrial specialties. [Pulcra SA]

Dehydol PID 6. CAS 34938-91-8; Laureth-6; emulsifier, wetting agent for industrial, cosmetic, pharmaceutical applications, in high electrolyte concs. [Pulcra SA]

Dehydol PLS 1. Ethoxylated C12-14 fatty alcohol (1 EO); surfactant. [Pulcra SA]

Dehydol PTA 7. CAS 61791-28-4; Talloweth-7; surfactant. [Pulcra SA]

Dehydol TA 11. CAS 61791-28-4; Talloweth-11; raw material for detergents, toilet cubes; emulsifier for tech. applications [Henkel KGaA; Pulcra SA]

Dehydran 520. Fatty acid ester; defoamer for polymerization, esp. during monomer stripping in mfg. of suspension PVC; biodeg. [Henkel/Functional Prods.]

Dehydran 1019. Nonsilicone defoamer; defoamer for production and processing of polymer dispersions. [Henkel/Functional Prods.]

Dehydran 1293. Modified polysiloxane; defoamer for water-based industrial coatings, printing inks, wood lacquers. [Henkel/Functional Prods.]

Dehydran P 12. Nonsilicone defoamer; for monomer recovery in production of synthetic rubber; antifoaming agent for synthetic polymer disps., water-based paints, adhesives, glues. [Henkel/Functional Prods.]

Dehydrite®. A registered trade name for magnesium perchlorate trihydrate; a drying agent. §

dehydroacetic acid. (DHA; 3-acetyl-6-methyl-2H-pyran-2,4(3H)-dione, ion(1-); 2H-pyran-2,4(3H)-dione, 3-acetyl-6-methyl-; 3-acetyl-4-hydroxy-6-methyl-2-pyrone) Cylcic ketone; $C_8H_8O_4$; CAS 520-45-6; 771-03-9; EINECS 212-227-4; fungicide, bactericide, plasticizer, chemical intermediate, medicated toothpastes.

Dehydrocholin. CAS 81-23-2; A proprietary preparation of dehydrocholic acid; a laxative. [British Drug Houses] *

Dehydrophen 65. Alkylaryl polyglycol ether; detergent, wetting, dishwashing agent. [Henkel KGaA] *

Dehydrophen PNP 4. CAS 9016-45-9; 2311-27-5; Nonoxynol-6; detergent, dispersant, emulsifier, wetting agent for solv. cleaners; emulsifier/stabilizer for paint additives; as deicing fluid; sludge dispersant in petrol. products [Pulcra SA]

Dehydrophen POP 4. CAS 9002-93-1; Octoxynol-4; emulsifier, detergent, dispersant; used in wax-based washing formulations, emulsion cleaners. [Pulcra SA]

Dehymuls. Mixture of higher molecular esters mainly mixed ester of penthaerithrityl fatty acid ester; w/o type ointment and creams. [Henkel Chemicals Ltd] *

Dehymuls E. Sorbitan sesquioleate, pentaerythritol cocoate, stearyl citrate, beeswax, aluminum stearate; w/o emulsifier; suitable for cosmetics. [Henkel/Cospha; Henkel Canada; Henkel KGaA]

Dehymuls F. Microcrystalline wax, pentaerythritol cocoate, stearyl citrate, glyceryl oleate, aluminum stearate, propylene glycol; w/o emulsifier for creams and emulsions. [Henkel/Cospha; Henkel Canada; Henkel KGaA]

Dehymuls HRE 7. CAS 61788-85-0; PEG-7 hydrogenated castor oil; emulsifier for w/o emulsions for personal care products [Henkel KGaA]

Dehymuls K. Petrolatum, decyl oleate, sorbitan sesquioleate, pentaerythritol cocoate, stearyl citrate, beeswax, mineral oil, ceresin, aluminum stearate; SE base for mfg. of cosmetic and pharmaceutical preparations of the w/o type. [Henkel; Henkel KGaA]

Dehymuls SML. CAS 1338-39-2; EINECS 215-663-3; Sorbitan laurate; w/o emulsifier and coemulsifier for cosmetic and pharmaceutical applications [Henkel; Henkel KGaA]

Dehymuls SMO. CAS 1338-43-8; EINECS 215-665-4; Sorbitan oleate; emulsifier and coemulsifier for ointments and creams of the w/o type. [Henkel; Henkel KGaA]

Dehymuls SMS. CAS 1338-41-6; EINECS 215-664-9; Sorbitan stearate; emulsifier for w/o; used in the cosmetic and pharmaceutical industry. [Henkel; Henkel KGaA]

Dehymuls SSO. CAS 8007-43-0; EINECS 232-360-1; Sorbitan sesquioleate; w/o emulsifier and coemulsifier for waxes and oils. [Henkel; Henkel KGaA]

Dehypon Conc. Ethoxylated, propoxylated fatty alcohols and anionics; wetting agent, emulsifier and dispersant; for carpet cleaners; high detergency, low foam. [Henkel; Henkel Canada; Henkel KGaA]

Dehypon LS-24. Ethoxylated, propoxylated lauryl alcohol; detergent and wetting agent; low foaming. [Henkel Canada; Henkel KGaA]

Dehypon LT 054. Fatty alcohol polyglycol ether; surfactant for bottle washing. [Henkel KGaA]

Dehyquart A. CAS 112-02-7; EINECS 203-928-6; Cetrimonium chloride; emulsifier for emulsion polymerization; softener, conditioner, bactericide, fungicide, and odor inhibitor in personal care products; antistat for hair and fibers. [Henkel/Cospha; Henkel/Functional Prods.; Henkel Chemicals Ltd]

Dehyquart AU-36. Bis (acyloxyethyl) hydroxyethyl methylammonium methosulfate, 15% IPA; highly biodeg. surfactant; raw material for production of softeners; good rewetting. [Pulcra SA]

Dehyquart C Crystals. CAS 104-74-5; EINECS 203-232-2; Lauryl pyridinium chloride; emulsifier in creams and lotions; hair conditioners, skin creams; antistat for hair and fiber; bactericide, fungicide, corrosion inhibitor, sequestrant; conditioner used in personal care products [Henkel/Cospha; Henkel/Functional Prods.; Henkel Chemicals Ltd]

Dehyquart CDB. Cetyl dimethyl benzylammonium chloride in liquid form; cationic emulsifier used in hair cosmetics

---

and antistatics. [Henkel Chemicals Ltd] *

Dehyquart D. CAS 17342-21-1; Lauryl pyridinium bisulfate; germicide, wetting agent with anticorrosive effect, emulsion breaker; acid stable. [Henkel/Cospha; Henkel/Functional Prods.]

Dehyquart DAM. CAS 107-64-2; EINECS 203-508-2; Distearyl dimethyl ammonium chloride; emulsifier for plastics industry; conditioning component for hair care preparations; antistat. [Henkel/Functional Prods.; Henkel Chemicals Ltd]

Dehyquart E. Hydroxycetyl hydroxyethyl dimonium chloride; antistat, conditioner for hair care preparations; compatible with anionic systems. [Henkel/Cospha; Henkel Canada]

Dehyquart LDB. CAS 139-07-1; EINECS 203-351-5; Lauralkonium chloride; bactericide and fungicide for disinfectants; emulsifier; external antistat for plastics. [Henkel/Functional Prods.]

Dehyquart LT. CAS 112-00-5; EINECS 203-927-0; Laurtrimonium chloride; emulsifier for plastics industry, wetting agent, antistat, bactericide, demulsifier, deodorant, conditioning component for hair care products [Henkel/Functional Prods.; Henkel Chemicals Ltd]

Dehyquart SP. CAS 58069-11-7; Quaternium-52; emulsifier, conditioning, softening and antistatic agent used in personal care products; metal corrosion inhibitor. [Henkel/Cospha; Henkel Canada; Henkel Chemicals Ltd]

Dehyton®AB-30. Fatty amine derivative with betaine structure, in liquid form; for liquid shampoos, especially baby and special shampoos. [Henkel Chemicals Ltd] *

Dehyton® G. CAS 68647-53-0; Cocoamphocarboxyglycinate; surfactant for mild and conditioning shampoos, bath products, baby shampoos, skin cleansers, foam baths. [Henkel Canada; Henkel KGaA; Pulcra SA]

Dehyton® K. Betaine in liquid form; additive for shampoos. [Henkel Chemicals Ltd] *

Dehyton® KE. Cocamidopropyl betaine; surfactant base. [Pulcra SA]

Dehyton® PAB-30. CAS 683-10-3; Lauryl betaine; high foaming detergent for mild shampoos; solubilizer for lauryl sulfates in conc. shampoos; thickener. [Pulcra SA]

Dehyton®PG. CAS 68650-39-5; Disodium cocoamphodiacetate; foamer, wetting agent for shampoos, foam baths, cosmetics requiring high foaming, mildness. [Pulcra SA]

Dehyton®PK. CAS 61789-40-0; Cocamidopropyl betaine; high foaming, conditioning detergent for mild shampoos; solubilizer for lauryl sulfates in conc. shampoos; thickener. [Pulcra SA]

Dehyton®PLG. CAS 14350-96-0; Sodium lauroamphoacetate; mild, high foaming surfactant for shampoos, foam baths, cosmetics. [Pulcra SA]

Dekol®. Protective colloids for dyeing cellulose fibers and mixtures with vat, sulfur, and naphthol dyes. [BASF AG; BASF plc]

Dekol®N. Lignin sulfonate; protection and glossing agent for furs. [BASF AG]

Dekrysil. A proprietary preparation of 4, 6-dinitrocresol. §

Delac MOR. CAS 102-77-2; N-Oxydiethylene-benzo-thiazole-2-sulfenamide; the most delayed action accelerator offered by Uniroyal; activated by thiurams, dithiocarbamates, BIK, guanidines and aldehyde-amines; nondiscoloring and nonstaining to rubber stocks and materials in contact with them; used in tire treads, carcass,

mechanicals and wire jackets. [Uniroyal] *

Delac NS. CAS 95-31-8; N-t-Butyl-o-benzothiazole-2-sulfenamide; a delayed action accelerator very safe at processing temperatures but producing high modulus stocks at curing temperatures; activated by thiurams, dithiocarbamates, aldehyde amines, guanidines and BIK; nondiscoloring and nonstaining to rubber stocks and materials in contact with them; used in tire treads, carcass, mechanicals and wire jackets. [Uniroyal] *

Delac S. CAS 95-33-0; N-Cyclohexyl-2-benzothiazolesulfenamide; an all-purpose delayed action accelerator which combines superior scorch safety with shorter curing cycles; used in tire tread, carcass, camelback and mechanical goods. [Uniroyal] *

Delafila. Slate powder. [Delabole Slate]

Delaglas®A. PMMA extruded sheet. [Asahi Chem. Industry]

Delalot's alloy. An alloy containing 80% copper, 18% zinc, 2% manganese, and 1% calcium phosphate.

Delalutin. CAS 630-56-8; Hydroxyprogesterone caproate; progestin. [Bristol-Myers Squibb Co Inc] *

Delan-Col. CAS 3347-22-6; Suspension concentrate containing 600 g dithianon per liter; fungicide used for control of scab in fruit apples and pears. [ICI Chem. & Polymers Ltd]

Delanium. A proprietary trade name for carbon and graphite materials highly resistant to all chemicals except some oxidizing agents. [Powell Duffryn Quarries Ltd] *

Delaphos. Zinc orthophosphate; anticorrosive pigment for paints, particularly primer fomulations. [Pasminco Europe Ltd/ISC Alloys Div]

Delaprism®. PMMA prismatic sheet. [Asahi Chem. Industry]

Delatestryl. CAS 315-37-7; Testosterone enanthate; androgen. [Bristol-Myers Squibb Co Inc] *

Delaville. Zinc dust (minimum 95% metallic); zinc rich anticorrosive paints and reducing agent in chemical processes. [Pasminco Europe Ltd/ISC Alloys Div]

Delax. Nongripping laxative. [The Boots Co plc] *

Delchowyte. A decolorizing carbon, prepared from peat.

Delegol®, Delegol-T. Disinfectant on a phenol base; veterinary medicine. [Bayer AG]

Delestrogen. CAS 979-32-8; Estradiol valerate; estrogen. [Bristol-Myers Squibb Co Inc] *

Delf® HD Aerosol Adhesive. Synthetic rubber and resin; universal heavy duty adhesive. [D L Forster Ltd] *

Delf®MP Aerosol Adhesive. Synthetic rubber and resin in blend of chlorinated and ketonic hydrocarbon solvents; general purpose adhesive in aerosol form for use in the furniture industry. [D L Forster Ltd]

Delf® Clene. 2-Butoxyethanol and surfactants; general-purpose water-based cleaner for the furniture industry. [D L Forster Ltd]

Delf®Drape. Isopropyl alcohol and surfactants; crease and wrinkle remover for curtain and furniture fabrics. [D L Forster Ltd]

Delf® Fabric Protector. CAS 71-55-6; 1,1,1-Trichloroethane, surfactants, butane propane; stain protector for general fabrics. [D L Forster Ltd]

Delfloc® 50. Cationic polymer solution; retention aid and flocculant for the paper industry. [Hercules]

Delf®Silicone Aerosol. Silicone oil, aliphatic hydrocarbon, butane, propane; universal release and lubricating agent; reduces friction on a wide range of surfaces. [DL

---

* Trade name not verified as available   † Trade name verified as obsolete

Forster Ltd]

Delft blue. A pigment; a mixture of indigo and ultramarine.

Delhi rustless iron. A corrosion resisting alloy containing 18% chromium, 1.5% silicon, and not more than 0.08% carbon.

Delial®. Suntan products combining sun protection with skin care. [Bayer AG]

Delicron. Elastomeric precision impression material; for double-mix technique. [Bayer AG] †

Delight. Spray dried light density granular detergent builder containing disilicate. [Kemira Kemi Ltd]

Delimon. A proprietary preparation of 1-phenyl-2,3-dimethyl-4-(2-phenyl-3-methyl-hydroxazino-methyl)-pyrazolone-(5)-hydrochloride, paracetamol and salicylamide; an analgesic. [Consolidated Chemicals Ltd] *

Delinal. CAS 1420-03-7; Propenzolate hydrochloride; anticholinergic. [Merrell Dow Pharmaceuticals Inc] *

Delios®. C8-10 triglycerides; solv. for flavors and oil-sol. food additives; surface treating agent for dried fruits, confectionery, dietary foodstuffs. [Grünau]

Dellatol® BBS. CAS 3622-84-2; n-Butylbenzene sulfonamide; plasticizer for polyamide 6, 66, 11, and 12 and copolyamides; also for flexibilizing cellulose derivatives, especially flame-retardant cable coatings based on cellulose acetate and cellulose acetobutyrate. [Bayer AG]

Delmate. CAS 13698-49-2; Delmadinone acetate; progestin; anti-androgen; anti-estrogen. [Syntex Laboratories Inc] *

Delnet. A nonwoven fabric made fom high density polyethylene, polypropylene and polypropylene copolymers; nonadherent facing for surgical dressings,feminine hygiene products, disposable press cloth, breathable laminate, fusible adhesive, reinforcement for paper, plastic foam and other nonwoven fabrics. [Hercules] *

Delo. Lubricating oil. [Chevron] *

Deloxan. Organo functional polysiloxanes. [Degussa Ltd]

Deloxil. An emulsifiable concentrate containing 190 g bromoxynil and 190 g ioxynil per liter; a post-emergence contact herbicide for cereal crops. [Hoechst UK]

Delpet®. PMMA resin [Asahi Chem. Industry]

Delrin®. A proprietary trade name for a stiff strong engineering plastic of the acetal resin type; excellent fatigue resistance; used as a replacement for die cast parts in gears, bearings and housings. [DuPont UK]

Delrin® 100, 500. CAS 105-57-7; Acetal homopolymer resin; engineering thermoplastic with high strength and stiffness; also offers high fatigue endurance, natural lubricity, corrosion resistance, and resilience where recovery and toughness under wide range of temp. and humidity conditions required; 100 grade is most visc. in series; used in machinery, agricultural equipment, interior automotive door handles, clock mechanisms, ballcock valves, videocassettes, and other molding applications [DuPont]

Delrin® 100AF, 500AF. Acetal homopolymer resin with Teflon fibers; low friction/low wear grade; see Delrin 100. [DuPont]

Delrin® 100ST, 500T. CAS 105-57-7; Acetal resin; toughened grades. [DuPont]

Delrin® 107, 507. CAS 105-57-7; Acetal resin; uv-stabilized grades for improved weatherability. [DuPont]

Delrin® 150SA, 550SA. CAS 105-57-7; Acetal resin; low die deposit grades for improved extrusion. [DuPont]

Delrin® 570. CAS 105-57-7; Acetal homopolymer resin, glass fiber-filled; high stiffness grade; see Delrin 100. [DuPont]

Delrin® 900. CAS 105-57-7; Acetal homopolymer resin; high flow grade; see Delrin 100. [DuPont]

Delrin® AF Blend. CAS 105-57-7; Acetal homopolymer; thermoplastic for use in moving parts where low friction and long wear are important; for bushings, gears, slides, housings, guides, low friction components, elec. components. [DuPont; Polymer Corp.]

Delsene® 50 DF. CAS 83601-81-4; Carbendazim; systemic fungicide. [DuPont UK]

Delsene® M Flowable. CAS 83601-81-4, 12427-38-2; A suspension concentrate containing 50 g carbendazim and 320 g maneb per liter; systemic fungicide for cereals. [DuPont UK]

Delta. Two-piece can inks; for two-piece cans, side wall printing of pilfer proof closures and fast cure sheet fed metal containers, [Coates Coatings Ltd]

delta acid. (F-acid; Cassella's Acid F) β-Naphthylamine-7-sulfonic acid; $(C_{10}H_6(NH)_2SO_3H)$; used as an intermediate for azo dyes.

Deltacast. A range of specially designed release agents for the foundry industry, most usually the pressure die casting of aluminum. [Acheson Colloids]

Delta Cortef. CAS 53-03-2; A proprietary preparation of prednisone. [Upjohn Ltd] *

Delta Cortelan. A proprietary preparation for the acetate of prednisone. [Glaxo Laboratories] *

Deltacortone. CAS 53-03-2; Prednisone; for treatment of certain endocrine and nonendocrine disorders responsive to corticosteroid therapy. [Merck & Co Inc] †

Deltacortril. CAS 50-24-8; A proprietary preparation of prednisolone; a corticosteroid. [Pfizer International]

Delta-Dome. CAS 53-03-2; Prednisone; glucocorticoid. [Miles Pharmaceuticals] *

Deltaforge. A range of lubricating dispersions of graphite or other lubricating pigments in water; designed especially for lubrication of dies or for coating of billets in the metal forging process. [Acheson Colloids]

Delta-Genacort. CAS 50-24-8; A proprietary preparation of prednisolone; an anti-inflammatory agent. [Fisons plc, Pharmaceutical Div] *

Deltaglaze. A range of products designed to protect titanium, steel and super alloys during heating and during the forging process. [Acheson Colloids]

Deltalin. Ergocalciferol; vitamin. [Eli Lilly & Co]

Delta Metals®. The trademark for a variety of metals, metallic alloys, and metal articles. §

Deltasone. CAS 53-03-2; Prednisone; glucocorticoid. [Upjohn]

Deltastab. CAS 52-21-1; Proprietary preparations of prednisolone, e.g. the acetate. [The Boots Co plc]

Delta-Therm® NDT-402. Rigid polyurethane systems; for dispensing froth foams through pressurized equip. for sandwich panels, thin wall refrigeration, coolers/freezers, trucks. [Flexible Prods.]

Delta-Therm® NDT-300. Rigid polyurethane systems; for dispensing froth foams through pressurized equip. for water heaters. [Flexible Prods.]

Deltyl. A proprietary trade name for a plasticizer; it is a fatty acid ester. §

Delweve. Nonwoven fabric made from polypropylene by the film extrusion-embossing-orientation process; excellent

---

‡ Trade name and manufacturer not verified   § Trade name without identified manufacturer

backing material, skirt and cambric liners for furniture, reinforcement for films, plastic foam, needlepunched nonwovens and wall coverings, bale coverings. [Hercules] *

Dema. CAS 60-54-8; A proprietary brand of tetracycline. [USV Pharmaceutical Corp] ‡

Demavet. CAS 67-68-5; Dimethyl sulfoxide in aqueous solution. [Ciba plc] *

Demelloy. Soft solder and brazing alloys for semiconductor industry. [Degussa Ltd]

Demerol. CAS 50-13-5; Meperidine hydrochloride; analgesic. [Sterling Drug Inc] *

Demetox. CAS 919-86-8; Insecticide containing demeton-s-methyl. [ICI Chem. & Polymers Ltd]

Demix®7730. Catalytically disproportionated tall oil product; emulsifier for SBR production. [Arizona]

Demser. Metirosine; used for the treatment of pheo-chromocytoma. [Merck & Co Inc]

Demulen. A proprietary preparation of mestranol and ethynodiol diacetate; an oral contraceptive. [G D Searle & Co Ltd]

Demulen 50. A proprietary preparation of ethinylestradiol and ethynodiol diacetate; an oral contraceptive. [G D Searle & Co Ltd]

Demulsifier 3837. Polymer; demulsifier for metalworking applications; for use at alkaline pH and without the use of any other chemical. [Hoechst Celanese/Colorants & Surf.]

D.E.N. Epoxy novolac resins and solutions designed to provide high temperature service for epoxy-type applications; used for coatings and adhesives for abrasives. [Dow UK]

D.E.N. 431. Epoxy-novolac resin; used in high-performance adhesives, structural and elec. laminates, potting and molding compounds, coatings and castings for elevated temp. service, and filament wound pipe; used primarily in elec. applications and as an additive to conventional epoxies; low visc. [Dow]

denatonium benzoate NF. (N-[2-[(2,6-dimethylphenyl) amino]-2-oxoethyl]-N,N-diethylbenzenemathanaminium benzoate; benzyldiethyl [(2,6-xylylcarbomoyl) methyl]ammonium benzoate; lignocaine benzyl benzoate) Organic compound.; $C_{21}H_{29}N_2O \cdot C_7H_5O_2$; CAS 3734-33-6; EINECS 223-095-2; pharmaceutical aid (alcohol denaturant, flavor); added to toxic substances as a deterrent to ingestion. [Atomergic Chemetals; EM Industries; Macfarlan Smith Ltd]

DenClen. Denture cleanser (powder, tablets and liquid), for cleaning dentures. [Richardson-Vicks Inc] *

Dendrid. CAS 54-42-2; A proprietary preparation of idoxuridine; used as eye-drops. [Alcon Universal] *

Denesive. Coating, separating and impregnating media for paper, leather, cardboard, asbestos, synthetic plastics, metal and textiles; all containing synthetic resins and silicones. [Wacker-Chemie GmbH] *

De-Nol. A proprietary preparation of tripotassium dicitratobismuthate; used in the treatment of peptic ulcers. [Brocades Pharma]

De-NOx Catalyst MDN-100. Nitric acid plant exhaust gas; applicable temps. 300-500 C. [Mitsubishi Kasei] *

Denquel. Sensitive teeth toothpaste, to relieve the discomfort of sensitive teeth. [Richardson-Vicks Inc] *

Densil. Industrial biocides. [ICI Chem. & Polymers Ltd]

Densites. Mining explosives; contain ammonium nitrate,

sodium or potassium nitrate, and trinitrotoluene.

Densodrin®. Combination of special natural and synthetic oils; lightfast waterproof agents; also for vegetable-tanned, synthetic-tanned or retanned leather. [BASF AG; BASF plc]

Dentomat. Dental amalgamator. [Degussa Ltd]

Dentplus Special. CAS 7757-93-9; A proprietary trade name for dicalcium phosphate dihydrate; used as a thickening agent, cleaning agent, and carrier in toothpaste. [Hoechst UK] *

Denver clay. See bentonite.

Denver mud. See bentonite.

Denzox. CAS 1314-13-2; Dense zinc oxide; used for speciality ceramics. [Eagle Zinc]

Deodorant #4761-C, OS. Mixture of fragrance materials; heat-resistant deodorizer for residual odors in the finished product [Andrea Aromatics] *

Deoxidine. Metal treating compositions. [ICI Chem. & Polymers Ltd] †

Deoxidizing Tubes. Pure copper tubes containing a variety of deoxidizing agents for the treatment of copper and nickel alloys. [Foseco (F.S.) Ltd]

Deoxiphos 600. Liquid, phosphoric acid based cleaner; deoxidizer for aluminum and steel. [Invequimica & CIA SCA] *

Deoxylyte. Metal treating compositions. [ICI Chem. & Polymers Ltd] †

DEP. See diethyl phthalate.

Depakene. CAS 99-66-1; Valproic acid; anticonvulsant. [Abbott Laboratories]

Depakote. Divalproex sodium; anticonvulsant. [Abbott Laboratories]

depAndro 100 200. CAS 58-20-8; Testosterone cypionate; androgen. [O'Neal, Jones & Feldman Pharmaceuticals]*

Depat. O,O-Diethyl phosphoroamidothioate; an intermediate for phosphorus pesticides. [A/S Cheminova] *

Depen. See Trolovol. [Degussa AG] *

Depepsen. CAS 9010-01-9; Sodium amylosulfate; enzyme inhibitor. [G D Searle & Co] *

depGynogen. Estradiol cypionate; estrogen. [O'Neal, Jones & Feldman Pharmaceuticals] *

Dephosphex. Highly basic oxidizing supplementary flux for phosphorus removal from arc melted steel. [Foseco (F.S.) Ltd] †

Depixol. A proprietary preparation of flupenthixol decanoate; a neuroleptic. [Pfizer International]

Deplet. A proprietary preparation of potassium teclothiazide; a diuretic. [Nicholas Laboratories Ltd] *

Depmedalone 40, also Depmedalone 80. CAS 53-36-1; Methylprednisolone acetate; glucocorticoid. [O'Neal, Jones & Feldman Pharmaceuticals] *

Depocillin. A proprietary preparation of procaine penicillin; an antibiotic. [Brocades Pharma] †

Depo-Medrol. CAS 53-36-1; Methylprednisolone acetate; glucocorticoid. [Upjohn]

Depomedrone. CAS 53-36-1; A proprietary preparation of methyl prednisolone acetate. [Upjohn Ltd] *

Depo-Penicillin. CAS 6130-64-9; Penicillin G procaine; antibacterial. [Upjohn] *

Depo-Provera. CAS 71-58-9; A proprietary preparation containing medroxyprogesterone acetate. [Upjohn; Upjohn Ltd]

Depostat. A proprietary preparation of gestronal hexanoate in oily solution; used to treat benign prostatic hypertrophy

---

and endometrial carcinoma. [Schering Health Care Ltd]

Depot Glumorin. CAS 9001-01-8; A proprietary preparation of kallikrein bound to a high molecular weight steroid. [FBA Pharmaceuticals] ‡

Depot-Impletol. Pharmaceutical preparation; neural therapeutic and diagnostic agents. [Bayer AG]

Depot-Padutin®. A vasoactive enzyme of prolonged action. [Bayer AG]

Depovirin. CAS 58-20-8; Testosterone cypionate; androgen. [Hoechst-Roussel Pharmaceuticals Inc] *

depPredalone. CAS 52-21-1; Prednisolone acetate; glucocorticoid. [O'Neal, Jones & Feldman Pharmaceuticals] *

Depronal S.A. A proprietary preparation of dextropropoxyphene hydrochloride; an analgesic. [Warner] *

Depsoline. Textile auxiliary chemicals. [ICI Chem. & Polymers Ltd]

Dequacaine. A proprietary preparation containing dequalinium chloride and benzocaine; anesthetic antibacterial lozenges. [The Boots Co plc]

Dequadin. CAS 522-51-0; A proprietary preparation containing dequalinium chloride; antibacterial. [The Boots Co plc]

Dequalone. Dequadin and prednisolone. [Allen & Hanburys Ltd] *

Dequaspon. A proprietary preparation of gelatin sponge impregnated with dequadin; dental packing for hemorrhage. [Allen & Hanburys Ltd] *

Dequest. Phosphonates; used in scale prevention and corrosion inhibition for industrial water treatment and to control metal ions in aqueous systems. [Monsanto Co]

D.E.R. Epoxy resins; a range of solid, liquid, flexible and brominated thermosetting polymers; features adhesion, hardness, flexibility, toughness, dimensional stability, clarity and chemical resistance. [Dow UK]

D.E.R. 317. CAS 25928-94-3; Bisphenol-A epoxy resin; ideal for adhesive, casting, potting, encapsulation, and wet lay-up applications; after cure, yields highly crosslinked thermoset polymers; a high-visc., fast-reacting resin for adhesive applications [Dow]

D.E.R. 362. CAS 25928-94-3; Bisphenol A epoxy resin; low visc. liq. epoxy resin for applications requiring high filler loading, excellent wet-out, and high performance; crystallization resist.; no diluents or solvs.; used for adhesives, flooring, encapsulants, coatings, composites. [Dow]

D.E.R. 383. CAS 25928-94-3; Epoxy resin; liq. epoxy resin for reduced visc. and extended pot life of catalyzed formulations; for filament winding, potting and encapsulation, etc.; yields higher filler loadings, better process control and handling, and reduced resin/hardener waste. [Dow]

D.E.R. 511-A80. Brominated epoxy resin. [Dow]

D.E.R. 642U. CAS 25928-94-3; Epoxy resin; for powd. coatings. [Dow]

D.E.R. 732. CAS 25928-94-3; Epichlorohydrin-polyglycol reaction product; flexible epoxy resin; used as additives to base epoxy systems; increases flexibility in adhesives, construction and civil engineering applications (aggregates, seamless floors, machine grouts), industrial maintenance coatings, and structural prepegs. [Dow]

Deracyn. CAS 57938-82-6; Adinazolam mesylate; antidepressant. [Upjohn] *

Derakane®. Vinyl ester resins formulated for the reinforced plastics industry; used to make articles by molding,

spray-up and filament winding, in sheet molding compounds and in coatings. [Dow UK]

Derakane®411-35. Epoxy-based vinyl ester resin; thermoset; provides superior toughness and corrosion resist.; used in chem. processing industry, pulp and paper mills; FDA compliance. [Dow]

Derakane®411C-50. Vinyl ester resin; resin transfer molding grade. [Dow]

Derakane®470-45. A proprietary polyvinyl ester resin possessing good chemical resistance to chlorinated solvents. [Dow UK]

Derakane®510-40. A proprietary polyvinyl ester resin used in fire-retardant laminates; contains 20% bromine. [Dow UK]

Derakane®510B-700PAT. Vinyl ester resin; formulated for use in the military marine market for the production of high-quality lightweight navy craft. [Dow]

Derakane® 8084. Elastomer-modified vinyl ester resin; offers increased adhesive strength, superior resist. to abrasion and mech. stress, exceptional toughness, chem. resist.; for FRP applications [Dow]

Deraspan. Insulating panels using Styrofoam brand plastic foam as the core material and a variety of outside layers such as aluminum, plywood, etc; for cold storage warehouses, dairy coolers, cold rooms for food processing, etc. [Dow UK] *

Derifil. Chlorophyllin copper complex, obtained by replacing the methyl and phytyl ester groups with alkali and magnesium with copper; deodorant. [Rystan Company Inc]*

Deriphat 151C. Lauraminopropionic acid; wetting agent, detergent, emulsifier, corrosion inhibitor; personal care and hard surface cleaners; general-purpose surfactant; high foaming, substantive. [Henkel/Cospha; Henkel Canada]

Deriphat 154. CAS 61791-56-8; Disodium N-tallow-β iminodipropionate; detergent, solubilizer for personal care products, hard surface cleaning, textiles, emulsion polymerization; good substantivity. [Henkel/Cospha; Henkel Canada]

Deriphat 160. CAS 3655-00-3; EINECS 222-899-0; Disodium N-lauryl β-iminodipropionate; detergent, solubilizer, primary emulsifier used in org. and inorg. compounds; emulsion polymerization and stabilization; wetting agent; mild surfactant for hair and skin products [Henkel/Cospha; Henkel/Functional Prods.; Henkel Canada]

Deriphat 160C. CAS 26256-79-1; Sodium-N-lauryl beta-iminodipropionate; detergent, solubilizer, stabilizer; used in petrol. processing, emulsion polymerization, foaming cleaners, personal care products [Henkel/Cospha; Henkel/Functional Prods.; Henkel Canada]

Derma. Synthetic organic dyestuffs for use in the leather industry. [Sandoz Products Ltd]

Dermacort. CAS 50-23-7; Hydrocortisone; glucocorticoid. [Rowell Laboratories Inc] *

Dermaffine®. CAS 143-28-2; EINECS 205-597-3; Oleyl alcohol; raw material for cosmetics and pharmaceuticals. [Laserson & Sabetay]

Dermafill. Impregnating agents in aqueous solution; for leather. [Colour-Chem Ltd] *

Dermalac. Nitrocellulose lacquers and emulsions; for leather. [Colour-Chem Ltd] *

Dermalcare® 1673. Sodium laureth sulfate, disodium laureth sulfosuccinate, laureth-6 carboxylic acid, cocamidopropyl betaine, ammonium chloride; high performance base for prep. of ultra-mild face and skin cleanser formulations   [Rhone-Poulenc Surf.; Rhone-Poulenc France]

Dermalcare® GMS-165. Glyceryl stearate, PEG 100 stearate; emulsifier for o/w creams and lotions; high electrolyte tolerance. [Rhone-Poulenc Surf.]

Dermalcare® GMS/SE. Glyceryl stearate SE; self-emulsifying emulsifier for creams and lotions. [Rhone-Poulenc Surf.]

Dermalcare® NI. Cetearyl alcohol, ceteareth-20; broad tolerance emulsifier for o/w systems, visc. builder for cosmetic creams, lotions, ointments. [Rhone-Poulenc Surf.; Rhone-Poulenc UK; Rhone-Poulenc France]

Dermalcare® POL. Cetearyl alcohol, ceteth-20, and glycol stearate; lubricant SE wax, emulsifier for lotions and creams; effective over broad pH range. [Rhone-Poulenc Surf.]

Dermalcare® SPS. CAS 8002-23-1; Cetyl esters; cosmetic grade emulsifier and emulsion stabilizer for creams, lotions, antiperspirants, creme rinse conditioners, personal care products; substitute for natural spermaceti; acid and alkali stable. [Rhone-Poulenc Surf.]

Dermalex. A proprietary preparation of squalene, hexachlorophene and allantoin; a protective skin lotion. [The Dermalex Co Ltd] *

Dermane. CAS 111-01-3; Synthetic squalane. [Universal Preserv-A-Chem Inc] ‡

Dermasome® A. Lecithin and allantoin; cosmetic ingred. encapsulated in lipid spheres; for skin care products [Microfluidics]

Dermasome® E. Lecithin and tocopheryl acetate; cosmetic ingred. encapsulated in lipid spheres; for skin care products [Microfluidics]

Dermasome® H. CAS 9067-32-7; Lecithin and sodium hyaluronate; cosmetic ingred. encapsulated in lipid spheres; for skin care products [Microfluidics]

Dermasome® MT. CAS 8002-43-5; EINECS 232-307-2; Lecithin; cosmetic ingred. encapsulated in lipid spheres; for skin care products [Microfluidics]

Dermasome® P. Panthenol and lecithin; cosmetic ingred. encapsulated in lipid spheres; for skin care products [Microfluidics]

Dermasome® S. Squalane and lecithin; cosmetic ingred. encapsulated in lipid spheres; for skin care products [Microfluidics]

Dermasome® SC. Lecithin and soluble collagen; cosmetic ingred. encapsulated in lipid spheres; for skin care products [Microfluidics]

Dermasome® SOD. Lecithin and superoxide dismutase; cosmetic ingred. encapsulated in lipid spheres; for skin care products [Microfluidics]

Dermasome® TRF. Tissue/skin respiratory factors and lecithin; cosmetic ingred. encapsulated in lipid spheres; for skin care products [Microfluidics]

Dermasome® U. Urea and lecithin; cosmetic ingred. encapsulated in lipid spheres; for skin care products [Microfluidics]

Dermasome® V. Aloe vera gel and lecithin; cosmetic ingred. encapsulated in lipid spheres; for skin care products [Microfluidics]

Dermasulph. A proprietary preparation of polythionates. sulfur skin ointment. [Crookes Laboratories] *

Dermatol. A basic bismuth salt of gallic acid, $C_6H_2(OH)_3 \cdot CO \cdot Bi(OH)_2$; used medicinally for the treatment of wounds and skin diseases, also as a remedy for perspiring feet.

Dermidex®. Lidocaine 1.2% w/w, aluminum chlorhydroxyallantoinate 0.25%, chlorbutanol 1% w/w, cetrimide 0.5% w/w; antiseptic, anesthetic, anti-irritant healing aid and emollient. [Seton Healthcare Group plc]

Dermoblock MA. CAS 134-09-8; Menthyl anthranilate. [Alzo]

Dermoblock OS. CAS 118-60-5; Octyl salicylate; uv light absorber for sunscreens. [Alzo]

Dermogesic. Benzocaine, calamine and metacresol; for topical relief of inflammatory and pruritic conditions of the skin. [Merck & Co Inc] †

Dermol. Bismuth chrysophanate, Bi $(C_{15}H_9O_4)_2Bi_2O_3$; used in the treatment of skin diseases as a 5-20% ointment.

Dermol 89. CAS 71566-49-9; Octyl isononanoate; emollient for skin care and make-up products; partial replacement for silicone oils; antitackiness aid in antiperspirants; resin plasticizer for hair sprays. [Alzo]

Dermol 105. CAS 60209-82-7; Isodecyl neopentanoate; dry emollient, low visc. oil; SPF booster. [Bernel]

Dermol 108. CAS 34962-91-9; Isodecyl octanoate; emollient for cosmetics; antitackiness aid. [Alzo]

Dermol 126. Laureth-2 benzoate; emollient, bloomer, wetting agent; compatible with surfactants. [Bernel]

Dermol 185. CAS 58958-60-4; EINECS 261-521-9; Isostearyl neopentanoate; emollient, pigment binder, freeze/thaw stabilizer. [Bernel]

Dermol 334. Isodecyl octanoate, octyl isononanoate, diethylene glycol diisononanoate, diethylene glycol dioctanoate; Carbopol dispersing aid; dry feel emollient. [Alzo]

Dermol 489. CAS 72269-52-4, 106-01-4; Diethylene glycol dioctanoate, diethylene glycol diisononanoate; emollient for cream and lotion formulations; dispersant for Carbopol powds. [Alzo]

Dermol BES. CAS 109-38-6; Butoxyethyl stearate [Alzo]

Dermol DISD; CAS 127358-81-0; Diisostearyl dimer dilinoleate; emollient for creams, lotions, and makeup; anti-irritant in formulations. [Alzo]

Dermol G-76. CAS 125804-12-8; Glycereth-7 benzoate; emollient for creams, lotions, bath products, liq. soaps, hydro-alcoholic sol'ns.; softener and moisturizer for skin. [Alzo]

Dermol GL-7A. CAS 57569-76-3; Glycereth-7 triacetate; emollient, solubilizer. [Alzo]

Dermol ICSA. CAS 138208-68-1; Isocetyl salicylate; emollient; solv. for benzophenone in sunscreen formulations. [Alzo]

Dermol L45. CAS 125804-13-9; Glycereth-4.5 lactate; emollient, humectant for hydro/alcoholic formulations, aftershave, body splashes. [Alzo]

Dermol MO. Glycereth-7 diisononanoate, diethylene glycol dioctanoate, diethylene glycol diisononanoate; emollient for skin care formulations and lip products; replacement for mineral oil. [Alzo]

Dermol OO. CAS 29710-25-6; Octyl oxystearate; prevents defatting of skin from harsh surfactants and detergents; for cosmetic and cleansing formulas; binder in pressed powds. [Alzo]

Dermolan GLH. CAS 138314-11-1; Glycereth-7.5 hydroxystearate; emollient in creams and lotions; self emulsifier; visc. builder for shampoos; conditioner for shampoos. [Alzo]

---

* Trade name not verified as available   † Trade name verified as obsolete

Dermonistat. CAS 22832-87-7; A proprietary preparation of miconazole nitrate; an antifungal agent. [Janssen Pharmaceutical Ltd] *

Dermoplast. A proprietary preparation of benzocaine, benzethonium chloride, menthol, hydroxyquinalone benzoate, and methyl paraben in the form of an aerosol; used as a soothing skin spray. [Wyeth Laboratories] †

Dermovate. CAS 25122-46-7; A proprietary preparation of clobetasol propionate; used in the treatment of eczema and psoriasis. [Glaxo Laboratories] *

Dermoxyl®. CAS 94-36-0; Benzoyl peroxide; for treatment of acne. [ICN Pharmaceuticals Inc]

Derosal WDG. CAS 83601-81-4; Carbendazim; systemic fungicide. [Hoechst UK]

De Rossi's stain. A microscopic stain. It consists of two solutions: a) tannic acid 25 g, distilled water 100 cc ; b) fuchsin 0.25 gram, phenol 5 g, alcohol 10 g, and distilled water 100 g.

Derris Dust. CAS 83-79-4; Contains rotenone; garden insecticide. [ICI Garden Products]

Derris Dust. CAS 83-79-4; Rotenone; insecticidal dust for use on crop and noncrop areas. [Vitax Ltd]

Derussole. CAS 1333-86-4; Carbon black dispersions; for simple and dust-free dyeing of paints, lacquers, paper, cardboard, plastics, synthetic fibers, printing inks, and mineral binders. [Degussa AG] *

Desamidocollagen. CAS 9007-34-5; Soluble collagen; moisturizer and film-former for toiletries. [Henkel/Cospha; Henkel Canada] *

Desavin. Plasticizer for use in coating materials based on unsaponifiable binders as well as in plastics dispersions and film coatings. [Bayer AG]

Descale. Acid detergent for milkstone removal. [Ciba plc]

Deseril. CAS 361-37-5; A proprietary preparation of methysergide; for treatment of migraine. [Sandoz] *

Desferal. A proprietary preparation of desferrioxamine mesylate; used to treat hemochromatosis and acute iron poisoning. [Ciba plc]

Desibyl Kapseals. A proprietary preparation of dried whole bile. [Parke-Davis] *

Desicchlora. A proprietary name for a perchlorate of barium, a drying agent to replace calcium chloride, sulfuric acid, and potassium hydroxide; absorbs 20% of its weight of water. §

Desimpal. An active oxygen compound; used in the textile industry during peroxide bleaching in order to prevent precipitation of hardening silicate when stabilizing with water glass. [Degussa AG] *

Desmalkyd®. Oil-modified polyurethanes; for the formulation of paints and varnishes as well as printing inks. [Bayer AG]

Desmocap®. Polyurethane coating raw material; for automotive engineering, mech. and plant engineering, industrial coatings, building, wood and furniture finishing, coatings for plastics and paper. [Bayer AG; Bayer plc]

Desmocast®. Polyols for elastomers. [Bayer AG]

Desmocoll®. High molecular hydroxyl polyurethanes; used for solution based adhesives with particular suitability for high grade footwear sole bonding and production of plastics laminates. [Bayer AG; Bayer plc]

Desmocoll® 110. High molecular weight linear polyester urethane; for use in adhesives; used primarily in contact bonding with heat activation in the footwear, laminating, and pkg. industries. [Bayer; Miles]

Desmocoll® 526. Linear polyester PU based on MDI; elastomeric PU for use in adhesives for bonding plastics, rubber, leather, wood, textiles, paper; suitable for contact bonding at room temperature [Bayer; Miles]

Desmoderm®. One-component polyurethane finishes; for polyurethane coated materials. [Bayer AG]

Desmoderm® Foil. Microporous polyurethane foil. [Bayer plc] †

Desmodur®. Crosslinking agent for adhesives based on Baycoll, Baypren, Desmocoll, and Dispercoll and PUR coatings; isocyanates for foams and elastomers. [Bayer AG; Bayer plc]

Desmodur® HL. TDI/HDI-based polyisocyanate; used for fast curing two-component coatings for furniture and metal; improved light stability. [Bayer AG]

Desmodur® IL. TDI-based polyisocyanate; used for extremely fast-curing two-component coatings for wood and metal. [Bayer AG]

Desmodur® L. TDI-based polyisocyanate; for fast-drying topcoats and primers for metal, plastic and several substrates; recommended for interior use. [Bayer AG]

Desmodur® R, RF. Bonding agent. [Bayer plc]

Desmodur® SL. 66% solution of a polyisocyanate; for use in combination with Leguval to increase adhesion on a wide variety of materials and to produce hard, tack-free surfaces. [Bayer AG]

Desmodur® VL. MDI-based polyisocyanate; used for high solids and solvent-free coatings, sealants and related products, cationic electrodeposition. [Bayer AG]

Desmodur® KA-8331. Polyisocyanate polyether PU based on MDI; solv.-free, fast-curing moisture-cure adhesive for bonding textiles, leather, wood, metal, and plastics. [Bayer; Miles] *

Desmodur® L75A. Adduct of toluene diisocyanate and polyol; cross-linking agent that improves resistance to heat, greases, oils, plasticizers, org. solvs., and the adhesion to many substrates; primarily used in adhesives. [Miles; Polysar]

Desmodur® RE. Triphenyl methane 4,4´,4´´-triisocyanate in ethyl acetate; room temperature crosslinking agent for adhesives based on Desmocoll and Baycoll polymers, nat. rubber, and synthetic rubbers. [Miles; Polysar]

Desmodur® RFE. Tris (p-isocyanato-phenyl)-thiophosphate in ethyl acetate; room temperature crosslinking agent for adhesives based on Desmocoll and Baycoll polymers, natural and synthetic rubbers. [Miles; Polysar]

Desmodur® VKS-2, VKS-4, VKS-18. CAS 101-68-8; Polymethylene polyphenyl isocyanate; room temperature crosslinking agent for adhesives based on Desmocoll and Baycoll polymers, natural and synthetic rubbers. [Miles; Polysar]

Desmoflex®. PU elastomer; for building, furniture, shoes, automotive and mech. engineering, sports surfacing, sporting goods, textiles, elec. industry, domestic appliances. [Bayer AG]

Desmolac®. Polyurethane coating raw material; for automotive engineering, mech. and plant engineering, industrial coatings, building, wood and furniture finishing, coatings for plastics and paper. [Bayer AG; Bayer plc]

Desmopan®. Thermoplastic polyurethane elastomer with various levels of shore hardness; for processing by injection molding and extrusion. [Bayer AG; Bayer plc]

Desmopan® 150. CAS 9009-54-5; Thermoplastic ester PU; injection molding grades with high mech. strength, good

---

‡ Trade name and manufacturer not verified   § Trade name without identified manufacturer

flow, short cycles, easy mold release; used for articles subjected to wear, e.g., castors, shoe heels. [Bayer AG]

Desmopan®385. CAS 9009-54-5; Thermoplastic ester PU; injection molding and extrusion grade with high mech. strength, increased resist. to hydrolysis, improved low-temp. flexibility; used for extruded articles and heavy-duty engineering components. [Bayer AG] *

Desmopan®585. CAS 9009-54-5; Thermoplastic ester PU; soft extrusion and injection molding grades with good resist. to microbial attack; used for parts where exposure to high mech. loads is combined with risk of attack from microorganisms, e.g., sports shoe soles, cable sheathing, animal marking tags; 585 is extremely wear resist. grade. [Bayer AG] *

Desmopan® 786. CAS 9009-54-5; Thermoplastic ether carbonate PU; exhibits very good resist. to hydrolysis and microbial attack; provide dependable performance in harshest outdoor environments; 786 for injection molding and extrusion; used for fire hoses. [Bayer AG]*

Desmopan® KA 8333. CAS 9009-54-5; Thermoplastic ether PU; injection molding grades with very good resist. to hydrolysis and microbial attack, very good low-temp. flexibility; used where impact resist. and flexibility at very low temps. is required, e.g., cable, ski boots. [Bayer AG]*

Desmophen®. Flexible and rigid polyurethane foams; flexible foam for cushioning applications, upholstered furniture, mattresses, automotive engineering, textiles, pkg.; rigid foam for heating and refrigeration engineering, building industry, sporting goods, automotive engineering; also for industrial coatings. [Bayer AG; Bayer plc; Miles]

Desmorapid®. Catalyst for adhesives based on Baycoll, Desmodur and Desmocoll. [Bayer AG]

Desmotherm®. Polyurethane coating raw material; for automotive engineering, mech. and plant engineering, industrial coatings, building, wood and furniture finishing, coatings for plastics and paper. [Bayer AG; Bayer plc]

Desodora. Skin protective soap with substances against bacterial attack and fungal infections of the skin; for protection and cleansing of the skin under environmental stress. [Dynamit Nobel Wien GmbH] *

Desogen. A proprietary preparation of dodecanoyl N methylaminoethyl(phenyl carbamyl methyl)dimethyl ammonium chloride. throat lozenges. [Ciba plc] *

Desogrip. Trade name for resin based adhesive in polychloroprene digression; used for attaching soles and uppers for shoe manufacturing. [Seal Ltd]

Desomeen. Ethoxylated fatty amines. [Witco/Organics]

desomorphine. CAS 427-00-9; (Permonid) 7,8-Dihydro-6-deoxymorphine.

DeSonate 50-S. Sodium dodecylbenzene sulfonate, 53% active; detergent component. [Witco/Organics]

DeSonate 60-S. Sodium dodecylbenzene sulfonate, 60% active; detergent component. [Witco/Organics]

DeSonate AOS. C_{14}-C_{16} sodium alpha olefin sulfonate; high foaming detergent component for personal care products, light duty detergents and industrial cleaners. [Witco/Organics]

DeSonate SA. Dodecylbenzene sulfonic acid; component for manufacture of many detergents. [Witco/Organics]

DeSonate SA-H. Alkylbenzene sulfonic acid; emulsifier, chemical intermediate (non biodegradable). [Witco/Organics]

Desonic®30C. Ethoxylated castor oil; emulsifier, lubricant,

dye leveler, antistatic agent and dispersant for various textile applications and the pulp and paper industry. [Witco/Organics]

Desonic® DA-4. CAS 26183-52-8; Deceth-4; wetting and rewetting agent for industrial applications. [Witco/Organics]

Desonic®DA-6. CAS 26183-52-8; Deceth-6; wetting agent for built scour systems. [Witco/Organics]

Desonic® N Series (i.e. 4N, 9N etc). Ethoxylated nonyl phenols, 1-100 moles ethylene oxide; emulsifiers, wetting agents, detergents and dispersants. [Witco/Organics]

Desonic® S Series. Octylphenol ethoxylates, various degrees of ethoxylation; emulsifiers, wetting agents, detergents and dispersants. [Witco/Organics]

DeSonol A. CAS 2235-54-3; EINECS 218-793-9; Ammonium lauryl sulfate; detergent and shampoo component. [Witco/Organics]

DeSonol AE. CAS 67762-19-0; Ammonium laureth (3) sulfate; component for detergents, shampoos and personal care products. [Witco/Organics]

DeSonol S. Sodium lauryl sulfate; detergent and shampoo component. [Witco/Organics]

DeSonol SE. Sodium laureth (3) sulfate; component for detergents, shampoos and personal care products. [Witco/Organics]

DeSonol SE-2. Sodium laureth (2) sulfate; component for detergents, shampoos and personal care products. [Witco/Organics]

DeSonol T. CAS 139-96-8; EINECS 205-388-7; TEA lauryl sulfate; mild shampoo component. [Witco/Organics]

Desophos®. Phosphate esters. [Witco/Organics]

Desoplas. Trade name for resin based adhesive in polyurethene dispersion; used for attaching soles and uppers for shoe manufacturing. [Seal Ltd]

DeSotan®SMO. CAS 1338-43-8; EINECS 215-665-4; Sorbitan monooleate; lipophilic emulsifier, fiber lubricant and softener. [Witco/Organics]

DeSotan®SMO-20. CAS 9005-65-6; Polysorbate 80; hydrophilic emulsifier and wetting agent. [Witco/Organics]

DeSotan®SMT. Sorbitan monotallate; lipophilic emulsifier, fiber lubricant and softener. [Witco/Organics]

DeSotan® SMT-20. PEG-20 sorbitan tallate; hydrophilic emulsifier and wetting agent. [Witco/Organics]

Destral. Nonselective herbicides; for weedkilling. [Borax Consolidated Ltd]

Destral. Mixture of 2,4-D, dalapon and diuron; used for total weed control in non crop areas. [ABM Chemicals Ltd]

Destral BR. Nonselective herbicides; weedkilling. [Borax Consolidated Ltd] *

Desulfex. Fluxes for removal of sulfur from molten iron or steel. [Foseco (F.S.) Ltd]

Desyrel. CAS 25332-39-2; Trazodone hydrochloride; antidepressant. [Mead Johnson & Co] *

DET. *See* diethyl toluamide.

DETA. *See* diethylenetriamine.

Detac. Denaturing materials for use in water washed spray booths. [Brent Chemicals International plc]

Detaclad®. For the chemical industry. [DuPont UK]

Detarex. Chelating agents of the aminopoly/carboxylic acid type; sequestering agents for trace metal control. [W R Grace Ltd]

Detarol. Ethylene diamine tetra acetic acid salts; chelating agents for trace metal control. [W R Grace Ltd]

---

* Trade name not verified as available   † Trade name verified as obsolete

Deteclo. A proprietary preparation ot chlortetracycline, tetracycline and democlocycline hydrochlorides; an antibiotic. [Lederle Laboratories] *

Deterflo A 210. Modified ethoxylated alcohol; low foaming degreasing sol'ns. for surfaces requiring strongly acid or alkaline detergents, at low temp. and under pressure; stable to acids and bases; biodeg. [Ceca SA]

Detergent 8®. Blend of an alkanolamine, glycol ethers, and an alkoxylated fatty alcohol; biodeg. detergent, wetting agent for cleaning circuit boards, electronic parts, phosphate-sensitive labware, delicate industrial parts, nuclear reactor cavities. [Alconox]

Detergyl. Textile auxiliary chemicals. [ICI Chem. & Polymers Ltd]

Deterpal. Specialty blends of anionic surfactants; used for industrial detergents, paints. [Atochem UK/Ceca]

Deterpal 832. Alkyl ether sulfate/solvent blend; degreaser, emulsifier. [Ceca SA]

Deterpal LC. Alkylsulfate/solvent blend; dispersant and antisettling agent for alkyd paints. [Ceca SA]

Dethlac. CAS 62-73-7; An insecticidal lacquer, containing dichlorvos; a slow release fly killer. [Gerhardt Pharmaceuticals] *

Dethmor. CAS 81-81-2; Warfarin; a rodenticide. [Gerhardt Pharmaceuticals] *

Detigon. CAS 511-13-7; Chlophedianol hydrochloride; an antitussive. [Bayer AG] †

Detigon Linctus. A proprietary preparation containing chlophedianol citrate and potassium guaiacol sulfonate; a cough linctus. [FBA Pharmaceuticals] ‡

Det-O-Jet®. Highly alkaline detergent containing potassium hydroxide, silicate of soda, and sodium hypochlorite; low sudsing biodeg. detergent with good wetting, emulsifying, and penetrating properties, for ultrasonic and mechanical washers, hospital and lab ware, optical and electronic components, pharmaceutical apparatus, industrial parts. [Alconox]

Dettol. A proprietary trade name for a germicide containing chloroxylenols and terpineol; very powerful in action and yet non-poisonous. §

Deva. A range of precious metal alloys; used for dentistry and dental engineering. [Degussa]

Devarda's alloy. An alloy of 45% aluminum, 50% copper, and 5% zinc.

Devrinol. CAS 15299-99-7; Suspension concentrate containing 450 g/l napropamide; amide herbicide for oilseed rape and fruit. [Embetec Crop Protection Ltd]

Devrinol T. CAS 15299-99-7, 1582-09-8; Suspension concentrate containing 140 g napropamide and 140 g trifluralin per liter; amide herbicide for oilseed rape and fruit. [Embetec Crop Protection Ltd]

Deward Steel. A proprietary nonshrinking steel containing 1.55% manganese, 0.3% molybdenum, and 0.9% carbon. §

Dewrance metal. See Durance's metal.

DEWT L. Blend of chromate and an organic; corrosion control aid for closed recirculating water systems. [Drew Ind. Div.]

Dexacidin® Ophthalmic Ointment. Neomycin and polymyxin B sulfates and dexamethasone in a white petrolatum base; for steroid-responsive inflammatory ocular conditions for which a corticosteroid is indicated and where bacterial infection or a risk of bacterial ocular infection exists. [Iolab Corp]

Dexacidin® Ophthalmic Suspension. Neomycin and polymyxin B sulfates and dexamethasone; for steroid-responsive inflammatory ocular conditions for which a corticosteroid is indicated and where bacterial infection or a risk of bacterial ocular infection exists. [Iolab Corp]

Dexacillin. CAS 26774-90-3; Epicillin; antibacterial. [Bristol-Myers Squibb Co Inc] *

Dexamist Ear Spray. Nonpressurized pump action spray containing dexamthethasone, neomycin sulfate and glacial acetic acid; for treatment of otitis externa in dogs. [Stafford-Miller] ‡

Dexa-Rhinaspray. A proprietary preparation of neomycin, dexamethasone and tramazoline. [Boehringer Ingelheim Ltd]

Dexawin. Racephenicol; antibacterial. [Sterling Drug Inc] *

Dexedrine Tablets. Dexamphetamine sulfate; used in the treatment of narcolepsy, also indicated in children with refractory hyperkinetic states. [SmithKline Beecham]

Dexel. Plastic molding materials, cellulose derivatives or synthetic resins; for industrial use. [Courtaulds Fibres Ltd]

Dexil. Core breakdown agent. [Foseco (F.S.) Ltd]

Dexine 521. A polyisobutylene material with a good resistance to chemical attack from oxidizing liquors up to 110 C; used for lining and covering metal tanks. [Dexine Rubber Co Ltd] *

Dexine 656. A natural rubber based compound with good abrasion resistance used for lining metal tanks. [Dexine Rubber Co Ltd] *

Dexine 687. A natural rubber based material with good resistance to chemicals especially sodium hypochlorite. [Dexine Rubber Co Ltd] *

Dexine 759. A Hypalon (qv) based lining and covering material with very good resistance to chemical attack. It can be used with sulfuric acid at concentrations up to 95% [Dexine Rubber Co Ltd] *

Dexine 779. A polyurethane lining and covering material with a very good resistance to abrasion. [Dexine Rubber Co Ltd] *

Dexlar®. Staple cement; for the automotive industry. [DuPont UK]

Dexon. Polyglycolic acid; surgical aid. [Davis & Geck] *

Dexone. CAS 50-02-2; Dexamethasone; glucocorticoid. [Rowell Laboratories Inc] *

Dexonite. A trade name for a proprietary hard rubber molded material for electrical insulation; a proprietary trade name for ebonite. [Dexine Rubber Co Ltd] *

Dexoplas. A proprietary trade name for a butadiene-styrene plastics material used for constructing corrosion resistant fittings. [Dexine Rubber Co Ltd] *

Dexten. A proprietary preparation of dexamphetamine resinate. [Nicholas Laboratories Ltd]

dextran. (macrose) Polymers of glucose with chain-like structures and m.w. to 200,000; $(C_6H_{10}O_5)_n$; CAS 9004-54-0; EINECS 232-677-5; blood plasma substitute or expander, confections, lacquers, oil-well drilling muds, filtration gel, food additive. [Accurate Chem. & Scientific; Am. Biorganics; Pharmacia AB; Schweizerhall; Spectrum Chem. Mfg.; U.S. Biochemical]

Dextran. A proprietary trade name for α1-6 polyglucose or polyanhydroglucose. [Fisons plc, Pharmaceuticals Div]*

Dextranomer. Dextran crosslinked with epichlorohydrin; a promoter of wound healing.

Dextraven. A proprietary preparation of dextrans in normal

---

‡ Trade name and manufacturer not verified   § Trade name without identified manufacturer

saline; used to restore blood volume. [Fisons plc, Pharmaceuticals Div] *

dextrin. (dextrine; starch gum; amylin; gommelin; artificial gum; vegetable gum) Gum produced by incomplete hydrolysis of starch; CAS 9004-53-9; EINECS 232-675-4; thickener, adhesives, sizing paper and textiles, substitute for natural gums, food industry, printing inks, in penicillin manufacture, fuel in pyrotechnic devices. [Am. Maize Prods.; Avebe BV; Grain Processing; Nat'l. Starch & Chem.; A.E. Staley Manufacture]

dextrine. See dextrin.

dextrin-maltose. See glucose syrup.

Dextrinozole. See ozole.

Dextroform. A condensation product of dextrin and formaldehyde; used as an antiseptic.

Dextrol OC-20. CAS 51811-79-1; Complex org. phosphate ester free acid; detergent, wetting agent, emulsifier; for pesticides, emulsion polymerization; corrosion inhibitor; dedusting agent for alkaline powd. [Dexter]

Dextrol OC-70. Phosphated aliphatic ethoxylate; detergent, wetting agent, dispersant, emulsifier; for alkyd paints, solv.-based coatings, colorant systems, pesticides, PVA and acrylic polymerization; corrosion inhibitor; dispersant for magnetic oxide in aromatic solvs. [Dexter]

Dextrone X. CAS 1910-42-5; Soluble concentrate containing 200 g/l paraquat; a pre-emergence bipyridilium herbicide to control weeds in field crops and ornamentals. [Chipman Ltd]

Dextrone X. CAS 1910-42-5; Soluble concentrate containing 200 g/l paraquat; a pre-emergence bipyridilium herbicide to control weeds in field crops and ornamentals. [ICI Agrochemicals Professional Products]

dextronic acid. See gluconic acid.

dextrose. See glucose.

Dextrostix®. A proprietary test strip impregnated with a buttered mixture of glucose oxidase, peroxidase, and a chromogen system; used to estimate blood glucose. [Bayer AG]

Dextrozyme. A balanced mixture of glucoamylase and pullulanase obtained from selected strains of *Aspergillus niger* and *Bacillus acidopullulyticus* by submerged fermentation; used in the starch syrup industry for the saccharification of liquified starch (maltodextrins) in the production of high dextrose syrups. [Novo Nordisk]

Dexuron. Suspension concentrate containing 300 g diuron and 100 g paraquat per liter; used to control weeds around trees and shrubs. [Chipman Ltd]

Dexytal. A proprietary preparation of sodium amylobarbitone and dexamphetamine sulfate; an antidepressive agent. [Eli Lilly & Co] †

D.F. 118. A proprietary preparation of dihydrocodeine bitartrate; an analgesic. [British Drug Houses] *

DHA. See dihydroxyacetone.

DI-43. Polymeric lubricity additive for draw and iron can body lubricants. [Ferro/Keil]

Diabase Developer®. Azoic bases used in combination with various Diathol Grounder prepares; prior to coupling the bases must be diazotized to form diazo salt. [Mitsubishi Kasei] *

Diabeta. CAS 10238-21-8; Glyburide; antidiabetic. [Hoechst-Roussel Pharmaceuticals Inc]

Diabetic Cough Mixture. A proprietary preparation containing codeine phosphate, pholcodine, butethamate citrate, ipecacuanha liquid extract and squill liquid extract; a cough mixture. [Rybar Laboratories Ltd] *

diabetin. See levulose.

Diabiformin. A proprietary preparation of chlorpropamide, metformin hydrochloride; an oral hypoglycemic agent. [Pfizer International]

Diabinese. CAS 94-20-2; A proprietary preparation of chlorpropamide; an oral hypoglycemic agent. [Pfizer International]

Diabiphage. A proprietary preparation of chlorpropamide, metformin hydrochloride; an oral hypoglycemic agent. [Pfizer International]

Diablack® A. CAS 1333-86-4; Carbon black; for rubber industry. [Mitsubishi Kasei] *

Diacelliton Dye®. Disperse dyes for acetate and nylon. [Mitsubishi Kasei] *

Diacetin. Glycerol diacetate, $(CH_3COOCH_2)_2CHOH$; auxiliary for use in foundries. [Bayer AG]

diacetone alcohol. (4-hydroxy-4-methyl-2-pentanone) $(CH_3)_2C(OH)CH_2COCH_3$; CAS 123-42-2; EINECS 204-626-7; Solvent for nitrocellulose, cellulose acetate, oils, resins, waxes, fats, dyes, tars, lacquers, dopes, coatings, wood preservatives, rayon, artificial leather, metal cleaning; laboratory reagent; hydraulic fluids; textile stripping agent. [Allchem Industries; Atochem N. Am.; BP Chem. Ltd; Hoechst Celanese; Shell; Union Carbide]

1,2-diacetoxyethane. See ethylene glycol diacetate.

diacetyl. (2,3-butanedione; biacetyl) $CH_3COCOCH_3$; CAS 431-03-8; EINECS 207-069-8; Aroma carrier in food products. [Aldrich; Penta Mfg.]

diacetyldioxime. See dimethyl glyoxime.

diacetyl tannin. See acetannin.

Diacid Dye®. Acid colors for wool, silk, and nylon; produces bright shades. [Mitsubishi Kasei] *

Diaclear®MA-3000 4L. Super high molecular weight polyacrylamide; flocculant for industrial waste water treatment and sedimentaiton, filtration, and centrifugal processes. [Mitsubishi Kasei] *

Diaclear®MK-166. Polyacrylamide; flocculant for industrial waste water treatment, dewatering and filtration of suspended organic sludge. [Mitsubishi Kasei] *

Diaclear® MO-3000 H. Super high molecular weight polyacrylamide; flocculant for enhanced oil recovery. [Mitsubishi Kasei] *

Diacotton Dye®. Direct dyes for vegetable fiber and viscose rayon. [Mitsubishi Kasei] *

Diacron Dye. Reactive hot-type dyes, esp. suitable for printing and continuous dyeing method. [Mitsubishi Kasei] *

Diacryl. Range of methacrylate monomers used in adhesive and coating resins. [Akzo Chemie UK Ltd]

Diacryl Dye®. Colors for dyeing polyacrylonitrile fiber. [Mitsubishi Kasei] *

Diacryl P Series®. Colors for dyeing cationic dyeable polyester. [Mitsubishi Kasei] *

Diacupro Dye®. Direct dyes whose fastness to light and washing can be improved considerably after treatment with copper salt. [Mitsubishi Kasei] *

Diadavin®. Cleaning and stain removing agent; for textile finishing. [Bayer AG; Bayer plc]

Diadem Chrome. Acid mordant wool dyestuffs. [Akzo Chemie] *

Diadol 13. Higher alcohol; surfactant for shampoo, light duty liq. detergents. [Mitsubishi Kasei] *

Diadol 18G. EINECS 248-470-8; Isostearyl alcohol; base for

---

oily cosmetic preps.; superfatting agent. [Mitsubishi Kasei] *

Diadur. Zirconia-alloyed aluminum oxide; used for abrasive industries, super refractories, sandblasting and for safes. [Lonza AG]

Diafen. CAS 132-18-3; Diphenylpyraline hydrochloride; antihistaminic. [3M Pharmaceuticals] †

Diaflex. Intraperitoneal dialysis solutions. [The Boots Co plc] *

Diaformer Z-AT. Methacryloyl ethyl betaine/methacrylates copolymer; polymer for hair care products (aerosol or pump hairspray, spritz, mousse; good compatibility with propellants. [Mitsubishi Petrochem.; Sandoz]

Diaginol. Sodium acetrizoate. [May & Baker Ltd] *

Diahold A-503. Methacrylates/acrylates copolymer amine salt; fixative and film-forming polymer for aerosol and pump hairsprays, setting lotions, spritzes; forms hard, glossy films; excellent hair holding performance. [Mitsubishi Petrochem.; Sandoz]

Diahope®-006. Activated carbon; for waste water treatment, water purification, solvent. purification, chems. purification. [Mitsubishi Kasei] *

Diahope®-S60. Activated carbon; for sugar refining, water works, food additives purification. [Mitsubishi Kasei] *

Diaion® CR10. Chelating resin for recovery of metal, waste water treatment, purification of brine for NaOH, membrane process, etc. [Mitsubishi Kasei] *

Diaion®HP 10. Synthetic adsorbent for vitamins, antibiotics, enzymes, steroids, fatty acids, perfumes, other bioactive substances; removal of phenol and surfactants; decolorization applications [Mitsubishi Kasei] *

Diaion® SA10A. Strongly basic anion exchange resin for demineralization of water, catalyst, separation of amino acid, recovery of metal. [Mitsubishi Kasei] *

Diaion® SK1B. Strongly acidic cation exchange resin for softening/demineralization of water, separation and recovery of metal, refining of chems., sugar, dextrose, and amino acid, dehydration of org. solv., catalyst. [Mitsubishi Kasei] *

Diaion®WK10. Weakly acidic cation exchange resin (methacrylic type) for recovery of metal, dealkalization, iron removal, refining of cane sugar, prep. of antibiotics, medicines, amino acids. [Mitsubishi Kasei] *

Diaion®WK20. Weakly acidic cation exchange resin (acrylic type) for water treatment, recovery of metal, dealkalization, iron removal, prep. of antibiotics, medicines, amino acids. [Mitsubishi Kasei] *

Diak. Curing agent. [DuPont UK]

Diak No. 4. CAS 13253-82-2; 4,4-Methylenebis (cyclohexylamine)carbamate [R.T. Vanderbilt]

Diakon. Acrylic molding and extrusion powders. [ICI Chem. & Polymers Ltd]

Dial. Diallylbarbituric acid; a powerful sedative and hypnotic.§

Dialead®. Coal tar pitch carbon fiber; for thermoset plastics, cements, metal, rubber applications [Mitsubishi Kasei] *

Dialen 6. Alpha olefin; comonomer for polyolefin; chemical intermediate. [Mitsubishi Kasei] *

diallyl maleate. CH$_2$:CHCH$_2$OCOCH:CHCOOCH$_2$-CH:CH$_2$; CAS 999-21-3; EINECS 213-658-0; polymers and copolymers, insecticides. [Aceto; Ashland]

diallyl phthalate. (DAP) C$_6$H$_4$(COOCH$_2$CH:CH$_2$)$_2$; CAS 131-17-9; EINECS 205-016-3; primary plasticizer; intermediate. [Allchem Industries; Arco; BP Chem. Ltd; C.P.

Hall; OxyChem; Rogers]

Dialose. CAS 7491-09-0; Docusate potassium; stool softener. [Stuart Pharmaceuticals] †

Dialose Plus. Contains casanthranol and docusate potassium (casanthranol is a purified mixture of the anthranol glycosides derived from *Cascara sagrada*; laxative and stool-softener. [Stuart Pharmaceuticals] †

Dialpha. Fungal α-amylaze dispersion in food grade filler; flour additive. [Diaflex Ltd] *

Dialume. CAS 21645-51-2; Aluminum hydroxide; antacid. [Armour Pharmaceutical Co] *

Dialuminous Dye®. Direct dyes with good to excellent lightfastness. [Mitsubishi Kasei] *

dialuramide. (Murexan). Uramil, C$_4$H$_5$N$_3$O$_3$.

dialyl. Methylamine and lithium citrate.

dialyzed iron. A solution containing ferric oxide and acetic acid, made by dialyzing ferric acetate; used medicinally.

Diamalt. Malt extract. [Abril Industrial Waxes] †

Diamex. Malt extract. [ABM Chemicals Ltd] *

diamidogen sulfate. *See* hydrazine sulfate.

Diamin(e)®. Substansive dyestuffs; used for textile dyeing and printing. [Cassella AG]

Diamine. Range of cationic surfactants of the diamine type, in liquid, solid or paste form; widespread use in industry and in household and personal care formulations, though mostly in the form of derived quaternaries and various salts. [Hoechst AG] *

Diamine B11. CAS 90640-45-2; EINECS 292-564-1; N-C16-22 alkyl propylene diamine; emulsifier, corrosion inhibitor; chemical intermediate producing ethoxylates for surfactants, acetates for emulsifiers and dispersants. [Berol Nobel AB]

Diamine BG. CAS 61791-55-7; EINECS 263-189-0; n-Tallow-propylene diamine; emulsifier, corrosion inhibitor; chemical intermediate, surfactant producing ethoxylates for detergent applications, and hydrochlorides/acetates for emulsifiers and dispersants. [Berol Nobel AB]

Diamine® H Extra. 1,6-Hexanediamine; intermediate for production of AH salt. [BASF AG]

Diamine HBG. CAS 68603-64-5; EINECS 271-696-6; Hydrogenated tallow propylene diamine; emulsifier, corrosion inhibitor, surfactant; chemical intermediate producing ethoxylates for detergent applications, hydrochlorides/acetates for emulsifiers, dispersants. [Berol Nobel AB]

Diamine KKP. CAS 61791-63-7; EINECS 263-195-3; n-Coco propylene diamine; emulsifier, corrosion inhibitor, surfactant; chemical intermediate producing ethoxylates for detergent applications, hydrochlorides/acetates for emulsifiers, dispersants, and alginates. [Berol Nobel AB]

Diamine OL. CAS 7123-62-8; EINECS 230-528-9; n-Oleyl propylene diamine; emulsifier, corrosion inhibitor, surfactant; chemical intermediate producing ethoxylates for detergent applications, hydrochlorides/acetates for emulsifiers and dispersants. [Berol Nobel AB]

diamine sulfate. *See* hydrazine sulfate.

1,2-diaminocyclohexane. (1,2-cyclohexanediamine) C$_6$H$_4$N$_2$; CAS 694-83-7; EINECS 211-776-7. [Aldrich; Du Pont; Milliken; Pacific Anchor]

3,3´-diaminodiphenyl sulfone. (3,3´-sulfonyldianiline) C$_{12}$H$_{12}$N$_2$O$_2$S; CAS 599-61-1. [BASF; Mitsui Toatsu; Schweizerhall]

4,4´-diaminodiphenyl sulfone. (4,4´-DDS; 4,4´-

---

‡ Trade name and manufacturer not verified § Trade name without identified manufacturer

sulfonyldianiline; bis(4-aminophenyl) sulfone) $(NH_2C_6H_4)_2SO_2$; CAS 80-08-0; EINECS 201-248-4; Curing agent. [Crown Metro]

Diaminogen®. A range of dyestuffs. [Cassella AG] *

2,6-diaminopyridine. (2,6-pyridinediamine) Aromatic amine; $C_5H_7N_3$; CAS 141-86-6; EINECS 205-507-2. [Cilag AG; Janssen Chimica; Reilly Industries; Schweizerhall]

di(α-amino-β-thiolpropionic acid). See cystine.

Diamite Epoxy Brushkote. Two-component solvent-release, epoxy compound used as a heavy-duty coating for protecting floors, walls and equipment against attrition and chemical attack. [Metalcrete Mfg Co]

Diamite Epoxy Flooring. 100% solids epoxy; used for repairing and surfacing floors subject to heavy traffic. [Metalcrete Mfg Co]

Diamond D 21. Partially hydrogenated vegetable oil (soybean, cottonseed), mono/diglycerides, glycerol-lacto ester of fatty acids, lecithin; kosher; shortening for cakes. [Van Den Bergh Foods]

Diamond D 31. CAS 8016-70-4; EINECS 232-410-2; Partially hydrogenated soybean oil; kosher; all-purpose deep fat frying filler fat for cookies, whipped toppings. [Van Den Bergh Foods]

Diamond cement. A cement containing 8 parts isinglass, 1 part gum ammoniacum, 1 part galbanum, and 4 parts alcohol; used for mending china and glass.

Diamond Fiber. A proprietary trade name for a vulcanized fiber ; a laminated acid-treated cotton cellulose. §

Diamondite. An alloy of 95.65% tungsten with 3.91% carbon.

Diamond Quality®. Unrefined castor oil USP; emollient, pigment wetter, cosolv., lubricant for cosmetics, makeup, antiperspirant sticks. [CasChem]

diamond white. See oil white.

Diamonine. Textile auxiliary chemicals. [ICI Chem. & Polymers Ltd]

Diamox. CAS 59-66-5; A proprietary preparation of acetazolamide; a diuretic. [Lederle Laboratories]

Diampron. NN-Di-(3-amidinophenyl)urea diiesthionate. [Rhone-Poulenc Rorer Ltd]

Dianabol. A proprietary preparation of methandienone; an anabolic agent. [Ciba plc] *

Dianal. Thermoplastic acrylic resins. [British Traders & Shippers Ltd]

Diane. Diphenylolpropane used as the phenolic reactant in resin manufacture. [Chemische Werke, Albert] *

Dianette. 2 mg Cyproterone acetate, 35 mcg ethinylestradiol [Schering Health Care Ltd]

Dianix Dye®. Disperse dyes with good fastness for dyeing synthetic fibers, esp. polyester; excellent dispersing props. [Mitsubishi Kasei] *

Dianol®. Raw materials for the manufacture of saturated and unsaturated polyester resins, glassmat binder resins, alkyd resins, microcellular and elastomeric polyurethanes. [Akzo Chemie UK Ltd]

Dianol® 220. CAS 32492-61-8, 901-44-0; Ethoxylated bisphenol A diol; reactive modifier for sat. and unsat. polyesters, vinyl esters, and PU resin formulations. [Akzo]

Dianol® FSD. Modified amine condensate in alcohol; provides fulling qualities and scouring efficiency for processing woolen and worsted fabrics.   [Rhone-Poulenc Surf.]

Dianthine. A trademark for a range of dyes. [National Lead Co] *

dianthine B. See erythrosin.

Diaparene. CAS 25155-18-4; Methylbenzethonium chloride; anti-infective, topical. [Sterling Drug Inc] *

Diapen. CAS 1538-09-6; A proprietary preparation of penicillin G benzathine; an antibiotic. [Pfizer International] †

Diaphan oil. A mixture of methylhexalin and sodium oleate; used in the preparation of transparent soaps.

Diapid. CAS 50-57-7; antidiuretic; vasoconstrictor. [Sandoz Pharmaceuticals]

Diapol® WMB® 1808. SBR carbon black masterbatch; for rubber industry. [Mitsubishi Kasei] *

Diaresin Dye®. Solvent dyes for coloring thermoplastics and thermosets; provides high brilliancy and transparency with good fastness props. [Mitsubishi Kasei] *

Diasone Sodium Enterab. CAS 144-75-2; Sulphoxone sodium; antibacterial. [Abbott Laboratories] *

diastase. (diastatic enzyme) Enzyme that converts starch into sugar; CAS 9000-92-4; EINECS 232-567-7; enzyme in textile industry used for starch desizing; degrades starch, dextrins, and oligosaccharides to maltose. [Spectrum Chem. Mfg.; U.S. Biochemical]

diastatic enzyme. See diastase.

Diastatin. CAS 1400-61-9; A proprietary preparation of nystatin; an antifungal. [Pfizer International] †

Diastix®. A proprietary test strip impregnated with glucose oxidase and peroxidase, plus potassium iodide; used to detect glycosuria. [Bayer AG]

Diathol Grounder®. Azoic coupling component used on cotton and rayon as prepare prior to combining with Diabase Developer to form insol. dye; also combined directly with diazotized bases to produce org. pigments. [Mitsubishi Kasei] *

diatomaceous earth. (kieselguhr; diatomite; infusorial earth) Mineral material consisting chiefly of the siliceous frustules and fragments of various species of diatoms; CAS 7631-86-9; Filtration, clarifying, decolorizing; insulation absorbent; mild abrasive; catalyst carrier; anticaking agents in fertilizer; conditioning agent for dusts, soil. [Celite; L.A. Salomon; Spectrum Chem. Mfg.]

diatomite. See diatomaceous earth.

Diavite. Vitamin mixture dispersed in food grade filler; flour additive. [Diaflex Ltd] *

Diax. A proprietary product used as a diastatic ferment. §

diazabicycloundecene. (DBU; 1,8-diazabicyclo [5.4.0] undec-7-ene (1,5-5). [2,3,4,6,7,8,9,10-octapyrimidol [1,2-a]azepine]) $C_9H_{16}N_2$; CAS 6674-22-2; EINECS 229-713-7. [Air Prods.; BASF; Fluka; Schweizerhall]

Diazamine. Direct dyes. [ICI Chem. & Polymers Ltd] †

diazepam. (7-chloro-1,3-dihydro-1-methyl-5-phenyl-2H-1,4-benzodiazepin-2-one) $C_{16}H_{13}ClN_2O$; CAS 439-14-5; EINECS 207-122-5; medicine (tranquilizer). [Hoffmann-La Roche SA]

Diazine. Direct duestuffs for cotton and artificial silk. [J C Bottomley] *

Diazinon Liquid. Insecticide. [DowElanco Ltd]

Diazitol. CAS 333-41-5; Diazonon insecticide.

Diazitol Liquid. Organophosphorus insecticide. [Ciba plc]*

Diazol. Direct dyes. [ICI Chem. & Polymers Ltd]

Diazol. CAS 333-41-5; Active ingredient: diazinon; organophosphorous agricultural insecticide with acaricidal properties. [Makhteshim Chemical Works Ltd]

Diazone. Fast dyestuffs for cotton. [J C Bottomley] *

---

* Trade name not verified as available   † Trade name verified as obsolete

Diazopon SS-837. CAS 9016-45-9; Nonyl phenol ethoxylate [Rhone-Poulenc Surf.]

Diazopon SS-837. Polyoxyethylated alkyl phenol (nonionic), a water-soluble surfactant with dispersing and solubilizing properties; anticrock agent for naphthol dyeings; soaping agent for naphthol-dyed stock, yarn and fabrics; leuco-vat-ester dyeing assistant for use in the acid bath to inhibit pigment agglomeration on the fiber. [ISP] *

diazoresorcin. CAS 550-82-3; (azoresorcin, resazoin). Resazurin, $C_{12}H_9NO_4$.

diazoxide. 7-Chloro-3-methyl-1,2,4-benzo-thiadiazine 1,1-dioxide.

Diazyme® L-200. CAS 9032-08-0; Glucoamylase; enzyme for hydrolysis of starch dextrins to glucose; for brewing and fermentation applications [Solvay Enzymes]

DIBA. See diisobutyl adipate.

dibasic lead carbonate. See lead carbonate (basic).

Dibenyline. CAS 63-92-3; A proprietary preparation of phenoxybenzamine hydrochloride; vascular antispasmodic. [SmithKline Beecham] *

Dibenyline Capsules. CAS 63-92-3; Phenoxybenzamine hydrochloride (α-adrenergic receptor antagonist); for short term management of hypertensive episodes associated with phaeochromocytoma. [SmithKline Beecham] *

Dibenzo GMF. Dibenzoyl-p-quinone-dioxime; a non-sulfur vulcanizing agent for natural, SBR, butyl and EPDM rubber; used to impart heat resistance to tyre curing bags, gaskets and wire insulation, also used as a coagent with peroxide curatives. [Uniroyal][as]

dibenzopyran, tricyclic. See xanthene.

dibenzosuberone. (10,11-dihydro-5H-dibenzo[a,d] cyclohepten-5-one) $C_{15}H_{12}O$; CAS 1210-35-1; EINECS 214-912-3. [Lonza; Penta Mfg.; Sandoz; Schweizerhall]

dibenzoyl peroxide. See benzoyl peroxide.

Dibenzyline. CAS 63-92-3; A proprietary preparation containing phenoxybenzamine hydrochloride. [SmithKline Beecham]

Dibexin. A proprietary preparation of Vitamin B. [Parke-Davis] *

Dibistin. CAS 2508-72-7, CAS 154-69-8; A proprietary preparation of antazoline hydrochloride and tripelennamine hydrochloride; antihistamine. [Ciba plc] *

Dibotin. CAS 834-28-6; A proprietary preparation of phenformin hydrochloride; an oral hypoglycemic agent. [Bayer AG]

Dibrogan. Dibromopropamidine isethionate/promethazine cream. [Rhone-Poulenc Rorer Ltd]

Dibrom. CAS 300-76-5; Naled insecticide. [Chevron] *

Dibromin. Dibromobarbituric acid; used as an antiseptic.

1,4-dibromobutane. (tetramethylene dibromide) $Br(CH_2)_4Br$; CAS 110-52-1; EINECS 203-775-5. [Humphrey; Janssen Chimica]

β-dibromo hydrin. See 2,3-dibromo-1-propanol.

2,3-dibromo-1-propanol. (β-dibromo hydrin; allyl alcohol dibromide) $CH_2BrCHBrCH_2OH$; CAS 96-13-9; EINECS 202-480-9; Intermediate in prep. of flame retardants, insecticides, and pharmaceuticals. [Aldrich; ICI Am.; Schweizerhall]

dibutoxyethyl phthalate. (n-butyl glycol phthalate; ethyldibutoxy phthalate) $C_6H_4(COOC_2H_4OC_4H_9)_2$; CAS 117-83-9; plasticizer for PVC, polyvinyl acetate, and other resins.

dibutyl-1,2-benzene dicarboxylate. See dibutyl phthalate.

2,6-di-t-butyl-p-cresol. See BHT.

di-t-butyl dicarbonate. (BOC anhydride; di-t-butyl pyrocarbonate) $[CH_3)_3COCO]_2O$; CAS 244-24-99-5; EINECS 246-240-1. [Aldrich; Fluka; Schweizerhall]

dibutyl fumarate. (fumaric acid, dibutyl ester) (C_4H_9)-$OOCCH:CHCOO(C_4H_9)$; CAS 105-76-9; Monomeric plasticizers, copolymers, intermediate. [Penta Mfg.]

dibutyl maleate. (DBM; 2-butenedioic acid, dibutyl ester) $C_4H_9OOCCH:CHCOOC_4H_9$; CAS 105-76-0; EINECS 203-328-4; copolymers, plasticizers, intermediate. [Aristech; Pentagon Chem. Ltd; Penta Mfg.; Unitex]

di-t-butyl peroxide. (DTBP; t-butyl peroxide; bis (1,1-dimethylethyl) peroxide) Organic peroxide; $(CH_3)_3COO$-$C(CH_3)_3$; CAS 110-05-4; EINECS 203-733-6; polymerization catalyst for resins (e.g., olefins, styrene, styrenated alkyds, silicones); ignition accelerator for diesel fuel; organic synthesis; intermediate. [Akzo; Atochem N. Am.; Witco/Argus]

2,6-di-t-butylphenol. (2,6 DTBP) Orthoalkylated aromatic; $[(CH_3)_3C]_2C_6H_3OH$; CAS 128-39-2; EINECS 204-884-0; Intermediate; antioxidant, stabilizer. [Allchem Industries; Ethyl; Penta Mfg.; PMC Spcialties; Schenectady]

dibutyl phthalate. (1,2-benzenedicarboxylic acid, dibutyl ester; dibutyl-1,2-benzene dicarboxylate; butyl phthalate) Aromatic diester of butyl alcohol and phthalic acid; $C_6H_4(COOC_4H_9)_2$; CAS 84-74-2; EINECS 201-557-4; plasticizer in nitrocellulose, lacquers, elastomers, explosives, nail polishes; solvent for perfumes, oils; perfume fixative; textile lubricating agent. [Aldrich; Aristech; BP Chem. Ltd; Chisso Am.; Daihachi Chem. Ind.; Eastman; C.P. Hall; Mitsubishi Gas; Unitex]

di-t-butyl pyrocarbonate. See di-t-butyl dicarbonate.

dibutyltin dilaurate. $(C_4H_9)_2Sn(OCOC_{10}H_{20}CH_3)_2$; CAS 77-58-7; EINECS 201-039-8; Stabilizer for vinyl resins, lacquers, elastomers; catalyst for urethane and silicones. [Air Prods.; Atochem N. Am.; Ferro/Bedford; KMZ Chem. Ltd; Witco]

dibutyltin maleate. (DBM) $[(C_4H_9)_2Sn(OOCCH)_2]_x$; CAS 15535-69-0; Stabilizer for PVC resins; condensation catalyst. [Atochem N. Am.; Ferro/ Bedford]

dicalcium phosphate (CTFA). See calcium phosphate, dibasic.

Dicalite 14, 14B, and 14W. Proprietary trade names for diatomaceous silica fillers; used for heat insulating and as a filler for plastics, etc. §

dicarboxylic acid C₃. See malonic acid.

dicarboxylic acid C₈. See suberic acid.

Dicarburetted Hydrogen. See olefiant gas.

Dicestal. CAS 97-23-4; Dichlorophen. [May & Baker Ltd] *

Dichan 100. CAS 3129-91-7; Dicyclohexylamine nitrite; vapor-phase corrosion inhibitor for ferrous metals; designed for items enclosed in pkg., e.g., hot water heating systems, nuclear reactor heat-exchange units, gas recovery systems, jet aircraft engine compressors, internal combustion engines, welding electrodes, double-walled pipes. [Olin]

Dichevrol. Dielectric oil. [Chevron] *

Dichlofuanide. N-Dimethylamino-N´-phenyl-N´-(fluorodichloromethylthio) sulfamide; a proprietary paint fungicide. [Bayer AG]

dichlone. (dichloronaphthoquinone) CAS 117-80-6; A fungicide used as a seed dressing, insecticide, organic catalyst.

---

‡ Trade name and manufacturer not verified   § Trade name without identified manufacturer

Dichloramine T. N, N-Dichloro-p-toluenesulfonamide; $CH_3C_6H_4SO_2NCl_2$; A disinfectant, antibacterial.

dichloramine-M. Methyldiphenylmethyl-dichloramine.

Dichloraminet. See Dichloramine T.

o-dichlorobenzene. (1,2-dichlorobenzene; orthodichlorobenzene) $C_6H_4Cl_2$; CAS 95-50-1; EINECS 202-425-9; Manufacture of 3,4-dichloroaniline; solvent for wide range of org. materials and for oxides of nonferrous metals; degreasing hides and wool; industrial odor control; herbicide, insecticide, soil fumigant. [Ashland; Mitsubishi Gas; Monsanto; PPG Industries]

p-dichlorobenzene. (1,4-dichlorobenzene) $C_6H_4Cl_2$; CAS 106-46-7; EINECS 203-400-5; Moth repellent, general insecticide, germicide, space odorant, manufacture of 2,5-dichloroaniline, dyes, intermediates, pharmaceuticals, agriculture (fumigating soil). [Ashland; Mitsubishi Gas; Monsanto; PPG Industries]

1,2-dichlorobenzene. See calcium phosphate, dibasic.

dichlorodifluoromethane. (difluorodichloromethane; Fluorocarbon-12) CCl_2F_2; CAS 75-71-8; Refrigerant, plastics, blowing agent, low-temp. solvent, leak detecting agent, freezing of foods by direct contact. [Air Prods.; Atochem N. Am.; ICI Spec.; PCR]

Dichloroditane. A proprietary trade name for p-dichlorodiphenylmethane. [Bakelite Corporation] *

1,2-dichloroethane. See ethylene dichloride.

dichloroisocyanuric acid potassium salt. See potassium dichloroisocyanurate.

dichloroisocyanuric acid sodium salt. See sodium dichloroisocyanurate.

dichloromethane. See methylene chloride.

dichlorophenyl urea. See diuron.

1,2-dichloropropane. See propylene dichloride.

Dichlor-Stapenor®. An antistaphylcoccal penicillin. [Bayer AG]

Dichlosuric. CAS 58-93-5; Hydrochlorothiazide; for the treatment of edema and hypertension. [Merck & Co Inc]†

Dichlotride. CAS 58-93-5; Hydrochlorothiazide; for the relief of edema and hypertension. [Merck & Co Inc]

Dickül varnish. See stand oil.

Dicloxin. CAS 13412-64-1; Dicloxacillin sodium; antibacterial. [Bristol Laboratories] *

Dicofen. CAS 122-14-5; Emulsifiable concentrate of 500 g fenitrothion per liter; an organophosphorus insecticide. [Pan Britannica Industries Ltd]

Diconal Tablets. A proprietary preparation of dipipanone hydrochloride and cyclizine hydrochloride; an analgesic and antienetic combination. [The Wellcome Foundation Ltd]

Dicontal®New. Broad spectrum insecticide; used for grape crops. [Bayer AG]

Dicosal. A range of fire extinguishing powders to extinguish metal fires including burning uranium. [Degussa AG] *

Dicotox. A selective weedkiller. [May & Baker Ltd]

Dicotox Extra. CAS 1702-17-6; 2,4-D; translocated herbicide for cereals and established grassland. [Rhone-Poulenc Environmental Prods. Ltd]

Dicrodamine. Series of cationic surfactants in the form of fatty alkyl propylene diamines or diamine salts, in which the alkyl group is coconut, tallow, hydrogenated tallow or oleic in origin; intermediates for di-quaternaries and polyethoxylates; dispersants for pigments; corrosion inhibitors; drawing aids for copper wire and tubing; flexible hardeners for epoxy resins; auxiliaries for oil and petro-

leum industries; bitumen emulsions. [Croda Chem. Ltd]*

Dicron 45Sc. Organophosphorus insecticide. [Ciba plc] *

Dicrylan. Finishing agent. [Ciba plc] *

Dicrylan 270. A proprietary trade name for an aqueous emulsion of an acrylic resin for increasing the abrasion resistance of crease-resistant fabrics. [Ciba plc] *

Dicumoxane. A proprietary preparation of cumetharol; an anticoagulant. [MCP Pharmaceuticals] *

dicumyl peroxide. (di-α-cumyl peroxide; cumene peroxide; diisopropylbenzene peroxide) Organic peroxide. $[C_6H_5C(CH_3)_2O]_2$; CAS 80-43-3; polymerization catalyst and vulcanizing agent; crosslinking agent for olefinic polymers. [Akzo; Atochem; Hercules; Mitsui Petrochem. Ind.; R.T. Vanderbilt]

Di-Cup. CAS 80-43-3; A range of dicumyl peroxide preparations; vulcanizing and polymerization agent. [Hercules]

Dicurane 500 FW. CAS 15545-48-9; Suspension concentrate containing 500 g chlorotoluron per liter; a contact urea herbicide for cereal crops. [Ciba-Geigy Agrochemicals]

Dicurane Duo 495FW. A liquid formulation containing 106 g bifenox and 389 g chlorotoluron per liter as a suspension concentrate; a residual herbicide for the control of weeds in winter wheat. [Ciba-Geigy Agrochemicals]

dicyandiamide. (cyanoguanidine) $NH_2C(NH)(NHCN)$; CAS 461-58-5; EINECS 207-312-8; Fertilizers; nitrocellulose stabilizer; organic synthesis, esp. of melamine, barbituric acid, and guanidine salts; explosives; catalyst for epoxy resin; pharmaceuticals; fireproofing compounds; stabilizer in detergents; modifier for starch production [Allchem Industries; Andrulex Trading Ltd]

dicyclohexyl carbodiimide. (DCC; N,N´-dicyclohexyl carbodiimide) $C_6H_{11}N:C:NC_6H_{11}$; CAS 538-75-0; EINECS 208-704-1; Chemical intermediate, coupling agent in peptide synthesis. [Janssen Chimica; Schweizerhall]

dicyclohexyl phthalate. (DCHP) $C_6H_4(COOC_6H_{11})_2$; CAS 84-61-7; Plasticizer for nitrocellulose, ethylcellulose, chlorinated rubber, PVAc, PVC, and other polymers. [Morflex; Novachem; Schweizerhall; Unitex]

Dicycloverine. CAS 77-19-0; (dicyclomine) β-Diethylaminoethyl cyclohexylcyclohexanecarboxylate.

Dicynene. CAS 2624-44-4; A proprietary preparation of ethamsylate; an antihemorrhagic agent. [Delandale Laboratories Ltd]

Didi-Col. An insecticide. [ICI Chem. & Polymers Ltd]

Didigram. An oil insecticide. [Plant Protection] *

Didimac. An insecticide. [ICI Chem. & Polymers Ltd]

Didin Fluid. Nitrification inhibitor. [Omex Agriculture Ltd]

didodecyl 3,3´-thiodipropionate. See dilauryl thiodipropionate.

DIDP. See diisodecyl phthalate.

Didrate. CAS 34195-34-1; Hydrocodone bitartrate; antitussive. [Penick Corp] *

Didrex. CAS 5411-22-3; A proprietary preparation of benzphetamine hydrochloride; an antiobesity agent. [Upjohn; Upjohn Ltd]

Didronel. CAS 7414-83-7; Etidronate disodium; regulator; pharmaceutic aid. [Norwich Eaton Pharmaceuticals Inc]

die-casting alloys. These are usually zinc-base alloys containing 86% zinc, 7-10% tin, 4-7% copper, 0.5-1% aluminum. Some alloys have a tin base, and a typical one contains 90% tin, 4% copper, and 6% antimony.

Di-el. A material used as an insulating material for high

---

voltage engineering.

Dieline. Sym-dichlorethylene, CHCI = CHCI; used as a solvent for cellulose acetate, rubber, and oils.

Dielmoth. A moth proofing agent for wool. [Shell UK] *

Diene®. Polybutadiene rubber; for high impact polystyrene and tires. [Firestone Syn. Rubber] *

Diene® 35AC. CAS 9003-17-2; Stereospecific polybutadiene rubber; used as the backbone rubber for graft polymers; impact modifier for PS and ABS to produce beverage cups, plates, dinnerware, football helmets, coolers, pkg., toys, furniture, automotive grills, dashboard components, consoles, headlight and interior light fixtures. [Firestone Syn. Rubber]

Diene® 70AC. CAS 9003-17-2; Polybutadiene rubber, alkyl lithium polymerized; features excellent color and low gel; imparts superior toughness and lower gloss when used as a grafted impact modifier for thermoplastic resins. [Firestone Syn. Rubber]

Dienol. CAS 84-17-3; Colloidal manganese.

Diepoxy. An epoxy mold material; used for dental molds. [Kemtron International Inc]

Dieselect. Lubes. [Chevron] *

Dieselmotive. Lubes. [Chevron] *

Dieselube. Lubes. [Chevron] *

Di-esterex N. A proprietary trade name for a rubber vulcanizing accelerator stated to be 60% of the dinitrophenyl ester of mercaptobenzthiazole and 40% of the acetate of diphenylguanidine. §

diethanolamine. (DEA; 2,2´-iminobisethanol; di(2-hydroxyethyl) amine; 2,2´-iminodiethanol) Aliphatic amine; $(HOCH_2CH_2)_2NH$; CAS 111-42-2; EINECS 203-868-0; gas scrubbing; rubber chemicals intermediate; manufacture of surfactants for textiles, herbicides, petrol. demulsifier; emulsifier and dispersant in agriculture., cosmetics, pharmaceuticals; textile lubricants; humectant; softening agent; in organic synthesis. [Hüls AG]

diethanolamine lauric acid amide. See lauramide DEA.

Diethoxol. CAS 111-90-0; EINECS 203-919-7; Monoethyl ether of diethylene glycol. [ICI Chem. & Polymers Ltd] †

diethylamine. $(C_2H_5)_2NH$; CAS 109-89-7; EINECS 203-716-3; rubber chemicals, textile specialties, selctive solvent, dyes, flotation agents, resins, pesticides, polymerization inhibitor, pharmaceuticals, petroleum chemicals, electroplating, corrosion inhibitors. [Air Prods.; Allchem Industries; Atochem; BASF; Union Carbide]

diethyl 1,2-benzenedicarboxylate. See diethyl phthalate.

diethylene diamine. See piperazine.

1,4-diethylene dioxide. See 1,4-dioxane.

diethylene ether. See 1,4-dioxane.

diethylene glycol. (DEG; dihydroxydiethyl ether; diglycol; 2,2´-oxybisethanol) Aliphatic diol; $CH_2OHCH_2O-CH_2CH_2OH$; CAS 111-46-6; EINECS 203-872-2; production of polyurethane and unsat. polyester resins, triethylene glycol; textile softener; solvent for nitrocellulose, dyes and oils; dehydration of natural gas, elasticizers, and surfactants; humectant for tobacco, casein, and synthetic sponges. [BASF; BP Chem. Ltd; Du Pont; Eastman; Hoechst Celanese; Mitsui Petrochem. Ind.; Mobil; Olin; OxyChem; Shell; Texaco; Union Carbide]

diethylene glycol butyl ether acetate. (butoxyethanol acetate) $C_4H_9OCH_2CH_2OCH_2CH_2OAc$; Solvent. [Arco; Eastman; OxyChem]

diethylene glycol dimethacrylate. CAS 2358-84-1; Crosslinker for rubber vulcanization, moisture barrier films and coatings, photopolymer printing plates and letterpress inks, conversion coatings and adhesives. [CPS; Sartomer]

diethylene glycol laurate. See PEG-2 laurate.

diethylene glycol monooleate. See PEG-2 oleate.

diethylene glycol stearate. See PEG-2 stearate.

diethylene oxide. See 1,4-dioxane.

diethylene oximide. See morpholine.

diethylenetriamine. (DETA; aminoethylethandiamine; 2,2|-iminodiethylamine; bis(2-aminoethyl) amine) $C_4H_{13}N_3$; $NH_2C_2H_4NHC_2H_4NH_2$; CAS 111-40-0; EINECS 203-865-4; solvent for sulfur, acid gases, various resins, dyes; saponification agent for acidic materials; fuel component. [Allchem Industries; Janssen Chimica; Nayler Chem. Ltd; Tosoh; Union Carbide]

diethyl ether. See ethyl ether.

di(2-ethylhexyl) adipate. See dioctyl adipate.

di(2-ethylhexyl) phthalate. See dioctyl phthalate.

diethylin. The diethyl ether of glycerol.

diethyl ketone. (3-pentanone) $CH_3CH_2COCH_2CH_3$; CAS 96-22-0; EINECS 202-490-3; medicine, organic synthesis. [BASF; Janssen Chimica; Penta Mfg.; Union Carbide]

diethyl phthalate. (DEP; ethyl phthalate; phthalic acid, diethyl ester; diethyl 1,2-benzenedicarboxylate) Aromatic diester of ethyl alcohol and phthalic acid; $C_{12}H_{14}O_4$; $C_6H_4(CO_2C_2H_5)_2$; CAS 84-66-2; EINECS 201-550-6; solvent for nitrocellulose and cellulose acetate; plasticizer, wetting agent, insectiad preps.; in perfumery as solvent and fixative; plasticizer in solid rocket propellants. [BP Chem. Ltd; Daihachi Chem. Ind.; Eastman; Hüls Am.; Morflex; Penta Mfg.; Unitex]

diethyl toluamide. (DEET; m-toluic acid diethylamide; N,N-diethyl-m-toluamide; DET) $CH_3C_6H_4CON(C_2H_5)_2$; CAS 134-62-3; EINECS 205-149-7; insect repellants, resin solvent, film formers. [Du Pont; Honeywill & Stein Ltd; Morflex]

Di-Farmon. Herbicide containing mecoprop and dicamba. [ICI Chem. & Polymers Ltd]

Di-Farmon. Soluble concentrate of 21 g dicamba and 319 g mecoprop per liter; used for weed control in cereals and grassland. [Farm Protection Ltd]

Difflam Cream. CAS 132-69-4; Cream containing 3% benzydamine hydrochloride. [3M Pharmaceuticals] †

Difflam Oral Rinse. CAS 132-69-4; Solution containing 0.15% benzydamine hydrochloride. [3M Pharmaceuticals] †

Diflucan®. CAS 86386-73-4; Fluconazole; antifungal. Also known as Candigyn (France), Flucam (Japan), Flucan (Turkey), Flucagyn (Germany), Flugyn (Argentina), Fungata (Austria), Fungustatin (Greece), Triflucan (Argentina, Brazil, Uruguay), Trifluco (France), Trizol (Great Britain, Ireland). [Pfizer International]

difluorodichloromethane. See dichlorodifluoromethane.

Difolatan. CAS 2939-80-2; Captafol fungicide. [Chevron] *

diformyl. $(CHO)_2$; CAS 107-22-2; Glyoxal.

Difusor. Peritoneal dialysis solutions. [The Boots Co plc] *

digallic acid. See tannic acid.

Diganox. CAS 20830-75-5; A proprietary preparation of digoxin. [L Wilcox & Co Ltd] ‡

Di-Gel®. Liquid: simethicone, aluminum hydroxide, magnesium hydroxide; tablets: simethicone, calcium carbon-

---

‡ Trade name and manufacturer not verified   § Trade name without identified manufacturer

ate, magnesium hydroxide; antacid, antigas. [Schering Corp]

Digestan®Non-Gamma 2 Amalgam. Silver amalgam without a gamma 2 phase in powder and capsule form. [Bayer AG]

Digibind. CAS 20830-75-5; A proprietary formulation of digoxin (specific antibody fab fragments); for treatment of potentially life threatening digoxin or digitoxin intoxication. [The Wellcome Foundation Ltd]

Digifortis Kapseals. A proprietary preparation of digitalis. [Parke-Davis] *

Digiglusin. Digitalis; cardiotonic. [Eli Lilly & Co] †

Digilanid. A proprietary preparation of a glycosidal complex of lanatosides; used for treatment of heart failure. [Sandoz] *

digitalis. The dried leaves of the flowering plants of *Digitalis purpurea;* used as a cardiac stimulant.

diglycine. See glycylglycine.

Diglycolamine® Agent (DGA®). 2-(2-aminoethoxy) ethanol; solv. for removal of CO_2 or H_2S from gases, for recovery of aromatics from refinery streams, for prep. of foam stabilizers, wetting agents, emulsifiers, and condensation polymers. [Texaco]

diglycol laurate. See PEG-2 laurate.

diglycol oleate. See PEG-2 oleate.

diglycol stearate. See PEG-2 stearate.

Diglyme. CAS 111-96-6; The dimethyl ether of diethylene glycol; a solvent for polystyrene, PVC/PVA copolymer (*qv*) and polymethyl methacrylate. [ICI Chem. & Polymers Ltd] †

Dihalo. Swimming pool water disinfectant. [Great Lakes Europe]

dihydro-2(3H)-furanone. See butyrolactone.

dihydroisophorone. CAS 873-94-9; 3,5,5- Triethylcyclohexanone; a high boiling point ketone solvent for surface coatings.

Dihydrojasmone. 2-Pentyl-3-methyl-2-cyclopenten-1-one. [Quest Int'l. UK Ltd]

dihydroxyacetone. (DHA; 1,3-dihydroxydimethyl ketone; 1,3-dihydroxy-2-propanone) Aliphatic ketone; $HOCH_2COCH_2OH$; CAS 96-26-4; enzyme; used as cosmetic stain for suntan lotion and reagent in chemical synthesis;humectant, plasticizer, fungicides. [EM Industries; Gist-Brocades Food Ingreds.; Janssen Chimica; Penta Mfg.; Spectrum Chem. Mfg.]

1,4-dihydroxyanthraquinone. See quinizarin.

1,2-dihydroxybenzene. See pyrocatechol.

3,4-dihydroxybenzoic acid. See protocatechuic acid.

2,4-dihydroxybenzophenone. See benzophenone-1.

2,2´-dihydroxy-4,4´-dimethoxybenzophenone. See benzophenone-6.

1,3-dihydroxydimethyl ketone. See dihydroxyacetone.

di(2-hydroxyethyl) amine. See diethanolamine.

dihydroxyethyl sulfide. See thiodiglycol.

1,6-dihydroxyhexane. See hexamethylene glycol.

2,2´-dihydroxy-4-methoxybenzophenone. See benzophenone-8.

1,6-dihydroxynaphthalene. See 1,6-naphthalenediol.

2,3-dihydroxynaphthalene. See 2,3-naphthalenediol.

2,7-dihydroxynaphthalene. See 2,7-naphthalenediol.

1,2-dihydroxypropane. See propylene glycol.

1,3-dihydroxy-2-propanone. See dihydroxyacetone.

2,6-dihydroxypurine. See xanthine.

2,4-dihydroxypyrimidine. See uracil.

diiodoform. (tetraiodoethylene.; ethylene periodide) Used in medicine.

diisobutyl adipate. (DIBA; bis(2-methylpropyl) hexanedioate; diisobutyl hexanedioate; hexanedioic acid, diisobutyl ester) Diester of isobutyl alcohol and adipic acid; $[C_2H_4COOCH_2CH(CH_3)_2]_2$; CAS 141-04-8; used as a plasticizer. [Aceto]

diisobutyl hexanedioate. See diisobutyl adipate.

diisobutyl ketone. (2,5-dimethyl-4-heptanone; Isovalerone) $(CH_3)_2CHCH_2COCH_2CH(CH_3)_2$; CAS108-83-8; EINECS 203-620-1; Solvent for nitrocellulose, rubber, synthetic resins; lacquers, coatings, organic synthesis, roll-coating inks, stains. [Allchem Industries; Eastman; Hüls AG; Union Carbide]

2,4-diisocyanatotoluene. See toluene diisocyanate.

diisodecyl phthalate. (DIDP) $C_6H_4(COOC_{10}H_{21})_2$; CAS 68515-49-1; Used as a plasticizer. [BASF; C.P. Hall; OxyChem]

diisononyl phthalate. CAS 68515-48-0; Plasticizer. [BASF; Chemisphere Ltd; Chisso Am.; C.P. Hall]

diisopropanolamine. (DIPA; 1,1´-iminobis-2-propenol) Aliphatic amine; $HN(CH_2CHOHCH_3)_2$; CAS 110-97-4; emulsifying agents for polishes, textile specialties, leather compounds, insecticides, cutting oils, aq. paints. [BASF]

diisopropylamine. $[(CH_3)_2CH]_2NH$; CAS 108-18-9; EINECS 203-558-5; intermediate, catalyst. [Air Prods.; Aldrich; Atochem N. Am.; BASF; Union Carbide]

1,3- and 1,4- diisopropyl benzene. (3- and 4-isopropylcumene) $C_{12}H_{18}$; CAS 99-62-7 and 100-18-5; EINECS 202-773-1 and 202-826-9; Solvent; intermediate. [Koch]

diisopropylbenzene peroxide. See dicumyl peroxide.

Dijex. Proprietary preparations containing either aluminum hydroxide and magnesium carbonate as a co-dried gel or aluminum hydroxide and magnesium hydroxide; antacids. [The Boots Co plc]

Dikar. Fungicide and miticide. [Rohm & Haas]

2,5-diketopyrrolidine. See succinimide.

Dilabil. CAS 81-23-2; Dehydrocholic acid; choleretic. [Sterling Drug Inc] *

Dilangio. CAS 14286-84-1; A proprietary preparation of bencyclane fumarate; a vasodilator. [Pfizer International] †

Dilantin. CAS 57-41-0; Phenytoin; anticonvulsant. [Parke-Davis]

Dilasoft. Softening and filling agent with hydrophilic properties; for synthetic fibers and blends with cellulosics. [Sandoz Products Ltd] *

Dilatin NA Liquid. Based on unchlorinated aromatic hydrocarbons; a proprietary preparation used in the dyeing of 100% polyester goods. [Sandoz] *

Dilatol®. Vasoactive preparation. [Bayer AG]

Dilaudid. CAS 71-68-1; Hydromorphone hydrochloride; analgesic. [Knoll Pharmaceutical Co]

dilauryl thiodipropionate. (didodecyl 3,3´-thiodipropionate; thiodipropionic acid, dilauryl ester) Diester; diester of lauryl alcohol and 3,3´-thiodipropionic acid; $(C_{12}H_{25}OOCCH_2CH_2)_2S$; CAS 123-28-4; antioxidant, additive for high-pressure lubricants and greases, plasticizer and softening agent, antioxidant for edible fats and oils. [Am. Cyanamid; Evans Chemetics; Morton Int'l.; Witco/Argus]

Dilectene. A proprietary trade name for an aniline-formalde-

---

* Trade name not verified as available   † Trade name verified as obsolete

hyde synthetic resin. §

Dilecto. A proprietary trade name for a phenol-formaldehyde synthetic resin laminated product. §

Dilexo. Copolymer dispersions for paints, adhesives and sealants. [RWE-DEA Chemicals UK Ltd]

dilinoleic acid. (9,12-octadecadienoic acid, dimer; dimer acid) 36-Carbon dicarboxylic acid formed by the catalytic dimerization of linoleic acid; $C_{36}H_{64}O_4$; CAS 6144-28-1; 61788-89-4; lubricant, corrosion inhibitor, mildness additive in household detergents, plastics, and protective coatings. [Arizona; Henkel/Emery; Sherex; Union Camp; Witco/Humko]

Dillex. Gripe mixture. [The Boots Co plc] *

Dilor-G. Dyphylline, guaifenesin USP; indicated as a bronchodilator-expectorant for treating bronchial asthma, emphysema, bronchitis, pneumonitis, and other related bronchopulmonary insufficiency conditions. [Altana Inc] ‡

Dilor Elixir. CAS 479-18-5; Dyphylline; indicated for relief of acute bronchial asthma and for reversible bronchospasm associated with chronic bronchitis and emphysema. [Altana Inc] ‡

Dilosyn. CAS 1229-35-2; A preparation of methdilazine hydrochloride. [British Drug Houses] *

Diluex₩, FG. CAS 1337-76-4; Attapulgite clay; absorbent carrier for pesticides in dusts and wettable powds. [Floridin] *

Dilver. An alloy containing 42% nickel; used in filament lamps, as it has the same coefficient of expansion as glass.

Dimacide. Acid dyes. [ICI Chem. & Polymers Ltd] †

Dimagel. A proprietary preparation containing dimagnesium aluminum trisilicate; an antacid. [Lewis Laboratories] ‡

Dimagel-Belladonna. A proprietary preparation containing dimagnesium trisilicate and belladonna alkaloids; an antacid. [Lewis Laboratories] ‡

dimagnesium orthophosphate. See magnesium phosphate, dibasic.

Dimanin. Quaternary ammonium compound; used for disinfection. [Bayer AG]

Dimanin A Special. Control of algae, molds and bacteria in greenhouses and gardens. [Bayer AG]

Dimatos. See infusorial earth.

dimazon. CAS 83-63-6; Diacetylaminoazotoluene, $C_{18}H_{19}N_3O_4$; a red dye used in ointment or as a dusting powder.

Dimdac. Dimethyldiallylammonium chloride. [Synthetic Chemicals Ltd]

Dimension. Range of nylon/PPE alloys including reinforced grades with glass and talc; applications include wheel covers, wheel hubs, water meter components, resonators, mirror housings and electrical components. [Allied Colloids Ltd]

Dimension®. For the medical industry. [DuPont UK]

Dimension D-9300 BK. Modified nylon 6/PPE alloy; blow molding grade with ultra-high melt visc., load-bearing at elevated temps., good impact resist.; used for automotive applications [Allied-Signal Engineered Plastics]

Dimension Master®. High stability films; for the electrical industry. [DuPont UK]

dimer acid. See dilinoleic acid.

Dimetane. CAS 980-71-2; Brompheniramine maleate; antihistaminic. [Wyeth Laboratories] *

dimethicone. CAS 9006-65-9; A polydimethylsiloxane

$CH_3[Si(CH_3)_2O]Si(CH_3)_3$; used in ointments and topical drug ingredients.

Dimethoate. O,O-Dimethyl-S-(N-methylcarbamoylmethyl) phosphorodithioate; an insecticide and acaricide with contact and plant systemic activity suitable for protection against a broad range of insects and mites. [A/S Cheminova] *

Dimethoate Bayer. CAS 60-51-5; Insecticide with limited systemic action for control of aphids, spider mites, whiteflies, mealy bugs, scale insects, fruit flies, leaf-eating caterpillars and leafminers on pome and stone fruit, grapes, sugar beet, cotton, citrus fruit, tobacco, vegetables. [Bayer AG]

2,6-dimethoxy phenol. (pyrogallol 1,3-dimethyl ether) $C_8H_{10}O_3$; CAS 91-10-1; EINECS 202-041-1. [Janssen Chimica; Penta Mfg.; Schweizerhall; Spectrum Chem. Mfg.]

dimethoxypropane. (acetone dimethyl acetal) Protein precipitant; $(CH_2)_2C(OCH_3)_2$; CAS 77-76-9; EINECS 201-056-0; chemical intermediate for pharmaceuticals; tissue dehydrating agent. [Aldrich; Janssen Chimica; Penta Mfg.; Schering Berlin Polymers; Sherex]

dimethoxytetrahydrofuran. (2,5-dimethoxy tetrahydrofuran) $OCH(OCH_3)C_2H_4CHOCH_3$; CAS 696-59-3; EINECS 211-797-1; biocide. [Chemie Linz UK Ltd]

dimethylacetic acid. See isobutyric acid.

dimethyl adipate. (dimethyl hexanedioate; methyl adipate) Diester of methyl alcohol and adipic acid; $CH_3O_2C-(CH_2)_4CO_2CH_3$; CAS 627-93-0; EINECS 211-020-6. [Du Pont; Morflex; UCB SA]

dimethylamine hydrochloride. $(CH_3)_2NH.HCl$; CAS 506-59-2; EINECS 208-046-5. [Janssen Chimica; Penta Mfg.; Schweizerhall]

p-dimethylaminobenzaldehyde. (4-dimethylaminobenzaldehyde) $C_9H_{11}NO$; CAS 100-10-7; EINECS 202-819-0; Dyes, medicine, reagent. [Aceto; BASF; Penta Mfg.; Schweizerhall]

dimethylaminoethanol. See dimethylethanolamine.

dimethylaminoethyl chloride hydrochloride. (DMC; 1-chloro-2-dimethylaminoethane hydrochloride; 2-chloro-N,N-dimethylethylamine hydrochloride; N-(2-chloroethyl)dimethylamine hydrochloride) $(CH_3)_2NCH_2CH_2Cl \cdot HCl$; CAS 4584-46-7; EINECS 224-970-1; Manufacture of antihistamines and other pharmaceuticals; organic intermediate for introduction of β-dimethylaminoethyl radical. [ICI Spec.; Janssen Chimica; Lonza]

n,n-dimethylaniline. (aniline, N,N-dimethyl) C_8H_{11}; $C_6H_5N(CH_3)_2$; CAS 121-69-7; EINECS 204-493-5; Dyes, intermediates, solvent, manufacture of vanillin, stabilizer (acid acceptor), reagent. [Allchem Industries; BASF; Du Pont; ICI Am.]

dimethyl anthranilate. (9,10-dimethyl anthranilate; N-methyl methyl anthranilate) $CH_3OOCC_6H_4NHCH_3$; CAS 85-91-6, 781-43-1; EINECS 212-308-4; Perfumes, flavoring, drugs. [Am. Bio-Synthetics; Bell Flavors & Fragrances; Penta Mfg.]

dimethylarsinic acid sodium salt. See sodium cacodylate.

dimethylbenzene. See xylene.

dimethyl 1,2-benzenedicarboxylate. See dimethyl phthalate.

α-(5,6-dimethylbenzimidazolyl)cyanocobamide. See vitamin B_{12}.

dimethyl butanedioate. See dimethyl succinate.

dimethyl carbinol. See isopropyl alcohol.

---

‡ Trade name and manufacturer not verified   § Trade name without identified manufacturer

dimethyldithiocarbamic acid, sodium salt. *See* sodium dimethyldithiocarbamate.

dimethylethanolamine. (DMAE; 2-dimethylaminoethanol; deanol) $(CH_3)_2NCH_2CH_2OH$; CAS 108-01-0; EINECS 203-542-8; Synthesis of dyestuffs, pharmaceuticals, textile auxiliaries. [Air Prods.; Atochem N. Am.; BASF; Nippon Nyukazai; Texaco; Union Carbide]

dimethyl ether. (methane, oxybis-; oxybismethane; methyl ether) Organic compound.; CH_3OCH_3; CAS 115-10-6; EINECS 204-065-8; Aerosol propellant.

n,n-dimethylethylamine. (N-ethyldimethylamine) $(CH_3)_2NC_2H_5$; CAS 598-56-1; EINECS 209-940-8.

1,1-dimethylethylhydroperoxide. *See* t-butyl hydroperoxide.

dimethyl ethynyl carbinol. *See* methyl butynol.

dimethyl formamide. (DMF) C_3H_7NO; $HCON(CH_3)_2$; CAS 68-12-2; EINECS 200-679-5; Solvent in vinyl resins and acetylene, butadiene, acid gases; polyacrylic fibers; catalyst in carboxylation reactions, organic synthesis, carrier for gases. [Air Prods.; Ashland; BASF; Du Pont; ICI Spec.; Mallinckrodt; Mitsubishi Gas; Nissan Chem. Ind.; UCB SA]

dimethyl glyoxime. (diacetyldioxime; butane dioxime) $CH_3C(:NOH)C(:NOH)CH_3$; CAS 95-45-4; EINECS 202-420-1; Analytical chemistry, esp. as reagent for nickel, biochemical research. [Atomergic Chemetals; Pfaltz & Bauer]

2,5-dimethyl-4-heptanone. *See* diisobutyl ketone.

dimethyl hexanedioate. *See* dimethyl adipate.

5,5-dimethyl hydantoin. *See* DM hydantoin.

dimethylhydroxybenzene. *See* xylenol.

5,5-dimethyl-2,4-imidazolidinedione. *See* DM hydantoin.

dimethyl lauramine. (lauryl dimethylamine; dodecyl-dimethylamine) Tertiary aliphatic amine; $CH_3(CH_2)_{10}$-$CH_2N(CH_3)_2$; CAS 112-18-5; 67700-98-5; EINECS 203-943-8; liq. cationic detergent; corrosion inhibitor; acid-stable emulsifier. [Ethyl; Lonza; Mason; Sherex]

dimethylmethane. *See* propane.

3,7-dimethyl-2,6-octadienal. *See* citral (cis and trans).

3,7-dimethyl-6-octenal. *See* citronellal.

3,7-dimethyl-6-octen-1-ol. *See* citronellol.

dimethylphenol. *See* xylenol.

dimethyl-phenylene-green. The tetramethyl derivative of Phenylene blue, $C_{16}H_{20}N_3Cl$; dyes silk and other fabrics.

dimethyl phthalate. (DMP; dimethyl 1,2-benzene-dicarboxylate; 1,2-benzenedicarboxylic acid dimethyl ester; phthalic acid dimethyl ester) Diester of methyl alcohol and phthalic acid; $C_6H_4(CO_2CH_3)_2$; CAS 131-11-3; EINECS 205-011-6; solvent for resin, plasticizer for cellulose acetate and nitrocellulose lacquers; plastics, rubber; coating agents; safety glass. [Allchem Industries; BASF; Daihachi Chem. Ind.; Eastman; Hüls Am.; Morflex; UCB SA; Unitex]

2,2-dimethyl-1,3-propanediol. *See* neopentyl glycol.

dimethyl succinate. (dimethyl butanedioate; methyl succinate) $CH_3OOC(CH_2)_2COOCH$; CAS 106-65-0; EINECS 203-419-9; Light and heat stabilizer for polyolefins, ABS polymer systems, flexible PVC, food pkg. [Du Pont; Penta Mfg.; Schweizerhall]

dimethyl sulfoxide. (DMSO; methyl sulfoxide) $(CH_3)_2SO$; CAS 67-68-5; EINECS 200-664-3; Solvent for polymerization; analytical reagent; industrial cleaners, pesticides, paint stripping, hydraulic fluids, medicine (anti-inflammatory), veterinary medicine, plant pathology and nutrition, pharmaceuticals, spinning synthetic fibers. [Aldrich; Atochem N. Am.; Gaylord]

Dimilin. CAS 35367-38-5; Diflubenzuron; insecticides to counteract a number of harmful organisms occurring in agricultural, horticultural and forestry circles, fungus growth, sub-tropical cultures (weevils, cotton worm, many varieties of fruit insects, etc.); blocks development from the la [ICI Agrochemicals]

Dimilin. CAS 35367-38-5; Diflubenzuron; insecticides to counteract a number of harmful organisms occurring in agricultural, horticultural and forestry circles, fungus growth, sub-tropical cultures (weevils, cotton worm, many varieties of fruit insects, etc.); blocks development from the larvae to the adult insect stage; does not harm the environment. [Duphar BV]

Dimipressin. CAS 50-49-7; A proprietary preparation of imipramine; an antidepressant. [RP Drugs] *

Dimodan LS Kosher. Sunflower oil dist. monoglyceride, unsat.; food emulsifier; w/o emulsifier for low-calorie spreads, icing shortenings and cake shortenings. [Grindsted Prods.; Grindsted Prods. Denmark]

Dimodan O Kosher. Partially hydrogenated soybean oil glyceride; food emulsifier for margarine, icing shortenings, coffee whiteners. [Grindsted Prods.; Grindsted Prods. Denmark]

Dimodan PM. Hydrogenated lard or tallow dist. monoglyceride, sat.; food emulsifier; starch complexing agent; for margarine, cake shortenings, confectionery coatings; softener for bread; peanut butter stabilizer. [Grindsted Prods.; Grindsted Prods. Denmark]

Dimodan PV, PV 300 Kosher. Hydrogenated soybean oil dist. monoglyceride, unsat.; food emulsifier, starch complexing agent, antisticking agent; crumb softener for bread; aerating agent in cake mixes and frozen desserts. [Grindsted Prods.; Grindsted Prods. Denmark]

Dimodan PVP Kosher. CAS 97593-29-8; Hydrogenated palm oil dist. monoglyceride; food emulsifier, starch complexing agent, aerating agent; for margarine, cake shortening, etc.; crumb softener for bread; peanut butter stabilizer. [Grindsted Prods.; Grindsted Prods. Denmark]

Dimodan S. CAS 97593-29-8; Lard dist. monoglyceride, unsat.; food emulsifier for margarine, cake shortenings, icing shortenings, coffee whiteners. [Grindsted Prods.]

Dimotane Plus Elixir. Brompheniramine maleate and pseudoephedrine hydrochloride; for symptomatic relief of allergic rhinitis. [Wyeth Laboratories]

Dimotane Tablets. CAS 980-71-2; Brompheniramine maleate; antihistamine for treatment of allergic conditions such as hay fever and urticaria. [Wyeth Laboratories]

Dimotapp Elixir. A proprietary preparation containing brompheniramine maleate, phenylephrine hydrochloride, and phenylpropanolamine hydrochloride; a cough linctus. [Wyeth Laboratories] *

Dimul DDM K. Monoglycerides; food emulsifier. [Witco/Humko]

Dimundite. *See* Camite. §

Dimycin. Streptomycin and dihydrostreptomycin for veterinary purposes. [Glaxo Laboratories] *

Dimyril. A proprietary preparation containing isoaminile citrate; a cough suppressant. [Fisons plc, Pharmaceuticals Div] *

Dinacrin. CAS 54-85-3; Isoniazid; antibacterial. [Sterling

---

* Trade name not verified as available   † Trade name verified as obsolete

Drug Inc] *

Dinamene. Selective weedkiller. [Murphy Chemical Co Ltd]†

Dindevan. CAS 83-12-5; A proprietary preparation of phenindione; an anticoagulant. [Duncan Flockhard Ltd]

Dingler's green. A pigment; a chromium phosphate.

Diniternal. See Dinitra.

Dinitra. 2:4-Dinitrophenol. §

3,5-dinitrobenzoic acid. $C_7H_4N_2O_6$; CAS 99-34-3; EINECS 202-751-1. [Exchem Industries Ltd; Nobel; Schweizerhall]

dinonyl phenol. Mixture of dialkyl substituted phenols; $(C_9H_{19})_2C_6H_3OH$; CAS 1323-65-5; Solvent. [Allchem Industries; Texaco]

Dinoram. Alkyl propylene diamines based on coco, tallow, and oleyl amines; bitumen emulsifiers and adhesion agents, production of cationic emulsions of oils and waxes, chemical synthesis intermediates, treatment of pigments and fillers for paints, epoxy curing agents, bactericides. [Atochem UK/Ceca]

Dinoramox. Ethoxylated alkyl propylene diamines; emulsifiers, corrosion inhibitors, wetting agents. [Atochem UK/Ceca]

3,3′-dioctadecyl thiodipropionate. See distearyl thiodipropionate.

Dioctyl. CAS 577-11-7; EINECS 209-406-4; A proprietary trade name for dioctyl sodium sulfosuccinate. §

dioctyl adipate. (DOA; bis(2-ethylhexyl) hexanedioate; di(2-ethylhexyl) adipate; hexanedioic acid, bis (2-ethylhexyl) ester) Diester of 2-ethylhexyl alcohol and adipic acid; $[CH_2CH_2COOCH_2CH(C_2H_5)C_4H_9]_2$; CAS 103-23-1; EINECS 203-090-1; used as a plasticizer, commonly blended with general purpose plasticizers (DOP, DIOP); solvent, aircraft lubricants. [BASF; Chisso; Eastman; Hüls AG; Inolex; Monsanto]

dioctyl phthalate. (DOP; di(2-ethylhexyl) phthalate; di-s-octyl phthalate; 1,2-benzenedicarboxylic acid dioctyl ester) Diester of 2-ethylhexyl alcohol and phthalic acid; $C_6H_4[COOCH_2CH(C_2H_5)C_4H_9]_2$; CAS 117-81-7; EINECS 204-211-0; plasticizer for many resins and elastomers. [Aristech; BASF; Chemisphere Ltd; Chisso; Daihachi Chem. Ind.; Eastman; C.P. Hall; Hüls AG; Mitsubishi Gas; UCB SA]

dioctyl sodium sulfosuccinate. (DSS; sodium dioctyl sulfosuccinate; sodium di(2-ethylhexyl) sulfosuccinate; docusate sodium) Sodium salt of the diester of 2-ethylhexyl alcohol and sulfosuccinic acid; $C_8H_{17}OOCCH_2CH(SO_3Na)COOC_8H_{17}$; CAS 577-11-7; 1369-66-3; EINECS 209-406-4; food additive (processing aid in sugar industry); stabilizer for hydrophilic colloids; wetting agent, dispersant, emulsifier in cosmetic, pharmaceutical, and industrial applications; adjuvant in tablet formation; solubilizer. [Alco; Am. Cyanamid; Brotherton Ltd; Eastern Color & Chem.; EM Industries; Hart Prod. Corp.; Henkel/Organic Prods.; Hodag; McIntyre; Mona; Sherex; Witco]

Dioderm. CAS 50-23-7; White aqueous cream containing 0.1% hydrocortisone BP; used for topical treatment of eczema, dermatitis, and all types of inflammatory, pruritic and allergic skin conditions. [Dermal Laboratories Ltd]

Diodoquin. A proprietary preparation of diiodohydroxyquinoline; an anti-amebic drug. [G D Searle & Co Ltd] *

Diofan®. Vinylidene chloride polymer dispersions; for production of chemically resist. and heat sealable pkg. materials impervious to water vapor, gas and aromas; binders for textile coatings; moisture-resist. coatings for building materials. [BASF AG; BASF plc]

Diofan®D. Polyvinylidene chloride dispersion. [BASF plc]*

Dioform. CAS 540-59-0; (1,2-dichloroethylene; acetylene dichloride; dichloracetylene) ClHC:CHCl; used as a general solvent for organic materials, dye extraction, perfume, lacquers, thermoplastics, organic synthesis.

Diogyn. CAS 50-28-2; Estradiol; estrogen. [Pfizer Inc] †

Diogyn E. CAS 57-63-6; Ethinyl estradiol; estrogen. [Pfizer Inc] †

Diogynets. CAS 50-28-2; Estradiol; estrogen. [Pfizer Inc] †

Diolpate®. Saturated polyesters; used for PVC and rubber compounding, surface coating industry. [Kemira Polymers]

Dioltech 311. Molybdate-based; corrosion and scale control agent for open recirculating cooling water systems. [Drew Ind. Div.]

Di-On. Herbicide. [Agan Chemical Manufacturers Ltd] †

Dion®. Polyester resins. [Reichhold]

Dion®FR 6308. Polyester resin; flame retardant for corrosion applications in filament winding and compression molding/SMC/BMC process; used for molded parts (industrial grating, computer housings, etc.); high physical strengths, MIL-R-7575 compliance; good fume handling corrosion resistance, high heat resistance; ambient or elev. temp. gel and cure. [Reichhold] *

Dion®FR 6657. Polyester resin; flame retardant, promoted resin for transportation and construction applications; only for use with alumina trihydrate. [Reichhold] *

Dion® VER 9100 NP. Bisphenol epoxy vinyl ester resin; thermoset with good corrosion resist., inherent strength and impact resist.; for filament winding centrifugal casting processes. [Reichhold/Reactive Polymers] *

Dion® VER. Vinyl ester resins. [Reichhold]

Dional 11, 113. Fluorinated hydrocarbon; solv. for dry cleaning. [Hoechst Celanese]

Dion®Cor-Res. Synthetic resins. [Reichhold]

Dionil®. A range of fatty acid amide polyglycol ethers; special detergent, soil suspending/leveling and protective colloid action; some of the range have an additional superfatting effect. [Hüls AG]

Dionil® OC. CAS 26027-37-2; 31799-71-0; PEG-3 oleamide; detergent for light and heavy-duty detergents, dishwashing agents, cosmetic preps.; component in textile auxliaries; refatting agent. [Hüls Am.; Hüls AG]

Dionil®SD. Fatty acid amide polyglycol ether; superfatting and preparation agent. [Hüls Am.; Hüls AG]

Dion-Iso®. Common metals and their alloys; for metal building materials, transportable buildings, railway tracks, nonelectric cables and wires, small hardware items, pipes and tubes, safes. [Reichhold]

Dionosil. Propyliodone suspensions or powders. [Glaxo Laboratories] *

Diorez®. Saturated polyester polyols; used for polyurethane systems, microcellular elastomers, thermoplastic polyurethane fabrics and surface coatings. [Kemira Polymers]

Diorez®SC. Polyesters, acrylics, epoxies; surface coatings intermediates. [Kemira Polymers]

Diosal. Sodium diiodosalicylate.

Diotroxin. L-Thyroxine and L-triiodo-thyronine mixture. [Glaxo Laboratories] *

---

‡ Trade name and manufacturer not verified   § Trade name without identified manufacturer

Diox DR 22. CAS 13463-67-7; EINECS 236-675-5; A trade name for a rutile (titanium dioxide) type white pigment with a blue tone and good dispersability. [U.S. Industrial Chem.] *

1,4-dioxane. (diethylene ether; 1,4-diethylene dioxide; diethylene oxide) Ether; $OCH_2CH_2OCH_2CH_2$; CAS 123-91-1; stabilizer for chlorinated hydrocarbons; solv. for adhesives, dyes, cellulose, lacquer, wax, pharmaceuticals, and coatings. [Ashland; BASF; CPS; Ferro/Grant; Mallinckrodt; Union Carbide]

Dioxatrine. CAS 5633-14-7; Benzetimide hydrochloride; anticholinergic. [Janssen Pharmaceutica Inc] *

dioxine. CAS 1746-01-6; (Gambine R). A dyestuff; nitrosodihydroxynaphthalene, $C_{10}H_7NO_3$; dyes iron mordanted fabrics, green, and chrome mordanted materials, brown.

Dioxitol. A colorless, slightly hygroscopic liquid with mild odor; used as a solvent in paints, lacquers, textile printing inks, and stains; it is effective as a metal degreasing agent and is used in production of safety glass; it is also used as a coupling agent for cutting oils and emulsifiable oils; and in production of plasticizers and also as an extraction agent for essences and perfumes. [Shell UK]

dioxogen. A 3% solution of hydrogen peroxide, H_2O_2; used as a disinfectant.

dioxopurine. See xanthine.

Dioxyanthranol. Anthrarobine, $C_{14}H_{10}O_3$; used externally for skin diseases.

dioxybenzone. See benzophenone-8.

2,4-dioxypyrimidine. See uracil.

DIPA. See diisopropanolamine.

Dipanol. Terpene rubber reclaimimg oil. [Crowley Chem.]

Dipar. CAS 834-28-6; A proprietary preparation of phenformin hydrochloride; a hypoglycemic agent. [Hoechst UK] *

Dipasic. A proprietary preparation of aminosalicylate of isonicotinic acid hydrazide; an antituberculous drug. [Geistlich Sohne AG] *

Dipel®. Bacillus thuringiensis; a bacterial insecticide for control of caterpillars. [English Woodlands Ltd]

dipentaerythritol. $(CH_2OH)_3CCH_2OCH_2C(CH_2OH)_3$; CAS 126-58-9; used as intermediate in manufacture of alkyds and drying oils, paints and coatings. [Allchem Industries; Honeywill & Stein Ltd]

Dipentek. A technical grade of dipentaerythritol. §

dipentene. (cinene; limonene, inactive; cajeputene) $C_{10}H_{16}$; CAS 138-86-3; EINECS 205-341-0; Solvent for oleoresinous products, rosin, ester gum, etc.; rubber compounding. and reclaiming; dispersant for oils, resins and combinations, pigments, driers, paints, enamels, lacquers; general wetting; printing inks, perfumes, flavors, waxes, polishes. [Arizona; Hercules; Penta Mfg.; SCM Glidco Organics; Veitsiluoto Oy]

Dipentene No. 122. Terpene liquid; solvent and antiskinning agent in paints. [Hercules] *

Di-Petronate Series. Range of diluted sodium petroleum sulfonates; emulsifiers, dispersing and wetting agents for use when lower viscosity and easier handling than the Petronate Series is required. [Witco Chemical Ltd] *

Dipex. A proprietary mold lubricant for rubber; a water-soluble sodium sulfonate obtained from petroleum-acid sludges. §

Diphasol. Dyeing and printing assistant. [Ciba plc] *

Diphen 60-B. A phenol-urea formaldehyde resin. §

diphenal. The sodium salt of diaminodihydroxybiphenyl; used as a photographic developer.

diphenatrile. See diphenylacetonitrile.

Diphentoin. CAS 630-93-3; Phenytoin sodium; anticonvulsant. [SmithKline Beecham] *

diphenyl. (1,1'-biphenyl; bibenzene; phenylbenzene) $C_6H_5C_6H_5$; CAS 92-52-4; Used as a heat transfer agent, fungistat for agricultural use, in organic synthesis. [Aldrich; Coalite Chem. Div; Koch; Monsanto; Sybron]

Diphenyl. Direct dyes. [Ciba plc] *

diphenylacetonitrile. (diphenatrile) $(C_6H_5)_2CHCN$; CAS 86-29-3; EINECS 201-662-5; Preparation of diphenylacetic acid, synthesis of antispasmodics, herbicide. [Andeno BV; R.W. Greeff; Janssen Chimica; Schweizerhall]

diphenyl ether. See diphenyl oxide.

diphenylmethane-4,4'-diisocyanate. See MDI.

diphenyloxazole. (DPO; 2,5-diphenyloxazole) $C_{15}H_{11}NO$; CAS 92-71-7; EINECS 202-181-3; Primary fluor used as scintillation counters or in wavelength shifters. [Du Pont Medical Prods.; Packard Instrument BV; Penta Mfg.; Spectrum Chem. Mfg.]

diphenyl oxide. (phenyl ether; diphenyl ether) $(C_6H_5)_2O$; CAS 101-84-8; EINECS 202-981-2; Perfumery, soaps; heat-transfer medium; chemical intermediate for halogenation, acylation, alkylation, etc. [Monsanto; Penta Mfg.]

Diphone. A range of sulfone chemicals; for use in continuous tin-plating processes, and as monomers for engineering plastics. [Yorkshire Chemicals plc] *

Diphyl®, Diphyl®T. Heat transfer media for indirect heating and cooling in the chemical, petrochemical, plastics, man-made fibers, fat and soap industries. [Bayer AG; Bayer plc]

Dipidolor. CAS 302-41-0; A proprietary preparation of piritramide; an analgesic. [Janssen Pharmaceutical Ltd]*

Diplosal. CAS 552-94-3; Salicyl-salicylic acid, $HO \cdot C_6H_4\text{-}COOC_6H_4COOH$; used in medicine in chronic and acute rheumatism.

Diplovax. Live oral poliovirus vaccine; immunizing agent. [Pfizer Inc] †

Dipolymer. Coumarone-indene resin.

dipotassium carbonate. See potassium carbonate.

dipotassium disulfite. See potassium metabisulfite.

dipotassium hydrogen orthophosphate. See potassium phosphate, dibasic.

dipotassium orthophosphate. See potassium phosphate, dibasic.

dipotassium phosphate. See potassium phosphate, dibasic.

Dippel's animal oil. See bone oil.

Dippel's oil. See bone oil.

dipping metal. A jeweler's alloy containing 48 parts of copper and 15 parts of zinc.

Diprane®. Polyester polyurethane systems; used for mining, quarrying, construction and engineering. [Kemira Polymers]

Diprivan. Injectable anesthetic. [ICI Chem. & Polymers Ltd]

Diprofarn. CAS 5907-38-0; Dipyrone; analgesic; antipyretic. [Farmitalia (Farmaceutici Italia)] *

dipropylene glycol. (di-1,2-propylene glycol; 1,1|-oxybis-2-propanol) Mixture of diols; $CH_3CHOHCH_2OCH_2CHOH\text{-}CH_3$; CAS 110-98-5; EINECS 203-821-4; polyester and alkyd resins, reinforced plastics, plasticizers, solvents.

---

* Trade name not verified as available

† Trade name verified as obsolete

[Aldrich; Olin; Texaco]

Diprosin A-100. Disproportionated rosin; polymerization emulsifier for synthetic rubbers and plastics. [Toho Chem. Industry]

Diprosin K-80. CAS 61790-50-9; Disproportionated rosin potassium soap; polymerization emulsifier for synthetic rubbers and plastics. [Toho Chem. Industry]

Diprosin N-70. CAS 61790-51-0; Disproportionated rosin sodium soap; polymerization emulsifier for synthetic rubbers and plastics. [Toho Chem. Industry]

Diprosono. CAS 5593-20-4; Betamethasone dipropionate; glucocorticoid. [Schering Corp]

Dipsal. CAS 7491-14-7; Dipropylene glycol salicylate; uv absorbent for sunscreens; suitable as inhibitor for uv degradation of polymers and dyestuffs; does not deteriorate in contact with perspiration; emollient for toiletries, alcohol lotions, vegetable or mineral-type products, and pharmaceutical specialties; suitable for hair applications; useful to reduce deterioration and discoloration of polymers. [Scher]

Dipterex®. CAS 52-68-6; Trichlorfon; primarily a stomach poison insecticidal spray used for control of mangold fly, fruit fly and pests of maize, alfalfa, and cotton. [Bayer AG]

Dipterex®80. CAS 52-68-6; Organophosphorus insecticide as a soluble powder containing 80% w/w trichlorfon; used to control mangolf fly on beet, cabbage white, leaf minor, and other caterpillars on brassicas. [Bayer plc]

Dirame. CAS 13717-04-9; Propiram fumarate; analgesic. [Miles Pharmaceuticals] *

diresorcinolphthalein. See fluorescein.

Diresul. Synthetic organic dyestuffs for use in the textile industry. [Sandoz Products Ltd]

Direx® 4L. CAS 330-54-1; Diuron suspension; flowable herbicide for control of many weeds and grasses in a variety of crops. [Griffin]

Dirocide. CAS 1642-54-2; Diethylcarbamazine citrate; anthelmintic. [Bristol-Myers Squibb Co Inc] *

Diroval. CAS 7779-90-0; Zinc phosphate. [James M. Brown Ltd]

Dirubin. CAS 1344-28-1; Refined corundum and crystalline aluminum oxide; for production of ramming mixes, shape bricks, and crucibles for the lining of high temperature furnaces; molding material for precision casting molds and the casting of aggressive steels; raw material for the electroslag remelting process, separating agent for annealing process. [Hoechst AG]

Disadine. Antiseptic for topical use, presented as a dry powder spray. [ICI Chem. & Polymers Ltd]

Disalcid™. CAS 552-94-3; Salsalate; tablets and capsules; analgesic; anti-inflammatory. [3M Pharmaceuticals]

Disalol. Phenylsalicylsalicylate, $HO \cdot C_6H_4 \cdot CO \cdot O \cdot C_6H_4COOC_6H_5$.

Disamide. CAS 671-88-5; A proprietary trade name for disulfamide; an oral diuretic. [British Drug Houses] *

Discase. CAS 9001-09-6; Chymopapain; enzyme. [Travenol Laboratories Inc] *

Discelite. A proprietary trade name for a diatomaceous silica filler. §

Discharge Agent DP. Reducing agent for discharge and discharge resist printing on polyester, acetate, triacetate, and their blends with polyamide. [BASF]

Discharge Lake R and RR. See paranitraniline red.

Disco 727. Nonionic wax; for use with cationic fixatives in textiles. [Callaway]

Discodye 1148. Natural org. compound; retarding agent for cotton and cotton/polyester blends. [Callaway]

Discofix DBA. Fixative for improvment of home laundry, fastness of fiber reactive, sulfur and indigo dyes. [Callaway]

Discol 715. Cationic water-sol. polymer; for indigo dye fixation in textiles. [Callaway]

Discol 1457. Silicone defoamer; for textile dyeing and finishing. [Callaway]

Discol DFW. CAS 577-11-7; EINECS 209-406-4; Dioctyl sodium sulfosuccinate; fast wetting agent for denim finishing and continuous carpet dyeing. [Callaway]

Discolite. $NaHSO_2 \cdot CH_2O \cdot 2H_2O$, a reducing agent for stripping in dyeing. §

Discoloc 70-A. Cationic resinous polymer dye; fixing agent and color control aid; for indigo dye fixation. [Callaway]

Discolube 473-A. Cationic lubricant improving beaming efficiency after indigo dyeing. [Callaway]

Discopen 216. Nonrewetting penetrant for rainwear-type finishes; dec. at temps. > 250 F. [Callaway]

Discosoft 1043-S. Fatty blend; softener and sanforizing lubricant that does not promote color bleeding in finish bath; excellent resist. to scorching and yellowing. [Callaway]

Discoterge 326-D. Crypto-anionic surfactant; surfactant for use in desizing, alkaline scouring, and peroxide bleaching. [Callaway]

Discozone MAC. Conc. antiozonant/softener for textiles; easy handling in automated feed systems. [Callaway]

Disfico. A trade name tor a vulcanized fiber used for electrical insulation. §

Disflamoll®. Phosphate plasticizers. [Bayer plc]

Disflamoll®DPK. CAS 26444-49-5; Diphenylcresyl phosphate; flame retardant plasticizer for plasticized PVC products; used in air ducts, tarpaulins, driving and conveyor belts, imitation leather, coatings, hoses and extruded goods, cable sheathing and insulation, soles and injection molded items. [Bayer AG; Miles; Polysar]

Disflamoll® DPO. Diphenyloctyl phosphate; flame retardant plasticizer for type PVC applications, dip, rotationally, extruded and injection molded parts, mechanical foam. [Bayer AG; Miles; Polysar]

Disflamoll® TCA. Trichloroethyl phosphate; monomeric plasticizer. [Bayer AG]

Disflamoll®TKP. Tricresyl phosphate; see Disflamoll DPK. [Bayer AG; Miles; Polysar]

Disflamoll® TOF. Trioctyl phosphate; plasticizer with extreme low temp. flexibility (to -65 C) [Bayer AG; Miles; Polysar]

Disflamoll®TP. CAS 115-86-6; Triphenyl phosphate; flame retardant; gelatinizing and plasticizing agent for collodion cotton; plasticizer w/o gelatinizing properties for acetyl cellulose; reduces flamm. of NC and acetyl cellulose-based plastic compounds and lacquer films; mfg. of photographic film materials, surface coatings. [Bayer AG; Miles; Polysar]

Disipal. A proprietary preparation of orphenadrine hydrochloride; used in the treatment of Parkinsonism. [Brocades Pharma]

Disipal (as HCl). CAS 4682-36-4; Orphenadrine citrate; relaxant; antihistaminic. [3M Pharmaceuticals] †

disodium dithionite. See sodium hydrosulfite.

disodium edetate. See disodium EDTA.

---

‡ Trade name and manufacturer not verified   § Trade name without identified manufacturer

disodium EDTA. (disodium edetate; disodium dihydrogen ethylene diamine tetraacetate; ethylenediaminetetraacetic acid, disodium salt; edetate disodium) CAS 139-33-3; Substituted diamine; $C_{10}H_{16}N_2O_8 \cdot 2Na$; food preservative, chelating and sequestering agent; anticoagulant; pharmaceutic aid. [W.R. Grace/Hampshire; R.W. Greeff]

disodium hydrogen phosphate. See disodium phosphate (anhydrous).

disodium hydrogen phosphate, dihydrate. See disodium phosphate, dihydrate.

disodium IMP. See disodium inosinate.

disodium inosinate. (disodium IMP; sodium inosinate) A 5'-nucleotide derived from seaweed or dried fish; $C_{10}H_{13}N_4O_8P \cdot 2Na$; CAS 4691-65-0; Flavor potentiator in foods. [Penta Mfg.; Schweizerhall]

disodium orthophosphate. See disodium phosphate (anhydrous).

disodium phosphate (anhydrous). (DSP-O; sodium phosphate, dibasic; disodium hydrogen phosphate; disodium orthophosphate) Inroganic salt; phosphoric acid, disodium salt; Na_2HPO_4; CAS 7558-79-4; 10140-65-5; 7782-85-6; EINECS 231-448-7; controls pH in mildly alkaline solutions; food products, water treatment, animal feed, textiles, pharmaceuticals, chemicals, fertilizers, detergents. [Albright & Wilson; Monsanto; Rhone-Poulenc Basic; U.S. Biochemical; Whiting, Peter Ltd]

disodium phosphate, dihydrate. (DSP-2; sodium phosphate, dibasic dihydrate; disodium hydrogen phosphate, dihydrate) Sodium phosphate; $Na_2HPO_4 \cdot 2H_2O$; CAS 10028-24-7; EINECS 231-448-7; controls pH in mildly alkaline solutions. [Albright & Wilson; Monsanto; Rhone-Poulenc Basic; U.S. Biochemical]

disodium tartrate. See sodium tartrate.

Disolite. A disinfectant used in the mushroom industry. [Coventry Chemicals Ltd]

Dispargen. A form of colloidal mercury.

Disparit B. CAS 79-01-6; Trichlorethylene, C_2HCl_3; a disinfecting cleaning compound. §

Dispello. A proprietary preparation of salicylic acid, zinc chloride and hypophosphorous acid; a corn cure. [Ayrton Saunders plc] ‡

Disperbyk®. Solution of an alkylolammonium salt of a higher molecular weight polycarboxylic acid; wetting, dispersing additive to prevent settling and flooding of pigments; for solv. and aq. systems, stains, wood preservatives, anticorrosive primers, wash primers, nitrocellulose primers, fillers, antifouling paints, emulsion paints; emulsifier. [Byk-Chemie USA]

Disperbyk®-181. Solution of an alkanolammonium salt of a polyfunctional polymer; wetting additive for emulsion paints based on polymethacrylates and copolymers, vinyl esters, styrene copolymers, water-reducible paint systems, solv.-based paint systems; improves color development, gloss. [Byk-Chemie USA]

Dispercab. CAS 9004-36-8; Cellulose acetate butyrate dispersions. [Tennant-KVK Ltd]

Dispercap. CAS 9004-39-1; Cellulose acetate propionate dispersions. [Tennant-KVK Ltd]

Dispercel. CAS 9004-70-0; Nitrocellulose dispersions. [Tennant-KVK Ltd]

Dispercoll C-74. Poly-2-chlorobutadiene aq. disp.; CR latex for formulation of water-based contact cements, laminating and mastic adhesives for the automotive, construc-tion, furniture, footwear, and pkg. industries; slight to med. cryst. [Bayer; Miles]

Dispercryl. Acrylic copolymer dispersions. [Tennant-KVK Ltd]

Disperfin. CAS 1309-37-1; Transparent iron oxide pigment pastes. [Brockhues AG]

Disperkyd. Alkyd copolymer dispersions. [Tennant-KVK Ltd]

Dispermid. Polyamide copolymer dispersions. [Tennant-KVK Ltd]

Dispersant 1084. Proprietary blend; pitch dispersant for pulp/paper industry. [Hart Chem. Ltd.]

Dispersant LF-88. Low foaming pulp machine pitch dispersant. [Hart Chem. Ltd.]

Disperse-Ayd. Dispersing agents; for paints, inks, etc. [Cornelius Chemical Co Ltd] *

Disperse-Ayd 1. Proprietary reaction prod of high molecular weight surfactants with a long oil alkyd; broad spectrum dispersant, wetting agent, stabilizing aid for solv.-thinned coatings; wets and deflocculates most org. and inorg. pigments. [Daniel Prods.]

Disperse-Ayd 6. Synergistic blend of wetting and dispersing agents; wetting, dispersing, and deflocculating agents for use with carbon black pigments in solv.-thinned coatings. [Daniel Prods.]

Disperse-Ayd 15. Modified thermoplastic acrylic in propylene glycol monomethyl ether acetate; pigment dispersing vehicle for high performance solv.-thinned coatings. [Daniel Prods.]

Disperse-Ayd W-22. Surfactant blend; pigment dispersant for water-thinned coatings. [Daniel Prods.]

Dispersite. A proprietary trade name for a dispersion of rubber, rubber-like and film-forming resins in water. §

Dispersogen A. CAS 9084-06-4; Sodium naphthalene formaldehyde sulfonate; dispersing and dyeing auxiliary; emulsifier, wetting agent, adjuvant for agricultural formulations. [Hoechst Celanese/Colorants & Surf.; Hoechst AG]

Dispersogen SI. Sodium polynaphthalene sulfonate and sodium C12-14 alkyl sulfate; surfactant for agricultural formulations. [Hoechst Celanese/Colorants & Surf.]

Dispersol. Condensation product of formaldehyde and sodium naphthalene sulfonate; dispersible powders and aqueous dispersions eg for dyestuffs, pigments, pest control. [ICI Chem. & Polymers Ltd]

Dispersol. Various liquid blends of dispersants for cooling water systems; used for open recirculating water systems. [Schaefer Technologies Inc]

Dispersol 103, 105. pitch dispersant for neutral to alkaline systems in paper/pulp processing; keeps particles suspended, prevents agglomeration; dispersant for clay slurries, pigments and coatings. [CNC Int'l. L.P.]

Disperstat. Antistatic agent. [Stephenson Thompson Textile Chemicals]

Dispervyn. Vinyl copolymer dispersions. [Tennant-KVK Ltd]

Dispex®. Polymeric dispersing agents; effective on minerals and inorganic pigments in aqueous systems; very widely used in aqueous paints, adhesives and ceramic production. [Allied Colloids Ltd]

Dispex® A40 and N40. Ammonium or sodium salt of polymeric carboxylate as a pale yellow liquid; nonfoaming pigment dispersing agent used in emulsion paints, sometimes in combination with polyphosphate dispers-

---

* Trade name not verified as available † Trade name verified as obsolete

ants. [Allied Colloids Ltd]

Dispex® G40 and GA40. Sodium salt of carboxylated polymer in the form of a pale yellow liquid; pigment dispersant for emulsion paints, especially sheen and gloss water based paints. [Allied Colloids Ltd]

Displasol DP. Quaternary ammonium compound blend; displaces acid dyes on nylon producing novel styling effects. [Am. Emulsions]

Disponil AAP 307. Alkylaryl polyglycol ether; coemulsifier for polyacrylates, acrylate-vinyl acetate copolymers, other applications; dispersant for emulsion paints. [Henkel/Functional Prods.]

Disponil AEP 5300. Ether phosphate, acid ester; emulsifier for rosin, vinyl acetate and acrylate systems. [Henkel/Functional Prods.]

Disponil AES 13. Sodium alkylaryl ether sulfate; emulsifier for vinyl acetate homopolymers, acrylate homo and copolymers, styrene acrylate copolymers, vinyl acetate-acrylate copolymers, VAE copolymers, PVDC latexes, vinyl chloride homo and copolymer latexes. [Henkel/Functional Prods.]

Disponil FES 32. CAS 9004-82-4; Sodium laureth sulfate; emulsifier for vinyl acetate copolymers, S/B latexes, vinyl chloride copolymers, acrylate homo- and copolymers. [Henkel/Functional Prods.]

Disponil FES 92E. Sodium laureth-12 sulfate; surfactant for low-irritation shampoos, emulsion polymerization. [Pulcra SA]

Disponil G 200. Isoeicosanol; surfactant; liq. fatty alcohol. [Henkel/Functional Prods.]

Disponil O 5. Cetoleth-5; emulsifier for emulsion polymerization; wetting agent. [Henkel/Functional Prods.]

Disponil SML 100 F1. CAS 1338-39-2; EINECS 215-663-3; Sorbitan laurate; surfactant for polymerization. [Henkel/Functional Prods.]

Disponil SML 104 F1. CAS 9005-64-5; Polysorbate 21; surfactant for polymerization. [Henkel/Functional Prods.]

Disponil SML 120 F1. CAS 9005-64-5; Polysorbate 20; surfactant for polymerization. [Henkel/Functional Prods.]

Disponil SMO 100 F1. CAS 1338-43-8; EINECS 215-665-4; Sorbitan oleate; surfactant for polymerization. [Henkel/Functional Prods.]

Disponil SMO 120 F1. CAS 9005-65-6; Polysorbate 80; surfactant for polymerization. [Henkel/Functional Prods.]

Disponil SMP 100 F1. CAS 26266-57-9; EINECS 247-568-8; Sorbitan palmitate; surfactant for polymerization. [Henkel/Functional Prods.]

Disponil SMP 120 F1. CAS 9005-66-7; Polysorbate 40; surfactant for polymerization. [Henkel/Functional Prods.]

Disponil SMS 100 F1. CAS 1338-41-6; EINECS 215-664-9; Sorbitan stearate; surfactant for polymerization. [Henkel/Functional Prods.]

Disponil SMS 120 F1. CAS 9005-67-8; Polysorbate 60; surfactant for polymerization. [Henkel/Functional Prods.]

Disponil SSO 100 F1. CAS 8007-43-0; EINECS 232-360-1; Sorbitan sesquioleate; surfactant for polymerization. [Henkel/Functional Prods.]

Disponil STO 100 F1. CAS 26266-58-0; EINECS 247-569-3; Sorbitan trioleate; surfactant for polymerization. [Henkel/Functional Prods.]

Disponil STO 120 F1. CAS 9005-70-3; Polysorbate 85; surfactant for polymerization. [Henkel/Functional Prods.]

Disponil STS 100 F1. CAS 26658-19-5; EINECS 247-891-4; Sorbitan tristearate; surfactant for polymerization. [Henkel/Functional Prods.]

Disponil STS 120 F1. CAS 9005-71-4; Polysorbate 65; surfactant for polymerization. [Henkel/Functional Prods.]

Disponil SUS IC 8. CAS 577-11-7; EINECS 209-406-4; Dioctyl sodium sulfosuccinate; wetting agent, coemulsifier for plastics industry. [Henkel/Functional Prods.]

Dispray. Disinfectant used for rapid disinfections of the skin before operations, injections or venepuncture. [ICI Chem. & Polymers Ltd]

Dissolvine. Sequestering and chelating agents. [Akzo Chemie UK Ltd]

Distaclor. CAS 70356-03-5; Cefaclor; for the treatment of respiratory tract infections including pneumonia, bronchitis, exacerbations of chronic bronchitis, pharyngitis, tonsillitis and sinusitis, otitis media and urinary tract infections. [Dista Products Ltd]

Distalgesic. A proprietary preparation of dextro-propoxyphene hydrochloride and paracetamol; an oral analgesic. [Dista Products Ltd]

Distamine. A proprietary preparation of penicillamine hydrochloride; used in the treatment of heavy metal poisoning and Wilson's disease. [Dista Products Ltd]

Distaquaine V-K. A proprietary preparation containing phenoxymethyl-penicillin potassium; an antibiotic. [Dista Products Ltd]

distearyl thiodipropionate. (3,3´-thiobispropanoic acid, dioctadecyl ester; 3,3´-dioctadecyl thiodipropionate; thiodipropionic acid, distearyl ester) Diester of stearyl alcohol and 3,3´-thiodipropionic acid; $(C_{18}H_{37}OO-CCH_2CH_2)_2S$; CAS 693-36-7; antioxidant, plasticizer, softening agent. [Am. Cyanamid; Evans Chemetics; Morton Int'l.; Witco/Argus]

Distec. Fatty acids and glycerides. [Akzo Chemie UK Ltd]†

Distillase® L-200. CAS 9032-08-0; Glucoamylase; industrial grade enzyme for fuel alcohol applications [Solvay Enzymes]

Distillex DS1. CAS 71-55-6; Recovered 1,1,1-trichloroethane. [Distillex Ltd]

Distillex DS2. CAS 79-01-6; Recovered trichloroethylene. [Distillex Ltd]

Distillex DS3. Recovered dichloromethane. [Distillex Ltd]

Distillex DS4. CAS 127-18-4; Recovered tetrachloroethylene. [Distillex Ltd]

Distillex DS5. CAS 76-13-1; Recovered 1,1,2-trichloro-1,2,2-trifluoroethane. [Distillex Ltd]

Distillex DS6. CAS 75-69-4; Recovered trichlorofluoromethane. [Distillex Ltd]

Distillex DS7. Recovered 75% tetrachloroethylene and 25%n-butanol v/v. [Distillex Ltd]

Distoline. A proprietary trade name for commercial oleic acid obtained from vegetable oils. §

disulfuric acid. *See* fuming sulfuric acid.

Disulphine Blue. A preparation of sulfan blue used as a visual diagnostic agent for circulatory disorders. [ICI Chem. & Polymers Ltd] †

Disyston® FE-10. CAS 298-04-4; Granular systemic insecticide containing 10% w/w disulfoton on Fuller's earth

---

‡ Trade name and manufacturer not verified   § Trade name without identified manufacturer

granules; used to control aphids and certain aphid-borne virus diseases on potatoes, carrots, celery, marrows, parsley, French and runner beans and Brussels sprouts. [Bayer plc]

Ditate-DS. Testosterone enanthate USP, extradiol valerate USP for injection; indicated in the treatment of moderate to severe vasomotor symptoms associated with the menopause. [Altana Inc] ‡

Ditensamine C, O and S. Cationic surfactants in the form of alkyl propylene diamines in which the alkyl group is coconut, oleic and tallow respectively; liquid or solid forms; synthesis intermediate, bitumen emulsions, corrosion inhibition. [Tensia SA] *

Dithane. CAS 8018-01-7; Protectant mancozeb fungicide. [Rohm & Haas UK]

Dithane. CAS 8018-01-7; Wettable powder or water dispersible granules containing mancozeb; protective fungicide for fruit, field crops and roses. [Pan Britannica Industries Ltd]

β,β′-dithiobisalanine. See cystine.

Dithiocream. Pale yellow aqueous cream containing dithranol BP; recommended for the topical treatment of sub-acute and chronic psoriasis including psoriasis of the scalp. [Dermal Laboratories Ltd]

dithiothreitol. (L-DTT; 1,4-dithio-L-threitol; Cleland reagent) $C_4H_{10}O_2S_2$; CAS 3483-12-3; EINECS 240-263-0; Reducing agent for proteins and enzymes, biochemical research. [Aldrich; Bio-Rad Labs; Biosynth AG; Schweizerhall; U.S. Biochemical]

dithizone. CAS 102-08-9; Diphenylthiocarbazone; used for the detection of heavy metals.

Dithrolan. Stiff yellow ointment containing 0.5% dithranol BP, 0.5% salicylic acid BP, in equal quantities of hard and soft paraffin; used for the topical treatment of quiescent psoriasis. [Dermal Laboratories Ltd]

ditrimethylolpropane tetraacrylate. Curing agent. [Sartomer]

Ditropan. CAS 1508-65-2; Oxybutynin chloride; anticholinergic. [Marion Merrell Dow Inc]

Diucardin. CAS 135-09-1; Hydroflumethiazide; antihypertensive; diuretic. [Wyeth Laboratories]

Diulo. CAS 17560-51-9; Metolazone; diuretic; antihypertensive. [G D Searle & Co] *

Diumide-K. Frusemide and controlled release potassium chloride; for edema. [Napp Laboratories Ltd] *

Diuresal. Tablets containing 40mg frusemide BP; injection containing 20mg/2ml frusemide BP and 50mg/5ml frusemide BP; diuretic for the management of edema of cardiac, renal or hepatic origin; pulmonary edema, toxemia of pregnancy, mild or moderate hypertension. [Lagap Pharmaceuticals Ltd]

diuretic salt. CAS 127-08-2; Potassium acetate.

Diurex. CAS 330-54-1; Active ingredient: diuron; 3-(3,4-dichlorophenyl)-1,1-dimethylurea; residual herbicide effective against a wide range of both broadleaf weeds and annual grasses. [Agan Chemical Manufacturers Ltd]

Diurex. CAS 14293-44-8; See Diurexan. [Degussa AG] *

Diurexan. CAS 14293-44-8; Xipamide; tablet form; diuretic/antihypertensive. [Degussa AG] *

Diuril. CAS 58-94-6; Chlorothiazide; for the treatment of edema and hypertension. (U.S.) [Merck & Co Inc]

Diurnal-Penicillin. CAS 6130-64-9; Penicillin G Procaine; antibacterial. [Upjohn] *

Diurol. Active ingredients: azolan plus diurex; multipurpose

herbicidal mixture which eradicates a wide spectrum of established weeds, while preventing further weed germination for extended periods. [Agan Chemical Manufacturers Ltd]

diuron. (3-(3,4-dichlorophenyl)-1,1-dimethylurea; N′-(3,4-dichlorophenyl)-N,N-dimethylurea; 1,1-dimethyl-3-(3,4-dichlorophenyl)urea) $C_6H_3Cl_2NHCON(CH_3)_2$; CAS 330-54-1; Pre-emergent herbicide, sugar cane flowering suppressant. [Griffin; Pacific Anchor; Rhone-Poulenc Agrochimie SA]

Diuron. Total and selective herbicide. [Rhone-Poulenc Environmental Prods. Ltd]

Diuron Bayer. CAS 330-54-1; Diuron; herbicide effective against emerging and young broad-leaved and grass weeds as well as mosses; suitable for both selective (sugar cane, cotton) and total weed control (orchards, vineyards). [Bayer AG]

Divergan® F, R. Crosslinked PVP for brewing industry; stabilizing agents for drinks (beer, wine, clear fruit juices). [BASF AG; BASF plc]

Diver's liquid. A liquid formed by absorbing ammonia in solid ammonium nitrate. It is capable of dissolving ammonium nitrate.

Divipan. CAS 62-73-7; Active ingredient: dichlorvos; one of the most useful fast-acting agricultural insecticides-acaricides. [Makhteshim Chemical Works Ltd]

Diwatex 30, 40. Sulfonated kraft lignin derivs.; dispersant for dyestuffs and pesticides. [Borregaard LignoTech]

Dixarit. CAS 4205-91-8; A proprietary preparation of clonidine hydrochloride; used for migraine and menopausal flushing. [Boehringer Ingelheim Ltd]

Dixie. See Catalpo. §

Dixie 5 and Dixie Special 102. Proprietary carbon black pigments. §

Dixie Clay®. CAS 1332-58-7; Hydrated aluminum silicate (kaolin clay); filler, extender or reinforcing pigment for paint, paper, rubber, ceramics, plastics, and specialities. [R T Vanderbilt Co Inc]

Dizene. Emulsifiable ortho-dichlorobenzene. [PPG Industries]

Diziktol. CAS 333-41-5; Diazinon; insecticide [Makhteshim Chemical Works Ltd]

DK-Ester. Sugar esters; food emulsifier. [Grünau]

DKP. See potassium phosphate, dibasic.

DLG-10, 20. CAS 557-05-1; Dispersible zinc stearate; sanding sealer aid in lacquers. [Hüls Am.] †

DLPA 375. Tablets of DL-phenylamine 375mg in a natural basis; used as a dietary suplement. [Larkhall Laboratories plc] *

DLS Base. Preneutralized emulsifiable amidoamine condensate; softener base imparting a dry, soft hand to cotton and other fabrics. [Clark]

D-Lube. Fuel additive. [Kalon Chemicals Ltd]

D.M. See Adamsite. §

DM-2. Mold repair/potting compound and general purpose adhesive; epoxide resin based mastic use as a heat resistant encapsulating material; used for encapsulation of electric components where solder connections are affected by heat, repair of mold porosity in rotational plastic molds. [Dynamold Inc]

DMAE. See dimethylethanolamine.

DMAMP-80. CAS 7005-47-2; 2-Dimethylamino-2-methyl-1-propanol; amine solubilizer for resins in aq. coatings; emulsifier for waxes; vapor-phase corrosion inhibitor;

---

* Trade name not verified as available † Trade name verified as obsolete

urethane catalyst; titanate solubilizer; raw material for synthesis. [Angus]

DMF. *See* dimethyl formamide.

DM hydantoin. (5,5-dimethyl hydantoin; 5,5-dimethyl-2,4-imidazolidinedione) Organic compound.; $C_5H_8N_2O_2$; CAS 77-71-4; intermediate for textiles and other applications. [Great Lakes; Janssen Chimica; Lonza]

DMI-689. A two-part urethane mixture, fast curing, 1:1 mix; hard copy replication system useful for replication of parts, measurement of dimensional accuracy, moldmaking and prototype tooling. [Dynamold Inc]

DM-Nitrophen™. Reagent. [Calbiochem Corp]

DMP. *See* dimethyl phthalate.

DMP. CAS 4744-10-9; 2,2-Dimethoxypropane; chemical intermediate for pharmaceuticals; dehydrating agent. [Schering Berlin Polymers]

DMP. Catalyst systems for polyurethane and epoxy resins. [Rohm & Haas]

DMR-503. A two-part silicone elastomer, fast curing, 1:1 mix, hand mixable; flexible putty useful for replication of parts, measurement of dimensional accuracy, moldmaking and prototype tooling. [Dynamold Inc]

DMR-504. A two-part silicone elastomer, fast curing, 1:1 mix, hand mixable; pourable flexible system useful for replication of parts, measurement of dimensional accuracy, moldmaking and prototype tooling. [Dynamold Inc]

DMS-4-828. Epoxy resin based mastic; moldable liquid shim materials used as a spacer between engines or skin of aircraft or ships; surface conforming structural epoxies used to fill gaps between metal parts and between structural members and composites such as graphite. [Dynamold Inc]

DMSO. *See* dimethyl sulfoxide.

DMSO. CAS 67-68-5; Dimethyl sulfoxide; aprotic solv. used as reaction medium in mfg. of pesticides, pharmaceuticals, dyes, inks; solv. in polymers, electronics, refining; chem. intermediate for pesticides. [Atochem N. Am./Fine Chems.]

D.N.T. CAS 121-14-2; Dinitrotoluene, $CH_3.C_6H_3$ $(NO_2)_2$; used in organic synthesis, toluidines, dyes, explosives.

DO-160. Chlorinated olefin. [Dover]

DOA. *See* dioctyl adipate.

Dobane (Detergent Alkylate). Linear alkylbenzenes which yield light colored, biodegradable sulfonates; used for general detergent applications, from household powders to light duty liquids. [Shell] *

Dobanic Acids JN and 83. Dark, viscous liquids which on neutralization give light colored sulfonates particularly suitable for the production of high performance, liquid detergents. [Shell] *

Dobanol. Colorless high purity liquids used as base materials for the manufacture of alcohol sulfates, alcohol ethoxylates, and alcohol ethoxysulfates; for production of detergents, wetting agents, dispersants, and emulsifiers. [Shell] *

Dobanol. Detergent alcohol. [Mitsubishi Petrochem.] *

Dobanol ethoxylates. Intermediates for the production of shampoo components, toiletry products, dishwashing liquids, and washing powders. [Shell] *

Dobanol ethoxysulfates. Aqueous or aqueous/ethanol solutions for various applications, such as components for toiletry products, light-duty liquid detergents, and high-performance liquid detergents. [Shell] *

Dobanox. Surfactant/emulsifier. [Shell UK] *

Dobatex. An anionic detergent. [Shell UK] *

Dobbin's reagent. Prepared by adding mercuric chloride solution to a solution of potassium iodide until a permanent precipitate is obtained. The solution is filtered and 1 gram of ammonium chloride added, then dilute caustic soda until a precipitate is formed; used for detecting traces of caustic alkalies in soap.

Dobell solution. An aqueous solution containing 1.5% sodium borate, 1.5% sodium bicarbonate, and 0.3% phenol and glycerin. An alkaline antiseptic.

Dobutrex. CAS 49745-95-1; Dobutamine hydrochloride; cardiotonic. [Eli Lilly & Co]

Doca Acetate. Desoxycorticosterone acetate; adrenocortical steroid. [Organon Inc] *

Dock-Ban. Soluble concentrate of 19.5 g dicamba, 245 g MCPA and 86.5 g mecoprop per liter; used for weed control in cereals and grassland. [Quadrangle Agrochemicals]

Docklene. Soluble concentrate of 336 g mecoprop, 84 g dicamba and 84 g MCPA per liter; used for weed control in cereals and grassland. [Schering Agrochemicals Ltd]

1-docosanol. *See* behenyl alcohol.

13-docosenamide. *See* erucamide.

Doctor metal. An alloy of 88% copper, 9.5% zinc, and 2.5% tin.

docusate sodium. *See* dioctyl sodium sulfosuccinate.

dodecahydrosqualene. *See* squalane.

1-dodecanethiol. *See* n-dodecyl mercaptan.

dodecanoic acid. *See* lauric acid.

dodecanoic acid, methyl ester. *See* methyl laurate.

1-dodecanol. *See* lauryl alcohol.

dodecene-1. (C_{12} α olefin; α-dodecylene; tetrapropylene) $H_2C:CH(CH_2)_9CH_3$; CAS 112-41-4; 6842-15-5; EINECS 203-968-4; Intermediate for surfactants and specialty industrial chemicals; flavors, perfumes, medicine, oils, dyes, resins. [Chevron; Ethyl; Shell]

dodecoic acid. *See* lauric acid.

dodecyl acrylate. *See* lauryl acrylate.

dodecyl alcohol. *See* lauryl alcohol.

dodecylbenzene sodium sulfonate. *See* sodium dodecylbenzenesulfonate.

dodecylbenzene sulfonic acid. (DDBSA) Substituted aromatic acid; $C_{18}H_{30}O_3S$; CAS 27176-87-0; 68411-32-5; 68584-22-5; 68608-88-8; 85536-14-7; EINECS 248-289-4; anionic detergent.

dodecylbenzenesulfonic acid, sodium salt. *See* sodium dodecylbenzenesulfonate.

dodecyldimethylamine. *See* dimethyl lauramine.

α-dodecylene. *See* dodecene-1.

n-dodecyl mercaptan. (DDM; m-lauryl mercaptan; 1-dodecanethiol; mercaptan C_{12}) n-$C_{12}H_{25}SH$; CAS 112-55-0; EINECS 203-984-1; Modifier in polymerization reactions, esp. for SBR; reducing initiator and chain transfer agent. [Atochem N. Am.; Phillips 66]

t-dodecyl mercaptan. (t-dodecanethiol) t-$C_{12}H_{26}S$; CAS 25103-58-6; EINECS 246-619-1. [Aldrich; Atochem N. Am.; Phillips 66]

1-dodecylpyridinium chloride. *See* laurylpyridinium chloride.

Dodicor 2565. Quaternary arylammonium chloride; metalworking surfactant; corrosion inhibitor for oil and gas industry; max. protection for zinc in acid cleaners. [Hoechst Celanese/Colorants & Surf.]

Dodigen. Range of cationic surfactants of the quaternary

---

‡ Trade name and manufacturer not verified   § Trade name without identified manufacturer

ammonium chloride type, in liquid, paste or solid form; antistatic agents, fabric conditioner/softener, fiber finishers, water-repellant and dewatering agents, wetting agents for oils, dispersants for pigments, flushing agents, foaming agents, spinning bath and viscous additives, corrosion inhibitors; flotation chemicals and anti-caking agents for rendering salts free-flowing, anchoring and wetting agents for tars and bitumens, surface coatings, lacquers, adhesives, disinfectants, hair cosmetics, auxiliaries for leather, textiles, rubber, and metals. [Hoechst UK] *

Dodine FL. CAS 2439-10-3; Dodine; a fungicide for the control of scab in apples and pears. [Truchem Ltd]

Dodine WP. CAS 2439-10-3; Dodine; a wettable powder fungicide for the control of scab in apples and pears. [Truchem Ltd]

Do Do. Bronchitis remedy. [Ciba plc] *

Doff. Range of insecticides and herbicides; for horticultural/ household use. [Doff Portland Ltd]

Dog Off. Quassia; animal repellent for outdoor crops. [Fieldspray]

Dolan. Acrylic fiber. [Hoechst UK]

Dolanit. Acrylic fiber. [Hoechst UK]

Dolasan. A proprietary preparation of dextropropoxyphene napsylate and aspirin; an analgesic. [Eli Lilly & Co] †

Dolasol TF. Nonionic surfactants and inorg. builders; industrial cleaner. [Zschimmer & Schwarz] *

Dolene. CAS 550-83-4; Propoxyphene hydrochloride; analgesic. [Lederle Laboratories USA] *

Doler Brass. A proprietary alloy; it is a silicon brass. §

Dolmatil. CAS 15676-16-1; Sulpiride. [Bristol-Myers Squibb Pharmaceuticals Ltd]

Dolo-Adamon®. A mixture of sodium phenyldimethyl pyrazolone methylaminomethane sulfonate (Noramidazophenum), codeine phosphate and 5-ethyl-5-crotyl barbituric acid (Crotarbital) and Adamon (qv); a strong pain relieving agent. [AG Chemische Fabrik] ‡

Dolobid. CAS 22494-42-4; Diflunisal; for the relief of pain. [Merck & Co Inc]

Dolofil. Dolomite. [Steetley Minerals Ltd]

Dolomol. A white insoluble powder, consisting mainly of magnesium stearate, with small amounts of magnesium oleate and palmitate; used as a dusting powder for skin.

Dolophine Hydrochloride. CAS 1095-90-5; Methadone hydrochloride; analgesic. [Eli Lilly & Co] *

Doloxene. A proprietary preparation of dextropropoxyphene hydrochloride; an analgesic. [Eli Lilly & Co]

Doloxene Compound-65. A proprietary preparation of dextropropoxyphene hydrochloride, phenacetin, aspirin, and caffeine; an analgesic. [Eli Lilly & Co] †

Doloxytal. A proprietary preparation of dexopropoxyphene hydrochloride and amylobarbitone; an analgesic and sedative. [Eli Lilly & Co] †

Dolviran®. An analgesic. [Bayer AG]

Domba oil. See laurel nut oil.

Dome-Acne. A proprietary preparation of sulfur and resorcinol acetate used in the treatment of acne vulgaris. [Dome/Hollister-Stier] †

Domeboro. CAS 8006-13-1; Aluminum acetate; astringent. [Miles Pharmaceuticals] *

Domestos. CAS 7681-52-9; Stabilized sodium hypochlorite. [Unilever] *

Domical. CAS 549-18-8; A proprietary preparation of amitryptiline hydrochloride; an antidepressant. [Berk

Pharmaceuticals Ltd] *

Dominate. A wettable powder containing a mixed culture of micro-organisms (*Anthrobacter* sp., *Aspergillus terreus, Bacillis subtilis, Bacillis thuringiensis, Bacteroides* sp., *Nocardia* sp., and *Pseudomonas* sp.); used to suppress growth of pathogenic soil fungi; applied to the soil; used for a wide variety of crops. [Westbridge Research Group]‡

Domoso. CAS 67-68-5; Dimethyl sulfoxide; anti-inflammatory. [Syntex Laboratories Inc] *

Donarit 1, Donarit 2, Donarit 3. Powdery ammon dynamites with the addition of aromatic nitro-compounds and explosive oil; particularly suitable for medium-hard rock; used in quarries, in agriculture and forestry under dry conditions. [Dynamit Nobel Wien GmbH] *

Donarite. An explosive containing 80% ammonium nitrate, 12% trinitrotoluene, 4% flour, 3.8% nitroglycerin, and 0.2% collodion wool.

Donnagel. A proprietary preparation containing kaolin, pectin, hyoscyamine sulfate, atropine sulfate and hyoscine hydrobromide; an antidiarrheal. [Wyeth Laboratories] *

Donnagel-PG. A proprietary preparation containing kaolin, pectin, hyoscyamine sulfate, atropine sulfate, hyoscyamine hydrobromide and opium; an antidiarrhoeal. [Wyeth Laboratories] *

Donnatal. A proprietary preparation containing hyoscyamine sulfate, atropine sulfate, hyoscine hydrobromide and phenobarbitone; an antacid. [Wyeth Laboratories] *

Donnazyme. A proprietary preparation containing hyoscyamine sulfate, atropine sulfate, hyoscine hydrobromide, phenobarbitone, pepsin, pancreatin and bile salts; an antacid. [Wyeth Laboratories] *

Dontalol®. Pharmaceutical preparation; mouthwash concentration. [Bayer AG]

Donut Pyro®. CAS 7758-16-9; Sodium acid pyrophosphate; leavening agent for baking, cereals. [Rhone-Poulenc Food Ingreds.]

Doom. A microbial insecticide in powder form containing viable spores of Bacillus popilliae, a specific pathogen which infects and kills Japanese beetle grubs; ready-to-use; for control of Japanese beetle grubs; only one application is needed as the living spores are self-perpetuating. [Fairfax Biological Laboratory Inc]

door plate brass. See brass.

DOP. See dioctyl phthalate.

Dopamet. CAS 555-30-6; A proprietary preparation of methyldopa; a drug used in the treatment of hypertension. [Berk Pharmaceuticals Ltd] *

Dopar. CAS 59-92-7; Levodopa; antiparkinsonian. [Norwich Eaton Pharmaceuticals Inc] *

Dopastat. CAS 62-31-7; Dopamine hydrochloride; adrenergic. [Parke-Davis] *

dope. The name given to various solutions of cellulose or cellulose compounds in acetone, amyl alcohol, amyl acetate, and other solvents; used for painting aeroplane wings, and other purposes.

Dopram Infusion. Clear colorless solution in flexible plastic bag containing doxapram hydrochloride 2 mg in 5% glucose intravenous infusion BP; ventilatory stimulant used principally for treatment of acute respiratory failure or following anesthesia. [Wyeth Laboratories]

Dopram Injection. CAS 62-31-7; Clear colorless solution

---

* Trade name not verified as available   † Trade name verified as obsolete

containing doxapram hydrochloride 20 mg per ml in water in 5 ml ampoules; ventilatory stimulant used principally for treatment of acute respiratory failure or following anesthesia. [Wyeth Laboratories]

Dorantamin. CAS 59-33-6; Pyrilamine maleate; antihistaminic. [Dorsey Pharmaceuticals] *

Dorbane. Danthron; laxative. [3M Pharmaceuticals] †

Dorbanex Capsules. Each capsule contains danthron BP 25 mg and poloxamer 188 200 mg [3M Pharmaceuticals]*

Dorbanex Forte. Each 5 ml spoonful contains danthron BP 75 mg and poloxamer 188 1000 mg [3M Pharmaceuticals] *

Dorbanex Liquid. Each 5 ml spoonful contains danthron BP 25 mg and poloxamer 188 200 mg [3M Pharmaceuticals]*

Dorcolor®. Spun-dyed man-made fibers. [Bayer AG]

Doré silver. A silver containing small amounts of gold.

Doriden. CAS 77-21-4; A proprietary preparation of glutethimide; a hypnotic. [Ciba plc] *

Dorin. Emulsifiable concentrate containing 125 g triadimenol and 375 g tridemorph per liter; used for control of mildew and rust in cereals. [Bayer plc]

Dorindan. Emulsifiable concentrate containing 125 g triadimenol and 375 g tridemorph per liter; used for control of mildew and rust in cereals. [Bayer plc]

Dorival. An analgesic, antipyretic drug. [Bayer AG]

Dorix®. Polyamide staple fiber; used for nonwovens, felts, insulation, coverings and mats. [Bayer AG]

Dorlastan®. Man-made fibers, filament yarn. [Bayer AG]

Dormakil. Fungicide. (Sold in UK for DowElanco) [ICI Chem. & Polymers Ltd]

Dormate. CAS 64-55-1; Mebutamate; antihypertensive. [Wallace Laboratories] *

Dormethan. Dextromethorphan hydrobromide; antitussive. [Dorsey Pharmaceuticals] *

Dormicum. See hypnovel. [Roche Products Ltd] *

Dormone. A selective weedkiller containing 465 g/liter 2,4-D as the diethanolamine salt; used for the control of broad leaved weeds on amenity areas, golf courses, playing fields etc. [Burts & Harvey]

Dormone. A selective weed killer containing 465 g/l 2,4-D as the diethanolamine salt; may be used for the control of broad leafed weeds on amenity areas, golf courses, playing fields, etc. [Rhone-Poulenc Environmental Prods. Ltd]

Dormonoct. CAS 61197-93-1; Loprazolam; used for short term treatment of insomnia and/or nocturnal waking. [Roussel Laboratories Ltd]

Dormopan®. Long-lasting hypnotic with immediate action. [Bayer AG]

Doroma®. A soporific. [Bayer AG]

Dorox. Aluminum alcoholates. [RWE-DEA Chemicals UK Ltd]

Dorsacaine. CAS 5987-82-6; Benoxinate hydrochloride; anesthetic. [Dorsey Pharmaceuticals] *

Dorsital. CAS 76-74-4; Pentobarbital; sedative. [Dorsey Pharmaceuticals] *

Dosaflo. CAS 19937-59-8; Suspension concentrate containing 500 g/l metoxuron; residual urea herbicide for the control of weeds in cereals and carrots. [ICI Chem. & Polymers Ltd]

Dosaflo. CAS 19937-59-8; Suspension concentrate containing 500 g/l metoxuron; residual urea herbicide for the control of weeds in cereals and carrots. [Farm Protection Ltd]

Dosulphin. A proprietary preparation of sulfaproxyline and sulfamerazine; an antibiotic. [Ciba plc] *

D.O.T.G. CAS 97-39-2; Di-o-tolylguanidine; a rubber vulcanization accelerator.

D.O.T.T. Di-o-tolylthiourea; $SC(NHC_6H_4CH_3)_2$; used as a metal pickling inhibitor.

Double Bond. Two component adhesive. [Stag Polymers & Sealants Ltd]

Double/Bubble. Package for two-part epoxies. [Hardman]

Double Green S.F. See methyl green.

double nickel salt. Nickel ammonium sulfate, $Ni(NH_4)_2$ $·(SO_4)_2.6H_2O$; used in the plating trade.

Double Shield. Marine antifouling; does not contain any tin compounds [Llewellyn Ryland Ltd]

Doublet. Herbicide. [May & Baker Ltd] *

Double Twitchell reagent. The barium salt of the sulfonated mixture of naphthalene and fatty acid; used in the decomposition of fats

Double White. CAS 1312-76-1; A proprietary trade name for a general purpose potassium silicate cement for acid conditions, e.g., as a bedding and jointing material for tiles. [Haworth (ARC) Ltd] ‡

Doucil. Chemical for use in industry. [Crosfield Chemicals Ltd]

Doverchlor 10. Chlorinated olefin. [Dover]

Doverguard® 152. Chlorinated paraffin; flame retardant for plastics, rubbers, adhesives, paints, fabric coatings. [Dover]

Doverguard® 700. Resinous chlorinated paraffin; flame retardant for plastics, rubbers, adhesives, paints, fabric coatings. [Dover]

Doverguard® 5761. Chlorinated paraffin; flame retardant for plastics, rubbers, adhesives, paints, fabric coatings. [Dover]

Doverguard® 8133. Resinous bromochlorinated paraffin; flame retardant for plastics, rubbers, adhesives, paints, fabric coatings. [Dover]

Doverguard® 8207-A. Liq. bromochlorinated paraffin; flame retardant for plastics, rubbers, adhesives, paints, fabric coatings. [Dover]

Doverguard® 8410. Liq. brominated paraffin; flame retardant for plastics, rubbers, adhesives, paints, fabric coatings. [Dover]

Doverlub 8136. Heavy-duty extreme-pressure sol. oil base providing excellent emulsion stability, good EP and antiwear props., good rust protection and lubricity [Dover]

Doverlub 8506. Chlorinated methyl ester; lubricity and EP agent for lubricants incl. synthetic coolant, cutting, drawing, and gear oils. [Dover]

Doverlub 8527. Chlorinated fatty acid; lubricity and EP agent for lubricants incl. synthetic coolant, cutting, drawing, and gear oils. [Dover]

Doverlub 8531. Chlorinated lard oil; lubricity and EP agent for lubricants incl. synthetic coolant, cutting, drawing, and gear oils. [Dover]

Doverlub 8621. Mixture of fatty compounds and chlorinated paraffin; lubricity and EP agent for lubricants incl. synthetic coolant, cutting, drawing, and gear oils. [Dover]

Doverphos®. Liquid and solid phosphites performing as antioxidants; provides process and service life stability in various polymers and intermediates in vinyl stabilizers.

---

‡ Trade name and manufacturer not verified   § Trade name without identified manufacturer

[Dover]

Doverphos® 4. CAS 26523-78-4; Trisnonylphenyl phosphite; heat stabilizer for PVC, ABS, polyolefins, some rubber products [Dover]

Doverphos® 4-HR. CAS 26523-78-4; Trisnonylphenyl phosphite, 0.75% triisopropanolamine; hydrolysis-resistant antioxidant for elastomer mfg. [Dover]

Doverphos® 6. CAS 25448-25-3; Triisodecyl phosphite; chelating agent with metal carboxylates as polymer additives, esp. for chlorinated polymers such as PVC and chlorinated PE; improves color, heat and light stability; antioxidant and EP additive for lubricants. [Dover]

Doverphos® 7. CAS 25550-98-5; Phenyl diisodecyl phosphite; chelating agent with metal carboxylates as polymer additives, esp. for chlorinated polymers such as PVC and chlorinated PE; improves color, heat and light stability. [Dover]

Doverphos® 8. CAS 26544-23-0; Diphenyl isodecyl phosphite; chelating agent with metal carboxylates as polymer additives, esp. for chlorinated polymers such as PVC and chlorinated PE; improves color, heat and light stability. [Dover]

Doverphos® 10. CAS 101-02-0; Triphenyl phosphite; improves color stability in polyesters, polyurethanes, and alkyd resins; also aids curing and hardening in epoxies. [Dover]

Doverphos® 10-HR. CAS 101-02-0; Triphenyl phosphite, 0.5% triisopropanolamine; improves color stability in polyesters, polyurethanes, and alkyd resins; also aids curing and hardening in epoxies. [Dover]

Doverphos®11. CAS 80584-85-6; Tetraphenyl dipropylene glycol diphosphite. [Dover]

Doverphos®12. CAS 80584-86-7; Poly (dipropylene glycol) phenyl phosphite; chelating agent with metal carboxylates as polymer additives, esp. for chlorinated polymers such as PVC and chlorinated PE; improves color, heat, and light stability. [Dover]

Doverphos® 49. Triisotridecyl phosphite; antioxidant and EP additive for lubricants. [Dover]

Doverphos® 53. CAS 3076-63-9; Trilauryl phosphite; antioxidant and EP additive for lubricants. [Dover]

Doverphos®75. Diphenyl isotridecyl phosphite [Dover]

Doverphos® 213. CAS 4712-55-4; Diphenyl phosphite. [Dover]

Doverphos® 251. Distearyl hydrogen phosphite. [Dover]

Doverphos® 253. Dioleyl hydrogen phosphite; antioxidant and EP additive for lubricants. [Dover]

Doverphos®269. Diisotridecyl hydrogen phosphite; antioxidant and EP additive for lubricants. [Dover]

Doverphos® 271L. Dilauryl hydrogen phosphite; antioxidant and EP additive for lubricants. [Dover]

Doverphos® 274. Dilauryl hydrogen phosphite. [Dover]

Doverphos® 298. Diisooctyl hydrogen phosphite; antioxidant and EP additive for lubricants. [Dover]

Doverphos®DIOP. CAS 36116-84-4; Diisooctyl phosphite. [Dover]

Doverphos® DPGDP. CAS 80584-85-6; Tetraphenyl dipropylene glycol diphosphite. [Dover]

Doverphos® DPIOP. CAS 26401-27-4; Diphenyl isooctyl phosphite. [Dover]

Doverphos® DPP. CAS 4712-55-4; Diphenyl phosphite. [Dover]

Doverphos® S-680. CAS 3806-34-6; Distearylpentaerythritol diphosphite; color stabilizer and melt flow aid for polymer processing. [Dover]

Doverphos® S-686, S-687. CAS 3806-34-6; Distearylpentaerythritol diphosphite; color stabilizer and melt flow aid for polymer processing. [Dover]

Doverphos®TIOP. CAS 25103-12-2; Triisooctyl phosphite. [Dover]

Doverphos® TLP. CAS 3076-63-9; Trilauryl phosphite. [Dover]

Doverphos®TTDP. Triisotridecyl phosphite. [Dover]

Dow 276-V2. α-methyl styrene derivative; a vinyl plasticizer. [Dow UK] *

Dow DBR. Dibenzoyl resorcinol; an ultraviolet absorber for plastics. [Dow UK] *

Dow V9. Alpha methyl styrene derivative; a vinyl plasticizer. [Dow UK] *

Dowanol®. A line of glycol ethers and glycol ether acetates used as solvents in a variety of unrelated industrial applications. [Dow UK]

Dowanol®DB. CAS 112-34-5; EINECS 203-961-6; Diethylene glycol n-butyl ether; solvent, coupling agent providing improved surface wetting, soil penetration in household, commercial and industrial cleaning products [Dow]

Dowanol® DM. CAS 111-77-3; Methoxydiglycol; solvent, coupling agent providing improved surface wetting, soil penetration in household, commercial and industrial cleaning products [Dow]

Dowanol® DPMA. PPG-2 methyl ether acetate; solvent, coupling agent providing improved surface wetting, soil penetration in household, commercial and industrial cleaning products [Dow]

Dowanol® DPnB. CAS 29911-28-2; Dipropylene glycol n-butyl ether; solvent, coupling agent providing improved surface wetting, soil penetration in household, commercial and industrial cleaning products [Dow]

Dowanol® EB. CAS 111-76-2; EINECS 203-905-0; Ethylene glycol n-butyl ether; solvent, coupling agent providing improved surface wetting, soil penetration in household, commercial and industrial cleaning products [Dow]

Dowanol® EPh. Ethylene glycol phenyl ether; solvent, coupling agent providing improved surface wetting, soil penetration in household, commercial and industrial cleaning products [Dow]

Dowanol® PM. CAS 107-98-2; Propylene glycol methyl ether; solvent, coupling agent providing improved surface wetting, soil penetration in household, commercial and industrial cleaning products [Dow]

Dowanol® PMA. CAS 108-65-6; Propylene glycol methyl ether acetate; solvent, coupling agent providing improved surface wetting, soil penetration in household, commercial and industrial cleaning products [Dow]

Dowanol® PnB. Propylene glycol n-butyl ether; solvent, coupling agent providing improved surface wetting, soil penetration in household, commercial and industrial cleaning products [Dow]

Dowanol® PPh. Propylene glycol phenyl ether; solvent, coupling agent providing improved surface wetting, soil penetration in household, commercial and industrial cleaning products [Dow]

Dowanol®PPM. Dipropylene glycol methyl ether; solvent, coupling agent providing improved surface wetting, soil penetration in household, commercial and industrial cleaning products [Dow]

Dowanol® TPM. PPG-3 methyl ether; solvent, coupling agent providing improved surface wetting, soil penetra-

---

tion in household, commercial and industrial cleaning products [Dow]

Dowclene. Industrial solvents, primarily for metal and suede. [Dow UK]

Dowco 179. CAS 2921-88-2; A proprietary preparation of chlorpyrifos; an insecticide. [Dow UK]

Dow Corning® 1-2531 Release Coating. Silicone resin coating; provides durable release coating for bakers pans, waffle irons; FDA approved for food contact use. [Dow Corning]

Dow Corning® 7 Compound. Silicone compound; lubricant, release agent for mold break-in; preservative and lubricant for rubber. [Dow Corning]

Dow Corning® 24 Emulsion. Silicone emulsion; food grade lubricant for manufacture of paper/paperboard in contact with food; FDA approved; water-dilutable. [Dow Corning]

Dow Corning® 190 Surfactant. Dimethicone copolyol; silicone surfactant, surface tensile depressant, wetting agent, emulsifier, foam builder, humectant, softener; used for producing flexible slab stock urethane foam; ingredient in personal care products; plasticizer for hair resins. [Dow Corning]

Dow Corning® 197 Surfactant. Silicone surfactant; for polyurethane foam industry. [Dow]

Dow Corning® 200 Fluid. Dimethicone; foam control agent for nonaq. systems, distillation, resin mfg., asphalt, oil refining, gas-oil separation. [Dow Corning]

Dow Corning® 203 Fluid. Silicone fluid; release film providing internal release and lubrication. [Dow Corning]

Dow Corning® 344 Fluid, 345 Fluid. CAS 69430-24-6; Cyclomethicone; lubricant, spreading agent, detackifier for skin cleansers. [Dow Corning]

Dow Corning® 556 Fluid. Phenyldimethicone; lubricant, emollient for skin care products [Dow Corning]

Dow Corning® 929. Amodimethicone, tallowtrimonium chloride and nonoxynol-10; imparts wet and dry combing ease to hair care formulations; also car polish ingred. [Dow Corning]

Dow Corning® 1500 Compd. Silica filled polydimethyl siloxane; defoamer for aq. or nonaq. systems, food processing, rendering, glycol scrubbing, cutting oils. [Dow Corning]

Dow Corning® 3225C Formulation Aid. Silicone glycol copolymer; surfactant for preparing w/o volatile silicone emulsions used in personal care products [Dow Corning]

Dow Corning® 7224. Amino functional silicone; hair conditioning ingred. [Dow Corning]

Dow Corning® ACH-303. Aluminum chlorhydrate; act. ingred. in antiperspirant and deodorant formulations. [Dow Corning]

Dow Corning® ACH7-308. Aluminum sesquichlorhydrate; act. ingreds. in antiperspirant formulations. [Dow Corning]

Dow Corning® AZG-368. Aluminum/zirconium chlorohydrate; act. ingred. for all forms of topical antiperspirants. [Dow Corning]

Dow Corning® AZG-370. Aluminum/zirconium glycine; act. ingred. for all forms of topical antiperspirants. [Dow Corning]

Dow Corning® Antifoam 1410. Silicone emulsion; foam control agent for inks, textile starching/sizing, cutting oils, resin mfg., gas processing, adhesives/coatings, waste water treatment, pesticide/fertilizer industries; for ex-

treme pH conditions. [Dow Corning]

Dow Corning® Antifoam A. CAS 8050-81-5; Simethicone; foam control agent for distillation, resin sizes, textile latex backing, paper, asphalt, lubricants, detergents, pesticides, edible oils, soaps, shampoos; also avail. in food grade. [Dow Corning]

Dow Corning® Antifoam C. CAS 8050-81-5; Simethicone; food grade foam control agent for food industry, paper coatings, pesticides, herbicides, fertilizers. [Dow Corning]

Dow Corning® FF-400. Silicone glycol copolymer; lubricant; heat stabilizer, antistat, fiber finish for textiles threads. [Dow Corning]

Dow Corning® FS-1265 Fluid. Dimethyl silicone fluid; foam control agent for nonaq. systems, aromatic/chlorinated solvs., gas-oil separation, dry cleaning, metal cleaning and degreasing, oil refining. [Dow Corning]

Dow Corning® Q1-6106. Coupling agent designed to promote adhesion of two dissimilar materials; in reinforced plastic systems, provides coupling of most reinforcing agents, e.g., glass, Kevlar to most polar org. polymers and engineering plastics, e.g., epoxies, urethanes, acrylics, polysulfones, PPS, melamines, polyimides, PC, and thermoplastic polyesters; adhesion promoter for plastics and metals. [Dow Corning]

Dow Corning® QF1-3593A. Dimethicone, trimethylsiloxysilicate; emollient with water repellency and low slip for skin care products [Dow Corning]

Dow Corning® Z-6020. CAS 1760-24-3; EINECS 212-164-2; N-2-aminoethyl-3-aminopropyl trimethoxy silane; coupling agent used for epoxies, phenolics, melamines, nylons, PVC, acrylics, polyolefins, polyurethanes, nitrile rubbers. [Dow Corning]

Dow Corning® Z-6030. CAS 2530-85-0; EINECS 219-785-8; 3-Methacryloxypropyl trimethoxysilane; coupling agent used for free-radical cross-linked polyester, rubber, polyolefins, styrenics, acrylics. [Dow Corning]

Dow Corning® Z-6032. N- [2- (Vinylbenzylamino) -ethyl] -3-aminopropyl trimethoxysilane; coupling agent used for most thermoset and thermoplastic resins. [Dow Corning]

Dow Corning® Z-6040. CAS 2530-83-8; EINECS 219-784-2; 3-Glycidoxy-propyl trimethoxysilane; coupling agent used for epoxies, urethane, acrylic, and polysulfide sealants. [Dow Corning]

Dow Corning® Z-6075. CAS 4130-08-9; Vinyltriacetoxy silane; coupling agent used for polyesters, polyolefins, EPDM, EPM (peroxide cured). [Dow Corning]

Dow Corning® Z-6076. CAS 2530-87-2; 3-Chloropropyl trimethoxy silane; coupling agent used for epoxy, styrenics, nylon. [Dow Corning]

Dowetch Deadline. Magnesium photoengraving plate; manufactured in different gauges and sizes to meet requirements of the newspaper and printing industry. [Dow UK]

Dowex. Ion exchange resins; used for water softening and recovering waste or undesirable materials from process streams. [Dow UK]

Dowex M-31. Catalyst for production of motor fuel oxygenates (MTBE, ETBE, and TAME). [Dow]

Dowex Monosphere. Ion exchange resins. [Dow UK]

Dowfax. A family of disulfonated anionic surfactants used in a variety of end-use applications such as cleaning products and bleaches, latex production and agricultural products. [Dow UK]

---

‡ Trade name and manufacturer not verified   § Trade name without identified manufacturer

Dowfax 2A0. Dodecyl diphenyloxide disulfonic acid; for use in acidic systems or for prep. of various salts. [Dow; Dow Europe]

Dowfax 2A1, 2EP. Sodium dodecyl diphenyloxide disulfonate; detergent, emulsifier, wetting agent, solubilizer, dispersant, spreading agent, penetrant for detergent formulation, emulsion polymerization, agriculture, electroplating, ore flotation, drilling muds; leveling agent for acid dyeing of nylon, dyeing assistant, emulsifier for dye carriers. [Dow; Dow Europe]

Dowfax 3B0. N-Decyl diphenyloxide disulfonate; for formulating cleaning products and agricultural products where salts other than NaCl are required. [Dow]

Dowfax 3B2. CAS 36445-71-3; Sodium n-decyl diphenyloxide disulfonate; detergent, emulsifier, wetting agent, solubilizer, dispersant, spreading agent, penetrant for detergent formulation, emulsion polymerization, agriculture, electroplating, ore flotation, drilling muds; leveling agent for acid dyeing of nylon, dyeing assistant, emulsifier for dye carriers; biodeg. [Dow; Dow Europe]

Dowfax 30C05, 30C10, 50C15. CAS 106392-12-5; EO/PO block copolymer; low foaming surfactant, defoamer base. [Dow Europe]

Dowfax 2A1. Sodium dodecylated oxydibenzene disulfonate; light colored free-flowing powder; anionic surfactant with high solubility, stability, coupling ability and surface activity in strong aqueous solutions of acids, alkalies and salts; moderate sudsing agent; used in metal cleaning including soak tank, steam and electrolytic systems; textiles; shampoos and cosmetics; emulsion polymerization; pulp and paper, mining, and food processing industries. [Dow UK]

Dowfax 9N2, 9N3 and 9N4. Nonionic surfactants of the nonylphenol ethoxylate type in the form of an almost colorless liquid; emulsifier, wetting agent, dry cleaning soap, antifoam; used in chemical intermediates; pesticides; metal cleaners; latex paints; dry cleaning formulations. [Dow UK]

Dowfax 9N5, 9N6 and 9N7. Nonionic surfactants of the nonylphenol ethoxylate type in the form of an almost colorless liquid; emulsifier, wetting agent and dry cleaning soap; used in pesticides; wax and polish; metal cleaners; metal working compounds; dry cleaning formulations. [Dow UK]

Dowfax 9N8, 9N9 and 9N10. Nonionic surfactant of the nonylphenol ethoxylate type in liquid form; detergent, emulsifier, wetting agent and penetrant; used in household, industrial, and institutional cleaners and specialties; textiles; pesticides; latex paints; pulp and paper; leather. [Dow UK]

Dowfax 9N12, 9N14/15 and 9N12W. Nonionic surfactant of the nonyl phenol ethoxylate type in liquid or semi-solid form; used for general detergency and wetting, involving elevated temperatures; used in light duty liquid detergents; cleaning specialties; soak tank and metal cleaners. [Dow UK]

Dowfax 8174. Alkylated disulfonated diphenyl oxide; detergent, emulsifier, wetting agent, solubilizer, dispersant, spreading agent, penetrant for detergent formulation, emulsion polymerization, agriculture, electroplating, ore flotation, drilling muds; leveling agent for acid dyeing of nylon, dyeing assistant, emulsifier for dye carriers; biodeg. [Dow]

Dowfax 8390. Sodium n-hexadecyl diphenyloxide disulfonate; detergent, emulsifier, wetting agent, solubilizer, dispersant, spreading agent, penetrant for detergent formulation, emulsion polymerization, agriculture, electroplating, ore flotation, drilling muds; leveling agent for acid dyeing of nylon, dyeing assistant, emulsifier for dye carriers; biodeg. [Dow; Dow Europe]

Dowfax XDS 8292.00. Sodium hexyl diphenyloxide disulfonate; lowest foaming and highest charge density product in Dowfax series; high solubilizing capabilities in acids, alkalies, and electrolytes; for cleaning, latex mfg., paints, adhesives, mineral and metal processing, textile applications [Dow]

Dowfax XDS 8390.00. Sodium n-hexadecyl diphenyloxide disulfonate; surfactant for emulsion polymerization. [Dow; Dow Europe]

Dowflake. CAS 10043-52-4; Calcium chloride. [Dow UK]

Dowfrost. CAS 57-55-6; EINECS 200-338-0; Inhibited propylene glycol; heat transfer fluid; used as a coolant in the manufacture of beer, wine, milk and other liquids; also used to freeze poultry and fish. [Dow UK]

Dowfroth. Flotation frothers; low-viscosity water-soluble liquids which produce highly selective foams; used by the mining industry in the recovery of minerals for ores. [Dow UK]

Dowgard. Coolant/antifreeze; designed to protect automobile radiators against overheating in summer and freezing in winter and contains additives to protect against foaming and corrosion. [Dow UK]

Dowicide. Phenolic-based antimicrobials used as active ingredients in disinfectant formulations, and also as preservatives in a variety of applications such as metal working fluids, adhesives and cosmetic preparations. [Dow UK]

Dowicide 1. A proprietary trade name for o-phenylphenol; an antiseptic and fungicide. [Dow] *

Dowicide 2. A proprietary trade name for 2:4:5-trichlorphenol; an antiseptic and fungicide. [Dow] *

Dowicide 3. A proprietary trade name for chloro-o-phenylphenol; an antiseptic and fungicide. [Dow] *

Dowicide 5. A proprietary trade name for brom-p-phenylphenol; a germicide. [Dow] *

Dowicide 6. A proprietary trade name for tetrachlorophenol; an antiseptic. [Dow] *

Dowicide 7. A proprietary trade name for pentachlorophenol; an antiseptic and fungicide. [Dow] *

Dowicide A. A proprietary trade name for sodium-o-phenylphenate; an antiseptic and germicide. [Dow] *

Dowicide B. A proprietary trade name for sodium 2 : 4 : 5-trichlorphenate; an antiseptic and germicide. [Dow] *

Dowicide C. A proprietary trade name for sodium-chloro-o-phenylphenate; an antiseptic and germicide. [Dow] *

Dowicide F. A proprietary trade name for sodium tetrachlorophenate; an antiseptic and germicide. [Dow]*

Dowicil®. Preservatives which provide microbial protection for various applications such as cosmetic and personal care formulations, household products, paints, adhesives, metal working fluids and latex emulsions. [Dow UK]

Dowicil® 75. 1-(3-chloroallyl)-3,5,7-triaza-1-azonia-adamantane chloride (act. ingred.) and sodium bicarbonate (stabilizer); preservative for aq. end products such as adhesives, latex emulsions, paints, metal-cutting fluids, drilling muds, biodeg. detergents, and paper coatings; antimicrobial activity. [Dow]

---

* Trade name not verified as available   † Trade name verified as obsolete

Dowlex. Linear low density polyethylene resins used for cast and blown films, injection molding, blow molding, and extrusion coating. [Dow UK]

Dowlex 2032. CAS 9002-88-4; LLDPE resin; general purpose liner resin with improved thermal stability for blown film extrusion (thin guage, med. stiffness film applications); FDA compliance. [Dow]

Dowlex 2035. CAS 9002-88-4; LLDPE; cast film extrusion resin. [Dow]

Dowlex 2042. CAS 9002-88-4; LLDPE; blown film extrusion resin. [Dow]

Dowlex 2500. CAS 9002-88-4; LLDPE; injection molding resin. [Dow]

Dowlex 3010. CAS 9002-88-4; LLDPE; extrusion coating resin. [Dow]

Dowmetal alloys. Aircraft alloys containing magnesium with small amounts of aluminum and manganese, sometimes with the addition of small quantities of copper, cadmium, and zinc. They have low specific gravity. *

Downright. Latex additives used to improve low and high temperature performance and durability in asphalt concrete applications. [Dow UK]

Dowper. CAS 127-18-4; Inhibited perchloroethylene; solvent for dry cleaning. [Dow UK]

Dow Plasticizer No. 5. Diphenyl mono ortho xenyl phosphate; a vinyl plasticizer. [Dow UK] *

Dow Plasticizer No. 55. Technical grade diphenyl mono ortho xenyl phosphate; a vinyl plasticizer. [Dow UK] *

Dowpon. Herbicide; for controlling grass species; used primarily in sugar cane, sugar beets, orchards and also in noncrop applications such as railroads and rubber plantations. [Dow UK] *

Dow Shield. CAS 57754-85-5; Soluble concentrate containing 200 g clopyralid per liter; a foliar herbicide for use on brassicas and field vegetables. [DowElanco Ltd]

Dowtherm 209. Heat transfer agent used as a temperature controlling liquid (coolant) for diesel-powered vehicles. [Dow UK]

Dowtherm A. A proprietary trade name for a product consisting of a mixture of biphenyl and biphenyl ether; used for heating industrial machinery to high temperatures (e.g. 200 C.) in place of steam. *

Doxaben. See Cardura. [Pfizer International]

Doxinate. CAS 577-11-7; Docusate sodium; stool softener; pharmaceutic aid. [Hoechst-Roussel Pharmaceuticals Inc] *

Doxylar®. CAS 24390-14-5; Capsules containing doxycycline hydrochloride BP equivalent to 100 mg doxycycline; clinically useful in treatment of a variety of infections caused by susceptible strains of Gram-positive and Gram-negative bacteria, e.g., pneumonia and other respiratory tract infections including acute and chronic bronchitis, genito-urinary infections and soft tissue infections. [Lagap Pharmaceuticals Ltd]

DP. 250. A polyester vinyl plasticizer. §

DP/4137-16. A proprietary fast-curing polyester resin with low viscosity; a fire retardant. [Synthetic Resins Ltd] *

D.P.G. Diphenylguanidine; a rubber vulcanization accelerator. §

DPNR. CAS 9006-04-6; Deproteinized NR; used for products with high elec. resistivity, low affinity for water; engineering grades for off-shore and underwater use, brake pads, vibration mounts, pharmaceutical products [H.A. Astlett]

DPO. See diphenyloxazole.

DPPG. CAS 41395-83-9; EINECS 255-350-9; Propylene glycol dipelargonate; emollient and oily rancidless additive for cosmetic and pharmaceutical preparations. [Gattefosse SA]

DPR® 40, 75, 400. cis 1,4-Polyisoprene of low molecular weight derived from virgin synthetic polyisoprene; used for potting, sealants, caulk, adhesives, and flexible rubber molds and ebonite stocks; cures to soft rubber; two-part compounds can be formulated for self-curing at room temperature; elevated temp. curable by sulfur, halogenated phenolics, peroxides, etc. [Hardman]

DPTT. CAS 120-54-7; Dipentamethylene thiuram hexasulfide; very active sulfur-bearing accelerator imparting heat resistance to sulfurless compounds; primary accelerator for Hypalon, butyl, and EPDM; vulcanizing agent for heat-resistant latex; nondiscoloring, nonstaining. [Akrochem]

Dracyl. Toluene, $C_6H_5CH_3$.

dracylic acid. See benzoic acid.

Dragendorf's reagent. (Kraut's reagent). Potassium iodobismuthate; used for testing alkaloids.

Dragon. Insecticide. [ICI Chem. & Polymers Ltd]

dragon gum. CAS 9000-65-1; Gum tragacanth.

Dragonmat. Insecticide vaporizing device. [ICI Chem. & Polymers Ltd]

dragon's blood. A red resin. The two varieties are Palm Dragon's Blood, obtained from the rattan palm, *Daemonorops draco*, of Sumatra and Borneo, and Socotra Dragon's Blood, from *Dracoena cinnabari*, of Socotra, and the West Indies; used as a pigment for the preparation of red lakes and varnishes.

Drainaid GL-73. Drainage aid for acidified pulp in paper industry. [Hart Chem. Ltd.]

Drakeol 5. Light mineral oil USP; CAS 8042-47-5; emollient. [Penreco]

Drakeol 7. Light mineral oil USP; CAS 8042-47-5; carrier, base ingred. in ointments, lotions, baby oils, sun tan lotions, makeup; solv. and emollient in creams, waterless hand cleaners; protective coating for foods; pigment dispersant, lubricant for plastics. [Penreco]

Drakeol 10. Light mineral oil USP; CAS 8042-47-5; plasticizer, lubricant for plastics. [Penreco]

Drakeol 19. Mineral oil USP; CAS 8042-47-5; primary plasticizer for ethyl cellulose; lubricant for textile/paper. [Penreco]

Draketex 50. Light mineral oil USP; CAS 8042-47-5; coning and finishing oil base for nylon and rayon production [Penreco]

Dralon®. Polyacrylic staple fiber and tow; used for outdoor textiles and acid proof protective clothing. [Bayer AG]

Dralon® T. Polyacrylic staple-fiber; used for industrial felts and filter media. [Bayer AG]

Dramamine. CAS 523-87-5; A proprietary preparation of dimenhydrinate; an antiemetic. [G D Searle & Co Ltd] *

Drapex. White mineral oil. [Pennzoil Products Co]

Drapex®. Epoxy plasticizer. [Argus Chemical Corporation]

Drapex® 4.4. Octyl epoxy tallate; plasticizer for vinyl compounds; compatible with primary plasticizers; used with Argus Ba/Cd stabilizers to synergistically improve heat and light stability; resistant to extraction and migration; food pkg. use. [Witco/Argus]

Drapex® 6.8. CAS 8013-07-8; EINECS 232-391-0; Epoxidized soybean oil; plasticizer for vinyl compounds;

---

‡ Trade name and manufacturer not verified   § Trade name without identified manufacturer

compatible with primary plasticizers; used with Argus Ba/ Cd stabilizers to synergistically improve heat and light stability; resistant to extraction and migration; food pkg. use. [Witco/Argus]

Drapex® 10.4. Epoxidized linseed oil; plasticizer for vinyl compounds; compatible with primary plasticizers; used with Argus Ba/Cd stabilizers to synergistically improve heat and light stability; resistant to extraction and migration; food pkg. use. [Witco/Argus]

Drapex®P-1. Polyester plasticizer; plasticizer for use in PVC compounds requiring durability, resist. to migration, high dielec. props., e.g., for elec. tape compounds, refrigerator gasketing, wall covering, automotive interiors, oil-resist. wire insulation. [Witco/Argus]

Drapolene. For treatment of nappy rash. [The Wellcome Foundation Ltd]

Drat. CAS 3691-35-8; An oil formulation containing 2.5 g/liter chlorophacinone, a powerful anticoagulent rodenticide; controls black rats, brown rats, house mice, long-tailed field mice, voles and musk rats. [Rhone-Poulenc Environmental Prods. Ltd]

drawing brass. See brass.

Draza®. CAS 2032-65-7; Pellet formulation containing 4% methiocarb; to control slugs and snails in any crop; reduces populations of leatherjackets; there is some evidence that cutworms, earwigs, and millipedes are controlled. [Bayer AG; Bayer plc]

Dreadnought powder. An explosive containing 73-77% ammonium nitrate, 14-17% sodium nitrate, 4-6% ammonium chloride, and 3-5% trinitrotoluene.

Dreft. A proprietary trade name for a washing material consisting of sodium lauryl sulfate. §

Drenamist. A proprietary preparation of adrenaline hydrochloride in an aerosol form; bronchial antispasmodic. [Nicholas Laboratories Ltd] *

Drene. A proprietary trade name for a shampoo containing sodium lauryl sulfate. §

Drenison. A proprietary preparation of flurandrenalone for dermatological use. [Eli Lilly & Co] *

Drenusil. A proprietary preparation of polythiazide; a diuretic. [Pfizer International] *

Dresden Thick Oil. A thick turpentine or oleo-resin; similar to Venice turpentine; used as a vehicle for colors for painting. §

Dresinate. Potassium and sodium soaps of rosin, modified rosin and tall oil derivatives; improves latex stability in rubber latices; used in formulating soluble cutting oils and drawing compounds; as modifiers of heavy-duty metal cleaners and other industrial cleaners. [Hercules]*

Dresinol. A range of resin dispersions; used to modify polymer properties in adhesives, in the production of pigments and resinated colors in the graphics and inks industry, in paint polymer emulsions and to help wet paint pigments. [Hercules] *

Dress All. Partially hydrogenated soybean oil, artificial flavor, TBHQ, artificial color; kosher; dressing oil for seafood, vegetables, frying, other prepared foods. [Van Den Bergh Foods]

Drewamine. Corrosion inhibitor for control of corrosion in the afterboiler section of steam generating systems by neutralizing the acidity of the condensate. [Drew Ind. Div.]

Drewbrom. Aq. solution of bromide ion; precursor for production of biocide used as disinfectant, sanitizer, bactericide, slimicide, and algicide in recirculating cool-

ing water systems, once-through cooling water, and waste water treatment systems. [Drew Ind. Div.]

Drewchlor®. Algicide and precursor for generation of chlorine dioxide for treating industrial recirculating cooling water systems incl. cooling towers, air washers and evaporator condensers. [Drew Ind. Div.]

Drewclean®26. Alkaline cleaner; cleaner for ion exchange resins and filter media fouled by oil, grease, other organics, clay or silt. [Drew Ind. Div.]

Drewcor®2130. Blended neutralizing amines and Mekor® voltile oxygen scavenger/metal passivator; corrosion inhibitor for protection against low pH and oxygen induced corrosion in condensate and steam lines. [Drew Ind. Div.]

Drewfax®0007. CAS 577-11-7; EINECS 209-406-4; Sodium dioctyl sulfosuccinate; wetting, penetrating, surface tensile reducing agent, dispersant for industrial coatings, adhesives, inks, pigments, textile, cosmetic, paper, metal, paint, rubber, plastics, petrol. and agricultural industries. [Drew Ind. Div.] *

Drewfax®412. Blend of silicone derivs. and glycol ethers; surfactant reducing surface tensile, wetting agent, flow and leveling agent, slip and mar aid for solv. and water based coatings and inks based on alkyds, epoxy, urethanes, acrylics. [Drew Ind. Div.]

Drewfax® 680. Silicone surfactant; surfactant, antifoam, flow and leveling agent, air release agent for solv.-based coatings and inks based on alkyd melamine, epoxy, urethanes, acrylics; esp. suited for wood coatings. [Drew Ind. Div.]

Drewfax® 818. Blend of silicone copolymers and glycol ether; surfactant, wetting agent, flow and leveling agent, slip and mar aid, anticrater for water-based coatings and inks based on acrylics, alkyds. [Drew Ind. Div.]

Drewfax® S-600. Silicone/silica derivs.; surfactant, antifoam, flow and leveling agent for solv. and water-based coatings and inks based on alkyds, epoxy, urethanes, acrylics; esp. suited for clear lacquers, pigmented coatings, uv coatings. [Drew Ind. Div.]

Drewfax®S-700. CAS 577-11-7; EINECS 209-406-4; Sodium dioctyl sulfosuccinate; surfactant, anticrater, dispersant, flow and leveling agent for high-solids and acrylic solv.-based inks and coatings. [Drew Ind. Div.]

Drewfloc®2270. High molecular weight emulsion polymer; flocculant and coagulant aid for dewatering industrial slurries, water clarification, and sludge conditioning applications [Drew Ind. Div.]

Drewgard® 120. Conc. polyphosphate; corrosion and deposit inhibitor in once-through and open recirculating cooling water systems and potable water distribution systems. [Drew Ind. Div.]

Drewgard® 189E. Isoquest-LT polymer, org. phosphorus and zinc; corrosion and deposit inhibitor for open recirculating cooling water systems. [Drew Ind. Div.]

Drewmulse®10K. Glyceryl mono-shortening from soya oil; lipophilic emulsifier used in food industry as dispersing aid, antistaling agent, antistick agent. [Stepan/PVO]

Drewmulse®200K. Glyceryl stearate; emulsifier, emollient, antistat, stabilizer, visc. builder, opacifier for creams, lotions, hair conditioners. [Stepan Europe]

Drewmulse® 900K. Glyceryl stearate; emulsifier for food industry as dispersing aid, antistaling agent, antistick agent. [Stepan/PVO]

Drewplast®017. Glycerol mono and diesters of fatty acids;

---

food-grade plastics additive; antistat for LDPE, lubricant for rigid and flexible PVC; good thermal props., stable to oxidative degradation. [Stepan/PVO]

Drewplast®030. Glycerol mono and diesters of fatty acids; food-grade plastics additive; antifog for PVC sheet and film; good thermal props., stable to oxidative degradation. [Stepan/PVO]

Drewplast®051. Glycerol mono and diesters of fatty acids; food-grade plastics additive; antifog for PVC sheet and film; cling agent for LDPE; lubricant for rigid and flexible PVC; good thermal props., stable to oxidative degradation. [Stepan/PVO]

Drewplex. Blends of natural and synthetic sludge conditioning agents; used for boiler water sludge conditioning, boiler water treatment. [Ashland Chemical Company]

Drewplus® L-108. Blend of mineral oils, emulsifiers, silica derivs.; defoamer for latex/rubber applications esp. monomer stripping, acrylic, PVAc, NBR, SBR, PVC. [Drew Ind. Div.]

Drewplus® L-123. Nonsilicone defoamer; defoamer for latexes. [Drew Ind. Div.]

Drewplus®L-162. Silicone defoamer; defoamer for latexes. [Drew Ind. Div.]

Drewplus®L-407. Modified polysiloxane copolymer emulsion; foam control agent for architectural paints and coatings (interior/exterior gloss and semigloss paints, waterborne industrial coatings, polymer emulsions, water-thinnable gravure and flexographic inks). [Drew Ind. Div.]

Drewplus®L-464. Blend of silica and organic solids; surfactant, defoamer, air release agent effective in water-reducible coatings such as polyurethanes, interior and exterior paints. [Drew Ind. Div.]

Drewplus®L-768. Blend of silicone fluid, silica derivs. and surfactants; defoamer for industrial/chemical processes, food/fermentation and agricultural applications, aq. and some nonaq. systems. [Drew Ind. Div.]

Drewplus® L-790. Blend of mineral oils, silica derivs., and emulsifiers; defoamer for food/fermentation and agricultural applications [Drew Ind. Div.]

Drewplus®L-813. Blend of dimethylpolysiloxane, silica, and emulsifiers; foam control agent for industrial food processing (starch slurries, fermentation, calcium chloride brines, adhesives, glues, vegetable and fruit processing, sugar, instant coffee, fruit juices, dehydrating and evaporating systems). [Drew Ind. Div.]

Drewplus® M-111. Blend of silica derivs. and surfactants; defoamer for gypsum, starch, cement, pigments, joint compound and adhesives that require a dry defoamer. [Drew Ind. Div.]

Drewplus® Y-200. Nonsilicone defoamer; defoamer for paints/coatings, inks, adhesives. [Drew Ind. Div.]

Drewplus® Y-381. Blend of organics and hydrocarbons; defoamer/antifoam for latex paints (acrylic, vinyl acrylic, S/B), coatings, ink, adhesives. [Drew Ind. Div.]

Drewpol® 3-1-O. CAS 9007-48-1; Polyglyceryl-3 monooleate; emulsifier for creams, makeup, lotions, conditioners; food emulsifier. [Stepan; Stepan/PVO]

Drewpol® 10-4-O. CAS 34424-98-1; Polyglyceryl-10 tetraoleate; emulsifier, emollient, antistat, stabilizer, visc. builder, opacifier for creams, makeup, lotions, hair conditioners; solubilizer for vitamins, flavors, medicaments. [Stepan; Stepan Canada; Stepan Europe]

Drewpone® 60K. CAS 9005-67-8; Polysorbate 60; food

emulsifier, foaming agent, dough conditioner. [Stepan/PVO]

Drewpone® 65K. CAS 9005-71-4; Polysorbate 65; food emulsifier, foaming agent, dough conditioner. [Stepan/PVO]

Drewpone® 80K. CAS 9005-65-6; Polysorbate 80; food emulsifier, foaming agent, dough conditioner. [Stepan/PVO]

Drewsorb®60K. CAS 1338-41-6; EINECS 215-664-9; Sorbitan stearate; food emulsifier. [Stepan/PVO]

Drewsperse® 611. CAS 9003-04-7; Sodium polyacrylate; pigment and paint dispersant. [Drew Ind. Div.] *

Drewsperse® S-825. High molecular weight polymer; wetting agent, dispersant, film-former for water-based acrylic, alkyd, styrene acrylic, epoxy, alkyd-melamine, or polyurethane coatings and inks; maximizes color development; stabilizes the dispersion against flocculation and settling. [Drew Ind. Div.]

Drewtrol®6955S. Polyphosphate, synthetic polymers, nat. org. compounds, sequestrants, alkali and antifoam; corrosion and deposit inhibitor for boiler water treatment. [Drew Ind. Div.]

Drexar 530. CAS 2163-80-6; MSMA; selective herbicide for post emergent weed control on lawns and ornamental turf. [Drexel Chemical Company] *

Driclor. CAS 7446-70-0; A colorless alcoholic solution containing aluminum chloride hexahydrate; used for the treatment of hyper-hydrosis of the axillas the hands and the feet. [Stiefel Laboratories (UK) Ltd] *

Dricoid® 200. CAS 11138-66-2; Xanthan gum; used for ice cream, lowfat frozen dessert, frozen yogurt, ice milk and novelties. [Kelco]

Dricoid® KB, KBC. CAS 9005-37-2; Propylene glycol alginate; used for ice cream, novelties, soft-serve yogurt. [Kelco]

Dricol. CAS 1421-68-7; A proprietary preparation of amidephrine mesylate; a nasal decongestant. [Bristol-Myers Squibb Co Inc] *

Dricold. CAS 124-38-9; Solid carbon dioxide pellets or slices; used as refrigerant. [ICI Chem. & Polymers Ltd]

drierite. CAS 7778-18-9; Anhydrous calcium sulfate; used for drying gases.

Driers. Those substances which are added during the process of boiling linseed oil, to accelerate its drying properties; absorbs the oxygen from the air and transfers it to the oil, thereby aiding its oxidation; term is used for the oxides of lead, manganese, and cobalt, which were formerly used as driers; more recently, the oxalate, acetate, and borate of manganese have been employed; at present, the metallic salts of fatty acids, such as lead and manganese linoleates are much used; the metallic resinates are also employed as driers. *Also see* boiled oil, terebine, and lead tungate.

Dri Film®DF1040. Polymethyl hydrosiloxane fluid; reactive silicone fluid forming water-repellent fluids with heat or heat/catalyst; used in textiles, particle treatment, magnesium oxide, and Calrod®units. [GE Silicones]

Driftol. Drift retardant. [Chevron] *

Drift Proof. Non-phytotoxic drift control agent, spreader-sticker and pesticide deposit builder. [W A Cleary]

Drikalite®. CAS 471-34-1, 1317-65-3; Calcium carbonate; extender pigment. [ECC]

Drikold. CAS 124-38-9; Sold carbon dioxide blocks or slices; refrigerating agent. [ICI Chem. & Polymers Ltd]

---

‡ Trade name and manufacturer not verified   § Trade name without identified manufacturer

drill rod brass. *See* brass.
Drimarene®. Specialty dye for aq. mediums, wood stains. [Sandoz]
Drimax®. Filter cake dewatering aids offering improvement in filter performance with substantial reductions in final cake moisture contents. [Allied Colloids Ltd]
Drimix®. Powder dispersion of liquids, dry liquid concentrates; free flowing power of liquid additives widely used in elastomer compounding. [Kenrich Petrochemicals]
Drinamyl. A proprietary preparation of dexamphetamine sulfate and amylobarbitone. [SmithKline Beecham] *
Drisdol. Ergocalciferol; vitamin [Sterling Drug Inc] *
Dri-Sil. Silicone water repellents. [Midland Silicones] ‡
Drisorb. Desiccant. [Production Chemicals Ltd]
Drisoy®. Modified soybean oils; used in paints. [Reichhold]
Dristan Inhaler. CAS 3595-11-7; Propylhexedrine; adrenergic. [Whitehall Laboratories]
Dristan Long Lasting Nasal Mist. CAS 2315-02-8; Oxymetazoline hydrochloride; adrenergic. [Whitehall Laboratories]
Dritan. Vegetable tannins. [Hodgson Chemicals Ltd]
Driton. Colloidal silica; antislip agent; prevents fiber slippage, dry hand; used only under alkaline conditions. [Sybron]
Drittel silver. An alloy of 67% aluminum and 33% silver.
Drivanil. A range of alkylene oxide addition products; used as a base for synthetic lubricants, base component for brake fluids, heat transfer medium, and a viscosity modifier for hydraulic fluids containing no mineral oil. [Hüls AG]
Driverit. CAS 75-09-2; Stabilized methylene chloride; solvent for metal degreasing; also suitable for the treatment of aluminum. [Hüls AG] *
Driverol MPL. Partial phosphate ester; corrosion inhibitor for the mineral oil industry. [Hüls AG]
Driverol OMM. An amide/anhydride mixture; corrosion inhibitor [Hüls AG]
Driveron. CAS 1634-04-4; Methyl tert-butyl ether; an anti-knock agent for motor fuels. [Hüls AG]
Drivolan. A range of dodecanedioic acid esters; base components for synthetic and semisynthetic lubricants. [Hüls AG]
Drivosol. Aerosol propellant. [Hüls AG]
Driwal. *See* Ceresit. §
Drolban. CAS 521-12-0; Dromostanolone propionate; antineoplastic. [Eli Lilly & Co] †
Droleptan. CAS 548-73-2; A proprietary preparation of droperidol; tranquilizer, premedicant, anti-emetic. [Janssen Pharmaceutical Ltd] *
Dromoran. A proprietary preparation of levorphanol tartrate; an analgesic. [Roche Products Ltd] *
Droncit®. Tapeworm drug for dogs and cats, also against *Echinococcus* spp. veterinary medicine. [Bayer AG]
Drontal. Multispectrum anthelmintic for dogs and cats. [Bayer AG]
Drontal Plus. Multispectrum anthelmintic for dogs. [Bayer AG]
Drossa. Protective hand lotion; for practice and laboratory. [Bayer AG] †
Drott. A proprietary pyroxylin plastic. §
Droxol 200. CAS 25322-68-3; PEG-200; chem. intermediate for textiles, lubricants, printing, solvs., cleaning formulations, humectants, visc. modifiers. [Henkel/Textiles]

Droxychrome. A photographic color developer. [May & Baker Ltd] *
Dry and Clear. CAS 94-36-0; Benzoyl peroxide; keratolytic. [Whitehall Laboratories] *
Dry-Blend® NCG Fiber. Nickel-coated graphite fiber in ABS/PC, ABS, or PC; for conductive plastics. [Am. Cyanamid]
Dry Flo®. CAS 9087-61-0; Aluminum starch octenyl succinate; body powds., antiperspirants, feminine hygiene sprays, foot powds. [Nat'l. Starch]
dry ice. CAS 124-38-9; Solid carbon dioxide. Its specific gravity is 1.56.
Dry Lightning. Inhibited sulfamic acid (dry); removes water-formed deposits. [Garvey Chemical Corp] *
Drymax®. Drier accelerator; for accelerating the drying of nonaqueous coatings. [Hüls Am.]
Drymet® 59. CAS 6834-92-0; EINECS 229-912-9; Sodium metasilicate anhyd.; alkaline silicate used for formulating specialty detergents and contributing buffering capacity and corrosion inhibition of soft metals and ceramic glazes. [Crosfield]
Dry Pexol® 200. Fortified pale rosin; dry size used with alum in paper and paperboard to produce resistance to water and aq. solutions [Hercules]
Dry Pexol® 243. Fortified dark rosin; dry size for lower brightness paper grades. [Hercules]
Dryptal. A proprietary preparation containing frusemide; a diuretic used for edema of cardiac, hepatic or renal angina. [Berk Pharmaceuticals Ltd] *
Dry Seed TRIGGRR. A dry powder containing trace minerals; used to enhance germination, maturation, and crop yields; applied to seed prior to planting; used for a wide variety of crops including corn, sorghum, wheat and vegetables. [Westbridge Research Group] ‡
Dry Size XL20C. Pale rosin, paraffin wax, ratio 80:20; dry size used with alum to impart high level of resistance to water and aq. solutions in specialized applications like molded pulp products; high levels can be used without brightness, foam, or other operating problems on foaming machine; food pkg. paper/paperboard. [Hercules]
Dryspersion®. Powder dispersions. [Kenrich Petrochemicals]
Drytech. Absorbant for aqueous liquids. [Dow UK]
Dryvax. Smallpox vaccine; immunizing agent. [Wyeth Laboratories] *
D.S.H.C. Dimethylsilanol hyaluronate; provides skin regeneration, strong hydrating action for cosmetic and health products such as milks, emulsions, creams, lotions. [Exsymol]
DSM. CAS 919-86-8; Preparation containing 50% demeton-S-methyl; insecticide. [L W Vass (Agricultural) Ltd] †
DSP-2. *See* disodium phosphate, dihydrate.
DSP-O. *See* disodium phosphate (anhydrous).
DSS. *See* dioctyl sodium sulfosuccinate.
D-S-S. CAS 577-11-7; Docusate sodium; stool softener; pharmaceutic aid. [Parke-Davis] *
D-S-S Plus. Contains casanthranol and docusate sodium; laxative, and stool-softener. [Parke-Davis] *
D-steel. A steel containing 1.1-1.4% manganese, 0.33% carbon and 0.12% silicon.
DSX 1514. Urethane associative thickener; rheology modifier for latex paints and adhesives; maximize high sheer visc. [Henkel] *
DTBP. *See* di-t-butyl peroxide.

---

* Trade name not verified as available   † Trade name verified as obsolete

DTIC-Dome. Dacarbazine; antineoplastic. [Miles Pharmaceuticals]

D-Trans. D-*trans* allethrin based intermediates. [McLaughlin Gormley King Co]

D.T.S. Dehydrothio-p-toluidinesulfonic acid.

Duact. Proprietary formulations containing acrivastine hydrochloride and pseudoephedrine hydrochloride; for temporary relief of symptoms associated with allergic rhinitis accompanied by nasal congestion. [The Wellcome Foundation Ltd]

Dualite M6001AE, M6017AE. Hollow composite microspheres (shell: PVDC/acrylonitrile copolymer; coating: calcium carbonate); low-dens. filler for use in plastics, coatings, adhesives, BMC, SMC, rubber compounding, paper mfg., etc.; M6001AE for moderate exposure to nonaggressive solvs.; M6017AE for prolonged exposure to aggressive solvs. [Pierce & Stevens]

Duallor. A range of precious metal alloys; for dentistry and dental engineering. [Degussa AG] *

Duasyn. Water soluble dyes for paper, toiletries and cleansers. [Hoechst AG]

Dubbin. Mixtures of waxes and tallow with coloring matter, sometimes with the addition of rosin; used to render leather waterproof, and to preserve it. §

Dubox®. For the electrical industry. [DuPont UK]

Duco. A proprietary trade name for pyroxylin lacquers, containing cellulose nitrate. §

Ducobee-Hy. CAS 13422-51-0; Hydroxocobalamin; vitamin. [Sterling Drug Inc] *

Dudley metal. An alloy of 98% tin, 1.6% copper, and 0.25% lead.

Duet®. For the agriculture industry. [DuPont UK]

Duette. Paint system with two tone decorative effect for walls and ceilings. [ICI Chem. & Polymers Ltd]

Dufox. A potato fungicide. [Murphy Chemical Co Ltd] †

Dugro. CAS 7681-76-7; Ronidazole; antiprotozoal. [Merck & Co Inc] *

DUK-880. Lenacil and phenmedipham; used for control of annual dicotyledons in sugar beet. [DuPont UK]

Dukatalon. CAS 85-00-7, 1910-42-5; Mixture of diquat and paraquat; herbicide. [Makhteshim Chemical Works Ltd]

Duke's metal. A heat-resisting alloy, containing 81% iron, 12% chromium, 4% cobalt, 1.5% carbon, and small quantities of manganese, tungsten, and silicon.

Dulceta. Textile auxiliary chemicals. [ICI Chem. & Polymers Ltd] †

dulcine. (sucrol, sucrene; 4-ethoxyphenylurea). Mono-p-phenetol-carbamide, $H_2NOCNHC_6H_4OC_2H_5$; CAS 150-69-6; used as a sweetening substance. It is 200 times sweeter than cane sugar, prohibited by the USA FDA for use in foods.

Dulcodos. A proprietary preparation containing bisacodyl and dioctyl sodium sulfosuccinate; a laxative. [Boehringer Ingelheim GmbH] *

Dulco-Lax. CAS 603-50-9; A proprietary preparation containing bisacodyl; a laxative. [Windsor Healthcare Ltd]

Dulenza. A proprietary viscose silk. §

Dullray. A heat-resisting alloy containing 60% iron, 34% nickel, and 5% chromium.

Dulux. A range of various interior/exterior paints and paint-related products. [ICI Chem. & Polymers Ltd]

Dumacene C13, NP707, NP7710 and NPX10. Alkylphenol ethoxylate nonionic surfactant; for textile scouring; wool washing. [Tensia SA] *

Dumet. A copper-clad nickel-iron alloy.

Duncaine. A proprietary trade name for Lignocaine. §

Dunclad CE. A proprietary trade name for a laminate of Penton (*qv*) and a synthetic rubber; offers a highly corrosion resistant lining at temperatures up to 125-130 C. [Dunlop Rubber] *

Dunclad VN. A laminate of unplasticized PVC and synthetic rubber designed to enable metal tanks to be lined using special adhesives, giving a highly corrosive resistant surface to prevent attack from oxidizing and other acids up to 85 C. [Dunlop Rubber] *

Dunlop 6593. A high grade neoprene compound giving high chemical and abrasion resistance at elevated temperatures. [Dunlop Rubber] *

Dunlop Grade 6167. A first quality butyl rubber compound which can be used up to 110 C. in corrosive conditions. [Dunlop Rubber] *

Dunlop PL. A laminate of polypropylene and synthetic rubber for tank lining. [Dunlop Rubber] *

Dunnite. An American explosive. The main constituent is picric acid.

Dunova®. Polyacrylic staple fiber; water absorbent used for sports wear. [Bayer AG]

Duo-Autohaler. Breath-actuated pressurized aerosol containing a suspension of isoprenaline hydrochloride BP 8 mg/ml, phenylephrine bitartrate 12 mg/ml; delivers 400 metered doses (0.16 mg isoprenaline hydrochloride, 0.24 mg phenylephrine bitartrate per dose). [3M Pharmaceuticals] †

Duocid. *See* Unasyn. [Pfizer International]

Duocrome®. Iridescent colors producing dual-color effects. [Mearl]

Duo-Decadron. Dexamethasone acetate and dexamethasone sodium phosphate; promotes and sustains corticosteroid activity for the relief of certain endocrine and nonendocrine disorders. [Merck & Co Inc]

Duofilm. Collodion based product containing salicyclic acid BP and lactic acid BP; topical application for the treatment of plantar and mosaic warts. [Stiefel Laboratories (UK) Ltd] *

Duofol T. Sulfated ester; wetting, rewetting agent, dispersant, penetrant, leveling agent, finishing agents for textile wet processing of cotton piece goods; stable to mild alkalies and org. acids. [Hart Chem. Ltd.]

Duogastrone. CAS 7421-40-1; A proprietary preparation of carbenoxolone sodium in a delayed release capsule; for the treatment of duodenal ulcers. [Berk Pharmaceuticals Ltd] *

Duolite. Ion exchange resins. [Rohm & Haas UK]

Duolith. *See* Cryptone. §

Duomac®. Series of cationic surfactants which consist of the acetate salts of alkyl propylene diamines, and which are soluble; used in pigment flushing, froth flotation and flocculation, particularly in mineral flotation; petroleum processing; leather; paper; pigments and surface coatings; ceramics. [Akzo Chemie UK Ltd] †

Duomat. Dental amalgamator. [Degussa Ltd]

Duomatic. Injector/drencher gun. [May & Baker Ltd] *

Duo-Medihaler™. Pressurized aerosol containing a suspension of isoproterenol hydrochloride, phenylephrine bitartrate; delivers 300 metered doses (0.16 mg isoproterenol hydrochloride, 0.24 mg phenylephrine bitartrate per dose); for asthma relief. [3M Pharmaceuticals]

---

‡ Trade name and manufacturer not verified   § Trade name without identified manufacturer

Duomeen. A range of coco, oleyl and stearyl diamines; used as waterproofing agents, bitumen emulsifiers, bitumen adhesion additives, anticorrosives and as agricultural sprays. [Harcros Australia] *

Duomeen®. Range of cationic surfactants of the alkyl propylene diamine type, possessing both primary and secondary amine groups; they form strongly bonded films on the surfaces of metal, textiles, plastics, etc.; used in petroleum; road emulsions; plastics; rubber; textiles; leather; herbicides; fungicides; rodent repellents; mineral flotation; paper; pigments, and surface coatings; water and sewage treatment; wax; sealant formulations; cement curing; metal working; emulsion for car underseals; carbon paper, and typewriter ribbon. [Akzo Chemie UK Ltd]

Duomeen® C. CAS 61791-63-7; N-Coco-1,3-diaminopropane; chemical intermediate; corrosion inhibitor, fuel oil additive, flotation agent; used in metals, textiles, plastics, herbicides; epoxy curing agent. [Akzo; Akzo Chem. BV]

Duomeen® CD. CAS 61791-63-7; Coco-1,3-diaminopropane; industrial surfactant; intermediate. [Akzo]

Duomeen® LT-4. 3-Tallowalkyl-1,3-hexahydropyrimidine; industrial surfactant. [Akzo]

Duomeen® OL. CAS 7173-62-8; EINECS 230-528-9; N-Oleyl-1,3-diaminopropane; industrial surfactant. [Akzo]

Duomeen®OTM. CAS 68715-87-7; N,N,N´-Trimethyl-N´-9-octadecenyl-1,3-diaminopropane; industrial surfactant. [Akzo]

Duomeen® T. CAS 61791-55-7; Tallow-1,3-diaminopropane; bitumen emulsifier, corrosion inhibitor, oil additive, antisettling agent; textile finishing agent, dispersant for inorg. pigments in paints, larvacidal oil additive. [Akzo; Akzo Chem. BV]

Duomeen®TDO. CAS 61791-53-5; N-Tallow-1,3-propanediamine dioleate; bitumen emulsifier, corrosion inhibitor, oil additive, antisettling agent; also in metal treatment as film and boundary lubricant, metal drawing additive; dispersant in paint industry. [Akzo; Akzo Chem. BV]

Duomeen® TTM. CAS 68783-25-5; N,N,N´-Trimethyl-N´-tallow-1,3-diaminopropane; industrial surfactant. [Akzo]

Duoquad®. Diamine quaternary ammonium chloride. [Akzo Chemie UK Ltd]

Duoquad® O-50. CAS 68310-73-6; N,N,N´,N´,N´-Pentamethyl-N-octadecenyl-1,3-diammonium dichloride, aq. IPA; industrial surfactant. [Akzo]

Duoquad® T-50. CAS 68607-29-4; N,N,N´,N´,N´-Pentamethyl-n-tallow-1,3-propanediammonium dichlorides, aq. IPA; detergent, corrosion inhibitor, metal cleaner; emulsifier for sec. oil recovery. [Akzo]

Duoteric. Surfactant blend. [Rhone-Poulenc UK]

Duothane. Two-component hot cure polyurethane adhesive. [Compounding Ingredients Ltd]

Duovac-C. B₁ type, B₁ strain, Newcastle and Connecticut type; for immunization of poultry. [Intervet Inc] †

Duovac-M. B₁ type, B₁ strain, Newcastle and mild Massachuttes type; for immunization of poultry. [Intervet Inc]

Duovac-Ma5. B₁ Type, B₁ strain Newcastle and Ma₅ Mass. type vaccine; for immunization of poultry. [Intervet Inc]

Duovent. A proprietary preparation of ipratropium bromide and fenoterol hydrobromide; a bronchodilator. [Boehringer Ingelheim Ltd]

Duphalac. CAS 4618-18-2; Lactulose; for treatment of obstipation and (pre-) hepatic coma. [Duphar BV] *

Duphaston. CAS 152-62-5; Dydrogesterone; progestative; counteracts complaints caused by hormonal disorders in women. [Duphar BV] *

Dupical. 4-Tricyclo(5.2.1.0. 2,6)-decylidene-8)-butanal. [Quest Int'l. UK Ltd]

Duplosan®. Translocated herbicide for cereals and grassland. [BASF plc]

Duplosan® CMPP. CAS 93-65-2; Mecoprop-P; selective herbicide for control of broadleaf weeds in cereals, meadows, pastures [BASF AG]

Duplosan®DP. CAS 28631-35-8; Dichlorprop-P; herbicide for control of broadleaf weeds in cereals [BASF AG]

Duplosan New System CMPP. CAS 93-65-2; Soluble concentrate containing 600 g/l mecoprop-P; for control of weeds in undersown cereals and grassland. [Rhone-Poulenc Crop Protection Ltd]

Duponol. Surfactants. [DuPont UK] *

Duponol LS. CAS 1847-55-8; EINECS 217-430-1; A proprietary trade name for sodium oleyl sulfate, a wetting agent. [Witco/Organics] *

Duponol ME. A proprietary trade name for sodium lauryl sulfate, a wetting agent. [Witco/Organics] *

Duponol WA. A proprietary trade name for a mixture of sodium salt of sulfated lauryl alcohol and lauryl alcohol. [Witco/Organics] *

Du Pont Accelerators. Accelerators for rubber vulcanization; No. 1: p-nitroso-dimethylaniline; No. 4: aniline; No. 5: formaldehyde-aniline; No. 6: methylene dianilide; No. 8: anhydro-formaldehyde-p-toluidine; No. 11: triphenylguanidine; No. 12: diphenylguanidine. [Du Pont] *

Du Pont Adjuvant. Nonionic spreader containing 900 g/l ethylene oxide condensate; wetting agent for use in herbicide, fungicide and insecticide sprays. [DuPont UK]

Du Pont Enrich®. For the agriculture industry. [DuPont UK]

Du Pont Linuron 50, 4L. CAS 330-55-3; Linuron; a residual urea herbicide for the control of weeds in field crops. [DuPont UK]

Du Pont Pakwrap®. Polymer. [DuPont UK]

Du Pont Permissible No. 1. A trademark for an explosive containing nitroglycerin, ammonium nitrate, wood pulp, and sodium chloride. [Du Pont] *

Dupranin CR. Quaternary ammonium compound; cationic; clear liquid; leveling agent for dyeing with cationic dyes (retarder). [Thor Chemicals (UK) Ltd]

Dupranin W. Alkyl amino polyglycol ether; nonionic; yellow liquid; leveling agent for acid, 1:1 metallic complex and after-chroming dyestuffs, recleaning agent for dyed and printed goods. [Thor Chemicals (UK) Ltd]

Duprene. A proprietary trade name for a synthetic rubber made by the polymerization of chloroprene; resistant to heat, oil, ozone, and most other chemicals. §

Durabolin. CAS 62-90-8; Nandrolone phenpropionate; androgen. [Organon Inc]

Durabond. Calcium/magnesium lignosulfonate; pellet binder in animal feeds; contributes some nutritive value. [Borregaard LignoTech]

Durabond 650, 655. Durable synthetic resin stiffener and hand modifier for finishing formulations. [Eastern Color & Chem.]

Duracillin. A proprietary preparation of procaine penicillin; an antibiotic. [Eli Lilly & Co] †

Duracore. High strength aluminum honeycomb cores; for military and aerospace sandwich panel applications.

---

* Trade name not verified as available   † Trade name verified as obsolete

[Fothergill Tygaflor Ltd] *

Duracreme. A proprietary preparation of propylene glycol, glycerine, sodium alginate, boric acid, and hexyl resorcinol; a spermicidal jelly. [LRC Products Ltd] *

Durad. Lubricant additives. [FMC]

Duradene® 706. Solution-polymerized B/S copolymer, nonstaining; used for mech. goods. [Firestone synthetic Rubber]

Duradene® 710. Solution-polymerized B/S copolymer, nonstaining; processing and extrusion aid to be used with other elastomers for such applications as shoe soling; also for asphalt modification and adhesives. [Firestone Syn. Rubber]

Duradene® 711. Solution SBR, staining; used where abrasion resist. is important and low oil levels are required; suitable for tire treads, conveyor belting, molded goods. [Firestone Syn. Rubber]

Duradene®750. Solution-polymerized B/S copolymer, 37.5 phr aromatic oil, staining; for tire tread service and body sidewall compding. [Firestone Syn. Rubber]

Duradene® 755. Solution-polymerized B/S copolymer, 20 phr naphthenic oil, slightly staining; for tire stocks. [Firestone Syn. Rubber]

Duradiene®. Styrene/butadiene solution SBR; for tires, molded goods and adhesives. [Firestone Syn. Rubber]*

Duraflex® 8410. CAS 9003-28-5; Polybutylene; base polymer or modifier in adhesive and sealant formulations; modifier for atactic PP, olefin polymers, EPDM, elastomers; for pkg., transportation, construction, nonrigid bonding adhesives/sealants; improved hot melt performance. [Shell]

Durafoam. High yield lightweight grout for void filling. [Foseco (F.S.) Ltd]

Duraform. A proprietary trade name for asbestos reinforced thermoplastics; they have greater stiffness, lower coefficient of expansion, higher heat distortion point, lower creep, higher tensile and flexural strengths than the basic resins. §

Duraguard. Polyester epoxy and epoxy polyester-based coating powders; RAL color range; used for electrostatic spray applications for decorative finishes, including wireworking, sheet metal fabrication and domestic wirework. [Plascoat Systems Ltd]

Dural. CAS 1344-28-1; Macro-crystalline regular aluminum oxide; used for abrasive industries, super refractories, sandblasting and for safes. [Lonza AG]

Dural. See Duralumin.

Duralac. Barium chromate jointing compound conforming to specification DTD 369A; for sealing of joints between dissimilar metals, protection of metals in contact with wood, synthetic resin compositions, leather, rubber, fabrics etc. [Llewellyn Ryland Ltd]

Duralam®. For the electrical industry. [DuPont UK]

Duralcon. Antistatic textile cohesive agent. [Stephenson Thompson Textile Chemicals]

Duralit. Dental stone. [Degussa Ltd]

Duralium. An alloy of aluminum with from 3.5-5.5% copper and small amounts of magnesium and manganese.

Duralkan K Concentrate. Amine-amide-formaldehyde condensation product; cationic; clear yellowish liquid; improves the wet-fastness of dyeings and prints with direct and reactive dyes; fiber-affinitive dyestuff fixing agent. [Thor Chemicals (UK) Ltd]

Duralloy. Dental silver alloys and mercury for mixing amal-

gams; for dentistry and dental engineering. [Degussa Ltd]

Duralon. A furan resin; vinyl plasticizer [U.S. Stoneware Co]*

Duraloy. Alloyed thermoplastics; used for vehicle bumpers, high impact applications. [Hoechst-Celanese] *

Duralum. An alloy of 79% aluminum, 10% copper, zinc, and tin.

Duralumin. Alloys of aluminum with from 3-5.5% copper, 0.5-1% manganese, 0.5% magnesium, and small quantities of silicon and iron. The alloy, containing 95.5% aluminum, 3% copper, 1% manganese, and 0.5%magnesium; resistant to sea water and dilute acids.

Duramite®. CAS 1317-65-3; Calcium carbonate; pigment with unique particle size distribution, easy dispersion; for mix-in application in carpet backing, roofing compounds, spackles, coatings. [ECC]

Duramycin. A proprietary preparation of diabekacin sulfate; an antibiotic. [Pfizer International] †

Duranalium. An aluminum alloy similar in composition and properties to hydronalium and B.S. sea-water alloy.

Durana metal. An alloy of 65% copper, 30% zinc, 2% tin, 1.5% iron, and 1.5% aluminum, has a golden yellow color.

Durance's metal. (Dewrance metal). A bearing metal, consisting of about 33% tin, 23% copper, and 45% antimony.

Durand's metal. An alloy of 66.6% aluminum, and 33.3% zinc.

Duranest. CAS 36637-18-0; Etidocaine; anesthetic. [Astra Pharmaceutical Products Inc] *

Duranic. An alloy containing aluminum with from 2-5% nickel and 1.5-2.5% manganese.

Duranit. A range of styrene/butadiene copolymers; used as reinforcing resins in rubber mixes and as reinforcing dispersion for natural and synthetic latex. [Hüls AG]

Duranite. A proprietary trade name for a fast-baking synthetic enamel. §

Duranox. Amitrole + atrazine + diuron; a liquid mixture of herbicides for weed control. [Agri-Technics Ltd]

Duranthrene® Dyes. A trade name for certain British dyestuffs. §

Duraplex®. Synthetic resins, common metals and their alloys. [Reichhold]

Duraplus® 1. Emulsion polymer; emulsion for ultra-high wear floor polishes with exceptional buffability. [Rohm & Haas]

Durapro. CAS 21256-18-8; Oxaprozin; anti-inflammatory. [Wyeth Laboratories] *

Duraquin. CAS 7054-25-3; Quinidine gluconate; cardiac depressant. [Parke-Davis] *

DuraSoft. Phemfilcon A; contact lens material. [Wesley-Jessen] *

Durasol Acid Blue B. The British brand of Alizarin saphirol B; a dyestuff. §

Durastat® AS-5760. Highly loaded static dissipative conc. for LDPE, LLDPE. [PPG/Polymer Prods.]

Durastat®AS-5814-2. Antistat conc. for dry foods which can be let-down in HDPE in extrusion process; prevents dust buildup during extrusion and molding. [PPG/Polymer Prods.]

Durastat® AS-5903-3. Highly loaded flame retardant and static dissipative film extrusion conc. which can be let-down in polyethylene at 8-20%. [PPG/Polymer Prods.]

---

‡ Trade name and manufacturer not verified　　　　　§ Trade name without identified manufacturer

Durastic. A bitumen compound stated to be composed of high grade bitumen freed from organic acids and used as a protective coating. §

Durastrength 200. Acrylic impact modifier; impact modifier for PVC, exterior durable building products [Atochem]

Duratex. CAS 68334-00-9; Partially hydrogenated cottonseed oil; lubricant for tablets; tablet release; candy dusting and coating for hydroscopic materials. [Van Den Bergh Foods]

Duration® 12 Hour Nasal Spray. CAS 2315-02-8; Oxymetazoline hydrochloride; adrenergic. [Schering Corp]

Durax®. CAS 95-33-0; N-Cyclohexyl-2-benzothiazole-sulfenamide; primary delayed action accelerator used in both natural and synthetic rubbers; safe at processing temperatures, fast at curing temperatures. [Goodyear]

Durax®. CAS 95-33-0; N-Cyclohexyl-2-benzothiazole-sulfenamide. [R.T. Vanderbilt]

Durazol. Direct dyestuffs. [ICI Chem. & Polymers Ltd] †

Durbar Bronze. A proprietary trade name for an alloy of copper with 24% lead and 4% tin. §

Durbar Hard Bronze. A proprietary trade name for an alloy of copper with 10% tin and 20% lead. §

Durcoton. A proprietary phenol-formaldehyde resin impregnated textile. §

Durecol. A proprietary glycero-phthalic synthetic resin. §

Durehete 900. A proprietary trade name for steel containing 1% chromium and 1/2% molybdenum. It is suitable for studs and bolts for service at temperatures up to 900 F. (482 C.). [Samuel Fox & Co Ltd] ‡

Durehete 950. A proprietary trade name for a steel containing 1% chromium, 1/2% molybdenum and 1/4% vanadium for studs and bolts for service at temperatures up to 950 F. (510 C.). [Samuel Fox & Co Ltd] ‡

Durehete 1050. A proprietary trade name for a steel containing 1% chromium, 1% molybdenum and 3/4% vanadium for bolting materials capable of operating at metal temperatures up to 1050 F. [Samuel Fox & Co Ltd] ‡

Durel. Polyarylate; used for microwave cookers, vehicle light lens and traffic light lens; customized product. [Hoechst AG]

Durelast. Moisture cure polyurethane adhesive. [MacPherson Polymers]

Durelast®. Polyurethane prepolymer; used for adhesives, sealants and surface coatings. [Kemira Polymers]

Dur-Em® 114. Glyceryl oleate; emulsifier for personal care products, foods (icings, cakes, margarine, vegetable dairy systems). [Van Den Bergh Foods]

Dur-Em® 117. Glyceryl stearate; textile lubricant and finishing agent; emulsifier for cosmetic and pharmaceutical creams and lotions, foods; lubricant for thermoplastics; dispersant for inorg. pigments. [Van Den Bergh Foods]

Dur-Em® GMO. Glyceryl oleate; solubilizer, dispersant, lubricant, wetting aid, penetrant for foods, personal care products, dry cleaning bases, paints, and insecticides. [Van Den Bergh Foods]

Durenat®. Long-acting sulfonamide. [Bayer AG]

durene. CAS 95-93-2; 1,2,4,5-Tetramethylbenzene, $C_6H_2(CH_3)_4$; used in organic synthesis, plasticizers, polymers, fibers.

Duresco. See lithopone.

Durethan®. Polyamide 6 and 6/6, copolyamide and amorphous polyamide; engineering thermoplastic for the manufacture of injection moldings; some grades also for film production; high stiffness and hardness, good impact strength; glass fiber reinforced and glass sphere and mineral filled grades; polymer and elastomer modified formulations; used for electrical engineering/electronics, power tools, domestic appliances, automotive industry, mechanical engineering, furniture industry, sporting goods, toys, packaging sector, semifinished products [Bayer AG]

Durethan®A 30 S. CAS 32131-17-2; Nylon 66; engineering plastic (elec., mechanical, precision, automotive), household appliances, building construction, furniture manufacture, and packaging applications; general-purpose grade for injection moldings (fast cycles, easy release). [Bayer AG]

Durethan®B 30 S, B 31 SK. CAS 25038-54-4; Nylon 6; B 30 S and B 31 SK are used for injection moldings with fast cycles; B 31 SK has a higher impact strength. [Bayer; Miles]

Durethan®B 35 F, B 38 F, B 40 F. CAS 25038-54-4; Nylon 6; extrusion grades for use in flat and blown film. [Bayer AG] *

Durethan®BKV. Glass-filled polyamide 6. [Bayer AG] *

Durethan®BKV 30 H. CAS 25038-54-4; Nylon 6 with 30% glass fiber and heat stabilizer; used for injection moldings exposed to elevated temps.; stiffness, and hardness. [Bayer; Miles]

Durethan®BKV 115. CAS 25038-54-4; Nylon 6, 15% glass-reinforced. [Miles]

Durethan®BM 30 X. CAS 25038-54-4; Nylon 6, 30% glass fiber/min.-reinforced; engineering plastic (elec., mechanical, precision, automotive), household appliances, building construction, furniture manufacture, and packaging applications; general-purpose grade for injection moldings (fast cycles, easy release). [Bayer; Miles]

Durethan® KL1-2402/30. CAS 25038-54-4; Nylon 6, impact-modiifed, 30% mineral-reinforced; injection molding grade for applications requiring impact strength, stiffness, heat resist. and dimensional stability, e.g., automotive fans, shrouds, mirror housings, power tools, furniture components, appliances, impeller blades, motor gears. [Miles]

Durethan®RM KU 2-2501/30. CAS 25038-54-4; Nylon 6, 30% glass-reinforced, reduced moisture; reduced moisture resin for elec. housings, automotive parts, consumer products, office chair seats, lawn/garden equip., power tool housings, hydraulic cylinder components. [Miles]

Durethane. A proprietary range of polyurethane thermoplastics used in injection molding. [Bayer AG] *

Durex. A proprietary trade name for an alloy of 83% copper, 10% tin, and 4-5% carbon; also a proprietary trade name for phenol-formaldehyde synthetic resin. §

Durex white. CAS 513-77-9; EINECS 208-167-3; Barium carbonate, $BaCO_3$; used in treatment of brines in chlorine-alkali cells to remove sulfates, rodenticide, barium salts, ceramic flux, optical glass, case-hardening baths.

Durez® 115. Two-stage phenolic resin; general-purpose thermoset resin for applications incl. automotive transmission parts and braking systems; for compression, transfer, and injection molding applications [OxyChem/Durez]

Durez® 123. Phenolic resin; impact-grade resin; used in compression, transfer, and injection molding; high shock resistance. [OxyChem/Durez]

Durez® 152. Phenolic resin; elec. asbestos-free resin for molded products requiring heat resistance and dimensional stability; used in compression, transfer, and injection molding. [OxyChem/Durez]

Durez® 18420. Phenolic resin; specialty molding material used in compression molding. [OxyChem/Durez]

Durez® 24150. Alkyd resin; molding resin for arc-track resistance (retains elect. properties at elevated temps.); used in transfer and injection moldings. [OxyChem/Durez]

Durez® 25000. Two-step phenolic molding compound; general purpose molding material for injection molding; good balance of mech. and elec. props. [OxyChem/Durez]

Durez® 32633. Two-step phenolic molding compound, glass-filled; molding compound for applications requiring high strength, dimensional stability, and heat resist.; for small motor and gear housings, brush holders, commutators, underhood automotive parts. [OxyChem/Durez]

Durfax® 60. CAS 9005-67-8; Polysorbate 60; food emulsifier; personal care products; preshave beard lubricant and softener products [Van Den Bergh Foods]

Durfax® 65. CAS 9005-71-4; Polysorbate 65; food emulsifier; pesticide dispersant. [Van Den Bergh Foods]

Durfax® 80. CAS 9005-65-6; Polysorbate 80; food emulsifier; personal care products; antifog agent in plastics and aerosol furniture polish. [Van Den Bergh Foods]

Durfax®EOM. CAS 51158-08-8; PEG-20 glyceryl stearate; food emulsifier; dough conditioner; lubricant and fabric softener. [Van Den Bergh Foods]

Durferrit. Carburizing, nitriding, annealing, hardening, tempering and heat transfer salts; used for heat treatment of metal in the machine, machine tool, motor vehicle, aircraft and other metal working industries. [Degussa Ltd]

Durham's stain. A microscopic stain. It contains a saturated solution of stannous chloride and a 15% solution of tannic acid in equal parts, with a few drops of an alcoholic solution of methylene blue.

Duricef. CAS 66592-87-8; Cefadroxil; antibacterial. [Mead Johnson & Co]

Durichlor. A hydrochloric acid resisting alloy containing 81% iron, 14.5% silicon, 3.5% molybdenum, and 1% nickel.

Duridine. Tetramethylphenylamine, $C_6H(CH_3)_4 \cdot NH_2$.

Durifan AR30, BK 30. Silicic acid dispersions; antislipping agents, especially for linings made of natural and synthetic fibers. [Thor Chemicals (UK) Ltd]

Durimet Alloys. Proprietary alloys for acid resistance. Alloy A is stated to contain iron with 25% nickel, no chromium, and 5% silicon; B contains iron with more nickel than A, and with chromium content about one-third of the nickel, and 5% silicon; alloy D contains iron with 15% nickel and smaller amounts of chromium and silicon. §

Durine. A formalin preparation.

Duriron. An acid resisting alloy, which is a silicon-iron alloy. It contains 15.5% silicon, 82% iron, 0.66% manganese, 0.83% carbon, and 0.57% phosphorus.

Durisol. A trademark. *See* Permali. §

Durkex 500. CAS 68334-28-1; Partially hydrogenated vegetable oil (cottonseed, soybean); kosher; used where high stability is needed; for coating/spraying (dried fruit, crackers), frying/roasting (nuts), antidusting (gravies, soups), color/flavor carrier, lubricant, moisture barrier, mineral oil replacement, ingred. in prepared foods. [Van Den Bergh Foods]

Durkex Durola. Partially hydrogenated canola oil (kosher); for coating/spraying (crackers, snacks); food ingred. where enhanced nutritional profile is desired. [Van Den Bergh Foods]

Durko. CAS 68334-28-1; Partially hydrogenated vegetable oil (soybean, cottonseed), mono and diglcyerides; kosher; shortening used in yeast-raised sweet goods. [Van Den Bergh Foods]

Durlac®100W. Glyceryl stearate lactate; emulsifier for food; confectionery gloss enhancer; starch gelling agent in industrial processes. [Van Den Bergh Foods]

Durlite F. CAS 68334-28-1; Partially hydrogenated vegetable oil (soybean, cottonseed); kosher; multipurpose shortening for vegetable dairy, bakery processed food. [Van Den Bergh Foods]

Durlite Gold MBN II. Partially hydrogenated vegetable oil (soybean, cottonseed), lecithin, TBHQ, beta carotene, natural flavors; kosher; shortening system for bakery items, replacement for cholesterol- and tropical oil-containing shortening systems. [Van Den Bergh Foods]

Dur-Lo®. Mono- and diglycerides with BHA and citric acid; food emulsifier for fat-reduced foods; fat replacement or reduction in sour dressings and other vegetable dairy systems, bakery cake mixes. [Van Den Bergh Foods]

Duro cement. A cement used in the manufacture of acid towers. It contains 96% silica and 4% sodium silicate.

Durocide. Biocides. [Durham Chemicals Ltd]

Durodi Steel. A proprietary trade name for a nickel-chromium-molybdenum steel. §

Durofer. Nontoxic carburizing process. [Degussa Ltd]

Duroftal. A proprietary synthetic resin. §

Duroglass. A proprietary borosilicate resistance glass for chemical use. §

Duroil. Self emulsifiable oil. [Stephenson Thompson Textile Chemicals]

Durola Select. Partially hydrogenated canola oil; kosher; shortening for cookies, crackers, biscuits; filler fat for cream centers, whipped toppings. [Van Den Bergh Foods]

Durolastik. Waterproofing, roofing materials. [Weatherguard/Marbleloid Products Inc] *

Duroloy®. Self-lubricating ball bearings with thermoplastic polyimide retainers; for photodiode mfg. equip. [Rogers]

Durolube. Textile yarn lubricant. [Stephenson Thompson Textile Chemicals]

Duromel. CAS 68334-00-9; Partially hydrogenated cottonseed oil; kosher; high performance fat for vegetable dairy, confectionery, bakery. [Van Den Bergh Foods]

Duromel B108. CAS 68334-00-9; Partially hydrogenated cottonseed oil; kosher; coating fat for bakery coatings. [Van Den Bergh Foods]

Duromine. CAS 122-09-8; Each capsule contains phentermine 15 mg or 30 mg as ion-exchange resin complexes. [3M Pharmaceuticals] †

Duromorph. CAS 57-27-2; A proprietary preparation of morphine in a long acting form; a hypnotic. [LAB Ltd] ‡

Duronze. A proprietary trade name for a high-silicon copper alloy. §

Durophen. A trade name for a series of plasticized phenolic resins widely used in baking finishes. [Chemische Werke, Albert] *

Durophen 127-B. An ammonia condensed phenol formaldehyde resin melting at 55 C. [Chemische Werke, Albert]*

‡ Trade name and manufacturer not verified   § Trade name without identified manufacturer

Durophen 170W. A butylated diane formaldehyde condensate cooked with trimethylene glycol maleate (65% solids). [Bush Beach Ltd] *

Durophen 218V. A butylated diene formaldehyde castor oil resin. [Chemische Werke, Albert] *

Durophen 287W. A butylated phenol urea formaldehyde resin sold at 58% solids. [Chemische Werke, Albert] *

Durophen 309W. A butylated xylenol formaldehyde resin containing butyl glyceryl adipate. [Chemische Werke, Albert] *

Durophen 330V. A butylated diene formaldehyde resin cooked with glyceryl phthalate and synthetic fatty acids. [Chemische Werke, Albert] *

Durophet. Each capsule contains levo- and dextro-amphetamine (ratio 1:3) 7.5 mg or 20 mg as ion-exchange resin complexes. [3M Pharmaceuticals] †

Duroplaz 610, 810, 911. Proprietary trade names for phthalate esters of straight chain alcohols. §

Duroprene®. A registered trade name for a product obtained by the exhaustive chlorination of natural rubber; can be molded, and is soluble in benzene, coal-tar naphtha, and carbon tetrachloride; it is resistant to chemical action and is used in paints and varnishes. §

Duroseal. CAS 637-12-7; Modified aluminum stearate, for waterproofing. [Durham Chemicals Ltd]

Durosehl. Epoxy and polyurethane resin systems; for casting and laminated structures, engineering patterns and toolmaking, electrical encapsulations. [Solochart Ltd] *

Durosil. Precipitated silica; low surface area reinforcing silica for rubber compounds [Degussa]

Duroslip. Antistatic textile fibers lubricant. [Stephenson Thompson Textile Chemicals]

Durosoft. Fiber softener. [Stephenson Thompson Textile Chemicals]

Durosol. Self emulsifiable oil. [Stephenson Thompson Textile Chemicals]

Durostabe. Stabilizers for PVC. [Durham Chemicals Ltd] *

Duroterm. Investment material; for casting precious metal alloys; used in dentistry. [Bayer AG]

Durotex. Yarn strengthening agent. [Stephenson Thompson Textile Chemicals]

Durotex 7603. Antimicrobial for latex carpet backing, dry film preservative. [Morton Int'l./Plastics Additives]

Durotint. Textile fugitive tints. [Stephenson Thompson Textile Chemicals]

Durowynd. Hydrophylic wetting agent. [Stephenson Thompson Textile Chemicals]

Durox. A mullite (qv) made by fusing kyanite and alumina; a proprietary trade name for an ammonium dynamite; an explosive. §

Duroxyn. Epoxide resins esterified with fatty acids; for chemical resistant paints. [Resinous Chemicals Ltd] †

Durrax. Hydroxyzine hydrochloride; tranquilizer. [Dermik Laboratories Inc] *

Dursban®. CAS 2921-88-2; Chlorpyrifos; an organophosphate insecticide for control of ticks, mosquitoes, and other insects. [DowElanco Ltd]

Dursban® 2E. CAS 2921-88-2; Chlorpyrifos; Insecticide containing 2 lb/gal of chlorpyrifos; for broad spectrum control of chinch bugs, sod web worms, ants, and earwigs. [Dow; W.A. Cleary]

Dursban®14G. See suSCon Blue. (Indonesia) [Incitec Ltd]

Durtan®60. CAS 1338-41-6; EINECS 215-664-9; Sorbitan stearate; food emulsifier, gloss enhancer for chocolate

coatings; dispersant for inorganics used in thermoplastics. [Van Den Bergh Foods]

Duspatal. CAS 3625-06-7; Mebeverine; spasmolytic, counteracts intestinal spasms. [Duphar BV] *

Duspatalin. CAS 3625-06-7; Mebeverine; spasmolytic, counteracts intestinal spasms. [Duphar BV] *

Dustallay. Wetting system for reducing dust in mines. [Foseco (F.S.) Ltd]

Dustex. Lignosulfonate; biodeg. environmentally safe dust binder and road stabilizer. [Borregaard LignoTech]

Dutch camphor. Obtained from the wood of the Japanese camphor laurel, *Cinnamomun camphora*.

Dutch metal. An alloy of 80% copper and 20% zinc.

Dutch oil. See ethylene dichloride.

Dutch pink. A pigment. It is a yellow color made by absorbing quercitron on barytes or alumina.

Dutch varnish. A solution of rosin in turpentine.

Dutch white. A pigment consisting of 1 part white lead with 3 parts heavy spar.

Du-Ter. CAS 76-87-9; Fungicide containing fentin hydroxide; for prevention of potato blight and disease control in sugar beet. [ICI Chem. & Polymers Ltd] *

Du-Ter. CAS 76-87-9; Fentin hydroxide; for control of potato blight and disease control in sugar beet. [Chiltern Farm Chemicals Ltd]

Du-Ter®. CAS 76-87-9; Triphenyltin hydroxide; flowable fungicide for pecans, potatoes, sugarbeets; restricted use. [Griffin]

Duthane. A proprietary polyester-based polyurethane elastomer cross-linked with diols. §

Dutral. A proprietary trade name for an ethylene-propylene synthetic rubber copolymer suitable for tank linings, seals, hose, cables. [Shell] *

Dutral-Co. A proprietary range of ethylene-propylene copolymers. [Montedison UK Ltd] *

Dutral-Ter. A proprietary range of ethylene-propylene-diene terpolymers (EPDM). [Montedison UK Ltd] *

Dutrex. Hydrocarbon oils. [Shell UK] *

Dutrex 20, 25. Extender-plasticizer oil; vinyl resin additive and plasticizer [Shell]

Dutrex Process and Extender Oils. Aromatic extracts produced during the refining of lubricating oils; by suitable selection and blending, a range of Dutrex grades with varying viscosities are obtained possessing good solvent characteristics and excellent polymer compatibility. [Shell] *

Duvadilan. CAS 395-28-8; Isoxsuprine; promotes blood flow rate. [Duphar BV] *

Duvadilan Retard. CAS 395-28-8; Isoxsuprine; promotes blood flow rate. [Duphar BV] *

Duvoid. CAS 590-63-6; Bethanechol chloride; cholinergic. [Norwich Eaton Pharmaceuticals Inc] *

Duxalid. A proprietary trade name for a synthetic resin. §

Duxite. An explosive containing nitro-glycerin, collodion cotton, sodium nitrate, wood meal, and ammonium oxalate; also the name applied to a resin from lignite.

Duxol. A proprietary trade name for a synthetic resin. §

DV. CAS 84-17-3; Dienestrol; estrogen. [Merrell Dow Pharmaceuticals Inc] *

DV-1801. CAS 115047-92-2; Behenyl polyethoxyethyl methacrylate [Rhone-Poulenc Surf.]

DV-1936. CAS 1462-55-1; 2-(Dodecylthio) ethanol. [Rhone-Poulenc Surf.]

DV-2301. CAS 6281-42-1; Aminoethyl ethylene urea

[Rhone-Poulenc Surf.]

D-Visor. Deodorant for industrial processes and products. [CPL Group Ltd]

D-Wax. Wax inhibitors and removers. [Baker Performance Chemicals Ltd]

D.X.L. High boiling tar acids. [Coalite Fuels & Chemicals Ltd]*

DY 023. CAS 2210-79-9; Cresyl glycidyl ether; diluent for epoxy [Ciba-Geigy/Plastics] *

Dyafac PEG 6DO. CAS 9005-07-6; PEG-12 dioleate; surfactant for textile use. [Henkel/Textiles]

Dyamul. A range of dyebath auxiliaries, including detergents, buffers, solvents, lubricants and combinations of these; assistants in textile processing and dyeing. [Yorkshire Chemicals plc] *

Dyapol. Dispersants, levelers, and retardants for disperse dyeing. [Yorkshire Pat-Chem; Yorkshire Chemicals plc] *

Dyazide Tablets. Combination of triamterene and hydrochlorothiazide; potassium conserving diuretic for use as antihypertensive. [SmithKline Beecham] *

Dybln®. For the chemical industry. [DuPont UK]

Dycastal. Shrinkage prevention in aluminum diecastings. [Foseco (F.S.) Ltd]

Dy-Chek. Four step process for detecting surface flaws in metallic and nonmetallic components. [Foseco (F.S.) Ltd]

Dycill. CAS 13412-64-1; Dicloxacillin sodium; antibacterial. [SmithKline Beecham]

Dyclone. CAS 536-43-6; Dyclonine hydrochloride; anesthetic. [Astra Pharmaceutical Products inc] *

Dycote. Foundry die coatings. [Foseco (F.S.) Ltd]

Dycron. CAS 1344-28-1; Crystalline aluminum oxide; for lapping of gearings and hydraulics, motors, electronics, glass etc. [Dynamit Nobel Wien GmbH] *

Dye Retarder #1. CAS 61789-77-3; EINECS 263-087-6; Dicoco dimethyl ammonium chloride; retardant for acrylics; oil and solv. emulsifier. [Sherex/Div. of Witco]

Dyeset® 100 Conc. Resin polymer; dye fixative for direct dyes for exhaust last rinse application for textile finish formulation. [Sybron]

Dyetone®. CAS 7789-38-0; Aq. solution of sodium bromate; dye oxidant for vat and sulfur dyes. [Olin]

Dyeweld SUPR. Fixative for acid dyes; for exhaust application after dyeing. [Sybron]

Dyflor 2000. Polyvinylidenefluoride (PVDF); for injection, extrusion and blow molding, manufacture of chemical apparatus, mechanical engineering, cable industry, electronics, manufacture of medical equipment. [Dynamit Nobel Wien GmbH]

Dyflor L90. A proprietary polyvinyl fluoride; processed in dispersion form. §

Dyfonate. CAS 944-22-9; Fonofos; soil and seed insecticide. [Farm Protection Ltd]

Dykor 204. CAS 24937-79-9; Polyvinylidene fluoride; primer coating containing mica to enhance adhesion at metal/coating interface; protective coating for lining chem. processing vessels, acid etching in mfg. of electronic components, bleaching operations in pulp/paper mills, corrosion protection. [Whitford]

Dylark® 132. CAS 9011-13-6; SMA-based engineering copolymer; resin offering excellent clarity. [Arco Chemical Co]

Dylene. A proprietary polystyrene. [Arco Chemical Co]

Dylite® D195B. CAS 9003-53-6; Expandable PS; for dynamic cushioning applications, insulation, foam drink cups, pkg. [Arco Chemical Co]

Dylite® R2595B EPS. CAS 9003-53-6; Expandable PS with 25% recycled content; resin for cushioning, insulation, protective pkg. applications. [Arco Chemical Co]

Dylon. Colloidal graphite in aqueous or solvent carriers; high temperature parting agent, release compounds, with good electrical and thermal conductivity. [Dylon Industries Inc] *

Dylonite. Resin impregnated impervious graphite; for heat exchangers, tubes and pumps with excellent resistance to high temperature corrosive chemicals. [Dylon Industries Inc] *

Dylox®. CAS 52-68-6; Trichlorfon; insecticidal spray used for control of mangold fly, fruit fly and pests of maize, alfalfa, and cotton. [Bayer AG]

Dylux®. [DuPont UK]

Dymacryl. 100% Acrylic copolymers, color stable inorganic pigments and surface penetrating agents; available in 10 colors plus clear; beautifies and protects above-grade masonry surfaces against damage caused by water absorption; used on architectural concrete, precast and poured concrete, GFR concrete, brick, natural stone, stucco and unglazed tile. [Dampney Company Inc] *

Dymax® Light-Weld® Adhesives. Aerobic adhesives which cure by exposure to ultraviolet light; used for medical devices, electronic assembly glass and plastics. [Dymax Corporation]

Dymax® Light-Welder™. Ultraviolet light sources including flood lights, focus beam lights, spot lights and a combination spot light/dispenser. [Dymax Corporation]

Dymax® Multi-Cure® Adhesives. Aerobic adhesives which cure by exposure to ultraviolet activator or heat; used for magnet, metal, fiber optics, assembly adhesives and coatings for electronics. [Dymax Corporation]

Dymel®. CAS 115-10-6; Dimethyl ether propellant. [DuPont UK]

Dymelor. CAS 968-81-0; A proprietary preparation of acetohexamide; an oral hypoglycemic agent, antidiabetic. [Eli Lilly & Co]

Dymerex. Partially dimerized rosin acids; used to reinforce specialty adhesive products, in specialty protective coatings. [Hercules] *

Dymetrol®. Fiber/polymer. [DuPont UK]

Dymsol® 38C. A proprietary anionic, biodegradable polymerization emulsifier for improving the processing characteristics of SBR, nitrile rubber and neoprene. [Henkel Inc] *

Dymsol® 2031. Sulfonated fatty acid, sodium salt; primary or sec. emulsifier for emulsion polymerization. [Henkel/Functional Prods.; Henkel-Nopco]

Dymsol® PA. Sulfated ester; primary emulsifier for emulsion polymerization. [Henkel/Functional Prods.]

Dynacal. CAS 1305-78-8; Fused calcium oxide; additive for the melting of high purity metallurgical products; raw material for the electroslag remelting process; auxiliary material for metallurgical desulfurization processes. [Hüls AG]

Dynacast. Fusion cased bricks; for the lining of slab pusher furnaces, billet pusher furnaces, ingot pusher furnaces, forging furnaces, rocker bar furnaces, roller hearth furnaces, soaking pit furnaces, tundishes of continuous casting plants, steel degassing plants; high temperature furnaces. [Hüls AG]

---

‡ Trade name and manufacturer not verified   § Trade name without identified manufacturer

Dynacerin®. Jojoba oil substitute, liquid wax ester; oily component for emulsions with emollient properties in pharmaceuticals and cosmetics. For ointments, creams, liquid emulsions, external suspension, skin lotions, bubble baths and shampoos. [Hüls AG]

Dynacerin®660. CAS 17673-56-2; Oleyl erucate; jojoba oil substitute; emollient for cosmetic preparations. [Hüls Am.]

Dynacet®. CAS 91723-32-9; Acetylated monoglycerides of edible fatty acids; food emulsifier; coating agent for shortenings, toppings, meat and sausages. [Hüls Am.; Hüls AG]

Dynacoll. Reactive polyesters; adhesive raw material for the formulation of reactive hot melts. [Hüls AG]

Dynaflex. Two-part polyurethane sealant; used as a security sealant for applications in institutional and correctional complex installations wherever the sealant may be exposed to physical abuse. [Pecora Corporation]

Dynaflock. CAS 1302-42-7; Sodium aluminate; caustic product for paper industry, water treatment, ceramic industry and building industry. [Hüls AG]

Dynaflush. Nonchlorinated, nonflamm., noncarcinogenic, non-ozone depleting solv. for flushing and cleaning residues from equip.; used in the urethane foam industry. [Dynaloy]

Dyna-Form. Formaldehyde; soil sterilant and fumigant for glass houses. [Fargro Ltd]

Dynaglaze. Polishes and valeting products for cars. [Spectra Brands plc]

Dynagrout. CAS 1302-42-7; Sodium aluminate; reactive agent for forming injection gels (grout), for the mining and building industry. [Hüls AG]

Dynagunit. Alkali aluminate formulations; liquid concrete setting accelerator for the mining and building industry. [Hüls AG]

Dynamag. CAS 1309-48-4; Electromagnesia (magnesium oxide); raw material for ramming mixes and shape bricks for the lining of high temperature furnaces; raw material for welding powder and the coating of electrodes. [Hüls AG]

Dynamar® Brand Specialities. A line of speciality chemicals used in the manufacture/processing of elastomer and plastics compositions. [3M]

Dynamar®FC. Fluorochemical; mold release agent for hot molding operations; esp. useful with silicone, ethylene acrylic and fluorosilicone elastomers. [3M]

Dynamar®PPA-790. Polymer processing additive providing enhanced release and flow in molding and extrusion of elastomer-based compounds [3M]

Dynamask. Dry film solder mask. [Morton Int'l. Ltd]

Dynamine. A rubber vulcanization accelerator. It is diphenylguanidine.

dynamite. (Kieselguhr dynamite) Principal ingredient is nitroglycerin; industrial explosive detonated by blasting caps.

dynamite acid. CAS 7697-37-2; EINECS 231-714-2; Concentrated nitric acid; used for making 96% mixed acids (34% nitric acid + 62% sulfuric acid).

dynamite glycerin. Glycerin of specific gravity 1.263, containing 98-99%. It contains no lime, sulfuric acid, chlorine, or arsenic.

Dynamites. Explosives first patented by Nobel, which consist of nitroglycerine rendered shock-stable by absorption onto Kieselguhr or some other absorbent.

Dynamullit. CAS 12141-46-7; Fused mullite (aluminum silicate); for production of shape bricks for high-charged zones in heating and melting furnaces; molding material for precision casting molds. [Hüls AG]

Dynamutilin. CAS 55297-96-6; Tiamulin fumarate; antibacterial. [Bristol-Myers Squibb Co Inc]

Dynamutilin Aqueous Solution. Tiamulin hydrogen fumarate in aqueous solution. [Bristol-Myers Squibb Pharmaceuticals Ltd] †

Dynamutilin Water Soluble Powder. Tiamulin hydrogen fumarate in powder base. [Bristol-Myers Squibb Pharmaceuticals Ltd] †

Dynamyte. CAS 88-85-7; Dinoseb; herbicide for use on beans, small grains, forage, cereal crops. [Draxel Chemical Company] ‡

Dynamyxin. Sulphomyxin; antibacterial. [Pfizer Inc] †

Dynapen. CAS 13412-64-1; Dicloxacillin sodium; antibacterial. [Bristol Laboratories]

Dynapol®H. Low-molecular, saturated polyesters; for the production of high-quality baking enamels (industrial enamels, enamels for household appliances, enamels for vehicles (metallic finishes)). [Dynamit Nobel Wien GmbH] *

Dynapol®H. Low-molecular, saturated polyesters; used for the production of high-quality baking enamels; industrial enamels, enamels for household appliances and metallic finishes for vehicles. [Hüls UK Ltd]

Dynapol®L. High molecular copolyesters; raw material for coil coating, can coating, metal decorating. [Dynamit Nobel Wien GmbH] *

Dynapol®L. High-molecular, copolyesters. raw material for coil coating, can coating, metal decorating. [Hüls UK Ltd]

Dynapol® L 205. Thermoplastic polyester; for corrosion resist. primers and stamping enamels. [Hüls Am.]

Dynapol®LH. Medium and low-molecular, saturated polyesters; particularly suitable for coatings coming in contact with foodstuffs; raw material for coil coating, metal decorating and adhesives. [Dynamit Nobel Wien GmbH]*

Dynapol® LH. Low-molecular, saturated polyesters; particularly suitable for coatings coming into contact with foodstuffs; raw material for coil coating, metal decorating, and adhesives. [Hüls UK Ltd]

Dynapol® P. Thermoplastic copolymers; suitable where protection against corrosion and good resistance to aggresive chemicals are required. [Dynamit Nobel Wien GmbH] *

Dynapol®P. Thermoplastic copolymers; used where protection against corrosion and good resistance to aggressive chemicals are required. [Hüls UK Ltd]

Dynapol®S. Thermoplastic, linear, saturated copolyesters; adhesive raw materials for hot melts. [Dynamit Nobel Wien GmbH] *

Dynapol® S. Thermoplastic, linear saturated copolyesters; adhesive raw material for hot melts. [Hüls UK Ltd]

Dynapor. Phenolic foaming resins; flame inhibiting and rot-proof insulating material for ceilings in flat roof construction, for wall and facade paneling or pipe sheatings. [Hüls AG]

Dynasan®. Fatty acid esters; for use as lubricants and retarding agents in tablets, crystallization accelerator in suppositories and chocolate, consistency regulators in ointments, creams and lotions, basic material for fatty powders. [Hüls AG]

---

* Trade name not verified as available   † Trade name verified as obsolete

Dynasan® 110. CAS 621-71-6; Tricaprin; emollient, lubricant, consistency regulator for cosmetic applications such as sticks, creams, lotions, powds.; for prep. of margarine, confectionery, milk products, fruit diets. [Hüls Am.]

Dynasan® 112. CAS 538-24-9; Trilaurin; consistency regulator for cosmetic creams, lotions; lubricant, powd. base, stick component [Hüls Am.]

Dynasan® 114. CAS 555-45-3; Trimyristin; binder, lubricant for tablets and compressed confectioneries; consistency regulator for creams and lotions; lubricant, powd. base, stick component. [Hüls Am.]

Dynasan® 116. CAS 555-44-2; Tripalmitin; lubricant, consistency regulator in cosmetic powds., cakes, and production of tablets; powd. and makeup base; stick component. [Hüls Am.]

Dynasan® 118. CAS 555-43-1; EINECS 209-097-6; Tristearin; crystallization accelerator in chocolate; lubricant in cosmetic powds., cakes, and production of tablets. [Hüls Am.; Hüls AG]

Dynasan® P60. Hydrogenated palm oil; consistency regulator for creams, lotions, makeup, decorative cosmetics; stabilizer for hindering oiling out of an emulsion. [Hüls Am.]

Dynasil®. Silicic acid esters; binders for the foundry industry and for inorganic zinc dust coatings. [Hüls AG]

Dynasil® 40. Polydiethoxysiloxane; silicone dioxide source. [Hüls Am.]

Dynasil® A. CAS 78-10-4; Tetraethoxysilane; coupling agent, chem. intermediate, blocking agent, release agent, lubricant, primer, reducing agent. [Hüls Am.]

Dynasil® CA. CAS 18407-94-8; Tetrakis (2-ethoxyethoxy) silane; coupling agent, chem. intermediate, blocking agent, release agent, lubricant, primer, reducing agent [Hüls Am.]

Dynasil® CM. Tetrakis (2-methoxyethoxy) silane; coupling agent, chem. intermediate, blocking agent, release agent, lubricant, primer, reducing agent [Hüls Am.]

Dynasil® M. CAS 681-84-5; Tetramethoxysilane; coupling agent, chem. intermediate, blocking agent, release agent, lubricant, primer, reducing agent [Hüls Am.]

Dynasolve 100. CAS 68-12-2; Dimethylformamide-based solv.; nonselective solv. [Dynaloy]

Dynasolve 150. CAS 109-99-9; Tetrahydrofuran-based solv.; flamm. solv. for room temperature use only. [Dynaloy]

Dynasolve 165. Blend of solvs. and org. chems.; room temperature decapsulation solv. for dissolving cast epoxy resins. [Dynaloy]

Dynasolve 699. CAS 96-48-0; Butyrolactone-based solv.; solv. for use at room temperature for thin acrylic coatings; used hot (120-150 C) for acrylic adhesives and epoxy powd. coatings. [Dynaloy]

Dynasolve M-30. Nonchlorinated, nonflamm. solv. for general cleaning operations; esp. for curing of uncured polymers such as epoxies, urethanes and silicones. [Dynaloy]

Dynasperse A. CAS 8061-51-6; Sodium lignosulfonate; primary dye dispersant, good heat stability and milling properties, low staining and very low dye reduction. [Borregaard LignoTech]

Dynaspinell. Fused spinel (magnesium aluminate); raw material for the production of pyrometer sheaths, crucibles, refractory bricks, ramming mixes. [Hüls AG]

Dynastite. An explosive containing 94% potassium chlorate, and 6% barium nitrate, dipped in nitro-toluene.

Dynasylan®. Organo-functional silane adhesion promoters, alkyl silanes; bonding agent between inorganic surfaces and organic polymers; binders for the foundry industry. [Dynamit Nobel Wien GmbH] *

Dynasylan®. Organo-functional silane adhesion promoters, alkyl silanes; bonding agent between inorganic surfaces and organic polymers; binders for the foundry industry. [Hüls UK Ltd]

Dynasylan® AMEO. CAS 919-30-2; EINECS 213-048-4; Aminopropyltriethoxysilane; coupling agent, chem. intermediate, blocking agent, release agent, lubricant, primer, reducing agent. [Hüls Am.]

Dynasylan® AMEO-P. CAS 919-30-2; EINECS 213-048-4; Aminopropyltriethoxy silane; organo-silicon compounds for bonding between org. and inorg. components of a system. [Hüls Am.]

Dynasylan® AMMO. CAS 13822-56-5; EINECS 237-511-5; Aminopropyltrimethoxysilane; coupling agent, chem. intermediate, blocking agent, release agent, lubricant, primer, reducing agent. [Hüls Am.]

Dynasylan® BDAC. CAS 13170-23-5; Di-t-butoxy-diacetoxysilane; coupling agent, chem. intermediate, blocking agent, release agent, lubricant, primer, reducing agent. [Hüls Am.]

Dynasylan® BSA. CAS 10416-59-8; Bis(trimethyl-silyl)acetamide; coupling agent, chem. intermediate, blocking agent, release agent, lubricant, primer, reducing agent. [Hüls Am.]

Dynasylan® BSM. Alkyl-alkoxy-silane mixtures; impregnating material for buildings. [Dynamit Nobel Wien GmbH]*

Dynasylan® BSM. Alkyl-alkoxy-silane mixtures; impregnating material for buildings. [Hüls UK Ltd]

Dynasylan® CPTEO. CAS 5089-70-3; 3-Chloropropyl-triethoxysilane; coupling agent, chem. intermediate, blocking agent, release agent, lubricant, primer, reducing agent. [Hüls Am.]

Dynasylan® CPTMO. CAS 2530-87-2; 3-Chloro-propyltrimethoxysilane; coupling agent, chem. intermediate, blocking agent, release agent, lubricant, primer, reducing agent. [Hüls Am.]

Dynasylan® DAMO. CAS 1760-24-3; EINECS 212-164-2; N-2-aminoethyl-3-aminopropyltrimethoxysilane; coupling agent, chem. intermediate, blocking agent, release agent, lubricant, primer, reducing agent. [Hüls Am.]

Dynasylan® DAMO-P. CAS 1760-24-3; EINECS 212-164-2; Aminoethylamino propyltrimethoxy silane; coupler for epoxy, phenolic, melamine, nylons, PVC, acrylics, polyolefins, PU and nitrile rubber. [Hüls Am.]

Dynasylan® DAMO-T. CAS 1760-24-3; EINECS 212-164-2; Aminoethylamino propyltrimethoxy silane, tech. grade; coupler for epoxy, phenolic, melamine, nylons, PVC, acrylics, polyolefins, PU and nitrile rubber. [Hüls Am.]

Dynasylan® ETAC. CAS 17689-77-9; Ethyltriacetoxy-silane; coupling agent, chem. intermediate, blocking agent, release agent, lubricant, primer, reducing agent. [Hüls Am.]

Dynasylan® GLYMO. CAS 2530-83-8; EINECS 219-784-2; 3-Glycidoxypropyltrimethoxysilane; CAS 2530-83-8; coupling agent, chem. intermediate, blocking agent, release agent, lubricant, primer, reducing agent. [Hüls Am.]

---

‡ Trade name and manufacturer not verified   § Trade name without identified manufacturer

Dynasylan®HMDS. Hexamethyldisilazane; CAS 999-97-3; coupling agent, chemical intermediate, blocking agent, release agent, lubricant, primer, reducing agent. [Hüls Am.]

Dynasylan® IBTMO. CAS 18395-30-7; Isobutyl-trimethoxysilane; CAS 18395-30-7; coupling agent, chemical intermediate, blocking agent, release agent, lubricant, primer, reducing agent. [Hüls Am.]

Dynasylan®IMEO. CAS 58068-97-6; N-[3-(Triethoxysilyl)-propyl] 4,5-dihydroimidazole; coupling agent, chemical intermediate, blocking agent, release agent, lubricant, primer, reducing agent. [Hüls Am.]

Dynasylan®MEMO. CAS 2530-85-0; EINECS 219-785-8; 3-Methacryloxypropyltrimethoxysilane; coupling agent, chemical intermediate, blocking agent, release agent, lubricant, primer, reducing agent. [Hüls Am.]

Dynasylan® MTES. CAS 2031-67-6; EINECS 217-983-9; Methyltriethoxysilane; coupling agent, chemical intermediate, blocking agent, release agent, lubricant, primer, reducing agent. [Hüls Am.]

Dynasylan® MTMO. CAS 4420-74-0; EINECS 224-588-5; 3-Mercaptopropyltrimethoxysilane; coupling agent, chemical intermediate, blocking agent, release agent, lubricant, primer, reducing agent. [Hüls Am.]

Dynasylan® MTMS. CAS 1185-55-3; EINECS 214-685-0; Methyltrimethoxysilane; coupling agent, chemical intermediate, blocking agent, release agent, lubricant, primer, reducing agent. [Hüls Am.]

Dynasylan®OCTEO. CAS 2943-75-1; EINECS 220-941-2; Octyltriethoxysilane; coupling agent, chemical intermediate, blocking agent, release agent, lubricant, primer, reducing agent. [Hüls Am.]

Dynasylan® TCS. CAS 10025-78-2; Trichlorosilane; coupling agent, chemical intermediate, blocking agent, release agent, lubricant, primer, reducing agent. [Hüls Am.]

Dynasylan® TRIAMO. CAS 35141-30-1; Trimethoxysilylpropyldiethylene triamine; coupling agent, chemical intermediate, blocking agent, release agent, lubricant, primer, reducing agent. [Hüls Am.]

Dynasylan®VTC. CAS 75-94-5; Vinyltrichlorosilane; CAS 75-94-5; coupling agent, chemical intermediate, blocking agent, release agent, lubricant, primer, reducing agent. [Hüls Am.]

Dynasylan® VTEO. CAS 78-08-0; EINECS 201-081-7; Vinyl triethoxysilane; coupling agent, chemical intermediate, blocking agent, release agent, lubricant, primer, reducing agent. [Hüls Am.]

Dynasylan® VTMO. CAS 2768-02-7; EINECS 220-449-8; Vinyltrimethoxysilane; coupling agent, chemical intermediate, blocking agent, release agent, lubricant, primer, reducing agent. [Hüls Am.]

Dynasylan®VTMOEO. CAS 1067-53-4; EINECS 213-934-0; Vinyl tris (methoxyethoxy) silane; coupling agent, chemical intermediate, blocking agent, release agent, lubricant, primer, reducing agent. [Hüls Am.]

Dynat W. A proprietary brand of mechanically comminuted rubber from Malaysia. §

Dynatex GTZ. CAS 9006-04-6; Low ammonia NR latex; suitable for all latex applications where excellent stability and good preservation parameters are desired. [Guthrie Latex]

Dynatherm. Electromagnesia; fused magnesia for heating elements. [Hüls AG]

Dynatred. Polyurethane sealant; used for sealing horizontal joints in parking decks, plazas, warehouse floors or other areas subject to heavy foot and vehicular traffic, particularly where slope exceeds 1%. [Pecora Corporation]

Dynatrol. Polyurethane sealant; used in general construction for caulking vertical expansion joints in walls and sealing around door and window frames. [Pecora Corporation]

Dynaweld. Urethane polymeric sealant adhesive; used for adhering pre-formed sections of Urexpan NR-200 to concrete or epoxy substrates and to each other. [Pecora Corporation]

Dynazirkon. CAS 1314-23-4; Zirconium oxide; raw material for the production of high refractory, slag resistant crucibles and moldings for the melting of high temperature alloys; raw material for high wear resisting casting nozzles in continuous casting plants. [Hüls AG]

Dyne. Iodophor/acid pipe line cleaner and sterilizer. [Ciba plc]

Dynemate 200. Hypochlorite/acid pipe line cleaner and sterilizer. [Ciba plc]

Dynobel. An explosive containing potassium perchlorate, nitroglycerin, ammonium oxalate, wood meal, and a little collodion cotton. §

Dynomel. Melancil and urea melamine resins; for decorative laminates and surfacing of wood-based panels. [Dynochem UK Ltd]

Dy´on. Blue colorant, concentrated liquid for coloring spray solutions (especially pesticide sprays) so that operator can see the spray coverage. [Regal Chemical Company]

Dyon. Marker dye. [Rhone-Poulenc Environmental Prods. Ltd]

Dyphene. Alkyl phenolic resins; used in adhesives, rubber, elastomers, bonding, abrasives and friction materials. [PMC Specialities Group Inc]

Dypur. Insulating pouring/feeding bush containing a ceramic foam filter. [Foseco (F.S.) Ltd]

Dyqex®. Sodium lignosulfonate; anionic dye dispersant extender. [Georgia-Pacific] *

Dyrene®. CAS 101-05-3; Anilazine; broad spectrum fungicide used for tobacco, potatoes, cereals and ornamentals. [Bayer AG]

Dyrenium. CAS 396-01-0; Triamterene; diuretic. [SmithKline Beecham]

Dysoid. A bearing bronze containing 62% copper, 18% lead, 10% tin, and 10% zinc.

Dytac. CAS 396-01-0; A proprietary preparation of triamterene; a diuretic. [SmithKline Beecham] *

Dytek® A. CAS 15520-10-2; 2-Methylpentamethylenediamine; epoxy curing agent; also used in polyurethanes, wet strength resins, scale and corrosion inhibitors, motor oil and gasoline additives, polyamide plastics, films, adhesives, and inks. [DuPont] *

Dytel®. Leak detective dye. [DuPont UK]

Dytherm®. Expandable copolymer; for rigid foams providing excellent resist. to temps. to 250 F; end uses include automotive door liners, instrument panels, solar panels, cores for composite structures. [Arco Chemical Co]

Dytide. A proprietary preparation of triamterene and benzthiazide; a diuretic. [SmithKline Beecham] *

Dytransin. CAS 1553-60-2; Ibufenac; analgesic; anti-inflammatory. [The Boots Co plc] *

Dytron® XL. Thermoplastic rubber; for electrical applications, e.g., wire and cable. [Advanced Elastomer Sys-

---

* Trade name not verified as available   † Trade name verified as obsolete

tems]

Dyvax®. Polymer. [DuPont UK]

DZ910. Dissociated zircon consisting of spheres of free silica with zirconia dendrites embedded therein; used for manufacture of ceramic pigments and refractory compositions. [Ferro] *

E

E45 Cream. A proprietary preparation of white soft paraffin, light liquid paraffin and wool fat used as a skin cream. [The Boots Co plc] †

E-3810. CAS 25928-94-3; Epoxy; low-cost, general purpose encapsulant for applications requiring min. thermal shock resistance, elec. performance. [ICI Fiberite]

E-3824. CAS 25928-94-3; Glass-filled epoxy; low pressure molding compound for encapsulation in electronics industry; good thermal shock resistance and wet elec. properties. [ICI Fiberite]

E-9405. CAS 25928-94-3; Glass fiber-filled epoxy; molding compound for encapsulation (delicate elec. components, large coils); thermal shock, moldability, retains elec. properties under humid conditions. [ICI Fiberite]

E alloy. An alloy of 76% aluminum, 20% zinc, 2.5% copper, 0.2% iron, 0.5% manganese, 0.5% magnesium, and 0.2% silicon.

Earex®. Arachis oil 33 1/3%, almond oil 33 1/3%, rectified camphor oil 33 1/3%; used for softening and removal of ear wax. [Seton Healthcare Group plc]

Early Impact. A suspension concentrate containing 150 g carbendazim and 94 g flusilazole per liter; systemic fungicide for use on cereals. [DuPont UK]

earth archil. Archil (*qv*), contaminated with mineral matter; used for the preparation of litmus.

earth wax. *See* ozokerite.

earth wax. *See* ceresin.

Ease Release 200 Series. General purpose release agent for release of most types of casting and molding systems, e.g., PU elastomers, PU foam, epoxy resin, polyester, RTV silicones, rubber, and thermoplastic polymers; effective on aluminum, chrome; RTV silicone, epoxy, polyester, rubber, and steel molds. [George Mann]

Ease Release 2040 Series. Semi-permanent release coating providing long lasting release for urethane, epoxy, rubber, aluminum, and steel molds, molding presses, coating machinery; effective with urethanes, rubbers, epoxies, hot-melt adhesives, polyesters, and other polymers and gums. [George Mann]

Ease Release 2191. Silicone compound blend; release agent for casting wax models and molds, injection molding, assembly lube; provides multiple releases. [George Mann]

Easigel. Rheological additive for use in organic liquid systems of widely differing type and polarity. [Akzo Chemie UK Ltd]

Easisperse. Easy dispersing pigments; used for printing inks and paints. [Manox Ltd] *

Easprin. CAS 50-78-2; Aspirin; analgesic; antipyretic; antirheumatic. [Parke-Davis]

Eastbond. Hot melt adhesives. [Eastman]

East India gum. (Bombay gum). A variety of gum arabic, pale amber or pinkish in color.

East Indian Balsam of Copaiba. CAS 8030-55-5; A name given to Gurjun balsam (the oleo-resin from the stems of *Diptero-carpus* species).

Eastman® 910. A proprietary trade name for a cyanoacrylate adhesive which sets with the application of pressure. Variants are: 910 EM for vinyls; 910 FS for quicker setting; 910 MHT for applications involving high temperatures. [Eastman Chemical Int'l. AG] *

Eastman® AQ-38S. Polymer; adhesives and sealants raw materials. [Eastman]

Eastman® C-11 Ketone. Ketone. [Eastman] *

Eastman® DTBHQ. CAS 88-58-4; 2,5-Di-t-butylhydroquinone; antioxidant for rubber, polyesters. [Eastman]

Eastman® HQMME. CAS 150-76-5; Hydroquinone monomethyl ether; antioxidant for monomers. [Eastman]

Eastman® Inhibitor DOBP. 4-Dodecyloxy-2-hydroxybenzophenone; industrial grade uv absorber/stabilizer for polyethylene and PP; suitable for unsat. polyesters, PS, PVC, CAB, NC, urethane, and acrylic surface coatings; also used as a screening agent in protecting rubber, flourescent pigments, polishes, and papers from the degrading effects of uv light. [Eastman] *

Eastman® Inhibitor OABH. Oxalyl bis(benzylidenehydrazide); stabilizer used in polyolefins in contact w/ copper or copper-containing alloys; copper deactivator. [Eastman] *

Eastman® Inhibitor OPS. A proprietary trade name for p-octylphenyl salicylate; a uv light inhibitor suitable for polyolefins. [Eastman] *

Eastman® Inhibitor Poly TDP 2000. CAS 63123-11-5; Thiodipropionate polyester. [Eastman] *

Eastman® Inhibitor RMB. CAS 136-36-7; Resorcinol monobenzoate; industrial grade uv absorber/stabilizer for cellulosic plastics and PVC formulations; uv absorber. [Eastman]

Eastman® MTBHQ. CAS 1948-33-0; Mono-tert-butylhydroquinone; antioxidant for rubber, monomers. [Eastman]

Eastman® P4C5B-030. CAS 9003-07-0; Polypropylene homopolymer. [Eastman] *

Eastman® Poly TDP 2000. CAS 63123-11-5; Thiodipropionate polyester; sec. antioxidant used with phenolic primary antioxidants in polypropylene and other polyolefins. [Eastman]

Eastman® Yellow. The sodium salt of glucosephenylosazone p-carboxylic acid; a yellow coloring

---

matter used as a corrective filter in photography. [Eastman] *

Eastobrite®OB-1. CAS 1533-45-5; 2,2´-(1,2-Ethenediyldi-4,1-phenylene)bisbenzoxazole; optical brightener; fluorescent whitening agent for use in linear polyester, PET, nylon fibers. [Eastman]

Eastoflex B1020. Propylene/butene copolymer; adhesives and sealants raw materials. [Eastman]

Eastoflex E1003. Propylene/ethylene copolymer; adhesives and sealants raw materials. [Eastman]

Eastoflex P1010. Amorphous polyolefin homopolymer; adhesives and sealants raw materials. [Eastman]

Eastotac H-100. Hydrocarbon tackifying resin; tackifier for adhesives and sealants applications. [Eastman]

Eastozone 32. N, N´-Dimethyl-N, N´-di-(1 methylpropyl) p-phenylenediamine; a proprietary antioxidant. [Eastman]*

Easy Cleen. Mild alkaline degreasing agent containing a blend of emulsifiers, solvents and solubilizing agents; general-purpose cleaner. [Vitax Ltd]

Easy-Flo. Fluxes for silver alloy brazing. [Johnson Matthey plc]

Easypoxy; Two-component epoxy adhesives [Conap]

eau de Brouts. Petitgrain water.

eau de Goudron. Tar water.

EB1500-1AR. Polyester, thermoset; flame retardant, abrasion-resist. elec. encapsulation grade general-purpose resin; for compression, transfer, and injection molding. [Cuyahoga Plastics]

EB3000-2. Polyester, thermoset, glass-reinforced; flame retardant thermoset for compression, transfer, and injection molding. [Cuyahoga Plastics]

Ebal. CAS 4748-78-1; EINECS 225-268-8; p-Ethylbenzaldehyde; additive for resins; intermediate for pharmaceuticals, fragrances. [Mitsubishi Gas]

Ebecryl®. Acrylic pre-polymers which are radiation curable. [UCB (Chem) Ltd]

Ebecryl®110. Oxyethylated phenol acrylate monomer; low visc., low odor monomer used in uv light and electron beam cure products [UCB Radcure]

Ebecryl®150. Bisphenol A deriv. diacrylate monomer; for applications where high reactivity and low volatility are required; esp. useful in inks and varnishes. [UCB Radcure]

Ebecryl®600. Bisphenol A epoxy diacrylate; low color grade featuring very fast cure, high gloss, excellent solv. resist.; for overprint varnishes, lithographic and screen inks, coatings for paper, paperboard, wood chipboard, and rigid plastics, laminating adhesives, paper upgrading. [UCB Radcure]

Ebecryl® 629. Epoxy novolac acrylate in 33% monomer blend; provides heat and solv. resist. to solder resists, marking inks, adhesion on metallized substrates, low shrinkage coatings, heat-resist. applications. [UCB Radcure]

Ebecryl®1360. Silicone hexaacrylate; additive contributing slip, substrate wetting and flow props. to overprint varnishes, clear coatings on paper, plastics and metals. [UCB Radcure]

Ebert and Merz's α-acid. CAS 92-41-1; 2, 7-Naphthalenedisulfonic acid, $C_{10}H_6(SO_3H)_2$; used as an intermediate for dyes.

Ebert and Merz's β-acid. CAS 581-75-9; 2,6-Naphthalenedisulfonic acid.

E.B. Golden Glitter, Neutral Glitter. Pearlescent material and aq. acrylic resinous binders; for decorative textile printing. [Eastern Color & Chem.]

Ebner's fluid. A mixture of 2.5 cc hydrochloric acid, 2.5 g sodium chloride, 100 cc water, and 500 cc alcohol; used for decolorizing in bacteriological work.

Ebonestos. A trade name for a series of proprietary molded products for electrical and heat insulation. §

ebonized monel. A monel metal with a fine finish produced by an oxidizing process.

ebonite. (vulcanite, hardened rubber). A material prepared by vulcanizing rubber with up to 75% sulfur or metallic sulfides, with the addition of chalk, gypsum, or other filling and coloring substances; used as an insulating material in the manufacture of electrical parts, filler for low-cost rubber products.

Ebontex. A proprietary trade name for an emulsified asphalt used for waterproofing tanks. §

ebony black. A blackish-brown dyestuff mixed with a blue dyestuff; used for dyeing cotton from a bath containing sodium sulfate and sodium carbonate, and half-wool.

Eborex. A proprietary preparation containing about 65-70% sodium fluosilicate; it is a light fluosilicate for use as an insecticide. §

E-BR® 8405. CAS 9003-17-2; Emulsion polybutadiene; nonstaining; used for tire carcass, sidewalls, mech. goods; exhibits high tack, good dynamic and low temp. props. [Ameripol Synpol]

E-BR® 8471. CAS 9003-17-2; Carbon black masterbatch emulsion polybutadiene, 62.5 parts high aromatic oil, 82.5 parts N-339 carbon blk., antiozonant; used for conveyor and power belting, extruded goods, retread rubber. [Ameripol Synpol]

Ebrok. A proprietary trade name for a bituminous plastic. §

EC-25®, -25K. Propylene glycol mono- and diesters of fats and fatty acids, mono- and diglycerides, partially hydrogenated soybean oil, lecithin, BHA, citric acid; food emulsifier for cakes, mixes. [Van Den Bergh Foods]

E.C.A. Cresylic acids. [Murphy Chemical Co Ltd] †

Eca. Oil additive packages. [Exxon UK] *

Ecco MP®-2004. CAS 97-23-4; 20% Dihydroxy dichlorodiphenyl methane; textile fungicide, mildewproofing agent; compat. with aluminum/wax type water-repellents. [Eastern Color & Chem.]

Eccoblanc W-55-Q. Single-pkg. pigment wh. printing compound containing its own binding system; useful in discharge printing; produces excellent whites with good durability to laundering and drycleaning; for screen printing and roller printing. [Eastern Color & Chem.]

Eccobond®. High strength, high performance adhesives for industrial and electronic use; thermally conductive, fire retardant, and clear formulations; elec. conductive adhesives also avail. [Emerson & Cuming]

Eccobond® 114. A proprietary one part filled epoxy adhesive. [Emerson & Cuming]

Eccobond®SF40. A proprietary low density two-component epoxy-based adhesive and rigid filler. [Emerson & Cuming]

Eccobond® Adhesive Special #2. Resin/solv. adhesive material; solv. system with initial tack and somewhat rapid evaporation; for textiles and paperboard. [Eastern Color & Chem.]

Eccobond® Paste 99. A proprietary one part thixotropic epoxy adhesive of high thermal conductivity; used in

---

‡ Trade name and manufacturer not verified   § Trade name without identified manufacturer

heat sink applications. [Emerson & Cuming]

Eccobrite RB. CAS 588-59-0; Stilbene; whitening agent for cotton and acetates. [Eastern Color & Chem.]

Eccoclean CR-46. Self-emulsifying solv. cleaner; added solvency for extra cleaning power; useful on difficult cleaning problems. [Eastern Color & Chem.]

Eccocoat®. Coatings for electronics, dip coats for resistors and capacitors, conformal coatings for circuit boards, spray or brush coating for general purpose applications [Emerson & Cuming]

Ecco Defoamer Heavy. Silicone defoamer; antifoam for a wide variety of systems. [Eastern Color & Chem.]

Ecco Defoamer NSD. Nonsilicone self-emulsifying hydrocarbon; antifoam for dyeing processes. [Eastern Color & Chem.]

Ecco Defoamer S. Petrol./silicone; defoamer for mills waste water. [Eastern Color & Chem.]

Eccodye Colors. Complete range of colors for oil-phase print systems containing alkyd resins as binding agents. [Eastern Color & Chem.]

Ecco Fast Binder 1500. Pigment binder for machine and rotary screen printing to give soft, durable prints with wet and dry crock resist. [Eastern Color & Chem.]

Eccofix 101-40. Resinous substantive dye fixative for direct and fiber reactive dyestuffs. [Eastern Color & Chem.]

Eccoflo® HiK. Free-flowing and pour-in-place powds. with adjusted surface current suppressants; for dielec. applications. [Emerson & Cuming]

Eccofloat. A proprietary polyester-resin-bound syntactic foam used to fill voids in submarine hulls. [Emerson & Cuming] *

Eccofloat EG35. A proprietary epoxy resin-bound syntactic foam material used in deep-sea applications. [Emerson & Cuming] *

Eccofloat HG452. A proprietary polyester-resin-bound low-density float material for use in deep-sea applications. [Emerson & Cuming] *

Eccofloat PC61. A proprietary polyester-resin-bound castable material for use in deep-sea applications. [Emerson & Cuming] *

Eccofloat PP22 and 24. Proprietary grades of polyester-bound syntactic foam which can be packed in situ to fill voids and to make buoys. [Emerson & Cuming] *

Eccofloat SP 12, 20. A proprietary polyester-bound low-density syntactic foam for use where buoyancy is required in harbor and off-shore applications. [Emerson & Cuming] *

Eccofloat SS40. A proprietary polyurethane rubber-bound material used in the making of deep-sea diving suits. [Emerson & Cuming] *

Eccofloat UG 36. A proprietary polyurethane-bound semi-flexible non compressible material. [Emerson & Cuming]*

Eccofloat US 35. A proprietary polyurethane - bound material - flexible, compressible and usable down to about 1000 ft depth of water. [Emerson & Cuming] *

Eccofloat Encapsulant 1421. A proprietary epoxy-resin-bound encapsulant used to protect under-sea components. [Emerson & Cuming] *

Eccofoam® FPH. Rigid, high temp. resist. polyurethane foam-in-place resin for dielec. applications. [Emerson & Cuming]

Eccofoam® PP. A proprietary group of hydrocarbon resin closed-cell foams used in high-frequency electrical appli-

cations. [Emerson & Cuming] *

Eccoful DL Conc. Modified amide ethoxylate; detergent, scouring and fulling agent for wool and wool blends; softener. [Eastern Color & Chem.]

Eccogel F. Proteinaceous compound; gives stiffness and hand to suit desired finishes on various tapes and narrow fabrics. [Eastern Color & Chem.]

Eccolube L-54. Self-dispersing lubricant for narrow fabrics and a yarn lubricant for braiding and weaving; resists yel. [Eastern Color & Chem.]

Eccopel 10. Semidurable solv. phase water repellent for use in a complete solv. system; used in drycleaning establishments for the reprocessing of outerwear garments to produce water repellency. [Eastern Color & Chem.]

Eccopuff. Printing system providing a raised, three-dimensional type of print for novelty printing; applicable to paper, textiles, nonwoven, and decorative tiles in building applications. [Eastern Color & Chem.]

Ecco Resin 234. Complexed acrylic emulsion compound; pigment binder for variety of fabric substrates; nonyellowing; gives max. softness of hand. [Eastern Color & Chem.]

Ecco Rez 3070. Cyclic methylated high solids resin; stiffener for nylon and synthetics; requires catalysts. [Eastern Color & Chem.]

Ecco Rez M-300-7. CAS 9003-08-1; High solids melamine-formaldehyde resin; produces stiffness and excellent durability to fabrics in resin finishing baths; requires catalyst. [Eastern Color & Chem.]

Eccoro®. Imidazolines; corrosion resistance of metals. [Eastern Color & Chem.]

Eccoscour CB. Sodium alkylaryl sulfonate; detergent, wetting agent, and emulsifier for textile scouring and dyeing applications. [Eastern Color & Chem.]

Eccoscour D-7. Sulfate alkyl phenol ethoxylate; surfactant and scouring agent for textile fabric and yarns; prevents redeposition of soils, dyestuffs, or pigments. [Eastern Color & Chem.]

Eccoseal®. Epoxy impregnant and casting resins. [Emerson & Cuming] *

Eccoshield®. Adhesives, coatings, caulks, sealants, tapes, and gaskets for EMI/RFI shielding applications and static dissipation and elec. ground planning. [Emerson & Cuming]

Eccosil®. Silicone adhesives and casting resins. [Emerson & Cuming]

Eccosil® 2CN. RTV silicone product; industrial grade for water clear repairable potting, embedment of circuits. [Emerson & Cuming] *

Eccosil® 1776. RTV silicone product; one-part general purpose adhesive, caulk, sealant; easy-to-use, nonflowing flexible, moisture-resistant elec. grade for glass ceramics, metal, plastics. [Emerson & Cuming] *

Eccosoft C-200. Fatty material; substantive softener for use on cottons and cellulosic materials, blends of synthetics and cellulosics; shows minimal yellowing. [Eastern Color & Chem.]

Eccosol 150. Self-emulsifying mineral oil lubricant for fine wire drawing. [Eastern Color & Chem.]

Eccosolv C-14. Self-emulsifying solvent cleaner for cleaning and degreasing of machine parts. [Eastern Color & Chem.]

Eccosorb® 269E. A proprietary epoxy coating; brushed onto surfaces, it increases their electrical loss in the S to

---

* Trade name not verified as available   † Trade name verified as obsolete

K bands of the high frequency range. [Emerson & Cuming]

Eccosorb®AN. Lightweight, flexible foam sheet, broadband microwave absorber. [Emerson & Cuming]

Eccosorb®MF. A proprietary range of magnetically-loaded epoxy resins. [Emerson & Cuming]

Eccosorb® Coating 268E. A proprietary epoxy coating; brushed onto surfaces, it increases their electrical loss in the L-band of the high frequency range. [Emerson & Cuming]

Eccospheres. Trade name for small hollow glass or silica spheres of diameter ranging from 10-250 microns; used as a loading material for plastics to impart lightness and reduced permittivity which it does by virtue of the large airspace. [Emerson & Cuming] *

Eccostat. Antistat for textiles. [Eastern Color & Chem.]

Eccostat C. Antistatic agent, softener for synthetic and natural fibers and blends. [Eastern Color & Chem.]

Eccoterge 200. PEG 400 ester; wetting agent, emulsifier, dispersant for solvs. in aq. systems. [Eastern Color & Chem.]

Eccoterge ASB. Amine alkylaryl sulfonate; detergent, wetting agent, emulsifier, scouring agent for textile applications. [Eastern Color & Chem.]

Eccoterge MV Conc. Fatty amino condensate; detergent, emulsifier, wetting and scouring agent for textile use. [Eastern Color & Chem.]

Eccotherm®TC-11. General-purpose silicone grease with moderate thermal conductivity; for circuit assembly. [Emerson & Cuming]

Eccowax UL-100. Substituted fatty amide condensate; wax conc. for preparation of stock softener solution or softener products at 20-27% solids; requires use of acetic acid in order to produce a cationic softener system; for cotton, synthetics and cotton/syn. blends. [Eastern Color & Chem.]

Eccowet®W-50. Sodium aliphatic ester sulfonate; wetting agent, penetrant, dispersant, solubilizer, emulsifier, detergent for textiles, metal processing, disinfectants, paints, pigments, wallpaper, rubber cements, adhesives, drycleaning detergents, topical pharmaceuticals, cosmetics. [Eastern Color & Chem.]

Ecco White® FW-5. Distyryl-phenyl; optical brightener for nylon and blends. [Eastern Color & Chem.]

Ecdel® 9965. Copolyester ether elastomer; offers clarity, toughness, and chem. resist. for flexible pkg. applications incl. medical applications. [Eastman] *

Ecdel® 9967. Copolyester ether elastomer; offers clarity, toughness, and chem. resist. for flexible pkg. applications incl. medical applications. [Eastman] *

Ecepox®PB1 and PB2. Epoxidized esters used as plasticizers and stabilizers for PVC compounds. Registered trade names. §

Èchappe silk. A name for floss or waste silk.

Echicaoutchin. A low-grade gutta-like material from *Alstonia scholaris.*

Echo. Thiadiazole derivatives used as crosslinking agents; used in the vulcanization of halogen-containing polymers such as polyepichlorohydrins and chlorinated polyethylene; typical applications are extruded and molded hose and tubing. [Hercules]

Echo. Soluble concentrate containing 240 g/l metluidide; grass growth suppressant. [ICI Agrochemicals Professional Prods.]

echurin. A mixture of picric acid and nitro-flavin.

Eclabron. CAS 5634-38-8; Guaithylline; bronchodilator; expectorant. [Rhone-Poulenc Rorer Pharmaceuticals] *

Eclipse. Granular or powder organic based fertilizers; a steady release, lower nitrogen fertilizer for horticultural crops, parks and gardens. [Humber Fertilizers plc]

Ecobinder®. Polymer emulsion; nonsolvent-containing binder for emulsion paint. [Scott Bader]

Ecobond. Environmentally friendly phosphating processes. [Brent Chemicals International plc]

Ecolac. A proprietary trade name for an air drying lacquer and adhesive for plastics. §

Ecolo. Compounded plastic material. [Mitsubishi Petrochem.] *

Ecolotec. Resin binder for sand cores, hardened with carbon dioxide gas. [Foseco (F.S.) Ltd]

Ecomytrin®. CAS 1402-82-0, CAS 1404-04-2; A proprietary preparation of amphomycin and neomycin; antibiotic skin cream. [Bayer AG]

Econacort. A proprietary preparation containing oconazole nitrate and hydrocortisone; used for the treatment of fungal and inflammatory skin disorders. [Bristol-Myers Squibb Pharmaceuticals Ltd]

Econocat. Catalysts. [Courtaulds Engineering Ltd]

Econochlor. CAS 56-75-7; Chloramphenicol; antibacterial; antirickettsial. [Alcon Laboratories Inc] *

Econogel. High performance rheological additive for paints, inks, sealants, mastics, etc.; a range of metal-based drying agents. [Akzo Chemie UK Ltd]

Economy Flor-Dri. CAS 1337-76-4; Attapulgite clay; absorbent [Floridin] *

Econopred. CAS 52-21-1; Prednisolone acetate; glucocorticoid. [Alcon Laboratories Inc] *

Ecoro. A proprietary packaging material of polypropylene loaded with calcium carbonate to ease disposal by incineration. [Mitsubishi Petrochem.] *

Ecostatin Cream. CAS 68797-31-9; Econazole nitrate in cream base. [Bristol-Myers Squibb Pharmaceuticals Ltd]

Ecostatin Lotion. CAS 68797-31-9; Econazole nitrate in lotion base. [Bristol-Myers Squibb Pharmaceuticals Ltd]

Ecostatin Pessaries. CAS 68797-31-9; Econazole nitrate vaginal pessaries. [Bristol-Myers Squibb Pharmaceuticals Ltd]

Ecostatin Powder Solution. CAS 68797-31-9; Econazole nitrate in aerosol spray. [Bristol-Myers Squibb Pharmaceuticals Ltd]

Ecosyl. Concentrated biological silage additive. [ICI Chem. & Polymers Ltd]

Ecothene EC 101. CAS 9002-88-4; HDPE homopolymer; post-consumer homopolymer usable straight, as a blend, or in coextruded structures; for blow molding, thermoforming, injection molding, and film extrusion applications. [Quantum/USI]

Ecotrin. CAS 50-78-2; Aspirin; analgesic; antipyretic; antirheumatic. [Menley & James Laboratories] *

ECP-170. Emulsifiable chlorinated paraffin; designed for metal displacement applications which leave residual films removable with water/alkali cleaners. [Ferro/Keil]

ecru silk. Silk which has lost about 3-4% of its weight of sericin or silk gum.

Ectimar®. Broad spectrum antimycotic; used for the treatment of dermatomycosis in veterinary medicine. [Bayer AG]

---

‡ Trade name and manufacturer not verified   § Trade name without identified manufacturer

Eczederm. Calamine in a soft emollient base; for the treatment of eczema. [Quinoderm Ltd]

Edaplan LA 400. Silicone-free leveling agent for lacquers and varnishes. [Münzing Chemie GmbH]

Edaplan LA 411. Silicone compound; leveling agent for lacquers, varnishes, aq. coatings, powd. coatings. [Münzing Chemie GmbH]

Edaplan VP LA 420. Silicone-free polymer; deaerator and leveling agent for coil coating systems. [Münzing Chemie GmbH]

Edasil. Natural ion exchanger for improvement and conditioning of the soil. [Süd-Chemie AG] *

edathamil. See edetic acid.

Edecril. CAS 58-54-8, 6500-81-8; Ethacrynic acid and sodium ethacrynate; potent diuretic for the relief of certain edemas. (Japan) [Merck & Co Inc]

Edecrin®. CAS 58-54-8, 6500-81-8; Ethacrynic acid and sodium ethacrynate; potent diuretic for the relief of certain edemas. [Merck & Co Inc]

Edecrina. CAS 58-54-8, 6500-81-8; Ethacrynic acid and sodium ethacrynate; potent diuretic for the relief of certain edemas. (Sweden) [Merck & Co Inc]

Edecrin Sodium. CAS 6500-81-8; Ethacrynate sodium; diuretic. [Merck & Co Inc] *

Edelfeka. A nickel-containing silver-copper-cadmium alloy.

Edelresanol. A proprietary synthetic resin. §

edelweiss. See oil white.

Edelwit. CAS 1305-62-0; Hydrated lime Ca(OH)$_2$; used in chemical industries, drinking water treatment, waste water treatment. [BV Nekami] *

Edenol 74. A proprietary alkyl-epoxy stearate type plasticizer for plastisols. [Henkel Chemicals Ltd, Hental & Cie]

Edenol 302. Propylene glycol dicaprylate/dicaprate; emollient oil and cosolv. for personal care products, bath oils, aerosols, antiperspirants; stable vehicle for pigmented cosmetics. [Henkel/Cospha; Henkel Canada]

Edenol B35. A proprietary alkyl-epoxy stearate type plasticizer for plastisols. [Henkel Chemicals Ltd, Hental & Cie]

Edenol B316. Epoxidized linseed oil; plasticizer used in rigid PVC. [Henkel Chemicals Ltd, Hental & Cie]

Edenol D72. A proprietary alkyl-epoxy stearate type plasticizer for plastisols. [Henkel Chemicals Ltd, Hental & Cie]

Edenol D82. CAS 8013-07-8; EINECS 232-391-0; An epoxidized soybean oil; plasticizer used in both rigid and plasticized PVC. [Henkel Chemicals Ltd, Hental & Cie]

Edenol HS 235. A proprietary alkyl-epoxy stearate type plasticizer for plastisols. [Henkel Chemicals Ltd, Hental & Cie]

Edenor ITS. Isotridecyl stearate. [Henkel/Cospha]

Edenor PTO. CAS 19321-40-5; Pentaerythritol tetraoleate. [Henkel/Cospha]

Eder's solution. A solution of mercuric chloride and ammonium oxalate used in photometric determinations.

edetate disodium. See disodium EDTA.

Edetate sodium. See tetrasodium EDTA.

Edeta®. Complexing agents for cosmetics industry. [BASF AG]

edetic acid. (EDTA (CTFA); N,N´-1,2-ethanediylbis[N-(carboxymethyl) glycine]; ethylene diamine tetraacetic acid; edathamil; Versene acid) Substituted diamine; (HOOCCH$_2$)$_2$NCH$_2$CH$_2$N(CH$_2$COOH)$_2$; CAS 60-00-4; EINECS 200-449-4; chelating agent; stabilizer; antioxidant in foods; pharmaceutic aid. [Allchem Industries; Allied Colloids; W.R. Grace/Hampshire; Protex SA;

Showa Denko]

Edicol. Coloring matter for use in foodstuffs, pharmaceuticals, and cosmetics. (Sold under licence from ICI). [ICI Chem. & Polymers Ltd]

Edifas. Vegetable derivatives; used in food manufacturing as binding, stabilizing and emulsifying agents. [Imperial Chemical Industries plc]

Edimet. Polymethyl methacrylate. §

Edinol. It contains p-aminosaligenin, acetone sulfite, potassium hydroxide, and potassium bromide; a photographic developer.

Edistir®. A proprietary range of polystyrene molding granules. [Montedison UK Ltd] *

Edistir®FA. CAS 9003-53-6; PS; general purpose, high flow grade for injection molding. [Montedipe Srl]

Edistir®N 1280, N 1281. CAS 9003-53-6; PS; high thermal resist. and high mech. props.; for extrusion; also suitable for injection molding. [Montedipe Srl]

Edistir® RC. CAS 9003-53-6; HIPS; glossy grade with balanced props. for injection molding, extrusion. [Montedipe Srl]

Edistir®RK. CAS 9003-53-6; HIPS; flame-retardant grade for injection molding, extrusion. [Montedipe Srl]

Edistir® RKV. CAS 9003-53-6; Semiexpandable HIPS; flame-retardant grade for injection molding. [Montedipe Srl]

Edistir® RV 8. CAS 9003-53-6; Semiexpandable HIPS; med.-high impact, high flow grade for injection molding. [Montedipe Srl]

Edistir® SR 550, SRL 550. CAS 9003-53-6; HIPS; high impact, improved flow grade for injection molding, thin sheet extrusion. [Montedipe Srl]

Edistir® UT/1. CAS 9003-53-6; HIPS; very high impact, thermal resist. grade for injection molding. [Montedipe Srl]

Edistir®UT/SF. CAS 9003-53-6; HIPS; high impact, thermal resist. grade for injection molding. [Montedipe Srl]

Edit®. For the agriculture industry. [DuPont UK]

Edolan®. Resist agent for wool/polyamide by the exhaust process. [Bayer AG]

Edrisal. A proprietary preparation of amphetamine sulfate, aspirin, and phenacetin. [SmithKline Beecham] *

Edrul®. Loop diuretic. [Bayer AG]

EDTA (CTFA. See edetic acid.

EDTA Na4. See tetrasodium EDTA.

Edunine. Textile auxiliary chemicals. [ICI Chem. & Polymers Ltd]

Edward's speculum. A zinc and arsenic bearing bronze containing 63.3% copper, 32.2% tin, 2.9% zinc, and 1.6% arsenic.

EE. See ethoxyethanol.

EEA. See ethoxyethanol acetate.

EE Acetate. CAS 111-15-9; Ethylene glycol monoethyl ether acetate. [Eastman] *

Eel antifriction metal. An alloy of 75% lead, 15% antimony, 6% tin, 1.5% cadmium, 0.5% arsenic, and 0.1% phosphorus.

EES. CAS 41342-53-4; Erythromycin ethylsuccinate; antibacterial. [Abbott Laboratories]

EE Solvent. CAS 110-80-5; Ethylene glycol monoethyl ether. [Eastman] *

Efcortelan. CAS 50-23-7; A proprietary preparation of hydrocortisone for dermatological use. [Glaxo Laboratories]

Efcortelan-N. A proprietary preparation of hydrocortisone and neomycin; used in dermatology as an antibacterial agent. [Glaxo Laboratories] *

Efcortelan Soluble. CAS 50-23-7; A proprietary preparation of hydrocortisone; for injection in cases of shock and adrenal crisis. [Glaxo Laboratories] *

Efcortesol. CAS 50-23-7; A proprietary preparation of hydrocortisone as the 21-disodium phosphate ester; used to treat adrenal insufficiency and shock. [Glaxo Laboratories]

Efetaal. 1-Ethoxy-1-phenylethoxyethane. [Quest Int'l. UK Ltd]

Effersyllium. Psyllium husk; laxative. [Stuart Pharmaceuticals] †

Effesay. Sulfonated alcohols and detergents. [ABM Chemicals Ltd] *

Effico Tonic. A proprietary preparation of thiamine hydrochloride, nicotinamide, tincture of gentian and caffeine hydrate; a tonic. [Pharmax Ltd]

Efica. Additives for the coatings industry. [Stoller Chemicals Ltd]

Efudex. CAS 51-21-8; Fluorouracil; antineoplastic. [Hoffmann-LaRoche Inc] *

Efudix. CAS 51-21-8; A proprietary preparation of fluorouracil; antineoplastic agent. [Roche Prods. Ltd] *

Efuranol. CAS 113-52-0; A proprietary preparation of imipramine hydrochloride; an antidepressant. [Pfizer International] †

Efweko. CAS 1333-86-4; Carbon black chips in nitrocellulose or plastics; used for dyeing lacquers, gravure and flexographic printing inks. [Degussa AG] *

Egalex. Liquid dyeing assistant; for textile industry. [Tensia SA] *

Egalisal. A fiber protection agent; used in the textile industry for dyeing wool. [Degussa AG] *

Egalon Colours, Egalon Auxiliaries, Egalon Thinners. Pigments and auxiliaries for nitrocellulose finishing. [Bayer plc]

EGDS. *See* glycol distearate.

Eglantine. A name which has been applied to both isobutyl benzoate and to isobutyl-phenol acetate; used in perfumery.

EGO-4. Antisettling additive. [United Catalysts]

Egyptianized clay. Clay rendered more plastic by the addition of tannin.

2-EH. *See* 2-ethylhexanol.

EHIDA Kit. Etifenin; diagnostic aid. [Amersham Corp] *

Ehrhard's metal. An alloy of 89% zinc, 4% copper, 4% tin, and 3% lead.

Ehrlich-biondi stain. A microscopic stain containing 100 cc of a saturated solution of Orange G, 30 cc of a saturated solution of acid fuchsin, and 50 cc of a saturated solution of methyl green.

Ehrlich's diazo reagent. For indole: 4 g p-dimethylaminobenzaldehyde in 380 cc alcohol and 80 cc concentrated hydrochloric acid. One volume of the solution to be tested is used with 1 volume of the reagent, a positive color being red.

Ehrlich's hematoxylin. A microscopic stain. It consists of 30 grains haematoxylin, 100 cc absolute alcohol, 100 cc glycerin, 30 grains ammonium alum, and 100 cc distilled water.

Eisler's bronze. A bronze containing 5.9% tin; used for art castings.

Ekaline. Aliphatic polyglycol ether; versatile dispersing, leveling, and scouring agent. [Sandoz Prods. Ltd] *

Ekammon. A proprietary preparation of aspirin, and vitamins C and K; an analgesic. [Octel Chemicals Ltd] †

Ekanda rubber. A rubber obtained from the shrub, *Raphionacme utilis* in Angola.

Ekatin. Insecticide containing thiometon for aphid control. [ICI Chem. & Polymers Ltd] †

E-Kote 3042. A trade name for a silver filled air drying epoxy coating material soluble in isobutyl ketone. [Allied Prods. Corporation] ‡

Ektapro® EEP Solvent. CAS 763-69-9; Ethyl 3-ethoxypropionate; high-performance retarder solv. for formulating enamels, lacquers, topcoats, and primers; urethane-grade quality; polymerization solv. [Eastman]

Ektar®FB PG003. PP, 30% glass fiber-reinforced; for eloc./electronic applications [Eastman]

Ektasolve®. Glycol ethers and esters. [Eastman]

Ektasolve® DB. CAS 112-34-5; EINECS 203-961-6; Butoxydiglycol; solvent [Eastman] *

Ektasolve® DB Acetate. Diethylene glycol butyl ether acetate; solvent [Eastman] *

Ektasolve® DE. CAS 111-90-0; EINECS 203-919-7; Ethoxydiglycol; solvent [Eastman] *

Ektasolve® DE Acetate. Diethylene glycol ethyl ether acetate; solvent [Eastman] *

Ektasolve® DE-HG. Diethylene glycol monoethyl ether, ethylene glycol. [Eastman] *

Ektasolve® DM. CAS 111-77-3; Diethylene glycol monomethyl ether; evaporating solv. used in brushing lacquers and dye stains; useful in wood stains, printing inks, and dye pastes for textiles; coalescing aid for PVAc latex paints; used in stamp pad and stencil inks; diluent for hydraulic brake fluids. [Eastman]

Ektasolve® DP. CAS 6881-94-3; Diethylene glycol monopropyl ether; evaporating, water-misc. solv. used in sol'n. and water-dilutable coatings; act. for many coating materials incl. NC, acrylic copolymers, epoxy resins, chlorinated rubber, and alkyd resins; strong coupling agent with some resin systems in water-dilutable coatings. [Eastman]

Ektasolve® EB. CAS 111-76-2; EINECS 203-905-0; Butoxyethanol; solv. for alkyd, phenolic, maleic, and cellulose nitrate resins; excellent retarder for lacquers, improving gloss and flow-out, blush resistance, and reducing the formation of orange peel; useful in formulating hot-spray, brushing, flow-coat, and aerosol lacquers. [Eastman]

Ektasolve® EB Acetate. CAS 112-07-2; Butoxyethanol acetate; high-boiling; useful as coalescing aid for latex paints; used in multicolor lacquers, lacquer emulsions; retarder in high-low lacquer thinners, printing inks, and epoxy coatings; used as solv. in silk-screen, stamp pad, and stencil inks and component of varnish removers. [Eastman]

Ektasolve® EEH. Ethylene/diethylene glycol 2-ethylhexyl ether. [Eastman] *

Ektasolve® EP. CAS 2807-30-9; Ethylene glycol monopropyl ether; slow evaporating solv. used in coatings; useful in waterborne coating systems; coupling solv. for resin/water systems; controls visc. of waterborne resins; effective for NC, acrylic, epoxy, polyamide, and alkyd resins; retarder in coating systems. [Eastman]

---

‡ Trade name and manufacturer not verified   § Trade name without identified manufacturer

Ektasolve® PM Acetate. CAS 108-65-6; Propylene glycol methyl ether acetate; Retarder solv. [Eastman]

Ektebin®. CAS 14222-60-7; Prothionamide; an antitubercular preparation. [Bayer AG]

Ektogan. (ektogen, zinc perhydrol, zinconal). A preparation of zinc oxide, containing from 40-60% zinc peroxide; an antiseptic used for dressing wounds and burns, also as an astyptic.

Ektogen. See Ektogan.

Elacid CLR. Elastin partial hydrolysate with elastin specific protein structures in weakly acid hydrophilic medium; products for aging and inelastic skin. [Dr. Kurt Richter; Henkel/Cospha]

Elacid Richter. O/w hair conditioner conc. for damaged hair; antistat. [Dr. Kurt Richter; Henkel/Cospha]

elainic acid. See oleic acid.

Elanone. CAS 24678-13-5; Lenperone; antipsycotic. [Wyeth Laboratories] *

Elaol. Stated to be dibutyl phthalate; a proprietary plasticizer. §

Elaol 1. A trade name for a plasticizer made from C_4 to C_6 paraffin fatty acids and hexanetriol. §

Elaol 2. A trade name for a plasticizer made from C_6 to C_9 paraffin fatty acids and hexanetriol. §

Elaol 3. A trade name for a plasticizer made from C_4 to C_6 paraffin fatty acids and pentaerythritol. §

Elaol 4. A trade name for a plasticizer made from C_6 to C_9 paraffin fatty acids and pentaerythritol. §

Elaol VI. Flame retardant hydraulic fluid; for use as a safety precaution (fire and consequential damage) in mining. [Bayer AG]

Elaqua XX. Urea; diuretic. [ICN Pharmaceuticals Inc] †

Elargol. A silver finish for mica and plastics. [Octel Chemicals Ltd] †

Elase. A proprietary preparation of fibrinolysin and desoxyribonuclease; dermatological stimulant. [Parke-Davis] *

Elastalloy® 6713. Contains Kraton® polymer. [GLS Plastics]

Elastan®. Polyol-isocyanate formulations; systems for sports ground surfaces, play grounds, tennis courts, artificial lawns. [BASF AG]

Elas-Tein AS-20. Elastin, ethyl ester in ethanol; cosmetic ingred. [Maybrook]

elastic asbestos. See mountain cork.

elastic bitumen. See elaterite.

Elasti-glass. A proprietary trade name for a vinyl copolymer used for belts, braces, raincoats, tobacco pouches, etc.§

Elastinhydrolysate, Liq. Hydrolyzed animal elastin; filmformer, moisturizer; enhances system stability. [Henkel/Cospha; Henkel Canada] *

Elastite. A sulfurized oil rubber substitute; also a proprietary flooring block made from asphalt, fiber, and fillers. §

Elastoblend 8480. Emulsion polybutadiene/SBR (40/60), 55 parts N-234 carbon blk., 20 parts high aromatic oil; masterbatch for high grade truck retreads for both precure and mold cure. [Ameripol Synpol]

Elastobond. Elastomeric bonding bitumen. [Feb Ltd]

Elastocarb Tech Light, Tech Heavy. CAS 546-93-0; Magnesium carbonate. inorg. filler providing flame retardancy and smoke suppression to elastomers, plastics, and thermosets incl. EPDM, PP, PE, PVC; used in wire and cable compounds, conduit/tubing, film and sheet [Morton Int'l.]

Elastocarb UF. CAS 546-93-0; Magnesium carbonate; precipitated ultra-fine grade filler for use in conduit, wire and cable compounds, aircraft interior components, other transportation uses; reduces smoke dens. and increases physical properties. [Morton Int'l./Plastics Additives]

Elastocell®. Microcellular polyurethane suspension components. [BASF]

Elastocoat®. PU; casting and coating systems. [BASF AG]

Elastoflex®. Soft PU; foam systems. [BASF AG]

Elastofoam® I. Polyol-isocyanate formulations; for polyurethane flexible integral skin foam systems for decorative parts subject to heavy duty. [BASF AG]

Elastoid 1300. High build elastomeric coatings based on multiphase synthetic rubber copolymers; tough, flexible, corrosion and weather resistant, abrasion and impact resistant; for heavy-duty protection in aggressive chemical, industrial and marine environments; extremely low water vapor transmission rate; used on metal, concrete, foam insulation; unaffected by water immersion. [Dampney Company Inc] *

Elastolac. A proprietary trade name for a shellac derivative; water and alcohol soluble. §

Elastolit® D. Polyol-isocyanate formulations; systems for rigid integral skin foams for production of desks, chairs, cabinets, window profiles, cable guides, molded parts, housings. [BASF AG]

Elastolith. A synthetic resin. See Bakelite. §

Elastollan® 1154D. Thermoplastic polyether PU elastomer; offers excellent low temp. props., hydrolysis and fungus resist.; for injection and blow molding and extrusion. [BASF] *

Elastollan® C-59D. Thermoplastic polyester PU elastomer; exhibits very good hydrolytic stability, excellent oil, fuel, and solv. resist.; for injection and blow molding and extrusion. [BASF] *

Elastollan® S-60D. Thermoplastic polyester PU elastomer; exhibits good hydrolytic stability, good ooil, fuel, and solv. resist.; for injection and blow molding and extrusion. [BASF] *

Elastomag® 100. CAS 1309-48-4; Magnesium oxide; chemical thickener for polyester resins; anticaking agent; used in synthetic rubber compounding, adhesives, fuel oil additives, and as acid acceptor for specialty plastics. [Akrochem]

Elastopal®. PU elastomers; for casting. [BASF AG]

Elastopan®. PU; shoe foam systems. [BASF AG]

Elastopor®. Hard PU; foam systems. [BASF AG]

Elastopreg®. Thermoplastic fiber composite; Semifinished products of glass mat-reinforced thermoplastics. [Elastogran]

Elastorid®. Wood dust-filled PP sheets; carrier for inner trim of passenger cars. [BASF AG]

Elastosil®. RTV-1 silicone rubbers; adhesives, sealing, coating, sealants in the electrical and electronics industry, seals in the automobile industry. [Wacker-Chemie GmbH]

Elastosil® LR 3001. CAS 63394-02-5; Liq. silicone rubber; used for anode caps; flame retardant. [Wacker Silicones]

Elastosil® LR 3003/20. CAS 63394-02-5; Liq. silicone rubber; features high mech. strength; for precision-molded components, o-rings, gaskets; approved for health care, food contact applications. [Wacker Sili-

---

* Trade name not verified as available   † Trade name verified as obsolete

cones]

elaterite. (elastic bitumen, mineral caoutchouc, Helenite). A fossil resin, resembling asphaltum, found in some of the lead mines in Derbyshire. It contains 6-7% mineral matter, and is slightly soluble in ether.

Elaterite, Artificial. A proprietary product made from liquid bitumen and vegetable oils, then treatment with heat and pressure with sulfur chloride, saltpeter, and sulfur; used for water proofing and insulation. §

Elaterium. The dried sediment from the juice of *Ecballium olaterium*. The active principle is elaterin.

Elavil. CAS 549-18-8; Amitriptyline hydrochloride; antidepressant with sedative properties. (U.S.) [Merck & Co Inc]

Elayl. *See* olefiant gas.

Elbasol. Solvent dyestuffs; for coloration of non aqueous solvents. [Holliday Dyes & Chemicals Ltd]

Elbelan. 2:1 Metal complex dyestuffs; for wool, nylon and blends containing one or both of these fibers. [Holliday Dyes & Chemicals Ltd] †

Elbelene. Dyes for polypropylene. [Holliday Dyes & Chemicals Ltd] †

Elbenyl. Anionic dyestuffs; dyes specially selected for their suitability for dyeing nylon. [Holliday Dyes & Chemicals Ltd]

Elbeplast. Mainly azo and anthraquinone-based solvent dyes; for coloration of plastics. [Holliday Dyes & Chemicals Ltd]

Elbestret. For desulfurization of gas. [Holliday Dyes & Chemicals Ltd]

Elcema®F150, G250, P100. CAS 9004-34-6; Cellulose NF; anticaking agent, tabletting aid for pharmaceutical industry. [Degussa]

Elcitonin®. Agent for bone metabolism. [Asahi Chem. Industry]

Elcomet. A proprietary trade name for a steel containing chromium, silicon, copper, and nickel. §

Eldecort. CAS 50-23-7; Hydrocortisone; glucocorticoid. [ICN Pharmaceuticals Inc]

Eldisine. CAS 59917-39-4; Vindesine sulfate; antineoplastic. [Eli Lilly & Co]

Eldopaque. CAS 123-31-9; Hydroquinone; depigmentor. [ICN Pharmaceuticals Inc]

Eldoquin. CAS 123-31-9; Hydroquinone; depigmentor. [ICN Pharmaceuticals Inc]

Electrafil® F-4/CN/40. Nylon, nickel-coated carbon fiber filled; static dissipative and conductive thermoplastic [Akzo Engineering Plastics] *

Electrafil®F-1700/CF/10/A. CAS 25134-01-4; PPO, modified, 10% PAN carbon fiber-reinforced; static dissipative and conductive thermoplastic [Akzo Engineering Plastics] *

Electrafil®G-1/SS/5. Nylon, stainless steel fiber filled; static dissipative and conductive thermoplastic [Akzo Engineering Plastics] *

Electrafil®G-50/SS/10. PC, 10% stainless steel fiber filled; static dissipative and conductive thermoplastic [Akzo Engineering Plastics] *

Electrafil®G-1204/SS/3. CAS 9003-56-9; ABS, 3% stainless steel fiber filled; static dissipative and conductive thermoplastic [Akzo Engineering Plastics] *

Electrafil®G-1704/SS/5. CAS 25134-01-4; PPO, modified, 5% stainless steel fiber filled; static dissipative and conductive thermoplastic [Akzo Engineering Plastics] *

Electrafil® G-1854/SS/7. CAS 26062-94-2; PBT, 7% stainless steel fiber filled; static dissipative and conductive thermoplastic [Akzo Engineering Plastics] *

Electrafil® J-1/30/CF/7/H; CAS 32131-17-2; Nylon 66, 30% PAN carbon fiber-reinforced; static dissipative and conductive thermoplastic [Akzo Engineering Plastics] *

Electrafil® J-2/CF/30; CAS 9008-66-6; Nylon 6/10, 30% PAN carbon fiber-reinforced; static dissipative and conductive thermoplastic [Akzo Engineering Plastics] *

Electrafil®J-3/CF/30. CAS 25038-54-4; Nylon 6, 30% PAN carbon fiber-reinforced; static dissipative and conductive thermoplastic [Akzo Engineering Plastics] *

Electrafil® J-4/CF/30. Nylon 6/12, 30% PAN carbon fiber-reinforced; static dissipative and conductive thermoplastic [Akzo Engineering Plastics] *

Electrafil® J-7/20/EC. Nylon, carbon black filled; static dissipative and conductive thermoplastic [Akzo Engineering Plastics] *

Electrafil® J-30/CF/20. CAS 9003-53-6; PS, 20% PAN carbon fiber-reinforced; static dissipative and conductive thermoplastic [Akzo Engineering Plastics] *

Electrafil® J-50/20/CF/10. PC, 10% PAN carbon fiber-reinforced; static dissipative and conductive thermoplastic [Akzo Engineering Plastics] *

Electrafil® J-60/CF/30. CAS 9003-07-0; PP, 30% PAN carbon fiber-reinforced; static dissipative and conductive thermoplastic [Akzo Engineering Plastics] *

Electrafil® J-80/CF/10/TF/10. CAS 105-57-7; Acetal, 10% PAN carbon fiber-reinforced, 10% PTFE; static dissipative and conductive thermoplastic [Akzo Engineering Plastics] *

Electrafil® J-100/CF/30. CAS 9009-54-5; Polyurethane, thermoplastic, 30% PAN carbon fiber-reinforced; static dissipative and conductive thermoplastic [Akzo Engineering Plastics] *

Electrafil® J-1100/CF/30. PES, 30% PAN carbon fiber-reinforced; static dissipative and conductive thermoplastic [Akzo Engineering Plastics] *

Electrafil® J-1105/CF/30. PEEK, 30% PAN carbon fiber-reinforced; static dissipative and conductive thermoplastic [Akzo Engineering Plastics] *

Electrafil® J-1106/CF/30. CAS 61128-46-9; Polyetherimide, 30% PAN carbon fiber-reinforced; static dissipative and conductive thermoplastic [Akzo Engineering Plastics] *

Electrafil® J-1200/CF/10. CAS 9003-56-9; ABS, 10% PAN carbon fiber-reinforced; static dissipative and conductive thermoplastic [Akzo Engineering Plastics] *

Electrafil®J-1300/CF/30/TF/15. CAS 9016-75-5; PPS, 30% PAN carbon fiber-reinforced, 15% PTFE; static dissipative and conductive thermoplastic [Akzo Engineering Plastics] *

Electrafil® J-1400/CF/20. ETFE, 20% PAN carbon fiber-reinforced; static dissipative and conductive thermoplastic [Akzo Engineering Plastics] *

Electrafil® J-1500/CF/20. CAS 25135-51-3; Polysulfone, 20% PAN carbon fiber-reinforced; static dissipative and conductive thermoplastic [Akzo Engineering Plastics] *

Electrafil® J-1700/CF/10. CAS 25134-01-4; PPO, modified, 10% PAN carbon fiber-reinforced; static dissipative and conductive thermoplastic [Akzo Engineering Plastics] *

Electrafil® J-1701/CF/10/FR. PPE, modified, 10% PAN carbon fiber-reinforced; flame retardant static dissipative

---

‡ Trade name and manufacturer not verified   § Trade name without identified manufacturer

and conductive thermoplastic [Akzo Engineering Plastics] *

Electrafil®J-1800/CF/30. CAS 25038-59-9; PET, 30% PAN carbon fiber-reinforced; static dissipative and conductive thermoplastic [Akzo Engineering Plastics] *

Electrafil®J-1850/CF/30. CAS 26062-94-2; PBT, 30% PAN carbon fiber-reinforced; static dissipative and conductive thermoplastic [Akzo Engineering Plastics] *

Electrafil® JM-61/CF/10. CAS 9003-07-0; PP, 10% PAN carbon fiber-reinforced; static dissipative and conductive thermoplastic [Akzo Engineering Plastics] *

Electrafil® M-1526/EC. Nylon, carbon black filled; static dissipative and conductive thermoplastic [Akzo Engineering Plastics] *

Electrafil®PC-50/EC. PC, carbon black filled; static dissipative and conductive thermoplastic [Akzo Engineering Plastics] *

Electrafil®PE-90/EC. CAS 9002-88-4; HDPE, carbon black filled; static dissipative and conductive thermoplastic [Akzo Engineering Plastics] *

Electrafil® PP-60/CC/20/EC. CAS 9003-07-0; PP, carbon black and calcium carbonate filled; static dissipative and conductive thermoplastic [Akzo Engineering Plastics] *

Electrafil® TR-1900/EC. CAS 9002-84-0; TPE, carbon black filled; static dissipative and conductive thermoplastic [Akzo Engineering Plastics] *

Electran. Reagents for electrophoresis. [BDH Chemicals Ltd]

Electrathane. Antistatic castable polyurethane; passive static discharge, charged transfer roller (photo copiers). [PEI Precision Elastomers Inc] ‡

electraurol. A form of colloidal gold.

electric bronze. An alloy of 87% copper, 7% tin, 3% zinc, and 3% lead.

electricidal. Electro-colloidal iridium.

electriridol. Colloidal iridium.

Electrisil. A proprietary silicone rubber composition used for insulating conductors. [GE Silicones] *

Electrisil 758. A proprietary flame retardant silicone rubber compound used to insulate high-voltage cables. [GE Silicones] *

Electrisil 9025. A proprietary silicone rubber compound used in applications where radiation and high temperatures may be encountered. [GE Silicones] *

Electrit. A trademark for goods of the abrasive and refractory class, the essential constituent of which is crystalline alumina. §

Electroclear. Clear lacquers used for metal protection. [ICI Chem. & Polymers Ltd]

electrocuprol. A form of colloidal copper.

Electrodag®. Dispersions of conducting pigment in resin; used for silk screen printable conducting inks for both flexible and rigid circuitry; shielding coatings; conducting coatings; heat generating coatings and inks; conducting impregnants. [Acheson Colloids]

Electrodag®112. Graphite pigment, acrylic binder in water; EMC shielding coating for plastics; protects sensitive electronic equip. [Acheson]

Electrodag® 415; Silver pigment, PVC binder in SB-1 or MIBK; EMC shielding coating for plastics; protects sensitive electronic equip. [Acheson]

Electrodag® 415C. Silver pigment, acrylic binder in SB-1; EMC shielding coating for plastics; protects sensitive electronic equip. [Acheson]

Electrodag®437i. Copper pigment, acrylic binder in SB-1, SB-8, or SB-10; EMC shielding coating for plastics; protects sensitive electronic equip. [Acheson]

Electrodag® 438. Copper/silver pigment, acrylic binder in SB-1, SB-8, or SB-10; EMC shielding coating for plastics; protects sensitive electronic equip. [Acheson]

Electrodag®439. Nickel pigment, acrylic binder in SB-1, SB-8, or SB-10; EMC shielding coating for plastics; protects sensitive electronic equip. [Acheson]

Electrodag® 442. Nickel pigment, polyester binder in MEK:xylol (1:1); EMC shielding coating for steel frames; protects sensitive electronic equip. [Acheson]

Electrodag® 550. Nickel pigment, acrylic binder in SB-8 or SB-10; EMC shielding coating for plastics; protects sensitive electronic equip. [Acheson]

Electrodag®24501. Silver pigment, urethane binder in PM acetate; EMC shielding coating for composites/aluminum; protects sensitive electronic equip. [Acheson]

Electrodyn. Sprayers. [ICI Chem. & Polymers Ltd]

electro-filtros. A diaphragm material. It consists of grains of pure crystalline silica cemented together with a fused siliceous binding substance; used in electrolytic processes.

Electrofine®S-70. Chlorinated paraffin; flame retardant for plastics, textiles, paper and cardboard in conjunction with antimony trioxide; improves hardness, gloss and resist. to acids and bases in paints. [Atochem]

Electro-fused Cement. See fused cement.

electro-granodized iron and steel. A process for forming a rust-preventing coat on iron and steel. An alternating current plates a continuous coating of zinc phosphate.

Electrolon. See Carbora. §

electrolyilc chlorogen (E.C.). A chlorinated soda prepared by the electrolysis of brine.

electrolyte acid. See sulfuric acid.

electromartiol. A form of colloidal iron.

Electromate. Zinc plating specialty chemicals. [Stowlin Ltd]

electromercurol. A form of colloidal mercury.

Electronite. A safety explosive containing 75% ammonium nitrate, 5% barium nitrate, with wood meal and starch.

Electronite No. 2. An explosive consisting of 95% ammonium nitrate, and 5% wood meal and starch.

electropalladiol. A form of colloidal palladium.

electroplatinol. A form of colloidal platinum.

electrorhodiol. A form of colloidal rhodium.

Electrorubin. A trademark for abrasive and refractory materials; the essential constituent is crystalline alumina. §

Electrose. A proprietary trade name for a shellac plastic. §

electroselenium. A form of colloidal selenium.

electrotype metal. An alloy of 93% lead, 4% antimony, and 3% tin.

Electro-Wash®. General-purpose cleaning agent, degreaser for removal of encrusted dirt, grease, oxidation and contaminants; for use with elec. interfaces, contacts and connectors. [Chemtronics]

Electrox. CAS 1314-13-2; Photoconductive zinc oxides. [Durham Chemicals Ltd]

electrozone. A similar preparation to Chloros (sodium hypochlorite solution); a disinfectant.

Electrum. See nickel silvers.

Electrundum. A trademark for materials of the abrasive type and consisting essentially of alumina. §

Elefac I-205. Octyldodecyl neopentanoate; dry emollient,

SPF booster, pigment wetter and binder. [Bernel]

Elektra. Organic brightener system; acid copper electroplating (decorative). [Harshaw Chemicals Ltd]

Elektron®. Alloys, flues, and hardeners. [Magnesium Elektron Ltd]

Elektron®. A trademark used in connection with certain magnesiuro alloys containing up to about 10% of various alloying constituents, such as aluminum, zinc and manganese; used in cast and wrought forms for aero engines and other purposes. §

Elemite. A proprietary trade name for a wetting agent and detergent; a combination of sulfonated oils and solvents.§

elephant bronze. An alloy of 85% copper, 10.5% tin, 2.75% zinc, 1.5% lead, and 0.1-0.2% phosphorus.

elephant-S bronze. An alloy of 80.5% copper, 10.2% tin, 9% antimony, and 0.1-0.3% phosphorus.

Elestol. A proprietary preparation of chloroquine phosphate, prednisolone and aspirin; an anti-inflammatory agent. [FBA Pharmaceuticals] ‡

Eleudron-Solution. Sulfonamide; used for coccidiosis in poultry; veterinary medicine. [Bayer AG]

Elexar®. A proprietary range of thermoplastic rubbers designed for use in the cable industry. [Shell] *

Elexar® 8421. Thermoplastic elastomer; UL recognized, 105 C flexible cord material for insulation, jacketing, and molding applications; excellent abrasion and impact resist., heat aging chars. [Shell]

Elfacos®. A range of nonionic emulsifiers and stabilizers used in water-in-oil and oil-in water cosmetic formulations. [Akzo Chemie UK Ltd.]

Elfan® 200. Sodium lauryl sulfate (C12) as a fine white powder; anionic surfactant used in toothpastes. [Akzo Chemie Nederland BV]

Elfan® 240 and 240S. Sodium lauryl sulfate (C12/C14), either natural (240) or based on a synthetic fatty alcohol (240S); supplied as a transparent to white paste; detergent and emulsifier used in shampoos, light duty detergents and cleaning pastes. [Akzo Chemie Nederland BV]

Elfan®240M and 240M/S. Monoethanolamine lauryl sulfate (C12/C14), either natural (240M) or based on a synthetic fatty alcohol (240M/S); supplied as a clear, yellowish, medium viscous liquid; detergent and emulsifier for shampoos and bubble baths. [Akzo Chemie Nederland BV]

Elfan®240T and 240T/S. CAS 139-96-8; EINECS 205-388-7; Triethanolamine lauryl sulfate (C12/C14), either natural (240T) or based on a synthetic fatty alcohol (240T/S); supplied as a yellowish clear liquid; detergent for shampoos and bubble baths. [Akzo Chemie Nederland BV]

Elfan®280. Sodium coconut fatty alcohol sulfate (C12-C18); supplied as a white powder or paste; detergent raw material for light duty detergents, all-purpose washing agents and hand cleansers. [Akzo Chemie Nederland BV]

Elfan® 680. Sodium oleyl-cetyl alcohol sulfate; supplied as a yellowish- brown paste; detergent for heavy and light duty detergent powders, washing and cleaning pastes. [Akzo Chemie Nederland BV]

Elfan® A432. Amphoteric surfactant supplied as a clear yellowish liquid; used for baby shampoos; bubble baths; strong acid and alkaline cleaning detergents. [Akzo Chemie Nederland BV]

Elfan® KT550. Anionic surfactant in which the cation is sodium and the anion is composed of 50% coconut/50% tallow fatty alcohol sulfate (C12-C18); supplied as a white paste; detergent raw material for heavy and light duty detergents, and washing powders. [Akzo Chemie Nederland BV]

Elfan® NS 242, NS 243S, NS 252 S. Anionic surfactant in liquid or paste form; used for shampoos, bubble baths, dishwashing liquids, light duty liquids, washing pastes, car shampoos. [Akzo Chemie UK Ltd]

Elfan®NS 243 S Mg. Anionic surfactant with magnesium as counter ion giving excellent mildness qualities; used for shampoos, bubble baths, special care and baby products. [Akzo Chemie UK Ltd.]

Elfan®NS 682 KS. Anionic surfactant supplied as a yellowish paste; used for shampoos, bubble baths, dishwashing liquids, light duty liquids, washing pastes, car shampoos. [Akzo Chemie UK Ltd]

Elfan® OS 46. Sodium alpha-olefine sulfonate (C14/C16); supplied as a yellowish liquid; anionic surfactant for shampoos, bubble baths, dishwashing detergents, liquid and paste-form cleaners. [Akzo Chemie UK Ltd]

Elfan®WA Series. Dodecylbenzene sulfonate as triethanolamine or sodium salt or in acid form; supplied as liquid, paste or powder; anionic surfactants used in heavy and light duty, all-purpose and dishwashing detergents, scouring and other powder formulations [Akzo Chemie Nederland BV]

Elfanol®510. Sodium sulfosuccinic acid monoester of a fatty acid alkylolamide; creamy colored paste; for washing and cleaning pastes; hand cleansers. [Akzo Chemie UK Ltd]

Elfanol® 616. Sodium sulfosuccinic acid monoester of an ethoxylated fatty alcohol; yellowish viscous liquid; used for shampoos, bubble baths, baby baths, liquid hand cleansers. [Akzo Chemie UK Ltd]

Elfanol® 850. Sodium sulfosuccinic acid monoester of an ethoxylated fatty acid; yellow-brown, nearly clear liquid; used for baby baths, shampoos, bubble baths, liquid hand cleansers. [Akzo Chemie UK Ltd]

Elfanol®883. CAS 577-11-7; EINECS 209-406-4; Sodium dioctylester of sulfosuccinic acid; colorless to slight yellow liquid; wetting agent for technical processes. [Akzo Chemie UK Ltd]

Elfapur®N50. Nonylphenol ethoxylate nonionic surfactant in the form of a clear, nearly colorless liquid; used for dishwashing detergents for automatic machines; emulsifiers for fats and mineral oils. [Akzo Chemie Nederland BV]

Elfapur® N70. Nonylphenol ethoxylate nonionic surfactant in the form of a clear, nearly colorless liquid; low foaming dishwashing detergents for automatic machines, industrial and solvent cleansers, fat-dissolving pastes, all-purpose washing pastes [Akzo Chemie Nederland BV]

Elfapur® N90, N120 and N150. Nonylphenol ethoxylate nonionic surfactants in liquid or paste form; wide range of detergents, cleansers, car shampoos, fat-dissolving pastes, all purpose washing pastes. [Akzo Chemie Nederland BV]

Elftex® 675. CAS 1333-86-4; Carbon black; for coloring plastics; very good uv protection. [Cabot; Cabot Carbon Ltd]

Elfugin. Wash-fast antistatic agent; for synthetic fibers. [Sandoz Prods. Ltd] *

---

‡ Trade name and manufacturer not verified   § Trade name without identified manufacturer

elhuyarite. A red allophane mineral.

Elianite I. An acid-resisting alloy, containing 82% iron, 15% silicon, and 0.6% manganese.

Elianite II. An acid-resisting alloy, consisting of 81% iron, 15% silicon, 0.5% manganese, 2.2% nickel, 0.8% carbon, and 0.06% phosphorus.

Eliminal. Aluminum-removing flux for copper alloys. [Foseco (F.S.) Ltd]

Elimite® . CAS 52645-53-1; Permethrin; scabicide. [Allergan, Inc]

Elintaal. 1-Ethoxy-1-(3,7-dimethyl-1,6-octadienyloxy)-ethane. [Quest Int'l. UK Ltd]

Elinvar. A nickel steel containing 36% nickel, 46% iron, 12% chromium, 4% tungsten, and 1-2% manganese.; used for the more delicate parts of watches.

Elite Fast. Anionic dyestuffs (neutral dyeing); for dyeing shades of wool with good wash fastness. [Holliday Dyes & Chemicals Ltd]

Elityran. A proprietary preparation of throid extract. [Bayer AG]

Eljon. Disazo yellow and orange pigment; azo yellow and red pigments; toners; used for printing inks, paints, plastics and artists colors. [European Colour (Pigments) Ltd]

Elkalub. Oils, oil additives, and greases. [Fernox Manufacturing Co Ltd]

Elkem Microsilica. Raw and processed amorphous silica (condensed silica fume); for refractories, polymers, insulation, fluid cracking catalysts and a range of chemical and mineral uses. [Elkem Chemicals Inc] *

Elkonite. A copper-tungsten alloy used for making welding dies. It has a Brinnell hardness of 225, a compression strength ot 208,000 lb per sq in, and is not annealed at red heat.

Elkosin. A proprietary trade name for sulfasomidine. §

ellagitannin. A variety of tannin found in divi-divi, knoppern, and myrobalans.

ellagite. A mineral. It is a variety of natrolite.

Elliott's Lawn Sand. CAS 10028-22-5; Ferrous sulfate; used for moss control in turf. [Thomas Elliott Ltd]

Elliott's Moss Killer. CAS 10028-22-5; Ferrous sulfate; used for moss control in turf. [Thomas Elliott Ltd]

Elmarid. An alloy of 89% tungsten, 4.5% cobalt, 5.9% carbon, and 0.4% iron.

Elner's German silver. A nickel silver containing 57.4% copper, 26.6% zinc, 13% nickel, and 3% iron.

Elocril. CAS 18181-70-9; Iodofenphos; an organophosphorus insecticide. [Ciba-Geigy Agrochemicals]

Elotex. Redispersible homo or copolymer powders (powdered emulsions); for building products. [Ebnother AG]*

Eloxal. A proprietary trade name for an anodized aluminum.§

Eloxyl. CAS 94-36-0; Benzoyl peroxide; keratolytic. [ICN Pharmaceuticals Inc] †

ELP-3. Epoxy terminated polysulfide polymer; modifier for epoxy resins; used in concrete adhesives, chemically resist. linings or coatings, bonding to metallic substrates with oily finishes. [Morton Int'l./Polymer Systems]

Elsner's reagent. A basic zinc chloride solution obtained by dissolving 500 g zinc chloride and 20 g zinc oxide in 425 cc water and warming. A solvent for silk.

Elspar. Asparaginase; for the treatment of acute lymphocytic leucemia. [Merck & Co Inc]

Eltaga® . Fiber. [DuPont UK]

Eltesol® . Aromatic sulfonates for laundry detergents.

[Albright & Wilson Ltd]

Eltesol® 4009, 4018. Xylene sulfonic acid modified with methanol and sulfuric acid; catalysts for curing cold-setting phenol-formaldehyde and phenol-furane resins used in the foundry industry as binders for sand in the production of molds and cores. [Albright & Wilson UK]

Eltesol® 4402, 4403 and FDA 55/8. Alkylaryl sulfonate anionic in acid form; used widely in the foundry industry for curing cold setting resins. [Albright & Wilson Ltd, Detergents Div, Marchon]

Eltesol® 5400 Series. Phenol sulfonic acid condensates; surfactant for leather processing. [Albright & Wilson Am.]

Eltesol® 7200 Series. Dihydroxy diphenyl sulfonates; surfactant for leather processing. [Albright & Wilson Am.]

Eltesol® AC60. CAS 37475-88-0; EINECS 253-519-1; Ammonium cumene sulfonate; surfactant, hydrotrope for agricultural applications. [Albright & Wilson UK]

Eltesol® ACS 60. Alkylaryl sulfonate anionic; pale yellow liquid; for manufacture of liquid detergent formulations. [Albright & Wilson Ltd, Detergents Div, Marchon]

Eltesol® AX 40. CAS 26447-10-9; EINECS 247-710-9; Ammonium xylene sulfonate; hydrotrope, cloud pt. depressant used in the detergent mfg.; solubilizer, coupler. [Albright & Wilson UK]

Eltesol® CA 65. CAS 28631-63-2; Cumene sulfonic acid; catalyst for foundry resins; descaling agent for metal cleaning; anti-stress additive and plating aid in electroplating bath; curing aid in the plastics industry; raw material in the mfg. of dyes and pigments; detergents industry. [Albright & Wilson UK]

Eltesol® MGX. Magnesium xylene sulfonate; hydrotrope for liq and spray-dried detergent formulations. [Albright & Wilson UK] *

Eltesol® PSA 65. CAS 1333-39-7; Phenol sulfonic acid; catalyst for foundry resins; descaling agent for metal cleaning; anti-stress additive and plating aid in electroplating bath; curing aid in the plastics industry; raw material in the mfg. of dyes and pigments; detergents industry; pharmaceutical chemicals and disinfectants. [Albright & Wilson UK]

Eltesol® PT 93. Potassium toluene sulfonate; hydrotrope, cloud pt. depressant used in the detergent mfg.; solubilizer, coupler. [Albright & Wilson UK]

Eltesol® PX 40. Potassium xylene sulfonate; hydrotrope, solubilizer, coupling agent, and visc. modifier in liq. formulations; cloud pt. depressant in detergent formulations. [Albright & Wilson UK]

Eltesol® PX 93. Potassium xylene sulfonate; hydrotrope, cloud pt. depressant used in the detergent mfg.; solubilizer, coupler. [Albright & Wilson UK]

Eltesol® SC 93. CAS 32073-22-6; Sodium cumene sulfonate; hydrotrope for hard surface cleaners. [Albright & Wilson UK]

Eltesol® ST 40. CAS 657-84-1; Sodium toluenesulfonate; hydrotrope, solubilizer, coupling agent, and visc. modifier in liq. formulations; cloud pt. depressant in detergent formulations. [Albright & Wilson UK]

Eltesol® SX 30. CAS 1300-72-7; EINECS 215-090-9; Sodium xylene sulfonate; hydrotrope, cloud pt. depressant used in the detergent mfg.; solubilizer, coupler. [Albright & Wilson UK]

Eltesol® TA 65. 65% Toluene sulfonic acid and 1.4% sulfonic acid aq. solution; curing agent for resins in foundry cores, plastics, coatings; intermediate; catalyst in foundry and chemical industries; hardening agent in plastics; activator for nicotine insecticides; descaling agent in metal cleaning; in electroplating baths. [Albright & Wilson UK]

Eltesol® TPA. Formulated product; tin-plating additive. [Albright & Wilson UK]

Eltesol® TSX. CAS 70788-37-3; p-Toluene sulfonic acid monohydrate BP; catalyst for org. synthesis, synthetic resins, mfg. of p-cresol, toluene derivs., pharmaceutical products, dyestuffs; chemical intermediate. [Albright & Wilson UK]

Eltesol®XA. CAS 25321-41-9; EINECS 246-839-8; Xylene sulfonic acid aq. solution; catalyst in foundry and chemical industries; hydrotrope for agricultural formulations. [Albright & Wilson UK]

Eltex. High density polyethylene. [Laporte Industries Ltd]

Eltex P. Polypropylene. [Laporte Industries Ltd]

Eltroxin. A proprietary preparation of thyroxine sodium; thyroid hormone. [Glaxo Laboratories]

Eludril Mouthwash. Antibacterial/antifungal mouthwash. 0.1% chlorhexidine, 0.1% chlorbutol, 0.5% chloroform; dilute 10 ml: half glass warm water; used for gingivitis, apthous/dental ulcers and mouth and throat infection; does not stain teeth or composites. [Concept Pharmaceuticals Ltd] *

Eludril Spray. Aerosol spray containing chlorhexidine 0.05% and amethecaine 0.015%; antibacterial/antifungal for apthous/dental ulceration and for mouth and throat infections. [Concept Pharmaceuticals Ltd] *

Elugent™. Detergent. [Calbiochem Corp]

Elvace®. Vinyl acetate and ethylene emulsion copolymers and terpolymers. [Reichhold]

Elvace® 1870. VAE copolymer emulsion; adhesive base with improved plasticizer response, quick tack, and adhesion compared with homopolymer. [Reichhold] *

Elvacite®. Acrylic resins. [DuPont UK]

Elvaloy®. Resin modifiers. [DuPont UK]

Elvaloy®EP-4043, HP441. Ethylene/acrylate/carbon monoxide terpolymer. [DuPont]

Elvamide®. Nylon multipolymer resins. [DuPont UK]

Elvanol®. Polyvinyl alcohol resins. [DuPont UK]

Elvanol®20-25. CAS 9002-89-5; Fully hydrolyzed polyvinyl alcohol; for use in adhesive, textile and paper applications [DuPont]

Elvanol®71-30. CAS 9002-89-5; Fully hydrolyzed polyvinyl alcohol; film-forming binder used in adhesives, paper, paperboard sizing and coatings, textiles, films, building products, hoses, gaskets, emulsification in emulsions and latexes, additive for concrete, cement, and food. [DuPont]

Elvanol®90-50. CAS 9002-89-5; Fully hydrolyzed polyvinyl alcohol; provides high film strength and binding power in low visc. systems; used in paper and paperboard coating and sizing, adhesives; pigment binder; food additive application [DuPont]

Elvaron®. CAS 1085-98-9; A wettable powder containing 50% w/w dichlofluanid; fungicide used to control botrytis on strawberries, raspberries, loganberries, blackberries, blackcurrants, redcurrants, gooseberries, outdoor grapes, tomatoes under cover, tulips and peonies; also controls cane spot and stamen blight; reduces raspberry mildew and gives some reduction of spur blight on raspberries, mildew and blackspot on roses, downy mildew on strawberries, leaf spot on blackcurrants and gooseberries. [Bayer AG; Bayer plc]

Elvax®. EVA polymer resins. [DuPont UK]

Elvax® 40-W. EVA copolymer, antiblock; resin useful as base polymers in hot melt and solv.-thinned coatings and adhesives; modifiers to improve properties such as processability, thermoplasticity, adhesion, and abrasion resistance; food pkg. applications. [DuPont]

Elvax® 260. EVA copolymer resin; high-visc. resin, toughness, flexibility, adhesion, and barrier properties for coating and adhesive applications, and flocked or fabric-laminated counters. [DuPont]

Elvax® 310. EVA copolymer resin; wax-compatible resin; enhances properties of hot melt blends containing microcryst. waxes; used in heat-sealable barrier coatings and hot melt adhesives; also for blending with waxes and modifying resins for formulations; giving fiber- or film-tearing bonds between paper substrates and nonporous pkg. materials (e.g., aluminum foil, PP and Mylar polyester films, K cellophane). [DuPont]

Elvax® 550, 560. EVA copolymer resin; high molecular weight and high melt visc. grades used in paper roll wrap and carpet seaming tape. [DuPont]

Elvax® 4260. EVA/acid terpolymer; high molecular weight resin used in hot melt systems requiring improved adhesion to polar, nonporous substrates; in coatings, provides superior hot tack, improved grease resistance, and optimum barrier properties. [DuPont]

Elvax®4320. EVA/acid terpolymer; intermediate molecular weight resin higher in visc. than Elvax 4310, intermediate in performance between Elvax 4310 and 4355; combinable with Elvax 4355 or 4260 to optimize performance at a desired visc. level. [DuPont]

Elvax® D. Proprietary dispersions of ionomers and vinyl resins. [DuPont UK] *

Elverite. A proprietary trade name for charcoal iron used for crushing mills. §

Elveron®. Polymer. [DuPont UK]

Elvon®. Polymer. [DuPont UK]

EM-1. Lard oil; additive improving wetting and lubricity in industrial applications. [Ferro/Keil]

EM-550. Fatty compound; Lubricity and detergency additive for compounding water-sol. machining fluids; excellent film strength. [Ferro/Keil]

EM-600. CAS 61791-00-2; PEG 600 monotallate; surfactant, emulsifier, wetting agent, detergent for industrial applications, sol. cutting oils and drawing compounds [Ferro/Keil]

EM-980. Mixed fatty acid diethanolamide containing excess diethanolamine; solubilizer, visc. builder, detergency and lubricity aid used in synthetic and semi-synthetic metalworking fluids, floor cleaners, paint strippers, buffing compounds; contributes to corrosion resistance; sec. emulsifier for sol. oils. [Ferro/Keil]

EMA. Ethylene-maleic anhydride copolymers resins; dispersing agents, film formers and chemical intermediates for use in capsule walls, liquid detergents and drilling muds, thickening agents in textile print pastes and cosmetics. [Monsanto Co]

Emac SP2205. Ethylene methyl acrylate copolymer; excellent heat stability; for coatings and laminations, compounding (impact modification, compatibilizer), as tie layer for LDPE, HDPE, PET, PP, PVDC, EVA, PVC, PC in films. [Chevron]

Emaillit. Bitumin on a solvent base; primer. [Vedag GmbH]*

---

‡ Trade name and manufacturer not verified

§ Trade name without identified manufacturer

Emalex 103. CAS 9004-95-9; Ceteth-3; emulsifier, cleaner, dispersant for cosmetics, esp. creams and lotions; esp. suited for creamy hair conditioners [Nihon Emulsion]

Emalex 200di-IS. PEG-4 diisostearate; oil-phase cosmetic ingred. [Nihon Emulsion]

Emalex 200 di-L. CAS 9005-02-1; PEG-4 dilaurate; oil-phase cosmetic ingred., emulsifier for creams, milky lotions, hair conditioners; cleaner, superfattening agent, thickener, reforming agent. [Nihon Emulsion]

Emalex 200 di-O. CAS 9005-07-6; PEG-4 dioleate; oil-phase ingred., emulsifier, dispersant with good spreadability for cosmetics, creams, milky lotions, foundations. [Nihon Emulsion]

Emalex 200 di-S. CAS 9005-08-7; PEG-4 distearate; pearling agent, hydrophobic component, reforming agent, emulsifier, thickener for cosmetics. [Nihon Emulsion]

Emalex 218. CAS 9004-96-0; PEG-3 oleate; surfactant for cosmetics; produces stable emulsions. [Nihon Emulsion]

Emalex 300 di-IS. PEG-6 diisostearate; emulsifier for cosmetic emulsions, dispersant, reforming agent. [Nihon Emulsion]

Emalex 400A. CAS 9004-99-3; PEG-3 stearate; oil-phase cosmetic ingred., pearlescent, emulsifier, dispersant, emulsion stabilizer, thickener. [Nihon Emulsion]

Emalex 400 di-IS. PEG-8 diisostearate; emulsifier for cosmetic emulsions, dispersant, reforming agent. [Nihon Emulsion]

Emalex 508. CAS 9004-98-2; Oleth-8; emulsifier, dispersant, solubilizer for cosmetics; suitable for hair tonics and hair care products as solubilizer for perfumes. [Nihon Emulsion]

Emalex 600 di-IS. PEG-12 diisostearate; emulsifier for cosmetic emulsions, dispersant, reforming agent. [Nihon Emulsion]

Emalex 640. CAS 9005-00-9; Steareth-40; emulsifier, dispersant, thickener for cosmetics, creams, milky lotions. [Nihon Emulsion]

Emalex 709. Laureth-9; emulsifier, penetrant, wetting agent, cleaner, dispersant for cosmetics, creams, milky lotions; paint and itch relieving effect for ointments. [Nihon Emulsion]

Emalex 805. CAS 9004-99-3; PEG-5 stearate; oil-phase cosmetic ingred., pearlescent, emulsifier, dispersant, emulsion stabilizer; thickener for cleansing foam. [Nihon Emulsion]

Emalex 1605. CAS 9004-95-9; Ceteth-5; emulsifier, dispersant, solubilizer, cleaner, wetting agent for cosmetics, creams, milky lotions, skin lotions. [Nihon Emulsion]

Emalex 1805. CAS 52292-17-8; Isosteareth-5; emulsifier, dispersant, solubilizer, cleaner, wetting agent for cosmetics, creams, milky lotions, skin lotions. [Nihon Emulsion]

Emalex 2405. PEG-5 decyltetradecyl ether; emulsifier, dispersant, solubilizer, cleaner for cosmetics, creams, milky lotions, skin lotions. [Nihon Emulsion]

Emalex 2505. PEG-5 decylpentadecyl ether; emulsifier, solubilizer, cleaner, thickener for cosmetics, creams, milky lotions, skin lotions. [Nihon Emulsion]

Emalex 6300 DI-ST. CAS 9005-08-7; PEG-150 distearate; thickener, stabilizer for shampoos, hair conditioners, cleansing foams. [Nihon Emulsion]

Emalex 6300 M-ST. CAS 9004-99-3; PEG-150 stearate; oil-phase cosmetic ingred., thickener for shampoos and hair conditioners. [Nihon Emulsion]

Emalex BHA-30. Beheneth-30; emulsifier, dispersant, solubilizer for cosmetics; produces moist, spreading emulsion. [Nihon Emulsion]

Emalex C-20. CAS 61791-12-6; PEG-20 castor oil; emulsifier, solubilizer, dispersant in cosmetics, medical pharmaceuticals. [Nihon Emulsion]

Emalex CC-10. Cetyl caprate; lipophilic base for cosmetics [Nihon Emulsion]

Emalex CC-16. CAS 540-10-3; EINECS 208-736-6; Cetyl palmitate; lipophilic base for cosmetics [Nihon Emulsion]

Emalex CC-18. CAS 2778-96-3; EINECS 220-476-5; Stearyl stearate; lipophilic base for cosmetics [Nihon Emulsion]

Emalex CC-168. CAS 59130-69-7; EINECS 261-619-1; Cetyl octanoate; lipophilic base for cosmetics [Nihon Emulsion]

Emalex CS-5. Choleth-5; emulsifier, solubilizer, dispersant, thickener for cosmetics, creams, milky lotions, skin lotions; emollient for hair care products; gloss aid for creams and milky lotions. [Nihon Emulsion]

Emalex CWS-3. PEG-3 cetyl ether stearate; SE emulsifying cosmetic ingred.; hydrophobic component and reforming agent for cosmetics and industrial areas. [Nihon Emulsion]

Emalex DEG-di-IS. PEG-2 diisostearate; oil-phase ingred. for cosmetic emulsions. [Nihon Emulsion]

Emalex DEG-di-L. CAS 9005-02-1; PEG-2 dilaurate; oil-phase cosmetic ingred., emulsifier for creams, milky lotions, hair conditioners; cleaner, superfattening agent, thickener, reforming agent. [Nihon Emulsion]

Emalex DEG-di-O. CAS 9005-07-6; PEG-2 dioleate; oil-phase ingred., emulsifier, dispersant with good spreadability for cosmetics, creams, milky lotions, foundations. [Nihon Emulsion]

Emalex DEG-m-S. CAS 9004-99-3; PEG-2 stearate; oil-phase cosmetic ingred., pearlescent, emulsifier, dispersant, emulsion stabilizer, thickener. [Nihon Emulsion]

Emalex DISG-2. Diglyceryl diisostearate; emulsifier for cosmetics and foods. [Nihon Emulsion]

Emalex DISG-3. Polyglyceryl-3 diisostearate; emulsifier for cosmetics and foods. [Nihon Emulsion]

Emalex DSG-2. Polyglyceryl-2 distearate; emulsifier for cosmetics and foods. [Nihon Emulsion]

Emalex EG-2854-IS. PEG-4 sorbitol triisostearate; oil-phase base for emulsions; emulsifier for w/o emulsions. [Nihon Emulsion]

Emalex EG-2854-O. PEG-4 sorbitol tetraoleate; oil-phase base for emulsions; emulsifier for w/o emulsions. [Nihon Emulsion]

Emalex EG-2854-S. PEG-4 sorbitol tristearate; oil-phase base for emulsions. [Nihon Emulsion]

Emalex EG-di-L. CAS 624-04-4; Ethylene glycol dilaurate; oil-phase cosmetic ingred., emulsifier for creams, milky lotions, hair conditioners; cleaner, superfattening agent, thickener, reforming agent. [Nihon Emulsion]

Emalex EG-di-O. CAS 928-24-5; Ethylene glycol dioleate; oil-phase ingred., emulsifier, dispersant with good spreadability for cosmetics, creams, milky lotions, foundations. [Nihon Emulsion]

Emalex EG-di-S. CAS 627-83-8; EINECS 211-014-3; Ethylene glycol distearate; pearling agent, hydrophobic

---

* Trade name not verified as available   † Trade name verified as obsolete

component, reforming agent, emulsifier, thickener for cosmetics. [Nihon Emulsion]

Emalex EGS-A. Ethylene glycol monostearate; oil-phase cosmetic ingred., pearlescent, emulsifier, dispersant, emulsion stabilizer, thickener; clouding agent for shampoos/hair conditioners. [Nihon Emulsion]

Emalex ET-2020. CAS 9005-64-5; Polysorbate 20; emulsifier, solubilizer for cosmetics, medical pharmaceuticals. [Nihon Emulsion]

Emalex ET-8020. CAS 9005-65-6; Polysorbate 80; emulsifier, solubilizer for cosmetics, medical pharmaceuticals. [Nihon Emulsion]

Emalex ET-8040. PEG-40 sorbitan oleate; emulsifier, solubilizer for cosmetics and medical pharmaceuticals. [Nihon Emulsion]

Emalex GM-5. CAS 51158-08-8; PEG-5 glyceryl stearate; emulsifier, solubilizer, thickener for cosmetics, creams, milky lotions, hair conditioners, facial cleansers. [Nihon Emulsion]

Emalex GMS-55FD. Glyceryl monostearate, SE; surfactant for food, cosmetic and medical pharmaceutical applications; stabilizer, superfatting agent, reforming agent for cleansing foams. [Nihon Emulsion]

Emalex GMS-A. Glyceryl monostearate; surfactant for food, cosmetic and medical pharmaceutical applications. [Nihon Emulsion]

Emalex GMS-ASE. Glyceryl monostearate, SE; surfactant for food, cosmetic and medical pharmaceutical applications. [Nihon Emulsion]

Emalex GWIS-115. PEG-15 glyceryl monoisostearate; emulsion stabilizer, dispersant, emulsifier, solubilizer for cosmetics. [Nihon Emulsion]

Emalex GWIS-303. PEG-3 glyceryl triisostearate; oil-phase cosmetic ingred. for creams and milky lotions. [Nihon Emulsion]

Emalex GWO-303. PEG-3 glyceryl trioleate; oil-phase cosmetics ingred. [Nihon Emulsion]

Emalex GWS-204. PEG-4 glyceryl distearate; oil-phase cosmetics ingred. [Nihon Emulsion]

Emalex GWS-303. PEG-3 glyceryl tristearate; oil-phase cosmetics ingred.; for lipsticks, ointments. [Nihon Emulsion]

Emalex HC-5. CAS 61788-85-0; PEG-5 hydrogenated castor oil; oil-phase ingred., w/o emulsifier, dispersant for cosmetics and medical products [Nihon Emulsion]

Emalex J.J. O-V. CAS 61789-91-1; Jojoba oil; cosmetics ingred. [Nihon Emulsion]

Emalex K.T.G. CAS 65381-09-1; Caprylic/capric acid triglyceride; oil-phase cosmetic ingred.; emollient. [Nihon Emulsion]

Emalex LWIS-2. PEG-2 lauryl ether isostearate; SE emulsifying cosmetic ingred.; oily base for creams, milky lotions, hair conditioners. [Nihon Emulsion]

Emalex LWS-3. PEG-3 lauryl ether stearate; SE emulsifying cosmetic ingred.; medical ointment base producing good emollient and spreading props. [Nihon Emulsion]

Emalex MSG-2. CAS 37349-34-1; Diglyceryl monostearate; emulsifier for cosmetics and foods. [Nihon Emulsion]

Emalex MTS-30E. CAS 2116-84-9; Phenyl trimethicone; oil-phase cosmetic ingred. for alcoholic milky-wh. lotions. [Nihon Emulsion]

Emalex N-83. Cocamide DEA; thickener or bubbling agent in hair shampoos, body shampoos, facial cleansing foams. [Nihon Emulsion]

Emalex NN-7. CAS 120-40-1; EINECS 204-393-1; Lauramide DEA; cleaner in hard water. [Nihon Emulsion]

Emalex NN-15. CAS 7545-23-5; EINECS 231-426-7; Myristamide DEA; thickener or bubbling agent in hair shampoos, body shampoos, facial cleansing foams. [Nihon Emulsion]

Emalex NP-2. CAS 9016-45-9; Nonoxynol-2; wetting agent, emulsifier, cleaner, dispersant, foaming agent, solubilizer for cosmetics. [Nihon Emulsion]

Emalex OD-5. Octyldodeceth-5; emulsifier, dispersant, solubilizer, cleaner, wetting agent for cosmetics, creams, milky lotions. [Nihon Emulsion]

Emalex OE-6. CAS 9004-96-0; PEG-6 oleate; cosmetics surfactant. [Nihon Emulsion]

Emalex OP-25. Octoxynol-25; emulsifier, wetting agent, cleaner, dispersant, foaming agent, solubilizer for cosmetics. [Nihon Emulsion]

Emalex O.T.G. CAS 538-23-8; Glyceryl trioctanoate; oil-phase cosmetic ingred.; emollient. [Nihon Emulsion]

Emalex PEIS-3. CAS 56002-14-3; PEG-3 isostearate; emulsifier, dispersant, solubilizer for cosmetics; emulsion stabilizer, skin fitness reformer for milky lotions, hair conditioners. [Nihon Emulsion]

Emalex PG-di-IS. Propylene glycol diisostearate; oil-phase cosmetic ingred. [Nihon Emulsion]

Emalex PG-di-L. Propylene glycol dilaurate; cosmetic ingred. [Nihon Emulsion]

Emalex PG-di-O. CAS 85049-34-9; EINECS 285-203-4; Propylene glycol dioleate; cosmetic ingred. [Nihon Emulsion]

Emalex PG-di-S. CAS 6182-11-2; Propylene glycol distearate; oil-phase cosmetic ingred. [Nihon Emulsion]

Emalex PGML. Propylene glycol monolaurate; surfactant for food and cosmetics; emulsion stabilizer for creams, milky lotions, hair conditioners. [Nihon Emulsion]

Emalex PGO. CAS 1330-80-9; EINECS 215-549-3; Propylene glycol monooleate; surfactant for food and cosmetics. [Nihon Emulsion]

Emalex PGS. CAS 1323-39-3; EINECS 215-354-3; Propylene glycol monostearate; surfactant for food and cosmetics. [Nihon Emulsion]

Emalex RWIS-105. PEG-5 hydrogenated castor oil isostearate; oil-phase component, emulsifier for w/o emulsions, dispersant for hydrophobic components. [Nihon Emulsion]

Emalex RWIS-305. PEG-5 hydrogenated castor oil triisostearate; solubilizer for oil-phase ingreds. in cosmetics; emulsifier for w/o emulsions; reforming agent, dispersant for hydrophobic components. [Nihon Emulsion]

Emalex RWL-120. PEG-20 hydrogenated castor oil laurate; emulsifier, dispersant, reforming agent for cosmetic creams, milky lotions, hair conditioners. [Nihon Emulsion]

Emalex SG-37. Caprylic/capric/stearic triglyceride; oil-phase cosmetic ingred.; excellent waxing material with high emulsifying ability. [Nihon Emulsion]

Emalex SPE-100S. CAS 1338-41-6; EINECS 215-664-9; Sorbitan stearate; oil-phase cosmetic ingred., surfactant; for creams, milky lotions, hair conditioners. [Nihon Emulsion]

Emalex SPE-150S. Sorbitan sesquistearate; oil-phase cosmetic ingred., surfactant; for creams, milky lotions, hair conditioners. [Nihon Emulsion]

‡ Trade name and manufacturer not verified   § Trade name without identified manufacturer

Emalex SPIS-100. CAS 71902-01-7; Sorbitan isostearate; oil-phase cosmetic ingred., surfactant; for creams, milky lotions, hair conditioners. [Nihon Emulsion]

Emalex SPIS-150. Sorbitan sesquiisostearate; oil-phase cosmetic ingred., surfactant; for creams, milky lotions, hair conditioners. [Nihon Emulsion]

Emalex SPO-100. CAS 1338-43-8; EINECS 215-665-4; Sorbitan oleate; oil-phase cosmetic ingred., surfactant; for creams, milky lotions, hair conditioners. [Nihon Emulsion]

Emalex SPO-150. CAS 8007-43-0; EINECS 232-360-1; Sorbitan sesquioleate; oil-phase cosmetic ingred., surfactant; for creams, milky lotions, hair conditioners. [Nihon Emulsion]

Emalex SWS-4. PEG-4 stearyl ether stearate; SE emulsifying cosmetic ingred.; hydrophobic component and reforming agent for cosmetics and industrial areas. [Nihon Emulsion]

Emalex TEG-di-IS. PEG-3 diisostearate; cosmetic ingred. [Nihon Emulsion]

Emalex TEG-di-L. CAS 9005-02-1; PEG-3 dilaurate; oil-phase cosmetic ingred., emulsifier for creams, milky lotions, hair conditioners; cleaner, superfattening agent, thickener, reforming agent. [Nihon Emulsion]

Emalex TPIS-303. PEG-3 trimethylolpropane triisostearate; oil-phase cosmetics ingred., emulsifier, solubilizer; offers smooth, clear appearance producing transparent cosmetic products [Nihon Emulsion]

Emalex TPM-303. PEG-3 trimethylolpropane trimyristate; emulsifier, solubilizer, oil-phase ingred. for cosmetics; reforming agent for creams, milky lotions, hair conditioners. [Nihon Emulsion]

Emalex TPS-203. PEG-3 trimethylolpropane distearate; oil-phase cosmetics ingred.; for hair creams, lipsticks, ointments, cold creams. [Nihon Emulsion]

Emalex TPS-303. PEG-3 trimethylolpropane tristearate; oil-phase cosmetics ingred.; for hair creams, ointments, cold creams; superfatting agent for creamy hair conditoners, cleansing foams. [Nihon Emulsion]

Emaline. Conventional and high build bituminous coatings. [Sigma Coatings] *

Ema Resins. EMA resins. [Monsanto plc]

Emaweld®. Electromagnetic welding systems for assembling thermoplastic components. [Ashland]

Embacel. Kieselguhr for gas chromatography. [May & Baker Ltd] *

Embacide. Sheep dip. [May & Baker Ltd] *

Embacoid. A cement for cine films. [May & Baker Ltd] *

Embadot. Photographic developer. [May & Baker Ltd] *

Embafix. Photographic fixer. [May & Baker Ltd] *

Embafume. CAS 74-83-9; Methyl bromide fumigant. [May & Baker Ltd] *

Embalith. Photographic developer. [May & Baker Ltd] *

Embamix. CAS 7681-11-0; Potassium iodide mixtures. [May & Baker Ltd]

Embanox. A food grade antioxidant. [May & Baker Ltd]

Embanox®. Antioxidant for food applications. [Rhone-Poulenc Food Ingreds.]

Embaphase. Stationary phases for gas chromatography. [May & Baker Ltd] *

Embark. Soluble concentrate containing 240 g/l metluidide; grass growth suppressant. [Gordon International Corp]

Embaspeed. Rapid working photographic developer. [May & Baker Ltd] *

Embatex. A cellulose acetate coated fabric. [May & Baker Ltd] *

Embathion. An insecticide. [May & Baker Ltd] *

Embatype. Photographic developer. [May & Baker Ltd] *

Embazin. A proprietary preparation of sulfaquinoxaline sodium; a veterinary coccidiostat. [May & Baker Ltd] *

Embedyne. Chlorodyne. [May & Baker Ltd] *

embelic acid. An acid obtained from the fruit of *Embelia ribes*; used in medicine for worms.

Embequin. Diiodohydroxyquinoline. [May & Baker Ltd] *

Embesafe. Plastic coated glass bottles. [May & Baker Ltd]*

Embesol. Photographic developer. [May & Baker Ltd] *

Emblet® M. Biaxially oriented polyester film, one side chemically treated. [SNIA (UK) Ltd]

Embond 55. Polymer reinforced resin/alcohol; needle-punched carpet flooring adhesive. [Marley Floors Ltd]

Embond 66. Tackified synthetic rubber latex; carpet flooring adhesive. [Marley Floors Ltd]

Embond 125. Tackified acrylic emulsion; wood and cork flooring adhesive. [Marley Floors Ltd]

Embond 168. Tackified acrylic emulsion; flooring adhesive. [Marley Floors Ltd]

Embond 169. Tackified acrylic/EVA copolymer emulsion; flooring adhesive. [Marley Floors Ltd]

Embond 212. Tackified synthetic rubber latex; flooring adhesive. [Marley Adhesives Ltd]

Embond 401. Two part solvent-free epoxide base; flooring adhesive. [Marley Floors Ltd]

Embond 560. Neoprene solution in flammable organic solvents; contact adhesive. [Marley Floors Ltd]

Embond Surface Tackifier; Modified acrylic emulsion; flooring adhesive. [Marley Floors Ltd]

embrithite. A mineral, $Pb_3Sb_2S_6$.

Embrol. Photographic developer. [May & Baker Ltd] *

Embutox. A selective weedkiller. [May & Baker Ltd] *

Emcast 1510, 1511. CAS 25928-94-3; One-part epoxy; uv-curable, semirigid epoxy for electronic encapsulation, dipping, and adhering; cures rapidly to tough, hard, resistant material with good impact. [Electronic Materials]

Emcast 1550, 1551. CAS 25928-94-3; One-part epoxy; uv-curable, flexible epoxy for electronic encapsulation, dipping, and adhering; cures rapidly to tough, resistant materials with good impact. [Electronic Materials]

Emcocel® 90M. CAS 9004-34-6; Microcryst. cellulose NF/BP; tablet binder, disintegrant for pharmaceuticals; features low frability, inherent lubricity, enhanced compression of other excipients. [Mendell]

Emcol. Surfactant for cosmetics, toiletries, pharmaceutical, processing, agricultural and other industries. [Baxenden Chemicals Ltd]

Emcol® 4. CAS 122-19-0; EINECS 204-527-9; Stearalkonium chloride; antistat, substantive conditioner, emollient. [Witco/Organics]

Emcol® 1655. CAS 68425-43-4; Cocamidopropyl dimethylamine propionate; cosmetics and toiletry surfactant used as antistat, conditioner, emollient, foaming and substantive agent. [Witco/Organics]

Emcol® 3780. CAS 55819-53-9; EINECS 259-837-7; Stearamidopropyl dimethylamine lactate; cosmetics and toiletry surfactant used as antistat, conditioner, emollient, foaming and substantive agent. [Witco/Organics]

Emcol® 4100M, 4150 and 4161-L. Sodium alkanolamide sulfosuccinate in liquid or soft paste form; for hair and

---

* Trade name not verified as available   † Trade name verified as obsolete

carpet shampoos; 4161-L has low irritancy and is suitable for baby products; 4150 is used in emulsion polymerization and ore flotation. [Witco Chemical Ltd] *

Emcol® 4161L. CAS 43154-85-4; EINECS 256-120-0; Disodium oleamido MIPA sulfosuccinate; dispersant, wetting, foam booster/stabilizer, detergent, conditioner, and emulsifying agent for bubble bath, shampoos, cleansers for cosmetics and toiletries. [Witco/Organics; Witco SA]

Emcol® 4300. CAS 39354-47-5; Disodium C12-15 pareth sulfosuccinate; dispersant, wetting, foaming, detergent, emulsifying agent for bubble bath, shampoo, cosmetics and toiletries; emulsifier for acrylic, vinyl acetate, vinyl acrylic polymerization. [Witco/Organics; Witco Chemical Ltd]

Emcol® 4350. Triethanolamine fatty alcohol ethoxylate sulfosuccinate in liquid form; used for low irritancy hair shampoos and foam baths. [Witco Chemical Ltd] *

Emcol®4500. CAS 577-11-7; EINECS 209-406-4; Sodium dioctyl sulfosuccinate; dispersant, detergent, wetting, foaming, emulsifying agent; for cosmetics, toiletries, textiles, industrial processing slurries. [Witco/Organics; Witco SA; Witco Chemical Ltd]

Emcol® 4600. Sodium bistridecyl sulfosuccinate in liquid form; dispersing agent; dry cleaning. [Witco Chemical Ltd] *

Emcol®4776. Sodium dihexyl sulfosuccinate in liquid form; used for dry cleaning. [Witco Chemical Ltd] *

Emcol®4910. Sodium lauryl/propoxy sulfosuccinate; emulsifier for acrylic, vinyl acrylic polymerization. [Witco/Organics]

Emcol® 5430. Cocamidopropyl betaine; detergent, foam booster/stabilizer, visc. modifier, wetting agent. [Witco/Organics]

Emcol® CC-42. CAS 9076-43-1; PPG-40 diethylmonium chloride; pigment dispersant, particle suspension aid, emulsifier, solv., conditioner, antistat, lubricant, corrosion inhibitor for toiletries, cosmetics, germicides, synthetic fibers and plastics, textiles, industrial processes; ore flotation additive. [Witco/Organics; Witco Chemical Ltd]

Emcol® CC-55. Polypropoxy quaternary ammonium acetate; general nonirritant quaternary cationic surfactant used especially where anionic compatability is required. [Witco Chemical Ltd] *

Emcol® DOSS. CAS 577-11-7; EINECS 209-406-4; Sodium dioctyl sulfosuccinate; emulsifier for acrylonitrile polymerization; improves surface wetting. [Witco/Organics]

Emcol® E-607L. CAS 6272-74-8; EINECS 228-464-1; Lapyrium chloride; emollient, emulsifier, conditioner, foamer, cleanser, substantive agent, deodorant for cosmetics, toiletries, industrial applications; hair conditioner. [Witco/Organics; Witco SA]

Emcol®E-607S. Steapyrium chloride; emollient, emulsifier, conditioner, foamer, cleanser, substantive agent, deodorant for cosmetics, toiletries, industrial applications; hair conditioner. [Witco/Organics; Witco SA]

Emcol® ISML. Isostearamidopropyl morpholine lactate; cosmetics and toiletries surfactant used as antistat, conditioner, emollient, foaming and substantive agent; nonirritating base for cream rinses and conditioning shampoos. [Witco/Organics]

Emcol® K8300. CAS 43154-85-4; EINECS 256-120-0; Disodium oleamido-MIPA sulfosuccinate; dispersant, particle suspension aid, wetting agent, foam booster/stabilizer, detergent; emulsifier in emulsion polymerization of acrylic, styrene acrylic, vinyl acrylic. [Witco/Organics; Witco SA; Witco Chemical Ltd]

Emcol®LO. Lauramine oxide; foam booster/stabilizer, visc. modifier. [Witco/Organics]

Emcompress®. CAS 7789-77-7; Dibasic calcium phosphate dihydrate USP/BP; excipient for production of pharmaceutical tablets by direct compression process. [Mendell]

EmCon E-5. CAS 8001-17-0; Egg oil; emollient, moisturizer for hair and skin care products; w/o emulsifier; superfatting agent, humectant, mold release agent; occlusive agent. [Fanning]

EmCon Limnanthes Alba. Meadowfoam seed oil; skin/hair conditioner, occlusive agent for shampoos, makeup, face, body and hand creams and lotions, baby products, lipsticks, cleansing products [Fanning]

EmCon TEA TREE. CAS 68647-73-4; Tea tree oil; fragrance component, antimicrobial. [Fanning]

EmCon W. CAS 8006-95-9; Wheat germ oil; skin/hair conditioner, occlusive solv.; for hair conditioners, shampoos, lipsticks, cleansers, moisturizing creams and lotions, skin care products [Fanning]

Emcor. CAS 1337-76-4; Ultra-short fiber reinforcement based upon attapulgite (magnesium aluminum silicate) used in nonasbestos friction compounds; used for nonasbestos disc pads, drum linings, truck and railroad blocks, friction papers, gasketing. [Engelhard]

Emcosoy®. CAS 68513-95-1; Soy polysaccharides; tablet disintegrant for direct compression preparation [Mendell]

Emcyt. CAS 52205-73-9; Estramustine phosphate sodium; antineoplastic. [Hoffmann-LaRoche Inc] *

Emdex®. CAS 50-99-7; Dextrose (95%), isomaltose (2%), gentiobiose (2%), maltose (1%), maltotriose (< 0.1%), panose (< 0.5%); vehicle for direct compression of pharmaceutical tablets. [Mendell]

Emdite. A proprietary trade name for a 50% w/w aqueous solution of ethyl-ammonium ethyl-dithiocarbonate, an alternative to hydrogen sulfide in qualitative inorganic analysis. [British Drug Houses]

Emdithene. Basically a range of PU resins specifically formulated for the protection of light electrical and electronic components, but offering better retention of flexibility than normal PU types under thermal cycling conditions at higher temperatures. A [Robnorganic Systems Ltd] *

Emerald. Extender, wetting agent and antitranspirant for use in pesticides. [Intracrop Ltd]

emerald bronze. An alloy of 50% copper, 49.7% zinc, and 0.3% aluminum.

Emercide®1199. Liquid preservative system for cosmetics. [Henkel/Cospha; Henkel Canada]

Emeressence®1150. CAS 105-95-3; Ethylene brassylate; musk chemical for fragrance or odor masking applications. [Henkel/Emery] *

Emeressence® 1151. Ethylene dodecanedioate; musk chemical for fragrance or odor masking applications. [Henkel/Emery] *

Emeressence®1160 Rose Ether. CAS 122-99-6; EINECS 204-589-7; Phenoxyethanol; cosmetic preservative; effective against gram negative microorganisms. [Henkel/

---

‡ Trade name and manufacturer not verified   § Trade name without identified manufacturer

Emery] *

Emeressence® 1174 Fir Balsam. Aromatic with sweet balsamic odor reminiscent of Canadian fir. [Henkel/Emery] *

Emerest®2301. CAS 112-62-9; EINECS 203-992-5; Methyl oleate; base for industrial lubricants; mold release agent, defoamer, flotation agent, plasticizer for cellulosic plastics, needle lubricants; when sulfated is useful as wetting, rewetting, and dye leveling agent in textile and leather industries. [Henkel/Emery]

Emerest®2302. Propyl oleate; base for industrial lubricants; mold release agent, defoamer, flotation agent, plasticizer for cellulosic plastics, needle lubricants; when sulfated is useful as wetting, rewetting, and dye leveling agent in textile and leather industries. [Henkel/Emery; Henkel/Textile]

Emerest® 2308. CAS 31556-45-3; EINECS 250-696-7; Tridecyl stearate; lubricant used in sewing thread mfg. and fiber finish applications where high heat stability is desired. [Henkel/Emery; Henkel/Textile]

Emerest®2310. Isopropyl isostearate; low visc. emollient, lubricant for bath oils, creams, lotions, shampoos; binder for pressed powd. [Henkel/Cospha; Henkel Canada]

Emerest®2314. CAS 110-27-0; EINECS 203-751-4; Isopropyl myristate; cosmetic emollient; sewing thread lubricant. [Henkel/Cospha; Henkel Canada]

Emerest®2316. CAS 142-91-6; EINECS 205-571-1; Isopropyl palmitate; lubricant used for synthetic fibers in applications where low friction is essential; emollient in cosmetic formulations; high purity. [Henkel/Cospha; Henkel Canada]

Emerest®2324. CAS 646-13-9; Isobutyl stearate; lubricant for textile applications. [Henkel/Emery; Henkel/Textile]

Emerest®2325. CAS 123-95-5; EINECS 204-666-5; Butyl stearate; emollient in creams and lotions; dye solubilizer in lipsticks; lubricant. [Henkel/Cospha; Henkel Canada]

Emerest®2328. CAS 142-77-8; EINECS 205-559-6; Butyl oleate; lubricant. [Henkel/Emery]

Emerest®2350. Glycol stearate; emulsifier, opacifying and pearlescing agent, thickener, stabilizer used in liq. cosmetic and detergent compounds [Henkel/Cospha; Henkel Canada]

Emerest®2355. CAS 627-83-8; EINECS 211-014-3; Glycol distearate; emulsifier, opacifier, pearlescent, thickener, stabilizer used in liq. detergent and cosmetic products [Henkel/Cospha; Henkel Canada]

Emerest® 2380. CAS 1323-39-3; EINECS 215-354-3; Propylene glycol stearate; auxiliary emulsifier, opacifier, pearlescent; for lotions, makeup, textile processing. [Henkel/Cospha; Henkel/Textile; Henkel Canada]

Emerest® 2384. CAS 68171-38-0; Propylene glycol isostearate; solubilizer for fragrances in low alcohol or oil preparations; emollient for personal care products [Henkel/Cospha; Henkel Canada]

Emerest® 2388. CAS 41395-83-9; EINECS 255-350-9; Propylene glycol dipelargonate; lubricant, low visc. emollient for preshaves, bath oils, creams and lotions, textile processing. [Henkel/Cospha; Henkel/Textile; Henkel Canada]

Emerest® 2400. Glyceryl stearate; emulsifier for hand creams, cosmetics, textiles, industrial lubricants, polishes, agriculture; lubricant softener for textiles; opacifier and pearling agent. [Henkel/Cospha; Henkel Canada]

Emerest® 2410. Glyceryl isostearate; emollient, lubricant,

pearling agent, and w/o emulsifier for creams and lotions, textile applications; excellent oxidation and color stability. [Henkel/Cospha; Henkel/Textile; Henkel Canada]

Emerest® 2419. CAS 25637-84-7; EINECS 247-144-2; Glyceryl dioleate; emulsifier, lubricant, rust preventive additive, mold release agent, solv. for dyes and pigments; used in leathers, lubricants and softeners. [Henkel/Cospha; Henkel Canada]

Emerest® 2421 (*See* Witconol 2421). CAS 37220-82-9; Glyceryl oleate; emulsifier for cosmetics and industrial applications, in mold release agents, anti-icing fuel additive, rust preventative; in textiles as a lubricant component in synthetic fiber spin finishes; vehicle for agricultural insecticides. [Henkel/Emery; Henkel/Cospha; Henkel Canada]

Emerest® 2423. CAS 122-32-7; EINECS 204-534-7; Triolein; lubricant, w/o emulsifier for metals, leather, textiles, cosmetics, pharmaceuticals; called synthetic olive oil; sulfated form used as softener in leather and textile industries. [Henkel/Cospha; Henkel/Textile; Henkel Canada]

Emerest® 2452. Polyglyceryl-3 diisostearate; emulsifier, solubilizer, dye and pigment wetter; emollient; thickener; solv.; for creams, lotions, lip products [Henkel/Cospha; Henkel Canada]

Emerest® 2485. CAS 14450-05-6; EINECS 238-430-8; Pentaerythritol tetrapelargonate; primary lubricant base or modifier in lubricant formulations used in metal working and synthetic fiber processing. [Henkel/Emery]

Emerest®2610. CAS 9004-99-3; PEG-20 stearate; emulsifier for glyceryl stearate in nonionic textile lubricants and softeners; thickener; antigellant in starch solutions [Henkel/Emery; Henkel/Textile]

Emerest®2617. CAS 9004-96-0; PEG-150 oleate; strongly hydrophilic emulsifier, stabilizer, lubricant; for agricultural formulations. [Henkel/Emery; Henkel/Cospha; Henkel Canada]

Emerest®2620. CAS 9004-81-3; PEG-4 laurate; emulsifier, coupling agent, defoamer in water base coatings, visc. control additive; visc. depressant in vinyl plastisols; agriculture, textiles. [Henkel/Emery; Henkel/Cospha; Henkel/Textile; Henkel Canada]

Emerest® 2622. CAS 9005-02-1; PEG-4 dilaurate; coemulsifer and lubricant in SE textile and industrial oils, agriculture, mold release agent, visc. control agent. [Henkel/Emery; Henkel/Cospha; Henkel Canada]

Emerest® 2624. CAS 9004-96-0; PEG-4 oleate; lubricant component in textile processing; softener/lubricant for leather during tanning; emulsifier for mineral oils, fatty oils, and solvs. for cutting oils, solvs. in metal cleaners and degreasers, w/o emulsifier for consumer pesticide aerosols. [Henkel/Emery; Henkel/Cospha; Henkel/Textile; Henkel Canada]

Emerest® 2625. CAS 56002-14-3; PEG-4 isostearate; emulsifier; component in fiber lubricants, processing aids, and conc. liq. fabric softeners. [Henkel/Cospha; Henkel Canada]

Emerest®2630. CAS 9004-81-3; PEG-6 laurate; hydrophilic emulsifier; lubricant component and scrooping agent for textile fibers and yarns; visc. control agent for plastisols; agricultural formulations. [Henkel/Cospha; Henkel Canada]

Emerest® 2634. PEG-6 pelargonate; emulsifier, wetting agent, emollient, textile softener, lubricant, defoamer,

---

stabilizer, visc. control agent, pigment wetting, mold release agent; agricultural formulations, textile processing. [Henkel/Emery; Henkel/Cospha; Henkel/Textile; Henkel Canada]

Emerest® 2636. CAS 9004-99-3; PEG-6 stearate; waxy emulsifier for oils and fats in industrial lubricants, agricultural formulations; softener and lubricant in textiles and leather. [Henkel/Emery]

Emerest® 2647. PEG-8 sesquioleate; emulsifier, wetting agent, emollient, textile softener, lubricant, defoamer, stabilizer, visc. control agent, pigment wetting, mold release agent, agricultural formulations. [Henkel/Emery; Henkel/Cospha; Henkel/Textile; Henkel Canada]

Emerest® 2660. CAS 9004-96-0; PEG-12 oleate; dye leveling agent in textiles; emulsifier in specialty lubricants, agricultural formulations; detergent; acid washing of printed circuit boards. [Henkel/Emery; Henkel/Cospha; Henkel/Textile; Henkel Canada]

Emerest® 2704. CAS 9005-02-1; PEG-4 dilaurate; emulsifier, lubricant, dispersant for bath oils; visc. control agent for creams and lotions; for cosmetic and industrial applications. [Henkel/Cospha; Henkel Canada]

Emerlube® 5919. Ethoxylated vegetable oil; lubricant for PP yarns, carpet backings. [Henkel/Emery; Henkel/Textile]

Emerox® 1110. CAS 123-99-9; Azelaic acid; detergent intermediate. [Henkel/Emery]

Emersist 7210. Acid dye leveling agent for nylon; optimum leveling and enhanced color yield. [Henkel/Textiles]

Emersoft 7700. Softener conc. for resin finishing of cotton and polyester/cotton blends; also as plasticizer for starch-finished goods. [Henkel/Textiles]

Emersol® 110. CAS 57-11-4; EINECS 200-313-9; Stearic acid; detergent intermediate; opacifier in cosmetics, soaps, emulsifiers, chemical specialties. [Henkel/Emery]

Emersol® 143. CAS 57-10-3; EINECS 200-312-9; Palmitic acid; detergent intermediate; opacifier in cosmetics, soaps, emulsifiers, chemical specialties. [Henkel/Emery]

Emersol® 210. CAS 112-80-1; EINECS 204-007-1; Oleic acid; detergent intermediate for personal care, emollient, household and industrial applications. [Henkel/Emery]

Emersol® 315. CAS 60-33-3; EINECS 200-470-9; Linoleic acid; detergent intermediate for personal care, emollient, household and industrial applications. [Henkel/Emery]

Emersol® 871. CAS 2724-58-5; EINECS 220-336-3; Isostearic acid; detergent intermediate for personal care, emollient, household and industrial applications. [Henkel/Emery]

Emersol® 6333 NF. CAS 112-80-1; EINECS 204-007-1; Low-linoleic content oleic acid USP/NF; food grade fatty acid; also as binder, defoamer and lubricant for pesticides; detergent intermediate in personal care, emollients, household/industrial detergents. [Henkel/Emery]

Emersol® 6349. CAS 57-11-4; EINECS 200-313-4; Stearic acid; food grade fatty acid; also as binder, defoamer and lubricant for pesticides; detergent intermediate in personal care, emollients, household/industrial detergents. [Henkel/Emery]

Emersol® 7021. CAS 112-80-1; EINECS 204-007-1; Oleic acid; food grade kosher fatty acid; detergent intermediate in personal care, emollients, household/industrial detergents. [Henkel/Emery]

Emerstat® 6660. Antistat for fiber lubricants. [Henkel/Emery]

Emerwax® 1251. Synthetic ester wax; for lipsticks, makeup, nail care products, etc. [Henkel/Cospha; Henkel Canada]

Emerwax® 1253. Beeswax substitute; visc. agent for cosmetic formulations incl. sticks, cold creams, makeup; beeswax substitute. [Henkel/Cospha; Henkel Canada]

Emerwax® 1266. Cetearyl alcohol and ceteareth-20; BP type emulsifying wax; o/w emulsifier for pharmaceuticals, creams, lotions, antiperspirants, hair care products, depilatories. [Henkel/Cospha; Henkel Canada]

Emery® 400. CAS 57-11-4; EINECS 200-313-4; Stearic acid; detergent intermediate. [Henkel/Emery]

Emery® 515. Tallow/coconut fatty acid blend; detergent intermediate. [Henkel/Emery]

Emery® 610. Soya fatty acid; detergent intermediate. [Henkel/Emery]

Emery® 621. Coconut fatty acid; detergent intermediate in personal care, emollients, household/industrial detergents. [Henkel/Emery]

Emery® 650. CAS 143-07-7; Lauric acid; detergent surfactant. [Henkel/Emery]

Emery® 654. CAS 544-63-8; EINECS 208-875-2; Myristic acid; detergent surfactant. [Henkel/Emery]

Emery® 657. CAS 124-07-2; EINECS 204-677-5; Caprylic acid; detergent surfactant. [Henkel/Emery]

Emery® 912. CAS 56-81-5; CP/USP glycerin; skin softener, solubilizer, visc. modifier, flavor enhancer, moisturizer, solv., humectant, thickener, and solubilizer in cosmetics, drug vehicles, food applications, glass, ceramics, agriculture, and adhesives. [Henkel/Emery]

Emery® 1202. CAS 112-05-0; EINECS 203-931-2; Pelargonic acid; detergent intermediate. [Henkel/Emery]

Emery® 1650. Anhyd. lanolin USP; emulsifier, emollient, conditioner, lubricant for cosmetics, sun care products, textiles. [Henkel/Cospha; Henkel/Textile; Henkel Canada]

Emery® 1730. Mineral oil/fraction of lanolin alcohols and sterols blend; liq. absorption base, penetrant, w/o emulsifier, emollient, stabilizer for creams, lotions, makeup, shampoos; provides hypoallergenic skin penetration. [Henkel/Cospha; Henkel Canada]

Emery® 2203. Methyl tallowate; detergent intermediate; solv. for pesticides and herbicides. [Henkel/Emery]

Emery® 2204. Hydrogenated methyl tallowate; detergent intermediate. [Henkel/Emery]

Emery® 2209. CAS 67762-39-4; Methyl caprylate-caprate; detergent intermediate; solv. for pesticides and herbicides. [Henkel/Emery]

Emery® 2214. CAS 124-10-7; EINECS 204-680-1; Methyl myristate; detergent intermediate; solv. for pesticides and herbicides. [Henkel/Emery]

Emery® 2216. CAS 112-39-0; EINECS 203-966-3; Methyl palmitate; detergent intermediate; solv. for pesticides and herbicides. [Henkel/Emery]

Emery® 2218. CAS 112-61-8; EINECS 203-990-4; Methyl stearate; detergent intermediate; solv. for pesticides and herbicides. [Henkel/Emery]

Emery® 2219. CAS 112-62-9; EINECS 203-992-5; Methyl oleate; detergent intermediate; solv. for pesticides and herbicides. [Henkel/Emery]

Emery® 2224. Methyl soyate; solv. for pesticides and herbicides. [Henkel/Emery]

Emery®2230. Methyl sunflowerate; solv. for pesticides and herbicides. [Henkel/Emery]

Emery®2231. Dist. methyl canolate; solv. for pesticides and herbicides. [Henkel/Emery]

Emery® 2232. Methyl canolate; solv. for pesticides and herbicides. [Henkel/Emery]

Emery®2253. CAS 61788-59-8; EINECS 262-988-1; Methyl coconate; detergent intermediate; solv. for pesticides and herbicides. [Henkel/Emery]

Emery® 2255. Methyl palm kernelate; detergent intermediate. [Henkel/Emery]

Emery® 2270. Methyl laurate; detergent intermediate; solv. for pesticides and herbicides. [Henkel/Emery]

Emery® 2301. CAS 112-62-9; EINECS 203-992-5; Methyl oleate; solv. for pesticides and herbicides. [Henkel/Emery]

Emery®2895 Foamaster Soap L. CAS 8052-48-0; Sodium tallowate; defoamer for dry agricultural formulations. [Henkel/Emery]

Emery®2900. Dimer ester; synthetic lubricant basestock for industrial applications. [Henkel/Emery]

Emery®2914. CAS 1732-10-1; Dimethyl azelate; synthetic lubricant. [Henkel/Emery]

Emery®2957. Diester; synthetic lubricant basestock. [Henkel/Emery]

Emery®3304. CAS 661-19-8; EINECS 211-546-6; Behenyl alcohol (5-15% C18, 5-20% C20, ≈70% C22); chemical intermediate for detergent mfg. [Henkel/Emery]

Emery® 3310. Oleyl/cetyl alcohol (25-33% C16, 60-70% C18); chemical intermediate for detergent mfg. [Henkel/Emery]

Emery® 3312. CAS 143-28-2; EINECS 205-597-3; Oleyl alcohol (2-8% C16, 87-93% C18); chemical intermediate for detergent mfg. [Henkel/Emery]

Emery® 3317. CAS 143-28-2; EINECS 205-597-3; Oleyl alcohol (2-10% C16, 87-95% C18); chemical intermediate for detergent mfg. [Henkel/Emery]

Emery® 3320. Tallow alcohol (25-35% C16, 60-67% C18); chemical intermediate for detergent mfg. [Henkel/Emery]

Emery® 3321. Hexyl alcohol; chemical intermediate for detergent mfg. [Henkel/Emery]

Emery® 3322. CAS 111-87-5; EINECS 203-917-6; Octyl alcohol; chemical intermediate for detergent mfg. [Henkel/Emery]

Emery® 3323. Decyl alcohol; chemical intermediate for detergent mfg. [Henkel/Emery]

Emery® 3324. CAS 111-87-5; EINECS 203-917-6; Octyl alcohol; chemical intermediate for detergent mfg. [Henkel/Emery]

Emery® 3326. CAS 112-53-8; Lauryl alcohol (44-50% C12, 14-20% C14, 8-10% C16, 8-12% C18); chemical intermediate for detergent mfg. [Henkel/Emery]

Emery® 3332. CAS 112-53-8; Lauryl alcohol; chemical intermediate for detergent mfg. [Henkel/Emery]

Emery®3334. CAS 112-72-1; EINECS 204-000-3; Myristyl alcohol; chemical intermediate for detergent mfg. [Henkel/Emery]

Emery®3336. CAS 36653-82-4; EINECS 253-149-0; Cetyl alcohol; chemical intermediate for detergent mfg. [Henkel/Emery]

Emery®3343. CAS 112-92-5; EINECS 204-017-6; Stearyl alcohol; chemical intermediate for detergent mfg. [Henkel/Emery]

Emery® 3357. CAS 112-53-8; Lauryl alcohol (40-48% C8, 51-59% C10); chemical intermediate for detergent mfg. [Henkel/Emery]

Emery® 5353 Lomar PW. Condensed naphthalene sulfonate; dispersant for agricultural formulations. [Henkel/Emery]

Emery®5366 (Lomar PWA Liq.). Ammonium naphthalene sulfonate; dispersant for agricultural formulations. [Henkel/Emery]

Emery® 5370 Sellogen W. CAS 9084-06-4; Sodium alkyl naphthalene sulfonate; wetting agent for agricultural formulations. [Henkel/Emery]

Emery®6220. EO/PO block polymer; surfactant for agricultural formulations. [Henkel/Emery]

Emery®6221 Monolan 2500. EO/PO block polymer; surfactant for agricultural formulations. [Henkel/Emery]

Emery® 6686. CAS 25322-68-3; PEG 600; chemical intermediate for coatings, adhesives, lubricants, metalworking, paper mfg., petrol. production, ceramics, printing, electronics, solvs., cleaners, latex paints, mold release agent, rubber. [Henkel/Emery]

Emery®6701. Trimethylolpropane tripelargonate; lubricant with good heat stability for textile processing. [Henkel/Emery; Henkel/Textile]

Emery® 6717. Internal-external lubricant for mfg. of glass fibers. [Henkel/Textiles]

Emery®6744. Cocamidopropyl betaine; surfactant, foamer for personal care products, shampoos, baby preparations, skin cleansers; also for textile applications. [Henkel/Cospha; Henkel/Textile; Henkel Canada]

Emery® 6750 Nopcosperse AD-6 Liq. Dispersant for agricultural dry flowables, wettable powds., and aq. suspensions. [Henkel/Emery]

Emery® HP-2050. Anhyd. lanolin; emulsifier, emollient, conditioner, moisturizer, pigment dispersant for personal care products, pharmaceuticals. [Henkel/Cospha; Henkel Canada]

Emery's L-110. A proprietary form of azelaic acid; used as a softener for alkyd resins. §

Emery's L-114. A proprietary mixture of low molecular weight aliphatic acids in which pelargonic acid, C_8H_7COOH, predominates; used in the oil modification of alkyd resins. §

Emeside. CAS 77-67-8; A proprietary preparation of ethosuximide; an anticonvulsant. [LAB Ltd] ‡

Emetrol. A proprietary preparation of laevulose, dextrose and orthophosphoric acid; an anti-emetic. [Pharmax Ltd]*

Emgard®2033. Blend; emulsifier for methyl oleate, agricultural formulations. [Henkel/Emery]

Emgard® 2063. Formulated product; spreading and penetrating agent for agricultural herbicide formulations. [Henkel/Emery]

Emge. Magnesium hyposulfite in ampoules and tablets.

EMI-24. 2-Ethyl-4-methylimidazole; a curing agent for epoxy resins used in low proportions thus improving chemical resistance. §

Emid®6500. CAS 68140-00-1; Cocamide MEA (1:1); thickener, foam stabilizer for shampoos, hair coloring products, liq. detergents, and rug cleaners. [Henkel/Cospha; Henkel Canada]

Emid®6515. Cocamide DEA (1:1); emulsifier, foam booster and stabilizer; inhibits redeposition of soils; thickener, superfatting agent for shampoos, bubble baths, cleans-

---

ers, liq. detergents; antiredeposition agent for soils on textiles. [Henkel/Cospha; Henkel/Textile; Henkel Canada]

Emid® 6519. CAS 120-40-1; EINECS 204-393-1; Lauramide DEA (1:1); foam booster, stabilizer, visc. modifier for personal care products, soaps, bath additives. [Henkel/Cospha; Henkel Canada]

Emid® 6545. CAS 93-83-4; EINECS 202-281-7; Oleamide DEA (1:1); foam suppressant in dye carrier and solv. emulsions; emulsifier for mineral oils for antistatic fiber processing aids and yarn lubricants. [Henkel/Emery; Henkel/Textile]

Emid® 6590. Lauramide DEA and propylene glycol; foam booster, visc. modifier for personal care products, bath additives, liq. soaps. [Henkel/Cospha; Henkel Canada]

EMI-X®. Thermoplastic compounds having electrical conduction properties. [LNP; ICI Chemicals & Polymers Ltd]

EMI-X® DC-1008. PC with 40% carbon fiber reinforcement; highly conductive composite effectively shielding electromagnetic and/or radio frequency interference; used in avionics housings, business machine enclosures, other electronic devices. [LNP]

EMI-X® OC-1008. CAS 9016-75-5; PPS with 40% carbon fiber reinforcement; highly conductive composite effectively shielding electromagnetic and/or radio frequency interference; used in avionics housings, business machine enclosures, other electronic devices. [LNP]

EMI-X® PC-1008. CAS 25038-54-4; Nylon 6 with 40% carbon fiber reinforcement; highly conductive composite effectively shielding electromagnetic and/or radio frequency interference; used in avionics housings, business machine enclosures, other electronic devices. [LNP]

EMI-X® PDX-83393. PC with 25% nickel; conductive attenuating composite effectively shielding electromagnetic and/or radio frequency interference; used in avionics housings, business machine enclosures, other electronic devices. [LNP]

EMI-X® PDX-A-88128. CAS 9003-56-9; ABS composite; attenuating composite for EMI/RFI shielding applications. [LNP]

EMI-X® PDX-D-87815. PC composite; attenuating composite for EMI/RFI shielding applications. [LNP]

EMI-X® PDX-O-91074. CAS 9016-75-5; PPS composite; attenuating composite for EMI/RFI shielding applications. [LNP]

EMI-X® PDX-P-90305. CAS 25038-54-4; Nylon 6 composite; attenuating composite for EMI/RFI shielding applications. [LNP]

EMI-X® PDX-R-89496. CAS 32131-17-2; Nylon 6/6 composite; attenuating composite for EMI/RFI shielding applications. [LNP]

EMI-X® PDX-W-88341. CAS 26062-94-2; PBT composite; attenuating composite for EMI/RFI shielding applications. [LNP]

EMI-X® RC-1008. CAS 32131-17-2; Nylon 6/6 with 40% carbon fiber reinforcement; highly conductive composite effectively shielding electromagnetic and/or radio frequency interference; used in avionics housings, business machine enclosures, other electronic devices. [LNP]

EMI-X® WC-1008. Thermoplastic polyester with 40% carbon fiber reinforcement; highly conductive composite effectively shielding electromagnetic and/or radio fre-

quency interference; used in avionics housings, business machine enclosures, other electronic devices. [LNP]

Emka DDBSA. Dodecylbenzene sulfonic acid; base for compounding detergents and emulsifiers for textile processing. [Emkay]

Emkabase CA. Emulsifier for mineral oil; mixed with 90 parts mineral oil to yield a water-disp. substantive product for textile processing. [Emkay]

Emkabase ODC-2. Alkylaryl sulfonate blend; emulsifier for orthodichlorobenzene. [Emkay]

Emkabond UR. Fiber reactive laminating adhesive for bonding urethane foam to fabric. [Emkay]

Emka Catalyst P-35. Org. amine; catalyst giving excellent bath stability for thermosetting resins in textile industry. [Emkay]

Emkacide GS-2. Disinfectant, cleaner, deodorizer, and fungicide with softening props. [Emkay]

Emka Defoam AA. Solvent type defoamer for textile processing. [Emkay]

Emka Defoam BC, NC. Nonsilicone; foam and froth control for industrial wastes, textile and paper processing. [Emkay]

Emka Defoam DP. Silicone emulsion; defoamer; may be cut 50-50 with water. [Emkay]

Emkadixol. Lubricants. [ICI Chem. & Polymers Ltd]

Emkafix RXC. Quaternary ammonium deriv.; dispersant which improves fixation of direct dyes. [Emkay]

Emkafol D. Highly sulfonated fatty acid ester; wetting and rewetting agent with excellent leveling props. for textile processing; also activates enzymes for starch removal. [Emkay]

Emkafume FA. Nondurable type gas fading inhibitor for acetate colors. [Emkay]

Emkagen 49. Amino condensate and alkylaryl sulfonate blend; detergent for print washing in textile applications. [Emkay]

Emkagen 49AM. Ammoniated detergent based on long chain alkylaryl sulfonate; detergent, wetting agent, emulsifier for heavy duty cleaning and scouring, industrial cleaning; resistant to hard water. [Emkay]

Emkagen BT. Amino-condensate blend; detergent for wool scouring. [Emkay]

Emka Graphite Remover. Effective cleaner for graphited fabrics. [Emkay]

Emkal BNS. Butyl naphthalene sodium sulfonate; dispersant, detergent, dyeing assistant, scouring, and wetting agent, emulsifier used in textile industry. [Emkay]

Emkal BNS Acid. Butyl naphthalene sulfonic acid; wetting, dispersing agent, detergent, dye assistant, scouring agent, emulsifier. [Emkay]

Emkal NNS, NNS Acid. Nonyl naphthalene sodium sulfonate; dispersant, detergent, dyeing assistant, scouring, and wetting agent, emulsifier used in textile industry. [Emkay]

Emkal NOBS. Sodium nonyl benzene sulfonate; dispersant, detergent, dyeing assistant, scouring, and wetting agent, emulsifier used in textile industry. [Emkay]

Emkalane WL. Amino condensate; scouring agent for wool and wool blends. [Emkay]

Emkalar Base C50L. Emulsifier; base for liq. dye carriers by mixing with ortho-phenyl phenol. [Emkay]

Emkalite BAC. Anionic sulfonate blend; wetting agent, partial stripping agent for cationic dyes from acrylic

---

‡ Trade name and manufacturer not verified   § Trade name without identified manufacturer

fibers; helps produce level redyes. [Emkay]

Emkalon KLA. Polyethoxylated condensate; lanolin base softener. [Emkay]

Emkalon TN. Polyethoxylate; lubricant, softener; highly compat. [Emkay]

Emkalube F-11. Bright oil which forms an emulsion in water; used as a trough oil in twisting and as a needle lubricant (textile industry). [Emkay]

Emkane Acid. Alkylaryl sulfonic acid; detergent, dyeing assistant, scouring and wetting agent; emulsifier used in the textile industry; biodeg. [Emkay]

Emkane HAD. Alkylaryl sulfonate; detergent, dyeing assistant, scouring and wetting agent, esp. for cleansing stubborn soils; emulsifier used in the textile industry; biodeg. [Emkay]

Emkane HAX. Amine-neutralized alkylaryl sulfonate; double-strength detergent, dyeing assistant, scouring and wetting agent; emulsifier used in the textile industry; biodeg. [Emkay]

Emkanet B. Stiff resin finish for veils and nets (textiles). [Emkay]

Emkanol NC, NCD 25, 35, 45, 55. Sulfonates; penetrants, wetting agents for mercerizing caustic solutions [Emkay]

Emkanyl 85. Leveling agent for dyeing of nylon hosiery. [Emkay]

Emkanyl BRX. Nylon dyeing assistant for acid dyes; prevents barré in woven fabrics; leveling agent. [Emkay]

Emkapel DE. Durable type water repellent for textile industry. [Emkay]

Emkapene AV, AVX. Low-foaming penetrant, emulsifier, scum preventatives for use in sulfur and vat dyeing; compat. with sodium sulfide. [Emkay]

Emkapene RW. Mixt. of sulfonated oils and pine oil; leveling agent and dye assistant. [Emkay]

Emkaperm. Permanent stiffening finish which withstands several washings (textiles). [Emkay]

Emkapol PO-18. CAS 143-18-0; EINECS 205-590-5; Potassium oleate; detergent, soap, emulsifier; stabilizer for natural latex; biodeg. [Emkay]

Emkapon 4S, DS, SS, TS. Amide sulfonates; general textile detergents and emulsifiers. [Emkay]

Emkapon BC. Wool scouring agent containing optical bleach. [Emkay]

Emkapon DAC, DAC-50. Detergent/emulsified solv. blend; detergent for cleaning obstinate soils, grease, stains, wax, water repellents; recommended as Dacron scour. [Emkay]

Emkapon Jel 500 Conc. Fatty methyl taurate; dispersant and surfactant for textile processing. [Emkay]

Emkapon ML. Modified alkylaryl sulfonate; scouring and dyeing agent for synthetics. [Emkay]

Emkapruf ABR, FL. Conc. flame resist. finishes for textiles. [Emkay]

Emkarate. Lubricant. [ICI Chem. & Polymers Ltd]

Emkaron GA-1. Sulfated fatty acid ester; low-foaming dyeing assistant, dispersant, wetting agent for textiles; resistant to acid and alkaline media. [Emkay]

Emkarox. Synthetic lubricants. [Imperial Chemical Industries plc]

Emkasan QA-50. Quaternary ammonium compound; algicide, germicide, disinfectant, and deodorant for textile and industrial use. [Emkay]

Emkasene 800. CAS 139-89-9; Trisodium hydroxyethyl ethylene diamine triacetate aq. solution; chelating agents for sequestering calcium, magnesium, and ferric ions in solution [Emkay]

Emkaset. Modified formaldehyde resin, thermosetting; stiff, crease-resist., shrink-resist.; for textiles. [Emkay]

Emkasize CF. Sizing agent for cotton and jute. [Emkay]

Emkasol DE. Modified oleyl-lauryl alcohol sulfate; dispersant, wetting agent, detergent, dye bath stabilizer. [Emkay]

Emkasorb. Used in finishing of nylon and other synthetics to increase the absorbency. [Emkay]

Emkastat MLT. Antistat lubricant, soil releasing agent; hard wax base which protects delicate fabrics. [Emkay]

Emkatan K. Synthetic softener and lubricant used as a finish for cotton and rayon knit goods and as a lubricant on dyed skeins to improve winding and eliminate chafing and abrasions; good antistat. [Emkay]

Emkatard. Dispersant for vat colors (textiles). [Emkay]

Emkaterge B. Synthetic detergents with alkyl terpenes; wetting agent, base product for compounding graphite remover. [Emkay]

Emkatex 11, 21. POE fatty acid; emulsifier for water kerosene formula. [Emkay]

Emkatex 49-P. Synthetic detergents; color dispersant, dyebath stabilizer. [Emkay]

Emkatex AA. Alkylaryl sulfonate; detergent, wetting agent, penetrant for textile applications. [Emkay]

Emkatex DX, DXP. Phosphated high molecular weight alcohol; dispersant, leveling agent, penetrant, scouring agent, and textile dye bath stabilizer; for kier, bleaching processes; also increases absorbency of flame proofing finishes. [Emkay]

Emkatex NE. Synthetic detergents, emulsifiers, dispersants; textile auxiliary. [Emkay]

Emkatint BRN. Optical bleach and brightener for nylon, arnel, and orlon. [Emkay]

Emkatol M. Graphite remover for nylon lace. [Emkay]

Emka Transfer Remover. Cleaner for redying nylon hosiery. [Emkay]

Emkawate AS. weighter finish with nonslip props.; minimizes fraying and ravelling (textiles). [Emkay]

Emkazyme. Mixt. of amylolytic enzymes and catalysts, solubilizing resins, starches, etc.; for textile applications. [Emkay]

emodin. (Frangula emodin; frangulic acid; 1,3,8-trihydroxy-6-methylanthraquinone) CAS 15687-27-1; $C_{14}H_4O_2$-$(OH)_3CH_3$; used in medicine as a cathartic.

Emol Keleet. A purified fuller's earth.

Empal. Aqueous solution containing 25% w/v MCPA as mixed sodium and potassium salts; for use as an agricultural herbicide. [Universal Crop Protection Ltd]

EM-PB. Lard oil; additive improving wetting and lubricity in industrial applications. [Ferro/Keil]

Empee® FR 42 LM. CAS 9002-88-4; HDPE, flame-retarded; used in wire and cable markets or typ. PE applications requiring flame retardance. [Monmouth Plastics]

Empee® PE-112. CAS 9002-88-4; LDPE, flame-retarded; UL-recognized grade for injection molded parts, certain elec. applications. [Monmouth Plastics]

Empee® PE-113. CAS 9002-88-4; LDPE, flame-retarded; for wire and cable insulation. [Monmouth Plastics]

Empee® PO Conc. 61. Imparts flame retardancy to low an dhigh dens. polyethylene. [Monmouth Plastics]

Empee®PP-301. CAS 9003-07-0; PP, flame-retarded; UL-

---

* Trade name not verified as available   † Trade name verified as obsolete

recognized resin for injection molded parts, extruded sheet, shapes, certain elec. applications. [Monmouth Plastics]

Empee®PP-459. CAS 9003-07-0; PP homopolymer, flame-retarded; high impact, UL-recognized resin for injection molding. [Monmouth Plastics]

Empee® PP-560. CAS 9003-07-0; PP copolymer, flame-retarded; for extrusion, wire insulation. [Monmouth Plastics] *

Empee®PP Conc. 33. CAS 9003-07-0; PP, flame retardant; imparts flame retardancy to most PP homopolymers; for use in fiber and film applications. [Monmouth Plastics]

Empee®PS-921. CAS 9003-53-6; PS, flame-retarded; UL-recognized resin for injection molded transparent crystal parts. [Monmouth Plastics]

Emperor alloy. A nickel-chromium alloy. It will resist a temperature of 1750 C-1800 C F.

Emperor brass. An aluminum bronze. It consists of 60% copper, 20% aluminum, and 20% zinc.

Empetal. 4-(4-Methyl-3-penten-1-yl)-3-cyclohexen-1-carboxyaldehyde. [Quest Int'l. UK Ltd]

Emphos. Series of anionic surfactants of the phosphate ester type showing a range of properties eg surface tension lowering, wetting, foaming, according to composition; antistats; dry cleaning; emulsion polymerization; industrial alkaline cleaners; papermaking; pesticides; textile processing; extreme pressure lubricants corrosion inhibitors; release agents; moisture barrier agents; oil well fluids; pigment dispersion. [Witco Chemical Ltd]*

Emphos CS-136. Nonoxynol-6 phosphate; lubricant; antistat; emulsifier for cutting fluids, PVAc, and acrylic film formation; detergent for hard surfaces, metal cleaners, and dry cleaning systems; waterless hand cleaner component. [Witco/Organics]

Emphos CS-1361. Sodium nonoxynol-9 phosphate; antistat; emulsifier for transparent gels; detergent for hard surfaces, metal cleaners, and drycleaning systems; particle dispersant for aq. systems; coupling agent; wetting agent. [Witco/Organics]

Emphos PS-21A. Alcohol ethoxylate phosphate ester; detergent base, emulsifier, foaming and wetting agent, lubricant, dispersant; used for detergent industry and industrial surfactants. [Witco/Organics]

Emphos TS-230. Phenol ethoxylate phosphate ester, acid form; lubricant; improves lubricity and load-bearing of water-based lubricants; industrial corrosion inhibitor, defoamer, metal processing surfactant for synthetic oils; low-foaming hydrotrope for alkaline cleaners. [Witco/Organics]

Empicol. An aliphatic alcohol sulfate and aliphatic alcohol alkyl ether sulfate; wetting agent. [Albright & Wilson Ltd]

Empicol® 0031/T. CAS 68585-44-4; DEA-lauryl sulfate; detergent used in personal care products [Albright & Wilson UK]

Empicol® 0045. Sodium lauryl sulfate; raw material and foaming agent for toothpaste, shampoos, foam baths. [Albright & Wilson UK]

Empicol® 0045V. Sodium lauryl sulfate; surfactant for toiletries. [Albright & Wilson Am.]

Empicol®0216. Fatty alcohol ethoxy phosphate ester; foam stabilizer, conditioner, and antistatic agent for toiletries; detergent used in textile processing. [Albright & Wilson UK]

Empicol® 0303. CAS 151-21-3; Sodium lauryl sulfate BP;

surfactant in toothpaste and pharmaceutical preparations; emulsion polymerization. [Albright & Wilson UK]

Empicol®0585/A. Sodium ethylhexyl sulfate; surfactant for toiletries; low foam degreasing agent for textiles. [Albright & Wilson Am.]

Empicol®0758. Sodium decyl sulfate; surfactant for toiletries. [Albright & Wilson Am.]

Empicol®0775. Sodium lauryl/tallow sulfate; surfactant for toiletries. [Albright & Wilson Am.]

Empicol®9060X. Formulated product; pearling/opacifying conc. for toiletries. [Albright & Wilson Am.]

Empicol® AL30. CAS 68081-96-9; EINECS 218-793-9; Ammonium lauryl sulfate; detergent for shampoos, carpet shampoos, leather processing. [Albright & Wilson/ Australia]

Empicol®ALL. Ammonium lauryl triethoxy sulfate; shampoo ingred. [Albright & Wilson Australia]

Empicol® BSD 52. Sodium/magnesium laureth sulfate; surfactant for toiletries. [Albright & Wilson Am.]

Empicol®CHC 30. Sodium cetyl oleyl sulfate; multipurpose textile detergent. [Albright & Wilson UK] *

Empicol®DA. CAS 68585-44-4; DEA-lauryl sulfate; ingred. in personal care products, automobile cleaners. [Albright & Wilson Australia]

Empicol® DLS. DEA-lauryl sulfate; detergent used in liq. and lotion shampoos. [Albright & Wilson UK] *

Empicol® EAA. CAS 67762-19-0; Ammonium laureth sulfate; surfactant for toiletries. [Albright & Wilson Am.]

Empicol®EAB. CAS 67762-19-0; Ammonium laureth sulfate; detergent and base used in the cosmetic industry. [Albright & Wilson UK]

Empicol® EAC. CAS 67762-19-0; Ammonium laureth sulfate; surfactant for toiletries. [Albright & Wilson Am.]

Empicol®EGB, EGC. CAS 67702-21-4; Magnesium laureth sulfate; detergent and toiletry raw material. [Albright & Wilson UK]

Empicol®EL. MEA-lauryl sulfate; detergent used in the mfg. of liq. shampoos. [Albright & Wilson UK] *

Empicol® EMB. CAS 977067-77-8; MEA laureth sulfate; detergent raw material for shampoos. [Albright & Wilson UK]

Empicol® ESA. CAS 68585-34-2; Sodium laureth sulfate; detergent and toiletry raw material. [Albright & Wilson UK]

Empicol® ESB. CAS 9004-82-4; 68585-34-2; Sodium laureth sulfate; shampoo ingred.; mild detergent for textile and leather processing; dispersant/emulsifier for emulsion and suspension polymerization. [Albright & Wilson Australia]

Empicol®ESC/AU. CAS 68585-34-2; Sodium laureth sulfate; surfactant used as a base in personal care products [Albright & Wilson Australia]

Empicol® ETB. CAS 270028-82-6; TEA-laureth sulfate; detergent and toiletry raw material. [Albright & Wilson UK]

Empicol® HL25. CAS 2044-56-6; Lithium lauryl sulfate; active shampoo detergent ingred. [Albright & Wilson UK]

Empicol®L Series. Alkyl sulfate; dispersant/emulsifier for emulsion and suspension polymerization, agricultural formulations; foaming agent for carpet backing; air entraining agent for construction; flotation aid. [Albright & Wilson UK]

Empicol® LM. CAS 68955-19-1; Sodium lauryl sulfate; detergent, foaming agent for toothpaste, shampoos,

---

‡ Trade name and manufacturer not verified   § Trade name without identified manufacturer

shaving products [Albright & Wilson UK]

Empicol®LMV/T. CAS 68955-19-1; Sodium lauryl sulfate; detergent for cream shampoos, shaving products [Albright & Wilson UK]

Empicol® LQ33/T. CAS 4722-98-9; 68908-44-1; MEA-lauryl sulfate; surfactant in the mfg. of personal care products; emulsifier in the mfg. of rubber latices and for resins; bactericidal detergents. [Albright & Wilson UK]

Empicol® LS30. CAS 151-21-3; Sodium lauryl sulfate; wetting, dispersing, emulsifying, and foaming agent for industrial processes, detergent/cleaner formulations. [Albright & Wilson UK] *

Empicol® LX. CAS 68585-47-4; Sodium lauryl sulfate; emulsifier in the mfg. of plastics, resins, and synthetic rubbers; foaming agent for rubber foams, personal care products, pharmaceuticals, and carpet and upholstery shampoos; lubricant in mfg. of molded rubber goods. [Albright & Wilson; Albright & Wilson UK]

Empicol®LXV. CAS 151-21-3; Sodium lauryl sulfate; emulsifier in the mfg. of plastics, resins, and synthetic rubbers; foaming agent for rubber foams, personal care products, toothpaste, and carpet and upholstery shampoos; lubricant in mfg. of molded rubber goods. [Albright & Wilson; Albright & Wilson UK]

Empicol® LY28/S. CAS 151-21-3; Sodium lauryl sulfate; coprecipitant in mfg. of photographic film; emulsion polymerization in the plastic and rubber industries; foaming agent in carpet processes. [Albright & Wilson UK] *

Empicol® LZ. CAS 68955-19-1; Sodium lauryl sulfate; detergent, wetting and foaming agent in personal care products, pharmaceuticals; emulsifier in mfg. of rubbers, plastics, and resins by emulsion polymerization; foaming agent in mfg. of foam rubber goods; lubricant used in plastic goods. [Albright & Wilson UK]

Empicol® LZG 30. CAS 151-21-3; Sodium lauryl sulfate; surfactant, foaming and wetting agent used in industrial processes; emulsifier in mfg. of synthetic rubbers, plastics, and resins by emulsion polymerization. [Albright & Wilson UK] *

Empicol® LZP. CAS 151-21-3; Sodium lauryl sulfate; detergent, wetting and foaming agent in personal care products, pharmaceuticals; emulsifier in mfg. of rubbers, plastics, and resins by emulsion polymerization; foaming agent in mfg. of foam rubber goods; lubricant used in plastic goods. [Albright & Wilson UK] *

Empicol® LZV. CAS 68955-19-1; Sodium lauryl sulfate; detergent and foaming agent for shampoos, toothpaste. [Albright & Wilson UK]

Empicol®MD. Sodium lauryl ethoxy sulfate as a pale straw liquid; foaming agent with good hard water performance and low viscosity and irritancy; used in baby and medical shampoos, foam baths, hand cleaners and fine fabric washing. [Albright & Wilson Ltd, Detergents Div, Marchon] *

Empicol® ML 26/F. CAS 68081-97-0; Magnesium lauryl sulfate; detergent used in the mfg. of shampoos and toothpaste. [Albright & Wilson UK]

Empicol® SDD. CAS 37354-45-5; Disodium lauryl ethoxy sulfosuccinate; mild raw material for toiletries and detergents. [Albright & Wilson UK]

Empicol®SLL. CAS 36409-57-1; Disodium lauryl sulfosuccinate; mild raw material for toiletries and detergents. [Albright & Wilson UK]

Empicol®STT. Disodium cetearyl sulfosuccinate; mild raw

material for toiletries and detergents. [Albright & Wilson UK]

Empicol® TA40. CAS 68908-44-1; TEA-lauryl sulfate; ingred. for personal care products [Albright & Wilson Australia]

Empicol®TAS30. CAS 68955-20-4; Sodium tallow sulfate; collector in the beneficiation of minerals by ore flotation; detergent active raw material; cofoaming agent for latex foam compounds [Albright & Wilson UK]

Empicol®TL40. CAS 139-96-8; EINECS 205-388-7; TEA-lauryl sulfate; detergent used in liq. and lotion shampoos; foam booster for fire fighting. [Albright & Wilson UK]

Empicol® XC35. Blend; pearlescent base for shampoo, bubble bath, liq. soap; formulations should contain preservative. [Albright & Wilson UK]

Empicol®XM 17. Formulated product; conc. detergent for carpet shampoo. [Albright & Wilson UK]

Empicol®XPA. Formulated blend; pearlized and opacified conc. for toiletries. [Albright & Wilson UK]

Empicryl®. Pigment dispersant; for a wide variety of water based paints. [Albright & Wilson Ltd, Detergents Div] *

Empicryl®6045. Alkyl methacrylate polymer in hydrocarbon oil; visc. index improver for formulation of high visc. index hydraulic fluids; emulsifier, dispersant for emulsion polymerization; reactive diluent for adhesives and coatings. [Albright & Wilson UK]

Empicryl® APD. Maleic anhydride/diisobutylene copolymer, disodium salt, aq. solution; pigment dispersant for water based paints, pigment stabilizer. [Albright & Wilson UK] *

Empicryl® DH122. Polyalkyl methacrylate copolymer in hydrocarbon oil; visc. index improver, pour pt. depressant, and low temp. sludge dispersant for formulation of multigrade crankcase oils. [Albright & Wilson UK] *

Empicryl®PPT38. Alkyl methacrylate copolymer in hydrocarbon oil; pour pt. depressant for formulation of lubricants incl. multigrade crankcase oils in conj. with hydrocarbon-based polymeric visc. index improvers. [Albright & Wilson UK] *

Empicryl®PT1334. Alkyl methacrylate polymer in hydrocarbon oil; visc. index improver/pour pt. depressant for formulation of multigrade crankcase oils. [Albright & Wilson UK] *

Empicyl. Lubricating oil additives. [Albright & Wilson Ltd]

Empigen. Tertiary amine and imidazoline derivates. [Albright & Wilson Ltd, Detergents Div]

Empigen® 5083. CAS 61788-90-7; EINECS 263-016-9; Coco dimethylamine oxide; foam booster/stabilizer and visc. modifier for shampoos, foam baths, cleaners; improves conditioning in shampoos; solubilizer for liq. bleach products [Albright & Wilson UK]

Empigen®5089. CAS 112-00-5; EINECS 203-927-0; Laurtrimonium chloride; bactericide for disinfectant and sanitizer formulations for household, institutional, agriculture, food processing applications, antiseptic detergents in pharmaceuticals; algicide for swimming pools; wood preservatives. [Albright & Wilson UK]

Empigen® 5107. Alkyl dimethylamine betaine; foaming agent/stabilizer, antistat, solubilizer for shampoos, foam baths, latex foam compounds, oil production, alkaline industrial hard surface cleaners; wetting and coupling agent for cleaners, traffic film removers. [Albright & Wilson UK]

Empigen® 5509. Cocamino hydroxy sulfobetaine; foam

---

* Trade name not verified as available   † Trade name verified as obsolete

booster, conditioner for shampoos, skin cleansers. [Albright & Wilson UK]

Empigen® AB. CAS 67700-98-5; Dimethyl lauramine; intermediate for mfg. of high quality derivs. such as quaternary ammonium compounds, betaines, amine oxides; catalyst for PU foam, resin curing agent, corrosion inhibitor, and flotation aid. [Albright & Wilson UK]

Empigen® AF. C12-16 alkyl dimethyl amine; intermediate for amine deriv.; resin curing agents; corrosion inhibitor; ore flotation chemicals. [Albright & Wilson UK]

Empigen® AG. C12-16 alkyl dimethyl amine; intermediate for amine deriv.; resin curing agent; corrosion inhibitor; ore flotation chemicals. [Albright & Wilson UK]

Empigen® AH. CAS 112-75-4; Dimethyl myristamine; intermediate for mfg. of high quality derivs. such as quaternary ammonium compounds, betaines, amine oxides; catalyst for PU foam, resin curing agent, corrosion inhibitor, and flotation aid. [Albright & Wilson UK]

Empigen® BAC50. Benzalkonium chloride NF; bactericide for disinfectant and sanitizer formulations for household, institutional, agriculture, food processing applications, antiseptic detergents in pharmaceuticals; algicide for swimming pools; wood preservatives; masonry biocides; permanent retarders in dyeing of acrylic fibers; phase transfer catalyst. [Albright & Wilson UK]

Empigen® BB. CAS 66455-29-6; Lauryl betaine; foam booster/stabilizer, emulsifier, dispersant, wetting agent, thickening agent, conditioner used for shampoos, detergents, latex foam compounds for carpet backing, industrial applications; formulation of film removers; stable over wide pH range. [Albright & Wilson Am.; Albright & Wilson UK]

Empigen® BCB50. CAS 68989-00-4; Benzalkonium chloride; disinfectant for dairy, food processing, restaurant, brewing, and bottling industries; retarder in dyeing of acrylic fibers. [Albright & Wilson UK]

Empigen® BS. Cocamidopropyl dimethylamine betaine; foaming agent/stabilizer and antistat for shampoos, foam baths, latex foam compounds, oil production; wetting and coupling agent for cleaners, traffic film removers. [Albright & Wilson UK]

Empigen® BS/H. Cocamidopropyl dimethylamine betaine; foaming agent/stabilizer and antistat for shampoos, foam baths, latex foam compounds, oil production; wetting and coupling agent for cleaners, traffic film removers. [Albright & Wilson UK]

Empigen® CDL60. CAS 68608-66-2; Disodium lauroamphodiacetate; detergent for nonirritating shampoos, skin cleansing, personal care products [Albright & Wilson UK]

Empigen® CDR10. Coconut imidazoline betaine; surfactant used in personal care products, household and industrial cleaners, textile processing. [Albright & Wilson UK]

Empigen® CDR40. CAS 68334-21-4; Sodium cocoamphoacetate; mild detergent for shampoos, skin cleansers, personal care products, textile processing. [Albright & Wilson UK]

Empigen® CHB40. Alkyl trimethyl ammonium bromide; bactericide. [Albright & Wilson UK]

Empigen® CM. Tallow trimethyl ammonium methosulfate; antistat and conditioning agent for personal care products; emulsifier and antistat for industrial use; retarder for dyeing of acrylic fibers. [Albright & Wilson UK]

Empigen® FKC75L. Alkyl diamido amine lactate; raw material used in fabric softener formulations. [Albright & Wilson UK]

Empigen® FKH75L. Dialkyl diamido amine lactate; surfactant. [Albright & Wilson UK]

Empigen® FRC75S. Tallow alkyl imidazoline methosulfate; textile conditioning agent, antistat, and lubricant for fibers; raw material for fabric softener formulations. [Albright & Wilson UK]

Empigen® OB. CAS 1643-20-5; 70592-80-2; Lauramine oxide; coactive, detergent, antistat, foam booster/stabilizer and visc. modifier for personal care products, surgical scrubs, fire fighting foam concs., foamed rubbers, bleach additive; solubilizer. [Albright & Wilson; Albright & Wilson UK]

Empigen® OC. Alkyl dimethyl amine oxide; surfactant. [Albright & Wilson Am.]

Empigen® OH25. CAS 3332-27-2; EINECS 222-059-3; n Myristyl dimethyl amine oxide; foam booster/stabilizer and visc. modifier for shampoos, foam baths, detergents; improves conditioning in shampoos; solubilizer for liq. bleach products [Albright & Wilson; Albright & Wilson UK]

Empigen® OS/A. CAS 68155-09-9; Cocamidopropyl dimethyl amine oxide; foam booster/stabilizer and visc. modifier for shampoos, foam baths, detergents; improves conditioning in shampoos. [Albright & Wilson; Albright & Wilson UK]

Empigen® OY. CAS 59355-61-2; PEG-3 lauramine oxide; detergent, antistat, foam booster/stabilizer for foamed rubbers, fire fighting, bleach additive. [Albright & Wilson; Albright & Wilson UK]

Empigen® XDR302. Sodium cocoamphoacetate with sodium lauryl sulfate; used in personal care products [Albright & Wilson UK]

Empilan®. Fatty acid esters, emulsifiers and foam stabilizers. [Albright & Wilson Ltd] *

Empilan® 0004. Castor oil EO/PO condensate; emulsifier and lubricant for formulation of cutting oils, grinding fluids, and textile lubricants; mfg. of brake fluids. [Albright & Wilson UK]

Empilan® 2020. Lauryl alcohol ethoxylate; stabilizer for latexes produced by emulsion polymerization. [Albright & Wilson UK] *

Empilan® 2502. CAS 8051-30-7; Cocamide DEA (1:1); foam booster/stabilizer. [Albright & Wilson UK]

Empilan® 7132. Fatty amine alkoxylate; surfactant. [Albright & Wilson UK]

Empilan® AM Series. Amine ethoxylates; emulsifier for agriculture emulsifiable and suspension concs.; antistat for plastics. [Albright & Wilson UK]

Empilan® BD. Blend of quaternary ammonium germicide with nonionic detergent; clear pale yellow stable liquid; bactericidal detergent used in washing of dishes, pans, glassware, dishcloths, etc. [Albright & Wilson Ltd, Detergents Div, Marchon] *

Empilan® BQ 100. CAS 9004-96-0; PEG-8 oleate; emulsifier for insecticides, spindle oils, industrial wetting agent, scouring agent, antifoam, PVC antistat; in laundry works, glue mfg., paper coating, hand cleaning jellies, turbine oil additive. [Albright & Wilson UK]

Empilan® CDE. Cocamide DEA (1:1); foam boosting/stabilizing agent, solubilizer, detergent for use in personal care and detergent products; antistat in plastics; softener for leather processing. [Albright & Wilson UK]

Empilan® CIS. CAS 68440-05-1; Cocamide MIPA; foam

---

‡ Trade name and manufacturer not verified   § Trade name without identified manufacturer

stabilizer, detergent, shampoo additive. [Albright & Wilson UK]

Empilan® CME. CAS 68140-00-1; Cocamide MEA; detergent, foam booster/stabilizer in detergent systems; stabilizer for hair and carpet shampoos; base for mfg. of ethoxylated alkylolamides; visc. modifier. [Albright & Wilson UK]

Empilan® EGMS. Glycol stearate; opacifier/pearling and emulsifying/stabilizing agent in shampoos; emollient. [Albright & Wilson UK]

Empilan® GMS LSE32. Glyceryl stearate SE; food grade emulsifier. [Albright & Wilson UK]

Empilan® GMS NSE32. Glyceryl stearate; food grade emulsifier. [Albright & Wilson UK]

Empilan® K Series. Alkyl ethoxylates; emulsifier, wetting agent for agricultural emulsifiable and suspension concs., spinning oils in textile processing; emulsifier, dispersant for emulsion polymerization; rubber compounding aid; antistat for plastics. [Albright & Wilson UK]

Empilan® KA10/80. CAS 68439-45-2; C10-12 alcohol ethoxylate; scouring and wetting agent for textiles, emulsifier, dye leveling and dispersing agent, detergent, in metal processing, cutting oils, paper industry, paints, insecticides and pesticides; biodeg. [Albright & Wilson UK]

Empilan® KB 2. CAS 68002-97-1; Laureth-2; emulsifier, foam booster, superfatting agent, used in detergents and emulsifying systems; mortar plasticizer. [Albright & Wilson UK]

Empilan® KCA Series. C12-13 alcohol ethoxylates; surfactant. [Albright & Wilson UK]

Empilan® KCB Series. C11 alcohol ethoxylates; surfactant. [Albright & Wilson UK]

Empilan® KCL Series. C12-15 alcohol ethoxylates; surfactant. [Albright & Wilson UK]

Empilan® KCMP 0703/F. CAS 68551-13-3; Fatty alcohol alkoxylate; emulsifier, detergent, wetting agents for industrial and domestic applications. [Albright & Wilson UK]

Empilan® KCP Series. C14-15 alcohol ethoxylates; surfactant. [Albright & Wilson UK]

Empilan® KCX Series. C13-15 alcohol ethoxylates; surfactant. [Albright & Wilson UK]

Empilan® KI Series. C13 alcohol ethoxylates; wetting agent for leather processing. [Albright & Wilson UK]

Empilan® KL 6. Ceteleth-6; emulsifier for mineral oils in cosmetics, detergent and wetting agents in textile industry. [Albright & Wilson UK]

Empilan® KM 11. CAS 68439-49-6; Ceteareth-11; emulsifier, foam control agent in synthetic heavy duty detergents, soap additive; dispersant for textile processing. [Albright & Wilson UK]

Empilan® KS Series. C9-11 alcohol ethoxylates; surfactant. [Albright & Wilson UK]

Empilan® LDE. CAS 120-40-1; EINECS 204-393-1; Lauramide DEA (1:1); foam booster/stabilizer, solubilizer, thickener, detergent used in shampoos and liq. detergent formulations; antistat for plastics. [Albright & Wilson UK]

Empilan® LIS. CAS 142-54-1; EINECS 205-541-8; Lauramide MIPA; foam booster/stabilizer for liq. and powd. detergents, shampoos; visc. modifier, base for mfg. of ethoxylated deriv. [Albright & Wilson UK]

Empilan® LME. CAS 142-78-9; EINECS 205-560-1; Lauramide MEA; foam booster/stabilizer in detergent systems; stabilizer for hair and carpet shampoos; base for mfg. of ethoxylated alkylolamides; visc. modifier. [Albright & Wilson UK]

Empilan® LP10. CAS 61791-08-0; Ethoxylated cocamide; foam and detergent booster. [Albright & Wilson UK]

Empilan® MAA. CAS 61791-08-0; PEG-6 cocamide; liq. detergent additive as foam stabilizer, detergent booster, solubilizer. [Albright & Wilson UK]

Empilan® NP9. CAS 9016-45-9; Nonoxynol-9; wetting agent, detergent, emulsifier, solubilizer; for agricultural emulsifiable and suspension concs., leather processing, emulsion and suspension polymerization; mortar plasticizer; plastics antistat. [Albright & Wilson UK]

Empilan® OPE9.5. Octyl phenol ethoxylate; emulsifier for agricultural emulsifiable and suspension concs. [Albright & Wilson UK]

Empilan® P7061. EO/PO condensate; wetting agent for agricultural emulsifiable and suspension concs.; compounding aid for pharmaceuticals. [Albright & Wilson UK]

Empilan® SM Series. Sorbitan ester ethoxylates; emulsifier for agricultural emulsifiable concs. [Albright & Wilson UK]

Empimin®. Alcohol ethoxy sulfates. [Albright & Wilson Ltd, Detergents Div] *

Empimin® 3060. Surfactant blend; high expansion fire fighting foam conc. [Albright & Wilson UK]

Empimin® 3116. Sulfosuccinate blend; flotation aid for phosphate rock. [Albright & Wilson UK]

Empimin® BMA. Formulated surfactant; surfactant used as air entraining agent for cementitious mixes, foaming agent for concrete, mortar plasticizer. [Albright & Wilson UK]

Empimin® BMB. Formulated surfactant; plasticizer/air entraining agent for construction industry. [Albright & Wilson UK]

Empimin® BMC. Formulated product; air entraining agent for mortar/cement. [Albright & Wilson UK]

Empimin® KSN27. CAS 9004-82-4; Sodium laureth sulfate (3 EO); detergent raw material for high-quality liq. detergents, textile and leather processing; emulsifier, dispersant for emulsion polymerization; biodeg. [Albright & Wilson UK]

Empimin® LAM30/AU. CAS 75422-21-0; Ammonium alkyl ether sulfate; wetting and foaming agent used in mfg. of plasterboard, geological drilling operations. [Albright & Wilson Australia]

Empimin® LR28. CAS 151-21-3; Sodium lauryl sulfate; foaming agent in rubber latex systems; emulsifier for emulsion polymerization. [Albright & Wilson UK]

Empimin® LSM30. Sodium alkyl ether sulfate; foaming agent for slurries in plasterboard mfg.; foam-boosting additive in latices for carpet backing; wetting agent; ingred. in alkaline liq. cleaners. [Albright & Wilson UK]

Empimin® MA. CAS 3006-15-3; Sodium dihexyl sulfosuccinate; emulsifier, dispersant, wetting agent for emulsion polymerization, agricultural emulsifiable concs. [Albright & Wilson UK]

Empimin® MH. Disodium cocoyl sulfosuccinamate; foaming agent for carpet backing. [Albright & Wilson UK]

Empimin® MHH. Disodium N-lauryl sulfosuccinamate as a pale cream liquid; foaming agent for rubber latices eg in carpet manufacture. [Albright & Wilson Ltd, Detergents

---

* Trade name not verified as available   † Trade name verified as obsolete

Div, Marchon] *

Empimin® MK/B. Disodium cetyl stearyl sulfosuccinamate; foaming agent for carpet backing. [Albright & Wilson UK]

Empimin® MKK. CAS 14481-60-8; EINECS 238-479-5; Disodium N-stearyl sulfosuccinamate as a pale cream soft paste or spray dried powder; foaming agent for rubber latexes e.g., in carpet manufacture. [Albright & Wilson Ltd, Detergents Div, Marchon] *

Empimin® MSS. Diammonium N-lauryl sulfosuccinamate as a pale cream liquid; foaming agent for rubber latexes eg in carpet manufacture. [Albright & Wilson Ltd, Detergents Div, Marchon] *

Empimin® MTT. Disodium N-oleyl sulfosuccinamate as a pale cream liquid; foaming agent for rubber latices e.g., in carpet manufacture. [Albright & Wilson Ltd, Detergents Div, Marchon] *

Empimin® OP70. CAS 1369-66-3; Sodium dioctyl sulfsuccinate; emulsifier, dispersant, wetting agent used for emulsion polymerization, oil slicks, textiles, agrochem. [Albright & Wilson UK]

Empimin®OT. CAS 577-11-7; EINECS 209-406-4; Sodium dioctyl sulfosuccinate; dispersant, emulsifier, detergent, wetting agent for o/w emulsions, agrochem., emulsion polymerization, filler and extender dispersions, leather processing. [Albright & Wilson UK]

Empimin® SDS. CAS 84501-49-5; Sodium decyl sulfate; emulsifier, dispersant, detergent, and wetting agent for industrial and institutional cleansers, mfg. of pigments, alkaline cleansers; dust suppression. [Albright & Wilson Australia]

Empimin® SQ25. CAS 68585-34-2; Sodium alkyl triethoxysulfate; dispersant, wetting and foaming agent in institutional, household, and industrial cleaners. [Albright & Wilson Australia]

Empiphos. A synthetic detergent comprising aliphatic phosphate. [Albright & Wilson Ltd] *

Empiphos DF Series. Phosphate esters; emulsifier for agricultural emulsifiable concs. [Albright & Wilson UK]

Empiquaternary An alkyl dimethylbenzyl ammonium chloride. [Albright & Wilson Ltd] *

Empirin. Proprietary formulations containing aspirin or aspirin and codeine phosphate; tablets; a general analgesic and anti-inflammatory agent. [The Wellcome Foundation Ltd]

Empiwax. Oil/water emulsifier; pharmaceutical and toilet preparations, ointments; penicillin cream bases. [Albright & Wilson Ltd, Detergents Div]

Empiwax SK. Cetearyl alcohol and sodium lauryl sulfate; SE wax as o/w emulsifier for pharmaceutical and toilet preparations and ointments. [Albright & Wilson UK]

Emplets Potassium Chloride. CAS 7447-40-7; EINECS 231-211-8; Potassium chloride; replenisher. [Parke-Davis] *

Emplex. CAS 25383-99-7; EINECS 246-929-7; Sodium stearoyl lactylate; starch and protein complexing agent for bakery products; emulsifier, conditioning agent, softener for processed foods. [Am. Ingredients/Patco]

Empol® 1004. Dimer acid, hydrogenated; surfactant for industrial applications, lubricants. [Henkel/Emery]

Empol® 1010. Dimer acid; polymer grade; surfactant for industrial applications, lubricants. [Henkel/Emery]

Empol®1040. Trimer acid; surfactant for industrial applications, lubricants. [Henkel/Emery]

Empol®1061. Dimer acid; surfactant for industrial applica-

tions, lubricants. [Henkel/Emery]

Empracet Codeine Phosphate. A proprietary formulation of codeine phosphate and acetaminophen; for relief of mild to moderately severe pain. [The Wellcome Foundation Ltd]

Emralon. Dispersions of PTFE in resin; used for dry film lubricants and parting agents for rigid and flexible substrates. [Acheson Colloids]

Emralon 304. CAS 9002-84-0; Colloidal PTFE in mixt. of alcohol/esters/aromatics; lubricant additives for dry gear and chain lubes, aerosols, assembly and thread lubes. [Acheson]

Emralon 8301-01; CAS 9002-84-0; Colloidal PTFE in water; lubricant additive for dry gear and chain lubes, aerosols, assembly and thread lubes. [Acheson]

EMS 209. Microsilica with a proprietary coating and treatment; incorporated in extruded PVC pipe to increase elastic modulus and pipe stiffness, increase impact resistance and to increase volumetric output from the extruder. [Elkem Chemicals Inc] *

Emsac Concrete Additive. Microsilica (condensed silica fume) based dry or slurried products formulated with dispensing agents such as those commonly used in the concrete industry meeting the requirements of ASTM C494; used to improve the strength and durability of ordinary Portland cement concretes, grouts and mortars. [Elkem Chemicals Inc] *

Emsodur. Cubic grit produced to close dimensional tolerances (0.5-3 mm); highly resistant to abrasion; for deflashing of thermoset molded parts, rubber parts, deflashing of small electric components, tool cleaning, varnish removal from sensitive surfaces, other surface treatments. [EMS-Am. Grilon; EMS-Grilon (UK) Ltd]

Emsodur Micro. Special deflashing grit with a cube size of 200-400 microns; for deflashing of the smallest electronic components. [EMS-Am. Grilon]

Emsorb® 2500. CAS 1338-43-8; EINECS 215-665-4; Sorbitan oleate; coupler, emulsifier, lubricant, and softener for textile fibers, leather, cosmetics, agriculture, household products; for formulating petrol. oils and waxes, natural fats and waxes, and alkyl esters. [Henkel/Emery; Henkel/Textile]

Emsorb® 2502. CAS 8007-43-0; EINECS 232-360-1; Sorbitan sesquioleate; coupler, coemulsifier for o/w systems; emulsifier for w/o systems; household aerosols, cosmetics, agriculture, industrial and textile oils. [Henkel/Emery; Henkel/Textile]

Emsorb® 2503. CAS 26266-58-0; EINECS 247-569-3; Sorbitan trioleate; coupler, emulsifier, lubricant, and softener for textile fibers and leather, cosmetics, agriculture, household products; for formulating petrol. oils and waxes, natural fats and waxes, and alkyl esters; coemulsifier for mineral oil. [Henkel/Emery; Henkel/Textile]

Emsorb® 2505. CAS 1338-41-6; EINECS 215-664-9; Sorbitan stearate; coupler, hydrophobic emulsifier; coemulsifier for industrial oils, household products, agriculture, and cosmetics; textile lubricant; paper and textile processing. [Henkel/Emery]

Emsorb® 2507. CAS 26658-19-5; EINECS 247-891-4; Sorbitan tristearate; coupler, emulsifier, lubricant, and softener for textile fibers and leather, cosmetics, agriculture, household products; for formulating petrol. oils and waxes, natural fats and waxes, and alkyl esters; coemulsifier for mineral oil. [Henkel/Emery]

---

‡ Trade name and manufacturer not verified   § Trade name without identified manufacturer

Emsorb® 2510. CAS 26266-57-9; EINECS 247-568-8; Sorbitan palmitate; coupler, emulsifier for cosmetic, agriculture and household products; fiber-to-metal lubricant. [Henkel/Emery]

Emsorb® 2515. CAS 1338-39-2; EINECS 215-663-3; Sorbitan laurate; coupler, emulsifier; used in household specialities, industrial oils, agriculture, cosmetics, and emulsion polymerization; antifoam properties. [Henkel/Emery; Henkel/Textile]

Emsorb® 2516. CAS 71902-01-7; Sorbitan isostearate; auxiliary emulsifier, solubilizer, corrosion inhibitor in lubricants, metal protectants and cleaners, emulsion polymerization. [Henkel/Emery]

Emsorb® 2518. CAS 68238-87-9; Sorbitan diisostearate; auxiliary emulsifier, solubilizer, corrosion inhibitor in lubricants, metal protectants and cleaners, emulsion polymerization. [Henkel/Emery]

Emsorb® 2720. CAS 9005-64-5; Polysorbate 20; o/w emulsifier, solubilizer, visc. modifier; used in creams, lotions, shampoos, conditioners, liq. soaps. [Henkel/Cospha; Henkel Canada]

Emsorb® 2721. CAS 9005-64-5; PEG-80 sorbitan laurate; wetting agent, dispersant, mild cleanser for baby products; anti-irritant. [Henkel/Cospha; Henkel Canada]

Emsorb® 2722. CAS 9005-65-6; Polysorbate 80; emulsifier, coemulsifier for cosmetics; dispersant for pigments in makeup; solubilizer for oils, flavors, fragrances. [Henkel/Cospha; Henkel Canada]

Emsorb® 2726. PEG-40 sorbitan diisostearate; emulsifier, solubilizer for flavors in mouthwashes, lipstick, perfume, for perfumes, germicides, and other polar substances in aq. systems. [Henkel/Cospha; Henkel Canada]

Emsorb® 2728. CAS 9005-67-8; Polysorbate 60; o/w emulsifier for cosmetics, hair straighteners, shaving products, sun care products; binder in powds.; with Emsorb 2505 to stabilize wax emulsions. [Henkel/Cospha; Henkel Canada]

Emsorb® 6900. CAS 9005-65-6; PEG-20 sorbitan oleate; dispersant for pigments in coatings; solubilizer for oils and fragrances; hydrophilic emulsifier; coemulsifier for petrol. oils, fats, solvs., and waxes in cosmetics, household products, industrial lubricants, and textile dye carriers; emulsifier for tobacco sucker control concs. [Henkel/Emery; Henkel/Textile]

Emsorb® 6901. CAS 9005-65-6; PEG-5 sorbitan oleate; o/w emulsifier and lubricant in industrial lubricants, textile lubricants and softeners, metal treatment, paints, emulsion polymerization, agriculture; color dispersant in plastics. [Henkel/Emery; Henkel/Textile]

Emsorb® 6903. CAS 9005-70-3; PEG-20 sorbitan trioleate; o/w emulsifier for petrol. oils, fats, waxes, and alkyl esters; lubricant for metals, textiles, leather; in sol. oils for metal processing and finishing; glass fiber lubricants; automotive lubricant additives; agricultural formulations. [Henkel/Emery; Henkel/Textile]

Emsorb® 6906. CAS 9005-67-8; PEG-4 sorbitan stearate; w/o emulsifier used in household formulations; fiber-to-metal lubricant for synthetic and cellulosic fibers and yarns. [Henkel/Emery]

Emsorb® 6908. PEG-16 sorbitan tristearate; o/w emulsifier, lubricant, softener; for textile processing and finishing compounds, agricultural formulations. [Henkel/Emery]

Emsorb® 6909. CAS 9005-67-8; PEG-4 sorbitan stearate; emulsifier for hydraulic fluids, metal treatment, emulsion polymerization, paints; color dispersants for plastics. [Henkel/Emery]

Emsorb®6913. CAS 9005-70-3; PEG-20 sorbitan trioleate; emulsifier for agricultural formulations. [Henkel/Emery]

Emsorb®6915. CAS 9005-64-5; PEG-20 sorbitan laurate; o/w emulsifier and solubilizer of petrol. oils, solvs., and fats; used in cosmetic creams and lotions; visc. modifier in shampoos; emulsifier for dye carriers, antistatic scrooping agent in primary spin finishes, and fiber processing aid in textile industry; agricultural formulations. [Henkel/Emery; Henkel/Textile]

Emsorb® 6917. PEG-16 sorbitan trioleate; o/w emulsifier, lubricant for metals, agriculture, textiles, leather, glass fiber, automotive additives. [Henkel/Emery]

Emthox® 2730. PEG-75 cocoa butter glycerides; emollient, humectant, emulsifier for personal care products [Henkel/Cospha; Henkel Canada]

Emthox® 5882. Laureth-4; dispersant for bath oil; emulsifier for creams and lotions; wetting agent; for eye make-up, deodorants, hair coloring products [Henkel/Cospha; Henkel Canada]

Emthox® 5885. CAS 68439-49-6; Ceteareth-20; emulsifier, solubilizer for cosmetics, conditioners, depilatories, hair straighteners, sun care products [Henkel/Cospha; Henkel Canada]

Emthox® 5967. Laureth-12; emollient, thickener in shampoos; emulsifier in creams and lotions. [Henkel/Cospha; Henkel Canada]

Emtryl. CAS 551-92-8; Dimetridazole; a veterinary antiprotozoan. [RMB Animal Health Ltd]

Emulamid TO-21. CAS 68092-28-4; Tall oil fatty alkanolamide; emulsifier, lubricant, corrosion inhibitor, wetting agent for water-extendable metalworking coolants. [Mayco Oil & Chem.]

Emulan. Mink oil; emollient with high spreading coefficient for skin and hair care formulations requiring good oxidative stability. [Emulan]

Emulan®. Emulsifiers; used for impregnation, lubrication, polishing and cleaning; anticorrosive emulsifier. [BASF plc]

Emulan® A. CAS 9004-96-0; Oleic acid oxyethylate; o/w emulsifier for mineral oils and metal polishing emulsions. [BASF AG]

Emulan®AF. Fatty alcohol oxyethylate; emulsifier for paraffin wax and mineral oil emulsions. [BASF AG]

Emulan® EL. Castor oil oxyethylate; emulsifier for nonaq. solvs. in water processing, light chemicals industry, for emulsifiable concs. for crop protection. [BASF AG]

Emulan® OC. Fatty alcohol oxyethylate; o/w emulsifier for dry bright emulsions, waxes; dispersant, stabilizer for emulsions. [BASF AG]

Emulan®PO. Alkylphenol oxyethylate; emulsifier for cleaners, formwork release oils, drilling and cutting oils; flotation agent. [BASF AG]

Emulan® PO. Nonionic surfactant of the alkylphenol ethoxylate type as a colorless to pale yellow clear oil; used for emulsification, mainly in combination with other emulsifiers; used in cold cleaners and other solvent based cleaners; drilling oil. [BASF AG] *

Emulcid. Hexahydro-1,3,5-tris (2-hydroxyethyl)-s-triazine; liquid; wet-state bactericide/fungicide for metalworking fluids, detergents, cooling waters, and for industrial plant cleaning. [Thor Chemicals (UK) Ltd]

Emuldan HV 40 Kosher, HV 52 Kosher. Hydrogenated

---

* Trade name not verified as available   † Trade name verified as obsolete

vegetable oil monodiglycerides; food emulsifier and stabilizer. [Grindsted Prods.]

Emulgade 1000 NI. Cetearyl alcohol and ceteareth-20; emulsifying agent, SE o/w base for creams, lotions. [Henkel/Cospha; Henkel Canada; Henkel KGaA]

Emulgade C. Emulsifying wax NF; self-emulsifying raw material for emulsions, hair conditioners, creams, lotions. [Henkel/Cospha; Henkel Canada]

Emulgade EO-10. Oleth-25 and cetyl alcohol; self-emulsifying base for preparation of creams and emulsions. [Pulcra SA]

Emulgade F. Cetearyl alcohol, PEG-40 castor oil, and sodium cetearyl sulfate; SE raw material for cosmetic creams and fluid emulsions; emulsifier for cosmetics, pharmaceuticals. [Henkel/Cospha; Henkel Canada]

Emulgator E 2149. Steareth-7, stearyl alcohol; emulsifier and stabilizer for cosmetic and pharmaceutical o/w emulsions. [Goldschmidt; Goldschmidt AC]

Emulgator E 2155. Stearyl alcohol, steareth-7, steareth-10; emulsifier and stabilizer for cosmetic and pharmaceutical o/w emulsions. [Goldschmidt; Goldschmidt AG]

Emulgator E 2568 SE. CAS 68439-49-6; Ceteareth-25; emulsifier and stabilizer for cosmetic and pharmaceutical o/w emulsions. [Goldschmidt; Goldschmidt AG]

Emulgator U4. CAS 9016-45-9; Nonoxynol-4; surfactant, emulsifier, wetting agent for liq. and powd. detergents, industrial preparations, textile, leather, pulp/paper, drycleaning, metal cleaners. [Unger Fabrikker AS]

Emulgeant 710. Polyethoxy ether; emulsifier for aliphatic and aromatic solvs., paraffin; in degreasers. [Ceca SA]

Emulgeen P. CAS 7492-30-0; EINECS 231-314-8; A proprietary trade name for potassium ricinoleate. [S & D Chemicals Ltd] *

Emulgeen S. CAS 5323-95-5; EINECS 226-191-2; A proprietary trade name for sodium ricinoleate. [S & D Chemicals Ltd] *

emulgen. A jelly-like mass used for the rapid emulsification of oils and resins. It contains tragacanth, gum arabic, pittoporad, glycerin, alcohol, and water.

Emulphogene®. Redesignated Rhodasurf® [Rhone-Poulenc Surf.]

Emulphopal HC. Proprietary; waterless hand cleaner base. [Stepan Canada]

Emulphor®. Redesignated Alkamuls® or Rhodasurf® [Rhone-Poulenc Surf.]

Emulpon. Emulsifiers, dispersants, solubilizers for oils, solvents and waxes; used for textile, agriculture and cosmetic industries. [Witco Chemical Ltd] *

Emulsamin. A proprietary trade name for menthol diurethane, a wetting agent and detergent. §

Emulsene. Emulsifying agents. [Bush Boake Allen Ltd]

Emulsiderm. Pale blue/green liquid emulsion containing 0.5% benzalkonium chloride BP, 25% liquid paraffin BP, 25% isopropyl myristate BPC; an aid in the treatment of dry skin conditions, especially those associated with eczema, ichthyosis or xeroderma; permits rehydration of the keratin by replacing lost lipids and its antiseptic properties assist in overcoming secondary infection. [Dermal Laboratories Ltd]

Emulsifier 4. Dialkyl quaternary ammonium chloride in water/IPA; emulsifier for nonpolar hydrophobes, e.g., mineral seal oil, waxes, silicones; used for auto spraywax, carnauba spraywax, mop treatment emulsions, stainless steel cleaners, vinyl dressings. [Exxon/ Tomah]

Emulsifier 632/90%. Modified alkyl phenol ethoxylate; low foam surfactant, dispersant for fats, oils, and waxes. [Ethox]

Emulsifier K 30 40%. Sodium alkane sulfonates based on n-paraffin; emulsifier for emulsion polymerization; effective over wide pH range; antistat; biodeg. [Bayer AG; Miles/Organic Prods.]

Emulsifier L.W. Cyclohexylammonium oleate. [Laporte Industries Ltd] *

Emulsifier WHC. Formulated emulsifier; emulsifier and detergent used in waterless hand cleaner formulations and degreasers. [Stepan; Stepan Canada]

Emulsil. Food grade silicone products; antifoams and release agents. [Siliconas Hispania SA]

Emulsilac S. CAS 25383-99-7; EINECS 246-929-7; Sodium stearoyl lactylate; emulsifier, dough conditioner, freeze/thaw stabilizer for food industry. [Witco/Humko]

emulsin. A ferment; decomposes the glucoside, amygdalin, into grape sugar, benzaldehyde, and hydrocyanic acid.

Emulsion C-340. CAS 8015-86-9; Carnauba wax emulsion; film-former for most substrates incl. fabric, metal, wood, leather, and painted surfaces; for spraywaxes, wood water repellent, protective coatings. [Exxon/Tomah]

Emulsi-Phos. Food grade phosphate; emulsifier specially designed for use in process cheese, cheese food and cheese spreads. [Monsanto Co] *

Emulsogen. Emulsifiers for oils and waxes. [Hoechst UK]

Emulsogen 2144. Ethoxylated rosin; low foaming emulsifier, wetting agent for metalworking fluids, high acidic conditions. [Hoechst Celanese/Colorants & Surf.]

Emulsogen CP 136. Calcium dodecylbenzene sulfonate/ castor oil ethoxylate; surfactant for agricultural formulations. [Hoechst Celanese/Colorants & Surf.]

Emulsogen EL-050. CAS 61791-12-6; PEG-5 castor oil; emulsifier additive for chlorinated paraffins, triglycerides; surfactant for textile processing. [Hoechst Celanese/ Colorants & Surf.]

Emulsogen HEL-050. CAS 61788-85-0; PEG-5 hydrogenated castor oil; surfactant for textile processing. [Hoechst Celanese/Colorants & Surf.]

Emulsogen IC. Calcium dodecylbenzene sulfonate/oleyl alcohol ethoxylate; surfactant for agricultural formulations. [Hoechst Celanese/Colorants & Surf.]

Emulsogen M. Fatty alcohol ethoxylate; mineral oil emulsifier, wetting agent; for metalworking fluids. [Hoechst Celanese/Colorants & Surf.]

Emulsynt GDL. CAS 27638-00-2; EINECS 248-586-9; Glyceryl dilaurate; emulsifier, thickener, and emollient for creams and lotions. [Van Dyk]

Emultex®. Vinyl acetate homo and copolymer dispersions; used for adhesives, surface coating and textiles. [Harlow Chemical Co Ltd]

Emultex. Liquid emulsifier; for sizing oils. [Tensia SA] *

Emultex 307, 328. A proprietary range of unplasticized vinyl acetate homopolymer emulsions stabilized with polyvinyl alcohol and used in the manufacture of adhesives. [Harlow Chemical Co Ltd]

Emultex AC431. A proprietary vinyl acetate-acrylic ester copolymer emulsion; used for general-purpose emulsion paints. [Harlow Chemical Co Ltd]

Emulvin®. Emulsifying agents. [Bayer AG; Bayer plc]

Emulvin® W. Aromatic polyglycol ether; emulsifier, stabilizer, wetting agent for latices. [Miles; Polysar]

---

‡ Trade name and manufacturer not verified   § Trade name without identified manufacturer

Emulzome. Hydrogenated polyisobutene, stearyl heptanoate, mineral oil; for cosmetic creams and milks. [Exsymol]

EMU®Powd. 120 FD. Styrene-based copolymer; binder for production of paints and impregnation/coating materials for paper; for modifying cement-containing dry mortar. [BASF AG]

Emvelop®. CAS 68334-00-9; Hydrogenated vegetable oil; for production of sustained-release tablet formulations. [Mendell]

E-Mycin®. CAS 114-07-8; Erythromycin; antibacterial. [Upjohn] *

E-Mycin E. CAS 41342-53-4; Erythromycin ethylsuccinate; antibacterial. [Upjohn] *

Emzyami No. 1. Phytoenzymatic complex, anti-free radical protector with anti-solar and anti-bacterial activity; helps prevent skin from aging. [Alban Muller]

Emzyami No. 1. Phytoenzymatic complex, anti-free radical protector with anti-solar and anti-bacterial activity; helps prevent skin from aging. [Alban Muller] *

Enadel. CAS 24166-13-0; A proprietary preparation of cloxazolam; an antianxiety agent. [Pfizer International]

enamel white. See lithopone.

Enanth. Nylon 7.

enanthic acid. See heptanoic acid.

enanthic anhydride. See heptanoic acid.

enanthyl alcohol. See heptylalcohol.

Enathene® EA 705-009. Ethylene n-butyl acrylate copolymer; for adhesives for case and carton sealing, bookbinding, laminates, sealants, caulks, textile coatings. [Quantum/USI]

Enathene® EA 720-009. Ethylene n-butyl acrylate copolymer; for extrusion coating; excellent heat seal response at low temps.; good thermal stability at high processing temps.; for coatings and laminations with oriented PP, PET, papers, clay-coated board and nylon substrates for specialty pkg.; FDA-approved. [Quantum/USI]

Enavid 5mg. A proprietary preparation containing norethynodrel and mestranol. [G D Searle & Co Ltd] *

Enavid-E. A proprietary preparation containing norethynodrel and mestranol. [G D Searle & Co Ltd] *

Enbucrilate. Butyl 2-cyanoacrylate; histoacryl; a surgical tissue adhesive. §

Encapsulated MgO. CAS 1309-48-4; Magnesium oxide. [Olin] *

Encapsulation. Thermosetting material; for molded products. [Sumitomo Bakelite Co Ltd] *

Encelac®. Org. and inorg. pigment formulations in ester-sol. nitrocellulose and dibutyl phthalate; for pigmenting nitrocellulose lacquers. [BASF AG]

Encem Steel. A proprietary trade name for a nickel-chromium-molybdenum steel. §

Encephabol®. Pyrithioxin. [E Merck]

Enceprint®. Org. and inorg. pigment formulations in a low-visc. alcoholic nitrocellulose containing plasticizers; for pigmenting of flexographic and gravure inks. [BASF AG]

Encore. Suspension concentrate containing 125 g isoproturon and 250 g pendimethalin per liter; used for annual weed control in winter wheat, rye and barley. [Cyanamid of Great Britain Ltd]

Endanil. Nylon dyes and pigments for plastics. [ICI Chem. & Polymers Ltd]

Endcor. A wide range of high performance, corrosion-resistant coating systems. Types include: acrylic, alkyd, chlorinated rubber, epoxy, pretreatments, primers, polyurethane, vinyl and zinc rich; general industrial maintenance for metal and masonry surfaces in a wide variety of exposure conditions, including weathering and chemical attack, immersion service. [Dampney Company Inc]*

Endegal. Polyester dyes and pigments for plastics. [ICI Chem. & Polymers Ltd] †

Endermol. A compound ointment vehicle, containing hydrocarbons of the paraffin series, and stearic acid amide. §

Endobil. Iodoxamic acid; diagnostic aid. [Bracco Industria Chimica SpA] *

Endocaine. A proprietary trade name for pyrrocaine. §

Endocrocine. The orange-yellow coloring matter isolated from *Nephioniopsis endocrocea*, a lichen growing in Japan. It is a hydroxyanthraquinone, $C_{16}H_{10}O_7$.

Endojodin®. CAS 123-47-7; Prolonium iodide; an injectable iodine preparation. [Bayer AG]

Endomirabil. Iodoxamic acid; diagnostic aid. [Bracco Industria Chimica SpA] *

endotryptase. A proteolytic enzyme.

Endox. CAS 1918-00-9, 7085-19-0; Soluble concentrate of 112 g dicamba and 265 g mecoprop per liter; herbicide used for weed control in cereals and grassland. [Farmers Crop Chemicals Ltd]

Endoxan. CAS 50-18-0; Cyclophosphamide; tablets and vials (dry substance); cystostatic. [Degussa AG] *

Endoxana. See Endoxan. [Degussa AG] *

Endrate. CAS 139-33-3; Edetate disodium; chelating agent; pharmaceutic aid. [Abbott Laboratories] *

Endrine. A proprietary preparation of ephedrine, menthol, camphor, and eucalyptol; a nasal decongestant. [Wyeth Laboratories] *

Endspray. Mixture of fentin hydroxide and metoxuron; used for control of potato blight and haulm desicant. [Pan Britannica Industries Ltd]

Endura®. Flame-retardant concs. [PPG/Polymer Prods.]

Endura. Epoxy and polyurethane primers. [Feb Ltd]

Enduracrete. A trowel-applied polyurethane screed with good impact and abrasion resistance; especially suited to hygienic demands; resistant to steam cleaning. [Feb Ltd]

Enduraflex. Two-component pitch-modified polyurethane pouring grade sealant for industrial floors. [Feb Ltd]

Enduraflor. Self-smoothing epoxy system. [Feb Ltd]

Enduragloss. Purified rosin ester gum/2-octyldodecyl myristate; neutral oil improving adhesion, gloss, conditioning in lipsticks, foundtion, eye shadow stick, hair creams. [U.S. Cosmetics]

Enduraguard EP. A pitch epoxy coating incorporating nonslip aggregate; produces high strength, extremely hard wearing coating. [Feb Ltd]

Endurakote. Nonstaining water-dispersible epoxy coating. [Feb Ltd]

Enduralay. A trowel-applied nontainting epoxy system suitable for use in food processing, brewing, and beverage industries. [Feb Ltd]

Enduralith. Premixed polymer-modified cementitious floor topping suitable for medium to heavy duty floor areas. [Feb Ltd]

Enduratop. A high strength polymer modified overlay for areas requiring resistance to impact and abrasion. [Feb Ltd]

En-Dur-Lon. Rubber base coating, contains no abrasive; antislip or decorative coating. [W J Ruscoe Co] †

---

* Trade name not verified as available   † Trade name verified as obsolete

Enduro Alloys. Proprietary corrosion resisting alloys of iron with chromium, or with nickel and chromium; Enduro A contains iron with from 16.5-18.5% chromium; Enduro KA2 contains iron with 17-20% chromium and 7-10% nickel; Enduro S has iron with 12.5-14.5% chromium. §

Endurol. Vat dye colors. [James Robinson & Co Ltd]

Enduron. CAS 135-07-9; A proprietary preparation of methylclothiazide; a diuretic. [Abbott Laboratories]

Endyne. CAS 31842-01-0; Indoprofen; analgesic; anti-inflammatory. [Adria Laboratories Inc] *

Enelbin®. Antirheumatic and antiarthritic pharmaceuticals; antiphlogistic paste. [Cassella AG]

Enelchem Products. Lead pigments, lead oxides, lead silicates, calcium carbonates, etc. [Rheox Inc]

Enerade® 3045. Oxyalkylated polyol; emulsion breaker base. [Rhone-Poulenc Oil Field Chem.]

Enerade® 7101, 7102. Oxyalkylated nonyl phenolic resin; emulsion breaker base. [Rhone-Poulenc Oil Field Chem.]

Energex. A broad range carbohydrase preparation produced by submerged fermentation of a selected strain of the aspergillus niger group; added to animal feed in order to boost feed digestability and improve feed utilization. [Novo Nordisk]

Eneril. CAS 103-90-2; A proprietary preparation of paracetamol; an analgesic. [Nicholas Laboratories Ltd]*

Enervite®. Fish oils, fish liver oils, cod liver oil B. vet C; dietary supplement for animals; conditioning oil for animals. [Seven Seas Ltd]

English bearing metal. Antifriction and fitting metal. It contains usually 53% tin, 33% lead, 10.5% antimony, and 2.5% copper.

English metal. A jeweler's alloy containing 88 parts tin, 2 parts copper, 2 parts brass, 2 parts nickel, 1 part bismuth, 8 parts antimony, and 2 parts tungsten.

English white bearing metal. An antifriction metal, containing 77% tin, 15% antimony, and 8% copper.

engobe. A fusible mixture of clay, telspar, and silica; used for the manufacture of glazes on pottery.

engravers acid. See nitric acid.

Enhance®. For the agriculture industry. [DuPont UK]

Enhance. Floor polishes. [Evans Vanodine International Ltd]

Enhance. Nonionic wetting agent containing 90% alkylphenol ethoxylate; spreader for use in herbicide, fungicide and insecticide sprays. [Midkem Agrochemicals]

Enide. CAS 957-51-7; Diphenamid; used for weed control in horticultural crops. [ICI Agrochemicals]

Enisyl. CAS 657-27-2; Lysine hydrochloride; amino acid. [Person & Covey Inc]

Enkade. CAS 66794-74-9; Encainide hydrochloride; cardiac depressant. [Mead Johnson & Co] *

Enlax. A proprietary preparation; it is a phenolphthalein preparation. §

Enmag. Magnesium ammonium phosphate fertilizer. [Scottish Agricultural Industries plc]

ENR 25. CAS 9006-04-6; Natural rubber; modified by epoxidation reaction to achieve greater oil resist., gas permeability, and damping chars.; used for specialty tires, milking inflations, hoses, footwear. [Guthrie Latex]

ENSA-6. Ethoxylated naphthol sulfonic acid; brightening and dispersing agent for tin plating. [Hart Chem. Ltd.]

Ensecote S. A proprietary trade name for a modified epoxy resin coating material for high temperature stoving or spraying. §

Enso DTO 10 - 30. Distilled tall oils with a rosin acids content of 10-30%; used for alkyd resins, liquid soaps, emulator, latex. [Enso-Gutzeit OY] †

Ensol 2. Tall oil fatty acids with 2% rosin acids; used for alkyd resins, liquid soaps. [Enso-Gutzeit OY] †

Ensoline. Self-emulsifying mineral oils; lubricant for synthetic fibers. [Atochem UK/Ceca]

Ensoline 203 AS. Antistatic oiling of all fibers. [Ceca SA]

Enso Rosin. Tall oil rosin; used for paper sizes. [Enso-Gutzeit OY] †

Entacyl. CAS 142-88-1; A proprietary preparation of piperazine adipate; an antihelmintic. [British Drug Houses] *

Entair. Capsules and syrup containing theophylline and guaiphenesin. Reduces secretion of mucus in chronic bronchitis. (182) .

Entair-A. A proprietary preparation of theophylline, guaiphenisin, and ephedrine hydrochloride; a bronchial antispasmodic. [British Drug Houses] †

Entair Expectorant. A preparation of diphenhydramine hydrochloride and guaifenesin. [British Drug Houses] *

Entamide. A proprietary trade name for diloxamide; an amebicide. [The Boots Co plc] †

Entarex. NTA salts;. chelating agents for trace metal control. [W R Grace Ltd]

Enterfram. A proprietary preparation containing framycetin sulfate and light kaolin; an antidiarrheal. [Fisons plc, Pharmaceuticals Div] *

Entericin. CAS 50-78-2; Aspirin; analgesic; antipyretic; antirheumatic. [Bristol-Myers Squibb Co Inc]

Enteromide. A proprietary trade name for the calcium salt of sulfaloxic acid; an antibiotic. [Consolidated Chemicals Ltd] *

Enterosan. A proprietary preparation of diiodo-hydroxyquinoline chlorodyne, tincture of belladonna and kaolin; an antidiarrheal. [Mayfair Chemicals] *

Entramin. CAS 121-66-4; 2-Amino-5 nitrothiazole premix. [May & Baker Ltd] *

Entramin A. Acetamidonitrothiazole premix. [May & Baker Ltd] *

Entrosalyl. CAS 54-21-7; A proprietary preparation of sodium salicylate in an enteric coated tablet; antirheumatic drug. [Cox Pharmaceuticals] †

Entrosalyl Standard. CAS 54-21-7; A proprietary preparation of enteric coated sodium salicylate. [Cox Pharmaceuticals] †

Entrox. CAS 1314-13-2; Zinc oxide, coated to improve incorporation into rubber. [Durham Chemicals Ltd] *

Enusin Colours. Pigment colors based on casein binders, for aniline and semi-aniline finishing. [Bayer plc]

Envacar. CAS 5714-04-5; A proprietary preparation of guanoxan sulfate; an antihypertensive. [Pfizer International]

Envex®1001. Polyimide, thermoplastic; for ultrasonic transducers, probes, tips; flow meter gears. [Rogers]

Envirez. Unsaturated polyester resins formulated for low smoke, reduced smoke toxicity and low volatility combined with fire retardancy; used for transportation (subways, aircraft etc.) and construction materials requiring both fire retardancy and reduc [Ashland Chemical Company] *

Envirocats. A range of catalysts used in alkylation, oxidation, acetylation, and sulfonylation processes. [Contract Chemicals (Knowsley) Ltd]

---

‡ Trade name and manufacturer not verified

§ Trade name without identified manufacturer

Envirosafe. Cleaner and degreaser. [Kalon Chemicals Ltd]

Enviroseal. Water based water repellent; used for concrete water proofing. [PCR]

Enviroset. Binder system for fuel agglomerates. [Foseco (F.S.) Ltd]

Enzactin. CAS 102-76-1; Triacetin; antifungal. [Wyeth Laboratories] *

Enzeon. Chymotrypsin; enzyme. [Sterling Drug Inc] *

Enzypan. A proprietary preparation of pepsin, pancreatin, and bile; digestive enzyme supplement. [Norgine Ltd]

Enzytol. A 10% aqueous solution of choline borate.

E.O.D. 2-Octyldodecyl erucate; neutral oil improving adhesion, gloss, conditioning in lipsticks, foundtion, eye shadow stick, hair creams. [U.S. Cosmetics]

eosin. CAS 15086-94-9; (eosin A, eosin B, eosin C, eosin A extra, eosin DH, eosin GGF, eosin G Extra, eosin 3J, eosin JJS, eosin G, eosin KS, eosin yellowish, water soluble eosin, acid eosin, eosin 4J extra). The alkali salts of tetrabromo-fluoresceine, $C_{20}H_6O_5Br_4Na_2$; dyes wool and silk yellowish red; used as a microscopic stain and a fluorescent tracer dye; red writing ink, cosmetic products, coloring motor fuel.

eosin J. See erythrosin.

Eosin YS. A British brand of Eosin. §

Eosolate. Silver acetylguaiacoltrisulfonate; an antiseptic.

EP-1. O,O-Diethyl phosphorodithioic acid; used as an intermediate for organophosphorus insecticides. [A/S Cheminova] *

EP-2. O,O-Diethyl phosphorochlorodithioate; mainly used in the production of organophosphorus insecticides. [A/S Cheminova] *

Epal®. Linear primary alcohols. [Ethyl Corp]

Epal® 6. CAS 111-27-3; Hexanol; detergent and emulsifier intermediate. [Ethyl Corp]

Epal® 8. CAS 111-87-5; EINECS 203-917-6; Octanol; detergent and emulsifier intermediate. [Ethyl Corp]

Epal® 10. CAS 112-30-1; Decanol; detergent and emulsifier intermediate. [Ethyl Corp]

Epal® 12. CAS 112-53-8; Dodecanol; detergent and emulsifier intermediate. [Ethyl Corp]

Epal® 14. CAS 112-72-1; EINECS 204-000-3; Tetradecanol; biodeg. detergent/emulsifier intermediate. [Ethyl Corp]

Epal® 16NF. CAS 36653-82-4; EINECS 253-149-0; Hexadecanol NF; USP grade biodeg. detergent/emulsifier intermediate. [Ethyl Corp]

Epal® 18NF. CAS 112-92-5; EINECS 204-017-6; Octadecanol NF; USP grade biodeg. detergent/emulsifier intermediate. [Ethyl Corp]

Epal® 20+. CAS 68921-61-5; EINECS 272-778-1; C18-C32 linear and branched alcohols (66%), C24-C40 hydrocarbons (40%); biodeg. detergent/emulsifier intermediate. [Ethyl Corp]

Epal® 108. Octanol, decanol, hexanol (55:41:4); biodeg. detergent/emulsifier intermediate. [Ethyl Corp]

Epal® 610. C8 alcohol (42%), decyl alcohol (54%), hexanol (4%); biodeg. detergent/emulsifier intermediate. [Ethyl Corp]

Epal® 810. C8 alcohol (45%), decanol (54%); biodeg. detergent/emulsifier intermediate. [Ethyl Corp]

Epal® 1012. Decyl alcohol (75%), dodecanol (23%); biodeg. detergent/emulsifier intermediate. [Ethyl Corp]

Epal® 1214. Dodecanol (66%), tetradecanol (27%), hexadecanol (6%); biodeg. detergent/emulsifier inter-

mediate. [Ethyl Corp]

Epal® 1412. Tetradecanol (58%), dodecanol (40%); intermediate for surfactants, plasticizers, lubricant additives, thioesters, specialty chems.. [Ethyl Corp]

Epal® 1618. Hexadecanol (47%), octadecanol (50%); biodeg. detergent/emulsifier intermediate. [Ethyl Corp]

Epanutin. CAS 630-93-3; A proprietary preparation of phenytoin sodium; an anticonvulsant. [Parke-Davis] *

ephedrine. CAS 299-42-3; An alkaloid obtained from the Chinese plant Ma Huang. It is α-phenyl-β-methylaminopropanol, $C_6H_5 \cdot CHOH \cdot CH \cdot CH_3NH \cdot CH_3$; used as a mydriatic and as an atropine substitute in ophthalmology.

ephos. A basic phosphate, containing 60-65% tricalcium phosphate.

Ephynal. CAS 59-02-9; A proprietary preparation of vitamin E; a vasodilator. [Roche Products Ltd] *

Epibond® 1217-A/B. Quick-set adhesive paste for aircraft industry; rapid cure; excellent adhesion to steel, Al, wood, plastics. [Ciba-Geigy/Furane]

Epibond® 1544-A/B. Self-extinguishing adhesive for thermoplastics in aircraft industry; low flow; bonds dissimilar substrates. [Ciba-Geigy/Furane]

epichlorhydrin. (chloromethyl oxirane; chloropropylene oxide) CAS 106-89-8; Major raw material for epoxy and phenoxy resins, manufacture of glycerol, curing propylene-based rubbers; solvent for cellulose esters and ethers; high wet-strength resins for paper industry.

epichlorohydrin elastomer. (chloropropylene oxide elastomer) Epichlorohydrin elastomer; $(CHCH_2ClCH_2O)_x$; elastomer used for fuel pump diaphragms, pipe gaskets, hose for fuel, oil, and gas, vibration isolators, motor mounts, rolls, adhesives, sponge goods, air conditioning hose and seals. [Ciba-Geigy; Conap; Hardman; Key Polymer; Morton Int'l.; Reichhold; Rhone-Poulenc/Perf. Resins; Sartomer; Shell; Union Carbide]

Epiclon. Anydyride epoxy curing agents. [Anchor Chemical Group plc]

EPICS. Multiparameter computer and laser instrument; used for counting and sorting types of biological cells. [Coulter Electronics Ltd]

Epi-Cure® 87. Modified aliphatic amine adduct; curing agent for applications where relatively short cure periods at room temperature required; end uses incl. tooling gel coats, laminating compounds, adhesives. [Shell]

Epi-Cure 8515. A proprietary amidoamine curing agent for epoxy resins. [Alma Paint & Varnish Co] ‡

Epidermin in Oil. Polyvalent tissue complex in oil medium; product for aging and sensitive skin. [Dr. Kurt Richter; Henkel/Cospha]

Epidermin Water-Soluble. Polyvalent tissue complex, hydro-alcohol solubilized; product for aging and sensitive skin. [Dr. Kurt Richter; Henkel/Cospha]

Epiflex. A proprietary trade name for an epoxy resin expansion jointing material. [Haworth (ARC) Ltd] ‡

Epifoam. Muco-adherent, white, odorless foam containing hydrocortisone acetate and pramoxine hydrochloride; for treatment of perineal trauma. [Stafford-Miller] ‡

Epifrin®. CAS 51-43-4; Epinephrine USP; antiglaucoma agent. [Allergan, Inc]

Epiglaubite. An impure calcium phosphate.

epihydrin. Propylene oxide, C_3H_6O.

Epikem. Nonstaining epoxy adhesive and grouting system used in the food processing and brewing industries. [Feb

---

* Trade name not verified as available   † Trade name verified as obsolete

Ltd]
Epikote. A series of epoxy resins whose characteristics may be modified by hardeners and other additives. [Shell]
Epikote DX-209-B-80, DX-210-B-80. A proprietary 80% solution of epoxy resin in methyl ethyl ketone; used with Epikure 3400 as a curing agent. [Shell] *
Epikote DX-231-B-91. A proprietary solution containing 91% epoxy resin in methyl ethyl ketone for use in work involving carbon film; it is cured with Epikure 3400. [Shell] *
Epikure 3400. A proprietary curing agent for epoxy resins. [Shell] *
Epilim. A proprietary preparation of sodium valproate; an anticonvulsant. [Reckitts Colours Ltd] *
Epilink. A range of water-based, solvent-free and solvent-based curing agents for use with epoxy resins. [Akzo Chemie NV]
E-Pilo-1®, -2®, -3®, -4®, -6®. Pilocarpine hydrochloride and epinephrine bitartrate; indicated for the treatment of chronic primary open-angle glaucoma; may be used in combination with carbonic anhydrase inhibitors, hyperosmotic agents, or beta-adrenergic blocking agents. [Iolab Corp]
Epilok. Epoxy curing agents. [Akzo Chemie] *
Epilon. A proprietary trade name for an epoxy resin cement. [Haworth (ARC) Ltd] ‡
Epinal. Epinephryl borate; adrenergic. [Alcon Laboratories Inc]
Epinalin. A proprietary preparation of adrenaline and ephedrine in solution. §
Epiphassol. A viscous oily liquid containing naphthenesulfonic acids. It is a similar preparation to Kontakt, and is used for cleaning cotton fabrics.
Epiphen. Epoxy resins. [Monomer-Polymer & Dajac Laboratories Inc]
Epires. Epoxy resin. [Akzo Chemie UK Ltd]
Epirez 501. CAS 2426-08-6; Butyl glycidyl ether; a proprietary reactive diluent for epoxy resin systems. [Alma Paint & Varnish Co] ‡
Epirez 502. A proprietary aliphatic diepoxide. [Alma Paint & Varnish Co] ‡
Epirez 520C. A proprietary epoxy resin of the bisphenol A type. [Alma Paint & Varnish Co] ‡
Episol. Chlorodiethylaminoethoxyphenylbenzthiazole (Halethazole); an antifungal agent. [Crookes Laboratories]
Episol. Dyes for leather. [James Robinson & Co Ltd]
EPlstatic® 100. Tribasic lead sulfate; heat stabilizer for flexible and rigid PVC compounds [Eagle-Picher] *
EPlstatic® 110. Basic lead silicosulfate; heat stabilizer for use in low-temp. flexible PVC. [Eagle-Picher] *
Epitar. An epoxy resin additive. [Midland-Yorkshire Tar Distillers] ‡
Epitate. An epoxy resin additive. [Midland-Yorkshire Tar Distillers] ‡
EPlthal 120. CAS 1344-40-7; Basic lead phthalate; heat stabilizer in 90 and 105 C PVC wire and cable compounds [Eagle-Picher] *
Epitone. 7 and 8-Acetyl-5-isopropyl-2-methylbicyclo (2.2.2)oct-2-ene. [Quest Int'l. UK Ltd]
Epitrate. CAS 51-42-3; Epinephrine bitartrate; adrenergic. [Wyeth Laboratories] *
Epivax. A proprietary vaccine used against canine distemper. [Coopers Animal Health Ltd] *

EP Lead. Specially-prepared lead naphthenate; used in formulating extreme pressure gear oils and greases. [Hüls Am.]
Eplink. Epoxy curing agents. [Akzo Chemie UK Ltd]
EPM rubber. (ethylene-propylene copolymer; EPR) CAS 9010-79-1; Rubber for injection molded and extruded goods (elec. components, wire insulation, o-rings, brake components); modifying PP and other plastics. [Hüls AG]
Epocap®. Two-part epoxy compounds. [Harcros UK, Hardman]
Epocap®16129 A/B. CAS 25928-94-3; Epoxy; flexible, low visc., unfilled system with excellent crack resist. and moisture protection for stress sensitive components; excellent penetration for cost effective sand potting. [Hardman]
Epocap® 16358 A/B. CAS 25928-94-3; Epoxy; rigid, high heat temp. distort., filled, flame retardant system with excellent thermal conduct., very low exotherm; excellent high temp. elec. props. [Hardman]
Epocast 1610. Syntactic lightweight honeycomb structure reinforcement compound [Ciba-Geigy Plastics UK] *
Epocrete. Two-part epoxy concrete materials. [Hardman]
Epocure. Epoxy curing agents. [Hardman]
Epodil. Glycidyl ether reactive dilutents for epoxy resins. [Anchor Chemical Group plc]
Epodur. High solids and solventless epoxy maintenance coatings for long-term corrosion protection under extreme exposure conditions; used for buried pipelines, marine and offshore equipment, heavy construction, chemical process equipment, water and sewage works equipment, structural steel, tank linings. [Dampney Company Inc] *
Epodyl. A proprietary preparation of triethyleneglycol diglycidyl ether; an anticancer agent. [ICI Chem. & Polymers Ltd]
Epok. Elastomeric sealant coating. [BP Chemicals Ltd]
Epolast. Two-part epoxy compounds. [Hardman]
Epolene®. Polyolefin waxes. [Eastman]
Epolene® C-10. CAS 9002-88-4; Polyethylene wax; nonemulsifiable low-density wax used in hot melt adhesives and coatings for papers and packaging materials, as paraffin modifiers, in slush and cast molding, rubber compounding, and in rigid and flexible vinyl compounds; high gloss coatings, low water-vapor-transmission rates, grease- and blocking-resistance. [Eastman]
Epolene® C-13. CAS 9002-88-4; Polyolefin resin, low molecular weight; wax designed for use with Epolene waxes or blends containing lower molecular weight materials; used as petrol.-wax modifiers to increase blend visc., improve grease resistance, blocking temp., scuff resistance; and gloss of paraffin; additive for inks and as compounding resins for hot melt adhesives. [Eastman]
Epolene®C-16. Polyolefin resin; nonemulsifiable wax used as hot melt coatings for paper (glossy barrier coatings, readily heat sealable to paper products, metal foils, and plastic films); in petrol.-wax coatings, and in wax-copolymer coatings for increased scuff resistance, gloss stabilization, and hot tack; tolerates high levels of inorganic fillers without drastic increases in melt visc. [Eastman]
Epolene®E-14. Oxidized polyethylene homopolymer resin; low-density and low soften. pt. emulsifiable wax, imparts slip resistance to floor polish films. [Eastman] *
Epolene® N-15. CAS 9003-07-0; PP homopolymer resin;

---

‡ Trade name and manufacturer not verified   § Trade name without identified manufacturer

high melting pt. and hardness; modifier for petrol. waxes to increase resistance to blocking, scuffing, and abrasion, compound resin for hot melt adhesives. [Eastman]

Epolene® N-20, N-21. CAS 9002-88-4; Polyethylene homopolymer wax; nonemulsifiable wax with improved resist. to solvs. and oils, and good hardness; used in cosmetics, hot-melt adhesives, dispersing aids for color concs., cable filling composition, slip additives for printing inks; modifiers for hot-melt highway marking. [Eastman] *

Epoleon® N-7C. Glycine betaine, org. and salt of org. acids, amine compounds, essential oils; odor neutralizing agent for restaurant and household garbage, wastewater treatment, sludge treatment, landfill applications, scrubber systems. [Epoleon Corp. of Am.]

Epoleon® N-100. Glycine betaine, sodium citrate, sodium dihydrogen phosphate, PEG lauryl ether; odor neutralizing agent for scrubber systems, waste water treatment. [Epoleon Corp. of Am.]

Epolite 1301. CAS 25928-94-3; Two-part epoxy resin; general purpose, chip-resistance, epoxy surface coat (fast tack free time); used in models, masters, stretch dies, checking and drill fixtures, etc.; nonsag properties; cures readily under high humidity conditions; mix ratio: 100 parts resin/10 parts hardener by wt. [Hexcel]

Epolite 1302. CAS 25928-94-3; Two-part epoxy resin and hardener; abrasion- and impact-resistance epoxy, metallic black surface coat with rapid cure; for automotive gates, router and trim fixtures, foundry patterns and core boxes, check and inspection fixtures, draw and stretch dies; mix ratio: 100 parts resin/7 parts hardener by wt. [Hexcel]

Epolite 2300. Two-part resin laminating system; laminating system, max, wettability, low visc., high mechanical strength; used in checking and inspection fixtures and gauges, Keller models, spotting racks, router and trim, assembly and drill fixtures; foundry patterns, core boxes, and stretch dies and models; mix ratio: 8 parts resin/1 part hardener by wt. [Hexcel]

Epolite 2315. CAS 25928-94-3; Two-part epoxy system; room temperature curing laminating resin. [Hexcel]

Epolite 3300. CAS 25928-94-3; Two-part epoxy compound; mass casting epoxy compounds, variable hardness, high impact resistance; used in hammer forms, stretch dies, drop hammer, close tolerance masters or models; mix ratio: 100 parts resin/10-30 parts hardener by wt. [Hexcel]

Epolite 5363. CAS 25928-94-3; Two-part epoxy system; adhesive for thin film bonding, encapsulating. [Hexcel]

Epomarine. Two-part epoxy compounds for splash zone and for underwater application. [Hardman]

Epon®. Epoxy resins. [Shell]

Epon® 8280. A proprietary liquid epoxy resin for use in filled compounds. [Shell] *

Eponac®. A range of epoxy resins; used for varnish industry, electrotechnical field and building industry. [AMC SPREA S.p.A.]

Epon® Resin DPL-1911. CAS 25928-94-3; Rubber modified liq. epoxy resin; used for composites exhibiting enhanced fracture toughness. [Shell]

Epontol. Pharmaceutical preparation; for the induction of anesthesia and for intravenous anesthesia of short duration. [Bayer AG] †

Epophen. Epoxide resins. [Borden (UK) Ltd] *

Eposet. Paint brush bristle-setting two-component epoxies. [Hardman]

Eposet. Two-part epoxy adhesives. [Harcros UK]

Eposolve 299-R. Versatile clean-up solv. (non-red label) for cured and uncured epoxies. [Hardman]

Epotal® 181 D. Ethylene polymer dispersion; additive for other dispersions; increases blocking resist. of adhesives; improves slip in paper finishing. [BASF AG]

Epo-Tek® E-3081. Single-component, elec. conductive die attach adhesive; for bonding large chips and substrates [Epoxy Tech.]

Epotuf®. Epoxy resins and hardeners. [Reichhold]

Epotuf® 38-690. Water-reducible epoxy ester. [Reichhold]

Epotuf® Hardener 37-612. Polyamide; epoxy hardener; general purpose hardener, good resiliency and adhesion; used in elec. potting, encapsulating, and casting, flooring, surfacing, coatings, and adhesives. [Reichhold]

Epotuf® Hardener 37-621. Polyamide; epoxy hardener; solution form; used in coatings. [Reichhold]

Epoweld®. Two-part epoxy compounds; structural adhesives. [Harcros UK, Hardman]

Epoweld® 19157. CAS 25928-94-3; Two-part epoxy adhesive; for ski industry applications; excellent cohesive strength and good peel strength. [Hardman]

Epox-S. A proprietary trade name for an epoxidized triester vinyl plasticizer. [Ruco Div] *

Epoxidized X-70 and X-75. Polysiloxane resin in an aliphatic vehicle system; water repellant for masonry. [Nova Chemical Inc] *

Epoxol 7-4. CAS 8013-07-8; EINECS 232-391-0; Epoxidized soybean oil; auxiliary plasticizer, acid scavenger, stabilizer for PVC compounds; food pkg. materials. [Am. Chem. Services]

Epoxol 8-2B. Epoxidized butyl esters of linseed oil fatty acids; auxiliary plasticizer with low visc. low volatility, heat and light stabilization; acid scavenger; stabilizer for PVC compounds; food pkg. materials. [Am. Chem. Services]

Epoxol 9-5. Epoxidized linseed oils; stabilizing plasticizer; food pkg. materials. [Am. Chem. Services]

Epoxol 80, 130. A proprietary trade name for vinyl plasticizers manufactured from soya bean. [Swift & Co] ‡

Epoxol G-5. Linseed oil epoxy resin. §

Epoxy Adhesive. High strength general purpose epoxy adhesive. [Sealocrete Ltd]

epoxyethane. See ethylene oxide.

1,8-epoxy-p-menthane. See eucalyptol.

Epoxy Plus. Epoxy resins used for repairing concrete, grouting and bonding. [SBD Construction Prods. Ltd]

Epoxyprene 50. Epoxidized natural rubber. [Guthrie Latex]

1,2-epoxypropane. See propylene oxide.

2,3-epoxypropyl methacrylate. See glycidyl methacrylate.

Epoxy Putty Pack (EP-3/EHP-12). Epoxy putty resin and hardener; for filling and fairing; can be used underwater. [Wessex Resins & Adhesives Ltd]

epoxy resin. Derived from epichlorohydrin and diethylene glycol; OCH_2CHOCH_2; CAS 25928-94-3; surface coatings, adhesive, casting metal-forming tools and dies; encapsulation of elec. parts. [Asahi Chem Industry Co Ltd; Ciba-Geigy; Conap; Hardman; Key Polymer; Morton Int'l.; Reichhold; Rhone-Poulenc/Perf. Resins; Sartomer; Shell; Union Carbide]

EP Pastes. Aqueous dispersions of organic and inorganic pigments. [Reckitts Colours Ltd]

---

* Trade name not verified as available   † Trade name verified as obsolete

Eppy/N. Epinephryl borate; adrenergic. [Barnes-Hind Inc] *

EPR. See EPM rubber.

Eprolin. CAS 59-02-9; Vitamin E supplement. [Eli Lilly & Co]†

Eprylac. Acrylic adhesives, sealants and coatings. [BP Chemicals Ltd]

Epsikapron. CAS 60-32-2; A proprietary preparation of 6-aminocaproic acid; used in the treatment of fibrinolysis. [KabiVitrum AB] *

Epsilan-M. CAS 59-02-9; Vitamin E supplement. [Adria Laboratories Inc] *

Epsom Salts. CAS 7487-88-9; (salts, salts of England, hair salt, bitter salt). Magnesium sulfate, $MgSO_4 \cdot 7H_2O$.

epsom salts. See magnesium sulfate.

EPsyn® 40-A. EPDM terpolymer; general purpose rubber for med.-loaded extruded and molded compounds requiring short cure times; suggested for wire and cable covering, injection molding, sponge; excellent ozone and heat resist. in vulcanizates; sulfur-vulcanizable; very fast curing with conventional EPDM acceleration systems. [Copolymer Rubber]

EPsyn® 55. EPDM terpolymer; used as additive in SBR, nitrile, neoprene, and other highly sat. rubbers for improved ozone and heat resist.; sulfur-vulcanizable; ultra fast curing with conventional EPDM acceleration systems. [Copolymer Rubber]

EPsyn® 5508. EPDM terpolymer; used in wire and cable, sponge, molded applications, where dimensional stability is important; accepts very high loadings of fillers and extenders; aids processing in blends with other EPDMs; thermoplastic EPDM yielding high green strength; sulfur-vulcanizable; very fast curing with conventional EPDM acceleration systems. [Copolymer Rubber]

EPsyn® P-557. EPDM terpolymer, 50 phr paraffinic oil; general purpose rubber which attains high tensiles and low compression set; sulfur-curable; very fast curing with conventional EPDM accelerator systems. [Copolymer Rubber]

E.P. Tablets®. Paracetamol, caffeine, codeine; analgesic. [Pharmax Ltd]

Eptam 6E. CAS 759-94-4; Emulsifiable concentrate of 720 g EPTC per liter; used for pre planting weed control in potatoes. [Farm Protection Ltd]

Eptoin. CAS 57-41-0; Soluble phenytoin. [The Boots Co plc]

Epurite. A mixture of bleaching powder, iron sulfate, and copper sulfate; used for the production of oxygen, which gas is obtained by action of water.

Equadiol. A proprietary preparation of meprobamate and ethinylestradiol; used for control of menopause. [Wyeth Laboratories] *

Equagesic Tablets. A proprietary preparation of ethoheptazine citrate, meprobamate BP and aspirin; an analgesic with muscle relaxant properties; used mainly for the treatment of musculo-skeletal disorders. [Wyeth Laboratories]

Equal. CAS 22389-47-0; Aspartame; sweetener. [G D Searle & Co] *

equalized guano. Natural guanos, blended or mixed with ammonium salts, to obtain definite proportions of nitrogen and phosphorus; a fertilizer.

Equanil Tablets. CAS 57-53-4; A proprietary preparation of meprobamate; tranquilizer for the short-term treatment of anxiety. [Wyeth Laboratories]

Equaprin. A proprietary preparation containing meprobam-

ate and aspirin; an analgesic. [Wyeth Laboratories] *

Equatrate. A proprietary preparation of meprobamate and pentaerythrityl tetranitrate; used in treatment of angina pectoris. [Wyeth Laboratories] *

Equiben. CAS 26097-80-3; Cambendazole; anthelmintic. [Merck & Co Inc] *

Equionic. Sanitizer/detergent compounds. [Glover (Chemicals) Ltd] ‡

Equipoise. CAS 13103-34-9; Boldenone undecylenate; anabolic. [Bristol-Myers Squibb Co Inc] *

Equipose. CAS 10246-75-0; A proprietary preparation of hydroxyzine pamoate; a sedative. [Pfizer International]†

Equiproxen. CAS 22204-53-1; Naproxen; anti-inflammatory; analgesic; antipyretic. [Syntex Laboratories Inc] *

equisetic acid. CAS 499-12-7; (Citridic acid). aconitic acid, $C_3H_3(COOH)_3$.

Equivert. A proprietary preparation of buclizine hydrochloride; an antinauseant. [Pfizer International] †

Equivurm Plus. CAS 31431-39-7; A proprietary preparation containing mebendazole; veterinary antihelmintic (horses). [Janssen Pharmaceutical Ltd] *

Eqvalan. CAS 70288-86-7; Ivermectin (a mixture of ivermectin component B1a and ivermectin component B1b); antiparasitic. [Merck & Co Inc] *

Era 147. A proprietary trade name for a steel containing 0.22% carbon, 0.20% silicon, 0.04% sulfur, 0.04% phosphorus, 0.50% manganese, 5.00% chromium and 0.50% molybdenum; used for forging steel pressure vessels for service with hydrogen. [Hadfield, George & Co] ‡

Era 164. A proprietary trade name for a steel containing 0.20% carbon, 0.25% silicon, 0.04% sulfur, 0.04% phosphorus and 1.5% manganese; used in the manufacture of forged steel pressure vessels for use with intermediate pressures. [Hadfield, George & Co] ‡

Era CR1. A proprietary trade name for a steel containing 0.06% carbon, 0.30% silicon, 0.04% sulfur, 0.04% phosphorus, 1.0% manganese, 18.50% chromium and 9.00% nickel; used in the manufacture of forged steel pressure vessels with good corrosion resistance. [Hadfield, George & Co] ‡

Era CR15 (CB). A proprietary trade name for a steel containing 0.06% carbon, 0.50% silicon, 0.04% phosphrous, 1.00% manganese, 19.00% chromium, 10.00% nickel and 0.6% niobium. [Hadfield, George & Co] ‡

Eraclene. A registered trademark for low density polyethylene. [Anic Agricoltura Spa] ‡

Eradacin. Capsules containing acrosoxacin; an antibacterial agent used for the treatment of gonorrhea. [Winthrop Laboratories] *

Eradite. Sodium hyposulfite, $Na_2S_2O_4$.

Era® Dyes. These are registered trade names for certain British dyestuffs. §

Era metal. A steel containing 21% chromium and 7% nickel.

Eranol. A form of colloidal iodine.

erbia. See erbium oxide.

erbium. Rare-earth element; Er; CAS 7440-52-0; EINECS 231-160-1; nuclear controls, special alloys, room-temperature laser. [Atomergic Chemetals; Cerac; Noah Chem.; Rhone-Poulenc Basic]

erbium oxide. (erbia) Er_2O_3; CAS 12061-16-4; EINECS 235-045-7; Phosphor activator, infrared-absorbing glass. [Atomergic Chemetals; Cerac; Noah Chem.; Rhone-Poulenc Basic]

---

‡ Trade name and manufacturer not verified   § Trade name without identified manufacturer

Ercal. CAS 379-79-3; A proprietary name for ergotamine tartrate. [Blue Line Chemical Co] ‡

ercerhinol. A colloidal silver.

Ercusol®. Aqueous acrylic polymer dispersion. [Bayer AG; Bayer plc]

Erdmann's reagent. Made by adding 40 cc of concentrated sulfuric acid to 20 drops of a solution containing 10 drops of nitric acid (specific gravity 1.153) and 20 cc of water; used in testing for alkaloids.

Erganol. Dibenzyl ether; a softening agent for cellulose esters.

Ergodryl. A proprietary preparation of ergotamine tartrate, caffeine citrate, and diphenhydramine hydrochloride; used for migraine. [Parke-Davis] *

Ergol. Stated to be benzyl benzoate, $C_6H_5 \cdot COO \cdot CH_2 \cdot C_6H_5$; a proprietary plasticizer; used as a softening agent for cellulose esters. §

Ergomar. CAS 379-79-3; Ergotamine tartrate; analgesic. [Fisons Corp, Pharmaceuticals Div] *

ergometrine. (ergonovine; ergobasine; ergostetrine-ergotocine; ergotrate) CAS 60-79-7; $C_{10}H_{23}O_2N_3$; used in medicine.

Ergostat. CAS 379-79-3; Ergotamine tartrate; analgesic. [Parke-Davis] *

ergot. (Secale cornutum; rye ergot) A dark colored fungus, which attacks damp rye and other grasses, and when contained in flour, causes ergotism. It is a mixture of alkaloids; used in medicine as a vasoconstrictor.

Ergotamine. CAS 113-15-5; p-Hydroxyphenylethylamine; $C_{33}H_{35}O_5N_5$; used in medicine.

Ergotrate Maleate. CAS 129-51-1; Ergonovine maleate; oxytocic. [Eli Lilly & Co]

Eribate. CAS 6381-77-7; Sodium erythorbate. [PMP Fermentation Prods. Inc]

Ericon. A proprietary trade name for a phenol-formaldehyde synthetic resin. §

Erinofort. A proprietary cellulose acetate plastic. §

Erinoid. A proprietary trade name for casein-formaldehyde synthetic resin insulating material. [Mobil] *

Erio. Acid dyes. [Ciba plc] *

Eriochrome. Mordant dyes. [Ciba plc] *

Erioclarite. Pretreatment agents. [Ciba plc] *

Erional. Dyeing and printing assistant. [Ciba plc] *

Erionyl. Acid dyes for polyamide fabrics. [Ciba plc] *

Eriopon. Surface active agents. [Ciba plc] *

Erkantol®. Wetting agents and padding auxilliaries; used for textile finishing. [Bayer AG; Bayer plc]

Erlicki's solution. A hardening agent used in microscopy; consists of potassium dichromate 2.5 parts, calcium sulfate 1 part and water 100 parts.

Ermite. A proprietary trade name for a synthetic resin. §

Ertalon® LFX. Internally lubricated cast nylon; for unlubricated moving parts applications. [Polymer Corp.]

Ertalon®PETP. PETP; used for bearing and wear applications requiring higher dimensional stability. [Polymer Corp]

Ertalyte®. CAS 25038-59-9; PET polyester; thermoplastic with excellent wear resist., low coeff. of friction, superior dimensional stability; for bearings, bushings, seals, spacers, rollers, guides, insulators, food contact parts, pump components, valve parts. [Polymer Corp.]

Erthro. CAS 114-07-8; Erythromycin; antibacterial. [Abbott Laboratories] *

Ertilen. Antibiotic veterinary ethicals. [Ciba plc] *

erucamide. (erucic acid amide; 13-docosenamide; cis 13-docosenamide) Aliphatic amide; $CH_3(CH_2)_7CH=CH-(CH_2)_{11}CONH_2$; CAS 112-84-5; EINECS 204-009-2; foam stabilizer; solvent for waxes and resins, emulsions; slip/antiblock agent for polyethylene. [Akzo; Chemax; Croda; Syn. Prods.; Witco/Humko]

erucic acid amide. See erucamide.

Ervamine. A proprietary trade name for melamine-formaldehyde. §

Ervamix. A proprietary trade name for a fibrous glass reinforced polyester molding compound. §

Ervevax. RA27/3 live attenuated rubella virus vaccine; for vaccination of pre-pubertal children against rubella. [SmithKline Beecham] *

Ervol. White mineral oil. [Witco]

ERYC. CAS 114-07-8; Erythromycin; antibacterial. [Parke-Davis]

EryDerm. CAS 114-07-8; Erythromycin; antibacterial. [Abbott Laboratories] *

Erygel®. CAS 114-07-8; Erythromycin USP; anti-acne gel. [Allergan, Inc]

Erymax®. CAS 114-07-8; Erythromycin USP; anti-acne solution. [Allergan, Inc]

Erypar. CAS 643-22-1; Erythromycin stearate; antibacterial. [Parke-Davis] *

Ery-Ped. CAS 41342-53-4; Erythromycin ethylsuccinate; antibacterial. [Abbott Laboratories]

Erysilin. An erysipelas vaccine for swine. [Glaxo Laboratories] *

Ery-Tab. CAS 114-07-8; Erythromycin; antibacterial. [Abbott Laboratories]

erythorbic acid. (D-erythro-hex-2-enonic acid, γ-lactone; isoascorbic acid) Isomer of ascorbic acid; $C_6H_8O_6$; CAS 89-65-6; atioxidant (industrial, food, brewing), reducing agent in photography. [Ashland; Spice King]

Erythrin. (spirit eosin, methyl eosin, primrose soluble in alcohol, erythrin methyl eosin, eosin soluble in spirit). A dyestuff; the potassium salt of tetrabromofluorescein methyl ether; dyes silk bluish-red, with a reddish fluorescence.

Erythrocin. CAS 3847-29-8; A proprietary preparation containing erythromycin as the lactobionate; an antibiotic. [Abbott Laboratories]

Erythrocin Lactobionate-IV. CAS 3847-29-8; Erythromycin lactobionate for injection; antibacterial. [Abbott Laboratories] *

Erythrolar. CAS 643-22-1; Tablets containing erythromycin stearate BP equivalent to 250mg erythromycin and 500mg erythromycin; each 5ml of reconstituted suspension contains erythomycin ethylsuccinate equivalent to 250mg of erythromycin; used for the prophylaxis and treatment caused by erythromycin-sensitive organisms. [Lagap Pharmaceuticals Ltd]

Erythromid. CAS 114-07-8; A proprietary preparation containing erythromycin; an antibiotic. [Abbott Laboratories] *

Erythroped. CAS 41342-53-4; A proprietary preparation containing erythromycin (as the ethyl succinate); an antibiotic. [Abbott Laboratories] *

Erythrosiderite. A mineral, potassium iron chloride; $2KCl \cdot FeCl_3 \cdot H_2O$.

erythrosin. CAS 16423-68-0; (erythrosin B, erythrosin D, pyrosin B, iodeosin B, eosin Bluish, eosin J, Rose B, dianthine B, primrose soluble, soluble primrose). The sodium or potassium salt of tetraiodofluorescein, $C_{20}H_6O_5I_4Na_2$. Dyes silk and wool bluish- red; used for paper staining, as a sensitizer of silver bromide in the photographic industry, and in the production of orthochromatic dry plates.

erythrosin A. The sodium or potassium salt of triiodofluorescein.

Erythrosin G. (dianthine G, pyrosin G, iodeosin G) The sodium or potassium salt of diiodofluoresceine, $C_{20}H_8O_5I_2Na_2$. Dyes wool yellowish-red, with yellowish-red fluorescence.

ES-7CF/000. Polyethersulfone, 7% carbon fiber-reinforced; thermoplastic resin. [Compounding Tech.] *

Esaflon. SF6 gas. [Montedison UK Ltd] *

Esbatal. A proprietary preparation of benthanidine sulfate; an antihypertensive for the treatment of moderate to severe hypertension. [The Wellcome Foundation Ltd]

Esbenite. A material made from cellulose, powdered mica, and magnesium silicate.

Esbrid NSG-240A. Nylon composite with 50% high strength glass and ceramic fibers; thermoplastic molding composite used as replacements for die cast metal parts in business machines, home appliances, automotive applications. [Thermofil]

Escacure®KB1. Benzyldimethyl ketal; photoinitiator for wh. coatings, inks, photopolymers, electronic photoresists, polyester-styrene wood filler composites; optimum absorp. 250-350 nm. [Sartomer]

Escaid. Specialty hydrocarbon for mineral applications. metal extraction solvent. [Exxon Int'l.] *

Escalol® 507. CAS 21245-02-3; Octyl dimethyl PABA; topical sunscreen. [Van Dyk]

Escalol® 557. CAS 5466-77-3; Octyl methoxycinnamate; sunscreen. [Van Dyk]

Escalol® 567. CAS 131-57-7; EINECS 205-031-5; Benzophenone-3; sunscreen. [Van Dyk]

Escalol® 587. CAS 118-60-5; Octyl salicylate; sunscreen. [Van Dyk]

Escane. Detergent intermediates. [Exxon UK] *

Eschel. A fine-grained light colored smalt (qv).

Eschka mixture. A mixture of 2 parts by weight pure calcined magnesia and 1 part pure anhydrous sodium carbonate; used for the determination of sulfur in coal by heating the coal with the mixture, then adding hydrochloric acid and barium chloride, when barium; sulfate is precipitated.

Esco Extract. A synthetic tannin prepared from sulfonated heavy tar oils, by condensation with formaldehyde, and then forming the chromium salt. §

Escomer. Polyethylene wax. [Exxon Int'l.]

Escopol® R-020. CAS 68527-24-2; Reactive polymer. [Exxon] *

Escor. Polyethylene copolymer; low density polyethylene. [Exxon Int'l.]

Escorene. Linear low density PE, high density PE; specialty plastics. [Exxon Int'l.]

Escorez. Hydrocarbon resins from petroleum feedstocks. [Exxon UK]

Escorpal. CAS 3735-90-8; Phencarbamide; anticholinergic. [Bayer AG] †

EsCort. CAS 73771-04-7; Prednicarbate; glucocorticoid. [Hoechst-Roussel Pharmaceuticals Inc] *

Escorto. A proprietary brand of artificial silk. §

Escoweld. Liquid epoxy adhesives and grouts. [Exxon UK]*

Esdeform. Stain removers. [S & D Chemicals Ltd] *

Esdesol. A paint stripper. [S & D Chemicals Ltd] *

Esdogen. Wetting agent and solvent soaps. [S & D Chemicals Ltd] *

E-Series®Electronics Cleaner/Degreaser 2000. General-purpose cleaning agent, penetrant removing oil, grease, silicone, flux, and adhesive; for PCBs, motors and interfaces. [Chemtronics]

E-Series® Freez-It®. Refrigerant for locating thermal intermittent electronic components during soldering or testing; replaces CFC12 refrigerants; freezes adhesives for easy removal. [Chemtronics]

E-Series®Freez-It®Antistat. High-pressure circuit refrigerant for static-sensitive devices; also for heat sink protection of components during soldering and testing. [Chemtronics]

E-Series® Ultrajet®. Ultra-pure precision duster for removal of dust, lint, and metallic oxide particles. [Chemtronics]

eserine. Physostigmine, $C_{15}H_{21}N_3O_2$; an alkaloid obtained from the calabar bean.

Eshalit. A bakelite (phenol-formaldehyde resin). §

Esi-Cryl 1E10N. CAS 9002-88-4; Polyethylene emulsion (based on Epolene E-10); used for floor finishes, general coatings, inks, lubricants, textiles. [Emulsion Systems]

Esi-Cryl 20/20. Zinc-free polymer; produces excellent maintenance floor finishes with high performance and durability without the need of metal crosslinking for removability. [Emulsion Systems]

Esi-Cryl 40. Polyethylene wax emulsion; provides excellent blk. heel mark and scuff resist., high gloss, color stability to floor finish formulations. [Emulsion Systems]

Esi-Cryl 246. Metal crosslinked acrylic copolymer; used in detergent-resist., high gloss, long-term durability floor finishes. [Emulsion Systems]

Esi-Cryl 325N. CAS 9002-88-4; Polyethylene emulsion (based on AC-325); used for floor finishes, general coatings, inks, lubricants, textiles. [Emulsion Systems]

Esi-Cryl 1540A. CAS 9002-88-4; Polyethylene emulsion (based on AC-540 copolymer); used for adhesives, floor finishes, and general coatings. [Emulsion Systems]

Esi-Cryl Respond I. Acrylic copolymer; CAS 62180-77-2; used for floor finish formulations exhibiting superior gloss and durability. [Emulsion Systems]

Esi-Det 21M. Modified alkanolamide; detergent for hard surface cleaners; visc. builder for aq. systems; base for multipurpose cleaners, strippers, degreasers; high tolerance to alkaline builders, solvs. [Emulsion Systems]

Esi-Det CDA. 1:1 Cocamide DEA; detergent, foam stabilizer, thickener for personal care products, light duty dishwash and household cleaners, industrial products; 100% biodeg. [Emulsion Systems]

Esi-Det EP-20. Complex phosphate ester of POE ethanol; detergent for wide range of products; high compatibility. [Emulsion Systems]

Esidrex. CAS 58-93-5; A proprietary preparation of hydrochlorothiazide (6-chloro-3,4- dihydro-7-sulfamyl-1,2,4-benzothiadiazine 1,1-dioxide); a diuretic. [Ciba plc]

Esidrex-K. A proprietary preparation of hydrochlorothiazide and potassium chloride in slow release core; a diuretic. [Ciba plc] *

Esi-Graph 743. Acrylic quad polymer; graphics arts emulsion for gift wrap printing, general coatings. [Emulsion

---

‡ Trade name and manufacturer not verified   § Trade name without identified manufacturer

Systems]

Esi-Graph 745. Styrenated acrylic emulsion; graphics arts emulsion for general coatings, as extending vehicle. [Emulsion Systems]

Esi-Graph 1045. Acrylic copolymer emulsion; graphics arts emulsion for single vehicle corrugated news ink. [Emulsion Systems]

Esi-Terge 10. Cocamide DEA; foam stabilizer, thickener for household, cosmetic, industrial products [Emulsion Systems]

Esi-Terge 40% Coconut Oil Soap. CAS 8001-31-8; EINECS 232-282-8; Cochin oil soap; detergent for liq. hand soap; mild cleaning and coupling. [Emulsion Systems]

Esi-Terge 320. Phosphated nonylphenoxy polyethoxy ethanol; detergent, visc. aid, coupling and wetting agent for synthetic hard surface cleaners with high caustic conc.; high alkali stable. [Emulsion Systems]

Esi-Terge B-15. 2:1 Cocamide DEA; detergent for mild all-purpose cleaners, foam builder, and thickener, wetting agent; for household and industrial cleaners. [Emulsion Systems]

Esi-Terge HA-20. Modified amine condensate; self-coupling detergent for cleaners, detergent base; visc. builder; for low and high alkali cleaners. [Emulsion Systems]

Esi-Terge L-75. Blend of surfactants, chelating and coupling agents; base for high foaming cleaners for dishwashing, carwashing, all-purpose cleaning, butyl degreasers. [Emulsion Systems]

Esi-Terge N-100. PEG ether surfactant; rapid wetter for formulating mild acid cleaners as well as neutral and mildly alkaline products [Emulsion Systems]

Esi-Terge S-10. 1:1 Cocamide DEA; detergent, emulsifier, foam stabilizer, thickener, for liq. dishwashing and car washing detergents, household, industrial, and cosmetic products [Emulsion Systems]

Esi-Terge SXS. CAS 1300-72-7; EINECS 215-090-9; Sodium xylene sulfonate; solubilizer for light- and heavy-duty cleaners. [Emulsion Systems]

Esi-Terge T-60. TEA dodecylbenzene sulfonate; detergent, wetting agent, foam stabilizer, for car and dishwashing detergent, synthetic hand soap, household cleaners. [Emulsion Systems]

Eskabarb. CAS 50-06-6; Phenobarbital; anticonvulsant; hypnotic; sedative. [SmithKline Beecham] *

Eskacef. CAS 38821-53-3; A proprietary preparation of cephradine; an antibiotic. [SmithKline Beecham] *

Eskacillin. CAS 6130-64-9; Penicillin G Procaine; antibacterial. [SmithKline Beecham] *

Eskacillin 100. A proprietary preparation containing benzylpenicillin (as the potassium salt); an antibiotic. [SmithKline Beecham] *

Eskacillin 100 Sulpha. A proprietary preparation of benzylpenicillin potassium and sulfadimidine; an antibiotic. [SmithKline Beecham] *

Eskacillin 200. A proprietary preparation containing procaine penicillin; an antibiotic. [SmithKline Beecham]*

Eskacillin 200 Sulpha. A proprietary preparation of procaine penicillin and sulfadimidine; an antibiotic. [SmithKline Beecham] *

Eskadiazine. CAS 68-35-9; Sulfadiazine; antibacterial. [SmithKline Beecham] *

Eskalith. CAS 554-13-2; Lithium carbonate; antimanic.

[SmithKline Beecham]

Eskamel Cream. Combination of resorcinol and sulfur; for treatment of acne. [SmithKline Beecham] *

Eskimo. Lubes. [Chevron] *

Eskornade. A proprietary preparation of isopropamide, diphenylpyraline and phenylpropanolamine hydro chloride; a nasal decongestant. [SmithKline Beecham] *

Eskornade Spansule Capsules. Combination of isopropamide iodide, phenylpropanolamine hydrochloride and diphenylpyraline hydrochloride; oral nasal decongestant with antihistaminic and anticholinergic activity. [SmithKline Beecham] *

Esmaillite. A dope consisting of a mixture of cellulose acetate and volatile solvent, usually ethyl formate.

Esoderm. CAS 58-89-9; Lindane; for infestation by lice. [Napp Laboratories Ltd] *

Esophotrast. CAS 7727-43-7; Barium sulfate; diagnostic aid. [Armour Pharmaceutical Co] *

Esorb. CAS 59-02-9; Vitamin E supplement. [Wyeth Laboratories] *

Esperal®115RG. CAS 80-43-3; Dicumyl peroxide; for med.-temp. applications as a polymerization and crosslinking agent. [Witco/Argus]

Esperal® 120. CAS 78-63-7; 2,5-Dimethyl-2,5-di(t-butylperoxy)hexane; initiator in crosslinking of polymers and elastomers; melt flow modifier for PP; vulcanization agent for silicone rubber, fluoro elastomers, EPDM, EVA; crosslinking agent for LDPE. [Witco/Argus]

Esperal® 230. CAS 1068-27-5; 2,5-Dimethyl-2,5-di(t-butylperoxy)hexyne-3; initiator in crosslinking of polyolefins and other polymers; crosslinking agent for HDPE and LLDPE at temps. above 180 C. [Witco/Argus]

Esperase®16.0 L. Proteinase; enzyme; defoamer for built liq. detergents. [Novo Nordisk] *

Espercarb®438M-60. CAS 19910-65-7; 60% solution di-s-butyl peroxydicarbonate in odorless mineral spirits; initiator, crosslinking agent for polymers. [Witco/Argus]

Espercarb® 840. CAS 16111-62-9; Di-2-ethylhexyl peroxydicarbonate; initiator. [Witco/Argus]

Esperfoam®FR. CAS 1338-23-4, 37187-22-7; MEK peroxide and acetylacetone peroxide; initiator for rapid cures of polyester resins; DOT org. peroxide label is not required. [Witco/Argus]

Esperox®. Organic peroxide catalyst. [Witco/Argus]

Esperox® 10. CAS 614-45-9; t-Butyl peroxybenzoate; initiator for polymerization of ethylene and styrene, and for high temp. molding of polyesters. [Witco/Argus]

Esperox® 12MD. CAS 107-71-1; 50% solution of t-butyl peroxy acetate in odorless mineral spirits; initiator for polymerization of styrene, ethylene, acrylates, etc. [Witco/Argus]

Esperox® 13M. CAS 23474-91-1; 75% solution of t-butyl peroxycrotonate in odorless mineral spirits; initiator for polymerization applications. [Witco/Argus]

Esperox® 28. CAS 3006-82-4; t-Butyl peroxy 2-ethyl hexanoate; initiator recommended for polymerization of ethylene plus use in med. temp. molding of polyester resin systems. [Witco/Argus]

Esperox® 31M. CAS 927-07-1; t-Butyl peroxypivalate in mineral spirit diluent; initiator used in polymerization of ethylenically unsat. monomers. [Witco/Argus]

Esperox® 33M. CAS 26748-41-4; t-Butyl peroxyneodecanoate in mineral spirit diluent; efficient and reactive initiator for polymerization of ethylenically unsat.

---

* Trade name not verified as available

† Trade name verified as obsolete

monomers. [Witco/Argus]

Esperox® 41-25A. CAS 1931-62-0; 25% t-Butyl peroxy maleate disp.; for polymerization of various ethylenically unsat. resins and monomers [Witco/Argus]

Esperox®497M. CAS 22313-62-8; t-Butyl peroxy 2-methyl benzoate sol'n. in mineral spirit diluent; initiator for polymerization of ethylene and styrene, high temp. molding of polyester resin systems, and vulcanization of silicon rubber. [Witco/Argus]

Esperox® 545M. CAS 68299-16-1; t-Amyl peroxyneodecanoate in mineral spirit diluent; efficient and reactive initiator for polymerization of ethylenically unsat. monomer. [Witco/Argus]

Esperox®551M. CAS 29240-17-3; t-Amyl peroxypivalate in mineral spirit diluent; initiator used in polymerization of ethylenically unsat. monomers. [Witco/Argus]

Esperox® 570. CAS 686-31-7; 95% t-Amyl peroxy 2-ethylhexanoate; for polymerization of ethylene, acrylates, and unsat. polyester resins [Witco/Argus]

Esperox®740M. CAS 130097-36-8; 75% solution of cumyl peroxyneoheptanoate in odorless mineral spirits; initiator for polymerization of vinyl chloride. [Witco/Argus]

Esperox® 747M. 75% solution of t-amyl peroxyneoheptanoate in odorless mineral spirits; initiator for polymerization of ethylenically unsat. monomers. [Witco/Argus]

Esperox®750M. CAS 26748-38-9; 75% solution of t-butyl peroxyneoheptanoate in odorless mineral spirits; initiator for polymerization of ethylenically unsat. monomers. [Witco/Argus]

Esperox® 939M. CAS 26748-47-0; Cumyl peroxyneodecanoate in mineral spirit diluent; efficient and reactive initiator for polymerization of vinyl chloride. [Witco/Argus]

Esperox® 5100. CAS 4511-39-1; t-Amyl peroxybenzoate; initiator for polymerization of ethylene, styrene, acrylates, and curing of unsat. polyester resins. [Witco/Argus]

Esperox® C-496. CAS 34443-12-4; t-Butyl peroxy-2-ethylhexyl carbonate; initiator for polymerization of vinyl monomers, styrene, acrylates, and unsat. polyester resins. [Witco/Argus]

Esrakon. Tablets for use in start up of electro slag refining process for steel billets and slabs. [Foseco (F.S.) Ltd]

Essar (W). A proprietary trade name for a general purpose acid and alkali resistant furane resin cement for bedding in acid resisting tiles. §

Essence of Bigarade. Oil of bitter orange peel.

essence of Bitter almonds. Benzaldehyde

essence of Mirbane. CAS 98-95-3; (oil of Mirbane). Nitrobenzene, $C_6H_5NO_2$; formerly used for scenting soap.

Essential Oil of Bitter Almonds. Benzaldehyde, C_6H_5CHO.

Essex powder. An explosive, containing 22-24% nitroglycerin, 0.5-1.5% collodion cotton, 33-35% potassium nitrate, 33-35% wheat flour, and 5-7% ammonium chloride.

Esshete 1250. A steel containing 16% chromium, 10% nickel and 6% manganese; an austenitic creep resisting steel. [Samuel Fox & Co Ltd] ‡

Esshete CML. A proprietary trade name for a steel containing 1% chromium and 1% molybdenum possessing high creep strength and corrosion resistance; suitable for operating up to 1000 C. [Samuel Fox & Co Ltd] ‡

Esshete CRM2. A steel containing 2 1/4% chromium and 1% molybdenum and superior properties to CML; can be used up to 1100 CF. [Samuel Fox & Co Ltd] ‡

Esshete CRM5. A steel containing 5% chromium and 1% molybdenum suitable for tubes exposed to high temperature steam. [Samuel Fox & Co Ltd] ‡

Esskol®. Vegetable drying oils used in preparation of paints, varnishes. [Reichhold]

Estabex. Epoxidized soya bean oil and epoxy ester. [Akzo Chemie UK Ltd]

Estabex 2307, 2349. Epoxidized soya bean oils. §

Estabex 2386. Epoxidized monoesters. §

Estaflex. CAS 77-90-7; Acetyl tributyl citrate; plasticizer. [Akzo Chemie UK Ltd]

Estaloc 61000 Series. Mixture of thermoplastic polyurethane, reinforcement, polymeric alloys, additives; reinforced engineering thermoplastic [BFGoodrich/Spec. Polymers] *

Estalol®. Dibasic ester; for paper, foundry, electronics, textile industries. [Aceto]

Estane®. A proprietary range of thermoplastic polyurethane molding and extrusion compounds. [BFGoodrich] *

Estane® 5701 F1. CAS 9009-54-5; Polyester-urethane polymer, calcium stearate-dusted; thermoplastic polymer, flexibility, chemical and abrasion resistance, wide compatibility; used in binders, coated fabrics, adhesive blending resins. [BFGoodrich/Spec. Polymers] *

Estane®58092. CAS 9009-54-5; Polyester-urethane compound; firm, flexible, low-blocking, thermoplastic compound, highest chemical and abrasion resistance and properties; used in injection molding and extrusion of film, sheet, tubing, and shapes. [BFGoodrich/Spec. Polymers] *

Estane®58300. CAS 9009-54-5; Polyether base urethane compound; thermoplastic compound, resistance to hydrolysis, improved fungus resistance, low-temp. flexibility; used in extrusion and injection molding for hose and wire jacketing, molded junctions. [BFGoodrich/Spec. Polymers] *

Estasol. Dimethyl esters of adipic, glutaric, and succinic acids; low toxicity compound solvent; general purpose solvent coil coatings, foundry resins. [Chemoxy International Ltd]

Ester Copal. An ester gum obtained by the interaction between glycerin and copal.

Estergel. CAS 110-27-0; EINECS 203-751-4; Isopropyl myristate; pharmaceutic aid. [Merck & Co Inc] *

Esterkem. Three-component product based on modified unsaturated polyester resin with peroxide catalyst; adhesive and grout for ceramic tiles and paviors in heavy-duty chemical-resistant application. [Feb Ltd]

Estermone. Mixture of 2,4-D and dicamba; used to control weeds in turf. [Synchemicals Ltd]

Esterol. A proprietary brand of benzyl succinate; also a proprietary trade name for alkyd synthetic varnish and lacquer resins. §

Esterolane. Disperse dyes. [ICI Chem. & Polymers Ltd] †

Esterox. A fast, air-drying, modified epoxy sealer-dustproofer; for protecting interior concrete floors. [Secure Inc] *

Esterpol. A proprietary synthetic resin. §

Estersil. Ethyl and propyl salicylglycollic esters. [Johnsons of Hendon] *

Estigyn. CAS 57-63-6; Tablets of ethinylestradiol-estrogen.

---

‡ Trade name and manufacturer not verified   § Trade name without identified manufacturer

[British Drug Houses] *

Estinyl. CAS 57-63-6; Ethinyl estradiol; estrogen. [Schering Corp]

Estol. Textile lubricants. [Crosfield Chemicals Ltd] †

Estol 1407. Glyceryl oleate; antistat for polyethylene and polypropylene; antifog agent for LDPE and PP; for lubricants, cosmetic emollients. [Unichema]

Estol 1468. Glyceryl stearate; antistat for polyethylene and polypropylene; antifog agent for LDPE; for lubricants, cosmetic emollients. [Unichema]

Estol 1476. CAS 646-13-9; Isobutyl stearate; lubricant for PVC processing; cosmetic emollients. [Unichema]

Estol 1481. Cetostearyl stearate; lubricant for PVC processing; cosmetic emollients. [Unichema]

Estolan. Polyester polyols; used for textile lamination foams, packaging foams, microcellular formulated systems, adhesives and flexible coatings. [Harcros UK]

Estoral. Menthyl borate.

Estrace. CAS 50-28-2; Estradiol; estrogen. [Mead Johnson & Co]

Estracyt. CAS 2998-57-4; A proprietary preparation of estramustine; antineoplastic agent. [Roche Prods. Ltd]*

Estragol. p-Propenylanisol, $CH_2:CH \cdot CH_2 \cdot C_6H_4OCH_3$.

Estron. Acetate tow and yarn. [Eastman]

Estrovis. CAS 152-43-2; A proprietary preparation of quinestrol used for the suppression of lactation. [Warner]*

Etadurin 31. Amine copolymer aq. solution; thermosetting wet strength resin for paper mfg. [Akzo]

Etapuron FT. Quaternary compound; cationic; yellow liquid; softening agent for flame resistant finishes; compatible with most of the salt-based flame retardants. [Thor Chemicals (UK) Ltd]

Etard's reagent. Anhydrous chromium oxychloride, an oxidizing agent.

Eteleen. Trigallic acetal.

EternaBrite. Leafing aluminum pigments; metallic printing inks (silver and gold). [Silberline Mfg Co]

Eternite. A slate-like mass made from 6 parts Portland cement and 1 part asbestos fiber; used for roofing.

Ethacol. Pyrocatechol monoethyl ether.

Ethacure®100. Diethyl toluene diamine; high-performance curing agent for epoxy resins; used in filament winding, elec. encapsulation, prepregs, tooling, potting and casting, laminating, coating, molding, and adhesive applications. [Ethyl Corp]

Ethacure®300. Curing agent for polyurethane and epoxy resins. [Ethyl Corp]

Ethafoam. Polyethylene foam used principally in cushioning and packaging applications. [Dow UK]

Etha-Keratin ISO. AMP-isostearoyl hydrolyzed keratin; cosmetics ingred. [Brooks Industries]

ethal. CAS 36653-82-4; Cetyl alcohol, $C_{16}H_{33}OH$.

Ethal 326. Laureth-3; emulsifier for textile and industrial applications. [Ethox]

Ethal 368. PEG-3 C16-18 alcohol; emulsifier, lubricant. [Ethox]

Ethal 926. Laureth-9; detergent, emulsifier. [Ethox]

Ethal 3328. PEG-3 C12-18 alcohol; low HLB emulsifier and coupler. [Ethox]

Ethal BPA-6. PEG-6 bisphenol A; monomer for polyester and urethane coatings. [Ethox]

Ethal CSA-25. PEG-25 C16-18 alcohol; hydrophilic emulsifier, detergent for mild acidic or alkaline solutions and hot aq. systems. [Ethox]

Ethal DA-4. CAS 26183-52-8; PEG-4 decyl alcohol; wetting agent for aq. solutions; intermediate for mfg. of anionic surfactants. [Ethox]

Ethal DDP-7. Multipurpose emulsifier and detergent; excellent greaser cutter. [Ethox]

Ethal DNP-8. CAS 9014-93-1; PEG-8 dinonyl phenol; low foam emulsifier for solv. systems; coupling agent for introducing water into nonaq. systems. [Ethox]

Ethal EH-2. PEG-2 2-ethylhexanol; intermediate for anionic surfactants; component of low foam wetting systems. [Ethox]

Ethal LA-4. Laureth-4; general-purpose emulsifier, lubricant. [Ethox]

Ethal NP-1.5. CAS 9016-45-9; Nonoxynol-1.5; emulsifier, intermediate. [Ethox]

Ethal NP-6. CAS 9016-45-9; Nonoxynol-6; coupling agent for surfactant systems, for coupling water into oil systems. [Ethox]

Ethal OA-10. CAS 9004-98-2; Oleth-10; moderate HLB emulsifier for esters and oils. [Ethox]

Ethal TDA-3. CAS 24938-91-8; PEG-3 tridecyl alcohol; surfactant, coupling agent. [Ethox]

Ethal TDA-6. CAS 24938-91-8; PEG-6 tridecyl alcohol; wetting agent, detergent, foamer, dispersant, emulsifier. [Ethox]

Ethal TDA-18. CAS 24938-91-8; PEG-18 tridecyl alcohol; high-foaming detergent for elevated temp. applications. [Ethox]

Ethana®. CAS 71-55-6; 1,1,1-Trichloroethane; solvent. [Asahi Chem. Industry]

ethane, 1,2-dichloro-. See ethylene dichloride.

1,2-ethanediol. See glycol.

1,2-ethanediol dimethacrylate. See ethylene glycol dimethacrylate.

ethaneidyl) derivatives. See polysorbate 40.

ethane, 1,1,1-trichloro-. See trichloroethane.

Ethanite. A proprietary plastic made from ethylene dichloride and calcium polysulfide. §

ethanol. See ethyl alcohol.

ethanolamine. (MEA; 2-aminoethanol; 2-aminoethyl alcohol; monoethanolamine) $NH_2CH_2CH_2OH$; CAS 141-43-5; EINECS 205-483-3; Scrubbing acid gases, esp. in synthesis of ammonia; nonionic detergents for dry cleaning wool treatment, emulsion paints, polishes, agricultural sprays; chemical intermediate; pharmaceuticals; corrosion inhibitor; rubber accelerator. [BP Chem. Ltd; OxyChem; Texaco; Union Carbide]

Ethanox®323. Nonylphenol disulfide; antioxidant for disproportionation and bleaching rosins and rosin esters. [Ethyl Corp]

Ethanox®330. 1,3,5-Trimethyl-2,4,6-tris (3,5-di-tert-butyl-4-hydroxybenzyl) benzene; antioxidant and stabilizer for plastic, resin, rubber, and wax; food industry. [Ethyl Corp]

Ethanox®398. CAS 118337-09-0; 2,2´-Ethylidenebis (4,6-di-t-butylphenyl) fluorophosphonite; antioxidant for PP, LLDPE. [Ethyl Corp]

Ethanox®702. 4,4´-Methylenebis(2,6-di-tert-butylphenol); antioxidant for rubber, plastic, resin, adhesive, petrol. oil, and wax. [Ethyl Corp]

Ethanox®703. 2,6-Di-tert-butyl-α-dimethylamino-p-cresol; antioxidant for rubber, plastic, resin, adhesive, petrol. oil, and wax. [Ethyl Corp]

---

* Trade name not verified as available

† Trade name verified as obsolete

Etha-Soy ISO. AMP-isostearoyl hydrolyzed soy protein; cosmetic ingredient for skin and hair care products [Brooks Industries]

Ethavan. CAS 121-32-4; Ethyl vanillin; used by flavor manufacturers to replace part of the vanillin to give bouquet to the finished flavor or fragrance. [Monsanto Co] *

Ethazate. CAS 14323-55-1; Zinc diethyldithiocarbamate; fast cures at low temperature; medium precure rate activated by Oxaf; used for latex compounding; in latex foam dipped goods and fabric coatings. [Uniroyal] *

Ethazate 50D. A 50% active water dispersion of Ethazate; ready-to-use form for latex compounding. [Uniroyal] *

Ethene. See olefiant gas.

ethene, homopolymer. See polyethylene.

ethenol, homopolymer. See polyvinyl alcohol.

ethenyl acetate. See vinyl acetate.

ethenyl acetate, homopolymer. See polyvinyl acetate (homopolymer).

1-ethenyl-2-pyrrolidinone, homopolymer. See PVP.

ether. See ethyl ether.

Etheramine 13. 3-(Isotridecyloxy) 1-propaneamine; corrosion inhibitor, chemical intermediate. [Berol Nobel AB]

Etherdiamine 13. Branched aliphatic ether diamine; corrosion inhibitor, chemical intermediate. [Berol Nobel AB]

Ethereal Oil of Bitter Almonds. Benzaldehyde, $C_6H_5.CHO$.

Etherin. See olefiant gas.

ether, methylated. Ether prepared from methylated spirit.

Ethfac 104. Aliphatic phosphate ester; low foaming penetrant and emulsifier. [Ethox]

Ethfac 142W. Aliphatic phosphate ester; emulsifier; lubricant additive for process fluids. [Ethox]

Ethfac 391. Aliphatic phosphate ester; detergent for textile scouring; effective dispersant at elevated temps. [Ethox]

Ethfac 1018. Aliphatic phosphate ester; antistat and softener with excellent lubricating props. when neutralized. [Ethox]

Ethfac NP-110. Aromatic phosphate ester; hydrophilic detergent and emulsifier, esp. for liq. detergent formulations. [Ethox]

Ethfac PB-1. Aliphatic phosphate ester; hydrophilic surfactant with high caustic stability; for highly alkaline cleaners. [Ethox]

Ethfac PD-6. Aliphatic phosphate ester; higher tolerance to alkali than Ethfac 161; useful as component of alkaline cleaners. [Ethox]

Ethfac PP-16. Aromatic phosphate ester; low foaming hydrotrope for coupling nonionics into alkaline systems. [Ethox]

Ethibond. Suture, nonabsorbable surgical aid. [Ethicon Inc]*

Ethidium. Animal health insecticide. [Schering Agrochemicals Ltd]

Ethidol. The ethyl ester of iodo-ricinoleic acid; stated to be suitable for intraglandular injection. §

Ethiflex. Suture, nonabsorbable surgical aid. [Ethicon Inc]*

ethinyl trichloride. See trichloroethylene.

Ethiodan. CAS 99-79-6; Ethyl 4-iodophenylundec-10-enoate (iophendylate); X-ray contrast medium for myclography. [British Drug Houses] *

Ethiodol. Ethiodized oil for injection; indicated for use in radiographic exploration for hysterosalpinography and lymphography. [Altana Inc] ‡

Ethion. O,O,O´,O´-Tetraethyl-S,S´-methylene di(phos-phorodithioate); has both acaricidal and insecticidal properties; its acaricidal action is widely used in the abatement of cattle ticks; as an insecticide it is used on citrus, deciduous fruits, tea, cotton and ornamental plants. [A/S Cheminova] *

Ethnine. CAS 509-67-1; A proprietary preparation containing pholcodine; a cough linctus. [Allen & Hanburys Ltd]*

Ethobral. A proprietary preparation of phenobarbitone, butobarbitone and quinalbarbitone; a hypnotic. [Wyeth Laboratories] *

ethocel. See ethylcellulose.

Ethocel Standard Premium. CAS 9004-57-3; Ethyl cellulose NF; binders for pharmaceutical tabletting. [Dow]

Ethoduomeen. A range of fatty diamine ethoxylates; used as wax emulsions, polishes, fuel additives, antistatic additives, dye assistants, viscose spinning additives, algicides, bactericides, and disinfectants. [Harcros Australia]

Ethoduomeen® T/13. CAS 61790-85-0; PEG-3 tallow aminopropylamine; emulsifier used in making of bitumen emulsions; dispersant for waxes; for textiles, asphalt, agricultural emulsions; wetting agent, corrosion inhibitor. [Akzo; Akzo Chem. BV]

Ethoduomeen® T/20. CAS 61790-85-0; PEG-10 tallow aminopropylamine; emulsifier, dispersant, wetting agent used in coating preparation on paperboard; corrosion inhibitor. [Akzo; Akzo Chem. BV]

Ethoduomeen® T/25. CAS 61790-85-0; PEG-15 tallow aminopropylamine; corrosion inhibitor in water treatment chemicals in sec. oil recovery. [Akzo; Akzo Chem. BV]

Ethoduoquad® T/15-50. N,N,N´,N´-Penta(2-hydroxyethyl)-N-tallowalkyl-1,3-propane diammonium diacetate, aqueous isopropyl alcohol; industrial surface active agent for agriculture, textiles, protective coatings, inks, pigment dispersions, acid pickling baths, metalworking, electroplating, and plastics manufacture. [Akzo]

Ethofat®. Ethoxylated fatty acids; used for mineral flotation; metalworking; paper; leather; pigments and surface coatings; plastic foams; textiles. [Akzo Chemie UK Ltd]†

Ethofat® 18/14. CAS 9004-99-3; PEG-4 stearate; industrial surfactant. [Akzo]

Ethofat® 242/25. CAS 61791-00-2; 65071-95-6; PEG-15 tallate; emulsifier, detergent, dispersant. [Akzo]

Ethofat® O/20. CAS 9004-96-0; PEG-10 oleate; emulsifier, detergent, dispersant. [Akzo]

Ethofoil. A proprietary trade name for ethyl cellulose film. §

ethohexadiol. See ethyl hexanediol.

Ethokem. Cationic surfactant containing 870 g/l polytehanoxy alkylamine; wetting agent for phosphonoglycine herbicides. [Midkem Agrochemicals]

Ethokem C/12. CAS 61791-31-9; Cationic surfactant adjuvant containing bis-2 hydroxyethyl cocamine; wetting agent for phosphonoglycine herbicides and ammonium sulfate. [Midkem Agrochemicals]

Ethomeen®. A range of fatty amine ethoxylates; used as wax emulsions, polishes, fuel additives, antistatic additives, dye assistants, viscose spinning additives, algicides, bactericides and disinfectants. [Akzo]

Ethomeen® 18/12. CAS 10213-78-2; PEG-2 stearamine; emulsifier, dispersant used in textile processing. [Akzo]

Ethomeen® 18/15. CAS 26635-92-7; PEG-5 stearamine; emulsifier, dispersant for textile processing. [Akzo]

Ethomeen® 18/20. CAS 26635-92-7; PEG-10 stearamine;

---

‡ Trade name and manufacturer not verified   § Trade name without identified manufacturer

emulsifier, dispersant used in textile processing. [Akzo]

Ethomeen® 18/25. CAS 26635-92-7; PEG-15 stearamine; emulsifier, dispersant used in textile processing. [Akzo]

Ethomeen® 18/60. CAS 26635-92-7; PEG-50 stearamine; emulsifier, dispersant used in textile processing; prevents premature coagulation of latex rubber [Akzo]

Ethomeen® C/12. CAS 61791-14-8; 61791-31-9; PEG-2 cocamine; emulsifier, dispersant used in textile processing. [Akzo; Akzo Chem. BV]

Ethomeen® C/15. CAS 61791-14-8; PEG-5 cocamine; emulsifier, dispersant for textile processing, dyeing assistant, desizing assistant, softener, antistat. [Akzo; Akzo Chem. BV]

Ethomeen® C/20. CAS 61791-14-8; PEG-10 cocamine; emulsifier, dispersant used in textile processing. [Akzo]

Ethomeen® C/25. CAS 61791-14-8; PEG-15 cocamine; emulsifier, dispersant used in textile processing. [Akzo; Akzo Chem. BV]

Ethomeen® S/12. CAS 61791-24-0; PEG-2 soyamine; emulsifier, dispersant used in textile processing. [Akzo; Akzo Chem. BV]

Ethomeen® S/15. CAS 61791-24-0; PEG-5 soyamine; emulsifier, dispersant for textile processing. [Akzo; Akzo Chem. BV]

Ethomeen® S/20. CAS 61791-24-0; PEG-10 soyamine; emulsifier, dispersant used in textile processing. [Akzo]

Ethomeen® S/25. CAS 61791-24-0; PEG-15 soyamine; emulsifier, dispersant used in textile processing. [Akzo; Akzo Chem. BV]

Ethomeen® T/12. CAS 61791-44-4; PEG-2 tallow amine; emulsifier, dispersant used in textile processing. [Akzo; Akzo Chem. BV]

Ethomeen® T/15. CAS 61791-26-2; PEG-5 tallow amine; emulsifier, dispersant for textile processing. [Akzo; Akzo Chem. BV]

Ethomeen® T/25. CAS 61791-26-2; PEG-15 tallow amine; emulsifier, dispersant used in textile processing. [Akzo; Akzo Chem. BV]

Ethomid®. A range of ethoxylated amides; used as lubricants. [Harcros Australia]

Ethomid® HT/23. CAS 68155-24-8; PEG-13 hydrogenated tallow amide; emulsifier, dispersant, detergent, dye leveling agent; for silicone finishing agents, sizing lubricants. [Akzo]

Ethomid® HT/60. CAS 68155-24-8; PEG-50 hydrogenated tallow amide; surfactant, emulsifier, sec. stabilizer for emulsion systems; dispersant, detergent. [Akzo; Akzo Chem. BV]

Ethomid® O/17. CAS 26027-37-2; PEG-7 oleamide; emulsifier, dispersant, detergent. [Akzo]

Ethomulsion. A proprietary trade name for ethyl cellulose lacquer emulsion. §

Ethoquad®. Ethoxylated quaternary ammonium salts; used as electroplating bath additives and as bacteriostats. [Harcros Australia]

Ethoquad® 18/12. CAS 3010-24-0; 28724-32-5; PEG-2 stearmonium chloride and IPA; antistat, emulsifier, dyeing assistant, leveling agent, antifoam used in textile industry, as electroplating bath additives. [Akzo]

Ethoquad® 18/25. CAS 28724-32-5; PEG-15 stearmonium chloride; antistat, emulsifier, dyeing assistant, leveling agent, antifoam used in textile industry, as electroplating bath additives. [Akzo]

Ethoquad® C/12. CAS 70750-47-9; PEG-2 cocomonium

chloride and IPA; antistat, emulsifier, dyeing assistant, electroplating bath additive. [Akzo]

Ethoquad® C/12 Nitrate. CAS 71487-00-8; PEG-2 cocomethyl ammonium nitrate, IPA; industrial surfactant for agricultural, textiles, protective coatings, inks, pigment dispersions, acid pickling baths, metalworking, electroplating, plastics mfg. [Akzo]

Ethoquad® C/25. CAS 61791-10-4; PEG-15 cocomonium chloride; antistat, emulsifier, dyeing assistant, leveling agent, antifoam used in textile industry, as electroplating bath additives. [Akzo]

Ethoquad® CB/12. CAS 61789-68-2; PEG-2 cocobenzonium chloride, IPA; industrial surfactant for agriculture, textiles, protective coatings, inks, pigment dispersions, acid pickling baths, metalworking, electroplating, plastics mfg. [Akzo]

Ethoquad® O/12. CAS 18448-65-2; PEG-2 oleamonium chloride and IPA; antistat, emulsifier, dyeing assistant, electroplating bath additive. [Akzo]

Ethoquad® T/12. CAS 67784-77-4; PEG-2 tallowalkyl methyl ammonium chloride, ethanol; industrial surfactant for agriculture, textiles, protective coatings, inks, pigment dispersions, acid pickling baths, metalworking, electroplating, plastics mfg. [Akzo]

Ethoquad® T/13-50. PEG-3 tallow alkyl ammonium acetate, aq. IPA; industrial surfactant for agriculture, textiles, protective coatings, inks, pigment dispersions, acid pickling baths, metalworking, electroplating, plastics mfg. [Akzo]

Ethosperse®. Polyoxyethylene ethers. [Lonza Inc]

Ethosperse® CA-2. CAS 9004-95-9; Ceteth-2; o/w emulsifier, thickener, stabilizer for hair care products, antiperspirants. [Lonza Inc]

Ethosperse® G-26. CAS 31694-55-0; Glycereth-26; emulsifier, humectant for cosmetic, pharmaceutical and industrial uses. [Lonza Inc]

Ethotal. Emulsifier dispersant, solubilizer; used for textiles; agricultural; cosmetics. [Witco Chemical Ltd] *

Ethox 25-R-8. Block polymer, 75% EO; dispersant, emulsifier and intermediate for esters. [Ethox]

Ethox 1122. CAS 31621-91-7; PEG-9 pelargonate; high cohesion lubricant for synthetic fiber production and processing. [Ethox]

Ethox 1212. Ethoxylated coconut glyceride; lubricant for textiles and metals. [Ethox]

Ethox 1358. Aliphatic phosphate ester; potassium salt as antistat and lubricant. [Ethox]

Ethox 1372. Polyoxyalkylene fatty amine; low foam dye leveler for acid dyes. [Ethox]

Ethox 2156. Short chain triglyceride ester; lubricant for textiles and metals. [Ethox]

Ethox 2423. PPG-25-laureth-25; low foam detergent, wetting and rewetting agent for textiles, emulsifier. [Ethox]

Ethox 2610. PPG 1025 ditallate; oil-in-oil dispersant and lubricant; for silicone and nonsilicone defoamers; low foam additive for various formulations. [Ethox]

Ethox 2659. Ethoxylated styrenated phenol; dispersant for pigments and org. emulsions; higher HLB than Ethox 2938. [Ethox]

Ethox 2684. Aliphatic phosphate ester; coupling agent for nonionic surfactants in alkali systems. [Ethox]

Ethox 3113. PPG-2 bisphenol A; monomer for polyester resins. [Ethox]

Ethox CAM-2. CAS 61791-14-8; PEG-2 cocamine; emulsi-

---

* Trade name not verified as available   † Trade name verified as obsolete

fier, dispersant, textile dyeing assistant, lubricant; intermediate for amphoterics. [Ethox]

Ethox CO-5. CAS 61791-12-6; PEG-5 castor oil; emulsifier, lubricant; dispersant for pigments and clays. [Ethox]

Ethox COA. Cocamide DEA; foaming agent, detergent, and dispersant. [Ethox]

Ethox DL-5. CAS 9005-02-1; PEG 200 dilaurate; lubricant and emulsifier for low foaming applications. [Ethox]

Ethox DO-9. CAS 9005-07-6; PEG-8 dioleate; emulsifier for oils and solvs.; used for industrial lubricants. [Ethox]

Ethox DT-15. PEG-15 tallow diamine; retarder for acid dyes; dispersant for use in acidic solutions [Ethox]

Ethox DTO-9A. CAS 61791-01-3; PEG-8 ditallate; emulsifier for oils and solvs.; used for industrial lubricants. [Ethox]

Ethox HCO-16. CAS 61788-85-0; PEG-16 hydrogenated castor oil; heat-stable emulsifier for natural and synthetic oils, lubricant. [Ethox]

Ethox HO-50. CAS 57171-56-9; PEG-50 sorbitol hexaoleate; emulsifier and lubricant for heat-stable systems. [Ethox]

Ethox L-61. Block polymer, 10% EO; low foam detergent, emulsifier, lubricant; intermediate for esters and polyesters. [Ethox]

Ethox LF-1226. Polyoxyalkylene glycol ether; detergent for removal and dispersion of oils and waxes; intermediate for anionic systems. [Ethox]

Ethox MA-8. PEG-8 monomerate; cost-effective emulsifier, lubricant. [Ethox]

Ethox MI-9. CAS 56002-14-3; PEG-9 isostearate; emulsifier and lubricant. [Ethox]

Ethox ML-5. CAS 9004-81-3; PEG-5 laurate; lubricant, emulsifier, detergent, softener; coupling agent; visc. control agent for plastisol resins. [Ethox]

Ethox MO-9. CAS 9004-96-0; PEG-9 oleate; lubricant, emulsifier for natural and synthetic oils, detergent, softener. [Ethox]

Ethox MS-8. CAS 9004-99-3; PEG-8 stearate; lubricant, wax and oil emulsifier, detergent, softener; for aq. processing. [Ethox]

Ethox OAM-308. PEG-30 oleyl amine; emulsifier, dispersant, textile dyeing assistant and lubricant; leveling agent for dyeing nylon; antiprecipitant for cross dyeing. [Ethox]

Ethox PPG 1025 DTO. PPG-20 ditallate; oil-in-oil dispersant and lubricant. [Ethox]

Ethox SAM-10. CAS 26635-92-7; PEG-10 stearamine; moderate HLB emulsifier; component of corrosion inhibitors. [Ethox]

Ethox SO-9. PEG-8 sesquioleate; all-purpose oil emulsifier. [Ethox]

Ethox TAM-2. CAS 61791-44-4; PEG-2 tallow amine; emulsifier, dispersant, textile dyeing assistant, lubricant; used for preparation of high pressure wax dispersions. [Ethox]

Ethox TO-8. CAS 61791-00-2; PEG-8 tallate; lubricant, emulsifier, detergent, softener, dispersant. [Ethox]

Ethoxol. A proprietary glycol ether. [ICI Chem. & Polymers Ltd]

ethoxyethanol. (EE; ethylene glycol ethyl ether; 2-ethoxyethanol; Cellosolve) Ether alcohol; CH_3CH_2O-CH_2CH_2OH; CAS 110-80-5; EINECS 203-804-1; solvent for nitrocellulose, lacquer, lacquer thinners, dyeing and printing textiles, varnish removers, cleaning solutions. [Arco; Ashland; OxyChem; Union Carbide]

ethoxyethanol acetate. (EEA; ethylene glycol ethyl ether acetate; 2-ethoxyethanol acetate; 2-ethoxyethyl acetate) Ester of ethoxyethanol and acetic acid; $HOCH_2CH_2OC_2H_5$; CAS 111-15-9; solvent; automobile lacquers; retards evaporation and imparts high gloss. [Allchem Industries; Arco; OxyChem; Union Carbide]

2-ethoxyethanol acetate. *See* ethoxyethanol acetate.

2-ethoxyethyl acetate. *See* ethoxyethanol acetate.

3-ethoxy-4-hydroxybenzaldehyde. *See* ethyl vanillin.

Ethoxylan® 1685. CAS 61790-81-6; PEG-75 lanolin; emollient, emulsifier, dispersant, foam stabilizer, resin plasticizer for cosmetic and pharmaceutical preparations, textile processing. [Henkel/Cospha; Henkel/Textile; Henkel Canada]

Ethoxyol® 1707. Polysorbate 80, cetyl acetate, acetylated lanolin alcohol; SE emollient with lubricating and penetrating properties; auxiliary emulsifier, solubilizer, pigment wetting agent for cosmetics, liq. soaps. [Henkel/Cospha; Henkel Canada]

Ethrane. CAS 13838-16-9; Enflurane; an inhalation anesthetic. [Abbott Laboratories] *

Ethrel. CAS 16672-87-0; Growth regulator containing ethephon. (Sold in UK on behalf of Amchem Prods. Inc.) [ICI Chem. & Polymers Ltd]

Ethrel C. Soluble concentrate containing 480 g 2-chlorethylphosphonic acid per liter; plant growth regulator. [ICI Agrochemicals]

Ethrel-E. Growth regulator. [A H Marks & Co Ltd] *

Ethrel-R. Defoliant . [A H Marks & Co Ltd] *

Ethril. CAS 643-22-1; Erythromycin stearate; antibacterial. [Bristol-Myers Squibb Co Inc] *

Ethrine. CAS 509-67-1; Linctus of pholcodine. [Allen & Hanburys Ltd] *

Ethsorbox L-20. CAS 9005-64-5; PEG-20 sorbitan laurate; emulsifier for oils, solvs., and fats; lubricant for cotton and rayon. [Ethox]

Ethsorbox O-20. CAS 9005-65-6; PEG-20 sorbitan oleate; emulsifier for industrial and textile lubricants. [Ethox]

Ethsorbox S-20. CAS 9005-67-8; PEG-20 sorbitan stearate; waxy lubricant, softener. [Ethox]

Ethsorbox TO-20. CAS 9005-70-3; PEG-20 sorbitan trioleate; lubricant and emulsifier for industrial process fluids. [Ethox]

Ethsorbox TS-20. CAS 9005-71-4; PEG-20 sorbitan tristearate; lubricant, softener for textile goods; emulsifier for fats and oils. [Ethox]

Ethulon. Ethyl cellulose film for tracing and industrial purposes. [May & Baker Ltd] *

Ethulose. A proprietary preparation of alcohol and levulose; a parenteral source of calories. [Geistlich Sohne AG] *

ethyl acetate. (acetic ether; acetic acid, ethyl ester; vinegar naphtha) Ester of ethyl alcohol and acetic acid; $CH_3COOC_2H_5$; CAS 141-78-6; EINECS 205-500-4; general solvent in coatings and plastics, organic synthesis, smokeless powders, artificial leather, photographic films and plates, pharmaceuticals, synthetic fruit essences; cleaning textiles. [Allchem Industries; BP Chem. Ltd; Chisso; Eastman; Hoechst Celanese; Lonza AG; Mallinckrodt; Monsanto; Union Carbide]

ethyl acetone. *See* methyl propyl ketone.

ethyl alcohol. (EtOH; ethanol; alcohol (CTFA); absolute alcohol) Undenatured ethyl alcohol; CH_3CH_2OH; CAS 64-17-5; EINECS 200-578-6; solvent; extraction medium; manufacture of acetaldehyde, denatured alcohol, pharmaceuticals (tonics, colognes), perfumery, organic

‡ Trade name and manufacturer not verified   § Trade name without identified manufacturer

synthesis; octane booster in gasoline; topical antiseptic; alcoholic beverages. [Eastman; Georgia-Pacific Resins; Gist-Brocades SpA; Grain Processing; Quantum/USI; Union Carbide; Vista]

ethyl-p-aminobenzoate hydrochloride. See benzocaine.

Ethylan. Nonionic ethoxylates of fatty alcohols, acids, amines, phenols and alkylolamides; used for detergency, low foam wetting, latex stabilization, emulsification, cosolvency and petroleum recovery. [Harcros UK]

Ethylan® 44. CAS 9016-45-9; Nonoxynol-4; emulsifier for paraffin hydrocarbons, silicones, mineral oil, alkyd resins, etc.; formulation of degreasers, hand cleaning gels. [Harcros UK]

Ethylan® 172. Ceteleth-3; emulsifier for mineral oils and hydrophobic waxes. [Harcros UK]

Ethylan® A2. CAS 9004-96-0; PEG-4 oleate; emulsifier for kerosene and mineral oil, antifoam agent. [Harcros UK]

Ethylan® A10. CAS 9004-96-0; PEG 1000 oleate; surfactant, emulsifier, wetting agent, detergent for areas where low toxicity is important, e.g., oil slick dispersants. [Harcros]

Ethylan®BD10. Low foam biodeg. surfactant, wetting agent for rinse aids, machine dishwashing, metal cleaning, bottle washing. [Harcros]

Ethylan® BV. CAS 9016-45-9; Nonoxynol-14; foam stabilizer and booster, solubilizer, emulsifier used in pesticides, perfumes, emulsion polymerization. [Harcros UK]

Ethylan® CD109. Isodeceth-9; wetting agent, emulsifier, detergent; sudsing agent for liq. detergents; scouring and dye leveling agent; coemulsifier for aromatic solvs., waxes, toxicants; detergent sanitizer; essential oil solubilizer. [Harcros UK]

Ethylan® CD123. C12 fatty alcohol ethoxylate; emulsifier, wetting agent, detergent; coemulsifier for mineral oils and waxes, alkyd resins, paraffin hydrocarbons; biodeg. [Harcros UK]

Ethylan® CD913. Synthetic lower fraction primary alcohol EO condensate (2.9 EO); detergent, o/w and w/o emulsifier, wetting agent, solubilizer; emulsifier for mineral oils and waxes, alkyd resins, paraffinic hydrocarbons. [Harcros UK]

Ethylan® CDP2. Linear middle fraction fatty alcohol ethoxylate (2 EO); surfactant, intermediate for toiletry grade sulfates and other specialty surfactants; coemulsifier for mineral oils, waxes, alkyd resins, paraffinic hydrocarbons. [Harcros UK]

Ethylan®D252. Synthetic primary alcohol ethoxylate (2 EO); o/w and w/o emulsifier for mineral oils and waxes, alkyd resins, paraffin hydrocarbons. [Harcros UK]

Ethylan®FO30. Fish oil ethoxylate; surfactant, emulsifier, wetting agent, detergent for areas where low toxicity is important, e.g., oil slick dispersants; also in antifoams. [Harcros UK]

Ethylan®FO30. CAS 9004-96-0; PEG 1000 oleate; surfactant, emulsifier, wetting agent, detergent for areas where low toxicity is important, e.g., oil slick dispersants. [Harcros]

Ethylan® GEL2. CAS 9005-64-5; Polysorbate 20; w/o emulsifier, solubilizer esp. with sorbitan esters; used in cosmetics, agriculture, perfumes, fiber and textile lubricants, textile antistats, polymer additives, suspension and emulsion polymerization. [Harcros UK]

Ethylan®GEO8. CAS 9005-65-6; Polysorbate 80; emulsifier for cosmetics, pharmaceuticals, agrochem. formula-tions, textile lubricants, plastic additives, emulsion and suspension polymerization; solubilizer for perfume, flavors, essential oils. [Harcros UK]

Ethylan®GEO81. CAS 9005-65-6; Polysorbate 81; emulsifier for cosmetics, agriculture, plastic additives, textile fiber lubricants and softeners, suspension and emulsion polymerization; solubilizer for perfume, flavors, essential oils. [Harcros UK]

Ethylan® GEP4. CAS 9005-66-7; Polysorbate 40; w/o emulsifier, solubilizer esp. with sorbitan esters; used in cosmetics, agriculture, perfumes, fiber and textile lubricants, textile antistats, polymer additives. [Harcros UK]*

Ethylan® GES6. CAS 9005-67-8; Polysorbate 60; general purpose, low toxicity emulsifier for cosmetics and agrochem.; textile lubricant; plastics additive; emulsion and suspension polymerization; solubilizer for perfume, flavors, essential oils. [Harcros UK]

Ethylan® GL20. CAS 1338-39-2; EINECS 215-663-3; Sorbitan laurate; emulsifier for cosmetics, pharmaceuticals, agriculture, plastic antifog, textile fiber lubricant/ softener, suspension and emulsion polymerization. [Harcros UK]

Ethylan® GO80. CAS 1338-43-8; EINECS 215-665-4; Sorbitan oleate; emulsifier for cosmetics, pharmaceuticals, agriculture, plastic antifog, textile fiber lubricant/ softener, suspension and emulsion polymerization. [Harcros UK]

Ethylan®GPS85. CAS 9005-70-3; Polysorbate 85; emulsifier for cosmetics, agriculture, plastic additives, textile fiber lubricants and softeners, suspension and emulsion polymerization; solubilizer for perfume, flavors, essential oils. [Harcros UK]

Ethylan® GS60. CAS 1338-41-6; EINECS 215-664-9; Sorbitan stearate; emulsifier for cosmetics, pharmaceuticals, agriculture, plastic antifog, textile fiber lubricant/ softener, suspension and emulsion polymerization. [Harcros UK]

Ethylan® GT85. CAS 26266-58-0; EINECS 247-569-3; Sorbitan trioleate; emulsifier for cosmetics, pharmaceuticals, agriculture, plastic antifog, textile fiber lubricant/ softener, suspension and emulsion polymerization. [Harcros UK]

Ethylan®LD. Cocamide DEA; foam stabilizer, emulsifier for hand cleaning gels, hard surface cleaners, shampoos; plastics antistat. [Harcros UK]

Ethylan®NP 1. CAS 9016-45-9; Nonoxynol-1.5; defoamer, oil emulsifier. [Harcros UK]

Ethylan® TN-10. CAS 61791-14-8; PEG-10 cocamine; wetting agent for acid or alkaline metal cleaners, stripping of surface coatings; oil emulsifier with anticorrosive properties, antistat for synthetic fibers with PS; for cosmetics mfg. [Harcros UK]

Ethylan® TT-15. CAS 61791-44-4; PEG-15 tallow amine; wetting agent for metal cleaning and stripping of surface coatings, fiber antistat, used in cosmetics. [Harcros UK]

Ethylbutylcarbinol. Sec-Heptyl alcohol.

Ethyl Cadmate. CAS 14239-68-0; Activated cadmium diethyldithiocarbamate; primary accelerator for NR and synthetic rubbers; used with a thiazole; gives heat resistance and low compression set properties. [R T Vanderbilt Co Inc]

ethyl carbinol. See n-propyl alcohol.

ethylcellulose. (cellulose, ethyl ether; ethocel) Ethyl ether of cellulose; CAS 9004-57-3; hot-melt adhesives and

---

coatings for cables, etc.; extrusion wire insulation, protective coatings, pigment-grinding bases, food additive; tablet binder in pharmaceuticals. [Aqualon; Colorcon; FMC; Hercules]

Ethyl Chloride BP. CAS 75-00-3; Ethyl chloride BP; used for local anesthesia. [Bengue & Co Ltd] *

Ethylcyanine. Lepidinquinolineethylcyanine bromide, a dyestuff; a similar dye to Cyanine, using ethyl bromide instead of amyl iodide; used as a substitute for cyanine in dyeing.

ethyldibutoxy phthalate. See dibutoxyethyl phthalate.

Ethyl Di-Icinol. CAS 110-90-0; EINECS 203-919-7; 2-(2-ethoxyethoxy)ethanol; solv. for use in protective coatings, inks, cleaning products, agricultural chems.; aids wetting, penetration, and soil removal; coupling solv. [ICI Australia]

N-ethyldimethylamine. See n,n-dimethylethylamine.

ethyl dimethyl methane. See isopentane.

Ethylene Glycol Distearate VA. CAS 627-83-8; EINECS 211-014-3; Glycol distearate; opacifier and pearling agent for hair care [Goldschmidt]

Ethylene Glycol Monostearate VA. Glycol stearate; stabilizer and thickener for creams and lotions; pearlizing and opacifying agent. [Goldschmidt]

ethylene alcohol. See glycol.

ethylene chloride. See ethylene dichloride.

ethylene diacetate. See ethylene glycol diacetate.

ethylenediamine. (1,2-diaminoethane) $NH_2CH_2CH_2NH_2$; CAS 107-15-3; EINECS 203-468-6; Fungicide, manufacture of chelating agents, dimethylolethylene-urea resins, chemical intermediate, solvent, emulsifier, textile lubricants, antifreeze inhibitor. [Aldrich; Allchem Industries; BASF; Texaco; Union Carbide]

ethylene diamine tetraacetic acid. See edetic acid.

ethylene diamine tetraacetic acid, sodium salt. See tetrasodium EDTA.

ethylene dichloride. (ethane, 1,2-dichloro-; Dutch oil; 1,2-dichloroethane) Halogenated aliphatic hydrocarbon; $Cl—CH_2CH_2—Cl$; CAS 107-06-2; production of vinyl chloride, trichloroethylene, vinylidene chloride, trichloroethane; lead scavenger in gasoline; paint, varnish remover; metal degreasing; soaps; wetting/penetrating agents; organic synthesis; ore flotation; solvent; fumigant. [Albright & Wilson Am.; Ashland; BASF; BP Chem. Ltd; Ethyl; Georgia Gulf; Norsk Hydro AS; OxyChem; PPG Industries]

ethylene dimethacrylate. See ethylene glycol dimethacrylate.

ethylene glycol. See glycol.

ethylene glycol bisthioglycolate. See glycol dimercaptoacetate.

ethylene glycol butyl ether. See butoxyethanol.

ethylene glycol butyl ether acetate. See butoxyethanol acetate.

ethylene glycol diacetate. (glycol diacetate; ethylene diacetate; 1,2-diacetoxyethane) Ester; $CH_3COO-CH_2CH_2OOCCH_3$; CAS 111-55-7; EINECS 203-881-1; extraction solv., foundry resins, perfume fixative, solv. for coatings. [Chemoxy Int'l. plc; CPS; Eastman]

ethylene glycol dimethacrylate. (1,2-ethanediol dimethacrylate; 1,2-bis(methacryloyoxy) ethane; ethylene dimethacrylate) $CH_2:C(CH_3)COOCH_2CH_2O-COC(CH_3):CH_2$; CAS 97-90-5; EINECS 202-617-2; Crosslinker and modifier for ABS, acrylic sheet and rods,

PVC, ion exchange resins, glaze coatings, dental polymers, paper processing aids, rubber modifier, adhesives, optical polymers, leather finishing, moisture barrier films. [Akzo; CPS; Sartomer]

ethylene glycol distearate. See glycol distearate.

ethylene glycol ethyl ether. See ethoxyethanol.

ethylene glycol ethyl ether acetate. See ethoxyethanol acetate.

ethylene glycol methyl ether. $CH_3OC_2H_4OH$; CAS 109-86-4; solvent. [Arco; Ashland; OxyChem; Union Carbide]

ethylene glycol monobutyl ether. See butoxyethanol.

ethylene oxide. (oxirane; epoxyethane) C_2H_4O; CAS 75-21-8; EINECS 200-849-9; Manufacture of ethylene glycol and higher glycols, surfactants, acrylonitrile, ethanolamines; petroleum demulsifier; fumigant; rocket propellant; industrial sterilant (medical plastic tubing); fungicide. [BASF; Hoechst Celanese; Hüls AG; Mitsubishi Petrochem.; Mitsui Petrochem. Ind.; Mitsui Toatsu Chem.; OxyChem; Shell; Union Carbide]

ethylene-propylene copolymer. See EPM rubber.

ethylene succinic acid. See succinic acid.

ethylene tetrachloride. See perchloroethylene.

ethylene thiourea. (ETU 2-imidazolidinethione; imidazoline-2-thiol; 2-mercaptoimidazoline) $NHCH_2CH_2NHCS$; CAS 96-45-7; EINECS 202-506-9; Electroplating baths; intermediate for antioxidants, insecticides, fungicides, synthetic resins, vulcanization accelerators, dyes. [Faesy & Besthoff; Ore & Chem. Corp.]

ethyl ether. (diethyl ether; ether) $CH_3CH_2OCH_2CH_3$; CAS 60-29-7; EINECS 200-467-2; Organic synthesis; smokeless powders; industrial solvent; analytical chemistry; anesthetic; extractant. [Aldrich; Exxon; Hüls AG; Mallinckrodt; Quantum/USI; Spectrum Chem. Mfg.]

Ethyl Ether Anhydrous A.C.S. CAS 60-29-7; Ethyl ether; reagent grade. [Quantum/USI] *

Ethyl Ether USP/ACS. CAS 60-29-7; Ethyl ether; laboratory reagent, surface antiseptic and cleaning agent, in liniments, as analgesic, expectorant; denaturant for alcohol formulations; chem. reaction media. [Quantum/USI] *

ethyl 3-ethoxypropionate. $C_2H_5OC_3H_5O_2C_2H_5$; CAS 763-69-9; solvent; intermediate for vitamin B_1. [Eastman; Union Carbide]

Ethylflo 162. Polyalphaolefin; synthetic lubricant for automotive crankcase oils, hydraulic fluids, gear and transmission fluids, compressor lubricants, metalworking lubricants, and personal care items. [Ethyl Corp]

Ethylflo 180. Polyalphaolefin; synthetic lubricant for automotive crankcase oils, hydraulic fluids, gear and transmission fluids, compressor lubricants, metalworking lubricants, and personal care items. [Ethyl Corp]

ethylformic acid. See propionic acid.

Ethyl Green. (Methyl Green). A dyestuff; zinc double chloride of ethylhexamethylpararosaniline bromide, $C_{27}H_{35}N_3Cl_3BrZn$. Dyes wool mordanted with sodium thiosulfate and sulfuric acid or zinc acetate, and silk and cotton mordanted with tannin a bluish-green.

ethyl hexanediol. (ethohexadiol; 2-ethyl-1,3-hexanediol; ethyl hexylene glycol) Aliphatic alcohol; $CH_3CH_2-CH_2CHOHCHCH_2CH_3CH_2OH$; CAS 94-96-2; EINECS 202-377-9; Insect repellent, cosmetics, vehicle and solvent in printing inks, medicine, chelating agent for boric acid. [Hüls Am.; Union Carbide]

2-ethyl-1,3-hexanediol. See ethyl hexanediol.

2-ethylhexanol. (2-EH; 2-ethylhexyl alcohol; octyl alcohol;

---

‡ Trade name and manufacturer not verified   § Trade name without identified manufacturer

alcohol C$_8$) CH$_3$(CH$_2$)$_3$CHC$_2$H$_5$CH$_2$OH; CAS 104-76-7; EINECS 203-234-3; Plasticizer for PVC resins; defoaming agent, wetting agent, organic synthesis, solvent mix for nitrocellulose; penetrant for plasticizing inks, etc. [Aristech; Ashland; BASF; BP Chem. Ltd; Eastman; Shell]

2-ethyl-1-hexanol. See isooctyl alcohol.

2-ethylhexoic acid. (butylethylacetic acid) C$_4$H$_9$CH-(C$_2$H$_5$)COOH; CAS 78-42-2; Paint and varnish driers (metallic salts); esters as plasticizers. [Aldrich; Ashland; BASF; Eastman; Neste UK; Union Carbide]

2-ethylhexyl acetate. CH$_3$COOCH$_2$CHC$_2$H$_5$C$_4$H$_9$; CAS 103-09-0; Solvent for nitrocellulose, resins, lacquers, baking finishes. [Eastman; Hüls AG; MTM Speciality Chem. Ltd; Penta Mfg.]

2-ethylhexyl acrylate. See octyl acrylate.

2-ethylhexyl alcohol. See 2-ethylhexanol.

ethyl hexylene glycol. See ethyl hexanediol.

2-ethylhexyl hexadecanoate. See octyl palmitate.

2-ethylhexyl palmitate. See octyl palmitate.

ethyl 4-hydroxybenzoate. See ethylparaben.

ethyl p-hydroxybenzoate. See ethylparaben.

2-ethyl-2-hydroxymethyl-1,3-propanediol. See trimethylolpropane.

Ethyl Icinol. CAS 110-80-5; 2-Ethoxyethanol; solv. for use in protective coatings, inks, cleaning products, agricultural chems.; aids wetting, penetration, and soil removal; coupling solv. [ICI Australia]

ethyl lactate. (lactic acid, ethyl ester) Ester; CH$_3$CHOH-COOC$_2$H$_5$; CAS 97-64-3; dipped latex products, coatings, specialty finishes, inks, diluents, adhesives, intermediates, photoresist solvs., screen printing of electronic parts, flavors and fragrances, solvent resins. [CPS; Farleyway Chem. Ltd; Penta Mfg.]

ethyl methacrylate. (ethyl 2-methyl-2-propenoate; ethyl-α-methyl acrylate) Ester of ethyl alcohol and methacrylic acid; H$_2$C:CCH$_3$COOC$_2$H$_5$; CAS 97-63-2; EINECS 202-597-5; polymers; chemical intermediates. [Rohm & Haas]

2-ethyl, 4-methyl imidazole. C$_2$H$_5$C$_3$N$_2$H$_2$CH$_3$; CAS 931-36-2; EINECS 213-234-5; Curing epoxy resin systems. [Schweizerhall]

ethyl methyl ketone. See methyl ethyl ketone.

ethyl methyl ketone peroxide. See methyl ethyl ketone peroxide.

ethyl 2-methyl-2-propenoate. See ethyl methacrylate.

Ethyl Namate®. Sodium diethyldithiocarbamate. [R.T. Vanderbilt]

Ethylol. CAS 64-17-5; EINECS 200-578-6; Essentially ethyl alcohol containing approximately 6% isopropyl alcohol; denatured with 3% butyl alcohol and 1% lead-free petrol; solvent in paint and lacquers, printing inks, foundries, dyes, industrial detergents, explosives, polishes, degreasers, rust removers, manufacture of xanthates and esters. [Sasolchem] *

Ethylol Colored. Mineralized methylated spirits; solvent. [Sasolchem]

ethyl orthosilicate. See ethyl silicate.

ethylparaben. (ethyl 4-hydroxybenzoate; ethyl p-hydroxybenzoate; ethyl parahydroxybenzoate) Ester of ethyl alcohol and p-hydroxybenzoic acid; C$_9$H$_{10}$O$_3$; CAS 120-47-8; EINECS 204-399-4; pharmaceuticals; preservative. [Inolex; Nipa Labs; Penta Mfg.]

ethyl parahydroxybenzoate. See ethylparaben.

ethyl parathion. O,O-Diethyl-O-(4-nitrophenyl) phosphorothioate; used as an insecticide for the protection of field crops, vegetables and fruit. [A/S Cheminova] *

ethyl phthalate. See diethyl phthalate.

ethyl safranate. Ethyl dehydrocyclo geranate. [Quest Int'l. UK Ltd]

Ethyl Selenac. Selenium diethyl dithiocarbamate; rubber accelerator for NR, SBR and IIR; vulcanizing agent; effective in low sulfur and sulfurless heat-resistant compounds. [R T Vanderbilt Co]

ethyl silicate. (ethyl orthosilicate; tetraethyl orthosilicate) (C$_2$H$_5$)$_4$SiO$_4$; CAS 78-10-4; Intermediate for manufacture of ethyl silicate products; produces binders; chemical-and heat-resistant paints, cements, weatherproofing; protective coatings. [Akzo; Aldrich; R.W. Greeff; Hüls Am.; PCR; Wacker Silicones]

Ethyl Tellurac®. Tellurium diethyldithiocarbamate; ultra accelerator for NR, SBR, NBR, EPDM; used with thiazole modifiers; produces high modulus vulcanization; particulary act. in IIR compounds [R.T. Vanderbilt]

Ethylthiurad. CAS 97-77-8; Tetraethylthiuram disulfide; sulfur-bearing accelerator and vulcanizing agent; cure modifier for neoprene. [Monsanto Co] *

ethyl toluenesulfonamide. (n-ethyl-p-toluenesulfonamide) Mixture of isomers of aromatic amides; C$_2$H$_5$NH-SO$_2$C$_6$H$_4$CH$_3$; CAS 80-39-7; 1077-56-1; plasticizer for shellac, cellulose acetate, and protein materials; used in NC lacquers, cellulose acetate compositions, PVAc emulsion adhesives, synthetic polyamides; food pkg. applications. [ICI Spec.; Rit-Chem]

n-ethyl-p-toluenesulfonamide. See ethyl toluenesulfonamide.

ethyltrichlorosilane. (trichloroethylsilane) C$_2$H$_5$SiCl$_3$; CAS 115-21-9; EINECS 204-072-6; Intermediate for silicones. [Hüls Am.; PCR]

Ethyl Tuads®. CAS 97-77-8; Tetraethylthiuram disulfide; accelerator and vulcanizing agent for rubber. [R T Vanderbilt Co]

Ethyl Tuex. CAS 97-77-8; Tetraethylthiuram disulfide; a nondiscoloring and nonstaining accelerator; sharp curing range with normal to high sulfur; used in natural, butyl and nitrile rubbers, steam hose and calendered air-cured sheeting. [Uniroyal] *

ethyl vanillin. (3-ethoxy-4-hydroxybenzaldehyde; benzaldehyde, 3-ethoxy-4-hydroxy-) Substituted phenolic; C$_9$H$_{10}$O$_3$; CAS 121-32-4; EINECS 204-464-7; flavoring agent. [Boehringer Mannheim GmbH; Penta Mfg.; Schweizerhall]

Ethyl Zimate®. CAS 14324-55-1; Zinc diethyldithiocarbamate; ultra accelerator; primary accelerator in NR and SBR; requires a thiazole modifier for safe processing and wide cure range; nondiscoloring in light colored stocks; stabilizer in thermoplastic rubbers and hot melts; 1:1 combination (dry wt.) with Zetax suggested for latex foam acceleration; accelerator for butyl latex vulcanization. [R.T. Vanderbilt]

Etingal® A. Phosphoric acid ester; antifoam for papermaking. [BASF AG]

Etingal® L. Ethyl ether deriv. of fatty acid; foam controller for papermaking. [BASF AG]

Etingal® S. Phosphoric acid ester; foam breaker for papermaking, leather, furs. [BASF AG]

Etocas. Emulsifiers and solubilizers for oils, solvents and waxes; used for cosmetics; metalworking fluids; textiles;

---

* Trade name not verified as available   † Trade name verified as obsolete

insecticides, herbicides; household products. [Croda Chem. Ltd]

EtOH. See ethyl alcohol.

Etophen 102. CAS 9016-45-9; Nonoxynol-2; detergent, dispersant, emulsifier, wetting agent for household and industrial detergents, textiles, paper, leather, and ceramic industries. [Zschimmer & Schwarz]

Etophen 114. CAS 9016-45-9; Nonoxynol-14; washing and cleansing agent. [Zschimmer & Schwarz]

E-Toplex. CAS 59-02-9; Vitamin E supplement. [USV Pharmaceutical Corp] ‡

Etrofolan®. CAS 2631-40-5; 2-Isopropyl-phenyl-N-methylcarbamate; insecticide effective against leafhoppers and bugs. [Bayer AG]

Etronite. The trade name for a synthetic resin-paper product; used for electrical insulation. §

Etrynit. CAS 2921-92-8; Propatyl nitrate; vasodilator. [Sterling Drug Inc] *

ETU 2-imidazolidinethione. See ethylene thiourea.

Eubeco. Vitamin B complex for injection. [Allen & Hanburys Ltd] *

eucalyptol. (cineol; cajeputol; 1,8-epoxy-p-menthane; 1,3,3-trimethyl-2-oxabicyclo[2.2.2]octane) $C_{10}H_{18}O$; CAS 470-82-6; EINECS 207-431-5; Pharmaceuticals (cough syrups, expectorants), flavoring, perfumery. [Penta Mfg.; Quest Int'l.; Ungerer]

Eucopine. A pine disinfectant containing substituted phenolic germicides. [William Pearson Ltd] *

Eudemine. CAS 364-98-7; A proprietary preparation of diazoxide; used in the treatment of hypertension and hypoglycemia. [Allen & Hanburys Ltd]

Euderm®. Finely disperse acrylic resin; used in the leather industry for tightening the grain and as a base coat in pigment finishing. [Bayer AG]

eugenic acid. See eugenol.

eugenol. (4-allyl-2-methoxyphenol; caryophyllic acid; eugenic acid) $C_{10}H_{14}O_2$; CAS 97-53-0; EINECS 202-589-1; perfumes, essential oils, medicine (analgesic), production of isoeugenol for the manufacture of vanillin, flavoring. [Firmenich; Lucta SA; Penta Mfg.; Schweizerhall; Spectrum Chem. Mfg.; Ungerer]

eugenyl methyl ether. See methyl eugenol.

Euglycin. A proprietary trade name for metahexamide. §

Eugynon 30. Proprietary preparations of 250 mcg levonorgestrel and 30 mcg ethinylestradiol; oral contraceptives. [Schering Health Care Ltd]

Euka-drya. A Dutch cellulose rayon.

Eukanol®. Covering pigment dispersions with casein binder for finishing; used in leather industry. [Bayer AG; Bayer plc]

Eukesolar® Dyes. Metal complex dyes for the leather and fur industry. [BASF AG]

Eukesol® Binder S. Alcoholic polymer solution; binder and thickener for applying pigment finishes in leather and fur industry. [BASF AG]

Eukinase. A powder obtained by the desiccation of the pancreatic juice of swine, and contains mainly trypsin.

Eulan® 33. General purpose textile finishing agent, fast to washing and processing; for the permanent protection of wool, hair, feathers and brush filaments against moths, carpet beetles (Anthrenus) and black carpet beetles (Attagenus). [Bayer AG; Bayer plc]

Eulan® BLS; Mothproofing agent; used for protection of carpets and furnishing fabrics against moths and carpet beetles. [Bayer AG; Bayer plc]

Eulan® WA. Mothproofing agent; auxiliary for the protection of goods made from wool or wool blends against moths and carpet beetles (Anthrenus) where demands made on fastness to washing are not exacting, e.g., carpets and furnishing fabrics. [Bayer AG; Bayer plc]

Eulysin® WP. pH regulator for dyeing wool or nylon with acid and premetallized acid dyes. [BASF]

Eumulgin B2. CAS 68439-49-6; Ceteareth-20; emulsifier for ointments, creams, low visc. emulsions, cosmetics, pharmaceuticals. [Henkel/Cospha; Henkel Canada; Henkel KGaA; Pulcra SA]

Eumulgin C4. CAS 61791-08-0; PEG-5 cocamide; foaming agent for detergents; emulsifier. [Henkel; Henkel KGaA]

Eumulgin EP2. Ceteleth-2; emulsifier for mineral oil, hydrocarbons, fats, metalworking oils, textile auxiliaries. [Henkel/Cospha; Henkel Canada; Henkel KGaA; Pulcra SA]

Eumulgin EP 2L. Ethoxylated oleyl/cetyl alcohol; emulsifier component of pronounced cold behavior. [Henkel KGaA]

Eumulgin HRE 40. CAS 61788-85-0; PEG-40 hydrogenated castor oil; solubilizer and o/w emulsifier for cosmetic and pharmaceutical preparations [Henkel KGaA]

Eumulgin L. PPG-2-ceteareth-9; emulsifier, solubilizer for aq. or hydroalcoholic media; for skin and hair care preparations [Henkel/Cospha; Henkel KGaA]

Eumulgin M8. CAS 9004-98-2; Oleth-10 and oleth-5; emulsifier, solubilizer for pesticides and cosmetics. [Henkel/Cospha; Henkel KGaA]

Eumulgin O5. CAS 9004-98-2; Oleth-5; emulsifier, solubilizer for pesticides, perfumes, cosmetics, floor polishes, hair dressings, creams, lotions, mineral oil, terpenes; dispersant improving color acceptance of emulsion paints. [Henkel/Cospha; Henkel Canada; Henkel KGaA]

Eumulgin PA 10. CAS 61791-44-4; PEG-10 tallowamine; surfactant. [Pulcra SA]

Eumulgin PA 12. CAS 61791-14-8; PEG-12 cocamine; surfactant. [Pulcra SA]

Eumulgin PA 30. PEG-30 oleamine; surfactant. [Pulcra SA]

Eumulgin PC 2. CAS 61791-08-0; PEG-2 cocamide; detergent, foam stabilizer, solubilizer for liq. detergent systems. [Pulcra SA]

Eumulgin PK 23. CAS 61791-29-5; PEG-23 cocoate; surfactant. [Pulcra SA]

Eumulgin PLT 4. CAS 9004-96-0; Ethoxylated oleic acid; emulsifier for paraffinic waxes and compounds, mineral oils; dyeing assistant, lubricant, antistat, emulsifier for textile industry. [Pulcra SA]

Eumulgin PPG 40. PPG-40; surfactant. [Pulcra SA]

Eumulgin PRT 36. CAS 61791-12-6; PEG-36 castor oil; detergent, emulsifier, dispersant, solubilizer for conc. pesticides, metal, leather, cosmetics, toiletries, pharmaceuticals, textile, and polymer industries. [Pulcra SA]

Eumulgin PST 5. CAS 9004-99-3; PEG-5.2 stearate; surfactant. [Pulcra SA]

Eumulgin PWM2. CAS 9004-98-2; Oleth-2; w/o emulsifier, dispersant, lipophilic cosolv.; fragrance grade; broad pH and electrolyte tolerance. [Pulcra SA]

Eumulgin RO 40. CAS 61791-12-6; PEG-40 castor oil; o/w emulsifier, solubilizer for perfume oils; for personal care creams and lotions. [Henkel/Cospha; Henkel Canada; Henkel KGaA]

Eumulgin SML 20. CAS 9005-64-5; Polysorbate 20; solubilizer and emulsifier for cosmetics and pharmaceuticals.

---

‡ Trade name and manufacturer not verified   § Trade name without identified manufacturer

[Henkel; Henkel KGaA; Pulcra SA]

Eumulgin SMO 20. CAS 9005-65-6; Polysorbate 80; solubilizer and emulsifier for cosmetics and pharmaceuticals. [Henkel KGaA]

Eumulgin SMS 20. CAS 9005-67-8; Polysorbate 60; solubilizer and emulsifier for cosmetics and pharmaceuticals. [Henkel KGaA]

Eumulgin WM5. CAS 9004-98-2; Oleth-5; raw material for foam-controlled laundry detergents. [Henkel KGaA; Pulcra SA]

Eumydrin. Methylatropine nitrate; anticholinergic. [Sterling Drug Inc] *

Eunatrol. CAS 143-19-1; Pure sodium oleate; ore flotation, waterproofing textiles, emulsifier of oil/water systems.

euosmite. A resin found in the lignite in Bavaria.

Eupad. An antiseptic consisting of bleaching powder and boric acid in equal parts.

Euparen®. CAS 1085-98-9; Dichlofluanid; fungicide with specific action against Botrytis. [Bayer AG]

Euparen®M. Tolylfluanid; broad spectrum fungicide; effective against Botrytis. [Bayer AG] *

Euperlan®. Mixture of fatty alcohol ether sulfates with special additives; for production of pearly sheen shampoos. [Henkel Chemicals Ltd] *

Euperlan®MPK 850. Sodium laureth sulfate, magnesium laureth sulfate, sodium laureth-8 sulfate, magnesium laureth-8 sulfate, sodium oleth sulfate, magnesium oleth sulfate, glycol stearate, PEG-3 distearate, cocamide MEA, laureth-10; fatty alcohol ether sulfates with pearlescents for use as a pearly gloss conc. for baby shampoos, bath products [Henkel/Cospha; Henkel Canada]

Euperlan®PK 810. Glycol distearate, sodium laureth sulfate, cocamide MEA, laureth-9; pearlescent base for lotion shampoos and bath products [Henkel/Cospha; Henkel Canada]

Euperlan®PK 900. PEG-3 distearate, sodium laureth sulfate; pearlescent base for lotion shampoos, bath products [Henkel/Cospha; Henkel Canada]

Euperlan®PK 3000. Glycol distearate, laureth-4, cocamidopropyl betaine; surfactant for pearlescent shampoos, bubble bath creams; brilliant, dense pearly gloss. [Henkel/Cospha; Henkel Canada]

Euphoramin. A proprietary preparation of meprobamate and methylamphetamine hydrochloride. [Rybar Laboratories Ltd] *

Euphyllin®. A registered trademark for a mixture of equal parts of primary and secondary theophylline-ethylenediamine, $C_2H_4(NH_2)_2 \cdot C_7H_8N_4O_2 \cdot C_2H_4(NH_2)_2 \cdot 2C_7H_8N_4O_2$; used in medicine. [Byk Gulden Lomberg Chem] *

Eupinal. A proprietary preparation of cabeine iodide. §

Eupolen®. Formulations of org. and inorg. pigments with low molecular weight polyethylene; used for mass dyeing of polyolefins and other polymers. [BASF AG]

Eurax-Hydrocortisone. A proprietary preparation of hydrocortisone and crotamiton; for dermatological use. [Ciba plc] *

Eurecryl. Cyanoacrylates; rapid curing, solvent free adhesives. [Schering Industrial Chemicals Ltd]

Euredur. Epoxy resin curing agents. [Schering Industrial Chemicals Ltd]

Eureka 102. CAS 8002-33-3; EINECS 232-306-7; Sulfated castor oil; emulsifier, detergent; grinding aid in pigment

disps.; plasticizer in finish coatings; topping oil for suede leather. [Atlas Refinery]

Eureka 392. CAS 61790-35-0; Sulfated tall oil fatty acid; emulsifier, carrier for refined oils, base for solv. fat liquor systems; for detergent formulations; penetrant, fiber lubricant for textile and leather processing. [Atlas Refinery]

Eureka 400-R. Sulfonated fish oil; emulsifier, lubricant, softener, fatliquor used in leather processing, fibers. [Atlas Refinery]

Eureka 800. Sulfonated natural oil; fatliquor for leathers; nonyellowing softener. [Atlas Refinery]

Eureka 800-R. Sulfonated animal oil; fiber lubricant; fat liquor for leather tanning. [Atlas Refinery]

Eureka 1014-M. Sulfated neatsfoot oil, sulfated synthetic sperm oil; fat liquor for finer upper leathers; imparts whiteness, softness and fullness. [Atlas Refinery]

Eureka 1067-A. Fish oil, sulfated fish oil, hydrocarbon oils; economical fat liquor for split leather. [Atlas Refinery]

Eureka Alloy. A trademark for a copper-nickel alloy used for electrical resistance wires. §

Eureka E-2. Fatty acid diamine condensate deriv.; emulsifier for raw oils, fiber lubricant, fabric softener. [Atlas Refinery]

Eurelon. Thermoplastic polyamides; binders for flexo and rotogravure inks, overprint varnishes and thixotropic agents for alkyd resins. [Schering Industrial Chemicals Ltd; Sherex Chemical Co Inc]

Euremelt. Polyamides and ethylene vinyl acetate hot melt adhesives. [Schering Industrial Chemicals Ltd]

Eurepox. Epoxy resins. raw material base for paints, surface coatings, resin mortars, two component adhesives, casting resins and laminates. [Schering Industrial Chemicals Ltd]

Euresol. Resorcinol monoacetate; antiseborrheic; keratolytic. [Knoll Pharmaceutical Co] *

Euresyst. Epoxy resin systems; binders for printing inks and injection resins for the building industry. [Schering Industrial Chemicals Ltd]

Euretek®. Vinyl plastisol adhesion promoter. [Sherex/Div of Witco]

Euretek® 540. Adhesion promoter for PVC plastisol; noncritical, conventional bake applications. [Schering Berlin Polymers] *

Eurobin. Chrysarobin triacetate.

European Elastin 10. CAS 73049-73-7; Hydrolyzed elastin; substantive protein adding to skin elasticity; protective colloid effect; film-former; moisturizer; for skin nourishing creams, shave creams, sun tan products, hair care and treatment products [Maybrook]

Europolymer. Thermoplastic polyurethanes for injection molding, and as adhesives and coatings. [Avalon Chemical Co Ltd] *

Europrene. A registered trademark for general purpose butadiene styrene copolymers coded as follows: CIS 1-4 cis-polybutadiene; SS high styrene copolymers; N butadiene acrylonitrile copolymers; AR acrylic; SOL butadiene styrene. [EniChem Elastomers Ltd]

Europrene AR. Polyacrylic rubbers. [EniChem Elastomers Ltd]

Europrene CIS. Polybutadiene rubbers. [EniChem Elastomers Ltd]

Europrene Lattice. Synthetic latexes. [EniChem Elastomers Ltd]

---

* Trade name not verified as available   † Trade name verified as obsolete

Europrene N. Nitrile rubbers. [EniChem Elastomers Ltd]

Europrene NEOCIS. Polybutadiene rubbers. [EniChem Elastomers Ltd]

Europrene SOL. Solution butadiene styrene rubbers. [EniChem Elastomers Ltd]

Europrene SOL T. Thermoplastic rubbers. [EniChem Elastomers Ltd]

Eurotex. Polyaminoamide resins, used as adhesion promoters for PVC plastisols. [Schering Industrial Chemicals Ltd]

Eurylon. Maize starch, high in amylose; used for food, textiles, board and paper industries. [Roquette (UK) Ltd]

Eusapon®. Wetting agents for leather and fur processing. [BASF AG]

Eusin®. Brilliant, transparent pigment dispersions based on casein binders for aniline and aniline effect finishing; used in the leather industry. [Bayer AG; Bayer plc]

Eusol. A solution containing 0.54% hypochlorous acid, 1.28% calcium biborate, and 0.17% calcium chloride. It is made by shaking Eupad (qv) in water, and filtering.

Eusolvan. A proprietary solvent stated to consist of ethyl lactate. §

Eutannin. A mixture of gallic acid and milk sugar. An intestinal astringent.

Eutanol G. CAS 5333-42-6; Octyldodecanol; lubricant, emollient for cosmetics and pharmaceuticals; carrier for oil-sol. activ ingreds.; pigment dispersant. [Henkel/Cospha; Henkel Canada]

Eutanol G16. CAS 36311-34-9; EINECS 252-964-9; Isocetyl alcohol; emollient for personal care products; carrier for oil-sol. active ingreds.; dispersant for pigments. [Henkel/Cospha; Henkel Canada]

Euthatal. CAS 57-33-0; Pentobarbitone sodium solution. [RMB Animal Health Ltd]

Euthroid. Liotrix; a mixture of liothyronine sodium and levothyroxine sodium in a ratio of 1:1 in terms of biological activity, or in a ratio of 1:4 in terms of weight; thyroid hormone. [Parke-Davis]

Euthylen®. Formulations of org. and inorg. pigments in PE with low molecular weight constituents; for coloring polyolefins. [BASF AG]

Eutonyl. CAS 306-07-0; Pargyline hydrochloride; antihypertensive. [Abbott Laboratories] *

Euvalerol B. A proprietary preparation of valerian root with phenobarbitone; a sedative. [Allen & Hanburys Ltd] *

Euvinyl®C. Formulations of org. and inorg. pigments in a VC copolymer; for mass coloring of rigid PVC, plasticized PVC, PVC pastes and VC copolymers. [BASF AG]

Euviprint®. Formulations of org. and inorg. pigments in a VC copolymer; for pigmenting gravure inks and surface coatings for plasticized PVC and aluminum foil and other pkg. materials. [BASF AG]

Euvitol. CAS 2240-14-4; A proprietary preparation of fencamfamin hydrochloride. [Allen & Hanburys Ltd] *

Euxyl. A range of preservatives used in cosmetics and toiletries. [Sterling-Winthrop Group Ltd]

EVA. Ethylene vinyl acetate. A flexible polythene-like polymer for molding and extrusion; adheres well to metals.

Evac-Q-Mag. CAS 3344-18-1; Magnesium citrate; laxative. [Adria Laboratories Inc] *

Evac-Q-Tabs. CAS 77-09-8; Phenolphthalein; laxative. [Adria Laboratories Inc] *

Evadyne Tablets. CAS 5585-73-9; A proprietary preparation of butriptyline hydrochloride; an antidepressant.

[Wyeth Laboratories]

Eval. Ethylene vinyl alcohol resins. [Quantum Chemical Corp]

Eval®E105. Ethylene vinyl alcohol copolymer; thermoplastic barrier resin used in pkg. materials, gas fill pkg., pkg. for oily foods, edible oils, mineral oils, agricultural pesticides, org. solvs.; suitable for film extrusion, sheet coextrusion, blow molding coextrusion, tube profile coextrusion, extrusion coating, and laminating procedures; EP-E105 grade used for sheets and films. [Eval Co. of Am.]

Eval® E151. Ethylene vinyl alcohol copolymer resin; melt phase forming, solid phase pressure forming, blow molding, tubes. [Eval Co. of Am.]

Eval® G115. Ethylene vinyl alcohol copolymer resin; cast film, specialty applications. [Eval Co. of Am.]

Eval® K102. Ethylene vinyl alcohol copolymer resin; solid phase pressure forming applications. [Eval Co. of Am.]

Evanacid®3CS. CAS 99-68-3; Carboxymethyl mercaptosuccinic acid; metal chelate; metal deactivator for the stabilization of glyceride oils. [Evans Chemetics]

Evangard® 18MP. CAS 31778-15-1; Octadecyl 3-mercaptopropionate. [Evans Chemetics]

Evanol. Proprietary; cosmetic cream base for depilatories, hair relaxing creams, and curl kits. [Evans Chemetics]

Evans' cement. A metallic cement made by adding cadmium amalgam (74% mercury) to mercury.

Evanstab® 12. CAS 123-28-4; Dilauryl thiodipropionate; antioxidant for polyethylene, PP, and polyolefins, ABS; stabilizer for polyolefins, oils and fats, food application; plasticizer for rubber products; lubricating oil additive; synthetic lubricant; chemical preservative in fats and oils. [Evans Chemetics]

Evanstab® 13. CAS 10595-72-9; Ditridecyl thiodipropionate; sec. antioxidant in ABS, polyolefins, and other polymer systems. [Evans Chemetics]

Evanstab® 14. CAS 16545-54-3; Dimyristyl thiodipropionate; sec. antioxidant for polyolefins. [Evans Chemetics]

Evanstab® 18. CAS 693-36-7; Distearyl thiodipropionate; sec. antioxidant for use in polyolefins; used in food-pkg. materials and edible fats and oils. [Evans Chemetics]

Evasperse. Ethylene vinyl acetate dispersions. [Runnymede Dispersions Ltd]

Evatane. Ethylene-vinyl acetate copolymers; used for hot melt adhesives. [Atochem UK Ltd; Atochem Inc]

Event. Adjuvant containing 90% alkyl and aryl ether phosphates; wetting agent for pesticide and liquid fertilizer mixtures. [Ideal Manufacturing Ltd]

Everbrite. A proprietary trade name for an alloy of 60% copper, 30% nickel, 3% iron, 3% silicon, and 3% chromium. §

Eveready Prestone. The trademarks applied to ethylene glycol antifreeze. §

Everflex® 81L. CAS 108-05-4; Vinyl acetate copolymer emulsion; paint and coating emulsion for use as binder for clay coating of paper and paperboard; factory finishes, ceiling tile, wall board, and textile treatments. [W R Grace/Organics] *

Everflex®515L. Vinyl emulsion in anionic/nonionic emulsifier system; pigment binder for use in surface coatings, e.g., interior flat wall paints and primer sealers. [W R Grace/Organics]

Everflex®E. Vinyl acrylic emulsion; general purpose paint/coatings emulsion for interior and exterior paints with well-balanced properties; latex, interior flat, primer seal-

---

‡ Trade name and manufacturer not verified   § Trade name without identified manufacturer

ers, exterior masonry. [W R Grace/Organics] *

Everflex® SP-1084. CAS 9003-20-7; Vinyl acetate copolymer emulsion; high molecular weight emulsion used as concrete additive to improve adhesion and compressive strength. [W R Grace/Organics]

Evergreen. Lawn fertilizer combined with selective weedkiller. [Fisons plc, Horticulture Div] *

Everitt's salt. Potassium ferrous ferro-cyanide, $K_2Fe \cdot Fe(CN)_6$.

Everlastic. A proprietary trade name for a textile incorporating Duprene. §

Everlube. Family of oven cured dry film lubricant coatings containing solid lubricants and suitable binder system; used for fasteners, slides, pins, clips and numerous applications requiring dry lubrication and corrosion protection particularly under high loads. [E/M Corporation]

Everseal. A bituminastic liquid applied as a corrosion-resisting material.

Eversoft Plastex. A low-freezing explosive containing a nitrated mixture of glycerin and ethylene glycol, ammonium nitrate, sodium chloride, wood meal, trinitrotoluene, and nitro-cotton.

Eversoft Sea Mex. A low-freezing explosive containing a nitrated mixture of glycerin and ethylene glycol, ammonium nitrate, sodium chloride, and wheat flour.

Eversoft Tees Powder. A low-freezing explosive consisting of a nitrated mixture of glycerin and ethylene glycol, ammonium nitrate, wood meal, and sodium chloride.

Eversun. Sun protection products, to protect skin while promoting a tan. [Richardson-Vicks Inc] *

Evidorm. A proprietary preparation of hexobarbitone and cyclobarbitone calcium; a hypnotic. §

Evipal. CAS 56-29-1; Hexobarbital; sedative. [Sterling Drug Inc] *

Evipan Sodium®. CAS 358-52-1; Hexapropymate; pharmaceutical preparation for intravenous anesthesia. [Bayer AG]

Evolve. Hydrocarbon or oxyhydrocarbon and aqueous-based solvents and equipment for industrial cleaning processes. [ICI Chem. & Polymers Ltd]

Evo-stik 873 Super. A proprietary synthetic rubber latex adhesive having a high solids contents. [Evode Plastics Ltd]

Evramycin. A proprietary preparation containing triacetyloeandomycin; an antibiotic. [Wyeth Laboratories] *

Ewer and Pick's acid. 1, 6-Naphthalene-disulfonic acid, $C_{10}H_6(SO_3H)_2$.

Ewo. Micronized white barytes. [Sachtleben Chemie GmbH]

EW-POL 8021. Aryl polyglycol ether; surface-active plasticizer, thickener for PVAc adhesive dispersions. [Henkel/Functional Prods.]

Exact-S®. CAS 75-18-3; Dimethyl sulfide; coking suppressor for ethylene production and for steel mill furnace walls; odorant for natural gas; presulfiding agent for catalysts in refinery processes. [Gaylord Chem.]

Exaltone. A trade name for muskone (cyclopentadecanone), the perfuming principle of natural musk. §

Exameen 824 3724. CAS 139-08-2; EINECS 205-352-0; Myristalkonium chloride; germicidal quaternary for hard surface disinfection and sanitization; algicide and slimicide for swimming pool and industrial water treatment. [Clough]

Exameen 2125 M 3704. Myristalkonium chloride,

quaternium-14; germicidal quaternary for bacteriological control in disinfectant and sanitizer formulations for hospitals, nursing homes, public institutions, and industry. [Clough]

Exameen 3500 3714. Benzalkonium chloride; germicidal quaternary for hard surface disinfection and sanitization; algicide and slimicide in swimming pool and industrial water treatment. [Clough]

EXC-33. Fluorochemical copolymer; release agent for rubber, plastic, and metal industries. [Releasomers]

Excelite. See Metalite; also an American trade name for a thermosetting fibrous plastic. §

Excellerex. A proprietary rubber vulcanization accelerator; it is an aniline derivative. §

Excello. An electrical resistance alloy containing 85% nickel, 14% chromium, 0.5% iron, and 0.5% manganese; also the name for a carbon black used in rubber mixings.

Excelo. A proprietary trade name for a hot die steel containing 2.5% tungsten, 1.5% chromium, 0.35% vanadium, and 0.55% carbon. §

Excelon. A proprietary trade name for acrylic denture material. §

Excelsior. A proprietary trade name for carbon black. §

Exchem GO-1. 2-Ethylhexyl nitrate; cetane improver. [Exchem Industries Ltd]

Exelderm. CAS 61318-91-0; Topical antifungal preparation containing sulconazole nitrate. [ICI Chem. & Polymers Ltd]

Exell. CAS 61791-44-4; Wetting agent containing 64% polyethoxylated tallow amine; spreader for use with phosphonoglycine and contact herbicides. [Truchem Ltd]

Exem. Anionic, nonionic and special anionic and nonionic blends of emulsifier; emulsifier for agricultural pesticides. [Makhteshim Chemical Works Ltd] †

Exgraphite. Expandable acid treated graphite. [Foseco (F.S.) Ltd] †

Exkin. Oxime antiskinning agents; for prevention of skin formation in solvent-based coating compositions. [Hüls Am.]

Exlax. A proprietary phenolphthalein preparation. §

Exl-die Steel. A proprietary trade name for a non-shrinking steel containing 1.15% manganese, 0.5% chromium, 0.5% tungsten, and 0.9% carbon. §

Exna. Benzthiazide; diuretic; antihypertensive. [Wyeth Laboratories] *

Exobloc BF-1000. Modified closed-cell silicone foam; fire-resist., nontoxic foam for critical elec. component insulation, vibration damping/gap filler; HVAC line insulation; high and low temp. gasketing. [Bisco Prods.]

Exocerol® OM. Blowing system based on citric and azodicarbonamide; chemical blowing agent for foaming of thermoplastics in injection molding. [Boehringer Ingelheim; Henley]

Exoderil. CAS 65473-14-5; Naftifine hydrochloride; antifungal. [Sandoz Pharmaceuticals] *

Exolan Cream. Pale yellow aqueous cream containing 1% 1:8:9 triacetoxyanthracene; used for the topical treatment of sub-acute and chronic psoriasis including psoriasis of the scalp. [Dermal Laboratories Ltd]

Exolit® IFR-23. Ammonium polyphosphate and synergists; halogen-free flame retardant for PE, PP, EVA, elastomers. [Hoechst Celanese/Spec. Chem.]

Exolon. A trademark for abrasive articles consisting essen-

---

*   Trade name not verified as available   † Trade name verified as obsolete

tially of silicon carbide. §

Exoryl. Water based acrylic adhesives. [Harcros UK]

Exosurf. A proprietary preparation of colfosceril palmitate; for the prophylaxis or treatment of neonatal respiratory distress syndrome. [The Wellcome Foundation Ltd]

EXP-28. Silicone fluid; paint and protective coatings additive improving performance props. [Genesee Polymers]

EXP-49. CAS 78-62-6; Dimethyldiethoxy silane; intermediate for blocking hydroxyl and amino groups in org. synthesis reactions; also for preparing hydrophobic and release materials and enhancing flow of powders. [Genesee Polymers]

EXP-51. CAS 1825-62-3; Trimethylethoxy silane; intermediate useful for blocking hydroxyl or amino groups in order to perform reactions on multifunctional org. compounds or polymers; also for deactivating glass surfaces used in gas chromatographic applications. [Genesee Polymers]

EXP-58. Silicone wax copolymer; lubricant, mold release agent and internal lubricant for plastics. [Genesee Polymers]

Expancel® 091. Wet, unexpanded microspheres used as blowing agents. [Expancel]

Expancel® 551 DE. Dry, expanded microspheres used as resilient lightweight fillers. [Expancel]

expanded graphite. A substance prepared by covering flake graphite with an oxidizing agent, to produce a film of graphitic acid, then heating strongly to cause the particles to become distended.

Expandex®5PT. CAS 3999-10-8; 5-Phenyltetrazole; blowing agent for foaming plastics and elastomers at elevated temps.. [Uniroyal]

Expandex® 175. 5-Phenyltetrazole, barium salt; high temp. chemical blowing agent for engineering resins. [Uniroyal]

expanding solder. An alloy of 37.5% lead, 6.75% bismuth, and 56.25% antimony. It expands on cooling, and is used for fixing metal into holes.

Expansyl Spansule. A proprietary preparation of trifluoperazine hydrochloride, diphenylpyraline hydrochloride, and ephedrine sulfate. [SmithKline Beecham]*

Expectal®. Cold remedy. [Bayer AG]

Explosive D. Ammonium picrate.

explosive gum. Nitroglycerin gelatinized with collodion cotton. It contains 96% of the former, and 4% of the latter compound.

Explotab®. CAS 9063-38-1; Sodium starch glycolate NF/BP; tablet disintegrant for formulations prepared by direct compression or wet granulation techniques. [Mendell]

Exprol. Photographic developer. [May & Baker Ltd] *

Expulin. A proprietary preparation of pholdocine, ephedrine hydrochloride, chlorpheniramine maleate, glycerin, and menthol; a cough medicine. [Galen Ltd] *

Exsel®. Selenium sulfide lotion USP; for seborrheic dermatitis, tines versicolor. [Allergan, Inc]

Exsyproteines 2%, 4%. CAS 9007-58-3; Hydrolyzed animal elastin; cosmetic ingred. for moisturizers, oily skin treatments, anti-aging creams. [Exsymol]

Extend®. For the medical industry. [DuPont UK]

Extendopel. Hydrophobic complex; fluorocarbon extender with no effect on oil repellency; improves water repellency. [Sybron]

Extendospheres® Metalite® Zinc. Zinc clad hollow microspheres used as filler to build high-mil coatings at less weight, and in zinc-rich coatings for structural steel, marine, automobile underbodies, tank linings, metal coils, etc. [PQ Corp.]

Extendospheres® SG. Hollow microspheres used as low sp.gr. extender in resin matrices and concrete applications to reduce weight, add thermal insulation and increase impact strength. [PQ Corp.]

Exterol Ear Drops. CAS 124-43-6; Viscous solution containing 5% w/w urea hydrogen peroxide; an aid in the removal of hardened ear wax. [Dermal Laboratories Ltd]

Extil. A proprietary preparation containing carbinoxamine maleate and pseudoephedrine hydrochloride; a cough suppressant. [British Drug Houses] *

Extil Compound Linctus. A proprietary preparation containing noscapine, pseudoephedrine, and carbinoxamino maleate; a cough linctus. [British Drug Houses]*

Extir. A proprietary range of expandable polystyrene beads. [Montedison UK Ltd] *

Extlat®. CAS 9015-54-7; Hydrolyzed animal protein; for food industry. [Asahi Chem. Industry]

Extol. A proprietary trade name for a sulfated compound with solvents used as a detergent. §

Extra Bond. Polyvinyl acetate liquid; used for wood and general builders glue. [Wessex Resins & Adhesives Ltd]

extract of yeast. See yeast extract.

Extrakt 52. Surfactant blend; for bath additives, hair shampoos, liq. body cleaners. [Zschimmer & Schwarz]

Extramycin®. Bactericidal broad-spectrum antibiotic. [Bayer AG]

extra white metal. An alloy of 50% copper, 30% nickel, and 20% zinc. It is a nickel silver (German silver).

Extrox. CAS 1314-13-2; Coated zinc oxide. [Harcros UK]

Extrudoil®. Lubricants which reduce friction between workpeice and tool or die in exturding or cold forming operations. [Witco/Allied-Kelite]

Extrusil. CAS 10101-39-0; Calcium silicate; extender for rubber compounds. [Degussa]

Extrusion-Plus. Nonchlorinated and nonsulfonated oil; amber colred liquid; extensively used by automotive fastener and aerospace fastener manufacturers. [Rustlan Chemical Co] ‡

Exxal®. Industrial alcohols. [Exxon UK]

Exxal® 6. CAS 68526-79-4; Hexyl alcohol. [Exxon] *

Exxal® 7. CAS 70914-20-4; Isoheptyl alcohol. [Exxon] *

Exxal® 8. CAS 68526-83-0; Isooctyl alcohol. [Exxon] *

Exxal® 9. CAS 68526-84-1; Isononyl alcohol. [Exxon] *

Exxal® 10. CAS 68526-85-2; Decyl alcohol. [Exxon] *

Exxal® 12. CAS 68526-86-3; Dodecyl alcohol. [Exxon] *

Exxal® 13. CAS 68526-86-3; Tridecyl alcohol. [Exxon] *

Exxal® 16. CAS 36311-34-9; EINECS 253-149-0; Hexadecanol. [Exxon] *

Exxal® 18. CAS 2745-8-931; Octyldecanol. [Exxon] *

Exxal® 20. CAS 5333-42-6; Octyldodecanol. [Exxon] *

Exxal® 26. CAS 70693-05-9; Undecylpentadecanol. [Exxon] *

Exxal® L1315. CAS 67762-41-8; Linear C13-C15 alcohol. [Exxon] *

Exxate®. Acetates of higher oxo-alcohols. solvents for high solids paint formulations. [Exxon Int'l.; Exxon UK]

Exxate® 600. Oxo-hexyl acetate [Exxon] *

Exxate® 800. CAS 108419-32-5; C8 alkyl acetate. [Exxon]*

Exxate® 900. CAS 108419-33-6; C9 alkyl acetate. [Exxon]*

Exxate® 1000. CAS 108419-34-7; C10 alkyl acetate.

---

‡ Trade name and manufacturer not verified   § Trade name without identified manufacturer

[Exxon] *

Exxate® 1300. CAS 108419-35-8; C13 alkyl acetate. [Exxon] *

Exxelor. Polymer modifiers. [Exxon UK]

Exxon® Bromo XP-50. Brominated copolymer of isobutylene and paramethylstyrene; elastomer with high damping props., good environmental and aging resist., ozone and uv stability; suggested for tire applications (sidewalls, treads, innerliners, carcasses, bladders), mech. goods (gaskets, diaphragms), hoses, belting, adhesives. [Exxon]

Exxon® Butyl 065. Isobutylene-isoprene elastomer, 0.05-0.20% zinc dibutyl dithiocarbamate (stabilizer); used for flat belts, coated materials, rubber-covered rolls, hose, mats, molded and extruded products, o-rings, shock and vibration products, inner tubes, water barrier applications, elec. goods; products exhibit shock absorp., weather resist., flex, tear, and abrasion resist., impermeability, chem. resist., excellent elec. props.; vulcanized with sulfur, resin, GMF. [Exxon] *

Exxon® Butyl 077. CAS 9010-85-9; Isobutylene-isoprene elastomer; for use in applications requiring FDA compliance. [Exxon] *

Exxon® Butyl 268. Isobutylene-isoprene elastomer, 0.05-0.20% zinc dibutyl dithiocarbamate (stabilizer); used for inner tubes and extrusions; products exhibit impermeability to gases, ozone weathering, heat, chem., tear, and abrasion resist., good elec. props., low compression set, excellent color retention; vulcanized with sulfur, resin, GMF. [Exxon] *

Exx-Print. Solvents for offset printing inks. [Exxon UK]

Exxsol®. High quality dearomatized aliphatic solvent. [Exxon Int'l.]

Exxsol® D-40, D-60, D-80, D-110, D-130. CAS 64742-47-8; Dearomatized aliphatic; solvent. [Exxon] *

Exxsol® Heptane. CAS 142-82-5, 64742-89-8; Heptane; solvent. [Exxon] *

Exxsol® Hexane. CAS 110-54-3, 64742-49-0; Hexane; solvent. [Exxon] *

Exxsol® Isopentane. CAS 78-78-4; Isopentane, dearomatized; solvent. [Exxon] *

Exxtraflex. Polyolefin film. [Exxon Int'l.] *

Exyphen. A proprietary preparation of brompheniramine maleate, guaifenesin, phenylephrine hydrochloride, and phenyl propanolamine hydrochloride; a cough medicine. [H N Norton & Co Ltd] ‡

Exzyme. Protein digesting enzyme powder. [PMP Fermentation Prods. Inc]

Eymid. Polyamides. [Ethyl SA]

Eymyd® Prepreg. Fluorine-containing thermoplastic polyimides prepregged onto unidirectional fibers or woven fabric; structural parts for aircraft, engines, and rockets for use up to 700 F. [Ethyl Corp]

Eymyd® Resin L-20N. Polyamic acid solution in 1-methyl-2-pyrrolidinone; forms a thermoplastic fluorinated polyimide after solv. evaporation and curing; produces coatings with excellent thermal oxidative stability, adhesion, frictional wear, and elec. props.; applications incl. erosion barrier coating, corrosion-resist. coating, thermal protective coating, tool coating, adhesive, fiber sizes. [Ethyl Corp] *

Eypel®. Polyphosphazene polymers. [Ethyl Corp]

Eypel® A Acoustic Barrier Sheet. Poly (aryloxyphosphazene) elastomer; fire-resistant acoustic mass sheeting [Ethyl Corp]

Eypel® F. Polyfluoroalkoxyphosphazene compound with incorporated peroxide curative; fluoroelastomer for use where low temp. flexibility, flex fatigue resist., and chem. resist. are required, e.g., o-ring, other sealing applications; broad temp. use (65-350 F), excellent fatigue and solv. resist. [Ethyl Corp]

EZA®. Zeolite A; detergent builder; solv.; anticaking agent for detergents and desiccants. [Ethyl Corp]

E-Z Mix. Series of dry liquid concentrates; for rubber and vinyl. [C P Hall] *

EZ Mold Lubricant. Glycol surfactant; mold release lubricant for natural and synthetic rubber compounds [TSE Industries]

---

* Trade name not verified as available   † Trade name verified as obsolete

F

F12. *See* Freon. §

F238. CAS 31717-87-0; Emulsifiable concentrate of 400 g dodemorph per liter; used for control of mildew in ornamental nursery stock. [BASF plc]

F309. Iodophor concentrate. [Evans Vanodine International Ltd]

F-310. Gasoline additive. [Chevron] *

F-500, -3600, etc. CAS 21645-51-2; Aluminum hydroxide compressed gels; antacid actives. [Reheis]

F-1000®. CAS 21645-51-2; Aluminum hydroxide; antacid active with good resuspending props. [Reheis]

F-1000 Dried Gel. CAS 21645-51-2; USP grade aluminum hydroxide; antacid powder. [Reheis]

F-2000. CAS 21645-51-2; Aluminum hydroxide compressed gel; antacid. [Reheis]

F-2000 Dried Gel. CAS 21645-51-2; USP grade aluminum hydroxide; antacid powder. [Reheis]

F-2000 FR. Brominated epoxy; exhibits excellent performance in engineering plastics (PBT, PET, nylon), styrenics (ABS, PC), and thermosets (epoxy, phenolic, unsat. polyesters); high thermal stability and thermal aging. [AmeriHaas; Dead Sea Bromine]

F-2001. Brominated epoxy oligomer; flame retardant additive for thermosets (epoxy, phenolic, unsat. polyesters); good uv stability. [AmeriHaas; Dead Sea Bromine]

F-2100 Dried Gel. CAS 21645-51-2; USP grade aluminum hydroxide; antacid powder. [Reheis]

F-2200. Brominated epoxy resin; flame retardant resin for potting, wet lay-up, and pre-preg laminates, adhesives, molding compounds, coatings. [AmeriHaas; Dead Sea Bromine]

F-2200 Dried Gel. CAS 21645-51-2; USP grade aluminum hydroxide; antacid powder. [Reheis]

F-2300. Med. molecular weightbrominated epoxy oligomer; flame retardant additive for use in thermoplastics (PBT, PC/ABS, ABS, HIPS); high thermal and uv stability. [AmeriHaas; Dead Sea Bromine]

F-2300H. High molecular weightbrominated epoxy polymer; flame retardant additive for use in thermoplastics (PBT, PET), thermoplastic elastomers, alloys (PC/ABS), etc.; high thermal and uv stability, good melt flow. [AmeriHaas; Dead Sea Bromine]

F-2400. High molecular weightbrominated polymer; flame retardant for use in thermoplastics (PBT, PET, polyamides, PU), alloys (PC/ABS), styrenics (ABS, HIPS); high thermal stability and thermal aging, high uv stability. [AmeriHaas; Dead Sea Bromine]

F-2400E. High molecular weightbrominated polymer; flame retardant for use in thermoplastics (PBT, PET, PU), alloys (PC/ABS), styrenics (ABS, HIPS), etc.; high thermal and uv stability, thermal aging. [AmeriHaas; Dead Sea Bromine]

FA-1. CAS 25928-94-3; Epoxy resin adhesive, gyro-grade; for bonding and sealing aluminum and other metals; mix ratio: 100 pbw adhesive/3.2 pbw of Activator BA-4. [Bacon]

FA-8. CAS 25928-94-3; Epoxy resin adhesive, gyro-grade; for bonding and sealing beryllium and other metals; mix ratio: 100 pbw adhesive/13.5 pbw Activator BA-5. [Bacon]

FA-14. CAS 25928-94-3; Two-part modified epoxy compound; adhesive coil for impregnation of electronic components with high percentage of fine wires where complete penetration and freedom from voids is important; for casting and coating applications; able to penetrate cracks and crevices; bonding for components made from fused beryllium oxide where low surface tensile compound needed; suitable for gyro use in contact with poly(bromotrifluoroethylene) oil. [Bacon]

Fabahistin. Mebhydrolin napadisylate; a pharmaceutical preparation also known as Incidal; day-time antihistamine. [Bayer AG]

Fabelnyl. Polyamide injection molding compounds. [Rhone-Poulenc UK]

Fabrene®. Fibers. [DuPont UK]

Fabrethane. A proprietary one-component foamable polyurethane, 100% solids. [W R Grace & Co] *

Fabrex. Preformed insulating refractory shapes for various applications. [Foseco (F.S.) Ltd]

Fabrifil. A proprietary trade name for a macerated cotton fabric filler. §

Fabriglide. Fabric lubricant. [Dow Corning]

Fabrikoid. A trademark for a fabric coated with pyroxylin. §

Fabroil. A proprietary trade name for a synthetic resin. §

Fabrol. Mucolytic. [Ciba plc] *

Fabrolite. A synthetic resin of the phenol-formaldehyde type. [Associated Electrical Industries (GEC)] ‡

Facet®. Quinclorac; for post-emergence control of barnyard grass and some other weeds in rice. [BASF AG]

F-acid. *See* delta acid.

Facteka. A proprietary trade name for a rubber substitute. §

factice. A polymerization product of natural fatty oils and sulfur or sulfur chloride, a processing promoter for an economical processing of rubber; used for extruded articles. [Rhein-Chemie Rheinau] *

factis. A term applied to rubber substitutes prepared from oils.

Factoprene NS. A proprietary hard factice. [Hubron Rubber Chemicals] *

Factoprene Z. A proprietary soft factice. [Hubron Rubber

---

‡ Trade name and manufacturer not verified   § Trade name without identified manufacturer

Chemicals] *

Factorate. Antihemophilic factor [Armour Pharmaceutical Co] *

Factrel. Gonadorelin hydrochloride from sheep, pig or other species; luteinizing hormone-releasing factor hydrochloride; gonad-stimulating principle. [Wyeth Laboratories] *

Faexin extract. The fatty acids of yeast.

Fagacid. A product derived from beech wood tar. It is an antiseptic agent used in the preparation of soaps and plasters.

Fahlun diamonds. (tin brilliants). Lead-tin alloys, containing about 40% lead; used for theater jewelery.

Fahralloy. A proprietary trade name for chromium-nickel-iron alloys. §

Fairey Metal. A proprietary trade name for an alloy of aluminum with copper and smaller amounts of magnesium. §

Fairprene. A proprietary trade name for a chloroprene polymer for fabric coating. §

Fairy Ring Destroyer. CAS 26644-46-2; Emulsifiable concentrate containing 190 g/l triforine; a systemic insecticide. [Synchemicals Ltd]

Faktex. A proprietary trade name for a yellow rubber substitute which can be dispersed in water for addition to latex. §

Faktogel. Range of factices for use in the rubber industry. [Bayer AG]

Falapen. A proprietary preparation containing benzylpenicillin; an antibiotic. [British Drug Houses] *

Falcodyl. A proprietary preparation of pholcodine and ephedrine hydrochloride; a bronchial antispasmodic. [H N Norton & Co Ltd] ‡

FAL Cypermethrin 10. CAS 66841-24-5; Emulsifiable concentrate containing 100 g cypermethrin per liter; a pyrethroid insecticide. [Fine Agrochemicals Ltd]

Falkaloid. A proprietary trade name for a soft oil modified alkyd resin. §

Falkyd. A proprietary trade name for a soft oil modified alkyd resin. §

Falmonox. CAS 5560-78-1; Teclozan; anti-amebic. [Sterling Drug Inc] *

FAM 30. Iodophor. [Evans Vanodine International Ltd]

Famel. CAS 61-76-7; Proprietary preparations containing phenylephrine hydrochloride; nasal decongestants. [The Boots Co plc]

Famel Syrup. A trademark for a syrup comprising purified beech wood creosote rendered water soluble by means of lactic acid in combination with calcium lactophosphate, aconite and codeine; used for treatment of infections of the lungs. See Beatin. [Optrex] *

Famid. Carbamate insecticide. [Ciba plc] *

Famodan MS Kosher. CAS 1338-41-6; EINECS 215-664-9; Sorbitan stearate; food emulsifier for fat crystal modification and bloom retarders; for cocoa butter substitutes, compound coatings, imitation dairy systems. [Grindsted Prods.]

Famodan TS Kosher. CAS 26658-19-5; EINECS 247-891-4; Sorbitan tristearate; fat crystal modifier preventing fat bloom in cocoa butter substitutes and compound coatings; improves texture of frostings and icings. [Grindsted Prods.]

Famosan. Agricultural disinfectant. [The Wellcome Foundation Ltd] *

Famous. CAS 8016-70-4; EINECS 232-410-2; Partially hydrogenated soybean oil, stabilized. [Karlshamns]

Fanal. Specialty colorants. [BASF plc]

Fanasil. CAS 2447-57-6; A proprietary preparation of sulfadoxine; sulfonamide antibiotic. [Roche Products Ltd] *

Fancol 707. CAS 9035-85-2; PPG-30 cetyl ether; skin conditioner, emollient for permanent waves, shampoos, blushers, indoor tanning preparations. [Fanning]

Fancol Acel. CAS 61788-48-5; EINECS 262-979-2; Acetylated lanolin; emollient, superfatting agent, lipophilic spreading agent for oils, creams, lotions, hair grooms, ointments, pharmaceuticals. [Fanning]

Fancol ALA. Cetyl acetate and acetylated lanolin alcohol; cosmetic, toiletry conditioner, emollient, pigment binder, spreading agent, oil coupler, solv., glossing and plasticizing agent for hairsprays. [Fanning]

Fancol ALA-10. Polysorbate 80, cetyl acetate, acetylated lanolin alcohol. skin/hair conditioner, sec. emulsifier, pigment wetter, solubilizer, emollient, humectant, superfatting agent for antiperspirants, lotions and creams, sunscreens. [Fanning]

Fancol C. Petrolatum, lanolin, lanolin alcohol; w/o emulsifier, moisturizing lubricant, emollient, fatting agent, water absorbent. [Fanning]

Fancol CA. CAS 36653-82-4; EINECS 253-149-0; Cetyl alcohol; surfactant, emulsifier, emulsion stabilizer, opacifier, skin conditioner/emollient, visc. booster, foam booster; for makeup, hair conditioners, cleansers, moisturizing creams and lotions. [Fanning]

Fancol CAB. Petrolatum, lanolin, lanolin alcohol; w/o emulsifier, moisturizing lubricant, emollient, fatting agent, water absorbent. [Fanning]

Fancol CB. CAS 8002-31-1; Cocoa butter; skin conditioner, occlusive solv., skin protectant for OTC drug products, makeup, moisturizing creams and lotions, suntan preparations. [Fanning]

Fancol CH. CAS 57-88-5; Cholesterol; film-former with lubricating, protective and anti-irritant props., aids cell regneration, emulsifier for w/o formulations, precursor for production of vit. D; for cosmetics, pharmaceuticals, hair/skin care products. [Fanning]

Fancol CH-24. CAS 27321-96-6, 9004-95-9; Choleth-24 and ceteth-24; surfactant, emulsifier for hair dyes, bubble bath, moisturizing creams and lotions, cleansing products. [Fanning]

Fancol CO-30. CAS 61791-12-6; PEG-30 castor oil; surfactant, emulsifier for cleansing products, hair conditioners, wave sets, face, body and hand creams, and lotions. [Fanning]

Fancol DL. CAS 16485-10-2; DL-Panthenol; conditioning agent for hair care products, makeup, moisturizing creams and lotions. [Fanning]

Fancol Gingko Extract. Ginko extract; biological extract for use in lotions, milks, creams, shampoos, bubble baths, soaps, gels. [Fanning]

Fancol HCO-25. CAS 61788-85-0; PEG-25 hydrogenated castor oil; surfactant, emulsifier for bath oils, tablets and salts, aftershave lotions, skin fresheners. [Fanning]

Fancol HL. CAS 8031-44-5; EINECS 232-452-1; Hydrogenated lanolin; emollient, moisturizer, lubricant, plasticizer, chemical intermediate, humectant, mold release agent for pharmaceuticals, cosmetics, industrial applications. [Fanning]

Fancol HL-20. CAS 68648-27-1; PEG-20 hydrogenated

---

*   Trade name not verified as available

† Trade name verified as obsolete

lanolin. solubilizer, superfatting agent, gelling agent for cosmetics, pharmaceuticals, makeup, nail polish, night creams, microemulsions. [Fanning]

Fancol HON. CAS 8028-66-8; Honey; biological additive, flavoring, skin conditioner, humectant for shampoos, face, body and hand creams and lotions, bath products, hair conditioners, cleansing products, moisturizing creams and lotions. [Fanning]

Fancol Karite Butter. CAS 68424-60-2; Shea butter; ointment base, anti-irritant for skin, skin conditioner, occlusive agent, solv. for suntan preparations, body lotions, winter sports products, wrinkle creams, soaps, shave foams, shampoos, balsams. [Fanning]

Fancol Karite Extract. CAS 68424-59-9; Shea butter extract; emollient with excellent spreadability for suntan preparations, skin toners, lipsticks, eye liners, ointments, suppositories. [Fanning]

Fancol LA. CAS 8027-33-6; EINECS 232-430-1; Lanolin alcohol; emollient, thickener, emulsifier, stabilizer, plasticizer, superfatting agent, dye dispersant, chemical intermediate, lubricant, humectant, mold release agent, conditioner for cosmetics, pharmaceuticals, soaps, industrial applications. [Fanning]

Fancol LA-5. CAS 61791-20-6; Laneth-5; emollient, emulsifier, moisturizer, spreading agent, coupler for shampoos, skin care products, cosmetics. [Fanning]

Fancol LAO. CAS 8012-95-1, 9027-33-6; Mineral oil and lanolin alcohol; conditioner, surfactant, stabilizer, moisturizer, humectant, penetrant, emollient, plasticizer, and primary emulsifier for use in cosmetics and pharmaceuticals; plasticizer in aerosol formulas. [Fanning]

Fancol Menthol. CAS 89-78-1; Menthol; denaturant, flavoring agent, fragrance component for mouthwashes, aftershaves, cleansing products, face, body and hand creams/lotions, skin care products. [Fanning]

Fancol OA-95. CAS 143-28-2; EINECS 205-597-3; Oleyl alcohol; plasticizer, emulsion stabilizer, antifoam and coupling agent, aerosol lubricant, petrol. additive, pigment dispersant; rust preventive; detergent, release agent, cosolvent, softener, tackifier; spreading agent used for metalworking, petrochemicals, pulp and paper, paints and coatings, plastics, and polymers, food applications, pharmaceuticals, cosmetics; chemical intermediate. [Fanning]

Fancol SA. CAS 112-92-5; EINECS 204-017-6; Stearyl alcohol; emulsion stabilizer, opacifier, skin conditioner, emollient, visc. booster, foam booster for makeup, hair conditioners, moisturizing creams and lotions, cleansers. [Fanning]

Fancol SA-15. CAS 25231-21-4; PPG-15 stearyl ether; skin conditioner, emollient for bath oils, tablets and salts, fragrance products, lipsticks, deodorants, cleansers. [Fanning]

Fancol TOIN. CAS 97-59-6; Allantoin; skin conditioner for makeup, cleansers, face, body and hand creams and lotions, moisturizing creams, skin care products. [Fanning]

Fancol WGFA. CAS 68484-43-1; Wool grease fatty acid; emulsifier with high adhesion to various substrates, oil binding props. for use in lubricating greases, polishes, anti-corrosion compounds. [Fanning]

Fancor D. Blend of animal fats and petroleum-derived materials; long term rust preventive; EP and slip agent for wire drawing compounds, slushing oils, cutting oils,

lubricants; crystallization inhibitor for waxes; dispersant; in leather stuffing greases, fat liquors; waterproofing agent in polishes; rope dressing; impregnant for paper industry. [Fanning]

Fancor IPL. CAS 63393-93-1; EINECS 264-119-1; Isopropyl lanolate; hydrophilic emollient, moisturizer, w/o auxiliary emulsifier, stabilizer and opacifier, wetting and dispersing agent for pigments and talc, aids slip and gloss of stick cosmetics, plasticizes wax systems, mold release, superfatting agent; binder for pressed powds. [Fanning]

Fancor LFA. CAS 68424-43-1; EINECS 270-302-7; Lanolin fatty acids; emollient, stabilizer, emulsifier, corrosion inhibitor for personal care and pharmaceutical products; used in industrial leather treating, coatings, polishes, corrosion inhibitors, lubricants. [Fanning]

Fancor Lanwax. CAS 68201-49-0; EINECS 269-220-4; Natural lanolin wax ester; plasticizer in wax crayons; water repellent, humectant, conditioner, corrosion inhibitor, emollient, lubricant for cosmetics, toiletries, pharmaceuticals, industrial applications; extender and crystallization inhibitor for natural waxes in industrial applications; emulsifier. [Fanning]

Fancorsil A. CAS 69430-24-6, 9006-65-9; Cyclomethicone and dimethicone; hair/skin conditioner, emollient, solv.; used in hair conditioners, makeup, moisturizing creams and lotions. [Fanning]

Fancorsil P. Modified silicone hydrocarbon; skin/hair conditioning agent, emollient; excellent afterfeel and lubricity. [Fanning]

Fancorsil SLA. Dimethicone copolyol adipate; skin/hair conditioning agent, emollient; provides soft and velvety afterfeel. [Fanning]

Fancoscour PO, VC. Detergent for textile scouring. [Reilly-Whiteman]

Fanfare. Suspension concentrate containing 450 g isoproturon and 19 g isoxaben per liter; used for annual weed control in cereals. [Ciba-Geigy Agrochemicals]

Fangerine. Tablets containing phenolphthalein, chocolate, and sucrose; a laxative. §

Fanghidi Sclofani. A yellow powder of volcanic origin, consisting chiefly of sulfur, with small quantities of iron, calcium, and manganese.

Fansidar. Proprietary combination antimalarial product containing sulfadoxine and pyrimethamine. [Roche Products Ltd] *

Fantan. Phenylcinchonoylurethane.

Fantasit. A proprietary casein plastic. §

Fanwax G. Stearyl alcohol, ceteareth-20; o/w self-emulsifying wax for cosmetics and toiletries, skin/hair lotions, antiperspirants, depilatories, creme rinses, opacified hair dyes, bleaches. [Fanning]

Fanwax P. CAS 8005-44-5, 9005-67-8; Cetearyl alcohol, polysorbate 60; o/w self-emulsifying wax for lotions, creams, hair and skin care products, creme rinses, antiperspirants. [Fanning]

Fanzil. CAS 2447-57-6; Sulfadoxine; antibacterial. [Hoffmann-LaRoche Inc] *

Far-Go/Avadex BW. Trichloroallyl diisopropylthiocarbamate; herbicide for wild oats. [Monsanto Co] *

Fargro Chlormequat. CAS 999-81-5; Soluble concentrate containing 460 g/l chlormequat; plant growth regulator. [Fargro Ltd]

farina. Flour, or potato starch.

---

‡ Trade name and manufacturer not verified   § Trade name without identified manufacturer

farine. A term applied in the West Indies to a product obtained by grating fresh cassava root, draining away the juice from the wet pulp, and then heating the residue. It is also known as Cassava meal.

Faringets. A proprietary preparation of myristyl benzalkonium iodine chloride. [Bayer AG]

farinose. Starch cellulose, or the outer covering of the starch granule.

Farlite. A proprietary trade name for a phenol-formaldehyde synthetic resin laminated product. §

Farmacel. CAS 999-81-5; Growth regulator for wheat and oats containing chlormequaternary [ICI Chem. & Polymers Ltd]

Farmacel. CAS 999-81-5; Soluble concentrate containing 460 g/l chlormequat; plant growth regulator for wheat and oats. [Farm Protection Ltd]

Farmacel 645. CAS 999-81-5; Soluble concentrate containing 645 g/l chlormequat; plant growth regulator. [Farm Protection Ltd]

Farmaneb. Fungicide for prevention of potato blight. [ICI Chem. & Polymers Ltd]

FAR Mark I through X. Elastomeric coatings made up of asphalt, neoprene, acrylic and Kraton rubber imbedded in polyester fabric; used for cold-applied liquid or fully adhered membrane type roof systems. [Flex-Shield] *

Farmatin. CAS 76-87-9; Fentin hydroxide; used for control of potato blight. [Farm Protection Ltd]

Farmon. Range of liquid herbicides of different formulations. [ICI Chem. & Polymers Ltd]

Farmon 2,4-D. CAS 1702-17-6; 2,4-D; Translocated herbicide for cereals and established grassland. [Farm Protection Ltd]

Farmon 2,4-DB Plus. 2,4-DB + MCPA; Translocated herbicide for cereal crops. [Farm Protection Ltd]

Farmon 2,4-DP+MCPA. Soluble concentrate of 358 g dichlorprop and 177 g MCPA per liter; used for control of weeds in cereals and grass. [Farm Protection Ltd]

Farmon Blue. Nonionic wetting agent containing 900 g/l alkylphenol ethoxylate; spreader for use in herbicide, fungicide, and insecticide sprays. [Farm Protection Ltd]

Farmon Condox. Soluble concentrate of 112 g dicamba and 265 g mecoprop per liter; used for weed control in cereals and grassland. [Farm Protection Ltd]

Farmon MCPA 50. CAS 94-74-6; MCPA; herbicide for cereals and grassland. [Farm Protection Ltd]

Farmon MCPB Plus. Soluble concentrate containing 41 g MCPA and 244 g MCPB per liter; for control of weeds in undersown cereals and grassland. [Farm Protection Ltd]

Farmon Mini Slug Pellets. CAS 108-62-3; Pellets containing 6% w/w metaldehyde; snail and slug bait. [Farm Protection Ltd]

Farmon PDQ. Soluble concentrate of 80 g diquat and 120 g paraquat per liter; used for weed control in field crops. [Farm Protection Ltd]

Farmon TCA. CAS 76-03-9; TCA; for control of weeds in field crops. [Farm Protection Ltd]

Farnesol. A sesquiterpene alcohol prepared from nerolidol (*qv*); a perfume.

Farrant's medium. A microscopic medium. It consists of a mixture of equal parts of glycerin and arsenious acid, to which is added powdered gum arabic, allowed to stand and filtered.

Farronic. A heat-resisting nickel-copper alloy.

Fascat®2000. Esterification catalyst; effective where catalyst removal is required; source of stannous tin. [Atochem]

Fascat®2004. CAS 7772-99-8; Anhydrous stannous chloride; esterification catalyst for mfg. of plasticizers and polyesters. [Atochem]

Fascat®4400. CAS 7646-78-8; Anhydrous tin tetrachloride; strong Lewis acid catalyst for variety of reactions incl. acetylation and alkylation of aromatic compounds, chloromethylation, reaction of epichlorohydrin with alcohols to form glycidyl ethers. [Atochem]

Fascol. Hexachlorophane anthelmintic. [The Wellcome Foundation Ltd] *

Fasigyn. CAS 19387-91-8; A proprietary preparation of tinidazole; an antiprotozoal. [Pfizer International]

Fasinex. Cattle and sheep flukicide. [Ciba plc]

Fastac. CAS 67375-30-8; Alphacypermethrin. contact insecticide. [Shell UK]

Fasteeth. Denture adhesive powder, for securing dentures. [Richardson-Vicks Inc] *

Fasteeth Extra Hold. To secure hard-to-hold lower dentures. [Richardson-Vicks Inc] *

Fastex. A proprietary trade name for a specially stabilized and purified rubber latex supplied in concentrations of 40 and 60%. §

Fastin. CAS 1197-21-3; Phentermine hydrochloride; appetite suppressant. [SmithKline Beecham]

F.A.S.T. Lube System. Airless spray lubrication systems for the metalforming industry. [Franklin Oil Corporation (Ohio)]

Fastocaine. A proprietary preparation of lignocaine hydrochloride and noradrenaline; a local anesthetic. [Leo Laboratories] *

Fastusol®. Substantive dyes for paper coloring. [BASF AG]

Fatal Flip. Trade name for total release fogger insecticide; for use by both professional and consumer trade to control many flying and crawling insects. [Colonial Products Inc]

Fat, Chocolate. *See* vegetable butter.

Fatsco. Sodium arsenate 3%; ant poison for sweet eating ants, kills roaches, moles, mice, woodchucks etc. [Fatsco] ‡

Faturan. *See* Bakelite. §

Faunolen. Canine contagious hepatitis vaccine. [The Wellcome Foundation Ltd] *

Faversham powder No. 2. An explosive containing ammonium nitrate, potassium nitrate, trinitrotoluene, and ammonium chloride.

Favierite No. 1. An explosive consisting of 88% ammonium nitrate and 12% dinitro-naphthalene.

Favierite No. 2. An explosive containing 90% of No. 1 and 10% ammonium chloride.

Favour. Mixture of metalaxyl and thiram; protectant fungicide against downy mildew in lettuce. [Ciba-Geigy Agrochemicals]

Fax. Edible fats. [Marfleet Refining Co] *

Faxola. Cooking oil. [Marfleet Refining Co] *

FB 48. CAS 1330-43-4; Sodium borate. fertilizers. [U.S. Borax & Chem.]

FBC CMPP. Selective herbicide. [Schering Agrochemicals Ltd]

FBC Fly Dip. Insecticide. [Schering Agrochemicals Ltd]

FBC MCPA. Selective herbicide. [Schering Agrochemicals Ltd]

FBC Pirimicarb 50. Insecticide. [Schering Agrochemicals

Ltd]

FBC Protectant Fungicide. Fungicide. [Schering Agrochemicals Ltd]

FBC Slug Destroyer. Mulluscicide. [Schering Agrochemicals Ltd]

FBC Winter Dip. Insecticide. [Schering Agrochemicals Ltd]

FC 113. Cleaning solvent. [ICI Chem. & Polymers Ltd]

feathered tin. Granulated tin.

Feb Brickclean. Cleaner and degreasing solution for the rapid removal of cement mortar stains, grime, oil, grease, algae, moss, and wax from brickwork, concrete, asbestos, terrazzo, and ceramics. [Feb Ltd]

Febclean. Hand cleaner to remove grease, fresh paint, mastic, ink, glue, bitumen, rubber cement, tile adhesive and food stains without the need for water. [Feb Ltd]

Febclear Super. Surface coating providing resistance to oil, dusting, abrasions and splash contact from acid and alkali and common chemicals. [Feb Ltd]

Febco. Concrete floor hardening and dust proofing liquid based on sodium silicate. [Feb Ltd]

Febcrete AEA. Resin-based air entraining agent for concrete. [Feb Ltd]

Febcure. Curing compounds. [Feb Ltd]

Febdura Standard. Premixed material for concrete repair. [Feb Ltd]

Febexpan. Pre-mixed shrinkage compensating grouting compounds. [Feb Ltd]

Febface. Single-place polyurethane mold coating providing an abrasion-resistant protective coating to formwork; unaffected by alkalis, petrol, oil and grease; provides impervious barrier to water and damp. [Feb Ltd]

Febfast PG. A fast setting, high early strength epoxy resin pouring grade mortar. [Feb Ltd]

Febflex One Coat. One-coat solvent-based bitumen roofing compound; for coating and renovating slate, galvanized iron, asbestos, cement, lead, zinc, asphalt, and timber. [Feb Ltd]

Febflor. Powder floor smoothing compound for preparing substrates for final floor covering. [Feb Ltd]

Febflow Accelerating. Concrete plasticizer, water reducing admixture with set accelerating properties for protection of concrete. [Feb Ltd]

Febflow Retarding. Non-air entraining, water reducing, set retarding admixture for concrete. [Feb Ltd]

Febfoam. Liquid foaming agent producing aerated concrete or cement. [Feb Ltd]

Febglaze. A lightly filled pigmented, solventless, epoxy resin-based compound; for surface application to achieve chemical resistant coatings for concrete storage tanks, etc. [Feb Ltd]

Febgrout. A water-reducing, expanding admixture for cement and mortar grouts. [Feb Ltd]

Febguard Bonding Agent. Adhesion promoter between concrete substrates and repair mortars. [Feb Ltd]

Feb Hybit. High solids protective coating based on emulsified bitumen; dries to tough water and vapor-proof film; used for general waterproofing and dampproofing of concrete, roofing felt, metal protection, insulation, etc. [Feb Ltd]

Feb Hyseal Slurry. A surface brush or trowel-applied waterproofing cement slurry; completely watertight even under conditions of high hydrostatic pressures. [Feb Ltd]

Febkol Elastomer 110 and 122. Proprietary brands of polysulfide sealant. [Feb Ltd] *

Febkol Plastomer 555. A proprietary polysulfide/epoxy sealant and adhesive. [Feb Ltd] *

Febmast GP. General-purpose pre-mixed trowel-applied sealing and bedding compound. [Feb Ltd]

Febmix Admix. Mortar plasticizer. [Feb Ltd]

Febol Standard. Shutter-applied surface retarder for concrete. [Feb Ltd]

Febond PVA. An integral PVA bonding admixture; multi-use adhesive, sealer, primer and bonding agent for bonding common building materials to themselves or to each other. [Feb Ltd]

Febond SBR. A styrene-butadiene synthetic rubber latex bonding compound for concrete and mortar; for water resistant renderings and general purpose floor screeding. [Feb Ltd]

Febox. Multi-use epoxy patching mortar. [Feb Ltd]

Feb Oxide. Coloring pigments for cement; contains no inert fillers or earth colors. [Feb Ltd]

Febpitch. Epoxy providing a waterproof membrane for use in concrete ducts, vats, tanks, retaining walls. [Feb Ltd]

Febplast Ready Mixed. A premixed decorative ceiling and wall finish; also as caulking and taping compound. [Feb Ltd]

Febplate. A proprietary range of epoxy mortars. [Feb Ltd]

Febproof. An integral liquid waterproofing admixture for use in mortar and concrete. [Feb Ltd]

Febrail No. 1. Epoxy grout specially formulated for British Rail for fixing rail shoulders into precast concrete railway sleepers. [Feb Ltd]

Febrilix. CAS 103-90-2; A proprietary preparation of paracetamol; an analgesic. [The Boots Co plc]

Febrok. Penetrates the concrete surface and reacts with free lime to form insoluble pore blocking compounds; results in a dust-free, durable and abrasive resistant surface. [Feb Ltd]

Febseal. Building sealant and caulking compound. [Feb Ltd]

Febset NF. Two-part, multipurpose epoxy material for bedding and repair work for concrete. [Feb Ltd]

Febsilicon. Colorless waterproofer containing silicone; external surface-applied waterproofer for permeable materials such as brickwork, cement rendering, concrete block, asbestos sheeting. [Feb Ltd]

Febspeed. Calcium chloride-based liquid rapid hardener; accelerates setting times of concrete and mortar to enable paths, floors, drives to be open for early use. [Feb Ltd]

Febstik. Contact adhesive for bonding plastic laminates, aluminum sheet, chipboard, plywood, plaster board and natural stone. [Feb Ltd]

Febstrike. Release agent for application to mold and shutter faces prior to casting concrete; approved for use with potable water. [Feb Ltd]

Feb Supercrete. Two-component multi-use polyester repair and patching compound for concrete. [Feb Ltd]

Feb Sylane. Based on alkyl alkoxy silane; water repellent for treatment of concrete. [Feb Ltd]

Febtex. Cold mix textured finish for internal use on plasterboard, cement rendering, sealed plaster, concrete. [Feb Ltd]

Febtile. Nonslip ceramic tile adhesive. [Feb Ltd]

Febtite Liquid. Waterproofing admixture for waterproofing cement, mortar. [Feb Ltd]

Febtone. Coloring agent that also plasticizes and improves water and frost resistance of concrete and mortar. [Feb

---

‡ Trade name and manufacturer not verified

§ Trade name without identified manufacturer

Ltd]

Febweld. A permanent high bond strength epoxy adhesive for internal or external bonding of renderings, granolithic toppings and concrete to concrete; used in repair of chloride contaminated concrete. [Feb Ltd]

Feb Wintamix. Mortar admixture for winter use. [Feb Ltd]

Fecap. A proprietary preparation of ferrous fumarate and folic acid; an iron supplement. [Pfizer International]

Fe-Cap. A proprietary preparation of ferrous glycine sulfate; used in the treatment of anemia caused by deficiency of iron. [MCP Pharmaceuticals] *

Fe-Cap C. A proprietary preparation of ferrous glycine sulfate and ascorbic acid; used in the treatment of anemia. [MCP Pharmaceuticals] *

Fe-Cap Folic. A proprietary preparation of ferrous glycine sulfate and folic acid; used in the treatment of anemia during pregnancy. [MCP Pharmaceuticals] *

Fecraloy. A proprietary trade name for an alloy of iron, chromium, aluminum and yttrium under development for use in sintered form as a catalyst to assist in the control of atmospheric pollution by reducing the emission of carbon monoxide and other fumes from [Atomic Energy Establishment] *

Feculose. The name given to various commercial starch esters.

Fedralite. A proprietary trade name for a Vinsol resin-treated laminated paper. §

Fedrazil Tablets. A proprietary formulation of pseudo-ephedrine and chlorcyclizine hydrochlorides; for temporary relief of nasal and sinus congestion associated with colds and hay fever. [The Wellcome Foundation Ltd]

Feedercalc. Micro computer programs to facilitate rapid reliable calculation of optimum feed requirements for all types of iron and steel castings. [Foseco (F.S.) Ltd]

Feedex. Moldable exothermic feeding compound. [Foseco (F.S.) Ltd]

Feedmate. Boiler water treatment. [Grace Dearborn Ltd]

Feedol. Feeding compounds for nonferrous metals. [Foseco (F.S.) Ltd]

Feen-A-Mint®. CAS 77-09-8; Pills: Yellow phenolphthalein, docusate sodium; gums: yellow phenolphthalein; laxative. [Schering Corp]

Fefol. A proprietary preparation of ferrous sulfate and folic acid; a hematinic. [SmithKline Beecham] *

Fefol Spansule Capsule. Ferrous sulfate and folic acid; for prophylaxis of iron and folic acid deficiency during pregnancy. [SmithKline Beecham] *

Fefol-Vit Spansule Capsule. Ferrous sulfate and folic acid with thiamine mononitrate, riboflavine, pyridoxine hydrochloride, nicotinamide, and ascorbic acid; for prophylaxis of iron and folic acid deficiency during pregnancy particularly when inadequate diet calls for supplementary vitamins B and C. [SmithKline Beecham] *

Fefol Z Spansule Capsule. Ferrous sulfate, folic acid, and zinc; for prophylaxis of iron and folic acid deficiency during pregnancy, when inadequate diet calls for supplementary zinc. [SmithKline Beecham] *

Fehlings solution. An alkaline solution of potassio-tartrate of copper. It is prepared in two solutions. a) Consisting of copper sulfate, and b) a solution of Rochelle salt (potassium sodium tartrate) and caustic soda; used for the identification and determination of sugars.

Fehling's solution, neutral. A solution made by adding 25 cc of a solution containing 2 g copper (7.86 g $CuSO_4 \cdot$

$7H_2O$) per liter to 25 cc of a solution containing 3.292 g sodium carbonate and 20 g Rochelle salt per liter. It is stated to be a sensitive reagent for the detection of sugars.

Fekta®RT. Algicide for swimming pools. [Goldschmidt] *

Felamine. A proprietary preparation of cholic acid and hexamine. [Sandoz] *

Felden. CAS 36322-90-4; A proprietary preparation of piroxicam; an anti-inflammatory, analgesic. [Pfizer International]

Feldene. CAS 36322-90-4; A proprietary preparation of piroxicam; an anti-inflammatory, analgesic. [Pfizer International]

Feliniffa P. Feline panleucopenia vaccine. [RMB Animal Health Ltd]

Feliniffa RC. Feline rhinotracheitis and calicivirus vaccine. [RMB Animal Health Ltd]

Felixite. An explosive. It is a 42-grain powder, and contains metallic nitrates, nitro-hydrocarbons, and 3% petroleum jelly.

Fellozine. CAS 58-33-3; Promethazine hydrochloride; anti-emetic; antihistaminic. [O'Neal, Jones & Feldman Pharmaceuticals] *

Felsinosima. A trade name for cultures of *Bacillus felsineus.*§

Felspar. (Potassium Felspar, Orthoclase). A potassium-aluminum silicate, $(K_2O \cdot 3SiO_2.) + (Al_2O_3 \cdot 3SiO_2)$; used in the manufacture of porcelain, as a building material, and as a fertilizer.

Felton 3T. Blend; detergent for felt of paper mfg. machines. [Toho Chem. Industry]

Felvinone. 7 and 8-Acetyl-5-isopropyl-2-methylbicyclo (2.2.2)oct-2-ene. [Quest Int'l. UK Ltd]

Felzodox. CAS 1314-13-2; Zinc oxide. [Pigment & Chemical Inc]

Femergin. CAS 379-79-3; A proprietary preparation of ergotamine tartrate; used for migraine. [Sandoz] *

Femerital. A proprietary preparation containing nifuratel and ambucetamide. [MCP Pharmaceuticals] *

Femin-9. Vitamin, mineral and fatty acid preparation; used for premenstrual syndrome. [Marfleet Refining Co] *

Feminone. CAS 57-63-6; Ethinyl estradiol; estrogen. [Upjohn] *

Feminor 21. A proprietary preparation of mestranol (16 pink tablets) and mestranol and norethynodrel (5 white tablets); oral contraceptive. [LRC Products Ltd] *

Feminor Sequential. A proprietary preparation of mestranol (15 pink tablets) or mestranol and norethynodrel (5 white tablets); oral contraceptive. [LRC Products Ltd] *

Femipausin. A proprietary preparation containing methyltestosterone and ethinylestradiol. [Hoechst UK] *

Femodene. 75 mdg gestodene and 30 mcg ethinylestradiol [Schering Health Care Ltd]

Femulen. CAS 297-76-7; A proprietary preparation of ethynodiol diacetate; used as an oral contraceptive. [G D Searle & Co Ltd] *

Fenafix. A proprietary trade name for a series of modified vinyl-pyrrolidone resins; used to modify the properties of other vinyl films and also to improve the adhesion of difficult surfaces. [Fine Dyestuffs & Chemicals Ltd] *

Fenamisal. Phenyl aminosalicylate.

fenasprate. 4-Acetamidophenyl o-acetylsalicylate.

Fenbid Spansule Capsul. CAS 15687-27-1; Ibuprofen; sustained release anti-inflammatory agent indicated in

---

* Trade name not verified as available   † Trade name verified as obsolete

rheumatoid arthritis and other painful conditions. [SmithKline Beecham] *

Fenitex. CAS 122-14-5; Fenitrothion; insecticide. [Makhteshim Chemical Works Ltd]

Fenitrothion EC. Organophosphorus insecticde. [Ciba plc]*

Fennite. A fungicide. [Fisons plc, Horticulture Div] *

Fenocil. Bromacil + pentachlorophenol; used for total weed control in noncrop areas. [Chemsearch (UK) Ltd]

Fenocin. CAS 132-98-9; A proprietary preparation of penicillin V potassium; an antibiotic. [Pfizer International]

Fenocin Forte. CAS 132-98-9; A proprietary preparation of penicillin V potassium; an antibiotic. [Pfizer International]

Fenoil®. Range of chemicals for exploration, production and enhanced recovery of oil and natural gas. [Bayer AG; Bayer plc]

Fenolac. Cyclized rubber used in adhesives and bonding agents. §

Fenolite. An Italian synthetic resin material for use in electrical insulation.

Fenopon AC-78. Anionic surfactant in which the cation is sodium and the anion is a coconut ester of isethionate; powder form; low salt content detergent with good foaming, lathering, and dispersing properties; used in detergent bars, dentifrices, shampoos, bubble baths, and other cosmetics. [ISP] *

Fenopon CD. Ammonium ethoxy sulfate in liquid form; high foaming surfactant for high electrolyte aqueous systems; air entraining properties used in the concrete industry; foaming agent for light weight cements; frothing agent for gypsum wallboard. [GAF Great Britain] *

Fenopon CN-42. Sodium N-cyclohexyl-N-palmitoyl taurate in paste form; low foaming detergent, dispersing agent, stabilizer; used in mechanical dishwashing detergents; industrial cleaners; synthetic rubber emulsions. [ISP] *

Fenopon CO. Sodium or ammonium nonylphenol ethoxy sulfate in liquid form; high foaming detergent with wetting, dispersing, emulsifying, antistatic and lime soap dispersion properties; used for scrub soaps; car washes; rug and hair shampoos; vinyl polymerization; petroleum wax; plastics and synthetic fibers. [ISP] *

Fenopon EP. Ammonium nonylphenol ethoxy sulfate in liquid form; versatile primary emulsifier and stabilizing agent used in emulsion copolymers. [GAF Great Britain]*

Fenopon SE. Sodium alkylaryl polyether sulfonate in liquid form; detergent base with good foaming and rinsability; used for cosmetic and pharmaceutical products and emulsion polymerization. [ISP] *

Fenopon T-33 and T-43. CAS 137-20-2; EINECS 205-285-7; Sodium N-methyl N-oleyl taurate as a clear liquid or a slurry; anionic surfactant used in textiles, rug shampoos, cleaning and detergent formulations. [GAF Great Britain] *

Fenopon T-51. CAS 137-20-2; EINECS 205-285-7; Sodium N-methyl N-oleyl taurate in the form of a readily soluble gel; used for textile processing; latex emulsion stabilizer. [GAF Great Britain] *

Fenopon T-77. CAS 137-20-2; EINECS 205-285-7; Sodium N-methyl N-oleyl taurate in powder form; wetting and dispersing agent used in dry blending; industrial and herbicidal wetting powders; textiles; rug shampoos; cleaning and detergent formulations. [GAF Great Britain] *

Fenopon TC-42. Sodium N-coconut acid N-methyl taurate in the form of a slurry; chemically stable detergent with foaming, lathering and dispersing properties; used in detergent bars; shampoos; bubble baths; cosmetics. [ISP] *

Fenopon TK32. CAS 61791-41-1; Sodium N-methyl N-tall oil acid taurate in liquid form; dispersing and suspending agent used as a precipitation inhibitor for salts of barium, calcium and strontium, and for scale prevention in oil well tubing and flow lines. [ISP] *

Fenopon TN-74. Sodium N-methyl N-palmitoyl taurate in powder form; dispersing and suspending agent used for dry blending into detergent formulations and in herbicides and insecticides. [GAF Great Britain] *

Fenopron. A proprietary preparation of fenoprofen calcium; an anti-inflammatory drug. [Dista Products Ltd]

Fenostil. A proprietary preparation of dimethindine-hydrogen maleate. [Zyma (UK) Ltd] ‡

Fenostil Rotard. CAS 3614-69-5; A proprietary preparation of dimethindene maleate; an antipruritic drug. [Zyma (UK) Ltd] ‡

Fenotec. Ester-phenolic resin binders for cold-set bonding of sand cores. [Foseco (F.S.) Ltd]

Fenoval. Phenoval.

Fenox. CAS 61-76-7; A proprietary preparation of phenylephrine hydrochloride; a nasal spray. [The Boots Co plc]

Fentachol. Tranquilizing and anticholinergic preparations. [Allen & Hanburys Ltd] *

Fentazin. CAS 58-39-9; A proprietary preparation of perphenazine; a sedative. [Allen & Hanburys Ltd]

Fenton's metal. A bearing metal. It contains about 80% zinc, 15% tin, and 5% copper.

Fenton's reagent. Hydrogen peroxide and a ferrous salt; used for the oxidation of polyhydric alcohols.

Fentro. Insecticide. [Murphy Chemical Ltd] †

Fenyrane. 2,4-Dimethyl-6-phenyldihydropyrane. [Quest Int'l. UK Ltd]

Feosol. Ferrous sulfate; hematinic. [Menley & James Laboratories] *

Feospan. A proprietary preparation of dried ferrous sulfate; a hematinic. [SmithKline Beecham] *

Feospan Spansule Capsule. Ferrous sulfate; for prevention and treatment of iron deficiency. [SmithKline Beecham] *

Feospan Z Spansule Capsule. Ferrous sulfate and zinc; for prevention and treatment of iron deficiency and when supplementary zinc is needed. [SmithKline Beecham] *

Feostat. CAS 141-01-5; Ferrous fumarate; hematinic. [O'Neal, Jones & Feldman Pharmaceuticals] *

F.E.P. A proprietary name for Teflon 100. [DuPont UK] *

Ferad. Bismuth and boron additives for malleable cast iron. [Foseco (F.S.) Ltd] †

Feraloy. A nickel-steel-chromium alloy, having a specific gravity of 8.15, and melting at 1480 C.

Feravol. A proprietary preparation of ferrous gluconate and folic acid; a hematinic. [Carlton Laboratories (UK) Ltd]*

Ferbelan. Preparations of B vitamins with iron. [British Drug Houses] *

Fergapol. Artificial and synthetic resins; for adhesives, paints, mastics and sealants. [Ferguson & Menzies Ltd]*

Fergatac. Artificial and synthetic resins; for adhesives, paints, mastics and sealants. [Ferguson & Menzies Ltd]*

Fergon. CAS 299-29-6; A proprietary preparation of ferrous gluconate; a hematinic. [Bayer AG]

Fer-In-Sol. Ferrous sulfate; hematinic. [Mead Johnson & Co] *

Ferlosa. A proprietary range of synthetic pulp for the paper and cement industry. [Montedison UK Ltd] *

Ferlucon. A proprietary preparation of ferrous gluconate and aneurine hydrochloride; a hematinic. [British Drug Houses] *

fermentation amyl alcohol. See fusel oil.

Fermenticide. An antiseptic compound. [Bush Boake Allen Ltd]

Fermenzyme®L-200. CAS 9032-08-0; Glucoamylase from Aspergillus niger; enzyme for fermentation of fuel alcohol. [Solvay Enzymes]

Fermet alloy. An alloy of 74.5% iron, 18% nickel, 2.2% manganese, 0.7% tungsten, 0.3% copper, and 0.35% carbon.

Fermin. CAS 68-19-9; A proprietary preparation of cyanocobalamin in an oral form; a hematinic. [Albion Group Ltd] ‡

Fermine. A proprietary trade name for methyl phthalate. §

Fernacol. A fungicide. [ICI Chem. & Polymers Ltd]

Fernambuco wood. See redwoods.

Fernasan. A nonmercurial seed dressing. [ICI Chem. & Polymers Ltd]

Fernasul. A lime and sulfur fungicide. [Plant Protection] *

Fernesta. A selective weed killer. [Plant Protection] *

Fernex. CAS 23505-41-1; Insecticide containing pirimiphos-ethyl. [ICI Chem. & Polymers Ltd]

Fernex. CAS 23505-41-1; Granules containing 10% w/w pirimiphos-ethyl; an organophosphorus insecticide to control scarid and phorid flies. [Fargro Ltd]

Fernico Alloy. See Kovar Alloy. §

Fernide. A foliage fungicide. [ICI Chem. & Polymers Ltd]

Fernimine. A selective weed killer. [ICI Chem. & Polymers Ltd]

Fernol. Textile flame retardants; for textile finishing. [Yorkshire Chemicals plc] *

Fernox. Water treatment chemicals including cleansers, descalers, corrosion proofers and inhibitors, and antifreezes. [Fernox Manufacturing Co Ltd]

Fernoxone. A selective weed killer. [ICI Chem. & Polymers Ltd]

Fero-Gradumet. Ferrous sulfate; hematinic. [Abbott Laboratories]

Ferox-Celotex. A proprietary trade name for Celotex (qv) which has to be treated to resist attack by fungi and termites. §

Ferozon. A disinfectant. §

Ferquatac. Artificial and synthetic resins, emulsions of artificial and synthetic resins; tackifiers, and modifiers for water-based adhesives, coatings, mastics, and sealants. [Ferguson & Menzies Ltd] *

Ferralium® Alloy 255. Super stainless steel alloy; high tensile strength. [Haynes Int'l.]

Ferrax. Flowable concentrate containing 400 g ethirimol, 30 g flutriafol and 10 g thiabendazole per liter; fungicide seed treatment for barley. [Bayer AG; Bayer plc;Dow Elanco Ltd;]

Ferrical. A proprietary preparation of ferrous iron, calcium, aneurine hydrochloride, and phenolphthalein; a tonic. [H N Norton & Co Ltd] ‡

ferric chloride. (iron chlorides; ferric chloride anhydrous; ferric trichloride) $FeCl_3$; CAS 7705-08-0; EINECS 231-729-4; Treatment of sewage and industrial wastes; etch-ing agent, mordant, disinfectant, pigment, feed additive. [Asahi Denka Kogyo; BASF; Eka Nobel AB; Mallinckrodt; Penta Mfg.; Rasa Ind.]

ferric chloride anhydrous. See ferric chloride.

ferric nitrate. (iron nitrate) $Fe(NO_3)_3 2 9H_2O$; CAS 10421-48-4; Dyeing, tanning, analytical chemistry. [Aldrich; General Chem.; Hoechst Celanese; Mallinckrodt; Sherman Chem. Ltd]

ferric oxide. (ferric oxide red; Iron oxide; red iron trioxide; ferrosoferric oxide) Fe_2O_3; CAS 1309-37-1; EINECS 215-168-2; Metallurgy, gas purification, paint and rubber pigment, in thermite, polishing compounds, mordant, laboratory reagent, catalyst, feed additive, electronic pigments for TV, permanent magnets, memory cores for computers, magnetic tapes. [BASF; Kerr-McGee; Miles]

ferric oxide red. See ferric oxide.

ferric sulfate. See iron sulfate (ic).

ferric trichloride. See ferric chloride.

ferric trisulfate. See iron sulfate (ic).

Ferri-Darotin. A sodium iron (iii) ethylenediamine tetracetic acid trihydrate and ammonium iron ethylenediamine tetracetic acid composition; used as a bleaching bath component for the development of color pictures, as an agent to act against chloros [Degussa AG] *

Ferrikalite. An artificial potassium ferric sulfate. §

Ferriplex. Iron chelate. [Rhone-Poulenc UK]

Ferriplus. Iron chelate. [Rhone-Poulenc UK]

Ferrisul. A trade name for ferric sulfate. §

ferrite. Nearly pure iron; phosphorus and sulfur may be present in minute quantities, but the carbon content is not more than 0.05%.

ferro-aluminum. It contains up to 20%. aluminum with iron, and is used in the preparation of iron and steel.

ferro-argentan. An alloy resembling silver and containing 70% copper, 20% nickel, 5.5% zinc, and 4.5% cadmium.

ferro-boron. An alloy of iron and boron, containing from 20-25% boron. It is added to steel.

Ferrocap. A proprietary preparation of ferrous fumarate and Vitamin B_1; a hematinic. [Consolidated Chemicals Ltd] *

Ferrocap F 350. A proprietary preparation of ferrous fumarate and folic acid; used in the treatment of anemia during pregnancy. [Consolidated Chemicals Ltd] *

ferro-carbon-titanium. An alloy of iron and titanium, containing carbon; used for making steel.

ferrochlor. A mixture of ferric chloride and calcium hypochlorite; used to clarify water.

ferro-chromium. A British Chemical Standard Alloy; low carbon No. 203 contains 69% and over chromium, 0.08% carbon, and 0.01% sulfur, while the high carbon alloy contains 71.4% and over chromium, 5.09% carbon, and 0.02%

ferro-cobalt. An alloy of 70% cobalt with iron; used for adding cobalt to steel.

ferrocobaltite. A mineral. It is a cobaltite containing iron.

Ferrocontin. Ferrous glycine sulfate in a controlled release tablet; for iron deficiency anemia. [Napp Laboratories Ltd] *

Ferrocrete. A brand of rapid hardening Portland cement of high strength. §

ferro-cupralium. An alloy of 75-80.5% copper, 11-12% aluminum, and 2-13% iron.

Ferro-Cure. For rubber compounding additives, such as antioxidants, vulcanizing agents and the like in Int Class 1. [Ferro] *

---

* Trade name not verified as available   † Trade name verified as obsolete

Ferrodic. A proprietary preparation containing ferrous carbonate and ascorbic acid; a hematinic. [Allen & Hanburys Ltd] *

ferrodur. A substance containing Nitrolim (qv); used for case hardening and tempering iron and steel.

ferroferrite. See magnetic iron ore.

Ferrofloc. CAS 7758-94-3; Water treatment chemical developed from ferrous chloride. [Rheox Inc]

Ferrogen. Ladle cleaning flux. [Foseco (F.S.) Ltd]

Ferrograd. A proprietary preparation of ferrous sulfate in a slow release form; a hematinic. [Abbott Laboratories] *

Ferrograd C. A proprietary preparation containing ferrous sulfate and ascorbic acid; a hematinic. [Abbott Laboratories]

Ferrograd Folic. A proprietary preparation of ferrous sulfate and folic acid; used in the treatment of anemia during pregnancy. [Abbott Laboratories]

ferro-magnesite. Obtained by burning magnesite with iron ore; used as a lining for furnaces.

ferro-manganese. An alloy of manganese with iron and carbon, usually made in the blast furnace. High-grade alloys contain about 78% manganese and 8% carbon. These alloys vary from 50-80% manganese, 10-42% iron, 2% silicon, and 5-8% carbon.

ferro-molybdenum. An alloy of iron with 80% molybdenum; used in the place of molybdenum in the manufacture of hard steel.

Ferromyn. CAS 10030-90-7; Ferrous succinate compounds; for the treatment and prophylaxis of iron deficient aneamia. [The Wellcome Foundation Ltd]

ferron. An alloy of 50% iron, 35% nickel, and 15% chromium. Also a building material prepared from the pickling liquor from steel mills. It consists of precipitated iron oxide.

ferro-nickel. (nickel iron). An alloy of iron and nickel, usually containing 25% nickel.

ferronite. A solid solution of about 0.27% carbon in β- iron.

ferro-phosphorus. An alloy of iron and phosphorus; used in steel making for thin castings.

Ferrophos Pigment. A refractory ferro-alloy developed for use in high performance specialty coatings; its primary use is in zinc-rich coatings where it works with zinc dust to provide good corrosion resistance and weldability characteristics; for use in weldable coil coatings and conductive paints for both EMI shielding and antistatic needs. [Occidental Chemical Corp]

ferrophytin. Neutral colloidal inositol hexaphosphate of iron containing about 7.5% iron and 6% phosphorus. A tonic and hematopoietic.

Ferrosil 14. CAS 12178-41-5; Ferroaluminum silicate; filler for abrasion-resistant plastic systems, caulks, sealants, polishes, abrasive compounds; extender pigment for primers and other coatings. [Kaopolite]

ferro-silicon. Alloys of iron and silicon made in the arc type electric furnace. They are graded upon the silicon content. The ordinary grades containing 25, 45-50, 75, and 95% silicon are used in steel works. The quality containing 95% silicon gene

ferro-silicon-aluminum. An alloy of iron, silicon, and aluminum, containing up to 15% silicon.

ferrosoferric oxide. See ferric oxide.

ferrostabil. A stabilized ferrous chloride preparation used in medicine in the form of tablets and suppositories.

Ferrotone. Mildly acidic scale remover. [Interox Chemicals Ltd]

Ferrotubes. An iron degassing and scavenging agent. [Foseco (F.S.) Ltd] †

ferro-tungsten. An alloy of iron and tungsten. It usually contains from 65-85% tungsten, and from 1-2% carbon; used in the steel industry.

ferro-uranium. Alloys of iron and uranium, containing from 30-50% uranium; used in steel making.

ferrous fumarate. $FeC_4H_2O_4$; CAS 141-01-5; EINECS 205-447-7; Dietary supplement. [Chemie Linz UK; Hüls AG; Nichia Kagaku Kogyo; Schweizerhall]

ferrous sulfate. See iron sulfate (ous).

ferro-vanadium. An alloy of iron and vanadium, containing from 20-40% vanadium. It is added to steel and iron.

ferro-vanadium (No. 205). A British Chemical Standard alloy. Vanadium 52.2%. (standard). It also contains carbon 0.16%. sulfur 0.03%. (not standard), silicon 1.18%. phosphorus 0.05%.

Ferrox. Trade name for yellow iron oxides used as paint pigments; consists of 98-99%. $Fe(OH)_3$, with calcium sulfate. §

Ferroxide. CAS 1309-37-1; Synthetic iron oxide pigments; ferric oxide; for surface coatings, plastics and cement coloring. [Mercian Minerals & Colours Ltd] ‡

ferroxyl reagent. A gelatin or agar-agar jelly containing phenolphthalein and potassium ferricyanide. When a piece of iron is placed in the jelly, colors are formed at the ends of the metal after a time. Iron ions give a color of Turnbull's blue with the potassium; ferricyanide and hydroxyl ions, a pink color with the phenophthalein; used to test corrosion in iron.

Ferrozell. A proprietary trade name for a synthetic resin. §

ferro-zirconium. A 20% zirconium alloy with iron; used to remove nitrogen and oxides from steel.

Ferrozoid. (Vestalin). Alloys. They are usually 28% nickel steels and are used as electrical resistances.

ferrozone. A saccharated iron and vanadium compound.

ferrugo. Ferric hydroxide, $Fe(OH)_3$.

ferrul. An alloy of 54.6% copper, 40% zinc, 5% lead, and 0.4% aluminum.

ferrum. The Latin name for iron.

Ferrutope. CAS 23383-11-1; Ferrous citrate Fe 59; radioactive agent. [Mallinckrodt Inc] *

Ferrux. Feeding compound for ferrous metals. [Foseco (F.S.) Ltd]

Ferry Alloy. A trademark for an electrical resistance alloy of 54% copper and the balance nickel, a material with a very low temperature coefficient of resistance. [Wiggin Alloys Ltd] ‡

Ferrybar. A proprietary preparation of iron and ammonium citrate, riboflavine, nicotinamide, and aneurine hydrochloride; used to prevent iron deficiency anemia. [Rybar Laboratories Ltd] *

Ferry metal. Alloys used for electrical resistance and containing 40-45% nickel and 55-60% copper. The name appears to be also applied to a bearing alloy and solder containing lead with 2% barium, 1% copper, and 0.25% mercury.

Fersaday. CAS 141-01-5; A proprietary preparation of ferrous fumarate; a hematinic. [Duncan Flockhard Ltd]

Fersamal. CAS 141-01-5; A proprietary preparation of ferrous fumarate; a hematinic. [Glaxo Laboratories] *

Fersolate. A proprietary preparation of ferrous, manganese and copper sulfates; a hematinic. [Glaxo Laboratories]*

Fertilox. CAS 1305-79-9; Calcium peroxide; used in agricul-

---

‡ Trade name and manufacturer not verified   § Trade name without identified manufacturer

ture and horticulture. [Interox Chemicals Ltd]

Fesovit. A proprietary preparation of ferrous sulfate, ascorbic acid, and B vitamins; used in the treatment of anemia. [SmithKline Beecham] *

Fesovit Spansule Capsule. Ferrous sulfate, thiamine mononitrate, riboflavine, pyridoxine hydrochloride, nicotinamide and ascorbic acid; hematinic with added vitamins for the prevention and treatment of iron deficiency, particularly in the elderly patient where inadequate diet calls for supplementary vitamins B and C. [SmithKline Beecham] *

Fesovit Z Spansule Capsule. Ferrous sulfate, thiamine mononitrate, riboflavine, pyridoxine hydrochloride, nicotinamide, ascorbic acid and zinc; oral iron and zinc preparation, with added vitamins, for the prevention of iron deficiency when inadequate diet calls for supplementary zinc and vitamins B and C. [SmithKline Beecham] *

Festoform. A solid preparation of formaldehyde, obtained by mixing an aqueous solution of formaldehyde with a soda soap solution; an antiseptic disinfectant and deodorizer.

Fetrilon® Combi. Easily sol. micronutrient mixed fertilizer with 9% magnesium; for all agricultural crops, vines, fruit, and hops. [BASF AG; BASF plc]

Fettel. Herbicide containing triclopyr, dicamba, and mecoprop. [ICI Chem. & Polymers Ltd]

Fettel. Emulsifiable concentrate of 78 g dicamba, 130 g mecoprop, and 72 g triclopyr per liter; used for weed control in cereals and grassland. [Farm Protection Ltd]

Feuille Morte. A jeweler's alloy, containing 70% gold and 30% silver.

Fevarin 50. CAS 54739-18-3; Fluvoxamine; for treatment of mental and physical complaints caused by disturbed moods and humors. [Duphar BV] *

Feximac. A proprietary preparation of buteximac in a cream base; used in the treatment of eczema. [Nicholas Laboratories Ltd] *

FFA-5, FFA-9. CAS 25928-94-3; Epoxy resin adhesive; offers strong bond; mix ratio: 100 pbw adhesive/150 pbw Activator BA-15; FFA-9 is fast setting, R.T. curing adhesive; bond is more brittle than FFA-5; mix ratio: 100 pbw adhesive/100 pbw Activator BA-11. [Bacon]

Fiba-Bond CI, W. Modified urea-formaldehyde resins; wet strength resins for paper industry; for controlled sizing and wet strength at lower use levels. [CNC Int'l. L.P.]

Fiber, Egyptian. See vulcanized fiber.

Fiberfil® J-1/30. CAS 32131-17-2; Nylon 6/6, glass fiber-filled; reinforced engineering thermoplastic for demanding components in automotive, consumer products, mech., and elec./electronic industries. [DSM]

Fiberfil® J-4/35. Nylon 6/12, glass fiber-filled; reinforced engineering thermoplastic for demanding components in automotive, consumer products, mech., and elec./electronic industries. [DSM]

Fiberfil® J-7/33. CAS 25038-54-4; Nylon 6, glass-filled; toughened nylon for automotive and consumer products. [DSM]

Fiberfil® J-7/33/IT. CAS 25038-54-4; Nylon 6, glass-filled; incrementally toughened nylon for automotive and consumer products. [DSM]

Fiberfil® J-17/30/VO. CAS 32131-17-2; Nylon 6/6, glass filled; flame retardant thermoplastic for elec./electronic,mech. and consumer/industrial applications.

[DSM]

Fiberfil® J-60/30/E8. CAS 9003-07-0; Polypropylene, glass fiber-filled; reinforced engineering thermoplastic for automotive, consumer, and mech. components where price/performance ratio is important and good chem. and moisture resist. are critical. [DSM]

Fiberfil® J-60/30/FR. CAS 9003-07-0; Polypropylene, glass-filled; flame retardant thermoplastic for elec./electronic,mech. and consumer/industrial applications. [DSM]

Fiberfil®M-1492. CAS 9003-07-0; Polypropylene, glass and mineral-filled; reinforced engineering thermoplastic for automotive, consumer, and mech. components where price/performance ratio is important and good chem. and moisture resist. are critical. [DSM]

Fiberfil®NY-7. CAS 25038-54-4; Nylon 6; toughened nylon for automotive and consumer products. [DSM]

Fiberfil®NY-7/VO. CAS 25038-54-4; Nylon 6; high impact flame retardant thermoplastic for elec./electronic,mech. and consumer/industrial applications. [DSM]

Fiberfil® NY-16/MF/40. CAS 32131-17-2; Nylon 6/6, mineral-filled; reinforced engineering thermoplastic for demanding components in automotive, consumer products, mech., and elec./electronic industries. [DSM]

Fiberfil® PP-60/TC/40. CAS 9003-07-0; Polypropylene, talc-filled; reinforced engineering thermoplastic for automotive, consumer, and mech. components where price/ performance ratio is important and good chem. and moisture resist. are critical. [DSM]

Fiberfrax®6000 RPS. CAS 1332-58-7; Kaolin ceramic fiber; reinforcement and filler for phenolic, epoxy, nylon, melamine and polyurethane systems. [Carborundum] *

Fiberglas®101C. Chopped glass strand; reinforcement for bulk molding compounds. [Owens-Corning Fiberglas] *

Fiberite. Thermosetting composites. [ICI Chem. & Polymers Ltd]

Fiberite 944. Polyimide; for missile fins, high temp. aerospace structures; self-extinguishing, low smoke generation; service temp. 316 C. [ICI Fiberite]

Fiberite 986. Bismaleimide; for high temp. structural aircraft; low odor; service temp. 218 C. [ICI Fiberite]

Fiberite 6070. Phenolic; for aircraft interiors; low smoke, self-extinguishing; designed for press molding; service temp. 177 C. [ICI Fiberite]

Fiberite 7669. CAS 25928-94-3; Epoxy; for aerospace structures; excellent tack and drape; service temp. 177 C. [ICI Fiberite]

Fiberite 7701. CAS 25928-94-3; Epoxy; for structural aircraft exteriors; self-extinguishing; solv. resist.; slef-adhesive to honeycomb core; press mold and vacuum bag processability; service temp. 93 C. [ICI Fiberite]

Fiberite 9002. Polyester; for aircraft ducting, radomes; self-extinguishing; vacuum bag, press, and autoclave processability; service temp. 163 C. [ICI Fiberite]

Fiberkal. CAS 1332-58-7; Fibrous calcined kaolin; reinforcement for plastic systems; asbestos replacement in some applications; filtration aid; texture paints. [Kaopolite]

Fiberlac. A proprietary trade name for cellulose nitrate lacquer. §

Fiberloc®803GR10. Glass-reinforced vinyl composite; for engineered applications incl. appliances, business equip., construction, elec. equip., heating, ventilating and air conditioning, marine products, plumbing and

---

water treatment, windows; high strength, stiffness, dimensional stability. [BFGoodrich/Geon Vinyl]

Fiberloid. See Viscoloid; a proprietary trade name for a cellulose nitrate plastic resistant to oils. §

Fiberlon. A proprietary trade name for a phenol-formaldehyde synthetic resin resistant to oils. §

Fiberod. Long, continuous, parallel reinforcing fibers set in thermoplastic resin systems; available in rod, bar, tape, and bidirectional fabric foam; thermoplastic molding compounds for compression, transfer and injection molding; used for automotive, appliances, equipment, sporting and other applications. [Polymer Composites]*

Fiberoid. An American grade of vulcanized fiber used for electrical insulation; also the name for a celluloid. §

Fiber Pare. A short length olefin fiber; asphalt reinforcement for highway paving, patching, seal coating, crack sealing, curb mix designs. [Hercules] *

Fiberstos. A silica-asbestos product used as an insulator.

Fiberstran®G-1/50. CAS 32131-17-2; Nylon 6/6, 50% long glass-reinforced; long fiber reinforced thermoplastic with high impact strength for demanding metal replacement applications in automotive, consumer products, mech. and elec/electronic industries. [DSM]

Fiberstran®G-3/50. CAS 25038-54-4; Nylon 6, 50% long glass-reinforced; long fiber reinforced thermoplastic with high impact strength for demanding metal replacement applications in automotive, consumer products, mech. and elec/electronic industries. [DSM]

Fiberstran®G-4/45. Nylon 6/12, 45% long glass-reinforced; long fiber reinforced thermoplastic with high impact strength for demanding metal replacement applications in automotive, consumer products, mech. and elec/electronic industries. [DSM]

Fibervorm. Wheat-fiber derivative; for treatment of complaints arising from inflation of the small intestine. [Duphar BV] *

Fibestos. A proprietary trade name for a cellulose acetate plastic. §

Fibra-Cel®. CAS 9004-34-6; Cellulose fibers; functional fillers, reinforcement, thickening aid, processing aid, conditioning agent for the rubber industry (shoe soles, v-belts, tires, gaskets), thermoset and thermoplastic resins (building products, pet food, asphalt, latex paints). [Celite]

Fibral. Polishing pads and lenses. [Carl Freudenberg] *

Fibralda. A proprietary trade name for untwisted cellulose acetate fibers. §

Fibravorma. Wheat-fiber derivative; for treatment of complaints arising from inflation of the small intestine. [Duphar BV] *

fibrino-plastic substance. See globulin.

fibrinoplastin. See globulin.

Fibro. A proprietary artificial silk product. It is a staple fiber (qv) made by the viscose process. §

Fibroc. A laminated product consisting of fiber impregnated with a phenol-formaldehyde resin. §

Fibron. A trademark for a surfacing material for resurfacing and treatment of floors resulting in a plastic finish. §

Fibro-Silk Powd. CAS 9009-99-8; Silk protein pigment. [Brooks Industries]

Fibrotex. A proprietary trade name for a roofing cement consisting of asbestos mixed with oil and gum. §

Fibrox. A silicon oxycarbide. It is a thermal insulator.

Fibrox 030 SC. Mineral wool fiber; mineral fiber for bitumen,

paper, gaskets, paints, friction products, plastics, rubber, coatings, caulks and other applications. [Industrial Fibers]

Fibrox 300. Mineral wool fiber; filler/reinforcer for bitumen, paper, gaskets, paints, friction products, plastics, rubber, coatings, caulks, and other applications. [Industrial Fibers]

Fibrredux. Fiber reinforced laminating resins and adhesives. [Ciba plc] *

Ficam. Public health insecticide. [Rhone-Poulenc Rorer Ltd]

Ficam. CAS 22781-23-3; Bendiocarb; public health insecticide. [Cambridge Animal & Public Health Ltd]

Ficel®. Blowing agents; azo polymerization initiators. [Sherex/Div of Witco]

Ficel® AC2. CAS 123-77-3; Azodicarbonamide; chem. blowing agent for sponge rubber, cushion vinyl floor covering, expanded vinyl coated fabrics, profiles/pipe, sealants. [Schering Berlin Polymers]

Ficel®AZDN-LF. 2,2-Azodiisobutyronitrile; polymerization initiator for a wide range of monomers. [Schering Berlin Polymers]

Fi-Chlor. Chlorcyanurate used for rendering wool shrink-resistant. [Fisons plc, Horticulture Div]

fichtelite. A hydrocarbon, $C_{18}H_{32}$, found in fossil coniferous resins.

Fi-Clor. Chlorinated cyanuric acid. [Schering Agrochemicals Ltd] *

Ficoid. CAS 152-97-6; A range of proprietary skin creams containing fluocortolone; used in the treatment of eczema and related dermatoses. [Fisons plc, Pharmaceuticals Div] *

Ficortril. CAS 50-03-3; A proprietary preparation of hydrocortisone acetate; a corticosteriod. [Pfizer International]

Ficote. Range of coated fertilizers. [Fisons plc, Horticulture Div] *

Fi-Cryl. Acrylic copolymer solutions and dispersions. [Fisons plc, Horticulture Div] *

Fiddle gum. Gum tragacanth.

Fielder®. For the agriculture industry. [DuPont UK]

Fierroso. See Acerado. §

Fi-Gard. Rubber and plastics additives. [Fisons plc, Horticulture Div] *

Filamid. Polymer-soluble dyes for spin coloration of polyamide. [Ciba plc]

Filastic. A proprietary rubber textile yarn in which the rubber latex impregnation takes place during spinning; the yarns contain 50% of rubber. §

Filcryl. A proprietary acrylic polymer for dental fillings. §

file bronze. An alloy of 64.4% copper, 18% tin, 10% zinc, and 7.6% lead.

Filester. Dyes for PET bottles. [Ciba plc]

Filex. CAS 25606-41-1; An aqueous concentrate containing propamocarb hydrochloride; a protective fungicide for use on all ornamentals and some edible crops against Pythium, Peronospora and Phytophthora. [Fisons plc]

filicic acid. An acid obtained from male fern. It is a vermicide.

Fi-Line. Dyeline chemicals for photocopying. [Fisons plc] *

Filite. A smokeless explosive. It is Ballistite (qv), drawn out into cords with the aid of a solvent.

Fillite. A proprietary inert silicate in the form of spheres; and as glass, it is used as a filling material for plastics. [Fillite (Runcorn) Ltd] *

Fillite Hollow Microspheres. Free flowing hollow alumina

‡ Trade name and manufacturer not verified   § Trade name without identified manufacturer

silica microspheres; used for concretes and various refractory cements, ceramic fillers, thermoplastics, undercoating filler material and in molded compounds. [Fillite USA Inc] *

Fillite Solid Microspheres - PFA. Solid alumina silica microspheres; used for concretes and various refractory cements, paints and pigment industry, ceramic fillers and in other molded compounds. [Horn's Crop Service Center] *

Fillmaster; Expanded polystyrene blockform for lightweight landfill civil engineering use; loading bearing fill for roads and embankments. [Vencel Resil Ltd]

Fillpak. Polyurethane foams; for sealing, as filler. [James Briggs & Sons Ltd] *

Filmex. Alcohol solvent; used in processing of ink, fabrics, latex, lacquers, pharmaceuticals, and chemical specialties, and in other industrial applications. [Quantum Chemical Corp]

Filmite. White oil preparations. [Murphy Chemical Co Ltd]†

Filmon®. Biaxially oriented nylon 6 film; heat shrinkable. [SNIA (UK) Ltd]

Film Plus. Corrosion inhibitors. [Baker Performance Chemicals Ltd]

Filofin. Pigment dispersions for polypropylene fibers. [Ciba plc]

Filon. Glass fiber reinforced polyester (GRP) sheeting; used for roofing and cladding. [BIP Chemicals Ltd] †

Filpro. Protein hydrolysates; suitable for any savory application: soups, frozen meals, snacks, convenience meals, meat pies and sausages. [PPF International Ltd]*

Filt-char. A proprietary brand of bone charcoal; used as a filtering medium. §

Filter-cel. A proprietary preparation of infusorial earth; used in filtering. §

Filtracarb. Carbonaceous materials for use as filtration media. [Coal Products Ltd]

Filtracite. Carbonaceous materials for use as filtration media. [Coal Products Ltd]

Filtram. Laminated filter drains. [ICI Chem. & Polymers Ltd]†

Filtrasorb® 100, 200. Activated carbon; for removal of dissolved organic chems. from municipal and industrial water. [Calgon Carbon] *

Filtrez. Rosin resin; used in printing inks, coatings and adhesives. [Monsanto Co] *

Filtrol. A trademark for a decolorizing substance consisting of fine silica with a little aluminum silicate. §

Filtros. Gum rosin; used in paints, varnishes and lacquers. [Monsanto Co] *

Filtrosol A. CAS 118-56-9; Homosalate; ultraviolet screen. [Norda Inc] ‡

Final Flip. Trade name for a pelleted rodenticide, single feed type, indoor or outdoor use, place pack size; contains chronic toxicant active material. [Colonial Products Inc]

Finalgon. A proprietary preparation of nonylic acid, vanillymide, and butoxyethyl nicotinate. [Boehringer Ingelheim GmbH] *

Final Touch. General degreasers. [Evans Vanodine International Ltd]

Finaplix. CAS 10161-34-9; Trenbolone acetate; anabolic (veterinary). [Roussel UCLAF, Fine Chemicals] *

Fine-Clad®. Synthetic resins for use in protective coatings. [Reichhold]

fine gold. A jeweler's alloy, containing 75% gold and 25%

silver.

fine silver. 99.9% pure silver.

Finesse®. Preparationonyl weed killer. [DuPont UK] *

Finestol. A phloroglucinol dye coupler. [Fisons plc] *

Finex-25. CAS 1314-13-2; Zinc oxide; provide excellent feel, covering power and skin adherance; uv absorber; effective as SPF booster. [Presperse]

Finex-25-020. CAS 1314-13-2, 9004-73-3; Zinc oxide, methicone; provide excellent feel, covering power and skin adherance; uv absorber; effective as SPF booster. [Presperse]

finings. The term applied to isinglass dissolved in an acid such as tartaric acid; used to clarify beer.

Finish®. For the agriculture industry. [DuPont UK]

Finistrol ESJ Concentrate. Polyethylene emulsion; nonionic; light brown viscous liquid; fat-free softener for cellulose fibers and synthetics, preferably polyamide and its blends. [Thor Chemicals (UK) Ltd]

Finistrol GZ. Polysiloxane emulsion; nonionic; thin white liquid; softener for use alone or in combination with synthetic resins or reactants; improves sewability of resin-finished materials. [Thor Chemicals (UK) Ltd]

Finitron. N-Ethyl perfluorooctanesulfonamide; insecticide. [Griffin]

Finizym. A fungal β-glucanase preparation produced by submerged fermentation of a selected strain of *Aspergillus niger*; used during fermentation and storage of the beer to prevent filtration difficulties and to prevent precipitation of β-glucans. [Novo Nordisk]

Finnish turpentine. *See* turpentine.

Finntitan. CAS 13463-67-7; EINECS 236-675-5; Titanium dioxide; used in paints, inks, rubbers, cosmetics, ceramics etc. [Cornelius Chemical Co Ltd] *

Fintex 572. Quaternary ammonium compound in liquid form; cationic surfactant used in fabric conditioning. [Berol Kemi (UK) Ltd] *

Fiolax. A trade name for a resistance glass. §

Fiorivert. 1-Ethoxy-1-phenylethoxyethane. [Quest Int'l. UK Ltd]

Fire PRF₂ 1000 FM. Two-component preaccelerated, precatalyzed phenol resorcinol-formaldehyde resin; flame-retardant, low-smoke thermoset resin for fabricating factory mutual-approved fume and smoke exhaust duct systems; mix ratio 45.7/34.3/20 (Part A/B/filler). [Indspec] *

fire-armour. Heat-resisting alloys containing 60-61% nickel, 18-20% chromium, 10-20% iron, 0-1.8% manganese, and 0.5% carbon.

Firebrake®. Flame retardant for plastics and coatings [Borax Consolidated Ltd]

Firebrake®ZB. CAS 12513-27-8; Zinc borate; flame retardant synergistic with antimony oxide or alumina trihydrate in most halogenated polyesters and vinyl esters; delays onset of oxidation; used in chlorinated polyesters, PVC plastisols, and epoxy systems; smoke suppressant; char promoter; improved elec. properties in polyester and nylon; promotes adhesion between plastics and metals; translucent; low toxicity, and nonhygroscopic. [U.S. Borax & Chem.]

Firecheck. Water-based elastomeric fire retardant protective coating, applied by brush or spray; for asbestos encapsulation, fire retardant weatherproofing, combustible substrates. [Liquid Plastics Ltd]

fireclay. Clay containing a considerable amount of free

---

* Trade name not verified as available   † Trade name verified as obsolete

silica.

Firecol. An alginate fire-fighting suspension. [Alginate Industries Ltd] *

Firecrete. A proprietary trade name for a calcined high alumina clay used as a refractory in furnaces. §

fire-damp. A gas, mainly consisting of methane, often found in coal mines.

FireGuard 910. Plenum compound; low smoke, low flame compound for plastics industry. [Teknor Apex]

FireMaster®. Fire protection products avail. in bulk, as blanket, board, putty, etc. for cable tray, duct, plastic pipe, and structural steel fire protection. [Thermal Ceramics]

Firemaster® 642. Proprietary; flame retardant. [Great Lakes]

Firemaster® HP-36. Halogenated phosphate ester; flame retardant. [Great Lakes]

Fire Retardant FR-8. Salt-type all-purpose fire retardant for textile fabrics and cellulosics. [Emco Services]

Firesaife. CAS 7722-76-1; Ammonium phosphate fire retardant. [Scottish Agricultural Industries plc]

FireShield® H. CAS 1309-64-4; Antimony trioxide; general purpose flame-retardant synergist for plastics, rubber, paper, and paint; polymerization catalyst for PET resins and fibers; also for electronics, glass, ceramics, petrol. refining, and chem. mfg. [Laurel Industries]

FireShield® HPM. CAS 1309-64-4; Antimony trioxide; catalyst, chem. intermediate, and flame retardant; suitable for sensitive electronic applications. [Laurel Industries]

FireShield® L. CAS 1309-64-4; Antimony trioxide; general purpose flame-retardant synergist for plastics, rubber, paper, and paint; polymerization catalyst for PET resins and fibers; also for electronics, glass, ceramics, petrol. refining, and chem. mfg. [Laurel Industries]

Firit. A foundry refractory coating. [Foseco (F.S.) Ltd]

Firmadent. Denture adhesive (powder, cream and liquid), for securing dentures. [Richardson-Vicks Inc] *

Firnagral. (Iranolin) A mineral drying oil extracted from aromatic petroleum residues; used to replace up to 30% of the linseed oil in putty. It is stated to give the putty a harder and smoother surface.

Firnis. Linseed oil and driers.

First Choice Electroless Palladium. Electroless plating solution for the autocatalytic chemical deposition of palladium metal; for plating of electronic components for its low contact resistance and high wear properties and for the metalization of ceramics, plastics, and other nonmetallics. [Callery Chemical Company] *

Firthite. A proprietary trade name for a material consisting of a mixture of tungsten and other carbides. §

fir wool oil. The oil of Scotch fir leaves.

Fischer-Langbein solution. A solution containing 450 g ferrous chloride, 500 g calcium chloride, and 750 cc water; used for the electro-deposition of iron, with a current density of up to 120 amps/ft^2. A temperature of 60-70 C is used.

Fischer's reagent. A test solution for sugars. It consists of 2 parts phenyl-hydrazine hydrochloride and 3 parts sodium acetate in 20 parts of water.

Fischer's salt. Potassium cobaltic nitrite, $K_3Co(NO_2)_6$; used for the detection and determination of potassium.

Fischer's yellow. See Fisher's salt.

fisetin. Tetrahydroxymethylanthraquinone, $C_{15}H_{10}O_6$; A yellow coloring from the wood of *Quebracho colorado*,

etc.

fish berry. CAS 124-87-8; The fruit of *Cocculus indicus* and *Anamirata paniculata*. It acts as a powerful fish poison. The active principle is picrotoxin, $C_{30}H_{34}O_{13}$.

fish gelatin. See isinglass.

Fisons 18-15, MCPB. A selective weedkiller. [Fisons plc, Horticulture Div] *

Fisons P.C.P. A weedkiller. [Fisons plc, Horticulture Div] *

Fi-Vi. A lightweight expanded PVC. [Fisons plc] *

Fixanal. Analytical chemicals accurately weighed and sealed, ready for rapid volumetric solution.

Fixaplus. X-ray fixer. [May & Baker Ltd] *

Fixapret®. Textile finishing aids for anti-crease, anti-shrink, and easy care finishing. [BASF AG; BASF plc]

Fixat. Inorganic molding sand binders for foundries, also in connection with lustrous carbon formers to obtain better castings. [Süd-Chemie AG] *

Fixatek. X-ray fixer. [May & Baker Ltd] *

fixed white. Commercial barium sulfate, $BaSO_4$.

Fixegal. Dispersing and levelling agent; used for dyeing in the textile industry (Degussa].*

Fixin. Aluminum lactate.

Fixinvar. An alloy having the same properties as Elinvar, but having greater stability.

Fixodent. Denture adhesive cream for securing dentures. [Richardson-Vicks Inc] *

Fixogene. Textile auxiliary chemicals. [ICI Chem. & Polymers Ltd]

Fixol. Adhesives. [Associated Adhesives] ‡

fixopone. See oil white.

Fix-Sol. A concentrated fixing and hardening solution. [Johnsons of Hendon] *

FK 140. Precipitated silica; for silicone rubber industry. [Degussa]

FK 300DS. Silica, precipitated; for paper and film industry. [Degussa]

FK 500LS. Precipitated silica; thickener for cosmetic creams and lotions; also in thermal insulation, paper, films, pesticides, pharmaceuticals. [Degussa]

FL7P. Liquid fertilizer. [Fisons plc, Horticulture Div] *

Flagyl. CAS 443-48-1; A proprietary preparation of metronidazole; for treatment of anaerobic and protozoal infections. [Rhone-Poulenc Rorer Ltd]

Flagyl Compak. A proprietary preparation comprising metronidazole, taken orally, and vaginal pessaries containing nystatin; used in the treatment of vaginitis. [Rhone-Poulenc Rorer Ltd]

Flagyl IV. CAS 69198-10-3; Metronidazole hydrochloride; antibacterial. [Rhone-Poulenc Rorer Ltd]

flake lead. See white lead.

flake litharge. Litharge made by the oxidation of lead.

flake white. (cremnitz, cremnitz white, kremnitz, crems white, blanc d'argent, silver white, London white, Nottingham white). CAS 1304-85-4; Lead whites. They are all carbonates of lead, and contain varying quantities of hydrated oxide of lead. Flake white is a variety of chamber white lead, obtained in flakey pieces by heating lead plates. A basic busmuth is also known as flake white. The term flake or pearl white is sometimes used for oxychloride of bismuth.

Flamarret®. CAS 1309-48-4; Magnesium oxide for fire protection. [Steetley Magnesia Products Ltd]

Flamco. Mold and core dressings. [Foseco (F.S.) Ltd]

Flamegard®908. Methylolthiourea-based compound; flame

---

‡ Trade name and manufacturer not verified　　　　　　　§ Trade name without identified manufacturer

retardant for nylon. [Sybron]

Flame guard. CAS 1344-28-1; Aluminum oxide trihydrate. [Alcoa Industrial Chemicals]

Flamenco®. Pigments of exceptional brilliance displaying twin colors; for cosmetics. [Mearl]

Flamenol. A proprietary trade name for a polyvinyl chloride synthetic resin. §

Flame Out #44. Decabromodiphenyl oxide/antimony trioxide 2:1 aq. disp.; general-purpose flame retardant for textile applications. [Emco Services]

Flame Out CO. Low-cost flameproofing agent for use on cottons; nonyellowing; imparts a soft hand. [Emco Services]

Flaming. A decolorizing agent for sugar juices.

Flammacerium. Silver cerium nitrate; for treatment of serious burns. [Duphar BV] *

Flammastik. Incombustible cable coatings applied by spray guns or spatula. [Degussa AG] *

Flammazine. Silver sulfadiazine; for treatment of burns. [Duphar BV] *

Flammentin ASN. Ammonium-phosphorus compound; clear solution; nonwater resistant flameproofing agent for textiles made from cellulose and wool; especially suitable for re-impregnation for decoration textile materials. [Thor Chemicals (UK) Ltd]

Flammentin PS. Reactive phosphorus-nitrogen compound; thermo-stable flameproofing agent for textiles of cellulose; especially for carrier fabrics to be coated and exposed to higher temperatures. [Thor Chemicals (UK) Ltd]

Flammex. Flame retardant. [The Wellcome Foundation Ltd]*

Flammocite. A safety explosive. It contains 44% ammonium nitrate, 16% sodium chloride, 14% sodium nitrate, 10% trinitro-toluene, 6% nitro-glycerin, 5% ammonium sulfate, and 5% cellulose.

Flamolin. Flame retarded polymers. [Quantum Chemical Corp]

Flamtard H. CAS 12027-96-2; Zinc hydroxystannate; flame retardant for PVC, polychloroprene, chlorosulfonated polyethylene, other halopolymers [Alcan]

Flamtard S. CAS 12036-37-2; Zinc stannate; flame retardant for PVC, polychloroprene, chlorosulfonated polyethylene, polypropylene, other halopolymers [Alcan]

Flandrac. A decolorizing agent for sugar juices.

Flar. A proprietary preparation containing a lactic ferment resistant to antibiotics, thiamine, riboflavine, pyridoxine, cyanocobalamin, nicotinamide, sodium pantothenate, inositol, folic acid and liver extract; used in the treatment of antibiotic side-effects. [Consolidated Chemicals Ltd]*

Flat-Ayd. Dispersed flatting agents; used for paints, inks etc. [Cornelius Chemical Co Ltd] *

Flat-Ayd®Bases. Dispersions of flatting agents in variety of vehicles; predispersed flatting bases for gloss control of coatings. [Daniel Products]

Flatting Agent OK412. Surface-treated precipitated silica; flatting agent for clear and pigmented solv.- and water-based coatings, printing inks; antisettling agent. [Degussa]

Flavaxin. CAS 83-88-5; Riboflavin; vitamin. [Sterling Drug Inc] *

flavaxin. See riboflavin.

Flavazol. A dyestuff, $C_{13}H_{12}N_2O_2$, prepared from p-toluidine and salicylic acid. It dyes yellow on chrome mordanted wool.

Flavelix. A proprietary preparation containing mepyramine maleate, ephedrine hydrochloride, ammonium chloride, and sodium citrate; a cough linctus. [Pharmax Ltd] *

Flaveosine. A dyestuff obtained by condensing m-acetaminodimethylaniline with phthalic anhydride. It dyes tannined cotton and wool reddish yellow, and silk golden yellow.

Flavinduline. (induline yellow). A dyestuff. Dyes tannined cotton yellow.

flavine. Three materials are known by this name: a) Diaminobenzophenone; b) a grade of quercitron bark extract; and c) diaminomethylacridinium chloride.

Flavocents. Concentrated flavor compositions. [PPF International Ltd] *

flavoline. 2-Phenyl-4-methylquinoline.

Flav-O-Lok. Free-flowing nonhygroscopic flavor powders; used in formulations in food, beverages and pharmaceuticals where liquid flavors cause difficulties. [Hercules] *

Flavomycin. Bambermycins; antibiotic complex, containing mainly moenomycin A and C; antibacterial. [Hoechst AG]

flavone. (2-phenylbenzopyran-4-one; 2-phenylchromone) One of a group of flavonoid plant pigments existing as colorless needles, insoluble in water and melting at 100 C. The flavones produce ivory and yellow colors in plants and flowers.

flavopurpurin. (alizarin No. 10CA, alizarin FA, alizarin GB, alizarin GI, Alizarin RG, alizarin SDG, alizarin X, alizarin VCA, alizarin CAF, DCA, JCA, VAR) Trihydroxy-anthraquinone, $C_{14}H_8O_5$. Dyes cotton mordanted with alumina, red.

Flavotint. Food color and flavor compounds. [PPF International Ltd] *

Flav-R-Keep FP-51. CAS 7758-29-4; Sodium tripolyphosphate, lemon juice solids, flavorings; food additive for meat, poultry, and seafood industries. [Rhone-Poulenc Food Ingreds.]

Flaxedil. CAS 65-29-2; A proprietary preparation of gallamine triethiodide; a muscle relaxant. [Rhone-Poulenc Rorer Ltd]

flax seed. Linseed.

flaxseed oil. See linseed oil.

flax wax. A wax associated with flax fiber and with the cortical tissues. The air-dried cortex contains as much as 10% by weight of wax. It is removed by extraction with a volatile solvent.

FLC-2. Aliphatic hydrocarbon resin; high softening point resin used as a fluid loss control additive. [Shell]

Flea-B-Gon. Flea killer. [Chevron] *

Flea Flip. A line of flea control products; insecticide for household, pet, outdoor use in the control of fleas, ticks and other insects; includes aerosols, concentrates and ready-to-use liquids. [Colonial Products Inc]

Flectol H. CAS 26780-96-1; Polymerized 1,2-dihydro-2,2,4-trimethyl-quinoline; antioxidant; resists effect of heat deterioration and normal aging, for dry rubber and latex. [Monsanto Co]

Flectol ODP. Octylated diphenylamine; antioxidant protecting rubber against heat deterioration and normal aging; for neoprene. [Monsanto Co]

Flectol Pastilles. CAS 26780-96-1; Polymerized 1,2-dihydro-2,2,4-trimethylquinoline; antioxidant used in tires, belts, hose retread rubber and general

---

* Trade name not verified as available   † Trade name verified as obsolete

mechanicals. [Monsanto Co] *

Fletcher's alloy. An alloy of 95.5% aluminum, 3% copper, 1% tin, 0.5% antimony, and 0.5% phosphor tin.

Fletchers Arachis Oil Enema. Arachis oil; a proprietary fecal softener for treatment of constipation. [Pharmax Ltd]

Fletcher's bearing alloys. Aluminum base alloys. One contains 92% aluminum, 7.5% copper, and O.25% tin, and another 90% aluminum, 7% copper, and 1% zinc.

Fletchers Enemette. Glycerol dioctylsulfosucunate; a proprietary micro enema for treatment of constipation, pre- and post-operative cleansing of the bowel, in obstetrics and prior to proctoscopy, sigmoidoscopy, or x-ray examination. [Pharmax Ltd]

Fletchers Phosphate Enema. Sodium acid and monosidum phosphates; a proprietary liquid evacuent enema for treatment of constipation, pre- and post-operative cleansing of the bowel, in obstetrics and prior to proctoscopy, sigmoidoscopy, or x-ray examination. [Pharmax Ltd]

Fleur. Intensive decorating colors with flux mask for enamel and earthenware. [Degussa Ltd]

Fleurelle®. Fibers. [DuPont UK]

Flex®. For the medical industry. [DuPont UK]

Flexade Regular. Lubricant, corrosion inhibitor; easily soluble in water; biodegradable; used for metal treatment. [Carboxyl Chemicals Ltd] *

Flexalyn. Diethylene glycol diabietate; a proprietary trade name for a plasticizer. §

Flexamine. A superflexing antioxidant containing 35% JZ; for use in heavy service truck tread and SBR treads, camelback, wire insulation, neoprene belting and molded soles; offers protection against copper and manganese. [Uniroyal] *

Flexan®130. Sodium polystyrene sulfonate; hair fixative for setting lotions, conditioners, blow drying aid; gloss and antistatic chars. [Nat'l. Starch]

Flexane®. Two-part urethane; for making flexible molds, cast parts and nonmarring holding fixtures, forming abrasion resistant and noise reduction linings, encapsulating parts. [ITW Devcon]

Flexane® 80. Two-part urethane elastomer; tough rubber compound for repairs for lining and protection applications, tooling and molding applications. [ITW Devcon]

Flexane® 94. Two-part urethane elastomer; tough rubber compound for repairs for lining and protection applications, tooling and molding applications. [ITW Devcon]

Flexbond. Vinyl acetate copolymers emulsions. [Air Prods. & Chems. Inc]

Flex Carbon. CAS 1333-86-4; A proprietary carbon black. §

Flexchlor. Chlorinated paraffins. [Witco]

Flexcote. Rubber/vinyl lacquer; used for decoration and protection of rubber articles. [W J Ruscoe Co] †

Flexcrete. Polymer-modified cementitious mortars (range) for repair and maintenance of (reinforced) concrete; used for high-rise housing, PRC housing, any damaged or defective concrete structures. [Liquid Plastics Ltd]

Flexcryl. Acrylic emulsions. [Air Prods. & Chems. Inc]

Flexel®1010. Thermoplastic compound; for wire and cable applications. [BFGoodrich/Geon Vinyl]

Flexeril. CAS 6202-23-9; Cyclobenzaprine hydrochloride; a muscle relaxant and tranquilizer. (U.S.) [Merck & Co Inc]

Flexiban®. CAS 6202-23-9; Cyclobenzaprine hydrochloride; muscle relaxant and tranquilizer. [Merck & Co Inc]

Flexible Fyrex®. Mixture of crystalline diammonium phosphate and monoammonium phosphate, penetrating and softening agents; inorganic flame retardant; improved surface wetting; improved hand in treated fabrics [Akzo]

Flexidor. CAS 82558-50-7; Suspension concentrate containing 500 g isoxaben per liter; used for control of annual dicotyledons in cereals, grass and fruit. [DowElanco Ltd]

Flexin. CAS 1343-98-2; Selected compound of silicic acid, cationic; antislip agent for fabrics and knitted goods. [Henkel Chemicals Ltd] *

Flexipol® FP-100(M). Flexible polyurethane systems; foams for acoustic, pkg., toys, padding cushioning applications [Flexible Prods.]

Flexipol®FSF-106. Flexible polyurethane systems; foams for toys, seats and cushions, padding cushioning and integral skin applications. [Flexible Prods.]

Flexipol®NDTP-311-1.8. Rigid polyurethane systems; for dispensing froth foams through low or high pressure pour machines, spray equip. or hand mix; for flotation and walk-in refrigeration applications. [Flexible Prods.]

Flexipol®NP-311-2. Rigid polyurethane systems; for dispensing froth foams through low or high pressure pour machines, spray equip. or hand mix; for carving and decorative molding applications. [Flexible Prods.]

Flexipol® NS-322-2. Rigid polyurethane systems; for dispensing froth foams through low or high pressure pour machines, spray equip. or hand mix; for flotation and spray insulation applications. [Flexible Prods.]

Flexobond 329. Two-part urethane system; non-TDI system curing to a tough rubber solid with good adhesion to many substrates; used as a fairing compound, adhesive, or encapsulant; for cure at ambient or elevated temps. [Bacon]

Flexobond 423. Two-part urethane system; non-TDI system curing to a hard solid with high clarity and low odor; used as an adhesive, encapsulant, or casting compound; for cure at ambient or elevated temps. [Bacon]

Flexocel. Twin component flexible polyurethane foam systems. [Baxenden Chemicals Ltd]

Flexol. A proprietary trade name for vinyl plasticizers. §

Flexol Plasticizer 3GH. Triethylene glycol-di-2-ethyl butyrate; a proprietary trade name for a plasticizer. §

Flexol Plasticizer 3GO. Triethylene glycol-di-2-ethyl hexoate; a proprietary trade name for a plasticizer. §

Flexomer DFDA-1137 Natural 7. Ethylene copolymer; resin with good resilience, toughness, flexibility, chem. resist., stress crack resist. for blow molding, injection blow molding, profile and sheet extrusion, thermoforming, small part injection molding applications where high flexibility is desired, e.g., hose and tubing, squeeze bottle, squeeze tube; blending resin with high pressure or LLDPE to improve their props.; FDA compliance. [Union Carbide]

Flexonyl. Pigments for aqueous flexographic inks. [Hoechst AG]

Flexoresin. Proprietary brands of glycol and glyceryl phthalates; also polymerized terpenes. §

Flexricin®9. CAS 26402-31-3; EINECS 247-669-7; Propylene glycol ricinoleate; wetting agent, dye solv., wax plasticizer, stabilizer for textile, household, and cosmetic applications, rewetting dried skins. [CasChem]

Flexricin®13. CAS 141-08-2; EINECS 205-455-0; Glyceryl ricinoleate; wetting agent, wax plasticizer, and mold release agent for rubber polymers, antifoam agent,

---

‡ Trade name and manufacturer not verified   § Trade name without identified manufacturer

household, and cosmetic applications, rewetting dried skins. [CasChem]

Flexricin® 15. CAS 106-17-2; EINECS 203-369-8; Glycol ricinoleate; wetting agent, plasticizer, textile, household, and cosmetic applications, rewetting dried skins; chemical intermediate. [CasChem] *

Flexricin® 17. Pentaerythritol ricinoleate; plasticizer; chemical intermediate. [CasChem]

Flexricin® 100. CAS 141-22-0; Ricinoleic acid deriv.; lubricant for textile, metalworking compounds; corrosion inhibitor intermediate; intermediate for water sol./disp. lubricants. [CasChem]

Flexricin® 115. N(β-Hydroxyethyl) ricinoleamide; lubricant/ antistat for plastics, metals; mold release, antiblocking agent for textile coatings; slip agent for varnishes and lacquers; also for elec. potting compounds, crayons, wax blends, high-temp. greases. [CasChem] *

Flexricin® 185. N,N´-Ethylene bis-ricinoleamide; lubricant/ antistat for plastics, metals; mold release, antiblocking agent for textile coatings; slip agent for varnishes and lacquers; also for elec. potting compounds, crayons, wax blends, high-temp. greases. [CasChem] *

Flexricin® P-1. CAS 141-24-2; EINECS 205-472-3; Methyl ricinoleate; low temp. lubricant plasticizer for rubber, phenolic and epoxy resins. [CasChem] *

Flexricin® P-3. Butyl ricinoleate; general purpose plasticizer, lubricant for NC. [CasChem] *

Flexricin® P-4. CAS 140-03-4; Methyl acetyl ricinoleate; all purpose plasticizer, lubricant for vinyls and lacquers. [CasChem] *

Flexricin® P-6. Butyl acetyl ricinoleate; lubricity additive for textile finishes; plasticizer. [CasChem] *

Flexricin® P-8. CAS 101-34-8; EINECS 202-935-1; Glyceryl (triacetyl ricinoleate); plasticizer for vinyl wire jacketing and semirigid vinyls; emollient; stabilizer for anhyd. pigmented systems. [CasChem] *

Flex-Shield. Elastomeric coatings made up of asphalt, neoprene, acrylic and Kraton rubber imbedded in polyester fabric; used for roof coating, maintenance cleaning products, degreaser, waxes, and wall coatings. [Flex-Shield] *

Flexsol 43. Deltafilcon A; contact lens material. [Alcon Laboratories Inc] *

Flexthane 610 EXP, 611 EXP, 620 EXP. Urethane hybrid polymer, aq. for coating and adhesive applications incl. graphic arts polymers, textile and vinyl coatings, leather finishes, plastic coatings, industrial coatings, film lamination. [Air Prods./Polymers] *

Flextron. A nontoxic, nonbiodegradable two-component, cold-curing polyurethane sealant; for expansion and construction joints in water retaining structures; sold in three viscosity grades, suitable for pouring or pressure gunning. [Berger Elastomers] *

Flexzone. A range of antiozonants which offer high protection against flexing, ozone, heat, and oxygen in rubber products. [Uniroyal] *

Flexzone 3C. CAS 101-72-4; N-Isopropyl-N´-phenyl-p-phenylene diamine; a proprietary antioxidant. [Uniroyal] *

Flexzone 6-H. N-Phenyl-N´-cyclohexyl-p-phenylene diamine; a proprietary antioxidant. [Uniroyal] *

Flicker Flake™. Aluminum pigments; used for coatings requiring a glitter or high sparkle effect. [Silberline Mfg Co]

Flint. CAS 330-55-2, 1582-09-8; Emulsifiable concentrate

containing 120 g linuron and 240 g trifluralin per liter; herbicide for winter cereals. [Ashlade Formulations Ltd]

flint. A form of silica, SiO_2.

flint alloy. A heat and corrosion-resisting alloy containing 83% iron, 12.5% chromium, 3% carbon, and 0.5% silicon.

flintcast. A white iron made in the electric furnace. It resists abrasion.

flint glass. (lead glass, crystal). A glass composed of lead and potassium silicates; used for hollow-ware, superior bottles, and optical work. See potash lead glass.

Flint Metal. A proprietary trade name for an alloy of iron with 4-4.5% nickel, 1.25-1.75% chromium, and 3-3.5% carbon. §

Flint SSD. Silver sulfadiazine; antiinfective, topical. [The Boots Co plc]

flinty zinc ore. Flinty calamine.

Fliselina. Nonwoven textile for the apparel industry. [Carl Freudenberg] *

Flit. A proprietary insecticide. §

Flixapret. Synthetic resins for textile finishing. [BASF plc] *

Flo-Aid. Anticaking agent; used to improve flow properties of fertilizers and explosives. [Crowley Chem.]

floatstone. Porous opal.

float tin. CAS siterite, occurring in the soil, and formed by the disintegration of tin rocks.

Flocculant T-9. Complex; cohesion and sedimentation agent for TiO_2 mfg. [Toho Chem. Industry]

Flochel. Poly-electrolyte flocculating agent. [Steetley Chemicals Ltd] *

Flo Chem Extra. Aq. fluorochemical; stain repellent and dry soil resist. finish for fabrics for aq. application. [Emco Services]

Flo-Cillin. CAS 6130-64-9; Penicillin G procaine; antibacterial. [Bristol-Myers Squibb Co Inc] *

Flock-Lok® 850. Flexible polyurethane flock adhesive; adhesive featuring excellent adhesion of flock to elastomers and plastics such as ABS; wide range of cure conditions; solv.-resist.; moisture-curable. [Lord]

Flock-Lok® 851. Flexible polyurethane flock adhesive; moisturing curing adhesive with excellent adhesion of flock to elastomers such as SBR and plastics such as ABS. [Lord]

Floclean 103. A proprietary preparation of polycarboxylic, alkyl sulfonic and organic acids; for removal of metallic salts from cellulosic reverse osmosis membranes. [Pfizer International] †

Flo-Con. Textile chemical. [Albright & Wilson Ltd, Phosphates & Speciality Business] *

Flocon 100. 35% Aqueous solution of a proprietary polyacrylate polymer; a scale control agent in desalination. [Pfizer International]

Flocsil. Activated silica. [Crosfield Chemicals Ltd]

Floex. A proprietary trade name for a wetting agent for paint, etc.; a condensation product of higher fatty alcohols. §

Flogel. Slurry explosives; for a wide range of surface blasting. [Hercules] *

Flolan. Epoprostenol sodium; inhibitor. [Burroughs Wellcome Co] *

Flolan. CAS 35121-78-9; A proprietary formulation of prostacyclin; for use as an alternative to heparin during renal dialysis. [The Wellcome Foundation Ltd]

Flolys®. CAS 50-99-7; Glucose syrup; sweetening agent for general food use. [Roquette (UK) Ltd]

---

* Trade name not verified as available   † Trade name verified as obsolete

Flomac. Granular desulfurizing agent for blast furnace iron and steel. [Foseco (F.S.) Ltd] †

Flo-Mo 5BMP. Free acid of a complex organic phosphate ester; emulsifier for phosphated and chlorinated pesticides; coupling agent for agricultural and industrial applications. [Witco/Organics]

Flo-Mo 80/20. Modified alcohol ethoxylate; agricultural adjuvant. [Witco/Organics]

Flo-Mo 1082. Formulated product; emulsifier for petroleum based agricultural sprays. [Witco/Organics]

Flo-Mo 1093. Formulated product; emulsifier for vegetable oil based agricultural sprays. [Witco/Organics]

Flo-Mo DEL, DEH. Formulated products; matched pair of emulsifiers for a broad range of pesticides. [Witco/Organics]

Flo-Mo Lowfoam. Modified alcohol ethoxylate; surfactant for the formulations of agricultural adjuvants. [Witco/Organics]

Flo-Mo Suspend. Formulated product; compatibility agent for use in agricultural formulations. [Witco/Organics]

Florafoam. Rigid foam (urethane and phenolic) for floral arrangements. [Baxenden Chemicals Ltd]

Floramat. Ethyl-2-t-butylcyclohexylcarbonate; fragrance raw material, for floral notes. [Henkel/Cospha]

Florane. 2-Heptyl tetrahydrofuran. [Quest Int'l. UK Ltd]

Floranid® N 32. Special slow-release nitrogeous fertilizer with 32% nitrogen for intensive horticultural crops, ornamentals, and lawns. [BASF AG]

Floranit. Wetting agent; for mercerizing and caustic lye padding. [Henkel Chemicals Ltd] *

Floraquin. A proprietary preparation of diiodohydroxyquinoline; vaginal antiseptic tablets. [G D Searle & Co Ltd] *

Floratex. Melded fabrics. [ICI Chem. & Polymers Ltd]

Florco®. CAS 1337-76-4; Attapulgite clay; floor absorbent for oil, grease, water, and other liqs.; offers lighter dens.; antislip agent on floors; used by automotive, steel, transportation, commercial, pet, food and beverage, institutional, and amusements industries, and for winter and home uses. [Floridin] *

Florco®-X. CAS 1337-76-4; Attapulgite clay; 8/30 mesh; absorbs oil and water spills; pet absorbent. [Floridin] *

Florel Fruit Eliminator. Etephon; used to eliminate messy fruit set from ornamental olives, carobs, apples and crab apples; applied at bloom; also used to control mistletoe in conifers and deciduous trees. [Lawn & Garden Products Inc]

Florence lake. (Vienna Lake, Paris Lake). Lakes produced from cochineal, by precipitating alkaline solutions of cochineal with alum, or with a mixture of alum and tin salts.

Florentine brown. (Vandyck red, hatchette brown). A pigment. It is copper ferrocyanide.

Florex®. CAS 1337-76-4; Granular attapulgite clay; agrochemical carrier oil refining. [Whitecourt Ltd]

Florex® Ag-Dri 6/30, LVM 8/16, RVM 8/16. CAS 1337-76-4; Attapulgite clay; absorbent and adsorbent [Floridin]*

Floricin. (florizine, derizine, dericin). A substance produced by heating castor oil to 300 C, and distilling 10% of it. There remains the product termed floricin, which solidifies at -20 C. It is also made by heating castor oil; with formaldehyde. It is miscible with ceresin and petroleum jelly and is used as a vehicle for menthol and oil of eucalyptus; unlike castor oil, it is insoluble in alcohol.

Florida phosphates. Mineral phosphates. There are two types, hard rock phosphates, containing 80% calcium phosphate, and soft clay phosphates, containing 40-60% calcium phosphate; used as fertilizers.

Floridin. CAS 50-59-9; A form of fuller's earth found in Florida, U.S.A. It removes the coloring matter from oils and waxes, and is used for this purpose.

Florigel® H-Y. CAS 1337-76-4; Attapulgite clay; absorbent and adsorbent [Floridin] *

Florinef. CAS 514-36-3; A proprietary preparation of fludrocortisone 21-acetate. [Bristol-Myers Squibb Pharmaceuticals Ltd] *

Florinef Acetate. CAS 514-36-3; Fludrocortisone acetate; adrenocortical steroid. [Bristol-Myers Squibb Co Inc]

Florisil. Chromatography absorbent. [Whitecourt Ltd]

Florite. A proprietary trade name for a carefully prepared and screened bauxite. §

Florizine. See Floricin.

Flor-Kleen. CAS 1337-76-4; Attapulgite clay; absorbent [Floridin] *

Florocyclene. Tricyclodecenyl propionate. [Quest Int'l. UK Ltd]

Florone. CAS 33564-31-7; Diflorasone diacetate; antiinflammatory. [Upjohn] *

Floropryl. CAS 55-91-4; Isoflurophate. cholinergic [Merck & Co Inc]

Florosal. Concrete additive. [Steetley Chemicals Ltd] *

Florox. CAS 94-36-0; Benzoyl peroxide dispersion in food grade filler; flour additive. [Diaflex Ltd] *

Flor Sherry. Dry yeast for wine production. [Ciba plc] *

Flosol. A proprietary trade name for colloidal barium silicofluoride for horticultural purposes. §

Flotox. CAS 7704-34-9; Sulfur pesticide. [Chevron] *

flour of sulfur. Powdered sulfur. It has been powdered by grinding, but is not so finely powdered as flowers of sulfur.

Flovan. Proofing agents. [Ciba plc] *

flowers of antimony. CAS 1309-64-4; Formed when antimony burns in air. It consists of antimony oxide, Sb_4O_6; used in medicinal preparations, and as a white pigment.

flowers of Benjamin. (flowers of Benzoin). Benzoic acid, C_6H_5COOH.

flowers of bismuth. Bismuth oxide, Bi_2O_3, obtained by burning bismuth metal at a red heat.

flowers of brimstone. See flowers of sulfur.

flowers of camphor. Camphor in crystalline form.

flowers of copper. The oxide of copper, CuO, produced when copper burns.

flowers of sulfur. (flowers of brimstone). CAS 7704-34-9; Consists of minute crystals of sulfur, obtained by chilling sulfur vapor.

flowers of Tin. Stannic oxide, SnO_2. A polishing powder.

flowers of zinc. See philosopher's wool.

flowers of zinc. See zinc oxide.

Flowfusor. Urological and topical irrigation solutions. [The Boots Co plc] *

Flow-Guard. Synthetic, amorphous, precipitated silica. [PPG Industries]

flow-powder. A mixture of white lead and salt which gives off chlorine on heating; used for the production of flow blues (cobalt blues) on ceramics. Cobalt chloride is formed, which, being volatile, gives blues of varying intensity.

Floxan SC-5211. Flocculant. [Henkel/Cospha]

---

‡ Trade name and manufacturer not verified   § Trade name without identified manufacturer

Floxapen. CAS 5250-39-5; Floxacillin; antibacterial. [SmithKline Beecham] *

Floxifral. CAS 54739-18-3; Fluvoxamine; for treatment of mental and physical complaints caused by disturbed moods and humours. [Duphar BV] *

Fluanxol®. Neuroleptic. [Bayer AG]

fluates. This term is used for fluosilicates. It is also the name for waterproofing compounds consisting of solutions of sodium silcate, or silicofluoride, and other silicofluorides, such as those of zinc, magnesium, and aluminum.

Fluax. Anti-influenza vaccine; for protection against influenza. [Merck & Co Inc] †

Flubenol. CAS 31430-15-6; A proprietary preparation containing flubendazole; veterinary antihelmintic. [Janssen Pharmaceutical Ltd] *

Flucagyn. See Diflucan. [Pfizer International]

Flucam. See Diflucan. [Pfizer International]

Flucan. See Diflucan. [Pfizer International]

Fluderma. CAS 2825-60-7; Formocortal; glucocorticoid. [Farmitalia (Farmaceutici Italia)] *

Fludor solder. An aluminum solder containing 56.5% tin, 40% zinc, 3% lead, 0.2% antimony, and 0.1% copper.

Fluf® **10-0-10.** Urea-formaldehyde suspension; fertilizer providing nitrogen and potassium to golf course turf. [W A Cleary]

Flugne 113. Trifluorotrichloroethane; a proprietary noninflammable solvent of low toxicity for cleaning precision equipment. [Rhone-Poulenc NV] *

Flugyn. See Diflucan. [Pfizer International]

Fluid EEZ 1000. Soybean oil with glycerol lacto esters of fatty acids, propylene glycol, mono- and diglycerides, TBHQ; kosher; fluid shortening for high quality cakes made without hydrogenated shortening; added emulsifiers for improved performance. [Van Den Bergh Foods]

Fluid Flex. Soybean oil, fatty acid glyceryl lactates, mono and diglycerides, TBHQ; kosher; fluid shortening for high quality cakes made without hydrogenated shortening; increases pliability and shelf life of soft tortillas. [Van Den Bergh Foods]

Fluidil. CAS 2259-96-3; Cyclothiazide; diuretic; antihypertensive. [Adria Laboratories Inc] *

Fluidiram. Fatty amine blends and aminated clays; anticaking agents for fertilizers. [Atochem UK/Ceca]

Fluilan. CAS 8006-54-0; Lanolin oil; w/o emulsifier; dispersant for pigments; conditioning agent; emollient, penetrant, superfatting agent for lipsticks, baby oils, brilliantines, cleansing lotions; plasticizer for hair spray resins; moisturizer in w/o emulsions; also for soaps, shampoos, dishwashing liqs., germicidal skin cleansers. [Croda Inc.; Croda Chem. Ltd.]

Fluilan AWS. CAS 68458-58-8; PPG-12-PEG-65 lanolin oil; emollient, solubilizer; plasticizer and film modifier for hair sprays. [Croda Inc.]

Fluisil®**S55K.** A low temperature silicone lubricant; used for refrigeration machinery. [Bayer AG]

Fluisol. A range of sulfated castor oils; for various applications in many industries. [Yorkshire Chemicals plc] *

Fluitex. Thinned (or fluidized) starches; used in textiles to strengthen and finish and in the paper industry. [Roquette (UK) Ltd]

Flukiver. CAS 57808-65-8; A proprietary preparation containing closantel; veterinary flukicide. [Janssen Pharmaceutical Ltd] *

Fluoderm. A proprietary preparation of fluorometholone, clioquinol, and chlorphenesin for dermatological use. [British Drug Houses] *

Fluogen. Influenza virus vaccine; immunizing agent. [Parke-Davis]

Fluolite. Textile auxiliaries. [ICI Chem. & Polymers Ltd]

Fluon®. A trademark for polytetrafluoroethylene (PTFE); a hard plastics material with a very water repellant surface which has a working temperature range from -200 to +280 C; has a low high frequency power factor and permittivity and outstanding nonstick properties; it is used in heat resistant glands, packings and bearings. [ICI Chem. & Polymers Ltd]

Fluon® **AD1, AD1L, AD1H.** CAS 9002-84-0; PTFE resin; general purpose aq. dispersion polymer for impregnation of packing yarns and glasscloth. [ICI Fluoropolymers]

Fluon®**CDI.** CAS 9002-84-0; PTFE resin; general purpose coagulated disp. polymer for thick-walled tubing and unsintered tape. [ICI Fluoropolymers]

Fluon®**G170.** CAS 9002-84-0; PTFE resin; general purpose molding powd. [ICI Fluoropolymers]

Fluonex™. CAS 356-12-7; Fluocinonide; topical corticosteroid. [ICN Pharmaceuticals Inc]

Fluonid®. CAS 67-73-2; Fluocinolone acetonide; cortocosteroid. [Allergan, Inc]

Fluorad® **FC-24.** Trifluoromethanesulfonic acid; catalyst and reactant increasing yields in polymerization of epoxies, styrenes, THF, in alkylation and acylation reactions; improves octane rating; used with nitric acid for higher yields of pharmaceuticals, explosives, dyes, and intermediates. [3M]

Fluorad® **FC-26.** Perfluorooctanoic acid; intermediate for preparation of monomers and surfactants. [3M/Industrial Chem. Prods.]

Fluorad® **FC-93.** Ammonium perfluoroalkyl sulfonate; wetting agent in etching solutions in semiconductor devices; foaming agent, leveling agent, corrosion inhibitor. [3M/Industrial Chem. Prods.]

Fluorad® **FC-95.** Potassium perfluoroalkyl sulfonate; wetting and foaming agents for coatings, etchants, plating baths, cleaning systems; corrosion inhibitor, leveling agent. [3M/Industrial Chem. Prods.]

Fluorad® **FC-118.** Ammonium perfluorooctanoate; surfactant for emulsion polymerization of fluorinated monomers. [3M/Industrial Chem. Prods.]

Fluorad® **FC-722.** Fluorochemical polymer in inert fluorocarbon solv. blend; conformal coating producing thin, transparent films possessing antimigration props.; as nonwetting barrier coating, as protective coating for elec. contacts and electronic components. [3M]

Fluorad®**FC-724.** Fluorochemical acrylate polymer in fluorinated inert vehicle; surface modifier producing transparent films with excellent antiwetting and antisticking props. against oils, silicones and photoresist solutions in mfg. of semiconductor devices; antimigration barrier; release coating. [3M]

Fluorad® **FC-740.** Fluoroaliphatic polymeric ester; oil well stimulation surfactant; foams hydrocarbon liqs.; active in low polarity org. solvs. [3M/Industrial Chem. Prods.]

Fluorad® **FX-8.** CAS 307-35-7; Perfluorooctanesulfonyl fluoride; intermediate for preparation of monomers and surfactants for textile treatment, paper sizes, inert fluids. [3M/Industrial Chem. Prods.]

Fluorad® **FX-13.** 2-(N-Ethylperfluorooctanesulfonamido) ethyl acrylate; monomer for polymerization reactions to

---

* Trade name not verified as available   † Trade name verified as obsolete

provide oil and water repellency, lubricity and release to polymers; used to mfg. textile treating resins, paper sizes, surfactants, and inert fluids. [3M]

Fluorad Surfactants. Fluorinated sufactants; used as wetting agents in paint and polishes. [3M UK plc]

Fluoral. CAS 7681-49-4; Sodium fluoride; dental caries prophylactic. [Oral-B Laboratories Inc] *

fluoram. Ammonium bifluoride.

Fluor-Amps. CAS 518-47-8; Fluorescein sodium 5ml amps 10 and 20%; diagnostic aid in the determination of circulation time, examination of opthalmic vasculature and differentiation of malignant and healthy tissue, visualization of gall bladder and bile duct before surgery and for all localization of brain tumors. [SAS Pharmaceuticals Ltd] ‡

Fluoranar. Coil coating compositions. [ICI Chem. & Polymers Ltd] †

Fluoraz. Modified structure of tetrafluorocthylene and propylene copolymers; used in producing sealing materials for chemical, petrochemical, and other applications where temperature and chemical resistance are of paramount importance. [Greene, Tweed & Co]

fluorchrome. Chromium fluoride, $CrF_3 \cdot 4H_2O$; a mordant.

Fluorel® FC-2120. Fluoroelastomer; incorporated cure gumstock developed for automotive fuel line hose; for extrusion, injection and transfer molding applications. [3M]

Fluorel® FC-2144. Fluoroelastomer; designed for molding of complex shapes requiring good compression set resist.; capable of higher filler loading at equiv. durometer; suitable for compression molding. [3M]

Fluorel® FC-2173. Specialty masterbatch of incorporated cure fluoroelastomer gum; best-flowing incorporated cure polymer with improved release over 2170 (no sacrifice in physical properties); used in molded goods, o-rings, hose. [3M]

Fluorel® FC-2211. Fluoroelastomer; used for modifying the visc. of other Fluorel grades to improve flow, sprayability, etc.; lowest visc. in series. [3M]

Fluorel® FT-2481. Fluoroelastomer terpolymer; gumstock with excellent heat and chem. resist.; developed for shaft seals, molded goods, diaphragms, and extrusions where improved chem. resist. is required; vulcanizable with amine and dihydroxy cure systems. [3M]

Fluorel® FX-9038. Fluoroelastomer; high fluorine, incorporated cure fluoroelastomer offering improved processability, mold release, and compression set; designed for molded goods applications requiring very high chem. resist. [3M]

fluoremetic. Antimony sodium fluoride, $SbF_3 \cdot NaF$.

Fluoresbrite. Fluorescent monodisperse carboxylated microspheres (polymer beads containing fluorescent dye); an identification tag and a size reference for agglutination tests, flow cytometry, instrument calibration, gel filtration, light scattering and phagocytosis. [Polysciences Inc]

fluorescein. (resorcinolphthalein; diresorcinolphthalein) $C_{20}H_{12}O_5$; CAS 2321-07-5; EINECS 219-031-8; Dyeing seawater for spotting purposes, tracer to locate impurities in wells, dyeing silk and wool, diagnostic aid in ophthalmology, indicator and reagent for bromine. [EM Industries; R.W. Greeff; Hilton Davis; Kraeber GmbH; U.S. Biochemical]

fluoresceine. (resorcinolphthalein; diresorcinolphthalein)

CAS 2321-07-5; It is obtained by heating phthalic anhydride with resorcinol, with green fluorescence; $C_{20}H_{12}O_5$; Its alkali salts dye silk and wool yellow, as does fluoresceine itself; it is used also as a tracer to locate impurities in wells, diagnostic aid in ophthalmology, indicator and reagent for bromine.

Fluorescent. Dyestuffs that fluoresce in daylight and/or uv light; for textiles, resin pigments, and various solvents. [Holliday Dyes & Chemicals Ltd]

fluorescent blue. (resorcin blue, iris blue). A dyestuff. It is the ammonium salt of tetrabromoresorufin. $C_{12}H_6Br_4N_2O_3$; dyes silk and wool blue, with brownish fluorescence.

Fluorescent Red 5B. A proprietary organic fluorescent-red dye used for coloring polystyrene polymethyl-methacrylate and unplasticized PVC. [Farbwerke Hoechst] *

Fluorescite. CAS 518-47-8; Fluorescein sodium; diagnostic aid. [Alcon Laboratories Inc]

Fluorfolpet. N-(Fluordichloromethylthio) phthalimid; fungicide applied as a paint. [Bayer AG]

fluorine. Nonmetallic halogen element; F; CAS 7782-41-4; Production of metallic and other fluorides, fluorocarbons, fluoridating compounds for drinking water and toothpaste. [Air Prods.; Allied-Signal; Solvay GmbH]

Fluorinert Liquids. Perfluorinated liquids; used in the electronics industry as a vapor phase soldering medium; in direct testing of electronic components and in direct cooling. [3M UK plc]

Fluorinse. CAS 7681-49-4; Sodium fluoride; dental caries prophylactic. [Oral-B Laboratories Inc] *

Fluor-I-Strip. CAS 518-47-8; Fluorescein sodium; diagnostic aid. [Wyeth Laboratories]

fluorobenzene. (phenyl fluoride) C_6H_5F; CAS 462-06-6; EINECS 207-321-7; Insecticide and larvicide intermediate, identification reagent for plastic or resin polymers. [Hoechst Celanese; ICI Am.; Schweizerhall]

Fluorocarbon-11. See trichlorofluoromethane.

Fluorocarbon-12. See dichlorodifluoromethane.

Fluorocomp® FC-101. CAS 9002-84-0; PTFE resin, 5% glass fiber-filled; long wearing, mech. stable composite for chem., elec., and mech. applications; for self-lubricating bearings, compressor rings, gaskets, seals, valve seats and liners, thin-wall rings. [ICI Fluoropolymers]

Fluorocomp® FC-144. CAS 9002-84-0; PTFE resin, 40% bronze-filled; long-wearing, self-lubricating filled composite for most mech. and some chem. applications; for bearings, rings gaskets, seals, valve seats, and liners. [ICI Fluoropolymers]

Fluorocomp® FC-174. CAS 9002-84-0; PTFE resin, 15% glass fiber-filled, 5% MoS_2; long wearing, mech. stable composite for chem., elec., and mech. applications; for self-lubricating bearings, compressor rings, gaskets, seals, valve seats and liners, thin-wall rings. [ICI Fluoropolymers]

Fluorocomp® FC-182. CAS 9002-84-0; PTFE resin, 55% bronze-filled, 5% MoS_2; long-wearing, self-lubricating filled composite for most mech. and some chem. applications; for bearings, rings gaskets, seals, valve seats, and liners. [ICI Fluoropolymers]

Fluoroether Grease 834. Solv./chem. resist. grease with superior high temp. props. [Wm F Nye]

fluoroform. (trifluoromethane; propellant 23; refrigerant 23) CAS 75-46-7; CHF_3. Refrigerant, intermediate in organic synthesis, direct coolant for infrared detector cells,

---

‡ Trade name and manufacturer not verified   § Trade name without identified manufacturer

blowing agent for infrared foams.

Fluoroglide. Fluoropolymer powders used as dry lubricants and additives. [Imperial Chemical Industries plc]

fluorol. CAS 7681-49-4; Sodium fluoride, NaF.

Fluorol® Dyes. Polycyclic dyes with intensive fluorescence; for coloring carburetor fuels, oils and lubricating greases. [BASF AG]

Fluorolene®. Polytetrafluoroethylene. [LNP] *

Fluorolube® GR-290, GR-362, GR-470, GR-544, GR-660. CAS 9002-83-9; Polychlorotrifluoroethylene; lubricating grease for use in chlorine and oxygen systems, metal-working, nuclear service, drilling, heat transfer media, damping fluids, plasticizers. [OxyChem]

Fluorolux. Coil coating compositions. [ICI Chem. & Polymers Ltd] †

Fluoromar. Fluroxene; anesthetic. [Anaquest, Div of BOC Inc] *

Fluoromelt® . CAS 9002-84-0; Melt processable fluorocompounds based on FEP, ECTFE, PFA, ETFE, and PVDF; color concs. for wire and cable applications. [ICI Fluoropolymers]

Fluoromelt® FP-F-FMX1. Foam conc. for blending with FEP base resin and subsequent extrusion onto wire. [ICI Fluoropolymers]

p-fluorophenol. (4-fluorophenol) C_6H_5FO; CAS 371-41-5; EINECS 206-736-0; Fungicide, intermediate for pharmaceuticals. [ICI Am.; PCR; Schweizerhall]

Fluoroplex®. CAS 51-21-8; Fluorouracil; for multiple actinic keratoses. [Allergan, Inc]

Fluor-Op® Ophthalmic Suspension USP 0.1%. CAS 426-13-1; Fluorometholone; indicated for the treatment of corticosteroid-responsive inflammation of the palpebral and bulbar conjunctiva, cornea and anterior segment of the globe. [Iolab Corp]

Fluorosint® 500. CAS 9002-84-0; PTFE with mica filler; low coeff. of thermal expansion, good elec. props., high chem. resist. and dimensional stability; for elec./electronic components, bearings, bushings, wear plates, thrust buttons. [Polymer Corp.]

Fluorotex. Fluorochemical; oil and water repellent for outerwear market. [CNC Int'l. L.P.]

fluorotrichloromethane. See trichlorofluoromethane.

Fluorouracil Roche. CAS 51-21-8; Proprietary preparation of fluorouracil; antineoplastic agent. [Roche Products Ltd] *

FluorSave. Reagent. [Calbiochem Corp]

fluorspar. (fluor, fluorite, Derbyshire spar). A mineral. It is calcium fluoride, $4[CaF_2]$; principal source of fluorine and its compounds by way of hydrogen fluoride, flux in open hearth steel furnaces and in metal smelting, ceramics, wood preservatives, optical equipment.

Fluothane. CAS 151-67-7; A proprietary preparation of halothane; a general anesthetic. [ICI Chem. & Polymers Ltd]

Fluowet 40 M. Fluoroaliphatic ethoxylate; surfactant for metalworking fluids; intermediate products; wetting agent for mordant baths, aq. and org. solv. systems. [Hoechst Celanese/Colorants & Surf.; Hoechst AG]

Fluowet OL. Ammonium salt of perfluorosulfate ester; surfactant for textile processing. [Hoechst Celanese/Colorants & Surf.]

Fluprim. Proprietary cough and cold remedy containing vitamin C, dextromethorphan, ephedrine, phenindamine, retinol and salicylamide. [Roche Products Ltd]*

Flurogestone Acetate. 9-Fluoro-11β, 17-di-hydroxy-pregn-4-ene-3, 20-dione, 17-acetate; a proprietary progestin. [G D Searle & Co Ltd] *

Flurothyl. Di-(2,2,2-trifluoroethyl)-ether. Indoklon.

Fluscorbin. A proprietary preparation of vitamin C, quinine dihydrochloride and sulfate, phenacetin and caffeine; a cold remedy. [Cox Pharmaceuticals] †

Flutec. Range of extremely inert, temperature stable, non-toxic, noninflammable liquids exhibiting excellent electrical insulating properties and good heat transfer characteristics. [Rhone-Poulenc UK]

Flutec PP1. A fluorinated hydrocarbon C_6F_{14}; perfluoro-n-hexane. §

Flutec PP2. A fluorinated hydrocarbon C_7F_{14}; perfluoro-methylcyclohexane. §

Flutec PP3. A fluorinated hydrocarbon C_8F_{16}; perfluoro-1,3-dimethylcyclohexane. §

Flutec PP9. A fluorinated hydrocarbon $C_{11}F_{20}$; perfluoro-1-methyldecalin. §

Fluvermal. CAS 31430-15-6; A proprietary preparation containing flubendazole; antihelmintic. [Janssen Pharmaceutical Ltd] *

Fluvirin. An inactivated influenza vaccine (surface antigen) containing highly purified hemaglutinin and neuraminidase antigens; for protection against influenza. [Evans Medical] *

Flux-Off® . Defluxing agent removing activated and nonactivated rosin flux, and dirt, grease, and molding compounds; for use on PC boards, elec. assemblies, other electronic components. [Chemtronics]

Fluxol. A hardwood pitch prepared from the distillation of hardwood; used as a rubber softener.

Fluzone. Influenza virus vaccine; immunizing agent. [Bristol-Myers Squibb Co Inc] *

FM 1132. Two-stage phenolic resin with macerated fabric as reinforcement; molding compound, good shock resistance and dielec. properties for impact, compression, and transfer molding. [ICI Fiberite]

FM 3510. Two-stage cotton fabric-filled phenolic resin; impact molding resin, good torque strength and molding surface. [ICI Fiberite]

FM 21288. One-stage glass roving-reinforced phenolic compound; high impact molding material. [ICI Fiberite]

F-MA 11® . CAS 21645-51-2, 546-93-0; Aluminum hydroxide, magnesium carbonate; antacid which minimizes constipative or laxative effects. [Reheis]

FMB 65-15 Quat, 65-28 Quaternary CAS 139-08-2; EINECS 205-352-0; Myristalkonium chloride; for formulation of mildewcides and swimming pool algicides. [Huntington Lab]

FMB 210-8 Quat, 210-15 Quaternary CAS 7173-51-5; Didecyldimonium chloride; for formulation of disinfectants, sanitizers, fungicides, water treatment microbicides, swimming pool algicides, mildewcides. [Huntington Lab]

FMB 302-8 Quaternary CAS 68424-95-3; Dicapryl/dicaprylyl dimonium chloride; for formulation of disinfectants, sanitizers, and fungicides. [Huntington Lab]

FMB 500-15 Quat U.S.P. Benzalkonium chloride; preservative in OTC drug products. [Huntington Lab]

FMB 504-5 Quaternary Dicapryl/dicaprylyl dimonium chloride, myristalkonium chloride; for formulation of disinfectants, sanitizers, fungicides, water treatment micro-

---

* Trade name not verified as available   † Trade name verified as obsolete

bicides, mildewcides. [Huntington Lab]

FMB 1210-5 Quat, 1210-8 Quaternary Dicetyldimonium chloride, myristalkonium chloride; for formulation of disinfectants, sanitizers, fungicides, water treatment microbicides, swimming pool algicides. [Huntington Lab]

FMB 3328-5 Quat, 3328-8 Quaternary Myristalkonium chloride, quaternium-14; for formulation of disinfectants, sanitizers, fungicides, water treatment microbicides, swimming pool algicides. [Huntington Lab]

FMB 6075-5 Quat, 6075-8 Quaternary n-Alkyl dimethyl benzyl ammonium chloride and n-alkyl dimethyl ethylbenzyl ammonium chloride; for formulation of disinfectants, sanitizers, fungicides, water treatment microbicldes, and swimming pool algicides. [Huntington Lab]

FML Forte Liquifilm®, Liquifilm . CAS 426-13-1; Fluorometholone; corticosteroid, anti-inflammatory. [Allergan, Inc]

FML-NEO Liquifilm® . Fluorometholone, neomycin sulfate; corticosteroid, anti-inflammatory. [Allergan, Inc]

FML-S Liquifilm® . Fluorometholone, sulfacetamide sodium; corticosteroid, anti-inflammatory, antibacterial. [Allergan, Inc]

FML S.O.P.®; CAS 426-13-1; Fluorometholone; corticosteroid, anti-inflammatory. [Allergan, Inc]

Foamacure. Defoaming agents; for aqueous coating compositions. [Hüls Am.]

Foamaster. Antifoams/defoamers. [Henkel Chemicals Ltd]

Foamaster 340. Nonsilicone defoamer; for pressure dyeing applications in textile industry. [Henkel/Textiles]

Foamaster 371-S. Silicone emulsion; defoamer for pressure and atmospheric processes in textile industry. [Henkel/Textiles]

Foamaster A. Polymerized alcohol; defoamer for use in degassing and monomer stripping operations involving synthetic latexes. [Henkel/Functional Prods.]

Foamaster AP. Low silicone hydrophobic silica; defoamer for use in degassing and monomer stripping operations involving synthetic latexes. [Henkel/Functional Prods.]

Foamaster NXZ. Low silicone, fatty acid base; defoamer for synthetic latexes, paints, adhesives, blade and roll coatings, emulsifiable latex stripping. [Henkel/Functional Prods.]

Foamaster Soap L. CAS 8052-48-0; Sodium tallowate; dry defoamer for wettable powd. and dry flowable pesticide formulations. [Henkel/Emery]

Foamaster VC. Low silicone hydrophobic silica; defoamer for use in degassing and monomer stripping operations involving synthetic latexes. [Henkel/Functional Prods.]

Foam-Coll 4C. CAS 68920-65-0; Potassium coco-hydrolyzed collagen; foaming protein, cosmetic ingred.; mild surfactant. [Brooks Industries]

Foam-Coll 4CT. CAS 68952-16-9; TEA coco-hydrolyzed collagen; foaming protein, cosmetic ingred.; mild surfactant. [Brooks Industries]

Foamer CD. Proprietary blend; foamer for air mist drilling, general detergents, gypsum board production, ether sulfates. [Harcros]

Foamex AD-50, AD-100, AD-300, J-275. Silicone defoamers. [Rhone-Poulenc Surf.]

FoamFlush™. N-Methyl-2-pyrrolidone, butyrolactone, other ingreds.; urethane remover for urethane foam manufacture. [ISP]

Foamid 117. CAS 68425-43-4; Cocamidopropyl dimethyl-amine propionate, water; surfactant, softener, emollient for creams, lotions, fingernail polish removers. [Alzo]

Foamkill® 30 Series. Organic and organo-silicone conc.; defoamer for food/pharmaceutical applications [Crucible] *

Foamkill® 30HP. Organo-silicone emulsion; antifoam/defoamer for waste treatment, pulp/paper, severe foaming situations. [Crucible]

Foamkill® 80J Series. Silicone compound; defoamer for food/pharmaceutical applications [Crucible] *

Foamkill® 400A. Organo-silicone emulsion; defoamer for pulp/paper applications incl. adhesives backings, latex paints and coatings, water-reducible inks, floor and ceiling coatings. [Crucible] *

Foamkill® 608. Nonsilicone; defoamer for paper coatings and adhesives and formulations sensitive to fish-eyeing, janitorial supply houses; readily emulsifiable. [Crucible]

Foamkill® 614. General purpose antifoam/defoamer containing very small amount of silicone; for inks, adhesives, and coatings; suitable for SBU, PVA, PVE, polyethylene, and miscellaneous copolymers. [Crucible]

Foamkill® 618 Series. Org. and organo-silicone conc.; defoamer for food/pharmaceutical applications incl. paper coatings and adhesives in contact with food. [Crucible] *

Foamkill® 634C. Nonsilicone; defoamer for food applications, canning trade, pasteurizer defoaming. [Crucible]

Foamkill® 639. Nonsilicone; defoamer for aq. systems, pulp/paper applications, acrylic, PVC, PVA, PVPC, and other coatings, paints, inks, antifoam formulating. [Crucible]

Foamkill® 639JOH. Organo-silicone; defoamer for aq. paint, ink, and coatings systems; excellent for vinyl acrylic paints and other severe foaming applications; may be used in grind or let-down. [Crucible]

Foamkill® 639Q. Organo-silicone; defoamer for aq. systems, paints, inks, adhesives and coatings, esp. for high-foaming acrylic or Joncryl resin systems; may be used in grind or let-down; high stability. [Crucible]

Foamkill® 649. Nonsilicone; highly compatible defoamer for aq. systems, adhesives, coatings, esp. acrylic systems, paints, inks, drawing and cutting fluids, paper applications; very little tendency to cause fish-eyeing. [Crucible]

Foamkill® 654NS. Nonsilicone; defoamer for drawing and cutting fluids; emulsifiable and compatible with synthetic or oil-based media. [Crucible]

Foamkill® 663J. Organo-silicone; defoamer for food/pharmaceutical applications incl. paper coatings and adhesives in contact with food, pulp/paper applications, water cooling towers; lubricant for fine papers; water-disp. [Crucible]

Foamkill® 810F. Dimethicone; defoamer for food/pharmaceutical applications incl. general aq. systems, paper coatings, and adhesives in contact with food, egg washing, cleaning/sanitizing solutions, cosmetics, pulp/paper applications. [Crucible]

Foamkill® 836B. Silicone emulsion; highly compatible defoamer for chemical compounding, synthetic or water-based metalworking lubricants, aq. systems, inks, janitorial applications, paints, coatings, paper industry, waste treatment, severe foaming applications; dilutable with water. [Crucible]

Foamkill® CMP. Bis-stearamide; efficient defoamer for paper reclaiming, caustic treatment, waste treatment,

---

‡ Trade name and manufacturer not verified   § Trade name without identified manufacturer

textile jet, or atmospheric dyeing. [Crucible]

Foamkill® EFT. Silicone emulsion; defoamer for waste treatment; good flash knockdown and staying power; for intermittent or continuous feed. [Crucible]

Foamkill® MS. Silicone emulsion; defoamer for waste treatment, atmospheric or pressure dyeing of textiles; for intermittent or continuous feed. [Crucible]

Foam-Kon 20. Thermoplastic structural foam concentrate; for high pressure foam systems; used in injection molding compounds (polyethylene, PP, ABS, SAN, PS, acetal, etc.). [LNP]

Foamole A. Linoleamide DEA (1:1); hair conditioner for shampoos, hair dyes; visc. booster, foam stabilizer for shampoos, liq. soaps. [Van Dyk]

Foamole B. CAS 68953-11-7; Minkamidopropyl dimethylamine; superfatting agent, conditioner for hair care products. [Van Dyk]

Foamole M. CAS 68140-00-1; Cocamide MEA (1:1); foam booster/stabilizer, thickener, emulsifier for creams and lotions, shampoos, bubble baths, other detergents. [Van Dyk]

Foamosul. Sodium foamaldehyde sulferylate; textile printing/emulsion polymerisation. [RV Chemicals Ltd]

Foamquat IAES. CAS 67633-63-0; EINECS 266-778-0; Isostearyl dimethylamidopropyl ethonium ethosulfate. [Alzo]

Foam-Soy C. Sodium cocoyl hydrolyzed soy protein; foaming cosmetic ingredient for skin and hair care products; mild surfactant. [Brooks Industries]

foam tannin. Tannin extracted from sumach, or galls, by means of ether.

Foam Tint. Pigment dispersions; for coloration of polyurethane foam. [Pacific Dispersions Inc] *

Foam-Wheat C. Sodium cocoyl hydrolyzed wheat protein; mild foaming protein for use in shampoos; gentle cleanser in facial makeup removers. [Brooks Industries]

Fob metal. See brass.

Focal. Flowable fungicide. [Schering Agrochemicals Ltd]

Focus. An analgesic, antipyretic drug. [Bayer AG]

Focus®. Cycloxydim; post-emergence graminicide against annual and perennial grasses; selective in broadleaf crops, e.g., sugar beet, cotton, soybean, vegetables, onions [BASF AG]

Fodel®. For the electrical industry. [DuPont UK]

Foerdite. An explosive consisting of 25% nitroglycerin, 1.5% collodion wool, 5% nitro-toluene, 4% dextrin, 3% glycerin, 37% ammonium nitrate, and 24% potassium chloride.

Foilcote. Overprint lacquers for aluminum, foil, carton boards, labels etc. [The Scottish Adhesives Co Ltd]

Foilgrip. Aluminum foil to paper laminating adhesives. [The Scottish Adhesives Co Ltd]

foil lead. An alloy of 86.5% lead, 12.5% tin, and 1% copper.

Folcovin. CAS 509-67-1; A proprietary preparation containing pholcodine; a cough linctus. [Rybar Laboratories Ltd] *

Folex 350. A proprietary preparation containing ferrous fumarate and folic acid; used in the treatment of anemia during pregnancy. [Rybar Laboratories Ltd] *

Folex-P. Foliar phosphate. [Omex Agriculture Ltd]

Foliac Super Red. A proprietary trade name for a graphite jointing compound. §

Folia-Feed. Foliar nutrient. [DowElanco Ltd]

Foliar 36 Extra. Foliar feed. [BASF plc]

Foliar Nitrophoska. Foliar feed. [BASF plc]

Foliar TRIGGRR. A liquid containing plant growth regulators (cytokinin and gibberellic acid) and trace minerals; used to increase crop yields and quality; applied to the foliage; used for a wide variety of crops including cotton, soybeans, wheat, fruits and vegetables. [Westbridge Research Group] ‡

folic acid. (pteroylglutamic acid; folacin; PGA) CAS 59-30-3; Part of the vitamin B complex; $C_{19}H_{19}N_7O_6$; used in medicine, nutrition, food additive.

Folicet. CAS 59-30-3; Folic acid; vitamin. [Mission Pharmacal Co] *

Folicin. Prophylatic folic acid and iron preparations; suitable for the prophylaxis and treatment of pregnancy anemia. [Paines & Byrne Ltd] *

Folicote Transpiration Minimizer. Refined wax, emulsifiers, preservatives, minimum 50% solids; FDA approved for use on edible crops; reduces water loss from plant foliage, winter protection, transplanting and transporting plants, christmas trees, wreaths, agricultural crops such as potatoes, corn, tobacco, transplants, stone and citrus fruits. [Aquatrols Corp of Am.] *

Folicur. CAS 107534-96-3; Tebuconazole; fungicide with systemic properties and broad spectrum activity against rusts, leaf spot diseases, e.g., *Septoria spp.* powdery mildew and several *Fusarium* species on cereals; whitemold, *Phoma* and various leaf spot diseases on oilseed rape. [Bayer AG]

Folidol®-E605. CAS 56-38-2; Parathion; insecticide and acaricide. [Bayer AG] *

Folidol® M. CAS 298-00-0; Parathion-methyl; insecticide with contact, stomach, and breathing poison action. [Bayer AG]

Folimat®. CAS 1113-02-66; Omethoate; systemic acaricide and insecticide. [Bayer AG; Bayer plc]

Folin-Dennis solution. A solution prepared by adding slowly 400 cc of a 0.7268% solution of silver nitrate to a solution containing 10 g mercuric cyanide and 180 g caustic soda, in 1,200 cc water; used for the determination of acetone.

Folin-McEllroy lactose reagents. a) A saturated picric acid solution (2 g picric acid in 100 cc water). b) A 20% sodium carbonate solution.

Folin-McEllroy sugar reagents. Qualitative. 100 g sodium pyrophosphate, $Na_4P_2O_7.10H_2O$, 30 g crystalline disodium monohydrogen phosphate, Na_2HPO_4, and 50 g dry sodium carbonate in 900 cc water. Dissolve and add 13 g copper sulfate dissolved in 200 cc water; Quantitative. a) acidified copper sulfate sol'n. containing 60 g copper sulfate in 900 cc water; dissolve and add 5 cc conc. sulfuric acid; make up to 1000 cc; b) Phosphate-carbonate-thiocyanate dry mixture prepared by mixing 100 g dry sodium carbonate, $Na_2CO_3H_2O$, and 30 g potassium thiocyanate mixed in a large mortar.

Folin's uranium acetate mixture. Uric acid reagent. It consists of 500 g ammonium sulfate, 5 g uranium acetate, and 6 cc glacial acetic acid dissolved in 650 cc water and made up to 1 liter.

Folin's Uric Acid Reagent. Add to 160 cc water, 50 cc syrupy phosphoric acid. Heat to 85 C., and add 100 g sodium tungstate. Boil for 1 hour under reflux. Place 25 g lithium carbonate in a beaker, add 50 cc syrupy phosphoric acid and 200 cc water. Boil 10 minutes, cool and add first solution; mix and dilute to 1000 cc.

---

* Trade name not verified as available   † Trade name verified as obsolete

Folio 575FW. Suspension concentrate containing 500 g chlorothalonil and 75 g metalaxyl per liter; a systemic fungicide for field crops. [Ciba-Geigy Agrochemicals]

Folithion®. CAS 122-14-5; Fenitrothion; broad spectrum insecticide for controlling biting and sucking insect pests. [Bayer AG]

follicle stimulating hormone. An extract of human post-menopausal urine containing primarily the follicle-stimulating hormone. Pergonal.

follotropin. A follicle-stimulating hormone.

Follutein. Gonadotropin, chorionic; gonad-stimulating principle. [Bristol-Myers Squibb Co Inc] *

Folosan. Horticultural fungicides. [Plant Protection] *

Folpan. CAS 133-07-3; Active ingredient: folpet; agricultural fungicide. [Makhteshim Chemical Works Ltd]

Folpet. N-(Trichlormethylthio) phthalimid; fungicide applied as a paint. [Bayer AG]

Folvite. CAS 59-30-3; Folic acid; vitamin [Lederle Laboratories USA] *

Fomac. A fungicide for rubber. [ICI Chem. & Polymers Ltd]†

Fomblin. A proprietary range of perfluoropolyether fluids. [Montedison UK Ltd] *

Fome-Cor. Light-weight, rigid board made of extruded polystyrene foam securely bonded between two layers of tough kraft linerboard; used in manufactured housing, automotive and graphic arts. [Monsanto Co] *

Fomescol. Nonionic surfactant. [Rhone-Poulenc UK]

fomitine. A liquid extract from the fungi *Fomes cinn'amomeus.*

Fomox®. Intumescent one- and two-component compounds for use as flame retardants in civil engineering, marine engineering, automotive engineering, and industrial construction. [Bayer AG]

Fomrez. Polyesters for flexible PU foam and PU elastomers. [Baxenden Chemicals Ltd]

Fomrez® 50. Trimethylol propane branched adipate polyester; used for flexible cellular PU foam for textile and industrial applications, dispersing agents, coatings and adhesives, elastomers. [Witco/Organics] *

Fomrez® 4393. Polyfunctional specialty polyether; used as semirigid foam crosslinkers, rigid foam additives, precursors for reactive diluents and chain extenders in urethane synthesis. [Witco/Organics]

Fomrez® A1228. Primary polyether triol based on glycerin; used in the mfg. of one-shot, high-resilient molded and slab foam, RIM, microcellular, and shoe sole systems. [Witco/Organics] *

Fomrez® ED400. Polyether diol; used to produce PU coatings, PU prepolymers for elastomers, adhesives, caulks, sealants, and dispersing agents. [Witco/Organics] *

Fomrez® EPD28. Polyether primary diol; faster reaction than conventional prepolymer grade diols; allows production of high vol. intricate elastomer parts where rapid mold turnover is required; also suggested as partial or total replacement for the polytetramethylene glycols. [Witco/Organics] *

Fomrez® ET190. Polyether triol; used in the mfg. of PU adhesives, coatings, foam, and elastomers; also used as crosslinkers and pigment dispersing vehicles. [Witco/Organics] *

Fongarid. CAS 57646-30-7; Furalaxyl; fungicide for ornamentals. [Ciba-Geigy Agrochemicals]

Fonoline® White, Yellow. Petrolatum USP; soft, low melting point for consumer use as petrol. jelly, ointments, industrial applications; as emollient, protective coating, binder, carrier, lubricant, moisture barrier, plasticizer, protective agent, softener. [Witco/Sonneborn] *

Fontaine's powder. An explosive consisting of potassium picrate and potassium chlorate.

foots. (foots oil) Matter deposited by oils on standing.

Foraflon® 1000 HD. CAS 24937-79-9; PVDF semicryst. homopolymer; thermoplastic for general use for extrusion and injection molding applications in the chemical engineering, pkg., and elec. wiring fields and for protection of other materials, e.g., paints, plastic films. [Atochem] *

Foral 85, 105 and AX. A range of hydrogenated rosin and rosin esters; tackifiers and polymer-modifying resins in adhesives and in hot-melt-applied decorative, pressure sensitive and heat-sealable coatings. [Hercules]

Foralkyl. Perfluoro alkyl iodides and derivatives. [Atochem UK Ltd]

Forane. CAS 26675-46-7; Isoflurane, an inhalation anesthetic. [Atochem SA]

Forane. Chlorinated fluorocarbons; aerosol propellant, foam blowing agent and refrigerant. [Pacific Chemical Industries Pty Ltd]

Foraperle. Leather finishing agents. [ICI Chem. & Polymers Ltd] †

Forbes metal. An alloy of 53.5% zinc and 46.5% copper.

Forbest 13. Polyester/fatty alkyls compound; wetting and dispersing agent for water-based varnishes, highly pigmented slurries; binding agent. [Lucas Meyer]

Forbest 50. Modified aryl-alkyl silicones; antifloating and antisilking agent for solv. systems, curtain coats. [Lucas Meyer]

Forbest 62B. Based on hydrocarbon polymers; slip agent to improve smoothness and abrasion resistance; recommended for pigmented air-dry and stoving enamels, clear coatings. [Lucas Meyer]

Forbest 410. CAS 9002-88-4; Polyethylene compound with reactive groups; leafing and stabilizing agent for leafing aluminum pigments; recommended for air-dry and stoving paints. [Lucas Meyer]

Forbest 780. Carboxylic reaction product; catalyst for gel formation in offset and heat-set printing inks, hydrocarbon-based structural varnishes. [Lucas Meyer]

Forbest 1000B. Fatty acid/paraffin oil blend; defoamer for solv. and aq. systems; used for solvent-based air-dry and stoving finishes, aq. systems, dip-coatings, emulsion paints, varnishes. [Lucas Meyer]

Forbest MW 23. CAS 63231-60-7; Microcryst. wax finely dispersed in C17 fatty acid; increases slip and abrasion resistance in clear and pigmented stoving enamels, e.g., industrial paints, coil-coating finishes, printing inks (flexo, letterpress). [Lucas Meyer]

Forbest WP. High molecular weight synthetic compound; wetting, dispersing, and antisettling agent for all air-dry and stoving solvent-based paints and systems with polar solvs. [Lucas Meyer]

Force. A malt preparation.

Forceval. A proprietary preparation of vitamins and minerals. [Unigreg Ltd]

Forcite. A trademark for various types of explosives. §

Fordath Resins. A proprietary range of phenolic and urea resins used in foundry work. [Fordath Engineering Co] ‡

Forest Bark. Chipped, ground or composted bark for use as

---

‡ Trade name and manufacturer not verified   § Trade name without identified manufacturer

a soil conditioner, planting aid or mulch. [ICI Garden Products]

For-Ester. CAS 1702-17-6; 2,4-D; translocated herbicide for cereals and grass. [Synchemicals Ltd]

Forex®. PVC foam panels, slightly expanded with density 700 kg/m^3, and 450 kg/m^3; possible applications include fair stands and exhibition booths, screen printing, signs and displays, lightweight construction in automotive and aviation industries and wall coverings for wet rooms. [Airex AG]

Forhistal Maleate. CAS 3614-69-5; Dimethindene maleate; antihistaminic. [Ciba-Geigy Corp] *

Forit. CAS 153-87-7; Oxypertine; antidepressant. [Sterling Drug Inc] *

Forlan. Petrolatum, lanolin alcohol, lanolin; absorption base, emollient. [RITA]

Forlan 200. Petrolatum, lanolin alcohol; absorption base for personal care products; enhance stability of emulsions, dispersions, and suspensions; epidermal emollient, moisturizer, lubricant. [RITA]

Forlan C-24. CAS 27321-96-6; Choleth-24; emulsifier, emulsion stabilizer, emollient, moisturizer, solubilizer, visc. modifier, pigment dispersant, plasticizer for cosmetics, pharmaceuticals. [RITA]

Forlan L; Synthetic lanolin; an ingredient that qualitatively approximates the composition of lanolin; enhances emulsion stability in both o/w and w/o systems, assists in the wetting and dispersion of pigments in facial makeup, lipstick and eye shadow preparations. [RITA] *

Forlan LM. Synthetic lanolin; An ingredient that qualitatively approximates the composition of lanolin. It enhances emulsion stability in both o/w and w/o systems, assists in the wetting and dispersion of pigments in facial makeup, lipstick and [RITA] *

Forlanit P. Sodium laureth phosphate; wetting agent, dispersant, flotation auxiliary for metal cleaners, galvanic baths, hot copper baths containing cyanides. [Henkel/Cospha; Henkel/Functional Prods.; Henkel Canada; Henkel KGaA]

Forlay. Selective weed killers. [Murphy Chemical Co Ltd] †

Formac 40. A solution polymer based on acrolein and formaldehyde with an active ingredient content of 40% by weight; used to control and eliminate algae, fungi and bacteria from industrial water. [Degussa Ltd]

Formacel®. For the chemical industry. [DuPont UK]

Formadermine. Guaiacol methylene ether.

formagen. A dental cement consisting of two parts : a) A liquid containing creosote, phenol, olive oil, and alcoholic formalin, and b) a powder consisting of aluminum silicate, magnesium and zinc carbonates, and lime.

formaldehyde. (oxymethylene; formalin; formic aldehyde; aldehyde C$_1$; methanal) HCHO; CAS 50-00-0; EINECS 200-001-8; Urea and melamine resins, polyacetal resins, phenolic resins, fertilizers, preservatives, reducing agent, corrosive inhibitor. [Aqualon; Du Pont; Farleyway Chem. Ltd; Georgia-Pacific Resins; Hoechst Celanese; Mallinckrodt; Mitsubishi Gas; Monsanto]

formaldehyde-ammonia 6 : 4. See hexamine.

formalin. (formol; formol-chloral). A 40% aqueous solution of formaldehyde which may contains 15% methyl alcohol to prevent the separation of polymerized compounds.

Formalite. A trade name for phenol-formaldehyde resin molded material for use in electrical insulation. §

formalith. A formaldehyde solution.

formamide. (amide C$_1$; methanamide) HCONH$_2$; CAS 75-12-7; EINECS 200-842-0; solvent, softener, intermediate in organic synthesis. [Aldrich; BASF; Fluka; Penta Mfg.]

formammidinesulfinic acid. See thiourea dioxide.

forman. Chloromethyl menthyl ether, C$_{10}$H$_{19}$O · CH$_2$C1.

Formanek's indicator. Alizarin green used as an indicator. It gives a violet color with pH 0.3, pink with pH .0, yellow with pH 12, and brown with pH 14.

formaniline. It is anhydro-formaldehyde aniline; a rubber vulcanization accelerator.

Formapex. A proprietary phenol-formaldehyde resin varnish. §

Formasal. A condensation product of formaldehyde and salicylic acid. §

Format. CAS 57754-85-5; Herbicide containing clopyralid. [ICI Chem. & Polymers Ltd]

Formax. CAS 141-53-7; Sodium formate. [May & Baker Ltd]*

Formel NF. A mixture of Isceon chlorofluorocarbons and chloromethanes; totally nonflammable; dielectric fluid for use in distribution transformers. [Rhone-Poulenc UK]

Formex. A proprietary trade name for an enamelled wire; the enamel is flexible and heat resisting up to about 185 C; contains a mixture of polyvinyl acetal and a thermosetting phenol-formaldehyde synthetic resin. §

Formica. A proprietary trade name for phenolic and urea resins. §

formic acid. (hydrogen carboxylic acid; methanoic acid) Organic acid; HCOOH; CAS 64-18-6; EINECS 200-579-1; Reducing agent, dyeing and finishing of textiles, leather treatment, chemicals, manufacture of fumigants, insecticides, refrigerants, solvs. for perfumes, lacquers; electroplating; silvering glass; ore flotation. [BASF; BP Chem. Ltd; Hoechst Celanese; Mallinckrodt; Norsk Hydro AS]

formic acid sodium salt. See sodium formate.

formic aldehyde. See formaldehyde.

formin. See hexamine.

Formit. See Bakelite. §

Formite. See Bakelite. §

Formitrol. A formaldehyde preparation.

Formkote. Dry film lubricant coating containing graphite and suitable binder system; high temperature titanium metal forming lubricant. [E/M Corporation]

Formodac. Chemical products for use as additives in the manufacture of concrete. [BP Chemicals Ltd]

formol. See formalin.

Formol 55. Urea formaldehyde precondensate; for further condensation with urea, melamine, phenol, or furfuryl alcohol for adhesives and foundry binders. [BASF AG]

formol-chloral. See formalin.

formolide. An antiseptic consisting of an aqueous solution of 15% alcohol, 2% boric acid, 4% sodium benzoate, and 1% formaldehyde.

formolites. Phenol-formaldehyde resins.

formolyptol. A formaldehyde preparation.

Formon®. Solder and braze compositions; for the electrical industry. [DuPont UK]

Formosa camphor. Camphor from China.

formose. A mixture of sugars obtained from formaldehyde by polymerization.

Formrez. Polyesters for flexible PU foam and PU elas-

---

* Trade name not verified as available   † Trade name verified as obsolete

tomers. [Baxenden Chemicals Ltd] †

Formrez®. A registered trade name for a polyester-based polyurethane elastomer cross-linked with a diamine. [Witco/Organics] *

Formrez® 11. Linear poly (diethylene adipate) polyester; used in PU industry for mfg. of prepolymers (isocyanate or hydroxyl-terminated), thermoplastic elastomers, coatings and adhesives, dispersing agents, microcellular shoe sole systems, millable gums, cast elastomers. [Witco/Organics] *

Formrez® L49-28. Primary polyether triol based on glycerin; for mfg. of one-shot urethane elastomers, RIM, microcellular, and shoe sole systems. [Witco/Organics]*

Formrez® T-279, T-280. Polyether triol based on trimethylol propane; used as urethane crosslinkers, precursors for reactive diluents, epoxy and melamine cure diluents. [Witco/Organics] *

Formula 90. Squash court plaster. [Prodorite Ltd]

Formula 111. Heavy-duty solv. for strong cleaning/degreasing on interfaces, metal contacts, electro-mech. assemblies, PCB connectors. [Chemtronics]

Formula 405. Pregnenolone succinate; nonhormonal sterol derivative. [Doak Pharmacal Co] ‡

Formula AC. Self-emulsifying mineral oil lubricant for wire drawing; gives superior antitarnish props. [Eastern Color & Chem.]

Formula S. Squash court plaster. [Prodorite Ltd]

Formusol®. CAS 149-44-0; Sodium formaldehyde sulfoxylate; used for textile printing, emulsion polymerization. [RV Chemicals Ltd]

Formusol® SA. Special acetaldehyde sulfoxylate. [RV Chemicals Ltd] †

Formvar. Polyvinyl formal resins; suitable for formulating structural adhesives, wash primers, can and drum linings and wood and knot sealers. [Monsanto Co]

Fornax. Finishing agents. [Ciba plc] *

fornitrol. A reagent used for the estimation of nitric acid, nitro-compounds, and nitrates.

Forociben Premix. Sulfonamide animal feed additive. [Ciba plc] *

Foroid. A proprietary preparation of sodium tetraiodophenolphthalein. §

Foron. Synthetic organic dyestuffs for use in textiles. [Sandoz Products Ltd]

Fortafix. A range of high temperature resistant adhesive cements and sealing compounds; used for bonding, sealing or insulating inorganic materials where heat resistance up to 1600 C is required. [Fortafix Ltd] *

Fortagesic. A proprietary preparation of pentazocine and paracetamol; an analgesic. [Sterling Research Laboratories] *

Fortex. Flavors for confectionery, foodstuffs, and beverages. [Bush Boake Allen Ltd]

Fortex®. CAS 63231-60-7; Hard microcryst. wax consisting of n-paraffinic, branched paraffinic, and naphthenic hydrocarbons; used in hot-melt coatings and adhesives, paper coatings, printing inks, plastic modification (as lubricant and processing aid), lacquers, paints, and varnishes, as binder in ceramics, for potting, filling, and impregnant; in elec./electronic components, in investment casting, rubber and elastomers (plasticizer, antisunchecking, antiozonant), as emulsion wax size, as fabric softener ingred., in emulsion and latex coatings, and in hand creams and lipsticks. [Petrolite]

Fortiflex®. High density polyethylene (HDPE); used for packaging, blow molded and injection molded containers, film for a wide variety of materials including foodstuffs; automotive and industrial parts; extruded pipe for potable water and natural gas transport; sheet industrial parts. [Solvay Polymers]

Fortiflex® A60-70-99, A60-70-119. CAS 9002-88-4; HDPE; blow molding grade with excellent stiffness and processability on reciprocating screw equip. for milk, water, juice, and fountain syrup bottles. [Solvay Polymers]

Fortiflex® B45-06R-09. CAS 9002-88-4; HDPE copolymer; high molecular weight ammunition grade for shot shell tubes. [Solvay Polymers]

Fortiflex® G36-24-149. CAS 9002-88-4; HDPE; sheet extrusion and thermoforming grade with excellent ESCR and thermal performance; for geomembranes. [Solvay Polymers]

Fortiflex® G38-70C. CAS 9002-88-4; HDPE; blk. conc. for use in NSF pipe grade resins. [Solvay Polymers]

Fortiflex® J36-25-142. CAS 9002-88-4; HDPE; film grade with high impact strength and tear; for blown film merchandise bags. [Solvay Polymers]

Fortiflex® K36-55-122. CAS 9002-88-4; HDPE; pipe grade; natural base resin suitable for pipe when blended with color conc. [Solvay Polymers]

Fortiflex® T50-200. CAS 9002-88-4; HDPE; injection molding grade with superior toughness and ESCR; for industrial parts. [Solvay Polymers]

Fortiflex® XF-855. CAS 9002-88-4; HDPE copolymer containing 25% recycled polyethyene; blow molding grade for applications requiring moderate ESCR. [Solvay Polymers]

Fortilene®. Polypropylene (PP); used for fiber applications including nonwovens and slit-tape; packaging, particularly injection molded containers, caps and closures, and blow molded bottles, film, industrial and automotive parts. [Solvay Polymers]

Fortilene® 1001. CAS 9003-07-0; PP homopolymer; extrusion grade resin with excellent thermal stability for sheet, strapping, profile extrusion. [Solvay Polymers]

Fortilene® 1602. CAS 9003-07-0; PP homopolymer; med. flow resin with good thermal stability for conventional injection molding of caps and closures, all purpose molding. [Solvay Polymers]

Fortilene® 1802. CAS 9003-07-0; PP homopolymer, antistat; controlled rheology resin with high flow, good impact for injection molding of caps and closures, thin-wall containers. [Solvay Polymers]

Fortilene® 2104. CAS 9003-07-0; PP homopolymer; film resin with good stretching and gauge control for bioriented film applications. [Solvay Polymers]

Fortilene® 3151. CAS 9003-07-0; PP homopolymer; fiber/filament grade for slit film, carpet backing applications. [Solvay Polymers]

Fortilene® 4104, 4109. CAS 9003-07-0; PP copolymer, antistat; blow molding grade with good clarity for hot-fill containers. [Solvay Polymers]

Fortilene® 4209. CAS 9003-07-0; PP copolymer; film grade with good processability, excellent clarity for biaxially oriented film. [Solvay Polymers]

Fortilene® 5801. CAS 9003-07-0; PP copolymer; injection molding resin with good impact for specialty closures, medical applications. [Solvay Polymers]

---

‡ Trade name and manufacturer not verified   § Trade name without identified manufacturer

Fortilene® 9000. CAS 9003-07-0; PP homopolymer; unstabilized sphere for compounding, automotive applications. [Solvay Polymers]

Fortimax. Reinforcing fillers for synthetic elastomers. [ICI Chem. & Polymers Ltd]

Fortisan. A proprietary trade name for a synthetic fiber made from regenerated cellulose. §

Fortral. CAS 64024-15-3; A proprietary preparation of pentazocine hydrochloride; an analgesic. [Sterling Research Laboratories] *

Fortress. Heat-resistant glass-fiber tissues. [ICI Chem. & Polymers Ltd]

Fortrex. An activated clay for reinforcing natural and synthetic rubber. [Croxton & Garry Ltd] *

Fortrol. CAS 21725-46-2; Suspension concentrate containing 500 g cyanazine per liter; a triazine herbicide. [Shell UK]

Fortron® 0205B4. CAS 9016-75-5; PPS resin; high temp. polymer with excellent thermal and elec. props., good solv. resist.; used for connectors, switches, coil bobbins, vapor phase soldering, automotive applications (engine compartment components, fuel line systems), applications for high temp. and corrosive environments (caps, gears, fittings, industrial parts); 0205B4 is a low visc., unfilled powd. [Hoechst Celanese/Engineering Plastics; Hoechst AG]

Fortunan. CAS 52-86-8; Haloperidol; pharmaceutical preparation for the treatment of schizophrenia. [M A Steinhard Ltd] *

Fosalsil. A proprietary trade name for a natural diatomaceous material made into bricks or used as a cement. §

Foscast. Range of preformed refractory shapes. [Foseco (F.S.) Ltd]

Foset. Acid-catalysed phenolic resin systems for cold-set bonding of sand cores. [Foseco (F.S.) Ltd] †

Fosfamide CPD-170. Complex alkanolamide phosphate ester; emulsifier for preparation of metalworking fluids and greases; imparts lubricity, anticorrosion and mild EP props.; grease additive. [Henkel/Cospha; Henkel Canada]

Fosfamide N. Fatty amido phosphate complex; detergent, foaming and wetting agent for textile, all-purpose, institutional, and metal cleaners. [Henkel/Cospha; Henkel Canada]

Fosferno. CAS 56-38-2; Parathion insecticide. [ICI Chem. & Polymers Ltd]

Fosfil. High alumina refractory material. [Foseco (F.S.) Ltd]

Fosfor. A proprietary preparation of phosphorylcolamine; a tonic. [Consolidated Chemicals Ltd] *

Fosfostilben. See Honvan. [Degussa AG] *

Foshell. Liquid phenol-formaldehyde novalac resins for coating sand by the warm process. [Foseco (F.S.) Ltd]†

Fosoil. Oil-based binders for cold-set bonding of sand cores. [Foseco (F.S.) Ltd] †

fossil flour. See infusorial earth.

fossil salt. Rock salt.

fossil wax. See ozokerite.

Fostap. Preformed highly insulating launder liners. [Foseco (F.S.) Ltd]

Fosterge LF. Alkylphenoxy POE acid phosphate; intermediate for foaming cleaners and solv. emulsion cleaners; emulsifier for pesticides; corrosion inhibitor. [Henkel/Cospha; Henkel Canada]

Fosterge R. Mono and dialkyl phosphoric acid; intermediate

for emulsifying, penetrating, and anticorrosion compounds. [Henkel/Cospha; Henkel Canada]

Fostex AMP. CAS 6419-19-8; EINECS 229-146-5; Amino trimethylene phosphonic acid; scale inhibitor and sequestrant in water treatment. [Henkel]

Fostex P. CAS 2809-21-4; Etidronic acid; scale inhibitor and sequestrant for water treatment. [Henkel] *

Fotofax. CAS 1314-13-2; Zinc oxide, high purity; for varistors, reprographics. [Manchem Ltd] *

fotosensin. A condensation product of phthalic acid and resorcinol containing small proportions of copper and iron. Small quantities increase the root and stem growth of plants.

Fototar. Coal tar; anti-eczematic. [ICN Pharmaceuticals Inc]

Fouadin. CAS 15489-16-4; Stibophen; trivalent antimony compound for the treatment of schistosomiasis. [Bayer AG] †

Fouane. Benzthiazide.

Foulagan. Wool fulling agent. [Ceca SA]

Foulon. Paper dyes. [ICI Chem. & Polymers Ltd] †

Foundrox. CAS 1309-37-1; Fe_2O_3; foundry grade red iron oxide for core and mold use in eliminating veining and other casting expansion defects. [DCS Color & Supply Co Inc] *

foundry clay. A clay containing from 80-90% silica and 15-18% alumina.

Foundry pattern metal. An alloy of 75% zinc and 25% tin.

Fouramine. Leather dyes. [ICI Chem. & Polymers Ltd]

Fourdrinier wire. A brass containing from 80-85% copper, and 15-20% zinc.

Four Sure Liquid. Foot-rot treatment compound for sheep. [Charles Tennant & Co Ltd]

Fovane. CAS 91-33-8; A proprietary preparation of benzthiazide; a diuretic. [Pfizer International] †

Foxstar. A liquid formulation containing 107 g bifenox, 286 g isoproturon and 143 g mecoprop per liter as a suspension concentrate; a post-emergence herbicide for the control of weeds in winter cereals. [Rhone-Poulenc Crop Protection Ltd]

FP-Vac. Fowl pox vaccine; for immunization of poultry. [Intervet Inc]

FR-11. CAS 12124-97-9; Ammonium bromide; flame retardant for paper, chipboard, and nonwashable textiles. [AmeriHaas; Dead Sea Bromine]

FR-20. CAS 1309-42-8; Magnesium hydroxide; flame retardant and smoke suppressant for ABS, PP, PS, rubbers. [AmeriHaas; Dead Sea Bromine]

FR 28. Flame retardant for cellulose. [U.S. Borax & Chem.]

FR-513. CAS 36483-57-5; Tribromoneopentyl alcohol; flame retardant for flexible and rigid PU; flame retardant intermediate. [AmeriHaas; Dead Sea Bromine]

FR-522. CAS 3296-90-0; Dibromoneopentyl glycol; flame retardant for unsat. polyesers, rigid PU and foams; flame retardant intermediate. [AmeriHaas; Dead Sea Bromine]

FR-612. CAS 615-58-7; Dibromophenol; flame retardant for epoxy resins, phenolic resins, polyester resins; flame retardant intermediate. [AmeriHaas; Dead Sea Bromine]

FR-613. CAS 118-79-6; 2,4,6-Tribromophenol; reactive flame retardant used mainly as an intermediate for polymeric flame retardants. [AmeriHaas; Dead Sea Bromine]

FR-705. CAS 87-83-2; Pentabromotoluene; flame retardant

---

* Trade name not verified as available   † Trade name verified as obsolete

for unsat. polyesters, polyethylene, PP, PS, SBR latex, textiles, rubbers. [AmeriHaas; Dead Sea Bromine]

FR-913. CAS 3278-89-5; Tribromophenyl allyl ether; aromatic flame retardant for expandable PS; synergist with hexabromocyclododecane. [AmeriHaas; Dead Sea Bromine]

FR-1025. CAS 594477-57-3; Poly (pentabromobenzyl) acrylate; polymeric flame retardant for engineering thermoplastics, PET, PBT, nylon, PP, and PS. [AmeriHaas; Dead Sea Bromine]

FR-1034. CAS 109678-33-3; Tetrabromodipentaerythritol; flame retardant for PP extruded fibers; processing aid for ABS and HIPS. [AmeriHaas; Dead Sea Bromine]

FR-1205. CAS 32534-81-9; Pentabromodiphenyl oxide; flame retardant for use in laminates (both epoxy and phenolic), unsat. polyesters, synthetic fibers, and flexible PU foams, suitable for textiles. [AmeriHaas; Dead Sea Bromine]

FR-1206. CAS 25637-99-4; Hexabromocyclododecane; fire retardant for wide range of plastics, textiles, adhesives, and coatings; esp. for styrene-based systems. [AmeriHaas; Dead Sea Bromine]

FR-1208. CAS 32536-52-0; EINECS 251-087-9; Octabromodiphenyl oxide; flame retardant for thermoplastics, e.g., ABS, HIPS, LDPE, PP random copolymer; recommended for injection moldings. [AmeriHaas; Dead Sea Bromine]

FR-1210. CAS 1163-19-5; EINECS 214-604-9; Decabromodiphenyl oxide; flame retardant used in thermoplastics and fibers, incl. HIPS, glass-reinforced thermoplastic polyester molding resins, LDPE extrusion coatings, PP (homo and copolymers), ABS, nylon, PBT, PET, PU, SBR latex, textiles, rubber. [AmeriHaas; Dead Sea Bromine]

FR-1360. A proprietary flame-retardant material comprising tribromoneopentyl alcohol used in the production of flexible polyurethane foam. [Dow UK] *

FR-1524. CAS 79-94-7; EINECS 201-236-9; Tetrabromobisphenol-A; reactive flame retardant used in the mfg. of epoxy, PC, ABS, phenolic, PS, and polyester resins, rubber; flame retardant intermediate. [AmeriHaas; Dead Sea Bromine]

FR-2124. Tetrabromobisphenol-A allyl ether; flame retardant for expandable PS. [AmeriHaas; Dead Sea Bromine]

FR-2406. Tris (2,3-dibromopropyl) phosphate; a proprietary fire-retardant additive used in acrylics, epoxies, latices, phenolics, polyesters, polystyrenes, polyvinyl chloride, rayon celluloses and polyurethanes. [Dow UK] *

Fracton. A refractory dressing. [Foseco (F.S.) Ltd]

Fractorite. An explosive containing 90% ammonium nitrate, 4% resin, 4% dextrin, and 2% potassium dichromate.

Fractorite B. An explosive consisting of 75% ammonium nitrate, 2.8% dinitro-naphthalene, 2.2% ammonium oxalate, and 20% ammonium chloride.

Fragarol. (Fragasol). The butyl ether of β-naphthol; a synthetic perfume.

Fragaroma. Perfumery products. [May & Baker Ltd] *

Fragasol. See Fragarol.

Fraissite. Benzyl iodide, $C_6H_5 \cdot CH_2 \cdot I$.

Framycort. A proprietary preparation of framycetin sulfate and hydrocortisone; used in dermatology as an antibacterial agent. [Fisons plc, Pharmaceuticals Div] *

Framygen. A proprietary preparation of framycetin sulfate;

a systemic antibiotic. [Fisons plc, Pharmaceuticals Div]*

Framyspray. An antibiotic aerosol. [Fisons plc, Pharmaceuticals Div] *

Francolor. Dyes, pigments and chemical auxiliaries. [ICI Chem. & Polymers Ltd]

frankincense. Olibanum, a gum resin.

Frankonite. (silitonite, tonsil). German bleaching earths obtained from deposits in Germany.

Franocide. CAS 1642-54-2; A proprietary preparation of diethylcarbamazine citrate used in veterinary work. [Coopers Animal Health Ltd] *

Franol. A proprietary preparation of theophylline, phenobarbitone, and ephedrine hydrochloride; a bronchial antispasmodic. §

Franol Plus. A proprietary preparation of throphylline, phenobarbitone, ephedrine sulfate, and thenyldiamine hydrochloride; used as a bronchospasm relaxant. [Winthrop Laboratories] *

Frantln. A proprietary anthelmintic given to unweaned lambs. [Coopers Animal Health Ltd] *

Frary metal. A calcium-barium-lead alloy containing up to 2% barium and 1% calcium; used as a bearing metal. It melts at 445 C.

Fraude's reagent. Perchloric acid, $HClO_4$.

Fredo. Calcium hydrosulfite.

Freeflo. Industrial cleaning compound. [The Wellcome Foundation Ltd] *

Freeman's nonpoisonous white lead. A pigment. It is a mixture of white lead, zinc white, baryta white, and magnesium carbonate.

Freeteem. Fluxes for uphill teemed killed steel ingots. [Foseco (F.S.) Ltd]

Freez-Gard® Formula FP-88E. Sodium hexametaphosphate, glassy, sodium chloride, and sodium erythorbate; food additive for meat, poultry, and seafood industries. [Rhone-Poulenc Food Ingreds.]

Freezine. See Preservaline.

Fre-Flex. Focofilcon A; 2-hydroxyethyl methacrylate polymer with methacrylic acid; contains 55% water; contact lens material. [Optech Inc] *

Frekote® 33. Releasing interface; aerosols and bulk. [Dexter/Frekote]

Frekote® 44-NC. Nonchlorofluorocarbon version of Frekote 44 [Dexter/Frekote]

Frekote® 800. Release agent for nat. and synthetic org. rubber compounds; for use on molds at temps. above 150 C. [Dexter/Frekote]

Frekote® EXITT®. Silicone-based; release agent for urethane elastomers, compression-molded resins, org. rubbers, EPDM rubber, and specialty molded resins; aerosol and bulk. [Dexter/Frekote]

Frekote® No. 1. Fluorocarbon; release agent for acrylics, epoxies, nylons, phenolics, polycarbonates, polypropylenes, polystyrenes, and rubber; applications incl. epoxy potting and encapsulating, prepreg fabricating, filament winding; dry lubricant for plastic gears, belts, threaded connectors, rubber, glass, wood, metals; aerosol and bulk. [Dexter/Frekote]

Frelen. Crosslinked polyethylene foam. [Carl Freudenberg]*

Fremy's salt. Potassium hydrogen fluoride, $KF \cdot HF$.

French cement. A mucilage of gum arabic mixed with powdered starch; used by naturalists, artificial flower makers, and confectioners.

French chalk. A variety of Steatite or Soapstone (qv). It is a

---

‡ Trade name and manufacturer not verified   § Trade name without identified manufacturer

hydrated silicate of magnesium, and is used for marking cloth, removing grease from silk, as a filler, and for other purposes.

French pine resin. *See* gum thus.

French polish. A polish for wood. It consists of shellac dissolved in alcohol.

French turpentine. (Bordeaux turpentine). The oleo-resin from *Pinus maritima*. It contains from 15-20% essential oil and 70-80% rosin.

French verdigris. Basic copper acetate.

French white. Powdered talc.

Frenokone. A selective weedkiller. [Plant Protection] *

Freon®. Fluorocarbons. [DuPont UK]

Frescile. 3-Methyl dodecanitrile. [Quest Int'l. UK Ltd]

Fresh Pak. Extra-thin polyethylene HD-film; for packaging or incorporation into laminates. [UCB nv Film Sector] *

Freudenberg Megulastik. Mechanical dampeners. [Carl Freudenberg] *

Freund's acid. 1-Naphthylamine-3: 6-disulfonic acid.

friable amber. *See* Gedanite.

Frick's alloys. Nickel silvers containing from 50-69% copper, 18-39% zinc, and 5-31% nickel.

friedelite. A mineral, Mn_2SiO_4.

Friedländer's stain. A microscopic stain. It contains 50 g of a saturated alcoholic solution of gentian violet, 10 g of glacial acetic acid, and 100 cc water.

Frigate. CAS 61791-44-4; Wetting agent containing 800 g/l tallow amine ethoxylate; spreader for use with phosphonoglycine herbicides. [Fermenta ASC Europe Ltd]

Frigen. Fluorinated hydrocarbon refrigerant. [Hoechst UK]

Frigesa® D, F, IC. Hydrocolloids; thickener and stabilizer for desserts, ice cream, dressings. [Grünau]

Frigid-Go® 2815, 2816. Moly low-temp. multipurpose synthetic grease. [Henkel/Emery]

Frimulsion. Hydrocolloid blends used to improve stability, appearance, texture, and mouthfeel of food; for food and beverage industries. [Hercules] *

Frishmuth's aluminum solder. Alloys. One contains 67% tin, 27% lead, and 3% aluminum. Another consists of 94% zinc, 4% aluminum, and 2% copper.

Frishout solder. An alloy of 46% tin, 23% zinc, 15% aluminum, 8% copper, and 9% silver.

Fritzsche's reagent. Dinitroanthraquinone, $C_{14}H_6N_2O_6$.

Froben. Potent nonsteroidal anti-inflammatory and analgesic agent. [The Boots Co plc]

Frodingham G.G.B.S. Ground granulated blastfurnace slag (GGBFS) for blending with Portland cement in the concrete mixer to produce Portland blastfurnace cement concrete; used for all classes and types of concrete to give long term durability; provides high chemical and sulfate resistance, low heat properties and reduces the risk of alkali aggregate reaction when required. [Frodingham Cement Co Ltd]

Frohde's reagent. An alkaloid reagent. It consists of 0.5 gram of sodium molybdate dissolved in 100 cc of concentrated sulfuric acid.

Frosted silver. *See* Dead silver.

Frost-Off. Antifreeze liquid; deicing fluid for automobile windshields and door locks. [Merix]

Frother 4171. A terpene-alcohol based flotation reagent; produces a more brittle froth than typical pine oil; for flotation of nonmetallic minerals. [Hercules] *

fructose. (levulose; D-fructose) Sugar occurring in fruit and honey; $C_6H_{12}O_6$; CAS 57-48-7 (D-form); EINECS 200-333-3; Used as a sweetener. [Am. Maize Prods.; Corn Prods.; Laevosan GmbH; Pfantsiehl Labs; A.E. Staley Manufacture]

Fruitonile. 2-Methyl decanitrile. [Quest Int'l. UK Ltd]

fruit sugar. *See* levulose.

Fruit Tree Grease. Mineral grease; banding grease to protect trees from wingless moths and other crawling insects. [Vitax Ltd]

Frumil. Frusemide and amiloride; for management of edema, in congestive cardiac failure, nephrosis, corticosteroid therapy, estrogen therapy, ascites associated with cirrhosis. [Berk Pharmaceuticals Ltd] *

Frusid. A proprietary preparation of frusemide; a diuretic. [DDSA Pharmaceuticals Ltd] *

Fruvit®. CAS 12071-83-9; Propineb; fungicide for control of downy mildews. [Bayer AG] *

FS Bordeaux Powder. Copper sulfate/lime complex; a protectant fungicide for the control of blight and canker. [Ford Smith & Co Ltd]

FS Derris. CAS 83-79-4; Emulsifiable concentrate or dustable powder containing rotenone; a contact insecticide for fruit, vegetables and greenhouse crops. [Ford Smith & Co Ltd]

FS Dricol 50. CAS 1332-40-7; Copper oxychloride; a protectant fungicide. [Ford Smith & Co Ltd]

FS Thiram 15% Dust. CAS 137-26-8; Thiram; fungicide with animal repellent properties. [Ford Smith & Co Ltd]

Fubol. Wettable powder containing mancozeb and metalaxyl; protectant fungicide for field crops and fruit. [Ciba-Geigy Agrochemicals]

Fucidin. A proprietary preparation containing sodium fusidate; an antibiotic. [Leo Laboratories] *

Fucidin V.P. A proprietary preparation of sodium fusidate and phenoxymethyl penicillin; an antibiotic. [Leo Laboratories] *

Fucidin-H. A proprietary preparation of sodium fusidate and hydrocortisone acetate; used in the treatment of dermatosis. [Leo Laboratories] *

Fudow. A proprietary range of phenolic molding materials. [Fudow] *

Fudowlite U. A proprietary range of urea formaldehyde molding materials. [Fudow] *

Fuelsaver®. CAS 2224-44-4, 1854-23-5; 4-(2-Nitrobutyl)morpholine, 4,4´-(2-ethyl-2-nitrotrimethylene) dimorpholine; preservative for diesel and other hydrocarbon fuels. [Angus]

Fugata. A proprietary preparation of ticonazole; an antifungal. [Pfizer International]

Fulacolor. Reactive clay used for carbonless copy paper. [Laporte Industries Ltd]

Fulbent. CAS 1302-78-9; Bentonite clay for binding, suspending and emulsifying. [Laporte Industries Ltd] *

Fulbond. A bonding agent for foundry sands. [Laporte Industries Ltd]

Fulbond. CAS 1302-78-9; Foundry bentonite. [Minas de Gádor SA]

Fulcat Catalysts. CAS 1318-93-0; Acid treated montmorillonite clays; catalysts for a range of organic reactions notably the alkylation of phenols. [Laporte Industries Ltd]

Fulcin. CAS 126-07-8; A proprietary trade name for a preparation of griseofulvin BP; an antifungal agent. [ICI Chem. & Polymers Ltd]

Ful-Glo. CAS 518-47-8; Fluorescein sodium; diagnostic aid.

---

* Trade name not verified as available   † Trade name verified as obsolete

[Barnes-Hind Inc] *

Fulgurite. An explosive containing 60% nitroglycerin and 40% wheaten flour and magnesium carbonate.

fuligo. Soot.

Fullasorb. Absorbent earth granules. [Laporte Industries Ltd]

Fullerite. A proprietary trade name for a slate powder used as a rubber filler. §

fuller's earth. A term applied to a sandy loam or argillaceous earth found in Surrey and Kent. It consists of aluminum and magnesium hydrosilicates. A deodorizer; used for clarifying oils, and used in cosmetics, rubber compounds, carrier for catalysts, filtering medium.

Fulmargin. A solution of colloidal silver.

Fulmenit. An explosive consisting of 86.5% ammonium nitrate, 5.5% trinitrotoluene, 2.5% paraffin oil, 1.5% charcoal, and 4% guncotton.

fulminating gold. A compound having the formula $2AuN_2H_3.H_2O$, prepared by the action of concentrated ammonia on gold hydroxide. It is explosive.

fulminating platinum. Explosive compounds formed by acting upon ammonium platinochloride with potassium hydroxide

fulminating silver. A compound of nitrogen and silver prepared by the action of ammonia on precipitated silver oxide. It is explosive.

Fulmont. Used for refining of vegetable, animal and mineral oils, fats and waxes. [Minas de Gádor SA]

Fulmont Activated Bleaching Earths. CAS 1318-93-0; Acid activated montmorillonite clays; color and impurity removal from edible oils, re-refining or purification of mineral oils. [Laporte Industries Ltd]

Fulton 404®. Acetal, 20% PTFE lubricant; thermoplastic compound for moving parts requiring excellent frictional properties, resistance to wear, and ease of fabrication; used in bearings and sliding-contact parts, e.g., gears, valve seats, stem packing, thrust washers, etc. in textile, printing, automotive, appliance, and furniture industries. [LNP]

Fulton White. A proprietary brand of lithopone (*qv*). §

Fulvicin-P/G . CAS 126-07-8; Griseofulvin; antifungal. [Schering Corp]

Fulvicin-U/F. CAS 126-07-8; Griseofulvin; antifungal. [Schering Corp] *

Fulvite. An artificial titanium monoxide.

fumagillin. An antibiotic produced by certain strains of *Aspergillus fumigatus*.

Fumarate 6-18. Diester of fumaric acid and C6-C18 alcohol; basic monomer for specialty polymers. [Henkel/Functional Prods.]

fumaric acid. HOOCCH:CHCOOH; CAS 110-17-8; EINECS 203-743-0; Modifier for polyester, alkyd and phenolic resins; paper sizing resins; plasticizers, rosin esters and adducts, alkyd resin coatings, upgrading natural drying oils, food additive, acidulant, flavoring agent, mordant, organic synthesis, inks. [Chemie Linz UK; Haarmann & Reimer; Lonza; Mitsubishi Gas; Monsanto; Schaefer Salt]

fumaric acid, dibutyl ester. *See* dibutyl fumarate.

Fume-Control. Fume suppressant for hydrochloric and sulfuric acids. [Crown Tech.]

Fumexol. Pretreatment agents. [Ciba plc] *

fuming nitric acid. Nitric acid containing some of the lower oxides of nitrogen.

fuming oil of vitriol. *See* fuming sulfuric acid.

fuming sulfuric acid. (oleum; nordhausen acid, pyrosulfuric acid) CAS 8014-95-7; It consists of sulfur trioxide dissolved in sulfuric acid. The commonest fuming acid contains 55% sulfuric acid, and 45% sulfur trioxide; used as a sulfating and sulfonating agent, dehydrating agent in nitrations, dyes, and explosives.

Fumite. General purpose insecticide in smoke form for greenhouse use. [ICI Garden Products]

Fumite Dicloran. CAS 99-30-9; Smoke fungicide (active ingredient dicloran); for use in enclosed areas against botrytis and rhizoctonia on protected crops. [Octavius Hunt Ltd]

Fumite Dicofol. CAS 115-32-2; Smoke acaricide (active ingredient dicofol); for use in enclosed areas against red spider and other mites on protected crops. [Octavius Hunt Ltd] †

Fumite Lindane. Smoke insecticide (active ingredient gamma HCH); broad spectrum insecticide for use in (enclosed) poultry, mushroom farms, ships' holds, public buildings, protected crops, against a wide range of pests. [Octavius Hunt Ltd]

Fumite Permethrin. CAS 52645-53-1; Smoke insecticide (active ingredient permethrin); for use in enclosed areas against whitefly and other pests of protected crops, cockroaches on stored produce, domestic insect pests. [Octavius Hunt Ltd]

Fumite Pirimiphos Methyl Smoke. CAS 29232-93-7; Smoke insecticide (active ingredient pirimiphos methyl); for use in enclosed areas against whitefly and other pests of protected crops, pests in grain silos etc. [Octavius Hunt Ltd]

Fumite Pirimiphos-Methyl Smoke. CAS 29232-93-7; Pirimiphos-methyl; contact fumigant and organophosphorus insecticide for greenhouse crops. [Octavius Hunt Ltd]

Fumite Propoxur. CAS 114-26-1; Smoke insecticide (active ingredient propoxur); for use in enclosed areas against whitefly and other pests of protected crops. [Octavius Hunt Ltd]

Fumite Ronilan. CAS 50471-44-8; Smoke fungicide (active ingredient vinclozolin); for use in enclosed areas against botrytis on protected crops. [Octavius Hunt Ltd]

Fumite TCNB. CAS 117-18-0; Smoke fungicide (active ingredient tecnazene); for use in enclosed areas against botrytis and mildew on protected crops. [Octavius Hunt Ltd] †

Fumite TCNB Smoke. CAS 117-18-0; Tecnazene; protectant fungicide and potato sprout suppressant. [Octavius Hunt Ltd]

Fumite Tecnalin. Smoke insecticide/fungicide (active ingredients gamma HCH/tecnazene); for use in enclosed areas against botrytis, whitefly and other insect pests on protected crops. [Octavius Hunt Ltd] †

Fumyl-O-Gas. Methyl bromide with amyl acetate; soil fumigant against soil-borne diseases. [Brian Jones & Associates Ltd]

Funduscein®-10 or -25 Injection. CAS 518-47-8; Fluorescein sodium; indicated for ophthalmic angiography and angioscopy in diagnostic examination of the fundus; in evaluation of the iris vasculature; to differentiate between viable and nonviable tissue; to assess aqueous flow; in differential diagnosis of malignant and nonmalignant tumors; and in determination of circulation time and

---

‡ Trade name and manufacturer not verified   § Trade name without identified manufacturer

adequacy of the circulation. [Iolab Corp]

Fungaflor. CAS 73790-28-0; Imazalil; used for control of mildew in greenhouse plants. [Hortichem Ltd]

Fungal Lactase 100,000. Fungal lactase; enzyme for hydrolyzing lactose in dairy products (milk, whey, cheese, yogurt), pharmaceuticals (digestive aids). [Solvay Enzymes]

Fungal Protease Conc. Protease; enzyme for hydrolysis of peptide bonds; for baking (improves grain, texture, loaf volume), meat tenderizer formulations; hydrolyzes and modifies plant and animal protein under acid conditions. [Solvay Enzymes]

Fungalysin™. Glucanase. [Calbiochem Corp]

Fungamyl. A purified fungal α–amylase preparation produced from a selected strain of *Aspergillus oryzae*; used in starch processing, brewing, alcohol production and baking industries. [Novo Nordisk]

Fungata. See Diflucan. [Pfizer International]

Fungex. A copper-containing fungicide. [Murphy Chemical Co Ltd] †

Fungi-Fluor. Stain solution and counterstain in an eight ounce kit for rapid identification of fungi; used for fungi identification in clinical cell cultures and tissue biopsy samples. [Polysciences Inc]

Fungilin Cream. CAS 1397-89-3; Amphotericin in cream base. [Bristol-Myers Squibb Pharmaceuticals Ltd]

Fungilin Lozenges. CAS 1397-89-3; Amphotericin in lozenge form. [Bristol-Myers Squibb Pharmaceuticals Ltd]

Fungilin Ointment. CAS 1397-89-3; Amphotericin in ointment form. [Bristol-Myers Squibb Pharmaceuticals Ltd]

Funginex. CAS 26644-46-2; Emulsifiable concentrate containing 66 g/l triforine; a systemic insecticide. [Synchemicals Ltd]

Fungitex 656. A proprietary trade name for a solution of an organo-mercuric complex; used as a durable mildew proofing agent for textiles. [Ciba plc] *

Fungitrol. Fungicides; used for nonaqueous coating compositions, caulking compounds, and other substrates. [Hüls Am.]

Fungitrol Tinox. Bis(tri-n-butyl)oxide; wood preservative and antifouling coating compositions. [Hüls Am.]

Fungizone for Infusion. CAS 1397-89-3; Amphotericin for intravenous infusion; an antibiotic. [Bristol-Myers Squibb Pharmaceuticals Ltd]

Fungo®. CAS 97-23-4; 15% Dichlorophen; fungicide, bactericide and algicide used as a moss, lichen, and mold killer. [Dax Products Ltd]

Fungus Fighter. Fungicide. [May & Baker Ltd] *

Fungustatin. See Diflucan. [Pfizer International]

Furac No. 3. The lead salt of dithiofuroic acid; a proprietary rubber vulcanization accelerator. §

Furacin. CAS 59-87-0; Nitrofurazone; anti-infective, topical. [Norwich Eaton Pharmaceuticals Inc] *

Furacin. Acid and solvent resisting cement for bedding and jointing tiles and bricks. [Prodorite Ltd]

Furadantin. CAS 67-20-9; Nitrofurantoin; antibacterial. [Norwich Eaton Pharmaceuticals Inc]

Furakem S. Two-component product based on furane resin; used for bedding, jointing tiles and paviors in highly corrosive environments; for operating temperatures to 180 C; resists most commonly used nonoxidizing acids, alkalis and up to 50% concentrated sulfuric acid. [Feb Ltd]

2-furaldehyde. See furfural.

Furamazone. CAS 3270-71-1; Nifuraldezone; antibacterial. [Norwich Eaton Pharmaceuticals Inc] *

Furamide. A proprietary preparation of diloxanide 2-furoate; an antiamebic. [The Boots Co plc]

Furanace. CAS 13411-16-0; Nifurpirinol; antibacterial. [Dainippon Pharmaceutical Co] *

2-furancarboxaldehyde. See furfural.

furan-2-carboxylic acid. See furoic acid.

Furanculine. Dried yeast.

2,5-furandione. See maleic anhydride.

2-furanmethanol. See furfuryl alcohol.

Fura-Tone® NC-1012. Furan resin; cured to hard infusible state resist. to solvs. and chems.; suggested for resin transfer molding. [Cardolite]

Furbac. CAS 95-33-0; n-Cyclohexyl - 2 - benzthiazyl sulfenamide; a proprietary accelerator. §

furfural. (2-furaldehyde; 2-furancarboxaldehyde; artificial oil of ants) Cyclic aldehyde; C_4H_3OCHO; CAS 98-01-1; EINECS 202-627-7; chemical intermediate for manufacture of derivatives. (furan, THF); solvent for petrol. lube, nitrocellulose; wetting agent; in manufacture of furfural-phenol plastics; vulcanization accelerator; insecticide, fungicide, germicide; reagent in analytical chemistry. [Aldrich; Allchem Industries; Great Lakes Sarl; QO]

furfural resins. Artificial resins obtained by the condensation of furfuraldehyde with phenols, cresols, or other similar bodies.

furfuralcohol. See furfuryl alcohol.

furfuryl alcohol. (2-furanmethanol; 2-furylcarbinol; furfuralcohol) $C_4H_3OCH_2OH$; CAS 98-00-0; EINECS 202-626-1; wetting agent, furan polymers, solvent for dyes and resins, flavoring. [Aldrich; Allchem Industries; Great Lakes Sarl; QO]

Furlong®. For the agriculture industry. [DuPont UK]

furnace-calamine. Masses consisting mainly of zinc oxide, formed during the smelting of zinciferous iron ores.

furnace black. See carbon black.

Furnex. CAS 1333-86-4; Carbon black. [Sevalco Ltd]

furoic acid. (furan-2-carboxylic acid; pyromucic acid) $C_5H_4O_3$; CAS 88-14-2; EINECS 201-803-0; Preservative, bactericide, furoates for perfume and flavoring, fumigant, textile processing, chemical intermediate. [R.W. Greeff; Penta Mfg.; QO]

Furotec. Furane resin binders; used for cold-set boning of sand cores. [Foseco (F.S.) Ltd]

Furoxone. CAS 67-45-8; Furazolidone; anti-infective, topical; antiprotozoal. [Norwich Eaton Pharmaceuticals Inc]*

Fursatil CS 12. A proprietary cold-setting resin based on urea/furane. [Fordath Engineering Co] ‡

Fursatil CS15. A faster-setting, lower strength variant of Fursatil CS 12. [Fordath Engineering Co] ‡

Fursatil CS25. A fast-setting, medium strength variant of Fursatil CS12 having a low fume level. [Fordath Engineering Co] ‡

Fursatil CS30. A proprietary phenol/furane cold-setting resin. [Fordath Engineering Co] ‡

Fursatil CS40. A proprietary plasticized urea cold-setting resin. [Fordath Engineering Co] ‡

Fursatil CS60, CS65. A proprietary phenol/urea cold-setting resin. [Fordath Engineering Co] ‡

Fursatil CS71. A proprietary modified phenolic cold-setting resin. [Fordath Engineering Co] ‡

Fursatil CS81. A proprietary modified urea formaldehyde

---

cold-setting resin. [Fordath Engineering Co] ‡

Furunculin. A yeast preparation

2-furylcarbinol. *See* furfuryl alcohol.

Fusabond® MB-110D. CAS 9002-888-4; Chemically modified LLDPE; anhydride-modified resin for use as compatibilizers in blends and alloys, polymeric coupling agents in reinforced or recycled PE or PP, adhesives and sealants. [DuPont]

Fusabond® MC-197D. CAS 24937-78-8; Chemically modified ethylene/vinyl acetate; anhydride-modified resin for use as compatibilizers in blends and alloys, polymeric coupling agents in reinforced or recycled PE or PP, adhesives and sealants. [DuPont]

Fusabond® MZ-109D. CAS 9003-07-0; Chemically modified polypropylene homopolymer; anhydride-modified resin for use as compatibilizers in blends and alloys, polymeric coupling agents in reinforced or recycled PE or PP, adhesives and sealants. [DuPont]

Fusabond® MZ-203D. CAS 9003-07 0; Chemically modified polypropylene impact copolymer; anhydride-modified resin for use as compatibilizers in blends and alloys, polymeric coupling agents in reinforced or recycled PE or PP, adhesives and sealants. [DuPont]

Fusafungine. An antibiotic produced by Fusarium lateritium 437.

Fusarex. CAS 117-18-0; Granules or dustable powder containing tecnazene; protectant fungicide and potato sprout suppressant. [ICI Chem. & Polymers Ltd]

Fusariol. A mercury-formaldehyde preparation. A seed preservative.

Fuscochlorin. A dark-green pigment from algae.

Fuscorhodin. A dark-red pigment from algae.

fused cement. (electro-fused cement). Terms applied to an aluminous cement with a high alumina content.

fusel oil. (fermentation amyl alcohol, potato oil, grain oil, Marc brandy oil). A by-product in alcoholic fermentation, especially in the preparation of potato spirit, and in the rectification of alcohol. It consists mainly of mixed C_2-C_5 alcohols; used in the manufacture of chemicals, explosives, as an intermediate, pharmaceuticals, synthetic rubber, varnishes, lacquers, solvent for resins and waxes, perfumery.

fusible salt. *See* microcosmic salt.

fusible salt of urine. *See* microcosmic salt.

fusidic acid. An antibiotic produced by a strain of *Fusidium*. Fucidin is the sodium salt.

Fusilade. CAS 69806-50-4; Emulsifiable concentrate of 125 g fluazifop-p-butyl per liter; used for grass weed control for broad-leaved crops. [ICI Chem. & Polymers Ltd]

Fusion®. For the agriculture industry. [DuPont UK]

Fussolon. A proprietary range of fluorocarbon and similar resins. [Daikin Kogyo Co] ‡

Fustic. (old fustic, Brazil wood, yellow wood, Cuba wood). $C_{13}H_{10}O_6$; the chips or extract from *Morus tinctoria*; natural dyestuff, the dyeing agents being morin, and maclurin (pentahydroxy-benzophenone); chiefly used for dyeing wool yellow.

fustin. The diazobenzene compound of maclurin (obtained from fustic).

Futura Flex. High performance protective and waterproofing coatings for harsh environments. [Baxenden Chemicals Ltd]

Futura Thane. High performance protective and waterproofing coatings for harsh environments. [Baxenden Chemicals Ltd]

Futurit. A proprietary plastic of the phenol-formaldehyde type. §

FW 18. CAS 1333-86-4; Channel-type carbon black; for high-quality film calendered systems. [Degussa]

FW 200 Beads and Powd. CAS 1333-86-4; Channel-type carbon black; for paints and coatings. [Degussa]

FX-512. Uv-activated epoxy curative for coating formulations. [3M]

FX Pastes. Acrylic dispersions of organic pigments. [Reckitts Colours Ltd]

Fyarestor. Flame retardant. [Witco]

Fybogel. A proprietary preparation of ispaghula husk with sodium bicarbonate and citric acid; a laxative. [Reckitts Colours Ltd] *

Fybranta. A proprietary preparation of bran and calcium phosphate; a laxative. [Norgine Ltd]

Fydulan. CAS 1194-65-6; Dichlobenil; direct, selective weed killing action in orchards, vineyards, flower beds, and parkland areas and along rail tracks, motorways and waterways. [Duphar BV] *

Fydulan. Mixture of dalapon and dichlobenil; used for total weed control in noncrop areas. [Chipman Ltd]

Fydumas. CAS 1194-65-6; Dichlobenil; direct, selective weed killing action in orchards, vineyards, flower beds, and parkland areas and along rail tracks, motorways and waterways. [Duphar BV] *

Fydusit. CAS 1194-65-6; Dichlobenil; direct, selective weed killing action in orchards, vineyards, flower beds, and parkland areas and along rail tracks, motorways, and waterways. [Duphar BV] *

Fyfanon. O,O-Dimethyl-S-(1,2-di(ethoxycarbonyl)ethyl) phosphorodithioate; a low toxic insecticide effective against insect pests on livestock, stored crops, agriculture, home and garden. [A/S Cheminova] *

Fyran J2K. Sulfamate-based compound; flame retardant for treatment of cellulosic fabrics; minimal effect on hand; buffered for nonyellowing. [Yorkshire Pat-Chem]

Fyrex®. Mixture of crystalline diammonium phosphate and monoammonium phosphate; inorganic flame retardant; produces essentially neutral solution. [Akzo]

Fyrol® 6. Diethyl N,N-bis(2-hydroxyethyl) aminomethylphosphonate; reactive flame retardant for rigid urethane foams; incorporated into foam structure by reacting as a polyol; foams exhibit good dimensional stability; for spray, froth, or pour-in-place applications [Akzo]

Fyrol® 38. Tri(β,β´-dichloroisopropyl)phosphate; scorch stabilized flame retardant for flexible and rigid polyurethane foams; reduces discoloration caused by high exotherm following processing of flexible polyether urethane foam [Akzo]

Fyrol® CEF. Tri(β-chloroethyl)phosphate; flame retardant well suited for transparent or pastel-shaded plastics and coatings [Akzo]

Fyrol® DMMP. CAS 756-79-6; Dimethyl methylphosphonate; flame retardant for applications where high phosphorus content, good solvency, and low visc. are desired; lowers visc. of epoxy resins and unsat. polyesters filled with hydrated alumina oxide [Akzo]

Fyrol® FR-2. Tri(β,β´-dichloroisopropyl)phosphate; flame retardant for flexible urethane foams [Akzo]

Fyrol® PBR. CAS 32534-81-9, 115-86-6; Pentabromodiphenyl oxide, triphenyl phosphate, organo phosphate

---

‡ Trade name and manufacturer not verified § Trade name without identified manufacturer

ester; flame retardant additive for flexible polyurethane foams; low propensity for discoloration caused by high exotherm in the processing of flexible polyether urethane foam [Akzo]

Fyrol® PCF. Tri(β-chloroisopropyl)phosphate; flame retardant. [Akzo]

Fyrquel® 150. Triaryl phosphate ester; fire-resist. hydraulic fluid for industrial applications incl. air compressors, glass, and metal furnace hydraulics, gas and steam turbine bearing lubrication, vacuum pumps. [Akzo]

Fyrquel® EHC. CAS 68937-41-7; Triaryl phosphate; fire-resist. electro-hydraulic control fluid. [Akzo]

Fytospore. Mixture of cymoxanil and mancozeb; fungicide for the control of potato blight. [ICI Chem. & Polymers Ltd]

Fyzol 11E. Adjuvant containing 99% highly refined mineral oil; wetting agent for contact carbamate herbicides. [Schering Agrochemicals Ltd]

G

G.500. A proprietary preparation containing hexamine mandelate and mothionine; a urinary antiseptic. [Octel Chemicals Ltd] †

G623. Silicone compound; general-purpose water-repellent coatings or sealants to prevent galvanic corrosion; lubricant for plastics and rubber. [GE Silicones]

G635. Methyl phenyl silicone compound; its sealing and dielec. props. make it useful for high voltage insulators, connectors, automobile and aircraft ignition systems and other electronic equip. [GE Silicones]

G-2162. PEG-25 propylene glycol stearate; surfactant. [ICI Am.; ICI Spec. Chem.]

G-3300. Alkylaryl sulfonate anionic surfactant in the form of a reddish-brown liquid; emulsifier, dispersant used for pigments in paints. [Atlas Chemical Industries (UK) Ltd]†

G-4252. CAS 9005-66-7; PEG-80 sorbitan palmitate; surfactant. [ICI Spec. Chem.]

G-4280. CAS 9005-64-5; PEG-80 sorbitan laurate; surfactant. [ICI Spec. Chem.]

GA3. See gibberellic acid.

Gaardocyclene. Tricyclodecenyl isobutyrate. [Quest Int'l. UK Ltd]

Gabbett's stain. A microscopic stain. It contains 2 g methylene blue, 25 cc sulfuric acid, with water up to 100 cc

Gabbro. A coarse, crystalline rock, composed mainly of lime-soda felspar.

Gabian oil. An inflammable mineral naphtha.

Gabraster. Polyester resins.

Gabrosa. Carboxymethyl cellulose for drilling muds. [Montedison UK Ltd] *

G-acid. 2-Naphthol-6 or 8-sulfonic acid; used as an azo dye intermediate.

Gadalan brands. Special catalysts for textile wash-and-wear finishes. [Thor Chemicals (UK) Ltd]

Gadorgel. CAS 1302-78-9; Bentonite; for drilling fluid and civil engineering to achieve high quality thixotropic muds. [Minas de Gádor SA]

Gadose. A grease prepared from cod-liver oil and lanolin; used as a basis for ointments.

Gad's cement. A mason's cement, consisting of 3 parts clay and 1 part ferric oxide.

Gaduol. An extract containing the alcohol-soluble constituents of cod-liver oil.

Gafac®. Redesignated Rhodafac®. [Rhone-Poulenc Surf.]

Gafamide CDD-518. Coconut oil diethanolamine condensate; adjuvant for nonionics, alkylaryl sulfonates and sulfated fatty alcohols; used in liquid manual-dishwashing formulations, drycleaning detergents, heavy-duty household detergents and industrial cleaners. [ISP] *

Gafen LB-400, LE-500 and LS-500. Organic phosphate esters in free acid form; supplied as liquids; extreme pressure additives for metal working fluids, with lubricant, emulsifier and rust inhibition properties; used in lubricating and rolling oils, cutting and hydraulic fluids. [ISP] *

Gafen LE-700, LP-700 and LK-500. Organic phosphate esters in free acid form; supplied as liquids; lubricity additives used in water-based cutting fluids with high concentration of inorganic rust inhibitors. [GAF Great Britain] *

Gafen LM-400. Organic phosphate ester in free acid form, supplied as a liquid; oil-soluble, water-dispersible lubricant for use as a rust inhibitor in aqueous cutting fluids, rolling oils, and hydraulic fluids. [GAF Great Britain] *

Gafen LM-600. Organic phosphate ester in free acid form, supplied as a liquid; oil-soluble emulsifier and metal lubricant used in rolling oils, cutting fluids, and hydraulic fluids. [ISP] *

Gaffix® VC-713. Vinylcaprolactam/PVP/dimethylaminoethyl methacrylate copolymer, ethanol; film-forming, fixative resin for use in mousses, gels, glazes, lotions and hairsprays. [ISP]

Gafgard 233 and 233E. Clear amber liquid blends of reactive monomers; radiation curable coatings formulated to provide abrasion resistant surfaces to plastics, paper, wood and other substrates. [ISP]

Gafgard 238. A general purpose, nonyellowing aliphatic urethane based oligomer; used in tough scuff resistant coatings for flexible substrates, including plastics, textiles and leather. [ISP]

Gafgard 245. Clear amber liquid blend of aliphatic oligomers and monomers; radiation curable coatings designed for high speed application/curing. [ISP]

Gafgard 277. Clear amber liquid blend of aliphatic oligomers and monomers; radiation curable coatings designed for flooring use. [ISP]

Gafgard 280. Clear amber liquid blend of aliphatic oligomers and monomers; radiation curable coating designed for curtain coating. [ISP]

Gafite. A range of thermoplastic polyester molding compounds; resins which can replace metals as well as other plastics in numerous applications, e.g., under-the-hood parts, electrical/electronic components, appliance parts, hardware, pumps, and hydraulic controls. [ISP]

Gafite LW. A range of thermoplastic polyester molding compounds; used for automotive exterior parts such as rear end panels, cowl vents, fender extensions and headlight housings. [ISP]

Gaflex. Thermoplastic polyester elastomers; uses include industrial tubing, fuel lines, hydraulic hoses, flexible couplings, fasteners, gaskets, seals, boots and bellows for mechanical drives, wire and cable jacketing, noise dampening devices and pump parts. [ISP]

Gafoam AD. Ammonium salt of ethoxylate sulfate; foaming agent primarily used as an air-drilling surfactant for oil and gas wells; also used in well cleanout and as a mobility control agent for carbon dioxide. [ISP]

Gafquat®734. CAS 53633-54-8; Polyquaternium-11; film-forming substantive polymer, conditioner for formulation of hair conditioners, rinses, sprays, shampoos, dyes, semipermanents, deodorants, antiperspirants, shaving preparations, antiseptics, toilet soaps, skin creams, sunburn remedies. [ISP]

Gafquat® HS-100. Polyquaternium-28. conditioning resin for personal care products; substantive, film-forming props.; for shampoos, conditioners, permanent waves, glazes, moisturizing creams. [ISP]

Gafstat AD-510 and AE-610. Free acids of complex phosphate esters, in liquid form; anionic surfactants in which the hydrophobe is aromatic; internal antistatic agents for plastics, with heat stability and low toxicity. Compatible with PVC, polyolefins, polystyrene and many other plastics. [GAF Great Britain] *

Gafstat AS-610 and AS-710. Free acids of complex phosphate esters, in liquid form; anionic surfactants in which the hydrophobe is aliphatic; internal antistatic agents for plastics, with heat stability and low toxicity; compatible with PVC, polyolefins, polystyrene, and many other plastics. [GAF Great Britain] *

Gaftuf. Impact modified polybutylene terephthalate compounds; used for hand tool housings, shrouds subjected to severe abuse, gasoline and brake clips etc. [ISP]

gagat. A variety of soft coal.

Gala. A proprietary casein plastic. §

galactan. (Gelose). A gum, $(C_6H_{10}O_5)_n$, from agar-agar.

Galactasol® Guar Derivs. CAS 39421-75-5; Hydroxypropyl guar; gums used to increase visc. in water and most brines; high visc. grades also as flocculants; some in series as stabilizers, suspending aids. [Aqualon]

β-galactosidase. (lactase) Derived from *E. coli*; CAS 9031-11-2; EINECS 232-864-1; enzyme-linked immunoassay, biochemical research.

Galag. Magnesium impregnated metallurgical coke for iron desulfurization. [Foseco (F.S.) Ltd] †

galagum. A mixture of modified polysaccharides. It gives a colloidal solution when boiled in water. It is a protective colloid, and is used in making baker's and flavoring emulsions, and in cosmetic and hair lotions.

Galahad A. A proprietary trade name for a steel containing 0.10% carbon, 0.50% silicon, 0.04% sulfur, 0.04% phosphorus, 0.60% manganese, 13.00% chromium, optional 1.50% nickel, and 0.25% molybdenum; used in the manufacture of forged steel pressure vessels with good corrosion resistance. [Hadfield, George & Co] ‡

Galalith. The trade name for a polymeric material, obtained by the action of formaldehyde upon casein. §

Galam butter. Shea butter.

Galaxy®. Bentazon, acifluorfen; for post-emergence control of annual broadleaf weeds in soybeans and peanuts [BASF AG]

Galben M. Benalaxyl + mancozeb; used for control of blight in potatoes. [DowElanco Ltd]

Galden. A proprietary range of perfluoropolyether fluids. [Montedison UK Ltd] *

Galenite. A pigment. It is a basic sulfate of lead. The name is sometimes applied to Galena.

Galenomycin. A proprietary preparation of oxytetracycline; an antibiotic. [Galen Ltd] *

Galettame silk. (Ricotti silk, Neri silk, Basinetto silk). The residue of the silk cocoon after reeling.

Galicar. An anti-friction bearing metal containing 83% tin.

Galipot. The resin from *Pinus maritima*.

gallacetophenone. CAS 528-21-2; Trihydroxyacetophenone, $C_8H_8O_4$; used medicinally as an antiseptic in skin diseases.

Gallal. Aluminum subgallate, $Al_4(C_7H_2O_5)_3$. An antiseptic and astringent.

Gallant. Herbicides based primarily on haloxyfop. [Dow UK]*

Gallatite. (Lactite, Lactoform, Cornalith, Ingalite, Lactorite, Sicalite, Proteolite). Casein preparations of a similar type to Galalith (*qv*).

Gallery One. Universal color dispersions for use in architectural coatings. [Engelhard]

gallic acid. (3,4,5-trihydroxybenzoic acid) $C_6H_2(OH)_3-CO_2H$; CAS 149-91-7; photography, writing ink, dyeing; manufacture of pyrogallol, tannins, paper; tanning agent; pharmaceuticals, engraving, lithography; analytical reagent. [Fuji Chem. Ind.; Mallinckrodt; Penta Mfg.; U.S. Biochemical]

gallic acid, propyl ester. *See* propyl gallate.

gallicin. Gallic acid methyl ester, $C_6H_2(OH)_3CO_2CH_3$; used by oculists as an antiseptic in conjunctivitis.

Gallion. Nonionic wetting agent containing ethylene oxide condensates; spreading agent for translocated phenoxy herbicides. [Intracrop Ltd]

Gallipoli oil. An olive oil used in the textile industries.

gallisin. That portion of commercial starch syrup which resists fermentation.

gallium. Metallic element; Ga; CAS 7440-55-3; EINECS 231-163-8; compounds used as semiconductors. [Aldrich; Atomergic Chemetals; Cerac; Eagle-Picher; Int'l. Gallium GmbH; Rhone-Poulenc Basic]

gallobromol. Dibromogallic acid, $C_6Br_2(OH_3)COOH$.

gallotannic acid. *See* tannic acid.

gallstone. A yellow pigment obtained from the gall bladder of oxen.

Galorn. A proprietary trade name for a casein plastic. §

Galoryl. Range of various surfactants. [CFPI]

Galt glass. Polyester resin bonded glass fiber moldings with a special surface giving good resistance to weather and chemicals.

galvanized iron. (zinced iron). Iron coated with metallic zinc.

galvanit. A plating powder consisting of a mixture of the salt of the metal to be deposited (silver for silver plating), and a more electro-positive metal.

galvano Lac. A mixture of celluloid varnish with powdered metal.

Galvoline. Magnesium ribbon anode. [Dow UK]

Galvomag. High current magnesium anode. [Dow UK]

Galvorod. Standard magnesium anode. [Dow UK]

Gamanase. A hemicellulase prepared by submerged fermentation of *Aspergillus niger*, particularly suited for applications where a rapid break-down of galactomannans to less viscous products is required. [Novo Nordisk]

Gamastan. Globulin, immune; immunizing agent. [Cutter Laboratories, Miles Laboratories Inc] *

Gambier. *See* cutch.

Gamboge. (Gummi gutta). A gum resin, the product of *Garcinia morella* of Siam. It is a yellow pigment used for water colors, also for coloring spirit and other varnishes.

Gamboge butter. A fat obtained from *Garcinia morella*.

Gamimune. Globulin, immune; immunizing agent. [Cutter Laboratories, Miles Laboratories Inc] *

gamma acid. 2-Amino-8-naphthol-6-sulfonic acid; used as an azo dye intermediate.

Gamma-BHC Dust. Insecticide. [Murphy Chemical Ltd] †

Gamma-Col. Suspension concentrate containing 800 g γ-HCH per liter; an organochlorine insecticide. [ICI Agrochemicals; ICI Chem. & Polymers Ltd]

Gammagee. Globulin, immunizing agent. [Merck & Co Inc]*

Gamma-HCH Dust. Garden insecticide. [Murphy Chemical Ltd] †

Gammalex. Captan + γ-HCH; insecticide and fungicide seed dressing for brassicas and oilseed rape. [ICI Agrochemicals; ICI Chem. & Polymers Ltd]

Gammalin. Insecticides. [ICI Agrochemicals; ICI Chem. & Polymers Ltd]

Gammar. Globulin, immune; immunizing agent. [Armour Pharmaceutical Co]

Gammasan. γ-HCH; liquid seed dressing. [ICI Agrochemicals; ICI Chem. & Polymers Ltd]

Gammatox. Livestock dips and sprays. [The Wellcome Foundation Ltd] *

Gammatrol. Nucleonic instruments for process control and measurement. [ICI Chem. & Polymers Ltd]

Gammexane. Insecticides. [ICI Chem. & Polymers Ltd]

Gammonativ. Normal immunoglobulin for intravenous use. [KabiVitrum AB] *

Gamonil®. CAS 23047-25-8; Lofepramin. [E Merck]

Ganex® Et-201. PVP/decene copolymer. [ISP]

Ganex® P-904. Butylated PVP; used in cosmetics and toiletries as moisture barrier, adhesive, protective colloid, and microencapsulating resin; as dispersant for pigments; as solubilizer for dyes; in petroleum industry as sludge and detergent dispersant; protective colloid in coatings; suspending aid in polymerization; dyeing assistant; antiredeposition agent in drycleaning; esp. as dispersant in aq. agric. chemicals or pigmented skin care products [ISP]

Ganex®V-216. CAS 32440-50-9; PVP/hexadecene copolymer; used in cosmetics and toiletries as moisture barrier, adhesive, protective colloid, and microencapsulating resin; as dispersant for pigments; as solubilizer for dyes; in petroleum industry as sludge and detergent dispersant. [ISP]

Ganex® V-220. CAS 28211-18-9; PVP/eicosene copolymer; used in cosmetics and toiletries as moisture barrier, adhesive, protective colloid, and microencapsulating resin; as dispersant for pigments; as solubilizer for dyes; in petroleum industry as sludge and detergent dispersant. [ISP]

Ganex®WP-660. Tricontanyl PVP; waterproofing polymer for personal care products; for sunscreens, skin care products, cosmetics, makeup, baby care products [ISP]

gangue. The earthy portion of an ore which leaves the metal when reduced.

Ganicin. Zinc-rich coatings. [DuPont UK]

ganister. A rock mineral with a composition corresponding to a pure silica with about 1/10 of its weight of clay; used in the manufacture of siliceous fire-bricks, and for lining furnaces.

Ganocide. CAS 5707-69-7; Fungicides containing drazoxolon. [ICI Chem. & Polymers Ltd]

Gansil. *See* Chloramine T. §

Gant. A barrier cream. [Croda Chem. Ltd] *

Gantanol. CAS 723-46-6; A proprietary preparation of sulfamethoxazole; an antibiotic. [Roche Laboratories; Roche Products Ltd]

Gantrez®. A range of poly(methylvinyl ether/maleic anhydride) copolymers; gelling agents, thickeners, stabilizers, explosive stabilizers, anticorrosion coatings, and suspending aids. [ISP]

Gantrez® AN-119. CAS 52229-50-2; PVM/MA copolymer; thickener, dispersant, stabilizer, gelling agent, coupler, protective colloid, suspending aid used in emulsion polymerization; adhesives; household detergents, liq. hand cleaners, acid bowl cleaners; produces clear films of high tensile and cohesive str. [ISP]

Gantrez® AN-8194. Stearylvinyl ether/maleic anhydride copolymer, toluene; hydrocarbon-sol. copolymer that forms waxy films with good water resistance and thermal properties; compatible with silicone release agents and modifies release level for pressure-sensitive labels and tapes; used in adhesives, coatings, and water-repellent coatings for porous and nonporous substrates. [ISP]

Gantrez® B-773. Poly(vinyl isobutyl ether), hexane; tacky polymer with excellent adhesion to plastic, metal, and coated surfs.; plasticizer and leveling agent for surface coatings. [ISP]

Gantrez® ES-225. Ethyl ester of PVM/MA copolymer, ethanol; copolymer forming clear, glossy films with substantivity and moisture resistance; used in hairsprays, mousses, gels, and lotions, coatings, polishes; emulsion stabilizer in creams and lotions. [ISP]

Gantrez® M-154. CAS 9003-09-2; Polyvinyl methyl ether; polymer functioning as tackifier, binder, and plasticizer; used in printing inks, textile sizes and finishes, latex modification. [ISP]

Gantrez® S-95. CAS 9011-16-9; PVM/MA copolymer; hydrolyzed low molecular weight polymer; water-sol. polyelectrolyte similar to Gantrez AN series; chelating agent. [ISP]

Gantrez® SP-215. Ethyl ester of PVM/MA copolymer; hair fixative for stiffer, harder holding products; up to 10% can be formulated in pump hair sprays. [ISP]

Gantrez®V-215. Methylvinyl ether/maleic anhdride copolymer, ethyl ester, ethanol SDA-40B. [ISP]

Gantrisin. *See* Gantanol. [Roche Products Ltd] *

Garamycin. CAS 1405-41-0; Gentamicin sulfate; a complex antibiotic substance, produced by *Micromonospora purpurea* nsp.; has three components, sulfates of gentamicin C_1, gentamicin C_2, and gentamicin C_{1A}; antibacterial. [Schering Corp]

Garbacryl. Lubricating oil additives. [Rhone-Poulenc UK]

Garbritol. Textile auxiliary. [Hickson & Welch Ltd] †

Garcinia oil. *See* Kokum Butter.

Garcrete. Refractory castable. [Foseco (F.S.) Ltd]

Gardamide. N-Methyl-N-phenyl-2-methyl butyramide. [Quest Int'l. UK Ltd]

Gardenal. CAS 50-06-6; Phenobarbitone. [Rhone-Poulenc Rorer Ltd]

Gardenal Sodium. CAS 57-30-7; A proprietary preparation of phenobarbitone sodium; a hypnotic. [Rhone-Poulenc Rorer Ltd]

---

‡ Trade name and manufacturer not verified   § Trade name without identified manufacturer

Gardeniol. Phenyl-methyl-carbinyl-acetate, $C_6H_5 \cdot$ CH(OOC $\cdot$ CH$_3$) $\cdot$ CH$_3$; a perfume.

Garden Lime. CAS 1317-65-3; Ground limestone; used to increase pH of acid soils. [Vitax Ltd]

Gardinol. Sodium salts of sulfonated higher fatty alcohols; wetting agents. [Hickson & Welch Ltd]

Gardinox. Textile auxiliaries. [Hickson & Welch Ltd]

Gardlite. A proprietary trade name for a synthetic resin, produced from toluene sulfonamide and formaldehyde.§

Gardoprim A 500FW. A suspension concentrate containing 100 g atrazine and 400 g terbuthylazine per liter; used for total weed control in forestry plantations. [Ciba-Geigy Agrochemicals]

Garganine. A madder extract.

Garj. A bituminous sandstone. It contains from 6-17% bitumen.

Garlon. CAS 69633-04-1; Herbicides based primarily on triclopyr. [Dow UK]

Garlon 2. CAS 69633-04-1; Emulsifiable concentrate containing 240 g/l triclopyr; herbicide to control perennial and woody weeds. [ICI Chem. & Polymers Ltd]

Garlon 4. CAS 69633-04-1; Emulsifiable concentrate containing 480 g/l triclopyr; herbicide to control perennial and woody weeds. [Chipman Ltd]

Garnex. Refractory insulating expendable tundish linings. [Foseco (F.S.) Ltd]

Garoflam. Flame retardant blends; for plastics and rubber. [Croxton & Garry Ltd] *

Garomix. Magnesium oxide and zinc oxide blend; for rubber goods. [Croxton & Garry Ltd] *

Garosorb. Desiccants; for plastics and rubbers. [Croxton & Garry Ltd] *

Garospers. Dispersed rubber chemicals; used for rubber goods. [Croxton & Garry Ltd] *

Garozinc. CAS 1314-13-2; Zinc oxides; used for rubber goods. [Croxton & Garry Ltd] *

Garpak. Rammable refractory. [Foseco (F.S.) Ltd]

Garseal. Refractory air-setting mortar. [Foseco (F.S.) Ltd]

Gartop. Refractory insulating powder or boards for covering the steel in continuous casting tundishes. [Foseco (F.S.) Ltd]

Gartube. Insulating shrouds fitted on the bottom of teeming ladles. [Foseco (F.S.) Ltd]

Garvox 3G. CAS 22781-23-3; Bendiocarb; a contact, systemic insecticide. [Schering Agrochemicals Ltd]

Gasbinda. CAS 1344-09-8; Sodium silicate binders for sand cores and molds using CO$_2$ hardening process. [Foseco (F.S.) Ltd]

gas black. (satin gloss black, hydrocarbon black, hydrocarbon gas black, silicate of carbon, jet black, ebony black). A carbon black made by the incomplete combustion of natural gas; used in rubber mixings.

Gasil. Micronized silica gel. [Crosfield Chemicals Ltd]

Gasil EBC, EBN. Proprietary compounds of silica used in the paint, resin and plastics industries, e.g., in the matting of electron beam-cured coatings. [Crosfield Chemicals Ltd]

Gaskoid. A proprietary trade name for a rubber jointing resistant to oil and petrol. §

gas oil. The name applied to all mineral oils intended for the preparation of gas, such as the light oils of brown coal tar, and shale oil; used as an absorption oil and in the manufacture of ethylene.

Gas Service Anywhere. Distributorship services in the field of liquified petroleum gas. [Quantum Chemical Corp]

Gastalar. A proprietary preparation containing aluminum hydroxide, magnesium carbonate and sorbitol; an antacid. [Armour Pharmaceutical Co] *

Gastex. A proprietary gas black used in rubber mixings. §

Gastracalm. An indigestion mixture. [The Boots Co plc]

Gastratox 6G Slug Pellets. CAS 108-62-3; Pellets containing 6% w/w metaldehyde; snail and slug bait. [Truchem Ltd]

Gastrils. CAS 21645-51-2, CAS 546-93-0; A proprietary preparation containing aluminum hydroxide and magnesium carbonate as a co-dried gel; an antacid. [Smith & Nephew Pharmaceuticals Ltd] *

Gastro Caloreen. A proprietary preparation of a polyglucose polymer used as a high-calorie food supplement. [Scientific Hospital Supplies] *

Gastrocote. A proprietary preparation of alginic acid, dried aluminum hydroxide gel, magnesium trisilicate and sodium bicarbonate; an antacid. [MCP Pharmaceuticals] *

Gastrografin. Sodium and meglumine salts of diatrizoate; x-ray contrast media. [Schering Health Care Ltd]

Gastrovite. A proprietary preparation of ferrous glycine sulfate, ascorbic acid and calcium gluconate; a hemintic. [MCP Pharmaceuticals] *

Gat 15. Rousselot special gelatin powder; developed as a substitute for gum arabic. [Sanofi Bio-Industries Ltd] ‡

Gatodan 415. Propylene glycol stearate and distilled monoglycerides; food emulsifier for cake and sponge improvers. [Grindsted Prods.; Grindsted Prods. Denmark]

Gaucho. CAS 105827-78-9; Imidacloprid; insecticidal seed dressing with excellent root systemic properties for control of sucking pests (virus vectors), e.g., aphids, thrips, leafhoppers, white files and some biting insects such as rice water weevil, Colorado potato beetle, wireworms, *Diabrotica* species, frit fly, beet fly and onion fly on various crops, particularly on rice, cereals, maize, cotton, potatoes, beets and vegetables. [Bayer AG]

Gauduin's fluid. A mixture of finely powdered cryolite, and a solution of phosphoric acid in alcohol. A soldering fluid.

gauging metal. An alloy similar to Delta metal, but containing iron.

Gaultheria oil. CAS 119-36-8; Oil of wintergreen; used in medicine, flavoring compounds and perfumery.

Gaultheriasalol. Methylsalicylo salicylate, HO $\cdot$ C$_6$H$_4 \cdot$ COO $\cdot$ C$_6$H$_4 \cdot$ COOCH$_3$.

Gaultheric acid. Methyl salicylate, CH$_3 \cdot$ C$_7$H$_5$O$_3$.

Gaviscon. A proprietary preparation of alginic acid, sodium alginate, magnesium trisilicate, aluminum hydroxide, and sodium bicarbonate; an antacid. [Reckitts Colours Ltd] *

Gazelle. Acid phosphates for the baking trade. [Albright & Wilson Ltd] *

GBL. CAS 96-48-0; γ-Butyrolactone; spinning and coagulating solv. for textiles, in nail polish removers, detergents, sunscreens, household cleaners, paint removers, agric. use, polymers, petrol. industry. Arco Chemical Co]

GDL. Glucono delta lactone. [Pfizer International]

GE 2557. A proprietary trade name for a polyester vinyl plasticizer. [GE Plastics] *

Geax. A trade name for a synthetic resin-paper product used as an electrical insulation. §

Geblitol. Sodium hydrosulfite; used as a disinfectant.

Gecet F100. Expandable engineering resin; foam bead material for use in sporting goods (water sports, boat components), automotive (upper instrument panels), furniture, building and construction, medical (sterilizable insulating containers and lab specimen shippers), shipping (high-str. dunnage), materials handling (functional pallets, cores in composite structures) applications; F100 offers widest range of mech. props., moderate heat resist.; end use temp. 220 F. [GE Plastics; Huntsman]

Gechophen. CAS 104-29-0; A trade name for chlorphenesin B.P. [British Drug Houses] *

Gedanite. (friable amber, soft amber). A resin found on the shores of the Baltic. It is a variety of amber, and melts at 150-180 C. It is also called soft amber. A variety of amber low in succinic acid.

Gedeflex®. A registered trademark for dibutyl, butylbenzyl and dioctyl phthalates. §

Gedelite®. A registered trademark for phenolic molding powders and resins. §

Gedge's metal. See Aich metal.

gedrite. A yellow resin found with Prussian amber. It is also the name for a mineral.

Gefarnil. CAS 51-77-4; A proprietary preparation of gefarnate; a gastrointestinal sedative. [Crookes Laboratories] *

Geko. Inorganic molding sand binders for foundries, also in connection with lustrous carbon formers to obtain better castings. [Süd-Chemie AG] *

Gel II. CAS 7681-49-4; Sodium fluoride; dental caries prophylactic. [Oral-B Laboratories Inc] *

Gelamite D. A nitroglycerin dynamite; used for construction and building, explosives, mining, petroleum and related industries. [Hercules] *

Gelaprime F. A nitroglycerin dynamite; used for construction and building, explosives, mining, petroleum and related industries. [Hercules] *

Gelatase. Bacterial proteolytic enzymes. [ABM Chemicals Ltd] *

gelatin. (gelatine) Obtained from partial hydrolysis of collagen derived from animal skin, connective tissues, and bones; CAS 9000-70-8; EINECS 232-554-6; photographic film, sizing, textile and paper adhesives, cements, capsules for medications, matches; clarifying agent; protective colloid in ice cream; stabilizer, thickener, texturizier in food. [Croda; DynaGel; G Fiske & Co Ltd; Hormel; Nitta Gelatin]

Gelatinastralite. An explosive consisting of ammonium nitrate, with some sodium nitrate, up to 20% dinitrochlorhydrin, and maximum amounts of 5% nitroglycerin, and 1% collodion cotton.

gelatin carbonite. An explosive consisting of 25% nitroglycerin, 0.7% collodion wool, 7% gelatin, 25% sodium chloride, and 42% ammonium nitrate.

Gelatine Donarit 1, 2, 3. Gelatinous ammon dynamites (ammon gelatins, ammon gelignites); main ingredients are explosive-oil, ammonium nitrate, aromatic nitrocompounds and cellulose nitrate; suitable for use at the surface as well as underground; very good water resistance [Dynamit Nobel Wien GmbH] *

Gelatine Donarit 2 E. Gelatinous explosive; main ingredients are explosive-oil, ammonium nitrate, nitrocellulose, etc. [Dynamit Nobel Wien GmbH] *

Gelatine Donarit S. An explosive for seismic prospecting which, on account of special sensitizers, can achieve best results even under high pressure; supplied in plastic tubes which can be coupled by means of screw threads to form longer charging units. [Dynamit Nobel Wien GmbH] *

gelatin, vegetable. Agar-agar.

gelato-glycerin. Glycerin jelly.

Gelcharg. High viscosity, low cost polymer for gelling water-gel explosives; thickener, water-blocking agent. [Hercules] *

Gelcosal®. Tar BP; strong coal tar solution and salicylic acid; aqueous gel; used to treat psoriasis and eczema in the chronic scaling phase. [Quinoderm Ltd]

Gelcotar®. Tar BP; strong coal tar solution in an aqueous gel base; used to treat psoriasis and eczema. [Quinoderm Ltd]

Gelcotar® Liquid. Strong coal tar solution and cade oil; a shampoo for treatment of psoriasis of the scalp, seborrhoeic dermatitis and dandruff. [Quinoderm Ltd]

Geleol. CAS 31566-31-1; Glyceryl stearate; emulsifier. [Gattefosse; Gattefosse SA]

Gelflex. Dimefilcon A; 2-hydroxyethyl methacrylate polymer with methyl methacrylate and ethylenebis dimethacrylate; contact lens material. [Dow Corning Ophthalmics Inc] *

Gel Flo. A slurry explosive; used in iron ore mining. [Hercules] *

Gelkyd. Thixotropic alkyds. [Cray Valley Ltd]

Gelline. A colloidal iodine gel.

Gelobel. A proprietary trade name for a gelatin dynamite; an explosive. §

Gelofusine. A proprietary preparation of gelatin, calcium chloride and sodium chloride; used in the treatment of shock. [Consolidated Chemicals Ltd] *

Gelosedine. Pharmaceutical preparation; gastrointestinal sedative and antispasmodic. [Bayer AG]

Gelosine. A mucilaginous material extracted from a Japanese algae. It is soluble in alcohol and water.

Gelot 64®. Glyceryl stearate and PEG-75 stearate; SE base for o/w cosmetic and pharmaceutical emulsions. [Gattefosse; Gattefosse SA]

Geloxite. An explosive consisting of 54-64 parts nitroglycerin, 4-5 parts nitro-cotton, 13-22 parts potassium nitrate, 12-15 parts ammonium oxalate, 0-1 part red ocher, and 4-7 parts wood meal.

Geloy® BG10. Acrylic-styrene-acrylonitrile resin; blow molding grade for applications requiring prolonged outdoor exposure. [GE Plastics]

Geloy® GY1020. Weatherable polymer [GE Plastics]

Geloy® GY1220. ASA/PVC alloy; weatherable polymer [GE Plastics]

Geloy® XP1001. Acrylic-styrene-acrylonitrile terpolymer resin; injection molding resin with excellent weatherability; applications incl. interior and exterior automotive/truck/RV trim and parts, providing color stability and property retention without painting. [GE Plastics]

Gel Power. Slurry explosives; underground non coal-type mines. [Hercules] *

Gelrite®. CAS 71010-52-1; Gellan gum; used for microbiological applications. [Kelco]

gel rubber. A term used for the residue of rubber left undissolved when raw rubber is treated with a solvent.

Gelsorb B. CAS 1337-76-4; Attapulgite clay. [Bromhead & Denison Ltd]

Gelucire 35/10. Saturated polyglycolized glycerides; excipient for hard gelatin capsules. [Gattefosse SA]

Gelusil. A proprietary preparation containing aluminum

---

‡ Trade name and manufacturer not verified   § Trade name without identified manufacturer

hydroxide and magnesium trisilicate; an antacid. [Warner] *

Gelva. Polyvinyl acetate resin emulsions/solutions; used to make pressure sensitive coatings and laminating adhesives for difficult-to-bond surfaces; used to formulate surface coatings and for specialty paper coatings. [Monsanto Co]

Gelvatol. Powdered resins. [Monsanto plc]

Gemex. Surfactants. [Akzo] *

Gemglo. A proprietary trade name for styrene and methacrylate. §

Geminimycin. CAS 1405-41-0; A proprietary preparation of gentamycin sulfate; an antibiotic. [Pfizer International] †

Gemlite. A proprietary trade name for a urea formaldehyde synthetic resin. §

Gemme. Crude turpentine.

Gemonil. CAS 50-11-3; Metharbital; anticonvulsant. [Abbott Laboratories] *

Gemstone. A proprietary trade name for a phenol-formaldehyde cast resin. §

Gemstone M.1.2. A proprietary trade name for a phenolic laminating resin. §

Gemtone®. Rich, lustrous pigments deriving color from both light interference and light absorption. [Mearl]

Genacort. CAS 50-23-7; A proprietary preparation of hydrocortisone for dermatological use. [Fisons plc, Pharmaceuticals Div] *

Genagen C-100. CAS 61791-29-5; PEG-10 cocoate; surfactant for textile processing. [Hoechst Celanese/Colorants & Surf.]

Genagen CA-050. CAS 61791-08-0; PEG-5 cocamide; cleansing skin protective component for detergents and cleaning agents. [Hoechst Celanese/Colorants & Surf.; Hoechst AG]

Genagen O-090. CAS 9004-96-0; PEG-9 oleate; surfactant for textile processing. [Hoechst Celanese/Colorants & Surf.]

Genagen P-070. CAS 9004-94-8; PEG-7 palmitate; surfactant for textile processing. [Hoechst Celanese/Colorants & Surf.]

Genagen PL-090. CAS 31621-91-7; PEG-9 pelargonate; surfactant for textile processing. [Hoechst Celanese/Colorants & Surf.]

Genagen S-080. CAS 9004-99-3; PEG-8 stearate; surfactant for textile processing. [Hoechst Celanese/Colorants & Surf.]

Genagen TA-080. CAS 61791-00-2; PEG-8 tallate; surfactant for textile processing. [Hoechst Celanese/Colorants & Surf.]

Genal P4300-CM. A proprietary heat resistant but asbestos-free phenolic molding compound. [GE Plastics] *

Genamid®; Polyamide resins, liquid reactable. [Cray Valley Ltd]

Genamid® 151. Amidoamine resin; epoxy curing agent offering superior wetting, excellent chem. resist., better internal plasticizer, fast cure time; 151 used for high solids coatings, castings, laminates, and adhesives. [Henkel] *

Genamin. Large range of cationic surfactants which may be primary, secondary or tertiary amines, diamines or amine salts; solid, liquid or paste forms; antistatic agents, fabric conditioner and softener, fiber finishers, water-repellant agents [Hoechst UK] *

Genamin C Grades. Ethoxylated cocamine (2-25 EO);

surfactant for agric. formulations. [Hoechst Celanese/Colorants & Surf.]

Genamin CTAC. CAS 112-02-7; EINECS 203-928-6; Cetrimonium chloride; cosmetic raw material for hair treatment preps.; antistat. [Hoechst Celanese/Colorants & Surf.]

Genamin DSAC. CAS 107-64-2; EINECS 203-508-2; Distearyldimonium chloride; cosmetic raw material for hair and skin treatment preps.; good wet combing, skin softening props. [Hoechst Celanese/Colorants & Surf.]

Genamin KDM-F. CAS 17301-53-0; Behentrimonium chloride; base, antistat, emulsifier for preparation of hair care products; conditioner. [Hoechst Celanese/Colorants & Surf.]

Genamin KSE. Distearyldimonium chloride, cetyl alcohol, ceteareth-15, ceteareth-3, PEG-3 distearate; SE base material, conditioner for hair treatment preps. [Hoechst Celanese/Colorants & Surf.]

Genamin KSL. PEG-5 stearyl ammonium lactate; gloss aid for hair care products [Hoechst Celanese/Colorants & Surf.]

Genamin T-020. CAS 61791-44-4; PEG-2 tallow amine; surfactant for agric. formulations, textile processing. [Hoechst Celanese/Colorants & Surf.]

Genamin TA Grades. CAS 61790-33-8; EINECS 263-125-1; Tallow fatty acid amine; surfactant for agric. formulations. [Hoechst Celanese/Colorants & Surf.]

Genamine C-020. Coconut fatty amine oxethylate; raw material for mineral oil additives, insecticides, pesticides, cosmetic bases, adhesives. [Hoechst Celanese]

Genaminox CS. CAS 61788-90-7; EINECS 263-016-9; Coco dimethyl amine oxide; foaming agent and stabilizer, thickener for personal care products; hair conditioner. [Hoechst AG]

Genaminox KC. CAS 61788-90-7; EINECS 263-016-9; Cocamine oxide; foam booster/stabilizer over wide pH range, thickener for shampoos, bath products; surfactant for textile processing. [Hoechst Celanese/Colorants & Surf.; Hoechst AG]

Genapol®. General surfactant properties, low foaming power and ability to reduce the foam of other surfactants; stable to acids, alkalies and most metal salts; drain aid in machine dish and bottle washing. [Hoechst UK] *

Genapol® 24-L-3. CAS 68439-50-9; C12-14 pareth-2.9; biodeg. detergent intermediate for sulfation for use in cosmetics, shampoos, light duty detergents; emulsifier, prewash spotter, agric. adjuvant, hydrocarbon-based cleaning systems. [Hoechst Celanese/Colorants & Surf.]

Genapol® 26-L-1. CAS 68551-12-2; C12-16 pareth-1; biodeg. detergent intermediate for sulfation for use in cosmetics, shampoos, light duty detergents; emulsifier, prewash spotter, agric. adjuvant, hydrocarbon-based cleaning systems. [Hoechst Celanese/Colorants & Surf.]

Genapol® 42-L-3. CAS 68439-50-9; C12-14 pareth-3; biodeg. detergent intermediate for sulfation for use in cosmetics, shampoos, light duty detergents; emulsifier, prewash spotter, agric. adjuvant, hydrocarbon-based cleaning systems. [Hoechst Celanese/Colorants & Surf.]

Genapol® 2299. Coconut alcohol ethoxylate; surfactant for textile processing. [Hoechst Celanese/Colorants & Surf.]

Genapol® AMS. TEA-PEG-3 cocamide sulfate; detergent, foaming agent used in top-grade cosmetics cleansers; lime soap dispersant; biodeg. [Hoechst Celanese/Colorants & Surf.; Hoechst AG]

Genapol® ARO. CAS 9004-82-4; Sodium laureth sulfate; raw material for cosmetics, detergents, and cleaning agents. [Hoechst Celanese/Colorants & Surf.; Hoechst AG]

Genapol® C-050. CAS 61791-13-7; Coceth-5; raw material for mfg. of textile, leather, paper auxs., detergents, emulsifiers, cosmetics, agric. [Hoechst Celanese/Colorants & Surf.]

Genapol® DA-040. CAS 26183-52-8; Deceth-4; surfactant for textile processing. [Hoechst Celanese/Colorants & Surf.]

Genapol® GC-050. CAS 61791-13-7; Coceth-5; surfactant for textile processing. [Hoechst Celanese/Colorants & Surf.]

Genapol® LRO Liq., Paste. CAS 9004-82-4; Sodium laureth sulfate; detergent, foaming agent used in cosmetic products, personal care products, agric., lime soap dispersant; biodeg. [Hoechst Celanese/Colorants & Surf.; Hoechst AG]

Genapol® O-020. CAS 9004-98-2; Oleth-2; raw material for mfg. of textile, leather, paper auxs., detergents, emulsifiers, cosmetics, agric. [Hoechst Celanese/Colorants & Surf.]

Genapol® PF 10. EO/PO block copolymer; surfactant for agric. formulations, textiles. [Hoechst Celanese/Colorants & Surf.]

Genapol® PGM Conc. Sodium laureth sulfate, glycol distearate, and cocamide MEA; pearl-luster conc. used in shampoos, bubble baths, cosmetics, liq. soaps, detergents; biodeg. [Hoechst Celanese/Colorants & Surf.; Hoechst AG]

Genapol® PL 120. EO/PO fatty alcohol adduct; low foaming surfactant. [Hoechst AG]

Genapol® PMS. CAS 627-83-8; EINECS 211-014-3; Glycol distearate; pearlescent agent for cosmetic washing agents and shampoo. [Hoechst Celanese/Colorants & Surf.; Hoechst AG]

Genapol® S-020. Stearyl alcohol polyglycol ether; raw material for mfg. of textile, leather, paper auxs., detergents, emulsifiers, cosmetics. [Hoechst Celanese/Colorants & Surf.] *

Genapol® T Grades. Tallow alcohol polyglycol ether (8-25 EO); detergent base and basic material for cosmetic and specialty chemical industries, agric. formulations; biodeg. [Hoechst Celanese/Colorants & Surf.; Hoechst AG]

Genapol® TS Powd. CAS 9005-08-7; PEG-3 distearate; detergent, pearlescent, opacifier for shampoos, bubble baths, shower preps. [Hoechst Celanese/Colorants & Surf.; Hoechst AG]

Genapol® TSM. PEG-3 distearate, sodium laureth sulfate, glycol distearate; detergent, opacifier for shampoos, bubble baths, shower preps. [Hoechst Celanese/Colorants & Surf.; Hoechst AG]

Genapol® UD-030. PEG-3 fatty alcohol; surfactant for oil and gas industry. [Hoechst Celanese/Colorants & Surf.]

Genapol® V 2908. C12-15 synthetic oxo-alcohol EO/PO block copolymer; surfactant for agric. formulations. [Hoechst Celanese/Colorants & Surf.]

Genapol® X-040. CAS 24938-91-8; Trideceth-4; surfactant for textile processing. [Hoechst Celanese/Colorants & Surf.]

Genapol® ZRO Liq., Paste. CAS 9004-82-4; Sodium laureth sulfate; raw material with good foaming and cleansing for cosmetics, detergents, cleaning agents. [Hoechst Celanese/Colorants & Surf.; Hoechst AG]

Genasco. A proprietary trade name for a bituminous softener. §

Genasprin. Acetylsalicylic acid tablets. [Fisons plc, Pharmaceuticals Div] *

Genatosan Skin Bar. A soap-free detergent bar. [Fisons plc, Pharmaceuticals Div] *

Gen-che. Zinc, manganese, copper, sulfur; chelated micronutrient for vegetable, tree and field crops. [Draxel Chemical Company] ‡

Genclor. Chlorinated polyvinyl chloride; adhesive. [ICI Chem. & Polymers Ltd]

Gendriv 162. Fiber recovery and retention aid for paper machine applications [Hercules]

Gendriv 492S. Formation aid for low shear or fast draining paper applications [Hercules]

Genelit. A spongy bronze-like bearing metal prepared from a very finely ground mixture of copper, tin, and graphite. When the mixture is heated the graphite burns away, the copper and tin melt together, leaving behind a porous mass which is able to absorb large quantities of lubricating oil.

Generol® 122. Soya sterol; emollient, auxiliary or primary w/o emulsifier, emulsion stabilizer, visc. modifier, solubilizer for cosmetics. [Henkel/Cospha; Henkel Canada; Henkel KGaA]

Generol® 122E5. PEG-5 soya sterol; emollient, primary and sec. emulsifier, conditioner, stabilizer and consistency modifier in o/w emulsions; substantive to hair and in shampoo. [Henkel/Cospha; Henkel Canada; Henkel KGaA]

Generon. A unit used to separate components. [Dow UK] *

Genesolv A Solvent. CAS 75-69-4; Trichlorofluoromethane; cleaning solvent for metal, plastic and glass; extractant. [Allied-Signal Inc] *

Genesolv D Solvent. CAS 76-13-1; Trichlorotrifluoroethane; cleaning solvent for electronic, electrical and other high value assemblies; carrier for specialty coatings; drying agent; dielectric fluid. [Allied-Signal Inc] *

Genetron® 11. CAS 75-69-4; Trichlorofluoromethane; centrifugal refrigerant; sec. coolant in low temp. systems; used in thermal insulation construction projects. [Allied-Signal]

Genetron® 12. CAS 75-71-8; Dichlorodifluoromethane; refrigerant for reciprocating and rotary type equip., household to industrial applications, centrifugal applications, sterilant gas, blowing agents, aerosols. [Allied-Signal]

Genetron® 13. CAS 75-72-9; Chlorotrifluoromethane; low temp. refrigerant used in low stage of cascade systems to provide evaporator temps. in range of -100 F. [Allied-Signal]

Genetron® 22. Hydrochlorofluorocarbon 22. refrigerant for residental, commercial and industrial applications; blowing agent in aerosols; intermediate to produce fluoropolymers. [Allied-Signal]

Genetron® 113. CAS 76-13-1; Trichlorotrifluoroethane; used in low capacity centrifugal chiller packaged units; operates with very low system pressures, high gas volumes. [Allied-Signal]

Genetron® 114. CAS 76-14-2; Dichlorotetrafluoroethane; used with centrifugal compressors for higher capacities or for lower evaporator temp. process applications; also for foam applications [Allied-Signal]

Genetron® 123. CAS 306-83-2; Dichlorotrifluoroethane; centrifugal refrigerant with low operating pressures. [Allied-Signal]

Genetron® 134a. Tetrafluoroethane; refrigerant; CRC subsitute for wide range of air conditioning and refrigeration systems in residential, commercial and industrial applications incl. automotive. [Allied-Signal]

Genetron® 141b. CAS 1717-00-6; 1,1-Dichloro-1-fluoroethane; blowing agent (replacement for CFCs) in foam applications, rigid board, flexible foam. [Allied-Signal]

Genetron® 500. Dichlorodifluoromethane (73.8%), hydrofluorocarbon 152a (26.2%) azeotrope; slightly higher vapor pressures; provides higher capacities from same compressor displacement. [Allied-Signal]

Genetron®502. Hydrochlorofluorocarbon 22 (48.8%), chlorofluorocarbon 115 (51.2%); for low evaporation temp. applications [Allied-Signal]

Genetron® 503. Hydrofluorocarbon 23 (40.1%), chlorotrifluoromethane (59.9%); used in low stage of cascade type systems where it provides gains in compressor capacity and in low temp. capability. [Allied-Signal]

Genetron Dry Refrigerants. Chlorofluorocarbon refrigerant fluids; used for refrigeration and air conditioning equipment, expansion agents for plastic foam applications, aerosol propellants. [Norplex] *

Genetron®HFC 23. Trifluoromethane; low temp. refrigerant replacing CFC-13 and R-503 in low stage of cascade systems. [Allied-Signal]

Genetron® HFC 125. Pentafluoroethane; for use in low temp. refrigeration applications [Allied-Signal]

Genetron® Refrigerant 32/125. Azeotropic mixt. of difluoromethane and pentafluoroethane; substitute for air conditioning and refrigeration applications [Allied-Signal]

Genetron® Refrigerant 125/143a. HFC-125 (45%) and HFC-143A (55%); for low temp. refrigeration applications [Allied-Signal]

Genexol. A proprietary preparation of triisopropylphenoxypolyethoxyethanol; a spermicide. [W J Rendell Ltd] *

Genie®. For the agriculture industry. [DuPont UK] *

Genisol. Medicated shampoo. [Fisons plc, Pharmaceuticals Div] *

Genitron. CAS 123-77-3; Azodicarbonamide; blowing agents and azo-initiators. [Schering Industrial Products Ltd]

Genklene. Industrial solvents. [ICI Chem. & Polymers Ltd]

Genoa oil. Fine olive oil.

Genochrome. A stabilized photographic color developer. [May & Baker Ltd] *

Genoform. Methyleneglycolsalicylic ester; used in medicine, and is said to split up in the intestines into formaldehyde, salicylic and acetic acids.

Genomoll P. Trichlorethyl phosphate plasticizer. [Hoechst UK]

Genoptic Liquifilm®. CAS 1405-41-0; Gentamicin sulfate; a complex antibiotic substance, produced by *Micromonospora purpurea nsp*; has three components, sulfates of gentamicin C_1, gentamicin C_2, and gentamicin C_{1A}; antibacterial. [Allergan, Inc]

Genoptic S.O.P.®; CAS 1405-41-0; Gentamicin sulfate; antibacterial. [Allergan, Inc]

Genotherm. Flexible PVC film. [Hoechst UK]

Genoxal. *See* Endoxan. [Degussa AG] *

Genoxide®. CAS 7722-84-1; A registered trade name for a special quality of hydrogen peroxide for medical purposes. [Laporte Industries Ltd] *

Gensil. Silicone antifoaming agents. [Bevaloid Ltd] *

Genster. A trademark for a manufacture of carbon. §

Gentacidin® Ophthalmic Solution and Ointment. CAS 1405-41-0; Gentamicin sulfate; indicated in the topical treatment of infections of the external eye and its adnexa caused by susceptible bacteria (e.g., conjunctivitis, keratitis and keratoconjunctivitis, corneal ulcers, blepharitis and blepharoconjunctivitis, acute meibomianitis, and dacryocystitis). [Iolab Corp]

gentamicin. An antibiotic produced by *Micromonospora purpurea*. Garamycin. Genticin is the sulfate.

Genthane SR. (GS 338). A proprietary trade name for a polyurethane based molding compound with a temperature range from -60 C to +160 C. §

genthelvite. A mineral. It is $(Zn,Fe,Mn)_8Be_6Si_6O_{24}S_2$.

gentian. The dried rhizome and roots of *Gentiana lutea*, or of other *Gentiana* species; used in medicine and liqueurs.

Gentiannie. The zinc double chloride of dimethyldiaminophenazthionium chloride, $C_{14}H_{14}N_3SCl$. Dyes mordanted cotton bluish-violet.

gentisin. (gentianic acid) The yellow pigment of *Gentiana lutea*. It is 1:3:7-trihydroxyxanthone-3-methyl ether.

Gentisone HC. A proprietary preparation of gentamicin sulfate and hydrocortisone acetate; used in the treatment of infections and inflammations of the ear. [Nicholas Laboratories Ltd] *

Gentran 40. Dextran 40; Blood flow adjuvant; plasma volume extender. [Travenol Laboratories Inc] *

Gentran 75. Dextran 75; a polysaccharide produced by the action of *Leuconostoc mesenteroides* on sucrose; average molecular weight: 75,000; plasma volume extender. [Travenol Laboratories Inc] *

Genu. A range of carrageenan and pectin powders; emulsifiers, stabilizers, thickeners and gelling agents. [Hercules]

Genuzan. A proprietary gum blend; thickener and suspender in food and beverage industries, for personal care products and cosmetics, and in the pharmaceutical and medical industry. [Hercules]

Geocillin. A proprietary preparation of carindacillin sodium; an antibiotic used in the treatment of genito-urinary tract infections. [Pfizer International]

Geolast®. Thermoplastic rubber; high oil resistance rubber. [Advanced Elastomer Systems]

Geolite. Polyurethane intermediate. [Union Carbide (UK) Ltd]

Geolith. Grout for combatting spontaneous combustion in coal mines. [Foseco (F.S.) Ltd]

Geon® 8700A. CAS 9002-86-2; Rigid vinyl compound; extrusion compound for interior building products; profile applications [BFGoodrich/Geon Vinyl]

Geon® 8720. CAS 9002-86-2; Vinyl compound; wire and cable sheath; coaxial cable, noncontaminating jacket.

---

* Trade name not verified as available

† Trade name verified as obsolete

[BFGoodrich/Geon Vinyl]

Geon® 8812, 8813. CAS 9002-86-2; Vinyl compound; general purpose flexible molding and extrusion compound [BFGoodrich/Geon Vinyl]

Geon® 8896. CAS 9002-86-2; Vinyl compound; wire and cable insulation; SPT cord. [BFGoodrich/Geon Vinyl]

Geon® 83457. CAS 9002-86-2; Vinyl compound; general purpose flexible molding and extrusion compound [BFGoodrich/Geon Vinyl]

Geon® 83718. CAS 9002-86-2; Vinyl compound; flexible molding and extrusion compound; weatherstrip. [BFGoodrich/Geon Vinyl]

Geon® 86100, 86101, and 86103. CAS 9002-86-2; Vinyl compound; low-gloss grade for coextrusion capstock. [BFGoodrich/Geon Vinyl]

Geon® 87239, 87241. CAS 9002-86-2; Rigid vinyl compound; tin-stabilized; injection molding compound [BFGoodrich/Geon Vinyl] *

Geon®87396. CAS 9002-86-2; Rigid PVC compound; high gloss, high impact general-purpose bottle compound for extrusion blow molding; for opaque bottles, nonfood contact applications such as pkg. for automotive and marine additives, shampoos, and charcoal lighter fluids. [BFGoodrich/Geon Vinyl] *

Geon® 87420. CAS 9002-86-2; Vinyl compound; medical grade injection molding compound with high flow, normal impact; FDA compliance. [BFGoodrich/Geon Vinyl] *

Geon® HTX-6110. Vinyl-based alloy; engineering thermoplastic for elevated temp. performance, inherent flame retardance, excellent chem. and corrosion resist., toughness, electrical performance, surface appearance, processability. [BFGoodrich/Geon Vinyl] *

Geon® HTX 92190. CAS 9002-86-2; PVC compound; interior grade, high heat deflection, low expansion, rigid cube profile extrusion compound [BFGoodrich/Geon Vinyl]

Geon®W015. CAS 9002-86-2; Vinyl compound; for wire and cable applications [BFGoodrich/Geon Vinyl]

Geopen. A proprietary preparation of carindacillin sodium; an antibiotic used in the treatment of genito-urinary tract infections. [Pfizer International]

Georgia Gulf 3132 Clear 02. CAS 9002-86-2; Vinyl compound rigid profile extrusion compound for profiles, electronic pkg., integrated circuit magazine tubes; high impact, good clarity. [Georgia Gulf]

Georgia Gulf 5006. CAS 9002-86-2; Vinyl compound rigid profile extrusion compound for window profiles, siding and sheet; excellent weatherability. [Georgia Gulf]

Georgia Gulf 5006 General. CAS 9002-86-2; Vinyl compound rigid profile extrusion compound for general purpose applications incl. profile, sheet and elec.; high gloss and tensile strength. [Georgia Gulf]

Georgia Gulf 9105. CAS 9002-86-2; Vinyl compound general purpose blow molding grade for pkg. of cosmetics, toiletries and household products; excellent combination of impact strength, clarity and chem. resist. [Georgia Gulf]

Georgia Gulf 9151. CAS 9002-86-2; Vinyl compound chem. resist. grade for blow molding bottles for chem. active products like pine oil cleaners; good impact strength. [Georgia Gulf]

Georgia Gulf 9175J. CAS 9002-86-2; Vinyl compound general purpose injection blow molding grade for pkgs. for hair dyes and shampoos; excellent clarity, flow and stability. [Georgia Gulf]

Georgia Gulf 9202. CAS 9002-86-2; Vinyl compound food grade for mfg. of large returnable water bottles; excellent processing, low odor and taste. [Georgia Gulf]

Georgia Gulf CL-7049. CAS 9002-86-2; Engineering vinyl; clear grade with med. flow for injection molding for FDA applications; excellent impact strength. [Georgia Gulf]

Georgia Gulf EH-71L. CAS 9002-86-2; PVC disp. resin; high molecular weight, low visc. grade for coating and molding applications [Georgia Gulf]

Georgia Gulf EX-240. CAS 9002-86-2; PVC disp. resin; high performance grade with elastomer-like props. for low temp. and fat fusion applications [Georgia Gulf]

Georgia Gulf HH-1900. CAS 9002-86-2; Engineering vinyl alloy; high heat resistant molding material with high flow. [Georgia Gulf]

Georgia Gulf HM-7054. CAS 9002-86-2; Engineering vinyl; special purpose molding material with high modulus and very good surface hardness; used for load bearing applications [Georgia Gulf]

Georgia Gulf SP-7107. CAS 9002-86-2; Engineering vinyl; special purpose molding material with high flow for applications requiring excellent thermal stability and very good mech. props. [Georgia Gulf]

Georgia Gulf UV-7160. CAS 9002-86-2; Engineering vinyl; uv-resistant, high flow super tough blend for enclosures. [Georgia Gulf]

Geoseal. Soil grouting resins. [Borden (UK) Ltd] *

Geostone®. Class IV dental stone; for high quality models and dyes. [Bayer AG]

Geostop. Accelerated anhydrite for mines roadways stoppings. [Foseco (F.S.) Ltd] †

Gepel. Reactive silane-modified siloxane; protective penetrant. [GE Silicones]

Geracryl. Patterned extruded acrylic sheet; for glazing, shower screens etc. [Cornelius Chemical Co Ltd] *

geraniol. (trans-3,7-dimethyl-2,6-octadien-1-ol) A terpene alcohol; $C_{10}H_{18}O$; CAS 106-24-1; EINECS 203-377-1; Perfumery, constituent of synthetic fragrances and synthetic linalool. [Penta Mfg.; Schweizerhall; SCM Glidco Organics]

geraniol acetate. See geranyl acetate.

geranium crystals. CAS 101-84-8; Diphenyl ether, $C_6H_5 \cdot O \cdot C_6H_5$; used in perfumery.

geranium oil. CAS 8000-46-2; Pale yellow or greenish liquid, having geraniol as its main constituent; used in perfumes and cosmetics.

geranyl acetate. (geraniol acetate) $(CH_3)_2C:CH-CH_2CH_2C(CH_3):CHCH_2OCOCH_3$; CAS 16409-44-2; EINECS 240-458-0; Perfumery, flavoring. [Firmenich; Int'l. Flavors & Fragrances; Penta Mfg.; SCM Glidco Organics]

Gerhardt's caustic. This consists of litharge boiled with potassium hydroxide until it is dissolved, and water is added.

Germaben® II. Diazolidinyl urea (30%), propylene glycol (56%), methylparaben (11%), and propylparaben (3%); broad-spectrum antimicrobial preservative for cosmetic products [Sutton Labs]

Germalgene. Trichlorethylene, C_2HCl_3.

Germall®115. CAS 39236-46-9; Imidazolidinyl urea; antimicrobial preservative for cosmetics. [Sutton Labs]

Germall® II. CAS 78491-02-8; Diazolidinyl urea; broad-spectrum antimicrobial preservative for cosmetics and

---

‡ Trade name and manufacturer not verified   § Trade name without identified manufacturer

toiletries. [Sutton Labs]

German silver. *See* nickel silvers.

German silver solder. Usually consists of 5 parts German silver, and 4 parts zinc.

German turpentine. *See* turpentine.

German yeast. Dried yeast.

Germanin®. Suramin; pharmaceutical preparation; antitrypanosomal preparation. [Bayer AG]

germanium. Nonmetallic element; Ge; CAS 7440-56-4; EINECS 231-164-3; solid state electronic devices, semiconducting applications, brazing alloys, phosphors, gold and beryllium alloys, infrared-transmitting glass. [Atomergic Chemetals; Cabot; Cerac; Eagle-Picher; New Metals & Chems. Ltd; Noah Chem.]

Germ-i-Tol. Benzalkonium chloride; pharmaceutic aid. [Hexcel Chemical Products] *

Germul. Vinyl acetate and vinyl acrylic emulsions; base for adhesives. [Atochem UK/Ceca]

Germul A 735. Crosslinking acrylic resin; textile hand modifier. [Ceca SA]

Geronol. Blend of nonionic and anionic surfactants; pesticide emulsifiers. [Geronazzo S.p.A]

Geronol AG-100/200 Series. Complex anionic-nonionic ethoxylate blends; emulsifiers for toxicants, esp. organophosphate insecticides and nonsaponifiable herbicides. [Rhone-Poulenc Surf.]

Geropon®ACR/4. Disodium laureth sulfosuccinate; emulsifier for emulsion polymerization, detergent base, foamer, foam stabilizer, dispersant. [Rhone-Poulenc France]

Geropon® AS-200. Sodium cocoyl isethionate, coconut acid, stearic acid; detergent, wetting agent, dispersant, suspending agent for textile wet processing, industrial/household detergents, cosmetics, agric. pesticides, leather, rubber, etc.; readily biodeg. [Rhone-Poulenc Surf.]

Geropon® CYA/60. CAS 577-11-7; EINECS 209-406-4; Sodium dioctyl sulfosuccinate; wetting agent, surface tension reducer, visc. depressant for emulsion polymerization of PVC. [Rhone-Poulenc Geronazzo]

Geropon® CYA/DEP. CAS 127-39-9; Sodium diisooctyl sulfosuccinate; wetting agent for textile industry, pesticides, emulsion polymerization, printing inks, water paints; stable in acid media. [Rhone-Poulenc Geronazzo]

Geropon® DOS. CAS 577-11-7; EINECS 209-406-4; Sodium dioctyl sulfosuccinate; wetting agent, emulsifier, demulsifier for textile wet processing, specialty cleaners, dewatering agent for flotation concs., oil spill clean-up blends; limited stability in alkaline or acidic media. [Rhone-Poulenc France]

Geropon®LSS. CAS 36409-57-1; Disodium lauryl sulfosuccinate; low-irritant detergent for shampoos and bubble baths. [Rhone-Poulenc Surf.; Rhone-Poulenc France]

Geropon®MLS/A. CAS 1561-92-8; Sodium methallyl sulfonate; dye improver reactive comonomer for acrylic fibers polymerization; reactive emulsifier or coemulsifier in latex emulsion polymerization. [Rhone-Poulenc Geronazzo]

Geropon®S-1585. Disodium octylphenoxy sulfosuccinate; surfactant. [Rhone-Poulenc France]

Geropon® SBFA-30. CAS 68815-56-5; Disodium laureth sulfosuccinate; skin protecting anti-irritant surfactant.

[Rhone-Poulenc Surf.]

Geropon®SBL-203. CAS 25882-44-4; EINECS 247-310-4; Disodium lauramido MEA-sulfosuccinate; improves flash foam of anionic systems; produces brittle, tack-free residue for carpet shampoos. [Rhone-Poulenc Surf.]

Geropon® SBR-3. CAS 40754-60-7; Disodium ricinoleamido MEA-sulfosuccinate; skin protecting anti-irritant surfactant. [Rhone-Poulenc Surf.]

Geropon® SS-L7DE. Sodium lauramido DEA sulfosuccinate; mild detergent, foam booster/stabilizer for liq. dish detergent and toiletry preps. [Rhone-Poulenc Surf.]

Geropon® T-22/A. CAS 137-20-2; EINECS 205-285-7; Sodium methyl oleoyl taurate; surfactant, dispersant for agric. formulations. [Rhone-Poulenc Surf.]

Geropon® T-33. CAS 137-20-2; EINECS 205-285-7; Sodium N-methyl-N-oleoyl taurate; detergent, dispersant, wetting agent for textile and general-purpose applications; dye assistant; for kier boiling, bleaching, wetting, finishing of textiles; in industrial detergents, rug shampoos, bottle washing compounds, metal cleaners, paper industry. [Rhone-Poulenc Surf.; Rhone-Poulenc France]

Geropon® TC-42. CAS 61791-42-2; Sodium N-methyl-N-cocoyl taurate; foamer, dispersant, detergent for detergent bars, shampoos, bubble baths, cosmetics; chemically stable. [Rhone-Poulenc Surf.; Rhone-Poulenc France]

Geropon® TK-32. CAS 61791-41-1; Sodium N-methyl-N-tallowyl taurate; detergent, suspending agent, dispersant; precipitation inhibitor for org. and inorg. salts of Ba, Ca, Sr; for petrol. industry. [Rhone-Poulenc Surf.]

Geropon® WS-25, WS-25-I. CAS 63217-13-0; Sodium dinonyl sulfosuccinate; rewetting agent for textile finishing, in application of resins, softeners, starches; wetting and dispersing agent for latex paints. [Rhone-Poulenc Surf.; Rhone-Poulenc France]

Geropon® X2152. Ammonium dioctyl sulfosuccinate; wetting agent, emulsifier, demulsifier, stabilizer in water treatment and petrol. processing; stable at neutral pH. [Rhone-Poulenc France]

Gesagard. CAS 7287-19-6; Wettable powder containing 50% w/w prometryn; a contact triazine herbicide. [Ciba-Geigy Agrochemicals]

Gesaprim 500FW. CAS 1912-24-9; A liquid formulation containing 500 g atrazine per liter as a suspension concentrate; a residual herbicide. [Ciba-Geigy Agrochemicals]

Gesatop. CAS 122-34-9; Suspension concentrate or wettable powder containing simazine; a triazine herbicide to control weeds and grasses in cane fruit, roses and some vegetables. [Ciba-Geigy Agrochemicals]

Gesilit. Safety explosives. No. 1 contains 30.75% nitroglycerin jelly, 5.25% dinitrotoluene, 7% sodium chloride, 18% sodium nitrate, and 39% dextrin. No. 2 consists of 30.75% nitroglycerin jelly, 5.25% dinitrotoluene, 22% ammonium nitrate, 21% sodium chloride, 21% dexrrin.

Gesteins-tremonit V. An explosive containing ammonium nitrate, nitroglycerin, vegetable meal, potassium perchlorate, and nitro-compounds.

Gesteins-Westfalit B and C. Explosives. They are ammonals containing dinitrobenzene, and dinitrotoluene respectively.

Gesterol 50. CAS 57-83-0; Progesterone; progestin. [O'Neal, Jones & Feldman Pharmaceuticals] *

---

* Trade name not verified as available   † Trade name verified as obsolete

Gestinal®. Amniotic fluid substitute and lubricant; veterinary medicine. [Bayer AG]

Gestone. CAS 57-83-0; Progesterone BP suitable for oral or injectable administration; used for the treatment of premenstrual syndrome and puerperal depression; also indicated for habitual and threatened abortion and dysfunctional uterine bleeding. [Paines & Byrne Ltd] *

Getah wax. See Java wax.

Getren®4/200. Silicone-free release agent for foundry and steel industry, plastics incl. unsat. polyester, epoxy, and PU resins; lubricant for tire production [Goldschmidt] *

Getren® FD 575. Release agent for the rubber industry; mainly for production of radiator hoses. [Goldschmidt] *

Gevodin. A proprietary preparation of famprofazone, paracetamol, isopropyl phenazone and caffeine; an analgesic. [Geistlich Sohne AG] *

Gevral. A proprietary preparation of multivitamins and minerals. [Lederle Laboratories] *

GFS®. Xanthan gum/galactomannan blend; gum for use in food preparations; stabilizer, suspending agent, thickener; provides rheological control to water- and milk-based foods. Kelco]

GH5. Granular fertilizer. [Fisons plc, Horticulture Div] *

gibberellic acid. (GA$_3$) C$_{19}$H$_{22}$O$_6$; CAS 77-06-5; EINECS 201-001-0; Agriculture and horticulture, malting of barley with improved enzymatic characteristics. [Am. Biorganics; Atomergic Chemetals; U.S. Biochemical]

Giemsa's stain. A microscopic stain for white blood corpuscles. It contains eosin, glycerin, and methanol.

Gigantem®. For the treatment of circulatory disorders. [Bayer AG]

Gilalgin. Alginate; dental impression material. [Giulini Corp]

Gildent. Stump material for dentures. [Giulini Corp]

gilding metal. A jeweler's alloy of 90% copper, and 10% zinc. Another alloy contains 70% copper, 17.5% brass, and 12.5% tin.

gilding solutions. These generally consist of solutions of gold chloride and potassium carbonate in water; used for the electro-deposition of gold.

gilsonite. An asphaltic material or solidified hydrocarbon; acid, alkali and waterproof coatings; black varnishes, lacquers, baking enamels; wire-insulation compounds; linoleum, floor tile; paving; insulation; diluent in low-grade rubber compounds; possible source of gasoline, fuel oil. [R.E. Carroll]

Gilsonite. Naturally occurring bitumen; for oil well cements and drilling fluids, printing inks, foundry sand additive, explosives, asphalt pavement sealer, bituminous paints. [Mercian Minerals & Colours Ltd] ‡

Gilsonite and Design. Uintaite resin. [Chevron] *

Gilstone. Synthetic gypsum. [Giulini Corp]

Gilumag. CAS 1309-42-8; Magnesium hydroxide. [Giulini Corp]

Gina. A proprietary preparation of propatylnitrate; used to treat angina pectoris. §

ginal. A purifier for sugars. It contains the sodium alginate (from seaweed).

gingelly oil. (gingili oil, teal oil, teel oil, til oil, beni oil, benne oil, beniseed oil). CAS 8008-74-0; Sesame oil obtained from the seeds of Sesamum indicum and of S. orientale; used in the manufacture of margarine and soap, and as a burning oil.

Gingicain. A proprietary preparation of amethocaine gentisate and chlorbutol in aerosol; a local anesthetic.

[Hoechst UK] *

gingili oil. See gingelly oil.

Ginsene. Ethylcyclohex-3-ene carboxylate. [Bush Boake Allen Ltd]

ginseng. The root of Panax quinque-folium; used in medicine

Gin-shi-bui-chi. A Japanese alloy of 30-50% silver with copper.

Gippon. Antiseptic paint. [J C Bottomley] *

Girard's reagent. Betaine hydrazide hydrochloride.

githagin. See struthiin.

Givgard DXN. 6-Acetoxy-2,4 dimethyl-m-dioxane; nonformaldehyde bactericidal and fungicidal agent for industrial use; used for all kinds of industrial water-based systems such as emulsions, suspensions and dispersions. [Givaudan SA] *

Givsorb® UV-1. CAS 57834-33-0; N^2-(4-Ethoxy-carbonylphenyl)-N′-methyl-N′-phenylformamidine; industrial uv absorber [Givaudan; Givaudan SA] *

Givsorb® UV-2. CAS 65816-20-8; N-(p-ethoxy-carbonylphenyl)-N′-ethyl-N′-phenylformamidine; industrial UV absorber, light stabilizer, and antioxidant for PU, PVC, polyolefins, ABS, nylon, and acetal resins; photostable, broad spectrum screening agent for protection against the adverse effects of both uv-B and uv-A radiation; used in surface coatings, polishes, dyestuffs, carpet treatments. [Givaudan; Givaudan SA] *

Gladiator. CAS 1698-60-8; Suspension concentrate containing 430 g chloridazon per liter; a pyridazinone herbicide for beet crops. [Tripart Farm Chemicals Ltd]

Gladixol®. Cardiac stimulant. [Bayer AG]

Glagerite. A Bavarian white clay.

Glanzan PHN Conc. Wax emulsion; nonionic; luster finish for yarns, woven and knitted fabrics made of cotton, regenerated cellulose as well as their blends. [Thor Chemicals (UK) Ltd]

Glascol®. A range of acrylic polymers in water; these resins are film formers, have pigment binding properties and dry to form hard, flexible, clear films; uses include paper and board coatings, printing inks, overprint varnishes, artists colors. [Allied Colloids Ltd]

Glascol®HN2. A proprietary aqueous acrylic copolymer, the sodium salt of which gives hard and brittle films soluble in water. [Allied Colloids Ltd] *

Glascol®HN4. A proprietary aqueous acrylic copolymer the sodium salt of which gives soft and flexible films soluble in water. [Allied Colloids Ltd] *

Glascol®PA6. A proprietary acrylic copolymer supplied in the form of low viscosity aqueous solutions; the ammonium salt gives tacky, pressure-sensitive films resistant to water. [Allied Colloids Ltd] *

Glascol®PA8. A proprietary acrylic copolymer supplied in the form of low viscosity aqueous solutions; the ammonium salt gives soft, tacky, pressure-sensitive films resistant to water. [Allied Colloids Ltd] *

Glascol®PN 8. A proprietary acrylic copolymer supplied in the form of low viscosity aqueous solutions; the sodium salt gives soft, tacky, pressure-sensitive films soluble in water. [Allied Colloids Ltd] *

Glasdag. A range of lubricating greases, oils and coatings based on graphite as the lubricating pigment; designed especially for the manufacture of glass containers. [Acheson Colloids]

Glaser's salt. Potassium sulfate and sulfite.

---

‡ Trade name and manufacturer not verified   § Trade name without identified manufacturer

Glasgro. Range of granular fertilizers. [Fisons plc, Horticulture Div] *

Glaskyd®. Thermoset polyester molding compounds for industrial applications requiring elec. insulation, high strength, rigidity at elevated temps. and corrosion resist. [Am. Cyanamid] *

Glass H. CAS 10124-56-8; Sodium hexametaphosphate. [FMC]

Glassclad® 18. 20% Sol'n. in t-butanol and diacetone alcohol; hydrophobic treatment and lubricant for glass and ceramics. [Hüls Am.]

Glass Guard. Viscous liquid applied to glass and metal for protection against acid-based cleaners; used for historical glass, aluminum window castings etc. [Nova Chemical Inc] *

glassite. Magnetic oxide of iron from precipitation of ferrous sulfate with caustic soda. Also called black rouge; used for buffing.

Glass Liquor. See soluble glass.

glass-maker's soap. Manganese dioxide, MnO_2.

Glassona. Cellulose-bonded fiberglass bandage. [Smith & Nephew Pharmaceuticals Ltd] *

glass silk. Glass wool.

Glass Sponge. A patented sponge-like product obtained by mixing glass wool with salt, heating, then dissolving out the salt. §

Glass, Water. See soluble glass.

Glauber's salt. CAS 7727-73-3; (Mirabilite). Sodium sulfate decahydrate, $Na_2SO_4 \cdot 10H_2O$; used in solar heat storage and air conditioning.

glauconic acids. Bluish-violet dyestuffs, obtained by the successive action of pyroracemic acid and formaldehyde on aromatic primary amines.

Glaucosil. The siliceous residue obtained by extracting greensand with mineral acids. It consists of practically pure silica; used as an absorbent for gases.

Glauramine. A proprietary solution of specially purified auramine; a powerful antiseptic. §

Glaurin. Diethylene glycol monolaurate; a proprietary trade name for a plasticizer. §

Glazamine. Textile auxiliary chemicals. [ICI Chem. & Polymers Ltd]

Glaze 'N Seal Concrete and Masonry Sealer. Solvent base sealer; a clear nonyellowing acrylic for application to exterior and interior concrete, stone and masonry type surfaces. [Glessner Corporation Inc (GGI Products) DBA] *

Glaze 'N Seal Waterbase Clear Concrete and Brick Sealer. Waterbase sealer; an acrylic emulsion for application to interior concrete, stone and masonry type surfaces; nonflammable. [Glessner Corporation Inc (GGI Products) DBA] *

glazier's salt. Potassium sulfate, K_2SO_4.

Glean® TP. Bromoxynil + chlorsulfuron + ioxynil; herbicide mixture for weed control in cereals. [DuPont UK]

Glekosa. See Chloramine T. §

Glendion. Polyether and polyester polyols. [Montedison UK Ltd] *

Gleptosil. Gleptoferron; hematinic. [Fisons plc, Pharmaceuticals Div] *

Glessite. A variety of amber melting at 250-300 C.

Gletvax. Porcine E coli vaccine. [The Wellcome Foundation Ltd] *

Glianimon®. Neuroleptic. [Bayer AG]

Glibadone. Glibenclamide; pharmaceutical preparation for the treatment of maturity onset diabetes. [M A Steinhard Ltd] *

Glibenese. CAS 29094-61-9; A proprietary preparation of glipizide; an oral hypoglycemic agent. [Pfizer International]

Glievor bearing metals. One alloy contains 76% lead, 14% antimony, 8% tin, and 2% iron, and another consists of 73% zinc, 9% antimony, 7% tin, 5% lead, and 4% copper.

glimmer. Mica.

Glissofluid® A 10, A 13. Aliphatic dicarboxylic acid ester; component for synthetic lubricants. [BASF AG; BASF plc]

Glissolube®. Polyalkyleneoxide derivs.; synthetic components for high-performance lubricants. [BASF AG; BASF plc]

Glissopal®. Intermediates and components for production of lubricating oil additives and lubricating oils. [BASF AG]

Glissosafe®. Hydraulic fluids for mining. [BASF AG; BASF plc]

Glissoviscal® B. Polyisobutylenes; thickener (visc. improver) for lubricating oils. [BASF AG; BASF plc]

glist. Mica.

Glitzi. Household pad. [Carl Freudenberg] *

Glizarin Binder. Textile finishing cross-linking agents. [BASF plc] *

GLOB®. Urethane polymer with sika powds.; urethane/carbide mixt. [TSE Industries]

Globe Granite. A stone similar to Ward's stone (qv).

globulin. (fibrino-plastic substance, paraglobin, paraglobulin, serum casein). Blood protein

Glocure. Benzoin ether. [Rhone-Poulenc UK]

Glofoam. Synthetic wax. [Rhone-Poulenc UK]

Glokem. Synthetic wax. [ABM Chemicals Ltd] *

Glokill. Heterocyclic biocide. [Rhone-Poulenc UK]

Glokill 77. Nonsurface active biocide. [Rhone-Poulenc UK]

Glokill PQ. Polymeric quaternary ammonium compound in the form of a pale yellow aqueous solution; slow acting nonfoaming biocide for water treatment where foam is a problem. [Rhone-Poulenc UK]

Glomeen. Chlorine releasing sterilant. [ABM Chemicals Ltd] *

Glo-Mold®. Elastomeric compound; mold cleaning compounds for preventive maintenance. [Glo-Mold] *

Glonoine oil. nitroglycerin.

Glopol. Dispersing agent for aqueous systems. [Rhone-Poulenc UK]

Glopol 461. Polymeric quaternary ammonium salt as a pale yellow slightly viscous liquid; cationic surfactant used as resin to impart electro-conductive properties to paper and ceramics. [Rhone-Poulenc UK]

Gloquat 1032. Dialkyl quaternary ammonium salt as a brown viscous liquid; cationic surfactant with hydrophobe, anticorrosive, dispersant and electrostatic properties; used in pigment dispersion, paints and coatings, car washes. [Rhone-Poulenc UK]

Gloquaternary Cationic surfactant, biocide. [Rhone-Poulenc UK]

Gloquaternary Alkylaryl trimethyl ammonium chloride. horticultural algicide and moss-killer. [Fargro Ltd]

Gloquaternary Alkylaryl trimethyl ammonium chloride. horticultural algicide and moss-killer. [Flowering Plants Ltd] †

Gloria®. White mineral oil. [Witco/Sonneborn]

---

* Trade name not verified as available   † Trade name verified as obsolete

glossite. An abrasive, the active material of which is stated to be black oxide of iron.

Glossova. Floor polishes. [Evans Vanodine International Ltd]

Glover's wool. See Tanner's wool.

Glowtein. Poultry feed additive, colorant. [Mitchell Cotts Chemicals Ltd]

Gloy. A trademark for an adhesive said to be a mixture of dextrin and starch, with magnesium chloride. §

Gluadin®AGP. Hydrolyzed wheat protein; pleasant skin feel for emulsions and surfactant preps. [Henkel/Cospha; Henkel Canada] *

Glucagon. A polypeptide hormone produced in the alpha cells of the islets of Langerhans in the pancreas; proprietary preparation. [Eli Lilly & Co]

glucal. A reducing compound, $C_6H_{10}O_4$, obtained by the reduction of β-aceto-bromo-glucose with zinc dust and acetic acid.

glucal. See calcium gluconate.

Glucam® E-10. CAS 68239-42-9; Methyl gluceth-10; humectant for personal care products; freezing pt. depressant; emollient in aq. and hydroalcoholic products; moisturizer; foam modifier in detergent and shampoo systems; solv. and solubilizer for topical pharmaceuticals; used in emulsions, toilet articles; adds gloss, conditioning. Amerchol Corp]

Glucam® E-20. CAS 68239-43-0; Methyl gluceth-20; humectant for personal care products; freezing pt. depressant; emollient in aq. and hydroalcoholic products; moisturizer; foam modifier in detergent and shampoo systems; solv. and solubilizer for topical pharmaceuticals. Amerchol Corp]

Glucam® E-20 Distearate. CAS 98073-10-0; Methyl gluceth-20 distearate; auxiliary o/w emulsifier, moisturizer, emollient and lubricant for cosmetics and pharmaceuticals; conditioner. Amerchol Corp]

Glucam®P-10. CAS 61849-72-7; PPG-10 methyl glucose ether; humectant for personal care products; freezing pt. depressant; emollient in aq. and hydroalcoholic products; moisturizer; foam modifier in detergent and shampoo systems; solv. and solubilizer for topical pharmaceuticals. Amerchol Corp]

Glucam® P-20. CAS 61849-72-7; PPG-20 methyl glucose ether; humectant for personal care products; freezing pt. depressant; emollient in aq. and hydroalcoholic products; moisturizer; foam modifier in detergent and shampoo systems; solv. and solubilizer for topical pharmaceuticals. Amerchol Corp]

Glucam® P-20 Distearate. PPG-20 methyl glucose ether distearate; skin moisturizer, conditioner, and emollient for cosmetics and pharmaceuticals; binder and plasticizer for pressed powds. Amerchol Corp]

Glucamate®DOE-120. PEG-120 methyl glucoside dioleate; thickener, emulsifier, solubilizer for shampoos. Amerchol Corp]

Glucamate® SSE-20. CAS 68389-70-8; PEG-20 methyl glucose sesquistearate; o/w emulsifier, solubilizer used with Glucate SS; effective at low concs. [Amerchol; Amerchol Europe]

Glucanal. Proprietary preparations of silver and anthraquinone glucosides. §

Glucanex. A β-glucanase preparation produced by a selected strain of *Trichoderma*; can be used in all cases where the aim is to improve the clarification and the filtrability of wines made from botrytized grapes. [Novo Nordisk]

Glucanex®L-300. β-Glucanase; enzyme for hydrolysis of β-glucans; for beer breweries. [Solvay Enzymes]

glucase. Maltase, an enzyme which converts maltose into glucose.

Glucate®DO. CAS 83933-91-3; Methyl glucose dioleate; w/o emulsifier, auxiliary emulsifier for o/w systems; conditioner, emollient, lubricant, plasticizer, and pigment dispersant. [Amerchol; Amerchol Europe]

Glucate®IS. Methyl glucose sesquiisostearate; primary w/o emulsifier for personal care products. Amerchol Corp]

Glucate® SS. CAS 68936-95-8; Methyl glucose sesquistearate; w/o emulsifier used with Glucamate SSE-20 to provide visc. stability, mildness. [Amerchol; Amerchol Europe]

Glucidex. A range of dried glucose syrups or maltodextrins; used in various food applications to provide, for example, bulk with or without sweetness. [Roquette (UK) Ltd]

Glucina. Beryllium oxide, BeO.

D-glucitol. See sorbitol.

Gluckauf. A German safety explosive containing ammoniuna nitrate, wood meal, dinitrobenzene, and copper oxalate.

Glucobay. Pharmaceutical preparation containing acarbose, a glucosidase inhibitor; used for diabetes mellitus. [Bayer AG]

Glucodin. A proprietary preparation of glucose and ascorbic acid; a food supplement. [Farley Health Products] ‡

Gluconal®CA A. CAS 299-28-5; Calcium gluconate anhydrous; pharmaceutical/food grade mineral source for human and veterinary pharmaceutical preps., dietary supplements, fortified foods and animal feed. [Akzo Chemie]

Gluconal®CA M. CAS 299-28-5; Calcium gluconate monohydrate; pharmaceutical/food grade mineral source for human and veterinary pharmaceutical preps., dietary supplements, fortified foods and animal feed. [Akzo Chemie]

Gluconal® CA M B. CAS 5743-34-0; Calcium borogluconate; pharmaceutical/food grade mineral source for human and veterinary pharmaceutical preps., dietary supplements, fortified foods and animal feed. [Akzo Chemie]

Gluconal® CO. Cobalt gluconate; pharmaceutical/food grade mineral source for human and veterinary pharmaceutical preps., dietary supplements, fortified foods and animal feed. [Akzo Chemie]

Gluconal®CU. CAS 527-09-3; Copper gluconate; pharmaceutical/food grade mineral source for human and veterinary pharmaceutical preps., dietary supplements, fortified foods and animal feed. [Akzo Chemie]

Gluconal®FE. CAS 299-29-6; Ferrous gluconate; pharmaceutical/food grade mineral source for human and veterinary pharmaceutical preps., dietary supplements, fortified foods and animal feed. [Akzo Chemie]

Gluconal® K. CAS 299-27-4; Potassium gluconate; pharmaceutical/food grade mineral source for human and veterinary pharmaceutical preps., dietary supplements, fortified foods and animal feed. [Akzo Chemie]

Gluconal® MG. Magnesium gluconate; pharmaceutical/food grade mineral source for human and veterinary pharmaceutical preps., dietary supplements, fortified foods and animal feed. [Akzo Chemie]

‡ Trade name and manufacturer not verified   § Trade name without identified manufacturer

Gluconal® MN. Manganese gluconate; pharmaceutical/ food grade mineral source for human and veterinary pharmaceutical preps., dietary supplements, fortified foods and animal feed. [Akzo Chemie]

Gluconal®NA. CAS 527-07-1; Sodium gluconate; pharmaceutical/food grade mineral source for human and veterinary pharmaceutical preps., dietary supplements, fortified foods and animal feed. [Akzo Chemie]

Gluconal® ZN. CAS 4468-02-4; Zinc gluconate; pharmaceutical/food grade mineral source for human and veterinary pharmaceutical preps., dietary supplements, fortified foods and animal feed. [Akzo Chemie]

gluconate. See chlorhexidine digluconate.

gluconic acid. (glyconic acid; glycogenic acid; dextronic acid) $CH_2OH(CHOH)_4COOH$; CAS 526-95-4; EINECS 208-401-4; pharmaceuticals and food products, cleaning and pickling metals, sequestrant, cleansers, catalyst in textile printing. [Akzo; Am. Biorganics; Faesy & Besthoff; Glucona; PMP Fermentation Prods.]

D-gluconic acid calcium salt. See calcium gluconate.

D-gluconic acid magnesium salt. See magnesium gluconate.

D-gluconic acid, monosodium salt. See sodium gluconate

D-gluconic acid potassium salt. See potassium D-gluconate.

gluconic acid sodium salt. See sodium gluconate.

Glucophage. A proprietary preparation of metformin hydrochloride; an oral hypoglycemic agent. [Rona Laboratories] ‡

Glucopon 225. CAS 68515-73-1; C8,10 alkyl polyglycoside; surfactant, detergent, wetting agent, surface tensile reducer, hydrotrope, dispersant for laundry detergents, liq. cleaners, hard surface cleaners, institutional and industrial cleaners; biodeg. [Henkel/Emery]

Glucopon 425. CAS 68515-73-1, 110615-47-9; C8-16 alkyl polyglycoside; surfactant, detergent, wetting agent, surface tensile reducer, hydrotrope, dispersant for laundry detergents, liq. cleaners, hard surface cleaners, institutional and industrial cleaners; biodeg. [Henkel/Emery]

Glucopon 600. CAS 110615-47-9; C12-16 alkyl polyglycoside; surfactant, detergent, wetting agent, surface tensile reducer, hydrotrope, dispersant for laundry detergents, liq. cleaners, hard surface cleaners, institutional and industrial cleaners; biodeg. [Henkel/Emery]

glucose. (dextrose; grape sugar; corn sugar) Sugar obtained from the hydrolysis of starch; $C_6H_{12}O_6$; CAS 50-99-7 (anhydrous), 5996-10-1 (hydrous); EINECS 207-757-8; Confectionery, foods, medicine, brewing, baking, canning. [Am. Biorganics; Amerchol NV; Avebe BV; Corn Prods.; Mallinckrodt; Mendell; U.S. Biochemical]

D-glucose, 4-O-β-D-galactopyranosyl-. See lactose.

glucose or sugar vinegar. A vinegar prepared by the conversion of starch substance into sugar, by the action of dilute acids, followed by fermentation and acetification.

glucose syrup. (dextrin-maltose). A partially hydrolyzed starch employed in brewing and in confectionery.

glucose syrup. See corn syrup.

Glucostix®. Reagent for blood sugar determination. [Bayer AG]

glucotannin. A tannin found in Chinese rhubarb. It is 1-galloyl-β-glucose.

Glucox. Glucose oxidase. [John & E Sturge Ltd, Selby] *

Glucquat® 125. Lauryl methyl gluceth-10 hydroxy-

propyldimonium chloride; substantive conditioner for hair and skin care products; humectant, moisturizer [Amerchol Corp]

Gluferate. A proprietary preparation of ferrous gluconate and ascorbic acid; a hematinic. [Wyeth Laboratories] *

Gluma. Dentin bonding agent for composite filling materials; used in dentistry. [Bayer AG]

Glumal. CAS 12607-92-0; Aceglutamide aluminum; anti-ulcerative. [Kyowa Hakko Kogyo Co Ltd] *

Glumorin. CAS 9001-01-8; A proprietary preparation of kallikrein. [FBA Pharmaceuticals] ‡

Glurenorm. CAS 33342-05-1; Tablets containing gliquidone; oral hypoglycemic agent used for treatment of noninsulin dependent diabetes. [Winthrop Laboratories] *

Glurub. A proprietary rubber-glue compound for rubber stiffening. §

Glutalys®. Maize gluten; used for animal feed. [Roquette (UK) Ltd]

glutamic acid. (L-glutamic acid; L-2-aminoglutaric acid; 2-aminopentanedioic acid) $C_5H_9NO_4$; CAS 56-86-0; EINECS 200-293-7; Medicine, biochemical research, salt substitute, flavor enhancer (L-form). [Am. Biorganics; R.W. Greeff; Penta Mfg.; U.S. Biochemical]

L-glutamic acid. See glutamic acid.

L-glutamic acid 5-amide. See L-glutamine.

glutamic acid, monosodium salt. See MSG.

L-glutamine. (L-glutamic acid 5-amide; 2-amino-4-carbamoylbutanoic acid) $C_5H_{10}N_2O_3$; CAS 56-85-9; EINECS 200-292-1; Medicine, culture media, biochemical research, feed additive. [Degussa; R.W. Greeff; Janssen Chimica; Penta Mfg.; Tanabe USA]

glutaraldehyde. $OCH(CH_2)_3CHO$; CAS 111-30-8; EINECS 203-856-5; Intermediate, fixative for tissues, for cross-linking protein and polyhydroxy materials, tanning of soft leathers. [Allchem Industries; BASF; Transol Chem. UK Ltd; Union Carbide]

Glutarol. CAS 111-30-8; Colorless evaporative solution containing 10% w/v glutaraldehyde; used for the topical treatment of warts, especially plantar warts. [Dermal Laboratories Ltd]

Glutarom® Range. Amino acid salt mixt. on carrier; natural flavor enhancer. [Grünau]

Glutol. Formaldehyde gelatin obtained by evaporating a solution of gelatin with one of formaldehyde; employed as an antiseptic powder for wounds; also a name applied to a proprietary synthetic resin. §

Glutolin. Methyl cellulose.

Glutril. CAS 26944-48-9; A proprietary preparation of glibornuride; used in the treatment of diabetes mellitus. [Roche Products Ltd] *

Glutrin. CAS 8061-52-7; Modified calcium lignosulfonates; binders for foundry and refractory brick manufacture. [Borregaard LignoTech]

Glyakol. Diglyceryl ether tetracetate.

Glybrom. Pyrabrom; antihistaminic. [Iolab Corp] †

Glycamyl. See glycerin of starch.

glycarbin. Glyceryl carbonate.

Glycene. A proprietary trade name for an alkyd synthetic resin used for dentures. §

Glyceria wax. A wax formed in the stem of cane grass, *Glyceria ranirgera*, of Australia. It melts at 82 C.

glycerin. (glycerol; 1,2,3-propanetriol; glycerine; glycyl alcohol) Polyhydric alcohol; $HOCH_2COHHCH_2OH$; CAS

56-81-5; EINECS 200-289-5; alkyd resins, dynamite, ester gums, pharmaceuticals, cosmetics, perfumery, lubricants, softener, bacteriostat, penetrant, solvent, humectant, plasticizer, emollient; antifreeze; in production of antibiotics. [Alba Int'l.; Asahi Denka Kogyo; Farleyway Chem. Ltd; Fina; Henkel/Emery; Lonza; Procter & Gamble; Unichema; Witco/Humko]

glycerin-formal. A condensation product of glycerol and formaldehyde; used as a solvent for lacquer, preservatives against molds and bacteria.

glycerite. See tannic acid.

Glycero-ester. A trade name for ester gum. §

Glycerogen. A mixture of polyhydric alcohols obtained by inversion of sugar to hexose, then reduction with hydrogen and vacuum distilled. The final product is 40% glycerine, 40% propylene glycol and 20% hexyl alcohols.

glycerol. See glycerin.

glycorol ester hydrolase. See lipase.

Glycero-piperaz. Basic piperazine glycerophosphate.

Glycerox. Water soluble emollients; used for cosmetics and toiletries. [Croda Chem. Ltd]

glyceryl laurate. (glyceryl monolaurate; dodecanoic acid, monoester with 1,2,3-propanetriol; dodecanoic acid, 2,3-dihydroxypropyl ester) Monoester of glycerin and lauric acid; $CH_3(CH_2)_{10}COOCH_2COHHCH_2OH$; CAS 142-18-7; EINECS 205-526-6; emulsifier, dispersant for food products, oils, waxes, solvents; antifoaming agent; drycleaning soap base. [Grindsted; Henkel/Emery; Lonza; Protameen]

glyceryl mono/distearate. Emulsifier for personal care products, food applications.

glyceryl monooleate. See glyceryl oleate.

glyceryl monoricinoleate. See glyceryl ricinoleate.

glyceryl monostearate. See glyceryl stearate.

glyceryl oleate. (glyceryl monooleate; monoolein; 9-octadecenoic acid, monoester with 1,2,3-propanetriol) Monoester of glycerin and oleic acid; $CH_3-(CH_2)_7CH=CH(CH_2)_7COOCH_2CCH_2OHHOH$; CAS 111-03-5; 37220-82-9; EINECS 203-827-7; 253-407-2; Emulsifier, coemulsifier, stabilizer, wetting agent, lubricant, and antistat; used in cosmetic, pharmaceutical, industrial, and food applications. [Calgene; Ferro/Keil; Grindsted; Henkel/Emery; ICI Spec.; Inolex; Karlshamns; Lonza; Mona; Patco; Stepan/PVO; Witco/Humko]

glyceryl ricinoleate. (12-hydroxy-9-octadecenoic acid, monoester with 1,2,3-propanetriol; monoricinolein; glyceryl monoricinoleate) Monoester of glycerin and ricinoleic acid; $C_3H_5(OOCC_{16}H_{32}OH)_3$; CAS 141-08-2; EINECS 205-455-0; Emulsifying agent. [CasChem; Lonza]

glyceryl stearate. (monostearin; glyceryl monostearate; Octadecanoic acid, monoester with 1,2,3-propanetriol) Monoester of glycerin and stearic acid; $CH_3-(CH_2)_{16}COOCH_2COHHCH_2OH$; CAS 123-94-4; 11099-07-3; 31566-31-1; 85666-92-8; 85251-77-0; EINECS 250-705-4; 234-325-6; 204-664-4; 286-490-9; Nonionic sec. o/w emulsifier for creams and lotions; visc. booster for emulsions. [Eastman; Goldschmidt; Grindsted; Hart Prod. Corp.; Henkel/Emery; ICI Spec.; Inolex; Karlshamns; Lanaetex; Lipo; Lonza; MTM Spec. Chem. Ltd; Patco; Van Dyk; Witco/Humko]

glyceryl triacetate. See triacetin.

glyceryl trinitrate. See nitroglycerin.

glyceryl trioleate. See triolein.

glycidyl acrylate. (acrylic acid 2,3-epoxypropyl ester) $H_2C:CHCOOCH_2CHCH_2O$; CAS 106-90-1; polyfunctional monomer; in coatings to improve adhesion to substrate and solv. resistance. [Estron]

Glycidyl Ether. 2,3-Epoxypropyl (n-butyl, allyl) ether; epoxy resin reactive diluent, raw material for curing agents of epoxy resins. [Yokkaichi Chemical Co Ltd] *

glycidyl methacrylate. (2,3-epoxypropyl methacrylate; methacrylic acid 2,3-epoxypropyl ester) $C_7H_{10}O_3$; CAS 106-91-2; EINECS 203-441-9; Polyfunctional monomer; in hydrogels for contact lenses and membranes, molding and casting compounds, impregnating paper, concrete, wood, coatings, printing inks, adhesives, sealants, elastomers. [Estron; Mitsubishi Gas; Sartomer]

glycine. (aminoacetic acid; glycocoll) Amino acid; H_2NCH_2COOH; CAS 56-40-6; EINECS 200-272-2; organic synthesis, nutrient, biochemical research, buffering agent, chicken-feed additive, reduces bitter taste of saccharin, retards rancidity in animal and vegetable fats. [Aldrich; Allchem Industries; Degussa; W.R. Grace/Hampshire; U.S. Biochemical]

glyco. See paraffin, liquid.

glyciobiarsol. Bismuth glycollylarsanilate.

glycobrom. The glyceryl ester of dibromohydrocinnamic acid.

glycocoll. See glycine.

Glycoderm. Sphingolipid liposomes with glycosaminoglycans; prep. for dry and cracked skin; for restoration of the lipid barrier of the stratum corneum. [Dr. Kurt Richter; Henkel/Cospha]

glycogenic acid. See gluconic acid.

glycol. (ethylene glycol; 1,2-ethanediol; ethylene alcohol) Aliphatic diol; $HOCH_2CH_2OH$; CAS 107-21-1; EINECS 203-473-3; antifreeze in cooling and heating systems; in hydraulic brake fluids; industrial humectant; solvent in paints, plastics, inks; softening agent for cellophane; stabilizer; in explosives, alkyd resins, elastomers, synthetic fibers and waxes; asphalt. [Ashland; BASF; Eastman; Hoechst Celanese; Mitsui Petrochem. Ind.; Mitsui Toatsu Chem.; Mobil; Olin; OxyChem; Shell; Texaco; Union Carbide]

glycol diacetate. See ethylene glycol diacetate.

glycol dimercaptoacetate. (ethylene glycol bisthioglycolate) $HSCH_2COOCH_2CH_2OOCCH_2SH$; Crosslinking agent for rubbers, accelerator in curing epoxy resins. [Evans Chemetics; Jansse Chimica]

glycol distearate. (EGDS; ethylene glycol distearate; octadecanoic acid, 1,2-ethanediyl ester) Diester of ethylene glycol and stearic acid; $CH_3(CH_2)_{16}COO-CH_2CH_2OCO(CH_2)_{16}CH_3$; CAS 627-83-8; EINECS 211-014-3; pearlescent and opacifier; thickener, intermediate, lubricant, emulsifier, emollient; for emulsion shampoos and foam baths.

glycolic acid. (hydroxyacetic acid) $HOCH_2COOH$; CAS 79-14-1; EINECS 201180-5; Leather dyeing and tanning; textile dyeing; cleaning, polishing and soldering compounds; copper pickling; adhesives; electroplating; petroleum demulsifier; chelating agent for iron; chemical milling; pH control. [Du Pont; R.W. Greeff; Hoechst Celanese]

glycoline. See paraffin, liquid.

Glycolube® 100. Polyol ester; internal lubricant for thermoplastics. [Lonza Inc]

---

‡ Trade name and manufacturer not verified   § Trade name without identified manufacturer

Glycolube® 140. Polyol ester; thermoplastic antistat. Lonza Inc]

Glycolube® P. Ester wax; external lubricant and antistat. Lonza Inc]

Glycolube® VL. CAS 8002-74-2; Synthetic wax; internal and external lubricant,release agent for thermoplastics; also used for textiles. Lonza Inc]

glyco metal. An alloy of 70-74% lead, 14-16% antimony, and 8-12% tin.

Glycomul®. Sorbitan esters. [Lonza Inc]

Glycomul® L. CAS 1338-39-2; EINECS 215-663-3; Sorbitan laurate; emulsifier for edible, cosmetic, industrial, pharmaceutical uses; antistat, antifog for PVC. Lonza Inc]

Glycomul® O. CAS 1338-43-8; EINECS 215-665-4; Sorbitan oleate; emulsifier for cosmetic, pharmaceutical, and industrial applications. Lonza Inc]

Glycomul® P. CAS 26266-57-9; EINECS 247-568-8; Sorbitan palmitate; emulsifier for cosmetic, pharmaceutical, and industrial applications. Lonza Inc]

Glycomul® S FG. CAS 1338-41-6; EINECS 215-664-9; Sorbitan stearate; emulsifier for food, cosmetic, household and industrial use. Lonza Inc]

Glycomul® SOC. CAS 8007-43-0; EINECS 232-360-1; Sorbitan sesquioleate; emulsifier for cosmetic, pharmaceutical, and industrial applications. Lonza Inc]

Glycomul® TO. CAS 26266-58-0; EINECS 247-569-3; Sorbitan trioleate; emulsifier for cosmetic, pharmaceutical, and industrial applications. Lonza Inc]

Glycomul® TS KFG. CAS 26658-19-5; EINECS 247-891-4; Sorbitan tristearate; emulsifier for food, cosmetic, household and industrial applications. Lonza Inc]

Glycon® G 100, G 300. CAS 56-81-5; Glycerin; humectant, bodying agent, moisture control agent for toothpaste, cosmetics, sugarless confections, controlled moisture foods and industrial applications. Lonza Inc]

Glycon® P-45. CAS 57-10-3; EINECS 200-312-9; 45% Palmitic acid; lubricant, defoamer, and component of other food additives. Lonza Inc]

Glycon® S-65. CAS 61790-38-3; EINECS 263-130-9; Hydrogenated tallow fatty acid; lubricant, defoamer, and component of other food additives. Lonza Inc]

Glycon® S-90. CAS 57-11-4; EINECS 200-313-4; 90% Stearic acid; lubricant, defoamer, and component of other food additives. Lonza Inc]

Glycon® TP. CAS 57-11-4; EINECS 200-313-4; Stearic acid, triple pressed; lubricant, defoamer, and component of other food additives. Lonza Inc]

glyconic acid. *See* gluconic acid.

Glyconol®. Amide wax; synthetic wax for textiles and other applications; gelling and visc. modifier for hot melts, wax blends, oil modifiers, metalworking lubricants. Lonza Inc]

Glyconyl. A photographic developer, the active constituent of which is p-hydroxyphenylglycine.

glycophenol. *See* saccharin.

glycosal. The mono-salicylic ester of glycerol.

Glycosin. *See* saccharin.

Glycosperse®. Polyethoxylated sorbitan esters. [Lonza Inc]

Glycosperse® HTO-40. PEG-40 sorbitan hexatallate; emulsifier for food, cosmetic, household or industrial applications. Lonza Inc]

Glycosperse® L-10. CAS 9005-64-5; PEG-10 sorbitan laurate; emulsifier for food, cosmetic, household or industrial applications. Lonza Inc]

Glycosperse® L-20. CAS 9005-64-5; 9062-73-1; Polysorbate 20; emulsifier for food, cosmetic, pharmaceutical, and industrial uses; flavor solubilizer and dispersant. Lonza Inc]

Glycosperse® O-5. CAS 9005-65-6; Polysorbate 81; flavor solubilizer and dispersant; emulsifier for cosmetic, pharmaceutical, and industrial use. Lonza Inc]

Glycosperse® O-20 FG, O-20 KFG. CAS 9005-65-6; Polysorbate 80; emulsifier for ice cream, frozen desserts; solubilizer and dispersant for shortenings; adjuvant for herbicides and plant growth regulators. Lonza Inc]

Glycosperse® P-20. CAS 9005-66-7; Polysorbate 40; emulsifier for food, cosmetic, pharmaceutical, and industrial uses; flavor solubilizer and dispersant. Lonza Inc] *

Glycosperse® S-20 FG, S-20 KFG. CAS 9005-67-8; Polysorbate 60; emulsifier for chocolate and confectionery coatings, icings, toppings, cakes, cream fillings, shortenings, desserts, dough conditioners. Lonza Inc]

Glycosperse® TS-20 FG, TS-20 KFG. CAS 9005-71-4; Polysorbate 65; emulsifier for ice cream, frozen desserts, cakes, whipped toppings, nondairy creamers, icings, fillings. Lonza Inc]

Glycosterine. A glycol glyceryl stearate; used to replace beeswax in certain polishes. §

glycothymoline. An antiseptic, usually containing thymol, eucalyptol, menthol, borates, bicarbonates, benzoates, and glycerin.

Glycowax® 765. CAS 110-30-5; EINECS 203-755-6; N,N'-Ethylene bisstearamide; synthetic wax for kraft brownstock pulp and paper defoaming, plastics processing, powd. metal lubrication, water treatment defoaming, other industrial applications. Lonza Inc]

Glycozone. A proprietary preparation claimed to contain 5% glyceric acid, and 90% glycerin; an antiseptic. §

glycylglycine. (diglycine) $NH_2CH_2CONHCH_2COOH$; CAS 556-50-3; EINECS 209-127-8. [Am. Biorganics; Penta Mfg.; Wchweizerhall; Spectrum Chem. Mfg.]

Glydant®. CAS 6440-58-0; DMDM hydantoin in water; preservative, broad spectrum antimicrobial for cosmetics and toiletries; effective against Gram-positive and Gram-negative bacteria, fungi, and yeast. Lonza Inc]

Glydexx N-10. CAS 26761-45-5; Glycidyl decanoate. [Exxon] *

Glydus. Herbicide. [Agan Chemical Manufacturers Ltd]

Glyecin. Ethylthiodiglycol. A solvent used in treatment of wool for dyeing.

Glyezin®. Dye solvents and fixing assistants for textile finishing. [BASF AG; BASF plc]

Glymin®. Leak detector fluid for double-walled storage tanks. [BASF AG]

glymol. *See* paraffin, liquid.

Glyoxal 40%. CAS 107-22-2; Glyoxal; used in textile, paper and oil field industries. [Cyanamid BV]

glyoxalin. *See* imidazole.

Glyphogan. CAS 1071-83-6; Glyphosate; herbicide. [Agan Chemical Manufacturers Ltd]

Glyprosol 20. Yeast glycoproteins; natural skin smoothing protein. [Brooks Industries]

Glyptal 2557, 2559. Proprietary trade names for polyester plasticizers. [GE Plastics] *

glyptal resins. Resinous products obtained by the interaction of glycerol and organic acids. A resin of this type is

---

made by reacting upon glycerol with oleic acid and phthalic anhydride; used in paints, varnishes, and lacquers.

Glyrol. CAS 56-81-5; Glycerin; pharmaceutical aid. [Iolab Corp]

glysal. The mono-glycol ester of salicylic acid.

Glysantin®. Antifreeze for combustion engines. [BASF AG; BASF plc]

Glysennid. Sennosides; laxative. [Sandoz Pharmaceuticals] *

glysobuzole. Isobusole.

Glytex. Mixture of isoxaben and methabenzthiazuron; soil-acting herbicide for winter cereals. [Bayer plc]

Glytex® 203. Polyol ester; heat-stable nylon and polyester lubricant, emulsifier, antistat. Lonza Inc]

Glytex® 513. Ester ethoxylate; nylon and polyester tire cord finish emulsifier, lubricant, antistat. Lonza Inc]

Glytex® 663. Hydantoin ester; nylon and polyester industrial yarn lubricant, emulsifier, antistat. Lonza Inc]

Glytex® 1085. Alcohol ester; textile spin finish lubricant, emulsifier, antistat. Lonza Inc]

Glythermin®. Glycol-based antifreeze thermal liqs. [BASF AG]

Glyvenol. CAS 10310-32-4; Tribenoside; sclerosing agent. [Ciba-Geigy Corp] *

GMD. Pigment dispersions; for coloration of aqueous and solvent-based trade sale paints. [Pacific Dispersions Inc] *

GMF. p-Quinonedioxime; a rapid and economical vulcanizing agent when used in conjunction with red lead; gives fast-curing high-modulus stocks; recommended for use in butyl curing bags, wire insulation and where a fast-curing high-modulus stock is desired. [Uniroyal] *

GMS Base. Glyceryl stearate; base for nonionic textile softeners. [Clark]

GMS/SE Base. Glyceryl stearate SE; base for nonionic textile softeners. [Clark]

Goa butter. See Kokum butter.

Gobapur Acide Pur. Sulfuric acid and oleum. [Rhone-Poulenc NV/CdF Chimie AZF] *

Gofrativ. CAS 7704-34-9; Sulfur; fungicide [Makhteshim Chemical Works Ltd]

Gofravik. CAS 7704-34-9; Sulfur; fungicide [Makhteshim Chemical Works Ltd]

Gohi Iron. A proprietary trade name for iron containing manganese, sulfur, phosphorus, copper (total less than 0.125%.). §

Gohsefimer. Modified polyvinyl alcohol. [British Traders & Shippers Ltd]

Gohsenol. Polyvinyl alcohol. [British Traders & Shippers Ltd]

Gohseran. Modified polyvinyl alcohol. [British Traders & Shippers Ltd]

gold. Metallic element; Au; CAS 7440-57-5; EINECS 231-165-9; infrared reflectors; electrical contact alloys; brazing alloys; laboratoryware; decorative arts; electronics; dental alloys; jewelry; colloidal dispersions for coloring glass, as nucleating agent, for specialized medical treatments. [Cerac; Degussa; Koch Chem. Ltd; Noah Chem.]

gold bronze. See copper.

gold button brass. See brass.

gold chloride. (gold trichloride; auric chloride) AuCl₃; CAS 13453-07-1; EINECS 236-623-1. [Degussa; Métaux Précieux SA; Spectrum Chem. Mfg.]

Gold Coinage. See Standard Gold.

goldenacorn. Nutmeg.

golden antimony sulfide. CAS 1315-04-4; (golden sulfuret of antimony, golden sulfide of antimony). Antimony pentasulfide, Sb_2S_5.

Golden Bear 102. Naphthenic distillate; oil used in high-loss systems, motor oils for cars with excessive oil consumption. [Witco/Golden Bear]

Golden Bear 2013-10. Solv.-refined naphthenic oil; used in industrial, metalworking, ink and similar formulations, for blending approved railroad and marine engine oils, motor oils designed for older automobiles where premium quality is not necessary. [Witco/Golden Bear]

Golden Bear 4013-10. Solv.-refined naphthenic oil; premium grade featuring extra low pour pts., outstanding oxidation stability to produce base stock for mfg. refrigeration oils, white oils, transformer oils, high-performance motor oils either by themselves or in combination with paraffinic oils. [Witco/Golden Bear]

Golden Dawn. Refined lanolins of pharmaceutical/cosmetic quality; emollients and w/o emulsifiers. [Westbrook Lanolin] *

Golden Fleece. Hypo-allergenic and super-refined lanolins; emollients and w/o emulsifiers in cosmetics and pharmaceuticals. [Westbrook Lanolin] *

Golden Hermetite. Golden gel, nonhardening; a high technology gasket jointing compound with excellent chemical and temperature resistance; high tack stabilizes gasket during assembly; clean to use. [Hermetite Products Ltd]*

Golden-Pea-Pro EN-15. Hydrolyzed golden-pea protein; cosmetic ingredient for skin and hair care products [Brooks Industries]

goldenseal. Hydrastis Canadensis; used as a source of yellow dye.

Golden Sulfide of Antimony. See golden antimony sulfide.

Golden Sulfuret of Antimony. See golden antimony sulfide.

golden syrup. (Drip Syrup). This is the product obtained when raw or brown sugar is dissolved and the solution clarified with animal charcoal and the white sugar crystallized from it.

Golden Wax. Mold release liquid wax for plastic molding. [Specialty Products Co] *

Gold Guard. Contact cleaner/lubricant for removal of oxides, dust and contaminants from precious metals and connectors. [Chemtronics]

gold size. Consists of a mixture of 1 part yellow ocher, 2 parts copal varnish, 3 parts linseed oil, 4 parts turpentine, and 5 parts boiled oil.

Gold Solders. Various alloys of gold, silver, and copper, sometimes with zinc. An ordinary gold solder contains 43% gold, 30% silver, 20% copper, and 7% zinc; consists of 66% gold, 22% copper.

gold trichloride. See gold chloride.

Goliath. CAS 13684-63-4; Emulsifiable concentrate containing 114 g/l phenmedipham; for weed control for beet crops. [ABM Chemicals Ltd]

Golpanol®. Electroplating additives. [BASF plc]

Golpanol® MBS. Oxidant and demetallizer for industrial cleaners. [BASF AG]

Goltix®. CAS 41394-05-2; A water dispersible granular formulation containing 70% w/w metamitron; used to control annual weeds in sugar beet grown on mineral and

---

‡ Trade name and manufacturer not verified   § Trade name without identified manufacturer

organic soils and red beet, fodder beet and mangolds grown on mineral soils. [Bayer AG; Bayer plc]

gommeline. See British gum.

Gon. Tablets for children. [Octel Chemicals Ltd] †

Gonacrine. Preparation of 2,8-diamino-10-methyl-acridinium chloride and diamino acridine. [May & Baker Ltd] *

Gonadotraphon FSH. Freeze-dried serum gonadotrophin; used to treat sterility in males due to defective spermatogenesis and is used to treat secondary amenorrhea and anovulatory sterility in females. [Paines & Byrne Ltd] *

Gonadotraphon LH. Sterile freeze-dried preparation of chorionic gonadotrophin; used in males for the treatment of delayed puberty, undescended testes, ectopic testes requiring surgery, oligospermia or aspermia with inactive testes; used in the female for the indication of ovulation, recurrent abortion, nenorrhazia due to persistent follicular phase and secondy amenorrhoea. [Paines & Byrne Ltd] *

Gonadotrophon. A group of proprietary preparations of human gonadotrophins. [Pharmax Ltd] *

Gondang wax. See Java wax.

gong metal. A brass containing 78% copper and 22% zinc.

Goniosol. CAS 9004-65-3; Hydroxypropyl methylcellulose; pharmaceutic aid. [CooperVision Inc] ‡

Gooch and Eddy reagent. A reagent used for precipitating magnesium as magnesium carbonate. It contains 180 parts concentrated ammonia, 800 parts water, and 900 parts absolute alcohol, the solution being saturated with ammonium carbonate.

Good Gulf. Gasoline. [Chevron] *

GoodLife. A powder organic based fertilizer comprising a composted organic base and chemical N, P & K to form the analysis; four types sold: all-pupose Fertilizer, flower garden fertilizer, vegetable fertilizer and lawn weed and feed fertilizer. [Humber Fertilizers plc]

Good-rite® 2528X10. CAS 1337-81-1; Vinyl pyridine latex; used for adhesives in tire cord chafer fabric and industrial rubber goods; produces wickproof chafer fabric in one pass; very stable high solids RFL dips; excellent adhesion retention. [BFGoodrich] *

Good-rite® 3150. 1,1´,1´´-(1,3,5-Triazine-2,4,6-triyltris ((cyclohexylimino)-2,1-ethanediyl)tris(3,3,5,5-tetra-methylpiperazinone); hindered amine light stabilizer. [BFGoodrich] *

Good-rite® GP-223. CAS 103-23-1; Dioctyl adipate; vinyl plasticizer. [BFGoodrich] *

Good-rite® GP-235. Octyldecyl adipate; vinyl plasticizer. [BFGoodrich] *

Good-rite® GP-236. Didecyl adipate; vinyl plasticizer. [BFGoodrich] *

Good-rite® GP-261. A phthalate vinyl plasticizer. [BFGoodrich] *

Good-rite®GP-265. CAS 119-07-3; Octyldecyl phthalate; a vinyl plasticizer. [BFGoodrich] *

Good-rite® GP-266. Didecyl phthalate; a vinyl plasticizer. [BFGoodrich] *

Good-rite® K-702. CAS 9003-01-4; Polyacrylic acid aq. sol'n.; detergent assistant, soap builder, particulate soil dispersant, sequesterant for calcium, magnesium, iron; scale inhibitor; for laundry, dishwash, consumer/institutional cleaning products [BFGoodrich]

Good-rite® K-752. CAS 9003-01-4; Polyacrylic acid aq. sol'n.; detergent assistant, soap builder, particulate soil

dispersant, sequesterant for calcium, magnesium, iron; scale inhibitor; for laundry, dishwash, consumer/institutional cleaning products [BFGoodrich/Spec. Polymers]

Good-rite®K-7058D. CAS 9003-04-7; Sodium polyacrylate; detergent assistant, soap builder, particulate soil dispersant, sequesterant for calcium, magnesium, iron; scale inhibitor; for laundry, dishwash, consumer/institutional cleaning products [BFGoodrich]

Good-rite Polyacrylates. Homopolymers and copolymers of acrylic acid and their salts in both liquid and dry powder forms; for water treatment, soap and detergents and dispersants. [BFGoodrich UK] *

Goodyear LPR-6632. SBR latex; for textile applications, carpet lamination and precoat, fabric backcoating; as sprayable binders, tape base saturants, binders for adhesive and coating applications [Goodyear] *

Go Pain. A proprietary analgesic preparation of salicylamide, potassium salicylate, calcium succinate, p-amino benzoic acid, vitamins B, and C, and aluminum hydroxide. [De Pree Co] ‡

Gopmann solder. An alloy of 49.1% tin, 20.3% zinc, and 26% lead.

Gordon Superflex D. A trade name for a graft of stereospecific rubber and polystyrene for high impact. Izod 1.35-1.65; elongation 27%; tensile strength 3800 psi; modulus 291,000 psi (flex). [PBI - Gordon Corp] †

Gossamer®. Polymer. [DuPont UK]

Gougeon Laminating Epoxy. Liquid epoxy resin and hardener; laminating resin specifically for use with glass cloth, carbon fibers, aramid and hybrids. [Wessex Resins & Adhesives Ltd]

Goulac. CAS 8061-52-7; Calcium lignosulfonate; binder for foundry, ceramic, and refractory products [Borregaard LignoTech]

Goulard powder. Lead acetate.

GP-II. Porofocon B; contact lens material. [Barnes-Hind Inc]*

GP-4. Amine functional silicone fluid; intermediate for synthesis of silicone/org. copolymers used in textiles, coatings car polishes; also in lubricant, coating and mold release formulations. [Genesee Polymers]

GP-4-E. Silicone emulsion; provides durability and detergent resistance in car polishes, vinyl conditioners. [Genesee Polymers]

GP66 Miracle Cleaner. Nonionic surfactant, synthetic detergent, wetting agents, emulsifiers, builders etc; a USDA A-1 certified cleaning compound used to clean forklifts, diesel engines, conveyors, robots, concrete floors, ovens, whitewalls, vinyl tops. [GP66 Chemical Corporation] *

GP-71-SS. Mercapto-modified dimethyl silicone copolymer fluid; plastic and rubber release agent; internal lubricant and release agent for sulfur and peroxide cure rubber; coreactant in vinyl polymerization; synthesis of org./silicone copolymers; heat stabilizer; in corrosion inhibitor coatings, inks. [Genesee Polymers]

GP-137. CAS 2897-60-1; 3-Glycidoxypropylmethyldiethoxy silane; coupling agent between inorg. fillers and epoxy, melamine, phenolic and urethane resins, PS, acrylic sealants, butyl rubber. [Genesee Polymers]

GP-165. Silicone resin emulsion; cures to solv.-resist. coating; as binder in aq. paints intended for application to substrates exposed to high temps.; component of water-based masonry water-repellent formulations. [Genesee

---

* Trade name not verified as available   † Trade name verified as obsolete

Polymers]

GP-180. Proprietary organosilicone resin conc. in 1,1,1-trichlorethane solv.; used as a dry film release coating for molding of plastics or rubber. [Genesee Polymers]

GP-187. Methyl silicone resin solution. in IPA; cures to a soft pliable consistency at ambient conditions, to a clear, hard film at elevated temps.; used in masonry water repellents, leather and fabric treating agents, mold release formulations; as binder for high temp. application paints and coatings. [Genesee Polymers]

GP-209. Dimethyl silicone EO/PO block copolymer; emulsifier, wetting agent, pigment dispersant, leveling agent, profoaming additive for PU foams, hard surface cleaners, polishes, cosmetic formulations; inverse sol. suggests use as defoamer for hot aq. surfactant sol'ns. [Genesee Polymers]

GP-210. Silicone emulsion; defoamer for hot and cold foaming systems, industrial applications, commercial cleaning compounds, latex stripping, adhesives, cutting oils, leather treating, sewage treatment, chemical processing, paints/coatings. [Genesee Polymers]

GP-217. Dimethylpolysiloxane EO block copolymer; wetting agent, emulsifier for water-based coatings, inks, polishes, hard surface cleaners; dispersant for clays, pigments; thread lubricant; leveling and flow control agent; profoaming additive in aq. systems. [Genesee Polymers]

GP-227. Organo-modified dimethylsilicone polymer; surfactant, w/o emulsifier for emulsions, auto and furniture polishes, vinyl conditioners; gloss aid; adds detergent resistance. [Genesee Polymers]

GP-262. Nonsilicone; defoamer for aq. systems, metalworking fluids, cleaner formulations, paints, coatings, paper/paperboard, waste water treatment; alkali resistant. [Genesee Polymers]

GP-310-I. Silicone aq. emulsion; dilutable antifoam for industrial processing in hot, cold, alkaline or aq. systems; for chemical processing, cleaning products, paints, paper and latex processing. [Genesee Polymers]

GP-7000. Methyl alkyl dimethyl silicone fluid; surface tensile reducer for nonaq. solv. and oil systems; wetting and leveling agent in inks, coatings, plastisols; antifoam, mold release agent; internal release agent for plastics and rubber; textile lubricant. [Genesee Polymers]

G-P-D. Colorant dispersions; for coloring of nonaqueous coating compositions. [Hüls Am.] *

G.P.V. CAS 87-08-1; A proprietary preparation of penicillin V. [Galen Ltd] *

GR Acid. α-Naphtholdisulfonic acid.

Graessorb. Sunscreen agents and ultra violet filters. [Nipa Laboratories Ltd.]

Grafene. A proprietary trade name for a lubricant containing graphite and oils. §

Grafil. High tensile carbon fibers for industrial use. [Courtaulds plc]

Grafita. A proprietary trade name for a lubricant containing graphite and grease. §

Grafitix (Anti-graffiti). Nonflammable solvent mixture manufactured in aerosol spray; used for removing of paint, felt, applied on stone, cement, bricks, wood, metal and fabrics. [S F C] *

Grafitix (Baent). Nonflammable solvent mixture water miscible, containing special wetting agent; specially designed for building trade applications; it strips all surfaces

covered with paint, rough coat and plastic facing coatings. [S F C] *

Grafitix (Ravalement). Aqueous product, basic reaction, low viscosity, completely soluble in water; used for cleaning and stripping the outside surfaces of buildings mainly in stone or cement soiled by atmospheric pollution. [S F C] *

Grafix. Graphic arts fixer. [May & Baker Ltd] *

grahamite. An asphaltic substance found in Mexico and Cuba. It is usually associated with mineral matter. It has a specific gravity of 1.17, melts at 175-230 C., up to 45% mineral matter, and is very soluble in carbon disulfide.

Grahams salt. See sodium hexametaphosphate.

grain alcohol. CAS 64-17-5; Ethyl alcohol, C_2H_5OH.

grain oil. See fusel oil.

grains D'Ambrette. Musk seeds.

grains of paradise. (Guinea Grains). Seeds of *Amomum melegueta*.

Grain Store Smoke. Insecticide. [DowElanco Ltd]

Gramazine. Herbicide. [ICI Chem. & Polymers Ltd]

Graminon® Plus. Bentazon, isoproturon, dichlorprop; for post-emergence control of grasses and broadleaf weeds in winter wheat and winter barley [BASF AG]

Gramixel. CAS 1910-42-5, 330-54-1; Herbicide containing paraquat and diuron. [ICI Chem. & Polymers Ltd]

Gramonol. CAS 1910-42-5, 330-55-2; Weedkiller containing paraquat and monolinuron. [ICI Chem. & Polymers Ltd]

Gramonol 5. CAS 330-55-2, 1910-42-5; Suspension concentrate containing 154 g monolinuron and 110 g paraquat per liter; for control of annual dicotyledons in cereals. [Hoechst UK]

Gramonol Five. CAS 330-55-2, 1910-42-5; Suspension concentrate containing 154 g monolinuron and 110 g paraquat per liter; for control of annual dicotyledons in cereals. [ICI Agrochemicals]

Gramoxone. CAS 1910-42-5; Paraquat; herbicide. [Schering Agrochemicals Ltd]

Gramoxone. CAS 1910-42-5; Paraquat weedkiller preparations. [ICI Chem. & Polymers Ltd]

Gramoxone X. CAS 1910-42-5; Soluble concentrate containing 200 g/l paraquat; a pre-emergence bipyridilium herbicide to control weeds in field crops and ornamentals. [ICI Agrochemicals; Schering Agriculture]

Gramp's solder. An alloy of 60.4% tin, 36.1% zinc, 3% copper, 0.25% lead, and 0.18% antimony. An aluminum solder.

Gram's iodine stain. This consists of 1 gram iodine, 2 g potassium iodide, and 300 cc distilled water.

Gram's stain. A microscopic stain. It contains gentian violet.

Gramuron. CAS 1910-42-5, 330-54-1; Herbicide containing paraquat and diuron. [ICI Chem. & Polymers Ltd]

Graneodin Ointment. Gramicidin and neomycin sulfate in ointment base. [Bristol-Myers Squibb Pharmaceuticals Ltd]

Graney bronze. An alloy of 76.5% copper, 9.2% tin, and 15.2% lead.

Granodine. Zinc phosphating solution for metal treatment. [ICI Chem. & Polymers Ltd]

granodized steel. Steel which has been treated with zinc phosphate to give the surface resistance to corrosion.

granol. Carbonized granulated peat.

Granolube. Metal treating compositions. [ICI Chem. & Polymers Ltd] †

---

‡ Trade name and manufacturer not verified   § Trade name without identified manufacturer

Granosan®. For the agriculture industry. [DuPont UK]

Granstock. Animal feed additive. [ICI Chem. & Polymers Ltd]

Granuform. CAS 30525-89-4; Free-flowing, dust-free, small white beads of 90.5±1% paraformaldehyde; used in the plastics industry for production of phenolic, urea, melamine and coating resins; in the chemical and pharmaceutical industries for chloromethylation processes; as disinfectant. [Degussa Ltd]

Granular Weedkiller. CAS 7775-09-9; Granules containing 30% w/w sodium chlorate; total weed control for paths, drives and noncrop areas. [Dimex Ltd]

Granulite®BF 6/16. Ground rubber from truck and passenger retread buffings, free of foreign material, fabric, and steel; used in pneumatic tires, auotmotive components, construction and paving materials, industrial rubber goods, building materials, sporting goods, plastic materials. [Baker Rubber]

Granulite®TR-10. Ground rubber from truck and passenger tread rubber, free of foreign material, fabric, and steel; used in pneumatic tires, auotmotive components, construction and paving materials, industrial rubber goods, building materials, sporting goods, plastic materials. [Baker Rubber]

Granulite®WTP-10. Ground rubber from whole passenger and/or truck tire or equivalent with fabric removed; free of foreign material; used in pneumatic tires, auotmotive components, construction and paving materials, industrial rubber goods, building materials, sporting goods, plastic materials. [Baker Rubber]

granulose. The inner part of the starch granule is known by this name.

grape sugar. *See* glucose.

Graphalloy, Silver. A trademark for a molded graphite impregnated with silver used in electrical brushes and similar appliances. §

graphite. (black lead; plumbago; mineral carbon) C; CAS 7782-42-5; EINECS 231-955-3; The allotropic form of carbon, occuring naturally in Madagascar, Ceylon, Mexico, Korea, Austria, USSR; can be synthetically produced by heating petroleum coke to ≈ 3000 C in an electric resistance furnace; reinforcing agent for pre-pregging, filament winding; lubricant additive for greases, engine oils, etc.; fillers, paints, coatings, self-lubricating bearings. [Cerac; Lonza; Sigri GmbH; Ucar Carbon]

graphite metal. An antifriction and fitting metal. It contains 15% tin, 68% lead, and 17% antimony.

graphitic carbon. The black shiny flakes of carbon present in pig iron.

graphitic temper carbon. The black amorphous carbon present in certain varieties of iron.

graphitites. Graphites which swell on moistening them with strong sulfuric acid, and then heating them to redness. They are not true graphites.

Graphitol. Synthetic organic pigment colors; used for printing inks and paints. [Sandoz Products Ltd]

Graphsize. Polyurethane based paper surface size. [Akzo]

Graphtol®. Pigment for coloring bar soaps, gels, powds., solids. [Sandoz]

Grappier cements. Hydraulic limes are slaked and passed through sieves. The hard lumps left on the sieve consist of unchanged limestone and calcium silicates. These are finely ground, and are then known as grappier cements.

Le Farge cement belongs to this class.

Graslam. Herbicide. [May & Baker Ltd] *

Grasselerator 101. A trademark for a rubber vulcanization accelerator; it is aldehyde ammonia. §

Grasselerator 102. CAS 100-97-0; Hexamethylenetetramine; a trademark for a rubber vulcanization accelerator.§

Grasselerator 508. Butylideneaniline; a proprietary rubber vulcanization accelerator. §

Grasselerator 552. A trademark for a rubber vulcanization accelerator. §

Grasselerator 833. A trademark for a liquid aldehyde amine condensation product; it is a low temperature rubber vulcanization accelerator, and has antioxidant properties. §

Grass Greenizit. Permanent grass colorant restoring green color to dormant or discolored grass. [W A Cleary]

Grasshopper. Compound fertilizer 8:1.5:1.5 plus 2,4-D dicamba and ferrous sulfate; lawn fertilizer. [ICI Garden Products]

Gravidox. CAS 58-56-0; Pyridoxine hydrochloride; vitamin. [Eli Lilly & Co] †

Gravulac. Gravure overlacquers. [The Scottish Adhesives Co Ltd]

Grazon 90. Emulsifiable concentrate containing 60 g clopyralid and 240 g triclopyr per liter; used for treatment of perennial weeds in grassland. [DowElanco Ltd]

Great Lakes BA-50. Bis (2-hydroxyethyl ether) of tetrabromobisphenol A; difunctional alcohol providing flame retardance; for unsat. polyester and epoxy thermoset resins; used for laminates for electronic circuit boards; corrosion resistance of systems useful in materials for tanks, ducts, and hoods. [Great Lakes]

Great Lakes BA-59P. CAS 79-94-7; EINECS 201-236-9; Tetrabromobisphenol A; flame retardant for thermoplastic and thermoset resin systems, epoxy systems; also used as reactive flame retardant for PC and additive for styrenic thermoplastics. [Great Lakes]

Great Lakes BE-51. Bis (allyl ether) of tetrabromobisphenol A; flame retardant used in EPS and foamed PS. [Great Lakes]

Great Lakes CD-75P. Hexabromocyclododecane; flame retardant used in thermoplastic and thermosetting polymers; textile treatments, latex binders, adhesives, unsat. polyesters, and coatings. [Great Lakes]

Great Lakes DBS. Dibromostyrene; flame retardant. [Great Lakes]

Great Lakes DE-71. CAS 32534-81-9; Pentabromodiphenyl oxide; high visc. flame retardant for thermosetting and thermoplastic resin systems; used for unsat. polyester, rigid and flexible urethane foams, epoxies, laminates, adhesives, and coatings. [Great Lakes]

Great Lakes DE-79. CAS 32536-52-0; EINECS 251-087-9; Octabromodiphenyl oxide; flame retardant for ABS, nylon, polycarbonate, and polyester thermoplastic polymers; additive for unsat. polyesters and epoxy thermoset resins. [Great Lakes]

Great Lakes DE-83R. CAS 1163-19-5; EINECS 214-604-9; Decabromobiphenyl oxide; halogenated flame retardant for thermoplastic, elastomeric, and thermoset polymer systems incl. HIPS, PBT, ABS, nylons, PP, LDPE, EPDM, unsat. polyesters, and epoxy resins. [Great Lakes]

Great Lakes FF 680. Bis (tribromophenoxy) ethane; flame

---

* Trade name not verified as available   † Trade name verified as obsolete

retardant for application where thermal stability at high processing temps. is important; for thermoplastic and thermoset systems, light-stable applications [Great Lakes]

Great Lakes FR-756. Disodium salt of tetrabromophthalate; flame retardant. [Great Lakes]

Great Lakes PDBS-10. Poly(dibromostyrene); flame retardant. [Great Lakes]

Great Lakes PDBS-80. Poly(dibromostyrene); flame retardant. [Great Lakes]

Great Lakes PE-68. Bis (2,3-dibromopropyl ether) of tetrabromobisphenol A; flame retardant used in PP, polyethylene, polybutylenes, and polyolefin copolymers; effective at low loading levels. [Great Lakes]

Great Lakes PH-73. CAS 118-79-6; 2,4,6-Tribromophenol; reactive intermediate for phenol-based reactions; flame retardant, antifungal agent, or chemical intermediate. [Great Lakes]

Great Lakes PHE-65. CAS 3278-89-5; Tribromophenol allyl ether; flame retardant. [Great Lakes]

Great Lakes PHT4. CAS 632-79-1; EINECS 211-185-4; Tetrabromophthalic anhydride; flame retardant in production of unsat. polyester resins and rigid PU polyols; cohardener for epoxy resins; cost efficient additive for latex emulsions; derivs. used as flame retardants in diverse application (wire coating, and wool, etc.). [Great Lakes]

Great Lakes PHT4-Diol. CAS 20566-35-2; EINECS 243-885-0; Tetrabromophthalatediol; reactive intermediate used to produce flame retardant rigid urethane foam; can replace chlorinated polyols; for PU elastomers, coatings, adhesives, and fibers. [Great Lakes]

Great Lakes PO-64P. Poly-dibromophenylene oxide; flame retardant which melts into most polymers to optimize physical properites; esp. for cryst. polymers (polyesters, polyamides); enhances flow into thin wall sections; permits higher regrind loading levels. [Great Lakes]

Great Lakes SP75. Stabilized hexabromocyclododecane; flame retardant. [Great Lakes]

green acid. (Green sulfonate, green sulfonic acid). Crude mixtures of sulfonic acids from refining petroleum sludge.

green cinnabar. See chromium oxide (ic).

Green Dot. A smokeless powder; double-base-type formulation minimizes charge weight and moisture absorption. [Hercules] *

green earth. (Veronese Green, Veronese Earth). A natural pigment. It contains ferrous iron, silica, magnesia, alumina, and lime; used as an absorbent for basic dyestuffs.

Greenfly & Blackfly Killer. Systemic insecticide. [Fisons plc, Horticulture Div] *

green gold. Alloys of 75% gold, with from 11-25% silver, and 4-12% cadmium.

green iodide of mercury. Mercurous iodide, HgI.

Greenkeeper. Range of turf fertilizers. [Fisons plc, Horticulture Div] *

Greenkeeper Mosskiller. Turf fertilizer with iron sulfate. [Fisons plc, Horticulture Div] *

Green Magic. Various granular fertilizer blends; fertilizers for lawns, gardens and flowers. [Horn's Crop Service Center] *

Greenmaster Autumn. CAS 10028-22-5; Ferrous sulfate; used for moss control in turf. [Fisons plc]

Greenmaster Extra. Granules containing MCPA and mecoprop; for control of weeds in amenity and roadside grass. [Fisons plc]

Greenmaster Mosskiller. CAS 10028-22-5; Ferrous sulfate; used for moss control in turf. [Fisons plc]

green mordant. Sodium thiosulfate; used as a mordant for fixing aniline greens on fiber.

green nickel oxide. See nickel oxide (ous).

green ocher. Usually a mixture of silica, clay, and ferrous hydroxide; used as a base for cheap lakes.

green oils. A fraction of oil obtained from shale by treating the distillate with sulfuric acid and sodium hydroxide and then again distilling.

Greenol. Liquid iron plant nutrient. [Chevron] *

green oxide of chromium. Chromic oxide, Cr_2O_3.

green powder. See methyl green.

Green Spar. See malachite.

green sulfonate. See green acid.

Green Sulphur. CAS 7704-34-9; Sulfur powder and green dye; garden fungicide. [Vitax Ltd]

green sulfuric acid. See green acid.

Green Tape. Co-fired ceramic multilayer materials. [DuPont UK]

green ultramarine. A pigment produced by heating kaolin, silica, sodium carbonate, sulfur, coal, and rosin, washing and grinding; used in water-colors, and is known as Lime green. When heated again with sulfur, it gives blue ultramarine.

Green Up. Mixture of 2,4-D and dicamba; used to control weeds in turf. [Synchemicals Ltd]

Green Up Autumn Liquid Lawn Feed. Liquid concentrate containing NPK 3:6:6 and 1% Fe; autumn feed for lawn areas. [Vitax Ltd]

Green Up Feed and Weed Plus Mosskiller. Dry powder containing NPK 8:4:4 plus 2,4-D, mecoprop and ferrous sulfate; combined fertilizer, weed and mosskiller. [Vitax Ltd]

Green Up Lawn Feed and Weed. Liquid concentrate containing NPK 14.5:3:3 and 2,4-D and dicamba; combined feed and weed for turf. [Vitax Ltd]

Green Up Lawn Feedn Weed Plus Moss Killer. Liquid concentrate containing NK 11:4 plus dichlorophen, mecoprop, dichlorprop, dicamba and benazolin; combined feed, weed and mosskiller. [Vitax Ltd]

Green Up Lawn Spot Weedkiller. CAS 1702-17-6, 7085-19-0; Trigger spray pack containing 2,4-D, mecoprop; selective spot weedkiller for lawn areas. [Vitax Ltd]

Green Up Liquid Lawn Feed. Liquid concentrate containing NPK 17:3.5:3.5; liquid fertilizer for turf. [Vitax Ltd]

Green-up Mossfree. CAS 10028-22-5; Ferrous sulfate; used for moss control in turf. [Synchemicals Ltd]

Green Up Mossfree. Soluble powder containing ferrous sulfate heptahydrate. [Vitax Ltd]

Green Up Weedfree Lawn Weedkiller. CAS 1702-17-6, 1918-00-9; Emulsifiable concentrate containing 2,4-D and dicamba; selective herbicide for use on turf. [Vitax Ltd]

Green Up Weedfree Spot Weedkiller for Lawns. CAS 1702-17-6, 1918-00-9; Aerosol containing 2,4-D and dicamba; selective spot weedkiller for lawns. [Vitax Ltd]

green vitriol. See iron sulfate (ous).

green wood spirits. Acetone alcohol.

Grefco. A proprietary trade name for a chrome ore cement used as a refractory. §

Greggio. Crude Sicilian sulfur.

---

‡ Trade name and manufacturer not verified

§ Trade name without identified manufacturer

Gregoderm. A proprietary preparation of neomycin sulfate, nystatin, polymyxin B sulfate and hydrocortisone; used in the treatment of skin disorders. [Unigreg Ltd]

Gregovite C. A vitamin C tablet. [Unigreg Ltd]

Grenacher's alum carmine. An aqueous solution containing 1-5% common or ammonia alum, boiled with 0.5-1% carmine, and filtered. A microscopic stain.

Grenacher's borax carmine. A microscopic stain. It contains 3 g carmine, 4 g borax, and 100 cc distilled water. After dissolving by heat, 100 cc of 70% alcohol added, and the solution filtered.

Grenadine S. *See* Acid Magenta. §

G Resin. A proprietary trade name for cumarone-indene resins. §

grey acetate. Crude acetate of lime, prepared with distilled pyroligneous acid. It contains from 80-82% calcium acetate, and 20% water.

grey antimony. Trigonal antimony obtained by allowing molten antimony to cool in a crucible.

grey cast iron. A cast iron containing much of its carbon in the uncombined state. A typical one contains 94% iron, 3.5% carbon, and 2.5% silicon. It has a specific gravity of 7.0 and melts at 1230 C.

grey forge pig. A pig iron usually containing less silicon than other grey irons.

grey gold. An alloy of gold and iron, sometimes with silver. One alloy contains 86% gold, 8.5% silver, and 5.5% iron, and an other 83% gold and 17% iron.

grey mixture. A mixture of 7 parts meal powder, with 100 parts saltpeter and sulfur; used in fireworks.

grey tin. A form of tin obtained by exposing the metal to low temperatures. It reverts back to the ordinary form when heated.

Grid®. For the agriculture industry. [DuPont UK]

Grifa. Lithium acetyl salicylate.

Griffith's white. *See* lithopone.

Griffith's zinc white. *See* lithopone.

Grifulvin V. CAS 126-07-8; Griseofulvin; antifungal. [Ortho Pharmaceutical Corp] *

Grignard's reagent. Magnesium reacts with alkyl and aryl halides, in the presence of ether, forming compounds of the type R.MgX.

Grilamid®. Nylon 12 (polyamide 12); granules for injection molding and extrusion; for precision engineering, components for optical and medical appliances, electrical/electronic industry, sports equipment, automotive industry; extrusion types for manufacture of tubes, sausage skins, films, cable sheathing. [EMS-Chemie AG]

Grilamid® ELY. Elastomeric polyamide 12; used for sports equipment. [EMS-Chemie AG]

Grilamid® ELY20NZ. Thermoplastic nylon 12 elastomer; general-purpose elastomer for injection molding and extrusion; used in ski boots, trekking boots, and applications requiring high flexibility. [EMS-Am. Grilon]

Grilamid® L16. Nylon 12 resin; extrusion grade [EMS-Am. Grilon]

Grilamid® L16G. Nylon 12 resin; with internal lubricant; high-flow injection molding resin for thin-section and long-flow-path components. [EMS-Am. Grilon] *

Grilamid® L25. Nylon 12 resin; general-purpose film extrusion grade without additives or stabilizers. [EMS-Am. Grilon]

Grilamid® TR55. Nylon 12 resin; transparent resin with superior chem. resist., strength, stiffness, and toughness

for fuel, air, or water filter bodies, domestic appliance parts, faucet and shower handles, elec./electronic components, eyeglass frames, automotive parts; agric. and food processing components, medical equip., fiber optic cable sheathing, instrument gauge lenses. [EMS-Am. Grilon; EMS-Chemie AG]

Grilamid® TR55LX. Nylon 12 resin; high flow grade for injection molding and extrusion; used for eyeglass frames, etc. [EMS-Am. Grilon]

Grilbond®. Blocked isocyanates; bonding agents for polyester-reinforced rubber goods in tire and belt industry. [EMS-Chemie AG]

Grilbond® PVC Bonding Agents. Modified polyurethanes and polyamines; bonding agents in PVC plastisols for underbody protection in the automotive industry. [EMS-Chemie AG]

Grilene Swiss Polyester. Polyethylene terephthalate fibers; used for nonwovens and sieves. [EMS-Chemie AG]

Grilesta®. Saturated polyester resins in granular form; used for polyester/epoxy and polyester TGIC powder coatings, particularly for short curing time or low curing temperatures; special products for transparent systems. [EMS-Chemie AG]

Grilesta® P 7205. Hybrid polyester resin; 50/50 hybrid with excellent flow, normal curing. [EMS-Am. Grilon] *

Grilesta® P 7304. Polyester resin; for outdoor durable systems; 93/7 TGIC-PES, normal curing. [EMS-Am. Grilon] *

Grilesta® P 7401. Hybrid polyester resin; for powd. coatings; 70/30-80/20 hybrid, TMA-free, fast curing. [EMS-Am. Grilon] *

Grillocin HY-77. Zinc ricinoleate, triethanolamine, zinc rosinate, isostearic acid, dipropylene glycol, sodium lactate, abietic acid, tocopherol; absorbs malodors from sol'ns. and surfs. [RITA]

Grilloten LSE87. CAS 25339-99-5; EINECS 246-873-3; Sucrose laurate; o/w emulsifier, solubilizer, foam booster, counter-irritant; stable over wide pH range. [RITA]

Grilloten LSE87K. Sucrose cocoate; o/w emulsifier, solubilizer, foam booster, counter-irritant; stable over wide pH range. [RITA]

Grilloten PSE141G. CAS 25168-73-4; EINECS 246-705-9; Sucrose stearate; o/w emulsifier, solubilizer; stable over wide pH range. [RITA]

Grilloten ZT40, ZT80, PSE 141G, LSE 87, LSE 87K. Sucrose ricinoleate; mild solvent-free surfactants; nontoxic, nonsensitizing and anti-irritant; both water and oil soluble; moisturizing and emulsifying properties do not affect formulation. [RITA] *

Grilon®. Nylon 6 (polyamide 6), granular form; used for injection molding, extrusion and blow molding for automotive, machinery, electrical/electronic, building industries, medical instruments, packaging films; also as staple fibers, monofilaments. [EMS-Chemie AG]

Grilon® A23GM. CAS 25038-54-4; Nylon 6 resin, internally lubricated; general-purpose injection molding resin for thin-section or long-flow-path components. [EMS-Am. Grilon]

Grilon® BT. Polyamide alloy, granular form; used for extrusion and injection molding of technical parts for machinery, mining, electrical/electronic, automotive and building industries. [EMS-Chemie AG]

---

* Trade name not verified as available   † Trade name verified as obsolete

Grilon® BT40Z. CAS 25038-54-4; Nylon 6 resin; impact-modified grade for injection molding, extrusion, and cast and blown film applications [EMS-Am. Grilon]

Grilon® C. Copolyamide, granular form; used for coextrusion of laminated blown films with high gloss and transparency ; also for flexible cable sheathing and monofilament, masterbatch base material. [EMS-Chemie AG]

Grilon® CA6E. Nylon 6/12 copolymer resin; for shrink wrap film applications [EMS-Am. Grilon]

Grilon® CF6S. Nylon 6/12 copolymer resin; film grade intended for shrink wrapping applications, incl. wrapping of food. [EMS-Am. Grilon]

Grilon® ELX. Elastomeric polyamide 6, granular form; used in ski, mountaineering and hiking boots, hydraulic and pneumatic tubing. [EMS-Chemie AG]

Grilon® ELX23NZ. Thermoplastic nylon 6 elastomer; general-purpose elastomer for injection molding and extrusion; used where cold impact resistance and high flexibility needed. [EMS-Am. Grilon]

Grilon® PV-15H. CAS 25038-54-4; Nylon 6 resin, 15% glass fiber reinforced, stabilized; injection-molding grade offering intermediate rigidity and strength characteristics. [EMS-Am. Grilon]

Grilon® PVN-15H. CAS 25038-54-4; Nylon 6 resin, 15% glass fiber reinforced, stabilized; impact-modified injection-molding grade offering high toughness. [EMS-Am. Grilon]

Grilon® R47HW. CAS 25038-54-4; Nylon 6 resin, plasticized, stabilized; higher molecular weight resin for injection molding and extrusion; used for tubes, rods, and profiles, flexible tubing for automotive and industrial applications, and for the mfg. of mandrels. [EMS-Am. Grilon]

Grilon® T. Nylon 6/6, granular form; used for injection molding in electrical/electronic, automotive, building and machinery industries. [EMS-Chemie AG]

Grilon® T300GM. CAS 32131-17-2; Nylon 6/6; unreinforced injection molding resin offering fast cycling, easy processability, excellent chem., solv., and fuel resist., wear and abrasion performance, elec. insulating props.; used for elec./electronic components, automotive connectors, furniture and window components, buttons, switches, housings for household appliance and power tool components. [EMS-Am. Grilon]

Grilon® TV-15H. CAS 32131-17-2; Nylon 6/6, 15% glass fiber-reinforced; heat-stabilized injection molding resin. [EMS-Am. Grilon]

Grilon® XE3106. Thermoplastic nylon 6 elastomer; very low hardness, highly flexible elastomer for injection molding and extrusion. [EMS-Am. Grilon]

Grilon® XE3222. Nylon 6-6/9 copolymer resin; for coextrusion for blown and cast film with good clarity and flexibility. [EMS-Am. Grilon]

Grilon® XE3303. Nylon 6/6-6/10 copolymer resin; for food pkg. and medical applications [EMS-Am. Grilon]

Grilonit®. Epoxy resins and hardeners, solid and liquid, unmodified and modified; used in civil engineeering, general surface protection, flooring systems, mortars, sealers, priming coats, crack injection systems, structural adhesives and coatings in the building industry. [EMS-Chemie AG]

Grilonit® Reactive Diluents. Liquid glycidyl ethers, mono- and difunctional; auxiliary agents to facilitate processing,

enhance wetting properties and increase flexibility in epoxy resin systems. [EMS-Chemie AG]

Grilpet®. Polyethylene terephthalate polyester, granular form; used for extrusion and injection molding of technical parts for electrical/electronic and automotive industries, sports and leisure articles, domestic appliances. [EMS-Chemie AG]

Grilpet® EV-30. CAS 25038-59-9; PET polyester resin, 30% glass fiber-reinforced; fast cycling resin for injection molding applications; produces moldings with high dimensional stability, low moisture absorp., high rigidity; suitable for elec./electronic, automotive, or structural applications [EMS-Am. Grilon] *

Grilpet® XE3060. CAS 26062-94-2; PBT polyester resin; high molecular weight extrudable resin with good mech. and thermal props., outstanding hydrolysis resist.; ideal for the outer layer of fiber optic loose buffer dual-tubes. [EMS-Am. Grilon] *

Griltex®. Copolyamide and copolyester granules and powders; for coating of fusible interlinings; excellent resistance to laundering and dry-cleaning; short fusing cycles at moderate temperatures; manufacture of hot-melt adhesive films, monofilaments; structural hot-melts for automotive, textile and machinery industries. [EMS-Chemie AG]

Grime Go. Solvent based hand cleaner liquid and paste. [Momar Industrial Services Ltd]

Grimm aluminum solder. An alloy of 69% tin, 29% lead, 1.5% zinc, and 0.75% silver.

grinding oils. Drying oils such as linseed, etc.; used for grinding pigments for paints.

grindstone. A sandstone consisting almost entirely of quartz.

Grindtek AML 60. Acetylated palm kernel glycerides; lubricant and plasticizer for plastics and coatings; cosolv. for polar additives in low polarity systems. [Grindsted Prods.] *

Grindtek FAL 1. CAS 25383-99-7; EINECS 246-929-7; Sodium stearoyl lactylate; o/w emulsifier. [Grindsted Prods.] *

Grindtek ML 90. CAS 142-18-7; EINECS 205-526-6; Glyceryl laurate; component in w/o and o/w creams, lubricant, antistat, antifogging agent in plastics; antimicrobial effects reported. [Grindsted Prods.]

Grindtek MM 90. Glyceryl myristate; component in w/o and o/w creams, lubricant, antistat, antifogging agent in plastics; antimicrobial effects reported. [Grindsted Prods.]

Grindtek MSP 32-6. Glyceryl stearate SE; component in w/o and o/w creams, lubricant, antistat, antifogging agent in plastics; antimicrobial effects reported. [Grindsted Prods.] *

Grindtek PGE 25. CAS 9007-48-1; Polyglyceryl-3 oleate; o/w emulsifier, antifogging agent for plastics. [Grindsted Prods.] *

Grindtek PGMS 90. CAS 1323-39-3; EINECS 215-354-3; Propylene glycol stearate; w/o emulsifier. [Grindsted Prods.] *

Grindtek SMS. CAS 1338-41-6; EINECS 215-664-9; Sorbitan stearate; w/o and o/w emulsifier. [Grindsted Prods.] *

Grindtek STS. CAS 26658-19-5; EINECS 247-891-4; Sorbitan tristearate; w/o and o/w emulsifier. [Grindsted Prods.] *

Griplet®. For the electrical industry. [DuPont UK]

---

‡ Trade name and manufacturer not verified   § Trade name without identified manufacturer

Grippal. A pharmaceutical preparation for prevention and treatment of chills and colds. [Bayer AG]

Grisactin. CAS 126-07-8; Griseofulvin; antifungal. [Wyeth Laboratories]

Grisounites. Coal mine explosives. They contain nitroglycerin, collodion cotton, ammonium nitrate, and sometimes magnesium sulfate in the place of ammonium nitrate. Others contain ammonium nitrate and nitro- naphthalene.

Grisoutite. An explosive consisting of 53% nitroglycerin, 14.5% kieselguhr, and 32.5% magnesium sulfate.

Grisovin. CAS 126-07-8; A proprietary preparation containing griseofulvin; an antibiotic. [Glaxo Laboratories] *

Gris-PEG®. CAS 126-07-8; Griseofulvin ultramicrosize USP; antifungal. [Allergan, Inc]

Grivory®. Partial aromatic polyamides, granular form; used for injection molding, blow molding and extrusion for automotive, leisure and sports equipment, electrical and telecommunication technology, packaging industry. [EMS-Chemie AG]

Grivory® XE3215. Amorphous, unreinforced engineering thermoplastic; impact-resist. resin for injection molding, extrusion, and blow molding; features high rigidity, dimensional stability, gloss, and easy colorability. [EMS-Am. Grilon]

Grivory® XE3290. Amorphous nylon, 35% glass fiber-reinforced; for injection molding applications; good dimensional stability. [EMS-Am. Grilon]

Grodex. Seed germination indicator. [May & Baker Ltd] *

GRO-HY. Nitrogen, phosphorus, potash plus trace elements as a slow-release fertilizer tablet; fertilizer for trees, shrubs and bushes. [Envhy Ltd] ‡

Grommett brass. See brass.

Grossmann reagent. An ammoniacal solution of dicyandiamidine sulfate, $(C_2H_6ON_4)_2.H_2SO_4$. A reagent for nickel.

Grossman's alloy. An alloy containing 87% aluminum, 8% copper, and 5% tin.

Gross solder. An alloy of tin, zinc, aluminum, lead and phosphorus.

Grotan. Range of biocides for the oil and metal working industries. [Sterling-Winthrop Group Ltd]

Groundhog. Aminotriazole + diquat + paraquat + simazine; used for total weed control in non crop areas. [ICI Agrochemicals Professional Products; ICI Chem. & Polymers Ltd]

ground-nut oil. A non-drying oil obtained from the groundnut. It consists chiefly of the glycerides of oleic and linolic acids. It is edible, and when hydrogenated is used in the manufacture of margarine. Lower qualities are employed in soap-making.

groundnut oil. See peanut oil.

grout. A mixture of cement and water. Sometimes including sand.

Groutcide 75. CAS 1897-45-6; Chlorothalonil; fungicide for Portland cement grout. [Henkel]

Growmore. Granular fertilizer containing NPK 7:7:7; general-purpose fertilizer. [Vitax Ltd]

gru-gru fat. The fat from the seeds of *Acrocomia sclerocarpa*, of the palm family.

Gruber solder. An alloy of 60% tin, 25% zinc, 2% aluminum, 10% copper, and 3% cadmium.

Grudekok. (Lignite char). Lignite from which about 30-40% water has been expelled; used in the manufacture of briquettes.

GT50A. A proprietary preparation of calciferol, aneurin, neostigmine, and carbachol, for use in osteoarthritis. [Geistlich Sohne AG] *

GTI. Lubricants for hot glass handling in the glass container industry. [Darmex Corp]

guadidium chloride. See guanidine hydrochloride.

guaiacol. (methylcatechol; Pyrocatechol methyl ether; o-Methoxyphenol; o-Hydroxyanisole) $OHC_6H_4OCH_3$; 90-05-1; EINECS 201-964-7; synthetic flavors, medicine (expectorant). [Penta Mfg.; Schweizerhall; Société Chimique Roche]

guaiacum resin. A resin obtained from the wood of *Guaiacum officinale* and *Guaiacum sanctum*.

Guaic. Guaiacol resin.

guaifenesin. (3-(2-methoxyphenoxy)-1,2-propanediol) $C_{10}H_{14}O_4$; CAS 93-14-1; medicine (expectorant). [R.W. Greeff; Henley; Peboc Ltd; Spectrum Chem. Mfg.]

Guaiol. Tiglic aldehyde, C_5H_8O.

guanidine hydrochloride. (aminoformamidine hydrochloride; guanidinium chloride; guadidium chloride) $NH_2C(:NH)NH_2 \cdot HCl$; CAS 50-01-1; EINECS 200-002-3. [Am. Biorganics; Dajac Labs; Fluka; Heico; U.S. Biochemical]

guanidine hydrothiocyanate. See guanidine thiocyanate.

guanidine nitrate. $NH_2C(:NH)NH_2 \cdot HNO_3$ CAS 506-93-4; EINECS 208-060-1; manufacture of explosives, disinfectants, photographic chemicals. [Dajac Labs; R.W. Greeff; Spectrum Chem. Mfg.]

guanidine thiocyanate. (guanidine hydrothiocyanate; guanidinium thiocyanate; guanidinium rhodanide) $NH_2C(:NH)NH_2.HSCN$; CAS 593-84-0; EINECS 209-812-1; potent protein denaturant used in isolation of intact DNA, RNA. [Dajac Labs; Eastman; Fulka; U.S. Biochemical]

guanidinium chloride. See guanidine hydrochloride.

guanidinium rhodanide. See guanidine thiocyanate.

guanidinium thiocyanate. See guanidine thiocyanate.

Guanimycin. A proprietary preparation of dihydro-streptomycin sulfate, sulfaguanidine and kaolin; an antidiarrhoeal. [Allen & Hanburys Ltd] *

guanine. (2-amino-1,7-dihydro-6H-purin-6-one; 2-aminohypoxanthine; 2-Amino-6-hydroxypurine) Natural purine; $C_5H_5N_5O$; CAS 73-40-5; EINECS 200-799-8; biochemical research, cosmetics. [R.W. Greeff; Henley; Janssen Chimica; Mearl; Penta Mfg.]

guano. (Bird Manure). It consists of deposits of excrements and skeletons of birds and animals; a fertilizer

Guanor Expectorant. A proprietary preparation of ammonium chloride, diphenhydramine hydrochloride, sodium citrate, chloroform, and menthol. [RP Drugs] *

Guara. The ground fruits of a species of *Caesalpinia*, from Central and South America; used as a tanning material.

Guaranine. See Theine.

Guardar. Acetic acid 2%, petroleum distillates 1%, polybutenes 1%, inert ingredients 96%; a liquid adjuvent for reducing pesticide use and for enhancing the "sticking" of foliar-applied materials to leaf surfaces and seeds. [SN Corp/Appropriate Technology Ltd]

Guardion. Chemicals and equipment for treating water, effluent and sewage. [ICI Chem. & Polymers Ltd]

Guardsep. Chemicals and equipment for treating water, effluent and sewage. [ICI Chem. & Polymers Ltd]

Guardsman. CAS 7784-25-0; Aluminum ammonium sul-

---

* Trade name not verified as available   † Trade name verified as obsolete

fate; bird and animal repellent. [Sphere Laboratories (London) Ltd]

Guardsman. Chemicals and equipment for treating water, effluent and sewage. [ICI Chem. & Polymers Ltd]

guar flour. *See* guar gum.

guar gum. (guar flour; gum cyamopsis) Natural material derived from the ground endosperms of *Cyamopsis tetragonolobus*; CAS 9000-30-0; EINECS 232-536-8; paper coating, cosmetics, pharmaceuticals; thickener, emulsifier in food products; as protective colloid, stabilizer; binding agent in tablets; flocculant in mining industry; coagulant aid in water treatment. [Aqualon; Hercules; Rhone Poulenc; Stan Chem Int'l. Ltd]

Guartec CAP. Water-sol. polymer for strength enhancement in paper. [Hercules]

Guartec CIP. Water-sol. polymer for strength enhancement in paper. [Hercules]

Guayale. (Durango). A rubber obtained from *Parthenium argentatum*.

Guettier metal. An alloy of 62% copper, 32% zinc, and 6% tin.

Guido's balsam. Liniment of opium.

Guignet's Green. Chromium hydroxide or hydrated chromium oxide. [Reading, Green & Marvell Ltd]

Guillaume Alloy. A proprietary trade name for an alloy of 66% iron and 34% nickel; has a low coefficient of expansion. §

Guillaume Metal. An alloy of 64% copper and 36% bismuth.§

Guinea Grains. *See* Grains Of Paradise.

Gulfad-C. Chemical additive for cements used in oil and gas wells. [Chevron] *

Gulfco. Lube oils and greases. [Chevron] *

Gulfcrest. Greases and motor fuel. [Chevron] *

Gulfcrown. Greases. [Chevron] *

Gulfcut. Cutting fluids. [Chevron] *

Gulfgem. Greases. [Chevron] *

Gulfknit. Lube oils. [Chevron] *

Gulfleet. Lube. [Chevron] *

Gulflex. Greases. [Chevron] *

Gulf Lite. Charcoal starter, patio torch fuel. [Chevron] *

Gulf Lubcote. Lube oil and anticorrosive. [Chevron] *

Gulflube. Motor oils. [Chevron] *

Gulf No-Rust. Lube oil and anticorrosive. [Chevron] *

Gulfpride. Motor oils. [Chevron] *

Gulfspin. Lube oil. [Chevron] *

Gulftene. Alpha olefins. [Chevron] *

Gulftex. Lubes. [Chevron] *

Gulftow. Lubes. [Chevron] *

Gulftronic. Electrostatic precipitators. [Chevron] *

Gulfwax. Paraffin wax. [Chevron] *

Gum D. A gelatin dynamite (French). It contains 69.5% nitroglycerin.

Gum E. A gelatin dynamite (French). It contains 49% nitroglycerin.

Gum MB. A gelatin dynamite (French). It contains 74% nitroglycerin.

gum, Amritsar. An acacia gum from *Acacia modesta*; used in calico printing.

gum arabic. (acacia gum; arabic gum; Arabin; calcium gummate) CAS 9000-01-5; EINECS 232-519-5; Gum acacia from *Acacia senegal* and other species; pharmaceuticals, adhesives, inks, textile printing, cosmetics, thickening agent, colloidal stabilizer, food products, bind-

ing agent in tablets, emulsifier. [Dr. Madis Labs; Meer; Penta Mfg.; TIC Gums]

gum Benguela. A semi-fossil copal used in varnishes.

gum Benjamin. CAS 9000-05-9; Benzoin.

gumbo. *See* bentonite.

gum camphor. *See* camphor.

gum, catechu. Catechu.

gum, Cowrie. Gum dammar.

gum cyamopsis. *See* guar gum.

gum dragon. Gum tragacanth.

gum, fiddle. Gum tragacanth.

gum Juniper. Gun sandarac.

gum lac. *See* lac.

gum Lini. A gum made from linseed by treatment with water, then treating the mass with 90% alcohol. The gum is soluble in water and is used as a substitute for gum arabic.

Gummeline. Dextrin.

gum rosin. *See* rosin.

gum thus. (French pine resin). Common frankincense, the crude turpentine from French pine-trees.

gum tragacanth. CAS 9000-65-1; A gum obtained from shrubs of the *Astragalus* family.

guncotton. (nitro-cotton, pyroxylin, collodion cotton). An explosive. It is nitro-cellulose, made by acting upon cotton with nitric and sulfuric acids.

Gun metal. An alloy containing from 89-91% copper, 8-11% tin, and 1-2% zinc. The alloy, containing 90% copper, 8% tin, and 2% zinc, has a specific gravity of 8.8, and melts at 1010 CC.

Gunner. CAS 52756-22-6; Emulsifiable concentrate of 200 g flamprop-M-isopropyl per liter; used for control of wild oats in cereal crops. [Quadrangle Agrochemicals]

Gunning's reagent. A 10% iodine solution in alcohol; used for the detection of acetone in urine.

gunpowder. (black powder). A mixture of saltpeter, carbon, and sulfur, in varying proportions. An average one contains 75% potassium nitrate, 10% sulfur, and 15% carbon. An explosive.

Guntapite. A monolithic refractory material used for lining and repairing linings of steelmaking vessels. [Pfizer International] †

Gurdynamite. *See* dynamite.

Gurjun balsam or oil. (wood oil). The oleo-resin from the stems of *Dipterocarpus* species.

Gurley's metal. An alloy of 86.5% copper, 5.4% zinc, 5.4% tin, and 2.7% lead.

Gurney's bronze. An alloy of 76% copper, 15% lead, and 9% tin.

Gurr. Biological stains and reagents. [BDH Chemicals Ltd]

Gusathion®. CAS 86-50-0; Azinphos-methyl; insecticide with broad spectrum of activity, for control of biting insects, especialy lepidopterous larvae, and sucking pests such as aphids, suckers, scale insects, etc. on a wide range of crops in all climatic zones; side-effect on mites. [Bayer AG]

Gusathion®A. CAS 2642-71-9; Azinphos-ethyl; insecticide with broad spectrum of activity against biting pests such as caterpillars, beetles and their larvae, as well as sucking pests such as aphids, thrips, leafhoppers, etc. on a wide range of crops; side-effect on mites. [Bayer AG]

Gusathion® MS. A wettable powder containing 25% w/w azinphos-methyl and 7.5% w/w demeton-S-methyl sulfone; used to control a wide range of biting and sucking

---

‡ Trade name and manufacturer not verified   § Trade name without identified manufacturer

pests on agricultural and horticultural crops. [Bayer plc]†

Gusto. CAS 13684-63-4; Emulsifiable concentrate containing 118 g/l phenmedipham; for weed control for beet crops. [Farm Protection Ltd]

Guthion®. CAS 86-50-0; Azinphos-methyl; insecticide with broad spectrum of activity, for control of biting insects, especialy lepidopterous larvae, and sucking pests such as aphids, suckers, scale insects, etc. on a wide range of crops in all climatic zones; side-effect on mites. [Bayer AG]

Guthrie's eutectic alloy. An alloy of 47% bismuth, 20% tin, 20% lead, and 13% cadmium.

gutta-percha. The coagulated latex of species of *Palaquium* and *Payena*, of Malay, Sumatra, and Borneo. The material is plastic when hot and can be molded. It was formerly used extensively for electrical insulaters, chemical containers etc. formerly used widely for golf ball covers; used in dentistry surgical accessories, and as an insulating medium in electrical devices.

gutta-shea. (Karite gum). A product resembling gutta-percha obtained from an African tree, *Bassia Parkii*. The gutta is separated from the fat (Shea butter).

Gutta-Siak. A low grade gutta-percha gum from *Payena Leerii* and other *Payena* species, of Siak, Sumatra. It is a mixed product with a high resin content.

gutta-sundik. A gutta-percha from *Payena* species.

gutta-susu. (Assam white, gutta gerip, gutta Singarip, Borneo rubber). A wild rubber mainly obtained from *Willughbeia firma*, of Borneo.

guvacine. An alkaloid derived from pyridine. It is 1,2,5,6-tetrahydropyridine-3-carboxylic acid. Its methyl ester is also called guvacine.

G Varnish. A proprietary trade name for varnish and lacquer resins made from glycerol and phthalic anhydride. §

Gynaflex. A proprietary preparation of noxytiolin and lignocaine hydrochloride; used in vaginitis. [Geistlich Sohne AG] *

Gyne-Lotrimin. CAS 23593-75-1; Clotrimazole; antifungal. [Schering-Plough HealthCare Products Inc]

gynocardia oil. (Chaulmoogra oil; hydrocarpus oil). It is obtained from the seeds of *Taraktogenon kurzii*; used in medicine for the treatment of leprosy and other infective skin diseases.

gynocardic acid. A term used for the acids contained in the oil expressed from the seeds of *Gynocardia odorata*.

Gyno-Daktarin. CAS 22832-87-7; A proprietary preparation of miconazole nitrate; used in the veterinary treatment of vaginal candidiasis. [Janssen Pharmaceutical Ltd] *

Gynogen LA 10. CAS 979-32-8; Estradiol valerate; estrogen. [O'Neal, Jones & Feldman Pharmaceuticals] *

Gynol. CAS 9016-45-9; Nonoxynol 9; spermaticide. [Ortho Pharmaceutical Corp]

gypsite. A deposit consisting of small grains of gypsum, disseminated through an earthy mass; used for the production of wall plastics.

Gypsona. Plaster of Paris bandage. [Smith & Nephew Pharmaceuticals Ltd] *

gypsum. (alabaster, selenite, annaline, terra alba, satinite, mineral white, light spar, satin spar, Lenzit). CAS 13397-24-5; A mineral. It is calcium sulfate, $CaSO_4 \cdot 2H_2O$; used in plasters, in Portland cement, paints, and as a filler for paper and cotton.

Gypsum-F. Source of calcium and sulfur for use on golf courses, ball parks, cemeteries, nurseries, industrial lawns, home lawns, and other turf applications; increases permeability of soils to lower sodium content. [W A Cleary]

Gyrane. 2-Butyl-4,6-dimethyldihydropyran. [Quest Int'l. UK Ltd]

---

* Trade name not verified as available   † Trade name verified as obsolete

H

H.11. A proprietary preparation of a polypeptide from male urine; used in the treatment of cancer. [Standard Laboratories] *

H-30. Hydrated alumina; for production of petrol, hydrotreating and cracking catalysts, aluminum sulfate, aluminum chloride, sodium aluminate, aluminum phosphate, TiO_2, zeolites for detergent industry; chemical heat sink (retards burning of host compounds). [La Roche Chem.]

H7250. CAS 1009-93-4; Hexamethylcyclotrisilazane; difunctional blocking agent; reagent for cyclosilylation. [Hüls Am.] *

H7301. Hexamethyldisilazane; adhesion promoter for photoresists incl. positive and negative photoresists on SiO_2 substrates; highly purified for electronic applications. [Hüls Am.]

H7310. CAS 107-46-0; Hexamethyldisiloxane; as cleaning, polishing and damping media; offers low toxicity, inertness. [Hüls Am.]

HA 819. Polyoxyethylene nonyl phenol; a water soluble material of medium chain length; used as a detergent. [Honeywell Atlas] ‡

Hachimycin. An antibiotic produced by *Streptomyces hachijoensis*; used in the treatment of trichomoniasis.

H-acid. 1,8-Diaminonaphthol-3, 6-disulfonic acid; used as an azo dye intermediate.

Hadranol. CAS 1702-17-6; Hormone based on 2,4-D ester; plant growth regulator. [Makhteshim Chemical Works Ltd]

Haelan. A proprietary preparation of flurandrenalone; used in the treatment of skin diseases. [Dista Products Ltd]

Haelan-C. Flurandrenolone with clioquinol BP; for the topical management of those dermatological disorders complicated by bacterial or fungal infection. [Dista Products Ltd] *

Haelan-X. A proprietary preparation of flurandrenolone for dermatological use. [Dista Products Ltd] *

Haelan Tape. Flurandrenolone; adjunctive therapy for chronic recalcitrant dermatoses that may respond to topical corticosteroids and particularly dry-scaling and localized lesions. [Dista Products Ltd]

Hemachates. Agates marked with red jasper.

Haemate-P. Antihemophilic factor. [Hoechst-Roussel Pharmaceuticals Inc] *

Haemostop. A proprietary preparation of naftazone; used to stop capillary bleeding by injection. [Consolidated Chemicals Ltd] *

Hagafilm. Water treatment chemicals. [Albright & Wilson Ltd, Phosphates & Speciality Business] *

Hagatreat. Water treatment chemicals. [Albright & Wilson Ltd, Phosphates & Speciality Business] *

Hager's reagent. Picric acid (1 gram) dissolved in 100 cc water.

Hagevap. Treatment for sea water evaporators. [Albright & Wilson Ltd]

Hahnmann's mercury. Black oxide of mercury.

Haine's solution. Copper sulfate, 8.314 g, is dissolved in 400 cc water, 40 cc glycerol, and 500 cc of 5% potassium hydroxide added. A test for sugar.

Hair Complex 20/70n. Placenta extract, B vitamins, sulfur-containing amino acids in water-alcohol medium; aq.-alcoholic lotions for regenerative hair care, oily scalps and dandruff. [Dr. Kurt Richter; Henkel/Cospha]

Hair Complex Aquosum. Herbs and B vitamins in water-alcohol medium; treatment of scalps with dandruff or greasiness; general hair protection. [Dr. Kurt Richter; Henkel/Cospha]

Hair salt. *See* epsom salts.

Halar E-CTFE Film. Copolymer of ethylene and chlorotrifluoroethylene; useful as release films. [Allied-Signal Inc]

Halar® 300. ECTFE; melt processable fluoropolymer offering excellent chem. resist., good elec. props., broad use temp. range, flame retardance, abrasion resist., and resist. to cobalt radiation; suitable for extrusion, injection molding, blow molding, rotomolding, and for application by fluidized bed or electrostatic coating processes; 300 grade has intermediate melt visc. for compression and blow molding, extrusion (wire coating, tubing, and film). [Ausimont; Fluorocarbon Co. Ltd.]

Halaurant. A proprietary preparation of halibut liver oil, orange juice and calciferol; a vitamin mixture. [Ayrton Saunders plc]‡

Halbase 10. Lead sulfate, tribasic; PVC stabilizer. [Halstab]*

Halberland metal. A brass containing 87% copper and 13% zinc.

Halcarb 20. CAS 598-63-0; Leadcarbonate, basic; PVC stabilizer. [Halstab] *

Halciderm. CAS 3093-35-4; A proprietary preparation of halcinonide used in the treatment of skin diseases. [Bristol-Myers Squibb Pharmaceuticals Ltd] *

Halcion. CAS 28911-01-5; Triazolam; sedative, hypnotic. [Upjohn]

Haldol. CAS 52-86-8; A proprietary preparation containing heloperidol; antipsychotic, anti-emetic. [Janssen Pharmaceutical Ltd] *

Haldol Decanoate. A proprietary preparation containing heloperidol decanoate; antipsychotic. [Janssen Pharmaceutical Ltd] *

Haldrate. CAS 1597-82-6; A proprietary trade name for the 21-acetate of paramethasone. [Eli Lilly & Co] †

---

‡ Trade name and manufacturer not verified § Trade name without identified manufacturer

Haldrone. CAS 1597-82-6; Paramethasone acetate; glucocorticoid. [Eli Lilly & Co] †

half-stuff. Refined wood cellulose obtained as sulfite pulp and in the form of thick sheets; used either alone or mixed with esparto pulp in the manufacture of paper.

Halibol. A proprietary preparation of halibut-liver oil with irradiated ergosterol. §

Haliborange. Orange flavored and colored chewable sugar coated tablets containing vitamin A 750 ug, vitamin D 5 ug, Vitamin C 25 mg; used as a vitamin supplement for adults and children. [Evans Medical] *

halite. See rock salt.

Haliverol. A proprietary preparation of halibut-liver oil with irradiated ergosterol. §

Hallco® C-491. Glyceryl diacetate. [C.P. Hall]

Hallco® C-918. Glyceryl acetate. [C.P. Hall]

Hallcomid®. Series of dimethyl amides; grinding aids, dispersing agents, insecticides, bio-agents. [C P Hall] *

Hallcomid® M-8-10. N,N-Dimethyl caprylamide-capramide; mutual solv. for polar and nonpolar ingred. in cosmetics, cleaners, corrosion inhibitor. [C.P. Hall]

Hallcomid® M-18-OL. N,N-Dimethyl oleamide; solubilizer, solv., dispersant, wetting agent. [C.P. Hall]

Hallcote®. Rubber lubricants. [C.P. Hall]

Hallcote® 573. CAS 471-34-1; Calcium carbonate; rubber dips and coatings. [C.P. Hall]

Hallcote® CSD. CAS 1592-23-0; EINECS 216-472-8; Calcium stearate-based slab dip; rubber dips and coatings. [C.P. Hall]

Hallcote® ZS 5050. CAS 557-05-1; Zinc stearate-based slab dip; rubber dips and coatings. [C.P. Hall]

Hallmark. Emulsifiable concentrate of 50 g lambda-cyhalothrin per liter; a pyrethroid insecticide. [ICI Agrochemicals]

Hal-Lub-D. Lead stearate, dibasic; PVC stabilizer, lubricant. [Halstab] *

Hal-Lub-N. CAS 1072-35-1; Lead stearate, normal; PVC stabilizer, lubricant. [Halstab] *

Halobrom. CAS 126-06-7; 1-Bromo-3-chloro-5,5-dimethylhydantoin; broad spectrum biocide for control of algae, bacterial and fungal slimes in swimming pools and industrial water systems; nonflam. [Dead Sea Bromine]

Haloflex. Vinyl acrylic-based polymer resin products. [ICI Resins BV]

Halog. CAS 3093-35-4; Halcinonide; anti-inflammatory [Bristol-Myers Squibb Co Inc] *

Halon. A proprietary polytetrafluoroethylene. [Allied-Signal Inc] †

Halotestin. CAS 76-43-7; Fluoxymesterone; androgen. [Upjohn] *

halothane. CAS 151-67-7; (2-Bromo-2-chloro-1,1, 1-trifluoroethane,fluothane) $CF_3CHBrCl$; used in medicine as an anesthetic.

Halothane M & B. CAS 151-67-7; A proprietary preparation of halothane; volatile anesthetic. [RMB Animal Health Ltd]

Halowax 1014. A proprietary trade name for hexachlornaphthalene. [Bakelite Corporation] *

Halowax 4000 B-2. A proprietary trade name for a chlorinated hydrocarbon vinyl plasticizer. [Union Carbide (UK) Ltd] *

Haloxil. Haloxon/oxyclozanide anthelmintic. [The Wellcome Foundation Ltd] *

Halphen reagent. A solution of sulfur (1% in carbon disulfide; used to test for cotton-seed oil. To 1 cc of oil add 1 cc of reagent and 1 cc of amyl alcohol, and heat. A red color is given with cotton-seed oil.

Halphos. Lead phosphite, dibasic; PVC heat/light stabilizer [Halstab] *

Halso® 99. CAS 25168-05-2; Monochlorotoluene; extender, diluent or substitute for other org. solvs.; for dye carrier, fuel oil additive, sludge solv., in paint thinners, metal parts cleaners, adhesives. [OxyChem]

Halstab 30. Complex lead salt; heat stabilizer for PVC. [Halstab] *

Halstab P-1. stabilizer/lubricant for PVC. [Halstab] *

Haltex 300. Alumina trihydrate; smoke suppressive and flame retardant filler for plastics; used for carpet backings, tub and shower stalls, wire and cable insulation, elec. uses, vinyl coated fabrics, rubber products. [Hitox]

Halthal. Lead phthalate, dibasic; PVC heat stabilizer. [Halstab] *

Halumin. An aluminum alloy with 1.48% copper, 2% nickel, 2.3% manganese, 0.47% iron, and 0.09% silicon. It is specially resistant to corroding agents.

Halycitrol. Vitamin emulsion with halibut liver oil; each ml is standardized to contain not less than 276 mcg (920 iu) Vitamin A and 1.9 mcg (76 iu) Vitamin D. [LAB Ltd] ‡

Hamameli tannin. A tannin originally isolated from Hamamelis virginica.

hambergite. A mineral, BeOH·BeBO₃.

Hamburg white. A pigment consisting of 1 part white lead and 2 parts heavy spar.

Hamilton metal. A brass containing 67% copper and 33% zinc. An alloy of 90% zinc, with small quantities of copper, lead, antimony, and phosphor tin, is also known as Hamilton metal.

hammer slag. A basic silicate of iron produced and used in the puddling process for iron.

Hammonia metal. An alloy of 64.5% tin, 32.2% zinc, and 3.2% copper.

Hamonite. An activated carbon, made from peat.

Hampamide B. CAS 4862-18-4; Nitrilotriacetamide aq. sol'n. [Evans Chemetics]

Hampden Steel. A proprietary trade name for a chromium tool steel containing 12.5% chromium, 0.25% nickel, 0.25% manganese, and 2.1% carbon. §

Hamp-Ene® 100. CAS 64-02-8; EINECS 200-573-9; Tetrasodium EDTA; general purpose chelating agent. [W R Grace/Hampshire]

Hamp-Ene® Acid. CAS 60-00-4; EDTA; chelating agent; used where sodium ion is undesirable. [W R Grace/Hampshire]

Hamp-Ene® Na2. Disodium EDTA (dihydrate); chelating agent. [W R Grace/Hampshire]

Hamp-Ene® Na3 Liq. CAS 150-38-9; Trisodium EDTA; chelating agent. [W R Grace/Hampshire]

Hamp-Ene® Na4. CAS 64-02-8; EINECS 200-573-9; Tetrasodium EDTA (dihydrate); chelating agent. [W R Grace/Hampshire]

Hamp-Ex® 80. CAS 140-01-2; Pentasodium pentetate; chelating agent for alkaline earth and heavy metal ions; peroxide bleaching. [W R Grace/Hampshire]

Hamp-Ex® Acid. CAS 67-43-6; Pentetic acid; chelating agent. [W R Grace/Hampshire] *

Hampfoam 35. CAS 61791-59-1; EINECS 263-193-2; Sodium cocoyl sarcosinate; surfactant used in alkaline industrial formulations, textile applications; forms stable small bubbles. [W R Grace/Organics; Chemplex Chems.]

Hamp-Ol®120. CAS 139-89-9; Trisodium HEDTA; general purpose chelating agent for control of iron at pH 6.5-9.5 as well as Ca and Mg. [W R Grace/Hampshire]

Hamp-Ol® Acid. HEDTA; chelating agent. [W R Grace/Hampshire] *

Hamp-Ol® Crystals. CAS 139-89-9; Trisodium HEDTA (dihydrate); chelating agent. [W R Grace/Hampshire]

Hamposyl®AL-30. CAS 68003-46-3; EINECS 268-130-2; Ammonium lauroyl sarcosinate; surfactant for shampoos, skin cleansers, bath gels; sec. emulsifier for emulsion polymerization. [W R Grace/Hampshire]

Hamposyl® C. CAS 68411-97-2; EINECS 270-156-4; Cocoyl sarcosine; detergent, wetting and foaming agent, foam stabilizer, emulsifier, anticorrosive agent, conditioner for hair and rug shampoos, cosmetics, skin cleansers; biodeg. [W R Grace/Hampshire; Chemplex Chems.]

Hamposyl® C-30. CAS 61791-59-1; EINECS 263-193-2; Sodium cocoyl sarcosinate; detergent, wetting and foaming agent, foam stabilizer, emulsifier, anticorrosive agent, conditioner for hair and rug shampoos, cosmetics, skin cleansers; biodeg. [W R Grace/Hampshire; Chemplex Chems.]

Hamposyl® L. CAS 97-78-9; EINECS 202-608-3; Lauroyl sarcosine; detergent, wetting and foaming agent, foam stabilizer, emulsifier, anticorrosive agent, conditioner for hair and rug shampoos, cosmetics, skin cleansers; biodeg. [W R Grace/Hampshire; Chemplex Chems.]

Hamposyl® L-30. CAS 137-16-6; EINECS 205-281-5; Sodium lauroyl sarcosinate; detergent, wetting and foaming agent, foam stabilizer, emulsifier, anticorrosive agent, conditioner for hair and rug shampoos, cosmetics, skin cleansers; biodeg. [W R Grace/Hampshire; Chemplex Chems.]

Hamposyl® M. CAS 52558-73-3; EINECS 258-007-1; Myristoyl sarcosine; detergent, wetting and foaming agent, foam stabilizer, emulsifier, anticorrosive agent, conditioner for hair and rug shampoos, cosmetics, skin cleansers; biodeg. [W R Grace/Hampshire; Chemplex Chems.]

Hamposyl® M-30. CAS 30364-51-3; EINECS 250-151-3; Sodium myristoyl sarcosinate; detergent, wetting and foaming agent, foam stabilizer, emulsifier, anticorrosive agent, conditioner for hair and rug shampoos, cosmetics, skin cleansers; biodeg. [W R Grace/Hampshire]

Hamposyl®O. CAS 110-25-8; EINECS 203-749-3; Oleoyl sarcosine; detergent, wetting and foaming agent, foam stabilizer, emulsifier, corrosion inhibitor, mold release agent, conditioner for hair and rug shampoos, cosmetics, skin cleansers; ceramic dispersant; biodeg. [W R Grace/Hampshire; Chemplex Chems.]

Hamposyl®S. CAS 142-48-3; EINECS 205-539-7; Stearoyl sarcosine; detergent, wetting and foaming agent, foam stabilizer, emulsifier, anticorrosive agent, conditioner for hair and rug shampoos, cosmetics, skin cleansers; biodeg. [W R Grace/Hampshire]

Hamposyl®TL-40. TEA lauroyl sarcosinate; provides mild, high lathering props. to skin cleansers. [W R Grace/Hampshire]

Hamposyl®TOC-30. TEA-oleoyl sarcosinate, TEA-cocoyl sarcosinate. biodeg. surfactant for hair and rug shampoos. [W R Grace/Hampshire]

Hampshire® DEG. CAS 139-41-3; Sodium dihydroxyethylglycinate; chelating agent used for for control of iron only in alkaline sol'ns. [W R Grace/Hampshire]

Hampshire®EDG. Ethanoldiglycine disodium salt; chelating agent used for the control of iron and chelates of calcium. [W R Grace/Hampshire]

Hampshire® NTA 150. CAS 5064-31-3; Trisodium NTA; chelating agents used in laundry detergents and specialty cleaning prods., water treatment, textiles, metal finishing, pulp and paper processing, and petrol. industry; nonphosphate detergent builder. [W R Grace/Hampshire]

Hampshire®NTA Acid. CAS 139-13-9; Nitrilotriacetic acid; general purpose chelating agent. [W R Grace/Hampshire]

Hampshire® NTA Na3. CAS 5064-31-3; Trisodium NTA monohydrate; chelating agent; detergent builder for low or nonphosphate formulations. [W R Grace/Hampshire]

HAN® 857. CAS 64742-06-9; Aromatic; solvent. [Exxon] *

Handi-Wrap. Plastic film for use as a household wrapping material. [Dow UK]

Handyfoam. One-part polyurethane foam in an aerosol; highly adhesive, expanding, filling and sealing material. [Feb Ltd]

Hang-ge. A drug. It is the root nodule of *Pinellia tuberifera*; used as an anti-emetic in Japan and China.

Hansa. Pigment powders; used for paints and inks. [Thomas Goldschmidt Ltd]

Hansa oil. A proprietary trade name for polymerized marine animal oil for soap manufacture. §

Hanus' iodine bromide solution. Iodine bromide (10 g), dissolved in 500 cc of glacial acetic acid; used for determining the iodine value of fats and oils.

Harco Foamstopper®. Blends of mineral oil and surfactants; used for adhesives and surface coatings. [Harlow Chemical Co Ltd]

hard aluminum. An alloy of 77% aluminum, 11% zinc, and 11% magnesium. Another alloy contains copper in the place of zinc.

Hardcote. Dressing for cores and molds. [Foseco (F.S.) Ltd]

Hard Cure. Balanced blend of sodium, potassium and meta silicates combined with surface tension reducing agents; for application to freshly placed concrete following finishing. [Secure Inc] †

hardened rosins. Metallic resinates prepared by heating rosin with metallic oxides, usually calcium, magnesium oxide and zinc oxide.

hardenite. A collective name for austenite and martensite of eutectoid composition.

hard-finish plaster. Plaster made from oven-burnt gypsum dipped in alum solution, and again calcined.

hard-head. The name by which the impurities obtained from the refining of tin are known.

Hardite X. A heat-resisting alloy. It contains 82-86% nickel, 10-13% chromium, and 2% manganese.

Hardite. Heat-resisting alloys containing 55-65% nickel, 15-18% chromium, 1-4% silicon, 1-2% manganese, the balance being iron.

hard jatoba. A Brazilian copal resin.

hard lead. (antimony lead). An alloy of lead with 10-30% antimony; used as a type metal.

hard metal. The name usually applied to a tin-copper alloy, containing 1 part tin, and 2 parts copper.

hard paraffin. *See* paraffin.

---

‡ Trade name and manufacturer not verified

§ Trade name without identified manufacturer

hard platinum. Platinum containing from about 5-30% iridium.

Hardset. Heat-curing hard rubber compound. [Hardman]

hard solder. An alloy of 86% copper, 9% zinc, and 4% tin.

hard zinc. An alloy of 92% zinc, 5% iron, and 3% lead.

Hargus Steel. A proprietary trade name for a die steel containing 1.0% manganese, 0.35% nickel, and 1.0% carbon. §

Harle's solution. A solution of sodium arsenite.

Harlington bronze. See Harrington Bronze.

Harmogen. A proprietary preparation of piperazine oestrone sulfate; used in the treatment of menopausal symptoms. [Abbott Laboratories] *

Harmomang A and B. Iron alloys containing carbon, manganese and molybdenum. Proprietary alloys.

Harmonia bronze. An alloy of 57% copper, 40% zinc, 1.8% iron, and 0.4% lead.

Harmony. Lube oil, hydraulic fluid. [Chevron] *

Harmony®. Mixture of metsulfuron-methyl and thifensulfuron-methyl; for control of annual dicotyledons in cereals. [DuPont UK]

Haro® Chem P28G. CAS 1072-35-1; Normal lead stearate; stabilizer for rigid and plasticized PVC applications. [Harcros UK] *

Haro® Chem PDF. Dibasic lead phosphite; stabilizer for rigid and plasticized PVC applications offers heat and light stability and good electrical properties. [Harcros UK] *

Haro® Chem PTS-E. Tribasic lead sulfate; stabilizer for rigid and plasticized PVC applications offers heat stability and good elec. properties. [Harcros UK] *

Haro® Mix CE-701. CAS 7439-92-1; Lead one-pack system; one-pack lead heat stabilizer systems for rigid PVC applications, UPVC pressure and nonpressure pipes. [Harcros UK] *

Haro® Mix CK-711. CAS 7439-92-1; Lead one-pack system; stabilizer/lubricant for extrusion of UPVC pressure or nonpressure pipes [Harcros UK] *

Haro® Mix MH-204. CAS 7439-92-1; Lead one-pack system; stabilizer/lubricant for extrusion of UPVC pressure pipes. [Harcros UK] *

Haro® Mix YK-110. Barium/cadmium/lead one-pack; stabilizer/lubricant for UPVC window profile extrusion. [Harcros UK] *

Haro® Mix ZC-028, ZC-029, ZC-030. Calcium/zinc one-pack; nontoxic one-pack heat and light stabilizer for rigid PVC applications (profiles). [Harcros UK] *

Haro® Mix ZT-514. Tin one-pack; stabilizer/lubricant for extrusion of UPVC pipes [Harcros UK] *

Haro® Wax L333. Lubricant blend for rigid and plasticized PVC applications, incl. food contact applications [Harcros UK] *

Harrier. Weedkiller containing mecoprop, 3,6-dichloropicolinic acid and ioxynil as potassium salts. [ICI Chem. & Polymers Ltd]

Harrington bronze. (Harlington bronze). An alloy of 55.75% copper, 42.5% zinc, 1% tin, and 0.75% iron. A white bearing metal.

Harringtonite. An Irish mineral containing 41.4% SiO_2, 30.2% Al_2O_3, 11.2% CaO, 5.2% Na_2O, and 12.5% H_2O.

Harris' hematoxyiin stain. A microscopic stain. It contains 1 g haematoxylin, 10 cc alcohol, 20 g alum, 0.5 g mercuric oxide, and 200 cc water.

Harrison's indicator. A small amount of starch boiled with a few cc's. of water, adding to it 100 cc of a freshly prepared 10% potassium iodide solution.

Hartamide 9137. CAS 93-83-4; EINECS 202-281-7; Oleamide DEA; coupling agent, emulsion stabilizer, lubricant and antistat. [Hart Chem. Ltd.]

Hartamide AD. Cocamide DEA (2:1); foam booster, solubilizer, emulsifier, coupling agent for detergent formulations. [Hart Prods. Corp.]

Hartamide LDA. CAS 120-40-1; EINECS 204-393-1; Lauramide DEA; detergent, foam stabilizer and thickener for liq. and powd. detergent systems, shampoos, bubble baths. [Hart Chem. Ltd.]

Hartamide LMEA. CAS 142-78-9; EINECS 205-560-1; Lauramide MEA; foam stabilizer for spray-dried powd. detergents and bubble bath preparations; visc. modifier for detergents, shampoos, bubble baths. [Hart Chem. Ltd.]

Hartamide OD. Cocamide DEA; detergent, foam stabilizer and visc. regulator for liq. and powd. detergent systems, shampoos, bubble baths. [Hart Chem. Ltd.]

Hartamine Series. Fatty acid imidazolines; corrosion inhibitors for gas and oil industry. [Hart Chem. Ltd.]

Hartasist 16. Silicone-based; defoamer for oil and gas industry. [Hart Chem. Ltd.]

Hartasist 46. Specialty surfactant; cold flow improver for oil deposit; heavy-duty cleaner. [Hart Chem. Ltd.]

Hartasperse DI-4900 Series. deinking agent for wash deinking systems; increases pulp brightness. [Hart Chem. Ltd.]

Hartbreak Series. Phenolic resins, polyol ester, alkylaryl sulfonate; demulsifiers for w/o emulsions in oil and gas industry. [Hart Chem. Ltd.]

Hartenol LAS-30. CAS 151-21-3; Sodium lauryl sulfate; detergent, wetting agent, emulsifier, foaming agent for light duty household detergents; biodeg. [Hart Prods. Corp.]

Hartenol LES 60. CAS 9004-82-4; Sodium laureth sulfate; detergent, wetting agent, dispersant, emulsifier, foaming agent for light duty household detergents; biodeg. [Hart Prods. Corp.]

hartin. A white resin found in the lignite in Austria.

hartite. A similar resin to Hartin, and found with it.

Hart Lawn Sand. CAS 10028-22-5; Ferrous sulfate; used for moss control in turf. [Maxwell Hart Ltd]

Hartmetall. See Camite. §

Hart Moss Killer. CAS 10028-22-5; Ferrous sulfate; used for moss control in turf. [Maxwell Hart Ltd]

Hartofix 2X. Cationic resin; fixative for direct and reactive dyes; improves fastness of most substantive colors to water, soaps, and detergents. [Hart Chem. Ltd.]

Hartofol 40. Sodium dodecylbenzene sulfonate; high foaming biodeg. surfactant for heavy-duty liq. detergents, textile processing, industrial cleaners, shampoos, automotive cleaners. [Hart Prods. Corp.]

Hartofol 60T. TEA dodecylbenzene sulfonate; surfactant for cosmetics, shampoos, liq. detergents, wool wash formulations; biodeg.; mild, high purity. [Hart Prods. Corp.]

Hartolan. CAS 8027-33-6; EINECS 232-430-1; Lanolin alcohols; spreading agent, dispersant, stabilizer, plasticizer, w/o emulsifier, conditioner, superfatting agent, moisturizer, and emollient for cosmetic and pharmaceutical systems. [Croda Inc.]

Hartolite. CAS 8027-33-6; EINECS 232-430-1; Lanolin alcohols fraction; w/o emulsifier, emollient, skin condi-

---

tioner, moisturizing agent. [Croda Chem. Ltd.]

Hartolon 5683. CAS 9002-88-4; Polyethylene and selected esters; nonyel. softener for 100% cotton toweling; provides static control and absorbency. [Hart Chem. Ltd.]

Hartolon NA. Sat. esters and alkyl condensate; softener, lubricant, antistat used in textiles. [Hart Chem. Ltd.]

Hartomer 4900. Sodium vinyl sulfonate; monomer for latex emulsion polymerization systems. [Hart Chem. Ltd.]

Hartomer GP 2164. Phosphate ester; surfactant for water-based adhesives and latex emulsion polymerization. [Hart Chem. Ltd.]

Hartomer LD 31. Polyacrylate; antiredeposition agent, dispersant, processing aid. [Hart Chem. Ltd.]

Hartomul PE-30. Blend of polyethylene and polyether derivs.; softener for use with resin finishes on lightweight fabrics; mold release agent for brake linings, urethane foam. [Hart Chem. Ltd.]

Hartonyl L531. Ethoxylated fatty amine; leveling aid for dyeing polyamides; migrating leveling agent and dispersant for acid and disperse dye stuffs; increases contrast and clarity between nylon fibers of different affinities; effective in continuous and winch dyeing of carpets. [Hart Chem. Ltd.]

Hartopol 25R2. Polyoxyalkylene glycol; low foaming surfactant for rinse aids and windshield washer fluids. [Hart Chem. Ltd.]

Hartopol L42. EO/PO block copolymer; low foaming rinse aid, defoamer, emulsifier, detergent, dispersant for many industries. [Hart Chem. Ltd.]

Hartopol LF-1. Polyoxyalkylene glycols; detergent, emulsifier, defoamer for low-foam dishwash, rinse aids, general household and industrial use. [Hart Chem. Ltd.]

Hartopol P65. Polyoxyalkylene glycol; dispersant, demulsifier, defoamer, emulsifier, detergent for many industries. [Hart Chem. Ltd.]

Hartosoft 171. Esters and quaternary amines; nonyel. softener with antistatic and lubricating props. for velours and toweling. [Hart Chem. Ltd.]

Hartosoft S5793. Amino functional silicone emulsion; softener providing lubricity to cotton and cotton blended knits and woven fabrics. [Hart Chem. Ltd.]

Hartotrope AXS. CAS 26447-10-9; EINECS 247-710-9; Ammonium xylene sulfonate; detergent, solubilizer and cloud pt. depressant for light duty and built liq. detergent systems. [Hart Chem. Ltd.]

Hartotrope KTS 44. Potassium toluene sulfonate; hydrotrope for high act. or built liq. detergent systems; for use where sodium ion undesirable. [Hart Chem. Ltd.]

Hartotrope STS-40, Powd. Sodium toluene sulfonate; detergent, solubilizer and cloud pt. depressant for light duty and built liq. detergent systems. [Hart Chem. Ltd.]

Hartotrope SXS 40, Powd. CAS 1300-72-7; EINECS 215-090-9; Sodium xylene sulfonate; detergent, solubilizer and cloud pt. depressant for light duty and built liq. detergent systems. [Hart Chem. Ltd.]

Hartowet 5917. Blend; wetting aid for fabrics; contains no phosphates. [Hart Chem. Ltd.]

Hartshorn oil. See bone oil.

Hartshorn spirit. See spirit of Hartshorn and volatile alkali.

Harvesan. Mercurial seed dressing. [The Boots Co plc] †

Harvestra®. For the agriculture industry. [DuPont UK]

Harvite. A proprietary trade name for a shellac compound.§

Hascrome. A proprietary alloy. It is a manganese-chromium-iron welding rod.

H.A. Solvent. Cyclohexyl acetate; a proprietary solvent for cellulose acetate and nitrate, rosin, rubber, oils, and metallic resinates. §

Hastelloy®. Nickel-base alloys; excellent corrosion resist. with moderate wear resist. [Haynes Int'l.]

hatchettine. See bitumen.

Havapen. CAS 983-85-7; A proprietary preparation containing penamecillin; an antibiotic. [Wyeth Laboratories] *

Hayem's solution. A solution of 5 g sodium sulfate, 1 gram sodium chloride, and 0.5 g mercuric chloride in 200 cc water; used in the examination of blood corpuscles.

Haylite No. 1. An English explosive. It contains 25-27% nitroglycerin, 0.5-1.5% collodion cotton, 19-21% potassium nitrate, 19-21% barium nitrate, 12-14% wood meal, 6-8% mineral jelly, and 10-12% ammonium oxal.

Haylite No. 3. An explosive. It consists of 9.5% nitroglycerin, 60% ammonium nitrate, 5% wood meal, 19.5% sodium chloride, and 5% ammonium oxalate.

Haymine®. Ephedrine, chlorpheniramine; antihistamine for treatment of hayfever. [Pharmax Ltd]

Haynes® 242. Ni-Mo-Cr alloy; age-hardenable alloy with high temp. strength, low thermal expansion chars., good oxidation resist. for service to 760 C; resist. to high-temp. fluorine and fluoride environments; for fluoropolymer and fluoroelastomer production. [Haynes Int'l.]

Haynes alloy No. 25. A proprietary trade name for an alloy of cobalt, nickel, tungsten, and chromium. Possesses exceptional mechanical properties up to 1800 F. §

Haynes metals. Alloys of from 10-75% iron, 20-30% chromium, and 5-25% cobalt. A harder alloy contains 45% cobalt, 40% tungsten, and 15% chromium, and a softer one, 62% cobalt, 28% tungsten, and 10% chromium; they are noncorrosive.

Haynon. A proprietary preparation of chlorpheniramine; an anti-allergic drug. [RP Drugs] *

Hayphryn. A proprietary preparation of phenylephrine hydrochloride and thenyldiamine hydrochloride; an anti-allergic nasal spray. [Winthrop Laboratories] *

Haysite. Thermoset polyester resin, fillers and glass; for electrical insulation, corrosion resistant; sold as custom molded parts, sheets, molding compound, SMC-BMC and pultruded shapes; MIL spec, NEMA and UL recognized materials. [Haysite Reinforced Plastics]

Haysite 14100. Electrical grade bulk molding compound for compression, transfer and injection molding. [Haysite Reinforced Plastics]

Haysite 24500. Corrosion-resist. bulk molding compound for compression, transfer and injection molding. [Haysite Reinforced Plastics]

Haysite 42000. Electrical grade sheet molding compound for compression molded parts requiring high strength and good elec. props. [Haysite Reinforced Plastics]

Haysite EHC-P. Fiberglass-reinforced polyester; general purpose economy pultrusion grade offering 155 C performance; for transformer and elec. applications. [Haysite Reinforced Plastics]

Haystellite. See Camite. §

Hazol. CAS 2315-02-8; A proprietary preparation of oxymetazoline hydrochloride; a nasal decongestant. [Allen & Hanburys Ltd] *

HB-40. Partially hydrogenated terphenyl; plasticizer for vinyl sheeting, films and fabric or paper coatings, and for vinyl protective coatings and adhesives. [Monsanto Co] *

HBD. See tributyltin oxide.

---

‡ Trade name and manufacturer not verified   § Trade name without identified manufacturer

HBR. A compressible closed-cell polyethylene foam rod; placed in a joint before applying a sealant and to assist the sealant in assuming the proper configuration; also serves as a bond breaker to prevent three-side adhesion of sealant to joint substance. [Hercules] *

H-B-Vax. Hepatitis B surface antigen; a vaccine for protection against hepatitis B. [Merck & Co Inc]

HC45. CAS 50-03-3; A proprietary preparation containing hydrocortisone acetate; a dermatological cream. [The Boots Co plc]

HC 200 Concentrate. Pyrethin; contact insecticide. [Certified Laboratories Ltd]

HC-913. Acrylate polymer in ethyl acetate solv.; paint primer conc. used as primer over galvanized metal, aluminum and plastics. [3M]

H Chrome Green 105. CAS 12001-99-9, 7787-59-9; Chromium hydroxide, bismuthoxychloride; inorg. colorant. [Presperse]

HD-Eutanol. CAS 143-28-2; EINECS 205-597-3; Oleyl alcohol; ultra pure grade used as as solubilizer for dyes and waxes; superfatting agent, emollient in cosmetic emulsions, creams, alcoholic lotions. [Henkel/Cospha; Henkel Canada]

HD-Ocenol 45/50. Oleyl/cetyl alcohol (25-33% C16, 60-70% C18); intermediate for surfactant mfg. [Henkel/Emery]

HD-Ocenol 90/95. CAS 143-28-2; EINECS 205-597-3; Oleyl alcohol (2-10% C16, 87-95% C18); emollient, superfatting agent, carrier for cosmetics; intermediate for surfactant mfg. [Henkel/Emery]

HD-Ocenol 110/130. Oleyl linoleyl alcohol (5-10% C16, 90-95% C18); intermediate for surfactant mfg. [Henkel/Emery]

HDPE. See polyethylene, high-density.

HDPE 04352N. CAS 9002-88-4; HDPE; injection molding resin. [Dow Plastics]

HDPE 25053-P. CAS 9002-88-4; HDPE copolymer resin; high molecular weight resin with outstanding hot-melt strength, high flex. mod., superior ESCR, and toughness; used for blow molding and sheet vacuum forming applications incl. gas tanks, drums, large industrial parts; FDA compliance. [Dow Plastics]

HDPE 32060C. CAS 9002-88-4; HDPE; blow molding resin. [Dow Plastics]

HDPE 35053. CAS 9002-88-4; HDPE; corrugated tubing resin. [Dow Plastics]

HDPE IP-10. CAS 9002-88-4; HDPE; improved processing resin designed for reusable cases. [Dow Plastics]

Head and Shoulders. CAS 13463-41-7; Pyrithione zinc; antibacterial; antifungal; antiseborrheic. [Procter & Gamble] *

Headland Charge. CAS 7085-19-0; Soluble concentrate containing 570 g/l mecoprop; for control of weeds in cereals and grassland. [SBC Technology Ltd] ‡

Headland Dephend. CAS 13684-63-4; Emulsifiable concentrate containing 114 g/l phenmedipham; for weed control for beet crops. [SBC Technology Ltd] ‡

Headland Dual. CAS 83601-81-4, 12427-38-2; A suspension concentrate containing 62 g carbendazim and 400 g maneb per liter; systemic fungicide for cereals. [WBC Technology Ltd]

Headland Guard. Mixture of organo copolymers and surfactants; spreader and sticking agent for agricultural chemicals. [WBC Technology Ltd]

Headland Inorganic Liquid Copper. CAS 1332-40-7; Suspension concentrate containing 435 g copper oxychloride per liter; a protectant fungicide. [WBC Technology Ltd]

Headland Intake. Mixture of organic acids and surfactants; wetting additive for a wide range of herbicide, fungicide, dessicant and growth regulator sprays. [WBC Technology Ltd]

Headland Relay. Mixture of dicamba, MCPA and mecoprop; used for weed control in cereals and grassland. [WBC Technology Ltd]

Headland Spirit. CAS 12427-38-2; Suspension concentrate containing 480 g maneb per liter; a dithiocarbamate fungicide to control blight, rusts and mildew. [WBC Technology Ltd]

Headland Swift. CAS 999-81-5; Soluble concentrate containing 750 g/l chlormequat; plant growth regulator. [WBC Technology Ltd]

Heart Shape Indigestion Tablets. A proprietary preparation of sodium bicarbonate, magnesium carbonate and calcium carbonate; an antacid. [Ayrton Saunders plc] ‡

Heavithane. A proprietary polyester-based polyurethane elastomer crosslinked by diols. [HMC Wheels] ‡

heavy carburetted hydrogen. See olefiant gas.

Hebron Pabracr. A suspension concentrate containing 80 g cetrimide and 80 g chlorpropham per liter. soil acting herbicide for lettuce. [Atlas Interlates Ltd]

H.E. cellulose. See hydroxyethylcellulose.

Heckel's solution. A solution containing sodium sulfite, benzoic acid, and water.

Hecla 35. A proprietary trade name for alloy steel containing 0.15% carbon, 0.25% silicon, 0.04% sulfur, 0.04% phosphorus and 0.70% manganese; used to manufacture forged steel pressure vessels. [Hadfield, George & Co]‡

Hecla 115. Alloy steel containing 0.35% carbon, 0.25% silicon, 0.04% sulfur, 0.04% phosphorus, 0.7% manganese and 1.00% nickel; used for manufacturing forged steel pressure vessels. [Hadfield, George & Co] ‡

Hecla 135, 138. An alloy steel containing 0.60% carbon, 0.30% silicon, 0.30% manganese, 2.00% chromium, 2.00% nickel and 0.45% molybdenum; used for the manufacture of forged high tensile steel pressure vessels. [Hadfield, George & Co] ‡

Hecla 138H. An alloy steel containing 0.40% carbon, 0.30% silicon, 0.04% sulfur, 0.04% phosphorus, 0.60% manganese, 0.70% chromium, 2.70% nickel, 0.50% molybdenum and 0.25% vanadium (optional); used for the manufacture of forged high tensile steel pressure vessels. [Hadfield, George & Co] ‡

Hecla 155. An alloy steel containing 0.12% carbon, 0.20% silicon, 0.04% sulfur, 0.04% phosphorus, 0.40% manganese, 2.30% chromium, 1.00% molybdenum; used for the manufacture of forged steel pressure vessels for use at higher temperatures. [Hadfield, George & Co] ‡

Hecla 174. An alloy steel containing 0.30% carbon, 0.30% silicon, 0.04% sulfur, 0.04% phosphorus, 0.50% manganese, 5.00% chromium, 0.90% vanadium; used for the manufacture of forged steel pressure vessels for hydrogen service and high tensile purposes. [Hadfield, George & Co] ‡

Hecla 180. An alloy steel containing 0.40% carbon, 0.30% silicon, 0.04% sulfur, 0.04% phosphorus, 0.60% manganese, 1.00% chromium, 3.2% nickel, 0.50% molybdenum and 0.25% vanadium; used for the manufacture of forged high tensile steel pressure vessels. [Hadfield, forged high tensile steel pressure vessels. [Hadfield, George & Co] ‡

Hecla 306. An alloy steel containing 0.30% carbon, 0.20% silicon, 0.04% sulfur, 0.04% phosphorus, 0.50% manganese, 3.00% chromium, 0.50% molybdenum and 0.20% vanadium; used for the manufacture of forged steel pressure vessels for hydrogen service. [Hadfield, George & Co] ‡

Hecla 307. An alloy steel containing 0.18% carbon, 0.20% silicon, 0.04% sulfur, 0.04% phosphorus, 0.50% manganese, 3.00% chromium, 0.50% molybdenum and 0.20% vanadium; used for the manufacture of forged steel pressure vessels for hydrogen service. [Hadfield, George & Co] ‡

Hecla powder. An explosive similar in composition to Giant powder (*qv*).

Hectabrite® AW. CAS 12173-47-6; Hectorite USP/NF; emulsifier, thickener, suspension agent in pharmaceutical, cosmetic and personal care prods. [Am. Colloid]

Hectalite® 200. CAS 12173-47-6; Hectorite USP/NF; viscosifier, suspension agent, binder for pharmaceutical, cosmetic and personal care prods. [Am. Colloid]

Hector bases. Basic substances obtained by the oxidation of thioureas with hydrogen peroxide; good vulcanization accelerators.

hectorite laponite. Sodium magnesium lithium fluoro silicate; a white, iron free suspending and gelling agent for aqueous systems.

hedgehog crystals. Crystals of ammonium urate found in urinary deposits.

Hedonal®. Range of herbicides containing 2,4-D, MCPA, dichlorprop, MCPP either alone or in combinations; growth regulator herbicide used for control of weeds in cereals. [Bayer AG]

Hedulin. CAS 83-12-5; Phenindione; anticoagulant. [Merrell Dow Pharmaceuticals Inc] *

Hegolit 3. A patented product which is a preparation of higher aliphatic alcohols, with a melting-point of 50 C; a plasticizer. §

Hégor. Shampoo, to wash and condition hair. [Richardson-Vicks Inc] *

Heidenhain's chrome hematoxyiin. A microscopic stain. It is produced by staining the object in 0.33% solution of haematoxylin in water, then soaking in 0.5% solution of potassium chromate.

Heiloy. A proprietary trade name for stainless steels used for dairy utensils. §

Helarion. CAS 108-62-3; A pelleted bait containing 6% w/w metaldehyde; used for control of slugs and snails. [Fisons plc]

Heleco. A proprietary pyroxylin preparation. §

Helenite. *See* Elaterite. §

Heliane. Vat dyes. [ICI Chem. & Polymers Ltd] †

Helicon. *See* aspirin. §

Helio®. Pigments; for surface coatings, printing inks, wallpaper, colored coated paper and lake producing industries. [Bayer AG]

heliochrysin. (sun gold). The sodium salt of tetranitro-α-naphthol.

Heliofil. Pigment dyestuffs in paste form for the dope dyeing of viscous continuous filaments and stable fibers. [Bayer AG]

Heliogen®. Phthalocyanine pigments; for superfast letterpress, offset, flexographic and gravure printing inks,

paints, lacquers, for coloring plastics, for production of artists chalks, paints and crayons. [BASF AG; BASF plc]

Heliolac. A nitrocellulose lacquer.

Heliophan. CAS 118-56-9; Homosalate; ultraviolet screen. [R W Greeff & Co Inc] *

Helizarin®. Pigment formulations for printing and dyeing textiles. [BASF AG; BASF plc]

Helmex. CAS 22204-24-6; A proprietary preparation of pyrantel pamoate; an antihelmintic. [Pfizer International]

Helmezine. Preparations of salts of piperazine. [Allen & Hanburys Ltd] *

Helothion. CAS 35400-43-2; Sulprofos; insecticide with stomach and contact action mainly against caterpillars of *Heliothis spp.* and *Spodoptera spp.* on cotton and other crops. [Bayer AG]

Heloxy® 7. Glycidyl ether of a mixture of C8-C10 alcohols; visc. reducing modifier for epoxy formulations used in flooring, casting, tooling, laminating, potting, coatings, etc. [Shell]

Heloxy® 61. CAS 2426-08-6; Butyl glycidyl ether; epoxy modifier; max. visc. reduction, min. loss of properties; increases impregnation of resin systems, and level of filler loading; used in elec., laminating, casting, tooling, flooring, and coatings. [Shell] *

Heloxy® 62. CAS 2210-79-9; o-Cresyl glycidyl ether; reactive diluent; increases level of filler loading in epoxy resins; used in flooring, low visc. casting, laminating, and decoupage. [Shell] *

Heloxy® 64. Nonyl phenyl glycidyl ether; epoxy functional modifier [Shell] *

Heloxy® 65. CAS 3101-60-8; p-tert-Butyl phenyl glycidyl ether; moderate reactive diluent; used in casting, tooling, laminating, and in flooring. [Shell] *

Heloxy® 67. CAS 2425-79-8; Diglycidyl ether of 1,4 butanediol; diluent; used in casting, laminating, tooling, potting, and elec. applications. [Shell] *

HEMA. *See* 2-hydroxyethyl methacrylate.

Hema-Combistix. A proprietary test strip comprised of four separate tests: (1) methyl red and bromothymol blue for pH; (2) a buffered mixture of glucose oxidase, peroxidase, o-toluidine, and a red dye for glucose; (3) buffered tetrabromophenol blue for glucose; (4) o-toluidine for blood; used to test urine. [B C Ames] *

Hemalum. A microscopic stain. One gram haematoxylin or its ammonium salt is dissolved in 50 cc of 90% alcohol, added to a solution of 50 g of alum in 1,000 cc water, and filtered.

Hemastix. A proprietary test-strip containing o-tolidine and an organic peroxide; used to detect the presence of blood in urine. [B C Ames] *

Hematest. A proprietary test tablet of o-toluidine and an organic peroxide; used to detect blood in feces and urine. [B C Ames] *

Hematin. CAS 15489-90-4; $C_{34}H_{32}N_4O_4FeOH$; used in biochemical research.

hematine paste and powder. *See* logwood.

Hemicellulase Conc. Fungal hemicellulase; enzyme for petroleum industry (breaker with guar gum), coffee (hydrolyzes coffee gums), flavors (extraction of essential oils and plant extracts). [Solvay Enzymes]

hemicelluloses. These are contained in plant cell walls. They are reserve celluloses and are readily converted into hexoses and pentoses by acids and by the enzyme cytase.

---

‡ Trade name and manufacturer not verified   § Trade name without identified manufacturer

hemlock oil. CAS 8021-28-1; Essential oil obtained by distilling *Tsuga canadensis* leaves; used in medicine.

Hemofil. Antihemophilic factor. [Hyland Therapeutics] *

hemoglobin; CAS 9008-02-0; Respiratory protein of the red blood cells that transfers oxygen from the lungs to the tissues and carbon dioxide from the tissues to the lungs; used in medicine.

Hemoplastin. (hemostatic serum). A fluid preparation from blood serum. It consists chiefly of prothrombin and thrombokinase.

Hemoplex. Vitamin B complex injection with liver. [Paines & Byrne Ltd] *

Hemostatic Serum. See Hemoplastin.

hemostatin. CAS 51-43-4; Adrenalin.

Hemoterge. Rinse and reference solution; used for cleaning and blanking of automatic hemoglobinometers. [Coulter Electronics Ltd]

Hemoxone. Soluble concentrate of 392 g dichlorprop and 210 g MCPA per liter; used for control of weeds in cereals and grass. [ICI Chem. & Polymers Ltd]

Hempel's solution. A solution made by mixing a solution of 120 g potassium hydroxide in 80 cc water with a solution of 5 g pyrogallol in 15 cc water; used for the determination of oxygen.

Hemrids. A preparation of phenylephrine hydrochloride, amethocaine hydrochloride, bismuth carbonate and tyloxapol; anesthetic suppositories. [Bayer AG]

henna. (Egyptian privet; flower of paradise) CAS 83-72-7; Derived from the leaves and roots of *Lawsonia inermis* or *L. alba*; used as a dye, and for staining the hair, in medicine as an antifungal agent.

HEP. N-(2-Hydroxyethyl)-2 pyrrolidone; solvent reaction intermediate, textile auxilliaries, cosmetic ingredient. [ISP] *

Hepacon. A proprietary name for a group of preparations containing liver extract and/or vitamins of the B complex. [Consolidated Chemicals Ltd] *

Hepacort Plus. A proprietary preparation of heparin, hydrocortisone acetate, and methyl parahydroxybenzoate. local anti-inflammatory cream. [Rona Laboratories] ‡

Hepaglobin®. Immunoglobulin. [Bayer AG]

Heparin Lock Flush. Heparin sodium; anticoagulant. [Abbott Laboratories] *

hepar sulfur. Potassium sulfide.

Hepatolite. Technetium Tc 99m Disofenin; diagnostic aid; radioactive agent. [DuPont-NEN Medical Products] *

Hep-B-Gammagee. Hepatitis B immune globin; post-exposure prophylaxis against hepatitis B. [Merck & Co Inc]

Hepicebrin. Hexavitamin capsules; combination vitamins. [Eli Lilly & Co] †

heptadecanoic acid, 16-methyl-. See isostearic acid.

heptahydrate: bitter salts. See magnesium sulfate.

heptaminol. 6-Amino-2-methylheptan-2-ol.

heptanal. Heptoic aldehyde, $CH_3(CH_2)_5CHO$.

heptane. (n-heptane; alkane C_7) Aliphatic hydrocarbon; $CH_3(CH_2)_5CH_3$; CAS 142-82-5; 64742-89-8; EINECS 205-563-8; standard for octane rating determinations; anesthetic; solvent; organic synthesis. [Ashland; Exxon; Humphrey; Phillips 66; Texaco]

heptanoic acid. (enanthic acid; carboxylic acid C_7; enanthic anhydride; n-heptylic acid; heptoic acid) $[CH_3(CH_2)_5CO]_2O$; CAS 111-14-8; EINECS 203-838-7; organic synthesis, production of lubricants for aircraft and brake fluids. [Atochem N. Am.; Exxon UK; Hoechst

Celanese; Penta Mfg.]

1-heptanol. See heptylalcohol.

2-heptanone. See methyl n-amyl ketone.

Hepteen Base. Heptaldehyde-aniline reaction product; fast-curing, high temperature accelerator with maximum processing safety; works well with basic furnace blacks used in natural rubber pure gum stocks. [Uniroyal] *

heptene. Heptaldehydeaniline; a proprietary rubber vulcanization accelerator §

heptoic acid. See heptanoic acid.

Heptokill. Insectidal formulation. [Mitchell Cotts Chemicals Ltd]

Heptomer. Gleptoferron; hematinic. [Fisons plc, Pharmaceuticals Div] *

Heptuss. CAS 129-03-3; Cyproheptadine; for the treatment of coughs and colds. [Merck & Co Inc] †

heptylalcohol. (1-heptanol; alcohol C_7; enanthyl alcohol) $C_7H_{16}O$; $CH_3(CH_2)_6OH$; CAS 111-70-6; EINECS 203-897-9; Organic intermediate, solvent, cosmetic formulations. [Atochem N. Am.; Penta Mfg.; Suchema AG]

n-heptylic acid. See heptanoic acid.

Her. Chemical resorcinol di (β-hydroxyethyl) ether; curative for urethane polymers. [Anderson Development Company] *

herapathite. Quinine sulfate periodide.

Herapath's salt. Quinine iodo-sulfate.

Herbatox®. Bentazon, isoproturon, dichlorprop; for post-emergence control of grasses and broadleaf weeds in winter cereals and spring wheat. [BASF AG]

Herbazin 50. CAS 122-34-9; A wettable powder containing simazine; Long term maintenance of weed-free pathways, bare ground and other areas requiring total weed control. [Fisons plc, Horticulture Div] *

Herbazin Plus. A wettable powder containing simazine and aminotriazole; a quick acting herbicide for control of existing weeds with long term persistence. [Fisons plc, Horticulture Div]

Herbazin Plus SC. CAS 61-82-5, 122-34-9; A suspension concentrate containing 180 g aminotriazole and 300 g simazine per liter; used for total weed control in non crop areas and fruit orchards. [Fisons plc]

Herbazin Special. Amitrole + atrazine + 2,4-D; used for total weed control in non crop areas. [Fisons plc]

Herbazin Total. CAS 1912-24-9; A granule containing atrazine; a persistent herbicide for the control of grasses and many annual and perennial broad-leaved weeds. [Fisons plc, Horticulture Div] *

Herbohn bronze. An alloy of 71% copper, 26% tin, and 3% zinc.

Herborane. 2-Butyl-4,4,6-trimethyl-1,3-dioxane. [Quest Int'l. UK Ltd]

Herbrak®. Herbicide for controlling the grass and broad-leaved weeds occurring in sugar and fodder beet; pre-drilling, pre-emergence or post-emergence applications. [Bayer AG]

Hercat 627. Dispersed rosin size; sizing agent for paper machine wet end processing. [Hercules]

Herclor. Epichlorohydrin elastomer. [Hercules Ltd] †

Hercobind DS. An oleoresin emulsion; resistant to rain and wind erosion; binds and agglomerates fine minerals for dust control. [Hercules] *

Hercobond®339. Water-sol. polymer for strength enhancement in paper. [Hercules]

Hercoflat®. Paint texturing/flatting agents. [Hercules Ltd]

---

* Trade name not verified as available   † Trade name verified as obsolete

Hercoflat® Texturing Pigments and Flatting Agent. CAS 9003-07-0; Special type of PP; easily dispersed in coating vehicles for textured, nonglare finishes; used in most finishes, maintains original texture in fully cured, baked systems. [Aqualon]

Hercoflav. Formulated flavors for foods. [Hercules Ltd] †

Hercoflex®600. Nonvolatile, low molecular weight plasticizers. [Hercules] *

Hercoflex®707. Nonvolatile, low molecular weight plasticizers. [Hercules] *

Hercoflex®707A. Nonvolatile, low molecular weight plasticizers. [Hercules] *

Hercoflex® 900. A polymeric plasticizer; contributes to clarity, green tack, permanence and low-temperature flexibility in polyvinyl acetate adhesives. [Hercules] *

Hercoflex®Plasticizer. Pentaerythritol ester; plasticizer for PVC with outstanding heat resistance; wire insulation for high-temp. service and government specification cable construction and high-quality plastisol formulations. [Aqualon]

Hercofloc. A series of high molecular weight synthetic water soluble polymers; flocculant polymers with many uses in water management. [Hercules] *

Hercofroth. A series of frothers; range includes pine oil, modified terpene alcohols, aliphatic-terpene alcohol, polypropylene glycol water-soluble based frothers; in flotation, frothers stabilize the air bubbles containing the unwetted particles into a froth that is easily removed from the liquid. [Hercules] *

Hercol 2. CAS 55-63-0; Nitroglycerin dynamite; used in construction and building, explosives, mining, petroleum and related industries. [Hercules] *

Hercol 2X. CAS 55-63-0; Nitroglycerin dynamite; used in construction and building, explosives, mining, petroleum and related industries. [Hercules] *

Hercolube®. A range of synthetic esters (polyol type) used as lubricants; functional fluids where exposure to both high and low temperature is a primary consideration. [Hercules] *

Hercolube® Synthetic Ester. Polyol esters; ready-to-use completely compound lubricating oils; wide-temp. variance gear oils designed to reduce energy consumption and maintenance in automotive and industrial uses; combine low channel point and superior thermal stability, low volatility at elevated temps. and high flash pt. [Aqualon]

Hercolyn®. CAS 127-25-3; Hydrogenated methyl abietate. [Hercules Ltd]

Hercolyn® D. The hydrogenated methyl ester of rosin; tackifying resin for adhesives, plasticizing and softening agent for chewing gum, used in flexographic, type T gravure and screen-process inks, as a plasticizing resin in cellulose-based coatings. [Hercules]

Hercomix. A blasting agent containing ammonium nitrate and fuel oil; suitable for use under dry borehole conditions. [Hercules] *

Hercon®2. CAS 55-63-0; Nitroglycerin dynamite; used for construction and building, explosives, mining, petoleum and related industries. [Hercules] *

Hercon®2X. CAS 55-63-0; Nitroglycerin dynamite; used for construction and building, explosives, mining, petroleum and related industries. [Hercules] *

Hercon®32. Cationic cellulose-reactive sizing emulsion for application in gypsum paper. [Hercules]

Hercon® 40. Cationic cellulose reactive sizing emulsions. [Hercules] *

Hercon® 48. Cationic cellulose reactive sizing emulsions. [Hercules] *

Herco®Pine Oil. Pine oil; various grades for household and industrial cleaners; disinfectant, antifoam agent, textile specialties. [Hercules]

Herco-Prills. CAS 6484-52-2; Ammonium nitrate prills containing a minimum of 33.5% nitrogen; used alone or for bulk blending of mixed fertilizers. [Hercules] *

Hercoprime. Adhesion promoters for powder coatings. [Hercules Ltd]

Hercopruf. Glycol-based freeze conditioning agents; prevents ice binding of wet minerals and ice build-up on conveyor belts. [Hercules] *

Hercose AP. A trademark for cellulose acetate propionate; used in lacquer manufacture. §

Hercose C. A trademark for cellulose acetobutyrate for use in lacquer manufacture. §

Hercosett®. Polyamide epichlorohydrin resin; for shrinkproofing of wool. [Hercules] *

Hercosett® 125. A water-soluble cationic resin containing reactive polyamide epichlorohydrin; imparts shrink resistance to wool. [Hercules] *

Hercosol. A trademark for a solvent made from pine oil. §

Hercosol TP-S. 65% solids solution of dark pine-tree-derived resin in a mixed-terpene hydrocarbon solvent; designed for rubber reclaimers who need an additional solvent or a second reclaiming solvent in their process. [Hercules] *

Hercosplit WR. CAS 55-63-0; A nitroglycerin dynamite; for construction and building industry, explosives, mining, petroleum and related industries. [Hercules] *

Hercotac®AD. A low molecular weight modified aromatic hydrocarbon resin; used mainly in pressure sensitive systems containing natural rubber and in hot-melts based on ethylene vinyl acetate copolymer. [Hercules]*

Hercotac®LA. A low molecular weight modified aliphatic hydrocarbon resin; used mainly in pressure sensitive systems containing natural rubber, and in hot melts based on ethylene vinyl acetate copolymer. [Hercules]*

Hercotuf. Powdered polypropylene. [Hercules Ltd]

Hercules®4 Defoamer. High-effiency defoamer for acid and alkaline papermaking systems, deinking systems. [Hercules]

Hercules®37M6-8. CAS 50-00-0; Formaldehyde sol'n.; for mfg. of synthetic resins by reaction with phenols, urea, melamines for molded goods, elec. insulation, binders, plywood adhesives, varnishes, wet-str. resins for paper and textiles; chem. intermediate. [Hercules] *

Hercules® 247. Release agent for tissue and towelling machines; improves wet and dry creeping. [Hercules]

Hercules®356 Defoamer. Defoamer for wet-end use in acid or alkaline papermaking systems. [Hercules]

Hercules®752 Size. Nonfortified dark paste rosin; dry size used with alum in paperboard and building prods. to produce high level of resistance to water and aq. sol'ns.; primarily for unbleached kraft southern pine pulps. [Hercules]

Hercules® 1098. Rosin soap emulsion; air entrainment aid for flotation save-alls. [Hercules]

Hercules® 2051GS Defoamer. Hydrocarbon oil-based; defoamer for use as drainage aid and foam killer in kraft pulpmill brownstock washing systems. [Hercules]

---

‡ Trade name and manufacturer not verified   § Trade name without identified manufacturer

Hercules® AR 150. PEG-15 rosinate; low foaming surfactant, detergent, emulsifier, wetting agent, suspending agent, dispersant for industrial cleaners. [Hercules]

Hercules® AR 160. PEG-16 rosinate; low foaming surfactant, detergent, emulsifier, wetting agent, suspending agent, dispersant for industrial cleaners and food related areas. [Hercules]

Hercules® AS. CAS 9004-70-0; Nitrocellulose; film-former; intermediate type used in place of Hercules SS in some applications. [Aqualon]

Hercules® K. CAS 9004-57-3; Ethylcellulose; for formulation of lacquers, varnishes, inks, foils, adhesives, plastics and for food contact, animal feed, and pharmaceutical goods; good toughness, flexibility at low temps.;, low flamm. [Hercules] *

Hercules® N. CAS 9004-57-3; Ethylcellulose; for formulation of lacquers, varnishes, inks, foils, adhesives, plastics and for food contact, animal feed, and pharmaceutical goods; good toughness, flexibility at low temps.;, low flamm. [Hercules] *

Hercules® RES A-2338. Synthetic resin aq. disp.; tackifier for acrylic-latex polymers used in label, tape, and construction adhesive applications; gives good oxidative and uv light stability in latexes. [Hercules]

Hercules® RS. CAS 9004-70-0; Nitrocellulose; film-former. [Aqualon]

Hercules® Improved Tech. PE. CAS 115-77-5; Pentaerythritol; used in production of alkyd resins, rosin esters, urethane resins, drying oils, synthetic lubricants, plasticizers, intumescent paints, plastics, stabilizers for plastics, explosives. [Hercules] *

Hercules metal. A bronze containing 85.5% copper, 10% tin, 2.5% aluminum, and 2% zinc. Another alloy contains 54% copper, 36% zinc, 7.5% iron, and 2.5% aluminum. The term is also used for an alloy of copper, nickel, and aluminum.

Hercules® Mono-PE. CAS 115-77-5; Pentaerythritol; used in production of alkyd resins, rosin esters, urethane resins, drying oils, synthetic lubricants, plasticizers, intumescent paints, plastics, stabilizers for plastics, explosives. [Hercules] *

Hercules® SS. CAS 9004-70-0; Nitrocellulose; film-former. [Aqualon]

Hercules® T. CAS 9004-57-3; Ethylcellulose; for formulation of lacquers, varnishes, inks, foils, adhesives, plastics and for food contact, animal feed, and pharmaceutical goods; good toughness, flexibility at low temps.; low flamm. [Hercules] *

Hercules® Type AS4. PAN-based carbon fiber; continuous, high strength fiber which has been surface treated and sized to improve props.; for use in weaving, prepregging, filament winding, pultrusion and molding compounds. [Hercules] *

Hercules® X Dry Size. Nonfortified pale dry rosin size; dry size used with alum in paperboard and building prods. to produce high level of resistance to water and aq. sol'ns. [Hercules]

Herculine FR. Surface explosive; surface charge for geophysical prospecting. [Hercules] *

Herculite. Molding plaster for molds used in making aluminum alloy castings. [Foseco (F.S.) Ltd] †

Herculon. Polypropyleneolefin fiber; for floor coverings, furniture and fixtures and textiles. [Hercules] *

Herculoy. A patented alloy; it is a silicon bronze containing

tin; high tensile strength; resists corrosion. Herculoy 418 contains copper with about 3% silicon and 0.5% tin. §

Hercures®. Aryl and alkylaryl resins; used for adhesives, hot melts, coatings, inks. [Hercules; Hercules Ltd]

Heresite. A proprietary trade name for a phenol-formaldehyde synthetic resin molding compound. [Heresite] *

Herkules. An alloy of 50% silver and 50% copper; used for fuse wire.

Hermann's fluid. (Platino-Aceto-Osmic Acid). A fixing agent used in microscopy. It contains 15 parts of a 1% platinum chloride solution, 4 parts of a 2% osmium tetroxide solution, and 1 part of glacial acetic acid.

Hermes. Suspension concentrate containing 480 g metoxuron and 30 g simazine per liter; post-emergence herbicide for winter cereals. [Atlas Interlates Ltd]

Hermite fluid. It contains magnesium oxide and hypochlorous acid, with from 4-5% available chlorine. A disinfectant.

Herolith. A synthetic resin of the phenol-formaldehyde type. *See* Bakelite. §

Heron. A paper-based grade of Tufnol industrial laminates. [Tufnol Ltd]

Herox®. Polymer. [DuPont UK]

Herpid. CAS 67-68-5; A proprietary preparation of dimethylsulfoxide; an antiviral solution used in the treatment of cutaneous herpes simplex and herpes zoster (shingles). [Boehringer Ingelheim Ltd]

Herplex Liquifilm®. CAS 54-42-2; Idoxuridine; antiviral. [Allergan, Inc]

Herrifex DS. A liquid containing 587.5 g per liter mecoprop as the potassium salt; used to control cleavers, common chickweed and a wide range of other broadleaved weeds in cereals, sports turf, grass seed crops and apple and pear orchards. [Bayer plc] †

Herrisol. A liquid containing 35.4% w/w dicamba (Banvel D), MCPA and mecoprop; used to control common chickweed, cleavers, knotgrass, redshank, scentless mayweed, corn spurrey and a wide range of other broadleaved weeds in cereals, grass crops and orchards. [Bayer plc]

Herschel's crystals. Hydrated calcium-tetrahydroxy-trisulfide, $Ca_3(OH)_4 \cdot S_3 \cdot 8H_2O$.

Hespan®. Hetastarch; a starch that is composed of more than 90% of amylopectin and that has been etherified to the extent that an average of 7 to 8 of the OH groups present in every 10 D-glucopyranose units of starch polymer have been converted into OCH_2CH_2OH groups; plasma volume extender. [DuPont Merck Pharmaceuticals]

hesperidene. (citrene, carvene) Limonene, $C_{10}H_{16}$.

Hest MS. CAS 17661-50-6; Myristyl stearate; bodying agent, emollient for creams and lotions; pearlescent in anionic systems; spermaceti replacement. [Heterene]

Hesthasulphid. A proprietary brand of sodium sulfide; used in leather manufacture. §

Hetacin-K. CAS 5321-32-4; Hetacillin potassium; antibacterial. [Bristol Laboratories] *

Hetamine 5L-25. CAS 55819-53-9; EINECS 259-837-7; Stearamidopropyl dimethylamine lactate; antistat, conditioner for hair; compat. with most anionic surfactants. [Heterene]

Hetan SL. CAS 1338-39-2; EINECS 215-663-3; Sorbitan laurate; lipophilic emulsifier. [Heterene]

Hetan SO. CAS 1338-43-8; EINECS 215-665-4; Sorbitan

---

*　Trade name not verified as available　　　　　　　† Trade name verified as obsolete

oleate; lipophilic emulsifier. [Heterene]

Hetan SS. CAS 1338-41-6; EINECS 215-664-9; Sorbitan stearate; lipophilic emulsifier. [Heterene]

Hetaplas®. [DuPont Merck Pharmaceuticals]

heteroauxin. See 3-indoleacetic acid.

Hetester 412. CAS 2778-96-3; EINECS 220-476-5; Stearyl stearate; emollient for stick cosmetics. [Heterene; Bernel]

Hetester 3236S. Myristyleicosyl stearate. [Bernel] *

Hetester FAO. C12-15 alcohols octanoate; unique skin feel ingred., wetting agent; used in personal care prods., hair and skin care prods., makeup, antiperspirants; emollient IPM replacement. [Heterene; Bernel]

Hetester HCA. CAS 27233-00-7; Glyceryl triacetyl hydroxystearate. gloss agent, film-former for lip oils, skin oils, lipsticks, eye makeup; high visc. and adherence. [Heterene; Bernel]

Hetester HCP. PPG-3 hydrogenated castor oil; emollient for lip and cosmetic prods. [Heterene; Bornel]

Hetester HSS. CAS 97338-28-8; Isocetyl stearoyl stearate; emollient with unusual skin feel, pigment dispersant for stick prods., emulsion systems; binding oil for pressed powds. [Heterene; Bernel]

Hetester ISS. Isostearyl stearoyl stearate; emollient for use in stick prods., pigmented emulsion-type systems; binding oil for use in pressed powds. [Heterene; Bernel]

Hetester MS. CAS 17661-50-6; Myristyl stearate; bodying agent in creams and lotions; in certain anionic systems, imparts pearling effects; replacement for spermaceti. [Heterene; Bernel]

Hetester PCA. Propylene glycol ceteth-3 acetate; emulsifier, pigment wetter, antichalking agent, emollient used in personal care prods., antiperspirants. [Heterene; Bernel]

Hetester PCP. Propylene glycol ceteth-3 propionate. [Bernel] *

Hetester PHA. Propylene glycol isoceteth-3 acetate; emulsifier, pigment wetter, antichalking agent, emollient used in personal care prods., antiperspirants. [Heterene; Bernel]

Hetester PMA. Propylene glycol myristyl ether acetate; emollient, solv., and plasticizer for anhyd. oil systems, emulsions. [Heterene; Bernel]

Hetester SSS. Stearyl stearoyl stearate; emollient used in cosmetic stick formulations, emulsion systems. [Heterene; Bernel]

Hetester TICC. Triisocetyl citrate; oily liq. emollient useful in stick and pigmented emulsion-type prods., skin care prods., cleansing creams, makeup; pigment dispersing properties. [Heterene; Bernel]

Hetlan AC. CAS 61788-49-6; EINECS 262-980-8; Acetylated lanolin alcohols; emollient for creams and lotions. [Heterene]

Hetoxamate 200 DL. CAS 9005-02-1; PEG-4 dilaurate; thickener, foam booster, foam stabilizer, emulsifier. [Heterene]

Hetoxamate 400 DS. CAS 9005-08-7; PEG-8 distearate; thickener, foam booster, foam stabilizer, emulsifier. [Heterene]

Hetoxamate FA-5. CAS 61791-00-2; PEG-5 tallate; detergent, emulsifier, lubricant, softener; coupling agent for cosmetics, textiles, leather, metal cleaning. [Heterene]

Hetoxamate LA-5. CAS 9004-81-3; PEG-5 laurate; detergent, emulsifier, lubricant, softener; coupling agent for cosmetics, textiles, leather, metal cleaning. [Heterene]

Hetoxamate MO-2. CAS 9004-96-0; PEG-2 oleate; detergent, emulsifier for personal care prods.; softener for leather. [Heterene]

Hetoxamate SA-5. CAS 9004-99-3; PEG-5 stearate; detergent, emulsifier, lubricant, softener, coupling agent for cosmetics, textiles, leather, metal cleaning. [Heterene]

Hetoxamide C-4. CAS 61791-08-0; PEG-5 cocamide. [Heterene]

Hetoxamine C-2. CAS 61791-14-8; PEG-2 cocamine; emulsifier, softener, antistat, water repellent, desizing agent in agriculture, waxes, oils, textile/leather, metal cleaning. [Heterene]

Hetoxamine O-2. PEG-2 oleamine; emulsifier, softener, antistat, water repellent, desizing agent in agriculture, waxes, oils, textile/leather, metal cleaning. [Heterene]

Hetoxamine S-2. CAS 61791-24-0; PEG-2 soyamine; emulsifier, softener, antistat, water repellent, desizing agent in agriculture, waxes, oils, textile/leather, metal cleaning. [Heterene] *

Hetoxamine ST-5. CAS 26635-92-7; PEG-5 stearamine; emulsifier, softener, antistat, water repellent, desizing agent in agriculture, waxes, oils, textile/leather, metal cleaning. [Heterene]

Hetoxamine T-2. CAS 61791-44-4; PEG-2 tallow amine; emulsifier, softener, antistat, water repellent, desizing agent in agriculture, waxes, oils, textile/leather, metal cleaning. [Heterene]

Hetoxide BN-13. CAS 35545-57-4; PEG-13 β-naphthol ether; emollient, emulsifier, visc. control agent, lubricant, pigment dispersant, perfume solubilizer, used in cosmetics, household, textile industry, metal treating and plating; intermediate. [Heterene]

Hetoxide BY-1.8. PEG-1.8 butynediol; metal plating emulsifier. [Heterene] *

Hetoxide C-2. CAS 61791-12-6; PEG-2 castor oil; emollient, emulsifier, solubilizer, pigment dispersant, detergent used in cosmetics, household, textile industry. [Heterene]

Hetoxide C-200-50%. CAS 61791-12-6; PEG-200 castor oil; lubricant, emulsifier, solubilizer. [Heterene]

Hetoxide DNP-4. CAS 9014-93-1; Nonyl nonoxynol-4; intermediate, emulsifier. [Heterene]

Hetoxide G-7. CAS 31694-55-0; Glycereth-7; emulsifier. [Heterene]

Hetoxide HC-16. CAS 61788-85-0; PEG-16 hydrogenated castor oil; emollient, emulsifier, solubilizer, pigment dispersant, detergent used in cosmetics, household, textile industry. [Heterene]

Hetoxide MPC. m,p-Cresol hydrophobe; solv. for lacquers and coatings. [Heterene]

Hetoxide NP-4. CAS 9016-45-9; Nonoxynol-4; detergent and emulsifier. [Heterene]

Hetoxide P-3. PEG-3 phenyl ether. [Heterene]

Hetoxol 15 CSA. CAS 68439-49-6; Ceteareth-15; detergent, emulsifier, leveling agent, intermediate for cosmetics, textiles, scouring agents, dyes, household prods., silicone emulsification, surfactants. [Heterene]

Hetoxol 916P. PEG-6 PPG-2.5 C9-C11 alcohols ether; surfactant for industrial applications. [Heterene]

Hetoxol C-24. Oleth-24 and ceteth-24; intermediate, emulsifier, wetting agent, solubilizer, coupling agent. [Heterene]

Hetoxol CA-2. CAS 9004-95-9; Ceteth-2; detergent, emul-

---

‡ Trade name and manufacturer not verified   § Trade name without identified manufacturer

sifier, leveling agent, intermediate, used for cosmetics, household formulations, silicone emulsification, textile processing. [Heterene]

Hetoxol CAWS. PPG-5, ceteth-20; intermediate, emulsifier, wetting agent, solubilizer, coupling agent. [Heterene]

Hetoxol CD-4. PEG-4 2-ethyl hexyl ether; intermediate, emulsifier, wetting agent, solubilizer, coupling agent. [Heterene]

Hetoxol CS. Cetearyl alcohol. [Heterene]

Hetoxol CS-4. CAS 68439-49-6; Ceteareth-4; detergent, emulsifier, leveling agent, intermediate for personal care prods., wax, oils, textiles, scouring agents, dyes, household prods., silicone emulsification, surfactdants. [Heterene]

Hetoxol D. Cetearyl alcohol and ceteareth-20; intermediate, emulsifier, wetting agent, solubilizer, coupling agent; for cosmetics, paper, textile industries. [Heterene]

Hetoxol G. Stearyl alcohol and ceteareth-20; intermediate, emulsifier, wetting agent, solubilizer, coupling agent. [Heterene]

Hetoxol IS-2. CAS 52292-17-8; Isosteareth-2. [Heterene]

Hetoxol J. Cetearyl alcohol and ceteareth-20; intermediate, emulsifier, wetting agent, solubilizer, coupling agent. [Heterene]

Hetoxol L-4. CAS 5274-68-0; Laureth-4; intermediate, emulsifier, wetting agent, solubilizer, coupling agent. [Heterene]

Hetoxol LS-9. Laureth-9, steareth-9; detergent, emulsifier, leveling agent, intermediate, used for cosmetics, household formulations, silicone emulsification, textile processing. [Heterene]

Hetoxol M-3. CAS 27306-79-2; Myreth-3; emulsifier and pigment dispersant in makeup; intermediate, wetting agent, solubilizer, coupling agent. [Heterene]

Hetoxol MP-3. CAS 63793-60-2; PPG-3 myristyl ether. [Heterene]

Hetoxol OA-3 Special. CAS 9004-98-2; Oleth-3; emulsifier and pigment dispersant for cosmetic applications. [Heterene]

Hetoxol OL-2. CAS 9004-92-2; Oleth-2; intermediate, emulsifier, wetting agent, solubilizer, coupling agent. [Heterene]

Hetoxol P. Emulsifying wax NF; emulsion base. [Heterene]

Hetoxol PLA. CAS 68439-53-2; PPG-30 lanolin ether; oily emollient, intermediate, emulsifier, wetting agent, solubilizer, coupling agent. [Heterene]

Hetoxol SP-15. CAS 25231-21-4; PPG-15 stearyl ether; oily emollient material in cosmetics. [Heterene]

Hetoxol STA-2. CAS 9005-00-9; Steareth-2; intermediate, emulsifier, wetting agent, solubilizer, coupling agent. [Heterene]

Hetoxol TD-3. CAS 24938-91-8; Trideceth-3; detergent, emulsifier, leveling agent, intermediate, used for cosmetics, household formulations, silicone emulsification, textile processing. [Heterene]

Hetoxol TDEP-15. PEG-10 PPG-15 tridecyl ether; surfactant for industrial use, automatic dishwashing formulations. [Heterene]

Hetphos OA-3. CAS 39464-69-2; Oleth-3 phosphate. [Heterene]

Hetphos SG. CAS 50643-20-4; PPG-5 ceteth-10 phosphate. [Heterene]

Hetquat S-20. CAS 122-19-0; EINECS 204-527-9; Stearalkonium chloride. [Heterene]

Hetron®. Chlorendic, bisphenol, furan and vinyl ester resins for chemical resistant applications; halogenated resins for fire retardant applications; used for fiberglass reinforced parts for the chemical resistant building construction, electrical and transportation markets. [Ashland Chemical Company] *

Hetron® 92. Unsat. polyester resin, unpromoted; for flame retardant reinforced thermosetting equip. manufactured by BMC/SMC molding or pultrusion processes; as halogen conc. resin for blending with non-flame retardant resins. [Ashland]

Hetron® 99P. Isophthalic polyester resin, halogenated, promoted. [Ashland]

Hetron® 197-3. Polyester resin, halogenated, unpromoted, thermoset; corrosion and heat resist., flame retardant resin for filament wound and hand lay-up tanks, pipes, pumps, stacks, scrubbers, and other equip. handling corrosive gases, vapors or liqs. [Ashland]

Hetron® 197AT. Unsat. styrene-containing polyester resin; Class I fire retardant, chemical- and heat-resistant, for high-temp. applications such as stacks and ducts, handling corrosive gases and vapors. [Ashland]

Hetron® 692. Polyester resin, halogenated; flame retardant resin for corrosion-resist. fiberglass-reinforced plastic equip. incl. press molded equip. [Ashland]

Hetron® 700DMA. Accelerated bisphenol A fumarate polyester resin; low-visc., chemical-resistant resin, good wet-out, excellent gel time, stability, and craze resistance; for corrosion control applications. [Ashland]

Hetron® 800. Furan resin; thermoset for corrosion-resist. fiberglass-reinforced plastic equip. incl. pipes, tanks, fume hoods, ducts, linings, and flake glass coatings where fire retardancy is not required. [Ashland]

Hetron® 922. Vinyl ester resin; used for hand lay-up, spray-up, and filament winding, flake glass, filled lining and coating compounds. [Ashland]

Hetron® FR 991. Vinyl ester resin; flame retardant, corrosion resist. resin for filament wound, hand lay-up and spray-up boats, tanks, pipes, duct, stacks, scrubbers, linings. [Ashland]

Hetsorb L-4. CAS 9005-64-5; Polysorbate 21; emulsifier, lubricant, thickener, corrosion inhibitor. [Heterene]

Hetsorb L-10. CAS 9005-64-5; PEG-10 sorbitan laurate; detergent, emulsifier, lubricant for cosmetics; corrosion inhibitor. [Heterene]

Hetsorb L-20. CAS 9005-64-5; Polysorbate 20; detergent, emulsifier, lubricant for cosmetics; corrosion inhibitor. [Heterene]

Hetsorb L-80-72%. CAS 9005-64-5; PEG-80 sorbitan laurate; surfactant for shampoos; used as counterirritant. [Heterene]

Hetsorb O-5. CAS 9005-65-6; Polysorbate 81; food emulsifier. [Heterene]

Hetsorb O-20. CAS 9005-65-6; Polysorbate 80; emulsifier, lubricant for cosmetics; corrosion inhibitor. [Heterene]

Hetsorb P-20. CAS 9005-66-7; Polysorbate 40; general purpose emulsifier. [Heterene]

Hetsorb S-4. CAS 9005-67-8; PEG-4 sorbitan stearate; emulsifier. [Heterene]

Hetsorb S-20. CAS 9005-67-8; Polysorbate 60; emulsifier for vitamins. [Heterene]

Hetsorb TO-20. CAS 9005-70-3; Polysorbate 85; general purpose emulsifier, thickener, lubricant, corrosion inhibitor. [Heterene]

---

* Trade name not verified as available   † Trade name verified as obsolete

Hetsorb TS-20. CAS 9005-71-4; Polysorbate 65; general purpose emulsifier, thickener, lubricant, corrosion inhibitor. [Heterene]

Hetsulf 40, 40X. Sodium dodecylbenzene sulfonate; wetting agent, emulsifier, dispersant, intermediate, detergent, liq. formulation syndet. [Heterene] *

Hetsulf 50A. CAS 1331-61-9; EINECS 215-559-8; Ammonium dodecylbenzene sulfonate; wetting agent, emulsifier, dispersant, for light duty detergent formulations. [Heterene] *

Hetsulf 60S. Sodium dodecylbenzene sulfonate; wetting agent, emulsifier, dispersant, base for formulated prods. [Heterene] *

Hetsulf 60T. Amine dodecylbenzene sulfonate; wetting agent, emulsifier, dispersant, for cosmetic, bath and shampoos uses. [Heterene] *

Hetsulf Acid. Sodium dodecylbenzene sulfonate; wetting agent, emulsifier, dispersant, intermediate, base for neutralized surfactant. [Heterene] *

Hetsulf IPA. MIPA-dodecylbenzenesulfonate; wetting agent, emulsifier, dispersant. [Heterene] *

Heusler alloy. An alloy of 66-68% copper, 18-22% manganese, 10-11% aluminum, and 0-4% lead.

Heveacrumb. A proprietary compressed rubber crumbled in oil. [RRI Malaya] ‡

Heveatex. The trade name to denote a series of preserved, concentrated, or processed Hevea rubber latexes. §

Hevikote. Two peak pitch epoxy protective coating. [Thomas Ness Ltd]

hex. See hexamine.

Hexa-Betalin. CAS 58-56-0; Pyridoxine hydrochloride; vitamin. [Eli Lilly & Co] †

hexabutyldistannoxane. See tributyltin oxide.

Hexacal. Fire retardant additives for rigid foams. [ICI Chem. & Polymers Ltd]

Hexacarb. Black dye for leather. [Pointing Ltd] *

Hexacert. USA certified foodstuff colorant. [Pointing Ltd] *

Hexacide. Insect and vermin killer. [Pointing Ltd] *

Hexacol. Foodstuff colorants of guaranteed specification. [Pointing Ltd] *

hexadecanoic acid. See palmitic acid.

1-hexadecanol. See cetyl alcohol.

1-hexadecanol lactate. See cetyl lactate.

hexadecylic acid. See palmitic acid.

n-hexadecyl lactate. See cetyl lactate.

1-hexadecylpyridinium chloride. See cetylpyridinium chloride.

n-hexadecyl stearate. See cetyl stearate.

Hexaderm. Fast dyes for leather. [Pointing Ltd] *

2,4-hexadienoic acid. See sorbic acid.

2,4-hexadienoic acid, potassium salt. See potassium sorbate.

Hexadrol. CAS 50-02-2; Dexamethasone; glucocorticoid. [Organon Inc]

Hexafoam. Isocyanate foams. [ICI Chem. & Polymers Ltd]†

hexaglycerol. See trimethylolpropane.

hexahydrobenzene. See cyclohexane.

hexahydrophenol. See cyclohexanol.

hexahydropyridine. See piperidine.

hexahydrothymol. See menthol.

hexahydrotoluene. See methyl cyclohexane.

hexahydroxy cyclohexane. See inositol.

Hexalan. Fast to light colors for wool. [Pointing Ltd] *

hexalin. See cyclohexanol.

Hexalin. CAS 108-93-0; Cyclohexanol. [DuPont] *

Hexallac. Dyes for cellulose lacquers. [Pointing Ltd] *

hexamethylamine. (HMTA; methenamine; aminoform; hexamethylene-tetramine) CAS 100-97-0; $C_6H_{12}N_4$; used in the curing of phenol formaldehyde and resorcinol formaldehyde, resins, rubber to textile adhesives, protein modifier, organic synthesis, pharmaceuticals.

hexamethyldisilane. $C_6H_{18}Si_2$; CAS 1450-14-2; EINECS 215-911-0; Starting material for prep. of trimethylsilyl alkali compounds. [R.W. Greeff]

hexamethyldisilazane. (HMDS; bis (trimethylsilyl) amine) Silica reaction product; $(CH_3)_3SiNHSi(CH_3)_3$; CAS 999-97-3; 68909-20-6; EINECS 213-668-5; chemical intermediate, chromatographic packings, silylating agent. [Aldrich; Dow Corning; Hüls Am.; Janssen Chimica; PCR]

hexamethylene. See cyclohexane.

hexamethylene glycol. (1,6-hexanediol; 1,6-dihydroxyhexane) $HO(CH_2)_6OH$; CAS 629-11-8; EINECS 211-074-0; intermediate in the production of nylon; manufacture of hexamethylenediamine, polyesters, polyurethanes; in gasoline refining; as plasticizer.

hexamethylene tetramine. (HMTA; methenamine; aminoform; urotropine; hexamine) $(CH_2)_6N_4$; CAS 100-97-0; EINECS 202-905-8; Curing of phenolformaldehyde and resorcinolformaldehyde resins, adhesives, fungicide, antibacterial. [Allchem Industries; Dajac Labs; R.W. Greeff; Mitsubishi Gas; OxyChem/Durez]

Hexamic Acid. CAS 100-88-9; Cyclamic acid. sweetener. [Abbott Laboratories] *

hexamine. (cystamin, cystogen, metramine, urotropine, formin, naphthamine, xametrin, vesaloin, urisol, uritone, hex, H.M.T., formaldehyde-ammonia 6:4, carin, ammonioformaldehyde, vesalvine). CAS 100-97-0; EINECS 202-905-8; Hexamethylenetetramine, $C_6H_{12}N_4$, curing of phenolformaldehyde and resorcinolformaldehyde resins; rubber to textile adhesive; ingredient of high explosive cyclonite; rubber accelerator; fungicide; corrosion inhibitor

6-hexanalactone, 2-oxepanone. See ε-caprolactone monomer.

hexanaphthene. See cyclohexane.

hexane. (n-hexane; alkane C_6) Aliphatic compound.; $CH_3(CH_2)_4CH_3$; CAS 110-54-3; 64742-49-0; EINECS 203-777-6; solvent, alcohol denaturant, paint diluent, polymerization reaction medium; filling for thermometers. [Ashland; BP Chem. Ltd; Exxon; Humphrey; Mitsui Petrochem. Ind.; Phillips 66; Texaco]

1,6-hexanediol. See hexamethylene glycol.

hexanhexol. Mannite, $C_6H_{14}O_6$.

hexanitrin. Mannitol-hexanitrate, $CH_2O(NO_2) \cdot (CHO \cdot NO_2)_4 \cdot CH_2ONO_2$.

hexanoic acid. See caproic acid.

1- or n-hexanol. See hexyl alcohol.

Hexapar. Detergent preparations. [Pointing Ltd] *

Hexaphos. CAS 10124-56-8; Sodium hexametaphosphate. [FMC]

Hexaplant Richter. Water, alcohol, fennel extract, hops extract, balm mint extract, mistletoe extract, matricaria extract, yarrow extract; polyvalent herbal extract in aq. alcohol; emollient for aq. and hydroalcoholic herbal cosmetics, skin and hair prods., emulsified preparations. [Dr. Kurt Richter; Henkel/Cospha]

Hexaplas. Phthalate plasticizers for PVC. [ICI Chem. &

---

‡ Trade name and manufacturer not verified   § Trade name without identified manufacturer

Hexaplus. Plasticizers. [ICI Chem. & Polymers Ltd] †

Hexaryl D60L. Triethanolamine alkylaryl sulfonate in liquid form; anionic surfactant. [Witco Chemical Ltd] *

Hexasol. Alcohol-soluble dyes. [Pointing Ltd] *

Hexatype. Dyes for doubletone printing inks. [Pointing Ltd]*

Hexavibex. CAS 58-56-0; Pyridoxine hydrochloride; vitamin. [Parke-Davis] *

Hexcel 164M. Nonexpanding two-part PU compound (TDI- and MOCA-free); universal cable plugging compound; forms open sheath moisture blocks and pressure dams in paper/pulp and plastic insulated telecommunications cables. [Hexcel]

Hexcel 174 "Rapid-Dam." Nonexpanding, two-part polyether PU compound (TDI-free); noncorrosive, cable plugging compound; for low cost moisture blocks and pressure dams in paper/pulp and plastic (PIC) insulated telecommunications cables. [Hexcel]

Hexcel 195 RE Hexagel. Re-enterable encapsulant; remains permanently soft and tacky; blocks out moisture, adhesion to cable and closures. [Hexcel]

Hexcelcure 160. Propoxylated amine; epoxy curative. [Zeeland]

Hexcelcure 169. Cyanoethylated amine; epoxy curative. [Zeeland]

Hexela. Cellulose acetate and nylon dyes. [Pointing Ltd] *

1-hexene. (C_6 linear alpha olefin; hexylene; butyl ethylene) $CH_3CH_2CH_2CH_2CH:CH_2$; CAS 592-41-6; Intermediate for surfactants and specialty industrial chemicals (flavors, perfumes, dyes, resins); polymer modifier. [Chevron; Ethyl; Hüls Am.; Phillips 66; Shell]

Hexetidine. CAS 141-94-6; Hexetidine; antimicrobial, antifungal agent for oral hygiene prods. [Angus]

hexil. (hexanitrodiphenylamine.; hexyl; hexite; dipicrylamine) CAS 131-73-7; $(NO_2)_3C_6H_2NHC_6H_2(NO_2)_3$; used as a booster explosive, analysis for potassium.

Hexnitrol. Leather stains. [Pointing Ltd] *

Hexo. Industrial detergent. [Crosfield Chemicals Ltd] †

Hexoil. Oil and varnish dyes. [Pointing Ltd] *

Hexomax. Immobilized invertase enzyme. [UOP Speciality Products]

Hexopal. A proprietary preparation of inositol nicotinate; a peripheral vasodilator. §

Hexoran. A sodium alkylaryl sulfonate; for scouring, wetting-out level bleaching, dyeing etc on all fibers. [Roehm Ltd]*

Hexoran A15. Sodium alkylaryl sulfonate; anionic surfactant in liquid form; used for scouring, wetting out, level bleaching, dyeing, etc.; for all fibers. [Roehm Ltd] *

Hexsotate. hexamethylene-tetramine sodio-acetate.

hexyl alcohol. (1- or n-hexanol; pentylcarbinol; amylcarbinol) Aliphatic alcohol; $CH_3(CH_2)_4CH_2OH$; CAS 111-27-3; 68526-79-4; EINECS 203-852-3; pharmaceuticals (antiseptics, perfume esters), solvent, plasticizer. [Ashland; Ethyl; Penta Mfg.; Vista]

hexylene. See 1-hexene.

hexylene glycol. (2-methyl-2,4-pentanediol; 4-methyl-2,4-pentanediol; α,α,α'-trimethyltrimethyleneglycol) Aliphatic alcohol; $(CH_3)_2COHCH_2CHOHCH_3$; CAS 107-41-5; EINECS 203-489-0; hydraulic brake fluids, printing inks, coupling agent and penetrant for textiles, cosmetics; ice inhibitor in carburetors; fuel and lubricant additive; emulsifier. [Ashland; BP Chem. Ltd; Mitsui Petrochem. Ind.; Penta Mfg.; Shell; Union Carbide]

Hexyl Jasmat®. Acetyl ethyl octanoate; floral, herbal, green, jasmine-like fragrance [BASF AG]

Hexyltan. Hexylresorcinal and tannic acid; for the treatment of burns. [Merck & Co Inc] †

Heyn's reagent. The double chloride of copper and ammonia; used to reveal ferrite in the micro-analysis of carbon steels.

HFC. Compounded plastic material. [Mitsubishi Petrochem.] *

Hibbo. A proprietary aluminum bronze containing iron. §

Hibiclens. CAS 18472-51-0; Chlorhexidine gluconate; an antimicrobial. [Imperial Chemical Industries plc]

Hibidil. CAS 55-56-1; An antiseptic and bactericidal solution containing chlorhexidine. [ICI Chem. & Polymers Ltd]

Hibiscrub. CAS 55-56-1; A proprietary preparation of chlorhexidine; an antiseptic. [ICI Chem. & Polymers Ltd]

Hibisol. CAS 55-56-1; Antiseptic hand rub containing chlorhexidine. [ICI Chem. & Polymers Ltd]

Hibispray. CAS 55-56-1; Antiseptic spray containing chlorhexidine. [ICI Chem. & Polymers Ltd]

Hibitane. A proprietary preparation of chlorhexidene digluconate; an antiseptic. [ICI Chem. & Polymers Ltd]

Hibosol. High boiling point solvents. [BP Chemicals Ltd]

Hibudine. A proprietary trade name for a synthetic rubber, probably derived from butadiene. §

Hi-Build. Paint filler. [ICI Chem. & Polymers Ltd]

Hi-Carbolon®. Carbon fiber and carbon fiber prepreg. [Asahi Chem. Industry]

Hi-Care® 1000. CAS 65497-29-2; Guar hydroxypropyltrimonium chloride; for skin care prods. [Rhone-Poulenc Surf.]

Hi-Cat. Cationic starch; used in the paper industry, wet end additive. [Roquette (UK) Ltd]

Hickstor. CAS 117-18-0; Tecnazene potato treatment. [Hickson & Welch Ltd]

Hickstor 6 + MBC. Carbendazim + tecnazene; protectant fungicide and sprout suppressant for stored potatoes. [Hickson & Welch Ltd]

Hicond-2000. Electrically conductive olefin; high melt strength compound suitable for profile and sheet extrusion, thermoforming and blow molding. [United Composites]

Hicore 90. A nickel-chromium-molybdenum case-hardening steel for heavy motor vehicles and other gears.

Hicoseen. A proprietary preparation of diethylaminoethyl phenylbutyrate, codeine phosphate and guaicol albuminate; a cough linctus. [Hommel Pharmaceuticals]*

HiD 9301. CAS 9002-88-4; HDPE resin; pressure pipe resin; excellent ESCR. [Chevron]

HiD 9602. CAS 9002-88-4; HDPE homopolymer resin; sheet resin used for toys, housewares, and other stiff sheet and thermoformed parts. [Chevron]

HiD 9632. CAS 9002-88-4; HDPE copolymer resin; film resin used for notion and millinery bags, coextrusions, insulation bags, shipping bags, multiwall liners. [Chevron]

HiD 9650. CAS 9002-88-4; HDPE copolymer resin; film resin for high strength and barrier films; used for notion and millinery bags, barrier pkg., coextrusions. [Chevron]

Hidosin. Disinfectant. [Thomas Goldschmidt Ltd]

Hiduminium. A registered trade name for a range of aluminum alloys. [High Duty Alloys (Properties) Ltd] *

Hi-Ex Foam. Fire fighting foam conc. [Henkel/Cospha]

HiFax AB 6023. CAS 9003-07-0; PP resin; acoustical barrier injection molding grade offering sound deadening, excellent processability, flexibility; used for automotive isola-

---

tors, fender seals, door seals. [Himont]

HiFax CA45A. CAS 9003-07-0; PP resin; semirigid, med. modulus injection molding grade with superior processability, excellent low temp. performance; used for automotive bumper fascias, valence panels, air dams, body side moldings, wheel flares, rub strips. [Himont]

HiFax CB 17AC. CAS 9003-07-0; PP resin; rigid, high modulus extrusion/blow molding grade with good dimensional stability, high heat resist.; used for thermoformed truck bumpers, blow molded truck wind skirts, roofing, and automotive blow molded bumper beams, extruded roof rack strips, blow molded rigid air ducts. [Himont]

HiFax ETA 3011. CAS 9003-07-0; PP resin; low modulus, soft grade for injection molding applications; excellent low temp. performance for automotive mud flaps, isolators, map pockets, bumper and caps, and cup holders. [Himont]

HiFax ETA 3095. CAS 9003-07-0; PP resin; semirigid, med. modulus injection molding grade with superior processability, excellent low temp. performance; used for automotive bumper fascias, valence panels, air dams, body side moldings, wheel flares, rub strips. [Himont]

HiFax ETA 5012. CAS 9003-07-0; PP resin; soft, low modulus grades for extrusion/blow molding; offers excellent low temp. performance, high heat oxidative stability; used for roofing and automotive air ducts, intake/outtake air hoses, seat back headrest covers. [Himont]

HiFax RTA 3263E. CAS 9003-07-0; PP resin; extruded roof rack strips, blow molded rigid air ducts; dens. 1.10 g/cc; melt flow 0.4 dg/min; tensile strength 22.8 MPa; tensile elong. 330%; Izod impact 748 J/m notch; tear strength 112 kN/m; Shore hardness D63; distort. temp. 131 F (264 psi). [Himont]

Hi-Fibre. Asbestos substitute. [Hill Brothers Chemical Co]

HiGel. CAS 637-12-7; Aluminum stearate; nonaqueous coating compositions. [Hüls Am.] †

Highlink 80®. Aliphatic dialdehyde; H_2S scavenger for oil and gas industry. [Hoechst Celanese]

Highsorb 40®. Aliphatic dialdehyde; H_2S scavenger for oil and gas industry. [Hoechst Celanese]

High Temperature Deodorant #4896, OS. Mixture of fragrance materials; masks odors released in the plant during high temp. processing (≈ 400 F). [Andrea Aromatics] *

high tensile brass. An alloy of 76% copper, 22% zinc, and 2% aluminum.

Hightensite. A trade name for a proprietary hard rubber composition for use in electrical insulation. §

High Trees Mixture B. Nonionic spreader; wetting agent for phosphonoglycine herbicide sprays. [Service Chemicals Ltd]

HiGlass BJ44A. CAS 9003-07-0; PP copolymer, glass-reinforced; high melt flow, high impact grade for underhood automotive, fan shrouds, headlamp retainers, lawn tractor grilles and other applications. [Himont]

HiGlass PF062-2. CAS 9003-07-0; PP homopolymer, glass reinforced; high melt flow series with 20 and 50% glass reinforcement; features superior resist. to warpage, excellent stiffness and high temp. performance, and a wide processing window; used for appliances, automotive parts, elec./electronic connectors. [Himont]

HiGlass PF072-1. CAS 9003-07-0; PP homopolymer, glass

reinforced; std. melt flow series with 10-40% glass reinforcement; features superior resist. to warpage, excellent stiffness and high temp. performance, wide processing window, improved impact and creep performance; used for dishwasher components, pump housings/components, plumbing parts, athletic equip. [Himont]

HiGlass SB 224-2; CAS 9003-07-0; PP copolymer, 20% glass-reinforced; std. melt flow series with 20-50% glass reinforcement; features excellent cold temp. impact resist.; uses incl. wheel well housings, housings for lawn and garden equip., splice closures. [Himont]

Hi-Gloss I. Solvent base sealer; an acrylic sealer with high solids for application to exterior and interior concrete, stone and masonry type surfaces for commercial use only. [Glessner Corporation Inc (GGI Products) DBA] *

Hi-heet. An American synthetic resin molded product for electrical insulation. §

Hiirogane. A blood-red colored metal prepared either by the treatment of copper with a solution of copper sulfate and verdigris, or by heating a copper alloy with a paste containing a salt of copper, borax, and water.

Hills Adult Balsam. CAS 509-67-1; A proprietary preparation containing pholcodine. [Windsor Healthcare Ltd]

Hills Adult Expectorant. CAS 93-14-1; A proprietary preparation containing guaifenesin. [Windsor Healthcare Ltd]

Hills Balsam Pastilles. A proprietary preparation containing benzoin tincture compound, capsicum oleoresin, peppermint oil, ipecacuanha liquid extract, tincture lobella syrup 1949 and menthol. [Windsor Healthcare Ltd]

Hills Junior Balsam. A proprietary preparation containing citric acid monohydrate and ipecacuanha liquid extract. [Windsor Healthcare Ltd]

Hills-McCanna Alloy No.45. An alloy containing 88% copper, 10.5% aluminum, and 1.5% iron.

Hinge. CAS 83601-81-4; A liquid formulation containing 500 g carbendazim per liter as a suspension concentrate; systemic fungicide. [Quadrangle Agrochemicals]

Hinochloa®. Mefenacet; herbicide effective against grasses (especially against *Echinochloa crus-galli* and some broad-leaved weeds in transplanted paddy rice; mainly used in combinations with other compounds. [Bayer AG]

Hinosan®. CAS 17109-49-8; Edifenphos; fungicide especially effective against *Pyricularia oryzae* on rice; also for control of other rice diseases. [Bayer AG]

Hiotrol. For commercial and residential agents for the control and reduction of odors associated with latrines, toilets, lavatories, locker and shower rooms, kennels, livestock pens, industrial malodors, sewage wastes, garbage, land fills and lagoons. [Ferro] *

Hioxyl. CAS 7722-84-1; A cream containing stabilized hydrogen peroxide 1.5%; used in the treatment of leg ulcers, minor wounds and infections. [Quinoderm Ltd]

Hipec®. Semiconductor protective coating. [Dow Corning]

Hipernick®. A registered trademark for an alloy of nickel and iron in equal parts; used for making cores of audio-frequency transformers. §

Hipersil®. A registered trademark for high permeability silicon steel; used in high-frequency communications equipment. §

Hipersolv. High performance solvents for HPLC. [BDH Chemicals Ltd]

Hi-Pflex® 100. CAS 471-34-1; Calcium carbonate, surface-treated; filler improving impact strength and flex. mod. in

---

‡ Trade name and manufacturer not verified   § Trade name without identified manufacturer

polymers; high-performance reinforcing agent for plastics; applications incl. PVC (pipe, wire and cable), PP (interior and exterior auto parts, toys, pallets, corrugated boxes), HDPE (pipe and other rigid applications, wire and cable and other flexible applications). [Pfizer]

Hi-pHase® 35. Dispersed rosin size; sizing agent for paper/paperboard. [Hercules]

Hi-pHorm 67. Rosin-based; surface treatment agent [Hercules]

Hipochem B-3-M. CAS 136-60-7; EINECS 205-252-7; Emulsified butyl benzoate; self-emulsifying carrier for polyester and triacetate fibers. [High Point]

Hipochem Carrier 761. Emulsified aromatic hydrocarbon; dye carrier for poly/wool blends. [High Point]

Hipochem Compatibilizer WMC. Surfactant blend; compatibilizer for cationic/disperse dyes. [High Point]

Hipochem EK-18. CAS 577-11-7; EINECS 209-406-4; Sodium dioctyl sulfosuccinate; wetting, rewetting, desizing, bleaching, scouring for textiles. [High Point]

Hipochem GM. Emulsified 1,2,4-trichlorobenzene; dyeing assistant; carrier for polyester in removal of trimers. [High Point]

Hipochem Jet Dye T. Emulsified trichlorobenzene; dyeing assistant; carrier for jet dyeing machine; aids in removal of trimers. [High Point]

Hipochem Jet Scour. Modified ethoxylated alcohol; low foaming scour for all fibers; good wetting; excellent removal of mineral oil, butyl stearate, greases. [High Point]

Hipochem M-51. Nonsurface active quaternary ammonium compound; dyeing assistant; migrating agent for dyeing basic dyeable fibers. [High Point]

Hipochem Migrator J. CAS 56-93-9; Benzyl trimethyl ammonium chloride; migrating agent for basic dyes. [High Point]

Hipochem MS-BW. Emulsified aromatic and aliphatic solvs.; low foaming detergent, solvent scour for removal of oils, grease, waxes, and fatty materials in textile operations. [High Point]

Hipochem MTD. Complex phosphate ester; wetting agent, leveling agent in caustic boil-off, bleach baths, other wet processing of piece goods, hosiery, knit goods; dispersant, antiredeposition aid. [High Point]

Hipochem PDO. Optically brightened detergent powd. for improving wash effectiveness on denim. [High Point]

Hipochem SRC. Self-emulsifying solv. systems; detergent; removes dyes, oil, grease, pigment, binders from pads and machinery. [High Point]

Hipofix 491. Methyl methylol resin; fixing agent for direct dyes on cellulosics. [High Point]

Hi-Point® 90. CAS 1338-23-4; MEK peroxide in dimethyl phthalate; catalyst/initiator for R.T. cures of polyesters. [Witco/Argus]

Hipolon New. Modified cationic surfactant; detergent, dye retarder, leveler. [High Point]

Hiposcour® 3-80. Modified ethoxylate; scouring agent for synthetics and cellulosics; detergent with non-redeposition props. for mineral oils, motor oils, grease, butyl stearate stains; biodeg. [High Point]

Hiposcour® BFS. Complex Varsol/surfactant blend; detergent scour for removal of oil and grease from cotton and blends; wetting agent; stable over broad pH range, to caustic; biodeg. [High Point]

Hipowet IBS. Complex modified ethoxylate; wetting and scouring agent, detergent, suspending agent; caustic stable; stable with enzymes. [High Point]

Hippuran I 131. CAS 133-17-5; Iodohippurate sodium I 131; diagnostic aid; radioactive agent. [Mallinckrodt Inc] *

hippuric acid. (benzaminoacetic acid; benzoylaminoacetic acid; benzoylglycocoll; benzoylglycin) $C_6H_5CONH-CH_2COOH$; CAS 495-69-2; EINECS 207-806-3; Organic synthesis; medicine. [Penta Mfg.; Schweizerhall; U.S. Biochemical]

Hippuryl Amide. N-Benzoyl-glycinamide; $C_6H_5CONH-CH_2CONH_2$; a white powder; a substrate in studies of papain action. [British Drug Houses] *

Hipputope. CAS 133-17-5; Iodohippurate sodium I 131; diagnostic aid; radioactive agent. [Bristol-Myers Squibb Co Inc] *

Hiprex. CAS 5714-73-8; Each tablet contains methenamine hippurate 1 g. [3M Pharmaceuticals] †

Hirathiol. A compound used as a substitute for ichthyol (qv).

Hirudoid. Heparidoid 0.3% in cream and gel bases; for treatment of varicose ulcers and concomitant symptons of varicose veins, superficial soft tissue injuries, bruising and hematomas. [Luitpold-Werk] *

Hi-Selon. Polyvinyl alcohol film. [British Traders & Shippers Ltd]

Hi-Sil. Synthetic, amorphous, precipitated silica. [PPG Industries]

Hismanal. CAS 68844-77-9; A proprietary preparation containing astemizole; nonsedative antihistamine, allergic rhinitis and conjunctivitis (hay fever), other allergic reactions. [Janssen Pharmaceutical Ltd] *

Hi-Sol®. Aromatic hydrocarbon solvents; used in coatings. [Ashland Chemical Company] *

Hi-Sol® 10, 15, 70. CAS 8030-30-6; Naphtha; aromatic solv. [Ashland]

Hispor 45WP. CAS 83601-81-4, 12427-38-2; Carbendazim + maneb; systemic fungicide for winter cereals. [Ciba-Geigy Agrochemicals]

Hispril. CAS 132-18-3; Diphenylpyraline hydrochloride; antihistaminic. [SmithKline Beecham] *

Histadyl E.C. A proprietary preparation containing codeine phosphate, ephedrine hydrochloride, phenylpyramine fumarate, ammonium chloride and chloroform; a cough linctus. [Eli Lilly & Co] †

Histalog. The hydrochloride of 3-(2-aminoethyl)pyrazole (ametazole hydrochloride). [Eli Lilly & Co] †

Histantin. CAS 1620-21-9; A proprietary preparation of chlorcyclizine hydrochloride; antihistamine. [The Wellcome Foundation Ltd]

Histazarin. 2:3-Dihydroxyanthraquinone.

histidine. (L-histidine; α-amino-β-imidazolepropionic acid) $HOOCCH(NH_2)CH_2C_3H_3N_2$; CAS 71-00-1; EINECS 200-745-3; Medicine, feed additive, biochemical research, dietary supplement. [Degussa; R.W. Greeff; Janssen Chimica; Penta Mfg.; Tanabe USA]

L-histidine. See histidine.

Histo-Acryl. A proprietary preparation of enbucrilate; a surgical tissue adhesive. §

Histogenol®. A combination of nucleic acid and sodium methyl arsenate; used in the treatment of tuberculosis. [High Duty Alloys (Properties) Ltd] *

Histryl. A proprietary preparation of diphenpyraline hydrochloride. [SmithKline Beecham] *

Histryl Spansule Capsule. CAS 132-18-3; Diphenylpyraline hydrochloride; antihistamine for use in hay fever

---

* Trade name not verified as available   † Trade name verified as obsolete

and other allergies. [SmithKline Beecham] *

Hitac 300. High viscosity amorphous polypropylene. [Crowley Chem.]

HiTEC®. Poly alpha olefins. [Ethyl Corp; Ethyl SA]

HiTEC®300. Lubricant additive for automotive and industrial gear oils. [Ethyl Corp]

HiTEC®800. Lubricant additive for diesel engine oils. [Ethyl Corp]

HiTEC®2900. Lubricant additive for antiwear hydraulic oils. [Ethyl Corp]

HiTEC® 4000. Performance additive for cleaner burning distillate fuel. [Ethyl Corp]

HiTEC® 4400. Performance additive for gasolines with enhanced detergency for optimum engine performance. [Ethyl Corp]

HiTEC®4700. Performance additive for gasoline and jet fuel with enhanced oxidation inhibition; lubricant additive for lubricants with improved oxidation inhibition. [Ethyl Corp]

Hi Temp EC-1000. Emulsion of aliphatic oils, salts and soaps with water; die casting and mold release agent, drilling and tapping fluid/coolant. [Hi Temp Lubricants Inc] *

Hi Temp EC-4000. Emulsion of aliphatic oils, salts and soaps with water; die casting mold release agent. [Hi Temp Lubricants Inc] *

Hi Temp EC-5000. Emulsion of aliphatic oils, salts and soaps with water; die casting mold release agent, die casting plunger lubricant. [Hi Temp Lubricants Inc] *

Hitenso. A proprietary trade name for a cadmium bronze. §

HiTint. Pigment dispersions; for coloration of solvent-based coatings. [Pacific Dispersions Inc] *

Hitox®. CAS 13463-67-7; EINECS 236-675-5; Titanium dioxide; buff-colored pigment developed as alternative to wh. titanium dioxide; used in alkyds, acrylic urethanes, high solids systems, water reducibles, water bases, powd. coatings, inks, and adhesives. [Hitox]

Hi-Zex. A trademark for Japanese high density polyethylene. [Mitsui Co Ltd] *

H-K Mastitis. CAS 5321-32-4; Hetacillin potassium; antibacterial. [Bristol Laboratories] *

HM-0230. Thermoplastic polyamide adhesive; offers good penetration of fibrous or porous substrates and rapid setting; as folding cement in shoe mfg. [H.B. Fuller] *

HM-0652. Thermoplastic polyamide adhesive; offers rapid setting and high temp. resistant bonds with fibrous or porous substrates; for mfg. of pleated air filters for automotives. [H.B. Fuller] *

HM-0814. Polyamide hot-melt adhesive; for fiber/foil tube winding operations; excellent adhesion to metals, foils, coated paper and other dense surfaces. [H.B. Fuller] *

HM-6300. Thermoplastic polyester hot-melt adhesive; for demanding production assembly applications; produces flexible bonds over broad temp. range with sufficient open time to bond large components or complex shapes. [H.B. Fuller] *

HMDS. See hexamethyldisilazane.

HMS Liquifilm®. CAS 2668-66-8; Medrysone; corticosteroid, anti-inflammatory. [Allergan, Inc]

H.M.T. See hexamine.

H.M.T.D. Hexamethylenetriperoxydiamine. A detonating explosive.

Hobane. CAS 1689-84-5, 1689-83-4; Bromoxynil and ioxynil; herbicide. [ICI Chem. & Polymers Ltd]

Hobane. CAS 1689-84-5, 1689-83-4; An emulsifiable concentrate containing 240 g bromoxynil and 160 g ioxynil per liter; a post-emergence contact herbicide for cereal crops. [Farm Protection Ltd]

Hochst New Blue. The calcium salt of the di- and trisulfonic acids of trimethyltriphenylpararosanilinetrisulfonic acid; a dyestuff; dyes wool blue from an acid bath. §

Hodag 20-L. CAS 9004-81-3; PEG-4 laurate; emulsifier, wetting agent, plasticizer for cosmetic, pharmaceutical and other uses. [Calgene]

Hodag 22-L. CAS 9005-02-1; PEG-4 dilaurate; emulsifier, wetting agent, plasticizer for cosmetic, pharmaceutical and other uses. [Calgene]

Hodag 40-O. CAS 9004-96-0; PEG-8 oleate; emulsifier, wetting agent, plasticizer for cosmetics, pharmaceuticals, other uses. [Calgene]

Hodag 40-R. CAS 9004-97-1; PEG-8 ricinoleate; emulsifier, wetting agent, plasticizer for general cosmetic, pharmaceutical, and other uses. [Calgene]

Hodag 40-S. CAS 9004-99-3; PEG-8 stearate; emulsifier, wetting agent, plasticizer for cosmetic, pharmaceutical and other uses. [Calgene]

Hodag 42-O. CAS 9005-07-6; PEG-8 dioleate; emulsifier, wetting agent, plasticizer for cosmetic, pharmaceutical and other uses. [Calgene]

Hodag 150-S. CAS 9004-99-3; PEG-6-32 stearate; emulsifier, wetting agent, plasticizer for cosmetic, pharmaceutical and other uses. [Calgene]

Hodag 602-S. CAS 9005-08-7; PEG-150 disearate. thickener and auxiliary emulsifier for creams and lotions, esp. amphoteric type shampoos. [Calgene]

Hodag Antifoam CO-350. Silicone antifoam; antifoam for nonaq. foaming systems; for fermentation, chem. processing, food, pharmaceutical, paint, adhesives, paper coatings, metalworking, lubricants, textile processing, petroleum, pulp and paper, cleaning compounds. [Calgene]

Hodag Antifoam F-1. CAS 8050-81-5; Simethicone; antifoam for aq. and nonaq. foaming systems; for fermentation, chem. processing, food, pharmaceutical, paint, adhesives, paper coatings, metalworking, lubricants, textile processing, petroleum, pulp and paper, cleaning compounds. [Calgene]

Hodag Antifoam FD-82. Silicone emulsion; antifoam for aq. foaming problems; for fermentation, chem. processing, food, pharmaceutical, paint, adhesives, paper coatings, metalworking, lubricants, textile processing, petroleum, pulp and paper, cleaning compounds. [Calgene]

Hodag C-100-L. CAS 136-99-2; EINECS 205-271-0; 1-Hydroxyethyl-2-lauric imidazoline; intermediate for quaternary ammonium compounds; strongly absorbed on textiles, paper and many metal surfs.; for agric., asphalt, cleaners, corrosion inhibitors, demulsifiers, flotation, metalworking, paints, pigment grinding, inks, textiles, wax emulsions. [Calgene]

Hodag C-100-O. 1-Hydroxyethyl-2-oleic imidazoline; intermediate for quaternary ammonium compounds; strongly absorbed on textiles, paper and many metal surfs.; for agric., asphalt, cleaners, corrosion inhibitors, demulsifiers, flotation, metalworking, paints, pigment grinding, inks, textiles, and wax emulsions. [Calgene]

Hodag C-100-S. CAS 95-19-2; EINECS 202-397-8; 1-Hydroxyethyl-2-stearic imidazoline; intermediate for quaternary ammonium compounds; strongly absorbed on textiles, paper and many metal surfs.; for agric., asphalt, cleaners, corrosion inhibitors, demulsifiers, flotation, metalworking, paints, pigment grinding, inks, textiles, wax emulsions. [Calgene]

Hodag C-100-T. CAS 61791-39-7; EINECS 263-171-2; 1-Hydroxyethyl-2-tall oil imidazoline; intermediate for quaternary ammonium compounds; strongly absorbed on textiles, paper and many metal surfs.; for agric., asphalt, cleaners, corrosion inhibitors, demulsifiers, flotation, metalworking, paints, pigment grinding, inks, textiles, wax emulsions. [Calgene]

Hodag CC-22. Propylene glycol dicaprylate/dicaprate; surfactant for food, cosmetics, and pharmaceutical industries; vehicle/diluent/carrier for vitamins, drugs, flavors, color, fragrance; emollient for makeup, bath and skin oils. [Calgene]

Hodag CSA-80. CAS 31394-71-5; PPG-26 oleate; nongreasy spreading agent, skin moisturizer, emollient, vehicle for skin and hair care prods., bath oils; carrier and dispersant for additives (hormones, vitamins, essnetial oils, germicides, etc.); foam depressing props. [Calgene]

Hodag CSA-101. Cetearyl alcohol, ceteth-20, glycol stearate. [Calgene]

Hodag DGL. CAS 9004-81-3; PEG-2 laurate; emulsifier. [Calgene]

Hodag DGO. CAS 9004-96-0; PEG-2 oleate; lubricant. [Calgene]

Hodag DGS. CAS 9004-99-3; PEG-2 stearate; emulsifier, opacifier, thickener for cosmetics, o/w emulsions; lubricant for stamping and drawing; protective coating for hygroscopic materials (tablts); opacifier/lubricant for paper industry; antitack agent. [Calgene]

Hodag DGS-C. PEG-2 stearate SE; self-emulsifying emulsifier, opacifier, thickener for cosmetics, o/w emulsions; lubricant for stamping and drawing; protective coating for hygroscopic materials (tablts); opacifier/lubricant for paper industry; antitack agent. [Calgene]

Hodag DOSS-70. CAS 577-11-7; EINECS 209-406-4; Dioctyl sodium sulfosuccinate; wetting agent, surface tension reducer. [Calgene]

Hodag EGMS. Glycol stearate; wetting agent, surface tension reducer. [Calgene]

Hodag GML. CAS 142-18-7; EINECS 205-526-6; Glyceryl laurate; emulsifier, opacifier, stabilizer for food, drug, and cosmetic industries. [Calgene]

Hodag GMO. Glyceryl oleate; emulsifier, opacifier, stabilizer for food, drug, and cosmetic industries. [Calgene]

Hodag GMP. Glyceryl palmitate; surfactant. [Calgene]

Hodag GMR. CAS 141-08-2; EINECS 205-455-0; Glyceryl ricinoleate; emulsifier, opacifier, stabilizer for food, drug, and cosmetic industries. [Calgene]

Hodag GMS. Glyceryl stearate; emulsifier, opacifier, stabilizer for food, drug, and cosmetic industries. [Calgene]

Hodag GTO. Glyceryl trioleate; emulsifier, opacifier, stabilizer for food, drug, and cosmetic industries. [Calgene]

Hodag Nonionic 1017-R. CAS 9003-11-6; Meroxapol 171. [Calgene]

Hodag Nonionic 1035-L. CAS 9003-11-6; Poloxamer 105; detergent, antifoam, wetting agent, emulsifier, antistat, demulsifier, visc. modifier, deduster, gelation aid, metalworking lubricants, dispersants. [Calgene]

Hodag Nonionic 1044-L. CAS 9003-11-6; Poloxamer 124; detergent, antifoam, wetting agent, emulsifier, antistat,

demulsifier, visc. modifier, deduster, gelation aid, metalworking lubricants, dispersants. [Calgene]

Hodag Nonionic 1064-L. CAS 9003-11-6; Poloxamer 184; detergent, antifoam, wetting agent, emulsifier, antistat, demulsifier, visc. modifier, deduster, gelation aid, metalworking lubricants, dispersants. [Calgene]

Hodag Nonionic 1088-F. CAS 9003-11-6; Poloxamer 238; detergent, antifoam, wetting agent, emulsifier, antistat, demulsifier, visc. modifier, deduster, gelation aid, metalworking lubricants, dispersants. [Calgene]

Hodag Nonionic 2017-R. CAS 9003-11-6; Meroxapol 172; detergent, antifoam, dispersant, demulsifier, dishwashing rinse visc. control agent. [Calgene]

Hodag Nonionic E-5. CAS 9016-45-9; Nonoxynol-5; detergent and wetting agent for cosmetics, insecticides and other formulations. [Calgene]

Hodag Nonionic GR-8. CAS 61791-12-6; PEG-8 castor oil; surfactant. [Calgene]

Hodag Nonionic GRH-25. CAS 61788-85-0; PEG-25 hydrogenated castor oil; surfactant. [Calgene]

Hodag Nonionic ID-5. Isodeceth-5; surfactant. [Calgene]

Hodag Nonionic L-4. Laureth-4; surfactant. [Calgene]

Hodag Nonionic S-2. CAS 9005-00-9; Steareth-2; surfactant. [Calgene]

Hodag Nonionic TD-15. CAS 24938-91-8; Trideceth-15; surfactant. [Calgene]

Hodag PB-285. CAS 9003-13-8; PPG-15 butyl ether. [Calgene]

Hodag PE-005. Alkylaryl phosphate ester; emulsifier, EP lube additive, antistat, corrosion inhibitor, surfactant. [Calgene]

Hodag PE-005-K. Potassium phosphate ester; solubilizer for nonionic surfactants in built liq. concs., esp. low-foam nonionics; also raises cloud pt. of rinse-aid formulations; stable over wide pH range; nondiscoloring on solid caustic. [Calgene]

Hodag PEG 200. CAS 25322-68-3; PEG-4; cosmetics and pharmaceuticals formulation; plasticizer for adhesives; inks; resins and coatings. [Calgene]

Hodag PEG 300. CAS 25322-68-3; PEG-6; cosmetics and pharmaceutical formulation; latex coagulating bath; plasticizer for adhesives, spray-on bandages; resins and coatings. [Calgene]

Hodag PEG 400. CAS 25322-68-3; PEG-8; cosmetic and pharmaceutical formulation; humectant, coupler for lotions; release agent for rubber; latex coagulating bath; in PVAc paints. [Calgene]

Hodag PEG 540. CAS 25322-68-3; PEG-6, PEG-32; cosmetic, pharmaceutical, and suppository formulation; humectant, plasticizer in adhesives; base for metal polishes; lubricant for paper sizes; inks; in alkyd resins and coatings. [Calgene]

Hodag PEG 600. CAS 25322-68-3; PEG-12; cosmetic and pharmaceutical formulation; resins and coatings. [Calgene]

Hodag PEG 1000. CAS 25322-68-3; PEG-20; cosmetic and pharmaceutical formulation; resins and coatings; imparts dimensional stability to paper wet strength resins, improves coatings gloss. [Calgene]

Hodag PEG 1450. CAS 25322-68-3; PEG-32; cosmetic and pharmaceutical formulation; resins and coatings. [Calgene]

Hodag PEG 3350. CAS 25322-68-3; PEG-75; cosmetic and pharmaceutical formulation; resins and coatings; humectant, plasticizer for adhesives; antistat for rubber conveyor belt; in shoe polish; lubricant for paper sizing; printing inks; tablet binder, lubricant. [Calgene]

Hodag PEG 8000. CAS 25322-68-3; PEG-150; cosmetic and pharmaceutical formulation; resins and coatings; antistat for rubber conveyor belting; in shoe polish; lubricant for paper size; printing inks; release agent for rubber; tablet binder/lubricant. [Calgene]

Hodag PGL-101. Decaglyceryl monolaurate. [Calgene]

Hodag PGL. Triglyceryl monolaurate. [Calgene]

Hodag PGML. Propylene glycol laurate; surfactant for food industry. [Calgene]

Hodag PGMP. Propylene glycol palmitate; surfactant. [Calgene]

Hodag PGMS. CAS 1323-39-3; EINECS 215-354-3; Propylene glycol stearate; surfactant for food industry. [Calgene]

Hodag PGO-61. CAS 9007-48-1; Hexaglyceryl monooleate [Calgene]

Hodag PGO-62. Hexaglyceryl dioleate. [Calgene]

Hodag PGO-101. CAS 9007-48-1; Decaglyceryl monooleate. [Calgene]

Hodag PGO-102. Decaglyceryl dioleate. [Calgene]

Hodag PGO-103. Decaglyceryl trioleate. [Calgene]

Hodag PGO-104 (formerly Hodag SVO-1047). CAS 34424-98-1; Decaglycerol tetraoleate. [Calgene]

Hodag PGO-108. Decaglyceryl octaoleate. [Calgene]

Hodag PGO-1010 (formerly Hodag SVO-10107). CAS 11094-60-3; Decaglycerol decaoleate. [Calgene]

Hodag PGS. CAS 1323-39-3; EINECS 215-354-3; Propylene glycol stearate; emulsifier for food processing. [Calgene]

Hodag PGS-61. CAS 37349-34-1; Hexaglyceryl monostearate. [Calgene]

Hodag PGS-62. Hexaglyceryl distearate. [Calgene]

Hodag PGS-101. CAS 37349-34-1; Decaglyceryl monostearate. [Calgene]

Hodag PGS-102. Decaglyceryl distearate. [Calgene]

Hodag PGS-103. Decaglyceryl tristearate. [Calgene]

Hodag PGS-104. Decaglyceryl tetrastearate. [Calgene]

Hodag PGS-108. Decaglyceryl octastearate. [Calgene]

Hodag PGS-1010. CAS 39529-26-5; EINECS 254-495-5; Decaglyceryl decastearate. [Calgene]

Hodag PGSH. Triglyceryl monoshortening. [Calgene]

Hodag PGSH-61. Hexaglyceryl monoshortening. [Calgene]

Hodag PGSH-62. Hexaglyceryl dishortening. [Calgene]

Hodag Polyglycol 5035. PPG-28-buteth-35. [Calgene]

Hodag PSML-20. CAS 9005-64-5; Polysorbate 20; food-grade emulsifier. [Calgene]

Hodag PSMO-20. CAS 9005-65-6; Polysorbate 80; emulsifier for food processing, industrial applications. [Calgene]

Hodag PSMP-20. CAS 9005-66-7; Polysorbate 40; emulsifier for food processing, industrial applications. [Calgene]

Hodag PSMS-20. CAS 9005-67-8; Polysorbate 60; emulsifier for food processing, industrial applications. [Calgene]

Hodag PSTS-20. CAS 9005-71-4; Polysorbate 65; emulsifier for food processing, industrial applications. [Calgene]

Hodag SML. CAS 1338-39-2; EINECS 215-663-3; Sorbitan laurate; emulsifier, oil additive, corrosion inhibitor; food-grade emulsifier. [Calgene]

Hodag SMO. CAS 1338-43-8; EINECS 215-665-4; Sorbitan oleate; emulsifier, oil additive, corrosion inhibitor; food-grade emulsifier. [Calgene]

Hodag SMP. CAS 26266-57-9; EINECS 247-568-8; Sorbitan palmitate; emulsifier, oil additive, corrosion inhibitor; food-grade emulsifier. [Calgene]

Hodag SMS. CAS 1338-41-6; EINECS 215-664-9; Sorbitan stearate; emulsifier for food processing. [Calgene]

Hodag SSO. CAS 8007-43-0; EINECS 232-360-1; Sorbitan sesquioleate. [Calgene]

Hodag STO. CAS 26266-58-0; EINECS 247-569-3; Sorbitan trioleate; emulsifier, oil additive, corrosion inhibitor; food-grade emulsifier. [Calgene]

Hodag STS. CAS 26658-19-5; EINECS 247-891-4; Sorbitan tristearate; emulsifier, oil additive, corrosion inhibitor; food-grade emulsifier. [Calgene]

Hodag SVO-9. CAS 9005-65-6; PEG-20 sorbitan oleate; emulsifier; kosher grade. [Calgene]

Hodag SVO-629. Hexaglyceryl distearate; food emulsifier. [Calgene]

Hodag SVO-1047. CAS 34424-98-1; Decaglyceryl tetraoleate; food emulsifier. [Calgene]

Hodag SVS-18. CAS 9005-67-8; PEG-20 sorbitan stearate; food emulsifier; kosher grade. [Calgene]

Hoe S 1816. EO/PO block copolymer; surfactant for agric. formulations. [Hoechst Celanese/Colorants & Surf.]

Hoe S 1984 (TP 2279). Calcium dodecylbenzene sulfonate/nonylphenol ethoxylate; surfactant for agric. formulations. [Hoechst Celanese/Colorants & Surf.]

Hoe S 2650. Dilaureth-4 dimonium chloride; surfactant for conditioners, shampoos. [Hoechst Celanese/Colorants & Surf.]

Hoe S 2713. Sodium dodecylbenzene sulfonate; surfactant for agric. formulations. [Hoechst Celanese/Colorants & Surf.]

Hoe S 2721. Polyglyceryl-2 sesquiisostearate; w/o emulsifier for personal care prods.; for prods. with high temp. stability. [Hoechst Celanese/Colorants & Surf.]

Hoe S 2749. TEA dodecylbenzene sulfonate; surfactant for agric. formulations. [Hoechst Celanese/Colorants & Surf.]

Hoe S 3618. EO/PO block copolymer bis-mono-phosphate ester; surfactant for agric. formulations. [Hoechst Celanese/Colorants & Surf.]

Hoe S 3680. CAS 8013-07-8; EINECS 232-391-0; Epoxidized soybean oil; surfactant for agric. formulations. [Hoechst Celanese/Colorants & Surf.]

Hoechst Wax C. Amide wax based on bis-stearoyl ethylenediamine; lubricant and release agent for polyolefins; improves processing in engineering thermoplastics such as polyamides and acetal. [Hoechst AG]

Hoechst Wax E. Glycol/butylene glycol montanate; internal/external lubricant and release agent for PVC and other polymers; for sheet, film and bottle applications, esp. for tin stabilized systems. [Hoechst AG; Hoechst Celanese]

Hoechst Wax OP. Butylene glycol montanate, calcium montanate; internal/external lubricant and release agent for PVC and other polymers; maintains clarity of finished product. [Hoechst AG; Hoechst Celanese]

Hoechst Wax PE 190. CAS 9002-88-4; Polyethylene wax; for mfg. of transparent PVC articles with high surface gloss. [Hoechst AG; Hoechst Celanese]

Hoechst Wax PE 520. CAS 9002-88-4; Polyethylene wax; carrier for pigment concs. for polyolefins; external lubri-

---

‡ Trade name and manufacturer not verified   § Trade name without identified manufacturer

cant for PVC. [Hoechst AG; Hoechst Celanese]

Hoechst Wax PED 191. Oxidized polyethylene wax; lubricant for PVC; processing aid for rigid PVC compounds; reduces sticking of tin-stabilized PVC melts to hot machine parts. [Hoechst AG]

Hoechst Wax PED 521. CAS 9002-88-4; Polyethylene wax; lubricant for PVC; reduces sticking of tin-stabilized PVC melts to hot machine parts. [Hoechst AG]

Hoechst Wax PP 230. CAS 9003-07-0; Polypropylene wax; external lubricant for PVC, carrier for pigments. [Hoechst AG]

Hoechst Wax S. CAS 68476-03-9; EINECS 270-664-6; Montan acid wax; lubricant and release agent for PVC and other polymers. [Hoechst AG; Hoechst Celanese]

Hoegrass. CAS 51338-27-3; Emulsifiable concentrate containing 378g diclofop-methyl per liter; used for control of weeds in grass. [Hoechst UK]

Hoenle's cement. A cement, consisting of 2 parts shellac, and 1 part Venice turpentine.

Holcote. Thixotropic ready-for-use water based coatings for sand molds and cores. [Foseco (F.S.) Ltd]

H₂old EP-1. CAS 9003-39-8; Vinylpyrrolidone terpolymer. [ISP]

Holdfast D. Soluble concentrate of 25 g dicamba and 250 g paclobutrazol per liter; used for weed control in cereals and grassland. [ICI Agrochemicals]

Holfos bronze. An alloy of copper with 11-12% tin, 0.25% lead, and 0.1-0.2% phosphorus.

Holite. A proprietary trade name for a synthetic resin for molding and laminating. §

Hollofil®. Fibers. [DuPont UK]

holmia. See holmium oxide.

holmium. Metallic element; Ho; CAS 7440-60-0; EINECS 231-169-0; getter in vacuum tubes, research in electrochemistry, spectroscopy. [Atomergic Chemetals; Cerac; Noah Chem.; Rhone-Poulenc Basic]

holmium oxide. (holmia) Ho₂O₃; CAS 12055-62-8; EINECS 235-015-3; refractories, special catalyst. [Atomergic Chemetals; Cerac; Noah Chem.; Rhone-Poulenc Basic]

Holocaine Hydrochloride. CAS 620-99-5; Phenacaine hydrochloride; anesthetic. [Abbott Laboratories] *

Holoxan. CAS 3778-73-2; Ifosfamide; vials (dry substance); cytostatic. [Degussa AG] *

Holtox. A suspension concentrate containing 250 g atrazine and 250 g cyanazine per liter; a residual herbicide. [Shell UK]

Homac. A proprietary synthetic resin. §

Homagenets Aoral. CAS 68-26-8; Vitamin A; vitamin (antixerophthalmic). [SmithKline Beecham] *

Homapin. CAS 80-49-9; Homatropine methylbromide; anticholinergic. [Mission Pharmacal Co] *

Homberg's metal. A fusible alloy, containing 3 parts bismuth, 3 parts tin, and 3 parts lead. It has a melting point of 122 C.

Homberg's phosphorus. Anhydrous calcium chloride, CaCl₂, which, when fused, and exposed to the sun, becomes phosphorescent in the dark.

Homberg's salt. Boric acid, H₃BO₃.

Hombifine®. Chemical products for industrial purposes, especially pigments for paints and colors, fibers, cosmetics and catalysts. [Sachtleben Chemie GmbH]

Hombisorp®. Chemical adsorbents for purifying liquids and gases. [Sachtleben Chemie GmbH]

Hombitan®. CAS 13463-67-7; EINECS 236-675-5; Titanium dioxide, TiO₂; white inorganic pigment for paints and coatings. [Sachtleben Chemie GmbH]

Hombitec®. Chemical products for industrial purposes, especially pigments for paints and colors, fibers, cosmetics and catalysts. [Sachtleben Chemie GmbH]

Home-Tet. Tetanus immune globulin; immunizing agent. [Savage Laboratories] †

homoanisic acid. See p-methoxyphenylacetic acid.

Homokol. A sensitizer for silver bromide plates. It is a mixture of quinoline red with an isocyanine dye.

homomenthyl salicylate. See homosalate.

homophan. See paratophan.

homosalate. (homomenthyl salicylate; metahomomenthyl salicylate; 3,3,5-trimethylcyclohexyl 2-hydroxybenzoate) Substituted phenolic compound; C₁₆H₂₂O₃; CAS 118-56-9; Sunscreening agent. [EM Industries; Penta Mfg.; Quest Int'l.; Witco/Humko]

Homo Size 7A. Blend; neutral sizing agent for paper mfg. (petrol. resin base). [Toho Chem. Industry]

Hondostab. Lead stabilizers. [British Traders & Shippers Ltd]

Hondurite. A proprietary molded composition, with cotton and a vulcanized binder, for electrical insulation. §

Honeycat. Catalysts for the control of air pollution; used as off-highway exhaust purification catalysts, industrial air pollution control catalysts. [Johnson Matthey plc]

Honey Wax. Mold release based wax for plastic molding. [Specialty Products Co] *

Honvan. CAS 522-40-7; Fosfestrol; tablets and ampoules; cytostatic. [Degussa AG] *

Honvol. See Honvan. [Degussa AG] *

hoof oil. See neatsfoot oil.

Hopcalite I. A mixture of 50% manganese dioxide, 30% copper oxide, 15% cobalt oxide, and 5% silver oxide; used as a catalyst to oxidize carbon monoxide.

Hopkin's-Cole reagent. To a liter of a saturated solution of oxalic acid, 60 g of sodium amalgam are added, and the mixture allowed to stand. It is then filtered and diluted with from 2-3 volumes of water; used for the detection of proteins.

Hopkin's-Cole Tyrosine C reagent. This contains mercuric sulfate dissolved in a solution of sulfuric acid.

Hopkin's lactic acid reagent. Thiophene.

Hopp II. A modified aqueous hop extract; standardized to 35% reduced iso-alpha acids; for addition to malt beverages after fermentation to standardize bitterness. [Pfizer International]

Hopper salt. Sodium chloride, which has been caused to crystalize in large hollow cubes which float when alum is added to the bath.

Hoppit. Quassia; animal repellent for outdoor crops. [Fieldspray]

Horco X. A proprietary trade name for a thermosetting polyvinyl butyral synthetic resin. §

Hordaflex. Chlorinated paraffins. [Hoechst UK]

Hordalub. Chlorinated paraffins. [Hoechst UK]

Hordamer. Polyethylene primary dispersions. [Hoechst UK]

Horizon®. For the agriculture industry. [DuPont UK]

Horizon/Horizont. CAS 107534-96-3; Tebuconazole; fungicide with systemic properties and broad spectrum activity against rusts, leaf spot diseases, e.g., Septoria spp., powdery mildew and several Fusarium species on cereals; whitemold, Phoma and various leaf spot dis-

---

* Trade name not verified as available   † Trade name verified as obsolete

eases on oilseed rape. [Bayer AG]

hormodin. See 3-indolebutyric acid.

Hormone Rooting Powder. Captan and hormone compound. [Murphy Chemical Ltd] †

Hormonin. Oestriol 0.27 mg, estradiol 0.6 mg and oestrone 1.40 mg; replacement therapy in estrogen deficiency. [G W Carnrick Co Ltd] ‡

Horna. Lead chromate yellow and molybdate red pigments. [Ciba plc]

Horn O' Plenty. Various granular fertilizer blends; farm and garden fertilizers. [Horn's Crop Service Center] *

Hortag Aquasulf. CAS 7704-34-9; Suspension concentrate containing 900 g/l sulfur; a protectant fungicide. [Avon Packers Ltd]

Hortag Carbotec. Carbendazim + tecnazene; protectant fungicide and sprout suppressant for stored potatoes. [Avon Packers Ltd]

Hortag Tecnacarb Dust. Carbendazim + tecnazene; protectant fungicide and sprout suppressant for stored potatoes. [Avon Packers Ltd]

Hortag Tecnazene Plus. CAS 117-18-0, 148-79-8; Dustable powder containing 6% w/w tecnazene and 1.8% thiabendazole; a protectant fungicide and potato sprout suppressant. [Avon Packers Ltd]

Hortag Thiram. CAS 137-26-8; Thiram; fungicide with animal repellent properties which is available as a dust, wettable powder or suspension. [Avon Packers Ltd]

Hortichem Spraying Oil. Emulsifiable concentrate containing 710 g/l petroleum oil; an insecticide and acaricide. [Hortichem Ltd]

Hortus. Fertilizers. [Scottish Agricultural Industries plc]

Hostacain. A proprietary preparation of butanilicaine phosphate, procaine phosphate, and adrenaline; a local dental anesthetic. [Hoechst UK] *

Hostacerin CG. Cetearyl alcohol, triceteareth-4 phosphate, PEG-6 oleamide, sodium C14-17 sec alkane sulfonate; self-emulsifying base material for mfg. of o/w creams. [Hoechst Celanese/Colorants & Surf.; Hoechst AG]

Hostacerin DGO. Polyglyceryl-2 sesquioleate; emulsifier for cosmetics and pharmaceuticals. [Hoechst Celanese/Colorants & Surf.; Hoechst AG]

Hostacerin DGS. Polyglyceryl-2-PEG-4 stearate; coemulsifier and thickener for cosmetic o/w emulsions. [Hoechst Celanese/Colorants & Surf.; Hoechst AG]

Hostacerin O-20. CAS 9004-98-2; Oleth-20; surfactant. [Hoechst Celanese/Colorants & Surf.]

Hostacerin T-3. CAS 68439-49-6; Ceteareth-3; emulsifier, superfatting agent, base for ointments, creams, liq. emulsions, shampoo additive. [Hoechst Celanese/Colorants & Surf.; Hoechst AG]

Hostacerin WO. Polyglyceryl-2 sesquiisostearate, beeswax, mineral oil, magnesium stearate, and aluminum stearate; emulsifier conc. for cosmetic w/o emulsions. [Hoechst Celanese/Colorants & Surf.; Hoechst AG]

Hostacor. Corrosion inhibitors. [Hoechst UK]

Hostacor 2098. Complex carboxylic acid; wetting agent, corrosion inhibitor in aq. systems, synthetic metalworking and cleaning formulations. [Hoechst Celanese/Colorants & Surf.]

Hostacor 2125. Carboxylic acid complex; corrosion inhibitor for synthetics on ferrous and nonferrous metals. [Hoechst Celanese/Colorants & Surf.]

Hostacor BBM. Boric acid/carboxylic acid amine condensate; emulsifier, wetting agent, and corrosion inhibitor for

semi-synthetic metalworking fluids. [Hoechst Celanese/Colorants & Surf.]

Hostacor H Liq. N. Arylsulfonamidocarboxylic acid; chemical intermediate for mfg. of corrosion inhibitors for aq. systems, metalworking fluids. [Hoechst Celanese/Colorants & Surf.]

Hostacor TP 2445. Boric acid/carboxylic acid amine condensate; high boron content corrosion inhibitor with surface active props.; wetting agent. [Hoechst Celanese/Colorants & Surf.]

Hostadrill Brands. Vinylamide/vinylsulfonic acid copolymers; filtration agents. [Hoechst AG]

Hostadur. A trade name for partially crystalline thermoplastic polyester based on ethylene terephthalate used for construction of rigid components, e.g., gears. [Hoechst UK] *

Hostaflam. Flame retardants for polymers. [Hoechst UK]

Hostaflex. Vinyl acetate modifying copolymer. [Hoechst UK]

Hostaflon®. Polytetrafluoroethylene resins and dispersions. [Hoechst AG]

Hostaflon® C2. A proprietary polychlorotrifluoroethylene. [Hoechst UK] *

Hostaflon® ET. A proprietary ethylene tetrafluoroethylene copolymer. [Hoechst UK] *

Hostaflon® TF 1101. CAS 9002-84-0; PTFE; ram extrusion grade. [Hoechst Celanese] *

Hostaflon® TF 1620. CAS 9002-84-0; PTFE suspension polymer; compression molding grade for isostatic compression molding; produces skived film with excellent elec. properties. [Hoechst AG] *

Hostaflon® TF 2071. CAS 9002-84-0; PTFE emulsion polymer; paste powd. for extrusion applications; dens. 450 g/l (bulk), 2.1 g/cc (sintered tube); Sintered tube: tensile strength 28 MPa; tensile elong. 300% (break). [Hoechst AG] *

Hostaflon® TF 5032. CAS 9002-84-0; Unmodified PTFE disp. polymer; good film-forming and penetration properties; for coatings on prepared metal or ceramic surfaces, and impregnation of absorb. materials (e.g., asbestos, glass fibers, carbon, sintered metal); nonstick coating on cookware, impregnation of glass fiber fabrics, production of PTFE-impregnated asbestos packing. [Hoechst AG]*

Hostaflon® TF 5537. CAS 9002-84-0; Modified PTFE disp. polymer; pigmented, aq. disp. for topcoats. [Hoechst AG] *

Hostaflot L Grades. Aliphatic dithiophosphates; flotation collector for sulfide minerals. [Hoechst AG]

Hostaform. Acetal copolymer molding resins. [Hoechst AG]

Hostaform C 2521. Acetal copolymer resin; std. grade for extrusion; also for injection molding of thick-walled parts. [Hoechst AG] *

Hostaform C 9021 ELS. Acetal copolymer resin, carbon blk.-modified; injection molding grade with very low surf. resist.; for applications in explosion-proof areas. [Hoechst AG] *

Hostaform C 9021 K. Acetal copolymer resin, mineral additive modified; low-friction low-wear injection molding grade modified with mineral additive for moldings with low dry sliding wear. [Hoechst AG] *

Hostaform S 27076. Acetal copolymer resin; high impact resistant, very flexible, vibration absorbing injection molding grade. [Hoechst AG] *

Hostalen®. A trademark for a high density polythene.

---

‡ Trade name and manufacturer not verified   § Trade name without identified manufacturer

[Hoechst UK] *

Hostalen®EP 4450. CAS 9002-88-4; UHMW polyethylene; flow enhanced polymer with high bearing props. [Hoechst Celanese] *

Hostalen® G. High density polyethylene. [Hoechst UK]

Hostalen® GB 6950. CAS 9002-88-4; HDPE copolymer with a narrow molecular weight distribution; high melt flow index, fast injection speeds, reduced dens., improved impact strength and ESCR behavior; used in paint pails, domestic ware, etc. [Hoechst AG] *

Hostalen® GM 5010 T2. CAS 9002-88-4; HDPE, HMW, carbon blk.; NSF-approved pipe grade for potable water applications. [Hoechst Celanese] *

Hostalen® GM 7745 HP. CAS 9002-88-4; HDPE, HMW; blow molding grade for storage tank liners, lawn and garden spray tanks, automated refuse carts, intermediate bulk containers; superior ESCR. [Hoechst Celanese]*

Hostalen®GUR 5121. CAS 9002-88-4; UHMW polyethylene; enhanced flow injection molding grade with excellent abrasion and chem. resist., high impact strength. [Hoechst Celanese]

Hostalen® OO. A trademark for isotactic polypropylene; a thermoplastic moderately rigid molding material. [Hoechst UK] *

Hostalen® PP 927. CAS 9003-07-0; PP homopolymer; stabilized extrusion grade resin; good mechanical properties at moderate draw ratios (10/1); for extrusion of bristles, ribbons, and general purpose applications. [Hoechst Celanese UK] *

Hostalit. Polyvinyl chloride polymers and copolymers. [Hoechst AG]

Hostalit Z. A blend of PVC and chlorinated polyolefin; a thermoplastic used for manufacturing pipes. [Hoechst AG]

Hostalub®. Lubricants for polymers. [Hoechst UK]

Hostalub®CAF 484 SB. Calcium stearate, synthetic paraffin wax, and oxidized polyethylene wax; lubricant for high line-speed extrusion of vinyl siding. [Hoechst Celanese]

Hostalub®FA 1. Amide wax; internal/external lubricant for PVC and thermosets; carrier for pigments. [Hoechst AG]

Hostalub® H 4. Modified hydrocarbon wax; external lubricant for PVC pipe and profile extrusion and for plasticized PVC. [Hoechst AG]

Hostalub® VP Ca W 2. CAS 68308-22-5; Calcium montanate; internal lubricant for PVC, polyamide, thermoplastic PU, other thermoplastics and thermosets. [Hoechst AG]

Hostalub® WE4. CAS 68476-38-0; Glyceryl montanate; lubricant and release agent for PVC and other polymers. [Hoechst AG; Hoechst Celanese]

Hostalub® XL 165. Blend of paraffinic wax components; external lubricant for rigid PVC extrusion, wire and cable applications. [Hoechst Celanese]

Hostalux KCB. Benzoxazole type; optical brightener, whitening agent used in all types of polymer processing; used in plastic films, press molding, and injection molding material fibers and bristles, paints and lacquers. [Hoechst Celanese]

Hostamer Brands. Vinylamide/vinylsulfonic acid copolymers; polymers for enhanced oil recovery, esp. at high temps. and salinity. [Hoechst AG]

Hostamid. A proprietary trade name for a group of nylons. [Hoechst UK] *

Hostamont. Montan waxes. [Hoechst AG]

Hostanox®. Antioxidants for polymers. [Hoechst UK]

Hostanox® 03. Benzene propanoic acid; antioxidant for HDPE, PP, PA, POM. [Hoechst Celanese] *

Hostanox® OSP 1. Phenol, 4,4´-thiobis 2-(1,1-dimethyl-ethyl) phosphite; antioxidant for HDPE, PP, PMMA, PA. [Hoechst Celanese] *

Hostanox® PAR 24. CAS 31570-04-4; Tris-(2,4-di-t-butylphenyl) phosphite; antioxidant for LLDPE, HDPE, PP. [Hoechst Celanese] *

Hostanox® SE 10. Dioctadecyl disulfide; antioxidant for HDPE, PP, PMMA, PA. [Hoechst Celanese] *

Hostapal 2345. Nonylphenol polyglycol ether carboxylate; surfactant for agric. formulations. [Hoechst Celanese/Colorants & Surf.]

Hostapal N-040. CAS 9016-45-9; Nonoxynol-4; detergent, wetting agent, emulsifier for general industrial applications, agric., textiles. [Hoechst Celanese/Colorants & Surf.]

Hostaperm. Pigments powders for prints and inks. [Hoechst UK]

Hostaphan. Polyethylene terephthalate films. [Hoechst UK]

Hostaphane. A proprietary trade name for polythylene terephthalate film. [Hoechst UK] *

Hostaphat. Phosphoric acid ester emulsifier. [Hoechst UK]

Hostaphat 2122. 2-Ethylhexyl phosphate; surfactant for textile processing. [Hoechst Celanese/Colorants & Surf.]

Hostaphat AR K. Nonylphenol ethoxylate phosphate ester, potassium salt; surfactant for agric. formulations. [Hoechst Celanese/Colorants & Surf.]

Hostaphat HI. Hexyl phosphate; surfactant for textile processing. [Hoechst Celanese/Colorants & Surf.]

Hostaphat KL 340N. Trilaureth-4 phosphate; o/w emulsifier for cosmetic purposes. [Hoechst Celanese/Colorants & Surf.; Hoechst AG]

Hostaphat KO 300. Trioleyl phosphate; emulsifier for cosmetic w/o emulsions. [Hoechst Celanese/Colorants & Surf.; Hoechst AG]

Hostaphat KO 380. Trioleth-8 phosphate; emulsifier for o/w emulsions for cosmetics industry. [Hoechst Celanese/Colorants & Surf.; Hoechst AG]

Hostaphat KW 340 N. Triceteareth-4 phosphate; emulsifier for creamy cosmetic emulsions based on hydrocarbons, fatty alcohols and fatty acids. [Hoechst Celanese/Colorants & Surf.; Hoechst AG]

Hostapon CAS. Mixture of sulfation products of fatty alcohol derivatives and fatty acid condensation products, supplied as a clear yellowish liquid; optimal foaming agent with very good skin compatability; used in cosmetics, e.g. cleansing agents for intimate hygiene, baby shampoos and foam baths. [Hoechst AG] *

Hostapon CT Paste. Sodium methyl cocoyl taurate; detergent used in shampoos; good foaming, skin compat. [Hoechst Celanese/Colorants & Surf.; Hoechst AG]

Hostapon IDC. Sodium N-palmityl N-cyclohexyl taurine; surfactant for textile processing. [Hoechst Celanese/Colorants & Surf.]

Hostapon KA Powd. Sodium cocoyl isethionate; detergent base for cosmetic industry, detergent bars; foamer, dispersant. [Hoechst Celanese/Colorants & Surf.; Hoechst AG]

Hostapon KTW New. CAS 70609-66-4; Sodium lauroyl taurate; detergent base for cosmetics, toothpastes.

---

* Trade name not verified as available   † Trade name verified as obsolete

[Hoechst Celanese/Colorants & Surf.; Hoechst AG]

Hostapon SO. CAS 137-20-2; EINECS 205-285-7; Sodium methyl oleoyl tauride; detergent, foamer, dispersant for cosmetic industry. [Hoechst Celanese/Colorants & Surf.]

Hostapon STT Paste. CAS 149-39-3; Sodium methyl stearoyl taurate; detergent for high quality cream shampoos. [Hoechst Celanese/Colorants & Surf.; Hoechst AG]

Hostapon T Powd. CAS 137-20-2; EINECS 205-285-7; Sodium methyl oleoyl taurate; cleaner for textile, leather, household prods., agric., cosmetics, etc. [Hoechst Celanese/Colorants & Surf.; Hoechst AG]

Hostapon TF. Anionic surfactant in which the cation is sodium and the anion is composed of unsaturated fatty acids condensed with methyl taurine; supplied as a clear, yellowish viscous liquid; foaming, washing and lime soap dispersing agent used in clear liquid and cream shampoos; foam baths; toothpastes. [Hoechst UK] *

Hostapor. Polystyrene foaming grades. [Hoechst UK] *

Hostapren. Chlorinated polyethylene. [Hoechst AG]

Hostaprime®HC 5. Furan dione propene polymer; coupling agent for PP, PA. [Hoechst Celanese] *

Hostaprint. Pigment preparations for plastics printing. [Hoechst UK]

Hostapur DOS Hi Conc. Dialkyl sodium sulfosuccinate; surfactant for textile processing. [Hoechst Celanese/Colorants & Surf.]

Hostapur DTC. Isotridecyl alcohol polyglycol ether carboxylate; surfactant for agric. formulations. [Hoechst Celanese/Colorants & Surf.]

Hostapur OS. Sodium olefin sulfonate with an alkylene radical (C15-C18) as a liquid or free-flowing powder; used for powdered detergents of all kinds, light duty detergents and cleaning agents, textile and leather auxiliaries, and upholstery and carpets. [Hoechst UK] *

Hostapur SAS. Series of three anionic surfactants of the alkane sulfonate type, where the chain length is C13-C18 and the cation is sodium; liquid, paste or flake form; wetting and foaming agent used in detergents, washing-up liquids, cleaning agents of all kinds, textiles and leather auxiliaries. [Hoechst UK] *

Hostaquick. CAS 23560-59-0; Emulsifiable concentrate containing 550 g heptenophos per liter; used for control of pests in greenhouses and a wide range of vegetable crops. [Hoechst UK]

Hostarex Grades. Sec. and tert. amines; anion/cation exchangers. [Hoechst AG]

Hostastat®. Polymer antistatic agents. [Hoechst AG]

Hostastat® FA 14. Ethoxylated alkylamine, coco deriv; antistat for polyolefins, polystyrene, ABS. [Hoechst Celanese] *

Hostastat® System E1956. Laurylamide in LDPE carrier; antistat conc. for PE, PP. [Hoechst Celanese] *

Hostatec. Polyether ketone. [Hoechst AG]

Hostathion. CAS 24017-47-8; Emulsifiable concentrate containing 420 g/l triazophos; an organophosphorus insecticide. [Hoechst UK]

Hostatint. Pigment preparations for tinting paint. [Hoechst UK]

Hostatron®. Foaming agents for polymers. [Hoechst UK]

Hostatron®System P1941. Carbonic acid-based endothermic blowing agents in proprietary carriers; blowing agent for PS, ABS, PE, PP. [Hoechst Celanese] *

Hostavin®. UV stabilizers for polymers. [Hoechst UK]

Hostavin® ARO 8. CAS 1843-05-6; EINECS 217-421-2; Benzophenone-12; uv absorber for plastics, esp. LDPE, HDPE, PP, polyisobutylene, cellulosics, PC, EVA copolymers, plasticized PVC. [Hoechst Celanese] *

Hostavin®N 30. Oligomeric HALS; uv stabilizer for PP, PE, elastomers, POM, PU. [Hoechst Celanese] *

Hostawet TDC. Ether carboxylate; emulsifier, wetting agent for metal cleaning formulations. [Hoechst Celanese/Colorants & Surf.]

Hostiren. Polystyrene resins. [Hoechst UK] *

Hotbac®. Polymer. [DuPont UK]

Hot Melt Wetness Indicator®. Dry adhesive particles for manufacturing use; to be melted and applied to textile fabrics. [Reichhold]

Hotspur. Herbicide for broad leafed weeds in cereals. [ICI Chem. & Polymers Ltd]

Hotspur. Mixture of clopyralid, fluroxypyr and ioxynil; used to control broad leaf weeds in cereals. [Farm Protection Ltd]

houillite. Anthracite.

House Plant Leaf Shine. Aerosol containing leafshine material; house plant spray. [Vitax Ltd]

House Plant Liquid Feed. Liquid concentrate containing NPK 5:2:2; house plant feed. [Vitax Ltd]

Houseplant Long Lasting Feed. Coated fertilizer. [Fisons plc, Horticulture Div] *

House Plant Pest Killer. Aerosol containing pyrethrum and resmethrin; house plant insecticide. [Vitax Ltd]

Howard's silver. Mercury fulminate; used for percussion caps.

Howflex. Plasticizer. [Laporte Industries Ltd]

Howsorb. Sorbitol syrup aqueous solution. [Laporte Industries Ltd] *

Howstik. Rubber/resin solutions in various solvents; contact adhesives for insulation, building, footwear, foam adhesives. [Howlett Adhesives Ltd]

Howtex. Ready-mixed water-based ceramic tile adhesives; for ceramic tiling, building, woodworking, laminating. [Howlett Adhesives Ltd]

Howtol. Cyclic alcohols. [Laporte Industries Ltd] *

Hoyle's metals. Bearing metals usually containing about 46% tin, 12% antimony, and 42% lead.

Hoyt Metal. A proprietary trade name for an antimonial lead containing 6-10% antimony. §

H P Acthar Gel. Corticotropin repository; hormone glucocorticoid; diagnostic aid. [Armour Pharmaceutical Co] *

HPC-9. CAS 80-15-9, 1338-23-4; Cumene hydroperoxide/MEK peroxide sol'n.; facilitates R.T. curing of vinyl ester of other unsat. polyester resins. [Witco/Argus]

HRF Ayerst Injection. Gonadorelin hydrochloride, a leutinizing hormone, releasing hormone; diagnostic use for evaluating the functional capacity and response of the gonadotropes of the anterior pituitary. [Wyeth Laboratories; Monmouth Pharmaceuticals Ltd]

HSB 1900. High styrene resin/SBR fluxed blend (100:100); used in shoe soles and household goods; gives easy processing, high durometer compounds; vulcanized with std. curing systems. [Mach-1 Compounding]

H-scale. A proprietary trade name for a synthetic pearl essence. §

HSZ-320NAA. Crystalline aluminosilicate; zeolite for catalyst and adsorbent applications. [Tosoh]

H.T. CAS 7758-23-8; Monocalcium phosphate monohy-

---

‡ Trade name and manufacturer not verified   § Trade name without identified manufacturer

drate; leavening agent. [Monsanto Co] *

HTH. CAS 7778-54-3; Calcium hypochlorite; dry chlorinating agent. [Olin]

HT Non-ionic Wetter. Spreader containing 90% alkylphenol ethylene oxide condensate; wetting agent for use in horticultural sprays. [Service Chemicals Ltd]

HT-Proteolytic Conc. Bacterial protease; enzyme for hydrolysis of proteins over neutral and alkaline pH range; for baking, proteins. [Solvay Enzymes]

H.T.S. A salt mixture used as a heat transfer medium. It contains approximately 40% sodium nitrite, 7% sodium nitrate, and 50% potassium nitrate by weight. The temperature limits are 290-1000 F.

HTSA #1. CAS 16260-09-6; Oleyl palmitamide; release agent providing slip, antiblocking to thermoplastics incl. PP film, nylon. [Hexcel]

HTSA #3. CAS 10094-45-8; Stearyl erucamide; release agent providing slip, antiblocking to thermoplastics incl. PP film, nylon. [Hexcel]

Huber 40C. CAS 1332-58-7; Kaolin clay; functional filler with med. brightness and particle size, moderate oil and water demand; for paint and plastics applications. [J.M. Huber]

Huber 65A. CAS 1332-58-7; Kaolin clay; air floated functional filler with med. brightness, med. particle size; for adhesives, paints, plastics. [J.M. Huber]

Huber 95. CAS 1332-58-7; Kaolin clay; water washed functional filler with med. brightness, fine particle size; for adhesives, paints, plastics, and inks. [J.M. Huber]

Huberbrite 1. CAS 7727-43-7; Barytes; pigment grade with high assay, high brightness, fine whiteness. [J.M. Huber]

Hubercarb® Q 6-20. CAS 471-34-1; Calcium carbonate; filler/extender for plastics, caulks/sealants, rubber, adhesives, glass, ceramics, paper, cleansers, paints/coatings, pesticides, asphalt, drilling mud, rice polishing, environmental cleanup. [J.M. Huber]

Hubercarb® W 2. CAS 471-34-1; Calcium carbonate; filler and extender with low moisture pickup; ideal for moisture-sensitive applications such as SMC, BMC, TMC, polyethylene film and single-component urethane caulks and for paints and coatings, rubber, adhesives, ceramics, paper, nonabrasive cleaners. [J.M. Huber]

Huber's reagent. A solution of ammonium molybdate and potassium ferrocyanide; used for the detection of free mineral acids.

Hubel's reagent. a) Iodine (50 g), dissolved in 1 liter of 95% alcohol. b) Mercuric chloride (60 g), dissolved in 1 liter of alcohol; used for obtaining the iodine value of fats and oils.

Hudroson. Metal cleaning and pretreatment systems. [Nickerson Chemicals Ltd]

Hugel A. A proprietary solution of the sodium salt of an acrylic copolymer; used for thickening natural and synthetic rubber latices. [Hubron Rubber Chemicals] *

Hugel AH. A proprietary viscous solution of an acrylic copolymer. [Hubron Rubber Chemicals] *

Hugel B. A proprietary trade name for emulsions of acrylic copolymer containing free carboxyl groups. [Hubron Rubber Chemicals] *

Hugel BC 10. A proprietary viscous solution, colorless and odorless, of the sodium salt of an acrylic polymer. [Hubron Rubber Chemicals] *

Hugel CH14. The sodium salt of a proprietary acrylic polymer; very viscous; takes the form of a short nonstringy yellow gel. [Hubron Rubber Chemicals] *

Hulot's solder. An alloy of 37.5% tin, 37.5% lead, and 25% zinc amalgam; used for soldering aluminum bronze.

Humagel. A proprietary preparation of paromycin sulfate, kaolin and pectin; an antidiarrheal. [Parke-Davis] *

Human albumin®. Serum fraction. [Bayer AG]

Humatin. CAS 1263-89-4; A proprietary preparation containing paromomycin (as the sulfate); an antibiotic. [Parke-Davis]

Humatrope. CAS 9002-72-6; Somatropin; growth regulator. [Eli Lilly & Co]

Humber. Granular or powder organic based fertilizers comprising organic base and chemical N, P and K to form the analysis; a steady release, lower nitrogen fertilizer for agricultural crops and grasslands. [Humber Fertilizers plc]

Humectant SD-35. CAS 667-83-4; Panthenyl ethyl ether; humectant. [Presperse]

Humectol®. Wetting, dispersion and leveling agents. [Cassella AG]

Humifen. Sulfonated aliphatic polyesters; Wetting agent for dyeing; yarn textile finishing; dry cleaning detergents; glass cleaners; wallpaper removers; battery separators. [GAF Great Britain] *

Humifen BA-77. Anionic surfactant in powder form; wetting, dispersing and antistatic agent for paints, printing inks, latex stabilization, leather, textile processing and agricultural chemistry. [GAF Great Britain] *

Humifen BX-78. Anionic surfactant in powder form; wetting, penetrating, dispersing and emulsification agent; used for cotton, rayon, dyestuffs, leather, agricultural chemicals, insecticides, paper, rubber latex and polymerization. [ISP] *

Humorsol. CAS 56-94-0; Demecarium bromide; cholinergic. [Merck & Co Inc] *

Humulin. Human insulin; a protein that has the normal structure of the natural antidiabetic principle produced by the human pancreas; antidiabetic. [Eli Lilly & Co]

Hünefeld solution. This contains 25 cc alcohol, 5 cc chloroform, 1.5 cc glacial acetic acid, and 15 cc turpentine; used for the detection of blood.

Huntsman 201. CAS 9003-53-6; Crystal PS. [Huntsman]

Huntsman 240. CAS 9003-53-6; HIPS, rubber-modified; shock resistance at ambient and low temps., gloss chars., good thermal properties; used for refrigerator and appliance parts. [Huntsman]

Huntsman 312. CAS 9003-53-6; Med. impact PS. [Huntsman]

Huntsman 331. CAS 9003-53-6; HIPS. [Huntsman]

Huntsman 351. CAS 9003-53-6; HIPS, ignition-retardant. [Huntsman]

Huntsman 474. CAS 9003-53-6; PS, rubber-modified; high-heat and impact polymer, high thermal properties, shock resistance at ambient and low temps., moldability; for small appliances, consoles, automotive parts (interior parts). [Huntsman]

Huntsman 765. CAS 9003-53-6; HIPS; color-stable grade. [Huntsman]

Huppert's reagent. A 10% aqueous solution of calcium chloride used for the detection of biliary pigments in urine.

Huron. An aluminum alloy containing from 3.5-6.6% copper and small amounts of manganese, magnesium, and chromium. The name is also used for a chromium steel containing 12.5% chromium, 1.0% vanadium, and 0.2%

---

* Trade name not verified as available   † Trade name verified as obsolete

carbon. It is used for dies.

Hurr nut. Myrobalans.

Husman metal. An alloy of 74% tin, 11% antimony, 10.6% lead, 4% copper, 0.22% iron, and 0.18% zinc.

Hu-Tet. Tetanus immune globulin; immunizing agent. [Hyland Therapeutics] *

HVA-2. CAS 3006-93-7; N,N´-m-Phenylenedimaleimide; a free radical regulator used as an auxiliary in the curing of Hypalon. [DuPont UK] *

HVP. CAS 977059-33-8; Hydrolyzed vegetable proteins; natural flavors for all types of processed foods to improve or impart a characteristic flavor. [Hercules] *

Hyalase. Hyaluronidase spreading agent. [Fisons plc, Pharmaceuticals Div] *

hyaluronic acid. (sodium hyaluronate) Natural mucopolysaccharide formed by bonding N-acetyl-D-glucosamine with glucuronic acid. $(C_{14}H_{21}NO_{11})N$; CAS 9004-61-9; EINECS 232-678-0; Moisturizer for skin care products. [Am. Biorganics; Croda; Solabia; Worthington Biochemical]

Hyamine® 10X. CAS 25155-18-4; Methylbenzethonium chloride; germicide, disinfectant, sanitizer in restaurant and pharmaceutical uses. [Lonza Inc]

Hyamine® 1622 50%. CAS 121-54-0; EINECS 204-479-9; Benzethonium chloride; germicide, disinfectant, sanitizer in restaurant and pharmaceutical uses; antistat, bacteriostat on fabrics; preservative for starch, glue, casein; cocatalyst for curing polyesters. [Lonza Inc]

Hyamine® 1622 Crystals. CAS 121-54-0; EINECS 204-479-9; Benzethonium chloride; bactericide, deodorant, preservative for veterinary and pharmaceutical prods. [Lonza Inc]

Hybaite. CAS 7775-14-6; Sodium hydrosulfite; sodium dithionite/alkali/buffers/clorites; for bleaching of paper pulps. [RV Chemicals Ltd]

Hyban. Soluble concentrate of 18.7 g dicamba and 300 g mecoprop per liter; used for weed control in cereals and grassland. [Agrichem (International) Ltd]

Hyb-lum. An alloy of aluminum in which the alloying elements, consisting of about 2%, are mainly nickel and metals of the chromium group; used for reflectors of therapeutic lamps.

Hybon® 2011. Continuous filament single-end glass strand rovings with proprietary sizing; reinforcement for polyester and vinyl ester resin systems in pultrusion applications. [PPG/Fiber Glass]

Hybri-Chem 100. Polyester/polyurethane hybrid system; for open mold casting or lamination; low exotherm; high heat distort. temp. [Hybri-Chem] *

Hy-Brite. Metal cleaner. [FMC]

Hybrite®. CAS 7775-14-6; Sodium hydrosulfite; sodium dithionite/alkali/buffers/chelates; used for bleaching of paper pulps. [RV Chemicals Ltd]

Hycal. A proprietary preparation of dextrose and related compounds; used as a high-calorie diet supplement. [SmithKline Beecham] *

Hycar® 1203X17. A proprietary 70 : 30 blend of Hycar medium-high acrylonitrile rubber and Geon PVC resin. [BFGoodrich] *

Hycar® 1204X5. A proprietary 100 : 70 : 120 pre-fluxed blend of Hycar medium-high acrylonitrile rubber, Geon PVC resin and a phthalate plasticizer. [BFGoodrich] *

Hycar® 1204X9. A proprietary 100 : 70 : 100 pre-fluxed blend of Hycar medium-high acrylonitrile rubber, Geon PVC

resin and a phthalate plasticizer, protected by nonstaining stabilizers. [BFGoodrich] *

Hycar® 1205X3. A proprietary 50 : 50 : 60 pre-fluxed blend of Hycar medium-high acrylonitrile rubber, Geon PVC resin and a phthalate plasticizer. [BFGoodrich] *

Hycar® 1273. A proprietary 70 : 30 blend of carboxy-modified Hycar acrylonitrile rubber and Geon PVC resin, possessing good resistance to abrasion. [BFGoodrich]*

Hycar® 1300X8. Reactive liq. polymers. [BFGoodrich/Spec. Polymers] *

Hycar® 1402 H82. A proprietary acryl-onitrile-butadiene copolymer in powder form, having a medium-high content of acrylonitrile. [BFGoodrich UK] *

Hycar® 1402 H83. A proprietary acrylonitrile-butadiene copolymer in powder form, having a high content of acrylonitrile. [BFGoodrich UK] *

Hycar® 1403 H84. A proprietary acrylonitrile-butadiene copolymer having a medium acrylonitrile content; supplied in powdered form and used when good behavior at low temperatures is required. [BFGoodrich UK] *

Hycar® 1552. Nitrile latex; used for adhesives, carpet backcoating, paper coatings and saturation, pigment binding, textile coatings; exhibits fast penetration. [BFGoodrich/Spec. Polymers] *

Hycar® 1577. Butadiene-acrylonitrile-styrene terpolymer latex; used for waterproof and greaseproof coatings, leather finishes, paints, paper coatings and saturation. [BFGoodrich/Spec. Polymers] *

Hycar® 2100. A proprietary range of polyacrylic-solution polymers used as pressure-sensitive adhesives. [BFGoodrich UK] *

Hycar® 2550H33. A proprietary reinforced styrene-butadiene copolymer rubber used in the manufacture of foam rubber. [BFGoodrich UK] *

Hycar® 2550H5. A proprietary aqueous, anionic dispersion of a cold-polymerized styrene-butadiene copolymer. [BFGoodrich UK] *

Hycar® 2550H55. A proprietary aqueous, anionic dispersion of a styrene-butadiene copolymer used in the foambacking of carpets. [BFGoodrich UK] *

Hycar® 2570H28 and 2570H29. A proprietary group of aqueous anionic dispersions of self-reactive styrene-butadiene copolymers; used in the making of carpet backings. [BFGoodrich UK] *

Hycar® 2570X5. A proprietary aqueous, anionic dispersion of a carboxy-modified styrene-butadiene copolymer reactive to heat; used for leather finishes and adhesives. [BFGoodrich UK] *

Hycar® 2671H49. A proprietary aqueous anionic dispersion of a heat-reactive carboxy-modified acrylic polymer used in the making of surgical rubber materials. [BFGoodrich UK] *

Hycar® 4021. A proprietary copolymer of ethyl acrylate having a small percentage of 2-chloro-ethyl vinyl ether. [BFGoodrich] *

Hycar® 4032. A proprietary polyacrylic rubber used to make rubber seals and gaskets. [BFGoodrich UK] *

Hycar® 4043. A proprietary acrylic rubber which remains flexible at -40 C, but which also gives good resistance to oil at high temperatures. [BFGoodrich] *

Hycar® 4201. A proprietary copolymer of ethyl acrylate having a small percentage of 2-chlorovinyl ether. [BFGoodrich] *

Hycar® 26345. Acrylic latex, heat-reactive; used for uphol-

---

‡ Trade name and manufacturer not verified   § Trade name without identified manufacturer

stery, drapery, mattress ticking coatings; foamable; durable. [BFGoodrich/Spec. Polymers] *

Hycar®ATBN, CTB, CTBN, VTBN. Reactive liq. polymer; liq. polymer which cures at room or elevated temps. to solid elastomer; improves crack resistance of epoxy, polyester systems, enhances fatigue performance of FRP; good elec. and wetting properties; used as modifier for epoxy structural adhesives; improves impact resistance of epoxy coatings; used in potting, encapsulation, cable fillers, moisture blocks, castable elastomers for roto-molding. [BFGoodrich/Spec. Polymers] *

Hycar® Reactive Liquid Polymer (RLP). RLP are homopolymers of butadiene or copolymers of butadiene/acrylonitrile; reactive groups are in both terminal positions of the polymer chain and, optionally, may have additional reactive groups pendent on the chain; three types (carboxyl, acrylated vinyl and secondary amine) available commercially; CT polymers in epoxy structural adhesives, encapsulants, coatings, composites, potting/encapsulation compounds, solid propellant binder; AT in epoxy adhesives, maintenance coatings encapsulation and geographical cable fillers, civil engineering/construction; VT in polyester BMC, SMC, anarobic adhesives, radiation curing compositions, fiberglass-reinforced plastic polyester composition [BFGoodrich UK] *

Hycathane. A proprietary polyurethane elastomer. [FPT Industries] *

Hycel M Series. Isocyanate/resin system; for rigid PU foam for thermal insulation, structural material, packing materials, etc. [Toho Chem. Industry]

Hycol. A coal tar black disinfectant composed of 30% coaltar acids, hydrocarbon oils and vegetable oil soap. [William Pearson Ltd] *

Hycolin. A general purpose hospital disinfectant containing 16% substituted phenolic germicides in a detergent base. [William Pearson Ltd] *

Hycon. Photographic developer. [May & Baker Ltd] *

Hycote. Spray paints; for matched colors and touch up paints. [James Briggs & Sons Ltd] *

Hydagen® B. α-Bisabolol; antiphlogistic active agent for emulsions, oils, lotions, and oral hygiene preps. [Henkel/Cospha; Henkel Canada]

Hydagen® C.A.T. CAS 77-93-0; Triethyl citrate; nonmicrobiocidal deodorant active agent. [Henkel/Cospha; Henkel Canada]

Hydagen®DEO. Triethyl citrate and BHT; nonmicrobiocidal act. ingred. for deodorant systems and personal care prods. [Henkel/Cospha; Henkel Canada]

Hydan. CAS 118-52-5; 1,3-Dichloro-5,5-dimethylhydantoin. [Rhone-Poulenc UK]

Hydan®. For the agriculture industry. [DuPont UK]

Hydantil. A proprietary preparation of methoin and phenobarbitone; an anticonvulsant. [Sandoz] *

Hydecat. Range of catalysts and absorbents used for purification. [ICI Chem. & Polymers Ltd]

Hydelta. CAS 50-24-8; Prednisolone; glucocorticoid. [Merck & Co Inc] *

Hydelta-TBA. CAS 7681-14-3; Prednisolone tebutate; glucocorticoid. [Merck & Co Inc] *

Hydeltrasol. CAS 125-02-0; Prednisolone sodium phosphate; for the treatment of certain endocrine and nonendocrine disorders responsive to corticosteroid therapy. (U.S.) [Merck & Co Inc]

Hydergine. A proprietary preparation of mesylates of

dihydroergocornine, dihydroergocristine and dihydroergokryptine; peripheral vasodilator. [Sandoz]

Hydex® 100 Gran. 206. CAS 50-70-4; Sorbitol; humectant, bodying agent, moisture control agent for toothpaste, cosmetics, sugarless confections, controlled moisture foods, industrial applications. [Lonza Inc; Lonza AG]

Hydon. Bromacil + picloram; used for total weed control in non crop areas. [Chipman Ltd]

Hydraffin. CAS 7440-44-0; Granulated carbon; used for treatment of drinking, process, and waste water. [Bayer AG]

Hydra-guard. Mixture of gamma-HCH and thiram; seed dressing for brassica crops. [Agrichem (International) Ltd]

Hydral® 710. Hydrated alumina; fire retardant, smoke suppressant improving arc track resist. in thermosets, thermoplastics, elastomerics; in fillers and coating pigments in fine printing papers for increased brightness; for vinyl compounding; as reinforcing pigment in adhesives; fine mild abrasive in waxes and polishes; filler for cosmetic powds., lotions; polishing agent in dentifrices. [Alcoa]

Hydralin. CAS 108-93-0; Methylcyclohexane.

Hydramyl. Pentane, C_5H_{12}.

Hydrangea Colourant. CAS 10043-01-3, 10045-89-3; Soluble powder containing aluminum sulfate hexahydrate and ferrous ammonium sulfate hexahydrate; hydrangea colorant. [Vitax Ltd]

Hydraphthal. A preparation containing 90% tetralin, 5% ammonium oleate, and 3% water. A wetting-out and scouring agent.

hydrargyrum. See mercury.

Hydrasal®. Zirconium, magnesium, rare-earth and other chemicals, powders and alloys; for use in industry, including metalworking. [Magnesium Elektron Ltd]

hydrated lime. See calcium hydroxide.

Hydraulan®. Brake fluids. [BASF AG; BASF plc]

hydraulic bronze. An alloy of 83% copper, 5% lead, 5% zinc, 5% tin, and 2% nickel. Another alloy contains 83% copper, 10.8% tin, 6% zinc, and 0.1% lead.

hydraulic cements or mortars. These are prepared by calcining mixtures of calcium carbonate with from 10-30% clay. Tricalcium silicate, $3CaO \cdot SiO_2$, and tricalcium aluminate, $3CaO \cdot Al_2O_3$, are formed.

hydraulic limes. Limes containing from 15-30% clayey matter (aluminum silicate). They are made by burning impure limestones at a low temperature. They slake in water, but show hydraulic properties.

hydraulic mortar. See hydraulic cements.

hydrazine. (hydrazine base; hydrazine anhydrous) H_2NNH_2; CAS 302-01-2; reducing agent, corrosion inhibitor, waste water treatment, electrolytic plating, retox reactions, polymerization catalyst; organic hydrazine derivatives.; rocket fuel. [Aldrich; Atochem SA; Fairmount; Miles; Olin; Spectrum Chem. Mfg.]

hydrazine anhydrous. See hydrazine.

hydrazine base. See hydrazine.

hydrazine sulfate. (diamine sulfate; diamidogen sulfate) $NH_2NH_2 \cdot H_2SO_4$; CAS 10034-93-2; EINECS 233-110-4; Chemical intermediate, condensation reactions, catalyst for making acetate fibers; analysis of minerals, slags, fluxes; determination of arsenic in metals; separation of polonium from tellurium; fungicide, germicide. [Fairmount; Janssen Chimica; Mallinckrodt; Otsuka Chem.;

---

* Trade name not verified as available   † Trade name verified as obsolete

Spectrum Chem. Mfg.]

Hydrea. A proprietary preparation of hydroxyurea, a carcino-chemotherapeutic agent. [Bristol-Myers Squibb Pharmaceuticals Ltd]

Hydrea Capsules. CAS 50-23-7; Hydrocortisone preparations. [The Boots Co plc]

Hydrenol D. Tallow alcohol (25-35% C16, 60-67% C18); wetting agent, emulsifier, emollient, consistency giving agent for skin creams and lotions; intermediate for surfactant mfg. [Henkel/Cospha; Henkel KGaA]

Hydrenox-M. CAS 135-09-1; A proprietary preparation of hydroflumethiazide; a diuretic. [The Boots Co plc]

Hydresol®. Condensates of naphthalene sulfonic or phenol-sulfonic acid and formaldehyde; superplasticizers for mineral binders, for improving workability of mineral-bonded mortar and concrete. [BASF AG]

Hydrex®. Polyester resins for coatings, casting and laminate application. [Reichhold]

Hydrholac. Nitrocellulose lacquer, used for leather finishing. [Rohm & Haas UK]

Hydrin® C. Epichlorohydrin/ethylene oxide copolymer; elastomer used for fuel pump diaphragms, pipe gaskets, hose for fuel, oil, and gas, vibration isolators, motor mounts, rolls, adhesives, sponge goods, boots, seals, o-rings, bladders; excellent resist. to ozone, oils, and fuels, and low temp. flexibility; vulcanized with ETU, TETA, HMDAC, Thiate E. [Zeon]

Hydrin® C-CG. Epichlorohydrin/ethylene oxide copolymer; cement grade of Hydrin C. [Zeon]

Hydrin® T. Ethylene oxide/epichlorohydrin/allyl glycidyl ether terpolymer; elastomer used for fuel pump diaphragms, pipe gaskets, hose for fuel, oil, and gas, vibration isolators, motor mounts, rolls, adhesives, sponge goods, boots, seals, o-rings, bladders; blendable with SBR and NBR; to improve low temp. performance, fuel, oil, and ozone resist.; vulcanized with sulfur, peroxide, or ETU. [Zeon]

β-hydrindone. See 2-indanone.

Hydrine. CAS 9004-99-3; 106-11-6; PEG-2 stearate; consistency stabilizer for ointments, cream lotions; opacifier in shampoos, liq. soaps. [Gattefosse; Gattefosse SA]

hydriodic acid. (hydroiodic acid) HI; CAS 10034-85-2; EINECS 233-109-9; Preparation of iodine salts, organic preparations, analytical reagent, disinfectant, pharmaceuticals. [Janssen Chimica]

Hydro. Fertilizers, chemicals, gases and plastics. [Norsk Hydro AS]

Hydro AWC. CAS 149-44-0; Sodium formaldehyde sulfoxylate; stripping agent for removing acid dyes from wool and nylon. [Henkel/Textiles]

Hydroace Series. Emulsifier for emulsion type fuel oils. [Toho Chem. Industry]

Hydroba. Jojoba oil derivatives, cosmetics, and toiletries additive. [A & E Connock (Perfumery & Cosmetics) Ltd]

Hydroblok. Cross-linked modified polyacrylamide. [Cyanamid BV]

Hydrobol. Water repellent finishes. [Ciba plc] *

hydrobromic acid. (hydrogen bromide in acetic acid) HBr · CH$_3$COOH; CAS 10035-10-6; EINECS 233-113-0; Analytical chemistry, solvent for ore minerals, manufacture of inorganic and some alkyl bromides, alkylation catalyst. [Allchem Industries; Associated Octel Co Ltd; EM Industries; Ethyl; Great Lakes]

hydrobuna. A name applied to hydrogenated rubber.

hydrocarbon-aldehyde resins. Resins obtained by the interaction of hydrocarbons such as naphthalene with formaldehyde in the presence of sulfuric acid.

hydrocarbon black. See gas black.

hydrocarbon cement. A cement made by mixing heated pitch or tar with from one to four times its volume of calcium or magnesium sulfate; used for paving or building purposes.

hydrocarbon gas black. See gas black.

hydrocarbon oil. See paraffin, liquid.

Hydrocell YP-30. Hydrolyzed yeast protein; cosmetic ingredient for skin and hair care prods. [Brooks Industries]

Hydrocerol®BIH. Blowing agent for foaming of thermoplastics in injection molding and extrusion; high gas yield at lower temps. [Boehringer Ingelheim; Henley]

Hydrocerol® CF 70. Blowing and nucleating agents for plastics; for injection molding and extrusion for fine-celled foam. [Boehringer Ingelheim; Henley]

Hydrocerol Compound. Blowing agent. [Henley Chemicals Inc]

Hydrocerol®LC. Citric acid esters; blowing agent for foaming of thermoplastics at higher processing temps.; esp. suitable for PC. [Boehringer Ingelheim; Henley]

Hydrocerol® TAF 50. Nucleating agent for extrusion processes for direct gassing PS and polyolefins. [Boehringer Ingelheim; Henley]

Hydro-Chem. Driers for water based coatings. [Mooney Chemicals Inc]

hydrochloric acid. HCl; CAS 7647-01-0; EINECS 231-595-7; Acidizing of petroleum wells, boiler scale removal, chemical intermediate, ore reduction, food processing, pickling and metal cleaning, industrial acidizing, general cleaning, alcohol denaturant, laboratory reagent. [Allied-Signal; Asahi Chem Industry Co Ltd; Asahi Denka Kogyo; Atochem; Dover; Du Pont; ICI SA; Miles; Nissan Chem. Ind.; OxyChem; PPG Industries; Rasa Ind.; Showa Denko; Vista; Witco/Argus]

hydrocinnamic acid. (3-phenylpropionic acid) C$_6$H$_5$-CH$_2$CH$_2$COOH; CAS 501-52-0; EINECS 207-924-5; Fixative for perfumes, flavoring. [R.W. Greeff; Janssen Chimica; Penta Mfg.]

Hydrocol®. Aqueous dispersion of iron oxides, ochers and composite pigments; used for coloration of cementitious products (mortars, blocks, bricks, slabs, panels, reconstructed stone, split blocks, etc.). [W Hawley & Son Ltd]

Hydrocol®. Drainage aids for pulp, paper and board manufacture. [Allied Colloids Ltd]

Hydrocoll AL-50, AL-55, EN-40, EN-55, EN-55-X, EN-SD, EN-SD-1M, EN-SD-10M. CAS 9015-54-7; Hydrolyzed animal protein; cosmetics ingred. [Brooks Industries]

Hydrocoll G-40. CAS 9000-70-8; Gelatin; cosmetics ingred. [Brooks Industries]

Hydrocoll G-55. Hydrolyzed gelatin; cosmetic ingred. [Brooks Industries]

Hydrocortone. CAS 50-23-7; Hydrocortisone; for the treatment of certain endocrine and nonendocrine disorders responsive to corticosteroid therapy. [Merck & Co Inc]

Hydro-Cure. Drying catalyst for water based paints. [Mooney Chemicals Inc]

Hydrodarco. CAS 7440-44-0; Activated carbon; used for water purification, purification of chemicals/pharmaceuticals, air purification and solvent recovery. [Am. Norit] *

Hydroderm. Hydrocortisone, neomycin and bacitracin; topical corticosteroid with topical antibiotics for the treatment

---

‡ Trade name and manufacturer not verified   § Trade name without identified manufacturer

of skin conditions. [Merck & Co Inc] †

Hydrodiuril. CAS 58-93-5; Hydrochlorothiazide; for the relief of edema and hypertension. [Merck & Co Inc]

Hydroferrox®. CAS 1309-37-1; Aqueous dispersion of Bayferrox iron oxides; used for coloration of cementitious products (mortars, blocks, bricks, slabs, panels, reconstructed stone, split blocks, etc.). [W Hawley & Son Ltd]

hydrofluoric acid. Hydrogen fluoride in aq. sol'n.; HF; CAS 7664-39-3; EINECS 231-634-8; aluminum production, fluorocarbons, pickling stainless steel, etching glass, acidizing oil wells, gasoline production, processing uranium. [Allied-Signal; Du Pont; Farleyway Chem. Ltd; General Chem.; Hoechst Celanese; Seimi Chem.]

Hydrofol Acid 1655. CAS 57-11-4; EINECS 200-313-4; Stearic acid; pharmaceutic aid. [Sherex/Div of Witco]

hydrogenated castor oil. (opalwax; castorwax; castor oil, hydrogenated) End product of controlled hydrogenation of castor oil; CAS 8001-78-3; in water-repellent coatings, candles, polishes, ointments, cosmetics; impregnant for paper, wood, cloth; as lubricant, mold release in manufacture of formed plastics and rubber goods. [Akzo; Arista; Sothern Clay Prods.]

hydrogenated vegetable oil. (vegetable oil, hydrogenated) End product of controlled hydrogenation of vegetable oil; CAS 68334-28-1; emulsifier for food processing; binder, lubricant in pharmaceutical tableting, pressed powders; cocoa butter replacement in cosmetics and pharmaceuticals; emollient, wax. [Arista; Jojoba Growers & Processors; Karlshamns; Lipo; A.E. Staley Manufacture]

hydrogen bromide in acetic acid. See hydrobromic acid.

hydrogen carboxylic acid. See formic acid.

hydrogenite. A mixture of 5 parts ferro-silicon, 90-95 parts silicon, 12 parts sodium hydroxide, and 4 parts slaked lime. When ignited it yields hydrogen.

hydrogen peroxide. Inorganic oxide; HOOH; CAS 7722-84-1; EINECS 231-765-0; bleaching and deodorizing textiles, wood pulp, hair, fur, etc.; plasticizers; refining and cleaning metals; visc. control for starch and cellulose derivatives. [Atochem N. Am.; Degussa; Du Pont; Farleyway Chem. Ltd; FMC; Mallinckrodt; Mitsubishi Gas]

hydrogen peroxide sodium carbonate adduct. See sodium percarbonate.

hydrogen rubeanide. The amide of dithiooxalic acid, $CS(NH_2)CS(NH_2)$.

hydrogen sulfate. See sulfuric acid.

hydroiodic acid. See hydriodic acid.

Hydrokeratin AL-30. CAS 69430-36-0; Hydrolyzed keratin; cosmetics ingred. [Brooks Industries]

Hydrokote®95. CAS 68334-28-1; Hydrogenated vegetable oil; specialty base used as replacement for cocoa butter in cosmetic and pharmaceutical applications. [Karlshamns]

Hydrol 100. Partially hydrogenated coconut oil; kosher; oil for whipped topping, coffee whitener, vegetable dairy systems, biscuit and crackers. [Van Den Bergh Foods]

Hydrolact. Lactase. [John & E Sturge Ltd, Selby] *

Hydrolactin 2500. Hydrolyzed milk protein; substantive skin and hair care conditioner, moisturizer; for shampoos, conditioner rinses, setting and waving lotions, moisturizing creams and lotions, night creams and lotions. [Croda Inc.]

Hydrolactol 70. Glyceryl stearate, propylene glycol stearate, glyceryl isostearate, propylene glycol isostearate,

oleth-25, ceteth-25; self-emulsifying base for fluid, semi-fluid lotions and o/w creams, mineral pigment formulations. [Gattefosse]

Hydrolete. Calcium hydride, CaH_2.

Hydrolin. CAS 7775-14-6; Sodium hydrosulfite sol'n.; for bleaching of clay, groundwood, and thermomech. pulp. [Olin]

hydrolith. A 90% calcium hydride which yields hydrogen on contact with water.

Hydro-Marc. Etafilcon A; contact lens material. [Vistakon Inc] *

Hydromedin. CAS 58-54-8, CAS 6500-81-8; Ethacrynic acid and sodium ethacrynate; potent diuretic for the relief of certain edemas. (Germany) [Merck & Co Inc]

Hydromet. Methyldopa and hydrochlorothiazide; for the treatment of hypertension. [Merck & Co Inc]

Hydromilk EN-20. Hydrolyzed milk protein; complete food containing all essential amino acids. [Brooks Industries]

Hydromol Cream. Arachis oil, isopropyl myristate, liquid paraffin, sodium pyrrolidone carboxylate, sodium lactate; an emollient cream for treatment of dry skin conditions. [Quinoderm Ltd]

Hydromol Emollient. Light liquid paraffin, isopropyl myristate; an emollient bath additive for dry skin conditions. [Quinoderm Ltd]

Hydromox. CAS 73-49-4; Quinethazone; diuretic. [Lederle Laboratories USA]

Hydromycin-D. A proprietary preparation of prednisolone and neomycin; used in dermatology as an antibacterial agent. [The Boots Co plc]

Hydron®. Sulfur vat dyestuffs; used for textile dyeing and printing. [Cassella AG]

Hydronal. Pure (+)-hydroxy citronellaldehyde. [Bush Boake Allen Ltd]

Hydronalium. An aluminum alloy containing from 7-9% magnesium and small amounts of silicon and manganese. It is resistant to sea-water, soap, and soda.

hydronaphthol. β-naphthol, $C_{10}H_7OH$.

Hydron Blue. A dark-blue vat dye, obtained by reducing nitroso-phenol with carbazole to form an indophenol, which is heated with sodium sulfide, and subsequently with sulfur; also used as a trademark for a range of other dyestuffs. [Cassella AG] †

hydrone. An alloy of 35% sodium and 65% lead, which generates hydrogen by action of water.

Hydronyx. A proprietary trade name for sodium sulfoxylate.§

Hydropalat 535. Amine neutralized polyester; suspending agent for water-reducible paints. [Henkel KgaA]

Hydropalat® A. A registered trade name for diethylhydrophthalate, a solvent. §

Hydropalat® B. A registered name for dibutylhydrophthalate, a solvent. §

Hydro Paste®. Water dispersible aluminum pigment; used for aqueous paints and coatings used for decorative metallic effects as well as a wide range of protective coatings applications. [Silberline Mfg Co Inc]

Hydrophilol ISO. CAS 68171-38-0; Propylene glycol isostearate. [Gattefosse SA]

Hydrophobol. Proofing agent. [Ciba plc] *

Hydropur®. Water reworking and treatment agents, flocculating agents. [Cassella AG]

hydroquinol. See hydroquinone.

hydroquinone. (1,4-benzenediol; p-dihydroxybenzene;

---

hydroquinol) Aromatic organic compound; $C_6H_6O_2$; $C_6H_4(OH)_2$; CAS 123-31-9; EINECS 204-617-8; Photographic developer (not for color film); dye intermediate, inhibitor; stabilizer in paints and varnishes; motor fuels and oils; antioxidant for fats and oils. [Aldrich; Allchem Industries; Eastman; Goodyear Tire & Rubber; Kraeber GmbH]

Hydro-resin A. A proprietary trade name for a water soluble resin. §

Hydros. CAS 7775-14-6; Sodium hydrosulfite; specially prepared for use as a reducing agent in vat color dyeing. [Albright & Wilson Ltd]

Hydros® 1. Sodium hydrosulfite and sodium pyrophosphate. [RV Chemicals Ltd]

Hydros® F. Sodium hydrosulfite; sodium dithionite/sodium metabisulfite; used for textile coloration/mineral bleaching/general bleaching. [RV Chemicals Ltd]

Hydro-Saluric®. CAS 58-93-5; Hydrochlorothiazide; for the relief of edema and hypertension. [Merck & Co Inc]

Hydrosan. Flocculating aid for water treatment. [Goldschmidt] *

Hydrosil®. Organofunctional silane coupling agents; for surface treatment of fillers, substrate primers, integral blend components of thermoplastic and thermoset polymers. [Hüls Am.]

hydrosol. An aqueous colloidal silver solution.

Hydrosol. Solubilized sulfur dyestuffs; used for textile dyeing and printing. [Hoechst AG]

Hydrosol®. Sulfur dyestuffs; used for textile dyeing and printing. [Cassella AG]

Hydrosoy 2000. CAS 977059-33-8; Hydrolyzed soy protein; conditioner for cosmetic prods. [Croda Inc.]

Hydrospray. Hydrocortisone, phenylpropanolamine, phenylephrine and neomycin; adjunctive local treatment of certain nasal disorders. [Merck & Co Inc] †

Hydrosulfite AWC. CAS 7775-14-6; Sodium hydrosulfite. [Henkel/Cospha]

Hydrosulphit®. CAS 7775-14-6; Sodium dithionite; reducing bleaches for wood pulps and wood-based old paper. [BASF AG]

hydrosulfite. (rongalite; hyraldite; decroline; redo; sodium dithionite; sodium hydrosulfite; blanchite; blankit); CAS 7775-14-6; Used in dyeing, and for decolorizing sugar syrups.; oxygen scavenger for synthetic rubbers.

Hydrosulfite A. A 10% solution of hydrosulfite NF, or hyraldite; used for testing dyed fabrics.

Hydrosulfite A.W. See Discolite. §

Hydrosulfite BASF. A 90% sodium hydrosulfite, $Na_2S_2O_4$.§

Hydrosulfite NF Conc. See Hyraldite C EXT. §

Hydrosulfite NF. (Rongalite Conc., Brittalite, Formosul). A condensation product of formaldehyde and sodium hydrosulfite; consists of a mixture of formaldehyde sodium bisulfite, $NaHSO_3 \cdot CH_2O$, and formaldehyde-sodium sulfoxylate, $NaHSO_2 \cdot CH_2O$; used as a discharger in calcio printing.

Hydrotek. Pigment dispersions; for coloration of aqueous coatings. [Pacific Dispersions Inc] *

Hydrotriticum 2000. CAS 70084-87-6; Hydrolyzed wheat protein; substantive conditioning agent, moisturizer for hair waving systems, shampoos, cream rinses, skin care prods. [Croda Inc.]

Hydrotriticum Powd. Hydrolyzed whole wheat protein. [Croda Inc.]

Hydrotriticum QL. Laurdimonium hydroxypropyl hydro-

lyzed wheat protein; film-forming conditioner for hair and skin care prods., e.g., waving systems, activated conditioner treatments, shampoos, styling mousses, hair coloring, wrinkle remover creams and lotions, liq. soap, facial scrubs, skin cleansers, bath prods. [Croda Inc.]

Hydrotriticum QM. Cocodimonium hydroxypropyl hydrolyzed wheat protein; film-forming conditioner for hair and skin care prods., e.g., waving systems, activated conditioner treatments, shampoos, styling mousses, hair coloring, wrinkle remover creams and lotions, liq. soap, facial scrubs, skin cleansers, bath prods. [Croda Inc.]

Hydrotriticum QS. Steardimonium hydroxypropyl hydrolyzed wheat protein; film-forming conditioner for hair and skin care prods., e.g., waving systems, activated conditioner treatments, shampoos, styling mousses, hair coloring, wrinkle remover creams and lotions, liq. soap, facial scrubs, skin cleansers, bath prods. [Croda Inc.]

Hydrotriticum WAA. Wheat amino acids; substantive moisturizer, humectant for hair and skin care prods., cleansers, antiwrinkle preps., sun screens; leaves soft and conditioned afterfeel. [Croda Inc.]

hydrous wool fat. See lanolin.

hydrous magnesium silicate. See talc.

Hydroxal. CAS 21645-51-2; A proprietary oral antacid consisting of a suspension of aluminum hydroxide. [Blue Line Chemical Co] ‡

hydroxyacetic acid. See glycolic acid.

o-hydroxyacetophenone. (2-hydroxyacetophenone) $HOC_6H_4COCH_3$; CAS 118-93-4; EINECS 204-288-0. [R.W. Greeff; Hoechst Celanese; Janssen Chimica; Penta Mfg.]

p-hydroxyacetophenone. (4-hydroxyacetophenone) $HOC_6H_4COCH_3$; CAS 99-93-4; EINECS 202-802-8. [R.W. Greeff; Hoechst Celanese; Janssen Chimica; Schweizerhall]

2-hydroxyacetophenone. See o-hydroxyacetophenone.

4-hydroxyacetophenone. See p-hydroxyacetophenone.

o-hydroxyanisole. See guaiacol.

p-hydroxybenzaldehyde. (4-hydroxybenzaldehyde) $C_7H_6O_2$; CAS 123-08-0; EINECS 204-599-1; pharmaceuticals. [Hoechst Celanese; Penta Mfg.; Schweizerhall; Spectrum Chem. Mfg.]

4-hydroxybenzaldehyde. See p-hydroxybenzaldehyde.

o-hydroxybenzaldehyde. See salicylaldehyde.

o-hydroxybenzamide. See salicylamide.

o-hydroxybenzoic acid. See salicylic acid.

2-hydroxybenzoic acid. See salicylic acid.

4-hydroxybenzoic acid butyl ester. See butylparaben.

4-hydroxybenzoic acid, propyl ester. See propylparaben.

1-hydroxybenzotriazole. $C_6H_5N_3O \cdot H_2O$; CAS 2592-95-2; EINECS 219-989-7; widely used additive to decrease racemization in the carbodiimide peptide coupling. [Aldrich; Janssen Chimica; Schweizerhall]

α-hydroxybenzyl phenyl ketone. See benzoin.

4-hydroxybutyric acid, γ-lactone. See butyrolactone.

4-hydroxycoumarin. $C_9H_6O_3$; CAS 1076-38-6; EINECS 214-060-2. [R.W. Greeff; Janssen Chimica; Penta Mfg.]

hydroxydimethylbenzene. See xylenol.

hydroxyethylcellulose. (cellulose, 2-hydroxyethyl ether; H.E. cellulose) Modified cellulose polymer containing hydroxyethyl side chains; CAS 9004-62-0; Thickener, suspending agent; stabilizer for vinyl polymerization; retards evaporation of water in mortars and cements; binder in ceramic glazes; used in paper and textile sizing.

---

‡ Trade name and manufacturer not verified   § Trade name without identified manufacturer

[Amerchol NV; Aqualon; Union Carbide]

2-hydroxyethyl methacrylate. (HEMA) $CH_2:C(CH_3)$-$COOCH_2CH_2OH$; CAS 868-77-9; EINECS 212-782-2; Acrylic resins, binder for nonwoven fabrics, enamels. [BP Chem.; Mitsubishi Gas; Rohm & Haas]

Hydroxylan. CAS 68424-66-8; EINECS 270-315-8; Hydroxylated lanolin; emulsifier, conditioner, emollient. [Fanning]

4-hydroxy-3-methoxybenzaldehyde. See vanillin.

2-hydroxy-4-methoxybenzophenone. See benzophenone-3.

2-hydroxy-4-methoxyphenyl. See benzophenone-3.

4-hydroxy-4-methyl-2-pentanone. See diacetone alcohol.

hydroxymimetite. A basic lead arsenate made artificially.

α-hydroxy-α-phenylacetophenone. See benzoin.

β-p-hydroxyphenylalanine. See tyrosine.

Hydroxyprolisilane C. Methylsilanol aspartate hydroxyprolinate; provides collagen restructuring for cosmetic and health prods. incl. stretch mark prevention creams, anti-aging formulations, acne preventives, eye contour creams, etc. [Exsymol]

2-hydroxy-1,2,3-propane tricarboxylic acid. See citric acid.

2-hydroxypropanoic acid. See lactic acid.

2-hydroxypropionic acid. See lactic acid.

hydroxypropyl alginate. See propylene glycol alginate.

2-hydroxypropylamine. See isopropanolamine.

hydroxypropyl methacrylate. $CH_3CHOHCH_2OO$-$CC(CH_3):CH_2$; CAS 27813-02-1; EINECS 248-666-3; monomer for acrylic resins, nonwoven fabric binders, detergent lube oil additives. [BP Chem.; Dajac Labs; Rohm & Haas]

1-hydroxy-2-pyridine. (2-hydroxypyridine; 2-pyridinol; 2-pyridone) C_5H_5NO; CAS 142-08-5; EINECS 205-520-3; Educt for the preparation of 1-(-alkenyl)-2-pyridones; useful in Diels-Alder reactions.

2-hydroxypyridine. See 1-hydroxy-2-pyridine.

8-hydroxyquinoline. (8-quinolinol; Oxine) C_9H_7NO; CAS 148-24-3; EINECS 205-711-1; Precipitating and separating metals, preparation of fungicides, chelating agent, disinfectant. [Penta Mfg.; Spectrum Chem. Mfg.; Superol BV; Tanabe USA]

N-hydroxysuccinic acid. (malic acid; apple acid) $COOHCH_2CH(OH)COOH$; CAS 6915-15-7; manufacture of esters and salts, wines; chelating agent, food acidulant, flavoring. [Allchem Industries; Haarmann & Reimer; Janssen Chimica; Schweizerhall]

α-hydroxytoluene. See benzyl alcohol.

β-hydroxytricarballylic acid. See citric acid.

Hydrozets. Hydrocortisone, bacitracin, tyrotricin, neomycin and benzocaine; for the effective relief of minor mouth and throat irritations. [Merck & Co Inc]

hydrozone. (Glycozone, Pyrozone). Trade names for hydrogen peroxide, H_2O_2; used as an antiseptic in dental practice.

hydryl. The product of the interaction between mercuric oxide and Orsudan (qv).

Hyflo NS, S. CAS 9006-04-6; Powd. NR, dusted with inert silica; used for adhesives, pharmaceutical and other tapes; Mooney visc. nonstabilized and stabilized resp. [H.A. Astlett]

Hyflo Super-Cel. CAS 7631-86-9; Diatomaceous earth; filter aids for industrial use (beer, dry cleaning, pharmaceuticals, chems., water, industrial waste). [Celite]

Hyflux. Soldering fluxes. [ABM Chemicals Ltd] *

Hyflux M. Hydrazine hydrobromide. [ABM Chemicals Ltd]*

Hy-glo Steel. A proprietary trade name for a stainless steel containing 17.0% chromium and 0.6% carbon. §

Hygrass. Soluble concentrate of 18.7 g dicamba and 300 g mecoprop per liter; used for weed control in cereals and grassland. [Agrichem (International) Ltd]

hygrol. Colloidal mercury.

Hygroplex HHG. Natural moisturizing factor in hydrophilic medium; for emulsified aq. and hydroalcoholic skin and hair moisturizing cosmetics. [Dr. Kurt Richter; Henkel/Cospha]

Hygroton. CAS 77-36-1; Chlorthalidone; diuretic. [Ciba plc; Parke-Davis] *

Hygroton K. A proprietary preparation of chlorthalidone and potassium chloride; a diuretic with a potassium supplement. [Ciba plc] *

Hyjet. Hydraulic fluid. [Chevron] *

Hyladerm®. CAS 9004-61-9; Hyaluronic acid. [Amerchol Corp]

hylastic. A high-manganese steel containing 1.6-1.8% manganese and 0.35% carbon.

Hylene®. A proprietary trade name for an organic diisocyanate used in the manufacture of polyurethane foam having a range of rigidities. [DuPont UK]

Hylite Color-Max. Fluorescent pigment/plasticizer dispersions; pourable colors for plastisols and other systems; offered custom matched to standard. [W J Ruscoe Co]

Hylorel. CAS 22195-34-2; Guanadrel sulfate; antihypertensive. [Upjohn] *

Hymec. CAS 7085-19-0; Soluble concentrate containing 570 g/l mecoprop; for control of weeds in cereals and grassland. [Agrichem (International) Ltd]

Hymod. Copper powder for sintering iron. [Steetley Chemicals Ltd] *

Hymolon CWC. Cocamide DEA; fulling agent for wool and household cleaners. [Hart Chem. Ltd.]

Hymono. Distilled monoglycerides; for various applications in the food industry e.g., margarines, shortenings, bakery and dairy. [Quest Int'l.]

Hymush. CAS 148-79-8; Wettable powder containing 60% w/w thiabendazole; a systemic fungicide. [Agrichem (International) Ltd]

Hyonic. Wetting agents. [Henkel Chemicals Ltd] *

Hyonic GL 400. Polyethoxy alkyl phenol; surfactant, co-emulsifier for specialty polymerizations. [Henkel/Functional Prods.]

Hyonic NP-40. CAS 9016-45-9; Nonoxynol-4; emulsifier for solv. cleaning compounds, waterless hand cleaners, agric. emulsifiable concs., silicone prods., detergent for petrol. oils; visc. reducer for plastisols; intermediate for production of anionic surfactants; metal cleaners; food contact applications. [Henkel/Functional Prods.; Henkel/Organic Prods.]

Hyonic OP-7. CAS 9002-93-1; Octoxynol-7; surfactant for textile use. [Henkel/Textiles]

Hyonic OP-55. scouring for textile desizing; emulsifier for waxes and oils preventing redeposition. [Henkel/Textiles]

Hyonic PE-100. CAS 9016-45-9; Nonoxynol-10; detergent, wetter, emulsifier, base, penetrant for household and industrial cleaners, paper toweling; latex stabilizer. [Henkel] *

hyoscine. CAS 51-34-3; Scopalamine, $C_{17}H_{21}NO_4$, an

---

* Trade name not verified as available   † Trade name verified as obsolete

alkaloid; used in medicine.

Hypacel. Bleaching agent for textiles. [RV Chemicals Ltd]†

Hypalon®. Chlorosulfonated polyethylene; synthetic elastomer; resist. to weather, ozone, oil, solvs., chemicals, heat, flame, and abrasion; used in jacketing and insulation for wire/cable, soles and heels, automotive components, coated fabrics, wh. tire sidewalls, sheet roofing, liners and covers for reservoirs and waste containment ponds, protective and decorative coatings, hose, linings for chemicals processing equipment, industrial rolls, seals, gaskets, and diaphragms. [DuPont] *

Hypalon® CP 826. Chlorinated, maleic anhydride-grafted PP resin; adhesion promoter for PP and blends; adhesion modifier for coatings, adhesives and inks. [DuPont]

Hypan® QT100. Acrylic multiblock copolymer; substantive to skin and hair; thickener; carrier for active substances; provides protective coating; primary emulsifier. [Kingston Tech.]

Hypan® SA100H. CAS 61788-40-7; Acrylic acid/acrylonitrogens copolymer; emulsifier and gellant for neutral pH; able to form conjugates with certain drugs for controlled delivery formulations. [Kingston Tech.]

Hypan® SR150H. CAS 136505-00-5; Acrylic acid/acrylonitrogens copolymer; thickener and gellant for aq. formulations, esp. highly conc. salt sol'ns., surfactants and drugs; emulsifier. [Kingston Tech.]

Hypan® SS201. CAS 123754-28-9; Ammonium acrylates/acrylonitrogens copolymer; gellant, emulsifier for cosmetics and related applications. [Kingston Tech.]

Hypan® SS500V. Tromethamine acrylates/acrylonitrogens copolymer; gellant, emulsifier for cosmetics and related applications. [Kingston Tech.]

Hypan® SS500W. TEA-acrylates/acrylonitrogens copolymer; gellant, emulsifier for cosmetics and related applications. [Kingston Tech.]

Hypaque Meglumine. Diatrizoate meglumine; diagnostic aid. [Sterling Drug Inc] *

Hypax. Oxidized wax/printing inks; anticorrosives. [Chemoxy International Ltd] †

Hyperab®. Specific human immunoglobulin. [Bayer AG; Miles Pharmaceuticals]

Hypercal. A proprietary preparation of rauwolfia alkaloids; an antihypertensive. [Carlton Laboratories (UK) Ltd] *

Hypercal B. A proprietary preparation of rauwolfia alkaloids and amylobarbitone; used in the control of hypertension. [Carlton Laboratories (UK) Ltd] *

Hyperdol. Reserpine and hydroflumethiazide; an antihypertensive. [The Boots Co plc]

HyperHep. Hepatitis B immune globulin; immunizing agent. [Cutter Laboratories, Miles Laboratories Inc] *

Hyper+Ion 1050. Coagulant and flocculant for potable water treatment, municipal wastewater treatment, industrial water and wastewater treatment, industrial applications. [General Chem.]

Hyper+Ion 2050A. Coagulant and flocculant for potable water treatment, municipal wastewater treatment, industrial water and wastewater treatment, industrial applications. [General Chem.]

Hyperit. This is the same as hyperol. §

Hyperkil Bait. Caciferol; a rodenticide. [Antec International Ltd]

Hyperlast®. Polyether polyurethane systems; used for oil and marine, electrical, engineering, and automotive applications. [Kemira Polymers]

Hypermer. Dispersants. [Imperial Chemical Industries plc]

Hypernick. See Hipernick. §

Hyperol. Proprietary name for a compound of hydrogen peroxide and urea, $CO(NH_2)_2 \cdot H_2O_2$; contains 35% hydrogen peroxide, which is obtained by dissolving in water or ether; one gram in 10 cc = a 10-volume strength solution of hydrogen peroxide. §

Hypersal®. Auxiliary agents used in impregnation with melamine and urea formaldehyde resins. [Cassella AG]

Hypersal. Functional additives for decorative laminates. [Hoechst AG]

Hypersol. Hyperdispersants for pigment dispersions. [Tennant-KVK Ltd]

Hyperstat. CAS 364-98-7; Diazoxide; antihypertensive. [Schering Corp]

Hypertensin. CAS 53-73-6; A proprietary preparation of angiotensin amide; a vasoconstrictor. [Ciba plc] *

Hyper-Tet. Specific human immunoglobulin. [Bayer AG; Miles Pharmaceuticals]

Hypertussis. Pertussis immune globulin; immunizing agent. [Cutter Laboratories, Miles Laboratories Inc] *

Hypnodil. A proprietary preparation containing metomidate hydrochloride; veterinary hypnotic (pigs). [Janssen Pharmaceutical Ltd] *

Hypnomidate. CAS 33125-97-2; A proprietary preparation containing etomidate; for induction of anesthesia. [Janssen Pharmaceutical Ltd] *

Hypnomidate Concentrate. A proprietary preparation containing etomidate hydrochloride; for induction of anesthesia. [Janssen Pharmaceutical Ltd] *

Hypnorm. A proprietary preparation containing fentanyl citrate and fluanisone; veterinary anesthetic (dogs, rabbits and guinea pigs). [Janssen Pharmaceutical Ltd] *

Hypnovel. CAS 59467-70-8; Proprietary preparation of midazolam; benzodiazepine with hypnotic and sedative properties. [Roche Products Ltd] *

hypo. CAS 7772-98-7; Sodium thiosulfate $Na_2S_2O_3$.

Hypol® FHP 2000. PU prepolymer derived from TDI; foamabile hydrophilic prepolymer offering water activation, high additive loading, controllable flexibility and texture, high water absorbence, low temp. exotherms, inherent flame retardancy. [W R Grace/Organics]

Hypon. A proprietary formulation of aspirin, caffeine and codeine; for symptomatic relief of pain and pyrexia. [The Wellcome Foundation Ltd]

Hyporit. CAS 7778-54-3; The trade name for calcium hypochlorite, containing 80% available chlorine; an antiseptic. §

HypoTears® Lubricating Eye Drops and Ointment. Polyvinyl alcohol 1%; for the temporary relief of burning and irritation due to dryness of the eye or to exposure to wind or sun; helps protect against further eye irritation. [Iolab Corp]

Hypovase. CAS 19237-84-4; A proprietary preparation of prazosin hydrochloride; for the treatment of hypertension, left ventricular failure and Raynaud's Disease. [Pfizer International]

HypRho® D. Specific human immunoglobulin. [Bayer AG]

Hypromellose. A surface-active agent. It is a partial mixed methyl and hydroxypropyl ether of cellulose.

Hyprone. Soluble concentrate of 18.7 g dicamba, 100 g MCPA and 194 g mecoprop per liter; used for weed control in cereals and grassland. [Agrichem (International) Ltd]

---

‡ Trade name and manufacturer not verified   § Trade name without identified manufacturer

HyPure A. CAS 7790-92-3; Hypochlorous acid; for bleaching, org. synthesis. [Olin]

HyPure C. CAS 7790-93-4; Chloric acid. [Olin]

HyPure K. CAS 7778-66-7; Potassium hypochlorite; for liq. bleach, water treatment, hard surface cleaners. [Olin]

HyPure L. CAS 13840-33-0; Lithium hypochlorite; for liq. bleach, water treatment, hard surface cleaners. [Olin]

HyPure N. CAS 7681-52-9; Sodium hypochlorite; for liq. bleach, water treatment, hard surface cleaners. [Olin]

Hyquat 70, 75. CAS 999-81-5; Chlormequat; plant growth regulator. [Agrichem (International) Ltd]

Hyraldite C Ext. Sodium formaldehyde sulfoxylate, $NaHSO_2 \cdot CH_2O \cdot 2H_2O$; used as a reducing agent in calico printing. §

Hysa. Derivative of Hybis polybutene; used in surfactant manufacture and as a corrosion inhibiting agent. [BP Chemicals Ltd]

Hysede. Mixture of γ-HCH, thiabendazole and thiram; seed dressing for brassica crops. [Agrichem (International) Ltd]

Hyskon. Dextran 70; a polysaccharide produced by the action of *Leuconostoc mesenteroides* on sucrose; average molecular weight: 70,000; plasma volume extender. [Kabi-Pharmacia]

Hysol® 1C Epoxi-Patch Kit. CAS 25928-94-3; Two-part epoxy system; general-purpose R.T. curing adhesive/sealant for bonding metal, wood, and most plastics; used in contact with meat during processing; easily sanded, cut, tapped, and machined; nontoxic when cured. [Dexter/Hysol]

Hysol® 6C Epoxi-Patch Kit. CAS 25928-94-3; Aluminum-filled two-part epoxy system; epoxy kit used for bonding, sealing, and repairing appliance condensers, radiators, other aluminum substrates. [Dexter/Hysol]

Hysol® 342. CAS 24937-78-8; EVA-based; hot-melt adhesive with fast set time and excellent hot tack; for packaging applications. [Dexter/Hysol]

Hysol® 2000. Vinyl acetate-based; hot-melt adhesive with unique combination of tackifiers allowing good adhesion of styrene-based plastics. [Dexter/Hysol]

Hysol® 7804. Polyamide; tough, elastomeric hot-melt adhesive with excellent impact resist. at low temps.; for demanding applications where substrates are exposed to temp. extremes; bonds to metals, plastics, wood, leather, fabric, nonwoven fabric, films and foils. [Dexter/Hysol]

Hysol® DoAll (1942). CAS 24937-78-8; EVA-based adhesive; medium setting adhesive for general purpose applications; excellent adhesion to wood and many plastics (nylon, PC, PVC, PS, ABS, acrylic). [Dexter/Hysol]

Hysol® EA9460. Two-part epoxy; outstanding combination of peel and shear strength, good adhesion to metals, ceramics, plastics, and FRP without primers; for structural bonding applications. [Dexter/Hysol]

Hysol® EE0067/HD3561 and Hysol EE0067/HD3615. Resin/hardener system with nonabrasive filler; low visc., low cost blk. casting system with nonabrasive filler for general purpose applications; mix ratios: 100/20 and 100/14 by wt. resp. [Dexter/Hysol]

Hysol® EO1016. CAS 25928-94-3; One-component epoxy system; casting compound for elec. insulation applications, thermal shock resistance and flameout properties; extremely stable at R.T.; not for casting masses larger than 50 g; heat curing. [Dexter/Hysol]

Hysol® ES4228. Thixotropic version of Hysol ES4128; for conformal coating or blobbing semiconductor chips; for use in potting; solid state watch circuits and chip carriers; mix ratio: 100/100 by wt. [Dexter/Hysol]

Hysol® MBI-02. A proprietary epoxy resin molding powder modified for use as a load-bearing material. [Dexter/Hysol] *

Hysol® MG1 Series. CAS 25928-94-3; One-component epoxy molding powd.; short flow compression grade; fast gel times, good dimensional stability and good release; not for molding delicate inserts; in thin wall cases, bobbins, insulators. [Dexter/Hysol]

Hysol® OS0100. General purpose LED lamp encapsulant; color stability and moisture resistance; mix ratio: 100/100 by wt. [Dexter/Hysol]

Hysol® PC18. One-component urethane; clear printed circuit coating for brush dip or spray application; tough abrasion resistance, and environmental protection. [Dexter/Hysol]

Hysol® Polyshot 1X. Hot melt adhesive for use on particle board, softwoods, and foam substrates. [Dexter/Hysol]

Hysol® XC7-W529. A proprietary flexible, one-component epoxy casting and potting compound possessing good thermal shock properties. [Dexter/Hysol] *

Hysorb. CAS 7758-29-4; Sodium tripolyphosphate. [FMC]

Hyspray. CAS 61791-44-4; Cationic surfactant containing 800 g/l polyethoxylated tallow amine; wetting and spreading agent for phosphonoglycine herbicide sprays. [Fine Agrochemicals Ltd]

Hystar®. Hydrogenated starch; for use in toiletries and special dietary food. [Lonza Ltd]

Hystar® 3375. Hydrogenated starch hydrolyzate; humectant, bodying agent, moisture control agent for toothpaste, cosmetics, sugarless confections, controlled moisture foods and industrial applications. [Lonza Inc]

Hystar® TPF. Hydrogenated starch hydrolyzate; lubricant; maintains optimal moisture control in industrial, pet food, tobacco, teat dip, and oral hygiene prods. [Lonza Inc]

Hystor 10. CAS 117-18-0; Granules containing 10% w/w tecnazene; protectant fungicide and potato sprout suppressant. [Agrichem (International) Ltd]

Hystrene® 1835. CAS 67701-05-7; Mixt. tallow/coconut acid (CTFA); chemical intermediate, emulsifier; used for personal care prods., soaps, waxes, textile auxiliary, and pharmaceuticals. [Witco/Humko]

Hystrene® 3022. Hydrogenated menhaden acid; chemical intermediate, emulsifier; used for personal care prods., waxes, greases, textile auxiliary, pharmaceuticals. [Witco/Humko]

Hystrene® 3675. 75% Dimer acid; corrosion inhibitor, intermediate; derivs. used as synthetic lube components, corrosion inhibitors for petrol. processing, as extenders and crosslinking agents for high polymeric systems; mildness additive in detergents. [Witco/Humko]

Hystrene® 4516. CAS 57-11-4; EINECS 200-313-4; Stearic acid; lubricant, textile auxiliary, emulsifier, plasticizer, intermediate, used in cosmetics, shampoos, pharmaceuticals. [Witco/Humko]

Hystrene® 5012. CAS 68938-15-8; Hydrogenated stripped coconut acid; chemical intermediate, emulsifier; used for personal care prods., soaps, lubricants, waxes, textile auxiliary, pharmaceuticals. [Witco/Humko]

Hystrene® 5016 NF. CAS 57-11-4; EINECS 200-313-4; Stearic acid, triple pressed; food grade acids used as

---

lubricants, release agents, binders, and defoamers, and in components for producing other food grade additives. [Witco/Humko]

Hystrene® 5460. CAS 68937-90-6; Trilinoleic acid; corrosion inhibitor, lubricant; intermediate for mfg. of soaps, emulsions, creams, lotions, ethoxylates, buffing compounds, lubricants. [Witco/Humko]

Hystrene® 5522. CAS 112-85-6; Behenic acid; chemical intermediate, emulsifier; used for personal care prods., soaps, waxes, textile auxiliary, pharmaceuticals. [Witco/Humko]

Hystrene® 7018 FG. CAS 57-11-4; EINECS 200-313-4; Stearic acid; food grade acids used as lubricants, release agents, binders, and defoamers, and in components for producing other food grade additives. [Witco/Humko]

Hystrene® 8016. CAS 57-10-3; EINECS 200-312-9; 80% Palmitic acid; intermediate for manufacture of soaps, emulsions, creams, lotions, ethoxylates, buffing compounds, lubricants. [Witco/Humko]

Hystrene® 9014. CAS 544-63-8; EINECS 208-875-2; Myristic acid (90%); lubricant, textile auxiliary, emulsifier, plasticizer, intermediate, used in cosmetics, shampoos, pharmaceuticals. [Witco/Humko]

Hystrene®9022. CAS 112-85-6; Behenic acid (90%); lubricant, textile auxiliary, emulsifier, plasticizer, intermediate, used in cosmetics, shampoos, pharmaceuticals. [Witco/Humko]

Hystrene® 9512. CAS 143-07-7; Lauric acid (95%); lubricant, textile auxiliary, emulsifier, plasticizer, intermediate, used in cosmetics, shampoos, pharmaceuticals. [Witco/Humko]

Hystrene® 9514. CAS 544-63-8; EINECS 208-875-2; Myristic acid (95%); chemical intermediate, emulsifier; used for personal care prods., waxes, textile auxiliary, pharmaceuticals. [Witco/Humko]

Hystrene®9718 NF FG. CAS 57-11-4; EINECS 200-313-4; Stearic acid NF (92%); food grade acids used as lubricants, release agents, binders, and defoamers; intermediate for producing food grade emulsifiers. [Witco/Humko]

HyStretch V-29. Heat-reactive, carboxylated saturated acrylic elastomer; sprayable, foamable, high compression recovery, low tack, durable, soft, inherent heat and light stability; can be heat sensitized; self-crosslinking at 275 F; accepts fillers, pigments, thickeners, and crosslinkers; used for textile, nonwoven, adhesive bindings, coatings, and finishes. [BFGoodrich] *

HyStretch V-43. Acrylates copolymer; heat-reactive, carboxylated sat. acrylic elastomer; sprayable, foamable, high compression and stretch recovery, low tack, durable, soft, low blocking; inherent heat and light stability; can be heat sensitized; self-crosslinking at 275 F; accepts fillers, pigments, thickeners, and crosslinkers; used for textile, nonwoven, adhesive bindings, coatings, and finishes. [BFGoodrich] *

Hysward. Mixture of dicamba, MCPA and mecoprop; used for weed control in cereals and grassland. [Agrichem (International) Ltd]

Hytak. Hot-melt adhesives for packaging and product assembly formulated from a variety of polymers, resins, waxes, plasticizers and antioxidants; used for carton sealing, bottle and can labelling, manufacture of disposable sanitary products, assembling of components in automotive and general industries and woodworking.

[Hytak Ltd] *

Hytakerol. CAS 67-96-9; Dihydrotachysterol; regulator. [Sterling Drug Inc] *

Hytane. CAS 34123-59-6; Suspension concentrate containing 500 g isoproturon per liter; used for annual weed control in cereals. [Ciba-Geigy Agrochemicals]

Hytec. CAS 117-18-0; Dustable powder containing tecnazene; protectant fungicide and potato sprout suppressant. [Agrichem (International) Ltd]

Hytec Super. CAS 117-18-0, 148-79-8; Dustable powder containing 6% w/w tecnazene and 1.8% thiabendazole; a protectant fungicide and potato sprout suppressant. [Agrichem (International) Ltd]

Hytemco. A proprietary trade name for an iron-nickel resistance alloy. §

HyTemp 4051. Acrylic ester copolymer; elastomer for hose, tubing, cable jacket, belting, and mechanical goods; offers service temperature of 0-400 F; excellent oil resistance. [Zeon]

HyTemp 4052. Acrylic ester copolymer; elastomer for hose, tubing, cable jacket, belting, and mechanical goods; service temperature -30 to 375 F. [Zeon]

HyTemp NPC-50. Quaternary ammonium compound; nonpost cure agent used with all HyTemp 4050 series elastomers. [Zeon]

Hyten®. Fibers. [DuPont UK]

Hyten M Steel. A proprietary trade name for a nickel-chromium-molybdenum steel. §

Hy-ten-sl. A proprietary trade name for an alloy of 66% copper, 19% zinc, 10% aluminum, and 5% manganese.§

Hytex. Printing ink alkyds. [Croda Resins Ltd]

Hythane. Polyurethane alkyds. [Croda Resins Ltd]

Hytherm. Heat resistant dyes for plastics. [Morton Int'l. Ltd]

Hytin. Mixture of fentin acetate and maneb; used for control of potato blight. [Agrichem (International) Ltd]

Hy-TL. CAS 148-79-8, 137-26-8; Mixture of thiabendazole and thiram; fungicide seed dressing. [Agrichem (International) Ltd]

Hytone. CAS 50-23-7; Hydrocortisone; glucocorticoid. [Dermik Laboratories Inc] *

Hytox. Germicidal detergent. [Unilever] *

Hytrel®. A group of polyester elastomers used as thermoplastic rubbers; they are graded for hardness as follows: Hytrel 4055, 92A; Hytrel 5550, 55D; Hytrel 6350, 63D. [DuPont UK]

Hytrol. 2,4-D, diuron, amitrole, simazine; total weedkiller with more than one seasons resistance; for use on garden paths and other noncrop areas. [Agrichem (International) Ltd]

Hytrol. Aminotriazole + 2,4-D + diuron + simazine; used for total weed control in non crop areas. [Farmers Crop Chemicals Ltd]

Hyvar® X. CAS 314-40-9; A wettable powder containing 80% w/w bromacil; used for control of weeds in cane fruit and noncrop areas. [DuPont UK]

Hyvar X.7; CAS 314-40-9; A wettable powder containing 80% w/w bromacil; used for control of weeds in cane fruit and noncrop areas. [Selectokil Ltd]

Hy-Vic. CAS 148-79-8, 137-26-8; Mixture of thiabendazole and thiram; fungicide seed dressing. [Agrichem (International) Ltd]

Hy-Vin. Homo and copolymeric polyvinyl chloride compounds (S-PVC/E-PVC); for production of plastic articles. [Norsk Hydro AS]

Hyvis. Polybutenes. [BP Chemicals Ltd]
Hyzod® AC-1000. PC sheet; features low flame spread, low smoke generation, low toxicity, high heat resist., excel- lent dimensional stability; for aircraft seat parts, tray tables, lavatories, cargo liners, window reveals, parti- tions, moldings. [Sheffield Plastics] *

I

IAA. *See* 3-indoleacetic acid.

Iachiol. Silver fluoride. AgF.

IA-IA alloy. An alloy of 60% copper and 40% nickel; used for electrical resistances.

IBA. *See* 3-indolebutyric acid.

Ibbal. p-Isobutylbenzaldehyde; intermediate for pharmaceuticals; fragrance. [Mitsubishi Gas]

Ibdu®. Slow release nitrogen fertilizer. [Mitsubishi Kasei] *

Ibex. Acid calcium phosphate for aerating flour. [Albright & Wilson Ltd]

IBR/IVP/P13. Live vaccines; for active immunization against infectious bovine rhinotracheitis, infectious pustulous vulvovaginitis/balanoposthitis and para-influenza-3 disease of cattle and for active interferonization. [Bayer AG]

Ibrin. Fibrinogen I 125; a preparation of human fibrinogen labelled with iodine-125; diagnostic aid; radioactive agent. [Amersham Corp] *

Ibugel. CAS 15687-27-1; Colorless gel containing 5% w/w ibuprofen BP; used for the treatment of backache, rheumatic and muscular pain, sprains, strains, and neuralgia. [Dermal Laboratories Ltd]

Ibular®. CAS 15687-27-1; Tablets containing 200mg and 400mg ibuprofen BP; indicated for its anti-inflammatory and analgesic effect in the treatment of rheumatoid arthritis (including juvenile rheumatoid arthritis or Still's disease), ankylosing spondylitis, osteoarthrosis and nonrheumatoid (sero-negative) anthropathies. [Lagap Pharmaceuticals Ltd]

IB-VAC. Infectious bronchitis vaccine; for immunization of poultry. [Intervet Inc] †

IB-VAC-H. Massachuttes Holland type bronchitis vaccine; for immunization of poultry. [Intervet Inc]

IB-VAC-M. Massachuttes type bronchitis vaccine; for immunization of poultry. [Intervet Inc] †

Icdal. Terephthalic acid resins; for insulation of heat-resistant electrical conductors. [Dynamit Nobel Wien GmbH]*

Iceberg®. CAS 12141-46-7; Aluminum silicate, anhyd., calcined; pigment used in paints, paper, board, rubber and plastics. [Burgess Pigment]

Icecap® K. CAS 12141-46-7; Aluminum silicate, anhyd., calcined; extender for TiO_2 in coatings where high Hegman grind is required; used in wire and cable, molded and extruded rubber and plastic prods., paper coatings, paints; good electricals, low compression set, low water absorp. [Burgess Pigment]

ice colors. Colors formed on the fiber by treating it with a phenol, and then with a diazotized amine, in the presence of ice. They are also known as ingrain colors.

Iceland spar. CAS 13397-26-7; Crystalline calcium carbonate.

Iceline. *See* Preservaline.

Ice Melt. CAS 10043-52-4; Calcium chloride pellets; used for snow and ice melting, dust control and tire weighting. [Standard Tar Products Co Inc]

Ice No. 2. Glyceryl stearate and polysorbate 80; stabilizer and emulsifier for the food industry; lubricant for textiles and plastics; fabric softener. [Van Den Bergh Foods]

Ice No. 12K. CAS 9005-71-4; 80/20 blend of mono- and diglycerides and polysorbate 65; emulsifier used with stabilizers in frozen desserts. [Van Den Bergh Foods]

Ichthadone. A proprietary preparation of ammonium ichthosulfonate. §

Ichthopaste. Zinc paste and ichthamol bandage. [Smith & Nephew Pharmaceuticals Ltd] *

Ichthymall. CAS 8029-68-3; Ichthammol; anti-infective, topical. [Mallinckrodt Inc] *

Ichthyocolla. *See* isinglass.

Ichthyol. CAS 8029-68-3; Ichthammol; anti-infective, topical. [Stiefel Laboratories Inc] *

Icinol. Polyalkylene glycols. [Imperial Chemical Industries plc]

Icinol DPM. CAS 34590-94-8; (2-Methoxy-methylethoxy) propanol; solv. for use in protective coatings, inks, cleaning prods., agric. chems.; aids wetting, penetration, and soil removal; coupling solv. [ICI Australia]

Icinol PM. CAS 107-98-2; 1-Methoxy 2-propanol; solv. for use in protective coatings, inks, cleaning prods., agric. chems.; aids wetting, penetration, and soil removal; coupling solv. [ICI Australia]

Icipen 300. A proprietary preparation of phenoxymethylpenicillin; an antibiotic. [ICI Chem. & Polymers Ltd] †

Icomeen® T-2. CAS 61791-44-4; PEG-2 tallow amine; wetting agent, penetrant, emulsifier, stabilizer, dispersant, antistat, lubricant, solubilizer. [BASF]

Icomeen® T-15. CAS 61791-44-4; PEG-15 tallow amine; surfactant. [BASF]

Iconol DA-4. CAS 26183-52-8; Deceth-4; wetting agent, detergent, emulsifier, dispersant, solubilizer for textile scouring and dyeing, industrial and institutional cleaners, household cleaning prods. and specialties. [BASF]

Iconol DA-6. CAS 26183-52-8; Deceth-6; wetting agent, detergent, emulsifier, dispersant, solubilizer for textile scouring and dyeing, industrial and institutional cleaners, household cleaning prods. and specialties. [BASF]

Iconol DA-9. CAS 26183-52-8; Deceth-9; wetting agent, detergent, emulsifier, dispersant, solubilizer for textile scouring and dyeing, industrial and institutional cleaners, household cleaning prods. and specialties. [BASF]

Iconol NP-30. CAS 9016-45-9; Nonoxynol-30; wetting

---

‡ Trade name and manufacturer not verified　　　　　§ Trade name without identified manufacturer

agent, penetrant, detergent, cleaning agent, emulsifier, latex stabilizer, dispersant for industrial, institutional and household cleaning prods.; emulsifier for emulsion polymerization, asphalt emulsions. [BASF]

Iconol NP-40. CAS 9016-45-9; Nonoxynol-40; wetting agent, penetrant, detergent, cleaning agent, emulsifier, latex stabilizer, dispersant for industrial, institutional and household cleaning prods.; emulsifier for emulsion polymerization, asphalt emulsions. [BASF]

Iconol NP-50. CAS 9016-45-9; Nonoxynol-50; wetting agent, penetrant, detergent, cleaning agent, emulsifier, stabilizer, dispersant, coemulsifier. [BASF]

Iconol NP-70. CAS 9016-45-9; Nonoxynol-70; wetting agent, penetrant, detergent, cleaning agent, emulsifier, latex stabilizer, dispersant for industrial, institutional and household cleaning prods.; emulsifier for emulsion polymerization, asphalt emulsions. [BASF]

Iconol NP-100. CAS 9016-45-9; 26027-38-3; Nonoxynol-100; wetting agent, penetrant, detergent, cleaning agent, emulsifier, latex stabilizer, dispersant for industrial, institutional and household cleaning prods.; emulsifier for emulsion polymerization, asphalt emulsions. [BASF]

Iconol OP-10. CAS 9002-93-1; Octoxynol-10; wetting agent, penetrant, detergent, cleaning agent, emulsifier, latex stabilizer, dispersant for industrial, institutional and household cleaning prods.; emulsifier for emulsion polymerization, asphalt emulsions. [BASF]

Iconol OP-30. CAS 9002-93-1; Octoxynol-30; emulsifier, wetting agent, dispersant, synthetic latex stabilizer, detergent in formulating industrial, institutional, and household cleaning prods.; primary emulsifier for acrylic and vinyl emulsion polymerization and for asphalt emulsion systems. [BASF]

Iconol OP-40. CAS 9002-93-1; Octoxynol-40; wetting agent, penetrant, detergent, cleaning agent, emulsifier, latex stabilizer, dispersant for industrial, institutional and household cleaning prods.; emulsifier for emulsion polymerization, asphalt emulsions. [BASF]

Iconol TDA-3. CAS 24938-91-8; Trideceth-3; wetting agent, detergent, emulsifier, dispersant, solubilizer for textile scouring and dyeing, industrial and institutional cleaners, household cleaning prods. and specialties. [BASF]

Iconol TDA-6. CAS 24938-91-8; Trideceth-6; wetting agent, detergent, emulsifier, dispersant, solubilizer for textile scouring and dyeing, industrial and institutional cleaners, household cleaning prods. and specialties. [BASF]

Iconol TDA-8. CAS 24938-91-8; Trideceth-8; wetting agent, detergent, emulsifier, dispersant, solubilizer for textile scouring and dyeing, industrial and institutional cleaners, household cleaning prods. and specialties. [BASF]

Iconol TDA-9. CAS 24938-91-8; Trideceth-9; wetting agent, degreaser, emulsifier, detergent. [BASF]

Iconol TDA-10. CAS 24938-91-8; Trideceth-10; wetting agent, detergent, emulsifier, dispersant, solubilizer for textile scouring and dyeing, industrial and institutional cleaners, household cleaning prods. and specialties. [BASF]

Iconol WA-1. Alkoxylated phenolic; surfactant; pigment dispersant for aq. systems. [BASF]

Iconol WA-4. Alkoxylated phenolic; surfactant; dispersant. [BASF]

Iconsim®. For the electrical industry. [DuPont UK]

ICR. Catalyst. [Chevron] *

Icthaband. A proprietary zinc paste and ichthammol bandage used in the treatment of eczema. [Seton Healthcare Group plc]

Ictotest. A proprietary test tablet of p-nitrobenzene diazonium, p-toluene sulfonate, salicylsulfonic acid and sodium bicarbonate; used for the detection of bilirubin in urine. [B C Ames] *

Ideal alloy. An alloy of 53.5% copper, 45% nickel, 0.66% iron, and 0.45% manganese.

Idemin. A proprietary preparation of meprobamate and benactyzine hydrochloride; antidepressant. [Horlicks] *

Idilon. See Carbora. §

Iditol. A proprietary shellac substitute. §

Idoklon. CAS 333-36-8; Flurothyl; stimulant. [Anaquest, Div of BOC Inc] *

idryl. CAS 206-44-0; Fluoranthrene, $C_{16}H_{10}$.

IE-40-A. PU elastomer; tough, abrasion-resist. engineering elastomer for R.T. hand batch processing; for rollers, pallets, prototypes, molds, fixtures, wheels, bumpers. [Innovative Engineering of Mich.]

Ifex. CAS 3778-73-2; Ifosfamide; antineoplastic. [Mead Johnson & Co]

Ifomide. See Holoxan. [Degussa AG] *

Igasurine. Impure brucine.

Igelit PCU. A proprietary manufacture of polyvinylchloride.§

Igepal® 131. β-Naphthol + 8-9 EO; surfactant. [Rhone-Poulenc France]

Igepal® BPA-6. Bisphenol A ethoxylate; monomer for coatings; reactive diluent; useful in alkyds and polyurethanes. [Rhone-Poulenc Surf.]

Igepal® CA-210. CAS 9002-93-1; Octoxynol-1.5; emulsifier for solv. cleaners, drycleaning, pesticides, floor polish; defoamer. [Rhone-Poulenc Surf.; Rhone-Poulenc France]

Igepal® CA-620 and CA-630. A range of nonionic surfactants; may be used in all phases of detergent compounding and aqueous processing in the textile and paper industries, in industrial metal cleaners, acid cleaners and floor cleaners, detergent-sanitizers and waterless hand cleaners. [Rhone-Poulenc Surf.] *

Igepal® CA-720. Oxtoxynol-12.5; a nonionic surfactant used as a hard surface detergent with aqueous solubility at high temperatures; used in hot spray, soak and steam-cleaning systems, electrolytic cleaning and metal pickling operations [Rhone-Poulenc Surf.] *

Igepal® CA-897. CAS 9002-93-1; Octoxynol-40; emulsifier, stabilizer for vinyl acetate and acrylate polymerization; dyeing assistant; emulsifier for fats and waxes. [Rhone-Poulenc Surf.; Rhone-Poulenc France]

Igepal® Cephene Distilled. CAS 122-99-6; EINECS 204-589-7; Phenoxyethanol; surfactant. [Rhone-Poulenc France]

Igepal® CO-210. CAS 9016-45-9; Nonoxynol-2 (1.5 EO); foam and emulsion stabilizer, detergent, coemulsifier, intermediate, defoamer, dispersant; for metalworking; biodeg. [Rhone-Poulenc Surf.; Rhone-Poulenc France]

Igepal® CO-430. CAS 9016-45-9; Nonoxynol-4; An oil soluble surfactant used as a coemulsifier with CO-850 and CO-880; plasticizer and antistatic agent for polyvinyl acetate; freeze-thaw stabilizer for latex emulsions, intermediate for the synthesis of high foaming, water soluble sulfate esters. [Rhone-Poulenc Surf.]

Igepal® CO-520. CAS 9016-45-9; Nonoxynol-5; an oil soluble surfactant and intermediate for anionic surfactants; de-icing fluid for jet aircraft fuels and automotive gasoline, added to home storage tanks to inhibit rusting, dispersant for petroleum oils. [Rhone-Poulenc Surf.]

Igepal® CO-530. CAS 9016-45-9; Nonoxynol-6; an oil soluble surfactant and intermediate for anionic surfactants; uses are similar to those given for CO-520 and it is also used as an emulsifier for silicones, agricultural compounds, and mineral oils. [Rhone-Poulenc Surf.]

Igepal® CO-610. CAS 9016-45-9; Nonoxynol-8; A water soluble surfactant widely used for detergency, wetting and emulsification applicable where low foaming is particularly important. [Rhone-Poulenc Surf.]

Igepal® CO-630. CAS 9016-45-9; Nonoxynol-9; nonionic surfactant; water soluble detergents, wetting agents and emulsifiers. [Rhone-Poulenc Surf.]

Igepal® CO-660. CAS 9016-45-9; Nonoxynol-10; nonionic surfactant, water soluble detergents, wetting agents and emulsifiers. [Rhone-Poulenc Surf.]

Igepal®CO-710. CAS 9016-45-9; Nonoxynol-10.5; nonionic surfactant, water soluble detergents, wetting agents and emulsifiers. [Rhone-Poulenc Surf.]

Igepal® CO-720. CAS 9016-45-9; Nonoxynol-12; nonionic surfactant, water soluble detergents, wetting agents and emulsifiers. [Rhone-Poulenc Surf.]

Igepal® CO-730. CAS 9016-45-9; Nonoxynol-15; nonionic surfactant, water soluble detergents, wetting agents and emulsifiers. [Rhone-Poulenc Surf.]

Igepal® CO-850. CAS 9016-45-9; Nonoxynol-20; A water-soluble detergent and wetting agent, emulsifier for fats, oils, waxes, solvents, demulsifier for crude petroleum oil emulsions. [Rhone-Poulenc Surf.]

Igepal® CO-880. CAS 9016-45-9; Nonoxynol-30; nonionic surfactant used as a detergent for the high temperature scouring of textiles in pressure equipment. [Rhone-Poulenc Surf.]

Igepal® CO-887. CAS 9016-45-9; An aqueous solution of CO-880; surfactant for emulsion polymerization and post-additive stabilization. [Rhone-Poulenc Surf.]

Igepal® CO-890. CAS 9016-45-9; Nonoxynol-40; A highly water-soluble emulsifier and stabilizer used in concentrated electrolyte solutions, synthetic latices, floor waxes and polishes. [Rhone-Poulenc Surf.]

Igepal® CO-970. CAS 9016-45-9; Nonoxynol-50; A very highly soluble surfactant effective at high temperatures and in concentrated electrolyte solutions, stabilizer for synthetic latexes, emulsifier and stabilizer in the preparation of floor waxes and polishes. [Rhone-Poulenc Surf.]

Igepal® CO-997. CAS 9016-45-9; Nonoxynol-100; emulsifier, stabilizer, wetting agent, dyeing assistant for plastics, latexes, floor polishes, etc. [Rhone-Poulenc Surf.; Rhone-Poulenc France]

Igepal® CTA-639W. Water-soluble surfactant with exceptional wetting and emulsifying properties; used in latex paints as a wetting agent, titanium oxide and as an emulsifier for oil or alkyd additives. [Rhone-Poulenc Surf.]

Igepal® DM-430. CAS 9014-93-1; Nonyl nonoxynol-7; emulsifier for agric., emulsion polymerization, leather industries. [Rhone-Poulenc Surf.; Rhone-Poulenc France]

Igepal® DM-530. CAS 9014-93-1; Nonyl nonoxynol-9; an emulsifier used in acid-based cleaners, aerosol, cosmetic, insecticide and wax emulsions, textile finishing

oils, dry-cleaning soaps, inks, lacquers and paints. [Rhone-Poulenc Surf.]

Igepal® DM-710. CAS 9014-93-1; Nonyl nonoxynol-15; a water-soluble surfactant for detergency and emulsification, especially where low foaming is desired; detergent for washing paper mill felts. [Rhone-Poulenc Surf.]

Igepal® DM-730. CAS 9014-93-1; Nonyl nonoxynol-24; highly water-soluble surfactant used in emulsion polymerization, latex stabilization and emulsifiable pesticide formulations. [Rhone-Poulenc Surf.]

Igepal® DM-970. CAS 9014-93-1; Nonyl nonoxynol-150; unique high-melting flake or solid nonionic surfactant used in household detergents and in industrial detergent formulations. [Rhone-Poulenc Surf.]

Igepal®OD-410. CAS 122-99-6; Phenoxydiglycol; a solvent for various resins (vinyl, phenolic, polyester, alkyd, nitrocellulose, cellulose acetate), ingredient of metal cleaners, paint strippers and other cleaning compounds and as an ink vehicle. [Rhone-Poulenc Surf.]

Igepal®RC-520. CAS 9014-92-0; Dodoxynol-6; A low foaming rewetting agent suitable for paper towels, tissues, and semichemical corrugating media. [Rhone-Poulenc Surf.]

Igepal® RC-620. CAS 9014-92-0; Dodoxynol-10; An all-purpose detergent and wetting agent; detergent in acid cleaning, detergent-sanitizing and grease cutting formulations, as an emulsifier and wetter in agricultural compounds, rug shampoos, and whitewall tire cleaners. [Rhone-Poulenc Surf.]

Igepon®. Redesignated Geropon®. [Rhone-Poulenc Surf.]

Igetaleim MA. Melamine resin; used for adhesives. [Sumitomo Bakelite Co Ltd] *

Igetaleim UA. Urea resin; used for adhesives. [Sumitomo Bakelite Co Ltd] *

Igewsky's reagent. A solution of 5% picric acid in absolute alcohol; used for etching in the micro-analysis of carbon steels.

Iglodine. A proprietary preparation of phenol and iodine; an antiseptic. [Ayrton Saunders plc] ‡

Ignicide. Phosphate base; for flameproof plywood and hardboard. [Stanley Smith & Co Plastics Ltd] †

Iguafen. Dispersing and soaping agent; solubilizing properties; retarder; antiprecipitant; used for textiles particularly acrylics and wool. [GAF Great Britain] *

IHSA I-125. Albumin, Iodinated I 125 serum; diagnostic aid; radioactive agent. [Mallinckrodt Inc] *

Ilcocillin. Antibiotic veterinary ethicals. [Ciba plc] *

Ildamen. CAS 15687-41-9; Oxyfedrin; tablets, ampoules and drops; therapy for ischemic heart disease. [Degussa AG] *

Iletin. *See* insulin.

Ilexan E. CAS 107-21-1; Inhibited ethylene glycol; leak indicating liquid, antifreeze agent and solar fluid. [Hüls AG]

Ilexan HT. CAS 25322-68-3; Polyethylene glycol; antifreeze agent stable at high temperatures for solar collectors. [Hüls AG]

Ilexan P. CAS 57-55-6; EINECS 200-338-0; Inhibited 1,2-propylene glycol; antifreeze agent and solar fluid for heating drinking water. [Hüls AG]

Ilexan S. A mixture of alkylbenzenes; heat transfer agent. [Hüls AG]

Ilinol. CAS 8001-26-1; Conjugated linseed oil. [Unilever] *

Illium. An acid-resisting alloy containing 60.65% nickel,

---

‡ Trade name and manufacturer not verified   § Trade name without identified manufacturer

21.07% chromium, 6.42% copper, 4.67% molybdenum, 2.13% tungsten, 1.04% silicon, 1.09% aluminum, 0.98% manganese, 0.76% iron, and 1.19% carbon and boron. [Stainless Foundry and Engineering Co.] *

Illosone. CAS 3521-62-8; Erythromycin estolate; antibacterial agent. [Dista Products Ltd]

ilmenite. (titaniferousilron ore, titanicilron, titaniferous iron). A mineral. It consists of about 52% titanic oxide, TiO_2, and 48% ferrous oxide, FeO.

Ilonium. A proprietary preparation of colophony, larch turpentine, turpentine oil, phenol and thymol; used in the treatment of skin disorders. [Ilon Laboratories]‡

Ilopan. CAS 81-13-0; Dexpanthenol; cholinergic. [Adria Laboratories Inc]

Ilosone. CAS 3521-62-8; Erythromycin estolate; antibacterial. [Eli Lilly & Co] *

Ilosone. CAS 3521-62-8; Erythromycin estolate BP; antibiotic. [Dista Products Ltd] *

Ilotycin. A proprietary preparation containing erythromycin, ethyl carbonate; an antibiotic. [Eli Lilly & Co]

Ilozyme. Pancrelipase; a concentrate of pancreatic enzymes standardized for lipase content; enzyme. [Adria Laboratories Inc]

Imacol. Low foaming lubricant; for prevention of creases and abrasions when wet finishing piece goods of synthetic fibers and blends with natural fibers. [Sandoz Products Ltd] *

Imadyl. CAS 53716-49-7; Proprietary preparation of carprofen; antirheumatic and analgesic. [Roche Products Ltd] *

Imavate. CAS 113-52-0; Imipramine hydrochloride; antidepressant. [Wyeth Laboratories] *

Imaverol. CAS 35554-44-0; A proprietary preparation containing enilconazole; veterinary dermatomycoses. [Janssen Pharmaceutical Ltd] *

Imbretil. Carbolonium bromide.

Imbrilon. CAS 53-86-1; A proprietary preparation of indomethacin; used in the treatment of arthritis. [Berk Pharmaceuticals Ltd] *

imesatin. β-Iminoisatin, $C_8H_6ON_2$.

Imferon. A proprietary preparation of iron dextran; a hematinic. [Fisons plc, Pharmaceuticals Div] *

Imicure EMI-24, 24S. Imidazoles accelerators. [Air Prods. & Chems. Inc]

imidazole. (glyoxalin) $C_3H_4N_2$; CAS 288-32-4; EINECS 206-019-2; Biological control of pests, esp. fabric-feeding insects; contact insecticide in an oil spray. [Aldrich; BASF; Janssen Chimica; Penta Mfg.]

Imidazoline 18. Series of organic nitrogenous bases; oleic or stearic acid reacted with various short chain amines to produce alkyl imidazolines; for dewatering of metal and other surfaces, anticorrosives, cationic emulsions for oils, pigment dispersing aids, antistatic agents. [Lakeland Laboratories Ltd]

imidazoline-2-thiol. See ethylene thiourea.

Imidrol. Cationic emulsifiers based on fatty imidazoline derivatives; used for the production of oil-based drilling muds and for lubrication in water-based muds. [Allied Colloids Ltd]

2,2′-iminobisethanol. See diethanolamine.

1,1′-iminobis-2-propenol. See diisopropanolamine.

2,2′-iminodiethanol. See diethanolamine.

2,2′-iminodiethylamine. See diethylenetriamine.

Imiodid. A substance obtained by heating p-ethoxyphenyl-

succinimide with a solution containing potassium iodide and iodine; a powerful antiseptic.

Imlar®. Vinyl coating; for the automotive industry. [DuPont UK]

Immadium. An alloy of manganese bronze containing aluminum.

Immedial®. Sulfur dyestuffs; used for textile dyeing and printing. [Cassella AG]

Immergan® A. Aliphatic sulfochloride; lightfast oil tanning agent for chamois and auxiliary tanning agent for very soft leathers. [BASF AG]

Immetal. Diiodoerucic acid isobutyl ester.

Immuglobin. Immune globulin; immunizing agent. [Savage Laboratories] †

Immukin. A proprietary preparation of recombinant human interferon gamma; used as an adjunct therapy to antibiotics to reduce the frequency of serious infections in patients with chronic granulomatous disease. [Boehringer Ingelheim Ltd]

Immuno-bed. Plastic embedding kit; for light microscopy immunohisto chemistry procedures. [Polysciences Inc]

Immu-Tetanus. Tetanus immune globulin; immunizing agent. [Parke-Davis] *

Imodium. CAS 53179-11-6; A proprietary preparation of loperamide; an antidiarrhoeal. [Janssen Pharmaceutical Ltd] *

imogen. Sodium diaminonaphtholsulfonate.

imogen sulfite. A photographic developer.

Impact. CAS 76674-21-0; Fungicide containing flutriafol. [ICI Chem. & Polymers Ltd]

Impact®. For the chemical industry. [DuPont UK]

Impact Excel. CAS 1897-45-6, 76674-21-0; Suspension concentrate containing 300 g chlorothalonil and 47 g flutriafol per liter; a systemic fungicide. [ICI Agrochemicals]

Impad. High density impact pads for continuous casting tundishes. [Foseco (F.S.) Ltd]

Imperacin. CAS 79-57-2; A proprietary preparation of oxytetracycline dihydrate; an antibiotic. [ICI Chem. & Polymers Ltd]

Imperial metal. An alloy of 80% copper and 20% nickel.

Impervite. See Ceresit. §

Impet® 330. CAS 25038-59-9; PET polyester, 30% glass-reinforced; injection molding resin for high-performance applications requiring toughness, rigidity, and exceptional dimensional stability; used for automotive applications (distributor housings, rotors, headlamp bezels, structural body parts); consumer electronics/appliances (motor housings, coffee makers, hair dryers), elec./electronics (connectors, terminal blocks, bobbins), furniture. [Hoechst Celanese] *

Implenal®. Mixt. of salts of dicarboxylic acids; finishing agent for chrome tanning, furs. [BASF AG]

Impletol®. Pharmaceutical preparation; neural therapeutic agent. [Bayer AG]

Imposil. Veterinary iron-dextran complex. [Fisons plc, Pharmaceuticals Div] *

Impra®. Organic wood preservatives and wood stainers; applied by brushing, spraying and dipping. [Weyl GmbH]

Impra-biolan. Ecological wood stainer for indoor and outdoor use; applied by brushing and spraying. [Weyl GmbH]

Impra-color. Wood stainer protecting against fungi and insect attack; for outdoor use; applied by brushing, spraying, and dipping. [Weyl GmbH]

Impra® Concentrates. Water-dilutable concentrates for protection of freshly sawn timber against sap stain and of logs against pinhole borers; applied by dipping and spraying. [Weyl GmbH]

Impra-elan. Fast-drying satin gloss joinery stainer; applied by brushing and spraying. [Weyl GmbH]

Imprafix®. Polyurethane products for coating and laminating textiles; fast to washing and to solvents; also suitable as binders for pigment printing. [Bayer AG]

Impralan®. Water-dilutable ecological stainer for indoor treatment of wood; applied by spraying and pouring. [Weyl GmbH]

Impraleum. A coaltar distillate for fencing; applied by brushing, spraying and dipping. [Weyl GmbH] †

Impralit. Wood preservative salts and fire retardant salts; for dipping, vacuum-pressure method. [Weyl GmbH]

Impranil®. Polyurethane products; used for coating and laminating textiles; fast to washing and to solvents. [Bayer AG]

Impregnant. Acrylic polymers; used for electrical moldings. [Monomer-Polymer & Dajac Laboratories Inc]

Impression. High temperature on glaze colors. [Degussa Ltd]

Impressional. Alginate impression material; used in dentistry. [Bayer AG]

Imprez. Petroleum resins. [ICI Chem. & Polymers Ltd]

Impriment Black 7101-A. Pigment dyestuffs. [Catawba-Charlab]

Imprimus®. Conc. pigment pastes for sheet offset printing inks. [BASF AG]

Improved Kelmar®. CAS 9005-36-1; Potassium alginate; used for dietetic and low-sodium foods, dry mixes, dental impression material, surgical impressions. [Kelco]

Impsonite. An asphaltic substance found in Arkansas and other places. It can be fused with difficulty, and is only slightly soluble in carbon disulfide.

Imron®. Urethane finish; for the automotive industry. [DuPont UK]

Imsil® A-10. Silica, amorphous; filler/extender in solv.-based or latex paints; thickener, carrier, free flow agent; powd. coating; paint flatting agent; used in protective and wire and cable coatings, PU elastomers, epoxy, phenolic, buffing, polishing and polyester compounds, elec. resistors, refractory prods., insulations, antiblock agents, toothpastes, agric. compositions, cosmetic, metal castings, injection thermoset moldings. [Unimin]

Imsol. Solvents. [ICI Chem. & Polymers Ltd] †

Imuran. CAS 446-86-6; A proprietary formulation of azathioprine; an immunosuppressive agent used in the management of organ transplant recipients and the treatment of patients with certain conditions considered to result from auto-immune phenomena. [The Wellcome Foundation Ltd]

Imwitor®. Emulsifier for the pharmaceutical industry, cosmetic industry, and food industry. [Dynamit Nobel Wien GmbH]

Imwitor®191. CAS 31566-31-1; 68308-54-3; Glyceryl stearate; coemulsifier, dispersant for personal care prods.; emulsifier in o/w and w/o emulsions; lubricants and binders used in the pharmaceutical industry; suspending agent, stabilizer, thickener; food emulsifier. [Hüls Am.; Hüls AG]

Imwitor®308. CAS 26402-26-6; Glyceryl caprylate; solubi-

lizer for pharmaceutical drugs; carrier/vehicle for drugs in capsules; coemulsifier for lipophilic materials. [Hüls Am.]

Imwitor®312. CAS 142-18-7; EINECS 205-526-6; Glyceryl laurate; coemulsifier for o/w emulsions; solubilizer, carrier for lipophilic drugs; superfatting agent for bath prods.; bacteriostatic effect. [Hüls Am.; Hüls AG]

Imwitor®369. CAS #91744-38-6; Monoglyceride citric ester; food emulsifier. [Hüls Am.; Hüls AG]

Imwitor® 370. CAS 91744-38-6; Glyceryl stearate citrate; food emulsifier; o/w emulsifier for very polar oils, fats and liq. wax esters in cosmetics. [Hüls Am.; Hüls AG]

Imwitor®375. Glyceryl citrate/lactate/linoleate/oleatel; o/w emulsifier for very polar oils and fats; for oil baths, cream baths. [Hüls Am.; Hüls AG]

Imwitor®742. CAS 26402-26-6; Caprylic/capric glycerides; plasticizer for hard fats, solv. for lipophilic ingreds.; emollient; coemulsifier, solubilizer, carrier for lipophilic drugs; dispersant, absorp. promoter; bacteriostatic effect. [Hüls Am.; Hüls AG]

Imwitor® 780 K. CAS 66085-00-5; Isostearyl diglyceryl succinate; w/o emulsifier for polar oils and fats, cosmetic and pharmaceutical preparations. [Hüls Am.; Hüls AG]

Imwitor®914. Glyceryl myristate; coemulsifier for o/w emulsions; solubilizer, carrier for lipophilic drugs; dispersant, consistency regulator in creams, lotions, anhyd. formulations. [Hüls Am.; Hüls AG]

Imwitor®928. CAS 61789-05-7; EINECS 263-027-9; Glyceryl cocoate; surfactant for pharmaceutical, cosmetic, and nutritional fields; as emulsifier, solubilizer, dispersion aid, plasticizer, lubricant, consistency regulator, skin and mucous membrane protectant, refatting agent, penetrant, carrier, adsorp. promoter. [Hüls AG]

Imwitor® 960. Glyceryl stearate SE; emulsion base for cosmetics; suspending agent in creams, lotions, emulsions, cosmetics. [Hüls Am.; Hüls AG]

Imwitor®988. CAS 26402-26-6; Glyceryl caprylate; surfactant for pharmaceutical, cosmetic and nutritional fields; as emulsifier, solubilizer, dispersion aid, plasticizer, lubricant, consistency regulator, skin and mucous membrane protectant, refatting agent, penetrant, carrier, adsorp. promoter. [Hüls AG]

Inacid. CAS 53-86-1; Indomethacin; analgesic and anti-inflammatory agent for the prompt and sustained relief of certain rheumatic conditions. (Spain) [Merck & Co Inc]

Inalium. An aluminum alloy containing 2% cadmium, 0.8% magnesium, and 0.4% silicon.

Inapassade, A. proprietary preparation of sodium para-aminosalicylic acid and isoniazid; antituberculous drug. [Geistlich Sohne AG] *

Inapsine. CAS 548-73-2; Droperidol; antipsychotic. [Janssen Pharmaceutica Inc] *

Incidal. Pharmaceutical preparation also known as Fabahistin; day-time antihistamine. [Bayer AG]

Incoblend. Flexible vinyl compound based on intrinsically conductive polymer (polyaniline); for applications requiring high elec. conductivity, such as EMI shielding. [Allied-Signal; Americhem; Zipperling Kessler]

Inco Chrome Nickel. Proprietary nickel alloys containing 12-14% chromium and 6-7% iron; they resist corrosion by the organic acids met within foodstuffs; the alloy containing 80% nickel, 14% chromium, and 6% iron melts at 1390 C. [Inco Alloys Int'l. Inc]

Incoloy Alloy 800. A trademark for an alloy of 20% chromium, 32% nickel and 48% iron; resistant to hydrogen/

---

‡ Trade name and manufacturer not verified   § Trade name without identified manufacturer

hydrogen sulfide corrosion. [Inco Alloys Int'l. Inc; Wiggin Alloys Ltd] *

Incoloy Alloy 825. A trademark for an alloy of 40% nickel, 21% chromium, 3% molybdenum, 2% copper, 1% titanium and the balance iron; resistant to corrosion in hot oxidizing acid conditions. [Inco Alloys Int'l. Inc; Wiggin Alloys Ltd] *

Incoloy Alloy 901. A trademark for an alloy of 12.5% chromium, 5.7% molybdenum, 2.9% titanium, 42% nickel and the balance iron. [Inco Alloys Int'l. Inc; Wiggin Alloys Ltd] *

Incoloy Alloy DS. A trademark for an alloy of 18% chromium, 2.3% silicon, 37% nickel and the balance iron. [Inco Alloys Int'l. Inc; Wiggin Alloys Ltd] *

Incomparable. Soldering fluxes. [Reade Metals & Minerals Corp] ‡

Inconel. A proprietary trade name for an alloy resistant to corrosion and heat, containing 80% nickel, 14% chromium, and 6% iron. [Inco Alloys Int'l. Inc] *

Inconel Alloy 600. A trademark for an alloy of 16% chromium, 7% iron and 77% nickel; good oxidation resistance at high temperatures. [Inco Alloys Int'l. Inc; Wiggin Alloys Ltd] *

Inconel Alloy 700. A trademark for an alloy of 15% chromium, 29% cobalt, 3% molybdenum, 2.25% titanium, 3.3% aluminum, and the balance nickel. [Inco Alloys Int'l. Inc; Wiggin Alloys Ltd] *

Inconel Alloy 718. A trademark for an alloy of 19% chromium, 3% molybdenum, 0.8% titanium, 5% niobium, 53% nickel, and the balance iron. [Inco Alloys Int'l. Inc; Wiggin Alloys Ltd] *

Inconel Alloy X-750. A trademark for an alloy of l5% chromium, 2.5% titanium, 0.9% aluminum, 0.9% niobium, 7% iron, and the balance nickel. [Inco Alloys Int'l. Inc; Wiggin Alloys Ltd] *

Incrocas 30. CAS 61791-12-6; PEG-30 castor oil; emulsifier, solubilizer, emollient, superfatting agent, lubricant for personal care prods., detergents, metalworking fluids, insecticides, herbicides, household prods.; lubricant, antistat, softener, dye leveling agent for textiles; also lime soap dispersant in alkaline scouring systems. [Croda Inc.] *

Incrodet TD7-C. Trideceth-7 carboxylic acid; mild surfactant for shampoos, bath gels, cleansers; neutralizer for trace caustic or thioglycolate residues present after hair permanents or relaxers; in neutralizing shampoos which follow ethnic hair straighteners; lime soap dispersant; emulsifier; stable at low and high pH. [Croda Inc.]

Incromate ALL. Almondamidopropyl dimethylamine lactate; conditioner providing slip for wet comb in hair prods. [Croda Inc.]

Incromate BAL. Babassamidopropyl dimethylamine lactate; conditioner with good foam for improved dry comb. [Croda Inc.]

Incromate CDL. CAS 68425-42-3; Cocamidopropyl dimethylamine lactate; surfactant, foamer, conditioner for personal care prods., conditioners; base for cationic emulsions. [Croda Inc.]

Incromate CDP. CAS 68425-43-4; Cocamidopropyl dimethylamine propionate; moderate foaming conditioner for clear rinses and shampoos; good detangling. [Croda Inc.]

Incromate IDL. CAS 55852-15-8; Isostearamidopropyl dimethylamine lactate; substantive conditioner improving

slip, wet comb and manageability of hair, feel in hand creams and lotions. [Croda Inc.]

Incromate ISML. Isostearamidopropyl morpholine lactate; substantive surfactant, conditioner, visc. builder for personal care prods. and cationic emulsions; improves slip, wet comb, and manageability in hair, feel in hand creams and lotions. [Croda Inc.]

Incromate Mink L. Minkamidopropyl dimethylamine lactate; foamer, conditioner for hair and skin care prods. [Croda Inc.]

Incromate ODL. Oleamidopropyl dimethylamine lactate; a cationic salt used in clear rinses, conditioners, conditioning shampoos, compatible in anionic systems. [Croda Chem. Ltd] *

Incromate OLL. Olivamidopropyl dimethylamine lactate; conditioner and foamer for hair and skin care prods. [Croda Inc.]

Incromate SDL. CAS 55819-53-9; EINECS 259-837-7; Stearamidopropyl dimethylamine lactate; visc. builder, opacifier, softener for hair shampoos, conditioners, fabrics; emulsifier for hand creams, cleansers, lotions; raw material. [Croda Inc.]

Incromate SEL. Sesamidopropyl dimethylamine lactate; conditioner and foamer for good slip, wet combing, and detangling. [Croda Inc.]

Incromate WGL. Wheat germamidopropyl dimethylamine lactate; conditioner with excellent slip, wet comb, and dry feel. [Croda Inc.]

Incromectant AMEA-100. CAS 142-26-7; EINECS 205-530-8; Acetamide MEA; clarifying detangling agent for shampoos, conditioners, cream rinses; humectant in creams and lotions. [Croda Inc.]

Incromectant AQ. Acetamidopropyl trimonium chloride; antistat for shampoos and conditioners; humectant; plasticizer for hair conditioning/setting polymers. [Croda Inc.]

Incromectant LAMEA. Acetamide MEA and lactamide MEA; humectant, moisturizing agent for hair and skin care prods. [Croda Inc.]

Incromectant LMEA. CAS 5422-34-4; EINECS 226-546-1; Lactamide MEA; clarifying detangling agent for shampoos, conditioners, and cream rinses; humectant in creams and lotions. [Croda Inc.]

Incromectant LMEA-70. Lactamide MEA; conditioning agents in cream rinses, humectants in creams and lotions. [Croda Surf. Ltd]

Incromectant LQ. Lactamidopropyl trimonium chloride; antistat for shampoos and conditioners; humectant; plasticizer for hair conditioning/setting polymers. [Croda Inc.]

Incromide ALD. Almondamide DEA (1:1); conditioner, visc. builder, foam stabilizer; for shampoos, bubble baths, soaps, bath prods. [Croda Inc.]

Incromide BAD. Babassuamide DEA (1:1); visc. builder, foam stabilizer, emulsifier; for shampoos, bubble baths, soaps, bath prods. [Croda Inc.]

Incromide BED. CAS 70496-39-8; Behenamide DEA (1:1); high melting pearling and opacifying agent used in stick preps. [Croda Inc.]

Incromide BEM. CAS 94109-05-4; Behenamide MEA (1:1); surfactant structural wax for antiperspirant and other stick prods. [Croda Inc.]

Incromide CA. Cocamide DEA; surfactant, foam stabilizer, emulsifier and thickener used in household, cosmetic, and industrial formulations. [Croda Inc.]

Incromide CM. CAS 68140-00-1; Cocamide MEA; foam stabilizer, thickener, opacifier for cosmetic, industrial, household formulations. [Croda Inc.]

Incromide L-90. CAS 120-40-1; EINECS 204-393-1; Lauramide DEA; foam stabilizer, thickener, detergent, and foaming agent in household, industrial and institutional cleaning comps., car washes, rug and upholstery cleaners, and personal care prods. [Croda Inc.]

Incromide LA. Linoleamide DEA; superfatting agent and thickener for personal care and household prods.; useful in increasing visc. of various sulfate and ether sulfate dilutions; conditioner and lubricant. [Croda Inc.]

Incromide LCL. CAS 142-78-9; EINECS 205-560-1; Lauramide MEA; surfactant. [Croda Inc.]

Incromide LLT. CAS 120-40-1; EINECS 204-393-1; Lauramide DEA. [Croda Inc.]

Incromide LM-70. CAS 120-40-1; EINECS 204-393-1; Lauramide DEA; foam stabilizer, thickener, detergent, and foaming agent in household, industrial and institutional cleaning comps., car washes, rug and upholstery cleaners, and personal care prods. [Croda Inc.]

Incromide Mink D. Minkamide DEA; visc. builder, foam stabilizer for shampoos, bubble baths, liq. soaps, bath prods. [Croda Inc.]

Incromide OD. CAS 93-83-4; EINECS 202-281-7; Oleamide DEA (1:1); low color visc. builder with good sol. in anionic surfactants; for shampoos, bubble baths, soaps, bath prods. [Croda Inc.]

Incromide OLD. Olivamide DEA (1:1); visc. builder, foam stabilizer; for shampoos, bubble baths, soaps, bath prods. [Croda Inc.]

Incromide SED. Sesamide DEA (1:1); visc. builder and foam stabilizer; for shampoos, bubble baths, soaps, bath prods. [Croda Inc.]

Incromide WGD. Wheat germamide DEA (1:1); visc. builder and foam stabilizer; for shampoos, bubble baths, soaps, bath prods. [Croda Inc.]

Incromine BB. CAS 60270-33-9; EINECS 262-134-8; Behenamidopropyl dimethylamine; emollient conditioner, lubricant, visc. builder, and moisturizer for hair care prods.; intermediate for hair conditioning agent. [Croda Inc.; Croda Surfactants Ltd]

Incromine CB. CAS 68140-01-2; EINECS 268-771-8; Cocamidopropyl dimethylamine; foaming agent and conditioner used in hair care prods.; intermediate for hair conditioning agent. [Croda Inc; Croda Surfactants Ltd]

Incromine IB. Isostearamidopropyl dimethylamine; used as an intermediate for hair conditioners and shampoo rinses. [Croda Surf. Ltd]

Incromine OPB. CAS 109-28-4; EINECS 203-661-5; Oleamidopropyl dimethylamine; emollient conditioner, lubricant and moisturizer for hair care prods.; intermediate for hair conditioning agent. [Croda Inc; Croda Surfactants Ltd]

Incromine PB. CAS 39669-97-1; Palmitamidopropyl dimethylamine; intermediate, substantive conditioner, thickener, and emulsifier used in hair care prods. [Croda Inc.]

Incromine SB. CAS 7651-02-7; EINECS 231-609-1; Stearamidopropyl dimethylamine; visc. builder, conditioner for hair care prods.; intermediate for hair conditioning agent for shampoo rinses. [Croda Inc; Croda Surfactants Ltd]

Incromine Oxide AL. Almondamidopropylamine oxide; visc. builder, cationic conditioner for hair care prods. [Croda Inc.]

Incromine Oxide B-30P. CAS 26483-35-2; Behenamine oxide; softener and conditioner for hair care prods.; visc. builder, emulsifier, lubricant, wetting, and foam stabilizer used in personal care prods. [Croda Inc.; Croda Surf. Ltd.]

Incromine Oxide BA. Babassuamidopropylamine oxide; foamer, foam stabilizer; for hair care prods., facial cleaners. [Croda Inc.]

Incromine Oxide C. CAS 68155-09-9; Cocamidopropylamine oxide, aq. solution; visc. builder, foam stabilizer, conditioner used in personal care prods. [Croda Inc.; Croda Surf. Ltd.]

Incromine Oxide I. Isostearamidopropylamine oxide; foam stabilizer, thickener, lubricant, and visc. builder used in cosmetic, household, and janitorial prods.; wetting agent for conc. electrolyte solutions [Croda Inc.; Croda Surf. Ltd.]

Incromine Oxide ISMO. Isostearamidopropyl morpholine oxide; foam stabilizer, thickener, lubricant, and visc. builder used in cosmetic, household, and janitorial prods.; wetting agent for conc. electrolyte solutions [Croda Inc.]

Incromine Oxide L-40. Lauramine oxide; foaming agent, foam stabilizer, degreaser; for shampoos and light-duty liqs. [Croda Inc.]

Incromine Oxide M. CAS 3332-27-2; EINECS 222-059-3; Myristamine oxide; surfactant, emulsifier, emollient, conditioner, visc. builder, foam booster used in personal care prods. [Croda Inc.; Croda Surf. Ltd.]

Incromine Oxide Mink. Minkamidopropylamine oxide; foaming agent, visc. builder. [Croda Inc.]

Incromine Oxide O. CAS 25159-40-4; EINECS 246-684-6; Oleamidopropylamine oxide; foam booster/stabilizer, thickener, lubricant, and visc. builder used in cosmetic, household, and janitorial prods.; wetting agent for conc. electrolyte solutions [Croda Inc.; Croda Surf. Ltd.]

Incromine Oxide OD-50. CAS 14351-50-9; EINECS 238-311-0; Oleyl dimethylamine oxide; foam stabilizer, visc. builder for cosmetic and household prods. [Croda Surf. Ltd.]

Incromine Oxide OL. Olivamidopropylamine oxide; thickener for clear systems. [Croda Inc.]

Incromine Oxide S. CAS 2571-88-2; EINECS 219-919-5; Stearamine oxide; conditioner, emulsifier, visc. builder for personal care prods. [Croda Inc.; Croda Surf. Ltd.]

Incromine Oxide SE. Sesamidopropylamine oxide; visc. builder, conditioner for hair care prods. [Croda Inc.]

Incromine Oxide WG. Wheat germamidopropylamine oxide; visc. builder, conditioner for hair care prods. [Croda Inc.]

Incronam 30. CAS 61789-40-0; Cocamidopropyl betaine; surfactant, emulsifier, coupling agent, visc. builder, foam detergent for personal care prods., chemical specialities, rug and upholstery shampoos, dishwashing compounds. [Croda Inc.; Croda Surf. Ltd.]

Incronam AL-30. Almondamidopropyl betaine; foam booster/stabilizer for skin prods. [Croda Inc.]

Incronam B-40. CAS 84082-44-0; Behenyl betaine; foaming surfactant, conditioner, lubricant used in personal care prods.; excellent slip, conditioning; good wetting and rinse aid. [Croda Inc.; Croda Surf. Ltd.]

---

‡ Trade name and manufacturer not verified   § Trade name without identified manufacturer

Incronam BA-30. Babassamidopropyl betaine; foam booster/stabilizer for shampoos. [Croda Inc.]

Incronam CD-30. CAS 68424-94-2; EINECS 270-329-4; Coco betaine; mild, high foaming surfactant for detergent systems. [Croda Surf. Ltd.]

Incronam I-30. Isostearamidopropyl betaine; foaming surfactant, conditioner, lubricant, visc. builder used in personal care prods. [Croda Inc.; Croda Surf. Ltd.]

Incronam Mink 30. Minkamidopropyl betaine; conditioner; foam booster/stabilizer, visc. builder. [Croda Inc.]

Incronam OD-50. CAS 871-37-4; EINECS 212-806-1; Oleyl betaine; high foaming surfactant, visc. builder for conditioning shampoos, clear rinses, cleansing creams and cosmetic lotions. [Croda Surf. Ltd.]

Incronam OL-30. Olivamidopropyl betaine; foam booster/stabilizer for skin care prods. [Croda Inc.]

Incronam OP-30. CAS 25054-76-6; EINECS 246-584-2; Oleamidopropyl betaine; visc. builder, high foaming surfactant for shampoos, bubble baths, cleansing lotions, hand cleaners, skin care. [Croda Inc.; Croda Surf. Ltd.]

Incronam SE-30. Sesamidopropyl betaine; foam booster/stabilizer, conditioner with good slip. [Croda Inc.]

Incronam WG-30. CAS 133934-09-5; Wheat germamidopropyl betaine; foam booster/stabilizer. [Croda Inc.]

Incropol CS-20. CAS 68439-49-6; Ceteareth-20; surfactant, emulsifier, lubricant, detergent for industrial and household prods.; coupling agent, antistat, fiber lubricant and solubilizer for personal care prods. [Croda Inc.]

Incropol L-23. Laureth-23; detergent, wetting agent, emulsifier, coupling agent, antistat, fiber lubricant, solubilizer for cosmetic formulations. [Croda Inc.]

Incroquat 100. Methyl bis (hydrogenated tallow amidoethyl) 2-hydroxyethyl ammonium chloride; nonfoaming, strongly substantive conditioner, antistat for hair care prods. [Croda Inc.]

Incroquat 248. Quaternium-72; economical hair conditioner. [Croda Inc.]

Incroquat AL-85. Almondamidopropalkonium chloride; conditioner, foamer for hair care prods. [Croda Inc.]

Incroquat B65C. Behenalkonium chloride, cetyl alcohol; substantive conditioner for hair care prods. [Croda Inc.]

Incroquat BA-85. Babassamidopropalkonium chloride; conditioner, foamer, antistat for hair care prods. [Croda Inc.]

Incroquat CR Conc. Cetearyl alcohol, PEG-40 castor oil, stearalkonium chloride; self-emulsifying wax, conditioner, softener, emollient, o/w emulsifier used in cream rinses, conditioners. [Croda Inc.]

Incroquat CTC-30. CAS 112-02-7; EINECS 203-928-6; Cetrimonium chloride; conditioner, antistat for hair care prods. [Croda Inc.]

Incroquat DBM-90. Dibehenyldimonium methosulfate; conditioner with good wetting and slip for hair care prods. [Croda Inc.]

Incroquat I-85. CAS 67633-59-4; Isostearaminopropalkonium chloride; conditioner with good foaming, slip and detangling for hair prods. [Croda Inc.]

Incroquat Mink-85. Minkamidopropalkonium chloride; conditioner, foamer, excellent slip for cosmetics, hair care prods. [Croda Inc.]

Incroquat O-50. CAS 37139-99-4; Olealkonium chloride; conditioner with good slip for hair care prods. [Croda Inc.]

Incroquat OL-85. Olivamidopropalkonium chloride; conditioner with good foaming and slip for hair prods. [Croda Inc.]

Incroquat S-75CG. Quaternium-27; low foaming conditioner with dry feel on hair, good bodying. [Croda Inc.]

Incroquat SDQ-25. CAS 122-19-0; EINECS 204-527-9; Stearalkonium chloride; surfactant used as ingred. in personal care prods., textile and paper; dispersant for pigments and dyestuffs; antistat for fibers and synthetics; hair conditioner. [Croda Inc.; Croda Universal Ltd.]

Incroquat SE-85. Sesamidopropalkonium chloride; conditioner, antistat with good foam and dry feel for hair care prods. [Croda Inc.]

Incroquat WG-85. Wheat germamidopropalkonium chloride; conditioner with good foam and slip for hair prods. [Croda Inc.]

Incroquat Behenyl BDQ/P. Propylene glycol and behenalkonium chloride; o/w emulsifier and conditioner. [Croda Inc.; Croda Universal Ltd.]

Incroquat Behenyl TMC. Cetearyl alcohol, behentrimonium chloride. [Croda Inc.] *

Incroquat Behenyl TMC/P. Propylene glycol and behenalkonium chloride; o/w emulsifier and conditioner. [Croda Inc.; Croda Universal Ltd.]

Incroquat Behenyl TMS. Behenalkonium methosulfate, cetearyl alcohol; self-emulsifying wax, conditioner, softener, emollient, o/w emulsifier used in hair and skin care prods. [Croda Inc.]

Incrosoft 100. Methyl bis (hydrogenated tallow amido ethyl) 2-hydroxyethyl ammonium chloride; fabric softener with good hand and antistatic properties; for home, commercial or industrial laundry applications, textile and finishing operations. [Croda Inc.]

Incrosoft 248. Quaternium-72; softener for dryer sheets. [Croda Inc.]

Incrosoft CF1-75. A range of alkyl imidazolinium methosulfates; fabric softeners for preparation of clear detergent liquid concentrates. [Croda Surf. Ltd] †

Incrosoft S-75. CAS 86088-85-9; Quaternium-27; softener base, lubricant, antistat and rewetting agent for fabrics and syns. [Croda Inc.]

Incrosoft T-75. CAS 130124-24-2; Quaternium-53; fabric softener, lubricant and antistat for home and commercial laundry prods. [Croda Inc.]

Incrosperse. Polymeric dispersants. [Croda Resins Ltd]

Incrosul LAFS. CAS 68890-92-6; Disodium laneth-5 sulfosuccinate; mild, low foaming, conditioning surfactant with good emulsifying props. used in personal care prods. [Croda Inc.]

Incrosul LMA. Diammonium lauramido-MEA sulfosuccinate; mild foaming surfactant for cosmetic prods. [Croda Inc.]

Incrosul LMS. CAS 25882-44-4; EINECS 247-310-4; Disodium lauramido MEA-sulfosuccinate; mild foaming agent and cleanser used in personal care prods. and carpet shampoos; lime soap dispersant. [Croda Inc.]

Incrosul LS. Disodium lauryl sulfosuccinate; mild foaming and conditioning surfactant, visc. modifier for cosmetic prods. [Croda Inc.]

Incrosul LSA. Diammonium lauryl sulfosuccinate; mild, high foaming surfactant used in personal care prods.; lime soap dispersant. [Croda Inc.]

Incrosul LTS. Disodium laureth sulfosuccinate; mild high foaming surfactant used in shampoos and general cleansing preparations; low eye irritation for baby shampoos. [Croda Inc.]

---

* Trade name not verified as available   † Trade name verified as obsolete

Incrosul OMS. Disodium oleamido MEA-sulfosuccinate; mild high foaming surfactant used in personal care prods.; good conditioner. [Croda Inc.]

Incrosul OTS. Disodium oleth-3 sulfosuccinate; mild, conditioning surfactant, foamer for shampoos, baby shampoos, bath gels, mild skin cleansers; anti-irritant for anionic surfactants. [Croda Inc.]

Incrosul TS. Disodium tridecyl sulfosuccinate; mild high foaming surfactant used in personal care prods. [Croda Inc.]

Indacin. CAS 53-86-1; Indomethacin; analgesic and anti-inflammatory agent for the prompt and sustained relief of certain rheumatic conditions. (Japan) [Merck & Co Inc]

Indalca. A thickening agent derived from natural gum; thickener for fabric-printing dyes. [Hercules] *

2-indanone. (β-hydrindone) C_9H_8O; CAS 615-13-4; EINECS 210-410-3. [Aldrich; Penta Mfg.; Schweizerhall]

Indanthren®. A range of vat dyes, having unsurpassed overall fastness properties on cotton and other vegetable fibers as well as on textiles made from regenerated cellulose. [Bayer AG; BASF plc]

Indazin. A proprietary trade name for a range of products used in the dyeing of textiles. [Cassella AG] *

Inderal. CAS 318-98-9; A proprietary preparation of propranolol hydrochloride. [ICI Chem. & Polymers Ltd]

Inderetic. Combined antihypertensive and diuretic containing propranolol hydrochloride and bendrofluazide. [ICI Chem. & Polymers Ltd]

Inderex. Proprietary preparation of propranol; for treating hypertension. [Imperial Chemical Industries plc]

India gum. (Persian gum). A gum resembling gum arabic.

Indian ocher. A native ferric oxide of North America.

Indian Red. (Venetian red, Venetian bole, rouge, colcothar, red bole, bole, Armenian bole, English red, angel red, chemical red, Pompeian red, Berlin red, iron minium, iron red, Persian red, raddle, reddle, red rudd, red ocher, red chalk, red earth, iron saffron, Spanish oxide, Turkey red oxide) Red pigments consisting mainly of ferric oxide, with varying amounts of natural argillaceous compounds. Some are natural products, burnt or unburnt, but the name is also applied to products made by; heating ferric sulfate; used as pigments in paints, rubbers, and plastics, and as polishing agents.

Indian yellow. (Piuri, Purree, Pioury). A pigment used in India, obtained from the urine of cows fed on mango leaves. The coloring principal is the magnesium or calcium salt of euxanthic acid, $C_{19}H_{16}O_{11}Mg \cdot 5H_2O$; used as a permanent water and oil color.

India tragacanth. See karaya gum.

indican. Potassim-indoxyl-sulfate.

indicator, universal. A mixture of indicators, usually methyl red, α-naptholphthalein, phenolphthalein, bromothymol blue, and cresol red. The color indicates the pH value when added to the solution.

indicolite. A mineral from Brazil. It is a blue tourmaline.

Indigal. Substrate for β-galactosidase. [U.S. Biochemical]

Indigo. CAS 482-89-3; Indigotine, $C_{16}H_{10}N_2O_2$. Natural indigo is obtained by steeping the leaves of indigo-bearing plants in water then Oxidizing the extract. Synthetic indigo is prepared by several methods; used for cotton, wool, silk, by steeping the material in a vat containing containing the leuco compound, then exposing to air.

Indigo Carmine. Indigotindisulfonate sodium; diagnostic aid. [Hynson Westcott & Dunning] *

Indilitans. An alloy of 36% nickel, 0.06% carbon, 0.68% manganese, 0.09% silicon, and remainder iron. The alloy has a low coefficient of thermal expansion.

Indio. Decorating colors for vitreous enamel. [Degussa Ltd]

indirubin. Indigo red; $C_{16}H_{10}N_2O_2$, a red coloring matter associated with indigo in amount usually from 1-5%. Java indigo often contains up to 10%. It is produced from the decomposition of indican.

Indisin®. A registered trademark currently awaiting re-allocation by its proprietors. [Cassella AG] *

indium. Metallic element; In; CAS 7440-74-6; EINECS 231-180-0; automobile bearings, electronic and semiconductor devices, brazing and soldering alloys, reactor control rods, electroplated coatings on aircraft bearings. [Aldrich; Atomergic Chemetals; Cerac; Noah Chem.]

indium oxide. In_2O_3; CAS 1312-43-2; EINECS 215-193-9; Manufacture of special glasses. [Atomergic Chemetals; Cerac; Noah Chem.]

Indocarbon. Black sulfur dyestuffs; used for textile dyeing and printing. [Cassella AG]

Indocid®, -R. CAS 53-86-1; Indomethacin; analgesic and anti-inflammatory agent for the prompt and sustained relief of certain rheumatic conditions. [Merck & Co Inc]

Indocin. CAS 53-86-1; Indomethacin; analgesic and anti-inflammatory agent for the prompt and sustained relief of certain rheumatic conditions. (U.S.) [Merck & Co Inc]

Indocybin. A proprietary trade name for psilocybin. §

Indofast®. High-quality organic pigments for the surface coatings industry, especially automotive coatings. [Bayer AG]

Indoil CPD 142 and CPD 143. A proprietary trade name for petroleum type vinyl plasticizers. [Standard Oil Co] *

Indoklon. A proprietary trade name for flurethyl. §

Indolar. Suppositories containing 100 mg indomethacin BP, capsules containing 25 mg, 50 mg, SR capsules containing 75 mg indomethacin BP; nonsteroidal analgesic and anti-inflammatory agent indicated in active rheumatoid arthritis, osteoarthritis, ankylosing spondylitis, acute gout; also indicated in periarticular disorders such as bursitis, tendinitis, synovitis, tenosynovitis and capsulitis and in inflammation, pain and edema following orthopedic procedures. [Lagap Pharmaceuticals Ltd]

3-indole acetic acid. (IAA; heteroauxin; β-indole acetic acid) $C_{10}H_9NO_2$; CAS 87-51-4; EINECS 201-748-2; Regulates cell membrane electron transport and proton flux. [Atomergic Chemetals; Biosynth Int'l.; Penta Mfg.]

β-indole acetic acid. See 3-indole acetic acid.

indole-α-aminopropionic acid. See DL-tryptophan.

3-indolebutyric acid. (IBA; 4-(3-indolyl)butyric acid; hormodin) $C_8H_6N(CH_2)_3COOH$; CAS 133-32-4; Plant hormone, esp. used in rooting plants. [Biosynth AG; Penta Mfg.; Pfaltz & Bauer; Spectrum Chem. Mfg.]

indoline. A basic navy blue dye.

Indomed. CAS 53-86-1; Indomethacin; anti-inflammatory. [Rowell Laboratories Inc] *

Indomee. CAS 53-86-1; Indomethacin; analgesic and anti-inflammatory agent for the prompt and sustained relief of certain rheumatic conditions. (Sweden) [Merck & Co Inc]

Indon. CAS 83-12-5; Phenindione; anticoagulant. [Parke-Davis] *

Indonex VG. A proprietary trade name for aromatic hydrocarbon vinyl plasticizer. [Standard Oil Co] *

Indopol. Polybutenes. [Amoco Chemical Co]

Indoptol. CAS 53-86-1; Indomethacin; eye drops for the prevention of aphakic cystoid macular edema. [Merck & Co Inc]

Indorm. CAS 362-29-8; A proprietary preparation of propiomazine; a hypnotic. [Wyeth Laboratories] *

Induclor. CAS 7778-54-3; Calcium hypochlorite. [PPG Industries]

Indulin®201. Tall oil fatty acid; emulsifier for soap solutions for high float emulsions. [Westvaco]

Indulin®AQS. Tall oil amino polycarboxylic acid; emulsifier for asphalt emulsions. [Westvaco]

Indulin®MQK. C21 dicarboxylic amido alkyl amine; asphalt emulsifier for quick set slurry seal applications. [Westvaco]

Indulin® SA-L. Sodium lignate; emulsifier, stabilizer, retarder for asphalt emulsions. [Westvaco]

Indulin®W-1. CAS 110152-58-4; Amine deriv. of pine lignin; emulsifier, stabilizer in asphalt emulsions, retarding agent in cement; also for oilfield chems. [Westvaco]

Indulin® XD-70. CAS 9016-45-9; Nonylphenol ethoxylate; coemulsifier for anionic or cationic slow-setting asphalt emulsions. [Westvaco]

Induline. A mixture of aryl-amino-azines, made by heating together aminoazobenzene, aniline, and aniline hydrochloride; used in making inks.

Indur. A proprietary trade name for a phenol-formaldehyde synthetic resin used for molding. §

Indurite. An explosive containing 40% guncotton, freed from lower nitrates, and 60% nitro-benzene. The name is also applied to a molding powder of the phenolformaldehyde condensation product type.

Indusoil. A proprietary product; it is a refined Talleol. §

Industrene® 104. CAS 112-80-1; EINECS 204-007-1; Oleic acid, low titer; chemical intermediate. [Witco/Humko]

Industrene® 120. CAS 463-40-1; EINECS 207-334-8; Linolenic acid; chemical intermediate. [Witco/Humko]

Industrene®126. Soya acid; chemical intermediate. [Witco/Humko]

Industrene®143. CAS 61790-37-2; Tallow acid; intermediate used in alkyd resins, rubber compounding, water repellents, polishes, soaps, abrasives, cutting oils, candles, crayons, emulsifiers; FG grades as lubricant, release agent, binder, defoamer in foods, intermediate for food emulsifiers. [Witco/Humko]

Industrene®223. CAS 68938-15-8; Hydrogenated coconut acid; chemical intermediate, emulsifier; used for personal care prods., waxes, textile auxiliary, pharmaceuticals. [Witco/Humko]

Industrene®224. Oleic-linoleic acid; chemical intermediate. [Witco/Humko]

Industrene® 365. CAS 67762-36-1; Mixt. caprylic/capric acid; intermediate used in alkyd resins, rubber compounding, water repellents, polishes, soaps, abrasives, cutting oils, candles, crayons, emulsifiers; FG grades as lubricant, release agent, binder, defoamer in foods, intermediate for food emulsifiers. [Witco/Humko]

Industrene®4516. CAS 57-10-3; EINECS 200-312-9; 45% Palmitic acid; intermediate for mfg. of soaps, emulsions, creams, lotions, ethoxylates, buffing compounds, lubricants. [Witco/Humko]

Industrene® 4518. CAS 57-11-4; EINECS 200-313-4; Single pressed stearic acid; intermediate used in alkyd resins, rubber compounding, water repellents, polishes,

soaps, abrasives, cutting oils, candles, crayons, emulsifiers; FG grades as lubricant, release agent, binder, defoamer in foods, intermediate for food emulsifiers. [Witco/Humko]

Industrene® 5016. CAS 57-11-4; EINECS 200-313-4; Double pressed stearic acid; intermediate used in alkyd resins, rubber compounding, water repellents, polishes, soaps, abrasives, cutting oils, candles, crayons, emulsifiers; FG grades as lubricant, release agent, binder, defoamer in foods, intermediate for food emulsifiers. [Witco/Humko]

Industrene®7018 FG. CAS 57-11-4; EINECS 200-313-4; 70% Stearic acid; lubricant, release agent, binder, defoamer for foods; intermediate for food-grade emulsifiers. [Witco/Humko]

Industrene®D. 40% Dimer acid; intermediate for lubricants, corrosion inhibitors for petrol. industry, extenders and cross-linking agents for high polymeric systems, in hot-melt adhesives, epoxy curing agents. [Witco/Humko]

Industrene®R. CAS 57-11-4; Hydrogenated rubber grade stearic acid; intermediate used in alkyd resins, rubber compounding, water repellents, polishes, soaps, abrasives, cutting oils, candles, crayons, emulsifiers; FG grades as lubricant, release agent, binder, defoamer in foods, intermediate for food emulsifiers. [Witco/Humko]

Industrial, cosmetic, or platy talc. See talc.

Industrial Dyne. Iodophor/acid pipeline cleaner and sterilizer. [Ciba plc] *

Industrial Dynemate. Hypochlorite pipeline cleaner/sterilizer. [Ciba plc] *

industrial gum. (tragasol, locust bean gum). Carob bean gum; an ingredient of mucilages. It is also used as a protective colloid.

Industrol® DW-5. Modified oxyethylated alcohol; biodeg., chlorine-stable, low-foaming detergent, dispersant, wetting agent, emulsifier for home/commercial machine dishwash, spray cleaners; defoamer for protein soils. [BASF]

Industrol® N3. CAS 9003-11-6; EO/PO block copolymer; detergent, dispersant, wetting agent for low to high temp. rinse aids where low foam and sheeting are important. [BASF]

Industrol®TFA-8. Fatty acid ethoxylate (8 EO); surfactant. [BASF]

Industrol®TO-16. CAS 61791-00-2; PEG-16 tallate; surfactant. [BASF]

Inertex. Protective powder for molten magnesium [Foseco (F.S.) Ltd]

Infacare. Baby bath, additive for bath. [Richardson-Vicks Inc] *

Infacol®. Dimethicone; used for treatment of infant colic. [Pharmax Ltd]

Infasoft. Shampoo, to wash and condition hair. [Richardson-Vicks Inc] *

Infavina. Vitamin B complex; tonic. [Merck & Co Inc] †

InFilm. Kaolin clay or metakaolinitic aluminum silicate produced by calcining kaolin clay; sold in powder form or as a slurry in ethylene glycol; antiblocking agents for plastic films. [ECC International Ltd]

Inflamase®Forte 1% Ophthalmic Solution. CAS 125-02-0; Prednisolone sodium phosphate; recommended for moderate to severe inflammations, particularly when unusually rapid control is desired; in stubborn cases of anterior segment eye disease, systemic adrenocortical

---

* Trade name not verified as available   † Trade name verified as obsolete

hormone therapy may be required; when the deeper ocular structures are involved, systemic therapy is necessary. [Iolab Corp]

Inflamase®Mild 1/8% Ophthalmic Solution. CAS 125-02-0; Prednisolone sodium phosphate; indicated for treatment of steroid-responsive inflammatory conditions of the palpebral and bulbar conjunctiva, cornea, and anterior segment of the globe (e.g., allergic conjunctivitis, acne rosacea, superficial punctate keratitis, iritis, etc.); corneal injury form chemical, radiation or thermal burns, or penetration of foreign bodies. [Iolab Corp]

Influvac. Influenza virus vaccine; for protection against virus infection. [Duphar BV] *

Infrotto®. Analgetic and antirheumatic pharmaceuticals. [Cassella AG]

infusible white precipitate. Dimercuridiammonium chloride.

infusorial earth. CAS 61790-53-2; (celite, fossil flour, kieselguhr, Tripolite, mountain flour). The siliceous remains of diatoms; used as a non-conducting material for boilers, an absorbent for liquids and liquid manures, in the preparation of dynamite, extender in paints, rubbers, and plastic products, ceramics, paper coatings, anticaking agents, fertilizers, asphalt compositions, drilling mud thickener.

infusorial earth. See diatomaceous earth.

Ingalite. See Gallatite.

Ingotol. Nonferrous chill and ingot dressing. [Foseco (F.S.) Ltd]

INH. CAS 54-85-3; Isoniazid; antibacterial. [Ciba-Geigy Corp]

Inhalit. A proprietary preparation of menthol, eucalyptol and pumilio pine oil; a nasal decongestant. [Ayrton Saunders plc]‡

Inhibitor. Gel time/pot life extenders for use with unsaturated polyester resins. [Akzo Chemie UK Ltd]

Inhibitor 60S. corrosion inhibitor for acid cleaning, pickling, industrial cleaning, consumer prods. [Exxon/Tomah]

Inhibitor RT 212. Fatty acid alkanolamide; anticorrosion additive for metal cleaning and metal working; lubricant in drilling and cutting oils. [Zschimmer & Schwarz] *

Inidal. Mebhydrolin napadisylate; daytime antihistamine. [Bayer AG]

Inipol. Fatty diamines and derivatives; cationic surfactant for organic solvent media for paint and pigment dispersion, corrosion inhibitors, lubricity enhancers. [Atochem UK/Ceca]

Initial®. For the agriculture industry. [DuPont UK]

Initiator BK. Silyl ether of benzopinacol in triethylphosphate (30%) and toluene (5%); initiator for radical polymerization processes with vinyl monomers, acrylic compounds or olefins; also suitable for BMC and SMC polyester prods. [Miles; Polysar]

Injacom. A proprietary vitamin injection for animals. [Roche Products Ltd] *

Injex. Iron/dextran injection. [The Wellcome Foundation Ltd]*

Inklurit®. Urea-formaldehyde precondensation product; water-sol., in conjunction with emulsifiers and hardening agent for enclosing and mfg. water-immiscible substances, e.g., liq. petrol. prods. [BASF AG]

Inkovar 335. Modified hydrocarbon resins; designed specifically for compatibility with cellulosic polymers. [Hercules] *

Inkovar 617. Modified hydrocarbon resins; designed for low energy heatset ink vehicles and flushed colors. [Hercules] *

Inkrustin®. Fluxes for wet and dry galvanizing, tinplating and leadcoating; as a soldering machine flux in the mfg. of tin cans. [BASF AG]

Innovace®. CAS 76095-16-4; Enalapril maleate; for the treatment of hypertension and congestive heart failure. [Merck & Co Inc]

Innovex. Polyethylene. [BP Chemicals Ltd]

Inochrome. Premetallized dyes. [ICI Chem. & Polymers Ltd]

Inocor. CAS 60719-84-8; Amrinone; cardiotonic. [Sterling Drug Inc] *

Inoculin. Inoculating compound for cast iron. [Foseco (F.S.) Ltd]

Inoderme. Leather dyes. [ICI Chem. & Polymers Ltd]

Inopak. Prepacked mold inoculants for use in production of grey and ductile iron castings. [Foseco (F.S.) Ltd]

inositol. (hexahydroxy cyclohexane; myo-Inositol; meso-Inositol) Cyclic polyol; constituent of body tissue; $C_6H_6(OH)_6 \cdot 2HOH$; CAS 87-89-8; EINECS 201-781-2; medicine, nutrition; intermediate. [Allchem Industries; R.W. Greeff; Mitsui Toatsu Chem.; Remy & Co; U.S. Biochemical]

Inotab. Tableted mold inoculants for use in production of grey and ductile iron castings. [Foseco (F.S.) Ltd] †

Insect Bite Cream (AS). A proprietary preparation of antazoline, benzocaine and cetrimide; an antihistamine cream. [Ayrton Saunders plc]‡

Insect Spray for House Plants. Contact fertilizer aerosol. [Fisons plc, Horticulture Div] *

insect wax. See Chinese wax.

Insektigun. Bioallethrin + permethrin; used for control of flies in agricultural premises. [Spraydex Ltd]

Insidon. CAS 909-39-7; A proprietary preparation of opipramol dihydrochloride; a sedative. [Ciba plc] *

Insol-U & RM. Urea-formaldehyde concentrate. [Georgia-Pacific]

Insomnol. A proprietary preparation of methylpentynol; a hypnotic. §

Instant Gasket. RTV silicone gasketing material; replaces cork, felt paper, asbestos, rubber and metal gaskets; gasket made in situ squeezed from the tube using unique applicator nozzle. [Hermetite Products Ltd] *

Instant Ocean. A dry, granular mixture of salts for preparation of synthetic seawater; used in aquariums for marine animals, culture medium and corrosion testing. [Aquarium Systems Inc]

Instoms. Indigestion tablets. [Fisons plc, Pharmaceuticals Div] *

instrument bronze. An alloy of 82% copper, 13% tin, and 5% zinc.

Insulgard. Polycarborate sheet; for conservatory and glasshouse glazing. [Insulgard Ltd]

insulin. CAS 9004-10-8; A hormone isolated from the pancreas; used in the treatment of diabetes.

Insullac. A copal spirit varnish with the resin acids neutralized; an insulating material.

Insulmag®. CAS 1309-48-4; Magnesium oxide for electrical cable filling. [Steetley Magnesia Products Ltd]

Insural. Insulating refractory shapes for use with aluminum alloys. [Foseco (F.S.) Ltd]

Insurok. A proprietary trade name for a phenol-formalde-

---

‡ Trade name and manufacturer not verified   § Trade name without identified manufacturer

hyde synthetic resin used for molding compounds and laminated products. §

Intal Compound. A proprietary preparation of disodium cromoglycate and isoprenaline sulfate; a bronchial inhalant. [Fisons plc, Pharmaceuticals Div] *

Integrin. CAS 153-87-7; A proprietary preparation of oxypertine; a tranquillizer. §

Integuard. Plastics materials in the form of films, sheets, strips or labels; tamper-evident plastic films for sealing and packaging applications. [Courtaulds plc]

Intensaïn®. INN: carbochromene; cardiac preparations. [Cassella AG]

Interacton. A proprietary preparation of cocarboxylase, glutaminase, histaminase, Co-enzyme A, ascorbic and allyl sulfide. [FAIR Laboratories]‡

Intercept. CAS 9016-45-9; Nonoxynol 9; spermaticide. [Ortho Pharmaceutical Corp] *

Intercide. Biocide/fungicide for PVC. [Akzo Chemie UK Ltd]

interferon. A protein formed by the interaction of animal cells with viruses. It is capable of conferring resistance to virus infection on animal cells .

internol. See paraffin, liquid.

Interol. See paraffin, liquid.

Interox H48. Magnesium monoperoxyphthalate hexahydrate low temperature bleach. [Interox Chemicals Ltd] *

Intersept®. Synergistic blend of substituted ammonium salts of alkylated phosphoric acids admixed with a free alkylated phosphoric acid; broad-spectrum biostat. [Interface Research] *

Interstab®. Lead and mixed metal heat stabilizers for PVC. [Akzo Chemie UK Ltd]

Interstab®BZ-4828A. Liq. metal soap stabilizer for flexible PVC [Akzo]

Interstab® BZ-4836. Barium-zinc-phosphite complex; liq. metal soap heat and light stabilizer for flexible PVC [Akzo]

Interstab® FR930. Fire retardant for polypropylene and polystyrene; lubricants for PVC. [Akzo Chemie UK Ltd]

Interstab®LT-4805R. Barium-zinc; heat and light stabilizer for flexible PVC [Akzo]

Interwood®. Machines, tools, parts and fittings for woodworking. [Interwood Ltd]

Intimate Contact. Silicone release film comprising super clear polyester coated with silicone release; release film for optical applications. [Custom Coating & Laminating Corp] *

Intob. Mold inoculants for iron castings. [Foseco (F.S.) Ltd]†

Intracarrier®ATM. Blend of esters and aromatic hydrocarbon derivs. with nonionic emulsifier; biodeg. dyeing accelerant for disperse dyeing of synthetic fibers and their blends; carrier for polyester and triacetate; moderate foaming; high flash pt. [Crompton & Knowles]

Intradex. A proprietary preparation of dextran in saline or dextrose solution; an intravenous infusion used to restore blood volume. [Glaxo Laboratories] *

Intraflodex. A proprietary preparation of dextran in saline or dextrose; used to improve capillary circulation. [Glaxo Laboratories] *

Intrafomil® AK. Nonsilicone; defoamer for textile operations, esp. jet equip. [Crompton & Knowles]

Intral®. Mineral iron sulfide (pyrite); 48-50% S, 44-45% Fe; used for manufacture of amber glass. [Metallgesellschaft AG]

Intralan®Salt HA. Condensed sodium alkylaryl sulfonate;

low foaming dispersing and leveling agent for disperse dyes and dyeing of synthetic fibers; esp. for beam, package and jet dyeing. [Crompton & Knowles]

Intralan® Salt N. EO condensate; dye assistant, leveling agent, penetrant for textiles. [Crompton & Knowles] *

Intralgin Gel. Benzocaine BP 2% w/w and salicylamide 5% w/w in an alcoholic gel vehicle. [3M Pharmaceuticals] †

Intralipid. Fat emulsion for parenteral use; high-energy source for intravenous nutrition. [KabiVitrum AB] *

Intraphasol COP. Hydrocarbons, solubilized, aliphatic and sulfonic acid salts; emulsifier, wetting agent assistant for dyeing of fabrics and carpets. [Crompton & Knowles] *

Intraquest®TA Solution CAS 64-02-8; EINECS 200-573-9; Tetrasodium EDTA; sequestering agent for scouring, bleaching, dyeing, and other wet finishing textile operations; softens water by complexing calcium, magnesium, and divalent heavy metal ions over broad pH range; detergent builder for scouring. [Crompton & Knowles]

Intrasoft®OCN. Fatty acid condensate; softener, antistat for textile fibers, esp. acrylic, polyamide, cellulosics; aq. solutions are stable to dilute acids, salts, hard water, nonionics and cationics. [Crompton & Knowles]

Intrassist® LA-LF. Quaternary compound; nonretarding, low-foaming leveling agent for Sevron®cationic dyes on acrylic fibers. [Crompton & Knowles]

Intratex® A. Complex amino condensate; nonfoaming leveling agent for dyeing wool. [Crompton & Knowles] *

Intratex® CA-2. Aliphatic ethoxylates; compatibilizer. [Crompton & Knowles]

Intratex®DD. Surfactant blend; leveling agent for disperse, direct and acid dyes with penetrating, emulsifying, and compatibilizing props.; for cotton and blends with polyester and nylon. [Crompton & Knowles]

Intratex®N. Phenolic condensate; dyebath auxiliary, leveling and aftertreating agent for acid dyes on nylon carpet and apparel. [Crompton & Knowles]

Intraval. CAS 71-73-8; A proprietary preparation of thiopentone sodium; used as a short duration anesthetic or for induction of anesthesia. [May & Baker Ltd]

Intraval Sodium. CAS 71-73-8; Thiopentone sodium. [Akzo Chemie UK Ltd]

Intravon® JU. EO condensate; detergent, emulsifier, dispersant, wetting agent, penetrant for textile, household and cosmetic applications; antiprecipitant for dyeing acrylic blends with anionic/cationic dyes in a one-bath method. [Crompton & Knowles]

Intravon®SOL-N. Aliphatic solv./surfactant blend; low odor solv.-based scouring detergent for all fibers, in one-bath dyeing/scouring processes; oil solubilizer; good wetting and penetration, moderate foamer; stable to alkali. [Crompton & Knowles]

Intrex. Functional surfactant blend. [Rhone-Poulenc UK]

Intrex DW81. Corrosion inhibitor. [Rhone-Poulenc UK]

Intrex HA70. Organic phosphoric acid derivative as an amber liquid; hydrotroping agent used in the solubilizing of other surface active agents in highly alkaline cleaner bases; base for acid cleaners for metals, etc. [Rhone-Poulenc UK]

Intron. Interferon; a protein formed by the interaction of animal cells with viruses capable of conferring resistance to virus infection on animal cells; antineoplastic; antiviral. [Schering Corp]

Intropin®. CAS 62-31-7; Dopamine hydrochloride; adrenergic. [DuPont Merck Pharmaceuticals]

---

* Trade name not verified as available   † Trade name verified as obsolete

Invaderm. Dyeing auxiliaries. [Ciba plc] *

Invaderm C9B. A sodium aryl disalphonate in powder form used as an anionic level dyeing assistant for leather. [Ciba plc] *

Invadine. A proprietary trade name for a wetting agent containing a sodium alkylphenylene sulfonate. §

Invalon. Dyeing and printing assistant. [Ciba plc] *

Invar. An alloy of 36% nickel and 64% steel (0.2% carbon in steel). It has very little expansion on heating, and is used for delicate instruments for measuring.

invariant. An alloy of 47% nickel and 53% iron. Has similar properties to Permalloy.

Invaro Steel. A proprietary trade name for a tool steel containing 1.15% manganese, 0.5% tungsten, and 0.9% carbon. §

Invasol. Fatliquors. [Ciba plc] *

Invenol. CAS 339-43-5; Carbutamide.

Invephos 20. Liquid chemical containing calcium to be diluted with water; used to produce a microcrystalline, corrosion resisting, paint bording, zinc phosphate coating on iron, steel, and zinc. [Invequimica & CIA SCA] *

Invephos 21C. Liquid chemical containing calcium to be diluted with water; corrosion preventing chemical used in pretreatment for electropainting, produces microcrystalline coatings on iron and steel. [Invequimica & CIA SCA]*

Inveres EVH. Polyvinyl acetate homopolymer; sizing agent, adhesives for wood and paper; available in a wide variety of viscosities. [Invequimica & CIA SCA] *

Inveres K-82. Highly plasticized homopolymer of polyvinyl acetate; binder for carpet backing. [Invequimica & CIA SCA] *

Inversal. CAS 539-21-9; A proprietary preparation of ambazone. [FBA Pharmaceuticals] ‡

Inversine. CAS 826-39-1; Mecamylamine hydrochloride; for the treatment of hypertension. [Merck & Co Inc]

Inversol 140. Complex polyglycol ester; inversely sol. lubricity agent for metalworking liqs.; reduces surface tensile and coefficient of friction. [Ferro/Keil]

invertase. See sucrase.

invertin. See sucrase.

invert sugar. CAS 8013-17-0; A mixture of molecular proportions of glucose and fructose, obtained in the hydrolysis of cane sugar by acids; used to improve wines, also in the manufacture of liqueurs, fruit preserves, and honey substitutes.

Invicta Duo 495FW. A liquid formulation containing 160 g bifenox and 400 g isoproturon per liter as a suspension concentrate; a residual herbicide for the control of weeds in winter cereals. [Farm Protection Ltd]

Invicta Duo 495FW. A liquid formulation containing 160 g bifenox and 400 g isoproturon per liter as a suspension concentrate; a residual herbicide for the control of weeds in winter cereals. [Farmers Crop Chemicals Ltd]

Iocare® Balanced Salt Solution. A sterile physiological solution containing sodium chloride, potassium chloride, calcium chloride, magnesium chloride, sodium acetate and sodium citrate; used for irrigation during various surgical procedures of the eyes, ears, nose and/or throat. [Iolab Corp]

Iodal. A substance prepared by the action of iodine upon a mixture of alcohol and nitric acid.

Iodanisol. o-Iodoanisole. An antiseptic.

Iodantifebrin. Iodoacetanilide.

iodeosin B. See erythrosin.

Iodesin. Tetraiodofluoresceine, $C_{20}H_8 I_4O_5$.

Iodex. A proprietary ointment containing 4% organically-combined iodine in a petroleum jelly base. [SmithKline Beecham] *

Iodicyl. Diiodosalicylic acid methyl ester.

Iodin. Iodized arachis oil.

iodine. Nonmetallic halogen element; I_2; CAS 7553-56-2; EINECS 231-442-4; dyes, alkylation and condensation catalyst, iodides, iodates, antiseptics, germicides, x-ray contrast media, food and feed additive, stabilizers, photographic film, water treatment, pharmaceuticals, medicinal soaps. [Aldrich; Andeno BV; Atomergic Chemetals; Cerac; Mallinckrodt]

iodine-potassium iodide solution. See Lugol's solution.

Iodized salt. Common salt, NaCl, containing a very small amount of sodium iodide. Goiter in certain districts is associated with lack of iodide in the water supply.

iodobenzene. C_6H_5I; CAS 591-50-4; EINECS 209-719-6. [R.W. Crcoff]

Iodobio 45. TEA hydro-iodide; slendering product for cosmetics. [Alban Muller]

iodoeosin. CAS 16423-68-0; EINECS 240-474-8; Erythrosin or Pyrosin; used as an indicator. V

iodoform. CAS 75-47-8; EINECS 200-874-5; Triiodomethane, CHI_3.

iodoglobin. Diiodotyrosine, $HO \cdot C_6H_2I_2 \cdot CH_2 \cdot CH(COOH)$-$NH_2$.

iodol. Tetraiodopyrrole, C_4I_4NH; used externally as a disinfectant.

iodomethane. See methyl iodide.

Iodosorb. Cadexomer iodine powder. [ICI Chem. & Polymers Ltd] †

Iodotope I-125. CAS 7790-26-3; Sodium iodide I 125; diagnostic aid; radioactive agent. [Bristol-Myers Squibb Co Inc] *

Iodotope I-131. CAS 7790-26-3; Sodium iodide I 131; antineoplastic; diagnostic aid; radioactive agent. [Bristol-Myers Squibb Co Inc]

Iodotope Therapeutic. CAS 7790-26-3; Sodium iodide I 131; antineoplastic; diagnostic aid; radioactive agent. [Bristol-Myers Squibb Co Inc]

Iodoval. Monoiodoisovalerylurea, $(CH_3)_2CH \cdot CHI \cdot CO \cdot NH \cdot CO \cdot NH_2$.

Iodozol. Diiodo-p-phenolsulfonic acid, $C_6H_2I_2(OH)SO_3H \cdot 3H_2O$.

Iodron. Bulk milk tank sanitizer. [The Wellcome Foundation Ltd] *

iohydrin. (ilothion). Diiodoisopropyl alcohol, $CH_2I \cdot CH(OH) \cdot CH_2I$.

Ionac ECP-88. A proprietary acrylic polymer used in the formulation of coatings applied by electrostatic spray. [Ionac Chemical Co] *

Ionamin. CAS 122-09-8; A proprietary preparation of phentermine in a resin complex; used in the treatment of obesity. [Pennwalt Corp] *

Ionex. Lube oil additives. [Shell UK] *

Ionil. CAS 69-72-7; Salicylic acid; keratolytic. [Galderma Laboratories]

Ionol. A proprietary trade name for a phenolic antioxidant possessing a symmetrical structure thus giving a low power loss at high frequencies when used in dielectrics. [Shell UK]

Ionol CP. Antioxidants. [Shell UK]

Ionol CPA-Feed. Antioxidants. [Shell UK]

---

‡ Trade name and manufacturer not verified   § Trade name without identified manufacturer

Ionol K65. Antioxidants. [Shell UK]

ionomer resins. A name given to thermoplastic resins containing both covalent and ionic bonds; carboxyl groups are located along the polymer chain by copolymerization to provide the anionic portion of the ionic cross links; metal ions constitute the cationic part of the links; sodium, potassium, magnesium, and zinc are examples of ions used. See also Surlyn.

Ionox. Range of antioxidants for polymeric systems, thermoplastic rubber and fatty acid distillates. [Shell UK]

Ionpure. Amorphous water-sol. inorg. preservative for cosmetics and plastic pkg. [U.S. Cosmetics]

iopropane. Iodopropanol.

Iopydone. 3,5-Diiodo-4-pyridone.

Iosan. Iodophor teat dip and udder cream; used for control of mastitis. [Ciba plc]

Iosol. CAS 552-22-7; Thymol iodide

Iotect. Iodine indicator. [May & Baker Ltd] *

Iotek 7010. Ionomer resin, zinc cation; thermoplastic for molding and extrusion of sporting goods, footwear, automotive parts, foams (buoys, ski lift seat cushions), perfume bottle stoppers, ignition tubes for explosives, impact modifiers. [Exxon]

Iotek 8000. Ionomer resin, sodium cation; thermoplastic for molding and extrusion of sporting goods, footwear, automotive parts, foams (buoys, ski lift seat cushions), perfume bottle stoppers, ignition tubes for explosives, impact modifiers. [Exxon]

Iothion. See iohydrin.

Iotox. Soluble concentrate containing 72 g ioxynil and 214 g mecoprop per liter; used for weed control in turf. [Rhone-Poulenc Environmental Prods. Ltd]

Iotril. Active ingredient: ioxynil octanoate; 4-cyano-2,6-diiodophenyl octanoate; selective post-emergence herbicide which controls a wide range of annual broadleaf weeds in cereals, onions, leeks, and sugar cane. [Agan Chemical Manufacturers Ltd]

Iotrilex. Ioxynil octanoate; herbicide. [Agan Chemical Manufacturers Ltd]

Iovist. CAS 79770-24-4; Aqueous solutions of iotrolan; x-ray contrast media. [Schering Health Care Ltd]

IPA. See isopropyl alcohol.

ipecac. (ipecacuanha). The dried root of Uragoga ipecacuanha; used in medicine; production of emetine

Ipecacuanha. See ipecac.

Ipecine. Emetine, an alkaloid.

Ipesandrine. A proprietary preparation of alkaloids of opium and ipecacuanha, and ephedrine hydrochloride; a cough linctus. [Sandoz] *

Ipexon. A proprietary preparation of guaifenesin, ephedrine hydrochloride, sucrose, and glycerine; a cough linctus. [British Drug Houses] *

Iphaneine. Pure isobutyl phenyl acetate. [Bush Boake Allen Ltd]

IPM. See isopropyl myristate.

ipomic acid. Sebacic acid, $CO_2H(CH_2)_8CO_2H$.

Iporka. Foamable urea resins. [BASF plc] *

IPP. See isopropyl palmitate.

Ipral. CAS 738-70-5; A proprietary preparation containing trimethoprim; an antibiotic. [Bristol-Myers Squibb Pharmaceuticals Ltd]

Ipropran. CAS 14885-29-1; Ipronidazole; antiprotozoal. [Hoffmann-LaRoche Inc] *

I.P.S. CAS 67-63-0; EINECS 200-661-7; Isopropyl alcohol.

[Foseco (F.S.) Ltd]

Ipsilene. A disinfecting gas made by heating ethyl chloride and iodoform under pressure.

Ipso. Suspension concentrate containing 450 g isoproturon and 19 g isoxaben per liter; used for annual weed control in cereals. [DowElanco Ltd]

IR. See polyisoprene.

Irabond. Primer and adhesive systems for irathane coatings and linings; for use on a variety of substrates to enhance the performance of irathane. [Irathane International Ltd]‡

Iragcet. Solvent soluble dyes. [Ciba plc] *

Irasolve. Solvents for use in cleaning/dissolving irathane; for assistance in surface preparation and cleaning of equipment. [Irathane International Ltd] ‡

Irathane. Elastomeric polyurethane coating and lining materials; for protection from severe abrasion and corrosion problems in mines, power plants, processing and offshore. [Irathane International Ltd] ‡

Ircogel. Metallo-organic complexes high in calcium content, having a particle size of 0.01 micron; thixotropic agents for PVC plastisols and organosols for cold dipping and general use to prepare nondrip compounds. [Stoller Chemicals Ltd]

Ircogel® 900. Calcium org.; thixotropic, antisag, and flow control agent for use in coating plastisols and organosols, cloth coating plastisols, PVC sealants, polysulfide sealants, polymercaptan sealants; exhibits wetting or dispersant effect on fillers. [Lubrizol]

Ircogel® 905. Org. thixotrope for urethane and high solids, solv.-based coatings. [Lubrizol]

Ircosperse 2170. Imidazoline in xylene; dispersant. [Lubrizol]

Irenat®. Thyroid depressant. [Bayer AG]

Iretol. 1,2,3-Trihydroxy-5-methoxybenzene, $C_7H_8O_4$; used in perfumery.

Irgaclarol. Wetting and scouring agents. [Ciba plc] *

Irgacure. Uv curing agents. [Ciba plc] *

Irgaderm. Dyes for leather finishing. [Ciba plc] *

Irgaferm BC Champagne. Dry yeasts for wine production. [Ciba plc] *

Irgafin. Predispersed pigments for plastics polymers. [Ciba plc] *

Irgafiner. Predispersed pigments for polyolefins. [Ciba plc]*

Irgafos. Co-stabilizers for plastics. [Ciba plc] *

Irgalan. Metal complex dyes. [Ciba plc] *

Irgalevone. Dyeing and printing assistants. [Ciba plc] *

Irgalite. Pigment dispersions for emulsion paints and other organic pigments. [Ciba plc]

Irgalite Blue GST. Beta form copper phthalocyanine blue pigment (CI Pigment Blue 15) with excellent texture, dispersibility, gloss, and high strength for letterpress and lithographic inks. [Ciba plc] *

Irgalite C-20. Pigment preparations. [Ciba plc] *

Irgalite Dispersed. Pigment plasticizer dispersions. [Ciba plc] *

Irgalite M-20. Pigment preparations. [Ciba plc] *

Irgalite MPS. Multipurpose pigment stainers for paints. [Ciba plc] *

Irgalite PDS. Predispersed pigment powders for paints. [Ciba plc] *

Irgalite PR. Predispersed pigment powders for rotogravure. [Ciba plc] *

Irgalite Yellow BGW. Metaxylidide bis-arylamide yellow

---

* Trade name not verified as available　　　　† Trade name verified as obsolete

pigment (CI Pigment Yellow 13) with excellent dispersibility and improved gloss and transparency for letterpress and lithographic inks. [Ciba plc] *

Irgalite Yellow F4G. A proprietary monoazo pigment of the arylamide type; greenish hue which makes it suitable for use in letterpress, offset litho, flexographic, and gravure printing inks. [Ciba plc] *

Irgalon. Pretreatment agent. [Ciba plc] *

Irganol. Acid dyes for wool [Ciba plc] *

Irganox. A trademark for a range of speciality antioxidants of the hindered phenol type developed initially for the stabilization of polyolefins for use at high frequencies. [Ciba plc] *

Irgapadol. Dyeing and printing assistants. [Ciba plc] *

Irgaphor. Rubber masterbatch pigments. [Ciba plc]

Irgaplastol M-20. Pigment preparations. [Ciba plc] *

Irgapyrol. Flame proofing agents. [Ciba plc] *

Irgarol. Paint additives. [Ciba plc] *

Irgasan. Bacteriostats. [Ciba plc] *

Irgasan DP 300. CAS 3380-34-5; Triclosan; disinfectant. [Ciba-Geigy Corp] *

Irgasol®. Dyeing and printing assistant. [Ciba plc] *

Irgasperse. Pigment dispersions for nonaqueous decorative paints. [Ciba plc]

Irgastab®. Stabilizers for polymers. [Ciba-Geigy/Additives; Ciba plc] *

Irgatan. Synthetic tanning agents for leather. [Ciba plc] *

Irgatron. Premetallized dyes for polyamides. [Ciba plc] *

Irgawax. Plastics lubricants. [Ciba plc] *

Irgazin. A range of organic pigments derived from iso-indolinone and dioxazine; typical colors are Yellow 2GLT, Yellow 3RLT, Orange RLT, Red 2BJT and Violet BLT. [Ciba plc]

Irgazin C-20, M-20. Pigment preparations. [Ciba plc] *

Irgoferm CM Montrachet. Dry yeasts for wine production. [Ciba plc] *

iridin. (irisin). The powdered extract of iris.

Iridite®. Chromate coatings; corrosion protective coating for plated and unplated metals; base coat for painting of nonferrous metals; approved for governmental and industrial spec. finishing of zinc plate, cadmium plate, hot-dipped galvanized steel, zinc die-cast alloys. [Witco/Allied-Kelite]

iridium. CAS 7439-88-5; EINECS 231-095-9; Metallic element of the platinum group; combined with platinum as an alloy for ammonia fuel-cell catalyst, electric contacts and thermocouples, commercial electrodes, resistance wires and laboratory ware.

iridium steel. A German steel containing 4% cobalt, 16% tungsten, 3.5% chromium, 0.67% vanadium, 0.8% molybdenum and 0.6% carbon.

Iridosmine. (osmiridium). An alloy of iridium and osmium containing 40-77% iridium, and 20-50% osmium. If there is more iridium, the alloy is called Nevyanskite, and Siserskite if the content of osmium is high.

Irigenin. 5,7,3´-trihydroxy-6,4´,5´-trimethoxyisoflavone.

Irilac®. Clear coatings providing protection against corrosion, fingerprinting, tarnish, and abrasion on all metals. [Witco/Allied-Kelite]

iriphan. (triphan). The strontium salt of atophan.

Irish moss extract. See carrageenan.

Irish pearl moss. Caragheen moss, a gelatinous seaweed, *Chondrus crispus*. It contains carrageenin allied to pectin. It is employed as a substitute for isinglass, as a size for thickening colors in calico printing, and for stiffening silk.

Irish Peat Wax. (Montana wax, montanin wax). Waxes extracted from Irish peat are sold under these names. They resemble montan wax.

irisin. See iridin.

Irisol®. Spirit soluble dyestuffs; for the surface coatings, printing ink, office supplies, and plastics industries. [Bayer AG]

Irlux®. For the chemical industry. [DuPont UK]

Irofol C. A proprietary preparation of ferrous sulfate, folic acid, sodium ascorbate, and vitamin C; a hematinic. [Abbott Laboratories] *

iron. (carbonyl-iron powder) Metallic element; Fe; CAS 7439-89-6; EINECS 231-096-4; steel and other alloys (e.g., cast and wrought iron); source of hydrogen by reaction with steam. [Aldrich; Cerac; ISP; Noah Chem.]

Iron A. A British chemical standard. It is a cast iron containing 0.734% combined carbon, 1.989% silicon, 0.047% sulfur, 0.049% phosphorus, 0.688% manganese, 0.042% arsenic, 0.052% titanium, and 2.387% graphitic carbon.

Iron B. A British chemical standard. It is a cast iron containing 0.39% combined carbon, 0.031% sulfur, 0.026% phosphorus, 0.031% arsenic, 0.108% titanium, and 2.67% graphitic carbon.

Iron D. (Phosphoric D). A British chemical standard containing 11.8% phosphorus.

Iron D2. A British chemical standard; grey phosphoric cast iron in the form of fine turnings. It contains 1.31% silicon, 1.07% phosphorus, and 1.64% manganese. The approximate quantities of other elements are 2/5% graphitic carbon, 0.8% combined carbon, and 0.03% sulfur.

Iron G. A standard cast iron containing 1.82% graphite carbon, 0.86% combined carbon, 1.3% silicon, 0.41% manganese, 0.125% sulfur, and 0.45% phosphorus.

Iron L. (nickel-chromium-copper-Austenitic Iron L). A British Chemical Standard containing 3.06% total carbon, 2.26% silicon, 1.01% manganese, 0.119% phosphorus, 13.45% nickel, 3.96% chromium, 4.73% copper, 0.031% sulfur, the remainder being iron.

Ironac. An acid-resisting alloy of iron and silicon. It contains 13% silicon, 84% iron, 0.77% manganese, 1.08% carbon, and 0.78% phosphorus.

iron-andradite. Synonym for Skiagite.

Iron, Armco. See Armco Iron. §

iron black. Antimony precipitated as a fine powder, by action of zinc upon an acid solution of an antimony salt; imparts the appearance of polished steel to paper mache and plaster of Paris.

iron chlorides. See ferric chloride.

iron flint. An opaque variety of quartz containing iron.

iron froth. A spongy variety of hematite.

iron-leucite. A mineral; $KFe\cdots Si_3O_8$.

iron liquor. See pyrolignite of iron.

Iron Man. Vitamin-fortified tonic. [Richardson-Vicks Inc] *

iron nitrate. See ferric nitrate.

Iron ore A, hematite type. A standard iron ore containing 58.19% iron, 0.056% phosphorus, 8.14% silica, and 0.066% sulfur; used as a standard for checking analyses of iron ore.

iron-ore cement. Cements in which a large proportion of the alumina is replaced by ferric oxide.

iron-orthoclase. Synonym for Ferriorthoclase.

iron oxide. See ferric oxide.

---

‡ Trade name and manufacturer not verified

§ Trade name without identified manufacturer

iron persulfate. *See* iron sulfate (ic).

iron putty. A mixture of ferric oxide and boiled linseed oil; used for joints in iron pipes.

Iron pyrolignite. *See* pyrolignite of iron.

iron sulfate (ic). (ferric sulfate; ferric trisulfate; iron persulfate; iron tersulfate) Fe(SO$_4$)$_3$; CAS 10028-22-5; EINECS 233072-9; pigments, reagent, etching aluminum, disinfectant, textile dyeing and printing, flocculant in water and sewage purification, soil conditioner, polymerization catalyst, metal pickling, chelated iron products, intermediate. [Boliden Intertrade; Faesy & Besthoff; Rhone-Poulenc]

iron sulfate (ous). (ferrous sulfate; iron vitriol; copperas; green vitriol; sal chalybis) FeO$_4$S · 7H$_2$O; CAS 7720-78-7; iron oxide pigment, other iron salts, ferrites, water and sewage treatment, catalyst esp. for synthetic ammonia, fertilizer, feed additive, flour enrichment, reducing agent, herbicide, wood preservative, process engraving. [EM Industries; J.M. Huber; Mallinckrodt]

iron tersulfate. *See* iron sulfate (ic).

iron vitriol. (green vitriol, copperas, green copperas). CAS 7782-63-0; Ferrous sulfate, FeSO$_4$ · H$_2$O.

Irox. A proprietary trade name for a synthetic yellow iron oxide. §

Irribral. CAS 14286-84-1; A proprietary preparation of bencyclane fumarate; a vasodilator. [Pfizer International] †

Isanol®. CAS 71-36-3; EINECS 200-751-6; 60% iso- and 40% n-butanol; solv. for production of adhesives and surface coating resins. [BASF AG]

Isarit. A bleaching earth.

Isatin. Intermediate for production of pharmaceuticals and dyes, stabilizer for plastics industry. [BASF AG]

Isceon. A range of halogen derivations of aliphatic hydrocarbons used as refrigerants and propellants. [ISC Chemicals Ltd]

isinglass. Sodium silicate solution; used for the clearing of liquids, such as beer and wine. *Also see* Thao.

Ismelin. CAS 55-65-2; A proprietary preparation of guanethidine; an antihypertensive. [Ciba plc] *

Ismelin-Navidrex K. A proprietary preparation of guanethidine sulfate, cyclopenthiazide, and potassium chloride in a slow release core. [Ciba plc] *

Ismotic. CAS 652-67-5; Isosorbide; diuretic. [Alcon Laboratories Inc]

isoamyl acetate. CH$_3$COOCH$_2$CH$_2$CH(CH$_3$)$_2$; CAS 123-92-2; flavoring, perfumes, solvent for nitrocellulose, masking undesirable odors. [Aldrich; Penta Mfg.; Spectrum Chem. Mfg.]

Isoante. Urethanes. [Dow UK]

isoanthraflavic acid. 2,7-Dihydroxyanthraquinone.

isoascorbic acid. *See* erythorbic acid.

Iso-Autohaler. Breath-actuated pressurized aerosol containing a suspension of isoprenaline sulfate BP 4 mg/ml; delivers 400 metered doses (0.08 mg per dose). [3M Pharmaceuticals] †

1,3-isobenzofurandione. *See* phthalic anhydride.

Isobond. Isocyanate binders. [Dow UK]

Isobrite®. Zinc brightener systems providng full bright plate with greater tolerance for high temp. [Witco/Allied-Kelite]

Isobu-M-AMD. N-(Isobutoxymethyl)acrylamide. [Cyanamid BV] *

isobutad. A mineral rubber or bitumen.

isobutane. (2-methylpropane; propane, 2-methyl-; isobutane) Hydrocarbon gas; CH(CH$_3$)$_3$; CAS 75-28-5; EINECS 200-857-2; Aerosol propellant. [Air Prods.; Phillips 66]

isobutane. *See* isobutane.

isobutanol. *See* isobutyl alcohol.

isobutyl acetate. CH$_3$COOCH$_2$CH(CH$_3$)$_2$; CAS 110-19-0; EINECS 203-745-1; Solvent for nitrocellulose; in thinners, sealants, topcoat lacquers; perfumery; flavoring agent. [BASF; Eastman; Hoechst Celanese; Janssen Chimica; Union Carbide]

isobutyl alcohol. (isobutanol; isopropylcarbinol; 2-methyl-1-propanol) Alcohol; (CH$_3$)$_2$CHCH$_2$OH; CAS 78-83-1; EINECS 201-148-0; flavors and fragrances; organic synthesis; latent solvent; intermediate; paint removers; fluorometric determinations; liq. chromatography. [BASF; CPS; Eastman; Hoechst Celanese; Neste UK; Shell; Union Carbide]

Isobutyl Niclate®. CAS 15317-78-9; Nickel diisobutyl-dithiocarbamate; antioxidant/antiozonant for protection in epichlorohydrin. [R.T. Vanderbilt]

isobutyric acid. (dimethylacetic acid; 2-methylpropanoic acid; isopropylformic acid) Organic acid; (CH$_3$)$_2$CH-COOH; CAS 79-31-2; EINECS 201-195-7; manufacture of esters for solvents, flavors, perfume bases, disinfecting agent, varnish, deliming hides, tanning agent. [Eastman; Hoechst Celanese; Hüls Am.]

Isoclad. Water-based, elastic anticorrosive cladding for all ferrous and nonferrous metals; used for any corrosive environment including industrially polluted areas. [Liquid Plastics Ltd]

Isocon. Refined, semirefined and polymeric isocyanates; curing coponents for use with Propocon polyether systems. [Harcros UK]

Iso-Cornox. Selective weedkillers. [The Boots Co plc]

Iso-Cornox. CAS 7085-19-0; Soluble concentrate containing 570 g/l mecoprop; herbicide for control of weeds in cereals and grassland. [Schering Agrochemicals Ltd]

Isocracking. Catalyst. [Chevron] *

Isocreme. Compounded lanolin-derived sterols. [Croda Chem. Ltd]

Isocure. A tertiary amine vapor-cured phenolic urethane resin system; liquid resins and catalysts used as binders for foundry core and mold production. When mixed with sand, these resins and catalysts act as an adhesive for foundry sands; these resin-bonded sand articles are used in the production of ferrous and nonferrous castings. [Ashland Chemical Company] *

isocyanuric acid. (s-triazine-2,4,6-triol; cyanuric acid; 2,4,6-trihydroxy-1,3,5-triazine) Ketone isomer of cyanuric acid; N=C(OH)N=C(OH)NCOH; CAS 108-80-15; EINECS 203-618-0; used to stabilize chlorine sol'ns. in swimming pools; bleaches, sanitizers. [Allchem Industries; Nissan Chem. Ind.; Schaefer Salt & Chem.; 3-V]

isocyanuric chloride. *See* trichloroisocyanuric acid.

Isodamp® C-1002. Thermoplastic; highly damped thermoplastic material for military tank ammunition racks. [E-A-R]

isodecyl methacrylate. Monomer; CH$_2$=C(CH$_3$)-COOC$_{10}$H$_{21}$; CAS 29964-84-9; pressure-sensitive adhesives, coatings for leather, textiles, paper, nonwoven fiber, polymer modifier and stabilizer, visc. index improver, dispersion for plastics and rubber, floor waxes, potting compounds, sealants. [CPS; Rohm & Haas;

---

* Trade name not verified as available   † Trade name verified as obsolete

Sartomer]

Isoderm®. Comprehensive range of nitrocellulose seasons for leather, aqueous or organic systems. [Bayer AG]

isodurindine. Tetramethylaniline, $C_6H(CH_3)_4.NH_2$.

Isofoam. Twin component rigid polyurethane foam systems. [Baxenden Chemicals Ltd]

isoform oxiosol. p-Iodoxyanisol.

Isogel. A proprietary preparation derived from the husks of mucilaginous seeds; a laxative. [Allen & Hanburys Ltd]*

Iso Isotearyle WL 3196. CAS 41669-30-1; EINECS 255-485-3; Isostearyl isostearate; cosmetics ingred. [Gattefosse SA]

Isoject-Streptomycin Injection. CAS 3810-74-0; Streptomycin sulfate; antibacterial. [Pfizer Inc]

Isol. CAS 107-41-5; An oil forming a permanent emulsion with hot or cold water; used for oiling textiles.

Isol R. Antiskinning agent for paints and varnishes. [Akzo Chemie UK Ltd]

Isolan®, Isolan® K. 1:2 Metal complex dyestuffs; used for dyeing wool and polyamide fibers. [Bayer AG]

Isolan®Gl 34. CAS 91824-88-3; Polyglyceryl-4 isostearate; w/o emulsifier for vegetable oils; low odor, easy to use; forms very stable emulsions; for personal care industry. [Goldschmidt]

Isolan®GO 33. CAS 9007-48-1; Polyglyceryl-3 oleate; w/o emulsifier for vegetable oils and natural triglycerides; for personal care industry. [Goldschmidt]

Isolantite. A proprietary trade name for a ceramic material made from steatite with binders. §

Isolene®40, 75, 400. cis 1,4-Polyisoprene of low molecular weight derived form natural rubber; polymer used in adhesives, sealants, caulks, and lubricants. [Hardman]

L-isoleucine. (2-amino-3-methylpentanoic acid) $CH_3-CH_2CH(CH_3)CH(NH_2)COOH$; CAS 73-32-5; EINECS 200-798-2; A natural protein amino acid; medicine, nutrition, biochemical research. [Degussa; R.W. Greeff; Tanabe USA; U.S. Biochemical]

Isolevin. A proprietary trade name for the hydrogen tartrate of isoprenaline. §

Isolierstahl. See Bakelite. §

Isolit. Dental separating agent. [Degussa Ltd]

Isoloss®LS. Polyurethane elastomer, cellular; high-density foam for difficult mech. energy control problems; good shock absorp. and vibration isolation performance; for gaskets, motor mounts, cushion pads, bumpers, springs, laminates, pressure pads, athletic paddings. [E-A-R]

Isomist. A proprietary preparation of isoprenaline sulfate; a bronchial antispasmodic. [Nicholas Laboratories Ltd] *

Isomol. Thixotropic ready-for-use spirit based coatings for all types of iron, steel, and nonferrous castings. [Foseco (F.S.) Ltd]

Isonal. Shading dyestuffs complementing the Isolan range. [Bayer plc]

Isonaphthol. CAS 101-68-8; β-Naphthol, $C_{10}H_7OH$.

Isonate® 125M. CAS 101-68-8; MDI; processing aid, intermediate for the production of cast, RIM, and thermoplastic PU elastomers, adhesives, binders, coatings, and sealants. [Dow]

Isonate®2125M. CAS 101-68-8; MDI; for PU industry for adhesives, binders, coatings, and sealants. [Dow]

Isonex. CAS 54-85-3; A proprietary preparation of isoniazid; used for the treatment of tuberculosis. [Pfizer International]

Isonex Forte. CAS 54-85-3; A proprietary preparation of isoniazid; used for the treatment of tuberculosis. [Pfizer International]

Isoniazid. CAS 54-85-3; Isonicotinoylhydrazine.

isonicotinic acid. (4-picolinic acid; pyridine-4-carboxylic acid) $C_6H_5NO_2$; CAS 55-22-1; EINECS 200-228-2. [Raschig; Reilly Industries]

Isonol. Polyols used in the manufacture of polyurethane products. [Dow UK]

Isonox®103. CAS 128-39-2; 2,6-Di-t-butylphenol; antioxidant for control of formation of gums and peroxides in fuels and oils; stabilizer for aviation turbine fuels. [Schenectady]

Isonox®129. CAS 35958-30-6; 2,2´-Ethylidenebis (4,5-di-tert.-butylphenol); antioxidant and thermal stabilizer for polymers; food pkg. application; used in PP, polyethylene, PVC, PS, ABS, hydrocarbon resins, EVA-modified compounds. [Schenectady]

Isonox® 132. CAS 17540-75-9; 2,6-Di-tert-butyl-4-sec-butylphenol; antioxidant used in polyols and rubber systems. [Schenectady]

isooctadecanoic acid. See isostearic acid.

isooctanol. See isooctyl alcohol.

isooctyl alcohol. (isooctanol; 2-ethyl-1-hexanol; alcohol C_8) $CH_3(CH_2)_3CH(C_2H_5)CH_2OH$; CAS 68526-83-0; 26952-21-6; EINECS 248-133-5; In plasticizers; intermediate for nonionic detergents and surfactants, hydraulic fluids; resin, solvent, emulsifier, antifoaming agent. [CDF Chimie Nederland BV]

isooctylmercaptoacetate. See isooctyl thioglyconate.

isooctyl thioglyconate. (isooctylmercaptoacetate) $HSCH_2COOCH_2C_6H_{15}$; CAS 25103-09-7; EINECS 246-613-9; Antioxidants, fungicides, oil additives, plasticizers, insecticides, stabilizers, polymerization modifiers, stabilizer for tin-sulfur compounds, stripping agent for polysulfide rubber. [Bock, Bruno Chemische Fabrik KG]

Isopar®. High purity isoparaffinic solvent. [Exxon Int'l.]

Isopar® C. CAS 64742-48-9; C7-8 isoparaffin; solvent. [Exxon] *

Isopar® E. CAS 64742-48-9; C8-9 isoparaffin; solvent. [Exxon] *

Isopar® G. CAS 64742-48-9; C10-11 isoparaffin; solvent. [Exxon] *

Isopar®H. CAS 64742-48-9; C11-12 isoparaffin; solvent. [Exxon] *

Isopar®L. CAS 64742-48-9; C11-13 isoparaffin; solvent. [Exxon] *

Isopar®M. CAS 64742-47-8; C13-14 isoparaffin; solvent. [Exxon] *

Isopaste. Foundry core paste that cures fast without heat [Ashland Chemical Company] *

isopentane. (2-methylbutane; ethyl dimethyl methane) Aliphatic hydrocarbon; $CH_3CH_2CH(CH_3)_2$; CAS 78-78-4; EINECS 201-142-8; solvent; manufacture of chlorinated derivatives; blowing agent for polystyrene. [Exxon; Phillips 66]

Isophane. Insulin preparation. [The Boots Co plc] *

isophorone. (3,3,5-trimethyl-2-cyclohexen-1-one) $C_9H_{14}O$; CAS 78-59-1; EINECS 201-126-0; In solvent mixtures for finishes, for polyvinyl and nitrocellulose resins, pesticides, stoving lacquers. [Allchem Industries; Atochem; BP Chem. Ltd; Hüls AG; Union Carbide]

Isoplac. A proprietary insulation. §

Iso-Planotox. Selective weedkiller. [May & Baker Ltd] *

Isoplast. Thermoplastic engineering resins. [Dow UK]

‡ Trade name and manufacturer not verified   § Trade name without identified manufacturer

Isoplast 101. CAS 9009-54-5; Thermoplastic PU; amorphous engineering resin with crystalline props.; for extrusion and injection molding; offers high impact, low moisture sensitivity, excellent chem. and solv. resist., high abrasion resist.; opaque; sp.gr. 1.19; melt flow 8 g/10 min (5 kg, 224 C); tensile strength 48 MPa (break); tensile elong. 6% (yield), 160% (break); flex. strength 68 MPa; Izod impact 1175 J/m notch; Rockwell hardness > R100. [Dow Plastics]

Isoplast 101LGF40 Nat, 101LGF60 Blk. CAS 9009-54-5; Thermoplastic PU, long glass fiber-reinforced; amorphous engineering resin with crystalline props.; offers high mod., strength, impact, excellent chem. and solv. resist. [Dow Plastics]

Isoplast 302. CAS 9009-54-5; Thermoplastic PU; amorphous engineering resin with crystalline props. [Dow Plastics]

Isopoxy. Insulating varnish. [Schenectady-Midland Ltd]

isoprene rubber. See polyisoprene.

Isoprep®. Cleaners for use on ferrous and nonferrous metals and on nonconductive substrates. [Witco/Allied-Kelite]

isopropanol. See isopropyl alcohol.

isopropanolamine. (MIPA; 1-amino-2-propanol; monoisopropanolamine; 2-hydroxypropylamine) Aliphatic amine; $H_2NCH_2CHOHCH_3$; CAS 78-96-6; EINECS 201-162-7; solubilizer, neutralizer, emulsifying agent; plasticizers, insecticides. [Ashland; BASF; Mitsui Toatsu Chem.]

Isopropenylbenzene. See α-methylstyrene monomer.

isopropyl alcohol. (IPA; isopropanol; 2-propanol; dimethyl carbinol) Aliphatic alcohol; $(CH_3)_2CHOH$; CAS 67-63-0; EINECS 200-661-7; solv. for essential oils, alkaloids, gums, resins, cellulose derivatives; coatings; deicing agent for liq. fuels; lacquers; extraction processes; dehydrating agent; denaturing ethyl alcohol; in cosmetics; manufacture of acetone, glycerol, isopropyl acetate. [Arco; Eastman; Exxon; Mallinckrodt; Mitsui Toatsu Chem.; Shell; Union Carbide]

Isopropylan® 33. Isopropyl palmitate, lanolin oil; binder in talc and pearl powd. systems; plasticizer, emollient, and moisturizer. [Amerchol Corp]

isopropylbenzene. See cumene.

isopropylcarbinol. See isobutyl alcohol.

isopropyl chloroformate. $(CH_3)_2CHOOCCl$; CAS 108-23-6; Chemical intermediate for free-radical polymerization initiators, organic synthesis. [Atochem N. Am.; BASF; PPG Industries]

6-isopropyl-m-cresol. See thymol.

3- and 4-isopropylcumene. See diisopropyl benzene, 1,3- and 1,4-.

isopropylformic acid. See isobutyric acid.

isopropyl n-hexadecanoate. See isopropyl palmitate.

4,4'-isopropylidenediphenol (CTFA). See bisphenol A.

isopropyl mercaptan. (2-propanethiol) $(CH_3)_2CH(HS)$; CAS 75-33-2; EINECS 200-861-4; Standard for petrol. analysis; intermediate. [Atochem N. Am.; Phillips 66]

2-isopropyl-5-methylphenol. See thymol.

isopropyl myristate. (IPM; 1-methylethyl tetradecanoate; tetradecanoic acid, 1-methylethyl ester; myristic acid isopropyl ester) Ester of isopropyl alcohol and myristic acid; $CH_3(CH_2)_{12}COOCH(CH_3)_2$; CAS 110-27-0; EINECS 203-751-4; cosmetic creams, topical medicinals. [Amerchol; Goldschmidt; Henkel/Emery; Inolex;

Lanaetex; Stepan; Unichema]

isopropyl palmitate. (IPP; isopropyl n-hexadecanoate; hexacanoic acid, 1-methylethyl ester; 1-methylethyl hexandecanoate) Ester of isopropyl alcohol and palmitic acid; $CH_3(CH_2)_{14}COOCH(CH_3)_2$; CAS 142-91-6; EINECS 205-571-1; emollient, emulsifier in lotions, creams, and similar cosmetics. [Amerchol; Goldschmidt; Henkel/Emery; Inolex; Stepan; Unichema]

Isoptin. CAS 152-11-4; Verapamil hydrochloride; anti-anginal; cardiac depressant (anti-arrhythmic). [Knoll Pharmaceutical Co]

Isopto Alkaline. A proprietary preparation of hypromellose for ocular use. [Alcon Universal] *

Isopto Atropine. Atropine sulfate solution for ocular use; anticholinergic. [Alcon Laboratories Inc]

Isopto Carbachol. CAS 51-83-2; Carbachol solution for ocular use; cholinergic. [Alcon Laboratories Inc]

Isopto Carpine. CAS 54-71-7; A proprietary preparation of pilocarpine hydrochloride solution for ocular use. [Alcon Universal] *

Isopto Cetamide. Sulfacetamide sodium solution; an ocular antiseptic, antibacterial. [Alcon Laboratories Inc]

Isopto Eserine. CAS 57-64-7; A proprietary preparation of physostigmine salicylate solution for ocular use. [Alcon Universal] *

Isopto Frin. CAS 61-76-7; A proprietary preparation of phenylephrine hydrochloride solution for ocular use. [Alcon Universal] *

Isopto Homatropine. CAS 51-56-9; A proprietary preparation of homatropine hydrobromide solution for ocular use. [Alcon Universal] *

Isopto Hydrocortisone. CAS 50-23-7; A proprietary preparation of hydrocortisone for ocular use. [Alcon Universal]*

Isopto Hyoscine. A proprietary preparation of hyoscine hydrobromide solution for ocular use. [Alcon Universal]*

Isopto P-ES. A proprietary preparation of pilocarpine hydrochloride and physostigmine salicylate solution for ocular use. [Alcon Universal] *

Isopto Plain. A proprietary preparation of hypromellose for ocular use. [Alcon Universal] *

Isordil. CAS 87-33-2; Isosorbide dinitrate; vasodilator. [Ives Laboratories Inc] *

Isordil Tablets. CAS 87-33-2; Isosorbide dinitrate; for treatment of angina pectoris, congestive heart failure. [Wyeth Laboratories; Monmouth Pharmaceuticals Ltd]

Isordil Tembids Capsules. CAS 87-33-2; Isosorbide dinitrate; for prophylaxis of angina pectoris. [Wyeth Laboratories; Monmouth Pharmaceuticals Ltd]

Isoset®. High performance, two-component adhesive systems for woodworking, panels, beams, plywood, laminating, coating, contact and pressure-sensitive adhesives, fiber bonding. [Ashland]

Isoset®WD3-A322 Emulsion Resin. Proprietary, reactive, water-based polymer for crosslinking with CX-hardener at ambient temperature; structural wood laminations, Type 1 waterproof/no-creep performance. [Ashland Chemical Company] *

Isoset® WD3-CM402 Emulsion Resin. Proprietary, reactive, water-based polymer for crosslinking with CX-hardener at ambient temperature; structural sandwich composite, metal or plastic faces to porous, i.e., wood cores. [Ashland Chemical Company] *

iso soap. A solid sulfonic derivative of castor oil, soluble in

---

* Trade name not verified as available   † Trade name verified as obsolete

hot water; used as a bleaching, washing, and dressing agent in the textile industries.

isostearic acid. (heptadecanoic acid, 16-methyl-; Isooctadecanoic acid; 16-methylheptadecanoic acid) Mixture of branched chain 18 carbon aliphatic acids; $C_{18}H_{36}O_2$; CAS 2724-58-5; EINECS 220-336-3; cosmetics, chemicals, dispersant, softener in rubber compounds, food packaging, suppositories, ointments. [Henkel/Emery; Nissan Chem. Ind.; Unichema; Union Camp]

isotachiol. Silver silico-fluoride, Ag_2SiF_6.

Isotagetone. 2,7-Dimethyloct-5-en-4-one. [Bush Boake Allen Ltd]

Isotagetone 50. 2,7-Dimethyloct-5-en-4-one in isopropyl myristate; tagette, chamomile odor for use in fragrances. [Bush Boake Allen Ltd]

Isotense. CAS 84-36-6; A proprietary preparation of syrosingopine; an antihypertensive. [Nicholas Laboratories Ltd] *

Isoterge. Detergent solutions; used for cleaning of automatic blood cell analyzers. [Coulter Electronics Ltd] †

Isotex® 100. CAS 92-88-6; 4,4´-Biphenol; for liq. crystal polymers. [Schenectady]

Isothan Q-75. Lauryl isoquinolinium bromide; anti-infective. [Onyx Chemical Co] *

Isoton. Diluents based on normal saline; used for blood cell counting and sizing; analysis medium for electrical sensing zone particle size analyzers. [Coulter Electronics Ltd]

Isotox. Insecticide seed treater. [Chevron] *

isovaleraldehyde. (3-methylbutyraldehyde; isovaleral; isovaleric aldehyde) $(CH_3)_2CHCH_2CHO$; CAS 590-86-3; EINECS 209-691-5; Flavoring, perfumes, pharmaceuticals, synthetic resins. [Aldrich; Hercules; Hoechst Celanese; Pfaltz & Bauer]

isovalerone. See diisobutyl ketone.

Isovanat. A proprietary preparation of vanadium in isotonic and isobaric solution. §

Isovue. CAS 60166-93-0; Iopamidol; diagnostic aid. [Bristol-Myers Squibb Co Inc]

Issolin. A phenol-formaldehyde resin, which is soluble in alcohol. §

Issolith. See Bakelite. §

Istin. CAS 15663-27-1; A proprietary preparation of cisplatin; an antineoplastic. [Pfizer International]

Istizin. Danthron; laxative. [Sterling Drug Inc] *

Isuprel Hydrochloride. Isoproterenol hydrochloride; adrenergic. [Sterling Drug Inc] *

Italcor. CAS 1344-28-1; Aluminum oxide. [Winchem Ltd]

ITP®. Thermosetting resin compositions. [Reichhold]

itrol. Silver citrate, $C_6H_5O_7Ag_3$.

Itrumil. A proprietary trade name for the sodium derivation of iodothiouracil. §

I T Talc. CAS 14807-96-6; Hydrous magnesium calcium silicate (industrial talc); used as fillers and extenders for paint, ceramics, plastics, rubber etc. [R T Vanderbilt Co Inc]

lupital® F10. Acetal copolymer; outstanding thermal and color stability, resist. to dimensional change, fatigue, wear, and corrosive environments. [Mitsubishi Gas] *

Ivaleur. A proprietary trade name for pyroxylin (cellulose nitrate). §

Ivarbase 98. Polysorbate 80, cetyl acetate, acetylated lanolin alcohol; cosmetics ingred. [Brooks Industries]

Ivarbase 101. Mineral oil, lanolin alcohol. easy-to-use o/w emulsifier for cosmetics applications. [Brooks Industries]

Ivarbase 3210. Cetyl acetate, acetylated lanolin alcohol. light greaseless emollient for cosmetics. [Brooks Industries]

Ivarbase 3230. Mineral oil, PEG-30 lanolin, cetyl alcohol; absorb. base, cosmetics ingred.; makes very stable milks and creams. [Brooks Industries]

Ivarbase 3240. Isopropyl lanolate, lanolin oil, oleyl alcohol; absorp. base, cosmetics ingred. [Brooks Industries]

Ivarbase 3250. Isopropyl palmitate, lanolin oil; rich light emollient for cosmetics use. [Brooks Industries]

Ivarlan 3100. Lanolin oil; cosmetic ingred. [Brooks Industries]

Ivarlan 3310. CAS 8027-33-6; EINECS 232-430-1; Lanolin alcohol; cosmetics ingred.; distilled grade. [Brooks Industries]

Ivarlan 3400. CAS 61790-81-6; PEG-75 lanolin; cosmetics ingred. [Brooks Industries]

Ivarlan 3406. CAS 61790-81-6; PEG-60 lanolin; cosmetics ingred. [Brooks Industries]

Ivarlan 3420. CAS 68458-88-8; PPG-12-PEG-50 lanolin; cosmetics ingred. [Brooks Industries]

Ivarlan 3450. CAS 68648-27-1; PEG-20 hydrogenated lanolin; cosmetics ingred. [Brooks Industries]

Ivarlan AWS. CAS 68458-58-8; PPG-12-PEG-65 lanolin oil; cosmetics ingred. [Brooks Industries]

Ivarlan C-24. Choleth-24, ceteth-24; cosmetics ingred. [Brooks Industries]

Ivarlan HL. CAS 8031-44-5; EINECS 232-452-1; Hydrogenated lanolin; cosmetics ingred. [Brooks Industries]

Ivarlan Light. Lanolin USP; cosmetic ingred. [Brooks Industries]

Ivarlan OH. CAS 68424-66-8; EINECS 270-315-8; Hydroxylated lanolin; cosmetic ingred. [Brooks Industries]

Ivax. A proprietary preparation of neomycin sulfate, sulfaguanidine and light kaolin; an antidiarrheal. [The Boots Co plc] †

Iversal®. 4-Amiddinohydrazonocyclohexa-2,5-diene-1-one thiosemicarbazone monohydrate; used for infections of the mouth and throat. [Bayer AG]

Iversal® A cum anesthetico. Pharmaceutical preparation; for the prophylaxis and treatment of mouth and throat infections. [Bayer AG]

Ivex® 10. Zinc undecylenate and undecylenic acid; antifungal powds. for foot care and personal hygiene prods. [CasChem]

Ivomec. CAS 70288-86-7; Ivermectin; a mixture of ivermectin component B1a and ivermectin component B1b; antiparasitic. [Merck & Co Inc] *

ivoride. A casein product used as an electrical insulation.

Ivorin-Profalon. Herbicide for beans and potatoes. [Hoechst UK] *

Ivosit. Selective contact herbicide. [Hoechst UK] *

Iwox. Oxidized microcrystalline wax; emulsion-based wax polish. [Industrial Waxes Ltd]

Ixan. Polyvinylidene chloride. [Laporte Industries Ltd]

Ixan®. Copolymers of vinylidene chloride and vinyl chloride or methylacrylate base; used for coextruded film, coatings on plastic film, coatings on paper. [Solvay Polymers]

Ixef® 1022. Polyarylamide; characterized by shock-absorbing properties, low creep, high dens. and modulus; used for parts subjected to heavy vibrations in automotive

---

‡ Trade name and manufacturer not verified   § Trade name without identified manufacturer

(body parts, under-the-hood parts), electromech. and electronic; (elec. motors, alternators, machine tools, high fidelity equip. housings). [Solvay Polymers]

Ixol. Halogenated polyols. [Laporte Industries Ltd]

ixolite. A resin found in Austria.

Ixper. Inorganic peroxides. [Interox Chemicals Ltd]

Izal. A distillate from coke residues. It is a proprietary disinfectant. §

J

J-13. CAS 14807-96-6; Jet-milled talc; cosmetic raw material. [U.S. Cosmetics]

Jabclad. Molded expanded polystyrene panels; for insulation panel applied externally on masonry walls for subsequent rendering. [Vencel Resil Ltd] *

Jabdec. Roof insulation board. [Vencel Resil Ltd]

Jabdie. Laminate of 12 mm fiberboard with expanded polystyrene; used for insulation of flat roofs under felt and mastic asphalt weatherproofing. [Vencel Resil Ltd] *

Jablina Insulating Panels. Laminate of expanded polystyrene with aluminum/paper facings; used for insulation lining for factory buildings. [Vencel Resil Ltd]

Jablite. Expandable polystyrene. [Vencel Resil Ltd] *

Jablite Cavity. Expanded polystyrene boards; used for partially filling the cavity in masonry wall construction. [Vencel Resil Ltd]

Jablite Flooring. Expanded polystyrene boards; used for underfloor thermal insulation. [Vencel Resil Ltd]

Jablite Insulation Board. Expanded polystyrene boards; for insulation of walls and floors, flat roof insulation. [Vencel Resil Ltd] *

Jablite Thermacel. Expanded polystyrene beads; for insulation of masonry cavity walls, infill for 'bean bags'. [Vencel Resil Ltd] *

Jablite Thermoclik. Laminate of bitumen roofing felt with expanded polystyrene; for insulation of flat roofs, used under felt and mastic asphalt weatherproofing. [Vencel Resil Ltd] *

Jablite Wall Lok; Expanded polystyrene tongue and grooved panels for partial cavity full in house building. [Vencel Resil Ltd]

Jaborandi. The native name for several drugs of a sudorific and salivating character, obtained from the leaves and twigs of various species of *Pilocarpus*.

Jabroc. *See* Permali. §

Jacana metal. An alloy of 70% lead, 20% antimony, and 10% tin.

J-acid. CAS 78-02-5; 6-Amino-1-naphthol-3-sulfonic acid; used as an azo dye intermediate.

Jacoby metal. An alloy of 85% tin, 10% antimony, and 5% copper.

Jacquemart's reagent. An aqueous solution of mercuric nitrate with nitric acid; used as a test for ethyl alcohol.

Jacutin®. Lindan. [E Merck]

Jadit. A proprietary preparation of buclosamide and salicylic acid; a skin fungicide. [Hoechst UK] *

Jadit H. A proprietary preparation containing buclosamide, salicylic acid and hydrocortisone; used in the treatment of inflammatory and fungal skin diseases. [Hoechst UK]*

Jagalux. UV and electron-beam curing resins. [Ernst Jager GmbH]

Jagalyd. Alkyd, epoxy ester synthetic resins. [Ernst Jager GmbH]

Jagapol. Polyester resins. [Ernst Jager GmbH]

JagDril CC. Semi-synthetic polymeric viscosifier for completion, workover, and low solids drilling operations. [Rhone-Poulenc/Water Soluble Polymers]

Jagotex. Acrylic resins. [Ernst Jager GmbH]

Jagotex Esi-Cryl. Acrylic resins for floor polishes, etc. [Ernst Jager GmbH]

Jaguar. An emulsifiable concentrate containing 22.2 g benazolin, 55.6 g bromoxynil, 27.8 g ioxynil and 413 g mecoprop per liter; a post-emergence herbicide for cereal crops and grass. [Schering Agrochemicals Ltd]

Jaguar® 413. CAS 39421-75-5; Hydroxypropyl guar gum; thickener for alcohol sol'ns. [Rhone-Poulenc/Water Soluble Polymers]

Jaguar® C. CAS 9000-30-0; Guar gum. [Rhone-Poulenc/Water Soluble Polymers]

Jaguar® C-13S, C-14S. CAS 65497-29-2; Guar hydroxypropyltrimonium chloride; thickener and conditioner for hair and skin care products, shampoos, creme rinses, lotions, creams. [Rhone-Poulenc Surf.]

Jaguar® C-162. CAS 71329-50-5; Hydroxypropyl guar hydroxypropyltrimonium chloride; conditioner for hair and skin care products, conditioning shampoos, bath gels, liq. soaps. [Rhone-Poulenc Surf.]

Jaguar® Guar Gum. CAS 9000-30-0; Guar gum; hydrocolloid for food applications. (baking, cereal, dairy/cheese, processed foods, beverages). [Rhone-Poulenc/Water Soluble Polymers]

Jaguar® HP 8. CAS 39421-75-5; Hydroxypropyl guar; polymer providing thickening, lubricating props., visc., suspension, and slip to aq. or hydroalcoholic cosmetic systems (shampoos, creme rinses, lotions, creams). [Rhone-Poulenc Surf.]

Jaguar® HP-11. CAS 39421-75-5; Hydroxypropyl guar. [Rhone-Poulenc Surf.]

Jaguar® HP 60. CAS 39421-75-5; Hydroxypropyl guar; polymer providing thickening, lubricating props., visc., suspension, and slip to aq. or hydroalcoholic cosmetic systems (shampoos, creme rinses, lotions, creams). [Rhone-Poulenc/Water Soluble Polymers]

jalap. The roots and tubers of certain convolvulaceous plants which yield purgative resins.

jalcase. A steel with a high resistance to wear with a soft core. It has forging properties.

Jamaica wood. *See* logwood.

Janimine. CAS 113-52-0; Imipramine hydrochloride; antidepressant. [Abbott Laboratories] *

---

‡ Trade name and manufacturer not verified § Trade name without identified manufacturer

janthone. A synthetic perfume obtained by condensing citral or lippial with mesityl oxide. It has a violet odor.

Janus. CAS 330-55-2, 1582-09-8; Liquid mixture of linuron and trifluralin; herbicide for winter cereals. [Atlas Interlates Ltd]

Jaon. CAS 299-27-4; Potassium gluconate; replenisher. [Adria Laboratories Inc] *

Japan Agar. See agar-agar.

Japan camphor. CAS 76-22-2; (Laurel Camphor). Ordinary camphor, $C_{10}H_{16}O$, which separates from the essential oil of *Laurus camphora.*

Japan drier. See Terebine.

Japan earth. See cutch.

Japanese acid clay. (Kambara earth). A clay having the formula $Al_2O_3 \cdot 6SiO_2 \cdot xH_2O$ (x is larger than 6). It has powerful adsorptive and decolorizing properties. The dried clay has strong dehydrating action.

Japanese bell metal. An alloy of 60.5% copper, 18.5% tin, 12% lead, 6% zinc, and 3% iron.

Japanese bronze. An alloy of from 81-83% copper, 10% lead, 4.6% tin, and 0-1.8% zinc.

Japanese gelatin. See agar-agar.

Japanese Silver. An alloy of 50% aluminum and 50% silver.

Japanese wax. See Chinese wax.

Japan isinglass. See agar-agar.

Japan sago. The starch from *Cycas revoluta.*

Japan tallow. (sumach wax, vegetable wax). Japan wax, derived from *Rhus succedanea, Rhus vernicifera,* and *Rhus sylvestric;* used for candles, floor waxes, polishes, substitute for beeswax, food packaging.

Japan varnishes. These are obtained by blending asphalt varnishes with dark colored copal or amber varnishes.

Japan wax. See Japan tallow.

Japidermic. A microbial insecticide in powder form containing viable spores of *Bacillus popilliae,* a specific pathogen which infects and kills Japanese beetle grubs; ready-to-use; for control of Japanese beetle grubs; only one application is needed as the living spores are self-perpetuating. [Fairfax Biological Laboratory Inc]

JAQ Powdered Quaternary CAS 139-08-2; EINECS 205-352-0; Myristalkonium chloride; for formulation of disinfectants, sanitizers, and swimming pool algicides. [Huntington Lab]

jara jara. 2-Methoxynaphthalene; used in perfumery.

Jarcal. CAS 10043-52-4; Food grade calcium chloride. [Jarchem Industries Inc]

Jarfix 391. Cellulose fixing agent; cationic auxiliary used in the Jarofast system of cationic dyeing for cellulosic fibers. [James Robinson & Co Ltd]

Jargonelle pear essence. A solution of isoamyl acetate in ethyl alcohol; used for flavoring confectionery.

Jarische's ointment. An ointment containing pyrogallic acid.

Jarofast. System name for the batchwise application of solubilized sulfur dyes by a cationic dyeing system, using a cationic auxiliary (Jarofix 391); the system is used in the garment dyeing industry for cellulosic fibers. [James Robinson & Co Ltd]

Jarosol. Solubilized sulfur dyestuffs; used in the Jarofast system of cationic dyeing for cellulosic fibers. [James Robinson & Co Ltd]

Jarozyme 491. Cationic reduction enzyme; enzyme used in the Jarofast system of cationic dyeing for cellulosic fibers to achieve the washdown effect. [James Robinson & Co Ltd]

Jarytherm. Heat transfer fluids. [Atochem SA]

Jascitile. 3-Methyloctanitrile. [Quest Int'l. UK Ltd]

Jasilyn. 4-Acetoxy-3-pentylterahydropyran. [Quest Int'l. UK Ltd]

Jasmacyclat. Methylcyclooctylcarbonate; fragrance raw material for floral notes. [Henkel/Cospha]

Jasmacyclene. Tricyclo decenyl acetate; a perfumery specialty. [Quest Int'l. UK Ltd]

Jasmal. That fraction of the essential oil of jasmine flowers distilling at 100 C.

Jasmatone. 2-n-Hexyl cyclopentanone. [Quest Int'l. UK Ltd]

Jasmolide. A perfumery chemical. [PPF International Ltd]*

Jasmopyrane. 4-Acetoxy-3-pentylterahydropyran. [Quest Int'l. UK Ltd]

Jasmorange®. 2-Methyl-3(4-methylphenyl) propanal; fragrance (fruity, balsamic, green, floral, aldehydic) [BASF AG]

Jatex. A proprietary brand of pure concentrated rubber latex, 60%. §

Jatob duro. A hard copal obtained from Ceara and Northern Bahia, Brazil; used in varnishes.

Jatob lagrima. (Trapoc resin). A rather soft copal from the Jatoba tree in Brazil; used in spirit varnish.

Jatob resin. Brazilian copals from *Hymenoea courbaril* and *Hymenoea parvifolia.* There are hard and soft qualities. The soft is called jatob, tean, and trapoc .

Java olives. See olives of Java.

Java wax. (Sumatra wax, Gondang wax, Kondang wax, Getah wax). A wax obtained from the bark of the gondang (wild fig) tree, *Ficus variegata.*

Javelin. Suspension concentrate containing 62.5 g diflufenican and 500 g isoproturon per liter; used for control of weeds in winter cereals. [Rhone-Poulenc Crop Protection Ltd]

Jaydalene. A proprietary soldering paste consisting of o-phosphoric acid with a base which vaporizes without decomposition, e.g., aniline, etc. §

Jayflex®. Plasticizers. [Exxon Int'l.]

Jayflex® 77. CAS 71888-89-6; Diisoheptyl phthalate; plasticizer [Exxon] *

Jayflex® 210. CAS 64742-53-6; Proprietary; secondary plasticizer [Exxon] *

Jayflex® 215. CAS 64742-14-9; Proprietary; secondary plasticizer [Exxon] *

Jayflex® 911. Dinonyl undecyl linear phthalate; plasticizer [Exxon] *

Jayflex® 3209. Proprietary adipate; plasticizer [Exxon] *

Jayflex® 7911. Diheptyl, nonyl, undecyl linear phthalate; plasticizer [Exxon] *

Jayflex® DHP. CAS 68515-50-4; Dihexyl phthalate; plasticizer [Exxon] *

Jayflex® DIDP. CAS 68515-49-1; Diisodecyl phthalate; plasticizer [Exxon] *

Jayflex® DINA. CAS 33703-08-1; Diisononyl adipate; plasticizer [Exxon] *

Jayflex® DINP. CAS 68515-48-0; Diisononyl phthalate; plasticizer [Exxon] *

Jayflex® DIOP. CAS 27554-26-3; Diisooctyl phthalate; plasticizer [Exxon] *

Jayflex® DOA. CAS 103-23-1; Dioctyl adipate; plasticizer [Exxon] *

Jayflex® DTDP. CAS 68515-47-9; Ditridecyl phthalate; plasticizer [Exxon] *

Jayflex®DUP. CAS 3648-20-2; Diundecyl linear phthalate; plasticizer [Exxon] *

Jayflex® L9P. Linear phthalate; high permanence plasticizer for extrusion and molding film, sheet and coated fabric for automotive, weather stripping, pool liners, membranes, tarps, specialty wire and cable applications. [Exxon]

Jayflex®TINTM. CAS 53894-23-8; Triisononyl trimellitate; plasticizer [Exxon] *

Jayflex®TOTM. CAS 3319-31-1; Trioctyl trimellitate; plasticizer [Exxon] *

Jayflex® UDP. CAS 68515-47-9; Undecyl dodecyl phthalate; plasticizer [Exxon] *

Jazz®. Fiber. [DuPont UK]

JB-4. Plastic embedding kit; used for light microscopy. [Polysciences Inc]

J-Black-20. Silicone color masterbatch. [Dow Corning STI]

Jectoflo. Lime based fluxes for desulfurisation of steel by deep ladle injection. [Foseco (F.S.) Ltd]

Jectomag. Magnesium based powders for sulfur removal from blast furnace iron. [Foseco (F.S.) Ltd]

Jectothane. A proprietary polyester-based polyurethane thermoplastic injection-molding compound. §

Jeffamine®. A series of polyoxyalkylene-derived di-and triamines; epoxy curing agents, polymer flexibilizers, and specialty polyamides. [Texaco]

Jeffamine® BuD-2000. Urea condensate of POP polyamine; epoxy modifier; nonreactive additive used in conc. of 5-20 phr to provide enhancement of metal-to-metal adhesion, thermal shock properties; results in increased elongation, higher impact and tensile strength, and lowered modulus, while heat deflection values are only slightly affected; reactive with formaldehyde to produce polymeric materials. [Texaco]

Jeffamine®D-2000. POP diamine; epoxy curing agent and modifier usable alone or in combination. [Texaco]

Jeffamine®DU-700. Urea condensate of POP polyamine; epoxy curing agent usable alone or in combination. [Texaco]

Jeffox. A series of poly(oxyethylene) glycols and poly(oxypropylene) glycols and triols; flexibilizers, humectants, and intermediates. [Texaco]

Jefron. Polyferose; a chelate complex of iron and a polymerized derivative of sucrose; hematinic. [Merrell Dow Pharmaceuticals Inc] *

jelly rock. See Wilkinite.

Jel-O-Mer®. Liquid alkyd resins and solutions. [Reichhold]

Jel-O-Mer® 46-902. Flow control agent. [Reichhold]

Jelonet. Paraffin gauze dressing. [Smith & Nephew Pharmaceuticals Ltd] *

Jelutong. (Pontianac, Fluvia, Gambia). A resinous latex yielded by *Dyera costulata*. It contains from 19-24% of rubber and 75-80% of resin, and is used for mixing with rubber and for other purposes.

Jenner's stain. A microscopic stain for white blood corpuscles. It consists of a) a solution of water-soluble, yellowish eosin, 0.5 g, in 100 cc methyl alcohol, and b) a solution of methyl blue, 0.5 g, in 100 cc methyl alcohol. For use 25 cc of a) are mixed with 20 cc of b).

Jeppel's oil. See bone oil.

Jer Dri WRN. Wax/zirconium salt complex; semidurable water repellent; effective on all fibers with excellent high temp. resist. [Sybron]

Jerotex P. Methylated urea-formaldehyde resin; high conc.

thermoset resin used as low cost stiffening agent for synthetic fabrics. [Sybron]

Jersey lily white. A pigment; a lithopone.

Jessate. Ethyl-2-hexylacetoacetate. [Quest Int'l. UK Ltd]

Jesuit's balsam. The oleo-resin copaiba.

Jesuit's bark. Cinchona bark.

Jet. A mineral that is a fossilized wood, and falls between lignite and coal; used for ornaments.

Jet Amine DC. CAS 61791-63-7; Coco diamine; emulsifier, fuel oil and gasoline additives, corrosion inhibitors, mineral flotation. [Jetco]

Jet Amine DE-13. CAS 22023-23-0; Tridecyl ether diamine; emulsifier, corrosion inhibitor. [Jetco]

Jet Amine DE 810. Octadecyl ether diamine; emulsifier, corrosion inhibitor. [Jetco]

Jet Amine DMCD. CAS 61788-93-0; EINECS 263-020-0; Dimethyl coco amine; intermediate for quats., surfactants, agriculture, and detergent formulations. [Jetco]

Jet Amine DMOD. CAS 14727-68-5; Dimethyl oleyl amine; intermediate for quats., surfactants, agriculture, and detergent formulations. [Jetco]

Jet Amine DMSD. CAS 61788-91-8; Dimethyl soya amine; intermediate for quats., surfactants, agriculture, and detergent formulations. [Jetco]

Jet Amine DMTD. CAS 68814-69-7; Dimethyl tallow amine; intermediate for quats., surfactants, agriculture, and detergent formulations. [Jetco]

Jet Amine DO. CAS 7173-62-8; Oleyl diamine; emulsifier, fuel oil and gasoline additives, corrosion inhibitors, mineral flotation. [Jetco]

Jet Amine DT. CAS 61791-55-7; Tallow diamine; emulsifier, fuel oil and gasoline additives, corrosion inhibitors, mineral flotation. [Jetco]

Jet Amine M2C. CAS 61788-62-3; Methyl dicocamine; intermediate for quats., surfactants, agriculture, and detergent formulations. [Jetco]

Jet Amine PC. CAS 61788-46-3; EINECS 262-977-1; Cocamine; corrosion inhibitor, flotation agent, emulsifier, mold release agent, lube oil additive, fertilizer anticaking agent, fabric finishing. [Jetco]

Jet Amine PE 1214. CAS 68511-41-1; Dodecyl/tetradecyl ether amine; emulsifier, corrosion inhibitor. [Jetco]

Jet Amine PHT. CAS 61788-45-2; EINECS 262-976-6; Hydrogenated tallow amine; corrosion inhibitor, flotation agent, emulsifier, mold release agent, lube oil additive, fertilizer anticaking agent. [Jetco]

Jet Amine PO. CAS 112-90-3; EINECS 204-015-5; Oleamine; corrosion inhibitor, flotation agent, emulsifier, mold release agent, lube oil additive, fertilizer anticaking agent. [Jetco]

Jet Amine PS. CAS 61790-18-9; Soyamine; corrosion inhibitor, flotation agent, emulsifier, mold release agent, lube oil additive, fertilizer anticaking agent. [Jetco]

Jet Amine PT. CAS 61790-33-8; EINECS 263-125-1; Tallowamine; corrosion inhibitor, flotation agent, emulsifier, mold release agent, lube oil additive, fertilizer anticaking agent. [Jetco]

Jet Amine TET. CAS 68911-79-5; Tallow tetramine; gasoline detergent, corrosion inhibitor, in petrol. products, dispersion agents for mineral pigments in org. vehicles, asphalt emulsifier. [Jetco]

Jet Amine TP. Tallow pentamine; gasoline detergent, corrosion inhibitor, in petroleum products, dispersion agents for mineral pigments in organic vehicles, asphalt emulsi-

---

‡ Trade name and manufacturer not verified   § Trade name without identified manufacturer

fier. [Jetco]

Jet Amine TRT. CAS 61791-57-9; Tallow triamine; gasoline detergent, corrosion inhibitor, in petrol. products, dispersion agents for mineral pigments in org. vehicles, asphalt emulsifier. [Jetco]

jet black. See gas black.

Jetfil 700C. CAS 14807-96-6; Talc. [Luzenac Am.]

Jet-Flex®101. Weatherable engineering plastic; for camper tops, spas, extrusion profile; stabilized for long-term outdoor weathering; can be coextruded with ABS substrates for thermoforming applications. [Multibase]

Jet Jel®. Flocculant for settling and clarifying reserve mud pits; biodeg. [Rhone-Poulenc/Water Soluble Polymers]

Jet-Lube J-75®. Lead-free drill steel lubricant for percussion rock drilling, blast hole drilling, road construction, logging, mining, coal drilling, pneumatic and tract drilling. [Jet-Lube]

Jet Quat 2C-75. CAS 61789-77-3; EINECS 263-087-6; Dicoco dimethyl ammonium chloride; bactericide, textile softener, asphalt emulsifier, petrol. processing. [Jetco]

Jet Quat C-50. CAS 61789-18-2; EINECS 263-038-9; Coco trimethyl ammonium chloride; bactericide, textile softener, asphalt emulsifier, petrol. processing; home and personal care products. [Jetco]

Jet Quat DT-50. CAS 68607-29-4; Methyl quaternary of tallow diamine; bactericide, textile softener, asphalt emulsifier, petrol. processing. [Jetco]

Jet Quat S-50. CAS 61790-41-8; EINECS 263-134-0; Soya trimethyl ammonium chloride; bactericide, textile softener, asphalt emulsifier, petrol. processing. [Jetco]

Jet Quat T-50. CAS 8030-78-2; Tallow trimethyl ammonium chloride; bactericide, textile softener, asphalt emulsifier, petrol. processing. [Jetco]

Jeunite. Copper fungicide. [Murphy Chemical Co Ltd] †

Jeweller's rouge. The finest calcined ferric oxide or hematite.

Jeyes disinfectant. A disinfectant containing creosote, rosin, caustic soda, and water. It forms emulsions with water.

Jicwood. See Permali. §

JL 43155AS. Silver pigment, epoxy/phenolic binder in MEK:toluene (1:1); EMC shielding coating for metals; protects sensitive electronic equip. [Acheson]

JL 43176; Silver pigment, epoxy/amine binder; EMC shielding coating for metals; protects sensitive electronic equip. [Acheson]

Jodomiron. CAS 440-58-4; Iodamide; diagnostic aid. [Bracco Industria Chimica SpA] *

Jogral. CAS 61791-44-4; Cationic surfactant containing 800 g/l tallow amine ethoxylate; wetting agent for phosphonoglycine herbicide sprays. [Ideal Manufacturing Ltd]

jojoba oil. (oils, jojoba) Oil from the seeds of the Jojoba desert shrub (*Simmondsia chinensis*); CAS 61789-91-1; emollient, conditioner, and lubricant for cosmetics and toiletries. [Arista; R.W. Greeff; Jojoba Growers & Processors; Lipo]

Jojobeads. Jojoba oil derivatives; used for cosmetics & toiletries. [A & E Connock (Perfumery & Cosmetics) Ltd]

Jonylon. Nylon 6 and 66 molding compounds. [BIP Chemicals Ltd]

Jordapon® CI-50 Disp. Sodium cocoyl isethionate; mild foaming surfactant for shampoos, bubble baths, creams, and lotions. [PPG/Specialty Chem.]

Jordaquat®350. Benzalkonium chloride; used for products requiring bacteriostatic, germicidal, and algicidal activity; also for static elimination at low use levels; used to disinfect hard surfs., sanitize food contact surface and fabrics, algae control in water systems. [PPG/Specialty Chem.]

Jothion. A proprietary preparation of iodopropanol. §

JR-228, JR-228-1. Two-component copolymers, one of which is coreacted with a high molecular weight epoxide; semiflexible adhesive systems; superior bonds to thermoplastics, elastomers, glass, and metals (outstanding adhesion to PC, polyesters, nylon, ABS, PVC, and acrylics). [Bacon]

J-Red-10. Silicone color masterbatch. [Dow Corning STI]

J-Red-12FS. Fluorosilicone color masterbatch. [Dow Corning STI]

JR Surfacer. Water soluble epoxy; for tool/pattern making applications and filler. [J R Technology Ltd]

J Slip NS-77. Modified silica; antislip agent; prevents fiber slippage; very effective on polyester; can be used over a wide pH range. [Sybron]

J-Soft 111E. Fatty amide; softener producing luxurious soft hand with no effect on crocking. [Sybron]

JSR-10. A proprietary ABS material possessing high impact strength. [Japan Syn. Rubber] *

JSR-12. A proprietary ABS material possessing high impact strength. [Japan Syn. Rubber] *

JSR-21. A proprietary ABS resin capable of giving good surface finish in molding operations. [Japan Syn. Rubber] *

Jubilee®. For the agriculture industry. [DuPont UK]

Juglone. 5-Hydroxynaphthoquinone.

Julin's chloride. Hexachlorobenzene, C_6Cl_6.

Jupital. CAS 1897-45-6; Chlorothalonil; a fungicide for a wide range of agricultural crops. [Fermenta ASC Europe Ltd]

Justice®. For the agriculture industry. [DuPont UK]

justite. A substance approximating to the formula, $(Ca \cdot Mg \cdot Fe \cdot Zn \cdot Mn)_3 Si_2O_7$, found in furnace slag. The name was formerly applied to the mineral Koenite.

jutahycica. (jutahy.) A copal from Brazil. It is obtained from the roots of *Hymenoea courbaril* and *Hymenoea parvifolia*.

jutahycica resins. (paragum, resina animé). Brazilian copal resins from *Hymenoea courbaril* and *Hymenoea parvifolia*.

Juvelith. See Bakelite. §

J Wet 19A. Nonrewetting wetting agent for flame retardant textile finishes. [Sybron]

JZF. CAS 74-31-7; N,N′-Diphenyl-p-phenylenediamine; an antioxidant for use in rubber, polyethylene, petroleum and vegetable oils and animal fats and oils; in natural rubber it protects against copper and manganese and gives protection against outdoor flexing and static weather cracking; protects against thermal oxidation in polyethylene, inhibits gum formation and degradation at elevated temperature in petroleum oils. [Uniroyal] *

---

* Trade name not verified as available   † Trade name verified as obsolete

K

K 129. CAS 1318-93-0; White montmorillonite; binder and plasticizer for ceramic formulations; ion exchange builder in detergents; thixotropic agent for liq. soaps; flocculant for water treatment. [Kaopolite]

K 129-H. CAS 1302-78-9; White bentonite; economical thickener and suspending agent for liq. abrasive cleaning compounds, water treatment; retention aid for paper. [Kaopolite]

K154. Aluminium-based clear masonry waterproofing solution; treats masonry, slates, tiles and all porous substrates. [Liquid Plastics Ltd]

K285. Absorbable dusting powder. [The Boots Co plc]

K.A. alloy. An aluminum alloy resembling duralumin.

kabaite. A mineral wax of the ozokerite type.

Kabikinase. CAS 9002-01-1; A proprietary preparation of streptokinase; a fibrinolytic agent. [KabiVitrum AB] *

kachin. A photographic developer. Its active constituent is pyrocatechol.

K-acid. 5-Amino-4-hydroxynaphthalene-1,7-disulfonic acid; $C_{10}H_4NH_2OH(SO_3H)_2$; used as an azo dye intermediate.

kadaya gum. See karaya gum.

Kadel® E-1000. Polyketone; high performance thermoplastic for high temp. applications in chemical processing, aviation/aerospace composites, advanced elec./electronic uses, oil drilling, self-lubricating sleeve bearings, antifriction parts, seals, backup rings, extruded shapes for use in hot corrosive media, wire and cable coatings, films, structural parts; excellent solv. and hydrolytic resist., low smoke and toxicity. [Amoco Chemical Co]

kaempferol. The coloring matter of the blue flowers of *Delphinium consolida*. It is a trihydroxyflavonol; $C_{15}H_{10}O_6$.

Kafil. CAS 52645-53-1; Insecticide containing permethrin. [ICI Chem. & Polymers Ltd]

Kagolin 5.8FG. Amitrole + atrazine + diuron; a granular mixture of herbicides for weed control. [Ciba-Geigy Agrochemicals]

Kagoo Oil. See Korung oil.

kainite. CAS 1318-72-5; A salt found in the Stassfurt deposits, consisting mainly of potassium magnesium sulfate and magnesium chloride, $K_2Mg(SO_4)_2 \cdot MgCl_2 \cdot 6H_2O$. The crude material consists of a mixture of kainite and rock salt, and contains 23% of pot; used in chemicals and fertilizers.

Kairoline. N-Ethyltetrahydroquinoline.

Kaiserling solution. A solution used for preserving tissue. It contains 3 g potassium acetate, 1 g potassium nitrate, 75 cc water, and 30 cc formaldehyde.

Kaladex. PEN film. [Imperial Chemical Industries plc]

Kalammon. A fertilizer containing 17% nitrogen and 30% calcium carbonate.

Kalar® 5214. CAS 9010-85-9; Cross-linked butyl composition; produces nonsagging butyl-based sealants, e.g., automotive windshield tape, hot melt sealant; as base for butyl mastics. [Hardman]

Kalar® 5263. CAS 9010-85-9; More highly cross-linked butyl rubber composition; used in elastic sealants requiring more resistance to flow or sag; as green strength enhancer for uncured butyl rubber compositions. [Hardman]

Kalbord. Insulating feeder head liners supplied in the form of flexible boards. [Foseco (F.S.) Ltd]

Kalcrete. Refractory castable for use in foundry ladles. [Foseco (F.S.) Ltd]

Kalene® 800. CAS 9010-85-9; Low molecular weight flowable butyl rubber derived from virgin butyl rubber; polymer used in sealants, coatings, elec. encapsulating compounds, and conformal coatings; gray; often used in solv. systems, e.g. toluene and Varsol 18. [Hardman]

Kaleoilris. A proprietary filling compound. §

Kalex 220 Crystal. CAS 64-02-8; EINECS 200-573-9; Tetrasodium EDTA; chelating agent. [Hart Chem. Ltd.]

Kalex® 13361. Two-component filled urethane compound; potting and encapsulating compound designed to withstand high humidity and temp. exposure, excellent thermal shock resist. [Hardman]

Kalex® 15036. Two-part urethane adhesive; extra fast setting, semiflexible, high shear strength adhesive with good peel strength. [Hardman] *

Kalex® 20171. Two-component urethane system; room temperature curing, low visc. elec. potting and encapsulation compound with very good elec. props., excellent hydrolytic stability, low elevated temp. wt. loss and low water absorp. [Hardman]

Kalex Acids. CAS 60-00-4; EDTA; sequestrant for preparation of amine or alkali metal salts; used where sodium ion is undesirable. [Hart Chem. Ltd.]

Kalex HMP. CAS 10124-56-8; Sodium hexametaphosphate; sequestering agent for calcium and magnesium ions for certain sensitive textile applications. [Hart Chem. Ltd.]

Kalex Liq. 50%. CAS 64-02-8; EINECS 200-573-9; Tetrasodium EDTA; general purpose chelating agent; complexes Ca, Mg, other common metals over wide pH range; for pulp/paper processing. [Hart Chem. Ltd.]

Kalex OH. CAS 139-89-9; Trisodium HEDTA; sequesters Ca, Mg, iron @ pH 8.0-10.5. [Hart Chem. Ltd.]

Kalex Penta. CAS 140-01-2; Pentasodium DTPA; sequestrant used when slightly higher chelate stability is required and when other strong complexing agents are

---

present; for pulp and paper, textile bleaching operations. [Hart Chem. Ltd.]

Kaliammon saltpeter. A potassium ammonium nitrate prepared by mixing equivalent molecular proportions of solid potassium chloride and ammonium nitrate in the presence of a little water; a fertilizer.

Kalidone®. CAS 4810-50-8; Potassium PCA; moisturizing agent for dermatological soap, shampoo, after-sun lotion, shower gel, nutritive and regenerative creams, hair comb-out balm. [UCIB]

Kalif. A proprietary copper-lead bearing alloy; melts at 952 C, with tensile strength of 10,000 psi at 21 C. §

Kalipol. Polyphosphate solution. [Albright & Wilson Ltd, Phosphates & Speciality Business]

Kalipol 18. A proprietary potassium polyphosphate solution used in the manufacture of liquid detergents. [Albright & Wilson Ltd]

Kalite. A proprietary form of chalk prepared by a special process whereby it has a very small particle size, and the particles are coated with a calcium soap; a rubber filler.§

Kaliuzoto. A fertilizer containing nitrogen, potassium, and organic matter manufactured from residual molasses.

kalkeisenolivin. Synonym for iron monticellite.

Kalkor. Disposable refractory plugs for plug-bottom ingot molds. [Foseco (F.S.) Ltd]

Kalleonicit. Pharmaceutical preparation also known as Nico Padutin; combination therapy for physiological vasodilation. [Bayer AG]

Kalle's acid. 1-Naphthylamine-2,7-disulfonic acid.

Kallodent®. Trade name for a methyl methacrylate thermoplastic material used for molding dentures. §

Kallodoc. Acrylic powder for artificial teeth and eyes. [Imperial Chemical Industries plc]

Kalmex. Thermit compound for feeding ingots and castings; also used in continuous casting tundishes to facilitate start up of casting. [Foseco (F.S.) Ltd]

Kalmin. Insulating riser sleeves and shapes. [Foseco (F.S.) Ltd]

Kalminex. Exothermic sleeves with highly insulating residual structures for lining feeder heads on iron and steel castings. [Foseco (F.S.) Ltd]

Kaloempang beans. See olives of Java.

Kalorex. Lightweight exothermic sideliner tiles for steel ingots. [Foseco (F.S.) Ltd]

Kalpack. Refractory ramming compound for securing nozzles in foundry ladles. [Foseco (F.S.) Ltd]

Kalpad. Moldable and preformed insulating materials to assist in feeding of steel and iron casting sections. [Foseco (F.S.) Ltd]

Kalpur. Insulating feeding sleeve with ceramic foam filter. [Foseco (F.S.) Ltd]

Kalrez®. Fluoroelastomer parts. [DuPont UK]

Kalseal. Refractory air-setting mortar used in foundry ladles. [Foseco (F.S.) Ltd]

Kalsert. System for applying feeder sleeves by insertion into pre-formed cavities in molds of cores. [Foseco (F.S.) Ltd]

Kaltas. Solvent-based bituminous binder; used for roads. [Vedag GmbH] *

Kaltek. Highly insulating disposable ladle lining. [Foseco (F.S.) Ltd]

Kalten. Combined hypertensive and diuretic containing atenolol, hydrochloride thiazide and aniloride hydrochloride. [ICI Chem. & Polymers Ltd]

Kaltop. Exothermic antipiping compounds in board form.

[Foseco (F.S.) Ltd]

Kalvan. A proprietary trade name for calcium carbonate for use in rubber to give wear resistance; has an ultra fine particle size. §

Kalzana. A proprietary calcium-sodium lactate. §

kalzose. A casein preparation containing calcium.

Kamala. See Kamela.

Kamax. Imidized methyl methacrylate molding powder. [Rohm & Haas UK]

Kamax T-150. Acrylic-imide copolymer; amorphous thermoplastic; injection molding resin featuring high heat resist., excellent optics, outdoor durability, excellent stiffness, easy processing; used for lighting, automotive, optical, pkg., medical, and appliance applications. [AtoHaas]

Kambe wood. See redwoods.

Kamela. (Kamala). A dyestuff obtained from the seeds or fruits of *Mallotus phillpenis* or *Rottlera tinctoria*; used in India as medicine, and for dyeing silk orange.

Kamillosan. Pharmaceutical preparation containing the active principle of *Matricaria chamomilla*. [Degussa AG]*

Kamoran. Actaplanin; a complex of glycopeptide-type antibiotics; growth stimulant. [Eli Lilly & Co]

Kane-Ace. MBS impact modifier for PVC. [European Vinyls Corporation Ltd]

Kane's salt. A salt prepared by dissolving mercuric nitrate in a boiling solution of ammonium nitrate.

Kanfotrex. A proprietary preparation of kanamycin, amphymycin and hydrocortisone used in dermatology as an antibacterial agent. [Bristol-Myers Squibb Co Inc]

Kanga butter. See lamy butter.

Kanigen. Chemical nickel plate. [Albright & Wilson Ltd] *

Kanja butter. See lamy butter.

Kankerex. CAS 21908-53-2; Mercuric oxide, HgO; for control of canker in apples and pears. [Universal Crop Protection Ltd]

Kannasyn. A proprietary preparation containing kanamycin (as the sulfate); an antibiotic. §

kanten. A variety of agar-agar from red tegusa seaweed of Japan.

Kanthal Alloy. A trademark for an iron alloy containing aluminum, cobalt, and chromium, and having a high degree of resistance to heat, a low electrical conductivity, and good hot and cold working properties. §

Kantmelt. Nonmelting industrial grease. [Specialty Products Co] *

Kantrex. CAS 25389-94-0; A proprietary preparation containing kanamycin sulfate; an antibiotic. [Bristol-Myers Squibb Co Inc]

Kantrexil. A proprietary preparation containing kanamycin sulfate, pectin, bismuth carbonate and activated attapulgite; an antidiarrheal. [Bristol-Myers Squibb Co Inc]

Kantrim. CAS 25389-94-0; Kanamycin sulfate; antibacterial. [Bristol Laboratories] *

Kantstik Q Powd. Refined synthetic wax ester; internal lubricant for injection molding; improves plastic flow in hard-to-reach mold areas; valuable in heavily pigmented molding resin mix; used in butyrate, PP, nylon, glass-filled nylon, PC, SAN, styrene, polyethylene, ABS, high-impact PS, PVC. [Specialty Prods.]

Kaochlor. A liquid potassium product sold only on prescription. [Hercules] *

Kaodene. Preparation for the treatment of diarrhea. [The Boots Co plc] *

---

* Trade name not verified as available

† Trade name verified as obsolete

Kaogel. A proprietary preparation containing kaolin and pectin; an antidiarrheal. [Parke-Davis] *

Kaokote. Partially hydrogenated vegetable oil (cottonseed, soybean), sorbitan stearate, polysorbate 60; kosher; coating fat for no-tempering low-liquor type coatings and centers. [Van Den Bergh Foods]

Kaokote F. CAS 68334-28-1; Partially hydrogenated vegetable oil (cottonseed, soybean); kosher; high performance fat for center fat, vegetable dairy systems. [Van Den Bergh Foods]

Kaola. CAS 68334-28-1; Partially hydrogenated vegetable oil (soybean, palm kernel); kosher; oil for vegetable dairy systems, candies, ice cream bar coatings, nut roasting. [Van Den Bergh Foods]

kaolin. (Bolus alba; China clay) Native hydrated aluminum silicate; $Al_2O_3 \cdot 2SiO_2 \cdot 2H_2O$; CAS 1332-58-7; EINECS 296-473-8; filler and coatings for paper, rubber, refractories, ceramics; in anticaking preps., paint; adsorbent for clarification of liqs. [Burgess Pigment; Dry Branch Kaolin; ECC Int'l.; J.M. Huber; Kapolite; Southeastern Clay; R.T. Vanderbilt]

kaolinase. A purified kaolin.

Kaomax. Partially hydrogenated soybean oil, sorbitan tristearate; kosher; high performance fat for no-tempering confectioners coating, vegetable dairy systems; steep melting shortening, oil migration inhibitor. [Van Den Bergh Foods]

Kaomel. CAS 68334-28-1; Partially hydrogenated vegetable oil (cottonseed, soybean); kosher; high performance fat for no-tempering confectioners coatings, vegetable dairy systems; steep melting shortening, oil migration inhibitor. [Van Den Bergh Foods]

Kaomycin. A proprietary preparation containing neomycin sulfate and kaolin; an antidiarrheal. [Upjohn Ltd] *

Kaon. A range of potassium products sold only on prescription; for pharmaceutical and medical uses. [Hercules] *

Kaon-Cl. CAS 7447-40-7; EINECS 231-211-8; Potassium chloride; replenisher. [Adria Laboratories Inc] *

Kaopectate. A proprietary preparation containing kaolin, bentonite and pectin; an antidiarrheal. [Upjohn; Upjohn Ltd]

Kaopolite® 1152. CAS 12141-46-7; Aluminum silicate, anhyd.; mild polishing props. for auto polishes, plastics; gentle abrasive that speeds cleaning; antiblocking agent for plastic film. [Kaopolite]

Kaopolite® AB. CAS 12141-46-7; Hydrated aluminum silicate; ultra-fine particle, soft silicate to moderate the polishing action of other abrasives; for auto polishes. [Kaopolite]

Kaopolite® SF. CAS 12141-46-7; Aluminum silicate, anhyd.; gentle abrasive for auto polish, metals, plastics, household, dentifrices; antiblocking agent in plastic film. [Kaopolite]

Kaoprem-E. Partially hydrogenated soybean oil with sorbitan tristearate; kosher; coating fat for no-tempering confectioners coating. [Van Den Bergh Foods]

Kaorich Beads. CAS 68334-28-1; Vegetable oil (cottonseed, soybean), hydrogenated; emulsifier for food processing. [Van Den Bergh Foods]

Kaorich Gold. CAS 68334-28-1; Partially hydrogenated vegetable oil (soybean, cottonseed), artificial butter flavor, artificial color (β carotene); kosher; shortening system for breading mixes and other dry mixes. [Van Den Bergh Foods]

Kaovax. A proprietary preparation of succinylsulfathiazole, and kaolin; an antidiarrheal. [H N Norton & Co Ltd] ‡

Kaowool®. CAS 1332-58-7; Kaolin clay-based; refractory bulk fibers for solving heat problems in furnaces (metal, petrochem., kilns in ceramic industry, boilers in utility industry). [Thermal Ceramics]

kapak. A material made from the mineral rubber elaterite; used in rubber mixings.

Kapazang oil. A fat obtained from the seeds of *Hodgsonia heteroclita*.

Kapex. Exothermic shapes to cover the surface of blind risers for iron and steel castings. [Foseco (F.S.) Ltd]

kapithamia piscum. *See* wood-apple gum.

Kapitol. Captan + nuarimol; systemic fungicide for apple and pear trees. [DowElanco Ltd]

kapok. A cotton-like down produced in the seed-pods of the kapok tree; used in upholstery, life jackets, insulation, and pillows.

Kapsol. A proprietary trade name for a methoxy-ethyloleate; used as a plasticizer. §

Kapsovit. Multi-vitamin capsules. [Allen & Hanburys Ltd] *

Kapton®. A proprietary trade name for polyimide resin in the form of film. [DuPont UK]

Kapur Kachri. *See* sanna.

Karafac 78. Conc. surfactant; neutralized (potassium salt) version of Karaphos XFA. [Clark]

Karakane. An alloy of 62.5% copper, 25% tin, 9.4% zinc, and 3.1% iron.

Kara Lube AL-Conc. Formulated product; fiber lubricant and dyestuff dispersant; leveling and migrating agent for direct dyeing of polyester/cotton blends; retarder/leveler for dyeing modacrylic fiber. [Rhone-Poulenc Surf.]

Karalube DKL. Fatty derivs./surfactant blend; lubricant for jet machine scouring and dyeing of synthetic fabrics. [Rhone-Poulenc Surf.]

Karamate. CAS 8018-01-7; Wettable powder or water dispersible granules containing mancozeb; protectant fungicide for fruit, field crops, and roses. [Rohm & Haas UK]

Karamide 121. Cocamide DEA (1:1); heavy-duty detergent, thickener, foam stabilizer in cleaning compounds, shampoos, textile scours; as lubricant, antistat. [Clark]

Karamide CO9A. Cocamide DEA (2:1); detergent, base for floor cleaners, all-purpose cleaners. [Clark]

Karamide ST-DEA. CAS 93-82-3; EINECS 202-280-1; Stearamide DEA; thickener, emulsifier for vegetable oil, mineral oil, microcrystalline wax. [Clark]

Karaphos HSPE. Blended phosphate ester, free acid; hydrotrope and wetter in heavy-duty detergents; high caustic-stable wetter and penetrant in textile formulations; as antistat; as rust preventative in metal cleaners; very low foaming; excellent alkali stability. [Clark]

Karaphos SWPE. Blended phosphate ester, free acid; fast wetting agent for detergents, cleaners, textile scours, penetrants, metal prep. compounds, antistats. [Clark]

Karasoft YB-11. Conc. self-emulsifying softener base; produces soft, dry, fluffy hand when applied to cotton, blends, and synthetics. [Clark]

Karasurf AS-26. Ammonium 2-ethyl hexanol sulfate; wetter, penetrant, hydrotrope, solubilizer for industrial cleaners, fire fighting foams, etc. [Clark]

Karate. λ-cyhalothrin; pyrethroid insecticide. [ICI Chem. & Polymers Ltd]

Karate. CAS 3691-35-8; An oil formulation containing 2.5%

---

‡ Trade name and manufacturer not verified   § Trade name without identified manufacturer

of chlorophacinone, an anticoagulant rodenticide; a bait to control black rats, brown rats, house mice and voles. [Lever Industrial Ltd]

Karathane. CAS 39300-45-3; Emulsifiable concentrate of 350 g dinocap per liter; used to control mildew in fruit, chrysanthemums and roses. [Rohm & Haas UK]

karaya gum. (Sterculia gum; India tragacanth; Kadaya gum) A hydrophilic polysaccharide from genus *Sterculia*; pharmaceuticals, textile coatings, ice cream and other food products, adhesives, protective colloids, stabilizers, thickeners, emulsifiers. [Meer; Penta Mfg.; Rhone-Poulenc Perf. Resins; TIC Gums]

Karaya Paste. A proprietary preparation of sterculia in isopropyl alcohol; used as a dressing around colostomies. [Abbott Laboratories] *

Karbolite. A Russian artificial resin made from phenols, formaldehyde, and naphtholsulfonic acid. §

Karboresin. Hydrocarbon-modified printing ink resins. [Hoechst AG]

Karbos. A carbonaceous decolorizer and filtering medium.

Karetnja. A bituminous insulation. It consists mainly of asphalt, with an aluminum stearate binder.

Karite gum. See gutta-shea.

Karlex 12006 BKFR, 12018 BKFR. Polycarbonate, 20% glass fiber-reinforced; flame retardant thermoplastic [Ferro/Engineering Thermoplastics] *

Karlex 40002 NA. PC/PET alloy; high heat deflection temp. [Ferro/Engineering Thermoplastics] *

Karlex 40003 NA. PC blend; high-impact. [Ferro/Engineering Thermoplastics] *

Karma metal. An alloy of 80% nickel and 20% chromium. It is a heat-resisting alloy and melts at 1415 C.

Karmarsch metal. An alloy containing 88.8% tin, 7.4% antimony, and 3.7% copper.

Karmex®. CAS 330-54-1; Diuron weedkiller. [DuPont UK]

Karmex. CAS 330-54-1; Diuron; residual urea herbicide. [Rohm & Haas UK]

Karolith. A casein preparation similar to Galalith; used for the manufacture of buttons and other objects.

Karox AO-30. CAS 61788-90-7; EINECS 263-016-9; Cocamine oxide; high foaming, mild surfactant, wetter, emulsifier, and coupling agent with excellent alkali tolerance. [Clark]

Karox LO. Dimethyl lauramine oxide; high foaming, mild detergent, coupling agent, wetter, emulsifier with good alkali and bleach stability. [Clark]

Karvol. A proprietary preparation containing various volatile oils; a decongestant. [The Boots Co plc]

Kasal®. Sodium aluminum phosphate, basic; food additive for baking, cereals, dairy and cheese. [Rhone-Poulenc Food Ingreds.]

Kasil. CAS 1312-76-1; Potassium silicates supplied in varying SiO_2:K_2O ratios; alkaline builder for liquid detergents and cleaners, binder for phosphors in cathode ray tubes. [PQ Corp] *

Kasof. CAS 7491-09-0; Docusate potassium; stool softener. [Stuart Pharmaceuticals] †

Kastone®. For the chemical industry. [DuPont UK]

Kastor. A proprietary preparation of senna and sodium potassium tartrate; a laxative. [Ayrton Saunders plc]‡

Katadolon. CAS 56995-20-1; Flupirtine; capsules and suppositories; analgesic. [Degussa AG] *

Katalabu gum. A gum of Nigeria, from *Acacia sieberiana*; an adhesive.

Katalon. CAS 85-00-7; Diquat; herbicide. [Makhteshim Chemical Works Ltd]

Katanol. A trademark for a range of mordants used in the dyeing of textiles. [Cassella AG] *

Katapol®. Redesignated Rhodameen®. [Rhone-Poulenc Surf.]

Katapone VV-328. Quaternary ammonium chloride; water soluble, acid-corrosion inhibitor for steel, copper, and aluminum. [ISP] *

katarsit. A calcium sulfite pellet for use as a dechlorinating agent for water.

Katavel oil. The oil from *Hydnocarpus wightiana*.

Katb₆l - ki - gond. See wood-apple gum.

Katchung oil. CAS 8002-03-7; Peanut oil.

katchung oil. See peanut oil.

katechu. See cutch.

Katemul IG-70. CAS 118777-77-8; Isostearamidopropyl dimethylamine glycolate; emulsifier, conditioner and softener for skin and hair products. [Scher]

Katemul IGU-70. CAS 129541-36-2; Isostearamidopropyl dimethylamine gluconate; emulsifier, conditioner and softener for skin and hair products. [Scher]

Katharin. CAS 56-23-5; A proprietary trade name for carbon tetrachloride; used as a grease remover. §

Kathon®. Biocide; preservative for coatings, floor polish and adhesives. [Rohm & Haas]

Kathon® 886. 5-Chloro-2-methyl isothiazolones; used as biocides and preservatives in a wide range of industrial applications. [Rohm & Haas]

Kathon® 893. N-Octyl isothiazolones; used as biocides in paint, leather, textiles, paper and plastics. [Rohm & Haas]

Kathro. CAS 57-88-5; Semi-refined cholesterol. [Croda Chem. Ltd] *

Katigen®. A registered trademark currently awaiting re-allocation by its proprietors to cover a range of dyestuffs. [Cassella AG] *

Katioran® AF. Fatty acid hydroxylalkylamide and ethoxylated fatty alcohol; emulsifier and thickener for hair care products; makes hair soft and manageable and prevents electrostatic charges. [BASF AG]

Katonium. A proprietary preparation of ammonium and potassium polystyrene sulfonate; a diuretic. §

Katorin. CAS 299-27-4; A proprietary preparation of potassium gluconate; a potassium supplement. [The Boots Co plc]

Katzenstein bearing metal. See bearing metals.

Kauk Catalyst. A proprietary trade name for a spherical catalyst of 5 mm diameter consisting of potassium salts and vanadium on a porous silica carrier. V_2O_5 content 6.5 %; used for converting SO_2 into SO_3. §

Kauramin®. Melamine-formaldehyde based; glues for product of weather-resist. chipboard and plywood; impregnating resin for decorative and overlay papers. [BASF AG]

Kauranat®. MDI-based; for production of weather-resist. chipboard. [BASF AG]

Kauresin®. Phenol-formaldehyde based; glues for production of weather-resist. plywood. [BASF AG]

Kaurit®. Urea-formaldehyde based; glues for production of veneered board, furniture; impregnating resins for decorative papers. [BASF AG]

Kauropal®. Assistants for glue and impregnating resins. [BASF AG]

---

* Trade name not verified as available   † Trade name verified as obsolete

kava-kava resin. A mixture of resins and resin acids from the dried roots of *Piper methysticum*.

Kawasaki Hakkinko. A proprietary Japanese steel containing 0.19% carbon, 1.8% silicon, 1.0% manganese, 17.0% nickel, 25.0% chromium, and 0.2% molybdenum; offers resistance to hydrogen embrittlement. §

Kayamer. Phosphate acrylate/methacrylate monomers. [British Traders & Shippers Ltd]

Kayarad. Monomeric/oligomeric acrylate monomers. [British Traders & Shippers Ltd]

Kay Ciel. CAS 7447-40-7; EINECS 231-211-8; Potassium chloride; replenisher. [Berlex Laboratories Inc] *

Kaydol. White mineral oil. [Witco]

Kaydox. 1,4-Dichlorobenzene paste. [Murphy Chemical Co Ltd] †

Kayexalate. Sodium polystyrene sulfonate; ion- exchange resin. [Sterling Drug Inc] *

Kaylene. A proprietary preparation of colloidal aluminum silicate. §

Kaylene-ol. A proprietary preparation of colloidal aluminum silicate with liquid paraffin. §

Kaynitro. Concentrated nitrogen/potash fertilizer. [ICI Chem. & Polymers Ltd] †

Kayphobe-ABO. CAS 1332-58-7; Kaolin. [Kaopolite]

K-Bond. Aluminum tripolyphosphate dihydrate or metaphosphate; hardener for waterglass; used in inorganic coatings (interior and exterior heat resistance and nonflammable). [Bromhead & Denison Ltd]

K-Contin. CAS 7447-40-7; EINECS 231-211-8; A proprietary preparation of potassium chloride in a controlled release tablet; used as a potassium supplement in diuretic therapy. [Napp Laboratories Ltd] *

K-Cop. Copper-ammonium complex; agricultural fungicide. [Griffin]

K de Krizia. A fragrance from Milan's foremost name in fashion design. [Richardson-Vicks Inc] *

Keene's alloy. *See* nickel silvers.

Keene's Cement. The name for a number of different plasters prepared by various manufacturers; usually obtained from plaster of Paris, dipped into a solution of alum or aluminum sulfate, and recalcining. §

Kefadol. CAS 42540-40-9; Cefamardole nafate; indicated in the treatment of infections of the respiratory tract, genitourinary tract, bones and joints, bloodstream (septicemia), skin and soft tissue, gall bladder and peritoneum and pelvic inflammatory disease in women; when due to susceptible microorganisms. [Dista Products Ltd]

Keffekilite. A fuller's earth.

Keffekill. Synonym for Sepiolite.

Keflex. CAS 15686-71-2; A proprietary preparation containing cephalexin monohydrate; an antibiotic. [Eli Lilly & Co]

Keflin. Cephalothin sodium; an antibiotic. [Eli Lilly & Co]

Keftab. Cefalexin hydrochloride; antibiotic. [Eli Lilly & Co]

Kefzol. Cephazolin sodium; an antibiotic. [Eli Lilly & Co]

Keical-Ace. CAS 10101-39-0; Ultra lightweight calcium silicate insulation; insulation for reactor, heat exchanger, vessel, tank, pipe. [Mitsubishi Kasei] *

Kekuna oil. Bakoly oil (candlenut oil).

Kelacid®. CAS 9005-32-7; EINECS 232-680-1; Alginic acid; used as gelling agent, emulsifier and stabilizer in food, pharmaceutical, and industrial applications; stabilizer in paper and textile industry. [Kelco]

Kelburon. An elastomer-modified polypropylene used for automotive applications (bumpers, dashboards). [DSM

NV] *

Kelco-Crete™. Welan gum; for cementitious applications. [Kelco]

Kelco-Gel® Gellan Gum. CAS 71010-52-1; Purified gellan gum; high molecular weight anionic polysaccharide; gelling agent for use in foods, pet foods, personal care products, industrial applications. [Kelco]

Kelcoloid® D. CAS 9005-37-2; Propylene glycol alginate; used as gelling agent, emulsifier and stabilizer in food, pharmaceutical, and industrial applications; stabilizer in paper and textile industry. [Kelco]

Kelcoloid® DH, DSF. CAS 9005-37-2; Propylene glycol alginate; used for emulsions and low pH systems. [Kelco]

Kelcoloid® HV, LV. CAS 9005-37-2; Propylene glycol alginates; used for emulsions and low pH systems. [Kelco]

Kelcoloid® O, S. CAS 9005-37-2; Propylene glycol alginate; beef foam stabilizer. [Kelco]

Kelcosol®. CAS 9005-38-3; Algin; used as gelling agent, emulsifier and stabilizer in food, pharmaceutical, and industrial applications; stabilizer in paper and textile industry. [Kelco]

Keldax®. Ethylene interpolymer resin. [DuPont UK]

Keleastoi. A proprietary trade name for a ricinoleate type of vinyl plasticizer. [Spencer Kellogg & Sons] *

Kelecin®. CAS 8002-43-5; EINECS 232-307-2; Industrial lecithin; emulsifying and dispersion agent, paint, mastics, feed and rubber. [Reichhold]

kelenmethyl. A mixture of ethyl and methyl chlorides; an anesthetic.

Kel-F 3700 Elastomer. A proprietary synthetic rubber resistant to high temperatures. [3M]

Kel-F 81 Plastic. Family of thermoplastic extrusion and molding materials that provide excellent chemical resistance over a broad temperature range; polychlorotrifluoroethylene. [3M]

Kelferron. A proprietary preparation of ferrous glycine sulfate; a hematinic. [MCP Pharmaceuticals] *

Kelfizina. CAS 152-47-6; Sulfalene; antibacterial. [Abbott Laboratories] *

Kelfizine W. A proprietary preparation of sulfametopyrazine; an antibiotic. [Farmitalia] *

Kelflo®. CAS 11138-66-2; Xanthan gum; liquid feed supplements. [Kelco]

Kelfo®. Xanthan gum/limestone blend; gum for use in animal feed; thickener, stabilizer. [Kelco]

Kelfolate. A proprietary preparation of ferrous glycine sulfate and folic acid; a hematinic. [MCP Pharmaceuticals]*

Kelgin. Sodium alginate, food grade. [Kelco Int'l. Ltd]

Kelgin® F. CAS 9005-38-3; Algin, refined; used as gelling agent, emulsifier and stabilizer in food, pharmaceutical, and industrial applications; stabilizer in paper and textile industry. [Kelco]

Kelgin® HV, LV, MV. Sodium alginates; used for paper coating and sizing, textile printing, dyeing and sizing, alkaline carpet dyeing and wallpaper adhesives. [Kelco]

Kelgin® QH, QL, QM. Sodium alginates; dispersible products for paper coating and textile applications. [Kelco]

Kelgin® XL. Sodium alginates; used for paper coating and sizing, textile printing, dyeing and sizing, alkaline carpet dyeing and wallpaper adhesives. [Kelco]

Kelgum. CAS 9005-38-3; A linseed oil rubber substitute.

Kelgum®. CAS 11138-66-2; Xanthan gum; used for desserts, gelled confectioneries, gels. [Kelco]

Kelig. CAS 8061-51-6; Modified sodium lignosulfonates; metal complexing agents for micronutrient formulations, industrial cleaners and water treatment formulations. [Borregaard LignoTech]

Kelig 32. Lignosulfonate and modified sugar acids; sequestrant, dispersant for the lubricant removal from zinc phosphate coatings, oil well cement retardant; used for cooling water treatment compounds, alkaline cleaners, scale control. [Borregaard LignoTech]

Kelig FS. Lignosulfonate; carrier for micronutrient metals in liq. formulations; inhibits leaf burn. [Borregaard LignoTech]

Kellin®. Drying oils. [Reichhold]

Kellite. A proprietary trade name for a synthetic resin. §

Kel-Lite™ CM. CAS 11138-66-2; Xanthan gum; used for cakes, cookies, muffins, waffles, pancakes. [Kelco]

Kellox®. Oxidized fish oils. [Reichhold]

Kelly's paint. A benzoated collodion containing tincture of benzoin, glycerin, and collodion; used for painting on abrasions of the skin.

Kelmar®. CAS 9005-36-1; Potassium alginate; gellant, emulsifier, and stabilizer in food and indust. application; gum, bodying agent for creams and lotions, dental impression materials; used for water holding in foods and industry. [Kelco]

Kelmer®. Industrial oils and greases; lubricants; dust absorbents; wetting and binding compositions; fuels; illuminants; candle wicks. [Reichhold]

Kelocyanor. A proprietary preparation of cobalt tetracemate in a glucose solution; an antidote for cyanide poisoning. [Rona Laboratories] ‡

Kelo-form. A proprietary trade name for ethyl aminobenzoate. §

kelp. (varec). Seaweed or the ash from seaweed; a source of iodine.

Kelpak. Seaweed concentrate; soil improver. [Omex Agriculture Ltd]

kelpchar. A decolorizing carbon obtained from seaweed by carbonizing in two stages and extracting with water and dilute hydrochloric acid.

Kelpol®. Paints, varnishes lacquers; preservatives against rust and deterioration of wood; colorants; mordants; raw natural resins; metals in foil and powder form for painters, decorators, printers, and artists. [Reichhold]

Kelpol® 835-M-50. Vinyl toluene monomer-modified alkyd. [Reichhold]

Kelpoxy®. Epoxy resins; for use in adhesives and coatings. [Reichhold]

Kelpoxy® G202-100. Elastomer modified epoxy conc. [Reichhold]

Kelprox. Thermoplastic EPDM-rubber; for automotive, appliances, industrial worms. [DSM NV] *

kelp salt. A mixture of potassium chloride, with some alkaline sulfates and carbonates, formed in the preparation of potassium chloride from kelp.

Kelrinal. CM rubber; used as a high-grade vulcanized rubber in the wire and cable industry and in the automotive industry. [DSM NV] *

Kelset®. Alginate; used as gelling agent, emulsifier and stabilizer in food, pharmaceutical, and industrial applications; stabilizer in paper and textile industry; self-gelling gum. [Kelco]

Kelsol®. Modified vegetable oil used as medium for tinting

color in paints. [Reichhold]

Kelsol® 3922-G-80. Waterborne long oil alkyd. [Reichhold]

Kelstar. Pigment dispersions; for coloration of non-polyurethane plastics. [Pacific Dispersions Inc] *

Keltan. Range of EPDM terpolymers having differing Mooney viscosities; used in the automotive industry for wires and cables, in the building industry and in domestic appliances and technical products. [DSM NV] *

Keltan 312. EPDM, nonstaining stabilizer; improved processability; very fast curing. [Copolymer Rubber]

Keltan 512. EPDM, nonstaining stabilizer; suitable for continuous curing; very fast cure rate. [Copolymer Rubber]

Keltan 4703. EPDM terpolymer, nonstaining stabilizer; high loading capacity, excellent mixing and extrusion behavior; very suitable for sponge articles, profiles; ultra fast curing. [Copolymer Rubber]

Keltan 4802. EPDM terpolymer, nonstaining stabilizer; good collapse resist., very good mech. and elastomeric props.; for solid profiles, radiator hoses, sealing rings, door seals for washing machines; very fast curing. [Copolymer Rubber]

Keltan TP. Thermoplastic rubber; used in the automotive, medical and pharmaceutical industries, for wires and cables, in household equipment and shoes. [DSM NV]*

Keltex®. Sodium alginates; used for textile printing, silver recovery. [Kelco]

Keltex® S. Specially designed sodium alginate; buffered product for textile printing. [Kelco]

Keltex® HV. Alginate; print paste thickener. [Kelco]

Kelthane. CAS 115-32-2; Dicofol; an acaricide. [Rohm & Haas UK]

Kelthix. Thixotropic resins; used for paint. ß

Keltone®. Sodium alginate; used for hot and cold water dessert gels, dental impression material. [Kelco]

Keltone® HV, LV. Sodium alginate; used for puddings, pie fillings, bakery fillings, chiffons, sauces and dessert gels. [Kelco]

Keltose®. Calcium alginate and ammonium alginate; gellant, binder, emulsifier, and stabilizer in food and industrial application. [Kelco]

Keltrol®. CAS 11138-66-2; Xanthan gum, food grade; used for bakery fillings, flavor emulsions, canned foods, dry mixes, frozen foods, juice drinks, pourable and spoonable dressings, relishes, gravies, syrups, baked goods, batters, puddings, toothpaste, shampoo, pharmaceuticals. [Kelco]

Keltrol®. Drying oils used in paints, varnishes, lacquers, and enamels. [Reichhold]

Keltrol® 1001-M-60. Vinyl toluene monomer-modified alkyd. [Reichhold]

Keltrol® F. CAS 11138-66-2; Food-grade xanthan gum; stabilizer for foods; thickener and emulsion stabilizer in creams and lotions; binder in toothpaste; suspending agent for fruit pulp. [Kelco]

Kelvar®. Varnishes, enamels. [Reichhold]

Kelvar® 3758-M-85. High solids alkyd. [Reichhold]

Kelvis®. CAS 9005-38-3; Sodium alginate, refined; gellant, emulsifier, and stabilizer in food and industrial application. [Kelco]

Kelzan®. CAS 11138-66-2; Xanthan gum; used for textured paint, carpet printing, cleaners and polishes, water-based lubricants, coatings, ceramic glazes, pigment and dye suspensions, agricultural products, metal working

---

* Trade name not verified as available   † Trade name verified as obsolete

products, foam dyeing/printing, coal slurries; other systems requiring suspension. [Kelco]

Kelzan® , D, M, XC Polymer. CAS 11138-66-2; Industrial grade xanthan gum; foam stabilizer, flocculant suspending, gelling agent, rheology modifier, lubricant for industrial applications incl. abrasives, adhesives, herbicides, fertilizers, ceramics, cleaners, emulsions, gels, mining, thixotropic paints, paper, petrol., pigments; viscosifier for drilling fluids; Kelzan is the std. industrial grade; D grade is for use with galactomannens; M grade is for use where lower salt content is required; XC Polymer is esp. useful as an additive to oil well drilling mud. [Kelco/Oil Field Products]

Kelzan® AR. CAS 11138-66-2; Xanthan gum; for industrial uses in highly alkaline systems. [Kelco]

Kelzan® S. CAS 11138-66-2; Dispersible xanthan gum product; gum providing suspension of solids slurries and rheological control for aq. systems. [Kelco]

Kemadrin Tablets/Injection. CAS 1508-76-5; A proprietary formulation of procyclidine hydrochloride; for treatment of Parkinsonism including the postenchephalitic, arteriosclerotic and idiopathic types. [The Wellcome Foundation Ltd]

Kemamide® B. CAS 3061-75-4; EINECS 221-304-1; Behenamide; lubricant, slip, antiblock, and mold release agent for plastics, crayons, petrol. products, asphalts, inks, metals, textiles; mold release agent for thermoplastic resins in injection molding; defoamer and water repellent in industrial/household application; corrosion inhibitor; pigment grinding aid and dyestuff dispersant in paints, enamels, varnishes, and lacquers; intermediate for textile emulsifiers and softeners; foam stabilizer in household detergents. [Witco/Humko]

Kemamide® E. CAS 112-84-5; EINECS 204-009-2; Erucamide; lubricant, slip, antiblock, and mold release agent for plastics, crayons, petrol. products, asphalts, inks, metals, textiles; mold release agent for thermoplastic resins in injection molding; defoamer and water repellent in industrial/household application. [Witco/Humko]

Kemamide® E-180. CAS 10094-45-8; Stearyl erucamide; lubricant additive, friction modifier for high-temp. plastics applications. [Witco/Humko]

Kemamide® E-221. Erucyl erucamide; lubricant additive, friction modifier for high-temp. plastics applications. [Witco/Humko]

Kemamide® O. CAS 301-02-0; EINECS 206-103-9; Oleamide, tech.; lubricant, slip, antiblock, and mold release agent for plastics, crayons, petrol. products, asphalts, inks, metals, textiles; mold release agent for thermoplastic resins in injection molding; defoamer and water repellent in industrial/household application. [Witco/Humko]

Kemamide® P-181. CAS 16260-09-6; Oleyl palmitamide; lubricant additive, friction modifier for high-temp. plastics applications. [Witco/Humko]

Kemamide® S. CAS 124-26-5; EINECS 204-693-2; Stearamide; lubricant, slip, antiblock, and mold release agent for plastics, crayons, petrol. products, asphalts, inks, metals, textiles; mold release agent for thermoplastic resins in injection molding; defoamer and water repellent in industrial/household application. [Witco/Humko]

Kemamide® S-180. Stearyl stearamide; lubricant additive, friction modifier for high-temp. plastics applications. [Witco/Humko]

Kemamide® S-221. Erucyl stearamide; lubricant additive, friction modifier for high-temp. plastics applications. [Witco/Humko]

Kemamide® W-20. CAS 110-31-6; EINECS 203-756-1; Ethylene dioleamide; lubricant, slip, antiblock, and mold release agent for plastics, crayons, petrol. products, asphalts, inks, metals, textiles; mold release agent for thermoplastic resins in injection molding; defoamer and water repellent in industrial/household application; internal and external lubricants in ABS, PS, polyethylene, PP, PVC, nylon, cellulose acetate, PVAc, and phenolic resins; defoamer in paper industry blk. liquoring, fabric dyeing, latex systems; metal processing, asphalts. [Witco/Humko]

Kemamide® W-39. CAS 110-30-5; EINECS 203-755-6; Ethylene distearamide; lubricant, slip, antiblock, and mold release agent for plastics, crayons, petrol. products, asphalts, inks, metals, textiles; mold release agent for thermoplastic resins in injection molding; defoamer and water repellent in industrial/household application. [Witco/Humko]

Kemamine® AS-650. Nitrogen derivs.; antistat for polyolefins, styrenics, and other plastics, esp. film applications; lubricity aid, mold release aid, pigment dispersant. [Witco/Humko]

Kemamine® BQ-2802C. CAS 16841-14-8; EINECS 240-865-3; Behenalkonium chloride; antistat, textile softener, dyeing aid, corrosion inhibitor, emulsifier; used in personal care products, e.g., creams, lotions, shampoos, hair conditioners. [Witco/Humko]

Kemamine® BQ-2982B. CAS 90730-68-0; Erucalkonium chloride. [Witco/Humko]

Kemamine® BQ-9701C. CAS 61789-73-9; Dihydrogenated tallow benzylmonium chloride. [Witco/Humko]

Kemamine® BQ-9702C. CAS 61789-72-8; EINECS 263-081-3; Dimethyl hydrogenated tallow benzyl ammonium chloride; germicide, sanitizer, slimicide, antistat, textile softener, dyeing aid, corrosion inhibitor, emulsifier; also for personal care products. [Witco/Humko]

Kemamine® BQ-9742C. CAS 61789-75-1; EINECS 263-085-5; Tallowalkonium chloride; antistat, textile softening, dyeing aid, corrosion inhibitor, emulsifier, leveling agent, shampoo and cream rinse conditioner. [Witco/Humko]

Kemamine® D-190. Arachidyl-behenyl 1,3-propylenediamine; gasoline detergent, bactericide, corrosion inhibitor in petrol. production, epoxy hardener, dispersant, asphalt emulsifier. [Witco/Humko]

Kemamine® D-650. N-Coconut 1,3-propylenediamine; gasoline detergent, bactericide, corrosion inhibitor in petrol. production, epoxy hardener, dispersant, asphalt emulsifier. [Witco/Humko]

Kemamine® D-970. CAS 68603-64-5; EINECS 271-696-6; N-hydrogenated tallow 1,3-propylenediamine; gasoline detergent, bactericide, corrosion inhibitor in petrol. production, epoxy hardener, dispersant, asphalt emulsifier. [Witco/Humko]

Kemamine® D-974. CAS 68439-73-6; EINECS 270-416-7; Tallowaminopropylamine; gasoline detergent, bactericide, corrosion inhibitor in petrol. production, epoxy hardener, dispersant, asphalt emulsifier. [Witco/Humko]

Kemamine® D-989. CAS 7173-62-8; EINECS 230-528-9; N-Oleyl 1,3-propylenediamine; gasoline detergent, bactericide, corrosion inhibitor in petrol. production, epoxy

---

‡ Trade name and manufacturer not verified   § Trade name without identified manufacturer

hardener, dispersant, asphalt emulsifier. [Witco/Humko]

Kemamine® D-999. Soyaminopropylamine; gasoline detergent, bactericide, corrosion inhibitor in petrol. production, epoxy hardener, dispersant, asphalt emulsifier. [Witco/Humko]

Kemamine® P-150, P-150D. 50% Arachidyl-behenyl primary amine (P-150Ddist.); emulsifier, flotation agent, dispersing and flushing agent, intermediate, used in metalworking oils, as fuel oil additive; mold release for rubber and plastics; lubricant and spinning aid in metalworking oils. [Witco/Humko]

Kemamine® P-650D. CAS 61788-46-3; EINECS 262-977-1; Cocamine, dist.; emulsifier, flotation agent, dispersing and flushing agent, intermediate, used in metalworking oils, as fuel oil additive; mold release for rubber and plastics; lubricant and spinning aid in metalworking oils. [Witco/Humko]

Kemamine® P-880, P-880D. CAS 143-27-1; EINECS 205-596-8; Palmityl primary amine (tech. and dist. resp.); emulsifier, flotation agent, dispersing and flushing agent, intermediate, used in metalworking oils, as fuel oil additive; mold release for rubber and plastics; lubricant and spinning aid in metalworking oils. [Witco/Humko]

Kemamine® P-970. CAS 61788-45-2; EINECS 262-976-6; Hydrogenated tallow amine (tech.); emulsifier, flotation agent, dispersing and flushing agent, intermediate, used in metalworking oils, as fuel oil additive; mold release for rubber and plastics; lubricant and spinning aid in metalworking oils. [Witco/Humko]

Kemamine® P-989D. CAS 112-90-3; EINECS 204-015-5; Oleamine, dist.; emulsifier, flotation agent, dispersing and flushing agent, intermediate, used in metalworking oils, as fuel oil additive; mold release for rubber and plastics; lubricant and spinning aid in metalworking oils. [Witco/Humko]

Kemamine® P-990, P-990D. CAS 124-30-1; EINECS 204-695-3; Stearamine (tech. and dist. resp.); emulsifier, flotation agent, dispersing and flushing agent, intermediate, used in metalworking oils, as fuel oil additive; mold release for rubber and plastics; lubricant and spinning aid in metalworking oils. [Witco/Humko]

Kemamine® Q-2802C. CAS 26597-36-4; Dibehenyldimonium chloride; antistat, textile softener, dyeing aid, corrosion inhibitor, emulsifier; used in personal care products, e.g., creams, lotions, shampoos, hair conditioners. [Witco/Humko]

Kemamine® Q-6502C. CAS 61789-77-3; EINECS 263-087-6; Dimethyl dicoconut ammonium chloride; germicide, sanitizer, slimicide, antistat, textile softener, dyeing aid, corrosion inhibitor, emulsifier; also for personal care products. [Witco/Humko]

Kemamine® Q-9702C. CAS 61789-80-8; EINECS 263-090-2; Quaternium-18; antistat, textile softener, dyeing aid, corrosion inhibitor, emulsifier; used in personal care products, e.g., creams, lotions, shampoos, hair conditioners. [Witco/Humko]

Kemamine® Q-9743C. Tallowtrimonium chloride; antistat, textile softening agent, dyeing aid, corrosion inhibitor, emulsifier. [Witco/Humko]

Kemamine® S-970. CAS 61789-79-5; EINECS 263-089-7; Hydrogenated ditallow amine; industrial use, grease, corrosion inhibitor. [Witco/Humko]

Kemamine® T-1902D. Dist. dimethyl-90% arachidyl-behenyl tert. amine; chemical intermediate for quaternary ammonium derivs. used for cosmetics and textiles; acid scavenger in petrol. products; epoxy hardener, catalyst in mfg. of flexible PU foams. [Witco/Humko]

Kemamine® T-6501. CAS 61788-62-3; Methyl dicoconut tert. amine; chemical intermediate for quaternary ammonium derivs. used for cosmetics and textiles; acid scavenger in petrol. products; epoxy hardener, catalyst in mfg. of flexible PU foams. [Witco/Humko]

Kemamine® T-6502D. CAS 61788-93-0; EINECS 263-020-0; Dist. dimethyl cocamine; chemical intermediate for quaternary ammonium derivs. used for cosmetics and textiles; acid scavenger in petrol. products; epoxy hardener, catalyst in mfg. of flexible PU foams. [Witco/Humko]

Kemamine® T-9701. CAS 61788-63-4; Dihydrogenated tallow methylamine; chemical intermediate for quaternary ammonium derivs. used for cosmetics and textiles; acid scavenger in petrol. products; epoxy hardener, catalyst in mfg. of flexible PU foams; corrosion inhibitor; gasoline additive. [Witco/Humko]

Kemamine® T-9742D. CAS 68391-07-1; Dimethyl hydrogenated tallow amine; intermediate for mfg. of quaternary ammonium chlorides; corrosion inhibitor; gasoline additive. [Witco/Humko]

Kemamine® T-9902. CAS 124-28-7; EINECS 204-694-8; Dimethyl stearamine; intermediate for quats; used in cosmetics and textiles; acid scavenger in petrol. products. [Witco/Humko]

Kematal. Acetal copolymer (POM); used for domestic equipment. i.e., jug kettle bodies, components for washing machines, food preparation equipment, plumbing (taps, cistern valves, and basins) [Hoechst AG]

Kemester® 104. CAS 112-62-9; EINECS 203-992-5; Methyl oleate; emulsifier, emollient for cosmetics; lubricant for leather; carrier for agricultural spray products. [Witco/Humko]

Kemester® 143. Methyl tallowate; intermediate in production of superamides, in metalworking lubricants, as solv.; lubricant, plasticizer for cosmetics, leather, rubber products. [Witco/Humko]

Kemester® 226. Methyl soyate; intermediate in production of superamides, in metalworking lubricants, as specialized solv.; opacifier, visc. control agent. [Witco/Humko]

Kemester® 1000. CAS 122-32-7; EINECS 204-534-7; Triolein; emollient used in cosmetics, textiles, leather, metalworking lubricants; base for sulfation. [Witco/Humko]

Kemester® 1418. CAS 17661-50-6; Myristyl stearate; for cosmetics/pharmaceuticals. [Witco/Humko]

Kemester® 2000. CAS 111-03-5; Glycerol oleate; emollient, emulsifier, stabilizer, wetting agent for textiles, personal care products. [Witco/Humko]

Kemester® 2050. CAS 1120-28-1; Methyl eicosenate; intermediate in production of alkanolamides, in metalworking lubricants, as specialized solvs.; foam depressant and nutrient in fermentation. [Witco/Humko]

Kemester® 3681. Dioctyl dilinoleate; lubricant for crankcase, turbine, compressor, gear, and metalworking formulations; also cosmetic emollient, pearling and bodying agent. [Witco/Humko]

Kemester® 4000. CAS 142-77-8; EINECS 205-559-6; Butyl oleate; emollient, wetting agent; plasticizer for textiles, leathers, elastomers, personal care products. [Witco/Humko]

---

* Trade name not verified as available   † Trade name verified as obsolete

Kemester® 4516. CAS 112-61-8; EINECS 203-990-4; Methyl stearate; intermediate in production of alkanolamides, in metalworking lubricants, as specialized solvs.; foam depressant and nutrient in fermentation. [Witco/Humko]

Kemester® 5221SE. CAS 9004-99-3; PEG-2 stearate SE; emollient, emulsifier, plasticizer, lubricant for cosmetics, rubber, textiles. [Witco/Humko]

Kemester® 5415. CAS 646-13-9; Isobutyl stearate; emollient, lubricant for textiles, metalworking fluids, personal care products. [Witco/Humko]

Kemester® 5500. Glyceryl stearate; emollient, emulsifier; stabilizer, plasticizer, lubricant for cosmetic, paper, textile, and industrial uses. [Witco/Humko]

Kemester® 5510. CAS 123-95-5; EINECS 204-666-5; Butyl stearate; emollient for cosmetics; lubricant, plasticizer. [Witco/Humko]

Kemester® 5654. Ditridecyl adipate; lubricant for crankcase, turbine, compressor, gear, and metalworking formulations; also cosmetic emollient, pearling and bodying agent. [Witco/Humko]

Kemester® 5721. CAS 31556-45-3; EINECS 250-696-7; Tridecyl stearate; emollient, dye carrier, textile lubricant, cosmetics/pharmaceuticals. [Witco/Humko]

Kemester® 5822. CAS 25339-09-7; EINECS 246-868-6; Isocetyl stearate; emollient, plasticizer for cosmetics. [Witco/Humko]

Kemester® 6000. Glyceryl stearate; cosmetic and industrial emulsifier, emollient; plasticizer for elastomers. [Witco/Humko]

Kemester® 9022. CAS 929-77-1; Methyl behenate; intermediate in production of superamides, in metalworking lubricants, as solv. [Witco/Humko]

Kemester® BE. CAS 18312-32-8; EINECS 242-201-8; Behenyl erucate; industrial lubricant, cosmetic emollient. [Witco/Humko]

Kemester® CP. CAS 540-10-3; EINECS 208-736-6; Cetyl palmitate; industrial lubricant, cosmetic emollient. [Witco/Humko]

Kemester® DMP. CAS 131-11-3; Dimethyl phthalate; emollient used in cosmetics, textiles, metalworking lubricants. [Witco/Humko]

Kemester® EE. CAS 27640-89-7; Erucyl erucate; industrial lubricant, cosmetic emollient. [Witco/Humko]

Kemester® EGDL. CAS 624-04-4; Glycol dilaurate; for cosmetics/pharmaceuticals. [Witco/Humko]

Kemester® EGDS. CAS 627-83-8; EINECS 211-014-3; Glycol distearate; intermediate in production of superamides, in metalworking lubricants, specialized solv.; industrial lubricant; opacifier, pearling additive, thickener for cosmetics and pharmaceuticals. [Witco/Humko]

Kemester® EGMS. Glycol stearate; intermediate in production of superamides, in metalworking lubricants, specialized solv.; industrial lubricant; opacifier, pearling additive, thickener for cosmetics and pharmaceuticals. [Witco/Humko]

Kemester® GDL. CAS 27638-00-2; EINECS 248-586-9; Glyceryl dilaurate; for cosmetics/pharmaceuticals. [Witco/Humko]

Kemester® GMS (Powd.). Glyceryl stearate; industrial lubricant; cosmetic emollient. [Witco/Humko]

Kemester® HMS. CAS 118-56-9; Homosalate; uv absorber; sunscreen agent; used in cosmetic skin preps. [Witco/Humko]

Kemester® JO. Erucyl arachidate; emollient for cosmetics. [Witco/Humko]

Kemester® MM. CAS 3234-85-3; EINECS 221-787-9; Myristyl myristate; lubricant for crankcase, turbine, compressor, gear, and metalworking formulations; also cosmetic emollient, pearling and bodying agent. [Witco/Humko]

Kemester® S20. CAS 1338-39-2; EINECS 215-663-3; Sorbitan laurate; for cosmetics/pharmaceuticals. [Witco/Humko]

Kemester® S40. CAS 26266-57-9; EINECS 247-568-8; Sorbitan palmitate; for cosmetics/pharmaceuticals. [Witco/Humko]

Kemester® S60. CAS 1338-41-6; EINECS 215-664-9; Sorbitan stearate; for cosmetics/pharmaceuticals. [Witco/Humko]

Kemester® S65. CAS 26658-19-5; EINECS 247-891-4; Sorbitan tristearate; for cosmetics/pharmaceuticals. [Witco/Humko]

Kemester® S80. CAS 1338-43-8; EINECS 215-665-4; Sorbitan oleate; for cosmetics/pharmaceuticals. [Witco/Humko]

Kemester® S85. CAS 26266-58-0; EINECS 247-569-3; Sorbitan trioleate; for cosmetics/pharmaceuticals. [Witco/Humko]

Kemester® THFO. Tetrahydrofurfuryl oleate; emollient used in cosmetics, textiles, metalworking lubricants. [Witco/Humko]

Kemflorseal. Epoxy die material and epoxy floor and wall product; used for dental molds, floor sealer, wall decorating and stain resistance. [Kemtron International Inc]

Kemgo. A proprietary trade name for inks for use with heat.§

Kemick. Heat-resisting paint. [ICI Chem. & Polymers Ltd] †

Kemira Phlogopite Mica. CAS 12001-26-2; Mica; filler for plastics, surface coatings, sound-deadening compounds etc. [Cornelius Chemical Co Ltd] *

Kemite. A ceramic material. Labstone is the name given to the material when used for laboratory bench tops.

Kemix. Paste form dispersions of activators and accelerators used in elastomer cures. [Kenrich Petrochemicals]

Kemlet metal. An alloy consisting mainly of zinc, with aluminum and copper.

Kemlex 10007 NAL1. Polyacetal, PTFE-lubricated. [Ferro/ Engineering Thermoplastics]

Kemmat. A range of glycol esters; used as emulsifiers for cosmetic and pharmaceutical creams and lotions, thickeners for shampoos and emulsifiers for mineral oils. [Harcros Australia] *

Kemmest. A range of glycol esters; used as emulsifiers in the cosmetic and pharmaceutical industry. [Harcros Australia] *

Kemonic. A range of lauryl alcohol ethoxylates; used as biodegradable detergents and emulsifiers. [Harcros Australia] *

Kemopol. Propylene oxide condensates of ethylene oxide/ glycol adducts; used as a detergent base for machine dishwashing and cattle antibloat. [Harcros Australia] *

Kemotan. A range of polysorbates; used as emulsifiers in the food and cosmetic industry. [Harcros Australia] *

kemp. The shorter fibers of mohair.

Kempol. A proprietary trade name for vulcanizable vegetable oil polymers. §

Kempore® 60/14FF. CAS 123-77-3; Azodicarbonamide-

---

‡ Trade name and manufacturer not verified   § Trade name without identified manufacturer

based; blowing agent for dynamic foaming processes, e.g., injection molding, extrusion, and calendering. [Uniroyal]

Kempoxy. An epoxy flooring material; floor resurfacer, for chemical resistance floors. [Kemtron International Inc]

Kempy wool. Wool prepared from sheep badly fed, or subjected to exposure. It dyes badly.

Kemset Epoxy. Three-pack epoxy mortar based on nonsolvented resin and silica aggregate; high compressive, flexural and tensile strengths. [Feb Ltd]

Kemstrene® 96.0%. CAS 56-81-5; Refined glycerin USP; humectant, solv.; used in cosmetics, liq. soaps, confections, inks, and lubricants; intermediate used in polyester and PU formulations. [Witco/Humko]

Kemtop. Epoxy die material and epoxy floor and wall product; used for dental molds, floor sealer, wall decorating and stain resistance. [Kemtron International Inc]

Kemwax. CAS 110-30-5; EINECS 203-755-6; Ethylene bisstearamide; used as plastics lubricants and slip agents in plastics. [Harcros Australia] *

Kenacort. CAS 124-94-7; Triamcinolone; glucocorticoid. [Bristol-Myers Squibb Co Inc]

Kenacort Diacetate Syrup. Triamcinolone diacetate; glucocorticoid. [Bristol-Myers Squibb Co Inc]

Kena® Formula FP-28, FP-85. Food additive for meat, poultry, and seafood industries. [Rhone-Poulenc Food Ingreds.]

Kenalog Injection. CAS 76-25-5; Triamcinolone acetonide in suspension in aqueous vehicle. [Bristol-Myers Squibb Pharmaceuticals Ltd]

Kencolor®. Dispersions of pigments in silicone oils and gums for silicone elastomer compositions; pigment and catalyst dispersions used in the coloring and curing of silicone elastomers. [Kenrich Petrochemicals]

Kendex 0220. Petroleum hydrocarbon crude petrolatum wax (black); used as a fuel in fire logs or for further refining to petrolatum. [Witco] *

Kendex 0834. Petroleum hydrocarbon heavy petroleum resins; used as a viscosity modifier in lubrication applications, plasticizer extender. [Witco] *

Kendex 0842. Petroleum hydrocarbon-cylinder stock; a base oil for lubrication formulations or plasticizer/extender. [Witco] *

Kendex 0847. Petroleum hydrocarbon150 bright stock; base oil in the formulation of motor oil or lubricants; a plasticizer/extender in the rubber industry. [Witco] *

Kendex 0866. Petroleum hydrocarbon-extract; a plasticizer/extender; can be used in adhesives. [Witco] *

Kendex 0898. Petroleum hydrocarbon-intermediate petroleum resins; viscosity modifier for lubricants; plasticizer/extender; additive stock for quench oils. [Witco] *

Kendex OCTG. Petroleum; corrosion inhibitor; specialty anticorrosion formulation for ferrous metals for tubular goods and other machined parts. [Witco] *

Kendurit. Skin protective soap with granules; for cleansing of dirty hands. [Dynamit Nobel Wien GmbH] *

Kenflex®. The condensation products of alkyl naphthalenes; vinyl plasticizer. [Kenrich Petrochemicals]

Kenflex® A. A dimethyl naphthalene derivative; elastomer and vinyl plasticizer and softener. [Kenrich Petrochemicals]

Kenlastic®. Elastomeric dispersion forms of activators and accelerators used in elastomer compounding; provides faster Banbury incorporation times; eliminates dust.

[Kenrich Petrochemicals]

Ken-Mag®. CAS 1309-48-4; Magnesium oxide; for the polymer industry. [Kenrich Petrochemicals]

Kenmix®. Paste dispersions of activators and accelerators; used in elastomer cures. [Kenrich Petrochemicals]

Kenplast®. Distilled aromatic and cumylphenol derived plasticizers; nonreactive and reactive diluents in epoxies, primary plasticizer in urethanes. [Kenrich Petrochemicals]

Kenplast® ES-2. Cumyl-phenyl acetate; plasticizer for urethanes; high flash pt. reactive diluent for epoxy, reducing odor and irritation; ideal for polyamide-cured epoxy floorings and coal tar pipe coatings; comonomer and impact modifier for phenolics; improves adhesion of vinyl plastisols coating metals and polar plastics. [Kenrich Petrochemicals] *

Kenplast® ESB. Cumyl-phenyl benzoate; primary plasticizer for PVC and PVC/nitrile; process aid for semirigid PVC extrudates; solvates conductive polyester and acrylic inks to improve impact. [Kenrich Petrochemicals]*

Kenplast® ESI. Biscumylphenyl trimellitate; plasticizer, process aid for extrusion of PVC, urethanes; lubricant and process aid for filled PS, PC; reactive diluent and flow promoter for epoxy powd. coatings; prevents oxidative depolymerization in nylons. [Kenrich Petrochemicals] *

Kenplast® ESN. Cumylphenyl neodecanoate; primary plasticizer for PVC, PVC/nitrile; process aid for semirigid PVC extrusions; solvates and impact modifies conductive polyester and acrylic inks. [Kenrich Petrochemicals]*

Ken-React®. Titanate, zirconate and aluminate coupling agents; typical composition: LICA 01 - titanium IV neoalkenolato, tris neodecanolato-O, NZ 01 - zirconium IV neoalkenolato, tris neodecanolato-O, KA 322 - diisopropyl (oleyl) aceto acetyl aluminate; coupling agents, adhesion promoters, composite improvers. [Kenrich Petrochemicals]

Ken-React® 7 (KR 7). Isopropyl dimethacryl isostearoyl titanate (monoalkoxy), IPA; see Ken-React. [Kenrich Petrochemicals]

Ken-React® 9S (KR 9S). Isopropyl tridodecylbenzenesulfonyl titanate (monoalkoxy), IPA; see Ken-React. [Kenrich Petrochemicals]

Ken-React® 12 (KR 12). Isopropyl tri(dioctylphosphato) titanate (monoalkoxy), IPA; see Ken-React. [Kenrich Petrochemicals]

Ken-React® 26S (KR 26S). Isopropyl 4-aminobenzenesulfonyl di(dodecylbenzenesulfonyl) titanate (monoalkoxy), IPA; see Ken-React. [Kenrich Petrochemicals]

Ken-React® 33DS (KR 33DS). Alkoxy trimethacryl titanate (monoalkoxy), IPA; see Ken-React. [Kenrich Petrochemicals]

Ken-React® 38S (KR 38S). Isopropyl tri(dioctylpyrophosphato) titanate (monoalkoxy), IPA; see Ken-React. [Kenrich Petrochemicals]

Ken-React® 39DS (KR 39DS). Alkoxy triacryl titanate (monoalkoxy), IPA; see Ken-React. [Kenrich Petrochemicals]

Ken-React® 41B (KR 41B). Tetraisopropyl di (dioctylphosphito) titanate (coordinate type), IPA; see Ken-React. [Kenrich Petrochemicals]

Ken-React® 44 (KR 44). Isopropyl tri(N ethylamino-

---

* Trade name not verified as available   † Trade name verified as obsolete

ethylamino) titanate (monoalkoxy), IPA; see Ken-React. [Kenrich Petrochemicals]

Ken-React® 46B (KR 46B). Tetraoctyloxytitanium di (ditridecylphosphite) (coordinate type), IPA; see Ken-React. [Kenrich Petrochemicals]

Ken-React® 55 (KR 55). Tetra (2, diallyoxymethyl-1 butoxy titanium di (di-tridecyl) phosphite (coordinate type), IPA; see Ken-React. [Kenrich Petrochemicals]

Ken-React® 133DS (KR 133DS). Titanium dimethacrylate oxyacetate (oxyacetate chelate type), IPA; see Ken-React. [Kenrich Petrochemicals]

Ken-React® 134S (KR 134S). Titanium di (cumylphenylate) oxyacetate (oxyacetate chelate type), IPA; see Ken-React. [Kenrich Petrochemicals]

Ken-React® 138D (KR 138D). KR 138S and 2-dimethylamino methyl propanol (quat type), IPA; see Ken-React. [Kenrich Petrochemicals]

Ken-React® 138S (KR 138S). Titanium di (dioctyl-pyrophosphate) oxyacetate (oxyacetate chelate type), IPA; see Ken-React. [Kenrich Petrochemicals]

Ken-React® 158FS (KR 158FS). Titanium di (butyl, octyl pyrophosphate) di (dioctyl, hydrogen phosphite) oxyacetate (oxyacetate chelate type), IPA; see Ken-React. [Kenrich Petrochemicals]

Ken-React® 212 (KR 212). Di (dioctylphosphato) ethylene titanate (A, B ethylene chelate type), IPA; see Ken-React. [Kenrich Petrochemicals]

Ken-React® 238S (KR 238S). Di (dioctylpyrophosphato) ethylene titanate (A, B ethylene chelate type); see Ken-React. [Kenrich Petrochemicals]

Ken-React® 262ES (KR 262ES). Di (butyl, methyl pyrophosphato) ethylene titanate di (dioctyl, hydrogen phosphite) (A, B ethylene chelate type), IPA; see Ken-React. [Kenrich Petrochemicals]

Ken-React® OPP2 (KR OPP2). Dicyclo (dioctyl) pyrophosphato, titanate (cycloheteroatom type), IPA; see Ken-React. [Kenrich Petrochemicals]

Ken-React® OPPR (KR OPPR). Dicyclo (dioctyl) pyrophosphato dioctyl titanate (cycloheteroatom type), IPA; see Ken-React. [Kenrich Petrochemicals]

Kensol 10. Petroleum naptha; light naptha used as fuel in sporting and camping equipment or as a solvent. [Witco]*

Kensol 13. Petroleum naptha; light petroleum naptha used in oil well deparaffinizing and as a solvent in blending asphalt. [Witco] *

Kensol 30. Petroleum hydrocarbonregular mineral spirits or Stoddard solvent; wide range of applications as a commercial solvent and as a parts cleaner or diluent. [Witco]*

Kensol 48T. Petroleum hydrocarbonlow odor petroleum distillate fraction; used as a solvent in various commercial formulations or as an oil to roll aluminum foil. [Witco]*

Kensol 50T. Petroleum hydrocarbonlow odor middle petroleum distillate; oil used in aluminum rolling applications and as a diluent in commercial formulations. [Witco] *

Kensol 51. Petroleum hydrocarbonmiddle distillate or a light mineral seal oil; used as a lubricant in the rolling of aluminum and as a diluent in commercial formulations. [Witco] *

Kensol 53. Petroleum hydrocarbonmineral seal oil; formulation of commercial products, coke absorber oil, compounding base stock, cuting oil. [Witco] *

Kensol 61. Petroleum hydrocarbonmineral seal oil; oil for compounding various commercial products, hydraulic oil base stock, light oil for diluents. [Witco] *

Kensol 80. Petroleum hydrocarbonV M & P type naphtha; solvent in paint or as a solvent in commercial products. [Witco] *

Kensol KM Metal Cleaner. Liquid metal polish; water-based suspension containing ammonia, oxalic acid and methanol among other ingredients; used primarily as an all-metal cleaner for commercial metal maintenance on architectural and ornamental metalwork as well as household use on brass, chrome, stainless steel, etc. [Kensol Corporation]

Kensol KV Rust Retarder. Hydrocarbon mixture, blend of petroleum products; primarily used as a rust preventative on polished steel and stainless steel on bank security equipment. [Kensol Corporation]

Kensol KX Oxide Resistor. Hydrocarbon mixture, blend of petroleum products; used primarily as an all-metal preservative for commercial metal maintenance on architectural and ornamental metalwork as well as household use to protect against rust and tarnish. [Kensol Corporation]

Kentish rag. A siliceous limestone; used as an adulterant of Portland cement.

Kentite. An explosive containing from 32-35% ammonium nitrate, 32-35% potassium nitrate, 16-18% ammonium chloride, and 14-16% trinitrotoluene.

Ken-Zinc®. CAS 1314-13-2; Zinc oxide dispersion; for use as an activator in elastomer compounding. [Kenrich Petrochemicals]

Kephos. Nonaqueous phosphating metal pretreatment. [ICI Chem. & Polymers Ltd]

Keramine H. Liquid protein ampules, to strengthen hair. [Richardson-Vicks Inc] *

Keramino 25. CAS 68238-35-7; Keratin amino acids; cosmetics ingred. [Brooks Industries]

keramyl. A solution of hydrofluosilicic acid, H_2SiF_6.

keraphen. Sodium tetraiodo-phenolphthalein.

Kera-Quat WKP. CAS 68915-25-3; Cocodimonium hydroxypropyl hydrolyzed keratin; substantive protein, moisturizer for hair and skin care products (shampoos, conditioners, styling products, liq. hand soaps, creams, lotions). [Maybrook]

Kerasol. Soluble keratin; proteinic conditioner for hair and nail care products. [Croda Inc.]

Kerasol. Tetraiodophenolphthalein, $C_{20}H_{10}I_2O_4$.

Kera-Tein 1000. CAS 69430-36-0; Hydrolyzed keratin; moisturizer for skin and hair care products; protective colloid effect; provides cystine to hair; minimizing damage from harsh treatments. [Maybrook]

Kera-Tein AA. CAS 68238-35-7; Keratin amino acids; substantive/protective protein, penetrant, moisturizer for skin and hair care products; source of cystine. [Maybrook]

Kera-Tein H. Keratin amino acids and sodium chloride; cosmetics ingred. [Maybrook]

Keratite. A name applied to a vulcanite.

Keratol. A cellulose waterproofing compound.

Kerb 50W. CAS 23950-58-5; Propyzamide; a residual herbicide in a wettable powder form for a wide range of agricultural crops. [Rohm & Haas UK]

Kerb 50W. CAS 23950-58-5; Propyzamide; a residual herbicide in a wettable powder form for a wide range of agricultural crops. [Pan Britannica Industries Ltd]

Kerb Propyzamide 50. CAS 23950-58-5; Propyzamide; a

---

residual herbicide in a wettable powder form for a wide range of agricultural crops. [Kommer-Brookwick Ltd]

Kerecid. CAS 54-42-2; A proprietary preparation of idoxuridine; an ocular antiseptic. [SmithKline Beecham]*

Kericompost. Compost for houseplants. [ICI Garden Products]

Kerigrow. Compound fertilizer 6:4:4; liquid fertilizer for houseplants. [ICI Garden Products]

Keriguards. Pellets containing fertilizer and insecticide in combination. [ICI Chem. & Polymers Ltd]

Kerimid 500. A proprietary polyamide-imide polymer in solution form. [Rhône-Poulenc NV]

Kerimid 501. A proprietary polyamide-imide polymer in film form. [Rhône-Poulenc NV]

Kerimid 502. A proprietary polyamide-imide polymer in the form of a green paste, comprising a thermosetting polymer and an aluminum powder filler. [Rhône-Poulenc NV]

Kerimid 503. A proprietary polyamide-imide polymer in film form, colored green. [Rhône-Poulenc NV]

Kerimid 601. A proprietary thermosetting polyimide used in the manufacture of glass-fiber laminates. [Rhône-Poulenc NV]

Keriroot. Hormone rooting powder. [ICI Chem. & Polymers Ltd]

Kerishine. Leaf glosser for houseplants. [ICI Garden Products]

Kerispikes. Compound fertilizer 6:4:4; food spikes for houseplants. [ICI Garden Products]

Kerispray. Contains pirimiphos-methyl with synergized pyrethrins; pesticide spray for houseplants. [ICI Garden Products]

Keristicks. Capilliary sticks for houseplants. [ICI Chem. & Polymers Ltd]

Kern's hydraulic bronze. An alloy of 78% copper, 12% tin, and 10% zinc.

Kerobit®. Mineral oil fraction antioxidants. [BASF plc]

Kerobit® BPD. CAS 101-96-2; N,N´-Di-s-butyl-p-phenylene diamine; antioxidant for crude oil distillates. [BASF AG]*

Kerofluid®. Additive for prevention of ice formation in carburetors and jet fuels. [BASF AG; BASF plc]

Keroflux® 5323, 5486, H, M. Low molecular weight wax; cold flow improver for middle distillates. [BASF AG; BASF plc]

kerogen. CAS 8032-30-2; The organic matter contained in shale. It amounts to from 20-27% and gives on distillation water, ammonia, gas, and oil; after fractionation and refinement the oil can yields 18% gasoline 30% kerosene, 27% gas oil, 15% light lube oil, and 10% heavy lube oil.

Kerokorn®. Corrosion inhibitors for fuels. [BASF AG]

Kerol. Disinfectant fluid. [The Wellcome Foundation Ltd] *

Kerolite. Space heater fuel. [Chevron] *

Keromask. A proprietary cosmetic preparation containing titanium oxide and ochre pigments. [Innoxa (England) Ltd] *

Keromet. Mineral oil fraction metal deactivator. [BASF plc]

Keromet MD 60, MD 80. CAS 94-91-7; N,N´-Disalicylidene-1,2-propane diamine; chelating agent and metal deactivator for fuels and lubricating oils. [BASF AG] *

Keronyx. A proprietary trade name for a casein plastic material used for the manufacture of combs, etc. §

Keropon®. Antifouling agents for crude oil processing plants; stabilizers for storage of petrol. distillates. [BASF AG]

Keropur®. Additives for cleaning and maintaining the inlet systems of internal combustion engines. [BASF AG; BASF plc]

kerosene. (paraffin oil, astral oil, coal oil) CAS 8003-20-6; A refined distillate of petroleum, 150-300 C; used as an illuminating oil, acetylating agent, starting point for production of acetic anhydride and acetate esters.

Kerosol 200. A low odor hydrocarbon fraction; aliphatic solvent. [Sasolchem]

Kerostat® 5009. Salts of amino carboxylic acid; antistatic additive for fuels. [BASF AG] *

Kessco®. Esters; emollients, lubricants, emulsifiers. [Stepan]

Kessco® 653. CAS 540-10-3; EINECS 208-736-6; Cetyl palmitate; synthetic spermaceti wax, thickener, visc. booster for pharmaceutical and cosmetic products; w/o emulsifier, lubricant for metalworking fluids. [Stepan; Stepan Canada]

Kessco® 874. Pentaerythritol tetracaprylate/caprate; high temp. stable lubricant base for textile and metalworking applications. [Stepan; Stepan Canada]

Kessco® 887. CAS 68956-08-1; Trimethylolpropane tricaprylate/caprate; high temp. stable lubricant base for textile and metalworking applications. [Stepan; Stepan Canada]

Kessco® BS. CAS 123-95-5; EINECS 204-666-5; Butyl stearate; biodeg. replacement for mineral oil; used as lubricants in textile spin finish, coning oils, carding, dye bath. [Stepan; Stepan Canada]

Kessco® EGAS. Glyceryl stearate, stearamide AMP; pearlescent, bodying agent for shampoos, liq. hand soaps; imparts soft, smooth skin feel to formulations. [Stepan; Stepan Canada]

Kessco® EGDS. CAS 627-83-8; EINECS 211-014-3; Glycol disterate; pearlescent, emollient, emulsifier. [Stepan; Stepan Canada]

Kessco® EGMS. Glycol stearate; pearlescent, bodying agent, emulsion stabilizer. [Stepan; Stepan Canada]

Kessco® GDL. CAS 27638-00-2; EINECS 248-586-9; Glyceryl dilaurate; surfactant for free-flowing lotions. [Stepan; Stepan Canada]

Kessco® GDS 386F. CAS 1323-83-7; Glyceryl distearate; emulsifier, opacifier, bodying agent. [Stepan; Stepan Canada]

Kessco® GMC-8. Glyceryl caprylate/caprate; solubilizer and emulsifier for vitamins, flavors, and medicaments. [Stepan; Stepan Canada]

Kessco® GML. CAS 142-18-7; EINECS 205-526-6; Glyceryl laurate; primary emulsifier for w/o emulsions. [Stepan; Stepan Canada]

Kessco® GMO. CAS 111-03-5; Glyceryl oleate; w/o emulsifier, emollient, spreading agent, pigment dispersant; lubricant for textiles, metalworking compounds; sperm oil replacement. [Stepan; Stepan Canada]

Kessco® GMS. Glyceryl stearate; emulsifier, opacifier, bodying agent; lubricant for textiles, metalworking compounds. [Stepan; Stepan Canada]

Kessco® IBS. CAS 646-13-9; Isobutyl stearate; lubricant for textiles, metalworking compounds; slip aid, wetting agent for pigmented lipsticks, bath oils, nail polish and removers, skin cleaners, creams, lotions. [Stepan; Stepan Canada]

Kessco® ICS. CAS 25339-09-7; EINECS 246-868-6;

---

* Trade name not verified as available

† Trade name verified as obsolete

Isocetyl stearate; emollient for makeup formulations. [Stepan; Stepan Canada]

Kessco® IPM. CAS 110-27-0; EINECS 203-751-4; Isopropyl myristate; biodeg. replacement for mineral oil; used as lubricants in textile spin finish, coning oils, carding, dye bath; emollient, solubilizer, vehicle for makeup, shaving preps., bath oils, hair preps. [Stepan; Stepan Canada]

Kessco® IPP. CAS 142-91-6; EINECS 205-571-1; Isopropyl palmitate; biodeg. replacement for mineral oil; used as lubricants in textile spin finish, coning oils, dye bath; emollient, solubilizer, vehicle for makeup, shaving preps., bath oils, hair preps. [Stepan; Stepan Canada]

Kessco® Octyl Isononanoate. CAS 71566-49-9; Octyl isononanoate; emollient with dry, nonoily skin feel, for creams, lotions, makeup, lipstick, antiperspirants. [Stepan]

Kessco® Octyl Palmitate. CAS 29806-73-3; EINECS 249-862-1; Octyl palmitate; biodeg. replacement for mineral oil; used as lubricants in textile spin finish, coning oils, carding, dye bath; gloss aid, emollient for makeup, hair grooms, creams, lotions; binder for pressed powds. [Stepan; Stepan Canada]

Kessco® PEG 200 DL. CAS 9005-02-1; PEG-4 dilaurate; lubricant, emulsifier, softener for textile and metalworking applications. [Stepan; Stepan Canada]

Kessco® PEG 200 DO. CAS 9005-07-6; PEG-4 dioleate; surfactants for cosmetics, pharmaceuticals, food, agriculture, plastic, and other industries; thickener, solubilizer. [Stepan; Stepan Canada]

Kessco® PEG 200 DS. CAS 9005-08-7; PEG-4 distearate; surfactants for cosmetics, pharmaceuticals, food, agriculture, plastic, and other industries; thickener, solubilizer. [Stepan; Stepan Canada]

Kessco® PEG 200 ML. CAS 9004-81-3; PEG-4 laurate; surfactants for cosmetics, pharmaceuticals, food, agriculture, plastic, and other industries; thickener, solubilizer. [Stepan; Stepan Canada]

Kessco® PEG 200 MO. CAS 9004-96-0; PEG-4 oleate; surfactants for cosmetics, pharmaceuticals, food, agriculture, plastic, and other industries; thickener, solubilizer. [Stepan; Stepan Canada]

Kessco® PEG 200 MS. CAS 9004-99-3; PEG-4 stearate; surfactants for cosmetics, pharmaceuticals, food, agriculture, plastic, and other industries; thickener, solubilizer. [Stepan; Stepan Canada]

Kessco® PGML. Propylene glycol laurate; emollient, emulsifier. [Stepan; Stepan Canada]

Kessco® PGMS. CAS 1323-39-3; EINECS 215-354-3; Propylene glycol stearate; emulsifier with m.p. near body temp. [Stepan; Stepan Canada]

Kest. A proprietary preparation of magnesium sulfate and phenolphthalein; used in the treatment of constipation. [Berk Pharmaceuticals Ltd] *

Kester. Fatty acid esters. [Croda Chem. Ltd] *

Kester Wax® K 48. CAS 136097-97-7; Ester wax. [Koster Keunen]

Ketaject. CAS 1867-66-9; Ketamine hydrochloride; anesthetic. [Bristol Laboratories] *

Ketalar. CAS 1867-66-9; Ketamine hydrochloride; anesthetic. [Parke-Davis]

Ketaset. CAS 1867-66-9; Ketamine hydrochloride; anesthetic. [Bristol Laboratories] *

Ketavet. CAS 1867-66-9; Ketamine hydrochloride; anesthetic. [Bristol Laboratories] *

Ketjenblack® EC-310 NW. CAS 1333-86-4; Carbon black aq. disp. electroconductive carbon black for rubber compounding. [Akzo]

Ketjencat. Fluid cracking catalyst. [Akzo]

Ketjenflex. Toluene sulfonamides. [Akzo]

Ketjenlube® 115. Butanol ester of α-olefin dicarboxylic acid copolymer; lubricant for load carrying applications; provides wear reduction. [Akzo]

Ketjensept. Chloramine T. [Akzo]

Ketjensil® SM 405. CAS 1344-00-9; Sodium-aluminum silicate, precipitated; reinforcing filler for silicone rubber applications; filler in disp. paints for partial replacement of TiO_2. [Akzo] *

Keto-Diastix. A proprietary test-strip used to detect ketones and glucose in urine. [B C Ames] *

α-ketoglutaric acid. (2-oxopentanedioic acid) HOOC-$CH_2CH_2COCOOH$; CAS 328-50-7. [Am. Biorganics; Penta Mfg.; Schweizerhall; U.S. Biochemical]

β-ketoglutaric acid. (ADA; acetonedicarboxylic acid) $HOOCCH_2CH_2COCOOH$; organic synthesis. [Dajac Labs; Janssen Chimica; Lonza; Penta Mfg.]

ketohexamethylene. See cyclohexanone.

Ketomax. Immobilized glucose isomerase. [UOP Speciality Products]

ketone base. Tetramethyldiaminobenzophenone; an intermediate for dyes.

ketone musk. CAS 81-14-1; Dinitro-tert-butylxylyl methyl ketone; an artificial musk perfume.

Ketonone. A proprietary trade name for benzoic acid derivatives used as plasticisers for cellulose acetate and cellulose nitrate. §

Ketonone B. A proprietary trade name for butylbenzoyl benzoate; a plasticizer. §

Ketonone E. A proprietary trade name for ethyl o-benzoyl benzoate; a plasticizer. §

Ketonone M. A proprietary trade name for methyl o-benzoyl benzoate; a plasticizer. §

Ketonone M.O. A proprietary trade name for methylethyl benzoyl benzoate; a plasticizer. §

α-ketopropionic acid. See pyruvic acid.

keto resins. Artificial resins obtained by the polymerization of aldehyde ketone condensation products.

Ketostix. A proprietary test strip of buffered sodium nitroprusside and glycine; used for the detection of ketones in urine, serum or milk. [B C Ames] *

γ-ketovaleric acid. See levulinic acid.

Ketovite. A complete vitamin supplement for restricted or synthetic diets; Ketovite tablets used in conjunction with Ketovite liquids will provide a complete vitamin supplement for use in conditions such as phenylketonuria, disorders of carbohydrate or amino acid metabolism. [Paines & Byrne Ltd] *

Ketrax. Anthelmintic containing levamisole. [ICI Chem. & Polymers Ltd]

Kevadon. CAS 50-35-1; Thalidomide; sedative; hypnotic. [Merrell Dow Pharmaceuticals Inc] *

Kevlar®. Aramid fiber. [DuPont UK]

Kevlar® 29, 49. Aramid; high strength, lightweight, flexible material; used for aircraft/aerospace, boat hulls, prosthetics, footwear, sporting goods, ropes and cables, fiber optics, bullet-resist. vests, fabrics, brakes and other friction products, in radial tires, as asbestos replacement, as reinforcement for mech. rubber goods; avail. as

---

‡ Trade name and manufacturer not verified

§ Trade name without identified manufacturer

continuous filament yarn, staple, engineered short fiber, pulp, spun yarn, needlepunched felt, paper, woven fabrics, cord, narrow webbing. [DuPont]

key alloy. A nickel-silver containing 60-65% copper, 20-26% zinc, 12% nickel, 1-2% lead, and 0-0.4% iron.

Keycide® X-10. CAS 56-35-9; Tributyltin oxide, stabilized; antimildew additive, antimicrobial for PVAc latex paints for packaged stability and mildew-resistant applied films. [Witco/Organics]

Keydime. Ketene dimer emulsion. [Tenneco Malros Ltd] *

Keykote®. Iron, zinc, or manganese phosphate coatings; protective coating for ferrous and nonferrous metals. [Witco/Allied-Kelite]

Keystone. Adhesives. [Associated Adhesives] ‡

Keytrol. A total weedkiller containing aminotriazole, atrazine and 2,4-D in a wettable formulation; provides broad spectrum control of grassy and broad-leaf weed species, including deep-rooted perennials. [Burts & Harvey] *

KF Polymer® C-1000. CAS 24937-79-9; PVDF; fluid-bed powd. coating. [Kureha Chem. Industry]

KF Polymer® T-850. CAS 24937-79-9; PVDF; for injection molding and extrusion of pipe, sheet, plate, film, filament, wire, insulation. [Kureha Chem. Industry] *

KF Polymer® T-1300. CAS 24937-79-9; PVDF; for extrusion, compression and blow molding of pipe and bottles. [Kureha Chem. Industry] *

KF Polymer® U-1000. CAS 24937-79-9; PVDF; powd. coating for anticorrosive coating applications. [Kureha Chem. Industry] *

KF Polymer® W-1000. CAS 24937-79-9; PVDF; extrusion molding for color masterbatch. [Kureha Chem. Industry]*

khaki. A coloring matter produced on the fiber. The material is dipped in chrome alum, ferrous sulfate, and pyrolignite of iron, and then passed through a solution of sodium silicate.

Khaki Yellow C. Sulfur dyestuffs. §

khari salt. A native salt of India consisting chiefly of sodium sulfate; used for curing skins.

Kidnamin. A proprietary preparation of essential aminoacids; used as a dietary supplement. [KabiVitrum Ltd]‡

kidney cotton. Peruvian cotton.

kiel compound. An insulating material containing rubber, sulfur, and mineral oil. It sometimes contains pumice and beeswax.

Kienmeyer's amalgam. An amalgam consisting of 2 parts mercury, 1 part tin, and 1 part zinc; used as a coating for frictional electrical machines.

kien oil. Turpentine oil obtained by the dry distillation of resinous wood.

Kieralon®. Textile scouring agents. [BASF plc]

Kieralon® C. Detergent blend; detergent for scouring; used in neutral and high strength caustic. [BASF/Fibers]

Kieralon® ED. Blend; low-foaming detergent, emulsifier, dispersant for scouring cotton, synthetic fibers, and blends. [BASF]

Kieralon® NB-ED. Nonionic/anionic formulation; low foaming detergent with emulsifying and dispersing props.; for desizing, scouring, caustic boil-off, bleaching processes. [BASF/Fibers]

Kieralon® TX-199. Emulsion; low foaming wetting/scouring additive for pretreatment, bleaching of all fabrics; for use in high turbulence equip. [BASF/Fibers]

Kieralon® TX-410 Conc. Polyoxyethylated blend; afterscouring agent, wetting agent, detergent for dyed and printed fabrics; stable to hard water, moderate concs. of acids and alkalies. [BASF/Fibers]

Kieselguhr. See infusorial earth.

Kieselguhr dynamite. (Gurdynamite, dynamite). Ordinary dynamite, consisting of nitroglycerin absorbed by kieselguhr.

kieselguhr. See diatomaceous earth.

Kil. Insecticides. [Fisons plc, Horticulture Div] *

Kilfoam. Silicone, emulsifiers, and stabilizer emulsion; defoamer for textile and aq. industrial processing, dye and finish baths; emulsifier for paints. [Arol Chem. Prods.]

Kilianite. A proprietary synthetic resin product. §

killed spirits. A solution of zinc chloride. Prepared by dissolving zinc in commercial hydrochloric acid until action ceases.

Killgerm® Py-Kill W. CAS 7696-12-0; Emulsifiable concentrate containing 22 g/l tetramethrin; for control of files in livestock houses. [Killgerm Chemicals Ltd]

Killgerm® Ratak Cut Wheat Rat Bait. CAS 56073-07-5; Difenacoum; a ready-to-use anticoagulant rodenticide. [Killgerm Chemicals Ltd]

Killgerm® Sewarin P. CAS 81-81-2; 0.025% Warfarin on pin-head oatmeal, with sugar and mold inhibitor; rat and mouse killer. [Killgerm Chemicals Ltd]

Killgerm® Sol Odamask H. Mixture of odor absorbents, fixatives and odor masking agents; industrial deodorant in sewage works, on refuse tips, maggot farms, etc. [Killgerm Chemicals Ltd]

Killgerm® ULV 400. Pyrethin; contact insecticide. [Killgerm Chemicals Ltd]

Killgerm® ULV 500. Mixture containing phenothrin and tetramethrin; for control of flying insects in agricultural premises. [Killgerm Chemicals Ltd]

Kilmet. Selective weedkillers. [May & Baker Ltd] *

Kilnet. Selective weedkiller. [May & Baker Ltd] *

Kinel 5502. A proprietary polyimide casting, potting, and encapsulating resin. [Rhône-Poulenc NV]

Kinel 5514. A proprietary polyimide molding composition reinforced with glass fiber. [Rhône-Poulenc NV]

Kinel 5517. A proprietary free-sintering self-lubricated, heat-resistant polyimide molding powder. [Rhône-Poulenc NV]

Kinetic No 12. See Freon. §

Kinetine. A combination of quinine and bectine.

Kinetite. A mixture of the jelly formed by dissolving guncotton in nitrobenzene, with potassium chlorate or potassium nitrate a sulfur; an explosive.

Kinevac. CAS 25126-32-3; Sincalide; choleretic. [Bristol-Myers Squibb Co Inc]

Kingston bronze. An alloy of 85% copper, 12% zinc, 2.5% tin, and 0.05% iron.

Kinite. A proprietary trade name for steel containing 12.5-14.5% chromium, 1.5% carbon, 1.1 cent molybdenum, 0.7% cobalt, 0.55% silicon, 0.5% manganese, and 0.4% nickel.

kino. (kino gum). The dried juice obtained from incisions in the trunk of *Pterocarpus marsupium*. It resembles catechu, and is used in dyeing and medicine.

kino, Australian. (kino, eucalyptus; kino, Botany Bay). The dried exudation of *Eucalyptus* species.

kino, Bengal. (kino, madras; Dhak gum). Butea gum, from *Butea frondosa*.

---

* Trade name not verified as available † Trade name verified as obsolete

kino, Botany Bay. See kino, Australian.

kino gum. See kino.

Kirnol® Range. Fatty acid mono and diglycerides; general purpose food emulsifiers. [Grünau]

kish. Crystalline graphite formed in blast furnace slag during iron smelting.

Kite. A paper based grade of Tufnol industrial laminates. [Tufnol Ltd]

Kiton. Acid wool dyestuffs. [Clayton Aniline Co Ltd] *

Kittool fiber. A fiber obtained from the leaves of a Ceylon palm, *Caryota urens*; used in the manufacture of brushes.

Kival. Insecticide. [May & Baker Ltd] *

Klavikordal. CAS 55-63-0; Nitroglycerin; vasodilator. [Rhone-Poulenc Rorer Pharmaceuticals] *

Klea. CFC ozone-benign replacement refrigerant. [ICI Chem. & Polymers Ltd]

Klearfac® 870. Block copolymer phosphate ester; surfactant. [BASF]

Klearfac® AA270. Alcohol alkoxylate phosphate ester, free acid; emulsifier, solubilizer, dedusting agent, hydrotrope, metal cleaner. [BASF]

Klearol. White mineral oil. [Witco]

Klebcil. CAS 25389-94-0; Kanamycin sulfate; antibacterial. [SmithKline Beecham] *

Kleen-Dent. Tooth polishing agent. [Reheis Inc]

Kleenite. Denture cleanser powder; for cleaning dentures. [Richardson-Vicks Inc] *

Kleenmold. Graphite lubricants for use in the glass industry, principally as mold releases. [Specialty Products Co] *

Kleenup. Weed and grass killer. [Chevron] *

Kleerox® HCS. Org. stabilizer for hydrogen peroxide bleach liquors in textile industry. [Rhone-Poulenc Surf.]

Klee's salt. Acid potassium oxalate, $KHC_2O_4 \cdot H_2O$.

Klegecell R30. CAS 9002-86-2; High performance PVC foam; rigid closed-cell foam with tridimensional grid structure, high thermal stability; structural sandwich core material providing low weight, insulation and stiffness. [Polimex]

Kleinenberg's fat mixture. A solution of cacao butter and spermaceti in castor oil; used as an embedding material in microscopy.

Kleinenberg's fixative. Used in microscopy. It consists of 100 cc of a saturated aqueous solution of picric acid, 3 cc of sulfuric acid, and 300 cc of water.

Kleinenberg's stain. A microscopic stain. It consists of a saturated solution of alum and calcium chloride in alcohol (70%. diluted with 6 times its volume of alcohol (70%.) to which is added an alcoholic solution of hematoxylin until the color is violet blue.

Klein's reagent. A saturated solution of cadmium borotungstate, $2(Cd_2H_2W_8O) \cdot 7(WO_3)B_2O_3 \cdot H_2O$; used for the separation of minerals.

Klenal. Industrial cleaner. [Specialty Products Co] *

Kleptose®. Betacyclodextrin; encapsulating agent for pharmaceutical and chemical industries. [Roquette (UK) Ltd]

Klerat. Rodenticide. [ICI Chem. & Polymers Ltd]

Klere-Seal®. Alkylalkoxysilane solution; reacts with minerals in concrete and masonry to form a chemical bond preventing the intrusion of water and water-borne salts into the substrates. [Pecora Corporation]

Kloben®. For the agriculture industry. [DuPont UK]

K-Lor. CAS 7447-40-7; EINECS 231-211-8; Potassium chloride; replenisher. [Abbott Laboratories] *

Kloref. A proprietary preparation containing betaine hydrochloride and potassium bicarbonate. [Cox Pharmaceuticals]

Kloro 6001. Chlorinated paraffin; very good heat stability and resist. to hydrolysis; used for sol. oils, synthetics, and semi-synthetics. [Ferro/Keil]

Kluberlubrication. Specialty lubricant and grease. [Carl Freudenberg] *

Klucel®. CAS 9004-64-2; Hydroxypropylcellulose; emulsion stabilizer, emulsification aid, whipping aid, suspending agent, thermoplastic resin and thickener. [Hercules] *

Klucel® E, G, H, J, L, M. CAS 9004-64-2; Hydroxypropylcellulose, standard grades; surface active thickener, stabilizer, film-former, suspending agent, protective colloid for coatings, adhesives, extrusions, moldings, paper, paint removers, encapsulations, inks. [Aqualon]

Klucel® EF. CAS 9004-64-2; Hydroxypropylcellulose, premium grade; stabilizer, film-former, suspending agent, protective colloid; esp. for nonaerosol hairspray formulations. [Aqualon]

Klucine. A proprietary waterproofing compound §

K.L.X. CAS 68334-28-1; Partially hydrogenated vegetable oil (cottonseed, soybean); icing stabilizer; adds solids to shortening; fat encapsulatin. [Van Den Bergh Foods]

K-Lyte. CAS 298-14-6; Potassium bicarbonate; pharmaceutic necessity. [Mead Johnson & Co]

K-Lyte/C1. CAS 7447-40-7; EINECS 231-211-8; Potassium chloride; replenisher. [Mead Johnson & Co]

KM Ammonium Metavanadate. 77-78% V_2O. [Kerr-McGee Chemical Corp] *

KM Ammonium Perchlorate. CAS 7790-98-9; 99.1-99.7% $NH_4C_2O_4$, depending on grade; manufactured in ordnance and industrial grades. [Kerr-McGee Chemical Corp] *

KMC. Diisopropyl naphthalene; high boiling solvent. [Collinda Ltd]

KMD-50. N,N′-Disalicylidene-1,2-propanediamine, xylene; metal deactivator for gasoline, distillate fuels, and other petrol. products. [Ferro/Keil]

KM Fly Ash. Damp bulk form. [Kerr-McGee Chemical Corp]*

KM Manganese Dioxide. CAS 1313-13-9; AB and SB battery active grades 90% minimum MnO_2 for use in Leclanche, alkaline and zinc chloride dry cell batteries. [Kerr-McGee Chemical Corp] *

KM Muriate of Potash. White agricultural grade.; 62% minimum K_2O in granular, coarse and standard grades. [Kerr-McGee Chemical Corp] *

K-Monel. A proprietary alloy containing nickel and copper in approximately the same ratio as in monel metal with the addition of 4% aluminum. §

KM Pebble Lime. CAS 1305-78-8; 90% minimum available CaO, in mill run, coarse, medium, fine and crushed grades. [Kerr-McGee Chemical Corp] *

KM Phosphate Rock. 68, 70, 72 and 75% BPL (bone phosphate of lime expressed as $Ca_3(PO_4)_2$) grades. [Kerr-McGee Chemical Corp] *

KM Potassium Chloride. CAS 7447-40-7; EINECS 231-211-8; High purity white industrial grade; 98.3% KCl (62% minimum K_2O equivalent). [Kerr-McGee Chemical Corp] *

KM Potassium Perchlorate. CAS 7778-74-7; 99.7% $KClO_4$, industrial and military grades. [Kerr-McGee Chemical Corp] *

---

‡ Trade name and manufacturer not verified   § Trade name without identified manufacturer

KM Sodium Chlorate. CAS 7775-09-9; Technical grade; 99.5% minimum in bulk; drummed may contain 0.25-0.5% anticaking agent; can be supplied in dry bulk, solution and with salt added to custom specifications. [Kerr-McGee Chemical Corp] *

KM Sodium Perchlorate. CAS 7601-89-0; 60-64% NaClO$_4$, aqueous solution. [Kerr-McGee Chemical Corp] *

KM Vanadium Pentoxide. CAS 1314-62-1; Fused flake and fine granular, 98% minimum V$_2$O$_5$. [Kerr-McGee Chemical Corp] *

Knapp's solution. Mercurous chloride (10.8 grains) are treated with potassium cyanide solution until the addition of caustic soda causes no precipitate. Caustic soda solution (100 cc of specific gravity 1.145), added, and the whole diluted to 1 liter; used for the estimation of glucose.

Knauerit 2. A plaster-shooting (mud-capping) explosive of greatest brisance and high velocity of detonation, developed for high performance; used without stemming for pop shots; from the cartridge of Knauerit 2 slices of adequate thickness are cut to be closely pressed against the boulder. [Dynamit Nobel Wien GmbH] *

Knauerit S. An explosive plaster charge of best adhesive strength which can be formed by hand to fit the shape of the underground; advantageous to use Knauerit S for demolition work, e.g., for cutting off iron constructions, bridge girders, rails, etc. [Dynamit Nobel Wien GmbH] *

Knave. Granular mixture of disulfoton and quinalphos; an organophosphorus insecticide. [Hortichem Ltd]

Kneiss alloy. An alloy of 42% lead, 40% zinc, 15% tin, and 3% copper; used for machine bearings; another alloy contains 50% zinc, 25% tin, and 25% lead.

Knight's patent zinc white. See lithopone.

Knit-Soft 30 NCPM. Blended softener to be pad or exhaust applied; excellent napping lubricant. [Yorkshire Pat-Chem]

Knittex. Finishing agent. [Ciba plc] *

Knot Out. CAS 82558-50-7; Suspension concentrate containing 125 g isoxaben per liter; used for control of annual dicotyledons in cereals, grass and fruit. [Synchemicals Ltd]

Koate®. Antihemophilic factor. [Bayer AG; Miles Pharmaceuticals]

kochenite. A fossil resin resembling amber.

Kochlin's Bearing Bronze. An alloy of 90% copper and 10% tin. §

Koch's acid. 1-Naphthylamine-3, 6, 8 trisulfonic acid; used as an azo dye intermediate.

Kocide® 20/20. Copper hydroxide plus nutritional zinc; wettable powd. agricultural fungicide, bactericide, and nutritional. [Griffin]

Kocide® Copper Sulfate Pentahydrate Crystals. CAS 7758-99-8; Copper sulfate, pentahydrate; fungicide to control plant diseases, in fertilizers to correct copper deficiencies in soils. [Griffin]

Kodabond® Copolyester 5116. Terephthalate-based copolyester; heat-seal layer in pkg. applications; applied as coextrusion or an extrusion coating; for coating on paper, polyesters, copolyesters, PC, polyvinylidene chloride. [Eastman]

Kodaflex® DBP. CAS 84-74-2; Dibutyl phthalate; plasticizer used in coatings industry as primary plasticizer-solv. for nitrocellulose lacquers; for rubbers and CAB, ethyl cellulose, PVAc, and synthetic resins; solv. for oil-sol. dyes, insecticides, peroxides, and org. compounds; anti-foamer and fiber lubricant in textile mfg. [Eastman]

Kodaflex® DBS. CAS 109-43-3; Dibutyl sebacate; plasticizer [Eastman]

Kodaflex® DEP. CAS 84-66-2; Diethyl phthalate; plasticizer; wetting agent in grinding pigments; pigment-disp. medium in cellulose acetate solutions and plastics, and solv. for natural resins and polymers; PVC products due to relatively high volatility. [Eastman]

Kodaflex® DIBP. CAS 84-69-5; Diisobutyl phthalate; plasticizer [Eastman]

Kodaflex® DIDA. CAS 27178-16-1; Diisodecyl adipate; plasticizer [Eastman]

Kodaflex® DIDP. CAS 68515-49-1; Diisodecyl phthalate; plasticizer. [Eastman]

Kodaflex® DMEP. CAS 117-82-8; Di-(2-methoxyethyl) phthalate; plasticizer. [Eastman]

Kodaflex® DMP. CAS 131-11-3; Dimethyl phthalate; plasticizer with high solv. power for cellulose acetate extrusion compounds; compatible with ethyl cellulose, CAB, PS, PVAc, polyvinyl butyral, and PVC; used in NC-based printing inks. [Eastman]

Kodaflex® DOA. CAS 103-23-1; Dioctyl adipate; plasticizer providing flexibility at low temps. to vinyl products; used in unfilled garden hose, clear sheeting, elec. insulation. [Eastman]

Kodaflex® DOP. CAS 117-81-7; Dioctyl phthalate; all-purpose plasticizer used with PVC resins incl. film and sheeting for upholstery, clothing, food pkg., paper coatings, molded vinyl products, elec. wire insulation; compatible with PS, methyl methacrylate, chlorinated rubber, NC, and CAB; low odor, relatively low toxicity, and low volatility. [Eastman]

Kodaflex® DOTP. Dioctyl terephthalate; primary plasticizer used with PVC resins, in PVC plastisols, rubber; application incl. wire coatings, automotive and furniture upholstery; compatible with acrylics, CAB, cellulose nitrate, polyvinyl butyral, styrene, oxidizing alkyds, nitrile rubber. [Eastman]

Kodaflex® DOZ. CAS 103-24-2; Dioctyl azelate; plasticizer. [Eastman]

Kodaflex® HS-3. Butyl octyl phthalate; primary plasticizer for polyvinyl homopolymer and copolymer resins, vinyl plastisols, expanded vinyl foams, and rotational molding and dip coating. [Eastman]

Kodaflex® HS-4. 60% Dioctyl phthalate, 5% dibutyl phthalate; high-solvating primary plasticizer; used in formulating vinyl plastisols having processing chars. required to be mechanically frothed; for PVC formulations, rotational molding, slush molding, dip coating, and filament coating application. [Eastman]

Kodaflex® OIDP. Octyl isodecyl phthalate; plasticizer. [Eastman]

Kodaflex® TEG-EH. CAS 94-28-0; Triethylene glycol di-2-ethylhexanoate; plasticizer [Eastman]

Kodaflex® TOTM. CAS 3319-31-1; Trioctyl trimellitate; primary plasticizer used in vinyl film and vinyl-coated fabrics. [Eastman]

Kodaflex® Triacetin. CAS 102-76-1; Glyceryl triacetate; low-toxicity plasticizer for vinyl compounds; used in adhesives, resinous and polymeric coatings, paper, and paperboard for food contact; water-insol. hydroxyethyl cellulose films. [Eastman]

Kodaflex® TXIB. CAS 6846-50-0; 2,2,4-Trimethyl-1,3-pentanediol diisobutyrate; primary plasticizer used in surface coatings, vinyl floorings, moldings, and vinyl products; compatible with film-forming vehicles used in lacquers for wood, paper, and metals; primary plasticizer for PVC plastisols for rotocasting and slush molding; used in PVC organosols processed by extrusion and injection molding. [Eastman]

Kodaloid. A proprietary trade name for a cellulose nitrate; it is made in the form of sheets. §

Kodapak®. Transparent cellulose acetate film; used for making packets. [Eastman]

Kodapak® 5214A. CAS 25038-59-9; PET polyester homopolymer; light weight material resistant to breaking, bursting, and shattering; improved barrier properties; for bottles for carbonated soft drinks, fruit juices, foods, etc. [Eastman]

Kodapak® PET Copolyester 13339. CAS 25038-59-9; PET polyester; for use in packaging and other applications. [Eastman] *

Kodar®. Copolyester thermoplastic. [Eastman]

Kodar® A150 Copolyester. Copolyester; produces extruded sheet with optical clarity, toughness at low temps., and high tear strength and elongation. [Eastman]

Kodar® PETG Copolyester 6763. CAS 25038-59-9; Amorphous glycol-modified PET; offers sparkling clarity in film and sheet form, easy thermoformability, toughness, sterilizability with ethylene oxide or gamma rays in medical applications; used for medical containers, thermoformed food containers and lids, blister pkg. for cosmetics, pharmaceuticals, heavy hardware, electronic parts, FDA compliance. [Eastman]

Kodar® Thermx Copolyester 6761. Crystallizable copolyester; for thermoformed dual ovenable trays providing superior temp. resist.; FDA compliance. [Eastman]

Kodel. Polyester yarn and fiber. [Eastman]

Kodofil. Polyester fiberfill. [Eastman]

Kodolite. Polyester fiberfill. [Eastman]

Kodosoff. Polyester fiberfill. [Eastman]

Koenig solder. An alloy of 60% tin, 30% aluminum, and 10% antimony.

Koerzit. An alloy for permanent magnets containing 1.1% carbon, 3.5% manganese, 36% cobalt, 4.8% chromium, the remainder being iron.

Koerzit, I, II, III. Proprietary cobalt steels containing 10, 20, and 30% cobalt respectively. §

Kogasin III. Nonaromatic hydrocarbon oil, approximately C15-C20. [Sasolchem]

Kohacool L-400. Alkylether sulfosuccinate; wetting and foaming agent for nonirritating hair shampoos, bubble baths, hair conditioners and lotions. [Toho Chem. Industry]

Koka Seki. A variety of pumice stone found in the Nujima Islands, near Tokio; used as a building material.

Koken. A proprietary synthetic resin. §

koko. The leaves of *Celastrus buxifolia*; used in Natal as a sumac substitute for tanning.

kokowai. A variety of rouge used by the Maori.

kokum butter. (Garcinia oil, concrete oil of mangosteen). A fat obtained from the seeds of *Garcinia indica* or *G. purpurea*. It is composed of stearine, myristicine, and oleine.

Koladex. "Pick-me-up" tablets containing natural kola nut. [LAB Ltd] ‡

Kolanticon. A proprietary preparation of aluminum hydrox-

ide, magnesium oxide, dicyclomine hydrochloride, and dimethicone; used as a gastrointestinal sedative. [Richardson-Vicks Inc] *

Kolantyl. A proprietary preparation containing dicyclomine hydrochloride, aluminum hydroxide, magnesium hydroxide, and methyl cellulose; an antacid. [Richardson-Vicks Inc] *

Kolantyl-NV. A proprietary preparation of dicyclomine hydrochloride, aluminum hydroxide, magnesium hydroxide, and magnesium trisilicate; an antacid. [Richardson-Vicks Inc] *

kola nut. (Kola Seeds). The seeds of *Cola acuminata* and *C. vera*.

kola seeds. See kola nut.

Kolax. An explosive of the same type as carbonite.

Kolene. Metal cleaning salts. [Degussa Ltd]

kol-kol gum. A gum of Nigeria, from *Acacia senegal*.

Kollercast. Moldable synthetic resins; for industrial purposes, e.g., flooring. [Scott Bader] *

Kollercure®. Epoxy curing agents. [Scott Bader]

Kollerdur® L 90. Polyurethane elastomer, IPDI based; evaporation curing high performance coatings for flexible substrates, e.g., foams, fabrics, leather, plastics, and clear overprint coatings. [Scott Bader]

Kollerdur® M0122. PU resin in methoxy propyl acetate/ xylene; aliphatic type; one-pack moisture cure coating for high performance industrial finishes; superior hardness, color retention. [Scott Bader]

Kollerdur® MO 118. One-pack polyurethane resin; moisture curing systems to produce primers, sealers, and coatings or formulated to produce aggregate-filled floors, roofing systems. [Scott Bader]

Kollernox®. Epoxy resins. [Scott Bader]

Kollidon® CL. Crospovidone; tablet-disintegrating agent, suspension stabilizer [BASF AG; BASF plc]

kolm. A variety of bituminous coal found in Sweden. The ash contains from 1-3% of uranium oxide, U_3O_8.

Kombat. Carbendazim + mancozeb; systemic fungicide for cereals. [Hoechst UK]

Kombé arrow poison. Strophanthus, the seed of *Strophanthus hispidus*.

Komed HC. CAS 50-03-3; Hydrocortisone acetate; glucocorticoid. [Barnes-Hind Inc] *

Komeen®. Copper-ethylenediamine complex; aquatic herbicide for hydrilla control in golf course, ornamental, and fish ponds, potable water reservoirs, fresh water lakes, fish hatcheries. [Griffin]

Kommoid. A sulfurized corn oil rubber substitute.

Kompak. Granular product for use on cast iron to produce a compacted graphite structure. [Foseco (F.S.) Ltd] †

Kompolite. Flooring materials. [Weatherguard/Marbleloid Products Inc] *

Konakion. Phytomenadione; a preparation of Vitamin K. [Roche Laboratories; Roche Products Ltd]

Konator. Dental equipment. [Degussa Ltd]

Kondang wax. See Java wax.

Kondremul. Mineral oil; laxative; pharmaceutical aid. [Fisons Corp, Pharmaceuticals Div] *

Konel. Proprietary nickel-cobalt-iron alloys containing about 2.5% titanium; they are high temperature resisting alloys and possess high tensile strength at elevated temperatures. §

Konform® AR. Acrylic resin; conformal coating providing insulation against high voltage arcing and corona shorts.

---

‡ Trade name and manufacturer not verified   § Trade name without identified manufacturer

[Chemtronics]

konilite. A silica in powder form.

Konker®. Vinclozolin, carbendazim; fungicide with systemic activity and contact effect for use in rape, sunflower, strawberries, vegetables [BASF AG]

konnan bark. Obtained from *Cassia fistula* of Southern India; a tanning material.

kon oil. (Kusum oil). Macassar oil obtained from the seeds of *Schleichera trijuga*. It has a saponification value of 215-230, an iodine value of 48-69, and an acid value of 6-35.

Konservan SN. Organic tin compound; nonionic dispersion; preservative for textiles. [Thor Chemicals (UK) Ltd]

Konstrastin. Basic zirconium basic acetate.

Konstructal. An aluminum alloy containing 1% copper or 8% zinc.

Kontakt. A purified form of the Twitchell reagent; used for the hydrolysis of fatty glycerides.

Kontrastin. Zirconium oxide, ZrO_2.

Konut. CAS 8001-31-8; EINECS 232-282-8; Coconut oil; kosher; oil for ice cream, coatings, nut roasting, corn popping. [Van Den Bergh Foods]

Konyne. Factor IX Complex; hemostatic. [Cutter Laboratories, Miles Laboratories Inc] *

konzentrole. A term used for essential oils free from terpenes and sesquiterpenes; used for flavoring.

Koolkat. Furane resin cold-set binders for sand cores. [Foseco (F.S.) Ltd]

Kopan. *See* Bakelite. §

Kopol®. Synthetic resins, soluble plastic resins. [Reichhold]

Koppert Moss Killer. A soap concentrate; used to kill mosses in turf. [Koppert (UK) Ltd]

Koppeschaar solution. A bromine solution of N/10 strength.

Kopr-Kote®. Copper flake graphite; anti-seize compound protecting metal parts for seizure, galling, rust, corrosion, and heat-freeze. [Jet-Lube]

Korad A. Acrylic film. [Rohm & Haas] *

Koraid PSM. Alumino-silicate, modified; suspension aid for pigments and abrasive particles in water-based systems without increasing visc.; used for paints, polishes, inks, agricultural, and pharmaceutical formulations. [Kaopolite]

Korantin. Additives for oil/water emulsions. [BASF plc]

Koraton. A proprietary trade name for a synthetic resin. §

Koreforte®. Reinforcing resins for natural and synthetic rubbers. [BASF AG]

Koreon. A basic chromium sulfate, $Cr(OH)SO_4$; used in the tanning industry.

Koresin®. Tackifier for natural and synthetic rubbers. [BASF AG; BASF plc]

Korestab®. Antioxidant for natural and synthetic rubbers. [BASF AG]

Koretack®. Tackifier for natural and synthetic rubbers. [BASF AG; BASF plc]

Korever®. Vulcanization resins. [BASF AG]

Korlan. CAS 299-84-3; Active ingredient: ronnel; insecticide used on cattle for the control of ticks, flies, maggots, and lice. [Dow UK] *

Korlite. Natural zeolite silicate mineral; absorbent for org. compounds, odors, sulfur dioxide gases and ammonia; builder in laundry detergents, cleaning compounds; decolorizer for org. liqs.; specialty filler and water treatments for lead and heavy metal removal. [Kaopolite]

Koro. A proprietary trade name for an alloy of 98% copper and 2% nickel. §

Korogel. A proprietary trade name for a soft Koroseal (*qv*).§

Korolac. A proprietary trade name for solutions of Koroseal (*qv*); used in acid-resisting tank linings. §

Koron. Refractory coatings for use on ingot molds, bottom plates, and slag pots. [Foseco (F.S.) Ltd]

Koronit. A German explosive.

Koronium Bromide. CAS 10476-81-0; The trade name for strontium bromide, $SrBr_2 \cdot 6H_2O$. §

Koroplate. A proprietary synthetic paint in which Koroseal (*qv*) is the base; extremely resistant to acid fumes. §

Koroseal. A proprietary trade name for a rubber-like thermoplastic varying in hardness, from soft jellies to hard rubber; detained by treating highly polymerized vinyl chloride with plasticizers at high temperatures and cooling; it can be worked like rubber when hot but requires rather higher temperatures; resistant to light, water, oils, and most other chemicals; used for impregnating and coating paper, fabrics and metals for the manufacture of tubing for corrosive materials and cable sheathing. §

Korpad. Antisplash pads to minimise splash defects during direct teeming of steel ingots. [Foseco (F.S.) Ltd]

Korspray. Foundry solvent (I.P.S.). [Foseco (F.S.) Ltd]

Kortaid. Saturated fatty acids. [Akzo Chemie UK Ltd.]

Korthix. CAS 1302-78-9; Bentonite, refined; thixotropic agent for water-based paints, inks, polishes, adhesives, and for household products. [Kaopolite]

Korthix H-NF. CAS 1302-78-9; Bentonite; bacteria-controlled grade for cosmetics and pharmaceuticals. [Kaopolite]

korung oil. (Kagoo oil). Pongam oil, obtained from the fruits of *Pongamia glabra*.

Kosmos®. Tin catalysts. [Thomas Goldschmidt Ltd]

Kosmos® 10. Stannous octoate, dioctyl phthalate; tin-org. catalyst for the mfg. of polyether PU foam; accelerates the gel reaction and intensifies the activation of the blowing reaction. [Goldschmidt AG]

Kosmos® 21. Dialkyl-tin-mercaptide; catalyst for PU polymerization, for molded PU foams, e.g., microcellular foams, rigid foams, high resilient foams, and RIM. [Goldschmidt AG] *

Kosmos Black, 3XB, BB, and F4. A proprietary gas black used in rubber mixings; also used as a black pigment. §

Kostil. A proprietary range of styrene acrylonitrile molding granules. [Montedison UK Ltd] *

Kotamite®. CAS 471-34-1, 1317-65-3; Calcium carbonate; coated pigment with easy dispersion in plastic compounds, e.g., polyolefins, rigid and flexible PVC; for wire and cable insulation compounds, improved impact props. in PP. [ECC]

Kotebond. Etch primers and finishes. [Brent Chemicals International plc]

Kovar® Alloy. A registered trademark for an alloy of iron with 23-30% nickel, 17-30% cobalt and 0.6-0.8% manganese; used for glass to metal seals. §

Kowet 12. CAS 26264-06-2; Calcium dodecylbenzene sulfonate [Rhone-Poulenc Surf.]

Kowet 3300. CAS 26264-05-1; Isopropylamine dodecylbenzene sulfonate. [Rhone-Poulenc Surf.]

KP-2. Designed to impart detergency and lubricity while minimizing the buildup of hard film deposits; plasticizer for dry rust inhibitor films. [Ferro/Keil]

KP-23. CAS 109-38-6; A proprietary trade name for a

---

* Trade name not verified as available   † Trade name verified as obsolete

plasticizer consisting of butoxyethyl stearate. §

KP 90. A proprietary trade name for a vinyl plasticizer of the epoxy type. §

KP-140®. CAS 78-51-3; Tributoxyethyl phosphate; plasticizer; leveling agent in floor polish formulations allowing uniform coverage, eliminates high and low spots in gloss, and preventing streaking, crazing, powd., and film contracting; flame retardant for plastics or synthetic rubbers of lower flammability; imparts low temp. flexibility to plastics or acrylonitrile rubbers; reduces visc. in plastisols. [FMC]

KP 201. CAS 84-61-7; Dicyclohexyl phthalate; a proprietary trade name for a vinyl plasticizer. §

KP 555. Bis(dimethylbenzyl) ether; a proprietary trade name for a vinyl plasticizer. §

K-Pool. Copper-TEA complex; algicide for swimming pools. [Griffin]

KR01 K-Resin Polymer. CAS 9003-55-8; Styrene-butadiene copolymer; resin for injection molding and extrusion processing; for crystal clear and warp resistant parts; in housings, blister packs, extruded tubes, molded boxes with integral hinges, toys, and where impact PS, oriented PS sheet, cellulosics, and rigid PVC have been used. [Phillips]

KR04 K-Resin Polymer. CAS 9003-55-8; Styrene-butadiene copolymer, antiblock; extrusion grade resin; clarity, toughness, rigidity in extruded parts, and detail on fast production cycles; for blending with general purpose PS; thermoformed blister packs, disposable containers, and where impact PS, oriented PS sheet, polyesters, cellulosics, and rigid PVC have been used; resin can be tinted or colored in a variety of transparent and opaque shades. [Phillips]

kraft paper. A paper produced by the sulfate pulp process. It is strong and inexpensive.

Kraft's metal. A fusible alloy containing 5 parts bismuth, 3 parts lead, and 1 part tin. It melts at 104 C.

Kraton®. Thermoplastic rubbers; used for footwear and adhesives. [Shell]

Kraton® D 1101. Linear styrene-butadiene-styrene block copolymer; thermoplastic rubber requiring no vulcanization; for formulating adhesives for the building/construction trade, hot-melt adhesives; FDA compliance. [Shell]

Kraton® D 1107. Linear styrene-isoprene-styrene block copolymer; thermoplastic rubber; FDA compliance. [Shell]

Kraton® D 1116. Branched styrene-butadiene copolymer; thermoplastic rubber for use in solv.-based construction mastic adhesives; high strength. [Shell]

Kraton® D 1320X. Branched styrene-isoprene multiarm copolymer; radiation crosslinkable thermoplastic rubber providing good tack in hot-melt pressure-sensitive adhesives even after cure. [Shell]

Kraton® D 2103. Thermoplastic elastomer (S-B-S block copolymer); injection molding and extrusion compound for medical/food, sporting goods, and misc. molded items; 2103 has FDA compliance. [Shell]

Kraton® FG 1901X. Maleic anhydride-functionalized styrene-ethylene-butylene-styrene block copolymer; thermoplastic rubber with excellent thermal, oxidative, and uv stability for toughened engineering thermoplastics, coextruded tie layers; FDA compliance. [Shell]

Kraton® G 1650. Linear styrene-ethylene-butylene-styrene block copolymer; thermoplastic rubber for use in adhesives, sealants, and coatings that must withstand weathering and high processing temps.; FDA compliance. [Shell]

Kraton® G 1701X. Styrene-ethylene-propylene diblock copolymer; thermoplastic rubber. [Shell]

Kraton® G 1726X. Linear styrene-ethylene-butylene-styrene block copolymer (70% S-EB diblock); thermoplastic rubber for use in adhesives, sealants, and applications where resist. to degradation is necessary and where softness and processability are more important than tensile strength; blendable with Kraton G 1652 to give desired props.; FDA compliance. [Shell]

Kraton® G 2701. Thermoplastic rubber compound (S-EB-S block copolymer blend); for use in extruded film, blow-molded containers, injection molded products, general rubber products; also to improve processing behavior of other thermoplastics rubbers and as impact modifier for high melt flow PP homopolymers; FDA compliance. [Shell]

Kraton® G 7430. Thermoplastic elastomer (S-EB-S block copolymer blend); injection molding and extrusion compound for automotive parts, sporting goods, and misc. molded items. [Shell]

Kraton® RP6404. S-I-S block copolymer; provides low visc., color-stable adhesives with high adhesion to various substrates used in hot melt assembly. [Shell]

Kremnitz. See flake white.

Kremser white. The purest form of white lead.

Krems white. See flake white.

Krenite. CAS 25954-13-6; A 48% water soluble liquid formulation of fosamine ammonium; applied to unwanted brush in late summer or autumn prevents bud break leading to death of treated plants the following spring. [Burts & Harvey; DuPont UK]

Krenite. CAS 25954-13-6; Soluble concentrate of 480 g fosamine-ammonium per liter; used for control of woody weeds in noncrop and forestry areas. [Selectokil Ltd]

Krennerite. See white tellurium.

Kresamin. (Cresamol). A mixture of 25% tricresol with ethylenediamine, $H_2N \cdot CH_2 \cdot CH_2 \cdot NH_2$. A powerful antiseptic.

Kresatin. m-Cresol acetate.

Kreside. Cresylic creosote [Sasolchem]

K-Resin Polymer KR01. Transparent and shatter resistant styrene-butadiene block copolymer containing at least 70 weight percent polymerized styrene; an injection molding grade for use where higher stiffness and warpage resistance is required; used for dust covers, point-of-purchase displays, molded boxes and containers, lids and office supplies. [Phillips]

K-Resin Polymer KR03. Transparent and shatter resistant styrene-butadiene block copolymer containing at least 70 weight percent polymerized styrene; an injection molding grade for use where higher impact resistance is required; used for overcaps, molded boxes and containers, toys, medical devices and tool handles. [Phillips] *

K-Resin Polymer KR04. Transparent and shatter resistant styrene-butadiene block copolymer containing at least 70 weight percent polymerized styrene; a resin for blending with general purpose polystyrene for sheet extrusion and thermoforming applications such as disposable cups and containers, blister packages and portion packaging. [Phillips]

K-Resin Polymer KR05. Transparent and shatter resistant

---

‡ Trade name and manufacturer not verified § Trade name without identified manufacturer

styrene-butadiene block copolymer containing at least 70 weight percent polymerized styrene; a resin for nonblended sheet extrusion and thermoforming, for blow molding (both extrusion and injection) and for profile extrusion; uses include blister packages, bottles, jars, medical devices and extruded tubes and profiles. [Phillips] *

K-Resin Polymer KR10. Transparent and shatter resistant styrene-butadiene block copolymer containing at least 70 weight percent polymerized styrene; a resin for blown or cast film production; uses include medical packaging, shrink wrap, overwrap, skin packaging, produce wrap, windows for envelopes and boxes and twist wrap. [Phillips] *

Kresival. A German preparation. It contains the water-soluble calcium salts of the sulfonic acids of the cresols.

Kresolin. See Kresopolin.

Kresopolin. (Kresolin). Preparations of crude carbolic acid; disinfectants.

Kricinol 35. CAS 7492-30-0; EINECS 231-314-8; Potassium ricinoleate; mold release. [Climax Fluids Additives]

Kriegr-o-dip. A proprietary trade name for liquid dyes for plastics. S-standard chemical dye; A-for cellulose acetate. W-powder dye for use in hot water. V-for polystyrene. §

Kristalex. α-Methylstyrene copolymer hydrocarbon resins; used for hot-melt product assembly adhesives and light colored caulking compounds. [Hercules]

Kristel Gold II. Partially hydrogenated vegetable oil (soybean, cottonseed), lecithin, artificial color and flavor; kosher; shortening system for danish pastries, sweet rolls, coffee cakes; higher melt pt. for high shear applications. [Van Den Bergh Foods]

Krist-o-kleer. A proprietary trade name for a plasticizer containing 50% dextrose and 50% levulose. §

Kristol. White oils. [Carless Refining & Marketing Ltd]

Krokoloy. A proprietary trade name for a steel containing 14% chromium with some cobalt. §

Kromaplast. Blended dry pigments; used in plastics. [Ampacet Corporation]

Kroma Red. CAS 1309-37-1; A precipitated red iron oxide for color pigment use. [Pfizer International] †

Kromax. An electrical resistance alloy of 80% nickel and 20% chromium.

Kromore. An alloy of nickel with 15% chromium; used for the heating elements in wire-wound electric furnaces. It has a specific resistance of 98 micro-ohms cm at 0 C.

Kromosperse. Pigment dispersing agent; for use in nonaqueous and aqueous coating compositions. [Hüls Am.]

Kronagold. Family of lubricating oils; gear oils, hydraulic oils, compressor oils, etc. [E/M Corporation]

Kronaplate. Family of lubricating greases; bearings, gears, cams, slides, etc. [E/M Corporation]

Krona-Syn. Synthetic lubricating fluids; compressor fluids, high temperature oven chain lubricants. [E/M Corporation]

Kronitex®. Flame retardant plasticizers. [FMC]

Kronitex® 25. CAS 68937-41-7; Triaryl phosphate; flame retardant plasticizer for PC/ABS, engineering resins, PVC, phenolic laminates, cellulosics. [FMC]

Kronitex® 50. CAS 68937-41-7; Triaryl phosphate; flame retardant plasticizer for PVC, flexible polyurethanes, synthetic rubber, belting. [FMC]

Kronitex® 100. CAS 68937-41-7; Triaryl phosphate; flame retardant plasticizer for PVC; aids fusion; in plastisols, visc. stability; compatibilizing plasticizer; catalyst carrier and pigment vehicle for PU; processing aid in rubber belting and mech. goods; flame retardant and processing aid in engineering resins. [FMC]

Kronitex® 200. CAS 68937-41-7; Triaryl phosphate; flame retardant plasticizer for cellulose polymers, PVC. [FMC]

Kronitex® 1840. CAS 68937-41-7; Triaryl phosphate; thixotropic flame retardant gel for use in coating of fiberglass and other media requiring high loading and low volatility. [FMC]

Kronitex® 3600. Alkylaryl phosphate ester; plasticizer with improved low temp. flexibility, high flame retardance, and low smoke evolution; ideal for vinyl film and sheeting, wire and cable insulation, coated fabrics, plastisols. [FMC]

Kronitex® PB-460. Brominated aromatic phosphate ester; flame retardant for engineering thermoplastics incl. PC, modified PPO, PBT, PET, ABS, and blends. [FMC]

Kronitex® TBP. CAS 126-73-8; Tributyl phosphate; primary plasticizer and solv.; antifoam for paints, pigment dispersant. [FMC]

Kronitex® TCP. CAS 68952-35-2; Tricresyl phosphate; general purpose flame retardant for vinyl compounds; low air, oil, and water loss; processing aid by improving flux rate of compounds containing slow-solvating plasticizers; rapid gelation and fusion rate make it useful in plastisols; plasticizer for NC lacquers and coatings; plasticizer and processing aid for rubbers; flame retardant sheeting. [FMC]

Kronitex® TOF. Trioctyl phosphate; low temp. plasticizer for use in PVC, rubber, paints, coatings; highly efficient solv. [FMC]

Kronitex® TPP. CAS 115-86-6; Triphenyl phosphate; flame retardant plasticizer for engineering resins, cellulosics. [FMC]

Kronitex® TXP. CAS 25155-23-1; Trixylenyl phosphate; flame retardant with better milling action in filled PVC compounds; for superior electrical compounds (wire and cable application). [FMC]

Kronos®. CAS 13463-67-7; EINECS 236-675-5; Titanium dioxide pigments; used for paint, paper, glass, ceramics, plastics and ink. [Rheox Inc]

Kronos®. CAS 13463-67-7; EINECS 236-675-5; Titanium dioxide pigment; used to impart opacity and brightness; for paints, inks, plastics, paper etc. [NL Chemicals (UK) Ltd] *

Kronos® 1000 (formerly Titanox 1000). CAS 13463-67-7; EINECS 236-675-5; Titanium dioxide, anatase; pigment for fibers, paper and coatings. [Kronos] *

Kronos® 2020 (formerly Titanox 2020). CAS 13463-67-7; EINECS 236-675-5; Titanium dioxide, rutile; pigment for coatings. [Kronos] *

Kronos® 2073 (formerly Titanox 2703). CAS 13463-67-7; EINECS 236-675-5; Titanium dioxide, rutile; pigment primarily for plastics. [Kronos] *

Kronos® 3020 (formerly Titanox 3020). CAS 13463-67-7; EINECS 236-675-5; Titanium dioxide; nonpigmentary product for ceramics, glass/glass fibers, glazes, vitreous enamels, welding rods. [Kronos] *

Krovar®. A wettable powder containing 40% w/w bromacil and 40% w/w diuron; used for total weed control in noncrop areas. [DuPont UK]

---

* Trade name not verified as available   † Trade name verified as obsolete

Krovar. A wettable powder containing 40% w/w bromacil and 40% w/w diuron; used for total weed control in noncrop areas. [Selectokil Ltd]

Kruppin. An electrical resistance alloy containing 28% nickel and the rest iron.

Kryalith. A proprietary trade name for a synthetic cryolite. §

Krylene® 606. A registered trademark for a cold polymerized, alum coagulated non-staining butadiene-styrene rubber. §

Krylene® 608. A registered trademark for a cold polymerized styrene butadiene rubber. §

Krynac® 19.65. CAS 9003-18-3; Butadiene-acrylonitrile copolymer, nonstaining, cold polymerized; NBR used for low temp. oil well specialties, belt covers, and idler rolls for low-temp. service, o-rings, seals, gaskets, hydraulic hose; vulcanized with sulfur, sulfur donor, or peroxide. [Miles; Polysar]

Krynac® 20H35. CAS 9003-18-3; Butadiene-acrylonitrile copolymer, nonstaining, hot polymerized; excellent low temp. flexibility; vulcanized with sulfur, sulfur donor, or peroxide. [Miles; Polysar]

Krynac® 27.50. A proprietary acrylonitrile rubber. [Miles; Polysar] *

Krynac® 34.35, 34.50. A proprietary cold-polymerized gel-free oil-resistant butadiene/acrylonitrile rubber. [Miles; Polysar] *

Krynac® 34.60 SP. A proprietary nitrile rubber capable of withstanding temperatures up to 135 C. [Miles; Polysar]*

Krynac® 34.80. A proprietary cold-polymerised gel-free oil-resistant butadiene/acrylontrile rubber. [Miles; Polysar]*

Krynac® 34.140. CAS 9003-18-3; Butadiene-acrylonitrile copolymer, nonstaining, cold polymerized; NBR used for plasticizer masterbatches, as visc. modifier, for o-rings, lip seals, gaskets; vulcanized with sulfur, sulfur donor, or peroxide. [Miles; Polysar]

Krynac® 823X2. A registered trademark for a copolymer of acrylonitrile and butadiene containing a medium level of bound acrylonitrile. [Miles; Polysar] *

Krynac® 833. A proprietary isoprene acrylonitrile rubber containing 31.0% bound acrylonitrile; Mooney viscosity is 70. [Miles; Polysar] *

Krynac® 843. Modified butadiene-acrylonitrile copolymer, 50 phr DOP, nonstaining, cold polymerized; NBR used for mechanical goods, plasticizer masterbatch, soft roll covers; vulcanized with sulfur, sulfur donor, or peroxide. [Miles; Polysar]

Krynac® 850. A trademark for a vinyl-modified nitrile rubber. [Miles; Polysar] *

Krynac® 881 and 882. Proprietary names for synthetic rubbers of the ethylacrylate type. [Miles; Polysar] *

Krynac® 882X1. A registered trademark for a low temperature resistant acrylic rubber for oil seals. [Miles; Polysar]*

Krynac® NV 850. NBR/PVC fluxed blend (50:50), nonstaining, cold polymerized; used for flame-resist. belting, footwear, fire hose covers, cable jackets, cellular products; excellent resist. to ozone, weathering, oils, and fuels; vulcanized with sulfur, sulfur donor, or peroxide. [Miles; Polysar]

Krynac® PXL 34.17. CAS 9003-18-3; Crosslinked butadiene-acrylonitrile copolymer, nonstaining, hot polymerized; impact modifier for PVC. [Miles; Polysar]

Krynac® X 1.46. Lightly carboxylated NBR-org. acid terpolymer, nonstaining, cold polymerized; used for spinning cots, roll covers, automotive seals, mech. goods, indus-

trial footwear; offers high abrasion resist., tensile and tear strength; vulcanized with sulfur, sulfur donor, or peroxide in combination with a metal oxide. [Miles; Polysar]

Krystalex. Hydrogenated hydrocarbon resins; for adhesives and hot melts. [Hercules] *

Krystallazurin. A fungicide consisting of ammoniacal copper sulfate.

krystallos. Quartz.

Krystaltite Film. PVC film useful as a shrink film for consumer applications. [Allied-Signal Inc]

Krytox®. A range of fluorinated greases used as lubricants in aircraft and missiles. [DuPont UK]

Krytox® GPL 206. Synthetic lubricant. [Miller-Stephenson]*

KS-052P. Thermoplastic olefin resin; for air quenched blown film extrusion of biohazard bags, medical pkg., trash bags; excellent puncture and low temp. impact resist. [Himont]

K-Slag. Potassium basic slag [Fisons plc, Pharmaceuticals Div] *

K.S. magnet steel. A cobalt steel containing 35% cobalt. It is suitable for short magnets.

K.S. powder. A 42-grain powder; an explosive.

KT-012P. Polyolefin resin; for air quenched blown and cast film processes for pkg. and shrink wrap; excellent heat sealing, clarity and gloss, good moisture barrier props. [Himont]

K.Tab. CAS 7447-40-7; EINECS 231-211-8; Potassium chloride; replenisher. [Abbott Laboratories] *

K-Tea. Copper-TEA complex; algicide for use in golf course, ornamental, and fish ponds, potable water reservoirs, fresh water lakes, and fish hatcheries. [Griffin]

KTPP. See potassium tripolyphosphate.

Kuhne phosphor bronze. An alloy of 78% coppper, 10.6% tin, 10.45% lead, 0.57% phosphorus, and 0.26% nickel.

Kukident. Denture cleanser for cleaning dentures; also denture adhesive for securing dentures. [Richardson-Vicks Inc] *

kukkersite. An oil shale of Esthonia, of specific gravity 1.2-1.4. It contains about 55% volatile matter, and when distilled at 500 C. yields from 70-80 gallons per ton of oil of specific gravity 0.92-0.93.

kuk-seng. A Chinese drug. It is the dried root of *Sophora fiavescens*, and contains the alkaloid matrine.

Kumulan®. Nitrothal-isopropyl, sulfur; for control of powdery mildew in apples. [BASF AG]

Kumulus® DF, FL. CAS 7704-34-9; Sulfur; for control of diseases and spider mites in fruit, vines, vegetable, ornamentals, and agricultural crops. [BASF AG]

Kumulus® S. CAS 7704-34-9; Wettable powder containing 80% w/w sulfur per liter; a protectant fungicide/foliar feed. [BASF plc; BASF AG]

Kunstharz HW. An ammonia condensed phenol formaldehyde resin melting at 55 C. [Chemische Werke, Albert]*

kunststein. An artificial stone made from magnesite.

kuoxam. A cellulose solvent prepared by dissolving 50 grams of copper sulfate in 300 cc water and adding ammonia until all the copper hydroxide is precipitated. The precipitate after filtration is dissolved in 25% ammonia solution.

Kupferdermasan. A salicyl-copper soap preparation containing 2% copper; a bactericide.

Kuracap. 2-Mercaptobenzthiazole + dibenzthiazole disulfide; a proprietary accelerator. [Nipa Laboratories Ltd] *

---

‡ Trade name and manufacturer not verified   § Trade name without identified manufacturer

Kurade. Accelerator for rubber. [Akzo Chemie] *

kurchi. The root bark of *Holarrhena antidysenteriea*; a febrifuge.

Kuro Bishi® . Coke; general use for carbide, ferro-alloy, copper and nickel refining. [Mitsubishi Kasei] *

Kurofan. Polyvinylidene chloride. [BASF plc]

Kurofan D. Polyvinylidene chloride dispersion. [BASF plc]*

Kurom 1. A jewellery alloy of copper with tin and cobalt.

Kuromoji oil. An oil from *Lindera* species; used in Japan for perfuming soaps and oils and contains α-phellandrene, nerolidol, linalol, and geraniol.

Kuron. CAS 93-72-1; Herbicide containing silvex as the active ingredient; herbicide used in ponds and other still water for the control of aquatic weeds, as well as control of brush on rangeland; also used industrially on railroads or under power lines for the control of weeds and brush. [Dow UK] *

Kurrodur. A proprietary trade name for an alloy of copper with 0.75% nickel and 0.5% silicon. §

Kurrol salts. Alkaline metaphosphates insoluble in water, but soluble in pyrophosphate solutions. They are produced by heating sodium trimetaphosphate or ethyl sodium phosphate.

Kusum oil. *See* kon oil.

kutch. *See* cutch.

Küttner silk. An artificial silk prepared by the viscose process.

K-Van. Potassium metavanadate. [Kerr-McGee Chemical Corp]

K-White. Aluminium tripolyphosphate; nontoxic white anticorrosive pigment used for anticorrosive coatings. [Bromhead & Denison Ltd]

Kwik Dri. Aliphatic hydrocarbon solvent; used in coatings, in fabric drycleaning, and in cold degreasing. [Ashland Chemical Company]

Kwikfill. Polyester two-part resin filler; for mending damaged body panels, filling dents, etc. [Hermetite Products Ltd] *

Kwik-Green. Nitrogen, sulfur, iron and zinc; used on turf, shrubs, trees, and potted plants to promote deep rich green foliage. [Lawn & Garden Products Inc]

Kymene® . Cationic wet-strength resins, including polyamide, polyamine, epoxide, and urea-formaldehyde resins; imparts strength to wet paper and paperboard; used primarily as internal additives in papermaking processes, but also as cationizing agents for starch added internally and as insolubilizing agents for starch in size press and pigmented coatings. [Hercules; Hercules Ltd]

Kymene® 109. Wet-strength resin for use in bleached or unbleached pulps and secondary fiber systems in paper industry. [Hercules]

Kymene® 435. Cationic, urea-formaldehyde resin; high-solids wet strength resin for paper and paperboard under acid pH conditions. [Hercules]

Kymene® 557H. Cationic, wet-strength resin; high-efficiency under acid or alkaline papermaking conditions; wet-strength resin, retention aid, and starch-cationizing agent, in tissue, toweling, and sanitary wadding grades, to wet-strength corrugating media, in linerboard and bag paper, surface applications, etc. [Hercules]

Kymene® 557LX. Environmentally compatible wet-strength resin for paper/paperboard applications. [Hercules]

Kynar® 301 F. CAS 24937-79-9; PVDF, crystalline high molecular weight; for solv.-based coatings; produces films with high resist. to gamma radiation and transparency to uv radiation. [Atochem N. Am./Plastics]

Kynar® 460. CAS 24937-79-9; PVDF; tough engineering resin with high abrasion resist. and stability in harsh thermal, chem., and uv environments; readily melt processable in molding and extrusion; used for coatings incl. corrosion-resist. coatings for chem. process equip.; long-life decorative finishes on building panels, film, filter cloth, instrumentation, and control equip. linings, membranes, static mixers, pipe and fittings, pumps, stock shapes, valves, elec./electronic jacketing. [Atochem N. Am./Plastics]

Kynar® 700 Series. CAS 24937-79-9; PVDF; *see* Kynar 460; 700 series avail. in wide range of visc. [Atochem N. Am./Plastics]

Kynar® 7200, 7201. Vinylidene fluoride-tetrafluoroethylene copolymer; easily melt processed by extrusion or molding into film, sheet, tube, monofilament, cable jackets, etc.; esp. useful for fiber optic applications. [Atochem N. Am./Plastics]

Kynar® Flex® 2800, 2801. CAS 24937-79-9; PVDF copolymer; used in wire and cable construction and other uses requiring high flexibility and improved impact resist. [Atochem N. Am./Plastics]

Kynar® Flex® 2850. Polyvinylidene fluoride-hexafluoropropylene copolymer; for various applications esp. wire and cable jacketing. [Atochem N. Am./Plastics]

Kynar® Flex® 2900. CAS 24937-79-9; Polyvinylidene fluoride copolymer; for wire and cable applications requiring extremely low smoke emission and low flame spread. [Atochem N. Am./Plastics]

Kynite. An explosive containing 24-26% nitroglycerin, 2-3% wood pulp, 32-321/2% starch, 31-34% barium nitrate, and 0-0.5% calcium carbonate.

Kynol. A highly cross-linked amorphous phenolic polymer. It resists temperatures up to 2500 C.

Kyolox BAT. Alkyl polyglycol ether; nonionic; colorless viscous liquid; washing, wetting and dispersing agent for all fibers and temperature ranges; used for kier boiling, bleaching, milling, and for after-soaping of dyed and printed materials. [Thor Chemicals (UK) Ltd]

Kypfarin. CAS 81-81-2; Warfarin. [Mechema Chemicals Ltd] *

kyrock. A rock asphalt consisting of sand with about 7% bitumen.

Kysite. A proprietary trade name for a phenol-formaldehyde synthetic resin with a fiber filler. §

Kytamer® PC. Chitosan PCA; water-sol. polymer. [Amerchol Corp]

KZ 55. Tetra (2,2 diallyloxymethyl) butyl, di (ditridecyl) phosphito zirconate, IPA; coupling agent. [Kenrich Petrochemicals]

KZ OPPR. Cyclo (dioctyl) pyrophosphato dioctyl zirconate, methyl naphthalene; coupling agent. [Kenrich Petrochemicals]

KZ TPP. Cyclo [dineopentyl (diallyl)] pyrophosphato dineopentyl (diallyl) zirconate, IPA; coupling agent. [Kenrich Petrochemicals]

KZ TPPJ. Cycloneopentyl, cyclo (dimethylaminoethyl) pyrophosphato zirconate, di mesyl salt; coupling agent. [Kenrich Petrochemicals]

K-Zinc. CAS 1314-13-2; Zinc oxide formulation; flowable nutrient for rice seed dressing. [Griffin]

---

* Trade name not verified as available   † Trade name verified as obsolete

L

L-55R® Acid Neutralizer. Aluminum-magnesium hydroxy carbonate; used for acid neutralization in the production of polyolefin resins. [Reheis Inc]

L-66. Sulfur chlorinated base; EP agent for use in threading and tapping operations, cutting and grinding oils. [Ferro/Keil]

labdanum. (ladanum). A resinous substance obtained from various species of *Cistus*; a stimulant expectorant.

Labitan. *See* Cormelian. [Degussa AG] *

Labiton. A proprietary preparation of thiamine hydrochloride, extract of Kola nuts, syrup and glycerophosphoric acid. [LAB Ltd] ‡

Labophylline. A proprietary preparation of theophylline and lysine; a respiratory and cardiac stimulant. [LAB Ltd] ‡

Laboprin. CAS 50-78-2, CAS 56-87-1; A proprietary preparation of aspirin and lysine; an analgesic. [LAB Ltd] ‡

Labosept. CAS 522-51-0; A proprietary preparation of dequalinium chloride; an antibacterial lozenge taken orally. [LAB Ltd] ‡

Labrafac Hydro WL 1219. Caprylic/capric triglycerides PEG-4 esters; hydrophilic oil for pharmaceutical and cosmetic formulations. [Gattefosse; Gattefosse SA]

Labrafac Lipophile WL 1349. CAS 65381-09-1; Caprylic/capric triglyceride. [Gattefosse]

Labrafil ISO. Isostearic ethoxylated glycerides; hydrophilic oil. [Gattefosse SA]

Labrafil M 1944 CS. CAS 97488-91-0; Apricot kernel oil PEG-6 esters; hydrophilic oil for pharmaceutical and cosmetic formulations. [Gattefosse; Gattefosse SA]

Labrafil M 1969 CS. Peanut oil PEG-6 esters; hydrophilic oil for pharmaceutical and cosmetic formulations. [Gattefosse; Gattefosse SA]

Labrafil M 1980 CS. Olive oil PEG-6 esters; hydrophilic oil for pharmaceutical and cosmetic formulations. [Gattefosse; Gattefosse SA]

Labrafil M 2125 CS. CAS 85536-08-9; Corn oil PEG-6 esters; hydrophilic oil for pharmaceutical and cosmetic formulations. [Gattefosse; Gattefosse SA]

Labrafil M 2130 BS. Hydrogenated palm/palm kernel oil PEG-6 esters; hydrophilic wax for pharmaceutical and cosmetic formulations. [Gattefosse; Gattefosse SA]

Labrafil M 2130 CS. Palm kernel oil, palm oil, PEG-6, and hydrogenated palm/palm kernel oil PEG-6 esters; hydrophilic wax for pharmaceutical and cosmetic formulations. [Gattefosse; Gattefosse SA]

Labrafil M 2735 CS. Triolein PEG-6 esters; hydrophilic hydrogenated oil; excipient for pharmaceutical and cosmetic formulations. [Gattefosse; Gattefosse SA]

Labrasol. CAS 85536-07-8; PEG-8 caprylic/capric glycerides; hydrophilic oil; excipient, solubilizer for pharmaceutical and cosmetic formulations; surfactant for microemulsions; wetting agent; penetration enhancer. [Gattefosse; Gattefosse SA]

Labrocol®. CAS 32780-64-6; Tablets containing 100 mg, 200 mg, and 400 mg labetalol hydrochloride; for the treatment of all forms of hypertension, and all grades of hypertension (mild, moderate, and severe) when oral antihypertensive therapy is desirable; also for treatment of angina pectoris when coexisting with hypertension. [Lagap Pharmaceuticals Ltd]

LABS-100. Linear alkylbenzene sulfonic acid; detergent intermediate. [Zohar Detergent Factory]

LABS 100/H.V. CAS 25496-01-9; EINECS 247-036-5; Tridecylbenzene sulfonic acid; higher visc. detergent intermediate. [Zohar Detergent Factory]

Labstix®. A proprietary test-strip used for the detection of pH, protein, glucose, ketones, and blood in urine. [Bayer AG]

Labstone. *See* Kemite.

lac. (gum lac, lacca, button lac, sheet lac, shellac). The resinous excretion of the lac insect, *Laccifer lacca*, cultivated in India, Burma, and Siam. The insects living on the twigs become surrounded with the lac, and in this form it is known as stick-lac; used as lacquer, a sealer coat under varnish, coat for floors and furniture.

Lacanite. A proprietary trade name for a shellac compound.§

lacca. *See* lac.

Laccain. A phenol-formaldehyde resin made with the aid of hydroxy acids, such as tartaric acid.

lac-dye. (lack-lack, dyer's lac). The coloring matter derived from lac (*qv*). Stick-lac contains 10% of coloring matter; used for dyeing wool mordanted with aluminum or tin salts.

L-Acid. *See* Laurent's acid.

Lacimoid. A proprietary trade name for a synthetic resin; used in laminated form for walls, etc. §

lac, Japanese. The lac obtained from *Rhus vernicifera*. It is a natural varnish or lacquer, and contains 85%, of urushic acid.

Lackmoid. The blue coloring matter obtained by heating resorcinol with sodium nitrite; used as an indicator in alkalimetry.

lacmus. Chemically pure litmus.

Lacolene. Aliphatic hydrocarbon solvent; used in coatings, adhesives and printing inks. [Ashland Chemical Company]

lacolin. *See* sodium lactate.

Lacorene. A proprietary trade name for a polystyrene molding resin. §

---

‡ Trade name and manufacturer not verified   § Trade name without identified manufacturer

Lacqran. A proprietary trade name for an ABS (*qv*) molding resin. §

Lacqrene 550. A proprietary polystyrene used in extrusion and injection molding. [Aquitaine-Organico] ‡

Lacqrene 635, 811, 835 and 836. Proprietary polystyrenes used to produce extrusions of differing tensile strengths. [Aquitaine-Organico] ‡

Lacqrene 740. An impact and heat resistant polystyrene suitable for use at 90 C. [Aquitaine-Organico] ‡

Lacqrene E. Antistatic polystyrene. [Aquitaine-Organico] ‡

Lacqsan 125 and 125L. Proprietary copolymers of styrene and acrylonitrile used in extrusion and injection molding. [Aquitaine-Organico] ‡

Lacqsan E. Antistatic styrene acrylonitrile. §

Lacqtene 1070 MN20, 1200 MN26. A proprietary low-density polyethylene used in injection molding. [Aquitaine-Organico] ‡

lacquer. Shellac dissolved in alcohol, and colored with saffron, annatto, or dragon's blood.

Lacril®. CAS 9004-65-3; Hydroxypropyl methylcellulose; ocular lubricant. [Allergan, Inc]

Lacri-Lube® NP, S.O.P®. White petrolatum, mineral oil; ocular lubricant. [Allergan, Inc]

Lacrinite. A proprietary casein-phenolformaldehyde product. §

lactase. An enzyme which converts lactose into d-glucose and d-galactose; used in biomedical research.

lactase. See β-galactosidase.

lacteol. (lactigen, lactilloids, lactobacilline, lactone). Preparations of lactic acid bacilli.

lactic acid. (2-hydroxypropanoic acid; 2-hydroxypropionic acid; milk acid) Organic acid; $CH_3CHOHCOOH$; CAS 50-21-5; EINECS 200-018-0; cultured dairy products, as acidulant, chemicals (salts, plasticizers, adhesives, pharmaceuticals), mordant in wool dyeing, food additive, manufacture of lactates. [Penta Mfg.; Pfanstiehl Labs; Purac Biochem BV]

lactic acid, ethyl ester. See ethyl lactate.

Lacticare. Contains lactic acid and sodium pyrrolidone carboxylate in an oil-in-water viscous lotion base; for the symptomatic relief of hyperkeratotic and other chronic dry skin conditions caused by low humidity or the use of detergents. [Stiefel Laboratories Inc; Stiefel Laboratories (UK) Ltd]

Lacticol 336. Specialty algin blend; milk-soluble stabilizer for milk-based systems. [Kelco Int'l. Ltd]

lactigen. See lacteol.

Lactile®. Sodium lactate, sodium PCA, hydrolyzed animal protein, fructose, urea, niacinamide, inositol, sodium benzoate, lactic acid; humectant, moisturing agent for creams and lotions. [Goldschmidt; Goldschmidt AG]

lactilloids. See lacteol.

Lactimon®. Solution of a partial amide and alkylammonium salt of a higher molecular weight unsat. polycarboxylic acid and a polysiloxane copolymer; wetting and dispersing additive to prevent settling of pigments; for solv. or solv.-free coating systems. [Byk-Chemie USA]

Lactin. CAS 63-42-3; Lactose, a sugar.

lactine. See vegetable butter.

Lactinium. A German preparation of neutral aluminum lactate; used an astringent and disinfectant.

Lactite. See Gallatite.

lactitis. A casein preparation containing borax and lead acetate. It is an artificial ivory.

Lactobacilline. See lacteol.

Lactodan B 30. Glyceryl stearate lactate; food emulsifier, improves aeration and foam stabilization. [Grindsted Prods. *

lactoflavin. See vitamin B_2.

lactoform. See Gallatite.

Lactoid. A casein preparation.

lactol. (lactonaphthol). The lactic acid ester of β-naphthol, $CH_3CHOH \cdot COO \cdot C_{10}H_7$; an intestinal astringent.

lactolin. (antimonin). The double salts of antimony lactate with alkalis, alkaline earths, and zinc salts. A convenient means for the transport of lactic acid. Also used as a substitute for tartar emetic in dyeing.

Lactolith. A casein preparation.

Lactoloid. A proprietary casein product. §

lactonaphthol. See lactol.

lactone. See lacteol.

Lactoprene. A patented vulcanizable synthetic rubber made from emulsified methyl or ethyl acrylate copolymerized with small quantities of a polyfunctional monomer such as butadiene, isoprene or allyl maleate; the copolymer is compounded with sulfur and accelerator and cured. §

lactorite. See Gallatite.

Lactosan. Dairy hygiene detergent sterilizer. [The Wellcome Foundation Ltd] *

lactose. (d-glucose, 4-O-β-d-galactopyranosyl-; milk sugar; saccharum lactis) Disaccharide; $C_6H_7O(OH)_4O-C_6H_7O(OH)_4$; CAS 63-42-3; pharmacy; infant foods; baking and confectionary; manufacture of penicillin, yeast; adsorbent in chromatography. [Dajac Labs; Penta Mfg.; Schweizerhall; Simonis BV]

lactose molasses. Molasses obtained from the preparation of milk sugar.

Lactozym. A β-galactosidase (lactase) preparation produced by submerged fermentation of a selected strain of the yeast Kluyveromyces fragilis; for sweet milk products, production of low lactose milk for persons suffering from lactose malabsorption; the milk can be consumed either directly or after condensing/drying. [Novo Nordisk]

lactulose. A sugar used in the treatment of hepatic coma and chronic constipation. It is 4-o-β -d-galactopyranosyl-d-fructose.

Ladalrod. Aluminum and aluminum alloy rod; used in alloying steel. [Reynolds Metal Co]

Laddok®. Bentazon, atrazine; herbicide for selective post-emergence control of broadleaf weeds [BASF AG]

Ladelloy. Ferro-alloy ladle additions. [Foseco (F.S.) Ltd]

Ladropen. Flucloxacillin; for treatment of infections caused by gram positive organisms. [Berk Pharmaceuticals Ltd]*

laevo-glucose. See levulose.

Laevuflex. A proprietary preparation of laevulose; used as a parenteral calorie supplement. [Geistlich Sohne AG] *

Lafil WL 3254. Polyglyceryl isostearostearate. [Gattefosse SA]

Lakeland AMA. Monosodium salt of an alkylamine dicarboxylate, as a clear liqud; used for traffic film removers and hard surface cleaners. [Lakeland Laboratories Ltd]

Lakeland AMA LF. Salt-free low-foam version of Lakeland AMA; low foam cleaners. [Lakeland Laboratories Ltd]

Lakeland C. Series of cationic emulsions of polyethylene wax and other polymers; used in textiles as softeners and lubricants. [Lakeland Laboratories Ltd]

Lakeland CAB. Amido betaine based on coconut alkyl

---

* Trade name not verified as available   † Trade name verified as obsolete

chain; viscosity modifier, solubilizer and foaming aid in cosmetic and toiletry formulations. [Lakeland Laboratories Ltd]

Lakeland CTA/N. Amino betaine based on coconut alkyl chain; amphoteric for production of mild shampoos, foam baths, hand cleaners; high foaming and detergency properties. [Lakeland Laboratories Ltd]

Lakeland N. Series of nonionic emulsions of polyethylene wax and other polymers; used in floor polish for nonslip, buffability, water resistance; in textiles as needle lubricants; in paper for scuff resistance, gloss, hardness, and water resistance. [Lakeland Laboratories Ltd]

Lakeland PA, PAE, PPE. Series of anionic surfactant phosphate esters in which the bases are nonylphenol ethoxylates, alcohol ethoxylates and alcohols; wetting agents, detergents, emulsifiers and lubricants which are stable to acid, alkaline and electrolyte conditions; used for heavy duty industrial cleaners; kier boiling of cottons; metal cleaning; various emulsification applications. [Lakeland Laboratories Ltd]

Lakewax. Series of nonionic emulsions of polyethylene wax and other polymers; used in paints and printing inks for scuff or rub resistance, slip properties, matt finish, water resistance. [Lakeland Laboratories Ltd]

Lalicopharsol. Mild emollient and emulsifier; used for fine cosmetics; pharmaceuticals; nonirritative barrier creams. [Solaver SA] *

Lalitecsol. Mild emollient and emulsifier; used for fine cosmetics; nonirritative barrier creams. [Solaver SA] *

Lamalgin. A thickening agent used for textile printing. [Degussa AG] *

Lambrex. Emulsion explosives (slurries), in which none of the ingredients is classified as an explosive; cap sensitivity is obtained by the mixing process; characterized by greatest handling safety, and can be supplied in cartridges or can be mixed in a pump truck on site and pumped into the boreholes. [Dynamit Nobel Wien GmbH] *

Lambrit (Anfo-explosives). Primed from the bottom of the borehole by Gelatine Donarit 1 and detonating fuse. [Dynamit Nobel Wien GmbH] *

Lamecreme LPM. Hydrogenated palm oil glycerides, cetyl alcohol, TEA-isostearyl hydrolyzed collagen; self-emulsifying raw material for emulsions, hair conditioners, creams, lotions. [Henkel/Cospha; Henkel Canada]

Lamecreme SA 7. POE stearoylether; emulsifier for cosmetics. [Grünau] *

Lamefin. A softening agent and lubricant; used by the textile industry for the after treatment of all types of fibers. [Degussa AG] *

Lamefix. Printing oils and fixation accelerators used for textile printing. [Degussa AG] *

Lamefix 680. Fatty acid polyglycol ether and PEG; accelerator for HT-fixation of polyester and triacetate. [Grünau]*

Lameform TGI. Polyglyceryl-3 diisostearate; emulsifier, emollient for w/o emulsions with good stability. [Henkel/Cospha; Henkel KGaA]

Lamefrost® Range. Emulsifier/stabilizer blends; emulsifier and stabilizer for ice cream. [Grünau]

Lamegin® DW 8000 HW. Diacetyl tartaric acid ester of mono/diglycerides; dough conditioner for bread and rolls. [Grünau; Henkel/Functional Prods.]

Lamegin® EE. CAS 68990-58-9; EINECS 273-612-0; Acetylated hydrogenated tallow glyceride; emulsifier

and plasticizer for cosmetic, food, and edible coatings. [Grünau]

Lamegin® GLP 10, 20. CAS 68990-06-7; Hydrogenated tallow glyceride lactate; emulsifier and plasticizer for cosmetics, foods, and drugs. [Grünau]

Lamegin® ZE 30, 60. CAS 68990-59-0; EINECS 273-613-6; Hydrogenated tallow glyceride citrate; emulsifier for cosmetic, margarine and meat industry. [Grünau]

Lamegum. A thickening agent used for textile printing. [Degussa AG] *

Lamemul® Range. Mono and diglycerides; food emulsifiers for baking additives, antistaling effect. [Grünau]

Lamephan. A softening agent and lubricant; used by the textile industry for the after treatment of all types of fibers. [Degussa AG] *

Lamepon. Dispersing and leveling agent, used for dyeing and stabilizer for peroxide bleaching in the textile industry. [Degussa AG] *

Lamepon PA-TR. CAS 68918-77-4; TEA-abietoyl hydrolyzed animal protein; mild detergent for shampoos, cleansers; skin compatible. [Henkel/Cospha; Henkel Canada; Henkel KGaA]

Lamepon S. CAS 68920-65-0; Potassium coco-hydrolyzed animal protein; mild detergent for personal care products, cleansers. [Henkel/Cospha; Henkel Canada; Henkel KGaA]

Lamepon ST40. CAS 68952-16-9; TEA-coco-hydrolyzed animal protein; mild detergent for shampoos, cleansers. [Henkel/Cospha; Henkel Canada; Henkel KGaA]

Lamepon UD. CAS 68951-92-8; Potassium undecylenoyl hydrolyzed animal protein; detergent for hair preparations, shampoos, skin cleansers for damaged skin, antidandruff shampoos. [Henkel/Cospha; Henkel Canada; Henkel KGaA]

Lameprint. A thickening agent used for textile printing. [Degussa AG] *

Lamequat L. Lauryldimonium hydroxypropyl hydrolyzed collagen; conditioning component for hair and body care preps. [Henkel/Cospha; Henkel Canada]

Lamequick® Range. Fat/emulsifier compounds; whipping base for toppings, desserts, other aerated foods. [Grünau]

Lamesoft LMG. Glyceryl laurate, TEA-coco-hydrolyzed collagen; refatting agent and thickener for foam bath and shampoos. [Henkel/Cospha; Henkel Canada]

Lamesorb® Range. Sorbitan esters of fatty acids/polysorbates; food emulsifier for o/w and w/o emulsions. [Grünau]

Lamex 173/FR. A self extinguishing type of the above resins complying with BS 476 Part I (Class II). [Croda Resins Ltd] †

Lamex 185. A trade name for a flexible amine preaccelerated polyester resin used for motor car body repairs. [Croda Resins Ltd] †

Lamex 186. A trade name for a rigid amine preaccelerated polyester resin used for motor car body repairs. [Croda Resins Ltd] †

Lamicoid. A proprietary trade name for a phenol-formaldehyde synthetic resin with a mica filler used for laminated products. §

Lamictal Tablets. CAS 84507-84-1; A proprietary formulation containing lamotrigine; anti-epileptic. [The Wellcome Foundation Ltd]

Laminac EPX-176. A self extinguishing (flame resistant)

---

‡ Trade name and manufacturer not verified   § Trade name without identified manufacturer

polyester resin; suitable for manufacture of reinforced plastics in the transportation industry. §

Lamitex. A proprietary trade name for a hard vulcanized fiber. §

Lampit® . Tetra hydro-3-methyl-4-(5-nitrofurfurylidene-amino)-1,4-thiazine 1,1-dioxide; for treatment of South American trypanosomiasis (Chagas disease). [Bayer AG]

Lamprene. CAS 2030-63-9; A proprietary preparation of clofazimine; an antileprotic. [Ciba plc] *

Lampronol. Leather dyes and finishes, pigments for plastics, dyes for printing inks. [ICI Chem. & Polymers Ltd]

Lamy butter. (Kanja butter, Kanga butter, Sierra Leone butter). The fat obtained from the seeds of *Pentadesma butyracea* of West Africa.

Lanacron. Wool dyestuffs. [Clayton Aniline Co Ltd] *

Lanain. *See* lanolin.

Lanaire. Scouring preparations. [Crosfield Chemicals Ltd]

Lanalin. *See* lanolin.

Lanamine® . CAS 10525-14-1; Mixed isopropanolamines myristate; high foaming mild detergent for shampoos, shaving soaps. [Amerchol Corp]

Lanaperl. Acid dyestuffs. [Hoechst AG]

Lanapex. Textile auxiliary chemicals. [ICI Chem. & Polymers Ltd]

Lan-Aqua-Sol 100. CAS 8039-09-6; PEG-75 lanolin; emulsifier for cosmetic and pharmaceutical emulsions; emollient, superfatting agent, conditioner for skin and hair care products, household detergents; solubilizer, wetting agent, dispersing aid. [Fanning]

Lanasol. Reactive dyes for wool and silk. [Clayton Aniline Co Ltd] *

Lanbritol. Self-emulsifying waxes. [Hickson & Welch Ltd]

Lancare. Household detergents and toiletries. [Harcros UK]

Lancer. CAS 37924-13-3; Flamprop-methyl; herbicide for wild oat control. [ICI Chem. & Polymers Ltd]

Lancosol. Mild emollient and emulsifier; used for fine cosmetics; nonirritative barrier creams. [Solaver SA] *

Landemul. Demulsifier components for oil field production chemicals. [Harcros UK]

Landromil. Ticlatone; antibacterial; antifungal. [Dorsey Pharmaceuticals] *

Lanesin. *See* lanolin.

Lanesta G. CAS 97404-50-7; Glyceryl lanolate; emollient, emulsifier; forms stable w/o emulsions. [Westbrook Lanolin]

Lanesta S. CAS 63393-93-1; EINECS 264-119-1; Isopropyl lanolate; nonsticky emollient rapidly absorbed by skin; lubricant; for lipsticks, nailcare, sunscreen preps., toilet soap; binder for pressed powds. [Westbrook Lanolin]

Lanestren. Dyestuffs for wool/polyester fiber blends. [BASF plc] *

Laneto 27. CAS 61790-81-6; PEG-27 lanolin; emulsifier, emollient, conditioner, moisturizer, stabilizer and solubilizer in makeup, lipstick, skin care products, bath products, shampoos, soap, shave creams, ointments, sun preps., veterinary products. [RITA]

Laneto AWS. CAS 68458-88-8; PPG-12-PEG-50 lanolin; auxiliary emulsifier, moisturizer, emollient for personal care products; plasticizer for hair spray, resins. [RITA]

Lanette. Self-emulsifying waxes. [Hickson & Welch Ltd]

Lanette 14. CAS 112-72-1; EINECS 204-000-3; Myristyl alcohol; emollient, consistency agent for cosmetic and pharmaceutical o/w and w/o creams, emulsions, sticks.

[Henkel/Cospha; Henkel Canada]

Lanette 16. CAS 36653-82-4; EINECS 253-149-0; Cetyl alcohol; emollient, consistency agent for cosmetic and pharmaceutical o/w and w/o creams, emulsions, sticks. [Henkel/Cospha; Henkel Canada]

Lanette 18 DEO. CAS 112-92-5; EINECS 204-017-6; Stearyl alcohol; emollient, consistency agent for cosmetic and pharmaceutical o/w and w/o creams, emulsions, sticks. [Henkel/Cospha; Henkel Canada]

Lanette E. CAS 59186-41-3; Sodium cetearyl sulfate; emulsifier and wetting agent for o/w emulsions, personal care products. [Henkel/Cospha; Henkel/Functional Prods.; Henkel Canada; Henkel KGaA]

Lanette N. Cetearyl alcohol and sodium cetearyl sulfate; emulsifying raw material for o/w creams and emulsions. [Henkel/Cospha; Henkel Canada]

Lanette O. Cetearyl alcohol; emollient, base, consistency factor for ointments, creams, emulsions. [Henkel/Cospha; Henkel KGaA]

Lanette SX. Cetearyl alcohol (90%) and sodium lauryl sulfate (10%); emulsifier; self-emulsifying base for mfg. of o/w ointment, creams, and emulsions. [Henkel/Cospha; Henkel Canada]

Lanette Wax. A proprietary trade name for a mixture of cetyl and stearyl alcohols. §

Lanette Wax Ester. A proprietary trade name for palmitic acid ester of cetyl and stearyl alcohols used in emulsions.§

Lanette Wax SX. A proprietary trade name for an emulsified mixture of cetyl and stearyl alcohols. §

Lanexol AWS. CAS 68458-88-8; PPG-12-PEG-50 lanolin; emollient, conditioner, superfatting agent, foam stabilizer, and lubricant for alcoholic and aq. compositions, plasticizer for hair sprays, o/w emulsifier, solubilizer. [Croda Inc.

Lanfrax® 1776. CAS 68201-49-0; EINECS 269-220-4; USP lanolin wax fraction; emulsifier, emollient; waxing agent, o/w auxiliary emulsifier; slip reducing agent in floor finishing compounds; w/o emulsion stabilizer and thickener. [Henkel/Cospha; Henkel Canada]

Langford Clay. CAS 1332-58-7; Hydrated aluminum silicate (kaolin clay); low cost reinforcer and inert filler for paint, paper, rubber, ceramics, plastics, and specialties. [R T Vanderbilt Co Inc]

Lanichol. *See* lanolin.

Laniol. *See* lanolin.

Lanital. A proprietary trade name for a casein textile fiber made by dissolving casein in a dilute alkaline solution and extruding the viscous compound in the form of thin filaments; these are treated with acid and rendered insoluble by means of formaldehyde. §

Lanitop. Medigoxin; used for congestive heart failure, cardiac arrhythmias. [Roussel Laboratories Ltd] †

Lankrocell® D15L. Blend of org. surfactants; mechanical foam promoter for PVC for carpet and floor backing applications. [Harcros UK]

Lankroflex®. Range of epoxidized fatty acid esters and oils; plasticizers for polymers. [Harcros UK]

Lankroflex® ED3. Octyl epoxy stearate; plasticizer. [Harcros UK]

Lankroflex® ED6. A proprietary epoxy plasticizer. [Harcros UK]

Lankroflex® GE. CAS 8013-07-8; EINECS 232-391-0; Epoxidized soya bean oil; plasticizer/stabilizer for epoxy.

---

* Trade name not verified as available   † Trade name verified as obsolete

[Harcros UK] *

Lankroflex® L. Epoxidized linseed oil; plasticizer/stabilizer for epoxy. [Harcros UK] *

Lankrol. Sulfated oils. [Harcros UK]

Lankrolan. Textile auxiliaries; shrink proofing agent for wool. [Harcros UK]

Lankroline. Sulfated oils and pigment finishes. [Harcros UK]

Lankrolyte. Sequestering agents. [Harcros UK]

Lankromark® . Range of mixed metal carboxylates, organotin compounds and organophosphites; stabilizers and antioxidants for polymers. [Harcros UK]

Lankromark® BM271. Butyltin carboxylate; transparent heat and light stabilizer for rigid and plasticized PVC applications [Harcros UK] *

Lankromark® BT050. Butyl thiotin; transparent heat stabilizer for rigid and plasticized PVC applications. [Harcros UK] *

Lankromark® BT120A. Dibutyltin mercaptide; stabilizer for PVC processing; excellent heat stability and clarity; for rigid sheeting, profiles, injection molding. [Harcros UK]*

Lankromark® DLTDP. CAS 123-28-4; Dilauryl thiodipropionate; stabilizer for polyolefins and other polymers; synergist with phenolic antioxidant or organophosphite. [Harcros UK] *

Lankromark® DP6404Z. Barium/zinc; self-lubricating stabilizer for suspension PVC for demanding semirigid formulations. [Harcros UK] *

Lankromark® DSTDP. CAS 693-36-7; Distearyl thiodipropionate; stabilizer for polyolefins and other polymers; synergist with phenolic antioxidant or organophosphite. [Harcros UK] *

Lankromark® LC68. Barium/cadmium; nonlubricating stabilizers for suspension PVC resins; LC310 grade also for emulsion resins. [Harcros UK] *

Lankromark® LC90. Cadmium/zinc; stabilizer/activator for chemically blown PVC; fast action. [Harcros UK] *

Lankromark® LC475. Cadmium-containing; stabilizer for suspension PVC resins; high efficiency, self-lubricating. [Harcros UK] *

Lankromark® LE65. CAS 101-02-0; Triphenyl phosphite; stabilizer for rigid and flexible PVC, PU; epoxy curing agent. [Harcros UK] *

Lankromark® LE109. CAS 26523-78-4; Tris (nonyl phenyl) phosphite; stabilizer for nontoxic rigid and flexible PVC; antioxidant for ABS/MBS, polyolefins, SBR. [Harcros UK] *

Lankromark® LE230. Substituted benzophenone; stabilizer for polyolefins, ABS, MBS, and SBR. [Harcros UK]*

Lankromark® LE285. CAS 1843-05-6; EINECS 217-421-2; 2-Hydroxy-4-octoxybenzophenone; uv absorber for PVC and polyolefins. [Harcros UK] *

Lankromark® LE296. CAS 131-57-7; EINECS 205-031-5; 2-Hydroxy-4-methoxybenzophenone; uv absorber for PVC and other polymers. [Harcros UK] *

Lankromark® LZ121. Barium/zinc; nonlubricating stabilizer for PVC emulsion resins. [Harcros UK] *

Lankromark® LZ187. Barium/zinc; stabilizer/activator for chemically blown PVC; slow action. [Harcros UK] *

Lankromark® LZ495. Calcium/zinc; nonlubricating stabilizer for PVC suspension and emulsion resins. [Harcros UK] *

Lankromark® LZ616. Barium/zinc; high efficiency self-lubricating stabilizer for PVC suspension resins; also for emulsion resins (rotational casting). [Harcros UK] *

Lankromark® LZ693. Barium/zinc; high efficiency nonlubricating stabilizers for PVC suspension and emulsion resins. [Harcros UK] *

Lankromark® LZ1034. Organic zinc/epoxy blend; plasticizer for stabilization of rigid PVC bottles. [Harcros UK]*

Lankromark® OT450. Octyl thiotin; transparent heat stabilizer for rigid and plasticized PVC applications; suitable for food contact applications. [Harcros] *

Lankromul. Oil spill dispersants. [Harcros UK] *

Lankroplast® . Lubricants in thermoplastic processing, antifogging agents for PVC film and viscosity modifiers for PVC pastes. [Harcros UK]

Lankroplast® L542. Tackifier for use in highly filled calendered flexible PVC formulations. [Harcros UK] *

Lankroplast® V2012. Visc. modifier for PVC plastisols. [Harcros UK] *

Lankropol®. Sulfated and sulfonated esters and acids; used as wetting agents, emulsifiers in textiles, paint and emulsion polymers. [Harcros UK]

Lankropol® KMA. CAS 6001-97-4; Sodium dihexyl sulfosuccinate, ethanol; emulsifier, wetting agent esp. in solutions of electrolytes; solubilizer for soaps, emulsion polymerization aid. [Harcros UK]

Lankropol® KO2. CAS 577-11-7; Sodium dioctyl sulfosuccinate, ethanol; wetting agent, emulsifier for emulsion polymerization. [Harcros UK]

Lankropol® KO Special. Sodium diisooctyl sulfosuccinate; clear pale straw liquid containing mineral oil; emulsifier and water carrier in solvents; used in hydrocarbon solvents, dry cleaning detergent formulations, and as a dewatering aid. [Henkel Chemicals Ltd] *

Lankropol® KPH70. CAS 577-11-7; Sodium dioctyl sulfosuccinate in propylene glycol/water; wetting agent, emulsifier, co-dispersant. [Harcros]

Lankropol® KSB 22. Monoester sulfosuccinate as a clear liquid; primary emulsifier in latex production. [Henkel Chemicals Ltd] *

Lankropol® KSG 72. Monoester sulfosuccinate as a clear liquid; low irritancy toiletry intermediate for shampoos, bubble baths, etc. [Henkel Chemicals Ltd] *

Lankropol® ODS. Disodium N-octadecyl sulfosuccinamate in liquid or paste form; foaming agent used for latex foam systems in carpet backing. [Henkel Chemicals Ltd] *

Lankrosol. Alkylaryl sulfonates, modified nonionics, phosphate esters; used as hydrotropes, wetters in electrolyte solutions. [Harcros UK]

Lankrosol HS101. Potassium salt of phosphate ester condensate; low foam hydrotrope for use in highly built liqs.; stable to acids, alkalis, electrolytes. [Harcros UK]

Lankrosol SXS. Anionic surfactant as a clear, pale straw liquid; solubilizing and viscosity modification agent used in high active liquid detergents and hard surface cleaners; also used for dye leveling in nylon dyeing. [Henkel Chemicals Ltd] *

Lankrosperse. Dispersing agents. [Harcros UK]

Lankrostat® 16. Antistats for plasticized PVC. [Harcros UK]*

Lankrostat® LME. Antistat for polyolefins, PS, crystal PS. [Harcros UK] *

Lankrothane. Polyurethane derivates. [Harcros UK]

Lannate®. CAS 16752-77-5; Methomyl insecticide. [DuPont UK]

Lanocerin®. CAS 68201-49-0; EINECS 269-220-4; Lanolin

wax; w/o emulsifier, emollient, conditioner used in cosmetics. [Amerchol Corp]

Lanogel® 21. CAS 61790-81-6; PEG-27 lanolin; emollient, emulsifier, dispersant, wetting agent, solubilizer, foam stabilizer, used in cosmetics, personal care products, pharmaceuticals, facial tissues, antiperspirants, germicidal hand soaps. [Amerchol Corp]

Lanogel® 31. Ethoxylated (40 mol) lanolin; nonionic surfactant, more water soluble than Lanogel 21. [Amerchol Corp]

Lanogel® 41. Ethoxylated (75 mol) lanolin; 50% aqueous gel, water souble, suitable for aqueous systems; conditioner in soaps and shampoos. [Amerchol Corp]

Lanogel® 61. Ethoxylated (85 mol) lanolin; nonionic 50% gel soluble in water and alcohol; conditioner in hydroalcoholic systems. [Amerchol Corp]

Lanogene®. Lanolin oil; emollient, moisturizer, and emulsifier which imparts oil sol. and spreading properties. [Amerchol Corp]

Lanoiac. Lanolin paint. [Croda Chem. Ltd] *

Lanoid. See Mulsoid. §

Lanol 14 M. CAS 59686-68-9; Myreth-3 myristate; cosmetic emulsifier. [Seppic]

Lanol 1688. EINECS 261-619-1; Cetearyl octanoate; cosmetic emulsifier. [Seppic]

Lanol P. CAS 4219-49-2; Glycol palmitate; cosmetic emulsifier. [Seppic]

Lanolex L-40. CAS 61790-81-6; PEG-50 lanolin; reforming agent, emollient for shampoos, hair conditioners. [Nihon Emulsion]

Lanolic Acid. Distilled lanolin fatty acids; cosolvent, emollient for skin and hair cosmetics, makeup, shaving foams and creams. [Croda Chem. Ltd]

lanolin. (anhydrous lanolin; wool wax; wool fat; lanalin; lanain; laniol; lanesin; lanichol) Derivative. of unctuous fatty sebaceous secretion of sheep consistg. of complex mixt. of esters of high molecular weight aliphatic, steroid, or triterpenoid alcohol and fatty acid; CAS 8006-54-0 (anhyd.), 8020-84-6 (hyd.); EINECS 232-348-6; ointments, leather finishing, soaps, face creams, facial tissues, hair set and sun-tan preps. [Amerchol; Croda; Henkel/Emery; Lanaetex; RITA; Stevenson Bros.; Westbrook Lanolin]

lanolin alcohol. (alcohols, lanolin; wool wax alcohol) Mixture of organic alcohols obtained from hydrolysis of lanolin; CAS 8027-33-6; EINECS 232-430-1; W/o emulsifier, stabilizer, softener, emollient, gelling agent, thickener, plasticizer, moisturizer for absorption bases, cosmetics, cleansing preps. [Chemmark; Croda; Henkel/Emery; Heterene]

Lanoline. A formulation of lanolin; an emollient to soothe and soften the skin. [The Wellcome Foundation Ltd]

Lanoplast. A proprietary cellulose acetate. §

Lanoquat® 1756. Quaternium-33, ethyl hexanediol; conditioner; emulsifier for skin moisturizers; substantive to hair; provides lubricating, conditioning, antistatic properties. [Henkel/Cospha; Henkel Canada]

Lanosol. Lanolized mineral oil. [Croda Chem. Ltd] *

Lanosterol. Lanosterol. [Croda Chem. Ltd]

Lanotein AWS 30. Propylene glycol, hydrolyzed animal protein, PPG-12-PEG-65 lanolin oil; conditioner, film former, lubricant, humectant, and emollient used in hair products. [Fanning]

Lanoxicaps. CAS 20830-75-5; A proprietary formulation of digoxin; for treatment of heart failure, atrial fibrillation, atrial flutter, and paroxysmal atrial tachycardia. [The Wellcome Foundation Ltd]

Lanoxin. CAS 20830-75-5; Proprietary formulation of digoxin; tablets, injections, oral solutions; for management of chronic cardiac failure and certain supraventricular arrhythmias, particularly atrial fibrillation and atrial flutter. [The Wellcome Foundation Ltd]

Lanoxine-PG. CAS 20830-75-5; A proprietary preparation of digoxin. [The Wellcome Foundation Ltd] †

Lanpharsol. Mild emollient and emulsifier; used for fine cosmetics; nonirritative barrier creams. [Solaver SA] *

Lanpol. Polyoxyethylated lanolin fatty acids. [Croda Chem. Ltd]

Lanpolamide 5. PEG-5 lanolinamide, PEG-5 lanolate; w/o emulsifier forming stable emulsions; emulsion stabilizer, corrosion inhibitor for aerosols. [Croda Inc.

Lanstar; Antistatic agents, emulsifiers; used for polystyrene; synthetic fibers; textile auxiliaries. [Lanstar Ltd] †

Lanstar NP2 and NP4. Nonylphenol ethoxylate nonionic surfactant in the form of an oil-soluble liquid; foam depressors; emulsifiers. [Lanstar Ltd] *

Lanstar NP40, NP50 and NP100. Nonylphenol ethoxylate nonionic surfactant in wax form; water-soluble surfactants effective at high temperatures and in concentrated electrolyte solutions; used in emulsion polymerization; latex stabilization; waxes and polishes. [Lanstar Ltd] *

Lanstar NP100/50. Nonylphenol ethoxylate in aqueous solution; nonionic surfactant effective at high temperatures and in concentrated electrolyte solutions; used in emulsion polymerization; latex stabilization; waxes and polishes. [Lanstar Ltd] *

Lanstar PCH, PC2 and PCO. Calcium petroleum sulfonate in liquid form; emulsifier, dispersing and wetting agent used in cutting oil, lube oil and fuel oil additives; rust preventatives; leather and textile industry. [Lanstar Ltd]*

Lanstar PS. Sodium petroleum sulfonate in liquid form; emulsifier, dispersing and wetting agent for cutting oil, lube oil and fuel oil additives; rust preventatives; leather and textile industry. [Lanstar Ltd] *

Lanstar PSW. Sodium petroleum sulfonate in liquid form; water-soluble anionic surfactant used in ore flotation; building industry; demulsification. [Lanstar Ltd] *

Lantecsol. Mild emollient and emulsifier for technical uses; rust inhibitor; bind material; for printing ink; rust inhibitors; petroleum; rubber adjuvants; lubricants; soaps; leather protection; painting industry. [Solaver SA] *

lanthana. See lanthanum oxide.

Lanthanol LAL. Sodium alkyl sulfoacetate in powder or flake form; anionic surfactant used in shampoos and bubble baths. [KWR Chemicals Ltd] *

lanthanum. Metallic element; La; CAS 7439-91-0; EINECS 231-099-0; Lanthanum salts, electronic devices, pyrophoric alloys, rocket propellants, reducing agent catalyst for conversion of nitrogen oxides to nitrogen in exhaust gases, phosphors in x-ray screens. [Aldrich; Atomergic Chemetals; Cerac; Rhone-Poulenc Basic]

lanthanum chloride. (lanthanum chloride anhydrous) LaCl$_3$; CAS 10099-58-8; EINECS 233-237-5; Anhydrous trichloride used to prepare the rare-earth metal. [Atomergic Chemetals; Cerac; Spectrum Chem. Mfg.]

lanthanum chloride anhydrous. See lanthanum chloride.

lanthanum nitrate. (lanthanum nitrate hexahydrate) La(NO$_3$)$_3$·6H$_2$O; CAS 10099-59-9; Antiseptic, gas

---

* Trade name not verified as available   † Trade name verified as obsolete

mantles. [Atomergic Chemetals; Cerac; Rhone-Poulenc Basic]

lanthanum nitrate hexahydrate. See lanthanum nitrate.

lanthanum oxide. (lanthana; lanthanum trioxide; lanthanum sesquioxide) La_2O_3; CAS 1312-81-8; EINECS 215-200-5; Calcium lights, optical glass, technical ceramics, cores for carbon-arc electrodes, fluorescent phosphors, refractories. [Atomergic Chemetals; Cerac; Rhone-Poulenc Basic; Seimi Chem.]

lanthanum sesquioxide. See lanthanum oxide.

lanthanum trioxide. See lanthanum oxide.

Lantrol® 1673. Lanolin oil; emollient and moisturizer for makeup, creams, lotions, hair care products, bath oils, medicinal preps., pigment dispersant. [Henkel/Cospha; Henkel Canada]

Lantrol® AWS 1692. CAS 68458-58-8; PPG-12-PEG-65 lanolin oil; emollient, plasticizer, solubilizer, and conditioner for hair products, shaving lotions, antiperspirants, body colognes. [Henkel/Cospha; Henkel Canada]

Lantrol® HP-2073. CAS 8006-54-0; Lanolin oil; emulsifier, emollient, conditioner, moisturizer, pigment dispersant for personal care products, pharmaceuticals. [Henkel/Cospha; Henkel Canada]

Lantrol® PLN. CAS 8039-09-6; Ethoxylated lanolin; cleaning and wetting agent; solubilizer for perfumes and germicides; conditioner for shampoos; superfatting agent for soaps; o/w emulsifier for nonfatty preps.; plasticizer for aerosols and hair sprays. [Pulcra SA]

Lanvis. CAS 154-42-7; A proprietary preparation of thioguanine; used in the treatment of acute myelogenous leukemia and acute lymphoblastic leukemia. [The Wellcome Foundation Ltd]

Lanxide. Ceramic matrix composites. [DuPont UK]

Lapis Smiridis. Emery. §

Lapix. A proprietary trade name for a flux used in steel molding; contains carbon and clay. §

Lapofloc. Water treatment chemicals. [Laporte Industries Ltd]

Laponite®. Synthetic clay products resembling hectorite with thixotropic gelling properties; used as gels in toothpastes and cosmetics; in surface coatings and sprays, conferring antistatic properties; used as retentive medium due to absorbancy. [Laporte Industries Ltd]

Laponite® B. Sodium magnesium fluorosilicate; imparts shear sensitive structure to waterborne formulations. [Laporte/Southern Clay]

Laponite® D. CAS 53320-86-8; Sodium magnesium silicate; used in conjunction with other thickeners for imparting a shear sensitive structure to clear gel and conventional toothpastes. [Laporte/Southern Clay]

Laponite® S. Sodium magnesium fluorosilicate, tetrasodium pyrophosphate; imparts a shear sensitive structure to waterborne formulations incl. paints and cleansers; may be coated onto paper or other surfaces to give elec. conductive films. [Laporte/Southern Clay]

Laponite® XLG. CAS 53320-86-8; Sodium magnesium silicate; inert base/carrier for act. ingreds.; suspending agent; promotes thixotropy giving stable suspensions; thickens cosmetic, toiletry creams, lotions, toothpaste products. [Laporte/Southern Clay]

Lapotan. CAS 10043-01-3; Basic aluminum sulfate. [Laporte Industries Ltd]

Lapudrine. CAS 15537-76-5; A proprietary preparation containing chlorproguanil hydrochloride; an antimalarial drug. [ICI Chem. & Polymers Ltd]

Laquanol. Water thinnable resins. [Croda Resins Ltd] †

Laractone®. CAS 52-01-7; Tablets containing 25 mg and 100 mg spironolactone BP; used in congestive heart failure, hepatic cirrhosis with ascites and edema, malignant ascites, nephrotic syndrome, diagnosis and treatment of primary hyperaldosteronism. [Lagap Pharmaceuticals Ltd]

Laraflex®. CAS 22204-53-1; Tablets containing 250 mg and 500 mg naproxen BP; for treatment of rheumatoid arthritis, osteoarthritis, ankylosing spondylitis, juvenile rheumatoid arthritis, acute gout and acute musculoskeletal disorders (such as sprains and strains, direct trauma, lumbrosacral pain, cervical spondylitis, tenosynovitis and fibrositis). [Lagap Pharmaceuticals Ltd]

Laratrim®. Tablets: 80 mg trimethoprim BP, 400 mg sulfamethoxazole BP and 160 mg trimethoprim BP, 800 mg sulfamethoxazole BP; suspension: 40 mg trimethoprim BP, 200 mg sulfamethoxazole BP in each 5 ml of suspension and 80 mg trimethoprim BP, 400 mg sulfamethoxazole BP in each 5 ml; used as an antibacterial agent effective against a wide range of Gram-positive and -negative organisms; for treatment of respiratory, genito-urinary, gastro-intestinal tracts and skin infections and other bacterial infections. [Lagap Pharmaceuticals Ltd]

Largactil. CAS 69-09-0; Chlorpromazine hydrochloride; a sedative. [Rhone-Poulenc Rorer Ltd]

Largon. CAS 1240-15-9; Propiomazine hydrochloride; sedative. [Wyeth Laboratories] *

Lariam. CAS 51773-92-3; Mefloquine hydrochloride; antimalarial. [Hoffmann-LaRoche Inc] *

Larnol. Rust preventative. [Croda Chem. Ltd] *

Larocin. CAS 26787-78-0; Proprietary preparation of amoxycillin; antibacterial. [Roche Products Ltd] *

Larodopa. CAS 59-92-7; A proprietary preparation of levodopa used in the treatment of Parkinson's disease. [Roche Laboratories; Roche Products Ltd]

Larodur®. Self-crosslinking, heat-curable polyacrylate resins; for single-coat baking finishes with good resist. to light, heat and chems.; for painting domestic appliances. [BASF AG]

Laroflex®. Vinyl chloride copolymers; for alkali- and acid-resist., lightfast, weather-resist. finishes on metal, concrete; binders for special gravure inks. [BASF AG]

Laromer®. Dipropylene glycol diacrylate; reactive diluent for radiation-curable systems [BASF AG] *

Laromer® POEA. CAS 48165-04-6; Phenoxyethyl acrylate; reactive diluent for radiation-hardenable systems [BASF AG] *

Laromer® TPGDA. CAS 68901-05-3; Tripropylene glycol diacrylate; reactive thinner for radiation-curing systems [BASF AG] *

Laromid®. Polyamides; hardeners for epoxy resins. [BASF AG]

Laromin®. Polyamines; hardeners for epoxy resins. [BASF AG]

Laropal®. Lightfast aldehyde and ketone resins; for surface coatings, production of pigment pastes, for flexographic and gravure inks. [BASF AG]

Larostat® 88. CAS 68308-67-8; Modified soyadimethylethyl ammonium ethosulfate; noncorrosive mold release agent; antistat; surface active. [PPG/Specialty Chem.]

Larostat® 143. Oleyldimethylethyl ammonium ethosulfate;

---

‡ Trade name and manufacturer not verified   § Trade name without identified manufacturer

surface active antistat. [PPG/Specialty Chem.

Larostat® 377 DPG. Lauric myristic dimethylethyl ammonium ethosulfate, dipropylene glycol; noncorrosive mold release agent; antistat; useful in polyurethanes; surface active. [PPG/Specialty Chem.

Larostat® 451. Stearyldimethylethyl ammonium ethosulfate; noncorrosive release agent forming a hard film; imparts gloss and antistatic properties; post treatment in polystyrenes and fiberglass; surface active. [PPG/Specialty Chem.

Larostat® JMR. Surface active antistat, lubricant. [PPG/Specialty Chem.

Larotid. CAS 26787-78-0; Amoxicillin; antibacterial. [SmithKline Beecham] *

Laroxyl. CAS 549-18-8; A proprietary preparation of amitriptyline hydrochloride; an antidepressive agent. [Roche Products Ltd] *

Larvacide. CAS 76-06-2; A proprietary trade name for chloropicrin; used as an insecticide. §

Larvex. A solution of sodium fluosilicate; a proprietary clothes-moth remedy. §

Larylin. Pharmaceutical preparations for cough and cold. [Bayer AG]

L.A.S. CAS 85536-07-8; PEG-8 caprylic/capric glycerides; nontoxic excipient for creams, lotions; surfactant for microemulsions. [Gattefosse SA]

Lasan. CAS 1143-38-0; Anthralin; antipsoriatic. [Stiefel Laboratories Inc] *

Laser. Emulsifiable concentrate containing 200 g cycloxydim per liter; used to control weeds in grass. [BASF plc]

Lasilium C. Lactoyl methylsilanol elastinate; provides hydrating, anti-inflammatory action, tissue regeneration, slimming action for hand creams, slimming products, toothpastes, cosmetic and health products. [Exsymol]

Lasilso. CAS 6834-92-0; EINECS 229-912-9; Sodium metasilicate. [Laporte Industries Ltd] *

Lasix. CAS 54-31-9; A proprietary preparation of furosemide; a diuretic. [Hoechst UK]

Lasix + K. A proprietary preparation of frusemide and potassium chloride; a diuretic containing a potassium supplement. [Hoechst UK] *

Lasma®. CAS 58-55-9; Theophylline; used for treatment of asthma. [Pharmax Ltd]

Lasonil®. Pharmaceutical preparation; for the treatment of sprains, contusions, bruises and varicose conditions. [Bayer AG]

Lasoven. A pharmaceutical preparation for the treatment of superficial venous thrombosis, superficial hematomas, sports injuries. [Bayer AG]

Lasso. 2-Chloro-2´,6´-diethyl-N-(methoxymethyl) acetanilide; pre-emergence herbicide. [Monsanto Co]

Lastex. Proprietary rubber latex threads spun with fiber. §

Lastil. Fungicide for bonded cork. [BDH Chemicals Ltd]

Lastilac. A proprietary molding compound. §

Latekoll®. Polyacrylic derivs.; thickener for polymer dispersions and latexes; for paints, adhesives, sealants, production of fiber webs. [BASF AG]

Latene. A solution of trimene base in rubber latex; a proprietary rubber vulcanization accelerator. §

Latensol AP8. A nonionic surfactant. [BASF plc] *

Latex Foam. A proprietary trade name for a type of cellular sponge rubber made from latex by a special method. §

Lathanol® LAL. Sodium lauryl sulfoacetate; emulsifier,

wetting agent, detergent, foaming agent, thickener used in cosmetics and personal care products. [Stepan; Stepan Canada; Stepan Europe]

Latibon®. CAS 544-17-2; Calcium formate; as feed additive and as silage additive. [Bayer AG]

Latkem. Two-component adhesive incorporating natural rubber latex for use in applications requiring high resilience and flexural strengths; excellent adhesion to steel and glass. [Feb Ltd]

Latol 4. Low-rosin grade fatty acid derived from tall oil; raw material for mfg. of emulsifiers used in disinfectants, cleaners and detergents; air-drying and baking alkyds; gloss oils and varnishes; toy enamels; metallic driers; core oils; masonry cements; flotation reagents; metal cleaners. [Climax Fluids Additives]

Latol 1550. Monocylcic C-21 dicarboxylic acid; biodeg. synthetic surfactant, coupling agent for surfactants; forms high solids fluid soaps; hydrotrope; for all-purpose household/industrial cleaners, disinfectant cleaners, laundry products; lubricity additive in metalworking formulations; curing agent for epoxy systems; also in concrete additives and adhesives. [Climax Fluids Additives]

Latol MOD. Blown tall oil; primary emulsifier for oil muds. [Climax Fluids Additives]

Laur 101A. CAS 63394-02-5; Silicone rubber disp.; used to render glass objects shatter resist.; fully compounded., ready to use; requires heat cure; long pot life. [Laur Silicone Rubber Compding.

Laur 676U. CAS 63394-02-5; Silicone rubber; used for o-rings; oil-resist., high tear strength, low compression set; peroxide-curable. [Laur Silicone Rubber Compding.

Laur Q-1330. CAS 63394-02-5; Silicone disp. base; used for coating of cloth and for elec. sleeving. [Laur Silicone Rubber Compding.

Laural. Fatty alcohol sulfates and ether sulfates; used for detergents, toiletries. [Atochem UK/Ceca]

Lauramide 11. Cocamide DEA; foam booster, thickener, superfatting agent. [Zohar Detergent Factory]

lauramide DEA. (lauric diethanolamide; N,N-bis(2-hydroxyethyl)dodecanamide; diethanolamine lauric acid amide) Mixture of ethanolamines of lauric acid; $CH_3(CH_2)_{10}CON(CH_2CH_2OH)_2$; CAS 120-40-1; 52725-64-1; EINECS 204-393-1; foam booster/stabilizer, detergency and visc. builder, emulsifier, wetting agent for personal care products, household and institutional detergents. [Chemron; Karlshamns; Mona; Norman, Fox; Sandoz; Scher; Sherex; Stepan; Witco/Humko]

Lauramide EG. Ethylene glycol stearate; opacifying agent. [Zohar Detergent Factory]

Laurel PEG 400 DT. CAS 61791-01-3; PEG-8 ditallate; emulsifier and solubilizer for mineral oils, fats, solvs.; for latex paints, metalworking fluids, industrial lubricants, textile specialties. [Reilly-Whiteman]

Laurel PEG 400 MO. CAS 9004-96-0; PEG-8 oleate; emulsifier for metalworking, paints, solvs., textile chem. specialties. [Reilly-Whiteman]

Laurel PEG 400 MT. CAS 61791-00-2; PEG-8 tallate; emulsifier for sol. oils, industrial lubricants; softener component for textile industry. [Reilly-Whiteman]

Laurel R-50. CAS 8002-33-3; EINECS 232-306-7; Sulfated castor oil; penetrant, lubricant, emulsifier used in detergent cleaners, metalworking lubricants, paint, textile lubricants, low-irritation and ethnic hair preps. and dyes, skin cleaners and lotions. [Reilly-Whiteman]

---

* Trade name not verified as available

† Trade name verified as obsolete

Laurel SBT. CAS 42808-36-6; Sulfated butyl tallate; industrial lubricant, emulsifier in textile formulations, rewetting agent for corrugated medium. [Reilly-Whiteman]

Laurel SD-101. Cocamide DEA (2:1); detergent, visc. builder, foamer for hard surface cleaners, laundry products, textile scouring, metal cleaners. [Reilly-Whiteman]

Laurel SD-400. CAS 93-83-4; EINECS 202-281-7; Oleamide DEA (2:1); emulsifier, visc. builder in metalworking fluids. [Reilly-Whiteman]

Laurel SD-520T. ErucamideTEA (2:1); surfactant, EP lubricant for metalworking formulations. [Reilly-Whiteman]

Laurel SMR. Sulfated methyl rapeseed ester; fiber-to-metal lubricant for textile fibers; improved wetting, high smoke pt. [Reilly-Whiteman]

Laurel SRO. CAS 617788-68-9; Sulfated rapeseed oil; used in metalworking lubricants and extreme pressure lubricants. [Reilly-Whiteman]

laurel nut oil. (Domba oil, Alexandrian laurel oil, poonseed oil, Tacamahac oil, Njamplung oil, Calaba oil, Dilo oil, N Dilo oil, Pinnay oil, U dilo oil). Calophyllum oil, from the nuts of *Calophyllum* species; used for medicinal and illuminating purposes, poisonous.

laurel oil. Bayberry oil, obtained from the berries of the laurel tree, *Laurus nobilis*.

Laurelphos 39. Phosphate ester, free acid; surfactant for oil and water-based metalworking lubricants and synthetic cutting fluids; EP lubricant; neutralized with caustic as wetting agent. [Reilly-Whiteman]

Laurelphos 400. CAS 39464-69-2; Aromatic phosphate ester, free acid; lubricant additive, emulsifier, rust inhibitor for metalworking lubricants. [Reilly-Whiteman]

Laurelphos RH-44. CAS 52623-95-7; Aliphatic phosphate ester, free acid; detergent, hydrotrope, visc. builder for strong alkali formulations; deduster for powd. alkalis. [Reilly-Whiteman]

Laurelterge 837, 1390. Detergent for textile scouring. [Reilly-Whiteman]

Laureltex 308, 308 LF. Detergent for textile scouring. [Reilly-Whiteman]

Laurel wax. (myrtle wax, bayberry tallow). Myrtle berry wax, obtained from *Myrica cerifera*.

Laurent's acid. (L-acid). l-Naphthyl-amine-5-sulfonic acid; $C_{10}H_6(OH)SO_3H$; sed as an azo dye.

Laurent's aluminum solder. The hard solder contains 63-74% zinc and 19-30% tin. The soft solder contains 60-70% zinc, 16-27% tin, and 12% lead.

Laurent's naphthalldinic acid. *See* Laurent's acid.

Laureol. *See* vegetable butter.

Laurex®. Higher fatty alcohols. [Albright & Wilson Ltd, Detergents Div] *

Laurex®. A proprietary trade name for a series of primary fatty alcohols. [Marchon France SA] *

Laurex® CS; Cetearyl alcohol BP; mfg. of surfactants; raw material for ethoxylation, sulfation, etc.; stabilizer in emulsion polymerization; lubricant in rigid PVC, also for pharmaceutical creams, hand lotions, bath oils, shaving creams. [Albright & Wilson Am.; Albright & Wilson UK]

lauric acid. (n-dodecanoic acid; dodecanoic acid; dodecoic acid) Fatty acid; $CH_3(CH_2)_{10}COOH$; CAS 143-07-7; EINECS 205-582-1; alkyd resins, wetting agents, soaps, detergents, cosmetics, insecticides, food additives. [Akzo; Henkel/Emery; Mirachem Srl; Unichema; Witco/Humko]

lauric diethanolamide. *See* lauramide DEA.

Lauridit. Foam stabilizers and skin protecting additives; for liquid and powder formulations. [Akzo Chemie Nederland BV] *

Laurodin. Bactericidal preparations of laurolinium acetate. [Allen & Hanburys Ltd] *

Laurox®. CAS 105-74-8; Dilauroyl peroxide; initiator for elevated-temp. polyester cures, for production of PVC resins, and acrylates. [Akzo] *

lauroyl chloride. (dodecanoyl chloride) $CH_3(CH_2)_{10}COCl$; CAS 112-16-3; EINECS 203-941-7; Surfactant, polymerization initiator, antienzyme agent, foamer; synthesis of lauroyl peroxide, sodium lauroyl sarcosinate, other sarcosinates. [Atochem N. Am.; Hüls AG; PPG Industries]

lauroyl sarcosine. (N-methyl-N-(1-oxododecyl) glycine) N-lauryl derivative. of N-methylglycine; $CH_3(CH_2)_{10}CON$-CH_3CH_2COOH; CAS 97-78-9; EINECS 202-608-3; detergent, corrosion inhibitor, foam booster and stabilizer, wetting agent, lubricant, emulsifier used in dentifrices, personal care, and household cleaning products, pharmaceuticals, metal processing and finishing, metalworking and cutting oils. [W.R. Grace/Hampshire; R.T. Vanderbilt]

Laur Red #10 Silicone Pigment. Iron oxide pigment in silicone rubber base; features excellent handling, improved heat stability, good dispersion. [Laur Silicone Rubber Compding. *

Laurydol. CAS 105-74-8; Lauroyl peroxide. [Akzo Chemie UK Ltd]

Laurydone®. CAS 30657-38-6; Lauryl PCA; emulsifier in oily or aq. continuous phase; lipophilic moisturizer for skin care products, hair care products, toiletries, makeup; exceptionally soft touch. [UCIB]

lauryl acrylate. (dodecyl acrylate) Monomer; $CH_2=CH$-$COOC_{12}H_{25}$; CAS 2156-97-0; EINECS 218-463-4; Uv-curable reactive diluent in inks and coatings, adhesives, visc. index improver, finishing aid for leather. [CPS; Sartomer]

lauryl alcohol. (1-dodecanol; C_{12} linear primary alcohol; dodecyl alcohol) Fatty alcohol; $CH_3(CH_2)_{10}CH_2OH$; CAS 112-53-8; 68526-86-3; EINECS 203-982-0; manufacture of sulfuric acid esters which are used as wetting agents; synthetic detergents, lube additives, pharmaceuticals, rubber, textiles, perfumes, flavoring agents. [Ethyl; M. Michel; Procter & Gamble; Vista]

lauryl dimethylamine. *See* dimethyl lauramine.

m-lauryl mercaptan. *See* n-dodecyl mercaptan.

laurylpyridinium chloride. (1-dodecylpyridinium chloride) Quaternary ammonium compound.; $C_5H_5NClC_{12}H_{25}$; CAS 104-74-5; EINECS 203-232-2; Cationic detergent, dispersing and wetting agent; ingredient of fungicides and bactericides. [Schweizerhall]

Lausofan. A hexamethylene ketone; used for destroying insects of the vermin type.

Lautal. A proprietary trade name for an aluminum alloy containing 4% copper and 2% silicon. §

Lauth's violet. (thionine). Aminoiminoiminodiphenyl sulfide, $C_{12}H_9N_3S$; used as a microscopic stain.

Lavacol. Rubbing alcohol; rubefacient. [Parke-Davis] *

Lavalloy. A proprietary trade name for a ceramic product made from mullite and alumina. §

lavarock. A heat-treated steatite.

lavasul. A mixture of 40% coke dust with sulfur; used for tank linings.

---

‡ Trade name and manufacturer not verified   § Trade name without identified manufacturer

Lavema. CAS 115-33-3; Oxyphenisatin acetate; laxative. [Sterling Drug Inc] *

lavender drops. Compound tincture of lavender.

Laventin® CW. Alkanol polyglycol ether; water-free wetting agent and detergent used in textile industry. [BASF AG]

lavite. A heated steatite product.

Lavrex. Fatty alcohols. [Surfachem Ltd]

Lawinit 100. Emulsion explosives (slurries), in which none of the ingredients is classified as an explosive; cap sensitivity is obtained by the mixing process; characterized by greatest handling safety; can be supplied in cartridges or can be mixed in a pump truck on site and pumped into the boreholes; special explosive for blasting avalanches. [Dynamit Nobel Wien GmbH] *

Lawn Food. Lawn fertilizer. [Fisons plc, Horticulture Div] *

Lawn Plus. Fertilizer and selective weedkiller. [ICI Chem. & Polymers Ltd]

Lawnsman. Range of lawn aids for garden use such as fertilizers, weedkillers, and a spreader. [ICI Garden Products]

Lawnsman Mosskiller. Mosskiller for lawns. [ICI Chem. & Polymers Ltd]

Lawnsman Spring Feed. Spring/summer lawn food. [ICI Chem. & Polymers Ltd]

Lawnsman Weed and Feed. Fertilizer combined with selective weedkiller. [ICI Chem. & Polymers Ltd]

Lawnsman Winterizer. Autumn lawn feed. [ICI Chem. & Polymers Ltd] *

Lawn Spot Weeder. Selective weedkiller aerosol. [Fisons plc, Horticulture Div] *

Lawnturf®. Artificial turf made from polyvinylidene chloride filament, etc. [Asahi Chem. Industry]

Lawn Weed Gun!; Contains 2,4-D and dicamba; ready-for-use herbicide spray for control of common lawn weeds. [ICI Garden Products]

Lawn Weedkiller. Selective weedkiller . [Murphy Chemical Ltd] †

Lawn Weeds Killer. Selective weedkiller. [Fisons plc, Horticulture Div] *

lawsone. CAS 481-39-0; Hydroxynaphthoquinone, $C_{10}H_6O_3$, the coloring matter of henna leaves.

LAX-1C. Aloin, P.E. cascara sagrada, P.E. belladonna (total alkaloids 0.04 mg), P.E. rhubarb, ginger; a laxative for temporary relief of simple constipation. [Pharmaceutical Basics Inc] *

LAX-2C. Aloin, P.E. cascara sagrada, P.E. belladonna (total alkaloids 0.08 mg), P.E. rhubarb, ginger; a laxative for temporary relief of simple constipation. [Pharmaceutical Basics Inc] *

LAX-3. Aloin, podophyllum resin, belladonna extract (total alkaloids 0.1 mg), strychnine sulfate; a laxative for temporary relief of simple constipation. [Pharmaceutical Basics Inc] *

LAX-4. White phenolphthalein, P.E. cascara sagrada, pepsin, aloin, podophyllum resin, diastase of malt, ginger; a laxative for temporary relief of simple constipation. [Pharmaceutical Basics Inc] *

LAX-42. Yellow phenolphthalein, P.E. cascara sagrada, pepsin, aloin, podophyllum resin, diastase of malt, ginger; a laxative for temporary relief of simple constipation. [Pharmaceutical Basics Inc] *

Laxans. See Purgen. §

Laxatin. See Purgen. §

Laxatol. See Purgen. §

Laxatoline. See Purgen. §

Laxen. See Purgen. §

Laxiconfect. See Purgen. §

Laxin. See Purgen. §

Laxoberal. A proprietary preparation containing sodium picosulfate; a laxative. [Windsor Healthcare Ltd]

Laxoin. See Purgen. §

Laxophen. See Purgen. §

Layor Carang. See agar-agar.

Laysa. Acoustic absorbers. [Carl Freudenberg] *

Laysa Plan. Acoustic absorbers. [Carl Freudenberg] *

Laytex. A proprietary insulating material derived from rubber latex. §

LCA-1. CAS 25928-94-3; Filled gyro-grade epoxy resin adhesive; adhesive with low coefficient of linear expansion; for bonding and sealing aluminum and other metals; mix ratio: 1.07 phr with Activator BA-4. [Bacon]

LCA-20. CAS 25928-94-3; Filled gyro-grade epoxy resin adhesive; semiflexible adhesive for sealing joints between dissimilar metals difficult to seal with regular gyro-grade adhesives; mix ratio: 27.4 phr with Activator BA-40. [Bacon]

LCA-127. CAS 25928-94-3; Epoxy adhesive; thermally conductive, elec. insulating. [Bacon]

LCP-20CF/000. Liq. crystal polymer, 20% carbon fiber. [Compounding Tech.] *

LDPE. See polyethylene, low-density.

L-DTT. See dithiothreitol.

lead. Metallic element; Pb; CAS 7439-92-1; EINECS 231-100-4; storage batteries, gasoline additive, radiation shielding, cable covering, ammunition, chemical reaction equip., solder and fusible alloys. [Asarco; Cerac; Noah Chem.]

lead acetate. (acetic acid, lead salt; sugar of lead; salt of Saturn) Inorganic salt; $[CH_3COO]_2Pb$; CAS 301-04-2; dyeing of textiles, waterproofing varnishes, insecticides, lead driers, chrome pigments, hair dye, weighting silks. [Am. Biorganics; Cerac; Chemetall GmbH; Hoechst Celanese; Mallinckrodt]

lead ashes. The skimmings formed during the melting of lead. It consists mainly of oxide.

lead bronze. An alloy of from 70-90% copper, 6-16% lead, and 4-13% tin.

lead carbonate (basic). (white lead; dibasic lead carbonate; lead subcarbonate) $2PbCO_3 \cdot Pb(OH)_2$; CAS 598-63-0; EINECS 215-290-6; exterior paint pigment, ceramic glazes. [Halstab]

leaded bronze. An alloy of 88.5% copper, 10% zinc, and 1.5% lead. Another source gives the following proportions: 80% copper, 10% tin, and 10% lead. It has a melting point of 945 C.

leaded gun metal. An alloy of 85.5% copper, 2% zinc, 9.5% tin, and 3% lead.

leaded zinc oxides. Pigments containing 20-35% lead sulfate and 60-80% zinc oxide.

lead flake. See white lead.

lead fluoborate. $B_2F_8 \cdot Pb$; CAS 13814-96-5; salt for electroplating lead. [Atochem N. Am.; Atomergic Chemetals; Hoechst Celanese; M&T Harshaw]

lead metasilicate. See lead silicate.

lead molybdate. $PbMoO_4$; CAS 10190-55-3; Analytical chemistry, pigments. [AAA Molybdenum Prods.; Atomergic Chemetals; Cerac]

lead nitrate. $Pb(NO_3)_2$; CAS 10099-74-8; EINECS 233-245-

9; Lead salts, mordant for dyeing and printing textiles and staining mother of pearl, matches, oxidizer in dye industry, sensitizer for photography, explosives, tanning, process engraving, lithography. [Blythe, William Ltd; Cerac; Mallinckrodt; Noah Chem.; Spectrum Chem. Mfg.]

lead ocher. Lead monoxide, PbO, found naturally.

Leadoxe. Lead dioxide; polysulfide sealant. [Mechema Chemicals Ltd] *

lead oxide. (lead suboxide; lead oxide, black; litharge, leaded) Pb_2O; In storage batteries. [Chemson GmbH; Hammond Lead Prods.; Nihon Kagaku Sangyo]

lead oxide, black. See lead oxide.

lead silicate. (lead metasilicate) $O_3Si \cdot Pb$; CAS 10099-76-0; Ceramics, fireproofing fabrics, paints, electrode position process in the automatic industry. [Eagle-Picher; Hammond Lead Prods.]

lead solder. An alloy of 50% lead and 50% tin used for soldering lead.

lead stearate. (stearic acid, lead salt) $Pb(C_{18}H_{35}O_2)_2$; CAS 1072-35-1; varnish and lacquer drier; in extreme pressure lubricants; lubricant in extrusion processes; stabilizer for vinyl polymers; corrosion inhibitor for petroleum; component of greases, waxes, and paints. [Ore & Chem. Corp.; Syn. Prods.; R.T. Vanderbilt]

lead subcarbonate. See lead carbonate (basic).

lead suboxide. See lead oxide.

lead tungate. A preparation obtained from lead acetate and tung oil; used as a drier in the preparation of paints.

lead vinegar. A basic lead acetate, $Pb(C_2H_3O_2)_2 \cdot PbO \cdot 2H_2O$.

lead water. A 1% solution of basic lead acetate.

Leak Detector. Aqueous special soap solutions; bubble forming fluids that detect and indicate leaks in any pressurized gas piping. [Dylon Industries Inc] *

lean coal. Coal of the poorest quality.

Leantin. A proprietary trade name for a bearing alloy of lead and tin. §

Leatherlubric. A proprietary trade name for a sulfonated sperm oil. §

Lebanon No. 34. A proprietary trade name for an alloy of 20% chromium, 30% nickel, 3.25% molybdenum, 5% copper, and 3.25% silicon; stated to be resistant to hydrochloric and sulfuric acid. §

Lebanon No. 48. A proprietary trade name for an alloy of 30% chromium, 30% nickel, 0.4% carbon, and the remainder iron. §

Lebaycid®. CAS 55-38-9; Fenthion; insecticidal spray used for control of biting and sucking pests and particularly effective against fruit flies, leafhoppers, cereal bugs and rice stem borers. [Bayer AG]

Lebbin Salt. A proprietary mixture containing nitrite; used for curing meat. §

lecin. An iron albuminate, stated to be an easily assimilable form of iron. It contains 20% albumin and 0.6% iron.

lecipon. A 10% lecithin in a soluble form.

Lecitase. A commercial preparation of phospholipase A-2 (phosphatide-2-acyl-hydrolase, E.C.3.1.1.4), manufactured from porcine pancreatic glands; used for the hydrolysis of lecithins of both animal and vegetable origin for improvement of emulsifying power. [Novo Nordisk]

Lecithan. A trade name for lecithin. §

lecithcerebrin. Lecithin prepared from brain substance.

lecithin. (yolk powder) Mixture of the diglycerides of stearic, palmitic and oleic acids linked to the choline ester of

phosphoric acid; found in plants and animals; $C_8H_{17}O_5NRR'$, R and R' are fatty acid groups; CAS 8002-43-5; EINECS 232-307-2; edible surfactant and emulsifier for food use, pharmaceuticals, cosmetics, leather treatment, textiles. [Am. Lecithin; Central Soya; W.A. Cleary; Duphar BV; Lucas Meyer GmbH; Spice King; U.S. Biochemical]

Lecithin L-Range. CAS 8002-43-5; EINECS 232-307-2; Lecithin; food emulsifier and coemulsifier for baking additives. [Grünau]

Lecithin Water Dispersible CLR. CAS 8002-43-5; EINECS 232-307-2; Hydrolyzed soya lecithin; mild refatting agent, emollient for aq. skin and hair care preparations, face cleansers, shampoos. [Dr. Kurt Richter; Henkel/Cospha]

lecithmedullan. A lecithin preparation from bone marrow.

lecithol. An emulsion of brain lecithin containing 1.5% lecithin.

Leclo. CAS 7758-95-4; Lead chloride. [Mechema Chemicals Ltd] *

Lectricon. A proprietary trade name for alkyl ammonium compounds dispersed in mixed aromatic/aliphatic oils with film strength agents; for electrostatically spraying onto form work for the casting of concrete. [British Solvent Oils] *

Lectro Cast. A proprietary trade name for an alloy of iron with 2.75% nickel, and 0.7% chromium. §

lecutyl. A combination of lecithin and copper cinnamate, containing 1.5% copper.

Ledac. CAS 301-04-2; Lead acetate. [Mechema Chemicals Ltd] *

Ledate®. Lead dialkyldithiocarbamate; accelerators for vulcanization of rubber. [R T Vanderbilt Co Inc]

Ledca. CAS 598-63-0; Lead carbonate. [Mechema Chemicals Ltd] *

Leddel alloy. An alloy of 86% zinc, 9.5% antimony, and 4.5% copper. Another alloy used for bearings consists of 87.5% zinc, 6.25% copper, and 6.25% aluminum.

Ledeburite. Austenite-cementite eutectic.

Ledebur's metal. Bearing metals. One contains 85% zinc, 10% antimony, and 5% copper. Another consists of 77% zinc, 17.5% tin, and 5.5% copper.

Ledercillin VK. CAS 132-98-9; Penicillin V potassium; antibacterial. [Lederle Laboratories USA] *

Ledercort. CAS 124-94-7; A proprietary preparation of triamcinolone; a steroid. [Lederle Laboratories] *

Lederfen. CAS 36330-85-5; Proprietory preparation containing fenbufen; nonsteroidal anti-inflammatory. [Lederle Laboratories] *

Ledermix. Combination dental kit consisting of a paste and cement with democlocyline and triamcinolone acetonide and a hardener; dental treatment combining antibiotic and anti-inflammatory agents. [Lederle Laboratories] *

Ledermycin. A proprietary preparation containing demethylchlortetracycline; an antibiotic. [Lederle Laboratories] *

Lederplex. A proprietary preparation of vitamin B complex. [Lederle Laboratories] *

Lederspan. CAS 5611-51-8; A proprietary preparation of triamcinolone hexacetonide; a steroid injection. [Lederle Laboratories] *

Ledfo. CAS 811-54-1; Lead formate. [Mechema Chemicals Ltd] *

Ledmin LPC. CAS 104-74-5; EINECS 203-232-2; A propri-

---

‡ Trade name and manufacturer not verified   § Trade name without identified manufacturer

etary trade name for lauryl pyridinium chloride; an emulsifier for waxes giving bacteriostatic polishes. [Hoechst UK] *

Ledni. CAS 10099-74-8; Lead nitrate. [Mechema Chemicals Ltd] *

Ledrite Brass. A proprietary trade name for a leaded brass containing 60-63% copper and 2.5-3.7% lead. §

Leecure B Series. CAS 7637-07-2; Boron trifluoride-based; hardeners, epoxy curing agents providing water, heat and chem. resist. compounds with high physical strength and elec. props.; used in electronic potting compounds, elec. varnishes, adhesives, filament winding, fiberglass composites, prepregs. [Leepoxy Plastics]

Leefex. A hop defoliant. [Plant Protection] *

Leegen®. A blend of a sulfonated petroleum product and selected mineral oil on an inert carrier; an effective dry foam plasticizer and processing aid for NR, SBR, IR, and IIR rubbers. [R T Vanderbilt Co Inc]

lees. Yeast and various suspended matters of the must produced during the fermentation of grape juice.

Leffmann and Beam's glycerol reagent. For Reichert-Meissl number. It is a mixture of 180 cc pure glycerol and 20 cc of 50% sodium hydroxide solution.

Legion. Lube oil. [Chevron] *

Legion. CAS 83601-81-4, 12427-38-2; A suspension concentrate containing 50 g carbendazim and 320 g maneb per liter; systemic fungicide for cereals. [Tripart Farm Chemicals Ltd]

Legumex Extra. A solution concentrate containing 27 g benazolin, 237 g 2,4-DB and 42.3 g MCPA per liter; a post-emergence herbicide. [Schering Agrochemicals Ltd]

Legupren®. A foam system based on Legural (unsaturated polyester resins) and a blowing agent for the system; for combination with light fillers to produce Legupren-based lightweight concrete for use in building construction. [Bayer AG]

Leguval®. Unsaturated polyester resins for the production of SMC, BMC, and DMC, etc.; used for the manufacture of glass reinforced corrugated sheet, light domes and light dome supports, cable distribution boxes, covers, light housings, bulk containers, boats, swimming pools, and vehicle bodies. [Bayer AG]

Leinsaat oils. Linseed oil obtained by extracting the residues from the presses.

Lekutherm®. A range of epoxy resins and hardeners; for the production of adhesives, for use in the construction of models, form plates and core boxes used in foundries, for jigs and tool construction. [Bayer AG]

Lemarquand's alloy. An alloy made from 75% copper, 14% nickel, 2.0% cobalt oxide, 1.8% tin, and 7.2% zinc. It is stated to be very resistant to oxidation.

Lembergite. An artificial hydrous aluminum-sodium silicate.

Lemberg's solution. A solution of aluminum chloride and extract of log wood. It colors calcite mineral surfaces violet.

Lembor. Lemon-scented cream, light duty hand cleanser; suitable for both factory and office washrooms. [Borax Consolidated Ltd] *

Lemco. A proprietary meat extract. §

Lemnian earth. (Lemnos earth). A red, yellow, or grey, earthy substance consisting of a hydrated silicate of aluminum, found at Lemnos. It is an ocher, and is used as a pigment.

Lem-O-Fos®. CAS 7758-29-4; Sodium tripolyphosphate; food additive for meat, poultry, and seafood industries. [Rhone-Poulenc Food Ingreds.

Lemol. Lemon oil substitute for flavoring. [Bush Boake Allen Ltd]

Lemolac. A proprietary preparation; a very light mercurous chloride. §

lemon chrome. See barium chromate.

Lemon Delph. Cleansing milk and skin freshener, for daily care of the skin. [Richardson-Vicks Inc] *

lemon lac. (orangelac, garnet lac). Terms referring to the color of lac, determined to some extent upon the tree from which it is obtained.

Lemon Plus. Throat drops. [Richardson-Vicks Inc] *

Lendorm. CAS 57801-81-7; Brotizolam; hypnotic. [Boehringer Ingelheim Pharmaceuticals Inc] *

Leneta. Paint testing panels and opacity charts; for testing of paints and inks. [Cornelius Chemical Co Ltd] *

Lenetol. Textile auxiliary chemicals. [ICI Chem. & Polymers Ltd]

Lenium. A proprietary preparation of selenium sulfide and bithionol; a skin cleanser. [Bayer AG] †

Lenka. Strongly alkaline detergent powder. [Shell UK] *

Lensine. Alkali detergent powders. [Shell UK] *

Lentagran. CAS 35512-33-9; Wettable powder containing 45% w/w pyridate; for annual weed control for cereals, oilseed rape and maize. [Ciba-Geigy Agrochemicals]

lentana. A red earthy ironstone of the type commonly called Laterite.

Lentard. Insulin zinc; antidiabetic. [Bristol-Myers Squibb Co Inc] *

Lente Iletin. Insulin zinc; antidiabetic. [Eli Lilly & Co]

Lenticillin. Procaine penicillin/benzathine penicillin. [May & Baker Ltd]

Lentizol. CAS 549-18-8; A proprietary preparation of amitryptyline hydrochloride; an antidepressant. [Warner] *

Lentopen. CAS 6130-64-9; Penicillin G procaine; antibacterial. [Wyeth Laboratories] *

Lentrax. Procaine and benzathine penicillin injection. [RMB Animal Health Ltd]

Lenzing P84. Aromatic polyimide; high performance nonflamm., thermally stable resin with excellent chem. resist.; used for production of needle felts for high temp. filtration, as reinforcement for PTFE, as compounding. ingred. for injection moldable polymers; avail. as fiber, powd., solution (in DMF or N-methylpyrrolidone), high-dens. molded part (P84-HD), low-dens. insulating material (P84-LD), med.-dens. rigid molded part (P84-MD). [Lenzing AG]

Lenzit. See gypsum.

lenzol. A pale cedar-wood oil of known viscosity and refractive index; used for oil immersion in microscopy.

Leo K. CAS 7447-40-7; EINECS 231-211-8; A proprietary preparation of potassium chloride; used in potassium replacement therapy. [Leo Laboratories] *

Leomin AN. Alkyl phosphonate; surfactant for textile processing; antistat for fiber mfg. and processing, plastics processing. [Hoechst Celanese/Colorants & Surf.; Hoechst AG]

Leomin FANF. Fatty alkyl quaternary ammonium salt; surfactant for textile processing. [Hoechst Celanese/Colorants & Surf.

Leona®. CAS 32131-17-2; Nylon 66 resin [Asahi Chem.

---

* Trade name not verified as available   † Trade name verified as obsolete

Industry]

Leonil. See Nekal.

Leonil DB Powd. Sodium diisobutyl naphthalene sulfonate; wetting agent and dyeing auxiliaries for textiles, agric. [Hoechst Celanese/Colorants & Surf.; Hoechst AG]

Leonil EBL. Oxidizing agents in detergents; fast desizing and wetting agent for cellulosic fibers and cellulosic/synthetic blends. [Hoechst AG]

Leonil OS. CAS 577-11-7; EINECS 209-406-4; Sodium dioctyl sulfosuccinate; surfactant for agric. formulations. [Hoechst Celanese/Colorants & Surf.

Leophen® BN. Alcohol sulfonate; foam-free wetting agent in mercerizing textiles. [BASF AG; BASF plc]

Lepandin. Highly dispersed pyrogenic silica; used for non-slip finishing. [Degussa AG] *

Lepro. Lead peroxide. [Mechema Chemicals Ltd] *

Lepton®. Polymer dispersions; binders for pigment and other finishes in leather and fur industry. [BASF AG]

Leptovax-Plux. Vaccine. [The Wellcome Foundation Ltd] †

Lerbek. Coccidiostat for poultry, turkeys and rabbits. [RMB Animal Health Ltd]

Lergoban. CAS 132-18-3; Each tablet contains diphenylpyraline hydrochloride BP 5 mg in a porous plastic matrix. [3M Pharmaceuticals] †

Lerifond. Phenolic resin. [Winchem Ltd]

Leriphen. Phenolic resin. [Winchem Ltd]

Lerite. Phenolic molding resin. [Winchem Ltd]

Leromoll. Plasticizer with minimum volatility and good resistance to alkalis and acids; for use in combination with unsaponifiable binders in the formulation of chemical-resistant coatings. [Bayer AG]

Le Sage cement. (plaster cement). A natural cement obtained from nodules found at Boulogne.

Lesan. Fungicide; used for control of soil borne diseases on ornamentals. [Bayer AG]

lethalbine. A lecithin albuminate, containing 20% lecithin.

Lethane. Insecticide concentrates supplied in petroleum distillate; used in industrial insecticide sprays and mosquito larvicides. [Rohm & Haas UK]

Lethi. Lead thiosulfate. [Mechema Chemicals Ltd] *

Lethidrone Injection. A proprietary formulation of nalorphine hydrobromide; for treatment of overdosage of opioids. [The Wellcome Foundation Ltd]

leucaniline. Triamino-diphenyl-tolyl methane, $(H_2N \cdot C_6H_4)_2CH \cdot C_7H_6 \cdot NH_2$.

Leucarsone. 4-Carbaminophenyl arsonic acid. [May & Baker Ltd] *

leucaurin. Trihydroxytriphenylmethane, $CH(C_6H_4 \cdot OH)_3$.

Leuchtol. A proprietary synthetic resin. §

l-leucine. (α-amino-γ-methylvaleric acid; α-amino-isocaproic acid) $C_6H_{13}NO_2$; CAS 61-90-5; EINECS 200-522-0; Nutrient and dietary supplement, biochemical research. [Degussa; R.W. Greeff; Nippon Rikagakuyakuhin; Tanabe USA; U.S. Biochemical]

leucobenzaurin. Di-p-hydroxy-triphenylmethane, $CH \cdot C_6H_5(C_6H_4OH)_2$.

leucogen. Sodium hydrogen sulfite, $NaHSO_3$; used in bleaching and paper making.

Leucol. CAS 91-22-5; An impure quinoline.

Leucoline. CAS 91-22-5; Quinoline.

Leucomycin®. Macrolide antibiotic. [Asahi Chem. Industry]

leuconine. (leukonin). An antimony preparation containing 98% of sodium metantimoniate. Recommended as a substitute for tin oxide for enameling.

Leucophor KNR. Fluorescent whitener for commerical/industrial laundry detergents, rug/upholstery cleaners, fabric softeners, laundry bleach, whitening soap, brightening polymers and plastics; esp. for cellulosics. [Sandoz]

Leucopure EGM Powd. Fluorescent whitener for commerical/industrial laundry detergents, rug/upholstery cleaners, fabric softeners, laundry bleach, whitening soap, brightening polymers and plastics; esp. for synthetics and wool. [Sandoz]

Leucovorin. A proprietary preparation of calcium folinate; used as an antidote to methotrexate. [Lederle Laboratories]

leukarion. See oil white.

Leukeran Tablets. CAS 305-03-3; A proprietary formulation of chlorambucil; for treatment of chronic lymphocytic leukemia, Hodgkins disease, certain forms of non-Hodgkins lymphoma, Walderstroms macroglobuliremia and advanced ovarian adrenocarcinoma. [The Wellcome Foundation Ltd]

Leukomycin®. CAS 56-75-7; Chloramphenicol; broad-spectrum antibiotic; veterinary preparation. [Bayer AG]

Leukon. A proprietary synthetic resin molding powder. §

leukonin. See leuconine.

Leukonöl LBA-2. Emulsifier for emulsion polymers; wetting agent for alkaline systems to pH 13. [Münzing Chemie GmbH]

Leukotan® 974. Acrylic syntan; retanning agent for leather. [Rohm & Haas] *

Leukotrop® W. Textile assistant for producing white discharge prints on cellulose fibers. [BASF AG; BASF plc]

Leuna Gas. A proprietary trade name for compressed propane. §

Leunaphos. A mixture of phosphate, nitrate, and sulfate of ammonia. fertilizer. §

Leuna saltpeter. A double salt of ammonium sulfate and nitrate; a fertilizer similar to Chilean nitrate in its action.

Leutalux. Inorganic and organic fluorescent substances; fluorescent screens for electron tubes and fluorescent lamps. [Leuchtstoffwerk GmbH] *

Levacast®. Reactive PUR systems for coating leather and other substrates. [Bayer AG]

Levacell®. Acid and direct powder and liquid dyestuffs for coloring paper; anionic direct liquid dyestuffs for the production of printing inks for tissue paper. [Bayer AG]

Levaderm®. Anionic dyestuffs for finishing and for drum dyeing and dyeing on the Multima machine. [Bayer AG]

Levafil. Dope dyes for polyamide fibers. [Bayer plc]

Levafix®. Reactive dyes; for dyeing cotton as well as textiles made from regenerated cellulose. [Bayer AG; Bayer plc]

Levaflex®. Thermoplastic rubber. [Bayer plc] †

Levaform®. A silicone release agent; used for the rubber, plastics, man-made fibers, metal, food and pharmaceutical industries. [Bayer AG; Bayer plc]

Levagard®. Flame retardants for rigid PU foams. [Bayer AG]

Levagel®. Polyurethane gels and their raw materials; ready-to-use gels covered with film or fabric; used in medical sector, e.g., for wheelchair cushions, mattress overlays to prevent pressure sores. [Bayer AG]

Levair. Lithium salts mixed with inhibitors compounded in aqueous solution; for dehumidifiers and chiller. [Leverton-Clarke Ltd]

Levair®. Sodium aluminum phosphate, acidic; leavening agent for baking, cereals. [Rhone-Poulenc Food

---

‡ Trade name and manufacturer not verified   § Trade name without identified manufacturer

Ingreds.]

Levalaine. Scouring agent. [Stephenson Thompson Textile Chemicals]

Levalan N. 1:2 Metal complex dyestuffs; used for dyeing wool and polyamide with great coloring strength, good solubility and outstanding fastness properties. [Bayer AG]

Levalin®. A range of auxilliaries for economical, continuous processes for dyeing wool, cellulosics and synthetic fibers. [Bayer AG; Bayer plc]

Levamisole. CAS 14769-73-4; (-)-2,3,5,6-Tetrahydro-6-phenylimidazo 2,1-β thiazole; used as an anthelmintic.

Levanox®. Aqueous preparations of organic pigments. [Bayer AG]

Levanyl®. Aqueous preparations of organic pigments. [Bayer AG; Bayer plc]

Levapon®. Degreasing agents, emulsifiers, cleaning and wetting agents used in the textile industry. [Bayer AG; Bayer plc]

Levapren®. EVA copolymer; includes electric wire and cable insulations that must withstand high temperatures or present good flame-retardant behavior. [Bayer AG; Bayer plc]

Levapren® 400. EVA copolymer; synthetic rubber for tech. moldings and extrudates, lamp seals, cable sheathings and insulations, cellular rubber goods, footwear soles, waterproof sheeting, hot-melt and pressure-sensitive hot-melt adhesives, etc.; suited for cables. [Bayer; Miles]

Levapren® K. EVA copolymer; grades for adhesives applications. [Bayer AG]

Levasil®. Mixing liquid for dental investment compounds; nonslip finishing for textiles; spinning lubricant additive; ceramic coating of magnetic steel plates; special ceramic mortars; sprayable compositions for formation of a protective layer on ingot mold baseplates; vacuum molding of ceramic fiber articles; fine casting by the shell process; wafer polishing (semiconductor manufacturing). [Bayer AG]

Levasint®. A fluidized-bed coating powder comprising saponified ethylene/vinylacetate copolymer; uses include maintenance coatings on metal parts, especially for fencing, cable rigs, traffic furniture and applications in water area. [Bayer AG; Bayer plc]

Levasol®. Solubilizers and fixation auxiliaries for phthalogen dyes. [Bayer AG; Bayer plc]

Levcarb. Noncyanide heat treatment salts and other compounded dry salts; for metals heat treatment. [Leverton-Clarke Ltd]

Levegal®. Leveling auxiliaries for dyeing synthetic and cellulosic fibers; dye carriers for polyesters. [Bayer AG; Bayer plc]

Levelan P148. Nonionic surfactant of the nonylphenol ethoxylate type in liquid form; stabilizer for latex. [Henkel Chemicals Ltd] *

Levelan® P208. Nonyl phenol ethoxylate aq. dilution; wetting agent, scouring agent, antistat, dye leveling agent at high temp. and high electrolyte conc. in textile industry; latex stabilizer and emulsifier in emulsion polymer industry. [Harcros UK]

Levelan P208. Nonionic surfactant of the nonylphenol ethoxylate type in liquid form; wetting agent for use at high temperatures; stabilizer for latex; emulsion polymerization aid. [Henkel Chemicals Ltd] *

Levelan P307. Nonionic surfactant of the nonylphenol ethoxylate type in liquid form; wetting agent at high temperatures. [Henkel Chemicals Ltd] *

Levelan P357. Nonionic surfactant of the nonylphenol type in liquid form; primary emulsifiers for emulsion polymerization. [Henkel Chemicals Ltd] *

Levepox®. Liquid epoxy resins for solvent free coatings. [Bayer AG]

Levesol. Solubilizers and fixation auxiliaries for phthalogen dyes. [Bayer plc]

Levius. A proprietary preparation of sustained-release aspirin; an analgesic. [Farmitalia (Farmaceutici Italia)] *

Levn-Lite. Sodium aluminum phosphate; leavening agent. [Monsanto Co] *

Levochrom®. Cobalt-chrome-molybdenum alloy for dental model cast prosthetics. [Bayer AG]

Levodip®. Liquid cold-dip hardener for dental industry. [Bayer AG]

Levo-Dromoran. Levorphanol tartrate; analgesic. [Hoffmann-LaRoche Inc] *

Levofin®. Fine investing compound for model cast techniques; used in dentistry. [Bayer AG]

Levogel®. Duplicating material for model cast techniques; used in dentistry. [Bayer AG]

Levogen®. Aftertreating agents; used for improving the wet fastness properties of dyeings. [Bayer AG; Bayer plc]

Levogen® LF. Amphoteric dyeing auxiliary; used in the leather industry to enhance the surface dyeing. [Bayer AG]

Levophed. A proprietary preparation of noradrenaline acid tartrate; used to raise blood pressure. §

Levopress. Dental preparation; color-stable auto-polymerizate for dentures, orthodontic appliances, repairs, etc. [Bayer AG]

Levoprome. CAS 60-99-1; Methotrimeprazine; analgesic. [Lederle Laboratories USA] *

Levotan®. Anionic or weakly cationic polyurethane retanning materials for soft leather with a tight grain and good drumming or staining properties. [Bayer AG]

Levotherm®. Rapid-setting precision investing compound for model cast constructions with an exact fit; for dentistry. [Bayer AG]

Levothroid. CAS 55-03-8; Levothyroxine sodium; thyroid hormone. [USV Pharmaceutical Corp] ‡

Levoxin®. Corrosion inhibition in closed water-steam boilers. [Bayer AG]

levulic acid. See levulinic acid.

Levuline. A proprietary preparation used in the textile industry for finishing. §

levulinic acid. (γ-ketovaleric acid; acetylpropionic acid; 4-oxopentanoic acid; levulic acid) $CH_3COCH_2CH_2COOH$; CAS 123-76-2; EINECS 204-649-2; Intermediate for plasticizers, solvents, resins, flavors, pharmaceuticals, acidulant, and preservative; chrome plating; solder flux; stabilizer for calcium greases; control of lime deposits. [Chemie Linz UK Ltd; Otsuka Chem.; Penta Mfg.; QO; Schweizerhall]

levulose. CAS 57-48-7; (fruit sugar, diabetin, levo glucose, sucro-levulose; fructose) Naturally occurring sugar; $C_6H_{12}O_6$; used in food stuffs, medicine, and as a preservative.

levulose. See fructose.

Lewasorb®. Powdered ion exchange resins regenerated for immediate use; used for water purification. [Bayer AG; Bayer plc]

Lewatit® . Cation and anion exchange resins based on polymerization type synthetic resins of differing mesh size and/or macroporous structure and with active groups of various acidities and basicities or selective properties; used for water treatment, waste water treatment, adsorption, decolorizing, separation, decontamination, catalysis, nutrient carrier for hydroponics. [Bayer AG; Bayer plc]

Lewisite. [β-Chlorovinyldichloroarsine, CHCl : CH · AsCl₂ , a military poison gas; the name Lewisite is also used for a mineral, $5CaO \cdot 2TiO_2 \cdot 3SbO_3$. §

Lewis metal. An alloy of 1 part tin and 1 part bismuth having the property of expanding when cooling. It has a melting-point of 138 C and is used for sealing and holding die parts.

Lewisol 28. Maleic-modified glycerol ester of rosin; used for Type C gravure inks and for nitrocellulose-based paper and wood coatings. [Hercules] *

Lexaine® C. Cocamidopropyl betaine; visc. builder, foam booster, thickener in conditioners, specialty shampoos, personal care products, dishwash, sanitizers. [Inolex]

Lexaine® CSB-50. CAS 68139-30-0; Cocamidopropylhydroxysultaine; surfactant, foaming and wetting agent used in personal care products (shampoos, conditioners, bath products), industrial (heavy-duty alkaline cleaners, paint strippers, metal cleaners); stable in systems containing acids, alkali, and electrolytes; lime soap dispersant. [Inolex]

Lexaine® IS. Isostearamidopropyl betaine; mild surfactant, foam booster, and thickener for shampoos, bubble baths, foaming conditioners; stable in acid and alkaline systems. [Inolex]

Lexaine® LM. CAS 4292-10-8; Lauramidopropyl betaine; mild surfactant, foam booster for bath products, shampoos, liq. soaps, dishwash liqs. [Inolex]

Lexaine® O. CAS 25054-76-6; EINECS 246-584-2; Oleamidopropyl betaine; surfactant used in personal care products. [Inolex]

Lexamine 22. CAS 16889-14-8; EINECS 240-924-3; Stearamidoethyl diethylamine; conditioner, emulsifier for hair and skin care products; conditioning shampoo. [Inolex]

Lexamine C-13. CAS 68140-01-2; EINECS 268-771-8; Cocamidopropyl dimethylamine; emulsifier for creams and lotions; when neutralized as conditioners for hair care products. [Inolex]

Lexamine L-13. CAS 3179-80-4; EINECS 221-661-3; Lauramidopropyl dimethylamine; conditioner, emulsifier for hair and skin care products. [Inolex]

Lexamine O-13. CAS 109-28-4; EINECS 203-661-5; Oleamidopropyl dimethylamine; emulsifier for creams and lotions; when neutralized as conditioners for hair care products. [Inolex]

Lexamine S-13. CAS 7651-02-7; EINECS 231-609-1; Stearamidopropyl dimethylamine; emulsifier for creams and lotions; when neutralized as conditioners for hair care products. [Inolex]

Lexamine S-13 Lactate. CAS 55819-53-9; EINECS 259-837-7; Stearamidopropyl dimethylamine lactate; conditioner for hair care products. [Inolex]

Lexan®. BPA polycarbonates; used for plastic components for automotive, electrical, electronics, lighting, medical, packaging, audio, etc. [GE Plastics Ltd]

Lexan® 121. PC resin; low visc., general purpose grade for molding intricate, hard-to-fill parts. [GE Plastics]

Lexan® 141L. PC resin; lower visc. general purpose grade. [GE Plastics]

Lexan® 144LR. PC resin; injection blow molding grade; FDA compliance; applications incl. baby bottles and other liq. pkg. [GE Plastics]

Lexan® 151, 153, 154. PC resin; extrusion blow molding grade; for water bottle applications requiring high impact strength, glass-like transparency, high gloss, and high heat distort. temp.; avail. as general purpose grade (151), uv-stabilized (153), food additive grade (154). [GE Plastics]

Lexan® 241. PC resin; general purpose injection molding grades. [GE Plastics]

Lexan® 500, 503. PC resin, high modulus grade; for applications requiring high rigidity along with toughness and impact strength; offers dimensional stability, low mold shrinkage for production of precise parts; 503 grade is uv-stabilized for outdoor applications. [GE Plastics]

Lexan® 920. PC resin; low visc. flame retardant resin used for elec., electronic, appliance, aircraft, lighting, computer, communications, business equipment, and hardware applications. [GE Plastics]

Lexan® 3412. PC resin, 20% glass fiber-reinforced; provides higher mech. props., reduced mold shrinkage over unreinforced grades; able to produce very precise parts. [GE Plastics]

Lexan® 8040. PC film; high-clarity, heat-resistant film for appliances, business machines, and transportation applications; FDA-approved grade, smooth both sides, polish texture. [GE Plastics]

Lexan® FL400. PC; engineering structural foam for applications where load bearing capability at elevated temps. is required; for appliance, automotive, telecommunications, material handling, and business machine industries. [GE Plastics]

Lexan® GR1310. PC resin; medical grade featuring superior gamma resistance. [GE Plastics]

Lexan® HF1110, HF1130, HF1140. PC resin; high flow resins providing enhanced processability and longer flow lengths; inherent mold release; ideal for disposables or where life cycles are not demanding, and in applications where clarity, heat resist., and dimensional stabilty are important; HF1110 is std., general purpose grade; HF1130 is uv stabilized; HF1140 is FDA-compliant grade. [GE Plastics]

Lexan® HP1, HPS1. PC resin; resin formulated for the health care industry and meeting FDA and USP standards; inherent mold release; HPS series are gamma radiation stabilized; HP avail. in natural, transparent tints, and custom opaque colors. [GE Plastics]

Lexan® OQ1020. PC resin, optical quality grade; offers optical clarity and resin purity for the most demanding optical applications; OQ1020 for compact disc market. [GE Plastics]

Lexan® PK2040. PC resin; packaging grade [GE Plastics]

Lexan® PPC4501. Polyphthalate carbonate copolymer resin; engineered for high heat and moisture resistance; FDA grades. [GE Plastics]

Lexan® SP1010. Polycarbonate resin; superior flow and high impact grade for thin wall and intricate parts. [GE Plastics] *

Lexan® WR1210, WR1240. PC resin, PTFE-filled; wear-

---

‡ Trade name and manufacturer not verified   § Trade name without identified manufacturer

resistant grade providing greater lubricity and reduced friction and wear between moving parts; for business equip., elec./electronics applications incl. keyboard frames, swivel bases, paper drives, printers, camera internal parts; WR1210 is general purpose, WR1240 is FDA-compliant grade. [GE Plastics]

Lexate BPQ. Lauramidopropyl betaine, TEA-coco-hydrolyzed collagen, oleamidopropyl dihydroxypropyl dimonium chloride; blended detergent, conditioner, and protein used as economical base for shampoo, bath gel, liq. soaps, dishwash, bubble baths, cleansers. [Inolex]

Lexate CRC. Stearamidopropyl dimethylamine, glycol stearate, and ceteth-2; cream rinse conc. and conditioner, emulsifier. [Inolex]

Lexate PX. Petrolatum, lanolin, and ozokerite; o/w and auxiliary emulsifier, lanolin cream base, conditioner in soaps and shaving cream, emollient. [Inolex]

Lexate TA. Glyceryl stearate, IPM, and stearyl stearate; auxiliary lipophilic emulsifier in w/o emulsions, emollient in cosmetic creams and lotions; provides barrier properties and slip. [Inolex]

Lexate TL. Glyceryl stearate, butyl stearate, and stearyl stearate; superfatting agent in milled bar soap, auxiliary lipophilic emulsifier in w/o systems, emollient; enhances barrier properties in creams and lotions. [Inolex]

Lexe® . A registered trademark for a cellulose acetatebutyrate insulating tape; it is flame retardant, has low moisture absorption and has a high dielectric strength. §

Lexein® A200. CAS 72319-06-3; Myristoyl hydrolyzed collagen; film-forming collagen protein deriv.; resin modifier for hair sprays, makeups, protection skin lotions and creams. [Inolex]

Lexein® A220. TEA-myristoyl hydrolyzed collagen; shampoos, hair conditioners, skin care products, creams, lotions, shave creams, bar and liq. soaps; aq. sol'ns. above pH 6.0. [Inolex]

Lexein® A520. CAS 68918-77-4; TEA-abietoyl hydrolyzed collagen; sebum control additive; causes delay in refatting of the scalp when used in shampoos. [Inolex]

Lexein® CP-125. Oleamidopropyl dimethylamine hydrolyzed collagen; protein for shampoos, cream rinses, hair conditioners. [Inolex]

Lexein® S620S/Superpro 5A. TEA-coco-hydrolyzed animal protein and sorbitol; mild, high foaming cleansing agent, humectant, moisturizer for skin care products. [Inolex]

Lexein® X-250HP. CAS 9015-54-7; Hydrolyzed collagen; low odor and color version; cosmetic industry protein exhibiting substantivity; for hair and skin care products. [Inolex]

Lexell. Hydrocarbon or oxyhydrocarbon and aqueousbased solvents and equipment for industrial cleaning processes. [ICI Chem. & Polymers Ltd]

Lexemul® 503. Glyceryl stearate; emulsifier, stabilizer, thickener, opacifier in emulsions or surfactant systems, cosmetics, toiletries, pharmaceuticals. [Inolex]

Lexemul® 561. Glyceryl stearate and PEG-100 stearate; cosmetic grade emulsifier for o/w cream or lotion systems. [Inolex]

Lexemul® CS-20. Cetearyl alcohol, ceteareth-20; emulsifier used in personal care products, topical preps. [Inolex]

Lexemul® EGDS. CAS 627-83-8; EINECS 211-014-3; Glycol distearate; lubricant, opacifier and pearling agent for cosmetic surfactant systems, liq. hand soaps, light duty liqs. [Inolex]

Lexemul® EGMS. Glycol stearate; opacifier and pearling agent for personal care products, light duty liqs.; emulsifier for creams, lotions, topicals; secondary suspending agent in o/w systems. [Inolex]

Lexemul® GDL. CAS 27638-00-2; EINECS 248-586-9; Glyceryl dilaurate; lipid for creams and lotions formulated to produce a dry skin feel; melts just below body temp.; emollient; coupling agent for more lipophilic materials; emulsion stabilizer. [Inolex]

Lexemul® P. Propylene glycol stearate SE; for low visc. emulsions for personal care products. [Inolex]

Lexemul® PEG-200 DL. CAS 9005-02-1; PEG-4 dilaurate; emulsifier, emollient, lubricant for cosmetics, pharmaceuticals, metalworking fluids, paints, polishes and misc. industrial formulations. [Inolex]

Lexemul® PEG-400ML. CAS 9004-81-3; PEG-8 laurate; emulsifier, emollient, lubricant dispersant for creams, lotions, spreading bath oils, cosmetic, pharmaceutical and industrial applications. [Inolex]

Lexemul® T. Glyceryl stearate SE; for use as emulsifier, opacifier, stabilizer, and emollient in alkaline anionic systems, cosmetics. [Inolex]

Lexgard B. CAS 94-26-8; Butylparaben; cosmetics preservative. [Inolex]

Lexgard M. CAS 99-76-3; Methylparaben USP; cosmetics preservative. [Inolex]

Lexgard P. CAS 94-13-3; Propylparaben USP; cosmetics preservative. [Inolex]

Lexite Granular Carpet. All-solids epoxy fortified with quartz granules; used to provide a tough wearing surface over industrial and commercial floors to make them both skid-proof and reasonably easy to clean. [Metalcrete Mfg Co]

Lexol 3975. Isopropyl palmitate, isopropyl myristate, isopropyl stearate; emollient; replacement for IPM. [Inolex]

Lexol EHP. CAS 29806-73-3; EINECS 249-862-1; Octyl palmitate; emollient for nonocclusive creams and lotions, bath oils, antiperspirants, other cosmetic and topical formulations. [Inolex]

Lexol GT 855, GT 865. CAS 65381-09-1; Caprylic/capric triglyceride; emollient with nonoily skin feel; moisturizer; for creams, lotions, bath oils, lipstick, makeup; solv. for perfume and flavor ingreds.; vehicle for medicinals, antibiotics, vitamins; solubilizer; oxidative stability. [Inolex]

Lexol IPM-NF. CAS 110-27-0; EINECS 203-751-4; Isopropyl myristate; NF emollient oil for cosmetic applications; colorless and essentially odorless, low visc., low f.p., outstanding spreading, good sol. [Inolex]

Lexol IPP. CAS 142-91-6; EINECS 205-571-1; IPP; emollient in conditioning cosmetics; solubilizer for cosmetic and topical pharmaceuticals. [Inolex]

Lexol PG 800. CAS 56519-71-2; Propylene glycol dioctanoate; emollient with nonoily feel, oxidation stability; for creams, lotions, topicals, lipsticks, glossers, makeup bases, bath oils, aftershaves; carrier/vehicle for fragrance. [Inolex]

Lexol PG 855. Propylene glycol dicaprylate/dicaprate; emollient, solubilizer for cosmetic creams and lotions; carrier/vehicle for flavors, fragrances, pigmented cosmetics, antibiotics. [Inolex]

Lexol PG 900. CAS 41395-83-9; EINECS 255-350-9; Propylene glycol dipelargonate; emollient for bath oils, preshave lotions, aerosol systems, lipsticks, glosses, makeup bases; carrier for fragrances. [Inolex]

Lexol SS. CAS 2778-96-3; EINECS 220-476-5; Stearyl stearate; emollient for bath oils, creams, lotions. [Inolex]

Lexolube® 2J-237. Tetraethylene glycol dicocoate; lubricant for polyester filament yarns. [Inolex]

Lexolube® 2T-237. PEG-4 di 2-ethylhexanoate; lubricant for polyester and nylon textile and industrial yarns; softener for synthetic rubber and other elastomers. [Inolex]

Lexolube® 2X-109. Ditridecyl adipate; base stock for crankcase and compressor oils; fiber lubricant in yarns. [Inolex]

Lexolube® 3G-310. Trimethylolpropane trioleate; lubricant component for hydraulic and metalworking oils. [Inolex]

Lexolube® 4N-415. Pentaerythrityl tetra C8-C10; high temp. lubricant for filament yarn production and processing, tire cord lubricant. [Inolex]

Lexolube® B-108. Isodecyl stearate; lubricant for synthetic fibers, metalworking, and industrial applications. [Inolex]

Lexolube® B-109. CAS 31556-45-3; EINECS 250-696-7; Tridecyl stearate; drawing and heat setting lubricant for textile/industrial filament yarns, plastic extrusion, magnetic tapes. [Inolex]

Lexolube® BS-Tech. CAS 123-95-5; EINECS 204-666-5; Butyl stearate; general purpose lubricant for textile synthetic fibers, metalworking, coatings, inks, plastics, rubber industries. [Inolex]

Lexolube® T-110. CAS 22047-49-0; EINECS 244-754-0; 2-Ethylhexyl stearate; lubricant for textile, metalworking, plastics industries. [Inolex]

Lexone®. CAS 21087-64-9; Metribuzin; triazinone herbicide for weed control in potatoes. [DuPont UK]

Lexorez 1100-25. Linear poly (diethylene glycol adipate); sat., linear, aliphatic polyester polyol with all primary hydroxyls; used in adhesives, flexible coatings, and castable elastomers. [Inolex]

Lexorez 1101-50A. Crosslinked poly (diethylene glycol adipate); sat., crosslinked primary hyroxyl polyester producing high-quality flexible polyester slab foams for use in textile laminates, pkg., filter media, and in noncellular applications such as coatings, adhesives, and elastomers; where good flexibility and solv. resist. are required. [Inolex]

Lexorez 1130-30. Linear poly (1,6-hexanediol adipate); sat. aliphatic polyester polyols with all primary hydroxyls; raw material for formulation of urethane prepolymers, coatings, and thermoplastic urethanes which exhibit outstanding low temp. flexibility, high strength, and softness. [Inolex]

Lexorez 1131-190. Crosslinked poly (1,6-hexanediol adipate); sat. primary hydroxyl functional polyester polyol; used in the formulation of high-performance urethane elastomers. [Inolex]

Lexorez 1400-35. Linear poly (1,6-hexanediol neopentyl glycol adipate); sat., linear, aliphatic, slightly branched polyester with primary hydroxyl functionality; used in semiflexible coatings, solution adhesives, prepolymer systems; produces urethanes with excellent solv. resist. and weatherability. [Inolex]

Lexorez 1721-65P. Crosslinked poly (diethylene glycol, neopentyl glycol, 1,6-hexanediol adipate); branched sat. aliphatic polyester polyol with all primary hydroxyl functionality; used to produce high-quality coatings for fabrics and flexible foams which are resist. to HC solvs. [Inolex]

Lexorez 1821-50. Crosslinked poly (diethylene glycol adipate); slightly crosslinked, sat. polyester polyol with primary hydroxyl functionality; used for flexible foams, cast elastomers, solution coatings, and adhesives; produces urethanes with excellent flexibility and good solv. resist. [Inolex]

Lexorez 5901-55. Crosslinked poly (dipropylene glycol phthalate adipate); sat. mixed aromatic/aliphatic polyester polyol; used for cast systems and solution coatings where extreme service is required; ideal for structural RIM applications. [Inolex]

Lexotan. CAS 1812-30-2; Proprietory preparation of bromazepam; benzodiazepine with anxiolytic properties. [Roche Products Ltd] *

Lexquat® 2240. CAS 68039-13-4; Polymethacrylamidopropyl trimonium chloride; film former, hair fixative, skin protectant, humectant, conditioner, antistat; pH stable; exceptional low use levels. [Inolex]

Lexquat® AMG-BEO. Behenamidopropyl PG-dimonium chloride; mild conditioning surfactant, emulsifier for shampoos, bath gels, conditioners, shave products. [Inolex]

Lexquat® AMG-IS. Isostearamidopropyl PG-dimonium chloride; mild conditioning surfactant, emulsifier for shampoos, bath gels, conditioners, shave products. [Inolex]

Lexquat® AMG-M. Lauramidopropyl PG-dimonium chloride; conditioner, emulsifier for hair and skin products; emollient in bath products, liq. soaps. [Inolex]

Lexquat® AMG-O. Oleamidopropyl PG-dimonium chloride; conditioner, emulsifier, emollient for hair and skin products, bath gels; forms clear dilutable gels with other fatty quats. [Inolex]

Lexquat® AMG-WC. Cocamidopropyl PG-dimonium chloride; foaming conditioner, emulsifier for hair and skin products, bath gels. [Inolex]

Lexquat® CH. Polyquaternium-29; film former, hair fixative, skin protectant, humectant, conditioner, antistat; pH stable; exceptional low use levels. [Inolex]

Lextron. A proprietory preparation of liver extract and ferric ammonium citrate. [Eli Lilly & Co] †

Ley Cornox. Selective weedkillers. [The Boots Co plc] †

Lffler's methylene blue. A stain for bacteria. It consists of 100 parts of a solution of sodium hydroxide (1 in 10,000) and 30 parts of a saturated alcoholic solution of methylene blue.

LG Wax. A montan wax derivative used in the production of nonionic and ionic dry-bright emulsions for floor polishes and similar applications. [Bush Beach Ltd] *

LHS 40% Coconut Oil Soap. Highly refined coconut oil, caustic potash; soap. [Emulsion Systems]

Liancare. Janitorial and hygiene products. [MTM plc]

Libavius' fuming spirit. CAS 7646-78-8; A solution of stannic chloride, $SnCl_4$.

Libfer®. High purity iron EDDHA chelate in soluble powder and granule form for correcting iron deficiency in soils of adversely high pH. [Allied Colloids Ltd]

Libfer® SP. High purity iron EDDHA chelate. [Atlas Interlates Ltd]

Libigen. Gonadotropin, chorionic; gonad-stimulating principle. [Savage Laboratories] †

Libollite. A variety of asphaltum.

Libral Range. Chelated micronutrients. [Atlas Interlates Ltd]

Libraxin. A proprietary preparation of chlordiazepoxide and

---

‡ Trade name and manufacturer not verified   § Trade name without identified manufacturer

clidinium bromide; a gastrointestinal sedative. [Roche Products Ltd] *

Librebor®. An organo-boron complex for correcting boron deficiency in most crops. [Allied Colloids Ltd]

Librel®. Chelated metallic micronutrients designed for soil or foliar application to correct specific trace element deficiencies in all crops. [Allied Colloids Ltd]

Libreleaf®. An economical range of copper, iron and manganese lignosulfonate chelates for correcting specific trace element deficiencies in all crops for foliar application only. [Allied Colloids Ltd]

Librium. CAS 438-41-5; A proprietory preparation of chlordiazepoxide hydrochloride; a sedative. [Roche Laboratories; Roche Products Ltd]

Libsorb®. A nonionic wetting and spreading agent which increases the effectiveness of foliar applied pesticides and fertilizers. [Allied Colloids Ltd]

Libspray®. A range of complete foliar fertilizers containing chelated trace elements in powder and liquid form to supplement soil applied fertilizers. [Allied Colloids Ltd]

LICA 01. Neoalkoxy, trineodecanoyl titanate; coupling agents which also act as adhesion promoters, antioxidants, antistats, antifoaming agents, accelerators, blowing agent activators, catalysts, curatives, corrosion inhibitors, disp. aids, emulsifiers, flame retardants, foaming agents, hardeners, impact modifiers, internal lubes, metal primers, process aids, pigment intensifiers, peroxide activators, release agents, retarders, stabilizers, surfactants, suspension aids, thixotropes, wetting agents. [Kenrich Petrochemicals]

LICA 09. Neoalkoxy, dodecylbenzenesulfonyl titanate; see LICA 01. [Kenrich Petrochemicals]

LICA 12. Neoalkoxy, tri (dioctylphosphato) titanate; see LICA 01. [Kenrich Petrochemicals]

LICA 38. Neoalkoxy, tri (dioctylpyrophosphato) titanate; see LICA 01. [Kenrich Petrochemicals]

LICA 44. Neoalkoxy, tri (N ethylaminoethylamino) titanate; see LICA 01. [Kenrich Petrochemicals]

LICA 97. Neoalkoxy, tri (m-amino) phenyl titanate, phenyl glycol ether; see LICA 01. [Kenrich Petrochemicals]

LICA 99. Neopentyl (diallyl)oxy, trihydroxy caproyl titanate; coupling agent. [Kenrich Petrochemicals]

Licella yarn. A product made from narrow strips of wood-pulp paper; used as a jute substitute.

lichen starch. Lichenin, $C_6H_{10}O_5$; a carbohytrate derived from Iceland moss.

lichen sugar. Erythritol, $C_4H_6(OH)_4$.

Lichner's blue. A variety of smalt. It is a silicate of cobalt and potassium.

Lichtenbergs's metal. An alloy of 50% bismuth, 30% lead, and 20% tin, melting at 91.5 C.

Licomer. Polyethylene primary dispersions. [Hoechst UK]

Liconite. A rubber subsitute made from bitumen and oils. §

Licowet. Fluorinated surfactants. [Hoechst UK]

Licryl-55. Licryfilcon A; 2-hydroxyethylmethacrylate polymer with ethylene dimethacrylate; contains 55% water as 0.9% saline; contact lens material. [Liquid Crystal Lens Co] *

Licryl-70. Licryfilcon A; 2-hydroxyethylmethacrylate polymer with ethylene dimethacrylate; contains 70% water as 0.9% saline; contact lens material. [Liquid Crystal Lens Co] *

Licuado Instante. Nutritional additive to milk. [Richardson-Vicks Ltd] *

Lida-Mantle. CAS 137-58-6; Lidocaine; anesthetic. [Miles Pharmaceuticals] *

Lidarral. CAS 65009-35-0; Lidamidine hydrochloride; antiperistaltic. [William H Rorer Inc] *

Lidex. CAS 356-12-7; Fluocinonide; glucocorticoid. [Syntex Laboratories Inc]

lidocaine. (α-Diethylaminoaceto-2,6-xylidide) C_6H_3-$(CH_3)_2NHCOCH_2N(C_2H_5)_2$; CAS 137-58-6; medicine (local anesthetic). [Atomergic Chemetals; R.W. Greeff; Wyckoff]

Lieben solution. A solution of iodine in potassium iodide.

Lieberkuhn's jelly. The jelly formed by mixing egg serum with about one third of its volume of twice normal sodium hydroxide.

Liebmann and Studer's acid. 1-Naphthol-7-sulfonic acid (1:7).

Life. Oil analysis services. [Chevron] *

Ligantraal. Methyl-N-(2,4-dimethyl cyclohexene-3-ylidenemethyl)-anthranilate. [Quest Int'l. UK Ltd]

Ligdynite. A coal mine explosive consisting of 25-27% nitroglycerin, 27-29% sodium nitrate, 10-12% sodium chloride, and 30-33% wood meal.

light carburetted hydrogen. See marsh gas.

Lightfast. Mixed metal oxide pigments. [Miles]

Lightguard. A line of extruded, expanded polystyrene foam insulation products having a protective coating on one side. [Dow UK] *

lightning powder. A double base smokeless rifle powder.

light spar. See gypsum.

Light Water Foam. Fluorochemical surfactants in solution; premium grade fire fighting foam proportioned in water; used for flammable liquid fire fighting and protection. [3M UK plc]

light white. See oil white.

Lignasan®. For the agriculture industry. [DuPont UK]

lignin dynamites. Nitroglycerin absorbed by a mixture of wood pulp, and a nitrate, usually sodium nitrate; used as explosives.

lignin sulfonate. (lignosulfonate) A metallic sulfonate salt; dispersant for concrete, carbon black-rubber mixes, extender for tanning agents, oil-well drilling mud additives, ore flotation agents, production of vanillin, industrial cleaners, gypsum slurries, dyestuffs, and pesticides. [Borregaard Lignotech; Holmen GmbH; R.T. Vanderbilt; Westvaco]

lignite. (brown coal). It is a low rank of coal and is intermediate between sub-bituminous and peat.

lignite char. See Grudekok.

lignocaine benzyl benzoate. See denatonium benzoate NF.

Lignorit. CAS 8061-52-7; Unfermented calcium lignosulfonate; extender for U/F resins in particle board mfg. [Borregaard LignoTech]

lignorosin. Calcium lignosulfonate, a by-product in the manufacture of paper pulp. It is a dark-brown syrup, and is used as an assistant in mordanting wool with chrome.

Lignosite®. CAS 8061-52-7; Calcium lignosulfonate; dispersant, emulsifier and emulsion stabilizer. [Georgia-Pacific]

Lignosol. Calcium and sodium lignosulfonates; specialty chemicals for use in a variety of binding, dispersing, complexing, stabilizing or copolymerizing applications. [Borregaard LignoTech]

Lignosol B. CAS 8061-52-7; Calcium lignosulfonate; wet-

---

* Trade name not verified as available   † Trade name verified as obsolete

ting agent, emulsifier, dispersant, binder; used in refractories, bricks, construction, insecticides, herbicides, fungicides; soil stabilizer. [Borregaard LignoTech]

Lignosol DXD. CAS 8061-51-6; Sodium lignosulfonate; wetting agent and dispersant in industrial cleaners, insecticides, herbicides and fungicides, emulsifier, emulsion stabilizer for asphalt emulsions. [Borregaard LignoTech]

Lignosol SD-60. Kraft lignin; primary dye dispersant providing excellent heat stability, low staining and good milling and paste chars. [Borregaard LignoTech]

Lignosol SF. CAS 8061-52-7; Calcium lignosulfonate; dispersant for Portland cement and concrete, retanning agent. [Borregaard LignoTech]

Lignosol TS. CAS 8061-53-8; Ammonium lignosulfonate; wetting agent, emulsifier, dispersant, tanning extract, slurry water reducer and grinding aid in cement mfg. [Borregaard LignoTech]

Lignostab. Lignocaine hydrochloride injection. [The Boots Co plc] *

Lignostone. See Permali. §

lignosulfonate. See lignin sulfonate.

Lignovest. Modified polybutadiene. [Hüls AG]

lignum vitae. Guaiacum wood.

Ligro. A proprietary trade name for a crude pine fatty acid mixture. §

ligroin. CAS 8032-32-4; A term rather loosely applied. It usually denotes a refined distillate of petroleum oil having a boiling-point of 120-135 C., of specific gravity 0.73; used as a polishing oil, and as a turpentine substitute in varnishes.

ligulin. The dyestuff of privet berries.

ligurite. A mineral. It is a variety of sphene.

Ligustral. 2,3-Dimethyl-3-cyclohexene-1-carboxy-aldehyde. [Quest Int'l. UK Ltd]

Lilaflot. Formulated cationic flotation agents based on primary fatty amines, primary ether amines, the corresponding diamines and quaternary ammonium compounds; mineral flotation for potash; feldspar; quartz; iron, sulfide and phosphate ores; scheelite. [Keno Gard (UK) Ltd] ‡

Lilamin AC-59 P. Long-chain fatty nitrogen deriv. with C14-22 alkyl chain length; mold release agent, processing aid for mfg. of natural and synthetic rubber products. [Berol Nobel AB]

Lilamin LSP 33. CAS 85632-63-9; EINECS 288-048-0; Tallow dipropylene triamine; wetting and adhesion agent, flushing agent, pigment grinding and dispersion agent for pigment industry; curing agent for epoxy paints and coatings; corrosion inhibitor. [Berol Nobel AB]

Lilaminox M4. CAS 3332-27-2; EINECS 222-059-3; Tetradecyl dimethylamine oxide; detergency booster, thickener for household bleaches based on sodium hypochlorite; foaming agent for hair shampoos. [Berol Nobel AB]

Lilaminox M24. CAS 85408-49-7; EINECS 287-011-6; C12-14 alkyl dimethylamine oxide; foam booster/stabilizer, antistat, softener for hair shampoos; thickener for household bleaches based on sodium hypochlorite, hard surface cleaners. [Berol Nobel AB]

Lilestralis. p-t-Butyl-α-methyldihydrocinnamic aldehyde. [Bush Boake Allen Ltd]

Lilestraus. p-t-Butyl-α-methyldihydrocinnamic aldehyde; fresh, light, green floral, reminiscent of lily; used in fragrances.

Lilion 6. Nylon 6 monofilament; used for sewing threads, belting, filtration fabrics. [SNIA (UK) Ltd]

Lilion 66. Nylon 66 monofilament; used for sewing threads, belting, filtration fabrics. [SNIA (UK) Ltd]

lillite. An earthy mineral resembling glauconite, of Bohemia.

Lilvert. 1-Ethoxy-1-hexoxyethane. [Bush Boake Allen Ltd]

lily of valley, artificial. Terpineol, $C_{10}H_{17}OH$.

lilyolene. See paraffin, liquid.

LIM6045. CAS 63394-02-5; Liq. two-component silicone rubber; for use in liq. injection molding to produce elastomeric parts with high clarity; excellent tear strength, very high tensile strength; for mech. parts, sporting goods, health care equip., camera parts, coating metal rolls, baby bottle nipples. [GE Silicones] *

Limanol. A preparation of salt-marsh mud used for rheumatism.

Limao. Shellac substitute. [British Traders & Shippers Ltd]

Lima wood. See redwoods.

limbachite. A Serpentine mineral.

Limbaki. Stretch velour fabric. [ICI Chem. & Polymers Ltd]†

Limbitrol. CAS 50-48-6, CAS 58-25-3; A proprietary preparation of amitriptyline and chlordiazepoxide; an antidepressant. [Roche Laboratories; Roche Products Ltd]

Limbux®. A form of mechanically slaked lime; used in agriculture, building and construction, metallurgical and chemical industries. [ICI Chem. & Polymers Ltd] †

lime. See calcium oxide.

lime. CAS 1305-78-8; (burnt lime, quicklime, caustic lime). Calcium oxide, CaO, produced by calcining calcium carbonate. Sometimes a mixture of calcium and magnesium oxides is sold under this name.

lime citrate. See calcium citrate.

limeite. A cement containing rubber, tallow, and lime. The addition of vermilion causes the mixture to harden.

lime mortar. Mixtures of slaked lime and sand.

lime nitrate. (Lime saltpeter). Calcium nitrate, $Ca(NO_3)_2 \cdot 2H_2O$; used an a fertilizer.

lime nitrogen. CAS 156-62-7; Calcium cyanamide; $CaCN_2$; used as a fertilizer, nitrogen products, pesticide, hardening iron and steel.

Limeolivine. Calcium orthosilicate Ca_2SiO_4.

lime saltpeter. See lime nitrate.

lime soap. Calcium resinate.

limestone. (calcium carbonate, natural; agricultural limestone; lithographic stone) $CaCO_3$; CAS 1317-65-3; EINECS 215-279-6; Building stone, metallurgy (flux), manufacture of lime; source of CO_2; Portland and natural cement; removal of sulfur dioxide from stack gases and sulfur from coal. [ECC Int'l.; EM Industries; Genstar Stone Prods.; Georgia Marble; J.M. Huber; Mallinckrodt; Nichia Kagaku Kogyo; Pfizer; Tilcon Ltd; Whittaker, Clark & Daniels]

lime-sulfur dips. Sheep dips for the treatment of scab. They contain flowers of sulfur, lime, and water.

lime water. CAS 1305-62-0; A saturated solution of calcium hydroxide, $Ca(OH)_2$, in water.

Limex. Sodium phosphate and metal preservatives; used in water using equipment where lime scale and corrosion are a problem. [Delaware Chemical Corp] *

Limex G. CAS 10124-56-8; Sodium hexametaphosphates; used in water treatment to sequester calcium minerals. [Delaware Chemical Corp] *

---

‡ Trade name and manufacturer not verified § Trade name without identified manufacturer

Limit. N-(Acetylamino) methyl-2-chlor-N-2,6-diethyl-phenylacetamide; turf grass regulator. [Monsanto Co] *

Limit 33. Defoamer for paper industry; used at the size press to defoam starch and mixtures of starch. [Monsanto Co]*

Limo. (sablon, tartrate of lime). The raw materials obtained by the precipitation of tartaric acid in tartar works or wine distilleries.

limonene. (p-mentha-1,8-diene; R(+)-limonene; (+)-carvene. (R)-4-isopropenyl-1-methyl-1-cyclohexene) Terpene; $C_{10}H_{16}$; CAS 5989-27-5; EINECS 227-813-5; flavoring, fragrance, and perfume materials; solvents, wetting agent, resin manufacture [Allchem Industries; Int'l. Flavors & Fragrances; Langley Smith Ltd; Penta Mfg.; SCM Glidco Organics]

limonene, inactive. See dipentene.

Limpetite. Air-cured liquid-applied synthetic rubber; for protection of items subject to extreme conditions of corrosion, erosion and electrolytic action. [Protective Rubber Coatings (Limpetite) Ltd]

Linalux. Gloss paint finishes for marine use. [ICI Chem. & Polymers Ltd]

Linaqua®. Water-soluble drying oil-based chemical vehicle for paint. [Reichhold]

Lincocin. CAS 7179-49-9; Lincomycin hydrochloride; antibacterial. [Upjohn]

lincomycin. CAS 154-21-2; An antibiotic produced by *Streptomyces lincolnensis var. lincolnensis*. Lincocin and mycivin are the hydrochloride.

Linctifed Expectorant. A proprietary preparation of triprolidine hydrochloride, pseudoephedrine hydrochloride, codeine phosphate and guaifenesin; an expectorant for the symptomatic relief of upper respiratory tract symptoms associated with productive cough. [The Wellcome Foundation Ltd]

Lindenol. Pure α-terpineol. [Bush Boake Allen Ltd]

Lindex. Insecticide and fungicide seed dressing. [DowElanco Ltd]

Lindex-Plus. Flowable concentrate of 43 g fenpropimorph, 545 g γ-HCH and 73 g thiram per liter; combined insecticide and fungicide seed treatment. [DowElanco Ltd]

linear C8 alpha olefin. See octene-1.

Linearil. A linear alkylate. [Montedison UK Ltd] *

Linesman. Line marking paint. [Kalon Chemicals Ltd]

Linevol 79. Blend of primary alcohols, the notable characteristic being a high straight chain alcohol content; readily esterified by conventional means to give a high quality plasticizer. [Shell]

Linevol 911. Blend of primary alcohols, the notable characteristic being a high straight chain alcohol content; readily esterified by conventional means to give a high quality plasticizer. [Shell]

Linevol Phthalates. Plasticizers based on straight chain Linevol alcohols; they are characterized by exceptionally low volatility and excellent low temperature properties; a range of three phthalates allows a versatile plasticizing performance. [Shell]

Linex® 4L. CAS 330-55-3; Linuron suspension; flowable herbicide for control of grasses and broadleaf weeds in certain crops and noncrop areas. [Griffin]

Linfos. CAS 58-89-9, 2921-88-2; Lindane and chlorpyrifos; insecticide. [Makhteshim Chemical Works Ltd]

Lingraine. CAS 379-79-3; A proprietary preparation of ergotamine tartrate; used for migraine. §

lining metal. Alloys of 70-90% lead, 2-20% tin, and 5-20%

antimony; used for bearings.

Linituss. A proprietary preparation of linseed, liquorice and chlorodyne; a cough linctus. [Ayrton Saunders plc] ‡

Linklon. Crosslinkable polyolefin compounds. [Mitsubishi Petrochem.] *

Linklon-X. Crosslinkable polyolefin compounds. [Mitsubishi Petrochem.] *

linnaeite. See Linnalite.

linnalite. (linnaeite). Native cobalt sulfide, Co_3S_4.

Linnet. CAS 330-55-2, 1582-09-8; Emulsifiable concentrate containing 106 g linuron and 192 g trifluralin per liter; herbicide for winter cereals. [Pan Britannica Industries Ltd]

Linocin. CAS 7179-49-9; A proprietary preparation containing lincomycin hydrochloride; an antibiotic. [Upjohn Ltd]*

LinoCure. An alkyd oil urethane no-bake binder system; utilized in the production of foundry cores and molds as a resin binder system. [Ashland Chemical Company] *

linoleic acid. (9,12-octadecadienoic acid; linolic acid) Unsaturated fatty acid; $CH_3(CH_2)_4=CHCH_2CH=CH-(CH_2)_7COOH$; CAS 60-33-3; EINECS 200-470-9; manufacture of paints, coatings, emulsifiers, vitamins. [Arizona; CasChem; Henkel/Emery; Hercules; Langley Smith Ltd]

linolic acid. See linoleic acid.

linotype metal. An alloy of 13.5% antimony, 2% tin, and 84.5% lead.

linoxyn. See solidified linseed oil.

linseed meal. Ground oil-cake.

linseed oil. (oils, linseed; flaxseed oil) Expressed oil from the dried ripe seed of Linum usitatissimum; CAS 8001-26-1; paints, varnishes, oil cloth, putty, printing inks, core oils, linings and packings, alkyd resins, soap, pharmaceuticals. [Arista; Ferro/Bedford; Penta Mfg.; John L Seaton Ltd]

linseed oil, artist's. Raw linseed oil which has been allowed to stand for weeks, then treated with litharge, and finally bleached by exposure.

linseed oil, refined. Raw linseed oil which has been treated with a 1% solution of sulfuric acid.

linseed oil soap. A potash soap.

Linsol. Dyestuff solutions in oleic acid. [Morton Int'l. Ltd]

Linter's starch. A soluble starch prepared by mixing raw starch with 7.5% hydrochloric acid in water, allowing it to stand for several days, with stirring. The solution is decanted, and the starch washed with water and dried.

Lintex A10. A proprietary aqueous emulsion of a styrene copolymer internally plasticized with a copolymerizal ester; contains crosslinkable groups and is used as a binder for emulsion paints. [Hüls AG] *

Linurex. CAS 330-55-3; Active ingredient: linuron; 3-(3,4-dichlorophenyl)-1-methoxy-1-methylurea; selective herbicide for both pre- and post-emergence application. [Agan Chemical Manufacturers Ltd]

Linuron. CAS 330-55-3; Linuron; a residual herbicide for the control of weeds in field crops. [Pan Britannica Industries Ltd]

Linuron 15. CAS 330-55-3; Linuron; a residual herbicide for the control of weeds in field crops. [Farm Protection Ltd]

Linuron 450 FL. CAS 330-55-3; Linuron; a residual urea herbicide for the control of weeds in field crops. [Farmers Crop Chemicals Ltd] †

Lipacide CCO. Caprylolyl hydrolyzed collagen; mild surfactant with excellent lathering and wetting props., substan-

---

* Trade name not verified as available   † Trade name verified as obsolete

tivity to hair and skin; for frequent-use shampoos, bath gels, soaps, shaving creams. [Rhone-Poulenc; R.T. Vanderbilt]

Lipacide DPHP. CAS 41672-81-5; Dipalmitoyl hydroxyproline; mild surfactant with excellent lathering and wetting props., substantivity to hair and skin; for frequent-use shampoos, bath gels, soaps, shaving creams; used in cosmetic preps. for maintenance of skins physiological balance. [Rhone-Poulenc; R.T. Vanderbilt]

Lipacide PCO. Palmitoyl animal collagen amino acids; mild surfactant with excellent lathering and wetting props., substantivity to hair and skin; for frequent-use shampoos, bath gels, soaps, shaving creams; for first aid creams, sunburn lotions. [Rhone-Poulenc; R.T. Vanderbilt]

Lipacide PK. Palmitoyl animal keratin amino acids. mild surfactant with excellent lathering and wetting props., substantivity to hair and skin; for frequent-use shampoos, bath gels, soaps, shaving creams. [Rhone-Poulenc; R.T. Vanderbilt]

Lipacide UCO. Undecylenoyl animal collagen amino acids. mild surfactant with excellent lathering and wetting props., substantivity to hair and skin; for frequent-use shampoos, bath gels, soaps, shaving creams. [Rhone-Poulenc; R.T. Vanderbilt]

Lipal. A proprietary name for a range of polyoxyethylene esters and ethers. [PVO International Inc] *

Lipamide MEAA. CAS 142-26-7; EINECS 205-530-8; Acetamide MEA; lubricating humectant for used in personal care products; hair conditioner for shampoos, rinses, conditioners; antistat, foam modifier. [Lipo]

Lipamin®. Cationic emulsified oils; fat liquoring agents for leather. [BASF AG]

Lipamine SPA. CAS 7651-02-7; EINECS 231-609-1; Stearamidopropyl dimethylamine; raw material for cosmetics, toiletries, pharmaceuticals. [Lipo]

lipase. (glycerol ester hydrolase; triacylglycerol lipase) Enzyme; CAS 9001-62-1; EINECS 232-619-9; hydrolyzes fat to glycerol and fatty acids; manufacture of cheese; removal of fat spots in drycleaning; in analytical chemistry of fats. [Atomergic Chemetals; Gist-Brocades Food Ingreds.; U.S. Biochemical]

Lipaton. Styrene acrylic, and styrene butadiene copolymer dispersions. [Hüls AG]

Lipex 102. CAS 68424-60-2; Shea butter. [Karlshamns]

Lipinol O. An octyl fatty acid ester; viscosity depressant especially for rotational molding plastisols. [Hüls AG]

Lipinol T. Dibenzyltoluene; secondary plasticizer, lowers the viscosity of PVC plastisols to a lesser degree than Lipinol O. [Hüls AG]

Lipiodul Ultra-Fluid. A proprietary preparation of iodized oil fluid injection; x-ray contrast medium. [May & Baker Ltd]*

Lipivas. Halofenate; antihyperlipoproteinemic; uricosuric. [Merck & Co Inc] *

Lipo AMS. Almond meal; natural abrasive for facial scrubs, abrasive body scrubs and foot products. [Lipo]

Lipo APS 40/60. Apricot seed powd. natural abrasive for facial scrubs, body scrubs, foot products. [Lipo]

Lipo DGLS. CAS 9004-81-3; PEG-2 laurate SE; spreading agent, emulsifier, dispersant, lubricant, opacifier, emulsion stabilizer, emollient, visc. builder used in bath oils, creams, lotions; defoamer for process applications. [Lipo]

Lipo EGDS. CAS 627-83-8; EINECS 211-014-3; Glycol distearate; spreading agent, emulsifier, dispersant, lubricant, opacifier, emulsion stabilizer, emollient, visc. builder used in bath oils, creams, lotions; defoamer for process applications. [Lipo]

Lipo EGMS. Glycol stearate; opacifier, pearlizer for shampoos, detergents; w/o emulsifier; stabilizer for o/w systems. [Lipo]

Lipo GMS 450. Glyceryl stearate; general purpose emulsifier, emollient, opacifier and visc. builder in creams and lotions; food emulsifier. [Lipo]

Lipo LUFFA 30/100. Luffa; natural abrasive for facial scrubs, abrasive body scrubs and foot products. [Lipo]

Lipo PGMS. CAS 1323-39-3; EINECS 215-354-3; Propylene glycol stearate; emulsifier, stabilizer for o/w lotions, soft creams. [Lipo]

Lipo Polyol NC. Hydrogenated starch hydrolysate; humectant for creams, lotions, antiperspirants, depilatories, wavesets. [Lipo]

Lipo SS. CAS 68334-28-1; Hydrogenated vegetable oil; emollient for skin, sunscreens, lipsticks, balms; enhances gloss, reduces blooming. [Lipo]

Lipo WSF 35/60, 60/100. Walnut shell flour; natural abrasive for facial scrubs, abrasive body scrubs and foot products. [Lipo]

Lipobee 102. Synthetic beeswax; raw material for cosmetics, pharmaceuticals, and toiletries. [Lipo]

Lipocire A, CM, DM. Hydrogenated palm glycerides, hydrogenated palm kernel glycerides. lipstick bases. [Gattefosse]

Lipoclor. Chlorinated natural fats; additives for lubricating and cutting oils and raw material for the preparation of greasings for leather. [Caffaro SpA] *

Lipoclor S. Chorinated animal fats. [SNIA (UK) Ltd]

Lipocol C. CAS 36653-82-4; EINECS 253-149-0; Cetyl alcohol; emollient, consistency builder for creams, lotions, molded stick products. [Lipo]

Lipocol C-2. CAS 9004-95-9; Ceteth-2; emulsifier, defoamer, wetting agent, solubilizer, conditioning agent for personal care products and pigment disp. [Lipo]

Lipocol L. Lauryl alcohol; nonoily afterfeel emollient for creams, lotions, makeup. [Lipo]

Lipocol L-4. Laureth-4; surfactant for pigment dispersions, antiperspirants, depilatories, creams, lotions; antistat, emulsifier. [Lipo]

Lipocol O. CAS 143-28-2; EINECS 205-597-3; Oleyl alcohol; nontacky afterfeel emollient for bath oils, creams, lotions, makeup. [Lipo]

Lipocol O-2. CAS 9004-98-2; Oleth-2; surfactant for pigment dispersions, antiperspirants, depilatories, creams, lotions; antistat, emulsifier. [Lipo]

Lipocol S. CAS 112-92-5; EINECS 204-017-6; Stearyl alcohol; waxy afterfeel emollient, consistency builder for creams, lotions, molded sticks. [Lipo]

Lipocol S-2. CAS 9005-00-9; Steareth-2; surfactant for pigment dispersions, antiperspirants, depilatories, creams, lotions; antistat, emulsifier. [Lipo]

Lipocol SC-4. CAS 68439-49-6; Ceteareth-4; surfactant for pigment dispersions, antiperspirants, depilatories, creams, lotions; antistat, emulsifier. [Lipo]

Lipocol TD-3. CAS 24938-91-8; Trideceth-3; emulsifier, wetting and scouring agent, dispersant for essential oils; raw material for sulfation and phosphation. [Lipo] *

Lipocutin®. Water, lecithin, cholesterol, and dicetyl phosphate. [Henkel/Cospha; Henkel Canada] *

---

‡ Trade name and manufacturer not verified   § Trade name without identified manufacturer

Lipodan CDS Kosher. Hydrogenated cottonseed oil/dist. monoglycerides blend; peanut butter stabilizer. [Grindsted Prods.; Grindsted Prods. Denmark]

Lipodan CRE Kosher. Hydrogenated cottonseed/rapeseed oils and dist. monoglycerides blend; stabilizer for peanut butter and other oil-based systems. [Grindsted Prods.

Lipodan SET Kosher. CAS 68334-28-1; Hydrogenated vegetable oils (rapeseed, cottonseed, soybean); stabilizer for peanut butter and other oil-based systems. [Grindsted Prods.

Lipoderm®. Fat liquoring agents for leather. [BASF AG]

Lipoflavonoid. A proprietary preparation of choline bitartrate, inositol, dimethionine, ascorbic acid, lemon bioflavnoid, thiamine, riboflavine, nicotinamide, pyridoxine, panthenol and cyanocobalamin; used for neurosensory deafness. [Lewis Laboratories] ‡

Lipo Gantrisin. Sulfisoxazole acetyl; antibacterial. [Hoffmann-LaRoche Inc] *

Lipo-Hepin. Heparin sodium; anticoagulant. [3M Pharmaceuticals] †

Lipolan. CAS 8031-44-5; EINECS 232-452-1; Hydrogenated lanolin; auxiliary w/o emulsifier; emollient; conditioner; lubricant. [Lipo]

Lipolan. Carboxylated styrene/butadiene copolymer dispersions; binders for carpet backing, textile coating, fabric lamination, needle felt material, nonwoven fabrics, shoe toe-caps, and other textile finishes imparting a firm consistency. [Hüls AG]

Lipolan 31. CAS 68648-27-1; PEG-24 hydrogenated lanolin; o/w emulsifiers, solubilizer, emollient, conditioner used in cosmetics, toiletries and topical pharmaceuticals. [Lipo]

Lipolan R. Lanolin oil; emollient, spreading agent, conditioner, cosolv., plasticizer, and lubricant for personal care products; dispersant for pigments. [Lipo]

Lipolase. Lipolase hydrolyzed triglycerides into mono- and diglycerides, glycerol and free fatty acids, all of which are more soluble than the original fats; used in detergent formulations to remove fat-containing stains as those resulting from frying fats, salad oils, butter, fat-based sauces, soups, human sebum, or certain cosmetics, e.g., lipstick. [Novo Nordisk]

Lipo-Lutin. CAS 57-83-0; Progesterone; progestin. [Parke-Davis] *

Lipomulse 165. Glyceryl stearate and PEG-100 stearate; general purpose emulsifier, emollient, opacifier, and visc. builder in creams and lotions. [Lipo]

Liponate 2-DH. PEG-4 diheptanoate; emollient. [Lipo]

Liponate CL. CAS 35274-05-6; Cetyl lactate; emollient for desired feel and penetration in personal care products; thickener and visc. controller. [Lipo]

Liponate CRM. CAS 10401-55-5; Cetyl ricinoleate; glosser, emollient with dry afterfeel. [Lipo] *

Liponate DPC-6. CAS 68130-24-5; Dipentaerythrityl hexacaprylate/hexacaprate; nontacky emollient for treatment products. [Lipo]

Liponate EM. Ethyl morrhuate. [Lipo] *

Liponate GC. CAS 65381-09-1; Caprylic/capric triglyceride; emollient ester for adjusting rub-in and afterfeel of personal care products; thickener and visc. controller. [Lipo]

Liponate IPM. CAS 110-27-0; EINECS 203-751-4; IPM; emollient ester for adjusting rub-in and afterfeel of personal care products; thickener and visc. controller. [Lipo]

Liponate IPP. CAS 142-91-6; EINECS 205-571-1; IPP;

emollient ester for adjusting rub-in and afterfeel of personal care products; thickener and visc. controller. [Lipo]

Liponate MM. CAS 3234-85-3; EINECS 221-787-9; Myristyl myristate; emollient ester for adjusting rub-in and afterfeel of personal care products; thickener and visc. controller. [Lipo]

Liponate NPGC-2. CAS 70693-32-2; Neopentyl glycol dicaprylate/dicaprate; dry feel emollient for creams, lotions, cleansers, antiperspirants. [Lipo]

Liponate PB-4. CAS 61682-73-3; Pentaerythrityl tetrabehenate; emollient ester for adjusting rub-in and afterfeel of personal care products; thickener and visc. controller. [Lipo]

Liponate PC. Propylene glycol dicaprylate/dicaprate; emollient ester for adjusting rub-in and afterfeel of personal care products; thickener and visc. controller. [Lipo]

Liponate PE-810. Pentaerythrityl tetracaprylate/tetracaprate; emollient. [Lipo]

Liponate PO-4. CAS 19321-40-5; Pentaerythrityl tetraoleate; emollient ester for adjusting rub-in and afterfeel of personal care products; thickener and visc. controller. [Lipo]

Liponate PS-4. CAS 115-83-3; Pentaerythrityl tetrastearate; emollient ester for adjusting rub-in and afterfeel of personal care products; thickener and visc. controller. [Lipo]

Liponate SPS. CAS 8002-23-1; Cetyl esters (syn. spermaceti); emollient ester for adjusting rub-in and afterfeel of personal care products; thickener and visc. controller. [Lipo]

Liponate SS. CAS 2778-96-3; EINECS 220-476-5; Stearyl stearate; emollient ester for adjusting rub-in and afterfeel of personal care products; thickener and visc. controller. [Lipo]

Liponate TDS. CAS 31556-45-3; EINECS 250-696-7; Tridecyl stearate; emollient for creams and lotions. [Lipo]

Liponate TDTM. CAS 70225-05-7; Tridecyl trimellitate; nontacky emollient for treatment products, hair. [Lipo]

Liponic 70-NC. CAS 50-70-4; Sorbitol; humectant, plasticizer, softener, and lubricant; adds sweet taste and pleasant mouthfeel to oral hygiene products such as dentifrices and mouthwashes; oral dosage pharmaceutical, also for adhesives, leather, and paper coatings. [Lipo]

Liponic EG-1. CAS 31694-55-0; Glycereth-26; humectant in creams and lotions, lubricant, plasticizer for hair resins, foam stabilizer, pigment dispersant, hair conditioner, foam modifier, antistat; used in personal care products. [Lipo]

Liponic SO-20. Sorbeth-20; humectant and plasticizer for cosmetics and toiletries; nongreasy rich afterfeel with moderate lubricity. [Lipo]

Lipopeg 2-DL. CAS 9005-02-1; PEG-4 dilaurate; dispersant, emulsifier, spreading agent and lubricant in personal care products, bath oils. [Lipo]

Lipopeg 4-DO. CAS 9005-07-6; PEG-8 dioleate; dispersant, emulsifier, spreading agent and lubricant in personal care products, bath oils. [Lipo]

Lipopeg 4-DS. CAS 9005-08-7; PEG-8 distearate; dispersant, emulsifier, spreading agent and lubricant in personal care products, bath oils. [Lipo]

Lipopeg 4-L. CAS 9004-81-3; PEG-8 laurate; spreading agent, emulsifier, dispersant, lubricant for bath oils, creams and lotions. [Lipo]

---

* Trade name not verified as available   † Trade name verified as obsolete

Lipopeg 6000-DS. CAS 9005-08-7; PEG-150 distearate; dispersant, emulsifier, spreading agent, and lubricant in personal care products, bath oils. [Lipo]

Lipo-Peptide AME 30. Acetamide MEA, lauroyl hydrolyzed collagen, glycerin; moisturizer, humectant, emollient, softener, auxiliary emulsifier for skin products; conditioner, bodying agent, antistat for hair products; foam booster in shampoos, mousses, liq. hand soaps. [Maybrook]

Lipoproteol LCO. Sodium/TEA-lauroyl hydrolyzed collagen amino acid; mild additive with good foaming props. for rich lather shampoos, facial cleaners, infant shampoos, detergents. [Rhone-Poulenc; R.T. Vanderbilt]

Lipoproteol LK. Sodium/TEA-lauroyl hydrolyzed keratin amino acids; shampoo base. [Rhone-Poulenc; R.T. Vanderbilt]

Lipoquat R. CAS 112324-16-0; Ricinoleamidopropyl ethyldimonium ethosulfate; conditioner, antistat, emollient, glosser, softener for personal care products and anhyd. systems. [Lipo]

Liposiliol C. Dioleyl tocopheryl methylsilanol; aids tissue regeneration; for anti-aging formulations, oily cosmetics and sticks. [Exsymol]

Liposorb™. Absorbent. [Calbiochem Corp]

Liposorb L. CAS 1338-39-2; EINECS 215-663-3; Sorbitan laurate; emulsifier, thickener, lubricant, antistat, all-purpose lipophilic surfactant used with POE Liposorb series; also used in defoamers, aerosol w/o emulsions, corrosion inhibition. [Lipo]

Liposorb L-10. CAS 9005-64-5; PEG-10 sorbitan laurate; o/w emulsifier, lubricant, antistat, all-purpose hydrophilic surfactant; used for solubilizing oils and in conjunction with Liposorb esters. [Lipo]

Liposorb L-20. CAS 9005-64-5; Polysorbate 20; o/w emulsifier, lubricant, antistat, all-purpose hydrophilic surfactant; used for solubilizing oils and in conjunction with Liposorb esters. [Lipo]

Liposorb O. CAS 1338-43-8; EINECS 215-665-4; Sorbitan oleate; emulsifier, thickener, lubricant, antistat, all-purpose lipophilic surfactant used with POE Liposorb series; also used in defoamers, aerosol w/o emulsions, corrosion inhibition. [Lipo]

Liposorb O-5. CAS 9005-65-6; Polysorbate 81; hydrophilic surfactant used for solubilizing oils; emulsifier, lubricant, antistat. [Lipo]

Liposorb O-20. CAS 9005-65-6; Polysorbate 80; surfactant for food processing; flavor and color dispersant for pickles; defoamer for beet sugar, yeast processing; wetting agent for poultry defeathering; crystal control agent for salt. [Lipo]

Liposorb P. CAS 26266-57-9; EINECS 247-568-8; Sorbitan palmitate; lipophilic surfactant, emulsifier, thickener, lubricant, antistat; also in defoamers, aerosol w/o emulsions, corrosion inhibition. [Lipo]

Liposorb P-20. CAS 9005-66-7; Polysorbate 40; hydrophilic surfactant, solubilizer for oils, emulsifier, lubricant, antistat. [Lipo]

Liposorb S. CAS 1338-41-6; EINECS 215-664-9; Sorbitan stearate; emulsifier, thickener, lubricant, antistat, all-purpose lipophilic surfactant used with POE Liposorb series; also for defoamers, aerosol w/o emulsions, corrosion inhibition. [Lipo]

Liposorb S-4. CAS 9005-67-8; Polysorbate 61. [Lipo] *

Liposorb S-20. CAS 9005-67-8; Polysorbate 60; food emulsifier, defoamer. [Lipo]

Liposorb SQO. CAS 8007-43-0; EINECS 232-360-1; Sorbitan sesquioleate; emulsifier, thickener, lubricant, antistat, all-purpose lipophilic surfactant used with POE Liposorb series. [Lipo]

Liposorb TO. CAS 26266-58-0; EINECS 247-569-3; Sorbitan trioleate; emulsifier, thickener, lubricant, antistat, all-purpose lipophilic surfactant used with POE Liposorb series. [Lipo]

Liposorb TO-20. CAS 9005-70-3; Polysorbate 85; o/w emulsifier, lubricant, antistat, all-purpose hydrophilic surfactant; used for solubilizing oils and in conjunction with Liposorb esters. [Lipo]

Liposorb TS. CAS 26658-19-5; EINECS 247-891-4; Sorbitan tristearate; emulsifier, thickener, lubricant, antistat, all-purpose lipophilic surfactant used with POE Liposorb series. [Lipo]

Liposorb TS-20. CAS 9005-71-4; Polysorbate 65; food emulsifier, defoamer. [Lipo]

Lipostat Tablets. CAS 81131-70-6; Pravastatin sodium [Bristol-Myers Squibb Pharmaceuticals Ltd]

Lipotin 100, 100J, SB. CAS 8002-43-5; EINECS 232-307-2; Refined soy lecithin, filtrated, deodorized; wetting and dispersing agents preventing sedimentation in paints; raw material for textile and leather industry compounds. [Lucas Meyer]

Lipotin A. Phosphoamino compound.; emulsifier, wetting and dispersing agent for aq. systems, stabilizer for latex and emulsion paints, leather finishes, water-disp. and -reducible air-dry and stoving alkyds, offset printing inks, textile auxiliaries. [Lucas Meyer]

Lipotin H. CAS 8029-76-3; EINECS 232-440-6; Hydroxylated soy lecithin; wetting and dispersing agent for air-drying and stoving paints; raw material for textile and leather compounds; emulsifier. [Lucas Meyer]

Lipotriad. A proprietary preparation of tricholine citrate, inositol, DL-methionine, cyanocobalamin, thiamine hydrochloride, riboflavine, nicotin amide, pyridoxine hydrochloride, and pantothenyl alcohol. [Lewis Laboratories]‡

Lipovol A. CAS 8024-32-6; Avocado oil; conditioner, glosser, emollient imparting a light, nongreasy, silky afterfeel to skin and hair products; high film gloss and rapid spread; used in personal care products. [Lipo]

Lipovol ALM. CAS 8007-69-0; Sweet almond oil; conditioner, glosser, emollient imparting a light, nongreasy, silky afterfeel to skin and hair products; high film gloss and rapid spread; used in personal care products. [Lipo]

Lipovol G. CAS 8024-22-4; Grape seed oil; conditioner, glosser, emollient imparting a light, nongreasy, silky afterfeel to skin and hair products; high film gloss and rapid spread; used in personal care products. [Lipo]

Lipovol HS. CAS 8016-70-4; EINECS 232-410-2; Hydrogenated soybean oil. [Lipo]

Lipovol J. CAS 61789-91-1; Jojoba oil, refined; emollient, conditioner, and lubricant for cosmetics and toiletries; rapid spread and soft, nontacky afterfeel; used in skin and personal care products and oils, anhyd. and emulsified makeups. [Lipo]

Lipovol MOS-70. Tridecyl stearate, neopentyl glycol dicaprylate/dicaprate, tridecyl trimellitate; specialty esters exhibiting tactile props. of mineral oil. [Lipo]

Lipovol MOS-130. Tridecyl stearate, tridecyl trimellitate, dipentaerythrityl hexacaprylate/hexacaprate; specialty esters exhibiting tactile props. of mineral oil. [Lipo]

---

‡ Trade name and manufacturer not verified   § Trade name without identified manufacturer

Lipovol MOS-350. Dipentaerythrityl hexacaprylate/hexacaprate, tridecyl trimellitate, tridecyl stearate, neopentyl glycol dicaprylate/dicaprate; specialty esters exhibiting tactile props. of mineral oil. [Lipo]

Lipovol P. Apricot kernel oil; emollient used in cosmetics and pharmaceuticals; soft, nontacky afterfeel and high film gloss. [Lipo]

Lipovol PAL. CAS 8002-75-3; Palm oil. [Lipo]

Lipovol SAF. CAS 8001-23-8; Safflower oil; conditioner, glosser, emollient imparting a light, nongreasy, silky afterfeel to skin and hair products; high film gloss and rapid spread; used in personal care products. [Lipo]

Lipovol SES. CAS 8008-74-0; Sesame oil; emollient, solv., and vehicle used in cosmetics, toiletries, and pharmaceuticals; offers light, nontacky feel and enhances gloss and spread of pigmented sticks and pot products. [Lipo]

Lipovol SO. Hybrid safflower oil; conditioner, glosser, emollient imparting a light, nongreasy, silky afterfeel to skin and hair products; high film gloss and rapid spread; used in personal care products. [Lipo]

Lipovol SOY. CAS 8001-22-7; Soybean oil; natural emollient oil, lubricant, conditioner for luxury skin products, hair care products, makeup, fine soaps, bath oils, anhyd. systems. [Lipo]

Lipovol SUN. CAS 8001-21-6; Sunflower seed oil; emollient imparting a pleasant, nongreasy feel to skin and hair care products and makeups; adds conditioning, spread, and sheen to personal care products; used in preparation of margarine. [Lipo]

Lipovol WGO. CAS 8006-95-9; Wheat germ oil; emollient imparting perceptible afterfeel to skin and hair care products; gloss to anhyd. and emulsified makeups. [Lipo]

Lipowax D. Cetearyl alcohol, ceteareth-20; o/w self-emulsifying wax, used in skin and hair creams and lotions, personal care products. [Lipo]

Lipowax G. Stearyl alcohol and ceteareth-20; emulsifier for personal care products. [Lipo]

Lipowax NI. Cetearyl alcohol and ceteth-20; see Lipowax D. [Lipo]

Lipowax P. Cetearyl alcohol, polysorbate 60; o/w self-emulsifying wax for neutral and mildly acidic and alkaline pH systems. [Lipo]

Lipowax P-SPEC. Cetearyl alcohol, polysorbate 60; emulsifying wax for formulation of creams, lotions, and ointments. [Lipo]

Lipowax PR. Cetearyl alcohol, polysorbate 60, PEG-150 stearate, steareth-20; emulsifying wax for formulation of creams, lotions, and ointments. [Lipo]

Lipowitz's alloy. An alloy of 50% bismuth, 27% lead, 13% tin, and 10% cadmium. It has a specific heat of 0.0345 calories per gram at from 5-50 C., and melts at 65 C; used for automatic sprinklers and other purposes.

Lipoxol® 12000. CAS 25322-68-3; PEG-240; cosmetic ingred. for lipsticks, deodorant sticks, soap bars, powd. bases, creams, and pastes. [Hüls Am.; Hüls AG]

Lipozyme. An experimental preparation of an immobilized lipase; the fungal lipase is produced by submerged fermentation of a selected strain of *Mucor miehei* for interesterification, alcoholysis of oils and fats and synthesis of esters. [Novo Nordisk]

Liquaemin sodium. Heparin sodium; anticoagulant. [Organon Inc] *

Liquamar. CAS 435-97-2; Phenprocoumon; anticoagulant.

[Organon Inc] *

Liquamycin. CAS 60-54-8; Tetracycline; antiamebic; antibacterial; antirickettsial. [Pfizer Inc] †

LiquaPar® Oil. Isopropylparaben, isobutylparaben, butylparaben; preservative for cosmetics and topical pharmaceuticals. [Sutton Labs]

Liquapen. CAS 113-98-4; Penicillin G potassium; antibacterial. [Pfizer Inc] †

Liquemin. CAS 9005-49-6; Proprietary preparation of heparin; anticoagulant. [Roche Products Ltd] *

liqueur de Ferraile. See pyrolignite of iron.

liqueur de van Swieten. Consists of 1 part mercuric chloride, 100 parts alcohol, and 900 parts water.

Liquibor. Organic borate. [Manchem Ltd]

Liquibor 169. Potassium glycol borate; corrosion inhibitor for glycol-based brake fluids. [Rhone-Poulenc UK]

Liquibor 524. A borax-glycol condensation product which is used at a concentration of 1-2% as a corrosion inhibitor in synthetic hydraulic fluids. [Rhone-Poulenc UK]

Liquibrom. Cooling water biocide. [Great Lakes Europe]

Liquical. CAS 10043-52-4; Liquid calcium chloride. [Dow UK]

Liqui-Cee. CAS 134-03-2; Sodium ascorbate; vitamin. [Baxter Health Care] *

Liquid 99. Liquid cleanser. [British Nova Works Ltd]

Liquid Q4 Borders and Beds Fertiliser. Liquid concentrate containing NPK 5.3:5.3:10 plus trace elements; border and bedding plant fertilizer. [Vitax Ltd]

Liquid Absorption Base Type A, T. Mineral oil, lanolin alcohol; emollient for liq. make-up to improve dispersion and application properties of pigments; primary oil phase ingred. in o/w emulsions; type T is better solv. for oil-sol. dyes. [Croda Inc.]

liquidambar. (Copalm balsam). A balsam obtained from a large Mexican tree. It contains cinnamyl cinnamate and styrene. It is also erroneously called liquid storax.

Liquid Bases. Compounded lanolin-derived sterols. [Croda Chem. Ltd]

liquid bronzes. Varnishes in which bronze colors are suspended.

Liquid Code XLR. CAS 10090-54-7; Potassium pyroantimonate solution; crosslinker for guar and derivatized guar at pH 3-5. [Rhone-Poulenc/Water Soluble Polymers]

Liquid Copper Fungicide. Fungicide. [Murphy Chemical Ltd] †

Liquid Crystal CN/9. CAS 57-88-5; Cholesteric esters; carriers for nutrients in skin care products; provides decorative, functional and aesthetic effects to cosmetics. [Presperse]

liquid drier. A name given to a concentrated solution of calcium chloride.

Liquid Feed for Hanging Baskets. Liquid concentrate containing NPK 4:2:6 plus trace elements; liquid fertilizer for hanging baskets, tubs, planters and window boxes. [Vitax Ltd]

liquid gold. Contains about 10% of gold (as chloride), resin, lavender oil, and bismuth; used for painting china.

Liquid Growmore. Liquid concentrate containing NPK 7:7:7; general purpose liquid fertilizer. [Vitax Ltd]

Liquid Latex. CAS 9003-20-7; Polyvinyl acetate; used in oil well cementing applications for fluid loss control, rheology modification, and improved bonding. [Rhone-Poulenc/Water Soluble Polymers]

---

* Trade name not verified as available　　　　　† Trade name verified as obsolete

Liquid Lightning. Inhibited sulfamic acid; liquid; removes water-formed deposits. [Garvey Chemical Corp] *

Liquidow. CAS 10043-52-4; Liquid calcium chloride; for applications on unpaved roads to control dust. [Dow UK]

liquid paraffin. See mineral oil.

liquid pitch oil. See oil of tar.

liquid resins. (Polyterpene, Sulfate Resin, Talleol). Semi-resinous compounds obtained as by-products in the manufacture of wood pulp by the sulfite and sulfate processes of paper making.

liquid silver. See mercury.

liquid storax. Storax, a balsam.

Liquid Tomato Feed. Liquid concentrate containing NPK 4.5:4.5:9 plus Mg; fertilizer. [Vitax Ltd]

Liquifilm Forte® . Polyvinyl alcohol; ocular lubricant. [Allergan, Inc]

Liquifilm Tears® . Polyvinyl alcohol; ocular lubricant. [Allergan, Inc]

Liquigel® . CAS 21645-51-2, CAS 1309-42-8; Aluminum hydroxide and magnesium hydroxide fluid gel; suspension antacids. [Reheis Inc]

Liquigel-AM. Aluminum magnesium fluid gel. [Reheis Inc]

Liquigel-HO. High oxide fluid gel; antacid formulations. [Reheis Inc]

Liquimeth. Liquid methionine salt. [Degussa Ltd]

Liqui-Nox® . Blend; phosphate-free detergent for critical manual and ultrasonic cleaning applications. [Alconox]

Liquinure. Liquid fertilizer. [Fisons plc, Horticulture Div] *

Liquiwax DC-EFA/SS. Dicetyl dilinoleate; emollient to soften the skin without greasiness and oiliness. [Brooks Industries]

Liquiwax DIADD. Diisoarachidyl dodecanedioate; emollient to soften the skin without greasiness and oiliness. [Brooks Industries]

Liquiwax DICDD. Diisocetyl dodecanedioate; emollient to soften the skin without greasiness and oiliness. [Brooks Industries]

Liquiwax DIEFA. Diisoarachidyl dilinoleate; emollient to soften the skin without greasiness and oiliness. [Brooks Industries]

Liquor, Glass. See soluble glass.

liquorice. CAS 58-55-9; The dried root of *Glycyrrhiza glandulifera*.

Lisat. Stearyl lactyl-2-lactylates. [Thomas Goldschmidt Ltd]

Liskonum Tablets. CAS 554-13-2; Lithium carbonate; controlled release tablet for treatment of acute episodes of mania or hypomania and for the prophylaxis of recurrent manic-depressive illness. [SmithKline Beecham] *

Lissamine. Acid and direct dyestuffs. [ICI Chem. & Polymers Ltd]

Lissanol. Leather finishes. [ICI Chem. & Polymers Ltd]

Lissapol. Synthetic detergents. [ICI Chem. & Polymers Ltd]

Listab. Metal stearates. [Chemson Ltd]

Listerine. A proprietary antiseptic containing boric acid, benzoic acid, thymol, and essential oils of eucalyptus, gaultheria, and others. §

litalbin. Lecithin albuminate.

Litefax. Low density insulating sideliner tiles for small to medium steel ingots. [Foseco (F.S.) Ltd] †

Litex. A range of styrene-butadiene copolymers, emulsions and dispersions; used for paints and anticorrosion coatings. [Hüls AG]

Lithane. CAS 554-13-2; A proprietary preparation of lithium carbonate; an antidepressant. [Pfizer International]

litharge. CAS 1317-36-8; (Massicot). Pigments consisting of lead monoxide, PbO. Litharge is obtained in silver refining, and has a more reddish color than Massicot, which is made by roasting lead.

Litharge 33. High-purity lead oxide; acid acceptor, activator, and vulcanizing agent in rubber compounding. [Eagle-Picher] *

litharge, leaded. See lead oxide.

lithargrite. A mixture of oxide of lead and calcined magnesia; used as a rubber filler.

Lithene®. Liquid polybutadiene telomers and compounds; components in chlorinated paint, electrical encapsulants, specialty adhesives, rubber coagent, surface coatings. [Reichhold]

Lithex. Lead dithiobenzoate precipitated on an inert base such as clay; a proprietary rubber vulcanization accelerator. §

lithic acid. CAS 69-93-2; Uric acid, $C_5H_4N_4O_3$; used in organic synthesis.

lithic acid. See uric acid.

lithiopiperazine. A preparation of piperazine and lithium salts; used as a solvent for uric acid.

lithium. Metallic element; at. no. 3; Li; CAS 7439-93-2; EINECS 231-102-5; Scavenger and degasifier for stainless and mild steels in molten state; deoxidizer in copper and alloys; rocket propellants; pharmaceuticals. [Atomergic Chemetals; Cerac; FMC; Leverton-Clarke Ltd]

lithium hypochlorite. LiOCl; CAS 13840-33-0; EINECS 237-558-1; Laundry bleach, swimming pool chlorination. [FMC]

lithium molybdate. Li_2MoO_4; Steel coating, petroleum cracking catalyst. [AAA Molybdenum Prods.; Atomergic Chemetals; Cerac]

lithium nitrate. $LiNO_3$; CAS 7790-69-4; EINECS 232-218-9; Ceramics, pyrotechnics, salt baths, heat exchange media, refrigeration systems, rocket propellant. [Atomergic Chemetals; Cerac; Mallinckrodt]

lithium octadecanoate. See lithium stearate.

lithium stearate. (lithium octadecanoate; octadecanoic acid, lithium salt) Lithium salt of stearic acid; $CH_3(CH_2)_{16}COOLi$; CAS 4485-12-5; cosmetics, plastics, waxes, greases, lubricant in powd. metallurgy, corrosive inhibitor, flatting agent, high-temp. lubricant. [Chemetall GmbH; Schweizerhall; Syn. Prods.; Witco]

Lithobid. CAS 554-13-2; Lithium carbonate; antimanic. [Ciba-Geigy Corp]

litho-carbon. A material resembling asphalt, found in Texas.

lithoclastite. An explosive. It is a dynamite.

Lithoform. A pretreatment for zinc. [ICI Chem. & Polymers Ltd] †

lithographer's varnish. See stand oil.

lithographic stone. See limestone.

lithol. Ammonium ichtho-sulfonate.

Lithol® Pigments. Azo dye lakes for letterpress, offset, flexographic and gravure inks, for paints, for coloring plastics. [BASF AG]

lithomarge. (Lithocolla, leucoargilla). A compact clay found in rock fissures.

Lithonate. CAS 554-13-2; Lithium carbonate; antimanic. [Rowell Laboratories Inc] *

litho-oil. (stand oil, polymerized oil). Raw linseed oil heated in such a manner that practically no oxidation occurs,

---

‡ Trade name and manufacturer not verified   § Trade name without identified manufacturer

thus thickening the oil through polymerization alone.

lithophone. See lithopone.

lithopone. (Griffith's zinc white; Beckton white; Duresco; enamel white; Fulton white; Jersey lily white; Knights patent zinc white; Marbon white; Nevin; Orrs white; oleum white; pinolith; porcelain white; Rosss white; zinc baryta white; zincolith) Mixture of zinc sulfide (26-60%), barium sulfate, and zinc oxide; CAS 1345-05-7; white pigment in paints, paper; provides thixotropy, improves gloss and flow; it has been largely replaced by titanium dioxide in paints and rubber goods. [Ore & Chem. Corp.; Sachtleben GmbH]

Lithopone®. Zinc sulfide/barium sulfate, $ZnS/BaSO_4$; white pigment for paints and coatings. [Sachtleben Chemie GmbH]

Lithostar. Reprographic chemical. [Octel Chemicals Ltd] †

Lithostat. CAS 546-88-3; Acetohydroxamic acid; enzyme inhibitor. [Mission Pharmacal Co] *

Lithotabs. CAS 554-13-2; Lithium carbonate; antimanic. [Rowell Laboratories Inc] *

litmus. (lichen blue, lacmus, tournesol, turnsole). The coloring matter derived from different species of lichens; used as an indicator used in analytical chemistry where precision is not required, soil testing.

litnum bronze. An alloy of from 80-85% copper, 10-15% aluminum, and 4% iron.

Litrison. A proprietary multivitamin preparation containing vitamins B_1, B_2, B_6, and B_{12}, methionine, choline tartrate, dexpanthenol, biotin, folic acid, and vitamin E; used in liver disease. [Roche Products Ltd] *

littoral. A decolorizing agent for sugar juices.

liver of antimony. The name applied to the impure double sulfides of antimony, obtained by heating antimony sulfide, Sb_2S_3, with various metallic sulfides, more especially with the alkali and alkaline earth sulfides.

Liveroid. A proprietary liver extract. §

liver ore. See cinnabar.

liver sugar. Glycogen.

Lizetan Spray. Insecticidal spray; used for control of biting and sucking pests. [Bayer AG]

LLDPE. See polyethylene, linear low density.

Lloyd's Cream®. 10% w/w Diethylamine salicylate cream; for relief of rheumatics aches, pains, strains, sprains, fibrositis, lumbago, and sciatica. [Seton Healthcare Group plc]

Lloyd's reagent. A hydrous aluminum silicate prepared from fuller's earth. It absorbs alkaloids.

LMT. Meat tenderizer. [Atomergic Chemetals Corp]

LNA. Leucine aminopeptidase substrate. [Monomer-Polymer & Dajac Laboratories Inc]

loadstone. See magnetic iron ore.

Loalin. A proprietary trade name for a polystyrene molding compound. §

Loams. Natural mixtures of clay and sand; used for the manufacture of bricks and tiles. §

Lobak. A proprietary preparation of chlormezanone and paracetamol. §

Lobosol. Oxygenated solvents. [BP Chemicals Ltd] †

Lobra. Partially hydrogenated canola oil. [Karlshamns]

Locan. A proprietary preparation of amethocaine, amylocaine, bismuth subnitrate, and zinc oxide; anesthetic anal suppositories. [British Drug Houses] *

Locke's solution. A saline solution containing dextrose and used for injections.

Lockite. Polyolefin; continuous extruded strip high slip fendering for docks and similar applications. [Stanley Smith & Co Plastics Ltd]

Locobase Cream and Ointment. A proprietary base for creams and ointments. [Brocades Pharma]

Locoid. CAS 13609-67-1; A proprietary preparation of hydrocortisone 17-butyrate; used as a topical corticosteroid. [Brocades Pharma]

Locorten. CAS 2002-29-1; A proprietary preparation of flumethasone pivalate for dermatological use. [Ciba plc]*

Locorten-N. A proprietary preparation of flumethasone pivalate and neomycin sulfate; used in dermatology as an antibacterial agent. [Ciba plc] *

locoum. A gum-like mass prepared from starch paste and sugar.

Locron. Aluminum chlorohydrate; anhidrotic. [Hoechst-Celanese]

Loctite. Single component structural adhesives and thread locking materials possessing the unique property of setting when air is excluded; some are based on oxygenated methacrylic molecules of patented formulation. §

locust bean gum. (carob flour; carob seed gum; algaroba) Polysaccharide plant mucilage; ground seed of the ripe fruit of St. John's Bread (Ceratonia siliqua); CAS 9000-40-2; EINECS 232-541-5; in foods as stabilizer, thickener, emulsifier; cosmetics; sizing and finishes for textiles; pharmaceuticals; paints; fiber bonding agent in paper manufacture; drilling fluid; coffee, chocolate substitute. [Grindsted; Hercules]

Lodestone. See magnetic iron ore.

Lodine. CAS 41340-25-4; Etodolac capsules or tablets; for acute or long-term use in rheumatoid arthritis and osteoarthritis. [Wyeth Laboratories]

Lodol. Dephosphorizing agent for ladle treatment of rimming steels. [Foseco (F.S.) Ltd] †

Lo-Dose. CAS 61791-44-4; Cationic surfactant containing 800 g/l tallow amine ethoxylate; wetting agent for phosphonoglycine herbicide sprays. [Quadrangle Agrochemicals]

Lodosin. CAS 28860-95-9; Carbidopa; a decarboxylase inhibitor for concurrent use with levodopa in the treatment of Parkinson's disease. [Merck & Co Inc] †

Lodosyn. See Lodosin.

Loestrin 20. A proprietary preparation of ethinylestradiol and norethisterone acetate; an oral contraceptive. [Parke-Davis] *

Lo-ex. An aluminum alloy containing 14% silicon, 2% nickel, 1% copper, and 1% magnesium.

Lofenalac. Proprietary name for an infant milk feed low in phenylalanine; used in the dietary treatment of phenylketonuria. [Bristol-Myers Squibb Co Inc]

Loftine. See mulsoid.

Logas. For degassing copper base alloys. [Foseco (F.S.) Ltd]

LoGel. CAS 637-12-7; Aluminum stearate; for nonaqueous coating compositions. [Hüls Am.] †

logwood. (Campeachy wood, Jamaica wood, Hematine paste and powder, steam black). A natural dyestuff from the wood of Haematoxylon campechianum. It is sold as chips or extract. The wood contains haematoxylin, $C_{16}H_{14}O_6 \cdot 3H_2O$; used for dying black with a chrome mordant, for wool; with an iron mordant for silk; and with a chrome or iron and aluminum mordant for cotton.

---

* Trade name not verified as available   † Trade name verified as obsolete

Logynon. 6 Tablets containig 30 mcg ethinylestradiol and 50 mcg leonorgestrel, 5 tablets containing 40 mcg ethinylestradiol and 75 mcg levonorgestrel, 10 tablets containing 30 mcg ethinylestradiol and 125 mcg levonorgestrel. [Schering Health Care Ltd]

Lo-Hetaplas®. For the medical industry. [DuPont UK]

Lohys steel. A mild steel having a high magnetic permeability.

Lokandi. *See* Ventilago Madraspanta.

Lokset. Resin anchor systems. [Foseco (F.S.) Ltd]

LoLoss®. Natural polymeric viscosifier stabilized to prevent thermal degradation; fluid loss control agent/flocculant for drilled solids; encapsulates cuttings, stabilizes the borehole. [Rhone-Poulenc/Water Soluble Polymers]

Lolotint 97. CAS 1317-39-1; Cuprous oxide. [CP International Chemicals Ltd]

Lomag. Magnesium removing flux. [Foseco (F.S.) Ltd] †

loman steel. An abbreviation of low-manganese steel. It contains from 7-10% manganese.

Lomar® D. CAS 9084-06-4; Sodium naphthalene sulfonate; dispersant for disperse and vat dyes. [Henkel/Textiles]

Lomar® HP. Condensed potassium naphthalene sulfonate; dispersant; secondary emulsifier for emulsion polymerization. [Henkel/Functional Prods.

Lomar® LS. CAS 9084-06-4; Condensed sodium naphthalene sulfonate; dispersant, emulsifier for emulsion polymerization, dyestuff mfg., agric. formulations; leveling agent for dyeing fibers; low salt. [Henkel/Functional Prods.; Henkel/Textile]

Lomar® PWA. Condensed ammonium naphthalene sulfonate; visc. depressant; for molding and extruding operations in ceramics; dispersant for emulsion paints, agric. formulations; emulsifier for emulsion polymerization of synthetic elastomers; visc. reducer for pigment slurries; stabilizer. [Henkel/Emery; Henkel/Functional Prods.

LO/MIT™-1. Silver colored, nonthickness dependent, low emissivity, radiant barrier coating for energy conservation and light and heat reflection; high temperature tolerant coating. [Solec]

Lomodex. A proprietary preparation of dextrans in saline or dextrose solution; used to improve capillary circulation. [Fisons plc, Pharmaceuticals Div] *

Lomod® AE2020a, AE2040a, AE2060a. Engineering elastomer; automotive exterior grade [GE Plastics]

Lomod® FR5020a. Engineering elastomer; flame retardant grade offering outstanding elec. props., chem. resist., toughness; for mating seals, complex parts. [GE Plastics]

Lomod® FR30125a. Engineering elastomer; flame retardant grade offering outstanding elec. props., chem. resist., toughness; for mating seals, complex parts. [GE Plastics]

Lomod® HG3015a. Engineering elastomer; high gravity grade providing structural integrity, sound-damping capabilities and the ability to be molded into complex parts. [GE Plastics]

Lomod® ST3090a. Engineering elastomer; provides soft, tactile feel and flexibility for sporting goods, appliances, power tools, and personal care articles; adheres to other resins permitting multi-stiffness or multi-color parts. [GE Plastics]

Lomod® TE3040a. Thermoplastic elastomer; offers flexibility, toughness, resist. to chems., ozone, hydrocarbons

and temp. extremes; for hoses, tubing, boots, bellows, diaphragms, seals, safety equip., sporting goods, athletic footwear components. [GE Plastics]

Lomotil. A proprietary preparation of diphenoxylate hydrochloride and atropine sulfate; an antidiarrhoeal. [G D Searle & Co Ltd]

Lomotil with Neomycin. A proprietary preparation containing diphenoxylate hydrochloride, atropine sulfate and neomycin sulfate; an antidiarrhoeal. [G D Searle & Co Ltd] *

Lomudase. A proprietary preparation of chymotrypsin and isoprenaline sulfate; for treatment of bronchitis. [Fisons plc, Pharmaceuticals Div] *

Lomupren. A proprietary preparation of isoprenaline sulfate; a bronchial antispasmodic. [Fisons plc, Pharmaceuticals Div] *

Lomusol. A proprietary preparation of sodium cromoglycate; as a nasal spray in the treatment of allergic rhinitis. [Fisons plc, Pharmaceuticals Div] *

Lonacol. Copper-free fungicide; used for tomatoes, beans, potatoes, maize, onions and fruits. [Bayer AG]

Lonarit. An acetyl cellulose product; can be molded and colored. §

Londal. A proprietary trade name for certain aluminum alloys. §

Londax®. For the agriculture industry. [DuPont UK]

London paste. A paste made by adding a third of the weight of water to a mixture of equal parts of sodium hydroxide and powdered lime. A caustic.

London White. *See* Flake White.

longifene. CAS 569-65-3; The hydrochloride of buclizine.

Longlife Plus. Mixture of 2,4-D and dicamba; used to control weeds in turf. [ICI Agrochemicals Professional Products]

Longlife Turf Foods. Fertilizers containing long-lasting nitrogen for sports grounds and parks. [Scottish Agricultural Industries plc]

long oil varnishes. A classification of varnishes. They contain about 1 part of the solid constituents to 1 1/2 parts of drying oil. Short oil varnishes contain 1 part of solid to 1/2 part oil.

Loniten. CAS 38304-91-5; Minoxidil; antihypertensive. [Upjohn]

Lonsicar. CAS 409-21-2; Silicon carbide; used for antislip and wear resistant concrete or resin surfaces. [Lonza AG]

Lontrel. CAS 1702-17-6; Herbicides based primarily on clopyralid. [Dow UK]

Lontrel. Weedkiller containing 3,6-dichloropicolinic acid, dichlorprop, and MCPA as potassium salts; (Distributed in UK for Dow Chemical Co) [ICI Chem. & Polymers Ltd]

Lontrel Plus. Soluble concentrate containing 15 g clopyralid, 420 g dichlorprop and 175 g MCPA per liter; a translocated herbicide for use on cereals. (Sold in UK for DowElanco) [ICI Chem. & Polymers Ltd]

Lonza KS. CAS 7782-42-5; Graphite; relatively round particle shape, high elec. and thermal conductivity, good compressibility; for brake linings, plastics, carbon brushes, batteries, electrochemistry, pencils, hard metals, lubricants, ceramics, catalysts. [Lonza G+T]

Lonzaine®. Betaine amphoterics; used in shampoos and industrial cleaners. [Lonza AG]

Lonzaine® 12C. CAS 68424-94-2; EINECS 270-329-4; Coco betaine; conditioner, foaming agent, visc. modifier, irritation mitigant used in personal care products and

---

‡ Trade name and manufacturer not verified　　　　　　　§ Trade name without identified manufacturer

industrial applications; biodeg. [Lonza Inc]

Lonzaine® C. Cocamidopropyl betaine; foaming agent, conditioner, visc. booster, wetting agent, used in cosmetics, toiletries, detergents, metal finishing, textile finishing, etc.; biodeg. [Lonza Inc]

Lonzaine® CS. CAS 68139-30-0; Cocoamidopropylhydroxysultaine; conditioner used in personal care products and industrial applications. [Lonza Inc]

Lonza Insta-Pearl®. Proprietary blend; high intensity pearlescent, opacifier for hair and skin care products. [Lonza Inc]

Lonzest®. A range of sorbitan esters, polyoxyethylene sorbitan esters and polyethylene glycol esters; used for emulsion formation and stabilization in foods, pharmaceuticals and cosmetics. [Lonza AG]

Lonzest® 143-S. CAS 6221-95-0; EINECS 226-300-9; Myristyl propionate; emollient, penetrant, and spreading agent in cosmetics; perfume solv. in personal care products; humectant. [Lonza Inc]

Lonzoid. A proprietary cellulose acetate. §

Loobwax 0651. CAS 9002-88-4; Polyethylene wax; oxidized lubricant for high-shear applications, pipes, profiles, clear films. [Astor Wax] *

Loobwax 0761. Esterified, oxidized polyethylene wax; lubricant. [Astor Wax] *

Lopid. CAS 25812-30-0; Gemfibrozil; antihyperlipoproteinemic. [Parke-Davis] *

Lopox®. A registered trademark for epoxy resins. §

Lopresor. CAS 56392-17-7; A proprietary preparation of metoprolol tartrate; used in the treatment of angina pectoris. [Ciba plc] *

Lopresoretic. Beta-blocker. [Ciba plc] *

Lopressor. CAS 56392-17-7; Metoprolol tartrate; anti-adrenergic. [Ciba-Geigy Corp] *

Lopressor SR. Beta-blocker. [Ciba plc] *

Loprox. CAS 41621-49-2; Ciclopiroxolamine; antifungal. [Hoechst-Roussel Pharmaceuticals Inc] *

Lorate®. For the agriculture industry. [DuPont UK]

Lorelco. Prescription drug for the treatment of hypercholesterolemia. [Dow UK] *

Loretine. (sulfiolinic acid). Iodohydroxyquinolinesulfonic acid, $C_9H_4NI(OH)SO_3H$; used as a germicide.

Lorexane. Preparations of γ benzene hexachloride; an insecticide; antiparasitic hair lotion. [ICI Chem. & Polymers Ltd]

Lorfan. A proprietary preparation of levallorphan tartrate; used in cases of narcotics overdosage and in obstetrical anesthesia. [Roche Products Ltd] *

Lorival. A proprietary electrical insulation made from a synthetic resin of the phenol-formaldehyde type. §

Lorol. A mixture of alcohols produced by the reduction of coconut oil; sulfated fatty alcohols. [Hickson & Welch Ltd]

Lorol C6. Hexyl alcohol; chemical intermediate for detergent mfg. [Henkel/Emery]

Lorol C8. CAS 111-87-5; EINECS 203-917-6; Octyl alcohol; intermediate for surfactant mfg.; in rolling oils and other lubricants. [Henkel/Emery]

Lorol C8-C10 Special. CAS 112-53-8; Lauryl alcohol (40-48% C8, 51-59% C10); chemical intermediate for detergent mfg. [Henkel/Emery]

Lorol C10. Decyl alcohol; intermediate for surfactant mfg.; component in rolling oils, lubricants. [Henkel/Emery]

Lorol C12. CAS 112-53-8; Lauryl alcohol; intermediate for

surfactant mfg.; component in rolling oils, lubricants. [Henkel/Emery]

Lorol C12-C14. CAS 112-53-8; Lauryl alcohol (65-69% C12, 24-28% C14, 4-8% C16); intermediate for surfactant mfg.; component in rolling oils, lubricants. [Henkel/Emery]

Lorol C14. CAS 112-72-1; EINECS 204-000-3; Myristyl alcohol; intermediate for surfactant mfg.; component in rolling oils, lubricants. [Henkel/Emery]

Lorol C16. CAS 36653-82-4; EINECS 253-149-0; Cetyl alcohol; intermediate for surfactant mfg.; component in rolling oils, lubricants. [Henkel/Emery]

Lorol C18. CAS 112-92-5; EINECS 204-017-6; Stearyl alcohol; intermediate for surfactant mfg.; component in rolling oils, lubricants. [Henkel/Emery]

Lorol DA. Diethanolamine lauryl sulfate in liquid form; anionic surfactant used in shampoos, milder in action than other alkyl sulfates in the Lorol series. §

Lorol MA and MR. Monoethanolamine lauryl sulfate in liquid form; Lorol MR is a built product; anionic surfactant for shampoos. §

Lorol NH. CAS 2235-54-3; EINECS 218-793-9; Ammonium lauryl sulfate in liquid form; anionic surfactant having a mild action; used in good quality shampoos. §

Lorol Special. CAS 112-53-8; Lauryl alcohol (70-75% C12, 24-30% C14); chemical intermediate for detergent mfg. [Henkel/Emery]

Lorol TA and TAR. CAS 139-96-8; EINECS 205-388-7; Triethanolamine lauryl sulfate in liquid form; Lorol TAR is a built product containing alkylolamide foam booster; anionic surfactant used in shampoos. §

Lorol TN and TNR. Triethanolamine/ammonium lauryl sulfate in liquid form; Lorol TNR contains alkylolamides; anionic surfactant used in shampoos. §

Lorsban. CAS 2921-88-2; Insecticides containing chlorpyrifos; used to control ticks on cattle, mosquitoes, and other insects. [Dow UK] *

Losan. A proprietary preparation consisting of quinophan and sodium bicarbonate. §

Losetic. A proprietary procaine hydrochloride and adrenalin preparation. §

Losilphos. Ferophosphorus briquettes; used in steelmaking for their phosphoros content. [Monsanto Co] *

Losophane. Triiodo-m-cresol, $C_6H_3(I_3)OHCH_3$; used externally in skin diseases, as an antiseptic and astringent.

LoSOx. High activity catalyst for the abatement of carbon monoxide and unburned hydrocarbons in the presence of high levels of sulfur compounds; most suitable for operations where the oxidation of sulfur dioxide must be limited. [Engelhard]

Lotader® 3210. Acrylic terpolymer; for aluminum coating, coextrusion applications. [Atochem SA]

Lotader® 3700. Acrylic terpolymer; for modification of thermoplastic polymers. [Atochem SA]

Lotader® AX 8660. Ethylene-acrylic ester-glycidyl methacrylate terpolymer. [Atochem N. Am.] *

Lotader® P3-3200. Acrylic terpolymer; for heat adhesive film and powd., pipe coating applications. [Atochem SA]

Lotensile See Carbora. §

Lotioblanc. White lotion; astringent; protectant (topical). [Baxter Health Care] *

Lotrimin. CAS 23593-75-1; Clotrimazole; antifungal. [Schering Corp]

Lotryl® 17 BG 04. Ethylene-butyl acrylate copolymer; for

---

* Trade name not verified as available   † Trade name verified as obsolete

films (gloves), coextrusion. [Atochem SA]

Lotryl® 20 MA 08. Ethylene-acrylic ester (EMA) copolymer; for films, coextrusion, coating, hot-melt adhesives, compounds. [Atochem SA]

Lotryl® 28 BA 175. Ethylene-butyl acrylate copolymer; for hot-melt adhesives. [Atochem SA]

Lotusate. CAS 115-44-6; Talbutal; sedative. [Sterling Drug Inc] *

Louse Powder. Insecticide. [Schering Agrochemicals Ltd]

Lo-Vel. Amorphous silica flatting agents. [PPG Industries]

lovol. A mixture of aliphatic alcohols formed by the high pressure hydrogenation of coconut oil.

Low Crock 18-G-3. Modified org. resin emulsion; excellent binder props. for aq. pigment printing systems. [Eastern Color & Chem.

Low Crock FB-I. Finishing auxiliary for reduced bleeding/crocking of dyed and printed fabrics; minimal effect on fabric hand, minimum buildup on equip.; resin and catalyst stable. [Yorkshire Pat-Chem]

Lowerite. Strippable coating compositions. [Croda Chem. Ltd] *

Lowilite® 22. CAS 1843-05-6; EINECS 217-421-2; Benzophenone-12; uv absorber for PP, PE, PVC, and other polymers; excellent heat resist. at extrusion temps. [Lowi]

Lowilite® 26. CAS 3896-11-5; EINECS 223-445-4; 2-(2′-Hydroxy-3′-t-butyl-5′-methylphenyl)-5-chlorobenzotriazole; uv stabilizer for polyolefins, polyester resins and coatings. [Lowi]

Lowilite® 55. CAS 2440-22-4; EINECS 219-470-5; 2-(2′-Hydroxy-5′-methylphenyl) benzotriazole; uv absorber for polymers incl. polyester, PVC, PS, HIPS, rubber, PC, PMMA. [Lowi]

Lowilite® 62. α-Methylstyrene/N-(2,2,6,6-tetramethylpiperidinyl-4)maleimide/N-stearyl maleimide terpolymer; HALS protecting LDPE, HPDE, LLDPE, and PP against degradation by oxidation and photo-oxidation; very high thermal resist. to 320 C. [Lowi]

Lowilite® 77. CAS 52829-07-9; EINECS 258-207-9; Bis (2,2,6,6-tetramethyl-4-piperidinyl) sebacate; HALS for polyolefins, PS, HIPS, ABS, SAN, ASA; also suitable for PU, polyamides, acetal. [Lowi]

Lowinox® 001. CAS 128-39-2; 2,6-Di-t-butylphenol; antioxidant for oils. [Lowi]

Lowinox® 002. 4,4-Methylenebis(2,6-di-t-butylphenol); antioxidant for oils. [Lowi]

Lowinox® 22IB46. 2,2′-Isobutylidene-bis(4,6-dimethylphenol); antioxidant for rubber, latex. [Lowi]

Lowinox® 22M46. CAS 119-47-1; 2,2′-Methylenebis (4-methyl-6-t-butylphenol); antioxidant for rubber, latex, adhesives, ABS, polyacetate. [Lowi]

Lowinox® 44B25. 4,4′-Butylidene-bis(2-t-butyl-5-methylphenol); antioxidant for rubber, latex, adhesives, ABS, polyamide. [Lowi]

Lowinox® 44S36. CAS 96-69-5; 4,4′-Thiobis (2-t-butyl-5-methylphenol); antioxidant for latex, adhesives, plastics, cross-linked polymers. [Lowi]

Lowinox® 070. CAS 98-54-4; t-Butylphenol; antioxidant for oils and fuels [Lowi]

Lowinox® 241. 2,2′-Methylenebis-(4-methyl-6-t-butylphenol), tris(2,4-di-t-butylphenyl)phosphite; antioxidant for adhesives. [Lowi]

Lowinox® 242. CAS 31570-04-4; Sterically hindered polynuclear phenol and tris-(2,4-di-t-butylphenyl)phosphite;

antioxidant for hot-melts. [Lowi]

Lowinox® 243. Pentaerythrityl tetrakis-3-(3′,5′-di-t-butyl-4′-hydroxyphenyl)propionate, bis(2,4-di-t-butylphenyl) pentaerythritol diphosphite; antioxidant for adhesives, hot-melts, plastic, polyolefins, PS, PC, SAN. [Lowi]

Lowinox® 244. 4,4′-Butylidene-bis(2-t-butyl-5-methylphenol) and tris-(2,4-di-t-butylphenyl) phosphite; antioxidant for plastics, polyolefins, PS. [Lowi]

Lowinox® 245. 4,4′-Butylidene-bis(2-t-butyl-5-methylphenol) and distearyl-3,3′-thiodipropionate; antioxidant for plastics, polyolefins. [Lowi]

Lowinox® 246. 2,2′-Methylenebis(4-methyl-6-t-butylphenol) and distearyl-3,3′-thiodipropionate; antioxidant for APP. [Lowi]

Lowinox® 247. 2,6-Di-t-butyl-4-methylphenol and ditridecyl thiodipropionate; antioxidant for polyurethane. [Lowi]

Lowinox® 624. 2,4-Dimethyl-6-t-butylphenol; antioxidant for fuels. [Lowi]

Lowinox® ACP. Polymeric 2,2,4-Trimethyl-1,2-dihydroquinoline; antioxidant for rubber, latex, crosslinked polymers. [Lowi]

Lowinox® AH25. CAS 79-74-3; 2,5-Di-t-amylhydroquinone; antioxidant for adhesives. [Lowi]

Lowinox® BHT. CAS 128-37-0; BHT; antioxidant for rubber, adhesives, plastics, polyolefins, polystyrene. [Lowi]

Lowinox® CA 22. CAS 1843-03-4; 1,1,3-Tris-(2-methyl-4-hydroxy-5-t-butylphenyl) butane; phenolic antioxidant. [Lowi]

Lowinox® DLTDP. CAS 123-28-4; Dilauryl-3,3′-thiodipropionate; antioxidant for plastics, polyolefins. [Lowi]

Lowinox® DSTDP. CAS 693-36-7; Distearyl-3,3′-thiodipropionate; antioxidant for plastics, polyolefins. [Lowi]

Lowinox® MBMC. 2-t-Butyl-m-cresol; intermediate for antioxidants, synthetic musk. [Lowi]

Lowinox® MBPC. CAS 2409-55-4; 2-t-Butyl-p-cresol; intermediate for antioxidants, uv absorbers. [Lowi]

Lowinox® ODA. Octylated diphenylamine; antioxidant for rubber. [Lowi]

Lowinox® PO35. CAS 2082-79-3; Octadecyl-3-(3′,5′-di-t-butyl-4′-hydroxyphenyl) propionate; antioxidant for adhesives, hot-melts, plastics, polyolefins, polystyrene, PVC, pC, SAN. [Lowi]

Lowinox® PP35. CAS 6683-19-8; Pentaerythrityl tetrakis-3-(3′,5′-di-t-butyl-4′-hydroxyphenyl) propionate; antioxidant for adhesives, hot-melts, plastics, polyolefins, PS, PC, SAN. [Lowi]

Lowinox® PTBT. CAS 98-51-1; p-t-Butyl toluene; antioxidant. [Lowi] *

Lowinox® SDA. Styrenated diphenylamine; antioxidant. [Lowi] *

Lowinox® TBMX. CAS 98-19-1; t-Butyl-m-xylene; intermediate for perfumes and fragrances. [Lowi] *

Lowinox® TNPP. CAS 26523-78-4; Tris-nonylphenyl phosphite; antioxidant for rubber, adhesives, ABS. [Lowi]

Lowroff phosphor bronze. Alloys of 70-90% copper, 4-13% tin, 5-16% lead, and 0.5-1% phosphorus.

lox. Liquid oxygen; used in mining explosives and in rocket propellants.

loxa bark. Pale cinchona bark.

Loxiol G 52. C16-18 fatty alcohol; surfactant for polymerization. [Henkel/Functional Prods.

Loxiol G-70, G-71, G-72 and G-73. Proprietary names for a range of multifunctional polyesters of high molecular weight; used as additives to PVC compounds to reduce

---

‡ Trade name and manufacturer not verified   § Trade name without identified manufacturer

surface tackiness. [Henkel Chemicals Ltd] *

Loxiol P 1420. C16-18 fatty alcohol; surfactant for polymerization. [Henkel/Functional Prods.

Loxiol VPG 1354. CAS 112-92-5; EINECS 204-017-6; Stearyl alcohol; surfactant for polymerization. [Henkel/Functional Prods.

Loxiol VPG 1451. CAS 661-19-8; EINECS 211-546-6; Behenyl alcohol; surfactant for polymerization. [Henkel/Functional Prods.

Loxiol VPG 1743. CAS 36653-82-4; EINECS 253-149-0; Cetyl alcohol; surfactant for polymerization. [Henkel/Functional Prods.

Loxitane. CAS 1977-10-2; Loxapine; tranquilizer. [Lederle Laboratories USA] *

Loxon. CAS 321-55-1; A proprietary preparation of haloxon; a veterinary anthelmintic. §

Lozol. CAS 26807-65-8; Indapamide; antihypertensive; diuretic. [USV Pharmaceutical Corp] ‡

LP®-12. Polysulfide, mercaptan-terminated; used for building, aviation, automotive, insulating glass, and marine sealants; excellent resist. to oils, org. solvs., weathering, low and high temp., gas permeability, oxidation, ozone, moisture; cured using inorg. peroxides, org. oxidizing agents, urethane or epoxy resins. [Morton Int'l./Polymer Systems]

LP-100. Lead dioxide; catalyst/curing agent for polysulfide, low molecular weight butyl and polyisoprene rubber; oxidizer in mfg. of dyes and to control burning rate of incendiary fuses or pyrotechnics. [Eagle-Picher] *

LP®-977. Polysulfide polymer; for sealants (aircraft, automotive, construction, marine, insulating glass), fluid membranes, elec. potting, leather impregnation. [Morton Int'l./Polymer Systems]

LPS 1 Greaseless Lubricant. A blend of solvents, lubricants, moisture displacers and corrosion inhibitors to provide greaseless lubrication of delicate mechanisms to dry out sensitive electrical and electronic circuits; lubricant, penetrant, water displacer, cleans electrical equipment. [Holt Lloyd Corporation] *

LPS 2 General Purpose Lubricant. A blend of lubricants and corrosion inhibitors that provide general purpose lubrication, corrosion inhibition and penetration of seized mechanisms; lubricant, penetrant and quick release, corrosion inhibitor. [Holt Lloyd Corporation] *

LPS 3 Heavy Duty Rust Inhibitor. A blend of waxes, solvents, corrosion inhibitors and oils to provide long term protection of metal parts; corrosion inhibitors, chain lubricant and antiseize compound. [Holt Lloyd Corporation] *

LPS 500 Plus. A blend of waxes, oils and corrosion inhibitors used to lubricate and prevent corrosion in severe conditions; as a lubricant, corrosion inhibitor or antiseize compound. [Holt Lloyd Corporation] *

LPS Brake Cleaner. A solvent-based cleaner for brakes and brake components. [Holt Lloyd Corporation] *

LPS Electro Contact Cleaner. A premium blend of solvents which is safe to use on rubber paint, plastics and to clean and degrease sensitive electronic equipment; cleans electrical/electronic components and delicate mechanisms. [Holt Lloyd Corporation] *

LPS Engine Degreaser. A petroleum-base cleaner which will emulsify with water; engine degreaser. [Holt Lloyd Corporation] *

LPS Heavy-Duty Silicone Lubricant. A water-based dry film silicone lubricant; lubricant, safe for use on almost any surface, approved by USDA for use in federally inspected meat and poultry plants, approved for use in areas with incidental food contact. [Holt Lloyd Corporation] *

LPS Instant Cold Galvanize. A combination of solvents, resin binders, and 95% pure zinc metal which provides a galvanic coating similar to hot dipped galvanize; cold galvanizing compound that inhibits corrosion for up to three years, acts as a primer or finish coat and repairs damaged or worn hot-dipped galvanized coatings. [Holt Lloyd Corporation] *

LPS Instant Super Cleaner/Degreaser. A combination of chlorinated solvents to clean and degrease metal parts and electrical equipment; cleaner/degreaser for equipment. [Holt Lloyd Corporation] *

LPS Paint Remover. A blend of solvents, waxes, and stripping aids for removing all types of paint from metal surfaces; for removal of paint and varnishes. [Holt Lloyd Corporation] *

LPS Tap-All. An engineered blend of lubricant aids and corrosion inhibitors that can be used for the machining of all metals; does not contain sulfur or chlorine; as a lubricant when tapping, drilling, grinding, milling, threading, reaming, turning, boring, sawing or engraving all metals. [Holt Lloyd Corporation] *

LSD. CAS 50-37-3; Lysergic acid diethylamide; $C_{20}H_{25}N_3O$; hallucinogen.

Luaktin®. Antiskinning agents for oil-based paints and baking finishes; improved gloss and leveling. [BASF AG]

Lubafax. A proprietary formulation of propylene glycol and hydroxypropyl methylcellulose; a surgical lubricant for selected medical implements. [The Wellcome Foundation Ltd]

Lubasin®. Adhesives for textile printing. [BASF AG; BASF plc]

Lube-Booster® 1320, II. Lubricity additive for sol. oils; designed not to upset existing HLB balance of sol. oils. [Ferro/Keil]

Lube-Lok®. Family of oven-cured dry film lubricant coatings containing solid lubricants and suitable binder system; used for fasteners, slides, pins, clips, and numerous applications requiring dry lubrication and corrosion protection particularly under high loads. [E/M Corporation]

Lubestat. Water soluble, biodegradable lubricants; fiber lubricants, textile softeners, yarn processing aids. [Milliken] *

Lubestine. CAS 14807-96-6; Talc, asbestine substitute; extender for paints and as a general purpose filler. [Bromhead & Denison Ltd]

Lubit® 64. All-purpose textile lubricant for all fibers; promotes better fabric penetration. [Sybron]

Lubix. Die lubricant for semicontinuous casting. [Foseco (F.S.) Ltd]

Lubolid. Dry powder lubricants. [Dow Corning Ltd] *

Lubrajel® CG, DV, MS, TW. Polyglycerylmethacrylate and propylene glycol; autoclavable nondrying water-sol. lubricant for medical and surgical use. [Presperse]

Lubran 145. Alkyl methacrylate polymer; pour pt. depressant, visc. index improver for lubricating oils. [Toho Chem. Industry]

Lubran AD. Hydrocarbon wax/naphthalene condensate; pour pt. depressant, dewaxing aid for lubricating oils. [Toho Chem. Industry]

---

* Trade name not verified as available   † Trade name verified as obsolete

Lubrhophos® HR-719. CAS 70321-78-7; Aliphatic phosphate ester of ethoxylated butanediol, acid form; biodeg. low foaming emulsifier, EP agent, corrosion inhibitor, cleaning and lubricity aids for metalworking fluids. [Rhone-Poulenc Surf.

Lubrhophos® LB-400; CAS 39464-69-2; Org. phosphate ester of ethoxylated oleyl alcohol, free acid; lubricity and EP additive, rust inhibitor, wetting agent, emulsifier, detergent, dispersant; for lubricating and rolling oils, hydraulic and water-based cutting fluids, glass cutting and polishing lubricants. [Rhone-Poulenc Surf.; Rhone-Poulenc France]

Lubrhophos® LE-500. CAS 51811-79 1; Aromatic phosphate ester of ethoxylated nonyl phenol; EP agent, emulsifier, corrosion inhibitor, detergent, lubricant for metalworking fluids. [Rhone-Poulenc Surf.; Rhone-Poulenc France]

Lubrhophos® LE-700. CAS 51811-79-1; Aromatic phosphate ester of ethoxylated nonylphenol, free acid; high foaming lubricant, emulsifier, EP agent, corrosion inhibitor for water-based cutting fluids. [Rhone-Poulenc Surf.

Lubrhophos® LF-200. CAS 39464-67-0; Org. phosphate ester of ethoxylated dodecylphenol, free acid; lubricant, emulsifier for oil-based cutting fluids, hydraulic fluids, rolling oils, slushing compounds. [Rhone-Poulenc Surf.; Rhone-Poulenc France]

Lubrhophos® LM-400. CAS 39464-64-7; Aromatic phosphate ester of ethoxylated dinonylphenol, free acid; lubricant, emulsifier, EP agent, corrosion inhibitor, detergent for cutting fluids, hydraulic fluids, rolling oils. [Rhone-Poulenc Surf.

Lubrhophos® LP-700. CAS 39464-70-5; Complex org. phosphate ester of ethoxylated phenol, free acid; low foaming emulsifier, EP agent, corrosion inhibitor, detergent, lubricant for water-based metalworking fluids. [Rhone-Poulenc Surf.; Rhone-Poulenc France]

Lubrhophos® LS-500. CAS 9046-01-9; Phosphate acid ester of ethoxylated tridecanol; aliphatic hydrophobic base; lubricant, emulsifier for oil and water-based cutting fluids, hydraulic fluids, rolling oils. [Rhone-Poulenc Surf.; Rhone-Poulenc France]

Lubri-Bond®. Family of air curing dry film lubricant coatings containing solid lubricants and suitable binder system; for frictional surfaces requiring dry lubrication particularly on substrates which cannot tolerate oven cure. [E/M Corporation]

Lubricant EHS. Fatty acid ester; heat-stable boundary lubricant for microemulsions for heavy-duty machining of aluminum. [Hoechst Celanese/Colorants & Surf.

Lubricin 25. Processing aid for rigid PVC during calendering, injection molding of pipe, fittings, conduit, sheet and profiles and for rubber improving appearance of molded goods. [CasChem]

Lubricin N-1. Additive for use in motor fuels, penetrating oils, cutting, machining and metalworking fluids providing lubricity and detegent action and reducing rust, corrosion and metal wear. [CasChem]

Lubrico. A proprietary trade name for an alloy of 75% copper, 20% lead, and 5% tin. §

Lubricomp®. Lubricated thermoplastic compounds. [LNP; ICI Chemicals & Polymers Ltd]

Lubricomp® 189. CAS 9016-75-5; PPS composite, PTFE lubricated; for injection molding applications incl. hard-to-mold thin-walled parts. [LNP]

Lubricomp® DFL-4036. PC, 30% glass fiber-reinforced, 15% TFE-lubricated, flame-retardant; thermoplastic compound. for gear and bearing applications requiring low frictional properties, high PV values, and good wear resistance. [LNP]

Lubricomp® DL-4030. PC resin with 15% PTFE lubrication; internally lubricated composite offering excellent wear and frictional char. for gear, cam, and sliding applications. [LNP]

Lubrigel. CAS 1318-93-0; Organically modified refined montmorillonite clay; for use in the conversion of lubricant base stocks to clay-based greases. [Akzo Chemie UK Ltd]

Lubril CAT-X/VC. Fatty imidazoline deriv.; softener for treatment of glass; improves adhesion of hydrocarbon coating materials. [Rhone-Poulenc Surf.

Lubril PF-570. Aq. solution of cationic polymer; lubricant for fiberglass mfg. [Rhone-Poulenc Surf.

Lubril QC. High molecular weight hydrophilic polyester dispersion; imparts soil release and moisture transport props. to polyester filament fabrics; oil scavenger keeping oils and greases in suspension during scouring/dyeing; lubricant for fiber-to-fiber lubrication. [Rhone-Poulenc Surf.

Lubrimet® P 600, P 900. PPG; lubricant; solubilizer for dyestuffs and surfactants. [BASF AG]

Lubrisol. Natural oil; textile lubricant. [Scher]

Lubritab®. CAS 68334-00-9; Hydrogenated vegetable oil NF; lubricant for pharmaceutical tablets. [Mendell]

Lubrite B33. Bleaching of mineral oils, greases and waxes, including re-refined/recycled lubricating oil. [Minas de Gádor SA]

Lubrizol® 2152. CAS 61789-86-4; Calcium sulfonate; pigment dispersant and wetting agent for color concs., paints, coatings, inks; pigment flushing aid for org. and inorg. pigments; visc. stabilizer/reducer in plastisols. [Lubrizol]

Lubrizol® 2153. Succinimide; pigment dispersant and wetting agent for color concs., inks, plastisols and organosols. [Lubrizol]

Lubrol. Fatty alcohol ethoxylates with a neutral reaction; used as surfactants. [ICI Chem. & Polymers Ltd]

Lubrol 90. Ester; nonoily press release agent for rubber roll release problems in pulp/paper industry; surface lubricant for tabulator card stock and paper board. [CNC Int'l. L.P.

Lubrol N5. Nonylphenol ethoxylate nonionic surfactant in viscous liquid form; used for conversion of premium paraffin to self-emulsifiable solvent; in industrial cleaning and degreasing including metal surfaces. [ICI Chem. & Polymers Ltd] *

Lubrol N13. Nonylphenol ethoxylate nonionic surfactant in semisolid form; wetting and emulsifying agent used in pest control, oleines, metal treatment, and degreasing. [ICI Chem. & Polymers Ltd] *

Lucalen®. Ethylene copolymers with polar groups. [BASF AG; BASF plc]

Lucantin® CX. Citranaxanthine; for pigmentation of egg yolks. [BASF AG; BASF plc]

Lucaphos® 40, 48. CAS 7757-93-9; Dicalcium phosphate; for the feed industry. [BASF AG]

Lucarotin® 10%. CAS 7235-40-7; β-Carotene; for the feed industry. [BASF AG]

lucca oil. CAS 8001-25-0; Olive oil.

---

‡ Trade name and manufacturer not verified   § Trade name without identified manufacturer

Lucerno. An alloy containing 67.9% nickel, 27.5% copper, 2.4% iron, and 2.2% manganese.

Luchem AS-946. Methyl methacrylate/allyl methacrylate copolymer; antishrink additive. [Atochem] *

Lucidene. Range of aqueous acrylic and styrene-acrylic emulsions in printing inks and overprint lacquers. [Morton Int'l. Ltd]

Lucidol. CAS 94-36-0; Benzoyl peroxide. [Akzo Chemie UK Ltd]

Lucidol 75FP. CAS 94-36-0; Benzoyl peroxide with 25% water; active constituent in anti-acne creams and soaps and as a polymerization initiator for plastics production. [Atochem]

Lucidol-78. CAS 94-36-0; Benzoyl peroxide; initiator for bulk, sol'n., and suspension polymerization, and high-temp. and room temperature cure of polyester resins. [Atochem]

Lucidol GS. CAS 94-36-0; A range of various concentration benzoyl peroxide powders. [Akzo Chemie UK Ltd]

Lucidol RM. Solid organic peroxide; used as free radical initiator in polymerization. [Akzo Chemie UK Ltd]

Lucidril. A proprietary trade name for the hydrochloride of meclofenoxate. [Lloyd's Pharmaceuticals Ltd] *

Lucilite. Silica hydrogel. [Rhone-Poulenc UK]

Lucipal. CAS 94-36-0; Benzoyl peroxide compositions. [Akzo Chemie UK Ltd]

Lucirin® BDK. Uv initiator for radiation-curing finishes and putties. [BASF AG]

Lucite®. Proprietary trade names for a methyl methacrylate and acrylate synthetic resins; the material is more transparent than glass and has a very high refractive index. [DuPont UK]

Lucitone. See Vernonite. §

Lucobit®. Ethylene copolymer/bitumen; for the production of damp-proof courses in buildings above ground and underground construction, for elastic foundations, as protection against corrosion for steel tanks, for improving quality of road bitumen, for production of injection molded parts. [BASF AG; BASF plc]

Lucofen S A. CAS 151-06-4; A proprietary preparation of chlorphentermine hydrochloride; antiobesity agent. [Warner] *

Luconyl®. Concs. of org. and inorg. pigments using nonionic dispersants; for coloring polymer emulsion paints, inks, water-based wood glazes. [BASF AG]

Lucovyl. A proprietary polyvinyl chloride. [Pechiney-St Gobain] *

Lucryl®. CAS 9011-14-7; Polymethylmethacrylate; thermoplastic molding compounds; std. grades for injection molding or extrusion offer high transparency and brilliance, high surface hardness and scratch resist., excellent weather resist., mech. strength and rigidity; impact-modified grades; for injection molding and extrusion for construction industry, motorcar industry, lighting engineering, sanitary products, household appliances, advertising displays. [BASF AG; BASF plc]

Ludigol®. Mild oxidizing agent for textile finishing. [BASF AG; BASF plc]

Ludigol F. CAS 127-68-4; Sodium m-nitrobenzene sulfonate; used to accelerate the stripping of nickel from steel, brass or copper; to control the reaction rate in pickling nickel-chomium-iron alloys and for dissolving nickel and copper; additive in phosphate coating of metals. [Rhone-Poulenc Surf.]

Ludiomil. CAS 10347-81-6; Maprotiline hydrochloride; antidepressant. [Ciba plc]

Ludlum alloy. A heat-resisting alloy of iron with from 13-17% chromium, 1% silicon, 1% molybdenum, and 0.4% carbon.

Ludlum No. 602 Steel. A proprietary trade name for a steel containing 1.7% silicon, 0.7% manganese, 0.4% molybdenum, 0.12% vanadium, and 0.48% carbon. §

Ludopal®. Unsat. polyester resins; P grades for paraffin-containing polyester finishes, dissolved in styrene; P 6 M is styrene-free resin for peroxide-containing finishes; E grades for elastifying P grades; U 150 for paraffin-containing and paraffin-free uv finishes, filling compounds. [BASF AG]

Ludorum. CAS 15545-48-9; Suspension concentrate containing 500 g chlorotoluron per liter; a contact urea herbicide. [Tripart Farm Chemicals Ltd]

Ludox®. Colloidal silica. [DuPont UK]

Lufibrol®. Extraction assistants in pretreatment of textile goods. [BASF AG; BASF plc]

Lufibrol® E. Extraction agent for textile desizing and scouring. [BASF]

Lufibrol® FW. Bleaching agent, wool fiber protection agent for prevention of yellowing during dyeing. [BASF]

Lufibrol® NB-7. Extracting and dispersing agent for detaching and washing impurities in desizing, scouring, and caustic boil-off operations. [BASF]

Lufilen®. Org. and inorg. pigments in polyethylene; for polyolefin spin dyeing. [BASF AG]

Luftseide. (celta, soie nouvelle, tubulated silk). Artificial silks of the rayon type made with hollow central spaces. They are formed by adding gas-evolving materials to the viscous solution.

Lugacin®. 2 ml Snap-off ampoule containing a 2 ml solution of gentamicin sulfate BP equivalent to 80 mg gentamicin base and a 2 ml solution containing gentamicin sulfate BP equivalent to 20 mg gentamicin base; bactericidal antibiotic active against extremely broad spectrum of Gram-positive and Gram-negative pathogens including *E. coli, klebsiella, proteus, pseodomonas aeruginosa* and antibiotic-resistant strains of *staphlococcus aureus.* [Lagap Pharmaceuticals Ltd]

Lugalvan. Electroplating additive. [BASF plc]

Luganil® Dyes. Anionic dyes for leather and fur dyeing, for coloring cleaners, etc. [BASF AG; BASF plc]

Lugol's solution. Iodine-potassium iodide solution; 5 g of iodine are titrated with 5 g of potassium iodide, and 100 cc of water, and diluted to 1 liter.

Lugo's powder. Powdered cinchona bark.

Luhydran®. Water-dilutable binders for electrocoating finishes, baking finishes, wood and plastic coatings, anti-corrosion and roadmarking paints, printing inks, overprinting varnishes. [BASF AG]

Lukens Bone Wax. Made from natural bees wax, salicylic acid, and a natural oil; applied by surgeon in order to produce hemostasis in bone during surgery in which bone is cut. [Lukens International Corp] ‡

Lulea tar. A variety of Stockholm tar.

Lumarith® EC. A registered trademark for an ethyl cellulose thermoplastic resistant to oils. §

Lumarith® ER. A registered trademark for ethyl rubber. §

Lumattin®. Wax modified silica; matting agents in paint and varnish industry. [BASF AG]

Lumbang oil. A drying oil obtained by pressing the seeds of

---

* Trade name not verified as available   † Trade name verified as obsolete

Aleurites moluccana used in the manufacture of paints and soap.

Lumen Alloy 11-C. A proprietary trade name for an alloy containing copper with 10% aluminum and 1% iron. §

Lumen bronze. An alloy of 86% zinc, 10% copper, 4% aluminum, and 0.1% magnesium. A softer variety contains 88% zinc, 8% aluminum, and 4% copper.

Lumicon 68 Silver Amalgam/Powder. Minimum corroding silver alloy of standardized grain size distribution, mixable in any commercial mixer. [Bayer AG]

Lumicon Non Gamma 2. Silver alloy, practically corrosion-free; dental specialty. [Bayer AG] †

Lumifore/Gluma. Hybrid-filled light-curing plastic filling/dentin bonding system for dental industry. [Bayer AG]

Lumilux. Luminous pigments. [Hoechst UK]

Luminal® . Pharmaceutical preparation; anticonvulsant, hypnotic and sedative agents. [Bayer AG]

Luminalettes® . Pharmaceutical preparations; anticonvulsant, hypnotic and sedative agents. [Bayer AG]

Luminal Sodium. CAS 57-30-7; Phenobarbital sodium; anticonvulsant; hypnotic; sedative. [Sterling Drug Inc]

Luminous®. PMMA optical fiber. [Asahi Chem. Industry]

Lumite. An aluminum alloy containing 5.6% nickel and 1% iron.

Lumiten® E. Foam suppressants for aq. emulsion paints, cement systems, dispersion adhesives. [BASF AG]

Lumiten® I, N. Wetting agent for emulsion paints, for the modification of cement systems, and for adhesives. [BASF AG]

Lumitol®. Hydroxylic polyacrylate resins; used in combination with polyisocyanates to produce chem.-resist. and solv.-resist. polyurethane finishes having excellent mech. props. [BASF AG]

Lummer's solution. Hydrogen platinochloride, H_2PtCl_6 (3 g in 100 cc water), with 0.02 gram lead acetate; used for coating platinum electrodes with finely divided platinum.

Lumnite cement. An alumina cement consisting of about 40% alumina, 40% lime, 15% iron oxide, 5% silica, and magnesia. The material is made from bauxite, and is stated to be stronger than Portland cement.

Lumo WW 75. Sodium dodecylbenzene sulfonate; heavy and light duty detergent. [Zschimmer & Schwarz]

Lumorol 4153. Nonionic/anionic surfactants with phosphate and dissolving agents; cleansing agents. [Zschimmer & Schwarz]

Lumorol K 28. Disodium laurethsulfosuccinate, cocamidopropyl betaine, magnesium lauryl sulfate; detergent for cosmetics, shampoos, bath preps. [Zschimmer & Schwarz]

Lumosäure A. Dodecylbenzene sulfonic acid; for low pH cleansing agents. [Zschimmer & Schwarz]

Lumo Stabil S 80. Sodium dodecylbenzene sulfonate; washing and cleansing agents, industrial cleaners. [Zschimmer & Schwarz] *

lump-Lac. See lac.

lunar caustic. CAS 7761-88-8; Silver nitrate, $AgNO_3$, fused and cast into sticks or rods. An energetic caustic for wounds and sores.

Lunosol. A proprietary preparation; it is a colloidal silver chloride. §

luo-calcite. A name given to the calcium bicarbonate occurring in solution in natural waters.l

luo-chalybite. A name given to the ferrous bicarbonate occurring in solution in natural waters.

luo-diallogite. A name given to the manganese bicarbonate occurring in solution in natural waters.

luo-magnesite. A name given to the magnesium bicarbonate occurring in solution in natural waters.

lupeose. (mannotetrose). Stachyose, $C_{24}H_{12}O_{21}$, a tetrasaccharide. It occurs in the tubers of *Stachy tubifera*.

Luperco 101-P20. CAS 78-63-7; 20% disp. of 2,5-dimethyl-2,5-di (t-butylperoxy) hexane on a PP powd. carrier; crosslinking agent for elastomers and thermoplastic resins. [Atochem]

Luperco 230-XL. CAS 995-33-5; n-Butyl-4,4-bis (t-butylperoxy) valerate on inert filler; initiator for curing elastomers, and for high-temp. cure of polyester resins. [Atochem]

Luperco 231-XL. CAS 6731-36-8; 1,1-Di (t-butylperoxy) 3,3,5-trimethyl cyclohexane on inert filler ($CaCO_3$); initiator for curing elastomers, crosslinking agent for thermoplastic modification. [Atochem]

Luperco 233-XL. Ethyl-3,3-di (t-butylperoxy) butyrate on inert filler; initiator for curing elastomers and for polymer modification thermoplastic cross-linking. [Atochem]

Luperco 331-XL. CAS 3006-86-8; 1,1-Di (t-butylperoxy) cyclohexane on inert filler; initiator for high-temp. cure of polyester resins and for curing elastomers. [Atochem]

Luperco 500-40KE. CAS 80-43-3; Dicumyl peroxide on Burgess KE clay; see Luperco 101XL. [Atochem]

Luperco 801-XL. CAS 3457-61-2; t-Butyl cumyl peroxide on inert filler; initiator for high-temp. cure of polyester resins, curing elastomers, and polymer modification thermoplastic cross-linking. [Atochem]

Luperco 802-40KE. α-α-Bis (t-butylperoxy) diisopropylbenzene on Burgess clay; crosslinking agent for polymer modification, curing elastomers, high-temp. cure of polyester resins. [Atochem]

Luperco A. CAS 94-36-0; Benzoyl peroxide with an inorganic filler; used for polymerization catalysis, drying accelerator, oxidizing agent and bleaching applications. [Atochem North America Inc/Organic Peroxides Div]

Luperco AC. CAS 94-36-0; Benzoyl peroxide with an organic filler; used for polymerization catalysis, drying accelerator, oxidizing agent and bleaching applications. [Atochem North America Inc/Organic Peroxides Div]

Luperco AFR. CAS 94-36-0; A trademark for 50% benzoyl peroxide in plasticizer. [Wallace & Tiernan] ‡

Luperco AFR-250. CAS 94-36-0; Benzoyl peroxide; fire retardant initiator for high-temp. and room temperature cure of polyester resins. [Atochem]

Luperco AST. Benzoyl peroxide with silicone oil; initiator for curing elastomers. [Atochem]

Luperfoam 40. Aq. mixture of t-butylhydrazinium chloride and cupric chloride; chemical blowing agent for unsat. polyester resins. [Atochem]

Luperox. Solid organic peroxides. [Atochem North America Inc/Organic Peroxides Div]

Luperox 2,5-2,5. 2,5-Dihydroperoxy-2,5-dimethylhexane; initiator for bulk, sol'n., emulsion, and suspension polymerization. [Atochem]

Luperox 118. 2,5-Dimethyl-2,5-bis-(benzoylperoxy) hexane; initiator for vinyl polymerization. [Atochem]

Luperox 500R. CAS 80-43-3; Dicumyl peroxide; initiator for bulk, sol'n., and suspension polymerization, polymer modification thermoplastic crosslinking, curing elastomers, high-temp. cure of polyester resins. [Atochem]

Luperox 802. CAS 25155-25-3; α-α-Bis (t-butylperoxy)

---

‡ Trade name and manufacturer not verified   § Trade name without identified manufacturer

diisopropylbenzene; crosslinking agent for thermoplastic modification, curing elastomers, and high temp. cure of polyesters; initiator for vinyl polymerization. [Atochem]

Lupersol 10. CAS 26748-41-4; t-Butyl peroxyneodecanoate; initiator for vinyl polymerizations. [Atochem]

Lupersol 11. CAS 927-07-1; t-Butyl peroxypivalate in odorless mineral spirits; see Lupersol 10. [Atochem]

Lupersol 70. CAS 107-71-1; t-Butyl peroxyacetate in odorless mineral spirits; see Lupersol 10. [Atochem]

Lupersol 80. CAS 109-13-7; t-Butyl peroxyisobutyrate in odorless mineral spirits; see Lupersol 10. [Atochem]

Lupersol 101. CAS 78-63-7; 2,5-Dimethyl-2, 5-di(t-butyl-peroxy) hexane; see Luperox 500R. [Atochem]

Lupersol 130. CAS 1068-27-5; 2,5-Dimethyl-2 5-di(t-butylperoxy) hexyne-3; initiator for bulk, solution, and suspension polymerization, high-temp. curing of polyester resins, and cure of acrylic syrup. [Atochem]

Lupersol 188-M75. CAS 26748-47-0; α-Cumylperoxy neodecanoate in odorless mineral spirits; see Luperox 2,5-2,5. [Atochem]

Lupersol 219-M60. Diisononanoyl peroxide in odorless mineral spirits; crosslinking agent for bulk, solution, and suspension polymerization. [Atochem]

Lupersol 220-D50. 2,2-Di(t-butylperoxy) butane in DOP; initiator for bulk, solution, and suspension polymerization, high-temp. curing of polyester resins, and cure of acrylic syrup. [Atochem]

Lupersol 221. CAS 16066-38-9; Di(n-propyl) peroxydicarbonate; see Luperox 2,5-2,5. [Atochem]

Lupersol 223. CAS 16111-62-9; Di (2-ethylhexyl) peroxydicarbonate; see Luperox 2,5-2,5. [Atochem]

Lupersol 224. 2,4-Pentanedione peroxide; curing agent for unsat. polyester thermoset resins. [Atochem]

Lupersol 225. CAS 19910-65-7; Di (sec-butyl) peroxydicarbonate; initiator for bulk, sol'n., and suspension polymerization and cure of acrylic syrup. [Atochem]

Lupersol® 227. A 50% solution of diisobutyryl peroxide in mineral spirits; an organic peroxide for polymerizing. [Wallace & Tiernan] ‡

Lupersol 230. CAS 995-33-5; n-Butyl-4,4-bis (t-butylperoxy) valerate; initiator for bulk, solution, and suspension polymerization, for high-temp. cure of polyester resins, for curing elastomers, and for cure of acrylic syrup. [Atochem]

Lupersol 231. CAS 6731-36-8; 1,1-Di (t-butylperoxy) 3,3,5-trimethyl cyclohexane; initiator for bulk, sol'n., and suspension polymerization, curing elastomers, high-temp. cure of polyester resins, and cure of acrylic syrup. [Atochem]

Lupersol 233-M75. Ethyl-3,3-di (t-butylperoxy) butyrate in odorless mineral spirits; initiator for bulk, solution, and suspension polymerization, high-temp. cure of polyesters and acrylics. [Atochem]

Lupersol 256. 2,5-Dimethyl-2,5-bis(2-ethylhexanoyl-peroxy) hexane; see Lupersol 10. [Atochem]

Lupersol 288-M75. CAS 26748-47-0; α-Cumyl peroxyneoheptanoate in odorless mineral spirits. [Atochem]

Lupersol 331-80B. CAS 3006-86-8; 1,1-Di (t-butylperoxy) cyclohexane in butylbenzyl phthalate; initiator for bulk, solution, and suspension polymerization, high-temp. curing of polyester resins, and cure of acrylic syrup. [Atochem]

Lupersol 531-80B. 1,1-Di-(t-amylperoxy) cyclohexane in butyl benzyl phthalate; initiator for bulk, solution, and suspension polymerization, for high-temp. cure of polyester resins, and for cure of acrylic syrup. [Atochem]

Lupersol 533-M75. CAS 67567-23-1; Ethyl 3,3-di(t-amylperoxy)butyrate in odorless mineral spirits; initiator for curing of acrylic syrup; crosslinking agent for bulk, solution, and suspension polymerization, thermoplastic modification. [Atochem]

Lupersol 546-M75. CAS 68299-16-1; t-Amylperoxy neodecanoate solution in OMS; initiator for bulk, solution, and suspension polymerization, and cure of acrylic syrup. [Atochem]

Lupersol 553-M75. 2,2-Di (t-amylperoxy) propane in odorless mineral spirits; CAS 3052-70-8; initiator for high temp. cure of polyesters, cure of acrylic syrup; crosslinking agent for bulk, solution, and suspension polymerization, thermoplastic modification. [Atochem]

Lupersol 554-M50, 554-M75. CAS 29240-17-3; t-Amylperoxy pivalate sol'ns. in OMS; initiators for bulk, solution, and suspension polymerization and for cure of acrylic syrup. [Atochem]

Lupersol 555-M60. t-Amyl peroxyacetate in odorless mineral spirits; crosslinking agent for bulk, solution, and suspension polymerization, high temp. cure of polyester, and cure of acrylic syrup. [Atochem]

Lupersol 575. CAS 686-31-7; t-Amylperoxy-2-ethyl-hexanoate; initiator for bulk, solution, and suspension polymerization, for high-temp. cure of polyester resins, and for cure of acrylic syrup. [Atochem]

Lupersol 665-M50. 1,1-Dimethyl-3-hydroxybutyl peroxy-2-ethylhexanoate [Atochem]

Lupersol 688-T50. 1,1-Dimethyl-3-hydroxybutylperoxy-neoheptanoate in toluene; crosslinking agent for bulk, solution, emulsion, and suspension polymerization, and cure of acrylic syrup. [Atochem]

Lupersol DDM. A trademark for 60% methyl ethyl ketone peroxide in dimethyl phthalate. [Wallace & Tiernan] ‡

Lupersol DDM-9. CAS 1338-23-4; MEK peroxide; polymerization initiator for cure of promoted unsat. polyester resins and vinyl ester resins at ambient temps.; promoter, transition metal salt, activates decomposition of peroxide initiator; flexibility in useful conc. range and pot life. [Atochem]

Lupersol DEL. A trademark for 60% methyl ethyl ketone peroxide in dimethyl phthalate. [Wallace & Tiernan] ‡

Lupersol DSW. CAS 1338-23-4; A proprietary trade name for methyl ethyl ketone peroxide; a liquid fire resistant peroxide containing 11.5% active oxygen. [Kingsley & Keith Chemical Corp] *

Lupersol KDB. Di-t-butyl diperoxyphthalate in DBP; see Lupersol 10. [Atochem]

Lupersol P-31, P-33. CAS 3006-82-4, 3006-86-8; t-Butyl peroctoate and 1,1-di (t-butylperoxy) cyclohexane blend in mineral oil; initiator for high-temp. cure of polyester resins. [Atochem]

Lupersol PDO. t-Butyl peroxy-2-ethylhexanoate in DOP; see Lupersol 10. [Atochem]

Lupersol TAEC. OO-t-amyl O-(2-ethylhexyl) mono-peroxycarbonate; crosslinking agent for emulsion, bulk, solution, and suspension polymerization, high temp. cure of polyester, and cure of acrylic syrup. [Atochem]

Lupersol TBEC. OO-t-butyl O-(2-ethylhexyl) mono-peroxycarbonate; see Lupersol 231. [Atochem]

Lupersol TBIC-M75. OO-t-Butyl O-isopropyl mono-peroxycarbonate in odorless mineral spirits; *see* Lupersol 10. [Atochem]

lupetazin. Dimethylpiperazine, (C₂H₃(CH₃)NH) ₂; used as a uric acid solvent.

Luphen® D. Polyurethane disp.; for production of adhesives. [BASF AG]

Lupinit. A proprietary casein product. §

Lupolen®. CAS 9002-88-4; HDPE, LDPE, LLDPE, LDPE/EVA copolymer; various grades for injection molding, extrusion, blow molding, powd. grades, compression molding grades, semifinished products [BASF AG; BASF plc]

Lupolen 1800 H/M/S. A proprietary low-density polyethylene used in injection molding. [BASF plc] *

Lupolen 1810E. A proprietary low-density polyethylene used in the manufacture of milk containers. [BASF plc]*

Lupolen 1810H. A proprietary low-density polyethylene used in blow molding. [BASF plc] *

Lupolen 1812D and 1812EH. Proprietary polyethylenes used in the manufacture of wires and cables. [BASF plc] *

Lupolen 1814E. A proprietary low-density polyethylene used in the manufacture of milk containers. [BASF plc] *

Lupolen 1852E/H. A proprietary low-density polyethylene used in the manufacture of pipes. [BASF plc] *

Lupolen 2040EX and 2410DX. Proprietary low-density polyethylenes used in the manufacture of bags. [BASF plc] *

Lupolen 2410S. A proprietary low density polyethylene used in injection molding. [BASF plc] *

Lupolen 2424H and 2425K. Proprietary low-density polyethylenes used in the manufacture of transparent packaging materials. [BASF plc] *

Lupolen 2430H. A proprietary low density polyethylene used in blow molding. [BASF plc] *

Lupolen 2452 E. A proprietary low density polyethylene used in the extrusion of pipes. [BASF plc] *

Lupolen 3010 S. A proprietary low density polyethylene used in injection molding. [BASF plc] *

Lupolen 3020 D. A proprietary low density polyethylene used in blow molding. [BASF plc] *

Lupolen 3020 KX and 3025 KX. Proprietary low-density polyethylenes used as over-wrappings. [BASF plc] *

Lupolen 4261 AX. A proprietary high density polyethylene of high molecular weight. [BASF plc] *

Lupolen 5011 K. A proprietary high density polyethylene used in injection molding. [BASF plc] *

Lupolen 5052 C. A proprietary high density polyethylene used in the extrusion of pipes. [BASF plc] *

Lupolen 6011 K. A proprietary high density polyethylene used in injection molding. [BASF plc] *

Lupolen 804H and 1814H. Proprietary low-density polyethylenes used in the manufacture of liners and barriers. [BASF plc] *

Lupolen V-2524EX and V-3510K. Proprietary low-density polyethylene copolymers. [BASF plc] *

Lupolex. Linear low density polyethylene. [BASF plc]

Lupranat®. MDI/TDI; for production of PU flexible foam, semirigid and rigid foams, thermoplastic elastomers for the furniture, automotive, construction, pkg., elec., and shoe industry. [BASF AG]

Lupranol®. Polyether polyols; for the production of PU flexible foam, semirigid and rigid foam for the furniture, automobile, construction, pkg., elec., and shoe industry. [BASF AG]

Lupraphen. Polyester polyols; for the production of PU, ready-to-use PU systems, thermoplastic granules, textile coating systems, finished parts made of PU and casting elastomers; for the automotive, machine and equip. construction industries. [BASF AG]

Luprenal®. Polyacrylate resins; heat-curable resins employing added crosslinkers; used in combination with melamine or urea resins to give hard, tough, weather-resist. and chem. resist. coatings. [BASF AG]

Luprimol®. Improves hand of pigment prints on textiles. [BASF AG; BASF plc]

Luprintan®. Fixing assistants for printing on synthetic fibers. [BASF AG]

Luprintan® ATP. Printing auxiliary for synthetics. [BASF plc]

Luprintan® DCA. Fixation accelerator for printing acetate, triacetate and polyester with disperse dyes. [BASF]

Luprintol®. Auxiliaries for emulsion printing. [BASF plc]

Luprintol® PE. Aryl polyglycol ether; emulsifier for prep. of o/w emulsions, solv.-containing or solv.-free thickenings for textile pigment printing. [BASF AG]

Luprofil®. Org. and inorg. pigments/PP masterbatches; for PP spin dyeing. [BASF AG]

Lupromag®. Magnesium propionate; mineralized single feedstuff for animal feeding. [BASF AG]

Lupron. CAS 74381-53-6; Leuprolide acetate; antineoplastic. [TAP Pharmaceuticals]

Luprosil®. CAS 79-09-4; Propionic acid; preservative for mixed feedstuffs. [BASF AG]

Luprosil® NC. Ammonium propionate; preservative. [BASF AG]

Luprosil® Salt. CAS 4075-81-4; Calcium propionate; preservative for feed industry. [BASF AG]

Luprosil® Sodium Salt. CAS 137-40-6; Sodium propionate; preservative for feed industry [BASF AG]

Lurafix® Dyes. Wetting agent-free dyes for transfer printing inks. [BASF AG]

Luramide®. PA-sol. metal complex dyes and acid dyes for dyeing PA chips in aq. liquors. [BASF AG]

Luran®. Styrene acrylonitrile copolymer in granule form; used for moldings for domestic appliances and sanitaryware. [BASF plc]

Luran® 358 N. CAS 9003-54-7; SAN copolymer; very easy flow grade for injection moldings of containers, record player covers, TV filter panels, cassettes. [BASF AG]

Luran® 378 P G7. CAS 9003-54-7; SAN copolymer, 35% glass fiber-reinforced; high rigidity reinforced grade for injection molding and extrusion; used for chassis for assembling elec. and other parts, headlamp casings, parts for record players, disc drives, and tape reels. [BASF AG] *

Luran® 757R. A proprietary acrylonitrile styrene-acrylonitrile copolymer used in injection molding. [BASF plc] *

Luran® 776S. A proprietary acrylonitrile styrene-acrylonitrile used in extrusion and injection molding. [BASF plc] *

Luran® KR 2517. A proprietary acrylo-nitrile-styrene copolymer containing 35% glass fiber. [BASF plc] *

Luran® KR 2556. CAS 9003-54-7; SAN copolymer; high heat resist. grade for instrument covers and cassettes; used for injection molding and extrusion. [BASF AG] *

Luran® S. Acrylonitrile/styrene/acrylate polymer; for injec-

---

‡ Trade name and manufacturer not verified   § Trade name without identified manufacturer

tion molding or extrusion; esp. suitable for structural parts subject to outdoor exposure, e.g., letterboxes, parts for garden appliances, road signs, boat hulls, windsurfers, vehicle external parts, hot water drainage pipes, patio furniture, toy [BASF AG; BASF plc]

Lurantin® . Lightfast direct dyes for dyeing and printing cellulose fibers, nylon fibers and wool. [BASF AG; BASF plc]

Luranyl®. Polyphenylene ether and HIPS blend; std. grades for high-precision components which are tough and resist. to heat distort.; glass fiber-reinforced grades for pumps and fittings, hot water and drinking water; impact-resist. grades for car interior parts; instrument panels; halogen-free fire-retardant grades for elec. insulation bodies, housing for office machines, meters. [BASF AG]

Lurapret®. Antislip agents for textile finishing. [BASF AG]

Lurazol® Dyes. Anionic dyes for leather and fur dyeing, coloring cleaners. [BASF AG]

Luredox® BP, PO. CAS 7775-14-6; Sodium dithionite; for mfg. of cleaners. [BASF AG]

Luredur® . Acrylate copolymers and modified poly-acrylamides; dry strength agents for strengthening the structure of paper and board. [BASF AG; BASF plc]

Luresin® . Based on aq. sol'ns. of formaldehyde-free cationic polyamido-amine-epichlorohydrin resins; for in-creasing wet and dry strength of papers, esp. in the neutral pH range. [BASF AG]

Lurgi metal. An alloy of lead with 2% barium.

Luron® Binder. Albumen-type condensate in aq. colloidal solution; assistant for laminate printing ink, leather and fur processing. [BASF AG]

Lurotex®. Textile dyeing and finishing auxiliary. [BASF plc]

Lurotex® A-25. Hydrophilic agent, crease inhibitor for syn-thetic fiber piecegoods and hank yarns. [BASF]

Luscin. Coal dust replacement additives for iron foundry sands. [Foseco (F.S.) Ltd] †

Lusol. General purpose maintenance spray; for lubrication, penetrating seized parts, freeing rusted mechanisms, driving moisture out of electrical parts, corrosion preven-tion on metal surfaces. [Hermetite Products Ltd] *

Lusol®. Wire drawing compounds for drawing of intermedi-ate and fine copper wire, as well as rod breakdown. [Witco/Allied-Kelite]

Lusolvan® FBH. Diisobutyl ester of a dicarboxylic acid mixt.; coalescent for aq. polymer dispersions; for paints and textured finishes. [BASF AG]

Lustilac. A proprietary molding compound. §

Lustra-Cellulose. Artificial silk. §

Lustral. See Zoloft. [Pfizer International]

Lustralite®. Synthetic resin for industrial use. [Reichhold]

Lustralite® 44-444. Gloss enhancer. [Reichhold]

Lustran. See Zoloft. [Pfizer International]

Lustran ABS. Acrylonitrile-butadiene-styrene thermoplas-tics; molding and extrusion grades, refrigerator inner liners, business machine and appliance housings, auto-motive parts, and industrial components. [Monsanto Co]

Lustran SAN. Styrene-acrylonitrile molding resins; clear rigid thermoplastics for molding high quality articles requiring transparency, brilliant surface, chemical resis-tance and stiffness. [Monsanto Co]

Lustran Ultra ABS. ABS molding grades, superior molded part appearance and gloss; used for telephones, appli-ances, video cassettes, power tool housings, office equipment, toys, sporting goods, and lawn and garden

equipment. [Monsanto Co]

Lustranyl. Solvent soluble dyes. [Morton Int'l. Ltd]

Lustra-Pearl® . Mica, titanium dioxide; pearlescent pig-ments for cosmetic eye, face, lip, and body makeup. [Van Dyk]

Lustrasol®. Acrylic or styrene based resin solutions. [Reich-hold]

Lustre. Protein binders based on modified casein disper-sions; used for leather. [Colour-Chem Ltd] *

Lustrex. Polystyrene molding and extrusion rsins; used for packaging, toys, appliance, photographic, furniture, au-dio cassette, medical and houseware markets, thermoformed containers, trays, cups and lids. [Mon-santo Co] *

Lustron. See Celanese; also a proprietary trade name for polystyrene molding compounds. §

Lustrose. A proprietary compound used in the textile indus-try for sizing. §

Lusynton® A. Mixt. of org. and inorg. substances; corrosion inhibitor; chlorite stabilizer with resistance to hard water and acid with wetting and detergent effects. [BASF AG; BASF plc]

Lutalyse. Dinoprost tromethamine; oxytocic; prostaglandin. [Upjohn] *

Lutan® . Basic aluminum salts; for tanning leather and fur. [BASF AG]

Lutate. CAS 630-56-8; A proprietary preparation of hydroxyprogesterone caproate. [Savage Laboratories]†

Lutavit. Animal feed vitamins. [BASF plc]

Lutecin. See nickel silvers.

Lutensit® . Ionic surfactants. [BASF plc]

Lutensit® A-BO. Sodium dioctylsulfosuccinate in water-neopentyl glycol; surfactant for chemical industry, deter-gents, cleaners; wetting agent, dispersant. [BASF AG]

Lutensit® A-ES. Sodium alkyl-phenol ether sulfate as a yellow liquid; wetting and dispersing agent with many applications, sometimes in combination with nonionics; emulsifying agent for methylene chloride. [BASF plc] *

Lutensit® A-LBA. Anionic surfactant in yellow liquid form; wetting, dispersing, cleaning and emulsifying agent used in household rinsing and cleaning formulations, all-pur-pose cleaners, and various industrial emulsions. [BASF plc] *

Lutensit® A-PS. Sodium alkane sulfonate as a slightly yellow liquid; alkaline industrial, household and metal cleaners, steam jet cleaners and pickling baths. [BASF plc] *

Lutensit® AN 10. Mixts. of various nonionic alkoxylates and an anionic, acidic, readily neutralizable surfactant; de-greasing agent, cold cleaners. [BASF AG]

Lutensit® AS 2230, 2270. Sulfated natural alcohol polyglycol ether, sodium salt; foaming surfactant for chemical and cosmetic industry. [BASF AG] *

Lutensit® K-LC. Dimethyl C12/C14 fatty alkylbenzyl ammo-nium chloride in almost colorless aqueous solution; dispersing and wetting agent with bactericidal and fungi-cidal effects; used in conjunction with nonionics to pro-duce disinfectant cleaners for the food and beverage industries and for miscellaneous household and indus-trial uses. [BASF AG] *

Lutensit® K-OC. Benzalkonium chloride; biocidal, wetting and dispersing agents for production of disinfectant cleaners for beverages and foodstuffs, trading concerns and household. [BASF AG] *

---

* Trade name not verified as available   † Trade name verified as obsolete

Lutensol® . Low foaming nonionic surfactants; used for domestic machine dishwashing detergents and rinse aids, low foam industrial cleaners. [BASF plc]

Lutensol® A 7. PEG-7 C16C18 fatty alcohol; water-sol. detergent for production of detergents and in chemical processing industry. [BASF AG]

Lutensol® AO 3. PEG-3 straight chain synthetic C13-15 fatty alcohol; detergent, emulsifying and dispersing agent used in household and industrial detergents, chemical industry; > 80% biodeg. [BASF AG]

Lutensol® AO 10. PEG-10 straight chain synthetic C13-15 fatty alcohol; detergent, emulsifying and dispersing agent used in household and industrial detergents, chemical industry; > 80% biodeg. [BASF AG]

Lutensol® AP 6. PEG-6 alkylphenol; detergent, wetting, emulsifying and dispersing agent, used in cleaners, detergents, leather, fur, paper, paint and dye industries. [BASF AG]

Lutensol® AP 10. CAS 9016-45-9; Nonoxynol-10; detergent, wetting, emulsifying and dispersing agent, used in cleaners, detergents, leather, fur, paper, paint and dye industries. [BASF AG]

Lutensol® AP 20. CAS 9016-45-9; Nonoxynol-20; detergent, wetting, emulsifying and dispersing agent, used in cleaners, detergents, leather, fur, paper, paint and dye industries. [BASF AG]

Lutensol® AT 11. PEG-11 sat. C16-C18 alcohol; detergent, wetting, dispersing and emulsifying agent used in chemical industry, for household and industrial detergents. [BASF AG]

Lutensol® ED 140. EO-PO ethylene diamine compound.; antistat, detergent, dispersant, defoamer, wetting agent, gelling agent, solubilizer, emulsifier, demulsifier, lubricant, foam suppressor for household and industrial uses. [BASF AG] *

Lutensol® FA 12. Oleyl amine ethoxylate; emulsifier, dispersant, wetting and degreasing agent used in heavy-duty and other detergents, shampoos; > 80% biodeg. [BASF AG]

Lutensol® FSA 10. CAS 31799-71-0; Oleic acid amide ethoxylate; emulsifier, dispersant, detergent, wetting agent, used in fine detergents, hand cleaners, fur dressing; > 80% biodeg. [BASF AG]

Lutensol® LF 220, 221, 223, 224. Alkoxylated straight chain alcohol; surfactant for dishwashing powds. and rinse aids. [BASF AG] *

Lutensol® ON 30. PEG-3 C10 oxo-alcohol; surfactant for chemical industry and production of detergents. [BASF AG]

Lutensol® TO 3. PEG-3 C13 oxo-alcohol; surfactant for chemical industry and production of detergents. [BASF AG]

luteol. CAS 57-83-0; Hydroxychlorodiphenylquinoxaline; used as an indicator.

Lutetia. Pigments, for paints and inks. [ICI Chem. & Polymers Ltd]

Lutexal® TX-401. Synthetic thickening agent for textile applications. [BASF; BASF plc]

Lutexan. Natural yellow color. [Atomergic Chemetals Corp]

2,6-lutidine. (2,6-dimethylpyridine) C_7H_9N; CAS 108-48-5; EINECS 203-587-3; Pharmaceuticals, resins, dyestuffs, rubber accelerators, insecticides. [Aldrich; Janssen Chimica; Raschig; Reilly Industries]

Lutofan® . CAS 9002-86-2; Vinyl chloride disp.; for produc-

tion of laminating and heat-sealable adhesives and pkg. materials, binders for bonding fiber webs and for textile coating. [BASF AG; BASF plc]

Lutonal® . Polyvinyl ether. [BASF plc]

Lutonal® A. Polyvinyl ethyl ether; nonyellowing, very resilient, soft resin for odorless and taste-free finishes in the paint industry, for gravure and flexographic inks, for producing pressure-sensitive, building adhesives, and as tackifying soft resins. [BASF AG]

Lutonal® D. Polyvinyl ether dispersion. [BASF plc] *

Lutonal® LC. Polyvinyl isobutyl ether. [BASF plc] *

Lutrabond. Polyester nonwovens for roof covers. [Carl Freudenberg] *

Lutradur. Polyester nonwoven. [Carl Freudenberg] *

Lutrizole® . Dimetridazole, dimetridazole HCl; active ingred. for control of black head disease in turkeys and bloody diarrhea complaints in swine. [BASF AG]

Lutrol® . Pharmaceutical grades of polyethylene glycol. [BASF plc]

Lutrol® E 300. CAS 25322-68-3; PEG-6; emulsifier, emollient, lubricant, and solv. for liq. preps. [BASF AG] *

Lutrol® E 400. CAS 25322-68-3; PEG-8; emulsifier, emollient, lubricant, and solv. for liq. preps. [BASF AG] *

Lutrol® E 1500. CAS 25322-68-3; PEG-32; emulsifier, binder, solubilizer, resorption promoter for substances insol. or sparingly sol. in water. [BASF AG] *

Lutrol® E 4000. CAS 25322-68-3; PEG-75; emulsifier, binder, solubilizer, resorption promoter for substances insol. or sparingly sol. in water. [BASF AG] *

Lutrol® E 6000. CAS 25322-68-3; PEG-150; emulsifier, binder, solubilizer, resorption promoter for substances insol. or sparingly sol. in water. [BASF AG] *

Lutrol® OP-2000. CAS 31394-71-5; PPG-26 oleate. [BASF]*

Lutrol® W-3520. PPG-28-buteth-35. [BASF] *

Lutron. Waterless foundry molding material. [Foseco (F.S.) Ltd]

Lutron. Electroplating additive. [BASF plc]

Lutropur. Electroplating additive. [BASF plc]

Luvican M170. Polyvinyl carbazole. [BASF plc] *

Luviflex® VBM 35. CAS 26589-26-4; Acrylates/PVP copolymer; film former with excellent hydrocarbon compatibility for weatherproof hairstyles. [BASF AG; BASF plc]*

Luviform® ES 22. Ethyl ester of PVM/MA copolymer; film-forming binders for hair sprays and setting lotions. [BASF AG] *

Luviform® ES 42. Butyl ester of PVM/MA copolymer; film-forming binders for hair sprays and setting lotions. [BASF AG] *

Luviform® FA 119. CAS 9011-16-9; PVM/MA copolymer; stabilizing and binding agent for toothpastes, denture retaining agents, shampoos, etc. [BASF AG]

Luviquat® . Quaternary compounds. [BASF plc]

Luviquat® FC 370. CAS 29297-22-0; Polyquaternium-16 (methylvinylimidazolium chloride/vinylpyrrolidone copolymer (30:70 ratio) aq. solution); substantive cationic polymer used as conditioner in products for hair and skin care; film former; foam stabilizing and lubricating effects. [BASF AG] *

Luviquat® FC 550. CAS 29297-22-0; Polyquaternium-16 (methylvinylimidazolium chloride/vinylpyrrolidone copolymer (50:50 ratio) aq. solution); substantive cationic polymer used as conditioner in products for hair and skin care; film former; foam stabilizing and lubricating effects.

---

‡ Trade name and manufacturer not verified   § Trade name without identified manufacturer

[BASF AG] *

Luviquat® FC 905. CAS 29297-22-0; Polyquaternium-16; hair conditioners, rinses, shampoos, bleaches, dyes, liq. soaps, bath preps., skin care products. [BASF AG] *

Luviquat® HM 552. CAS 29297-22-0; Polyquaternium-16; conditioner and setting resin for hair care. [BASF AG] *

Luviquat® Mono CP. Hydroxyethyl cetyldimonium phosphate; conditioner for hair cosmetics, esp. cold waves. [BASF] *

Luviset. Resins for hairsprays. [BASF plc]

Luviset CA 66. CAS 25609-89-6; Vinyl acetate/crotonic acid copolymer; hair fixative for aerosols, hair sprays, setting lotions, conditioners. [BASF] *

Luviset CAP. Vinyl acetate/crotonic acid/vinyl propionate copolymer; film-forming agents for hair-sprays and fixing lotions; compatible in aerosols with propane/butane. [BASF] *

Luviset CAP X. Crotonic acid/vinyl acetate/vinyl propionate terpolymer; hair fixative for aerosols, hair sprays, setting lotions, conditioners. [BASF] *

Luviskol®. Cosmetic and technical grades of PVP, vinyl acetate copolymers. [BASF plc]

Luviskol® K12, K17, K30, K60. CAS 9003-39-8; PVP; film-forming agent, hair fixative, thickener, protective colloid, suspending agent, and dispersant for cosmetics, adhesives, paints, coatings, paper, detergents, glass fibers, inks, ceramics, nonpharmaceutical tableting, photographic films, crop protection agents. [BASF AG]

Luviskol® K80, K90. CAS 9003-39-8; PVP; film-forming agent, hair fixative, thickener, protective colloid, suspending agent, and dispersant for cosmetics, adhesives, paints, coatings, paper, detergents, glass fibers, inks, ceramics, nonpharmaceutical tableting, photographic films, [BASF AG]

Luviskol® VA 28 E. CAS 25086-89-9; PVP/VA copolymer (20:80 ratio), ethanol; film-forming agents for hair-sprays and fixing lotions. [BASF AG] *

Luviskol® VAP 343 E. PVP/VA/vinyl propionate copolymer (30:40:30 ratio), ethanol; film-forming agents for hair-sprays and fixing lotions; compatible in aerosols with propane/butane. [BASF] *

Luvisoft. Fabric softeners. [BASF plc]

Luvitol EHO. EINECS 261-619-1; Cetearyl octanoate; emollient oil component for cosmetics and pharmaceuticals. [BASF]

Luvitol HP. CAS 61693-08-1; Hydrogenated polyisobutene; emollient oil component for cosmetics and pharmaceuticals. [BASF] *

Luwax A. CAS 9002-88-4; LDPE homopolymer; additive for printing inks; flatting and antisettling agent in paints; dispersant and color enhancer; in floor plishes; lubricant for plastics processing. [BASF]

Luwax AF 30. CAS 9002-88-4; Micronized polyethylene wax; additive for printing inks, as flattening and antisettling agent in paints, dispersant in color concs.; improves hardness in wax compounds, black heel mark resist. in floor polishes; lubricant in plastics processing. [BASF]

Luwax AH 3. CAS 9002-888-4; HDPE wax; additive for printing inks, as flattening and antisettling agent in paints, dispersant in color concs.; improves hardness in wax compounds, black heel mark resist. in floor polishes; lubricant in plastics processing. [BASF]

Luwax EAS 1. CAS 9002-88-4; Polyethylene copolymer wax; additive for printing inks, as flattening and antiset-tling agent in paints, dispersant in color concs.; improves hardness in wax compounds, black heel mark resist. in floor polishes; lubricant in plastics processing. [BASF]

Luwax EVA 1. CAS 9002-88-4; Polyethylene copolymer wax; additive for printing inks, as flattening and antisettling agent in paints, dispersant in color concs.; improves hardness in wax compounds, black heel mark resist. in floor polishes; lubricant in plastics processing. [BASF]

Luwax OA 2. Oxidized polyethylene wax; additive for printing inks, as flattening and antisettling agent in paints, dispersant in color concs.; improves hardness in wax compounds, black heel mark resist. in floor polishes; lubricant in plastics processing. [BASF]

Luwipal®. CAS 9003-08-1; Melamine formaldehyde resins, butanol or methanol etherified; combined with alkyd, polyacrylate, and epoxy resins to give light-resist. and weather-resist. baking finishes, e.g., automotive finishes. [BASF AG]

Luxalloy. Dental silver alloys and mercury for mixing amalgams; for dentistry and dental engineering. [Degussa Ltd]

Luxate. Polyurethane intermediates for coatings, elastomers, adhesives, sealants, thermoplastics. [Olin]

Luxene. A proprietary synthetic resin of the phenol-formaldehyde class. §

Luxene 44. A proprietary denture compound made from a vinyl copolymer. §

Luxer®. Nonwoven fabric of synthetic filaments. [Asahi Chem. Industry]

Luxol. Solvent soluble dyes. [Morton Int'l. Ltd]

Luxor. A combination of hydrolyzed vegetable proteins and other ingredients; used as a natural flavor in vegetables, meat, chicken and seafood dishes. [Hercules] *

Luzenac 8170. CAS 14807-96-6; Talc; ultrabright appearance grade for polyolefins, PVC, elastomers, most other plastics. [Luzenac; R.T. Vanderbilt]

Luzerne. A proprietary trade name for a hard rubber. §

Luzidol. Benzoyl peroxide, $C_{14}H_{10}O_4$.

LX. Heat-reactive petroleum resins. [Neville Chemical Co]

LX-685®,125. Petrol. hydrocarbon resin; used in adhesives (hot melt pressure sensitive, mastic, pressure sensitive), coatings (alum., can and drum, emulsion, industrial, marine, traffic), inks (gelled varnishes, gravure, heat set, letterpress, lithographic), rubber (cements, mechanical goods, molded goods, tires), concrete-curing compounds, and caulking compounds. [Neville]

LX-782®. Petrol. hydrocarbon resin; used for inks, adhesives, coatings, rubbers, and concrete cures. [Neville]

lyargol. Silver proteinate.

Lycadex®. CAS 9050-36-6; Maltodextrin; fat replacers in foods. [Roquette (UK) Ltd]

Lycal®. Lightly calcined sea water magnesia. [Steetley Magnesia Products Ltd]

Lycanol. CAS 339-44-6; A proprietary trade name for glymidine. §

Lycasin. Hydrogenated glucose syrup; a sugar substitute, in foods (confectionery) and pharmaceuticals (syrups and lozenges). [Roquette (UK) Ltd]

lycetol. (tetradine). Dimethylpiperazine tartrate; used as a solvent for uric acid.

lycine. CAS 61-19-8; The base of *Lycium barbarum*. It is identical to betaine.

lycopodium. (club moss; vegetable sulfur) A pale yellow powder consisting of the spores of *Lycopodium*

---

clavatum; used as a dusting powder, in explosives; in the pharmaceutical industry as a coating.

Lycopon. A proprietary trade name for sodium hyposulfite.§

Lycra®. Fibers. [DuPont UK]

Lycresse®. Fibers. [DuPont UK]

Lyddite. (Melinite, Pertite, Shimose). Picric acid; an explosive.

Lydian stone. (lydite, touchstone). A siliceous slate containing about 84% silica, 5% alumina, and 1% ferric oxide; used for testing gold by rubbing it upon the stone and testing the streak of metal produced with acid.

lye. *See* potassium hydroxide.

lye glycerin. Glycerin obtained from soap liquor.

Lynex®. Polyphenylene ether-polyamide alloy [Asahi Chem. Industry]

Lynite. A trademark for an alloy of 88% aluminum, 10% copper, 1.5% iron, and 0.25% magnesium; specific gravity 2.95. §

Lynx. A cotton fabric grade of Tufnol industrial laminates. [Tufnol Ltd]

Lyocol. Low foaming scouring,leveling and dispersing agent; combined scouring and dyeing of polyester fibers; beam dyeing and jet application. [Sandoz Products Ltd]

Lyofix. Finishing agent. [Ciba plc] *

Lyofix 363. A proprietary trade name for a modified urea formaldehyde resin precondensate used in the crease-resistance finishing of cellulosic textiles. [Ciba plc] *

Lyofix F. A proprietary trade name for a melamine-formaldehyde resin precondensate used in finishing paper makers' felts. [Ciba plc] *

Lyogen. Textile leveling agent; used for wool, nylon, polyurethane and polyamide fibers; vat dyestuffs; acid, milling, chrome and direct dyestuffs. [Sandoz Products Ltd]

Lyonore. A proprietary trade name for a steel containing 0.2% copper with some chromium and nickel. §

Lyons sugar. Sucramine, the ammonium salt of saccharine.

Lyopan. A proprietary preparation of phenazone theophyllin and urethane. §

Lyophrin. A proprietary preparation of adrenaline hydrogen tartrate with sodium sulfate as a preservative; for treatment of glaucoma. [Alcon Universal] *

Lyovac Sodium Edecrin. CAS 6500-81-8; Ethacrynate sodium; diuretic. [Merck & Co Inc] *

Lypsyl. Lipsalve. [Ciba plc] *

Lyracamine. Acrylic dyes. [ICI Chem. & Polymers Ltd] †

Lyric®. For the agriculture industry. [DuPont UK]

Lyril®. For the agriculture industry. [DuPont UK]

Lysamine. Protein derived from vegetable matter; animal feedstuffs supplement. [Roquette (UK) Ltd]

Lysargine. A colloidal silver, containing 60% of silver; used as an antiseptic.

Lysase. Enzyme preparations; used for conversion of starch or starch hydrolysis products. [Roquette (UK) Ltd]

Lyse S, S III, S III Diff. Reagents which destroy red blood cells to leave white blood cells in suspension for analysis; used for white blood cell and hemoglobin determination. [Coulter Electronics Ltd]

lysidine. (piperazenyl). Tetrahydropyrazine.

l-lysine. (α, ε-diaminocaproic acid) $C_6H_{14}N_2O_2$; CAS 56-87-1; EINECS 200-294-2; Biochemical and nutritional research, pharmaceuticals, culture media, fortification of foods and feeds, nutrient supplement, animal feed additive. [Degussa; R.W. Greeff; U.S. Biochemical; Walton Pharmaceuticals Ltd]

l-lysine monohydrochloride. $NH_2(CH_2)_4CH(NH_2)COOH \cdot HCl$; CAS 70-53-1; EINECS 200-739-0. [Degussa; R.W. Greeff; Spectrum Chem. Mfg.; Tanabe USA]

Lysinex. A proprietary preparation of l-lysine hydrochloride and stanolone; an anabolic agent. [Lloyd, Hamol] *

Lysitol. *See* Lysol. §

Lysivane. CAS 1094-08-2; Ethopropazine hydrochloride. [May & Baker Ltd] *

Lysmeral®. 2-Methyl-3-(4-t-butylphenyl) propanol; fragrance (floral, powdery, fresh, lily-of-the-valley-like) [BASF AG]

Lysochlor. Chloro-m-cresol, $C_6H_3Cl\,(CH_3)OH$. A disinfectant. It is a similar preparation to Eusapyl (*qv*).

Lysodren. CAS 53-19-0; Mitotane; antineoplastic. [Bristol Laboratories] *

Lysoform. A solution of formaldehyde in alcoholic potash soap solution; a disinfectant much like Lysol. §

Lysol. High activity, wide spectrum disinfectant. [Coventry Chemicals Ltd]

Lysol. Consists of crude cresols mixed with a soft-soap solution; a disinfectant used in surgery, and for cleaning floors and walls. §

Lythol Oil. A commercial phenolated oil used as a disinfectant. §

Lytor & RM. Tall oil resin. [Georgia-Pacific]

Lytron. Range of aqueous styrene copolymer emulsions with controlled particle size; for use in paper coatings and opacifying liquid detergents. [Morton Int'l. Ltd]

---

‡ Trade name and manufacturer not verified   § Trade name without identified manufacturer

M

M33, MN3. Polyimide resins. [Rhone-Poulenc NV] *
M50. CAS 11113-70-5; Basic lead silico chromate; anticorrosive pigments for coatings. [Rheox Inc] *
M66. Magnesia alumina spinel; used with magnesia to produce bricks and monoliths for steel, nonferrous, glass and cement industries where generation of toxic hexavalent chrome from chrome-bearing refractories is a problem. [Alcoa]
M100. CAS 1308-38-9; Chromic oxide pigment. [British Chrome & Chemicals Ltd]
M9030. CAS 18769-78-3; Methyltridecylsilane; thermally stable fluid with superior metal-on-metal lubrication and wear chars. [Hüls Am.]
MA 20. Soap powder for launderettes. [Unilever] *
Maali resin. A yellowish-white resin resembling elemi of Samoa.
Maalox. A proprietary preparation containing magnesium aluminum hydroxide gel; an antacid. [Rorer] †
Macadamite. An alloy of 72% aluminum, 24% zinc, and 4% copper.
Macaja butter. Mocaya oil.
Macaloid. Rheological additives; used for paints, sealants, agricultural products, cosmetics and ceramics. [Rheox Inc]
Macassar nutmeg butter. (papua nutmeg butter). The fat from *Myristica argentae*.
Maceal. 7-Formyl-5-isopropyl-2-methyl bicyclo (2,2,2)-oct-2-ene. [Quest Int'l. UK Ltd]
mace butter. Nutmeg butter, obtained from the seeds of *Myristica officinalis*, of the East.
Mace Butter, American. *See* otoba butter.
Macerase®. Pectinase. [Calbiochem Corp]
MacFarland's alloy. Heat-resisting alloys. a) Contains 59% nickel, 30% chromium, and 11% copper. b) Contains 46% nickel, 43% chromium, and 11% copper.
Macgill metal. An alloy containing 88% copper, 7% nickel, 4.5% iron, and traces of tin and lead. It resists corrosion.
Machacon juice. An alkaline decoction of the juice of the root of a plant having the same name; used to coagulate rubber latex.
Machete. CAS 23184-66-9; Butachlor; herbicide. [Monsanto Co] *
machine bronze. Variable alloys containing 50-90% copper, 25% nickel, 0-30% tin, and 0.8% lead. Some alloys contain no nickel, and also contain zinc. Another alloy of this type consists of 83% copper, 16% zinc, and 1% tin.
Mach's metal. An alloy of aluminum with 2-10% magnesium.
Macht's metal. An alloy of 57% copper and 43% zinc; used for castings.

M-acid. 1-Amino-5-naphthol-7-sulfonic acid; $C_{10}H_5NH_2OH\text{-}SO_3H$; used as an azo dye intermediate.
Mackam 1C. Sodium cocoamphoacetate; surfactant for high alkaline cleansers, shampoos, baby shampoos. [McIntyre]
Mackam 1L. Sodium lauramphoacetate; surfactant for nonirritant shampoos, cleaners. [McIntyre]
Mackam 2C. Disodium cocoamphodiacetate; surfactant for baby shampoo, high alkaline cleaners. [McIntyre]
Mackam 2CT. Disodium caproamphodiacetate, sodium trideceth sulfate, and hexylene glycol; nonirritating shampoo. [McIntyre]
Mackam 2CY. Disodium caprylamphodiacetate; low-foaming alkaline cleaner. [McIntyre]
Mackam 2L. Disodium lauroamphodiacetate; surfactant for nonirritating shampoos, cleaners. [McIntyre]
Mackam 2W. Disodium wheat germamphodiacetate; mild surfactant for baby shampoos; wetting agent in caustic cleaners. [McIntyre]
Mackam 35. Cocamidopropyl betaine; foamer for shampoos, bubble baths, dishwash. [McIntyre]
Mackam 151C. CAS 84812-94-2; EINECS 284-219-9; Cocaminopropionic acid; mild surfactant, conditioner for shampoos. [McIntyre]
Mackam 151L. Lauraminopropionic acid; mild surfactant, conditioner for shampoos. [McIntyre]
Mackam 160C. Sodium lauriminodipropionate; surfactants for personal care and industrial applications. [McIntyre]
Mackam CB-35. CAS 68424-94-2; EINECS 270-329-4; Cocobetaine; surfactant, conditioner, visc. builder, foam booster for shampoos, skin cleansers, bath products, heavy duty cleaners. [McIntyre]
Mackam CET. CAS 693-33-4; EINECS 211-748-4; Cetyl betaine; surfactant, conditioner, visc. builder, foam booster for shampoos, skin cleansers, bath products, heavy duty cleaners. [McIntyre]
Mackam CSF. CAS 68919-41-5; Sodium cocoamphopropionate; surfactant for heavy duty cleanser, metal and all-purpose cleaner, nonirritating shampoos. [McIntyre]
Mackam HV. CAS 25054-76-6; EINECS 246-584-2; Oleamidopropyl betaine; hair conditioner, foamer, emollient, visc. builder. [McIntyre]
Mackam ISA. Isostearamidopropyl betaine; surfactant, conditioner, visc. builder, foam booster for shampoos, skin cleansers, bath products, heavy duty cleaners. [McIntyre]
Mackam J. Cocamidopropyl betaine; surfactant, conditioner, visc. builder, foam booster for shampoos, skin cleansers, bath products, and heavy duty cleaners. [McIntyre]

---

* Trade name not verified as available   † Trade name verified as obsolete

Mackam LMB. Lauramidopropyl betaine; surfactant for shampoos, bubble baths, dishwash formulations. [McIntyre]

Mackam MLT. Sodium lauroamphoacetate and sodium trideceth sulfate; surfactant for shampoos, cleaners, detergents. [McIntyre]

Mackam OB-30. CAS 871-37-4; EINECS 212-806-1; Oleyl betaine; visc. builder for alkaline cleanser. [McIntyre]

Mackam RA. CAS 86089-12-5; Ricinoleamidopropyl betaine; surfactant, conditioner, visc. builder, foam booster for shampoos, skin cleansers, bath products, heavy duty cleaners. [McIntyre]

Mackam TM. Dihydroxyethyl tallow glycinate; surfactant, conditioner for shampoos, thickener for alkaline oven cleaners, acid bowl cleaners. [McIntyre]

Mackam WGB. CAS 133934-09-5; Wheat germamidopropyl betaine; surfactant, conditioner, visc. builder, foam booster for shampoos, skin cleansers, bath products, heavy duty cleaners. [McIntyre]

Mackamide AME-75, AME-100. CAS 142-26-7; EINECS 205-530-8; Acetamide MEA; humectant, surfactant, thickener, foam booster/stabilizer for personal care and industrial applications. [McIntyre]

Mackamide C. Cocamide DEA (1:1); foam stabilizer and thickener for shampoos, industrial applications. [McIntyre]

Mackamide CMA. CAS 68140-00-1; Cocamide MEA; surfactant, thickener, foam booster/stabilizer for personal care and industrial applications. [McIntyre]

Mackamide ISA. CAS 52794-79-3; EINECS 258-193-4; Isotearamide DEA; lubricant, surfactant, thickener, foam booster/stabilizer for personal care and industrial applications. [McIntyre]

Mackamide L10. CAS 120-40-1; EINECS 204-393-1; Lauramide DEA; surfactant, thickener, foam booster/stabilizer for personal care and industrial applications. [McIntyre]

Mackamide LME. CAS 5422-34-4; EINECS 226-546-1; Lactamide MEA; conditioner for shampoos; surfactant, thickener, foam booster/stabilizer for personal care and industrial applications. [McIntyre]

Mackamide LMM. CAS 142-78-9; EINECS 205-560-1; Lauramide MEA; surfactant, thickener, foam booster/stabilizer for personal care and industrial applications. [McIntyre]

Mackamide LOL. Linoleamide DEA; surfactant, thickener, foam booster/stabilizer for personal care and industrial applications. [McIntyre]

Mackamide MO. CAS 93-83-4; EINECS 202-281-7; Oleamide DEA (1:1); surfactant, thickener, foam booster/stabilizer for personal care and industrial applications. [McIntyre]

Mackamide ODM. Oleamide DEA, DEA oleate; surfactant, thickener, foam booster/stabilizer for personal care and industrial applications. [McIntyre]

Mackamide OP. CAS 111-05-7; EINECS 203-828-2; Oleamide MIPA; surfactant, thickener, foam booster/stabilizer for personal care and industrial applications. [McIntyre]

Mackamide PK. Palm kernelamide DEA; surfactant, thickener, foam booster/stabilizer for personal care and industrial applications. [McIntyre]

Mackamide PKM. Palm kernelamide MEA; surfactant, thickener, foam booster/stabilizer for personal care and industrial applications. [McIntyre]

Mackamide R. CAS 40716-42-5; EINECS 255-051-3; Ricinoleamide DEA; emulsifier; softener; surfactant, thickener, foam booster/stabilizer for personal care and industrial applications. [McIntyre]

Mackamide S. CAS 68425-47-8; EINECS 270-355-6; Soyamide DEA (1:1); surfactant, thickener, foam booster/stabilizer for personal care and industrial applications; hair conditioner. [McIntyre]

Mackamide SMA. CAS 111-57-9; EINECS 203-883-2; Stearamide MEA; surfactant, thickener, foam booster/stabilizer for personal care and industrial applications. [McIntyre]

Mackamine CAO. CAS 68155-09-9; Cocamidopropylamine oxide; detergent, wetting agent for heavy-duty cleaners; hair conditioner, visc. builder, foam booster for personal care products. [McIntyre]

Mackamine CO. CAS 61788-90-7; EINECS 263-016-9; Cocamine oxide; detergent, wetting agent for heavy-duty cleaners; hair conditioner, visc. builder, foam booster/stabilizer for personal care products. [McIntyre]

Mackamine IAO. Isostearamidopropylamine oxide; detergent, wetting agent for heavy-duty cleaners; hair conditioner, visc. builder, foam booster for personal care products. [McIntyre]

Mackamine ISMO. Isostearamidopropyl morpholine oxide; detergent, wetting agent for heavy-duty cleaners; hair conditioner, visc. builder, foam booster for personal care products. [McIntyre]

Mackamine LAO. CAS 61792-31-2; EINECS 263-218-7; Lauramidopropylamine oxide; detergent, wetting agent for heavy-duty cleaners; hair conditioner, visc. builder, foam booster for personal care products. [McIntyre]

Mackamine LO. Lauramine oxide; detergent, wetting agent for heavy-duty cleaners; hair conditioner, visc. builder, foam booster for personal care products. [McIntyre]

Mackamine O2. CAS 14351-50-9; EINECS 238-311-0; Oleamine oxide; detergent, wetting agent for heavy-duty cleaners; hair conditioner, visc. builder, foam booster for personal care products. [McIntyre]

Mackamine OAO. CAS 25159-40-4; EINECS 246-684-6; Oleamidopropylamine oxide; detergent, wetting agent for heavy-duty cleaners; hair conditioner, visc. builder, foam booster for personal care products. [McIntyre]

Mackamine SAO. CAS 25066-20-0; Stearamidopropylamine oxide; detergent, wetting agent for heavy-duty cleaners; hair conditioner, visc. builder, foam booster for personal care products. [McIntyre]

Mackamine SO. CAS 2571-88-2; EINECS 219-919-5; Stearamine oxide; detergent, wetting agent for heavy-duty cleaners; hair conditioner, visc. builder, foam booster for personal care products. [McIntyre]

Mackamine WGO. Wheat germamidopropylamine oxide; detergent, wetting agent for heavy-duty cleaners; hair conditioner, visc. builder, foam booster for personal care products. [McIntyre]

Mackanate A-102. Disodium deceth-6 sulfosuccinate; mild surfactant for personal care products. [McIntyre]

Mackanate A-103. Disodium nonoxynol-10 sulfosuccinate; mild surfactant for personal care products. [McIntyre]

Mackanate AY-65TD. Diamyl sodium sulfosuccinate, tridecyl alcohol; surfactant, wetting agent for industrial applications. [McIntyre]

Mackanate CM. Disodium cocamido MEA-sulfosuccinate;

---

‡ Trade name and manufacturer not verified   § Trade name without identified manufacturer

base for rug cleaners, shampoos, bubble baths. [McIntyre]

Mackanate CP. CAS 68515-65-1; EINECS 271-102-2; Disodium cocamido MIPA-sulfosuccinate; mild surfactant for personal care products. [McIntyre]

Mackanate DC-30. Disodium dimethicone copolyol sulfosuccinate; mild surfactant for personal care products. [McIntyre]

Mackanate DC-30A. Diammonium dimethicone copolyol sulfosuccinate; mild surfactant for personal care products. [McIntyre]

Mackanate DOS-40. CAS 577-11-7; EINECS 209-406-4; Dioctyl sodium sulfosuccinate; surfactant, wetting agent for industrial applications. [McIntyre]

Mackanate EL. Disodium laureth sulfosuccinate; surfactant for shampoos and bubble baths. [McIntyre]

Mackanate LA. Diammonium lauryl sulfosuccinate; mild surfactant for personal care products, hand cleaners. [McIntyre]

Mackanate LM-40. CAS 25882-44-4; EINECS 247-310-4; Disodium lauramido MEA-sulfosuccinate; high foaming surfactant for personal care products. [McIntyre]

Mackanate LO. Disodium lauryl sulfosuccinate; mild surfactant for hand and skin cleaners, shampoos, bubble baths. [McIntyre]

Mackanate O-3. Disodium oleth sulfosuccinate; mild surfactant for personal care products. [McIntyre]

Mackanate OD-30. CAS 56388-43-3; EINECS 260-143-1; Disodium oleamido PEG-2 sulfosuccinate; surfactant for mild conditioning shampoo. [McIntyre]

Mackanate OM. Disodium oleamido MEA-sulfosuccinate; nonirritating high foaming surfactant for shampoos. [McIntyre]

Mackanate OP. CAS 43154-85-4; EINECS 256-120-0; Disodium oleamido MIPA-sulfosuccinate; mild, skin-protective surfactant for personal care products. [McIntyre]

Mackanate RM. Disodium ricinoleamido MEA-sulfosuccinate; mild, skin-protective surfactant for personal care products. [McIntyre]

Mackanate UM. Disodium undecylenamido MEA-sulfosuccinate; mild surfactant for personal care products. [McIntyre]

Mackanate WGD. Disodium wheat germamido PEG-2 sulfosuccinate; nonirritating emollient surfactant for personal care products. [McIntyre]

Mackazoline C. CAS 61791-38-6; EINECS 263-170-7; Cocoyl hydroxyethyl imidazoline; emulsifier; corrosion inhibitor for acid bowl cleaners, pickling systems; salts as antistats, water displacer. [McIntyre]

Mackazoline CY. CAS 37478-68-5; EINECS 253-521-2; Capryl hydroxyethyl imidazoline; emulsifier; corrosion inhibitor for acid bowl cleaners, pickling systems; salts as antistats, water displacer. [McIntyre]

Mackazoline L. CAS 136-99-2; EINECS 205-271-0; Lauryl hydroxyethyl imidazoline; emulsifier; corrosion inhibitor for acid bowl cleaners, pickling systems; salts as antistats, water displacer. [McIntyre]

Mackazoline O. Oleyl hydroxyethyl imidazoline; emulsifier; corrosion inhibitor for acid bowl cleaners, pickling systems; salts as antistats, water displacer. [McIntyre]

Mackechnie. Copper sulfate, pentahydrate, monohydrate and anhydrous. [McKechnie Chemicals Ltd] *

Mackechnie's bronze. Usually an alloy of 57% copper, 41% tin, 1% zinc, 1% iron, and 0.5% lead.

Mackenzie's amalgam. Bismuth (2 parts) and lead (4 parts) are melted separately in crucibles, and each poured into 1 part mercury. These amalgams are then rubbed together.

Mackenzie's metal. An alloy of 70% lead, 17% antimony, and 13% tin; also 68% lead, 16% antimony, and 16% bismuth. Electrotype metals.

Mackernium 006. CAS 26062-79-3; Polyquaternium-6; slip agent for liq. hand soaps, conditioner for shampoos without buildup. [McIntyre]

Mackernium 007. CAS 26590-05-6; Polyquaternium-7; slip agent for liq. hand soaps; conditioner for shampoos without buildup. [McIntyre]

Mackernium KP. CAS 37139-99-4; Oleakonium chloride; conditioner, lubricant, antistat for personal care products. [McIntyre]

Mackernium NLE. Quaternium-84; conditioner, lubricant, antistat for personal care products. [McIntyre]

Mackernium SDC-25. CAS 122-19-0; EINECS 204-527-9; Stearalkonium chloride; conditioner, lubricant, antistat for personal care products. [McIntyre]

Mackester EGDS. CAS 627-83-8; EINECS 211-014-3; Glycol distearate; emulsifier, lubricant, antistat, defoamer for metalworking, textile lubricants, plastics, paper; emulsifier, pearlescent, emollient for cosmetics. [McIntyre]

Mackester EGMS. Glycol stearate; emulsifier, lubricant, antistat, defoamer for metalworking, textile lubricants, plastics, paper; emulsifier, pearlescent, emollient for cosmetics. [McIntyre]

Mackester IDO. CAS 59231-34-4; EINECS 261-673-6; Isodecyl oleate; emulsifier, lubricant, antistat, defoamer for metalworking, textile lubricants, plastics, paper; emulsifier, pearlescent, emollient for cosmetics. [McIntyre]

Mackester IP. Glycol stearate, other ingreds.; emulsifier, lubricant, antistat, defoamer for metalworking, textile lubricants, plastics, paper; emulsifier, pearlescent, emollient for cosmetics. [McIntyre]

Mackester TD-88. Triethylene glycol dioctoate; emulsifier, lubricant, antistat, defoamer for metalworking, textile lubricants, plastics, paper; emulsifier, pearlescent, emollient for cosmetics. [McIntyre]

Mackine 101. CAS 68140-01-2; EINECS 268-771-8; Cocamidopropyl dimethylamine; intermediate for cationic surfactants, chemical specialties, hair conditioners, mild cleansers; corrosion inhibitor; salts as emulsifier for acid systems. [McIntyre]

Mackine 201. Ricinoleamidopropyl dimethylamine; intermediate for cationic surfactants, chemical specialties, hair conditioners, mild cleansers; corrosion inhibitor; salts as emulsifier for acid systems. [McIntyre]

Mackine 301. CAS 7651-02-7; EINECS 231-609-1; Stearamidopropyl dimethylamine; intermediate for cationic surfactants, chemical specialties, hair conditioners, mild cleansers; corrosion inhibitor; salts as emulsifier for acid systems. [McIntyre]

Mackine 321. CAS 55852-13-6; Stearamidopropyl morpholine; intermediate for cationic surfactants, chemical specialties, hair conditioners, mild cleansers; corrosion inhibitor; salts as emulsifier for acid systems. [McIntyre]

Mackine 401. CAS 67799-04-6; EINECS 267-101-1; Isostearamidopropyl dimethylamine; intermediate for cat-

---

* Trade name not verified as available   † Trade name verified as obsolete

ionic surfactants, chemical specialties, hair conditioners, mild cleansers; corrosion inhibitor; salts as emulsifier for acid systems. [McIntyre]

Mackine 421. Isostearamidopropyl morpholine; intermediate for cationic surfactants, chemical specialties, hair conditioners, mild cleansers; corrosion inhibitor; salts as emulsifier for acid systems. [McIntyre]

Mackine 501. CAS 109-28-4; EINECS 203-661-5; Oleamidopropyl dimethylamine; intermediate for cationic surfactants, chemical specialties, hair conditioners, mild cleansers; corrosion inhibitor; salts as emulsifier for acid systems. [McIntyre]

Mackine 601. CAS 60270-33-9; EINECS 262-134-8; Behenamidopropyl dimethylamine; intermediate for cationic surfactants, chemical specialties, hair conditioners, mild cleansers; corrosion inhibitor; salts as emulsifier for acid systems. [McIntyre]

Mackine 701. Wheat germamidopropyl dimethylamine; intermediate for cationic surfactants, chemical specialties, hair conditioners, mild cleansers; corrosion inhibitor; salts as emulsifier for acid systems. [McIntyre]

Mackine 801. CAS 3179-80-4; EINECS 221-661-3; Lauramidopropyl dimethylamine; intermediate for cationic surfactants, chemical specialties, hair conditioners, mild cleansers; corrosion inhibitor; salts as emulsifier for acid systems. [McIntyre]

Mackine 901. CAS 68188-30-7; Soyamidopropyl dimethylamine; intermediate for cationic surfactants, chemical specialties, hair conditioners, mild cleansers; corrosion inhibitor; salts as emulsifier for acid systems. [McIntyre]

Mackpro KLP. Quaternium-79 hydrolyzed keratin; conditioner for hair and skin care products. [McIntyre]

Mackpro MLP. Quaternium-79 hydrolyzed milk protein; conditioner for hair and skin care products. [McIntyre]

Mackpro NLP, NLP-Special. Quaternium-79 hydrolyzed animal protein; hair conditioner. [McIntyre]

Mackpro NLW. Quaternium-79 hydrolyzed wheat protein; conditioner for hair and skin care products. [McIntyre]

Mackpro NSP. Quaternium-79 hydrolyzed silk; conditioner for hair and skin care products. [McIntyre]

Mackpro SLP. Quaternium-79 hydrolyzed soy protein; conditioner for hair and skin care products. [McIntyre]

Mackpro WWP. Wheatgermamidopropyl dimethylamine hydrolyzed wheat protein; conditioner for hair and skin care products. [McIntyre]

Mack's cement. Prepared by adding calcined sodium sulfate or potassium sulfate to dehydrated gypsum.

Mackstat® DM. CAS 6440-58-0; DMDM hydantoin; broad spectrum cosmetic preservative for shampoos, skin cleansers, bath products, lotions and creams. [McIntyre]

Macocyn. CAS 79-57-2; A proprietary preparation of oxytetracycline; an antibiotic. [Pfizer International]

Macol® 1. CAS 9003-11-6; Poloxamer 181; defoamer, deduster, emulsifier, detergent, dispersant, dye leveler, gellant, antistat, solubilizer, dispersant, wetting agent, lubricant base for metalworking and textile lubricants, cosmetics, medical, paper, pharmaceutical, chemical intermediates. [PPG/Specialty Chem.]

Macol® 2. CAS 9003-11-6; Poloxamer 182; detergent, emulsifier, wetting agent, dispersant, antistat, defoamer, gellant, solubilizer, lubricant base for cosmetic, medical, paper, metalworking, pharmaceutical and textile industries. [PPG/Specialty Chem.]

Macol® 4. CAS 9003-11-6; Poloxamer 184; detergent, foam-

ing agent, emulsifier, wetting agent, dispersant, antistat, defoamer, gellant, solubilizer, lubricant base for cosmetic, medical, paper, pharmaceutical and textile industries. [PPG/Specialty Chem.]

Macol® 8. CAS 9003-11-6; Poloxamer 188; detergent, foaming agent, wetting agent, dispersant, antistat, gellant, solubilizer, lubricant base for cosmetic, medical, paper, pharmaceutical and textile industries. [PPG/Specialty Chem.]

Macol® 15. CAS 9003-11-6; Meroxapol 105; defoamer, detergent, emulsifier, pulp and paper additive, dispersant, lubricant, leveling aid, wetting agent. [PPG/Specialty Chem.]

Macol® 16. CAS 9003-11-6; Meroxapol 108; defoamer, deduster, detergent, emulsifier, dispersant, dye leveler, gellant, antistat. [PPG/Specialty Chem.]

Macol® 18. CAS 9003-11-6; Meroxapol 171; defoamer, wetting agent, deduster, demulsifier, detergent, dispersant, dye leveler, gellant, antistat; synthetic lubricant base fluid for metalworking and textile lubricants, chemical intermediates; pulp and paper additive. [PPG/Specialty Chem.]

Macol® 19. CAS 9003-11-6; Meroxapol 172; defoamer, wetting agent, deduster, demulsifier, detergent, dispersant, dye leveler, gellant, antistat; synthetic lubricant base fluid for metalworking and textile lubricants, chemical intermediates; pulp and paper additive. [PPG/Specialty Chem.]

Macol® 23. CAS 9003-11-6; Poloxamer 403; foaming agent, emulsifier, wetting agent, dispersant, antistat, gellant, solubilizer, lubricant base for cosmetic, medical, paper, pharmaceutical and textile industries. [PPG/Specialty Chem.]

Macol® 27. CAS 9003-11-6; Poloxamer 407; emulsifier, wetting agent, dispersant, antistat, defoamer, gellant, solubilizer, lubricant base for cosmetic, medical, paper, pharmaceutical and textile industries. [PPG/Specialty Chem.]

Macol® 32. CAS 9003-11-6; Meroxapol 251; wetting agent, dispersant, antistat, defoamer, gellant, solubilizer, lubricant base for cosmetic, medical, paper, pharmaceutical and textile industries. [PPG/Specialty Chem.]

Macol® 33. CAS 9003-11-6; Meroxapol 311; wetting agent, dispersant, antistat, defoamer, gellant, solubilizer, lubricant base for cosmetic, medical, paper, metalworking, pharmaceutical and textile industries. [PPG/Specialty Chem.]

Macol® 34. CAS 9003-11-6; Meroxapol 254; wetting agent, dispersant, antistat, defoamer, gellant, solubilizer, lubricant base for cosmetic, medical, paper, pharmaceutical and textile industries. [PPG/Specialty Chem.]

Macol® 35. CAS 9003-11-6; Poloxamer 105; wetting agent, dispersant, antistat, defoamer, gellant, solubilizer, lubricant base for cosmetic, medical, paper, pharmaceutical and textile industries. [PPG/Specialty Chem.]

Macol® 40. CAS 9003-11-6; Meroxapol 252; wetting agent, dispersant, antistat, defoamer, gellant, solubilizer, lubricant base for cosmetic, medical, paper, metalworking, pharmaceutical and textile industries. [PPG/Specialty Chem.]

Macol® 42. CAS 9003-11-6; Poloxamer 122; wetting agent, dispersant, antistat, defoamer, gellant, solubilizer, lubricant base for cosmetic, medical, paper, pharmaceutical and textile industries. [PPG/Specialty Chem.]

---

‡ Trade name and manufacturer not verified   § Trade name without identified manufacturer

Macol® 44. CAS 9003-11-6; Poloxamer 124; wetting agent, dispersant, antistat, defoamer, gellant, solubilizer, lubricant base for cosmetic, medical, paper, pharmaceutical and textile industries. [PPG/Specialty Chem.]

Macol® 46. CAS 9003-11-6; Poloxamer 101; wetting agent, dispersant, antistat, defoamer, gellant, solubilizer, lubricant base for cosmetic, medical, paper, pharmaceutical and textile industries. [PPG/Specialty Chem.]

Macol® 72. CAS 9003-11-6; Poloxamer 212; wetting agent, dispersant, antistat, defoamer, gellant, solubilizer, lubricant base for cosmetic, medical, paper, pharmaceutical and textile industries. [PPG/Specialty Chem.]

Macol® 77. CAS 9003-11-6; Poloxamer 217; wetting agent, dispersant, antistat, defoamer, gellant, solubilizer, lubricant base for cosmetic, medical, paper, pharmaceutical and textile industries. [PPG/Specialty Chem.]

Macol® 85. CAS 9003-11-6; Poloxamer 235; wetting agent, dispersant, antistat, defoamer, gellant, solubilizer, lubricant base for cosmetic, medical, paper, pharmaceutical and textile industries. [PPG/Specialty Chem.]

Macol® 101. CAS 9003-11-6; Poloxamer 331; wetting agent, dispersant, antistat, defoamer, gellant, solubilizer, lubricant base for cosmetic, medical, paper, pharmaceutical and textile industries. [PPG/Specialty Chem.]

Macol® 108. CAS 9003-11-6; Poloxamer 338; wetting agent, dispersant, antistat, defoamer, gellant, solubilizer, lubricant base for cosmetic, medical, paper, pharmaceutical and textile industries. [PPG/Specialty Chem.]

Macol® 123. Ceteareth alcohol, ceteareth-20, ceteareth-10; emulsifier for cosmetic and pharmaceutical applications. [PPG/Specialty Chem.]

Macol® 125. Stearyl alcohol, ceteareth-20; emulsifier for cosmetic and pharmaceutical applications. [PPG/Specialty Chem.]

Macol® 300. PPG-7 buteth-10; detergent for toilet cleaners, laundry detergents, emulsion polymerization, defoamers, metalworking fluids, hydraulic fluids. [PPG/Specialty Chem.]

Macol® 660. PPG-12 buteth-16; detergent for toilet bowl cleaners, laundry; defoamer, rubber lubricant, intermediate; hydraulic, heat transfer, and metal working fluids; mold release agent; emulsion polymerization. [PPG/Specialty Chem.]

Macol® CA-2. CAS 9004-95-9; Ceteth-2; emulsifier, detergent, wetting agent, dispersant, solubilizer, coupling agent for cosmetics, textile, metalworking, household, industrial and other applications. [PPG/Specialty Chem.]

Macol® CPS. Cetearyl alcohol, polysorbate 60, PEG-150 stearate, steareth-20. emulsifier for cosmetic and pharmaceutical applications. [PPG/Specialty Chem.]

Macol® CSA-2. CAS 68439-49-6; Ceteareth-2; detergent, wetting agent, emulsifier, dispersant, solubilizer, and coupling agent for cosmetics, textiles, metalworking lubricants, household products, and industrial applications. [PPG/Specialty Chem.]

Macol® DNP-5. CAS 9014-93-1; Nonyl nonoxynol-5; emulsifier, detergent, wetting agent, dispersant, solubilizer, coupling agent for cosmetics, textile, metalworking, household, industrial and other applications. [PPG/Specialty Chem.]

Macol® E-200. CAS 25322-68-3; PEG-4; chemical intermediate for fatty acid esters, lubricant bases in textile and metalworking, as components in pharmaceutical and cosmetic preps. [PPG/Specialty Chem.]

Macol® LA-4. Laureth-4; detergent, wetting agent, emulsifier, dispersant, solubilizer, stabilizer, coupling agent for cosmetics, textiles, metalworking lubricants, household products, industrial uses. [PPG/Specialty Chem.]

Macol® LF-110. Polyalkoxylated aliphatic ether; biodeg. low foam wetting aid and low surface tension surfactant; synthetic lubricant base fluid for metalworking, hard-surface cleaning, and metal cleaning and degreasing; used in cleaners and rinse aids. [PPG/Specialty Chem.]

Macol® NP-4. CAS 9016-45-9; Nonoxynol-4; emulsifier, detergent, wetting agent, dispersant, solubilizer, coupling agent for cosmetics, textile, metalworking, household, industrial and other applications. [PPG/Specialty Chem.]

Macol® OA-2. CAS 9004-98-2; Oleth-2; detergent, wetting agent, emulsifier, dispersant, solubilizer, stabilizer, coupling agent for cosmetics, textiles, metalworking lubricants, household products, industrial uses. [PPG/Specialty Chem.]

Macol® OP-3. CAS 9002-93-1; Octoxynol-3; emulsifier, detergent, wetting agent, dispersant, solubilizer, coupling agent for cosmetics, textile, metalworking, household, industrial and other applications. [PPG/Specialty Chem.]

Macol® P-500. PPG-9; defoamer for aq. systems, in mold release applications, lubricant bases for textile, paper, metalworking formulations, chemical intermediates for fatty acid esters, components for urethane resins. [PPG/Specialty Chem.]

Macol® SA-2. CAS 9005-00-9; Steareth-2; detergent, wetting agent, emulsifier, dispersant, solubilizer, stabilizer, coupling agent for cosmetics, textiles, metalworking lubricants, household products, industrial uses. [PPG/Specialty Chem.]

Macol® TD-3. CAS 24938-91-8; Trideceth-3; detergent, wetting agent, emulsifier, dispersant, solubilizer, stabilizer, coupling agent for cosmetics, textiles, metalworking lubricants, household products, industrial uses. [PPG/Specialty Chem.]

Macol® WSL-2000. Non-butanol functional fluid; lubricant in metalworking, textile, and hydraulic fluids. [PPG/Specialty Chem.]

Macor. Photographic chemicals. [Makhteshim Chemical Works Ltd]

Macquer's salt. CAS 7784-41-0; Potassium arsenate, KH_2AsO_4.

Macrodantin. CAS 67-20-9; Nitrofurantoin; antibacterial. [Norwich Eaton Pharmaceuticals Inc]

Macrodex. Dextran 70; a polysaccharide produced by the action of *Leuconostoc mesenteroides* on sucrose; average molecular weight: 70,000; plasma volume extender. [Kabi-Pharmacia]

macrogols. Polyethylene glycols.

macrogol stearate. Polyoxyl 40 stearate or polyoxyl 8 stearate.

Macrolex®. High quality dyestuffs; used for dyeing polystyrene, styrene copolymers, polymethacrylate rigid PVC and polycarbonate. [Bayer AG; Bayer plc]

Macromite. CAS 471-34-1; Calcium carbonate; very fine wet-ground product for paint and polymer industries. [ECC]

macrose. *See* dextran.

Macrosorb. Inorganic adsorbent. [Crosfield Chemicals Ltd]

---

* Trade name not verified as available   † Trade name verified as obsolete

Macrosorb. Range of inorganic matrices for separation and support. [Microporous Materials Ltd]

Macrospherical® 95. Aluminum chlorohydrate powder with a spherical shell; antiperspirant active. [Reheis Inc]

Macrynal. Hydroxy acrylic resins for use with polyisocyanates; used for automotive paints and industrial finishes. [Resinous Chemicals Ltd]

Maculanin. Potassium amylate.

Macuprax. Bordeaux/cufraneb fungicide. [McKechnie Chemicals Ltd] *

Madanite. A product used as a binding material made from 2 parts petroleum jelly and 1 part rubber.

Madar fiber. A bast fiber, known in India by this name. obtained from *Calotropis procera* and *C. gigantea*.

madder. The powdered root of the plant, *Rubia tinctorum*. The chief constituent is ruberythric acid. This acid is a glucoside, and is split up into alizarin and a sugar by the action of acids.

Maddrell salts. Alkaline meta-phosphates which are insoluble in water. They are made by heating monosodium phosphate at 245-250 C.

Madecassol. A proprietary preparation extract of *Centella asiatica* in an ointment; used for skin protection. [Rona Laboratories] ‡

madol oil. An oil obtained from the seeds of *Garcinia echinocarpa*.

Madopar. A proprietary preparation of levodopa and benserazide; used in the treatment of Parkinson's disease. [Roche Products Ltd] *

Madquat Q-6. CAS 5039-78-1; Methylchloride quaternary salt of dimethylaminoethyl methacrylate [Rhone-Poulenc Surf.]

Madribon. CAS 122-11-2; A proprietary preparation containing sulfadimethoxine; used as an anti-infective. [Roche Products Ltd] *

Madurit®. Melamine resin products; used for laminate and chipboard production, improvement of paper wet strength, bonding of glass fleeces. [Cassella AG]

Mafloc® 700. Powd. polymer; dewatering aid in vacuum filters and centrifuge systems; esp. effective for municpal and industrial wastewater sludge and potable water clarification. [PPG/Specialty Chem.]

Mafloc® 718. Emulsion polymer; used as sludge thickener, vacuum filter, centrifuge, plate and frame filter press, belt filter press in industrial and municipal waste dewatering applications. [PPG/Specialty Chem.]

Mafloc® 764. Polyacrylate polymer; flocculating power towards suspension of solids. [PPG/Specialty Chem.]

Mafo® 13. Potassium salt of complex n-stearyl amino acid; biodeg. emulsifier, wetting agent, corrosion inhibitor, suspending agent, solubilizer for difficult materials, dairy chain and metal-to-metal lubricant, emollient; for burnishing and polishing compounds. [PPG/Specialty Chem.]

Mafo® C. Cocamidopropyl betaine; biodeg. solubilizer, emollient, coupling agent, emulsifier, foam booster for shampoos. [PPG/Specialty Chem.]

Mafo® CB 40. CAS 68424-94-2; EINECS 270-329-4; Coco betaine; biodeg. foam booster, visc. builder, solubilizer, emollient, coupling agent, emulsifier for shampoos. [PPG/Specialty Chem.]

Mafo® CSB. CAS 68139-30-0; Cocamidopropyl hydroxysultaine; biodeg. solubilizer, emollient, coupling agent, emulsifier, foam booster for shampoos. [PPG/Specialty Chem.]

Mafo® LMAB. Lauramidopropyl betaine; biodeg. solubilizer, emollient, coupling agent, emulsifier, foam booster for shampoos. [PPG/Specialty Chem.]

Mafo® OB. CAS 871-37-4; EINECS 212-806-1; Oleyl betaine; biodeg. solubilizer, emollient, coupling agent, emulsifier, foam booster for shampoos. [PPG/Specialty Chem.]

Maftec®. CAS 1344-28-1; Alumina fiber; reinforcing agent for high temp. usage and composite materials, e.g., furnace lining. [Mitsubishi Kasei] *

Mafu®. CAS 62-73-7; DDVP; insecticidal aerosol used for control of stored product pests in filled and empty silos, store rooms, etc.; for flying insects, weevils, mites and bugs. [Bayer AC]

Mafura fat. A fat obtained from the seeds of *Mafureira oleifera*. It contains from 92-95% fatty acids; used in the manufacture of soaps.

Mag-40. Liquid chemical defoliant of cotton and a desiccant in silverskin onion. [Makhteshim Chemical Works Ltd] †

Magadi soda. An East African soda. It contains sodium carbonate and sodium bicarbonate.

Magala® 0.5E. CAS 61632-57-3; Di-n-butyl magnesium/triethyl aluminum complex (0.5:1); used in production of catalysts for polymerization of olefins or dienes and as an alkylating agent. [Akzo]

Magan. CAS 18917-89-0; A range of magnesium salicylate tablets sold only on prescription for medical and pharmaceutical uses. [Hercules] *

MagChem® 1060. CAS 1309-48-4; Magnesium oxide; hardburned screened grade for refractories, water treatment, acid neutralization, desilication, heavy metals removal. [Martin Marietta Magnesia Spec.] *

MagChem® 10B. CAS 1309-48-4; Magnesium oxide; hardburned screened grade specially processed to provide low boron level in fused elec. heating element insulation. [Martin Marietta Magnesia Spec.] *

MagChem® 20. CAS 1309-48-4; Magnesium oxide; lightburned reactive grade for mfg. of magnesium chems., construction products, lubrication oil additives, fuel additives, oil drilling chems., in neoprene and chlorinated polymers, sugar refining, uranium ore processing, acid neutralization, leather tanning, water treatment. [Martin Marietta Magnesia Spec.] *

Magcoke. Magnesium impregnated coke for the desulfurization and nodularization of cast iron. [Foseco (F.S.) Ltd]

Magecol. A proprietary lampblack suitable for rubber goods.§

Magicote Masonry Paint. Vinyl copolymer based, water thinned; used for brickwork, cement, stone, concrete, pebbledash, internal and external. [Berger Jenson & Nicholson Ltd] *

Magicote Non Drip and Liquid Gloss. Alkyd resin based, white spirit thinned; for internal and external application in the DIY market; non drip; needs no undercoat, liquid requires Magicote undercoat. [Berger Jenson & Nicholson Ltd] *

Magicote Solid Emulsion. Vinyl copolymer based, water thinned; available in Vinyl Matt and Vinyl Silk to apply by roller to interior walls and ceilings; supplied ready-for-use in roller tray. [Berger Jenson & Nicholson Ltd] *

Magicote Vinyl Matt. Vinyl copolymer based, water thinned; washable finish for interior walls and ceilings. [Berger

---

‡ Trade name and manufacturer not verified   § Trade name without identified manufacturer

Jenson & Nicholson Ltd] *

Magicote Vinyl Silk. Vinyl copolymer based, water thinned; washable finish for interior walls and ceilings. [Berger Jenson & Nicholson Ltd] *

Magisal. CAS 132-49-0; Magnesium acetylsalicylate; used for similar purposes as aspirin. §

magister of bismuth. Basic bismuth nitrate, $BiO \cdot NO_3 \cdot H_2O$.

magister of sulfur. Sulfur precipitated in an amorphous condition from solutions of hyposulfites or polysulfides, by acids.

magistery of lead. See white lead.

Maglite® D. CAS 1309-48-4; Magnesium oxide; used in cure of polychloroprene compounds to aid aging stability, absorbing HCl. [Marine Magnesium]

Maglite Y. CAS 1309-48-4; A trademark for magnesium oxide. [E Merck] †

maglstral. An impure copper sulfate containing ferric oxide, sodium sulfate, and sodium chloride.

Magmet. Magnetic metal particles; used in coatings for magnetic tapes and discs. [Hercules] *

Magna A. CAS 68334-28-1; Partially hydrogenated vegetable oil (cottonseed, soybean); kosher; high melting, high performance fat for no-tempering coatings, centers, vegetable dairy systems; shortening for bakery applications. [Van Den Bergh Foods]

Magnabrite® F. Magnesium aluminum silicate NF; stabilizing and suspending agent for cosmetics and pharmaceuticals. [Am. Colloid]

Magnacell. Nitrocellulose paste dispersions. [Tennant-KVK Ltd]

Magnacide. Biocides and algicides. [Baker Performance Chemicals Ltd]

Magnaclean. Cleaning chemicals and antifoulants. [Baker Performance Chemicals Ltd]

Magnaclear. Reverse emulsion breakers and water clarifiers. [Baker Performance Chemicals Ltd]

Magnacryl. Second generation acrylic adhesives and radiation curable (uv/EB) adhesives and coatings; for pressure sensitive and permanent bonding, specialty overcoating applications. [Beacon Chemical Company Inc] *

Magnaflo. Replaces magnesium sulfate in stainblocking treatment of carpets with Stain-Free. [Sybron]

Magnafloc®. Synthetic flocculants and coagulants designed to improve the separation and handling of finely divided solids in aqueous suspension; a very wide range of products is available in terms of chemical type, molecular weight and ionic character; used in sedimentation, filtration and centrifugation. [Allied Colloids Ltd]

Magnafloc® LT. High molecular weight polyacrylamides for use as coagulant aids in treatment of drinking water and waterworks sludges; synthetic organic coagulants as alum replacements. [Allied Colloids Ltd]

Magna Flow. Flow improvers and pour point depressants. [Baker Performance Chemicals Ltd]

Magnakyd. Alkyd paste dispersions. [Tennant-KVK Ltd]

magnalium. Alloys of magnesium and aluminum. One contains from 1-2% of magnesium. It is lighter than aluminum, and is as hard as brass. Another alloy contains 10% magnesium; is used for parts of machinery, for cooking utensils, and for opt ical mirrors.

Magnamite. Carbonized or graphitized polyacrylonitrile fiber; used to reinforce thermoset and thermoplastic resins, and as a replacement for steel, aluminum, tita-

nium and other metals. [Hercules] *

Magnapen. A proprietary preparation of ampicillin and flucloxacillin; an antibiotic. [SmithKline Beecham] *

Magnaphoscal®. Multiple phosphate (sodium, calcium, magnesium phosphate) granulated; for mineral feeds and mixed feeds. [Bayer AG]

Magnaplas PA213-BF83. Nylon 6, barium ferrite-filled; permanent magnetic material [Bay Resins]

Magnaset. High pigmented binderless paste dispersions. [Tennant-KVK Ltd]

Magnasoft. Silicone textile softener. [Union Carbide]

Magnasol. Pigment solution. [Tennant-KVK Ltd]

Magnasorb. Glycols/alkanolamines. [Baker Performance Chemicals Ltd]

Magnasperse. Paste dispersions. [Tennant-KVK Ltd]

Magnaspheres®. CAS 1309-42-8; Magnesium hydroxide for water treatment. [Steetley Magnesia Products Ltd]

Magna Tac. Specialty adhesives for commercial and industrial applications; for pressure sensitives, two-part epoxies, solvent- and water-based systems, hot melts and hat sizings. [Beacon Chemical Company Inc] *

Magnate. CAS 73790-28-0; Imazalil; fungicide. [Makhteshim Chemical Works Ltd]

Magnatreat. Oxygen and sulfide scavengers. [Baker Performance Chemicals Ltd]

magnesia. CAS 1309-48-4; Magnesium oxide; MgO; used in refractories, for steel furnace liningspolycrystalline ceramic for aircraft windshields, electrical insulation, pharmaceuticals and cosmetics.

magnesia. See magnesium oxide.

magnesia alba. A basic magnesium carbonate of variable composition.

magnesia bleaching liquid. Magnesium oxychloride, $Mg(OCl)_2$; used for bleaching.

magnesia-citrate mixture. Citric acid (20 g), is dissolved in 20% ammonium hydrate, and mixed with 1 liter of magnesia mixture (qv); used in the determination of phosphorus magnesia mixture.

magnesia magma. See magnesium hydroxide.

magnesia usta. See magnesium oxide.

magnesia white. Both magnesium oxide and magnesium carbonate are known by this name.

magnesium. Metallic element; Mg; CAS 7439-95-4; EINECS 231-104-6; aluminum alloys for structural parts, etc.; pyrotechnics; photography; production of iron, nickel, zinc, titanium, zirconium, steel; gasoline additive; magnesium compounds; cathodic protection; reducing agent; precision instruments; optical mirrors. [Aldrich; Norsk Hydro A/S; Pechiney Electrométallurgie]

magnesium acetate. $Mg(OOCCH_3)_2$; CAS 142-72-3; dye fixative in textile printing, deodorant, disinfectant, antiseptic. [EM Industries; Hoechst Celanese; Verdugt BV]

magnesium aluminum silicate. (aluminum magnesium silicate; aluminosilicic acid, magnesium salt) Complex silicate refined from naturally occurring minerals; $Al_2MgO_8Si_2$; CAS 1327-43-1; 12199-37-0; suspending agent, thickener; antacid. [Am. Colloid; Dry Branch Kaolin; ECC Int'l.; Kaopolite; R.T. Vanderbilt; Volclay Ltd]

magnesium base alloys. The magnesium in these alloys usually varies from 90-95%.

magnesium carbonate. (carbonic acid, magnesium salt (1:1); carbonic acid, magnesium salt (2:1)) Basic dehydrated magnesium carbonate or a normal hydrated magnesium carbonate; $CH_2O_3 \cdot Mg$; CAS 546-93-0; magnesium salts, heat insulation and refractory, rubber reinforcing agent, inks, glass, pharmaceuticals, dentifrices, cosmetics, table salt, antacid; used in foods as drying agent, color retention agent, anticaking agent. [Lonza; Magnesia GmbH; Mallinckrodt; Morton Int'l.; Whittaker, Clark & Daniels]

magnesium chloride. (magnesium chloride anhydrous) $MgCl_2$; CAS 7786-30-3; EINECS 232-094-6; source of magnesium; disinfectants, fire extinguishers, fireproofing wood, cement, refrigerating brines, ceramics, cooling drilling tools, textile sizes and lubricants, paper manufacture, dust control on roads, flocculating agent, catalyst. [Aldrich; Magnesia GmbH; Mallinckrodt; Schaefer Salt & Chem.]

magnesium chloride anhydrous. See magnesium chloride.

magnesium citrate, dibasic. (acid magnesium citrate) $MgHC_6H_5O_7 \cdot 5H_2O$; CAS 144-23-0; laxative, dietary supplement. [EM Industries; Magnesia GmbH; Mallinckrodt]

magnesium gluconate. (magnesium D-gluconate; D-gluconic acid magnesium salt) $(C_6H_{11}O_7)_2Mg$; CAS 3632-91-5; EINECS 222-848-2; mineral source for pharmaceutical and food products [Akzo; Atomergic Chemetals; Spectrum Chem. Mfg.]

magnesium hydrate. See magnesium hydroxide.

magnesium hydrogen phosphate. See magnesium phosphate, dibasic.

magnesium hydroxide. (magnesium hydrate; milk of magnesia; magnesia magma) Inorganic base; $Mg(OH)_2$; CAS 1309-42-8; EINECS 215-1703; intermediate for obtaining magnesium metal, sugar refining, medicine (antacid, laxative), residual fuel oil additive, sulfite pulp, uranium processing, dentifrices, in foods as drying agent, frozen desserts. [Aldrich; Climax Performance; Croxton & Garry Ltd; J.W.S. Delavau; Mallinckrodt; Morton Int'l.; Reheis; Solem Industries]

Magnesium-Monel. A proprietary trade name for an alloy of 50% magnesium and 50% monel metal. §

magnesium nitrate. $Mg(NO_3)_2 \cdot 2H_2O$; CAS 10377-60-3; pyrotechnics. [Blythe, William Ltd; EM Industries; Hoechst Celanese; Mallinckrodt]

magnesium octadecanoate. See magnesium stearate.

magnesium oxide. (magnesia; calcined magnesia; magnesia usta) MgO; CAS 1309-48-4; EINECS 215-171-9; Refractories, esp. for steel furnace linings, polycrystalline ceramic for aircraft windshields, elec. insulations, pharmaceuticals, cosmetics, inorg. rubber accelerator, paper manufacture; white color standard; reflector in optical instruments. [EM Industries; Harwick; Hüls Am.; Magnesia GmbH; Mallinckrodt; Morton Int'l.]

magnesium-perhydrol. See Biogen.

magnesium phosphate, dibasic. (dimagnesium orthophosphate; magnesium phosphate, secondary; magnesium hydrogen phosphate) $MgHPO_4 \cdot 3H_2O$; CAS 7757-86-0; stabilizer for plastics, food additive, medicine (laxative). [Ashland; Mallinckrodt; Rasa Ind.; Spectrum Chem. Mfg.]

magnesium phosphate, secondary. See magnesium phosphate, dibasic.

magnesium silicate. (silicic acid, magnesium salt (1:1)) Inorganic salt of variable composition; $3MgSiO_3 \cdot 5HOH$ (variable); $\approx MgO \cdot SiO_2 \cdot xH_2O$; CAS 1343-88-0; EINECS 215-681-1; Rubber filler, ceramics, glass, re-

fractories; absorbent for crude oil spills; animal and vegetable oils (bleaching agent); odor absorbent; filter medium; anticaking agent for foods. [Cyprus Industrial Min.; PQ Corp.; R.T. Vanderbilt]

magnesium stearate. (magnesium octadecanoate; octadecanoic acid, magnesium salt) magnesium salt of stearic acid; $[CH_3(CH_2)_{16}COO]_2Mg$; CAS 557-04-0; baby dusting powder; lubricants in making tablets; drier in paints and varnishes; stabilizer and lubricant for plastics; emulsifying agent for cosmetics. [EM Industries; Ferro/Grant; Magnesia GmbH; Mallinckrodt; Norac; Syn. Products; Witco]

magnesium sulfate. (sulfuric acid, magnesium salt (1:1); trihydrate; heptahydrate: bitter salts; epsom salts) Inorganic salt; H_2O_4S Mg; $MgSO_4$; CAS 7487-88-9; EINECS 231-298-2; fireproofing, textiles (warp sizing, dyeing, etc.), mineral waters, catalyst carrier, paper (sizing), cosmetic lotions. [Blythe, William Ltd; Heico; Mallinckrodt; PQ Corp.]

magnesium sulfonate. CAS 71786-47-5; Lube and fuel oil additive, rust preventive. [King Industries; Lubrizol; Witco/Sonneborn]

magnesium superoxyl. CAS 14452-57-4; Magnesium peroxide, MgO_2.

Magnesol. Hydrated, amorphous, synthetic magnesium silicate; used for adsorption, catalyst support, edible oil and fat reclamation, anticaking, deodorization. [Reagent Chemical & Research Inc] *

Magnetic Black. A proprietary trade name for a finely ground magnetic iron oxide for use as a pigment. §

magnetic iron ore. (Loadstone, lodestone, magnetite, black oxide of iron, ferroferrite). A black ore of iron. It is a ferroso-ferric oxide, $FeO.Fe_2O_3$, and contains over 72% iron.

magnetic oxide of iron. See magnetic iron ore.

magnetic pyrites. (Pyrrhotine, pyrrhotite, magnetic sulfide of iron). A mineral. In composition it approximates to ferrous sulfide, FeS, or ferroso-ferric sulfide, Fe_3S_4, but often contains nickel.

magnetic sulfide of iron. Magnetic iron ore.

magnetite. See magnetic iron ore.

magnet steel. Alloys of iron with from 5-50% cobalt, 5-18% nickel, 0-7% manganese, 1-12% tungsten, and 0-12% chromium.

Magnevist. Dimeglumine gadopentetate; MRI agent. [Schering Health Care Ltd]

Magnifin® H7A. CAS 1309-42-8; Magnesium hydroxide; halogen-free flame retardant/filler for plastics and rubber, esp. EPDM, EPM, XL-PE, XL-EVA; environmentally friendly. [Martinswerk GmbH]

Magnifin® H10. CAS 1309-42-8; Magnesium hydroxide; halogen-free, general purpose flame retardant/filler for plastics and rubber; environmentally friendly. [Martinswerk GmbH]

Magnifin® H10C. CAS 1309-42-8; Magnesium hydroxide; halogen-free flame retardant/filler for plastics and rubber, esp. PVC, EVA, EPDM, PE copolymer; environmentally friendly. [Martinswerk GmbH]

Magnilor. CAS 4936-47-4; Nifuratel. [The Wellcome Foundation Ltd] *

Magnisal. CAS 10377-60-3; Magnesium nitrate hexahydrate; used for agricultural (for curing magnesium deficiency via irrigation system, direct soil application or foliar spray) and technical (in metal, textile, ceramic and

---

‡ Trade name and manufacturer not verified   § Trade name without identified manufacturer

other industries) applications. [Haifa Chemicals Ltd] *

Magno. An electrical resistance alloy containing 95% nickel and 5% manganese.

Magnobond 3. Construction adhesive; bonds old to new concrete; cures underwater. [Magnolia Plastics]

Magnobond 6504. Construction adhesive; airport lighting sealer. [Magnolia Plastics]

Magno-Ceram. Three-part casting system consisting of resin, curing agent and ceramic grain; for injection and compression dies having low shrinkage and good dimensional stability; low thermal expansion enables embedment of reinforcement rods and cooling or heating coils. [Magnolia Plastics]

Magnocid. Magnesium oxychloride, $Mg(OH)(OCl)$. It has bleaching properties.

magnodat. See Biogen.

Magnolax. A proprietary preparation of magnesium hydroxide with liquid paraffin. §

magnolia metal. A bearing metal consisting mainly of antimony amd lead, and sometimes tin, with small quantities of iron and bismuth.

Magnolium. An alloy of 90% lead and 10% antimony.

Magnoloop I. Construction adhesive; traffic loop sealer. [Magnolia Plastics]

Magnox. CAS 1309-37-1; Magnetic iron oxides; used in coatings for magnetic inks, discs and tapes. [Hercules]*

Magnum®, F. Chlordazon, ethofumesate [BASF AG; BASF plc]

Magnum. 60 in. Membrane, spiral wound configuration; both TFC and Roga available; used for reverse osmosis water treatment. [Allied-Signal Fluid Systems]

Magnum. A family of acrylonitrile-butadiene-styrene resins (ABS) used in appliances, automotive, recreational, medicine and electronics applications. [Dow UK]

Magnum 240. CAS 9003-56-9; ABS resin; automotive low gloss grade. [Dow Plastics]

Magnum 275. CAS 9003-56-9; ABS resin; sheet extrusion grade. [Dow Plastics]

Magnum 445 HQ. CAS 9003-56-9; ABS resin; for automotive trim applications. [Dow Plastics]

Magnum 788HP. CAS 9003-56-9; ABS resin; high heat, low gloss grade. [Dow Plastics]

Magnum 2610. CAS 9003-56-9; ABS resin; for health care industry. [Dow Plastics]

Magnum 3661. CAS 9003-56-9; ABS resin; ignition-resist. grade. [Dow Plastics]

Magnum 4420. CAS 9003-56-9; ABS resin; ignition-resist. resin with excellent processability and toughness for computer and business equip. [Dow Plastics]

Magnum 9450P. CAS 9003-56-9; ABS resin; plateable grade. [Dow Plastics]

Magnum FG960. CAS 9003-56-9; ABS resin; pipe fitting grade. [Dow Plastics]

Magnuminium. A proprietary trade name for magnesium-aluminum alloys. §

Magnum-White. Magnesium hydroxide and calcium carbonate blend; fire retardant/smoke suppressant filler for PVC compounds and SBR latex formulations. [RMc Minerals]

Magocarb-33. CAS 546-93-0; Magnesium carbonate; flame retardant in plastics. [Kaopolite]

Magoh-S. CAS 1309-42-8; Magnesium hydroxide; fire retardant filler for plastics; extender pigment for flame retardant coatings. [Kaopolite]

Magotex. CAS 1309-48-4; Fused magnesium oxide; high thermal conduct. coeff., low oil absorp., good elec. resistivity, low moisture; for thermal conductive molding compounds, brake lining, elec. potting compounds, special refractories. [Kaopolite]

Magox® 98 HR. CAS 1309-48-4; Magnesium oxide; reactive tech. grade for chem. process areas where a rapid reaction rate is necessary, as in detergents and chem. neutralization. [Kaopolite]

Magox® Super Premium. CAS 1309-48-4; Magnesium oxide; adsorption, absorption agent and scavenger with high surface area for plastics, rubber industries, chemical neutralization, etc. [Kaopolite]

Magrex. Fluxes for magnesium alloys. [Foseco (F.S.) Ltd]

Magrods. CAS 1309-48-4; Magnesium oxide sticks. [The Chemical & Insulating Co Ltd]

Magspa. Fertilizer additive. [ICI Chem. & Polymers Ltd]

Magtran. CAS 7783-40-6; Magnesium fluoride optical qualities. [BDH Chemicals Ltd]

mahogany acid. Crude mixtures of sulfonic acid for the refining of petroleum sludge.

mahogany brown. A sienna which has been ignited, ground wet, and made up in the form of pieces, and dried.

Ma Huang. A Chinese drug. It contains the alkaloid ephedrine.

Maillechort. A nickel silver containing copper, zinc, nickel, and iron.

Maincote. Acrylic resin emulsions. [Rohm & Haas UK]

Mainstay®. CAS 8001-69-2; Cod liver oil BP; dietary supplement; for relief of joint aches, plains, and stiffness. [Seven Seas Ltd]

Maizena. A proprietary trademark for corn starch. §

maize oil. See corn oil.

Maizolith. A material resembling hard rubber obtained by the alkaline treatment of corn-stalk or corn-cob. §

Majamin. Sodium-β-tetralin sulfonate. Majamin-kalium, the potassium salt, and majammonium, the ammonium salt, are similar products. They are added to soap and soap powders to increase lathering power.

Majammonium. Ammonium-β-tetralin-sulfonate.

Majeptil. CAS 316-81-4; Thioproperazine. [May & Baker Ltd] *

majolica. A pottery enamelled with a tin oxide enamel.

majunga noir. A rubber yielded by *Landolphia perrieri*.

Makalot. A proprietary trade name for a phenol-formaldehyde synthetic resin; used for molding. §

Makon® 4. CAS 9016-45-9; Nonoxynol-4; detergent, emulsifier used in chemical specialties, cosmetic, agric., industrial and metal cleaners, textile, paper and petrol. industries. [Stepan; Stepan Canada; Stepan Europe]

Makon® 8240. CAS 61791-12-6; PEG-36 castor oil; wetting agent, lubricant, coupling agent, defoamer for metalworking fluids, corrosion inhibitors. [Stepan; Stepan Canada]

Makon® NI 10, NI 20, NI 30. Alkylphenol alkoxylate; emulsifier, dispersant for agric. microemulsions and emulsifiable concs. [Stepan Europe]

Makon NP6, NP10 and 4,8,12,14 and 30. Nonylphenol ethoxylate nonionic surfactant in liquid form; detergents, wetting agents and emulsifiers. [KWR Chemicals Ltd] *

Makon® OP-6. CAS 9002-93-1; Octoxynol-6; detergent, wetting agent, emulsifier for household/industrial cleaners. [Stepan Europe]

Makon OP6 and OP9. Octylphenol ethoxylate nonionic

---

* Trade name not verified as available   † Trade name verified as obsolete

surfactant in liquid form; detergents, wetting agents and emulsifiers. [KWR Chemicals Ltd] *

Makroblend® . PC/polyester blends, characterized by their chemical resistance and low temperature impact strength. [Miles]

Makroblend® DP4-1368. PC/PET blend, impact-modified, flame-retardant; good chem. resist. and low temp. toughness; for appliances, personal care products, elec. connectors, meter housings, instrument enclosures. [Miles]

Makroblend® UT 400. PC/PET polyester blend; unfilled, general purpose resin for injection molding; offers high melt flow for easy processability, dimensional stability and good surface appearance, high impact strength and toughness, good heat resist.; used for vacuum cleaner housings and nozzles, utility locks, protective face guards, tractor parts, instrument housings, radio housings, speaker grills. [Miles]

Makrofol® . Polycarbonate-based electrical insulating film with high dielectric strength and long-term thermal stability up to 130 C; used for dial plates and signalling and warning indicators on motor vehicles. [Bayer AG; Bayer plc]

Makrofol® BL 2-2. PC film; tech. film; light-diffusing, dazzle-free, greater scratch resist.; natural translucent; both sides very fine matte. [Bayer AG]

Makrofol® DE 1-1. PC film; tech. film with high clarity; natural color, both sides gloss. [Bayer AG] *

Makrofol® FR 6-2. PC film; tech. film, nonoptical grade; dazzle-free, greater scratch resist.; natural translucent, one side fine velvet, one side very fine matte. [Bayer AG]*

Makrolon® . Polycarbonate for the manufacture of injection moldings primarily for use in the electrical industry and for industrial mechanical applications. [Bayer AG; Bayer plc]

Makrolon® 1006 Tint. PC; optical grade. [Bayer; Miles]

Makrolon® 1143. PC; uv-stabilized extrusion grade with structural visc. [Bayer; Miles] *

Makrolon® 2400. PC; injection molding grade used in electronics, elec. engineering, lighting, photographic, optical equipment, mechanical process and precision engineering, office supplies, domestic ware, sports and safety applications; avail. in all major colors transparent, translucent, or opaque cylindrical grans.; sol. in a number of commercial solvs.; other organic compounds, e.g., benzene, acetone, and CCl$_4$ have a superficial swelling effect. [Bayer; Miles] *

Makrolon® 2600, 2800. PC; injection molding grade used in electronics, elec. engineering, lighting, photographic, optical equipment, mechanical process and precision engineering, office supplies, domestic ware, sports and safety applications. [Bayer; Miles]

Makrolon® 3100, 3200. PC; for injection molding and extrusion; see Makrolon 2400. [Bayer; Miles]

Makrolon® 3208. PC; FDA grade for extrusion applications; more resistant to hydrolysis than 3200; see also Makrolon 2400. [Bayer AG] *

Makrolon® 6355, 6455. PC; flame-retardant grade with release agent; amorphous engineering thermoplastic for automotive, transportation, building/construction, business machine, consumer, elec./electronic, medical, and optical applications. [Bayer; Miles]

Makrolon® 8325. PC, 20% glass-reinforced; high modulus grade. [Bayer; Miles]

Makrolon® AL-2647 1068 Tint. PC; auto lens grade. [Bayer; Miles]

Makrolon® FCR-2405, FCR-2407, FCR-2458. PC resin; amorphous engineering thermoplastic for automotive, transportation, building/construction, business machine, consumer, elec./electronic, medical, and optical applications; general purpose grades processable by injection molding and extrusion; FCR grades are general-purpose high-productivity with additives for easy release and uv stability. [Bayer; Miles]

Makrolon® GV. A trademark for glass filled polycarbonate resin. [Bayer; Miles] *

Makrolon® HMS-3118. PC; extrusion grade with FDA compliance. [Bayer; Miles]

Makrolon® LQ-2847, LQ-3147, LQ-3187. PC; optical grade. [Bayer; Miles]

Makrolon® LTG-3123. PC; lighting grade for HID lamps. [Bayer; Miles]

Makrolon® Rx-2530. PC; radiation-stabilized, high-flow resin offering transparency and reduced yellowing after radiation exposure; for injection molding applications requiring min. color change after sterilizing doses of gamma radiation. [Bayer; Miles]

Makrolon® SF-600. PC structural foam, 5% glass-reinforced; flame-retardant. [Bayer; Miles]

Maktion. CAS 950-37-8, 60-51-5; Methidathion and dimethoate; insecticide. [Makhteshim Chemical Works Ltd]

malabar tallow. A fat obtained from the seeds of Vateria indica. It melts at 37.5 C., has a saponification value of 188.7-189.3, and an iodine value of 37.8-39.6.

malacca primers. Primers for bonding materials such as rubber, wood, fabric, and metals. They usually have a rubber base.

malachite. (Mountain green, green verditer, copper rust, green spar). A mineral used as a pigment. It is a hydrated basic copper carbonate, CuCO$_3$ · Cu(OH)$_2$, and when ground is sold as mountain green and mineral green.

Malapaho. Oil of Panao, collected from Dipterocarpus vernicifluus.

Malatex. A proprietary preparation of propylene glycol, malic acid, benzoic acid and salicylic acid; a desloughing agent. [H N Norton & Co Ltd] ‡

malathion. (S-[1,2-bis (ethoxycarbonyl) ethyl]O,O-dimethyl phosphorodithioate; O,O-dimethyl phosphorodithioate of diethyl mercaptosuccinate) (CH$_3$O)$_2$P(S)SCH-(COOC$_2$H$_5$)CH$_2$COOC$_2$H$_5$; CAS 121-75-5; Insecticide useful on Mediterranean fruit fly. [Allchem Industries; Am. Cyanamid; Sariaf SpA]

Malathion 60. CAS 121-75-5; Diethyl 2-(dimethoxyphosphinothioyl-thio)succinate; broad spectrum organophosphorus insecticide. [Farm Protection Ltd]

Malathion Dust. Insecticide. [Murphy Chemical Ltd] †

Malathion Liquid. Insecticide. [Murphy Chemical Ltd] †

Malazide. Plant growth inhibitor. [Fisons plc, Horticulture Div] *

Maldene. Bonding agent. [GE Plastics ABS Ltd] *

Maldene 285. A proprietary copolymer of butadiene and maleic anhydride supplied as a 25% solution in acetone for use as an intermediate. [Unichem] *

Maldene 286. A copolymer similar to Maldene 285 except that it is a 25% solution of the partial ammonium salt in water. [Unichem] *

Maldene 288. A copolymer similar to Maldene 285 except

---

that the solids content in the solution is 35%. [Unichem]*

Maldene 289. A proprietary copolymer of butadiene and maleic anhydride partially ethyl-esterified and dissolved to 25% in ethyl alcohol. [Unichem] *

Maldene 292. A proprietary copolymer of butadiene and maleic anhydride supplied as a partially-butyl-esterified 25% solution in butanol. [Unichem] *

Maldene 293. A copolymer similar to Maldene 285 except that it is a 25% solution of the partial amide-ammonium salt in water. [Unichem] *

Maldene 300. A copolymer similar to Maldene 285 except that it is a 25% solution of the partial octyl ester in toluene. [Unichem] *

Maldene 631. A copolymer similar to Maldene 285 except that it is an 18% solution of a zinc-ammonium-complexed form of Maldene 286 in water. [Unichem] *

male fern. The rhizome of *Aspidium felix-mas*.

maleic acid. (maleinic acid; cis-butenedioic acid) HOOCCH:CHCOOH; CAS 110-16-7; EINECS 203-742-5; Organic synthesis (malic, succinic, aspartic, tartaric, propionic, lactic, malonic, and acrylic acids); dyeing and finishing of cotton, wool, silk; preservative for oils and fats. [Genl. Chem.; Penta Mfg.; Thor]

maleic anhydride. (2,5-furandione; toxilic anhydride; cis-butenedioic anhydride) $C_4H_2O_3$; CAS 108-31-6; EINECS 203-571-6; Polyester resins, alkyd coating resins; fumaric and tartaric acid manufacture; pesticides; preservative for oils and fats; pharmaceuticals. [Amoco; Aristech; Ashland; Atochem SA; Hüls AG; Mitsui Toatsu Chem.; Monsanto; OxyChem]

maleinic acid. See maleic acid.

Malenite. A material containing an antimony double salt, $SbF_3 \cdot Na_2SO_4$, in addition to sodium fluoride, and the sodium compound of dinitro-phenol or dinitro-o-cresol; used for impregnating wood.

malethamer. Maleic anhydride-ethylene polymer.

Malezafin 55 Plus. Emulsifiable concentrate of ethyl ester of 2,4-D acid and butoxyethanol ester of 2,4-DP acid; broadleaf herbicide for pasture land. [Invequimica & CIA SCA] *

Malezafin 57 LV. Emulsifiable concentrate of a mixture of butoxyethanol esters of 2,4-D and 2,4-DP acid; low volatile, broad leaf herbicide, brushkiller for pasture land. [Invequimica & CIA SCA] *

Malezafin LV-4. Emulsifiable concentrate of butoxyethanol ester of 2,4-D acid; low volatile, broadleaf herbicide for corn crops and pasture land. [Invequimica & CIA SCA]*

malic acid. See hydroxysuccinic acid.

Malix. Tablets containing 2.5 mg and 5 mg glibenclamide BP; indicated for the treatment of noninsulin dependent diabetes which is not adequately controlled by dietary measures alone. [Lagap Pharmaceuticals Ltd]

malladrite. Sodium fluosilicate, $Na_2 SiF_2$.

malleable iron. (wrought iron). Practically pure iron, through which are scattered particles of slag or oxide.

malleable nickel. Nickel commercially refined, and treated with a deoxidizing agent such as magnesium, and cast into ingots. It is suitable for hot or cold working.

Mallebrein. Aluminum chlorate; used as an antiseptic and astringent.

Mallet alloy. A brass containing 74.6% zinc and 25.4% copper.

Mallet bark. The bark of *Eucalyptus occidentalis*. It contains from 35-52% tannin.

Malloydium. An alloy of 61% copper, 23% nickel, 14% zinc, and 1% iron. It is stated to be acid-resisting.

Malogen CYP 200. CAS 58-20-8; Testosterone cypionate; androgen. [O'Neal, Jones & Feldman Pharmaceuticals]*

Malogen LA 200. CAS 315-37-7; Testosterone enanthate; androgen. [O'Neal, Jones & Feldman Pharmaceuticals]*

malonic acid. (dicarboxylic acid C_3; methanedicarbonic acid) $CH_2(COOH)_2$; CAS 141-82-2; EINECS 205-503-0; Intermediate for barbiturates and pharmaceuticals. [Aldrich; R.W. Greeff; Lonza; Penta Mfg.]

malonic methyl ester nitrile. See methyl cyanoacetate.

Malonoben. 2-(3,5 di-tert-butyl-4-hydroxybenzylidene) malononitrile; used as a pesticide.

Maloprim. A proprietary preparation of dapsone and pyrimethamine; used in the prophylaxis of malaria. [The Wellcome Foundation Ltd]

Maloran. Substituted urea herbicide. [Ciba plc] *

Malotte's alloy. An alloy of 46% bismuth, 20% lead, 34% tin. Melting-point is 203 F.

Malros. Saponified rosin size. [Tenneco Malros Ltd] *

maltase. See glucase.

maltha. A variety of mineral tallow or wax found in Finland. It is also the name applied to certain types of soft bitumen.

malthactite. A clay of the fuller's earth type.

malthite. A name for viscous bitumens.

Maltisorb. Polyhydric alcohol; a sugar substitute in powder or liquid form for foods (confectionery) and pharmaceuticals. [Roquette (UK) Ltd]

maltobiose. See maltose.

maltose. (malt sugar; maltobiose) Malt sugar, an isomer of cellobiose; $C_{12}H_{22}O_{11} \cdot H_2O$; CAS 69-79-4; nutrient, sweetener, culture media, stabilizer for polysulfides, brewing. [Am. Biorganics; Avebe BV; Penta Mfg.; Pfanstiehl Labs]

Maltrin® M040. CAS 9050-36-6; Maltodextrin; nonsweet, nutritive polymer useful for wet binding and anticaking; adds solution visc., good mouthfeel; for pharmaceuticals, foods; DE 4-7. [Grain Processing]

Maltrin® M200. CAS 68131-37-3; Corn syrup solids; dried glucose syrup with very good coating and binding properties; directly compressible; for pharmaceutical use; DE 20-23. [Grain Processing]

Maltrin® M510. CAS 9050-36-6; Agglomerated maltodextrin; flowable form; exhibits excellent dispersibility and dissolution; directly compressible binder and diluent; for pharmaceutical use; free-flowing gran.; DE 9-12. [Grain Processing]

Maltrin® QD M600. CAS 68131-37-3; Agglomerated corn syrup solids; agglomerated form of Maltrin M200; directly compresible binder and good carrier; for pharmaceutical use; free-flowing gran.; DE 20-23. [Grain Processing]

malt sugar. See maltose.

maltyl. Dry malt extract containing about 90% soluble carbohydrates.

maltzyme. Diastase, an enzyme.

maluminum. An alloy of 87% aluminum, 6.4% copper, 4.8% zinc, 1.4% iron, and traces of manganese, silicon, and lead.

Mammol. CAS 1304-85-4; Bismuth subnitrate; pharmaceutic necessity. [Abbott Laboratories] *

manaca. (Camganiba; geraticaca; vegetable mercury) The dried root of *Brunfelsia hopeana*, of Brazil. An extract is used as a diuretic and diaphoretic.

Manal. CAS 638-38-0; Manganese acetate. [Mechema Chemicals Ltd] *

Manalox. Aluminum organic compounds; rheology modifiers, ink industry, water repellent for masonry and damp proofing and water repellent for timber. [Rhone-Poulenc UK]

Manalox AG. A proprietary trade name for glycalox. §

Manalox AS. A proprietary trade name for sucralox. §

Mancarb Plus. A liquid formulation containing 80 g carbendazim, 150 g chlorothalonil and 200 g maneb per liter as a suspension concentrate; eradicant fungicide for use on cereals. [Ashlade Formulations Ltd]

Manchem. Metal carboxylates; drier for paint or printing ink, additives for fuel oil or diesel oil. [Rhone-Poulenc UK]

Mancobride Mancanese. Cobalt bromide solution. [Mechema Chemicals Ltd] *

Mancopper. It is an ethylene bisdithiocarbamate-mixed metal complex containing about 13.7% manganese and about 4% copper; used as a pesticide.

Mandelamine. CAS 587-23-5; Methenamine mandelate; antibacterial. [Parke-Davis]

mandelic acid. (phenylglycolic acid; a-phenylhydroxyacetic acid; benzoglycolic acid; amygdalic acid) $C_6H_5CH_2O-COOH$; CAS 90-64-2; organic synthesis, medicine (urinary antiseptic). [W.R. Grace Ltd; R.W. Greeff]

Mandelin's reagent. An alkaloidal reagent, consisting of 1 gram of ammonium vanadate dissolved in 200 cc of concentrated sulfuric acid.

Mandol. CAS 42540-40-9; Cefamandole nafate; antibacterial. [Eli Lilly & Co]

Mandops Barleyquat B. CAS 999-81-5; Soluble concentrate containing 620 g/l chlormequat; used to increase barley yields. [Mandops (UK) Ltd]

Mandops Bettaquat B. CAS 999-81-5; Soluble concentrate containing 620 g/l chlormequat; used to increase oats yields. [Mandops (UK) Ltd]

Mandops Halloween, Hele Stone. Mixture of chlormequat and di-1-p-menthene. plant growth regulator. [Mandops (UK) Ltd]

Mandops Narsty. CAS 7784-25-0; Aluminum ammonium sulfate; bird and animal repellent. [Mandops (UK) Ltd]

Mandops Podquaternary Chlormequat and di-1-p-menthene. plant growth regulator. [Mandops (UK) Ltd]

Mandops Spring Poquaternary CAS 999-81-5; Soluble concentrate containing 590 g/l chlormequat; used to increase yields of oilseed rape, pea and bean crops. [Mandops (UK) Ltd]

Mandurin. A proprietary preparation of hexamine mandelate; a urinary antiseptic. [Harker Stagg] ‡

Manex. Suspension concentrate containing maneb and zinc; a protectant fungicide against potato blight, mildew and rust. [Chiltern Farm Chemicals Ltd; Griffin; Quadrangle Agrochemicals; L W Vass (Agricultural) Ltd;]

Manfloc. CAS 1302-42-7; Sodium aluminate composition; retention aid for paper water treatment. [Manchem Ltd]*

Mangabeira rubber. A rubber from the small tree, *Hancornia speciosa* cultivated in Paraguay and Venezuela.

Mangal. Chemical complex of manganese and aluminum; drier for cobalt-free paint or printing ink. [Manchem Ltd]*

mangaloy. An alloy of nickel containing iron and manganese.

manganaluminum bronze. An alloy containing from 9-10% manganese, 85.5-86% copper, and 4 1/2-5% aluminum.

Manganar. Manganese arsenate.

manganated linseed oil. Linseed oil which has been boiled with manganese dioxide to increase its drying properties.

manganese. Mn; CAS 7439-96-5; EINECS 231-105-1; Ferroalloys, i.e., steel manufacture; improves corrosion resistance and hardness in nonferrous alloys; purifying and scavenging agent in metal production; manufacture of aluminum. [Atomergic Chemetals; Cerac; Chemetals; Kerr-McGee]

manganese acetate. (acetic acid manganese (2+) salt) $Mn(C_2H_3O_2)_2 \cdot 4H_2O$; CAS 638-38-0; Textile dyeing, oxidation catalyst, paint and varnish drier, fertilizer, food packaging, feed additive. [Atomergic Chemetals; Hoechst Celanese; Nihon Kagaku Sangyo; Spectrum Chem. Mfg.; Verdugt BV]

manganese-aluminum brass. An alloy of 56% copper, 40% zinc, 3% manganese, and 1% aluminum.

manganese binoxide. *See* manganese dioxide.

manganese black. *See* manganese dioxide.

manganese boron. An alloy of manganese and boron; used for making other alloys.

manganese brass. Variable alloys, usually containing 51-69% copper, 1-4% manganese, 0-3% iron, 29-40% zinc, 0-2% tin, 0-2% nickel, and sometimes aluminum.

manganese bronze. An alloy made by adding ferro-manganese or manganese to bronze. It usually contains from 82-83.5% copper, 8% tin, 5% zinc, 3% lead, and 0.5-2% manganese; an alloy containing 59% copper, 39% zinc, 1.5% manganese, and 0.5% iron.

manganese cupro nickel. This is usually an alloy containing from 65-83% copper, 15-30% manganese, and 2-8% nickel.

manganese dioxide. (manganese oxide; manganese binoxide; manganese black; manganese peroxide) MnO_2; CAS 1313-13-9; EINECS 215-202-6; oxidizing agent, depolarizer in dry cell batteries, pyrotechnics, matches, catalyst, laboratory reagent, scavenger and decolorizer, textile dyeing, source of metallic manganese (as pyrolusite). [Aldrich; Atomergic Chemetals; Eagle-Picher; Hoechst Celanese; Kerr-McGee; Nichia Kagaku Kogyo]

manganese German silver. An alloy of 80% copper, 15% manganese, and 5% zinc.

manganese gluconate. $Mn(C_6H_{11}O_7)_2 \cdot 2H_2O$; Mineral source for pharmaceutical and food prods, vitamin tablets, feed additive. [Akzo; Spectrum Chem. Mfg.]

manganese nickel. An alloy of from 51-82% copper, 14-31% manganese, and 3-16% nickel. An alloy containing 95% nickel and 5% manganese, and melting at 1420 C. is also known as manganese nickel.

manganese nickel brass. Alloys containing 51-65% copper, 5-40% zinc, 0-2.78% iron, 1.5-3.24% manganese, and 2-18% nickel.

manganese nickel silver. Alloys containing 50-70% copper, 9-40% zinc, 1-20% manganese, and 2-20% nickel.

manganese ore A. A standard manganese ore. It contains 51.3% manganese, 14.3% available oxygen, 6.5% silica, 1.3% iron, and 0.22% phosphorus.

manganese oxide. *See* manganese dioxide.

manganese peroxide. *See* manganese dioxide.

manganese silver. *See* Manganese German Silver.

manganese steel. An alloy of manganese and steel containing up to 20% manganese. Commercial manganese steel contains 11-14% manganese, 1-1.3% carbon,

---

‡ Trade name and manufacturer not verified   § Trade name without identified manufacturer

0.3% silicon. 0.05-0.08% sulfur, and 0.05-0.08% phosphorus.

manganese sulfate (ous). (manganous sulfate) $MnSO_4 \cdot 4H_2O$; CAS 7785-87-7; fertilizers, feed additive, paints and varnishes, ceramics, textile dyes, medicine, fungicide, ore flotation, catalyst in viscose process, synthetic manganese dioxide. [Aldrich; Chemetals; Mallinckrodt; Nihon Kagaku Sangyo]

manganese velvet brown. *See* umber.

Manganese Violet. (Nuremberg violet, mineral violet, permanent violet). Manganic phosphate; pigment used as a mineral color. [Reckitts Colours Ltd]

manganese white. A pigment. It is manganous carbonate, $MnCO_3$.

manganic. Nickel containing a small percentage of manganese.

Manganin. Alloys usually containing 70-86% copper, 4-25% manganese, and 2-12% nickel. One of the best varieties contains 83.6% copper, 13.6% manganese, 2.5% nickel, and 0.3% iron; used in electrical instruments where constant resistivity is important.

manganite. (Brown manganese ore). A mineral. It is a hydrated oxide of manganese, $Mn_2O_3 \cdot H_2O$. It is also a name for a war gas which consisted of a mixture of hydrocyanic acid and arsenic trichloride.

Mangan-Neusilber. A nickel silver containing manganese. It contains from 59-72% copper, 5-20% zinc, 10-18% nickel, and 2-20% manganese.

Mangano Steel. A proprietary trade name for a non-shrinking steel containing 1.6% manganese, 0.2% chromium and 0.95% carbon. §

mangano-titanium. This is usually an alloy of manganese with 30% titanium, and is used as a deoxidizer in making bronze and brass castings.

manganous sulfate. *See* manganese sulfate (ous).

Mangatrace. CAS 7773-01-5; Manganese chloride tetrahydrate; supplement. [Armour Pharmaceutical Co]*

Mangnamite. Graphite fiber; for structural strength applications. [Hercules] *

Mangol. A powder consisting of basic magnesium hypochlorite; used for testing alkaloids.

Mangonic. A manganese-nickel alloy containing about 3% manganese.

Mangoxe. CAS 1313-13-9; Manganese dioxide; polysulfide sealant. [Mechema Chemicals Ltd] *

Manguard®. CAS 12427-38-2; Water-dispersible granule containing 75% w/w maneb; for use as an agricultural fungicide. [Universal Crop Protection Ltd]

Manhardt's Aluminum Bronze. An alloy of 83.3% aluminum, 6.25% copper, 10.13% tin, 0.16% antimony, 0.05% magnesium, and 0.08% phosphorus. §

Manjak. (glance pitch). A bitumen found in Mexico, South America, and West Indies; used as a paint, as a roofing material, and in connection with drilling for oil.

manna. CAS 69-65-8; A sugary exudation occurring in the rising sap of *Fraxinus ornus*, and *F. rotundifolia*; the crude material contains from 12-13% water, 10-15% sugar, 32-42% mannitol, 40-41% mucilaginous substances, organic acids, and nitrogenous matter; used in medicine as a laxative.

Mannheim gold. A brass containing 80% copper and 20% zinc. A jeweler's alloy. The term is also applied to a bronze consisting of 83.7% copper, 9.3% zinc, 7% tin, with a little phosphorus.

Mannol. Ethylacetanilide; A febrifuge. It is also used as a softening agent for cellulose esters.

Mannolit. *See* Chloramine T. §

Manoblend. Mixtures of rubber chemicals and fillers [Manchem Ltd] *

Manobond. Cobalt boroacylate; rubber-to-metal bonding agents. [Rhone-Poulenc UK]

Manocat. Metal carboxylates; catalysts for unsaturated polyesters and polyurethane foams. [Rhone-Poulenc UK]

Manofast. Thiourea dioxide; reducing agent for dyes. [Rhone-Poulenc UK]

Manofil. Chemically treated fillers. for rubber and plastic manufacture. [Manchem Ltd] *

man oil. *See* bone oil.

Manomet. Metal carboxylates; stabilizers for PVC, pigment dispersing aids and kicker for PVC foams. [Rhone-Poulenc UK]

Manosec. Range of metal carboxylates based on synthetic organic acids; paint driers. [Rhone-Poulenc UK]

Manosil. Combinations of silica or silicates with speciality chemicals; for rubber, plastics. [Manchem Ltd] *

Manosperse. Dispersed rubber chemicals (various); used in rubber industry. [Rhone-Poulenc UK]

Manox. CAS 7704-34-9; Insoluble sulfur; used in rubber vulcanizing. [Rhone-Poulenc UK]

Manox. Iron blue pigments; for printing inks, paints and fungicides. [Manox Ltd] *

Manoxol. Sulfosuccinate surface active agents; wetting agent, dispersing agents. [Manchem Ltd] *

Manoxol MA. Sodium di(methyl amyl) sulfosuccinate in 60% water/alcohol solution; powerful wetting agents which can act as aids to dispersion, detergency, and emulsification; used for adhesives, asbestos, agriculture and horticulture, bleaching, clays, determination of anionic detergents, dry cleaning and laundry; dust control, electroplating, emulsion polymerization, etching, germicides, leather, metal technology, oil additives, paint/printing ink, papermaking, pharmaceutical, photography, resins, rubber, textiles, wax polishes, wallpaper removal. [Manchem Ltd] *

Manoxol OT, OT/P and OT/B. CAS 577-11-7; EINECS 209-406-4; Sodium dioctyl sulfosuccinate, available as solid, water/alcohol solution, powder (85% with 15% sodium benzoate), or specially pure grade for pharmaceutical use; powerful wetting agents which can act as aids to dispersion, detergency, and emulsification; used for adhesives, asbestos, agriculture and horticulture, bleaching, clays, determination of anionic detergents, dry cleaning and laundry; dust control, electroplating, emulsion polymerization, etching, germicides, leather, metal technology, oil additives, paint/printing ink, papermaking, pharmaceutical, photography, resins, rubber, textiles, wax polishes, wallpaper removal. [Manchem Ltd] *

Manoxol OT60. Sodium diisooctyl sulfosuccinate; used as a wetting and dispersing agent. [Harcros Australia] *

Manoxolot. CAS 577-11-7; EINECS 209-406-4; A proprietary trade name for dioctyl sodium sulfosuccinate. §

Manplex. Manganese zinc ethylenebisdithiocarbamate complex; for control of potato blight and rust, blight and mildew in winter wheat. [Kommer-Brookwick Ltd]

Manqueta. (Manquta). African names for a fossil gum resin, resembling copal.

Manro. Foam boosters and stabilizer; powder and liquid

---

* Trade name not verified as available　　　　　　† Trade name verified as obsolete

detergents; hair shampoos; hand cleaning jellies and cleaners. [Manro Products Ltd] *

Manro ALS. Ammonium primary alcohol sulfate as a clear, pale yellow liquid; foaming and wetting agent for hair, carpet and upholstery shampoos. [Manro Products Ltd]*

Manro BA and NA. Dodecylbenzene sulfonic acid; anionic surfactant supplied as a brown viscous liquid, low in free oil and inorganic content; raw material for the production of detergents and emulsifiers such as powder and liquid detergents, hand cleaning gels, machine degreasers and tank cleaners. [Manro Products Ltd] *

Manro BES. Range of anionic surfactants in liquid form; anion: sodium; cation: synthetic alcohol ethoxy sulfate; high foaming agents used in liquid detergent formulations, bubble baths and shampoos; industrial applications include use as a drilling aid. [Manro Products Ltd]*

Manro D Paste. Sodium cetyl/oleyl sulfate as a white/pale yellow stiff paste; wetting, dispersing and emulsifying agent used in textile scouring and kier boiling. [Manro Products Ltd] *

Manro DL28. Sodium primary alcohol sulfate as a water white liquid; foaming and wetting agent with good handling characteristics used in latex processing and emulsion polymerization. [Manro Products Ltd] *

Manro DS 35. Sodium primary alcohol sulfate as a pale yellow liquid; wetting and emulsification agent used in metal cleaner formulations etc. [Manro Products Ltd] *

Manro HA. Dodecyl benzene sulfonic acid, derived from propylene tetramer; anionic surfactant supplied as a brown viscous liquid, low in free oil and inorganic content; raw material for production of detergents and emulsifiers such as powder and and liquid detergents, hand cleaning gels, machine degreasers and tank cleaners. [Manro Products Ltd] *

Manro HCS. Anionic surfactant; supplied as a pale amber sparkling viscous liquid; used for emulsification of a wide range of chemicals, especially in emulsifiable solvent degreasing, e.g., for hand cleaning jellies or engine cleaners. [Manro Products Ltd] *

Manro KXS. Potassium xylene sulfonate; anionic surfactant in liquid form; cloud point depressant and solubilization agent, especially for heavy duty detergent formulations. [Manro Products Ltd] *

Manro MA 35. Disodium octadecyl sulfosuccinamate as a pale cream liquid; foaming agent in the manufacture of latex foams. [Manro Products Ltd] *

Manro ML33. Monoethanolamine primary alcohol sulfate in pale yellow liquid form; foaming and wetting agent for high quality hair shampoos and other toiletry products. [Manro Products Ltd] *

Manro NEC. Anionic surfactant as a colorless liquid; cation: sodium; anion: natural or Ziegler alcohol ethoxy sulfate; foaming and cleaning agent for liquid and lotion hair shampoos. [Manro Products Ltd] *

Manro NP. Range of nonionic surfactants of the nonylphenol ethoxylate type in liquid, paste or wax form; general purpose nonionic detergent bases; emulsifiers for agrochemicals and pesticides; emulsion polymerization. [Manro Products Ltd] *

Manro PTSA. Toluene sulfonic acid; anionic surfactant in liquid form; intermediate in detergent manufacture, also has descaling and catalyst properties; used in metal industries for resin bound sand castings; manufacture of esters, acetylation and resin production; hydrotrope pro-

duction for detergents. [Manro Products Ltd] *

Manro SBS. Anionic surfactant; supplied as a clear pale yellow liquid with low inorganic content and controlled levels of minor constituents; detergent and emulsifier, e.g., light duty liquid detergents, hard surface cleaners; scouring and wetting in textile industries; emulsifiable insecticides; plastics; rubber. [Manro Products Ltd] *

Manro SDBS. Anionic surfactant; supplied as a white/pale yellow paste with low inorganic content and controlled levels of minor constituents; used as a detergent and emulsifier eg light duty detergents, hard surface cleaners; scouring and wetting in textile industries; rubber; plastics; emulsifiable insecticides. [Manro Products Ltd]*

Manro SIOS. Sodium isooctyl sulfate as a pale yellow, clear mobile liquid; wetting agent with good alkali stability used in metal cleaner formulations and similar products. [Manro Products Ltd] *

Manro SLS28. Sodium primary alcohol sulfate as a very pale yellow liquid; foaming and wetting agent for carpet and upholstery shampoos ; foam rubber; emulsion polymerization in plastics and rubber industries. [Manro Products Ltd] *

Manro SLS45. Sodium primary alcohol sulfate as a white paste; foaming and wetting agent for hair shampoos. [Manro Products Ltd] *

Manro STS. Sodium toluene sulfonate; anionic surfactant in liquid form; used for reduction of slurry viscosity before spray drying in heavy duty detergent powder manufacture. [Manro Products Ltd] *

Manro SXS. CAS 1300-72-7; EINECS 215-090-9; Sodium xylene sulfonate; anionic surfactant in liquid form; cloud point depressant, coupling and stabilization agent for light and heavy duty detergents. [Manro Products Ltd] *

Manro TDBS. Anionic surfactant, pale amber viscous liquid; a detergent, mild to the skin; used in bubble baths, car shampoos and other detergents. Also used as an emulsifier in emulsification polymerization. [Manro Products Ltd.] *

Manro TL40. Triethanolamine primary alcohol sulfate in the form of a clear, pale amber liquid; foaming and wetting agent for high quality hair shampoos and other toiletry products. [Manro Products Ltd] *

Manro XSA. CAS 25321-41-9; EINECS 246-839-8; Xylene sulfonic acid; anionic surfactant in liquid form; intermediate in detergent manufacture, also has descaling and catalyst properties; used in metal industries for resin bound sand castings; manufacture of esters, acetylation and resin production; hydrotrope production for detergents. [Manro Products Ltd] *

Mansil. CAS 21738-42-1; A proprietary preparation of oxamniquine; an antihelmintic. [Pfizer International]

Mansonil®. Veterinary preparation; used against tapeworm infestation in ruminants, dogs, and cats. [Bayer AG]

Mansu. CAS 7785-87-7; Manganese sulfate. [Mechema Chemicals Ltd] *

Mantadil Cream. A proprietary formulation of chlorcyclizine hydrochloride and hydrocortisone acetate; for treatment of pruritic skin eruptions and other dermatoses. [The Wellcome Foundation Ltd]

Mantadine®. For the medical industry. [DuPont UK]

Mantin. Organotin/dithiocarbamate. [Ciba plc] *

Mantrilon®. Liq. manganese fertilizer; foliar fertilizer to prevent and cure deficiency of manganese in all agric.

crops, vines and fruit. [BASF AG]

Mantrilon® FL. Foliar feed. [BASF plc]

Manucol. Sodium alginate, food grade. [Kelco Int'l. Ltd]

Manucol DH. Sodium alginate; used for bakery glazes, filling creams, cheesecake toppings, cheese spreads, processed cheeses, instant puddings, and aspics. [Kelco Int'l. Ltd]

Manucol DM. Sodium alginate; used for ice creams, chilled desserts. [Kelco Int'l. Ltd]

Manucol Ester E/RK. CAS 9005-37-2; Propylene glycol alginate; used for fermented milks, fruit drinks, citrus concentrates. [Kelco Int'l. Ltd]

Manucol Ester EX/LL. CAS 9005-37-2; A proprietary trade name for propylene glycol alginate; a food grade emulsifying agent. [Alginate Industries Ltd] *

Manucreme. Hand cream, to soothe dry and rough hands. [Richardson-Vicks Inc] *

Manugel PTJ. Specialty algin blend; gelling powder for structured glace fruit, high solid fillings, bakery fillings, bakery jellies. [Kelco Int'l. Ltd]

Manutex. Sodium alginate, technical grade. [Kelco Int'l. Ltd]

Manzanate. Ethyl-2-methyl pentanoate. [Quest Int'l. UK Ltd]

Manzate®. CAS 12427-38-2; Suspension concentrate or wettable powder containing maneb; a dithiocarbamate fungicide to control blight, rusts, and mildew. [DuPont UK]

Manzate® 200 DF. CAS 8018-01-7; Wettable powder containing mancozeb; protectant fungicide for fruit, field crops and roses. [DuPont UK]

Maolate. CAS 886-74-8; Chlorphenesin carbamate; relaxant. [Upjohn]

Map₄. A coarse starch obtained from the fruit of *Inocarous edulis*.

Mapeg® 200 DL. CAS 9005-02-1; PEG-4 dilaurate; emulsifier, dispersant used in cosmetics, pharmaceuticals, metalworking and fiber lubricants, etc. [PPG/Specialty Chem.]

Mapeg® 200 DO. CAS 9005-07-6; PEG-4 dioleate; emulsifier, dispersant used in cosmetics, pharmaceuticals, metalworking and fiber lubricants, etc. [PPG/Specialty Chem.]

Mapeg® 200 DOT. CAS 61791-01-3; PEG-4 ditallate; surfactant, emulsifier for metalworking lubricants; emollient for hair preps., creams and lotions; solubilizer for bath oils and fragrances. [PPG/Specialty Chem.]

Mapeg® 200 DS. CAS 9005-08-7; PEG-4 distearate; emulsifier, dispersant used in cosmetics, pharmaceuticals, metalworking and fiber lubricants, etc. [PPG/Specialty Chem.]

Mapeg® 200 ML. CAS 9004-81-3; PEG-4 laurate; emulsifier, dispersant used in cosmetics, pharmaceuticals, metalworking and fiber lubricants, etc. [PPG/Specialty Chem.]

Mapeg® 200 MO. CAS 9004-96-0; PEG-4 oleate; emulsifier, dispersant used in cosmetics, pharmaceuticals, metalworking and fiber lubricants, etc. [PPG/Specialty Chem.]

Mapeg® 200 MOT. CAS 61791-00-2; PEG-4 tallate; emulsifier, dispersant used in cosmetics, pharmaceuticals, metalworking and fiber lubricants, etc. [PPG/Specialty Chem.]

Mapeg® 200 MS. CAS 9004-99-3; PEG-4 stearate; emulsifier, dispersant used in cosmetics, pharmaceuticals, metalworking and fiber lubricants, etc. [PPG/Specialty Chem.]

Mapeg® 1500 MS. CAS 9004-99-3; PEG-6-32 stearate;

emulsifier, dispersant used in cosmetics, pharmaceuticals, metalworking and fiber lubricants, etc. [PPG/Specialty Chem.]

Mapeg® 1540 DS. CAS 9005-08-7; PEG-32 distearate; emulsifier, dispersant used in cosmetics, pharmaceuticals, metalworking and fiber lubricants, etc. [PPG/Specialty Chem.]

Mapeg® 6000 DS. CAS 9005-08-7; PEG-150 distearate; emulsifier, dispersant used in cosmetics, pharmaceuticals, metalworking and fiber lubricants, etc. [PPG/Specialty Chem.]

Mapeg® CO-5. CAS 61791-12-6; PEG-5 castor oil. [PPG/Specialty Chem.]

Mapeg® CO-16H. CAS 61788-85-0; PEG-16 hydrogenated castor oil; surfactant, emulsifier, dispersant, wetting agent, emollient for hair preps., creams and lotions; solubilizer for bath oils and fragrances. [PPG/Specialty Chem.]

Mapeg® DGLD. CAS 9004-81-3; Diethylene glycol laurate; surfactant for formation of gels. [PPG/Specialty Chem.]

Mapeg® EGDS. CAS 627-83-8; EINECS 211-014-3; Glycol distearate; emulsifier, dispersant used in cosmetics, pharmaceuticals, metalworking and fiber lubricants, etc.; thickener, opacifier, pearling additive. [PPG/Specialty Chem.]

Mapeg® EGMS. Glycol stearate; emulsifier, dispersant used in cosmetics, pharmaceuticals, metalworking and fiber lubricants, etc.; thickener, opacifier, pearling additive. [PPG/Specialty Chem.]

Mapeg® S-40. CAS 9004-99-3; PEG-40 stearate; emulsifier, dispersant used in cosmetics, pharmaceuticals, metalworking and fiber lubricants, etc. [PPG/Specialty Chem.]

Mapeg® TAO-15. CAS 61791-00-2; PEG-660 tallate; emulsifier, dispersant used in cosmetics, pharmaceuticals, metalworking and fiber lubricants, etc. [PPG/Specialty Chem.]

Maphos® 17. Aromatic phosphate ester; emulsifier for emulsion polymerization; solubilizer. [PPG/Specialty Chem.]

Maphos® 33. Aliphatic phosphate ester; emulsifier, lubricant with anticorrosive/antifrictional props. for oil and water-sol. lubricant systems, e.g., greases, synthetic cutting oils, drawing compounds, chain-belt lubricants, gear oils, and rust preventatives. [PPG/Specialty Chem.]

Maphos® 60A. Complex org. phosphate acid ester; textile wetting, hard surface detergent; lubricant, anticorrosive, dispersant, hydrotrope, solubilizer, emulsifier; metalworking. [PPG/Specialty Chem.]

Maphos® 66H. Phosphate acid ester, neutralized; hard surface cleaning hydrotrope; antistat for solubilization of low foam and conventional surfactants in alkaline liqs.; metalworking. [PPG/Specialty Chem.]

Maphos® 78. Complex org. phosphate acid ester; hydrotrope, detergent. [PPG/Specialty Chem.]

Maphos® 8135. Aromatic phosphate ester; dispersant, hydrotrope, emulsifier, EP lubricant additive for greases, synthetic cutting oils, drawing compounds, and hard surface cleaners. [PPG/Specialty Chem.]

Maphos® FDEO. Phosphate ester; surfactant, hydrotrope, detergency aid, antistat for drycleaning and lubricant systems, emulsion polymerization. [PPG/Specialty Chem.]

Maphos® JA 60. Aliphatic phosphate ester; detergent, coupling agent; compat. with builders; for hard surface

---

cleaners, built detergents, metalworking fluids. [PPG/ Specialty Chem.]

Maphos® L 13. Aliphatic phosphate ester; surfactant, lubricant, anticorrosive, coupling agent for metalworking, dry cleaning, hard surface cleaning, dedusting, lubrication, emulsion polymerization. [PPG/Specialty Chem.]

Mapico. CAS 1309-37-1; Synthetic iron oxide. [Sevalco Ltd]

maple sugar sand. A by-product in the manufacture of maple sugar. The sap from the maple is evaporated in pans, and a precipitate forms when the water content is about 35%. This precipitate is maple sugar sand. The chief constituent is calcium malate (60-80%) and malic acid is easily prepared from it.

Mapo®. Tris[1-(2-methyl-aziridinyl) phosphine oxide; textile chemical; resin raw material, crosslinker, adhesion promoter. [Aceto]

Maprenal. Etherified melamine/formaldehyde resins; used for stoving finishes and acid curing lacquers. [Resinous Chemicals Ltd]

Maprenal®. Melamine and benzoguanamine resin products; for production of stoving (baking) lacquers. [Cassella AG]

Maprofix 563 and LK.USP. Sodium lauryl sulfate in powder or granular form; detergency, foaming and emulsification agent, food and dentifrice grade; also used in emulsion polymerization. [Millmaster-Onyx UK] *

Maprofix 60S and 60N. Sodium or ammonium lauryl 3EO sulfate in liquid form; foaming surfactant for light duty detergents. [Millmaster-Onyx UK] *

Maprofix ES-2. Sodium lauryl 2EO sulfate in liquid form; foaming and dispersing agent for shampoos, bubble baths and general cosmetics. [Millmaster-Onyx UK] *

Maprofix ESY. Sodium lauryl 1EO sulfate in liquid form; used for emulsification, foaming and wetting agent for emulsion polymerization, shampoos and gels. [Millmaster-Onyx UK] *

Maprofix MG. Magnesium lauryl sulfate in liquid form; wetting, emulsifying, and dispersing agent with low cloud point; used in nonalkaline shampoos and rug shampoos. [Millmaster-Onyx UK] *

Maprofix NH and NHL. CAS 2235-54-3; EINECS 218-793-9; Ammonium lauryl sulfate in liquid form; detergency and foaming agent with high buffering capacity pH 6-7; used in nonalkaline shampoos, for general cosmetic use and in industrial foams. [Millmaster-Onyx UK] *

Maprofix TAS. Sodium tallow alcohol sulfate in paste form; detergent and dispersing agent used in high temperature detergents and ore flotation. [Millmaster-Onyx UK] *

Maprofix TLS. CAS 139-96-8; EINECS 205-388-7; Triethanolamine lauryl sulfate in liquid form; detergent with good foam stability and low cloud point; used in clear shampoos, bubble baths, fine fabric detergent, industrial and household cleaners and rug and upholstery shampoos. [Millmaster-Onyx UK] *

Maprofix WA, WAC and WAQ. Sodium lauryl sulfate in liquid or paste form; detergency, foaming and dispersing agent for liquid and cream shampoos; general cosmetic uses; rug and upholstery shampoos; latices and paint pigments. [Millmaster-Onyx UK] *

Maprofix WAC-LA and LCP. Sodium lauryl sulfate in liquid form; high purity, low salt content and viscosity, light in color. wetting, emulsification and dispersing agent, polymerization grade. [Millmaster-Onyx UK] *

Mapromin. A proprietary trade name for a sulfated fatty alcohol used as a wetting agent. §

Mapron®. Soybean milk; food additive. [Mitsubishi Kasei]*

Maprosyl® 30. CAS 137-16-6; EINECS 205-281-5; Sodium n-lauroyl sarcosinate; detergent, wetting and foaming agent used in personal care and household detergent products; anticorrosive props. [Stepan; Stepan Canada]

Marabond 21. Lignosulfonate; low temp., low cost oil well cement retarder. [Borregaard LignoTech]

Marabout silk. A white silk which still contains its gum. It is dyed and used for the manufacture of imitation feathers.

Maracarb. Modified lignosulfonate; dispersant, humectant, chelating agent for mfg. of alkaline metal cleaners. [Borregaard LignoTech]

Maracarb N-1. CAS 8061-51-6; Sodium lignosulfonate; modifier, dispersant, and humectant in dyestuff pastes; industrial cleaners; plant foliar spray; chelates metal ions; slime control agent for paper mill systems. [Borregaard LignoTech]

Maracell XE. CAS 8061-51-6; Sodium lignosulfonate, partially desulfonated; sludge conditioner, chelating agent for prevention of scale formation in treatment of boiler water, industrial cleaners. [Borregaard LignoTech]

Maracon. Calcium and sodium lignosulfonates; concrete admixtures. [Borregaard LignoTech]

Maramul SS. Lignosulfonate; emulsifier for asphalt emulsions. [Borregaard LignoTech]

Maranil DBS. Dodecylbenzene sulfonic acid; intermediate for mfg. of liq., powd., or paste detergents, emulsifiers, textile auxiliaries; acid catalyst. [Henkel/Functional Prods.; Pulcra SA]

Maranil Powd. A. Sodium dodecylbenzene sulfonate; base for mfg. detergents, dishwashes, cleaning agents; wetting agent; emulsifier for PVC copolymers, carboxylated S/B latexes. [Henkel/Functional Prods.]

maranta. Arrowroot starch.

Maranyl®. Nylon molding and extrusion compounds. [ICI Chem. & Polymers Ltd]

Maranyl® A125. CAS 32131-17-2; Nylon 6/6, unfilled, lubricated; engineering material with high mech. props., abrasion, thermal, chem. and creep resist.; for injection molding or extrusion applications. [ICI Advanced Materials]

Maranyl® A175S. CAS 32131-17-2; Nylon 6/6, 30% glass fiber-reinforced, heat stabilized, lubricated; engineering material with high mech. props., abrasion, thermal, chem. and creep resist.; for injection molding or extrusion applications. [ICI Advanced Materials]

Maranyl® A360. CAS 32131-17-2; Toughened nylon 6/6, 24% glass fiber-reinforced, flame retardant; engineering material with high mech. props., abrasion, thermal, chem. and creep resist.; for injection molding or extrusion applications. [ICI Advanced Materials]

Maranyl® TA505HS. CAS 32131-17-2; Toughened nylon 6/6, heat stabilized; engineering material with high mech. props., abrasion, thermal, chem. and creep resist.; for injection molding or extrusion applications. [ICI Advanced Materials]

Marasperse 52 CP. CAS 8061-51-6; Polymerized sodium lignosulfonate; dispersant for disperse dyestuffs; low stain, high heat stability; pitch dispersant in paper mills. [Borregaard LignoTech]

Marasperse GFC. CAS 8061-52-7; Calcium lignosulfonate, sugar-free; additive to high strength concrete; esp. effective with silica fume concretes. [Borregaard LignoTech]

Marasperse N-22. CAS 8061-51-6; Sodium lignosulfonate, purified, oxidized; dispersant, o/w emulsion stabilizer, emulsifier; mfg. of disperse dyes for dyeing acetate and polyesters; dispersant and sequestering agent in cooling water treatments; agric. chemical formulations; gypsum board additive; industrial cleaners. [Borregaard LignoTech]

Marax. A proprietary preparation of hydroxyzine hydrochloride, ephedrine sulfate and theophylline; an anti-asthmatic. [Pfizer International]

Marbalettes®. Socket pellets for extraction wounds in dental industry. [Bayer AG]

Marble Dust. CAS 471-34-1, 1317-65-3; Calcium carbonate; coarse ground filler for putties, glazes and mild abrasive compounds. [ECC]

Marbleloid. Flooring materials. [Weatherguard/Marbleloid Products Inc] *

Marblemite. CAS 1317-65-3; Calcium carbonate; high brightness filler for high loading and min. black specks in cultured marble applications. [ECC]

Marblette. A proprietary trade name for a phenol-formaldehyde cast resin. §

Marbo. A proprietary trade name for a chlorinated rubber. §

Marbolith. See Bakelite. §

Marbon B. A proprietary trade name for a cyclo-rubber. §

Marbon Latex. Marmix reactive SBR and ABS latexes. [GE Plastics ABS Ltd] *

Marbon Resins. ABS polymers; for injection molding, sheet extrusions. [GE Plastics ABS Ltd] *

Marbon White. A proprietary brand of Lithopone (qv). §

Marcain. A proprietary preparation of bupivacaine and adrenaline; long acting local anesthetic. [British Drug Houses] *

Marcaine. CAS 14252-80-3; Bupivacaine hydrochloride; anesthetic. [Sterling Drug Inc] *

marcasite. (white iron pyrites, coxcomb pyrites, radiated pyrites). A mineral. It is disulfide of iron, FeS_2. The term is also occasionally applied to bismuth.

Marcasol. See markasol.

marchies. See margines.

Marchon® DC 1102. Formulated product; shale and cuttings wash cleaner for oilfield industry. [Albright & Wilson UK]

Marcoumar. CAS 435-97-2; A proprietary preparation of phenprocoumon; an anticoagulant. [Roche Products Ltd] *

marcs. The name given to the residue from wine factories, consisting of the stems and skins of grapes; used for making verdigris.

Mareepa. Kernels of the fruits of the cokerite palm of British Guiana.

Maretin. CAS 1491-41-4; Naftalofos; anthelmintic. [Bayer AG]

Marevan. CAS 129-06-6; A proprietary preparation of warfarin sodium; an anticoagulant. [Duncan Flockhard Ltd]

Marexine-CA. FC-126 strain, frozen turkey herpes virus marek's vaccine; for immunization of poultry. [Intervet Inc] †

Marezine Injection. Cyclizine ractate; antinauseant. [Burroughs Wellcome Co] *

Marezzo marble. An artificial marble from oxychloride cement; used for building.

Marfanil-Prontalbin®. Sulfonamide powder; used against wound infections in veterinary medicine. [Bayer AG]

Margalite. A phenol-formaldehyde resin product; used in the manufacture of varnishes and insulators. §

margarodite. A mica having an appearance similar to talc.

margines. (marchies). The residues obtained from the manufacture of olive oil.

margol. A mixture of volatile fatty acids; used as a flavoring material for margarine, to give it the taste of butter.

margosa bark. Indian azadirach.

Margosan-O. CAS 11141-17-6; Active ingredient is azadirachtin, a tetranortriterpinoid; growth regulator on many insects in various life stages, due to hormonal disruption preventing normal metamorphosis; repellancy and/or antifeedency through olfactory or gustatory rejection by various flying and crawling pests; currently restricted to nonfood crop use. [Vikwood Botanicals Inc]*

Margraff alloy. An alloy consisting of 58% copper, 28% tin, and 14% zinc.

Maricol. Magnesium ricinoleate, $(HO.C_{17}H_{32}.CO_2)_2Mg$

Marignac's salt. Potassium-stanno-sulfate.

marine acid. Hydrochloric acid, HCl.

marine fiber. A fiber obtained by dredging the shallow water of a gulf in South Australia. It is a hydrated lignocellulose.

marine oil. A mixture of blown rape oil and a mineral oil; used for marine engines.

Marine Plasma Extract. Hydrolyzed marine protein, brown algae extract, marine plasma fluid, iodine, silicon, iron, sodium, potassium, magnesium, calcium, ascorbic acid, niacinamide; sea plasma extract containing essential elements for development of ocean life. [Brooks Industries]

marine salt. Sodium chloride, NaCl.

marine soap. A soap made from coconut oil, which is soluble in fresh and sea water.

Marinol. Dronabinol; anti-emetic. [Unimed Inc] *

Mark 80. Organic additive system (non coumarin); semibright nickel electroplating (Duplex). [Engelhard Technologies Ltd]

Mark® 1330. A proprietary sulfur-containing organotin stabilized for use in PVC for injection molding and pipe extrusion. [Argus Chemical Corporation]

Mark® 1414. A proprietary organotin mercaptide stabilizer for PVC used in the extrusion of rigid pipe. [Argus Chemical Corporation]

Mark® 2140. CAS 95823-35-1; Pentaerythrityl hexyl-thiopropionate; stabilizer for use in polyolefins and other polymeric systems; synergistic with primary antioxidants; color which reduces or eliminates the need for phosphite; used at 0.1-0.3 phr in PP, 0.05-0.10 phr in HDPE, 0.25-0.5 phr in elastomers, and 0.5-1.0 phr in S.B. latex. [Witco/Argus]

Mark® 4700. Ba-Zn salt; non-Cd heat stabilizer for PVC filled and clear compounds. [Witco/Argus]

Mark® 5089. CAS 96328-09-1; Pentaerythritol alkyl thiodipropionate; antioxidant/stabilizer for plastics. [Witco/Argus]

Mark® 5095. Lauryl/stearyl thiodipropionate; antioxidant for polyolefins and other polymeric systems including synthetic rubber; also for pharmaceuticals, cosmetics, industrial oils, greases, lubricants. [Witco/Argus]

Mark-A-Leak AW. High foaming leak detector for use under a ll weather conditions, including freezing temps. [Climax Fluids Additives]

markasol. (marcasol). Bismuth boro-phenate; used as a substitute for iodoform.

---

* Trade name not verified as available   † Trade name verified as obsolete

Marksman. CAS 330-55-2, 1582-09-8; Mixture of linuron and trifluralin; herbicide for winter cereals. [Farmers Crop Chemicals Ltd]

Markstat® AL-12. Quaternary ammonium chloride deriv. of polyalkoxy tert. amines; antistat additive for PU films and thermoplastics. [Witco/Argus]

Markus alloy. See nickel silvers.

Markwet NR-25. Ethoxylated alcohol blend; textile penetrant and detergent imparting nonrewetting chars.; wetting agent and detergent in alkaline, acid, and neutral media; does not promote color transfer of dispersed dyes. [Ivax Industries]

Markwet WL-12. Surfactant; wetting agent, dispersant for dyestuffs, pigments and dirt; solubilizer; stable in alkaline and mild acid media. [Ivax Industries]

Marlamid®. A range of fatty acid alkanolamides; stabilizes foam, increases soil-suspending power and has a superfatting effect; intermediates for textile dressing agents and softeners and starting materials for fabric softeners. [Hüls AG]

Marlamid® A 18. CAS 141-21-9; EINECS 205-469-7; Stearamidoethyl ethanolamine; base for textile auxiliary agents and softeners. [Hüls Am.; Hüls AG]

Marlamid® D 1885. CAS 93-83-4; EINECS 202-281-7; Oleamide DEA; foam stabilizer, thickener, superfatting agent for liq. detergents, shampoos. [Hüls Am.; Hüls AG]

Marlamid® DF 1218. CAS 61790-63-4; Cocamide DEA; foam stabilizer, thickener, superfatting agent for liq. detergents, shampoos. [Hüls Am.; Hüls AG]

Marlamid® DF 1818. CAS 68425-47-8; EINECS 270-355-6; Soyamide DEA; foam stabilizer, thickener, superfatting agent for liq. detergents, shampoos. [Hüls Am.; Hüls AG]

Marlamid® KL. Cocamidopropyl lauryl ether; pearlescent surfactant, foam stabilizer, thickener, opacifier for liq. and paste detergents, shampoos. [Hüls Am.; Hüls AG]

Marlamid® KLP. Cocamidopropyl lauryl ether, sodium laureth sulfate; pearlescent base for shampoos, bubble baths, liq. soaps. [Hüls Am.; Hüls AG]

Marlamid® M 1218. CAS 68140-00-1; Cocamide MEA; foam stabilizer, thickener in household, personal, industrial detergents, superfatting agent. [Hüls Am.; Hüls AG]

Marlamid® M 1618. CAS 68153-63-9; Tallowamide MEA; foam stabilizer, thickener for household, personal, and industrial detergents. [Hüls Am.; Hüls AG]

Marlamid® PG 20. Cocamide MEA, glycol ditallowate; pumpable pearlescent base for shampoos, bubble baths, liq. soaps. [Hüls Am.; Hüls AG]

Marlate 2-MR Emulsifiable Insecticide. CAS 72-43-5; 24% Methoxychlor in an emulsifiable solvent; for control of insects on livestock, agricultural premises, forest and shade trees, agricultural crops, ornamentals and flowers and for mosquito control. [Kincaid Enterprises Inc] *

Marlate 50 WP. CAS 72-43-5; 50% Methoxychlor wettable powder; for control of insects in stored grain and for livestock, vegetables and fruits. [Kincaid Enterprises Inc] *

Marlate 300 Flowable. CAS 72-43-5; 30% Methoxychlor (3.0 lbs/gal); seed treatment for insect infestations. [Kincaid Enterprises Inc] *

Marlate 400 Flowable Concentrate. CAS 72-43-5; 40.5% Methoxychlor (4.0 lbs/gal); used for elms, forage and field crops, vegetables and seed treatment. [Kincaid Enterprises Inc] *

Marlate Methoxychlor Insecticide. CAS 72-43-5; 50% Methoxychlor wettable powder; primarily for home use for flowers, gardens, trees and ornamentals. [Kincaid Enterprises Inc] *

Marlazin®. A range of fatty amine polyglycol ethers; used in alkali-resistant and acid-resistant industrial cleaners and dyeing auxiliaries. [Hüls AG]

Marlazin® KC 21/50. CAS 61789-71-7; EINECS 263-080-8; Cocoalkonium chloride; bactericide; production of disinfectant cleaning agents. [Hüls Am.; Hüls AG]

Marlazin® KC 30/50. CAS 61789-18-2; EINECS 263-038-9; Cocotrimonium chloride; quaternary for production of hair conditioning agents. [Hüls Am.; Hüls AG]

Marlazin® L 10. PEG-10 lauramine; surfactant for production of low-foaming acidic cleaners, textile auxiliaries; very high acid resist. [Hüls Am.; Hüls AG]

Marlazin® OL 2. CAS 26635-93-8; PEG-2 oleamine; detergent for industrial cleaners, acid cleaners, textile auxiliaries; resistant to acids and alkalies. [Hüls Am.; Hüls AG]

Marlazin® S 10. CAS 26635-92-7; PEG-10 stearamine; detergent for industrial cleaners, acid cleaners, textile auxiliaries; resistant to acids and alkalies. [Hüls Am.; Hüls AG]

Marlazin® T 10. CAS 61791-44-4; PEG-10 tallowamine; detergent for industrial cleaners, textile and dyeing auxiliaries; resistant to acids and alkalies. [Hüls Am.; Hüls AG]

Marlex 1708. A proprietary trade name for a low density polyethylene suitable for the extrusion of heavy duty film; Type 1 Class A Grade 4 resin; density 0.917; melt index 0.8. [Pacific Petroleums (Quebec)] ‡

Marlex® BMN 55500. A high density polyethylene; used for injection molding of thin wall containers, toys and overcaps. [Phillips] *

Marlex® BMN TR-880. A high density polyethylene; used for injection molding of milk cases, tote boxes, automotive and industrial components and high quality housewares. [Phillips] *

Marlex® CL-50. A crosslinkable high density polyethylene; used for rotational molding of trash containers, chemicals and sewage tanks, seats, boats, camper tops. [Phillips] *

Marlex® CL-100. A crosslinkable high density polyethylene; used for rotational molding of trash containers, industrial and agricultural chemical storage tanks, small engine, snowmobile and automotive fuel tanks. [Phillips] *

Marlex® CL-200. CAS 9002-88-4; Polyethylene resin; rotational molding resin used in agric., chemical, and sewage tanks, automotive fuel tanks, military pkg., trash containers; crosslinks during molding; excellent environmental stress cracking resistance and impact strength at low temps. [Phillips]

Marlex® EHM 6003. A high density polyethylene; used for blow molding of large chemical tanks and parts such as trash cans requiring high stiffness; for sheet extrusion and thermoforming of sheet, tote boxes, deep draw thermoformed parts. [Phillips] *

Marlex® EHM 6006. A high density polyethylene; used for blow molding of containers for bottling products such as milk and distilled water, fruit juices, etc. [Phillips] *

Marlex® EHM 6007. A high density polyethylene; used for blow molding of lightweight containers for bottling products such as milk and distilled water. [Phillips] *

Marlex® EMN TR-885. A high density polyethylene; used for

---

‡ Trade name and manufacturer not verified   § Trade name without identified manufacturer

injection molding of thin walled containers where higher production rate and rigidity is required. [Phillips] *

Marlex® ER9-0002. A reinforced high density polyethylene; used for sheet extrusion and thermoforming for structural automotive applications, housings, tote boxes. [Phillips]*

Marlex® ER9-0020. A reinforced high density polyethylene; used for sheet extrusion and thermoforming for structural automotive applications, seating, shrouds, housings. [Phillips] *

Marlex® HGH-050. A polypropylene homopolymer; used for injection molding of appliance parts and chemical equipment. [Phillips] *

Marlex® HGL-050-01. A polypropylene homopolymer; used for extrusion of soda straws. [Phillips] *

Marlex® HGL-050-01 (Antistatic). A polypropylene homopolymer; used for injection molding of food containers, housewares and toys. [Phillips] *

Marlex® HGL-120-01 (Antistatic). A polypropylene homopolymer; used for injection molding of thin wall containers, medical supplies and housewares. [Phillips] *

Marlex® HGL-200 (Antistatic). A polypropylene homopolymer; used for injection molding of thin wall containers, medical supplies and housewares. [Phillips] *

Marlex® HGL-350 (Antistatic). Controlled rheology polypropylenes; used for injection molding of thin wall containers, housewares and closures. [Phillips] *

Marlex® HGN-020-01. A nucleated polypropylene; used for extrusion blow molding for drugs and toiletries; for extrusion of profiles, sheet and solid phase pressure forming. [Phillips] *

Marlex® HGN-120-01 (Nucleated). A polypropylene homopolymer; used for injection molding of closures and food containers. [Phillips] *

Marlex® HGN-200 (Nucleated). A polypropylene homopolymer; used for injection molding of thin wall containers, medical supplies and housewares. [Phillips] *

Marlex® HGN-200A. A controlled rheology polypropylene homopolymer; used for injection molding of pill vials and medical jars. [Phillips] *

Marlex® HGN-350 (Nucleated). Controlled rheology polypropylenes; used for injection molding of thin wall containers, housewares and closures. [Phillips] *

Marlex® HGX-010. A polypropylene homopolymer; used for extrusion, blow molding and fiber extrusion for drugs, toiletries and strappings. [Phillips] *

Marlex® HGX-030. CAS 9003-07-0; PP homopolymer resin; for monofilament, slit film filament, and staple applications; low water, carry over, processing stability. [Phillips]

Marlex® HGX-040. A polypropylene homopolymer used for general extrusion and slit film and monofilament extrusion for woven carpet backing and bags, rope and cordage; Woven carpet backing and bags, rope and cordage. [Phillips] *

Marlex® HGX-330 (Controlled Rheology). A polypropylene; used for multifilament extrusion for multifilament staple. [Phillips] *

Marlex® HGZ-050-02. A polypropylene homopolymer; used for injection molding of food containers, housewares and toys. [Phillips] *

Marlex® HGZ-120-02. CAS 9003-07-0; PP homopolymer resin; for injection molding of thin-walled containers, syringes, medical supplies, and closures; hardness and

abrasion resistance, high rigidity, and processing stability. [Phillips]

Marlex® HGZ-120-04. A polypropylene used for multifilament extrusion; used for multiultifilament staple. [Phillips] *

Marlex® HGZ-200. A polypropylene homopolymer; used for injection molding of thin wall containers, medical supplies and housewares. [Phillips] *

Marlex® HGZ-350. Controlled rheology polypropylenes; used for injection molding of thin wall containers, housewares and closures. [Phillips] *

Marlex® HHM-4515. A high density polyethylene; used for injection molding of 5-gallon shipping containers, institutional seating, fuel tanks and closures. [Phillips] *

Marlex® HHM 4903. CAS 9002-88-4; HDPE resin; for blow molding, sheet, and thermoforming of large industrial containers, fuel tanks, automotive parts, housings, trays, etc.; ease of processing, stress cracking resistance, surface appearance, and melt strength. [Phillips]

Marlex® HHM 5202. A high density polyethylene; used for blow molding of bleach and detergent containers and chemical packaging; for sheet extrusion and thermoforming of tote boxes, trays, industrial housings, shrouds. [Phillips] *

Marlex® HHM TR-130. A medium density polyethylene; used for film extrusion of merchant bags, produce bags, trash bags and multiwall bag liners. [Phillips] *

Marlex® HHM TR-140. CAS 9002-88-4; Polyethylene resin; film grade resin for produce, merchant, and trash bags, multi-wall bag liners; excellent impact and tear strength, good moisture barrier properties, nonblocking chars. [Phillips]

Marlex® HHM TR-144. A high density polyethylene; used for film extrusion of merchant bags, produce bags, multiwall bag liners and trash bags. [Phillips] *

Marlex® HHM TR-210. A high density polyethylene; used for wire and cable coating for primary insulation for telephone conductors. [Phillips] *

Marlex® HHM TR-226. A high density polyethylene used for wire and cable coating; foam skin insulation on telephone singles. [Phillips] *

Marlex® HHM TR-230 Black. A high density polyethylene used for wire and cable coating; telephone cable jacketing. [Phillips] *

Marlex® HHM TR-232 Black. A high density polyethylene; used for wire and cable coating for power cable jacketing. [Phillips] *

Marlex® HHM TR-250 Black. A high density polyethylene; used for wire and cable coating for aerial cable jacketing of drop wire, line wire and tree wire. [Phillips] *

Marlex® HHM TR-400. CAS 9002-88-4; Med.-dens. polyethylene resin; base resin for oil field and industrial pipe and fittings. [Phillips]

Marlex® HHM TR-418 (Black, Orange). A polyethylene; used for pipe extrusion and injection molding of pipe fittings for gas distribution pipe, potable water pipe, engineered pipe. [Phillips] *

Marlex® HLM-020. A polypropylene homopolymer; used for injection blow molding for drugs, toiletries, cosmetics and spices. [Phillips] *

Marlex® HLN-120-01. A polypropylene homopolymer; used for injection molding for closures and food containers. [Phillips] *

Marlex® HLN-200 (Antistatic, Nucleated). A polypropylene

---

* Trade name not verified as available   † Trade name verified as obsolete

homopolymer used for injection molding; thin wall containers, medical supplies and housewares. [Phillips] *

Marlex® HLN-350 (Antistatic, Nucleated). Controlled rheology polypropylenes used for injection molding; thin wall containers, housewares and closures. [Phillips] *

Marlex® HMN-938. A high density polyethylene used for rotational molding; agricultural and industrial containers, FDA approved drums and tanks, trash containers. [Phillips] *

Marlex® HMN 4550. CAS 9002-88-4; HDPE; injection molding resin with high stress crack resist, impact strength, good warpage resist.; for agric./industrial containers, food handling containers, seating, fuel tanks. [Phillips]

Marlex® HMN 5060. A high density polyethylene used for injection molding; industrial containers, fuel tanks, closures and feeder tubs. [Phillips] *

Marlex® HMN 5580. A high density polyethylene; used for injection molding of pails, housewares, closures and crates requiring good toughness. [Phillips] *

Marlex® HMN 6060. A high density polyethylene; used for injection molding of trays, industrial parts, beverage crates and safety helmets. [Phillips] *

Marlex® HMN 54140. A high density polyethylene; used for injection molding of industrial containers, crates and boxes, large frozen food containers. [Phillips] *

Marlex® HMN TR-942. A high density polyethylene; used for rotational molding for agricultural and industrial containers and food handling containers. [Phillips] *

Marlex® HMX-020-01 (Lubricant). A polypropylene homopolymer used for injection blow molding; drugs, toiletries, cosmetics, and spices. [Phillips] *

Marlex® HNS-080. CAS 9003-07-0; PP homopolymer resin; for high clarity blown, quenched, or cast film applications, including soft goods, stationery, bakery goods, candy wrapping; high slip and med. anti-block chars., cleanliness, sparkling clarity, low haze, and high gloss. [Phillips]

Marlex® HXM 50100. CAS 9002-88-4; HDPE resin; for sheet and thermoforming including large formed parts, cattle feeders, pallets, and boats; excellent stress cracking resistance, melt strength, impact strength even at low temps. [Phillips]

Marlex® RGX-020. A polypropylene random copolymer used for extrusion blow molding; syrup bottles, food containers and toiletries. [Phillips] *

Marlex® RGX-020 (Antistat). A polypropylene random copolymer; used for extrusion blow molding for syrup bottles, food containers and toiletries. [Phillips] *

Marlex® RMN-020C. A polypropylene random copolymer; used for injection blow molding for drugs, toiletries, cosmetics, and spices. [Phillips] *

Marlex® RMX-020. CAS 9003-07-0; PP random copolymer; clearer than homopolymer grades. [Phillips]

Marlex® TR.610. A proprietary trade name for high-density polyethylene; for use as a wire and cable insulation. [Pacific Petroleums (Quebec)] ‡

Marlex® TR.885. A proprietary trade name for high density polyethylene; for injection molding thin walled containers; density 0.965 gm/cc;melt index 30. [Pacific Petroleums (Quebec)] ‡

Marley Bitumen Paint Primer. Bitumen/solvent solution; metal and concrete paint. [Marley Floors Ltd]

Marleybond PVA. Polyvinyl acetate emulsion; for multi use PVC. [Marley Floors Ltd]

Marley Carpet Cleaner. Neutral detergent solution; for carpet cleaning. [Marley Floors Ltd]

Marley Cement Accelerator. CAS 10043-52-4; Calcium chloride solution; cement admixture. [Marley Floors Ltd]

Marley Cement Colorant. Pigment dispersion; cement admixture. [Marley Floors Ltd]

Marley Cement Dustproofer. CAS 1344-09-8; Sodium silicate solution; cement/concrete surface treatment. [Marley Floors Ltd]

Marley Cement Plasticiser. Natural resin solution; cement admixture. [Marley Floors Ltd]

Marley Cement Waterproofer. CAS 143-18-0; EINECS 205-590-5; Potassium oleate solution; cement admixture. [Marley Floors Ltd]

Marley Cork Tile Adhesive. Tackified acrylic emulsion; cork tile adhesive. [Marley Floors Ltd]

Marley Exterior Water Repellent. Silicone, solvent solution; wall treatment. [Marley Floors Ltd]

Marley Floor Cleaner. Detergents and alkalies; floor cleaner. [Marley Floors Ltd]

Marley Floor Gloss. Metal cross linked resin and wax emulsion; floor polish. [Marley Floors Ltd]

Marley Floor Primer. Modified synthetic polymer emulsion; subfloor treatment. [Marley Floors Ltd]

Marley Homelay Adhesive. Modified bitumen emulsion; flooring adhesive. [Marley Floors Ltd]

Marley Interior Waterproofer. Polyurethane, solvent solution; wall, floor and ceiling treatment. [Marley Floors Ltd]

Marley Mastic. Bitumen solvent solution; filler for roofs. [Marley Floors Ltd]

Marley Patchit®. Mineral filled synthetic latex; filler/leveller for floors. [Marley Floors Ltd]

Marley Roofbond®. Bitumen solvent solution; roofing felt adhesive. [Marley Floors Ltd]

Marley Roofseal®. Modified bitumen emulsion; roof treatment. [Marley Floors Ltd]

Marley Stick and Lift. Modified acrylic emulsion; flooring adhesive. [Marley Floors Ltd]

Marley Superwax. Acrylic and soft wax emulsion; floor polish. [Marley Floors Ltd]

Marley Universal Flooring Adhesive. Tackified acrylic emulsion; flooring adhesive. [Marley Floors Ltd]

Marlican®. CAS 67774-74-7; Straight-chain dodecylbenzene; detergent intermediate, solubilizer; secondary plasticizer; reduces visc. of PVC pastes; solv. for carbonless copy papers; biodeg. [Hüls AG]

Marlie's alloy. An alloy containing 10% iron, 35% nickel, 25% brass, 20% tin, and 10% zinc, which has been quenched in a mixture of acids.

Marlinat®. A range of sodium salts of sulfosuccinic acid esters; highly effective wetting agents for the textile, paint and paper industries; used for the preparation of cleaners for sensitive textiles. [Hüls AG]

Marlinat® 242/28. CAS 9004-82-4; Sodium laureth (2) sulfate; strongly foaming base surfactant for detergents, shampoos, liq. soaps. [Hüls Am.; Hüls AG]

Marlinat® CM 40. Laureth-5 carboxylic acid; surfactant for mild detergents, shampoos, foam baths, cleaners. [Hüls Am.; Hüls AG]

Marlinat® DF 8. CAS 577-11-7; EINECS 209-406-4; Dioctyl sodium sulfosuccinate; highly act. wetting agent for textile, paint and paper industries used in cleaners, cosmetic preparations. [Hüls Am.; Hüls AG]

Marlinat® DFK 30. CAS 151-21-3; Sodium lauryl sulfate;

---

‡ Trade name and manufacturer not verified   § Trade name without identified manufacturer

base surfactant for hair shampoos, foam baths, shower foams, liq. soaps. [Hüls Am.; Hüls AG]

Marlinat® DFL 40. CAS 139-96-8; EINECS 205-388-7; TEA lauryl sulfate; finely porous foaming surfactant. [Hüls AG]

Marlinat® DFN 30. CAS 2235-54-3; EINECS 218-793-9; Ammonium lauryl sulfate; base surfactant for hair shampoos, foam baths, shower foams, liq. soaps. [Hüls Am.; Hüls AG]

Marlinat® KT 50. Sodium tallow sulfate, sodium coco-sulfate; surfactant for production of hand-washing pastes. [Hüls Am.; Hüls AG]

Marlinat® SL 3/40. Disodium laureth sulfosuccinate; base surfactant for hair shampoos, foam baths, shower foams, liq. soaps. [Hüls Am.; Hüls AG]

Marlinat® SRN 30. Disodium lauramido MEA-sulfosucci-nate, sodium C12-14 olefin sulfonate; base surfactant for carpet and upholstery cleaners. [Hüls Am.; Hüls AG]

Marlipal® . A range of fatty alcohol polyglycol ethers; nonionic surfactants used as bases for detergents and dish-washing preparations, having dispersing, wetting, detergent, cleaning, soil-suspending and homogenizing properties. [Hüls AG]

Marlipal® 1/12. PEG methyl ether; surfactant for production of methyl-terminated fatty acid esters. [Hüls AG]

Marlipal® 24/20. CAS 68439-50-9; Laureth-2; dispersant, wetting agent, emulsifier, detergent for washing, clean-ing, soil suspension, textile pretreating and dyeing. [Hüls Am.; Hüls AG]

Marlipal® 24/939. C12-14 alcohol ethoxylate blend; dispers-ant, wetting agent, emulsifier, detergent for washing, cleaning, soil suspension, textiles. [Hüls AG]

Marlipal® 124. CAS 9002-92-0; Laureth-4; surfactant for cosmetics, textile auxiliary agents; solubilizer for oils and perfumes. [Hüls Am.; Hüls AG]

Marlipal® 1012/4. CAS 26183-52-8; Deceth-4; wetting sur-factant, esp. for hard surface cleaning. [Hüls Am.

Marlipal® 1618/6. CAS 68439-49-6; Ceteareth-6; dispers-ant; production of powd. detergents; binding agent and base material for solid cleaning agents (toilet sticks); coating material for foam suppressant, enzymes, etc.; dyeing auxiliaries. [Hüls AG]

Marlipal® 1850/5. CAS 9004-98-2; Oleth-5; surfactant for mfg. of washing powds., textile auxiliaries. [Hüls Am.; Hüls AG]

Marlipal® KF. CAS 66455-15-0; Deceth-6; surfactant, deter-gent raw material, wetting agent, and additive used in dishwashing agents, glass and hard surface cleaners, in mfg. of tablet soaps; degreaser for textiles and leather. [Hüls Am.; Hüls AG]

Marlipal® MG. CAS 9002-92-0; Laureth-7; solubilizer for act. ingreds. and oils in cosmetics. [Hüls Am.; Hüls AG]

Marlipal® O11/30. C11-oxo alcohol ethoxylate (3 EO); wetting agent for hard surface cleaners, textile pretreat-ment. [Hüls AG]

Marlipal® O13/20. C13-oxo alcohol ethoxylate (2 EO); dispersant, wetting agent, detergent, cleaning, soil sus-pending and homogenizing agent, emulsifier for textiles. [Hüls AG]

Marlipal® SU. CAS 54045-08-8; Cetoleth-25; surfactant for mfg. of washing powds., dyeing assistants; plasticizer for soap production; yields finely porous foam. [Hüls Am.; Hüls AG]

Marloid® CAS. Specialty algin blend; milk-soluble product

for use in milk-based systems. [Kelco]

Marlon® . A range of alkylbenzenesulfonates; used for the manufacture of detergents and cleaners. [Hüls AG]

Marlon® A 365. CAS 38411-30-3; Sodium dodecylbenzene sulfonate; surfactant for production of detergents, clean-ing agents, textile auxiliaries; biodeg. [Hüls Am.; Hüls AG]

Marlon® A350. Anionic surfactant in liquid form; high quality products with low salt content used in liquid detergents and cleaning materials for domestic and industrial use. [Hüls UK Ltd]

Marlon® A360, A365, A375. Anionic surfactant in paste form; high quality products with low salt contents used for paste detergents and cleaning materials for domestic and industrial use. [Hüls UK Ltd]

Marlon® A390, A396, ARL. Anionic surfactant in solid form; high quality products with low salt contents used for powder detergents and cleaning materials for domestic and industrial use. [Hüls UK Ltd]

Marlon® AMX. Amine dodecylbenzene sulfonate; base material for production of drycleaning detergents, de-greasing agents for metal industry, floor cleaners; biodeg. [Hüls Am.; Hüls AG; Hüls UK Ltd]

Marlon® ARL. Sodium dodecylbenzene sulfonate and so-dium toluene sulfonate; detergent component, foaming, wetting agent; for powd. detergents and scouring powds. [Hüls Am.; Hüls AG]

Marlon® AS3. CAS 85536-14-7; Dodecylbenzene sulfonic acid; intermediate for mfg. of anionic surfactants, deter-gents, sulfonates, textile auxiliaries; biodeg. [Hüls Am.; Hüls AG; Hüls UK Ltd]

Marlon® PF 40. Sodium C13-17 alkane sulfonate, sodium laureth sulfate; mild, high foaming detergent for domestic and industrial applications; biodeg. [Hüls Am.; Hüls AG]

Marlon® PS 30. Sodium C13-17 alkane sulfonate; surfactant for production of liq., conc., mild cleaning agents, hair shampoos, foam baths, textile auxiliaries. [Hüls Am.; Hüls AG]

Marlophen® . A range of nonylphenol polyglycol ethers with wetting, detergent, dispersing and homogenizing prop-erties; used as textile and paper auxiliaries, as wetting agents for coal, rock dusts and pigments and as an additive for concrete manufacture. [Hüls AG]

Marlophen® 81N. CAS 85005-55-6; Nonoxynol-1; wetting agent, detergent, dispersant; homogenizing capacity; polymer remover in floor care. [Hüls AG]

Marlophen® 85. CAS 9002-93-1; Octoxynol-5; detergent, wetting agent, dispersant with washing and homogeniz-ing capacity. [Hüls Am.; Hüls AG]

Marlophen® 810. CAS 9002-93-1; Octoxynol-10; wetting agent for acidic, neutral, and alkaline cleaning agents; production of textile auxiliaries. [Hüls Am.; Hüls AG]

Marlophen® 810N. CAS 9016-45-9; 37205-87-1; Nonoxy-nol-10; wetting agent, detergent, dispersant; homog-enizing capacity; for acidic, neutral and alkaline clean-ers, textile auxiliaries. [Hüls AG]

Marlophen® 830N. CAS 9016-45-9; 37205-87-1; Nonoxy-nol-30; wetting agent, detergent, dispersant; homog-enizing capacity; binding agent and base material for solid cleaning agents such as toilet sticks. [Hüls AG]

Marlophen® DNP 16. CAS 9014-93-1; Nonyl nonoxynol-16; raw material for textile and paper auxiliaries, dispersant. [Hüls Am.; Hüls AG; Hüls UK Ltd]

Marlophen® P 1. Phenol ethoxylate (1 EO); solubilizer,

---

solvent. [Hüls Am.; Hüls AG]

Marlophen® X. Alkylphenol polyglycol ether; wetting agent for use in binders for coal dust, pigments, and in concrete mfg. [Hüls AG; Hüls UK Ltd]

Marlophor. A range of partial phosphate esters in the form of acids and salts; used as low foam wetting agents for alkaline solutions, starting materials for the manufacture of acid cleaners, water and oil-soluble wetting agents for the textile and paper industries, antistatic agents, and drycleaning detergents. [Hüls AG]

Marlophor® CS-Acid. CAS 76483-21-1; Isopropyl phosphate ester; base compound for formulating acid cleaners for glass, metal, and ceramics, flame retardants, mercerizing wetting agents and antitstats; anticorrosive and rust-removing props. [Hüls Am.; Hüls AG]

Marlophor® DS-Acid. CAS 68439-39-4; n-Butyl phosphate ester; surfactant used as base material for acid cleaners for glass, metal, and ceramics, flame retardants, mercerizing wetting agents and antistats; anticorrosive and rust-removing props. [Hüls Am.; Hüls AG]

Marlophor® FC. Sodium alkylpolyglycol ether phosphate in liquid form; low foam wetting agent for detergents for dishwashers and industrial cleaning. [Hüls UK Ltd]

Marlophor® HS-Acid. CAS 39407-03-9; n-Octyl phosphate ester; emulsifier for silicone oils. [Hüls Am.; Hüls AG]

Marlophor® IH-Acid. CAS 12645-31-7; 2-Ethylhexyl phosphate ester; low foam wetting agent for weakly to med. strong alkaline range in textile pretreatment and finishing. [Hüls Am.; Hüls AG]

Marlophor® LN-Acid. Trilauryl phosphate; wetting agent for textile, paper, and personal care industries; antistat, drycleaning detergent. [Hüls Am.]

Marlophor® MD. Organic phosphate ester surfactant; wetting agent (oil and water soluble) and antistat; used in paper; textiles; natural and synthetic fibers; dry cleaning detergents. [Hüls UK Ltd]

Marlophor® MO 3-Acid. CAS 39464-66-9; Laureth-3 phosphate; surfactant for production of textile and dyeing auxiliaries, antistats, drycleaning detergents. [Hüls Am.; Hüls AG]

Marlophor® ND-Acid. Isoalkylphosphate ester/alkylphenol PEG ether blend; detergent; wetting agent; antistat; component for textile auxiliary agents, drycleaning formulations, paper industry. [Hüls Am.; Hüls AG]

Marlophor® T10-Acid. Diceteareth-10 phospate; special liq. formulation for production of drycleaning detergents with antistatic props., textile and dyeing auxiliaries. [Hüls Am.; Hüls AG]

Marlopon. A range of alkylbenzenesulfonates; used for the manufacture of cosmetic detergents and dishwashing agents. [Hüls AG]

Marlopon® ADS 50. CAS 26545-53-9; EINECS 247-784-2; DEA dodecylbenzene sulfonate; detergent raw material for liq. phosphate-containing detergents and cleaners; biodeg. [Hüls Am.; Hüls UK Ltd]

Marlopon® AMS 60. Amine dodecylbenzene sulfonate/nonionic blend; surfactant used for detergents, dishwashing agents, industrial cleaners, car shampoos, hand cleaners. [Hüls Am.; Hüls AG]

Marlopon® AT. CAS 29381-93-9; TEA-dodecylbenzene sulfonate; surfactant for mfg. of cosmetic detergents, dishwashing agents. [Hüls Am.; Hüls UK Ltd]

Marlosoft® A 18 M. Blend of fatty acid alkanolamide acetate and fatty acid polyglycol ester; for production of cold

water-sol. textile softeners. [Hüls AG]

Marlosoft® IQ 75. Imidazolinium methosulfate; base for fabric softeners. [Hüls Am.]

Marlosoft® IQ 90. Quaternized tallow fatty imidazolinium methosulfate; surfactant for production of fabric softeners. [Hüls AG]

Marlosol® . A range of polyglycol esters of fatty acids; starting materials for preparing agents used in the synthetic fiber industry. [Hüls AG]

Marlosol® 183. CAS 9004-99-3; PEG-3 stearate; raw material for finishing agents in the synthetic fiber industry; emulsifier. [Hüls Am.; Hüls AG]

Marlosol® BS. CAS 9005-08-7; PEG-12 distearate; superfatting agent, visc. enhancer for hair shampoos, cosmetic preps., fabric softeners, synthetic fiber finishing. [Hüls Am.; Hüls AG]

Marlosol® FS. CAS 9005-07-6; 52668-97-0; PEG-12 dioleate; superfatting agent, visc. enhancer for hair shampoos, cosmetic preps., fabric softeners, synthetic fiber finishing. [Hüls Am.; Hüls AG]

Marlosol® OL2. CAS 9004-96-0; PEG-2 oleate; raw material for prep. agents for synthetic fiber industry. [Hüls AG]

Marlosol® R70. CAS 61791-12-6; PEG-70 castor oil; thickener, conditioner for toilet sticks. [Hüls Am.; Hüls AG]

Marlosol® RF3. PEG-3 rapeseed fatty acid ester; preparation agent. [Hüls AG]

Marlosol® TF3. CAS 61791-00-2; PEG-3 tallate; raw material for prep. agent for synthetic fiber industry. [Hüls AG]

Marlotherm. A range of benzyltoluenes; suitable for use as heat transfer mediums. [Hüls AG]

Marlowet® . A wide range of emulsifiers; emulsifiers for mineral oils, hydrocarbons, waxes, oleic acids, solvents, pesticides, cold cleaners, spindle oils, textile lubricants, furniture polishes and leather dressings, etc. [Hüls AG]

Marlowet® 1072. C12-14 alcohol polyglycol ether carboxylic acid; emulsifier for metalworking, water-miscible cooling lubricants, drilling oils, textile auxiliaries. [Hüls Am.; Hüls AG]

Marlowet® 4536. Nonylphenol polyglycol ether carboxylic acid; emulsifier for water-miscible cooling lubricants and drilling oils, production of textile auxiliaries. [Hüls AG]

Marlowet® 4538. C13 oxo-alcohol polyglycol ether carboxylic acid; emulsifier for metalworking, water-miscible cooling lubricants, drilling oils, textile auxiliaries. [Hüls Am.; Hüls AG]

Marlowet® 4539. C9 oxo-alcohol polyglycol ether carboxylic acid; surfactant for metalworking, water-miscible cooling lubricants, drilling oils. [Hüls Am.; Hüls AG]

Marlowet® 4702. CAS 52668-97-0; C18 fatty acid polyglycol ester; emulsifer for mineral oils, spindle oils, metalworking, textile lubricants. [Hüls Am.; Hüls AG]

Marlowet® 4800. CAS 68439-49-6; C16-18 alcohol polyglycol ether; emulsifier for waxes, car and furniture polishes, textile lubricants, leather care. [Hüls Am.; Hüls AG]

Marlowet® 4900. Nonylphenol PEG ether; emulsifier for mineral oils used in metalworking and in cold cleaners. [Hüls Am.; Hüls AG]

Marlowet® 5311. CAS 97999-44-5; Isononanol phosphate ester; emulsifier used in paint removers; wetting agent for chlorinated hydrocarbons; production of textile auxiliaries. [Hüls Am.; Hüls AG]

Marlowet® 5324. CAS 39464-70-5; Phenol polyglycol ether phosphate ester; emulsifier for mineral oils; lubricant

---

‡ Trade name and manufacturer not verified   § Trade name without identified manufacturer

auxiliaries; metal processing. [Hüls Am.; Hüls AG]

Marlowet® 5400. CAS 26635-93-8; Alkylamine polyglycol ether; emulsifier for mineral oils, paraffin oils, car wash rinses, furniture polishes, textile auxiliaries. [Hüls Am.; Hüls AG]

Marlowet® 5440. CAS 95-38-5; Substituted imidazoline; emulsifier for mineral oils, corrosion protection, car wash rinses. [Hüls Am.; Hüls AG]

Marlowet® 5459. Fatty acid DEA; corrosion inhibitor for metalworking. [Hüls Am.; Hüls AG]

Marlowet® BL. CAS 9002-92-0; C12 alcohol polyglycol ether; emulsifier for mineral oils, spindle oils, textile lubricants, bitumen. [Hüls Am.; Hüls AG]

Marlowet® FOX. CAS 68439-49-6; 68920-66-1; Ceteareth-28; emulsifier for oleic acid and waxes, textile lubricants, car and furniture polishes, leather care. [Hüls Am.; Hüls AG]

Marlowet® ISM. CAS 37205-87-1; Nonylphenol polyglycol ether; emulsifier for solvs., for pesticides, cold cleaners. [Hüls Am.; Hüls AG]

Marlowet® LVS. C18 fatty acid ester of ethoxylated castor oil; emulsifier for vegetable oils; used in metalworking, leather auxiliaries, release agents. [Hüls Am.; Hüls AG]

Marlowet® NF. Mixt. of carboxylic acid polyglycol esters; emulsifier for leather, auxiliaries, textile lubricants. [Hüls Am.

Marlowet® OCM. Fatty acid alkanolamide polyglycol ether; anticorrosive agent for water-misc. cooling lubricants. [Hüls AG]

Marlowet® PW. CAS 68439-49-6; C16-18 alcohol polyglycol ether; surfactant, emulsifier for paraffin, lanolin waxes, textile and paper impregnation, mold release agents, furniture polishes. [Hüls Am.; Hüls AG]

Marlowet® R 11. CAS 61791-12-6; Ethoxylated castor oil; emulsifier for animal and vegetable oils and neutral fats, leather auxiliaries. [Hüls Am.; Hüls AG]

Marlowet® R 40. CAS 61791-12-6; PEG-40 castor oil; emulsifier for fatty acids, solvs., cosmetic oils; textile lubricants, dyeing auxiliaries, pesticides, creams; biodeg. [Hüls Am.; Hüls AG]

Marlowet® WOE. CAS 9004-98-2; Oleth-5; emulsifier for paraffinic mineral oils, textile lubricants. [Hüls Am.; Hüls AG]

Marlox® . A range of alkylene oxide addition products; components of low-foaming detergents and cleaners particularly for low-foaming dishwashing agents and for low-foaming textile finishing agents and processing aids. [Hüls AG]

Marlox® 3000. Butyl glycol/alkylene oxide addition product; surfactant for mfg. of textile auxiliaries and finishing agents. [Hüls Am.; Hüls AG]

Marlox® FK 14. Propylene glycol capreth-4; detergent, antistat, foam controller; for low-foaming detergents and cleaners, dishwash, industrial cleaners, textile auxiliaries. [Hüls Am.; Hüls AG]

Marlox® FK 64. CAS 68154-97-2; PPG-6 deceth-4; detergent, antistat, foam controller; for low-foaming detergents and cleaners, dishwash, industrial cleaners, textile auxiliaries. [Hüls Am.; Hüls AG]

Marlox® L 6. CAS 9064-14-6; PPG-7 lauryl ether; detergent, antistat, foam controller, textile auxiliary agent. [Hüls Am.; Hüls AG]

Marlox® MO 124. CAS 68439-51-0; PPG-4 laureth-2; component for low foaming detergents and cleaners, auto-

matic dishwasher formulations, industrial cleaners, textile auxiliaries. [Hüls Am.; Hüls AG]

Marlox® MS 48. CAS 69227-21-0; C12-18 fatty alcohol alkylene oxide addition product; surfactant for mfg. of textile auxiliaries and finishing agents. [Hüls Am.; Hüls AG]

marls. Natural mixtures of clay and chalk (aluminum silicate and calcium carbonate); used in the manufacture of cements. The term is also applied to friable earths which are devoid of chalk, such as those of Staffordshire.

Marme's reagent. Consists of 10 parts cadmium iodide, CdI_2, and 20 parts potassium iodide, KI, dissolved in 80 parts water; used for testing for alkaloids.

Marmite. A yeast extract; it is a food preparation resembling meat extract. §

Marmo Bardiglio de Bergamo. Vulpinite, a variety of anhydrite, mixed with silica; used for ornamental purposes.

Marphos. CAS 7664-38-2; Phosphoric acid. [GE Plastics ABS Ltd] *

Marplan. CAS 59-63-2; A proprietary preparation of isocarboxazid; an antidepressant. [Roche Laboratories; Roche Products Ltd]

Marquat Pigments. Cadmium, cobalt and titanium pigments; for coatings industry. [Degussa]

Marseilles soap. Olive oil soap.

Marshal 10G. CAS 55285-14-8; Carbosulfan; a systemic carbamate insecticide. [Rhone-Poulenc Crop Protection Ltd]

Marshal/suSCon. CAS 55285-14-8; 100 g/kg Carbosulfan; for termite control in reforestation programs with Eucalyptus species; control of Hylobius weevils in pine plantations. [Incitec Ltd]

Marshal/suXon. See Marshal/suSCon. (France) [Incitec Ltd]

marsh gas. CAS 74-82-8; (Light Carburetted Hydrogen). Methane, CH_4.

Marsilid. CAS 54-92-2; A proprietary preparation of iproniazid; an antidepressant. [Roche Products Ltd] *

Marsipol. Leather finishes. [ICI Chem. & Polymers Ltd] †

martensite. A solid solution of carbon in iron, and is a characteristic constituent of steel which has been tempered at a temperature a little above the transformation point.

Martifin. CAS 21645-51-2; Ground and finely precipitated aluminum hydroxide; used as a filler and coating pigment for the paper and cardboard industry. [Lonza AG]

Martinal. CAS 21645-51-2; Ground and finely precipitated aluminum hydroxide; used as a flame retardant for plastics and rubber. [Lonza AG]

Martinal® OL-111 LE. CAS 21645-51-2; Aluminum hydroxide; finely precipitated filler, flame retardant; suitable for automatic bulk handling. [Martinswerk GmbH]

Martinal® ON-4608. CAS 21645-51-2; Aluminum hydroxide; med. particle size filler/flame retardant for thermoset plastics, carpetbacking latexes. [Martinswerk GmbH]

Martinal® OS. CAS 21645-51-2; Aluminum hydroxide; filler/flame retardant for use where coarse fillers are needed, e.g., artificial marble, polymer concrete, resin flooring, chipboard. [Martinswerk GmbH]

martinite. A mineral, $Ca_5H_2(PO_4)_4$.

Martino's alloys. Alloys containing 17.25% pig iron, 3-4.5% ferro-manganese, 1.5-2% chromium, 5.25-7.5% tungsten, 1.25-2% aluminum, 0.5-0.75% nickel, 0.75-1%

---

* Trade name not verified as available   † Trade name verified as obsolete

copper, and 65-70% wrought iron; used for drilling and cutting tools.

Martin's cement. A similar cement to Keene's, except that potassium carbonate solution is used instead of alumina.

Martin steel. (Open Hearth Steel). Steel obtained in the Martin process by melting from 75% of cast iron in a reverberatory furnace with the necessary quantity of wrought iron to obtain the required amount of carbon.

Martipol. CAS 1344-28-1; Speciality aluminum oxide product; used as polishing aluminas for the ceramic industry. [Lonza AG]

Martisorb. CAS 1344-28-1; Speciality aluminum oxide product; used for the purification of water. [Lonza AG]

Martoxin. CAS 1344-28-1; Speciality aluminum oxide product; used as a coating pigment for carbonfree self-copying paper. [Lonza AG]

Marvaloy 750. Acrylic-modified styrene; offers high clarity, favorable economics for injection molding and extrusion applications; used for specialty medical, pharmaceutical, food, and cosmetic pkg., advertising displays, high-str. toys. [Marval Industries]

Marvanbrite CF. Heterocyclic stilbene deriv.; nonyel. optical brightener for cellulosic, wool, silk, nylon and surgical whites. [Marlowe-Van Loan]

Marvanfix® ATA. Nonformaldehyde fixing agent for improved wetfastness of direct dyes on cellulosic fibers. [Marlowe-Van Loan]

Marvanfix® C. Quaternary ammonium resin complex; textile fixative; improves wetfastness on direct and reactive dyes. [Marlowe-Van Loan]

Marvanlube® 92. Silicone emulsion; textile lubricant for preboarding, finish boarding, hosiery. [Marlowe-Van Loan]

Marvanlube® BHC. Complex polymeric; low foaming dyebath lubricant for hosiery and piece good dyeings (nylon hosiery, nylon/spandex body suits, cotton tights). [Marlowe-Van Loan]

Marvanol® Aftertreat 2AF. Emulsion polymer; dyeing assistant; exhaust pigment aftertreat for garments, hosiery. [Marlowe-Van Loan]

Marvanol® BAN. Sulfated vegetable oil; dye dispersant; minimizes barré. [Marlowe-Van Loan]

Marvanol® Carrier BB. CAS 136-60-7; EINECS 205-252-7; Self-emulsifiable butyl benzoate; carrier for atmospheric and pressure dyeing. [Marlowe-Van Loan]

Marvanol® Defoamer AM-2. Conc. silicone emulsion; foam inhibitor and depressant for textile applications. [Marlowe-Van Loan]

Marvanol® GAW. Ethoxylated org. deriv.; textile leveling agent; compatibilizer for basic and acid dyes. [Marlowe-Van Loan]

Marvanol® LSL. Ethoxylated surfactant blend; compatibilizer for basic and acid dyes on synthetics. [Marlowe-Van Loan]

Marvanol® Penetrant 35. Amine neutralized sulfonic acid; detergent, wetting agent, textile dyeing, scouring agent, finishing, leveling and retarding agents for acid dyes. [Marlowe-Van Loan]

Marvanol® Pretreat GD-P. Polymeric resin; dyeing assistant; exhaust pigment pretreat for garments, hosiery. [Marlowe-Van Loan]

Marvanol® RD2-1852. Sulfated ester blend; dyeing assistant; dyebath lubricant for direct dyes on cotton. [Marlowe-Van Loan]

Marvanol® REAC A-213. Polymer salt; low foaming dispers-

ing agent, textile scouring agent for fiber reactives; improves dye fixation. [Marlowe-Van Loan]

Marvanol® SBO (60%). Sulfated butyl oleate, sodium salt; detergent, wetting and leveling agent, emulsifier, dyeing assistant, lubricant. [Marlowe-Van Loan]

Marvanol® SCO (50%). CAS 8002-33-3; EINECS 232-306-7; Sulfated castor oil; detergent, wetting agent, dyeing assistant and lubricant used in finishing operations; leveling agent for cotton. [Marlowe-Van Loan]

Marvanol® Scour 2 Base. Alcohol ether condensate; multifunctional biodeg. textile scour. [Marlowe-Van Loan]

Marvanol® SPO (60%). Sulfated propyl oleate, sodium salt; detergent, wetting and leveling agent, dyeing assistant, lubricant for textiles. [Marlowe-Van Loan]

Marvanquest 1022. Org. blend; nonsilicate/hydrogen peroxide bleach stabilizer; prevents calcium deposits. [Marlowe-Van Loan]

Marvanscour® KW. Solvs. and detergents; detergent for prescour and afterscour. [Marlowe-Van Loan]

Marvanscour® LF. Phosphate ester; low-foaming wetting agent and dyeing assistant for jet dyeing cotton and blends. [Marlowe-Van Loan]

Marvansoft 1771. Silicone emulsion; softener for textile finishing, printing cellulosics and synthetics. [Marlowe-Van Loan]

Marvantex RBDS. Inorg. salts; one-bath scour, bleaching agent, and dye. [Marlowe-Van Loan]

Marvylan. Polyvinylchloride; used in the plastics processing industry, applications in building construction (tubes and pipes, profiles and cables); in the packaging industry (bottles); as synthetic leather (bags, wall covering, clothing); in hoses etc. [DSM NV] *

Marweld M-17. Epoxy resin, liquid sealant, two component, flexible; pipe and flange thread sealant for sealing threads in flanged ductile iron and cast iron water pipes, EPA approved. [RJ Manufacturing Inc] *

Marzine. Proprietary formulations of cyclizine hydrochloride or cyclizine lactate; tablets and injections; for treatment of nausea and vomiting. [The Wellcome Foundation Ltd]

Marzine RF Tablets. CAS 298-57-7; A proprietary formulation of cinnarizine; for prophylaxis of motion sickness. [The Wellcome Foundation Ltd]

Mascot Clearing. CAS 15310-01-7; Benodanil; a systemic fungicide. [Rigby Taylor Ltd]

Mascot Cloverkiller. CAS 7085-19-0; Soluble concentrate containing 300 g/l mecoprop; for control of weeds in cereals and grassland. [Rigby Taylor Ltd]

Mascot Contact Turf Fungicide. CAS 50471-44-8; Vinclozolin; a protectant fungicide for turf. [Rigby Taylor Ltd]

Mascot Gauntlet. CAS 1912-24-9; Atrazine; a residual herbicide. [Rigby Taylor Ltd]

Mascot Highway. CAS 61-82-5, 122-34-9; Aminotriazole + simazine; used for total weed control in non crop areas. [Rigby Taylor Ltd]

Mascot Moss Killer. CAS 97-23-4; Dichlorophen; fungicide, bactericide and algicide. [Rigby Taylor Ltd]

Mascot Selective. Soluble concentrate containing 60 g 2,4-D and 200 g mecoprop per liter; used to control weeds in grassland. [Rigby Taylor Ltd]

Mascot Super Selective. Soluble concentrate of 15 g dicamba, 100 g MCPA and 200 g mecoprop per liter; used for weed control in cereals and grassland. [Rigby

---

Taylor Ltd]

Mascot Systemic Turf Fungicide. CAS 83601-81-4; A liquid formulation containing 500 g carbendazim per liter as a suspension concentrate; systemic fungicide. [Rigby Taylor Ltd]

Mascot Ultrasonic. Glyphosate + simazine; a translocated and residual herbicide for the control of grasses and weeds. [Rigby Taylor Ltd]

Masil® 173. Silicone fluid dispersed in halogenated org. solv.; defoamer for chem. processing, refiners, solv. cleaning; esp. for defoaming light hydrocarbon solvs. [PPG/Specialty Chem.]

Masil® 264. Alkyl methyl polysiloxane; processing aid, lubricant, mold release; for aerosol pkg., ink/printing industry. [PPG/Specialty Chem.]

Masil® 280. Dimethicone copolyol; antistat, wetting agent for personal care products, etc. [PPG/Specialty Chem.]

Masil® 1066C. Dimethicone copolyol; lubricant and antistat for plastics, textiles, metal processing; wetting and leveling char.; antifog for glass cleaners. [PPG/Specialty Chem.]

Masil® 756. Tetrabutoxypropyl methicone; cosmetic ingred. providing high gloss to hair; nonoily emollient for skin care. [PPG/Specialty Chem.]

Masil® EM 100. Dimethylpolysiloxane fluids aq. emulsion; emulsifier, release aid in molding, extrusion, laminating, and casting for rubber, plastics, and metals; for leather, glass, and vinyl cleaners, polishes, textile softeners, textile/fiber lubricants; recommended for glass hard surface cleaner applications, and imparts nonsmearing, low gloss, and ease of wipe properties to such formulations. [PPG/Specialty Chem.]

Masil® EM 350. Dimethylpolysiloxane fluids aq. emulsion; food grade release emulsion used in mfg. of articles in contact with food. [PPG/Specialty Chem.]

Masil® EM 100,000. Dimethylpolysiloxane fluids aq. emulsion; emulsifier, release aid in molding, extrusion, laminating, and casting for rubber, plastics, and metals; for leather, glass, and vinyl cleaners, polishes, textile softeners, textile/fiber lubricants. [PPG/Specialty Chem.]

Masil® SF 5. Dimethicone; release aid, defoamer for nonaq. processes, esp. in the petrol., foods, and printing inks industries; internal lubricant for plastics, rubber, and metal; also in furniture and auto-wax polishes, household and personal care products; textile lubricant; lower visc. fluids recommended for cosmetic applications; higher visc. fluids (> 10,000 cSt) for formulating lubricants and mold releases in mfg. of plastics, rubber parts. [PPG/Specialty Chem.]

Masil® SF 201. Vinyl-terminated silicone polymer; functional polymer for compounding silicone elastomers for use as encapsulants in elec./electronic industry. [PPG/Specialty Chem.]

Masil® SF 500. Dimethicone; release aid, defoamer for nonaq. processes, esp. in the petrol., foods, and printing inks industries; internal lubricant for plastics, rubber, and metal; also in furniture and auto-wax polishes, household and personal care products. [PPG/Specialty Chem.]

Masil® SF 500,000. Dimethicone; release aid, defoamer for nonaq. processes, esp. in the petrol., foods, and printing inks industries; internal lubricant for plastics, rubber, and metal; also in furniture and auto-wax polishes, household and personal care products. [PPG/Specialty

Chem.]

Masil® SF 1,000,000. Dimethicone; release aid, defoamer for nonaq. processes, esp. in the petrol., foods, and printing inks industries; internal lubricant for plastics, rubber, and metal; also in furniture and auto-wax polishes, household and personal care products. [PPG/Specialty Chem.]

Masil® SF-MH. CAS 9004-73-3; Methicone; reactive fluid for modification of polyester and methacrylic resins, as waterproofing and impregnating agents for textiles, paper, leather and hydrophobizing agents for powds., silicas, and other fillers. [PPG/Specialty Chem.]

Masil® SF-V. CAS 69430-24-6; Cyclomethicone; volatile silicone fluid imparting silky light feel and spreadability to cosmetics (hair care products, skin creams and lotions, antiperspirants, deodorants, suntan preps.). [PPG/Specialty Chem.]

Masil® SFR 70. CAS 31692-79-2; Dimethiconol; reactive fluid; raw material in compounding silicone RTV systems, textile and paper coatings, plasticizer/processing aid for silicone elastomers, hydrophobizing silica, in water repellent formulations. [PPG/Specialty Chem.]

Maslip® 500. Proprietary formula containing no chlorinated or sulfonated compounds; lubricant base for metalworking fluids which require extreme-pressure qualities for heavy-duty work. [PPG/Specialty Chem.]

Masmoran. CAS 68-88-2; A proprietary preparation of hydroxyzine; an ataractic. [Pfizer International]

Masocare. Iodophors. [Evans Vanodine International Ltd]

Masodine. Iodophors. [Evans Vanodine International Ltd]

Masol. Cement mixture for pump packing in mines roadways. [Foseco (F.S.) Ltd]

Masonry Stain and Seal. Methylmethacrylate acrylic, pigments and polysiloxane resins in an aliphatic and aromatic solvent vehicle system; for semi-transparent stain for masonry. [Nova Chemical Inc] *

Masoten® . Veterinary preparation; for the control of ectoparasites in fish. [Bayer AG]

Massa Estarinum. Neutral hard fats based on mixtures of triglycerides; used for preparation of suppositories. [Hüls AG]

Massa Estarinum® CM. Hydrogenated palm glycerides, hydrogenated palm kernel glycerides; consistency regulator for decorative cosmetics, sticks, pencils, powds., glosses, and eye shadows. [Hüls Am.]

Massaranduba. (Balata rans, brittle balata). A pseudo gutta-percha derived from the sap of the Brazilian cow tree, *Mimusops elata*.

massecuite. The boiled mass of beet sugar syrup. It is a semi-solid mass formed during the evaporation of the sugar juice, and consists of sugar crystals and a thick syrup. It contains from 3.5-7% water.

Massicot. See litharge.

Masterblok. Solid phenolic ester in panels; used for master patterns, N/C trials, models, fixtures and vac forming tools. [J R Technology Ltd]

Masterbond. Cellulose acetate laminating adhesives. [The Scottish Adhesives Co Ltd]

Master Bond AC82. One-part nickel conductive adhesive coating; for EMI/RFI shielding applications. [Master Bond]

Master Bond EP11HT. CAS 25928-94-3; One-part epoxy; heat-resist. version of EP11; service temp. -60 to 350 F. [Master Bond]

Master Bond EP21HT. CAS 25928-94-3; Two-part epoxy; high temp. resist. version of EP21; excellent chem. resist.; service temp. -60 to 400 F. [Master Bond]

Master Bond EP30HT. CAS 25928-94-3; Two-part epoxy; high temp. resist. version of EP30; exceptional adhesion to glass, ceramics, wood; excellent long-term durability; service temp. -60 to 400 F. [Master Bond]

Master Bond EP34CA. CAS 25928-94-3; Two-part epoxy; heat-resist. laminating resin for filament winding of structural FRP components; service temp. -60 to 450 F. [Master Bond]

Master Bond EP75. CAS 25928-94-3; Two-part epoxy, graphite-filled; elec. conductive adhesive/sealant; service temp. -60 to 300 F. [Master Bond]

Master Bond Supreme 3HT. CAS 25928-94-3; One-part epoxy; high peel strength, heat-resist. version of EP3; service temp. -60 to 350 F. [Master Bond]

Master Bond Supreme 11HT. CAS 25928-94-3; Two-part epoxy; heat-resist. version of Supreme 11; service temp. -60 to 400 F. [Master Bond]

Mastercarb. Mineral filled concentrates in polyolefins; used for plastics moldings, sheet and film. [Collinda Ltd]

Mastercolor. Masterbatches of thermoplastics; color and additive concentrates used in plastics. [Ampacet Corporation]

Masterflam. Halogenic and nonhalogenic masterbatches for making self-extinguishing thermoplastics [VAMP srl]*

Masteril. A proprietary preparation of drostanolone propionate; for treatment of mammary carcinoma. [Syntex Laboratories Inc] *

Masterone. CAS 521-12-0; Dromostanolone propionate; antineoplastic. [Syntex Laboratories Inc] *

Master Sil 701. Silicone; general purpose bonding/sealing agent featuring high temp. resist.; used esp. with formed-in-place gaskets. [Master Bond]

Masterwood. Wood-filled thermoplastic polymers; used for plastics moldings and extrusions. [Collinda Ltd] †

mastic. The name for an important resin obtained from *Pistachia lentiscus*, from various parts of the Mediterranean coast; used in the manufacture of spirit varnishes. The term Mastic is also applied to a mixture of asphalt rock and Trinidad pitch.

Masticillin® C, M. Preparations containing penicillin, streptomycin or penicillin sulfonamide; used for the treatment of mastitis in veterinary medicine. [Bayer AG]

Mastrite. Iodophor mastitis treatment. [The Wellcome Foundation Ltd] *

Masuron. A proprietary trade name for a cellulose acetate plastic. §

Matalex. X-ray developer system. [May & Baker Ltd] *

matali. Rubber obtained from the roots of various species of *Apocynaceae*.

Match. CAS 21725-46-2; Suspension concentrate containing 500 g cyanazine per liter; a triazine herbicide. [Shell UK]

maté. (Jesuit's tea, St. Bartholomew's tea). Paraquay tea, the dried leaves and shoots of an evergreen, *Ilex paraguayensis*. It contains caffeine and theobromine.

Mateflex. Polyethylene resin; used for tennis courts. [Sumitomo Bakelite Co Ltd] *

Mater-Bi®. Chemically modified corn starch alloys; thermoplastic of vegetable origin with synthetic substances; biodegradable plastic for pkg. films, bags, 6-pack yokes, disposable diapers, hospital and sanitary products. [Novamont N. Am.]

Matexil. Textile auxiliary chemicals. [ICI Chem. & Polymers Ltd]

Mathesius Metal. A proprietary trade name for an alloy of lead with calcium and strontium in small amounts. §

matico-camphor. A camphor from the Peruvian matico.

Matikus. CAS 56073-10-0; Rodenticide containing brodifacoum. [ICI Chem. & Polymers Ltd]

Matrikerb. Mixture of clopyralid and propyzamide; a soil and leaf herbicide for winter oilseed rape. [Pan Britannica Industries Ltd]

Matrikerb. Clopyralid + propyzamide; a post-emergence herbicide for use on winter oilseed rape. [Rohm & Haas UK]

Matrimid® 5292 System. Bismaleimide resin with O,O′-diallyl bisphenol A hardener; used for advanced composites, high temp. adhesives, laminating, casting, filament winding. [Ciba-Geigy] *

Matrix®. For the agriculture industry. [DuPont UK]

Matrix alloy. An alloy of 48% bismuth. 28.5% lead. 14.5% tin, and 9% antimony. It expands on cooling and is used to hold tools in position.

Matromycin. CAS 2751-09-9; A proprietary preparation of troleandomycin; an antibiotic. [Pfizer International]

Matromycin-T. CAS 2751-09-9; A proprietary preparation of troleandomycin; an antibiotic. [Pfizer International] †

Matromycin Tao. A proprietary preparation of oleandomycin phosphate; an antibiotic. [Pfizer International] †

matteucinol. 6, 8-dimethyl-5, 7-dihydroxy-4′-methoxy-flavanone.

Mattheylec. Conductive adhesives and coatings; metallising preparations for the electrical and electronic industries. [Johnson Matthey plc] †

Mattina®. Matte pigments displaying intense color with subtle pearlescent glow for soft luster or matte effects; recommended for pressed powd. makeups. [Mearl]

Matt salt. Acid ammonium fluoride, $NH_4F \cdot HF$.

Matulane. CAS 366-70-1; Procarbazine hydrochloride; antineoplastic. [Hoffmann-LaRoche Inc] *

maturex. A purified alfa-acetolactate decarboxylase (ALDC) produced by a strain of *Bacillus subtilis*; added at the beginning of an alcohol fermentation (e.g. for beer), it prevents the formation of diacetyl by catalyzing the decarboxylation of alfa-acetolactate to acetoin.

maucherite. A mineral, Ni_3As_2.

mawele. A millet of East Africa.

Maxahibit TT-50. CAS 64665-57-2; Sodium tolyltriazole; corrosion inhibitor for copper, brass, nonferrous metals in water treatment, cooling waters, engine coolants, cleaners, metalworking fluids. [Climax Fluids Additives]

Maxair™ Inhaler. Pirbuterol acetate; inhalation aerosol for asthma. [3M Pharmaceuticals.

Maxepa®. Fish oil concentrate; used for triglyceride lowering. [Seven Seas Ltd]

Maxhete. A steel containing nickel, chromium, tungsten, copper, and silicon.

Maxibolin. CAS 965-90-2; Ethylestrenol; anabolic. [Organon Inc] *

maxi braun®. Self-tanning product. [Bayer AG]

Maxicam. CAS 34552-84-6; Isoxicam; anti-inflammatory. [Parke-Davis] *

Maxicrop Moss Killer & Conditioner. CAS 10028-22-5; Ferrous sulfate; used for moss control in turf. [Maxicrop International Ltd]

---

‡ Trade name and manufacturer not verified   § Trade name without identified manufacturer

Maxidex. A proprietary preparation of dexamethasone, benzalkonium chloride, and phenylmercuric nitrate; anti-inflammatory eye drops. [Alcon Universal] *

Maxidex Ointment. CAS 2392-39-4; Dexamethasone sodium phosphate; glucocorticoid. [Alcon Laboratories Inc]

Maxiflor® . CAS 33564-31-7; Diflorasone diacetate; cortocosteroid, anti-inflammatory. [Allergan, Inc]

Maxigard. A multicomponent nitrite borate cooling water treatment; diesel engine water treatment used to prevent corrosion and mineral deposits as well as corrosion due to cavitation in medium and high speed diesel engines. [Ashland Chemical Company] *

Maxilon. Modified basic dyes. [Ciba plc] *

Maxilvry Steel. A proprietary high nickel-chromium steel containing copper; stated to be corrosion resisting, and particularly resistant to attack by cider. §

Maxim. CAS 83601-81-4; A liquid formulation containing 500 g carbendazim per liter as a suspension concentrate; systemic fungicide. [Farmers Crop Chemicals Ltd]

MaxiMate. CAS 83601-81-4, 12427-38-2; A suspension concentrate containing 62 g carbendazim and 400 g maneb per liter; systemic fungicide for cereals. [Farmers Crop Chemicals Ltd] †

Maximite. An explosive. It is similar to cordite.

Maxitrol. A proprietary preparation of dexamethasone, neomycin, polymyxin B; used as eye-drops. [Alcon Universal] *

maxium metal. Castings of magnesium metal.

Maxolon. CAS 54143-57-6; A proprietary preparation of metoclopramide hydrochloride; an anti-emetic. [SmithKline Beecham] *

Maxon. Polyglyconate; surgical aid. [Davis & Geck] *

Mayari iron. An iron made from Cuban ores. Small amounts of vanadium and titanium are present which give strength to the metal obtained from these ores.

Mayari steel. A Cuban low nickel-chromium steel.

Mayclene. Selective herbicide. [Engelhard Technologies Ltd]

Mayco Base 1351. Sulfurized lard oil; heavy-duty EP lubricant additive for cutting oils, heavy-duty gear oils, way lubricants for machine tools; for use with highly paraffinic and re-refined oils. [Mayco Oil & Chem.

Mayer's albumen. A fixing agent used in microscopy. It consists of 50 cc white of egg, 50 cc glycerin, and 1 g sodium salicylate.

Mayer's solution. Mercuric iodide dissolved in aqueous potassium iodide. An alkaloidal reagent.

Mayphos 45. Org. phosphate ester, free acid; EP lubricant additive, metal wetting agent, surfactant; noncorrosive to ferrous and nonferrous metals; prevents water-induced plugging of diatomaceous earth filter. [Mayco Oil & Chem.

Maypon. Salts of collagen polypeptide fatty acid condensate; used as hair care additives. [Harcros Australia] *

Maypon 4C. CAS 68920-65-0; Potassium coco-hydrolyzed collagen; detergent used in personal care products, general purpose cleansers. [Inolex]

Maypon 4CT. CAS 68952-16-9; TEA-coco-hydrolyzed collagen; visc. builder, foam modifier. [Inolex]

Maypon UD. CAS 68951-92-8; Potassium undecylenoyl hydrolyzed collagen; detergent used in personal care products, antifungal properties. [Inolex]

maytee. Fenugreek seeds.

May-Tein C. CAS 68920-65-0; Potassium cocoyl hydrolyzed collagen; mild surfactant, conditioner, softener, moisturizer, lubricant, antistat, anti-irritant for hair relaxers, shampoos, conditioners, liq. soaps, depilatories; biodeg. [Maybrook]

May-Tein CT. CAS 68952-16-9; TEA-cocoyl hydrolyzed collagen; mild surfactant, conditioner, anti-irritant for shampoos, conditioners, baby products, mousses, makeup removers, shave creams, liq. soaps; biodeg. [Maybrook]

May-Tein KK. Potassium cocoyl hydrolyzed keratin; mild biodeg. surfactant providing protection and gentle cleaning to skin and hair products, hand, body and face soaps, bath products; substantive, foaming protein, softener, conditioner, antistat; anti-irritant in depilatories and hair relaxers. [Maybrook]

May-Tein KT. Sodium cocoyl hydrolyzed keratin; mild biodeg. surfactant providing gentle cleansing to skin and hair; source of cystine for damaged hair; substantive, foaming protein, lubricant, antistat; anti-irritant for harsh ingreds. [Maybrook]

May-Tein R. Potassium cocoyl rice protein; biodeg. protein surfactant for hair and skin care products; provides gentle cleaning, conditioning, foaming, substantivity, lubricity, antistatic props.; for shampoos, hair relaxers, makeup removers, creams, bath products. [Maybrook]

Mazak. Die casting alloy. [Pasminco Europe Ltd/ISC Alloys Div]

Mazam. A dextrin of high molecular weight. §

Mazamide® 68. Cocamide DEA (2:1); biodeg. thickener, emulsifier, foam builder; for hard surface cleaners, dishwash, shampoos, metalworking fluids, waterless hand cleaners, automotive specialties. [PPG/Specialty Chem.]

Mazamide® 1214. CAS 120-40-1; EINECS 204-393-1; Lauramide DEA (2:1); biodeg. thickener, emulsifier, foam builder; for hard surface cleaners, dishwash, shampoos, metalworking fluids, waterless hand cleaners, automotive specialties. [PPG/Specialty Chem.]

Mazamide® C-2. PEG-3 cocamide MEA; biodeg. foam builder/stabilizer for cosmetic and pharmaceutical shampoos. [PPG/Specialty Chem.]

Mazamide® CFAM. CAS 68140-00-1; Cocamide MEA (1:1); biodeg. foam stabilizer, visc. builder for cosmetic and pharmaceutical shampoos. [PPG/Specialty Chem.]

Mazamide® L-5. PEG-6 lauramide DEA; emulsifier, lubricant, rust inhibitor, buffing compound, detergent, foam builder, stabilizer for cosmetic and pharmaceutical creams, lotions, bath oils, shampoos; biodeg. [PPG/Specialty Chem.]

Mazamide® L-298. CAS 120-40-1; EINECS 204-393-1; 2:1 Lauramide DEA; foam builder and stabilizer, emulsifier, dispersant, visc. builder, solubilizer for hard surface cleaners, dishwashing, shampoos, metalworking fluids, automotive specialties, fiber and hair conditioners, dry cleaning, agric. sprays, leather/fur preparations, emulsifiable waxes, rust inhibitors, polishes, paint removers, rug shampoos, fuel oil additives, textile detergents; biodeg. [PPG/Specialty Chem.]

Mazamide® LLD. Linoleamide DEA (2:1); biodeg. thickener, emulsifier, foam builder; for hard surface cleaners, dishwash, shampoos, metalworking fluids, waterless hand cleaners, automotive specialties. [PPG/Specialty Chem.]

---

* Trade name not verified as available

† Trade name verified as obsolete

Mazamide® O 20. CAS 93-83-4; EINECS 202-281-7; 2:1 Oleamide DEA; biodeg. thickener, emulsifier, foam builder, corrosion inhibitor, dispersant; for hard surface cleaners, dishwash, shampoos, metalworking fluids, waterless hand cleaners, automotive specialties. [PPG/Specialty Chem.]

Mazamide® SMEA. CAS 111-57-9; EINECS 203-883-2; Stearamide MEA (1:1); biodeg. emulsifier, pearlescent for syndet soap bars. [PPG/Specialty Chem.]

Mazamide® SS 20. 2:1 Linoleamide DEA; biodeg. thickener, emulsifier, foam builder, corrosion inhibitor; for hard surface cleaners, dishwash, shampoos, metalworking fluids, waterless hand cleaners, automotive specialties. [PPG/Specialty Chem.]

Mazanor. CAS 22232-71-9; Mazindol; anorexic. [Wyeth Laboratories]

Mazawax® 163R. Cetearyl alcohol and polysorbate 60; emulsifier for pharmaceutical and cosmetic applications; base; emolliency and thickening properties. [PPG/Specialty Chem.]

Mazawet® 36. Surfactant; low foaming wetting agent, detergent, and dispersant; for machine dishwashing rinse additives, penetrant for textile finishing and scouring operations. [PPG/Specialty Chem.]

Mazawet® DOSS 70. CAS 577-11-7; EINECS 209-406-4; Dioctyl sodium sulfosuccinate; fast wetting surfactant, emulsifier, dispersant for drycleaning, paper and textile processing, control agent in coal dedusting, antifog for glass cleaners. [PPG/Specialty Chem.]

Mazclean EP. Proprietary emulsifier; microemulsifier package for use in preparing terpene-based cleaning formulations. [PPG/Specialty Chem.]

Mazeen® 173. CAS 102-60-3; EINECS 203-041-4; Tetrahydroxypropyl ethylenediamine; insecticide and herbicide emulsifier, antistat and rewetting agent, grease additive, textile lubricant; emulsifier for lubricants, inks, and cosmetics; chelant in electroless deposition formulations for electronics industry; crosslinker for rigid polyurethane. [PPG/Specialty Chem.]

Mazeen® C-2. CAS 61791-14-8; PEG-2 cocamine; emulsifier, rewetting agent, lubricant, coupler used in insecticides and herbicides, grease additives, textile lubricants, water-based inks, cosmetics; plastics antistat; visc. modifier and rust inhibitor in acid media for metalworking. [PPG/Specialty Chem.]

Mazeen® DAPI. CAS 67799-04-6; EINECS 267-101-1; Isostearylpropyl dimethylamine. [PPG/Specialty Chem.]

Mazeen® DAPL. CAS 68140-01-2; EINECS 268-771-8; Cocamidopropyl dimethylamine. [PPG/Specialty Chem.]

Mazeen® S-2. CAS 61791-24-0; PEG-2 soyamine; emulsifier, rewetting agent, lubricant, coupler used in insecticides and herbicides, grease additives, textile lubricants, water-based inks, cosmetics; plastics antistat. [PPG/Specialty Chem.]

Mazeen® S-13. CAS 7651-02-7; EINECS 231-609-1; Stearamidopropyl dimethylamine. [PPG/Specialty Chem.]

Mazeen® S-15. CAS 61791-24-0; PEG-15 soyamine; emulsifier, rewetting agent, lubricant, coupler used in insecticides and herbicides, grease additives, textile lubricants, water-based inks, cosmetics; plastics antistat. [PPG/Specialty Chem.]

Mazeen® SHCFA. CAS 68140-01-2; EINECS 268-771-8; Cocamidopropyl dimethylamine. [PPG/Specialty Chem.]

Mazeen® T-2. CAS 61791-44-4; PEG-2 tallow amine; emulsifier, rewetting agent, lubricant, coupler used in insecticides and herbicides, grease additives, textile lubricants, water-based inks, cosmetics, metalworking; plastics antistat; visc. modifier, rust inhibitor in acid media. [PPG/Specialty Chem.]

Mazide 25. CAS 123-33-1; Maleic hydrazide, 250 g/liter; a tree growth regulator to control shoots on the trunk and suckers around the base of street trees. [Synchemicals Ltd]

Mazide Selective. Soluble concentrate of 6 g dicamba, 200 g maleic hydrazine and 75 g MCPA per liter. plant growth regulator for grass verges. [Synchemicals Ltd]

Mazin®. CAS 12427-38-2, 1314-13-2; Wettable powder containing maneb 80% w/w and zinc oxide; protective fungicide against potato blight. [Universal Crop Protection Ltd]

Mazol® 80 MG K. CAS 51158-08-8; PEG-20 glyceryl stearate; emulsifier, dough conditioner for cakes, icings, whipped toppings. [PPG/Specialty Chem.]

Mazol® 165C. Glyceryl stearate and PEG-100 stearate; emulsifier blend for cosmetic and pharmaceutical o/w emulsions; acid-stable. [PPG/Specialty Chem.]

Mazol® 300 K. CAS 111-03-5; Glyceryl oleate, kosher; GRAS dispersant for oil or solv. systems; antifoam for sugar and protein processing; coemulsifier with T-Maz 60 or 80. [PPG/Specialty Chem.]

Mazol® 1400. CAS 65381-09-1; Caprylic/capric triglyceride; carrier for flavors, fragrances, vitamins, antibiotics, pigmented cosmetics, medicinals. [PPG/Specialty Chem.]

Mazol® GMO. CAS 111-03-5; Glyceryl oleate; GRAS dispersant for oil or solv. systems; antifoam for food processing; base for cosmetic creams, lotions, ointments; w/o emulsifier with emolliency, thickening properties; plasticizer; lubricant; antifog for PVC; emulsifier, coupling agent for metalworking applications. [PPG/Specialty Chem.]

Mazol® GMR. CAS 141-08-2; EINECS 205-455-0; Glyceryl ricinoleate; base for cosmetic creams, lotions, ointments; w/o emulsifier with emolliency, thickening properties; plasticizer. [PPG/Specialty Chem.]

Mazol® GMS. Glyceryl stearate; lubricant, emulsifier, plasticizer, and thickener for foods, drugs, and cosmetics. [PPG/Specialty Chem.]

Mazol® PETO. CAS 19321-40-5; Pentaerythritol tetraoleate; emulsifier, coupling agent for metalworking applications. [PPG/Specialty Chem.]

Mazol® PGMS. CAS 1323-39-3; EINECS 215-354-3; Propylene glycol stearate; coemulsifier for edible oil and shortenings, dispersing aid for nondairy creamers. [PPG/Specialty Chem.]

Mazol® PGO-31 K. CAS 9007-48-1; Triglyceryl oleate, kosher; emulsifier for food products, in cosmetics, toiletries, pharmaceuticals, lubricants, mold release compounds, plasticizers; solubilizer for essential oils and flavors. [PPG/Specialty Chem.]

Mazol® PGO-104. CAS 34424-98-1; Decaglyceryl tetraoleate; emulsifier for food products, cosmetics, toiletries, pharmaceuticals, lubricants, mold release compounds, in plasticizers for synthetic fabrics and plastics. [PPG/Specialty Chem.]

Mazon® 18A. Proprietary surfactant; visc. booster for aq.

---

‡ Trade name and manufacturer not verified   § Trade name without identified manufacturer

systems. [PPG/Specialty Chem.]

Mazon® 41. Ammonium salt of alkylphenol ethoxylate; surfactant for cleaning formulations. [PPG/Specialty Chem.]

Mazon® 60T. TEA dodecylbenzene sulfonate; high foaming formulated detergent, emulsifier for shampoos, dishwashing, car wash, textile, hard surface cleaners, metal cleaners and other formulations. [PPG/Specialty Chem.]

Mazon® 85. Modified dodecylbenzene sulfonic acid; self-emulsifying degreaser, detergent, emulsifier for cleaning applications; used with hydrocarbon solvs. [PPG/Specialty Chem.]

Mazon® 1045A. POE sorbitol fatty acid ester; emulsifier for pesticide, herbicide, metalworking, die-cast lubricant formulations, and emulsion polymerization; humectant, emollient. [PPG/Specialty Chem.]

Mazon® RI 4A. Amine-based; ferrous corrosion inhibitor developed to replace sodium nitrite in metalworking fluids. [PPG/Specialty Chem.]

Mazox® CAPA. CAS 68155-09-9; Cocamidopropylamine oxide; surfactant, conditioner, emollient, emulsifier, foam booster, visc. builder, lime soap dispersant for cosmetics, toiletries, household and industrial uses. [PPG/Specialty Chem.]

Mazox® CDA. CAS 7128-91-81; EINECS 230-429-0; Palmitamine oxide; surfactant, conditioner, emollient, emulsifier, foam booster, visc. builder for cosmetics, toiletries, household and industrial uses; textile softener. [PPG/Specialty Chem.]

Mazox® KCAO. Potassium dihydroxyethyl cocamine oxide phosphate; foam booster/stabilizer, detergent, emollient, lime soap dispersant for dishwash, heavy-duty detergents, caustic sol'ns.; stable in caustic soda; compat. with most surfactants, builders, many solvs. [PPG/Specialty Chem.]

Mazox® LDA. Lauramine oxide; surfactant, conditioner, emollient, emulsifier, foam booster, visc. builder for cosmetics, toiletries, household and industrial uses. [PPG/Specialty Chem.]

Mazox® MDA. CAS 3332-27-2; EINECS 222-059-3; Myristamine oxide; surfactant, conditioner, emollient, emulsifier, foam booster, visc. builder for cosmetics, toiletries, household and industrial uses. [PPG/Specialty Chem.]

Mazox® ODA. CAS 14351-50-9; EINECS 238-311-0; Oleamine oxide; surfactant, conditioner, emollient, emulsifier, foam booster, visc. builder for cosmetics, toiletries, household and industrial uses; textile softener. [PPG/Specialty Chem.]

Mazox® SDA. CAS 2571-88-2; EINECS 219-919-5; Stearamine oxide; surfactant, conditioner, emollient, emulsifier, foam booster, visc. builder for cosmetics, toiletries, household and industrial uses; textile softener. [PPG/Specialty Chem.]

Maztreat® 246. Organic defoamer; defoamer for metalworking formulations. [PPG/Specialty Chem.]

Maztreat® BOM. Additive for boiler water treatment. [PPG/Specialty Chem.]

Maztreat® CA Powd. Additive for cooling water treatment. [PPG/Specialty Chem.]

Mazu® 10 P Mod 11. Organic defoamer; defoamer for yeast fermentation. [PPG/Specialty Chem.]

Mazu® 43 C. Organic defoamer; defoamer for enzyme, effluent wastewater applications. [PPG/Specialty Chem.]

Mazu® 142. Organic defoamer; defoamer for whitewater, screen room, bleach plant. [PPG/Specialty Chem.]

Mazu® 252. Organic defoamer; defoamer for brownstock, adhesives, latex paints. [PPG/Specialty Chem.]

Mazu® 5118. Organic defoamer; defoamer for delayed coking. [PPG/Specialty Chem.]

Mazu® DF 100S. Silicone compound; food-grade defoamer for fermentation, vegetable oils. [PPG/Specialty Chem.]

Mazu® DF 200SP. CAS 8050-81-5; Simethicone; food grade defoamer for fermentation, vegetable oils; also is formulated to meet the specific needs of pharmaceutical industry. [PPG/Specialty Chem.]

Mazu® DF 205SX. Silicone emulsion; industrial defoamer formulated with low visc. for ease of handling and pumping; for leather finishing, metalworking, carpet cleaning, waste treatment. [PPG/Specialty Chem.]

Mazu® DF 243. Silicone emulsion; emulsifier; defoamer for adhesive, water-based paints; soap mfg., antifreeze, hot aq. systems, insecticides, textile, paper, petrol., vinyl latex binders and emulsions, waste treatment; formulated for rigid dilution requirements. [PPG/Specialty Chem.]

M.B. General chemicals and pharmaceuticals. [May & Baker Ltd] *

MB 450. CAS 68586-07-2; Monoethanolamine borate; corrosion inhibitor. [Werner G. Smith]

M & B 693. CAS 144-83-2; Sulfapyridine. [May & Baker Ltd]*

M.B.A. N,N´-Methylenebisacrylamide. [Cyanamid BV] *

MBS. A terpolymer of methylmethacrylate, butadiene and styrene. Its properties include rigidity, hardness, high impact strength, heat resistance and good clarity.

MBT. CAS 149-30-4; 2-Mercaptobenzothiazole; very act., nondiscoloring org. accelerator. [Akrochem]

MBT®. CAS 149-30-4; 2-Mercaptobenzothiazole; intermediate; medium temperature general-purpose accelerator for natural and synthetic rubbers; very active above 240 F (116 C); moderately low activation temperature [Uniroyal]

MBT. See 2-mercaptobenzothiazole.

MBTS. CAS 120-78-5; Benzothiazyl disulfide; nonstaining org. accelerator; very act. at temps. above 280 F. [Akrochem]

M B T S. CAS 120-78-5; Benzothiazyl disulfide; a general-purpose accelerator similar to MBT but with a higher activation temperature for greater processing safety; very active above 287 F (142 C) used in both natural and synthetic rubbers. [Uniroyal] *

MC®. CAS 25038-54-4; Type 6 cast nylon; used for seals, wear applications. [Polymer Corp]

MC580. Silicone resin; semirigid thermosetting transfer molding compound with high temp. and thermal shock resist.; for encapsulating high power electronic devices. [GE Silicones]

MC 2508. General-purpose mold cleaner for urethane and epoxy residues. [George Mann]

MCA. See monochloroacetic acid.

McGill Metal. A proprietary trade name for a group of copper-aluminum-iron alloys, one of which contains 89% copper, 9% aluminum, and 2% iron. §

M-Clean D. CAS 75-09-2; Methylene chloride vapor. cleaner and degreaser. [Occidental Chemical Corp]

McLube 1700. Fluorocarbon; solv.-based release coating for epoxies and thermosets, and natural, nitrile, SBR and silicone rubber. [McLube]

---

* Trade name not verified as available

† Trade name verified as obsolete

McLube 1777. CAS 9002-84-0; TFE polymer disp.; release coating for hot molds used to form rubber and plastic parts; antistick coating for cure of hose and other extrusions and on tools and process equip. [McLube]

McLube 1829. Fluorochemical/resin mixt.; release coating for rubber molding processes; antitack coating for uncured rubber. [McLube]

McNamee Clay®. CAS 1332-58-7; Kaolin clay; low cost reinforcer and inert filler for paint, paper, rubber, ceramics, plastics and specialilties. [R T Vanderbilt Co Inc]

MCP. See calcium phosphate (monobasic).

MCPB. 3-phenoxybutyric acid; used as a systemic fungicide active against chocolate spot disease in broad beans.

MD 50. CAS 737-31-5; Diatrizoate sodium; diagnostic aid. [Mallinckrodt Inc] *

MD 60. Diatrizoate meglumine; diagnostic aid. [Mallinckrodt Inc] *

MDI. (methylene di-p-phenylene isocyanato; methylene bisphenyl isocyanate; diphenylmethane-4,'4-diisocyanate) $CH_2(C_6H_4NCO)_2$; CAS 101-68-8; Prep. of polyurethane resin; bonding rubber to rayon and nylon. [Allchem Industries; BASF; ICI Polyurethanes; Miles]

MEA. See ethanolamine.

Mearl Film. Irridescent film; for packaging, laminating, visual effect. [Cornelius Chemical Co Ltd] *

Mearlin. CAS 13463-67-7, 12001-26-2; Titanium dioxide coated mica pearl pigments; for plastics, paints, automotive finishes. [Cornelius Chemical Co Ltd] *

Mearlin®. CAS 13463-67-7, 12001-26-2; Titanium dioxide/mica; for pearlescent, metallic, antique, and iridescent effects in powd. coating systems. [Mearl] *

Mearlite® GBU. CAS 7787-59-9; Bismuth oxychloride; pearl pigments and pastes used for high opacitiy and smooth luster providing frosted effects in makeups, lipsticks, blushers, etc. [Mearl]

Mearlmaid® AA. Water, guanine, isopropyl alcohol, methylcellulose; pearl colors for nail enamels, makeup, lotions, hair care products, natural cosmetics. [Mearl]

Mearlmica® SVA. CAS 12001-26-2; Org. surface-treated mica; imparts smooth texture, soft feel, soft luster or matte effect to cosmetics (pressed powd. makeup, eye shadows). [Mearl]

Measac. Ethanolamine sesquisulfite. [Albright & Wilson Ltd]*

Measurin. CAS 50-78-2; Aspirin; analgesic; antipyretic; antirheumatic. [Sterling Drug Inc] *

meat-sugar. CAS 87-89-8; Inositol, $C_6H_6(OH)_6$.

Mebadin. A proprietary trade name for dehydroemetine. §

Mebaral. CAS 115-38-8; Mephobarbital; anticonvulsant; sedative. [Sterling Drug Inc] *

Mebatreat. CAS 31431-39-7; A proprietary preparation containing mebandazole; Veterinary antihelmintic (cats and dogs). [Janssen Pharmaceutical Ltd] *

Mebenvet. CAS 31431-39-7; A proprietary preparation containing mebendazole; veterinary antihelmintic (game birds, poultry). [Janssen Pharmaceutical Ltd] *

Mebryl. A proprietary preparation of embramine hydrochloride; an anti-allergic. [SmithKline Beecham] *

Mecadox. CAS 6804-07-5; Carbadox; an antibacterial used in swine. [Pfizer International]

Mecca balsam. An oleoresin obtained from *Balsamodendron gileadense*, of Arabia. It is known in India as Balsan-katel, and is imported there under the name of Duhnul-balasan. Balm of Gilead is another name for

Mecca balsam.

Meccarb. Silicone carbide. [Winchem Ltd]

Meclan Cream. CAS 73816-42-9; Meclocycline sulfosalicylate; antibacterial. [Ortho Pharmaceutical Corp] *

Meclomen. CAS 6385-02-0; Meclofenamate sodium; anti-inflammatory. [Parke-Davis]

Meco. A cupro-nickel alloy.

meconium. Opium.

Mecpa. Selective weedkillers. [Murphy Chemical Co Ltd] †

Mecufix. An adhesive for polyester film. [May & Baker Ltd]*

Meculon. Metallized polyester film. [May & Baker Ltd] *

Medac Cream. Acne cream. [Fisons plc, Pharmaceuticals Div] *

medal bronze. An alloy of from 92-97% copper, 1-8% tin, and 0-2% zinc.

medal metal. An alloy of 84% copper and 16% zinc.

Medang Losoh oil. An oil from the wood of *Cinnamonum parthenoxylon*. It consists mainly of safrole.

MedGel. CAS 637-12-7; Aluminum stearate; for nonaqueous coating compositions. [Huls Am.] †

Medialan KA. CAS 61791-59-1; EINECS 263-193-2; Sodium cocoyl sarcosinate; detergent used in cream shampoos, foaming agent, mild toiletries, hair shampoos. [Hoechst Celanese/Colorants & Surf.

Medialan KF. TEA-palm kernel sarcosinate; detergent used in preparation of clear, liq. shampoos and body lotions. [Hoechst Celanese/Colorants & Surf.

Medialan LD. CAS 137-16-6; EINECS 205-281-5; Sodium lauroyl sarcosinate; surfactant for dental care products. [Hoechst Celanese/Colorants & Surf.; Hoechst AG]

Medicaire. For relief of nasal congestion. [The Wellcome Foundation Ltd]

Medihaler-Epi™. CAS 51-42-3; Pressurized aerosol containing a suspension of epinephrine bitartrate; delivers 300 metered doses (0.28 mg per dose); for asthma relief. [3M Pharmaceuticals]

Medihaler Ergotamine. CAS 379-79-3; Ergotamine tartrate; analgesic. [3M Pharmaceuticals] †

Medihaler-Iso™. CAS 6700-39-6; Pressurized aerosol containing a suspension of isoproterenol sulfate; delivers 300 metered doses (0.08 mg per dose); adrenergic; for asthma relief. [3M Pharmaceuticals]

Medihaler-iso Forte. Pressurized aerosol containing a suspension of isoprenaline sulfate BP 20 mg/ml; delivers 400 metered doses (0.4 mg per dose). [3M Pharmaceuticals] †

Medihaler-Tetracaine. Tetracaine; anesthetic. [3M Pharmaceuticals] †

Mediker. Shampoo, to wash and condition hair. [Richardson-Vicks Inc] *

Meditar. A proprietary preparation of coal tar; a dermatological product. [Brocades Pharma] †

Medium 7. A preparation consisting of dry sweet whey, sodium caseinate, disodium phosphate and soluble growth factors derived from *Saccharomyces cerevisiae*; used to prepare culture medium for the growth of thermophilic lactic acid bacteria. [Pfizer International] †

Medium 10. A preparation consisting of dry sweet whey, nonfat dry milk, disodium phosphate and soluble growth factors derived from *Saccharomyces cerevisiae*; used to prepare culture medium for the growth of thermophilic lactic acid bacteria. [Pfizer International] †

Medium VS. A preparation consisting of dry sweet whey, sodium caseinate, sodium citrate and soluble growth

---

‡ Trade name and manufacturer not verified   § Trade name without identified manufacturer

factors derived from *Saccharomyces cerevisiae*; used to prepare culture medium for the growth of thermophilic lactic acid bacteria. [Pfizer International] †

Medley®. For the agriculture industry. [DuPont UK]

Medo. EINECS 215-293-2; Oil-based soap containing cresylic acid; pruning compound and canker cure for garden trees. [Vitax Ltd]

medol. A combination of cresols and iodine; recommended as an antiparasitic in skin troubles.

Medolit. A phenol-formaldehyde condensation product. It is a resinous material, and is recommended as a shellac varnish substitute.

Medomet. CAS 555-30-6; A proprietary preparation of methyldopa; an antihypertensive. [DDSA Pharmaceuticals Ltd] *

Medomin. A proprietary preparation of heptabarbitone; a hypnotic. [Ciba plc] *

Medro - Cordex. A proprietary preparation of methylprednisolone and aspirin; an antirheumatic agent. [Upjohn Ltd] *

Medrol ADT Pak, also Medrol Dosepak. CAS 83-43-2; Methylprednisolone; glucocorticoid. [Upjohn]

Medrol Enpak. CAS 53-36-1; Methylprednisolone acetate; glucocorticoid. [Upjohn] *

Medrol Stabisol. CAS 5015-36-1; Methylprednisolone sodium phosphate; glucocorticoid. [Upjohn] *

Medrone. CAS 83-43-2; A proprietary preparation of methylprednisolone. [Upjohn Ltd] *

Medrone Medules. CAS 83-43-2; A proprietary preparation of methylprednisolone. [Upjohn Ltd] *

Medrone Veriderm. CAS 53-36-1; A proprietary preparation of methylprednisolone acetate for dermatological use. [Upjohn Ltd] *

Medusa. A waterproofing compound mixed with cement.

Meehanite. A proprietary trade name for a close-grained, pearlitic, sorbitic iron with properties superior to cast iron; has good casting and machining properties. §

Meena Harma. A name given to an opaque variety of bdellium gum-resin.

MEF®-LD. Moldable polyethylene foam. [Asahi Chem. Industry]

Mefarol®. Disinfectant based on a quaternary ammonium compound; veterinary medicine. [Bayer AG]

Mefoxin. CAS 35607-66-0; Cefoxitin; a beta-lactam antibiotic effective against a wide range of Gram-positive and Gram-negative bacteria. [Merck & Co Inc]

Mefranal. 3-Methyl-5-phenyl-1-pentanal. [Quest Int'l. UK Ltd]

Mefrasol. 3-Methyl-5-phenyl-1-pentanol. [Quest Int'l. UK Ltd]

Megace. CAS 595-33-5; Megestrol acetate; antineoplastic. [Mead Johnson & Co] *

Megaclor. CAS 1181-54-0; A proprietary preparation containing clomocycline; an antibiotic. [Pharmax Ltd] *

Meganite. An explosive containing nitroglycerin, and dinitrocellulose, to which has been added a nitro mixture to ensure complete combustion.

Meganox Plus. An emulsifiable concentrate containing 100 g aminotriazole, 250 g atrazine and 100 g MCPA per liter; used for total weed control in non crop areas. [Agri-Technics Ltd]

Megaperm 4510. A magnetic alloy containing 45% nickel, 45% iron, and 10% manganese.

Megaperm 6510. A magnetic alloy containing 65% nickel, 25% iron, and 10% manganese.

Megaphen®. CAS 69-09-0; Chlorpromazine hydrochloride; pharmaceutical preparation; classical broad-spectrum neuroleptic. [Bayer AG]

Megapoly (formerly Heaveaplus MG). Graft polymer of natural rubber latex with methyl methacrylate; adhesive and bonding agent, in water-based adhesives, latex gloves, pressure-sensitive tapes, surgical tapes, footwear, flooring, composite hoses. [H.A. Astlett]

Megapren C 150. A proprietary chloroprene rubber used in the manufacture of moldings and extrusions resistant to sunlight, weathering and ozone. [Rhein-Chemie Rheinau] *

Megapren Si 10, 20, 30, and 60. A proprietary range of materials based on silicone rubber. [Rhein-Chemie Rheinau] *

Megapren U225. A proprietary polyurethane rubber. [Rhein-Chemie Rheinau] *

Megasil. Silicone dielectric materials. [Midland Silicones]

Megilp. A mixture of linseed oil and mastic varnish; used in artist's oil paints.

Megimide. A proprietary preparation of bemegride. [Nicholas Laboratories Ltd] *

Meglumine. CAS 6284-40-8; N-Methylglucamine.

Megomit. A mica product used as an electrical insulator.

Megum. Bonding agent for the bonding of rubber to metals and other materials under vulcanizing conditions. [Chemetall GmbH] *

MEK. *See* methyl ethyl ketone.

Mekad. Antiskinning agent for paints. [Octel Chemicals Ltd]†

Meketone. Methyl ethyl ketone; solvent in paint and lacquer thinners, natural and synthetic resins, gums and rubbers, printing inks, PVC cloth manufacture, cleaning agent for metal surfaces, adhesives and cements; refining and dewaxing of mineral and lubricating oils. [Sasolchem] *

Mekon® White. CAS 63231-60-7; Microcryst. wax; release agent; used in hot-melt coatings and adhesives, paper coatings, printing inks, plastic modification (as lubricant and processing aid), lacquers, paints, varnishes, as binder in ceramics, for potting/filling in elec./electronic components, in investment casting, rubber and elastomers (plasticizer, antisunchecking, antiozonant), as emulsion wax size in papermaking, as fabric softener ingred., in emulsion and latex coatings, and in cosmetic hand creams and lipsticks. [Petrolite]

Mekor®. Volatile oxygen scavenger/metal passivator, corrosion inhibitor for steam generating systems. [Drew Ind. Div.

Mekor® 70. Corrosion inhibitor which chemically removes oxygen from feedwater, boiler water and condensate in steam generation systems; passivates iron and copper surfaces. [Drew Ind. Div.]

MEK peroxide. *See* methyl ethyl ketone peroxide.

Mekure T1. Copper chrome arsenate; wood preservative. [Mechema Chemicals Ltd] *

Mekure T2. Copper chrome arsenate; wood preservative. [Mechema Chemicals Ltd] *

MEL 80-P. CAS 9003-08-1; Melamine-formaldehyde resin; produces excellent wash fastness on dacron and chintz; has embossing and dimensional stability. [CNC Int'l. L.P.]

Melacos. Bottle washing detergents. [ICI Chem. & Polymers Ltd] †

---

* Trade name not verified as available   † Trade name verified as obsolete

Melafix DM. A proprietary trade name for a melamine formaldehyde resin; used as a shrink-resistant in wool. [Ciba plc] *

Melalith. A steatite-porcelain product.

melamine/formaldehyde resin. (melamine resin; 1,3,5-triazine,2,4,6-triamine, polymer with formaldehyde) Amino resin; reaction product of melamine and formaldehyde; $(CH_3H_6N_6CH_2O)_x$; CAS 9003-08-1; thermosetting resin. [Akzo; Am. Cyanamid; Astro Industries; Bakelite GmbH; BASF; Monsanto; Rhone-Poulenc Water Treatment; Sybron]

melamine resin. See melamine/formaldehyde resin.

Melamit 200. A proprietary melamine formaldehyde cellulose molding powder. [Bush Beach Ltd] *

Melampyrite. Dulcitol, $CH_2(OH)(CH \cdot OH)_4 \cdot CH_2OH$.

Melanate. CAS 22204-88-2; Tramadol hydrochloride; analgesic. [Upjohn] *

Melanex. A proprietary trade name for metahexamide. §

melaniline. CAS 102-06-7; Diphenyl-guanidine, NH: C(NH $\cdot C_6H_5)_2$; used as a basic accelerator for rubber.

melanoid. A colloidal bituminous paint material; used as a preservative paint for metal which is in contact with corrosive gases.

Melatix. A proprietary melamine formaldehyde molding compound. [Nisshin Boseki] *

Melax. Proprietary, polymeric based conditioning and sealing agents for fiberglass molds and other porous surfaces; used for fiberglass tools. [Axel Plastics Research Laboratories Inc]

Melclif® . Zirconium, magnesium, rare-earth and other chemicals, powders and alloys; for use in industry, including metalworking. [Magnesium Elektron Ltd]

melco. A synthetic milk made from the peanut.

Melcril 4079. A trademark for tetrahydrofurfuryl acrylate. [Danbert Chemical Co] ‡

Melcril 4083. A trademark for an ethylene glycol acrylate phthalate. [Danbert Chemical Co] ‡

Melcril 4085. A trademark for a benzyl acrylate. [Danbert Chemical Co] ‡

Melcril 4087. CAS 48165-04-6; A trademark for a phenoxy ethyl acrylate. [Danbert Chemical Co] ‡

Melcril 5919. A trademark for a melamine acrylate. [Danbert Chemical Co] ‡

Meld. Broad spectrum systemic fungicide for use in cereals. [BASF plc]

Meldin® 3000A. Polyimide resin; melt processable resin featuring wear resist.; for bearings, thrust washers, piston rings. [Furon]

Meldin® 3000D. Polyimide resin; melt processable resin featuring wear resist., nonhardened matting surface; for bearings, thrust washers. [Furon]

Meldin® 3000F. Polyimide resin; melt processable resin featuring high modulus, high dimensional stability; for picker finger copier, elec./electronic parts. [Furon]

Meldin® 3000G. Polyimide resin; melt processable resin featuring high modulus and strength; for gears, bearing retainers. [Furon]

Meldin® 3000H. Polyimide resin; melt processable resin featuring high modulus and elec. conductivity; for gears, piston rings. [Furon]

Melhi N, NS and NLM. Acidic, medium-hard, dark amber-colored resins obtained as coproduct during processing of rosin to modified forms; used in low-cost hot-melt adhesives and construction mastic adhesives. [Her-

cules] *

Melibiase. An enzyme which splits melibiose into glucose and galactose.

Meligrin. A condensation product of dimethyl-hydroxy-quinoline and methyl- phenylacetamide.

Melilot. p-Methylacetophenone, $C_6H_4(CH_3) \cdot CO \cdot CH_3$. It imparts the honey-like fragrance of sweet clover, and is used for perfuming soap.

Melimax. For acetonaemia in cattle. [The Wellcome Foundation Ltd] *

Melinar. PET polymer. [ICI Chem. & Polymers Ltd]

Melinex® . Polyethylene terephthalate film; an extremely tough material used for cable lapping, motor insulation and capacitors, valve diaphragms, and conveyor belting. [ICI Chem. & Polymers Ltd]

Melinex® 393. Polyester film; super clear, adhesion pre-treated one side for printing; good handelability; for metallizing, facestock, overlaminate applications. [ICI Films]

Melinex® 505. Polyester film; super clear, adhesion pre-treated two sides for printing; good handleability; for facestock and overlaminate labels. [ICI Films]

Melinex® 994. Polyester film; opaque white, high gloss, adhesion pretreated one side for printing; excellent handleability; for facestock labels. [ICI Films]

Melinite. The French name for Lyddite (qv). It consists of 70% picric acid and 30% collodion cotton.

melinose. A mineral. It is wulfenite.

Meliodent® . Heat-curing denture resin based on methyl methacrylate; used in dentistry. [Bayer AG]

melioform. A ruby-red liquid containing 25% formaldehyde and 15% aluminum acetate; used as a disinfectant.

Melioran. Sodium alkyl sulfate; used for textiles, flotation. [Atochem UK/Ceca]

Melioran F6. Alcohol sulfate in paste form; powerful detergent with wetting-out, dispersing, level dyeing and fiber protection properties used in the processing of wool, cotton, linen, silk, artificial and synthetic fibers. [Rewo Chemicals Ltd] *

Meliose. Fructose containing corn syrup; used for foods, e.g., soft drinks and confectionery. [Roquette (UK) Ltd]

Melit. Melamine resins.

Melite® . Zirconium, magnesium, rare-earth and other chemicals, powders and alloys; for use in industry, including metalworking. [Magnesium Elektron Ltd]

Mellaril. CAS 130-61-0; Thioridazine hydrochloride; antipsychotic; sedative. [Sandoz Pharmaceuticals]

Mellaril-S. CAS 50-52-2; Thioridazine; antipsychotic; sedative. [Sandoz Pharmaceuticals] *

Mellavax. Calf salmonellosis vaccine. [The Wellcome Foundation Ltd] *

Melleril. CAS 50-52-2; A proprietary preparation of thioridazine; a sedative. [Sandoz] *

Mellinese. CAS 94-20-2; A proprietary preparation of chlorpropamide; an oral hypoglycemic agent. [Pfizer International]

Mellite. Stabilizers for PVC. [Harcros UK] *

Mellol. Extra pure 2-phenylethyl alcohol. [Bush Boake Allen Ltd]

Melmac. A proprietary trade name for a melamine-formaldehyde synthetic resin and adhesive. §

Melmag® . Zirconium, magnesium, rare-earth and other chemicals, powders and alloys; for use in industry, including metalworking. [Magnesium Elektron Ltd]

---

‡ Trade name and manufacturer not verified   § Trade name without identified manufacturer

Melmex. Melamine formaldehyde compounds; used for molding. [BIP Chemicals Ltd]

Melnox® . Zirconium, magnesium, rare-earth and other chemicals, powders and alloys; for use in industry, including metalworking. [Magnesium Elektron Ltd]

Melocol. A proprietary trade name for a polyamide-formaldehyde product. §

Meloids. Throat lozenges. [The Boots Co plc] *

Melolam. Melamine/formaldehyde resins. [Ciba plc] *

Melolanel. Melancil and urea melamine resins; used for laminates and surfacing of wood-based panels. [Dynochem UK Ltd]

Melopas. Melamine/formaldehyde resins/molding compounds. [Ciba plc]

Meloprufe® . Zirconium, magnesium, rare-earth and other chemicals, powders and alloys; for use in industry, including metalworking. [Magnesium Elektron Ltd]

Melox® . Zirconium, magnesium, rare-earth and other chemicals, powders and alloys; for use in industry, including metalworking. [Magnesium Elektron Ltd]

Meloxide® . Zirconium, magnesium, rare-earth and other chemicals, powders and alloys; for use in industry, including metalworking. [Magnesium Elektron Ltd]

Melpure® . Zirconium, magnesium, rare-earth and other chemicals, powders and alloys; for use in industry, including metalworking. [Magnesium Elektron Ltd]

Melrasal® . Zirconium, magnesium, rare-earth and other chemicals, powders and alloys; for use in industry, including metalworking. [Magnesium Elektron Ltd]

Melsed. CAS 72-44-6; A proprietary preparation of methaqualone; a hypnotic. [The Boots Co plc]

Melsedin. CAS 340-56-7; A proprietary preparation of methaqualone hydrochloride; a hypnotic. [The Boots Co plc]

Melsprea®. A range of melamine molding compounds; used for tablewear and ashtrays. [AMC SPREA S.p.A.]

Meltan® . Zirconium, magnesium, rare-earth and other chemicals, powders and alloys; for use in industry, including metalworking. [Magnesium Elektron Ltd]

Meltatox®. CAS 31717-87-0; Dodemorph-acetate; for control of powdery mildew in roses and other ornamentals [BASF AG]

Meltron® . Zirconium, magnesium, rare-earth and other chemicals, powders and alloys; for use in industry, including metalworking. [Magnesium Elektron Ltd]

Melurac. A proprietary trade name for a melamine-urea-formaldehyde laminating synthetic resin. §

Melweld® . Zirconium, magnesium, rare-earth and other chemicals, powders and alloys; for use in industry, including metalworking. [Magnesium Elektron Ltd]

Melwhite® . Zirconium, magnesium, rare-earth and other chemicals, powders and alloys; for use in industry, including metalworking. [Magnesium Elektron Ltd]

Membrettes. CAS 57-83-0; Progesterone; progestin. [Wyeth Laboratories] *

Memilite. See Randanite.

Memosil. Addition-cured silicone based transparent bite registration material; used in dentistry. [Bayer AG]

Menest. Oestrogens, esterified; estrogens. [SmithKline Beecham]

Menformom A. CAS 53-16-7; Estrone; estrogen. [Savage Laboratories] †

Meningovax-C. Meningococcal vaccine; for immunization against infection caused by strains of Group C meningo-

cocci. [Merck & Co Inc] †

menispermin. An extract of Canadian moon-seed.

Menopax. A proprietary preparation of ethinylestradiol, carbomal and bromvaletone; used in the treatment of menopausal symptoms. [Nicholas Laboratories Ltd] *

Menopax Forte. A proprietary preparation of ethinylestradiol, methyltestosterone, carbromal, bromvaletin and mephenesin; used in the treatment of menopausal disorders. [Nicholas Laboratories Ltd] *

p-mentha-1,8-diene. See limonene.

menthol. (hexahydrothymol; 5-methyl-2-(1-methylethyl) cyclohexanol; racemic menthol) Diterpene; CH_3-$C_6H_9(C_3H_7)OH$; CAS 89-78-1; Perfumery, cigarettes, liqueurs, flavoring agent, chewing gum, chest rubs, cough drops, nasal inhalers. [Haarmann & Reimer; Janssen Chimica; Penta Mfg.; Quest Int'l.; Robeco]

Meothrin. CAS 39515-41-8; Fenpropathrin; a contact pyrethroid based acaricide. [Shell UK]

Mephaneine. Pure methyl phenylacetate. [Bush Boake Allen Ltd] †

Mephenytoin. CAS 50-12-4; 5-Ethyl-3-methyl-5-phenylhydantoin.

Mephetol. Herbicides. [ICI Chem. & Polymers Ltd]

Mephetol Extra. Soluble concentrate of 20.8 g dicamba, 333 g dichlorprop and 200 g MCPA per liter; used for weed control in cereals and newly sown grass. [BritAg Industries Ltd]

Mephine. CAS 6190-60-9; A proprietary preparation of mephentermine sulfate; used to raise blood pressure. [Wyeth Laboratories] *

Mephosol. A proprietary preparation of mephenesin and salicylamide; an antirheumatic. [Crookes Laboratories]*

Mephyton. Phytonadione; for the treatment of certain coagulation disorders. [Merck & Co Inc]

Mepilin. A proprietary preparation of ethinylestradiol and methyltestosterone. for treatment of senile and menopausal conditions. [British Drug Houses] *

Mepred. CAS 53-36-1; Methylprednisolone acetate; glucocorticoid. [Savage Laboratories] †

Meprobase-200. Tridihexethyl chloride and meprobamate; possibly effective as adjunctive therapy in peptic ulcer and in the irritable bowel syndrome, especially when accompanied by anxiety or tension. [Pharmaceutical Basics Inc] *

Meprobase-400. Tridihexethyl chloride and meprobamate; possibly effective as adjunctive therapy in peptic ulcer and in the irritable bowel syndrome, especially when accompanied by anxiety or tension. [Pharmaceutical Basics Inc] *

Mepron. CAS 63-68-3; Protected methionine. [Degussa]

Meprospan. CAS 57-53-4; Meprobamate; sedative. [Wallace Laboratories]

Meptid Injection. CAS 54340-58-8; 100 mg Meptazinol/1 ml ampoule; used for the treatment of moderate to severe pain, particularly post-operative pain, obstetric pain, and the pain of renal colic. [Wyeth Laboratories; Monmouth Pharmaceuticals Ltd]

Meptid Tablets. CAS 54340-58-8; 200 mg Meptazinol; used for the short-term treatment of moderate to severe pain. [Wyeth Laboratories; Monmouth Pharmaceuticals Ltd]

Meraneine. Pure geranyl acetate. [Bush Boake Allen Ltd]

Meranol. CAS 106-24-1; High purity geraniol (trans-3,7-dimethyl octa-2,6-dien-1-ol); high-quality rose petal odor used in fragrances. [Bush Boake Allen Ltd]

---

* Trade name not verified as available   † Trade name verified as obsolete

Merantine. Level dyeing wool dyestuffs. [Holliday Dyes & Chemicals Ltd]

Merbentyl. CAS 67-92-5; A proprietary preparation containing dicyclomine hydrochloride; an antispasmodic. [Richardson-Vicks Inc] *

Merbron R. Machine cleaner for aftercleaning of dyed polyester, removal of trimers from fibers or equip. [Sybron]

mercaptan C3. See n-propyl mercaptan.

mercaptan C12. See n-dodecyl mercaptan.

2-mercaptoacetic acid. See thioglycolic acid.

2-mercaptobenzimidazole. (2-benzimidazolethiol) $C_7H_6N_2S$; CAS 583-39-1; EINECS 209-502-6. [Aceto]

2-mercaptobenzothiazole. (MBT; 2(3H)-benzothiazole-thione; 2-benzothiazolethiol) CHCHCHCHCC-SC(SN)N; CAS 149-30-4; EINECS 205-736-8; Vulcanization accelerator for rubber (requires stearic acid for full activation); tire treads; mechanical specialties, fungicide, corrosion inhibitor, petrol production; extreme pressure additive in greases. [Allchem Industries; Am. Cyanamid; Monsanto; Uniroyal]

2-mercaptoethanol. (thioethylene glycol) $HSCH_2CH_2OH$; CAS 60-24-2; EINECS 200-464-6; Solvent for dyestuffs, intermediate for producing dyestuffs, pharmaceuticals, rubber chemicals, flotation agents, insecticides, plasticizers, reducing agent, biochemical reagent, PVC staiblizers, agricultural chemicals, textile auxiliary. [BASF; Morton Int'l.]

2-mercaptoimidazoline. See ethylene thiourea.

mercaptosuccinate. See malathion.

Merce Assist ADB. Low foaming mercerization penetrant. [Sybron]

Mercerisin OR. Sodium salt of sulfonation product; anionic; yellow-brown liquid; wetting agent for mercerizing. [Thor Chemicals (UK) Ltd]

mercerized cotton. Cotton which has been immersed in a solution of sodium hydroxide. It has a lustrous appearance.

Mercer's liquor. A solution containing potassium ferricyanide; used for etching.

Merclor D. A proprietary trade name for a solution of sodium hypochlorite NaOCl; a bleaching agent. §

mercolloid. A colloidal mercury sulfide.

Mercoloy. A proprietary trade name for a nickel bronze containing 60% copper, 25% nickel, 10% zinc, 1% tin, 2% lead, and 2% iron. §

Mercresin. A proprietary trade name for a solution of 0.1% o-hydroxy-phenyl-mercuric chloride and 0.1% sec-amyl-tricresol in 50% alcohol, 10% acetone, and water; a germicide. §

Mercuhydrin. A proprietary trade name for meralluride. §

mercurammonium chloride. (fusible white precipitate; ammonio-mercuric-chloride). Mercuri-diammonium chloride, $N_2H_6Hg.Cl_2$.

mercuric potassium iodide. See Mayer's solution.

mercuric acetate. See mercury acetate (ic).

mercuric chloride. See mercury chloride (ic).

mercurichrome. A fluorescein derivative of mercury; used as a bactericide.

mercuricide. Lithio-mercuric-iodide, $3LiI+HgI_2$; used as a germicide.

mercuric oxide. See mercury oxide (ic), red and yellow.

mercuric sulfate. See mercury sulfate (ic).

mercuric sulfide. See mercury sulfide (ic), red and black.

mercuriocoleols. A double stearate of cholesterol and mercury.

mercuriol. A mercury amalgara with aluminum.

mercury. (quicksilver; hydrargyrum; liquid silver) Metallic element; Hg; CAS 7439-97-6; EINECS 231-106-7; Amalgam, catalyst, electrical apparatus, cathodes for production of chlorine and caustic soda, thermometers, barometers. [Aldrich; Atomergic Chemetals; Cerac; Cox Chem. Ltd; Spectrum Chem. Mfg.]

mercury acetate (ic). (mercuric acetate) $(CH_3COO)_2Hg$; CAS 1600-27-7; EINECS 216-491-1; Catalyst for organic synthesis, pharmaceuticals. [Atomergic Chemetals; Cerac; Noah Chem.; Thor]

mercury bichloride. See mercury chloride (ic).

mercury blende. See cinnabar.

mercury chloride (ic). (mercuric chloride; mercury bichloride) $HgCl_2$; CAS 7487-94-7; EINECS 231-299-8; manufacture of calomel, other mercury compounds; disinfectant, organic synthesis, analytical reagent, metallurgy, tanning, catalyst, sterilant, fungicide, insecticide, wood preservative, embalming fluids, textile printing, photography, dry batteries. [Aldrich; Atomergic Chemetals; Spectrum Chem. Mfg.; Thor]

mercury oxide (ic), red and yellow. (mercuric oxide; red precipitate; yellow precipitate) HgO; CAS 21908-53-2; EINECS 244-654-7; Red: chemicals, paint pigment, perfumery, cosmetics, pharmaceuticals, ceramics, dry batteries, polishes, analytical reagent, antifouling paints, fungicide, antiseptic; yellow: antiseptic, mercury compounds. [Cerac; Noah Chem.; Spectrum Chem. Mfg.; Thor UK]

mercury persulfate. See mercury sulfate (ic).

mercury sulfate (ic). (mercuric sulfate; mercury persulfate) $HgSO_4$; CAS 7783-35-9; Calomel and corrosive sublimate, catalyst in the conversion of acetylene to acetaldehyde, extracting gold and silver from roasted pyrites, battery electrolyte. [Atomergic Chemetals; Noah Chem.; Spectrum Chem. Mfg.]

mercury sulfide (ic), red and black. (mercuric sulfide; red: vermilion; quicksilver vermilion; artificial cinnabar) HgS; CAS 1344-48-5; EINECS 215-696-3; Pigment. [Atomergic Chemetals; Cerac; Noah Chem.]

Merecol® FA. Water-sol. gum blend; stabilizer for salad dressing, bakery, confectionery, flavors, emulsions, flavored beverages with pulp. [Meer]

Merecol® I. Water-sol. gum blend; stabilizer, gelling agent for fillings and dips; substitute for gelatin in kosher marshmallows. [Meer]

Meretestate. CAS 58-22-0; Testosterone; androgen. [Sterling Drug Inc] *

Merfusan. Horticultural product. [May & Baker Ltd] *

Mergal. Fungicides and bactericide. [Hoechst UK]

Mergamma. Mixture of γ-HCH and phenylmercury acetate; fungicide seed dressing for cereals. [ICI Chem. & Polymers Ltd]

Mergital OC 30E. Ethoxylated oleo-cetyl alcohol (30 EO); surfactant. [Pulcra SA]

Mergital ST 30/E. CAS 9004-99-3; PEG-30 stearate; surfactant. [Pulcra SA]

Merigraph. A range (18 varieties available) of viscous, uv light sensitive liquid photopolymers; used to make printing plates for letterpress, flexo and dry offset printing processes. [Hercules] *

Merital. CAS 32795-47-4; Nomifensine maleate; antide-

---

‡ Trade name and manufacturer not verified   § Trade name without identified manufacturer

pressant. [Hoechst-Roussel Pharmaceuticals Inc] *

Meritol. A photographic developer. [Johnsons of Hendon]*

Merix. A line of aerospace, automotive, industrial, paper, photography, rubber, safety and sporting goods chemicals. [Merix]

Merlon. Polycarbonate resins characterized by their toughness, heat resistance, dimensional stability and clarity. [Miles] *

Merocets. CAS 123-03-5; A proprietary preparation of cetylpyridinium chloride; throat lozenges. [Richardson-Vicks Inc] *

Merolan. Fungicide containing dithiocarbonate. [ICI Chem. & Polymers Ltd]

Merpafol. CAS 2939-80-2; Active ingredient: captafol; nonsystemic agricultural fungicide. [Makhteshim Chemical Works Ltd]

Merpan. CAS 133-06-2; Active ingredient: captan; broad-spectrum agricultural fungicide. [Makhteshim Chemical Works Ltd]

Merpectogel. CAS 62-38-4; Phenylmercuric nitrate; pharmaceutical aid. [Poythress Laboratories Inc] ‡

Merpelan AZ. Wettable powder formulation to control weeds pre-emergence. [Bayer AG]

Merpentine. A proprietary trade name for a sodium alkyl naphthalene sulfonate product used as a wetting agent.§

Merpol® . Surface active agent. [DuPont UK]

Merpol® 100. Octylphenol ethoxylate; surfactant, emulsifier, wetting agent, detergent, solubilizer, emulsion stabilizer for metal cleaning, industrial and household detergents, agric.; stable to acids, bases, heat, freezing. [DuPont]

Merpol® A. Ethoxylated phosphate; low foaming wetting agent, surface tensile reducer for chemical mfg., cosmetics, metal processing, paper, petrol., inks, plastics, soaps, synthetic fibers, textiles; stable to acids, bases, heat to 100 C, freezing. [DuPont]

Merpol® DSR. A nonionic softener for use in conjunction with acrylic type soil release agents in durable press finishes on textiles. [DuPont UK]

Merpol® SE. Long-chain fatty alcohol ethoxylate; low foaming wetting and rewetting agent, emulsifier, dispersant for textiles, paper, o/w emulsions, asbestos; stable to acids, bases, heat, freezing. [DuPont]

Merpol® SH. CAS 24938-91-8; Long-chain alcohol ethoxylate; detergent, wetting agent for textiles, hard surface cleaning, paper, metal processing; biodeg.; stable to acids, bases, heat, freezing to 5 C. [DuPont]

Merquinox. Fungicide for industrial applications. [Octel Chemicals Ltd] †

Mersalyl BDH. Injection and tablets; a proprietary diuretic (mercurial). [British Drug Houses] *

Merse® 7F. Catalyst for polyester weight reduction work, for removal of insolubilized WD size; used with caustic. [Sybron]

Mersil. Turf fungicide. [May & Baker Ltd] *

Mersilene. Nonabsorbable surgical suture. [Ethicon Inc] *

Mersize. Paper sizing agent; used in paper mill to reduce sizing costs, improve resistance to ink, water and other aqueous solutions. [Monsanto Co] *

Mersol. A proprietary trade name for a solvent. §

Mersolat H 30. Sodium alkane sulfonates based on n-paraffin; biodeg. detergent base, wetting agent for mfg. electrolyte-resistant textile and leather auxiliaries, alkaline and acid detergents, floor cleaners, bubble baths, disinfectants, car shampoos. [Miles/Organic Products

Mersolat® CI. Chloralkane sulfonate; raw material used in the manufacture of special products, especially fat-liquoring agents. [Bayer AG]

Mersolat® H. Sodium alkane sulfonate; detergent for use in the production of special purpose materials, especially electrolyte-resistant wetting agents for textiles, leather auxiliaries, acid and alkaline cleansers. [Bayer AG; Bayer plc]

Mersolat® W 40. Sodium alkane sulfonates based on n-paraffin; biodeg. detergent base, wetting agent for mfg. electrolyte-resistant textile and leather auxiliaries, alkaline and acid detergents, floor cleaners, bubble baths, disinfectants, car shampoos. [Bayer AG; Miles/Organic Prods.

Mertan. Polyurethane water emulsions, polyurethane solutions, granule form; thermoplastic polyurethane (heel tops, soles etc.). [Merquinsa] *

Mertec. A proprietary trade name for a chlorinated rubber-base paint. §

Merthioiate. A proprietary preparation of thiomersal (sodium ethyl-mercurithio-salicylate); a skin antiseptic. [Eli Lilly & Co]

Meruvax. Rubella virus vaccine live; immunizing agent. [Merck & Co Inc] *

Meruvax II. Rubella vaccine; immunization against German measles. [Merck & Co Inc]

Mervan. CAS 22131-79-9; Alclofenac; anti-inflammatory. [Continental Pharma] *

Mesamoll® . Alkyl sulfonic ester of phenol; universal plasticizer used in PVC calendering, extrusion, injection molding, dip coating, high-pressure foam, rotational and compression moldings, film for linings and food pkg., shower curtains, floorcoverings, imitation leather, tarpaulins, protective clothing, cable insulation, structure profiles, tubing, shoes, tech. articles, toys; also used with PS, joint sealants, natural, S/B, nitrile/butadiene, chlorinated and butyl rubber; cleansing agent for PU processing equip. [Bayer AG; Bayer plc; Miles; Polysar]

Mesamoll®-Verdingrin. Paraffin sulfonated ester; a vinyl plasticizer. [Bayer AG]

Mesantoin. CAS 50-12-4; Mephenytoin; anticonvulsant. [Sandoz Pharmaceuticals]

Mesgamma. Combined insecticide fungicide and seed dressing. [Plant Protection] *

Mesicerin. Tri-θ -hydroxymesitylene, $C_6H_3(CH_2OH)_3$.

Mesidine. CAS 88-05-1; 2, 4, 6-trimethylaniline; $(CH_3)_3C_6H_2NH_2$; used as a dyestuff intermediate.

Mesitol® . Range of aftertreating and resist agents; used for polyamide, polyamide/cellulosic blends and half wool. [Bayer AG; Bayer plc]

Mesontoin. A proprietary preparation of methoin; an anticonvulsant. [Sandoz]

Mestinon. CAS 101-26-8; A proprietary preparation containing pyridostigmine bromide; cholinestetase inhibitor. [Roche Products Ltd] *

Mesurol® . CAS 2032-65-7; Methiocarb; versatile product formulated for different uses; especially as a molluscicide against slugs and snails as well as a seed dressing for repelling depredating birds; also as insecticide/acaricide against foliar-feeding caterpillars and sucking pests on various crops. [Bayer AG]

Met. Coated titanium anodes and cathodes. [ICI Chem. & Polymers Ltd]

Meta. Metaldehyde produced by the polymerization of

---

acetaldehyde.

metabisulfite. Sodium metasulfite, $Na_2S_2O_5$.

Metablen. Acrylic modifiers and processing aids for PVC. [British Traders & Shippers Ltd]

Metablen® C-301. Impact modifier for PVC blow molding compounds; improves oil resist. of bottles; used for olive oil bottles. [Metablen BV; Metco N. Am.]

Metachloron® A4 Liq. Organo metallic compound; color bleed assistant for direct dyes on wet cotton and rayon goods. [Crompton & Knowles]

Metachrome. Chrome dyestuffs. [J C Bottomley] *

Metacide. CAS 298-00-0; Parathion-methyl; spray for control of insect pests especially on cotton. [Bayer AG]

metacinnabarite. A mineral having the same composition as cinnabar, but black in color.

Metacon. Rust remover. [Croda Chem. Ltd] *

Metacrylene. A proprietary trade name for a styrene-methyl methacrylate copolymer. §

Metacrylene BS. A proprietary trade name for an MBS terpolymer. §

Metacure® T-1. CAS 1067-33-0; Dibutyl tin diacetate; catalyst for use in production of PU coatings, adhesives, and sealants. [Air Prods. & Chems. Inc]

Metacure® T-5. Tin-based; catalyst for production of PU coatings, adhesives, and sealants; increased pot life over Metacure T-12. [Air Prods. & Chems. Inc]

Metacure® T-9. Stannous octoate; catalyst for use in production of PU coatings, adhesives, and sealants; uniform activity and excellent stability. [Air Prods. & Chems. Inc]

Metacure® T-12. CAS 77-58-7; Dibutyl tin dilaurate; catalyst for production of PU coatings, adhesives, and sealants; formulated to remain liq. > 18 C for easier handling [Air Prods. & Chems. Inc]

Metacure® T-45. Potassium octoate; catalyst for PU; catalyzes the trimerization of isocyanates and polyol-isocyanate reaction. [Air Prods. & Chems. Inc]

Metacure® T-120; Organotin; catalyst for one-part moisture cure and two-part isocyanate coatings; compatible with amine co-catalysts. [Air Prods. & Chems. Inc]

Metacure® T-125; Organotin; catalyst for PU coatings, adhesives, and sealants; suitable for one-part moisture cure and two-part reactions. [Air Prods. & Chems. Inc]

Metacure® T-131; Organotin; catalyst for PU coatings, adhesives, and sealants; provides delayed action catalysis of isocyanate/polyol reaction. [Air Prods. & Chems. Inc]

metaethyl. A mixture of methyl and ethyl chlorides; used as an anaesthetic.

Metaferrin. An iron albumin compound containing 10% iron; used in the treatment of anaemia. §

Metaform. A packing material for packing stuffing-boxes, and consisting of powdered white metal, graphite, cylinder oil, and asbestos fiber. §

Metafos. Glassy phosphates. [Peridot Chemicals Inc]

Metaglo Super K. Metallic print binder for screen and machine printing on all types of fabrics; durable to laundering and drycleaning. [Eastern Color & Chem.

Metagon. Sodium hexametaphosphate. [Albright & Wilson Ltd, Phosphates & Speciality Business]

metahomomenthyl salicylate. See homosalate.

Metahydrin. CAS 133-67-5; Trichlormethiazide; diuretic; antihypertensive. [Merrell Dow Pharmaceuticals Inc] *

metal argentum. An alloy consisting of 85 1/2% tin and 14 1/2% antimony.

Metalclad. High performance cladding overpaint for a wide

range of cladding panels. [Feb Ltd]

Metal Deactivator S. CAS 94-91-7; N,N'-Disalicylidene-1,2-propane diamine; copper chelating agent for refinery industry. [Hart Chem. Ltd.]

metal-furnace slag. A slag formed and used in the preparation of copper metal. It is essentially a silicate of iron, and contains about 4% copper.

Metalite. Proprietary trade names for aluminum oxide abrasives. §

Metall. General purpose primer for all metal surfaces; suitable for wide range of substrates prior to the application of other polyurethane based systems. [Feb Ltd]

Metallic 9500. Conductive additive for coatings or composites; used in thermoplastics for EMI shielding and ESD protection for elec. products. [DJ Enterprises]

Metallichrome. A finish for motor bodies. [ICI Chem. & Polymers Ltd] †

Metallic Sodium. Pharmaceutical processing. [Foseco (F.S.) Ltd]

metalline. An alloy consisting of 35% cobalt, 30% copper, 10% iron, and 25% aluminum; used in jeweller's work.

Metallyte 70-2, 70-4, 70-U, 80-2, 80-4, 80-U, 140-2, 140-4, 140-U. CAS 9003-07-0; 1 side coextruded oriented PP film, 1 side metal, heat sealable; high opacity film with adhesion promoting layer on one side and broad seal range surface on the other. [Mobil] *

Metalset. Epoxy resin base adhesives and cements. [Smooth-On Inc] *

metal soaps. Salts of the heavy metals with fatty acids.

Metalyn 582. Ester of pentaerythritol and tall oil fatty acid; synthetic lubricant additive for formulations requiring high thermal and hydrolytic props. at elevated temps.; for rolling oils. [Climax Fluids Additives]

Metalyn 582. An ester of fatty acids designed for metal-rolling and working lubricants that require particular thermal and hydrolytic properties at elevated temperatures; used as an ingredient in wood preserving agents, as a secondary plasticizer for vinyl copolymer resin, as a plasticizer, softener and tackifier for rubber (nitrile) compounding and in various low cost adhesives, metal working lubricants. [Hercules] *

Metamsustac. A proprietary preparation of methyl-amphetamine hydrochloride, in a slow release tablet; an antiobesity agent. [Pharmax Ltd] *

Metamucil. A proprietary preparation containing psyllium mucilloid, dextrose, benzyl benzoate, sodium bicarbonate, potassium biphosphate and citric acid; a laxative. [G D Searle & Co Ltd] *

Metandren. CAS 58-18-4; Methyltestosterone; androgen. [Ciba-Geigy Corp] *

metanilic acid. (m-sulfanilic acid; m-aminobenzenesulfonic acid) $C_6H_4(NH_2)SO_3H$; CAS 121-47-1; Azo dye manufacture (sodium salt), sulfa drug synthesis. [Enichem Am.]

Metanium Ointment. Titanium dioxide 20%, titanium peroxide 5%, titanium salicylate 3%, titanium tannate 0.1% in a siliconized excipient; used for the prevention and treatment of nappy rash and other macerated skin conditions. [Bengue & Co Ltd] *

Metaprel. CAS 5874-97-5; Metaproterenol sulfate; bronchodilator. [Dorsey Pharmaceuticals] *

Metaquest. Sequestering agents. [Fisons plc] *

Metasal. Multipurpose fire extinguishing powders for extinguishing fires of incandescent solid, liquid and gaseous materials. [Degussa AG] *

Metasap® 537. CAS 637-12-7; Aluminum stearate; process aid, lubricant for PVC, polyolefins, PS, and ABS. [Syn. Prods.]

Metasil A, A+, Extra A+, B, C, E. A pure diatomaceous silica, each powder having different retentive properties; used for the coarse filtration of suspensions down to very fine filtrations giving brilliant filtrates on most liquids e.g. syrups, oils, chemicals, water, beer, vinegar, etc. [Stella-Meta Filters]

Metasil AL. Metasil A reinforced with hydrate of aluminum; used for the production of sparkling water, e.g., mineral water manufacturers. [Stella-Meta Filters]

Metasil ALAG. Metasil AL carrying a surface coating of silver; used for the production of sparkling water for drinking purposes from low quality water supplies. [Stella-Meta Filters]

Metasil D. Polystyrene micro beads; special barrier type precoat material. [Stella-Meta Filters]

Metasil DA. CAS 7440-44-0; Carbon; used for decolorizing, dechlorinating and taste removal. [Stella-Meta Filters]

Metasil MQC. Hydrated silicate; used for the adsorption of oxides in rolling oil coolants. [Stella-Meta Filters]

Metasil Purasil. Perlite (volcanic rock) type; used for general clarification of liquids not too strongly acid or alkaline, e.g. swimming pool. [Stella-Meta Filters]

Metasil R. Chopped rayon fiber; special barrier type precoat material. [Stella-Meta Filters]

Metasil SA, SB. Pure cellulose; used for the filtration of products where silica is unsuitable e.g. miscella oils, removal of oil from condensate water. [Stella-Meta Filters]

Metasil W/2. CAS 7440-44-0; Carbon; used for the filtration of strong alkali liquids. [Stella-Meta Filters]

Metaspirine. A proprietary preparation consisting of acetylsalicylic acid and caffeine. §

Metasystox 55. CAS 8022-00-2; An emulsifiable concentrate containing demeton-S-methyl 580 g per liter; used to control aphids, red spider mites and certain other pests on arable crops, fruit, market garden and greenhouse crops and ornamentals. [Farm Protection Ltd]

Metasystox® I. CAS 919-86-8; Demeton-S-methyl; systemic insecticide for control of sucking insects on a wide range of crops. [Bayer AG]

Metasystox R; CAS 301-12-2; A liquid concentrate of oxydemeton-methyl 570 g per liter; used to control aphids, red spider mites and certain other pests on arable crops, fruit, market garden and greenhouse crops and ornamentals; the choice of the grower who prefers a product with a less penetrating odor.; for all crops for which Metasys. [Farm Protection Ltd]

Metasystox® R. CAS 301-12-2; Emulsifiable concentrate containing 570 g/l oxydemeton-methyl; a systemic organophosphorus insecticide used to control aphids, red spider mites and certain other pests. [Bayer plc]

Metatone. A proprietary preparation containing thiamine hydrochloride, and calcium, potassium, sodium, manganese, and strychnine glycerophosphates; a tonic. [Parke-Davis] *

Metazene®. Mixt. of fatty alcohol esters of methyl methacrylic acid; odor counteractant. [Pestco]

Metco 450. A nickel/aluminum composite material for building up metal surfaces by spraying. Similar materials are Metco 451 (nickel-chromium), Metco 44 (nickel base/

chromium) and Metco 51 (aluminum bronze). §

Meteor. Mixed metal oxide color pigments used in plastics, coatings, and other applications. [Engelhard]

Meteor Plus. Fine particle size mixed metal oxide color pigments used in plastics, coatings and other applications. [Engelhard]

meteorite. An alloy of aluminum with from 1-2% zinc and 1-4% phosphorus.

metformin. N'N'-Dimethyldiguanide.

Metglas®. Amorphous metal alloys. [Allied-Signal]

Methacrol. Fabric and yarn lubricants. [DuPont UK]

β-methacrylic acid. See crotonic acid.

Methadose. CAS 1095-90-5; Methadone hydrochloride; analgesic. [Mallinckrodt Inc] *

Metha-Meridiazine. Trisulfapyrimidines (oral suspension); mixture of sulfadiazine, sulfamerazine and sulfamethazine; antibacterial. [McNeil Pharmaceuticals] *

methanal. See formaldehyde.

methanamide. See formamide.

methanedicarbonic acid. See malonic acid.

methane, oxybis-. See dimethyl ether.

methanoic acid. See formic acid.

methanol. See methyl alcohol.

Methaplex. Methylmethacrylate acrylic in an aromatic and aliphatic solvent vehicle system; water repellant and sealer for masonry. [Nova Chemical Inc] *

Methar 30. Selective herbicide for crabgrass control on grasses. [W A Cleary]

Metharbitone. CAS 50-11-3; 5,5-Diethyl-l-methyl barbituric acid.

Methasan. CAS 137-30-4; Zinc dimethyldithiocarbamate; vulcanization accelerator. [Monsanto Co] *

Methasol. Dyes for printing inks. [ICI Chem. & Polymers Ltd]

Methazate. CAS 137-30-4; A proprietary preparation of zinc dimethyl dithiocarbamate. [Naugatuck (US Rubber)] *

Methergin. A propetary preparation of methylergometrine maleate; used in obstetrics to stimulate uterine contraction. [Sandoz] *

Methergine. CAS 57432-61-8; Methylergonovine maleate; oxytocic. [Sandoz Pharmaceuticals]

methethyl. A mixture of ethyl and methyl chlorides; used as a local anaesthetic.

Methic. Basic dyestuffs. [ICI Chem. & Polymers Ltd]

DL-methionine. (2-amino-4-(methylthio)butyric acid; 2-amino-4-(methylmercapto)butyric acid) $CH_3SCH_2CH_2$-$CH(NH_2)COOH$; CAS 59-51-8; EINECS 200-432-1; Pharmaceuticals, feed additive, vegetable oil enrichment, single-cell protein. [Degussa; Penta Mfg.; U.S. Biochemical]

Methisul. CAS 144-82-1; A proprietary preparation of sulfamethisole; an antibiotic. [RP Drugs] *

Methocel. CAS 9004-67-5; Methylcellulose; used in adhesives, cosmetics, foods, paints, pharmaceuticals and textiles. [Dow UK]

Methocel® 40-202. CAS 9004-65-3; Hydroxypropyl methylcellulose; cold water-disp. grade for personal care applications. [Dow]

Methocel® A4C, A4M. CAS 9004-67-5; Methylcellulose; food gums used as thickener, stabilizer, emulsifier, adhesive, and gellant. [Dow]

Methocel® A4MP. CAS 9004-67-5; Methylcellulose; for personal care prods. [Dow]

Methocel® A15-LV. CAS 9004-67-5; Methylcellulose; food gums used as thickener, stabilizer, emulsifier, adhesive,

---

* Trade name not verified as available   † Trade name verified as obsolete

and gellant; also for tablet coating applications. [Dow]

Methocel® E3 Premium. CAS 9004-65-3; Hydroxypropyl methylcellulose; for tablet coating applications. [Dow]

Methocel® E4M. CAS 9004-65-3; Hydroxypropyl methylcellulose; food gums used as thickener, stabilizer, emulsifier, adhesive, and gellant. [Dow]

Methocel® E5. CAS 9004-65-3; Hydroxypropyl methylcellulose; food gums used as thickener, stabilizer, emulsifier, adhesive, and gellant. [Dow]

Methocel® F4M. CAS 9004-65-3; Hydroxypropyl methylcellulose; food gums used as thickener, stabilizer, emulsifier, adhesive, and gellant. [Dow]

Methocel® K35. CAS 9004-65-3; Hydroxypropyl methylcellulose; food gums used as thickener, stabilizer, emulsifier, adhesive, and gellant. [Dow]

Methocel® K100MP. CAS 9004-65-3; Hydroxypropyl methylcellulose; for personal care products. [Dow]

Methocidin. A proprietary preparation of hydroxymethylgramicidin ephedrine and cetylpyridinium chloride; throat lozenges. [Rona Laboratories] ‡

Meth-O-Gas. CAS 74-83-9; Methyl bromide. [Great Lakes Europe]

Methokill. Insecticidal formulation. [Mitchell Cotts Chemicals Ltd]

Methoklone. CAS 75-09-2; Stabilized methylene chloride. [ICI Chem. & Polymers Ltd]

Methomex. CAS 16752-77-5; Methomyl; insecticide. [Makhteshim Chemical Works Ltd]

Methotrexate. CAS 59-05-2; Preparations containing methotrexate; antineoplastic agent. [Lederle Laboratories]

Methoxone. CAS 24017-47-8; Emulsifiable concentrate containing 420 g/l triazophos; an organophosphorus insecticide. [ICI Chem. & Polymers Ltd]

methoxychlor. (DMDT; 1,1´-(2,2,2-trichloroethylidene)-bis[4-methoxybenzene]; methoxy DDT; 2,2-bis (p-methoxyphenol)-1,1,1-trichloroethane) Cl_3CCH-$(C_6H_4OCH_3)_2$; CAS 72-43-5; Insecticide effective against mosquito larvae and houseflies; recommended for use in dairy barns. [Kincaid Enterprises]

o-methoxyphenol. See guaiacol.

p-methoxyphenylacetic acid. (4-methoxyphenylacetic acid; homoanisic acid; p-methoxy-α-toluic acid) $CH_3OC_6H_4CH_2COOH$; CAS 104-01-8; EINECS 203-166-4; pharmaceuticals, other organic compounds. [Penta Mfg.; Schweizerhall]

4-methoxyphenylacetic acid. See p-methoxyphenylacetic acid.

p-methoxy-alpha-toluic acid. See p-methoxyphenylacetic acid.

Methozin. See Antipyrine. §

Methral. CAS 2119-75-7; Fluperolone acetate; glucocorticoid. [Pfizer Inc] †

methyl acetate. (acetic acid, methyl ester) Ester of methyl alcohol and acetic acid; CH_3COOCH_3; CAS 79-20-9; EINECS 201-185-2; paint remover compounds, lacquer solv., intermediate, synthetic flavoring; solv. for nitrocellulose, acetylcellulose; manufacture of artificial leather. [Akzo; Hoechst Celanese; Penta Mfg.]

methylacetic acid. See propionic acid.

methyl acetone. CAS 78-93-3; A crude fraction of wood distillation. Its principal constituents are acetone, methyl alcohol, and methyl acetate;$CH_3CO \cdot CH_2CH_3$; used as a solvent for nitrocellulose,cellulose acetate, rubber, gum,

resins; paint and varnish removers; extracting perfumes.

methyl acrylate (monomer). (2-propenoic acid methyl ester; acrylic acid methyl ester) Ester of methyl alcohol and acrylic acid; $CH_2:CHCOOCH_3$; CAS 96-33-3; EINECS 202-500-6; acrylic polymers; amphoteric surfactants; chemical intermediate; leather finish resins; textile and paper coatings; plastic films. [BASF; Hoechst Celanese]

methyl acrylate polymer. [BASF; Hoechst Celanese]

methyl adipate. See dimethyl adipate.

methylal. CAS 109-87-5; (formal; dimethoxymethane). Methylene-dimethyl ether $CH_2(O \cdot CH_3)_2$; used as a solvent; in organic synthesis; perfumes, adhesives, protective coatings; special fuel.

methyl alcohol. (methanol; wood alcohol; carbinol) CH_3OH; CAS 67-56-1; Industrial solvent; manufacture of formaldehyde, acetic acid, dimethyl terephthalate; chemical synthesis; antifreeze; solv. for nitrocellulose, polyvinyl butyral, shellac, rosin, manila resin, dyes; source of hydrogen for fuel cells; denaturant for ethanol. [Air Prods.; Albright & Wilson; CPS; Du Pont; Eastman; Hoechst Celanese; Mitsui Toatsu Chem.; Nissan Chem. Ind.; Norsk Hydro A/S; Quantum/USI]

methylaminoacetic acid. See sarcosine.

methyl-o-aminobenzoate. See methyl anthranilate.

methyl amyl alcohol. (MIBC; methylisobutyl carbinol; 4-methylpentanol-2) $(CH_3)_2CHCH_2CH(CH_3)OH$; CAS 108-11-2; Solvent for dyestuffs, oils, gums, resins, waxes, nitrocellulose, ethylcellulose; organic synthesis; froth flotation; brake fluids. [Allchem Industries; Ashland; Shell; Union Carbide]

methyl n-amyl ketone. (2-heptanone) $CH_3CH_2CH_2$-$CH_2CH_2COCH_3$; CAS 110-43-0; Industrial solvent for nitrocellulose lacquers, synthetic flavoring and perfumery. [Ashland; Eastman; Union Carbide]

methyl anthranilate. (methyl-o-aminobenzoate; Neroli oil, artificial; 2-aminobenzoic acid, methyl ester) $H_2NC_6H_4$-CO_2CH_3; CAS 134-20-3; EINECS 205-132-4; Flavoring, fragrance, perfume, cosmetics, pomades; intermediate for pharmaceuticals and dyes. [Bell Flavors & Fragrances; Haarmann & Reimer; PMC Specialties]

methyl benzene. (retinaphtha, phenyl methane; toluene) $C_6H_5CH_3$; CAS 108-88-3; Used as aviation gasoline and high octane blending stock; solvent for paints and coatings.

methylbenzene. See toluene.

methylbenzenesulfonic acid, sodium salt. See sodium toluenesulfonate.

4-methylbenzenesulfonic acid. See toluene sulfonic acid.

methyl benzoate. (benzoic acid methyl ester; Niobe oil) Ester; $C_6H_5COOCH_3$; CAS 93-58-3; EINECS 202-259-7; Perfumery; solvent for cellulose esters and ethers, resins, rubber; flavoring. [Morflex; Pentagon Chem. Ltd; Penta Mfg.; Schweizerhall; Sybron]

2-methylbenzoic acid. See o-toluic acid.

3-methylbenzoic acid. See m-toluic acid.

4-methylbenzoic acid. See p-toluic acid.

2-methylbenzophenone. (phenyl tolyl ketone) $C_{14}H_{12}O$; Perfume additive (fixative). [Janssen Chimica; Spectrum Chem. Mfg.]

4-methylbenzoyl chloride. See p-toluoyl chloride.

methyl bromide. (bromomethane) CH_3Br; CAS 74-83-9; Soil and space fumigant; disinfestation of potatoes, tomatoes, other crops; organic synthesis; extraction

‡ Trade name and manufacturer not verified   § Trade name without identified manufacturer

solvent for vegetable oils. [Akzo; Ethyl; Great Lakes]

2-methyl-1,3-butadiene, homopolymer. *See* polyisoprene

2-methylbutane. *See* isopentane.

methyl butynol. (2-methyl-3-butyn-2-ol; dimethyl ethynyl carbinol) Tertiary acetylenic alcohol; $(CH_3)_2COHCCH$; CAS 115-19-5; EINECS 204-070-5; corrosion inhibitor; reactive intermediate in manufacture of pharmaceuticals, plastics, rubbers, fragrances, agriculture.; as solv., acid inhibitor, viscosity reducer, and stabilizer in vinyl plastisols, platinum catalyst blocker for silicones. [Air Prods.; BASF]

2-methyl-3-butyn-2-ol. *See* methyl butynol.

methyl caprate. (methyl decanoate; methyl caprinate) $CH_3(CH_2)_8COOCH_3$; CAS 110-42-9; EINECS 203-766-6; intermediate for detergents, emulsifiers, wetting agents, stabilizer, resins, lubricants, plasticizer. [Penta Mfg.; Procter & Gamble]

methyl caprinate. *See* methyl caprate.

methyl catechol. (guaiacol) $C_6H_4(OH)(OCH_3)$, CAS 90-05-1; The chief constituent of beech-wood creosote; used in medicine as an expectorant; synthetic flavors.

Methyl Cellosolve. A proprietary name for the monomethyl ether of ethylene glycol; a colorless and nearly odorless liquid boiling at 124.5 C; has the lowest boiling-point and greatest rate of evaporation of all available glycol ethers; it is a solvent for cellulose acetate, nitrocellulose and hydrocarbons. §

methylcellulose. (cellulose methyl ether) Methyl ether of cellulose; CAS 9004-67-5; protective colloid in water-based paints to prevent flocculation of pigment; film and sheeting; binder in ceramic glazes; leather tanning; dispersing, thickening, and sizing agent; food additive; adhesive; paper greaseproofing; pharmaceuticals. [Allchem Industries; Aqualon; Courtaulds Water Soluble Polymers; Shin-Etsu Chem.]

methyl chloride. (chloromethane; monochloromethane) CH_3Cl; CAS 74-87-3; Catalyst carrier in low-temp. polymerization, tetramethyl lead, silicones, refrigerant, fluid for thermometric/thermostatic equipment, methylating agent in organic synthesis, extractant and low-temp. solvent, herbicide, topical anesthetic. [Air Prods.; Mitsui Toatsu Chem.; OxyChem]

methylchloroform. *See* trichloroethane.

Methyl Cumate® . CAS 137-29-1; Copper dimethyldithiocarbamate; accelerator for high-speed vulcanization of SBR, IIR, EPDM. [R.T. Vanderbilt]

methyl cyanoacetate. (malonic methyl ester nitrile) $NCCH_2COOCH_3$; CAS 105-34-0; EINECS 203-288-8; Organic synthesis, pharmaceuticals, dyes. [R.W. Greeff; Hüls Am.; Lonza]

methyl cyclohexane. (hexahydrotoluene) C_7H_{14}; CAS 108-87-2; EINECS 203-624-3; Solvent for cellulose ethers, organic synthesis. [Janssen Chimica; Penta Mfg.; Phillips 66]

methyl decanoate. *See* methyl caprate.

Methyl Di-Icinol. CAS 111-77-3; 2-(2-Methoxyethoxy)ethanol; solv. for use in protective coatings, inks, cleaning products, agric. chems.; aids wetting, penetration, and soil removal; coupling solv. [ICI Australia]

methyl dodecanoate. *See* methyl laurate.

methylene bichloride. *See* methylene chloride.

methylene bisacrylamide. $(CH_2:CHCONH)_2CH_2$; CAS 110-26-9; EINECS 203-750-9; Organic intermediate, crosslinking agent. [Am. Cyanamid; Bio-Rad Labs;

Fluka; Schweizerhall]

methylene bisphenyl isocyanate. *See* MDI.

methylene chloride. (dichloromethane; methylene dichloride; methylene bichloride) Halogenated organic compound.; CH_2Cl_2; CAS 75-09-2; EINECS 200-838-9; Paint removers, solvent degreasing, plastics processing, blowing agent in foams, solvent extraction, solvent for cellulose acetate, aerosol propellant, pharmaceutic aid. [Ashland; Atochem N. Am.; Farleyway Chem. Ltd; ICI Specialties; Mallinckrodt; Mitsui Toatsu Chem.; OxyChem]

methylene dichloride. *See* methylene chloride.

methylene di-p-phenylene isocyanate. *See* MDI.

Methyl Ester L. Lard methyl ester; lubricant and wetting agent for industrial lubricants, e.g., sol. cutting and drawing compounds, motor oils, rolling oils; also for textiles and cosmetics. [Climax Fluids Additives]

Methyl Ester S. Soya methyl ester; lubricant additive; substitute for triglycerides. [Climax Fluids Additives]

methyl ether. *See* dimethyl ether.

1-methylethyl hexandecanoate. *See* isopropyl palmitate.

methyl ethyl ketone. (MEK) ethyl methyl ketone; 2-butanone; 2-oxobutane) Aliphatic ketone; $CH_3CO-CH_2CH_3$; CAS 78-93-3; EINECS 201-159-0; solvent in nitrocellulose coatings and vinyl films, paint removers, cements, adhesives, organic synthesis; manufacture of smokeless powder; cleaning fluids; priming, catalyst carrier; acrylic coatings. [Atochem N. Am.; BP Chem. Ltd; Exxon; Hoechst Celanese; Mallinckrodt; Shell; Texaco; Union Carbide]

methyl ethyl ketone peroxide. (MEK peroxide; ethyl methyl ketone peroxide; 2-butanone peroxide) $C_2H_5C(OOH)-(CH_3)OOC(OOH)(CH_3)C_2H_5$; CAS 1338-23-4; EINECS 215-661-2; Initiator/catalyst for cure of unsaturated polyester resins. [Akzo; Atochem; Cook Composites & Polymers; Norac; Witco/Argus]

methyl ethyl ketoxime. $CH_3C(:NOH)CH_2CH_3$; CAS 96-29-7; EINECS 202-496-6; antiskinning agent for paint industry. [Akzo; Allied-Signal; KMZ Chem. Ltd]

1-methylethyl tetradecanoate. *See* isopropyl myristate.

methyl eugenol. (4-allyl-1,2-dimethoxybenzene; 4-allyl veratrole; eugenyl methyl ether) $(CH_3O)_2C_6H_3-CH_2CH:CH_2$; CAS 93-15-2; insect attractant, flavoring. [Firmenich; Penta Mfg.; Schweizerhall]

methylglycinate. *See* sodium lauroyl sarcosinate.

N-methylglycine. *See* sarcosine.

methyl-glycocoll. CAS 107-97-1; (sarcosine; methylaminoacetic acid) $CH_3 \cdot NH \cdot CH_2 \cdot CO_2H$; used in the synthesis of foaming antienzyme compounds for toothpaste, cosmetics, and pharmaceuticals.

methyl glycocoll. *See* sarcosine.

methyl glycol. *See* propylene glycol.

2-methyl glyoxaline. *See* 2-methyl imidazole.

methyl green. (Paris green, light green, double green SF, green powder, methyl aniline green). A dyestuff. It is the zinc double chloride of heptamethyl-pararosaniline chloride, $C_{26}H_{33}Cl_4N_3Zn$. Dyes silk green from a soap bath.

16-methylheptadecanoic acid. *See* isostearic acid.

methyl hexadecanoate. *See* methyl palmitate.

Methylhexalin® . A trade name for a mixture of three isomeric methylcyclohexanols; a solvent for fats, resins, oils, and waxes. §

methylhydroquinone. *See* toluhydroquinone.

methyl 2-hydroxybenzoate. *See* methyl salicylate.

---

* Trade name not verified as available   † Trade name verified as obsolete

methyl hydroxycellulose. Binder in plasters, adhesive, and troweling compounds.

methyl 12-hydroxy-9-octadecenoate. *See* methyl ricinoleate.

Methyl Icinol. CAS 109-86-4; 2-Methoxyethanol; solv. for use in protective coatings, inks, cleaning products, agric. chems.; aids wetting, penetration, and soil removal; coupling solv. [ICI Australia]

2-methyl imidazole. (2MZ; 2-methyl glyoxaline) CHCHNC(CH₃)NH; CAS 693-98-1; EINECS 211-765-7; Dyeing auxiliary for acrylic fibers, plastic foams. [Allchem Industries; BASF; Janssen Chimica]

methyl iodide. (iodomethane) CH₃I; CAS 74-88-4; EINECS 200-819-5; Organic synthesis, microscopy, testing for pyridine. [Akzo; Andeno BV, Burlington Bio-Medical; Fiarmount; R.W. Greeff]

methylisobutyl carbinol. *See* methyl amyl alcohol.

methyl lardate. Solvent, carrier for agricultural spray products; defoaming component in metalworking, paper deinking, pharmaceutical fermentation; wetting agent, lubricant ingred. for metalworking and lubricating oils. [Anar]

methyl laurate. (methyl dodecanoate; dodecanoic acid, methyl ester) Ester of methyl alcohol and lauric acid; CH₃(CH₂)₁₀COOCH₃; CAS 111-82-0; 67762-40-7; EINECS 203-911-3; intermediate for detergents, emulsifiers, wetting agents, stabilizers, lubricants, plasticizers, textiles, flavoring. [Henkel/Emery; Procter & Gamble; Stepan]

Methyl Ledate. CAS 19010-66-3; Lead dimethyldithiocarbamate; ultra accelerator recommended for NR, SBR, IIR, IR, and BR rubbers; used for ultra accleration, high speed, high temperature vulcanization. [R T Vanderbilt Co Inc]

methyl methacrylate (monomer). Methyl ester of methacrylic acid; CH₂:C(CH₃)COOCH₃; CAS 80-62-6; EINECS 201-297-1; monomer for polymethacrylate resins, impregnation of concrete. [Aldrich; Allchem Industries; Cyro Industries; Degussa; Mitsubishi Gas; Rohm & Haas; Transol Chem. UK Ltd]

methyl methacrylate polymer. (acrylite; methyl methacrylate resin) (C₅H₈O₂)ₙ; CAS 9011-14-7; Thermoplastic acrylic resin used in coatings, barrier coatings for PS, vinyl topcoats, product finishes, printing inks. [Aristech; Cyro Industries; Sybron]

methyl methacrylate resin. *See* methyl methacrylate polymer.

N-methyl methyl anthranilate. *See* dimethyl anthranilate.

5-methyl-2-(1-methylethyl) cyclohexanol. *See* menthol.

5-methyl-2-(1-methylethyl) phenol. *See* thymol.

p-methyl morpholine. (4-methyl morpholine) CH₂CH₂O-CH₂CH₂NCH₃; CAS 109-02-4; EINECS 203-640-0; Catalyst in polyurethane foams; extraction solvent; stabilizer for chlorinated hydrocarbons; self-polishing waxes; corrosion inhibitor; pharmaceuticals.

4-methyl morpholine. *See* p-methyl morpholine.

methyl mustard oil. Methyl isothiocyanate, CH₃ · NCS.

methyl myristate. (methyl tetradecanoate; tetradecanoic acid, methyl ester) Ester of methyl alcohol and myristic acid; CH₃(CH₂)₁₂COOCH₃; CAS 124-10-7; EINECS 204-680-1; intermediate for myristic acid detergents; emulsifiers, wetting agents, stabilizers, resins, lubricants, plasticizers, textiles, animal feeds; standard for gas chromatography; flavoring. [Henkel/Emery; Stepan]

Methyl Namate®. Sodium dimethyldithiocarbamate aq. solution; water treatment chemical; readily forms water-insol. salts with heavy metals such as cadmium, copper, chromium, and nickel; clarification agent for wastewater from plating, photo finishing, ore beneficiation processes. [R.T. Vanderbilt]

α and β-methylnaphthalene. C₁₀H₇CH₃; CAS 90-12-0 and 91-57-6; EINECS 201-966-8 and 78-3; Carrier for polyester/wool blended fabrics; beta form used in insecticides. [Allchem Industries; Crowley Tar Prods.; Koch]

Methyl Niclate®. CAS 15521-65-0; Nickel dimethyldithiocarbamate; antioxidant for epichlorohydrin and peroxide vulcanized elastomers. [R.T. Vanderbilt]

methyl octadecanoate. *See* methyl stearate.

Methylon. A range of substituted phenolic condensates; adhesion promoters for polysulfide sealants, chemical-resistant drum and can linings. [Cornelius Chemical Co Ltd] *

S( ) methyloxirane. *See* propylene oxide.

methyl palmitate. (methyl hexadecanoate; hexadecanoic acid, methyl ester) Ester of methyl alcohol and palmitic acid; CH₃(CH₂)₁₄COOCH₃; CAS 112-39-0; EINECS 203-966-3; chemical intermediate, chemical synthesis; lubricant in mineral, cutting, lamination, textile oils, and rust inhibitors; textile and leather application. [Henkel/Emery; Penta Mfg.; Stepan]

Methyl Parathion. O,O-Dimethyl-O-(4-nitrophenyl) phosphorothioate; used extensively in cotton producing areas. [A/S Cheminova] *

2-methyl-2,4-pentanediol. *See* hexylene glycol.

4-methyl-2,4-pentanediol. *See* hexylene glycol.

4-methylpentanol-2. *See* methyl amyl alcohol.

2-methylphenol. *See* o-cresol.

3-methylphenol. *See* m-cresol.

4-methylphenol. *See* p-cresol.

methyl phenyl polysilocane. *See* phenyl dimethicone.

2-methylpropane. *See* isobutane.

2-methylpropanoic acid. *See* isobutyric acid.

2-methyl-1-propanol. *See* isobutyl alcohol.

2-methyl-1-propene, homopolymer. *See* polyisobutene.

methyl propyl ketone. (MPK; 2-pentanone; ethyl acetone) CH₃COCH₂CH₂CH₃; CAS 107-87-9; EINECS 203-528-1; Solvent, substitute for diethyl ketone, flavoring. [Ashland; Janssen Chimica; Penta Mfg.]

methyl pulvate. Vulpic acid, C₁₉H₁₄O₅

4-methylpyridine. *See* γ-picoline.

methyl-2-pyrone. *See* dehydroacetic acid.

N-methyl-2-pyrrolidone. (NMP) CH₃NCH₂CH₂CH₂-CO; CAS 2555-05-7; Solvent for resins, acetylene, etc.; pigment dispersant; petroleum processing; spinning agent for PVC; intermediate. [Aldrich; Ashland; BASF; Chemoxy Int'l. plc; ISP; Janssen Chimica]

α-methylquinoline. *See* quinaldine.

methyl ricinoleate. (12-hydroxy-9-octadecenoic acid, methyl ester; methyl 12-hydroxy-9-octadecenoate; castor oil acid, methyl ester) Ester of methyl alcohol and ricinoleic acid; CH₃(CH₂)₅COHHCH₂CH=CH(CH₂)₇-COOCH₃; CAS 141-24-2; EINECS 205-472-3; plasticizer, lubricant, cutting oil additive, wetting agent. [Penta Mfg.; Reilly-Whiteman]

methyl salicylate. (methyl 2-hydroxybenzoate; sweet birch oil; oil of wintergreen) Ester of methyl alcohol and salicylic acid; C₆H₄OHCOOCH₃; CAS 119-36-8; EINECS 204-317-7; flavor in foods, beverages, topical

---

‡ Trade name and manufacturer not verified   § Trade name without identified manufacturer

analgesic, odorants; uv absorber in sunburn lotions. [Allchem Industries; R.W. Greeff; Penta Mfg.; Quest Int'l.; Schweizerhall]

Methyl Selenac. CAS 144-34-4; Selenium dimethyl dithiocarbamate; rubber accelerator for NR, SBR, and IIR; vulcanizing agents; effective in low sulfur and sulfurless heat resistant compounds. [R T Vanderbilt]

methyl stearate. (methyl octadecanoate; octadecanoic acid, methyl ester) Ester of methyl alcohol and stearic acid; $C_{19}H_{38}O_2$; $CH_3(CH_2)_{16}COOCH_3$; CAS 112-61-8; EINECS 203-990-4; intermediate for stearic acid detergents, emulsifiers, wetting agents, stabilizers, resins, lubricants, plasticizers. [Ferro/Keil; Penta Mfg.; Witco/Humko]

α-methylstyrene monomer. (isopropenylbenzene; 2-phenylpropene; 2-phenylpropylene) $C_6H_5C(CH_3):CH_2$; CAS 98-83-9; EINECS 202-705-0; Polymerization monomer, esp. for polyesters. [Allied-Signal; Ashland; Honeywill & Stein Ltd; Mitsui Petrochem. Ind.; Mitsui Toatsu Chem.]

methyl styryl ketone. See benzylidene acetone.

methyl succinate. See dimethyl succinate.

methyl sulfoxide. See dimethyl sulfoxide.

methyl tetradecanoate. See methyl myristate.

methyltheobromine. See caffeine.

methyl toluene. See xylene.

Methyl Tuads®. CAS 137-26-8; Tetramethylthiuram disulfide; ultra accelerator; for NR and synthetic rubbers (esp. IIR, CR); accelerator and vulcanizing agent; cure modifier for Neoprene (retards G types; accelerates vulcanization of W types); nondiscoloring in light stocks. [R.T. Vanderbilt]

Methyl Zimate® . CAS 137-30-4; Zinc dimethyldithiocarbamate; ultra accelerator for NR and synthetic rubbers; latex accelerator; act. over wide temp. range; generally requires thiazole modifier for safe processing and wide curing range; nondiscoloring in light stocks. [R.T. Vanderbilt]

Meticortelone Acetate. CAS 52-21-1; Prednisolone acetate; glucocorticoid. [Schering Corp] *

Meticorten. CAS 53-03-2; Prednisone; glucocorticoid. [Schering Corp]

Meti-Derm. CAS 50-24-8; Prednisolone; glucocorticoid. [Schering Corp] *

Metilar. A proprietary trade name for the 21-acetate of paramethazone. [Syntex Laboratories Inc] *

Metillure. An acid-resisting alloy consisting of 17% silicon, 81% iron, 0.9% manganese, 0.25% aluminum, 0.6% carbon, and 0.17% phosphorus.

Metiloil® . Methyl esters of ricinoleic acid and other fatty acids in castor oil; used for tanning and in textiles and for the manufacture of sulfonated emulsifying agents. [Aquitaine-Organico] ‡

Metol. Mono-methyl-p-amino-m-cresol sulfate, $(C_6H_3(OH)CH_3(NH)CH_3)_2H_2SO_4$; used as a developing agent used in photography.

Metolat FC 355. Low foaming wetting agent for color paints, tinting pastes, org./inorg. pigments; improves color acceptance in emulsion paints. [Münzing Chemie GmbH]

Metolat FC 515. Dispersing for extenders and pigments, emulsion paints. [Münzing Chemie GmbH]

Metolat LA 524. Grinding and wetting agent for lacquers and varnishes; wetting agent for organically modified bentonite. [Münzing Chemie GmbH]

Metolat TH 75. Emulsifier for mineral oil, esp. for cutting oils. [Münzing Chemie GmbH]

metolhydroquinone. (metol-quinone). It contains Metol (qv), hydroquinone, sodium phosphate, sodium sulfite, and potassium carbonate; used as a photographic developer.

Metopirone. CAS 54-36-4; A proprietary preparation of metyrapone; used to test pituitary gland function. [Ciba plc] *

Metosyn. CAS 356-12-7; A proprietary preparation of fluocinonide; a steroid skin cream. [ICI Chem. & Polymers Ltd]

Metox. CAS 364-62-5; Metoclopramide; pharmaceutical preparation for the treatment of dyspepsia, flatulence and hiatus hernia. [M A Steinhard Ltd] *

Metozin. See Antipyrine. §

Metprep. Alkaline and solvent paint strippers. [Brent Chemicals International plc]

Metra. CAS 50-58-8; Phendimetrazine tartrate; appetite suppressant. [O'Neal, Jones & Feldman Pharmaceuticals] *

Metral® . For the electrical industry. [DuPont UK]

metramine. See hexamine.

Metraspray. Tetracaine; anesthetic. [3M Pharmaceuticals]†

Metrax. Industrial detergent. [Crosfield Chemicals Ltd] †

Metrazol. (cardiazol; pentamethylene-tetrazole) $C_6H_{10}N_4$; CAS 54-95-5; used in medicine.

Metreton. CAS 125-02-0; Prednisolone sodium phosphate; glucocorticoid. [Schering Corp] *

Metro IV. CAS 443-48-1; Metronidazole; antiprotozoal. [Am. McGaw] *

Metrolyl® . CAS 443-48-1; Tablets containing 200 mg and 400 mg metronidazole BP; suppositories containing 500 mg and 1 g metronidazole BP; injection 0.5% w/v in 100 ml bottles or minibags (500 mg metronidazole per 100 ml); used for treatment of infections; prevention of postoperative infections due to anaerobic bacteria; treatment of acute ulcerative gingivitis and acute dental, pericoronitis and apical infections; trichomonas infections, amebiasis, and giardiasis. [Lagap Pharmaceuticals Ltd]

Metro-nite. A refined natural mineral composed of calcium carbonate, and carbonates and silicates of magnesium; used in the paint industry as a pigment.

Metropad. Aq. pigment dispersions; nonresinated systems for pad dyeing applications; excellent resin/catalyst stability. [Yorkshire Pat-Chem]

Metrosol AZ. Detergent for textile scouring. [Reilly-Whiteman]

Metrotect. A proprietary bitumen paint for protecture treatment of iron-work, etc. §

Metrotex. Aq. pigment dispersions; for textile printing applications; contains no formaldehyde derivs. [Yorkshire Pat-Chem]

Metrotex Colors. Aq. pigment dispersions containing no formaldehyde derivatives; for textile printing. [Yorkshire Pat-Chem]

Metro Tiles. Industrial floor tiles for heavy abrasion and impact. [Prodorite Ltd] †

Metrotonin. A mixture stated to contain sparteine and acetyl-choline; used in medicine.

Metruien M. A proprietary preparation containing ethynodiol and mestranol. [G D Searle & Co Ltd] *

---

* Trade name not verified as available   † Trade name verified as obsolete

Metrulen. A proprietary preparation containing ethynodiol and mestranol. [G D Searle & Co Ltd] *

Metso. CAS 6834-92-0; EINECS 229-912-9; Sodium metasilicate. [Crosfield Chemicals Ltd]

Metso. Sodium metasilicate (anhydrous or pentahydrate) or sodium orthosilicate; alkaline component of heavy duty household, institutional and industrial cleaning compounds, metal cleaning. [PQ Corp] *

Metubine Iodide. CAS 7601-55-0; Metocurine iodide; blocking agent. [Eli Lilly & Co]

Meturon® 4L. CAS 2164-17-2; Fluometuron suspension; flowable herbicide controlling annual grasses and broadleaf weeds in cotton and sugarcane. [Griffin]

Mevacor. CAS 75330-75-5; Lovastatin; for the reduction of elevated total and LDL cholesterol levels in patients with primary hypercholesterolemia. [Merck & Co Inc]

Mevantraal. Methyl-2-methyl pentylidene anthranilate. [Quest Int'l. UK Ltd]

Mevasine. CAS 826-39-1; Mecamylamine hydrochloride; an oral antihypertensive agent. [Merck & Co Inc] †

Mevilin-L. A freeze-dried preparation of a living attenuated virus of the Schwarz strain which has been grown in chick embryo fibroblast-tissue cultures; active immunization against measles. [Evans Medical] *

Mevinacor. CAS 75330-75-5; Lovastatin; for the reduction of elevated total and LDL cholesterol levels in patients with primary hypercholesterolemia. [Merck & Co Inc]

Mewlon. Polyvinyl acetate fiber filaments. [British Traders & Shippers Ltd]

Mexapol. Floor polishes. [Evans Vanodine International Ltd]

Mexate. CAS 59-05-2; Methotrexate; antineoplastic. [Bristol Laboratories] *

Mexican onyx. A variety of calcite.

Mexican turpentine. *See* turpentine.

Mexico seeds. Castor oil seeds.

Mexitil. CAS 5370-01-4; A proprietary preparation of mexiletine hydrochloride; an anti-arrhythmic agent. [Boehringer Ingelheim Pharmaceuticals Inc; Boehringer Ingelheim Ltd]

Mexphalte. Trademark for varieties of bitumen; used for road dressing and other purposes. §

Meyerhofferite. CAS 12007-56-6; Calcium borate; an artificially prepared mineral. §

Meyer's solution. Mercury-potassium iodide solution, obtained by dissolving 13.35 g mercuric chloride and 49.8 g potassium iodide separately in water, mixing the solutions and diluting to 1 liter.

Meyprofix® 509 (redesignated Polycare® 509). Vinyl acetate, isobutyl maleate, vinyl neodecanoate copolymer; hairstyling fixative for superior holding and curl retention props. [Rhone-Poulenc UK]

Mezlin. CAS 51481-65-3; Mezlocillin; antibacterial. [Miles Pharmaceuticals]

M.F.C. Materials for chromatography. [British Drug Houses]*

MG2/MG4. Range of granular fertilizers. [Fisons plc, Horticulture Div] *

MGA. CAS 2919-66-6; Melengestrol acetate; antineoplastic; progestin. [Upjohn] *

M-Gard. Wood preservative. [Mooney Chemicals Inc]

MGH-93. CAS 1309-42-8; Magnesium hydroxide; plastics additive permitting higher processing temps. for olefins, PP, nylons; absorbs more heat than hydrated alumina; does not generate poisonous gases during combustion;

dilutes smoke produced by decomposing polymers. [RMc Minerals]

Mgoa rubber. Commercial name for the rubber from *Muscarenhasia elastica* of East Africa.

MHA. Methionine hydroxy analog; calcium feed supplement for poultry and other animal feeds. [Monsanto Co] *

miazine. CAS 289-95-2; Metadiazine; used in medicine and biochemical research.

MIBC. *See* methyl amyl alcohol.

mica. (Muscovite mica; laminated talc; glimmer glist) Silicate minerals; CAS 12001-26-2; filler/extender for plastics, rubber, coatings, and pearlescent pigment applications; for making fireproof window-panes and lamp-chimneys, electrical equipment; oil well drilling muds; binder and reinforcement in lipsticks. [Feldspar; Franklin Industrial Min.; Mearl; Mykroy/Macalex Ceramics; Norwegian Talc UK; Nyco Min.; Van Dyk]

Micabond. A proprietary trade name for a material consisting of mica, shellac, and resin. §

Micacoat®. CAS 12001-26-2; Chemically coupled muscovite mica, coarse and fine grinds: SiO_2, Al_2O_3, K_2O, Fe_2O_3, Na_2O, CaO, TiO_2, MnO_2, P, S; used for polymer composites and high performance coatings. [NYCO® Minerals Inc]

Micafil B. A proprietary synthetic resin-varnish-paper product used in electrical insulation. §

Micafil G. A proprietary electrical insulator made in the form of tubes from shellac, coated paper, and mica. §

Micafil S. A proprietary trade name for a shellac varnish-paper product used as an electrical insulation. §

Micafolium. A general name for electrical insulators made from mica splittings and paper.

Mica-Kote. A proprietary trade name for a roofing felt made from asphalt impregnated felt. §

micanite. A mica material built up of small plates of mica with an insulating material such as shellac, or on a foundation of paper or cloth; used as an electrical insulating material.

micanite cloth. (mica Cambric, toile micanite). Products used for electrical insulation, made from mica splittings on a cotton-cambric backing.

Micarta. A trade name for a range of varnished paper and fabric products using natural and synthetic resin varnishes; used for electrical insulation. §

micarta folium. A similar product to micafolium.

mica silk. An electrical insulating tape made from mica splittings on a silk cloth.

Micatin. CAS 22832-87-7; Miconazole nitrate; antifungal. [Ortho Pharmaceutical Corp] *

Michel XO-24. Modified amine; antistat additive, integral release agent for linear polyethylene and PVC. [M. Michel]

Michel XO-150-12. CAS 3913-02-8; Isododecyl alcohol. [M. Michel]

Michel XO-150-16. CAS 36311-34-9; EINECS 252-964-9; Isocetyl alcohol. [M. Michel]

Michel XO-150-20. CAS 5333-42-6; Octyldodecanol. [M. Michel]

Michel XO-150-1620. CAS 70693-04-8; EINECS 248-470-8; Isostearyl alcohol. [M. Michel]

Michler's hydrol. CAS 119-58-4; p-p Bisdimethylaminobenzhydrol, HO · CH[$C_6H_4N(CH_3)_2$]$_2$; used as a dye intermediate and in organic synthesis.

Michler's ketone. CAS 90-94-8; Tetramethyldiamino-

---

‡ Trade name and manufacturer not verified   § Trade name without identified manufacturer

benzophenone, $[(CH_3)_2N \cdot C_6H_4]_2CO$; used for making dyestuffs, especially auramine derivatives.

Mi-Col. A mildew fungicide. [Plant Protection] *

Micolette Micro-enema. Sodium lauryl sulfoacetate, sodium citrate, glycerol, potassium sorbate, citric acid, sorbitol; for treatment of chronic or acute constipation in the rectum and sigmoid colon. [Cusi] ‡

Micoren. A proprietary preparation of cropropamide and crotethamide; a respiratory stimulant. [Ciba plc] *

Micracet. Pigment preparations. [Ciba plc]

Micral® 855. Alumina trihydrate; flame retardant/smoke suppressant for wire and cable jacketing and insulation, injection molded and extruded polyolefins. [J.M. Huber/Solem]

Micral® 932. Alumina trihydrate; halogen-free smoke suppressor/flame retardant for wire and cable insulation, injection-molded polyolefins, coatings, adhesives, rubber goods, paper filler and coating, PVC, EPDM, EPR, ABS, XLPE, and compression-molded thermosets. [J.M. Huber/Solem]

Micral® 1000. Alumina trihydrate; economical smoke suppressor/flame retardant with high brightness; for wire and cable insulation, injection molded polyolefins, coatings, adhesives, rubber goods, PVC, EPDM, EPR, XLPE, EVA, and compression molded thermosets. [J.M. Huber/Solem]

Micralax. A proprietary preparation containing sodium citrate, sodium alkyl sulfoacetate and sorbic acid; an enema. [SmithKline Beecham] *

Micralax Micro-enema. Combination of sodium citrate, sodium alkylsulfoacetate, sorbic acid together with glycerin, sorbitol and purified water; micro-enema for relief of constipation. [SmithKline Beecham] *

Micro-K. CAS 7447-40-7; EINECS 231-211-8; Potassium chloride; replenisher. [Wyeth Laboratories] *

Micro P Extender. Amorphous mineral silicate; lightweight resin extender, filler for aircraft, military, appliances, business equip., construction, consumer products, elec./electronic, land transport, and marine applications. [DJ Enterprises]

micro-asbestos. An Austrian asbestos of short fiber unsuitable for the ordinary uses of asbestos.

Microbator PC-78. A tamed and stabilized nontoxic chlorine dioxide complex concentrate formulated for use as an additive deodorant without free chlorine release; effectively arrests malodors caused by viruses, fungi, bacteria, and coliform densities when added to coolants, cutting oils, industrial sumps, sludge pits, cooling towers, waste water, marine holding stations and ships bilge areas, animal housing; rest room surfaces. [Punati Chemical Corp] ‡

Microbiotone. A peptic digest of beef tissue that is a water soluble granular product; used in various fermentations, veterinary biologicals and diagnostics as a nutrient for faster growth of various organisms. [Am. Labs]

Microbloc® . CAS 14807-96-6; Surface-treated talc with proprietary coating; antiblock for plastic film industry; enhanced compat. with polyolefins for improved film clarity and low COF. [Pfizer]

Microcal. CAS 10101-39-0; Calcium silicate. [Crosfield Chemicals Ltd]

Microcal ET. CAS 10101-39-0; Series of synthetic calcium silicates of controlled particle size used as extenders in emulsion paint. [Crosfield Chemicals Ltd]

Microcarb. CAS 63-25-2; Carbaryl; contact insecticide and worm killer. [Micro-Biologicals Ltd]

Microcarb T. Carbaryl + pyrethins; insecticide spray for the control of fleas and flies in poultry houses. [Micro-Biologicals Ltd]

Microcatalase® . CAS 9001-05-2; Bacterial catalase derived from *Micrococcus lysodeikticus*; enzyme which removes residual hydrogen peroxide after antimicrobial treatment of milk. [Solvay Enzymes]

Micro-Cel® A. CAS 10101-39-0; Synthetic calcium silicate; functional filler used as carriers, grinding aids, anticaking agent, and conditioner in agric. chemicals; toxicants; carriers for liq. seed inoculants; inert carriers to convert sticky visc. liq. to dry liqs. [Celite]

Micro-Cel® T-38. CAS 10101-39-0; Calcium silicate; extender pigment, opacifier contributing good hiding power to emulsion paints; flatting properties for flat wall paints. [Celite]

Micro-Chek® 11 P. CAS 26530-20-1; 2-n-Octyl-4-isothiazolin-3-one; antimicrobial, mildewcide for PVC, polyurethane, other polymers for use in roofing membranes, exterior automotive trims, awnings, tarpaulins, pond liners, marine upholstery, shower curtains, outdoor furniture. [Ferro/Bedford]

Microcillin. α-Carboxybenzylpenicillin sodium; injectable penicillin especially against gram-negative "problem organisms". [Bayer AG] †

microcosmic salt. (Phosphorus salt, fusible salt of urine, fusible salt, essential salt of urine). Sodium-ammonium-hydrogen phosphate, $Na(NH_4)H \cdot PO_4 \cdot 4H_2O$.

Microcult® GG. Reagent for detecting pathogenic urinary germs, *Candida* species and *Neisseria* in relevant media. [Bayer AG]

Microdol (Extra). A trade name for ground dolomite. §

Micro-Dry® . Aluminum chlorohydrate powder; antiperspirant active. [Reheis Inc]

Microduct® . CAS 9050-36-6; Maltodextrin; carrier for fragrances; emollient oils, bath products; food additive. [CasChem]

Micro-fine® . Fibers. [DuPont UK]

MicroForm B. CAS 1302-78-9; Dry, modified bentonite clay; microfloc coagulant [Hercules]

MicroForm BCS. CAS 1302-78-9; Slurried, modified bentonite clay; microfloc coagulant [Hercules]

Microgen Plus. Mixture of formaldehyde and γ-HCH; used for control of pests in poultry houses. [Micro-Biologicals Ltd]

Microgynon 30. A proprietary preparation of 30 mcg ethinylestradiol and 15 mcg levonorgestrel; an oral contraceptive. [Schering Health Care Ltd]

Microhoba. Jojoba oil derivatives. [A & E Connock (Perfumery & Cosmetics) Ltd]

Microlan. Powdered lanolin. [Croda Chem. Ltd] *

Microlith-A. Pigment preparations. [Ciba plc] *

Microlube A. CAS 8002-74-2; Paraffin wax emulsion; for surface application to paper and other materials; improves water resist. [Hercules]

Microlut. CAS 797-63-7; Norgestrel; progestin. [Schering AG] *

micromeritol. An alcoholic solution of yerba buena.

Micromet. A slowly soluble sodium metaphosphate. [Albright & Wilson Ltd] *

Micromid 1022. Polyamide resin disp.; for use in water-based adhesive and release coating formulations.

[Union Camp]

Micromite. CAS 122-14-5; Emulsifiable concentrate of 500 g fenitrothion per liter; an organophosphorus insecticide. [ICI Agrochemicals]

Micromulse WIO. Designed to form stable w/o emulsions or o/w microemulsions. [Norman, Fox]

Micronal®. Aq microcapsule dispersions; for production of the donor face of non-carbon copying papers. [BASF AG]

Micronase. CAS 10238-21-8; Glyburide; antidiabetic. [Upjohn] *

Micronor. CAS 68-22-4; Norethindrone; progestin. [Ortho Pharmaceutical Corp]

Micropaque. CAS 7727-43-7; X-ray grade barium sulfate. [Nipa Laboratories Ltd.

MicroPflex 1200. CAS 14807-96-6; Surface-modified microtalc. [Pfizer]

Micropil. Silica for industrial use. [Crosfield Chemicals Ltd]

Micropil. Range of spherical inorganic oxide particles for containment and adsorption. [Microporous Materials Ltd]

Micropoly 520. CAS 9002-88-4; Polyethylene; improves body, texture, visual attributes (luster, coverage, opacity) for wide variety of product formulations. [Presperse]

Micropore. A proprietary surgical tape of rayon with a hypoallergenic adhesive. [3M] *

microporite. An insulating concrete made from ground silica and lime, hardened by treatment with steam.

Micropur. Tablets containing 0.1 mg silver for water sterilization (domestic use). [The Boots Co plc]

Micropure® Ultra. CAS 2555-05-7; N-Methyl-2-pyrrolidone. [ISP]

Microsan. A copper fungicide. [Mechema Chemicals Ltd] *

Microseal. Process of deposition of thin dry solid film lubricant coating by impingement; mold release for plastic molds, bearings, sliding surfaces and vacuum lubrication. [E/M Corporation]

Microsil. Rubber reinforcing agent. [Crosfield Chemicals Ltd]

Microsized. Micron sized salt. [Akzo Salt]

Microsperse. CAS 1333-86-4; Range of carbon black dispersions in water; used in paper, concrete, ink, paint etc industries. [Breamhurst Ltd] ‡

Micro-Step® H-301. Sulfonate/nonionic blend; microemulsifier for broad range of pesticides. [Stepan; Stepan Canada; Stepan Europe]

Microstix®. Reagent for detecting pathogenic urinary germs, Candida species and Neisseria in relevant media. [Bayer AG]

Micro-supplex®. Fibers. [DuPont UK]

Microtex GTZ. [Guthrie Latex]

Microthene®. Polyolefin powders. [Quantum Chemical Corp]

Microthene® FA 150-00. Polyolefin; powd. coating resin. [Quantum/USI] *

Microthene® MA 530-060. CAS 9002-88-4; Polyethylene; for rotational molding. [Quantum/USI] *

Microthene® MN 701-00. CAS 9002-88-4; LDPE; powd. coating resin. [Quantum/USI] *

Microthene® MP 625U. CAS 9002-88-4; Polyethylene; for rotational molding. [Quantum/USI] *

Microthene® MU 760-00. Polyethylene, 19% vinyl acetate incorporated; for rotational molding. [Quantum/USI] *

Micro-triever®. [DuPont UK]

Microtuff 1000. CAS 14807-96-6; Surface-treated talc; filler for polyolefins. [Pfizer]

Microval Tablets. Levonorgestrel; oral contraceptive. [Wyeth Laboratories]

Micro-White® 07 Slurry. CAS 1317-65-3; Calcium carbonate; ultrafine wet-ground product for applications requiring high gloss and opacity in coatings and inks. [ECC]

Micro-White® 10 Codex. CAS 1317-65-3; Calcium carbonate; food-grade meeting purity requirements for food and food contact applications (flour, cake mixes, cereals, chewing gum, crackers). [ECC]

Micro-White® 15. CAS 1317-65-3; Calcium carbonate; ultrafine wet-ground product for paint and polymer industries. [ECC]

Micro-White® 25. CAS 1317-65-3; Calcium carbonate; easy dispersing pigment for paint, rubber, plastics, floor covering and other industries. [ECC]

Micro-White® 40. CAS 1317-65-3; Calcium carbonate; fine ground filler for economy paint. [ECC]

Micro-White® 100. CAS 1317-65-3; Calcium carbonate; wh. general-purpose filler for paints, rubber goods, putties, and joint compounds. [ECC]

Microx. CAS 1314-13-2; Zinc oxide, high surface area. [Harcros UK]

Micryston. A proprietary name for a group of hormone preparations used for replacement therapy. [LAB Ltd] ‡

mictine. 1-Allyl-6-amino-3-ethylpymidine-2,4-dione.

Mictral. Granules containing nalidixic acid, sodium citrate, citric acid and sodium bicarbonate; antibacterial agent for the treatment of cystitis and lower urinary tract infections. [Winthrop Laboratories] *

Midamide. CAS 17440-83-4; Amiloride hydrochloride; a potassium-conserving agent. [Merck & Co Inc] †

Midamor. CAS 17440-83-4; Amiloride hydrochloride; a potassium-conserving agent. [Merck & Co Inc]

Midas®. For agriculture. [DuPont UK]

Midas Gold®. Colorants in solid form and dispersions. [Reichhold]

middle oils. (Carbolic oils). That fraction of coal tar distilling at 170-230 C.

Midicel. CAS 80-35-3; A proprietary preparation containing sulfamethoxypyridazine. [Parke-Davis] *

Midrid. Isometheptene mucate 65 mg, dichloralphenazone 100 mg and paracetamol 325 mg; for treatment of migraine and tension headache. [G W Carnrick Co Ltd]‡

Midstream. CAS 85-00-7; Diquat; a granular contact herbicide and pre-harvest crop desiccant. [ICI Chem. & Polymers Ltd]

Midvale Alloys. Heat-resisting alloys; ATV alloy contains from 33-39% nickel, 10-12% chromium, 1.1-1.8% manganese, and the balance iron; BTG alloy contains 60-62% nickel, 10-11% chromium, 1.2-1.5% manganese, and the balance iron. §

Midvaloy H.R. A proprietary nickel-tungsten-chromium alloy; resists corrosion. §

Mifaslug. CAS 108-62-3; Pellets containing 6% w/w metaldehyde; moluscicide with many crop uses; snail and slug bait. [Farmers Crop Chemicals Ltd]

Mifatox. CAS 919-86-8; An emulsifiable concentrate containing 580 g demeton-S-methyl per liter; a systemic organophosphorus insecticide. [Farmers Crop Chemicals Ltd]

Migafar. Proofing agent. [Ciba plc] *

Migafar AL. A proprietary trade name for an aqueous

---

‡ Trade name and manufacturer not verified   § Trade name without identified manufacturer

emulsion of fatty products suitable for the removal of chafe marks from dyed fabrics. [Ciba plc] *

Migassist® NYL. Leveler for acid dyes. [Sybron]

Migatex. Finishing agents. [Ciba plc] *

Mi-Gee Brand. CAS 75-11-6; Methylene iodide; used for gem and mineral testing. [Geoliquids] *

Migen. House dust mite vaccine. [SmithKline Beecham] †

Might-Weld Multi-Cure Structural Adhesives. Acrylic structural adhesives, ultraviolet curing structural adhesives; used for magnet, metal, glass and fiber optics bonding, assembly adhesives and coatings for electronics. [Dymax Corporation] †

Mighty Soft. Dough softener. [Eastman]

Migil. Migraine remedy. [The Wellcome Foundation Ltd] †

Miglyol® . Special oils/neutral oils, triglycerides; used for production of oily suspensions, suppositories, ointments and creams. [Hüls AG]

Miglyol® 808. CAS 538-23-8; Tricaprylin; good skin spreading/penetrating props. [Hüls Am.]

Miglyol® 812. CAS 65381-09-1; Caprylic/capric triglyceride; dispersant, lubricant, anticaking agent, carrier, solv., solubilizer, suspending agent for cosmetics, dietetic products. [Hüls Am.; Hüls AG]

Miglyol® 818. Caprylic/capric/linoleic triglyceride; emollient for topicals. [Hüls Am.]

Miglyol® 829. Caprylic/capric/diglyceryl succinate; emollient, suspending agent for cosmetic and pharmaceutical topicals with dens. above 1.0. [Hüls Am.]

Miglyol® 840. Propylene glycol dicaprylate/dicaprate; emollient, dispersant, lubricant, suspending agent, solubilizer; act. ingred. for cosmetics and pharmaceuticals; carrier/vehicle and solv. for injection products, topical ointments, creams, lotions, suppositories; dietetic products. [Hüls Am].

Miglyol® Gel. Caprylic/capric triglyceride and stearalkonium hectorite; stabilizer; improves consistency and thermostability in cosmetic and pharmaceutical creams; emulsion stabilizer; as a base for w/o and o/w creams when combined with suitable emulsifiers; does not melt at even 100 C. [Hüls Am.] *

mignonette-geranium oil. An oil obtained by distilling geraniol over mignonette flowers.

Mignonette Green. See May Green. §

Migraine Dolviran®. CAS 379-79-3; Analgesic combination with ergotamine tartrate. [Bayer AG]

migra iron. A special pig iron for high quality castings obtained by a special heat treatment before casting, which results in a remarkably fine grain.

Migraleve. A proprietary preparation of buclizine dihydrochloride, paracetamol, codeine phosphate and dioctylsodium sulfosuccinate tablets; used in the treatment of migraine. [Int'l. Chemical Co Ltd] *

Migrane-Dolviran. Pharmaceutical preparation; used for the relief of migraine and migrainous headaches. [Bayer AG]

Migranil 858. Metallic salt; antimigrant for pigment dyeing. [Sybron]

Migrassist® D. Benzyl trimethyl ammonium chloride; migrator/dye leveling agent for cationic dyes; nonretarding. [Sybron]

Migril. A proprietary preparation of ergotamine tartrate, cyclizine hydrocloride and caffeine; used in the treatment of migraine. [The Wellcome Foundation Ltd]

Mikacion Dye® . Reactive cold-type dyes. [Mitsubishi Kasei]*

Mikamycin. An antibiotic produced by *Streptomyces mitakaensis*. Mikamycin B is Ostreogrycin B.

Mikawhite® . Fluorescent whitening agent for acetate, polyester, polyacrylic fiber. [Mitsubishi Kasei] *

Mikolite. A proprietary material; it is a vermiculite which has been expanded by calcination giving a very fine product; used in paints. §

Mikrobin. Sodium-p-chlorobenzoate; used as a preservative for wines.

Mikron. Organic and inorganic pigments. [Tennant-KVK Ltd]

Mila. Skin cream, milk cleanser and lotion, for daily care of the skin. [Richardson-Vicks Inc] *

Milanol. Basic bismuth trichloro-butyl-malonate.

Milcap. CAS 23947-60-6, 2939-80-2; Fungicide containing ethirimol and captafol for use on wheat. [ICI Chem. & Polymers Ltd]

Mil-Col. CAS 5707-69-7; Drazoxolon; mildew fungicide and seed dressing. [ICI Agrochemicals; ICI Chem. & Polymers Ltd]

Milcurb. CAS 5221-53-4; Fungicide containing dimethirimol. [ICI Chem. & Polymers Ltd]

mild alkali. Sodium carbonate, Na_2CO_3.

Mildison. CAS 50-23-7; A proprietary preparation of hydrocortisone; used as a topical corticosteroid. [Brocades Pharma]

mild lime. Calcium carbonate (chalk), $CaCO_2$, is known in agriculture as mild lime.

Mildothane. CAS 23564-05-8; Suspension concentrate containing 500 g/l thiophanate-methyl; a systemic insecticide. [DowElanco Ltd]

Mildothane. CAS 23564-05-8; Suspension concentrate containing 500 g/l thiophanate-methyl; a systemic insecticide. [Hortichem Ltd]

Mildothane Turf Liquid. CAS 23564-05-8; Suspension concentrate containing 500 g/l thiophanate-methyl; a systemic insecticide. [Rhone-Poulenc Crop Protection Ltd]

Mildvac-C. Connecticut type bronchitis vaccine; for immunization of poultry. [Intervet Inc] †

Mildvac-M. Mild Massachusetts type bronchitis vaccine; for immunization of poultry. [Intervet Inc]

Mildvac Ma5 (Mild Mass.Type). For immunization of poultry. [Intervet Inc]

Milgard. Cleansing lotion, for babies and sensitive skin. [Richardson-Vicks Inc] *

Milgo. CAS 23947-60-6; Fungicide containing ethirimol. [ICI Chem. & Polymers Ltd]

Milid. See Proglumide.

milk acid. See lactic acid.

Milkamino 20. Milk amino acids; substantive moisturizer for cosmetics field; low salt. [Brooks Industries]

milk glass. A soda or flint glass rendered opaque by the addition of a mineral phosphate.

milk of lime. Slaked lime and water in a thin cream.

milk of magnesia. CAS 1309-42-8; A suspension of magnesium hydroxide, $Mg(OH)_2$.

milk protein. See casein.

milkstone. A mixture of milk salts and protein obtained from milk.

milk sugar. CAS 63-42-3; Lactose, $C_{12}H_{22}O_{11}$

milk sugar rennet. See Pegnin.

milk tree wax. Cow tree wax.

---

* Trade name not verified as available   † Trade name verified as obsolete

Millad® . Trade name for family of sorbitol-based clarifying agent additives to improve the transparency of polyolefins, especially polypropylenes; for polypropylene random copolymers and homopolymers used for housewares, protective packaging, cases, medical devices, blow molded bottles and sheet. [Milliken]

Millad® 5L71-10. 10% Conc. of Millad 3905 in LDPE; clarifying agent for use in LLDPE. [Milliken]

Millad® 3905. Additive to enhance clarity and aesthetics of polyolefins, esp. PP, LLDPE, some HDPE. [Milliken]

Millad® HBPA. Hydrogenated bisphenol A; used in prep. of alkyd, polyester, and epoxy resins where good color stability and improved weatherability are important; for casting, laminating, coatings and fiber production. [Milliken]

Millaloy. A proprietary trade name for a nickel-chromium steel containing 4% nickel, 1.5% chromium, and 0.4% carbon. §

Millamine® 5260. Cycloaliphatic diamine (93% 1,2-diaminocyclohexane, 6% methylpentamethylenediamine); epoxy curing agent offering outstanding heat distort. temp., chem. resist. and physical props. in cured system; also as corrosion inhibitor, PU crosslinker, intermediate for polyamides and other chems. [Milliken]

Millathane® 66. Polyester urethane rubber; millable urethane which can be processed and cured by conventional rubber equip. and techniques; excellent resist. to abrasion, ozone, fuels, and oils, very good hot tear strength; peroxide curable. [TSE Industries]

Millathane® 88. Polyether urethane rubber; millable urethane can produce transparent or brightly colored parts with excellent abrasion resist., low temp. flexibility; very good uv stability; peroxide curable. [TSE Industries]

Mill Creek. Natural products, a line of hair and skin care products containing natural ingredients including keratin, aloe vera, jojoba, henna, elastin, apricot and chamomile. [Richardson-Vicks Inc] *

Milldride® . Alkenyl succinic anhydrides, dicarboxylic anhydrides; epoxy curing agents, starch modifiers, additives for motor oil and transmission fluid. [Milliken]

Milldride® DDSA. Dodecenyl succinic anhydride (C12-branched); curing agent for epoxy resins, corrosion inhibitor for nonaq. lubricating oils, intermediate for prep. of alkyd or unsat. polyester resins, intermediate in chem. reactions. [Milliken]

Milldride® HDSA. Hexadecenyl succinic anhydride; curing agent for epoxy resins, corrosion inhibitor for nonaq. lubricating oils, intermediate for prep. of alkyd or unsat. polyester resins, intermediate in chem. reactions. [Milliken]

Milldride® HHPA. Hexahydrophthalic anhydride; epoxy curing agent; also for prep. of alkyd and polyester resins where good color stability is important; for casting, laminating, embedding, coating, and impregnating elec. components. [Milliken]

Milldride® MHHPA. Methyl hexahydrophthalic anhydride; epoxy curing agent; for casting and impregnation applications where low visc., light color and good heat resist. are required. [Milliken]

Milldride® nDDSA. n-Dodecenyl succinic anhydride (C12-linear); curing agent for epoxy resins, corrosion inhibitor for nonaq. lubricating oils, intermediate for prep. of alkyd or unsat. polyester resins, intermediate in chem. reactions. [Milliken]

Milldride® nDSA. n-Decenyl succinic anhydride; curing agent for epoxy resins, corrosion inhibitor for nonaq. lubricating oils, intermediate for prep. of alkyd or unsat. polyester resins, intermediate in chem. reactions. [Milliken]

Milldride® ODSA. Octadecenyl succinic anhydride; curing agent for epoxy resins, corrosion inhibitor for nonaq. lubricating oils, intermediate for prep. of alkyd or unsat. polyester resins, intermediate in chem. reactions. [Milliken]

Milldride® OSA. CAS 26680-54-6; Octenyl succinic anhydride; curing agent for epoxy resins, corrosion inhibitor for nonaq. lubricating oils, intermediate for prep. of alkyd or unsat. polyester resins, intermediate in chem. reactions. [Milliken]

Milldride® TDSA. Tetradecenyl succinic anhydride; curing agent for epoxy resins, corrosion inhibitor for nonaq. lubricating oils, intermediate for prep. of alkyd or unsat. polyester resins, intermediate in chem. reactions. [Milliken]

Millektrol. A German sodium carbohydrate preparation; used as a styptic.

Millidet. An electric blasting cap with a bronze or aluminum shell; used in quarries, open-pit mines, underground mining operations, coal stripping, shafts, tunnels and heavy construction projects. [Hercules] *

Millifoam. Foaming agent for plasterboard. [Millmaster-Onyx UK]

milling silver. A nickel silver. It is an alloy of 56% copper, 27.5-31% zinc, 12-16% nickel, and 0.5-1% lead.

Millithix® 925. CAS 32647-67-9; Dibenzylidene sorbitol; thixotrope and gellant for use in unsat. polyester and vinyl ester resins; also as clear antiperspirant. [Milliken]

Millon's mase. Hydroxy-dimercuro ammonium hydroxide, $OH \cdot Hg_2NH_2O$; used for coloring porcelain.

Millon's reagent. Mercury dissolved in an equal weight of nitric acid (specific gravity 1.41), and the solution diluted to twice its volume. After standing, the liquid is decanted from the precipitate; used as a test for proteins.

Millophyline. A proprietary preparation of diethylaminoethyltheophylline and camphorsulfonate; used as a cardiac and respiratory stimulant; also as a bronchial antispasmodic; for veterinary use. [Dales Pharmaceuticals Ltd] *

Mills Plastic. A proprietary trade name for a vinylidene chloride synthetic resin. §

Milontin. CAS 86-34-0; A proprietary preparation of phensuximide; an anticonvulsant. [Parke-Davis]

Milowite. A proprietary amorphous silica for paint, polishing, and chemical trades; very white in color, and 90% is below 0.01 mm particle size. §

Milstem. CAS 23947-60-6; Liquid ethirimol seed dressing. [ICI Chem. & Polymers Ltd]

Milton. Sterilizing fluid and crystals, to sterilise baby bottles. [Richardson-Vicks Inc] *

Miltopan. Blend of alkylarylsulfonates and solvents, anionic; all purpose detergent. [Henkel Chemicals Ltd] *

Miltown. CAS 57-53-4; Meprobamate; sedative. [Wallace Laboratories]

Milvan Steel. A proprietary trade name for a high speed tool steel containing 19% tungsten, 4% chromium, and 2% vanadium. §

Milwaloy. A proprietary trade name for a chromium-vanadium steel. §

---

‡ Trade name and manufacturer not verified   § Trade name without identified manufacturer

Mimico. See Carbora. §

Minadex. Orange flavored, green syrup containing in each 5 ml the following vitamins and minerals; vitamin A 650 iu, vitamin D 65 iu, iron (as green ferric ammonium citrate) 12 mg, calcium glycerophosphate 11.25 mg, potassium glycerophosphate 1.125 mg, manganese sulfate, 0.5 mg, copper sulfate 0.5 mg; a vitamin and mineral supplement and appetite restorative for children and adults, particularly during and after illness. [Evans Medical] *

Minamino. A proprietary preparation of vitamins, minerals, and aminoacids; used as a dietary supplement. [Consolidated Chemicals Ltd] *

minargent. An alloy of copper, nickel, and aluminum; used as a silver substitute.

minargentatum. An alloy of 56.82% copper, 39.77% nickel, 2.84% tungsten, and 0.57% aluminum.

Mindel® A-670. Polysulfone, ABS-modified; platable, hot water-resist., tough, dimensionally stable resin; used for automotive applications where decorative parts pass through bake cycles or see service near heat sources; also for potable water use, food service applications, plumbing parts; FDA and NSF compliance. [Amoco Chemical Co]

Mindel® B-310. CAS 25135-51-3; Amorphous polysulfone/crystalline polymer blend, glass-reinforced; offers dimensional stability, low shrinkage and warpage; designed for molded elec. connectors; flame-retardant. [Amoco Chemical Co]

Mindel® B-322. CAS 25135-51-3; Amorphous polysulfone/crystalline polymer blend; offers dimensional stability, low shrinkage and warpage, high insulation resist., superior hydrolytic stabiilty; used for connectors, items for service in hot-humid atmospheres. [Amoco Chemical Co]

Mindel® M-800. CAS 25135-51-3; Polysulfone, mineral-filled; improved ESCR; for applications requiring good dimensional stability, excellent creep resist., and toughness; elec. insulation applications. [Amoco Chemical Co]

Mindel® S-1000. CAS 25135-51-3; Polysulfone-based proprietary resin; offers good hydrolytic stability, high heat distort. temp., chem. resist., steam and hot water resist., autoclavability, excellent elec. props., dimensional stability; used for food processing, filtration, food service, medical and hospital applications where sterilization necessary; FDA and NSF compliance; avail. in flame-retardant and glass-filled grades. [Amoco Chemical Co]

Minder. Biocide. [Rhone-Poulenc UK]

Minder. Adjuvant containing 94% rape oil; wetting agent for phosphonoglycine herbicide sprays. [Stoller Chemicals Ltd]

Mindererus's spirit. A solution of ammonium acetate.

Mindust Series. Blend; dedusting and wetting agent for use in coal prep. [Hart Chem. Ltd.

mineral acid. An inorganic acid.

mineral black. A pigment. It is a shale found naturally, and contains 70% silica and 30% carbonaceous matter.

mineral brown. See umber.

mineral butters. A term formerly used for several of the metallic chlorides, such as those of antimony, arsenic, bismuth, and zinc.

mineral caoutchouc. See elaterite.

mineral carbon. CAS 7782-42-5; Anthracite.

mineral carbon. See graphite.

Mineral Cotton. See slag wool.

mineral fat. See petroleum jelly.

mineral flour. A Florida clay used in rubber mixings.

mineral glycerin. See paraffin, liquid.

mineral grey. The ash from lapislazuli after the extraction of ultramarine.

Mineral Gum. See soluble glass.

Mineral Jelly. See petroleum jelly.

Mineral Jelly No. 5. Petrolatum; preblended base for cosmetic mfg.; odorless; tasteless; misc. with cosmetic ingred. used in oil-base formulations. [Penreco]

mineral khaki. A mineral color produced on the fiber by impregnating cotton with a mixture of ferrous and chromic acetates, drying, and then steaming. Mixtures of basic ferric and chromate acetates are formed on the material, which are fixed by passing the fiber through solutions of sodium carbonate and sodium hydroxide.

mineral lake. A basic chromate of tin, prepared by adding potassium chromate solution to stannous chloride solution; used for coloring paper, and in oil painting.

mineral oil. (heavy or light mineral oil; paraffin oil; liquid paraffin) Liq. mixture of hydrocarbons obtained from petroleum; CAS 8012-95-1; 8042-47-5; EINECS 232-384-2; cathartic; laxative; protectant; lubricant. [Carless Refining & Marketing Ltd; Exxon; Magie Bros. Oil; Mobil; Penreco; Witco/Golden Bear, Sonneborn]

mineral pitch. See bitumen.

mineral pulp. See agalite.

mineral rubber. Bitumens of the gilsonite type.

mineral syrup. See paraffin, liquid.

mineral tallow. See bitumen.

mineral umber. See umber.

mineral wax. See ceresin and ozokerite.

mineral white. See gypsum.

mineral wool. See slag wool.

mineral yeast. Torula, a yeast-like organism; used for fodder production.

Minex 2. Anhyd. sodium potassium aluminosilicate; filler for paint, plastics, adhesives and sealants, rubber, abrasives. [Unimin]

Minflo. Mining flotation reagents; depressants which selectively separate gangue materials in mineral flotation, increase the purity of the concentrate, increase efficiency of the collector, and can act as filter aids to agglomerate fines and prevent filter plugging. [Hercules]*

Minihist. CAS 59-33-6; Pyrilamine maleate; antihistaminic. [Ives Laboratories Inc] *

Minipress. CAS 19237-84-4; A proprietary preparation of prazosin hydrochloride; for the treatment of hypertension, left ventricular failure and Raynaud's Disease. [Pfizer International]

Mini Slugit Pellets. CAS 108-62-3; Metaldehyde slug killer. [Murphy Chemical Ltd] †

Minite. A similar explosive to Kohlen carbonite (qv), without barium nitrate.

Minitran™. CAS 55-63-0; Nitroglycerin transdermal delivery system; anti-arrhythmic. [3M Pharmaceuticals.

minium. See red lead.

minium tego. A high dispersed red lead marketed in Germany.

Minizide. A proprietary preparation of prazosin hydrochloride and polythiazide; an antihypertensive. [Pfizer Inter-

---

* Trade name not verified as available

† Trade name verified as obsolete

national]

mink oil. (oil of mink) Oil obtained from the subdermal fatty tissues of the mink; emollient for skin and hair care formulations, makeup removers. [Croda; Lanaetex; Geo Pfaus Sons]

Minlon®. Nylon 66 reinforced with 40% mineral filler; used as an engineering thermoplastic resin. [DuPont UK]

Minocin. CAS 13614-98-7; Minocycline hydrochloride; antibacterial. [Lederle Laboratories USA] *

Minofor. Alloys used by jewelers. They contain from 9-64% antimony, 20-84% tin, 2-10% copper, and 1-10% zinc.

Minol. See paraffin, liquid.

Minolite Antigrisouteuse. A Belgian explosive containing 72% ammonium nitrate, 23% sodium nitrate, 3% trinitrotoluene, and 2% trinitronaphthalene.

Mintacol Solubile. Pharmaceutical preparation; miotic. [Bayer AG] †

mint camphor. Menthol.

Mintezol. CAS 148-79-8; Thiabendazole; anthelmintic. [Merck & Co Inc]

Mintite. A patent finish for rubber surfaces; consists of powdered mica. §

Mint-O-Mag. Milk of magnesia; antacid; laxative. [Bristol-Myers Squibb Co Inc] *

Minugel. CAS 1337-76-4; Colloidal attapulgite clay; for suspension systems. [Whitecourt Ltd]

Min-U-Gel® 100. CAS 1337-76-4; Attapulgite clay; absorbent and adsorbent; gelling and suspending agent [Floridin] *

Min-U-Gel® 200. CAS 1337-76-4; Colloidal attapulgite clay; suspending agent in suspension fertilizers and other agric. suspensions; absorbent, adsorbent. [Floridin] *

Min-U-Gel® AR. CAS 1337-76-4; Attapulgite clay; thixotropic thickener for asphalt cutbacks [Floridin] *

Min-U-Gel® CW. CAS 1337-76-4; Attapulgite clay; stabilizer for coal water slurries [Floridin] *

Min-U-Gel® LF. CAS 1337-76-4; Colloidal attapulgite clay; thickener, stabilizer for liq. animal feeds containing immiscible materials (fat emulsions, vitamins, limestone powd.); suspending agent. [Floridin] *

Minulet. Ethinylestradiol and gestodene; oral contraceptive. [Wyeth Laboratories]

Min-U-Sil. Micronized natural silica; used for silicon rubber, surface coatings. [Cornelius Chemical Co Ltd] *

Minyak Kerung. An oleo-resin obtained from *Dipterocarpus* species of Malay. It is obtained as a viscous liquid.

Miochol® 1:100 Intraocular. CAS 60-31-1; Acetylcholine chloride; used to obtain complete miosis of the iris in seconds after delivery of the lens in cataract surgery; in penetrating keratoplasty, iridectomy and other anterior segment surgery where rapid, complete miosis may be required. [Iolab Corp]

Mio-Pressin. A proprietary preparation of rauwolfia serpentina, protoveratrine, and phenoxybenzamine hydrochloride; an antihypertensive. [SmithKline Beecham]*

Miostat. CAS 51-83-2; Carbachol; cholinergic. [Alcon Laboratories Inc]

MIPA. See isopropanolamine.

Mipcin® . CAS 2631-40-5; 2-Isopropylphenyl methylcarbamate; insecticide, esp. for leafhoppers and plant hoppers on paddy rice. [Mitsubishi Kasei] *

Mipolam. A proprietary trade name for polyvinyl chloride which, when plasticized, has properties resembling soft

rubber; used in molding and for sheathing cables. §

Mirabilite. See Glauber's salt.

Miracare® 2MCA. Disodium cocoamphodiacetate, sodium lauryl sulfate and hexylene glycol; used in formulating nonirritating shampoos. [Rhone-Poulenc Surf.; Rhone-Poulenc France]

Miracare® 2MCA-SF. Disodium cocoamphodipropionate, sodium lauryl sulfate, hexylene glycol; surfactant for nonirritating and non-eye-sting shampoos. [Rhone-Poulenc Surf.]

Miracare® ANL. Sodium C14-16 olefin sulfonate, sodium laureth sulfate, lauramide DEA; high foaming surfactant conc. for high-performance shampoos and skin cleansers. [Rhone-Poulenc Surf.; Rhone-Poulenc France]

Miracare® BC-10. PEG-80 sorbitan laurate, cocamidopropyl betaine, sodium trideceth sulfate, sodium lauroamphoacetate, PEG-150 distearate, sodium laureth-13 carboxylate; conc. for prep. of baby shampoo, bubble bath, bath gel and liq. hand soap products requiring mildness [Rhone-Poulenc Surf.; Rhone-Poulenc France] *

Miracare® BT. Disodium lauroamphodiacetate and sodium trideceth sulfate; surfactant for nonirritating and non-eye-sting shampoos. [Rhone-Poulenc Surf.]

Miracare® CT100. Stearyl alcohol and cetrimonium bromide; emulsifier for creme rinse/conditioner; base for permanent wave foam neutralizers. [Rhone-Poulenc Surf.]

Miracare® M1. Sodium lauryl sulfate, stearamide MEA, glycol stearate and cocamide MEA; built pearlized base for cream shampoos having high flash foam. [Rhone-Poulenc Surf.]

Miracare® MHT. Sodium lauroamphoacetate and sodium trideceth sulfate; surfactant for nonirritating and non-eye-sting shampoos, foaming skin cleansing products. [Rhone-Poulenc Surf.]

Miracare® MS-1. PEG-80 sorbitan laurate, sodium trideceth sulfate, PEG-150 distearate, disodium lauraminopropionate, cocamidopropyl hydroxysultaine, sodium laureth-13 carboxylate; conc. for prep. of baby shampoo, bubble bath, bath gel and liq. hand soap products requiring mildness. [Rhone-Poulenc Surf.]

Miracare® NWC. Sodium laureth sulfate, cocamide DEA, TEA-lauryl sulfate; biodeg. base for shampoos and bubble baths. [Rhone-Poulenc Surf.]

Miracare® SCS. CAS 122-19-0; EINECS 204-527-9; Stearalkonium chloride in compatible emulsifier base; formulated base for simplified production of cream rinse conditioners. [Rhone-Poulenc Surf.]

Miracare® XL. DEA-lauryl sulfate, DEA-lauraminopropionate, sodium lauraminopropionate, propylene glycol; high foaming base for prep. of hair and body shampoos. [Rhone-Poulenc Surf.]

Miracil D. A proprietary trade name for the hydrochloride of lucanthone. §

Miracle Man. Waterproofing cleaner and penetrant. [Rhone-Poulenc UK]

Miraculoy. A proprietary trade name for a steel containing 1.25% nickel, 0.65% chromium, 0.4% molybdenum, and 1.55% manganese. §

Miradon. CAS 117-37-3; (Anisindione; 2-p-anisyl-1,3-indandione); $C_{16}H_{12}O_3$; anticoagulant.

Miraflon® . Fluorocarbon rubber. [Asahi Chem. Industry]

Mirage. CAS 67747-09-5; Prochloraz; fungicide. [Makhteshim Chemical Works Ltd]

---

‡ Trade name and manufacturer not verified

§ Trade name without identified manufacturer

Miralite. A light aluminum alloy which can be cast or rolled, and drawn into wire. It contains 12% copper and 2% tin.

Miramant. A cutting alloy of heat resisting metals with a definite fraction of stable and hard carbides, especially molybdenum and tungsten carbides in eutectic proportions.

Mira metal. Acid-resisting alloys. One contains 74.7% copper, 16-3% lead, 6.8% antimony, 0.91% tin, 0.62% zinc, 0.43% iron, and 0.24% nickel. Another consists of 75% copper, 16% lead, 8% tin, and 1% nickel.

Miramine® C. CAS 67784-90-1; Coco hydroxyethyl imidazoline; emulsifier, corrosion inhibitor, softener, antistat, for textiles, asphalt, plastics, petrol. industry, cutting oils; water repellent treatment of cement, concrete, and plaster; antifungal agent for wood; tar emulsion breaker; slime control additive in paperboard. [Rhone-Poulenc Surf.]

Miramine® CODI. CAS 68140-01-2; EINECS 268-771-8; Cocamidopropyl dimethylamine; emulsifier; base for emulsions for creams, lotions, and hair rinses; conditioner, antistat, visc. builder. [Rhone-Poulenc Surf.]

Miramine® GS. CAS 95-19-2; EINECS 202-397-8; Stearyl hydroxyethyl imidazoline; wetting agent for foam wax stripper, degreaser, cleaning formulations; stable to acid and alkali. [Rhone-Poulenc France]

Miramine® HPS-B. CAS 61791-39-7; EINECS 263-171-2; Tall oil hydroxyethyl imidazoline. [Rhone-Poulenc Surf.]

Miramine® O. CAS 21652-27-7; Oleyl hydroxyethyl imidazoline; emulsifier, corrosion inhibitor, softener, antistat, wetting and flocculating agent, lubricant for textiles, asphalt, car wax emulsions, cleaners, paint mfg., agric. applications, synthetic coolants; dispersant for clay and pigments in solv. systems; tar emulsion breaker. [Rhone-Poulenc Surf.]

Miramine® SODI. CAS 7651-02-7; EINECS 231-609-1; Stearamidopropyl dimethylamine; emulsifier, conditioner; produces cationic emulsions. [Rhone-Poulenc Surf.; Rhone-Poulenc France]

Miramine® TO. CAS 61791-39-7; EINECS 263-171-2; Tall oil hydroxyethyl imidazoline; emulsifier, corrosion inhibitor in oil burning systems, pickling bath operations, asphalt emulsions; dispersant for clay and pigments in solv. systems. [Rhone-Poulenc Surf.]

Miranate® B. CAS 67990-17-4; Sodium butoxyethoxy acetate; low foaming wetting agent, emulsifier, lubricant for wax strippers, degreasers, metalworking fluids, other cleaner formulations; compat. with high concs. of electrolytes; stable to acid and alkali media. [Rhone-Poulenc Surf.; Rhone-Poulenc France]

Miranate® LEC. CAS 70632-06-3; Sodium laureth-13 carboxylate; auxiliary detergent for shampoo systems, lime soap dispersant; emulsifier, lubricant for metalworking fluids. [Rhone-Poulenc Surf.]

Miranol. Liquid foaming agent with outstanding emulsifying properties. [Foseco (F.S.) Ltd] †

Miranol® 2CIB. CAS 68647-53-0; Disodium cocoamphodiacetate; detergent for high foaming, nonirritating shampoos, skin cleansers, cosmetics, industrial detergents; emulsifier, solubilizer, coupling agent for heavy-duty liq. cleaners. [Rhone-Poulenc Surf.; Rhone-Poulenc France]

Miranol® BM Conc. CAS 686-8-66-2; Disodium lauroamphodiacetate; surfactant for nonirritating shampoos, skin cleansers. [Rhone-Poulenc Surf.; Rhone-Poulenc France]

Miranol C2M. Imidazoline based amphoteric with whole coconut as fatty radical; clear aqueous solution; nonirritating surfactants, emulsifiers, solubilizers, stabilizers; used in shampoos, cleaners, pharmaceutical applications. [Venture Chemical Products Ltd] *

Miranol® C2M Anhyd. Acid. CAS 68919-40-4; Cocoamphodipropionic acid; detergent, wetting agent, high foaming surfactant for disp. on caustic soda and on powd. mixes, etc., leveling agent in tin plating from acid baths; metal cleaning; industrial cleaning. [Rhone-Poulenc Surf.; Rhone-Poulenc France]

Miranol® C2M Conc. NP-PG. CAS 68650-39-5; Disodium cocoamphodiacetate, propylene glycol; surfactant. [Rhone-Poulenc Surf.]

Miranol C2M-SF. Imidazoline based amphoteric with whole coconut as fatty radical; clear amber liquid; for industrial cleaner formulations. [Venture Chemical Products Ltd]*

Miranol® C2M-SF 70%. CAS 68604-71-7; Disodium cocoamphodipropionate; detergent used for heavy-duty liq. cleaning compounds, steam cleaners, nonirritating shampoos, medicated cosmetics. [Rhone-Poulenc Surf.; Rhone-Poulenc France]

Miranol CM. Imidazoline based amphoteric with whole coconut as fatty radical; light amber liquid; emulsifies grease, suspends particulate soil. [Venture Chemical Products Ltd] *

Miranol® CM Conc. NP. CAS 68608-65-1; Sodium cocoamphoacetate; detergent, wetting and foaming agent, sequestrant, emulsifier, dispersant, germicidal, visc. builder; for extra heavy duty cleaners, steam, pressure, metal, all-purpose cleaners; biodeg. [Rhone-Poulenc Surf.; Rhone-Poulenc France]

Miranol® CM-SF Conc. CAS 68919-41-5; Sodium cocoamphopropionate; coemulsifier for emulsion polymerization; emulsifier, wetting agent for industrial, institutional and household cleaners; biodeg. [Rhone-Poulenc Surf.; Rhone-Poulenc France]

Miranol® CS Conc. CAS 68604-73-9; EINECS 271-705-0; Sodium cocoamphohydroxypropylsulfonate; detergent, wetting agent, corrosion inhibitor, emulsifier, sequestrant, foaming agent for shampoos, cold water fabrics, household and industrial cleaners; biodeg. [Rhone-Poulenc Surf.; Rhone-Poulenc France]

Miranol DM. Imidazoline based amphoteric with stearic fatty radical, in the form of a creamy white paste; fiber softener, wool lubricant. [Venture Chemical Products Ltd] *

Miranol® DM Conc. 45%. CAS 68608-63-9; Sodium stearoamphoacetate; viscosifier, lubricant, softener, conditioner for cosmetics, textiles, industrial, institutional and household cleaners. [Rhone-Poulenc Surf.; Rhone-Poulenc France]

Miranol® Ester PO-LM4. CAS 96726-23-9; Pentaerythrityl tetralaurate; emulsifier, emollient, conditioner, antistat for personal care products. [Rhone-Poulenc Surf.]

Miranol® FA-NP. CAS 68608-65-1; Sodium cocoamphoacetate; surfactant for extra-heavy duty liq. cleaning compounds, steam, pressure, metal, and all-purpose cleaners. [Rhone-Poulenc Surf.; Rhone-Poulenc France]

Miranol® FB-NP. CAS 68650-39-5; Disodium cocoamphodiacetate; surfactant for nonirritating shampoos, skin cleansers, medicated cosmetics, med.-duty liq. cleaners. [Rhone-Poulenc Surf.; Rhone-Poulenc France]

Miranol® FBS. CAS 68604-71-7; Disodium cocoamphodipropionate; surfactant for extra heavy duty liq. cleaning compounds, e.g., steam, pressure, metal, and all-purpose cleaners. [Rhone-Poulenc Surf.; Rhone-Poulenc France]

Miranol® H2C-HA. CAS 3655-00-3; EINECS 222-899-0; Disodium N-lauryl iminodipropionate; foamer and wetting agent used in alkaline cleaners, fire fighting compounds. [Rhone-Poulenc Surf.]

Miranol® H2M Conc. CAS 68608-66-2; Disodium lauroamphodiacetate; surfactant for nonirritating shampoos, skin cleaners; biodeg. [Rhone-Poulenc Surf.; Rhone-Poulenc France]

Miranol® H2M-SF Conc. CAS 68610-43-5; Disodium lauroamphodipropionate; surfactant for nonirritating shampoos and skin cleaners, esp. aerosols; biodeg. [Rhone-Poulenc Surf.; Rhone-Poulenc France]

Miranol® HM Conc. CAS 68608-66-2; Sodium lauroamphoacetate; detergent, wetting and foaming agent, sequestrant, emulsifier, dispersant, germicidal for shampoos, dishwashing, paints; biodeg. [Rhone-Poulenc Surf.; Rhone-Poulenc France]

Miranol® HM-SF Conc. CAS 61901-02-8; Sodium lauroamphopropionate; emulsion polymerization. [Rhone-Poulenc Surf.] *

Miranol® J2M Conc. CAS 68608-64-0; Disodium capryloamphodiacetate; emulsifier, caustic soda wetting agent, food washing and peeling, industrial, institutional and household cleaners, wax stripper, emulsion polymerization of synthetic rubbers and resins; biodeg. [Rhone-Poulenc Surf.; Rhone-Poulenc France]

Miranol® J2M-SF Conc. CAS 68815-55-4; Disodium caprylamphodipropionate; salt free version of Miranol J2M Conc.; emulsifier, wetting agent, coupling agent, solubilizer; for dispersion on caustic soda and on powd. mixes, metal cleaning, industrial cleaning; higher tolerance for alkalies and/or electrolytes; biodeg. [Rhone-Poulenc Surf.]

Miranol® JAS-50. CAS 68877-55-4; Capryloamphopropionate; emulsifier, wetting agent for industrial, institutional and household cleaners. [Rhone-Poulenc Surf.]

Miranol® JB. CAS 68608-64-0; Disodium capryloamphodiacetate; wetting agent used in caustic soda based cleaners used for food washing and peeling; also in wax stripper formulations. [Rhone-Poulenc Surf.; Rhone-Poulenc France]

Miranol® JBS. CAS 68815-55-4; Disodium capryloamphodipropionate; low foaming surfactant for medicated shampoos and skin cleansers; emulsifier, wetting agent for industrial cleaners. [Rhone-Poulenc Surf.; Rhone-Poulenc France]

Miranol JEM. Imidazoline based amphoteric with caprylic and ethylhexoic fatty radicals; clear aqueous solution; used in bottle washing, wax stripping, degreasing, etc. [Venture Chemical Products Ltd] *

Miranol® JS Conc. CAS 68610-39-9; Sodium caprylamphohydroxypropylsulfonate; emulsifier, wetting agent, corrosion inhibitor used in pickling acids, low foam, acid and alkali stable industrial cleaners; biodeg. [Rhone-Poulenc Surf.; Rhone-Poulenc France]

Miranol L2M-SF. Imidazoline based amphoteric with linoleic fatty radical, in aqueous solution form; used for wax-type polishes and floor finishes, etc. [Venture Chemical Products Ltd] *

Miranol® L2M-SF Conc. CAS 68991-88-8; Disodium

tallamphodipropionate; high foaming detergent, emulsifier for oil, wash and wax products, heavy-duty detergents, metalworking fluids. [Rhone-Poulenc Surf.]

Miranol® OS-D. CAS 68610-38-8; Sodium oleoamphohydroxypropylsulfonate; detergent, shampoos. [Rhone-Poulenc Surf.]

Miranol® S2M-SF Conc. CAS 68815-45-2; Disodium caproamphodipropionate; low wetting surfactant for medicated shampoos, cleansers; biodeg. [Rhone-Poulenc Surf.; Rhone-Poulenc France]

Miranol® SM Conc. CAS 68608-61-7; Sodium caproamphoacetate; wetting agent, foaming agent, detergent used in medicated and germicidal shampoos and hand soaps, rug and upholstery shampoos, in emulsion polymerization; biodeg. [Rhone-Poulenc Surf.; Rhone-Poulenc France]

Miranol SM. Imidazoline based amphoteric with capric fatty radical; clear aqueous solution; used for medicated, germicidal, rug and upholstery shampoos, hand soaps and surgical soaps. [Venture Chemical Products Ltd] *

Miranol® TBS. CAS 68991-88-8; Sodium tallamphodipropionate [Rhone-Poulenc Surf.]

Mirapol® 9, 95, 175. Polyquaternium-27; conditioner, antistat, emollient for hair and skin products; improves thickening of Bentonite clays in aq. systems. [Rhone-Poulenc Surf.]

Mirapol® 1941. Acrylates/steareth-20 methylacrylate copolymer; pH sensitive thickening agent, emulsifier for liq. detergents, shampoos, and cosmetics; optimum thickening at slightly alkaline pH. [Rhone-Poulenc Surf.]

Mirapol® 550. CAS 26590-05-6; Polyquaternium-7; polymer used in hair and skin care products to impart lubricity, slip, detangling, and luster to hair and a smooth, soft feel to skin [Rhone-Poulenc Surf.; Rhone-Poulenc France]

Mirapol® A-15. CAS 68555-36-2; Polyquaternium-2; softening, conditioning, lubricant, antistat, surface modifying agent used in cream rinses, conditioning-type shampoos, textile processing. [Rhone-Poulenc Surf.; Rhone-Poulenc France]

Mirapol® AD-1. CAS 90624-75-2; Polyquaternium-17; conditioner and antistat for personal care products. [Rhone-Poulenc Surf.]

Mirapol® AZ-1. CAS 90624-76-3; Polyquaternium-18; conditioner and antistat for personal care products. [Rhone-Poulenc Surf.] *

Mirasheen® 202. Glycol stearate, lauramide DEA, cocamidopropyl betaine, glycerin; pearl conc. for cold blend cosmetic formulations, liq. soaps; contains visc. building, foam boosting, and mild conditioning agents. [Rhone-Poulenc Surf.; Rhone-Poulenc France]

Mirasol. A proprietary trade name for alkyd varnish and lacquer resins. §

Mirataine® A2P-TS-30. Disodium N-tallow aminodipropionate; foamer, wetting agent with tolerance to alkali; for use in alkaline cleaners. [Rhone-Poulenc Surf.]

Mirataine® BB. CAS 86438-78-0; Lauramidopropyl betaine; mild substantive surfactant, visc. builder and foam booster, wetting agent for shampoos and dishwashing liqs.; conditioner, antistat, emollient; as solubilizer, visc. builder, foam booster with lauryl sulfates; stable to acid and alkali media. [Rhone-Poulenc Surf.; Rhone-Poulenc France]

Mirataine® BET-C-30. CAS 70851-07-9; Cocamidopropyl betaine; mild foaming agent, conditioner, detergent,

---

‡ Trade name and manufacturer not verified   § Trade name without identified manufacturer

emulsifier, foam booster/stabilizer, visc. builder for shampoos, liq. soaps, facial cleansers, bath gels, bubble baths. [Rhone-Poulenc Surf.]

Mirataine® BET-CS. See Mirataine® CBS.

Mirataine® BET-O-30. CAS 25054-76-6; EINECS 246-584-2; Oleamidopropyl betaine; conditioner, detergent, emulsifier, foam booster/stabilizer, visc. builder for shampoos, liq. soaps, facial cleansers, bath gels, bubble baths. [Rhone-Poulenc Surf.; Rhone-Poulenc France]

Mirataine® BET-P-30. CAS 693-33-4; EINECS 211-748-4; Cetyl betaine; foaming agent, conditioner, detergent, emulsifier, foam booster/stabilizer, visc. builder for shampoos, liq. soaps, facial cleansers, bath gels, bubble baths. [Rhone-Poulenc Surf.] *

Mirataine® BSC. CAS 70851-08-0; Cocamidopropyl hydroxysultaine. [Rhone-Poulenc Surf.]

Mirataine® CBS, CBS Mod. CAS 70851-08-0; Cocamidopropyl hydroxysultaine; mild high foaming surfactant, visc. builder, foam booster, emulsifier, wetting agent for shampoo formulations, liq. bubble baths, industrial cleaners. [Rhone-Poulenc Surf.; Rhone-Poulenc France]

Mirataine® H2C-HA. CAS 14960-06-6; Sodium lauriminodipropionate; high foaming surfactant, foam booster, wetting agent, dispersant for shampoos and skin cleansers, and hard surface cleaners. [Rhone-Poulenc Surf.; Rhone-Poulenc France]

Mirataine® HC-Acid. CAS 1462-54-0; Lauraminopropionic acid. [Rhone-Poulenc Surf.]

Mirataine® JC-HA. Alkyl iminopropionate; low foam hydrotrope for spray cleaners, rinse additives, mech. scrubbing systems, highly built alkaline systems, hard surf. cleaners, circulating systems. [Rhone-Poulenc Surf.]

Mirataine® T2C-30. CAS 61791-56-8; Disodium tallowiminodipropionate; detergent, solubilizer, moderate foaming surfactant used in textile, leather, metalworking, industrial and personal care products. [Rhone-Poulenc Surf.]

Mirataine® TM. CAS 61791-25-1; Dihydroxyethyl tallow glycinate; wetting agent, viscosifier for industrial and household cleaners; conditioner for shampoos; HCl thickener. [Rhone-Poulenc Surf.; Rhone-Poulenc France]

Miravon. Flow modifier. [Rhone-Poulenc UK]

Miravon B12DF. Ethylene/propylene oxide adducts; biodeg. surfactant, defoamer for dishwash rinse aids, dairy cleaning, metal degreasing, hard surf. cleaners, biocides. [Rhone-Poulenc France; Rhone-Poulenc UK]

mirbane oil. (nitrobenzene; oil of mirbane) $C_6H_5NO_2$; CAS 98-95-3; used in the manufacture of aniline, solvent for cellulose ethers, modifying esterification of cellulose acetate, ingredient of metal polishes; manufacture of benzoline, quinoline, and azobenzene.

Mirion. A proprietary preparation; it is iodo-hexamine. §

Mirrolac. Calendering varnishes. [The Scottish Adhesives Co Ltd]

mirror bronze. A copper and tin alloy, containing 28-35% tin. It sometimes contains a little nickel.

Mirvale. CAS 101-21-3; Chlorpropham; potato sprout depressant. [Ciba-Geigy Agrochemicals]

Mischzinn. A tin alloy. Theoretically it is an eutectic containing 63% tin and 37% lead, but in practice the tin is at least 55%. antimony and copper must be more than 3.5 and 0.5% respectively, and zinc is present in traces.

miscible carbon disulfide. A mixture of carbon disulfide with castor oil, caustic potash, denatured alcohol, and water. An insecticide for destroying the Japanese beetle in the soil without serious damage to the plant.

Misco. An alloy of 57.5% iron, 15% chromium, 25% nickel, 1.5% silicon, 0.5% manganese, and 0.5% carbon.

Miscrome. A proprietary trade name for a corrosion-resisting alloy of iron with 28% chromium. §

Missile. CAS 13457-18-6; Pyrazophos; systemic organophosphorus fungicide. [Hoechst UK]

Mission Prenatal. CAS 59-30-3; Folic acid; vitamin. [Mission Pharmacal Co]

mistletoe rubber. A rubber obtained from the fruit of certain Loranthaceae as parasites on the coffee tree.

Mist-O-Matic. A liquid fungicide seed dressing. [DowElanco Ltd]

Mist-o-matic Ferrax. Fungicide seed treatment. [Murphy Chemical Ltd] †

Mistral. CAS 67306-03-0; Emulsifiable concentrate of 750 g fenpropimorph per liter; used for control of mildew in cereals. [Rhone-Poulenc Crop Protection Ltd]

Mistral CT. A suspension concentrate containing 250 g chlorothalonil and 187 g fenpropimorph per liter; a systemic fungicide for winter wheat. [Rhone-Poulenc Crop Protection Ltd]

Mistron CB. CAS 14807-96-6; Talc; surface treated ultrafine platy talc; offers excellent impact props. to polyolefins, good film props. to films, improved tear and aging props. in rubber. [Luzenac Am.]

Mistron Vapor-R. CAS 14807-96-6; Talc; ultrafine plty talc as reinforcing filler for elastomers; processing aid for elastomers, plastics; rheological control agent for resins, plastisols, adhesives; nucleating agent for plastic foams. [Luzenac Am.]

Mistron ZSC. CAS 14807-96-6; Talc, zinc stearate-coated; flatting agent in solv.-based coatings, rheological control in resins, plastisols and adhesives, nucleating agent in plastic foams, mold release agent, rubber dusting aid. [Luzenac Am.]

Mitaban. CAS 33089-61-1; Amitraz; scabicide. [Upjohn] *

Mitac 20. CAS 33089-61-1; Amitraz; acaricide and insecticide for use on fruit trees and hops. [Schering Agrochemicals Ltd]

Mitas® . Compound seasonings; for food industry. [Asahi Chem. Industry]

Mitchalloy A. A proprietary trade name for an alloy of iron with 2.5% nickel and 0.9% chromium. §

Mitec® GP105A. Isocyanate adduct; yellowing general-purpose hardener for chem.-resist., anticorrosion, concrete, wood finishes, PU finishes. [Mitsubishi Kasei] *

Mithracin. A proprietary preparation of mithramycin; an antineoplastic. [Pfizer International]

Mitigan. CAS 115-32-2; Active ingredient: dicofol; specific miticide. [Agan Chemical Manufacturers Ltd]

mitigated caustic. A fused mixture of 1 part silver nitrate with 2 parts potassium nitrate.

Mitin. Moth proofing agents. [Ciba plc] *

Mitine. A base for ointments prepared from an emulsion which is superfatted with a non-emulsifying fat. Wool fat is used as the fat, and milk as the serum-like liquid to the extent of 50%.

Mitoxana. See Holoxan. [Degussa AG] *

Mitrelle. Polyester yarn. [ICI Chem. & Polymers Ltd]

Mitschlich's ammoniacal salt. A double compound of

---

mercuroxy-ammonium nitrate, and mercuriammonium nitrate ($NH_2 \cdot Hg_2O)NO_3 \cdot (NH_2Hg)NO_3 \cdot H_2O$.

Mitsubishi 4300J. CAS 9003-07-0; Polypropylene resin; for injection molding applications. [Mitsubishi Kasei] *

Mitsubishi BT002. CAS 9002-888-4; HDPE; for blow molding of large bottles. [Mitsubishi Kasei] *

Mitsubishi ET008. CAS 9002-888-4; HDPE; for monofilament (rope, net, screen), inflation film (wrapper for foods). [Mitsubishi Kasei] *

Mitsubishi F101A. CAS 9002-888-4; LDPE; for film (heavy duty bags), blow molding (large articles). [Mitsubishi Kasei] *

Mitsubishi JS050. CAS 9002-888-4; HDPE; for injection molding of pails, bottles, caps. [Mitsubishi Kasei] *

Mitsubishi L300. CAS 9002-888-4; LDPE; for lamination applications. [Mitsubishi Kasei] *

Mitsubishi UF421. CAS 9002-888-4; LLDPE; for high-clairy film applications. [Mitsubishi Kasei] *

Mitsubishi Kasei GF-PET 6010G15. CAS 25038-59-9; PET, 15% glass-reinforced; UL94HB grade. [Mitsubishi Kasei] *

Mitsubishi Kasei PBT 5008. CAS 26062-94-2; PBT; high flow grade for injection molding, esp. thin-walled products. [Mitsubishi Kasei] *

Mitsubishi Kasei PBT 5010F1. CAS 26062-94-2; PBT, inorg. filled; reinforced grade for warp resist. and heat resist. applications. [Mitsubishi Kasei] *

Mitsubishi Kasei PPS 704G40. CAS 9016-75-5; PPS, 40% glass reinforced; UL94V-O grade. [Mitsubishi Kasei] *

Mitsubishi Yuka-ECX. Electroconductive polymer. [Mitsubishi Petrochem.] *

Mitsubishi Yuka-SPX. Soft polyolefin. [Mitsubishi Petrochem.] *

Mittel AEP. Ethyl-p-toluenesulfonate; a proprietary softening agent for cellulose esters. §

Mittel KP. Cresyl-p-toluenesulfonate; a softening agent for cellulose esters. [sm]

Mittel L. A solvent resembling turpentine.

Mivacron. A proprietary formulation of micacurium chloride; a muscle relaxant used in general anesthesia to facilitate endotracheal intubation and to provide skeletal muscle relaxation during surgery. [The Wellcome Foundation Ltd]

Mix. Stabilizing agent for liquid fertilizer-pesticide application. [Draxel Chemical Company] ‡

Mixad. Sand conditioner. [Foseco (F.S.) Ltd]

mixed acid; See nitrating acid.

mixed ether. An ether containing two different alkyl radicals, as in ethyl-methyl ether, $C_2H_5 \cdot O \cdot CH_3$.

mixed metal. A term used for alloys of cerium, lanthanum, and praseodymium.

mixed vitriol. (Salzburg Vitriol). Cupric-ferrous sulfate, $CuSO_4 \cdot 3FeSO_4 \cdot 28H_2O$.

Mix-Kit. Package device for two-part compounds. [Hardman]

Mixol. A timber insecticide. [ICI Chem. & Polymers Ltd] †

Mixxim® BB/50. Blend; uv light absorber for processing engineering resins (nylon, PC, PET, PBT, PPO) and PVC, styrene, and acrylics; highly effective where long term permanent uv light stability is required. [Fairmount]

Mixxim® BB/100. CAS 103597-45-1; Bis[2-hydroxy-5-t-octyl-3-(benzotriazol-2-yl)phenyl]; methane; uv light absorber for processing engineering resins (nylon, PC, PET, PBT, PPO) and PVC, styrene, and acrylics; highly

effective where long term permanent uv light stability is required. [Fairmount]

Mixxim® HALS 57. Tetrakis (2,2,6,6-tetramethyl-4-piperidyl)-1,2,3,4-butane tetracarboxylate; hindered amine light and heat stabilizer for polyolefins such as PP, polyethylene, PS, ABS, PVC, PU, and engineering plastics. [Fairmount] *

Mixxim® HALS 63. [1,2,2,6,6-Pentamethyl-4-piperidyl/ β,β,β´,β´-tetramethyl-3,9-(2,4,8,10- tetraoxaspiro (5,5) undecane) diethyl; -1,2,3,4-butane tetracarboxylate; high molecular weight hindered amine light and heat stabilizer for PP, polyethylene, PS, ABS, engineering plastics, and elastomers, esp. in PP monofilament, tapes, molded and extruded products, polyethylene blown film. [Fairmount] ^

Mixxim® HALS 68. [2,2,6,6-Tetramethyl-4-piperidyl/ β,β,β´,β´-tetramethyl-3,9-(2,4,8,10-tetraoxaspiro (5,5) undecane) diethyl; -1,2,3,4-butane tetracarboxylate; high molecular weight hindered amine light and heat stabilizer for PP, polyethylene, PS, ABS, engineering plastics, and elastomers, esp. in PP monofilament, tapes, molded and extruded products, polyethylene blown film. [Fairmount] *

MKP. See potassium phosphate.

M-M-R® II. Measles, mumps and rubella virus vaccine live; for combined protection against measles, mumps and German measles. [Merck & Co Inc]

M-M-VAX. Measles and mumps vaccine; for protection against measles and mumps infection. [Merck & Co Inc]

MN powder. (Maxim-Nordenfelt powder). An American guncotton powder gelatinized with ethyl acetate.

M.N.T. CAS 1321-12-6; Mononitrotoluene, $CH_3 \cdot C_6 \cdot H_4 \cdot NO_2$; used as an intermediate in the preparation of trinitrotoluene.

moac. Very finely divided mica.

Moban. CAS 15622-65-8; Molindone hydrochloride; antipsychotic. [Endo Laboratories Inc]

Mobil 1240. CAS 9003-53-6; Crystal PS; high heat grade for injection molding and extrusion. [Mobil/Polystyrene] *

Mobil 2120. CAS 9003-53-6; Crystal PS; general purpose extrusion grade. [Mobil/Polystyrene] *

Mobil 5350. CAS 9003-53-6; HIPS; super high impact extrusion grade. [Mobil/Polystyrene] *

Mobil 5600. CAS 9003-53-6; HIPS; ESCR-resist. extrusion grade. [Mobil/Polystyrene] *

Mobil 8020. CAS 9003-53-6; PS; ignition-resist. structural foam for injection molding. [Mobil/Polystyrene] *

Mobil MX 4354. CAS 9003-53-6; HIPS; injection molding grade for durable goods, appliance parts, and any large part requiring good flow and gloss chars. [Mobil/Polystyrene]

Mobilrap. Pallet wrap stretch films including one-side cling films, in thicknesses varying from 17 to 35 microns; for wrapping around pallets for increased product protection during transport and warehousing. [Mobil Plastics Europe] *

Mobilrapper. Pallet wrap with stretch film system, which can be handled by one person, offering a variety of stretch films; for wrapping pallets for increased product protection during transport and warehousing. [Mobil Plastics Europe] *

Mobilsol® . Modifying diluents for epoxy resins; they give flexibility. [Mobil] *

Moca® . Polymer. [DuPont UK]

---

‡ Trade name and manufacturer not verified   § Trade name without identified manufacturer

Mocap 10G. CAS 13194-48-4; Granules containing ethoprophos; used to control cyst nematodes and wireworms in potatoes. [Rhone-Poulenc Crop Protection Ltd]

Mocasco Iron. A proprietary trade name for a nickel-chromium-molybdenum cast iron containing 1-1.35% nickel, 0.25-0.3% chromium, and 0.75% molybdenum. §

mocaya oil. (Macaja butter). Paraguay palm oil obtained from the kernels of *Acrocomia scelerocarpa*.

mocha-stone. Agates of white or brown chalcedony from India, with markings due to oxides of iron and manganese.

mock epsoms. Needle crystals of sodium sulfate, Na_2SO_4.

mock lead. Both tungsten ore found in Cornwall and zinc blend are known by this name.

mock silver. An alloy of 84% aluminum, 10% tin, 5.5% copper, and 0.1% phosphorus.

mock vermilion. Lead chromate.

M.O.D. Octyldodecyl myristate; emollient; rancidless additive for cosmetics and pharmaceuticals. [Gattefosse]

Mod Acid. Modified toluene sulfonic acid production; hydrotrope, solv., intermediate, catalyst. [Ruetgers-Nease]

Modaflow. Resin modifier; additive for improving flow, leveling and adhesion properties of surface coatings. [Monsanto Co] *

Modane. A danthron preparation; laxative tablets. [Hercules] *

Modane Soft. CAS 577-11-7; Docusate sodium; stool softener; pharmaceutic aid; surfactant. [Adria Laboratories Inc]

Modar. Thermosetting acrylic resins. [ICI Chem. & Polymers Ltd]

Modar 814. Modified acrylic resin; features low visc., rapid cure, fast cycle times, optimal fire retardancy, low smoke, low combustion toxicity; used in RTM, hand lay-up, spray-up, and filament winding applications [ICI Acrylics]

Modar 826HT. Modified acrylic resin; features low profile, nonshrink, optimal fire retardancy/low smoke, low combustion toxicity, optimum surface finish; for pultrusion. [ICI Acrylics]

Modar 865. Modified acrylic resin; features low visc., rapid cure, fast cycle times, highest strength, chem. resist., fire retardant/low smoke; used primarily in RTM, pultrusion, and filament winding applications [ICI Acrylics]

Modarez APVC 8. Acrylic polymer; modifier for rigid PVC for mfg. of bottles and hollow casings; improves surface and gloss. [Protex]

Modarez MFP Powd. Acrylic polymer on silica support; leveling agent for powd. paints, varnishes and coatings; improves spreading props., increases wettability of surfaces. [Protex]

Moddite. An explosive. It is a variety of Cordite, but is made with a nitro-cellulose partially soluble in ether alcohol.

Modecate. Fluphenazine decanoate; used in the treatment of psychotic disorders. [Bristol-Myers Squibb Pharmaceuticals Ltd]

Moderator. A soluble concentrate containing 300 g atrazine and 12.5 g imazapyr per liter; used for total weed control in non crop areas. [Chipman Ltd]

Modic. Adhesive polyolefin. [Mitsubishi Petrochem.] *

Modicol L. PEG fatty ester; chemical and mech. stabilizer, coagulant, wetting, visc. control, and dispersing agent, used in latex and resin emulsions. [Henkel/Functional Products]

Modicol S. Sulfated fatty acid; stabilizer; in rubber industry to stabilize natural and synthetic latexes during compounding, storage, and applic; ensures mechanical and chemical stability; prevents premature coagulation during high-speed agitation, acidification, or pigmentation. [Henkel/Functional Prods.]

Modified Butacite. A proprietary trade name for thermosetting polyvinyl butyral synthetic resin. §

modified soda. A mixture of sodium carbonate and bicarbonate used as a cleaning agent in laundries.

Modified Vinylite X. A proprietary trade name for thermosetting polyvinyl butyral synthetic resin. §

Modinal T. A proprietary trade name for a wetting agent consisting of a long chain alcohol sulfate. §

Moditen. A proprietary preparation of fluphenazine hydrochloride; a sedative. [Bristol-Myers Squibb Pharmaceuticals Ltd] *

Moditen Enanthate. CAS 2746-81-8; A proprietary preparation of fluphenazine enanthate; used in the treatment of psychotic disorders. [Bristol-Myers Squibb Pharmaceuticals Ltd] *

Moditen Tablets. Fluphenazine hydrochloride in tablet form. [Bristol-Myers Squibb Pharmaceuticals Ltd]

Modrenal. CAS 13647-35-3; Capsules containing trilostane; used in the treatment of adrenal cortical hyperfunction. [Winthrop Laboratories] *

Moducren. Timolol maleate, hydrochlorothiazide and amiloride hydrochloride; for the treatment of hypertension. [Merck & Co Inc]

Modulan®. CAS 61788-48-5; EINECS 262-979-2; Acetylated lanolin; conditioner, emollient, softener, lubricant for cosmetic and pharmaceutical products. [Amerchol Corp]

Modulex. CAS 1333-86-4; A trademark for a carbon black for use as a pigment. §

Moduret 25. Hydrochlorothiazide and amiloride hydrochloride; for the treatment of hypertension. [Merck & Co Inc]†

Moduretic. CAS 17440-83-4, CAS 58-93-5; Amiloride hydrochloride and hydrochlorothiazide; potassium-conserving diuretic for the treatment of edema and hypertension. [Merck & Co Inc]

Moellon R. Synthetic anhyd. moellon from pure refined herring oil; leather additive, fiber lubricant, softener; nonyel. for use on pastel colored leathers. [Atlas Refinery]

Mofix. Oxime/triazine herbicide. [Ciba plc] *

Mo-Flo. Liquid acid drain cleaner. [Momar Industrial Services Ltd]

Mogadon. CAS 146-22-5; A proprietary preparation of nitrazepam; a hypnotic. [Roche Products Ltd] *

Mogador gum. See Morocco gum.

Mogul® L. CAS 1333-86-4; Carbon black; for coloring resistive plastics. [Cabot; Carbot Carbon Ltd]

mohair. A material made from the hair of the Angora goat; used in fabrics for clothing, draperies and upholstery.

Mohawk Steel. A proprietary trade name for a hot die steel containing about 14% tungsten, 3.5% chromium, 0.7% vanadium, and 0.45% carbon. §

Mohr's salt. CAS 10045-89-3; Ferrous ammonium sulfate, $FeSO_4 \cdot (NH_4)_2 \cdot SO_4 \cdot 6H_2O$; used in analytical chemistry and metallurgy.

Molaschar. A decolorizing carbon used for sugar juices.

molascuit. A cattle food. It is the fine fiber of the sugar cane

---

* Trade name not verified as available   † Trade name verified as obsolete

or begasse, with cane molasses absorbed by it.

molasocarb. A decolorizing black made from molasses.

molasses. CAS 68476-78-8; The non-crystallizable residue from sugar. Cane molasses contain 55% sugar, 20% water, and 9% ash, whilst beet molasses consists of 50% sugar, 10% salts, 20% water, 10% nitrogenous matter, and 10% non-nitrogenous matter; used as a cattle food; in food, raw material for various alcohols, acetone, citric acid and yeast propagation.

molassine meal. A mixture of molasses and peat moss; used as a cattle food.

Molco. Spirit based mold and core coatings. [Foseco (F.S.) Ltd]

Moldabaster®, Moldabaster S. Plaster of Paris adjusted to work with Moldano, normal and quick setting; dental speciality. [Bayer AG]

Moldag 200. CAS 1317-33-5; Colloidal MoS$_2$ in 150 solv. refined paraffinic petrol. oil; lubricant additive for aerosols, machine oils, assembly and thread lubes. [Acheson]

Moldano®. Blue hard plaster of high Brinell hardness; dental specialty. [Bayer AG]

Moldaroc®. Yellow, extra hard plaster, for especially resistant models; dental specialty. [Bayer AG]

Moldasil. Condensation-cured silicone putty material for use in the dental laboratory. [Bayer AG]

Moldastone. Thixotropic class IV extra-hard dental stone for especially resistant models; used in dentistry. [Bayer AG]

Moldasynt. Synthetic class IV extra-hard dental stone for especially resistant models; used in dentistry. [Bayer AG]

Moldcote. Spirit based mold and core coatings. [Foseco (F.S.) Ltd]

Moldensite. A proprietary synthetic resin product used for electrical insulation. §

Moldesite®. A range of phenolic molding compounds; used for tool machines, motor industry, pottery, electrotechnical industry, sanitary field and electrical field. [AMC SPREA S.p.A.]

Moldex. Crosslinkable polyethylene resin; molding compounds. [Sumitomo Bakelite Co Ltd] *

MoldPro 613. Mold release for epichlorohydrin, EPDM, fluoro elastomers, TPE. [Witco/Humko]

MoldPro 759. Internal mold release and flow agent for nylon 6/6, nylon 6. [Witco/Humko]

MoldPro 830. Internal mold release agent for styrene butadiene block copolymer. [Witco/Humko]

Mold Wiz Ext. A series of nonsilicone, nonwax, nonstearate polymeric based mold release agents, solvent or water based; used for reinforced plastics composites, injection molding, polyurethane foam and natural and synthetic rubber. [Axel Plastics Research Laboratories Inc]

Mold Wiz Int. A series of polymeric based additive lubricants/release agents used as processing aids; used for reinforced plastic composites, melamine/phenolic/urea laminates and overlays, thermoplastic injection molding, natural and synthetic rubbers and urethane elastomers. [Axel Plastics Research Laboratories Inc]

moler. A Danish diatomaceous earth containing 82.6% silica, 5.33% aluminum oxide, a small proportion of ferric oxide, and organic matter. It is very light in weight, and is used in the manufacture of heat-insulating materials.

Molera. A heat insulator obtained by mixing fine clay with cork dust, and firing.

Molipaxin. CAS 25332-39-2; Trazodone hydrochloride; antidepressive. [Roussel Laboratories Ltd]

Mol-Iron. Ferrous sulfate; hematinic. [Schering Corp] *

Molivate. CAS 25122-57-0; A proprietary preparation of clobetasone butyrate; used in the treatment of eczema. [Glaxo Laboratories] *

Mollescal® C Conc. Bactericidal assistant for soaking hides in leather and fur industry. [BASF AG; BASF plc]

Mollifex. Embedding aid in microscopy. [BDH Chemicals Ltd]

Mollin. A base for ointments. It is a soft soap containing 17 percent of uncombined fat.

Mollit. A proprietary trade name for a polystyrene synthetic resin. §

Mollit B. CAS 614-33-5; A proprietary trade name for glyceryl tribenzoate. §

mollphorus. A glycerin substitute consisting of raw and invert sugar.

Molochite®. A mixture of mullitic aluminum silicate and amorphous siliceous glass produced by the calcination of kaolin clay; refractory aggregate for producing kiln furniture, investment casting molds and refractory bricks and shapes. [ECC International Ltd]

Moloie. A proprietary trade name for a manganese-molybdenum steel. §

Molsidain(e). *See* Corvaton. (Spain, Argentina, Uruguay) [Cassella AG]

Molsidolat. *See* Corvaton. (Italy, Austria) [Cassella AG]

Moltopren®. Color pastes; used for coloring ester and ether-based polyurethane foams. [Bayer AG]

Molybdate Red. CAS 12656-85-8; A lead chromate pigment consisting of mixed crystals of lead chromate, lead sulfate, and a small proportion of lead molybdate. Its color varies from reddish-orange to scarlet.

molybdenum. Metallic element; Mo; CAS7439-98-7; EINECS 231-107-2; alloying agent in steels and cast iron; pigments for printing inks, paints, ceramics; catalyst; solid lubricants; missile and aircraft parts; reactor vessels; cermets; die-casting copper-base alloys; special batteries. [AAA Molybdenum Prods.; Atomergic Chemetals; Cerac; Climax Molybdenum]

molybdenum anhydride. *See* molybdenum trioxide.

molybdenum dioxide. MoO$_2$; CAS 18868-43-4. [AAA Molybdenum Prods.; Atomergic Chemetals; Climax Molybdenum]

molybdenum disulfide. (molybdenum sulfide; molybdic sulfide) MoS$_2$; CAS 1317-33-5; EINECS 215-263-9; lubricant in greases, oil dispersions, resin-bonded films, dry powders, etc.; hydrogenation catalyst. [AAA Molybdenum Prods.; Climax Molybdenum; Dow Corning/ E/M Corp.]

molybdenum nickel. An alloy of 75% molybdenum and 25% nickel; used in the manufacture of saws.

Molybdenum Permalloy. A proprietary trade name for an alloy of 81% nickel, 17% iron, and 2% molybdenum; has a higher permeability than Standard Permalloy (*qv*). §

molybdenum steel. A variable alloy. It usually contains from 0.06-1.73% carbon and 0.23-15% molybdenum.

molybdenum sulfide. *See* molybdenum disulfide.

molybdenum trioxide. (molybdenum anhydride; molybdic oxide; molybdic acid hydride) MoO$_3$; CAS 1313-27-5; EINECS 215-204-7; Source of Mo; reagent for analytical chemistry; agriculture; manufacture of metallic Mo; cor-

---

‡ Trade name and manufacturer not verified   § Trade name without identified manufacturer

rosive inhibitor; ceramic glazes; enamels; pigments; catalyst. [AAA Molybdenum Prods.; Atomergic Chemetals; Cerac; Climax Molybdenum]

molybdic acid hydride. *See* molybdenum trioxide.

molybdic oxide. *See* molybdenum trioxide.

molybdic sulfide. *See* molybdenum disulfide.

molybdosodalite. A variety of sodalite containing nearly 3% MoO_3. It is green in color.

Molydag. Dispersions of molybdenum disulfide in resin/ solvent systems; used for dry film lubrication, or additives for motor oils and special machine greases. [Acheson Colloids]

Molydag 204. CAS 1317-33-5; MoS_2 in 500 solv. refined paraffinic petrol. oil; lubricant additive for assembly and thread lubes. [Acheson Colloids]

Molydag 206. CAS 1317-33-5; Colloidal MoS_2 in water; lubricant additive in dry gear and dry chain lubes, aerosols, penetrating lubes, machine oils, assembly lubes, thread lubes. [Acheson Colloids]

Molydag 208. CAS 1317-33-5; MoS_2 in 2500 solv. refined bright stock petrol. oil; lubricant additive for greases. [Acheson Colloids]

Molydag 210. CAS 1317-33-5; Colloidal MoS_2 in anhyd. IPA; lubricant additive for dry gear and chain lubes, aerosols, assembly and thread lubes. [Acheson Colloids]

Molydag 211. CAS 1317-33-5; Colloidal MoS_2 in trichloroethane; lubricant additive for dry gear and chain lubes, aerosols, assembly and thread lubes. [Acheson Colloids]

Molydag 214. CAS 1317-33-5; Colloidal MoS_2 in mineral spirits; lubricant additive for dry chain lubes, aerosols, penetrating lubes, assembly lubes. [Acheson Colloids]

Molykote®. Silicone, MoS_2 lubricant base; environmental lubricants. [Dow Corning]

Molyte. A trade name for a patented mixture of calcium and molybdenum oxides with a flux. §

Molyvan. A series of proprietary molybdenum compounds; friction reducers used as antiwear and extreme pressure agents in lubricants; can be used as antioxidants. [R T Vanderbilt Co Inc]

momea. (Mimea). A hemp preparation made in Tibet.

momordicine. Elaterin, $C_{20}H_{28}O_5$.

Mona NF-10. low-foaming detergent, wetting agent, solubilizer for spray, soak tank, in-place pipeline cleaners, floor scrubbing formulations. [Mona Industries]

monacetin. CAS 102-76-1; Glyceryl monoacetate, $CH_2OH \cdot CHOH \cdot CH_2OOCCH_3$; used for tanning; solvent for dyes; food additive; gelatinizing agent in explosives.

Monachit. An explosive containing 12% trinitro-xylene, 1% charcoal, and 1% collodion cotton.

Monacrin. CAS 134-50-9; Aminacrine hydrochloride; antiinfective, topical. [Sterling Drug Inc] *

Monafax. Range of surfactants, each of which is a mixture of mono- and di-phosphate esters derived from ethylene oxide based surfactants; mainly viscous liquids which are the acid form of the compound; can be converted to salts of alkali, metal, amine or ammonia by mixing with the desired base; used for emulsion polymerization; industrial cleaners e.g., soak tank cleaners, all-purpose and steam cleaners; herbicide and insecticide emulsifiers; metalworking lubricants; dry cleaning detergents. [D F Anstead Ltd] *

Monafax 785. Nonoxynol-9 phosphate; emulsifier, lubricant,

antistat, detergent, corrosion inhibitor for emulsion polymerization, agric., metalworking lubricants, alkaline cleaners, industrial use; antisoil redeposition for dry cleaning. [Mona Industries]

Monafax 1214. Deceth-4 phosphate; hypochlorite-stable surfactant, surface tensile reducer, wetting agent for mildew removers, tile cleaners, bowl cleaner, tire cleaners, bleaching of paper pulp and textiles, dairy cleaners, hard surface cleaners. [Mona Industries]

Monalube 29-78. Alkanolamide; corrosion inhibitor, lubricant for metalworking lubricants, fiber lubricants, textile specialties, wire and deep metal drawing. [Mona Industries]

Monalube 780. Modified alkanolamide; lubricant for copper drawing; corrosion inhibitor. [Mona Industries]

Monalux CAO. CAS 68155-09-9; Cocamidopropylamine oxide. [Mona Industries]

Monamate C-1142. CAS 68515-65-1; EINECS 271-102-2; Disodium cocamido MIPA sulfosuccinate; foaming/ cleaning surfactants for personal care and household products. [Mona Industries]

Monamate CPA. Sulfosuccinate half ester of an alkanolamide in liquid form; high foaming agent which produces a dense rich lather and imparts a soft and silky feel to the skin and hair and has low irritancy; used in mild shampoos, bubble baths, skin cleaners; rug shampoos; detergent formulations. [D F Anstead Ltd] *

Monamate LA-100. Disodium lauryl sulfosuccinate; high foaming, low irritation surfactant for personal care and household products. [Mona Industries]

Monamate LNT-40. Ammonium lauryl sulfosuccinate; high foaming, low irritation surfactant used in personal care products; biodeg. [Mona Industries]

Monamate OPA. Sulfosuccinate half ester of an alkylolamide as a light yellow liquid; produces flash foam and a rich dense lather with good rinsing and viscosity control; used in shampoos, e.g., baby and family types, gel face cleaners. [D F Anstead Ltd] *

Monamate OPA-100. CAS 56388-43-3; EINECS 260-143-1; Disodium oleamido PEG-2 sulfosuccinate; high foaming, nonirritating surfactant for shampoos, bubble baths, soap bars. [Mona Industries]

Monamate RMEA-40. Disodium ricinoleamido MEA-sulfosuccinate. [Mona Industries]

Monamid® 7-100. Cocamide DEA (1:1); foam booster/ stabilizer for industrial and household detergents; biodeg. [Mona Industries]

Monamid® 15-70W. 1:1 Linoleamide DEA; thickener, visc. builder, hair and fiber conditioner; biodeg. [Mona Industries]

Monamid® 150-ADY. 1:1 Linoleamide DEA; thickener for aq. systems; w/o emulsifier, corrosion inhibitor for sol. oils; biodeg. [Mona Industries]

Monamid® 150-CW. CAS 136-26-5; EINECS 205-234-9; 1:1 Capramide DEA; flash foamer, foam stabilizer for cosmetic preps.; also as emulsifier, solubilizer, visc. control agent; biodeg. [Mona Industries]

Monamid® 150-LMWC. CAS 120-40-1; EINECS 204-393-1; Lauramide DEA; foam booster/stabilizer for industrial and household detergents; biodeg. [Mona Industries]

Monamid® 150-MW. CAS 7545-23-5; EINECS 231-426-7; 1:1 Myristamide DEA; nonirritating foam stabilizer, emulsifier for aq. or nonaq. cosmetics and toiletries; thickener for systems containing sodium ions; biodeg. [Mona

---

* Trade name not verified as available   † Trade name verified as obsolete

Industries]

Monamid® 718. CAS 93-82-3; EINECS 202-280-1; Stearamide DEA; emulsifier, thickener, opacifier for creams and lotions; biodeg. [Mona Industries]

Monamid® CMA. CAS 68140-00-1; 1:1 Cocamide MEA; foamer, thickener for liq. and powd. detergents; biodeg. [Mona Industries]

Monamid® LIPA. CAS 142-54-1; EINECS 205-541-8; 1:1 Lauramide MIPA; foamer, thickener for liq. and powd. detergents; biodeg. [Mona Industries]

Monamid® LMA. CAS 142-78-9; EINECS 205-560-1; 1:1 Lauramide MEA; foamer, thickener for liq. and powd. detergents; biodeg. [Mona Industries]

Monamid® R31-42. Lauramide DEA and propylene glycol; foam stabilizer for cosmetic preps.; biodeg. [Mona Industries]

Monamid® S. CAS 111-57-9; EINECS 203-883-2; 1:1 Stearamide MEA; emulsifier, thickener for kerosene, mineral oils; biodeg. [Mona Industries]

Monamide. CAS 68140-00-1; Cocamide MEA; foam booster, thickener, superfatting agent. [Zohar Detergent Factory]

Monamine. Foam booster and stabilizers, emulsifiers, detergents, wetting agents, corrosion inhibitors, viscosity builders, lubricants, dispersants; used for cosmetics; shampoos, dry cleaning detergents; metal cleaners, toiletries; rust inhibitors, metal cutting fluids, emulsifiable waxes; agricultural sprays, fuel oil additives, leather and fur preparations. [D F Anstead Ltd] *

Monamine 779. Cocamide DEA and DEA-laureth sulfate; foaming agent, visc. builder, detergent, soil suspending agent, solubilizer, wetting and penetrating agent used in shampoos, bubble baths, household and industrial cleaners, germicides, uv absorbers; biodeg. [Mona Industries]

Monamine AA-100. 1:2 Distilled cocamide DEA and diethanolamine; detergent, emulsifier, thickener; biodeg. [Mona Industries]

Monamine ACO-100. Lauramide DEA and diethanolamine; detergent, emulsifier, thickener; biodeg. [Mona Industries]

Monamine ADY-100. Linoleamide DEA and diethanolamine; detergent, emulsifier, thickener; biodeg. [Mona Industries]

Monamine LM-100. Lauramide DEA and diethanolamine; detergent, wetting agent, emulsifier, thickener, corrosion inhibitor; biodeg. [Mona Industries]

Monamine T-100. CAS 68155-20-4; Tallamide DEA and diethanolamine; detergent, wetting agent, foam booster, thickener for household detergents; biodeg. [Mona Industries]

Monamulse 653-C. Alkanolamide, modified; emulsifier, solubilizer, degreaser for solvs. such as mineral spirits, Stod., kerosene, pine oil; used for cleaners for engine blocks, garage floor, truck bodies, silk screens, mechanics hand cleaners. [Mona Industries]

Monamulse CI. Imidazoline, modified; corrosion inhibitor improving water resistance in greases, oil-based lubricant systems, and on cast iron; improves emulsion stability; penetrant. [Mona Industries]

Monaprin. Pigments. [Imperial Chemical Industries plc]

Monaquat ISIES. CAS 67633-57-2; EINECS 266-778-0; Isostearyl ethylimidonium ethosulfate; antistat, lubricant, softener, corrosion inhibitor used in cosmetic industry, in industrial and textile applications; biodeg. [Mona Industries]

Monaquat TG. Bishydroxyethyl dihydroxypropyl stearaminium chloride; surfactant used in personal care products; conditioner for hair rinses; antistat and foaming used in fabric laundering and softening products; thickener for acid bowl cleaners, naval gels. [Mona Industries]

Monarch® 1100. CAS 1333-86-4; Carbon black; for coloring plastics. [Cabot; Cabot Carbon Ltd]

Monase. A proprietary trade name for the acetate of etryptamine. §

Monaspor. Antibacterial. [Ciba plc] *

Monastat 1195. Formulated conc.; antistat/cleaner for glass and plastic, e.g., TV screens, computers, medical diagnostic equip., safety goggles. [Mona Industries]

Monastral. Insoluble phthalocyamine pigments. [ICI Chem. & Polymers Ltd]

Monaterge. Excellent detergency and wetting; low to moderate foaming; alkaline stability; wide range of alkaline cleaners, e.g., floor, wall, drain; liquid laundry detergents. [D F Anstead Ltd] *

Monaterge 85 HF. Cocamide DEA, DEA-acrylinoleate, and DEA-dodecylbenzene sulfonate; high foaming detergent and wetting agent, hydrotrope; stable in high electrolyte systems. [Mona Industries]

Monateric 805. Cocoamphodiacetate, disodium cocamido MIPA-sulfosuccinate; high foaming, mild surfactant for hair conditoners, skin care products. [Mona Industries]

Monateric 810-A-50. Caprylic/capric carboxylic propionate, imidazoline-derived, salt-free; surfactant for industrial detergents, cleaners, cosmetics; hydrotrope, coupling agent, and/or solubilizer; corrosion inhibitor in metalworking systems, oil well flooding, and aerosol pkgs. [Mona Industries]

Monateric 811. CAS 68815-55-4; Disodium capryloamphodipropionate; corrosion inhibitor, detergent, wetting agent in noncorrosive cleaners and industrial detergents; biodeg. [Mona Industries]

Monateric 811. Amphoteric surfactant based on alkyl imidazolines, in amber liquid form; has corrosion inhibiting properties with detergent and surface active properties; used in a broad range of noncorrosive cleaners and industrial detergents. [D F Anstead Ltd] *

Monateric 951A. Disodium lauroamphodiacetate; flash foamer for shampoos, hand and body cleansers; biodeg. [Mona Industries]

Monateric 985A. Sodium lauroamphoacetate, sodium trideceth sulfate; high foaming, mild shampoo base for adult and baby products; biodeg. [Mona Industries]

Monateric 1000. CAS 68815-55-4; Disodium capryloamphodipropionate; corrosion inhibitor, detergent, wetting agent for metal cleaning, cutting fluids, synthetic lubricants; biodeg. [Mona Industries]

Monateric 1000. Amphoteric surfactant based on alkyl imidazolines; light amber liquid; detergent and wetting agent with corrosion inhibiting properties, suggested primary surfactant for metal cleaning, soak tank cleaners, etc. [D F Anstead Ltd] *

Monateric 1188M. CAS 3655-00-3; EINECS 222-899-0; Disodium lauryl beta-iminodipropionate; high foaming surfactant, hydrotrope for household and industrial hard surface cleaners, shampoos, bubble bath, mild skin cleansers, down hole foamers, air drilling; textile wetter;

---

‡ Trade name and manufacturer not verified   § Trade name without identified manufacturer

biodeg.; stable to acid and alkali. [Mona Industries]

Monateric 1202. Dihydroxyethyl tallow glycinate; detergent, substantive conditioner for hair care products; coupling agent, visc. control agent. [Mona Industries]

Monateric 1203. Sodium hydrogenated tallow dimethyl glycinate; surfactant, substantive conditioner for hair care products. [Mona Industries]

Monateric ADA. Cocamidopropyl betaine; high foaming surfactant used in air drilling, foam drilling, foam blanketing, air entraining agent for cement, gypsum, wallboard; for use in presence of brine and oil; biodeg. [Mona Industries]

Monateric CA-35. CAS 68919-41-5; Sodium cocoamphopropionate; detergent, wetting agent, emulsifier, dispersant, foaming agent used in cosmetic, household, and industrial products; coupling agent, solubilizer; biodeg.; stable to acid, alkali, electrolytes. [Mona Industries]

Monateric CA-35%. Amphoteric surfactant based on coconut imidazoline, in amber liquid form; detergent, wetting, emulsifying and dispersing agent with good foam and lather; used in cosmetics, e.g., shampoos, bubble baths, skin cleaners, hair dye formulations, and in high foam floor and metal cleaners. [D F Anstead Ltd] *

Monateric CAB. A cocamido betaine as a clear yellow liquid containing 4.8% sodium chloride; used in bubble bath and shampoo formulations. [D F Anstead Ltd] *

Monateric CAM-40. CAS 68919-41-5; Sodium cocoamphopropionate; surfactant. [Mona Industries]

Monateric CDL. Cocoamphodiacetate, sodium laureth sulfate; mild, high foaming shampoo base. [Mona Industries]

Monateric CDTD. Disodium cocoamphodiacetate, sodium trideceth sulfate; surfactant for industrial detergents, cleaners, cosmetics; hydrotrope, coupling agent, and/or solubilizer; corrosion inhibitor in metalworking systems, oil well flooding, and aerosol pkgs. [Mona Industries]

Monateric CDX-38. Disodium cocoamphodiacetate; detergent, foaming agent, mild base surfactant for shampoos and skin cleansers; biodeg. [Mona Industries]

Monateric CDX38. An imidazoline derived dicarboxylic acid amphoteric in the form of a light amber viscous liquid; good flash foam and lathering properties; used in baby shampoos, daily use shampoos and skin cleansers. [D F Anstead Ltd] *

Monateric CEM-38. Disodium cocoamphodipropionate; detergent, wetting agent, emulsifier, dispersant, solubilizer, hydrotrope used in liq. detergent systems, heavy-duty detergents, acid bowl cleaners, cosmetics; biodeg. [Mona Industries]

Monateric CEM-38%. Sodium salt of a dicarboxethyl fatty acid derived from imidazoline, in liquid form, clear to hazy amber in color; detergent, wetting, emulsifying and solubilizing agent used in liquid medium duty all-purpose detergents, toilet bowl cleaners and aluminum brighteners. [D F Anstead Ltd] *

Monateric CM-36S. Sodium cocoamphoacetate; foaming agent, emulsifier, high foaming detergent, wetting agent, solubilizer, conditioner, coupling agent, fulling agent used in cosmetic and textile industries; biodeg. [Mona Industries]

Monateric CSH-32. Disodium cocoamphodiacetate; high foaming, nonirritating surfactant for baby shampoos; biodeg. [Mona Industries]

Monateric CSH 32. Sodium salt of a dicarboxymethyl fatty acid derived from imidazoline, supplied as a clear yellow liquid; mild detergent properties used in shampoo formulations, particularly baby and daily use types. [D F Anstead Ltd] *

Monateric CyNa-50. Sodium capryloamphopropionate; detergent, emulsifier, coupling agent, solubilizer, wetting agent, low to moderate foaming surfactant used in conc. electrolyte systems, bottle washing, wax strippers, steam cleaners, industrial cleaning, textile processing, acid pickling baths; biodeg.; stable to acid, alkali, electrolytes. [Mona Industries]

Monateric Cy Na-50%. An amber liquid based on a capryl imidazoline; used in bottle washing, steam cleaners, wax strippers, all-purpose cleaners, degreasers, Kier boiling, mercerization and acid-pickling. [D F Anstead Ltd] *

Monateric ISA-35. CAS 68630-96-6; EINECS 271-929-9; Sodium isostearoamphopropionate; surfactant used in cosmetic and industrial products; conditioner, lubricant, thickener; biodeg. [Mona Industries]

Monateric ISA-35%. Amphoteric surfactant based on isostearic imidazoline; supplied as an amber flowable gel; used in shampoos such as protein, low pH and daily use types. [D F Anstead Ltd] *

Monateric L30. Sodium lauroamphoacetate. [Mona Industries]

Monateric LF. An amber liquid, low foaming detergent with high acid and alkaline stability; low or high temperature alkaline and acid cleaners, automatic car wash detergents, steam cleaners and truck body and aircraft cleaners. [D F Anstead Ltd] *

Monateric LMAB. Lauramidopropyl betaine; high foaming shampoo base. [Mona Industries]

Monateric LMM-30. Sodium lauroamphoacetate; high foaming detergent for nonirritating shampoos, skin cleansers, other cosmetics; biodeg. [Mona Industries]

Monateric TA-35. Sodium tallamphopropionate; surfactant for industrial detergents, cleaners, cosmetics; hydrotrope, coupling agent, and/or solubilizer; corrosion inhibitor in metalworking systems, oil well flooding, and aerosol pkgs. [Mona Industries]

Monateric TDB-35. CAS 61791-56-8; Disodium tallow beta-iminodipropionate; hydrotrope and detergent for high electrolyte systems such as heavy-duty liq. cleaners and wax strippers. [Mona Industries]

Monatrope 1250. Sodium alkanoate; surfactant hydrotrope for formulating alkaline built liq. detergent concs.; coupling agent for nonionic and other surfactants in high concs. of electrolytes; for household and industrial detergents, spray washes, textiles, hypochlorite detergents/sanitizers, dishwash liqs. [Mona Industries]

Monawet 1240. Disodium nonoxynol-10 sulfosuccinate; wetting agent for emulsion polymerization, adhesives, paints. [Mona Industries]

Monawet MB-45. CAS 127-39-9; Diisobutyl sodium sulfosuccinate; wetting, dispersing, emulsifying, penetrating and solubilizing agent used in emulsion polymerization of S/B for rug backing, paper coating, water treatment. [Mona Industries]

Monawet MB-45. Sodium diisobutyl sulfosuccinate as a clear colorless liquid; anionic surfactant for styrene/butadiene emulsion systems for rug backing. [D F Anstead Ltd] *

Monawet MM-80. Dihexyl sodium sulfosuccinate, 15%

---

water, 5% IPA; wetting agent, detergent for emulsion and suspension polymerization, rug backing, paper coating, textiles, paint, agric., cosmetic, detergent, mining, water treatment, electroplating baths, and food industries; electrolyte tolerant. [Mona Industries]

Monawet MM-80. Sodium dihexyl sulfosuccinate as a clear colorless liquid; anionic surfactant for pesticidal sprays, shampoos, detergents, latex paints, coatings and textile fibers. [D F Anstead Ltd] *

Monawet MO. CAS 577-11-7; EINECS 209-406-4; Series of anionic surfactants composed of sodium dioctyl sulfosuccinate in liquid form; used for textiles, e.g., cotton cloth desizing, printing and dyeing processes, wool carbonizing; agriculture, e.g., liquid fertilizers, insecticides and fungicides; cosmetics, e.g., creams and lotions, bath oils, shampoos. [D F Anstead Ltd] *

Monawet MO-65-150. CAS 577-11-7; EINECS 209-406-4; Dioctyl sodium sulfosuccinate, anhyd.; wetting, penetrating and spreading agent, emulsifier used in oil well cleaning, drycleaning detergents, solv. cleaners and strippers, lubricants, agric., paints, mining, water treatment, cosmetic applications. [Mona Industries]

Monawet MO-70. CAS 577-11-7; Dioctyl sodium sulfosuccinate, 20% water, 10% diethylene glycol butyl ether; wetting, dispersing, emulsifying, penetrating and solubilizing agent used in emulsion and suspension polymerization, adhesives, paints, textile, fertilizer, mining, water treatment, fire fighting, cosmetic, food industries. [Mona Industries]

Monawet MO-84R2W. CAS 577-11-7, 57-55-6; Dioctyl sodium sulfosuccinate, 16% propylene glycol; wetting agent for general use, agric., paints, mining, water treatment, cosmetics applications. [Mona Industries]

Monawet MT Series. CAS 2673-22-5; EINECS 220-219-7; Anionic surfactants containing ditridecyl sodium sulfosuccinate in liquid form; used for vinyl chloride suspensions and styrene emulsions for coatings, paints. MT-80H2W is used in nonaqueous pigment dispersions for printing inks. [D F Anstead Ltd] *

Monawet MT-70. Ditridecyl sodium sulfosuccinate, 12% water, 18% hexylene glycol; wetting, dispersing, emulsifying, penetrating and solubilizing agent used in emulsion and suspension polymerization, paints, coatings, indirect food additives. [Mona Industries]

Monawet SNO-35. Tetrasodium dicarboxyethyl stearyl sulfosuccinamate; wetting agent, solubilizer, emulsifier, dispersant, visc. depressant, mild detergent used in polymerization, paints, coatings, textile, cosmetic, agric. products; biodeg. [Mona Industries]

Monawet SNO-35. Tetrasodium N-(1,2 dicarboxyethyl) N-alkyl(C18) sulfosuccinamate as a clear light amber liquid; mild detergent with good wetting and calcium tolerance; emulsifier; used in industrial detergents, cosmetic and textile products, specialty emulsions or dispersions for polymerization systems. [D F Anstead Ltd] *

Monawet TD-30. Disodium deceth-6 sulfosuccinate; surfactant, emulsifier, foaming agent used in emulsion polymerization, cosmetic and textile industries. [Mona Industries]

Monawet TD-30. Half ester of sulfosuccinic acid based on an ethoxylated fatty alcohol. Light yellow liquid; wetting, foaming and emulsifying agent with low inorganic electrolyte content; used in emulsion polymerization, textile wet processing and a broad range of cosmetic and fine

fabric detergent formulations. [D F Anstead Ltd] *

Monazoline C. CAS 61791-38-6; EINECS 263-170-7; Cocoyl hydroxyethyl imidazoline; wetting agent, emulsifier for nonpolar liq., detergent, thickener, corrosion inhibitor, antistat, softener, bactericide used in paint and textile industries; also dispersant for clay and pigments, in acid dairy cleaners, in oil well acidifying and secondary recovery operations; biodeg. [Mona Industries]

Monazoline C, CY, O and T. 1-Hydroxyethyl-2 alkylimidazoline where the alkyl portion is caprylic, coconut, oleic and tall oil respectively; amber liquids; cationic surfactants with wetting, emulsifying, detergency, thickening, corrosion inhibiting, antistatic, softening and bactericidal properties; used in agricultural sprays; gravel to asphalt bonding; dairy cleaners; rinse aids; paints, sealants, inks; oil well recovery; sludge dispersion; corrosion control; plastics; metal treatment; textiles, e.g., softening; ore flotation; leather processing. [D F Anstead Ltd]*

Monazoline CY. CAS 37470-60-5; EINECS 253-521-2, Capryl hydroxyethyl imidazoline; wetting agent, emulsifier for nonpolar liq., detergent, thickener, corrosion inhibitor, antistat, softener, bactericide used in paint and textile industries; biodeg. [Mona Industries]

Monazoline IS. CAS 68966-38-1; EINECS 273-429-6; Isostearyl hydroxyethyl imidazoline; corrosion inhibitor and lubricant. [Mona Industries]

Monazoline O. Oleyl hydroxyethyl imidazoline; wetting agent, emulsifier for nonpolar liq., detergent, thickener, corrosion inhibitor, antistat, softener, bactericide used in paint and textile industries; also dispersant for clays and pigments, in agric. preparations, acid dairy cleaners; biodeg. [Mona Industries]

Monazoline T. CAS 61791-39-7; EINECS 263-171-2; Tall oil hydroxyethyl imidazoline; wetting agent, emulsifier for nonpolar liq., detergent, thickener, corrosion inhibitor, antistat, softener, bactericide used in paint and textile industries; also dispersant for clays and pigments, aids gravel-to-asphalt bonding, rinse aid for automatic car washes, printing ink additive, protective metal coatings; biodeg. [Mona Industries]

Monceren®. CAS 66063-05-6; Dustable powder containing 12.5% w/w pencycuron; a phenylurea fungicide to control black scurf in potatoes. [Bayer plc]

Moncler Derma. Facial cream, gel, lotion and stick, to improve complexion. [Richardson-Vicks Inc] *

Mond 70 alloy. An alloy of 70% nickel, 26% copper, and 4% manganese.

Mond gas. A combustible gas produced by passing air and steam over heated coal or peat. It consists of a mixture of carbon monoxide, hydrogen, and nitrogen.

Mondur. Toluene diisocyanate/polymethyl diisocyanate. [Miles]

Monece®. Fibers. [DuPont UK]

Monel Alloy 400. An alloy of 30% copper, 1% manganese, 2.5% (max) iron and the balance nickel; a general engineering alloy with good resistance to corrosion by sea water, sulfuric, hydrochloric and phosphoric acids. [Wiggin Alloys Ltd] ‡

Monel Alloy 414. Monel 400 with a high carbon content; improved machining properties. [Wiggin Alloys Ltd] ‡

Monel Alloy K-500. An alloy of 30% copper, 3% aluminum, 0.5% titanium and the balance nickel. [Wiggin Alloys Ltd]‡

Monel metal. An alloy. The cast metal usually contains from

---

‡ Trade name and manufacturer not verified   § Trade name without identified manufacturer

68-70% nickel, 28% copper, 2% iron, 1% silicon, and 0.25% manganese; and the forged alloy consists of 68% nickel, 28% copper, 2% iron, 1.5 [Huntington Alloys, Inco Alloys Int.] ‡

monesia. A South American vegetable extract, said to be obtained from the bark of *Lucuma glycyphloea*; used as an astringent used in diarrhea.

monesin. A saponin-like substance extracted from the bark of the South African plant, *Chrysophyllum viridifolium*; used in medicine.

monetite. A calcium phosphate, CaHPO$_4$, found in guano.

Monex. CAS 97-74-5; Tetramethylthiuram monosulfide; a nonstaining and nondiscoloring delayed action accelerator; has a short sharp curing range with normal to high sulfur in natural rubber; used in natural, SBR, butyl, nitrile and neoprene rubbers for wire insulation, druggist sundries, mechanicals, sponge and footwear. [Uniroyal] *

Monistat-Derm. CAS 22832-87-7; Miconazole nitrate; antifungal. [Ortho Pharmaceutical Corp] *

Monit. CAS 16051-77-7; Isosorbide mononitrate tablets; for prophylaxis of angina pectoris; (Sold in UK by Stuart Pharmaceuticals). [ICI Chem. & Polymers Ltd]

Monitan. Polysorbate 80; pharmaceutic aid. [Ives Laboratories Inc] *

Monite. A proprietary plastic. §

Monitor. CAS 10265-92-6; Methamidophos insecticide. [Chevron] *

Monnex. A nontoxic compound obtained by reacting urea with potassium carbonate; used as a fire extinguisher. [ICI Chem. & Polymers Ltd]

Mono Ammonium Phosphate. CAS 7722-76-1; Mono ammonium phosphate technical grade; used for other chemicals, water treatment, fire fighting, cosmetic products, food industry, pharmaceutical products. [Rhone-Poulenc NV] *

Mono Ammonium Phosphate (Agricultural Grade). CAS 7722-76-1; Mono ammonium phosphate; used for fertilizers, other chemicals. [Rhone-Poulenc NV] *

Mono-Baycuten®. Broad-spectrum angifungal. [Bayer AG]

Monobed. Ion exchange resins. [Rohm & Haas]

Monobel. (A2 Monobel). A trademark for a smokeless powder. It is a mixture of 9-11 parts nitroglycerin, 56-61 parts ammonium nitrate, 8-10 parts wood meal, 0.5-1.5 parts magnesium carbonate, and 18.5-21.5 parts potassium chloride; used in mines. §

mono-tert-butyl hydroquinone. *See* t-butyl hydroquinone.

monocalcium phosphate, monohydrate. *See* Calcium phosphate (monobasic).

Monocast® MC 901. Heat-stabilized nylon; produced by direct polymerization of monomer within a mold at atmospheric pressure; used in paper and textile, construction, mining, metalworking, and material handling industries for bearings, valve seats, seals, gears, wheels, guides, tooling fixtures, insulators, wear parts, etc.; good machinability. [Polymer Corp.]

monochloroacetic acid. (MCA; chloroacetic acid) ClCH$_2$COOH; CAS 79-11-8; EINECS 201-178-4; Herbicide, preservative, bacteriostat; intermediate in production of carboxymethylcellulose, ethyl chloroacetate, glycine, synthetic caffeine, sarcosine, thioglycolic acid, EDTA, 2,4-D, 2,4,5-T. [Allchem Industries; Atochem; Eka Nobel AB; Hoechst Celanese]

monochlorobenzene. (chlorobenzene; phenyl chloride) C$_6$H$_5$Cl; CAS 108-90-7; Phenol, solvent, pesticide inter-

mediate, heat transfer.

monochloromethane. *See* methyl chloride.

Monocid. CAS 61270-78-8; Cefonicid sodium; antibacterial. [SmithKline Beecham]

Mono-Coat® . A nontransferring, semipermanent mold release that gives multiple releases with no transfer to the molded part [Chem-Trend]

Monocortin. CAS 1597-82-6; Paramethasone acetate; glucocorticoid. [Syntex Laboratories Inc] *

Monocron. CAS 6923-22-4; Active ingredient: monocrotophos; contact and systemic agricultural insecticide belonging to the enolphosphates. [Makhteshim Chemical Works Ltd]

Monodral. A proprietary preparation containing penthienate methobromide; an antispasmodic. [Richardson-Vicks Inc] *

monoethanolamine. *See* ethanolamine.

monoformin. The formyl derivative of glycerin, C$_3$H$_5$-(OH)$_2$(OCHO).

monogermane. CAS 7782-65-2; Germanium tetrahydride, GeH$_4$.

monoglyme. CAS 110-71-4; Ethylene glycol dimethyl ether; CH$_3$OCH$_2$CH$_2$OCH$_3$; used as a solvent.

monoisopropanolamine. *See* isopropanolamine.

Mono-Kay. Phytonadione; vitamin. [Abbott Laboratories] *

Monol. Calcium permanganate, Ca(MnO$_4$)$_2$ · 4H$_2$O; used in the textile industry.

Monolan®. Ethylene and/or propylene oxide condensates; used in low foam nonionics, lubricants, dishwashing, and cosolvency. [Harcros UK]

Monolan® 8000 E/80. CAS 9003-11-6; EO/PO block polymer; low foam wetting agent and detergent, dispersant; primary emulsifier in emulsion polymerization. [Harcros UK]

Monolan® P222. EO-PO block polymer; low foaming detergent, antifoam for detergents, wetting agents, synthetic lubricants. [Harcros UK]

Monolan® PT. Complex EO-PO copolymer; low foam detergent and rinse aid, pigment dispersant; antifoam; dye leveling agent; car shampoos, window cleaners, yarn lubricants. [Harcros UK]

Monolastex Smooth. Fast drying water-based external or internal elastomeric coating; brush or roller applied; used for outside/inside of housing, offices, prefabricated units. [Liquid Plastics Ltd]

Mono-Line. A monolithic refractory material used for lining steelmaking vessels. [Pfizer International] †

Monolite® . Insoluble lake colors and pigments for paints, inks and plastics. [ICI Chem. & Polymers Ltd]

Mono-Lube® . Custom formulations for use in tire industry; processing aids for tire manufacture. [Chem-Trend]

Monomax AH90 B. Distilled monoglycerides; emulsifier/stabilizer for ice cream, margarine, shortening, peanut butter, confectionery, other foods; chewing gum plasticizer; starch complexing agent in pastas. [Australian Bakels]

Monomuls® Range. Fatty acid mono and diglycerides; general purpose food emulsifiers. [Grünau]

Monomuls® 90-L12. CAS 142-18-7; EINECS 205-526-6; Glyceryl laurate; coemulsifier, refatting agent and thickener for personal care products, bath additives, shampoos. [Henkel/Cospha; Henkel Canada]

Monomuls® 90-O18. CAS 111-03-5; Glyceryl oleate; emulsifier, stabilizer, refatting agent, thickener for cosmetics,

---

food and drugs. [Henkel/Cospha; Henkel KGaA]

monoolein. *See* glyceryl oleate.

Monopar. CAS 3784-99-4; Stilbazium iodide; anthelmintic. [Burroughs Wellcome Co] *

monopentaerythritol. *See* pentaerythritol.

Monophane. Insulin preparation. [The Boots Co plc] *

Monoplas 279. A proprietary dialkyl (C$_7$ -C$_9$) phthalate plasticizer. §

Monoplex® . Specialty plasticizers; for vinyl and synthetic lubricants. [C P Hall]

Monoplex® 5. A proprietary trade name for dibenzyl sebacate; a vinyl plasticizer. [C.P. Hall] *

Monoplex® DDA. CAS 27178-16-1; Diisodecyl adipate; lubricant; plasticizer. [C.P. Hall]

Monoplex® DIOA. Diisooctyl adipate; plasticizer. [C.P. Hall]

Monoplex® DOA. CAS 103-23-1; Di-2-ethylhexyl adipate; plasticizer for PVC food pkg. film. [C.P. Hall]

Monoplex® DOS. Di-2-ethylhexyl sebacate; plasticizer for elec. PVC compounds, low temp. sheet, film. [C.P. Hall]

Monoplex® NODA. n-Octyl, n-decyl adipate; plasticizer. [C.P. Hall]

Monoplex® S-73. Epoxidized octyl tallate; PVC stabilizer, plasticizer. [C.P. Hall]

Monoplex® S-75. Epoxidized glycol dioleate; PVC stabilizer, plasticizer. [C.P. Hall]

monopol oil. *See* Turkey red oils.

monopol soap. (avirol). Sulfonated oils similar to Turkey red oil; used as wetting-out agents.

monopotassium orthophosphate. *See* potassium phosphate.

Monopoxy. Single-component epoxy resin adhesive. [Hardman]

monoricinolein. *See* glyceryl ricinoleate.

Monoset. Pre-packed ultra rapid hardening cementitious mortars and concretes; used for raising of manhole covers and frames, repairs to motorways, repairs in tidal conditions and wherever a minimum downtime is needed. [Ronacrete Ltd]

Monosiliol C. Methylsilanol tri PEG-8 glyceryl cocoate; aids tissue regeneration; for anti-aging formulations, cosmetic and health emulsions, oils, milks, and soaps. [Exsymol]

monosodium dihydrogen phosphate. *See* sodium phosphate.

monosodium L-glutamate monohydrate. *See* MSG.

Monosorb. CAS 7558-80-7; Absorbed sodium phosphate. [FMC]

monostearin. *See* glyceryl stearate.

Monosulfiram. Tetra-ethylthiuram monosulfide; used as a parasiticide.

Monosulph. Anionic surfactants. [Henkel Chemicals Ltd] *

Monotard. Insulin zinc; antidiabetic. [Bristol-Myers Squibb Co Inc] *

Monotard Human. Insulin human; a protein that has the normal structure of the natural antidiabetic principle produced by the human pancreas; antidiabetic. [Bristol-Myers Squibb Co Inc] *

Monotheamin Pulvules. CAS 573-41-1; A proprietary preparation of theophylline monoethanolamine. [Eli Lilly & Co] †

Monotheamin and Amytal Pulvules. A proprietary preparation of theophylline monoethanolamine and amylobarbitone. [Eli Lilly & Co] †

Mono Thiurad. CAS 97-74-5; Tetramethylthiuram

monosulfide; vulcanization accelerator. [Monsanto Co]*

Monox. A product containing mainly silicon monoxide, with some silicon, silicon dioxide, and small quantities of silicon carbide. It is obtained by heating sand with silicon, carborundum, or coke, in the electric furnace; good thermal and electrical insulator.

Monphytol. A proprietary preparation of boric acid, chlorbutol, methyl salicylate, salicylic acid and undecylenic esters; antifungal skin powder. [LAB Ltd] ‡

Monsanto Salt. A proprietary trade name for o-chloro-p-toluene sodium sulfonate, Cl · C$_7$H$_6$ · SO$_3$ · Na. §

Monsell's salt. Basic ferric sulfate, Fe$_4$O(SO$_4$)$_5$; used in medicine.

Montago. A German light alloy with a specific gravity lower than aluminum.

Montana gold. An alloy of 89% copper, 10.5% zinc, and 0.5% aluminum.

Montana wax. *See* Irish peat wax.

Montane 20. CAS 1338-39-2; EINECS 215-663-3; Sorbitan laurate; emulsifier. [Seppic]

Montane 40. CAS 26266-57-9; EINECS 247-568-8; Sorbitan palmitate; emulsifier. [Seppic]

Montane 60. CAS 1338-41-6; EINECS 215-664-9; Sorbitan stearate; emulsifier. [Seppic]

Montane 65. CAS 26658-19-5; EINECS 247-891-4; Sorbitan tristearate; emulsifier. [Seppic]

Montane 80. CAS 1338-43-8; EINECS 215-665-4; Sorbitan oleate; emulsifier. [Seppic]

Montane 481. Sorbitan oleate, beeswax, and stearic acid; cosmetic ingred. [Seppic]

montanine. A liquid containing 31% hydrofluosilicic acid; Recommended as a disinfectant for the walls of breweries and distilleries. It is obtained from by-products in the pottery industry.

montanin wax. *See* Irish peat wax.

Montanox 20 DF. CAS 9005-64-5; Polysorbate 20; emulsifier. [Seppic]

Montanox 40 DF. CAS 9005-66-7; Polysorbate 40; emulsifier. [Seppic]

Montanox 60 DF. CAS 9005-67-8; Polysorbate 60; emulsifier. [Seppic]

Montanox 61. CAS 9005-67-8; PEG-4 sorbitan stearate; emulsifier. [Seppic]

Montanox 65. CAS 9005-71-4; Polysorbate 65; emulsifier. [Seppic]

Montanox 70. CAS 66794-58-9; PEG-20 sorbitan isostearate; emulsifier. [Seppic]

Montanox 80 DF. CAS 9005-65-6; Polysorbate 80; emulsifier. [Seppic]

Montanox 81. CAS 9005-65-6; Polysorbate 81; emulsifier. [Seppic]

Montanox 85. CAS 9005-70-3; Polysorbate 85; emulsifier. [Seppic]

montan pitch. The residue from the production of montan wax. The crude material gives an ash of 1.7%, has an acid value of 3, and a saponification value of 6.

Montax. A proprietary trade name for a filler for rubber, etc.; a mixture of hydrated magnesium carbonate and silica.§

Monteban. CAS 55134-13-9; Narasin; coccidiostat; growth stimulant. [Eli Lilly & Co]

Monteine LCK-32. CAS 68920-65-0; Potassium cocohydrolyzed animal protein; surfactant for cosmetics. [Seppic]

Monteine LCQ. Cocamidopropyldimethylaminohydroxy-

---

‡ Trade name and manufacturer not verified   § Trade name without identified manufacturer

propyl hydrolyzed animal protein; raw material for shampoos and hair conditioners. [Seppic]

Monteine LCT. CAS 68952-16-9; TEA-coco-hydrolyzed animal protein; surfactant for shampoo and general cosmetic use. [Seppic]

Monterey 30% Iron. Iron sulfate; a granular material used to correct iron deficiency; used on turf, flower beds, vegetables, etc. [Lawn & Garden Products Inc]

Monterey Bayleton. CAS 43121-13-3; Triadimefon; a fungicide for the control of diseases in turf and the control of powdery mildew, rusts and blight of ornamental plants. [Lawn & Garden Products Inc]

Monterey Bloom Popper. A 8-32-7 liquid fertilizer formulated with humic acid for better foliar uptake; high phosphate levels help in the blooming and fruit products of a plant and the humic acid helps in the uptake of that nutrient. [Lawn & Garden Products Inc]

Monterey Foliar Nutrient 11-4-6. Nitrogen, phosphoric acid, potash and chelated zinc, iron and manganese; used as a foliar nutrient on turf, ornamentals, trees, vegetables and house plants. [Lawn & Garden Products Inc]

Monterey Herbicide Helper. Petroleum distillate and alkylphenoxy polyethoxy ethanols; a spreader/penetrant that makes weed killers work faster and better. [Lawn & Garden Products Inc]

Monterey Insulate. Aids in the prevention of frost damage; used on ornamentals, vegetables, citrus, fruit trees, etc.; applied to frost conditions and repeated at 7-21 day intervals. [Lawn & Garden Products Inc]

Monterey Iron Chelate 10%. Used to correct iron deficiencies, either as a soil drench or as a foliar spray; may be used on ornamentals, vegetables, fruit trees, etc. [Lawn & Garden Products Inc]

Monterey Kryocide. CAS 15096-52-3; Cryolite; an easy-to-use insecticide for the control of many worm pests on vegetables, grapes, fruit tree and ornamentals; used either as a spray or dust. [Lawn & Garden Products Inc]

Monterey Liqui-Cop. Copper-Count-N; a liquid copper used as a dormant copper spray on stone fruits and citrus; used as a replacement for Bordeaux mixtures. [Lawn & Garden Products Inc]

Monterey Perc-O-Late Plus. Surfactants plus nitrogen, zinc, iron and manganese; used for water penetration on turf, flower beds, potted plants, etc. [Lawn & Garden Products Inc]

Monterey Signal. A blue colored dye to put into the spray tank to let you know where you are spraying; avoid skips and overlaps; breaks down in sunlight. [Lawn & Garden Products Inc]

Monterey Stimulator 12. Humic acid derived from completely organic sources; aids the plant in the uptake of nutrients from the soil or through foliar application; helps plants utilize the fertilizer you give them. [Lawn & Garden Products Inc]

Monthier's blue. A colored compound obtained by the oxidation of the precipitate formed by the action of ammoniacal ferrous chloride upon potassium ferrocyanide, $(Fe_2)_2(Fe(CN)_6)_3 \cdot 6NH_3 \cdot 9H_2O$.

Monthyle. CAS 111-60-4; Glycol stearate; emulsifier, stabilizer for ointments, cream lotions. [Gattefosse; Gattefosse SA]

Montigel. CAS 1302-78-9; Bentonites with specific swelling properties for various technical purposes, such as for

iron ore pelletisations and for sealing of sanitary land fill. [Süd-Chemie AG] *

montmorillonite. CAS 1302-78-9; A mineral, $Al_2O_3 \cdot SiO_2$. It is a colloidal clay similar to bentonite, and is often combined with alkalies or alkaline earths; used for decolorizing oils.

Montothene G50. A proprietary trade name for an ethylene vinyl acetate copolymer; translucent, nontoxic; good mechanical properties; used for film, injection and blow molded articles and extrusions. [Armour Pharmaceutical Co] *

Montreal potash. Commercial potassium carbonate.

Moogrol. A proprietary preparation; a mixture of the acids of the chaulmoogric series; used as a therapeutic agent in leprosy. §

Moore Floc. Organic and inorganic flocculants, cationic, anionic, alum replacement and caustic replacement products; used for flocculating fine dissolved or suspended solids in potable water plants, color and collodial and turbidin solids, algae precipitating aid. [Benzsay & Harrison Inc]

Moorland. A proprietary preparation for use as an antacid. [The Boots Co plc]

Moplefan. A proprietary range of polypropylene films. [Montedison UK Ltd] *

Moplen. Polypropylene; a flexible hard, tough hydrocarbon thermoplastic used for molding domestic ware and for electrical purposes. [Montedison UK Ltd] *

Morat white. (Moudan white). A white pigment. It is a clay found in Switzerland.

Moreau marble. A marble prepared by immersing soft amorphous limestone in a bath of zinc sulfate, and drying.

Morell's solution. A disinfecting solution containing arsenious acid, caustic soda, and a small quantity of phenol, dissolved in water.

Morestan®. CAS 2439-01-2; A wettable powder containing 25% w/w quinomethionate; fungicide with protective and eradicative action against powdery mildews on pome, stone and small fruits, cucurbits and ornamentals; as acaricide effective against eggs and mobile stages of mites and spider mites on pome and stone fruit, citrus, vegetables, ornamentals and coffee. [Bayer AG]

Morestan. CAS 2439-01-2; Wettable powder containing 25% w/w quinomethionate; for control of red spider mites, including organophosphorus strains, on apples, gooseberries, strawberries, and marrows and American gooseberry mildew (partial control of leafspot) on gooseberries, powdery mildew on marrows and willow anthracnose. [Hortichem Ltd]

Morfast. Liquid dyes for surface coatings. [Morton Int'l. Ltd]

Morfax. CAS 95-32-9; 4-Morpholinyl-2-benzothiazole disulfide; rubber accelerator for NR and synthetic rubbers; provides good curing activity; suggested for tires and mechanical goods requiring maximum strength and quality. [Goodyear]

Morflex 100. CAS 27554-26-3; A proprietary trade name for diisooctyl phthalate; a vinyl plasticizer. §

Morflex 125. CAS 119-07-3; n-Octyl n-decyl phthalate; a proprietary plasticizer. [Pfizer International] †

Morflex 150. CAS 84-61-7; Dicyclohexyl phthalate; plasticizer for adhesives, nitrocellulose lacquers. [Morflex]

Morflex 175. CAS 119-07-3; A proprietary trade name for octyl decyl phthalate; a vinyl plasticizer. §

Morflex 190. CAS 85-70-1; n-Butyl phthalyl-n-butyl glycolate; vinyl plasticizer. [Morflex]

Morflex 210. Diethyl hexyl phthalate; a proprietary plasticizer. [Pfizer International] †

Morflex 240. CAS 84-74-2; Dibutyl phthalate; a proprietary plasticizer. [Pfizer International] †

Morflex 310. Di-2 ethyl hexyl adipate; a proprietary plasticizer. [Pfizer International] †

Morflex 325. n-Octyl n-decyl adipate; a proprietary plasticizer. [Pfizer International] †

Morflex 330. Didecyl adipate; a vinyl plasticizer. [Pfizer International] †

Morflex 410. Di-2-ethyl hexyl azelate; a proprietary plasticizer. [Pfizer International] †

Morflex 510. Tri-2-ethyl hexyl trimellitate; a proprietary plasticizer. [Pfizer International] †

Morflex 525. Tri (n-octyl n-decyl) trimellitate; a proprietary plasticizer. [Pfizer International] †

Morflex 530. Triisodecyl trimellitate; a proprietary plasticizer. [Pfizer International] †

Morflex 560. CAS 1528-49-0; Tri-n-hexyl trimellitate; plasticizer with high efficiency, good permanence. [Morflex]

Morflex 1129. CAS 1459-93-4; Dimethyl isophthalate; chemical intermediate. [Morflex]

Morflex MSC. CAS 1337-33-3; Monostearyl citrate; chelating agent, surface lubricant. [Morflex]

Morflex P50. A proprietary n-alkyl phthalate plasticizer. [Pfizer International] †

Morhal resin. A resin obtained from *Vatica lanceoefolia*, of India.

Morhulin. Cod liver oil and zinc oxide; for treatment of wounds, scalds and dermatitis. [Napp Laboratories Ltd]*

Morillol®. 1-Octene-3-ol; fragrance and flavoring. [BASF AG]

Morkit®. CAS 84-65-1; Anthraquinone; seed treatment to reduce rook feeding damage until plant emergence in arable and vegetable crops. [Bayer AG]

Morland's salt. (Cr(NH$_3$)$_2$(SCN)$_4$)HNH. The guanidinium salt of the same complex as Reinecke's salt, formed as a by-product in the preparation of the latter salt.

Mornidine. CAS 84-04-8; A proprietary trade name for pipamazine. §

Moroccan olive oil. Argan oil from *Arganum sideroxylon*.

Morocco gum. (Mogador gum, brown barberry gum). A variety of gum acacia in the form of tears.

Morocide. Fungicide for mildew. [Hoechst UK] *

Moroline. Petrolatum, white; pharmaceutic aid; protectant. [Schering Corp]

Moronal. Aluminum-formaldehyde-sulfite; used as an antiseptic and astringent.

Morpan. Cationic surfactant, biocide. [Rhone-Poulenc UK]

Morpan BC. Aqueous solution of benzalkonium chloride; Bactericide, algicide and antistatic used in fields such as food; brewing; hospitals; farming; veterinary; general sterilizing. [ABM Chemicals Ltd].

Morpan NBB. A proprietary trade name for a 50% active composition of octyldimethylbenzylammonium bromide; a low forming flocculating agent used as a filtering aid. §

morpholine. (tetrahydro-1,4-oxazine; tetrahydro-2H-1,4-oxazine; diethylene oximide) Heterocyclic organic compound.; C$_4$H$_8$ONH; CAS 110-91-8; EINECS 203-815-1; rubber accelerator; solvent; additive to boiler water; optical brightener for detergents; corrosion inhibitor; organic intermediate. [Air Prods.; BASF; Nippon

Nyukazai; PMC Specialties; Texaco]

Morplas. Heat resistant dyes for plastics. [Morton Int'l. Ltd]

Mor-Rex® I-920. CAS 9050-36-6; Maltodextrin; nondispersive thinner for water-based lime muds. [Grain Processing]

morrhua oil. *See* cod liver oil.

Morsep. Cetrimide, vitamin A and vitamin D$_2$; used for diaper rash. [Napp Laboratories Ltd] *

Morstrip. Paint stripper. [Kalon Chemicals Ltd]

Mortegg Emulsion. Emulsifiable concentrate containing 600 g/l tar oils; fungicidal winter wash for fruit. [DowElanco Ltd]

Mortha. A proprietary preparation of morphine hydrochloride and tetrahydroaminacrine hydrochloride; an analgesic. [Octel Chemicals Ltd] †

Morto. Weedkiller and potato haulm destroyer. [Murphy Chemical Co Ltd] †

Morton's fluid. A solution containing iodine, potassium iodide, and glycerin.

Morwet EFW. Wetting agents and dispersants; used in pesticide formulations. [Witco/Organics]

mosaic gold. (mock gold, cat's gold, bronzing powder, tin bronze) Flaky yellow form of disulfide of tin, SnS$_2$; a substance also called mosaic gold is made from an amalgam of tin, mercury, with ammonium chloride and sulfur. formerly used for gilding and imitating bronze

mosaic silver. An alloy of tin and bismuth.

Moss Gun! Ready-to-use mosskiller in spray form. [ICI Garden Products]

Mosskil. Lawn fertilizer with iron sulfate. [Fisons plc, Horticulture Div] *

moss starch. Lichenin, (C$_6$H$_{10}$O$_5$)$_x$.

Mos-Tox. Moss eradicant. [May & Baker Ltd] *

Mota. Tablets of metaldehyde.

mother of pearl sulfur. (nacreous sulfur). A form of monoclinic sulfur obtained by heating sulfur with benzene at 140 C. It is unstable.

Motilium. CAS 57808-66-9; A proprietary preparation containing domperidone; antiemetic. [Janssen Pharmaceutical Ltd] *

Motipress. Fluphenazine hydrochloride and nortriptyline hydrochloride; used for the treatment of depressive and anxiety states. [Bristol-Myers Squibb Pharmaceuticals Ltd]

Motipress Motival Tablets. Fluphenazine hydrochloride and nortriptyline hydrochloride in tablet form. [Bristol-Myers Squibb Pharmaceuticals Ltd]

Motival. A proprietary preparation of fluphenazine hydrochloride and nortriptyline hydrochloride; a tranquilizer. [Bristol-Myers Squibb Pharmaceuticals Ltd] *

Motrin. CAS 15687-27-1; Ibuprofen; anti-inflammatory. [Upjohn]

Motrin-A. Ibuprofen aluminum; anti-inflammatory. [Upjohn]*

Motung Steel. A proprietary trade name for a high speed steel containing 7.5-8.5% molybdenum, 1.25-2% tungsten, 3.5-4.5% chromium, 0.9-1.5% vanadium, 0.8 carbon, 0.2-0.4% manganese, and 0.25-0.5% silicon. §

mou-iéou. (Pi-yu). Chinese vegetable tallow.

Mould Release Agent N 32. Brushable mold coating on a plastic base. [Chemetall GmbH] *

Mouldrite. Thermosetting molding powders. [BIP Chemicals Ltd] †

mountain blue. (azure blue, mineral blue, mopper blue,

---

‡ Trade name and manufacturer not verified   § Trade name without identified manufacturer

Hamburg blue, English blue). A basic copper carbonate; a pigment used by painters.

mountain butter. A hydrated aluminum sulfate found in fibrous masses.

mountain cork. (elastic asbestos). An asbestos which floats on water.

mountain flax. *See* amianthus.

mountain flour. *See* infusorial earth.

mountain green. (mineral green) CAS 12002-03-8; A pigment prepared by precipitating a boiling solution of alum and copper sulfate with a hot solution of sodium or potassium sulfate.

mountain leather. (mountain paper). Thin, tough types of asbestos.

mountain milk. An earth similar to infusorial earth; used as a rubber filler.

mountain paper. *See* mountain leather.

mountain tallow. *See* bitumen.

mountain wood. A variety of asbestos.

Mountford's paint. A waterproof paint. It consists of asbestos, ground in water, potassium or sodium aluminate, and potassium or sodium silicate sometimes with oil, and zinc white.

Mourey's aluminum solder. An alloy of 82% zinc and 18% aluminum.

Mouse Killer. Rodenticide. [Murphy Chemical Ltd] †

Mouser. CAS 56073-10-0; Contains brodifacoum; ready-for-use bait box, mouse killer. [ICI Garden Products]

Moussett's alloy. An alloy of 60% copper, 27.5% silver, 9.5% zinc, and 3% nickel.

Movelat. A proprietary preparation of adrenocortical extract, salicylic acid and mucopolysaccharide polysulfuric acid ester; used in the treatment of arthritis. [Luitpold-Werk]*

Movin B. Agents for the antimicrobial finishing of textiles. [Bayer AG]

Mowchem. CAS 53780-34-0; A grass growth regulator containing 240 g/l mefluidide; suppresses most grasses for up to 8 weeks; for grassed areas not subject to heavy wear. [Rhone-Poulenc Environmental Prods. Ltd]

Mowilith®. Vinyl acetate homo and copolymer, acrylic and styrene, acrylic copolymer dispersions; used for adhesives and surface coatings. [Harlow Chemical Co Ltd]

Mowilith. Polyvinyl acetate dispersion and solids. [Hoechst AG]

Mowiol®. Polyvinyl alcohol (partially/fully hydrolyzed polyvinyl acetate); for dispersion manufacture, paper manufacture, adhesives soluable film. [Harlow Chemical Co Ltd]

Mowiol. Polyvinyl alcohol. [Hoechst UK] *

Mowital. Polyvinyl butyl resin. [Hoechst UK]

Mow-It-Less. Grass seed mixture. [ICI Garden Products]

Moxam. CAS 64953-12-4; Moxalactam disodium; anti-infective. [Eli Lilly & Co]

Mozanon®. Sodium alginate; fungicide for prevention of TMV infection on tobacco. [Mitsubishi Kasei] *

MP-2. O,O-Dimethyl phosphorochloridothioate; mainly used in the production of organophosphorus insecticides. [A/S Cheminova] *

MP-1. O,O-Dimethyl phosphorodithioic acid; used as an intermediate for organophosphorus insecticides. [A/S Cheminova] *

MP-10CF-4CC/15T. CAS 25134-01-4; Modified PPO, 10% carbon fiber, 4% conductive carbon, 15% PTFE. [Compounding Tech.] *

M-P-A. Thixotropic agent; used as component of paint for protective coating of a substance. [Rheox Inc]

MPDiol Glycol. CAS 2163-42-0; 2-Methyl-1,3-propanediol; used in unsat. polyesters, gel coats, sat. polyester and alkyd coatings, plasticizers. [Arco Chemical Co]

MPEM. O,O-Dimethyl-S-methoxycarbonylmethyl phosphorodithioate; used as an intermediate for phosphorus pesticides. [A/S Cheminova] *

MPI DMSA Kidney Reagent. CAS 304-55-2; Succimer; diagnostic aid. [Medi-Physics Inc] *

MPI Indium DTPa In III. Pentetate indium disodium In 111; diagnostic aid; radioactive agent. [Medi-Physics Inc] *

MPI Indium Oxine In 111. Indium In 111 oxyquinoline; radioactive agent; diagnostic aid. [Medi-Physics Inc] *

MPI Krypton Kr 81m Gas Generator. Krypton, isotope of mass 81; radioactive agent. [Medi-Physics Inc] *

MPK. *See* methyl propyl ketone.

MPS 500. A chlorinated fatty acid ester; a vinyl plasticizer. [Occidental Chemical Corp] *

M-Quat® 32. Octadecyl diethanol methyl ammonium chloride; emulsifier, defoamer, coagulant. [PPG/Specialty Chem.]

M-Quat® 40. CAS 26062-79-3; Polyquaternium 6; homopolymer quaternary for hair and skin products; emollient. [PPG/Specialty Chem.]

M-Quat® 257. CAS 61789-80-8; Quaternium-18, isopropyl alcohol; quaternary for use as textile softener. [PPG/Specialty Chem.]

M-Quat® 522. CAS 67633-63-0; EINECS 266-778-0; Isostearamidopropyl ethyldimonium ethosulfate; conditioner for hair conditioners and shampoos. [PPG/Specialty Chem.]

M-Quat® 1033. CAS 68308-67-8; Soya ethyldimonium ethosulfate; conditioner for hair conditioners and shampoos; antistat for cleaners, rug shampoos; low foaming. [PPG/Specialty Chem.]

M-Quat® 2475. CAS 61789-77-3; EINECS 263-087-6; Dicocodimcnium chloride and isopropanol; emulsifier, defoamer, coagulant; for auto spray wax. [PPG/Specialty Chem.]

M-Quat® B-25. CAS 122-19-0; EINECS 204-527-9; Stearalkonium chloride; quaternary for hair conditioners. [PPG/Specialty Chem.]

M-Quat® Dimer 18. Hydroxypropyl bisstearyldimonium chloride; conditioner for hair and skin, perms, mousses; emulsifier. [PPG/Specialty Chem.]

M-Quat® JN. CAS 112324-16-0; Ricinolamidopropyl ethyldimonium ethosulfate; quaternary for shampoo, hair conditioners, lotions; substantive to hair. [PPG/Specialty Chem.]

M-Quat® JO-50. CAS 37139-99-4; Olealkonium chloride; conditioner, antistat for clear hair rinses. [PPG/Specialty Chem.]

MR-1, MR-1A, MR-17A, MR-17B. Proprietary nomenclature for allyl resins. §

MR-502K, MR-502P, MR-502Y. Dimethyl polysiloxane combination; magnetic rubber inspection material for nondestructive inspection of ferromagnetic parts used for inspection of thread, gear roots, inaccessible areas, coated areas; provides permanent record of the inspection. [Dynamold Inc]

MRV 1000. Polytetrafluoroethylene polymer; release agent for epoxy resins and any molded plastics; lubricants for drive belts, gears and most machinery. [Loes Enter-

---

prises Inc]

M-R-VAX II. Measles and rubella vaccine; combined protection against measles and German measles. [Merck & Co Inc]

MS 4. A silicone electrical insulating compound. [Midland Silicones] ‡

MS-26; Epoxy resin based mastic; moldable liquid shim materials used as a spacer between engines or skin of aircraft or ships; surface conforming structural epoxies used to fill gaps between metal parts and between structural members and composites such as graphite. [Dynamold Inc]

MS-122. CAS 116-14-3; Tetrafluorethylene telomer; release agent, dry lubricant for use on cold molds, esp. for epoxy potting/encapsulating, PU, nylon, acrylics, PP, PC phenolics, PS, foams, rubber molding. [Miller-Stephenson]*

MS-170 1,1,1 Trichloroethane Solv. CAS 71-55-6; 1,1,1-Trichloroethane. [Miller-Stephenson] *

MS-180 Freon® TF Solv. CAS 76-13-1; Trichlorotrifluoroethane. [Miller-Stephenson] *

MS Contin. CAS 64-31-3; Morphine sulfate; analgesic. [The Purdue Frederick Co]

MSG. (sodium glutamate (CTFA); glutamic acid, monosodium salt; monosodium L-glutamate monohydrate) Monosodium salt of L-form of glutamic acid; HOOC-CH₂CH₂CHNH₂COONa; CAS 142-47-2; flavor enhancer for foods. [Ajinomoto Co Inc; Allchem Industries; Asahi Chem Industry Co Ltd; Penta Mfg.; Schweizerhall]

M-Soft-1. Silicone-enriched cationic; softener for all fibers, yarns, fabrics; esp. for brushed or raised fabrics. [Zohar Detergent Factory]

M-Soft-10. Amino functional silicone plus cationic; softener producing an elastomeric silicone finish with excellent soft handle on all kinds of fibers, yarns, and fabrics. [Zohar Detergent Factory]

MSP. *See* sodium phosphate.

MSS 2,4-D Amine. CAS 1702-17-6; 2,4-D; translocated herbicide for cereals and established grassland. [Mirfield Sales Services Ltd]

MSS 2,4-D Ester. CAS 1702-17-6; 2,4-D; translocated herbicide for cereals and grass. [Mirfield Sales Services Ltd]

MSS 2,4-DB + MCPA. 2,4-DB + MCPA; translocated herbicide applied to cereals and undersown clovers. [Mirfield Sales Services Ltd]

MSS 2,4-DP. CAS 120-36-5; Soluble concentrate of 500 g dichlorprop per liter; used for control of weeds in barley, wheat and oats. [Mirfield Sales Services Ltd]

MSS 2,4-DP + MCPA. Mixture of dichlorprop and MCPA; used for weed control in cereals and turf. [Mirfield Sales Services Ltd]

MSS Aminotriazole. CAS 61-82-5; Amitrole; translocated herbicide. [Mirfield Sales Services Ltd]

MSS Chlormequat 40, 46, 60, 70. CAS 999-81-5; Chlormequat; plant growth regulator. [Mirfield Sales Services Ltd]

MSS CIPC. CAS 101-21-3; Chlorpropham; a carbamate herbicide and sprout depressant in stored potatoes. [Mirfield Sales Services Ltd]

MSS CMPP. CAS 24017-47-8; Emulsifiable concentrate containing 600 g/l triazophos; an organophosphorus insecticide. [Mirfield Sales Services Ltd]

MSS CMPP/DP. Soluble concentrate of 520 g dichlorprop and 130 g mecoprop per liter; used for control of weeds

in barley, wheat and oats. [Mirfield Sales Services Ltd]

MSS IPC 50. CAS 122-42-9; Wettable powder containing 50% w/w propham; used for weed control for beet crops and peas. [Mirfield Sales Services Ltd]

MSS MCPA 50. CAS 94-74-6; MCPA; herbicide for cereals and grassland. [Mirfield Sales Services Ltd]

MSS MH18. CAS 123-33-1; Maleic hydrazide; a plant growth regulator for grass and to reduce bud growth in trees, hedges and vegetables. [Mirfield Sales Services Ltd]

MSS Mircam. Soluble concentrate of 18.7 g dicamba and 300 g mecoprop per liter; used for weed control in cereals and grassland. [Mirfield Sales Services Ltd]

MSS Mircam Plus. Soluble concentrate of 19.5 g dicamba, 245 g MCPA and 86.5 g mecoprop per liter; used for weed control in cereals and grassland. [Mirfield Sales Services Ltd]

MSS Simazine/Aminotriazole 43FL. CAS 61-82-5, 122-34-9; A suspension concentrate containing 155 g aminotriazole and 275 g simazine per liter; used for total weed control in non crop areas and fruit orchards. [Mirfield Sales Services Ltd]

MSS Sugar Beet Herbicide. A suspension concentrate containing 37.5 g chlorpropham, 25 g fenuron and 150 g propham per liter; an herbicide for use on beet crops. [Mirfield Sales Services Ltd]

MST. CAS 64-31-3; Morphine sulfate in a controlled release tablet; for chronic severe pain. [Napp Laboratories Ltd]*

Mucaine Suspension. A proprietary preparation of oxethazaine, magnesium hydroxide and aluminum hydroxide gel; antacid mixture containing a topical anesthetic; used for the treatment of oesophagitis, particularly that associated with hiatus hernia, due to radiation therapy or the heartburn of late pregnancy. [Wyeth Laboratories]

muccocota gum. *See* ocota cocota gum.

Mucicarmin. A solution of 2 parts carmine, 1 part aluminum chloride, and 4 parts water; a staining solution.

Mucodyne. A proprietary preparation of carboxymethylcysteine; an expectorant. [Berk Pharmaceuticals Ltd] *

Mucogel®. CAS 21645-51-2, CAS 1309-42-8; Aluminum and magnesium hydroxide; antacid suspension for antacid therapy in gastric and duodenal ulcer gastritis, heartburn, gastric hyperacidity, indigestion. [Pharmax Ltd]

Mucomycin. CAS 992-21-2; A proprietary trade name for lymecycline. [British Drug Houses] *

Mucomyst. CAS 616-91-1; Acetylcysteine; mucolytic. [Mead Johnson & Co]

Mucopront. CAS 638-23-3; A proprietary preparation of carbocysteine; a mucolytic. [Pfizer International]

Mucron. Decongestant. [Ciba plc] *

mudar gum. A material obtained from *Calotropis giganteas.* It resembles gutta-percha, and contains about 20% of a rubbery material.

Mudge's speculum metal. An alloy of 69% copper and 31% tin.

muga silk. The product of the caterpillar, *Antheraca assama,* of Assam.

Mulch-Magic. A brown colored dye used on flower bed mulch to restore the faded color to the new freshly applied color. [Lawn & Garden Products Inc]

muldan. An orthoclase mineral.

mule gum. A name sometimes applied to Ceara rubber.

---

‡ Trade name and manufacturer not verified   § Trade name without identified manufacturer

Mulgofen. Tallow amine. water-soluble emulsifier for mineral oils; agricultural chemicals. [GAF Great Britain] *

Muller's fluid. A solution of phosphoric acid in alcohol; dichromate; and sodium sulfate dissolved in distilled water. A soldering fluid for brass and copper. The term is also used for a hardening agent used in microscopy. It consists of potassium

Mullex. A proprietary refractory material made from mixtures containing various proportions of clay and mullite. §

Mullfrax 301. A mullite-alumina product. [Carborundum] *

mullicite. A mineral. It is blue iron earth.

mullite. A refractory material formed by heating sillimanite to a temperature of 1550 C. It is also made from the minerals andalusite. dumortierite, and Indian cyanite, and by the electric fusion of alumina and silica.

Mulsifan CB. Beheneth-10; emulsifier for creams and lotions. [Zschimmer & Schwarz]

Mulsifan CPA. CAS 5274-68-0; Laureth-4; o/w emulsifier for cosmetic creams and lotions. [Zschimmer & Schwarz]

Mulsifan RT 1. Fatty acid polyglycol ester; emulsifier for mineral oils, textile processing, metal processing, drilling and cutting oils, chemo-tech. products. [Zschimmer & Schwarz]

Mulsifan RT 23. CAS 3055-95-6; Laureth-5; emulsifier for paraffin oils, wh. oils for formulation of lubricating agents, spin finishes, coning oils, emulsions for cosmetics. [Zschimmer & Schwarz]

Mulsifan RT 72. Coco fatty acid MEA ethoxylate; washing agent. [Zschimmer & Schwarz]

Mulsifan RT 141. CAS 9005-64-5; Polysorbate 20; solubilizer for perfumes and volatile oils. [Zschimmer & Schwarz]

Mulsifan RT 146. CAS 9005-65-6; Polysorbate 80; solubilizer for perfumes and volatile oils. [Zschimmer & Schwarz]

Mulsifan RT 203/80. CAS 68131-39-5; C12-15 pareth-12; solubilizer for perfumes and volatile oils. [Zschimmer & Schwarz]

Mulsivin. A proprietary preparation containing bromoform, codeine phosphate, tincture of aconite, belladonna alkaloids, ipecacuanha alkaloids, benzoin, balsam of tolu, storax and liquid paraffin; a cough linctus. [Rybar Laboratories Ltd] *

Multaglut. A mixture of persulfate and calcium phosphate; used to improve flour.

Multex. CAS 12141-46-7; Aluminum silicate; large particle size hard silicate for heavy-duty abrasives; water filtration aid. [Kaopolite]

Multi-W FL. CAS 83601-81-4, 12427-38-2; A suspension concentrate containing 50 g carbendazim and 320 g maneb per liter; systemic fungicide for cereals. [Pan Britannica Industries Ltd]

Multi Base. Powdered compost additive containing NPK 2.6:2.2:23 plus 3% Mg and trace elements; compost additive. [Vitax Ltd]

Multibase ABS 3075. CAS 9003-56-9; ABS; high-impact grade. [Multibase]

Multibase ABS 3525 CL. CAS 9003-56-9; ABS; clear grade. [Multibase]

Multibase ABS 3959. CAS 9003-56-9; ABS; plating grade. [Multibase]

multibrol. An organic combination of bromine consisting primarily of the sodium derivative of brom-oleate with a bromine content of 16%.

Multicel. Dry powder dyes; used in the paper industry. [Multicrom SA]

Multicet. Dry powder disperse dyes; used in the textile industry for dyeing nylon and acetate. [Multicrom SA]

Multicoild. Film cement. [May & Baker Ltd] *

Multicrack. Fluid cracking catalyst. [Akzo]

Multicrom. Dry powder inorganic and organic pigments; used in the manufacture of inks, paints and as dispersions for textile printing. [Multicrom SA]

Multicuer. Dry powder acid dyes; dyes developed for the leather industry. [Multicrom SA]

Multiflow. Resin modifier; additive for nonaqueous industrial coatings; improves flow, reduces pinholes and craters and improves substrate wetting. [Monsanto Co] *

Multigreen® II. Chelated micronutrients; water-soluble organic blend of metal chelates of iron, zinc, copper and manganese, to improve root growth, color and stress tolerance. [Regal Chemical Company]

MultiGuard. Nonnitrite, molybdate formula; for closed system treatment. [Garvey Chemical Corp] *

Multilind. Nystatin and zinc oxide; used for the treatment of fungal skin infections. [Bristol-Myers Squibb Pharmaceuticals Ltd]

Multilind Ointment. Nystatin and zinc oxide in ointment base. [Bristol-Myers Squibb Pharmaceuticals Ltd]

Multiluz. Dry powder direct dyes; light fast dyes used in the textile industry. [Multicrom SA]

Multimet. Mixed metal carboxylates; drier for paint or printing ink, PVC stabilizers. [Manchem Ltd] *

Multionic. Cooling water treatment; for corrosion and scale control. [Garvey Chemical Corp] *

Multisil. Precipitated silica; used in the rubber industry as a filler. [Multicrom SA]

MultiSperse. Boiler treatment; for scale and sludge control. [Garvey Chemical Corp] *

Multisperse CP. Highly effective silicate dispersant for prevention of silicate scale build-up in continuous bleaching equip. [BASF]

Multispray. Public health insecticide. [The Wellcome Foundation Ltd] *

Multistix® . A proprietary test strip for the detection of pH, protein, glucose, ketones, bilirubin, blood and urobilinogen in urine. [Bayer AG]

Multiter. Dry powder disperse dyes; used in the textile industry for dyeing polyester fibers. [Multicrom SA]

Multitherm IG-2. Paraffinic hydrocarbon; heat transfer fluid. [Multitherm Corporation] *

Multitherm PG-1. Naphthenic hydrocarbon; light white mineral oil USP; heat transfer fluid. [Multitherm Corporation] *

Multivac 4. Dental vacuum investor. [Degussa Ltd]

Multivite. A proprietary preparation of vitamins A, D, C and thiamine hydrochloride. vitamin supplement. [British Drug Houses] *

Multiwax® 180-M. CAS 63231-60-7; Microcryst. wax NF; plasticizer or modifier for polymeric coatings and adhesives; hot melt adhesives and coatings, chewing gum base, protective coatings; FDA approved. [Witco/Sonneborn] *

Multiwax® HS. CAS 63231-60-7; Microcryst. wax; wax used in foil/tissue laminations, heat-seal laminations, glassine laminations; FDA approved. [Witco/Sonneborn] *

Multronol. Polyether polyol. [Miles]

Mulukilivary. A gum-resin obtained from *Balsamodedron*

---

berryi; used as a myrrh substitute.

Mumetal. A patented nickel-iron-copper alloy having the highest permeability of all known commercial materials; its exceptional magnetic properties and low losses make it invaluable for cable loading, instrument transformers, relays, magnetic shields, etc. §

Mumpsvax. Mumps vaccine; for protection against mumps. [Merck & Co Inc]

Municol. Sodium alginate, food/pharmaceutical grade. [Kelco Int'l. Ltd]

munjeet. The root of *Rubia munjista*. It contains purpurin, and is an important Indian dyestuff.

Munjistin. (purpuroxanthic acid). Di-hydroxyanthra-quinonecarboxylic acid, $C_{15}H_8O_6$.

Muntz metal. A brass containing 60% copper and 40% zinc.

Murac. An insulating material stated to be made by treating the latex of *Sapotaceae* species.

Murald. Aldrin insecticides. [Murphy Chemical Co Ltd] †

Murcurite. A mercury fungicide. [Murphy Chemical Co Ltd]†

Murdiel. CAS 60-57-1; Dieldrin insecticides. [Murphy Chemical Co Ltd] †

Murex. A proprietary trade name for a manganese steel containing 3% manganese, 1% carbon, and 0.85% nickel. §

Murexan. *See* dialuramide.

murexide. (Naples red). An obsolete red basic dyestuff, obtained by the action of nitric acid upon guano, and subsequently treating the product with ammonia.

Murfite. Acaricide. [Murphy Chemical Ltd] †

Murfixtan. A mercury fungicide. [Murphy Chemical Co Ltd]†

Murfly. An insecticide. [Murphy Chemical Co Ltd] †

Murfotox. Organo-phosphorus insecticides. [Murphy Chemical Co Ltd] †

Murfume. Pesticidal smoke generators. [Murphy Chemical Co Ltd] †

Murfume Grain Store Smoke. γ-HCH; an organochlorine insecticide. [DowElanco Ltd]

muriate of potash. Potassium chloride, KCl.

muriate of soda. Sodium chloride, NaCl.

muriatic acid. *See* hydrochloric acid.

Murine Ear Drops. Carbamide peroxide, a compound of urea with hydrogen peroxide (1:1); anti-infective, topical. [Abbott Laboratories] *

Murine Plus. CAS 522-48-5; Tetrahydrozoline hydrochloride; adrenergic. [Abbott Laboratories] *

Muripsin. A proprietary preparation of glutamic acid hydrochloride and pepsin. [Norgine Ltd]

Muritan. Rotent control product (mice). [Bayer AG]

Murlin Premium Ladle Wash. A blend of finely divided, inorganic solids dispersed in an aqueous medium; used to coat and protect surfaces, tools and equipment coming in contact with molten aluminum. [Murlin Chemical Inc]

Murman's alloy. One contains 92% aluminum, 4.4% zinc, and 3.6% magnesium. Another consists of 72% aluminum, 14.5% zinc, and 13.5% magnesium.

Murnil® . Veterinary preparation; for activation of the skin metabolism of dogs and cats. [Bayer AG]

Murphex. Disinfestation products. [Murphy Chemical Co Ltd] †

Murphicol. Pesticide suspensions. [Murphy Chemical Co Ltd] †

Murphos. CAS 56-38-2; Parathion insecticides. [Murphy Chemical Co Ltd] †

Murvin 85. CAS 63-25-2; Carbaryl; contact insecticide and worm killer. [DowElanco Ltd]

muscarine. (Campanuline). A dyestuff. It is the dihydroxy derivative of Meldola's blue; $C_{18}H_{15}N_2Cl.O_2$; dyes cotton mordanted with tannin and tartar emetic blue; used for calico printing.

Muscatox. Paint-on-bait insecticide for fly control, e.g., in farms. [Bayer AG]

muscle sugar. Inositol, $C_6H_6(OH)_6$.

muscovite mica. *See* mica.

Mushet steel. (self-hardening steel). Steels containing from 0.7-1.2% carbon and 2-3% tungsten. They require no quenching or tempering.

mushroom sugar. CAS 69-65-8; Mannitol; $C_6H_8(OH)_6$; used as base for dietetic foods, diluent, determination of boron, pharmaceutical products, medicine, thickener, and stabilizer in food products.

Musiv gold. An alloy of from 66-70% copper and 30-34% zinc.

musk. CAS 300-54-9; The dried animal secretion of the musk deer. It has been practically superseded by synthetic compounds; used in cosmetics and perfumery, fragrances, and mothproofing agent.

musk B. *See* musk baur.

musk C. *See* ketone musk.

Musk R-1. 11-Oxahexadecanolide. [Quest Int'l. UK Ltd]

musk ambrette. CAS 83-66-9; Ambrottolide.

musk baur. (musk B, tonquinol). An artificial musk. It is 2 : 4 : 6-trinitro-l-methyl-3-tertiary-butyl-toluene, $C_6H(CH_3)$-$(NO_2)_3 \cdot C(CH_3)_3$; used for soap and toilet purposes.

Musketeer. Suspension concentrate containing 50 g ioxynil, 250 g isoproturon and 180 g mecoprop per liter; used for control of weeds in wheat and barley. [Hoechst UK]

Musol 20. Sol. yeast mucins; cosmetic ingred. for moisturizing. [Brooks Industries]

mussanin. An extract from *Albizzin anthelmintica*; used as a vermifuge.

mustard gas. (yperite, yellow cross gas) CAS 505-60-2; . Dichloro-diethyl-sulfide; used as a military poison gas

mustard oil. Ethyl-isothiocyanate, C_2H_5NCS.

Mustargen® . CAS 55-86-7; Mechlorethamine hydrochloride; intravenous indicated for the palliative treatment of Hodgkins disease (stages III and IV), lymphosarcoma, chronic myeloctic or chronic lympocytic leukemia, polycythemia vera, mycosis fungoides, and bronchogenic carcinoma. [Merck & Co Inc]

Mustine. A proprietary preparation of mustine hydrochloride; a carcino-chemotherapeutic agent. [The Boots Co plc]

Mustone. A Japanese chloroprene polymer.

Mutamycin. Mitomycin; antineoplastic. [Bristol Laboratories] *

Muthmann's liquid. CAS 79-27-6; Acetylene-tetra-bromide, $CHBr_2$ $CHBr_2$; used as a solvent.

muthol. *See* paraffin, liquid.

mutton tallow. *See* tallow.

M'Varavara. The roots of *Securidaca longipedunculata*; used medicinally.

M Violet 112. CAS 20202-66-3, 7787-59-9; Manganese violet, bismuthoxychloride; inorg. colorant. [Presperse]

MVP. Methacrylate; permanent structural adhesive for bonding metals, plastics and ceramics. [ITW Devcon]

MXM-7500. A proprietary epoxy putty used for filling in difficult radii and depressions in epoxy-fiber glass com-

---

‡ Trade name and manufacturer not verified   § Trade name without identified manufacturer

ponents. [Fiberite West Coast Corp] ‡

Myacide® AS Plus. CAS 52-51-7; 2-Bromo-2-nitropropane-1,3-diol; water-sol. antimicrobial, preservative for paper-making, adhesives, coatings, cooling towers, process waters, oilfield water-flooding operations, deodorizing, other aq. applications. [Angus; Boots Co. PLC]

Myacide® S-1. CAS 52-51-7; 2-Bromo-2-nitropropane-1,3-diol in dipropylene glycol methyl ether/water; water treatment antimicrobial for recirculating cooling towers and process water; preservative for oilfield drilling-fluid and water-flooding operations. [Angus; Boots Co. PLC]

Myacide® S-2. CAS 52-51-7; 2-Bromo-2-nitropropane-1,3-diol in dipropylene glycol methyl ether/water; water treatment antimicrobial for recirculating cooling towers and process water; preservative for oilfield drilling-fluid and water-flooding operations. [Angus; Boots Co. PLC]

Myacide® SP. CAS 1777-82-8; 2,4-Dichlorobenzyl alcohol; antifungal agent, preservative. [Inolex; Boots Co. PLC]

Myambutol. CAS 74-55-5; A proprietary preparation containing ethambutol; an antituberculous agent. [Lederle Laboratories]

Myanesin. CAS 533-06-2; A proprietary trade name for mephenesin carbamate; muscle relaxant and tranquilizer. [British Drug Houses] *

Mybasan. CAS 54-85-3; A proprietary trade name for isoniazid. §

My-B-Den. CAS 61-19-8; Adenosine phosphate; used in biochemical research.

Mybond®. Formulated natural and synthetic adhesives in aqueous, hot melt, solventless and solventborne forms for use in industry; for bonding a wide range of substrates for the paper and board, packaging, woodworking, building and decorating, textile, footwear, automotive, bookbinding, electrical and product assembly markets. [Mydrin Ltd]

Mycardol. CAS 78-11-5; A proprietary trade name for pentaerythritol tetranitrate. §

Mycelex. CAS 23593-75-1; Clotrimazole; antifungal. [Miles Pharmaceuticals]

Mycifradin. CAS 1405-10-3; A proprietary preparation of neomycin sulfate; an antibiotic. [Upjohn Ltd] *

Myciguent. CAS 1405-10-3; A proprietary preparation of neomycin sulfate; used in dermatology as an antibacterial agent. [Upjohn; Upjohn Ltd]

Mycil. CAS 2398-96-1; Proprietary preparations containing tolnaftate; used for treatment of fungal infections of the skin. [The Boots Co plc]

Mycil. CAS 104-29-0; A proprietary trade name for chlorphenesin; antifungal. [British Drug Houses] *

Mycivin. CAS 7179-49-9; A proprietary preparation containing lincomycin hydrochloride; an antibiotic. [The Boots Co plc]

Mycocide. A slime control agent. [Great Lakes Europe] †

mycodermine. A yeast preparation used in medicine.

Mycolactine. A proprietary preparation of dried yeast, ox bile, lactic principles, frangula, aloes and belladonna; a laxative. [L Wilcox & Co Ltd] ‡

Mycose. CAS 149-29-1; Trehalose, $C_{12}H_{22}O_{11}$.

Mycospor®. CAS 60628-96-8; Bifonazole; pharmaceutical preparation used as broad spectrum antifungal agent. [Bayer AG; Miles Pharmaceuticals]

Mycostatin. CAS 1400-61-9; Nystatin; antifungal. [Bristol-Myers Squibb Co Inc]

Mycota. A proprietary preparation of undecenoic acid and zinc undecenoate; used for treatment of fungal infections of the skin. [The Boots Co plc] †

Mycota. Antifungal cream, spray, powder for treatment of athlete's foot. [Seton Healthcare Group plc]

Mycotal. *Verticillium lecanii*; a fungal parasite to control aphids and whitefly. [Koppert (UK) Ltd]

Mycovac-L. Mycoplasma gallisepticum; for immunization of poultry. [Intervet Inc]

Mycozol. A proprietary preparation of chlorbutol, salicylic acid, benzoic acid and malachite green; for treatment of fungal infections of the skin. [Smith & Nephew Pharmaceuticals Ltd] *

Mydflex. A triethanolamine salicylate; topical analgesic. [Hercules] *

Mydfrin. CAS 61-76-7; Phenylephrine hydrochloride; adrenergic. [Alcon Laboratories Inc]

Mydochrome. Color reversal processing system. [May & Baker Ltd] *

Mydoneg. Color film processing system. [May & Baker Ltd]*

Mydoprint. Color paper processing system. [May & Baker Ltd] *

Mydriacyl. CAS 1508-75-4; Tropicamide; anticholinergic (ophthalmic). [Alcon Laboratories Inc]

Mydriasine. Atropine methyl-bromide, $C_{17}H_{23}NO_3 \cdot CH_3Br$; a mydriatic.

Mydrilate. CAS 5870-29-1; A proprietary preparation containing cyclopentolate hydrochloride; an antimuscarinic agent; eye drops. [Boehringer Ingelheim Ltd]

myelin. A white, fatty substance obtained from various animal and vegetable tissues.

Myer's naphthol green. (standard). It consists of 50 mg naphthol green B dissolved in 1,000 cc water; used for cholesterol determination

Myflam®. Formulated aqueous emulsion polymer coatings for use in the textile industry; coating and impregnating compounds imparting fire retardant and other technical performance requirements; for upholstery, drapes, blinds, carpet, bedding, automotive, aerospace, industrial, wallcoverings, apparel and flooring industries. [Mydrin Ltd]

Mykon. Wetting and detergent, dispersing and emulsifying properties; used for textiles; general purpose detergent and wetting agent. [Warwick International Ltd] *

Mykon 817. Phosphated alcohol anionic surfactant as a clear liquid; wetting and dispersing agent, scouring and bleach assistant; used primarily in the preparation and dyeing of cotton and synthetic fibers. [Warwick Chemical Ltd] *

Mykonaid. Dispersing and leveling agent; used for textiles especially in high temperature dyeing. [Warwick International Ltd] *

Mykroy/Mycalex. Machinable moldable ceramic material made from glass and mica powder. The material is completely inorganic and is supplied in sheet, rods or parts molded to specification; machinable with standard shop tools and no after-firing is required; for use in areas where dimensional stability is required over a broad range of temperatures; e.g. high voltage switch gears, arc barriers, asbestos-filled material replacement; thermal barrier for molding presses, thermal switch covers. [Mykroy/Mycalex] *

Mylanta. Antacid preparations. [ICI Chem. & Polymers Ltd]†

Mylar. A polyester film used for electrical insulation, cable lapping, magnetic tape; features very high tensile

---

* Trade name not verified as available   † Trade name verified as obsolete

strength. [ICI Chem. & Polymers Ltd] †

Mylar®. Polymer. [DuPont UK]

Myleran. Busulfan; antineoplastic. [Burroughs Wellcome Co] *

Myleran Tablets. CAS 55-98-1; A proprietary formulation of busulfan; for the palliative treatment of chronic myelogenous (myeloid, myelocytic, granulocytic) leukemia. [The Wellcome Foundation Ltd]

Mylicon. CAS 8050-81-5; Simethicone; antiflatulent. [Stuart Pharmaceuticals] †

Mylocon. A proprietary preparation containing methyl polysiloxane in flavored vehicle. [Parke-Davis] *

Mylol. An insect repellant. [The Boots Co plc]

Mylomide. A proprietary preparation of amylobarbitone and megemide; a sedative. [Nicholas Laboratories Ltd] *

Mylosar. CAS 320-67-2; Azacitidine; antineoplastic. [Upjohn] *

Mynah. A proprietary preparation of ethambutol and isoniazid; used in the therapy of tuberculosis. [Lederle Lab] *

Myocardol. CAS 78-11-5; A proprietary preparation of pentaerythritol tetranitrate; a vasodilator used in angina pectoris. [Bayer AG] †

Myochrysine. CAS 12244-57-4; Gold sodium thiomalate; a mixture of the mono- and di- sodium salts of gold thiomalic acid; antirheumatic. [Merck & Co Inc]

Myocrisin. A proprietary preparation of sodium aurothiomalate. [May & Baker Ltd] *

Myocrisin. Sodium aurothiomalate. [Rhone-Poulenc Rorer Ltd]

Myodil. Ethyl iodophenylundecylate. [Glaxo Laboratories] *

myo-Inositol. See inositol.

Myolgin. A proprietary preparation of aspirin, paracetamol, codeine phosphate, caffeine citrate and acetomenaphthone; an analgesic. [Cox Pharmaceuticals]

Myplabin®. Macrolide antibiotic. [Asahi Chem. Industry]

Mypolex®. For the chemical industry. [DuPont UK]

Myprozine. CAS 7681-93-8; Natamycin; antibacterial. [Am. Cyanamid]

myrabola oil. A German soap-making material. It is an oil stated to be a mixture of different fatty acids and their glycerides. It is obtained from fat waste.

Myras. Internal gear pump; for light oil and viscous liquid pumping. [Sihi Pumps (UK) Ltd]

Myrickite. A trade name for a chalcedony. §

Myringacaine Drops. CAS 90-03-9; Mercufenol chloride; anti-infective. [Upjohn] *

myristica. Nutmeg.

myristic acid. (tetradecanoic acid) Organic acid; $CH_3(CH_2)_{12}COOH$; CAS 544-63-8; EINECS 208-875-2; soaps, cosmetics, synthesis of esters for flavors and perfumes; component of food-grade additives. [Akzo; Aldrich; Henkel/Emery; Mirachem Srl; Unichema; Witco/Humko]

myristic acid isopropyl ester. See isopropyl myristate.

myristyl alcohol. (C14 linear primary alcohol; 1-tetradecanol; tetradecyl alcohol) Fatty alcohol; $CH_3(CH_2)_{12}CH_2OH$; CAS 112-72-1; EINECS 204-000-3; surfactant intermediate; organic synthesis; antifoam agent; perfume fixative for soaps and cosmetics; specialty cleaning preparations; emollient for cold creams. [Ethyl; W.R. Greeff; M. Michel; Vista]

myristyl lactate. (2-hydroxypropanoic acid, tetradecyl ester; tetradecyl 2-hydroxypropanoate) Ester of myristyl

alcohol and lactic acid; $CH_3COHHCOOCH_2(CH_2)_{12}$-$CH_3$; CAS 1323-03-1; EINECS 215-350-1; Imparts lubricity, sheen, etc. to cosmetic and pharmaceutical applications. [Dinoval Chem. Ltd]

Myritol. Skin compatible oil component and fattening agent; used for cosmetic preparations. [Henkel Chemicals Ltd]*

Myritol 318. CAS 65381-09-1; Caprylic/capric triglyceride; emollient for pharmaceutical and cosmetic preps. in emulsion form; oily component with solv. capacity; solubilizer. [Henkel/Cospha; Henkel Canada]

Myrj®. Polyexyethylene alkylesters. [ICI Am.]

Myrj® 45. CAS 9004-99-3; PEG-8 stearate; general purpose o/w emulsifiers for cosmetics, pharmaceuticals, etc. [ICI Spec. Chem.; ICI Surf. Belgium]

Myrj® 52. CAS 9004-99-3; PEG-40 stearate; pharmaceutical aid. [ICI Am.]

Myrj® 53. CAS 9004-99-3; PEG-50 stearate; pharmaceutical aid. [ICI Am.]

Myrj® 59. CAS 9004-99-3; PEG-100 stearate; o/w emulsifier for cosmetic and pharmaceutical applications. [ICI Spec. Chem.; ICI Surf. Belgium]

myrmekite. A quartz mineral.

myrobalans. The fruit of *Terminalia chebula*, of India. This is the chief variety of this product, but there are at least five varieties of the commercial article which are named after the district where they are marketed. They contain from 24-39% t

myrrh. The true myrrh is that known as Herabol myrrh, a gum resin obtained from various species of *Balsamodedron* and *Cammiphora*; used in perfumery, incense, and toiletries.

myrtol. A refined myrtle oil. It is an essential oil containing myrtenol, pinene, and cineol; used in medicine for bronchial and pulmonary infections and as an antiseptic.

Mysoline. CAS 125-33-7; Primidone; anticonvulsant. [Wyeth Laboratories]

Mysoline. CAS 125-33-7; A proprietary preparation of primidone; an anticonvulsant. [ICI Chem. & Polymers Ltd]

Myspamol. Proquamezine fumarate. [RMB Animal Health Ltd]

Mysteclin Capsules. Nystatin and tetracycline hydrochloride in capsule form. [Bristol-Myers Squibb Pharmaceuticals Ltd]

Mysteclin Syrup. Tetracycline and amphotericin; an antibiotic. [Bristol-Myers Squibb Pharmaceuticals Ltd]

Mysteclin Tablets. Tetracycline and nystatin; an antibiotic. [Bristol-Myers Squibb Pharmaceuticals Ltd]

Mystery gold. An alloy of 1 part platinum and 2 parts copper, with a little silver.

Mystic metal. An alloy of 88.7% lead, 10.8% antimony, 0.4% iron, and 0.1% bismuth.

mystin. A mixture of formaldehyde and sodium nitrite; used as a preservative.

Mystolene. Water repellents and proofers. [Catomance Group]

Mystox. Preservatives for textiles. [Catomance Group]

Mytab®. CAS 1119-97-7; EINECS 214-291-9; Myrtrimonium bromide; surfactant used as emulsifier and antistat in hair rinses; antimicrobial for cosmetics, topicals. [Zeeland]

Mytex. Formulated aqueous emulsion polymer coatings for use in the textile industry; coating and impregnating compound to impart technical properties such as fabric

---

‡ Trade name and manufacturer not verified   § Trade name without identified manufacturer

stabilization, seam slippage resistance, abrasion resistance, tuft lock, stain resistance, shower repellency and waterproofing, air and moisture permeability, thermal and electrical conductivity, blackout; for upholstery, soft furnishings, carpet, bedding, wallcovering, automotive, aerospace, footwear, apparel, label, motif and industrial markets. [Mydrin Ltd]

Mytolac. Acne lotion and cream, to clear acne blemishes. [Richardson-Vicks Inc] *

Mytrex. Neomycin sulfate and triamcinolone acetonide; indicated in the treatment of corticosteroid-responsive dermatoses with secondary infection. [Altana Inc] ‡

Mytrex F. Nystatin and triamcinolone acetonide; indicated for the treatment of cutaneous candidiasis. [Altana Inc]‡

Myuizone. p-Acetylaminobenzaldehyde thiosemicarbazone.

Myvacet®. Acetylated monoglycerides. [Eastman]

Myvacet® 5-07K. Acetylated hydrogenated vegetable glyceride, distilled; emulsifier, emollient; forms films with good moisture vapor barrier properties for nuts, dried fruits, sausages. [Eastman]

Myvacet® 7-07K. Acetylated hydrogenated vegetable glyceride; forms thin transparent film to serve as oxygen moisture barrier for food pkg. [Eastman]

Myvacet® 9-08K. Acetylated hydrogenated coconut oil glyceride, distilled; emulsifier, emollient for food processing. [Eastman]

Myvacet® 9-40. CAS 8029-92-3; Acetylated lard glyceride, distilled; emulsifier; food-grade lubricant and emollient; deaerator in some systems. [Eastman]

Myvacet® 9-45K. Acetylated hydrogenated soybean oil glyceride; emulsifier, lubricant, solv. for icings and shortenings. [Eastman]

Myvaplex®. Glyceryl monostearate. [Eastman]

Myvaplex® 600K. Glyceryl stearate, food grade; emulsifier in macaroni and cereal products; starch complexing agent, lubricant, processing aid for foods. [Eastman]

Myvaplex® 600PK. Glyceryl monostearate (from hydrogenated soybean oil); starch complexing agent, lubricant, processing aid for foods. [Eastman]

Myvatem® 30. Diacetyl tartaric acid ester of distilled monoglycerides (from edible tallow); emulsifier, dispersant for food, pharmaceutical, and cosmetic applications. [Eastman]

Myvatex®. Food emulsifiers. [Eastman]

Myvatex® 3-50K. Gyceryl stearate (distilled) and propylene glycol stearate (distilled); emulsifier in food applications. [Eastman]

Myvatex® 8-06K. Hydrogenated soybean oil monoglyceride; food emulsifier. [Eastman]

Myvatex® 40-06S K. Distilled propylene glycol esters, distilled monoglycerides, lactylic esters of stearic acid, and water; food emulsifier for cakes. [Eastman]

Myvatex® 90-10K. Hydrogenated rapeseed oil, hydrogenated cottonseed oil; food emulsifier. [Eastman]

Myvatex® Do Control K. Succinylated palm oil monoglycerides, palm oil monoglycerides, distilled; dough strengthener for yeast-raised bakery goods. [Eastman]

Myvatex® Mighty Soft®. Distilled monoglyceride prepared

from edible vegetable oil; softener, bodying agent, extension aid for yeast-raised bakery goods, ice cream, frozen desserts, pasta. [Eastman]

Myvatex® Monoset® K. Rapeseed oil monoglycerides, cottonseed oil monoglycerides, distilled; food emulsifier. [Eastman]

Myvatex® Super DO. Succinylated monoglycerides and distilled monoglycerides; emulsifier for baked goods. [Eastman] *

Myvatex® Texture Lite® K. Soybean oil monoglycerides, propylene glycol monoesters, sodium stearoyl lactylate, silicon dioxide; emulsifier for cakes, icings, cream fillings, whipped toppings, sauces. [Eastman]

Myverol®. Propylene glycol emulsifier. [Eastman]

Myverol® 18-00. Distilled hydrogenated animal glyceride; emulsifier for baked goods, confectionery products, cosmetics, dehydrated potatoes, etc. [Eastman] *

Myverol® 18-04K. CAS 67784-87-6; Hydrogenated palm oil glyceride; food emulsifier for candy, infant formula, margarine, peanut butter. [Eastman]

Myverol® 18-06K. Hydrogenated soy glyceride; food emulsifier for candy, chewing gum base, cake mixes, infant formula, whipped toppings. [Eastman]

Myverol® 18-07K. Hydrogenated cottonseed glyceride; food emulsifier for candy, chewing gum base, cake mixes, confectionery coatings, infant formula, whipped toppings. [Eastman]

Myverol® 18-30. CAS 61789-13-7; Tallow glyceride, distilled; emulsifier for baked goods, confectionery products, cosmetics, etc. [Eastman]

Myverol® 18-35. Palm oil glyceride, distilled; emulsifier for baked goods, confectionery products, cosmetics, etc. [Eastman] *

Myverol® 18-40. CAS 61789-10-4; Lard glyceride, distilled; emulsifier for baked goods, confectionery products, cosmetics, etc. [Eastman]

Myverol® 18-50K. CAS 61789-08-0; Hydrogenated vegetable glyceride, distilled; food emulsifier used in icings, cream fillings, cake mixes, shortenings. [Eastman]

Myverol® 18-85K. CAS 8029-44-5; Cottonseed glyceride; food emulsifier for icings, cream fillings, shortenings. [Eastman]

Myverol® 18-92, 18-92K. Sunflower seed oil glyceride; food emulsifier for icings, cream fillings, shortenings. [Eastman]

Myverol® 18-98. Safflower glyceride, distilled; emulsifier for diet margarines, icing and cream-fillings shortenings. [Eastman] *

Myverol® 18-99K. Low-erucic rapeseed oil monoglyceride, distilled; food emulsifier for icings, cream fillings, shortenings. [Eastman]

Myverol® P-06K. Distilled monoester from hydrogenated soybean oil and propylene glycol; food emulsifier, stabilizer, aerating agent. [Eastman]

Myverol® SMG VK. Hydrogenated palm oil or palm stearine succinylated monoglyceride; emulsifier for shortenings, dough strengthener, softener for bread baking. [Eastman]

2MZ. See 2-methyl imidazole.

N

N. CAS 1344-09-8; Sodium silicate solution (3.22 SiO$_2$:Na$_2$O ratio; 41 Be density); builder for laundry detergents and cleaners, adhesive for corrugated, spiral wound and laminated paper products, binder for foundry cores and molds, manufacture of silica gels, zeolites, catalysts. [PQ Corp] *

N33. An alloy of cast iron with additions of nickel, copper, and chromium.

N-521® Biocide. Tetrahydro-3,5-dimethyl-2H-1,3,5-thiadiazine-2-thione; fungicide and bactericide for use in leather, paint, glue, casein, starch, and paper manufacturing and processing. [Akzo]

N-948® Biocide. CAS 6317-18-6; Methylene bis (thiocyanate); industrial biocide controlling algae, bacteria, yeast and fungi. [Akzo]

Nacap®. CAS 2492-26-4; Sodium 2-mercaptobenzothiazole; metal deactivator; corrosion inhibitor for water, alcohol, and glycol systems; used in antifreeze; chemical intermediate. [R.T. Vanderbilt]

Nacconal. Sodium alkylbenzene sulfonate; detergent, wetting agent. [Stepan] *

Nacconol 35SL. Anionic surfactant in liquid form; used for liquid detergents. [KWR Chemicals Ltd] *

Nacconol® 40G. Sodium dodecylbenzene sulfonate; foamer, dispersant, wetting agent, detergent for agric., cement, dyeing, emulsion polymerization, textile, metal cleaning, metalworking, mining, paper industries; biodeg. [Stepan; Stepan Canada]

Nacconol 90F and 40F. Anionic surfactant in cream flake form; used for powdered detergents. [KWR Chemicals Ltd] *

Naclex. CAS 135-09-1; A proprietary preparation of hydroflumethiazide; a diuretic. [Glaxo Laboratories] *

Nacton. Oral anticholinergic. [SmithKline Beecham]

Nacton. Poldine methylsulfate; anticholinergic. [McNeil Pharmaceuticals] *

Nadavin®. Agents to increase the wet strength of papers. [Bayer AG; Bayer plc]

Nafcil. CAS 985-16-0; Nafcillin sodium; antibacterial. [Bristol Laboratories] *

Nafion®. Perfluorosulfonic acid membrane. [DuPont UK]

Naftin®. CAS 65473-14-5; Naftifine hydrochloride; antifungal. [Allergan, Inc]

Naftocit. Vulcanizing accelerators for latex and rubber processing. [Chemetall GmbH] *

Naftogran®. Polymer bounded vulcanizing accelerators for improved and nonpolluting processing in rubber chemicals. [Chemetall GmbH]

Naftolen. Plasticizers and extender oils for natural and synthetic rubber. [Chemetall GmbH] *

Naftolen R 100, 510, 530, 550, 570, X413, X414, X10, 134. Vinyl plasticizers. [Wilmington Chemical Co] *

Naftomix. PVC stabilizer and lubricant. [Chemson Ltd]

Naftonox. Antioxidants for the production and processing of natural and synthetic rubbers and their latexes, thermoplastics, and adhesives. [Chemetall GmbH] *

Naftopast. Solid dispersions for improved and nonpolluting processing in rubber compounds. [Chemetall GmbH] *

Naftovin. PVC stabilizer. [Chemson Ltd]

Naftozin. Stearic acids as processing auxiliaries in the production of compounds and as lubricants in the processing of thermoplastics. [Chemetall GmbH] *

Naganol. Veterinary preparation; for the control of trypanosomiasis of domestic animals. [Bayer AG]

Nageli's solution. A solution containing a mixture of zinc chloride and iodide; used as a disinfectant.

Naglusol. CAS 527-07-1; Sodium gluconate solution. [Jungbunzlauer Inc]

nahcolite. A native sodium bicarbonate.

Nalan. Water repellent. [DuPont UK] *

Nalcite. A proprietary trade name for a water softener containing an organic zeolite type exchanger material. §

Nalco® 131. Glycol-type surfactants; food grade antifoam for beet sugar and enzyme aq. operations. [Nalco]

Nalco® 2300. Silica/silicone; defoamer for coatings, latex and high solids systems. [Nalco]

Nalco® 2340. Silica-organic; nonsilicone antifoam for aq. resin and polymer processing; short term persistency. [Nalco]

Nalco® 8669. Silica-organic; nonsilicone antifoam for aq. fiberglass mat resin processing; effective over wide temp. range. [Nalco]

Nalcrom. Treatment for ulcerative colitis, etc. [Fisons plc, Pharmaceuticals Div] *

Nalfleet. Chemical treatments for the marine industry. [ICI Chem. & Polymers Ltd] *

Nalfloc. Water and effluent treatment chemicals. [ICI Chem. & Polymers Ltd] *

Nalfon. Fenoprofen calcium; anti-inflammatory; analgesic. [Eli Lilly & Co]

nalidixic acid. (1-ethyl-7-methyl-1,8-naphthyridin-4-one-3-carboxylic acid) C$_{12}$H$_{12}$N$_2$O$_3$; CAS 389-08-2; EINECS 206-864-7; Antibacterial for medicine. [Am. Int'l. Chem.]

Nalidone®. CAS 28874-51-3; Sodium PCA; moisturizer for dermatological soap, shampoo, after-sun gel, nutritive and regenerative creams and lotions, hair comb-out balm. [UCIB]

Nalorex®. CAS 16590-41-3; Naltrexone. [DuPont Merck Pharmaceuticals]

Nalutron. CAS 57-83-0; Progesterone; progestin. [Sterling

---

‡ Trade name and manufacturer not verified

§ Trade name without identified manufacturer

Drug Inc] *

Nalzin. CAS 55799-16-1; Zinc hydroxy phosphite; nontoxic anticorrosive pigment for paint. [Rheox Inc]

Nametal. Foil wrapped metallic sodium. [Foseco (F.S.) Ltd]

nancic acid. Lactic acid, $C_3H_6O_3$.

Nandel®. Fibers. [DuPont UK]

Nandrobolic. CAS 62-90-8; Nandrolone phenpropionate; androgen. [O'Neal, Jones & Feldman Pharmaceuticals]*

Nandrobolic LA. CAS 360-70-3; Nandrolone decanoate; androgen. [O'Neal, Jones & Feldman Pharmaceuticals]*

Nangawhite. A nonstaining antioxidant for rubber. [Rubber Regenerating Co] ‡

Nankor. A proprietary preparation of fenchlorphos; an insecticide. §

Nansa. Aklylaryl sulfonates. [Albright & Wilson Ltd, Detergents Div]

Nansa® 1042. CAS 68584-22-5; Dodecylbenzene sulfonic acid; intermediate used in manufacturing of detergents, laundry products, emulsifiers for emulsion polymerization. [Albright & Wilson UK] *

Nansa® 1106/P. Sodium dodecylbenzene sulfonate; surfactant for emulsion polymerization. [Albright & Wilson UK]*

Nansa® 1169/P. Alkylaryl sulfonate anionic; supplied as a golden liquid; surfactant for emulsion polymerization. [Albright & Wilson Ltd, Detergents Div, Marchon] *

Nansa® AS 40. CAS 1331-61-9; EINECS 215-559-8; Ammonium dodecylbenzene sulfonate; formulation of domestic and industrial liq. detergents. [Albright & Wilson UK]

Nansa® BMC. Blended anionic nonionic foaming agent, in yellow liquid form; air entraining agent for mortar and concrete. [Albright & Wilson Ltd, Detergents Div, Marchon] *

Nansa® EVM50. CAS 68953-96-8; Calcium dodecylbenzene sulfonate in aromatic solv.; emulsifier, dispersant for agrochemicals, textiles, surface coatings, polymerization, leather industries. [Albright & Wilson UK]

Nansa® HS80S. CAS 25155-30-0; Sodium dodecylbenzene sulfonate; formulation of detergents, hard surface and bottle cleaners; metal treatment and paper processing; scouring and wetting agent for textile industry; foamer and mortar plasticizer in building industry. [Albright & Wilson UK]

Nansa® LES42. Predominantly straight chain sodium dodecylbenzene sulfonate, lauryl ether sulfonate and magnesium xylene sulfonate blend giving an anionic in the form of a golden liquid; used for dishwashing detergents; hard surface cleaners. [Albright & Wilson Ltd, Detergents Div, Marchon] *

Nansa® LSS38/A. CAS 68439-57-6; EINECS 270-407-8; Sodium C14-16 olefin sulfonate; general purpose detergent base; wetting agent for agric. wettable powds., textile processing; emulsifier for enhanced oil recovery. [Albright & Wilson UK]

Nansa® MA30. Sodium dodecylbenzene sulfonate and ethoxylated nonionic blend; detergent base for dishwashing and hard surface cleansers; scouring agent for textiles; mortar plasticizer in manufacturing of masonry cement. [Albright & Wilson UK]

Nansa® SB 30. Branched sodium alkylbenzene sulfonate; surfactant. [Albright & Wilson UK]

Nansa® SBA. CAS 68608-88-8; Branched dodecylbenzene sulfonic acid; detergent intermediate. [Albright & Wilson UK]

Nansa® SL 30. Sodium dodecylbenzene sulfonate; surfactant used in detergent formulations. [Albright & Wilson UK] *

Nansa® SS 30. Sodium dodecylbenzene sulfonate; surfactant for detergent formulations; emulsifier for agric. emulsifiable concs.; curing/foaming agent for thermosets. [Albright & Wilson UK]

Nansa® SSA. CAS 68584-22-5; Dodecylbenzene sulfonic acid; detergent intermediate; in prep. of emulsifiers for emulsion polymerization. [Albright & Wilson UK]

Nansa® TDB. CAS 25496-01-9; EINECS 247-036-5; Tridecylbenzene sulfonic acid; detergent ingred. [Albright & Wilson UK] *

Nansa® TS 50. CAS 68584-25-8; TEA alkylbenzene sulfonate; detergent raw material. [Albright & Wilson UK]

Nansa UC. A range of detergent powders containing a sodium alkyl benzene sulfonate as the active ingredient.§

Nansa UCA/S and UCP/S. Spray-dried built powders based on straight chain alkylbenzene sulfonate; blue, free-flowing powder, dust free; anionic for built detergents. [Albright & Wilson Ltd, Detergents Div, Marchon] *

Nansa® YS94. CAS 68584-24-7; Isopropylamine dodecylbenzene sulfonate; emulsifier for solv.-based hand cleaners, agric. emulsifiable concs.; coupling agent for water in charge detergent systems; biodeg. [Albright & Wilson UK]

Nansen. A high-speed tungsten steel containing 18% tungsten.

napalite. A dark-red wax which melts at 42 C. It is a hydrocarbon and occurs naturally.

napalm. An aluminum soap consisting of a mixture of oleic naphthenic and coconut fatty acids. Makes petrol thicken and gel; used in flame throwers and fire bombs.

Napelec. Nondraining impregnant (obtained by mixing Napvis with water). [BP Chemicals Ltd]

Napeline. Benzaconine, $C_{32}H_{43}NO_{10}$.

Napental. CAS 57-33-0; Pentobarbital sodium; hypnotic; sedative. [SmithKline Beecham] *

Napgel. Antifreeze. [BP Chemicals Ltd]*

naphalane. A crude naphtha product containing soap. It is similar to naphthalane (qv).

Naphcon. CAS 550-29-2; Naphazoline hydrochloride; adrenergic. [Alcon Laboratories Inc]

naphtalin. See tar camphor.

naphtha. (coal tar naphtha; petroleum naphtha; benzin) Petroleum distillate; the less volatile portion obtained in redistilling benzine (qv), boiling from about 95-100 C; CAS 8030-30-6; as as a source of gasoline, special naphthas, petroleum chemicals, especially ethylene; thinners in paint, drycleaning fluid; blending with natural gas. [Ashland; Mobil; Monsanto; Norsk Hydro A/S; Texaco]

Naphthalamine. See naphthalidam.

naphthalene. (tar camphor) $C_{10}H_8$; CAS 91-20-3; EINECS 202-049-5; Intermediate (phthalic anhydride, naphthol, chlorinated naphthalenes, dyes, etc.), moth repellent, fungicide, smokeless powder, lubricant, synthetic resin, tanning, preservative, textile auxiliary, emulsion breaker, scintillation counter, antiseptic. [Allied-Signal; Aristech; Crowley Tar Prods.; Koch; Stan Chem Int'l. Ltd; Texaco]

1,6-naphthalenediol. (1,6-dihydroxynaphthalene) Polycyclic phenol; $C_{10}H_6(OH)_2$; CAS 575-44-6; EINECS 209-386-7. [Aceto]

---

* Trade name not verified as available   † Trade name verified as obsolete

2,3-naphthalenediol. (2,3-dihydroxynaphthalene) Polycyclic phenol; $C_{10}H_6(OH)_2$; CAS 92-44-4; EINECS 202-156-7; complexing reagent.

2,7-naphthalenediol. (2,7-dihydroxynaphthalene) Polycyclic phenol; $C_{10}H_6(OH)_2$; CAS 582-17-2; EINECS 209-478-7; reagent. [Aceto]

naphthalidam. (Naphthalidine).α-Naphthylamine, $C_{10}H_7.NH_2$.

naphthalidine. See naphthalidam.

Naphthalin. See tar camphor.

naphthalol. (betol). Naphthol salicylate.

naphthamine. See hexamine.

naphthazarin. 5, 6-dihydroxy-1, 4 naphthoquinone, $C_{10}H_4(OH)_2O_2$.

naphthenic acid. Cyclopentane carboxylic acid; CAS 1338-24-5; paint dryers, fungicides, metal catalysts, corrosion inhibitors, lubricants, fracturing fluids, cellulose preservatives, solvents, detergents, rubber reclaiming agent. [Crowley Chem.; Orange Chem. Ltd]

naphthionic acid. CAS 84-86-6; 1-Naphthylamine 4-sulfonic acid; used as an intermediate for azo dyes.

naphthite. Trinitro-naphthalene; used in explosives.

Naphthochrome. Wool dyestuffs. [Clayton Aniline Co Ltd]*

Naphthocyanine. Fast acid wool dyestuffs. [J C Bottomley]*

naphthocyanole. A homologue of pinacyanol (a red sensitizer for silver bromide plates), prepared by the condensation of β-naphtho-quinaldine ethiodide with formaldehyde in the presence of alcoholic potash; used as a red sensitizer for photographic plates.

naphthoformol. A product of α-naphthol and formaldehyde; used as a dusting powder.

Naphthol. Azoic dyestuffs. [Hoechst AG]

Naphthol Aristol. Iodo-naphthol, $C_{10}H_6I_2O_2$; used as an antiseptic.

Naphtholite. A proprietary trade name for a light petroleum distillate used as a solvent, etc. §

naphtholith. A bituminous shale.

Naphthopone E. Dispersing agent; used in naphthol AS developing baths for aftersoaping fast dyeings. [Bayer AG]

naphthoresorcin. 1, 3-Dihydroxy-naphthalene.

Naphthoride. A proprietary preparation of naphthalene tetrachloride. §

naphthosalol. See betol.

naphthosultone. The anhydride of 1-naphthol-8-sulfonic acid.

Naphtopon E. Dispersing agent in naphtol AS developing baths. [Bayer plc]

Naphtopone® E. Dispersing agent in naphtol AS developing baths for aftersoaping fast dyeings. [Bayer AG]

Napisan. Nappy sterilizer, to sanitize baby diapers. [Richardson-Vicks Inc] *

Naples yellow. (Paris yellow) CAS 13510-89-9; . A pigment. It is an antimonate of lead, $PbO \cdot Sb_2O_5$. A mixture of this body with carbonate and chromate of lead is also sold under this name cadmium sulfide, CdS, and a pale yellow ocher have been called by this term; used to stain glass, crockery, and porcelain.

Naplithin. Lithium-β-hydroxynaphthalene-α-monosulfonate.

Napliwi. A sand cemented together by the rubber latex from the roots of the Chondrilla plant in Russia. The roots are attacked by the larvae of certain insects, when the latex exudes and runs into the sandy soil where it coagulates.

The sand usually contains ≈ 2-2.5% rubber and 10% resin.

Naprosyn. CAS 22204-53-1; Naproxen; anti-inflammatory; analgesic; antipyretic. [Syntex Laboratories Inc]

Napryl. Polypropylene. [BP Chemicals Ltd] †

Napsalgesic. Oral analgesic. [Dista Products Ltd] *

Napsoft FL. Synthetic softener for cellulose fibers to be napped; produces a lush, rich suede finish. [CNC Int'l. L.P.]

Naptel. Polybutenes. [BP Chemicals Ltd]

Naptol. CAS 1954-81-4; Chloramben; residual herbicide for use in ornamentals. [Synchemicals Ltd]

Napvis. Polybutenes. [BP Chemicals Ltd]

Naqua. CAS 133-67-5; Trichlormethiazide; diuretic; antihypertensive. [Schering Corp]

Narcan®. CAS 357-08-4; Naloxone hydrochloride; antagonist. [DuPont Merck Pharmaceuticals]

narceine. CAS 131-28-2; The sodium bisulfite compound of p-sulfobenzeneazo-β-naphthol, $C_{16}H_{12}N_2O_7S_2Na_2$; used in calico printing.

Narceol. CAS 140-39-6; . It is p-tolyl acetate, $CH_3 \cdot C_6H_4 \cdot O \cdot COCH_3$; a synthetic perfume

Narcodeon. Double salts of narcotine and codeine with di- and poly-basic acids.

Narcotil. A mixture of ethyl and methyl chlorides; used as an anaesthetic. Methylene chloride, CH_2Cl_2; used as a local anesthetic, is also known by this name.

Narcotine. CAS 128-62-1; A proprietary trade name for noscapine. §

narcyl. Ethylnarceine hydrochloride.

Nardil. A proprietary preparation of phenelzine dihydrogen sulfate; an antidepressant. [Warner] *

Narex. A proprietary preparation of phenylephrine and chlorbutol; used as a nasal spray. [H N Norton & Co Ltd]‡

Nargentol. A protein compound of silver, containing 24% silver; an antiseptic.

Narki. An acid-resisting silicon-iron alloy.

Narlex EP-2. Dispersant; fluid loss control additive effective even at sat. level of calcium ions; for high pressure (500 psi), high temp. (175 C) requirements; bentonite extender. [Hart Chem. Ltd.

Narlex LD 42. Nonfoaming non-surface act. polymer; dispersant, plasticizer for pigments, clays, particulates; optimum performance > pH 6.5. [Hart Chem. Ltd.

Narphen. CAS 1239-04-9; A proprietary preparation of phenazocine hydrobromide; an analgesic. [Smith & Nephew Pharmaceuticals Ltd] *

Nasalcrom. CAS 15826-37-6; Cromolyn sodium; anti-asthmatic. [Fisons Corp, Pharmaceuticals Div] *

Naseptin. A proprietary preparation of chlorhexidene hydrochloride and neomycin sulfate. [ICI Chem. & Polymers Ltd]

Nasoflu. A proprietary live influenza vaccine. [SmithKline Beecham] *

NAS® 10. Styrene methyl methacrylate copolymer; featuring sparkling clarity, good scratch resist.; for cosmetics pkg., houseware, displays, office accessories. [Novacor]

NAS® 30. Styrene methyl methacrylate copolymer; offers sparkling clarity, good scratch resist., easy processing; exhibits less yellowing after gamma sterilization; for medical applications (catheter tubes, closures, IV flow regulators, thermoformed medical pkg.), meter covers, brush blocks, drafting instruments, bathroom accessories; FDA compliance. [Novacor]

---

‡ Trade name and manufacturer not verified   § Trade name without identified manufacturer

NAS® 50. Methyl methacrylate styrene copolymer; offers sparkling clarity, excellent scratch resist., and resist. to household detergents; used for salad bowls, tumblers, mugs, pitchers, cutlery handles; FDA compliance. [Novacor]

NaTa. CAS 76-03-9; TCA; for control of weeds in field crops. [Hoechst UK]

Natac®. Resin acids-amine resin soaps blend; tackifier for IIR, NR, SBR; NBR, SBR molding aid; slightly activates thiazole and thiuram accelerators. [Whitney & Oettler] *

Natacyn. CAS 7681-93-8; Natamycin; antibacterial. [Alcon Laboratories Inc]

Natalon. See Carbora. §

natamycin. An antibiotic produced by *Streptomyces natalensis*. Pimafucin. Pimaricin.

Natene. A trademark for Ziegler-type polyethylenes. [Pechiney-St Gobain] *

Nathin. Bread improvers. [Rhone-Poulenc UK]

Natipide. Liposomes. [Rhone-Poulenc UK]

Natirose. A proprietary preparation of glyceryl trinitrate, ethylmorphine hydrochloride and hyoscyamine hydrobromide; used in the treatment of angina pectoris. [L Wilcox & Co Ltd] ‡

Natisedine. A proprietary preparation of quinidine phenylethylbarbiturate; used as a cardiac sedative. [L Wilcox & Co Ltd] ‡

native paraffin. See ozokerite.

Natopherol. CAS 59-02-9; Vitamin E supplement. [Abbott Laboratories] *

natrena®. Calorie-free sweetener for hot and cold drinks, yogurt, low-fat curd cheese, desserts. [Bayer AG]

Natritope Chloride. CAS 7647-14-5; Sodium chloride Na 22; radioactive agent. [Bristol-Myers Squibb Co Inc] *

natrium. Latin name for sodium, Na.

natrium. See sodium

Natrolith. A water-softening material said to consist of granulated clay which removes lime and magnesium salts from hard water when used as a filter. §

Natrosol®. CAS 9004-62-0; Hydroxyethylcellulose; multipurpose nonionic cellulosic; thickener, protective colloid, binder, stabilizer and suspending agent. [Hercules] *

Natrosol® 250. CAS 9004-62-0; Hydroxyethylcellulose; thickener, protective colloid, binder, stabilizer, suspending agent for coatings, cosmetics, toiletries, textile printing pastes, inks, adhesives, electroplating, ceramics, textile and paper sizing, acid thickening for acidizing oil wells. [Aqualon]

Natrosol® FPS. Fluidized polymer suspension; latex paint thickener. [Aqualon]

Natrosol® Hydroxyethylcellulose. CAS 9004-62-0; Hydroxyethylcellulose; thickener for hair care products, creams and lotions, latex paints; protective colloid in emulsion polymerization; suspending aid in joint and tile cements; binder for welding rods; sol. in hot and cold water, DMSO; tolerates up to 70% polar org. solv. in water. [Aqualon]

Natrosol® Plus CS. Cetyl hydroxyethylcellulose; thickener and visc. stabilizer for aq. and surfactant systems in personal care field; binder, stabilizer, film-former, suspending agent. [Aqualon]

Natrovis® Water-Soluble Polymer. Hydroxypropyl hydroxyethylcellulose; water-sol. polymer; additive to hydraulic binder and latex-modified building materials, e.g., gypsum plasters, cement stuccos, mortars, masonry

cements; controls water balance, workability, adhesion, consistency, tackiness. [Aqualon]

Natrundum. See Carbora. §

Natsyn® 2200. Polyisoprene, solution polymerized; CAS 9003-31-0; nonstaining synthetic elastomer for light colored goods, adhesives, footwear, sponge products, tires, pharmaceutical goods, rubber bands, molded and mech. goods. [Goodyear] *

Natulan. CAS 671-16-9; A proprietary preparation of procarbazine; an antimitotic. [Roche Products Ltd] *

natural asphalt. See bitumen.

Natural Bone Ash, BCP 400. Calcium hydroxyapatite; used to coat and protect all surfaces contacted by molten nonferrous metals; used extensively in both aluminum and copper industries. [Murlin Chemical Inc]

Natural Extract AP. Trimethyl glycine; emollient, humectant, pigment, conditioner, lubricant for personal care formulations. [RITA]

natural rubber. Also see Polyisoprene. (NR) Used for adhesive, dipping, coating, foam, and molded materials, latex applications., foamed rubber, textile, medical, cement, asphalt; processing aid for blending with other rubber. [Firestone Syn. Rubber & Latex; Hardman; A. Schulman]

Naturechem® CAR. Cetyl acetyl ricinoleate; mild, noncomedogenic, nonoily emollient for skin care products. [CasChem]

Naturechem® CR. CAS 10401-55-5; Cetyl ricinoleate; mild, noncomedogenic, nonoily emollient for cosmetics. [CasChem]

Naturechem® EGHS. CAS 33907-46-9; EINECS 251-732-4; Glycol hydroxystearate; auxiliary emulsifier, emollient, thickener, opacifier for cosmetics, household products. [CasChem]

Naturechem® GMHS. CAS 1323-42-8; EINECS 215-355-9; Glyceryl hydroxystearate; auxiliary emulsifier, emollient, opacifier, bodying and thickening agent for cosmetics, household products. [CasChem]

Naturechem® GTH. CAS 27233-00-7; Glyceryl triacetyl hydroxystearate; mild, noncomedogenic emollient, wetting agent, stabilizer for pigmented products; imparts gloss. [CasChem]

Naturechem® GTR. CAS 101-34-8; EINECS 202-935-1; Glyceryl triacetyl ricinoleate; mild emollient, pigment wetter, cosolv.; softener for waxes and resins. [CasChem]

Naturechem® MAR. CAS 140-03-4; Methyl acetyl ricinoleate; light emollient; reduces greasiness of emollients such as mineral oil; cosolv. properties; solubilizer for benzophenone-3; superior freeze/thaw properties. [CasChem]

Naturechem® MHS. CAS 141-23-1; EINECS 205-471-8; Methyl hydroxystearate; opacifiier, pearlescent, emulsifier, visc. builder for surfactant systems. [CasChem]

Naturechem® OHS. Octyl hydroxystearate; emollient, softener for cosmetics; refatting additive for soaps, cleansers. [CasChem]

Naturechem® PGHS; CAS 33907-47-0; EINECS 251-734-5; Propylene glycol hydroxystearate; auxiliary emulsifier, dispersant, opacifier, thickener, emollient, stabilizer for cosmetics, household products. [CasChem]

Naturechem® PGR. CAS 26402-31-3; EINECS 247-669-7; Propylene glycol ricinoleate; wetting agent, stabilizer, pigment/dye dispersant providing emolliency, gloss,

---

* Trade name not verified as available					† Trade name verified as obsolete

plasticization to cosmetics, household products. [CasChem]

Naturechem® THS-200. PEG-200 trihydroxystearin; emulsifier, emollient, thickener, stabilizer for cosmetics, household products; stable over broad pH range. [CasChem]

Nature's Own Spray Helper. A spreader sticker made from cottonseed oil; used in fungicide sprays and others to prolong their life on the plant. [Lawn & Garden Products Inc]

Naturetin. CAS 73-48-3; Bendroflumethiazide; diuretic; antihypertensive. [Bristol-Myers Squibb Co Inc] *

Naturvue. Hefilcon B; 2-hydroxyethyl methacrylate polymer with 1- vinyl-2-pyrrolidinone and ethylene dimethacrylate; contact lens material. [Milton Roy Co] *

Naubuc. A nickel-silver. It contains 58% copper, 25% nickel, 16.25% zinc, and 0.75% iron.

Naugalube® 403. CAS 101-96-2; N,N′-Di-s-butyl-p-phenylenediamine; acrylic polymerization inhibitor; gasoline antioxidant/sweetener. [Uniroyal]

Naugalube® 438. CAS 101-67-7; Dioctyl diphenylamine; antioxidant in automatic transmission fluids, turbine oil, and syn. lubricants used in jet turbine engines; thermal stabilizer for automatic transmission fluid at high temps. [Uniroyal]

Naugalube® 470. N,N′-Di-isopropyl-p-phenylenediamine in methanol; acrylic polymerization inhibitor. [Uniroyal]

Naugalube® PDA. N-Phenyl-p-phenylenediamine; intermediate for lubricants. [Uniroyal]

Nauganlite. Alkylated phenol; a proprietary antioxidant. [Rubber Regenerating Co] ‡

Nauganlite Powder. A proprietary name for alkylated bisphenol. [Rubber Regenerating Co] ‡

Naugard® 10. CAS 6683-19-8; Tetrakis [methylene (3,5-di-t-butyl-4-hydroxyhydrocinnamate); methane; antioxidant effective against thermal oxidative degradation during long term heat aging; for polyolefins, styrenics, elastomers, adhesives, lubricants, and oils. [Uniroyal]

Naugard® 76. CAS 2082-79-3; Octadecyl 3,5-di-t-butyl-4-hydroxyhydrocinnamate; antioxidant for stabilizing polymeric substances such as polyolefins, styrenics, EPDM, and PVC; provides good thermal and color stability. [Uniroyal]

Naugard® 445. CAS 10081-67-1; 4,4′-Bis (α,α-dimethylbenzyl) diphenylamine; nondiscoloring antioxidant for polyolefins, styrenics, polyether polyols, hot-melt adhesives, lubricant additives, nylon and other polymers. [Uniroyal]

Naugard® 524. CAS 31570-04-4; Tris (2,4-di-t-butyl phenyl) phosphite; antioxidant used in thermoplastic and thermoset polymers where color and processing stability are critical. [Uniroyal]

Naugard® I-2. CAS 3081-14-9; N,N′-Bis(1,4-dimethylpentyl)-p-phenylenediamine; styrene polymerization inhibitor; column antifoulant. [Uniroyal]

Naugard® I-3. N-(1,4-Dimethylpentyl)-N′-phenyl-p-phenylenediamine; styrene polymerization inhibitor; column antifoulant. [Uniroyal]

Naugard® I-4. N-Phenyl,N′-isopropyl-p-phenylenediamine; styrene polymerization inhibitor; column antifoulant. [Uniroyal]

Naugard® I-6. N-Phenyl-N′-cyclohexyl-p-phenylenediamine; column antifoulant. [Uniroyal]

Naugard® J. CAS 74-31-7; N,N′-Diphenyl-p-phenylenedi-amine; acrylic polymerization inhibitor. [Uniroyal]

Naugard® NBC. CAS 13927-77-0; Nickel dibutyldithiocarbamate; nickel chelating uv stabilizer for polyolefins. [Uniroyal]

Naugard® PANA. CAS 90-30-3; Phenyl-α-naphthylamine; antioxidant for lubricants. [Uniroyal]

Naugard® XL-1. CAS 70331-94-1; 2,2′-Oxamido bis-[ethyl 3-(3,5-di-tert-butyl-4-hydroxyphenyl) propionate; antioxidant and metal deactivator; used in polymerization, processing, and in end use applications, wire and cable insulation, pipe and injection parts for automobiles and appliances; processing stabilizer for polyolefins, film, sheet, and blow molded bottles. [Uniroyal]

Naugawhite. Phenolic/bisphenolic; antioxidant for thermoplastics; hot melt adhesives. [Uniroyal]

Naugex SD-1. CAS 103-34-4; 4,4′-Dithiodimorpholine; a sulfur donor used as a partial or total replacement of sulfur for resistance to heat and ageing in NR, SBR, BR, NBR, and EPDM. [Uniroyal]

Nauli gum. An oleoresin from a tree found in the Solomon Islands. It contains 10% volatile oil, 8% resin, and 3% water-soluble matter containing anisic acid.

Navac. Vacuum processed metallic sodium. [Foseco (F.S.) Ltd]

naval bronze. An alloy of 88.1% copper, 9.74% tin, and 2.04% zinc.

Navane. A proprietary preparation of thiothixene hydrochloride; a psychotherapeutic agent. [Pfizer International]

Navane Hydrochloride. Thiothixene hydrochloride; antipsychotic. [Pfizer Inc]

Navidrex. A proprietary preparation of cyclopenthiazide and potassium chloride in slow release core; a diuretic. [Ciba plc] *

Navidrex-K. A proprietary preparation of cyclopenthiazide and slow- release potassium chloride; a diuretic. [Ciba plc] *

navy green paint. A mixture of barium sulfate, lead chromate, and an organic blue.

Naxchem CD-6M. Surfactant blend; detergent, foam booster/stabilizer for cosmetics, carpet shampoos, dishwash, laundry detergents, textile processing, pigment dispersions. [Ruetgers-Nease]

Naxchem Detergent CNB. Surfactant blend; detergent, foam booster/stabilizer for lotions and creams, carpet and upholstery shampoos, dishwash, laundry, textile processing, pigment disps. [Ruetgers-Nease]

Naxchem Dispersant K. Blend of alkanolamides and syndets; biodeg. dispersant, emulsifier for oil slicks, naphthas, kerosene, other solvs.; general cleaning of crude oil, marine and land transport, and manufacturing plants. [Ruetgers-Nease]

Naxchem Emulsifier 700. Blend of esters and surfactants; antistatic base for emulsifying low visc. pale mineral oils; for metal cutting, textile fiber lubricants. [Ruetgers-Nease]

Naxchem N-Foam 802. Blend of crosslinking and emulsifying agents; curing and foaming agent in mixtures with urea-formaldehyde resins to mask odors from landfills and for foam insulation; as cleaning compound when neutralized. [Ruetgers-Nease]

Naxel AAS-40S. Sodium dodecylbenzene sulfonate; biodeg. surfactant, foamer, wetting agent, detergent for household and industrial detergents, rug shampoos, textile wet processing, metal cleaners, dairy cleaners,

---

‡ Trade name and manufacturer not verified   § Trade name without identified manufacturer

cosmetics. [Ruetgers-Nease]

Naxel AAS-60S. TEA dodecylbenzene sulfonate; foaming agent for bubble baths, shampoos, household and industrial detergents, textile dyeing compounds, etc. [Ruetgers-Nease]

Naxel AAS-98S. Dodecylbenzenesulfonic acid; biodeg. detergent intermediate, wetting agent, emulsifier. [Ruetgers-Nease]

Naxel AAS-Special 3. Isopropylamine dodecylbenzene sulfonate; emulsifier for drycleaning, metal cleaning, emulsifiable solv. cleaners, fuel oil additives, mop treatments. [Ruetgers-Nease]

Naxel DDB 500. Dodecylbenzene; biodeg. detergent intermediate for production of dishwash, laundry, all-purpose and industrial cleaners, in specialty coatings and other industrial applications. [Ruetgers-Nease]

Naxell. Alkylaryl sulfonates. [Ruetgers-Nease Chemical Co]

Naxide 1230. CAS 61788-90-7; EINECS 263-016-9; Cocamine oxide; detergent, foam booster/stabilizer, visc. builder, conditioner for shampoos, hand cleaners, dishwash, light duty detergents, textiles, lubricants, paper coatings. [Ruetgers-Nease]

Naxolate WA-97. CAS 151-21-3; Sodium lauryl sulfate USP, BP; biodeg. detergent, wetting agent, foamer, emulsifier for cosmetic and household products including shampoos, bubble baths, rug shampoos, toothpaste, dishwash, laundry detergents. [Ruetgers-Nease]

Naxonac. Phosphate esters. [Ruetgers-Nease Chemical Co]

Naxonac 510. Nonyl phenol ether phosphate; detergent, wetting agent, emulsifier, lubricant, hydrotrope for heavy-duty and household detergents, waterless hand cleaners, solv. degreasers, emulsion polymerization, paint and wax strippers, electrolytic cleaners. [Ruetgers-Nease]

Naxonate® 4AX. CAS 26447-10-9; EINECS 247-710-9; Ammonium xylene sulfonate; hydrotrope, stabilizer, solubilizer used in formulating detergents, inks, electroplating baths, dyestuffs, polymers. [Ruetgers-Nease]

Naxonate® 4KT. Potassium toluene sulfonate; hydrotrope, stabilizer, solubilizer used in formulating detergents, inks, electroplating baths, dyestuffs, polymers. [Ruetgers-Nease]

Naxonate® 4L. CAS 1300-72-7; EINECS 215-090-9; Sodium xylene sulfonate; hydrotrope, stabilizer, solubilizer used in formulating detergents, inks, electroplating baths, dyestuffs, polymers. [Ruetgers-Nease]

Naxonate® 4ST. Sodium toluene sulfonate; hydrotrope, stabilizer, solubilizer used in formulating detergents, inks, electroplating baths, dyestuffs, polymers. [Ruetgers-Nease]

Naxonate® 5KT. Potassium toluene sulfonate; hydrotrope, stabilizer, solubilizer used in formulating detergents, inks, electroplating baths, dyestuffs, polymers. [Ruetgers-Nease]

Naxonate® 45SC. CAS 32073-22-6; Sodium cumene sulfonate; hydrotrope, stabilizer, solubilizer used in formulating detergents, inks, electroplating baths, dyestuffs, polymers. [Ruetgers-Nease]

Naxonate® SC. CAS 32073-22-6; Sodium cumene sulfonate; hydrotrope, stabilizer, solubilizer used in formulating detergents, inks, electroplating baths, dyestuffs, polymers. [Ruetgers-Nease]

Naxonate® SX. CAS 1300-72-7; EINECS 215-090-9; Sodium xylene sulfonate; hydrotrope, stabilizer, solubilizer used in formulating detergents, inks, electroplating baths, dyestuffs, polymers. [Ruetgers-Nease]

Naxonic NI-40. CAS 9016-45-9; Nonoxynol-4; wetting agent, dispersant, penetrant, emulsifier, detergent, solubilizer for textile wet processing, agric., cosmetic, industrial and household detergents, latex and polymers, wax/polishes; demulsifier for petrol. [Ruetgers-Nease]

Naxonol CO. Cocamide DEA (2:1); detergent, wetting agent, thickener for household and industrial detergents, shampoos, textile wet processing, leather and fur processing, metal cleaning, solv. emulsification. [Ruetgers-Nease]

NBC. CAS 13927-77-0; A proprietary name for nickel dibutyl dithiocarbamate. [DuPont UK] *

NC-4. Bis (p-ethylbenzylidene) sorbitol; nucleating agent improving optical and physical properties of PP for container, pkg. film, injection syringe and other applications. [Mitsui Toatsu]

N.C.T. Nitrocellulose tutular, a pyro-collodion powder, made from a gelatinized nitrocellulose, pressed in the form of rods. §

N.C.T.3 Alloy. *See* Alloy N.C.T.3. §

Ndilo oil. *See* Laurel nut oil.

NE. CAS 79-24-3; Nitroethane; intermediate, stabilizer for halogenated solvs., as fuels, explosives, and solvs. for coatings or industrial processes. [Angus]

Neacid. A pickling agent for gold- and silversmiths and in the jewelry industry. [Degussa]

Nealpon. *See* pantopon.

Neantina. Seed dressing for control of fungal diseases on cereals, rice, cotton and vegetables. [Bayer AG]

Neapolitan ointment. Mercury ointment.

neatsfoot oil. CAS 8002-64-0; A fixed oil obtained by boiling ox or cow's feet in water. It is also the name for a mixture of 1 part lard and 3 parts colza oil; used in the leather industry for fat liquoring and softening, lubricant, oiling wool.

Nebacortril. A proprietary preparation of neomycin sulfate, bacitracin and hydrocortisone; an antibiotic and anti-inflammatory. [Pfizer International]

Nebasulf. A proprietary preparation of neomycin sulfate, bacitracin and sulfacetamide; a topical antibacterial. [Pfizer International]

Nebcin. CAS 79645-27-5; Tobramycin sulfate (injection); antibacterial. [Eli Lilly & Co]

Nebony® 100, L-55. Aromatic petrol. hydrocarbon resin; used in wire/cable coatings, rubber (cements, mechanical goods, molded goods, tires), and caulking compounds. [Neville]

Nebs. CAS 103-90-2; Acetaminophen; analgesic; antipyretic. [Norwich Eaton Pharmaceuticals Inc] *

Nebulin. CAS 117-18-0; Tecnazene in liquid fogging solution; used for controlling sprouting and dry rot in stored potatoes. [Wheatley Chemical Co Ltd] *

Nebulin. CAS 117-18-0; Tecnazene in liquid fogging solution; for controlling sprouting and dry rot in stored potatoes. [Dean Agrochemicals Ltd]

Neburex. CAS 555-37-3; Active ingredient: neburon; selective herbicide for both pre- and post-emergence application. [Agan Chemical Manufacturers Ltd]

Necol. Cellulose lacquers, organic enamels, adhesives and plastic wood. [ICI Chem. & Polymers Ltd]

Nectandra bark. Bebeera bark.

---

* Trade name not verified as available   † Trade name verified as obsolete

Nedi. Polyvinyl acetate water soluble film and contract packing service. [Production Chemicals Ltd]

needle bronze. An alloy of 84.5% copper, 8% tin, 5.5% zinc, and 2% lead.

needle tin ore. Acute pyramidal crystals of cassiterite.

Neem bark. Indian azadirach.

Neem oil. *See* veepa oil.

Nefomolit. A proprietary trade name for a plastic made from mineral oil, formaldehyde, etc. §

Nefranutrin. A proprietary preparation of essential amino acids used in the dietary treatment of renal failure. [Geistlich Sohne AG] *

Nefrolan. Chlorexolone. [May & Baker Ltd] *

Neganol®. For the control of trypanosomiasis in domestic animals. [Bayer AG]

Negasunt®. Dusting powder against bacterial infection and fly-larvae (myiasis); veterinary preparation controlling fly maggots in wounds. [Bayer AG]

Negex. Chemical substances used in industry; negative expanders for lead acid batteries. [Associated Lead Manufacturers Ltd] *

NegGram. CAS 389-08-2; Nalidixic acid; antibacterial. [Sterling Drug Inc] *

Negram. CAS 389-08-2; A proprietary preparation containing nalidixic acid; a urinary antiseptic. §

Neguvon®. Insecticide against cattle grubs, mange and worm infestation, in particular for control of ectoparasites in the poultry house; veterinary medicine. [Bayer AG]

Neillite. A proprietary phenol-formaldehyde synthetic resin molding compound. §

Neisser's stain. A microscopic stain. a) Solution contains 0.1 g methylene blue, 2 cc alcohol, 5 cc glacial acetic acid, and 95 cc water. b) Solution contains 0.2 g bismarck brown in 100 cc boiling water.

Nekal. Anionic surfactants. [BASF plc]

Nekal® BA-77. Redesignated Rhodacal® BA-77. [Rhone-Poulenc Surf.]

Nekal® BX. Alkylaryl sulfonate anionic surfactant; wetting, dispersing, emulsifying and foaming agents with widespread industrial applications eg in the pigment, paint and paper industries; in leakage control in pipes, fittings and valves. [BASF plc] *

Nekal® BX-78. Redesignated Rhodacal® BX-78. [Rhone-Poulenc Surf.]

Nekal® SBS. Anionic surfactant in free acid form; wetting, dispersing, emulsifying and foaming agent with widespread industrial uses eg in the paper, paint and pigment industries; in leakage control in pipes, fittings and valves. [BASF plc] *

Nekal® WS-25 and WS-25-I. Redesignated Geropon® WS-25, WS-25-I. [Rhone-Poulenc Surf.]

Nekal® WT-27. Redesignated Geropon® WT-27. [Rhone-Poulenc Surf.]

Nekanil® 907. Low-ethoxylated alkyl phenol; detergent, wetting agent used in textile industry. [BASF AG; BASF plc]

Nelco. A proprietary trade name for whiting. §

Nelio Resin. A proprietary trade name for a purified wood resin. §

Nema. CAS 127-18-4; Tetrachloroethylene; anthelmintic. [Parke-Davis] *

Nemacin. TBZ and iodophor in dry granular form; used for controlling soil and seed borne diseases in onions. [Wheatley Chemical Co Ltd] *

Nemacur®. CAS 22224-92-6; Fenamiphos; nematicide with systemic action against root-knot, cyst-forming and free-living nematodes on bananas, cotton, soybeans, tobacco, citrus, vegetables and ornamentals. [Bayer AG]

Nemafax. CAS 23564-06-9; Thiophanate; a veterinary anthelmintic. [RMB Animal Health Ltd]

Nembutal. CAS 57-33-0; A proprietary preparation of pentobarbitone sodium; a hypnotic. [Abbott Laboratories]

Nembutal Sodium. CAS 57-33-0; Pentobarbital sodium; hypnotic; sedative. [Abbott Laboratories]

Neoamfo. Neomycin sulfate, thiostrepton and amphotericin in aqueous suspension. [Ciba plc] *

Neoarsycodyl. Sodium methyl-arsenate.

Neobacrin. A proprietary preparation of neomycin sulfate and zinc bacitracin; an antibiotic skin ointment. [Glaxo Laboratories] *

Neobee® 18. CAS 8001-23-8; Hybrid safflower oil; emollient oil for cosmetics and pharmaceuticals, solubilizer, solv. [Stepan/PVO; Stepan Europe]

Neobee® 20. Propylene glycol dicaprylate/dicaprate; emollient oil for cosmetics and pharmaceuticals, carrier for flavors and colors in foods; solubilizer, cosolv. [Stepan/PVO]

Neobee® 62. CAS 555-43-1; EINECS 209-097-6; Tristearin; solubilizer, stabilizer used in food applications. [Stepan/PVO]

Neobee® M-5. CAS 65381-09-1; Caprylic/capric triglyceride; diluent vehicle/carrier, solubilizer, cosolv. for flavoring, medicinals, colorings used in foods, beverages, cosmetic and pharmaceutical product, emollient. [Stepan/PVO; Stepan Europe]

Neobiotic. CAS 1405-10-3; Neomycin sulfate; antibacterial. [Pfizer Inc] †

Neobor. Disodium tetraborate pentahydrate; partially dehydrated borax to effect economy of transport and handling. [Borax Consolidated Ltd]

Neocaine-surrénine. A proprietary preparation consisting of ethocaine hydrochloride with adrenalin. §

Neo-Calglucon. Calcium glubionate; replenisher. [Dorsey Pharmaceuticals]

Neo-Cantil. A proprietary preparation of mepenzolate bromide and neomycin sulfate; an antidiarrheal. [MCP Pharmaceuticals] *

Neo-Cebitate®. CAS 6381-77-7; Sodium erythorbate; food additive for meat, poultry, and seafood industries. [Rhone-Poulenc Food Ingreds.]

Neocid. CAS 50-29-3; A proprietary preparation containing 5% DDT (*qv*); an insecticide. §

Neocidol Veterinary Powder. Organophosphorus wound dressing. [Ciba plc] *

Neocinchophen. Ethyl 6-methyl-2 phenylcinchonate.

Neo-Cobefrin. Levonordefrin; adrenergic. [Cook-Waite Laboratories Inc] *

Neo Cortef. A proprietary preparation of hydrocortisone acetate and neomycin sulfate; used in dermatology as an antibacterial agent. [Upjohn Ltd] *

Neocosal. An oil binder; used for cleaning bodies of water and ground surfaces that have been contaminated with oil. [Degussa AG] *

Neocrest. Saturated polyester resin products. [ICI Resins BV]

NeoCryl. Thermoplastic acrylic resins in solid form, homo and copolymers of acryl-methacrylic and styrene; used

---

‡ Trade name and manufacturer not verified   § Trade name without identified manufacturer

in various paints, e.g., marine paints, road paints and paints on plastic; flexo, gravure and screen painting inks; dry toner resins, aerosols and other applications. [Polyvinyl Chemie Holland BV]

Neocryl. Acrylic polymer resin products. [ICI Resins BV]

Neo-Cultol. Mineral oil; laxative; pharmaceutic aid. [Fisons Corp, Pharmaceuticals Div] *

Neo-Cytamen. CAS 13422-51-0; A proprietary preparation of hydroxycobalamin. [Duncan Flockhard Ltd]

Neodene® 6. CAS 592-41-6; Linear C6 alpha olefin; intermediate for biodeg. surfactants and specialty industrial chemicals. [Shell]

Neodene® 6/8. C6-8 alpha olefin; intermediate for biodeg. surfactants and specialty industrial chemicals. [Shell]

Neodene® 6/8/10. C6-10 alpha olefin blend; intermediate for biodeg. surfactants and specialty industrial chemicals. [Shell]

Neodene® 6/12. CAS 592-41-6, 112-41-4; C6/12 alpha olefin blend; intermediate for biodeg. surfactants and specialty industrial chemicals. [Shell]

Neodene® 8. CAS 111-66-0; Linear C8 alpha olefin; intermediate for biodeg. surfactants and specialty industrial chemicals. [Shell]

Neodene® 10-11/12-13. Decene-1, undecene, dodecene, tridecene blend; intermediate for biodeg. surfactants and specialty industrial chemicals. [Shell]

Neodene® 10-11/12-13/14. Decene-1, undecene, dodecene, tridecene, tetradecene blend; intermediate for biodeg. surfactants and specialty industrial chemicals. [Shell]

Neodene® 10. CAS 872-05-9; Linear C10 alpha olefin; intermediate for biodeg. surfactants and specialty industrial chemicals. [Shell]

Neodene® 10/12/1314. Alpha olefin/internal olefin blend; intermediate for biodeg. surfactants and specialty industrial chemicals. [Shell]

Neodene® 12. CAS 112-41-4; Linear C12 alpha olefin; intermediate for biodeg. surfactants and specialty industrial chemicals. [Shell]

Neodene® 12/1314. Dodecene-1, tridecene, tetradecene blend; intermediate for biodeg. surfactants and specialty industrial chemicals. [Shell]

Neodene® 14. CAS 1120-36-1; Linear C14 alpha olefin; intermediate for biodeg. surfactants and specialty industrial chemicals. [Shell]

Neodene® 16. CAS 629-73-2; Linear C16 alpha olefin; intermediate for biodeg. surfactants and specialty industrial chemicals. [Shell]

Neodene® 18. CAS 112-88-9; Linear C18 alpha olefin; intermediate for biodeg. surfactants and specialty industrial chemicals. [Shell]

Neodene® 20. CAS 3452-07-1; Linear C20 alpha olefin; intermediate for biodeg. surfactants and specialty industrial chemicals. [Shell]

Neodene® 810. CAS 111-66-0, 872-05-9; Octene-1, decene-1; intermediate for biodeg. surfactants and specialty industrial chemicals. [Shell]

Neodene® 1012. CAS 872-05-9, 112-41-4; Decene-1, dodecene-1 blend; intermediate for biodeg. surfactants and specialty industrial chemicals. [Shell]

Neodene® 1014. C10-14 alpha olefin blend; intermediate for biodeg. surfactants and specialty industrial chemicals. [Shell]

Neodene® 1112. CAS 68411-00-7; C11-12 internal olefins; intermediate for biodeg. surfactants and specialty industrial chemicals. [Shell]

Neodene® 1420. CAS 64743-02-8; C14-20 alpha olefin blend; intermediate for biodeg. surfactants and specialty industrial chemicals. [Shell]

Neodene® 1624. CAS 64743-02-8; C16-24 alpha olefin blend; intermediate for biodeg. surfactants and specialty industrial chemicals. [Shell]

Neodene® 2024. CAS 64743-02-8; C20-24 alpha olefin blend; intermediate for biodeg. surfactants and specialty industrial chemicals. [Shell]

Neodol® 1. CAS 112-42-5; Undecyl alcohol; detergent intermediate. [Shell]

Neodol® 1-3. Undeceth-3; detergent intermediate, emulsifier for general industrial usage, hard surface cleaning; biodeg. [Shell]

Neodol® 5. CAS 629-76-5; Pentadecyl alcohol; detergent intermediate. [Shell]

Neodol® 23. CAS 75782-86-4; C12-13 alcohols; detergent intermediate. [Shell]

Neodol® 23-6.5. CAS 66455-14-9; C12-13 pareth-7; detergent intermediate used in preparation of sulfates for highfoaming liq. detergents, household and industrial use; biodeg. [Shell]

Neodol® 25. CAS 63393-82-8; C12-15 alcohols; detergent, emulsifier intermediate. [Shell]

Neodol® 25-3. CAS 68131-39-5; C12-15 pareth-3; detergent intermediate used in preparation of sulfates for highfoaming liq. detergents; emulsifier; for cosmetic, industrial, dishwashing and liq. detergents; biodeg. [Shell]

Neodol® 45. CAS 75782-87-5; C14-15 alcohol; detergent intermediate. [Shell]

Neodol® 45-7. C14-15 pareth-7; detergent, wetting agent, emulsifier for general industrial and household detergent products; biodeg. [Shell]

Neodol® 91. CAS 66455-17-2; C9-11 alcohols; detergent intermediate. [Shell]

Neodol® 91-2.5; CAS 68439-46-3; C9-11 pareth-3 (2.5 EO); oil-sol. emulsifier and wetting agent, detergent intermediate, dispersant, surfactant; for general industrial usage, hard surface cleaning; biodeg. [Shell]

Neo-Duroterm® 3, 5 and 7. Investment compound system for precious metal casts; dental specialty. [Bayer AG]

Neo-Duroterm® L. Soldering investment material for the precious metal technique; dental specialty. [Bayer AG]

Neoferrum. CAS 20344-49-4; A proprietary preparation of ferric hydroxide; a hematinic. [Crookes Laboratories] *

Neoflex® 9. CAS 68527-05-9; Isononyl alcohol. [Shell]

Neoflex® 11. CAS 112-42-5; C11 primary alcohol; intermediate. [Shell]

Neogen. An alloy of 58% copper, 12% nickel, 27% zinc, 2% tin, and 0.5% aluminum. It is a nickel silver (German silver).

Neogest. A proprietary preparation containing 37.5 mcg levonorgestrel contained in 75 mcg norgestrel; an oral contraceptive. [Schering Health Care Ltd]

Neo Heliopan, Type 303. CAS 6197-30-4; Octocrylene; uv-B absorber for cosmetics, waterproof sunscreens. [Haarmann & Reimer GmbH]

Neo Heliopan, Type AV. CAS 5466-77-3; 2-Ethylhexyl p-methoxycinnamate; uv-B absorber for cosmetic applications, waterproof sunscreens. [Haarmann & Reimer GmbH]

Neo Heliopan, Type BB. CAS 131-57-7; EINECS 205-031-

5; 2-Hydroxy-4-methoxybenzophenone; uv-A and uv-B broad spectrum absorber for sunscreen formulations. [Haarmann & Reimer GmbH]

Neo Heliopan, Type Hydro. CAS 27503-81-7; 2-Phenylbenzimidazole-5-sulfonic acid; uv-B filter for sunscreen formulations. [Haarmann & Reimer GmbH]

Neo Heliopan, Type MA. CAS 134-09-8; Menthyl anthranilate; uv-A absorber for waterproof sunscreen formulations. [Haarmann & Reimer GmbH]

Neo Heliopan, Type OS. CAS 118-60-5; 2-Ethylhexyl salicylate; uv-B absorber for cosmetic applications, waterproof sunscreens; solubilizer for oxybenzone. [Haarmann & Reimer GmbH]

Neo-Hombreol. CAS 57-85-2; Testosterone propionate; androgen. [Organon Inc] *

Neolan. Metal complex dyes. [Ciba plc] *

Neoleptol. Triformyl-trimethylene-triamine.

Neoleukorit. A proprietary synthetic resin of the phenol-formaldehyde class. §

Neoloid. Castor oil; laxative. [Lederle Laboratories USA] *

Neolyn. Elastomeric alkyd type rosin-based resins with good grease resistance; used to impart flexibility and grease resistance to adhesives, is used for vinyl floor tiles and vinyl inks, in vinyl based and type T-gravure inks. [Hercules] *

Neolysol. Lysol made with chlorocresol; used as an antiseptic.

Neo-Medrone Acne Lotion. A proprietary preparation of sulfur, aluminum chlorhydroxide, methylpred nisolone acetate and neomycin sulfate; used in dermatology for acne. [Upjohn Ltd] *

Neomedrone Veriderm. A proprietary preparation of methylprednisolone and neolaycin; used in dermatology as an antibacterial agent. [Upjohn Ltd] *

Neomin. CAS 1405-10-3; A proprietary preparation containing neomycin sulfate; an antibiotic. [Glaxo Laboratories]*

Neomix. CAS 1405-10-3; Neomycin sulfate; antibacterial. [Upjohn] *

Neomodelon 100. Blend; neutral sizing agent for paper manufacturing. [Toho Chem. Industry]

Neo-Naclex. A proprietary preparation of bendrofluazide; a diuretic. [Duncan Flockhard Ltd]

Neo-Naclex-K. A proprietary preparation of bendrofluazide and potassium chloride; a diuretic. [Glaxo Laboratories]*

Neonalium. An alloy of aluminum with 6-14% copper, 1% nickel, and small amounts of other metals.

Neonite. A 33-grain sporting rifle powder. It contains 10% barium or potassium nitrate, 6% petroleum jelly, and insoluble nitrocellulose.

Neonite® EG60/6mm, EG61/12mm. CAS 25928-94-3; Epoxy molding compound; for elec. engineering, encapsulation of solenoids, mech. engineering (pump and valve parts, flanges). [Ciba-Geigy GmbH] *

Neopac. Acrylic dispersions. [ICI Chem. & Polymers Ltd]

Neopelline. An alkaloid, $C_{32}H_{45}NO_8$, obtained from *Aconitum napellus*.

Neopen® Dyes. Metal complex dyes; for inks, ball-pen pastes, copying toners. [BASF AG]

Neopen SS. A proprietary trade name for a wetting agent containing sodium abietene sulfonate. §

neopentyl glycol. (2,2-dimethyl-1,3-propanediol) $HOCH_2C(CH_3)_2CH_2OH$; CAS 126-30-7; EINECS 204-

781-0; Resin intermediate; insect repellent. [BASF; Eastman; Hüls AG; Mitsubishi Gas]

Neophax FA. A proprietary brown factice added to polychloroprene, nitrile rubber, Hypalon and polyurethane when maximum resistance to oil is required. [Hubron Rubber Chemicals] *

Neophenoquin. A proprietary preparation of lithium phenylcinchoninate. §

Neophryn. CAS 61-76-7; A proprietary trade name for the hydrochloride of phenylephrine. §

neopine. Hydroxy-codeine, $C_{18}H_{20}NO_3(OH)$; alkaloid from opium.

Neoplen. Foamed polyethylene slabstock and moldable beads; for cushion packaging of fragile items, antirattle stowage of tools in automotives. [BASF plc] *

Neopolen® E. CAS 9002-88-4; Polyethylene foam; for foamed moldings, packaging, building trade, maritime sector, automotive industry, recreation and leisure equip. [BASF AG; BASF plc]

Neopolen® P. CAS 9003-07-0; Polypropylene foam; for foamed moldings, packaging, building trade, maritime sector, automotive industry, recreation and leisure equip. [BASF AG; BASF plc]

Neopon. Surfactant for cosmetics, toiletries, pharmaceutical, processing, agricultural and other industries. [Baxenden Chemicals Ltd]

Neopon 33. Sodium olefin sulfonate and sodium ether sulfate as a clear amber liquid; foaming agent and detergent for cosmetic and household uses. [Witco Chemical Ltd] *

Neopon LAM. CAS 2235-54-3; EINECS 218-793-9; Ammonium lauryl sulfate, which may be based on natural or synthetic alcohols; liquid form; for hair and carpet shampoos. [Witco Chemical Ltd] *

Neopon LOA/F. Ammonium alcohol 2EO sulfate in liquid form; for hair shampoos. [Witco Chemical Ltd] *

Neopon LOS, LOS/F and LOS/NF. Sodium alcohol 2EO sulfate in liquid form; used for hair shampoos, foam baths, emulsion polymerization. [Witco Chemical Ltd] *

Neopon LOT/F. Triethanolamine alcohol 2EO sulfate in liquid form; for hair shampoos, foam baths, liquid detergents. [Witco Chemical Ltd] *

Neopon LS. Sodium lauryl sulfate, which may be based on natural or synthetic alcohols; liquid form; for hair and carpet shampoos; emulsion polymerization. [Witco Chemical Ltd] *

Neopon LT. CAS 139-96-8; EINECS 205-388-7; Triethanolamine lauryl sulfate which may be based on natural or synthetic alcohols; liquid form; for hair shampoos. [Witco Chemical Ltd] *

Neopralac. Pigments for textile printing. [ICI Chem. & Polymers Ltd] †

neoprene. See polychloroprene.

Neoprene Latex 115. CAS 126-99-8; Polychloroprene latex; stable disp., very resist. to deterioration by prolonged intensive mixing; for products requiring high-mod. properties; used for adhesives, bonded batts, coatings, saturants; exhibits high mech. strength, mech. and chem. stability; vulcanized with ZnO and thiocarbanilide or Thiuram E and Tepidone. [DuPont]

Neoprene Latex 400. Anionic colloidal system containing a copolymer of chloroprene and 2,3 dichloro-1,3-butadiene; general purpose disp. for products needing high modulus, abrasion and heat resistance, improved tear

---

‡ Trade name and manufacturer not verified   § Trade name without identified manufacturer

strength, and exceptional ozone and weathering resistance. [DuPont]

Neoprene Latex 750. Anionic colloidal system containing a med.-modulus highly crystallization-resistant copolymer of chloroprene and 2,3 dichloro-1,2-butadiene; general purpose latex for dipped goods requiring low modulus, wet/dry flexibility, and excellent crystallization resistance; adhesives. [DuPont]

Neoprene NPG 6856. CAS 126-99-8; Polychloroprene homopolymer; adhesive grade with improved rheology. [DuPont] *

Neo-protosll. A proprietary colloidal silver preparation; contains about 20% silver iodide, AgI, combined with a protein base, and is a germicide. §

Neopybuthrin. Synergized synthetic pyrethroids. [The Wellcome Foundation Ltd] *

Neorad. Radiation-curable oligomers and formulations. [ICI Resins BV]

Neoresit. A phenol-formaldehyde resin. §

NeoRez. Water and solvent based polyurethane resins; used in paints, inks and special coatings on all substrates, e.g.,. wood, metal, plastic, concrete, etc. [Polyvinyl Chemie Holland BV]

Neorez. Polyurethane resin products. [ICI Resins BV]

Neorode. Anti-fouling resin. [ICI Chem. & Polymers Ltd]

neosaccharin. See saccharin.

Neosar. CAS 50-18-0; Cyclophosphamide; antineoplastic; immunosuppressive. [Adria Laboratories Inc] *

Neoscan. CAS 41183-64-6; Gallium citrate Ga 67; diagnostic aid; radioactive agent. [Medi-Physics Inc] *

Neoscoa 203C. POE alkyl ether; detergent, scouring agent for cotton, wool, and synthetic fibers; deinking agent for paper. [Toho Chem. Industry]

Neosolve® AD-1. Wetting agent for asbestos. [M.S. Paisner]

Neosorb. Sorbitol powders or syrups [Roquette Corp]

Neosorexa. CAS 56073-07-5; Difenacoum; a ready-to-use anticoagulant rodenticide. [Sorex Ltd]

Neosote. The phenoloids of blast-furnace tar. It contains a small quantity of phenol, and a large amount of cresols.

Neospectra. CAS 1333-86-4; A proprietary trade name for carbon black. §

Neospinol 264. Blend of special anionics, nonionics, and mineral oil; dispersant for sulfur in rayon production; spinning oil for nylon. [Toho Chem. Industry]

Neosporin Products. Proprietary formulations of polymyxin B sulfate, neomycin sulfate and gramicidin and bacitracin zinc depending on the formulation; for short-term use in the treatment of topical infections (primary or secondary), due to susceptible organisms. [The Wellcome Foundation Ltd]

Neostar. Organic brightener system; for acid zinc electroplating. [Harshaw Chemicals Ltd] *

neosulfexine. Neohexal (hexamethylene-tetramine-salicyl-sulfonate).

Neosyl®. Silica for industrial use. [Crosfield Chemicals Ltd]

Neosyn. Synthetic tannins and auxiliaries. [Hodgson Chemicals Ltd]

Neo-Synephrine Hydrochloride. CAS 61-76-7; Phenylephrine hydrochloride; adrenergic. [Sterling Drug Inc] *

Neotac. Acrylic emulsion. [ICI Chem. & Polymers Ltd]

Neoteben®. CAS 54-85-3; Isoniazid; antitubercular. [Bayer AG]

Neotex. CAS 1333-86-4; Carbon black. [Sevalco Ltd]

Neothane. A proprietary polyester based polyurethane elastomer cross linked with diamine. [Goodyear] *

Neothene. Solvent; for gem and mineral testing. [Geoliquids] *

Neotrizine. Trisulfapyrimidines (oral suspension); mixture of sulfadiazine, sulfamerazine, and sulfamethazine; used as an antibacterial. [Eli Lilly & Co] †

Neotulle. A proprietary preparation of paraffin tulle with neomycin sulfate, zinc bacitracin, and polymixin B sulfate; wound dressing. [Fisons plc, Pharmaceuticals Div]*

NeoVac. Miscellaneous chemical compositions; for coatings and printing inks. [Polyvinyl Chemie Holland BV] *

Neovadine. Dyeing and printing assistant. [Ciba plc] *

Neovax. A proprietary preparation of neomycin sulfate, succinylsulfathiazole, and kaolin; an antidiarrheal. [H N Norton & Co Ltd] ‡

Neovit. A proprietary preparation of thiamine hydrochloride, riboflavine, pyridoxine hydrochloride, nicotinamide, manganese, sodium and potassium glycerophosphates; a tonic. [Rybar Laboratories Ltd] *

Neo-Voronit®. Fuberidazole, sodium-N-dimethyldithiocarbamate; nonmercurial liquid seed dressing for treatment of cereal seed against smuts and snow mold. [Bayer AG]

Neozapon® Dyes. Metal complex dyes with good sol. in polar solvs.; for production of flexographic and gravure inks. [BASF AG]

Neozone A. Phenyl-α-naphthylamine; a proprietary antioxidant for rubber. §

Neozone B. A proprietary trade name for meta-toluylene diamine; an antioxidant. [Imperial Chemical Industries plc; Du Pont (UK) Ltd] *

Neozone C. A proprietary antioxidant for rubber; resembles Neozone standard, but contains less m-toluylene diamine. §

Neozone D. Pure phenyl-β-naphthylamine; a proprietary antioxidant for rubber. §

Neozone E. A proprietary trade name for an antioxidant for rubber containing 75 parts of phenyl-β-naphthylamine and 25 parts of meta-toluylene diamine. [Imperial Chemical Industries plc; Du Pont (UK) Ltd] *

NEP. N-Ethyl-2-pyrrolidone; solvent, reaction intermediate, textile auxilliaries, cosmetic ingredient. [ISP] *

NEPD. CAS 597-09-1; 2-Nitro-2-ethyl-1,3-propanediol; chemical and pharmaceutical intermediate, in tire cord adhesives, as formaldehyde release agents, deodorants, antimicrobials. [Angus]

Nephril. CAS 346-18-9; A proprietary preparation of polythiazide; a diuretic. [Pfizer International]

Nephroflow. CAS 133-17-5; Iodohippurate sodium I 123; diagnostic aid; radioactive agent. [Medi-Physics Inc] *

Nepol® PP40. CAS 9003-07-0; PP, 40% Long glass fiber-reinforced; for injection molded parts requiring high stiffness and good impact resist. [Neste Composite Materials] *

N.E. powder. A 36-grain powder, containing metallic nitrates, and nitro hydrocarbons.

NEPS. Polystyrene expandable beads. [Asahi Chem. Industry]

Neptazane. CAS 554-57-4; Methazolamide; carbonic anhydrase inhibitor. [Lederle Laboratories USA] *

Nepton EXT. Durable water repellent and fluorocarbon extender; approved for Quarpel type water repellent treatments; nonsmoking; contains no solvs.; excellent resin/catalyst stability. [Yorkshire Pat-Chem]

---

* Trade name not verified as available   † Trade name verified as obsolete

Neptun® Bases. Dye bases for production of ball-pen pastes and typewriter ribbon inks. [BASF AG]

Neradol. Synthetic tannins generally prepared by the condensation of phenol-sulfonic acids with formaldehyde, under conditions that only water-soluble products are formed. §

Neral. Z-3,7-Dimethyl-2,6-octadiene-1-al; fragrance and flavoring; fresh, lemon-like, green, slightly lime-like. [BASF AG]

neral. β-Citral, $C_9H_{15}CHO$.

Neramine. Wood stain dyes and pigments for plastics. [ICI Chem. & Polymers Ltd]

Nerco. Solvents and detergents. [ABM Chemicals Ltd] *

Nercol. Soluble cutting oils. [ABM Chemicals Ltd] *

Nercolan. Blend of phenolic biocides. [Rhone-Poulenc UK]

Nercosol. Viscosity depressant, antifoaming; used for vinyl plastisols. [ABM Chemicals Ltd] *

Nerfinol. Textile finishing agent. [ABM Chemicals Ltd] *

Nergandin. An alloy containing 70% copper, 28% zinc, and 2% lead; used for condenser tubes.

Nericur Gel 5. CAS 94-36-0; 5% Benzoyl peroxide in aqueous gel. [Schering Health Care Ltd]

Nerisone Cream. 0.1% Difluocortolone valerate. [Schering Health Care Ltd]

Nerloate. Paint dryers. [ABM Chemicals Ltd] *

Nerolidol. Z,E-3,7,11-Trimethyl-1,6,10-dodecatriene-3-ol; fragrance and flavoring; sweetly floral, green, woody, lilly-like. [BASF AG]

nerolidol. CAS 7212-44-4; Methyl-vinyl-homo-geranyl carbinol; used as a perfume.

nerolin. CAS 93-18-5; (yara-yara; β-naphthol-methyl ester) $C_{10}H_7.O.CH_3$; β-Naphthol-ethyl ester is also known under this name.

nerolin II. (bromelia).β-Naphthol ethyl ester, $C_{10}H_7 \cdot O \cdot C_2H_5$; a synthetic perfume.

neroli oil. Oil obtained from the fresh flowers of the bitter orange; used as a perfume and flavoring substance.

Neroli oil, artificial. *See* methyl anthranilate

Nervan. Wetting agents and finishing oils. [Rhone-Poulenc UK]

Nervan CP. Anionic surfactant. [Rhone-Poulenc UK]

Nervanaid. Sequestering agent, metal chelate, filter aid powder. [Rhone-Poulenc UK]

Nervanase. Bacterial α-amylase. [Rhone-Poulenc UK]

nerve oil. *See* neatsfoot oil.

nervin. An extract of meat.

Nesacaine. CAS 3858-89-7; Chloroprocaine hydrochloride; anesthetic. [Astra Pharmaceutical Products Inc]

Nesfield's triple tablets. A water sterilizer. It consists of a) a tablet containing an iodide and an iodate; b) one containing citric or tartaric acid; and c) one containing sodium sulfite.

Nessler's reagent. CAS 7783-33-7; An alkaline solution of mercuric iodide in potassium iodide; used as a delicate test for ammonia.

Neste Polyethylene. Polyethylene of low, medium and high density as well as linear low pressure polyethylene; for packaging, distribution, household and construction films, insulation and jacketing of wire and cable, production of pipes, molded materials and paper coating. [Neste OY Chemicals & Unifos Kemi AB] *

Nestosyl Ointment. Benzocaine 2%, butyl aminobenzoate 2%, resorcin 2%, zinc oxide 10%, hexachlorophane 0.1% (in ointment form); topical anesthetic for the relief of local pain and irritation in lacerated skin conditions, also in hemorrhoids and anal pruritis. [Bengue & Co Ltd]*

Nethaprin Dospan. A proprietary preparation of etafedrine hydrochloride, bufylline, doxylamine succinate and phenylephrine hydrochloride; used in the treatment of bronchospasm. [Richardson-Vicks Inc] *

Nethaprin Expectorant. A proprietary preparation of etafedrine hydrochloride, bufylline, doxylamine succinate, and glyceryl guaiacolate; an expectorant. [Richardson-Vicks Inc] *

Netromycin. CAS 56391-57-2; Netilmicin sulfate; antibacterial. [Schering Corp] *

Nettolin. Humus complex-fertilizer on peat basis for the culture of wine, hops, fruit and vegetables, for flowers and lawns. [Süd-Chemie AG] *

Neudorfite. A resinous hydrocarbon found in Bavarian coal pits.

Neulactil. CAS 2622-26-6; Pericyazine; a sedative. [Rhone-Poulenc Rorer Ltd]

Neuphor® 100. An anionic rosin emulsion with 35% solids used as a sizing agent; used in paper and paperboard to impart resistance to water and aqueous solutions. [Hercules] *

Neuphor® 635. Emulsion size for use in manufacturing of paper/paperboard; improved stabilizing system. [Hercules]

Neurocil®. Pharmaceutical specialty; neuroleptic. [Bayer AG]

Neuro-Phosphates. A proprietary preparation of calcium glycerophosphate, sodium glycerophosphate, and strychnine; a tonic. [SmithKline Beecham] *

Neuro-Transentin. A proprietary preparation of adiphenine hydrochloride and phenobarbitone. [Ciba plc] *

Neustrene® 045. CAS 68002-72-2; Hydrogenated menhaden oil; textile lubricant, pharmaceutical intermediate, emulsifier, mold release agent, buffing compound. [Witco/Humko]

Neustrene® 059. CAS 67701-27-3; Hydrogenated tallow glycerides; used in manufacturing of alkali metal soaps, monoglycerides, textile auxiliaries, greases. [Witco/Humko]

Neustrene® 064. CAS 68002-71-1; Hydrogenated soybean oil; textile lubricant, pharmaceutical intermediate, emulsifier, mold release agent, buffing compound. [Witco/Humko]

Neutralaleisen. A Swedish silicon-iron alloy, which is stated to resist the action of acids.

neutral alum. A neutral basic alum obtained by the addition of sodium hydroxide to a solution of alum, until the precipitate produced is just redissolved.

Neutral Degras. CAS 68815-23-6; Fatty esters of wool grease; lubricant with EP and slip chars.; wire drawing compounds, slushing and cutting oils, lubricants; rust preventative; plasticizer and lubricant for adhesives; textile lubricant; inhibits crystallization of wax components; ink formulations; dispersant for other waxes; used in leather in stuffing greases; waterproofing agent. [Fanning]

Neutralite. An asphaltic material made in Germany.

neutral oils. The name given to the lightest lubricating oils from American petroleum. The term is also applied to refined coal-tar oils.

neutral phosphate. A fertilizer prepared by digesting mineral phosphate, bonemeal, or a mixture of both, with

---

‡ Trade name and manufacturer not verified

§ Trade name without identified manufacturer

small amounts of sulfuric acid. This renders the P_2O_5 more available. The product contains 20-25%. P_2O_5, and is neutral.

neutral red. (toluylene red). A dyestuff. It is the hydrochloride of dimethyldiaminotoluphenazine, $C_{15}H_{17}N_4Cl$; dyes cotton mordanted with tannin and tartar emetic, bluish-red; used as an acid-base indicator in the range of pH 6.8—8.0; red in acid, yellow-brown in alkali; also as a biological stain.

neutral tartar. Potassium tartrate.

Neutramag®. CAS 1309-42-8; A suspension of magnesium hydroxide for neutralising acidic effluents and reducing sludge volumes. [Steetley Magnesia Products Ltd]

Neutraphylline. A proprietary preparation of diprophyline; a bronchial antispasmodic. [Cox Pharmaceuticals] *

Neutrase. A bacterial proteinase made by submerged fermentation of a selected strain of Bacillus subtilis; can be used in any case where the aim is breakdown of proteinaceous matter for production of peptides and amino acids. [Novo Nordisk]

Neutrichrome. Premetallized dyes for wool and wool blends. [ICI Chem. & Polymers Ltd] †

Neutrigan®. Alkaline salts and salt mixtures; basifying, deacidifying, and masking agents for chrome, leather, and furs. [BASF AG]

Neutrogene. Azoic dyes. [ICI Chem. & Polymers Ltd] †

Neutrol® TE. CAS 102-60-3; EINECS 203-041-4; Tetrahydroxypropyl ethylenediamine; neutralizing agent for cosmetics industry. [BASF; BASF AG]

Neutrolactis. A proprietary preparation containing aluminum hydroxide gel, magnesium trisilicate, calcium carbonate and milk solids; an antacid. [Sandoz] *

Neutronyx®. Nonyl phenol polyglycol ethers. [Millmaster-Onyx UK]

Neutronyx® 656. CAS 9016-45-9; Nonoxynol-11; detergent, dispersant, wetting agent, emulsifier for household detergents, dishwashing, fine fabrics, metal cleaning and degreasing, industrial cleaning, sanitizers, insecticides, herbicides; silicone emulsifier; lime dispersant; stable in acid, alkali and hard water. [Stepan; Stepan Canada]

Nevada silver. See nickel silvers.

Nevastain. An alloy of 86% iron, 9.5% chrondum, 4% silicon, and 0.43% carbon; it is non-corrosive.

Nevastain® A. Hindered phenolic compound; antioxidant used in mastic adhesives, rubber goods such as mech. and molded goods, caulking compounds, cement, and antiskinning agents. [Neville]

Nevastain R.A. A stainless steel alloy containing iron with approximately 16% chromium, 1% copper, 1% silicon, 0.4% manganese, 0.03% phosphorus and sulfur, and 0.1% carbon (maximum).

Nevchem® 70. Alkylated petrol. hydrocarbon resin; resin used in adhesives (hot melt pressure sensitive, mastic); coatings (aluminum, emulsion, industrial, marine, paper, traffic, wire/cable), inks, rubber (cements, mechanical and molded goods, tires), concrete-curing compounds, caulking compounds. [Neville]

Nevex® 100. Modified hydrocarbon resin; enhances compatibility of wax and ethylene-vinyl acetate systems; in adhesives (hot melt, hot melt pressure sensitive, mastic, pressure sensitive), coatings (can & drum, industrial, paper), rubber (cements, mechanical and molded goods, tires). [Neville]

Nevidene. A proprietary trade name for a coumarone resin.§

Nevile and Winther's acid. CAS 84-87-7; α-Naphthol-4-sulfonic acid.

Nevillac® 10° XL. Hydroxy modified resin; plasticizer/tackifier in nitrile rubber; used in adhesives (hot melt, hot melt pressure sensitive, pressure sensitive), epoxy coatings, rubber cements, antiskinning agents. [Neville]

Neville. A proprietary trade name for coumarone-indene resins. [Neville Chemical Co] *

Nevillite. A proprietary trade name for a hydrocarbon. §

Nevin. A proprietary brand of lithopone (qv). §

Nevindene. A proprietary trade name for coumarone-indene resins. §

Nevinol. A proprietary coumarone plasticizing oil; a viscous liquid polymer practically nondrying at room temperature. §

Nevolin®. CAS 75-09-2; Methylene chloride; carrier and anti-evaporation agent for agrochemicals being applied by thermal fogging machine. [Fargro Ltd]

Nevoxy® EPX-L. Hydroxy modified resin; excellent color stability; esp. suited for use in wh. and pastel coatings; produces formulations with good prop. retention on aging. [Neville]

Nevpene® 9500. Modified hydrocarbon resin; aromatic producing an exceptional tackifier with modification; used in adhesives, coatings, rubber, and caulking compounds. [Neville]

nevraltein. A name for sodium-p-phenetidine-methane-sulfonate.

nevrosthénine. An alkaline solution of the glycero-phosphates of sodium, potassium, and magnesium.

Nevroz® 1420. Modified hydrocarbon resin; reactive unsat. chars.; replacement resins for rosin derivatives, and in inks. [Neville]

Nevtac® 80. Synthetic polyterpene resin; synthetic tackifier used in adhesives, coatings, rubber products, concrete-curing compounds, and caulking compounds. [Neville]

Nevyanskite. See iridosmine.

New 5C Cycocel. Chlormequat chloride and choline chloride; plant growth regulator for cereals, linseed, winter oilseed rape, ornamentals. [BASF plc; BASF AG]

New 5C Cycocel. Soluble concentrate containing 645 g chlormequat and 32 g choline chloride per liter; plant growth regulator for use in cereals and ornamentals. [Cyanamid of Great Britain Ltd]

Newagit. A trademark for abrasive and refractory materials consisting essentially of alumina. §

Newaloy. A proprietary trade name for a steel containing copper. §

New Brick. Blended organic and inorganic acids in combination with surfactants and wetting agents; cleaner for brick in new construction (mortar smears, dirt etc). [Nova Chemical Inc] *

New Brick (Heavy Duty). Blended organic and inorganic acids in combination with surfactants and wetting agents; cleaner for brick in new construction (mortar smears, dirt etc). [Nova Chemical Inc] *

Newcavac, -T. Newcastle disease, inactivated vaccine; for immunization of poultry. [Intervet Inc]

Newdamp Balancing Fluids. Blends of halogenated alkyl aryl hydrocarbons; controlled high dens., low visc. balancing fluids; replaces visc. damping fluid when balancing gyroscope floats; stability, low volatility, non-corrosiveness. [Bacon]

---

* Trade name not verified as available † Trade name verified as obsolete

New Formulation SBK Brushwood Killer. Liquid concentrate containing 2,4-D, mecoprop and dicamba; selective herbicide for use on coarse and woody weeds. [Vitax Ltd]

New Hickstor 6. CAS 117-18-0; Dustable powder containing 6% w/w tecnazene; protectant fungicide and potato sprout suppressant. [Hickson & Welch Ltd]

New Hystor. CAS 117-18-0; Granules containing 5% w/w tecnazene; protectant fungicide and potato sprout suppressant. [Agrichem (International) Ltd]

New Kotol. γ-HCH; an organochlorine insecticide. [Embetec Crop Protection Ltd]

New Legumex. Selective weedkillers. [Fisons plc, Horticulture Div] *

Newloy. An alloy of 64% copper, 35% nickel, and 1% tin.

New Murbetex. Pre-emergence herbicide. [Murphy Chemical Ltd] †

Newton's alloy. An alloy of 50% bismuth, 31.2% lead, and 18.7% tin. It is a fusible alloy, and melts at 94.5 C.

New Verdone. Selective weedkiller. [Plant Protection] *

New white lead. A sulfate of lead.

New-wrap. A proprietary viscose packing material. §

New Zealand dammar. A name given to Kauri copal resin, obtained from *Dammara Australis*.

Nex. CAS 1563-66-2; Carbofuran; a systemic insecticide in the form of 5% w/w granules for the soil treatment of beetles and nematodes. [Tripart Farm Chemicals Ltd]

Nez. A proprietary preparation of phenylephrine hydrochloride, paracetamol, caffeine and ascorbic acid; cold remedy. [Rybar Laboratories Ltd] *

NFB. CAS 7664-38-2; Phosphoric acids 80% and 85%; used in chemical polishing of aluminum. [Monsanto Co] *

NFT Fertilizer. Soluble fertilizer. [Fisons plc, Horticulture Div] *

Ngai camphor. A camphor obtained from *Blumea balsamifera*. It is closely related to borneol.

N-Hance® 3000. CAS 65497-29-2; Guar hydroxypropyltrimonium chloride; conditioner, viscosifier, substantive polymer for hair and skin care products, shampoos, lotions, liq. soaps. [Aqualon]

N'hangellite. An elastic bitumen.

NI-20GF. CAS 9008-66-6; Nylon 6/10 resin, 20% glass fiber-reinforced. [Compounding Tech.]

Niac. CAS 59-67-6; Niacin; vitamin. [O'Neal, Jones & Feldman Pharmaceuticals] *

Niacet. Metal acetates and propionates. [Niacet Corp]

Niacet Calcium Acetate Tech. CAS 62-54-4; Calcium acetate; sequestrant, thickener, pH control agent for petroleum and textile industries. [Niacet]

Niacet Sodium Acetate Anhyd. Tech. CAS 127-09-3; Sodiumacetate [Niacet]

niacin. (nicotinic acid; pyridine-3-carboxylic acid) CAS 59-67-6; Used in medicine as a cholesterol lowering agent, nutrition, feeds, enriched flours, dietary supplement.

niacin. See nicotinic acid.

Niagara blue. CAS 72-57-1; Trypan blue.

Niamid. CAS 51-12-7; A proprietary preparation of nialamide; an antidepressant. [Pfizer International]

Niamide. CAS 51-12-7; A proprietary preparation of nialamide; an antidepressant. [Pfizer International]

Niaproof® Anionic Surfactant 4. CAS 139-88-8; Sodium tetradecyl sulfate; detergent, wetting agent, penetrant, emulsifier used in adhesives and sealants, coatings, photo chemicals, emulsion polymerization, metal processing, electrolytic cleaning, pickling baths, plating,

pharmaceuticals, leather, textiles. [Niacet]

Niaproof® Anionic Surfactant 08. CAS 126-92-1; Sodium 2-ethylhexyl sulfate; detergent, wetting agent, penetrant, emulsifier used in textile mercerizing, metal cleaning, electroplating, photo chemicals, adhesives, emulsion polymerization, household and industrial cleaners, agric., pharmaceuticals; stable to high concs. of electrolytes. [Niacet]

Niax. Urethane catalyst. [Union Carbide]

Nibiol. CAS 4008-48-4; A proprietary preparation of nitroxoline; for veterinary use. [Dales Pharmaceuticals Ltd] *

Nibren. Flame-retardant impregnating, encapsulating and dipping waxes with high dielectric constants; for use in the manufacture of paper capacitors. [Bayer AG]

Nibren wax. CAS 1321-65-9; Chlorinated naphthalene.

Nibrite. Barrel bright nickel plating process. [Hanshaw Chemicals] ‡

Nicar. Nickel carbonate. [Mechema Chemicals Ltd] *

Nicaragua wood. See redwoods.

Nicat. Nickel catalysts and Raney-type nickel catalysts; hydrogenation catalysts. [Crosfield Chemicals Ltd]

Nicfo. CAS 3349-06-2; Nickel formate. [Mechema Chemicals Ltd] *

Ni-chillite. A proprietary trade name for a nickel-chromium-molybdenum cast iron. §

Nichroloy. Electrical resistance alloys containing 23-75% nickel, 7-20% chromium, 7-50% iron, and 1-3% manganese. by steam and dilute acids; They are used for electrical resistance wires, heat resistant products and chemical equipment.

Nichrome®. Alloys of 54-80% nickel, 10-20% chromium, 7-27% iron, 0-11% copper, 0-5% manganese, 0.3-4.6% silicon, and sometimes 1% molybdenum, and 0.25% titanium; used as electrical resistance metals; resists acids. §

Nichrosi. Alloys. a) Contains 25-30% chromium, 16-18% silicon, balance nickel. b) Contains 15-25% chromium, 16-18% silicon, balance nickel.

nickel. Ni; CAS 7440-02-0; EINECS 231-111-4; Electroplating; hydrogenation catalyst; in iron- and copper-based alloys. [Aldrich; Atomergic Chemetals; Cerac; Inco Europe; Spectrum Chem. Mfg.]

Nickel 200. CAS 7440-02-0; A grade of nickel containing 99.0% nickel. [Wiggin Alloys Ltd] ‡

Nickel 201. CAS 7440-02-0; A grade of nickel containing 99.0% nickel (min) and 0.02% carbon (max). [Wiggin Alloys Ltd] ‡

Nickel 204. CAS 7440-02-0; Nickel containing 4% cobalt. [Wiggin Alloys Ltd] ‡

Nickel 205. CAS 7440-02-0; Nickel containing 99.0% min. nickel and low carbon. [Wiggin Alloys Ltd] ‡

Nickel 211. CAS 7440-02-0; Nickel containing 5% manganese. [Wiggin Alloys Ltd] ‡

Nickel 212. CAS 7440-02-0; Nickel containing 2% manganese. [Wiggin Alloys Ltd] ‡

Nickel 213. CAS 7440-02-0; Nickel with improved machining properties; nickel 96% min., manganese 2 %; high carbon and silicon content. [Wiggin Alloys Ltd] ‡

Nickel 222. CAS 7440-02-0; Nickel containing 99.5% min. nickel, 0.06-0.09% magnesium and very low impurity levels. [Wiggin Alloys Ltd] ‡

Nickel 223. CAS 7440-02-0; Nickel containing 99.5% min. nickel, 0.035-0.065% magnesium and very low impurity

---

‡ Trade name and manufacturer not verified   § Trade name without identified manufacturer

levels. [Wiggin Alloys Ltd] ‡

Nickel 229. CAS 7440-02-0; Nickel containing 97.5% min. nickel, 1.8-2.2% tungsten, 0.35-0.65% magnesium and 0.02-0.04% aluminum. [Wiggin Alloys Ltd] ‡

Nickel 270. CAS 7440-02-0; Nickel 99.9% pure. [Wiggin Alloys Ltd] ‡

nickel acetate. (nickel acetate tetrahydrate) $Ni(OOCCH_3)_2$ · $4H_2O$; CAS 373-02-4, 6018-89-9; EINECS 206-761-7; textile mordant, catalyst. [Ashland; Atomergic Chemetals; Mallinckrodt; Nihon Kagaku Sangyo]

Nickeladium. A proprietary trade name for a nickel-vanadium cast steel. §

nickel aluminum bronze. An alloy of 10-40% nickel, 10-88% copper, and 2-30% aluminum. One alloy contains 20% tin.

Nickel Babbitt. A proprietary trade name for a tin-copper-nickel alloy used as a bearing metal for high speeds. §

nickel brass. A nickel silver. One alloy contains 55% copper, 43% zinc, and 2% nickel, and another 50% copper, 34% zinc, 15% nickel, and 0.1% aluminum.

nickel bronze. A nickel silver. It usually contains from 20-30% nickel, 50-86% copper, and 8-25% tin, but other alloys contain 11-18% zinc, and 0-18% lead.

nickel carbonate, basic. $NiCO_3$ · $2Ni(OH)_2$ · $4H_2O$; CAS 3333-67-3; EINECS 222-068-2; Electroplating, preparation of nickel catalysts, ceramic colors and glazes. [Ashland; Atochem N. Am.; Farleyway Chem. Ltd; M&T Harshaw; Nihon Kagaku Sangyo]

nickel chloride (ous). (nickelous chloride) $NiCl_2$ or $NiCl_2$ · $6H_2O$; CAS 7718-54-9; Electroplated nickel coatings, chemical reagent. [Ashland; Atochem N. Am.; Atomergic Chemetals; Mallinckrodt; Nihon Kagaku Sangyo]

nickelene. A name suggested for nickel silver (German silver) alloys.

nickel glance. A mineral, $Ni(AsS)_2$.

Nickelin. Electrical resistance alloys of nickel and copper, usually with zinc. One alloy contains 55.3% copper, 31%, 13%, 0.4% iron, and 0.2% lead; another 68% copper and 32% nickel; and a third 74.5% copper, 25% nickel, and 0.5% iron.

nickel iron. See ferro-nickel.

Nickel-linnaeite. See Polydymite.

nickel manganese bronze. An alloy containing 2.5% nickel, 53.4% copper, 39% zinc, 1.7% manganese, and 2.6% tin, with small quantities of aluminum and lead.

nickel-manganese-copper. An alloy of 73% copper, 24% manganese, and 3% nickel; used for electrical resistances.

nickel-molybdenum steels. Alloys containing 0.13-0.54% carbon, 0.12-4.4% molybdenum, and 1.8% nickel.

nickel nitrate (ous). (nickelous nitrate) $Ni(NO_3)_2$ · $6H_2O$; CAS 13138-45-9; nickel plating, preparation of nickel catalysts, manufacture of brown ceramic colors. [Mallinckrodt; Nihon Kagaku Sangyo; Noah Chem.]

Nickeloid. A proprietary trade name for a dual metal; it is zinc faced with nickel. §

nickel oreide. A nickel silver. It is an alloy of from 63-87% copper, 6-33% zinc, and 2-7% nickel.

nickelous chloride. See nickel chloride (ous).

nickelous nitrate. See nickel nitrate (ous).

nickelous oxide. See nickel oxide (ous).

nickel (II) oxide. See nickel oxide (ous).

nickel oxide (ous). (nickelous oxide; nickel (II) oxide; nickel protoxide; green nickel oxide) NiO; CAS 1313-99-1;

nickel salts, porcelain painting, fuel cell electrodes. [Atomergic Chemetals; Cerac; Nihon Kagaku Sangyo; Noah Chem.]

Nickeloy. An alloy of 1.5% nickel, 4% copper, and 94% aluminum.

nickel protoxide. See nickel oxide (ous).

nickel silvers. Ternary alloys of copper, nickel, and zinc, the standard of which is determined by the nickel con-tent. Albatra, Alfenide, Alpacca, Amberoid, Ambrac, American silver, Aphit, Argentan, Argentin, Argiroide, Argentan solder, Argyroide, Argyrolith, Aterit, Benedict plate, bismuth bronze, brazing solder, China silver, Carbondale silver, Craig gold, electroplate, electrum, Elner's German silver, Nevada silver, nickel brass, nickel bronze, nickelin, silverite, silveroid, Victoria silver, white button alloy, and white solder are all types of nickel silver. These alloys are used in the manufacture of tableware.

nickel silver solder. An alloy of 35% copper, 57% tin, and 8% nickel.

nickel steel. An alloy of nickel with steel, usually containing from 3-5% of nickel, but sometimes a larger amount. One alloy contains 30% of nickel, 1% manganese, and 1% chromium. Nickel steels are used for armor plates, ships' screws, boiler plates, cable wires, and gun barrels.

nickel sulfamate. (nickel (II) sulfamate dihydrate) $Ni(SO_3NH_2)_2$ · $2H_2O$; CAS 13770-89-3; EINECS 237-396-1; For nickel electroforming and plating; aerospace and electronics applications. [Atochem N. Am.; Atomergic Chemetals; M&T Harshaw; Witco/Allied-Kelite]

nickel (II) sulfamate dihydrate. See nickel sulfamate.

nickel zirconium. An alloy of 86.4% nickel, 6% aluminum, 6% silicon, and 1.5% zirconium; used for cutting tools.

Niclad. A proprietary trade name for a duplex metal in which nickel or nickel alloy is deposited on steel or iron. §

Niclate®. Nickel dialkyldithiocarbamate; rubber antioxidants offering antiozonant and antioxidant protection in epichlorohydrin; used for optimum aging properties. [R T Vanderbilt Co Inc]

Niclocide. CAS 50-65-7; Niclosamide; anthelmintic. [Miles Pharmaceuticals]

Nico. CAS 1313-99-1; Nickel oxide. [Mechema Chemicals Ltd] *

Nico-400. CAS 59-67-6; Niacin; vitamin. [Marion Merrell Dow Inc] *

Nicobid. CAS 59-67-6; Niacin; vitamin. [USV Pharmaceutical Corp] ‡

Nicocap. CAS 59-67-6; Niacin; vitamin. [ICN Nutritional Biochemicals Corp] *

Nicolane. A proprietary trade name for noxapine. §

Nicolar. CAS 59-67-6; Niacin; vitamin. [USV Pharmaceutical Corp] ‡

Nicolle's Carbol-thionin blue. A microscopic stain. It consists of a mixture of 10 cc of a saturated solution of thionin blue in 50% alcohol, and 100 cc of a 2% carbolic acid solution.

Nicolmelt®. Hot-melt adhesives. [Reichhold]

Nicon. An alloy of 70% iron and 30% nickel.

Nico Padutin. Pharmaceutical preparation also known as Kalleonicit; combination therapy for physiological vasodilation. [Bayer AG]

Nico-Padutin® forte. For the treatment of disorders of the arterial circulation. [Bayer AG]

---

* Trade name not verified as available   † Trade name verified as obsolete

Nicor. Mixture of nickel/cobalt oxide. [Mechema Chemicals Ltd] *

Nicorette. Nicotine polacrilex; methacrylic acid polymer with divinyl benzene, complex with nicotine; deterrent. [Merrell Dow Pharmaceuticals Inc] *

Nicoschwab. See Uba. §

Nicotine 40% Shreds. Insecticide smoke. [Murphy Chemical Ltd] †

nicotinic acid. (niacin; vitamin B; 3-picolinic acid; pyridine-3-carboxylic acid) $C_6H_5NO_2$; CAS 59-67-6; EINECS 200-441-0; for drugs and for the vitamin reinforcement of foods. [Degussa; R.W. Greeff; Lonza; Schweizerhall]

nicotinyl alcohol. CAS 100-55-0; 3-Pyridylmethanol.

Nicral alloys. Aluminum alloys containing varying percentages of nickel, chromium, and copper.

nicro-copper. An alloy of 98% copper and 2% nickel.

Nicrolan. Wool grease-based compositions; anticorrosive coating materials for metal protection. [Westbrook Lanolin] *

Nicroman. A proprietary trade name for a tool steel containing 1% chromium, 1.65% nickel, 0.35% copper, and 0.7% carbon; an oil-hardening hob steel. §

Nicron 325. CAS 14807-96-6; Platy talc; general purpose economical filler for industrial coatings and plastics that require max. filler loadings. [Montana Talc]

Nicron 665. CAS 14807-96-6; Platy talc; ultrafine, high purity talc with excellent reinforcing and dielec. properties; for extrusion wire and cable applications, additive for gloss control in paints and coatings. [Montana Talc]

Nicron JS 422. CAS 14807-96-6; Platy talc; filler providing mineral oil absorp. for low VOC and high solids paints/coatings applications. [Montana Talc]

Nicrosil. A proprietary trade name for an alloy of iron with 18% nickel and 4-6% silicon. §

Nicrosilal. An alloy of 71.2% iron, 18% nickel, 6% silicon, 2% chromium, 1.8% carbon, and 1% manganese. It is a non-magnetic grey cast iron which is resistant to staining and has great ductility.

Nicu steel. A nickel steel containing 2.13% nickel, 0.2% copper, 0.51% manganese, 0.03% sulfur, 0.03% silicon, and 0.006% phosphorus.

Nidazol. CAS 443-48-1; Metronidazole; pharmaceutical preparation for the treatment of trichomonal infestation. [M A Steinhard Ltd] *

Nidrin. CAS 21645-51-2, CAS 546-93-0; A proprietary preparation containing aluminum hydroxide and magnesium carbonate in a co-dried gel; an antacid. [Smith & Nephew Pharmaceuticals Ltd] *

Niello silver. (Russian tula; blue silver). An alloy of silver, copper, lead, and bismuth with a bluish color.

Niferex. A proprietary preparation of a polysaccharide-iron complex used in the treatment of anemia. [L Wilcox & Co Ltd] ‡

Niflor. Precision nickel/PTFE composite; precision thickness coating for hard, self-lubricating and corrosion resistant surfaces on engineered components. [Fothergill & Harvey plc] ‡

Nigagin. Methyl-p-hydroxybenzoate; a preservative.

Night of Olay Nightcare. Cream for nightly care of the skin. [Richardson-Vicks Inc] *

nigraniline. Aniline black, $C_{30}H_{25}N_5$.

nigre. The impure soap remaining after the good soap has been removed by running out. It contains iron soaps, caustic soda, and sodium chloride.

Nigrin. CAS 3930-19-6; Streptonigrin; antineoplastic. [Pfizer Inc] †

nigrol. The residue obtained after the removal of kerosene, gasoline, and light oil from petroleum naphtha.

Nigrosin. Water-soluble acid dyestuffs for a variety of special applications. [Bayer AG]

Nigrosin Bases. For shoe polish, printing ink, office supplies and plastics industries. [Bayer plc]

Nigroth Metal. A proprietary trade name for a heat-resisting nickel-chromium cast iron. §

nigrotic acid. Dihydroxysulfonaphthoic acid.

Ni-hard. Alloys of iron with from 4-5% nickel, 1.5% chromium, and varying amounts of silicon and carbon.

Nikal®. Nickel compound; antiseize compound for extreme temp. application; acid and chem. resist. [Jet-Lube]

Niklad. Electroless nickel systems for plating onto prepared substrates including stainless steel, steel alloys, cast iron, copper and alloys, aluminum, titanium, berylium, nickel alloys, nonconductors such as glass and ceramics. [Witco/Allied-Kelite]

Nikrome. Proprietary trade name for nickel-chromium steels. Nikrome M contains 2.25% nickel, 1% chromium, 0.45% molybdenum, and 0.4% carbon. §

Nikro-trimmer Steel. A proprietary trade name for a nickel-chromium steel containing 0.3% nickel, 0.55% chromium, and 0.85% carbon. §

Nilergex. CAS 1225-60-1; A proprietary preparation of isothipendyl hydrochloride. [ICI Chem. & Polymers Ltd]†

Nilevar. CAS 52-78-8; A proprietary preparation of norethandrolone; an anabolic steroid. [G D Searle & Co Ltd] *

Nilex. A proprietary 36% nickel steel used for pendulums, etc., on account of its low coefficient of expansion. §

Nilfom 2X. Nonsilicone defoamer; defoamer for atmospheric and pressure dyeing; compat. with all systems. [Am. Emulsions]

Nilfom DF-155. Nonsilicone defoamer; defoamer which does not contribute to flamm. [Am. Emulsions]

Nilo Alloy 36. A trademark for controlled expansion alloys; 36% nickel, balance iron. [Wiggin Alloys Ltd] ‡

Nilo Alloy 42. A trademark for controlled expansion alloys; 42% nickel, balance iron. [Wiggin Alloys Ltd] ‡

Nilo Alloy 48. A trademark for controlled expansion alloys; 48% nickel, balance iron. [Wiggin Alloys Ltd] ‡

Nilo Alloy 51. A trademark for controlled expansion alloys; 51% nickel, balance iron. [Wiggin Alloys Ltd] ‡

Nilo Alloy 475. A trademark for controlled expansion alloys; 47% nickel, 5% chromium, balance iron. [Wiggin Alloys Ltd] ‡

Nilo Alloy K. A trademark for controlled expansion alloys; 29% nickel, 17% cobalt, balance iron. [Wiggin Alloys Ltd] ‡

Nilo Alloy K45. A trademark for controlled expansion alloys; 32% nickel, 13% cobalt, 54% iron. [Wiggin Alloys Ltd] ‡

Nilo Alloy P50. A trademark for controlled expansion alloys; 50% nickel, 50% iron. [Wiggin Alloys Ltd] ‡

Nilodin. A proprietary trade name for the hydrochloride of lucanthrone. §

Nilomag Alloy 48. A trademark for magnetic alloys made by a powder metallurgy process. It is 48% nickel and the balance is iron. [Wiggin Alloys Ltd] ‡

Nilomag Alloy 51. A trademark for magnetic alloys made by a powder metallurgy process. It is 51% nickel and the balance is iron. [Wiggin Alloys Ltd] ‡

---

‡ Trade name and manufacturer not verified   § Trade name without identified manufacturer

Nilomag Alloy 471. A trademark for magnetic alloys made by a powder metallurgy process. It is 47% nickel, 50% iron, and 3% molybdenum. [Wiggin Alloys Ltd] ‡

Nilomag Alloy 475. A trademark for magnetic alloys made by a powder metallurgy process. It is 47% nickel, 5% chromium, and the balance is iron. [Wiggin Alloys Ltd] ‡

Nilomag Alloy K. A trademark for magnetic alloys made by a powder metallurgy process. It is 29% nickel, 17% cobalt, and the balance is iron. [Wiggin Alloys Ltd] ‡

Nimate. CAS 13770-89-3; Nickel sulfamate; for nickel electroforming of foils, perforated parts, moldings, and record stampers; also resizing of worn parts and nickel plating for aerospace and electronics applications. [Albright & Wilson UK] *

Nimocast Alloy 242. A patented alloy of 21% chromium, 10% cobalt, 10.5% molybdenum and the balance nickel. [Wiggin Alloys Ltd] ‡

Nimocast Alloy 713. 13.4% chromium, 4.5% molybdenum, 1% titanium, 6.2% aluminum, 2.3% niobium, balance nickel. [Wiggin Alloys Ltd] ‡

Nimocast Alloy 771. 77% nickel, 14% iron, 5% copper, 4% molybdenum. [Wiggin Alloys Ltd] ‡

Nimocast Alloy PE10. 20% chromium, 6% molybdenum, 6.5% niobium, 2.5% tungsten, balance nickel. [Wiggin Alloys Ltd] ‡

Nimocast Alloy PK24. An alloy consisting of 10% chromium, 15.2% cobalt, 3% molybdenum, 5.2% titanium, 5.5% aluminum, and the balance is nickel. [Wiggin Alloys Ltd] ‡

Nimol. An alloy of cast iron with 20% monel metal and 2-4% chromium. It is non-magnetic and has high resistance to corrosion by acid and sea water.

Nimonic Alloy 75. A trademark for an alloy of 20% chromium, 0.4% titanium, and the balance is nickel. [Wiggin Alloys Ltd] ‡

Nimonic Alloy 80A. A trademark for an alloy of 20% chromium, 2.3% titanium, 1.3% aluminum, and the balance is nickel. [Wiggin Alloys Ltd] ‡

Nimonic Alloy 90. A trademark for an alloy of 20% chromium, 17% cobalt, 2.5% titanium, 1.5% aluminum, and the balance is nickel. [Wiggin Alloys Ltd] ‡

Nimonic Alloy 93. A trademark for an alloy of 20% chromium, 17% cobalt, 2.5% titanium, 1.5% aluminum, and the balance is nickel; closer control maintained than Alloy 90. [Wiggin Alloys Ltd] ‡

Nimonic Alloy 105. A trademark for an alloy of 15% chromium, 20% cobalt, 5% molybdenum, 1.2% titanium, 4.7% aluminum, and the balance is nickel [Wiggin Alloys Ltd] ‡

Nimonic Alloy 115. A trademark for an alloy of 15% chromium, 15% cobalt, 3.5% molybdenum, 4% titanium, 5% aluminum, and the balance is nickel [Wiggin Alloys Ltd] ‡

Nimonic Alloy 118. A trademark for an alloy of 15% chromium, 15% cobalt, 3.5% molybdenum, 4% titanium, 5% aluminum, balance nickel; a fully vacuum-melted and cast version of Alloy 115. [Wiggin Alloys Ltd] ‡

Nimonic Alloy PE 11. A trademark for an alloy of 18% chromium, 5.2% molybdenum, 2.3% titanium, 0.8% aluminum, 38% nickel, and the balance is iron. [Wiggin Alloys Ltd] ‡

Nimonic Alloy PE 13. A trademark for an alloy of 22% chromium, 1.5% cobalt, 9% molybdenum, 18.5% iron, 0.6% tungsten, balance nickel. [Wiggin Alloys Ltd] ‡

Nimonic Alloy PK 31. A trademark for an alloy of 20% chromium, 14% cobalt, 4.5% molybdenum, 2.3% titanium, 0.4% aluminum, 5% niobium, and the balance is nickel. [Wiggin Alloys Ltd] ‡

Nimonic Alloy PK 33. A trademark for an alloy of 19% chromium, 14% cobalt, 7% molybdenum, 2% titanium, 2% aluminum, and the balance is nickel. [Wiggin Alloys Ltd] ‡

nimorazole. (4-2-(5-nitroimidazol-1-yl) ethyl; morpholine; nitrimidazine) A drug used in the treatment of trichomoniasis.

Nimotop®. CAS 66085-59-4; Nimodipine; vasodilator, cerebral therapeutic agent. [Bayer AG; Miles Pharmaceuticals]

Nimox. Nickel molybdenum oxides on alumina. [Laporte Industries Ltd] *

Nimrod. CAS 41483-43-6; An emulsifiable concentrate containing 250 g bupirimate per liter; a systemic fungicide to control powdery mildew. [ICI Agrochemicals; ICI Garden Products]

Nimrod T. CAS 41483-43-6, 26644-46-2; An emulsifiable concentrate containing 62.5 g bupirimate and 62.5 g triforine per liter; a systemic fungicide to control powdery mildew in ornamental plants. [ICI Agrochemicals; ICI Garden Products]

Ninate®. Alkyl benzene sulfonate; emulsifier. [Stepan] *

Ninate® 401. Calcium alkylbenzene sulfonate; emulsifier, dispersant used in pesticide formulations and in self-dispersing liq.; foaming agent. [Stepan; Stepan Canada; Stepan Europe]

Ninate 401. Anionic surfactant as a dark viscous liquid; emulsifier used in oil additives. [KWR Chemicals Ltd] *

Ninate® 401-A. Calcium alkylbenzene sulfonate; emulsifier for agric. formulations. [Stepan; Stepan Canada; Stepan Europe]

Ninate® 411. Amine dodecylbenzene sulfonate; emulsifier, solv. degreaser, drycleaning detergent, surface tens. reducer, defoamer; emulsifier used in emulsifiable kerosene formulations, agric. formulations, textiles, metalworking. [Stepan; Stepan Canada]

Ninate 411. Anionic surfactant as a light viscous liquid; emulsifier for solvent degreasers and dry cleaning detergents. [KWR Chemicals Ltd] *

Ninate 415. Anionic surfactant in light amber liquid form; emulsifier for solvent degreasers and dry cleaning detergents. [KWR Chemicals Ltd] *

Ninate® DS 70. CAS 577-11-7; EINECS 209-406-4; Sodium dioctyl sulfosuccinate; dispersant, wetting agent, emulsifier for agric. flowables, suspension and emulsifiable concs., gran. and powd. formulations. [Stepan Europe]

ninhydrin. CAS 485-47-2; (Triketol). Triketohydrindenehydrate. A colorimetric reagent for aminoacids.

Ninol®. Alkylolamides; thickeners, foam stabilizer, detergent. [Stepan] *

Ninol® 11-CM. Cocamide DEA, modified; detergent base for synthetic cleaners; emulsifier, lubricant, antistat for textile applications; emulsifier, corrosion inhibitor in cutting fluids, drawing compounds, metal cleaning. [Stepan; Stepan Canada]

Ninol® 30-LL. CAS 120-40-1; EINECS 204-393-1; Lauramide DEA; foam booster/stabilizer, visc. builder/modifier for liq. detergents, shampoos, hand soaps, bath products. [Stepan; Stepan Canada]

Ninol® 201. CAS 93-83-4; EINECS 202-281-7; Oleamide

DEA; emulsifier, corrosion inhibitor in industrial lubricant systems, cutting fluids, drawing compounds, metal cleaners; thickener for personal care and liq. detergent products; emulsifier, lubricant, antistat for textiles. [Stepan; Stepan Canada]

Ninol® AX. CAS 120-40-1; EINECS 204-393-1; Lauramide DEA; surfactant. [Stepan Canada]

Ninol® CMP. CAS 68140-00-1; Cocamide MEA; foam booster, visc. builder for liq. detergents, detergent blocks or bars. [Stepan; Stepan Canada]

Ninol® CNR. CAS 68140-00-1; Cocamide MEA; emollient, thickener, foam booster/stabilizer for shampoos, bubble baths, liq. soaps, shower gels, toilet soaps. [Stepan Europe]

Ninol® LMP. CAS 142-78-9; EINECS 205-560-1; Lauramide MEA; foam booster/stabilizer, thickener, emollient, detergent for dishwash, liq. detergents, detergent blocks or bars, shampoos, hand soaps, bath products [Stepan; Stepan Canada; Stepan Europe]

Ninol® M10. Monoisopropanolamide; thickener, emollient, anticorrosive agent, foam booster/stabilizer for shampoos, bubble baths, liq. soaps, shower gels, toilet soaps. [Stepan Europe]

Ninox. Foam stabilizer; shampoos; foam baths; detergents. [KWR Chemicals Ltd] *

Ninox® . Amine oxide; foam enhancers and thickeners. [Stepan] *

Ninox® FCA. CAS 68155-09-9; Cocamidopropylamine oxide; thickener, foam booster/stabilizer, detergent, antistat for scale-removing liqs., liq. soaps, cleaning foams, personal care products. [Stepan Europe]

Ninox® L. Lauramine oxide; thickener, foam booster/stabilizer, detergent, antistat for scale-removing liqs., liq. soaps, cleaning foams. [Stepan Europe]

Ninox® M. CAS 3332-27-2; EINECS 222-059-3; Myristamine oxide; thickener, foam booster/stabilizer, detergent, antistat for scale-removing liqs., liq. soaps, cleaning foams. [Stepan Europe]

Ninox® SO. CAS 2571-88-2; EINECS 219-919-5; Stearamine oxide; thickener, emollient, mild detergent for scouring pastes or creams; additive for hypochlorite. [Stepan Europe]

Niobe oil. See methyl benzoate.

niobium. (columbium) Metallic element; Nb; CAS 7440-03-1; EINECS 231-113-5; superconducting and magnetic alloys, cermets, missiles and rockets, cryogenic equipment, ferroniobium for alloy steels. [Atomergic Chemetals; Cabot; Cerac; Noah Chem.]

niobium oxide. (niobium pentoxide; niobium (V) oxide) Nb_2O_5; CAS 1313-96-8; EINECS 215-213-6; Intermediate, electronics. [Atomergic Chemetals; Cabot; Cerac; Noah Chem.]

Niong. CAS 55-63-0; Nitroglycerin. vasodilator [Rhone-Poulenc Rorer Pharmaceuticals] *

Nipabenzyl. CAS 94-18-8; Benzylparaben; preservative, bactericide, fungicide for pharmaceuticals, cosmetics, foods, medicinal preps., industrial applications. [Nipa Labs]

Nipabutyl. Butyl-4-hydroxybenzoate; a preservative. [Nipa Laboratories Ltd.

Nipabutyl Potassium. CAS 38566-94-8; Potassium butylparaben. preservative. [Nipa Labs]

Nipabutyl Sodium. CAS 36457-20-2; Sodium butylparaben; preservative, bactericide, fungicide for pharmaceu-

ticals, cosmetics, foods, medicinal preps., industrial applications. [Nipa Labs]

Nipacide® . Formulated biocides. [Nipa Laboratories Ltd.

Nipacide® BCP. CAS 120-32-1; o-Benzyl-p-chlorophenol; germicide in disinfectant cleaners for hospitals, schools, homes, etc.; readily degradable. [Nipa Labs]

Nipacide® BIT. 1,2-Benzisothiazolin-3-one. [Nipa Laboratories Ltd.]

Nipacide® BK. CAS 4719-04-4; Hydroxyethyl-s-triazine; preservative, bacteriostat, fungistat for sol. cutting fluids and coolants and other products. [Nipa Labs]

Nipacide® DP. 2,4-Didichlorophen. [Nipa Laboratories Ltd.]

Nipacide® DX. CAS 133-53-9; 2,4-Dichloro-m-xylenol. [Nipa Laboratories Ltd.]

Nipacide® F. 2,2-Dihydroxy-5,5-dichloro diphenyl monosulfide. [Nipa Laboratories Ltd.

Nipacide® MX. CAS 88-04-0; p-Chloro-m-xylenol; antimicrobial, preservative for disinfectant, algicide, slimicide, and water treatment pesticide products, polymer emulsions, adhesives, latex paints, metalworking cutting fluids. [Nipa Labs]

Nipacide® OPP. CAS 90-43-7; O-phenyl phenol; disinfectant, preservative for detergents, cooling lubricants, textile/leather finishing, adhesives, paper, citrus fruit, polishes, wax emulsions, ceramic glazes, soap solutions [Nipa Labs]

Nipacide® PC. 2,4-Di-4-chloro-3-methyl phenol. [Nipa Laboratories Ltd.]

Nipacide® PTAP. CAS 80-46-6; p-t-Amyl phenol; disinfectant. [Nipa Labs]

Nipacide® PX. CAS 133-53-9; 2,4-Di-4-chloro-3,5-xylenol. [Nipa Laboratories Ltd.]

Nipacombin SK. Compounded sodium-4-hydroxybenzoic acid; preservative for liq. antacid suspensions and other alkaline solutions [Nipa Labs]

Nipafax. Chemicals for thermal paper. [Nipa Laboratories Ltd.]

Nipagin A. CAS 120-47-8; Ethylparaben; preservative, bactericide, fungicide for pharmaceuticals, cosmetics, foods, medicinal preps., industrial applications. [Nipa Labs]

Nipagin A Potassium. CAS 36457-19-9; Potassium ethylparaben; a preservative. [Nipa Labs]

Nipagin A Sodium. CAS 35285-68-8; Sodium ethylparaben; a preservative, bactericide, fungicide for pharmaceuticals, cosmetics, foods, medicinal preps., industrial applications. [Nipa Labs]

Nipagin M. CAS 99-76-3; Methylparaben; preservative, bactericide, fungicide for pharmaceuticals, cosmetics, foods, medicinal preps., industrial applications. [Nipa Labs]

Nipagin M Potassium. CAS 26112-07-2; Potassium methylparaben; a preservative [Nipa Labs]

Nipagin M Sodium. CAS 5026-62-0; Sodium methylparaben; preservative, bactericide, fungicide for pharmaceuticals, cosmetics, foods, medicinal preps., industrial applications. [Nipa Labs]

Nipaguard® DMDMH. CAS 6440-58-0; DMDM hydantoin; antimicrobial for cosmetics and personal care products. [Nipa Labs]

Nipaguard® DME. A blend of esters of 4-hydroxybenzoic acid and DMDM hydantoin; used for preservation of cosmetics and toiletries. [Nipa Laboratories Ltd]

Nipaguard® MPS. A blend of esters of 4-hydroxybenzoic

---

‡ Trade name and manufacturer not verified   § Trade name without identified manufacturer

acid; used for preservation of cosmetics and toiletries. [Nipa Laboratories Ltd]

Nipaheptyl. n-Heptyl p-hydroxybenzoate; preservative, bactericide, fungicide for pharmaceuticals, cosmetics, foods, medicinal preps., industrial applications. [Nipa Labs]

Nipanox® BHT. CAS 128-37-0; BHT; antioxidant. [Nipa Labs]

Nipanox® S-1. Propyl gallate (20%), citric acid (10%) in propylene glycol; antioxidant for vegetable oil industry. [Nipa Labs]

Nipanox® Special. BHA, propyl gallate, citric acid in propylene glycol; antioxidant for foods; fat and oil stabilizer. [Nipa Labs]

Nipantiox. CAS 25013-16-5; Butylated hydroxyanisole. [Nipa Laboratories Ltd] †

Nipantiox 1-F. CAS 25013-16-5; BHA; antioxidant. [Nipa Labs]

NiPar 640. Nitroethane and 1-nitropropane blend; wetting additive for solv.-based coatings including inks, adhesives, etc.; provides improved film integrity, solv. release, and dry times. [Angus]

NiPar S-10. CAS 108-03-2; 1-Nitropropane; intermediate, solvs. for coatings. [Angus]

NiPar S-20. CAS 79-46-9; 2- Nitropropane; used in solv. blends for inks and coatings; esp. for NC, chlorinated rubber, vinyl, epoxy, acrylic, PU, polyamide systems; automotive finishes. [Angus]

Nipa salt. A material obtained by ignition of the plant *Nipa fructicans*; used to coagulate rubber latex.

Nipasept. Compounded esters of 4-hydroxybenzoic acid; a preservative. [Nipa Laboratories Ltd]

Nipasol M. CAS 94-13-3; Propylparaben; preservative, bactericide, fungicide for pharmaceuticals, cosmetics, foods, medicinal preps., industrial applications. [Nipa Labs]

Nipasol M Potassium. CAS 84930-16-5; Potassium propylparaben; preservative. [Nipa Labs]

Nipasol M Sodium. CAS 35285-69-9; Sodium propylparaben; preservative, bactericide, fungicide for pharmaceuticals, cosmetics, foods, medicinal preps., industrial applications. [Nipa Labs]

Nipastat. Methylparaben, butylparaben, ethylparaben, and propylparaben. preservative, bactericide, fungicide for pharmaceuticals, cosmetics, foods, medicinal preps., industrial applications. [Nipa Labs]

Nipol® 1000X132. CAS 9003-18-3; Butadiene-acrylonitrile copolymer, slightly staining antioxidant; used for hose, sheet packing, molded and extruded goods, oil seals, sundries, adhesives; excellent oil and fuel resist.; vulcanized with sulfur-accelerator, sulfur donor, or peroxide systems. [Zeon]

Nipol® 1001 CG. CAS 9003-18-3; Butadiene-acrylonitrile copolymer, slightly staining antioxidant; cement grade used for coated materials, sheet packing, molded and extruded goods, o-rings, oil seals, adhesives, plastic modification; excellent oil and fuel resist.; vulcanized with sulfur-accelerator, sulfur donor, or peroxide systems. [Zeon]

Nipol® 1022X59. CAS 9003-18-3; Precrosslinked butadiene-acrylonitrile copolymer; used in plastics compding. to impart rubbery properties to thermoplastics. [Zeon]

Nipol® 1072. CAS 9003-18-3; Carboxy-modified butadiene-acrylonitrile copolymer, nonstaining antioxidant; used for

heels and soles, rubber-covered rolls, molded and extruded goods, oil seals, adhesives; improved low temp. brittleness, hot tear and abrasion resist.; vulcanized with bivalent metal ions such as zinc oxide or with sulfur. [Zeon]

Nipol® 1203F60. NBR/PVC prefluxed blend (70:30), nonstaining antioxidant; used for coated materials, hose, flat belts, molded and extruded goods, shells and soles, rubber-covered rolls, wire and cable jacketing, matting, foam products, oil seals, adhesives; excellent resist. to ozone, weathering, fuels, oils, solvs., abrasion, heat; excellent colorability. [Zeon]

Nipol® 1204X22. NBR/PVC/DOP prefluxed blend (100/60/120); used for soft rolls, low durometer applications. [Zeon]

Nipol® 1411. CAS 9003-18-3; Butadiene-acrylonitrile copolymer, slightly staining antioxidant; used for friction products, plastic modification; vulcanized with sulfur-accelerator systems; finely divided powd.; nonsol. [Zeon]

Nipol® 2782. Block polymer; emulsifier for agric. formulations. [Stepan; Stepan Canada]

Nipol® AR-31. Acrylic rubber; offers resist. to high and low temps.; used in automobile gaskets, etc. [Zeon]

Nipol® AR-42. Acrylic rubber; low temp.-resist. rubber with shortened post-cure time; better compression set, low corrosion and improved extrusion moldability. [Zeon]

Nipol® AR-74. Acrylic rubber; elastomer with ultra low temp. resist.; used with soap-sulfur cure system; gives excellent low temp. properties for o-rings, seals, gaskets. [Zeon]

Nipol® DP5123P. Nitrile rubber; nonmigratory, nonextractable, nonvolatile modifier for plastics for improved oil and chem. resist., low temp. and heat aging properties, superior physical properties, increased abrasion resist. and enhanced rubber-like feel. [Zeon]

Nippon Ant and Crawling Insect Killer. CAS 52645-53-1, 7696-12-0; Aerosol containing permethrin and tetramethrin; residual contact insecticidal surface spray. [Vitax Ltd]

Nippon Ant Killer Liquid. Liquid bait material containing borax; used for control of black ants. [Vitax Ltd]

Nippon Ant Killer Powder. CAS 52645-53-1; Permethrin on talc dust carrier; contact residual insecticide. [Vitax Ltd]

Nippon Fly Killer Spray. CAS 52645-53-1, 7696-12-0; Aerosol containing permethrin and tetramethrin; domestic insecticide. [Vitax Ltd]

Nippon Ready For Use Ant and Crawling Insect Killer. CAS 52645-53-1; Trigger spray pack containing permethrin as a ready-for-use liquid. residual contact insecticidal surface spray. [Vitax Ltd]

Nipride. CAS 14402-89-2; A proprietary preparation of sodium nitroprusside; intravenous vasodilator. [Roche Laboratories; Roche Products Ltd]

Nipro (i). CAS 105-60-2; Caprolactam; used for nylon manufacturing. [Columbia Nitrogen Corporation] *

Nipro (ii). CAS 7783-20-2; Ammonium sulfate; fertilizer. [Columbia Nitrogen Corporation] *

Ni-resist. A corrosion and heat-resisting cast iron containing 12-15% nickel, 5-7% copper, and 1.5-4% chromium. §

Nirex. A proprietary trade name for an alloy of 80% nickel, 14% chromium, and 6% iron. §

Nirez® 7002. Terpene phenolic resin; resin for paste ink applications. [Arizona]

---

* Trade name not verified as available   † Trade name verified as obsolete

Nirez® 9007. Ester of selected tall oil fatty acids; specialty ink solv. for use in high solids, low VOC ink systems and in soybean oil systems to reduce visc. and adjust tack. [Arizona]

Nirez® 9011. Ester of selected tall oil fatty acids; specialty ink solv. for use in high solids, low VOC ink systems and in soybean oil systems to reduce visc. and adjust tack. [Arizona]

Nirolex Expectorant Linctus. A proprietary preparation containing guaifenesin, ephedrine sulfate and mepyramine maleate; a cough linctus. [The Boots Co plc]

Nirostaguss. A non-rusting and heat-resisting 34% chromium cast iron.

Nisapas. CAS 65-49-6, CAS 54-85-3; A proprietary preparation of sodium p-aminosalicylic acid and isoniazid; an antituberculous agent. [Antigen International Ltd] ‡

Nisentil. CAS 561-78-4; Alphaprodine hydrochloride; analgesic. [Hoffmann-LaRoche Inc] *

Nispan Alloy C-902. A trademark for an alloy of 42% nickel, 5.2% chromium, 2.4% titanium, 0.6% aluminum, and the balance iron. [Wiggin Alloys Ltd] ‡

Nisso PB. Hydroxyl terminated polybutadiene. [British Traders & Shippers Ltd]

niter. See saltpeter.

niter. See potassium nitrate.

niter cake. A residue from the manufacture of nitric acid consisting of a mixture of normal and acid, sodium sulfate.

niter spirit. See spirit of niter.

Nitofol®. CAS 10265-92-6; Methamidophos; broad spectrum systemic insecticide used for treatment of sucking and biting insects and spider mites. [Bayer AG]

Nitolac. A proprietary polyurethane flooring material. [Tercol, Shifnal, Shropshire] *

Nitoman. CAS 58-46-8; A proprietary preparation of tetrabenazine. [Roche Products Ltd] *

Nitracc. Fertilizers. [ICI Chem. & Polymers Ltd]

Nitrados. CAS 146-22-5; Nitrazepam; for relief of insomnia due to anxiety or stress. [Berk Pharmaceuticals Ltd] *

Nitral. CAS 10024-97-2; A trade name for moist nitrous oxide used as a bactericide. §

Nitraline. CAS 7697-37-2; Nitric acid used as a dairy pipeline cleaner. [Ciba plc]

Nitralloy. A nitrided aluminum-chromium-molybdenum steel.

Nitram. CAS 6484-52-2; Ammonium nitrate fertilizer. [ICI Chem. & Polymers Ltd]

Nitrammite. A native ammonium nitrate.

Nitrammomkalk. A mixture of ammonium and calcium nitrates in a granular form of Norwegian manufacture; used as a fertilizer.

Nitraniline N. Nitraniline mixed with sufficient sodium nitrite necessary for its diazotization.

p-nitraniline. See p-nitroaniline.

Nitraphen. 2:4-Dinitrophenol. §

Nitrapo. A product obtained from crude caliche by crystallization. It contains about 66% sodium nitrate, 29% potassium nitrate, and a little sodium chloride; used as a fertilizer.

Nitraprill. CAS 7727-37-9; 34.5% Nitrogen prilled fertilizer. [Kemira Ince Ltd]

nitrated oils. Thick syrupy liquids obtained by treating castor oil or linseed oil with a mixture of concentrated sulfuric and nitric acids; they form homogenous mixtures with nitrocellulose; dissolved in acetone, these oils form varnishes, which are used for enameling leather or similar material, and mixing paints.

nitrate of tin. A mixture of stannous and stannic chlorides; used by dyers.

nitrating acid. (mixed acid) A mixture of sulfuric acid and nitric acid; consists of 36% nitric acid and 61% sulfuric acid; used for nitrating in the manufacture of explosives and plastics.

nitrazol. See Paranitraniline red.

nitre. See potassium nitrate.

Nitrene 11230. Modified cocamide DEA (2:1); detergent, wetting agent, emulsifier, thickener, foam stabilizer for industrial and specialty cleaners, floor strippers and degreasers, household hard surface cleaners. [Henkel/Cospha; Henkel Canada]

Nitrene C Extra. Cocamide DEA (1:1); emulsifier, dispersant, wetting agent, foam booster/stabilizer, detergent, and visc. builder for industrial/household detergents, metalworking; intermediate for liq. detergents; clarifier for liq. soaps; solv. cleaners, drycleaning. [Henkel/Cospha; Henkel Canada]

Nitrex. CAS 67-20-9; A proprietary name for nitro furantoin. [Vale Chemical Co] *

nitric acid. (aqua fortis; engravers acid; azotic acid) HNO_3; CAS 7697-37-2; EINECS 231-714-2; Manufacture of ammonium nitrate for fertilizer and explosives, organic synthesis (dyes, drugs, explosives, cellulose nitrate, nitrate salts), metallurgy, photo-engraving, etching steel, ore flotation, urethanes, rubber chemicals, nuclear fuel. [Aceto; Air Prods.; Am. Cyanamid; Angus; Asahi Chem Industry Co Ltd; Du Pont; Miles; Monsanto; Nissan Chem. Ind.; Norsk Hydro A/S]

nitric ether. Ethyl nitrate, $C_2H_5NO_3$.

nitrided steel. Steel which has been treated with ammonia gas at a temperature of 950 C, whereby nitrogen is absorbed on the surface giving a hard, nonbrittle surface; steels containing form 0.5-2.0% aluminum and 0.5-4.0% of other elements are used.

Nitrile 10 D. CAS 1975-78-6; EINECS 217-830-6; Decane nitrile, distilled; chemical intermediate. [Berol Nobel AB]

Nitrile 12. CAS 2437-25-4; EINECS 219-440-1; Dodecane nitrile; chemical intermediate. [Berol Nobel AB]

Nitrile BG. Tallow nitrile; chemical intermediate. [Berol Nobel AB]

nitrile C4. See butyronitrile.

nitrilotriacetic acid. (NTA; N,N-bis(carboxymethyl)glycine; triglycollamic acid; triglycine) $N(CH_2COOH)_3$; CAS 139-13-9; EINECS 205-355-7; Chelating and sequestering agent; builder in synthetic detergents. [W.R. Grace/hampshire; R.W. Greeff]

2,2´,2´´-nitrilotris(ethanol). See triethanolamine.

1,1´,1´´-nitrilotris-2-propanol. See triisopropanolamine.

nitrimidazine. See nimorazole.

nitrite rubber. A product obtained by treating latex with a nitrite and coagulating with an acid.

Nitro-26. A nitrogeneous fertilizer. [Fisons plc] *

Nitro BT. Nitro blue tetrazolium. [Monomer-Polymer & Dajac Laboratories Inc]

p-nitroaniline. (4-nitroaniline; p-nitraniline) $NO_2C_6H_4NH_2$; CAS 100-01-6; EINECS 202-810-1; Intermediate for dyes, antioxidants; gasoline gum inhibitors; corrosion inhibitor. [Enichem Am.; Hoechst Celanese; Monsanto]

---

‡ Trade name and manufacturer not verified   § Trade name without identified manufacturer

nitro base. p-Nitrosodimethylaniline; an intermediate for dyes.

o-nitrobenzaldehyde. (2-nitrobenzaldehyde) $C_7H_5NO_3$; CAS 552-89-6; EINECS 209-025-3; used in the synthesis of dyes, pharmaceuticals, surface active agents. [Penta Mfg.; Schweizerhall]

p-nitrobenzaldehyde. (4-nitrobenzaldehyde) $C_7H_5NO_3$; CAS 555-16-8; EINECS 209-084-5; used in the synthesis of dyes, pharmaceuticals, surface active agents. [Penta Mfg.; Schweizerhall]

4-nitrobenzaldehyde. See p-nitrobenzaldehyde.

5-nitrobenzene-1,3-dicarboxylic acid. See 5-nitroisophthalic acid.

p-nitrobenzoic acid. (4-nitrobenzoic acid; p-nitrodracylic acid) $C_6H_4(NO_2)COOH$; CAS 62-23-7; EINECS 200-526-2; Organic synthesis. [Du Pont; Hüls Am.; Nobel Chem. Ltd; Schweizerhall]

4-nitrobenzoic acid. See p-nitrobenzoic acid.

nitrocalcite. (wall saltpeter). Calcium nitrate, $Ca(NO_3)_2$.

nitrocarbol. See nitromethane.

nitrocellulose. (cellulose, nitrate; nitrocotton; guncotton) Cellulose derivative; $C_{12}H_{16}(ONO_2)_4O_6$; CAS 9004-70-0; in lacquers, high explosives, rocket propellant; printing ink base; leather finishing, etc. [Allchem Industries; Aqualon; Asahi Chem Industry Co Ltd; Hercules; SNPE Chimie]

nitrocellulose varnishes. Varnishes used in the manufacture of artificial leather; they consist of nitrocelluloses dissolved in amyl acetate, and colored; used also for painting iron work.

Nitro-chalk. A proprietary fertilizer consisting of an intimate mixture of chalk and ammonium nitrate in the form of a fine powder. [ICI Chem. & Polymers Ltd]

nitrochloroform. See Chloropicrin.

Nitrocontin. Glyceryl trinitrate; for treatment of angina pectoris. [Napp Laboratories Ltd] *

nitrocotton. See nitrocellulose.

nitro-dextrin. A similar product to nitro-starch; used in explosives.

Nitrodisc. CAS 55-63-0; Nitroglycerin; vasodilator. [G D Searle & Co]

nitrodracrylic acid. CAS 62-23-7; p-Nitrobenzoic acid.

p-nitrodracylic acid. See p-nitrobenzoic acid.

nitro-erythrite. Erythrol tetranitrate.

nitroethane. Nitroparaffin; $CH_3CH_2NO_2$; CAS 79-24-3; EINECS 201-188-9; intermediate for synthesis; stabilizer for chlorinated solvs.; fuel additive; solvent. [Angus; W.R. Grace/Nitroparaffins; Spectrum Chem. Mfg.]

Nitro Fast. Specialty dye for automotive, shoe, and furniture polishes, oils, waxes, and solvs. [Sandoz]

Nitroferrite. It contains 93% ammonium nitrate, 2% trinitronaphthalene, 2% potassium ferricyanide, and 3% sugar; an explosive.

Nitroform. CAS 517-25-9; Tetranitromethane.

Nitrofuel® . CAS 75-52-5; Brand of racing nitromethane containing Nitroguard®; fuel for automotive racing and model engines. [Angus]

nitrogen monoxide. See nitrous oxide.

nitroglycerin. (glyceryl trinitrate; trinitroglycerin; glonoine oil; pyro-glycerin) $CH_2NO_3CHNO_3CH_2NO_3$; CAS 55-63-0; High explosive, production of dynamite, medicine (vasodilator), combating fires in oil wells, rocket propellants. [Hüls Am.; ICI Specialties]

5-nitroisophthalic acid. (5-nitrobenzene-1,3-dicarboxylic acid) CH_3NO_2; CAS 75-52-5; EINECS 200-876-6. [Pfister; Schweizerhall]

Nitrolac. A German nitrocellulose lacquer.

nitrolignin. Wood which has been nitrated.

Nitrolim. Commercial nitrolim contains 57-63% calcium cyanamide, 20% lime, 14% graphite, and 7-8% silica, iron oxide, and alumina; a fertilizer. §

Nitrolite. An explosive containing nitroglycerin.

nitromannite. $C_6H_8(ONO_2)_6$; CAS 15825-70-4; Mannitol hexanitrate; used as an explosive cap ingredient and in medicine.

nitromethane. (nitrocarbol) Nitroparaffin; CH_3NO_2; CAS 75-52-5; EINECS 200-876-6;stabilizer for chlorinated solvs.; chemical intermediate; solv. for cellulosic compounds, polymers, waxes; rocket fuel, gasoline additive; in coatings industry. [Angus; W.R. Grace/Nitroparaffins]

Nitro-muriatic Acid. See aqua regia.

Nitron. A base which forms a nitrate almost insoluble in water; used for the determination of nitric acid; also used as a rubber vulcanization accelerator; also a proprietary trade name for a cellulose nitrate plastic. §

Nitronet. CAS 55-63-0; Nitroglycerin; vasodilator. [Rhone-Poulenc Rorer Pharmaceuticals] *

Nitrophos® 20-20-0. Complex fertilizer with 20% nitrogen, 20% phosphate; for all agric. and horticultural crops. [BASF AG]

Nitrophoska® 10-15-20; Complex fertilizer with 10% N, 15% P_2O_5, 20% K_2O; for agric. and horticultural crops with low basal demand for nitrogen. [BASF AG]

nitrophosphate. A fertilizer sometimes wrongly called ammonium superphosphate. It is prepared by mixing calcium superphosphate with ammonium sulfate. Some mixtures contain ammonium phosphate and calcium sulfate.

Nitropore® ATA. Azodicarbonamide/DNPT blend; chemical blowing agent. [Uniroyal]

Nitropress. CAS 14402-89-2; Sodium nitroprusside; antihypertensive. [Abbott Laboratories] *

nitropropane. (1-nitropropane) Nitroparaffin; $CH_3CH_2CH_2NO_2$; CAS 108-03-2; EINECS 203-544-9; solv. for inks and coatings, cellulose acetate, vinyl resins, lacquers, synthetic rubbers, fats, oils, dyes; chemical intermediate; rocket propellant; gasoline additive. [Angus; W.R. Grace/Nitroparaffins]

2-nitropropane. (sec-nitropropane) Nitroparaffin; $CH_3CH(NO_2)CH_3$; CAS 79-46-9; EINECS 201-209-1; solv. for coatings and inks, cellulose acetate, vinyl resins, lacquers, synthetic rubbers, fats, oils, dyes; chemical intermediate; rocket propellant; gasoline additive. [Angus; Ashland; W.R. Grace/Nitroparaffins]

sec-nitropropane. See 2-nitropropane.

nitropropiol. Sodium-o-nitro-phenyl-propiolate.

Nitrosin Saltpetre. A proprietary preparation containing nitrite; used for curing meat. §

Nitrospan. CAS 55-63-0; Nitroglycerin; vasodilator. [USV Pharmaceutical Corp] ‡

nitro-starch. (xyloidin). CAS 9056-38-6; A nitric ester of starch, probably the octonitrate $C_{12}H_{12}(NO_2)_8O_{10}$; used for blasting explosives, either alone, or by mixing 10% of it with a mixture of sodium nitrate and carbonaceous material.

Nitrostat. CAS 55-63-0; Nitroglycerin; vasodilator. [Parke-Davis]

nitrosulfate. Ferric sulfate $Fe_2(SO_4)_3$; used as a mordant for

---

* Trade name not verified as available

† Trade name verified as obsolete

dyeing.

nitrosyl silver. Silver hyponitrite; $Ag_2N_2O_2$.

nitrosyl sulfuric acid. (lead chamber crystals) CAS 7782-78-7; Nitro-sulfonic acid, $NO_2(SO_2.OH)$.

nitrous ether. CAS 109-95-5; Ethyl nitrite, $C_2H_5NO_2$.

nitrous gas or air. Nitrogen dioxide, NO_2.

nitrous oxide. (nitrogen monoxide) N_2O; CAS 10024-97-2; Anesthetic in dentistry and surgery, propellant gas in food aerosols, leak detection. [Air Liquide Hellas SA; Air Prods.; Monsanto; Nissan Chem. Ind.; Showa Denko]

nitrous sulfuric acid. A solution of nitrosyl-sulfuric acid (Weber's acid), in sulfuric acid.

Nitroxan. It is stated to be a compound of barium metaplumabate and barium manganate; the ammonia is oxidized to nitric acid, which is retained as barium nitrate. A catalyst used for the conversion of ammonia into nitric acid.

nitroxoline. (nibiol) 8-Hydroxy-5-nitroquinoline.

Nitto Nitoflon. Fluorocarbon plastic products, tapes, films, tubes and moldings; used to modify nonadhesive fluoroplastic surfaces into bondable surfaces; used for electric applications and electronics, food production, paper and pulp, etc. [Nitto Electric Industrial Co Ltd] *

Nitto SPV. Surface protection adhesive films and sheets; protects the surface of stainless steel, aluminum, decorated laminates, etc., from damage in transportation, storage and fabrication; used for architectural structures, rolling stock, household articles. [Nitto Electric Industrial Co Ltd] *

nivan. See oil white.

Nivaquine. Chloroquin sulfate; an antimalarial. [Rhone-Poulenc Rorer Ltd]

Nivar. A proprietary trade name for Invar (qv). §

Nivebaxin. A proprietary preparation of neomycin sulfate, bacitracin zinc, polymyxin B sulfate, calcium phosphate, glycine and starch. [The Boots Co plc]

Nivembine. Chloroquine/di-iodohydroxyquinoline tablets. [Rhone-Poulenc Rorer Ltd]

Nivemycin. CAS 1405-10-3; A proprietary preparation containing neomycin sulfate; an antibiotic. [The Boots Co plc]

Nix Creme Rinse. CAS 52645-53-1; A proprietary formulation of permethrin; for treatment of infestation with pediculus humanus variety capitis (the head louse) and its nits (eggs). [The Wellcome Foundation Ltd]

Nix Dermal Cream. CAS 52645-53-1; A proprietary formulation of permethrin; for treatment of infestation with Sarcoptes scabiei. [The Wellcome Foundation Ltd]

Nix Dermal Cream. CAS 52645-53-1; A proprietary formulation of permethrin; for treatment of infestation with Sarcoptes scabiei. [The Wellcome Foundation Ltd] *

Nixenoid. See Viscoloid. §

Nixon C/A. A proprietary brand of cellulose acetate. §

Nixon C/N. A proprietary brand of cellulose nitrate. §

Nixon E/C. A proprietary brand of ethyl cellulose. §

Nixonite. A proprietary trade name for a cellulose acetate plastic. §

Nixonoid. A proprietary trade name for a cellulose nitrate plastic. §

Nix Stix L-515. Silicone dispersion; mold release for rubber and urethane products. [Dwight Prods.

Nix Stix X-9021. Mold release spray for urethane and rubber products. [Dwight Prods.

Nizin. Sulfanilate zinc; antibacterial. [Broemmel Pharmaceuticals] *

Nizoral. CAS 65277-42-1; A proprietary preparation containing ketoconazole; antifungal. [Janssen Pharmaceutica Inc; Janssen Pharmaceutical Ltd]

Njamplung oil. See Laurel nut oil.

Njatuo tallow. A fat obtained from Palaquium oblongifolium.

Njave butter. (Nari oil, Noumgou oil, Adjab fat). A fat obtained from the seeds of Mimusops njave or djave, also from Bassia toxisperma and Bassia djave. The nuts are known as Abeku, Bako, or Mahogany nuts.

NL-10GF. Nylon 6/12 resin, 10% glass fiber-reinforced; outstanding mech. properties, chem. resist., dimensional stability, low moisture absorp. [Compounding Tech.

NLA-10. CAS 84-74-2; Dibutylphthalate; a vinyl plasticizer. [National Lead Co] *

NLA-20. Di-2-ethylhexylphthalate; a vinyl plasticizer. [National Lead Co] *

NLA-30. CAS 68515-49-1; Diisodecylphthalate; a vinyl plasticizer. [National Lead Co] *

NLA-40. Didecylphthalate. [National Lead Co] *

N-Labstix. A proprietary test-strip used to detect pH, protein, glucose, ketones, blood and nitrite in urine. [B C Ames] *

NL Chem. Industrial chemicals, specialty chemicals, namely organic and inorganic chemicals generally including polymers and argochemicals. [Rheox Inc]

NLF-32. A mixed adipate; a vinyl plasticizer. [National Lead Co] *

NLF-33. A modified adipate; a vinyl plasticizer. [National Lead Co] *

NM. CAS 75-52-5; Nitromethane; intermediate, stabilizer for halogenated solvs., as fuels, explosives, and solvs. for coatings or industrial processes. [Angus]

NM-55®. CAS 75-52-5; Blend of nitromethane; fuel for automotive racing and model engines; solvent. [Angus]

NM-AMD. N-Methylolacrylamide. [Cyanamid BV] *

NMP. CAS 76-39-1; 2-Nitro-2-methyl-1-propanol; chemical and pharmaceutical intermediate, in tire cord adhesives, as formaldehyde release agents, deodorants, antimicrobials. [Angus]

NMP. CAS 872-50-4; 2-Methyl-2-pyrrolidone; used as coatings solv., in stripping and cleaning of paints and varnishes, industrial cleaning, mold cleaning, petrochem. processing, agric. solv., polymer solv. [Arco Chemical Co]

NMP. See N-methyl-2-pyrrolidone.

NMP Conc. CAS 76-39-1; 2-Nitro-2-methyl-1-propanol; chemical intermediate, formaldehyde donor, textile reactant; reduces formaldehyde on finished cloth. [Angus]

NN-10GF. CAS 32131-17-2; Nylon 6/6, 10% glass fiber-reinforced; tough, strong, abrasion-resist. resins resist. to most industrial chem. and solvs. [Compounding Tech.

NN-20CF. CAS 32131-17-2; Nylon 6/6 resin, 20% carbon fiber-reinforced; used for applications requiring ultra high strength, fatigue endurance, thermal conductivity, wear resist. and/or elec. conductivity; suitable for aerospace industry applications. [Compounding Tech.

No. 1 White. CAS 471-34-1, 1317-65-3; Calcium carbonate; general-purpose pigment for fineness is of secondary importance. [ECC]

No. 3 White. CAS 471-34-1; Calcium carbonate; wh. general-purpose filler for low-cost paints, rubber goods, mild abrasive compounds, and putties. [ECC]

---

‡ Trade name and manufacturer not verified   § Trade name without identified manufacturer

Noatac. Tackifier resins. [Georgia-Pacific] *

Nobble. Aluminum sulfate + copper sulfate + sodium tetraborate; used for slug and snail control in field and glasshouse crops. [Fieldspray]

Nobecutane. A proprietary preparation of acrylic resin dissolved in acetic esters used in aerosol form as a plastic wound dressing. [Astra Chemicals Ltd] ‡

Nobel Ardeer powder. A dynamite containing 33% nitroglycerin, 49% magnesium sulfate, 13% kieselguhr, and 5% potassium nitrate.

Nobel Polarite. It is a mixture of potassium perchlorate and ammonium nitrate, with trinitro-toluene, a little starch, and wood meal; an explosive.

Nobese. CAS 154-41-6; Phenylpropanolamine hydrochloride; adrenergic. [O'Neal, Jones & Feldman Pharmaceuticals] *

Noblen. Polypropylene. [Montedison UK Ltd] *

Nobrium. CAS 2898-12-6; A proprietary preparation of medazepam; a tranquilizer. [Roche Products Ltd] *

Noctec. CAS 302-17-0; Chloral hydrate; a sedative. [Bristol-Myers Squibb Pharmaceuticals Ltd]

Noctec Capsules. CAS 302-17-0; Chloral hydrate in soft gelatin capsule. [Bristol-Myers Squibb Pharmaceuticals Ltd]

No-Del. A proprietary preparation of allyl isothiocyanate, ethyl nicotinate, methyl salicylate, eugenol, oil of turpentine, and cholesterol in a hydrophillic base. [Rybar Laboratories Ltd] *

Nodulant. Magnesium/iron tablets for production of S.G. cast irons. [Foseco (F.S.) Ltd]

Nofome 2510. Nonsilicone water-based; textile defoamer for atmospheric and pressure applications. [Sybron]

Nogos. Organophosphorus insecticde. [Ciba plc] *

Noheet metal. (Tempered lead). An alloy of 98.4% lead, 1.4% sodium, 0.11% antimony, and 0.08% tin.

Noil. An alloy of 80% copper and 20% tin.

Noiret's aluminum solder. An alloy of 80% zinc and 20% tin.

Nokol. Coal dust replacement additives for iron foundry green sands. [Foseco (F.S.) Ltd] †

Nolibond® 1. Polyamide-imide resin; high-temp. adhesive for aeronautic, aerospace, electrotech. and mech. industries to produce metal or composite assemblies that require high tensile and shear strength. [Rhone-Poulenc] *

Nolicoat FE 71008. Polyamide-imide resin; used as protective varnish or formulated to give pigmented coatings; provides self-lubricating, high heat resist., and decorative properties. [Shell] *

Noltam. CAS 54965-24-1; Tamoxifen citrate; for treatment of breast cancer and anovulatory infertility. [Lederle Laboratories] *

Noludar. A proprietary preparation of methyprylone; a hypnotic. [Roche Products Ltd] *

Nolvadex. CAS 10540-29-1; A proprietary preparation of tamoxifen used in the treatment of breast cancer. [ICI Pharma]

no-mag. A nonmagnetic alloy of 77% iron, 12% nickel, 6% manganese, 3% carbon, and 2% silicon. It has a high specific resistance, is close-grained, and of good mechanical properties.

No-max. A proprietary trade name for a high speed molybdenum-tungsten steel for cutting tools. §

Nomaze. Aerosol nasal decongestant. [Fisons plc, Pharmaceuticals Div] *

Nomel®. Zirconium, magnesium, rare-earth and other chemicals, powders and alloys; for use in industry, including metalworking. [Magnesium Elektron Ltd]

Nomelle®. Fibers. [DuPont UK]

Nomelt®. High-temp. greases with excellent water resist., shear stability, minimal oil separation, outstanding resist. to shrinkage; suitable for high-temp. applications such as coke oven door hinges, roll-out table bearings in steel mills. [Witco/Allied-Kelite]

Nomex®. Nylon fiber specially fabricated to withstand exposure to 500 F; does not melt or drip. [DuPont UK]

Nonad Tulle. Paraffin gauze dressing. [Allen & Hanburys Ltd] *

Nonaid. Nonionic surfactant. [Rhone-Poulenc UK]

Nonal 206. Nonylphenol ethoxylate; detergent, penetrant, emulsifier, scouring agent, wetting agent, dispersant; for textiles. [Toho Chem. Industry]

Nonanol. Plasticizer, containing 3:5:5-trimethyl hexanol. [ICI Chem. & Polymers Ltd] †

Nonanol N. Mixture of isomeric primary nonyl alcohols; starting material for production of plasticizers for PVC and vinyl chloride copolymers; low volatile leveling agent for baking finishes; antifoam. [BASF AG]

Nonasol 3922. Blend of sulfates, sulfonates, and amide; detergent base for car wash products, etc. [Hart Chem. Ltd.

Nonasol N4AS. CAS 67762-19-0; Ammonium laureth sulfate; detergent base for dishwash, hard surface cleaners, shampoos; wetting agent; biodeg. [Hart Chem. Ltd.

Nonasol N4SS. CAS 9004-82-4; Sodium laureth sulfate; detergent base for high foaming liq. detergents, dishwash, hard surface cleaners, shampoos; wetting agent; biodeg. [Hart Chem. Ltd.

noncarbinol. See n-decyl alcohol.

Noncorrodite. A proprietary trade name for a chromium steel. §

Nonex. Surfactants. [BP Chemicals Ltd]

Nonex C5E. CAS 61791-29-5; PEG 500 cocoate; lubricant for textile applications. [Hart Chem. Ltd.

Nonex DL-2. CAS 9005-02-1; PEG 200 dilaurate; coemulsifier and lubricant in industrial and textile oils. [Hart Chem. Ltd.

Nonex DO-4. CAS 9005-07-6; PEG 400 dioleate; emulsifier, solubilizer for oils, fats, solvs. [Hart Chem. Ltd.

Nonex O4E. CAS 9004-96-0; PEG 400 oleate; emulsifier for mineral and animal oils. [Hart Chem. Ltd.

Nonex S3E. CAS 9004-99-3; PEG 300 stearate; lubricant and softener for textiles. [Hart Chem. Ltd.

Nonflammable Decobest DA. CAS 131-17-9; Diallyl phthalate; for decorative laminates. [Sumitomo Bakelite Co Ltd] *

nongo. A gum resembling tragacanth, obtained from *Albizzia brownei*, of Uganda.

non-gran metal. A bronze containing 87% copper, 11% tin, and 2% zinc.

Nonidet. Ethylene oxide condensate. [Shell UK] *

Non Ionic Emulsifier T-9. Proprietary; low foam emulsifier for oils and emulsion systems; 100% biodeg. [Werner G. Smith]

Nonipol 20. CAS 9016-45-9; POE nonyl phenyl ether; penetrant, wetting agent, spreader-sticker; base material for detergents; emulsifier for agric. pesticides and emulsion polymerization, org. solv., machine oils, liq.

---

* Trade name not verified as available   † Trade name verified as obsolete

paraffins. [Sanyo Chem. Industries]

No-Nox. Motor fuel. [Chevron] *

Nonox. Group of antioxidants. [ICI Chem. & Polymers Ltd]

Nonox ZA. 4-Isopropylamine diphenylamine; a proprietary antioxidant. [ICI Chem. & Polymers Ltd] †

non-pareil metal. An alloy of 78% lead, 17% antimony, and 5% tin.

Nonsepara®. Waterproof greases for high-moisture applications where temps. do not exceed 77 C; used in spring loaded reservoirs, grease cups, guns, pressurized automatic lubrication systems. [Witco/Allied-Kelite]

Nontoxol. Froth flotation reagents containing hydrocarbons. [Coal Products Ltd]

nonyl phenol. A mixture of isomeric monoalkyl phenols; $C_9H_{19}C_6H_4OH$; CAS 25154-52-3; nonionic surfactant (nonbiodegradable), lube oil additives, stabilizers, petroleum demulsifiers, fungicides, antioxidants for plastics and rubber. [Allchem Industries; Ashland; Berol Nobel AB; GE Specialty; Hüls AG; Mitsui Toatsu Chem.; Texaco]

Nopalcol. Nonionic surfactants. [Henkel Chemicals Ltd] *

Nopalcol 1-L. CAS 9004-81-3; PEG-2 laurate; general purpose emulsifier, plasticizer, lubricant, wetting agent, dispersant, binding and thickening agent for emulsion polymerization, dry cleaning, leather, mineral oil emulsions, paper industry, wall-tile mastics, solv. emulsions. [Henkel/Functional Prods.]

Nopalcol 1-TW. CAS 68153-64-0; PEG-2 tallowate; emulsifier, plasticizer, lubricant, wetting agent, defoamer, binding and thickening agent, used in cosmetics, dry cleaning, leather, textile industries. [Henkel/Functional Prods.]

Nopalcol 2-DL. CAS 9005-02-1; PEG-4 dilaurate; general purpose emulsifier, plasticizer, lubricant, wetting agent, dispersant, binding and thickening agent for emulsion polymerization, dry cleaning, leather, mineral oil emulsions, paper industry, wall-tile mastics, solv. emulsions. [Henkel/Functional Prods.]

Nopalcol 4-C. CAS 61791-29-5; PEG-8 cocoate; general purpose emulsifier, plasticizer, lubricant, wetting agent, dispersant, binding and thickening agent for emulsion polymerization, dry cleaning, leather, mineral oil emulsions, paper industry, wall-tile mastics, solv. emulsions. [Henkel/Functional Prods.]

Nopalcol 4-DTW. PEG-8 ditallowate; emulsifier. [Henkel/Functional Prods.]

Nopalcol 4-L. CAS 9004-81-3; PEG-8 laurate; general purpose emulsifier, plasticizer, lubricant, wetting agent, dispersant, binding and thickening agent for emulsion polymerization, dry cleaning, leather, mineral oil emulsions, paper industry, wall-tile mastics, solv. emulsions. [Henkel/Functional Prods.]

Nopalcol 4-O. CAS 9004-96-0; PEG-8 oleate; emulsifier; dispersant for leather pigments; paper coating defoamer, plasticizer and leveling agent. [Henkel/Functional Prods.]

Nopalcol 4-S. CAS 9004-99-3; PEG-8 stearate; general purpose emulsifier, plasticizer, lubricant, wetting agent, dispersant, binding and thickening agent for emulsion polymerization, dry cleaning, leather, mineral oil emulsions, paper industry, wall-tile mastics, solv. emulsions. [Henkel/Functional Prods.]

Nopalcol 6-DO. CAS 9005-07-6; PEG-12 dioleate; general purpose emulsifier, plasticizer, lubricant, wetting agent,

dispersant, binding and thickening agent for emulsion polymerization, dry cleaning, leather, mineral oil emulsions, paper industry, wall-tile mastics, solv. emulsions. [Henkel/Functional Prods.]

Nopalcol 6-R. CAS 9004-97-1; PEG-12 ricinoleate; general purpose emulsifier, plasticizer, lubricant, wetting agent, dispersant, binding and thickening agent for emulsion polymerization, dry cleaning, leather, mineral oil emulsions, paper industry, wall-tile mastics, solv. emulsions. [Henkel/Functional Prods.]

Nopalcol 10-COH. CAS 61788-85-0; PEG-20 hydrogenated castor oil; general purpose emulsifier, plasticizer, lubricant, wetting agent, dispersant, binding and thickening agent for emulsion polymerization, dry cleaning, leather, mineral oil emulsions, paper industry, wall-tile mastics, solv. emulsions. [Henkel/Functional Prods.]

Nopalcol 30-TWH. PEG-60 hydrogenated tallowate; general purpose emulsifier, plasticizer, lubricant, wetting agent, dispersant, binding and thickening agent for emulsion polymerization, dry cleaning, leather, mineral oil emulsions, paper industry, wall-tile mastics, solv. emulsions. [Henkel/Functional Prods.]

Nopalcol 200. CAS 25322-68-3; PEG 200; general purpose emulsifier, plasticizer, lubricant, wetting agent, dispersant, binding and thickening agent for emulsion polymerization, dry cleaning, leather, mineral oil emulsions, paper industry, wall-tile mastics, solv. emulsions. [Henkel/Functional Prods.]

Nopalcol 400. CAS 25322-68-3; PEG 400; general purpose emulsifier, plasticizer, lubricant, wetting agent, dispersant, binding and thickening agent for emulsion polymerization, dry cleaning, leather, mineral oil emulsions, paper industry, wall-tile mastics, solv. emulsions. [Henkel/Functional Prods.]

Nopalcol 600. CAS 25322-68-3; PEG 600; general purpose emulsifier, plasticizer, lubricant, wetting agent, dispersant, binding and thickening agent for emulsion polymerization, dry cleaning, leather, mineral oil emulsions, paper industry, wall-tile mastics, solv. emulsions. [Henkel/Functional Prods.]

Nopco® 1179. Fatty amido condensate; wool and worsted fulling and scouring; general purpose detergent. [Henkel/Functional Prods.]

Nopco® 2031. Sulfonated fatty product, sodium neutralized; emulsifier for emulsion polymerization. [Henkel/Functional Prods.]; Henkel-Nopco]

Nopco® 2272-R. Highly sulfated fatty ester; wetting agent in paper towels, latex impregnation and dipping, in metal treating and textile dyeing. [Henkel/Functional Prods.]; Henkel-Nopco]

Nopco® NXZ. Defoamer for synthetic latex emulsions, paint and adhesives from SBR, PVAc, acrylic, water-sol. resins. [Henkel/Cospha]

Nopco® PD#1-D. Defoamer for adhesives, paints, joint compounds, plaster; off-wh. powd.; water-wettable. [Henkel] *

Nopcocastor. CAS 8002-33-3; EINECS 232-306-7; Sulfated castor oil; emulsifier; superfatting agent for cosmetics. [Henkel/Functional Prods.]

Nopcochex RA. Fatty amide condensate; oil-sol. corrosion preventer. [Henkel/Cospha]

Nopcocide® N-40-D. CAS 1897-45-6; Chlorothalonil; mildewcide, fungicide for trade sales paints. [Henkel] *

Nopcocide® N-96. CAS 1897-45-6; Tetrachloroiso-

---

‡ Trade name and manufacturer not verified   § Trade name without identified manufacturer

phthalonitrile; broad spectrum microbicide for control of fungi in latex exterior and interior emulsion paints, solv.-based paints. [Henkel] *

Nopcocide® N-96-S. CAS 1897-45-6; 2,4,5,6-Tetrachloro-isophthalonitrile; antimicrobial for marine antifouling coatings. [Henkel] *

Nopco® Colorsperse 188-A. Ethoxylated fatty acid; pigment wetting and dispersing. [Henkel-Nopco]

Nopcofloc. Water treatment chemical. [Henkel Chemicals Ltd] *

Nopco® Foamaster. Antifoams/defoamers. [Henkel Chemicals Ltd] *

Nopcogen. Speciality surfactant. [Henkel Chemicals Ltd] *

Nopcogen 14-S. CAS 93-82-3; EINECS 202-280-1; Stearamide DEA; textile softener. [Henkel-Nopco]

Nopcogen 16-L. CAS 2016-56-0; Lauric polyamine condensate; emulsifier and textile finishing agent. [Henkel-Nopco]

Nopcogen 22-O. Oleyl imidazoline; emulsifier for mineral oil, kerosene, wetting agent, corrosion inhibitor, used in textile, asphalt, paper, agric. industries, car wax formulations, acid detergents. [Henkel/Functional Prods.]

Nopcolan SHR3. Thiosulfate, organic; shrink resister for wool. [Henkel]

Nopcolene. Leather processing aid. [Henkel Chemicals Ltd]*

Nopcolube. Textile processing aid. [Henkel Chemicals Ltd]*

Nopcone. Textile processing aid. [Henkel Chemicals Ltd] *

Nopcosant. CAS 9084-06-4; Sodium salt of condensed naphthalene sulfonate; dispersant for NR, SR latexes, pigments; used for paints, cements, sealants. [Henkel]*

Nopcosant L. Sulfated naphthalene; NR, SR latex dispersant for use in aq. systems, paints, paper coatings, textiles, and leather. [Henkel] *

Nopcosize. Speciality sizes. [Henkel Chemicals Ltd] *

Nopcosperse 28-B. Textile dispersant for pigments providing high grinding efficiency and superior color yield. [Henkel/Textiles]

Nopcostat 237. Fluorochemical-compatible polypropylene finish for carpet fiber. [Henkel/Textiles]

Nopcosulf CA-60, -70. CAS 8002-33-3; EINECS 232-306-7; Sulfated castor oil; softener in finishing starches, gums; plasticizer for starches; furniture base polish. [Henkel-Nopco]

Nopcosulf TA-30. Sulfated tallow; softener for cotton goods. [Henkel/Functional Prods.]

Nopcosulph. Anionic surfactant. [Henkel Chemicals Ltd] *

Nopcosurf CA. CAS 8002-33-3; EINECS 232-306-7; Sulfated castor oil; softener in finishing starches, gums; plasticizer for starches; furniture polish base. [Henkel/Functional Prods.]

Nopcotan. Leather processing aid. [Henkel Chemicals Ltd]*

Nopcote. Paper processing aid. [Henkel Chemicals Ltd] *

Nopcote C-104. CAS 1592-23-0; EINECS 216-472-8; Calcium stearate disp.; lubricant for paper coatings. [Henkel] *

Nopcotex. Textile softener lubricants. [Henkel Chemicals Ltd] *

Nopcowax. Synthetic waxes. [Henkel Chemicals Ltd] *

Nopcowax 22-DS. CAS 110-30-5; EINECS 203-755-6; Ethylene bisstearamide; high melting synthetic wax; binder, thickener for latex formulation, coatings, adhesives; used in powd. metallurgy as internal lubricant. [Henkel] *

Nopco Worsted Oil 12. Textile processing aid. [Henkel Chemicals Ltd] *

nopinene. See -pinene.

Nor-Q D. CAS 68-22-4; Norethindrone; progestin. [Syntex Laboratories Inc]

Nora. Rubber flooring and soling. [Carl Freudenberg] *

Noradran. A proprietary preparation of ephedrine hydrochloride, theophylline and papaverine hydrochloride; a bronchial antispasmodic. [Norma] *

noradrenaline. CAS 51-41-2; (-)-2-Amino-1-(3,4-dihydroxyphenyl)ethanol; levophed; $V_8H_{11}NO_3$.

Noraflor. Spun-bonded flooring for carpets. [Carl Freudenberg] *

Noralastic. Acoustic flooring. [Carl Freudenberg] *

Noralen. Elastic roof cover material. [Carl Freudenberg] *

Noram. Fatty primary, secondary and tertiary monoamines derived from coco, tallow and oleyl alkyl chains; amine salts are cationic emulsifiers, also intermediate in chemical synthesis, mineral flotation, anticaking of fertilizers, corrosion inhibitors. [Atochem UK/Ceca]

Noramac. Acetate salts of fatty mono amines; bactericides, flotation, flushing agents for pigments, polishes, hydrophobic treatment of textiles, wood, metals, etc. [Atochem UK/Ceca]

Norament. Special designed floors. [Carl Freudenberg] *

Noramid. Flooring material. [Carl Freudenberg] *

Noramium. Quaternary ammonium compounds derived from fatty amines with coco, tallow or oleyl alkyl chains; fabric softeners, organophilic clays, biocides, antistatics. [Atochem UK/Ceca]

Noramox. Ethoxylated fatty monoamines; emulsifiers, corrosion inhibitors, dispersing and wetting agents, textile auxiliaries, fuel additives. [Atochem UK/Ceca]

Noraplan. Flooring material. [Carl Freudenberg] *

Noratex. A proprietary preparation of talc, kaolin, zinc oxide, cod liver oil and wool fat used as a protective skin cream. [H N Norton & Co Ltd] ‡

Norbide. A proprietary trade name for an amorphous boron carbide; an abrasive. §

Norbo. A proprietary trade name for phenolic resin. §

Norcast 142 Systems. CAS 25928-94-3; Unfilled 100% solids epoxy resin; casting and impregnating materials [R.H. Carlson] *

Norcast 154FR. CAS 25928-94-3; Epoxy resin; flame retardant casting resin [R.H. Carlson] *

Norcast 1460-1. CAS 25928-94-3; One-component epoxy resin system; dipping compound [R.H. Carlson] *

Norcast 3220G-1. CAS 25928-94-3; One-component epoxy system; dip and brush coating [R.H. Carlson] *

Norcast 3258. CAS 25928-94-3; Epoxy; potting/encapsulating system [R.H. Carlson] *

Norcast 3705. CAS 25928-94-3; Two-part epoxy system; variable flexibility system for use as casting resin or high peel strength adhesive; high moisture resistance and ability to withstand physical and thermal shock; mix ratio range from 40-200 parts B/100 parts A; visc. 10,000 cps (mixed). [R.H. Carlson] *

Norcast 4914-1. CAS 25928-94-3; Epoxy, silver-filled; adhesive [R.H. Carlson] *

Norcure 131. Aliphatic amine; curing agent (used with DER 741) for compatibility and long pot life at room temperature; fast cure; mix ratio: 15 phr with DER 741. [R.H. Carlson] *

Norcure 3298. Epoxy hardener; used as primary hardener

---

* Trade name not verified as available   † Trade name verified as obsolete

for Resin 220-0320 for soft, flexible epoxy system; easily repaired when silica filled; mix ratio: 2/1 resin/hardener. [R.H. Carlson] *

Norcuron. CAS 50700-72-6; Vecuronium bromide; blocking agent (neuromuscular). [Organon Inc]

Nordel®. Ethylene-propylene synthetic rubber. [DuPont UK]

Nordel® 2744. A trademark for a fast curing EPDM hydrocarbon rubber possessing high green strength. [DuPont UK] *

Nordhausen acid. See fuming sulfuric acid.

Nordot 101F. Urethane adhesive; solv.-free/water-free adhesive for installing flooring and sport surfaces; for wood, vinyl, rubber, concrete and asphalt. [Syn. Surfaces]

Nordox. CAS 1317-39-1; Cuprous oxide, paint grade, red, micro milled; active ingredient in antifouling paints. [Nordox Industrier AS] *

Noreplast. A proprietary trade name for laminated thermosetting, plastic thermosetting and thermoplastic synthetic resins. §

Norepol. A proprietary trade name for vulcanizable vegetable polymers. §

Noreseal. A proprietary trade name for a cork substitute made from low cost domestic raw materials; stated to be equal in strength to cork. §

Norfer. CAS 141-01-5; A proprietary preparation of ferrous fumarate; used in the treatment of anemia. [H N Norton & Co Ltd] ‡

Norflex™ Injectable. CAS 4682-36-4; Each 2 ml ampule contains orphenadrine citrate 30 mg/ml; muscle relaxant; antihistaminic. [3M Pharmaceuticals]

Norflex™ Tablets. CAS 4682-36-4; Each tablet contains orphenadrine citrate 100 mg in a slow-release base; muscle relaxant. [3M Pharmaceuticals]

Norfox® 1101. CAS 61789-30-8; Potassium cocoate; flash foamer, emulsifier for shampoo bases, liq. hand soaps; lubricant for conveyors; coupling agent for heavily built liq. alkali systems such as steam cleaners and whitewall tire cleaners. [Norman, Fox]

Norfox® Agent 2A-2S. Air entraining agent, wetting agent for production of mortar, stucco, and plastic cement. [Norman, Fox]

Norfox® ALPHA XL. CAS 68439-57-6; EINECS 270-407-8; Sodium C14-16 alpha olefin sulfonate; base for shampoos, hand soaps, bath products, home and janitorial cleaners, dishwash, and light duty liqs. [Norman, Fox]

Norfox® Coco Betaine. Cocamidopropyl betaine; mild substantive surfactant, flash foamer; base for mild shampoos; electrolyte tolerance over wide pH range. [Norman, Fox]

Norfox® DC. CAS 61791-31-9; Cocamide DEA and diethanolamine; intermediate for detergent manufacturing, liq. dishwash, cosmetics; suds stabilizer and dedusting agent for dry products. [Norman, Fox]

Norfox® F-221. CAS 93-83-4; EINECS 202-281-7; Oleamide DEA; invert emulsifier with hydraulic properties; wetting agent, penetrant; salt tolerant when dissolved in oil phase over a wide temp. range; for herbicides, drilling fluids for petrol. industry, penetrating oils, specialty lubricants. [Norman, Fox]

Norfox® IM-38. CAS 62449-33-6; N-Oleyl imidazolinium hydrochloride; high-foaming wetting agent and emulsifier; base for automotive rinse-wax formulations, corrosion inhibition compounds. [Norman, Fox]

Norfox® SLES-60. CAS 9004-82-4; Sodium laureth sulfate,

14% denatured alcohol; base for shampoos and light duty liq. formulators; flash foam enhancer for automotive, household, personal care, and industrial cleaners; in manufacturing of gypsum wallboard, gas drilling of deep wells. [Norman, Fox]

Norfox® SLS. CAS 151-21-3; Sodium lauryl sulfate; biodeg. foamer and wetting agent for household, industrial and personal care products, shampoos, hand and body soaps, fabric care products. [Norman, Fox]

Norfox® Sorbo T-60. CAS 9005-67-8; PEG-20 sorbitan stearate; emulsifier for frozen desserts, salad dressings, cake mixes, icings; dough conditioner. [Norman, Fox]

Norfox® T-60. TEA dodecylbenzene sulfonate; detergent, wetting agent and foamer for agric. and industrial/household use, light duty detergents, hard surface cleaners, shampoos, wool and fine fabric washing. [Norman, Fox]

Norfox® TLS. CAS 139-96-8; EINECS 205-388-7; TEA lauryl sulfate; biodeg. mild ingred. for shampoo, light duty liqs., fine fabric detergents. [Norman, Fox]

Norfox® Unimulse OW. Polymer/surfactant; emulsifier for oily liqs. into oil and water emulsions. [Norman, Fox]

Norfox® X. Cocamide DEA; modified; used with acidic builders; wash-wax formulations, glass and appliance cleaners. [Norman, Fox]

Norfroth. Mixed propylene glycol isobutyl ethers and surfactants; ore flotation chemicals. [Chemoxy International Ltd]

Norgesic™ Forte Tablets. Orphenadrine citrate, aspirin, caffeine; analgesic. [3M Pharmaceuticals.]

Norgesic™ Tablets. Orphenadrine citrate, aspirin, caffeine; analgesic. [3M Pharmaceuticals]

Norgeston. 30 mcg Levonorgestrel. [Schering Health Care Ltd]

Norgine. The sodium-ammonium salt of laminaric acid from seaweed, Laminaria digitata and Saccharinus digitatus; used in the treatment of textiles.

Norgotin. A proprietary preparation of ephedrine, amethocaine, chlorohexidine acetate, and propylene glycol; used as eardrops. [Norgine Ltd] †

Norinyl-1, Norinyl-2. A proprietary preparation of norethisterone and mestranol; used as an oral contraceptive. [Syntex Laboratories Inc]

Norisodrine Aerotrol. Isoproterenol hydrochloride; adrenergic. [Abbott Laboratories]

Norisodrine Sulfate. CAS 6700-39-6; Isoproterenol sulfate; an adrenergic. [Abbott Laboratories]

Noristerat. Norethisterone oenanthate in oily solution for intramuscular administration. [Schering Health Care Ltd]

Norit, C, PK, R, RO. CAS 7440-44-0; Activated carbon; used for purification and decolorization of food products, pharmaceutical products, water (potable, process and waste), recovery of gold (CIP and CIL) and air purification (removal of H_2S, mercury, SO_2 etc) [R W Greeff & Co Inc]

Norithene. CAS 7440-44-0; Air conditioning carbons. [Norit UK Ltd] *

Norlestrin. A proprietary preparation of norethisterone acetate and ethinyl-estradiol; for the treatment of menstrual disorders. [Parke-Davis]

Norlestrin 21. A proprietary preparation of norethisterone acetate and ethinyl estradiol; used as an oral contraceptive. [Parke-Davis] *

Norlig. Calcium and sodium lignosulfonates; specialty

---

‡ Trade name and manufacturer not verified   § Trade name without identified manufacturer

chemicals for use in a variety of binding, dispersing, complexing, stabilizing, or copolymerizing applications. [Borregaard LignoTech]

Norlig 11 DA. CAS 8061-52-7; Calcium lignosulfonate; binder for pesticide/herbicide carrier substitutes; aids in subsequent dispersion. [Borregaard LignoTech]

Norlig 415. Modified sodium-calcium lignosulfonate; dispersant and water reducer for gypsum wallboard manufacturing; stucco dispersant giving low retardation. [Borregaard LignoTech]

Norlig A. CAS 8061-52-7; Calcium lignosulfonate; dispersant, binder, soil and dust stabilizer, carbon black, briquetting and pelletizing of coal and charcoal, ceramic additive. [Borregaard LignoTech]

Norlutate. CAS 51-98-9; Norethindrone acetate; progestin. [Parke-Davis] *

Norlutin. CAS 68-22-4; Norethindrone; progestin. [Parke-Davis]

Norlutin A. A proprietary preparation of norethisterone acetate; used in the control of menstrual irregularity. [Parke-Davis] *

Normacol. A proprietary preparation containing sterculia; a laxative. [Norgine Ltd]

normal powder. A gelatinized gun-cotton powder.

Normasol®. CAS 7647-14-5; 0.9% Sodium chloride solution; used for irrigation of eyes, wounds, and burns. [Seton Healthcare Group plc]

Normax. A proprietary preparation containing dioctyl sodium sulfosuccinate and 1-8-dihydroxyanthraquinone; a laxative. [Horlicks] *

Nor-Mer 020. CAS 25928-94-3; Epoxy system; uv-curable adhesive [R.H. Carlson]

Normet. Metallized biaxially oriented polypropylene film for use in manufacture. [Quantum Chemical Corp]

Normison Capsules. CAS 846-50-4; Temazepam; benzodiazepine hypnotic. [Wyeth Laboratories]

Normodyne. CAS 32780-64-6; Labetalol hydrochloride; anti-adrenergic. [Schering Corp]

No-Roma. A proprietary preparation of paraformaldehyde, sodium carboxy-methyl cellulose, sodium hydroxide and methylene blue; a medical deodorant. [Salt] *

Norox® **MCP.** MEK peroxides and cumyl hydroperoxide in plasticizers; polymerization initiator for room temperature cure of unsat. polyester and vinyl ester resins; low peak exotherm, longer gel time. [Norac]

Noroxin. CAS 70458-96-7; Norfloxacin; urinary antiseptic. (U.S.) [Merck & Co Inc]

Norpace. CAS 22059-60-5; Disopyramide phosphate; cardiac depressant. [G D Searle & Co]

Norpar®. High purity normal paraffin solvent. [Exxon Int'l.

Norpar® **12.** CAS 64771-72-8; Normal paraffin; solvent. [Exxon] *

Norplex laminates. Woven fiberglass fabric impregnated with thermoset resin, with thin copper foil on one or two sides, pressed under high temperature and heat to form plastic laminated sheets; substrate material for printed circuit boards. [Norplex] *

Norpramin. CAS 58-28-6; Desipramine hydrochloride; an antidepressant. [Merrell Dow Pharmaceuticals Inc] *

Norpramine. CAS 113-52-0; A proprietary preparation of imipramine hydrochloride; an antidepressant. [H N Norton & Co Ltd] ‡

Norprop. Biaxially oriented polypropylene film for use in manufacture. [Quantum Chemical Corp]

Norsed. A proprietary preparation of cyclobarbitone and amylobarbitone; a hypnotic. [H N Norton & Co Ltd] ‡

Norsil 711. Methyl phenol-based silicone; rubber for use at -115 to 204 C; allows for greater flexibility in high temp. applications. [R.H. Carlson] *

Norsil 1000. Optical coupling compound; reduces internal reflections and refractures in optical equip.; optically clear gel coupling in fiber optics. [R.H. Carlson] *

Norsil RTV 811. CAS 63394-02-5; Silicone rubber; for broad range of mech. and elec./electronic applications; potting and encapsulation, cast-in-place gasket and molds. [R.H. Carlson] *

Norsil SG 131. Dimethyl silicone compound; semitransparent compound used as release agent at -40 to 400 F; lubricant for rubber, plastic and leather parts; insulates and repels water from elec. connections. [R.H. Carlson]*

Norsil SG 169. Dimethyl silicone compound, polyethylene-filled; release agent, lubricant for plastic and rubber surfaces; very low dielec. constant. [R.H. Carlson] *

Norsolene®. Coumarone and polyindene resins. [CdF Chimie] *

Norsomix®. A trademark for polyester compounds. §

Norsophen®. Phenolic resins and compounds; composites offering strength, durability, excellent fire performance, high heat resist.; used for aerospace, construction, mass transit, mining, and automotive applications. [Norold Composites]

Norsorex. Polynorbomene. [Sartomer]

Nortech. Pre-colored plastic resins, polyolefin foam concentrates, specialty filled plastic resins and polylefin resins. [Quantum Chemical Corp]

Northovan. Sodium-o-vanadate. §

Nortron. CAS 26225-79-6; Emulsifiable concentrate of 200 g ethofumesate per liter; used for weed control in field crops. [Schering Agrochemicals Ltd]

Nortron Leyclene. Bromoxynil + ethofumesate + ioxynil; herbicide mixture for new grass lays. [Schering Agrochemicals Ltd]

Nortuff RA 1700-MO. CAS 9003-07-0; PP resin; extrusion grade [Quantum/USI] *

Nortuff RA 7020-KO. CAS 9002-88-4; Polyethylene; extrusion and injection molding grade [Quantum/USI] *

Nortuff RC 1700-MO. CAS 9003-07-0; PP resin; injection molding grade [Quantum/USI] *

Noruben. CAS 555-37-3; Neburon; herbicide. [Agan Chemical Manufacturers Ltd]

Norunil. CAS 330-55-3; Linuron; herbicide. [Agan Chemical Manufacturers Ltd]

Norust. Specialty blends of fatty amines and derivatives; corrosion inhibitors, fuel oil additives. [Atochem UK/Ceca]

Norval. CAS 577-11-7; EINECS 209-406-4; A proprietary preparation o-dioctyl-sodium sulfosuccinate in gelatine; a fecal softener. [Horlicks] *

Norvas. See Norvasc. [Pfizer International]

Norvasc®. CAS 88150-42-9; Amlodipine; used for hypertension and angina. Also known as Amlocor (Norway), Amlogard (Costa Rica), Amlor (Benelux, France), Amlovar (Finland, Norway, Sweden), Ampec (Germany), Darcoxin (Greece), Isitn (Great Britain, Ireland), Norvas (Argentina, Germany, Mexico, Venezuela). [Pfizer International]

Norvinyl. Polyvinyl chloride homo and copolymers (S-PVC); used as a main product or as an extender for production

---

* Trade name not verified as available   † Trade name verified as obsolete

of plastic articles. [Norsk Hydro AS]

Norvinyl DX 550. CAS 9002-86-2; Crosslinkable plasticized PVC compound; for manufacturing high temp. resist. cables and wires. [Hydro Plast]

Norwegian saltpeter. See air saltpeter.

Noryl® . Modified polyphenylene oxide; used for plastic components for automotive, electrical, electronics, lighting, medical, packaging, audio etc. [GE Plastics Ltd]

Noryl® 731. CAS 25134-01-4; Modified PPO resin; thermoplastic general purpose injection molding grade resin, excellent dimensional stability; heat resistance to 265 F, good impact strength, low moisture resistance; used in computers, business equip., automotive, elec., electronics, construction, telecommunications, appliances, and other industries. [GE Plastics]

Noryl® 1402B. Engineered blow molding grade for demanding automotive exterior body parts, fluid handling equip., medical products, motor casings. [GE Plastics]

Noryl® BN25. CAS 25134-01-4; Modified PPO resin; engineered blow molding grade with high heat resist., impact strength, dimensional stability, moisture resist., elec. insulating properties; used for computers, appliances, business equip., automotive, fluid handling, telecommunications equip. [GE Plastics]

Noryl® EM5101. CAS 25134-01-4; Modified PPO resin; energy management grade for automotive interior applications; superior processability, improved flow chars. [GE Plastics]

Noryl® EN185. CAS 25134-01-4; PPO resin; extrusion grade; combines excellent impact and heat resistance, UL recognition, formability and processability; used for electronics housings, beverage cases, transportation components. [GE Plastics]

Noryl® FN150. CAS 25134-01-4; Modified PPO foam resin; engineering structural foam resin for business machine housings and their structural bases, weather-resistant elec. enclosures, and lightweight structural components for the transportation industry which resist the greenhouse effect. [GE Plastics]

Noryl® FN215X. CAS 25134-01-4; Modified PPO structural foam resin; high mechanical strength, heat- and impact-resistance; used in business machine housings and structural bases, weather-resistant elec. enclosures, and lightweight structural components for transportation. [GE Plastics]

Noryl® GTX810. CAS 25134-01-4; PPO resin; unreinforced engineering thermoplastic resin; offers excellent chem. resist., mech. performance, high heat resist., dimensional stability; for automotive body panels, exterior hardware, and other applications requiring high heat resist., paintability. [GE Plastics]

Noryl® HM3020. CAS 25134-01-4; Modified PPO resin; high modulus grade [GE Plastics]

Noryl® HS1000X. CAS 25134-01-4; Modified PPO resin; injection molding grade thermoplastic resin for high-strength, thin-wall applications. [GE Plastics]

Noryl® N190X. CAS 25134-01-4; Modified PPO resin, flame retardant; thermoplastic injection molding grade for business machines, small appliance housings, current carrying parts, and telecommunication internal componentry; highly stable under load at elevated temps.; UL listed. [GE Plastics]

Noryl® PC180X. CAS 25134-01-4; Modified PPO resin; injection molding grade; computer and business equip-

ment grade thermoplastic resin for portable computer applications. [GE Plastics]

Noryl® PN235. CAS 25134-01-4; Modified PPO resin; thermoplastic, platable, injection molding grade that offers improved dimensional stability; lighter weight, more economical than metals; peel strength, low creep, and low temp. impact resistance; in automobile grilles, wheel covers, and headlamp bezels, appliance range knobs, and decorative trim areas. [GE Plastics]

Noryl® PX0722. CAS 25134-01-4; Modified PPO resin; automotive grade thermoplastic resin with elec. insulating capabilities for instrument panels, speaker grilles, glove box doors, connectors, fuse blocks, instrument panel components, mirrors, exterior trim, grilles, and wheel covers; PX0722 for interior trim. [GE Plastics]

Noryl® SE1GFN2. CAS 25134-01-4; Modified PPO resin, 20% glass reinforced; thermoplastic injection molding grade, broad UL recognition, excellent mechanical and dimensional chars.; used in elec. applications, connectors, structural parts, television components, liq. handling, business machines. [GE Plastics]

Noryl® SE100. CAS 25134-01-4; Modified PPO resin, flame retardant; thermoplastic injection molding grade for applications requiring moderate heat deflection and excellent dimensional stability. [GE Plastics]

Nosifeed 40. Feed additive (growth promoter) for swine and poultry. [Mitsubishi Kasei] *

Nosiheptide. It is a peptide obtained from cultures of *Streptomyces actuosus* 40037 or the same substance obtained by any other means; a veterinary antibiotic.

Nosiheptide. feed additive (growth promoter) for swine and poultry. [Mitsubishi Kasei] *

No-Swab. Mold coating containing solid lubricants and suitable binder system; molds for glass containers in I.S. machines for extension of mold swabbing cycles. [E/M Corporation]

Notak. Products for nondestructive detection of surface flaws in iron and steel components. [Foseco (F.S.) Ltd]

Notrilen®. Antidepressant. [Bayer AG]

Nottingham white. See flake white.

Nouraid. Polyunsaturated fatty acids. [Akzo Chemie UK Ltd.]

Nourycryl. Range of methacrylate monomers for use in adhesive and coating bases. [Akzo Chemie UK Ltd]

Nourydrier. Range of metal-based drying agents. [Akzo Chemie UK Ltd]

Nourymix. Additive concentrate for polyolefins. [Akzo Chemie UK Ltd]

Nouryset® . Allyl-based monomers; for use in optical and ophthalmic resins. [Akzo Chemie UK Ltd]

Nouryset® 156. Styrene proprietary product; used in the manufacturing of high-quality optical materials, e.g., high minus lenses. [Akzo] *

Nouryset® 200 HV 250. Prepolymerized diethylene glycol bis (allyl carbonate); CAS 142-22-3; polymer CAS 25656-90-0; offers lower shrinkage on curing; used for optical applications. [Akzo] *

Nova 2001. Emulsion stripper. [British Nova Works Ltd]

Nova PC-1000BK. PC resin. [Nova Polymers]

Nova-T. Polyethylene IUCD made radio-opaque by barium sulfate, with silver-cored copper wire. [Schering Health Care Ltd]

Novablend® 501. CAS 9002-86-2; PVC compound; general-purpose rigid injection grade; provides an economical alternative in opaque injection molded parts; also for

---

‡ Trade name and manufacturer not verified   § Trade name without identified manufacturer

extruded profiles. [Novatec Plastics & Chem.] *

Novablend® 5555. CAS 9002-86-2; PVC compound, impact-modified; rigid extrusion blow molding compound providing excellent processing and thermal stability; for water bottle applications. [Novatec Plastics & Chem.]

Novabloo. Toilet bowl cleanser. [British Nova Works Ltd]

Novabold. Toilet bowl cleanser. [British Nova Works Ltd]

Novacarb. CAS 1333-86-4; Carbon black; black conc. and impact modifier for ABS and PVC. [Nova Polymers]

Novacare. Floor maintainer. [British Nova Works Ltd]

novacetoform. Aluminum acetate.

Novacleer. Window cleanser. [British Nova Works Ltd]

Novacorn. An emulsifiable concentrate containing 240 g bromoxynil and 160 g ioxynil per liter; a post-emergence contact herbicide for cereal crops. [Farmers Crop Chemicals Ltd]

Novacote. Intermediates for plastics, for use in wrapping and packaging. [Imperial Chemical Industries plc]

Novacrete. Floor sealer. [British Nova Works Ltd]

Novacross. Sanitizing fluid. [British Nova Works Ltd]

Novacryl. Acrylic sealer and primer. [British Nova Works Ltd]

novaculite. A quartz rock used as an abrasive.

Novacut. Traffic lane cleaner. [British Nova Works Ltd]

Novadelox. CAS 94-36-0; Benzoyl peroxide based; bleaching agent for baking flour. [Akzo Chemie UK Ltd]

Novadine. Iodophor sanitizer for brewery plant; for sanitizing fermenting vessels, lager tanks, maturation tanks, bright beer tanks, fillers and mains. [Harshaw Chemicals Ltd]*

Novafed. Used primarily in the treatment of coughs, colds, and upper respiratory conditions. [Dow UK] *

Novafil. Polybutester; surgical aid. [Davis & Geck] *

NovaFlo. Liquid rosin size; internal sizing agent to give water resist. to paper and paperboard. [Georgia-Pacific] *

Novafrost. Chewing gum remover. [British Nova Works Ltd]

Nova Furnipol. Furniture polish. [British Nova Works Ltd]

Novagem. Furniture polish. [British Nova Works Ltd]

Novagrip. Penetrating seal. [British Nova Works Ltd]

Nova Highgloss. Paste wax polish. [British Nova Works Ltd]

Novahistine. Used primarily in the treatment of coughs, colds and upper respiratory conditions. [Dow UK] *

novalak resins. Synthetic resins of the formaldehyde-phenol type.

Novalar. impact modifier for ABS, PVC, vinyl chloride copolymers, PC, PU, epoxies, PBT, acrylics; improves impact strength, ductility, low temp. properties, eliminates brittleness, upgrades scrap. [Nova Polymers] *

Novalast 5000, 9000. Thermoplastic elastomer; low-cost material with good compression set and heat aging properties. [Nova Polymers]

Novaldin. CAS 5907-38-0; Dipyrone; analgesic; antipyretic. [Sterling Drug Inc] *

Novalene. Impact modifier for homopolymer PP and HDPE; upgrades scrap PP homopolymer to copolymer; improves ductility and low temp. properties; eliminates brittleness. [Nova Polymers]

Novalift. Kitchen grease remover. [British Nova Works Ltd]

Nova Lime-Lite. Floor maintainer. [British Nova Works Ltd]

Novalite. Carpet shampoo. [British Nova Works Ltd]

Nova Long-Life. Floor dressing. [British Nova Works Ltd]

Novaloy 6521. A proprietary one component epoxy resin supplied in powder form for the coating of electronic components. [Rogers Corp]

Novaloy® 9000. ABS/PVC alloy; engineering alloy for injection molding, profile and sheet extrusion; provides excellent processing, high heat distort. temps., excellent retention of properties on aging; used for appliance housings, elec. boxes, communications, lawn and garden applications, recreational products. [Novatec Plastics & Chem.

Novamid® 1010C. CAS 25038-54-4; Nylon 6; general purpose injection molding grade suitable for fast cycling injection molding; low visc. [Mitsubishi Kasei] *

Novamid® 1020CA2. CAS 25038-54-4; Nylon 6; extrusion molding grade for co-extrusion film applications; suitable for both air and water-cooled blown film; med. visc. [Mitsubishi Kasei] *

Novamid® 2020A, 2420A. Nylon copolymer; extrusion molding grade offering excellent transparency, high flexibility; for monofilament applications; med. visc. [Mitsubishi Kasei] *

Novamyl. A maltogenic amylase produced by a strain of *Bacillus subtilis*; added to flour it modifies the starch in the breadmaking process so that retrogradation is less likely to occur and staling is retarded 2 days or more. [Novo Nordisk]

Novantrone. CAS 65271-80-9; Injection containing mitozantrone; antineoplastic agent. [Lederle Laboratories] *

Novanyl. A range of acid dyes; for dyeing of polyamide fibers. [Yorkshire Chemicals plc] *

Nova One. Liquid cleanser. [British Nova Works Ltd]

Nova One Plus. Bactericidal liquid cleanser. [British Nova Works Ltd]

Novapel. Leather finishes. [ICI Chem. & Polymers Ltd]

Novaphalte. Asphalt sealer. [British Nova Works Ltd]

Nova Pine Fluid. Disinfectant. [British Nova Works Ltd]

Novapint. Universal pigment dispersions. [Tennant-KVK Ltd]

Novaplaste. Leather dyes. [ICI Chem. & Polymers Ltd] †

Novapol. Liquid wax polish. [British Nova Works Ltd]

Novapol® GF-0218-F. CAS 9002-88-4; LLDPE with process stabilizer; features toughness, strength, ease of extrusion; for blending. [Novacor Ltd.

Novapol® GI-2024-A. CAS 9002-88-4; LLDPE; features med. flow, easy processability, improved dispersion, good ESCR; for additive and color concs., masterbatches, compding., housewares, trash cans, industrial containers, toys. [Novacor Ltd.

Novapol® HB-L455-A. CAS 9002-88-4; HDPE with process stabilizer; features excellent processability on high speed, multiple head blow molding lines; for household and industrial chem. bottles, automotive fluid bottles, toys. [Novacor Ltd.

Novapol® LC-0517-A. CAS 9002-88-4; LDPE; features excellent adhesion, high draw down, excellent sealability; for dry soup pouches, pharmaceutical pkg., potato chip bags, snack food pkg. [Novacor Ltd.

Novapol® LE-0220-A. CAS 9002-88-4; LDPE; features easy processing, high melt strength; for profile extrusions, tape and sheet extrusion, foam, concs. [Novacor Ltd.]

Novapol® LF-0223-B. CAS 9002-88-4; LDPE, med. slip and antiblock; features good clarity, stiffness and strength; for consumer pkg., food wrap, bakery bags, liners, blends with LLDPE. [Novacor Ltd.]

Novapol® LF-Y819-D. CAS 9002-88-4; LDPE, high slip and antiblock; features high strength and toughness, excel-

---

lent processability and shrink film chars.; for industrial pkg., liners, shrink film, blends with LLDPE. [Novacor Ltd.

Novapol® PF-0118-B. CAS 9002-88-4; LLDPE, med. slip and antiblock; features toughness and strength; for trash bags, liners, general pkg. applications. [Novacor Ltd.

Novapol® PR-0636-UG. CAS 9002-88-4; Linear MDPE with process and uv stabilizer; rotational molding grade featuring high impact strength, very good ESCR; for toys, custom molding, carts. [Novacor Ltd.]

Novaquik. Barrier seal. [British Nova Works Ltd]

Novaract. Defoamer. [British Nova Works Ltd]

Novarex® 7022A. PC; injection molding grade, esp. for thin-walled parts; good moldability; low visc. [Mitsubishi Kasei]

Novarex® 7025G10. PC, 10% short glass fiber-reinforced; suitable for injection molding where good dimensional stability is required; med. visc. [Mitsubishi Kasei] *

Novarex® 7025NB. PC, nonflamm.; self-extinguishing grade. [Mitsubishi Kasei] *

Novasan. Deodorant blocks. [British Nova Works Ltd]

Novasheen. Heavy duty seal. [British Nova Works Ltd]

Novashield. Floor dressing. [British Nova Works Ltd]

NovaSize, Dark Fortified. Tall oil rosin; internal sizing agent to give water resistance to paper and paperboard. [Georgia-Pacific] *

Novasolve. Wax remover. [British Nova Works Ltd]

NovaSperse. Rosin aq. disp.; internal sizing agent to give water resist. to paper/paperboard. [Georgia-Pacific] *

Nova Starbright. Emulsion dressing remover. [British Nova Works Ltd]

Novastet. Antistatic treatment. [British Nova Works Ltd]

Nova Stripper Super. Seal remover. [British Nova Works Ltd]

Nova Supercote. Floor dressing. [British Nova Works Ltd]

Nova Superphalte. Buffable emulsion dressing. [British Nova Works Ltd]

Nova Supratreet. Carpet shampoo. [British Nova Works Ltd]

Novata 299, A, AB, B, BBC, BC, BCF, BD, C, D, E. Cocoglycerides; suppository bases; consistency giving agent for ointments, creams, and stick preps. [Henkel; Henkel KGaA]

Novatec. Polyolefin resins. [British Traders & Shippers Ltd]

Novatec. CAS 83915-83-7; Lisinopril; for the treatment of hypertension. [Merck & Co Inc]

Novatec 240H. Extrudable adhesive polyolefin resin; for barrier and HDPE bonding. [Mitsubishi Kasei] *

Novatec-AP. Extrudable adhesive polyolefin resin; for coextrusion of film, bottle, tube, and sheet. [Mitsubishi Kasei]*

Novathion. Dimethyl-O-(3-methyl-4-nitrophenyl) phosphorothioate; all-round low toxic insecticide for forest protection, agriculture and public health; used especially where long term effect is desired. [A/S Cheminova] *

Novatreet. Carpet shampoo. [British Nova Works Ltd]

Nova Tri-Power. Degreasant and cleanser. [British Nova Works Ltd]

Novaways. All purpose cleanser. [British Nova Works Ltd]

Novazole. CAS 26097-80-3; Cambendazole; anthelmintic. [Merck & Co Inc] *

Novazyd. Lisinopril, hydrochlorothiazide; for the treatment of hypertension in patients for whom combination therapy is appropriate. [Merck & Co Inc]

No Vein Compound. CAS 1309-37-1; A proprietary blend of red iron oxides; for use in the foundry industry to eliminate veining and other casting expansion defects. [DCS Color & Supply Co Inc] *

Noveloid. Proprietary products of cellulose esters and ethers. §

Novemol. A trade name for a wetting agent containing sulfonated terpene alcohols. §

Novester. Disperse dyes. [ICI Chem. & Polymers Ltd] †

Novex. Low density polyethylene. [BP Chemicals Ltd]

Novex. Benzalbisdimethyldithio-carbamate. It has a specific gravity of 1.365, is a white crystalline powder, and melts at 175 C; used as a rubber vulcanization accelerator.

Novidium. Homidium chloride. [May & Baker Ltd] *

Novitane. A proprietary polyurethane elastomer. [BFGoodrich] *

Novite. A proprietary trade name for an alloy of iron with about 1.5% nickel and 0.5% chromium. §

novobiocin. An antibiotic produced by *Streptomyces niveus* and *Streptomyces spheroides*; Streptonivicin. Albamycin is the calcium salt and biotexin and cathomycin are the sodium salts.

Novocain. Procaine hydrochloride; anesthetic. [Sterling Drug Inc] *

Novocoll® NC222 US. Isocyanate-terminated polyester urethane adhesive; laminating adhesive giving high adhesion on aluminum and transparent pkg. films. [Pierce & Stevens]

Novodur®. Acrylonitrile/butadiene/styrene copolymer; engineering thermoplastic for the manufacture of injection moldings; good toughness, strength, stiffness and resistance to chemicals; excellent surface quality; problem-free processing; range of standard grades, increased heat resistant, glass fiber-reinforced, flame-retardant, electroplating, and extrusion grades; used for housings, covers, automotive components, domestic appliances, radio, TV and phono equipment, office machinery, photographic and film sectors, toys. [Bayer AG; Bayer plc]

Novodur® L3FR. CAS 9003-56-9; ABS; flame-retardant grade. [Bayer AG]

Novodur® P2H-AT. CAS 9003-56-9; ABS; standard injection molding grade. [Bayer AG] *

Novodur® P2HE. CAS 9003-56-9; ABS; extrusion grade. [Bayer AG] *

Novodur® P2HGV. CAS 9003-56-9; ABS, glass fiber-reinforced. [Bayer AG] *

Novodur® P2T, P2T-AT. CAS 9003-56-9; ABS graft polymer; injection molding grade polymers with higher heat deflection temps.; *see also* Novodur. [Bayer AG]*

Novodur® PMTM. CAS 9003-56-9; ABS; increased heat distort. grade. [Bayer AG] *

Novofix. Photographic fixer. [May & Baker Ltd] *

Novogel® ST. Aluminum stearate, mineral oil. [Rhone-Poulenc]

Novol. CAS 143-28-2; EINECS 205-597-3; Super refined oleyl alcohol; emollient, emulsion stabilizer, superfatting agent, pigment suspending aid, used in cosmetics, personal care products; lipsticks, sunscreens, antiperspirants, bath oils. [Croda Inc.

Novol. CAS 143-28-2; EINECS 205-597-3; Cosmetic quality oleyl alcohol. [Croda Chem. Ltd]

novolac resin. (novolak resin; 2-stage phenolic resin; phenol-formaldehyde (novolak) resin) Thermoplastic phenol-formaldehyde type resin; molding materials; bonding agent in brake linings, abrasive grinding wheels,

---

‡ Trade name and manufacturer not verified   § Trade name without identified manufacturer

electrical insulation; reinforcing agent and modifier for nitrile rubber; air-drying varnishes; bonding materials.

novolak resin. See novolac resin.

Novolen®. A range of polypropylenes and polypropylene copolymers supplied in granular form; used for injection molding of technical parts and packaging, extrusion into special soft film, extrusion into fibers and tapes, blow molded bottles. [BASF AG; BASF plc]

Novoline. Photographic developer. [May & Baker Ltd] *

Novolith. Photographic developer. [May & Baker Ltd] *

Novomatic. Photographic developer. [May & Baker Ltd] *

Novonasco. A proprietary trade name for a wetting agent containing modified sodium alkylnaphthalenesulfonate.§

Novon® 2020-6001. Biodeg. polymer based on corn/potato starch; water-disp. free-foam extrusion polymer; for biodeg. interior cushioning materials. [Novon Prods.

Novon® 3001. Biodeg. polymer based on corn/potato starch; water-disintegrating injection molding grade for biodeg. golf tees. [Novon Prods.

Novon® M0121. Biodeg. polymer based on corn/potato starch; injection molding grade for ultrasound needle guides and other single-use medical devices which deform in steam autoclave, solution sterilizing, or exposure to water. [Novon Prods.

Novon® M0289. Biodeg. polymer based on corn/potato starch; injection molding grade for biodeg. stakes for golf course resodding. [Novon Prods.

Novoperm. Pigments powders; used for paints and inks. [Hoechst AG]

Novoplas. A proprietary trade name for an organic polysulfide synthetic elastic material. §

Novoprotin. A proprietary preparation of crystalline vegetable albumen. §

Novor. Crosslinker for rubber. [Durham Chemicals Ltd] *

Novor 950. CAS 101-68-8; Methylene bis(4-phenylisocyanate)-based; vulcanizing agent for natural and synthetic rubbers giving exceptional reversion and high-temp. aging resist., reduced emission levels. [Akrochem; Rubber Consultants]

Novotak. Photographic developer. [May & Baker Ltd] *

Novotriad. Triple sulfonamides. [May & Baker Ltd] *

Novozone®. A German registered trade name for a magnesium peroxide, MgO_2, prepared for medical purposes; an antiseptic used internally, and externally as an ointment, for wounds and gatherings. §

Novozym. A proteolytic enzyme prepared by submerged fermentation of a selected strain of Bacillus licheniformis; used in the manufacture of effective denture cleaners. [Novo Industri A/S] †

Novulatin. Ethynerone; antifertility agent. [Merck & Co Inc]†

Novulatum. Ethynerone; antifertility agent. [Merck & Co Inc]†

Noxamine. Ethoxylated fatty amine oxides; used for detergents, toiletries. [Atochem UK/Ceca]

Noxamium. Ethoxylated quaternary ammonium compounds; used for detergents, paints, crude oil demulsifier. [Atochem UK/Ceca]

Noxyflex. CAS 15599-39-0; A proprietary preparation of noxythiolin; used to irrigate the bladder. [Geistlich Sohne AG] *

noxythiolin. N-Hydroxymethyl-N´-methylthiourea.

Nozinan. CAS 60-99-1; A proprietary preparation of methotrimeprazine; sedative. [May & Baker Ltd] *

Nozolex. Free-flowing refractory fillers used in sliding gate nozzles. [Foseco (F.S.) Ltd]

NP-10. Aromatic plasticizer; chemically inert, non-saponifiable grade, with low reactivity; used in adhesives (mastic, pressure sensitive), rubber (cements, mechanical and molded goods, tires), and caulking compounds. [Neville]

NPG® Glycol. CAS 126-30-7; Neopentyl glycol; resin intermediate. [Eastman]

NPH Iletin. Insulin, isophane; antidiabetic. [Eli Lilly & Co]

N.P.L. alloy. An alloy of 94.5% aluminum, 2.5% nickel, and 1.5% magnesium.

NPP-1. O,O-Di-n-propyl phosphorodithioic acid; used as a high grade intermediate for organophosphorus insecticides and other products. [A/S Cheminova] *

NPR 3911. Neoprene latex; low visc. polymer which can accept large quantities of fillers while retaining good workability; for formulation of mastics and low-cost facing adhesives. [DuPont]

NPR 5587. Neoprene latex; used where rapid bond development is required, as a quick-break adhesive. [DuPont]

NS®. Used for higher PV applications. [Polymer Corp]

NSAE Powder. Sodium alkyl-naphthalene sulfonate; anionic surfactant in powder form; wetting and dispersing agent with high electrolyte tolerance, e.g., agricultural dispersing agent for sulfur. [Millmaster-Onyx UK] *

N.S.B. A proprietary preparation of noxytiolin, and vinyl pyrrolidone/vinyl acetate copolymer in an aerosol; wound dressing. [Geistlich Sohne AG] *

N-Serve. A line of nitrogen stabilizers based primarily on nitropyrin. [Dow UK]

N.S. fluid. A mixture of sodium chloride, aluminum chloride, and iron chloride.

NT-15GF/000. CAS 32131-17-2; Toughened nylon 6/6, 15% glass fiber-reinforced. [Compounding Tech. *

NTA. CAS 5064-31-3; Sodium nitrilotriacetate powder and 40% solution; for domestic and industrial laundry detergents, hard surface cleaning formulations, boiler treatment, textile auxiliary. [Monsanto Co] *

NTA. See nitrilotriacetic acid.

Nuact. Loss-of-dryness inhibitors; prevents loss of drying of nonaqueous coating compositions. [Hüls Am.

Nuade. Slip and mar agents; for water-borne baking finishes. [Hüls Am.] †

Nuba. A proprietary trade name for a thermoplastic coal-tar pitch and a cumarone-indene resin for paints. §

Nubain. CAS 23277-43-2; Nalbuphine hydrochloride; analgesic; antagonist. [Endo Laboratories Inc]

Nubex. Building remedial treatments. [Tenneco Organics Ltd] *

Nubrite. A bright nickel plating process. [Hanshaw Chemicals] ‡

Nubun. A proprietary trade name for a synthetic rubber latex insulation for power and communication cables; it is made from a special modification of buna S synthetic rubber. §

Nucleant. An aluminum alloy grain refiner. [Foseco (F.S.) Ltd]

Nucol. Wide range of high build, conventional and modified chlorinated rubber and vinyl coatings. [Sigma Coatings]*

Nucoline. See vegetable butter.

Nucrel®. Ethylene-methacrylic acid copolymer. [DuPont UK]

Nuelin Liquid. CAS 8000-10-1; Each 5 ml contains theophylline sodium glycinate 120 mg equivalent to theophylline hydrate BP 60 mg. [3M Pharmaceuticals] *

Nuelin-SA. CAS 58-55-9; Each tablet contains 175 mg anhydrous theophylline in a slow-release formulation. [3M Pharmaceuticals] †

Nuelin SA-250. CAS 58-55-9; Each tablet contains 250 mg anhydrous theophylline in a slow-release formulation. [3M Pharmaceuticals] †

Nuelin Tablets. CAS 58-55-9; Each tablet contains theophylline BP 125 mg in microcrystalline form. [3M Pharmaceuticals] †

Nufol. Nitrogenous fertilizers. [ICI Chem. & Polymers Ltd] †

Nuglas. Tubes molded with glass reinforcement; for electrical, structural, and thermal applications, jigs, fixtures and tooling, lightweight support structures for antennas, lighting, and surveillance. [Fothergill Tygaflor Ltd] *

Nujol. Mineral oil; laxative; pharmaceutic aid. [Schering Corp]

Nulacin. A proprietary preparation containing milk, doxtin, maltose base with magnesium trisilicate, calcium carbonate, magnesium oxide, and oil of peppermint; an antacid. [Horlicks] *

Nulogyl. CAS 6506-37-2; A proprietary preparation of nimorazole; used in the treatment of trichomonas infections. [Bristol-Myers Squibb Co Inc]

Nuloid. See Bakelite. §

Nulomoline. A solution of partly inverted sugar; used for some purposes as a substitute for glycerine.

Numoquin. Ethyl-hydrocupreine.

Numorphan. CAS 357-07-3; Oxymorphone hydrochloride; analgesic. [Endo Laboratories Inc] *

Numorphan Oral. A proprietary trade name for hydromorphinol; analgesic. [British Drug Houses] *

Numotac. CAS 2576-92-3; Each tablet contains isoetharine hydrochloride 10 mg in a porous plastic matrix. [3M Pharmaceuticals] †

NuoCide®. CAS 1897-45-6; Tetrachloroisophthalonitrile; fungicide and mildewcide for both water and solvent-based coatings, and as a marine antifouling agent in marine coatings. [Hüls Am.

Nuocure. Metal salts of organic acids; driers for water-borne coating compositions, catalyst for polyurethane foam. [Hüls Am.

Nuocure 28. Catalyst for polyurethane foams. [Harcros UK]

Nuodex. Metal salts of organic acids, fungicides and bactericides, bodying agents, dispersing agents; driers for coating compositions, curing catalysts, biocides for coating compositions, industrial biocides, thickeners, pigment dispersants. [Hüls Am.

Nuodex 84. CAS 2492-26-4; 50% aqueous solution of the sodium salt of 2-mercaptobenzothiazole; antimicrobial preservative for fibrous substrates, adhesives and aqueous emulsions. [Hüls Am.] †

Nuodex 87. Biocidal wall washing compound. [Harcros UK]

Nuodex 100. Compositions containing the dodecyldimethylbenzylammonium salt of naphthenic acid; antimicrobial preservatives for textiles, shoe linings and plasticized PVC compositions. [Hüls Am.] †

Nuodex 321 Extra. Mercurial biocide. [Harcros UK]

Nuodex NA. A pigment dispersant; used in vinyl and acrylic lacquer formulations. [Hüls Am.] †

Nuolate. Metal salts of tall oil fatty acids; driers for nonaqueous coating compositions. [Hüls Am.] †

Nuophene. CAS 97-23-4; Dihydroxy-dichloro-diphenyl-methane; fungicide for use on textiles, cordage, and hair felt. [Hüls Am.] †

Nuoplaz. Plasticizers; used for plasticizing PVC and other polymers. [Hüls Am.

Nuosept. Antimicrobial preservatives; used for preservation of aqueous coating compositions, pigment dispersions, inks, adhesives, caulks, metalworking fluids, paper coatings, drilling muds. [Hüls Am.

Nuosperse. Dispersing agents; used for coating compositions. [Hüls Am.

Nuostabe. Stabilizers; for heat and light stabilization of PVC compositions. [Hüls Am.] †

Nuostabe 1317. A proprietary trade name for a Ba/Cd/Zn stabilizer for PVC. [Durham Raw Materials] *

Nuostabe 1374. A proprietary trade name for a non sulfide staining stabilizer for PVC used in blown foams. [Durham Raw Materials] *

Nuostabe 1605. A stabilizer for PVC based on a barium/cadmium/zinc complex. [Durham Raw Materials] *

Nuosyn. Metal salts of organic acids; driers for nonaqueous coating compositions. [Hüls Am.] †

Nupel®. Rosin emulsion sizes. [Reichhold]

Nupercainal. A proprietary preparation of cinchocaine hydrochloride; an anesthetic ointment. [Ciba plc] *

Nupercaine. A proprietary preparation of cinchocaine hydrochloride; surface anesthetic. [Ciba plc] *

Nupercaine Hydrochloride. CAS 61-12-1; Dibucaine hydrochloride; anesthetic. [Ciba-Geigy Corp] *

Nuprin. CAS 15687-27-1; Ibuprofen; anti-inflammatory and analgesic. [Bristol-Myers Squibb Co Inc]

Nuprin. CAS 729-99-7; A proprietary trade name for sulfamoxole. §

Nupyrin. See aspirin. §

Nurac. Stated to be diphenyl-guanidine; also been said to be thiocarbanilide; a proprietary rubber vulcanization accelerator. §

Nuram. Nitrogenous fertilizers. [ICI Chem. & Polymers Ltd]†

Nuremberg gold. An alloy of 90% copper, 7.5% aluminum, and 2.5% gold. A jeweler's alloy.

Nurofen. CAS 15687-27-1; Proprietary preparations containing ibuprofen; analgesics. [The Boots Co plc]

Nurolon. Nonabsorbable surgical suture. [Ethicon Inc] *

Nusat. A satin finish nickel plating process. [Engelhard Technologies Ltd]

Nu-Seals Aspirin. CAS 50-78-2; Aspirin in an enteric coated form; an analgesic. [Eli Lilly & Co] †

Nu-Seals Potassium Chloride. CAS 7447-40-7; EINECS 231-211-8; Potassium chloride in enteric coated tablets. [Eli Lilly & Co] †

Nu-Seals Sodium Salicylate. CAS 54-21-7; Sodium salicylate used as an analgesic and antipyretic. [Eli Lilly & Co]†

Nu-Set. Bristle-setting cement. [Hardman]

Nusolv ABP-62. CAS 25640-78-2; Alkyl biphenyl mixture; solv. possessing excellent solvency, chem. and thermal stability, nonvolatility; for aerosols, adhesives, electronic parts manufacturing and cleaning, industrial cleaning, inks and paper coatings, metal cleaning, paints, plastics, sealants, tile manufacturing. [Ridge Tech.]

Nusolv ABP-103. CAS 69009-90-1; Bis(methylethyl)-1,1´-biphenyl; solv. possessing excellent solvency, chem. and thermal stability, nonvolatility; for aerosols, adhesives, electronic parts manufacturing and cleaning, industrial cleaning, inks and paper coatings, metal cleaning, paints, plastics, sealants, tile manufacturing. [Ridge

---

‡ Trade name and manufacturer not verified   § Trade name without identified manufacturer

Tech.]
Nustar®. For agriculture. [DuPont UK]
nutmeg butter. CAS 8007-12-3; Expressed oil of nutmeg.
nut oil. A term used in China for tung oil (Chinese wood oil). The same name is also used for walnut and arachis oils.
Nutracort. CAS 50-23-7; Hydrocortisone; glucocorticoid. [Galderma Laboratories]
Nutralys. Animal feedstuff. [Roquette (UK) Ltd]
Nutramigen. A proprietary artificial lactose-free infant milk food. [Bristol-Myers Squibb Co Inc]
Nutramin. Flour improver; vitamin enrichment additive. [Akzo Chemie UK Ltd]
Nutramon. Aqueous ammonia. [ICI Chem. & Polymers Ltd]†
Nutranel. Protein, fat, carbohydrate, vitamins, minerals, and trace elements; liquid feed. [Roussel Laboratories Ltd] †
Nutraphos. Nutrients for effluent treatment. [Scottish Agricultural Industries plc]
Nutrapon AL 1. CAS 67762-19-0; Ammonium lauryl ether (1) sulfate; surfactant for shampoos, bubble baths, and other cosmetic products below pH 7. [Clough]
Nutrapon AL 60. CAS 67762-19-0; Ammonium laureth-3 sulfate; surfactant for shampoo concs., bubble baths, dishwashing, and light duty detergents. [Clough]
Nutrapon B 1365. Sodium lauryl sulfate and ethylene glycol stearate; shampoo and detergent blend with pearlizing agent for pearlescent formulations; for shampoos, bubble baths, liq. hand soaps. [Clough]
Nutrapon BM 3960. Sodium laureth-3 sulfate; surfactant for clear liq. shampoos, bubble baths, other cosmetics. [Clough]
Nutrapon DE 3796. DEA-lauryl sulfate; mild bubble bath and shampoo conc. [Clough]
Nutrapon DL 3891. CAS 151-21-3; Sodium lauryl sulfate; high foaming surfactant for personal care and industrial formulations. [Clough]
Nutrapon ES-60 3568. CAS 9004-82-4; Sodium laureth (3) sulfate; mild surfactant for personal care and industrial products; flash foam in hard water. [Clough]
Nutrapon ESY 2299. Sodium laureth-1 sulfate; surfactant for shampoos, bubble baths, cosmetics; high tolerance to hard water. [Clough]
Nutrapon FA-50 0066. Ammonium deceth sulfate; high foaming surfactant for specialty applications, e.g., secondary oil recovery. [Clough]
Nutrapon HA 3841. CAS 2235-54-3; EINECS 218-793-9; Ammonium lauryl sulfate; surfactant for personal care products and detergent cleaners. [Clough]
Nutrapon KF 3846. Sodium laureth-3 sulfate; buffered sulfate for use in mild shampoos, bubble baths, shower gels; flash foam in hard water. [Clough]
Nutrapon KPC 0156. CAS 9004-82-4; Sodium laureth-3 sulfate; surfactant for shampoos, bubble bath, dishwashing detergents. [Clough]
Nutrapon PP 3563. CAS 2235-54-3; EINECS 218-793-9; Ammonium lauryl sulfate; used in nonalkaline shampoos, bubble baths, mild detergents and cleaners below pH 7.0. [Clough]
Nutrapon RS 1147. Sodium lauryl sulfate, sodium cocoyl sarcosinate; surfactant for rug and upholstery shampoo; high foaming, min. wetting, low cloud pt.; leaves dry residue for easy removal. [Clough]
Nutrapon TK 3603. Sodium lauryl sulfate, glycol stearate; pearlized base for shampoos and liq. hand soaps.

Nutrapon TLS-500. CAS 139-96-8; EINECS 205-388-7; TEA-lauryl sulfate; mild ingred. in cosmetic and industrial formulations. [Clough]
Nutrapon TW 3987. TEA-lauryl sulfate, sodium lauryl sulfate; surfactant for shampoos, bubble baths, liq. hand soaps. [Clough]
Nutrapon W 1367. CAS 151-21-3; Sodium lauryl sulfate; high foaming surfactant for personal care and industrial formulations. [Clough]
Nutrapon WAQE 2364. CAS 151-21-3; Sodium lauryl sulfate; high foaming surfactant for personal care and industrial formulations. [Clough]
NutraSweet. CAS 22389-47-0; Aspartame; sweetener. [G D Searle & Co] *
Nutregen. A proprietary gluten-free wheat starch used in the dietary treatment of coeliac disease. [Energen] *
Nutrifos. CAS 7722-88-5; Food grade tetrasodium pyrophosphate; used in meat curing. [Monsanto Co] *
Nutrilan® FPK, H, M. CAS 9015-54-7; Hydrolyzed animal protein; substantive to skin and hair. [Henkel KGaA/Cospha] *
Nutrilan® I-50. CAS 9015-54-7; Hydrolyzed animal protein; protein for shampoos, conditioning rinses, hair care products. [Henkel/Cospha; Henkel Canada]
Nutrilan® Keratin W. CAS 69430-36-0; Hydrolyzed animal keratin; protective protein for hair care products. [Henkel/Cospha; Henkel Canada] *
Nutrilan® L. CAS 9015-54-7; Hydrolyzed animal protein; protein for shampoos, conditioning rinses, hair care products, cold waves, bath and shower preps. [Henkel/Cospha; Henkel Canada]
Nutrilife®. Enzyme-protein compounds; synergistic effects with emulsifiers for baking processes. [Grünau]
Nutrimalt® Range. CAS 8002-48-0; Malt extract; taste and volume improver in baked goods. [Grünau]
Nutrisoft® 55. Distilled monoglyceride; food emulsifier for baking additives, confectionery; antistaling effect. [Grünau]
Nutrol 100. CAS 9002-93-1; Octoxynol-9; wetting agent, dispersant for metal and acid cleaners, pesticides. [Clough]
Nutrol 600. CAS 9016-45-9; Nonoxynol-9; detergent, emulsifier, wetting agent, dispersant. [Clough]
Nutrol 611. CAS 9016-45-9; Nonoxynol-8; detergent, emulsifier, wetting agent, dispersant. [Clough]
Nutrol 622. CAS 9016-45-9; Nonoxynol-4; emulsifier, light duty detergent, moderate foaming agent. [Clough]
Nutrol 640. CAS 9016-45-9; Nonoxynol-15; emulsifier, light duty detergent, moderate foaming agent. [Clough]
Nutrol 656. CAS 9016-45-9; Nonoxynol-11; emulsifier, light duty detergent, moderate foaming agent. [Clough]
Nutrol Betaine MD 3863. Cocamidopropyl betaine; surfactant, foaming agent, foam stabilizer, wetting agent for shampoos, bubble baths, liq. hand soaps. [Clough]
Nutrol Betaine OL 3798. Cocamidopropyl betaine; foamer, foam stabilizer, wetting agent for industrial and household cleaners, dishwashing, liq. hand soaps. [Clough]
Nutrol S-60 5350. Ammonium nonyl phenoxy polyethoxy sulfate; emulsifier for use in polymers; high foaming detergent, wetting agent for textiles. [Clough]
Nutrol SXS 5418. CAS 1300-72-7; EINECS 215-090-9; Sodium xylene sulfonate; hydrotrope, coupling agent, solubilizer. [Clough]

---

* Trade name not verified as available

† Trade name verified as obsolete

Nuvacon. A proprietary preparation of megestrol acetate and ethinyl estradiol; oral contraceptive. [British Drug Houses] *

Nuvamide. Triple sulfonamide with neomycin. [May & Baker Ltd] *

Nuvan. CAS 62-73-7; Emulsifiable concentrate of 500 g dichlorvos per liter; a fumigant organophosphorus insecticide. [Ciba-Geigy Agrochemicals]

Nuvan Fly Spray. Organophosphorus/pyrethrin insecticide. [Ciba plc] *

Nuvanol. Organophosphorus insecticide. [Ciba plc]

Nuvan Top Aerosol. Organophosphorus insecticde. [Ciba plc] *

Nuvis. Thixotropic bodying agents; for control of sag and flow of nonaqueous coating compositions. [Hüls Am.]

Nuxtra. Metal salts of synthetic organic acids; driers for nonaqueous coating compositions. [Hüls Am.]

NY-10GF. CAS 25038-54-4; Nylon 6, 10% glass fiber-reinforced; versatile thermoplastic. [Compounding Tech.]

NY-30CF. CAS 25038-54-4; Nylon 6 resin, 30% carbon fiber-reinforced; offers balance of strength to weight for use in aerospace and automotive industries. [Compounding Tech.]

Nyacol®. Colloidal dispersions of silica (SiO_2), antimony pentoxide, alumina and other metal oxides; binders for investment casting molds and fibrous refractories; polishing agent for semiconductor wafers, rigid disks, etc., antislip coating for paper, textile fibers, etc., flame retardant for plastics (antimony oxide dispersion). [PQ Corp]

Nyacol® A-1530. CAS 1314-60-9; Colloidal disp. of antimony pentoxide in water; flame retardant additive to latex emulsions; durable treatment for fabrics, nonwovens, fiberfill, paper, fiberglass, vinyls; suitable for FR adhesives. [PQ Corp.]

Nyad® 1250. CAS 13983-17-0; Wollastonite; reinforcer for polyvinylsiloxane used for dental impressions. [NYCO® Minerals Inc]

Nyad® Wollastonite. High aspect ratio and fine particle size calcium metasilicate minerals: CaO, SiO_2, Fe_2O_3, Al_2O_3, MnO, MgO, TiO_2; used for plastics, coatings, refractories, fire resistant board, adhesives, rubber and elastomers and polymer concrete. [NYCO® Minerals Inc]

Nyala. Acidulant; used for chemically leavened bread. [Albright & Wilson Ltd]

Nybex 12034 BKFR. CAS 25038-54-4; Nylon 6, 10% glass fiber-reinforced; flame retardant thermoplastic [Ferro/Engineering Thermoplastics] *

Nybex 12056 BKFR. CAS 25038-54-4; Nylon 6, 45% glass fiber-reinforced; flame retardant thermoplastic [Ferro/Engineering Thermoplastics] *

Nybex 13001 BKC. CAS 25038-54-4; Nylon 6, 30% carbon fiber-reinforced; conductive grade. [Ferro/Engineering Thermoplastics] *

Nybex 15011 NA. CAS 25038-54-4; Nylon 6, 30% mineral/glass fiber-reinforced; low warpage grade. [Ferro/Engineering Thermoplastics] *

Nybex 17000 NAX. CAS 25038-54-4; Nylon 6, 30% glass-reinforced; EMI shielding grade. [Ferro/Engineering Thermoplastics] *

Nybex 22008 BKUT. CAS 32131-17-2; Nylon 66, 33% glass-reinforced, plasticized; automotive grade. [Ferro/Engineering Thermoplastics] *

Nybex 42002 BKHS. CAS 25038-54-4; Nylon 6, 13% glass fiber-reinforced, heat-stabilized; blow molding grade. [Ferro/Engineering Thermoplastics] *

Nybex 52000 NA. CAS 32131-17-2; Nylon 66 alloy; moisture-resist. grade. [Ferro/Engineering Thermoplastics] *

Nycoa® 438. CAS 25038-54-4; Nylon 6; high flow, fast-setting, injection molding grade resin for very fast cycle operation in difficult-to-fill molds; NSF-approved for use in molded pipe fittings (potable water). [Nylon Corp. of Am.]

Nycoa® 446. CAS 25038-54-4; Nucleated nylon 6; high visc., nucleated nylon for extrusion of tubing, rods, sheet, and film with fast set-up time. [Nylon Corp. of Am.]

Nycoa® 500. CAS 32131-17-2; Nylon 6/6; std. grade with good moldability, toughness, and chem. resist.; for molding and extrusion. [Nylon Corp. of Am.]

Nycoa® 528. CAS 32131-17-2; Nylon 6/6, nucleated; nucleated grade with controlled crystallinity, excellent moldability and cycle, increased stiffness and abrasion resist. [Nylon Corp. of Am.]

Nycoa® 567. CAS 25038-54-4; Plasticized nylon 6; high impact strength extrusion and molding resin, flexibility and toughness; used in injection molded shoe heel lifts, extruded softhand fishing line and monofilament. [Nylon Corp. of Am.]

Nycoa® 714. CAS 25038-54-4; Plasticized nylon 6 copolymer; softness and flexibility for extrusion and molding for specialty products, e.g., molded athletic shoe soles, fishing line, etc. [Nylon Corp. of Am.]

Nycoa® 870. CAS 25038-54-4; Heat-stabilized nylon 6; extrusion resin for jacketing (wire, cable) including thermoplastic-insulated building wires and gasoline-resistant types. [Nylon Corp. of Am.]

Nycoa® 1417. CAS 25038-54-4; Impact-modified nylon 6 resin; med. high impact resin for applications where notch sensitivity of standard nylons produces inconsistent part performance. [Nylon Corp. of Am.]

Nycoa® 4015. CAS 25038-54-4; Nylon 6, 15% glass-reinforced; reinforced grade with excellent heat distort., high rigidity and strength. [Nylon Corp. of Am.]

Nycoa® 5015. CAS 32131-17-2; Nylon 6/6, 15% glass-reinforced; reinforced grade with excellent heat distort., high rigidity and strength. [Nylon Corp. of Am.]

Nycoat®. Family of chemically coupled minerals; e.g. alumina trihydrate, barytes and celestite (varies depending on substrate); for specialty applications: plastics, high performance coatings, adhesives, rubber and elastomers. [NYCO® Minerals Inc]

Nycor® Barytes. 200 + 325 mesh barium sulfate: barium sulfate, silica, ferric oxide, manganese and lead; for coatings, rubber and elastomers, friction products, refractories and sound deadening compounds. [NYCO® Minerals Inc] *

Nycor® Celestite. 200 + 325 mesh strontium sulfate: strontium sulfate, barium sulfate, calcium carbonate and alumina; for coatings, rubber and elastomers, friction products, refractories and sound deadening compounds. [NYCO® Minerals Inc] *

Nycor® R. CAS 13983-17-0; High aspect ratio wollastonite, untreated; for use in coatings, refractory, friction, construction board and plastics. [NYCO® Minerals Inc]

Nydrane. CAS 501-68-8; A proprietary preparation of beclamide; an anticonvulsant. [Rona Laboratories] ‡

Nydrazin. CAS 54-85-3; Isoniazid; antibacterial. [Bristol-Myers Squibb Co Inc] *

---

‡ Trade name and manufacturer not verified   § Trade name without identified manufacturer

Nydur. Polyamide resins (Nylon 6). impact modified with low temperature impact, as well as abrasion, heat and chemical resistance. [Miles]

Nyebar. A solution of a low surface energy fluorocarbon polymer in a fluorinated solvent; provides a nonwettable surface which controls or prevents the migration or creep of lubricants or other fluids. [Wm F Nye] *

NyeTact 520. 6-Ring polyphenylether in trichlorotrifluoroethane; elec. connector lubricant. [Wm F Nye]

Nyflake®. CAS 12001-26-2; Muscovite mica; coarse and fine grinds: SiO_2, Al_2O_3, K_2O, Fe_2O_3, Na_2O_3, CaO, TiO_2, MnO_2, P, S; for plastics, coatings, oil well drilling mud, adhesives. [NYCO® Minerals Inc]

Nyglas®. Chemically coupled ground glass; soda lime glass, platey, various particle sizes: SiO_2, Na_2O, CaO, MgO, Al_2O_3, K_2O, Fe_2O_3; for plastics and coatings. [NYCO® Minerals Inc]

Nykon. Corrosion resistant rheological additive; used for greases. [Rheox Inc]

Ny-Kon® I. Nylon 6/12, < 5% MoS_2 lubricant. [LNP]

Ny-Kon® P. CAS 25038-54-4; Nylon 6, < 5% MoS_2 lubricant. [LNP]

Ny-Kon® Q. CAS 9008-66-6; Nylon 6/10, < 5% MoS_2. [LNP]

Ny-Kon® R. CAS 32131-17-2; Nylon 6/6, < 5% MoS_2. [LNP]

Ny-Kon® V. High-impact nylon, < 5% MoS_2. [LNP]

Nylander's reagent. An alkaline solution of bismuth subnitrate and Rochelle salt obtained by dissolving 40 g Rochelle salt and 20 g bismuth subnitrate in 1,000 cc of 8% caustic soda; used for the detection of glucose in urine by boiling 5 parts of the glucose solution with 1 part of the reagent when reduction occurs and a black precipitate is produced.

Nylatron®. Nylon and molybdenum disulfide; molding compounds for bearing and wear applications (bearings, bushings, wear pads, etc.) [Polymer Corp]

Nylatron® 1018 HS. CAS 32131-17-2; Nylon 6/6 resin, 33% fiberglass-reinforced, heat-stabilized; injection molding resin. [DSM]

Nylatron® 1024 HS. CAS 32131-17-2; Nylon 6/6 resin, heat-stabilized; general purpose, heat-stabilized injection molding resin with internal lubricant. [DSM]

Nylatron® GS. Nylon, MoS_2-lubricated; wear resistance, low surface friction, and high strength and rigidity; used in bearings, valve seats, thrust washers, wear surfaces, rollers, gears, forming dies, tooling fixtures, etc.; avail. in rod, disc., strip, etc. [DSM]

Nylatron® GS-63. CAS 25038-54-4; Nylon 6 resin with molybdenum disulfide lubricant, uv-stabilized; injection molding resin containing carbon black for uv stabilization. [DSM] *

Nylatron® NSB-90. CAS 32131-17-2; Nylon 6/6 resin; injection molding resin specially formulated for use in bearing and wear applications. [DSM]

Nylax. A proprietary preparation containing bisacodyl, phenolphthalein and senna leaf powder; a laxative. [The Boots Co plc]

Nylocrom. Dry powder acid dyes; specially developed acid dyes used in the textile industry for dyeing nylon. [Multicrom SA]

Nylofixan. Aromatic sulfonate, black-tanning agent; for polyamides. [Sandoz Products Ltd] *

Nyloflex. Flexographic printing plates. [BASF plc]

Nylok® 170. Amino-funcitonal clacined clay; filler for reinforcement of polyamide resin systems. [J.M. Huber/Clay Div.]

Nylomine. Dyestuffs for nylon and polyamide fibers. [ICI Chem. & Polymers Ltd]

nylon. A family of polyamide polymers char. by presence of amide group CONH; CAS 63428-83-1; thermoplastic resin for tire cord, hosiery, wearing apparel, brush bristles, cordage, fish lines, tennis rackets, rugs, artificial turf, parachutes, composites, sails, film, gears/bearings, insulation, surgical sutures, metal coating, fuel tanks. [Ashley Polymers; Bamberger Polymers; BASF; DSM; EMS-Am. Grilon; Hoechst Celanese; Hüls Am.; ICI GmbH; LNP; Miles; Monsanto]

nylon 6. (poly[imino(1-oxo-1,6-hexanediyl)]; poly(iminocarbonylpentamethylene)) Polyamide. $[NH(CH_2)_5CO]_x$; CAS 25038-54-4; tire cord; fishing lines; tow ropes; hose manufacture; woven fabrics. [Snia UK]

nylon 6/6. (poly[imino (1,6-dioxo-1,6-hexanediyl) imino-1,6-hexanediyl; poly(hexamethyleneadipamide)) Polyamide; polymeric amide formed by the reaction of adipic acid with hexylenediamine. $[NH(CH_2)_6NHCO-(CH_2)_4CO]_n$; CAS 32131-17-2; thermoplastic resin for injection molding, extrusion. [Asahi Chem Industry Co Ltd; Snia UK]

nylon 12. (azacyclotridecane-2-one polyamide; poly(laurolactam)) Polyamide derived from 12-aminododecanoic acid; CAS 25038-74-8; 24937-16-4; thermoplastic resin for injection molding and extrusion applications. [Atochem SA; Daicel-Hüls; Hüls AG]

Nylon N-012. CAS 32131-17-2, 9004-73-3; Nylon 6/6, methicone; imparts soft, lubricious feel to cosmetic formulations, anhyd. systems, emulsions, and powds. [Presperse]

Nylon Resist NCO. Nylon dye resist conc. for level and even dyeings; retards dyeing rate. [Eastern Color & Chem.

Nyloprint. Flexible printing plates. [BASF plc]

Nylosan®. Synthetic organic acid dyestuffs; specialty dye for aq. mediums, textiles, fertilizers. [Sandoz; Sandoz Products Ltd]

Nyloset Finish. Amide resin; nylon builder and softener. [Scher Chemicals Inc] *

Nylosolv. Environmentally friendly solvent for printing plates. [BASF plc]

Nylox. Adhesive composition for nylon, etc. [Hardman]

Nyogel. An inorganically gelled series of greases based upon synthetic lubricating oils; for specialty lubrication of delicate machinery and engineered components. [Wm F Nye] *

NyoGel® 744. Fluorocarbon-filled low friction grease for cams and sliding parts. [Wm F Nye]

NyoSil. Halogenated silicone oil; potentiometer lubricant for wide variety of wipter/substrate combinations. [Wm F Nye]

Nypel. Reinforced and unreinforced recycled nylon 6 resins utilizing select feedstocks to combine quality, performance and processability at a lower price. [Allied-Signal Inc]

Nypene. A proprietary trade name for a polyterpene hydrocarbon resin used in adhesives, paints, and varnishes. §

Nyrim. Reaction injection molding nylon, prepared by in-mold polymerization of caprolactam (feedstock for nylon) to a nylon-6 block polymer with utilization of a catalyst 2.0 agents; used for industrial, agriculture, and automotive applications. [DSM NV] *

Nyspheres®. Fine particle hollow glass spheres, light-

---

* Trade name not verified as available   † Trade name verified as obsolete

weight: silica, alumina, iron oxides, calcium, magnesium, and alkalis; for plastics, coatings, adhesives, cement products, refractories. [NYCO® Minerals Inc]

Nystadermal. Nystatin and triamcinolone acetonide; used in the treatment of fungal and eczematous skin disorders. [Bristol-Myers Squibb Pharmaceuticals Ltd]

Nystan Cream. CAS 1400-61-9; Nystatin in cream base. [Bristol-Myers Squibb Pharmaceuticals Ltd]

Nystan Pastille. CAS 1400-61-9; Nystatin in a gelatin pastille formulation. [Bristol-Myers Squibb Pharmaceuticals Ltd]

nystatin. CAS 1400-61-9; Produced by *Streptomyces noursei*; $C_{47}H_{75}NO_{17}$; used in medicine as an antifungal antiobiotic, feed additive.

Nystavescent. CAS 1400-61-9; A proprietary preparation of nystatin in an effervescent pessary; used in the treatment of vaginal candidiasis. [Bristol-Myers Squibb Pharmaceuticals Ltd]

Nystex. CAS 1400-61-9; Nystatin; indicated in the treatment of cutaneous or mucocutaneous mycotic infections caused by *Candida albicans* and other candida species. [Altana Inc] ‡

NYsyn® 30-5. Acrylonitrile-butadiene copolymer, cold polymerized; for applications requiring easy processing and excellent low temp. flexibility, e.g., industrial and automotive hose and seals, cable jackets, molded and extruded mech. goods. [Copolymer Rubber]

NYsyn® 33-5HM. Acrylonitrile-butadiene copolymer, cold polymerized; nonblooming; general purpose polymer with good oil resist.; used for molded products; compounds exhibit high modulus, tensile strength and tear; fast cure. [Copolymer Rubber]

NYsyn® 305V. NBR/PVC (70:30); recommended for hose, shoe soles, wire and cable compounds, close cell sponge, colored products. [Copolymer Rubber]

NYsynblak® 9010. Acrylonitrile-butadiene copolymer black masterbatch, cold polymerized (NYsyn 35-8 with 50 parts N-550 carbon black); used for extruded and molded goods requiring med.-high solv. and oil resist. [Copolymer Rubber]

NYsynblak® DN 120. Acrylonitrile-butadiene copolymer black masterbatch, cold polymerized (NYsyn 33-3 with 50 parts N-234 carbon black); used for extruded and molded goods requiring med.-high solv. and oil resist.; also as black conc. impact modifier for ABS. [Copolymer Rubber]

Nytal® 100. CAS 14807-96-6; Hydrous magnesium silicate; reinforcing filler for PVC, vinyl asbestos tile, polyester in match molded articles, body patching compounds, PP, nylon, phenol formaldehyde, polyethylene, ceramic wall tile and artware; improves stiffness; auxiliary flux in vitreous ceramic bodies. [R.T. Vanderbilt]

NZ 01. Neoalkoxy trisneodecanoyl zirconate, IPA; coupling agents which sometimes also act as adhesion promoters, antioxidants, antistats, antifoaming agents, accelerators, blowing agent activators, catalysts), curatives, corrosion inhibitors, disp. aids, emulsifiers, flame retardants, foaming agents, grinding aids, hardeners, impact modifiers, internal lubes, metal primers, process aids, pigment intensifiers, peroxide activators, release agents, retarders, stabilizers, surfactants, suspension aids, thixotropes, wetting agents. [Kenrich Petrochemicals]

NZ 09. Neoalkoxy tris (dodecyl) benzene sulfonyl zirconate, IPA; see NZ 01. [Kenrich Petrochemicals]

NZ 12. Neoalkoxy tris (dioctyl) phosphato zirconate, IPA; see NZ 01. [Kenrich Petrochemicals]

NZ 33. Neopentyl (diallyl) oxy, trimethacryl zirconate, IPA; coupling agent. [Kenrich Petrochemicals]

NZ 38. Neoalkoxy tris (dioctyl) pyrophosphato zirconate, IPA; see NZ 01. [Kenrich Petrochemicals]

NZ 39. Neopentyl (diallyl) oxy, tri(9,10 epoxy stearoyl) zirconate; coupling agent. [Kenrich Petrochemicals]

NZ 44. Neoalkoxy tris (ethylene diamino) ethyl zirconate, IPA; see NZ 01. [Kenrich Petrochemicals]

NZ 89. Neopentyl (diallyl) oxy, trimercapto-phenyl zirconate; coupling agent. [Kenrich Petrochemicals]

NZ 90. Neopentyl (diallyl) oxy, tri(dodecyl)benzene-sulfonyl zirconate, IPA; coupling agent. [Kenrich Petrochemicals] *

NZ 97. Neoalkoxy tris (m-amino) phenyl zirconate, phenyl glycol ether; see NZ 01. [Kenrich Petrochemicals]

---

‡ Trade name and manufacturer not verified   § Trade name without identified manufacturer

O

O9810. Octamethylcyclotetrasiloxane; as cleaning, polishing and damping media; offers low toxicity, inertness. [Hüls Am.]

O9816. CAS 107-51-7; Octamethyltrisiloxane; as cleaning, polishing and damping media; offers low toxicity, inertness. [Hüls Am.]

OA 40-30. Chlorinated paraffin; for industrial applications. [Witco/Argus]

OA-100A. Hydrocarbon-derived; pour pt. depressant exhibiting exceptional shear stability; for formulating motor oils, hydraulic oils, paraffinic or paraffinic/naphthenic-based elec. insulating oils. [Witco/Argus]

OA-154. Alkanolamide; rust inhibitor, emulsifier for coolant or water-based products. [Witco/Argus]

OA-252. Sulfurized fatty acids; for formulating coolant or water-based systems. [Witco/Argus]

OA-252. Sulfurized fatty acid; additive for industrial oils and greases. [Witco/Argus]

OA-270. Sulfurized methyl ester; additive for industrial oils and greases. [Witco/Argus]

OA-300. Sulfurized sperm oil replacement; additive for industrial oils and greases. [Witco/Argus]

OA-502. Nonylated diphenylamine; antioxidant for industrial and automotive oil, grease, and fluid; stabilizer in hydraulic brake fluid. [Witco/Argus]

OA-505. Diphenylamine; high-temp. antioxidant for formulating high-temp. greases, automotive, diesel, and turbine lubricants; synergistic with other antioxidants and additives, enhancing thermal and oxidation stability. [Witco/Argus]

OA-700. Chlorinated ester; lubricant imparting exceptional metal wetting characteristics to industrial oils. [Witco/Argus]

OA-770. Sulfur-chlorinated additive; additive for industrial oils and greases. [Witco/Argus]

OA-951. Chlorinated fatty acid; for formulating coolant or water-based systems. [Witco/Argus]

Oak Draw 720, 728, 830A. A line of soluble oil, semi-synthetic and synthetic metal forming lubricants; for drawing, stamping, bending, forming, piercing. [Oak International Inc]

Oakite Defoamant. O/w emulsion containing org. esters, alcohols, silicone, hydrocarbons, and stabilizers; foam control agent for industrial applications, e.g., paper mill stock systems, gas dehydration units, amine scrubbing units, propane deasphalting units. [Oakite Prods.]

Oakite Ladd. Blend of org. defoamer, detergent, solvs.; low-temp. cleaning/foam controlling additive for acidic or alkaline detergents, spray washing. [Oakite Prods.]

Oak Kool 625, 632A, 648. A line of soluble oil, semi-synthetic and synthetic machining and grinding coolants; for CNC machining, screw machining, turning. [Oak International Inc]

oak moss resin. A resin obtained from lichens; used as a fixative in perfumery.

Oak Oils. Petroleum oils; for drawing, stamping, grinding, and machining. [Oak International Inc] †

oak red. A coloring matter, phlobaphene, $C_{28}H_{22}O_{11}$, obtained by the hydrolysis of quercitannic acid.

Oak Syncrolube. Synthetic (oil free) lubricant; for stamping and machining. [Oak International Inc] †

Obanol 516. Polyoctyl polyamino ethyl glycine and POE alkylphenol ether; germicide, disinfectant, deodorant, fungicidal cleaning aid. [Toho Chem. Industry]

Obazoline 662Y. Imidazoline deriv.; antistat and softener for synthetic fibers; base material for shampoos and hair rinse. [Toho Chem. Industry]

Obermayer's reagent. A solution of ferric chloride in concentrated hydrochloric acid (4 g ferric chloride dissolved in 1 liter of concentrated hydrochloric acid); used for the detection of indoxyl in urine, indigo being formed if this substance is present.

Obermine Black & Yellow. CAS 1197-21-3; Phentermine hydrochloride; appetite suppressant. [O'Neal, Jones & Feldman Pharmaceuticals] *

Obinese. A proprietary preparation of chlorpropamide and metformin hydrochloride; an oral hypoglycemic agent. [Pfizer International]

obsidene. A plastic residue from the distillation of petroleum.

Obsidian. See pumice stone.

OBTS. CAS 102-77-2; N-Oxydiethylene-2-benzothiazole sulfenamide; primary accelerator for natural, SBR, nitrile, and other general-purpose rubbers. [Akrochem]

Obturin. Soluble fluoresceine.

Occidine. It contains copper sulfate, iron sulfate, sulfur, naphthalene, and calcium carbonate; used as a fungicide.

Occlusin. Light cured dental filling composite. [ICI Chem. & Polymers Ltd] †

Occultest. A proprietary test tablet of o-toluidine, strontium peroxide, calcium acetate, tartaric acid, and sodium bicarbonate; used for the detection of blood in urine. [BC Ames] *

Ocenol. The mixture of fatty alcohols derived from sperm oil; also a proprietary trade name for technical oleic acid. §

ocher. (yellow ocher, oxide yellow, Chinese yellow) CAS 1309-37-1; A natural pigment consisting of hydrated oxides of iron and manganese, mixed with clay and sand. The term is frequently restricted to a pale, yellowish-brown variety.

---

* Trade name not verified as available   † Trade name verified as obsolete

ochermatite. A mineral, 3(3Pb(AsO$_4$)$_2$ · PbCl$_2$)4Pb$_2$MoO$_5$.

ochran. A yellow bole (or clay-earth).

ocota cocota gum. (Muccocota gum). Names applied in West Africa to varieties of copal.

octabenzone. See benzophenone-12.

OctaBoost® 620. Alumina controlled zeolite; catalyst providing superior octane performance and higher gasoline yields in refining processes. [Akzo]

octacosactrin. A corticotrophic peptide.

9,12-octadecadienoic acid. See linoleic acid.

9,12-octadecadienoic acid, dimer. See dilinoleic acid.

octadecanamide. See stearamide.

n-octadecanoic acid. See stearic acid.

octadecanoic acid butyl ester. See butyl stearate.

octadecanoic acid, 1,2-ethanediyl ester. See glycol distearate.

octadecanoic acid, lithium salt. See lithium stearate.

octadecanoic acid, magnesium salt. See magnesium stearate.

octadecanoic acid, monoester with 1,2,3-propanetriol. See glyceryl stearate.

octadecanoic acid, potassium salt. See potassium stearate.

octadecanoic acid, sodium salt. See sodium stearate.

octadecanoic acid, zinc salt. See zinc stearate.

1-octadecanol. See stearyl alcohol.

n-octadecanol. See stearyl alcohol.

9-octadecenamide. See oleamide.

9-octadecenoic acid. See oleic acid.

9-octadecenoic acid, 1,2,3-propanetriyl ester. See triolein.

9-octadecenoic acid, sodium salt. See sodium oleate.

9-octadecen-1-ol. See oleyl alcohol.

N-9-octadecenyl hexadecanamide. See oleyl palmitamide

octadecylbenzenemethanaminium chloride. See stearalkonium chloride.

N-octadecyl-13-docosenamide. See stearyl erucamide.

octadecyl methacrylate. See stearyl methacrylate.

Octaflex. A proprietary preparation of octaphen resinacrylate and methacrylate polymers; plastic wound dressing. [Octel Chemicals Ltd] †

Octamine® Flake, Powd. Octylated diphenylamine; a solid amine-type antioxidant, gives minimum discoloration with maximum protection; effective in natural, SBR, BR, neoprene, and nitrile rubbers [Uniroyal]

1,8-octanedicarboxylic acid. See sebacic acid.

octanedioic acid. See suberic acid.

n-octanoic acid. See caprylic acid.

1-octanol. See capryl alcohol.

Octave. CAS 67747-09-5; Wettable powder containing 50% w/w prochloraz; a broad-spectrum fungicide for cereal crops. [Fisons plc]

octene-1. (Linear C$_8$ alpha olefin) CH$_3$(CH$_2$)$_5$CH:CH$_2$; CAS 111-66-0; EINECS 203-893-7; Intermediate for surfactants and specialty industrial chemicals. [Air Prods.; Aldrich; Chevron; Ethyl; Shell; Texaco]

Octoate Z. CAS 136-53-8; Zinc di-2-ethylhexoate; rubber activator used in soluble cure systems in place of stearic acid and partial replacement of zinc oxide for natural and synthetic rubbers. [R T Vanderbilt]

Octocure 456. CAS 7704-34-9; Sulfur aq. disp.; rubber accelerator; also for vulcanization processes in aq. latex compounds. [Tiarco]

Octocure 553. CAS 1314-13-2; French process zinc oxide

aq. disp.; pigment and accelerator for vulcanization of rubber; well-suited for use in gelled latex compounds. [Tiarco]

Octocure ZDB-50. CAS 136-23-2; Zinc dibutyldithiocarbamate; latex and rubber accelerator. [Tiarco]

Octocure ZDE-50. CAS 14323-55-1; Zinc diethyldithiocarbamate; latex and rubber accelerator. [Tiarco]

Octocure ZDM-50. CAS 137-30-4; Zinc dimethyldithiocarbamate; latex and rubber accelerator. [Tiarco]

Octocure ZMBT-50. CAS 155-04-4; Zinc mercaptobenzothiazole; latex and rubber accelerator. [Tiarco]

Octoguard FR-01. CAS 1163-19-5; EINECS 214-604-9; Decabromodiphenyloxide aq. disp.; flame retardant for water-based polymer compounds such as latex adhesives, binders, coatings, and foams. [Tiarco]

Octoguard FR-10. CAS 1309-64-4; Antimony trioxide; flame retardant for water-based polymer compounds such as latex adhesives, binders, coatings, and foams. [Tiarco]

Octoguard FR-15. CAS 1309-64-4, CAS 1163-19-5; Antimony trioxide/decabromodiphenyloxide aq. disp. (1:5 ratio); flame retardant for water-based polymer compounds such as latex adhesives, binders, coatings, and foams. [Tiarco]

octoic acid. See caprylic acid.

Octoil. CAS 117-81-7; Dioctyl phthalate; a proprietary trade name for a plasticizer. §

Octoil S. Dioctyl sebacate; a proprietary trade name for a plasticizer. §

Octojet 104. CAS 1333-86-4; Carbon black disp.; medium jetness carbon black. [Tiarco]

Octolite 544. Phenylenediamine; antioxidant and metal inhibitor. [Tiarco]

Octolite 561. Bisphenol aq. dispersion; antioxidant suitable for all white or light-colored latex compounds where color is critical; also for rubber intended for repeated or continuous contact with food. [Tiarco]

Octolite AO-28. Polymeric hindered phenol/thioester; antioxidant. [Tiarco]

Octomer DBM. CAS 105-76-0; EINECS 203-328-4; Dibutyl maleate; plasticizer, intermediate. [Tiarco]

Octomer DIBM. CAS 14234-82-3; Diisobutyl maleate; plasticizer, intermediate. [Tiarco]

Octomer DIOM. CAS 1330-76-3; Diisooctyl maleate; plasticizer, intermediate. [Tiarco]

Octomer DOM. Dioctyl maleate; plasticizer, intermediate. [Tiarco]

Octonativ. Substance for injection; factor VIII concentrate for the treatment of hemophilia. [KabiVitrum AB] *

Octopirox. CAS 68890-66-4; Piroctone olamine; antiseborrheic. [Hoechst AG] *

Octopol NB-47. CAS 136-30-1; Sodium dibutyldithiocarbamate; ultra accelerator for SBR and natural rubber latex compounds; in polymerization of chloroprene rubber; esp. for latex compounds where copper staining of zinc salt dithiocarbamates is a problem. [Tiarco]

Octopol SDE-25. Sodium diethyldithiocarbamate; natural rubber latex preservative; precipitant for heavy metals in waste water treatment. [Tiarco]

Octopol SDM-40. Sodium dimethyldithiocarbamate; polymerization shortstop in SBR rubber; precipitant for heavy metals in waste water treatment. [Tiarco]

Octoran. Low foaming nonionic surfactant of the alkylphenol ethoxylate type in liquid form; wetting, detergency,

---

‡ Trade name and manufacturer not verified   § Trade name without identified manufacturer

emulsification and antistatic agent. [Roehm Ltd] *

Octorez. Chemical modified rosins. [Tenneco Malros Ltd] *

Octosol. Modified rosin soap. [Tenneco Malros Ltd] *

Octosol 449. CAS 143-18-0; EINECS 205-590-5; Potassium oleate; foaming agent, stabilizer, emulsifier, dispersant; primary frothing aid in gelled latex foam compounds. [Tiarco]

Octosol 474. CAS 112-03-8; EINECS 203-929-1; Octadecyl trimethyl ammonium chloride; emulsifier for cationic or cationic/anionic emulsion systems. [Tiarco]

Octosol 562. CAS 112-00-5; EINECS 203-927-0; Lauryl trimethyl ammonium chloride; gel sensitizer for latex foam rubber. [Tiarco]

Octosol A-1. CAS 58353-68-7; Disodium N-[3-(dodecyloxy)propyl; sulfosuccinamate; emulsifier, dispersant, wetting agent, foaming agent for frothed latex compounds and adhesives; suspending agent in emulsion polymerization; textile softener. [Tiarco]

Octosol A-18. CAS 14481-60-8; EINECS 238-479-5; Disodium N-octadecyl sulfosuccinamate; emulsifier, dispersant, foaming agent for latex compounds, cleaners; textile softener; suspending agent in emulsion polymerization; stable in acid and alkaline aq. systems. [Tiarco]

Octosol A-18-A. CAS 68128-59-6; Diammonium N-octadecyl sulfosuccinamate; emulsifier, stabilizer, foaming agent for acrylic latex frothed compounds, formulations where reduced sodium ion content is desirable; suspending agent in emulsion polymerization. [Tiarco]

Octosol ALS-28. CAS 2235-54-3; EINECS 218-793-9; Ammonium lauryl sulfate; stabilizer, emulsifier, dispersant, foaming agent for industrial aq. systems; generates stable, high foam. [Tiarco]

Octosol HA-80. Sodium dihexyl sulfosuccinate; emulsifier for latex emulsion polymerization. [Tiarco]

Octosol IB-45. CAS 127-39-9; Sodium diisobutyl sulfosuccinate; emulsifier for latex emulsion polymerization. [Tiarco]

Octosol SLS. CAS 151-21-3; Sodium lauryl sulfate; stabilizer, frothing aid, emulsifier, dispersant for latex compounds; low cloud pt., low salt content. [Tiarco]

Octosol TH-40. CAS 23386-52-9; EINECS 245-629-3; Sodium dicyclohexyl sulfosuccinate; surfactant for emulsion polymerization, mfg. of carboxylated latexes. [Tiarco]

Octosperse TS-10. Silicone emulsion; defoamer. [Tiarco]

Octotint 103. CAS 1309-37-1; Yellow iron oxide; color dispersion. [Tiarco]

Octotint 138. CAS 13463-67-7; EINECS 236-675-5; Anatase titanium dioxide. [Tiarco]

Octovit Tablets. Vitamin A, thiamine mononitrate, riboflavine, nicotinamide, pyridoxine, cyanocobalamin, ascorbic acid, cholecalciferol, tocopherol, calcium, ferrous sulfate, magnesium and zinc; a multivitamin/mineral product indicated where supplementation with vitamins and minerals may be of benefit. [SmithKline Beecham] *

Octowax 321. CAS 8002-74-2; Refined paraffin wax emulsion; heat-stable, nondiscoloring wax emulsion with 125-130 F melting range; for use in paper and wood coating, sizing, textile lubrication, as processing aid in latex foams; antiozonant characteristics. [Tiarco]

Octowax 518. CAS 9002-88-4; HDPE emulsion; wax emulsion for use in paper and wood coating, sizing, textile lubrication, as processing aid in latex foams; antiozonant

characteristics. [Tiarco]

Octowet 40. CAS 577-11-7; EINECS 209-406-4; Sodium dioctyl sulfosuccinate; wetting agent, emulsifier, penetrant for textile washing and dyeing operations, agriculture, mining, paper, printing. [Tiarco]

Octowet 70A. Ammonium dioctyl sulfosuccinate; high speed wetting agent, solubilizer, penetrant for textile processing, agriculture, mining, paper, printing. [Tiarco]

octyl acrylate. (acrylic acid, 2-ethylhexyl ester; 2-ethylhexyl acrylate) $C_{11}H_{20}O_2$; CAS 103-11-7. [BASF; Hoechst Celanese; Sartomer; Union Carbide]

octyl alcohol. See capryl alcohol.

octyl alcohol. See 2-ethylhexanol.

octyl palmitate. (2-Ethylhexyl palmitate; 2-Ethylhexyl hexadecanoate) Ester of 2-ethylhexyl alcohol and palmitic acid; $CH_3(CH_2)_{14}COOCH_2(CH_2CH_3)CH(CH_2)_3CH_3$; CAS 29806-73-3; EINECS 249-862-1; ester for sunscreens, antispersipirants, bath oils, liq. make-up; imparts gloss; binder for pressed powds.; solubilizer for benzophenone-3. [Inolex; Van Dyk]

ocuba wax. Ocuba fat, from *Myristica ocuba.*

Ocufen Liquifilm®. Flurbiprofin sodium; anti-inflammatory. [Allergan, Inc]

Oculinum®. Botulinum toxin; neuromuscular blocker. [Allergan, Inc]

Ocusert. CAS 92-13-7; Pilocarpine; antiglaucoma agent; cholinergic. [Ciba-Geigy Corp] *

Ocusol. A proprietary preparation of sodium sulfacetamide, zinc sulfate and cetrimide; used as eye-drops. [The Boots Co plc]

Odoron. Disinfectants. [Evans Vanodine International Ltd]

Odylen®. Veterinary preparation; for external use against ectoparasites. [Bayer AG]

OEstradin. A proprietary preparation of ethinylestradiol, phenobarbitone, sodium bromide, and glyceryl trinitrate. [H N Norton & Co Ltd] ‡

Oestroform. Injection: estradiol benzoate; tablets: estradiol-estrogenic hormone. [British Drug Houses] *

OFA. Fuel additive. [Chevron] *

OFHC Copper. Oxygen-free, high conductivity copper; nickel 0.0006%, bismuth 0.001%, cadmium 0.0001%, lead 0.001%, mercury 0.0001%, oxygen 0.001% (max), phosphorus 0.003%, selinium 0.001%, sulfur 0.0018%, tellurium 0.001%, zinc 0.0001%, iron 0.0005%; 0.00 to 70 total max (tin, antimony, arsenic, bismuth, manganese, selenium, tellurium). [Amax Inc] *

Oflox®. CAS 82419-36-1; Ofloxacin; antibacterial. [Allergan, Inc]

Oftanol®. CAS 25311-71-1; Isofenphos; soil-applied insecticide used for control of insects in rice crops and pear sucker. [Bayer AG]

Oftentral. A proprietary preparation containing miconazole nitrate, polymixin B sulfate; for eye infections in cats and dogs. [Janssen Pharmaceutical Ltd] *

OGA. Gasoline additive. [Chevron] *

Ogen. Estropipate; estrogen. [Abbott Laboratories]

Ogtac 85 V. CAS 3033-77-0; Glycidyl trimethyl ammonium chloride; intermediate for cationic surfactants; modifier for synthetic polymers such as starch, cellulose, gelatins, polyacrylic acids, epoxy resins; used for production of emulsion layers on photographic plates. [Chem-Y GmbH]

Ogwin. A mixture of lime and starch used to increase the rate of sedimentation of solids in water.

---

* Trade name not verified as available   † Trade name verified as obsolete

O-hi-o. A proprietary trade name for a die steel containing 12% chromium, 1.55% carbon, 0.85% vanadium, 0.4% cobalt, and 0.8% manganese. §

OHlan®. CAS 68424-66-8; EINECS 270-315-8; Hydroxylated lanolin; primary w/o emulsifier, auxiliary emulsifier and stabilizer, pigment wetting and dispersing agent, emollient and conditioner in personal care products, absorp. bases, pharmaceuticals. [Amerchol; Amerchol Europe]

Ohmal. A resistance alloy similar in composition to manganin. It usually contains 87.5% copper, 9% manganese, and 3.5% nickel.

ohmlac kapak. A refined elaterite.

Ohmoid. A proprietary trade nane for a phenol-formaldehyde synthetic resin laminated product used for electrical insulation. §

ohm oil. A mincral oil which has been treated or contains in solution an antioxidant, thereby stabilizing the oil and increasing its electrical resistivity.

oil asphalt. A thick fluid remaining after distilling crude petroleum; used for roofing materials, and for paving when mixed with natural asphalt.

oil babulum. *See* neatsfoot oil.

oil, Bari. *See* aix oil. §

oil blue. A pigment prepared by introducing copper filings into boiling sulfur, and after cooling, boiling the mass with sodium hydroxide to remove excess of sulfur. It is applicable as an oil color only.

Oil Die. A proprietary trade name for tool steel containing 1.6% chromium, 0.45% tungsten, and 0.9% carbon. §

Oil Gard. Blend of petroleum oil, viscosity index improvers; reduces oil consumption in automotive vehicles, increases oils' viscosity index. [Gard Corporation] *

oil of mink. *See* mink oil.

oil of Palma Christi. *See* castor oil.

Oil of Partridge berry. Oil of winter green.

Oil of Pennyroyal. (oil of Poley). CAS 8007-44-1; Oil of pulegium. It consists chiefly of pulegone, $C_{10}H_{18}O$.

oil of Peter. (oil of Petre). Rock oil, or a mixture of 1 part of oil rosemary, 4 parts turpentine, and 4 parts of Barbados tar.

oil of Petitgrain. The oil obtained from the leaves of the bitter orange tree.

oil of Pompillon. An ointment of poplar seeds, also green elder ointment.

oil of Portugal. Oil of sweet orange peel.

oil of Ptychotis. Oil of ajowan.

oil of rapeseed. *See* rapeseed oil.

oil of Rhodium. (oil of duty). The oil obtained from the root of *Genista canariensis*. Also a mixture of sandalwood oil, and otto of rose, or oil of rose geranium.

oil of spike. The volatile oil from *Lavandula spica*. It is also the name for a mixture of lavender oil and oil of turpentine, colored with alkanet.

oil of St. John Wort. The oil obtained by digesting the flowering tops of *Hypericum perforatum* in warm olive oil.

oil of sweet birch. Oil of Betula from *Betula lenta*.

oil of tar. (oil of liquid pitch). Creosote.

oil of tartar. Deliquescent potassium carbonate, K_2CO_3.

oil of tea. The oil obtained from the seeds of *Camellia species*.

oil of turpentine. (spirits of turpentine, essence of turpentine, turps). Derived from the pine, *Pinus palustris* and *P. Taeda*, and from the Scotch fir, *Pinus sylvestris*.

oil of verbena. The oil obtained from *Verbena triphylla*. Also the name for oil of lemongrass.

oil of wax. *See* wax butter.

oil of wheat. The oil obtained from bruised wheat.

oil of wheat germ. *See* wheat germ oil.

oil of wintergreen. Oil of gaultheria, obtained by distilling the leaves of *Gaultheria procumbens*, or from the bark of *Betula lenta*. The chief constituent is methyl salicylate, $C_6H_4(OH)COO \cdot CH_3$.

oil Paalsgaard. *See* Schou oil.

oil pulp. (thickener, fluid gelatin, viscom). A gelatinous material made by heating aluminum oleate with mineral oil. It is sold to give increased viscosity to mineral oils.

oil, Riviera. *See* aix oil. §

oil skin. A waterproof material made by impregnating cotton or other fabric with hardening oils.

oil shale. A sedimentary rock which yields from 12-60 gallons of shale oil per ton on distillation.

Oilatum Application. A proprietary preparation of arachis oil and polyvinyl pyrrolidine; used in dermatology. [Stiefel Laboratories (UK) Ltd] *

Oilatum Emollient. A proprietary preparation of liquid fraction of acetylated lanolin alcohols in a semi-dispersing emollient base; used in dermatology. [Stiefel Laboratories (UK) Ltd] *

Oilatum Soap. Contains unsaponified arachis oil BP and high molecular weight fatty acid salts; as an aid in the management of dry, pruritic skin conditions such as senile pruritus, atopic dermatitis and in the drying stage of eczema. [Stiefel Laboratories (UK) Ltd] *

Oildag. Colloidal graphite in 750 hydrotreated refined naphthenic petrol. oil; lubricant additive for chain lubes, aerosols, penetrating lubes, assembly lubes, thread lubes. [Acheson Colloids]

oiled silk. Thin silk fabric which has been impregnated with oil, usually linseed.

Oilfos. CAS 7558-80-7; Glassy sodium phosphate; controls viscosity of oil well drilling muds. [Monsanto Co] *

oils, cod liver. *See* cod liver oil.

oils, corn. *See* corn oil.

oils, jojoba. *See* jojoba oil.

oils, linseed. *See* linseed oil.

oils, olive. *See* olive oil.

Oilsol. Dyestuffs soluble in oils, waxes and plastics. [Morton Int'l. Ltd]

oil-soluble resins. Synthetic resins of the phenol-formaldehyde type obtained by a fixing process, using rosin in the mixing. They dissolve in hydrocarbon solution and are no longer soluble in alcohol.

oils, palm. *See* palm oil.

oils, palm kernel. *See* palm kernel oil.

oils, pine. *See* pine oil.

oils, safflower. *See* safflower oil.

oils, vegetable. *See* vegetable oil.

oils, walnut. *See* walnut oil.

oil-stone. *See* stink-stone.

Oil-Treet. Fuel oil conditioner and combustion catalyst. [Schaefer Technologies Inc]

oil, var. *See* aix oil. §

oil varnishes. Solutions of resins in linseed oil.

oil vulcanized. *See* Thinoline.

oil wax. *See* wax butter.

oil white. (light white, leukarion, albanol, diamond white, Edelweiss, snow white. anti-white lead, blenda, condor,

---

‡ Trade name and manufacturer not verified   § Trade name without identified manufacturer

fixopone, nivan). White lead substitutes. They consist chiefly of lithopone mixed with white lead or zinc white, also with whiting, gypsum, magnesia, or silica.

Ointment Base No. 3, 4, 6. White petrolatum USP; ointment base for eye and skin medications; carriers for medical materials. [Penreco]

oiticica. CAS 8016-35-1; Oil derived from the seeds of the Brazilian oiticica tree; used as a drying oil in paints and varnishes

Okerin. Antiozonant waxes, blends of petroleum waxes; for use in tire sidewall compounds and general rubber products. [Astor Chemical Ltd]

Oko®. Household insecticide for combating domestic flying insects. [Bayer AG]

Okol. A disinfectant consisting of an emulsion containing phenols. It is miscible with water.

Okstan M 62. Butyltin carboxylate, mercpatide-boosted; stabilizer for plastisol processing; provides heat and light stability, superior transparency, excellent color; very suitable for coating of canvas material and metal surfs. [Bärlocher GmbH]

Okstan M 69 S. Butyltin mercaptide; stabilizer for plastisol processing; provides superior stability and color properties; resist. to water staining; esp. low-odor; suitable for automotive dashboard panels. [Bärlocher GmbH]

Okstan X3. A trade name for a butyl tin mercaptide; a stabilizer for PVC with an exceptionally high heat stability. [Otto Bärlocher GmbH, Chemische Werke Munchen]*

Okstan XO. CAS 58229-88-2; A di-n-octyl tin mercaptide for stabilization of nontoxic PVC compounds. [Croda Resins Ltd] †

Olapon ND-9, SW. Detergent for textile scouring. [Reilly-Whiteman]

Olay Beauty Bar. A cleansing bar to clean, soften and smooth the skin. [Richardson-Vicks Inc] *

Olay Beauty Cleanser. Gentle, greaseless facial cleanser to leave skin soft and smooth. [Richardson-Vicks Inc] *

olcotrop leather. A leather made from a species of shark skin.

Old Plantation. CAS 6484-52-2; Ammonium nitrate; fertilizer. [Columbia Nitrogen Corporation] *

olea europaea oil. See olive oil.

oleamide. (9-octadecenamide; oleyl amide) Aliphatic amide; $CH_3(CH_2)_7CH:CH(CH_2)_7CONH_2$; CAS 301-02-0; EINECS 206-103-9; slip/antiblock agent for extrusion of polyethylene; wax additive; ink additive. [Akzo; Chemax; Chemron; Croda Universal; Henkel/Emery; Mona; Syn. Prods.; Witco/Humko]

Oleandocyn. A proprietary preparation of oleandomycin phosphate; an antibiotic. [Pfizer International] †

oleandomycin. An antibiotic produced by certain strains of *Streptomyces antibioticus.*

olefiant gas. (heavy carburetted hydrogen, dicarburetted hydrogen, elayl, ethene, etherin; ethylene) C_2H_4; CAS 74-85-1; Used in the manufacture of polyethylene, polypropylene, ethylene oxide, ethylene glycols, ethylene chlorohydrin, acetaldehyde, ethyl alcohol, polystyrene, styrene, polyvinyl chloride, etc.; as a refrigerant, welding and cutting of metals, in orchard sprays to accelerate fruit ripening.

oleic acid. (cis-9-octadecenoic acid; red oil; elainic acid; 9-octadecenoic acid) Unsaturated fatty acid; CH_3-$(CH_2)_7CH:CH(CH_2)_7COOH$; CAS 112-80-1; EINECS 204-007-1; soap base, manufacture of oleates, ointments, cosmetics, polishing compounds, lubricants, food-grade additives, Turkey red oil, driers; waterproofing textiles; oiling wool; pharmaceutic aid (solv.). [Akzo; Arizona; Henkel/Emery; Hercules; Unichema; Union Derivan SA; Witco/Humko]

oleic acid, potassium salt. See potassium oleate.

olein. CAS 122-32-7; Trioleyl derivative of glycerol, $C_3H_5(OC_{18}H_{33}O)_3$; term applied commercially to any liquid oil obtained from solid fats by pressure, to crude oleic acid, and to the potassium, sodium, or ammonium salt of the sulfonate of oleic acid; used as textile lubricants.

olein. See triolein.

olein of saponification. Commercial oleic acid prepared by the saponification of pure fats, and the separation from stearine, by pressing.

oleite. Sodium-sulfo-ricinoleate.

oleo. (premier jus). The oil expressed from beef-fat; used in margarine manufacture.

oleobismuth. An oil suspension of bismuth oleate.

Oleo-Coll LP. Lecithin, butyl stearate, coco-hydrolyzed animal protein, oleoyl sarcosine, sesame oil, lanolin alcohol; cosmetics ingred. [Brooks Industries]

Oleogen. See Parogen.

oleoguaiacol. Guaiacol oleate, $CH_3O \cdot C_6H_4 \cdot O \cdot CO \cdot C_{17}H_{23}$; used as an antiseptic.

Oleo Keratin ISO. AMP isostearoyl hydrolyzed keratin, isostearic acid, myristyl myristate, isopropyl palmitate; cosmetics ingred. [Brooks Industries]

oleoresins. Resins mixed with a volatile oil; they are semisolid and tacky at room temperature and soft and sticky at high temperatures.

Oleosol. See Oildag and Aquadag. [Acheson Colloids] *

Oleo-Soy C. Coco-hydrolyzed soy protein; cosmetic ingredient for hair and skin care products. [Brooks Industries]

Olepal ISO. CAS 56002-14-3; PEG-6 isostearate. [Gattefosse SA]

oleum. See fuming sulfuric acid.

oleum spirits. A petroleum distillate.

oleum white. A pigment. It is a sulfide of zinc mixed with a small proportion of barium sulfate (a lithophone); used as a rubber filler.

Olex. Fat soluble powdered flavors. [Bush Boake Allen Ltd]

oleyl alcohol. (9-octadecen-1-ol; cis-9-octadecen-1-ol) Unsaturated fatty alcohol; $CH_3(CH_2)_7CH=CH(CH_2)_8OH$; CAS 143-28-2; EINECS 205-597-3; in metal cutting oils, printing inks, textile finishing; as antifoam agent, plasticizer; manufacture of sulfuric esters used as surfactants, detergents, wetting agents; carrier for medicaments. [Croda; R.W. Greeff; Lanaetex; M. Michel; Ronsheim & Moore; Sherex]

oleyl amide. See oleamide.

oleyl palmitamide. (hexadecanamide, N-9-octadecenyl-; N-9-octadecenyl hexadecanamide) Substituted aliphatic amide; $CH_3(CH_2)_{14}CONHCH_2(CH_2)_7CH-CH(CH_2)_7CH_3$; CAS 16260-09-6; release agent providing slip, antiblocking to thermoplastics incl. PP film, nylon. [Croda Universal; Witco/Humko]

olibanum. (Indian frankincense, Salaigugl). A gum-resin obtained from *Boswellia* species. It contains from 8-18% of essential oil, 55-57% of resin, and 20-23% gum (carbohydrates).

Olicat® C. Polyisobutylene in mineral oil; tackiness agent,

---

lubricant in lubricating oils and greases. [Alox] *

Oligan. Shrink-resist additive. [Ciba plc] *

Olinor. Blend of hydrocarbons and sulfated fatty alcohols; softening agent for raising cotton goods. [Henkel Chemicals Ltd] *

olivenite. A mineral, $4CuO \cdot As_2O_5$.

olive oil. (oils, olive; olea europaea oil) Fixed oil obtained from the ripe fruit of *Olea europaea*; pale yellow to greenish liquid, nondrying with slight odor and taste; CAS 8001-25-0; EINECS 232-277-0; salad dressing and other foods; ointments, liniments, soaps, as lubricant and emollient, in cosmetics, pharmaceuticals. [Arista Industries; Croda; Penta Mfg.; Reilly-Whiteman]

olives of Java. (Kaloempang beans, Beligno seeds, Sterculia kernels). Seeds of *Sterculi foetida*, the source of Sterculia oil.

olivite. A substance having a rubber base; used as an acid-proofing material in pumps.

Olminal. A trade name for a commercial aluminum oleate; contains 1.5% aluminum. §

OLOA. Lubricating oil additive. [Chevron] *

Olobintin. A German preparation. It is a 10% solution of a mixture of various rectified turpentine oils.

olutkombul. The gelatinous sap of the plant *Abroma angustum*; used in India in dysmenorrhea.

Olympic. CAS 14807-96-6; Talc; for cosmetic applications incl. baby powds., creams and lotions, foot powds., dusting powds. [Cyprus Industrial Min.

Olympic Bronze. A proprietary trade name for an alloy of copper with 3% silicon and 1% zinc. §

Olympic Bronze G. A proprietary trade name for an alloy of copper with 22% zinc and 1% silicon. §

Omacide® P-BBP-5. 5% Zinc pyrithione, 80% butyl benzyl phthalate plasticizer; antimicrobial disp. for protection of PVC systems. [Olin]

Omacide® P-DIDP-5. 5% Zinc pyrithione, 65% diisodecyl phthalate plasticizer; antimicrobial disp. for protection of PVC systems. [Olin]

Omacide® P-ESO-5. 5% Zinc pyrithione, 75% epoxidized soybean oil plasticizer; antimicrobial disp. for protection of PVC systems. [Olin]

Omadine® MDS. Bispyrithione, magnesium sulfate; anti-dandruff agent for nonalkaline hair care products; antimicrobial agent for Gram-negative and Gram-positive bacteria; also inhibits the growth of fungi. [Olin] *

Omal. Trichloro-phenol.

Omarsan. A detergent and sterilizer. [PPF International Ltd]*

Ombrelub FC 533. Waterproofing agent for aq. systems, e.g., printing inks, cement, and concrete mixtures. [Münzing Chemie GmbH]

Omeril®. Anti-allergic agent. [Bayer AG]

omic acid. *See* osmium tetroxide.

Omnicryl. Aqueous color dispersions for use in water based ink systems and rubber latex applications. [Engelhard]

Omnilac. Non-nitrocellulose overlacquers. [The Scottish Adhesives Co Ltd]

Omnilube. A series of synthetic oils and greases based on polyalphaolefins and various organic compounds; a variety of oils used for OEM applications in appliances, power tools, computers, automotive electrical motors, and industrial maintenance applications. [Ultrachem Inc]*

Omnipaque. CAS 66108-95-0; Iohexol; diagnostic aid.

[Sterling Drug Inc] *

Omnipen. CAS 69-53-4; Ampicillin; antibacterial. [Wyeth Laboratories]

Omnipen-N. CAS 69-53-4; Ampicillin; antibacterial. [Wyeth Laboratories]

Omnisorb®. Cells. [Calbiochem Corp]

Omnivax. Anticlostridial vaccine. [The Wellcome Foundation Ltd] *

Omnopon. CAS 8002-76-4; A proprietary preparation of papavertum; an analgesic. [Roche Products Ltd] *

Omnopon-Scopolamine. Pre-anesthetic medication. [Roche Products Ltd] *

OMTS. CAS 102-77-2; N-Oxydiethylene-2-benzothiazole-sulfenamide; delayed-action accelerator for SBR, NR, and nitrile rubbers. [Akrochem]

Omyastab. A secondary stabilizer for chlorinated polyester resins. It is a colorless cyclic organic compound containing ether linkages which are compatible with the polyester resin.

Onadox-118. A proprietary preparation of aspirin and dihydrocodeine bitartrate; an analgesic. [British Drug Houses] *

Onamine 12. Dodecyl dimethylamine in liquid form; intermediate in the synthesis of surfactants, antioxidants, oil and grease additives. [Millmaster-Onyx UK] *

Onamine 14. Tetradecyl dimethylamine in liquid form; intermediate in the synthesis of surfactants, antioxidants, oil and grease additives. [Millmaster-Onyx UK] *

Onamine 16. Hexadecyl dimethylamine in liquid form; intermediate in the synthesis of surfactants, antioxidants, oil and grease additives. [Millmaster-Onyx UK] *

Onamine 18. Stearyl dimethylamine in liquid form; intermediate in the synthesis of surfactants, antioxidants, oil and grease additives. [Millmaster-Onyx UK] *

Onamine 65, 835 and 1214. Alkyl dimethylamine in liquid form; intermediate in the synthesis of surfactants, antioxidants, oil and grease additives. [Millmaster-Onyx UK]*

Onamine RO. 1-(2-hydroxyethyl) 2-n-heptadecenyl-2-imidazoline in liquid form; acid-stable emulsifier used in corrosion inhibition, solvent degreasing, defoaming, and demulsifying. [Millmaster-Onyx UK] *

Once. Hand cleanser. [Hardman]

Once. Decorative waterborne high opacity paint. [ICI Chem. & Polymers Ltd]

Oncol. CAS 82560-54-1; Insecticide containing benfuracarb. [ICI Chem. & Polymers Ltd]

Oncol 10G. CAS 82560-54-1; Benfuracarb; soil applied insecticide and nematicide for use on sugar beet. [Farm Protection Ltd]

Oncomouse®. For the medical industry. [DuPont UK]

Oncor. Basic lead silico pigments; anti-corrosive pigment for paint. §

Oncor 75. Flame retardant compositions. [Anzon Ltd] *

Oncovin. CAS 57-22-7; Vincristine. [Dista Products Ltd]

Oncovin. CAS 2068-78-2; Vincristine sulfate; antineoplastic. [Eli Lilly & Co]

Ondene®. Insecticide with fast killing activity and good penetrating action particularly for the control of sucking insects and bugs, cocoa flat bugs, leafhoppers and other pests; for use on cereals, pome and stone fruit, citrus fruit, rice, sugar cane. [Bayer AG]

Ondoita. A proprietary synthetic resin molding powder. §

Ongard. Smoke suppressants. [Anzon Ltd] *

---

‡ Trade name and manufacturer not verified   § Trade name without identified manufacturer

Oniachlor. Chlorinated isocyanates. [Sartomer]

Onion's alloy. A fusible alloy containing 50% bismuth, 30% lead and 20% tin.

Onslaught. CAS 330-55-2, 1582-09-8; Suspension concentrate containing 160 g linuron and 320 g trifluralin per liter; herbicide for winter cereals. [Quadrangle Agrochemicals]

Ontario Steel. A proprietary trade name for a non-shrinking steel containing 11% chromium, 0.75% molybdenum, 0.25% vanadium, 0.35% silicon, 0.30% manganese, and 1.45% carbon. §

on the ball. Superwash for golfballs, reducing skin friction and adding substantial distance to a golfball drive. [Merix]

O.N.V. Diphenylcarbamyl dimethyldithiocarbamate; a proprietary trade name for a rubber vulcanization accelerator. §

onyx. Consists chiefly of silica.

Onyx Classica OC-1000. Hydrated alumina; developed for synthetic onyx for synthetic marble industry; features high whiteness, purity, consistency, flame retardancy, compatibility. [Alcoa]

Onyxide® 75. Alkenyl(90% C18, 10% C16) dimethyl ethyl ammonium bromide in paste form; cationic surfactant algicide used in recirculating water systems; swimming pools; humidifiers. [Millmaster-Onyx UK] *

Onyxide® 172. Alkyl dimethyl ethyl benzyl ammonium cyclohexyl sulfamate in liquid form; antifungal agent and preservative used in latex emulsions; paints; adhesives; coated fabrics; cutting oils. [Millmaster-Onyx UK] *

Onyxide® 200. Hexahydro-1.3.5-tris (2-hydroxyethyl)-s-triazine; preservative for sol. cutting fluids and coolants; bactericide for oilfield drilling fluids, enhanced oil recovery operations. [Stepan; Stepan Canada]

Onyxide® 3300. Alkyl dimethylbenzyl ammonium saccharinate in powder form; germicide, conditioner and disinfectant with low skin and eye irritation; used in cosmetics and pharmaceuticals; hair preparations; detergent-sanitizers; disinfectants. [Millmaster-Onyx UK] *

onyx marble. A marble containing fossil shells.

onyx of Tecali. A variety of alabaster. The color varies from milk-white to pale yellow and pale green.

Onyxol. Alkanolamides. [Millmaster-Onyx UK]

Onyx Premier WP-31. Alumina trihydrate; filler for cultured onyx; provides highest whiteness, excellent translucency; flame retardant, smoke suppressant. [Alcan]

O.O.D. CAS 22801-45-2; 2-Octyldodecyl oleate; neutral oil improving adhesion, gloss, conditioning in lipsticks, foundation, eye shadow stick, hair creams. [U.S. Cosmetics]

Oolitic limestone. A massive variety of calcium carbonate; used for building purposes.

Opacicoat. CAS 471-34-1, 1317-65-3; Calcium carbonate; ultrafine ground coated pigment with unique particle size distribution; permits higher loadings in polymer systems while still maintaining properties. [ECC]

Opacimite. CAS 471-34-1, 1317-65-3; Calcium carbonate; finely ground product with an engineered particle size distribution for use in paints and polymers. [ECC]

opacite. Stannic chloride, $SnCl_4$.

Opacode. An edible ink for pharmaceutical or food use. [Colorcon Ltd] *

Opacolor. Coloring system for sugar coated confectionery pieces. [Colorcon Ltd] *

Opadry. Complete film coating system in a dry form for reconstitution; Aqueous film coating, organic film coating and enteric coating. [Colorcon Ltd] *

Opalite. A proprietary trade name for an amorphous silica.§

opal jasper. It is silica, resembling jasper.

Opalon. A proprietary trade name for phenol-formaldehyde cast resins. §

Opalon 740. A trade name for a graft copolymer used as a semi rigid wire insulation. §

Opalux. Coloring system for sugar coated tablets; sugar coating for pharmaceutical products. [Colorcon Ltd] *

opalwax. Hydrogenated castor oil.

Opaspray. Coloring system for film coated tablets; aqueous film coating colorant, organic film coating colorant. [Colorcon Ltd] *

Opatint. A multipurpose liquid color dispersion formulated for the coloring of all types of food and confectionery. [Colorcon Ltd] *

Opazil. Acid and alkaline activated bentonites for the adsorption of detrimental substances in the papermaking process. [Süd-Chemie AG] *

Opazil®. Activated bentonite-based; for absorption of interfering substances in papermaking. [BASF AG]

Opbipol. Polymeric optical fiber cable. [BASF plc]

Opera. Suspension concentrate containing 150 g terbutryn and 200 g trifluralin per liter; for weed control in winter cereals. [Tripart Farm Chemicals Ltd]

Operidine. A proprietary preparation of phenoperidine hydrochloride. Analgesic supplement in anesthesia. [Janssen Pharmaceutical Ltd] *

Opex® 80. Dinitro; chemical blowing agent. [Uniroyal]

Ophorite. An ignition powder for projectiles. It consists of magnesium powder and potassium chlorate.

Ophthaine. CAS 5875-06-9; Proparacaine hydrochloride; an anesthetic. [Bristol-Myers Squibb Co Inc] *

Ophthaine Solution. Proxymetacaine hydrochloride injection in aqueous solution. [Ciba plc] *

Ophthalgan. Glycerin; a pharmaceutic aid. [Wyeth Laboratories] *

Ophthalmadine. Idoxuridine 0.1% eye drops and 0.5% eye ointment; treatment of ocular *Herpes Simplex*. [SAS Pharmaceuticals Ltd] ‡

Ophthetic®. Proparacaine hydrochloride; anesthetic. [Allergan, Inc]

Opilon. A proprietary preparation of thymoxamine hydrochloride; a vasodilator. [Warner] *

Opino®. Venous therapeutic agent. [Bayer AG]

opium. CAS 8008-60-4; The dried juice from the unripe capsules of *Papaver somniferum*. It contains morphine, codeine, narcotine, narceine, thebane, papaverine, and meconin.

Opobyl (Pills). Desiccated liver 50 mg, sodium tauroglycocholate 50 mg, aqueous extract of boldo 10 mg, podophyllin 2 mg, alcoholic extract of euonymus 2 mg, aloes 20 mg per pill; Hepatic insufficiency, biliary stasis, habitual constipation caused by intestinal sluggishness. [Bengue & Co Ltd] *

Opogard 500. Suspension concentrate containing 150 g terbuthylazine and 350 g terbutryn per liter; weed germination inhibitor. [Ciba-Geigy Agrochemicals]

Opol. Chemical polishers for plastics. [Laporte Industries Ltd] *

opopanax. The dried juice from the roots of *Pastinaca opopanax*; used in perfumery, and medicinally as an

---

* Trade name not verified as available   † Trade name verified as obsolete

antispasmodic.

OPPalyte® . Range of polypropylene films; used in the packaging of food and nonfood products and special industrial applications. [Mobil Plastics Europe] *

OPPalyte® 233 TW, 278 TW, 350 TW. CAS 9003-07-0; Oriented PP film; nonheat sealable; high opacity expanded core with modified OPP skin layers both sides for durability and WVTR; treated both sides for ink and adhesive anchorage. [Mobil] *

OPPalyte® 250 ASW, 350 ASW. Oriented PP film, 1 side acrylic coated, 1 side PVDC coated, heat sealable; high opacity pearlescent core; acrylic coating provides excellent machinability, gloss, aroma barrier, and superior surface for solv. and water-based inks; PVDC coating provides strong seals, hot tack, moderate oxygen barrier, and excellent adhesion to cold seal; useful in single wall or lamination, and overwrap applications. [Mobil] *

Oppanol® . Polyisobutylene in the form of viscose liquids or rubbery solid crumb, depending on molecular weight; used in sheet or film form for waterproofing or corrosion resistant coatings, to modify properties of polyolefin polymers and waxes used in film or impregnation; for sealants and adhesives. [BASF plc]

Oppanol® B. Polyisobutylene; for the adhesive and sealant industry, elec. insulating oils, bases for chewing gums, for production of damp-proof courses containing fillers in construction industry. [BASF AG]

Oppanol® D. Polyisobutylene dispersion. [BASF plc] *

Oppasin® . Inorganic and organic. pigment concs.; for coloring rubber compounds and solutions; optical brighteners for textile fibers. [BASF AG]

OPS® . Biaxially oriented polystyrene sheet and film. [Asahi Chem. Industry]

Opsan. Antibacterial eye drops. [The Boots Co plc]

Op-Sulfa 30. Sulfacetamide sodium; an antibacterial. [Broemmel Pharmaceuticals] *

optannin. Basic calcium tannate.

Optbas. Green, effervescent tablets which are dissolved in water to produce an antiseptic eye lotion; each tablet contains adrenaline BP 0.6% w/w, phenylephrine hydrochloride BP 0.05% w/w and acriflovine BPC 1963 0.005% w/w; for relief of tired strained eyes, watering and bloodshot eyes, inflammation and catarrhal conjunctivitis. [Evans Medical] *

Optemet. Prefabricated disposable pouring trumpets for uphill teemed steel ingots. [Foseco (F.S.) Ltd] †

Opthaine. A proprietary preparation of proxymetacaine hydrochloride solution; an ocular anesthetic. [Bristol-Myers Squibb Pharmaceuticals Ltd] *

Op-Thal-Zin. Zinc sulfate; astringent (ophthalmic). [Alcon Laboratories Inc] *

optical bronze. An alloy of 89% copper, 6.5% zinc, and 4.5% tin.

Opticite. Polystyrene-based packaging labelling material. [Dow UK]

Opticlean® L-1000. Bacterial protease; enzyme for hydrolysis of protein; excellent stability in presence of oxygen bleaches; for use in highly alkaline laundry detergents. [Solvay Enzymes]

Opticorten. Steroid veterinary ethicals. [Ciba plc] *

Opticrom. CAS 15826-37-6; Cromolyn sodium; an antiasthmatic. [Fisons Corp, Pharmaceuticals Div]

Opticrom. Eye drops. [Fisons plc, Pharmaceuticals Div] *

Optiflex. Bovine somata tropine. [Dista Products Ltd]

Optigard®. A water blocking agent and lubricant used as an additive in the manufacture of optical and fiber cables. [Dow Corning]

Optigel. Thixotropic clay gellant. [Production Chemicals Ltd]

Optigel WM. Bentonite, cellulose gum; thixotrope for aq. cosmetic/toiletry systems, putty, caulk, cleaning compounds, waxes, polishes. [United Catalysts]

Optima 23B. CAS 68334-28-1; Partially hydrogenated vegetable oil (cottonseed, soybean); center fat for confectionery and bakery applications. [Van Den Bergh Foods]

Optima C. Activated bleaching earth; used for refining of vegetable, animal and mineral oils, fats and waxes. [Minas de Gádor SA]

Optimase® APL. Alkaline protease; enzyme for hydrolysis of proteins; used in moderately alkaline laundry detergents, silver recovery from x-ray and photographic film. [Solvay Enzymes]

Optimase® M-440. Alkaline protease; enzyme for hydrolysis of proteins; used in moderately alkaline laundry detergents, silver recovery from x-ray and photographic film. [Solvay Enzymes]

Optimase® PAG. Protease amylase; enzyme for hydrolysis of proteins; used in moderately alkaline laundry detergents, silver recovery from x-ray and photographic film. [Solvay Enzymes]

Optimask®. For the electrical industry. [DuPont UK]

Optimem. Plastic (or other polymeric material) membranes used for separation and related liquid treatments. [ICI Chem. & Polymers Ltd]

Optimine. CAS 3978-86-7; Azatadine maleate; antihistaminic. [Schering Corp] *

Optinol. Carriers used for dyeing polyester. [Yorkshire Pat-Chem; Yorkshire Chemicals plc]

Option® . For the agriculture industry. [DuPont UK]

Optipol® . Polymeric optical fiber with PC core and special cladding; high temp. resist. and flexibility suited for use in harsh conditions with relatively short transmission lengths; for machine and appliance automation, traffic signals, auto instrument illumination, process control engineering. [Bayer AG] *

Optivest. Dental chrome investment. [Degussa Ltd]

Optiwhite®. CAS 12141-46-7; Aluminum silicate, thermo-optic; pigment retaining whiteness and hiding when embedded in binders; for use in paper coating, paints, rubber, and plastics to extend TiO_2 or other costly pigments; excellent dispersion, good electricals, high hiding, improved film properties. [Burgess Pigment]

Optiwite Series. Optical whites for use on lycra, nylon, acetate, and wool or blends. [CNC Int'l. L.P.]

Optocillin® . Broad-spectrum penicillin combination with penicillinase-fast component. [Bayer AG]

Optodent®. Artificial teeth. [Bayer AG]

Optognath®. Artificial teeth. [Bayer AG]

Optosil®, Optosil Hard. Elastomeric impression material and ancillaries; dental speciality for practice and laboratory. [Bayer AG] †

Optosil® Liquid. Elastomeric isolating film on silicone basis; used for dental techology. [Bayer AG]

Optosil® Plus. Organic silicone polymer; used for dental impressions. [Bayer AG]

Optox. Melted lead oxide granules of high purity; basic raw material for optical glass. [Chemson Polymer Additiv GmbH] *

---

‡ Trade name and manufacturer not verified   § Trade name without identified manufacturer

Optran. High grade optical chemicals. [BDH Chemicals Ltd]

Optrex. Proprietary preparations containing distilled witch hazel; eye treatment preparations. [The Boots Co plc]

Opulets. A proprietary preparation of atropine sulfate in gelatin; used as a long acting mydriatic as an aid in diagnosis, and in treatment of uveitis. [Pharmax Ltd] *

ORA. Refinery process stream additive. [Chevron] *

Orabase. A proprietary preparation containing sodium carboxy-methyl cellulose, pectin, and gelatin in a liquid paraffin polyethylene base. [Bristol-Myers Squibb Pharmaceuticals Ltd] *

Oracet. Solvent soluble dyes. [Ciba plc]

Oracle®. Isoproturon and metsulfuron-methyl; used for annual weed control in wheat and barley. [DuPont UK]

Ora-Cop. Poultry nutrition supplement. [Mineral Research & Development Corp]

Oragrafin Calcium. CAS 1151-11-7; Ipodate calcium; diagnostic aid. [Bristol-Myers Squibb Co Inc] *

Oragrafin Sodium. CAS 1221-56-3; Ipodate sodium; diagnostic aid. [Bristol-Myers Squibb Co Inc] *

Orahesive. A proprietary preparation containing sodium carboxymethyl cellulose, pectin, and gelatin. [Bristol-Myers Squibb Pharmaceuticals Ltd] *

Oralcer®. A proprietary preparation of clioquinol (35 mg) and ascorbic acid (6 mg); used in the treatment of mouth ulcers. [Vitabiotics Ltd]

Oraldene. CAS 141-94-6; A proprietary preparation of hexetidine; an antiseptic mouth wash. [Warner] *

oralith. Organic pigments.

Oramide DL 200 AF. Cocamide DEA; cosmetic emulsifier. [Seppic]

Oramorph. CAS 64-31-3; A proprietary preparation of morphine sulfate; for oral administration. [Boehringer Ingelheim Ltd]

Oranabol. CAS 145-12-0; A proprietary trade name for oxymesterone. §

Orange III. A coloring for all types of food and confectionery. [Colorcon Ltd] *

orange lac. See lemon lac.

orange lead. (orange mineral, orange red, sandix, Saturn red). A red lead obtained by calcining powdered white lead. It is a better red lead than crystal minium.

orange tungsten. Saffron bronze, tungsten-sodium tungstate, $Na_2WO_4 \cdot W_2O_5$.

Oranit. See Nekal.

oranium bronze. (Dirigold). An aluminum bronze. It contains from 87-97% copper and 3-11% aluminum.

Orap. CAS 2062-78-4; Pimozide; antipsychotic. [McNeil Pharmaceuticals] *

Orasol. Solvent soluble dyes. [Ciba plc]

Orasone. CAS 53-03-2; Prednisone; glucocorticoid. [Rowell Laboratories Inc] *

Ora-Testryl. CAS 76-43-7; Fluoxymesterone; androgen. [Bristol-Myers Squibb Co Inc] *

Oratrast. CAS 7727-43-7; Barium sulfate; diagnostic aid. [Armour Pharmaceutical Co] *

Oratrol. CAS 120-97-8; Dichlorphenamide; carbonic anhydrase inhibitor. [Alcon Laboratories Inc] *

Oravue. CAS 41473-08-9; Iopronic acid; diagnostic aid. [Bristol-Myers Squibb Co Inc] *

Orbenin. CAS 61-72-3; A proprietary preparation containing cloxacillin; an antibiotic. [SmithKline Beecham] *

Orbicin. A proprietary preparation of diabekacin sulfate; an antibiotic. [Pfizer International] †

Orbigastril. A proprietary preparation of oxyphencyclimine and meprobromate; an antispasmodic and tranquilizer. [Pfizer International] †

Orbinamon. A proprietary preparation of thiothixene hydrochloride; a psychotherapeutic agent. [Pfizer International]

Orbitol. Photographic developer. [May & Baker Ltd] *

Orca. A French synthetic resin prepared from acrolein; used for electrical insulation. §

Orchard Herbide. CAS 61-82-5, CAS 330-54-1; Amitrole + diuron; used for total weed control in fruit orchards. [Hoechst UK]

Orchidee. Gold bearing on glaze colors for porcelain, bone china, and earthenware. [Degussa Ltd]

Orchidee. (Sanfoin). The isoamyl ester of salicylic acid or o-oxy-benzoic acid, $C_6H_4(OH)COOC_5H_{11}$; used in perfumery.

orchil. See archil.

orchindone. Isobutyl salicylate; used in perfumery.

orcin. CAS 504-15-4; (dihydroxytoluene; methylresorcinol; orcinol) $C_6H_3 \cdot CH_3(OH)_2$; Reagent for beet sugar, lignin, and pentoses.

ordeal bark. The bark of *Erythrophloeum guincense*.

ordeal bean. (split nut). Calabar bean, a source of eserine.

Ordnance 204, 500. Oil-based, water-soluble cutting oil (extreme pressure). Multipurpose oil-based soluble cutting oil for all metal removing operations where physical and chemical extreme pressure assistance is required to assist functional cooling properties on many metals where a soluble oil is preferred. [Sumner Oil Industries] ‡

ordonezite. A mineral; $2(ZnSb_2O_6)$.

Ordoval. A synthetic tannin made from sulfonated anthracene. §

Ordoval G.2G. See Neradol. §

ore-furnace slag. A slag obtained when roasted copper sulfide ore is mixed with oxidized ores and slag and fused for the production of coarse metal. It often contains unfused quartz and less than I% copper, and is mainly a silicate of iron.

Oregon balsam. The true oleoresin is obtained from *Pseudotsuga mucronata*, but another product sold under the same name consists of a mixture of rosin and turpentine.

Oreide. A yellow alloy resembling gold. It usually contains from 80-90% copper, 10-14.5% zinc, and 0-4.5% tin.

Orelite. See Metalite. §

orellin. A yellow coloring matter found in annatto. It is probably an oxidation product of bixin, another coloring matter of annatto.

Oreoselone. Oreoselin, $C_{14}H_{12}O_4$.

Oresmasin. Pigments for textiles. [Ciba plc] *

Oretic. CAS 58-93-5; Hydrochlorothiazide; diuretic. [Abbott Laboratories]

Oreton. CAS 58-22-0; Testosterone; androgen. [Schering Corp] *

Oreton Methyl. CAS 58-18-4; Methyltestosterone; androgen. [Schering Corp] *

Oreton Propionate. CAS 57-85-2; Testosterone propionate; androgen. [Schering Corp] *

Orevac® 9309. EVA terpolymer; for cast film coextrusion with PE/PA, ionomers/PA. [Atochem] *

Orevac® 18211. EVA terpolymer; for sheet coextrusion with PS, EVOH, PE, adhesive emulsions (aluminum). [Ato-

---

chem] *

Orevac® 18302. CAS 9002-88-4; PE; for blown film, and bottle blow molding coextrusion; for PE/PA, PS/EVOH/PE adhesives for barrier materials, thermo adhesive film. [Atochem] *

Orevac® PP-C. CAS 9003-07-0; PP, grafted; for blown film coextrusion, blow molding coextrusion, steel pipe coating, underfloor heating pipes. [Atochem] *

Orgamide R. A proprietary trade name for nylon 6. §

Organidin. Glycerol, iodinated; iodinated dimers of glycerol; expectorant. [Wallace Laboratories]

Organol. Dyes for the petroleum industry. [ICI Chem. & Polymers Ltd]

Organopol®. A range of retention aids used in papermaking specific to DIP/TMP/CTMP and highly contaminated stocks where normal retention aids have no effect. [Allied Colloids Ltd]

Organosilane Si203, Si208. For penetrating sealers that protect concrete and wood from water and chloride ion intrusion. [Degussa]

organosiloxane. See silicone.

Orgatrax. Hydroxyzine hydrochloride; a tranquilizer. [Organon Inc] *

Orglas. GRP linings and coatings. [Prodorite Ltd]

Orgozon CC 1118. Fatty phosphate ester, potassium salt; detergent and emulsifier for use in mild to strong alkaline conditions. [Clough]

Orgozon Conc. 0680. Phosphate ester, free acid form; detergent, emulsifier for use in mild to strong alkaline conditions; may be neutralized with sodium hydroxide, potassium hydroxide, etc. [Clough]

Oriental powder. An explosive used in fireworks. It is a mixture of gamboge and potassium nitrate.

Oriental sweet gum. See Storax.

Orimeten. Treatment for breast cancer. [Ciba plc] *

Orimune. Poliovirus vaccine live oral; immunizing agent. [Lederle Laboratories USA] *

Orinase. CAS 64-77-7; Tolbutamide; antidiabetic. [Upjohn]

Orinase Diagnostic. CAS 473-41-6; Tolbutamide sodium; diagnostic aid (diabetes). [Upjohn] *

Oriodide-131. CAS 7790-26-3; Sodium iodide I 131; antineoplastic; diagnostic aid; radioactive agent. [Abbott Laboratories] *

Orisan. Seed dressing for control of fungal diseases on cereals, rice, cotton and vegetables. [Bayer AG] *

Orisulf. CAS 526-08-9; A proprietary preparation of sulfaphenazole; an antibiotic. [Ciba plc] *

Orkan. Floor cleaning degreaser. [Gansow UK Ltd]

Orlean. See annatto.

Orlest 28. A proprietary preparation of ethinylestradiol and norethisterone acetate; an oral contraceptive. [Parke-Davis] *

Orlon®. Fiber. [DuPont UK] †

Ormolu. One alloy contains 58% copper, 25.3% zinc, and 16% tin, and another consists of 90.5% copper, 3% zinc, and 6.5% tin.

Ornalith. See Bakelite. §

Ornamental Weeder. CAS 1582-09-8; Trifluralin; a preemergence granular material to use on ornamentals, shrubs, trees, roses, and flower beds; control grasses and many broadleaf weeds. [Lawn & Garden Products Inc]

ornithine. CAS 70-26-8; α-δ-Diamino-valeric acid, $(H_2N)CH_2 \cdot CH_2 \cdot CH_2 \cdot CH(NH_2)COOH$.

Ornithite. A tricalcic phosphate, $Ca_3P_2O_8 \cdot 2H_2O$.

Ornitrol®. CAS 1249849; 20,25 Diazacholestenol dihydrochloride; chemosterilant for pigeon control. [Avitrol Corporation]

Oroglas DR. An impact modified polymethacrylate; used for injection molding, extension. [Rohm & Haas]

Oromid®. Nylon 6 and 66 granules; engineering plastic. [SNIA (UK) Ltd]

Oronal LCG. Sodium coceth sulfate, PEG-40 glyceryl cocoate; mild surfactant for cosmetics. [Seppic]

Oronite. Lubricating oil additive. [Chevron] *

Oropon. Composition of the enzymes of pancreas absorbed in sawdust or kieselguhr, and intimately mixed with ammonium chloride or boric acid; a bate for leather. [Rohm & Haas]

Orotan. Pigment dispersing agent; used for coatings. [Rohm & Haas]

Orovite. Oral vitamins B and C. [SmithKline Beecham]

orpiment. (yellow sulfide of arsenic) A mineral, As_2S_3.

orris camphor. Essential oil of orris.

orris root. The rhizome of Iris florentina.

Orr's white. See lithopone.

orseille. See archil.

Ortegol®. Polyurethane crosslinking agents. [Thomas Goldschmidt Ltd]

Ortegol® 204. Delayed reaction crosslinking additive for production of high resilience PU slabstock foams. [Goldschmidt AG] *

Orthene. Acephate insecticide. [Chevron] *

Orthenex. Pesticide. [Chevron] *

Ortho. Pesticides. [Chevron] *

orthoboric acid. See boric acid.

Orthochrom. Leather finishes. [Rohm & Haas]

Orthochrome T. p-Toluquinaldine p-toluquinoline-ethylcyanine bromide. A red sensitizer for silver bromide plates.

ortho-chrysotile. A mineral; $2[Mg_3Si_2O_5(OH)_4]$

Orthocide. CAS 133-06-2; Captan fungicide. [Chevron] *

Ortho-Clear. Leather finishes. [Rohm & Haas]

Ortho-Gro. Liquid plant food. [Chevron] *

Ortho-Klor. Insecticide killer. [Chevron] *

Ortholate. 2-t-Butyl cyclohexyl acetate. [Quest Int'l. UK Ltd]

Ortholeum. Grease stabilizer and lubricant assistant. [DuPont UK]

Ortholite. Leather finishes. [Rohm & Haas]

Orthomatic. Lawn sprayer. [Chevron] *

Orthophen® 278. CAS 80-46-6; Amylphenol; intermediate for chemical specialties; also in mfg. of photographic chemicals, oil demulsifiers, phenolic resins, agricultural surfactants, antiskinning agents. [Atochem] *

orthophosphoric acid. See phosphoric acid.

orthophosphorous acid. See phosphorous acid.

Orthorix. Lime-sulfur fungicide. [Chevron] *

Orthosil. A proprietary trade name for an anhydrous sodium orthosilicate; a detergent. §

Orthotrol. Drift retardant. [Chevron] *

orthotungstic acid. See tungstic acid.

Orthoxicol. A proprietary preparation containing methoxyphenamine hydrochloride, codeine phosphate, and sodium citrate; a cough linctus. [Upjohn Ltd] *

Orthoxine. CAS 93-30-1; A proprietary preparation of methoxyphenamine; a bronchial antispasmodic. [Upjohn Ltd] *

Orth's stain. A microscopic stain. It contains 1 g lithium

---

‡ Trade name and manufacturer not verified   § Trade name without identified manufacturer

carbonate and 2.5 g carmine in 100 cc water.

Ortizon. See Hyperol. §

Ortol. Dyestuffs for wool and polyamide. [BASF plc] *

Ortol. A photographic developer. Methyl-o-amino-phenol, $C_6H_4(OH)(NHCH_3)$ (2 mols), combined with hydroquinone (1 mol.), forms the basis of this developer.

Ortolan. A proprietary insulation. §

Ortolan® Dyes. 1:2 Metal complex dyes for dyeing and printing of wool and nylon fibers [BASF AG; BASF plc]

Ortosol. It consists of 10% chlorobenzene, 88% o- and m-dichlorobenzenes, and 2% p-dichlorobenzene. A mixture for dry cleaning.

Orudis. CAS 22071-15-4; Ketoprofen used in the treatment of arthritis. [Rhone-Poulenc Rorer Ltd]

Oruvail. CAS 22071-15-4; Ketoprofen in a pH sensitive controlled release system; used for treatment of arthritis. [Rhone-Poulenc Rorer Ltd]

Orvus WA. Sodium lauryl sulfate; a proprietary trade name for a wetting agent. §

Orzan. A range of ammonium and sodium lignin sulfonates; used as cement grinding aids and as thinners for drilling muds. [Harcros Australia] *

Orzol. White mineral oil. [Witco]

OS-2. Emulsifiable ester; forms stable emulsions for buffing, lapping, drawing, grinding compounds; rust preventative. [Werner G. Smith]

Osbil. CAS 3115-05-7; Lobenzamic acid. [May & Baker Ltd]*

Oscodal. A proprietary preparation of vitamin cod-liver oil product. §

Oscrete. Range of concrete admixtures. [Oils & Soaps Ltd]

Osimol. Dispersing and leveling agent; used for dyeing. [Degussa AG] *

osmic acid. (perosmic acid anhydride; perosmic oxide) CAS 20816-12-0; Osmium tetroxide, OsO_4; used as a microscopic stain, in photography, as a oxidation catalyst in organic synthesis.

Osmiridium. See Iridosmine.

Osmitrol. CAS 69-65-8; Mannitol; diagnostic aid; diuretic. [Travenol Laboratories Inc] *

osmium. Metallic element; Os; CAS 7440-04-2; EINECS 231-114-0; hardener for iridium and platinum, pen points, instrument pivots, catalyst. [Aldrich; Atomergic Chemetals; Degussa; Noah Chem.]

osmium tetroxide. (osmic acid; perosmic acid anhydride; perosmic oxide) OsO_4; CAS 20816-12-0; EINECS 244-058-7; microscopic staining, photography, oxidation catalyst in organic synthesis. [Aldrich; Atomergic Chemetals; Degussa; Janssen Chimica; Spectrum Chem. Mfg.]

osmo-calamine. Colloidal calamine.

Osmoglyn. CAS 56-81-5; Glycerin; pharmaceutic aid. [Alcon Laboratories Inc]

osmo-kaolin. A preparation of kaolin obtained by a patented electro-osmosis process. It has a high covering power and clinging properties, and is used in toilet powders.

osmondite. The stage in the transformation of austentite, at which the solution in dilute sulfuric acid reaches its maximum rapidity.

osmo-sil. A dye absorbent. It is a very pure form of silica, SiO_2.

Osnol. A proprietary preparation of liquid paraffin with vitamin D. §

OSO® 440. CAS 1309-37-1; Synthetic yel. iron oxide;

pigment with good suspension and stabilization properties for coatings, automotive finishes, appliance enamels, paints, plastics, building products, rubbers, inks. [Hitox]

OSO® 1905. CAS 1309-37-1; Synthetic red iron oxide; pigment with good suspension and stabilization properties for coatings, automotive finishes, appliance enamels, paints, plastics, building products, rubbers, inks. [Hitox]

Ospolot. CAS 61-56-3; A proprietary preparation of sulthiame; an anticonvulsant. [FBA Pharmaceuticals] ‡

Ospolot®, Ospolot Mite. Pharmaceutical preparation; anticonvulsant specially for temporal lobe epilepsy. [Bayer AG]

osram. An alloy of osmium and tungsten.

ossein. A variety of gelatin prepared from bones.

ossein. See collagen.

os sepiae. (white fish-bone). The calcareous shell lying within the back of the cuttle fish. It consists mainly of calcium carbonate.

Ossivite. A proprietary preparation of bonemeal and vtamins A and D. [Wyeth Laboratories] *

Ostamer. Polyurethane foam; prosthetic aid. [Merrell Dow Pharmaceuticals Inc] *

Ostan. A trade name for pure sodium and potassium hydroxides in disk form (5 mm diameter); suitable for analytical work and convenience in weighing. §

Ostelin. A proprietary preparation; it is a vitamin cod-liver oil product. §

Ostreocin. A proprietary trade name for a mixture of Ostreogrycins B and G. §

ostreogrycin. Antimicrobal substances produced by *Streptomyces ostreogriseus* (specific substances are designated by a terminal letter; thus, Ostreogrycin B). Ostreocin is a mixture of Ostreogrycins B and G.

Ostrilan. Sulfonamide quinoline hormone veterinary ethical. [Ciba plc] *

Oswego. A proprietary trademark for cornflour. §

Osyrol. 3,7-Dimethyl-7-methoxyoctan-2-ol. [Bush Boake Allen Ltd]

Otamidyl. Antibacterial/antifungal ear drops. [May & Baker Ltd] *

ote seeds. (acoomo seeds). Seeds of *Myristica angolensis*, of Nigeria. The seeds yield Kombe fat.

otita. A solution containing 1.8 grains quinine dihydrochloride per fluid ounce in glycerin and water.

otoba butter. (Otoba wax, American mace butter). Otoba fat, obtained from the fruit of *Myristica otoba*.

otoba fat. See otoba butter.

otoba wax. See otoba butter.

Otopred. A proprietary preparation of chloramphenicol, prednisolone, thiomersal; used as anti-infective ear-drops. [Typharm] *

Otoryl. Compound ear drops, veterinary. [May & Baker Ltd]*

Otosporin. A proprietary preparation of polymyxin B sulfate, neomycin sulfate, and hydrocortisone; used as anti-infective ear-drops. [The Wellcome Foundation Ltd]

Ototrips. A proprietary preparation of polymixin B, bacitracin, trypsin; used as anti-infective ear-drops. [Consolidated Chemicals Ltd] *

Otox. Ear drops. [The Boots Co plc] *

Otrivine. Hay fever formula nasal decongestant CCP. [Ciba plc] *

Otrivine-Antisitin. Nasal decongestant. [Ciba plc] *

---

* Trade name not verified as available   † Trade name verified as obsolete

Otrivin Hydrochloride. CAS 1218-35-5; Xylometazoline hydrochloride; used as an adrenergic (vasoconstrictor). [Ciba-Geigy Corp] *

Ottasept. For chemical compositions having antiseptic, germicidal and fungicidal properties such as para-, chloro-, meta- xylenol, in Class 18. [Ferro] *

Ouabaine Arnaud. CAS 630-60-4; A proprietary preparation of ouabaine; used in cases of threatened heart failure. [L Wilcox & Co Ltd] ‡

Oulu 102. CAS 61790-12-3; EINECS 263-107-3; Tall oil fatty acid; raw material. [Veitsiluoto Oy]

Oulu 331. Tall oil rosin; raw material. [Veitsiluoto Oy]

Oulumer 70. Tall oil rosin; tackifier for pressure-sensitive and hot-melt adhesives; raw material for high softening pt. rosin esters; also in printing inks. [Veitsiluoto Oy]

Oulupale XB 100. Pale rosin; tackifier with excellent thermal stability for disposable hot melts in hygiene sector, pressure-sensitive adhesives, bookbinding adhesives; compat. with EVA, SB, SIS, and NR elastomers and waxes. [Veitsiluoto Oy]

Oulures. Printing ink resins. [Veitsiluoto Oy]

Oulutac 20 D. Modified tall oil rosin aq. disp.; plasticizer for other resin disps.; gives very soft, water-resist. tacky film; good storage and mech. stability and resist. against sl. freezing. [Veitsiluoto Oy]

Oulutac 20 EP. Triethylene glycol ester of tall oil rosin; adhesive tackifier and plasticizer. [Veitsiluoto Oy] *

Oulutac 30. Ethylene glycol ester of tall oil rosin; adhesive tackifier. [Veitsiluoto Oy] *

Oulutac 30 D. Aqueous emulsion of tall oil rosin based ethylene glycol ester; adhesive tackifier. [Veitsiluoto Oy]*

Oulutac 80. Glycerol ester of tall oil rosin; adhesive tackifier. [Veitsiluoto Oy] *

Oulutac 80 D. Aqueous dispersion of tall oil rosin based glycerol ester; adhesive tackifier. [Veitsiluoto Oy] *

Oulutac 80 D/HS. Glyceryl rosinate aq. disp.; tackifier for natural and synthetic rubber-based latex adhesives; good mech. stability and resist. to sl. freezing. [Veitsiluoto Oy]

Oulutac 90. Pentaerythritol ester of tall oil rosin; adhesive tackifier. [Veitsiluoto Oy] *

Oulutac 90 D. CAS 8050-26-8; Pentaerythrityl tall oil rosinate; tackifier for natural and synthetic rubber-based latex adhesives; good mech. and storage stability and resist. to sl. freezing. [Veitsiluoto Oy]

Oulutac 90 D. Aqueous dispersion of tall oil rosin based pentaerythritol ester; adhesive tackifier. [Veitsiluoto Oy]*

Oulutac 105. Pentaerythritol ester of tall oil rosin; adhesive tackifier. [Veitsiluoto Oy] *

ounce metal. A bronze consisting of 85% copper, 5% tin, 5% zinc, and 5% lead.

Ouralpatti. See Ventilago Madraspanta.

Oust. Descaler products. [Dylon International Ltd]

ouvarovite. A mineral. It is a lime-chrome-garnet.

Ovaban. CAS 595-33-5; Megestrol acetate; an antineoplastic. [Schering Corp] *

Ovac. A blend of the thiazole derivatives of two specially selected aldehyde-amines; a proprietary trade name for a rubber vulcanizing accelerator. §

Ovacryl. Floor polishes. [Evans Vanodine International Ltd]

Overnite. A slug killer. [Murphy Chemical Co Ltd] †

Ovicide. Miscible tar oil winter wash. [ICI Chem. & Polymers Ltd]

Ovigest. For the weakly lamb. [The Wellcome Foundation Ltd] *

Ovitelmin. CAS 31431-39-7; A proprietary preparation containing mebendazole; veterinary antihelmintic (sheep and goats). [Janssen Pharmaceutical Ltd] *

Ovol. A proprietary preparation of dicyclomine hydrochloride and dimethicone; used in the treatment of infant colic. [Carter-Wallace Ltd] *

ovolecithin. See lecithin.

Ovran. A proprietary preparation of ethinylestradiol and levonorgestrel; an oral contraceptive. [Wyeth Laboratories]

Ovranette. A proprietary preparation of ethinyl estradiol and levonorgestrel; an oral contraceptive. [Wyeth Laboratories]

O-V Statin. CAS 1400-61-9; Nystatin; antifungal. [Bristol-Myers Squibb Co Inc] *

Ovucire WL 2944. Hemi-synthetic glycerides. [Gattefosse SA]

Ovulen 1 mg. A proprietary preparation of ethynodiol diacetate and mestranol; an oral contraceptive. [G D Searle & Co Ltd] *

Ovulen 50. A proprietary preparation of mestranol, ethynodiol diacetate; an oral contraceptive. [G D Searle & Co Ltd] *

Oxaban®-A. CAS 51200-87-4; Dimethyl oxazolidine; cosmetic preservative, antimicrobial. [Angus]

Oxaban®-E. CAS 7747-35-5; EINECS 231-810-4; 7-Ethyl bicyclooxazolidine; antibacterial for cosmetics and toiletries. [Angus]

Oxaf. CAS 155-04-4; Zinc-2-mercaptobenzothiazole; a medium temperature accelerator widely used in latex compounding; also used in proofing, wire and druggist sundries where fast curing and a minimum of odor are required. [Uniroyal] *

Oxaf 50D. A 50% water active dispersion; ready-to-use form for latex compounding. [Uniroyal] *

oxalan. Oxaluramide, $C_3H_5N_3O_3$.

oxalantin. Leucoturic acid, $C_6H_6N_4O_6$.

oxalate blasting powder. A safety explosive powder containing 71% niter, 14% charcoal, and 15% ammonium oxalate.

oxalic acid. (oxalic acid dihydrate) HOOCCOOH · H_2O; CAS 6153-56-6; EINECS 205-634-3; Automobile radiator cleanser, metal and equipment cleaning, purifying agent, intermediate, leather tanning, catalyst, laboratory reagent, stripping agent, textile bleaching, rare-earth processing, printing and dyeing auxiliary. [Ashland; General Chem.; Hoechst Celanese; KMZ Chem. Ltd; Mallinckrodt; Mitsubishi Gas]

oxalic acid dihydrate. See oxalic acid.

oxalic acid potassium salt. See potassium oxalate.

oxalic acid tin (II) salt. See tin (II) oxalate.

oxalic ether. Ethyl oxalate (C_2H_5 · COO)$_2$.

Oxalid. CAS 129-20-4; Oxyphenbutazone; anti-inflammatory; antirheumatic. [USV Pharmaceutical Corp] ‡

oxalumina. An abrasive consisting of small crystals of alumina.

Oxanid. CAS 604-75-1; Oxazepam; pharmaceutical preparation for the treatment of schizophrenia. [M A Steinhard Ltd] *

oxazolidine. (2-oxazolidine) CAS 497-25-6; 51200-87-4; 7747-35-5; 6542-37-6; Preservative, antibacterial agent for water-based paints, latexes, emulsions, metalwork-

---

‡ Trade name and manufacturer not verified   § Trade name without identified manufacturer

ing fluids; for oilfield water-flooding operations; corrosion inhibitor; crosslinking agent, catalyst for resin systems. [Angus]

Oxetal 500/85. Fatty alcohol polyglycol ether (5 EO); detergent, dispersant, emulsifier, wetting agent for household and industrial use, textile, paper, and leather industries. [Zschimmer & Schwarz]

Oxetal D 104. CAS 26183-52-8; Deceth-4; detergent, dispersant, emulsifier, wetting agent for household and industrial use, textile, paper, and leather industries. [Zschimmer & Schwarz]

Oxetal ID 104. Isodeceth-4; detergent, dispersant, emulsifier, wetting agent for household and industrial use, textile, paper, and leather industries. [Zschimmer & Schwarz]

Oxetal O 108. Cetoleth-8; detergent, dispersant, emulsifier, wetting agent for household and industrial use, textile, paper, and leather industries. [Zschimmer & Schwarz]

Oxetal TG 111. CAS 61791-28-4; Talloweth-11; detergent, dispersant, emulsifier, wetting agent for household and industrial use, textile, paper, and leather industries. [Zschimmer & Schwarz]

Oxetal VD 20. Laureth-2; washing and cleansing agent. [Zschimmer & Schwarz]

Oxicebral. A proprietary preparation of vincamine; a cerebral vasodilator. [Pfizer International]

Oxi-Chek. For antioxidants for plastics and elastomers. [Ferro] *

oxidized linseed oil. See solidified linseed oil.

oxidized oils. See blown oils.

Oxilube. Polyoxyalkylene diols and derivatives; base components for compressor oils, greases, gear oils, and aviation turbine lubricants. [Shell UK]

Oximony. A proprietary red oxide of iron used as a rubber pigment. §

oxine. See 8-hydroxyquinoline.

oxirane. See ethylene oxide.

Oxi-tan. A proprietary tanning compound. §

Oxitex. Textile fiber lubricant. [Shell UK]

Oxitol. A colorless, slightly hygroscopic liquid with mild odor; used as a solvent in paints, varnishes, inks and stains; also effective as an extraction agent for antibiotics and as a degreasing solvent. [Shell UK]

Oxivent. CAS 30286-75-0; Oxitropium bromide; an antimuscarinic bronchodilator. [Boehringer Ingelheim Ltd]

2-oxobutane. See methyl ethyl ketone.

Oxolin. A patented material made from oxidized oil, jute fiber, and sulfur; it is a rubber substitute. §

Oxone. (Oxolin). Compressed sodium peroxide, Na_2O_2; used in washing powders, scouring powders, metal cleaners, hair wave neutralizers, general oxidizing reactions. [DuPont] ‡

Oxonite. An explosive. It is made from 54% of nitric acid (specific gravity 1.5), and 46% of picric acid.

2-oxopentanedioic acid. See α-ketoglutaric acid.

4-oxopentanoic acid. See levulinic acid.

2-oxopropionic acid. See pyruvic acid.

Oxsoralen-Ultra. CAS 298-81-7; Methoxsalen; pigmentation agent. [ICN Pharmaceuticals Inc]

Oxucide. CAS 144-29-6; Piperazine citrate; anthelmintic. [Sterling Drug Inc] *

Oxxef. Latamoxef; semi-synthetic broad-spectrum beta-lactam antibiotic. [Merck & Co Inc] †

oxyalizarin. Purpurin, $C_{14}H_5O$.

oxyammonia. Hydroxylamine, NH_2OH.

oxyanthracene. Anthrol; $C_{14}H_{10}O$.

oxybenzone. See benzophenone-3.

oxybismethane. See dimethyl ether.

1,1′-oxybis-2-propanol. See dipropylene glycol.

oxybuprocaine. 2-Diethylaminoethyl 4-amino-3-butoxy-benzoate. Novesine is the hydrochloride.

oxycamphor. Campholenic acid; $C_9H_{15}.CO_2H$.

Oxycel. A proprietary preparation of oxidized cellulose; used to stop hemorrhage. [Parke-Davis] *

oxychloride. It is a solution of sodium hypochlorite, containing 10-12% available chlorine; a disinfectant.

oxychloride of tin. See tin salts.

Oxycholine. Muscarine, $C_5H_{15}O_3N$.

oxycinchophen. 3-Hydroxy-2-phenyl-cinchonic acid.

oxycymol. Carvacrol, $C_{10}H_{14}O$.

oxydase. An oxidizing enzyme.

Oxydasine. Consists mainly of a 0.05% solution of vanadic acid; an antiseptic recommended for wounds.

oxydislin. A compound of silicon having the formula Si_2H_2O, prepared by treating calcium silicide with cold diute alcoholic hydrochloric acid in the dark. It is a white solid spontaneously inflammable in air.

Oxydon. CAS 79-57-2; A proprietary preparation containing oxytetracycline; an antibiotic [RP Drugs] *

Oxydpech. Oxide pitch obtained as a residue from the distillation of the fatty acids obtained from the oxidation of paraffin.

Oxydurit®. Anionic product (oxidation agents/dispersants); auxiliary in the dyeing and printing of textiles. [Cassella AG]

oxygen. Nonmetallic gaseous element; O; CAS 7782-44-7; EINECS 231-956-9; copper smelting, steel production; in manufacture of ammonia, methyl alcohol, acetylene; oxidizer for rocket propellants; resuscitation, heart stimulant; decompression chambers; spacecraft; chemical intermediate; in oxidation of municipal/industrial wastes. [Air Prods.; Norsk Hydro A/S; Showa Denko]

oxygenated oil. Olive oil through which chlorine has been passed for several days.

oxygenated paraffin. See Parogen.

oxygen cubes. Made by mixing sodium peroxide and bleaching powder together and compressing into tablets. They contain 100 parts of bleaching powder (33-35% available chlorine), and 39 parts of sodium peroxide. On contact with water, oxygen is evolved.

oxygenite. A mixture of perchlorates or nitrates with a combustible substance. When ignited the mixture produces oxygen, and the material is used for this purpose.

oxygen powder. Sodium peroxide, Na_2O_2.

oxygen, solid. See oxygen cubes.

Oxy-Gro. CAS 1305-79-9; Calcium peroxide; for agricultural and horticultural uses. [Interox Chemicals Ltd]

Oxyguard. 2, 6-Di-t-butyl-p-cresol; a proprietary antioxidant. [Naugatuck (US Rubber)] *

oxyhemoglobin. Commercial form of hemoglobin.

Oxyliquit. A blasting explosive. It is formed by rapidly mixing liquid air, rich in oxygen, with powdered charcoal, petroleum residues, or cotton wool.

oxylith. A compressed powder. It is a mixture of sodium peroxide and bleaching powder, which evolves oxygen on treatment with water.

Oxylone. CAS 426-13-1; Fluorometholone; glucocorticoid.

---

* Trade name not verified as available   † Trade name verified as obsolete

[Upjohn] *

Oxymaster. CAS 79-21-0; Peracetic acid for sewage treatment. [Interox Chemicals Ltd]

oxymel. Clarified honey (80%.) mixed with acetic acid (10%.) and water (10%.).

oxymethylene. See formaldehyde.

oxymuth saca. A colloidal suspension of bismuth oxyhydrate.

Oxymycin. CAS 79-57-2; Oxyletracycllne; antibacterial. [O'Neal, Jones & Feldman Pharmaceuticals] *

Oxyneurine. CAS 107-43-7; Betaine, $C_5H_{11}NO_2$.

Oxynone. CAS 537-65-5; 2, 4-Diamino-diphenylamine; a proprietary trade name for a rubber antioxldant. §

Oxyper. CAS 497-19-8; EINECS 207-838-8; Sodium carbonate peroxyhydrate. [Interox Chemicals Ltd]

Oxyphenine. Direct cotton dyestuffs. [Clayton Aniline Co Ltd] *

Oxypon 288. Olive oil PEG-10 esters; solubilizer and refatting agent for cosmetics. [Zschimmer & Schwarz]

Oxypon 306. Mink oil PEG-13 esters; solubilizer and refatting agent for cosmetics. [Zschimmer & Schwarz]

Oxypon 328. PEG-26 jojoba acid, PEG-26 jojoba alcohol; refatting agent for cosmetics. [Zschimmer & Schwarz]

Oxypon 365. Avocado oil PEG-11 esters; refatting agent for cosmetics. [Zschimmer & Schwarz]

Oxypon 2145. PEG-15 glyceryl isostearate; emollient, superfatting agent for cosmetics. [Zschimmer & Schwarz]

oxytoluol. Cresol, $CH_3C_6H_4OH$.

Oxytracyl. Antibiotic veterinary ethical. [Ciba plc] *

oxytri. Polymerized β-hydroxy-trimethylene sulfide.

Oxytril. Selective weedkiller. [May & Baker Ltd] *

Oxytril CM. An emulsifiable concentrate containing 200 g bromoxynil and 200 g ioxynil per liter; a post-emergence contact herbicide for cereal crops. [Rhone-Poulenc Crop Protection Ltd]

Oxytril P. Selective herbicide [Murphy Chemical Ltd] †

Oxzone. CAS 7722-84-1; Hydrogen peroxide. [ABM Chemicals Ltd] *

Ozasol. Offset printing plates. [Hoechst AG]

Ozatec. Photoresists for acid plating and etching. [Hoechst AG]

ozocerite. See ozokerite.

ozogen. Hydrogen peroxide, H_2O_2.

Ozokerine. Trade name applied to varieties of soft paraffin. Also see petroleum jelly. §

ozokerite. (ozocerite; mineral wax; fossil wax) CAS 8021-55-4; Hydrocarbon wax derived from mineral or petroleum sources; electrical insulation, rubber products, paints, leather products, printing inks, floor and furniture polishes, cosmetics, ointments; substitute for carnauba and beeswax. [ISP]

ozole. A term applied to volatile aromatic odors contained in certain dextrins and plant extracts. Dextrinozole is the name applied to the body giving the scent of commercial dextrin.

P

P3. A proprietary degreasing material; a mixture of water-glass and trisodium phosphate in solid form; used for cleaning metals, glass, and textiles; stated to have no corrosive action on aluminum, aluminum alloys, tin, zinc, and brass. §

P3-Almeco. Chemicals for aluminum anodizing process. [Nickerson Chemicals Ltd]

P3-Armourbond. Heat resistant anticorrosive coatings. [Nickerson Chemicals Ltd]

P3-Carclin. Detergent-based vehicle wash. [Nickerson Chemicals Ltd]

P3-Croni. Spray booth paint denaturants. [Nickerson Chemicals Ltd]

P3-Ferroclene. Derusting chemical system. [Nickerson Chemicals Ltd]

P3-Ferromede. Metal protection system. [Nickerson Chemicals Ltd]

P3-Maxan. Paint strippers. [Nickerson Chemicals Ltd]

P3-Rodine. Acid inhibitors. [Nickerson Chemicals Ltd]

P3-Stripalene. Paint strippers. [Nickerson Chemicals Ltd]

P3-Suncorrite. Anti-corrosive coatings. [Nickerson Chemicals Ltd]

P®-10 Acid. CAS 141-22-0; Ricinoleic acid; chemical inter-mediate; imparts lubricity and rust-proofing to sol. cutting oils; basis for grease, soaps, resin plasticizers, and ethoxylated derivs. [CasChem]

P-11. CAS 25928-94-3; Epoxy resin compound; highly-filled gyro grade potting compound; absence of volatiles, usable in BFC [poly(bromotrifluoroethylene)] damping fluids, unaffected by poly(chlorotrifluoroethylene) oil; excellent machinability for general purpose applications; mix ratio: 3.20 phc with Activator BA-1. [Bacon]

P 13 N. A trademark for a range of polyimide varnishes and coatings. [TRW Inc] *

P-51. CAS 25928-94-3; Unfilled epoxy compound; general-purpose potting compound, room temperature cure to clear, slightly flexible material; for encapsulation of elec-tronic components for use up to 250 F; mix ratio: 19.5 phc with Activator BA-21. [Bacon]

P-51. CAS 1072-35-1; A proprietary name for dibasic lead stearate used as a heat stabilizer for PVC. [BFGoodrich UK] *

P-56. CAS 25928-94-3; Epoxy resin compound; thermally conductive potting compound; P-56 settles but offers max. thermal conductivity for settled portion; mix ratio: 71 phc with Activator BA-22. [Bacon]

P-60-10 and P-60-20 Cold Molding Compounds. Two-component polysulfide system; molding material, excel-lent transfer of pattern detail for flexible molds (fine detail with low dimensional tolerance); for ceramic applications (reproduces high quality plaster molds from flexible blocks or cases to increase production and reduce breakage); mix ratio: 100 parts P-60 or P-60-S Base A/ 3 parts Activator C; blk. liq.; Shore hardness A 20 and 10 resp. [Perma-Flex Mold]

P-80F. CAS 25928-94-3; Highly filled epoxy compound; instrument grade potting compound; contains different ceramic filler than P-80C; preferable for new applications due to marginally better properties; mix ratio: 6.0 phr BA-42 activator. [Bacon]

P-85. CAS 25928-94-3; Highly filled epoxy compound; low cost, heat-curing potting compound for casting applica-tions requiring ease of handling, long work life, short cure time; offers low coeff. of thermal expansion, high heat distilled temp., good elec. properties; in electronic appli-cations involving high voltages; mix ratio: 8.0 phc with Activator BA-60. [Bacon]

P 210-D. Acrylic copolymer; processing aid for PVC; im-proved material handling, faster fusion, increased pro-duction output. [Novacor]

P-289. CAS 1072-35-1; A proprietary name for a PVC plasticizer of the lead stearate type. [Haagen Chemie BV] *

P-400 Series. High bake phenolic coatings; for immersion service in acids, ammonias, petrol. products, dairy and food products, industrial chems.; for tank linings, heat transfer equip., piping, fans, blowers, duct work, exhaust hoods, tank cars, process equip. [Heresite Protective Coatings]

P 506. Styrene methyl methacrylate copolymer; features sparkling clarity, scratch resistance, easy processing; for appliances, displays, medical devices, toys, office ac-cessories. [Novacor]

P0820. Propyltris(trimethylsiloxy)silane; low visc. silicone fluid with high control over volatility, visc., dens. and ref. index, and unique solv. properties; used as lubricant. [Hüls Am.]

P-2003-K and P-2020-T. Low-density polyethylenes, of Soviet origin. §

PA-57. CAS 9006-04-6; Lightly cross-linked natural rubber extended with a light-colored, nonstaining oil; processing aid for extrusions, calendering, and open steam curing. [Akrochem]

PA-80. CAS 9006-04-6; Lightly cross-linked natural rubber extended with a light-colored, nonstaining oil; processing aid when blended with natural rubber, SBR, neoprene, or nitrile rubber; for extrusions, calendering, and open steam cure. [Akrochem]

PA-111. CAS 32131-17-2; Nylon 6/6; general purpose molding compound for mech. parts, gears, bearings, etc.; UL recognized. [Bay Resins]

PA-111CF30. CAS 32131-17-2; Nylon 6/6, 30% carbon fiber-reinforced. [Bay Resins]

PA-111G13. CAS 32131-17-2; Nylon 6/6, 13% glass fiber-reinforced; molding compound of exceptional strength and stiffness; recommended for replacement of alloy castings, gears, brackets, etc. [Bay Resins]

PA-121. CAS 32131-17-2; Nylon 6/6, impact-modified; molding compound with increased Impact resistance and toughness; used for clips, automotive components, power tool housings, recreational parts, etc. [Bay Resins]

PA-211. CAS 25038-54-4; Nylon 6 homopolymer; molding compound for mech. components, fittings, gears, handles, and UL elec. applications such as switches, receptacles, and plugs. [Bay Resins]

PA-211G13. CAS 25038-54-4; Nylon 6, 13% glass fiber-reinforced; molding compound for max. strength and stiffness; used for automotive under-hood components such as timing chain covers, vacuum reservoirs, painted exterior body parts, door and window hardware, gears, and connectors. [Bay Resins]

PA-211N40. CAS 25038-54-4; Nylon 6, 25% mineral- and 15% glass fiber-reinforced; molding compound offering balance of engineering properties, low warpage, resistance to sink-mark formation; for painted exterior body parts, door and window hardware. [Bay Resins]

PA-221. CAS 25038-54-4; Nylon 6 copolymer, impact-modified; molding compound recommended for applications requiring improved toughness over homopolymer; used for trim clips, fasteners, elec. connectors, high-impact fact masks, safety helmet headbands, medical clamps, power tool components. [Bay Resins]

Paalsgaard oil. See schou oil.

Pabagel. Aminobenzoic acid; uv screen. [Galderma Laboratories] *

Pabanol. Aminobenzoic acid; uv screen. [ICN Pharmaceuticals Inc] *

Pabracort. CAS 50-03-3; A proprietary preparation of hydrocortisone acetate; a nasal decongestant spray. [Pharmax Ltd] *

Pabrinex. Vitamins B and C preparation; used in the case of severe depletion or malabsorption of vitamins B and C particularly in alcoholism, after acute infections, postoperatively and in psychiatric states. [Paines & Byrne Ltd] *

P.A.C. A proprietary manufacture of formaldehyde. §

Pacatal. A proprietary trade name for the hydrochloride or the acetate of pecazine; a sedative. [Warner] *

Pacer. Soluble concentrate containing 360 g chlormequat and 180 g 2-chloroethylphosphoric acid per liter; plant growth regulator for use in winter wheat. [Farm Protection Ltd]

pacherite. A mineral, $BiVO_4$.

Pacific Sea Kelp Glycolic Extract B-1063. Source of minerals, vitamins, and amino acids; used in hair care cosmetics adding luster to hair and helping to keep scalp healthy. [Bell Flavors & Fragrances]

Pacitron. CAS 73-22-3; L-Tryptophan; for treatment of mild to moderate depressive illness. [Berk Pharmaceuticals Ltd] *

Packfong. A nickel silver. It contains from 26-44% copper, 16-37% zinc, and 32-41% nickel. One alloy contains 40.4% copper, 25.4% zinc, 31.6% nickel, and 2.6% iron.

Packman. Closed fill system pack opener. [Schering Agrochemicals Ltd]

paco. (pacos). A Peruvian term for ferruginous earth containing small quantities of metallic silver.

pacos. See Paco.

Pacvac. Liquid ring vaccuum pump, baseplate mounted complete with liquid/air separator; used for vacuum applications where the customer requires a pump set where only electrical pipe connections have to be made. [Sihi Pumps (UK) Ltd]

Pacwet. Dust suppressant (especially in coal mines both above and underground). [Pacific Chemical Industries Pty Ltd] *

Padac®. β-Lactamase substrate. [Calbiochem Corp]

Paddox. Herbicide containing the sodium/potassium salts of MCPA, mecoprop, and dicamba. [ICI Chem. & Polymers Ltd]

Paddox. Soluble concentrate of 18 g dicamba, 237 g MCPA, and 80 g mecoprop per liter; used for weed control in cereals and grassland. [Farm Protection Ltd]

Padimate O. CAS 21245-02-3; Octyl dimethyl PABA; sunscreen agent. [Lipo]

Padukrein®. Enzyme for vascular therapy. [Bayer AG]

Padutin®. Kallidinogenase; vasodilator [Bayer AG]

Padutin® 100. Kallidinogenase; for the treatment of male infertility due to reduced number and mobility of spermatozoa. [Bayer AG]

Paedo Sed. A proprietary preparation of dichloralphenazone and paracetamol; a hypnotic. [Pharmax Ltd] *

Pafra. Synthetic emulsion adhesives [Pafra Ltd] *

Pagid. Asbestos and nonasbestos friction material; used for drum brake linings for trucks and passenger cars, disc brake pads for passenger cars, clutch facings for trucks and passenger cars, rolled material for industrial application, slip bearings for tooling machines; repair, servicing and maintenance of vehicle brakes. [Caramba Chemie GmbH] *

painter's naphtha. A petroleum distillate. It has a boiling point of 105-200 C.

Paladac. A proprietary preparation containing vitamin A palmitate, calciferol, thiamine, riboflavine, pyridoxine, nicotinamide, and ascorbic acid. [Parke-Davis] *

pala gum. (Indian gutta-percha). A product from a Ceylon tree. The coagulated juice resembles gutta-percha.

palaite. A mineral; it is a hydrated manganese phosphate.

Palamid®. Masterbatches of organic and inorganic dyes in polyamide; for polyamide spin dyeing (fibers and filaments). [BASF AG]

Palamoll®. Plasticizers for plasticized PVC products resistance to oil, gasoline and bitumen with little migration tendency. [BASF AG; BASF plc]

Palamoll® 632. A polyester of adipic acid and propanediol; a plasticizer for PVC. [BASF plc] *

Palamoll® 644 and 646. A polyester of adipic acid and butanediol; a plasticizer for PVC. [BASF plc] *

Palamoll® 645 and 647. Propriety polyadipates having viscosities of 60 mPa·s and 10,000 mPa·s at 20 C respectively; used as polymeric plasticizers. [BASF plc]*

Palamoll® 855. A proprietary polymeric plasticizer having a viscosity of 5000 mPa·s. [BASF plc] *

Palanil®. Dyestuffs for polyester fibers. [BASF plc]

Palanil® Carrier. Auxiliaries for dyeing polyester fibers. [BASF plc] *

Palanil® Dyes. Disperse dyes for dyeing polyester fibers and

---

for printing acetate, triacetate, polyester, polyamide, and acrylic fibers. [BASF]

Palanil® P Dyes. Disperse dyes with special finish for printing polyester, acetate, triacetate, and acrylic fibers, esp. in combination with synthetic thickening agents of the Lutexal type. [BASF]

Palanil® T Dyes. Disperse dyes esp. for the thermosol dyeing of polyester fibers and polyester/cellulosic fiber blends. [BASF]

Palanthrene®. Vat dyes for dyeing cellulosic fibers, textile printing. [BASF]

Palanthrene® T. Specialty vat dyes meeting demands imposed in the thermosol, pad-steam dyeing of polyester/cellulosic fiber blends. [BASF]

palao amarillo. A rubber obtained from the Mexican *Euphorbia fulva*. It has a high resin content.

Palapent. CAS 57-33-0; Pentobarbital sodium; hypnotic; sedative. [Bristol-Myers Squibb Co Inc]

Palapreg®. Unsaturated polyester resins; SMC/BMC resins and resin systems for applications in elec., automotive, building, and sanitary industries; light-hardening, glass fiber-reinforced, semifinished grades available. [BASF AG; BASF plc]

Palatal®. Unsaturated polyester resins and vinyl ester resins; for glass fiber-reinforced and unreinforced applications, e.g., building sector, containers, pipes, silos, boat and ship building, motor vehicle sector, elec. industry, chem. industry, mech. engineering, space and air travel, consumer goods, furniture, polyester concrete, agglomerated marble, fillers, buttons, pourable sealing compounds, and laminates. [BASF AG; BASF plc]

Palatal® KR 1397. A proprietary unsaturated polyester based on chlorendic acid dissolved in monostyrene; used in the manufacture of articles made of glass-reinforced plastics. [Dexine Rubber Co Ltd] *

Palatal® P5. A proprietary unsaturated polyester resin dissolved in monostyrene; it has low viscosity and is used in the manufacture of articles made of glass-reinforced plastics. [BASF plc] *

Palatal® P8, P50T and P52TL. Proprietary polyester resins used in the manufacture of articles made of glass reinforced plastics. [BASF plc] *

Palatal® S333. A proprietary polyester resin used in the manufacture of articles made of glass-reinforced polyesters. [BASF plc] *

Palatase. A fungal lipase produced by fermentation of a selected strain of *Aspergillus niger*; used for production of certain Italian cheese types and other specialty cheeses in which a modest lipolysis is desired. [Novo Nordisk]

Palatex® NB-2. Crease inhibitor during dyeing of cotton fabrics in becks and jets. [BASF]

Palatin® Fast Dyes. Sulfo-group containing 1:1 metal complex dyes for dyeing and printing wool, polyamide, and silk fibers. [BASF; BASF AG; BASF plc]

palatinit. A mixture of sodium hydrosulfite (blankit) and zinc dust; used as a bleaching agent.

Palatinol. Phthalate plasticizers. [BASF plc]

Palatinol 1C. CAS 84-69-5; Diisobutyl phthalate; a nonvolatile softening agent for cellulose esters. [BASF AG] §

Palatinol® 11. Undecyl phthalate; plasticizer for plasticized PVC [BASF AG]

Palatinol® 711. Phthalic acid ester consisting of C7-11 alcohols; plasticizer with low volatility for plasticized PVC

products with low brittle temps. [BASF AG]

Palatinol® A. CAS 84-66-2; Diethyl phthalate. plasticizer for cellulose lacquers; dissolves cellulose nitrate, ester gum, coumarone, etc. [BASF] *

Palatinol® AH. Di-2-ethyl benzylphthalate and dioctylphthalate; PVC plasticisers for general application. [BASF plc] *

Palatinol® C. CAS 84-74-2; Dibutyl phthalate; plasticizer for nitrocellulose dyes, surface coatings, for leather/fur industry [BASF AG]

Palatinol® D10. CAS 27554-26-3; Diisooctyl phthalate; plasticizer for PVC. [BASF plc] *

Palatinol® DBP. CAS 84-74-2; Di-n-butylphthalate; plasticizer for PVAc-based coatings and adhesives; used to bond paper and in lamination of vinyl wall coverings to gypsum board; NC lacquers and cellophane coatings; manufacturing of smokeless powds. for military use, processing aid for nitrile rubbers, carrier solv. for pigments and dyes, solv. for printing ink vehicles, fixative for perfumes, modifier of phenolic laminates, component of floor waxes. [BASF] *

Palatinol® DIDP. CAS 68515-49-1; Diisodecylphthalate; primary monomeric plasticizer for vinyls; used in high-temp. processing in application such as wire and cable formulations, automotive formulations. [BASF] *

Palatinol® DN. CAS 68515-48-0; Diisononyl phthalate; plasticizer for PVC. [BASF] *

Palatinol® DOA. CAS 103-23-1; Di (2-ethylhexyl) adipate; plasticizer used in PVC, NC, and rubber; food contact applications; plastisols; dip coating formulations. [BASF]*

Palatinol® DOP. CAS 117-81-7; Di (2-ethylhexyl) phthalate; general-purpose plasticizer for vinyl compositions; solvating power for PVC. [BASF] *

Palatinol® K. Dibutylglycol phthalate; plasticizer for PVC. [BASF plc] *

Palatinol® M. CAS 131-11-3; Dimethyl phthalate. [BASF]*

Palatinol® N. CAS 68515-48-0; Diisononyl phthalate; primary plasticizer for PVC, plastisols; used for film for the building trade and engineering. [BASF] *

Palatinol® TOTM. CAS 3319-31-1; Tri (2-ethylhexyl) trimellitate; primary, monomeric plasticizer for PVC homopolymer and copolymer resins; used where extreme low volatility of a plasticizer is required; used in vinyl compositions, in high temp. wire insulation and critical automotive applications. [BASF] *

Palatinol® Z. CAS 68515-49-1; Diisodecyl phthalate; plasticizer for PVC. [BASF plc] *

Palatone. A proprietary trade name for an acrylic denture material. §

Palau. A platinum substitute. It is an alloy of gold and palladium, usually 80% gold and 20% palladium. Another alloy termed Palau contains 60% nickel, 20% platinum, 10% palladium, and 10% vanadium.

Pale 4. Polymerized castor oil; pigment wetting/dispersing agent; plasticizer for resins, gums, polymers; lubricant, penetrant; coupling solv.; adhesion promoter; for cellulose lacquers, inks, adhesives, industrial lubricants, polishes, caulks, leather dressing, hydraulic fluids, rubber compounding, gasket cement. [CasChem]

pale acid. Nitric acid containing less than 0.1% nitrogen oxides.

pale catechu. See cutch.

Palegal® A. Dispersant and leveling agent for polyester

---

dyeing. [BASF]

Palegal® NB-SF. Nonfoaming leveling agent, auxiliary for high temp. dyeing of polyester fibers. [BASF]

Palenine. Leveling and stripping agent for textiles. [Catawba-Charlab]

pale oils. A name applied to a distillate from the residue of petroleum which has been treated with acid and soda, washed or filtered to a certain degree of refining or color. They have a light and medium viscosity; used lubricants for rapid motion machines.

Palette 70. On glaze colors for porcelain. [Degussa Ltd]

Paleva. CAS 50-78-2, 103-90-2; A proprietary preparation of aspirin and paracetemol; an analgesic. [Concept Pharmaceuticals Ltd] *

Palfium. CAS 357-56-2; A proprietary preparation of dextromoramide; an analgesic. [MCP Pharmaceuticals]*

Paliocrom. Special effect copper phthalocyanine pigment. [BASF plc]

Paliogen®. Org. pigments; for high-quality paints, coloring plastics, tin plate inks, artists' paints and crayons. [BASF AG]

Paliotan®. Co-finish pigments; for high-quality industrial finishes and polymer dispersions. [BASF AG]

Paliotol®. Org. pigments; for special surface coatings, for coloring plastics, for production of artists' paints and crayons, for tin plate inks. [BASF AG]

Palite. Chloromethyl chloroformate, $COCl \cdot OCNH_2Cl$; a military poison gas.

palladium. Metallic element; Pd; CAS 7440-05-3; EINECS 231-115-6; alloys for electrical relays and switching systems in telecommunications equip.; air craft spark plugs; protective coatings. [Aldrich; Atomergic Chemetals; Degussa; Noah Chem.]

palladium asbestos. An asbestos coated with palladium used in gas analysis for the absorption of hydrogen.

palladium black. A finely divided palladium used as a catalyst in the hydrogenation of oils.

palladium chloride anhydrous. See palladium chloride (ous).

palladium chloride (ous). (palladium chloride anhydrous; palladous chloride; palladium dichloride) $PdCl_2$; CAS 7647-10-1; EINECS 231-596-2; Analytical chemistry, electroless coatings for metals, photography, leak detection in gas lines, indelible inks, catalyst. [Aldrich; Atomergic Chemetals; Dajac Labs; Degussa; Spectrum Chem. Mfg.]

palladium dichloride. See palladium chloride (ous).

palladium gold. (white gold). An alloy of 90% gold and 10% palladium. An alloy of 40% copper, 31% gold, 19% silver, and 10% palladium, is also known by this name.

palladium monoxide. See palladium oxide.

palladium nitrate (ous). (palladous nitrate) $Pd(NO_3)_2$; CAS 10102-05-3; EINECS 233-265-8; Analytical reagent, catalyst. [Atomergic Chemetals; Degussa]

palladium oxide. (palladium monoxide) PdO; CAS 1314-08-5; EINECS 215-218-3; Reduction catalyst in organic synthesis. [Atomergic Chemetals; Degussa]

palladium red. Ammonio-chloride of palladium; a red pigment.

palladous chloride. See palladium chloride (ous).

palladous nitrate. See palladium nitrate (ous).

Pallgrip. Adhesives for use as an antislip agent on bags to be stacked. [Norsk Hydro AS] *

Palliag. A range of precious metal alloys; for dentistry and dental engineering. [Degussa Ltd]

Pallinal®. Metiram, nitrothal-isopropyl; for control of powdery mildew in fruit, vegetables, hops, ornamentals [BASF AG]

Pallitop®. Nitrothal-isopropyl, metiram; for control of powdery mildew in apples, pears [BASF AG; BASF plc]

Pallitop® S. Nitrothal-isopropyl, sulfur; contact fungicide against powdery mildew in apples [BASF AG]

palm butter. See palm oil.

palmerite. A mineral that is an aluminum-potassium phosphate.

palm grease. See palm oil.

palmine. See vegetable butter.

palmitic acid. (hexadecanoic acid; cetylic acid; hexadecylic acid) Saturated fatty acid; $CH_3(CH_2)_{14}COOH$; CAS 57-10-3; EINECS 200-312-9; manufacture of metallic palmitates, soaps, lube oils, waterproofing, food-grade additives. [Akzo; Ashland; Henkel/Emery; Unichema; Witco/Humko]

palmitic acid, hexadecyl ester. See cetyl palmitate.

palmitin. Commercial palmitic acid is incorrectly called by this name.

palmityl alcohol. See cetyl alcohol.

palm kernel oil. (oils, palm kernel) Oil obtained from seeds of *Elaeis guineensis;* CAS 8023-79-8. [Alba Int'l.; Karlshamns; Penta Mfg.; Stevenson Bros.]

palm oil. (oils, palm; palm butter; palm grease) Natural oil obtained from *Elaeis guineensis;* CAS 8002-75-3; yellow borwn buttery edible solid; soap manufacture, pharmacy, food shortening; cutting tool lubricant; cosmetics; softener in rubber processing; cotton goods finishing; mold release agent; emollient. [Alba Int'l.; Karlshamns; Penta Mfg.; Stevenson Bros.]

palm pitch. A pitch obtained by the treatment of palm oil with sulfuric acid.

palm wax. A yellow wax from *Ceroxylon andicola;* used as a beeswax substitute.

Palohex. CAS 6556-11-2; Inositol niacinate; vasodilator. [Sterling Drug Inc] *

Palomar®. Phthalocyanine pigments for the surface coatings industry, especially for automotive coatings. [Bayer AG]

Palorium. A platinum substitute. It is a white alloy of gold and platinum only distinguished from platinum with difficulty. It is a ductile, homogeneous alloy with a melting-point of 1310 C, and it remains stronger than platinum on heating.

Palormone. Aqueous solution containing 50% w/v 2,4-D as the amine salt; for use as an agricultural herbicide. [Universal Crop Protection Ltd]

Paludrine. A proprietary preparation of proguanil hydrochloride; an antimalarial. [ICI Chem. & Polymers Ltd]

Palusol. Intumescent fireboard based on sodium silicate; used for fire and smoke seals at fire resistant doors and service penetrations of fire walls. [BASF plc]

Pamak. A range of tall oil fatty acids; used in chemicals and chemical processing, construction and building, floor coverings, paints and coatings, rubber, soaps, detergents and household products. [Hercules]

Pamelor. Nortriptyline hydrochloride; antidepressant. [Sandoz Pharmaceuticals]

Pamergan. Promethazine/pethidine preparations. [May & Baker Ltd] *

---

‡ Trade name and manufacturer not verified   § Trade name without identified manufacturer

Pameton. Tablets containing paracetamol and DL-methionine; analgesic for minor painful conditions; particularly useful where the possibility of misuse or overdosage exists. [Winthrop Laboratories] *

Paminal. A proprietary preparation of hyoscine methobromide and phenobarbitone. [Upjohn Ltd] *

Pamine. CAS 155-41-9; A proprietary preparation of methscopolamine bromide; a gastrointestinal sedative. [Upjohn; Upjohn Ltd]

Pamolyn. CAS 61790-12-3; EINECS 263-107-3; Tall oil fatty acids. [Hercules Ltd]

Pamolyn 100, 100 FG, 100 FGK. CAS 112-80-1; EINECS 204-007-1; Oleic acids from vegetable sources; used as plasticizers, emulsifiers and textile and wet-processing aids, in the food and drink industry, in the manufacture of personal care products and cosmetics. [Hercules]

Pamolyn® 125. CAS 112-80-1; EINECS 204-007-1; Oleic acid; for conversion to soaps and sulfonates for use as textile processing aids, automotive additives, mold lubricants, surfactants, agents for production of synthetic rubber. [Climax Fluids Additives]

Pamolyn 200, 240. CAS 60-33-3; EINECS 200-470-9; Technical grade linoleic acid derived from vegetable sources; used in caulking and sealant compositions; used to produce oleoresinous printing ink vehicles, for making epoxy resin ester coatings and pale color-retentive fast drying alkyds. [Hercules]

Pamolyn 300. CAS 60-33-3; EINECS 200-470-9; Conjugated linoleic acid derived fom vegetable sources; used as epoxy ester resin intermediates for adhesives and sealants, as chemical intermediates for conjugated double-bond reactions, as modifiers of stirenated, vinylated and methacrylated alkyds. [Hercules]

Pamolyn 327B. A partially bodied, conjugated fatty acid; used as a replacement for G-H-viscosity dehydrated castor oil in short to medium oil alkyds and copolymer alkyd resins. [Hercules]

Pamolyn 380. CAS 60-33-3; EINECS 200-470-9; Conjugated linoleic acid derived fom vegetable sources; used as epoxy ester resin intermediates for adhesives and sealants, as chemical intermediates for conjugated double-bond reactions, as modifiers of styrenated, vinylated, and methacrylated alkyds. [Hercules]

Panabath. Panacide preparation for water baths. [BDH Chemicals Ltd]

Panablock. Biocide. [BDH Chemicals Ltd]

Panacide. CAS 97-23-4; Dichlorophen; fungicide, bactericide and algicide used as a moss-killer. [BDH Chemicals Ltd]

Panaclean. Biocidal cleaner. [BDH Chemicals Ltd]

Panacryl. Cationic dyestuffs; for dyeing acrylic fibers, modacrylic fibers and basic dyeable polyesters and nylons. [Holliday Dyes & Chemicals Ltd]

Panacur. CAS 43210-67-9; Fenbendazole; an anthelmintic. [Hoechst-Roussel Pharmaceuticals Inc] *

Panadeine CO. A proprietary preparation of paracetamol and codeine phosphate; an analgesic. §

Panadol. CAS 103-90-2; A proprietary preparation paracetamol; an analgesic. §

Panadonin. A preparation of *Adonis davurica*, a Japanese herb; advocated as a substitute for digitalis.

Panagran. Biocide. [BDH Chemicals Ltd]

Panalane® L-14E. CAS 61693-08-1; Hydrogenated polyisobutene; specifically designed for cosmetics applications; outstanding feel and moisturizing ability [Amoco Chemical Co]

Panama bark. Quillaia bark.

Panama crimson. A coloring matter obtained from the leaves of a vine called china.

Panar. CAS 8049-47-6; A proprietary preparation containing pancreatin. [Armour Pharmaceutical Co] *

Panasand. Biocide. [BDH Chemicals Ltd]

panase. A combination of digestive enzymes of the pancreas, derived from the pancreatic glands of the pig.

Panasol. Aromatic solvents. [Amoco Chemical Co]

Panasorb. A proprietary preparation of paracetamol in a sorbitol base; an analgesic. §

Panaspray. Biocide. [BDH Chemicals Ltd]

Panastat. Biocide. [BDH Chemicals Ltd]

Panatest. Bacterial test kit. [BDH Chemicals Ltd]

Panazyme. Fungal protease. [Rhone-Poulenc UK]

Pancil T. CAS 26530-20-1; Paste containing 1% w/w octhilinone; fruit tree canker paint. [Rohm & Haas UK]

Pancoxin. A proprietary preparation containing sulfaquinoxaline; a veterinary coccidiostat. §

Pancoxin. A proprietary preparation of amprolium; an antiprotozoan for veterinary use. §

Pancoxin. A proprietary preparation of ethopabate; a veterinary anti-protozoal. §

pancreas diastase. Amylopsin; an enzyme.

Pancrease. Pancrelipase; a concentrate of pancreatic enzymes standardized for lipase content; enzyme. [McNeil Pharmaceuticals] *

Pancreatic Lipase 250. CAS 9001-62-1; Lipase; enzyme for hydrolysis of triglycerides to glycerol and fatty acids; used for development of flavors. [Solvay Enzymes]

pancreatokinase. A mixture of Eukinase (*qv*), and pancreatin.

Pancreol. A trypsin preparation; used for bating skins. [Hodgson Chemicals Ltd]

pancreozymin. A hormone obtained from duodenal mucosa.

Pancrex. A pancreatic deficiency supplement; used in the treatment of cystic fibrosis and chronic pancreatic steatorrhea. [Paines & Byrne Ltd] *

Pancrex V. CAS 8049-47-6; A proprietary preparation of concentrated pancreatin; used in the treatment of cystic fibrosis. [Paines & Byrne Ltd] *

Pancrolin. Enzyme preparation; used for leather processing. [Hodgson Chemicals Ltd]

Pandex. A selenium preparation for rubber vulcanization.

Panelyte. A proprietary trade name for phenol-formaldehyde laminated products and paper, fabric, wood veneer, fiber glass, and asbestos base thermosetting plastics for structural work. §

Panets. CAS 103-90-2; Pediatric paracetamol preparations. [The Boots Co plc] *

Panex. Diacetyltartaric ester of edible mono-diglycerides; used as emulsifier/ antistaling additive for bread fats, mayonnaise and sauces. [Harcros Australia] *

Panilax®. A trade name for materials made from aniline-formaldehyde synthetic resin; they are thermoplastic but have a softening point about 100 C. §

Panitrin. Papaverine nitrate in acetyl diethyl-amide.

Panmycin. CAS 60-54-8; A proprietary preparation of tetracycline; an antibiotic. [Upjohn Ltd] *

Panmycin Hydrochloride. CAS 64-75-5; Tetracycline hydrochloride; anti-amebic; antibacterial; antirickettsial.

---

* Trade name not verified as available   † Trade name verified as obsolete

[Upjohn] *

Panmycin Syrup. CAS 60-54-8; Tetracycline; anti-amebic; antibacterial; antirickettsial. [Upjohn]

Panodan 235, FDP-K, SD, SD-K. Monoglyceride diacetyl tartaric acid ester; food emulsifier. [Grindsted Products]

Panogen M. CAS 151-38-2; 2-Methoxyethylmercury acetate; cereal seed treatment. [Embetec Crop Protection Ltd]

Panok. CAS 103-90-2; A proprietary trade name for paracetamol. §

Pan-O-Lite. Sodium aluminum phosphate; food leavening agent. [Monsanto Co] *

Panolog. Triamcinolone acetonide, mycostatin, neomycin sulfate, and thiostrepton in cream, ointment, or lotion base. [Ciba plc] *

PanOxyl Aquagel. CAS 94-36-0; White viscous gels in three grades containing 2.5%, 5%, and 10% benzoyl peroxide; for use in the topical treatment of acne vulgaris. [Stiefel Laboratories Inc; Stiefel Laboratories (UK) Ltd]

PanOxyl Wash. CAS 94-36-0; A viscous lotion containing benzoyl peroxide; an antibacterial cleanser for use in the topical treatment of acne vulgaris. [Stiefel Laboratories (UK) Ltd] *

pan scale. The calcium sulfate, containing some sodium chloride, which settles out during the crystallization of salt from brine. It is sold as salt lick for cattle, also for manuring purposes.

Pansorbin®. Cells. [Calbiochem Corp]

Pantal. An aluminum alloy containing 0.8-2.0% magnesium, 0.4-1.4% manganese, 0.5-1.0% silicon, and 0.3% titanium. It resists corrosion, and has a tensile strength of from 18-33 kgmm^2.

Pantalast® 1120. VC-EVA copolymer; engineered material; low smoke, low flamm., good abrasion and high temp. resistance, low HCl generation, good weathering properties; for demanding wire/cable applications; nonmigratory; available in standard colors. [Pantasote Polymers] *

Pantarol. A proprietary product for the protection of metal; it is applied by brushing, spraying, and dipping; resists the action of light, sea air, steam, acid fumes, but is destroyed by concentrated acids and alkalis. §

Pantene. Hair tonic, shampoo and conditioner to keep hair healthy and attractive. [Richardson-Vicks Inc] *

Pantene Grooming Lotion. Grooming lotion. [Roche Products Ltd] *

Panteric. CAS 8049-47-6; A proprietary preparation containing pancreatin (triple B.P. strength). [Parke-Davis] *

Panther. Suspension concentrate containing 50 g diflufenican and 500 g isoproturon per liter; used for control of weeds in winter cereals. [Rhone-Poulenc Crop Protection Ltd]

Panthoderm. CAS 81-13-0; Dexpanthenol; cholinergic. [USV Pharmaceutical Corp] ‡

Pantholin. CAS 137-08-6; Calcium pantothenate; vitamin. [Eli Lilly & Co] †

Pantolit. A proprietary synthetic resin of the phenol-formaldehyde type. §

Pantopaque. CAS 99-79-6; Iophendylate; diagnostic aid. [Alcon Laboratories Inc] *

pantopon. (omnopon, nealpon; papaveretum). Mixtures of the soluble hydrochlorides of opium alkaloids.

Pantosept. A German antiseptic in which the active agent is hypochlorous acid.

pantothenyl alcohol. (N-pantoyl-3-propanolamine; pantothenylol) CAS 81-13-0; The alcohol corresponding to pantothenic acid with vitamin activity; used in biochemical research, as a food additive and dietary supplememnt.

Pantrop® retard. Psychotropic agent. [Bayer AG]

Panturon. A proprietary preparation containing atropine sulfate, papaverine hydrochloride phenobarbitone, aluminum hydroxide gel, kaolin, magnesium carbonate, magnesium trisilicate, and oil of peppermint; an antacid. [Norma] *

Panwarfin. CAS 129-06-6; Warfarin sodium; anticoagulant. [Abbott Laboratories] *

paoferro. The inner bark of the Brazilian ironwood tree; used as an antidiabetic.

papain. (vegetable pepsin; papayotin; papoid) Pepsin-like enzyme; a vegetable digestive ferment obtained from the unripe fruit of the papaw tree; CAS 9001-73-4; EINECS 232-627-2; meat tenderizer, tobacco, pharmaceutical; antihazing agent for beer; cosmetics, leather, textiles. [EM Industries; Dr. Madis Labs; Meer; Spice King; Stan Chem Int'l. Ltd]

Papain Conc. Protease; enzyme for hydrolysis of proteins; for brewing (stabilizes and chillproofs beer); meat tenderizers; pharmaceutical (digestion aids); protein modification. [Solvay Enzymes]

papaw. The seeds of *Asimina tribola*; an emetic.

papayotin. *See* papain.

Paperad. Finely divided aluminum trihydrate paper pigment; used in papers. [Reynolds Metal Co]

paper-clay. *See* bentonite.

paperhanger's alum. Aluminum sulfate; used for sizing paste.

Paperine. A starch product used in paper manufacture.

Paper-Pac®. Paper manufacturing additive for neutral sizing. [Sachtleben Chemie GmbH]

paper-spar. A mineral that is a variety of calcite.

Papi 27. Polymeric MDI; for PU industry for molded elastomers; for use in structural foam molding formulations, appliances. [Dow]

Papi 4901. Polymeric MDI; for PU industry for molded flexible foam. [Dow]

Papite. Acrolein with stannic chloride; a tear gas.

papoid. *See* papayotin.

paposite. A mineral that is a ferric sulfate.

Pappenheim's stain. (Pyronin stain). A microscopic stain. It consists of 1 part of a concentrated solution of pyronin and 3 parts of a concentrated solution of methyl green.

paprika. Cayenne pepper.

Papyrus. A proprietary casein product. §

Par. *See* Catalpo. §

Parabar. Oxidation inhibitors. [Exxon UK] *

Parabis. CAS 80-05-7; A pure form of bisphenol A; used as an intermediate in the production of polycarbonate and polysulfone resins. [Dow UK]

Parable. Soluble concentrate of 100 g diquat and 100 g paraquat per liter; used for weed control in field crops. [ICI Agrochemicals]

Parabolix® 100. Degreaser, destaticizer for electronics, parabolic light fixtures. [Merix]

Paracodol. A proprietary preparation of paracetamol and codeine phosphate; an analgesic. [Fisons plc, Pharmaceuticals Div] *

Paracol. Herbicide containing diuron and paraquat. [ICI

Chem. & Polymers Ltd]

Paracol. Wax and wax-rosin emulsions with excellent shelf, shear and chemical stability; used in paper and paperboard and to impart resistance to water and aqueous solutions; also in many types of wet or dry-formed building products. [Hercules]

Paracol® 403A6. Paraffin wax-pale rosin emulsion; sizing agent for surface sizing of paper and paperboard where all-wax emulsions might cause excessive slipperiness; improves sizing and finish of writing and printing papers; used in food-pkg. grades of paper/paperboard. [Hercules]

Paracol® 404C. CAS 63231-60-7; Refined microcrystalline wax emulsion; sizing agent for improving surface properties of paper and paperboard; detackifier; antiblocking agent. [Hercules]

Paracol® 800N. CAS 8002-74-2; Paraffin wax emulsion; sizing particleboard; superior pumping, mechanical, and spraying stability. [Hercules]

Paracol® 810N. CAS 8002-74-2; Paraffin wax emulsion; sizing agent for particleboard, hardboard, and insulation board; mechanical, pumping, and spraying stability for trouble-free operation in metering pumps and spray nozzles. [Hercules]

Paracol® 1886. CAS 8002-74-2; Paraffin-based aq. emulsion; acid-breaking-type emulsion used in specialized internal sizing applications. [Hercules]

Paracol® M161. CAS 8002-74-2; Paraffin wax emulsion; for surface sizing to improve sheet water resistance and finish where sheet brightness is of concern. [Hercules]

paracon. A generic name for polyester elastomers. See Paraplex. §

Paracort. CAS 53-03-2; Prednisone; glucocorticoid. [Parke-Davis] *

Paracortol. CAS 50-24-8; Prednisolone; glucocorticoid. [Parke-Davis] *

paracoumarone resin. (cumar resin, cumar gum, coumarone resin, benzo-furane resin, cumar). A synthetic resin produced from coal-tar distillates; used in the production of varnishes, polishes, artificial leathers.

Paracril® 1880, 1880LM. NBR, hot polymerized, slightly staining antioxidant; improved low temp. performance, good heat resistance and resilience; used for many military applications. [Uniroyal]

Paracril® 2813. NBR, cold polymerized, nonstaining antioxidant; used for continuously cured extruded goods including sponge and PVC blends; FDA compliance; excellent processing; fast cure rate. [Uniroyal]

Paracril® BJLT M-30. NBR, low temp. polymerized, nonstaining stabilizer; easy and safe processing, low mold fouling, low water absorp.; ideal balance of low temp. flexibility and oil resistance for most automotive applications; used in seals, o-rings, shoe soles, molded and extruded goods. [Uniroyal]

Paracril® OZO. NBR/PVC fluxed blend (70/30), nonstaining stabilizer; used for shoe soles, rolls, wire and cable jacketing, hose tubes and covers, sponge, belting, and gasketing; excellent ozone resistance, good fuel and oil resistance, high abrasion resist, excellent color stability; easy to mix. [Uniroyal]

Paracure. Dispersions and emulsions of various chemical additives, primarily rubber latex antioxidants, accelerators and curatives; chemical additives for rubber latex compounding. [Testworth Laboratories Inc]

Paradene®. A proprietary trade name for coumarone-indene resins. [Neville] *

Paradene® No. 2. Petrol. hydrocarbon resin; used for inks, adhesives, coatings, rubbers, and concrete cures. [Neville]

Paradione. CAS 115-67-3; Paramethadione; anticonvulsant. [Abbott Laboratories]

Paradol. A range of fat liquors; for modification of leather handle. [Yorkshire Chemicals plc] *

Paradol. See Neradol. §

Paradow. CAS 106-46-7; p-Dichlorobenzene; for commercial and industrial solvents, deodorants and sanitary products. [Dow UK] *

Paradura. A proprietary trade name for a phenolic synthetic resin for varnish and lacquers. §

Paradyne. Fuel oil additives; flow improvers. [Exxon Int'l.; Exxon UK] *

Paraffagar. A proprietary preparation of liquid paraffin and agar-agar. §

paraffin. (paraffin wax; hard paraffin; petroleum wax, crystalline) Hydrocarbon; solid mixture of hydrocarbons obtained from petroleum; characterized by relatively large crystals; CnH_{2n+2}; CAS 8002-74-2; EINECS 232-315-6; Candles, paper coating, protective sealant for food products; lubricants; hot-melt carpet backing, floor polishes, cosmetics, chewing gum base; raising melting point of ointments. [EM Industries; Exxon; Humphrey; Jonk BV; Koster Keunen; Phillips 66; Texaco; Vista]

paraffin, liquid. A mixture of liquid hydrocarbons obtained by the distillation of the liquid remaining after the lighter hydrocarbons have been removed from petroleum. It is decolorized and purified.

paraffin oil. See mineral oil.

Paraffinum Molle. See petroleum jelly.

paraffin wax. See paraffin.

Parafil. Fibre reinforced thermoplastic rope/cable. [ICI Chem. & Polymers Ltd] †

Parafilm. A proprietary trade name for a rubber composition.§

Parafix. Photographic fixer. [May & Baker Ltd] *

Paraflex. CAS 95-25-0; Chlorzoxazone; relaxant. [McNeil Pharmaceuticals] *

Paraflow. Dewaxing oils and pour point depressants for lubricants and power transmission fluids. [Exxon Int'l.] *

Para-Flux® 4156. Aromatic process oil; plasticizer for CR compounds [C.P. Hall]

Paraform. Molded insulation shapes. [The Chemical & Insulating Co Ltd]

paraform. See paraformaldehyde.

paraformaldehyde. (para-formaldehyde; polyoxymethylene; paraform) A polymer of formaldehyde in which n = 8-100; $HO(CH_2O)_nH$; CAS 30525-89-4; EINECS 200-001-8; fungicides, bactericides, disinfectants, adhesives, hardener and waterproofing agent for gelatin contraceptive creams. [Andrulex Trading Ltd; Degussa; Hoechst Celanese; Mitsubishi Gas; Mitsui Toatsu Chem.]

Paraglas. Cast acrylic sheet; used for production of light domes, illuminated signs, wash basins and bathtubs, safety coverings for machines, showcases, graduated dials, models etc. [Degussa Ltd]

paraglobin. See globulin.

paraglobulin. See globulin.

Paragon-15. Polysiloxane resin in an aliphatic vehicle

---

* Trade name not verified as available   † Trade name verified as obsolete

system; water repellant for masonry. [Nova Chemical Inc] *

Paragon II. Propylene glycol, DMDM hydantoin, methylparaben, propylparaben; patented dual action cosmetics preservative with bactericidal and fungicidal properties. [McIntyre]

Paragon Steel. A proprietary trade name for a non-shrinking steel contains 1.55% manganese, about 0.6% chromium, and 0.25% vanadium. §

Paragrid. Synthetic reinforcement and support materials for coil engineering applications. [ICI Chem. & Polymers Ltd] †

Paragutta. A patented insulating compound for use in the manufacture of submarine telegraph and telephone cables; made from deproteinized rubber obtained by the heat treatment of rubber latex to hydrolyze the protein and washing. This rubber has reduced water-absorbing properties and is mixed with gutta-percha or balata and suitable waxes to produce paragutta. §

Parahypon. A proprietary formulation of aspirin, codeine phosphate and caffeine; an analgesic. [The Wellcome Foundation Ltd]

Parake. A proprietary preparation of paracetamol and codeine phosphate; an analgesic. [Galen Ltd] *

Paral. CAS 123-63-7; Paraldehyde; hypnotic; sedative. [O'Neal, Jones & Feldman Pharmaceuticals] *

Paralac. Solvated solutions of synthetic rubber and resins; used for contact adhesives and coatings. [Testworth Laboratories Inc]

Paralene. A range of synthetic tanning agents; for leather industry. [Yorkshire Chemicals plc] *

Paralgin. A proprietary preparation of paracetamol, caffeine and codeine phosphate; an analgesic. [H N Norton & Co Ltd] ‡

Paralink. Synthetic reinforcement and support materials for coil engineering applications. [ICI Chem. & Polymers Ltd] †

Paraloid® BTA-702. Methacrylate/butadiene styrene; impact modifier [Rohm & Haas] *

Paraloid® EXL 2607. Acrylic and MBS additives for engineering plastics; toughening polymers particularly for automotive and leisure segments. [Rohm & Haas]

Paraloid® EXL-3330. Butyl acrylate-based; toughener for PC, PBT and PET. [Rohm & Haas] *

Paraloid® HT-510. Polyglutarimide acrylic copolymer; heat distort. temp. modifier for PVC [Rohm & Haas] *

Paraloid® K-120N. CAS 9011-14-7; Methyl methacrylate polymer; hard acrylic resin with excellent exterior durability and abrasion resistance; processing aid for vinyl topcoats. [Rohm & Haas]

Paraloid® KM-318F. Acrylic; impact modifier for vinyl foam applications such as sheet, profile and pipe; controls cell structure. [Rohm & Haas] *

Paraloop. Synthetic reinforcement and support materials for coil engineering applications. [ICI Chem. & Polymers Ltd] †

Paralux. Lube oils. [Chevron] *

Paramax. A proprietary preparation of paracetamol with metoclopramide; for symptomatic treatment of migraine. [SmithKline Beecham] *

Paramel. A range of synthetic resins; used for tanning of leather. [Yorkshire Chemicals plc] *

Paramel. Modified melamine formaldehyde resin; used in the paper industry. [Cyanamid BV]

Paramet Ester Gum. A proprietary trade name for rosin-glycerol synthetic resin for lacquer and varnish manufacture. §

Paramid. Lube oils. [Chevron] *

Paramin DF. CAS 1698-60-8; Chloridazon; a pyridazinone herbicide for beet crops. [BASF plc]

Paramins. Additives for the petroleum industry. [Exxon UK]

Paramol-118. A proprietary preparation of paracetamol and dihydrocodeine bitartrate; an analgesic. [British Drug Houses] *

Paramos. Benzalkonium chloride; algicide and moss killer for paths and flower pots. [Chemsearch (UK) Ltd]

Paramount. Lube oils. [Chevron] *

Paramount B. CAS 68990-82-9; Partially hydrogenated palm kernel oil; kosher; center fat for confectioner's coatings, vegetable dairy systems, candy centers, icings, cosmetic/pharmaceutical applications. [Van Den Bergh Foods]

Paramul® SAS. Stearamide DIBA-stearate; emulsifier, pearlizing agent, opacifier. [Bernel]

paranitraniline red. (azophor red, para red, nitrazol, discharge lake R and RR). p-Nitro-benzeneazo β-naphthol, $C_{16}H_{11}N_3O_3$; used for dyeing cotton, and in the preparation of lakes for paper-staining.

Paranol. A proprietary trade name for a phenol-formaldehyde synthetic resin. §

Paranox. Detergents and dispersants. [Exxon Int'l.] *

Parapak. Industrial oil additives. [Exxon UK] *

para palm oil. (para butter). Pinot oil, a semi-drying oil from the seeds of *Euterpe oleracea.*

Parapel® HC-85. Linoleamidopropyl ethyldimonium ethosulfate, dimethyl lauramine isostearate; emulsifier compatible with most anionic surfactants; hair conditioner. [Bernel]

Parapel® LAM-100. CAS 5422-34-4; EINECS 226-546-1; Lactamide MEA; hair conditioner, humectant; compatible with thioglycolate. [Bernel]

Parapel® LIS. CAS 70729-87-2; Dimethyl lauramine isostearate; shampoo and conditioner additive; silky, velvety emollient. [Bernel]

Paraplast 8100. Inorganic powd.; used to cast mandrels or cores used in production of hollow plastic articles (reinforced plastic ducts, filament-wound pressure vessels, etc.); offers heat resistance to 275 F. [Hexcel]

Paraplex®. Series of polymeric plasticizers; used for flexible vinyl and rubber. [C P Hall]

Paraplex® G-25 100%. High molecular weight polyester sebacate; plasticizer for elec. tapes, high temp. insulation, coaxial cable, upholstery, coated fabric. [C.P. Hall]

Paraplex® G-50. Polyester adipate; pigment grinding medium; PVC plasticizer; for insulation, upholstery, window channels, liners, gaskets, coated fabrics. [C.P. Hall]

Paraplex® G-60. CAS 8013-07-8; EINECS 232-391-0; Epoxidized soybean oil; polymeric type plasticizer for coating formulations based on PVC and copolymers, NC, chlorinated rubber and paraffin; permanence in surface coating films under severe exposure, high plasticizing efficiency; good flexibility at low temp.; stabilizes materials with acid-producing components. [C.P. Hall]

Paraplex® G.62. CAS 8013-07-8; EINECS 232-391-0; Epoxidized soya bean oil; vinyl plasticizer and vinyl stabilizer. [Rohm & Haas]

Parapoid. Extreme pressure additives for gear lubricants. [Exxon Int'l.] *

Parapol. Liquid polyisobutylene. [Exxon]

Paraquat + Plus. Herbicide. [Chevron] *

para red. See paranitraniline red.

pararosaniline. CAS 569-61-9; Triamino-triphenyl-carbinol; HOC($C_6H_4NH_2$)$_3$; dyes wool and silk, purplered, and cotton with mordants.

Paraset. Printing ink distillates. [Carless Refining & Marketing Ltd]

Parasiticine. A fungicide containing 57% copper sulfate and sodium carbonate and bicarbonate.

Parasol 17. A wide distillation range white spirit substitute with 17% aromatic content. [Sasolchem]

Paratac. Additive for lubricating oils and greases. [Exxon Int'l.] *

paratartaric acid. CAS 133-37-9; Racemic tartaric acid, $C_4H_6O_6$.

Paratect Bolus. CAS 26155-31-7; Morantel tartrate; a cattle antihelmintic. [Pfizer International]

Paratemp. CAS 10101-39-0; Asbestos-free calcium silicate; thermal insulator. [The Chemical & Insulating Co Ltd]

Paratex. A proprietary preparation of thiamine hydrochloride, calcium and strychnine glycerophosphates, and sodium hypophosphate; a tonic. [Albion Group Ltd] ‡

Paratherm NF. CAS 8042-47-5; White mineral oil; nonfouling heat transfer fluid. [Paratherm]

parathion. (O,O-diethyl O-p-nitrophenyl phosphorothioate; diethoxy, nitro-phenoxy phosphorothioate; ethyl parathion) CAS 56-38-2; A powerful insecticide.

Para-Thor-Mone. A proprietary preparation of parathyroid hormone. [Eli Lilly & Co] †

Paratie. Synthetic reinforcement and support materials for coil engineering applications. [ICI Chem. & Polymers Ltd] †

Paratol. Natural and synthetic rubber latex based adhesives and coatings; cohesive, pressure-sensitive and synthetic resin emulsions; used for numerous adhesive bonding applications. [Testworth Laboratories Inc]

Paratone. Viscosity index improver. [Exxon UK] *

para toner. Paranitraniline red; used as a toner for lakes.

Paratophan. (homophan, methyl-atophan). 6-Methyl-2-phenylquinoline-4-carboxyic acid.

Paratulle®. CAS 8002-74-2; White soft paraffin; tulle primary wound dressing in the treatment of burns, cuts, wounds and abrasions. [Seton Healthcare Group plc]

Paraweb. Fiber-reinforced thermoplastic webbing for civil engineering applications and used as cargo slings, show fencing, windbreaks. [ICI Chem. & Polymers Ltd] †

Paraxin. CAS 56-75-7; Chloramphenicol.

Parazol. Crude dinitrodichlorobenzene. It contains m-dinitro-p-dichlorobenzene, o-dinitro-p-dichlorobenzene, and p-dinitro-p-dichlorobenzene; used as a high explosive.

Parazolidin. A proprietary preparation of phenylbutazone and paracetamol; used in the treatment of arthritis. [Ciba plc] *

Par Clay®. CAS 1332-58-7; Hydrated aluminum silicate (kaolin clay); mineral filler used as filler, extender or reinforcing agent for paint, paper, rubber, ceramics, plastics and specialities. [R T Vanderbilt Co Inc]

Parco. Wear reducing processes. [Brent Chemicals International plc]

Parco® 58-C-55. Vinyl acrylic copolymer emulsion; used in paints (interior wall, gloss and semigloss exterior, masonry); clarity, high gloss. [Thibaut & Walker] *

Parcolene. Chemical treatment for use after "bonderizing" and "parkerizing" treatments. [Brent Chemicals International plc]

Parcryl® 250. 100% Acrylic emulsion; used in caulking compounds and masonry paints; wet adhesion, high gloss. [Thibaut & Walker] *

Parcryl® 311. 100% Acrylic emulsion; extremely flexible polymer for high pigment loading; for roof and tennis-court coatings; wet adhesion, high gloss. [Thibaut & Walker] *

Pardale. A proprietary preparation of paracetamol, codeine phosphate, and caffeine; an analgesic for veterinary use. [Dales Pharmaceuticals Ltd] *

Paredrine. CAS 306-21-8; Hydroxyamphetamine hydrobromide; adrenergic. [SmithKline Beecham] *

pareira. The dried root of Chondrodendron tomentosum.

Parel®. Propylene oxide/allyl glycidyl ether copolymer; elastomer with outstanding balance of low and high temp. resistance; used for dynamic mounts and high flex applications. [Zeon]

Parel 58. Sulfur vulcanizable, elastomeric copolymer of polypropylene oxide and alkyl glycidyl ether; used for flexible wing seals in supersonic aeroplanes, for motor mounts and other noise suppression applications requiring long term durability and as a blending rubber in tubes and tires. [Hercules] *

Parenamine. CAS 9015-54-7; Protein hydrolysate; replenisher. [Sterling Drug Inc] *

Parencillin. CAS 6130-64-9; Penicillin G procaine; antibacterial. [Merrell Dow Pharmaceuticals Inc] *

Parenol. Consists of 65% soft paraffin, 15% wool fat, and 20% distilled water.

Parenol liquid. Consists of 70% liquid paraffin, 5% white bees wax, and 25% distilled water.

Parentrovite. Parenteral vitamins B and C. [SmithKline Beecham]

Parenzyme. CAS 9002-07-7; Trypsin, crystallized; enzyme (proteolytic). [Merrell Dow Pharmaceuticals Inc] *

Parez. Modified resins of melamine and formaldehyde; used in the paper industry. [Cyanamid BV]

Parez 631NC. Cationic modified polyacrylamide; used in the paper industry. [Cyanamid BV]

Parfenac. CAS 2438-72-4; A proprietary preparation of bufexamac used as a skin cream. [Lederle Laboratories]*

Parian cement. A cement which is similar to Keene's cement, except that a solution of borax is used instead of alum.

parianite. An asphaltum from the pitch lake at Trinidad.

Paricin® 1. CAS 141-23-1; EINECS 205-471-8; Methyl hydroxystearate; lubricant/processing aid for butyl rubber; wax firming agent in cosmetics and specialty inks; source of hydroxystearic acid for glycerin-free lithium greases. [CasChem] *

Paricin® 6. Butyl acetoxystearate; oxidation-stable plasticizer for vinyls. [CasChem] *

Paricin® 8. Glyceryl tri(acetoxystearate); lubricant plasticizer for vinyls, esp. high-temp. wire jacketing; heat and oxidation-stable; grinding med. for pigment dispersions; plasticizer for nitrocellulose. [CasChem] *

Paricin® 9. CAS 33907-47-0; EINECS 251-734-5; Propylene glycol hydroxystearate; wax modifier, emollient; phys. properties near to those of spermaceti wax. [CasChem] *

---

* Trade name not verified as available   † Trade name verified as obsolete

Paricin® 13. CAS 1323-42-8; EINECS 215-355-9; Glyceryl hydroxystearate; wax modifier, emollient; phys. properties near to those of beeswax. [CasChem] *

Paricin® 15. CAS 33907-46-9; EINECS 251-732-4; Ethylene glycol hydroxystearate; wax modifier, emollient; phys. properties near to those of candelilla wax. [CasChem] *

Paricin® 18. Stearyl 12-hydroxystearate; saturated, low melting point wax [CasChem] *

Paricin® 210. N-Stearyl 12-hydroxystearamide; lubricant/antistat for plastics, metals; mold release, antiblocking agent for textile coatings; slip agent for varnishes and lacquers; also for elec. potting compounds, crayons, wax blends, high-temp greases; release agent in contact with food. [CasChem]

Paricin® 220. N (2-Hydroxyethyl) 12-hydroxystearamide; internal mold release agent, lubricant for polyolefins, PVC, styrenics. [CasChem]

Paricin® 285. N,N´-Ethylene bis 12-hydroxystearamide; internal lubricant, mold release, slip additive for PVC. [CasChem]

Parilene. A proprietary trade name for polyparaxylyene; a plastics material used for film manufacture for electrical purposes. §

Paris green. See emerald green and methyl green.

Paris salts. A disinfectant containing 50 parts zinc sulfate, 50 parts ammonia alum, 1 part potassium permanganate, and 1 part lime.

Parixistil. CAS 68-88-2; A proprietary preparation of hydroxyzine; an ataractic. [Pfizer International] †

Parkerised Steel. A patented process for the treatment of steel with iron and manganese phosphates to give the surface resistance to corrosion. §

Parker's cement. See Roman cement.

Parlay. Growth regulator for growers of grass seed. [ICI Chem. & Polymers Ltd]

Parlodel. CAS 22260-51-1; Bromocriptine mesylate; enzyme inhibitor. [Sandoz Pharmaceuticals]

Parlodion. A trademark for a shredded form of pure collodion. §

Parlon. Chlorinated rubber available in six viscosity grades; used as film-formers in adhesives, in corrosion resistant coatings for wood, metal, and concrete, for wood floor finishes and sealers, in inks, etc. [Hercules] *

Parlon P. A range of chlorinated polypropylenes; forms clear, hard, protective films that are resistant to chemicals, salt solutions and water; used as film formers in adhesives and sealants, the construction and building industry, for lumber and wood products, for floor coverings, etc. [Hercules] *

Parmentine. A mixture of glycerin, gelatin, dextrine, sodium sulfite, and zinc sulfate; used for sizing and finishing cotton, wool, and silk.

Parmetol. Range of preservatives; used in the paint, ink, and adhesives industries. [Sterling-Winthrop Group Ltd]

Parmid®. CAS 54143-57-6; Tablets: metoclopramide hydrochloride BP equivalent to 10 mg anhydrous metoclopramide; syrup: each 5 ml containing metoclopramide hydrochloride BP equivalent to 5 mg anhydrous metoclopramide; injection: each 1 ml of solution contains metoclopramide hydrochloride BP equivalent to 5 mg of the anhydrous metoclopramide; for treatment of digestive disorders, nausea and vomiting, migraine, post-operative hypotonia, post-operative syndrome. [Lagap

Pharmaceuticals Ltd]

Parmr. The trade name for a blown bitumen residue; it is a mineral rubber for use in the rubber industry; Grade I melts from 190-310 F and Grade II at above 300 F. §

Parnate Tablets. CAS 13492-01-8; Tranylcypromine sulfate (nonhydrazine monoamine oxidase inhibitor); for treatment of depressive illness particularly where phobic symptons are present. [SmithKline Beecham]

Paroa-caxy Oil. The seed-oil of *Penta-clethra flamentosa*.§

Parodyne. See Antipyrine. §

Parogen. (oleogen, oxygenated paraffin, vasogen). Consists of 2 parts liquid paraffin, 2 parts oleic acid, and 1 part ammoniated alcohol (5%).

Paroidin. CAS 9002-64-6; Parathyroid; regulator. [Parke-Davis] *

Paroil®. Chlorinated paraffin, liquid; used in cutting oils, industrial lubricant, flame retardant and plasticizer in plastics. [Dover]

Paroil® 10, 145, 1061. Chlorinated paraffin; lubricant additive. [Dover]

Parol 70. Mineral oil tech.; CAS 8042-47-5; used for animal feed dedusting, food pkg. materials, meat pkg., household cleaners and polishes. [Penreco]

Par-o-lac. An impregnating compound.

Paroleine. See paraffin, liquid.

Paroline. See paraffin, liquid.

Parolite. CAS 24887-06-7; Zinc formaldehyde sulfoxylate; stripping agent for removing dyes from wool and nylon. [Henkel/Textiles]

paromomyicn. CAS 7542-37-2; An antibiotic produced by *Streptomyces rimosus forma paromomycinus*.

Paroven. A proprietary preparation of hydroxyethylrutosides; used in the treatment of varicose veins. [Zyma (UK) Ltd] ‡

Parraynite. A trade name for a rubber compound which is used by X-ray operators to protect them from injury by exposure to the rays. §

parrot coal. Cannel coal.

Parr's alloys. Anti-corrosion alloys. One alloy contains 80% nickel. 15% chromium, and 5%, copper. Another alloy contains 66.6% nickel, 18% chromium, 8.5% copper, 3.3% tungsten, 2% alumimium, and 1% manganese.

Parsidol. CAS 1094-08-2; Ethopropazine hydrochloride; antiparkinsonian. [Parke-Davis] *

parsley camphor. (camphre de Persil). Crystallized apiole.

Parsol® 1789. CAS 70356-09-1; Butyl methoxydibenzoylmethane; uv-A absorber for production photostability; used in sunscreens. [Bernel; Givaudan]

Parsol® MCX. CAS 5466-77-3; Octyl methoxycinnamate; uv absorber, sunscreening agent in the wavelength range of 2900-3200 Å which causes sunburn and skin damage, stimulates tanning. [Bernel; Givaudan]

Parsolin. Organophosphorus insecticide. [Ciba plc] *

Parson's Alloy. A proprietary trade name for an alloy of 56% copper, 41.5% zinc, 1.2% iron, 0.7% tin, 0.1% manganese, and 0.46% aluminum; specific gravity 8.4. §

Parstelin. A proprietary preparation of tranylcypromine, and trifluoperazine. [SmithKline Beecham] *

Parstelin Tablets. Combination of tranylcypromine sulfate and trifluoperazine hydrochloride; for treatment of depressive illness complicated by anxiety particularly where phobic symptons are present. [SmithKline Beecham] *

Partagon. A German preparation. It consists of rods contain-

---

ing silver chloride and - sodium-silver chloride.

Partinium. An aluminum alloy; varies in composition, and often contains tungsten, copper, tin, zinc, and magnesium; one alloy contains 96% aluminum, 2.4% antimony, 0.8% tungsten, 0.64% copper, and 0.16% tin; another alloy consists of 88.5% aluminum, 7.4% copper, 1.7% zinc, 1.3% iron, and 1.1% silicon. §

PartsPrep™ Degreaser. CAS 872-50-4; N-Methyl-2-pyrrolidone and surfactants; degreaser formulations. [ISP]

Parvol. Contraceptive jelly, containing polyoxyethyleneothylcresol. [British Drug Houses] *

Parvol. A range of auxiliary products; for use as assistants in tanning processes. [Yorkshire Chemicals plc] *

parvoline. Dimethylethylpyridine, $C_9H_{13}N$.

Parylene N. A plastic material used to make thin film membranes, 2-1000 Angstroms thick. [Union Carbide (UK) Ltd] *

Pasilex. CAS 12141-46-7; Precipitated aluminum silicate; used as filler for the paper industry. [Degussa AG] *

Pasonex. A proprietary preparation of aminosalicylate calcium, isoniazid, pyridoxine hydrochloride and menadione; an antitubercular. [Pfizer International] †

Passini's solution. An aqueous solution of mercury and sodium chlorides and glycerine; used to preserve animal tissue.

Passow's slag cement. Prepared by blowing into liquid slag as it issues from the blast furnace, when it becomes granulated. It is then finely ground.

pastaccio. A residue from the manufacture of calcium citrate. It consists of vegetable cellulose with some hydrocarbons.

Pasturol. Soluble concentrate of 25 g dicamba, 200 g MCPA, and 400 g mecoprop per liter; broad-spectrum herbicide used for weed control in cereals and grassland. [Farmers Crop Chemicals Ltd]

Patafol. Fungicide containing ofurace for the control of potato blight. (Sold in UK for Chevron Chemical Co) [ICI Chem. & Polymers Ltd]

Patafol Plus. Manganese zinc ethylenebisdithiocarbamate and ofurace mixture; for control of potato blight. (Sold in UK for Chevron Chemical Co.) [ICI Chem. & Polymers Ltd]

patava oil. (batana oil). Coumou oil, a semi-drying oil obtained from the kernels of the Brazilian palm tree, *Oenocarpus batava.*

patchouli. An Indian herb, *Pogostemon patchouly*; used in perfumery.

Patco® 3. Blend of Emplex sodium stearoyl lactylate and Verv® calcium stearoyl-2-lactylate; conditioner/softener; starch and protein complexing agent for use in yeast-leavened bakery products [Am. Ingredients/Patco]

Patcote® 305. Silicone emulsion, 10% filled; defoamer for general food and industrial applications. [Am. Ingredients/Patco]

Patcote® 306. 20% Filled silicone emulsion; defoamer for general food and industrial applications; FDA compliance. [Am. Ingredients/Patco]

Patcote® 309. Nonsilicone aq. emulsion; defoamer for food and industrial use; FDA compliance. [Am. Ingredients/Patco]

Patcote® 315. 10% Silicone emulsion; defoamer for food and industrial applications; FDA compliance. [Am. Ingredients/Patco]

Patcote® 337. Nonsilicone; defoamer for use in alcohol production via grain fermentation; FDA compliance. [Am. Ingredients/Patco]

Patcote® 500. Silicone-containing; process defoamer, dispersing aid for latex trade sales. [Am. Ingredients/Patco]

Patcote® 512. Silicone-containing; defoamer for use in urethane-modified resins. [Am. Ingredients/Patco]

Patcote® 525. Silicone-containing; defoamer for use in water-reducible alkyd and industrial acrylic systems. [Am. Ingredients/Patco]

Patcote® 555K. 100% filled silicone; kosher grade; defoamer for food and nonfood processing; FDA compliance. [Am. Ingredients/Patco]

Patcote® 803. Nonsilicone; defoamer for acrylic and terpolymer emulsions for trade sales; FDA compliance. [Am. Ingredients/Patco]

Patcote® 811. Nonsilicone; defoamer for graphic arts water-based acrylic systems. [Am. Ingredients/Patco]

patent bark. Commercial quercetin.

patent black. An acid dyestuff. It is a substitute for logwood.

patent red. *See* cinnabar.

patents. The small portion of very white flour obtained from wheat. It is poor in proteins, and is used for fancy breads.

patent zinc white. A pigment made by adding a soluble sulfide to a zinc chloride or zinc sulfate solution, filtering off the precipitate, drying it, and then calcining it. It has the composition, $5ZnS \cdot ZnO$.

Pat-Fix RBD. Fixative for direct dyes and prints. [Yorkshire Pat-Chem]

Pathclear. Contains aminotriazole, diquat, paraquat and simazine; long-acting weedkiller for paths, drives, and patios. [ICI Garden Products]

Path Gun!; Ready-for-use herbicide spray. [ICI Chem. & Polymers Ltd]

Pathilon. Tridihexethyl chloride; anticholinergic. [Lederle Laboratories USA] *

Pathocil. CAS 13412-64-1; Dicloxacillin sodium; antibacterial. [Wyeth Laboratories]

Pathozone. CAS 62893-20-3; Cefoperazone sodium; an antibiotic used for mastitis in dairy cows. [Pfizer International]

Path Weeds Killer. Total weedkiller. [Fisons plc, Horticulture Div] *

patina. The green film which forms on copper and bronze moldings. It consists of basic copper carbonate or other basic copper salts.

Pationic 122A. CAS 29051-57-8; Sodium capryl lactylate; emulsifier/foam booster for facial cleansers and personal care products; microbial inhibitor properties. [RITA]

Pationic 138C. CAS 13557-75-0; Sodium lauroyl lactylate; detergent, conditioner, foam booster, visc. builder, lipophilic emulsifier, cleansing agent used in shampoos, facial cleansers. [RITA]

Pationic CSL. CAS 5793-94-2; Calcium stearoyl-2-lactylate; w/o emulsifier and protein complexer for cosmetics. [RITA]

Pationic ISL. CAS 66988-04-3; EINECS 266-533-8; Sodium isostearoyl-2-lactylate; surfactant, emulsifier for cosmetics; perfume solubilizer; substantive conditioner for hair and skin. [RITA]

Pationic SSL. CAS 25383-99-7; EINECS 246-929-7; Sodium stearoyl 2-lactylate; emulsifier, visc. builder, protein complexer for cosmetics products [RITA]

Patlac® IL. CAS 42131-28-2; Isostearyl lactate; surfactant,

---

* Trade name not verified as available   † Trade name verified as obsolete

emollient for cosmetics. [Am. Ingredients/Patco; RITA]

Patlac® LA. CAS 50-21-5; EINECS 200-018-0; Lactic acid; moisture binder, humectant. [Am. Ingredients/Patco; RITA]

Patlac® NAL. CAS 72-17-3; EINECS 200-772-0; Sodium lactate; pH buffer, humectant, stabilizer, component of stratum corneium; for food, pharmaceutical, and cosmetic industries. [Am. Ingredients/Patco; RITA]

Pat-Lube Series. Lubricants for jet and beck dyeing offering fiber-to-metal and fiber-to-fiber lubricity. [Yorkshire Pat-Chem]

Patogen 311. Low foam wetting agent/scour for bleach baths. [Yorkshire Pat-Chem]

Patogen 353. Wetting and scouring agent for bleach baths, dye baths. [Yorkshire Pat-Chem]

Patogen AO-30. CAS 68155-09-9; Cocamidopropyl dimethylamine oxide; fugitive wetter, foam booster, detergent, emulsifier, foaming agent for textile applications. [Yorkshire Pat-Chem]

Patogen P-10 Acid. Alkyl phosphate; scouring agent for textiles. [Yorkshire Pat-Chem]

Patoran. Substituted urea herbicide. [Ciba plc] *

Patoran® FL. CAS 3060-89-7; Metobromuron; for pre-emergence weed control in potatoes, soybeans, tobacco, tomatoes [BASF AG]

Pat-Quest CS. Chelating/sequestering agent for textile bleaching and scouring operations; excellent caustic stability. [Yorkshire Pat-Chem]

Patrol. CAS 67306-00-7; Emulsifiable concentrate of 750 g fenpropidin per liter; a systemic fungicide. [ICI Agrochemicals]

Pat-Soft 1442. Softener and lubricant for textile fabrics; excellent heat stability; recommended for finishing prior to printing. [Yorkshire Pat-Chem]

pattern metal. An alloy of 83% copper, 10% zinc, 4% tin, and 3% lead.

Pattinson's white lead. A pigment. It is basic lead chloride, $PbCl_2.2Pb(OH)_2$.

Pattonex. CAS 3060-89-7; Active ingredient: metobromuron; 3-(4-bromophenyl)-1-methoxy-1-methylurea; residual herbicide for use as a selective weedkiller. [Agan Chemical Manufacturers Ltd]

Pattrex. Pattern plaster. [Foseco (F.S.) Ltd]

Pattrit. Casting plaster. [Foseco (F.S.) Ltd] †

Pat-Wet LF-55. Low foaming wetting agent with non-rewetting properties for textile processing. [Yorkshire Pat-Chem]

Pat-Wet SP. Alkyl phosphate; low foaming wetting agent, bath stabilizer for textile dyeing; excellent for sulfur and indigo dye baths. [Yorkshire Pat-Chem]

Pavabid. Papaverine hydrochloride; relaxant. [Marion Merrell Dow Inc]

Pavacol D. A proprietary preparation containing pholcodine, balsam of tolu, oil of clove, tincture of capsicum, oil of peppermint, alcohol and chloroform; a cough linctus. [Boehringer Ingelheim Ltd]

Pavacol Diabetic Cough Syrup. A proprietary preparation containing agents as in Pavacol without carbohydrates; a cough linctus. [Octel Chemicals Ltd] †

pavlin. Fraxin, $C_{16}H_{18}O_{10}$; a substance which occurs in the bark of the common ash.

Pavulon. CAS 15500-66-0; Pancuronium bromide; blocking agent. [Organon Inc] *

Pavy's solution. A modified Fehling's solution used for the determination of sugar.

Paxadon. CAS 58-56-0; Pyridoxine hydrochloride; pharmaceutical preparation for the treatment of pregnancy sickness and peripheral neuritis. [M A Steinhard Ltd] *

Paxalgesic. Dextropropoxyphene with paracetamol; pharmaceutical preparation for the treatment of mild to moderate pain. [M A Steinhard Ltd] *

Paxalol. CAS 525-66-6; Propranolol; pharmaceutical preparation for the treatment of cardiac arrhythmias. [M A Steinhard Ltd] *

Paxane. CAS 17617-23-1; Flurazepam; pharmaceutical preparation for the treatment of insomnia. [M A Steinhard Ltd] *

Paxbestos. Asbestos products bonded with hydraulic cement and impregnated with bitumen; used as insulating materials.

Paxipam. CAS 23092-17-3; Halazepam; sedative. [Schering Corp] *

Paxofen. CAS 15687-27-1; Ibuprofen; pharmaceutical preparation for the treatment of rheumatic pain. [M A Steinhard Ltd] *

Paxofenac. Diclofenac; pharmaceutical preparation for the treatment of musculo-skeletal disorders and gout. [M A Steinhard Ltd] *

Paxolax. CAS 603-50-9; Bisacodyl; pharmaceutical preparation for the treatment of constipation. [M A Steinhard Ltd] *

Paxolin. A synthetic resin bonded paper product used for insulating purposes.

Payne's grey. An oil and water color prepared from black alizarin, madder, and indigo.

Payne's solution. A solution of sodium hypobromite.

Payzone. A proprietary preparation of nitrovin hydrochloride; a veterinary growth promoter. §

Pazo. An insecticide. [Murphy Chemical Co Ltd] †

PBI Slug Pellets. CAS 108-62-3; Pellets containing 6% w/w metaldehyde; snail and slug bait. [Pan Britannica Industries Ltd]

PBI Spreader. Nonionic wetting agent for use in agrochemical sprays. [Pan Britannica Industries Ltd]

P.B.N. Phenyl β-naphthylamine; a rubber antioxidant. §

PBT-1100. CAS 26062-94-2; PBT; molding compound for engineering parts requiring heat resistance, dimensional stability, rigidity, str., chem. resistance; used for elec., automotive, appliance, and telecommunications parts. [Bay Resins]

PBT-1100G15. CAS 26062-94-2; PBT, 15% glass-reinforced. [Bay Resins]

PBT-1300. CAS 26062-94-2; PBT, highly impact-modified. [Bay Resins]

PBT-1700. CAS 26062-94-2; PBT, flame retardant. [Bay Resins]

PC-000/5T. PC, 5% PTFE-lubricated; improved surface wear characteristics, reduced static and dynamic coefficients of friction. [Compounding Tech.]

PC-20CF. PC resin, 20% carbon fiber-reinforced; offers exceptional combination of dimensional stability and mech. strength and stiffness; used for wide range of precision electro-mech. parts and compounds. [Compounding Tech.]

PC-20GF/15T. PC, 20% glass fiber-reinforced, 15% PTFE-lubricated; improved mech. strength and stiffness, greater chem. resistance, thermal resistance, superior accuracy in molding, dramatically increased wear resis-

---

‡ Trade name and manufacturer not verified　　　　　　　§ Trade name without identified manufacturer

tance; used for precision electro-mech. components and gears. [Compounding Tech.]

PC-1100. PC; general purpose molding compound offering high precision, dimensional stability, and toughness. [Bay Resins]

PC-1100G10. PC, 10% glass-reinforced; molding compound offering additional rigidity for frames, cases, appliance housings. [Bay Resins]

PC-1100H30. PC, 30% carbon fiber-reinforced; electrically dissipative molding compound with excellent mech. properties; for business machine components. [Bay Resins]

PC-1244. Defoamer; used to suppress foaming tendencies in lubricating oil formulations. [Monsanto Co] *

PC-1344. Defoamer in nonaqueous hydrocarbon and solvent systems. [Monsanto Co] *

PC-1700G10FR. PC, 10% glass-reinforced, flame-retardant; molding compound for elec./electronic, business machine applications. [Bay Resins]

PCA 4-1. Polyether (PTMEG) based urethane liq. polymer; prepolymer producing elastomers with good aging properties when subjected to heat, humidity, fungus, and other environments, good abrasion and tear resistance, low-temp. flexibility, toughness, ease of processing. [Polyurethane Corp. of Am.] *

PCA 301. Isocyanate terminated polyester based PU prepolymer; yields cured elastomers with abrasion and tear resistance, high load bearing capacity; for solid industrial tires, rollers, sheet goods applications; ideal for pouring large parts. [Polyurethane Corp. of Am.] *

PCI. Corrosion inhibitors used in deicers. [Georgia-Pacific]

PCMX. For chemical compositions having antiseptic and germicidal properties, in Class 18. [(Ferro Corporation)]‡

PCP. See pentachlorophenol.

PC resin. See polycarbonate resin.

PDS. Polydioxanone; surgical aid. [Ethicon Inc] *

PDX-82427. PEEK with 30% nickel; conductive attenuating composite effectively shielding electromagnetic and/or radio frequency interference; used in avionics housings, business machine enclosures, and other electronic devices. [LNP]

PDX-84367. CAS 61128-46-9; PEI; antistatic grade composite for protection of mod. sensitive elec. components from low voltages. [LNP]

PDX-84368. PC; antistatic grade composite for protection of mod. sensitive elec. components from low voltages. [LNP]

PDX-84369. CAS 26062-94-2; PBT; antistatic grade composite for protection of mod. sensitive elec. components from low voltages. [LNP]

PE. See pentaerythritol.

PE 100 228 FH-VP. Halogen-free flame-retardant compound; extrusion compound for processing at ≈185 C. [Zipperling Kessler]

PE 1017. CAS 9002-88-4; LDPE homopolymer resin; extrusion resin; high-speed coating and laminating resin. [Chevron]

PE 4517. CAS 9002-88-4; LDPE homopolymer resin; coating and laminating resin for carton stock coating and photographic paper. [Chevron]

PE 5222. LDPE copolymer resin with 6% EVA and high slip and antiblock additives; film resin offering good heat seal; used for frozen foods and ice bags. [Chevron]

PE 5554-H. CAS 9002-88-4; LDPE homopolymer resin with med. slip and antiblock additive; high-clarity film resin. [Chevron]

PE 5861. CAS 9002-88-4; LDPE homopolymer resin with high slip and med. antiblock additive; film resin for garment applications. [Chevron]

peach black. A variety of carbon black similar to lampblack.

peach wood. See redwoods.

peanut oil. (arachis oil; groundnut oil; Katchung oil) Refined fixed oil obtained from seed kernels of one or more cultivated varieties of *Arachis hypogaea*; CAS 8002-03-7; edible oil; substitute for olive oil; vehicle for medicine; in manufacture of margarine, soaps, paints; heat transfer medium in laboratory. [Croda; Karlshamns; Penta Mfg.]

peanut ore. A mineral. It is a variety of wolframite.

Pea Pro-Tein BK. CAS 9008-99-8; Hydrolyzed pea protein; substantive protein, film-former, moisturizer for skin and hair car products (shampoos, conditioners, creams, lotions, liq. hand soaps); anti-irritant in anionic formulations. [Maybrook]

Pearex-L®. CAS 9032-75-1; Fungal pectinase; enzyme used to prevent haze formation in pear and apple processing. [Solvay Enzymes]

Pearl. CAS 8008-20-6; Kerosene. [Chevron] *

Pearl I, II, III. CAS 7787-59-9; Bismuth oxychloride. [Presperse]

pearl alum. A specially prepared aluminum sulfate used in the paper industry.

pearl ash. CAS 3811-04-9; 584-08-7; A variety of potassium carbonate, K_2CO_3.

Pearl Dust®. CAS 584-08-7; A registered trade name for a form of potassium carbonate, K_2CO_3; used as a filler. §

Pearlex GC 0311. Proprietary blend; pearlizing/opacifying agent for shampoos, bubble baths, liq. hand soaps at low addition levels. [Clough]

Pearl-Glo®. CAS 7787-59-9; Bismuth oxychloride; pearlescent pigment powds. and disps. for cosmetic eye, face, lip, and body makeup. [Van Dyk]

pearl-hardening. Calcium sulfate, $CaSO_4$; used as a loading for paper.

pearlite. Iron carbide eutectoid, consisting of alternate masses of ferrite and cementite.

pearl powder. Bismuth oxychloride.

pearl spar. A double carbonate of magnesium and calcium, $Mg.Ca(CO_3)_2$.

Pearlstick. Polyurethane pellets; raw materials for adhesives. [Merquinsa]

Pearlstick 45-05/40. Linear polycaprolactone-polyurethane; features very high rate of crystallization; additive for adhesives for wood and automotive industry and plastics; plasticizer for EVA, PVC, ASA hot-melts for impregnation and textile coating. [Merquinsa]

Pearlstick 46-10/06. CAS 9009-54-5; Linear polyurethane; features high rate of crystallization; for two-component adhesives used in footwear, wood, pkg. [Merquinsa]

Pearlstick 65-05. Linear polycaprolactone-polyurethane; features very high rate of crystallization; additive for two-component adhesives used in wood and automotive industries; plasticizer in EVA, PVC, ASA hot melts for impregnation and textile coating. [Merquinsa]

Pearl Supreme UVS. CAS 7787-59-9; Bismuth oxychloride. pearlescent pigment. [Presperse]

pearl white. (flake white). Bismuth oxychloride, BiOCl, (blanc de perle). A basic bismuth nitrate, $Bi(OH)_2NO_3$, is

---

also known as pearl white. The term is sometimes used in connection with a white lead which has been tinted with Paris blue or indigo.

pear oil. CAS 123-92-2; 628-63-7; Isoamyl acetate, $CH_3.COO.C_5H_{11}$; used in the manufacture of fruit essences for flavoring confectionery.

Pearsall. Flame retardant. [Witco]

Pearsol. Synthetic phenolic germicides in a terpeneol vegetable oil soap base; for general disinfection/antisepsis; a substitute for chloroxylenol solution. [William Pearson Ltd] *

Pearson's cerate. Consists of 4 parts lead plaster, 1 part beeswax, and 3 parts almond oil.

Pearson's solution. A solution of dried sodium arsenate 1 per 100 to 1 per 1,000.

Peat. The partially decayed remains of plants; used as fuel.

peat coal. An intermediate between peat and lignite.

Peaweed. Suspension concentrate containing 152 g prometryn and 304 g terbutryn per liter; for weed control in peas, beans and potatoes. [Farmers Crop Chemicals Ltd; Pan Britannica Industries Ltd]

Pebax®. Polyether block amides; for sparking applications, gears. [Atochem Inc] *

Pebax® 2533 SA 00, 2533 SD 00, 2533 SN 00. Polyether block amide polymers; SA indicates additive-free; SD is uv stabilized with mold release additive; SN is uv stabilized; extrusion/molding grade polymer used in sport, automobile, elec./electronics, medical, agriculture, and mechanical handling industries; also as a basis for hot-melt adhesive formulations. [Atochem N. Am./Plastics]

Pebax® 4011 MA 00. Polyether block amide polymer, antistatic, hydrophilic grade; molding grade polymer used in sport, automobile, elec./electronics, medical, agriculture, and mechanical handling industries; also as a basis for hot-melt adhesive formulations. [Atochem N. Am./Plastics]

Pebax® 4033 SN 70, 5533 SN 70. Polyether block amide polymers, uv stabilized with antistatic fillers; extrusion and molding grade polymer. [Atochem N. Am./Plastics]

Pebax® 6333 SA 00, 6333 SD 00, 6333 SN 00. Polyether block amide polymers; SA indicates additive-free; SD is uv stabilized with mold release additive; SN is uv stabilized; see Pebax 2533 SA 00. [Atochem N. Am./Plastics]

pecan oil. Oil obtained from the seed of the North American walnut, Juglans niger.

Peceol Isostearique. CAS 32057-14-0; Glyceryl isostearate; w/o coemulsifier; pigment dispersant; additive for lipsticks; superfatting agent for emulsified preparations [Gattefosse SA]

Pecosil CAS-36. Octyldodecyl/dimethicone copolyol citrate. [Phoenix]

Pecosil DAS 36. Dilinoyl dimethicone copolyol citrate. [Phoenix]

Pecosil GSA 36. Octyldodecyl dimethicone copolyol adipate. [Phoenix]

Pecosil OS-100B. Dimethicone propylethylenediamine behenate. [Phoenix]

Pecosil OS-100DA. Dimethicone propylethylenediamine dilinoleate. [Phoenix]

Pecosil OS-100HS. Dimethicone propylethylenediamine hydroxystearate. [Phoenix]

Pecosil OS-100L. Dimethicone propylethylenediamine laurate. [Phoenix]

Pecosil OS-100M. Dimethicone propylethylenediamine myristate. [Phoenix]

Pecosil OS-100U. Dimethicone propylethylenediamine undecylenate. [Phoenix]

Pecosil PS-100. Dimethicone copolyol phosphate; imparts silky feel to aq. formulations, skin creams/lotions, mousses, hair conditioners, shaving products, bath products and gels. [Phoenix]

Pecosil PS-100K. Potassium dimethicone copolyol phosphate. [Phoenix]

Pecosil SG-20. Octyldodecoxy dimethicone. [Phoenix]

Pecosil SSP. Hydrolyzed soy protein/dimethicone copolyol phosphate copolymer. [Phoenix]

Pectamol. A proprietary preparation containing oxeladin citrate, in flavored vehicle containing menthol, chloroform, and glycerin; a cough linctus. [British Drug Houses] *

pectase. A clotting enzyme, which produces vegetable jellies.

pectin. (citrus pectin) Polysaccharide; purified carbohydrate product obtained from the dilute acid extract of the inner portion of the rind of citrus fruits or from apple pomace; CAS 9000-69-5; EINECS 232-553-0; jellies, foods, cosmetics, drugs, protective colloids, emulsifying agents, dehydrating agents. [Hercules; Penta Mfg.; Pomosin GmbH; Spice King; U.S. Biochemical]

pectinase. Enzyme from Rhizopus sp.; CAS 9032-75-1; EINECS 232-885-6; biochemical research; juice and jelly industries. [Gist-Brocades Food Ingreds.; U.S. Biochemical]

Pectinase AT. CAS 9032-75-1; Fungal pectinase; enzyme for cranberry juice depectinization. [Solvay Enzymes]

Pectinex. A purified enzyme preparation produced from a selected strain of Aspergillus niger, used in the case where the aim is breaking down of soluble and insoluble pectins with varying degrees of esterification, for reduction of viscosity, clarification, maceration of plant tissue and depectinization. [Novo Nordisk]

Pecutrin®. Vitaminized mineral salt mixture; for individual dosing and as a feed additive for all animals kept for use. [Bayer AG]

PediaCare 1. Dextromethorphan hydrobromide; antitussive. [McNeil Consumer Products Co]

Pediaflor. CAS 7681-49-4; Sodium fluoride; dental caries prophylactic. [Ross Laboratories] *

Pediamycin. CAS 41342-53-4; Erythromycin ethylsuccinate; antibacterial. [Ross Laboratories] *

Pedvax® HIB. Hemophilus B conjugate vaccine (meningococal protein conjugate); indicated for route immunization against invasive disease caused by hemophilus influenza type B in infants/children 2 to 71 months of age. [Merck & Co Inc]

Peerless®. CAS 1332-58-7; Kaolin clay; filler and extender. [R T Vanderbilt Co Inc]

Peerless alloy. A heat-resisting alloy containing 78.5% nickel, 16.5% chromium, 3% iron, and 2% manganese.

Peerless® No. 1. CAS 1332-58-7; Secondary kaolin clay; filler used in adhesives, wallboard, paint, paper, fertilizer, roofing gran., crayons, powd. soaps, pharmaceuticals, ceramics; sanitaryware, artware, generalware, floor tile, elec. and chemical porcelain, and special refractories; imparts more plasticity to cast piece. [R.T. Vanderbilt]

PEG-4. (2,2´-[oxybis(2,1-ethanediyloxy)bisethanol]; CAS 25322-68-3 (generic); 112-60-7; EINECS 203-989-9; $H(OCH_2CH_2)_nOH$, avg. n = 4; lubricant for rubber molds,

---

‡ Trade name and manufacturer not verified   § Trade name without identified manufacturer

textile fibers, metalworking; in food and food pkg.; in cosmetics and hair preparations; pharmaceutic aid; in gas chromatography; in paints, paper coatings, polishes, ceramics.

PEG 200. See PEG-4.

PEG-2 laurate. (diethylene glycol laurate; diglycol laurate; PEG 100 monolaurate) CAS 141-20-8; EINECS 205-468-1; $CH_3(CH_2)_{10}CO(OCH_2CH_2)_nOH$, avg. n = 2; W/o emulsifier, dispersant, antistat, defoamer for textile, paper processing, cutting oils, polishes, emulsion cleaners, rubber latex, wool lubricants, paints. [Henkel/Emery; Inolex; Karlshamns; Lonza; Mona; Stepan; Witco/Humko]

PEG (1-4) lauryl ether sulfate, sodium salt. See sodium laureth sulfate.

PEG 100 monolaurate. See PEG-2 laurate.

PEG 100 monostearate. See PEG-2 stearate.

PEG-2 oleate. (diethylene glycol monooleate; diglycol oleate) CAS 106-12-7; EINECS 203-364-0; $CH_3(CH_2)_7CH$-$CH(CH_2)_7CO(OCH_2CH_2)_nOH$, avg. n = 2; emulsifier, dispersant, antistat for cosmetic, textile, paper processing, cutting oils, polishes, emulsion cleaners, rubber latex, wool lubricants; leather softener. [Henkel/Emery; Inolex; Karlshamns; Lipo; Lonza; Mona; Witco/Humko]

PEG-4 sorbitan laurate. See polysorbate 21.

PEG-20 sorbitan laurate. See polysorbate 20.

PEG-5 sorbitan oleate. See polysorbate 81.

PEG-20 sorbitan oleate. See polysorbate 80.

PEG-4 sorbitan stearate. See polysorbate 61.

PEG-20 sorbitan stearate. See polysorbate 60.

PEG-20 sorbitan trioleate. See polysorbate 85.

PEG-20 sorbitan tristearate. See polysorbate 65.

PEG-2 stearate. (diethylene glycol stearate; diglycol stearate) CAS 106-11-6; 9004-99-3 (generic); 85116-97-8; EINECS 203-363-5; 285-550-1; $CH_3(CH_2)_{16}CO$-$(OCH_2CH_2)_nOH$, avg. n = 2; Emulsifier, plasticizer, lubricant, wetting agent, binding and thickening agent, dispersant, antistat, opacifier, pearlescent, stabilizer used in cosmetics, dry cleaning, leather, textile industries, paper processing, rubber. [Henkel/Emery; Inolex; Karlshamns; Lipo; Lonza; Stepan; Witco/Humko]

pegmatite. A felspathic rock, similar to Cornish stone.

pegnin. A preparation of lactose and rennet, which yields a finely divided curd from cows' milk; used in infant food.

Pegol® L-10. EO/PO block copolymer; emulsifier for agric. formulations, cutting and grinding fluids, asphalt emulsions. [Rhone-Poulenc Surf.]

Pegosperse®. Polygycol esters and glycol esters. [Lonza Inc]

Pegosperse® 50 DS. CAS 627-83-8; EINECS 211-014-3; Glycol distearate; emulsifier, opacifier, stabilizer for suspensions and dispersions; emollient, lubricant and pigment dispersant in pharmaceuticals and cosmetics; thickener, wetting agent and plasticizer in hair products. [Lonza Inc]

Pegosperse® 50 MS. CAS 9004-99-3; Glycol stearate; dispersant, emulsifier for o/w emulsions for industrial use. [Lonza Inc]

Pegosperse® 100 L. CAS 9004-81-3; PEG-2 laurate; emulsifier for o/w emulsions; dispersant; for industrial use. [Lonza Inc]

Pegosperse® 100 O. CAS 9004-96-0; PEG-2 oleate; emulsifier for o/w emulsions; dispersant; for industrial use. [Lonza Inc]

Pegosperse® 100 S. CAS 9004-99-3; PEG-2 stearate; emulsifier for o/w emulsions; dispersant; for industrial use. [Lonza Inc]

Pegosperse® 200 DL. CAS 9005-02-1; PEG-4 dilaurate; emulsifier, dispersant, opacifier, visc. control agent, defoamer for cosmetics, household products, textiles, plastics, water treatment. [Lonza Inc]

Pegosperse® 400 DOT. CAS 61791-01-3; PEG-8 ditallate; emulsifier, dispersant, opacifier, visc. control agent, defoamer for cosmetic, household products, textiles, paper, water treatment. [Lonza Inc]

Pegosperse® 1750 MS. CAS 9004-99-3; PEG-40 stearate; emulsifier for o/w emulsions; dispersant; for industrial use. [Lonza Inc]

Pegosperse® PMS CG. CAS 1323-39-3; EINECS 215-354-3; Propylene glycol stearate; emulsifier, dispersant, opacifier, visc. control agent, defoamer for cosmetics, household products, textiles, plastics, water treatment. [Lonza Inc]

pegu. See cutch.

pegu catechu. See cutch.

Pekafill. Resin based light-curing small particle size hybrid composite filling material; used in dentistry. [Bayer AG]

Pekafix. Cold curing denture resin on the basis of methyl methacrylate; dental preparation. [Bayer AG] †

Peka Glas. A proprietary safety glass. §

Pekalux. Resin based micro-filled light-curing composite filling material; used in dentistry. [Bayer AG]

Pekatop. Heat-curing denture resin on the basis of methyl methacrylate; dental preparation. [Bayer AG] †

Pekatray®. Cold curing plastic for the preparation of individual impression trays; dental specialty. [Bayer AG]

Peladow. CAS 10043-52-4; Calcium chloride. [Dow]

Pelamag. Salt coated magnesium granules for desulfurizing blast furnace iron; (sold under licence from Dow Chemical Company). [Foseco (F.S.) Ltd]

Pelamagsalt. Coated magnesium granules; used for desulfurization of blast furnace iron. [Dow]

Pelargene. 2,4-Dimethyl-6-phenyldihydropyrane. [Quest Int'l. UK Ltd]

Pelargone. A trade name for nylon 9. §

Pelaspan. Expandable polystyrene resin used in manufacture of loose-fill packing. [Dow UK]

Pelaspan 333FR. A flame retardant expandable polystyrene. [Dow UK]

Pelaspan GP. General purpose expandable polystyrene. [Dow]

Pelaspan Mold-a-Pac. For packaging products using Pelaspan PAC loose fill coated with an adhesive to form a resilient molded cushion. [Dow UK]

Pelaspan PAC. Expanded polystyrene; loose-fill packaging material. [Dow UK]

Pelentan. CAS 548-00-5; A proprietary trade name for ethyl biscoumacetate. §

pelionite. A coal of the cannel type.

Pellethane®. A wide range of polyurethane elastomers, 'rubber-plastics' used cured or uncured fabricated into various shapes and forms by conventional methods. [Dow UK] *

Pellethane® 2102-55D. PU elastomer (polyester polycaprolactone resin); can be used uncured or cured; tough, high-performance, wear-resistance material providing clarity, chem. resistance, flexibility for automotive (fascia, brake cable jacketing, body side trim, etc.),

---

health care; (tubing, catheter components, transdermal patches) applications, in film or sheets as bladders, liners, or belting; in fabricated products for casters, athletic shoes, bushings, wheels, extruded profiles. [Dow Plastics]

Pellethane® 2354-45DGA. PU elastomer; automotive grade; see Pellethane 2102-55D. [Dow Plastics]

Pellethane® 2363-55D. PU elastomer; health care resin; see Pellethane 2102-55D. [Dow Plastics]

Pclonit D. Contains aluminum, has a strong shoving effect and is used where coarse fragmentation is required. [Dynamit Nobel Wien GmbH] *

Pels. CAS 1310-73-2; Sodium hydroxide. [PPG Industries]

Pcltex. Ferro-chromium lignosulfate, modified; wetting agent, emulsifier, dispersant; oil well drilling mud thinner. [Borregaard LignoTech]

Pen A. CAS 69-53-4; A proprietary preparation of ampicillin; an antibiotic. [Pfizer International]

Pen A/N. CAS 69-52-3; Ampicillin sodium; antibacterial. [Pfizer Inc]

Penacolite® B-18-S. CAS 65876-95-1; Resorcinol-formaldehyde resin; dry bonding agent formulated for rubber-to-wire adhesion; for tires, industrial belts, and retreads. [Indspec]

Penacolite® R-2170. CAS 24969-11-7; Resorcinol-formaldehyde resin; bonding agent for dipping formulas for bonding synthetic industrial fabrics (aramid, polyester and glass) to rubber. [Indspec]

Penapar VK. CAS 132-98-9; Penicillin V potassium; antibacterial. [Parke-Davis] *

Penaryl A. Amyldiphenyl; a proprietary trade name for a plasticizer. §

Penaryl B. Diamyldiphenyl; a proprietary trade name for a plasticizer. §

Penbritin. CAS 69-53-4; A proprietary preparation containing ampicillin; an antibiotic. [SmithKline Beecham] *

Penchlor. A proprietary acid-proof cement made from cement powder and sodium silicate solution; used for lining tanks. §

penchlorol. See pentachlorophenol.

pendare. A name for Venezuelan chicle.

pendecamaine. A surface-active agent present in tegobetaines. It is N,N-dimethyl-(3-palmitamidopropyl)-glycine betaine.

Pendramine. See Trolovol. [Degussa AG] *

Penecort®. CAS 50-23-7; Hydrocortisone; corticosteroid. [Allergan, Inc]

Peneteck. Mineral oil tech.; CAS 8042-47-5; emollient. [Penreco]

Penetral NA 20. Low-foam wetting agent for textile mercerizing; stable in strongly alkaline solution [Ceca SA]

Penetrol. A proprietary preparation containing various volatile oils; a decongestant. [The Boots Co plc]

Penetrol 2-EHS. Ammonium 2-ethylhexyl sulfate; solubilizer and wetting agent esp. at high pH and temp. [Clark]

Penetrol. It is a sulfonated oxidation product of petroleum; as insecticide against aphides

Penetrol. A compound used as a textile detergent.

Penetron OT-30. 2-Ethylhexyl sulfosuccinate; penetrant, wetting agent, surface tens. depressant for textiles, paper, paint, plastic, rubber and metal industries. [Hart Products Corp.]

Penicals. A proprietary preparation containing phenoxymethylpenicillin (as the calcium salt); an antibiotic

[Leo Laboratories] *

Penicals 333. A proprietary preparation containing phenoxymethylpenicillin calcium; an antibiotic. [Leo Laboratories] *

penicillin. CAS 1406-05-9; The name given to the antibiotic principle of the mold *penicillium notatum*. The material is now prepared by special fermentation processes in large quantities.

Penicillinase. An enzyme obtained from cultures of *Bacillus cereus* which hydrolyzes benzylpenicillin to penicilloic acid; it antagonizes the antibacterial action of penicillin; used in biomedial research.

Penicillin V Pulvules. A proprietary preparation containing phenoxymethyl penicillin; an antibiotic. [Eli Lilly & Co]

Penicillin V Sulpha. A proprietary preparation of phenoxymethyl penicillin sulfadiazine, sulfamerazine, and sulfadimidine; an antibiotic. [Eli Lilly & Co]

Penidural. Oral drops or oral suspension containing benzathine penicillin; for the treatment of mild to moderately severe penicillin-sensitive infections, particularly in children. [Wyeth Laboratories]

Penisem. CAS 113-98-4; Penicillin G potassium; antibacterial. [SmithKline Beecham] *

Penitriad. Trisulfonamide/potassium penicillin V preparations. [May & Baker Ltd] *

Penjectin. Procaine penicillin intramammary injection. [May & Baker Ltd] *

Pennad 150. CAS 100-37-8; Diethylaminoethanol; intermediate, emulsifier, catalyst in urethane foams, curing agent, corrosion inhibitor. [Atochem] *

Pennchem® Mortar. Two-component, silica-filled, vinyl ester resin-based mortar; for field mixing for setting brick and tile or for masonry construction. [Atochem]

Penncozeb. CAS 8018-01-7; Wettable powder containing mancozeb; protectant fungicide for fruit, field crops, and roses. [Shell UK]

Pennfloat® 3-2277. Mercaptan; intermediate. [Atochem] *

Pennodorant® 1013. CAS 110-01-0; Tetrahydrothiophene (99% min.); odorant for natural gas to permit detection of leaks. [Atochem] *

Pennstop® 1866. CAS 3710-84-7; N,N-Diethylhydroxylamine; free radical scavenger used by the rubber industry as an emulsion polymerization inhibitor; vapor phase inhibitor for olefin or styrene monomer recovery systems; in-process inhibitor for production of styrene, divinyl benzene, butadiene, isoprene; intermediate in the synthesis of silicone rubber and photographic developers. [Atochem] *

pennyroyal oil. (pulegium oil; hedeoma oil) CAS 8007-44-1; Yellow to reddish essential oil; used in the manufacture of pulegone, flavoring alcoholic beverages.

Pennzone B 0685. CAS 109-46-6; 1,3-Dibutylthiourea [Atochem N. Am.] *

Pennzone E 0686. CAS 105-55-5; 1,3-Diethylthiourea [Atochem N. Am.] *

Penotrane. CAS 14235-86-0; A proprietary preparation containing hydrargaphen. [Octel Chemicals Ltd] †

Penreco 1520, 3070. Petrolatum, tech.; used in rubber processing aids, carbon papers, buffing and polishing compounds, corrosion preventatives, general purpose lubricants, printing inks, solder pastes. [Penreco]

Penreco 2251 Oil. CAS 64742-14-9; Petroleum distillates; high purity hydrocarbon processing solv., foam control agent, in waterless hand cleaners, agricultural sprays,

---

‡ Trade name and manufacturer not verified   § Trade name without identified manufacturer

polishes, fruit and vegetable processing, cleaning oils. [Penreco]

Penreco Amber. Petrolatum USP; emollient, base for cosmetic and pharmaceutical preparations; waterproofing agent for butcher paper; lubricant, water repellent, moisture barrier for textile and paper; carrier for modeling clays, soldering paste and flux; pigment carrier for carbon paper; binder and conditioner for crayons. [Penreco]

Penreco Red. Tech. petrolatum; tech.-grade base; base and binder for polishes. [Penreco]

Pensa's rubber. A rubber substitute made from coal tar, petroleum tar, oil of turpentine, and boric or phosphoric acids.

Pensil® 100. CAS 63394-02-5; One-component silicone rubber sealant; sealant offering high strength and excellent primerless adhesion to building substrates; seals against spread of smoke and fire; autobondable. [GE Silicones]

penta. See pentachlorophenol.

Penta G.P. 79. Rust preventatives. [Croda Chem. Ltd] *

Pentabor. Sodium pentaborate. [Borax Consolidated Ltd] *

pentabromodiphenyl oxide. CAS 32534-81-9; flame retardant. [Ethyl; Great Lakes]

pentabromotoluene. (2,3,4,5,6-pentabromotoluene) $C_6Br_5CH_3$; CAS 87-83-2; flame retardant for unsaturated polyesters, polyethylene, PP, PS, SBR latex, textiles, rubbers. [Great Lakes]

2,3,4,5,6-pentabromotoluene. See pentabromotoluene.

Pentac Aquaflow. CAS 2227-17-0; Emulsifiable concentrate of 480 g dienochlor per liter; an acaricide. [DowElanco Ltd]

pentachlorophenol. (PCP; penta; penchlorol) Obtained from chlorinated of phenol; C_6HCl_5O; CAS 87-86-5; EINECS 201-778-6; fungicide, bactericide, algicide, herbicide; preservation of wood, starches, dextrins, glues. [Penta Mfg.; Vulcan]

Pentacizers. Proprietary trade names for plasticisers. §

pentacosactride. A corticotrophic peptide.

pentaerythritol. (PE; 2,2-bis(hydroxymethyl)-1,3-propanediol; monopentaerythritol; tetramethylolmethane) $C_5H_{12}O_4$; CAS 115-77-5; EINECS 204-104-9; In synthetic resins, paints, varnishes. [Aqualon; Degussa; Hoechst Celanese; Mitsubishi Gas; Mitsui Toatsu Chem.; Penta Mfg.; Perstorp Polyols]

pentaerythritol tetrastearate. See pentaerythrityl tetrastearate.

Penta-erythritol Tetrastearate (PET). A proprietary release agent used in injection-molding processes. [DuPont UK] *

pentaerythrityl tetrastearate. (pentaerythritol tetrastearate) Tetraester of pentaerythritol and stearic acid; $C_{77}H_{148}O_8$; CAS 115-83-3; polishes, coatings, textile finishes. [Hercules; Lipo; Lonza]

pentaerythrityl triacrylate. (2-propenoic acid-2-(hydroxymethyl)-2-(((1-oxo-2-propenyl) oxy) methyl)-1,3-propanediyl ester) ($H_2C:CHCO_2CH_2)_3CCH_2OH$; CAS 3524-68-3; Crosslinking agent used in adhesives, coatings, inks, textile products, photoresists, castings, modifiers for polyester, fiberglass, or polymers. [Sartomer]

Pentagan. Chlormequat + choline chloride; plant growth regulator. [Agan Chemical Manufacturers Ltd]

pental. (trimethylethylene) C_5H_{10}; CAS 57-33-0; Colorless, volatile liquid; used in organic synthesis, high-octane fuel manufacture.

Pental 8D. A range of rosin esters; used to produce chewing gum base. [Hercules] *

Pental 28. A range of rosin esters; used to modify nitrocellulose based coatings and lacquers. [Hercules] *

Pental 802A. A range of rosin esters; used in heat set lithographic inks. [Hercules] *

Pental A. A range of rosin esters; a tackifying agent for natural rubber and SBR based pressure sensitive adhesives. [Hercules] *

Pental G, X. A range of rosin esters; used in heat set lithographic inks. [Hercules] *

Pentalan. Pentaerythritol ester of woolwax fatty acids. [Croda Chem. Ltd]

pentaline. CAS 76-01-7; Pentachlorethane, $CHCl_2 \cdot CCl_3$; used as a solvent for oil and grease in oil cleaning; for separation of coal from impurities by density difference.

Pentalyn. Pentaerythritol esters of rosin, modified rosins, dibasic acid- and phenolic-modified rosins; synthetic resins used in heat-set offset inks, letterpress printing inks, flexographic inks. [Hercules]

Pentalyn 255, 261, 856. Alkali-soluble resins used in emulsion floor polishes. [Hercules]

Pentalyn 344, C, K. Tackifying and reinforcing resins for adhesives and sealants. [Hercules]

Pentalyn 802A, 802A Pale, 833, G, X. Used in overprint varnishes. [Hercules]

Pentalyn A. Used in the production of chewing gum base. [Hercules]

Pentalyn H. Tackifying and reinforcing resin for adhesives and sealants; also in the production of chewing gum base. [Hercules]

pentamethyleneamine. See piperidine.

Pentamid Cl2. Fatty acid polydiethanolamide; metal-working corrosion inhibitor. [Pentagon Urethanes Ltd] *

Pentamid KH. Methyl N-octadecyl terephthalamate; gelling agent for high-temperature greases. [Pentagon Urethanes Ltd] *

Pentamin BDMA etc. N,N-Dimethyl benzylamine; polyurethane catalyst, epoxy curing agent. [Pentagon Chemicals Ltd] *

n-pentane. (amyl hydride; alkane C_5) Aliphatic; $CH_3(CH_2)_3CH_3$; CAS 109-66-0; EINECS 203-692-4; solvent; artificial ice manufacture, low-temp. thermometers, solvent extract processes, blowing agent in plastics (expandable polystyrene), pesticide. [Ashland; Phillips 66]

n-pentanoic acid. See n-valeric acid.

2-pentanone. See methyl propyl ketone.

3-pentanone. See diethyl ketone.

Pentanox 4X, 24X. Alkyl dimethyl amine oxides; thickening and foaming agents. [Pentagon Chemicals Ltd]

Pentaphane. A proprietary trade name for a film made from a chlorinated polyether (polymerized 3,3-bis(chloromethyl)oxetane). [British Cellophane] ‡

Pentaphen® 67. CAS 80-46-6; p-t-Amylphenol; intermediate for chemical specialties; in germicidal formulations; also in manufacturing of photographic chemicals, oil demulsifiers, phenolic resins, agric. surfactants, antiskinning agents. [Atochem]

pentapotassium tripolyphosphate. See potassium tripolyphosphate.

Pentaquest Extra 0685. CAS 140-01-2; Pentasodium pentetate; chelating agent for iron chelation up to pH 11.5. [Clough]

---

* Trade name not verified as available   † Trade name verified as obsolete

Pentaquest OPAC 0201. CAS 67-43-6; DTPPA; scale and corrosion inhibitor for aq. systems; stable to hydrogen peroxide. [Clough]

Pentaquest OPNA 0256. CAS 140-01-2; Pentasodium DTPPA; scale and corrosion inhibitor for aq. systems; stable to peroxide. [Clough]

pentasodium triphosphate. See sodium tripolyphosphate.

Pentasol. CAS 71-41-0; 87-86-5; A mixture of pure amyl alcohols containing 75% primary alcohol and 25% secondary alcohol, and is obtained from the pentane fraction of gasoline; and is used as a varnish and lacquer solvent.

Pentasol. CAS 577-11-7; EINECS 209-406-4; Sodium dioctyl sulfosuccinate; wetting agent. [Pentagon Chemicals Ltd]

Pentateric 24B, B, BLG. Alkyl betaines; used for toiletries. [Pentagon Chemicals Ltd]

Pentek. A proprietary trade name for a technical grade of pentaerithritol used in synthetic resins and in the paint and varnish industry. §

Pentelex. Photographic developer. [May & Baker Ltd] *

Pentex® 40. Alkylaryl sulfonate; wetting agent, penetrant, emulsifier; textile and industrial processing; improves dye uniformity; aids rewetting of leather, paper, and paper-mill felts. [Rhone-Poulenc/Latex & Spec. Polymer]

Penthrane. CAS 76-38-0; A proprietary preparation of methoxyfluorane; inhalation anesthetic used in obstetrics. [Abbott Laboratories] *

Penthrinit. It is a plastic mixture of 80% pentaerythritol-tetranitrate and 20% nitroglycerine; an explosive.

Penthrit. CAS 78-11-5; Pentaerythritol tetranitrate.

Pentids. CAS 113-98-4; Penicillin G potassium; antibacterial. [Bristol-Myers Squibb Co Inc] *

Pentine 1185 5432. Isopropylamine dodecylbenzene sulfonate; emulsifier for emulsion degreasers, dry cleaning soaps, agric. emulsifier. [Clough]

Pentine Acid 5431. CAS 27176-87-0; Dodecylbenzene sulfonic acid; base for dishwashing and laundry detergents, industrial and institutional cleaners. [Clough]

Pentol. Timber fungicides. [Plant Protection] *

Pentonate DB. Fatty alcohol benzoate; substitute for isopropyl myristate; emollient. [Pentagon Chemicals Ltd] *

Pentonite. Sodium benzoate/sodium nitrite mixture; corrosion inhibitor for aqueous systems. [Pentagon Urethanes Ltd] *

Pentonium 50, 80. Alkyl dimethyl benzalkonium chlorides, alkyl trimethyl ammonium chlorides; biocides. [Pentagon Chemicals Ltd] *

Pentopan. A pentosanase; added to the flour to improve dough handling properties during baking. [Novo Nordisk]

Pentostam. Injection of pentavalent sodium stibogluconate; for the treatment of Leishmaniasis. [The Wellcome Foundation Ltd]

Pentothal. CAS 71-73-8; A proprietary preparation of thiopentone sodium; intravenous anesthetic. [Abbott Laboratories] *

Pentothal Sodium. CAS 71-73-8; Thiopental sodium; anesthetic; anticonvulsant. [Abbott Laboratories] *

Pentovls. CAS 1169-79-5; A proprietary preparation containing quinestradol. [Warner] *

Pentoxyl M. Perfumery speciality. [Bush Boake Allen Ltd]

Pentrex. Phenolic-modified and maleic modified esters of rosin; Pentrex G and X rosin esters are used in heat set lithographic inks; Pentrex 28 is used in nitrocellulose-based coatings and lacquers. [Hercules] *

Pentrexyl. CAS 69-53-4; A proprietary preparation of ampicillin; an antibiotic. [Bristol-Myers Squibb Co Inc]

Pentritol. CAS 78-11-5; Pentaerythritol tetranitrate; vasodilator. [USV Pharmaceutical Corp] ‡

Pentrium. A proprietary preparation of chlordiazepoxide and pentaerythritol tetranitrate; used in the treatment of angina pectoris. [Roche Products Ltd] *

Pentrone. Anionic surfactant. [Rhone-Poulenc UK]

Pentrone ON. Sodium 2-ethylhexyl sulfate in liquid form; anionic surfactant for dispersing and alkaline wetting agent used in electroplating and lye peeling. [Rhone-Poulenc UK]

Pentrone S. Range of anionic surfactants of the disodium mono- alkyl polyalkylene sulfosuccinate type; supplied as liquids; dispersing, foaming, and emulsification agents with low toxicity; used in emulsion polymerization, surgical scrubs, and shampoos. [Rhone-Poulenc UK]

Pentrosan. Dispersing and solubilizing agents. [Rhone-Poulenc UK]

Pentryate 80. CAS 78-11-5; Pentaerythritol tetranitrate; vasodilator. [O'Neal, Jones & Feldman Pharmaceuticals] *

pentyl. Synonym for the amyl group, C_5H_{11}.

pentylcarbinol. See hexyl alcohol.

Pentylol. Mixture of amyl alcohol isomers; solvent. [Sasolchem]

Pen-Vee. CAS 87-08-1; Penicillin V; antibacterial. [Wyeth Laboratories] *

Pen-Vee Drops. CAS 5928-84-7; Penicillin V benzathine; antibacterial. [Wyeth Laboratories]

Pen Vee Dural. A proprietary preparation containing phenoxymethylpenicillin potassium and benzathine penicillin; an antibiotic. [Wyeth Laboratories] *

Pen-Vee K. CAS 132-98-9; Penicillin V potassium; antibacterial. [Wyeth Laboratories]

Pen-Vee Suspension. CAS 5928-84-7; Penicillin V benzathine; antibacterial. [Wyeth Laboratories] *

Penzold's reagent. A solution of diazo-benzosulfonic acid and potassium hydroxide; a reagent for sugar in urine.

Pep. Polyester promoters. [Air Prods. & Chems. Inc]

Pepcid®. CAS 76824-35-6; Famotidine; for the treatment of gastric and duodenal ulcer and hypersecretory conditions such as Zollinger-Ellison syndrome. [Merck & Co Inc]

Pepcidine. See Pepcid®. [Merck & Co Inc]

Pepdine. See Pepcid®. [Merck & Co Inc]

Pepdul. See Pepcid®. [Merck & Co Inc]

Pepper Dust. Powdered pepper; animal deterrent. [Vitax Ltd]

peppermint camphor. CAS 89-78-1; Menthol, $C_{10}H_{19}OH$; used in medicine.

Pep Set. Resins and catalysts associated with the production of foundry cores and molds; when applied to foundry sand, provide a binder system useful in the production of foundry cores and molds; these binder systems fall into the large category of no-bake binders used within the foundry industry and cure at room temperature. [Ashland Chemical Company] *

pepsin. (pepsinum) A digestive enzyme of gastric juice; it decomposes albuminous bodies into peptone; CAS 9001-75-6; EINECS 232-629-3; medicine (digestive fer-

---

‡ Trade name and manufacturer not verified   § Trade name without identified manufacturer

ment); substitute for rennet in cheesemaking. [Am. Biorganics; EM Industries; G Fiske & Co Ltd; R.W. Greeff; Worthington Biochemical]

pepsinum. See pepsin.

Peptacol- 10. A proprietary preparation of homatropine methylbromide and phenobarbitone. [Pharmax Ltd] *

Peptard. CAS 6835-16-1; Each tablet contains 1-hyoscyamine sulfate 0.2 mg in a porous plastic matrix. [3M Pharmaceuticals] †

Peptavlon. CAS 5534-95-2; Pentagastrin; for diagnostic testing of gastric secretions. [Imperial Chemical Industries plc]

Peptein® 2000®. CAS 9015-54-7; Hydrolyzed animal protein; conditioner for cosmetic applications; esp. substantive to hair. [Hormel] *

Peptein® VgW. Hydrolyzed wheat protein; for cosmetic hair and skin care products [Hormel]

Peptizer 566. Naphthenic oil/sulfonate ester blend. [C.P. Hall]

Peptizer 7010. Mineral oil/sulfonate blend. [C.P. Hall]

Peptoil. Petroleum oil; spray adjuvant for enhancing herbicide activity and defoliation performance with foam eliminator. [Draxel Chemical Company] ‡

Peptorub. Pre-plasticized comminuted rubber.

Peptrex. Rubber peptizer and rubber. [Miles]

Peradinol. Textile dyeing auxiliaries. [Fine Dyestuffs & Chemicals Ltd] *

Peralfan T Concentrate. Combination of volatile solvents of specific activity with nonionic emulsifiers and anionic substances; yellowish liquid; spotting, dispersing and scouring agent to remove stubborn soilings. [Thor Chemicals (UK) Ltd]

Perapret®. Additives for textile resin finishing. [BASF plc]

Perapret® PE-2. Softener and lubricant improving abrasion and sewability in durable press finishing. [BASF]

perborax. Sodium perborate, $NaBO_3 \cdot 4H_2O$, a washing and bleaching agent.

perborin. Sodium perborate, $NaBO_3$, a constituent of washing powders.

perborol. See perborax.

Perbunan®. Acrylonitrile-butadiene rubber. [Bayer plc; Miles Inc]

Perbunan® N. Acrylonitrile butadiene rubber; used for rubber goods that must withstand oils, greases, and petrol and must have good resistance to heat, ageing and abrasion, e.g., seals, roller covers, hoses, cable sheathings, plant linings, conveyor belting. [Bayer AG]

Perbunan® N 1807 NS. NBR; low permeability to gases and good physiological properties; used in technical moldings e.g., seals, sleeves, diaphragms, bellows, valves, vibration dampers, footwear soles; roll covers, printing blocks and blankets, belting, fabric proofings (e.g., for flexible silos), technical hoses, cable sheathings, brake and clutch linings; open and closed cell sponge rubber; punching blocks, gloves, adhesives; nonstaining. [Bayer; Miles]

Perbunan® N/VC70. NBR/PVC fluxed blend (70% Perbunan N 2807 NS/30% PVC); nonstaining stabilizer; elastomer used in rubber goods with excellent resistance to mineral oil, grease, fuels, superior resistance to ozone and weathering; applications including hose, cable jackets, diaphragms, roll covers, thermal blown insulation, spinning aprons. [Bayer; Miles]

Percarbamid. CAS 7722-84-1; A hydrogen peroxide prepa-

ration; used in the cosmetics and pharmaceuticals industry. [Degussa AG] *

perchlorethylene. CAS 127-18-4; Tetrachloroethylene, C_2Cl_4.

perchloric acid. $HClO_4$; CAS 7601-90-3; EINECS 231-512-4; Analytical chemistry, catalyst, manufacture of esters, ingredient of electrolytic bath in deposition of lead, electropolishing, explosives. [Spectrum Chem. Mfg.]

perchloroethylene. (tetrachloroethylene; tetrachloroethene; ethylene tetrachloride) $Cl_2C=CCl_2$; CAS 127-18-4; EINECS 204-825-9; Dry-cleaning solvent; vermifuge; drying agent; degreasing metals. [Asahi-Penn; Ashland; Atochem; General Chem.; ICI Spec.; OxyChem; PPG Industries]

Perchloron. CAS 7778-54-3; A technical calcium hypochlorite containing 68.1% available chlorine.

Percist. Insecticidal formulation. [Mitchell Cotts Chemicals Ltd]

Perclene. CAS 127-18-4; Perchloroethylene; cleaner and degreasant. [Occidental Chemical Corp]

Perclene TG. CAS 127-18-4; Perchloroethylene; transformer grade. [Occidental Chemical Corp]

Percodan®. CAS 50-78-2, CAS 76-42-6; Aspirin and oxycodone. [DuPont Merck Pharmaceuticals]

Percogesic. Acetaminophen and phenyltoloxamine; analgesic for enhanced pain relief. [Richardson-Vicks Inc] *

Percol®. Polyacrylamide-based high effienciency retention aids for paper manufacture. [Allied Colloids Ltd]

Percolaye. Attapulgite or sepiolite clay; used for refining of minerals and chemicals. [Bromhead & Denison Ltd]

Percorten Acetate. Desoxycorticosterone acetate; adrenocortical steroid. [Ciba-Geigy Corp] *

Percorten M Crystules. A proprietary preparation of deoxycortone pivalate; for treatment of adrenocortical insufficiency. [Ciba plc] *

Percorten Pivalate. Desoxycorticosterone pivalate; adrenocortical steroid. [Ciba-Geigy Corp] *

Percresan. A mixture of cresols, soap, and water; used as a disinfectant in 1-2% solution.

Percumyl D. CAS 80-43-3; Dicumyl peroxide. [British Traders & Shippers Ltd]

Perduren. An organic polysulfide synthetic rubber.

Perdynamine. A compound of albumen and hemoglobin.

Perecot. A copper fungicide. [ICI Chem. & Polymers Ltd]

Peregal®. Leveling agent when dyeing cellulose fibers with vat dyes. [BASF AG]

Peregal® O. Polyoxyethylated fatty alcohol (nonionic); dyeing assistant for use with basic and direct colors; assistant for the dyeing, leveling and stripping of vat dyes; leveling agent in acetate printing. [ISP]

Peregal® OK. Methylpolyethanol quaternary amine (cationic); used as a vat dye retarder. [ISP]

Peregal® ST. CAS 9003-39-8; Polyvinylpyrrolidone; stripping assistant for cotton and rayon yarns or fabrics that have been dyed or printed with vat, sulfur or direct colors; rag stripping assistant in high grade paper. [ISP]

pereiro bark. The bark of Geisso-spermum vellosii. A Brazilian febrifuge.

Pereman. A copper fungicide. [Plant Protection] *

Perenol El. Polyvinyl isobutyl ether; defoamer for petrol. solv.-based paints. [Henkel KgaA]

Perenox. A copper fungicide. [ICI Chem. & Polymers Ltd]

Perenyi's fluid. (chromo-nitric acid). It contains 3 parts of 92% alcohol, 4 parts of 10% nitric acid, and 3 parts of

---

* Trade name not verified as available   † Trade name verified as obsolete

0.5% chromic acid. A fixing agent used in microscopy; the objects are treated with alcohol after fixing.

Perfecta. Petrolatum USP. [Witco]

Perfection®. CAS 7758-16-9; Sodium acid pyrophosphate; food grade leavening acid for baking, cereals. [Rhone-Poulenc Food Ingreds.]

Perfekthion®. CAS 60-51-5; Dimethoate; systemic insecticide for control of sucking and biting insects [BASF AG]

Perfix. High speed photographic fixer. [May & Baker Ltd] *

Perflex. Partially hydrogenated vegetable oil (soybean, cottonseed), propylene glycol mono and diesters of fats and fatty acids, mono and diglcyerides, lecithin, BHA; kosher; multipurpose shortening for cakes, nonstandard yeast-leavened products [Van Den Bergh Foods]

Perflex. A proprietary trade name for unstretched vinylidene chloride. §

PerforMax® 403. Polymeric; deposit inhibitor for open recirculating cooling water systems. [Drew Ind. Div.]

perfumery oil. Refined petroleum, of specific gravity 0.880-0.885; used in perfumery.

Perfusamine. CAS 85068-76-4; lofetamine hydrochloride; diagnostic aid; radioactive agent. [Medi-Physics Inc] *

Pergacid. Dyes for paper. [Ciba plc] *

Pergalen. A proprietary preparation of sodium apolate and benzyl nicotinate; for resolution of bruises and local trauma. [Hoechst UK] *

Pergamin. (Pergamyn). A grease-proof paper made from cellulose pulp.

Pergamyn. *See* Pergamin.

Pergantine. Pigment dispersions for paper. [Ciba plc] *

Pergaprint. Crosslinking agents for starch. [Ciba plc] *

Pergascript. Chemicals for carbonless paper. [Ciba plc] *

Pergasol. Dyes for paper. [Ciba plc] *

pergenol. A mixture of sodium perborate and bitartrate. It gives hydrogen peroxide on the addition of water.

Perglanz-Konzen-Trat B48 and B30. Blend of amphoterics and anionics, in the form of a white liquid; sheen additive for shampoos, bath, and shower preparations [(Th Goldschmidt Ltd).] ‡

Perglanzmittel GM 4006. Nonionics and fatty alcohol ether sulfate; pearlescent for hair shampoos and bath additives. [Zschimmer & Schwarz] *

Perglow. A bright nickel plating process. [Hanshaw Chemicals] ‡

perglycerol. An aqueous solution of sodium lactate; used as a substitute for glycerol for medical and cosmetic purposes.

Pergonal. A proprietary trade name for a follicle-stimulating hormone. §

Pergopak. Organic fillers. [Ciba plc] *

Pergopak. Urea formaldehyde resins. [Lonza Inc]

Pergut®. Chlorinated rubber; used for the formulation of coatings with high water, chemical, and low-temperature resistance as well as for use in printing inks. [Bayer AG; Bayer plc]

Perhydrate. *See* Hyperol. §

Perhydrit. *See* Hyperol. §

Perhydrol. Hydrogen peroxide, H_2O_2, one volume of 30% hydrogen peroxide giving 100 volumes of oxygen; used for bleaching, also as a disinfectant.

perhydrol of magnesia. *See* Biogen.

perhydrosqualene. *See* squalane.

Peri acid. CAS 82-75-7; 1-Naphthylamine-8-sulfonic acid; used as an azo dye intermediate.

Periactin. CAS 41354-29-4; Cyproheptadine hydrochloride; for relief of allergic and pruritic conditions and migraine headache; also an appetite stimulant. [Merck & Co Inc]

Periactin-Vita. Cyproheptadine hydrochloride and multivitamins; an appetite stimulant which helps prevent hypovitaminosis which may be associated with poor eating habits or an inadequate diet. [Merck & Co Inc]

Perichthol. A proprietary preparation of ammonium ichthosulfonate. §

Peri-Colace. Contains casanthranol and docusate sodium; laxative and stool-softener. [Mead Johnson & Co]

Perifenil. A proprietary preparation of pheneizine and pentaerythritol tetranitrate. [Warner] *

Perigen. CAS 52645-53-1; Formulations of permethrin. [The Wellcome Foundation Ltd] *

Perikol. A sensitizer for silver bromide plates, prepared by treating the addition product of toluquinaldine and the ethyl ester of toluene-sulfonic acid with alcoholic potassium hydroxide.

perilla oil. An oil obtained from the seed of the Asiatic mint, *Perilla ocymoides*; used as a drying oil in substitution for linseed oil.

periodic acid. $HiO_4 \cdot 2H_2O$; CAS 10450-60-9; oxidizing agent, increases wet strength of paper, photographic paper. [Atomergic Chemetals; Blythe, William Ltd; EM Industries; Janssen Chimica; Spectrum Chem. Mfg.]

periodic acid potassium salt. *See* potassium periodate.

Periograf. CAS 1306-06-5; Durapatite; prosthetic aid. [Sterling Drug Inc] *

Peripress. CAS 19237-84-4; A proprietary preparation of prazosin hydrochloride; for the treatment of hypertension, left ventricular failure and Raynaud's Disease. [Pfizer International]

Peristaltin. A cascara preparation containing the water-soluble glucosides extracted from the bark of *Rhamnus purshiana*; $C_{14}H_{18}O_8$; it stimulates peristalsis without drastic purgative action

Peritrate. CAS 78-11-5; Pentaerythritol tetranitrate; vasodilator. [Parke-Davis]

Perium. CAS 7681-80-3; Pentapiperium methylsulfate; anticholinergic. [William H Rorer Inc] *

Perizin®. For the control of ectoparasites in bees (varroa). [Bayer AG]

Perkacit® CBS. CAS 95-33-0; N-Cyclohexyl-2-benzothiazole sulfenamide; accelerator for rubber [Akzo] *

Perkacit® CDMC. CAS 137-29-1; Copper dimethyldithiocarbamate; accelerator for rubber [Akzo] *

Perkacit® DCBS. N,N´-Dicyclohexyl-2-benzothiazole sulfenamide; accelerator for rubber [Akzo] *

Perkacit® DOTG. CAS 97-39-2; N,N-Di-o-tolylguanidine; accelerator for rubber [Akzo] *

Perkacit® DPG. CAS 102-06-7; N,N-Di-phenylguanidine; accelerator for rubber [Akzo] *

Perkacit® DPTT. CAS 120-54-7; Dipentamethylenethiuram tetrasulfide; accelerator for rubber [Akzo] *

Perkacit® ETU. Ethylene thiourea (2-mercapto-imidazoline); accelerator for rubber [Akzo] *

Perkacit® MBS. 2-(Morpholinothio) benzothiazole; accelerator for rubber [Akzo] *

Perkacit® MBT. CAS 149-30-4; 2-Mercaptobenzothiazole; accelerator for rubber [Akzo] *

Perkacit® MBTS. CAS 120-78-5; Dibenzothiazole disulfide; accelerator for rubber [Akzo] *

Perkacit® NDBC. CAS 13927-77-0; Nickel dibutyl dithiocar-

---

‡ Trade name and manufacturer not verified   § Trade name without identified manufacturer

bamate; accelerator for rubber [Akzo] *

Perkacit® SDMC. Sodium dimethyldithiocarbamate; accelerator for rubber industry; shortstopper for polymer production [Akzo] *

Perkacit® TBBS. CAS 95-31-8; Butyl 2-benzothiazole sulfenamide; accelerator for rubber [Akzo] *

Perkacit® TDEC. Tellurium diethyl dithiocarbamate; accelerator for rubber [Akzo] *

Perkacit® TETD. CAS 97-77-8; Tetraethylthiuram disulfide; accelerator for rubber [Akzo] *

Perkacit® TMTD. CAS 137-26-8; Tetramethylthiuram disulfide; accelerator for rubber [Akzo] *

Perkacit® TMTM. CAS 97-74-5; Tetramethylthiuram monosulfide; accelerator for rubber [Akzo] *

Perkacit® ZBEC. Zinc dibenzyl dithiocarbamate; accelerator for rubber [Akzo] *

Perkacit® ZDBC. CAS 136-23-2; Zinc di-n-butyl dithiocarbamate; accelerator for rubber [Akzo] *

Perkacit® ZDEC. CAS 14323-55-1; Zinc diethyl dithiocarbamate; accelerator for rubber [Akzo] *

Perkacit® ZDMC. CAS 137-30-4; Zinc dimethyl dithiocarbamate; accelerator for rubber [Akzo] *

Perkacit® ZMBT. CAS 155-04-4; Zinc-2-mercaptobenzothiazole; accelerator for rubber [Akzo] *

Perkadox®. Organic peroxides; crosslinking agents [Akzo Chemie UK Ltd]

Perkadox® 14. CAS 25155-25-3; Di-(2-t-butylperoxyisopropyl)benzene; low reactivity peroxide useful as a finishing initiator at high temps. for styrenics; synergist for some halogen-containing flame retardants; also for cross-linking of olefin copolymers, EPDM, SBR, Neoprene, Hypalon. [Akzo] *

Perkadox® 16. CAS 15520-11-3; Bis(4-t-butylcyclohexyl) peroxydicarbonate; ultrafast initiator for polyester cure above 180 F; pultrusion, matched die molding; short ambient temp. compound shelf life. [Akzo] *

Perkadox® 16-W40-GB5. CAS 15520-11-3; Di-(4-t-butylcyclohexyl) peroxydicarbonate; initiator for polymerization of vinyl chloride, acrylates, methacrylates, etc.; recommended for manufacturing of PVC for elec. applications. [Akzo]

Perkadox® 20. CAS 3034-79-5; Di(2-methylbenzoyl) peroxide [Akzo] *

Perkadox® 26-fl. CAS 53220-22-7; Dimyristyl peroxydicarbonate [Akzo] *

Perkadox® 30. CAS 1889-67-4; 2,3-Dimethyl-2,3-diphenylbutane [Akzo] *

Perkadox® 58. CAS 10192-93-5; 3,4-Dimethyl-3,4-diphenylhexane [Akzo] *

Perkadox® BC. CAS 80-43-3; Dicumyl peroxide; high-temp. initiator used as a flame retardant synergist; also as cross-linking agent for a variety of natural and synthetic rubbers and olefins. [Akzo] *

Perkadox® GS. Initiator for the polymerization of vinyl chloride, vinyl acetate, acrylates, etc. [Akzo Chemie UK Ltd]

Perkadox® RM. Solid organic peroxides; used as free radical initiators in polymerization. [Akzo Chemie UK Ltd]

Perkadox® SE-8. CAS 762-16-3; Dioctanoyl peroxide [Akzo] *

Perkaglycerol. An aqueous solution of potassium lactate; used as a substitute for glycerol for medical and cosmetic purposes.

Perkalink® 300. CAS 101-37-1; Triallyl cyanurate; co-agent to improve efficiency of peroxide-induced crosslinking of rubber; sensitizer for radiation-cured compounds. [Akzo] *

Perkalink® 301. Triallyl isocyanurate; co-agent to improve efficiency of peroxide-induced cross-linking of rubber; sensitizer for radiation-cured compounds. [Akzo] *

Perkalink® 400. CAS 3290-92-4; Trimethylolpropane trimethacrylate; co-agent to improve efficiency of peroxide-induced cross-linking of rubber; sensitizer for radiation-cured compounds. [Akzo] *

Perkalink® 401. CAS 97-90-5; Ethylene glycol dimethacrylate; co-agent to improve efficiency of peroxide-induced cross-linking of rubber; sensitizer for radiation-cured compounds. [Akzo] *

Perkasil® KS 207. Aluminum-magnesium-sodium silicate; reinforcing filler for rubber industry; medium reinforcing properties. [Akzo] *

Perkasil® KS 300. Precipitated silica; highly dispersible reinforcing filler with excellent processing; for rubber industry [Akzo] *

Perkasil® KS 404. Precipitated silica; highly dispersible reinforcing filler with excellent transparency; for rubber industry [Akzo] *

Perkasil® VP 406. Precipitated silica; reinforcing filler for silicone rubber applications. [Akzo] *

Perkin's Base. p-Tolylaminoditolyl-p-toluquinone diimine.

Perklone. CAS 127-18-4; Perchloroethylene; a solvent for dry cleaning. [ICI Chem. & Polymers Ltd]

Perlankrol®. Anionic alcohol-, alcohol ether-, amide ether- and phenol ether- sulfates; used in toiletries, as foam boosters, emulsifiers, cement and gypsum foamers. [Harcros UK]

Perlankrol® ADP3. Sodium laureth (3) sulfate; biodeg. surfactant; base for preparation of high foaming shampoos and toiletries. [Harcros UK]

Perlankrol® ATL40. CAS 139-96-8; EINECS 205-388-7; TEA lauryl sulfate; biodeg. surfactant for preparation of high foaming shampoos and toiletries; emulsifier for emulsion polymerization. [Harcros UK]

Perlankrol® DAF25. CAS 2235-54-3; EINECS 218-793-9; Ammonium lauryl sulfate; foaming agent, base; formulation of toiletries and carpet shampoos. [Harcros UK]

Perlankrol® DSA. CAS 151-21-3; Sodium lauryl sulfate; foaming agent for synthetic latexes, emulsion polymerization aid; base for preparation of high foaming shampoos and toiletries; wetting agent; industrial detergent additive. [Harcros UK]

Perlankrol® ESD. Synthetic primary alcohol ether sulfate, sodium salt; biodeg. high foam additive for liq. household and industrial detergents; emulsifier for emulsion polymerization. [Harcros UK]

Perlankrol® ESK32. Sodium primary alcohol ethoxylate sulfate; foam booster/stabilizer for detergent formulations, foam cleaning, fire fighting foams, plasterboard production [Harcros]

Perlankrol® PA Conc. CAS 30416-77-4; Alkyl phenol ether sulfate, ammonium salt; foam booster/stabilizer, detergent base, frothing agent used in liq. detergent formulations; emulsifier for cresylic acid, emulsion polymerization. [Harcros UK]

perlate salt. See salt perlate.

Perlatum 400, 410, 410 CG, 420, 510. Wh. petrolatum USP; basic ingredient in cosmetics, ointments, lubricants; dust

---

* Trade name not verified as available   † Trade name verified as obsolete

control agents, and lubricants in baking industry, animal feed, pharmaceuticals, food processing. [IGI Petroleum Spec.] *

Perlatum 415, 415 CG, 425. Petrolatum USP; basic ingredient in cosmetics, ointments, lubricants; dust control agents, and lubricants in baking industry, animal feed, pharmaceuticals, food processing. [IGI Petroleum Spec.] *

Perlex. CAS 7787-59-9; Bismuth oxychloride pearls; used for decorative cosmetic products. [Morton Int'l. Ltd] †

Perlextra. See Perlex. [Morton Int'l. Ltd] †

Perlit®. Water repellents based on silicone, paraffin or fatty acid condensation products; used for textile finishing. [Bayer AG; Bayer plc]

perlite. A eutectic product resulting from an alloy of ferrite and cementite in steel.

Perlon. Monofil; used for fishing lines, zip fasteners, woven fabric and industrial uses. [Bayer AG]

Perlygel®. CAS 94-36-0; A trademark for benzoyl peroxide.§

Permabond. OPP laminating adhesives. [The Scottish Adhesives Co Ltd]

Permador. A precious metal alloy; for dentistry and dental engineering. [Degussa AG] *

Perma-Flex Blak-Stretchy®. Polysulfide system; two-component system for making room temperature setting rubbery, flexible, extremely elastic molds and patterns, for precision casting where max. stretch and good dimension control are important; in plaster casting and foundry pattern material; mix ratio: 100A/15B by wt.; the addition of ≈ 0.25% Pink or Yel. Curative C will speed up final set and give a more rubbery bounce to Blak Stretchy. [Perma-Flex Mold]

Perma-Flex Blak-Tufy®. Polysulfide system with carbon black pigment; three-component room temperature vulcanizing molding synthetic rubber reinforced with blk. reinforcing pigment with good toughness, low visc. and pouring properties in uncured form; in plaster shops for waste molds, intermediate working molds for prototype reproductions and prior to heavy duty thermoplastic Koroseal mold fabrication; mix ratio: 100A/20B/2C by wt. [Perma-Flex Mold]

Perma-Flex Blu-Sil. Silicone-type cold molding compound; two-component system, sets to flexible rubbery solid; for molds to cast Perma-Flex CMC U-2 PU, hot melt polyethylenes, epoxy resins, and acrylic compounds formed cold, and later cured at elevated temps.; mix ratio: 10A/25B by wt. [Perma-Flex Mold]

Perma-Flex Green-Sil. Silicone cold molding compound; two-component, high strength, low visc., room temperature curing elastomer for general mold making applications, e.g., wheels, gaskets, liners; for limited reprods. when casting polyester, epoxy, and PU rigid foam resins, and for electronic applications such as potting and encapsulating; mix ratio: 100A/5B by wt. [Perma-Flex Mold]

Permafuse. Modified phenolic adhesives; for bonding friction materials to their backing, e.g., brake shoes, clutch facings, transmission linings etc. [The Permafuse Corp]*

permalba. A composite pigment consisting mainly of barium sulfate. An artist's color.

Perma-Leaf. Aluminium pigment paste; used for protective coatings. [Reynolds Metal Co]

Permalens. Perfilcon; contact lens material. [CooperVision

Inc] ‡

Permali. A proprietary trade name for laminated products containing wood or paper impregnated with synthetic resin; some are made from thin wood coated with synthetic resin solution and compressed under heat, others are impregnated under pressure, solvent removed and then compressed. §

Permalloy®. A trademark for alloys of nickel and iron containimg more than 30% nickel; they are prepared by certain heat treatment and show unusual magnetic properties, giving a high initial permeability; one of the best alloys contains 78.5% nickel and 21.5% iron; another contains 78.5% nickel, 18% iron, 3% molybdenum and 0.5% manganese; a typical analysis gives 78.23% nickel, 21.35% iron, 0.04% carbon, 0.03% silicon, 0.035% sulfur, 0.22% manganese, 0.37% cobalt, 0.1% copper and traces of phosphorus. §

Permalon. A proprietary trade name for stretched vinylidene chloride. §

Permalose. Textile auxiliary chemicals. [ICI Chem. & Polymers Ltd]

Permalux. Neoprene accelerator. [DuPont UK]

Permalyn® 7085. Polyhydroxy ester of rosin; thermoplastic resin for use in hot-melt adhesives where light color and resistance to thermal degradation are required; improved oxidation resistance [Hercules]

Perma-Mold®. Release agents especially formulated for rigid urethane insulation. [Chem-Trend]

Permanax 6PPD. N-1,3-Dimethylbutyl-N´-phenyl-p-phenylene diamine; staining antiozonant for rubber. [Akzo]*

Permanax BL, BLN. Acetone/diphenylamine condensate; staining antioxidant for rubber. [Akzo] *

Permanax BLW. Acetone/diphenylamine condensate on inert carrier; Staining antioxidant for rubber. [Akzo] *

Permanax CNS. Nonstaining antioxidant for rubber. [Akzo]*

Permanax CR. Staining antiozonant and antioxiant forchloroprene rubber. [Akzo] *

Permanax DPPD. CAS 74-31-7; Diphenyl-p-phenylene diamine; staining antiozonant for rubber. [Akzo] *

Permanax HD. Heptylated diphenylamine; Staining antioxidant for rubber. [Akzo] *

Permanax HD (SE). Heptylated diphenylamine, Self-emulsifying grade; staining antioxidant for rubber. [Akzo] *

Permanax IPPD. CAS 101-72-4; N-Isopropyl-N´-phenyl-p-phenylene diamine; staining antiozonant for rubber. [Akzo] *

Permanax OD. Octylated diphenylamine; staining antioxidant for rubber. [Akzo] *

Permanax OZNS. Nonstaining antiozonant and antioxidant for rubber. [Akzo] *

Permanax TQ. Polymerized 2,2,4-trimethyl-1,2-dihydroquinoline; staining antioxidant for rubber. [Akzo] *

Permanax WSL. 2,4-Dimethyl-6-(1-methyl cyclohexyl) phenol; nonstaining antioxidant for rubber. [Akzo] *

Permanax WSL Pdr. 2,4-Dimethyl-6-(1-methyl cyclohexyl) phenol on inert carrier; nonstaining antioxidant for rubber. [Akzo] *

Permanax WSO. High molecular weight phenolic compound nonstaining antioxidant for rubber. [Akzo] *

Permanax WSP. 2,2´-Methylene-bis(6-(1-methyl-cyclohexyl)-p-cresol); nonstaining antioxidant for rubber. [Akzo] *

Permanax WSP (PQ). 2,2´-Methylene-bis(6-(1-methyl-cyclohexyl)-p-cresol); nonstaining antioxidant for use in

---

‡ Trade name and manufacturer not verified   § Trade name without identified manufacturer

polyethylene. [Akzo] *

Permanent. Pigment powders; used in paints and inks. [Hoechst AG]

Permanent Encapsulant 185N. Urethane compound; fast-setting, low exotherm castable elastomer for encapsulating telecommunications cable to provide moisture and mechanical protection and elec. insulation to cable splices; nonexpanding. [Hexcel]

Permanite. A cobalt steel which has very high magnetic properties.

Permapen. CAS 1538-09-6; A proprietary preparation of penicillin G benzathine; an antibiotic. [Pfizer International] †

Permaplex. An ion exchange membrane. [The Permutit Co]*

Permasect. CAS 52645-53-1; Permethrin; a pyrethroid insecticide. [Mitchell Cotts Chemicals Ltd]

Permasep®. Polymer. [DuPont UK]

Perma Shield. An elastomeric coating; suitable for application to concrete, brick, concrete block, stucco, metal, wood, sheetrock, masonite and plywood as a protective and decorative finish. [Secure Inc] †

Perma-Slik. Family of air curing dry lubricant coatings containing solid lubricants and suitable binder system; for frictional surfaces requiring dry lubrication particularly on substrates which cannot tolerate oven cure. [E/M Corporation]

Perma Sta. Ethylene gycol base; antifreeze. [Chemcentral Corp]

Permatag. Insecticidal cattle eartag. [Mitchell Cotts Chemicals Ltd]

Permathin. Tetrafilcon A; contact lens material. [UCO Optics Inc] *

Permatol A. A proprietary trade name for a preservative for wood; it contains pentachlorphenol in oil. §

Permax. A nickel steel containing 76% nickel, of French manufacture. It has magnetic properties.

Permax. CAS 66104-23-2; Pergolide mesylate; for neurological disorders. [Eli Lilly & Co]

Permethyl 99A. CAS 13475-82-6; Isododecane. cosolubilizer for nonhydrocarbon materials; for mascara, eyeliner, antiperspirant and where residual film is not desirable; solv. for debris on skin. [Presperse]

Permethyl 101A. CAS 4390-04-9; Isohexadecane. cosolubilizer for nonhydrocarbon materials; in eyeliners, mascaras, sun care and skin products; cleanser for eye and face makeup. [Presperse]

Permethyl 102A. CAS 93685-79-1; Isoeicosane. cosolubilizer for nonhydrocarbon materials; for skin care and sun care products; plasticizer for mascara. [Presperse]

Permethyl 104A. CAS 9003-29-6; Polyisobutene (isooctahexacontane); cosolubilizer for nonhydrocarbon materials; for lipsticks and glossers to improve wear and impart sheen; in skin care and sun care products [Presperse]

Permidan. Dimethylaminopyrazolone, $C_5H_9N_3O$.

Perminal. Textile auxiliary chemicals. [ICI Chem. & Polymers Ltd]

Perminvars. Proprietary alloys having exceptional magnetic properties; particularly suited for use in electrical communication circuits; one alloy contains 45% nickel, 25% cobalt, and 30% iron, and has a high initial permeability.§

Permitil. Fluphenazine hydrochloride; antipsycotic.

[Schering Corp]

Permobel. Alkyd-based top coat for automobiles. [ICI Chem. & Polymers Ltd]

Permonite. An explosive used in mines. It is a mixture of potassium perchlorate and ammonium nitrate, with trinitro-toluene, a little starch, and wood meal.

Permutite. An artificially made zeolite, prepared by igniting together china clay (aluminum silicates), and (sometimes) quartz or sand, with alkali carbonates; used for removing calcium and magnesium salts, sodium and potassium salts, and manganese and iron, from water. §

Permyl B-100. For stabilizers for halogenated hydrocarbon resins in US Class 6. [Ferro] *

Pernambuco wood. See redwoods.

Pernax. An artificial gutta-percha made from rubber, wax, and rosin. §

Pernivit. A proprietary preparation of nicotinic acid and acetomenaphthone. [British Drug Houses] *

Pernomol. A proprietary preparation of chlorobutol, phenol, camphor, tannic acid and spirit chilblain paint. [LAB Ltd]‡

Perone. CAS 7722-84-1; A proprietary trade name for pure hydrogen peroxide. §

Peronoid. A trade name for a mixture of copper sulfate and lime; a fungicide. §

Peropal®. CAS 41083-11-8; Azocyclotin; acaricide used for control of mobile stages of spider mites on pome and stone fruit, grapes, citrus, vegetables, etc. [Bayer AG]

perosmic acid anhydride. See osmium tetroxide.

perosmic oxide. See osmium tetroxide.

Perovskite. See calcium titanate.

Perox. Dyes for plastics. [Morton Int'l. Ltd]

Peroxal. CAS 7722-84-1; A trade name for hydrogen peroxide, H_2O_2. §

peroxidase. An enzyme found in most plant cells and some animal cells; CAS 9003-99-0; EINECS 232-668-6; promotes the oxidation of various substrates such as phenols, aromatic amines, etc. by means of hydrogen peroxide. [Am. Int'l. Chem.; Sigma; Spectrum Chem. Mfg.]

Peroxide RH-2. A proprietary trade name for a high melting, stable, aromatic organic peroxide; used as a polymerization catalyst. §

Peroxidol®. Epoxidized oils. [Reichhold]

Peroximon. A proprietary range of organic peroxides. [Montedison UK Ltd] *

Peroximon® DC 40 MG. CAS 80-43-3; Dicumyl peroxide/EPM masterbatch; vulcanizing and crosslinking agent for ethylene-propylene elastomers, polyethylene, EVA copolymers, nitrile and PU elastomers, SBR, PVC, neoprene; for rubber car parts, injection molded articles, elec. cable insulation, conveyor belts, shoe soles. [Akrochem]

Peroximon® DC-40. CAS 80-43-3; Dicumyl peroxide. [Akrochem]

Peroximon® S-164/40P. CAS 6731-36-8; 1,1-Di(t-butylperoxy)-3,3,5-trimethylcyclohexane. [Akrochem]

Peroxol. CAS 7722-84-1; Disinfectant solutions containing hydrogen peroxide, sometimes mixed with other disinfectants. §

peroxydisulfuric acid dipotassium salt. See potassium persulfate.

peroxydol. Sodium perborate, $NaBO_3 \cdot 4H_2O$; an antiseptic, deodorant, and bleaching agent.

Peroxyl. CAS 7722-84-1; Hydrogen peroxide. [May & Baker

---

* Trade name not verified as available   † Trade name verified as obsolete

Ltd] *

Perpentol. A tetralin preparation used for cleaning wool.

Perrindo®. High quality organic pigments for the surface coatings industry, especially for automotive coatings. [Bayer AG]

Persadox. CAS 94-36-0; Benzoyl peroxide; keratolytic. [Galderma Laboratories] *

Persantin. CAS 58-32-2; A proprietary preparation of dipyridamole; a vasodilator. [Boehringer Ingelheim Ltd]

Persantine. CAS 58-32-2; Dipyridamole; vasodilator. [Boehringer Ingelheim Pharmaceuticals Inc]

Persellig T. Leveling agent for dispersed dyes on polyamide and polyester. [Ceca SA]

Persian balsam. Compound tincture of benzoin.

Persian berry carmine. (Dutch Yellow). A pigment consisting of the aluminum and calcium lakes of the Persian berry coloring matters.

Persian yellow. Nitro-tolueneazonitrosalicylic acid, $C_{14}H_{10}N_4O_7$. Dyes chromed wool yellow; also used in cotton printing giving yellow shades with chromium acetate.

Persiderm®. Preparation for fashionable writing and luster effects on suede; used in leather industry. [Bayer AG]

Persiderm® Black. Aniline pigment for the lustering of black suedes. [Bayer plc]

persionin. An acetone extract of cud-bear.

Persistol®. Hydrophobic and oleophobic agents for textile finishing. [BASF AG; BASF plc]

Persistol® E. Zircon paraffin emulsion; for rendering suede furs water-repellent. [BASF AG]

persodine. A mixture of ammonium and potassium persulfates.

Persoftal®. Softeners for textiles; selected brands simultaneously impart an antistatic effect. [Bayer AG]

Persoftal® PE Special. Oligomer binder; used in polyester dyeing. [Bayer AG; Bayer plc]

Persoz's reagent. Zinc oxide (2 g), is added to a solution of zinc chloride (10 g), in 10 cc water. It dissolves silk, and detects silk in the presence of wool.

Perspex. Acrylic methyl methacrylate resins in sheet form. [ICI Chem. & Polymers Ltd]

Perstoff. Trichloromethyl-chloroformate, $CICO \cdot OCCl_3$; a posion gas. §

Perstorp Phenolic Moulding Compound. Toilet seats, pan handles, meter cases, automotive, domestic accessories, law bowls, electrical accessories. [Perstorp Ferguson Ltd] *

Perstorp Urea Moulding Compound. Electrical accessories, toilet seats, closures. [Perstorp Ferguson Ltd] *

Persulon. Fungicide; used for control of powdery mildew on cereals, fruit, vegetables and ornamentals. [Bayer AG]

Persyst. CAS 919-86-8; Emulsifiable concentrate containing 580 g demeton-S-methyl per liter; a systemic organophosphorus insecticide. [Ashlade Formulations Ltd]

Perthane. Trademark for an agricultural insecticide based on diethyldiphenyldichloroethane, supplied as a wettable powder or emulsifiable concentrate; controls insects on plants and livestockp; also used as a moth protection for textiles. §

Pertinit. A proprietary synthetic resin of the urea-formaldehyde type. §

Pertite. An Italian explosive; the main constituent is picric acid.

Pertofran. CAS 50-47-5; Desipramine; antidepressant.

[Ciba plc] *

Pertofrane. CAS 58-28-6; Desipramine hydrochloride; antidepressant. [USV Pharmaceutical Corp] ‡

Pertscan-99m. CAS 23288-60-0; Sodium pertechnetate Tc 99m; radioactive agent. [Abbott Laboratories] *

Pertusa. A proprietary preparation containing ephedrine hydrochloride, tincture of belladonna, liquid extract of ipecacuanha, syrup of tolu, honey, citric acid and sodium benzoate; a cough linctus. [The Boots Co plc]

Perugen. (synthetic Peru balsam). A synthetic Peru balsam made by mixing benzyl benzoate with storax, benzoic, and tolu balsams.

Peruol®. A registered trademark currently awaiting reallocation by its proprietors to cover a range of pharmaceuticals. [Cassella AG] *

Peruscabin. Benzyl benzoate, $C_6H_5 \cdot CO_2 \cdot CH_2 \cdot C_6H_5$. It is the active constituent of Peru balsam, and is used in the same manner, and for the same purposes as Peruol.

Peru silver. See Chinese silver.

Peruvian balsam. The oleoresin of *Myroxylon Pereirae*, of Central America.

Peruvian bark. Cinchona bark.

Peruvin. Cinnamyl alcohol, $C_6H_5CH : CHCH_2OH$.

Pervon. Fully reacted polyurethane/pitch coating. [Sigma Coatings] *

Pescola oil. An oil used in the tanning industry.

Pest-B-Gon. Roach bait. [Chevron] *

Pestex. An insecticide. [Fisons plc, Horticulture Div] *

Pestilizer®. Phosphate esters; compatibilizers for liquid fertilizers. [Stepan] *

Pestilizer® B Series. Amphiphilic alkyl sulfosuccinate; emulsifiers for liq. fertilizers; high stability in electrolytic systems. [Stepan Europe] *

Petalin. Distilled tall oils and fatty acids. [Surfachem Ltd]

Petameth. Racemethionine; acidifier. [O'Neal, Jones & Feldman Pharmaceuticals] *

Petcat R-9. CAS 1309-64-4; Antimony trioxide; catalyst in PET polyester production [Laurel Industries]

Pethidine Roche. A proprietary preparation of pethidine; narcotic analgesic. [Roche Products Ltd] *

Pethilorfan. A proprietary preparation of pethidine hydrochloride and levallorphan tartrate; an analgesic. [Roche Products Ltd] *

Petiole. Phenyl ethyl isopropyl ether. [Quest Int'l. UK Ltd]

Petlon®. Thermoplastic PET polyesters; modified polyethylene terephthalate grades which offer good dimensional stability, high rigidity, high heat resistance, good electrical properties and chemical resistance. [BASF AG; Miles Inc]

Petra. Range of polyethylene terephthalate polymers, including glass reinforced compounds; used for automotive ignition and carburetor components; electrical connectors, bobbins, relays and switches; intricate mechanical parts that maintain dimensional stability under load, or at temperatures above 150 G, or in the presence of moisture. [Allied-Signal Inc]

Petra® 130. CAS 25038-59-9; PET polyester, 30% glass-reinforced; for automotive parts, e.g., mirror housing assembly, chair shells; features strength, rigidity. [Allied-Signal Engineered Plastics]

Petra® 130FR. CAS 25038-59-9; PET polyester, 30% glass-reinforced, flame-retarded; for elec. connectors; features processability, dimensional stability, rigidity. [Allied-Signal Engineered Plastics]

---

‡ Trade name and manufacturer not verified   § Trade name without identified manufacturer

Petra® 230. CAS 25038-59-9; PET polyester, 35% mineral/ glass-reinforced; for automotive grill reinforcement, band saw pulleys; features strength and stiffness, min. warpage. [Allied-Signal Engineered Plastics]

Petra® 242. CAS 25038-59-9; PET polyester, 40% mineral/ glass-reinforced, impact-modified; for automotive roof rack; impact modified; features rigidity, toughness, paintability, strength. [Allied-Signal Engineered Plastics]

Petrac® 165. Petrol. wax; lubricant used in rigid PVC compounds to control external lubrication and fusion. [Syn. Prods.]

Petrac® 215. Oxidized polyethylene wax; lubricant used to control external lubrication of rigid PVC compounds and provide high gloss surface [Syn. Prods.]

Petrac® 270. CAS 57-11-4; EINECS 200-313-4; Tech. stearic acid; external lubricant in flexible PVC processing; chemical intermediate for metallic stearates, esters, etc.; dispersant, plasticizer, activator, lubricant in rubber compounding; thicknener for greases. [Syn. Prods.]

Petrac® CP-11. CAS 1592-23-0; EINECS 216-472-8; Calcium stearate; mold release agent, lubricant, pigment suspension aid, and flow control agent for plastics, paints, inks, waterproofing of cement and clay tile, lubrication and glossing in paper coatings; plastics application. [Syn. Prods.]

Petrac® Eramide®. CAS 112-84-5; EINECS 204-009-2; Erucamide; slip, release, antitack, and/or internal mold release agent; used in PP for extrusion of sheets, in injection molding; antistat; in polyvinyls for films and sheeting; in polyethylene it imparts slip and antiblock characteristics in film application; internal mold release agent in molded products; lamination of polyethylene to cellophane and in polyethylene extrusion coatings; withstands high processing temps.; food contact applications [Syn. Prods.]

Petrac® GMS. Glyceryl stearate; lubricant and mold release agent for plastics industry, pigment suspension in paints and inks, waterproofing agent for cement and clay tile, lubricant and glossing aid in paper coatings; used in rigid PVC pipe, foundry resins. [Syn. Prods.]

Petrac® MG-20 NF. CAS 557-04-0; Magnesium stearate NF; dry lubricant and anticaking agent used in food products, and filling of pharmaceutical capsules; antistick properties in tableting; improves stability, smoothness, and texture of cosmetic emulsions, creams, ointments; improves texture and water-repellency in baby and medical powds. [Syn. Prods.]

Petrac® PHTA. CAS 61790-38-3; EINECS 263-130-9; Partially hydrogenated tallow fatty acids; used in industrial products including synthetic lubricants, bar soaps, cosmetics, rubber tires. [Syn. Prods.]

Petrac® Slip-Eze. CAS 301-02-0; EINECS 206-103-9; Oleamide; slip, release, antitack, and/or internal mold release agent; polyethylene film and sheeting; injection or extruded molded products; polyvinyl film; PVC plastisol systems where "quick slip" characteristics are required; food pkg. materials. [Syn. Prods.]

Petrac® Vyn-Eze®. CAS 124-26-5; EINECS 204-693-2; Stearamide; slip, antitack, antiblock, and/or internal mold release agent; polyvinyl molded products and PVC plastisol systems; polyethylene film; food pkg. materials. [Syn. Prods.]

Petrac® ZN-41. CAS 557-05-1; Zinc stearate USP; lubricant

and mold release agent in PS, melamine, U-F, phenol-formaldehyde, and polyester molding resins; dusting agent for uncured rubber slabs; suspending and flattening agent in solv.- and water-based paints; anticaking agent used in extinguishers; lubricant in powd. metallurgy. [Syn. Prods.]

Petralol. See paraffln, liquid.

Petralon. A German name for a preparation of wood tar; an antiseptic.

Petramin. Special disperse dyes for dyeing nickel-modified polypropylene fibers. [Bayer plc]

Petrasul. A synthetic stone material made from asbestos and Portland cement, which is sulfur-impregnated; the sulfur content varies from 15-35% and the material can be colored; suitable for counter-tops, or similar purposes. §

Petre. CAS 7757-79-1; Potassium nitrate, KNO_3; used in pyrotechnics. §

Petrex. A proprietary trade name for a polybasic acid used in synthetic resin manufacture, the essential constituent of which is 3-isopropyl-6-methyl-3:6 endo-ethylene-Δ_4-tetrahydrophthalic anhydride. §

Petrex 7-75T. A solution of an alkyd-type resin derived from a terpene polybasic acid; used in coatings for cellophane. [Hercules] *

Petrin. CAS 1607-17-6; Pentrinitrol; vasodilator. [Parke-Davis] *

Petrinex. Propylene glycol mixtures. [ICI Chem. & Polymers Ltd]

Petro. See paraffin, liquid.

Petro 11. Alkyl naphthalene sodium sulfonate; hydrotrope and surfactant. [Witco/Organics] *

Petro 22. Modified alkyl napthalene sodium sulfonate; low foam surfactant for various cleaning formulations. [Witco/Organics]

Petro AG Special. Alkyl naphthalene sulfonate; water soluble anticaking agent for detergents, fertilizers and various salts. [Witco/Organics]

Petro BAF. Linear alkyl naphthalene sodium sulfonate; hydrotrope and surfactant for detergent and cleaning formulations. [Witco/Organics] *

Petro P. Alkyl napthalene sodium sulfonate; wetting agent for pesticide and fertilizer formulations. [Witco/Organics]*

Petro S. Alkyl napthalene sodium sulfonate; soil conditioner. [Witco/Organics] *

Petro ULF. Modified alkyl naphthalene sodium sulfonate; low foam surfactant for various cleaning applications. [Witco/Organics] *

Petro WP. Modified alkyl naphthalene sodium sulfonate; wetting agent for industrial cleaning formulations and pesticide formulations. [Witco/Organics] *

Petroacid. A proprietary trade name for a mixture of fatty acids obtained from petroleum distillates. §

petrobenzol. A petroleum distillate solvent; it has a boiling point of 61-96 C.

Petroclastite. (Petroklastite). An explosive. It contains potassium nitrate, sulfur, coal tar pitch, and potassium dichromate.

Petrofibe 201, 210, 215, 235. Petrolatum; industrial grades for applications requiring good moisture resistance, corrosion protection and lubricity. [IGI Petroleum Spec.] *

Petrofracteur. An explosive consisting of 10% nitrobenzene, 67% potassium chlorate, 20% potassium nitrate,

---

* Trade name not verified as available

† Trade name verified as obsolete

and 3% antimony pentasulfide.

Petrogils. Drilling muds. [Rhone-Poulenc UK]

petrol. A product of the distillation of petroleum. The term is synonymous with gasoline and petroleum spirit. Other names for the same product are naphtha, petroleum naphtha or mineral naphtha, benzoline, benzine, and carburine.

Petrolagar. A proprietary emulsion of liquid paraffin and agar-agar. §

Petrolane. Liquefied gas for general purposes. [Quantum Chemical Corp]

petrolatum. (petroleum jelly; petrolatum amber; petrolatum white; mineral jelly; paraffinum molle) Petroleum hydrocarbons; semisolid mixture of hydrocarbons obtained from petroleum; consists of the yellow, semi-solid, purified residue left when petroleum is distilled; CAS 8009-03-8 (NF); 8027-32-5 (USP); EINECS 232-373-2; specific gravity 0.87-0.90; laxative; textile lubricant; dispersant; as ointment base in pharmaceuticals and cosmetics; leather grease; shoe polish; rust preventives. [Exxon; Harcros; Magie Bros. Oil; Mobil; Penreco; Stevenson Bros.; Witco/Sonneborn]

petrolatum amber. *See* petrolatum.

petrolatum white. *See* petrolatum.

Petrolax. *See* paraffin, liquid.

Pétrole Hahn. Hair tonic, shampoo and conditioner to keep hair healthy and attractive. [Richardson-Vicks Inc] *

petrolenes. (malthenes). Constituents of bitumens which are soluble in hexane.

petroleum ether. (canadol; light ligrin; gasoline, solene). A distillate from petroleum oil. It consists essentially of pentane and hexane, and is a solvent for resins.

petroleum jelly. *See* petrolatum.

petroleum naphtha. CAS 8030-30-6; A term very loosely applied. It often denotes the first fraction of boiling-point up to 150 C., obtained from the distillation of crude petroleum oil, but is sometimes applied to any low boiling petroleum product.

petroleum naphtha. *See* naphtha.

petroleum pitch. CAS 8052-42-4; Asphalt.

petroleum spirit. (light petroleum;benzine; naphtha) CAS 8032-32-4; Both benzoline and naphtha are sold under this name. They are used as motor spirits, and for drycleaning cloths.

petroleum wax. Petroleum hydrocarbon derived from petroleum; CAS 8002-74-2; lubricant for formulating PVC and elec. wire and cable compounds; protectant for elastomers. [Astor Wax; Mobil]

petroleum wax, crystalline. *See* paraffin.

Petrolia. *See* paraffin, liquid.

Petrolig. Dispersant for oil well drilling muds. [Borregaard LignoTech]

petroline. A fraction of petroleum distillation of boiling-point 120-150 C., of specific gravity 0.722-0.737; used for defatting, or cleaning. The term is also used for a volatile oil yielded by asphalt, when it is distilled with water.

Petrolit. A German explosive containing potassium chlorate and mineral oil. §

Petrolite® C-400. Salt of an oxidized polyethylene; wax used as a gelling agent for solvs. in the formulation of paste products, e.g., shoe or floor polishes. [Petrolite]

Petrolite® C-7500. Oxidized synthetic wax; used in the formulation of emulsions having hard films with good gloss and dry-bright polish properties; oil-binding proper-

ties make it useful in solv.-based polishes and release agents. [Petrolite]

Petromix 9. Balanced blend of sodium sulfonates, auxiliary fatty acid soaps and coupling agents with antifoaming agents; strong emulsification agent used in textile processing oils, cutting oils, metal degreasers. [Witco Chemical Ltd] *

Petromor. Petroleum sulfonates. [Burmah-Castrol Ltd] *

Petronate® L, HL, K, CR and S. Series of anionic surfactants in which the cation is sodium and the anion is petroleum sulfonate; emulsifiers, dispersing and wetting agents; CR is rust preventative; used in printing inks; dry cleaning soaps; leather oils; lubricating grease; metal working; mineral dressings; paint; petroleum additives; emulsion breaking; rubber; textile processing oils. [Witco/Sonneborn; Witco Chemical Ltd]

Petronate® RP. Blend of sodium and calcium petroleum sulfonates; anionic surfactant used in rust preventative formulations. [Witco Chemical Ltd]

Petronauba® C. Oxidized microcryst. wax; used in the formulation of polishes and emulsions; carnauba substitute. [Petrolite]

Petrone A4 and A6C. Amine salt of alkylaryl sulfonic acid; brown viscous liquid; solubilization, antistat, dewatering, formulation, pigment dispersion, emulsification; corrosion inhibition properties. [ABM Chemicals Ltd] *

Petronol. *See* paraffin, liquid.

Petropul. A proprietary trade name for a synthetic resin. §

Petro-Rez 801. Cycloaliphatic hydrocarbon resin; tackifier for rubber compounding applications including molded mech. goods, extrusions, shoe soling, elec. insulation, flooring; processing aid for mineral-filled elastomers; extender in black loaded compounds. [Akrochem]

Petrosio. *See* paraffin, liquid.

Petrostep. Petroleum sulfonates; for enhanced oil recovery. [Stepan] *

Petrostep A-70. Branched chain alkylate sulfonic acid, anionic surfactant properties, viscous amber liquid; an emulsifier intermediate for speciality products. [KWR Chemicals Ltd] *

Petrosul® H-50. Sodium petrol. sulfonate; surfactant and corrosion inhibitor; used as motor and fuel oil additives, rustproofing formulations; also in dry cleaning solvs., leather processing, printing inks, oil well drilling fluids. [Penreco]

Petrosul® M-50, M-60, M-70. Sodium petrol. sulfonate; surfactant, emulsifier, and corrosion inhibitor for metalworking fluids; also for dry cleaning solvs., leather processing, textile oils, printing inks, oil well drilling fluids. [Penreco]

Petrothene®. Polyethylene. [Quantum Chemical Corp]

Petrothene® GA 501. CAS 9002-88-4; LLDPE, butene comonomer; film extrusion resin for heavy duty shipping sacks, liners, consumer pkg.; excellent puncture resistance, elong., and heat-seal strength [Quantum/USI]

Petrothene® GA 564. CAS 9002-88-4; LLDPE; resin for injection molding of industrial containers, trash cans, large parts; features high rigidity, good low temp. impact strength, high ESCR. [Quantum/USI]

Petrothene® GA 808-090. CAS 9002-88-4; High molecular weight LLDPE resin; wire and cable compound [Quantum/USI] *

Petrothene® HD 5903B. CAS 9002-88-4; High molecular weight HDPE; wire and cable compound [Quantum/

---

‡ Trade name and manufacturer not verified   § Trade name without identified manufacturer

USI]*

Petrothene® LB 5003-00. CAS 9002-88-4; HDPE; blow molding resin [Quantum/USI] *

Petrothene® LB 6001-00. CAS 9002-88-4; HDPE; sheet and profile extrusion grade [Quantum/USI] *

Petrothene® LF 6030-00. CAS 9002-88-4; HDPE copolymer; injection molding resin with good toughness and warp resistance for spray pump parts and overcaps; FDA compliance. [Quantum/USI]

Petrothene® LP 5102-00. CAS 9002-88-4; MMW-HDPE; film resin with high stiffness and strength, excellent appearance, good draw down for merchandise bags; FDA compliance. [Quantum/USI]

Petrothene® LS 3150-00. CAS 9002-88-4; HDPE copolymer; injection molding resin with excellent impact resistance, good flow forr containers, closures, housewares; FDA compliance. [Quantum/USI]

Petrothene® LT 5704-00. CAS 9002-88-4; HDPE, high load melt index; blow molding resin [Quantum/USI] *

Petrothene® NA 155-000. CAS 9002-88-4; LDPE; film resin with good stiffness, optics for industrial films with improved clarity. [Quantum/USI]

Petrothene® NA 204-000. CAS 9002-88-4; LDPE resin; extrusion coating and injection molding resin [Quantum/USI]

Petrothene® NA 341-000. CAS 9002-88-4; LDPE homopolymer; high clarity film resin with good melt strength for clarity bundling. [Quantum/USI]

Petrothene® PA 436. CAS 9002-88-4; MDPE, butene comonomer; resin for rotational molding of agric. and chem. storage containers; excellent ESCR, low temp. impact strength, warp resistance. [Quantum/USI]

Petrothene® PP 1510-HC. CAS 9003-07-0; PP impact copolymer; heat-stabilized resin with chem. resistance for film, profile, tubing, coating applications. [Quantum/USI]

Petrothene® PP 2004-MR. CAS 9003-07-0; PP homopolymer; extrusion grade resin with excellent elong., high melt flow for film and filament. [Quantum/USI]

Petrothene® PP 7300-KF. CAS 9003-07-0; PP random copolymer; resin with excellent clarity, impact strength, stiffness for general purpose blow molding, sheet extrusion. [Quantum/USI] *

Petrothene® PP 8000-GK. CAS 9003-07-0; PP homopolymer; injection molding resin with toughness, excellent surface appearance for general purpose containers, housewares. [Quantum/USI]

Petrothene® PP 8770-HU. CAS 9003-07-0; PP copolymer; heat-stabilized super-impact injection molding resin for tool boxes, pet carriers, toys. [Quantum/USI]

Petrothene® XL. Crosslinkable polyethylene resins. [Quantum Chemical Corp]

Petrowet® R. Sodium alkyl sulfonate; wetting agent, detergent, dispersant, penetrant, foamer for petrol., metalworking, textile, and paper industries, chem. manufacturing, oil well servicing; industrial formulations and cleaning. [DuPont]

peucedanin. Imperatorin. It occurs in the root of masterwort.

Pevafix. An adhesive for polyvinyl alcohol film. [May & Baker Ltd] *

Pevalon. Polyvinyl alcohol film. [May & Baker Ltd] *

Pevidine. CAS 25655-41-8; A proprietary preparation of povidone iodine; used as a skin disinfectant; marketed for veterinary use only. [Berk Pharmaceuticals Ltd] *

Pevikon. Polyvinyl chloride homo and copolymers (E-PVC); used in organosols and plastisols for production of plastic articles. [Norsk Hydro AS]

pewter. A variable alloy of from 73-89% tin, 1.6-6.7 antimony, 1-6.8% copper, and 0-20.5% lead, and sometimes zinc.

Pexalyn. Polar, acidic hydrocarbon-based synthetic resins; Pexalyn A500 and A600 are used as tackifier resins for solvent and emulsion adhesives and as modifier resins for ethylene/vinyl acetate copolymer and wax-based hot-melt coatings and adhesives. [Hercules] *

Pexate. Metal resinates. [Hercules Ltd] †

Pexid. CAS 6724-53-4; Perhexiline maleate; vasodilator. [Merrell Dow Pharmaceuticals Inc] *

Pexid. CAS 6724-53-4; A proprietary preparation of perhexiline maleate; used in the treatment of angina pectoris. [Richardson-Vicks Inc] *

Pexite. Wood rosins; used in wax modification, as a chemical intermediate, in solder fluxes and in wax based coatings. [Hercules] *

Pexol® 50, Dark Fluid. Fortified rosin size; sizing agent for paper/paperboard producing resistance to water and aq. sol'ns.; for use on unbleached kraft grades. [Hercules]

Pexol® 245. Fortified dark paste rosin size; sizing agent for paper/paperboard producing high level of water resistance; used in unbleached pulps and secondary fiber pulps used in multi-ply board. [Hercules]

Peyron's chloride. CAS 15663-27-1; Diaminodichloroplatinum.

Peyton powder. It is a nitrocellulose and nitroglycerin powder, containing 20% ammonium picrate; an explosive.

PF-10GF/15T. CAS 25135-51-3; Polysulfone resin, 10% glass fiber-reinforced, 15% PTFE-lubricated; improved mech. properties, reduced surface wear characteristics, wide temp. service range; used for applications. with moving parts requiring retention of properties, esp. creep resistance. [Compounding Tech.]

PF-20GF. CAS 25135-51-3; Polysulfone resin, 20% glass fiber-reinforced; features low creep and shrinkage; used for elec. connectors, business machine components, coffee maker bodies, microwave ovenware, sterilizer trays. [Compounding Tech.]

Pfeilringspalter. A catalyst used in the decomposition of fats. It is prepared by treating a mixture of hydrogenated ricinoleic acid and naphthalene with sulfuric acid.

Pferrico. CAS 1309-37-1; A cobalt treated iron oxide for magnetic media use. [Pfizer International] †

Pferrisperse. CAS 1309-37-1; A high-solids iron oxide pigment slurry. [Pfizer International] †

Pferritan. A zinc ferrite compound for high temperature color pigment use. [Pfizer International] †

Pferrocal. A steel-clad calcium wire effective in the production of high quality specialty steels. [Pfizer International]†

Pferromet. A metallic iron particle for magnetic tapes and disks. [Pfizer International] †

Pferrox. CAS 1309-37-1; A gamma ferric oxide for magnetic tape and disk applications. [Pfizer International] †

Pfico₂-Hop. A nonisomerized carbon dioxide extract of hops; used by adding to the brew kettle during boiling in the manufacture of malt beverages, to add bitterness to this product. [Pfizer International]

Pfico₂-Isohop. A modified aqueous hop extract produced

from a liquid carbon dioxide base concentrate and standardized to 35% isomerized alpha acids; for addition to malt beverage after fermentation to standardize bitterness. [Pfizer International]

Pfico₂-Redihop. A modified aqueous hop extract produced from a liquid carbon dioxide base concentrate and standardized to 35% reduced alpha acids; for addition to malt beverage after fermentation to standardize bitterness. [Pfizer International]

Pfiklor. CAS 7447-40-7; EINECS 231-211-8; Potassium chloride; replenisher. [Pfizer Inc] †

Pfinodal. A copper-nickel-tin alloy used in the fabrication of electronic connectors. [Pfizer International] †

Pfizer, -E, -EM. CAS 643-22-1; A proprietary preparation of erythromycin stearate; an antibiotic. [Pfizer International] †

Pfizercycline. CAS 60-54-8; A proprietary preparation of tetracycline; an antibiotic. [Pfizer International] †

Pfizerpen. CAS 113-98-4; Penicillin G potassium; antibacterial. [Pfizer Inc]

Pfizerpen A. CAS 69-53-4; Ampicillin; antibacterial. [Pfizer Inc] †

Pfizerpen-AS. CAS 6130-64-9; Penicillin G procaine; antibiotic; antibacterial. [Pfizer Inc]

Pfizerpen Vk. CAS 132-98-9; Penicillin V potassium; antibacterial. [Pfizer Inc]

Pfizerquine. CAS 54-05-7; A proprietary preparation of chloroquine; an antimalarial. [Pfizer International]

P-Flakes; Hydrogenated palm oil. [Karlshamns]

Phaltan. Fulpet fungicide. [Chevron] *

Phamosan. Skin protection ointments, lotions and soaps; for protection and care for the skin under environmental stress. [Dynamit Nobel Wien GmbH] *

Phanodorm®. Cyclobarbitone calcium; hypnotic and sedative. [Bayer AG]

Phanteine. CAS 115-95-7; Pure linalyl acetate. [Bush Boake Allen Ltd] †

Phantol. Pure linalol. [Bush Boake Allen Ltd] †

Phantom. CAS 23103-98-2; Granules containing 50% w/w pirimicarb; used for control of aphids. [Bayer plc]

Pharmagel. CAS 9000-70-8; A proprietary trade name for pure gelatin. §

Pharmakon. CAS 1314-13-2; Zinc oxide BP; pharmaceutical. [Manchem Ltd] *

Pharmasorb. CAS 1337-76-4; Pharmaceutical grade attapulgite (hydrous magnesium aluminum silicate); used for tablet or liquid antidiarrheals, as tableting aid, pharmaceutical carrier, inert, in cosmetics. [Engelhard]

Pharmaton. A proprietary preparation containing multivitamins and minerals; also contains ginseng. [Windsor Healthcare Ltd]

Pharmatone. A hydrolyzed pork tissue that is spray dried; over 90% protein; water soluble; used in veterinary biologicals and food supplements, for its nutrient and high nitrogen content. [Am. Labs]

Pharmolin. CAS 1332-58-7; Fine particle sized pharmaceutical grade hydrous kaolin; used as an extender in pharmaceuticals, tableting, cosmetics. [Engelhard]

Phasal®. CAS 554-13-2; Sustained-release tablets containing 300mg lithium carbonate BP; for treatment of acute manic or hypomanic episodes; prophylaxis in manic-depressive disorders, recurrent mania or recurrent depression. [Lagap Pharmaceuticals Ltd]

Phase II®. SMC resin for automotive fabrication. [Ashland]*

Phase Alpha®. Resins for automotive fabrication. [Ashland]*

phaseomannite. CAS 87-89-8; Inositol, $C_6H_{12}O_6$.

Phathalogen. Phthalocyanine dyes for blue and green shades for cotton and regenerated cellulose. [Bayer AG]

Phazyme. Pink, sugar coated tablet containing specially activated simethicone; deflatulent. [Stafford-Miller] ‡

P.H.D. A proprietary trade name for a plasticising oil. §

Phe-Mer-Nite. CAS 62-38-4; A phenylmercuric nitrate; pharmaceutic aid. [SmithKline Beecham] *

Phemerol Chloride. CAS 121-54-0; EINECS 204-479-9, Benzethonium chloride; antiinfective, topical; pharmaceutic aid. [Parke-Davis] *

Phemox. Mercury fungicide. [Murphy Chemical Co Ltd] †

Phenac. A proprietary trade name for a phenolic synthetic resin for varnish and lacquer. §

phenacemide. CAS 63-98-9; (Phenylacetyl)urea.

phenacetin. CAS 62-44-2; Acetyl-p-phenetidine, $C_6H_4(O \cdot C_2H_5) NH \cdot CO \cdot CH_3$.

Phenaldine. CAS 102-06-7; Diphenylguanidine; a proprietary trade name for a rubber vulcanization accelerator.§

phenald resins. A general term for phenol-formaldehyde resins.

Phenalgin. See Ammonal. §

Phenaphen. CAS 103-90-2; Acetaminophen; analgesic; antipyretic. [Wyeth Laboratories] *

phenazone. See antipyrine. §

phenedin. See phenacetin.

Phenegol. The mercury-potassium salt of nitro-p-phenolsulfonic acid. A bactericide.

Phenergan. CAS 58-33-3; Promethazine hydrochloride; antiemetic, antihistaminic. [Rhone-Poulenc Rorer Ltd]

Phenester. A proprietary trade name for a synthetic resin of the coumarone-indene type. §

phenethyl mustard oil. p-Ethylphenyl-thiocarbimide, SCN $\cdot C_6H_4 \cdot (C_2H_5)$.

Phenetidine. Amino-phenyl ethyl ether, $C_6H_4(OC_2H_5)NH_2$; an antipyretic.

phenetole. (ethoxy benzene; phenyl ethyl ether) $C_6H_5 \cdot O \cdot C_2H_5$; CAS 103-73-1.

Phenex. An aldehyde-amine; a proprietary trade name for a rubber vulcanization accelerator. §

phenic acid. (phenic alcohol). Phenol, C_6H_5OH; CAS 108-95-2; used in rubber manufacture, solvents, in phenolic resins.

phenic alcohol. See phenic acid.

Phenin. CAS 62-44-2; A proprietary preparation of phenacetin. §

phenindione. (2-phenylindane-1,3-dione; phenylindanedione) $C_{15}H_{10}O_2$; CAS 83-12-5; Used as a blood coagulant.

Phenistix. A proprietary preparation of ferric ammonium sulfate, magnesium sulfate and cyclohexylsulfamic acid impregnated on a test strip; used for the detection of phenylketonuria and ingestion of salicylates. [BC Ames]*

Phenitol. CAS 62-38-4; Phenylmercuric nitrate; pharmaceutic aid. [Alcon Laboratories Inc] *

Phenmerzyl Nitrate. CAS 62-38-4; Phenylmercuric nitrate; pharmaceutic aid. [Merrell Dow Pharmaceuticals Inc] *

Phenobarbitone Spansule Capsules. CAS 50-06-6; Phenobarbitone in sustained release form; anticonvulsant indicated in treatment of epilepsy. [SmithKline Beecham] *

Phenocitrain. A proprietary preparation of phenolic ana-

---

‡ Trade name and manufacturer not verified

§ Trade name without identified manufacturer

logues, citral and lignocaine; used in the treatment of varicose ulcers. [Whaley Pharmaceuticals] *

Phenodip. Liquid, compounded esters of 4-hydroxybenzoic acid; preservative. [Nipa Laboratories Ltd] †

Phenodorm. Pharmaceutical preparation; soporific. [Bayer AG] †

Phenodur. Heat hardening phenol/formaldehyde resins; used in industrial paints, brake and clutch linings and abrasives. [Hoechst AG]

Phenodur. Heat hardening phenol/formaldehyde resins; used for industrial paints, brake and clutch linings and abrasives. [Resinous Chemicals Ltd]

Phenoform. See Bakelite. §

Phenolax. CAS 77-09-8; Phenolphthalein; laxative. [Upjohn] *

phenol camphor. (carbolic camphor). Phenol with camphor (*Phenol cum camphorae B.P.*).

Phenolene, Phenolene Supra. Tin plating bath brighteners. [Ciba plc] *

phenol-formaldehyde (novolak) resin. See novolac resin.

phenolic resin. (phenol-formaldehyde; one step: A-stage resin, one-stage resin, resole; two step: novolac resin, novolak resin, two-stage resin) See also phenol-formaldehyde resin, novolac resin. Thermosetting resin from condensation of phenol or substituted phenol with aldehydes such as formaldehyde, acetaldehyde, and furfural; thermosetting resin for applications. requiring heat resistance, exhaust duct systems, wiring devices, switch gears, ovens, toasters, pot handles, elec. devices, coil bobbins, coatings, laminates. [3M; Akzo; Arakawa; Arizona; Asahi Yukizai Kogyo; Bakelite GmbH; BP Chem. Ltd; Georgia-Pacific; Hüls AG; PMC Specialties; QO; Raschig]

Phenoline. A disinfectant identical with Lysol. It is a cresol made soluble in water by saponification.

Phenolite. A proprietary trade name for a phenol-formaldehyde synthetic resin laminated product. §

phenolphthalein. $C_{20}H_{14}O_4$; CAS 143-74-8; The lactone of dioxy-triphenyl-carbolic-carboxylic acid; used as an indicator in alkalimetry.

phenol red. Phenol-sulfonphthalein; $(C_6H_4OH)_2CO-CO_2C_6H_4$; used as an acid-base indicator; diagnostic reagent medicine, laboratory reagent.

Phenonip. Phenoxyethanol, methylparaben, ethylparaben, propylparaben, butylparaben; fully active liq. preservative system with low toxicity and wide spectrum activity, esp. against pseudomonads; for cosmetics and pharmaceuticals. [Nipa Labs]

Phenopreg. A proprietary trade name for phenolic impregnated fabrics and papers. §

Phenoro. Proprietary preparation of canthaxanthin and β-carotene. photoprotective agent. [Roche Products Ltd]*

phenosalyl. A mixture of phenol, salicylic acid, menthol, and lactic acid; an antiseptic.

Phenosept. Mixture of propylene phenoxetol and p-chloro-m-xylenol; an antiseptic. [Nipa Laboratories Ltd]

Phenoweld. Phenolic adhesives for nylon, etc. [Hardman]

Phenox. A proprietary trade name for phenylmercury hydroxide. §

Phenoxetol. CAS 122-99-6; EINECS 204-589-7; Phenoxyethanol; antimicrobial preservative for cosmetics and pharmaceuticals. [Nipa Labs]

phenoxin. Carbon tetrachloride, CCl_4.

Phenoxine. CAS 562-09-4; Chlorphenoxamine hydrochloride; relaxant; anti-Parkinsonian. [Merrell Dow Pharmaceuticals Inc] *

phenoxyacetic acid. $C_6H_5OCH2COOH$; CAS 122-59-8; EINECS 204-556-7; Intermediate for dyes, pharmaceuticals, pesticides, other organics, fungicides, flavoring, laboratory reagent, precursor in antibiotic fermentations esp. penicillin V. [Chemie Linz UK; Great Lakes; Penta Mfg.; Schweizerhall]

Phenoxylene 50. CAS 94-74-6; MCPA; herbicide for cereals and grassland. [Schering Agrochemicals Ltd]

Phenoxylene Plus. A selective weed killer. [Fisons plc, Horticulture Div] *

phenoxy resin. High,molecular weight plastics copolymer of Bisphenol A and epichlorohydrin; $[OC_6H_4C(CH_3)_2-C_6H_4OCH_2CH(OH)CH_2]_n$; coatings and adhesives; blow-molded containers, pipe, ventilating ducts, and other molded parts. [Aldrich; Union Carbide]

Phensedyl. CAS 58-33-3; Promethazine hydrochloride compound; cough linctus. [Rhone-Poulenc Rorer Ltd]

Phenurone. CAS 63-98-9; Phenacemide; anticonvulsant. [Abbott Laboratories]

Phenyform. $[C_6H_4(OH)CH_2OH]_xCH_2O$; An antiseptic powder prepared from phenol and formaldehyde; used as an indicator and for denaturing purposes.

phenylacetic acid. (α-toluic acid) $C_6H_5CH_2COOH$; CAS 103-82-2; EINECS 203-148-6; Perfume, precursor in manufacture of penicillin G, fungicide, flavoring, laboratory reagent. [Calaire Chimie SNC; Penta Mfg.; Schweizerhall]

β-phenylacrylic acid. See cinnamic acid.

phenylacrylyl chloride. See cinnamoyl chloride.

phenylalanine. CAS 63-91-2; α-Amino-β-phenyl-propionic acid, $C_6H_5 \cdot CH_2CH(NH_2)COOH$; used in medicine and nutrition.

phenylamine. (aniline) $C_6H_5NH_2$; CAS 62-53-3.

phenyl benzene. See biphenyl.

phenylcarbinol. See benzyl alcohol.

phenyl chloride. See chlorobenzene.

phenyl cyanide. See benzonitrile.

phenyl dimethicone. (methyl phenyl polysilicane; polyphenylmethyl siloxane) Silicone polymer; CAS 2116-84-9; 63148-58-3.

m-phenylenediamine. (1,3-benzenediamine; 1,3-phenylenediamine; 3-aminoaniline) Aromatic amine; $C_6H_4(NH_2)_2$; CAS 108-45-2; EINECS 203-584-7; manufacture of dyes, hair dyes; rubber curing agent; in photography; as reagent for gold and bromine. [Du Pont]

p-phenylenediamine. (1,4-benzenediamine; 1,4-phenylenediamine) Aromatic amine; $C_6H_4(NH_2)_2$; CAS 106-50-3; EINECS 203-404-7; azo dye intermediate; photographic developing agent; intermediate in manufacture of antioxidants, accelerators for rubber, synthetic fibers; dyeing hair and fur. [Du Pont; Hoechst Celanese; Janssen Chimica]

1,3-phenylenediamine. See m-phenylenediamine.

1,4-phenylenediamine. See p-phenylenediamine.

phenyl ether. See diphenyl oxide.

phenyl fluoride. See fluorobenzene.

phenylformic acid. See benzoic acid.

phenyl-γ acid. 2-Phenylamino 8-naphthol-6-sulfonic acid.

phenylglycolic acid. See mandelic acid.

α-phenylhydroxyacetic acid. See mandelic acid.

2-phenyl imidazole. CAS 670-96-2; Epoxy curing agent for printed circuit boards, molding compounds, potting; ac-

---

* Trade name not verified as available   † Trade name verified as obsolete

celerator for dicyandiamide and anhydrides. [BASF; Janssen Chimica; Schweizerhall]

phenylmercuric acetate. (PMAC; phenylmercury acetate. (acetato)phenylmercury; acetoxyphenylmercury) Metallo-organic compound.; $C_6H_5HgOCOCH_3$; CAS 62-38-2; EINECS 200-532-5; fungicide, herbicide, mildewcide for paints; slimicide in paper mills. [Allchem Industries; Atomergic Chemetals; W.A. Cleary; EM Industries]

phenylmercuric oleate. (phenylmercury oleate) C_6H_5-HgOOC$(CH_2)_7$CH=CHC$_8H_{17}$; Mildewproofing agent for paints; fungicide, germicide. [Noah Chem.; Thor]

phenylmercury oleate. See phenylmercuric oleate.

phenyl methane. See methyl benzene.

phenylmethane. See toluene.

phenylmethanol. See benzyl alcohol.

phenylmethanone. See benzophenone-3.

phenylmethyl acetate. See benzyl acetate.

phenyl-peri acid. CAS 82-76-8; 1-Phenyl-naphthylamine-8-sulfonic acid.

o-phenylphenol. ((1,1′-biphenyl)-2-ol; 2-phenyl phenol; o-xenol) Substituted aromatic compound.; $C_{12}H_{10}O$; CAS 90-43-7; EINECS 201-993-5; Intermediate; dyes, germicides, fungicides, rubber chemicals; laboratory reagents; food pkg.; disinfectant. [Coalite Chem. Div.; Nipa Labs]

2-phenyl phenol. See o-phenylphenol.

2-phenylpropene. See α-methylstyrene monomer.

3-phenylpropenoic acid. See cinnamic acid.

3-phenylpropionic acid. See hydrocinnamic acid.

2-phenylpropylene. See α-methylstyrene monomer.

5-phenyltetrazole. HNN=NN=C(C_6H_5); CAS 3999-10-8; Blowing agent for foaming plastics and elastomers at elevated temps. [Hüls Am.]

phenyl tolyl ketone. See 2-methylbenzophenone.

phenyltrichlorosilane. $C_6H_5Cl_3Si$; CAS 98-13-5; Intermediate for silicones; laboratory reagent. [Aldrich; Hüls Am.; PCR]

phenyltrimethoxysilane. (trimethoxysilyl)benzene) $C_6H_5Si(OCH_3)_3$; CAS 2996-92-1; EINECS 221-066-9; Coupling agent, release agent, lubricant, blocking agent, chemical intermediate. [Hüls Am.; Janssen Chimica; PCR]

Pheny-PAS-Tebamin. CAS 133-11-9; Phenyl aminosalicylate; antibacterial. [The Purdue Frederick Co] *

pheophytin. The brownish derivative obtained by treating chlorophyll with acid.

Pherocon Adox-O Adoxamone. Pheromones; for control of tortrix moth in tree fruit. [DowElanco Ltd]

Pherocon Archemone. Pheromones; for control of tortrix moth in tree fruit. [DowElanco Ltd]

Pherocon CM Codelemone. Pheromones; for control of codling moth in apples. [DowElanco Ltd]

Pherocon GFun Funemone. Pheromones; for control of plum tree moth in plums. [DowElanco Ltd]

Philacid 0810. CAS 67762-36-1; Caprylic/capric acid (C8-10); intermediate for manufacturing of synthetic specialty detergents, esters for use as lubricants and emollients. [United Coconut Chem.]

Philacid 0818. Whole distilled coconut fatty acid (C8-18); intermediate for manufacturing of soap, synthetic detergents, fatty amines. [United Coconut Chem.]

Philacid 1200. CAS 143-07-7; Lauric acid (C12); intermediate for manufacturing of toilet soaps, synthetic deter-

gents, cosmetics and pharmaceuticals. [United Coconut Chem.]

Philacid 1214. Lauric/myristic acid (C12-14); intermediate for manufacturing of soaps, detergents. [United Coconut Chem.]

Philacid 1400. CAS 544-63-8; EINECS 208-875-2; Myristic acid (C14); intermediate for manufacturing of toilet soaps, cosmetics, esters for use as flavors and perfumes. [United Coconut Chem.]

philanized cotton. Cotton material which has been treated with concentrated nitric acid to convert it into a wool-like fabric.

Philcohol 1200. CAS 112-53-8; Lauryl alcohol (C12); intermediate for manufacturing of synthetic detergents (e.g., lauryl sulfates, ethoxylated lauryl alcohol, lauryl ether sulfates, lauric alkanolamides, etc.). [United Coconut Chem.]

Philcohol 1214. CAS 68425-37-6; EINECS 270-351-4; C12-14 coconut fatty alcohol; intermediate for surfactant manufacturing. [United Coconut Chem.]

Philcohol 1400. CAS 112-72-1; EINECS 204-000-3; C14 alcohols; intermediate for surfactant manufacturing. [United Coconut Chem.]

Philcohol 1600. CAS 36653-82-4; EINECS 253-149-0; C16 alcohols; intermediate for surfactant manufacturing. [United Coconut Chem.]

Philcohol 1618. C18-18 alcohols; intermediate for surfactant manufacturing. [United Coconut Chem.]

Philcohol 1800. CAS 112-92-5; EINECS 204-017-6; C18 alcohols; intermediate for surfactant manufacturing. [United Coconut Chem.]

philosopher's wool. CAS 1314-13-2; (flowers of zinc). The zinc oxide produced in a flocculent condition by burning zinc.

Phisomed. CAS 18472-51-0; A proprietary preparation of chlorhexidine gluconate in a detergent base used as a skin cleanser. [Winthrop Laboratories] *

phlobaphenes. Red or brown coloring matter from barks, usually oak bark.

phloba-tannins. Tannins which give the phlobaphene reaction.

phloroglucine. See phloroglucinol.

phloroglucinol. (phloroglucine; 1,3,5-trihydroxybenzene) $C_6H_6O_3$; CAS 108-73-6; EINECS 203-611-2; Analytical chemistry, decalcifying agent for bones, preparation of pharmaceuticals and dyes, resins, preservative for cut flowers, textile dyeing and printing. [ISK Europe SA; Schweizerhall; Spectrum Chem. Mfg.]

Phloroglucinol. CAS 108-73-6; 1,3,5-Trihydroxybenzene; intermediate for pharmaceuticals; coupler in dyeline printing; photographic chemicals; available in anhyd. and dihydrate forms. [Schering Berlin Polymers]

phlorol. CAS 90-00-6; o-Ethylphenol, $C_6H_4 \cdot C_2H_5$ (OH).

phlorone. (metaphlorone) p-Xyloquinone, $C_6H_2(CH_3)_2O_2$.

Phobol. Finishing agent. [Ciba plc] *

Phobotex FTN. A proprietary trade name for a fat modified melamine resin with outstanding water-repellancy; fast to repeated washing in soap and soda; provides a durable water repellant finish for textiles. [Ciba plc] *

Phobotone. Proofing agent. [Ciba plc] *

Phoenix alloy. An electrical resistance alloy containing 25% nickel and 75% iron.

Phoenixite. A proprietary pyroxylin product. §

Phoenix powder. An explosive containing 28-31% nitro-

---

glycerin, 0-1% nitro-cotton, 30-34% potassium nitrate, and 33-37% wood meal.

Pholin's alloy. An alloy of 77% tin, 19% bismuth, and 4% copper.

Pholtex. Each 5 ml spoonful contains pholcodine BP 15 mg and phenyltoloxamine 10 mg as ion-exchange resin complexes in a thixotropic vehicle. [3M Pharmaceuticals] †

Phono Bronze. A proprietary trade name for certain copper alloys containing about 1.25% tin and small ammounts of silicon and cadmium. §

Phoran. See Camite. §

Phos-Ad 100. CAS 126-73-8; Tributyl phosphate; defoamer conc.; for drilling fluids. [Chemron]

Phosai. CAS 7722-76-1; Monammonium phosphate fertilizer. [Scottish Agricultural Industries plc]

Phosal. CAS 8002-43-5; EINECS 232-307-2; Lecithin fractions. [Rhone-Poulenc UK]

Phosal 25 SB. Phosphatidylcholine; natural emulsifying aid, hair conditioner, oil restorer, source of essential fatty acids for cosmetics (shampoos, creams, oil baths, soap additive, hair rinses); so-called vitamin F. [Seppic]

Phosal 53 MCT. Phosphatidylcholine; production of liposomes for dermatology and cosmetics (improves skin moisturization and penetration); solubilizer/dissolving intermediary for lipophilic substances; source for dietetics. [Seppic]

Phosbrite®. Chemical brightening solutions. [Albright & Wilson Ltd, Phosphates & Speciality Business]

Phosbrite® 172. Patented product for chemically polishing aluminum, producing a reflective surface on a wide range of alloys. [Albright & Wilson Am.] *

Phos-Chek. Fire retardant, ammonium polyphosphate; catalyst used in intumescent coatings for paints, used to help meet flammability requirements for rigid and flexible polyurethane foams. [Monsanto Co] *

Phosclene. Metal finishing cleaners. [Albright & Wilson Ltd, Phosphates & Speciality Business] *

Phosclere. Stabilizers and antioxidants. [Harcros UK]

Phos-copper. A proprietary welding alloy composed essentially of copper with from 5-10% phosphorus; melts at 700 C, becoming extremely fluid at 750 C. §

Phosfetal 201 K. Calcium alkyl polyglycol ether phosphate; cosmetics surfactant. [Zschimmer & Schwarz]

Phosfetal 205. Alkyl polyglycol ether phosphate, acid; cleansing agent. [Zschimmer & Schwarz]

Phosfetal 603. Alkylaryl polyglycol ether phosphate, acid; cleansing agent. [Zschimmer & Schwarz]

Phos Flex. Mixed triaryl phosphate ester fire retardant plasticizers containing halogen and possessing better flame retardant properties in PVC than tricresyl phosphate. [Akzo Chemie UK Ltd]

phosgene. (carbonyl chloride; carbon oxychloride; chloroformyl chloride) $COCl_2$; CAS 75-44-5; used in the synthesis of isocyanates, polyurethane, and polycarbonate resins; organic carbonates and chloroformates; pesticides and herbicides; dye manyfacture.

Phosphaljel. CAS 13530-50-2; Aluminium phosphate; antacid. [Wyeth Laboratories] *

phospham. Phosphorus imidonitride, PN_2H.

phosphammite. Native ammonium phosphate.

Phosphanol Series. Phosphate ester; emulsifier for textile spinning oils, polymerization; antistat; anticorrosive agent. [Toho Chem.]

Phosphazote. It is an intimate mixture of superphosphate and urea contanig 4-11% nitrogen and 10-14% P_2O_5; a fertilizer.

Phosphin. 3-Amino-9-p-aminophenylacridine; used in the treatment of malaria.

PhosPho 642. CAS 8029-76-3; EINECS 232-440-6; Hydroxylated lecithin; surfactant, suspending agent for nail polish, hair/skin conditioner. [Fanning]

PhosPho E-100. CAS 8002-43-5; EINECS 232-307-2; Lecithin; emulsifier, suspending agent, solubilizer, superfatting agent for face and hand creams. [Fanning]

PhosPho F-97. CAS 8002-43-5; EINECS 232-307-2; Lecithin; for makeup, cleansers, body creams and lotions, shampoos, moisturizing creams. [Fanning]

PhosPho H-00. Hydrogenated soya lecithin; for liposome applications. [Fanning]

PhosPho H-150. Hydrogenated lecithin; for liposome applications. [Fanning]

PhosPho LCN-TS. CAS 8002-43-5; EINECS 232-307-2; Lecithin; phospholipid used as surfactant/emulsifier and skin conditioning agent for makeup, cleansers, body creams, shampoos, moisturizing creams, face creams, cosmetics, pharmaceuticals. [Fanning]

PhosPho PL-50. Phospholipids; emulsifier, refatting agent, source of vitamin F; for soft gelatin capsules, lotions, creams, oil baths, natural cosmetics. [Fanning]

PhosPho S-85. CAS 8002-43-5; Phospholipids; wetting agent for makeup, cleansers, body creams and lotions, shampoos, moisturizing creams. [Fanning]

PhosPho T-20. CAS 8002-43-5; Phospholipids; hydrophilic cosmetic ingredient for makeup, cleansers, body creams and lotions, shampoos, moisturizing creams. [Fanning]

Phosphocol P32. CAS 7789-04-0; Chromic phosphate P32; radioactive agent. [Mallinckrodt Inc] *

Phospho-gélose. A German clarifier for sugar juice. It consists of 70% phosphate of lime, and 30% kieselguhr.

Phospholan®. Alkylaryl sulfonates, modified nonionics, phosphate esters; used as hydrotropes and wetters in electrolyte solutions. [Harcros UK] *

Phospholan® ALF5. Tallow fatty alcohol phosphate ester on sodium carbonate carrier; foam limiter for use with anionic surfactants. [Harcros UK]

Phospholan® KPE4; CAS 36863-52-2; Arylethoxy phosphate, potassium salt; hydrotrope for solubilization of low foaming surfactants into highly built liqs.; industrial and domestic detergents and cleaners; fertilizer and toxicant formulations. [Henkel/Cospha; Harcros UK]

Phospholan® PHB14. Phosphate ester, free acid; hydrotrope and compatibility agent for agrochemical formulations; solubilizers anionic and nonionic surfactants into high electrolyte concs. [Harcros UK]

Phospholan® PTP7. Phosphate ester; surfactant offering excellent stability in high concs. of acid, alkali and electrolyte. [Harcros]

Phospholine Iodide. Ecothiopate iodide; for treatment of chronic open-angle glaucoma, chronic angle-closure glaucoma after iridectomy, accommodative esotropia. [Cusi] ‡

Phospholipid EFA. Linoleamidopropyl PG-dimonium chloride phosphate; patented component of skin and hair formulations; emulsifier, antimicrobial, moisturizer; solubilizer and fixative for fragrances; suitable for hypoallergenic products [Mona Industries]

Phospholipid PTC. Cocamidopropyl PG-dimonium chlo-

---

* Trade name not verified as available   † Trade name verified as obsolete

ride phosphate; patented biodeg. substantive, mild foamer, hydrotrope, visc. builder, wetter, surface tension reducer, bactericide for personal care products, baby products, feminine washes, ophthalmic preparations, disinfectant cleansers. [Mona Industries]

Phospholipid PTD. CAS 83682-78-4; Lauramidopropyl PEG-dimonium chloride phosphate; bactericidal, conditioner, antistat, detergent, foamer, emulsifier, solubilizer, dispersant, thickener and wetting agent for personal care, household, pharmaceutical, veterinary products, fire fighting foams, petrol. production, photographic processes, agric., mining, textiles. [Mona Industries]

Phospholipid PTS. Stearamidopropyl PG-dimonium chloride phosphate; patented mild substantivity agent, skin conditioner, thickener for personal care products; forms stable, low pH, smooth and elegant cosmetic emulsions. [Mona Industries]

Phospholipid SV. Stearamidopropyl PG-dimonium chloride phosphate and cetyl alcohol; patented substantive emulsifier, skin conditioner for skin and personal care products; moisturizer. [Mona Industries]

Phospholipon® 90/90G. Phosphatidylcholine; emulsifier for dermatology and cosmetics, solubilizer for parenteralia, raw material for liposomes; source for drugs and dietetics. [Seppic]

Phospholipon® CC. CAS 3436-44-0; 1,2-Dicaproyl-sn-glycero(3) phosphatidylcholine. [Seppic]

Phospholipon® MC. CAS 18194-24-6; 1,2-Dimyristoyl-sn-glycero(3) phosphatidylcholine. [Seppic]

Phospholipon® PC. CAS 2644-64-6; 1,2-Dipalmitoyl-sn-glycero(3) phosphatidylcholine. [Seppic]

Phospholipon® SC. CAS 816-94-4; 1,2-Distearoyl-sn-glycero(3) phosphatidylcholine. [Seppic]

Phospholutein. See lecithin.

phosphomolybdic acid. (PMA; phospho-12-molybdic acid) $H_3PMo_{12}O_{40} \cdot xH_2O$; CAS 11104-88-4; 51429-74-4; reagent for alkaloids; pigments; catalyst; fixing agent in photography; additive in plating processes; imparts water resistance to plastics, adhesives, and cement. [AAA Molybdenum Prods.; Atomergic Chemetals; Noah Chem.]

Phosphomort. Organo-phosphorous insecticides. [Murphy Chemical Co Ltd] †

phosphonic acid. See phosphorous acid.

phosphor bronze. A bearing metal. It is an alloy of from 70-97% copper, 3-13% tin, 0-16% lead, and 0.1-1.0% phosphorus and sometimes a little zinc. The alloys for casting usually contain 85-92% copper, 7-13% tin 0.3-1.0 phosphoros, and traces of lead and zinc.

phosphor copper. An alloy of copper with from 5-15% of phosphorus; used as an addition to other metals, and in the manufacture of phosphor bronze.

phosphor resin. A resin prepared from triphenyl phosphamide by heating and passing carbon dioxide through the compound.

phosphor steel. An alloy of steel with phosphorus.

phosphoric acid. (orthophosphoric acid) Inorganic acid; H_3PO_4; CAS 7664-38-2; EINECS 231-633-2; in manufacture of inorganic phosphates, fertilizers, detergents, chemical polishing, priming metals, petroleum refining; acid catalyst; as acidulant and flavor, antioxidant and sequestrant in food; pharmaceutic acid; in dental cements; as analytical reagent. [Albright & Wilson; Ashland; Farleyway Chem. Ltd; FMC; Mallinckrodt;

Mitsui Toatsu Chem.; Monsanto; Rasa Ind.; Rhone-Poulenc Basic]

o-phosphoric acid, triethyl ester. See triethyl phosphate.

phosphoric acid, triphenyl ester. See triphenyl phosphate.

phosphoric acid, trisodium salt. See trisodium phosphate, anhydrous.

phosphoric anhydride. See phosphorus pentoxide.

phosphoric chloride. See phosphorus pentachloride.

phosphoric ether. Triethyl phosphate, $OP(OC_2H_5)_3$.

phosphoric oxide. See phosphorus pentoxide.

phosphoric perchloride. See phosphorus pentachloride.

phosphorus acid. (orthophosphorous acid; phosphonic acid) H_3PO_3; CAS 10294-56-1; 13598-36-2; EINECS 237-066-7; Intermediate for manufacture of diphosphonic acids and phosphite salts used as pesticides, chelates, and plastic additives; restricts color formation in esterification and condensation reactions (in small quantities); chemical reducing agent. [Albright & Wilson; Janssen Chimica; Lonza; Rasa Ind.; Rhone-Poulenc Basic; Witco/Argus]

phosphorus ether. Triethyl phosphite, $(OC_2H_5)_3P$.

phosphor tin. An alloy of tin and phosphorus, containing up to 10% phosphorus.

phosphorus chloride. See phosphorus trichloride.

phosphorus oxychloride. (phosphoryl chloride) Cl_3OP; CAS 10025-87-3; EINECS 233-046-7; manufacture of insecticides, phosphate esters, pharmaceuticals; gasoline additives; dopant for semiconductor grade silicon, tricresyl phosphate, and fire-retarding agents. [Albright & Wilson; Aldrich; Cerac; FMC; Hoechst Celanese; Rhone-Poulenc Basic]

phosphorus pentachloride. (phosphoric chloride; phosphoric perchloride) PCl_5; CAS 10026-13-8; Chlorinating and dehydrating agent, catalyst. [Aldrich; Atomergic Chemetals; Hoechst Celanese; Noah Chem.]

phosphorus pentoxide. (phosphoric anhydride; phosphoric oxide) P_2O_5; CAS 1314-56-3; EINECS 215-236-1; Manufacture of chemicals, dessicant (drying agent), surfactants, condensing agent in organic synthesis, sugar refining, lab reagent, fire extinguishing, special glasses. [Albright & Wilson; Aldrich; Atochem N. Am.; Cerac; Hoechst Celanese; Rasa Ind.; Rhone-Poulenc Basic]

phosphorus trichloride. (phosphorus chloride) PCl_3; CAS 7719-12-2; EINECS 231-749-3; Source of phosphorus in manufacture of phosphite and phosphonate esters; chlorinating agent in manufacture of organic acid chlorides; analysis; catalyst. [Albright & Wilson; Aldrich; Atomergic Chemetals; FMC; Rhone-Poulenc Basic]

phosphorustriphenyl. See triphenyl phosphine.

phosphoryl chloride. See phosphorus oxychloride.

Phosphoteric® T-C6. Sodium dicarboxyethylcoco phosphoethyl imidazoline; patented surfactant, hydrotrope; synergizes detergency with ethoxylated nonionics; improves wetting, penetrating, and detergency; for high electrolyte industrial cleaners. [Mona Industries]

Phosphotope. CAS 7558-80-7; Sodium phosphate P 32; antineoplastic; antipolycythemic; diagnostic aid; radioactive agent. [Bristol-Myers Squibb Co Inc] *

phosphotungstic acid. (phosphotungstic acid hydrate; tungstophosporic acid; phosphowolframic acid) $H_3P(W_3O_{10})_4 \cdot H_2O$; CAS 12067-99-1; EINECS 235-087-6; Reagent in analytical chemistry and biology;

‡ Trade name and manufacturer not verified   § Trade name without identified manufacturer

manufacture of organic pigments; additive in plating industry; imparts water resistance to plastics, adhesives, cement; catalyst for organic reactions; photographic fixative; textile antistat. [Atomergic Chemetals; Noach Chem.; Spectrum Chem. Mfg.]

phosphotungstic acid hydrate. See phosphotungstic acid.

phosphowolframic acid. See phosphotungstic acid.

Phosteem. Metal treating compositions. [ICI Chem. & Polymers Ltd] †

Phostin. Tin-plating process. [Albright & Wilson Ltd, Phosphates & Speciality Business] *

Phostoxin. CAS 20859-73-8; Aluminum phosphide; used for gasing of rabbits and moles. [Rentokil Ltd]

Photal. A photographic developer.

Photine. Fluorescent whitening agents for paper textiles and detergents. [Hickson & Welch Ltd]

Photocure 51. Benzil dimethyl ketal; uv photoinitiator. [Aceto]

Photoglaze. Uv and eb cure coatings of varying viscosities and chemical composition; used to protect and preserve the original appearance of a wide range of materials, e.g., vinyls, plastics, paper, wood, even metallized surfaces, giving good resistance to abrasion, chemical stains and scuffing. [Lord Corporation (UK) Ltd]

Photomer®. Radiation curing chemicals. [Harcros UK]

Photomer® 3005. Acrylated epoxidized oil; radiation-cured chem. for eb coatings, uv inks and varnishes. [Harcros UK] *

Photomer® 5029. Polyester acrylate; radiation-curing chem. for eb coatings, uv varnishes. [Harcros UK] *

Photomer® 6360. Polyurethane acrylate, aliphatic; radiation-curing chem. for eb coatings, uv inks and varnishes. [Harcros UK] *

Photophor®. A trade name for a calcium phosphide, Ca_2P_2; used for signal fires. §

photoxylin. See celloidin.

Photozinc. CAS 1314-13-2; Photoconductive zinc oxide. [Pigment & Chemical Inc]

phoxim-methyl. O-α-cyanobenzylideneamino OO-dimethyl phosphorothioate; used as a pesticide.

PHPH. See biphenyl.

Phtalofix® FN. Premordant; used for Phtalogen K dyestuffs. [Bayer AG]

Phtalogen®. Phthalocyanine dyes; used to produce blue and green shades of outstanding fastness on cotton and regenerated cellulose. [Bayer AG]

Phtalogen® K, N1. Heavy metal donor for phtalogen dyestuffs; used in dyeing and textile printing. [Bayer AG]

Phtalotrop® B. Resist agent; for phtalogen resist printing. [Bayer AG]

Phthalamaquin. Quinetolate; relaxant. [Penick Corp] *

phthalic acid, diethyl ester. See diethyl phthalate.

phthalic acid dimethyl ester. See dimethyl phthalate.

phthalic anhydride. (1,3-isobenzofurandione) $C_8H_4O_3$; CAS 85-44-9; EINECS 201-607-5; Alkyd resins, plasticizers, hardener for resins, polyester, insecticides, laboratory reagent; manufacture of phthaleins, phthalates, benzoic acid. [Aldrich; Aristech; Atochem SA; BASF; Exxon; Mitsubishi Gas; Mitsui Toatsu Chem.; OxyChem; Stepan; UCB SA]

Phthalofyne. See Whipcide. §

Phthalopal®. Oil-free phthalate resins; for spirit-based and NC finishes; in hydrolyzed form, binder for aq. flexographic inks and ballpen inks. [BASF AG]

Phycon 15, 18, 25. Modified polybasic buffered acids; acid buffers for dyeing and printing operations. [Am. Emulsions]

Phycon LPH. Proprietary blend; pH control agent for stain-resist finished nylon carpet. [Am. Emulsions]

Phylatol. Biocidal preparation. [BDH Chemicals Ltd]

Phyldrox-G. A proprietary preparation of theophylline sodium glycinate, ephedrine hydrochloride, and phenobarbitone; a bronchial antispasmodic. [Carlton Laboratories (UK) Ltd] *

Phyllocontin. CAS 317-34-0; A proprietary preparation of aminophylline in a controlled release tablet; used in the treatment of bronchospasm. [Napp Laboratories Ltd] *

Phyllocontin. CAS 317-34-0; Aminophylline; relaxant. [The Purdue Frederick Co]

Phyllol. CAS 97-53-0; Pure eugenol. [Bush Boake Allen Ltd]

Phyomone. Growth promoting hormone. [ICI Chem. & Polymers Ltd]

Physeptone Linctus. CAS 1095-90-5; A proprietary preparation containing methadone hydrochloride; a cough linctus. [The Wellcome Foundation Ltd]

Physeptone Tablets and Injection. CAS 1095-90-5; A proprietary preparation of methadone hydrochloride; an analgesic. [The Wellcome Foundation Ltd]

physic nut oil. See purging nut oil.

physiological salt solution. (normal salt solution, isotonic salt solution, surgical solution). This consists of 8.5 g of sodium chloride in 1,000 cc of distilled water. It is sterilised and used for intravenous injection.

physostigmine. CAS 57-47-6; (eserine; calabarine) $C_{15}H_{21}N_3O_2$, an alkaloid from the calabar bean; used as an anticholinesterase in medicine.

Physostol. CAS 57-47-6; A sterile solution of 1% of an eserine salt in olive oil; used in the treatment of certain eye diseases.

Phytex®. A proprietary preparation of boratannic complex in alcohol in ethylacetate solvent; used for fungal infections in dermatology. [Pharmax Ltd]

phytic acid. CAS 83-86-3; (phytinic acid). Inositolhexaphosphoric acid, $C_6H_6(OPO_3H_2)_6$; used for chelating heavy metals in processing of animal fats and vegetableoils, corrosion inhibitor, metal treating, in the treatment of hard water.

phytin. Calcium magnesium salt of inositolhexaphosphoric acid; contains about 22% phosphorus, 12% calcium, and 1.5% magnesium. A powerful nerve and general tonic.

Phyt'iod. Ethylic esters of the iodized fatty acid of poppy seed oil (ethiodized oil); slendering product for cosmetics. [Alban Muller]

Phytodermine. Antifungal preparations. [May & Baker Ltd]*

Phytoforol. CAS 59-02-9; Vitamin E capsules. [British Drug Houses] *

phytol. CAS 150-86-7; A primary alcohol, $C_{20}H_{39}OH$, obtained by the decomposition of chlorophyll, the coloring matter of plants.

PI-20GF/000. CAS 61128-46-9; Polyetherimide, 20% glass fiber-reinforced. [Compounding Tech.] *

PIB. Isoprenaline and atropine; used for bronchospasm. [Napp Laboratories Ltd] *

P.I.B. An abbreviation for polyisobutylene.

PIB. See polybutene.

Pibiter. A proprietary range of molding products based on saturated thermoplastic polyesters. [Montedison UK Ltd] *

---

* Trade name not verified as available   † Trade name verified as obsolete

Picaltal®. Mixture of aromatic sulfonic acids; nonswelling, complex-forming acid in chrome tanning. [BASF AG]

Picamar. Propylpyrogallol dimethyl ether, $C_{11}H_{16}O_3$; used in perfumery.

Picco® 5000, 6000. Aromatic hydrocarbon resin derived from petroleum; used in higher-vinyl acetate content ethylene/vinyl acetate copolymer-based hot melt adhesives; used in caulking compounds, sealants, and concrete curing resins, and in news ink and oil-based lithographic inks. [Hercules]

Piccodiene®. Aliphatic hydrocarbon resins, composed mainly of polydicyclopentadienes; used in concrete curing compounds, in metallic paints and varnishes and as a compounding ingredient and process aid in many rubber goods. [Hercules]

Piccolastic®. A range of styrene monomer hydrocarbon resins; used for pressure sensitive, hot melt and solvent adhesives and as epoxy resin extenders. [Hercules]

Piccolastic® A-5. Proprietary trade name for vinyl plasticizers based upon polymerized styrene and homologues. [Hercules]

Piccolyte®. A proprietary trade name for thermoplastic terpene resins. [Hercules]

Piccolyte® A. Polyterpene hydrocarbon resins derived from α-pinene; used in adhesives for tapes and labels, in construction adhesives, in packaging adhesives, as tackifiers for styrene-butadiene rubber and styrene-block-copolymer rubber. [Hercules]

Piccolyte® C. A series of polyterpene hydrocarbon resins derived from d-limonene; used as tackifiers for natural rubber based pressure sensitive adhesives and can sealants used in the production of chewing gum base §

Piccolyte® HM110. A hydrocarbon-modified terpene resin; a tackifier resin of particular value in adhesives requiring high shear and high tack. [Hercules] *

Piccolyte® S. A range of polyterpene hydrocarbon resins derived from β-pinene; used in natural rubber-based pressure sensitive adhesives and styrene-butadiene rubber can sealants and in textile dry sizes. [Hercules]

Piccomer® XX. A range of low molecular weight resins produced from petroleum-derived monomers; binder and plasticizer resins in putty and sealants, as felt saturants in felt-based sheet goods, in waterproofing treatments for paperboard, as softeners for elastomers in rubber compounding. [Hercules]

Picconol® A100. A range of aliphatic hydrocarbon resin emulsions; tackifiers in adhesives and in the production of waterproof finishes. [Hercules]

Picconol® A200. Anionic terpene resin emulsions based largely on low molecular weight thermoplastic terpene resins produced from β-pinene or δ-limonene monomers; used in water-based laminating, case-sealing adhesives, in laminating paper, and as tackifiers for natural rubber latex. [Hercules] *

Picconol® A300. Anionic terpene resin emulsions based largely on low molecular weight thermoplastic terpene resins produced from β-pinene or d-limonene monomers; used in water-based laminating, case-sealing adhesives, in laminating paper and as tackifiers for natural rubber latex. [Hercules] *

Picconol® A400. A low molecular weight anionic pure monomer resin emulsion; used with other aqueous thermoplastic and/or elastomeric systems to produce excellent adhesives and coatings. [Hercules] *

Picconol® A500 A600. A range of aromatic hydrocarbon resin emulsions; used as adhesives, laminants and tackifiers. [Hercules] *

Piccopale®. Range of aliphatic hydrocarbon resins manufactured from petroleum-derived monomers; used in high ethylene/vinyl acetate hot-melt and natural rubber-based adhesives, in can coatings for packaging, in varnishes to impart gloss and flowability, in paper saturation, and as waterproofing agents for paper and textiles. [Hercules]

Piccopyn. A range of phenolic-modified terpene resins; used as laminating agents and modifiers in ethylene/vinyl acetate copolymer-based packaging and tray forming adhesives, in specialty coatings, to modify specific polyurethanes, as tackifiers and process aids in specialty rubber compounding. [Hercules] *

Piccotac®. A range of aliphatic hydrocarbon resins produced from mixed monomers of petroleum origin; developed especially for adhesives, particularly pressure sensitive and hot-melt types. [Hercules]

Piccotex®. Vinyltoluene copolymer hydrocarbon resins; used in hot-melt product assembly adhesives, for wax based coatings, in transparentizing paper, as dry sizing agents for textiles. [Hercules]

Piccotoner. Styrene acrylic copolymer hydrocarbon resins; used as toner resins for dry reproduction inks. [Hercules]*

Piccoumaron. A proprietary trade name of terpene varnish and lacquer resins. §

Piccoumarone Resins. Proprietary trade name for coumarone-indenes. §

Piccovar® AB. Aliphatic hydrocarbon resin; used as reinforcing agents in adhesives with high-temperature requirements, in heat set printing ink applications, in specialty coatings, in rubber compounding. [Hercules]

Piccovar® AP. Aromatic hydrocarbon resins used as plasticizers, softeners and tackifiers; suitable for use in various adhesive systems based on natural rubber, styrene-butadiene rubber, and poly-chloroprene. [Hercules]

Piccovar® L. Alkylated aromatic hydrocarbon resins; nonpolar, low molecular weight resins used as saturants, plasticizers, and tackifiers in adhesives and hot melts. [Hercules]

pichurim camphor. A substance resembling laurel camphor obtained from pichurim beans.

picked turkey gum. CAS 9000-01-5; (white sennaar gum). The best variety of gum acacia.

Picket. CAS 52645-53-1; Contains permethrin; garden insecticide. [ICI Garden Products]

pickle alum. Aluminum sulfate, $Al_2(SO_4)_3$; used for packing and preserving.

pickle green. A commercial variety of Scheele's green.

α-picoline. (2-methylpyridine; 2-picoline) $C_5H_4N(CH_3)$; CAS 109-06-8; EINECS 203-643-7; Organic intermediate for pharmaceuticals, dyes, rubber chemicals, solvent, source for vinyl pyridine, laboratory reagent. [Lonza; Nepera; Schweizerhall]

β-picoline. (3-picoline; 3- methylpyridine) CAS 108-99-6; EINECS 203-636-9; Solvent in synthesis of pharmaceuticals, resins, dyestuffs, rubber accelerators, insecticides; preparation of nicotinic acid, nicotinic acid amide, waterproofing agents; laboratory reagent. [Nepera; Schweizerhall]

---

‡ Trade name and manufacturer not verified

§ Trade name without identified manufacturer

γ-picoline. (4-methylpyridine; 4-picoline) NCHCHC-(CH$_3$)CHCH; CAS 108-89-4; EINECS 203-626-4; Solvent in synthesis of pharmaceuticals, resins, dyestuffs, rubber accelerators, pesticides, waterproofing agents; laboratory reagent; manufacture of isoniazid; catalyst; curing agent. [Lonza; Schweizerhall]

4-picoline. See γ-picoline.

3-picolinic acid. See nicotinic acid.

4-picolinic acid. See isonicotinic acid.

Pi-cone. A proprietary lithopone containing 15% titanium oxide, 25% zinc sulfide, and 60% precipitated barium sulfate; stated to have a much higher covering power than ordinary lithopone. §

picrasmin. Quassin, C$_{10}$H$_{12}$O$_3$.

picrate powder. The name given to explosive powders, in which the main constituent is the potassium or ammonium salt of picric acid.

picric acid. (trinitrophenol; nitroxantthic acid; carbazotic acid; phenoltrinitrate) C$_6$H$_2$.OH(NO$_2$)$_3$; CAS 88-89-1; used for making explosives; matches, and dyes. A solution is employed in the treatment of burns, erysipelas, eczema, and gonorrhea.

picric powder. An explosive consisting of ammonium picrate and potassium nitrate.

picro-aniline blue. A stain used in microscopy. It is prepared by adding aniline blue to a saturated solution of picric acid in 92% alcohol until the liquid becomes deep blue-green in color.

picrocarmine. A microscopic stain obtained by mixing 1 gram carmine in 10 cc water and 3 cc strong ammonia solution, and adding the mixture to 200 cc of a saturated solution of picric acid.

picronigrosine. An alcoholic solution of picric acid and nigrosin; a microscopic stain.

picrontric acid. See picric acid.

picro-sulfuric acid. A liquid made by adding to 100 volumes water, 2 volumes sulfuric acid, and about 0.25% picric acid; used in microscopy as a fixing agent.

pictet crystals. White crystals, SO$_2$ · xH$_2$O, formed when liquid sulfur dioxide evaporates.

Pictet's fluid. Liquid carbon dioxide; used for freezing machines.

Pictet's liquid. A mixture of liquid carbon dioxide and sulfur dioxide; used for producing low temperature.

pictolin. A mixture of liquid carbon dioxide and sulfur dioxide.

Pictressin Tannate. Argipressin tannate; antidiuretic. [Parke-Davis] *

Pidolidone® . CAS 98-79-3; PCA; amino acid, cellular penetration vector for amino acids or mineral salts; used in aq. phases of skin and hair care formulations. [UCIB]

Pielanase. Enzyme. [ABM Chemicals Ltd] *

Pierrot metal. An alloy consisting mainly of zinc, with smaller ammounts of copper, tin, antimony, and lead.

Pif-Paf. Insecticide preparations. [The Wellcome Foundation Ltd] *

pig iron. Crude form of cast iron; product of blast furnace ore reduction cast into ingots called pigs.

pig lead. Lead is obtained from galena by heating in a reverberatory furnace with a silica flux, then heated with coke, and sometimes lime in a cupola furnace. The lead drawn off is called pig lead.

Pigmentar. A proprietary trade name for standardized pine tar prepared for use as a rubber softener in rubber compounding. §

Pilagan Liquifilm® . CAS 148-72-1; Pilocarpine nitrate; antiglaucoma agent. [Allergan, Inc]

pilasonite. A mineral, Bi$_3$Te$_2$.

Piliogrip Adhesive System for Styructural Bonding. 100% Reactive urethane structural adhesives designed for bonding thermosets, thermoplastics and metals; SMC to SMC for automotive and truck body assemblies; SMC to metal for automotive body assemblies. [Ashland Chemical Company] *

Pilocar® Ophthalmic Solution. CAS 54-71-7; Pilocarpine hydrochloride; indicated for the treatment of primary open-angle galucoma and also to lower intraocular pressure prior to surgery for acute angle-closure glaucoma; may be used in combination with other miotics, beta-adrenergic blocking agents, carbonic anhydrase inhibitors, hyperosmotic agents, or epinephrine. [Iolab Corp]

Pilofrin. CAS 148-72-1; Pilocarpine nitrate; cholinergic. [Allergan Pharmaceuticals Inc] *

Pilomiotin. CAS 54-71-7; Pilocarpine hydrochloride; cholinergic. [CooperVision Inc] ‡

Pilot. CAS 76578-14-8; Quizalofop-ethyl; selective herbicide for control of grasses in mustard, rape and beet crops. [Schering Agrochemicals Ltd]

Pilot. Rolling oils for aluminum. [Schering Agrochemicals Ltd]

Pilot SXS-40. CAS 1300-72-7; EINECS 215-090-9; Sodium xylene sulfonate; coupling agt., hydrotrope, solubilizer, solv., stabilizer; used in liq. cleaners, organic polymers and dyestuffs, petrol. industry, pulping, animal glues. [Pilot]

Pimafucin. CAS 7681-93-8; A proprietary preparation of natamycin; used in the treatment of fungal infections. [Brocades Pharma] †

Pimaricin. CAS 7681-93-8; A proprietary trade name for natamycin. §

Pimel® . Photosensitive polyimide. [Asahi Chem. Industry]

pimelic ketone. See cyclohexanone.

pimelite. A mineral. It is meerschaum containing nickel.

pimple metal. A term used for a type of copper metal produced from the coarse metal obtained from sulfide ores which have been fused with an excess of copper oxide in their purification.

pinachrom. p-Ethoxy-quinaldine-p-methoxy-quinoline-ethyl-cyanine-bromide. A red sensitizer for silver bromide plates.

pinacoline. CAS 75-97-8; Methyl-t-butyl-ketone, CH$_3$ · CO · C(CH$_3$)$_3$.

pinacyanol. A red sensitizer for silver bromide plates obtained by treating quinaldinium salts with formaldehyde followed by alkali.

pinaflavol. A basic dye used as a green sensitizer in photography.

pinakol. A pyrogallol photographic developer in which part of the alkali usually employed is replaced by sodium amino-acetate.

Pinaverdol. A green sensitizer for silver bromide plates.

pinchbeck. An alloy of from 83-93% copper and 6-17% zinc. A brass.

Pincoffin. Commercial alizarin. §

pine gum. (pine resin, white pine resin, Cypress pine resin). Names applied to Australian sandarac resin, obtained from *Callitris quadrivalis* and *C. calcarata*.

α-pinene. (2-pinene; 2,6,6-trimethylbicyclo(3.1.1)-2-hept-

---

* Trade name not verified as available   † Trade name verified as obsolete

2-ene) Terpene hydrocarbon; $C_{10}H_{16}$; CAS 80-56-8; 7785-70-8; EINECS 232-077-3; solvent for protective coating, synthesis of camphene, pine oil, odorant, lube oil additives, flavoring, insecticides. [Aldrich; Arizona; Hercules; SCM Glidco Organics; Veitsiluoto Oy]

β-**pinene.** (6,6-dimethyl-2-methylenebicyclo [3.1.1] heptane; nopinene) Terpene hydrocarbon; $C_{10}H_{16}$; CAS 127-91-3; polyterpene resins; intermediate for perfumes and flavorings. [Arizona; Penta Mfg.; SCM Glidco Organics]

pine oil. (oils, pine; yarmor) CAS 8002-09-3; Volatile oil obtained from distillation of the species *Pinus*; the name was originally applied to turpentine oils obtained from pine trees; the term is used in U.S. to designate turpentine obtained by distilling pine wood; it is also used for the lighter oils of pine tar, a refined rosin oil, and a by-product; odorant, disinfectant, penetrant, wetting agent, preservative, laboratory reagent, fragrances. [Allchem Industries; Arizona; Hercules; Langley Smith Ltd; Penta Mfg.]

Pineotrene K. Detergent for textile scouring. [Reilly-Whiteman]

pine resin. See pine gum.

pine tar. CAS 8011-48-1; Sticky, viscous, dark brown to black liquid; used in ore flotation, paintts and varnishes, roofing materials, medicine.

piney tallow. Malabar tallow. an edible fat obtained from the seeds of *Vateria indica*, of East Indies.

pinguin. Alantol, $C_{10}H_{16}O$, obtained from the roots of *Inula elecampane*.

pink salt. Ammonium-stannic-chloride; $SnCl_4 \cdot 2NH_4Cl$; formerly used as a mordant for dyes.

Pinnacle. Mixture of imazamethabenz-methyl and isoproturon; used for control of weed grasses in winter cereals. [Cyanamid of Great Britain Ltd]

pinolin. A name for rosin oil (the first distillate from rosin, boiling at from 78-250 C.).

pinolith. See lithopone.

Pinwire brass. An alloy of from 66-73% copper and 27-34% zinc.

Pioloform. Polyvinyl butyral resins. [Wacker Chemicals Ltd]

Pioneer. Single-stage centrifugal glandless circulator; for domestic and industrial central heating circulator. [Sihi Pumps (UK) Ltd]

Pioneer. 12-20-20 Compound fertilizer. [Kemira Ince Ltd]

Pioneer alloy. An alloy of 20% copper, 38% nickel, 4% silicon, 3% molybdenum, 2% tungsten, and the remainder iron.

Pioxol. A proprietary trade name for pemoline. §

Pipanol. A proprietary preparation of benzhexol hydrochloride; used in Parkinsonism. §

pipeclay. Aluminum silicate; an abrasive.

piperazenyl. See lysidine.

piperazidine. See piperazine.

piperazine. (piperazine anhydrous; diethylene diamine; pyrazine hexahydride; piperazidine) $C_4H_{10}N_2$; CAS 110-85-0; EINECS 203-808-3; Corrosion inhibitor, anthelmintic, insecticide, accelerator for curing polychloroprene. [Allchem Industries; BASF; Janssen Chimica]

piperazine anhydrous. See piperazine.

piperazine calcium edetate. A chelate produced by reacting ethylene-diamine-NNN'N'-tetra-acetic acid with calcium carbonate and piperazine. Perin.

piperidine. (hexahydropyridine; pentamethyleneamine) $C_5H_{11}N$; CAS 110-89-4; EINECS 203-813-0; Solvent and intermediate, curing agent for rubber and epoxy resins, catalyst for condensation reactions, ingredient in oils and fuels, complexing agent. [Aldrich; Janssen Chimica; Nepera; Schweizerhall]

piperonyl butoxide. (4,5-methylenedioxy-2-propyl-benzyldiethylene glycol butyl ether) $C_{19}H_{30}O_5$; CAS 51-03-6; EINECS 200-076-7; Synergist in insecticides in combinations with pyrethrins. [Burlington Bio-Medical; Wellcome Foundation Ltd]

Piportil. CAS 37517-26-3; Pipotiazine palmitate; antipsychotic. [Ives Laboratories Inc] *

Piportil Depot. CAS 37517-26-3; A proprietary preparation of pipothiazine palmitate; depot tranquilizer. [May & Baker Ltd] *

pip pip. An abbreviated name for piperidinium-pentamethylene-dithiocarbamate. An accelerator for rubber vulcanization.

Pipracil. CAS 59703-84-3; Piperacillin sodium; antibacterial. [Lederle Laboratories USA] *

Piprelix. Worm expellent. [The Boots Co plc]

Pipricide. Veterinary anthelmintic. [The Wellcome Foundation Ltd] *

Pipril. CAS 61477-96-1; Injection containing piperacillin; antibiotic. [Lederle Laboratories] *

Piptal. A proprietary preparation containing pipenzolate methobromide; a gastrointestinal sedative. [MCP Pharmaceuticals] *

Piptalin. A proprietary preparation of pipenzolate bromide and simethicone; used as a gastro-intestinal sedative. [MCP Pharmaceuticals] *

Piral. Pyrogallol, $C_6H_3(OH)_3$.

Piria's naphthionic acid. α-Naphthylamine-4-sulfonic acid.

Piriex. A proprietary preparation containing chlorpheniramine maleate, ammonium chloride and sodium citrate; a cough linctus. [Allen & Hanburys Ltd] *

Pirimor. CAS 23103-98-2; Wettable powder containing 50% w/w pirimicarb; for control of aphids. [ICI Chem. & Polymers Ltd]

Piriton. CAS 2438-32-6; A proprietary preparation of chlorpheniramine maleate; antihistaminic. [Allen & Hanburys Ltd]

Piror. Biocide. [Union Carbide]

Pirsch-Baudoin's alloy. An alloy of 71% copper, 16.5% nickel, 1.75% cobalt oxide, 2.5% tin, and 7% zinc.

Pistol. CAS 13684-63-4; Formulated phenmedipham. [Rhone-Poulenc UK]

Pistol 400. CAS 13684-63-4; Emulsifiable concentrate containing 118 g/l phenmedipham; for weed control for beet crops. [Quadrangle Agrochemicals]

Pitayin. CAS 56-54-2; Quinidine, $C_{20}H_{24}N_2O_2$; an alkaloid used in medicine.

pitch barm. A cement made from casein, water-glass, and caustic lime.

pitch mineral. See bitumen.

Pit-ite No. 2. An explosive consisting of 23-25% nitroglycerin, 28-31% potassium nitrate, 33-36% wood meal, and 7-9% ammonium oxalate.

Pitocin. CAS 50-56-6; A proprietary preparation of oxytocin; used to induce labor. [Parke-Davis] *

Pitralon. It is a wood tar derivative; a German antiseptic.

Pitressin. CAS 9034-50-8; Vasopressin (injection); antidiuretic hormone. [Parke-Davis] *

Pitressin Tannate. A proprietary preparation of vasopressin

---

‡ Trade name and manufacturer not verified   § Trade name without identified manufacturer

tannate; antidiuretic hormone. [Parke-Davis] *

Pittabs. CAS 7778-54-3; Calcium hypochlorite. [PPG Industries]

pittaccal. Eupittonic acid, $C_{25}H_{26}O_9$.

Pittclor. CAS 7778-54-3; Calcium hypochlorite. [PPG Industries]

Pitteliene. A mixture of coal tar and oil; used as an insecticide.

Pitti. See Ventilago Madraspanta.

pittinite. A mineral. It is a variety of gummite.

pittylen. A mixture of pine tar with formaldehyde; used in skin diseases.

Pivaloxycyclene. Tricyclodecenyl dimethyl propanoate. [Quest Int'l. UK Ltd]

Pivarose. 2-Phenylethyl pivalate. [Quest Int'l. UK Ltd]

Pivatil. CAS 33817-20-8; Pivampicillin; a broad spectrum antibiotic ester of ampicillin less effected by the presence of food in the gut. [Merck & Co Inc] †

Pivofax. A dried yeast.

Pix®, ULV. CAS 24307-26-4; Mepiquat chloride; bioregulator for reduction of undesired vegetative growth of cotton, better boil retention, earlier maturity; improves yield and market quality of garlic and onions [BASF AG]

Pixol. A form of wood tar soluble in water, made from tar and soap; used as a disinfectant.

pix solubilis. (soluble pitch). A soluble modification of the tar obtained by sulfonating the tar obtained from peat.

Pixtonet. A trade name for a slate substiute; used as an electrical insulator. §

PJ1 Chain Lube, Blue Label. Keeps o-rings soft and pliable to maintain chain's seal against wear-causing dirt and moisture. [PJ1 Corporation] ‡

PJ1 Octane Plus. Together with racing fuel additives, it boosts the power of lower octane gasolines with complete safety; gas stabilizers keep the gas in tank from going stale and prevent gum and varnish build-up in the carburetors and fuel system. [PJ1 Corporation] ‡

PJ1 Super Cleaner. Cleans and degreases all metal parts, also disperses water and works on all electrical parts that need to be cleaned or dried. [PJ1 Corporation] ‡

PK-10GF/000. PEEK, 10% glass fiber-reinforced. [Compounding Tech.] *

PK-20CF/000. PEEK, 20% carbon fiber-reinforced. [Compounding Tech.] *

PKWF. Printing ink distillates and oils. [Haltermann Ltd]

Placadol. A proprietary preparation of papaverine hydrochloride, homatropine methylbromide, and codeine phosphate. [Int'l. Synthetics Ltd] *

Placentaliquid Oil-Soluble. Extract of unborn bovine placentas; products for aging skin. [Dr. Kurt Richter; Henkel/Cospha]

Placet alloy. An alloy of 60% nickel, 20% iron, 15% chromium, and 5% manganese; an electrically resistant alloy.

Placidyl. CAS 113-18-8; Ethchlorvynol; sedative. [Abbott Laboratories]

Planavin. A trademark for a nitrogenous weed-killer. [Shell Chemie GmbH] *

Planell Oil. Squalene, squalane, glycolipids, phytosterol, tocopherol; natural plant lipid extract for use as cosmetic emollient; maintains normal skin function by augmenting skin's own naturally occurring lipid membrane. [Brooks Industries]

Planet. Nonionic wetting agent containing 85% alkyl polyglycol ether and fatty acid; for use in agrochemical sprays. [Ideal Manufacturing Ltd]

Planetol. Photographic preparations. [May & Baker Ltd] *

Planidets. A proprietary preparation of dibromopropamidine embonate, chlorphenoctium amsonate, and butyl aminobenzoate; used in the treatment of mouth ulcers. [May & Baker Ltd] *

Planocaine. Procain hydrochloride. [May & Baker Ltd] *

Planocaine. A proprietary preparation of ethocaine hydrochloride. §

Planochrome. Mercurochrome. [May & Baker Ltd] *

Planofix. Pre-harvest fruit drop inhibitor. [May & Baker Ltd]*

Planotox. Selective weedkiller. [May & Baker Ltd] *

Plantaren 600 CS UP. C12-14 fatty alcohol glycoside; surfactant for dishwashing agents, laundry detergents, cleaners. [Henkel KGaA]

Plantaren 1200. C12-16 alkyl polyglycoside; mild surfactant, foamer, cleanser for low-irritation personal care products, hair shampoos, bath and shower gels, foam baths, facial cleansers; biodeg. [Henkel/Cospha; Henkel Canada]

Plantaren 2000. Decyl polyglucose; mild surfactant, foamer, cleanser for low-irritation personal care products, hair shampoos, bath and shower gels, foam baths, facial cleansers; biodeg. [Henkel/Cospha; Henkel Canada]

Plantaren CG 60. C8-10 fatty alcohol glycoside; solubilizer, wetting agent for alkali bottle washing, neutral and acidic cleaners. [Henkel KGaA]

Plantex®. Fiber. [DuPont UK]

plant indican. Indican, $C_{26}H_{31}NO_{17}$.

Plantvax. CAS 5259-88-1; Oxycarboxin; a systemic fungicide. [ICI Agrochemicals]

Plantvax 20. CAS 5259-88-1; Oxycarboxin; a systemic fungicide for the control of rust in wheat. [Uniroyal Chemical Ltd]

Plantvax 75. CAS 5259-88-1; Oxycarboxin; a systemic fungicide for the control of rust in ornamental plants. [Fargro Ltd]

Plaquenil. CAS 747-36-4; A proprietary preparation of hydroxychloroquin sulfate; an antimalarial. §

Plasadd™. Granulated concentrate designed to improve the physical properties of polyolefins. [Cabot Plastics Ltd]

Plasblak® masterbatches. Granulated concentrates of carbon black in plastics; used for coloring of thermoplastic resins and for imparting resistance to weathering degradation. [Cabot Plastics Ltd]

Plasblak® EV 1755. EVA masterbatch with 40% SRF carbon blk.; blk. masterbatch offering adequate weathering properties, good opacity, easy processing for molding and extrusion applications; compatible with LDPE, HDPE, PP, PS. [Cabot Plastics Ltd.]

Plasblak® EV 3524. 50% Carbon blk. (avg. particle size 27 nm) in LDPE copolymer carrier; blk. masterbatch for molding and extrusion of articles for pkg. (boxes, lids), electronic equip. cases (cassettes, VCRs), radios, TVS, domestic appliances; designed for coloring PS, ABS, nylon 6, and PET. [Cabot Plastics Ltd.] *

Plasblak® PE 1851. 50% SRF carbon blk. (avg. particle size 60 nm) in LDPE carrier with calcium carbonate extender as antiblocking agent; blk. masterbatch for film, molding, and extrusion applications; compatible with LDPE, HDPE, PP, and ethylene copolymers. [Cabot Plastics Ltd.]

Plasbumin. Albumin human; blood volume supporter. [Cut-

---

ter Laboratories, Miles Laboratories Inc] *

Plas-Chek. For plasticizers for vinyl halide resins, in Int Class 1. [Ferro] *

Plascoat Plasinter. Polyethylene coating powders; used for coating domestic wirework, display stands, tools, clips, WRC approved, insulation of electrical components. [Plascoat Systems Ltd]

Plascoat PPA. Polyolefin alloy; thermoplastic coating powder; used for coating wire dishwasher baskets, process valves, pipework and units, electrical switchgear, seat frames and the lining of fire extinguishers. [Plascoat Systems Ltd]

Plascoat PPA 31 series. Resistant to aqueous chemicals at temperatures up to 100 C. [Plastic Coatings Ltd] *

Plasdeg™. Granulated concentrate designed to impart controlled degradability to polyolefin-based products. [Cabot Plastics Ltd]

Plasdone®. CAS 9003-39-8; Povidone; pharmaceutic aid (dispersing and suspending agent). [ISP]

Plasdone® C. CAS 9003-39-8; A range of pyrogen-free polyvinylpyrrolidones; solubilizer, stabilizer, protective colloid for veterinary pharmaceuticals which minimizes toxic side effects and reduces irritation at site of infection. [ISP]

Plasdone® K. CAS 9003-39-8; A range of polyvinyl-pyrrolidones; tablet binder and coating agent, cohesive agent, stabilizer and protective colloid, detoxicant for many poisons and irritants, drug vehicle and retardant, film forming agent in medicinal aerosols. [ISP]

Plasgon. A proprietary trade name for plastic gasket and joint cement. §

Plasgrey®. Granulated concentrates of carbon black in plastics; used in small amounts as colorants for naturally uncolored plastics. [Cabot Plastics Ltd]

Plaslube® AC-80/TF/10. Acetal copolymer, 10% PTFE; internally lubricated engineering thermoplastic. [Akzo Engineering Plastics] *

Plaslube® G-1/30/SI/2. CAS 32131-17-2; Nylon 6/6, 2% silicone; internally lubricated engineering thermoplastic. [Akzo Engineering Plastics] *

Plaslube® G-3/40/MS/5. CAS 25038-54-4; Nylon 6, 5% molybdenum disulfide; internally lubricated engineering thermoplastic. [Akzo Engineering Plastics] *

Plaslube® G-50/20/TF/10. PC, 10% PTFE; internally lubricated engineering thermoplastic. [Akzo Engineering Plastics] *

Plaslube® J-1/30/MS/5. CAS 32131-17-2; Nylon 6/6, 5% molybdenum disulfide; internally lubricated engineering thermoplastic. [Akzo Engineering Plastics] *

Plaslube® J-1/33/TF/13/SI/2. CAS 32131-17-2; Nylon 6/6, 13% PTFE, 2% silicone; internally lubricated engineering thermoplastic. [Akzo Engineering Plastics] *

Plaslube® J-1/CF/15/TF/20. CAS 32131-17-2; Nylon 6/6, 15% PAN carbon fiber, 20% PTFE; internally lubricated engineering thermoplastic. [Akzo Engineering Plastics]*

Plaslube® J-3/30/MS/5. CAS 25038-54-4; Nylon 6, 5% molybdenum disulfide; internally lubricated engineering thermoplastic. [Akzo Engineering Plastics] *

Plaslube® J-4/30/TF/15. Nylon 6/12, 15% PTFE; internally lubricated engineering thermoplastic. [Akzo Engineering Plastics] *

Plaslube® J-4/CF/30/TF/10. Nylon 6/12, 30% PAN carbon fiber, 10% PTFE; internally lubricated engineering thermoplastic. [Akzo Engineering Plastics] *

Plaslube® J-50/20/TF/10. PC, 10% PTFE; internally lubricated engineering thermoplastic. [Akzo Engineering Plastics] *

Plaslube® J-50/30/SI/2. PC, 2% silicone; internally lubricated engineering thermoplastic. [Akzo Engineering Plastics] *

Plaslube® J-50/30/TF/10. PC, 10% PTFE; internally lubricated engineering thermoplastic. [Akzo Engineering Plastics] *

Plaslube® J-50/CF/10/TF/15/FR. PC, 10% PAN carbon fiber, 15% PTFE; flame retarded, internally lubricated engineering thermoplastic. [Akzo Engineering Plastics]*

Plaslube® J-77/30/TF/15. CAS 32131-17-2; Nylon 6/6 resin, 30% chopped fiber reinforcement, 15% PTFE lubricant; see Plaslube G-1/30/MS/5. [Akzo] *

Plaslube® J-80/20/TF/15. Acetal copolymer, 15% PTFE; internally lubricated engineering thermoplastic. [Akzo Engineering Plastics] *

Plaslube® J-80/CF/10/TF/10. Acetal copolymer, 10% PAN carbon fiber, 10% PTFE; internally lubricated engineering thermoplastic. [Akzo Engineering Plastics] *

Plaslube® J-1300/30/TF/15. CAS 9016-75-5; PPS, 15% PTFE; internally lubricated engineering thermoplastic. [Akzo Engineering Plastics] *

Plaslube® J-1300/CF/20/MS/10/TF/15. CAS 9016-75-5; PPS, 20% PAN carbon fiber, 10% molybdenum disulfide, 15% PTFE; internally lubricated engineering thermoplastic. [Akzo Engineering Plastics] *

Plaslube® NY-1/MS/5/TF/30. CAS 32131-17-2; Nylon 6/6, 5% molybdenum sulfide, 30% PTFE; internally lubricated engineering thermoplastic. [Akzo Engineering Plastics] *

Plaslube® NY-1/SI/5. CAS 32131-17-2; Nylon 6/6, 5% silicone-lubricated; internally lubricated engineering thermoplastic. [Akzo Engineering Plastics] *

Plaslube® NY-1/TF/10. CAS 32131-17-2; Nylon 6/6, 10% PTFE; internally lubricated engineering thermoplastic. [Akzo Engineering Plastics] *

Plaslube® NY-4/TF/10. Nylon 6/12, 10% PTFE; internally lubricated engineering thermoplastic. [Akzo Engineering Plastics] *

Plaslube® PC-50/TF/10. PC, 10% PTFE; internally lubricated engineering thermoplastic. [Akzo Engineering Plastics] *

plasmin. CAS 9001-90-5; (fibrinolysin) The proteolytic enzyme derived from the activation of plasminogen; it dissolves fibrin and facilitates the solution of clots formed in the bloodstream.

plasminogen. The specific substance derived from plasma which, when activated, has the property of lysing fibrinogen, fibrin, and other proteins.

Plasmosan. CAS 9003-39-8; A proprietary trade name for povidone. §

Plastacele. A proprietary trade name for a plasticized cellulose acetate compound. §

Plastamid. Polyamides. [Croda Resins Ltd] †

Plastammone. An explosive containing ammonium nitrate, glycerin, mononitro toluene, and nitro-semicellulose.

Plastamol. Plasticizers for PVC. [BASF plc] *

Plastazote. Crosslinked foamed polyethylene. [BXL Plastics Ltd] *

Plastech™. A compound based on a modified thermoplastic resin; designed for technical applications. [Cabot Plastics Ltd]

---

‡ Trade name and manufacturer not verified   § Trade name without identified manufacturer

Plastech™ EP 8126. CAS 9003-07-0; PP copolymer, rubber-modified; rubber masterbatch modifying impact resistance. [Cabot Plastics Ltd.]

Plastech™ PP 3344. CAS 9003-07-0; 40% mineral-filled PP homopolymer compound; designed for manufacturing of injection molded parts for automotive industry, e.g., dashboards, door moldings, lamp housings, under-the-hood applications such as fans, fan housings. [Cabot Plastics Ltd.]

plaster cement. Cements made from gypsum.

plaster of Paris. A partially dehydrated gypsum, $2(CaSO_4)$ · H_2O. It is made from gypsum by heating the latter from 212-400 F, when 3 parts of the water of crystallization is given off.

Plasteryl. An explosive. It is a mixture of 99.5% trinitrotoluene and 0.5% resin.

Plast-E-Tint. Pigment dispersions; for coloration of plastisols and organisols. [Pacific Dispersions Inc] *

Plasthall®. Series of monomeric and specialty plasticizers; used for rubber and flexible vinyl. [C P Hall]

Plasthall® 6-10P. Plasticizer for PVC. [C.P. Hall]

Plasthall® 8-10 TM-E. n-Octyl, n-decyl trimellitate; plasticizer for polychloroprene compounds. [C.P. Hall]

Plasthall® 83SS. Dibutoxyethoxyethyl sebacate substitute; plasticizer for rubber industry. [C.P. Hall]

Plasthall® 100. Isooctyl tallate; plasticizer. [C.P. Hall]

Plasthall® 200. CAS 117-83-9; Dibutoxyethyl phthalate; plasticizer. [C.P. Hall]

Plasthall® 201. Dibutoxyethyl glutarate; plasticizer. [C.P. Hall]

Plasthall® 203. Dibutoxyethyl adipate; plasticizer. [C.P. Hall]

Plasthall® 205. Dibutoxyethyl azelate; plasticizer. [C.P. Hall]

Plasthall® 207. Dibutoxyethyl sebacate; plasticizer. [C.P. Hall]

Plasthall® 220. Dibutoxyethoxyethyl phthalate; plasticizer. [C.P. Hall]

Plasthall® 224. Dibutoxyethoxyethyl glutarate; plasticizer. [C.P. Hall]

Plasthall® 226. Dibutoxyethoxyethyl adipate; plasticizer. [C.P. Hall]

Plasthall® 325. Butoxyethyl oleate; plasticizer. [C.P. Hall]

Plasthall® 503. CAS 142-77-8; EINECS 205-559-6; Butyl oleate; textile surface finisher, softener, thread lubricant and antistat; plasticizer. [C.P. Hall]

Plasthall® 4141. PEG-3 caprate-caprylate; plasticizer for rubber goods; lubricant for aluminum can industry. [C.P. Hall]

Plasthall® BSA. CAS 3622-84-2; N,N-butyl benzene sulfonamide; plasticizer for emulsion adhesives, pkg., caulk, printing ink, surface coatings. [C.P. Hall]

Plasthall® DBS. CAS 109-43-3; Dibutyl sebacate; plasticizer. [C.P. Hall]

Plasthall® DBZZ. Dibenzyl azelate; plasticizer. [C.P. Hall]

Plasthall® DIBA. CAS 141-04-8; Diisobutyl adipate; plasticizer. [C.P. Hall]

Plasthall® DIBZ. Diisobutyl azelate; plasticizer. [C.P. Hall]

Plasthall® DIDA. CAS 27178-16-1; Diisodecyl adipate; plasticizer. [C.P. Hall]

Plasthall® DIDG. Diisodecyl glutarate; plasticizer; lubricant additive. [C.P. Hall]

Plasthall® DIOA. Diisooctyl adipate; plasticizer. [C.P. Hall]

Plasthall® DIODD. Diisooctyl dodecanedioate; lubricant

additive; plasticizer. [C.P. Hall]

Plasthall® DOA. CAS 103-23-1; Dioctyl adipate; plasticizer. [C.P. Hall]

Plasthall® DODD. Dioctyl dodecanedioate dioate; lubricant additive; useful as textile surface finishes, softeners, thread lubricants and/or antistats; plasticizer. [C.P. Hall]

Plasthall® DOP. CAS 117-81-7; Di-2-ethylhexyl phthalate. [C.P. Hall]

Plasthall® DOS. Dioctyl sebacate; plasticizer. [C.P. Hall]

Plasthall® DOSS. Dioctyl sebacate substitute; plasticizer for rubber industry. [C.P. Hall]

Plasthall® DOZ. CAS 103-24-2; Dioctyl azelate; plasticizer. [C.P. Hall]

Plasthall® ESO. CAS 8013-07-8; EINECS 232-391-0; Epoxidized soybean oil; plasticizer. [C.P. Hall]

Plasthall® HA7A. Polyester adipate; plasticizer. [C.P. Hall]

Plasthall® NODA. n-Octyl, n-decyl adipate; plasticizer. [C.P. Hall]

Plasthall® P-550. Polyester glutarate; plasticizer for PVC applications; flexibilizing, permanent, nonmigrating; adhesive for film backing and varieties of tape. [C.P. Hall]

Plasthall® P-7035. Polyester glutarate; plasticizer for flexible PVC. [C.P. Hall]

Plasthall® P-7068. Polyester phthalate; plasticizer. [C.P. Hall]

Plasthall® R-9. Octyl tallate; transfer aid on correctable ribbon; penetration and tack agent for computer ribbons, carbon paper; plasticizer. [C.P. Hall]

Plasthall® TIOTM. CAS 53894-23-8; Triisooctyl trimellitate; plasticizer. [C.P. Hall]

Plasthall® TOTM. CAS 3319-31-1; Trioctyl trimellitate; plasticizer. [C.P. Hall]

Plastibase. Ointment base containing polyethylene and mineral oil. [Bristol-Myers Squibb Pharmaceuticals Ltd]

Plastic. Solvent dyestuffs; for coloration of plastics. [Holliday Dyes & Chemicals Ltd] †

Plastic A. CAS 614-33-5; A proprietary trade name for glyceryl tribenzoate. §

Plastic X. Tricresyl phosphate; a proprietary plasticizer with specific gravity from 1.177-1.18. §

Plasticalk. A proprietary trade name for a plastic resin used as an adhesive and filler. §

plastic bronze. An alloy of 64% copper, 30% lead, 5% tin, and 1% nickel.

Plasticede. A proprietary trade name for a plasticizer for clay; it contains tannins and lignin. §

Plasticizer 9. Glycerol ether alcohol; plasticizer for naphtha-resistance and benzene-resistance nitro lacquers. [BASF AG]

Plasticizer 13. The neutral ester of p-oxybenzoic acid and 2-ethyl hexanol; a plasticizer for polyamides. [BASF plc]

Plasticizer 28P. CAS 117-81-7; Dioctyl phthalate; a vinyl plasticizer. §

Plasticizer CEL. Phthalic acid glycol ester; plasticizer for cellulose acetate coatings, nitrocellulose lacquers; low volatility, good light stability; also as a desensitizer for peroxides. [Bayer AG]

Plasticizer E. A proprietary trade name for a chlorinated paraffin plasticizer. §

Plasticizer REO. Mixture of paraffinic/naphthenic process oils and petroleum sulfonates; plasticizer, process aid for synthetic and natural rubber compounds; peptizer; reclaiming agent; improves filler dispersion in dry compounds. [Akrochem]

---

* Trade name not verified as available   † Trade name verified as obsolete

Plastic Magnet. Thermosetting material; for molded products. [Sumitomo Bakelite Co Ltd] *

plastic metal. An alloy of 80.5% tin, 9.5% copper, 8.6% antimony, and 1.4% iron.

Plasticote. Vinyl/resin lacquer; used for decoration of vinyl articles. [W J Ruscoe Co] †

Plastic Plant Product. See plastic wood.

Plastic Steel. Paste containing 80% powdered steel and 20% epoxy resin capable of being hardened. [ITW Devcon]

plastic sulfur. Prepared by heating sulfur to 225 C; used to a limited extent as a material for preparing molds for electrotyping.

plastic wood. (plastic plant product). A material prepared by cooking vegetable matter with neutral salt solution followed by mechanical treatment to break down intercellular binding material. The name is also used to describe wood cellulose in solution with certain additional solvent such as ether or acetone; used for filling up holes in wood or other materials.

Plastifix. Polychloroprene. [Bayer AG]

Plastigel®. Chemical thickeners for polyester SMC, BMC, TMC. [Plasticolors]

Plastigen® G. Carbamide resin; very lightfast, non-hydrolyzable resin for nitrocellulose and chlorinated rubber finishes. [BASF AG]

Plastikon. A proprietary trade name for a rubber putty for filling and adhesive purposes. §

Plastilit® 3060. Polypropylene glycol alkylphenol ether; plasticizer for polymer dispersions in the paint, construction, chemical, adhesive and sealant industries. [BASF AG]

Plastisorb. Uv and weather-resistant colorants and additives for polymer systems. [Plasticolors]

Plastisperse®. Low dusting, high loaded, dry concs. for thermoplastic elastomers, bulk molding compounds. [Plasticolors]

Plastite. A name applied to a vulcanite.

Plastitoy. Vinyl/resin lacquer; used for decoration of vinyl toys. [W J Ruscoe Co] †

Plastitube. A proprietary trade name for cellulose acetate tubing. §

Plastodent. Dental wax. [Degussa Ltd]

Plastogen. A mixture of an oil soluble sulfonic acid of high molecular weight with a paraffin oil; an effective plasticizer in all elastomers; used in sponge and low durometer stocks. [King Industries]

Plastokyd. A range of alkyds for industrial paints and inks. [Croda Resins Ltd]

Plastokyd SC. A range of alkyds and polyesters modified with silicone for heat resistance and improved exterior exposure. [Croda Resins Ltd]

Plastol®. Low-visc. polymer plasticizers for PVC; for products which must have moderate resistance to chemicals. [BASF AG]

Plastolin I. A proprietary solvent that is benzyl acetate, $C_6H_5 \cdot CH_2 \cdot OOC \cdot CH_3$. §

Plastomag® 170. CAS 1309-48-4; Oil-dispersed magnesium oxide; chemical thickener for polyester resins; anticaking agent; used in synthetic rubber compounding, adhesives, fuel oil additives, and as acid acceptor for specialty plastics. [Morton Int'l./Plastics Additives]

Plastomenite. A German explosive powder made by incorporating 1 part nitro-lignin with 5 parts fused dinitrotoluene, and granulating. It may contain barium nitrate.

Plastomoll®. Adipic acid esters; for low-temp. resistance and light-stable nitro lacquers; plasticizers for plasticized PVC products with low brittle temps. [BASF AG; BASF plc]

Plastomoll® 34. An aliphatic acid ester mixture; vinyl plasticizer [BASF plc] *

Plastomoll® BMB. CAS 3622-84-2; N-Butylbenzene sulfonamide; a plasticizer for polyamides. [BASF plc] *

Plastomoll® DMA. Adipic ester of hydrogenated higher cyclic alcohols; vinyl plasticizer. [BASF plc] *

Plastomoll® DOA. CAS 103-23-1; Di-2-ethylhexyl adipate; PVC plasticizer. [BASF plc] *

Plastomoll® NA. CAS 33703-08-1; Diisononyl adipate; PVC plasticizer. [BASF plc] *

Plastomoll® TAH. Thiodibutyric acid ester of a synthetic octyl alcohol; a vinyl plasticizer. [BASF plc] *

Plastomoll® WH. Ester of an aliphatic dicarboxylic acid; vinyl plasticizer. [BASF plc] *

Plastone. Methylene diphenyldiamine mixed with a small amount of stearic acid or with naphthalene or naphthalene oil; a proprietary rubber vulcanization accelerator.§

Plastone A. A proprietary trade name for a phenolic molding compound containing cotton seed hull. §

Plastone B. A proprietary trade name for an inorganic and phenolic molding compound. §

Plastopal®. Urea-formaldehyde resins, butanol or methanol etherified; binders for coatings and paints. [BASF AG]

Plastopal® 11. A modified urea-formaldehyde condensation product in the form of a 65% butanol-white spirit solution. [BASF plc]

Plastoprene. A proprietary trade name for an isomerized rubber resin used for printing inks and chemical resistant coatings. [Croda Resins Ltd]

Plastorit. Natural coalescance of quartz, chlorite and mica; used for surface coatings including anticorrosive paints, textured and powder. [Mercian Minerals & Colours Ltd]‡

Plastose. Proprietary synthetic resin molding powders. §

Plastosol. An antiseptic liquid plaster containing copper, guaiacol-sulfonate, and penetrodine dissolved in a volatile organic solvent.

Plastosperse. A range of pigments dispersed in plasticisers for use in plastics. [Collinda Ltd] †

Plastosperse 40. 40% Carbon black dispersed in a plasticizer. [Collinda Ltd] †

Plastpak. Priming preparations and plasticizing additive. [ICI Chem. & Polymers Ltd]

Plastplate. A proprietary trade name for molded plastics plated with chromium, copper, gold, or nickel. §

Plastrotyl. An explosive. It is a plastic product prepared from trinitrotoluene, resin, collodion cotton, and crude dinitrotoluene. Sometimes larch turpentine is used.

Plastules with Folic Acid. A proprietary preparation of ferrous sulfate, folic acid, yeast and liver extract; a hematinic. [Wyeth Laboratories] *

Plastules with Liver. A proprietary preparation of ferrous sulfate, yeast and liver extract; a hematinic. [Wyeth Laboratories] *

Plastyrol. A proprietary trade name for acrylated, styrenated and vinylated alkyd resins for fast air drying paints. [Croda Resins Ltd]

Plastyrol E6X. A styrenated epoxide ester resin; a rapid air drying resin. [Croda Resins Ltd]

---

‡ Trade name and manufacturer not verified

§ Trade name without identified manufacturer

Plasvata®. Thrombolytic agent. [Asahi Chem. Industry]

Plasvita. Formaldehyde casein; tablet disintegration agent. [Dynamit Nobel Wien GmbH] *

Plasvita® TSM. Methylene casein; tablet disintegration agent; high capillary activity, low swelling effect. [Hüls AG]

Plaswite® LL 7014. 70% Rutile titanium dioxide in LLDPE carrier; wh. masterbatch for high quality LLDPE film pigmentation for applications requiring good dispersion, thermal and light stability, and which need to run at fast LLDPE extrusion rates; compatible with LDPE, LLDPE, HDPE, PP, EVA. [Cabot Plastics Ltd.]

Plaswite® LL 7105. 70% Titanium dioxide/lithopone blend in LLDPE carrier; wh. masterbatch for high quality LLDPE film pigmentation for applications requiring good dispersion, thermal stability, and which need to run at fast LLDPE extrusion rates; compatible with LDPE, LLDPE, HDPE, PP, EVA. [Cabot Plastics Ltd.]

Plaswite® masterbatches. Granulated concentrates of titanium dioxide in plastics; used for coloring of thermoplastic resins. [Cabot Plastics Ltd]

Plaswite® PS 7174. 50% Rutile titanium dioxide in crystal PS carrier, with additional toner; wh. masterbatch for manufacturing of various articles including thin-walled containers such as yogurt pots, disposable drinking cups, domestic appliances, toys, pkg. materials; compatible with PS. [Cabot Plastics Ltd.]

Platalargan. An alloy of aluminum and silver with some platinum. It is similar to Alargan, except that it contains platinum; used as a platinum substitute.

Platamid. Polyamides for extrusion and injection molding.

Platamid® Series. Hot melt adhesive produced from a copolyamide base; for extrusion of fusible film, monofilament, netting, web, and multifilament; for thin film fusible coatings; for application by disp. techniques used in textile industry; for fusion bonding of textiles, leather, wood, glass, and metals, and misc. applications including automotive interiors, belts, color concs., hats, hose construction, rainwear, shoes, television tubes, wire coating, etc. [Atochem] *

Plataril. Monofilament; for grass trimmer and fishing line. [Atochem Inc] *

plate pewter. An alloy of 90% tin, 6% antimony, and 2% each bismuth and copper.

plate powder. Bone ash of which calcium phosphate, $Ca_3(PO_4)_2$, forms 80% is sold as a non-mercurial plate powder under the name of white rouge.

plate sulfate. The double sulfate, $K_2SO_4 \cdot Na_2SO_4$, is called plate sulfate. It crystallizes from hot water, a flash of light accompanying the separation of each crystal.

platina. An alloy of 53.5% tin, and 46.5% copper.

platine. A brass containing 43% copper and 57% zinc.

Platine-autitre. A proprietary trade name for a platinum substitute containing 65-83% silver and platinum. §

Platinite. A proprietary trade name for a nickel steel containing 46% nickel, and 0.15% carbon; has a low coefficient of expansion, and can be sealed in glass. §

platinized asbestos. Loosely fiberd asbestos moistened with a concentrated solution of platinum chloride, dried, dipped into ammonium chloride solution, again dried, and brought to a red heat. It usually contains 8-8.5% platinum; and is used in the manufacture of sulfuric anhydride in the contact process for sulfuric acid.

Platino. An alloy of 11% platinum and 80% gold. It is resistant to fused potassium nitrate, and to alkalis.

platino-aceto-osmic acid. *See* Hermann's fluid.

platinoid. An alloy of 60% copper, 24% zinc, 2% tungsten, and 14% nickel; used as the material which connects filaments with outside wires of electric lamps. It has the same coefficient of expansion as glass.

Platinol. CAS 15663-27-1; Cisplatin; antineoplastic. [Bristol Laboratories] *

Platinor. An alloy of 2 parts platinum, 5 parts copper, 1 part silver, and 1 part nickel.

platinum. Pt; CAS 7440-06-4; EINECS 231-116-1; Metallic element; catalyst, laboratory ware, rayon and glass fiber manufacture, jewelry, dentistry, electrical contacts, thermocouples, surgical wire, bushings, electroplating, electric furnace windings, chemical reaction vessels, permanent magnets. [Aldrich; Degussa; Handy & Harman; Noah Chem.]

platinum black. CAS 7440-06-4; Finely divided platinum metal; used as a catalyst; to absorb hydrogen, oxygen.

platinum bronze. An alloy consisting of 90% nickel, 9% tin, and 1% platinum.

platinum gold. Variable alloy containing from 12-81% copper, 9-58% platinum, 0-70% gold, 0-37% silver, and 0-4% zinc.

platinum grey. *See* zinc grey.

platinum iridium. An alloy usually containing 90% platinum and 10% iridium; used in jewelry, electrical contacts, fuse wire, and hypodermic needles, and in other applications where high corrosion resistance is required.

platinum silver. An alloy containing 66.6% silver and 33.3% platinum.

platinum, soft. *See* soft platinum.

platinum solder. An alloy usually consisting of 73% silver and 27% platinum.

platinum substitute. An alloy of 72% nickel 23.6% aluminum, 3.7% bismuth, and 0.7% gold.

platinum yellow. A barium chloroplatinate or other alkaline chloroplatinate; used as a coating for fluorescent screens in X-ray work.

platnam. An alloy consisting of 56% nickel, 31% copper, 12% lead. 0.48% iron, and 0.32% aluminum.

platnik. An alloy of nickel and platinum. A platinum substitute.

Platol II. Spray-on coating for arc furnace electrodes. [Foseco (F.S.) Ltd] †

Platone. A peptone powder that is water soluble; used in the electroplating industry. [Am. Labs]

Plegine. CAS 50-58-8; Phendimetrazine tartrate; appetite suppressant. [Wyeth Laboratories]

Plesmet. A proprietary preparation of ferrous glycine sulfate and folic acid; used in the treatment of anemia of pregnancy. [Napp Laboratories Ltd] *

Plessite. A German explosive containing potassium chlorate and mineral oil. It is also the name for the mineral - Gersdorffite (NiAsS).

Plessy's Green. $Cr(PO_3)_3$, CAS 7789-04-0; Chromic phosphate, used as a pigment.

Plexar. Adhesive; for multilayer film and plate between polyolefins and polar plastics (polyamide, polyester) and metals. [DSM NV]

Plexar®. For polyolefin based extrudable adhesives; for use in multilayer barrier packaging, reactive polyolefin adhesives for use in boding polar plastics to nonpolar polyolefins, and modified polyolefin adhesive that form reaction bonds with polar plastics and metals. [Quantum Chemical Corp]

Plexar® PX 108. 9% EVA-based resin adhesive; tie-layer resin for bonding various substrates for barrier pkg. in coextruded bottles, films, and sheet. [Quantum/USI]

Plex-Hormone. A proprietary preparation of methyltestosterone, deoxycortone acetate, ethinyloestradiol and vitamin E; used to treat male androgen deficiency. [Consolidated Chemicals Ltd] *

Plexiglas® DR. Impact-modified acrylic resin; molding pellets for injection molding and extrusion; good long-term outdoor performance. [Rohm & Haas]

Plexiglas® HFI-10. Impact-modified acrylic resin; high impact molding pellets with improved injection molding char.; good long-term outdoor performance. [Rohm & Haas]

Plexiglas® V045. Acrylic resin; injection molding and extrusion pellets with high heat resistance, environmental stability; transparent, translucent, and opaque pellets. [Rohm & Haas]

Plexiglas® VS. Acrylic resin; max. flow, lowest heat resistance for applications where ease of injection molding is the major factor. [Rohm & Haas]

Plexiglo. A proprietary trade name for a polish and cleaner for transparent plastics. §

Plexigum. Pure acrylic resins; for surface coatings, inks, firing lacquers, heat seal lacquers, road marking paints, cementitious coatings. [Cornelius Chemical Co Ltd] *

Pleximon. Polymeric/oligomeric methacrylate based resins; thickeners, dental etc. [Cornelius Chemical Co Ltd] *

Plexisol. Acrylic resin solution; for paints, inks etc. [Cornelius Chemical Co Ltd] *

Plexol. Oil additives. [Rohm & Haas UK]

Plexophor. Textile sequestering agents. [Sandoz Products Ltd]

Plextol. Pure acrylic resin emulsions; for paints, concrete add mixtures, textile finishes, leather dressings, wood coatings, etc. [Cornelius Chemical Co Ltd] *

Plialite. A proprietary product that is rubber resin. §

Plictran. CAS 13121-70-5; A line of miticides based primarily on cyhexatin. [Dow UK]

Plictran. CAS 13121-70-5; Acaracide containing cyhexatin; (Sold in UK on behalf of Dow Chemical Co). [ICI Chem. & Polymers Ltd]

Plimmer and Paine's stain. A microscopic stain. It contains 10 g tannic acid, 18 g aluminum chloride, 18 g zinc chloride, 1.5 g rosaniline hydrochloride, and 40 cc 60% alcohol.

Plimmer's salt. Sodium antimony tartrate.

plinthite. A red clay found in Ireland.

Pliobond®. A broad line of solvent and water borne adhesives based on various polymer systems. Includes contact adhesives, heat reactive systems, pressure sensitives and elastomeric sealants; used for bonding forms, metals, plastics, wood products, fiberglass, rubber in various construction, consumer, industrial and automotive applications. [Ashland Chemical Company]*

Pliobond Adhesives. Rubber/solvent base adhesives; adhesive for bonding and sealing a variety of substrates (metal, rubber, wood, glass, ceramic, leather, cork, canvas, fiberglass and aluminum). [W J Ruscoe Co] *

Plio-Caulk. Butyl and acrylic caulking compounds; supplied 1/10 gallon cartridges; used to caulk and seal windows, doors, siding, stone and masonry. For interior and exterior use. [Ashland Chemical Company] *

Pliocord® LVP-4668. Vinyl pyridine-styrene-butadiene latex; used for cord adhesion in tires, conveyor belts, hose, etc.; exhibits excellent adhesion between fabric and rubber; used alone or with resorcinol-formaldehyde dip formulations. [Goodyear] *

Plioflex. A proprietary trade name for polyvinyl chloride. §

Plioflex® 1006. SBR, hot polymer, FA emulsifier, nonstaining stabilizer; general purpose grade. [Goodyear] *

Plioflex® 1028. SBR, FA emulsifier; food grade. [Goodyear]*

Plioflex® 1905. SBR resin masterbatch (80:20 SBR/high styrene resin), FA emulsifier, nonstaining stabilizer; used in sponge products, rug underlay, shoe soles, household goods; gives high styrene reinforcement; vulcanized with standard curing systems including sulfur and peroxide. [Goodyear] *

Plioform. A proprietary type of rubber plastic obtained by the action of halogenated acids on rubber. §

Pliogrip®. Industrial and structural adhesives. [Ashland]

Pliolite. Modified isomerized rubber, rubber derivatives and rubber-like resins. [Rhone-Poulenc UK]

Pliolite® 7103, 7104. Carboxylated styrene-acrylic aq. disp.; designed to replace solv.-borne resins in demanding specialty and OEM coating applications. [Goodyear] *

Pliolite® AC-80. Styrene/acrylate emulsion copolymer; high uv resistance; for high build maintenance coatings and architectural paints. [Goodyear] *

Pliolite® LPF-2108. SBR latex; for spread foam on carpet or fabric, molded foam for automotive and home furnishings, binders for mastic and adhesive formulations, asphalt modifier for construction. [Goodyear] *

Pliolite® S-6B, S-6F. CAS 9003-55-8; S/B polymer; reinforcers to increase hardness, stiffness, and abrasion resistance. [Goodyear] *

Pliolite® VT. Vinyl-toluene/butadiene emulsion copolymer; produces films with clarity, strength, hardness and chem. resistance; for traffic paints, block fillers, paper coatings, adhesives, laminates, printing inks, hot melts, abrasion-resistance coatings. [Goodyear] *

Pliolite® VTAC. Vinyl-toluene/acrylate emulsion copolymer; for texture coatings, interior and exterior masonry paints, intumescent fire-retardant paints. [Goodyear] *

Plio-Nail. High performance synthetic rubber-based adhesives; certified by PFS Corporation to meet the American Plywood Associations AFG-ol specification; used for joints, sub floors, siding, decorative panels, gypsum board, fixtures and a wide variety of materials. [Ashland Chemical Company] *

Plio-Seam. Elastomeric rubber sealant; provides a strong and durable seal for aluminum, steel, glass, wood, masonry, ceramics, fiberglass joints; remains flexible and tough; may be painted; available in clear, white, aluminum, black and architectural bronze; used to install or repair downspouts, rain troughs, roof flashings, shower stalls and metal or wood structures; weather seals glass in metal or wood storm windows and doors. [Ashland Chemical Company] *

Plio-Tac. Neoprene contact adhesive supplied in aerosol spray cans; will not cavitate polystyrene foams; also used to bond wood, carpet, vinyl fabric, metal, rubber and many other materials; used for bonding insulation to various surfaces, craft assembly projects, rebonding carpet or vinyl flooring, counter or table top laminates,

---

‡ Trade name and manufacturer not verified   § Trade name without identified manufacturer

mounting signs and photos and attaching labels to various surfaces. [Ashland Chemical Company] *

Plio-Tac 38. Contact cement designed for DIY; meets CPSA guidelines; for bonding decorative laminates to wood and metal surfaces. Also bonds leather, fabrics, unglazed ceramics, wall boards, cove base and carpets to themselves and each other. [Ashland Chemical Company] *

Pliovic® DR-450, DR-453, DR-454, DR-600, DR-602, DR-652. Vinyl disp. resin; developed for plastisol and organosol applications. [Goodyear] *

Pliovic® M-50, M-70, M-70SC, M-90. Vinyl homopolymer resin; blending resin for modifying plastisol compounds resulting in lower visc., improved visc. stability, increased hardness, minimized surface gloss. [Goodyear]*

Pliovic® WO-1, WO-2, WO-3, WO-S. Vinyl homopolymer disp. resin; used in plastisol and organosol compounds for automotive air and oil filters, dipped goods including gloves, coatings for awnings, rainwear, wall coverings, laminating plastisols for fabric-to-fabric adhesion, foams for garments, upholstery, and carpet backings, weatherstripping, rootcast and slush molded toys, boots, and novelties. [Goodyear] *

Plioway® EC1. Acrylic resin; film-forming resin for odorless interior paints, exterior masonry and concrete coatings. [Goodyear] *

Plitex. A proprietary trade name for a wood and phenolic resin. §

Plombit. A German acid-resisting material made from hard rubber, oleic acid, sulfuric acid, and sulfur.

Plondrel. CAS 5131-24-8; Fungicide containing ditalimfos; for the control of powdery mildew and scab. (Sold in UK on behalf of Dow Chemical Co). [ICI Chem. & Polymers Ltd]

plumbago. See graphite.

plumbago grease. A mixture of plumbago and tallow; used for lubricating.

plumber's solder. Usually a mixture of lead and tin, sometimes with a little antimony. Coarse solder contains 75% lead and 25% tin, and melts at 250 C. Ordinary solder (slicker solder) usually consists of 67% lead and 33% tin, and melts at 227 C. Plumber's fine or soft solder, contains 50% lead and 50% tine and melts at 188 C.

plumber's white alloy. An alloy of from 54-58% copper, 25-27% zinc, 13-17% nickel, 1-7% lead, and sometimes 1% tin and 1% iron.

plumbocalcite. See tarnowitzite.

plumboxan. A compound or solid solution of sodium manganate and sodium metaplumbate, of the composition, $Na_2MnO_4 \cdot Na_2PbO_3$. It gives up oxygen when treated with steam.

Plumbral. Copper/lead alloys flux. [Foseco (F.S.) Ltd]

Plumbrex. Flux for lead and alloys. [Foseco (F.S.) Ltd]

Plumbrit. Cleansing flux for lead alloys. [Foseco (F.S.) Ltd]†

plumose mica. A variety of muscovite.

Pluracol® 220. Polyether polyol; specialty polyol for low modulus foams and blown elastomers. [BASF] *

Pluracol® 355. Amine-based polyol; cross-linking agent for semiflexible urethane foams, coatings, adhesives, and polymers. [BASF] *

Pluracol® 450. POP deriv. of pentaerythritol; crosslinking agent for rigid urethane foams. [BASF] *

Pluracol® 581. Polymer reinforced (graft polyol) polyether;

high-resilience molding polyol for high firmness and strength in flexible foam applications; used in automotive molded, carpet underlay, semiflexible applications. [BASF] *

Pluracol® E200. CAS 25322-68-3; PEG-4; intermediate for preparation of nonionic surfactants; binder, base, coating, stabilizer, solv., vehicle, extender, and coupling agent for pharmaceutical, cosmetic, and toiletries; lubricant for metal applications, rubber industry; wood treatment; textile conditioning, antistat, and sizing agent, softener. [BASF]

Pluracol® E300. CAS 25322-68-3; PEG-6; see Pluracol E200; also dispersant in food tablets and preparations; plasticizer. [BASF]

Pluracol® E400. CAS 25322-68-3; PEG-8; see Pluracol E300. [BASF]

Pluracol® E400 NF. CAS 25322-68-3; PEG-8; chemical intermediate, base, coupler, thickener, lubricant, mold release agent, defoamer, softener, conditioner, antistat, sizing agent, dispersant for pharmaceutical, cosmetic, and oral care preparations, in metal polishing and cleaning formulations, rubber products, paper and wood products, textile processing, ink formulations. [BASF]

Pluracol® E600 NF. CAS 25322-68-3; PEG-12; see Pluracol E400 NF. [BASF]

Pluracol® E1000. CAS 25322-68-3; PEG-20; see Pluracol E400 NF. [BASF]

Pluracol® E1500. CAS 25322-68-3; PEG-6-32; see Pluracol E300. [BASF] *

Pluracol® E2000. CAS 25322-68-3; PEG-40; see Pluracol E400 NF. [BASF]

Pluracol® E4000. CAS 25322-68-3; PEG-75; see Pluracol E300. [BASF]

Pluracol® E4500. PEG; see Pluracol E400 NF. [BASF]

Pluracol® E6000. CAS 25322-68-3; PEG-150; see Pluracol E200. [BASF] *

Pluracol® E8000. CAS 25322-68-3; PEG; see Pluracol E400 NF. [BASF]

Pluracol® V-10. Polyoxyalkylene glycol polyol; thickening agent to control the visc. of water-glycol type, fire-resistant hydraulic fluids; resistance to shearing stresses and will not hydrolyze or degrade under use conditions; noncarbonizing and nongumming at high temps. [BASF]

Pluracol® W170. CAS 74623-31-7; PPG-5-buteth-7; component in demulsifying and wetting formulations; brake and metalworking fluids; rubber and fiber lubricant; textile applications; defoamer for hot and cold applications, food and chemical processing; cosmetic formulations. [BASF] *

Pluracol® W660. CAS 74623-31-7; PPG-12-buteth-16; see Pluracol W170. [BASF] *

Pluracol® W2000. CAS 74623-31-7; PPG-20-buteth-30; see Pluracol W170. [BASF] *

Pluracol® W3520N. CAS 9038-95-3; PPG; see Pluracol W170. [BASF]

Pluracol® W5100N. CAS 74623-31-7; PPG-33-buteth-45; see Pluracol W170. [BASF]

Plurafac®. Low foam nonionic surfactants. [BASF plc]

Plurafac® A-38. CAS 68439-49-6; Ceteareth-27; detergent, dispersant, wetting agent, emulsifier for heavy-duty detergents, metal cleaners, detergent tablets, electrolytic cleaning; biodeg. [BASF]

Plurafac® A-39. CAS 68439-49-6; Ceteareth-55; detergent, dispersant, wetting agent, emulsifier for heavy-duty de-

---

* Trade name not verified as available   † Trade name verified as obsolete

tergents, metal cleaners, detergent tablets, electrolytic cleaning; biodeg. [BASF]

Plurafac® B-25-5. Straight chain primary aliphatic oxyalkylated alcohol; detergent, dispersant, wetting agent, emulsifier for heavy-duty detergents, all-purpose liqs., detergent tablets, sanitizers, metal cleaners; biodeg. [BASF]

Plurafac® D-25. PPG-6 C12-18 pareth-11; detergent, dispersant, wetting agent, emulsifier, deduster for heavy-duty detergents, all-purpose liqs., detergent tablets, sanitizers, metal cleaners, hard surface cleaners; biodeg. [BASF]

Plurafac® LF 120. Alkoxylated fatty alcohol; surfactant for acid and alkaline low foaming cleaners. [BASF AG]

Plurafac® LF 220. Alkoxylated straight chain alcohol; low foaming surfactant for dishwashing powds., rinse aids. [BASF AG]

Plurafac® RA-20. Straight chain primary aliphatic oxyalkylated alcohol; detergent, dispersant, wetting agent, emulsifier, defoamer, deduster used in rinse aids and dishwashing products [BASF]

Pluraflo® E4A E5G, N5G. Formulated product; dispersant, wetting agent for pesticides. [BASF]

Plurasafe. Glycol hydraulic fluid. [BASF]

Pluriol®. Polyglycols. [BASF plc]

Pluriol® E 200. PEG; solubilizer, impregnating agent, humectant, mold release agent; flow improver, thermal and hydraulic fluid, organic intermediate; detergent and cleaner; dye and pigment dispersant; inks; textile and coatings industry; coloring ceramics; softener in paper industry; plasticizer in adhesives industry and in production of cellulose film; ceramics and metalworking lubricant. [BASF AG] *

Pluriol® E 4000. PEG; solubilizer, humectant; binder and hardener in personal care products; dispersing dyes and pigments; inks; textile and coating industry; coloring ceramics; paper industry softener; plasticizer in adhesives industry and production of cellulose film. [BASF AG] *

Pluriol® P 600. PPG; mold release agent, additive for oils and fluids; lubricant and antifoam for rubber; consistency improver and solubilizer; intermediate in industrial applications [BASF AG] *

Pluriol® P 2000. PPG; mold release agent, additive for oils and fluids; lubricant and antifoam for rubber; consistency improver and solubilizer; intermediate in industrial applications. [BASF AG] *

Plurivite. Vitamin preparations. [The Boots Co plc] *

Plurol Isostearique. Polyglyceryl-6 isostearate; emulsifier; cosurfactant for microemulsions. [Gattefosse SA]

Plurol Oleique WL 1173. Polyglyceryl-6 dioleate; emulsifier; cosurfactant for microemulsions. [Gattefosse SA]

Plurol Stearique WL 1009. Polyglyceryl-6 distearate; consistency agent and stabilizer for heated o/w emulsions; food emulsifier. [Gattefosse SA]

Pluronic® 10R5. CAS 9003-11-6; Meroxapol 105; emulsifier, wetting agent, binder, stabilizer, plasticizer, lubricant, solubilizer, dispersant, visc. control agent, defoamer, intermediate for hard surface detergents, rinse aids, automatic dishwashing, textile processing; cosmetics; pharmaceuticals, pulp, paper, and petrol. industries, agric. products, in iodophors, water treating systems, fermentation, cutting and grinding fluids. [BASF]

Pluronic® 10R8. CAS 9003-11-6; Meroxapol 108; see

Pluronic 10R5. [BASF]

Pluronic® 17R1. CAS 9003-11-6; Meroxapol 171; see Pluronic 10R5; also for foam control in paper sizing operations. [BASF]

Pluronic® 17R2. CAS 9003-11-6; Meroxapol 172; see Pluronic 10R5. [BASF]

Pluronic® 17R4. CAS 9003-11-6; Meroxapol 174; see Pluronic 10R5. [BASF]

Pluronic® 17R8. CAS 9003-11-6; Meroxapol 178; see Pluronic 10R5; also dry toilet bowl cleaners, dye levelers, solubilizer of drugs, stick type cosmetics; soap bars, dispersant in delnking operations. [BASF]

Pluronic® 25R1. CAS 9003-11-6; Meroxapol 251; see Pluronic 10R5; also foam control in paper sizing operations and antifreeze. [BASF]

Pluronic® 25R2. CAS 9003-11-6; Meroxapol 252; see Pluronic 10R5; also wetting and rinse aid; lubricant and leveling agent for paper coating. [BASF]

Pluronic® 25R4. CAS 9003-11-6; Meroxapol 254; see Pluronic 10R5. [BASF]

Pluronic® 25R5. CAS 9003-11-6; Meroxapol 255; see Pluronic 10R5; also thickener for cosmetic pastes and creams. [BASF]

Pluronic® 25R8. CAS 9003-11-6; Meroxapol 258; see Pluronic 10R5; also dry toilet bowl cleaners, dye levelers for fabrics; solubilizer for drugs; thickener for cosmetics; deinking operations; felt washing operations. [BASF]

Pluronic® 31R1. CAS 9003-11-6; Meroxapol 311; see Pluronic 10R5; also floating bath oils; foam control in antifreeze. [BASF]

Pluronic® 31R2. CAS 9003-11-6; Meroxapol 312; see Pluronic 10R5; paper coating color additive; deinking and felt washing operations. [BASF]

Pluronic® 31R4. CAS 9003-11-6; Meroxapol 314; see Pluronic 10R5. [BASF]

Pluronic® F38. CAS 9003-11-6; Poloxamer 108; wetting agent, emulsifier, demulsifier, foam and visc. control agent, dispersant, antistat, gelling agent, dyeing assistant, leveler, lubricant for agric., cosmetics, pharmaceuticals, metal cleaning, pulp/paper, textile scouring, water treatment. [BASF] *

Pluronic® F68. CAS 9003-11-6; Poloxamer 188; see Pluronic® F38. [BASF]

Pluronic® F68LF. CAS 9003-11-6; Poloxamer 108; see Pluronic® F38. [BASF]

Pluronic® F77. CAS 9003-11-6; Poloxamer 217; see Pluronic® F38 [BASF]

Pluronic® F87. CAS 9003-11-6; Poloxamer 237; see Pluronic® F38. [BASF] *

Pluronic® F88. CAS 9003-11-6; Poloxamer 238; see Pluronic® F38 [BASF]

Pluronic® F98. CAS 9003-11-6; Poloxamer 288; see Pluronic® F38 [BASF]

Pluronic® F108. CAS 9003-11-6; Poloxamer 338; see Pluronic F38. [BASF]

Pluronic® F127. CAS 9003-11-6; Poloxamer 407; see Pluronic F38. [BASF]

Pluronic® L31. CAS 9003-11-6; Poloxamer 101; see Pluronic® F38 [BASF]

Pluronic® L35. CAS 9003-11-6; Poloxamer 105; see Pluronic® F38 [BASF]

Pluronic® L42. CAS 9003-11-6; Poloxamer 122; see Pluronic® F38 [BASF]

Pluronic® L43. CAS 9003-11-6; Poloxamer 123; see

---

‡ Trade name and manufacturer not verified

§ Trade name without identified manufacturer

Pluronic® F38 [BASF]

Pluronic® L44. CAS 9003-11-6; Poloxamer 124; see Pluronic® F38 [BASF]

Pluronic® L61. CAS 9003-11-6; Poloxamer 181; see Pluronic® F38 [BASF]

Pluronic® L62. CAS 9003-11-6; Poloxamer 182; see Pluronic® F38 [BASF]

Pluronic® L62D. CAS 9003-11-6; Poloxamer 108; see Pluronic® F38 [BASF]

Pluronic® L62LF. CAS 9003-11-6; Poloxamer 108; see Pluronic® F38 [BASF]

Pluronic® L63. CAS 9003-11-6; Poloxamer 183; see Pluronic® F38 [BASF]

Pluronic® L64. CAS 9003-11-6; Poloxamer 184; see Pluronic® F38 [BASF]

Pluronic® L72. CAS 9003-11-6; Poloxamer 212; see Pluronic® F38 [BASF]

Pluronic® L81. CAS 9003-11-6; Poloxamer 231; see Pluronic® F38 [BASF]

Pluronic® L92. CAS 9003-11-6; Poloxamer 282; see Pluronic® F38 [BASF]

Pluronic® L101. CAS 9003-11-6; Poloxamer 331; see Pluronic® F38 [BASF]

Pluronic® L121. CAS 9003-11-6; Poloxamer 401; see Pluronic® F38 [BASF]

Pluronic® L122. CAS 9003-11-6; Poloxamer 402; see Pluronic® F38 [BASF]

Pluronic® P65. CAS 9003-11-6; Poloxamer 185; see Pluronic® F38 [BASF]

Pluronic® P75. CAS 9003-11-6; Poloxamer 215; see Pluronic® F38 [BASF]

Pluronic® P84. CAS 9003-11-6; Poloxamer 234; see Pluronic® F38 [BASF]

Pluronic® P85. CAS 9003-11-6; Poloxamer 235; see Pluronic® F38 [BASF]

Pluronic® P94. CAS 9003-11-6; Poloxamer 284; see Pluronic® F38 [BASF] *

Pluronic® P103. CAS 9003-11-6; Poloxamer 333; see Pluronic® F38 [BASF]

Pluronic® P104. CAS 9003-11-6; Poloxamer 334; see Pluronic® F38 [BASF]

Pluronic® P105. CAS 9003-11-6; Poloxamer 335; see Pluronic® F38 [BASF]

Pluronic® P123. CAS 9003-11-6; Poloxamer 403; see Pluronic® F38 [BASF]

Pluronic® PE 3100. PO/EO block polymer; for defoaming and controlled foam detergents. [BASF AG]

Plus. Range of fertilizers for garden use. [ICI Chem. & Polymers Ltd]

Plusbrite. A chemical for bright nickel plating. [Albright & Wilson Ltd] *

Plus-Gas C. Noncorrosive cutting fluids. [Foseco (F.S.) Ltd]†

Plus-Pac®. Palladium-based product; for recovery of platinum lost during the catalytic oxidation of ammonia to make nitric acid. [Johnson Matthey plc]

Pluviusin. A proprietary trade name for a synthetic resin of the urea type. §

Plyamul®. Polyvinyl acetate emulsions, adhesives. [Reichhold]

Plyamul® 40305-00. CAS 108-05-4; Vinyl acetate emulsion copolymer; adhesive base for high-speed applications. [Reichhold/Emulsion Polymers] *

Plymul 98-759. A range of polyvinyl acetate thermosetting

emulsions used for bonding cellulose to cellulose. [Reichhold] *

Plyophen. A proprietary trade name for a phenolic laminating resin and varnish. §

Plyothene. A proprietary phenolic molding material. [Reichhold] *

Ply·Pro 25. Tissue-converting adhesive. [Aqualon]

Plyron. Continuous fiber-reinforced thermoplastics. [ICI Chem. & Polymers Ltd]

Plysolene. Modified polyisobutylene sheet; waterproof membrane for insulation. [Plysolene Ltd]

Plytron. Semi-finished plastic. [ICI Chem. & Polymers Ltd]

PMA. See phosphomolybdic acid.

PMA 18, 60. CAS 62-38-4; Solubilized form and powdered form, respectively, of phenylmercuric acetate; preservative and fungicide for aqueous paints. [Hüls Am.] †

PMAC. See phenylmercuric acetate.

PMAS. Mercurial fungicide for prevention and control of pink and gray snow mold. [W A Cleary]

P.M.G. Metal. A proprietary trade name for an alloy of copper with 3-4% silicon, 2% iron, and 2% zinc. §

P.M.T. Alloy. A proprietary alloy made as a substitute for Admiralty gun metal; contains 88% copper, 2% zinc, and 10% silicon, manganese, and iron. §

Pneulec Core Gum. A proprietary product; it is a linseed oil and wood extract material, and is used as a binder for the sand for cores in metal casting. §

pneumatogen. A mixture of the peroxides of potassium and sodium.

Pneumovax. Pneumococcal vaccine, polyvalent; effective protection against certain pneumococcal infections. [Merck & Co Inc]

P.N.P. p-Nitro-phenol; used as fungicide in the rubber industry.

Poast®. CAS 74051-80-2; Sethoxydim; post-emergence graminicide against annual and perennial grasses. [BASF AG]

Pocan®. Thermoplastic polyester based on polybutylene terephthalate; for the manufacture of injection moldings with brilliant surface finish; uses include household appliances, office machines and electrical components. [Bayer AG; Bayer plc; Miles]

Pocan® B 1300. CAS 26062-94-2; PBT polyester; FDA-grade with med. visc. [Bayer AG]

Pocan® B 1305. CAS 26062-94-2; PBT polyester; heat-stabilized engineering plastic, easy mold release, med. visc. [Bayer AG] *

Pocan® B 1505. CAS 26062-94-2; PBT polyester; engineering thermoplastic, high heat deflection temps., stiffness and hardness, min. surface friction and cold flow, abrasion resistance, low moisture absorp., good dynamic fatigue strength, resistance to chemicals; and performance in fire tests; flow and fast cycling. [Bayer AG] *

Pocan® KU1-7033. CAS 26062-94-2; PBT polyester, 30% glass-reinforced, impact-modified; general-purpose grade; see Pocan B 1501. [Bayer AG] *

Pocan® S 1506. CAS 26062-94-2; PBT polyester, elastomer modified; high impact even at low temps. [Bayer AG] *

podophyllum. (devil's apple; Indian apple; mandrake root) The dried rhizome of Podophyllum peltatum; used in medicine.

POE. See PEG.

Poilite. A trade name for an asbestos cement product used

---

for building work. §

poison flour. Arsenious oxide, AS_4O_6.

Poivrette. The ground stones of olive fruit; used as an adulterant and toning agent in spices.

Pokalon. Polycarbonate film. [Lonza AG]

Polacure® 740M. Diamine curative; curative for high-performance elastomers. [Air Prods./Polyurethanes] *

Polamine® 250. PTMEG-diamine; curative for high-performance room temperature cast and cured elastomers for cast prototypes, coatings, adhesives, sealants, and spray systems; epoxy flexibilizer. [Air Prods./Polyurethanes] *

Polamine® 650, 1000, 2000. CAS 54667-43-5; Polytetramethyleneoxide-di p-aminobenzoate; curative for high-performance room temperature cast and cured elastomers for cast prototypes, coatings, adhesives, sealants, and spray systems; epoxy flexibilizer. [Air Prods./Polyurethanes] *

Polaqua. Water based primer/adhesive with low solids and containing no organic solvents; for extrusion and lamination of film to film, paper and foil adhesion promoter to films, paper and foil. [ADM Tronics Unlimited Inc]

Polar. Acid dyes for wool. [Ciba plc] *

Polaramine. Dexchlorpheniramine maleate; antihistaminic. [Schering Corp]

Polar Dynobel. (Polar Monobel No. 2; Polar Rex; Polar Saxonite; Polar Stomonal; Polar Super Clifite; Polar Thames Powder; Polar Viking). Proprietary low freezing explosives containing a mixture of nitrated glycerin and polyglycerin or glycerin and ethylene glycol, ammonium nitrate, sodium chloride, wood meal, etc. §

Polargel® NF. CAS 1302-78-9; Purified wh. bentonite USP/NF; thickener and suspending agent for cosmetics and pharmaceuticals. [Am. Colloid]

Polargel® HV. CAS 1302-78-9; Bentonite USP/NF; high visc. wh. montmorilonite used as disintegrant, binder, suspension agent and thickener. [Am. Colloid]

Polarin® Range. Glycerol and derivs.; solvs., humectants for food industry. [Grünau]

Polaris. Boiler and cooling water treatment. [Laporte Industries Ltd]

Polarite®. Surface-treated metakaolinitic aluminum silicate produced by the calcination of kaolin clay; surface-treating agents include substituted silanes and elastomers; functional fillers for rubbers and plastics. [ECC International Ltd]

Polarite® 420E(W). Low-profile additive for DMC and BMC applications; for low shrink compounds. [ECC]

Polarite® 420G(W). Low-profile additive for DMC and BMC applications; for zero shrink compounds. [ECC]

Polarite® 880E(W). Aluminum trihydrate; low profile flame retardant/smoke suppressant additive for BMC moldings. [ECC]

Polarwhite. A headless white paint. [J C Bottomley] *

Polathane. Polyester and polyether toluene diisothiocyanate prepolymers. [Air Prods. & Chems. Inc]

Polathane STE-73D, STE-83A, STE-90A, STE-95A. TDI-PTMEG; polyurethane prepolymer [Air Prods./Polyurethanes] *

Polathane STS-55. TDI-ester; polyurethane prepolymer [Air Prods./Polyurethanes] *

Polathane XPE-10, XPE-20, XPE-30. TDI-PTMEG; polyurethane prepolymer [Air Prods./Polyurethanes] *

Polawax®. Emulsifying wax NF; emulsifier, thickener, opacifier, suspending agent; stabilizer for o/w emulsions; for cosmetics, pharmaceuticals, hair straighteners, moisturizers, nail preparations, sunscreens, antibiotic creams and lotions, acne preparations, analgesic rubs. [Croda Inc.; Croda Surface Ltd.]

Polawax® A31. Emulsifying wax NF; emulsifier used in quick-breaking foams and mousses, cosmetics, pharmaceuticals. [Croda Inc.; Croda Surf. Ltd.]

Polectron 430. Vinylpyrrolidone/styrene copolymer; a stable opacifier especially used in high pH systems, e.g.,detergents and cold wave lotions; in surface coatings, provides adhesion, film toughness, color receptivity and noncorrosiveness. [ISP]

Poley oil. See oil of Pennyroyal.

Polidene®. Vinylidene chloride copolymer emulsions; used for paints, binders and membrane coatings. [Scott Bader]

Polidene® 33-001. VdC copolymer; paper coating; binder or coating for bonding of glass nonwoven. [Scott Bader]

Polidene® 33-004. VdC copolymer; fire retardant coatings, adhesives, and fiber impregnants; pigment binder. [Scott Bader]

Polidene® 33-055. Carboxylated VdC copolymer emulsion; highly resistance rustproof aq. primers for steel. [Scott Bader]

Polidene 528F. A proprietary polyvinlyidene chloride emulsion. [A E Staley Manufacturing Co] *

Polifil® C-10. CAS 9003-07-0; PP homopolymer, 10% calcium carbonate-reinforced; features high impact, stiffness, heat agint, good colorability, ESCR, low mold shrinkage; for appliances, elec. components, housewares, toys, automotive and utility products [Ralco Industries]

Polifil® CAS-40. CAS 9003-07-0; PP homopolymer, 40% calcium sulfate-reinforced; features higher whiteness level, reduced mold shrinkage, easier colorability and processability. [Ralco Industries]

Polifil® GFPP-10. CAS 9003-07-0; PP homopolymer,10% glass fiber-reinforced; provides high impact, stiffness, hardness, higher continuous use temp.; for appliances, elec. components, automotive and utility products [Ralco Industries]

Polifil® GFPPCC-10. CAS 9003-07-0; PP homopolymer,10% chemically coupled glass fiber-reinforced; offer superior strength, stiffness, improved high temp. performance, higher impact strength; for chem. resistance applications, appliances, elec. components, automotive, ignition and utility products [Ralco Industries]

Polifil® M-20. CAS 9003-07-0; PP homopolymer, 20% mica-reinforced; used where high modulus values are required; improved high stiffness values and low mold shrinkage. [Ralco Industries]

Polifil® RMC-10. CAS 9003-07-0; High impact PP, 10% calcium carbonate-reinforced; provides highest impact, and good stiffness, heat aging resistance, solv. resistance, ESCR, and surface quality; for automotive, appliances, elec. components, housewares, utility products [Ralco Industries]

Polifil® RMT-10. CAS 9003-07-0; High impact PP, 10% talc-reinforced; provides high impact strength, flex. modulus, stiffness and heat deflection resistance; used in automotive applications. [Ralco Industries]

Polifil® T-10. CAS 9003-07-0; PP homopolymer, 10% talc-reinforced; features high flex. modulus, stiffness and

---

‡ Trade name and manufacturer not verified   § Trade name without identified manufacturer

deflection temps, max. stiffness, low shrinkage, good colorability; used in underhood automotive, major appliances, elec. goods, housewares, utility products [Ralco Industries]

Poligen. Polymer emulsions dispersions. [BASF plc]

Poligen MMV. A proprietary aqueous dispersion of styrene/acrylic copolymers. [BASF plc] *

Poligen PE. CAS 9002-88-4; Polyethylene aq. disp.; dry-bright floor polish emulsions, release coats. [BASF AG; BASF plc]

Poligen WE 1. High molecular weight polyethylene wax aq. secondary emulsion; dry-bright floor polish emulsions. [BASF AG] *

Polimex TR. High performance expanded polymer foam core material; dimensionally stable at elevated temps.; for aircraft interior panels, antennas and radome structures, insulated freight containers, fuel tank insulation, sporting goods, marine structures, tooling. [Polimex]

Poliomyelitis Vaccine. Attenuated live poliomyelitis vaccine; immunization against poliomyelitis. [SmithKline Beecham] *

Polisax. Polyethylene sacks. [ICI Chem. & Polymers Ltd] †

polishing oil. A fraction of petroleum oil, having a boiling-point of 130-160 C., and a specific gravity of 0.74-0.77; used as a turpentine substitute.

Polish turpentine. *See* turpentine.

Politarp. Polyethylene sheets. [ICI Chem. & Polymers Ltd]

Politec. Cold box resin sand process. [Foseco (F.S.) Ltd]

Politint. Proprietary dyes for plastics. No. 1 for methylmethacrylate. No. 2 for cellulose acetate, cellulose acetate butyrate, and ethyl cellulose. §

Politol®. Sodium lignate; protein coagulant in purification of fats and oils. [Westvaco]

Pollack's cement. A cement glycerin, litharge, and red lead.

Pollinex. Hay fever pollen vaccine. [SmithKline Beecham]

Pollopas. Urea resin molding compounds; used for electrical engineering, sealing caps and sanitary equipments. [Dynamit Nobel Wien GmbH] *

Polnac. A range of polyester resins; for GFRP building coverings, buttons, electrical industry, various applications in GFRP, silos for fodder, varnishes for wood, pieces for industrial coachwork, prefab for the building industry, tanktrucks for raw material stockage; marble aggregates for floorings and decoration for furniture etc. [AMC SPREA S.p.A.]

Polocaine. CAS 1722-62-9; Mepivacaine hydrochloride; anesthetic. [Astra Pharmaceutical Products Inc]

Polomyx. Modified acrylate based coarse particle size suspension; architectural coating that is spray applied to gypsum board and other substrates and produces a textured, tone-on-tone or multi toned, seamless wallcovering. [Polomyx Industries Inc] *

Poloxalkol. A polymer of ethylene oxide, propylene oxide, and propylene glycol. §

Poloxyl Lanolin. A polyoxyethylene condensation-product of anhydrous lanolin. §

Poly C4M. Oil and grease additive. [Crowley Chem.]

polyacrylamide. (2-propenamide, homopolymer) Polyamide of acrylic monomers. [CH$_2$CHCONH$_2$]$_x$; CAS 9003-05-8; thickener, dispersant, antiprecipitant, solubilizer, binder, sizing, flocculating, suspending, crosslinking agent, filtering aid, lubricant; used in adhesives, agriculture, cement, coatings, cosmetics, detergents, latex manufacture, plaster, printing ink. [Aldrich; Allied

Colloids; Cyanamid BV; Calgon; Rhone-Poulenc Water Treatment]

polyacrylate. (acrylic resin) Solutions. of polyacrylic acid; dispersant for pesticides; anticaking agent; coatings additive; flocculant. [Polysar BV; Scott Bader Ltd]

polyacrylic acid, potassium salt. *See* potassium polyacrylate.

polyacrylic acid, sodium salt. *See* sodium polyacrylate.

Polyaldo® 2O10 KFG. Polyglyceryl-10 dioleate; food emulsifier, kosher grade; surfactant; polysorbate 80 replacer. [Lonza Inc]

Polyaldo® 2P10 KFG. Polyglyceryl-10 dipalmitate; food emulsifier, surfactant, gelling agent; replacement for polysorbate 60. [Lonza Inc]

Polyaldo® 2S6 KFG. Polyglyceryl-6 distearate; kosher food grade emulsifier; aerating and whipping agent. [Lonza Inc]

Polyaldo® DGDO KFG. CAS 11094-60-3; Polyglyceryl-10 decaoleate; kosher grade food dispersing agent. [Lonza Inc]

Polyaldo® HGDS KFG. Polyglyceryl-6 distearate; kosher food grade cake emulsifier. [Lonza Inc]

Polyaldo® TGMS KFG. CAS 37349-34-1; Polyglyceryl-3 stearate; kosher food grade emulsifier; aerating and whipping agent. [Lonza Inc]

Polyalk. A proprietary preparation of dimethicone and aluminum hydroxide gel; an antacid. [Galen Ltd] *

polyamide. *See also* nylon. A high molecular weight polymer in which amide linkages (CONH) occur along the molecular chain; may be either natural or synthetic; Natural polyamides include casein, soybean, and peanut proteins, zein; synthetic polyamides typified by various nylons; used for plastics, textile fibers, adhesives. [Aldrich; Arizona; Ashley Polymers; BASF; EMS-Am. Grilon; Georgia-Pacific; Hoechst Celanese; Hüls AG; Miles; Monsanto; SNIA UK; Union Camp]

Polyanthrene KS Liq. New. Polymeric aliphatic compound; formaldehyde-free fixing agent for improving washfastness of direct dyes on cellulose. [Crompton & Knowles]

Polybactrin. Antibacterial. [The Wellcome Foundation Ltd]

Poly bd® R-45HT. CAS 9003-17-2; Butadiene polymer, hydroxyl-terminated; for elec. potting and encapsulating applications including cable connectors, high voltage capacitors, shock absorbers, transformers, voltage regulators. [Atochem]

Polybead. Monodisperse latex polymer beads; an identification tag and a size reference for agglutination tests, flow cytometry, instrument calibration, gel filtration, light scattering, and phagocytosis. [Polysciences Inc]

Poly/Bed 812. Replacement for Shell's discontinued Epon 812; embedding kit for light microscopy. [Polysciences Inc]

Polybilt. Asphalt modifiers. [Exxon UK]

Polyblack. Nitriles. [BP Chemicals Ltd] †

Polyblends. Nitrile/PVC blends. [BP Chemicals Ltd] †

Polybond®. Polyolefins. [BP Chemicals Ltd]

Polybond® 1000. PP homopolymer, acrylic acid modified; thermoplastic for use as a chemical coupling agent, compatibilizing agent, metal adhesive and nucleating agent. [BP Performance Polymers]

Polybond® 1009. CAS 9002-88-4; HDPE homopolymer, acrylic acid modified; thermoplastic with good adhesion to a wide variety of substrates, excellent compatible with fillers and reinforcers; as chemical coupling agent for glass and mica-reinforced polyethylene; reduces moisture sensitivity in polyamides. [BP Performance Polymers]

Polybond® 1011. PP copolymer, high impact, acrylic acid modified; impact modifier in reinforced PP. [BP Performance Polymers]

Polybond® 2005. PP, acrylic acid-modified; additive for adhesion to metals and polar polymers; extrusion coating; FDA applications. [BP Performance Polymers]

Polybond® 2021. LLDPE, maleic anhydride; additive for adhesion to metals and polar polymers; film and sheet extrusion, injection molding [BP Performance Polymers]

Polybor® . CAS 12280-03-4; Disodium octaborate tetrahydrate; fire retardant for treatment of lumber. [U.S. Borax & Chem.; Borax Consolidated Ltd]

Polybrene. A proprietary trade name for hexadimethrine bromide. §

Polybut. Polybutenes. [British Traders & Shippers Ltd]

polybutadiene. (BR; butadiene rubber; cis-polybutadiene) (CH₂CH=CHCH₂)ₙ; CAS 9003-17-2; elastomer for tire industry, footwear, molded goods; blending ingredient in SBR; additive for plastics; coating resin in liq. form. [Ameripol Synpol; Asahi Chem Industry Co Ltd; BASF; Firestone Syn. Rubber & Latex; Goodyear; Phillips 66; Reichhold/Emulsion Polymers; Revertex Ltd]

1,2-polybutadiene. CAS 29406-96-0 (syndiotactic); Thermosetting resin for wire coating, EPDM peroxide-cured modifier, coatings, processing aid.

polybutene. (PIB; polybutylene; 1-butene, homopolymer) Polymer formed by polymerization of a mixture of iso- and normal butenes. [CH₂CH(C₂H₅)]ₙ; CAS 9003-28-5; Thermoplastic resin; used as tackifier, strengthener, and extender in adhesives, as plasticizer for rubber, as vehicle and fugitive binder for coatings, as cling additive for LLDPE stretch wrap films, as reactive intermediate for specialty chemicals. [Amoco; Ashland; BP Chem. Ltd; Harcros]

polybutylene. *See* polybutene.

Polycarbafil®. A trade name for flame retardant polycarbonate materials. §

polycarbonate resin. (PC resin) Thermoplastic resin for molded products, solutions in cast or extruded film, structural parts, tube and piping, prosthetic devices, optical parts, windows, computer and business equip., household appliances, compact disks, food contact and medical applications. [Aldrich; Ashland; LNP; Miles; Westlake Plastics]

Polycarboxylate AMC 60. Acrylic acid-maleic acid copolymer; high calcium dispersion power; co-builder in phosphate-free washing powds. [Hüls AG]

Polycare® 509. Vinyl acetate, isobutyl maleate, vinylneodecanoate copolymer; hair styling fixative resin providing superior hold and curl retention even under extremely humid conditions [Rhone-Poulenc Surf.]

Polycat® . Amine catalysts for polyurethane foam. [Air Prods. & Chems. Inc]

Polycat® 5. Amine-based catalyst for improved processing of flexible slabstock PU foam; promotes blow reaction. [Air Prods./Polyurethanes]

Polycat® 12. Catalyst for improved processing of flexible slabstock PU foam; for methylene chloride systems. [Air Prods./Polyurethanes]

Polycat® 58. Amine-based catalyst for PU flexible molded foam; for improved surface cure. [Air Prods./Polyurethanes]

Polycat® 77. Amine-based catalyst for improved processing of flexible slabstock PU foam; promotes gel reaction. [Air Prods./Polyurethanes]

Polycat® 77. Amine-based catalyst for PU flexible molded foam. [Air Prods./Polyurethanes]

Polycat® 91. Amine-based catalyst for improved processing of flexible slabstock PU foam; gives delayed action. [Air Prods./Polyurethanes]

Poly Check. Boiler water treatment. [Dearborn Chemicals Ltd] *

Poly-Chek. For additives for polymers, such as heat and light stabilizers, lubricants and the like, in Int Class 1. [Ferro] *

polychloroprene. (CR; neoprene; 2-chlorobutadiene 1,3; chloroprene rubber) CAS 126-99-8; Elastomer for molding, extrusion, and calendering for adhesive compounding, construction, automotive, hose and cable jackets, conveyor belts, closed cell sponge, etc.

Polychol 5. CAS 61791-20-6; Laneth-5; o/w emulsifier, dispersant, gellant, emollient, solubilizer for personal care products, hair straighteners, pharmaceuticals; bromo dye solv. [Croda Inc.; Croda Chem. Ltd.]

Polychrest salt. Potassium sulfate, K₂SO₄. The term is also applied to Rochelle salt.

polychrome blue of unna. A dyestuff prepared by the action of potassium carbonate on methylene blue; used in microscopy.

Polycillin. CAS 69-53-4; Ampicillin; antibacterial. [Bristol Laboratories]

Polycillin-N. CAS 69-52-3; Ampicillin sodium; antibacterial. [Bristol Laboratories]

Polycin 12. Urethane polyol [CasChem] *

Polycizer 162. CAS 117-81-7; Dioctyl phthalate; vinyl plasticizer. [Harwick Standard Chemical Co] ‡

Polycizer 332. CAS 103-23-1; Dioctyl adipate; vinyl plasticizer. [Harwick Standard Chemical Co] ‡

Polyclar® 10. Polyvinylpolypyrrolidone homopolymer; insol. cross-linked polymer; stabilizer for beverage clarification and stabilization; adsorbent in thin-layer and column chromatography. [ISP]

Polyclear. Textile reduction clearing additive. [Hoechst AG]

Polyclear 32-F. Clearing agent for use on polyester and polyester/cotton blends; removes loose dye from disperse dyes on polyester and nylon fabrics. [Henkel/Textiles]

Polycomp® 139. CAS 9016-75-5; PPS-filled PTFE composite; excellent wear properties; not for moderate to highly loaded applications due to low deformation resistance. [ICI Fluoropolymers]

Polycomp® 185. CAS 9016-75-5; PPS-filled PTFE composite; good deformation resistance with excellent wear properties, very low abrasiveness. [ICI Fluoropolymers]

Polycon. Photographic developer. [May & Baker Ltd] *

Polycon II. Silafocon A; contact lens material. [Syntex Ophthalmics Inc] *

Polycon S60 K. CAS 1338-41-6; EINECS 215-664-9; Sorbitan stearate; lipophilic food emulsifier used in emulsions where weaker water-binding properties and enhanced aeration are desired; crystallization promoter, surface film former. [Witco/Humko]

Polycon S80 K. CAS 1338-43-8; EINECS 215-665-4; Sorbitan oleate; lipophilic food emulsifier used in emul-

‡ Trade name and manufacturer not verified   § Trade name without identified manufacturer

sions where weaker water-binding properties and enhanced aeration are desired; crystallization promoter, surface film former. [Witco/Humko]

Polycon T60 K. CAS 9005-67-8; Polysorbate 60; hydrophilic food emulsifier, surfactant for formulating o/w emulsions. [Witco/Humko]

Polycon T80 K. CAS 9005-65-6; Polysorbate 80; hydrophilic food emulsifier, surfactant; used in pickle industry due to very low melting point. [Witco/Humko]

Polycote Pedigree. Benomyl + iodofenphos + metalaxyl; a fungicide and insecticide seed coating for seeds. [Seedcote Systems Ltd]

Polycote Prime. Powder mixture of iprodione, metalaxyl and triabendazole; a polymer seed coating for carrots. [Seedcote Systems Ltd]

Polycoupler IMP RFB X-353. Textile resin copolymer utilizing multifunctional groups of a hydantoin-acetylenic deriv. to achieve resin fixation. [CNC Int'l. L.P.]

polycroit. The coloring matter of saffron.

Polycrol. A proprietary preparation containing methylpolysiloxane, aluminum hydroxide gel and magnesium hydroxide; an antacid. [Nicholas Laboratories Ltd] *

Polycron. Disperse dyestuffs; for coloration of polyester fibers. [Holliday Dyes & Chemicals Ltd]

Polycryl. Range of acrylic polymer emulsions; for textile finishing. [Morton Int'l. Ltd]

Polycure. Polyester catalysts. [Mooney Chemicals Inc]

Polycycline Hydrochloride. CAS 64-75-5; Tetracycline hydrochloride; anti-amebic; antibacterial; antirickettsial. [Bristol-Myers Squibb Co Inc]

polydimethylsiloxane rubber. See silicone elastomer.

Polydis® TR 121. CAS 301-02-0; EINECS 206-103-9; Oleamide; slip agent for polyethylene films; lubricant and mold release for injection molding applications, processing thermoplastic resins, thermoplastic elastomers, and thermoset rubber systems; dispersant. [Struktol]

Polydis® TR 131. CAS 112-84-5; EINECS 204-009-2; Erucamide; slip agent for polyolefin films; release agent for polymer systems; dispersant in color concs. and printing inks; process aid/lubricant for thermoplastic elastomers, thermoplastic resins, thermoset rubber systems. [Struktol]

Polydur. Polyester resin molding compounds; for electronics and electrical engineering. [Dynamit Nobel Wien GmbH] *

polydymite. (nickel-linnaeite). A mineral, $(Ni \cdot Co)_4S_5$.

Polydyol 30-G. Dye carrier giving good color value, level and spot-free dyeings. [Eastern Color & Chem.]

Polyeite. A proprietary polyester laminating resin. [Reichhold] *

Poly-Em. Polyethylene wax emulsions. [Rohm & Haas UK]

Polyester 1606. Complex polyester; emulsion preventer/breaker base for oil treatment; base for oil-emulsion prevention. [Chemron]

Polyester N-95. Surfactant; emulsion preventer/breaker base for oil treatment; well stimulation additive. [Chemron]

polyestradiol phosphate. An oestrogen currently undergoing clinical trial as Leo 114 and Estradurin. It is a polyester of oestra-1,3,5(10)-triene-3, 17β-diol and phosphoric acid.

Polyestren®. Blended dyestuffs (dispersion/vat or sulfur/vat dyes); for dyeing and printing of blended textiles. [Cassella AG]

Poly-Eth. Plastic resins. [Chevron] *

Poly-Eth Hi-D. Plastic resins. [Chevron] *

polyethylene. (ethene, homopolymer) Polymer of ethylene monomers. $[CH_2CH_2]_x$; CAS 9002-88-4; EINECS 200-815-3; laboratory tubing; prostheses; elec. insulation; pkg. materials; kitchenware; tank and pipe linings; paper coatings; textile stiffeners. [Asahi Chem Industry Co Ltd; Ashland; Atochem SA; Eastman; EniChem UK; Exxon; LNP; Mitsubishi Petrochem.; Quantum/USI]

polyethylene glycol. *Also see* PEG. (PEG; polyglycol; polyether glycol) Condensation polymers of ethylene glycol; CAS 25322-68-3; EINECS 203-473-3; chemical intermediates, plasticizers, softeners, humectants, ointments, polishes, paper coating, mold lubricants, bases for cosmetics and pharmaceuticals, solvents, binders, metal and rubber processing, food additives, laboratory reagent. [BASF; BP Chem. Ltd; Calgene; Dow; Du Pont; Harcros; Henkel; Hüls; Inolex; Olin; Rhone-Poulenc Surf.; Texaco; Union Carbide]

poly (ethylene glycol adipate). Functional polyol for formulating polyurethane for solution. coatings, adhesives, and castable elastomers. [Werner G. Smith]

polyethylene, high-density. (HDPE) Blow-molded products, injection-molded items, film and sheet, piping, fibers, gasoline and oil containers. [Allied-Signal; BP Chem. Ltd; Chevron; Chisso; Du Pont; Exxon; Hüls Am.; OxyChem; Quantum/USI; Solvay Polymers; Westlake Plastics]

polyethylene, linear low density. (LLDPE) [BP Chem. Ltd; Neste Polyeten AB]

polyethylene, low-density. (LDPE) Packaging film, food packaging, paper coating, liners for drums, wire and cable coating, toys, cordage, waste bags, chewing gum base, squeeze bottles, electrical insulation. [Allied-Signal; Chevron; Eastman; Exxon; Hüls Am.; Neste Polyeten AB; Quantum/USI; Westlake Plastics]

polyethylene wax. Wax for polishes, plastics and rubber processing, printing inks, pigment masterbatches, hot melts; PVC lubricant. [Allied-Signal; Hoechst Celanese; Hüls AG; IGI Boler; Sartomer; Stevenson Bros.; Syn. Prods.]

Polyfax/Polysporin. Proprietary formulations of polymyxin B sulfate and bacitracin zinc; ointment and ophthalmic ointment; ointment: for treatment of primary infected dermatoses, secondary bacterial infection, accidental cuts, scratches and abrasions following surgical procedures and burns, etc.; for prophylactic use in burns, graft donor sites, surgical incisions; ophthalmic ointment: for treatment of bacterial infections of the eye and its adnexa, also pre- and post-operative use to prevent ocular infection following surgical procedures. [The Wellcome Foundation Ltd]

Polyfeed. Water soluble, chlorine free N-P-K fertilizer, with chelated micro-nutrients; for direct soil application, via irrigation system or foliar spray. [Haifa Chemicals Ltd] *

Polyfil® WC. Anhyd. organofunctional pigment; pigment reducing water vapor transmission, yielding excellent wet and dry elec. properties and good long-term stability in EPR and crosslinked polyethylene. [J.M. Huber/Clay Div.]

Polyfilm. Polyethylene film; used in packaging and industrial applications. [Dow UK] *

Polyfine MF15C. Thermoplastic resin; formulated as economical substitute for polyacetal resin in applications

---

* Trade name not verified as available   † Trade name verified as obsolete

where friction and wear properties are required; recommended for VTR tape reel, noiseless gear, bearing surface, etc.; for injection molding; MF15C offers high rigidity, high gloss for audio tape reel. [Advanced Web Prods.]

Polyflex. 6-Ethoxy-2,2,4-trimethyl-1,2-dihydroquinoline; a proprietary antioxidant. [Naugatuck (US Rubber)] *

Polyflex®. A registered trademark for a flexible polystyrene sheet and fiber. §

Polyflo. Oil additives. [Hoechst AG]

Polyflon. A proprietary brand of polytetrafluoroethylene (PTFE). [Daikin Kogyo Co] ‡

Polyfon. CAS 8061-51-6; A range of sodium lignin sulfonates; used as concrete retard additives and solids dispersion additives. [Harcros Australia] *

Polyfon® F. CAS 8061-51-6; Sodium lignosulfonate; dispersant for industrial applications from agric. to ceramics. [Westvaco]

Polyfusor Solutions. A range of sterile pyrogen-free solutions. [The Boots Co plc] *

Poly-G® 20-28. Polyether polyol (diol); for caulks, sealants, coatings, tire fill, clay pipe seal, high mod. RIM applications [Olin] *

Poly-G® 20-56. Polyether polyol (diol); for adhesives, caulks, sealants, coatings, castable elastomers, tire fill, clay pipe seal. [Olin] *

Poly-G® 30-56. Polyether polyol (triol); for adhesives, coatings. [Olin] *

Poly-G® 55-28. Polyether polyol (VHP diol); suitable for applications where reactivity is important; for adhesives, caulks, sealants, thermoplastic urethanes and hot melts, one-shot elastomers, RIM applications [Olin] *

Poly-G® 200. CAS 25322-68-3; PEG 200; chemical intermediate for production of surfactants for cleaners, textiles, paper, cosmetics; carrier for pharmaceuticals; also in cosmetics and personal care products, textiles, rubber mold releases, printing inks and dyes, metalworking fluids, foods, paints, paper, wood products, adhesives, agric. products, ceramics, elec. equipment, petrol. products, photographic products, resins. [Olin]

Poly-G Fluids. Polyalkylene glycols. [Olin UK]

Poly-G Polyols. Polyether polyols for urethane foam. [Olin UK]

polygalin. See struthiin.

polygallic acid. See struthiin.

Polygard. Tris (mixed mono and dinonyl phenyl) phosphite; antioxidant, processing and color stabilizer for PP, LDPE, LLDPE, HDPE, HIPS, ABS, PVC, PC, and EVA and polyamide hot-melt adhesives. [Uniroyal]

polygeline. A polymer of urea and polypeptides derived from denatured gelatin.

Polygeline™. CAS 9007-34-5; Modified collagen. [Calbiochem Corp]

polyglactin. A synthetic suture capable of being absorbed by the patient's body. It is a mixture of lactic acid polyester with glycolic acid.

polyglandin. A solution of the autacoid principles of the thyroid, parathyroid, ovary, testic, and pituitary gland substances.

Polyglobin®. Specific human immunoglobulin. [Bayer AG]

Polygloss. Flexographic printing inks; for printing flexible film. [Allied-Signal Inc/Sinclair and Valentine Division] *

Polyglucadyne. 1-3 and 1-6 β-glucans copolymer; patented material for protection of skin in cosmetics; macrophage stimulating factor. [Brooks Industries]

Polyglycol B-11-50. EO/PO random copolymer based on butanol; surfactant for textile processing. [Hoechst Celanese/Colorants & Surf.]

polyglycolic acid. A synthetic suture capable of being absorbed by the patient's body. Poly(oxycarbonylmethylene).

Polygon. CAS 7758-29-4; Sodium tripolyphosphate. [Albright & Wilson Ltd, Phosphates & Speciality Business] *

Polygrade. Degradable plastic materials; additive concentrates to cause degradation (after useful lifetime). [Ampacet Corporation]

Polyguide®. [DuPont UK]

polyhalite. (isobelite). A mineral which occurs in the Strassfurt deposits. It is a crystalline mixture of the sulfates of calcium, magnesium, and potassium, and is found with rock salt. It has the formula, $K_2SO_4 \cdot MgSO_4 \cdot 2CaSO_4 \cdot 2H_2O$.

Polyhall® 21J. CAS 9003-05-8; Very high molecular weight polyacrylamide; flocculant for settling coal floation residue and slime tailings; settling aid for phosphate slimes, gold and copper tailings; clarification and thickening aid. [Rhone-Poulenc/Water Soluble Polymers]

poly(hexamethyleneadipamide). See nylon 6/6.

polyhexanide. Poly-(1 - hexamethylenebiguanide hydrochloride); an anti-bacterial.

Polyhipe. Range of organic microporous matrices for immobilization. [Microporous Materials Ltd]

polyhydrite. A mineral. It is a silicate of iron.

polyimide, thermoplastic. Thermoplastic resin for compression molding, injection molding and film casting; structural composites, adhesives, film, insulation, coatings.

polyimide, thermoset. Thermoset resin for structural and nonstructural aerospace applications., bearings, automotive components, films for elec. equip.

poly(iminocarbonylpentamethylene). See nylon 6.

poly[imino (1,6-dioxo-1,6-hexanediyl) imino-1,6-hexanediyl. See nylon 6/6.

poly[imino(1-oxo-1,6-hexanediyl)]. See nylon 6.

polyisobutene. (polyisobutylene; 2-methyl-1-propene, homopolymer) Homopolymer of isobutylene. $[CH_2C(CH_3)HCH_2]_x$; CAS 9003-27-4; 9003-29-6; synthetic rubber; polymeric additive; thickener for lubricating oils; for the adhesive and sealant industry, elec. insulating oils, bases for chewing gums, for production of damp-proof courses containing fillers in construction industry. [BASF; Rit-Chem]

Polyisobutylene. See polyisobutene.

polyisoprene. (IR; isoprene rubber; 2-methyl-1,3-butadiene, homopolymer; cis-1,4-polyisoprene rubber) Thermoplastic polymer of isoprene. $[CH_2C(CH_3)CHCH_2]_x$; CAS 9006-04-6; 9003-31-0; elastomer for light colored goods, adhesives, footwear, sponge products, tires, pharmaceutical goods, rubber bands, molded and mech. goods. [Goodyear; A. Schulman]

Polykol. CAS 9003-11-6; Poloxamer 188; waxy, nonionic surfactant of the poly(oxypropylene)poly(oxyethylene) copolymer type; laxative. [Upjohn] *

poly(laurolactam). See nylon 12.

Polylite®. Polyester resins or polyurethane foam; also chemicals used in industry, science, photography, agriculture, horticulture, forestry, unprocessed artificial res-

---

‡ Trade name and manufacturer not verified   § Trade name without identified manufacturer

ins, unprocessed plastics, manures, fire extinguishing compositions, tempering and soldering preparations, etc. [Reichhold]

Polylite. Alkylated diphenylamine; closely related to octamine in chemical structure and may be considered a liquid form of octamine; has the same broad protective action with the same minimum of discoloration and staining; for use as an antioxidant and stabilizer in the manufacture of SBR and nitrile polymers and as an antioxidant in latex compounding. [Uniroyal] *

Polylite. A mineral; it is a variety of pyroxene. §

Polylite® 32-162. Isophthalic polyester resin, acrylic-modified; produces Corian®-like product with good stain resistance, high heat distort. temp.; for vanities, etc. [Reichhold]

polylithionite. A mineral. It is a variety of zinnwaldite containing lithium.

Polyloy® 6. Polyamide 6, granular form; used in automotive, machinery, electrical/electronics, building industries. [EMS-Chemie AG]

Polyloy® A. Polyamide 6/6, granular form; used in automotive, machinery, electrical/electronics, building industries. [EMS-Chemie AG]

Polylube 1105. Blend of fatty acid esters, polyoxyalkylene derivs.; lubricant, antistat for nylon. [Hart Chem. Ltd.]

Polylube ASTL. Polyoxyalkylene alcohol phosphate; antistat component for lubricant formulations for all fibers. [Hart Chem. Ltd.]

Polylube DDL. Blend of polyoxyalkylene alcohol derivs. and polyalkylene glycol ethers; heat-stable lubricant for glass extrusion applications. [Hart Chem. Ltd.]

Polylube Wax. Blend of polyoxyalkylene derivs.; wax for sizing operations; cohesive agent for short staple fiber in woolen systems. [Hart Chem. Ltd.]

Polymate. Cooling water treatment. [Dearborn Chemicals Ltd]

Polymekon. Silicone antifoam emulsions. [Thomas Goldschmidt Ltd]

Polymekon®. CAS 63231-60-7; Modified microcryst. wax; used in the formulation of inks and coatings and as binder, antislip and antimar agent. [Petrolite]

Polymel #7. Modified polyethylene wax; low molecular weight wax incorporated into rubber batches giving excellent mold release. [Frank B. Ross]

Polymene AZ. CAS 61791-12-6; Blend; for textile use. [Rhone-Poulenc Surf.]

Polymer C. Microcrystalline wax replacement. [Crowley Chem.]

Polymeric Sealant Gun. Thermal extruder for rubber-like sealants. [Hardman] *

polymerized oils. See blown oils.

Polymet®. [Polymer Corp]

polymethylmethacrylate. (2-propenoic acid, 2-methyl-, methyl ester, homopolymer) Polymer of methyl methacrylate; [CH₂CCH₃COOCH₃]ₙ; CAS 9011-14-7; thermoplastic used as main contituent of acrylic sheet, molding and extrusion compounds [Atochem SA; Shuman Plastics]

Polymica 200, 325, 400. CAS 12001-26-2; Wet-processed musscovite mica; filler for coating and performance polymer applications where high brightness, color, particle size, and consistency are important. [Franklin Industrial Minerals]

polymignite. A mineral, $4(Ca \cdot Ce \cdot Fe)O \cdot (Ti \cdot Zr)O_2 \cdot CaO \cdot$ Nb_2O_5.

Polymin®. Polyacrylamide-based, modified polyethylene imines; retention, dewatering, and flocculating agents. [BASF AG]

Polymist® F-5. CAS 9002-84-0; Polytetrafluoroethylene; additive in elastomers or plastics where improved lubricity and/or wear resistance are required. [Ausimont]

Polymoist Mask. CAS 9007-34-5; Collagen fiber material; moisturizer and vehicle for cosmetic active agents. [Henkel/Cospha; Henkel Canada]

Polymon. Soluble dyes for plastics. [ICI Chem. & Polymers Ltd]

Polymone. CAS 1702-17-6, 120-36-5; Aqueous solution containing 10% w/v 2,4-D and 40% w/v Dichlorprop; for use as an agricultural herbicide. [Universal Crop Protection Ltd]

Polymox. CAS 26787-78-0; Amoxicillin; antibacterial. [Bristol Laboratories]

Polymul. Range of polyethylene emulsions; variety of protective and decorative coatings. [Henkel Chemicals Ltd]*

polymyxin. Antimicrobial substances produced by *Bacillus polymyxa*. Specific substances are designated by a terminal letter, e.g., Polymyxin B sulfate; active against gram-negative bacteria.

Polyoil Hüls 110. CAS 9003-17-2; Polybutadiene resin; improves water resistance and dry characteristics of alkyds; used in anticorrosion and electrodeposition coatings. [Hüls Am.]

Polyox. A range of water soluble, high molecular weight polymers of ethylene oxide; used in adhesives, binders, pharmaceuticals, and lubricants. [Union Carbide (UK) Ltd]

polyoxymethylene. See paraformaldehyde.

Polyox® WSR 3333. CAS 25322-68-3; PEG-9M; water-sol. polymer. [Amerchol Corp]

Polyox® WSR N-10. CAS 25322-68-3; PEG-2M; thickener. [Amerchol Corp]

Poly-Pale®. Polymerized rosin and rosin esters; used for hot melt coatings and adhesives. [Hercules; Hercules Ltd]

Poly-Pale® Ester 10. Polymerized rosin esters; tackifying resin for solvent and emulsion pressure-sensitive adhesives and for hot-melt packaging adhesives. [Hercules]

Polypeg-E. Polyethylene glycol fatty ester. [Olin UK]

Polypel. Fertilizer. [Chevron] *

Polypenco® Cast Acrylic Rod. Cast acrylic; optical clarity, high tensile strength, good elec. properties; used in displays, signs, furniture components, lenses, elec./electronic parts. [Polymer Corp.]

Polypenco® Nylon 101. CAS 32131-17-2; Type 6/6 nylon; strong, stiff nylon for use in food processing, machinery, electronics, military and other industries for bearings, bushings, valve seats, seals, rollers, gears, insulators, fasteners, liners, tooling fixtures, forming dies, etc. [Polymer Corp.]

Polypenco® PEEK. PEEK; thermoplastic offering high continuous service temp. (480 F), low emission of smoke and toxic fumes on exposure to flame; for elec./electronic components, aircraft components, microwave, automotive applications, pump and valve parts, chem. processing. [Polymer Corp.]

Polypenco® Polycarbonate. PC; thermoplastic high impact. strength, good elec. properties; used in stand-off insulators, coil forms, optical and transparent structural

---

* Trade name not verified as available   † Trade name verified as obsolete

components. [Polymer Corp.]

Polypenco® Polysulfone. CAS 25135-51-3; Polysulfone; semi-transparent, heat, chem., and hydrolysis resistance, high-performance engineering thermoplastic; for medical tubing, food and beverage contact parts, dairy equip., aerospace components, circuit boards and connectors. [Polymer Corp.]

Polypenco® Q200.5. CAS 9003-53-6; Crosslinked PS; rigid insulating material; outstanding dielec. properties, good impact strength; used in UHF, VHF, and microwave insulators, communication and electronic equipment. [Polymer Corp.]

Polypentek. A proprietary trade name for polypentaerythritol. §

Polypeptide 10. CAS 9015-54-7; Hydrolyzed collagen; protective colloid, anti-irritant, substantivity agent, moisturizer for hair and skin care products, dish detergents; dye leveler and fiber protectant in textile industry. [Maybrook]

Polypeptide 37. CAS 9015-54-7; Hydrolyzed collagen; surfactant for personal care products [Inolex]

polyphenylmethyl siloxane. See phenyl dimethicone.

Polyphos. Sodium hexametaphosphate. [Olin UK]

polyphosphoric acid. (superphosphoric acid) $H_{n+2}P_nO_{3n+1}$; CAS 8017-16-1; Acid used in the manufacture of phosphates, phosphate esters, catalysts, fuel cell electrolytes, metal cleaning and brightening, organic reactions. [Albright & Wilson]

Polyplasdone®. Crospovidone; pharmaceutic aid. [ISP] *

Polyplasdone® XL. Polyvinylpolypyrrolidones; insoluble crosslinked polymer of N-vinyl-2-pyrrolidone used as a tablet disintegrant, complexing agent, detoxifier and antidiarrhea agent; adsorbent in thin-layer chromatography. [ISP]

Polyplate 90. CAS 1332-58-7; Kaolin clay; delaminated functional filler with high brightness, finer particle size; for adhesives, paints, plastics, and inks. [J.M. Huber]

Polypor. Silica gel. [Quantum Chemical Corp]

Poly-Pred Liquifilm®. Prednisolone acetate, neomycin sulfate, polymyxin B sulfate; corticosteroid, antibacterial. [Allergan, Inc]

Polyprene. Weber's name for rubber. §

Polypress. A proprietary preparation of prazosin hydrochloride and polythiazide; an antihypertensive. [Pfizer International]

Poly-Pro. Plastic resins. [Chevron] *

polypropene. See polypropylene.

polypropylene. (PP; 1-propene, homopolymer; propylene polymer; polypropene) Polymer of propylene monomers; three forms: isotactic (fiber-forming), syndiotactic, atactic (amorphous). $[CH_2(CH_3)CH]_x$; CAS 9003-07-0; 9010-79-1 (nucleated); Isotactic: fishing gear, ropes, filter cloths, laundry bags, protective clothing, blankets, fabrics, carpets, yarns. [Amoco; Aristech; Ashland; Chisso; Eastman; Exxon; Fina; Hüls; LNP; Mitsubishi Petrochem.; Mitsui Toatsu Chem.; Neste UK; Quantum/ USI; Shell; Solvay Polymers]

polpropylene glycol. Also see PPG. $HO(C_3H_6O)_nH$; Hydraulic fluids, rubber lubricants, antifoam agents, intermediates for urethane foams, adhesives, coatings, elastomers, plasticizers, paint formulations, lab reagent. [Aldrich; Arco; Ashland; BASF; BP Chem. Ltd; Calgene; Dow; Harcros; Hüls AG; Miles; Olin; PPG Industries; Rhone-Poulenc Surf.; Texaco; Witco]

Polyquart H. PEG-15 tallow polyamine; surfactant used in personal care products; hair conditioner, antistat. [Henkel/Cospha; Henkel Canada]

Polyquest. Boiler water treatment. [Grace Dearborn Ltd]

Polyquest 80. CAS 140-01-2; Pentasodium diethylenetriamine pentaacetate; sequestering agent for iron. [CNC Int'l. L.P.]

Polyrad. 5- and 11-mole ethylene oxide adducts of Amine D dehydroabietylamine; Polyrad rosin is used as a corrosion inhibitor and detergent in petroleum-processing agents and for inhibiting hydrochloric acid used in industrial and household cleaners. [Hercules] *

Polyram. Alkyl propylene polyamines based on coco and tallow alkyl chains; bitumen adhesion agents, intermediate for chemical synthesis. [Atochem UK/Ceca]

Polyram®-Combi, DF. CAS 9006-42-2; Metiram; for control of fungus diseases in fruit, hops, vines, vegetables, ornamentals [BASF AG]

Polysalt. Based on salts of polycarboxylic acids; dispersants for extenders, papercoating pigments; for stabilizing coating mixtures and slurries; grinding assistants for chalk. [BASF AG]

Polysar Bromobutyl 2030. Brominated isobutylene/isoprene copolymer; nonstaining stabilizer; low Mooney visc. version of Polysar Bromobutyl X2 for easier processing. [Miles; Polysar]

Polysar Butyl 100. CAS 9010-85-9; Isobutylene/isoprene copolymer rubber; nonstaining stabilizer; butyl rubber for med. and high voltage insulation, membranes for roofing and reservoir linings; vulcanizable with sulfur, quinoid, resin; slow curing. [Miles; Polysar]

Polysar Chlorobutyl 1240. Chlorinated isobutylene/isoprene copolymer, nonstaining stabilizer; low Mooney visc. version of Polysar Chlorobutyl 1255 for easier processing; vulcanized with zinc oxide, sulfur, quinoid, resin, peroxide. [Miles; Polysar]

Polysar EPDM 227. EPDM terpolymer; used alone to promote heat resistance or in blends to increase hardness while decreasing compound visc.; for injection molding and peroxide-cured heat-resistance compounds; slow cure rate. [Miles; Polysar]

Polysar EPDM 6463. EPDM, with 50 phr nonstaining paraffinic oil; oil-extended rubber for optimum heat stability and weather resistance; standard cure rate; used for heat-resistant applications, hose, elec. insulation, roofing membranes, and molded, extruded, and calendered goods. [Miles; Polysar]

Polysar EPDM XG 006. EPDM terpolymer extended with paraffinic oil; high visc. product designed to improve uv and color stability in low flex. mod., high flow polyolefin blends; for impact modification of polyolefins and for use in thermoplastic elastomers. [Miles; Polysar]

Polysar EPM XF 004. Ethylene-propylene copolymer; low visc. product designed to improve flow and gloss in polyolefin blends, for use in impact modification in polyolefins, for use in thermoplastic elastomers. [Miles; Polysar]

Polysar XL 30102. CAS 9010-85-9; Lightly crosslinked butyl terpolymer; used for preformed tape in automotive and architectural applications, solv. release sealants, pressure-sensitive adhesives, hot-applied sealants. [Polysar]

Polyseal. Polyurethane concrete sealer; suitable for use on any concrete floor that is subject to heavy abrasive traffic

---

‡ Trade name and manufacturer not verified   § Trade name without identified manufacturer

or is exposed to chemicals or oils. [Secure Inc] †

Poly/Sep 47. Mixture of 47 buffers; electrofocusing buffer for electrophoresis; generates a stable linear pH in polyacrylamide gels in the pH range of 2.5-10; avoids inconsistencies of conventional SCAM's. [Polysciences Inc]

Polyset 100. Acrylic triazine resin solution; stiffening agent for polyester. [CNC Int'l. L.P.]

Polysil. Silicone rubber gums and compounds. [Midland Silicones] ‡

Polysoft 35. Emulsion synthetic softener/lubricant for pre-cure, post-cure, and crease resistance finishes with excellent resistance to yellowing. [CNC Int'l. L.P.]

Polysoft B. CAS 9002-88-4; Polyethylene emulsion; textile softener improving physical properties and sewability in durable press applications [Sybron]

Polysoft CA. Polyethylene; antistatic agent, lubricant and softener compatible with resin finishes. [Scher Chemicals Inc] *

Poly-Solv®. Glycol ether solvents. [Olin UK]

Poly-Solv® DE (High Gravity). Diethylene glycol monoethyl ether (75%) and ethylene glycol (25%); solvent for brake fluids, hard-surface cleaners, leather dyeing, paints, coatings, printing inks, textile vat dyeing and printing, adhesives, antifreeze, floor waxes/polishes, insect repellents; solubilizer for dyes; plasticizer; extraction and crystallization solvent; vinyl chloride dispersant; coupler in cosmetic preparations. [Olin]

Poly-Solv® DM. CAS 111-77-3; Diethylene glycol monomethyl ether; solvent for brake fluids, hard-surface cleaners, leather dyeing, paints, coatings, printing inks, textile vat dyeing and printing, adhesives, antifreeze, floor waxes/polishes, insect repellents; solubilizer for dyes; plasticizer. [Olin]

Poly-Solv® DPM. Dipropylene glycol monomethyl ether; solvent for brake fluids, hard-surface cleaners, leather dyeing, paints, coatings, printing inks, textile vat dyeing and printing, adhesives, antifreeze, floor waxes/polishes, insect repellents; solubilizer for dyes; plasticizer. [Olin]

Poly-Solv® EE. CAS 110-80-5; Ethylene glycol monoethyl ether; solvent for brake fluids, hard-surface cleaners, leather dyeing, paints, coatings, printing inks, textile vat dyeing and printing, adhesives, antifreeze, floor waxes/polishes, insect repellents; solubilizer for dyes; plasticizer. [Olin]

Poly-Solv® EM. CAS 109-86-4; Ethylene glycol monomethyl ether; solvent for brake fluids, hard-surface cleaners, leather dyeing, paints, coatings, printing inks, textile vat dyeing and printing, adhesives, antifreeze, floor waxes/polishes, insect repellents; solubilizer for dyes; plasticizer. [Olin]

Poly-Solv® MPM. CAS 107-98-2; Monopropylene glycol monomethyl ether; solvent for brake fluids, hard-surface cleaners, leather dyeing, paints, coatings, printing inks, textile vat dyeing and printing, adhesives, antifreeze, floor waxes/polishes, insect repellents; solubilizer for dyes; plasticizer. [Olin]

Poly-Solv® TM. Triethylene glycol monomethyl ether; solvent for brake fluids, hard-surface cleaners, leather dyeing, paints, coatings, printing inks, textile vat dyeing and printing, adhesives, antifreeze, floor waxes/polishes, insect repellents; solubilizer for dyes; plasticizer. [Olin]

Poly-Solv® TPM. PPG-3 methyl ether; solvent for brake

fluids, hard-surface cleaners, leather dyeing, paints, coatings, printing inks, textile vat dyeing and printing, adhesives, antifreeze, floor waxes/polishes, insect repellents; solubilizer for dyes; plasticizer. [Olin]

Polysolvan 0. A solvent composed of the ester of isobutyl alcohol with glycollic and butyl glycollic acids.

Polysolvan E. A proprietary trade name for a solvent mixture comprising the acetates of propyl, isobutyl, and amyl alcohols. §

Polysolvan O. Coalescing solvent for adhesives. [Hoechst UK] *

Polysolvan SHS. A proprietary trade name for acetic acid esters of alcohols up to C_{11}. §

polysorbate 20. (POE (20) sorbitan monolaurate; PEG-20 sorbitan laurate; Sorbimacrogol laurate 300) Mixture of laurate esters of sorbitol and sorbitol anhydrides, with ≈ 20 moles ethylene oxide; CAS 9005-64-5 (generic); o/w emulsifier, solubilizer; used in agriculture., cosmetics, leather, metalworking, and textile industries.

polysorbate 21. (POE (4) sorbitan monolaurate; PEG-4 sorbitan laurate) Mixture of laurate esters of sorbitol and sorbitol anhydrides, with ≈ 4 moles ethylene oxide; CAS 9005-64-5 (generic); emulsifier for PVC polymerization, solubilizer for colorants, dye leveing agent.

polysorbate 40. (POE (20) sorbitan monopalmitate; sorbimacrogol palmitate 300; sorbitan, monohexadecanoate, poly(oxy-1,2-ethaneidyl) derivatives) Mixture of palmitate esters of sorbitol and sorbitol anhydrides, with ≈ 20 moles of ethylene oxide; CAS 9005-66-7; o/w emulsifier, solubilizer; used in agriculture., cosmetics, leather, metalworking, and textile industries.

polysorbate 60. (POE (20) sorbitan monostearate; PEG-20 sorbitan stearate; sorbimacrogol stearate 300) Mixture of stearate esters of sorbitol and sorbitol anhydrides, with ≈ 20 moles ethylene oxide; $C_{64}H_{126}O_{26}$; CAS 9005-67-8 (generic); industrial chemicals, solvent, emulsifier, pharmaceuticals, veterinary drug.

polysorbate 61. (POE (4) sorbitan monostearate; PEG-4 sorbitan stearate) Mixture of stearate esters of sorbitol and sorbitol anhydrides, with ≈ 4 moles ethylene oxide; CAS 9005-67-8 (generic); emulsifier, solubilizer, lubricant for textile use, household formulations, suppositories in pharmaceuticals.

polysorbate 65. (POE (20) sorbitan tristearate; PEG-20 sorbitan tristearate; sorbimacrogol tristearate 300) Mixture of stearate esters of sorbitol and sorbitol anhydrides, with ≈ 20 moles ethylene oxide; CAS 9005-71-4; o/w emulsifier, solubilizer; used in agriculture., cosmetics, leather, metalworking, and textile industries.

polysorbate 80. (POE (20) sorbitan monooleate; PEG-20 sorbitan oleate; Sorbimacrogol oleate 300) CAS 9005-65-6 (generic); 37200-49-0; Mixture of oleate esters of sorbitol and sorbitol anhydrides, with ≈ 20 moles ethylene oxide; pharmaceutic aid (surfactant); as emulsifier and dispersant in medicinal products; as defoamer and emulsifier in foods.

polysorbate 81. (POE (5) sorbitan monooleate; PEG-5 sorbitan oleate) CAS 9005-65-5 (generic); Mixture of oleate esters of sorbitol and sorbitol anhydrides, with ≈ 5 moles ethylene oxide; o/w emulsifier, solubilizer; used in agriculture., cosmetics, leather, metalworking, and textile industries.

polysorbate 85. (POE (20) sorbitan trioleate; PEG-20 sorbitan trioleate; sorbimacrogol trioleate 300) CAS

---

* Trade name not verified as available   † Trade name verified as obsolete

9005-70-3; Mixture of oleate esters of sorbitol and sorbitol anhydrides, with ≈ 20 moles ethylene oxide; used as a surfactant.

polysphaerite. A mineral, $(Pb \cdot Ca)_3 \cdot (PO_4)_2 (Pb \cdot Ca)_2 \cdot Cl(PO_4)$.

Polysphere 3000 SP. CAS 9003-53-6, 111-01-3; Polystyrene, squalane; binder for pressed powds.; lubricious, lusterous, high-grade filler for liq. and powd. formulations. [Presperse]

Polyspin MP-7-29. Polyoxyalkylene deriv.; heat stable lubricant for polypropylene filament. [Hart Chem. Ltd.]

Polyspin PA. Blend of synthetic oils and polyoxyethylene derivs.; spin finish for nylon 6 and polypropylene. [Hart Chem. Ltd.]

Polysporin. See Polyfax. [The Wellcome Foundation Ltd]

Polystab. Polymer additives. [Harcros UK]

Polystal®. Composites (profiles, tape, sheet, etc.) of thermosets and thermoplastics with glass or carbon fibers; for aeronautic and aerospace engineering, traffic engineering, elec. engineering/electronics, machine, plant and appliance engineering. [Bayer AG]

Polystat. Antistatic agent. [Crosfield Chemicals Ltd] †

Polystate C. CAS 9004-99-3; PEG-6 stearate; base for cosmetic lotions. [Gattefosse; Gattefosse SA]

Polystay AA-1. Anilino-phenyl methacrylamide; antioxidant in emulsion polymers, NBR, SBR, BR, ABS, and CR. [Goodyear] *

Polystep®. Various surfactants; for emulsion polymerization. [Stepan] *

Polystep® A-7. Sodium dodecylbenzene sulfonate, linear; emulsifier for emulsion polymerization, S/B, vinyl chloride, vinylidene chloride latexes. [Stepan; Stepan Canada]

Polystep® A-11; Isopropylamine dodecylbenzene sulfonate, branched; emulsifier, pigment dispersant; emulsion polymerization (S/B, vinyl chloride, vinylidene chloride latexes). [Stepan; Stepan Canada; Stepan Europe]

Polystep® A-13. Linear dodecylbenzene sulfonic acid; emulsion polymerizing surfactant; catalyst in acid catalyzed reactions. [Stepan; Stepan Canada]

Polystep® A-15-30K. CAS 27177-77-1; EINECS 248-296-2; Potassium dodecylbenzene sulfonate, linear; surfactant for styrene-butadiene, vinyl chloride, and vinylidene chloride latexes; thermal and hydrolytic stability. [Stepan; Stepan Canada]

Polystep® A-18. CAS 68439-57-6; EINECS 270-407-8; Sodium alpha olefin (C14, C16) sulfonate; surfactant for vinyl and vinylidene chloride, acrylic, styrene-acrylaic, SBR polymerization. [Stepan; Stepan Canada; Stepan Europe]

Polystep® B-1. CAS 9051-57-4; Ammonium nonoxynol-4 sulfate; emulsifier for emulsion polymerization. [Stepan; Stepan Canada]

Polystep® B-3. CAS 151-21-3; Sodium lauryl sulfate; emulsifier for emulsion polymerization, vinyl chloride, sol. acrylics. [Stepan; Stepan Canada; Stepan Europe]

Polystep® B-7. CAS 2235-54-3; EINECS 218-793-9; Ammonium lauryl sulfate; emulsion polymerization surfactant; latex foaming agent; water resistance in coatings. [Stepan; Stepan Canada]

Polystep® B-11. CAS 67762-19-0; Ammonium laureth sulfate (4 EO); emulsifier for emulsion polymerization (acrylics, styrene-acrylic, vinyl acrylics). [Stepan; Stepan Canada; Stepan Europe]

Polystep® B-12. CAS 9004-82-4; Sodium laureth sulfate (4 EO); emulsifier for polymerization (acrylics, styrene-acrylics, vinyl acrylics). [Stepan; Stepan Canada; Stepan Europe]

Polystep® B-25. Sodium decyl sulfate; emulsifier for emulsion polymerization (S/B, vinyl chloride, acrylic), high surface tension latex; hydrophilic. [Stepan; Stepan Canada; Stepan Europe]

Polystep® B-27. CAS 9014-90-8; Sodium nonoxynol-4 sulfate; emulsifier for acrylics, SBR, vinyl chloride, and butyl rubber. [Stepan; Stepan Canada; Stepan Europe]

Polystep® B-29. Sodium octyl sulfate; low-foaming emulsifier for vinyl chloride systems. [Stepan; Stepan Canada; Stepan Europe]

Polystep® C-OP3S. Sodium octoxynol-3 sulfate; emulsifier for controlling particle size in vinyl acetate specialty copolymers. [Stepan; Stepan Canada; Stepan Europe]

Polystep® F-1. CAS 9016-45-9; Nonoxynol-4; nonfoaming pigment dispersant; emulsifier for emulsion polymerization (styrene acrylic, acrylic, vinyl acrylic, S/B). [Stepan; Stepan Canada; Stepan Europe]

Polystep® F-95B. CAS 9016-45-9; Nonoxynol-34; emulsifier for acrylics and vinyl acetate; blended with other surfactants to increase the latex particle size. [Stepan; Stepan Canada; Stepan Europe]

polystyrene. (PS; styrene polymer; ethenylbenzene, homopolymer; benzene, ethenyl-, homopolymer) Polymer; grades: crystal, impact, expandable. $(C_8H_8)_x$; CAS 9003-53-6; thermoplastic resin for injection molding, extrusion of egg carton foam, pill bottles, pkg., appliances, electronics, toys, recreation, and construction; expandable polystyrene for insulation, protective pkg. [Amoco; Asahi Chem Industry Co Ltd; Ashland; Atochem SA; BASF; Chevron; Fina; Hüls AG; LNP; Mitsubishi Petrochem.; Mitsui Toatsu Chem.; Westlake Plastics]

Polystyrene 101. CAS 9003-53-6; Crystal polystyrene; heat-resistance extrusion and injection molding grade with high clarity, good dimensional stability; for medical molding, thick-wall housewares, sheet glazing, coextrusion. [Novacor Ltd.]

Polystyrene 220. CAS 9003-53-6; Crystal polystyrene; soft flow injection molding grade for housewares, drinkware, cutlery, toys, molded pkg., cosmetic and medical molding, as blending resin for transparent impact and molded pkg. [Novacor Ltd.]

Polystyrene 410. CAS 9003-53-6; Impact polystyrene; med. impact injection molding grade with excellent gloss, high flow; for cassettes, housewares, closures, hangers, toys, compact disc case insert. [Novacor Ltd.]

Polystyrene P 2122. CAS 9003-53-6; HIPS; extrusion and injection molding grade with superior gloss. [Novacor Ltd.]

Polystyrol. Polystyrene granules; used for extrusion (packaging) and injection molding (household appliances). [BASF plc]

Polystyrol 143E. A proprietary polystyrene having easy melt-flow and mechanical properties. [BASF plc] *

Polystyrol 165H. A proprietary polystyrene similar to Polystyrol 143E but possessing greater mechanical strength. [BASF plc] *

Polystyrol 168N. A proprietary polystyrene stabilized against ultraviolet light. [BASF plc] *

Polystyrol 427M. An impact-resistant polystyrene with good resistance to deformation at high temperatures. [BASF

---

‡ Trade name and manufacturer not verified   § Trade name without identified manufacturer

plc] *

Polystyrol 432F. A proprietary impact resistant styrene-butadiene copolymer. [BASF plc] *

Polystyrol 466 I, 472 D. A proprietary styrene/butadiene copolymer with high impact resistance. [BASF plc] *

Polystyrol 473 E. A proprietary styrene-butadiene copolymer offering high impact resistance and easy flow properties. [BASF plc] *

Polystyrol 475 K. A proprietary styrene butadiene copolymer offering high resistance to impact. [BASF plc] *

Polystyrol KR 253 and KR 2538. Proprietary styrene-butadiene copolymers offering very high resistance to impact at low temperatures. [BASF plc] *

Polystyrol KR 2536. A proprietary styrene/butadiene copolymer offering good resistance to impact and to deformation at high temperatures. [BASF plc] *

polysulfide. Synthetic polymer; elastomer for use in sealants, adhesives, potting and encapsulating compounds, gasoline and oil-loading hose, casting of molds, barrier coatings, binder in solid rocket fuel,. [Morton Int'l.]

polysulfone. Engineering thermoplastic; CAS 25135-51-3; amorphous thermoplastic with low flammability and smoke emission, good elec. properties; for injection molding and extrusion of food and chemical processing equip., elec./electronic components, medical and hospital parts requiring sterilization. [Aldrich; BASF; LNP]

PolySurf. Cetyl hydroxyethyl cellulose; hydrophobically modified hydroxyethylcellulose; thickener and visc. stabilizer for aq. and surfactant systems in personal care applications [Aqualon]

Polysystems. Urethane foam chemical systems. [Olin UK]

Polytac. Sealant, adhesive. [Crowley Chem.]

Polytac® 100. Rosin-based resinate; tackifier resin for adhesives. [Arizona]

PolyTalc 445. CAS 14807-96-6; Surface-modified platy talc; filler for polymer applications where color is critical; provides enhanced long term heat stability. [Pfizer]

Polytar. A proprietary preparation of liquid paraffin, tar crude oil, coal tar, and arachid oil extract of coal tar used in the treatment of skin diseases. [Stiefel Laboratories Inc; Stiefel Laboratories (UK) Ltd]

polytelite. A mineral, $(Pb \cdot Ag_2)_4Sb_2S_7$, with $(Zn \cdot Fe)_4Sb_2S_7$.

Polyterge PAT. Surfactant for removal and suspension of residual disperse dyes on polyester. [CNC Int'l. L.P.]

Poly-Tergent® 2A1 Acid. CAS 80260-73-2; Dodecyl diphenyl ether disulfonic acid; industrial cleaners for textile industry. [Olin]

Poly-Tergent® 2A1-L. Sodium dodecyl diphenyl ether disulfonate; industrial cleaners for textile industry. [Olin]

Poly-Tergent® 3B2. CAS 36445-71-3; Sodium decyl diphenyl ether disulfonate; surfactant for industrial and institutional detergents and textiles. [Olin]

Poly-Tergent® 3B2 Acid. Decyl diphenyl ether disulfonic acid; surfactant for industrial and institutional detergents and textiles. [Olin]

Poly-Tergent® 4C3. CAS 70191-76-3; Sodium alkyl diphenyl ether disulfonate; surfactant for paper, textiles, metalworking, industrial and household cleaners. [Olin]

Poly-Tergent® B-150. CAS 9016-45-9; Nonoxynol-4.5; surfactant for a wide variety of applications [Olin]

Poly-Tergent® CS-1. Carboxylated linear alcohol alkoxylate, sodium salt; builder surfactant, emulsifier, sequestrant for laundry detergents, hard surface cleaners, bottle washing, dairy and food service, alkaline, metal, trans-

portation cleaners; high temp. electrolyte and alkaline stability. [Olin]

Poly-Tergent® E-17A. EO/PO block polymer; foam control agent, solubilizer, dispersant, wetting agent, spreading agent for automatic dishwashing, rinse aids, industrial laundry, metal cleaning, water treatment, textiles, agric., paper processing. [Olin]

Poly-Tergent® P-17A. EO/PO block polymer; emulsifier, detergent, wetting agent, foam control agent, deduster, binder for rinse aids, automatic dishwashing, coatings, sizes, water treatment, egg washing, textiles, foods. [Olin]

Poly-Tergent® S-305LF. Alkoxylated linear alcohol; defoamer, wetting agent, dispersant for rinse aids, dairy cleaners, textiles, hard surface cleaners, automatic dishwashing, metal cleaning; biodeg. [Olin]

Poly-Tergent® SL-42. Alkoxylated linear alcohol; wetting agent, emulsifier, for prespots, laundry detergents, transportation cleaners, toilet bowl cleaners, glass cleaners, textiles, metal cleaners, degreasers, abrasive cleaners; biodeg. [Olin]

Poly-Tergent® SLF-18. Alkoxylated linear alcohol; low-foaming biodeg. detergent, dispersant, wetting agent, emulsifier, deduster; for commercial detergents, rinsing aid in machine dishwashing; used in removing protein-type soils. [Olin]

polytetrafluoroethylene. (PTFE, TFE; tetrafluoroethene homopolymer; tetrafluoroethylene polymer; polytetrafluoroethylene resin) Thermoplastic homopolymer. $[CF_2CF_2]_x$, $x \approx 20,000$; CAS 9002-84-0; as tubing or sheeting for chemical laboratory and process work; gaskets and pump packings; as elec. insulators esp. in high frequency applications.; filtration fabrics; protective clothing; prosthetic aid. [Janssens NV]

Polytetrafluoroethylene resin. See polytetrafluoroethylene.

polytetramethylene ether glycol. (PTMEG; PTMG; poly(oxy-1,4-butanediyl)-α-hydro-ω-hydroxy) Polyether glycol; $HO[(CH_2)_4O]_nH$; CAS 24979-97-3; 25190-06-1; for polyurethane formulation for automotive hose and gaskets, tires, industrial belts, tank and pipe liners, floor and roof coatings, medical devices. [BASF; QO]

polythene. The general term for a range of solid polymers of ethylene. *

polythiazide. (nephril; renese) CAS 346-18-9; 6-Chloro-3,4-dihydro-2-methyl-3-(2,2,2-trifluoroethylthio-methyl)benzo-1,2,4-thiadiazine-7-sulfonamide 1,1-dioxide. .

Polytrap®. A polymer used for entrapping solid and/or liquid materials. [Dow Corning]

Polytrap® Q5-6035. Cyclomethicone and acrylates copolymer; delivers volatile silicone fluid in powd. form; provides lubricious skin feel for skin care and cosmetics formulations. [Dow Corning]

Polytrap® Q5-6603. Acrylates copolymer; adsorbs high levels of lipopholic and certain hydrophilic liqs.; for skin care, sun care, and cosmetic formulations. [Dow Corning]

Polytrend. Colorant dispersions, polyester extender and pigment vehicle; for coloring plastic compositions, extender for unsaturated polyester compositions. [Hüls Am.]

Polytrim®. Trimethoprim sulfate, polymyxin B sulfate; ophthalmic ointment, ophthalmic solution, cream; for the treatment or prophylaxis of superficial bacterial infec-

---

* Trade name not verified as available   † Trade name verified as obsolete

tions of the skin and eye. [The Wellcome Foundation Ltd]

Polytrix. CAS 25322-68-3; PEG-115M and hyaluronic acid. [Amerchol Corp]

Polytron SMV 9081. Static dissipative alloy featuring durability, flame retardance, colorability, impact and chem. resistance, and processability; used for circuit cards, tote bins, paper handling equip.; rapid decay series can dissipate 5000 V to 0 V in < 2 s; also available as standard decay series and low dusting series. [BFGoodrich/Geon Vinyl] *

Polytron SMV 9804. Alloy; static dissipative alloy for injection molding; standard decay. [BFGoodrich/Geon Vinyl]

Polytrope. Rheological additives; used for unsaturated polyester resins. [Rheox Inc]

Polyurax. Urethane intermediates polyols. [BP Chemicals Ltd] †

polyurethane elastomer or rubber. ; Elastomer for extrusion, injection molding, and calendering; for sealants, caulks, adhesives, film and sheet, shoes, encapsulation of electronic components, automotive parts, flexible and rigid casting shapes. [Air Products; Crowley; DSM UK; Ferro/Bedford; BFGoodrich; Hardman; Miles; Morton Int'l.; Polyurethane Corp. of Am.; Soluol; UCB SA]

polyurethane foam. CAS 9009-54-5; foam insulation for thermal insulation, fabricated shapes, pipe covering, void filling, and cold storage applications. [Grace NV]

Polyval B. Vat dye stripper. [Ceca SA]

Polyvel CR-5F. CAS 78-63-7; 2,5-Dimethyl-2,5-di(t-butyl-peroxy) hexane; flow modifier and processing aid for PP. [Polyvel]

Polyvel CR-5P. Bis(t-butylperoxyisopropyl) benzene; flow modifier and processing aid for PP. [Polyvel]

Polyvel CR-5T. CAS 614-45-9; t-Butyl perbenzoate; additive for grafting and extrusion reactions. [Polyvel]

Polyvel CR-L10. CAS 1068-27-5; 2,5-Dimethyl-2,5-di(t-butylperoxy)hexyne-3; additive for roto molding and crosslinking. [Polyvel]

Polyvel PCL-20. CAS 80-43-3; Dicumyl peoxide; additive for crosslinking polyethylene. [Polyvel]

Polyvest. Polybutadienes used for treatment of minerals. [Hüls AG]

Polyvest 25. A polymeric filler activator; used for activating light-colored silicate fillers in EPDM compounds. [Hüls AG] *

Polyvest C70. A polymeric chalk filler activator; used for activating carbonate fillers in EPDM compounds. [Hüls AG] *

polyvinyl acetate (homopolymer). (PVAc; acetic acid, ethenyl ester, homopolymer; acetic acid vinyl ester polymers; ethenyl acetate, homopolymer) Homopolymer of vinyl acetate. [CH$_2$CHOOCOCH$_3$]$_x$; CAS 9003-20-7; resin with weathering resistance; used for paints; adhesives for food pkg., paper, wood, glass, and metals; primer sealers; dry wall cement; intermediate for conversion to polyvinyl alcohol and acetals; paper coating; component of lacquers, inks. [Aldrich; H.B. Fuller; Monsanto; Nat'l. Starch & Chem.; Wacker-Chemie GmbH]

polyvinyl alcohol. (PVA; PVAL; ethenol, homopolymer; PVOH) Polymer; [CH$_2$CHOH]$_x$; CAS 9002-89-5 (super and fully hydrolyzed); EINECS 209-183-3; in plastics industry in molding compounds, surface coatings, films resistant to gasoline, textile sizes; for elastomers (artificial sponges, fuel hoses); printing inks; pharmacuetical finishing; cosmetics; film and sheeting; ophthalmic lubri-

cant. [British Traders & Shippers; Dajac Labs; Honeywill & Stein Ltd]

polyvinyl butyral. (PVB; vinyl acetal polymers, butyrals; vinyl acetyl polymers, butyrates; polyvinyl butyral resin) Polymer produced by condensation of polyvinyl alcohol and butyraldehyde; (-CH$_2$CH(OOCCH$_3$)-)$_n$; CAS 63148-65-2; 9003-62-7; thermoplastic for extrusion, molding, coating, and casting processes; for adhesives, paints, lacquers, films, as sheet interlayer in safety glass and shatter-resistant protection in aircraft. [Cairn Chem Ltd; Monsanto; Wacker]

polyvinyl chloride. (PVC; chloroethene homopolymer; chloroethylene polymer) [CH$_2$CHClCH$_2$CHCl]$_n$; CAS 9002-86-2; EINECS 208-750-2; Rubber substitutes; elec. wire and cable coverings; pliable thin sheeting; film finishes for textiles; nonflamm. upholstery; raincoats; tubing; belting; gaskets; shoe soles. [Aldrich; Ashland; Atochem SA; Chisso; Georgia Gulf; BFGoodrich; Goodyear; Hüls Am.; Mitsui Toatsu Chem.; Norsk Hydro AS; OxyChem; Vista; Wacker-Chemie GmbH]

polyvinylpyrrolidone. *See* PVP.

Polyviol. Polyvinyl alcohol in a range of viscosities and degrees of saponification; used as protective colloid for emulsions, dispersions and suspensions, textile auxiliaries (finishes, impregnating agents, sizes); thickening agent for glues and adhesives; release agent in the processing of polyester resins. [Wacker-Chemie GmbH]

Polywax® 500. CAS 9002-88-4; Polyethylene homopolymer; release agent in carbon ink formulations; modifier for paraffin waxes; component in hot-melt coatings, adhesives, and chewing gum base; lubricant in plastics and rubber processing, elec. insulation, powd. coatings, and printing inks; antiblocking agent; binder. [Petrolite]

Polywet® ND-2. Functionalized oligomer, sodium salt; dispersant for pigments, mins., extenders, fillers in aq. systems; latex paints and enamels; coatings, adhesives, paper and paperboard; boiler water. [Uniroyal]

Polywet® Z1766. Sodium salt of a polyfunctional oligomer; dispersant for titanium dioxide, other pigments. [Uniroyal]

Poly-zole AZDN. Azo diisobutyronitrile; a vinyl polymerization catalyst; gives freedom from side reactions and is not readily poisoned. [National Polychemicals] ‡

Polyzote. A proprietary trade name for nitrogen-expanded synthetic resin plastics. §

Pomace. The residue from the extraction of apple juice in cider manufacture; a cattle food. §

Pomarsol®. CAS 137-26-8; Thiram; fungicidal spray used for control of scab, storage rots, skin and other diseases on pome and stone fruit, grapes and vegetables; also as a seed dressing. [Bayer AG]

Pombe. A beer made from Sorghum millet.

Pommetrol M. Mixture of chlorpropham and propham; a plant growth regulator for potato sprout growth suppression. [Sam Fletcher Agricultural Specialists]

Pomoloy. A proprietary trade name for a cast iron made by a special process. §

Ponderax PA. CAS 404-82-0, A proprietary preparation of fenfluramine hydrochloride in a sustained-release capsule; an appetite suppressant. [Servier Laboratories Ltd]*

pondermite. A mineral, Ca$_2$B$_6$O$_{11}$ · 4H$_2$O. A source of boric acid.

Ponder's stain. A microscopic stain. It consists of 0.02 g

---

‡ Trade name and manufacturer not verified   § Trade name without identified manufacturer

toluidine blue, 1 cc glacial acetic acid, and 2 cc absolute alcohol in 100 cc distilled water.

Pondicherry oil. (nut oil). Arachis oil.

Pondimin. CAS 404-82-0; Fenfluramine hydrochloride; anorexic. [Wyeth Laboratories] *

Pondinil Roche. CAS 17243-57-1; Proprietary preparation of mefenorex; anorexic. [Roche Products Ltd] *

ponite. A mineral that is a variety of rhodocrosite containing iron.

ponolith. (sunolith, superlith). Lithopone pigments.

Ponoxylan. Polynoxylin; for treatment of infective skin conditions including furuncolosis and pustular acne, burns. [Berk Pharmaceuticals Ltd] *

Ponsital® . 3-Chloro-10 [LCB]-γ-[N'-β'-(1-methyl-2-oxo-imidazolidyl-3)-ethyl-N-piperazinyl]-propyl[RCB]-phenothiazine dihydrochloride; a neuroleptic agent. [AG Chemische Fabrik] ‡

Ponstan. CAS 61-68-7; A proprietary preparation of mefenamic acid; an analgesic. [Parke-Davis] *

Ponstel. CAS 61-68-7; Mefenamic acid; anti-inflammatory; analgesic. [Parke-Davis]

Pontallor. A range of precious metal alloys; for dentistry and dental engineering. [Degussa Ltd]

ponticin. See rhaponticum.

Pontocaine. Tetracaine; anesthetic. [Sterling Drug Inc] *

Pontocaine Hydrochloride. CAS 136-47-0; Tetracaine hydrochloride; anesthetic. [Sterling Drug Inc] *

Pool-Chem; Consumer product line for control of water chemistry in swimming pools. [Puma Chemical Co Inc]*

poonac. Coconut cake, a cattle food. The term is also used for the residue from castor oil seeds after cold and hot pressing and solvent extraction; used for caulking timber.

poonahlite. A mineral that is a hydrated aluminum-calcium silicate.

Pope's solution. A solution of 1 part in 10,000 of a mixture of 10 parts 2 : 7-dimethyl-3 : 6-diamino-acridinium-methylo-chloride hydrochloride and 1 part crystal violet; an antiseptic for wounds.

poppy capsules or heads. The dried, immature fruit of Papaver somniferum.

populin. Benzoyl-salicin; $C_{20}H_{22}O_8$.

porcelain. A mixture of clay, quartz, and felspar. A normal mix consists of 50% clay, 25% quartz, and 25% felspar.

porcelain clay. Synonym for Kaolinite.

porcelain earth. See China clay.

porcelain white. See lithopone.

porcelanite. A fused clay and shale found in burned coal seams.

Porcelave. A proprietary trade name for a ceramic material.§

Porfiromycin. 6-Amino-8-carbamoyloxymethyl - 1, 1α, 2, 8, 8α, 8β-hexahydro - 8α - methoxy - 1,5 - dimethylazirono [2 ´, 3´ : 3, 4] pyrrolo[1, 2-α-indole-4,7-dione; an antibiotic.

Porocel. A proprietary trade name for a carefully prepared and screened bauxite. §

Porofor® . A range of chemical blowing agents for the production of cellular plastics primarily based on PVC, polyethylene, ABS and polystyrene. [Bayer AG; Bayer plc]

Porofor® ADC/E. CAS 123-77-3; Azodicarbonamide; chemical blowing agent for production of plastic foams; used in expanded UPVC pipe and sections; extrusion of foamed polyethylene. [Bayer; Miles; Polysar]

Poron® 4701-01. High-density microcellular PU; used for gaskets, seals, product protection, cushion pads, vibration mounts, motor mounts, RF shielding, PCB cushions, spacers, foam-backed tapes, athletic padding in automotive, elec./electronic industries, etc. [Rogers] *

Poron® S2000-80-24031. Cellular silicone material; for use at temp. and performance extremes; excellent flame resistance, low smoke generation, low toxicity of combustion by-products; uv and ozone resistance. [Rogers]

Porosil-Clarcel. Diatomite fillers; for painting, fertilizers coating and defluoration. [Atochem UK/Ceca] *

porous alum. Sodium aluminum sulfate (soda alum), $Al_2(SO_4)_3 \cdot Na_2SO_4 \cdot 24H_2O$.

porpezite. A native alloy of gold and palladium; commonly contains silver as well.

porphyry. A building stone having the same composition as felspar.

porporino. An alloy of mercury, tin, and sulfur; used for decorating purposes.

Portagen. A proprietary artificial infant food containing medium chain triglycerides and nonlactose carbohydrates, for use in cases of intolerance of fat and lactose. [Bristol-Myers Squibb Co Inc]

Portalac. CAS 4618-18-2; Lactulose; for treatment of obstipation and (pre-) hepatic coma. [Duphar BV] *

Portland arrowroot. The starch from Arum maculatum.

Portland cement. CAS 65997-15-1; Made by heating an intimate mixture of argillaceous and calcareous substances, such as lime and clay, and pounding the product. The material does not slake with water, and has energetic hydraulic properties.

Portland stone. CAS 1317-65-3; A limestone.

Portman 5C Chormequat. Chlormequat + choline chloride; plant growth regulator for use in cereals and ornamentals. [Portman Agrochemicals Ltd]

Portman Chlormequat 400, 460, 600,700. CAS 999-81-5; Chlormequat; plant growth regulator. [Portman Agrochemicals Ltd]

Portman Isotop. CAS 34123-59-6; Suspension concentrate containing 500 g isoproturon per liter; used for annual weed control in cereals. [Portman Agrochemicals Ltd]

Portman Propachlor 50FL. CAS 1918-16-7; Propachlor; a pre-emergence herbicide for various horticultural crops. [Portman Agrochemicals Ltd]

Portman Supaquat. Chlormequat + choline chloride; plant growth regulator for use in cereals and ornamentals. [Portman Agrochemicals Ltd]

Portsmouth Accelerator No. 3. Phenyl-o-tolyl-guanidine; a proprietary rubber vulcanization accelerator. §

Portugallo oil. Essential oil of orange peel.

Portuguese turpentine. See turpentine.

Portyn Kapseals. A proprietary preparation containing benzinolinium bromide; an antispasmodic. [Parke-Davis] *

Posicor. Quazinone; cardiotonic. [Hoffmann-LaRoche Inc]*

Posistac. CAS 53003-10-4; Salinomycin; an ionophorous antibiotic used as a growth stimulating nutritional aid in cattle and swine. [Pfizer International]

Poskine. (o-propionylhyoscine) A depressant of the central nervous system; proscopine is the hydrobromide.

Poskydal. Unsaturated polyester resins; used for the formulation of furniture finishes with and without paraffin wax as well as for use in fillers. [Bayer AG]

---

* Trade name not verified as available   † Trade name verified as obsolete

Pos O Print. Ammonium hydroxide - 26 Baumé-29.4% concentrate, 24 Baumé-25.5% concentrate, 23 Baumé-23.5% concentrate, 20 Baumé-17.7% concentrate; developing solution for blue prints in the engineering/drafting industries; developing solution for microfilm in the micrographic industry. [W D Service Company Inc]

Post-4. Anti-settling additives; used for paints. [Rheox Inc]

Post-Kite. Suspension concentrate containing 50 g ioxynil, 250 g isoproturon and 180 g mccoprop per liter; used for control of weeds in wheat and barley. [Schering Agrochemicals Ltd]

Potaba. CAS 138-84-1; Potassium p-aminobenzoate in tablet, capsule and powder form; approved in the management of Peyronie's disease and scleroderma. [Glenwood Inc]

potarite. A mineral (Pd · Hg).

potash. See potassium carbonate.

potash alum. CAS 10043-67-1; 7784-24-9; A double sulfate of potassium and aluminum, $Al_2(SO_4)_3 \cdot K_2SO_4 \cdot 24H_2O$; used in paper, matches, paints, tanning agents, waterproofing agents, purification of water, astringent, cement hardener.

potash bordeaux mixture. Contains 6 lb copper sulfate, 2 lb potassium hydroxide, and 50 gallons water.

pot-ashes. Impure potassium carbonate.

potash felspar. See felspar.

potash glass. A glass containing silicate of potassium.

potash-lead glass. A glass usually containing from 40-50% SiO_2, 28-53% PbO, 8-11% K_2O, and 1% of both Al_2O_3 and Fe_2O_3.

potash salts. See Abraum salts.

potash water-glass. A mixture of potassium silicates.

potassalumite. A mineral that is a potash alum.

potassic superphosphate. A manure made by combining calcium superphosphate with potash salts.

potassium. Metallic element; K; CAS 7440-09-7; EINECS 231-119-8; intermediate for potassium peroxide, heat-exchange alloys; laboratory reagent; component of fertilizers (as potassium chloride). [Aldrich; Atomergic Chemetals]

potassium acetate. CH_3COOK; CAS 127-08-2; EINECS 204-822-2; Dehydrating agent, textile conditioner, analytical reagent, medicine, cacodylic derivatives, crystal glass, synthetic flavors. [Am. Int'l. Chem.; EM Industries; General Chem.; Heico; Honeywill & Stein Ltd; Niacet; Schaefer Salt & Chem.]

potassium acid tartrate. See potassium bitartrate.

potassium alginate. (potassium polymannuronate; alginic acid, potassium salt) Potassium salt of alginic acid; CAS 9005-36-1; gellant, emulsifier, and stabilizer in food and indust. application; gum, bodying agent for creams and lotions, dental impression materials; used for water holding in foods and industry. [Atomergic Chemetals; Kelco Int'l.]

potassium amalgam. An alloy of potassium and mercury, formed by the combination of the elements.

potassium benzoate. (benzoin acid potassium salt) C_6H_5COOK; CAS 582-25-2; EINECS 209-481-3. [Am. Biorganics; Mallinckrodt; Pentagon Chem. Ltd; Schweizerhall; Verdugt BV]

potassium bichromate. See potassium dichromate.

potassium bitartrate. (cream of tartar; potassium acid tartrate) $KHC_4H_4O_6$; CAS 868-14-4; Baking powder, preparation of other tartrates, galvanic tinning of metals,

food additive. [Penta Mfg.; Spectrum Chem. Mfg.]

potassium bromate. $BrKO_3$; $KBrO_3$; CAS 7758-01-2; EINECS 231-829-8; Laboratory reagent, oxidizing agent, permanent wave compounds, dough conditioner, food additive. [Allchem Industries; Gist-Brocades Food Ingreds.]

potassium bromide. KBr; CAS 7758-02-3; EINECS 231-830-3; Photography, process engraving and lithography, special soaps, spectroscopy, infrared transmission, lab reagent. [Aldrich; Great Lakes; Mallinckrodt; Morton Int'l.]

potassium cadmium iodide. See Marme's reagent.

potassium carbonate. (carbonic acid, dipotassium salt; dipotassium carbonate; potash) Inorganic salt; K_2CO_3; CAS 584-08-7; EINECS 209-529-3; special glasses (optical, TV tubes), potassium silicate, fertilizer manufacture, dehydrating agent, pigments, printing inks, lab reagent, general purpose food additive, textile, dyeing. [Hüls; Mallinckrodt; OxyChem]

potassium chloride. Inorganic salt; KCl; CAS 7447-40-7; EINECS 231-211-8; Fertilizer; foods, pharmaceuticals; in photography; in buffer solutions, electrode cells. [Aldrich; Heico; Mallinckrodt; Morton Int'l.; Reheis]

potassium citrate. $C_6H_5O_7 \cdot 3K$; CAS 866-84-2; food additive; antacid. [Haarmann & Reimer; Mallinckrodt; Pfizer SA; Schweizerhall]

potassium cyanide. KCN; CAS 151-50-8; EINECS 205-792-3; Gold and silver extraction, analytical reagent, insecticide, fumigant, electroplating. [Atochem N. Am.; Degussa; Du Pont; W.R. Grace Ltd; Mallincrodt]

potassium dichloroisocyanurate. (dichloroisocyanuric acid potassium salt; potassium dichloro-s-triazinetrione) $C_3HCl_2N_3O_3 \cdot K$; CAS 2244-21-5; Bleaching compound., sanitizer/disinfectant, oxidizer in dishwashing compositions.

potassium dichloro-s-triazinetrione. See potassium dichloroisocyanurate.

potassium dichromate. (potassium bichromate; red potassium chromate) $K_2Cr_2O_7$; CAS 7778-50-9; EINECS 231-906-6; Oxidizing agent, analytical reagent, brass pickling, electroplating, pyrotechnics, explosives, matches, textile dyeing and printing, adhesives, tanning leather, wood stains, lithography, synthetic perfumes, pigments, alloys, ceramics, batteries. [Hoechst Celanese; Mallinckrodt]

potassium dihydrogen orthophosphate. See potassium phosphate.

potassium felspar. See felspar.

potassium d-gluconate. (D-gluconic acid potassium salt) $C_6H_{11}O_7K$; CAS 299-27-4; EINECS 206-074-2; Mineral source for pharmaceutical and food products [Akzo; R.W. Greeff]

potassium 2,4-hexadienoate. See potassium sorbate.

potassium hydrate. See potassium hydroxide.

potassium hydroxide. (caustic potash; potassium hydrate; lye) Inorganic base; KOH; CAS 1310-58-3; EINECS 215-181-3; Soap manufacture, bleaching, manufacture of chemicals, electrolyte, absorbent, liq. fertilizers, food additive, herbicides, electroplating, mercerizing, paint removers, reagent. [Hüls Am.; ICI Specialties; Olin; OxyChem]

potassium iodate. KIO_3; CAS 7758-05-6; EINECS 231-831-9; Analysis (testing for zinc and arsenic), iodometry, reagent, feed additive, dough conditioner, medicine

---

‡ Trade name and manufacturer not verified   § Trade name without identified manufacturer

(topical antiseptic). [Atomergic Chemetals; Blythe, Williams Ltd; R.W. Greeff; Mallinckrodt; Spectrum Chem. Mfg.]

potassium iodide. Inorganic salt; KI; CAS 7681-11-0; EINECS 231-659-4; reagent in analytical chemistry; photographic emulsions; animal feed additive; dietary supplement; in table salt; nylon stabilizer. [Aldrich; Atomergic Chemetals; R.W. Greeff; Mitsui Toatsu]

potassium lignosulfonate. Used as a dispersant. [Wesco Tech. Ltd.]

potassium metabisulfite. (disulfurous acid, dipotassium salt; potassium pyrosulfite; dipotassium disulfite) Inorganic salt; $K_2S_2O_5$; CAS 16731-55-8; EINECS 240-795-3; antiseptic; reagent; source of sulfurous acid; photographic developing agent; brewing, wine making; food preservative; bleaching agent. [Allchem Industries; Farleyway Chem. Ltd; Mallinckrodt]

potassium (meta)periodate. See potassium periodate.

potassium muriate. See muriate of potash.

potassium nitrate. (niter; nitre; saltpeter) KNO_3; CAS 7757-79-1; EINECS 231-818-8; Pyrotechnics, explosives, matches, fertilizer, reagent, glass manufacture, tempering steel, curing foods, oxidizer in solid rocket propellants. [Am. Biorganics; EM Industries; Mallinckrodt; San Yuan Chem. Co. Ltd.; Whiting, Peter Ltd]

potassium 9-octadecenoate. See potassium oleate.

potassium oleate. (potassium 9-octadecenoate; oleic acid, potassium salt) Potassium salt of oleic acid; $CH_3(CH_2)_7CH=CH(CH_2)_7COOK$; CAS 143-18-0; EINECS 205-590-5; Liq. soap for hand cleaners, tire mounting lubricant; emulsifier and corrosion control in paint strippers. [Emkay; Norman, Fox]

potassium oxalate. (potassium oxalate monohydrate; oxalic acid potassium salt) $(COOK)_2 \cdot H_2O$; CAS 583-52-8; analytical reagent, source of oxalic acid, bleaching and cleaning, removing stains from textiles, photography. [Am. Int'l. Chem.; General Chem.; Heico; Verdugt BV]

potassium oxalate monohydrate. See potassium oxalate.

potassium perchlorate. $KClO_4$; CAS 7778-74-7; EINECS 231-912-9; Explosives, oxidizing agent, photography, pyrotechnics and flares, reagent, oxidizer in solid rocket propellants. [Am. Int'l. Chem.; Eka Nobel AB; Mallincrodt; San Yuan Chem. Co. Ltd.]

potassium periodate. (potassium (meta)periodate; periodic acid potassium salt) KIO_4; CAS 7790-21-8; analysis, oxidizing agent. [Atomergic Chemetals; Cerac; Spectrum Chem. Mfg.]

potassium permanganate. $KMnO_4$; CAS 7722-64-7; EINECS 231-760-3; Oxidizer, disinfectant, deodorizer, bleach, dye, tanning, radioactive decontamination of skin, analytical reagent, medicine (antiseptic), manufacture of organic chemicals, air and water purification. [Am. Biorganics; Am. Int'l. Chem.; Blythe, William Ltd; Mallinckrodt; Schweizerhall]

potassium peroxydisulfate. See potassium persulfate.

potassium persulfate. (peroxydisulfuric acid dipotassium salt; potassium peroxydisulfate) $K_2S_2O_8$; CAS 7727-21-1; EINECS 231-781-8; Bleaching, oxidizing agent, reducing agent in photography, antiseptic, soap manufacture, analytical reagent, polymerization promoter, pharmaceuticals, starch modifier, flour-maturing agent, textile desizing. [Allchem Industries; Du Pont; FMC; Mallinckrodt; San Yuan Chem. Co. Ltd.; Transol Chem. UK Ltd]

potassium phosphate. (MKP; potassium phosphate, monobasic; potassium dihydrogen orthophosphate; monopotassium orthophosphate) Inorganic salt; KH_2PO_4; CAS 7778-77-0; EINECS 231-913-4; Food products, baking powder, nutrient solutions, water treatment, buffer and sequestrant, laboratory reagent. [Albright & Wilson; Aldrich; FMC; Heico; Monsanto]

potassium phosphate, dibasic. (DKP; dipotassium phosphate; dipotassium hydrogen orthophosphate; dipotassium orthophosphate) K_2HPO_4; CAS 7758-11-4; food and automotive industry; buffer in antifreezes; nutrient; humectant; pharmaceuticals. [Albright & Wilson; Aldrich; FMC; Heico; Monsanto; U.S. Biochemical]

potassium phosphate, monobasic. See potassium phosphate.

potassium phosphate, tribasic. (TKP; tripotassium phosphate) K_3PO_4; CAS 7778-53-2; EINECS 231-907-1; Detergent, water treatment, automotive products, fertilizer, foods as emulsifier. [Albright & Wilson; Ashland; FMC; Monsanto]

potassium polyacrylate. (polyacrylic acid, potassium salt) Potassium salt of polyacrylic acid; $(C_3H_4O_2)_x \cdot xK$; CAS 25608-12-2; dispersant for latex paints and coatings, pigments.

potassium polymannuronate. See potassium alginate.

potassium pyrophosphate. (TKPP; tetrapotassium pyrophosphate; potassium pyrophosphate, normal) $K_4P_2O_7 \cdot 3H_2O$; CAS 7320-34-5; soap and detergent builder, sequestering agent, peptizing and dispersing agent. [Atochem N. Am.; FMC; Monsanto]

potassium pyrophosphate, normal. See potassium pyrophosphate.

potassium pyrosulfite. See potassium metabisulfite.

potassium sodium tartrate. (potassium sodium tartrate tetrahydrate; Rochelle salt; Seignette salt) $KOCOCH(OH)CH(OH)COONa \cdot 4H_2O$; CAS 6381-59-5; EINECS 206-156-8; Baking powders, medicine (cathartic), silvering mirrors. [Aldrich; EM Industries; Mallinckrodt; Pfizer Spec. Chem.]

potassium sodium tartrate tetrahydrate. See potassium sodium tartrate.

potassium sorbate. (2,4-hexadienoic acid, potassium salt; sorbic acid, potassium salt; potassium 2,4-hexadienoate) $CH_3CH=CHCH=CHCOOK$; CAS 590-00-1; EINECS 246-376-1; As mold and yeast inhibitor. [Chisso Am.; Gist-Brocades Food Ingreds.; Hoechst Celanese; Pfizer Spec. Chem.; Protameen]

potassium stannate. $K_2SnO_3 \cdot 3H_2O$; CAS 12142-33-5; Textile dyeing and printing, alkaline tinplating bath. [Allchem Industries; Atochem N. Am.; Blythe, William Ltd; M&T Harshaw; Nihon Kagaku Sangyo]

potassium stearate. (octadecanoic acid, potassium salt; stearic acid, potassium salt) Potassium salt of stearic acid; $CH_3(CH_2)_{16}COOK$; CAS 593-29-3; in manufacture of textile softeners. [Original Bradford Soap Works; Witco]

potassium sulfate. K_2SO_4; CAS 7778-80-5; EINECS 231-915-5; Analytical reagent, medicine (cathartic), gypsum cements, fertilizer, manufacture of alum and glass, food additive. [Chisso; General Chem.; Heico; Mallinckrodt]

potassium tripolyphosphate. (KTPP; pentapotassium tripolyphosphate) $K_5P_3O_{10}$; CAS 13845-36-8; Detergents, paints, cleaners, specialty fertilizers, sequestrant. [Albright & Wilson; FMC]

---

* Trade name not verified as available   † Trade name verified as obsolete

potato gum. (almadina, euphorbia gum). Almeidina gum, stated to be derived from *Euphorbia rhipsaloides*, of West Africa. The latex contains about 10% rubber, 32% water, 51% resin, 1% protein, 6% insoluble matter, and gives an ash of 2.5%. The dry material contains 14.3% rubber and 75.8% resin.

potato oil or spirit. The alcohol obtained from potato starch.

Potato-Pro EN-15. Hydrolyzed potato protein; cosmetic ingredient for skin and hair care products. [Brooks Industries]

potato rubber. *See* potato gum.

Potazote. A French fertilizer containing 14% nitrogen, as ammonium chloride, and 20% potassium oxide as potassium chloride.

Potentite. *See* Tonite.

Potenzol V. Alkyl aryl polyether alcohol as an emulsifiable compound; wetting agent for herbicides [(Invequimica & CIA SCA).] ‡

P.O.T.G. Phenyl-o-tolyl-guanidine; used as a rubber vulcanization accelerator.

Potin. An alloy of 72% copper, 25% zinc, 2% lead, and 1% tin.

Potinjaune. *See* Potin.

Potkem. CAS 1312-76-1; Two-component products based on potassium silicate; as mortars for use with acid resistant bricks and paviors in heavy-duty chemical applications, e.g., brick chimney linings. [Feb Ltd]

pot metal. An alloy of lead and copper.

Potosi silver. *See* nickel silvers.

potstone. An impure steatite (*qv*).

Potter's clay. *See* pipeclay.

Potting Base. Powdered compost additive containing NPK 3.6:2.2:2.9 plus 3% Mg and trace elements; compost additive. [Vitax Ltd]

Pouckpong gum. (Touchpong gum). A rubber gum of British Guiana.

Poudre B. (Vieille powder). A French explosive. It is a smokeless powder made from a mixture of soluble and insoluble nitrocellulose, thoroughly gelatinized with a mixture of ether and alcohol, rolled into sheets, and cut into strips.

Poudre EF. A French explosive made from nitro-cellulose and binding material.

Poudre J. A French explosive containing 83% guncotton and 17% potassium bichromate.

Poudre Pyroxulée. A French sporting powder; consists of insoluble nitro cellulose, with 35% barium and potassium nitrates.

Poulenc 309. sym-Disodium-m-amino-benzoyl-m-amino-p-methyl-benzoyl-1-naphthyl-amino-4 : 6 : 8-trisulfonate-urea. §

Poutet's reagent. Consists of 1 cc mercury dissolved in 12 cc nitric acid, specific gravity 1.42; used for testing oils.

Povan. CAS 3546-41-6; Pyrvinium pamoate; anthelmintic. [Parke-Davis] *

Povidone. CAS 9003-39-8; Poly(vinylpyrrolidone) Plasmosan.

povidone. *See* PVP.

povidone-iodine. A complex produced by reacting iodine with poly(vinylpyrrolidone); betadine.

Powaspray Glymark. An inert white carrier marker for CDA use; an amenity and industrial weed control system that allows approved pesticides to be added to Glymark and applied at lower rates, e.g., glyphosate/Kerts flowable.

[Polymer Corp]

Powax. Additive to hot rinse tanks used after acid pickling of steel sheet and rod. [Foseco (F.S.) Ltd] †

Powder 19/04/15H Black 904. A proprietary polyethylene used in rotational molding and carpet-backing applications; can be used in contact with foodstuffs. §

Powder 215 Natural. A proprietary 400 micron powder used in flame-retardant rotational moldings. §

Powder 22/04/00A 400. A proprietary polyethylene powder of micron size having a low melting point; used for making interliners for fabrics and as carpet backing. §

Powder 26/04/00. A proprietary polyethylene powder possessing good rigidity; used in rotational molding. §

Powdered Aloe Vera (1:200) Food Grade. Aloe vera gel; for personal care products, suntan, sun treatment, burn gels, first aid creams, soaps, hair care, cosmetics, weight control, oral hygiene. [Tri-K Industries]

powdered hydrocyanic acid. The name applied to a calcium cyanide prepared from calcium carbide and hydrocyanic acid. It evolves hydrocyanic acid with moisture, hence the name. A fumigator.

powder of Algaroth. (basic chloride, English powder, powder of Algarotti). A mixture of antimony oxychloride, $SbOCl$, and antimony oxide, Sb_2O_3; used in the preparation of tartar emetic.

powder of Algarotti. *See* powder of Algaroth.

powellite. A mineral. It is calcium molybdate, $CaMoO_4$.

Power 64, 640, 700. CAS 999-81-5; Chlormequat; plant growth regulator. [Kommer-Brookwick Ltd]

Power Chlorothalonil 50. CAS 1897-45-6; Chlorothalonil; a fungicide for a wide range of agricultural crops. [Kommer-Brookwick Ltd]

Power Demo. CAS 23103-98-2; Granules containing 50% w/w pirimicarb; for control of aphids. [Kommer-Brookwick Ltd]

Power Diquat. CAS 85-00-7; Diquat; a contact herbicide and pre-harvest crop desiccant. [Kommer-Brookwick Ltd]

Power Drive. CAS 50471-44-8; Vinclozolin; a protectant fungicide for oilseed rape, peas and beans. [Kommer-Brookwick Ltd]

Power DSM. CAS 919-86-8; Emulsifiable concentrate containing 580 g demeton-S-methyl per liter; a systemic organophosphorus insecticide. [Kommer-Brookwick Ltd]

Power Ethephon 48. Soluble concentrate containing 480 g 2-chlorethylphosphonic acid per liter; plant growth regulator for cereal crops. [Kommer-Brookwick Ltd]

Power Flame. CAS 52756-22-6; Emulsifiable concentrate of 200 g flamprop-M-isopropyl per liter; used for control of wild oats in cereal crops. [Kommer-Brookwick Ltd]

Power Flamprop. CAS 52756-22-6; Emulsifiable concentrate of 200 g flamprop-M-isopropyl per liter; used for control of wild oats in cereal crops. [Kommer-Brookwick Ltd]

Power Gard. Proprietary blend of fuel additives; used for improved fuel consumption, reduced carbon deposits, cleaner carburator, reduced valve deposits. [Gard Corporation] *

Power Gro-Stop. Mixture of chlorpropham and propham; a plant growth regulator for potato sprout growth suppression. [Kommer-Brookwick Ltd]

Power MCPA. CAS 94-74-6; MCPA; herbicide for cereals and grassland. [Kommer-Brookwick Ltd]

---

‡ Trade name and manufacturer not verified   § Trade name without identified manufacturer

Power Non-ionic Wetter. Nonionic wetting agent containing 900 g/l alkylphenol ethylene oxide concentrate; for use in herbicides and pesticides. [Kommer-Brookwick Ltd]

Power Phosphine Pellets. CAS 20859-73-8; Aluminum phosphide; used for gasing of rabbits and moles. [Kommer-Brookwick Ltd]

Power Platoon. Soluble concentrate containing 155 g 2-chlorethylphosphonic acid and 305 g mepiquat chloride per liter; plant growth regulator for cereal crops. [Kommer-Brookwick Ltd]

Power Propiconazole. CAS 60207-90-1; Propiconazole; a systemic triazole fungicide for control of powdery mildew and rust in cereals. [Kommer-Brookwick Ltd]

Powerspire. CAS 60207-90-1; Propiconazole; a systemic triazole fungicide for control of powdery mildew and rust in cereals. [Kommer-Brookwick Ltd]

Power Spray Save. CAS 61791-44-4; Cationic surfactant containing 800 g/l tallow amine ethoxylate; wetting agent for phosphonoglycine herbicide sprays. [Kommer-Brookwick Ltd]

Powers Terebine. Drier solution; for addition to certain paints and varnishes to speed drying. [Llewellyn Ryland Ltd]

Power Swing. CAS 34123-59-6; Suspension concentrate containing 500 g isoproturon per liter; used for annual weed control in cereals. [Kommer-Brookwick Ltd]

Power Task. CAS 67306-03-0; Emulsifiable concentrate of 750 g fenpropimorph per liter; used for control of mildew in cereals. [Kommer-Brookwick Ltd]

Powmet. Metal powders and premixes. [McKechnie Chemicals Ltd] *

PP. *See* polypropylene.

PP-10GF/000. CAS 9003-07-0; PP, 10% glass fiber-reinforced. [Compounding Tech.] *

PP Captan 83. CAS 133-06-2; Captan; a dicarboximide fungicide. [ICI Agrochemicals]

P.P.D. Piperidine-pentamethylene-dithio-carbamate. A rubber vulcanization accelerator.

PPO. Polyphenylene oxide.

P.P.S. A proprietary polyphenylene sulfide; a cross-linkable aromatic thermoplastic with a high modulus used as a coating material capable of withstanding temperatures in the range 200-260 C. [LNP] *

PPSB Cutter®. Hemostatic agent. [Bayer AG]

PP-Vac. Pigeon pox; for immunization of poultry. [Intervet Inc]

PPX-30GF/000 HC. CAS 9003-07-0; PP, highly chemically coupled, 30% glass fiber-reinforced. [Compounding Tech.] *

Practolol. CAS 6673-35-4; A beta adrenergic receptor blocking agent. It is 4-(2-hydroxy-3-isopropyl-aminopropoxy)acetanilide.

Pradone Plus. Herbicide. [May & Baker Ltd] *

Praenitrona. CAS 588-42-1; A proprietary trade name for trolnitrate phosphate. §

Praepagen. Textile softening materials. [Hoechst AG]

Praestol® 186K. Polydimethyl-dialkyl ammonium chloride; polymer for use as wastewater coagulant in mining/mineral processing industries, water plants, biological wastewater facilities; filtration aid. [Stockhausen]

Praestol® A3010L. Acrylamide emulsion polymer; flocculating agent for effluent treatment (papermill, textile mill, food processing waste, petrochem. wastewater, steel mill effluent), mineral processing, brine clarification, sand/gravel washing, sugar juice clarification. [Stockhausen]

Praestol® K2001. CAS 7446-70-0; Basic aluminum chloride; coagulant for municipal and industrial water treatment; effective for turbidity reduction, phosphorus removal, water clarification, flotation, oil/water demulsification. [Stockhausen]

Pragmatar. A proprietary preparation of cetyl alcohol and coal tar distillate with sulfur and salicylic acid; used in the treatment of dandruff. [SmithKline Beecham] *

Pragmatar Ointment. Combination of cetyl alcohol-coal tar distillate, precipitated sulfur and salicylic acid; for treatment of dandruff, seborrheic conditions and common scaly skin disorders where skin is unbroken. [SmithKline Beecham] *

Pragmoline. A proprietary preparation of acetylcholine bromide. §

Praims. Cough drops. [Richardson-Vicks Inc] *

Prajmalium bitartrate. A preparation used in treatment of arrhythmia of the heart. N-Propylajmalinium hydrogen tartrate.

Pramidex. CAS 64-77-7; A proprietary preparation of tolbutamide; used in the treatment of diabetes. [Berk Pharmaceuticals Ltd] *

Prantal. CAS 62-97-5; Diphemanil methylsulfate; anticholinergic. [Schering Corp] *

Prapagen WK, WKL, and WKT. Cationic surfactants of the quaternary ammonium chloride type in liquid or paste form; antistatic agents, fabric conditioner and softener, fiber finishers, water-repellant agents and dewatering agents; wetting agents for oils, dispersants for pigments, flushing agents, foaming and wetting agents, spinning bath and viscous additives; flotation chemicals and anti-caking agents for rendering salts free-flowing, corrosion inhibitors; anchoring/wetting agents for tars and bitumen, surface coatings, lacquers, adhesives, disinfectants, hair cosmetics, auxiliaries for leather, textiles, rubber and metal industries. [Hoechst UK] *

prase. A mineral, SiO_2.

Praxilene. CAS 3200-06-4; Nafronyl oxalate; vasodilator. [Lipha SA] *

Prazepam. CAS 2955-38-6; 7-Chloro-1-(cyclopropyl-methyl)-1, 3-dihydro-5-phenyl-2H-1,4-benzo-diazepin-2-one; used as a muscle relaxant.

prazosin. CAS 19216-56-9; 1-(4-Amino -6, 7-dimethoxy-quinazolin- 2-yl) -4-(2-furoyl)piperazine; used as an anti-hypertensive.

PRC. Motor fuel additive. [Chevron] *

Preact. Galvanizing fluxes. [Mineral Research & Development Corp]

Prebane 500. CAS 886-50-0; Suspension concentrate containing 500 g/l terbutryn; for weed control in cereals. [Ciba-Geigy Agrochemicals]

Prebond. Pre-phosphate refining additive. [Brent Chemicals International plc]

Precef. CAS 60925-61-3; Ceforanide; antibacterial. [Bristol Laboratories] *

Prechem 90. Phosphate detergent; detergent, wetting agent, dispersant, soil suspending agent for textile preparation and dyeing; leveling agent; penetrant. [Ivax Industries]

Prechem NPX. Nonylphenoxy polyethoxy alcohol; detergent conc., wetting agent, emulsifier for textile wet pro-

---

* Trade name not verified as available   † Trade name verified as obsolete

cessing of cotton, wool and synthetics; aids dyebath stability. [Ivax Industries]

Precifac ATO. CAS 540-10-3; EINECS 208-736-6; Cetyl palmitate. [Gattefosse SA]

precipitated phosphate. Insoluble calcium phosphate.

Precirol ATO. CAS 8067-32-1; Glyceryl di/tripalmito stearate; additive for tablets, binder, lubricant, sustained release. [Gattefosse SA]

Precirol WL 2155. CAS 8067-32-1; Glyceryl ditristearate; additive for tablets manufacturing. [Gattefosse; Gattefosse SA]

Preconativ. Factor IX concentrate for the treatment of hemophilia B. [KabiVitrum AB] *

Predalone TBA. CAS 7681-14-3, Prednisolone tebutate; glucocorticoid. [O'Neal, Jones & Feldman Pharmaceuticals] *

predazzite. A mineral. It is a mixture of calcite and brucite.

Predef. CAS 338-98-7; Isoflupredone acetate; anti-inflammatory. [Upjohn] *

Predenema®. Prednisolone metasulfobenzoate; a proprietary liquid enema formulation for treatment of ulcerative colitis and procitis. [Pharmax Ltd]

Predfoam®. Prednisolone metasulfobenzoate; a proprietary foam enema formulation for treatment of ulcerative colitis and procitis. [Pharmax Ltd]

Pred Forte®. CAS 52-21-1; Prednisolone acetate; corticosteroid, anti-inflammatory. [Allergan, Inc]

Pred-G Liquifilm®. Prednisolone acetate, gentamicin sulfate; corticosteroid, antibacterial. [Allergan, Inc]

Pred-G S.O.P. Prednisolone acetate, gentamicin sulfate; corticosteroid, antibacterial. [Allergan, Inc]

Preditec®. For agriculture industry. [DuPont UK]

Pred Mild®. CAS 52-21-1; Prednisolone acetate; corticosteroid, anti-inflammatory. [Allergan, Inc] *

Predne-Dome. CAS 50-24-8; Prednisolone; glucocorticoid. [Miles Pharmaceuticals] *

Prednelan. CAS 50-24-8; A proprietary preparation of prednisolone. [Glaxo Laboratories] *

Prednesol. A proprietary preparation of prednisolone disodium phosphate. [Glaxo Laboratories]

Prednis. CAS 50-24-8; Prednisolone; glucocorticoid. [USV Pharmaceutical Corp] ‡

prednisolone. CAS 50-24-8; (11β, 17α-21-trihydroxy-pregna-1,4-diene-3,20-dione; 1,2-dehydrocortisone; metacortandralone) Analog of hydrocortisone; used in medicine

Prednisone. CAS 53-03-2; (17α-21-Dihydroxypregna-1,4-diene-3,11,20-trione-1,2-dehydrocortisone; metacortandracin). Analog of cortisone; used in medicine.

Predsol. A proprietary preparation of prednisolone disodium phosphate. [Glaxo Laboratories]

Predsol-N. A proprietary preparation of prednisolone sodium phosphate and neomycin sulfate. [Glaxo Laboratories]

Pree®. Metazachlor; for control of grasses and broadleaf weeds in maize [BASF AG]

Pre-Empt. Emulsifiable concentrate containing 46 g linuron, 54 g trietazine and 200 g trifluralin per liter; herbicide for winter cereals. [Schering Agrochemicals Ltd]

Prefera® Range. Sodium or calcium stearyl-2-lactylates; food emulsifier for improvement of fermentation tolerance, volume and texture of yeast-raised baked goods; antistaling effect. [Grünau]

Preferid. CAS 51333-22-3; A proprietary preparation of

budesonide; used as a topical corticosteroid. [Brocades Pharma]

Prefin Liquifilm. CAS 61-76-7; Phenylephrine hydrochloride; adrenergic. [Allergan Pharmaceuticals Inc]

Prefix D. CAS 1194-65-6; A granular formulation containing 6.75% dichlobenil; provides season-long weed control of both annual and perennial grasses and broad leaved weeds. [Burts & Harvey, Hoechst UK; Shell UK] *

Prefrin-A®. Pyrilaamine maleate, phenylephrine hydrochloride; adrenergic. [Allergan, Inc]

Prefrin Liquifilm®. Phenylephrine hydrochloride, polyvinyl alcohol; redness reliever, ocular lubricant. [Allergan, Inc]

Pregaday. A proprietary preparation of ferrous fumarate and folic acid; hematinic for use in pregnancy. [Duncan Flockhard Ltd]

Pregamal. A proprietary preparation of folic acid and ferrous fumarate; a hematinic. [Glaxo Laboratories] *

Pregeflo®. Pre-cooked starches; thickener for foods; binding agent in pharmaceuticals. [Roquette (UK) Ltd]

Pregfol. A proprietary preparation of folic acid and ferrous sulfate; a hematinic. [Wyeth Laboratories] *

Pregl's solution. A solution of potassium iodide and sodium iodate with a small amount of sodium chloride and bicarbonate.

Pregnyl. Gonadotropin, chorionic; gonad-stimulating principle. [Organon Inc]

pregolan. A chlorine compound giving 65-72% available chlorine.

pregrattite. A mineral that is a variety of paragonite.

Pregwood. A proprietary synthetic resin impregnated wood, made by impregnating and then subjecting the wood to heat and pressure. §

prehnite. (Jacksonite). A mineral; it is a silicate of aluminum and calcium, $2CaO \cdot Al_2O_3 \cdot 3SiO_2$.

Pre-Kite. Selective herbicide. [Schering Agrochemicals Ltd]

Prelude. CAS 67747-09-5; Prochloraz; protectant fungicide against seed-borne diseases. [Agrichem (International) Ltd]

Preludin. CAS 1707-14-8; Phenmetrazine hydrochloride; anorexic. [Boehringer Ingelheim Pharmaceuticals Inc] *

Premaline. Herbicide. [May & Baker Ltd] *

Premalox. Herbicide. [May & Baker Ltd] *

Premarin Tablets. Natural conjugated estrogens; for menopausal and postmenopausal estrogen therapy in women without a uterus for vasomotor symptoms such as sweating and flushes, allied disorders of the menopause such as atrophic vaginitis, kraurosis, vulvae, athropic urethritis; prophylaxis of osteoporosis in women at risk of developing fractures; palliation of selected cases of breast cancer. [Wyeth Laboratories]

Premarin Vaginal Cream. Conjugated estrogen (natural); for treatment of atrophic vaginitis and postmenopausal atrophic urethritis. [Wyeth Laboratories]

Premerge. CAS 88-85-7; Active ingredient: dinosebl; herbicide used for the control of broadleaf weeds in peas, soybeans, potatoes, and orchards. [Dow UK] *

Premier alloy. A heat-resisting alloy containing 61% nickel, 11% chromium, 25% iron, and 3% manganese.

Premix. Masterbatches in thermoplastics; for pigmentation of thermoplastics. [Cornelius Chemical Co Ltd] *

Prempak. Natural conjugated estrogen plus norgestrel; for menopausal and postmenopausal estrogen replacement therapy and allied disorders such as postmeno-

---

‡ Trade name and manufacturer not verified   § Trade name without identified manufacturer

pausal osteoporosis, atrophic vaginitis, postmenopausal atrophic urethritis. [Wyeth Laboratories] †

Prempak-C. Natural conjugated estrogens plus norgestrel; menopausal and postmenopausal estrogen replacement therapy in women with an intact uterus for vasomotor symptoms such as sweating and hot flushes, allied disorders of the menopause such as atrophic vaginitis, kraurosis, vulvas, atrophic urethritis; prophylaxis of osteoporosis in women at risk of developing fractures. [Wyeth Laboratories]

Prenite. A proprietary trade name for an asbestos sheet bonded with neoprene; a packing material. §

Prenol. 3-Methyl-2-butene-1-ol; fragrance and flavoring (fresh, herbal, green, fruity, slightly lavender-like) [BASF AG]

Prenomiser. A proprietary preparation of isoprenaline sulfate in aerosol form; a bronchial antispasmodic. [Fisons plc, Pharmaceuticals Div] *

Prent® . CAS 37517-30-9; Acebutolol; pharmaceutical preparation; cardioselective beta blocker. [Bayer AG]

prenylamine. CAS 503-60-6; N - (3,3-Diphenylpropyl)-α-methyl phenethylamine. Segontin and synadrin are the lactate.

Prepagen. Textile softening agents. [Hoechst UK] *

Prepagen WK. CAS 107-64-2; EINECS 203-508-2; A proprietary trade name for 75% distearyl dimethyl ammonium chloride in isopropanol; used as a softener in the laundry trade. [Hoechst UK] *

Pre-Par. CAS 26652-09-5; Ritodrine; uterospasmolytic, prevents premature birth. [Duphar BV] *

prepared bark. See quercitron.

prepared calamine. Obtained by calcining and powdering negative zinc carbonate or calamine, and freeing the product from gritty particles. It consists of zinc carbonate with some oxide of iron.

prepared chalk. (*Creta proeparata B.P.*) Washed chalk or whiting.

prepared cobalt oxide. Cobalt oxide, CoO, obtained by heating the black oxide, Co_2O_3; used in the ceramic industry.

preparing salt. Sodium stannate, $Na_2SnO_3 \cdot 3H_2O$; used as a mordant in dyeing and calico printing.

Prepcort. CAS 50-23-7; Hydrocortisone; glucocorticoid. [Whitehall Laboratories] *

Preperite. Compound used to remove rust from iron and steel and corrosion products from aluminum. [Chemicals International plc] ‡

Prepon®. Casting and modelling wax, blue, green and ivory; dental preparation. [Bayer AG]

PrepRite™ Coating Remover. N-Methyl-2-pyrrolidone, butyrolactone, and other ingreds. [ISP]

Presamine. CAS 113-52-0; Imipramine hydrochloride; antidepressant. [USV Pharmaceutical Corp] ‡

Preservaline. (Freezine, Iceline). Names for formaldehyde used as a preservative for milk.

Preservals® . Parabens; preservatives for cosmetics and pharmaceuticals. [Laserson & Sabetay]

Preservative. Mixture of antimicrobials; preservative; for cosmetic preparations. [Rewo Chemicals Ltd] *

Preservol. Creosote, partially emulsified with pyroligneous acid; used for preserving wood. §

Preservo-O-Sote. Wood preservatives. [Crowley Chem.]

Preservotabs. Water bath sanitizing tablets. [The Boots Co plc]

Presidal. A proprietary trade name for pentacynium methylsulfate. §

Presinol®, Presinol Mite. CAS 555-30-6; Methyldopa; antihypertensive. [Bayer AG]

Presite® . For agriculture industry. [DuPont UK]

Presol. Dampproofer and wood preserver solvents. [Carless Solvents Ltd] *

Presol W. Mercury fungicide solution. [Great Lakes Europe]†

Presomet. Black bitumous paint. [Thomas Ness Ltd] *

Press-Aid. CAS 8002-74-2, 9010-66-6; Synthetic wax, corn gluten protein; binder for pressed powds. [Presperse]

press-cake. The mill-cake formed by mixing the ingredients of gunpowder in the incorporating mill, is subjected to a high pressure to make press-cake.

pressed amber. See ambroid.

Pressimmune. A proprietary preparation of antihuman lymphocyte globulin; used in tissue transplants to produce immuno-suppression. [Hoechst UK] *

Press N Seal. Caulking tape in roll form; used to caulk windows, doors and other openings in construction buildings and homes. [Chemseco] *

Pressolith. (Sillimanith). Earthenware porcelain products.

Pressonex Bitartrate. CAS 33402-03-8; Metaraminol bitartrate; adrenergic. [Sterling Drug Inc] *

Pressphan. A German name for press-boards made from wood pulp; used as insulating materials.

Presszell. A German synthetic resin varnish-paper product used as an electrical insulator. §

Prestochlor. CAS 7778-54-3; Calcium hypochlorite. [P B & S Chemical Co Inc] ‡

Prestogen® K. Stabilizer for low silicate peroxide bleaching of cellulosic fibers and blends. [BASF; BASF plc]

Prestogen® TX-155. Multifunctional stabilizer for peroxide bleaching; high alkali stability. [BASF]

Prest-o-lite. A proprietary brand of acetylene gas compressed in cylinders. §

Prestone. Ethylene glycol antifreeze. [National Carbon Company Inc] *

Preston salts. (smelling salts). Consist of acid ammonium carbonate; NH_4HCO_3.

Presto Steel. A proprietary trade name for a steel containing 1.4% chromium. §

Presuren. A proprietary trade name for hydroxydione sodium succinate. §

pretamazium iodide. 4-(Biphenyl-4-yl)-3-ethyl-2-[4-(pyrrolidin-1-yl) styryl] thiazolium iodide; a preparation used in the treatment of enterobiasis.

Pretolone. Dyeing and printing assistant. [Ciba plc] *

Prevail 3050. Thermoplastic PU/ABS blend; engineering thermoplastic providing unique combination of flexibility and toughness; suggested for bumpers and fairings on commercial trucks, on snowmobiles, ATVs and campers, and in the automotive market for paintable soft fascia. [Dow Plastics]

Prevenol®. Wide range of microbicides/preservatives and corrosion inhibitors. [Bayer AG; Bayer plc]

Prevenol® A2. An organic inhibitor; for use in the formulation of fungicidal interior paints (emulsion and solvent based paints) excluding air-drying paints. [Bayer AG]

Prevenol® CI 5. Dibenzylsulfoxide; an organic inhibitor for use in cleansing acids and in the surface treatment of metals. [Bayer AG]

Prevenol® CI 7-100. CAS 29385-43-1; Tolyl triazole; corro-

---

sion inhibitor for copper, copper alloys and other metals; particularly suitable for antifreezes, coolants, cutting fluids and hydraulic fluids. [Bayer AG]

Preventol® CI 8. CAS 95-14-7; Benzotriazole; corrosion inhibitor for copper, copper alloys and other metals; particularly suitable for antifreezes, coolants, cutting fluids and hydraulic fluids. [Bayer AG]

Preventol® O Extra. CAS 90-43-7; o-Phenyl phenol; preservative for chemical and industrial products; raw material for disinfectants. [Bayer AG]

Prevex® BJA. Phenylene ether copolymer alloy; flame-retardant structural foam molding grade polymer for both thin- and thick-wall sections; used for CRT and printer housings, copier panels, instrument bases, card cages, and workstations. [GE Plastics]

Prevex® PMA. Polyphenylene ether resin; good heat resistance and dimensional stability; for small appliances, lawn care tools, power tools, industrial devices. [GE Plastics]

Prevex® PQA. Phenylene ether copolymer alloy; general purpose injection molding grade polymer; used for power tools, portable appliances, transportation components, hot water appliances, and liq. handling devices. [GE Plastics]

Prevex® VFA. Polyphenylene ether resin; good impact strength, dimensional stability for use in business machine applications, internal parts, other enclosures. [GE Plastics]

Prevex® VGA. Polyphenylene ether resin, flame-retardant; business machine applications, internal parts, other enclosures. [GE Plastics]

Prevex® W20. Phenylene ether polymer alloy; automotive polymer with high flow, high impact, and high heat resistance used for instrument panels and interior trim components. [GE Plastics]

Prevex® W30. Phenylene ether copolymer alloy; heat-resistant automotive polymer used for instrument panels and interior trim components. [GE Plastics]

Priadel. CAS 554-13-2; A proprietary preparation of lithium carbonate; used in the treatment of bipolar affective disorder. [Delandale Laboratories Ltd]

Priamid. Lather booster and stabilizer, emulsifier; antistatic agents; used for powdered detergents; toilet and shaving soap; shampoos; bubble baths; thermoplastics and synthetic fibers; liquid paste cleaners. [Unichema International] *

pribramite. A mineral. It is a variety of sphalerite.

Pricat 9900. Catalyst for hydrogenated of vegetable and animal fats and oils. [Hart Chem. Ltd.]

Pricerine 9071. CAS 56-81-5; Glycerin; polyol for low color alkyd resins. [Unichema]

pricite. (Bechilite). A mineral. It is a calcium borate; $3CaO \cdot 4BO_3 \cdot 6H_2O$.

priderite. A mineral. It is $(K,Ba)_{1.3} (Ti,Fe)_8O_{16}$.

Prifac 7912. Groundnut fatty acid; used in manufacturing of alkyd resins for stoving enamels, acid curing lacquers and NC lacquers; in wood and metal varnishes. [Unichema]

Prifac 7920. Tallow acid; raw material for surfactants, soaps, nitrogen derivs., buffing formulations. [Unichema]

Prifac 7951. Soybean fatty acid; used as raw material for high quality long oil alkyds for consumer paints, for med. and short oil alkyd resins for air and oven-drying lacquers with high flexibility. [Unichema]

Prifac 7960. Sunflower fatty acid; for high quality med. to fast drying alkyd resins, suitable for nonyellowing lacquer systems with excellent heat stability. [Unichema]

Prifac 9428. CAS 61790-38-3; EINECS 263-130-9; Hydrogenated tallow acid; chemical intermediate for surfactants, stabilizers, detergents, fabric softeners. [Unichema]

Prifrac 2901. CAS 124-07-2; EINECS 204-677-5; Caprylic acid; intermediate for ester production, synthetic lubes, latex stabilizers, substituted glycerides used as skin protectors. [Unichema]

Prifrac 2906. CAS 334-48-5; EINECS CAS 334-48-5; Capric acid; detergent raw material for soaps, industrial cleaners. [Unichema]

Prifrac 2920. CAS 143-07-7; Lauric acid; emulsifier for hot emulsion polymerization of NBR, and NR latex stabilization. [Unichema]

Prifrac 2940. CAS 544-63-8; EINECS 208-875-2; Myristic acid; intermediate for ester, detergent, and surfactant products [Unichema]

Prifrac 2960. CAS 57-10-3; EINECS 200-312-9; Palmitic acid; intermediate for surfactants, soap and cosmetic formulations. [Unichema]

Prifrac 2980. CAS 57-11-4; EINECS 200-313-4; Stearic acid; intermediate for surfactants, soap and cosmetic formulations. [Unichema]

Prifrac 2989. CAS 112-85-6; Behenic acid; intermediate for surfactants, soap and cosmetic formulations. [Unichema]

Prifrac 2990. CAS 112-86-7; Erucic acid. [Unichema]

prilocaine. CAS 721-50-6; N-(α-Propylaminopropionyl)-o-toluidine. Citanest.

Primacor. A family of adhesive polymers used for extrusion coating and layers in flexible packaging. [Dow UK]

Primacor 1320. Adhesive polymer; for blown film; used as adhesive layer or sealant layer in flexible pkg. structures; FDA compliance. [Dow]

Primacor 2912. Adhesive polymer; extrusion coating resin used for aseptic, medical, pharmaceutical and condiment pkg., dry mix, and moist and dry food pkgs., laminated tubes, lidding stock. [Dow]

Primacor 4990 Dispersion. CAS 9010-77-9; Ethylene acrylic acid copolymer disp.; disp. polymer for use as binder for nonwoven fibers including PP, polyester, glass, and nylon; provides soft fabrics having excellent antisoil redeposition properties. [Dow]

Primafloc. Polyelectrolyte coagulants, flocculants. [Rohm & Haas UK]

Primal®. For the prophylaxis and treatment of mouth and throat infections. [Bayer AG]

Primal. Acrylic emulsions; used for decorative and industrial coatings, binders for textile and nonwoven applications, floor polishes, leather, adhesives, cement modifiers. [Rohm & Haas]

Primalan. CAS 29216-28-2; A proprietary preparation of mequitazine; antihistamine. [May & Baker Ltd] *

Primallor. Gold casting alloy for dental applications. [Degussa]

Primapel. Soil retardent and leather chemicals. [Rohm & Haas UK]

Primaquine Bayer. 8-(4-Amino-1-methyl-butylamino)-6-methoxyquinoline; antimalarial. [Bayer AG]

Primasol® AMK. Padding auxiliary to prevent dye migration. [BASF; BASF plc]

---

‡ Trade name and manufacturer not verified   § Trade name without identified manufacturer

Primasol® FP. Aliphatic sulfonic acids, organic phosphate aq. solution; detergent, wetting agent, textile auxiliary, pigment padding, mercerizing and bleaching processes; migration inhibitor; low foaming; stable to electrolytes, alkalies. [BASF/Fibers; BASF AG]

Primasol® NB-NF. Phosphate ester; low foaming wetting and deaerating agent for dyeing, desizing, bleaching, pad bath application; modified surface tension to minimize foam formation. [BASF/Fibers]

Primatene Mist. CAS 51-43-4; Epinephrine; adrenergic. [Whitehall Laboratories]

Primatol AA. Triazine amino triazole. [Ciba plc] *

Primatol AD 85WP. Aminotriazole + atrazine + 2,4-D; used for total weed control in non crop areas. [Ciba-Geigy Agrochemicals]

Primatol AP. Triazine, pictoram total herbicide. [Ciba plc] *

Primatol SE 500FW. CAS 61-82-5, 122-34-9; A suspension concentrate containing 180 g aminotriazole and 300 g simazine per liter; used for total weed control in non crop areas and fruit orchards. [Ciba-Geigy Agrochemicals]

Primax. Textile and leather chemicals. [Rohm & Haas UK]

Primax. Surface modified, UHMWPE. [Air Prods. & Chems. Inc]

Primax UH-1060. CAS 9002-88-4; UHMWPE, modified; surface modified particles producing cast elastomer composites with high abrasion resistance and low coeff. of friction; imparts improved abrasion, corrosion and chem. resistance in coatings. [Air Prods./Polyurethanes]*

Primax UH-1080. CAS 9002-88-4; UHMWPE, modified; surface modified particles producing cast elastomer composites with high abrasion resistance and low coeff. of friction; imparts improved abrasion, corrosion and chem. resistance in coatings. [Air Prods./Polyurethanes]*

Primax UH-1250. CAS 9002-88-4; UHMWPE, modified; surface modified particles producing cast elastomer composites with high abrasion resistance and low coeff. of friction; imparts improved abrasion, corrosion and chem. resistance in coatings. [Air Prods./Polyurethanes]*

Primaxin. Imipenem and cilastatin sodium; broad-spectrum β-lactam antibiotic. [Merck & Co Inc]

Primazin®. Reactive dyes for dyeing and printing cellulosic fibers. [BASF; BASF plc]

Primazin Fixing Agent RP. Fixing agent for resist printing with fiber reactive dyes. [BASF]

Primecoat®. For the electrical industry. [DuPont UK]

Prime Flavours. Meat and savory flavors; for food products (soups, sauces etc) and processed meats. [Fries & Fries] *

Primene. Tert-alkylamines; stabilizers [Rohm & Haas UK]

Primicid. CAS 23505-41-1; Soil applied insecticide containing pirimiphos-ethyl. [ICI Chem. & Polymers Ltd]

Primobolan. CAS 434-05-9; A proprietary preparation of methenolone acetate; an anabolic agent. [Schering Chemicals Ltd] *

Primobolan Depot. CAS 303-42-4; A proprietary preparation of methenolone enanthate; an anabolic agent. [Schering Chemicals Ltd] *

Primodos. A proprietary preparation of ethinylestradiol and norethisterone acetate; used in the treatment of secondary amenorrhea. [Schering Chemicals Ltd] *

Primofax. CAS 56377-79-8; Nosiheptide; growth stimulant.

[Rhone-Poulenc UK] *

Primogyn Depot. CAS 979-32-8; A proprietary preparation of estradiol valerate; used in the treatment of amenorrhoea and prostatic carcinoma. [Schering Chemicals Ltd] *

Primolut Depot. A proprietary trade name for the caproate of hydroprogesterone. [Schering Health Care Ltd] *

Primolut N. 5 mg Norethisterone BP. [Schering Health Care Ltd]

Primor. Lubricating and industrial oils. [Burmah-Castrol Ltd]*

Primorol 1511. Branched chain C24-26-28 alcohol. [Henkel/Cospha]

Primotec. CAS 23505-41-1; Seed dressing containing pirimiphos-ethyl. [ICI Chem. & Polymers Ltd]

Primoteston Depot. CAS 315-37-7; A proprietary preparation of 250 mg testosterone oenanthate in clear nonaqueous solution; used in the treatment of male osteoporosis or sterility. [Schering Health Care Ltd]

Primperan. CAS 54143-57-6; A proprietary preparation of metoclopramide hydrochloride; an anti-emetic. [Berk Pharmaceuticals Ltd] *

primrose smokeless. A smokeless 42 grain powder.

primrose soluble. See erythrosin.

primuline base. p-Toluidine heated with sulfur.

Primus. Fiber-rich bar to help regularity. [Richardson-Vicks Inc] *

Primus®. For the automotive industry. [DuPont UK]

Prince's blue. A mineral. It is a blue variety of Sodalite (qv). The slabs are polished for ornamental purposes.

Prince's metal. (Prince Rupert's Metal). An alloy. It is a variety of brass containing from 61-83% copper and 17-39% zinc. Another alloy, also called Prince's metal, consists of 84.75% tin and 15.25% antimony.

Prince's metallic. See Prince's mineral.

Prince's mineral. (Prince's metallic). A clay containing about 40% oxides of iron.

Princillin. CAS 69-52-3; Ampicillin sodium; antibacterial. [Bristol-Myers Squibb Co Inc] *

Principen. CAS 69-53-4; Ampicillin; antibacterial. [Bristol-Myers Squibb Co Inc]

Principen/N. CAS 69-52-3; Ampicillin sodium; antibacterial. [Bristol-Myers Squibb Co Inc]

Prinil. CAS 83915-83-7; Lisinopril; for the treatment of hypertension. (Switzerland) [Merck & Co Inc]

Prinivil. CAS 83915-83-7; Lisinopril; for the treatment of hypertension. [Merck & Co Inc]

Prinsyl. Antiseptic. [Coventry Chemicals Ltd]

Printel. Aqueous dispersions of pigments for textile printing. [European Colour (Pigments) Ltd]

Printer's acetate. Aluminum acetate, $Al(C_2H_3O_2)_3$.

printer's iron liquor. A deep black solution of ferrous acetate, containing some ferric acetate. It contains about 10% iron.

Printex. CAS 1333-86-4; Carbon black; used for printing inks. [Degussa]

Printex 25 Beads and Powd. CAS 1333-86-4; Furnace black; esp. developed for uv inks, low uv absorption; glossy; improves shelf-life of inks. [Degussa]

Printex P. CAS 1333-86-4; Furnace black; for plastics formulations requiring highest uv absorp. [Degussa]

Printex U Beads and Powd. CAS 1333-86-4; Channel-type carbon black; general purpose industrial black coating systems [Degussa]

---

* Trade name not verified as available   † Trade name verified as obsolete

Printing Black for Wool. A dyestuff produced by the reduction of a mixture of 1 : 5- and 1 : 8-dinitro-naphthalene by means of glucose in alkaline solution in the presence of sodium sulfite. Dyes wool violet black from an acid bath. Employed in printing.

printing inks. Inks consisting of pigments incorporated with varnish made by heating linseed oil.

Printlok 1046. Vinyl acrylic; self-crosslinking pigment printing binder. [Catawba-Charlab]

Printogen. Printing oils and fixation accelerators used for textile printing. [Degussa AG] *

Printol® . Vegetable soaps, xylenols, pine oils; soluble phenolic disinfectant for hospital use. [Coventry Chemicals Ltd]

Printosol. Prinking ink distillates and oils. [Haltermann Ltd]

Printsolve™ Ink Remover. N-Methyl-2-pyrrolidone, dipropylene glycol methyl ether; ink remover for the pulp and paper industry, flexographic/rotogravure industries. [ISP]

Printwash. Mixture of esters, alcohols, toluene; for cleaning of print rollers. [Solrec Ltd] *

Prinza® Range. CAS 9000-30-0; Guar gum; thickener and stabilizer for food industry. [Grünau]

Prinzide. Lisinopril, hydrochlorothiazide; for the treatment of hypertension in patients for whom combination therapy is appropriate. [Merck & Co Inc]

Prioderm. CAS 121-75-5; A proprietary preparation of malathion; used in the treatment of infestation by lice. [Napp Laboratories Ltd] *

Priolene 6900. CAS 112-80-1; EINECS 204-007-1; Oleic acid; intermediate for ethoxylates, esters, nitrogen derivs., surfactants; used in soaps, personal care products, lubricant and metalworking fluids, for NR latex stabilization. [Unichema]

Priolube 1400. CAS 112-62-9; EINECS 203-992-5; Methyl oleate; synthetic lubricant-based fluids, metalworking, industrial applications. [Unichema] *

Priolube 1405. CAS 142-77-8; EINECS 205-559-6; n-Butyl oleate; plasticizer for chloroprene rubber. [Unichema]

Priolube 1407. Glyceryl oleate; lubricant for PVC processing; antifog agent for PVC film. [Unichema]

Priolube 1409. CAS 25637-84-7; EINECS 247-144-2; Glyceryl dioleate; lubricant for PVC processing; antifog for PVC film. [Unichema]

Priolube 1414. CAS 84988-79-4; EINECS 284-868-8; Isobutyl oleate; additive to improve ozone resistance of chloroprene rubber. [Unichema]

Priolube 1429. CAS 85049-34-9; EINECS 285-203-4; Propylene glycol dioleate; synthetic lubricant-based fluids, metalworking, industrial applications. [Unichema] *

Priolube 1435. CAS 122-32-7; EINECS 204-534-7; Glyceryl trioleate; synthetic lubricant-based fluids, metalworking, industrial applications. [Unichema] *

Priolube 1451. CAS 123-95-5; EINECS 204-666-5; n-Butyl stearate; plasticizer for butyl rubber; lubricant for PVC processing. [Unichema]

Priolube 1458. CAS 91031-48-0; EINECS 292-951-5; Isooctyl stearate; lubricant for PVC processing. [Unichema]

Priormatt. Synthetic emulsion paint; for decorative purposes on walls; for coating the interior of GRP boats. [Llewellyn Ryland Ltd]

Priplast 1431. Epoxidized oleate; plasticizer for PVC. [Unichema]

Priplast 1562. Plasticizer for nitrile rubber to be used for petrol. hose connections. [Unichema]

Priplast 3013. CAS 109-31-9; Di-n-hexyl azelate; plasticizer for PVC. [Unichema]

Priplast 3018. CAS 103-24-2; Di-2-ethylhexyl azelate; plasticizer for chloroprene and nitrile rubbers. [Unichema]

Priplast 3157. Polymeric plasticizer; plasticizer for rubber with excellent heal stability and good solv. extraction resistance. [Unichema]

Priplast 3191. Dimer-based polyester; used for flexible PU foams with outstanding hydrolytic stability, low fogging properties (suitable for automotive applications). [Unichema]

Pripol 1004. C44 dimer acid; modifier for nylon; polyester fibers; polyamide for hot melts; urethane elastomer. [Unichema]

Pripol 1009. Dimer acid; building block/modifier permitting production of higher molecular weight condensation polymers with flexibility, toughness, improved impact resistance, low moisture absorp., hydrolytic resistance; used in polyester fibers, polyamides, urethane elastomers. [Unichema]

Pripol 1025. Hydrogenated dimer acid; for hot melt adhesives; polyamide for thermographic inks. [Unichema]

Pripol 1040. Trimer acid; used in polyamino-amides for PVC plastisols to be used as car underbody coatings and in water-sol. alkyd resins. [Unichema]

Priscol. CAS 59-97-2; A proprietary preparation of tolazoline hydrochloride; a vasodilator. [Ciba plc] *

Priscoline Hydrochloride. CAS 59-97-2; Tolazoline hydrochloride; vasodilator (peripheral). [Ciba-Geigy Corp]

Prism. Gas separators; separates industrial gases through use of gas permeable hollow fiber membranes. [Monsanto Co]

Prism. Polyurethane RIM system. [Miles]

prismatic niter. Potassium nitrate, KNO_3, so-called from the form of the crystals.

Prisorine 3508. CAS 2724-58-5; EINECS 220-336-3; Isostearic acid; polysulfide rubber additive; results in use of less solv. or plasticizer, easier preparation, better cure retarding, better heat stability in final products [Unichema]

Pristane. Iso-octadecane, $C_{18}H_{38}$.

Pristene 180. Mixed tocopherols; food-grade antioxidant; natural source of vitamin E. [UOP]

Pristene R20. CAS 84604-14-8; Rosemary extract; food-grade natural flavoring and stabilizer. [UOP]

Pristerene 4904. CAS 57-11-4; EINECS 200-313-4; Stearic acid; intermediate for ethoxylates, esters, nitrogen derivs., personal product formulations; heat-stable. [Unichema]

Pristinamycin. An antibiotic produced by *Streptomyces pristina spiralis*.

Pristine. Natural mixed tocophenols; antioxidants. [UOP Inc]

Privine Hydrochloride. CAS 550-29-2; Naphazoline hydrochloride; adrenergic. [Ciba-Geigy Corp] *

Pro-Actidil Tablets. CAS 6138-79-0; Prolonged action triprolidine hydrochloride tablets; for the symptomatic relief of allergic conditions such as hayfever, urticaria and pruritus. [The Wellcome Foundation Ltd]

Proaid 9802. Nonstaining, nondiscoloring processing and dispersing aid for polychloroprene and EPDM. [Akrochem]

---

‡ Trade name and manufacturer not verified   § Trade name without identified manufacturer

Proaid 9814. Homogenizing agent and softening resin for use in most elastomers; improves processing. [Akrochem]

Proaid 9904. Specialty processing aid for use in CM (CPE) and CSM (Hypalon) polymers; gives some mold release properties; stabilizes hot air aging properties. [Akrochem]

Proban. Textile flame retardant. [Albright & Wilson Ltd, Phosphates & Speciality Business]

Pro-Banthine. CAS 50-34-0; Propantheline bromide; anticholinergic. [G D Searle & Co] *

Probe. CAS 20354-26-1; Wettable powder containing 75% methazole; for post-emergence weed control. [ICI Chem. & Polymers Ltd]

Probenecid. 4-(Di-n-propylsulfamoyl)benzoic acid; Benemid.

proberite. A mineral; 2(Na,Ca, $B_8O_9 \cdot 5H_2O$).

Probilin. CAS 77-09-8; A preparation of phenol phthalein. §

Probimer. Photo-crosslinkable synthetic resin. [Ciba plc]

Probimide. Polyimides. [Ciba plc]

Probucol. It is 4,4'-(iso-propylidenedithio) bis-2,6-di-t-butylphenol. An agent used to control the increase of cholesterol in the blood beyond normal limits.

procainamide. (N-(4-aminobenzoyl)-2-diethylamino-ethamine; 4-amylino-N-(2-diethylaminoethyl) benzamide). CAS 51-06-9; Pronestyl.

procaine hydrochloride. See benzocaine.

Procal. Vacuum-formed ceramic fiber insulation. [Foseco (F.S.) Ltd]

Procan. CAS 614-39-1; Procainamide hydrochloride; cardiac depressant. [Parke-Davis] *

procarbazine. (N-4-Isopropylcarbamoylbenzyl -N'-methylhydrazine) CAS 671-16-9; Natulan is the hydrochloride.

Procardia. CAS 21829-25-4; A proprietary preparation of nifedipine; for the treatment of angina and hypertension. [Pfizer International]

Procetyl 10. (PPG-10 cetyl ether) CAS 9035-85-2; Emollient, coupler, cosolvent, plasticizer, superfatting, wetting and spreading agent, penetrant; lubricant in cosmetics and personal care products; alcoholic and aq. alcoholic compositions. [Croda Inc.]

Procetyl AWS. (PPG-5 ceteth-20) CAS 9087-53-0; Emulsifier, plasticizer, coupler, humectant, dispersant, emollient, and fragrance solubilizer in aq. and aq. alcoholic systems, personal care products, antiperspirants, bath oils. [Croda Inc.; Croda Chem. Ltd.]

Procetyl AWS Modified. PPG-8-ceteth-2. [Croda Inc.]

Prochinor. Specialty blends of fatty amines and derivatives; demulsifiers for crude oil production and industrial oils. [Atochem UK/Ceca]

prochlorite. A mineral similar to chlorite.

prochlorperazine. CAS 58-38-8; 2-Chloro-10-[3-(4-methylpiperazin-1-yl)-propyl]pheno-thiazine. Compazine is the dimaleate or the edisylate.

Procilene. Polyester/cotton dyes. [ICI Chem. & Polymers Ltd]

Procinyl. Reactive disperse dyes. [ICI Chem. & Polymers Ltd] †

Procion. Reactive dyestuffs. [ICI Chem. & Polymers Ltd]

Procol®. An extensive range of flotation reagents for both sulfide and nonsulfide minerals; includes collectors, frothers, depressants and modifiers. [Allied Colloids Ltd]

Procom. Polypropylene compound. [ICI Chem. & Polymers Ltd]

Procond-101. CAS 9003-07-0; Electrically conductive PP copolymer; extrusion grade [United Composites]

Procor 75 AB-X. Two-side acrylic coated biaxially oriented PP film; for horizontal, vertical and general overwrap applications; excellent seal strength, moisture/flavor/odor barrier properties, wide sealing range [Mobil/Films]*

Pro-cort. CAS 50-23-7; Hydrocortisone; glucocorticoid. [Barnes-Hind Inc] *

Proctocort. CAS 50-23-7; Hydrocortisone; glucocorticoid. [Rowell Laboratories Inc] *

Proctofibe. Proprietary product of grain fiber, citrus fiber; used for diverticular disease, irritable colon, constipation. [Roussel Laboratories Ltd] †

Proctofoam HC. Muco-adherent, white, odorless aerosol foam containing hydrocortisone acetate and pramoxine hydrochlorite; topical antihemorrhoiddal. [Stafford-Miller] ‡

Procyclidine. 1-Cyclohexyl-1-phenyl - 3-pyrrolidinopropan-1-ol. Kemadrin is the hydrochloride.

procythol. A liver extract.

Prodag. CAS 7782-42-5; Graphite in water; lubricant additive for machine oils, assembly lubes. [Acheson Colloids]

Prodew 100. Sodium lactate, sodium PCA, sorbitol, hydrolyzed animal protein, proline; formulated moisturizer for cosmetics, soaps, hair care products; humectant. [Ajinomoto] *

Prodex. CAS 10265-92-6; Methamidophos; insecticide. [Makhteshim Chemical Works Ltd]

Pro-Diaban®. CAS 25046-79-1; Glisoxepide; oral antidiabetic preparation. [Bayer AG]

Prodoraqua. Liquid waterproofer and hardener. [Prodorite Ltd]

Prodorbond. A range of polymer modified heavy duty industrial floor finishes. [Prodorite Ltd]

Prodorcrete GT. A range of urethane resin based floor toppings for acid, alkali and chemical resistance. [Prodorite Ltd]

Prodorfilm. Series of special light stable wall coatings and potable linings for vessels and tanks. [Prodorite Ltd]

Prodorflor. Acid and chemical resisting epoxy resin based floor finish. [Prodorite Ltd]

Prodorglas. A range of stoved coatings for vessels and tanks to resist chemical corrosion. [Prodorite Ltd]

Prodorglaze. A range of textured decorative multi-colored wall coatings. [Prodorite Ltd]

Prodorguard. Epoxy resin based floor coatings for application to concrete floors and walls. [Prodorite Ltd]

Prodorite. An acid-resisting material; it is a concrete with a hardened pitch binder; the mineral part is carefully graded and mixed with the pitch; stated to be suitable for plants containing corrosive gases. §

Prodorlac. Heavy bituminous paint. [Prodorite Ltd]

Prodorshield. Self-leveling epoxy resin based floor topping. [Prodorite Ltd]

Prodox. Alkyl phenols; intermediate chemicals for antidegradents, agricultural chemicals and resins. [PMC Specialities Group Inc] *

Prodoxol. CAS 14698-29-4; A proprietary preparation of oxolinic acid; a urinary antiseptic. [Warner] *

Product AAS 90. Anionic surfactant in the form of a free flowing powder; for powder type detergents and cleaning

---

* Trade name not verified as available   † Trade name verified as obsolete

formulations where free flowing product is required. [Henkel Chemicals Ltd] *

Product MB320. Cyclohexylamine lauryl sulfate in the form of a solid block which is water and oil soluble; emulsifier for insecticides; printing ink manufature. §

Productol. EINECS 215-293-2; Cresylic acid. [PMC Inc]

Produkt 2058. Fatty alcohol/fatty acid ester; industrial surfactant. [Zschimmer & Schwarz]

Produkt GM 4210. MIPA-ammonium fatty alcohol ether sulfate, propylene glycol; industrial surfactant. [Zschimmer & Schwarz]

Produkt GS 5001. Fatty acid DEA, modified; for hair shampoos, bath additives, dishwash, cleansing agents. [Zschimmer & Schwarz] *

Pro-Etch. Patented glass etching system; for etching vehicle registration number onto glass car windows. [Hermetite Products Ltd] *

Profalon. Emulsifiable concentrate containing 200 g chlorpropham and 100 g linuron per liter; used to control weeds in bulb crops. [Hoechst UK]

Profax. Insulating refractory sideliner tiles for killed steel ingots. [Foseco (F.S.) Ltd]

Pro-fax® 6323. CAS 9003-07-0; PP homopolymer resin, heat-stabilized; high flow, max. stiffness, general-purpose resin used in automotive, hospital and institutional ware, housewares, closures by injection and fiber extrusion processes; UL rated at 110 C. [Himont] *

Pro-fax® 65F4-4. CAS 9003-07-0; PP, 40% talc-filled; max. stiffness and high temp. performance; UL (115 C continuous use); in automotive, appliances, industrial components. [Hercules]

Pro-fax® 65F5-4. CAS 9003-07-0; PP, 40% calcium carbonate-filled; best high flex modulus/impact balance, good colorability, surface finish; in housewares, small appliances. [Hercules]

Pro-fax® 7523. CAS 9003-07-0; PP copolymer resin, heat-stabilized; general-purpose resin for injection, fiber extrusion; improved low temp. impact, greater toughness allowing wider design latitude; UL rated at 110 C; used in totes, trays, automotive, furniture. [Himont] *

Pro-fax® 8523. CAS 9003-07-0; PP copolymer resin, heat-stabilized; high impact grade, toughness, low temp. impact resistance, long high temp. life; UL rated at 110 C; used in automotive parts, luggage, totes, bins, appliances. [Himont] *

Pro-fax® HB-301. CAS 9003-07-0; PP resin; ultra-high molecular weight polymer with consistent particle size; for manufacturing of porous products [Himont] *

Pro-fax® PC-072PM. CAS 9003-07-0; PP homopolymer modified for chemical coupling with glass; high heat performance; used in appliance, automotive, chemical processing equipment. [Himont] *

Pro-fax® PD-064. CAS 9003-07-0; PP homopolymer resin; film resin for biaxially oriented film. [Himont] *

Pro-fax® PF-101. CAS 9003-07-0; PP homopolymer resin; for thermoforming sheet; features clarity, low odor and taste, good processability; used for cookie trays, pkg. cups; available with antistat; FDA compliance. [Himont]*

Pro-fax® SA-747M. CAS 9003-07-0; PP random copolymer; features outstanding clarity, fast cycling, FDA compliance for housewares, food containers. [Himont] *

Pro-fax® SB-242. CAS 9003-07-0; PP resin; injection molding grade for housewares; very high flow; good impact/stiffness balance. [Himont] *

Pro-fax® SD-062. CAS 9003-07-0; PP copolymer resin; high-speed coating/laminating resin for films; used for fabric and paper coating; FDA compliance. [Himont] *

Pro-fax® SE-191. CAS 9003-07-0; PP copolymer resin; for wire and cable primary insulation. [Himont] *

Pro-fax® SV-256M. CAS 9003-07-0; PP random copolymer, clarifier; for blow molding applications requiring optimum clarity. [Himont] *

Pro-fax® Z-39S. CAS 9003-07-0; PP fiber resin; low gas fading for staple. [Himont] *

Proferdex. Iron dextran; hematinic. [Fisons plc, Pharmaceuticals Div] *

proflavine. CAS 92-62-6; 3 : 6-Diamino-acridine-sulfate, $C_{13}H_{11}N_3 \cdot H_2SO_4 \cdot H_2O$; used as an antiseptic.

Progacyl® ADG. High visc. thickener for continuous dyeing, random dyeing, and printing of tufted carpet using acid, premetallized acid, dispersed, direct, basic and nickel chelating dispersed colors; also for other industries where alkaline stability required. [Rhone-Poulenc Surf.]

Progacyl® CP-7. CAS 9000-30-0; Guar; polymeric thickener for flatbed screen printing of carpets, rotary screen printing. [Rhone-Poulenc Surf.]

Progacyl® CP-82; Synthetic hydrocolloid; print thickener for use on carpets. [Rhone-Poulenc Surf.]

Progallin LA. CAS 1166-52-5; Dodecyl gallate; antioxidant for cosmetics. [Nipa Labs]

Progallin P. CAS 121-79-9; Propyl gallate; antioxidant for cosmetics. [Nipa Labs]

Proganol. See Protargentum.

Pro-Gas (Gas Disclaimed). CAS 71-23-8; Propanol. [Chevron] *

Progasol® COG. Low-foaming detergent for scouring cotton, synthetics, and blends; for use in jet dyeing machines. [Rhone-Poulenc Surf.]

Progene. Liquid detergents. [Unilever] *

Progilite. A proprietary trade name for phenol-formaldehyde. §

Proglumide. 4-Benzamido N,N- dipropylglutaramic acid; used as an anti-gastrinic.

Proglycem. CAS 364-98-7; Diazoxide; antihypertensive. [Schering Corp] *

Progressite. An explosive containing 89% ammonium nitrate, 4.7% aniline hydrochloride, 6% ammonium sulfate, and 0.2% coloring matter.

Progynova. CAS 979-32-8; A proprietary preparation of estradiol valerate; used in the treatment of menopausal symptoms. [Schering Health Care Ltd]

Proheptazine. CAS 77-14-5; Hexahydro-1,3-dimethyl-4-phenyl-4-propionyloxy-azepine.

proidonite. A mineral, SiF_4.

Proil. Rust preventatives. [Croda Chem. Ltd] *

proiodin. A combination of iodine with protein, containing 4.4% iodine.

Project® 70 Stainless Type 316. Molybdenum bearing austenitic steel with increased percentages of nickel; higher tensile and creep strength at elevated temps.; for paper/pulp handling equip., process equip. for producing photographic chems., inks, rayon, rubber, textile bleaches and dyestuffs, high temp. equip. [Carpenter Tech.]

Prokayvit Oral. Tablets of acetome-naphthone (vitamin K). [British Drug Houses] *

Proklar. CAS 144-82-1; Sulfamethizole; antibacterial. [O'Neal, Jones & Feldman Pharmaceuticals] *

---

‡ Trade name and manufacturer not verified   § Trade name without identified manufacturer

Prokliman Ciba. A German product. It consists of tablets containing ovarial hormone, peristaltin, nitro-glycerin, pyramidone, and a sodium salt of caffeine.

Proladone. A proprietary preparation of oxycodone pectinate; an analgesic. [Crookes Laboratories] *

Prolan®. Preparations of gonadotropins; for the treatment of reproduction disorders in animals. [Bayer AG]

Prolaurin. A proprietary trade name for propylene glycol monolaurate. §

Proleaf. Horticultural foliar feed fertilizer. [Schering Agrochemicals Ltd]

Prolein. CAS 1330-80-9; EINECS 215-549-3; A proprietary trade name for propylene glycol monooleate. §

Prolene. Nonabsorbable surgical suture. [Ethicon Inc] *

Prolex. CAS 1918-16-7; Active ingredient: propachlor; 2-chloro-N-isopropyl-acetanilide; pre-emergence weed control of annual weeds. [Agan Chemical Manufacturers Ltd]

proline. (L-proline; 2-pyrrolidine carboxylic acid) A nonessential amino acid; $(CH_2)_3NHCHCOOH$; CAS 147-85-3 (L-form); EINECS 205-702-2; in moisturizers, biochemical and nutritional research, microbiological tests, culture media, dietary supplements, lab reagent. [Am. Biorganics; Degussa; R.W. Greeff; Nippon Rikagakuyakuhin; Penta Mfg.]

Pro-Line. Universal color dispersions for use in architectural coatings. [Engelhard]

L-proline. See proline.

Prolit. Bleaching earths (activated clays); for refining vegetable and mineral oils, vegetable and animal fats and in purifying solvents; as color developing agent for carbonless copying papers. [Caffaro SpA] *

Prolit. Activated bleaching earths; for the decolorizing of vegetable and mineral oils, animal and vegetable fats and solvents. [SNIA (UK) Ltd]

Prolith. Photographic developer. [May & Baker Ltd] *

Prolixin. Fluphenazine hydrochloride; antipsycotic. [Bristol-Myers Squibb Co Inc]

Prolixin Enanthate. CAS 2746-81-8; Fluphenazine enanthate; antipsycotic. [Bristol-Myers Squibb Co Inc]

Proloid. Thyroglobulin; thyroid hormone. [Parke-Davis] *

Prolongal®. Iron preparations; for prophylaxis and treatment of piglet anemia. [Bayer AG]

Proloprim. CAS 738-70-5; Trimethoprim; antibacterial. [Burroughs Wellcome Co] *

Proloprim Tablets. CAS 738-70-5; A proprietary formulation of trimethoprim; for treatment of initial episodes of uncomplicated urinary tract infections due to susceptible organisms. [The Wellcome Foundation Ltd] *

Prolugen. CAS 57-55-6; EINECS 200-338-0; Propylene glycol. [Octel Chemicals Ltd] †

Proluton Depot. 17-α-Hydroxyprogesterone hexanoate BP in clear oily solution. [Schering Health Care Ltd]

Promacetin. CAS 128-12-1; Acetosulfone sodium; antibacterial. [Parke-Davis] *

Promapar. CAS 69-09-0; Chlorpromazine hydrochloride; antiemetic; antipsycotic. [Parke-Davis] *

Promax® HV. CAS 68153-28-6; Soy protein; conc. with superior fat emulsification, emulsion stabilization and water binding capability; highly digestible protein-rich food ingredient for portion meats, poultry, seafood products, protein beverages. [Central Soya]

prometal. A variety of cast iron; used in the construction of furnace parts.

promethazine. CAS 60-87-7; 10-(2-Dimethylamino-n-propyl)phenothiazine; phenergen is the hydrochloride.

Promethus. A blasting powder. It contains potassium chlorate, manganese dioxide, iron oxide, mono-nitro-benzene, turpentine oil, and naphtha.

Prometrex. CAS 7287-19-6; Active ingredient: prometryne; 2,4-bis-(isopropylamino)-6-methylthio-1,3,5-triazine; selective pre- and post-emergence herbicide for the control of broadleaf and grass weeds in a variety of crops. [Agan Chemical Manufacturers Ltd]

Promicrol. Ultra fine grain photographic developer. [May & Baker Ltd] *

Prominal®. N-Methylethylphenyl-malonyl-urea (methylphenobarbitone); a hypnotic. [Bayer AG]

Promintic. A proprietary preparation of methyridine; a veterinary anthelmintic. §

Promodan SP. CAS 1323-39-3; EINECS 215-354-3; Propylene glycol stearate; food emulsifier. [Grindsted Prods.]

Promol. Release agent for rubber and plastics. [Rhone-Poulenc UK]

Promoloid. A fertilizer containing colloidal magnesium silicate; a Japanese product.

Promosoy® 20/60. CAS 68153-28-6; Soy protein; high digestible protein-rich food ingredient. [Central Soya]

Promotor 301. Metal compound, hydrocarbon solv; accelerator used in combination with peresters to achieve an optimal cure in hot press molding and a fast cure at room temperature and elevated temps. [Akzo] *

Promozyme. A heat stable debranching enzyme obtained from a novel species of *Bacillus* by submerged fermentation; belongs to the group of debranching enzymes known as pullulanases; used in the production of dextrose and maltose. [Novo Nordisk]

Promulgen® D. Cetearyl alcohol and ceteareth-20; gelling agent; o/w emulsifier, emollient, and stabilizer for cosmetics and pharmaceuticals; highly resistant to acidic and alkaline conditions. [Amerchol Corp]

Promulgen® G. Stearyl alcohol and ceteareth-20; gelling agent; o/w emulsifier, emollient, and stabilizer for cosmetics and pharmaceuticals; highly resistant to acidic and alkaline conditions. [Amerchol Corp]

Promyr. CAS 110-27-0; EINECS 203-751-4; Isopropyl myristate; emollient and solv. for cosmetics, toiletries, makeups; nongreasing rub in. [Amerchol Corp]

Promyristyl PM-3. CAS 63793-60-2; PPG-3 myristyl ether; low-visc. emollient for clear analgesic, deodorant, and fragrance sticks. [Croda Inc.]

Pronal 502, 502A. Polyalkylene glycol ester; defoaming agent for paper, latex, and water paint manufacturing. [Toho Chem. Industry]

Pronal EX-100. Silicone emulsion; antifoaming agent for petrochem. industry. [Toho Chem. Industry]

Pronalys. Analytical grade reagents. [May & Baker Ltd] *

Pronase®. Protease. [Calbiochem Corp]

Prondol Tablets. CAS 5560-72-5; A proprietary preparation of iprindole; an antidepressant. [Wyeth Laboratories]

Pronel Capsules. CAS 9000-70-8; A proprietary preparation of gelatin; used in the treatment of flaking fingernails. [Bioglan Laboratories] *

Pronestyl. CAS 614-39-1; Procainamide hydrochloride; cardiac depressant. [Bristol-Myers Squibb Co Inc]

Pronestyl Solution. CAS 614-39-1; Procainamide hydrochloride in aqueous solution for injection. [Bristol-Myers

---

* Trade name not verified as available   † Trade name verified as obsolete

Squibb Pharmaceuticals Ltd]

Pronestyl Tablets. CAS 614-39-1; Procainamide hydrochloride in tablet form. [Bristol-Myers Squibb Pharmaceuticals Ltd]

Pronova. CAS 9067-32-7; Sodium hyaluronate; used for medical, pharmaceutical, veterinary, botanical, microbiological, and cosmetic uses, as well as for bioreactor processes. [Protan]

Proofite. A proprietary trade name for a product similar to Aqualec. §

proof spirit. A term originally intended to denote alcohol, that was just strong enough to ignite gunpowder, when burnt upon it. It is alcohol containing 49.24 parts of alcohol to 50.76 parts of water by weight, or 100 volumes of alcohol to 81.82 volumes of water.

proof vinegar. A vinegar containing 66% acetic acid.

Propaderm. CAS 5534-09-8; A proprietary preparation of beclomethasone dipropionate; for dermatological use. [Glaxo Laboratories]

Propaderm A. A proprietary preparation of beclomethasone dipropionate and chlortetracycline hydrochloride; for dermatological use. [Glaxo Laboratories] *

Propaderm C. A proprietary preparation of beclomethasone dipropionate and clioquinol; for dermatological use. [Glaxo Laboratories] *

Propaderm N. A proprietary preparation of beclomethasone dipropionate and neomycin sulfate; for dermatological use. [Glaxo Laboratories] *

Propadrine. CAS 154-41-6; Phenylpropanolamine hydrochloride; adrenergic. [Merck & Co Inc] *

Propaesin. CAS 94-12-2; (Propocaine). The propyl ester of p-amino-benzoic acid, $NH_2 \cdot C_6H_4 \cdot COO \cdot C_3H_7$; a local anaesthetic.

Propafilm. Balanced biaxially oriented polypropylene film. [ICI Chem. & Polymers Ltd]

Propafoil. Metallized oriented polypropylene film. [ICI Chem. & Polymers Ltd]

Propain. Paracetamol, codeine phosphate, diphenhydramine, caffeine in yellow scored tablets; for treatment of headache, migraine, muscular pain, period pain and toothache; also for the relief of symptons of influenza and feverish colds. [Luitpold-Werk] *

Propaklone. Industrial solvent. [ICI Chem. & Polymers Ltd]

Propal. CAS 142-91-6; EINECS 205-571-1; Isopropyl palmitate; emollient and solv. for cosmetics, toiletries, makeups; nongreasing rub in. [Amerchol Corp]

propal. See proponal.

Propamine. Catalysts for polyurethane foams. [Harcros UK]

Propamine D. Tetramethyl ethylene diamine; a liquid catalyst miscible with water and organic liquids [Harcros UK]

propanal. (propionic aldehyde; propyl aldehyde) CH_3CH_2CHO; CAS 123-38-6; Used in the manufacture of propionic acid, polyvinyl and other plastics,, synthesis of rubber chemicals, and preservatives.

propane. (dimethylmethane; propyl hydride) $CH_3CH_2CH_3$; CAS 74-98-6; Hydrocarbon propellant; organic synthesis, household and industrial fuel, manufacture of ethylene, extractant, solvent, refrigerant, gas enricher. [Air Prods.; Fina; Phillips 66]

1,2-propanediol. See propylene glycol.

propane, 2-methyl-. See isobutane.

1-propanethiol. See n-propyl mercaptan.

2-propanethiol. See isopropyl mercaptan.

1,2,3-propanetriol. See glycerin.

1,2,3-propanetriol triacetate. See triacetin.

propanoic acid. See propionic acid.

propanoic acid, sodium salt. See sodium propionate.

propanoic acid, 3,3'-thiobis. See 3,3'-thiodipropionic acid.

propanol. Normal propyl alcohol, $CH_3 \cdot CH_2 \cdot CH_2OH$.

1-propanol. See n-propyl alcohol.

2-propanol. See isopropyl alcohol.

Proparacaine. A proprietary trade name for proxymetacaine. §

Propasol. Propyl oxide glycol ether solvent. [Union Carbide]

Propathene. Polypropylene, the lightest of the thermoplastics; features good rigidity and tensile strength which are retained at elevated temperatures, excellent resistance to chemicals with no tendency to environmental stress cracking; used in plastics manufacture and as molding and extrusion compounds. [ICI Chem. & Polymers Ltd]

Propcorn. Chemical products for the treatment of corn. [BP Chemicals Ltd]

2-propenamide, homopolymer. See polyacrylamide.

propene. See propylene.

1-propene, homopolymer. See polypropylene.

2-propenoic acid methyl ester. See methyl acrylate (monomer).

propenol. (allyl alcohol; propenyl alcohol) $CH_2 : CH \cdot CH_2OH$; CAS 107-18-6; Intermediate for pharmaceuticals and other organic chemicals, herbicide.

properidine. CAS 561-76-2; Isopropyl 1-methyl-4-phenylpiperidine-4-carboxylate; a narcotic analgesic.

Proper-Myl. A proprietary preparation of lyophilized yeasts, for injection. [Consolidated Chemicals Ltd] *

Propetal 241. Fatty alcohol EO/PO adduct; biodeg. detergent, wetting agent, emulsifier, antifoam; intermediate for preparation of low foaming detergents for household and industry, dishwash, sanitary and floor tile cleaners, industrial spray cleaners, metal pickling, paper, textile, leather. [Zschimmer & Schwarz]

Propetamphos. CAS 31218-83-4; Z-0-2-iso-propoxy-carbonyl-1-methylvinyl o-methyl ethyl phosphoramido = thioate; used as a pesticide.

propezite. A natural alloy of palladium and gold, containing 7% gold.

Propham. CAS 122-42-9; Propham; plant growth regulator for control of sprouting in stored potatoes and in some cases as herbicide against weeds in vegetables. [Bayer AG]

Propine®. CAS 64019-93-8; Dipivefrin hydrochloride; antiglaucoma agent. [Allergan, Inc]

Propiofan®. Polymer dispersions based on vinyl propionate; binder for exterior and interior paints, for composite thermal insulation systems; modifier for cement mortar; raw material for production of building adhesives, laminate adhesives and pkg. adhesives; binders for textile coating. [BASF AG; BASF plc]

Propiofan® D. Polyvinyl propionate dispersion. [BASF plc]*

propiolic acid. (o-nitro-phenyl-propiolic acid) $C_6H_4(NO_2)-C:C \cdot CO_2H$, CAS 471-25-0; Known commercially by this name; in the form of a thin paste.

Propione. CAS 96-22-0; Diethyl-ketone $(C_2H_5)_2 \cdot CO$; used as an hypnotic and anaesthetic.

propionic acid. (methylacetic acid; propanoic acid; ethylformic acid) C_2H_5COOH; CAS 79-09-4; EINECS 201-176-3; Esterifying agent; in production of cellulose propionates, etc.; as mold inhibitors and preservatives;

---

‡ Trade name and manufacturer not verified   § Trade name without identified manufacturer

in manufacture of ester solvs., fruit flavors, perfume bases; antifungal. [BASF; Eastman; Hoechst Celanese; Penta Mfg.; Union Carbide]

propionic acid, calcium salt. See calcium propionate.

Proplatinum. An alloy of 72% nickel, 23.6% silver, 3.7% bismuth, and 0.7% gold.

Proplex. Factor IX complex; hemostatic. [Hyland Therapeutics] *

Propocon. Formulated polyurethane systems for rigid, semi-rigid , microcellular foams; used in insulation, construction, automotive and shoe soling. [Harcros UK]

Propomeen 2HT-11. CAS 71060-61-2; Bis-hydrogenated tallowalkyl-2-hydroxypropyl amines; industrial surfactant. [Akzo]

Propomeen C/12. CAS 68516-06-3; Dipropylene glycol cocamine; industrial surfactant. [Akzo]

Propomeen T/12. CAS 68951-72-4; Dipropylene glycol tallowamine; industrial surfactant. [Akzo]

proponal. (propal, homobarbital). CAS 93-65-2; 142-91-6; Di-propyl-malonyl urea or dipropyl- barbituric acid, $(C_3H_7)_2 \cdot C(CO_2) \cdot (NH)_2CO$; used as a soporific in medicine.

Proponesin. A proprietary trade name for the hydrochloride of tolpronine. §

Propoquad® 2HT/11. CAS 68554-09-6; Di(hydrogenated tallowalkyl) (2-hydroxy-2-methylethyl) quaternary ammonium chloride, aq. IPA; industrial surfactant. [Akzo]

Propoquad® T/12; CAS 79770-97-1; Tallowalkylmethyl-bis(2-hydroxy-2-methylethyl)quaternary ammonium methylsulfates, diethylene glycol; industrial surfactant. [Akzo]

Pro-Portion. CAS 7681-49-4; Sodium fluoride; dental caries prophylactic. [Oral-B Laboratories Inc] *

Proposote. Creosote-phenyl-propionate; stimulating expectorant.

Propoxyol® 1695. PPG-5 lanolin wax glyceride; emollient, stabilizer, and pigment dispersant for anhyd. makeups; cosmetic additive. [Henkel/Cospha; Henkel Canada]

propyl acetate. $CH_3COOCH_2CH_2CH_3$; CAS 109-60-4; EINECS 203-686-1; flavoring agent, perfumery, solvent for nitrocellulose and other cellulose derivatives, natural and synthetic resins, lacquers, plastics, organic synthesis, lab reagent. [BASF; BP Chem. Ltd; Eastman; Hoechst Celanese; Union Carbide]

n-propyl alcohol. (1-propanol; ethyl carbinol) Aliphatic alcohol; $CH_3CH_2CH_2OH$; CAS 71-23-8; EINECS 200-746-9; organic synthesis; chemical intermediate; solvent for waxes, vegetable oil, natural and synthetic resins. [Eastman; Hoechst Celanese; Mallinckrodt; Union Carbide]

Propylan. Polyether polyols for flexible, semirigid and rigid polyurethane foams; used in upholstery, bedding, insulation, automotive, elastomers, coatings, and as pigment dispersing media. [Harcros UK]

Propylan A350. A proprietary amine initiated polyether used in the manufacture of polyurethane foam. [Harcros UK]*

Propylan G600. A proprietary polyoxypropylene triol of low molecular weight used in the production of rigid urethane foams and other urethane compositions, including elastomers. [Harcros UK] *

Propylan RF55. A proprietary modified sorbitol-based polyether for making rigid, flame-proof urethane foams. [Harcros UK] *

propyl carbinol. See butyl alcohol.

propyl cyanide. See butyronitrile.

propylene. (propene) $CH_3CH{:}CH_2$; CAS 115-07-1; EINECS 204-062-1; Chemical intermediate for manufacture of isopropyl alcohol, polypropylene, synthetic glycerol, acrylonitrile, propylene oxide, heptene, cumene, polymer gasoline, acrylic acid, vinyl resins, oxo chemicals. [Air Prods.; Amoco; BP Chem. Ltd; Chevron; Exxon; Fina; Mobil; OxyChem; Phillips 66; Shell; Texaco; Vista]

propylene chloride. See propylene dichloride.

propylene dichloride. (1,2-dichloropropane; propylene chloride) $CH_3CHClCH_2Cl$; CAS 78-87-5; EINECS 201-152-2; Intermediate for perchloroethylene, CCl_4; lead scavenger for antiknock fluids; solvents for fats, oils, waxes, gums, resins, cellulose esters and ethers; scouring compounds; metal degreasers; soil fumigant for nematodes. [Naphtachimie SA]

propylene glycol. (1,2-propanediol; 1,2-dihydroxypropane; methyl glycol) Aliphatic alcohol; $CH_3CHOHCH_2OH$; CAS 57-55-6; EINECS 200-338-0; solvent, emulsifier, production paints, resins, foods, drugs, antifreeze in breweries and dairies; substitute for ethylene glycol and glycerol; mold growth and fermentation inhibitor. [Aldrich; Arco; Asahi Denka Kogyo; Ashland; BP Chem. Ltd; Hüls AG; Olin; Texaco]

propylene glycol alginate. (hydroxypropyl alginate; alginic acid, ester with 1,2-propanediol) Mixture of propylene glycol esters of alginic acid. $(C_9H_{14}O_7)_8$; CAS 9005-37-2; food additive (human). [Kelco Int'l.; Meer]

propylene oxide. (1,2-epoxypropane; S(-)-methyloxirane) C_3H_6O; CAS 75-56-9; EINECS 200-87902; Intermediate; polyols for urethane foams, propylene glycols, surfactants, detergents, isopropanolamines, fumigant, synthetic lubricants, synthetic elastomers, solvent. [Arco; Ashland; Hüls AG; Texaco]

propylene polymer. See polypropylene.

Propylex. Polypropylene plastics in the form of sheets, bands, strips, films, plates, slabs, blocks, tubes, etc. [Courtaulds Fibres Ltd]

propyl gallate. (3,4,5-trihydroxybenzoic acid, propyl ester; n-propyl 3,4,5-trihydroxybenzoate; gallic acid, propyl ester) Aromatic ester of propyl alcohol and gallic acid; $(HO)_3C_6H_2COOCH_2CH_2CH_3$; CAS 121-79-9; EINECS 204-498-2; food and feed antioxidant, flavor and packaging material. [Aceto; Eastman; Nipa Labs; UOP]

propylhexedrine. CAS 101-40-6; 1-Cyclohexyl-2-methylaminopropane.

propyl hydride. See propane.

propyl p-hydroxybenzoate. See propylparaben.

Propyliodone. CAS 587-61-1; N-Propyl 3,5-di-iodo-4-pyridone-N-acetate; 3,5-diodo-1-propoxycarbonyl-methylpyrid-4-one; $I_2(O)C_5H_2NCH_2COOC_3H_7$; used in medicine

n-propyl mercaptan. (1-propanethiol; mercaptan C_3) n-C_3H_7SH; CAS 107-03-9; EINECS 203-455-5; Chemical intermediate, herbicide. [Atochem N. Am.; Phillips 66]

Propylol. Essentially normal propyl alcohol containing approximately 12% secondary butyl alcohol; solvent in paints and lacquers, foundries, deoiling waxes, grinding media, for glass forming, manufacture of hair lacquers, floor polishes, latex rubber production, disinfectants, hand cleaners, degreasers, rust removers, printing inks; production of xanthates and other mining chemicals. [Sasolchem]

---

* Trade name not verified as available   † Trade name verified as obsolete

propyl oleate. Ester of propyl alcohol and oleic acid; base for industrial lubricants; mold release agent, defoamer, flotation agent, plasticizer for cellulosic plastics, needle lubricants; when sulfated is useful as wetting, rewetting, and dye leveling agent in textile and leather industries. [Witco/Humko]

propylparaben. (propyl p-hydroxybenzoate; 4-hydroxy-benzoic acid, propyl ester; propyl parahydroxybenzoate) Organic ester; ester of n-propyl alcohol and p-hydroxybenzoic acid; $C_{10}H_{12}O_3$; CAS 94-13-3; EINECS 202-307-7; food preservatives, fungicide, mold control in sausage casings, pharmaceutic aid. [Allchem Industries; R.W. Greeff; Mipa Labs;Penta Mfg.]

propyl parahydroxybenzoate. See propylparaben.

n-propyltrichlorosilane. $C_3H_7Cl_3Si$; CAS 141-57-1; Intermediate for silicones. [Hüls Am.; PCR]

n-propyl 3,4,5-trihydroxybenzoate. See propyl gallate.

Propyl Zithate®. CAS 1000-90-4; Zinc isopropyl xanthate; rubber accelerator used in natural amd synthetic cements and doughs; nondiscoloring in presence of copper or iron. [R T Vanderbilt Co Inc]

propyphenazone. 4-Isopropyl-2, 3-dimethyl-1-phenyl- 5-pyrazolone; used as an analgesic.

propytal. See proponal.

Proresid. A proprietary preparation of mitopopozide; an antimitotic. [Sandoz] *

Prosan. Biocide. [Union Carbide]

Proscar®. Finasteride; for the treatment of benign prostatic hypertrophy. [Merck & Co Inc]

Proscopine. A proprietary trade name for the hydrochloride of poskine. §

Pro Seal. Silicone sealant, gasketing compounds; for making formed-in-place gaskets and for use within cut gaskets. [Novest Inc] ‡

Prosil®. Organosilinanes. [PCR]

Prosil® 178. CAS 18395-30-7; Isobutyltrimethoxysilane; coupling agent for hydrophobic treatment. [PCR] *

Prosil® 196. CAS 4420-74-0; EINECS 224-588-5; γ-Mercaptopropyl trimethoxy silane; coupling agent having both organic and inorganic reactivity; for acrylic, epichlorohydrin, nitrile, polysulfone, PS, PVC, urethane thermoplastics; thermoset acrylic, epoxy, nitrile/phenolic, phenolic, polybutadiene; and elastomerics. [PCR]*

Prosil® 220. CAS 919-30-2; EINECS 213-048-4; γ-Aminopropyl triethoxy silane, tech. grade; coupling agent enhancing and promoting chemical bonding between inorganic and organic molecules; for acetal, acrylic, epichlorohydrin, nitrile, NC, polyamide, PC, polyethylene, polyimide, polymethacrylate, PP, polysulfone, PS, PVC, urethane, and vinyl thermoplastics; thermoset acrylic (thermoset and latex), alkyd, epoxy, furan, melamine, nitrile/ phenolic, phenolic, polyester, vinyl butyral/phenolic; and elastomers. [PCR] *

Prosil® 248. CAS 2530-85-0; EINECS 219-785-8; γ-Methacryloxypropyl trimethoxy silane; coupling agent having reactive methacrylate and trimethoxysilyl groups; improves adhesion of organic thermoset resins to inorganic materials such as fiberglass, clay, quartz, and other siliceous surfaces; for ABS, acrylic, polyethylene, polyimide, polymethacrylate, PP, PS, silicone, SAN, and urethane thermoplastics; alkyd, DAP, epoxy, polybutadiene, polyester, and crosslinked polyethylene thermosets; and elastomers. [PCR] *

Prosil® 3128. CAS 1760-24-3; EINECS 212-164-2; n-(2-Aminoethyl)-3-aminopropyl trimethoxysilane; coupling agent for epoxies, phenolics, melamines, nylons, PVC, urethanes, acrylics. [PCR] *

Prosil® 5136. CAS 2530-83-8; EINECS 219-784-2; 3-Glycidoxypropyl trimethoxysilane; coupling agent for epoxies, urethanes, acrylics, polysulfides. [PCR] *

Prosil® 9202. CAS 2943-75-1; EINECS 220-941-2; n-Octyltriethoxysilane; coupling agent for hydrophobic treatment. [PCR] *

Prosil® HMDS. Hexamethyldisilazane; coupling agent; silica treatment for silicone elastomers, novolac photoresist adhesion promoter. [PCR] *

Prosobee. A proprietary artificial baby milk derived from soya; used in cases of intolerance to cows' milk. [Bristol-Myers Squibb Co Inc]

Prosol 525. Emulsified mineral oil with selected antistatic agents; antistatic oil for woolen and worsted processes. [Hart Chem. Ltd.]

Prosol 4692. Emulsified mineral oil; lubricant for garnetting operations; suitable for oil fibers. [Hart Chem. Ltd.]

prosopite. A mineral, $Ca(F \cdot OH)_2 \cdot Al_2(F \cdot OH)_3$.

Prosparol. A proprietary preparation of arachis oil and water emulsion; a high calorie food. [Duncan Flockhard Ltd] *

Prospect. Processing oils. [Carless Refining & Marketing Ltd]

Prospect®. For the agriculture industry. [DuPont UK]

Prostaphlin. CAS 7240-38-2; Oxacillin sodium; antibacterial. [Bristol Laboratories]

Prostearyl 15. CAS 25231-21-4; PPG-15 stearyl ether; emollient, lubricant for cosmetics, bath oils, sunscreens, hair products, aerosol antiperspirants, hand and body lotions; coupler for fragrances. [Croda Inc.]

Prostigmin. CAS 114-80-7; A proprietary preparation of neostigmine bromide; used in the treatment of myesthenia gravis and to counteract the effect of drugs resembling curare. [Roche Products Ltd] *

Prostin/15M. CAS 58551-69-2; Carboprost tromethamine; oxytocic. [Upjohn] *

Prostin F2 Alpha. Dinoprost tromethamine; oxytocic; prostaglandin. [Upjohn] *

Prostin VR Pediatric. Alprostadil; vasodilator. [Upjohn]

Protagon. See lecithin.

protan. See albutannin.

protargentum. (protargin, proganol). A compound of gelatin and silver. It contains 8% silver, and is used in aqueous solution in medicine.

Protargin. See Protargentum.

Protargolgranulat. A German product. It consists of 1 part protargol and 2 parts urea. It has the advantage of easy solubility.

Protars. A dry arsenical fungicide prepared from talc, lime, and arsenic oxide.

Protasan. Denture adhesive (powder, cream and liquid), for securing dentures. [Richardson-Vicks Inc] *

Protavic. Conductive resins. [Protex]

protease. CAS 9014-01-1; EINECS 232-752-2; Enzyme which breaks down protein; used in detergent compositions, food processing, tanning, protein hydrolysis, desizing textiles. [Am. Biorganics; PMP Fermentation Prods.; Schweizerhall; Solvay Enzymes GmbH; U.S. Biochemical]

Protectoid. A proprietary trade name for a cellulose acetate plastic in the form of a nonflammable film. §

---

‡ Trade name and manufacturer not verified   § Trade name without identified manufacturer

Protectol. A brown, syrupy liquid used for the protection of animal products such as hair, wool, silk, skin, and leather from the action of alkaline liquids. It contains sodium lignin sulfonate.

Protectol®. Biocides. [BASF plc]

Protectol® DMT. CAS 696-59-3; Dimethoxytetrahydrofuran; biocide for disinfectants [BASF AG]

Protectol® GDA, GT 50. CAS 111-30-8; Glutaraldehyde; biocide for aq. systems and cleaners. [BASF AG]

Protectol® GL 40. CAS 107-22-2; Glyoxal; biocide for aq. systems and cleaners. [BASF AG]

Protectol® KLC 50, 80. Benzalkonium chloride; biocidal surfactant for chemical industry, detergent manufacturing. [BASF AG]

Protectol® TOE. Thiadiazine deriv.; biocide for use in aq. systems. [BASF AG]

Protectyl. A solution containing 0.2% mercury, 1% salicylic acid, 3% glycerin, and 95.8% water; used as a disinfectant.

Protegin®. Mineral oil, petrolatum, ozokerite, glyceryl oleate, lanolin alcohol. emollient, emulsifier. [Goldschmidt; Goldschmidt AG]

Protegin® W, WX. Petrolatum, ozokerite, hydrogenated castor oil, glyceryl isostearate, polyglyceryl-3 oleate; SE w/o emulsifier, emollient, absorp. base for cosmetics and pharmaceuticals. [Goldschmidt; Goldschmidt AG]

proteids. The same as albuminoids.

Pro-Tein ES-20. CAS 68951-89-3; Hydrolyzed collagen, ethyl ester; substantive to hair; plasticizer for styling resins; improves film properties and gloss in cream rinses, hair conditioners, mousses; anti-irritant and protectant in facial toners, antiperspirants. [Maybrook]

Pro-Tein SA-20. CAS 68952-15-8; Lauroyl hydrolyzed collagen in ethanol; adds body, mitigates drying effects of alcohol in skin and hair care products (hair sprays, aftershaves, facial toners, antiperspirants); plasticizer for hair styling resins. [Maybrook]

Pro-Tein SM-20. CAS 72319-06-3; Myristoyl hydrolyzed collagen; cosmetic ingredient. [Maybrook]

proteinase. Bacterial protease. [Rhone-Poulenc UK]

Protein Grade®. Detergents. [Calbiochem Corp]

Proteodermin. Sol. proteoglycans; for skin care preparations for aging or sun damaged skin. [Dr. Kurt Richter; Henkel/Cospha]

Proteol. A combination of casein with formaldehyde; an antiseptic dusting powder.

Proteosilane C. Methylsilanol elastinate; aids tissue regeneration; for regenerative creams, anti-aging formulations, stretch mark preventives, other cosmetic and health emulsions, creams, alcoholic lotions. [Exsymol]

Prote-pon P 2 EHA-02-Z. Alkyl ether phosphoric acid; wetting agent, detergent, hydrotrope, emulsifier, rust inhibitor, EP lubricant for alkaline detergents, metal cleaners/lubricants, hard surface cleaners, textile scours/lubricants, emulsion polymerization, agric., and drycleaning formulations. [Protex]

Prote-sorb SML. CAS 1338-39-2; EINECS 215-663-3; Sorbitan laurate; emulsifier for food, cosmetic, household and industrial, agric., leather, metalworking, and textile industries. [Protex]

Prote-sorb SMO. CAS 1338-43-8; EINECS 215-665-4; Sorbitan oleate; emulsifier for food, cosmetic, household and industrial, agric., leather, metalworking, and textile industries. [Protex]

Prote-sorb SMP. CAS 26266-57-9; EINECS 247-568-8; Sorbitan palmitate; emulsifier for food, cosmetic, household and industrial, agric., leather, metalworking, and textile industries. [Protex]

Prote-sorb SMS. CAS 1338-41-6; EINECS 215-664-9; Sorbitan stearate; emulsifier for food, cosmetic, household and industrial, agric., leather, metalworking, and textile industries. [Protex]

Prote-sorb STO. CAS 26266-58-0; EINECS 247-569-3; Sorbitan trioleate; emulsifier for food, cosmetic, household and industrial, agric., leather, metalworking, and textile industries. [Protex]

Prote-sorb STS. CAS 26658-19-5; EINECS 247-891-4; Sorbitan tristearate; emulsifier for food, cosmetic, household and industrial, agric., leather, metalworking, and textile industries. [Protex]

Pro-Tex. Maneb (32.63%) and triphenyltin hydroxide (4.72%) solution; flowable fungicide for potatoes and sugar beets; restricted use. [Griffin]

Protex. A proprietary safety glass. §

Protexulate. Loose-fill mineral powder; for underground pipework for heat insulation and as a water barrier to stop corrosion. [Croxton & Garry Ltd] *

Prothar. Factor IX complex; hemostatic. [Armour Pharmaceutical Co] *

protheite. A mineral. It is a variety of pyroxene.

Prothera™. White petrolatum; skin cleanser and protectant. [ICN Pharmaceuticals Inc]

protocatechuic acid. (3,4-dihydroxybenzoic acid) $C_7H_6O_4$; CAS 99-50-3; EINECS 202-760-0. [Dinoval Chem. Ltd]

Protogest. CAS 9015-54-7; Protein hydrolysate veterinary. [The Wellcome Foundation Ltd] *

Proto-Lan 4R. Mineral oil, coco-hydrolyzed collagen, cetyl alcohol, myristyl myristate, ceteth-16, hydrogenated lanolin; emollient, moisturizer, auxiliary emulsifier, antistat, conditioner, base for skin and hair care products (hair conditioners, creams, lotions, lipsticks); provides lubricious film, water emulsifying/binding properties [Maybrook]

Proto-Lan 8. Lecithin, butyl stearate, coco-hydrolyzed animal protein, oleoyl sarcosine, sesame oil, lanolin alcohol. Nongreasy emollient, moisturizer, emulsifier, conditioner, antistat, lubricant for creams, lotions, ethnic formulations, conditioners, baby oils, bath oils, suntan products, shave creams, makeup, and face soaps. [Maybrook]

Proto-Lan 20. Hydrolyzed animal protein, PEG-75 lanolin, propylene glycol, ceteth-16, lanolin oil; substantive emollient, softener, conditioner, moisturizer, lubricant, base for skin and hair products (shampoos, conditioners, ethnic products, creams, lotions, face masks, and bath products). [Maybrook]

Proto-Lan 30. Propylene glycol, PPG-12-PEG-65 lanolin oil, hydrolyzed animal protein; substantive emollient, moisturizer, adjunct emulsifier for skin and hair care products (shampoos, conditioners, creams, lotions, bath products, antiperspirants, facial toners). [Maybrook]

Proto-Lan IP. CAS 2724-58-5, 977066-20-8; Isostearic acid, isostearoyl hydrolyzed collagen; nongreasy emollient, moisturizer for ethnic formulations, conditioners, hair treatments, lip care, skin care, nail polish, soap bars, and antiperspirants. [Maybrook]

Proto-Lan KT. Isostearic acid, sorbitan oleate, cocoyl hydrolyzed keratin; cosmetic ingredient. [Maybrook]

---

* Trade name not verified as available   † Trade name verified as obsolete

Protopam Chloride. CAS 51-15-0; Pralidoxime chloride; cholinesterase reactivator. [Wyeth Laboratories] *

Protopet. Petrolatum USP. [Witco]

Protophane. Insulin, isophane; antidiabetic. [Bristol-Myers Squibb Co Inc] *

Protormone. CAS 57-83-0; Veterinary injection of progesterone. [The Wellcome Foundation Ltd] *

Protovit. Proprietary preparation containing the B complex vitamins, vitamins A, C, D and E, biotin and nicotinamide. [Roche Products Ltd] *

Protropin. Somatrem; hormone. [Genentech Inc] *

Protrum K. CAS 13684-63-4; Emulsifiable concentrate containing 114 g/l phenmcdipham, for weed control for beet crops and strawberries. [Atlas Interlates Ltd]

Protugan. CAS 34123-59-6; Isoproturon; herbicide. [Agan Chemical Manufacturers Ltd]

Provent. CAS 58-55-9; A proprietary formulation of theophylline; for prevention and treatment of bronchial smooth muscle constriction in asthma and chronic bronchitis. [The Wellcome Foundation Ltd]

Proventil Inhaler. CAS 18559-94-9; Albuterol; bronchodilator. [Schering Corp]

Proventin. Protective agent for bleaching polyamide fibers. [Degussa]

Proventin 7. Antioxidant for use with nylon to prevent attack of peroxide compounds during bleaching and washing. [Henkel/Textiles]

Proventin 7. An active oxygen compound; used in the textile industry as specific protection agent for polyamide fibers during peroxide bleaching and in the dyeing process. [Degussa AG] *

Provera. CAS 71-58-9; Medroxyprogesterone acetate; progestin. [Upjohn]

Provera Dosepak. CAS 71-58-9; Medroxyprogesterone acetate; progestin. [Upjohn] *

Provil®. Addition-cured silicone based precision impression material in four different viscosities; for dentistry. [Bayer AG]

Pro-Viron. CAS 1424-00-6; 25 mg Mesterolone. [Schering Health Care Ltd]

provitamin A. See carotene.

Provocholine. CAS 62-51-1; Methacholine chloride; cholinergic. [Hoffmann-LaRoche Inc] *

Provol 50. CAS 52581-71-2; PPG-50 oleyl ether; emollient, superfatting agent, lubricant, pigment dispersant, coupler for personal care products, ethnic hair products; aids spreading and pigment dispersion in make-up systems. [Croda Inc.]

Provol. A compound used for electrical insulation. It is a mixture of pitch, bitumen or similar materials, and mineral matter.

Pro Weld. Two-part cold welding compound; for bonding aluminum, brass and steel castings. [Novest Inc] ‡

Proxel. Industrial microbiocides. [ICI Chem. & Polymers Ltd]

Proxitane. CAS 79-21-0; Peracetic acid. [Interox Chemicals Ltd]

Prox-onic 2EHA-1/02. PEG-2 2-ethylhexyl ether; detergent, wetting agent, emulsifier, dispersant, solubilizer, defoamer for textiles, metal cleaners, industrial, institutional and household cleaners, hand cleaners. [Protex]

Prox-onic BP-02 P. POP (2) bisphenol; aromatic dihydric alcohol for manufacturing unsaturated polyester and alkyd resins. [Protex]

Prox-onic CC-05. CAS 61791-29-5; PEG-5 cocoate; emul-

sifier, lubricant additive for metalworking, textiles, cosmetics, defoamers; visc. control agent in plastisols. [Protex]

Prox-onic CSA-1/04. CAS 68439-49-6; Ceteareth-4; emulsifier, emulsion stabilizer, detergent, wetting agent, dispersant, solubilizer, defoamer, dye assistant, leveling agent for cosmetics, textiles, metal cleaners, industrial, institutional and household cleaners, emulsion polymerization. [Protex]

Prox-onic DA-1/04. CAS 26183-52-8; Deceth-4; wetting agent, emulsifier, detergent, dispersant, solubilizer, de foamer for textiles, metal cleaners, industrial, institutional and household cleaners. [Protex]

Prox-onic DDP-09. CAS 9014-92-0; Dodoxynol-9; detergent, wetting agent for industrial and heavy-duty detergents. [Protex]

Prox-onic DNP-08. CAS 9014-93-1; Nonyl nonoxynol-8; emulsifier, detergent, wetting agent, dispersant, solubilizer, coupling agent for textiles, metalworking, household, industrial, agric., paper, paint and other industries. [Protex]

Prox-onic DT-03. PEG-3 tallow diamine; surfactant, emulsifier, lubricant additive, antistat, detergent for textile, metal, plastics, dyeing assistants, degreasers, corrosion inhibitor, agric.; intermediate for quats. [Protex]

Prox-onic EP 1090-1. Difunctional block polymer ending in primary hydroxyl groups; defoamer for metalworking, cosmetic, paper, textiles; base for low foaming surfactants, antifoams, dishwash, dispersing and wetting agents for paints, drilling muds, emulsifiers, petrol. demulsifiers, emulsion polymerization. [Protex]

Prox-onic HR-05. CAS 61791-12-6; PEG-5 castor oil; emulsifier, pigment dispersant, leveling agent, softener, rewetting agent, degreaser, antistat, emulsion stabilizer, lubricant for leather, paint, paper, plastics, textile, and cosmetics industries; solubilizer for perfumes. [Protex]

Prox-onic HRH-05. CAS 61788-85-0; PEG-5 hydrogenated castor oil; emulsifier, pigment dispersant, leveling agent, softener, rewetting agent, degreaser, antistat, emulsion stabilizer, lubricant for leather, paint, paper, plastics, textile, and cosmetics industries; solubilizer for perfumes. [Protex]

Prox-onic L 081-05. POE (5) linear alcohol ether; biodeg. low foam detergent, wetting agent, emulsifier for household, agric. and industrial cleaners; coupling agent and solubilizer for perfumes and organic additives. [Protex]

Prox-onic LA-1/02. Laureth-2; coupling agent, solubilizer, emulsion stabilizer for cosmetic and hair care products; with anionic surfactants for emulsion polymerization; in coning and textile spin finishes. [Protex]

Prox-onic MC-02. CAS 61791-14-8; PEG-2 cocamine; surfactant, emulsifier, lubricant additive, antistat, detergent for textile, metal, plastics, dyeing assistants, degreasers, corrosion inhibitor, agric.; intermediate for quaternaries. [Protex]

Prox-onic MG-020 P. CAS 61849-72-7; POP (20) methyl glucoside; cosolv., perfume fixative. [Protex]

Prox-onic MHT-015. CAS 61791-26-2; PEG-15 hydrogenated tallow amine; surfactant, emulsifier, lubricant additive, antistat, detergent for textile, metal, plastics, dyeing assistants, degreasers, corrosion inhibitor, agric.; intermediate for quats. [Protex]

Prox-onic MO-02. PEG-2 oleamine; surfactant, emulsifier, lubricant additive, antistat, detergent for textile, metal,

---

‡ Trade name and manufacturer not verified   § Trade name without identified manufacturer

plastics, dyeing assistants, degreasers, corrosion inhibitor, agric.; intermediate for quats. [Protex]

Prox-onic MS-05. CAS 26635-92-7; PEG-5 stearamine; surfactant, emulsifier, lubricant additive, antistat, detergent for textile, metal, plastics, dyeing assistants, degreasers, corrosion inhibitor, agric.; intermediate for quats. [Protex]

Prox-onic MT-02. CAS 61791-44-4; PEG-2 tallow amine; surfactant, emulsifier, lubricant additive, antistat, detergent for textile, metal, plastics, dyeing assistants, degreasers, corrosion inhibitor, agric.; intermediate for quats. [Protex]

Prox-onic NP-04. CAS 9016-45-9; Nonoxynol-4; surfactant, detergent, defoamer. [Protex]

Prox-onic OA-1/04. CAS 9004-98-2; Oleth-4; coupling agent, solubilizer, emulsion stabilizer for cosmetic and hair care products; with anionic surfactants for emulsion polymerization; in coning and textile spin finishes. [Protex]

Prox-onic OCA-1/06. Cetoleth-6; detergent, wetting agent, emulsifier, dispersant, solubilizer, defoamer for textiles, metal cleaners, industrial, institutional and household cleaners, hand cleaners. [Protex]

Prox-onic OL-1/05. CAS 9004-96-0; PEG-5 oleate; surfactant for cutting oils, degreasing solvs., metal cleaners, textiles, leather, cosmetics; dyeing assistant; emulsifier for mineral oils, fatty oils. [Protex]

Prox-onic OP-09. CAS 9002-93-1; Octoxynol-9; emulsifier for metal, textile processing, household and industrial cleaners, vinyl and acrylic polymerization. [Protex]

Prox-onic PEG-2000. CAS 25322-68-3; PEG-2M; low foam wetting in paper pulping; emulsifier for metal degreasing; bottle cleaner defoamer; binder for tobacco; polyurethane manufacturing; mold release agent; agric. [Protex]

Prox-onic PH-01. CAS 122-99-6; EINECS 204-589-7; Phenoxyethanol. [Protex]

Prox-onic PPG-900. PPG (900); agric. emulsifiable concs. [Protex]

Prox-onic SA-1/02. CAS 9005-00-9; Steareth-2; detergent, wetting agent, emulsifier, dispersant, solubilizer, defoamer for textiles, metal cleaners, industrial, institutional and household cleaners, hand cleaners. [Protex]

Prox-onic SA1-015/P. CAS 25231-21-4; POP (15) stearyl alcohol; surfactant. [Protex]

Prox-onic SML-020. CAS 9005-64-5; Polysorbate 20; emulsifier, solubilizer for petrol. oils, solvs., vegetable oils, waxes, silicones, etc.; for agric., cosmetic, leather, metalworking and textile industries. [Protex]

Prox-onic SMO-05. CAS 9005-65-6; Polysorbate 81; emulsifier, solubilizer for petrol. oils, solvs., vegetable oils, waxes, silicones, etc.; for agric., cosmetic, leather, metalworking and textile industries. [Protex]

Prox-onic SMO-020. CAS 9005-65-6; Polysorbate 80; emulsifier, solubilizer for petrol. oils, solvs., vegetable oils, waxes, silicones, etc.; for agric., cosmetic, leather, metalworking and textile industries. [Protex]

Prox-onic SMP-020. CAS 9005-66-7; Polysorbate 40; emulsifier, solubilizer for petrol. oils, solvs., vegetable oils, waxes, silicones, etc.; for agric., cosmetic, leather, metalworking and textile industries. [Protex]

Prox-onic SMS-020. CAS 9005-67-8; Polysorbate 60; emulsifier, solubilizer for petrol. oils, solvs., vegetable oils, waxes, silicones, etc.; for agric., cosmetic, leather, metalworking and textile industries. [Protex]

Prox-onic ST-05. CAS 9004-99-3; PEG-5 stearate; emulsifier, lubricant additive for metalworking, textiles, cosmetics, defoamers; visc. control agent in plastisols. [Protex]

Prox-onic TA-1/08. CAS 61791-00-2; PEG-8 tallate; emulsifier, detergent for degreasers and neutral or mildly alkaline detergents. [Protex]

Prox-onic TD-1/03. CAS 24938-91-8; Trideceth-3; intermediate for surfactants; emulsifier, detergent, foam builder, dispersant, wetting agent, solubilizer for mech. dishwash, alkaline cleaners, pulp and paper, textiles, corrosion inhibition. [Protex]

Prox-onic TM-06. POE (6) isolauryl mercaptan; wetting agent, surfactant, detergent for metal cleaning, household and industrial cleaners/scours, degreasers; emulsifier for herbicides/insecticides, greases, soils. [Protex]

Prox-onic UA-03. Isoundeceth-3; detergent, wetting agent, emulsifier, dispersant, solubilizer, defoamer for textiles, metal cleaners, industrial, institutional and household cleaners, hand cleaners. [Protex]

Proxy. A hydrogen peroxide solution.

Proxyl. A proprietary trade name for a pyroxylin denture material. §

Prozac. Fluxetine. [Dista Products Ltd]

Prozac. Fluoxetine hydrochloride; antidepressant. [Eli Lilly & Co]

Prozine. A proprietary preparation of meprobamate and promazine hydrochloride. [Wyeth Laboratories] *

Prozinex. CAS 139-40-2; Active ingredient: propazine; 2-chloro-4,6-bis-(isopropylamino)-1,3,5-triazine; selective pre-emergence herbicide. [Agan Chemical Manufacturers Ltd]

Prozone. Solvent for cleaning for electronic applications. [BP Chemicals Ltd]

PR Spray. Proprietary preparations for topical pain relief. [The Boots Co plc]

PR Tablets. Analgesic. [The Boots Co plc] *

Prulet. CAS 77-09-8; Phenolphthalein; laxative. [Mission Pharmacal Co]

prussic acid. CAS 74-90-8; (hydrocyanic acid; formonitrile) HCN; used in the manufacture of acrylonitrile, acrylates, adiponitrile; chelates; pesticides.

Pruteen. Single-cell protein used as a feed additive. [ICI Chem. & Polymers Ltd] †

Pruv. CAS 4070-80-8; Sodium stearyl fumarate NF; lubricant for pharmaceutical tablets. [Mendell]

Pryfon. CAS 25311-71-1; Isofenphos; insecticide for treatment of soil against subterranean termites. [Bayer AG]

Prystal. A formaldehyde-urea condensation product, a French proprietary rubber vulcanization accelerator; also a proprietary trade name for a cast, clear phenolic molding. §

Prystaline. A proprietary molding compound of the urea-formaldehyde type of synthetic resin. §

PS. See polystyrene.

PS3/PS4/PS5. Range of granular fertilizers. [Fisons plc, Horticulture Div] *

PS-10GF/000. Thermoplastic polyester, 10% glass fiber-reinforced. [Compounding Tech.] *

PS021. Tetrachlorophenyl-siloxane dimethyl siloxane copolymer, branched; thermal silicone fluid for metal-to-metal lubrication; base for high temp. grease lubricants; lubricant for instruments. [Hüls Am.]

PS-30GM/000. CAS 26062-94-2; PBT polyester, 30% milled glass-reinforced. [Compounding Tech.] *

---

* Trade name not verified as available   † Trade name verified as obsolete

PS034. Dimethicone; antifoam, leveling and flow control agent for coatings; heat exchangers, baths and thermostats; dielec. media; for cooling in transduction applications; ultrasonic applications. [Hüls Am.]

PS071. Ethylene oxide-modified polydimethylsiloxane; surfactant, wetting agent for photographic plates; antifog for glass and plastic optics; slip agent for flexographic and gravure inks. [Hüls Am.]

PS072. Ethylene oxide/propylene oxide-modified polydimethylsiloxane; surfactant, lubricant for fibers and plastics, metal-to-plastic wear interfaces; antitack and mar resistance aid for urethane coatings; slip agent in flexographic and gravure inks. [Hüls Am.]

PS073. Ethylene oxide-modified polydimethylsiloxane; surfactant, antifoam in water-based coatings. [Hüls Am.]

PS130. CAS 68607-75-0; Polyoctadecylmethylsiloxane; component in thread and fiber lubricants; process aid in melt spinning. [Hüls Am.]

PS140. CAS 68440-90-4; Polyoctylmethylsiloxane; lubricant for soft metals; rubber and plastic lubricant esp. when mated against steel or aluminum; for aluminum machining operations; process aid and plasticizer in polyolefin rubbers. [Hüls Am.]

PS181. CAS 63148-56-1; Polymethyl-3,3,3-trifluoropropylsiloxane; lubricant for EP applications, automotive and aerospace lubes, elec. contacts, precision timers; flotation medium for inertial guidance systems; in sonar lens, mech. vaccum pumps. [Hüls Am.]

PS187. Polymethyl-3,3,3-trifluoropropylsiloxane- (50%) dimethylsiloxane copolymer; lubricant for EP applications, automotive and aerospace lubes, elec. contacts, precision timers; flotation medium for inertial guidance systems; in sonar lens, mech. vaccum pumps. [Hüls Am.]

PSA. Phenol sulfonic acid; electrolyte for tin plating; catalyst for resins. [Hart Chem. Ltd.]

PSDF04. 1,1,5,5-Tetraphenyl-1,3,3,5-tetramethyltrisiloxane; diffusion pump fluid with excellent resistance to heat, oxidation and chems. [Hüls Am.]

PSDF05. 1,1,3,5,5-Pentaphenyl-1,3,5-trimethyltrisiloxane; diffusion pump fluid with excellent resistance to heat, oxidation and chems. [Hüls Am.]

P.S.E. No. 15 powder. An explosive. It is a mixture of ammonium perchlorate and rosin.

pseudo-alums. Double sulfates of aluminum and another metal containing a bivalent metal sulfate instead of a monovalent one. A type is $MnSO_4 \cdot Al_2(SO_4)_3 \cdot 24H_2O$. They are not isomorphous with the alums.

Pseudocollagen. High molecular weight matric oligosaccharides and sol. proteins; cosmetic ingredient forming moisture-retentive films on skin, leaving skinsoft and supple. [Brooks Industries]

pseudoephedrine. CAS 90-82-4; (+)-2-Methylamino-1-phenylpropan-1-ol (a stereoisomer of ephedrine).

pseudo-galena. See zinc blende.

Psoranide. CAS 67-73-2; Fluocinolone acetonide; glucocorticoid. [ICN Pharmaceuticals Inc] †

Psoriderm Bath Emulsion. Buff-colored liquid emulsion containing 40% special coal tar extract; an aid in the treatment of sub-acute and chronic psoriasis. [Dermal Laboratories Ltd]

Psoriderm Cream. Buff colored aqueous cream containing 6% special coal tar extract, 0.4% lecithin; used for the topical treatment of sub-acute and chronic psoriasis

including psoriasis of the scalp and flexures. [Dermal Laboratories Ltd]

Psoriderm Scalp Lotion. Amber colored foaming liquid containing 2.5% special coal tar extract, 0.3% lecithin; used for the topical treatment of psoriasis of the scalp. [Dermal Laboratories Ltd]

Psorion®. CAS 5593-20-4; Betamethasone dipropionate; topical corticosteroid for treatment of psoriasis. [ICN Pharmaceuticals Inc]

Psorox. Dermatological lotion and ointment. [Fisons plc, Pharmaceuticals Div] *

PTAL. CAS 140-87-0; EINECS 203-246-9; p-Tolualdehyde; additive for resins; intermediate for pharmaceuticals; fragrance. [Mitsubishi Gas]

P.T.F.E. Polytetrafluoroethylene.

PTFE, TFE. See polytetrafluoroethylene.

PTFE-19. CAS 9002-84-0, 13463-67-7; PTFE, titanium dioxide; binder for pressed powds.; excellent thermal and chem. resistance, good skin adhesion; imparts luxurious, lubricious feel. [Presperse]

PTMEG. See polytetramethylene ether glycol.

PTMG. See polytetramethylene ether glycol.

PTSA. See p-toluenesulfonamide.

PTSA 70. p-Toluene sulfonic acid; catalyst for resins. [Hart Chem. Ltd.]

PTZ® Phenothiazine Purified. Phenothiazine; antioxidant, monomer stabilizer. [ICI Polymer Additives]

Pulluzyme. Pullulanase. [Rhone-Poulenc UK]

Pulmadil Auto. CAS 31842-61-2; Breath-actuated pressurized aerosol containing a suspension of rimiterol hydrobromide 10 mg/ml; delivers 300 metered doses (0.2 mg per dose). [3M Pharmaceuticals] †

Pulmadil Inhaler. CAS 31842-61-2; Pressurized aerosol containing a suspension of rimiterol hydrobromide 10 mg/ml; delivers 300 metered doses (0.2 mg per dose). [3M Pharmaceuticals] †

Pulmo Bailly. Contains guaiacol 75 mg, phosphoric acid concentrated 85 mg, codeine 7 mg per 5 ml spoonful; cough sedative and expectorant. [Bengue & Co Ltd] *

Pulmolite. Technetium, Tc 99m aggregated albumin; used as a diagnostic aid and a radioactive agent. [DuPont-NEN Medical Products] *

Pulmovax. Pneumococcal vaccine; effective protection against certain pneumococcal infections. [Merck & Co Inc] †

Pulpex E and P. Polypropylene pulps; special fibrous additives that can be mixed in any proportion with natural pulps and can be handled on conventional equipment; very wide range of applications. [Hercules] *

Pulpzyme. A liquid xylanase preparation for reducing the need of bleaching chemicals in kraft pulp bleaching. [Novo Nordisk]

Pulsar. CAS 7778-54-3; Calcium hypochlorite; dry chlorinating agent. [Olin]

Pulsar. A solution concentrate containing 200 g bentazone and 200 g MCPA per liter; a post-emergence herbicide. [BASF plc]

Pulse®. Fish oil concentrate; dietary supplement of omega 3 fatty acids. [Seven Seas Ltd]

Pulse. A family of polycarbonate resin blends used for interior automotive applications. [Dow UK]

Pulse 600. Engineering thermoplastic; engineering thermoplastic providing easy processing, toughness, resistance to temp. extremes; suitable for snowmobile com-

---

‡ Trade name and manufacturer not verified   § Trade name without identified manufacturer

ponents that must withstand freezing conditions as adjacent parts become red-hot, cellular phones; exposed to heat inside closed automobiles; automotive grade. [Dow]

Pulse 1310. Engineering thermoplastic; engineering thermoplastic providing easy processing, toughness, resistance to temp. extremes; suitable for snowmobile components that must withstand freezing conditions as adjacent parts become red-hot, cellular phones exposed to heat inside closed automobiles; for durables. [Dow]

Pulse 1735. PC/ABS resin; uv-stable resin with high heat distort., easy processing, practical toughness for computer and business equip. [Dow Plastics]

Pultac. Strippable protective for paint spray walls, windows and floor grills. [Brent Chemicals International plc]

Pulvatex. A trademark for rubber (raw or partly prepared) for manufacture. §

Pumice Plus. Abrasive solvent base hand cleaner gel. [Momar Industrial Services Ltd]

pumice stone. (obsidian). A volcanic mineral (lava froth), consisting mainly of aluminum silicate; used as an abrasive.

Pumiline. A proprietary preparation of pine oil. §

Pump Repair Putty. Ceramic filled epoxy resin; used to repair, rebuild and protect pumps. [ITW Devcon]

Punch® C. Carbendazim + flusilazole; eradicant fungicide for use on cereals. [DuPont UK]

Punctilious® Ethyl Alcohol. CAS 64-17-5; EINECS 200-578-6; Ethanol; solv. and extraction medium. [Quantum/USI]

punicin. A coloring matter obtained from *Purpura capillus* and other shell fish.

Purac. CAS 7440-44-0; Decolorizing and absorptive activated carbon. [Lancashire Chemical Works Ltd]

Puraspec. Range of catalysts and absorbents used for purification. [ICI Chem. & Polymers Ltd]

Puratronic. High purity chemicals suitable for electronic materials; used in electronic device materials, crystal growing, epitaxy. [Johnson Matthey plc]

Purbeck stone. A limestone used in building.

Purdox. CAS 1344-28-1; A proprietary trade name for high purity recrystallized alumina. [Morgan Refractories Ltd]*

Purdurum. *See* Camite. §

P.U.R.E.-CMC. PU rubber elastomer (Part A); isocyanate (Part B); cold molding compound for making rubber molds, flexible blocks and cases, and for applications requiring unusual resistance to abrasion, impact, aging, and severe wear; the cured material has outstanding tear and tensile strength to withstand extensive flexing and use; for use with plaster, gypsum, and cements; mix ratio: 1:1 by wt. [Perma-Flex Mold]

Pure-Dent® B700. CAS 9005-25-8; Corn starch USP, NF; binder and diluent for granulations and tablets when used wet or dry; disintegrant; for pharmaceutical use. [Grain Processing]

Purez. Polyester polyols based on adipic acid. [Avalon Chemical Co Ltd]

Purganol. CAS 77-09-8; A phenol-phthalein preparation. §

Purgatin. (Purgatol). Anthrapurprin diacetate, $C_{14}H_5O_2$-$(OH)(O \cdot C_2H_3O)_2$. An aperient.

Purgatol. Consists mainly of the anhydrides and lactones of fatty acids; used for dressing hides. Also *see* Purgatin.

Purgen. CAS 77-09-8; Proprietary preparation of phenolphthalein, $C_6H_4(OC)C \cdot (C_6H_4 \cdot OH)_2O$, sometimes with

malic acid. §

purging nut oil. (physic nut oil). Curcas oil, from the seeds of *Jatropha Curcas*; used in soap-making, and for lubricating.

Purifloc. Polyacrylamides used as flocculants. [Dow UK]

Purinethol. CAS 50-44-2; A proprietary formulation of mercaptopurine; for remission induction, remission consolidation and maintenance therapy of acute leukemias. [The Wellcome Foundation Ltd]

Purisol. Chemicals and equipment for treating water, effluent and sewage. [ICI Chem. & Polymers Ltd]

Purity® 21. CAS 9005-25-8; Corn starch; binder, filler and disintegrant for cosmetic and pharmaceutical formulations, body powds., foot powds., dry shampoos, makeup, eye liner, mascara. [Nat'l. Starch]

Pur-Oba®. Natural wax ester; water-white jojoba oil for high-quality cosmetics. [Goldschmidt AG]

Purochem. Organotin bactericides. [Akzo Chemie] *

Puromix. Food phosphate mixtures. [Albright & Wilson Ltd, Phosphates & Speciality Business] *

Puron. Food grade phosphates. [Albright & Wilson Ltd, Phosphates & Speciality Business]

Purozone. A proprietary preparation consisting of an alcoholic solution of the sodium salts and acids of wood tar or similar materials; a disinfectant. §

Purple Copp. CAS 1317-39-1; Purple cuprous oxide. [Am. Chemet]

Purplecopp 97N Premium. CAS 1317-39-1; Cuprous oxide. [CP International Chemicals Ltd]

Pursennid. A proprietary preparation containing sennosides A and B (as calcium salts); a laxative. [Sandoz] *

Purton CFD. Cocamide DEA; foam stabilizer, thickener and superfatting agent for cosmetics, cleaners. [Zschimmer & Schwarz]

Purton SFD. Linoleamide DEA; foam stabilizer, thickener and superfatting agent for cosmetics, cleaners. [Zschimmer & Schwarz]

Purus. *See* Chloramine T. §

Purzaust® Catalysts. Catalysts for broad range of air purification applications [Allied-Signal]

Pusher. Secondary oil recovery polymer; an additive to water injected into petroleum reservoirs; decreases the mobility of the water flood in relation to the mobility of the crude. [Dow UK]

putrescine. CAS 110-60-1; Tetramethylene-diamine, $NH_2(CH_2)_4 \cdot NH_2$. A base found in ergot.

putty powder. An impure stannic oxide, SnO_2; used for polishing glass. It is also used sometimes in rubber mixing specific gravity of 6.6.

puzzuolana. A volcanic material found in various parts of Italy, especially Puzzuoli. It is employed for the conversion of pure lime into a hydraulic lime.

PV, PV Fast. Pigment powders for plastics. [Hoechst AG]

PVA; Abbreviation for polyvinyl alcohol.

PVA. *See* polyvinyl alcohol.

PVAc. *See* polyvinyl acetate (homopolymer).

Pvacote. Compounded polymeric emulsion; for spray coating of asbestos stripped rooms to bind residue fibers. [Howlett Adhesives Ltd] *

PVAL. *See* polyvinyl alcohol.

PVB. *See* polyvinyl butyral.

PVC. Abbreviation for polyvinyl chloride.

PVC. *See* polyvinyl chloride.

---

* Trade name not verified as available   † Trade name verified as obsolete

P V Carpine Liquifilm. CAS 148-72-1; Pilocarpine nitrate; cholinergic. [Allergan Pharmaceuticals Inc] *

PVC Deodorant #5417, OS. Mixture of fragrance materials; masks odors during processing and in finished product [Andrea Aromatics] *

PVF2. A polyvinyl fluoride. §

PVOH. See polyvinyl alcohol.

PVP. (polyvinylpyrrolidone; povidone; 1-ethenyl-2-pyrrolidinone, homopolymer) Polymer of 1-vinyl-2-pyrrolidone monomers. $(C_6H_9NO)_x$; CAS 9003-39-8; EINECS 201-800-4; film-forming agent, hair fixative, thickener, protective colloid, suspending agent, and dispersant for cosmetics industry, tech. applications.; drug vehicle and retardant; tablet binder, pharmaceutical excipient; in adhesives; in detergents. [Allchem Industries; BASF; ISP]

PVP-Iodine. CAS 25655-41-8; Povidone-iodine; anti-infective. [ISP] *

PX-10GF/000. Polyphenylene ether, 10% glass fiber-reinforced. [Compounding Tech.]

PX-104. CAS 84-74-2; Dibutyl phthalate; reagent grade plasticizer [Aristech] *

PX-109. CAS 68515-48-0; Diisononyl phthalate; reagent grade plasticizer [Aristech] *

PX-111. CAS 3648-20-2; Diundecyl phthalate; reagent grade plasticizer [Aristech] *

PX-120. CAS 68515-49-1; Diisodecyl phthalate; reagent grade plasticizer [Aristech] *

PX-126. Ditridecyl phthalate; reagent grade plasticizer [Aristech] *

PX-138. CAS 117-81-7; Dioctyl phthalate; reagent grade plasticizer [Aristech] *

PX-209. Diisononyl adipate; reagent grade plasticizer [Aristech] *

PX-238. CAS 103-23-1; Dioctyl adipate; reagent grade plasticizer [Aristech] *

PX-338. CAS 3319-31-1; Trioctyl trimellitate; reagent grade plasticizer [Aristech] *

PX-339. CAS 53894-23-8; Triisononyl trimellitate; reagent grade plasticizer [Aristech] *

PX-504. CAS 105-76-0; EINECS 203-328-4; Dibutyl maleate; reagent grade plasticizer [Aristech] *

PX-538. Dioctyl maleate; reagent grade plasticizer [Aristech] *

PX-800. CAS 8013-07-8; EINECS 232-391-0; Epoxidized soya oil; reagent grade plasticizer [Aristech] *

PX-914. Butyl octyl phthalate; reagent grade plasticizer [Aristech] *

Pybuthrin. Synergized pyrethrins. [The Wellcome Foundation Ltd] *

Pycamisan. A proprietary preparation containing sodium aminosalicylate, isoniazid; an antituberculous agent. [Smith & Nephew Pharmaceuticals Ltd] *

Pycasix. A proprietary preparation containing sodium aminosalicylate, isoniazid; an antituberculous agent. [Smith & Nephew Pharmaceuticals Ltd] *

Pycazide. CAS 54-85-3; A proprietary preparation containing isoniazid; an antituberculous agent. [Smith & Nephew Pharmaceuticals Ltd] *

PY Garden Insect Killer. Liquid concentrate containing pyrethrum and piperonyl butoxide; contact insecticide for use on crop and noncrop plants. [Vitax Ltd]

PY Garden Insecticide. Aerosol containing pyrethrins and piperonyl butoxide; contact insecticide for use on crop and noncrop areas. [Vitax Ltd]

Pylkrome. A composite alloy steel. The base is mild steel, and it has a corrosion-resisting surface of high chrome or high chromium-nickel-iron alloys.

Pylumin. Process for boning paint to aluminum and its alloys. [Brent Chemicals International plc]

Pylura. A proprietary preparation of adrenaline, benzyamine lactate and phenol; used in the treatment of hemorrhoids. [Pharmax Ltd] *

Pymafed. CAS 59-33-6; Pyrilamine maleate; antihistaminic. [Hoechst-Roussel Pharmaceuticals Inc] *

Pynol. Disinfectants. [Evans Vanodine International Ltd]

Pynosect 30. Mixture of pyrethrins and resmethrin; an insecticide for greenhouse and horticultural crops. [Mitchell Cotts Chemicals Ltd]

Pyoctanin. CAS 548-62-9; (Pyoktanin). The name given to different coal-tar colors. (1) Yellow pyoctanin (auramine, *[qv]*), and (2) Blue pyoctanin (methyl violet, *(qv)*); used in surgery as bactericides.

Pyoktanin. See Pyoctanin.

Pyopen. A proprietary preparation containing carbenicillin (as the disodium salt); an antibiotic. [SmithKline Beecham] *

PY Powder Garden & Household Insect Killer. Dust containing pyrethrins and piperonyl butoxide; contact domestic and garden insecticide. [Vitax Ltd]

Pyracur® FL. Chloridazon, metolachlor; pre-emergence herbicide for control of grasses and broadleaf weeds in sugar beet and fodder beet [BASF AG]

Pyradex® T. Chloridazon, triallate; pre-plant incorporated herbicide for control of broadleaf weeds and grasses in sugar and fodder beet [BASF AG]

Pyradiolin. A proprietary synthetic plastic used as a dielectric material in wireless telegraphy, and for other purposes; it is a modified pyroxylin plastic. §

Pyradur®. Chloridazon + metolachlor; pre-emergence herbicide for control of grasses and broad-leaved weeds in sugar and fodder beet crops. [BASF AG]

Pyra-Fog 100. Pyrethin; contact insecticide. [Chemsearch (UK) Ltd]

Pyralin®. Polyimide high temperature-resistant materials. [DuPont UK]

Pyralin; CAS 9004-70-0; A celluloid product available in transparent, translucent, opaque colored and colorless forms; resistant to hydrocarbons and oils.

pyraloxin. Oxidized pyrogallic acid, a dark brown powder obtained by oxidation with air and ammonia.

Pyralux®. Flexible materials. [DuPont UK]

Pyralvex. A proprietary preparation of anthraquinone glycosides and salicylic acid; used in the treatment of mouth ulcers. [Norgine Ltd]

Pyramid. Potassium silicates, granular sodium disilicates. [Crosfield Chemicals Ltd]

Pyramin®. CAS 1698-60-8; Chloridazon; for pre- and post-emergence control of weeds in sugar beet, fodder beet, Swiss chard, some ornamentals. [BASF AG; BASF plc]

Pyramol. Silica sols for industry. [Crosfield Chemicals Ltd]

Py-Ran. CAS 7758-23-8; Anhydrous monocalcium phosphate; used as acid component in leavening agents for self-raising flours, baking powders, cake mixes and corn meal. [Monsanto Co] *

Pyranet. Detergent sanitizer. [Crosfield Chemicals Ltd] †

Pyranol. CAS 1336-36-3; Sodium acetylsalicylate, $C_2H_3O \cdot OC_6H_4COONa$.

---

‡ Trade name and manufacturer not verified   § Trade name without identified manufacturer

Pyrasteel. A proprietary trade name for a heat-resisting alloy of iron with 25% nickel, 14% chromium, and 2.5-3.0% silicon. §

Pyratex. 2-Vinylpyridine-styrene-butadiene terpolymer; used to treat tire cord and other textiles to improve their adhesion to rubber and enhance the compatability of the textile-rubber system. [Bayer AG; Bayer plc]

Pyraton. A proprietary trade name for diacetone alcohol. §

Pyrax® A. CAS 12269-78-2; Pyrophyllite (hydrated aluminum silicate); inert filler/extender for NR and synthetic rubbers and latexes. [R.T. Vanderbilt]

Pyrax talcs A and B. Qualities of pure white talc mineral from deposits in America; used in rubber, textile, and ceramic industries.

pyrazinamide. CAS 98-96-4; Pyrazinoic acid amide; $C_5H_5N_3O$.

Pyrazine. See Antipyrine. §

pyrazine hexahydride. See piperazine.

Pyrazoline. See Antipyrine. §

Pyre-ML®. Wire enamel and insulating varnish; used in automotive industry. [DuPont UK]

Pyrene. Range of cleaners, phosphating process and paint shop care products. [Brent Chemicals International plc]

pyrene oil. (Begasses oil). An inferior olive oil.

pyrethrum powder. An insecticide made from the powdered flowers of some species of pyrethrum plants. The active principle is rotenone (*qv*).

Pyribenzamine Citrate. CAS 6138-56-3; Tripelennamine citrate; antihistaminic. [Ciba-Geigy Corp] *

Pyribenzamine Hydrochloride. CAS 154-69-8; Tripelennamine hydrochloride; antihistaminic. [Ciba-Geigy] *

Pyricit. Sodium fluoroborate, $NaBF_4$.

Pyricol. See Acopyrin. §

pyridine. Organic compound; $N(CH)_5$; CAS 110-86-1; EINECS 203-809-9; used in agrochemicals, pharmaceuticals, photographic materials, coatings, curing agents, rubber chemicals, plastics, antidandruff shampoos, textiles, dyestuffs. [Nepera; Penta Mfg.; Schweizerhall; Whitecourt Ltd]

pyridine-3-carboxylic acid. See nicotinic acid.

pyridine-4-carboxylic acid. See isonicotinic acid.

2,6-pyridinediamine. See 2,6-diaminopyridine.

2-pyridinol. See 1-hydroxy-2-pyridine.

Pyridium. Phenazophridine hydrochloride; analgesic. [Parke-Davis]

2-pyridone. See 1-hydroxy-2-pyridine.

Pyrido rubber. A polymerized acrolein-methyl-amine. A rubber-like material.

pyridoxine. CAS 65-23-6; (Vitamin B_6; 3-hydroxy-4,5-dimethylol-2-methylpyridine) A group of naturally occurring derivatives of pyridine with B_6 activity; $CH_3C_5HN-(OH)(CH_2OH)_2$; used in medicine and nutrition.

pyrimithate. (o-2-di-methylamino-6-methylpyrimidin-4-yl o,o-diethyl phosphorothioate) $C_{11}H_{20}N_3O_3PS$; CAS 5221-49-8; Used as an acaricide and an insecticide.

Pyrinex. CAS 2921-88-2; Active ingredient: chlorpyrifos; organophosphorous agricultural insecticide effective against a broad range of insects. [Makhteshim Chemical Works Ltd]

pyrites, cockscomb. See marcasite.

pyrites, coxcomb. See marcasite.

pyrites, radiated. See marcasite.

pyrites, white iron. See marcasite.

pyrithione zinc. See zinc pyrithione.

Pyro. CAS 64-17-5; EINECS 200-578-6; Denatured alcohol. [Quantum Chemical Corp]

pyro alcohol. Methyl alcohol, CH_3OH.

Pyroban G. Durable fire retardant for 100% polyester industrial products [CNC Int'l. L.P.]

pyro-bitumen. Bitumen which is insoluble in carbon tetrachloride.

Pyroblak. Oxide blackening process for iron and steel. [Brent Chemicals International plc]

Pyrobond. Wet blast anticorrosion additive. [Brent Chemicals International plc]

Pyrobor. Anhydrous sodium dimetaborate. [Kerr-McGee Chemical Corp]

Pyrobrite. Copper plating processes. [Albright & Wilson Ltd, Phosphates & Speciality Business] *

Pyrocast. A proprietary trade name for a nickel-chromium cast iron. §

pyrocatechin. (catechol; pyrocatechol) $C_6H_4(OH)_2$; CAS 120-80-9; used as an antiseptic, in photography, dyestuffs, electroplating; specialty inks, light stabilizers.

pyrocatechin. See pyrocatechol.

pyrocatechol. (1,2-benzenediol; pyrocatechin; 1,2-dihydroxybenzene) Phenol; $C_6H_4(OH)_2$; CAS 120-80-9; EINECS 204-427-5; in photography; dyeing furs; as reagent. [Aldrich; Coalite Chem. Div.; James River; Spectrum Chem. Mfg.]

pyrocatechol arsenic acid. o-Hydroxyphenyl arsenate; a reagent for alkaloids.

pyrocatechol methyl ether. See guaiacol.

Pyro-Chek® 68PB. Brominated polystyrene; flame retardant for plastics (esp. engineering plastics); suitable for thermosetting resins and polyolefins; synergist with antimony oxide. [Ferro]

Pyro-Chek® LM. Brominated polystyrene; flame retardant for styrenics. [Ferro]

Pyrochlor. Fire-resistant and wood preservative. [Hickson & Welch Ltd]

Pyrocide. Pyrethrins based intermediates. [McLaughlin Gormley King Co]

Pyroclean. Acid, alkaline, and emulsion cleaners. [Brent Chemicals International plc]

Pyroclense. Chassis and body cleaners for public transport and commercial haulage operators. [Brent Chemicals International plc]

pyrocollodion. A soluble nitrocellulose containing the highest practicable percentage of nitrogen, about 12.5%.

pyro cotton. A nitrated cellulose, not so fully nitrated as guncotton.

Pyrodialite. An explosive containing 80-88% potassium chlorate, 5-6% charcoal, 10-18% gas tar, and 3-4% sodium and ammonium bicarbonates.

Pyroforane. Halon fire extinguishing agents. [Atochem SA]

pyrofulmin. A yellow substance obtained by heating mercuric fulminate. It is probably a mixture of mercuric oxycyanide and oxide.

pyrogallic acid. CAS 87-66-1; Pyrogallol $C_6H_3(OH)_3$. It absorbs oxygen, and is used in gas analysis.

pyrogallic acid. See pyrogallol.

pyrogallol. (1,2,3-benzenetriol; pyrogallic acid; 1,2,3-trihydroxybenzene) Phenol; $C_6H_6O_3$; CAS 87-66-1; EINECS 201-762-9; protective colloid in preparation of metallic colloidal sol'n., photography, dyes, intermediate, synthetic drugs, medicine, lab reagent, reducing agent, antioxidant in lubricating oils. [Burlington Bio-

Medical; Fuji Chem. Ind.; Hoechst Celanese; Mallinckrodt; Schering Berlin Polymers]

Pyrogallol. CAS 87-66-1; 1,2,3-Trihydroxybenzene; chemical intermediate for electronics; photographic chemicals. [Schering Berlin Polymers]

pyrogallol 1,3-dimethyl ether. See 2,6-dimethoxy phenol.

Pyrogastrone. Tablets or liquid containing carbenoxolone sodium with alginate and antacids; used in the treatment of oesophagitis. [Winthrop Laboratories] *

pyrolignite of iron. (iron liquor, black mordant, black liquor, liqueur de ferraile). Ferrous acetate, $Fe(C_2H_3O_2)_2$, prepared by the action of pyroligneous acid upon iron turnings. The solution also contains ferric acetate; used in calico printing, and in dyeing, for the preparation of blue, violet, black, and brown colors.

pyrolignite of lime. CAS 62-54-4; Calcium diacetate.

Pyrolith. Fire retardant treatment for timber; for internal use only. [Hickson & Welch Ltd]

Pyromet® Alloy 625. Nonmagnetic nickel-base alloy; corrosion and oxidation-resistance ally with high strength and toughness; for heat shields, furnace hardware, gas turbine engine ducting, combustion liners, chem. plant hardware, seawater applications [Carpenter Tech.]

pyromic. A particularly pure form of induction melted nickel-chromium used for electric furnaces, heaters, ovens, etc.

pyromorphic phosphorus. Prepared by heating red phosphorus with a trace of iodine at 280 C, or in vacuo.

pyromucic acid. See furoic acid.

Pyronate. Alkylaryl petroleum sulfonate, completely hydrophilic, water soluble; dark in color; for demulsification; froth flotation of nonmetallic ores. [Witco Chemical Ltd]*

Pyronil. CAS 135-31-9; A proprietary trade name for the phosphate of pyrrobutamine. [Eli Lilly & Co] †

Pyronite. See Tetryl. §

Pyronium. A proprietary substitute for tin oxide in enamels; used in conjunction with tin oxide. §

Pyrophan. A combination of pyrogallol and dimethylamine.

pyroracemic acid. See pyruvic acid.

pyroretin. A brown resin found in lignite.

Pyros. A paramagnetic alloy consisting of nickel with 7% chromium, 5% tungsten, 3% manganese, and 3% iron. It is suitable for expansion pyrometers.

Pyroset. Organic phosphonium salt in water; flame retardants for textile industry. [Cyanamid BV]

pyrosine B. See erythrosin.

pyrosulfuric acid. See fuming sulfuric acid.

Pyroter CPI-40. PEG-40 hydrogenated castor oil PCA isostearate; emulsifier, solubilizer and thickener used in personal care products and detergents; low irritation, nontoxic. [Ajinomoto; Nihon Emulsion]

Pyroter GPI-25. Glycereth-25 PCA isostearate; moisturizer, emulsifier, solubilizer, dispersant and emollient. [Ajinomoto; Nihon Emulsion]

Pyrovatex. Proofing agent. [Ciba plc] *

Pyrox®. Mineral iron sulfide (pyrite); 48-50% S, 45-46% Fe; filler for resin-bonded grinding wheels, brake linings, etc. [Metallgesellschaft AG]

pyrrhol. CAS 109-97-7; (Pyrroline). Pyrrole, C_4H_5N; used in the manufacture of pharmaceuticals.

pyrrhotine. See magnetic pyrites.

pyrrhotite. See magnetic pyrites.

pyrrodiazole. Triazole, $C_2H_3N_3$.

2-pyrrolidine carboxylic acid. See proline.

2-pyrrolidone. (butyrolactam; pyrrolidone-2) CH_2CH_2-$CH_2C(O)NH$; CAS 616-45-5; EINECS 204-648-7; plasticizer and coalescing agent; solv. for veterinary medicine. [Allchem Industries; BASF; ISP; UCB SA]

pyrroline. See pyrrhol.

pyruvic acid. (2-oxopropionic acid; α-ketopropionic acid; acetylformic acid; pyroracemic acid) $CH_3COCOOH$; CAS 127-17-3; EINECS 204-824-3; Biochemical research. [Penta Mfg.; Schweizerhall; U.S. Biochemical]

Python. An emulsifiable concentrate containing 105 g aminotriazole, 207 g atrazine and 100 g 2,4-D per liter; used for total weed control in non crop areas. [CDA Chemicals Ltd]

pyxol. An emulsion of coal-tar acids with soap; used as a disinfectant.

---

‡ Trade name and manufacturer not verified   § Trade name without identified manufacturer

Q

Q-1300. N-Nitrosophenylhydroxylamine ammonium salt; polymerization inhibitor; chelating agent; antioxidant; germicides, fungicides, agricultural chems.; lustering agent for plating; corrosion inhibitor for metals; heat stabilizer for chlorosulfonated polyethylene; raw material for dye synthesis; analytical reagent. [Wako Pure Chem. Ind.]

Q-1301. CAS 15305-07-4; N-Nitrosophenylhydroxylamine aluminum salt; polymerization inhibitor; ideal for uv ink stabilizer. [Wako Pure Chem. Ind.; Wako Chem. USA]

QA-555. CAS 1337-76-4; Attapulgite clay; absorbent, adsorbent. [Floridin] *

Qaulineg. Color film processing system. [May & Baker Ltd]*

Qazul. Polyamide fibers for luggage. [ICI Chem. & Polymers Ltd] †

Q-Broxin. Drilling mud additive. [Georgia-Pacific]

QC-8800. Vinyl ester-based sheet molding compound; for compression molding of components requiring high structural strength and fatigue resist. [Quantum Composites]

Q-Cast. A castable refractory monolithic material for high temperature applications. [Pfizer International] †

Q-Cel® 300. Inorg. silicate hollow microspheres; extender/filler for plastics, fiberglass-reinforced plastics, cultured marble, cast polyester furniture and decorative parts, bowling ball cores, cast urethane and epoxy systems, autobody repair fillers, marine putties, PVC plastisol compounds; slurry explosives. [PQ Corp.]

Q-Crete. A monolithic refractory concrete for high temperature applications. [Pfizer International] †

Q-Gun. A gunnable refractory monolithic material for high temperature applications. [Pfizer International] †

QO® Furan. Furan; chemical intermediate in the mfg. of herbicides, pharmaceuticals, plastics, and fine chemicals. [QO Chem.]

QO® Furcarb® . Modified furan-phenolic resins; offers storage stability, low volatility; can be cured with acidic or basic catalysts; suitable for molding or extrusion; used in bonding carbon, graphite, basic refractory grains, silicon carbide, sand, and other aggregates to form shaped articles; applications incl. preformed carbon and graphite shapes, extruded anodes and electrodes, refractory shapes or brick, tap hole mixes, grinding wheels, blast furnace runner block, ramming mixes, crucibles, heat exchangers. [QO Chem.]

QO® Furfural. 2-Furaldehyde; chemical intermediate for mfg. of derivatives (furan and THF); solv. for separating sat. from unsat. compounds in petrol. lubricating oil, gas oil, and diesel fuel; extractive distillation of C4 and C5 hydrocarbons for the mfg. of synthetic rubber; decolorizing agent for wood rosin; solv. and processing aid for anthracene; ingred. in resins; reactive solv. and wetting agent in abrasive wheels and brake linings. [QO Chem.]

QO® Furfuryl Alcohol (FA®). CAS 98-00-0; Furfuryl alcohol; used in the production of foundry sand binders and corrosion-resistant resins; intermediate for esterification and etherification; impregnating sol'n. and carbon binder; wood adhesive component; solv. and temporary plasticizer for phenolic resins in the mfg. of cold-molded abrasive wheels; visc. reducer, cure promoter, and carrier in amine-cured epoxy resins. [QO Chem.]

QO® Polymeg® 650. Polytetramethylene ether glycol polyol; used in urethane elastomers, fibers, coatings, and adhesives, in the production of high-performance thermoset and thermoplastic elastomers, elastomeric polyesters, as polyester modifiers. [QO Chem.]

QO® Quacorr® Resin/Catalyst Systems. Modified furfuryl alcohol-based (furan) resins and liq. catalyst; produces laminates exhibiting excellent chem. resist.; inherently fire resist.; used to fabricate fiberglass-reinforced plastic equip. with outstanding corrosion resist., retention of strength at elevated temps., low flame spread, low smoke emission; applications incl. chem. process equip. such as tanks, process vessels, ducting, stacks, scrubbers, hoods, and pipe. [QO Chem.]

QO® Tetrahydrofuran (THF). CAS 109-99-9; Tetrahydrofuran; Industrial solv. with high solvency for broad range of materials and low boiling pt.; also used in the production of heterocyclic and open-chain compounds; chemical intermediate. [QO Chem.]

QO® Tetrahydrofurfuryl Alcohol (THFA®). CAS 97-99-4; Tetrahydrofurfuryl alcohol; High boiling solv. and carrier for pesticides; FDA approved for use in paper processing; chemical intermediate; also in industrial and consumer cleaners, leather and textile dyeing, epoxies, coatings, inks, paints, and adhesives; plasticizer and vinyl stabilizer carrier. [QO Chem.]

Q-Therm. High temperature fluids with excellent thermal conductivity, temperature range and dielectric properties; used for the design of new functional fluids and lubricants with low toxicity values, e.g., heat transfer fluids for process equipment [Anderson Development Company] *

Quab 151. 2,3-Epoxypropyl trimethyl ammonium chloride; cationic biopolymers (e.g. starch, cellulose, guar, gelatine, protein) and synthetic polymers (e.g., polyacrylic acid, acrylamide-acrylic acid copolymers, polyaminoamides, polyethylene imide). [Degussa]

Quab 188. CAS 3327-22-8; 3-Chloro-2-hydroxypropyl trimethyl ammonium chloride; for quaternization of com-

---

pounds with hydroxyl, amino, and other functional groups, especially corresponding polymers, production of cationic polyelectrolytes. [Degussa]

Quabond® 210. CAS 9003-20-7; High molecular weight polyvinyl acetate homopolymer aq. emulsion; provides good adhesion to metal, glass, wood, cork, leather, textile, paper, and plastic surfaces; stiffener and hand builder on cotton and cotton/polyester fabrics. [Rhone-Poulenc Surf.]

Quabond® 230. Acrylic copolymer aq. emulsion; self-cross-linking emulsion for use on synthetics and cotton fabrics. [Rhone-Poulenc Surf.]

Quad DSM. CAS 919 86-0; Emulsifiable concentrate containing 500 g demeton-S-methyl per liter; a systemic organophosphorus insecticide. [Quadrangle Agrochemicals]

Quad MCPA 50%. CAS 94-74-6; MCPA; herbicide for cereals and grassland. [Quadrangle Agrochemicals]

Quadban. Soluble concentrate of 19.5 g dicamba, 245 g MCPA and 86.5 g mecoprop per liter; used for weed control in cereals and grassland. [Quadrangle Agrochemicals]

Quadefome® MAB (redesignated Foamex MAB). [Rhone-Poulenc Surf.]

Quad-Fast. Surfactant containing di-I-p-menthene; coating agent for contact herbicides, pesticides, growth regulators, and foliar feeds. [Quadrangle Agrochemicals]

Quad-Keep. CAS 117-18-0; Dustable powder containing 3% w/w tecnazene; protectant fungicide and potato sprout suppressant. [Quadrangle Agrochemicals]

Quad Mini Slug Pellets. CAS 108-62-3; Pellets containing 6% w/w metaldehyde; snail and slug bait. [Quadrangle Agrochemicals]

Quadrafos. Polyphosphate; water softener. [Marlowe-Van Loan]

Quadrafos. Glassy phosphates. [Peridot Chemicals Inc]

Quadrangle Chlormequat 700. CAS 999-81-5; Soluble concentrate containing 700 g/l chlormequat; plant growth regulator. [Quadrangle Agrochemicals]

Quadrangle Cropspray 11E. Adjuvant containing 99% highly refined mineral oil; wetting agent for herbicide and fungicide sprays. [Quadrangle Agrochemicals]

Quadrangle Cyper. CAS 66841-24-5; Emulsifiable concentrate containing 100 g cypermethrin per liter; a pyrethroid insecticide. [Quadrangle Agrochemicals]

Quadrangle Super-Tin 4L. CAS 76-87-9; Suspension concentrate containing 480 g fentin hydroxide per liter; used for control of potato blight. [Quadrangle Agrochemicals]

Quadrilan® AT. Quaternized fatty amine ethoxylate as a clear amber liquid; antistatic agent for PVC and other polymers. [Harcros UK]

Quadrilan® BC. Benzalkonium chloride BP grade; bactericide, fungicide, germicide in disinfectants and detergent sanitizers, emulsifier in the dyeing industry; biodeg. [Harcros UK]

Quadrilan® MY 211. Specially developed cationic surfactant as a hazy amber liquid; used in alkaline spray cleaning concentrates for vehicle and chassis cleaning and crate washing. [Harcros UK]

Quadrol®. CAS 102-60-3; EINECS 203-041-4; Tetra (2-hydroxypropyl) ethylenediamine; polyol; chelating agent; intermediate used in resins, emulsifiers, surfactants, pharmaceuticals, herbicides, fungicides, insecticides, adhesives, and plasticizers. [BASF; BASF plc]

Quad Store. CAS 117-18-0; Granules containing 5% w/w tecnazene; protectant fungicide and potato sprout suppressant. [Quadrangle Agrochemicals]

Qualamox. CAS 26787-78-0; Amoxycillin injection. [RMB Animal Health Ltd]

Qualidot. Machine lith developer system. [May & Baker Ltd]*

Qualiflx. Photographic fixer. [May & Baker Ltd] *

Qualitol. Photographic developer. [May & Baker Ltd] *

Quallofil®. Fibers. [DuPont UK]

Quallofirm®. Fibers. [DuPont UK]

Qualloflex®. Fibers. [DuPont UK]

Quamectant AM-50. 6-(N-Acetylamino)-4-oxahexyl-trimonium chloride; substantive humectant with emollient feel; nonoily and nongreasy; skin moisturizer; hair care products. [Brooks Industries]

Quamilin. CAS 52645-53-1; Formulations for permethrin. [The Wellcome Foundation Ltd] *

Quantacure®. A range of photoinitiators and coinitiators for photo-curing formulations. [Octel Chemicals Ltd]

Quantril. CAS 63-12-7; Benzquinamide; antiemetic. [J B Roerig] *

Quantum. Protective coatings for concrete which provide mechanical strength, abrasion resistance, chemical corrosion resistance and weather resistance for floors, bridge and parking decks, garages, livestock pens, swimming pools, etc. [Quantum Chemical Corp]

Quantum. Refining catalysts. [Crosfield Chemicals Ltd]

Quantum®. For the agriculture industry. [DuPont UK]

Quantum. FCC catalysts. [Crosfield Chemicals Ltd]

Quantum (LOGO). A full line of petrochemicals and oleochemical preparations for industrial use. [Quantum Chemical Corp]

quartz. (silicon dioxide; silica) *Also see* silica; crystallized silicon dioxide; SiO_2; CAS 14808-60-7; EINECS 238-878-4; there are three types. a) crystalline, such as tridymite and cristobalite. b) crypto-crystalline, such as chalcedony, and c) hydrated silica or opal; electronic components; TV components. [Unimin; U.S. Silica; Westo Industrial Prods. Ltd]

quartz glass. CAS 60676-86-0; Fused silica glass.

quartzilite. A metallic carbide formed by action of silica and carbon at 2000-3000 C. It is suitable for electrical resistances.

quarzal. An alloy of aluminum with 15% copper, 6% manganese, and 0.5% silicon; used for the cylinders of internal combustion engines.

Quasar. FCC catalysts. [Crosfield Chemicals Ltd]

Quasilan. Pure MDI prepolymers; curing components for use with Propocon polyether systems. [Harcros UK]

Quat-Coll CDMA 40. Cocodimonium hydroxypropyl hydrolyzed animal protein; cosmetics ingred. [Brooks Industries]

Quat-Coll IP10-30. Hydroxypropyltrimonium gelatin; cosmetics ingred. [Brooks Industries]

Quat-Coll QS. CAS 11174-62-0; Steartrimonium hydrolyzed animal protein; cosmetics ingred. [Brooks Industries]

quaternary steels. Steels containing two special elements in addition to the iron and carbon.

Quat Keratin WKP. CAS 68915-25-3; Cocodimonium hydroxypropyl hydrolyzed animal keratin; cosmetics ingred. [Brooks Industries]

Quat-Pro E. CAS 111174-64-2; Triethonium hydrolyzed

---

‡ Trade name and manufacturer not verified   § Trade name without identified manufacturer

collagen ethosulfate; substantive protein for hair and skin care products (shampoos, leave-on conditioners, mousses, creams, lotions, face tonics, liq. soaps). [Maybrook]

Quat-Pro S, S-30. CAS 11174-62-0; Steartrimonium hydrolyzed animal protein; substantive film-former, moisturizer, conditioner, protectant for hair and skin care products (shampoos, conditioners, creams, lotions, liq. hand soaps, bath products). [Maybrook]

Quatramine. A range of quaternary ammonium salts; used in antiseptics, disinfectants, bactericides and algicides. [Harcros Australia] *

Quatrene 7670. Fatty amidoamine quaternary compound; wetting agent, demulsifier, and corrosion inhibitor for petrol. production. [Henkel/Cospha]

Quatrene C-5-6. Coco benzyl imidazolinium chloride; corrosion inhibitor for continuous treatment in oil and gas production, pipelines. [Henkel/Cospha]

Quatrene CB-50. Benzalkonium chloride; corrosion inhibitor for continuous treatment in oil and gas production, pipelines. [Henkel] *

Quatrex. Epoxy resins used for the electronics industry. [Dow UK]

Quatrex 152. Quaternized imidazoline; emulsifier, stabilizer for oilfield applications. [Chemron]

Quatrex 162. Surfactant; wetting agent, additive for reverse emulsion breakers, oil treating; emulsion preventer in acids; clay stabilizer. [Chemron]

Quatrex 182. Quaternized imidazoline; corrosion inhibitor conc.; surfactant, solubilizer for other corrosion inhibitor components; oilfield applications; emulsion preventer; clay stabilizer. [Chemron]

Quatrex 1010. CAS 25928-94-3; Epoxy resin; electronic grade resin for laminating applications. [Dow]

Quatrex 2410. Phenol-epoxy novolac; electronic grade characterized by extremely low hydrolyzable chlorine content. [Dow]

Quatrex 5010. One-component brominated epoxy resin; ignition-resist. grade with high glass transition temp. (180 C). [Dow]

Quatrex 6410. Brominated epoxy resin; electronic grade designed for use where max. bromine is needed to offset large ratios of hardener or nonbrominated resin. [Dow]

Quatrex CRC. Cetearyl alcohol, stearalkonium chloride, PEG-40 castor oil. conc. surfactant for cream rinses; fully compatible with esters, fatty alcohols, and proteins. [Chemron]

Quatrex CT-100. Stearyl alcohol, cetrimonium chloride; cream rinse conc., conditioner for quality salon and professional hair care products. [Chemron]

Quatrex CTAC. CAS 112-02-7; EINECS 203-928-6; Cetrimonium chloride; surfactant, conditioner for hair treatment applications. [Chemron]

Quatrex S. Soyamidopropalkonium chloride; substantivity and conditioning agent for conditioning shampoos, sprays, mousses, setting gels, conditioners. [Chemron]

Quatrex STC-25. CAS 122-19-0; EINECS 204-527-9; Stearalkonium chloride; personal care surfactant for after-shampoo cream rinses. [Chemron]

Quatrisoft Polymer LM-200. CAS 107987-23-5; Polyquaternium-24; stabilizer for emulsions; thickener for surfactant systems; substantive conditioner for hair and skin care products.. [Amerchol Corp]

Quat-Silk QTM-10. Hydroxypropyltrimonium hydrolyzed silk. cosmetics ingred. [Brooks Industries]

Quat-Soy CDMA-25. CAS 977039-11-4; Cocodimonium hydroxypropyl hydrolyzed soy protein; cosmetic ingredient for hair and skin care products. [Brooks Industries]

Quat-Soy LDMA-30. Lauryldimonium hydroxypropyl hydrolyzed soy protein; cosmetic ingredient for skin and hair care products. [Brooks Industries]

Quat-Veg Q-30. Hydroxypropyltrimonium hydrolyzed vegetable protein; cosmetic ingredient for skin and hair care products. [Brooks Industries]

Quat-Wheat CDMA-30. Cocodimonium hydroxypropyl hydrolyzed wheat protein; cosmetic ingred. for skin and hair care products. [Brooks Industries]

Quat-Wheat QTM-20. Hydroxypropyltrimonium hydrolyzed wheat protein; cosmetic ingred. for hair and skin care products. [Brooks Industries]

Quat-Wheat SDMA-30. Soya dimonium hydrolyzed wheat protein; substantivity agent for hair care products; enhances compability, softness, conditioning; esp. as pre-wrap before chemical treatment or as post-perm treatment. [Brooks Industries]

quebrachite. (quebrachitol). The monomethyl ether of laevo-inositol, $C_6H_6(OH)_5(OCH_3)$. It is found in the latex of *Hevea brasiliensis* (the rubber tree) to the extent of 1-2%.

quebracho. CAS 1401-55-4; *Loxopteryngium lorenzii.* The wood of this tree contains about 20% of tannin, and is used in the form of an extract for tanning.

Quecodur AE. Dimethylol ethylene urea; reactant for wash and wear finishes. [Thor Chemicals (UK) Ltd]

Quecodur B Granular. Urea-formaldehyde condensation product; used for textile crease-proofing and resilient finishing. [Thor Chemicals (UK) Ltd]

Quecodur CW Conc. Modified dimethylol dihydroxy ethylene urea; clear colorless liquid; used for low formaldehyde and chlorine resistant textile wash-and-wear finishes. [Thor Chemicals (UK) Ltd]

Quecophob HPA. Fluorocarbon resin emulsion; cationic; oleophobic and water-repelling agent; may be used in combination with synthetic resin precondensates, reactants and nonionic finishing agents. [Thor Chemicals (UK) Ltd]

Queen's metal. A jeweler's alloy. It is very variable in composition. It contains from 50-85% tin, 7-16% antimony, 0-16% lead, 0-3.5% copper, and 1-12% zinc.

Quelicin. CAS 71-27-2; Succinylcholine chloride; blocking agent. [Abbott Laboratories] *

Quellada. A proprietary preparation of γ-benzene hexachloride; antiparasitic for dermatological use. [Stafford-Miller] ‡

Quell Oil. Oil spill dispersants. [Harcros UK]

Quenty®. Functional cosmetics containing collagen. [Bayer AG]

Quenty® forty. Functional cosmetics containing Elastin-Bayer and collagen. [Bayer AG]

quercetin. $C_{15}H_{10}O_7$; CAS 117-39-5; Medicine; possible formation of epoxy resins. [EM Industries; Schweizerhall; U.S. Biochemical]

quercitron. A coloring matter, sold as chips, or as a coarse powder, obtained by grinding the bark of *Quercus tinctoria* and *Q. nigra.* The dyeing principle is quercitin or flavin, $C_{15}H_{10}O_7$; which forms yellow lakes with aluminum and tin salts; used for calico printing and wool dyeing.

---

* Trade name not verified as available   † Trade name verified as obsolete

Querton 14Br-40. CAS 1119-97-7; EINECS 214-291-9; Myrtrimoniumbromide; germicidal disinfectant, detergent for hospital and industrial cleaning and disinfection. [Berol Nobel AB]

Querton 16Cl-29. CAS 112-02-7; EINECS 203-928-6; Cetrimonium chloride; emulsifier, dispersant, antistat for hair conditiners, shampoos, detergent sanitizers. [Berol Nobel AB]

Querton 210Cl-50. CAS 7173-51-5; EINECS 270-331-5; Didecyl dimonium chloride; bactericide, fungicide for food processing industry, breweries, catering and hospitals; in detergent sanitizers. [Berol Nobel AB]

Querton 280. CAS 68391-03-7; EINECS 269-922-0; Trimethyl alkyl ammonium chloride; general purpose quaternary surfactant. [Berol Nobel AB]

Querton 441-BC. CAS 61789-72-8; EINECS 263-081-3; Hydrogenated tallow dimethylbenzyl ammonium chloride; agent for organophilic bentonites production. [Berol Nobel AB]

Querton 442. CAS 61789-80-8; EINECS 263-090-2; Dihydrogenated tallow dimethyl ammonium chloride; imparts soft feel and antistatic props. to textile softeners for commercial laundry and textile mfg. applications; also for hair conditioners, paper chemicals, mfg. of organoclays. [Berol Nobel AB]

Querton GCl-50. CAS 61789-71-7; EINECS 274-022-6; 1-Coco alkylguanidinium chlorides; bactericide, fungicide for detergent sanitizers. [Berol Nobel AB]

Querton KKBCl-50. CAS 61789-71-7; EINECS 263-080-8; Cocoalkyl dimethylbenzyl ammonium chloride; corrosion inhibitor for oil drilling; emulsifier and dispersant for sludge in oil drilling and waste water treatment. [Berol Nobel AB]

Quesbrom. CAS 126-06-7; 1-Bromo-3-chloro-5,5-dimethylhydantoin; 95% active tablet; an extremely effective microbiocial bactericide, fungicide and algicide used in cooling towers, once-thru and closed loop water system. [Ques Industries]

Queschlor. CAS 87-90-1; 1,3,5-Triazine-2,4,6(1H,3H,5H)-trione,1,3,5-trichloro-; a chlorinated isocyanurate containing 90% available chlorine; a high performance chlorine microbiocide used to control algae, bacteria, and fungi growth in recirculating water and cooling towers. [Ques Industries]

Quesfloc. A series of organic polymers suitable for waste water clarification; available in both dry and liquid compositions; used for industrial and municipal waste water (effluent) clarification. [Ques Industries]

Quesfloc F11283-1. A high molecular weight anionic polymer which performs as a coagulant aid and sludge conditioning agent in a variety of solid-liquid separation processes; in its dry form, it is a white, dustless free-flowing granular powder; used in conjunction with ferric chloride and alum for phoshorus removal; as a component in a dual anionic-cationic system for sledge dewatering operations. [Ques Industries]

Questal DI 0770. Disodium EDTA; chelating agent for use in mildly acidic dry formulations. [Clough]

Questal Extra Powd. Conc. 0780. CAS 64-02-8; EINECS 200-573-9; Tetrasodium EDTA; chelating agent for dry formulations. [Clough]

Questal FEC 0800. CAS 139-89-9; Trisodium HEDTA; chelating agent effective for iron to pH 12. [Clough]

Questal Special 0860. CAS 64-02-8; EINECS 200-573-9;

Tetrasodium EDTA; chelating agent over wide pH range. [Clough]

Questran. A proprietary preparation of cholestyramine chloride; used in the treatment of hypercholesterolemia. [Bristol-Myers Squibb Co Inc]

Questran. Cholestyramine resin; ion-exchange resin; antihyperlipoproteinemic. [Mead Johnson & Co] *

Questric Acid 5286. CAS 60-00-4; EDTA; chelating agent for use in dry formulations where sodium salt is undesirable. [Clough]

Quevenne's iron. Reduced iron.

Quiacryl. Acrylic water emulsions; for leather finishes, textile, coatings, adhesives. [Merquinsa]

Quiana®. Fibers. [DuPont UK]

Quick Cure®. For the electrical industry. [DuPont UK]

quickening liquid. A solution of mercuric nitrate or cyanide; used in electro-plating.

quicklime. See lime.

quicklime. See calcium oxide.

Quick-pach. A plastic fire-clay for making monolithic linings and quick repairs. §

Quickset® Extra. CAS 1338-23-4; MEK peroxide; high purity initiator for R.T. curing of polyester resins; increased reactivity with lower peroxide conc. [Witco/Argus]

quicksilver. See mercury.

quicksilver. Mercury, Hg. A pigment which consists of sulfide of mercury is also known under this name.

quicksilver vermilion. See mercury sulfide (ic), red and black.

Quicksol. Insulin preparation. [The Boots Co plc] *

Quickstir. Ceramic glaze and body stains offering easy dispersability; used for ceramic tile, sanitaryware, hobby ware, and procelain enamel. [Engelhard]

Quide. CAS 3819-00-9; Piperacetazine; antipsychotic. [Merrell Dow Pharmaceuticals Inc] *

Quidur. Urethane prepolymers; binder for cork-compositions, cross linkers for PU adhesives. [Merquinsa]

Quiess. Hydroxyzine hydrochloride; tranquilizer. [O'Neal, Jones & Feldman Pharmaceuticals] *

Quikote. Release agent. [Morton Int'l.]

Quikset. Blended metallic complex; accelerator for one-bath soil-release finishes on textiles. [CNC Int'l. L.P.]

Quilastic. Polyurethayne water emulsions, polyurethane solutions, granule form; raw materials for adhesives, leather finishes, synthetic leather, textile. [Merquinsa]

quillaic acid. An acid obtained from the inner bark of *Quillaja saponaria* (soap bark); an expectorant.

Quinaband. A proprietary bandage coated with zinc paste, calamine and iodochlorohydroxyeuinolone; used in the treatment of leg ulcers. [Seton Healthcare Group plc]

Quinaglute. CAS 7054-25-3; Quinidine gluconate; cardiac depressant. [Berlex Laboratories Inc] *

quinaldine. (chinaldine; α-methylquinoline) $C_{10}H_9N$; CAS 91-63-4; EINECS 202-085-1; Manufacture of dyes, pharmaceuticals, fine organic chemicals, acid-base indicators. [Allchem Industries]

Quinalspan. A proprietary preparation of quinalbarbitone sodium; a hypnotic. [Pharmax Ltd] *

Quinamin. CAS 804-63-7; Quinine sulfate; antimalarial. [Merrell Dow Pharmaceuticals Inc] *

Quindex. Fungicides containing copper 8-quinolinolate; for preservation of textiles, cordage, paper, adhesives and caulking compounds. [Hüls Am.] †

---

‡ Trade name and manufacturer not verified   § Trade name without identified manufacturer

quindoxin. Quinoxaline 1,4-dioxide. a growth promoter.

Quindo®. High-quality quinacridone pigments for the surface coatings and plastics industries. [Bayer AG]

Quine. CAS 804-63-7; Quinine sulfate; antimalarial. [Rowell Laboratories Inc] *

quineine. A standard solution of *Cinchona succiruba,* containing 7% of alkaloids, 5% of which is quinine.

quinicardine. CAS 56-54-2; $(C_{20}H_{24}O_2N_2)_2 \cdot H_2SO_4$; Quinidine sulfate.

Quinidex. CAS 50-54-4; Quinidine sulfate; used as a cardiac depressant. [Wyeth Laboratories] *

quinine sulfate. $(C_{20}H_{24}N_2O_2)_2 \cdot H_2SO_4 \cdot 2H_2O$; CAS 6119-70-6; EINECS 212-359-2; used as an antimalarial drug and a muscle relaxant. [Atomergic Chemetals; Dajac Labs; R.W. Greeff]

quinizarin. (1,4-dihydroxyanthraquinone) $C_{14}H_6O_2(OH)_2$; Antioxidant in synthetic lubricants, dyes. [BASF; Sandoz]

Quinn's Rubber. A patented rubber substitute made from rapeseed oil, petroleum, and chloride of sulfur. §

Quinocort. Potassium hydroxyquinoline sulfate and hydrocortisone in a cream base; used for inflamed dermatoses where infection is present. [Quinoderm Ltd]

Quinoderm® Cream. A proprietary preparation of benzoyl peroxide (10%) and potassium hydroxyquinoline sulfate in a cream base; used in the treatment of acne and acneform eruptions. [Quinoderm Ltd]

Quinoderm® Cream 5. Benzoyl peroxide (5%) and potassium hydroxyquinoline sulfate; aqueous cream; for treatment of acne, acneform eruptions where a lower strength of benzoyl peroxide is required [Quinoderm Ltd]

Quinoderm® Cream with Hydrocortisone. Benzoyl peroxide and potassium hydroxyquinoline sulfate with hydrocortisone; used in the treatment of inflamed acne. [Quinoderm Ltd]

Quinoderm® Lotio-Gel 5%. Benzoyl peroxide (5%) and potassium hydroxyquinoline sulfate; thixotropic gel; for treatment of acne, acneform eruptions where use on dry skin is required. [Quinoderm Ltd]

Quinoformine. A compound of quinic acid and hexamethylene-tetramine; used as a solvent for uric acid.

quinol. CAS 123-31-9; Hydroquinone (p-dihydroxy-benzene), $C_6H_4(OH)_2$; used as a photographic developer.

quinoline. (chinoline) C_9H_7N; CAS 91-22-5; EINECS 202-051-6; Preserving anatomical specimens, manufacture of quinolinol sulfate, niacin and copper-8-quinolinolate,

flavoring. [Allchem Industries; Expansia SA; Koppers Industries]

8-quinolinol. *See* 8-hydroxyquinoline.

Quinoped. A proprietary preparation of benzoyl peroxide and potassium hydroxyquinoline sulfate in a cream base; used in the treatment of fungal foot infections. [Quinoderm Ltd]

Quinosol. (Chinosol, Sunoxol). The potassium salt of oxyquinoline sulfonic acid, $C_9H_6 \cdot NO \cdot SO_3K$; used as an antiseptic.

Quinovasugar. Quinovitol, $C_6H_9O(OH)_3$.

Quinta-Pro Conc. Hydrolyzed collagen, triethonium hydrolyzed collagen ethosulfate, cationic collagen polypeptides, hydrolyzed keratin, collagen amino acids; protein blend for hair and skin care products (shampoos, conditioners, hair treatment products, creams, lotions, bath products, liq. hand soaps, shave products). [Maybrook]

Quintesse. Polyamide fibers for upholstery. [ICI Chem. & Polymers Ltd]

Quintiofos. O-Ethyl O-8-quinolyl phenylphosphonothioate; used as an insecticide.

Quintozene. CAS 82-68-8; Wettable powder containing 20% w/w quintozene; fungicide for various agricultural crops. [Rhone-Poulenc Environmental Prods. Ltd]

3-quinuclidinone hydrochloride. (1-azabicyclo [2.2.2] octan-3-one hydrochloride) $C_7H_{11}NO \cdot HCl$; CAS 1193-65-3; EINECS 214-776-5. [Janssen Chimica; Schweizerhall]

quisqueite. An asphaltum-like compound containing much sulfur. It is found in Peru.

Quiver. Suspension concentrate containing 200 g cyanazine and 350 g isoproturon per liter; a residual herbicide for winter cereals. [Shell UK]

Quixalud. CAS 8067-69-4; Halquinol in feed additive. [Ciba plc] *

Quorn. Mycoprotein. [Marlow Foods Ltd]

Quso® G27, G29, G35, G38, WR55, WR55-FG, WR83. Precipitated silica; thickener for pastes, creams, lotions in cosmetics and toiletries; suspending agent; improves free-flowing chars. of fine powds.; for pulp and paper defoaming. [Degussa]

Quso® WR55-FG. Precipitated silica; thickener for pastes, creams, lotions in cosmetics and toiletries; suspending agent; improves free-flowing chars. of fine powds.; for food grade defoamers. [Degussa]

Q-Vibe. A vibratable refractory monolithic material for high temperature applications. [Pfizer International] †

---

* Trade name not verified as available   † Trade name verified as obsolete

R

R-2 Crystals. A proprietary trade name for an ultra accelerator for latex, etc.; it is the reaction product of carbon disulfide with methylenedipiperidine. §

R-502, MR-502K, MR-502Y, MR-502P. Dimethyl polysiloxane combination; magnetic rubber inspection material for nondestructive inspection of ferromagnetic parts used for inspection of thread, gear roots, inaccessible areas, coated areas; provides permanent record of the inspection. [Dynamold Inc]

R-1000. Antacid compressed gel. [Reheis Inc]

R-1007. One-part silicone; RTV dispersion coating, ink. [McGhan NuSil]

Rabalon. Thermoplastic elastomer. [Mitsubishi Petrochem.] *

Rabro. A proprietary preparation containing liquorice, bismuth subnitrate, magnesium carbonate, sodium bicarbonate, alder buckthorn bark and calamus rhizome; an antacid. [Rybar Laboratories Ltd] *

racemic menthol. See menthol.

Rachromate-51. CAS 10039-53-9; Sodium chromate Cr 51; diagnostic aid; radioactive agent. [Abbott Laboratories]*

R-Acid. 2-Naphthol-3, 6-disulfonic acid; used as an azo dye intermediate; the disodium salt is used as a reagent in detecting nitrogen dioxide in the air.

Rackarock. A blasting explosive, consisting of 79% potassium chlorate and 21% nitrobenzene, mixed sometimes with picric acid or sulfur.

Rackarock Special. An explosive similar to Rackarock, but containing 12-16% picric acid.

Racumin®. CAS 5836-29-3; Coumetralyl; a ready-to-use anticoagulant rodenticide. [Bayer AG; Bayer plc]

Radar. CAS 60207-90-1; Propiconazole; a systemic triazole fungicide for control of powdery mildew and rust in wheat and barley. [Farm Protection Ltd]

Radar Propiconazole. CAS 60207-90-1; Propiconazole; a systemic triazole fungicide for control of powdery mildew and rust in wheat, barley, rye, sugar beet and oilseed rape. [Ciba-Geigy Agrochemicals; ICI Chem. & Polymers Ltd]

Radarsan. See Rawstol.

radauite. A mineral; a variety of labradorite.

Radel® A-100. Polyarylsulfone resin; excellent toughness, good chemical resist., hydrolytic stability; for medical devices, chemical processing, elec./electronic, food packaging, aviation applications; A-100 and A-200 have FDA clearance. [Amoco Chemical Co]

Radex. Insulating powders for ladles and continuous casting tundishes. [Foseco (F.S.) Ltd]

Radia® 7040. CAS 84988-74-9; EINECS 284-863-0; Butyl oleate; chemical intermediate, lubricant; chemical synthesis; carbon source in antibiotic culture broths; lubricity improvers in mineral oils; formulation of cutting, lamination, and textile oils, rust inhibitors; textile and leather industry. [Fina Chemicals]

Radia® 7051. CAS 85408-76-0; EINECS 287-039-9; Butyl stearate; chemical intermediate, lubricant, plasticizer; chemical synthesis; lubricant in mineral, cutting, lamination, and textile oils, rust inhibitors; textile and leather application. [Fina Chemicals]

Radia® 7060. CAS 67762-38-3; EINECS 267-015-4; Methyl oleate; chemical intermediate, lubricant; chemical synthesis; lubricity improvers in mineral oils; formulation of cutting, lamination, and textile oils; rust inhibitors; textile and leather industry. [Fina Chemicals]

Radia® 7108. CAS 85409-09-2; EINECS 287-075-5; Glyceryl C8-10 triester; chemical intermediate, chemical synthesis; lubricant in mineral, cutting, lamination, textile oils, and rust inhibitors; textile and leather application; also as emollient, plasticizer, solubilizer of act. components in cosmetics and pharmaceuticals. [Fina Chemicals]

Radia® 7110. CAS 85586-21-6; EINECS 287-824-6; Methyl stearate; chemical intermediate, chemical synthesis; lubricant in mineral, cutting, lamination, textile oils, and rust inhibitors; textile and leather application. [Fina Chemicals]

Radia® 7117. CAS 61788-59-8; EINECS 262-988-1; Methyl cocoate; emollient, plasticizer, lubricant for cosmetics, pharmaceuticals, plastics, lubricating oils, textile and leather additives, cutting oils for metallurgy; chemical intermediate. [Fina Chemicals]

Radia® 7120. CAS 112-39-0; EINECS 203-966-3; Methyl palmitate; chemical intermediate, chemical synthesis; lubricant in mineral, cutting, lamination, textile oils, and rust inhibitors; textile and leather application. [Fina Chemicals]

Radia® 7131. CAS 91031-48-0; EINECS 292-951-5; Isooctyl stearate; cosmetics emollient, solv.; plasticizer for PVC; lubricant for PS. [Fina Chemicals]

Radia® 7171. CAS 68604-44-4; EINECS 271-694-2; Pentaerythritol tetraoleate; lubricant, chemical intermediate; formulation of cutting, lamination, and textile oils; corrosion inhibitors; chemical synthesis. [Fina Chemicals]

Radia® 7176. CAS 91050-82-7; EINECS 293-029-5; Pentaerythritol tetrastearate; lubricant, chemical intermediate; formulation of cutting, lamination, and textile oils; corrosion inhibitors; chemical synthesis. [Fina Chemicals]

Radia® 7185. CAS 91031-43-5; EINECS 292-945-2; Ethyl stearate; chemical intermediate, chemical synthesis;

---

‡ Trade name and manufacturer not verified § Trade name without identified manufacturer

lubricant in mineral, cutting, lamination, textile oils, and rust inhibitors; textile and leather application; also as emollient, plasticizer, solubilizer of act. components in cosmetics and pharmaceuticals. [Fina Chemicals]

Radia® 7187. CAS 85049-36-1; EINECS 285-206-0; Ethyl oleate; chemical intermediate, chemical synthesis; lubricant in mineral, cutting, lamination, textile oils, and rust inhibitors; textile and leather application; also as emollient, plasticizer, solubilizer of act. components in cosmetics and pharmaceuticals. [Fina Chemicals]

Radia® 7190. CAS 110-27-0; EINECS 203-751-4; Isopropyl myristate; chemical intermediate, chemical synthesis; lubricant in mineral, cutting, lamination, textile oils, and rust inhibitors; textile and leather application; also as emollient, plasticizer, solubilizer of act. components in cosmetics and pharmaceuticals. [Fina Chemicals]

Radia® 7200. CAS 142-91-6; EINECS 205-571-1; Isopropyl palmitate; chemical intermediate, chemical synthesis; lubricant in mineral, cutting, lamination, textile oils, and rust inhibitors; textile and leather application; also as emollient, plasticizer, solubilizer of act. components in cosmetics and pharmaceuticals. [Fina Chemicals]

Radia® 7204. CAS 85049-34-9; EINECS 285-203-4; Propylene glycol dioleate; chemical intermediate, chemical synthesis; lubricant in mineral, cutting, lamination, textile oils, and rust inhibitors; textile and leather application; also as emollient, plasticizer, solubilizer of act. components in cosmetics and pharmaceuticals. [Fina Chemicals]

Radia® 7230. CAS 84988-79-4; EINECS 284-868-8; Isobutyl oleate; chemical intermediate, lubricant; chemical synthesis; carbon source in antibiotic culture broths; lubricity improvers in mineral oils; formulation of cutting, lamination, and textile oils, rust inhibitors; textile and leather industry. [Fina Chemicals]

Radia® 7231. CAS 85116-87-6; EINECS 285-540-7; Isopropyl oleate; chemical intermediate, lubricant; chemical synthesis; lubricity improvers in mineral oils; formulation of cutting, lamination, and textile oils; rust inhibitors; textile and leather industry. [Fina Chemicals]

Radia® 7241. CAS 85865-69-6; EINECS 288-668-1; Isobutyl stearate; cosmetics emollient, solv.; plasticizer for PVC; lubricant for PS. [Fina Chemicals]

Radia® 7266. CAS 91031-31-1; EINECS 292-932-1; Ethylene glycol distearate; chemical intermediate, chemical synthesis; lubricant in mineral, cutting, lamination, textile oils, and rust inhibitors; textile and leather application; also as emollient, plasticizer, solubilizer of act. components in cosmetics and pharmaceuticals. [Fina Chemicals]

Radia® 7331. CAS 85049-37-2; EINECS 285-207-6; Ethylhexyl oleate; chemical intermediate, lubricant; chemical synthesis; lubricity improvers in mineral oils; formulation of cutting, lamination, and textile oils; rust inhibitors; textile and leather industry. [Fina Chemicals]

Radia® 7345. CAS 72869-62-6; EINECS 276-951-2; Sorbitan tristearate; chemical intermediate, chemical synthesis; lubricant in mineral, cutting, lamination, textile oils, and rust inhibitors; textile and leather application; also as emollient, plasticizer, solubilizer of act. components in cosmetics and pharmaceuticals. [Fina Chemicals]

Radia® 7355. CAS 85186-88-5; EINECS 286-074-7; Sorbitan trioleate; chemical intermediate, chemical synthesis; lubricant in mineral, cutting, lamination, textile oils, and

rust inhibitors; textile and leather application; also as emollient, plasticizer, solubilizer of act. components in cosmetics and pharmaceuticals. [Fina Chemicals]

Radia® 7363. CAS 67701-30-8; EINECS 266-948-4; Glyceryl trioleate; lubricant, chemical intermediate; formulation of cutting, lamination, and textile oils; corrosion inhibitors; chemical synthesis; as carbon source in antibiotic culture broths. [Fina Chemicals]

Radia® 7370. CAS 68002-79-9; EINECS 268-093-2; Trimethylpropane trioleate; lubricant, chemical intermediate; formulation of cutting, lamination, and textile oils; corrosion inhibitors; chemical synthesis. [Fina Chemicals]

Radia® 7371. CAS 68002-79-9; EINECS 268-093-2; Trimethylol propane triester, unsaturated; chemical intermediate, chemical synthesis; lubricant in mineral, cutting, lamination, textile oils, and rust inhibitors; textile and leather application; also as emollient, plasticizer, solubilizer of act. components in cosmetics and pharmaceuticals. [Fina Chemicals]

Radia® 7500. CAS 97404-33-6; EINECS 306-797-4; Cetyl palmitate; chemical intermediate, chemical synthesis; lubricant in mineral, cutting, lamination, textile oils, and rust inhibitors; textile and leather application; also used for wax formulation due to its high melting point [Fina Chemicals]

Radia® 7501. CAS 85536-04-5; EINECS 287-484-9; Stearyl stearate; chemical intermediate, chemical synthesis; lubricant in mineral, cutting, lamination, textile oils, and rust inhibitors; textile and leather application; also used for wax formulation due to its high melting point [Fina Chemicals]

Radia® 7505. CAS 90193-76-3; EINECS 290-580-3; Distearyl phthalate; chemical intermediate, chemical synthesis; lubricant in mineral, cutting, lamination, textile oils, and rust inhibitors; textile and leather application; also as emollient, plasticizer, solubilizer of act. components in cosmetics and pharmaceuticals. [Fina Chemicals]

Radia® 7506. CAS 68955-45-3; EINECS 203-755-6; Ethylene bis-stearamide; chemical intermediate, chemical synthesis; lubricant in mineral, cutting, lamination, textile oils, and rust inhibitors; textile and leather application; also as emollient, plasticizer, solubilizer of act. components in cosmetics and pharmaceuticals. [Fina Chemicals]

Radia® 7510. CAS 91031-57-1; EINECS 292-960-4; Isononyl stearate; chemical intermediate, chemical synthesis; lubricant in mineral, cutting, lamination, textile oils, and rust inhibitors; textile and leather application [Fina Chemicals]

Radia® 7514. CAS 84539-90-2; EINECS 283-078-0; Pentaerythritol tetrabehenate; lubricant, chemical intermediate; formulation of cutting, lamination, and textile oils; corrosion inhibitors; chemical synthesis. [Fina Chemicals]

Radiamac 6149. CAS 61790-59-8; Hydrogenated tallow amine acetate; flotation reagent, anticaking aid for fertilizer, corrosion inhibitor, emulsifier. [Fina Chemicals]

Radiamac 6169. CAS 61790-57-6; EINECS 263-147-1; Coconut oil amine acetate; flotation reagent, anticaking aid for fertilizer, corrosion inhibitor, emulsifier. [Fina Chemicals]

Radiamine 6140. CAS 61788-45-2; EINECS 262-976-6;

---

* Trade name not verified as available   † Trade name verified as obsolete

Hydrogenated tallow amine; mineral flotation, corrosion inhibitor, pigment dispersant; cosmetics; lubricant and mold release for hard rubber, textile chemical, chemical synthesis; antistat and antifog additive for plastic foils. [Fina Chemicals]

Radiamine 6141. CAS 61788-45-2; EINECS 262-976-6; Hydrogenated tallow amine, distilled; mineral flotation, corrosion inhibitor, pigment dispersant; cosmetics; lubricant and mold release for hard rubber, textile chemical, chemical synthesis; antistat and antifog additive for plastic foils. [Fina Chemicals]

Radiamine 6160. CAS 61788-46-3; EINECS 262-977-1; Coconut oil amine; mineral flotation, corrosion inhibitor, pigment dispersant; cosmetics; lubricant and mold release for hard rubber, textile chemical, chemical synthesis; antistat and antifog additive for plastic foils. [Fina Chemicals]

Radiamine 6161. CAS 61788-46-3; EINECS 262-977-1; Coconut oil amine, distilled; mineral flotation, corrosion inhibitor, pigment dispersant; cosmetics; lubricant and mold release for hard rubber, textile chemical, chemical synthesis; antistat and antifog additive for plastic foils. [Fina Chemicals]

Radiamine 6164. CAS 124-22-1; Lauramine, distilled; mineral flotation, corrosion inhibitor, pigment dispersant; cosmetics; lubricant and mold release for hard rubber, textile chemical, chemical synthesis; antistat and antifog additive for plastic foils. [Fina Chemicals]

Radiamine 6170. CAS 61790-33-8; EINECS 263-125-1; Tallow amine; mineral flotation, corrosion inhibitor, pigment dispersant; cosmetics; lubricant and mold release for hard rubber, textile chemical, chemical synthesis; antistat and antifog additive for plastic foils. [Fina Chemicals]

Radiamine 6171. CAS 61790-33-8; EINECS 263-125-1; Tallow amine, distilled; mineral flotation, corrosion inhibitor, pigment dispersant; cosmetics; lubricant and mold release for hard rubber, textile chemical, chemical synthesis; antistat and antifog additive for plastic foils. [Fina Chemicals]

Radiamine 6172. CAS 112-90-3; EINECS 204-015-5; Oleyl amine; mineral flotation, corrosion inhibitor, pigment dispersant; cosmetics; lubricant and mold release for hard rubber, textile chemical, chemical synthesis; antistat and antifog additive for plastic foils. [Fina Chemicals]

Radiamine 6240. CAS 16789-79-5; Dihydrogenated tallow amine; intermediate for chemicals used in production of fabric softeners, household products, disinfectants. [Fina Chemicals]

Radiamine 6260. CAS 61789-76-2; Dicocamine; intermediate for wide range of chems. used as fabric softeners, household products, disinfectants. [Fina Chemicals]

Radiamine 6270. CAS 68783-24-4; Ditallowamine; intermediate for wide range of chems. used as fabric softeners, household products, disinfectants. [Fina Chemicals]

Radiamine 6310. CAS 7396-58-9; Di n-decylamine; intermediate for wide range of chems. used as fabric softeners, household products, disinfectants. [Fina Chemicals]

Radiamine 6343. CAS 61788-63-4; Dihydrogenated tallow methylamine; intermediate for wide range of chems. used as fabric softeners, household products, disinfectants. [Fina Chemicals]

Radiamine 6346. CAS 4088-22-6; Distearyl methylamine; intermediate for wide range of chems. used as fabric

softeners, household products, disinfectants. [Fina Chemicals]

Radiamine 6360. CAS 61788-62-3; Dicoco methylamine; intermediate for wide range of chems. used as fabric softeners, household products, disinfectants. [Fina Chemicals]

Radiamine 6540. CAS 68603-64-5; EINECS 271-696-6; Hydrogenated tallow propanediamine; corrosion inhibitor, dispersant, emulsifier, intermediate for chemical synthesis. [Fina Chemicals]

Radiamine 6560. CAS 61791-63-7; Coconut oil propane diamine; corrosion inhibitor, dispersant, emulsifier, intermediate for chemical synthesis. [Fina Chemicals]

Radiamine 6570. CAS 68439-73-6; Tallow propane diamine; corrosion inhibitor, dispersant, emulsifier, intermediate for chemical synthesis. [Fina Chemicals]

Radiamine 6572. CAS 68037-97-8; Oleyl propane diamine; corrosion inhibitor, dispersant, emulsifier, intermediate for chemical synthesis. [Fina Chemicals]

Radiamuls® Acetem 2021, 2134. Acetylated mono-diglycerides; alpha-tending emulsifier for food industry. [Fina Chemicals]

Radiamuls® Citrem 2931, 2932. Citrylated mono-diglycerides; anti-spattering aids for cooking margarines; emulsifier for meat industry; 2932 is self-emulsifying. [Fina Chemicals]

Radiamuls® CSL 2980. CAS 5793-94-2; Calcium stearoyl lactylate; food emulsifier; dough structure improver for baked goods. [Fina Chemicals]

Radiamuls® Datem 2001, 2008. Diacetyl tartaric monoglycerides; food emulsifier; 2008 is also a dough structure improver for bread and other baked goods. [Fina Chemicals]

Radiamuls® GTH 2375, GTH 2376. Glyceryl triheptanoate; butter tracers in food industry. [Fina Chemicals]

Radiamuls® Lactem 2950. Lactylated mono-diglycerides; alpha-tending emulsifier for food industry. [Fina Chemicals]

Radiamuls® MCT 2108. Glyceryl tricaprylate-caprate; edible oil; forms very thin films for coating confectionery and dried fruits; mold release aid for bakery, confectionery; lubricant for food processing equip.; visc. depressant, carrier of actives in oleoresins; lipid for dietetic foods. [Fina Chemicals]

Radiamuls® MG 2141. Glyceryl stearate SE; food emulsifier, plasticizer, foam inhibitor, homogenizer, rehydrator, anticaking agent. [Fina Chemicals]

Radiamuls® MG 2142. Glyceryl stearate; food emulsifier; lubricant for extruded foods and for food manufacturing equip. [Fina Chemicals]

Radiamuls® MG 2143. Vegetable glyceryl stearate; food emulsifier, plasticizer, homogenizer, foam inhibitor, rehydrator, anticaking agent. [Fina Chemicals]

Radiamuls® MG 2152. Glyceryl oleate; food emulsifier, plasticizer, homogenizer, foam inhibitor, rehydrator, anticaking agent. [Fina Chemicals]

Radiamuls® MG 2600. Glyceryl stearate; food emulsifier, plasticizer, homogenizer, foam inhibitor, rehydrator, anticaking agent. [Fina Chemicals]

Radiamuls® MG 2602. Glyceryl soyate; food emulsifier, plasticizer, homogenizer, foam inhibitor, rehydrator, anticaking agent. [Fina Chemicals]

Radiamuls® MG 2900. Glyceryl stearate; food emulsifier, plasticizer, homogenizer, foam inhibitor, rehydrator,

---

‡ Trade name and manufacturer not verified   § Trade name without identified manufacturer

anticaking agent. [Fina Chemicals]

Radiamuls® PG 2201. CAS 1323-39-3; EINECS 215-354-3; Propylene glycol stearate; food emulsifier. [Fina Chemicals]

Radiamuls® PG 2206. CAS 1330-80-9; EINECS 215-549-3; Propylene glycol oleate; food emulsifier. [Fina Chemicals]

Radiamuls® Poly 2248. Polyglyceryl monostearate; aerator and stabilizer for food products. [Fina Chemicals]

Radiamuls® Poly 2253. Polyglycerol polyricinoleate; w/o emulsifier for production of mold release aids for bakery and other foodstuffs, lubricants for food processing equip.; improves chocolate fluidity. [Fina Chemicals]

Radiamuls® Sorb 2145. CAS 1338-41-6; EINECS 215-664-9; Sorbitan stearate; food emulsifier, emulsion stabilizer; solubilizer and dispersant for flavors and other ingreds. [Fina Chemicals]

Radiamuls® Sorb 2147. CAS 9005-67-8; PEG-20 sorbitan stearate; food emulsifier, emulsion stabilizer; solubilizer and dispersant for flavors and other ingreds. [Fina Chemicals]

Radiamuls® Sorb 2157. CAS 9005-65-6; PEG-20 sorbitan oleate; food emulsifier, emulsion stabilizer; solubilizer and dispersant for flavors and other ingreds.; emulsifier for calf milk replacers. [Fina Chemicals]

Radiamuls® Sorb 2161. CAS 1338-41-6; EINECS 215-664-9; Sorbitan stearate, self-emulsifying; food emulsifier. [Fina Chemicals]

Radiamuls® Sorb 2166. CAS 1338-41-6; EINECS 215-664-9; Sorbitan stearate; food emulsifier. [Fina Chemicals]

Radiamuls® Sorb 2344. CAS 26658-19-5; EINECS 247-891-4; Sorbitan tristearate; food emulsifier. [Fina Chemicals]

Radiamuls® Sorb 2345. CAS 26658-19-5; EINECS 247-891-4; Sorbitan tristearate; food emulsifier, emulsion stabilizer; solubilizer and dispersant for flavors and other ingreds. [Fina Chemicals]

Radiamuls® SSL 2990. CAS 25383-99-7; EINECS 246-929-7; Sodium stearoyl lactylate; food emulsifier; dough structure improver for bread and other baked goods. [Fina Chemicals]

Radiaquat 6410, 6412. CAS 7173-51-5; Didecyl dimethyl ammonium chloride; bactericidal quaternary for hard surface cleaners, mildew preventers for commercial and industrial laundries. [Fina Chemicals]

Radiaquat 6412. CAS 7173-51-5; Didecyl dimethyl ammonium chloride; bactericidal quaternary for hard surface cleaners, mildew preventers for commercial and industrial laundries. [Fina Chemicals]

Radiaquat 6442. CAS 61789-80-8; EINECS 263-090-2; Dihydrogenated tallow dimethyl ammonium chloride; surfactant, softener for laundry applications. [Fina Chemicals]

Radiaquat 6444. CAS 112-02-7; EINECS 203-928-6; Palmityl trimethyl ammonium chloride; softener, antistat, detergent for textile, leather, detergent, and cosmetic industries; clay modifier. [Fina Chemicals]

Radiaquat 6462. CAS 61789-77-3; EINECS 263-087-6; Dicoco dimethyl ammonium chloride; detergent, antistat, softener, bactericide for detergents, textiles, fabric softeners, cosmetics. [Fina Chemicals]

Radiaquat 6470. CAS 68153-32-2; Ditallow dimethyl ammonium chloride; surfactant, softener for laundry applications. [Fina Chemicals]

Radiaquat 6471. CAS 68002-61-9; Tallow trimethyl ammonium chloride; detergent, antistat, softener for textile, leather, detergent and cosmetic products. [Fina Chemicals]

Radiaquat 6475, 6480. CAS 61789-80-8; EINECS 263-090-2; Dihydrogenated tallow dimethyl ammonium chloride; surfactant, softener for laundry products. [Fina Chemicals]

Radiasurf® 7125. CAS 68154-36-9; EINECS 268-910-2; Sorbitan laurate; emulsifier, descouring aid, antistat; anticorrosive agent for pipelines; cleaner for metallic surfaces; superfatting, bodying and antifog aid; pigment dispersant; detergent; emulsion of solv.; cutting oils; textile lubricant additive, concrete and paper additives; leather auxiliary; cosmetics and pharmaceuticals; plastics; pesticides; dry cleaning formulations. [Fina Chemicals]

Radiasurf® 7135. CAS 26266-57-9; EINECS 247-568-8; Sorbitan palmitate; emulsifier, descouring aid, antistat; anticorrosive agent for pipelines; cleaner for metallic surfaces; superfatting, bodying and antifog aid; pigment dispersant; detergent; emulsion of solv.; cutting oils; textile lubricant additive, [Fina Chemicals]

Radiasurf® 7137. CAS 9005-64-5; Polysorbate 20; emulsifier, descouring aid, antistat; anticorrosive agent for pipelines; cleaner for metallic surfaces; superfatting, bodying and antifog aid; pigment dispersant; detergent; emulsion of solv.; cutting oils; textile lubricant additive, [Fina Chemicals]

Radiasurf® 7140. CAS 85251-77-0; EINECS 286-490-9; Glyceryl stearate; emulsifier, wetting agent, defoamer, rust inhibitor, pigment grinding, antistat; internal lubricant for PVC. [Fina Chemicals]

Radiasurf® 7145. CAS 91031-74-2; EINECS 292-979-8; Sorbitan stearate; emulsifier, descouring aid, antistat; anticorrosive agent for pipelines; cleaner for metallic surfaces; superfatting, bodying and antifog aid; pigment dispersant; detergent; emulsion of solv.; cutting oils; textile lubricant additive, [Fina Chemicals]

Radiasurf® 7147. CAS 9005-67-8; Polysorbate 60; emulsifier, descouring aid, antistat; anticorrosive agent for pipelines; cleaner for metallic surfaces; superfatting, bodying and antifog aid; pigment dispersant; detergent; emulsion of solv.; cutting oils; textile lubricant additive, [Fina Chemicals]

Radiasurf® 7150. CAS 68424-61-3; EINECS 270-312-1; Glyceryl oleate; internal lubricant for PVC; biodeg. surfactant, wetting agent, emulsifier for cosmetics, pharmaceuticals, agriculture, chemical synthesis, explosives, polymers, glass fibers, surface coatings, textiles and leather. [Fina Chemicals]

Radiasurf® 7155. CAS 85186-88-5; EINECS 286-074-7; Sorbitan oleate; emulsifier, descouring aid, antistat; anticorrosive agent for pipelines; cleaner for metallic surfaces; superfatting, bodying and antifog aid; pigment dispersant; detergent; emulsion of solv.; cutting oils; textile lubricant additive, [Fina Chemicals]

Radiasurf® 7156. CAS 85711-45-1; EINECS 288-305-7; Pentaerythritol oleate; corrosion inhibitor for lubricating oils and greases; biodeg. surfactant, wetting agent, emulsifier for cosmetics, pharmaceuticals, agriculture, chemical synthesis, explosives, polymers, glass fibers, surface coatings, textiles and leather. [Fina Chemicals]

Radiasurf® 7157. CAS 9005-65-6; Polysorbate 80; emulsi-

---

* Trade name not verified as available   † Trade name verified as obsolete

fier, descouring aid, antistat; anticorrosive agent for pipelines; cleaner for metallic surfaces; superfatting, bodying and antifog aid; pigment dispersant; detergent; emulsion of solv.; cutting oils; textile lubricant additive, [Fina Chemicals]

Radiasurf® 7175. CAS 85116-93-4; EINECS 285-547-5; Pentaerythritol stearate; chemical intermediate, emulsifier, detergent, wetting agent, lubricant; chemical synthesis; cosmetics and pharmaceuticals; pearlescent shampoo formulations; detergency and cleaning products; cutting, lamination, and textile oils; rust inhibitor; pigment wetting, grinding, and improved gloss in paints, lacquers, printing inks; paper industry; lubricant and mold release in plastics; textile and leather processing. [Fina Chemicals]

Radiasurf® 7270. CAS 97281-23-7; EINECS 306-522-8; Ethylene glycol stearate; wetting aid, lubricant, opacifier, antistat, dispersant, w/o emulgent, scouring and detergent aid, defoamer, plasticizer, rust inhibitor; cosmetics and pharmaceuticals, lubricating and cutting oils, textile and leather aids, pigment grinding [Fina Chemicals]

Radiasurf® 7400. CAS 93455-78-8; EINECS 297-364-8; Diethylene glycol oleate; emulsifier, wetting agent, defoamer, rust inhibitor, pigment grinder, antistat. [Fina Chemicals]

Radiasurf® 7402. CAS 85736-49-8; EINECS 288-459-5; PEG-4 oleate; wetting aid, lubricant, opacifier, antistat, dispersant, o/w emulgent, scouring and detergent aid, defoamer, plasticizer, rust inhibitor, visc. modifier, antifog aid; cosmetics and pharmaceuticals; lubricating and cutting oils; textile and leather aids; pigment grinding aids in paints and printing inks; latex and emulsion paints; plastics; waxes and maintenance products; glass fiber; insecticides; silicones. [Fina Chemicals]

Radiasurf® 7403. CAS 85736-49-8; EINECS 288-459-5; PEG-8 oleate; wetting aid, lubricant, opacifier, antistat, dispersant, o/w emulgent, scouring and detergent aid, defoamer, plasticizer, rust inhibitor, visc. modifier, antifog aid; cosmetics and pharmaceuticals; lubricating and cutting oils. [Fina Chemicals]

Radiasurf® 7404. CAS 85736-49-8; EINECS 288-459-5; PEG-12 oleate; wetting aid, lubricant, opacifier, antistat, dispersant, o/w emulgent, scouring and detergent aid, defoamer, plasticizer, rust inhibitor, visc. modifier, antifog aid; cosmetics and pharmaceuticals; lubricating and cutting oils. [Fina Chemicals]

Radiasurf® 7410. CAS 85116-97-8; EINECS 285-550-1; Diethylene glycol stearate; wetter, lubricant, opacifier, antistat, dispersant, detergent, defoamer, plasticizer, rust inhibitor; for cosmetics, pharmaceuticals, textiles, leather, paints, inks, plastic, waxes, maintenance products, insecticides. [Fina Chemicals]

Radiasurf® 7414. CAS 97281-23-7; EINECS 306-522-8; PEG-12 stearate; wetting aid, lubricant, opacifier, antistat, dispersant, o/w emulgent, scouring and detergent aid, defoamer, plasticizer, rust inhibitor, visc. modifier, antifog aid; cosmetics and pharmaceuticals; lubricating and cutting oils. [Fina Chemicals]

Radiasurf® 7417. CAS 97281-23-7; EINECS 306-522-8; PEG-1500 stearate; wetting aid, lubricant, opacifier, antistat, dispersant, o/w emulgent, scouring and detergent aid, defoamer, plasticizer, rust inhibitor, visc. modifier, antifog aid; cosmetics and pharmaceuticals; lubricating and cutting oils. [Fina Chemicals]

Radiasurf® 7423. CAS 37318-14-2; EINECS 253-458-0; PEG-8 laurate; wetting aid, lubricant, opacifier, antistat, dispersant, o/w emulgent, scouring and detergent aid, defoamer, plasticizer, rust inhibitor, visc. modifier, antifog aid; cosmetics and pharmaceuticals; lubricating and cutting oils. [Fina Chemicals]

Radiasurf® 7443. CAS 85736-49-8; EINECS 288-459-5; PEG-8 dioleate; wetting aid, lubricant, opacifier, antistat, dispersant, o/w emulgent, scouring and detergent aid, defoamer, plasticizer, rust inhibitor, visc. modifier, antifog aid; cosmetics and pharmaceuticals; lubricating and cutting oils. [Fina Chemicals]

Radiasurf® 7444. CAS 85736-49-8; EINECS 288-459-5; PEG-12 dioleate; biodeg. surfactant, wetting agent, emulsifier for cosmetics, pharmaceuticals, agriculture, chemical synthesis, explosives, polymers, glass fibers, surface coatings, textiles and leather. [Fina Chemicals]

Radiasurf® 7453. CAS 97281-23-7; EINECS 306-522-8; PEG-8 distearate; wetting aid, lubricant, opacifier, antistat, dispersant, o/w emulgent, scouring and detergent aid, defoamer, plasticizer, rust inhibitor, visc. modifier, antifog aid; cosmetics and pharmaceuticals; lubricating and cutting oils. [Fina Chemicals]

Radiasurf® 7454. CAS 97281-23-7; EINECS 306-522-8; PEG-12 distearate; wetting aid, lubricant, opacifier, antistat, dispersant, o/w emulgent, scouring and detergent aid, defoamer, plasticizer, rust inhibitor, visc. modifier, antifog aid; cosmetics and pharmaceuticals; lubricating and cutting oils. [Fina Chemicals]

Radiasurf® 7473. CAS 97281-23-7; EINECS 306-522-8; PEG-8 stearate; biodeg. surfactant, wetting agent, emulsifier for cosmetics, pharmaceuticals, agriculture, chemical synthesis, explosives, polymers, glass fibers, surface coatings, textiles and leather. [Fina Chemicals]

Radiasurf® 7600. CAS 85251-77-0; EINECS 286-490-9; Glyceryl stearate; wetter, defoamer, rust inhibitor, pigment grinder, emulsifier, antistat. [Fina Chemicals]

Radiasurf® 7900. CAS 91052-47-0; EINECS 293-208-8; Glyceryl stearate; biodeg. surfactant, wetting agent, emulsifier for cosmetics, pharmaceuticals, agriculture, chemical synthesis, explosives, polymers, glass fibers, surface coatings, textiles and leather. [Fina Chemicals]

radicle vinegar. Acetic acid, glacial.

Radiflam A AE. Nylon; self-extinguishing; molding compound. [Radilon] *

Radilon A, A 32E. Nylon; molding and extrusion compound. [Radilon] *

Radilon A CP300. Nylon, reinforced and filled; molding compound. [Radilon] *

Radilon S, S 35FL/FLC. Nylon; for extrusion. [Radilon] *

Radilon S BHS200/201. Nylon; for molding. [Radilon] *

Radiocaps-131. CAS 7790-26-3; Sodium iodide I 131; antineoplastic; diagnostic aid; radioactive agent. [Abbott Laboratories] *

Radio-malt A proprietary preparation consisting of malt extract with irradiated ergosterol (radiostol); contains vitamins A, B, and D. §

Radiometal. A nickel-iron-copper alloy having high incremental permeability and low losses. It is largely used for radio transformers, relays, etc.

Radiopaque. A special barium sulfate.

Radiose. A French nitro-cellulose lacquer.

Radiostol. A proprietary preparation of calciferol, vitamin D_2. [British Drug Houses] *

---

‡ Trade name and manufacturer not verified   § Trade name without identified manufacturer

Radiostoleum. A proprietary preparation of vitamins A and D in oil. §

Radmolite. An electrical insulating material for supporting heating coils; largely consists of diatomaceous earth; has a slower rate of heat absorption; replaces fireclay. §

Radspor FT, 65 WP. CAS 2439-10-3; Dodine; a fungicide for the control of scab in apples and pears. [Truchem Ltd]

Radumine. A synthetic oxalic acid.

R.A.E. 57 alloy. An alloy of aluminum with 4% copper, 2% iron, and 0.5% magnesium; its specific gravity is 2.8.

raffinate. Material from which a soluble substance is extraced, e.g., petroleum from which higher hydrocarbons have been extracted.

raffinose. (gossypose; millitose; mellitriose) CAS 512-69-6; A sugar; $C_{18}H_{32}O_{16} \cdot 5HOH$; used in bacteriology, in the preparation of other saccharides.

Rainbow Custom Colored Mortars. CAS 1309-37-1; Blends of iron oxide colors and mortars; preblended colored mortar to produce more uniform color in the mortar joints of a building. [DCS Color & Supply Co Inc]*

Rainbow Ware. A proprietary synthetic resin of the urea-formaldehyde type. §

RAK®. Pheromones. [BASF AG]

Rakel's alloy. An aluminum bronze. It contains 87.5% copper, 10.5% aluminum, 1% manganese, and 1% lead or zinc.

Rakusol®. Org. and inorganic pigments in paraffin oil and glycerol ester; used for coloring plastics; less suitable for PVC. [BASF AG]

Ralox® 02. 4,4´-Methylene-bis-(2,6-di-t-butylphenol); nonstaining antioxidant. [Raschig]

Ralox® 46. CAS 119-47-1; 2,2´-Methylene-bis-(6-t-butyl-4-methylphenol); nonstaining antioxidant for plastics, rubber, latex, adhesives, hot melts, cables. [Raschig]

Ralox® 530. CAS 2082-79-3; Octadecyl-3-(3,5-di-t-butyl-4-hydroxyphenyl) propionate; nonstaining antioxidant for plastics, hot melts, cables, mineral oil processing (lubricants). [Raschig]

Ralox® 630. CAS 6683-19-8; Tetrakis [methylene (3,5-di-t-butyl-4-hydroxyhydrocinnamate)] methane; nonstaining antioxidant for plastics, hot melts, cables. [Raschig]

Ralox® BHT food grade. CAS 128-37-0; BHT; nonstaining antioxidant for plastics in contact with food, rubber, latex, adhesives, hot melts, mineral oil processing (lubricants), petrol., feedstuffs. [Raschig]

Ralox® TMQ-R. 2,2,4-Trimethyl-1,2-dihydroquinoline polymer; staining antioxidant for rubber processing. [Raschig]

Ralufon® 414. Cocamidopropyl betaine; surfactant. [Raschig]

Ralufon® DL. N,N-Dimethyl-N-lauryl-N-(3-sulfopropyl)-ammonium betaine; surfactant. [Raschig]

Ralufon® DT. N,N-Dimethyl-N-tallow-N-(3-sulfopropyl)-ammonium betaine; surfactant. [Raschig]

Ralufon® N. Nonylphenol polyethylene oxide sulfopropyl ether, potassium salt; surfactant. [Raschig]

Ralufon® TA. N,N-Dimethyl-N-stearic acid-amidopropyl-N-(3-sulfopropyl)-ammonium betaine; surfactant. [Raschig]

Raluquin®. 6-Ethoxy-2,2,4-trimethyl-1,2-dihydroquinoline; staining antioxidant. [Raschig]

Ramasit® KGT. Water repelling agent for textiles. [BASF; BASF plc]

Rambufaside. 14-Hydroxy-3-(4-O-methyl-α-l-rhamno-pyranosyloxy)-14β-bufa-4, 20 ,22-trienolide. 4´-o-Methylproscillaridin; a cardiac glucoside.

Ramenti ferri. Iron filings.

Ramet. A proprietary cutting material for steel alloys, cast iron, etc.; consists of tantalum carbide with nickel; melts at 4100 C. §

Rametin. A proprietary preparation of naphthalophos; a veterinary anthelmintic. §

ramie. (Chinese grass; rhea; rhea ramine;green ramie) A fiber obtained from *Boehmeria tenacissima*.

Ramix. A proprietary magnesite refractory. §

Ramos. A proprietary phenol-formaldehyde resin. §

Rampart. CAS 1563-66-2; Carbofuran; a systemic carbamate insecticide and nematode in the form of granules for soil treatment. [Sipcam UK Ltd; Universal Crop Protection Ltd]

Ramrod. CAS 1918-16-7; Propachlor; a pre-emergence herbicide for various horticultural crops. [Monsanto plc]

Ramtap. Fibrous tap hole plugs for electric arc and basic oxygen steel-making furnaces. [Foseco (F.S.) Ltd]

Rancho®. Herbicide for control of grass weeds and some broad-leaved weeds in irrigated rice crops. [Bayer AG]

Randanite. (Ceyssatite, Memilite). Varieties of hydrated silica or opal.

Raneoff® S. Silicone resin emulsion; durable water repellent. [Eastern Color & Chem.]

Ranestol. CAS 5714-82-9; Triclofenol piperazine; anthelmintic. [Parke-Davis] *

Raney nickel. CAS 7440-02-0; A form of finely divided nickel used in hydrogenating certain organic compounds; used as a catalyst for hydrogenation.

Ranide. CAS 22662-39-1; Rafoxanide; anthelmintic. [Merck & Co Inc] *

Ranotex. Textile softening agents. [Rhone-Poulenc UK]

Ransome's stone. An artificial stone made by mixing sand with sodium silicate and a little chalk, or other similar material. The product is molded to shape, and immersed in a solution of calcium chloride.

Rantudil®. Anti-inflammatory and antirheumatic agent. [Bayer AG]

Raolein 131. Triolein I 131; radioactive agent. [Abbott Laboratories] *

RAP; Polypropylene film, for use by industry in overwrap packaging of consumer items. [Quantum Chemical Corp]

Rapadex. Photographic developer. [May & Baker Ltd] *

rapeseed oil. (Brassica campestris oil; oil of rapeseed; Colza oil) Vegetable oil expressed from seeds of *Brassica campestris*; CAS 8002-13-9; metal lubricant additive; edible oil for salad dressings, margarine; soft soaps. [Arista Industries; Climax Performance; Penta Mfg.; Reilly-Whiteman; Werner G. Smith; Witco/Humko]

Rapiblend. Predispersed rubber chemicals. [Anchor Chemical Group plc]

rapic acid. A name which has been applied to the fatty acids of rape oil; identical to oleic acid.

Rapi-Cure BHC. Hydroxybutyl vinyl ether carbonate. [ISP]

Rapi-Cure CHMVE. Cyclohexane dimethanol, monovinyl ether. [ISP]

Rapi-Cure CHVE. 1,4-Cyclohexane dimethanol, divinyl ether. [ISP]

Rapi-Cure CVE. Cyclohexyl vinyl ether. [ISP]

Rapi-Cure DVE-3. Triethylene glycol divinyl ether; reactive diluent [ISP]

---

* Trade name not verified as available   † Trade name verified as obsolete

Rapi-Cure EHVE. CAS 103-44-6; 2-Ethylhexyl vinyl ether. [ISP]

Rapi-Cure HBVE. Hydroxybutyl vinyl ether. [ISP]

Rapid. CAS 23103-98-2; Pirimicarb; garden insecticide. [ICI Chem. & Polymers Ltd]

Ra-Pid-Gro. Plant food. [Chevron] *

Rapidogen. Dyestuffs; used for dyeing cotton by the warp sizing process. [Bayer AG]

Rapidosept®. A hand disinfectant; gentle to skin. [Bayer AG]

RapidPurge 2. Nonabrasive purging compound effective on all thermoplastics; for cleaning injection molding, hot runner tools, extrusion, multi-layer dies, blow molding equip. [RapidPurge]

Rapier. CAS 23950-58-5; Propyzamide; a residual herbicide for oil seed rape. [Farmers Crop Chemicals Ltd]

Rapifen. CAS 70879-28-6; A proprietary preparation containing alfentanil hydrochloride; analgesic supplement in anesthesia. [Janssen Pharmaceutical Ltd] *

Rapitard. A proprietary preparation of beef insulin crystals in a rapid acting solution of neutral pork insulin. [British Drug Houses] *

Rappor. CAS 13516-27-2; Guazatine; a fungicide seed dressing for wheat. [DowElanco Ltd]

Rappor Plus. CAS 13516-27-2, 73790-28-0; Guazatine + imazalil; a fungicide seed dressing for barley and oats. [DowElanco Ltd]

Raschit. CAS 59-50-7; p-Chloro-m-cresol; used as a preservative for latex.

Rasorite. Kernite (hydrated sodium borate), and in general boron ores and products. [Borax Consolidated Ltd] *

Rassamix CDA. Amitrole + atrazine + diuron; a mixture of herbicides for weed control. [Denoon CDS]

Rassapron. Amitrole + atrazine + diuron; a mixture of herbicides for weed control. [BP Oil Ltd]

Rastinon. CAS 64-77-7; A proprietary preparation of tolbutamide; an oral hypoglycemic agent. [Hoechst UK]*

Ratak. CAS 56073-07-5; Difenacoum; a ready-to-use anticoagulant rodenticide. [ICI Garden Products]

Rat Flip. Trade name for a line of rat and mouse bait products for both indoor and outdoor use; anticoagulant active ingredients are safer around domestic animals and pets; vitamin K is an antidote. [Colonial Products Inc]

Ratio. CAS 82558-50-7; Suspension concentrate containing 125 g isoxaben per liter; used for control of annual dicotyledons in cereals, grass and fruit. [Tripart Farm Chemicals Ltd]

Ratox. CAS 81-81-2; Warfarin-based rodenticide. [The Wellcome Foundation Ltd] *

Raudixin. A proprietary preparation of rauwolfia; an antihypertensive. [Bristol-Myers Squibb Pharmaceuticals Ltd]

Rau-Sed. CAS 50-55-5; Reserpine; antihypertensive. [Bristol-Myers Squibb Co Inc] *

Rauserfia. Rauwolfia serpentina; antihypertensive. [Penick Corp] *

Rautrax Sine K. A proprietary preparation of rauwolfia and hydroflumethiazide; used as an antihypertensive agent. [Bristol-Myers Squibb Pharmaceuticals Ltd] *

Rautrax Tablets. *Rauwofia serpentina* whole root, hydroflumethiazide, and potassium chloride in tablet form. [Bristol-Myers Squibb Pharmaceuticals Ltd] †

Rauverid. Rauwolfia serpentina; antihypertensive. [O'Neal, Jones & Feldman Pharmaceuticals] *

Rauwiloid. Each tablet contains selected alkaloid hydrochlorides of rauwolfia serpentina 2 mg [3M Pharmaceuticals] *

Rauxite. Proprietary trade name for urea-formaldehyde varnish and lacquer resins. §

Rauxone. A proprietary trade name for alkyd varnish and lacquer resins. §

Rauzene. A proprietary trade name for a phenolic varnish and lacquer resin. §

Rauzene Ester. A proprietary trade name for an ester gum.§

Raven. CAS 1333-86-4; Carbon black. [Sevalco Ltd]

Ravinil. PVC suspension and mass polymers. [European Vinyls Corporation Ltd]

Ravocaine Hydrochloride. CAS 550-83-4; Propoxycaine hydrochloride; anesthetic. [Cook-Waite Laboratories Inc] *

Ravoien. Plasticizer and rubber extender. [Burmah-Castrol Ltd] *

Ravolen 11(T). A proprietary trade name for a decolorized petroleum aromatic extract. [Burmah-Castrol Ltd] *

Rawstol. (Radarsan). Vermicides consisting of solutions of fluosilicic acid.

raw turkey umber. *See* umber.

raw umber. *See* umber.

Raxil. CAS 107534-96-3; Tebuconazole; cereal seed dressing with systemic properties for control of seed-borne diseases such as stinking smut, loose smuts and covered smut; highly effective at low dosage rates. [Bayer AG]

Raybar. CAS 7727-43-7; A proprietary preparation of barium sulfate. [Fleet Co Inc] *

Rayo. An electrical resistance alloy consisting of 85% nickel and 15% chromium.

Rayon. Semisynthetic fiber composed of regenerated cellulose which has been cogulated or solidified from a solution of cellulose xanthate, cellulose nitrate or from a solution of cellulose in ammoniacal copper oxide; used in nonwoven fabrics, surgical dressings, mechanical rubber goods, coated fabrics.

Rayox. CAS 13463-67-7; EINECS 236-675-5; A proprietary trade name for titanium dioxide, TiO_2. §

Razoxin. CAS 21416-87-5; Anti-cancer preparation containing razoxane. [ICI Chem. & Polymers Ltd]

RBC. A proprietary preparation of phenylmercuric nitrate, isobutyl p-aminobenzoate, N-butyl p-aminobenzoate, benzocaine, cholesterol and calaraine; antipruritic. [Rybar Laboratories Ltd] *

RC 7. Fluorocarbon; mold release agent and lubricant for silicone rubber molding operations, thermoset plastic molding. [Releasomers]

RCR Grey Squirrel Killer Concentrate. CAS 81-81-2; Warfarin; bait for grey squirrels. [Leo Fay Ltd]

RD10. Roadway dust suppression agent. [Foseco (F.S.) Ltd]

RDPE. Dimethyl ester solvent. [Orange Chemicals Ltd]

Reach® 101, 201, 501. Aluminum chlorohydrate; antiperspirant for increased wetness protection, especially for aerosols. [Reheis]

Reach® AZP-701, AZP-703. Aluminum zirconium tetrachlorohydrex glycine; enhanced efficacy antiperspirant. [Reheis]

Reacrone. Acrylic resins; for paints, inks and metal decorating. [Resinas Sinteticas SA] *

Reactal. Alkyds; for paints, lacquers and varnishes, inks and metal decorating. [Resinas Sinteticas SA] *

---

‡ Trade name and manufacturer not verified   § Trade name without identified manufacturer

Reactine® . CAS 83881-51-0; Cetirizine; antihistamine. [Pfizer International]

Reactint®. Polymeric colorants for flexible urethane foams. [Milliken]

Reactivan. A proprietary preparation of fencamfamin hydrochloride, thiamine, pyridoxine, cyanocobalamin, and ascorbic acid; a tonic. [E Merck] *

Reactobond. Chemically reactive oils for tube drawing. [Brent Chemicals International plc]

Reacton. Rare earth metals and compounds; for electronic materials, phosphors, magnetic materials, sputtering targets. [Johnson Matthey plc]

Ready-FeR. Iron chelate; used for iron deficiency in plants. [Makhteshim Chemical Works Ltd]

Reafor. Etherified amino resins; for paints, lacquers and varnishes, inks and metal decorating. [Resinas Sinteticas SA] *

Reafree. Saturated polyester; for paints, inks and metal decorating. [Resinas Sinteticas SA] *

Reagens-CF2. CAS 3327-22-8; 3-Chloro-2-hydroxypropyltrimethyl ammonium chloride. [Synthetic Chemicals Ltd]

Reakt. Colorless dye intermediates for non-carbon copying paper. [BASF AG]

realgar. CAS 56320-22-0; (ruby sulfur, red orpiment, red arsenic, red arsenic glass, ruby arsenic, arsenic orange). Arsenic disulfide, As_2S_2; used in the leather industry, paint, pyrotechnics, and taxidermy.

Realox®. CAS 1344-28-1; Reactive calcined alumina; thermally reactive alumina for ceramic applications requiring high density to be attained at lower firing temps. with a minimum of fluxing additives. [Alcoa]

Reamul. Aqueous dispersions (polyvinyl acetate, styrene, acrylic etc.); used for paints and inks. [Resinas Sinteticas SA] *

Réamur's alloy. An alloy of 70% antimony and 30% iron.

Rearguard. Suspension concentrate containing 64% w/v sulfur, 16% w/v maneb, 1% w/v copper oxychloride; for use as an agricultural fungicide. [Universal Crop Protection Ltd]

Reatane. Polyisocyanates; used for lacquers and varnishes. [Resinas Sinteticas SA] *

Reater. Modified polyester resins; used for lacquers and varnishes. [Resinas Sinteticas SA] *

Reatint. Trade name for family of polymeric, liquid, reactive colorants used to color polyurethane polymers; colorants for flexible polyether/polyester urethane foam used for carpet underlay, packaging, sponges, etc., and RIM polyurethane systems for automotive and consumer applications. [Milliken]

reaumerite. A compound, $(Ca \cdot Na_2)O \cdot 3SiO_2$, obtained by heating glass at its softening temperature.

Reax® 45A. CAS 8061-51-6; Sodium lignosulfonate with anionic wetting agent; wetting agent and dispersant for pesticides; recommended for use with chlorinated hydrocarbons and organophosphates. [Westvaco]

Reax® 80C. CAS 8061-51-6; Sodium lignosulfonate, modified; dispersant, suspending agent for dyestuffs, lead acid storage batteries. [Westvaco]

Réboulet's solution. An aqueous solution of calcium chloride, potassium nitrate, and alum; used to preserve anatomical specimens.

Recoil. A wettable powder containing 10% w/w oxadixyl and 56% w/w mancozeb; fungicide to control foliar and tuber blight in potatoes. [Schering Agrochemicals Ltd]

Recoil. A wettable powder containing 10% w/w oxadixyl and 56% mancozeb; a systemic fungicide for the control of foliar and tuber blight in potatoes. [Bayer plc]

Recoura's sulfate. A chromium hexahydrated sulfate, $Cr_2(SO_4)_3 \cdot 6H_2O$.

recovered grease. The oil used to lubricate wool during spinning is recovered from the washwater; used to manufacture a low-grade stearin.

ReCovr. CAS 154-69-8; Tripelennamine hydrochloride; veterinary antihistaminic. [Bristol-Myers Squibb Co Inc]*

Recresal. A proprietary preparation of acid sodium phosphate in tablets. §

Recrete NRC. Flowable replacement concrete used to replace damaged, defective or deteriorated surfaces. [Feb Ltd]

Rectified Spirit S.V.R. A specially rectified ethyl alcohol 68-69 over proof, containing 96-97% ethyl alcohol by volume; used in perfumery and in pharmaceutical extracts and tinctures.

Recupex. Flux for recovering nonferrous scrap. [Foseco (F.S.) Ltd]

Red 139. CAS 1309-37-1; Iron oxides, bismuthoxychloride; inorganic colorant. [Presperse]

red acid. Nitric acid of 40 Bé, or stronger. It contains dissolved nitrogen oxides.

red algar. Arsenic disulfide, As_2S_2.

Redalloy. A proprietary trade name for brass containing 85% copper, 14% zinc, and 1% tin. §

red antimony. (antimony blende, pyrantimonite, pyrostibnite, antimony cinnabar). A mineral; an oxysulfide of antimony; $Sb_2O_3 \cdot 2Sb_2S$; is also obtained by treating antimony chloride with sodium thiosulfate in aqueous solution; used as a pigment to replace ordinary cinnabar.

red argol. See argol.

red brass. A brass containing 90% copper and 10% zinc. Also Tombac (qv), which has been pickled in acid.

red charcoal. A wood charcoal made at low temperature. It contains hydrogen and oxygen.

red chromate of potash. Potassium bichromate, $K_2Cr_2O_7$.

red cobalt. A mineral that is erythrite.

Red Copp. CAS 1317-39-1; Red cuprous oxide. [Am. Chemet]

Redcopp 97N Premium. CAS 1317-39-1; Cuprous oxide. [CP International Chemicals Ltd]

red copper. See violet copper.

Redd Citrus Specialties. A range of essential citrus oils, natural citrus aroma and natural citrus specialty products; used for a wide range of applications in the food and beverage industries, personal care products, cosmetics, soaps, detergents and household products. [Hercules]*

reddingite. A mineral, $3(Mn \cdot Fe)O \cdot P_2O_5$.

Red Dot. A smokeless powder; designed for light and standard shotshell loads of all gauges; can also be used in specific handgun cartridges. [Hercules] *

red drops. (red lavender). Compound tincture of lavender.

Redeptin. CAS 1841-19-6; A proprietary preparation of fluspiriline; used in the treatment of schizophrenia. [SmithKline Beecham] *

Redeptin Injection. CAS 1841-19-6; Aqueous suspension of fluspirilene; long acting intramuscular injection for treatment of schizophrenia. [SmithKline Beecham] *

red gold. A jeweler's alloy containing 75% gold and 25% copper.

---

* Trade name not verified as available   † Trade name verified as obsolete

Red Hermetite. Red paste, semi hardening; used as gasket jointing compound to supplement all gaskets flanged or threaded applications; ensures leak-free joints in most environments. [Hermetite Products Ltd] *

Red Hot Pellets. CAS 1305-62-0; Pelleted calcium hydroxide. [Schaefer Salt & Chemical Co]

Rediclear. Hydrosulfite/dispersing agent blend. [RV Chemicals Ltd] †

Redicote. Specialized cationic bitumen emulsifiers. [Akzo]

red iodide of mercury. Mercuric iodide, HgI_2.

red iron trioxide. See ferric oxide.

Redisol. CAS 68-19-9; Cyanocobalamin; vitamineral [Merck & Co Inc] *

red lavender. See red drops.

red lavender spirit. See spirit of red lavender.

red lead. (red lead oxide, minium, Paris red, Saturn red; lead tetroxide). CAS 1314-41-6; A pigment. It is oxide of lead, Pb_3O_4, made by heating litharge, PbO. There are several kinds on the market distinguished by their color and amount of lead dioxide they contain; used in storage batteries, glass, potteriy and enameling, varnish, purification of alcohol, packing pipe joints, metal-protective paints, fluxes, and ceramic glazes.

red lead oxide. See red lead.

red liquor. (mordant rouge). A solution corresponding to the formula, $Al_2(C_2H_3O_2)_6$, which appears to consist of a diacetate of aluminum, and acetic acid. Red liquor is; largely used in dyeing and calico printing, especially for the production of red colors, for the manufacture of dense lakes, and for waterproofing woolen fabrics.

Redmanol. See Bakelite. §

red metal. A term usually applied to an alloy of 90% copper, and 10% zinc.

red mordant. See red liquor.

red nickel ore. A mineral that is niccolite or nickeline.

red oil. CAS 112-80-1; Commerical grade of oleic acid consisting of 70% oleic acid, 15% of linolic acid, and 15% of stearic acid; these oils are used for general lubrication.

Red oil. See oleic acid.

red oxide of chromium. Chromium trioxide, CrO_3.

red oxide of lead. See red lead.

red oxide of mercury. (red precipitate). CAS 21908-53-2; Mercuric oxide, HgO.

Redoxon. Vitamin C preparations. [Roche Products Ltd] *

red potassium chromate. See potassium dichromate.

red precipitate. See red oxide of mercury.

red precipitate. See mercury oxide (ic), red and yellow.

red prussiate of potash. Potassium ferricyanide; $K_3Fe(CN)_6$.

Redray. An electrical resistance alloy containing 85% nickel and 15% chromium.

red salts. Both crude sodium acetate and crude sodium carbonate, colored red by ferric oxide, are known as red salts.

red saunderswood. See redwoods.

red soda. A solution of red ink containing a little gum arabic and sodium carbonate; used as a marking ink for blue prints.

Red Star Powder. It is a 33-grain smokeless powder containing metallic nitrates, nitro-hydrocarbons, and petroleum jelly; used as an explosive.

red storax. (solid storax). An artificial product obtained by mixing poor storax with sawdust, and pressing the mixture; used for fumigating candles and powders.

Reduce®-150. Blend of sodium stearoyl lactylate, calcium sulfate, and sodium sulfite; dough conditioner for use in flour tortillas, pie crusts, and pizza shells. [Am. Ingredients/Patco]

reduced turpentine. A mixture of turpentine oil with petroleum.

reducin. Triamino-resorcinol, $C_6H(NH_2)_3 \cdot (OH)_2$; used in photography as a developer.

Reductone. CAS 7775-14-6; Sodium hydrosulfite solution; for continuous vat dyeing and afterscouring. [Olin]

Redul. Glycodiazine; an oral hypoglycemic agent. [Bayer AG]

Redurit. Electrically fused standard grade corundum; used for production of abrasives, abrasive paper, discs and cloth. [Hüls UK Ltd]

Reduxol Z. CAS 24887-06-7; Soluble zinc formaldehyde sulfoxylate. [RV Chemicals Ltd] †

Redux® 501. CAS 25928-94-3; Two-component epoxy adhesive; ambient or warm curing epoxy resin matrix system; mix ratio 100:15 pbw [Ciba-Geigy Plastics UK]*

red vermilion. See mercury sulfide (ic), red and black.

red vitriol. (Botryogen). A native ferroso-ferric sulfate from Sweden.

red wash. A zinc sulfate solution containing red coloring matter.

red water bark. Sassy bark.

redwoods. (red dye woods). These dye woods are divided into two classes: a) Soluble, which comprise Brazil, Pernambuco or Fernambuco wood, Peach wood, Lima wood, Sapan wood, Bimas redwood, and Nicaragua wood. All of them contain the coloring principle brazilin, $C_{16}H_{14}O_5$, which, by oxidation, is converted into brazilein, $C_{16}H_{12}O_5$, which give purple shades with chrome mordants, and crimson with alum. b) Insoluble redwoods, consisting of camwood or; came wood, barwood, saunderswood, santalwood, sandlewood, Bresille wood, and Caliatur wood. The dying principle is santaline. These woods have a limited application for dying wool with alumina, chrome, tin, or iron mordants.

Reed C-ABS-17415. ABS-based black concentrate. [Reed Plastics]

Reed C-NY-261. Nylon-based color concentrate; for use in nylon 6 and 6/6. [Reed Plastics]

Reed C-NY-4892. Nylon-based color concentrate with 20% carbon black; for use in nylon 6 and 6/6 especially for filled and toughened products; superior hiding power. [Reed Plastics]

Reed C-PBT-1338. PBT polyester-based color concentrate, 20% carbon black; concentrate for use in PBT polyester. [Reed Plastics]

ReedLite C-NY. Heavy metal-free nylon color concentrates; for automotive, electronic and mech. applications. [Reed Plastics]

ReedLite CPC. Heavy metal-free PC color concentrates; for elec., mech., automotive, appliance and business machine applications. [Reed Plastics]

Reese's alloy. An alloy used in dental work. It contains 87% tin, 8.6% silver, and 4.4% gold.

Rees's thionin stain. A microscopic stain. It consists of 1.5 g thionin and 10 cc alcohol in 100 cc of 5% solution of carbolic acid; used at the rate of 5 cc in 20 cc water.

Refagan. A proprietary preparation of salicylamide, phenacetin, caffeine and mebhydrolin napadisylate. [FBA Pharmaceuticals] ‡

---

‡ Trade name and manufacturer not verified   § Trade name without identified manufacturer

Refagan® N. Relief from colds. [Bayer AG]

refikite. A resin found in lignite.

Refine®. For the agriculture industry. [DuPont UK]

refined silver. A silver usually containing from 99.7-99.9% metal.

Refinex. CAS 1337-76-4; Powdered attapulgite clay; used for oil refining. [Whitecourt Ltd]

Refkon. A range of water-based coatings having high solids content. [Foseco (F.S.) Ltd]

Reflectafoam. Closed cell polyethylene backing with highly reflective aluminized polyester surface; used for preventing heat loss behind radiators and insulating airing cupboards. [Piccadilly Products Ltd] *

Reflite. A proprietary synthetic resin molding powder. §

reflorit. Picric acid; used in the disinfection of seed-corn.

reform phosphate. Rock phosphate which has been treated with small quantities of dilute acid to render it more porous, converting calcium carbonate into calcium hydrogen carbonate.

Refosporin®. CAS 56187-47-4; Cefazedone. [E Merck]

Refrax. Bricks made from recrystallized silicon carbide. A refractory material.

Refresh®. Polyvinyl alcohol, povidone; ocular lubricant. [Allergan, Inc]

Refresh® P.M. White petrolatum, mineral oil; ocular lubricant. [Allergan, Inc]

Refuse Trol. Liquid waste dispersant and deodorant. [Momar Industrial Services Ltd]

Regal® 400R. CAS 1333-86-4; Carbon black; for coloring higher dielec. PVC cable compounds. [Cabot; Cabot Carbon Ltd]

Regal Crown. Combination of growth stimulators to enhance plant growth by accelerating root growth. [Regal Chemical Company]

Regalite. Hydrogenated hydrocarbon resins. [Hercules Ltd]

Regalox. A sintered material comprising 88% alumina.

RegalStar. Herbicide to control crabgrass, goosegrass and other annual weeds; applied in early spring to turfgrass and cultivated nursery fields prior to weed seed germination. [Regal Chemical Company]

Regamycin. CAS 801-52-5; Porfiromycin; antibacterial; antineoplastic. [Upjohn] *

regenerated turpentine. A product of synthetic camphor manufacture; boils at 170 C.

Regenex. Copper and nickel alloy flux. [Foseco (F.S.) Ltd]†

Regent® 12XX. CAS 7758-23-8; Monocalcium phosphate, monohydrate; leavening agent for baking, cereal, beverages. [Rhone-Poulenc Food Ingreds.]

Reginal. Proofing agent. [Ciba plc] *

Reginol 2701. Surfactant blend; scouring agent for dyeing; biodeg. [Hart Prods. Corp.]

Regitine Mesylate. Phentolamine mesylate; antiadrenergic. [Ciba-Geigy Corp] *

Reglan. CAS 54143-57-6; Metoclopramide hydrochloride; antiemetic. [Wyeth Laboratories] *

Reglone. CAS 85-00-7; Diquat; a contact herbicide and preharvest crop desiccant. [ICI Chem. & Polymers Ltd]

Reglox. CAS 85-00-7; Diquat herbicide. [ICI Chem. & Polymers Ltd]

Regnis. Machine lubricant for glass forming machinery. [Specialty Products Co] *

Regonol. CAS 101-26-8; Pyridostigmine bromide; cholinergic. [Organon Inc]

Regulex. CAS 77-06-5; Gibberellic acid; a plant growth regulator; increases cropping in apples and pears. [ICI Chem. & Polymers Ltd]

Regulox K. Maleic hydrazine; a plant growth regulator for grass and to reduce bud growth in trees, hedges, and vegetables. [Rhone-Poulenc Environmental Prods. Ltd]

Regulus. Thermoplastic polyimide film; super heat resistant film for wire and cable, thermoplastic composites, pressure-sensitive adhesive tapes, primary insulation film. [Advanced Web Prods.]

regulus metal. This is usually a 5-12% antimony with lead.

regulus of antimony. Produced by heating antimony ore, Sb_2S_3. It contains about 10% of iron.

regulus of Venus. An alloy of copper and antimony, $SbCu_2$.

Regu-mate. CAS 850-52-2; Altrenogest; progestin. [Roussel UCLAF, Fine Chemicals] *

Regutol. CAS 577-11-7; Docusate sodium; stool softener; pharmaceutic aid. [Schering Corp] *

Rehydragel® Compressed Gel. CAS 21645-51-2; Aluminum hydroxide; adsorbent gel for pharmaceuticals (enhances suspensions, builds visc.); carrier for toxins in veterinary vaccines. [Reheis]

Rehydrol®. Aluminum chlorohydrate and propylene glycol; antiperspirant active. [Reheis Inc]

Reich's bronze. An aluminum bronze containing 85.2% copper, 7.52% iron, 6.6% aluminum, 0.5% manganese, and 0.15% lead.

Reicolit. A proprietary insulation. §

Reillex 202. Cross-linked poly-2-vinylpyridene. [Reilly Industries Inc]

Reillex 402 and 425. Cross-linked poly-4-vinylpyridene. [Reilly Industries Inc]

Reilline 2200 and 240. Linear poly-2-vinylpyridene. [Reilly Industries Inc]

Reilline 4200 and 450. Linear poly-4-vinylpyridene. [Reilly Industries Inc]

Reinecke's salt. CAS 13573-16-5; (ammonium tetrathiocyanodiaminochromate; ammonium reineckate) $NH_4[Cr(NH_3)_2(SCN)_4] \cdot HOH$, produced when ammonium cyanate is melted and ammonium bichromate added; used as a precipitating agent for organic bases and as a reagent for mercury.

Reiset's first base. Plato-diamine hydroxide, $Pt(NH_3 \cdot NH_3 \cdot OH)_2$.

Reiset's first chloride. Plato-diamine-chloride, $Pt(NH_3 \cdot NH_3 \cdot Cl)_2 \cdot H_2O$.

Reith alloy. An alloy of 75% copper, 10% tin, 10% lead, and 5% antimony.

Rela. CAS 78-44-4; Carisoprodol; relaxant. [Schering Corp]*

Reldan 50. CAS 5598-13-0; Chlorpyrifos-methyl; an organophosphate insecticide for the treatment of pests in stored grain and oilseed rape. [DowElanco Ltd]

Release Agent NL-1. Wax mixture, hydrocarbon solv. release agent for molded products; used as a first thin layer on the mold surface [Akzo] *

Release Agent NL-2. CAS 9002-89-5; Polyvinyl alcohol solution, alcohol water solv. film-forming release agent [Akzo] *

Release Agent NL-10. Wax mixture, paraffin oil solv. release agent for molded products; used as a first thin layer on the mold surface [Akzo] *

Releasil. Silicone release agents. [Dow Corning Ltd] *

Releez. Methyl oleate, methyl stearate, methyl palmitate,

---

methyl laurate, methyl myristate; asphalt release agent [Alzo]

Relief . Phenylephrine hydrochloride, polyvinyl alcohol; redness reliever, ocular lubricant. [Allergan, Inc]

Relimate . For the electrical industry. [DuPont UK]

Relipress . For the electrical industry. [DuPont UK]

Reloder 7. A smokeless powder; designed for rifle loads and 'benchrest' type reloads. [Hercules] *

Relugan . Resin or aldehyde tanning agents for leather. [BASF AG; BASF plc]

Relugan GT. An aqueous solution of glutaraldehyde; used as a tanning agent.

Remafin. Pigment masterbatches for plastics. [Hoechst UK]

Remazol. Reactive dyestuffs. [Hoechst UK]

Remcoil. Leather oils. [Hodgson Chemicals Ltd]

Remcopal. Nonionic surfactants (ethylene oxide derivatives); emulsifiers, wetting agents, antifoams. [Atochem UK/Ceca]

Remcopal 4. Laureth-4; emulsifier for beeswax; intermediate for polyethoxy ether sulfate. [Ceca SA]

Remcopal 6. CAS 9004-96-0; PEG-6 oleate; emulsifier for mineral oil and solvs.; stabilizer for polyurethane foams. [Ceca SA]

Remcopal 10. Cetoleth-10; base for emulsifiers and visc. spindle oils. [Ceca SA]

Remcopal 18. Cetoleth-18; retarder for colors; emulsifier for fatty alcohol and olein. [Ceca SA]

Remcopal 20. Laureth-20; emulsifier for fatty alcohol and olein. [Ceca SA]

Remcopal 25. Cetoleth-25; raw material for degreaser compounds; emulsifier for fatty alcohol, olein, and waxes. [Ceca SA]

Remcopal 29. CAS 9016-45-9; Nonoxynol-8.5; emulsifier. [Ceca SA]

Remcopal 40. CAS 61791-12-6; PEG-31 castor oil; emulsifier for olein. [Ceca SA]

Remcopal 40 S3. CAS 61791-12-6; PEG-40 castor oil; emulsifier for vitamins and essential oils. [Ceca SA]

Remcopal 121. Laureth-3; intermediate for sulfation. [Ceca SA]

Remcopal 207. CAS 9004-96-0; PEG-4.5 oleate; emulsifier for vitamins. [Ceca SA]

Remcopal 220. Cetoleth-25; emulsifier. [Ceca SA]

Remcopal 229. CAS 68439-49-6; Ceteareth-25; emulsifier for emulsion polymerization, washing detergents. [Ceca SA]

Remcopal 234. Cetoleth-4; surfactant. [Ceca SA]

Remcopal 238. CAS 68439-49-6; Ceteareth-20; surfactant for degreaser compounds. [Ceca SA]

Remcopal 258. Laureth-9; wetting agent. [Ceca SA]

Remcopal 273. Isodeceth-3; wetting agent for pigments and fillers. [Ceca SA]

Remcopal 306. CAS 9002-93-1; Octoxynol-5.5; wetting agent for carbon black, emulsifier for aromatic hydrocarbons, turpentine oil, and tallow. [Ceca SA]

Remcopal 334. CAS 9016-45-9; Nonoxynol-4; emulsifier, dispersant for pigments. [Ceca SA]

Remcopal 349. CAS 9016-45-9; Nonoxynol-8; emulsifier, dispersant for pigments. [Ceca SA]

Remcopal 666. CAS 9016-45-9; Nonoxynol-6; emulsifier for turpentine oil, heavy aromatic solvs.; wetting agent and dispersant for pigments. [Ceca SA]

Remcopal 3112. CAS 9016-45-9; Nonoxynol-2; antifoam, emulsifier. [Ceca SA]

Remcopal 3712. CAS 9016-45-9; Nonoxynol-12; emulsifier, wetting agent. [Ceca SA]

Remcopal 3820. CAS 9016-45-9; Nonoxynol-20; surfactant. [Ceca SA]

Remcopal 4000. CAS 61791-12-6; PEG-31 castor oil; surfactant. [Ceca SA]

Remcopal 4018. CAS 61791-12-6; PEG-23 castor oil; surfactant. [Ceca SA]

Remcopal 6110. CAS 9016-45-9; Nonoxynol-9; emulsifier, wetting agent. [Ceca SA]

Remcopal 21411. Laureth-11; emulsifier for light hydrocarbons, raw material shampoos. [Ceca SA]

Remcopal 21912 AL. Laureth-12; surfactant. [Ceca SA]

Remcopal 31250. CAS 9016-45-9; Nonoxynol-50; emulsifier for epoxy resin. [Ceca SA]

Remcopal 33820. CAS 9016-45-9; Nonoxynol-20; emulsifier for emulsion polymerization. [Ceca SA]

Remcopal D. Cetoleth-23; surfactant. [Ceca SA]

Remcopal HC 7. CAS 61788-85-0; PEG-7 hydrogenated castor oil; surfactant. [Ceca SA]

Remcopal HC 20. CAS 61788-85-0; PEG-20 hydrogenated castor oil; surfactant. [Ceca SA]

Remcopal HC 33. CAS 61788-85-0; PEG-33 hydrogenated castor oil; surfactant. [Ceca SA]

Remcopal HC 40. CAS 61788-85-0; PEG-40 hydrogenated castor oil; surfactant; solubilizer for essential oils and vitamins. [Ceca SA]

Remcopal HC 60. CAS 61788-85-0; PEG-60 hydrogenated castor oil; surfactant. [Ceca SA]

Remcopal L9. Laureth-9; emulsifier for heavy hydrocarbons, degreaser, antistat for synthetic fibers. [Ceca SA]

Remcopal L12. Laureth-10.5; surfactant. [Ceca SA]

Remcopal L30. CAS 9016-45-9; Nonoxynol-27; surfactant. [Ceca SA]

Remcopal LO 2B. Isodeceth-3; surfactant. [Ceca SA]

Remcopal LP. Laureth-9; surfactant. [Ceca SA]

Remcopal NP 30. CAS 9016-45-9; Nonoxynol-27; emulsifier. [Ceca SA]

Remcopal O11. CAS 9002-93-1; Octoxynol-11; solubilizer for essential oils; emulsifier. [Ceca SA]

Remcopal O12. CAS 9002-93-1; Octoxynol-12; solubilizer for essential oils; emulsifier. [Ceca SA]

Remcopal O9. CAS 9002-93-1; Octoxynol-9; solubilizer for essential oils; emulsifier. [Ceca SA]

Remcopal PONF. CAS 9016-45-9; Nonoxynol-11; emulsifier. [Ceca SA]

Remex. Desulfurizing flux for injection into steel melted in electric arc furnaces. [Foseco (F.S.) Ltd]

Remiderm. CAS 76-25-5, 8067-69-4; A proprietary preparation of triamcinolone acetonide and halquinol; used in dermatology as an antibacterial agent. [Bristol-Myers Squibb Pharmaceuticals Ltd] †

Remiderm Cream. CAS 76-25-5, 8067-69-4; Triamcinolone acetonide and halquinol in cream base. [Bristol-Myers Squibb Pharmaceuticals Ltd] †

Remiderm Ointment. CAS 76-25-5, 8067-69-4; Triamcinolone acetonide and halquinol in ointment base. [Bristol Myers Squibb Pharmaceuticals Ltd]

Remiderm Spray. CAS 76-25-5, 8067-69-4; Triamcinolone acetonide and halquinol in aerosol spray. [Bristol-Myers Squibb Pharmaceuticals Ltd] †

Remnos. CAS 146-22-5; A proprietary preparation of nitrazepam; a hypnotic. [DDSA Pharmaceuticals Ltd] *

Remotic. A proprietary preparation of halquinol and

---

‡ Trade name and manufacturer not verified   § Trade name without identified manufacturer

triamcinolone acetonide; used as a local antiinfective agent. [Bristol-Myers Squibb Pharmaceuticals Ltd]

Remotic Capsules. CAS 76-25-5, 8067-69-4; Triamcinolone acetonide and halquinol in castor oil. [Bristol-Myers Squibb Pharmaceuticals Ltd]

Remsed. CAS 58-33-3; Promethazine hydrochloride; antiemetic; antihistaminic. [Endo Laboratories Inc] *

Remsynol. Leather oils. [Hodgson Chemicals Ltd]

Remtal. CAS 1912-26-1, 122-34-9; Trietazine and simazine; selective herbicide for peas and beans. [Schering Agrochemicals Ltd]

Renacit® 4. Nonstaining antioxidant. [Bayer AG; Bayer plc]

Renacit® 7. CAS 133-49-3; Pentachlorothiophenol absorbed on clay; peptizing agent facilitating open mill and internal mixer mastication in rubber industry. [Miles; Polysar]

renaglandin. CAS 51-43-4; An extract of the suprarenal gland.

Renaleptine. See renalglandin.

Renault alloy. An aluminum alloy containing 88% aluminum, 10% zinc, and 2% copper.

Rencal. CAS 720-5-52-9; Phytate sodium; chelating agent. [Bristol-Myers Squibb Co Inc] *

Rendells. A proprietary preparation of nonoxynol used in the form of contraceptive pessaries. [W J Rendell Ltd] *

Rendrock. An explosive. It is a modification of Lithofracteur, consisting of 40% potassium nitrate, 40% nitroglycerin, 13% wood pulp, and 7.0% paraffin or pitch.

Renektan. Leather dyes. [ICI Chem. & Polymers Ltd]

Renese. CAS 346-18-9; A proprietary preparation of polythiazide; a diuretic. [Pfizer International]

Renex. Polyexyethylene alkyl or alkyl acryl esters. [ICI Chem. & Polymers Ltd]

Rengasil. CAS 31793-07-4; Pirprofen; anti-inflammatory. [Ciba-Geigy Corp] *

Renitec. CAS 76095-16-4; Enalapril maleate; for the treatment of hypertension and congestive heart failure. [Merck & Co Inc]

Reniten. CAS 76095-16-4; Enalapril maleate; for the treatment of hypertension and congestive heart failure. (Switzerland) [Merck & Co Inc]

rennet. CAS 9001-98-3; A clotting enzyme, which coagulates milk by precipitating the casein. It is the aqueous or alcoholic infusion of the dried stomach of the calf. It is used in making cheese, pharmaceuticals coagulation of casein for plastics, and as a food additive.

Rennilase. A milk clotting enzyme produced by a selected nonpathogenic strain of the fungus *Mucor miehei*; used in cheese-making for coagulation as an alternative to calf rennet. [Novo Nordisk]

rennin. A solid form of rennet.

Reno-M. Diatrizoate meglumine; diagnostic aid. [Bristol-Myers Squibb Co Inc]

Renol. Pigment and dyestuff materials for plastics. [Hoechst UK]

Renoquid. CAS 17784-12-2; Sulfacytine; antibacterial. [Parke-Davis] *

Renova. Solvent cleaner and degreaser. [British Nova Works Ltd]

Renovue-65, -DIP. Iodamide meglumine; diagnostic aid. [Bristol-Myers Squibb Co Inc]

Rentokil Deadline. CAS 28772-56-7; Bromadiolone; an anticoagulant rodenticide as a concentrated bait. [Rentokil Ltd]

Renyx. A proprietary trade name for an alloy of aluminum with nickel, copper, and silicon. §

Reochlor (LF and 54). Chlorinated paraffins; an extenderplasticizer for PVC [Ciba plc] *

Reoflam. A range of proprietary plasticizers used with PVC. [Ciba plc] *

Reofos. A range of synthetic organic phosphates. [Ciba plc]*

Reogen. A mixture of an oil soluble sulfonic acid of high molecular weight with a paraffin oil; plasticizer effective in all elastomers. [King Industries]

Reolube. A trade name for a range of synthetic organic phosphates. [Ciba plc] *

Reolube FAD. A proprietary trade name for a long chain fatty acid mixture, the principal components being C_{14}, C_{16}, and C_{18} acids. [Ciba plc] *

Reomet®. Metal treatment additives. [Ciba plc] *

Reomol. Plasticizers. [Ciba plc] *

Reomol 4PG. CAS 85-70-1; Butyl phthalyl butyl glycollate; a proprietary plasticizer. [Ciba plc] *

Reomol BCF. Butyl carbinol formal; a proprietary plasticizer. [Ciba plc] *

Reomol D79S. A mixture of heptyl and nonyl sebacates; a vinyl plasticizer §

Reomol DBS. CAS 109-43-3; Dibutyl sebacate; vinyl plasticizer. §

Reomol DCP. Dicapryl phthalate; a vinyl plasticizer §

Reomol DOS. Di-2-ethylhexyl sebacate; a vinyl plasticizer §

Reomol P. CAS 117-82-8; A proprietary trade name for dimethoxy ethyl phthalate; a chemical bonding agent for cellulose acetate staple fiber. [Ciba plc] *

Reomol PBPS. A sebacic acid polyester and a small proportion of nonpolymeric ester. [Ciba plc] *

Reomol TC9. A proprietary trade name for a chemical bonding agent for terylene and cellulose triacetate fibers.§

Reoplast. Epoxy plasticizers/stabilizers. [Ciba plc] *

Reoplex. Polyester plasticizers. [Ciba plc] *

Reoplex 200, 220, 300. Vinyl plasticizers of the polyester type. [Ciba plc] *

Reoplex 901. A trademark for a plasticizer for PVC sheeting intended for manufacture of surgical and electrical tapes. [Ciba plc] *

Reoplex 902. A plasticizer with good resistance to extraction by petroleum. [Ciba plc] *

Reostene. A nickel-iron alloy.

Repak. Calcium hypochloride. [PPG Industries]

Repelit. A proprietary synthetic resin varnish-paper product used for electrical insulation. §

Repello DC. Resin-wax blend; fabric water repellent. [Scher Chemicals Inc] *

Repel-O-Tex® QCJ. High molecular weight hydrophilic polyester emulsion; imparts soil release and moisture transport properties to polyester filament fabrics; oil scavenger keeping oils and greases in suspension during scouring/dyeing processes; fiber-to-fiber lubricant. [Rhone-Poulenc Surf.]

Replay RP 2177. CAS 9003-53-6; Impact polystyrene; contains controlled amts. of post-consumer recycled PS. [Huntsman]

Replay RP 2236. CAS 9003-53-6; Polystyrene; general purpose grade containing controlled amounts of post-consumer recycled PS. [Huntsman]

Replicast CS. Patented process for making castings in

---

* Trade name not verified as available   † Trade name verified as obsolete

which an expanded polystyrene pattern is coated with ceramic material then the pattern is burnt out. [Foseco (F.S.) Ltd] †

Replicast FM. Patented process for making castings using expanded polystyrene patterns in unbonded sand compacted under vacuum. [Foseco (F.S.) Ltd]

Reprodin®. Prostaglandin for cattle, horses, pigs and sheep. [Bayer AG]

Reproxal. Trimellitate plasticizers. [RWE-DEA Chemicals UK Ltd]

Repulse. CAS 1897-45-6; Chlorothalonil; a fungicide for a wide range of agricultural crops. [ICI Agrochemicals]

resacetophenone. (2, 4-Dihydroxy-aceto- phenone) CAS 89-84-9; $C_6H_3(OH)_2 \cdot CO \cdot CH_3$.

Resad. Polymer emulsions, polyvinyl acetate homopolymers and copolymers, acrylic and styrene acrylic polymers; used in adhesives, textile treatments and surface coatings. [Resadhesion Ltd]

Resamine. A proprietary trade name for formaldehyde resins. [Chemische Werke, Albert] *

Resan. See Bakelite. §

Resarit. Acrylic molding compound; for double and triple-walled sheets, rear lights, automotive parts, instrument covers, lampshades, condenser lenses, casings, covers for measuring instruments, etc. [Resart-IHM AG] *

Resarix SF. Scratch resistant coating system for surface treatment; for optical industry (lenses, magnifying glasses, lenses for sunglasses, scales) and for head protection (visors for crash helmets and astronaut helmets). [Resart-IHM AG] *

Resart. A range of melamine molding compounds; used for molded parts with tracking resistance, molded parts with high-grade dimensional stability and electrical components such as switches and relays. [Resart-IHM AG] *

Resartglas GS. Cast acrylic sheets LDII; used for roof windows, light domes, windscreens, caravan windows, displays, advertising gifts, furniture, for solarium equipment and solar beds - transparent to uva light. [Resart-IHM AG] *

Resart-PMMA XT. Standard extruded acrylic sheets 500 high impact; used for advertising aids (letters, displays, advertising transparencies), engineering components (housings, machine covers etc.), lighting fittings (cover for long-field light fittings, exterior fittings, etc.), roof hoods for caravans; drawing instruments, dome lights, door glazing, roof lights and caravan roof vents. [Resart-IHM AG] *

Resartherm. Glass fiber-reinforced polyester; used for electrical engineering, household appliances, electric tools, car ignition systems. [Resart-IHM AG] *

resazoin. See diazoresorcin.

Rescon. Disposable devices used for obtaining samples of molten steel. [Foseco (F.S.) Ltd] †

Rescue. Fuel additives. [UOP Inc]

Resectisol. CAS 69-65-8; Mannitol; diagnostic aid; diuretic. [Am. McGaw] *

reserpine. CAS 50-55-5; An alkaloid obtained from *Rauwolfia serpentina serpasil*; $C_{33}H_{40}N_2O_9$; used as an antihypertensive agent and as a tranquilizer.

Reserpoid. CAS 50-55-5; Reserpine; antihypertensive. [Upjohn] *

Resibon. Phenolic resins; used for metal decorating, adhesives, abrasives, thermal insulation refractories, interior can coatings. [Resinas Sinteticas SA] *

Resicart. Wet strength resins. [Ciba plc] *

Residuren. CAS 101-21-3; Chlorpropham; herbicide. [ICI Chem. & Polymers Ltd]

Residuren Extra. CAS 101-21-3, 330-54-1; Chlorpropham + diuron; an herbicide for treatment of grass in bulbs, peas, and beans. [Farm Protection Ltd]

Resigum. Hydrogenated rosin esters. [Barron Chemicals Ltd] ‡

Resilia. A proprietary trade name for a silico-manganese spring steel. §

Resilita. Polyamide resins; used for inks, paints, and adhesives. [Resinas Sinteticas SA] *

Resilla. A proprietary trade name for a special silicon-manganese spring steel. §

Resilon. CAS 409-21-2; Silicon carbide; used in the electrical industry. [Lonza AG]

Resimene. Melamine-formaldehyde resin; used as binders in adhesives, molded products, and foundry cores, paint coating ingredients and printing ink products. [Monsanto Co] *

resin. See colophony.

Resin 18. Poly-α-methylstyene. [Amoco Chemical Co]

Resin 164. CAS 9009-54-5; Polyether PU compound; nonexpanding compound for making low-cost moisture blocks and pressure dams in paper, pulp, and plastic insulated telecommunications cable (direct sheath injection method); service proven in all climates and environments; approved standard of major telephone companies. [Hexcel]

Resin 731D. Modified dehydrogenated (disproportionated) rosin; pale, oxidation-resistant, thermoplastic resin used in hot-melt-applied adhesives and coatings for paper and paperboard substrates, as tackifier and processing aid for rubber-based adhesives and molding compounds; for use in contact with food. [Hercules]

Resin 885, 3072. Mixed emulsifiers containing both disproportionated fatty acid and rosin acid; used in emulsion polymerization of elastomers calling for mixed emulsifier system. [Hercules]

Resin EX. Mixed glycol ester of rosin plus emulsifiers; self-emulsifying liquid rosin ester; plasticizer and tackifier in water-based adhesives. [Blueminster Ltd]

Resin H. Hydrogenated rosin. [Barron Chemicals Ltd] ‡

Resin M.S.2. Cyclohexanone condensation products. [Laporte Industries Ltd] *

Resin NC-11. Acidic resin derived from rosin; pale, noncryst., oxidation-resistant, thermoplastic resin used in formulating adhesive masses for surgical and industrial tapes, as resin component of various hot melt coating and adhesive compositions; for use in food contact. [Hercules]

Resin WP. A proprietary trade name for a melamine based thermosetting resin used for crease-resisting finishes. [Ciba plc] *

Resinall 153. Zincated modified rosin; high melting point resin with excellent sol. in low KB aliphatic solvs.; used in adhesives and rubber compounding; also as modifier for phenolic resins, printing inks and gloss oil. [Resinall Corp.]

Resinase. A liquid lipase containing preparation used to eliminate pitch/resin related problems in the paper industry. [Novo Nordisk]

resin blende. Zinc blende, ZnS, of a yellow color, is sometimes called by this name.

---

‡ Trade name and manufacturer not verified   § Trade name without identified manufacturer

resin essence. *See* rosin spirit.

Resin Ether L. A proprietary synthetic resin for use as a cellulose-lacquer plasticizer; nondrying, not susceptible to atmospheric oxidation, and has a low acid value. §

Resinette. A synthetic resin obtained from phenol and formaldehyde. §

Resinite. *See* Bakelite. §

resin lutea. (acaroid balsam). A name applied to yellow acaroid balsam, a yellow resin obtained from *Xanthorrhoea Hastile*.

Resinoid 1324. Phenolic resin, glass-reinforced; thermoset for injection and transfer molding; excellent creep resist. and retention of properties on prolonged exposure to elevated temps.; used for motor commutators, slip rings, automotive transmission parts, solenoid covers, circuit breaker housings, carburetor spacers. [Resinoid Engineering]

Resinoid 2002-4. Phenolic resin, fabric-reinforced; thermoset for transfer and compression molding; features good strength with excellent mech. shock resist.; used for automotive timing gears, aircraft and automotive pulleys, butterfly valve housings, explosion-proof flashlights, elec. connectors, conduit bushings; bulk. [Resinoid Engineering]

resin oil. That fraction of the distillation of resin (colophony), which distills over from 300-400 C. It consists principally of terpineol, $C_{10}H_7OH$; used as a lubricant.

resinol. A varnish substitute obtained by the dehydrogenation of petroleum, distillation, and Polymerization.

resinous silica. A variety of hydrated silica or opal.

Resinox. A proprietary synthetic resin molding powder of the phenol- formaldehyde type. §

Resin Release N. Mold release agent for fiberglass reinforced hand layup molding or casting. [Specialty Products Co] *

resins, acrolein. Resins obtained by the polymerization of acrolein by means of inorganic and organic bases or salts of iron and lead. Orea is a trade name for a resin of this type. Acrolein also condenses with phenols to form resins. *

resin spirit. *See* rosin spirit.

Resiosol. *See* Anusol. §

Resipol. Unsaturated polyester; used for paints, lacquers and varnishes, reinforced plastics. [Resinas Sinteticas SA] *

Resipol DL. Glyceryl dilactate; a proprietary trade name for a plasticizer. §

Resipol ML. Glyceryl monolactate; a proprietary trade name for a plasticizer. §

Resiren®. Sublimable disperse dyestuffs; used for heat transfer printing preferably on PES and other synthetics. [Bayer AG]

Resisco. A proprietary trade name for an alloy of 91% copper, 7% aluminum, and 2% nickel. §

Resista. A glass similar in composition to Pyrex glass. It contains 70% silica and 13.5% boric oxide.

Resistac. A proprietary trade name for an alloy of copper with 9% aluminum and 1% iron. §

Resistal. A heat-resisting alloy containing 63.5% iron, 16.6% nickel, 15% chromium, 4.5% silicon, and 0.3% carbon.

resistance bronze. A term for an alloy of from 84-86.5% copper, 11.5-13.5% manganese, and 2% iron.

Resista steel. An alloy of iron, nickel, and manganese, which is ductile at low temperatures.

Resistherm®. Raw materials used for the formulation of wire enamels with high thermal stability. [Bayer AG]

Resistin. An electrical resistance alloy containing 84-86% copper, 2% iron, and 11-13% manganese.

Resistoflex. A proprietary trade name for polyvinyl alcohol synthetic resins. §

Resistolac. Heat resistant cigarette carton lacquers. [The Scottish Adhesives Co Ltd]

Resistone. Cationic surfactant, biocide. [Rhone-Poulenc UK]

Resistone QD. Alkylaryl quaternary ammonium salt in pale yellow aqueous solution; bactericide and algicide, useful in alkaline media; used in static suppression. [Rhone-Poulenc UK]

Resistopen. Penicillin combination especially active against problem organisms. [Bayer AG] †

Resistox. An aldehyde-amine condensation product; a proprietary antioxidant for rubber. §

Resithren. Combinations of disperse and vat dyes; used for the one-bath dyeing of polyester/cellulosic blends. [Bayer AG]

Resitone QD. Alkylaryl quaternary ammonium salt in pale yellow aqueous solution; bactericide and algicide, useful in alkaline media; used in static suppression. [Rhone-Poulenc UK] *

Reslin. CAS 28434-01-7; Synergized bioresmethrin. [The Wellcome Foundation Ltd] *

Reslin S. Synergized pyrethroid/S-bioallethrin. [The Wellcome Foundation Ltd] *

Resmax. A proprietary preparation of hydroxyzine hydrochloride, ephedrine sulfate, and theophylline; an antiasthmatic. [Pfizer International] †

Resochin®. Chloroquine sulfate; for the treatment of chronic arthritis, certain skin diseases and malaria. [Bayer AG]

Resocoton. Mixtures of disperse and reactive dyestuffs; used for printing polyester/cotton blends. [Bayer AG]

Resogen® 35 Conc. Amino-aldehyde condensate; aftertreatment fixing agent for improving wetfastness of direct dyes on cellulose; good penetration, excellent lightfastness; low formaldehyde (< 1.5%). [Crompton & Knowles]

Resoglaz. A proprietary trade name for a polymerized styrene. §

Resolamin. Dyestuffs; for the one-bath dyeing of wool and polyester blends. [Bayer AG]

Resolin®. Disperse dyestuffs; used for polyester fibers. [Bayer AG; Bayer plc]

Resolin® P. A range of disperse dyestuffs; used for the dyeing of polyamide fibers. [Bayer AG]

Resoltex. Phenolic and melaminic resins; used for bonding of wood, foundry sands, and abrasives. [RWE-DEA Chemicals UK Ltd]

Resonium-A. A proprietary preparation of sodium polystyrene sulfonate; used in hyperkalaemia. §

Resopol. A proprietary polyester laminating resin. [DSM Resines France] *

Resorband. Phenolic resin adhesives. [Georgia-Pacific]

Resorbin®. A registered trademark currently awaiting reallocation by its proprietors to cover a range of pharmaceuticals. [Cassella AG] *

resorcin. (resorcinol; m-dihydroxybenzene; 3-hydroxyphenol) $C_6H_4(OH)_2$; CAS 108-46-3; used in the manufacture of resorcinaol-formaldehyde resins, dyes, cos-

---

* Trade name not verified as available   † Trade name verified as obsolete

metics, pharmaceuticals, as a crosslinking agent in the manufacture of neoprene.

resorcin. See resorcinol.

resorcinal. A mixture of equal parts of resorcinol and iodoform; used as an antiseptic dusting-powder.

resorcinol. (1,3-benzenediol; m-dihydroxybenzene; resorcin) Phenol; $C_6H_4(OH)_2$; CAS 108-46-3; EINECS 203-585-2; topical antiseptic; keratolytic agent; in tanning; in manufacture of resins, adhesives, hexylresorcinol, p-aminosalicylic acid, explosives, dyes; in cosmetics; dyeing and printing textiles. [Fairmount; Indspec; Janssen Chimica; Penta Mfg.; Whitecourt Ltd]

resorcinolphthalein. See fluorescein.

Resorufin. Hydroxyphenazone, $C_{12}H_7NO_3$.

Resotren. An amebacide. [Bayer AG] †

Resovin. A proprietary synthetic resin of the vinyl type. §

Resovyl. Textile auxiliary chemicals. [ICI Chem. & Polymers Ltd] †

Respenyl. CAS 93-14-1; A proprietary preparation containing guaiafenesin (guaiacol glyceryl ether); a cough suppressant. [Crookes Laboratories] *

Respiral. Cough drops. [Richardson-Vicks Inc] *

Respirot. Respiratory stimulant veterinary ethical. [Ciba plc]*

Responsar. Insecticide for control of all insect pests, e.g., cockroaches, beetles, flies, mosquitoes. [Bayer AG]

Respumit®. Antifoam; dyeing and printing auxilliary. [Bayer AG; Bayer plc]

Restor-E (Restoration Chemical Products). Commercial line of products for professional and do-it-yourself restoration of a variety of surfaces (wood, floors, soft goods, odor control, corrosion control) following fire, floods, etc. [Puma Chemical Co Inc] *

Restoration Cleaner. Blended organic and inorganic acids in combination with surfactants and wetting agents; for heavier duty concentrated cleaner for building exteriors. [Nova Chemical Inc] *

Restoration Cleaner (Heavy Duty). Blended organic and inorganic acids in combination with surfactants and wetting agents; used for heavier duty concentrated cleaner for building exteriors. [Nova Chemical Inc] *

Restoration Cleaner (Super Heavy Duty). Blended organic and inorganic acids in combination with surfactants and wetting agents; for heavier duty concentrated cleaner for building exteriors. [Nova Chemical Inc] *

Restoration Rinse. Blended organic and inorganic acids in combination with surfactants and wetting agents; for light duty cleaning for historical building exteriors. [Nova Chemical Inc] *

Restore-X Exterior Paint Remover. Green color, heavy-bodied liquid, nonflammable, water soluble, sodium hydroxide remover; surface preparation tool for removal of deteriorated, exterior paints and heavy-bodied stains. [Restech Industries Inc] *

Restore-X Weathered Wood Renewer. Blue color, heavy-bodied liquid, nonflammable, water soluble, sodium hydroxide remover; removes semi-transparent stain and the grey, weathered look from wood. [Restech Industries Inc] *

Resydrol. Water soluble synthetic resins; used for paints and printing inks. [Resinous Chemicals Ltd]

Resyn® 28-1310. CAS 25609-89-6; Vinyl acetate/crotonic acid copolymer; hair fixative; for hair sprays, setting lotions, conditioners. [Nat'l. Starch]

Resyn® 28-2913. CAS 55353-21-4; Vinyl acetate/crotonic acid/vinyl neodecanoate copolymer. hair fixative providing excellent holding power, manageability, gloss; for aerosol and pump hairsprays, setting lotions, spritzes, cosmetics. [Nat'l. Starch]

Retain PE-1001. CAS 9002-88-4; Recycle content polyethylene; for injection molding applications. [Dow Plastics]

Retain PE-5009. CAS 9002-88-4; Recycle content polyethylene; for overwrap film applications and blown film extrusion. [Dow Plastics]

Retain PS-4000. CAS 9003-53-6; Recycled content polystyrene resin; rubber modified; for injection molding applications. [Dow Plastics]

Retain RP-120. CAS 9002-88-4; Pigmented recycle content polyethylene; for film, injection molding and blow molding applications. [Dow Plastics]

Retaminol®. Paper auxiliaries; used to increase the filler and pigment yields in paper manufacture, to improve drainage speed and for backwater clarification. [Bayer AG]

Retard. Potassium salt of maleic hydrazide; growth retardant for trees, shrubs, ivy and grass. [Draxel Chemical Company] ‡

Retarder AK. CAS 85-44-9; Modified phthalic anhydride; nondiscoloring retarding agent to reduce scorching of rubber compounds at processing temps.; also acts as an activator for certain blowing agents. [Akrochem]

Retarder BA, BAX. CAS 65-85-0; EINECS 200-618-2; Predominantly benzoic acid; retarding agent for natural and synthetic rubbers and latexes; nonstaining; acts as an activator for certain blowing agents; processing aid with certain cis-polybutadiene rubbers; BAX is oil-treated. [Akrochem]

Retarder ESEN. CAS 85-44-9; Surface treated phthalic anhydride; a nondiscoloring retarder of vulcanization in all stock with all accelerators at processing temperature with a minimum retarding action at curing tmperatures. [Uniroyal] *

Retarder N. CAS 139-07-1; EINECS 203-351-5; Dimethyl lauryl benzyl ammonium chloride; retarder in dyeing of acrylics and polyesters; antistatic agent. [Hart Prods. Corp.]

Retarder OC. Quaternary base; retarding agent for dyeing of acrylic fibers; substantive. [Eastern Color & Chem.]

Retarder PX. CAS 85-44-9; Phthalic anhydride, oil treated; nondiscoloring retarding agent to reduce scorching of rubber compounds at processing temps. [Akrochem]

Retarder SAFE. Treated aromatic sulfonamide; nonstaining, nondiscoloring retarder for natural and synthetic rubber compounds; also useful for replasticizing of slightly scorched stocks by cold mill mixing. [Akrochem]

Retarder SAX. CAS 69-72-7; Tech. salicylic acid (90%) and light process oil treatment (10%); retarder; vulcanization inhibitor for SBR and natural rubber compounds; also as accelerator for W types of Neoprene; blowing agent activator in sponge rubber compounds [Akrochem]

Retarder V-48. Textile retarder; promotes leveling on fast striking cationic dyes. [Sybron]

Retardine. Retarder/leveling agent for dyeing cellulosics with vat, direct and sulfur dyes. [Henkel/Textiles]

Retardit A. Org. condensate; retarder and leveling agent offering good performance with improved economy. [Eastern Color & Chem.]

Retardol. Flame retardants. [Albright & Wilson Ltd]

Retargal. Leveling agent; for dyeing acrylic fibers with basic

---

‡ Trade name and manufacturer not verified   § Trade name without identified manufacturer

dyes. [Sandoz Products Ltd] *

Retariox. A proprietary trade name for an aldehyde-amine condensation product. §

Retcin. CAS 114-07-8; A proprietary preparation of erythromycin; an antibiotic. [DDSA Pharmaceuticals Ltd] *

Reten®. High molecular weight synthetic water-soluble polymers; used as thickeners, flocculants, antistatic agents, film formers, adhesives, slip agents, solids-suspending agents and crosslinking agents. [Hercules]

Reten® 157. Acrylamide-based copolymer; retention aid for paper industry; provides max. retention of fiber fines, paper chems., and fillers. [Hercules]

Reten® 763. Polymer solution; retention aid for fiber, fines, and fillers and provides drainage and flocculation of wh.-water solids in paper machine and flotation save-all operations; functions over a wide pH range in alum and alum-free systems; used as dry-creping aid, in kraft liquor clarification, and process water treatment; for use in manufacturing of paper/paperboard for contact with food (up to 0.2% resin). [Hercules]

Reten® 1232. Acrylamide-based copolymer; retention aid and flocculant. [Hercules]

retene. (methylisopropylphenanthrene) CAS 483-65-8; $C_{18}H_{18}$.

Retenema. CAS 2152-44-5; A proprietary enema containing betamethasone valerate; used in the treatment of proctitis. [Glaxo Laboratories] *

Reticusol. Hydrolyzed reticulin; moisturizer, conditioner for skin care products. [Croda Inc.]

Retilox. A proprietary range of organic peroxides suitable for polymer crosslinking. [Montedison UK Ltd] *

Retilox® F 40 MG. CAS 25155-25-3; α,α'-Bis(t-butylperoxy)-m/p-diisopropylbenzene/EPM masterbatch; curing and crosslinking agents for EPM, EPDM, polyethylene, silicone rubbers, NBR, EVA copolymers, SBR, chlorosulfonated polyethylene, PVC, PU rubbers, polybutadiene rubbers, neoprene, natural rubber, chlorinated polyethylene; for manufacturing of rubber articles, membranes for waterproofing, elec. cable insulation, conveyor belts, shoe soles. [Akrochem]

Retin-A. Tretinoin; keratolytic. [Ortho Pharmaceutical Corp]

retin asphalt. See retinite.

Retingan® R6, R7, R4B. Resin tanning materials; used for filling aftertreatment of chrome upper leather, particularly from cattle hides and sheepskins. [Bayer AG; Bayer plc]

retinite. (retin asphalt, walchowite). A fossil resin found in brown coal. It occurs in Derbyshire and in Walchow. The material found near Walchow is a polymeric resin made up chiefly of sesquiterpenes.

retinol. (codoil, rosinol, rosin oil). CAS 68-26-8; a) A product obtained by the distillation of rosin; b) vitamin A alcohol; $C_{20}H_{29}OH$; used as an antiseptic and a vehicle for ointments.

retinol. See vitamin A.

Retnolite. A proprietary trade name for a phenol-formaldehyde synthetic resin. §

Retrocure® G. CAS 136-85-6; Blend of 4- and 5-methylbenzotriazole (tolyltriazole); prevulcanization retarder for sulfur modified (G type) polychloroprene rubbers; also for NBR systems where MBTS/sulfur cure systems are used. [Akrochem]

Retrovir. CAS 30516-87-1; Proprietary formulations of zidovudine; for the management of patients with human immunodeficiency virus infection. [The Wellcome Foundation Ltd]

Retz alloy. An alloy of 75% copper, 10% lead, 10% tin, and 5% antimony.

reuniol. See roseol.

reussinite. A reddish-brown resin found in certain coal deposits.

Re-Vac II. Reovirus disease, inactivated vaccine; for immunization of poultry. [Intervet Inc]

Revacryl®. Acrylic and styrene acrylic copolymer dispersions; used for adhesives, surface coatings and textiles. [Harlow Chemical Co Ltd]

Revalon. A proprietary trade name for an alloy of 76% copper, 22% zinc, and 2% aluminum. §

Revatol. Reserving agent; for printing. [Sandoz Products Ltd] *

Revatol S. A proprietary trade name for sodium m-nitrobenzenesulfonate. §

Revertex®. Range of evaporated natural rubber latex concentrates; binders for reconstituted leatherboard, modifiers for asphalt, bitumen, etc. [Revertex Ltd]

Revlen. Esterifilcon A; butyl methacrylate polymer with butyl acrylate and ethylene dimethacrylate; contact lens material. [BioContacts Inc] ‡

Revolex. Antistatic formulations. [Rhone-Poulenc UK]

Revolite. A proprietary phenol-formaldehyde synthetic resin impregnated cloth. §

Revona. A proprietary trade name for a water-soluble aminoplast; a very effective pitch dispersant in paper making. [Ciba plc] *

Revultex®. Proprietary range of prevulcanized natural rubber latex concentrates; used for dipped goods, e.g., medical, balloons, etc. [Revertex Ltd]

Rewagit. CAS 1344-28-1; Crystalline aluminum oxide as blasting corundum; especially suited for descaling, derusting, roughening of work piece surfaces and blasting of austenitic steels. [Hüls UK Ltd]

Rewo-amid. Deodorant additive, perfume extender additive, versatile additive; foam booster - pearlizing agent, thickening agent, superfatting agent; for synthetic detergents, soaps, shampoos, bubble baths, cosmetics, toiletries. [Rewo Chemicals Ltd] *

Rewocid®. Undecylenic acid derivatives; surfactants. [Rewo Chemicals Ltd]

Rewocid® DU 185. Undecylenamide DEA and diethanolamine; detergent, emulsifier used as bacteriocide, thickener, foam stabilizer in shampoos. [Rewo GmbH]

Rewocid® SBU 185 P. Disodium undecylenamido MEA-sulfosuccinate; antidandruff shampoo surfactant, fungicide. [Rewo GmbH]

Rewocid® U 185. Undecylenamide MEA; fungicide, antimycotic agent; detergent; improves foam quality, stability, superfatting, increases viscosity. [Rewo GmbH]

Rewocid® UTM 185. Undecylenamidopropyltrimonium methosulfate; bactericide, fungicide for toiletries; conditioner for shampoos, antistat. [Rewo GmbH]

Rewocor. Anti-corrosive additive; for soluble cutting oils, etc. [Rewo Chemicals Ltd] *

Rewocoros. Corrosion inhibitors. [Rewo Chemicals Ltd]

Rewocoros B 2045. Alkenyl sulfosuccinic acid anhydride; corrosion inhibitor, tar adhesive agent. [Rewo GmbH]

Rewocoros B 3010. Alkenyl succinic acid, disodium salt; rust preventive additive in aq. media. [Rewo GmbH]

Rewocoros B 3032. Alkenyl succinic acid, TEA salt; rust preventive additive in aq. media. [Rewo GmbH]

---

* Trade name not verified as available   † Trade name verified as obsolete

Rewocoros BAC. Butyl ammonium caprylate; corrosion inhibitor. [Rewo GmbH]

Rewocoros RA 280. Oleic acid dibutylamide; corrosion inhibitor. [Rewo GmbH]

Rewocoros RAB 90. Modified boric DEA; low foaming corrosion inhibitor for sol. aq. metalworking oils, synthetic cold lubricants, water glycol hydraulic fluids, grinding lubricants. [Rewo GmbH]

Rewocoros TPAC 100. Tallow propylene diammonium caprylate; corrosion inhibitor for oils and boiler feed water systems. [Rewo GmbH]

Rewoderm® ES 90. CAS 68201-46-7; PEG-7 glyceryl cocoate; emulsifier for cosmetics, superfatting agent, solubilizer. [Rewo GmbH]

Rewoderm® LI 48. PEG-80 glyceryl tallowate; mild surfactant, thickener and superfatting agent for cosmetics. [Rewo GmbH]

Rewoderm® S 1333. Disodium ricinoleamido MEA-sulfosuccinate; detergent; used for skin protection, washing up liqs., personal care products; decreases irritancy of alkylbenzene sulfonate and other surfactants; emulsifier for emulsion polymerization. [Rewo GmbH]

Rewoderm® SPS. Disodium sitosteareth-14 sulfosuccinate. surfactant for shampoos, shower and foam baths, mild skin cleaners, baby baths, skin care products. [Rewo GmbH]

Rewolan®. Lanolin derivatives; surfactants. [Rewo Chemicals Ltd]

Rewolan® 5. CAS 68890-92-6; Disodium laneth-5 sulfosuccinate; detergent, moisturizer, superfatting agent for personal care products; skin protective agent. [Rewo GmbH]

Rewolan® AWS. PEG-75 lanolin oil; superfatting agent for personal care products. [Rewo GmbH]

Rewolan® LP. Isopropyl lanolate and lanolin; emollient for skin care, toiletries. [Rewo GmbH]

Rewolub KSM 80. Dicarboxylic acid diamide modified; surfactant for synthetic cooling oils, metalworking lubricants. [Rewo GmbH]

Rewolub TMP 275. Trimethylolpropane oleic acid ester; lubricant in metalworking fluids and textile auxs. [Rewo GmbH]

Rewomid®. Fatty acid alkylolamides. [Rewo Chemicals Ltd]

Rewomid® C 212. CAS 68140-00-1; Cocamide MEA; detergent, thickener, foam booster/stabilizer, superfatting agent used in detergent products; stabilizer of emulsions. [Rewo GmbH]

Rewomid® DL 203 S. Lauramide DEA and diethanolamine; foam booster/stabilizer, superfatting agent, and thickener for personal care products, floor cleaners, general purpose cleaners, textile lubricants. [Rewo GmbH]

Rewomid® DO 280 SE. CAS 93-83-4; EINECS 202-281-7; Oleamide DEA (1:1); detergent, thickener, foam booster/stabilizer, superfatting agent for cosmetic products; conditioner for shampoos, bath oils. [Rewo GmbH]

Rewomid® DO 280. CAS 93-83-4; Oleamide DEA and diethanolamine; detergent; products in the cosmetic, cleaning, and detergent industries. [Rewo GmbH]

Rewomid® F. Linoleamide DEA and diethanolamine; foam booster, superfatting agent and thickener for shampoos. [Rewo GmbH]

Rewomid® IPE 280. CAS 111-05-7; EINECS 203-828-2; Oleamide MIPA; detergent, emulsifier, foam stabilizer, thickener, superfatting agent; additive for skin protecting products. [Rewo GmbH]

Rewomid® IPL 203. CAS 142-54-1; EINECS 205-541-8; Lauramide MIPA; foam stabilizer, detergent for shampoos, shaving foams, and dishwashing liqs.; additive for solid and paste end products; improved washing power; stabilizer of emulsions. [Rewo GmbH]

Rewomid® IPP 240. Cocamide MIPA; detergent, foam stabilizer, thickener, additive for solid and paste end products; improved washing power; emulsion stabilizer. [Rewo GmbH]

Rewomid® L 203. CAS 142-78-9; EINECS 205-560-1; Lauramide MEA; detergent, thickener, foam booster/stabilizer, superfatting agent for detergent preparations; fixation of perfumes; stabilizer of emulsions. [Rewo GmbH]

Rewomid® R 280. CAS 106-16-1; EINECS 203-368-2; Ricinoleamide MEA; surfactant for synthetic soap bars; foam stabilizer, thickener. [Rewo GmbH]

Rewomid® S 280. CAS 111-57-9; EINECS 203-883-2; Stearamide MEA; detergent, thickener, foam booster, superfatting agent; stabilizer of emulsions; synthetic soap bars; anti-inflammatory agent. [Rewo GmbH]

Rewomin. Amine oxides; surfactants. [Rewo Chemicals Ltd]

Rewominox. Amine oxides; surfactants. [Rewo Chemicals Ltd]

Rewominox B 204. CAS 68155-09-9; Cocamidopropylamine oxide; foam booster, thickener for personal care products; antistat; hair conditioner; skin compatible. [Rewo GmbH]

Rewominox L 408. Lauramine oxide; foam booster, thickener for personal care products; hair conditioner, antistat. [Rewo GmbH]

Rewominox S 300. CAS 2571-88-2; EINECS 219-919-5; Stearamine oxide; foam booster, antistat, hair conditioning agent. [Rewo GmbH]

Rewominoxid. Foam booster; conditioner; fabric softener; for shampoos and foam baths. [Rewo Chemicals Ltd] *

Rewominoxid L 408. Lauramine oxide; foam booster for personal care products. [Rewo GmbH] *

Rewominoxid S 300. CAS 2571-88-2; EINECS 219-919-5; Stearamine oxide; foam booster, antistat, conditioner. [Rewo GmbH] *

Rewomul MG SE. CAS 977053-96-5; Glyceryl stearate SE; emulsifier for cosmetics. [Rewo GmbH]

Rewopal® BN 13. PEG-13 naphthole; wetting agent for electroplating baths. [Rewo GmbH]

Rewopal® C 6. CAS 61791-08-0; PEG-6 cocamide; dispersant, emulsifier, foam booster, wetting agent for calcium soap, personal care products; solubilizer for perfume oils. [Rewo GmbH]

Rewopal® HV 4. CAS 9016-45-9; Nonoxynol-4; emulsifier for mineral oils, petroleum, aliphatic hydrocarbons. [Rewo GmbH]

Rewopal® LA 3. Laureth-3; emulsifier for solvs. (cold water detergents), cosmetic oils (bath oils), mineral oils (textile auxiliary, metalworking), coupler, raw material for the production of ether sulfates; shampoos. [Rewo GmbH]

Rewopal® MPG 10. CAS 122-99-6; EINECS 204-589-7; Phenoxyethanol; solv., solubilizer for preservatives. [Rewo GmbH]

Rewopal® MT 65. Fatty alcohol/PEG methyl ether; low foaming detergent for strong acidic cleaners, textile auxiliaries; acid-stable. [Rewo GmbH]

---

‡ Trade name and manufacturer not verified   § Trade name without identified manufacturer

Rewopal® O 8. PEG-9 oleamide; detergent; wetting agent, o/w emulsifier, and dispersant for calcium soap; suitable for machine washing formulations. [Rewo GmbH]

Rewopal® PEG 6000 DS. CAS 9005-08-7; PEG-150 distearate; thickener for toiletries. [Rewo GmbH]

Rewopal® PG 280. CAS 627-83-8; EINECS 211-014-3; Glycol distearate; pearlizing agent for cosmetics. [Rewo GmbH]

Rewopal® RO 40. Castor oil ethoxylate; emulsifier for metalworking fluids, textile auxiliaries, agricultural chemicals. [Rewo GmbH]

Rewopal® TA 11. CAS 61791-28-4; Talloweth-11; detergent, wash-act. base, wetting agent, dispersant, emulsifier for waxes; biodeg. [Rewo GmbH]

Rewophat. Phosphate esters; surfactants. [Rewo Chemicals Ltd]

Rewophat EAK 8190. CAS 39464-66-9; Laureth-3 phosphate; corrosion inhibitor, emulsifier, dispersant, wetting agent; for metalworking fluids, textile auxiliaries and cleaners; antistat; high pressure additive; biodeg. [Rewo GmbH]

Rewophat NP 90. Nonylphenol polyglycol ether phosphate; emulsifier for emulsion polymerization; textile auxiliaries; antistat, raw material for industrial cleaners. [Rewo GmbH]

Rewophos EAK 8190. Phosphate ester alkyl polyglycol ether in liquid form; corrosion inhibitor, emulsifier, antistat for metal treatment agents. [Rewo Chemicals Ltd] *

Rewophos TD40. Phosphate ester alkyl polyglycol ether in liquid form; low foaming surfactant for metal treatment and antistatic applications. [Rewo Chemicals Ltd] *

Rewophos TD70 and OP80. Phosphate ester alkykl polyglycol ether in liquid form; surfactant/ hydrotrope used in metal treatment, antistatic applications and textile auxillary applications. [Rewo Chemicals Ltd] *

Rewopol. Anionic surfactants. [Rewo Chemicals Ltd]

Rewopol® B 1003. CAS 90268-48-7; EINECS 290-850-0; Disodium tallow sulfosuccinamate; foaming and antigelling agent for latex foam backings and coatings; emulsifier for emulsion polymerization; flotation agent. [Rewo GmbH]

Rewopol® B 2003. Tetrasodium dicarboxyethyl stearyl sulfosuccinamate; flotation reagent; emulsifier for emulsion polymerization; foaming agent for latex emulsion (carpet backing); antigelling agent, cleaning agent for paper mill felts. [Rewo GmbH]

Rewopol® CHT 12. Coco-EDTA-amide; sequestering agent, complexing surfactant for detergents. [Rewo GmbH]

Rewopol® CL 30. Laureth-3 carboxylic acid; surfactant for household cleaners and toiletries. [Rewo GmbH]

Rewopol® CLN 100. CAS 33939-64-9; Sodium laureth-11 carboxylate; surfactant for toiletries. [Rewo GmbH]

Rewopol® CT 65. Trideceth-7 carboxylic acid; acid-stable cleaner for household and industrial use, textile auxiliaries, mineral oil emulsions, tert. oil recovery; emulsifier, wetting agent for personal care products. [Rewo GmbH]

Rewopol® DLS. DEA-lauryl sulfate in liquid form; raw material for shampoos, detergents, etc. [Rewo Chemicals Ltd]

Rewopol® MLS 30. MEA-lauryl sulfate; surfactant for personal care products, foam baths, shampoos, liquid detergents; detergent raw material. [Rewo GmbH]

Rewopol® NEHS 40. Sodium octyl sulfate; low foaming wetting agent for alkaline cleaners, mercerizing, electroplating; hydrotrope. [Rewo GmbH]

Rewopol® NL 2-28. CAS 9004-82-4; Sodium laureth sulfate; surfactant for shampoos, shower gels, foam baths, liquid soaps, dishwashing liqs., emulsion polymerization, air entrainment agent, textile auxiliaries. [Rewo GmbH]

Rewopol® NLS 15 L. CAS 151-21-3; Sodium lauryl sulfate; emulsifier for emulsion polymerization. [Rewo GmbH]

Rewopol® NOS 5. Nonyl phenol polyglycol ether sulfate; emulsion polymerization surfactant especially for styrene. [Rewo GmbH]

Rewopol® SBC 212. Disodium cocamido MEA-sulfosuccinate; surfactant for foam cleaners, light duty detergents, and personal care products. [Rewo GmbH]

Rewopol® SBDB 45. CAS 127-39-9; Diisobutyl sodium sulfosuccinate; emulsifier for emulsion polymerization; stabilizer for disps.; pigment dispersant. [Rewo GmbH]

Rewopol® SBDC 40. CAS 23386-52-9; EINECS 245-629-3; Dicyclohexyl sodium sulfosuccinate; emulsifier for emulsion polymerization; stabilizer for disps.; pigment dispersant. [Rewo GmbH]

Rewopol® SBDD 65. Diisodecyl sodium sulfosuccinate; emulsifier for emulsion polymierzation; stabilizer for disps.; pigment dispersant. [Rewo GmbH]

Rewopol® SBDO 75. CAS 577-11-7; EINECS 209-406-4; Dioctyl sodium sulfosuccinate; wetting agent, solubilizer; emulsion polymerization. [Rewo GmbH]

Rewopol® SBF 12. Disodium lauryl sulfosuccinate; detergent raw material for personal care products; carpet and upholstery shampoos; toilet and syndet soaps. [Rewo GmbH]

Rewopol® SBFA 30. Disodium laureth sulfosuccinate; detergent raw material for personal care products, cleansing agents. [Rewo GmbH]

Rewopol® SBL 203. CAS 25882-44-4; EINECS 247-310-4; Disodium lauramido MEA-sulfosuccinate; detergent for cosmetics; aids spray-drying; carpet and upholstery shampoos; soaps. [Rewo GmbH]

Rewopol® SBMB 80. Diisohexyl sulfosuccinate; emulsion polymerization surfactant; stabilizer for disps.; pigment dispersant. [Rewo GmbH]

Rewopol® SBZ. Disodium PEG-4 cocamido MIPA-sulfosuccinate; surfactant for mild foam baths, shampoos, light duty detergents. [Rewo GmbH]

Rewopol® SK 275. Oleyl sarcosinic acid; emulsifier and corrosion inhibitor for mineral oils. [Rewo GmbH]

Rewopol® SLS. Sodium lauryl sulfate in paste form; raw material for shampoos, detergents, etc. [Rewo Chemicals Ltd]

Rewopol® SMS. Alkyl disodium sulfosuccinamate as a clear liquid; spreading and penetrating agent for latex emulsions. [Rewo Chemicals Ltd]

Rewopol® TLS 40. CAS 139-96-8; EINECS 205-388-7; TEA-lauryl sulfate; surfactant raw material for shampoos, foam baths, liquid detergents. [Rewo GmbH]

Rewopol® TLS 90 L. CAS 661-61-6; TIPA-lauryl sulfate; raw material for high active ingred. cosmetic products, foaming bath oils. [Rewo GmbH]

Rewopol® TMS and ODS. Alkyl disodium sulfosuccinamate in liquid or paste form; foaming agent for latex emulsions, emulsifier for emulsion polymerization. [Rewo Chemicals Ltd]

Rewopon® . Imidazolines/amines; surfactants. [Rewo Chemicals Ltd]

---

* Trade name not verified as available   † Trade name verified as obsolete

Rewopon® IM-BT. Fatty acid quaternary imidazoline in liquid/paste form; cationic surfactant which improves adhesion of bitumens and other binding and coating agents. [Rewo Chemicals Ltd]

Rewopon® IM OA. 1-Hydroxyethyl-2-alkyl-imidazoline; corrosion inhibitor, emulsifier, antistat. [Rewo GmbH]

Rewoquat B 10. CAS 7173-51-5; Didecyldimonium chloride; disinfectant for cleaners, e.g., for dairies and food industry. [Rewo GmbH]

Rewoquat B 50. Benzalkonium chloride; disinfectant for cleaners, dairy and food industries; algicide; textile dyeing auxiliary [Rewo GmbH]

Rewoquat CPEM. CAS 68989-03-7; PEG-5 cocomonium methosulfate; hair conditioner for shampoos, emulsifier in emulsion polymerization, antistat. [Rewo GmbH]

Rewoquat CR 3099. Difatty acid ester dimethyl ammonium methosulfate as a viscous liquid; softener, dry-cleaning agent, textile auxiliary, leather auxiliary. [Rewo Chemicals Ltd]

Rewoquat DQ 35. CAS 93572-63-5; EINECS 297-495-0; PEG-3 tallow propylenedimonium dimethosulfate; antistat and wetting agent. [Rewo GmbH]

Rewoquat RTM 50. Ricinoleamidopropyltrimonium methosulfate; conditioner for personal care products, liquid soaps; antistat. [Rewo GmbH]

Rewoquat W 75 H. CAS 91723-55-6; Quaternium-83; fabric softener. [Rewo GmbH]

Rewoquat W 75 PG. Quaternium-27, free of isopropanol; fabric softener. [Rewo GmbH]

Rewoquat W 222 LM. Ditallow amidoammonium methosulfate; fabric softener. [Rewo GmbH]

Rewoquat W 3690, W 3690 PG. Dioleyl imidazoline methosulfate; fabric softener. [Rewo GmbH]

Rewoquat W 7500. Quaternary imidazoline in liquid form; cationic surfactant used as a fabric softener. [Rewo Chemicals Ltd]

Rewoquaternary Cationic fabric softeners. [Rewo Chemicals Ltd]

Reworyl®. Alkylaryl sulfonates; surfactants. [Rewo Chemicals Ltd]

Reworyl® ACS. CAS 37475-88-0; EINECS 253-519-1; Ammonium cumene sulfonate; anionic surfactant in liquid form; hydrotrope. [Rewo Chemicals Ltd]

Reworyl® C. CAS 28631-63-2; Cumene sulfonic acid; anionic surfactant in liquid form; catalyst for synthetic resins; hydrotropes. [Rewo Chemicals Ltd]

Reworyl® K. Dodecylbenzene sulfonic acid; biodeg. raw material for anionic detergent systems. [Rewo GmbH]

Reworyl® KXS. Potassium xylene sulfonate; anionic surfactant in liquid form; hydrotrope. [Rewo Chemicals Ltd]

Reworyl® NCS. CAS 32073-22-6; Sodium cumene sulfonate; anionic surfactant in liquid form; hydrotrope. [Rewo Chemicals Ltd]

Reworyl® NKS 100. Sodium dodecylbenzene sulfonate; detergent and cleaner, textile auxiliaries. [Rewo GmbH]

Reworyl® NTS. Sodium toluene sulfonate; anionic surfactant in liquid form; hydrotrope. [Rewo Chemicals Ltd]

Reworyl® NXS 40. CAS 1300-72-7; EINECS 215-090-9; Sodium xylene sulfonate; hydrotrope for detergent systems. [Rewo GmbH]

Reworyl® T. p-Toluene sulfonic acid; anionic surfactant in liquid form; catalyst for foundry resins. [Rewo Chemicals Ltd]

Reworyl® TKS 90/L. TEA dodecylbenzene sulfonate;

biodeg. raw material for liquid detergents. [Rewo GmbH]

Reworyl® X. CAS 25321-41-9; EINECS 246-839-8; Xylene sulfonic acid; anionic surfactant in liquid form; catalyst for synthetic resins. [Rewo Chemicals Ltd]

Rewoteric® . Amphoteric baby shampoos; surfactants. [Rewo Chemicals Ltd]

Rewoteric® AM 2C NM. CAS 61791-32-0; Disodium cocoamphodiacetate; mild nonirritating raw material for personal care products. [Rewo GmbH]

Rewoteric® AM 2C SF. Disodium cocoamphodipropionate; high-foaming surfactant for baby and intimate hygiene products; excellent for high pH systems; coupler and detergent in high electrolyte formulations. [Rewo GmbH]

Rewoteric® AM 2L-40. CAS 14350-97-1; Disodium lauroamphodiacetate; mild nonirritating raw material for baby shampoos, mild shampoos, foam baths, intimate hygiene products. [Sherex/Div. of Witco]

Rewoteric® AM B-13. CAS 61789-40-0; Cocamidopropyl betaine; mild raw material, foam booster, visc. builder for personal care products (shampoos, skin cleansers), liquid soaps, all-purpose cleaners; lime soap dispersant; coemulsifier. [Sherex/Div. of Witco]

Rewoteric® AM B-14 LS. Cocamidopropyl betaine; low-salt surfactant for personal care products. [Sherex/Div. of Witco]

Rewoteric® AM B-15. Cocamidopropyl betaine; foam booster for shampoos, visc. builder, low-irritation skin cleanser, lime soap dispersant; aids in deposition of protein and cationic polymers on hair; coemulsifier. [Sherex/Div. of Witco]

Rewoteric® AM CA. Disodium lauroamphodiacetate and sodium laureth sulfate; raw material for mild foam baths, hair shampoos, baby preps. [Rewo GmbH]

Rewoteric® AM CAS. CAS 68139-30-0; Cocamidopropyl hydroxysultaine; high-foaming mild detergent for shampoos, bubble bath; coemulsifier. [Sherex/Div. of Witco]

Rewoteric® AM CAS-15. CAS 68139-30-0; Cocamidopropyl hydroxysultaine; surfactant used in personal care products and acid and alkaline cleaners. [Sherex/Div. of Witco]

Rewoteric® AM DM-35L. CAS 11140-78-6; Lauryl betaine; high-foaming mild surfactant for baby shampoos, hard surface cleaners, steam jet cleaners, and pickling baths. [Sherex/Div. of Witco]

Rewoteric® AM G30. Sodium lauroamphoacetate, sodium lauryl sulfate and hexylene glycol; surfactant for baby and child care cosmetic formulations. [Rewo GmbH]

Rewoteric® AM HC. CAS 13197-76-7; Lauryl hydroxysultaine; mild detergent foam booster, skin conditioning agent; hydrotrope; excellent for high pH systems. [Sherex/Div. of Witco]

Rewoteric® AM KSF 40. CAS 93820-52-1; Sodium cocoamphopropionate, salt-free; raw material, high-foaming surfactant, wetting agent used in personal care products (baby and intimate hygiene products) and industrial cleaners. [Rewo GmbH]

Rewoteric® AM LP. Sodium lauryliminodipropionate; high-foaming surfactant for baby and intimate hygiene products; excellent for high pH systems. [Sherex/Div. of Witco]

Rewoteric® AM R40. CAS 86089-12-5; Ricinoleamidopropyl betaine; mild cosmetic and medicated cleaning agents; baby and child care cosmetics. [Rewo GmbH]

Rewoteric® AM TEG. CAS 61791-25-1; Dihydroxyethyl

---

‡ Trade name and manufacturer not verified   § Trade name without identified manufacturer

tallow glycinate; high-foaming mild surfactant for acidic and alkaline cleaning products, conditioning shampoos; thickener. [Rewo GmbH]

Rewoteric® AM V. CAS 13039-35-5; EINECS 235-907-2; Sodium capryloamphoacetate; raw material, wetting agent, rust inhibitor for pickling baths, acid and alkaline cleaners. [Sherex/Div. of Witco]

Rewoteric® AM VSF. Sodium capryloamphopropionate, salt-free; wetting agent for alkaline and acidic cleaners. [Rewo GmbH]

Rewoteric® QAM 50. CAS 100085-64-1; EINECS 309-206-8; Cocobetainamido amphopropionate; surfactant for hard surface disinfectants, deodorants, cleaning and bacteriocidal products for use in industrial catering, hotels, hospitals, and households; biodeg. [Rewo GmbH]

Rewowax. Synthetic waxes. [Rewo Chemicals Ltd]

Rewowax CG. CAS 540-10-3; EINECS 208-736-6; Cetyl palmitate; spermaceti wax replacement; emulsifier for cosmetic creams and lotions. [Rewo GmbH]

Rex. A trademark for abrasive goods consisting essentially of alumina. §

Rex 95. A proprietary trade name for a cobalt steel containing 5% cobalt, 14% tungsten, and 4% chromium, 2% vanadium, and 0.5% molybdenum. §

Rex-blak. A special preparation of carbon black, containing carbon, rubber, and glue, in varying proportions; used in rubber mixings.

Rexene® 11S12. CAS 9003-07-0; PP homopolymer; super-clean grade for electronics, medical and food packaging, material handling equip. for cleanroom environments, tooling for healthcare devices. [Rexene Prods.]

Rexene® 13S10A. CAS 9003-07-0; PP copolymer; radiation-resist. improved clarity grade for injection molded parts in medical market, food packaging, see-through reusable household storage containers, electronics, consumer products [Rexene Prods.]

Rexene® 14S4A. CAS 9003-07-0; PP med.-impact copolymer; easy processing for injection molding. [Rexene Prods.]

Rexene® 18C3A. CAS 9003-07-0; PP impact copolymer; very high impact grade for underhood automotive applications, consumer products. [Rexene Prods.]

Rexene® PE 1903. Ethylene-vinyl acetate (9%) copolymer; excellent toughness and flexibility at low temps., good heat seal response; for laminating film, controlled atmosphere and heavy-duty packaging, medical tubing. [Rexene Prods.]

Rexene® PE 6010. CAS 9002-88-4; LDPE; good stress crack resist.; for blow molding of small bottles, containers, vials. [Rexene Prods.]

Rexene® PE 6076. CAS 9002-88-4; LDPE homopolymer; for blow molding of bottles with good clarity and excellent stiffness. [Rexene Prods.]

Rexene® PP 12R10A. CAS 9003-07-0; PP random copolymer; good contact clarity, impact resist., stiffness, high deflection temp.; for rigid containers, single-use food packaging, laboratoryware, medicine vials, hypodermic syringes, radiation sterilization resist. applications. [Rexene Prods.]

Rexene® PP 23S2. CAS 9003-07-0; PP copolymer; blow molding grade for bottles. [Rexene Prods.]

Rexene® PP 41E2. CAS 9003-07-0; PP homopolymer resin; for extrusion into highest quality biaxially oriented capacitor film with stiffness and excellent optical and barrier properties; high dielec. strength, low elec. loss. [Rexene Prods.]

Rexene® PP 9234. CAS 9003-07-0; PP copolymer; radiation-resist. grade for thermoformed trays, basins, and utensils for medical field; gamma sterilizable with little change in color, strength, or flexibility; FDA compliance. [Rexene Prods.]

Rexenite. A proprietary trade name for a cellulose acetate butyrate plastic. §

Rexhide. A rubber-glue stock material for use in rubber mixings.

Rexin X. Liquid mixed glycol ester of rosin; plasticizer and tackifier in adhesive compositions. [Bluemin ster Ltd]

Rexite. A blasting explosive, containing 6.5-8.5% nitroglycerin, 64-68% ammonium nitrate, 13-16% sodium nitrate, 6.5- 8.5%, trinitro-toluene, and 3-5% wood meal.

Rexobase BAT. Alkylaryl detergent; economical detergent and emulsifier which can be cut 50-50 with water for use as detergent and 50-50 with varsol to make a solv. scour. [Emkay]

Rexobase EN. Emulsifier for naphtha or toluol. [Emkay]

Rexobase HD. Degreaser for hide tanning; stable in brine solution [Emkay]

Rexobase XX. Amide condensate; emulsifier for varsol or Stod.; forms bright sol. oil used in water sol'ns. for the removal of stubborn soil and grease. [Emkay]

Rexobond. Resin gum; prepared weighter and body builder finish compatible with softeners and glues; excellent lining finish (textiles). [Emkay]

Rexoclean. Synthetic detergents, terpenes, and solvs.; complete scouring agent for stubborn oil and grease stains and for removal of nylon warp size. [Emkay]

Rexoclean 200N. Emulsified blend of solvs. and detergents; multipurpose cleaner for heavy duty cleaning including rubber marks and stains, in soap dispensers. [Emkay]

Rexodull CT. Multipurpose substantive delustrant with a titanium dioxide base. [Emkay]

Rexofos WS. Mixt. of water softeners and mild alkalies for softening and conditioning water. [Emkay]

Rexogel. Purified gelatin base product; economical stiff finish for textiles requiring no curing. [Emkay]

Rexogum GL. Blend of selected glues and gums; finishing agent for textile fabrics; gives a clear film without discoloration. [Emkay]

Rexol. A mixture of ammonium perchlorate, potassium chlorate, rosin, zinc or aluminum, and mineral oil or wax; an explosive.

Rexol 25/4. CAS 9016-45-9; Nonoxynol-4; detergent, dispersant, stabilizer; low foaming emulsifier for oils and petrol. solvs., emulsion polymerization; intermediate for anionic sulfonates; pigment dispersant. [Hart Chem. Ltd.]

Rexol 25/10. CAS 9016-45-9; Nonoxynol-10; detergent, dispersant, emulsifier, wetting agent; for paint, textiles, pulp/paper industries, emulsion polymerization; degreaser for leather; scouring agent for raw wool. [Hart Chem. Ltd.]

Rexol 35/3. Alcohol ethoxylate; emulsifier for o/w; intermediate for sulfation and phosphation. [Hart Chem. Ltd.]

Rexol 45/1. CAS 9002-93-1; Octoxynol-1; emulsifier, detergent, dispersant; coemulsifier for surfactant and solv. preparation blends. [Hart Chem. Ltd.]

Rexol 65/4. CAS 9014-92-0; Dodoxynol-4; ingred. for solv. cleaner and drycleaning formulations; emulsifier for ag-

---

* Trade name not verified as available   † Trade name verified as obsolete

ricultural oils; degreaser. [Hart Chem. Ltd.]

Rexol 2000 HWM. Amide ethoxylate, modified; antistatic emulsifier for mineral oils, manufacturing of textile lubricants; anticorrosive and cohesive properties [Hart Chem. Ltd.]

Rexol AE-1. Linear alcohol ethoxylate; intermediate for shampoo and detergent manufacturing. [Hart Chem. Ltd.]

Rexole 612. Plasticizer and softener for use in textile sizing formulations. [Emkay]

Rexolene. Sulfonated base dispersing agent for acetate dyes; dyeing assistant. [Emkay]

Rexoloid. Alkyd resin; clear resin requiring no curing; used as a ribbon finish on acetates; imparts stiff hand with high luster. [Emkay]

Rexolube. Lubricant and chafe eliminator for use in the diazotizing bath to eliminate chafes on developed black and navy colors. [Emkay]

Rexonic 1006. Linear alcohol ethoxylate; wetting and scouring agent for natural and synthetic fibers; leveling and dispersing aid for acid and disperse dyes; foaming agent for Kuster dyeing of carpets; compatible with dyestuffs over wide pH range; biodeg. [Hart Chem. Ltd.]

Rexonic N23-3. Linear alcohol ethoxylate; detergent, emulsifier, dispersant, and chemical intermediate used in solv. emulsion cleaners and dry cleaning detergents; biodeg. [Hart Chem. Ltd.]

Rexonic P-1. Propoxylated ethoxylated linear alcohol; low-foaming detergent, wetter, emulsifier, dispersant for natural and synthetic fibers, industrial and commercial low foam detergents; base for rinse aids and machine dishwashing formulations. [Hart Chem. Ltd.]

Rexonic RL. Alcohol ethoxylate; detergent, post-scouring agent. [Hart Chem. Ltd.]

Rexonit D. Protein-resin mixt.; heavy finish for rayon and acetate fabrics; compatible with salt weighters for finishing knit goods. [Emkay]

Rexopal 3928. Modified linear alcohol ethoxylate; low foaming wetting and scouring agent for natural and synthetic fibers; biodeg. [Hart Chem. Ltd.]

Rexopal SM-5. nonfoaming surfactant, wetting agent, detergent for prescouring. [Hart Chem. Ltd.]

Rexopene. Sodium alkylaryl sulfonate; wetting, dye leveling agent, penetrant, scouring assistant. [Emkay]

Rexophos 25/67. Alkylaryl phosphate ester; emulsifier for solvs. used over wide temp. range; surfactant for hard surface cleaners, alkaline metal cleaners. [Hart Chem. Ltd.]

Rexophos 4668. Linear alcohol phosphate ester; detergent; stable to highly alkaline cleaning sol'ns. [Hart Chem. Ltd.]

Rexophos BP-2. Linear alcohol phosphate ester; scouring agent and cleaner for textile industry; high alkaline stable. [Hart Chem. Ltd.]

Rexopon E. Amide amine condensate, modified; synthetic detergent; after wash for prints. [Emkay]

Rexopon SK. Silk degumming and scouring agent; used with soap or detergent to accelerate degumming and boil-off of silk goods. [Emkay]

Rexoscour SF. Scour and fulling agent for wool. [Emkay]

Rexoslip AS. Nonyellowing nonslip agent and weighter used primarily as satin finish. [Emkay]

Rexosolve 150. Sulfonated oils/emulsifier blend; detergent, scouring assistant; oil and grease remover. [Emkay]

Rexoteric XCE. Carboxyethylated coco amphoteric in liquid form; for alkaline industrial cleaners. [Jan Dekker BV] *

Rexoteric XCG. Specially designed amphoteric in liquid form; used in nonirritating conditioning shampoos, bath and toiletry products when high viscosity is required. [Jan Dekker BV] *

Rexoteric XCO. Carboxymethylated coco imidazoline derivative in liquid form; for high foaming, nonirritating shampoos, bath and toiletry products. [Jan Dekker BV]*

Rexoteric XJO. Carboxymethylated caprylic imidazoline derivative in liquid form; detergent and wetting agent for low foaming alkaline and acid cleaners. [Jan Dekker BV]*

Rexoteric XOO. Carboxymethylated oleic imidazoline derivative as a viscous liquid; imparts softening properties and is used especially in high viscosity hand cleaners and detergent formulations. [Jan Dekker BV] *

Rexoteric YCB. Carboxyethylated coco imidazoline derivative; high foaming detergent for use in all-purpose cleaners. Viscosity modulator, gel-builder and conditioner for low pH nonirritating shampoos, bath and toiletry products. [Jan Dekker BV] *

Rexoteric YCE. Carboxyethylated coco amine derivative; high foaming light to heavy duty cleaners in acid to alkaline pH range. [Jan Dekker BV] *

Rexoteric YJE. Carboxyethylated short chain amine derivative; moderately foaming light to heavy duty cleaners, giving good surface tension reduction over the whole pH range. [Jan Dekker BV] *

Rexoteric YOB. Carboxyethylated alkyl imidazoline derivative; specially designed for use in corrosion inhibitors and as a softener. [Jan Dekker BV] *

Rexoteric ZXCO. A blend of Rexoteric XCO and an anionic surfactant, in liquid form; ready-made base for nonirritating shampoos, bath and toiletry products. [Jan Dekker BV] *

Rexowax CNN. Water-disp. wax and resin; recommended as satin finish. [Emkay]

Rexowet 77. Heptadecyl sodium sulfate; fast wetting and penetrating agent for textiles; resistant to acid and alkali media. [Emkay]

Rexowet 500. Sulfonated dioctyl succinate; surface tension depressant; stable in acid, alkaline, and neutral baths. [Emkay]

Rexowet ASG-81. CAS 577-11-7; EINECS 209-406-4; Sodium dioctyl sulfosuccinate; wetting agent for textile use. [Emkay]

Rexowet CR. Sulfonated isopropyl oleate and cresylic acid; wetting agent and penetrant for cotton fabrics; scouring agent for removing mineral oil stains. [Emkay]

Rextox. A proprietary trade name for a material consisting of copper with a layer of cuprous oxide formed on the surface of the metal at high temperatures. §

Rextrude. A proprietary trade name for a cellulose-acetate-butyrate plastic. §

Reyalite. Aluminum alloys containing lithium. [Reynolds Metal Co] †

Reyalith. Aluminum alloys containing lithium. [Reynolds Metal Co] †

Reycomp. Laminated sheets of aluminum having plastic inner cores. [Reynolds Metal Co] †

Reydox. Aluminum and aluminum alloy castings. [Reynolds Metal Co]

Reynolds Wrap. Aluminum foil sheets and rolls; for consumer use in cooking, wrapping, etc. [Reynolds Metal

---

‡ Trade name and manufacturer not verified   § Trade name without identified manufacturer

Co]

Reynolon. Plastic film, laminated and unsupported; for consumer and food use. [Reynolds Metal Co]

Reynolon. PVC with various plasticizers and stabilizers; shrink film for presentation applications. [S Kempner Ltd]

REZ 300. Urea-formaldehyde resin; theromsetting resin paste producing superior crease resist. and shrinkage control on cottons and synthetics with a smooth, round hand. [CNC Int'l. L.P.]

Rezal®. Aluminum-zirconium tetrachlorohydrex glycine or pentachlorohydrex glycine; antiperspirant actives. [Reheis Inc]

Rezal® 36GP. Aluminum zirconium tetrachlorohydrex-glycine; antiperspirant active. [Reheis Inc]

Rezcat #2, #3. Highly reactive polyfunctional aziridine products; crosslinkers in polymeric systems containing carboxyl or hydroxyl functionalities; improves water and solv. resist. and increases adhesion to many substrates. [CNC Int'l. L.P.]

Rezex. Construction auxiliaries. [Crosfield Chemicals Ltd]

Rezifilm. CAS 137-26-8; Thiram; antifungal. [Bristol-Myers Squibb Co Inc] *

Rezistal. Corrosion and heat-resisting steels consisting of iron with up to 0.4% carbon, 1-5.5% silicon, 8-26% chromium, and 7-35% nickel.

Rezolin 164. Polyether PU compound; nonexpanding compound for making low-cost moisture blocks and pressure dams in paper, pulp, and plastic insulated telecommunications cable (direct sheath injection method); service proven in all climates and environments; approved standard of major telephone companies. [Hexcel]

Rez-O-Sperse® 3. Chlorinated paraffin (Paroil 170-HV) aq. emulsion; flame retardant for plastics, rubbers, adhesives, paints, fabric coatings, inks, carpet backings, paper coatings; plasticizer, tackifier. [Dover]

Rez-O-Sperse® A-1. Resinous chlorinated paraffin aq. disp.; flame retardant for plastics, rubbers, adhesives, paints, fabric coatings, inks, carpet backings, paper coatings; used where increased hardness is desired. [Dover]

Rezthane Series. Polyurethane emulsion coatings; topcoat, basecoat, water-repellent, low crock, abrasion-resist., oily and waxy textile finishes. [CNC Int'l. L.P.]

Rezyl. A proprietary trade name for an alkyd synthetic resin.§

rhamnose. $C_6H_{12}O_5$; CAS 3615-41-6; Synthetic sweetener. [Penta Mfg.; Pfanstiehl Labs; U.S. Biochemical]

rhaponticin. CAS 478-43-3; (rheic acid, rheumin, rhubarbaric acid, rhubarbarin). Chrysophanic acid, $C_{15}H_{10}O_4$, found in rhubarb root.

rhaponticum. (rhapontin; ponticin). The crystalline substance from the common English rhubarb.

rhapontin. See rhaponticum.

Rhapsodie®. Fibers. [DuPont UK]

Rhapsody®. For the agriculture industry. [DuPont UK]

rhatany. The dried root of *Krameria triandra* and *K. argentea*.

Rheaform. Clioquinol; anti-amebic; anti-infective, topical. [Bristol-Myers Squibb Co Inc] *

rheic acid. See rhaponticin.

Rhenalkote. Alkyd resins and polyesters; used in surface coating industry. [RWE-DEA Chemicals UK Ltd]

Rhenalyd. Alkyd resins and polyesters; used in surface coating industry. [RWE-DEA Chemicals UK Ltd]

Rhenania phosphate. (Vesta phosphate). Prepared by sintering together in a furnace at 1200-1300 C, a mixture of raw phosphate, limestone, and alkali silicate. The resulting product approximates to the formula $Ca_2KNa(PO_4)_2$.

Rhenanit V. A German explosive containing nitroglycerin, ammonium nitrate, vegetable meal, nitro-compounds, and potassium perchlorate.

Rhenappret B. Synthetic polymer solution; handle improvement agent for rayon and viscose staple fibers and synthetics. [Thor Chemicals (UK) Ltd]

Rheniforming. Catalyst. [Chevron] *

Rhenish dynamite. A solution of 75% nitroglycerin in naphthalene, 2% chalk or barium sulfate, and 23% kieselguhr.

Rhenital. Phenolic and melaminic resins; used for the bonding of wood, foundry sands and abrasives. [RWE-DEA Chemicals UK Ltd]

Rhenoblend. Polymer blends. [Bayer plc]

Rhenocure. Accelerators for polymers. [Bayer plc]

Rhenodiv. Separating agents. [Bayer plc]

Rhenodiv. A combination of surface active agents with film forming substances, partly enriched with corrosion inhibitors; prevents the sticking together of raw rubber, uncured rubber compounds, blanks and extrudates. [Rhein-Chemie Rheinau]

Rhenofit. Activators. [Bayer plc]

Rhenoflex. Chlorinated PVC; used for lacquer industry, adhesives industry, pyrotechnics. [Dynamit Nobel Wien GmbH] *

Rhenogran. A range of polymer-bound rubber chemicals with an activity content of 80%; for technical molded and extruded articles, tires and cable coverings. [Rhein-Chemie Rheinau]

Rhenogran. Predispersed polymer-bound rubber chemicals. [Bayer plc]

Rhenomag. CAS 1309-48-4; Magnesium oxide. [Bayer plc]

Rhenomag. CAS 1309-48-4; A range of magnesium oxide preparations of several qualities; acid acceptor and vulcanization activator used in the rubber industry. [Rhein-Chemie Rheinau]

Rhenopor. Blowing agents for the rubber and plastic industries. [Bayer plc]

Rhenopren. A polymeric processing promoter for the rubber industry; for technical molded and extruded articles. [Rhein-Chemie Rheinau] *

Rhenosin. A range of thermoplastic hydrocarbon resins; used as softening resins and homogenizers in the rubber industry; used for light and dark colored compounds, e.g., tires and conveyor belts. [Rhein-Chemie Rheinau]*

Rhenosin. Range of hydrocarbons for use as processing promoters in the rubber industry. [Bayer plc]

Rhenosorb. Dessiccants. [Bayer plc]

Rhenosorb. CAS 1305-78-8; Very finely divided calcium oxide; prevents porosity caused by moisture when cured continuously and without pressure; used for rubber compounds of all kinds. [Rhein-Chemie Rheinau]

Rheocin. CAS 139-44-6; Trihydroxystearin; thixotrope for aliphatic solv. systems, trade sales alkyds, stains, industrial alkyds, other coatings. [United Catalysts]

Rheocin. Antisettling and thickening agents for paints, varnishes, lubricants, adhesives, coatings, putties, and cosmetics. [Süd-Chemie AG] *

Rheolate. Rheological additives; used for water borne coatings, adhesives and paper coatings. [Rheox Inc]

---

* Trade name not verified as available   † Trade name verified as obsolete

Rheolube 350SBG-2. Mixed-soap petrol.-based; multipurpose bearing grease excellent for high speeds. [Wm F Nye]

Rheomacrodex. Dextran 40; a polysaccharide produced by the action of *Leuconostoc mesenteroides* on sucrose; average m.w. 40,000; blood flow adjuvant; plasma volume extender. [Kabi-Pharmacia]

Rheostene. A nickel-iron alloy. It has a specific resistance of 77 microhms per cm^3 at 0 C.

Rheotan I. An electrical resistance alloy containing 84% copper, 12% manganese, and 4% zinc.

Rheotan II. An alloy for electrical resistances. It contains 25% nickel, 52% copper, 5% iron, and 18% zinc. This alloy or a similar one has been called Rheostan.

Rheotemp 500. Polyester-based; bearing grease for high speed and wide temp. range (-65 to 350 F); for computer cooling fans, aircraft applications. [Wm F Nye]

Rheothik Polymer 80-11. Polysulfonic acid; thickener, lubricant, suspending agent, slip agent for aq. lubricants, synthetic cutting oils; thickens HF and other acids, alkaline systems; stable in strong acids. [Henkel/Cospha]

Rheotix AS. Antisettling and antisag agent for high solids/low VOC formulations; does not appreciably increase apparent visc. [United Catalysts]

Rheotol®. Polymerized alkyl phosphate in solvents; rheology modifier, wetting and dispersing agent. [R T Vanderbilt Co Inc]

Rheovis®. Associative thickening agents designed primarily for emulsion paints. [Allied Colloids Ltd]

Rheox. Rheological additives; used for paints, coatings, paper, plastic, ink, grease, cosmetics, adhesives, sealants and caulks. [Rheox Inc]

Rhesonativ. Human immunoglobulin for prophylaxis of rhesus immunization. [KabiVitrum AB] *

Rheumajecta. A proprietary preparation of sulfuryl-sulfokinase, cholinacetylase and catalase; used in the treatment of rheumatic diseases. [Enzypharm Biochemicals Ltd] ‡

rheumin. See rhaponticin.

Rheumon®. Percutaneous and antirheumatic agent. [Bayer AG]

Rheumox. Azapropazone dihydrate; capsules or tablets; indicated in the treatment of osteo and rheumatoid arthritis. [Wyeth Laboratories]

Rhevois. Associate rheology modifiers. [Allied Colloids Ltd]

Rhine metal. An alloy of 97% tin and 3% copper. It has a specific gravity of 7.35 and melts at 300 C.

Rhizopon A, AA. CAS 87-51-4; Indol-3-yl acetic acid; a root growth promoter in either powder or tablet form. [Fargro Ltd]

Rhizopon B. CAS 86-87-3; 1-Naphthylacetic acid; a plant growth regulator used to control suckering in fruit trees. [Fargro Ltd]

RH Maneb 80. Protectant fungicide. [Rohm & Haas UK]

Rhodacal® 70/B. Calcium dodecylbenzene sulfonate; emulsifier, dispersant for herbicides and pesticides. [Rhone-Poulenc Surf.; Rhone-Poulenc Geronazzo]

Rhodacal® 301-10. CAS 68439-57-6; EINECS 270-407-8; Sodium C14-16 olefin sulfonate; emulsion polymerization surfactant. [Rhone-Poulenc Surf.]

Rhodacal® 330. CAS 26264-05-1; Isopropylamine dodecylbenzene sulfonate; emulsifier, wetting agent, grease/pigment dispersant, lubricant, solubilizer, solv., penetrant, high foaming base for shampoos and cleaners,

metalworking fluids, agricultural pesticides, emulsion polymerization, drycleaning, latex paints, metal cleaning. [Rhone-Poulenc Surf.]

Rhodacal® 2283. Amine dodecylbenzene sulfonate; emulsifier for agricultural formulations. [Rhone-Poulenc Surf.; Rhone-Poulenc Geronazzo]

Rhodacal® ABSA. CAS 27176-87-0; Dodecylbenzenesulfonic acid. [Rhone-Poulenc Surf.]

Rhodacal® CA, 70%. Calcium dodecylbenzene sulfonate; biodeg. o/w emulsifier for agriculture, industrial applications; dispersant for polyester yarn dyeing. [Rhone-Poulenc Surf.; Rhone-Poulenc Surf. Canada]

Rhodacal® DDB-40. CAS 25155-30-0; Sodium dodecylbenzene sulfonate; high foaming emulsifier, dispersant, wetting agent for industrial, institutional and household cleaners, agricultural formulations. [Rhone-Poulenc Surf.]

Rhodacal® DDB 60T. TEA-dodecylbenzenesulfonate. [Rhone-Poulenc Surf.]

Rhodacal® DOV. TEA dodecylbenzene sulfonate; emulsion polymerization surfactant. [Rhone-Poulenc Surf.]

Rhodacal® DSB. CAS 28519-02-0; Sodium dodecyl diphenyloxide disulfonate; detergent, foamer, emulsifier, textile dye leveling agent, coupling agent, solubilizer; dispersant in metal and other industrial cleaners; co-emulsifier in emulsion polymerization; wetting agent, dispersant for agricultural formulations; alkaline and chlorine stable. [Rhone-Poulenc Surf.; Rhone-Poulenc France]

Rhodacal® IN. CAS 1322-93-6; Sodium isopropyl naphthalene sulfonate; dispersant and suspending agent for pesticide formulations. [Rhone-Poulenc Surf.]

Rhodacal® IPAM. CAS 26264-05-1; Isopropylamine dodecylbenzene sulfonate; emulsifier for drycleaning charge soaps, solv. degreasers; solubilizer in fuel oil; forms clear blends of water and kerosene. [Rhone-Poulenc Surf.]

Rhodacal® LA Acid. Dodecylbenzene sulfonic acid; biodeg. detergent, wetter, emulsifier, penetrant, foamer, intermediate used in liquid dishwashing detergents, cleaners, personal care products, and degreasers. [Rhone-Poulenc Surf.]

Rhodacal® Liquid. CAS 9084-06-4; Sodium polynaphthalene sulfonate [Rhone-Poulenc Surf.]

Rhodacal® N. CAS 9084-06-4; Sodium polynaphthalene sulfonate; process aid in paper, leather; dispersant for pulp/paper, metal cleaning; emulsifier, wetting agent for industrial cleaners, pesticides; moisture reducer in concrete; textile dye leveling agent. [Rhone-Poulenc Surf.]

Rhodacal® RM/77-D. Dinaphthalene-methane sulfonate sodium salt; dispersant, protective colloid, stabilizer of natural and synthetic elastomers; pigment grinding aid for paints. [Rhone-Poulenc Geronazzo]

Rhodacal® RM/210. CAS 26264-58-4; EINECS 247-561-6; Polynaphthalene-methane sulfonate, sodium salt; fluidizing and plasticizer agent for concrete and mortar. [Rhone-Poulenc Geronazzo]

Rhodacal® T. TEA dodecylbenzene sulfonate; base for clear liquid detergents and bubble baths. [Rhone-Poulenc Surf.]

Rhodafac® BG-510. CAS 108818-88-8; Aliphatic phosphate ester of ethoxylated isodecyl alcohols, free acid; detergent, emulsifier, wetting agent, dispersant for liquid industrial alkaline cleaners, hard surface detergents, soak-tank metal cleaning, steam cleaning, household

---

‡ Trade name and manufacturer not verified   § Trade name without identified manufacturer

cleaning; compatible in concentrated electrolyte systems. [Rhone-Poulenc Surf.; Rhone-Poulenc France]

Rhodafac® BP-769. CAS 39464-70-5; Aromatic phosphate ester of ethoxylated phenol; low foaming surfactant, hydrotrope for industrial, institutional and household cleaners with electrolyte content; good alkali stability. [Rhone-Poulenc Surf.]

Rhodafac® GB-520. CAS 68186-34-5; Sodium oleth-7 phosphate; emulsifier, lubricant, softener, textile finishing aid for wool and synthetic fibers, metalworking fluids. [Rhone-Poulenc Surf.; Rhone-Poulenc France]

Rhodafac® LO-529. CAS 37340-60-6; Sodium nonoxynol-6 phosphate; detergent, foamer, corrosion inhibitor, emulsifier for detergent concentrates, floor cleaners, agricultural concentrates, metalworking fluids. [Rhone-Poulenc Surf.; Rhone-Poulenc France]

Rhodafac® MC-470. CAS 42612-52-2; Sodium laureth-4 phosphate; detergent, emulsifier, visc. builder for creams and lotions, polymerization and stabilization of latexes; fatliquoring of leathers; metalworking fluids; textile antistat, lubricant, softener; emulsifier for mineral oils. [Rhone-Poulenc Surf.]

Rhodafac® MD-12-116. Aliphatic phosphate ester, free acid; low foaming emulsifier, dispersant, wetting agent with good rinsability; for industrial cleaners, pesticides, oil well cleanout; hydrotrope for metal cleaners, rinse aids. [Rhone-Poulenc Surf.]

Rhodafac® PA-15. Complex organic phosphate ester, free acid; hydrophilic detergent for liquid industrial cleaners; low foaming; stable to alkalies. [Rhone-Poulenc France].

Rhodafac® PE-510. CAS 68412-53-3; Nonoxynol-6 phosphate; detergent for drycleaning, waterless hand cleaners; intermediate for textile lubricants; emulsifier for emulsion polymerization (PVAc and acrylic films); emulsifier, dispersant for agricultural formulations; corrosion inhibitor; good electrolyte tolerance. [Rhone-Poulenc Surf.; Rhone-Poulenc France]

Rhodafac® PEH. CAS 12645-31-7; 2-Ethylhexyl phosphate; detergent, dispersant, and wetting agent in textile wet processing. [Rhone-Poulenc Surf.]

Rhodafac® PL-620. CAS 68649-29-6; C10-16 ethoxy, propoxy, phosphate; surfactant for well cleanout (asphaltene, paraffin and scale removal). [Rhone-Poulenc Surf.]

Rhodafac® RA-600. CAS 68130-47-2; Deceth-4 phosphate; detergent, emulsifier, wetting agent, foamer, dispersant for hard surface cleaners, industrial alkaline detergents, textile wet processing; coupler used in liquid alkali detergents. [Rhone-Poulenc Surf.; Rhone-Poulenc France]

Rhodafac® RB-400. CAS 39464-69-2; Oleth-4 phosphate; lubricant, antistat, emulsifier for fibers and metal. [Rhone-Poulenc Surf.]

Rhodafac® RD-510. CAS 39464-66-9; Laureth-4 phosphate; emulsifier, antistat, lubricant, solubilizer for fibers, metals, cosmetics, agricultural formulations. [Rhone-Poulenc Surf.]

Rhodafac® RE-610. CAS 68412-53-3; Nonoxynol-9 phosphate; detergent, emulsifier, wetting agent, dispersant, antistat, lubricant, dedusting agent for drycleaning, pesticides, emulsion polymerization, textile wet processing, metals, household and industrial detergents. [Rhone-Poulenc Surf.; Rhone-Poulenc France]

Rhodafac® RM-410. CAS 39464-64-7; Nonyl nonoxynol-7

phosphate; detergent, emulsifier, antistat for drycleaning, pesticides; rust inhibitor; textile wetting agent. [Rhone-Poulenc Surf.]

Rhodafac® RS-610. CAS 9046-01-9; Trideceth-6 phosphate; emulsifier for emulsion polymerization, waterless hand cleaners, pesticides; detergent for drycleaning formulations, textile wetting agent, lubricant for fiber and metal treatment, paper-mill felt washing; antistat for aerosols. [Rhone-Poulenc Surf.; Rhone-Poulenc France]

Rhodameen® HT-50. CAS 68783-22-2; PET-50 hydrogenated tallow amine; wetting agent, penetrant, emulsifier, stabilizer, dispersant, antistat, lubricant. [Rhone-Poulenc Surf.]

Rhodameen® O-12. PEG-2 oleamine; coemulsifier in emulsifiable concentrates (herbicides), w/o emulsions of solvs., waxes, and oils, catiaonic asphalt emulsions. [Rhone-Poulenc France]

Rhodameen® OA-860. PEG-30 oleamine; hydrophilic emulsifier, leveling agent; textile dyeing assistant; antiprecipitant for dyeing processes. [Rhone-Poulenc Surf.]

Rhodameen® OS-12. PEG-2 oleyl/stearyl amine; coemulsifier and antistat for cosmetics, textile and plastic processing. [Rhone-Poulenc France]

Rhodameen® PN-430. CAS 61791-26-2; PEG-5 hydrogenated tallowamine; emulsifier, corrosion inhibitor, lubricant for metalworking fluids, agricultural formulations; acid corrosion inhibitor for ferrous alloys. [Rhone-Poulenc Surf.; Rhone-Poulenc France]

Rhodameen® T-5. CAS 61791-26-2; PEG-5 hydrogenated tallow amine; emulsifier, dispersant; textile scouring and desizing assistant; softener and antistatic agent. [Rhone-Poulenc Surf.]

Rhodameen® VP-532/SPB. CAS 61791-26-2; PEG-8 hydrogenated tallowamine; wetting agent, penetrant, emulsifier, stabilizer, antistat, and lubricant; retarder, antiprecipitant, leveling agent in dyeing; dispersant for fiberglass mat manufacturing. [Rhone-Poulenc Surf.]

Rhodamine B. The hydrochloride of diethyl-m-aminophenol-phthalein, $C_{28}H_{31}Cl \cdot N_2O_3$; a dyestuff; dyes wool and silk bluish-red with fluorescence, also tannined cotton, violet-red. §

Rhodamine G and G Extra. Dyestuffs, consisting chiefly of triethyl-rhodamine, $C_{26}H_{27}Cl \cdot N_2O_3$; dyes wool, silk, and tannined cotton, red. §

Rhodamine S. The hydrochloride of dimethyl-m-aminophenol-succineine, $C_{20}H_{23}Cl \cdot N_2O_3$; a dyestuff; dyes cotton red, and is used for dyeing half silk goods, and for coloring paper pulp and wood. §

Rhodamox® CAPO. CAS 68155-09-9; Cocamidopropyl dimethylamine oxide; foaming agent/stabilizer, thickener, emollient for shampoos, bath products, dishwash, rug shampoos, fine fabric detergents, shaving creams, lotions, foam rubber, electroplating, paper coatings; used in toiletries for mildness. [Rhone-Poulenc Surf.]

Rhodamox® LO. CAS 1643-20-5; Lauryl dimethylamine oxide; foaming agent/stabilizer, thickener, emollient for shampoos, bath products, dishwash, fine fabric detergents, shaving creams, lotions, textile softeners, foam rubber, in electroplating, paper coatings; used in toiletries for mildness. [Rhone-Poulenc Surf.; Rhone-Poulenc France]

Rhodapex® 674/C. CAS 25446-78-0; Sodium trideceth sulfate. [Rhone-Poulenc Surf.]

---

* Trade name not verified as available

† Trade name verified as obsolete

Rhodapex® AB-20. CAS 67762-19-0; Ammonium laureth sulfate; emulsifier for emulsion polymerization and mineral oils; detergent. [Rhone-Poulenc France]

Rhodapex® CO-415. CAS 9051-57-4; Ammonium nonoxynol-4 sulfate. [Rhone-Poulenc Surf.]

Rhodapex® CO-433. CAS 68891-39-4; Sodium nonoxynol-4 sulfate; high foaming detergent, wetting agent, dispersant for dishwashing, scrub soaps, car washes, rug and hair shampoos; emulsifier for emulsion polymerization, petrol. waxes; antistat for plastics and synthetic fibers; lime soap dispersant. [Rhone-Poulenc Surf.; Rhone-Poulenc France]

Rhodapex® CO-436. CAS 9051-57-4; Ammonium nonoxynol-4 sulfate; high foaming detergent, wetting agent, dispersant for dishwashing, scrub soaps, car washes, rug and hair shampoos; emulsifier for emulsion polymerization, petrol. waxes; antistat for plastics and synthetic fibers; lime soap dispersant. [Rhone-Poulenc Surf.; Rhone-Poulenc France]

Rhodapex® EA. CAS 67762-19-0; Ammonium laureth (3) sulfate; high foaming emulsifier for industrial, institutional and household cleaners. [Rhone-Poulenc Surf.]

Rhodapex® EAY. CAS 67762-19-0; Ammonium laureth sulfate; detergent for shampoos, bubble baths. [Rhone-Poulenc Surf.]

Rhodapex® ES. CAS 9004-82-4; Sodium laureth (3) sulfate; emulsifier, high foaming base for shampoos, light duty detergents, bubble baths; polymerization surfactant. [Rhone-Poulenc Surf.]

Rhodapex® EST-30. CAS 25446-78-0; Sodium trideth sulfate; emulsifier, wetting agent, dispersant for baby shampoo, other personal care products, household, industrial, institutional and industrial formulations, emulsion polymerization of styrene systems, textile scouring, dishwash. [Rhone-Poulenc Surf.; Rhone-Poulenc France]

Rhodapex® F-85/SD. Sodium octylphenol ethoxy sulfate; surfactant. [Rhone-Poulenc France]

Rhodapon® 101-10. CAS 151-21-3; Sodium lauryl sulfate. [Rhone-Poulenc Surf.]

Rhodapon® BOS. CAS 126-92-1; Sodium 2-ethylhexyl-sulfate; biodeg. wetting agent, emulsifier, detergent, foamer with high electrolyte tolerance for industrial, institutional, and household cleaners; latex stabilizer, nickel brightener; metal treatment; textile and plywood manufacturing; fruit and vegetable washing. [Rhone-Poulenc Surf.; Rhone-Poulenc France]

Rhodapon® CAV. CAS 68299-17-2; Sodium isodecyl sulfate; wetting agent, emulsifier, detergent, foamer, rinse aid, visc. control agent; post-stabilizer in latex paints; metal treatment; textile and plywood manufacturing; fruit/vegetable washing; hard surface cleaners; emulsion polymerization of vinyl, acrylic, SBR, PVC. [Rhone-Poulenc Surf.]

Rhodapon® EC111. CAS 59186-41-3; Sodium cetyl/stearyl sulfate; emulsifier, detergent, flotation agent; collector in ore flotation; softener/lubricant for textiles; cosmetics, and toiletries. [Rhone-Poulenc Surf.; Rhone-Poulenc France]

Rhodapon® L-22, L-22/C. CAS 2235-54-3; EINECS 218-793-9; Ammonium lauryl sulfate; high foaming detergent, emulsifier for shampoo, bubble bath, pet shampoos, industrial and institutional cleaners, wool scouring, fire fighting foams, assistant for pigment dispersion; emulsion polymerization aid. [Rhone-Poulenc Surf.; Rhone-Poulenc France]

Rhodapon® LCP. CAS 151-21-3; Sodium lauryl sulfate; low cloud point emulsion polymerization surfactant. [Rhone-Poulenc Surf.]

Rhodapon® LM. CAS 3097-08-3; Magnesium lauryl sulfate; high-foaming emulsifier, wetting agent, dispersant, detergent for rug and upholstery shampoos, bubble baths, shampoos; food packaging applications. [Rhone-Poulenc Surf.]

Rhodapon® LT-6. CAS 139-96-8; EINECS 205-388-7; TEA lauryl sulfate; emulsifier, high foaming base for industrial and household detergents, shampoos, bubble baths; obtains creamy, mild lather. [Rhone-Poulenc Surf.]

Rhodapon® OLS. CAS 142-31-4; Sodium octyl sulfate; wetting agent; rinse aid for industrial, institutional and household cleaners; mercerizing agent for cotton goods; surfactant in electrolyte baths for metal cleaning; hard surface cleaning; neoprene dispersant; emulsifier for emulsion polymerization. [Rhone-Poulenc Surf.]

Rhodapon® OS. CAS 1847-55-8; EINECS 217-430-1; Sodium oleyl sulfate; specialty emulsifier for emulsion polymerization. [Rhone-Poulenc Surf.]

Rhodapon® TDS. CAS 3026-63-9; Sodium tridecyl sulfate; emulsifier, wetting agent, emulsion polymerization of vinyl chloride, styrene, and styrene/acrylic monomers; detergent formulations; base for shampoos, foaming bath oils, cosmetic emulsions. [Rhone-Poulenc Surf.]

Rhodaquat® M214B/99. CAS 1119-97-7; EINECS 214-291-9; Myrtrimonium bromide; conditioner with superior antistatic properties and light feel. [Rhone-Poulenc Surf.]

Rhodaquat® M242B/99. CAS 57-09-0; EINECS 200-311-3; Cetrimonium bromide; quaternary ammonium compound used in conditioners to increase antistatic and comb out effects. [Rhone-Poulenc Surf.]

Rhodaquat® M242C/29. CAS 112-02-7; EINECS 203-928-6; Cetrimonium chloride; surfactant with conditioning and emolliency effect on hair. [Rhone-Poulenc Surf.]

Rhodaquat® M270C/18. CAS 122-19-0; EINECS 204-527-9; Stearalkonium chloride. [Rhone-Poulenc Surf.]

Rhodasurf® 25-7. CAS 68131-39-5; C12-15 pareth-7; biodeg. detergent, wetting agent, emulsifier for household and industrial cleaners; coupler and solubilizer for perfumes and organic additives. [Rhone-Poulenc Surf.]

Rhodasurf® 91-6. CAS 68439-46-3; C9-11 pareth-6; biodeg. detergent, wetting agent, emulsifier for household and industrial cleaners; coupler and solubilizer for perfumes and organic additives. [Rhone-Poulenc Surf.]

Rhodasurf® 860/P. Isodeceth-6; wetting agent, foamer, emulsifier, detergent for textile processing, metal degreasing, herbicides, waxes, resins; stable to strong acidic and alkaline media. [Rhone-Poulenc France]

Rhodasurf® A-1P. CAS 68439-49-6; Ceteareth-23; surfactant. [Rhone-Poulenc France]

Rhodasurf® B-1. Laureth-16; surfactant. [Rhone-Poulenc France]

Rhodasurf® BC-420. CAS 24938-91-8; Trideceth-3; emulsifier, detergent, dispersant for petrol. oils, agricultural formulations; intermediate for manufacturing of high-foaming anionic surfactants. [Rhone-Poulenc Surf.]

Rhodasurf® C-2. CAS 9004-95-9; Ceteth-2; surfactant, emollient for creams and lotions. [Rhone-Poulenc Surf.]

Rhodasurf® DA-4. CAS 26183-52-8; Isodeceth-4; wetting

---

‡ Trade name and manufacturer not verified   § Trade name without identified manufacturer

agent, penetrant for pressure dyeing of fabrics, textile and industrial applications. [Rhone-Poulenc Surf. Canada]

Rhodasurf® DA-530. CAS 61827-42-7; Isodeceth-4; low foaming rapid wetting agent for industrial, institutional and household cleaners, textile compounding; scouring agent, emulsifier for defoamers. [Rhone-Poulenc Surf.]

Rhodasurf® E 400. CAS 25322-68-3; PEG-8; surfactant intermediate; binder and lubricant in compressed tablets; softener for paper, plasticizer for starch pastes and polyethylene films; coupling agent for skin care lotions. [Rhone-Poulenc Surf.]

Rhodasurf® L-4. CAS 68002-97-1; Laureth-4; emulsifier, thickener, wetting agent, pigment dispersant, lubricant, solubilizer for cosmetic and industrial emulsions; textile scouring agent, emulsion polymerization, metal cleaning, monomer systems, floor waxes, paper finishes, rubber; emollient for pharmaceuticals. [Rhone-Poulenc Surf.]

Rhodasurf® LA-3. CAS 68131-39-5; C12-15 pareth-3; detergent base and emulsifier for dishwash, personal care products, industrial applications; textile lubricant. [Rhone-Poulenc Surf.]

Rhodasurf® ON-870. CAS 9004-98-2; Oleth-20; high foaming emulsifier, stabilizer, dispersant, wetting agent, solubilizer for mineral oils, fatty acids, waxes; for industrial cleaners, metal cleaners, agriculture, paints, adhesives, textile, leather, cosmetic, pharmaceutical industries. [Rhone-Poulenc Surf.; Rhone-Poulenc France]

Rhodasurf® PEG 400. CAS 25322-68-3; PEG-9; intermediate, plasticizer, solv., lubricant, coupler in cosmetic lotions. [Rhone-Poulenc France]

Rhodasurf® PEG 3350. CAS 25322-68-3; PEG-75; mold release and antistat for rubber products; binder. [Rhone-Poulenc Surf.]

Rhodasurf® S-2. CAS 9005-00-9; Steareth-2; emollient, detergent, emulsifier for creams and lotions, facial cleansers, bath and toiletries. [Rhone-Poulenc Surf.]

Rhodasurf® T-95. CAS 24938-91-8; Trideceth-10; surfactant. [Rhone-Poulenc France]

Rhodasurf® TB-970. CAS 9005-00-9; Steareth-200; detergent, viscosifier, foam booster, dispersant, surfactant for most dry blending operations in detergent formulations, sanitizers; controls dissolution rate of solid or block type hard surface cleaners; textile wet processing aid; biodeg. [Rhone-Poulenc Surf.; Rhone-Poulenc France]

Rhodasurf® TDA-6. CAS 24938-91-8; Trideceth-6; coemulsifier, wetting agent, dispersant for cutting oils, low temp. wool scouring. [Rhone-Poulenc Surf. Canada]

Rhodasurf® TMP-3. Trimethylolpropane triethoxylate; monomer for coatings; reactive diluent; useful in alkyds and polyurethanes. [Rhone-Poulenc Surf.]

Rhodaterge® 206C. Proprietary; high foaming upholstery cleaner concentrate. [Rhone-Poulenc Surf.]

Rhodaterge® LD-50Q. Blend; detergent concentrate for dishwashing and general purpose cleaning; relatively mild to skin. [Rhone-Poulenc Surf. Canada]

Rhodaterge® SSB. Sodium lauryl sulfate, disodium lauryl sulfosuccinate, propylene glycol; surfactant. [Rhone-Poulenc Surf.]

Rhodaterge® WHC-347. Formulated blend; high foaming surfactant for formulating waterless hand cleaners; wetting agent. [Rhone-Poulenc Surf.]

Rhodeftal. Polyamide-imide wire enamel. [Rhone-Poulenc UK]

rhodeoretin. Jalapin, the chief constituent of Jalap resin.

Rhodester. Polyester resins. [Rhone-Poulenc NV] *

Rhodialite. A proprietary cellulose acetate. §

Rhodialux. Ultraviolet stabilizers for polymers. [Rhone-Poulenc UK]

Rhodialux A. CAS 131-57-7; EINECS 205-031-5; Benzophenone-3. [Rhone-Poulenc]

Rhodialux S. CAS 4065-45-6; EINECS 223-772-2; Benzophenone-4. [Rhone-Poulenc France]

Rhodia-Phos. CAS 7758-29-4; Sodium tripolyphosphate; used for detergents, water treatment, paper industry, food industry, textile industry, animal food. [Rhone-Poulenc NV] *

Rhodiarome. CAS 121-32-4; Ethyl vanillin; flavor for baking, cereal, diary, cheese, processed foods, beverages, confections. [Rhone-Poulenc Food Ingreds.; Rhone-Poulenc UK]

Rhodiasolv RPDE. Dibasic ester solvent. [Rhone-Poulenc UK]

Rhodiastab 50. Food grade PVC heat stabilizer. [Rhone-Poulenc UK]

Rhodiastab 83. Technical grade PVC heat stabilizer. [Rhone-Poulenc UK]

Rhodicare XC. CAS 11138-66-2; Xanthan gum; thickening and stabilizing hydrocholloid; emulsion/foam stabilizer; imparts visc. and suspends insol. additives in cosmetics and toiletry formulations. [Rhone-Poulenc Surf.; Rhone-Poulenc France]

Rhodigel®. CAS 11138-66-2; Xanthan gum, food grade; emulsion stabilizer, suspending agent, thickener for cosmetic and pharmaceutical applications. [Rhone-Poulenc Food Ingreds.; Rhone-Poulenc UK] *

Rhodigel® EZ. CAS 11138-66-2; Xanthan gum; used in cosmetic applications to suspend insol. additives; excellent pseudo plastic properties; in baking, cereals, meat, poultry, dairy/cheese, processed foods, beverages, and confections. [Rhone-Poulenc Food Ingreds.]

Rhodine. Acetyl salicylic acid. [Rhone-Poulenc UK]

rhodinol. CAS 6812-78-8; A terpene alcohol prepared from the oils of rose, geranium, and citronella. It is practically pure geraniol; used in perfume manufacture and as a flavoring agent.

rhodite. See rhodium gold.

rhodium. Metallic element; Rh; CAS 7440-16-6; EINECS 231-125-0; alloy with platinum for high temp. thermocouples, furnace windings, laborabory crucibles, spinnerets for rayon, electrical contacts, jewelry, catalyst, optical instrument mirrors, electrodeposited metal coatings, vacuum-deposited glass coatings. [Aldrich; Atomergic Chemetals; Degussa; Noah Chem.]

rhodium chloride. (rhodium trichloride) RhCl3; CAS 10049-07-7; EINECS 233-165-4; Manufacture of rhodium trifluoride. [Aldrich; Atomergic Chemetals; Degussa; Noah Chem.]

rhodium gold. (rhodite). A native alloy of from 57-66% gold and 34-43% rhodium.

rhodium trichloride. See rhodium chloride.

Rhodizite. A borate of lime, $3CaO \cdot 4B_2O_3$, imported from the West Coast of Africa.

Rhodocap. Cyclo dextrins. [Rhone-Poulenc UK]

Rhodoid. A proprietary plastic material made from cellulose acetate. §

rhodole. Intermediate products between fluoresceine-

---

* Trade name not verified as available

† Trade name verified as obsolete

phthalein and the rhodamines.

Rhodopas. Polyvinyl acetates. [Rhone-Poulenc NV] *

Rhodopas AX. Polyvinyl acetate/polyvinyl chloride copolymers. [Rhone-Poulenc NV]

Rhodopas X. Polyvinyl chlorides. [Rhone-Poulenc NV]

Rhodopol® 23. CAS 11138-66-2; Xanthan gum; thickener for agricultural formulations. [Rhone-Poulenc Surf.; Rhone-Poulenc Chemicals Ltd; Rhone-Poulenc France]

Rhodopol® XGD. CAS 11138-66-2; Xanthan gum; viscosifier for drilling. [Rhone-Poulenc/Water Soluble Polymers]

Rhodorsil®. Silicones. [Rhone-Poulenc UK]

Rhodorsil® 1505. Methylphenyl silicone resin; used for high-temp. paints; features excellent adherence to metal. [Rhone-Poulenc Silicones]

Rhodorsil® 10336. Methyl silicone resin; used for high-temp. paints; for high air-drying speed. [Rhone-Poulenc Silicones]

Rhodorsil® AF 422. Dimethylpolysiloxane emulsion; antifoam for aq. systems, agriculture, textile, chemical, rubber, metallurgy industries. [Rhone-Poulenc Silicones]

Rhodorsil® AF 426R. Silicone emulsion; antifoam for agricultural formulation. [Rhone-Poulenc Silicones]

Rhodorsil® RS 44, RS 48. CAS 63394-02-5; Silicone rubber; for molding, calendering, or extrusion applications; recommended for general purpose applications including sheet goods, gasketing, oil seals, misc. parts; peroxide-cured; no post-cure required. [Rhone-Poulenc Silicones] *

Rhodorsil® SC 5020. Silicone emulsion; antifoam for agricultural formulation. [Rhone-Poulenc Silicones]

Rhodoviol. Polyvinyl alcohol. [Rhone-Poulenc NV]

Rhometal. A complex nickel-iron alloy having high electrical resistivity and retaining its permeability up to very high frequencies; used for television transformers, special radio transformers, HF alternators, etc.

Rhon'Sec. Laboratory absorbent. [Rhone-Poulenc UK]

Rhoplex®. An acrylic resin for textile finishes. [Rohm & Haas UK]

Rhoplex® 2133. Acrylic polymer emulsion; for interior concrete and resilient tile sealers. [Rohm & Haas]

Rhoplex® AC-64. Acrylates copolymer; vehicle for exterior and interior latex paints. [Rohm & Haas]

Rhoplex® B-15. Acrylic latex; semireactive emulsion used for coated materials, adhesives; excellent cohesive and tensile strength, durability; will crosslink with melamine or urea/formaldehyde resins. [Rohm & Haas]

Rhoplex® K-3. Acrylates copolymer; provides softness, resilience and low temp. flexibility to treated fabrics; hand modifier, glass fabric adhesive, fabric finishing, pigment printing and dyeing, fabric backing. [Rohm & Haas]

Rhoplex® Multilobe 100. Acrylates copolymer; vehicle for exterior flat, sheen and semigloss paints. [Rohm & Haas]

Rhoplex® NT-2624. Acrylic polymer emulsion; provides excellent gloss and durability for floor polishes. [Rohm & Haas]

Rhotanium. A series of alloys, consisting mainly of gold (60-90%.) and palladium, and in some cases with a small proportion of rhodium. They are said to be more resistant to hot concentrated sulfuric acid and fused caustic soda than lead.

Rhovinal B. Polyvinyl butyrals. [Rhone-Poulenc NV] *

Rhovinal F. Polyvinyl formals. [Rhone-Poulenc NV] *

Rhubafuran. 2,4-Dimethyl-4-phenyltetrahydrofuran.

[Quest Int'l. UK Ltd]

rhubarbaric acid. See rhaponticin.

rhubarbarin. See rhaponticin.

rhyolite. A volcanic rock. It usually contains lime and iron.

Riamat. CAS 51-48-9; Thyroxine I 125; radioactive agent. [Mallinckrodt Inc] *

Riblene. A trade name for polyethylene. [ABCD Petrochimica] ‡

riboflavin. (vitamin B_2; 7,8-dimethyl-10-(1´d-ribityl) isoalloxazine; flavaxin; riboflavine) Organic compound.; $C_{17}H_{20}N_4O_6$; CAS 83-88-5; EINECS 201-507-1; crystalline pigment; dietary supplement; principal growth-promoting factor of vitamin B_2 complex (functions as flavor protein in tissue respiration). [Am. Biorganics; BASF; Bio-Rad Labs; EM Industries; Hoffmann-La Roche; Honeywill & Stein Ltd; Takeda USA]

D-ribosyl uracil. See uridine.

Ricaccel. Cure rate accelerator for chloroprene elastomers; imparts good storage life and gives outstanding cured properties [Ricon Resins]

rice paper. A paper made from plant pith, particularly in China and Japan.

Rice-Pro EN-20. CAS 977-59-33-8; Hydrolyzed rice protein; cosmetic ingredient for skin and hair care products. [Brooks Industries]

rice rubber. An elastic cellulose product made from Japanese rice.

Rice's bromide solution. A solution containing 125 parts bromine, 125 parts sodium bromide, and 1,000 parts water; used in the determination of urea.

Richard's aluminum solder. An alloy of 71.5% tin, 25% zinc, and 3.5% aluminum.

Richardson's speculum metal. An alloy of 65.3% copper, 30% tin, 2% silicon, 2% arsenic, and 0.7% zinc.

Riché gas. A gas obtained during the dry distillation of wood. It contains, on an average, about 60% carbon dioxide, 25% carbon monoxide, 15% methane, and a very small quantity of hydrogen.

ricin. CAS 9009-86-3; The name given to the toxic constituents of castor beans; used as a reagent for pepsin and trypsin.

Ricinion. CAS 61791-12-6; PEG-33 castor oil. solv., emulsifier for cosmetic/pharmaceutical emulsions. [Gattefosse]

ricinus oil. See castor oil.

Ricobond 1031, 1731, 1756. Polymeric; reactive adhesive promoter for compounding with elastomers to give increased adhesion to metal, elastomer, plastic, mineral, fabric, fiber, etc. [Ricon Resins]

Ricolite. See Bakelite. §

Ricon 100. SBR random copolymer; thermosetting liquid resin system; outstanding elec. properties, thermal stability, moisture and age resistance, adhesion, and chemical resistance; Ricon 100 used for elec. potting and impregnation of transformers, capacitors, motors laminates, molding compounds and castings, rubber modifiers, mica paper binder, nuclear heat shield; for harder cure zinc-rich coatings, and molding compounds. [Ricon Resins]

Ricon 130MA8. Polybutadiene maleic anhydride adducted resin; resin for elec. applications (low dielec. and moisture absorp. flexible epoxy formulations), room temperature cured elastomers for sealant and coating applications. [Ricon Resins]

---

‡ Trade name and manufacturer not verified   § Trade name without identified manufacturer

Ricon 159. CAS 9003-17-2; 1,2-Polybutadiene homopolymer resin; thermosetting resin used in laminates, radomes, mica sheets, bonding flexible mica parts; active homopolymer used for continuous production of laminate; forms prepregs which block easily; not recommended for B-stage applications. [Ricon Resins]

Ricon P30/Dispersion. 1,2-Vinylpolybutadiene disp. in Micro-Cel E; additive for carbon-reinforced natural rubber; nonextractable plasticizer and processing aid resulting in improved carbon black loading; crosslinkable peptizer. [Ricon Resins]

Ricoroof. Two-part room temperature curing elastomer; formulated for outdoor applications such as roofing, flashing repair, parking garage concrete sealants. [Ricon Resins]

Ricoseal. Two-part elastomer; room temperature curable soft sealant and potting compound; resist. to penetration by moisture, flexible at low temps., very impact energy absorbent. [Ricon Resins]

Ricotuff, Ricotuff L.V. Two-part anhydride/epoxy system built on the polybutadiene chain; strong adhesive for structural applications; excellent for elec. applications; for manufacturing of reinforced laminates; also suitable for radomes, coatings; superior elec. properties, better heat resist.; mix ratio 2:1 (A/B). [Ricon Resins]

Ric-Syn Wax. Hydrogenated castor oil. [United Catalysts]

Ridacto®. Activator. [Kenrich Petrochemicals]

Rid-A-Roach. A cockroach trap which contains an attractant to bring the roach to the sticky trap; a nonchemical method of control that is safe and easy-to-use. [Lawn & Garden Products Inc]

Ridaura. CAS 34031-32-8; Auranofin; antirheumatic. [SmithKline Beecham]

Ridene. CAS 3691-35-8; An oil formulation containing 2.5% of chlorophacinone; an anticoagulant rodenticide; bait to control black rats, brown rats, house mice and voles. [Ace Chemicals Ltd]

Ridoline. Metal treating compositions. [ICI Chem. & Polymers Ltd] *

Ridomil MBC 60WP. Carbendazin + metalaxyl; a fungicide mixture used to prevent the spread of fruit and vegetable storage diseases. [Ciba-Geigy Agrochemicals]

Ridomil Plus 50WP. Copper oxychloride + metalaxyl; a systemic fungicide mixture used to prevent mildew and root rot in a range of fruit and vegetables. [Ciba-Geigy Agrochemicals]

Ridosol. Metal treating compositions. [ICI Chem. & Polymers Ltd] †

Rifadin. CAS 13292-46-1; Rifampin; an antibiotic used in the treatment of pulmonary tuberculosis. [Dow UK] *

Rifamate. A combination product consisting of rifampin and isoniazid; used in the treatment of tuberculosis. [Dow UK] *

rifamycin. Antibiotics isolated from a *Streptomyces mediterranei*.

Rifleite. Nitrocellulose gelatinized by acetone; an explosive.

Rigidex. Polyolefins and derivatives; used for manufacture of plastics and plastic articles. [BP Chemicals Ltd]

Rigidex 3. High density polyethylene copolymer having a density of 0.946, melt index 0.3. [BP Chemicals Ltd]

Rigidex 9. High density polyethylene with a density of 0.960 and a melt index of 0.9. [BP Chemicals Ltd]

Rigidex X4RR. High density polyethylene having a density

of 0.946. [BP Chemicals Ltd]

Rigidite®. Advanced composites based on reactive resins and glass, carbon, and aramid fibers. [BASF AG]

Rigidoil. Oil spill clean up products. [BP Chemicals Ltd]

Rigilene. Polyethylene; sheet and block for dockside fendering. [Stanley Smith & Co Plastics Ltd]

Rigipore. Expandable polystyrene. [BP Chemicals Ltd]

Rikospray Balsam. Benzoin topical protective in aerosol form containing benzoin BPC 12.5% w/w (equivalent to dissolved solids of benzoin 9% w/w), prepared storax BPC 2.5% w/w. [3M Pharmaceuticals] †

Rikospray Silicone. Topical protective in aerosol form containing aluminum dihydroxyallantoinate 0.5% w/w, cetylpyridinium chloride BP 0.02% w/w, dimethicone 1000 BPC to 100%. [3M Pharmaceuticals] †

rilan wax. Waxy acids and vegetable oils hardened by hydrogenation. It is obtained from the higher alcohols of the fat series.

Rilata. Vulcanized bitumen linseed oil mixture containing 16% sulfur.

Rilsan®. Nylon 11 and nylon 12, molding goods, extrusion goods and fine powder coating for metal; used for medical and sporting goods. [Atochem Inc] *

Rilsan® BESHVO, BESVO. CAS 25035-04-5; Nylon 11; extrusion grade resin for tubular structures for applications (hoses, coiled air tubing, fuel lines); in blow molding of automotive gas tanks, fuel containers, and hollow products; low temp. toughness, resistance to hydraulic fluids and chemicals, dimensional stability, and resistance to cold flow; BESHVO is high visc. grade for large diameter pipe and complicated profile shapes. [Atochem N. Am./Plastics]

Rimactane. A proprietary preparation of rifampicin; an antibiotic and antituberculous agent. [Ciba-Geigy Corp]

Rimactazid. A proprietary preparation of rifampicin and isoniazid; used in the treatment of tuberculosis. [Ciba plc] *

Rimadyl. CAS 53716-49-7; Carprofen; anti-inflammatory. [Hoffmann-LaRoche Inc] *

Rimevax Injection. Live attenuated measles vaccine; immunization against measles. [SmithKline Beecham] *

Rimflex A/A. Thermoplastic elastomer; provides high mech. properties within a given hardness range; suitable to replace thermoplastic PU, partially cured thermoplastic vulcanizates, thermoset rubber, and high performance PVC compounds; processable via conventional thermoplastic techniques allowing total recyclability of regrind. [Syn. Rubber Tech.]

Rimifon. CAS 54-85-3; Proprietary preparation of isoniazid; antitubercular agent. [Roche Products Ltd] *

RIMline® 8711B/8700A. Two-component modified polyurethane system; for open or closed pour molding of low dens. structural RIM composites; for automotive interior trim, e.g., door panels, package trays, sunshades, seat backs. [ICI Polyurethanes]

RIMline® GMR-5000. Two-component modified polyurethane system; for structural reaction injection molding of fiber-reinforced composites; for radiator or battery supports or anywhere load-bearing thermal, environmental, and structural requirements are important. [ICI Polyurethanes]

RIMline® GMR-8711. Two-component PU system; low dens. structural reaction injection molding system providing design flexibility and tailored thickness and glass

---

* Trade name not verified as available   † Trade name verified as obsolete

content. [ICI Polyurethanes]

Rimplast® . Silicone-containing thermoplastic compositions; for injection molding and extrusion of polymers having interpenetrating networks. [Hüls Am.]

Rimthane. Multi-component systems used in the manufacture of polyurethane products via reaction injection molding. [Dow UK]

Rinatec. CAS 22254-24-6; A proprietary preparation of ipratropium bromide; an anticholinergic agent, a nasal spray used for the treatment and management of water rhinorrhea associated with perennial rhinitis. [Boehringer Ingelheim Ltd]

Ringer solution. An isotonic solution containing 0.7% sodium chloride, 0.03% potassium chloride, and 0.025% calcium chloride in water; used in physiological experiments.

Ringmaster. CAS 5259-88-1; Oxycarboxin; a systemic fungicide for the control of fairy rings in grass. [Rhone-Poulenc Environmental Prods. Ltd]

Rinoxin. A rodenticide. [Gerhardt Pharmaceuticals] *

Rinsan. Dairy hygiene circulation cleaner. [The Wellcome Foundation Ltd] *

Rintal®. Broad-spectrum anthelmintic for use in sheep, goat, camels, cattle, horses; veterinary medicine. [Bayer AG]

Rintal® Plus. Broad-spectrum anthelmintic plus boticide; used for horses. [Bayer AG]

Riopan. CAS 1317-26-6; Magaldrate; aluminum magnesium hydroxide sulfate hydrate; antacid. [Wyeth Laboratories] *

Rio Resin. A proprietary amorphous resin. [R T Vanderbilt Co Inc] †

Ripercol. A proprietary preparation containing levamisole hydrochloride; veterinary antihelmintic (cattle). [Janssen Pharmaceutical Ltd] *

Ripping ammonal. It contains 84-87% ammonium nitrate, 7-9% aluminum, 2-3% charcoal, and 3-4% potassium dichromate; used as an explosive.

Rippite. An English explosive containing 56-63% nitroglycerin gelatinized with a small quantity of collodion cotton, potassium nitrate, wood meal, castor oil, ammonium oxalate, with the addition of calcium or magnesium carbonate, and a petroleum jelly.

Rismavac-CR6. CV1 988, clone, CR-6 strain, frozen chicken herpes virus; for immunization of poultry. [Intervet Inc]

Riso. A material consisting of ammonium carbonate ground in a mixture of mineral and vegetable oils (cycline oil); used in the manufacture of sponge rubber.

Risolex. Tolclofos-methyl; an organophosphorus fungicide which gives protection against soil-borne diseases. [Schering Agriculture]

Rissicol. A castor oil powder, containing 49% castor oil and 36% of inorganic matter, mainly magnesia.

Ristin. A 25% solution of ethylene-glycol monobenzol ester.

Riston®. For the electrical industry. [DuPont UK]

Risunal. A proprietary preparation of β-diethylaminobutyric acid, aniline hydrochloride, isopropylphenazone, and ethyl and benzyl nicotinate; a rubefacient skin ointment. [Geistlich Sohne AG] *

Rita AZ. Guay-azulene sodium sulfonate; emollient, humectant, pigment, conditioner, lubricant for personal care formulations. [RITA]

Rita CA. CAS 36653-82-4; EINECS 253-149-0; Cetyl alcohol; thickener, opacifier, emollient for cosmetic formula-

tions. [RITA]

Rita EDGS. CAS 627-83-8; EINECS 211-014-3; Glycol distearate; emulsifier, emulsion stabilizer, emollient, pearlescent, opacifier, visc. builder for cosmetic systems. [RITA]

Rita EGMS, GMS. Glycol stearate; emulsifier, emulsion stabilizer, emollient, pearlescent, opacifier, visc. builder for cosmetic systems. [RITA]

Rita HA C-1. CAS 9067-32-7; Sodium hyaluronate; protein for use in skin and hair care preps. [RITA]

Rita KA. CAS 501-30-4; Kojic acid; emollient, humectant, pigment, conditioner, lubricant for personal care formulations. [RITA]

Rita SA. CAS 112-92-5; EINECS 204-017-6; Stearyl alcohol; thickener, opacifier, emollient for cosmetic formulations. [RITA]

Ritabate 20. Polysorbate 20; emulsifier for personal care products. [RITA]

Ritabate 60. CAS 9005-67-8; Polysorbate 60; emulsifier for personal care products. [RITA]

Ritabate 80. CAS 9005-65-6; Polysorbate 80; emulsifier for personal care products. [RITA]

Ritacet-20. CAS 68439-49-6; Ceteareth-20; emulsifier for personal care products. [RITA]

Ritaceti. CAS 8002-23-1; Cetyl esters; raw material for personal care products. [RITA]

Ritacetin. CAS 117-39-5; Quercetin; nutrient, humectant for hair and skin care formulations; hair repair agent; soothing to skin. [RITA]

Ritacetyl®. CAS 61788-48-5; EINECS 262-979-2; Acetylated lanolin; superfatting agent for soaps, shampoos; film-former for creams and lotions, water resistant films. [RITA]

Ritachlor 50%. Aluminum chlorohydrate; raw material for personal care products. [RITA]

Ritachol®. Mineral oil and lanolin alcohol; used in cosmetics and toiletries; stabilizer for emulsions, dispersions, and suspensions; primary or auxiliary emulsifier; epidermal moisturizer, lubricant, and emollient. [RITA]

Ritachol® 1000. Cetearyl alcohol, polysorbate 60, PEG-150 stearate, steareth-20; emulsifer for personal care products, pharmaceuticals, and household specialties. [RITA]

Ritachol® SS. CAS 2778-96-3; EINECS 220-476-5; Stearyl stearate; raw material for personal care products. [RITA]

Ritacollagen BA-1. CAS 9007-34-5; Soluble collagen; protein for use in skin and hair care preps. [RITA]

Ritaderm® . Petrolatum, lanolin, sodium PCA, polysorbate 85; emollient, moisturizer, and lubricant used in cosmetics. [RITA]

Ritahydrox. CAS 68424-66-8; EINECS 270-315-8; Hydroxylated lanolin; w/o emulsifier, hypoallergenic emollient. [RITA]

Ritalafa® . CAS 68424-43-1; EINECS 270-302-7; Lanolin acid; film-former, emollient; rewetting of makeup preparations. [RITA]

Ritalan® . Lanolin oil USP; moisturizer, plasticizer, penetrant, emollient; hypoallergenic, nonsensitizing skin lubricant. [RITA]

Ritalan® AWS. CAS 68458-58-8; PPG-12-PEG-65 lanolin oil; auxiliary emulsifier, moisturizer, emollient. [RITA]

Ritalan® C. IPP and lanolin oil; blending agent, emollient, epidermal penetrant, rewetting agent, and solubilizer for waxes and other oil-sol. or disp. materials; personal care

---

‡ Trade name and manufacturer not verified   § Trade name without identified manufacturer

products. [RITA]

Ritalanine. CAS 56-41-7; β-Alanine; protein for use in skin and hair care preps. [RITA]

Ritalastin EL-10. Hydrolyzed elastin; protein for use in skin and hair care preps. [RITA]

Ritalin Hydrochloride. CAS 298-59-9; Methylphenidate hydrochloride; stimulant. [Ciba-Geigy Corp] *

Ritaloe 1X. Aloe vera; emollient, humectant, pigment, conditioner, lubricant for personal care formulations. [RITA]

Ritamectant K2. Dipotassium glycyrrhizinate; humectant for personal care products. [RITA]

Ritamectant PCA. CAS 28874-51-3; Sodium PCA; humectant for personal care products. [RITA]

Ritapan CAP. CAS 137-08-6; Calcium pantothenate; nutrient, humectant for hair and skin care formulations; hair repair agent; soothing to skin. [RITA]

Ritapan D, DL. d-Panthenol; nutrient, humectant for hair and skin care formulations; hair repair agent; soothing to skin. [RITA]

Ritapan NAP. Sodium pantothenate; nutrient, humectant for hair and skin care formulations; hair repair agent; soothing to skin. [RITA]

Ritapan TA. Panthenyl triacetate; nutrient, humectant for hair and skin care formulations; hair repair agent; soothing to skin. [RITA]

Ritapeg 150 DS. CAS 9004-99-3; PEG-150 stearate; emulsifier for personal care products. [RITA]

Ritapeg 400 DS. CAS 9005-08-7; PEG-8 distearate; emulsifier for personal care products. [RITA]

Ritaphenone 3. CAS 131-57-7; EINECS 205-031-5; Benzophenone-3; raw material for personal care products. [RITA]

Ritaplast. Mineral oil and polyethylene; raw material for personal care products. [RITA]

Ritaplast R. Lanolin oil, polyethylene; raw material for personal care products. [RITA]

Ritaplast TN. C12-15 alcohols benzoate and polyethylene; raw material for personal care products. [RITA]

Ritapro 100. Cetearyl alcohol, steareth-20, and steareth-10; o/w emulsifier. [RITA]

Ritapro 165. Glyceryl stearate and PEG-2 stearate; o/w emulsifier for creams, lotions. [RITA]

Ritaquat Q. Steartrimonium hydrolyzed animal protein; protein for use in skin and hair care preps. [RITA]

Ritasilk. CAS 96690-41-4; Hydrolyzed silk protein; protein for use in skin and hair care preps. [RITA]

Ritasilk Powd. CAS 9009-99-8; Silk powd.; protein for use in skin and hair care preps. [RITA]

Ritasol. CAS 63393-93-1; EINECS 264-119-1; Isopropyl lanolate; emollient, spreading agent, water-resistant film former for lip preparations. [RITA]

Ritasynt IP. Glycol stearate and other ingreds.; foam booster, thickener, opacifier for cosmetics. [RITA]

Ritatin. CAS 58-85-5; Biotin; nutrient, humectant for hair and skin care formulations; hair repair agent; soothing to skin. [RITA]

Ritawax. CAS 8027-33-6; EINECS 232-430-1; Lanolin alcohol; emulsifier, emollient for skin preparations. [RITA]

Ritawax 5. CAS 61791-20-6; Laneth-5; emulsifier, solubilizer, emollient for creams and lotions. [RITA]

Ritawax AEO. Polysorbate 80, acetylated lanolin alcohol, cetyl acetate; emollient, lubricant, moisturizer, penetrant, solubilizer, dispersant, plasticizer for personal

care products. [RITA]

Ritawax ALA. Cetyl acetate and acetylated lanolin alcohol; emollient, lubricant, moisturizer, penetrant. [RITA]

Rit-Cizer #8. N-Ethyl o/p toluenesulfonamide; plasticizer for adhesives, paints, printing inks, epoxy resins, polyamide resins, phenolics, melamine resins. [Rit-Chem]

Rit-Cizer #9. o/p Toluene sulfonamide; plasticizer for thermosetting resins (melamine, urea, phenolics), nylon, casein, PVAc; imparts good gloss, improves wetting action. [Rit-Chem]

Riteflex®. Thermoplastic polyester elastomer. [Hoechst AG]

Riteflex® 347ZS. Thermoplastic polyester elastomer; processable by blow molding or extrusion; improved thermal stability for improved long-term prop. retention. [Hoechst Celanese] *

Riteflex® 540. Thermoplastic polyester elastomer; softest, most flexible grade; processable on conventional injection molding and extrusion equip.; recommended for colors and polymer modification. [Hoechst Celanese] *

Riteflex® BP 8929. Thermoplastic polyester elastomer alloy; high impact grade with excellent processability, especially good mold filling characteristics in injection molding applications. [Hoechst Celanese] *

Riteflex® BP 9057. Thermoplastic polyester elastomer alloy; most flexible grade with excellent low temp. impact properties; excellent processability and mold filling capabilities. [Hoechst Celanese] *

Rit-Ester B-100. Functional polyester; basic polymer or properties enhancer for adhesives and coatings; contains functional hydroxyl groups which can be reacted with melamines, isocyanates, and other resin systems. [Rit-Chem]

Ritha. An alkaline deposit found on the land in India; used as a soap substitute.

Ritoleth 2. CAS 9004-98-2; Oleth-2; o/w emulsifier, solubilizer; stable over wide pH range. [RITA]

Rit-O-Lite MHP-S. Formaldehyde/toluenesulfonamide polymer; modifier, adhesion promoter for coatings, lacquers, printing inks, adhesives; compatible with alkyds, acrylics, urethanes, nitrocellulose and vinyls. [Rit-Chem]

Rit-O-Lite MS-80. Promotes, modifies, and improves adhesion, gloss retention, solv. release, and lowers visc. of other resin systems. [Rit-Chem]

Ritox 35. Laureth-23; emulsifier for personal care products. [RITA]

Ritox 52. CAS 9004-99-3; PEG-40 stearate; emulsifier for personal care products. [RITA]

Ritox 721. CAS 9005-00-9; Steareth-21; emulsifier for personal care products. [RITA]

Rival. Glyphosate + simazine; a translocated and residual herbicide for the control of grasses and weeds. [Chipman Ltd]

Rivalit P. A German explosive containing nitro-glycerin, ammonium nitrate, vegetable meal, nitro-compounds, and potassium perchlorate.

rivet metal. An alloy of copper and tin, to which zinc is sometimes added.

Riviera Oil. *See* aix oil. §

Rivotril. CAS 1622-61-3; A proprietary preparation of clonazepam; an anticonvulsant. [Roche Products Ltd] *

RIX 90149. Modified cycloaliphatic amine adduct; curing agent for room temperature cure of epoxy resins; suggested end uses including industrial floor toppings, high

---

* Trade name not verified as available   † Trade name verified as obsolete

build glaze, sealer or gel coatings, general purpose castings and encapsulations. [Shell] *

Rizolex. Tolclofos-methyl; fungicide. [Schering Agrochemicals Ltd]

RJ-100. Styrene-allyl alcohol resins; paint coating ingredient. [Monsanto Co] *

R-MA 11®. CAS 21645-51-2, CAS 546-93-0; Aluminum hydroxide, magnesium carbonate; antacid which minimizes constipative or laxative effects. [Reheis]

RMD. High molecular weight polymers based on sodium acrylate; used in the mining industry. [Cyanamid BV]

RMR. High molecular weight polymers based on sodium acrylate; used in mining industry. [Cyanamid BV]

Roaccutane. Proprietary preparation of isotretinoin (13-cis-retinoic acid); systemic treatment for acne. [Roche Products Ltd] *

Roachban. A cockroach insecticide. [Murphy Chemical Co Ltd] †

Roach Stoppers. Blatticide. [Atomergic Chemetals Corp]

Roanoid. A proprietary urea-formaldehyde molding compound. §

roaster slag. A slag produced in the purification of copper metal. It contains from 17-40% copper as silicate and metal.

Ro-A-Vit. CAS 68-26-8; Proprietary preparation of vitamin A (retinol). [Roche Products Ltd] *

Robac. Principally rubber accelerators and vulcanizing agents; some activators and antioxidants; for vulcanization and protection of natural and synthetic rubbers. [Robinson Brothers Ltd]

Robac T.B.Z. Zinc thiobenzoate; a low temperature peptizing agent for natural rubber.

Robacure. Principally rubber accelerators and vulcanizing agents; some activators and antioxidants; for vulcanization and protection of natural and synthetic rubbers. [Robinson Brothers Ltd]

Robalate. CAS 13682-92-3; A proprietary preparation containing dihydroxy aluminum amino acetate; an antacid. [Wyeth Laboratories] *

Robane®. CAS 111-01-3; Squalane NF; moisturizer, emollient, lubricant, humectant; aids spread of topical agents over the skin, increases skin respiration, prevents insensible water loss, imparts suppleness to skin without greasy feel; cosmetics and pharmaceuticals. [Robeco]

Robaxin. CAS 532-03-6; Methocarbamol USP; tablets or injection; tablets for short-term symptomatic treatment of acute musculoskeletal disorders associated with painful muscle spasm; injection for treatment of acute painful muscle spasm due to musculoskeletal disorders or trauma. [Wyeth Laboratories]

Robaxisal Forte Tablets. Methocarbamol USP 400 mg and acetylsalicylic acid 325 mg; used for short-term management of pain and skeletal muscle spasm associated with musculoskeletal disorders such as lumbago, fibrositis, sprains, strains, etc. [Wyeth Laboratories]

Robengatope I-125. Rose bengal sodium I 125; radioactive agent. [Mallinckrodt Inc] *

Robengatope I-131. Rose bengal sodium I 131; diagnostic aid; radioactive agent. [Bristol-Myers Squibb Co Inc] *

Robertson alloy. Consists of 1 part gold, 3 parts silver, and 2 parts tin. It is mixed with mercury for use as a dental filler.

Robert's reagent. For proteins. It consists of 1 volume of pure nitric acid with 5 volumes of a 40%. (saturated)

solution of magnesium sulfate.

Robicillin Vk. CAS 132-98-9; Penicillin V potassium; antibacterial. [Wyeth Laboratories] *

Ro-Bile. Pancrelipase; a concentrate of pancreatic enzymes standardized for lipase content; enzyme. [Rowell Laboratories Inc] *

Robimycin. CAS 114-07-8; Frythromycin; antibacterial. [Wyeth Laboratories] *

Robinul Injection. CAS 596-51-0; Glycopyrrolate USP; anticholinergic. [Wyeth Laboratories]

Robinul Neostigmine Ampoules. Amber glass ampoules containing 500 mcg glycopyrrolate and 2.5 mg neostigmine methylsulfate BP; used for the reversal of residual nondepolarizing (competitive) neuromuscular block. [Wyeth Laboratories]

Robinul Powder. CAS 596-51-0; Glycopyrrolate USP; used for the treatment of plantar and palmar skin for idiopathic hyperhidrosis. [Wyeth Laboratories]

Robitet. CAS 60-54-8; Tetracycline; anti-amebic; antibacterial; antirickettsial. [Wyeth Laboratories] *

Robitussin. CAS 93-14-1; A proprietary preparation containing guaifenesin; a cough linctus. [Wyeth Laboratories] *

Robitussin A.C. A proprietary preparation containing codeine phosphate, guaiaphenesin and pheniramine maleate; a cough linctus. [Wyeth Laboratories] *

Robond. Acrylic polymer based adhesives. [Rohm & Haas]

Roburite I. An explosive containing 87.5% ammonium nitrate, 7% dinitro-benzene, 0.5% potassium permanganate, and 5% ammonium sulfate.

Roburite. An explosive used in mines. It consists of 86% ammonium nitrate and 14% chlorodinitro-benzene.

Roburite III. An explosive containing 87% ammonium nitrate, 11% dinitro-benzene, and 2% chloro-naphthalene.

Rocagel. Soil stabilization formulations. [Rhone-Poulenc UK]

Rocaltrol. Proprietary preparation of calcitriol; vitamin D metabolite used in correction of abnormalities of calcium and phosphate metabolism. [Roche Products Ltd] *

Roccal. A proprietary preparation of benzalkonium chloride; a preoperative skin cleanser. [Bayer AG]

Rocel. Plastics in the form of sheets, bands, strips, films, plates, slabs, blocks, tubes, etc. [Courtaulds plc]

Rocephin. CAS 73384-59-5; Proprietary preparation of ceftriaxone; broad-spectrum cephalosporin antibiotic. [Roche Laboratories; Roche Products Ltd]

Rochdale salt. See Rochelle salt.

Rochelle salt. See potassium sodium tartrate.

Rochelle salt. CAS 304-59-6; (Rochdale salt, tartrated soda, Seignette's salt). Sodium potassium-tartrate, $C_4H_4O_6Na \cdot K \cdot 4H_2O$; used to reduce the silver salts in the silvering of mirrors, and in medicine as a mild aperient. It is the active constituent of seidlitz powders.

rock ammonia. Ammonium carbonate $(NH_4)_2CO_3$.

rock asphalt. Limestone or other material and found naturally, impregnated with bitumen.

rock cork. A variety of asbestos. Also see rock wool.

rock crystal. Transparent and colorless quartz.

rock dammar. A variety of dammar resin derived from Hopea odorata of Burma.

Rocket® Ultra. Tridemorph, fenpropiomorph; systemic fungicide for control of cereal diseases [BASF AG]

rock salt. (halite). Sodium chloride, NaCl, from sea water.

Rocksil. A proprietary trade name for rockwool insulating

---

‡ Trade name and manufacturer not verified   § Trade name without identified manufacturer

materials; withstands 760 C. [Cape Insulation Cape Asbestos Co] *

rock tallow. See bitumen.

Rocktex. A proprietary trade name for rock wool. §

rock wool. (rock cork). A furnace product made from self-fluxing siliceous and argillaceous dolomite in which the basic and acidic constituents are present in proportions that their fluxing action is nearly balanced. The molten rock at 2800-3000 F is atomized by a blast of steam under pressure. The wool treated with a binder is sold as rock cork.

Roclys®. Glucose syrup; sweetening agent for general food use. [Roquette (UK) Ltd]

Rocol P.R. A proprietary silicone-based spray used for mold release. [Rocol Ltd] *

Rocol R.S.7. A proprietary nonsilicone-based wet film spray used for mold release. [Rocol Ltd] *

rocou. See annatto.

Rocryl. Specialty acrylic and methacrylic monomers. [Rohm & Haas UK]

Rocsol. Oxidized hydrocarbon waxes. [Croda Chem. Ltd]

R.O.D. 2-Octyldodecyl ricinoleate; neutral oil improving adhesion, gloss, conditioning in lipsticks, foundation, eye shadow stick, hair creams. [U.S. Cosmetics]

Rodea. CAS 40716-42-5; EINECS 255-051-3; Ricinoleic diethanolamide; emulsifier, thickener, solvating agent, corrosion inhibitor. [Rhone-Poulenc Surf.] †

Rodeo. Aquatic herbicide. [Monsanto Co] *

Rodinal. CAS 123-30-8; A photographic developer. The active constituent is p-aminophenol hydrochloride.

Rodo. Blend of essential oils (deodorants); neutralizes typical dry rubber odors and in emulsions acts as a deodorant in finished latex products. [R T Vanderbilt Co Inc]

rod wax. A wax-like mass deposited on the drill rods in many petroleum oil wells.

Roebaryt. A barium sulfate prepared for use in X-ray work.

Roesch's aluminum solder. An alloy of 50.2% zinc, 50% tin, 0.7% antimony, and 0.2% copper.

Roferose. CAS 50-99-7; Dextrose monohydrate; used for food, pharmaceutical, and industrial applications. [Roquette (UK) Ltd]

Roga. Cellulose acetate spiral wound membrane; reverse osmosis water treatment. [Allied-Signal Fluid Systems]*

Rogé Cavaills. Soap, bath additive, and shampoo; for cleansing care of skin and hair. [Richardson-Vicks Inc]*

Roghan. (Afridi wax). Obtained by boiling safflower oil for 2 hours, then putting it into vessels partly filled with water.

Rogitine. A proprietary preparation of phentolamine mesylate. [Ciba plc] *

Rogor E. Insecticide. [Schering Agrochemicals Ltd]

Rohafloc. Methacrylate based ionic/nonionic flocculants; for sludge dewatering, industrial waste treatment, paper sizing, etc. [Cornelius Chemical Co Ltd] *

Rohagit® SM V. Acrylic polymer; thickener. [Rohm Tech]

Rohamere® 8662. Thermoplastic acrylic polymer emulsion; forms a medium film; as finishing agent in textile, leather, paper applications. [Rohm Tech]

Rohatol® BV 382. Self-crosslinking acrylic polymer emulsion; for nonwoven bonding for household, hospital, hygiene applications, flocking adhesives for textiles. [Rohm Tech]

Rohatol® D 362. Thermoplastic acrylic polymer emulsion; for textile coating, protective foils, paper finishes. [Rohm

Tech]

Rohn alloys. Heat-resisting alloys containing nickel, chromium, and iron, sometimes with the addition of manganese and molybdenum.

Rohrbach's solution. A solution of barium and mercuric iodides (100 g barium iodide and 130 g mercuric iodide heated with 20 cc water to 150-200 C.). The solution is allowed to cool, when a double salt is deposited. The liquid is decanted. The salt is used in the separation of heavy metals by gravity as it has a specific gravity of 3.58.

Rohypnol. CAS 1622-62-4; Proprietary preparation of flunitrazepam; a benzodiazepine with hypnotic properties. [Roche Products Ltd] *

Roica®. Elastic polyurethane filament. [Asahi Chem. Industry]

RokLok® B-3. Polyurethane; binders to control water and broken strata in mines, tunnels, shafts and stream sealing applications; upon injection they permeate the strata and form expanded binder with excellent mech. properties [Flexible Prods.]

Rokon. CAS 149-30-4; 2-Mercaptobenzothiazole; corrosion inhibitor for coolants, greases and fuel; metal deactivator and extreme pressure agent for petroleum lubricants; chemical intermediate. [R T Vanderbilt Co Inc]

Rolafix. Photographic fixer. [May & Baker Ltd] *

Rolaids. CAS 539-68-4; Dihydroxyaluminum sodium carbonate; an antacid. [Parke-Davis] *

Rollit. Graphite powder; used as mandrel bar lubricant. [Lonza AG]

Rollofix X100. One-component expanding polyurethane foam comprising polyol, methyl di-isocyanate, freon and niax; used for filling cavities (domestic and industrial), bonding (sealing against noxious vapors and moisture). [Piccadilly Products Ltd] *

roll sulfur. Sulfur which has been melted and poured into molds.

Rol-man Steel. A proprietary trade name for a high-manganese steel containing 11-14% manganese and 1-1.4% carbon. §

Rolox. Two-part epoxy compounds. [Hardman]

Roman alum. Potash alum, $Al_2(SO_4)_3 \cdot K_2SO_4 \cdot 24H_2O$.

Roman bronze. An alloy of 90% copper and 10% tin.

Roman cement. (Parker's cement). A natural cement made by calcining the modules of argillaceous limestone mixed with calcareous spar, which occurs in London and other clays.

Romane. CAS 111-01-3; Squalane. [A & E Connock (Perfumery & Cosmetics) Ltd]

Romanite. See Roumanite.

Romanium. An alloy of 97.43% aluminum, 1.75% nickel, 0.25% copper, 0.25% antimony, 0.17% tungsten, and 0.15% tin.

Romensin. CAS 17090-79-8; Monensin. [Dista Products Ltd]

Romilar. Proprietary preparation of dextromethorphan; an antitussive. [Roche Products Ltd] *

Romite. It is ammonium nitrate mixed with a solid, melted hydrocarbon (paraffin or naphthalene), gelatinized with a liquid hydrocarbon (paraffin oil), and contains gelatinized potassium chlorate; an explosive.

Rompel's alloy. An antifriction metal, containing 62% copper, 10% zinc, 10% tin, and 18% lead.

Romperit G. A German explosive containing nitroglycerin, ammonium nitrate, vegetable meal, nitro-compounds,

---

* Trade name not verified as available   † Trade name verified as obsolete

and potassium perchlorate.

Rompun® . Sedative, analgesic, anesthetic, and muscle relaxant; for use in cattle. [Bayer AG]

Ronabond®. Pre-packed ready-to-use cementitious mortar; used in concrete repair, the laying of thin screeds and floors and the fixing of building components. [Ronacrete Ltd]

Ronafix. Styrene butadiene waterproof bonding additive; used in concrete repair, the laying of thin screeds and floors and the fixing of building components. [Ronacrete Ltd]

Ronase. CAS 1156-19-0; Tolazamide; antidiabetic. [Rowell Laboratories Inc] *

Ronaset® . Pre-packed high-strength mortars, concretes, and transition strip; used for bedding preformed bridge expansion joints, transition strips, motorway repairs, floors, and screeds. [Ronacrete Ltd]

Rondase. CAS 9001-54-1; A proprietary preparation of hyaluronidase; used to facilitate absorption of injected fluids. [British Drug Houses] *

Rondec. Contains carbinoxamine maleate and pseudoephedrine hydrochloride; antihistaminic and adrenergic. [Ross Laboratories] *

Rondec DM. Contains carbinoxamine maleate, dextromethorphan hydrobromide, and pseudoephedrine hydrochloride; antihistaminic; antitussive; adrenergic. [Ross Laboratories] *

Rondis. Disproportionated resin esters and salts; emulsifier in synthetic latex. [Akzo Chemie UK Ltd]

Rondomycin. CAS 3963-95-9; A proprietary preparation of methacycline hydrochloride; an antibiotic. [Pfizer International]

Ronfalin. Acrylonitrile butadiene styrene; plastic for general purposes and for special applications in the automotive and electrical industries: for toys, domestic appliances, telematic equipment and extruded products; uses include dashboard components and radiator grilles, switches; housings for hair driers and coffee makers, telephones, office machines, pipe and plate material. [DSM NV] *

Ronfaloy E. Blends/compounds with ABS; for use in the open air. [DSM NV] *

Ronfaloy V. Blends/compounds with ABS; for the telematic industry. [DSM NV] *

Ronfusil Steel. A proprietary trade name for a manganese-steel containing 12% manganese. §

Rongal®. Reducing agents for dyeing cellulose fibers with vat dyes. [BASF AG]

Rongalit®. Textile discharging and reducing agents. [BASF plc]

Rongalit® C. CAS 870-72-4; Sodium hydroxymethane sulfonate; reducing and discharge agents for textile printing. [BASF AG]

Rongalite C. A combination of the sodium salt of the unstable sulfoxylic acid and formaldehyde; a reducing agent used in the dye industry, and as a photographic developer.

Roniacol. CAS 100-55-0; Nicotinyl alcohol; vasodilator. [Hoffmann-LaRoche Inc] *

Ronia metal. A brass containing small quantities of cobalt, manganese, and phosphorus.

Ronicol. A proprietary preparation of nicotinyl tartrate; used in the treatment of circulatory disorders. [Roche Products Ltd] *

Ronilan®, DF, FL. CAS 50471-44-8; Vinclozolin; contact fungicide for use in vines, fruit, strawberries, vegetables, ornamentals, hops, etc. [BASF AG]

Ronilon. CAS 50471-44-8; Vinclozolin; a protectant fungicide for a wide range of vegetables and fruit. [BASF plc]

Ronnel. A proprietary preparation of fenchlorphos; an insecticide. §

Ronoxan. Antioxidant pastes for food. [Roche Products Ltd]*

Ronseal. Varnishes for general wood protection. [Sterling Ronseal Ltd]

Ronstar 2G. CAS 19666-30-9; Oxadiazon; a contact herbicide to control weeds and grasses in fruit and ornamental crops. [Embetec Crop Protection Ltd]

Ronstar TX. CAS 16118-49-3, 19666-30-9; Carbetamide + oxadiazon; a residual herbicide for pre-weed emergence control for container grown nursery plants. [Hortichem Ltd]

Roofcover LM. One-pack moisture-curing paint providing protection to roofing materials. [Feb Ltd]

Root Guard. CAS 333-41-5; Diazinon insecticide. [Murphy Chemical Ltd] †

Rooting Powder. CAS 86-87-3; Powder containing 1-naphthylacetic acid; rooting powder. [Vitax Ltd]

Root-Out. CAS 7773-06-0; EINECS 231-871-7; Ammonium sulfamate 98%; herbicide to control weeds and grasses in vegetables and ornamentals prior to planting; tree, weed and brushwood killer. [Dax Products Ltd]

Ropaque. Opaque polymer emulsion; opacifying agent for decorative coatings, paper coatings, etc. [Rohm & Haas]

Ro-Pel. Animal repellent. [Atomergic Chemetals Corp]

rose B. See erythrosin.

Roseclear. Contains bupirimate, pirimicarb and triforine; combined insecticide and fungicide for garden use. [ICI Garden Products]

Rose Ester. A trade name for trichloromethyl phenyl carbinyl acetate; m.p. 85-87 C; a rose scent used in the perfumery industry. [Nipa Laboratories Ltd] *

Rose Food. Granular fertilizer. [Fisons plc, Horticulture Div]*

rosein. An alloy of 44.4% nickel, 33.3% aluminum, 11.1% silver, and 11.1% tin. Another alloy contains 40% nickel, 30% aluminum, 20% tin, and 10% silver; used by jewelers.

Roseline. Finishing agent. [Ciba plc] *

Roselle fiber. A Malay fiber similar to jute.

Rosé oil. (Roshé oil, oil of geranium, oil of rose-geranium. oil of pelargonium, ginger grass oil, Turkish geranium oil, oil of palmarosa). Andropogon oils, obtained from a grass, *Andropogon nardus*. They contain geraniol, $C_{10}H_{18}O$.

roseol. (reuniol). Names applied to citronellol, C_9H_{17} · CH_2OH, or to mixtures of this body with geraniol; used in the manufacture of perfumes.

Rose Plus. Granular fertilizer containing magnesium. [ICI Garden Products]

rose quartz. CAS 14808-60-7; A mineral. It is a variety of quartz, SiO_2, which is stated to owe its color to manganese.

Rose's metal. Fusible bismuth alloys. a) Consists of 42% bismuth, 42% lead, and 16% tin. It melts at 79 C. b) Consists of 33.3% bismuth, 33.3% lead, and 33.3% tin. c) Consists of 48.9% bismuth, 27.5% lead, and 23.6% tin.

Rosette copper. This is obtained in thin films by throwing water on to the surface of molten copper and removing

---

the crusts formed.

rosin. (colophony; gum rosin; rosin gum) Residue from distilling off the volatile oil from the oleoresin obtained from *Pinus palustris* and other species of *Pinaceae*; CAS 8050-09-7; 8052-10-6; EINECS 232-475-7; manufacture of varnishes, paint driers, printing inks, cements, soap, sealing wax, wood polishes, paper, plastics, fireworks, sizes, rosin oil; waterproofing paper, walls; pharmaceutic aid. [Am. Cyanamid Arakawa; Arizona; Georgia-Pacific; Hercules BV; Meer; Natrochem; Veitsiluoto Oy]

Rosinal. Rubber softener and tackifier. [Crowley Chem.]

rosin grease. A combination of rosin oil (*qv*) with lime.

rosin gum. See rosin.

rosinjack. See zinc blende.

rosin oil. CAS 8002-16-2; An oil obtained by the distillation of rosin from 300-400 C; used as an adulterant for olive oil, also as a lubricant for iron bearings.

rosin spirit. (essence of rosin, resin spirit). The first distillate from rosin, 78-250 C. It is a complex mixture of hydrocarbons somewhat resembling turpentine; used as a substitute for turpentine.

Rosintene. Terpene-rosin polymer. [Crowley Chem.]

rosin tin. A yellow variety of the mineral cassiterite or tinstone.

Rosite® 3250A. Thermoset polyester BMC, 5% glass fiber-reinforced; outstanding elec. characteristics; for applications where molds designed for high shrinkage are employed. [Rostone]

Rosite® 4010ES. Thermoset polyester SMC, 22-25% glass fiber-reinforced; SMC for business equip. applications; high strength, dimensional control, and surface smoothness. [Rostone]

Rosite® 4030FS. Thermoset polyester SMC, 30% glass fiber-reinforced; zero shrink material providing surface smoothness for demanding applications, e.g., auto headlamp retainers. [Rostone]

Rosite® ESD Cond. C. Thermoset polyester BMC, 15% glass fiber-reinforced; electrostatic dissipative compound for molded paper guides in printers and plotters. [Rostone]

Roskens. Hand conditioner. [Fisons plc, Pharmaceuticals Div] *

Roskydal®. Unsaturated polyester resins; used for formulation of furniture waxes with and without paraffin wax. [Bayer AG; Bayer plc]

rosolene. A rosin oil obtained by the distillation of rosin.

Ross alloy. A bronze containing 68% copper and 32% tin.

Ross Japan Wax Substitute 525. Fish glycerides, tallow glycerides, oleostearine, microcrystalline wax; for coatings, textile finishes, cosmetics, lubricants, crayons, wax modeling, buffing compounds, pharmaceuticals. [Frank B. Ross]

Ross Spermaceti Wax Substitute 573. CAS 8002-23-1; Cetyl esters; for adhesives, cosmetic creams, lotions, and soaps, medicinals (ointments, salves), candles, textiles, coatings. [Frank B. Ross]

Ross's white. See lithopone.

Ross Wax #100. CAS 8002-74-2; Synthetic wax; mold release in rubber and plastics; external lubricant for PVC; increases melting point, opacity, gloss in wax blends, floor waxes, textiles, coatings, candles, hot melts, paints, inks, asphalt. [Frank B. Ross]

Ross Wax #145, 165. CAS 8002-74-2; High melting point fully refined petroleum wax; lubricant for formulating PVC and elec. wire and cable compounds. [Frank B. Ross]

Rota. pH indicator papers. [Rhone-Poulenc Laboratory Prods. Ltd]

Rotal. Electrolytic coloring of anodized aluminum. [Albright & Wilson Ltd]

Rotalin. CAS 330-55-2; Herbicide containing linuron. [ICI Chem. & Polymers Ltd]

Rotalin. CAS 330-55-3; Linuron; a residual urea herbicide for the control of weeds in field crops. [Farm Protection Ltd]

Rotax®. CAS 149-30-4; 2-Mercaptobenzothiazole; accelerator; primary accelerator for natural and synthetic rubbers; nonstaining and nondiscoloring; used in proofing compounds where lowest odor is desired; corrosion inhibitor in automotive chems., industrial cleaners; protects silverware from sulfur blackening. [R.T. Vanderbilt]

rotenone. CAS 83-79-4; A crystalline material found in the roots of derris, a plant grown in the rubber plantations of the Malay Peninsula; also found in the South American cube plant. It occurs up to 5.5% in derris and up to 7% in cube; $C_{23}H_{22}O_6$; used as an insecticide, in flea powders, fly sprays, and as mothproofing agents.

Roter. A proprietary preparation containing bismuth subnitrate, magnesium carbonate, sodium bicarbonate and frangula; an antacid used in the treatment of peptic ulcers. [FAIR Laboratories] ‡

Rotercholon. A proprietary preparation of turmeric, ox-bile extract, peppermint oil, fennel oil, caraway oil and methyl salicylate; used in the treatment of biliary disorders. [FAIR Laboratories] ‡

Rotersept. A proprietary aerosol spray containing chlorhexidine digluconate; used in the prophylaxis of puerperal mastitis. [FAIR Laboratories] ‡

Rotoval. Gravure printing inks; used for package printing. [Allied-Signal Inc/Sinclair and Valentine Division] *

Rotoxit. Resistant high-silicon-copper alloy; resistant to uric, fluosilicic, fatty acids, dilute hydrochloric, sulfuric, and acetic acids, 30% phosphoric acid, hydrogen peroxide, ammonia, lyes, and sulfates, but not to chromic, nitric, and lactic acids.

Rotra bark. The bark of *Rotra fotsy* and *R. meno*. The bark contains 12.6% tannin.

Rotten stone. A soft, friable aluminum silicate, containing a little organic matter; used as a polishing material. The term is also sometimes applied to Tripoli (*qv*).

Rotuba H. CAS 9004-35-7; Cellulose acetate; molding material. [Rotuba] *

Rouen white. A pigment. It is a clay found near Rouen.

rouge. CAS 1309-37-1; Good qualities of rouge consist of very fine iron oxide, Fe_2O_3, and are used as abrasives. The finest rouge is prepared from safflower.

Roumanite. (Romanite). The amber of Roumania, which much resembles the Prussian variety.

Roundup. CAS 1071-83-6; Glyphosate; herbicide. [Schering Agrochemicals Ltd]

Rousselot Gelatine. CAS 9000-70-8; Gelatin derived from animal tissue available in several mesh sizes; used for photographic emulsions, pharmaceutical hard, soft and micro capsules, edible gelatins for confectionery, meat, dairy and dessert industries. [Sanofi Bio-Industries Ltd]‡

Roussel's solution. A solution of sodium phosphate.

Roussin's black salt. A compound, $Fe_3H_2N_4O_4S_5$, obtained by adding a solution of ferrous or ferric chloride slowly to the mixed solution of potassium nitrite and ammonium sulfide, and then boiling.

Roussin's red salt. A salt, $Fe_2S_4N_2O_2Na_2 \cdot H_2O$, obtained by treating the sodium salt of Roussin's black salt with excess of acid after boiling, and then evaporating.

Roussin's salts. Salts of the type $KFe_4(NO)_7S_3$, obtained when nitric oxide is passed through a suspension of ferrous sulfide in a sulfide solution.

Roux's stain. A microscopic stain that contains 0.5 g gentian violet, 1.5 g methyl green, and 200 cc distilled water.

Rovace 571. CAS 9003-20-7; High molecular weight, highly branched vinyl acetate homopolymer, PVAL stabilized; emulsion for wood and packaging adhesives, textile hand stiffeners, masonry coatings; high wet tack, ease of compounding, machinability. [Rohm & Haas]

Rovace 2113. CAS 9003-20-7; High molecular weight vinyl acetate homopolymer; emulsion used in joint cement, carpet back size, acoustical coatings, textile sizing, corrugated adhesives; high filler load capacity, fast setting, water resistance, starch compatibility. [Rohm & Haas]

Rovace 9100. Vinyl/acrylic copolymer emulsion; binder for interior flat and semigloss paints. [Rohm & Haas]

Roval WP. CAS 36734-19-7; Iprodione; a fungicide with protectant activity for use in nursery stock and fruit trees. [Embetec Crop Protection Ltd]

Roval Dust. CAS 36734-19-7; Iprodione; a fungicide with protectant activity for lettuce and glass house crops. [Hortichem Ltd]

Roval Flo. CAS 36734-19-7; Iprodione; a fungicide with protectant activity for use in field crops, cereals, fruit trees and bulbs. [Rhone-Poulenc Crop Protection Ltd]

Roval Green. CAS 36734-19-7; Iprodione; a fungicide with protectant activity for use in turf and amenity grasses. [Rhone-Poulenc Environmental Prods. Ltd]

Rovamycin. Spyramycin. [May & Baker Ltd] *

Rovel. Weatherable polymers used to manufacture spas, truck toppers, automotive trim. [Dow UK]

Rovigon. Proprietary vitamin preparation containing vitamins A and E. [Roche Products Ltd] *

Rovimix. Vitamin supplements for animal feeds. [Roche Products Ltd] *

Rovisol. Water miscible vitamins for animals. [Roche Products Ltd] *

Rovral. Fungicide. [Rhone-Poulenc Rorer Ltd]

Rovral WP. Horticultural fungicide. [Embetec Crop Protection Ltd]

Roxadyl. CAS 40034-42-2; Rosozacin; antibacterial. [Sterling Drug Inc] *

Roxarsone. CAS 121-19-7; 4-hydroxy-3-nitrophenylarsonic acid; an antiprotozoal and growth promoter.

Roxite. A synthetic resin product used for electrical insulation. §

Roxon. A proprietary synthetic resin of the phenol-formaldehyde type. §

Roxotit. A copper-silicon acid-resisting alloy.

Royalac 133. A modified dithiocarbamate. produces excellent vulcanizates in EPDM polymers when used with normal sulfur levels; develops high degree of tensile strength at shore cure times while retaining equivalent compression set properties; the processing safety is reduced as compared to the standard Monex/MBT acceleration system. [Uniroyal] *

Royalac 136. A proprietary ultra-accelerator developed for use in EPDM polymers. [Uniroyal] *

Royalac 140. Zinc (hexadecyl-octadecyl) isopropyl dithiocarbamate. improves physical properties of EPDM diene blends; EPDM-NBR blends vulcanized with sulfur curing systems employing Royalac 139 and 140 compare favorably with CR compound in heat ageing, have essentially the same oil resistance, exhibit better lower temperature properties, are highly ozone resistant [Uniroyal] *

Royalcast® 3105. Castable thermoset plastic; provides rigidity of a structural plastic with advantages of liquid castability such as low tooling costs, production of large parts or parts with thick or varied wall cross-sections. [Uniroyal]

Royalene 301-T. EPDM terpolymer; used as an antiozonant in tire sidewalls and coverstrip and in blends with butyl rubber to improve heat and ozone resist. in inner tubes; vulcanizable with sulfur, peroxide, or radiation. [Uniroyal]

Royalene 306. EPDM terpolymer, nonstaining stabilizer; very good low temp. impact, excellent weathering properties; blends readily with polyolefins; also for wire and cable applications. [Uniroyal]

Royalene 521. EPDM terpolymer, nonstaining stabilizer; used for high quality molded mech. goods, press-cured sponge, wire and cable, coated fabric; suitable for mill mixing, calendering; exceptional flexibility. [Uniroyal]

Royalene 622. EPDM terpolymer, 40 phr nonstaining naphthenic oil, nonstaining stabilizer; accepts high filler loadings; high grn. strength, good shape retention in extrusions; used in extruded hose and molded mech. goods. [Uniroyal]

Royal Slo-Gro. Maleic hydrazine; a plant growth regulator for grass and to reduce bud growth in trees, hedges and vegetables. [Uniroyal Chemical Ltd]

Royaltherm® 1411. Silicone-modified EPDM elastomer; general-purpose elastomer used for automotive applications such as belts, diaphragms, protective boots, CV window gaskets, and for solar tubing, architectural gaskets, expansion joints; sulfur- or peroxide-curable. [Uniroyal]

Roydalox. A proprietary trade name for alumina porcelain with good resistance to corrosion and thermal shock; used in ball mills and grinding balls. [Doulton] ‡

Roydazide. CAS 12033-89-5; A proprietary trade name for silicon nitride; used for the manufacture of turbine blades and generally where a temperature of up to 1650 C is present. [Doulton] ‡

RPDE. Dibasic ester solvent. [Rhone-Poulenc UK]

RPM. Motor oil. [Chevron] *

RR 5. Semipermanent mold release agent for most thermosetting rubber and plastic materials; avail. in cold and hot formulations. [Releasomers]

RR 53 alloy. An aluminum alloy containing 91.85% aluminum, 2.25% copper, 1.3% nickel, 1.5% magnesium, 1.5% iron, 1.5% silicon, 0.1% titanium; used for die casting pistons.

RRV. Resorcylidene; a proprietary antioxidant for rubber. §

R-salt. The sodium salt of R-acid (qv).

RS Nitrocellulose. CAS 9004-70-0; Nitrocellulose; used as a clear, tough, fast drying film former. [Hercules] *

RT/Duroid® M. CAS 9002-84-0; PTFE, glass microfiber-

---

reinforced; offers high heat resist., dimensional stability, creep resist., low coeff. of friction, chemical, thermal and wear resist.; for seals, bushings, wear strips, gaskets, bearings, and piston rings. [Rogers]

RTF 762. CAS 63394-02-5; Two-component silicone rubber foam; cures at room temperature to a med.-dens. foam with the addition of a curing agent; applications including potting compound, cast-in-place thermal or elec. insulation, lightweight fabricated or molded rubber parts, sheet stock, die-cut as a gasket in industrial and construction applications. [GE Silicones] *

RTP 100 GB 10. CAS 9003-07-0; PP, 10% glass bead-reinforced, heat-stabilized; thermoplastic resin to exhibit optimum flow and warpage control; used in larger flat parts and thin-walled parts; bonds at knit lines are improved. [RTP]

RTP 200FR. CAS 32131-17-2; Nylon 6/6, flame-retardant nonreinforced; thermoplastic material exhibiting max. moldability, thermal stability, and flexibility for unfilled flame-retardant nylon 6/6; non-corrosive. [RTP]

RTP 201A. CAS 25038-54-4; Nylon 6, 10% glass fiber-reinforced; thermoplastic resin, improved strength, moduli, and deflective temp. over the base resin; improved moldability with less shrinkage, plus superior surface finish over nylon 6/6 materials; in automotive parts, electronic components, power tools. [RTP]

RTP 201B. CAS 9008-66-6; Nylon 6/10, glass-reinforced; thermoplastic resin, improvement in strength, moduli, and deflective temp. over base resin; very low water absorp. compared to other nylons; dimensional stability and high physical strengths. [RTP] *

RTP 201C. CAS 25035-04-5; Nylon 11, 10% glass fiber-reinforced; thermoplastic resin, lower water absorp., lower specific gravity, better chemical resistance, and better low temp. resistance than most glass-filled nylons; physical strengths are reduced. [RTP]

RTP 201D. Nylon 6/12, 10% glass fiber-reinforced; thermoplastic; better dimensional stability, toughness, and lower water absorp. than most glass-filled nylons; good moldability and ease of flow; in pumps, electronic applications, etc. [RTP]

RTP 301. PC, 10% glass fiber-reinforced; thermoplastic material; dimensional stability and improved rigidity, heat resistance, and strengths over the base resin; self-extinguishing characteristics maintained; in business machine parts, power tool components, gears, and cams. [RTP]

RTP 401. CAS 9003-53-6; PS, 10% glass fiber-reinforced; thermoplastic, improvements in strength, moduli, and dimensional stability over base resin; very rigid with low shrink; used in housings, automotive applications, etc. [RTP]

RTP 501. CAS 9003-54-7; SAN resin, 10% glass fiber-reinforced; thermoplastic; improved strengths over base resin and RTP 400 Series; balance of properties with strengths, low shrinkage, dimensional stability, and cost effectiveness; in tape hubs, housings, pumps, fan blades, etc. [RTP] *

RTP 601. CAS 9003-56-9; ABS resin, 10% glass fiber-reinforced; thermoplastic; improved properties over base resin; in automotive applications. [RTP]

RTP 701. CAS 9002-88-4; HDPE, 10% glass fiber-reinforced; thermoplastic; improved strengths and dimensional stability over base resin; flows very well and has <

1/10 shrinkage of base resin; used when molding large semiirregularly shaped parts of up to 25 lb in wt.; in bread trays, storage cases, and battery case applications. [RTP]

RTP 801. Acetal copolymer resin, 10% glass fiber-reinforced; thermoplastic material with improved dimensional control, strength, and stiffness over base resin; shrinkage reduced and elec. properties maintained; creep and wear resistance; in cams, gears, and bearings. [RTP] *

RTP 901. CAS 25135-51-3; Polysulfone, 10% glass fiber-reinforced ; strong, rigid, thermoplastic material used in applications requiring good heat resistance (up to 300 F); self-extinguishing; good elec. properties; used in elec., pump, plumbing, and food service applications. [RTP]

RTP 1001. CAS 26062-94-2; PBT, 10% glass fiber-reinforced; thermoplastic polyester having balance of properties (good strengths, elec. and thermal properties, low water absorp.); used in connectors and automotive applications requiring 300 F heat resistance. [RTP]

RTP 1105FR. CAS 25038-59-9; PET, 30% glass fiber-reinforced, flame-retardant; thermoplastic material; strengths, elec. properties, and smooth, molded surface. [RTP]

RTP 1201-80D. Thermoplastic PU elastomer, 10% glass fiber-reinforced; 80-D durometer thermoplastic material; abrasion and chemical resistance, controlled flexibility, and inherent toughness at high and low temps.; processed on conventional thermoplastic injection machines; in compressor rings, gears, nozzles, and recreational applications. [RTP] *

RTP 1301. CAS 9016-75-5; PPS, 10% glass fiber-reinforced; thermoplastic material; chemical and heat resistance, excellent strengths, flame retardant; used in low load applications up to 500 F, (hostile environment pump parts and connectors). [RTP]

RTP 1378. CAS 9016-75-5; Lubricated PPS; thermoplastic material combining strength, heat and wear resistance, chemical inertness, and flame retardance; can operate at temps. over 400 F. [RTP]

RTP 1401. PES, 10% glass fiber-reinforced; thermoplastic material; excellent physical properties including impact strength and tensile elongation; flame retardant; operates to temps. up to 400 F; used in connectors and aerospace applications. [RTP]

RTP 1501. Thermoplastic polyester elastomer, 10% glass fiber-reinforced; thermoplastic elastomer with strength, controlled resiliency, impact resistance, high and low temp. capabilities, and good chemical and elec. properties; combines the best characteristics of plastics and rubber; readily processed on standard injection machines; for sanding pads and seals applications. [RTP]

RTP 2301A. CAS 9009-54-5; Rigid thermoplastic polyurethane, 10% glass fiber-reinforced; offers dimensional stability and excellent molding characteristics [RTP]

RTP 2381A. CAS 9009-54-5; Rigid thermoplastic polyurethane, 10% PAN carbon fiber-reinforced; offers dimensional stability, moldability, good impact strength, and conductive properties [RTP]

RTP 3403-3. Liquid crystal polymer, 20% glass fiber-reinforced; offers outstanding heat and chemical resist., inherent flame retardance; excellent flowability. [RTP]

RTP 3405-3 TFE 15. Liquid crystal polymer, glass fiber-

---

reinforced, 15% PTFE; outstanding heat resist., chemical resist., intumescent flame retardance for wear-resist. applications. [RTP]

RTP 4001. Polyphthalamide, 10% glass fiber-reinforced; high heat distort., excellent strength and stiffness at room and elevated temps. [RTP]

RTP 4081. Polyphthalamide, 10% PAN carbon fiber-reinforced; offers high strength, stiffness, heat distort. temp. and elec. conductivity. [RTP]

RTP ESD-300 EM FR. PC, glass-reinforced, flame-retardant, static dissipative; flame retardant material with static dissipative characteristics; excellent toughness and flow. [RTP]

RTV 11. RTV silicone; general purpose encapsulant; with the addition of a curing agent, cures to a durable silicone compound at room temperature with service temp. of -60 to 204 C; 11 is easily pourable, self-leveling grade for general purpose potting; protects connectors, switches, components, and coils from dust, moisture, and shock. [GE Silicones]

RTV 31. RTV silicone; high-temp. encapsulant offering moderate strength for applications requiring -54 to 260 C continuous operating performance with intermittent performance to 316 C; 31 is a pourable grade for elec. potting and mech. protection. [GE Silicones]

RTV 133. One-component RTV silicone adhesive sealant; cures to a strong, durable, resilient silicone rubber at room temperature and on exposure to atmospheric moisture; offers high- and low-temp. performance; 133 is UL recognized; used as sealant on firewalls or as flame-retardant coating; also for coating electronic printed circuits, semiconductors, and for vertical applications. [GE Silicones]

RTV 511. CAS 63394-02-5; Two-part RTV silicone; low-temp. encapsulant offering moderate strength for extended low temp. performance (to -115 C); 511 is a pourable, self-leveling product for potting, encapsulating, and coating circuit assemblies and components between -115 and 204 C. [GE Silicones]

RTV 615. CAS 63394-02-5; Two-part RTV silicone rubber; encapsulant; addition-cure kits offering reliable deep-section cure even in closed assemblies; room temperature or heat curing; 615 is general purpose grade providing excellent elec. insulation and shock resist. at -54 to 204 C temp. range. [GE Silicones]

RTV 6156. Two-part methyl-phenyl silicone gel; dielec. gel which cures to form soft elastomer to preserve dielec. integrity and provide protection from vibration and shock for delicate electronic assemblies operating in harsh environments; solventless; supplied in 1:1 kit with curing agent; 6156 offers extended low-temp. performance; room temperature or heat curing. [GE Silicones] *

R Type Solvent®. Adipate polyester plasticizer; suitable for softening hot melt adhesive of packaging grade. [Nordson (UK) Ltd]

Rubalt. A proprietary compound of rubber, bitumen, and benzene; it is a waterproof, rust-proof, and acid-resisting paint, and is stated to be highly resistant to mineral acids, alkalis, chlorine, ammonia, and salt solution. §

rubber. CAS 9006-04-6; An elastic material contained in the latex of certain plants. The most important plant is *Hevea brasiliensis*, of South America, which yields the para rubber of commerce; used in tires, conveyor belts, foam rubber, electrical insulation.

rubber cements. These are made by dissolving rubber in suitable solvents, such as coal-tar naphtha or carbon disulfide. Sometimes rosin or turpentine is added. Rubber cement is often a mixture of rubber and sulfur dissolved in oil.

rubber formolite. A product obtained by the action of formaldehyde on a petroleum ether solution of a pale crepe rubber to which has been added concentrated sulfuric acid.

rubberite. An artificial rubber made from asphalt, oxidized oil, petroleum jelly, and sulfur.

Rubberlene. A white refined petroleum product; used as a solvent, and can be substituted for carbon disulfide for dissolving rubber. It boils at 145-300 F.

Rubbermakers Sulfur. CAS 7704-34-9; Sulfur; for rubber compounding. [Akrochem]

rubber-sulfur. Amorphous plastic sulfur, obtained from the Kobui sulfur mine, Japan.

Rubbone. A patented composition stated to be rubber resin prepared by oxidizing rubber catalytically; used in paints, varnishes, etc., in electrical insulation, and in the impregnation of coils. §

Rubelix. A proprietary preparation containing pholcodine and ephedrine hydrochloride; a cough linctus. [Pharmax Ltd] *

rubellan. A mineral that is a variety of mica.

Rubel metal. An alloy of 55% copper, 40% zinc, and 5% aluminum-iron-manganese-nickel alloy. Another alloy contains 51% copper, 40% zinc, and 5% aluminum-iron-manganese-nickel alloy, and 4% ferro-manganese.

Rubelogen. Rubella virus vaccine live; immunizing agent. [Parke-Davis] *

rubeosine. A nitrochlorofluorescein, obtained by the action of nitric acid upon aureosin.

ruberite. A name for red copper ore, Cu_2O.

Rub-erok. A proprietary trade name for hard rubber for electrical insulation. §

Rub-er-red. A red iron oxide of fine particle size, which is acid and alkali free, and contains no soluble salts; a proprietary pigment for rubber. §

rubianic acid. (ruberythric acid) $C_{22}H_{28}O_4$.

rubidium carbonate. Rb_2CO_3; CAS 584-09-8; EINECS 209-530-9; Special glass formulations. [Atomergic Chemetals; Cabot; Cerac; Noah Chem.]

rubidium chloride. RbCl; CAS 7791-11-9; EINECS 232-240-9; Analysis (testing for perchloric acid), source of rubidium metal. [Atomergic Chemetals; Cabot; Cerac; Noah Chem.]

rubidium iodide. RbI; CAS 7790-29-6; EINECS 231-198-1. [Atomergic Chemetals; Cabot; Cerac; Noah Chem.]

rubidium sulfate. Rb_2SO_4; CAS 7488-54-2; EINECS 231-301-7. [Atomergic Chemetals; Cabot; Cerac; Noah Chem.]

Rubinate® LF-168. CAS 101-68-8; Modified 4,4´-diphenylmethane diisocyanate; for manufacturing of high performance polyurethane elastomeric materials including reaction injection molding processed and cast elastomers, sealants, coatings, and adhesives. [ICI Polyurethanes] *

Rubini's essence. A saturated solution of camphor in alcohol.

rubio ore. A brown ore of iron, from Bilbao, in Spain.

Rubmag. A proprietary trade name for a light magnesium carbonate used for rubber reinforcing. §

---

‡ Trade name and manufacturer not verified   § Trade name without identified manufacturer

Rubout. Deodorants for rubber processing and products. [CPL Group Ltd]

Rubox. CAS 1314-13-2; Zinc oxide, rubber grade; used in the rubber industry. [Manchem Ltd] *

Rubramin PC. CAS 68-19-9; Cyanocobalamin; vitamin. [Bristol-Myers Squibb Co Inc]

Rubratope-57. Cyanocobalamin Co 57; diagnostic aid; radioactive agent. [Bristol-Myers Squibb Co Inc]

rubrax. A mineral rubber containing 98-99% of material soluble in chloroform, with an ash less than 0.5%.

rubrescin. An indicator prepared from resorcinol and chloral hydrate.

rubrica. A red pigment. It is a natural burnt ocher, containing varying quantities of iron.

Rubriment. A proprietary preparation of nicotinic acid benzyl ester and capsicin; an embrocation. [Horlicks] *

Rub-tex. A proprietary trade name for a hard rubber. §

ruby ore. Cuprous oxide, Cu_2O.

ruby powder. A sporting 42-grain powder containing 50% nitro-cellulose, metallic nitrate, 8% nitro-hydrocarbon, and 6% starch.

ruby tin. A red variety of the mineral Cassiterite, or tinstone.

Rucoflex® F-2014. Slightly branched, hydroxyl-terminated polyester polyol; specialty polyester designed for reaction with diisocyanates to produce crosslinked elastomeric PUs; used in solution laminating adhesives, solution coatings, prepolymers, and one-shot castables; low melting, soft solid which crystallizes very slowly at room temperature [Ruco Polymer]

Rucoflex® S-101 Series. Saturated, aliphatic, linear, hydroxyl-terminated polyester diol based on short chain glycols and polyester; polyester used for solution coatings and adhesives, thermoplastic elastomers, and castable prepolymers; PUs have high strength, solv., tear and abrasion resistance, and high resiliency in castables; melts at mod. temps. [Ruco Polymer]

Rucote 102. Polyester powd. coating resin; yields coatings with superior flow and flexibility; cured coatings feature outstanding weatherability, excellent flexibility and impact resist. [Ruco Polymer]

Rucothane 2010 L. PU resin latex; solv.-free, water disp. of high m.w., aliphatic-based, thermoplastic elastomer; polymer is extremely supple, exhibits good adhesion to a variety of substrates, and can be readily frothed by mechanical means; used as frothable interlayer/adhesive, as saturated and binder for nonwovens, and as blending resin for other latexes. [Ruco Polymer]

Rudol. White mineral oil. [Witco]

rufiopin. Tetrahydroxyanthraquinone, $C_{14}H_8O_6$.

Ruge's solution. A solution containing 1 cc glacial acetic acid, 2 cc formalin, and 100 cc distilled water.

Rulan®. Polymer. [DuPont UK]

Rumensin. Monensin, produced by *Streptomyces cinnamonensis*, and used as the sodium salt; an antibacterial and antifungal. [Eli Lilly & Co]

rumicin. Chrysophanic acid, $C_{15}H_{10}O_4$.

Runa. CAS 13463-67-7; EINECS 236-675-5; Rutile type titanium dioxide. [Laporte Industries Ltd] *

Runaway. Color remover. [Dylon International Ltd]

Ruolz alloys. See nickel silvers.

Ruselite. A proprietary trade name for an alloy of 94% aluminum, 4% copper, 2% chromium, and 2% molybdenum; it is stated to be resistant to corrosion. §

rusma. A mixture of arsenic sulfide (orpiment), As_2S_3, with

lime. It is made into a paste with water, and used for unhairing skins prior to tanning them.

Ruspini's solution. A styptic containing tannic acid, rose water, alcohol, and water.

Russian tallow. A mixture of beef and mutton fat.

Russian turpentine. The oleo-resin from *Pinus sylvestris* and *P. Ledebourii*.

Russian white lead. A white lead having the composition, $5PbCO_3 \cdot 2Pb(OH)_2 \cdot PbO$.

Russol. See paraffin, liquid.

Rustban. Protective coatings. [Exxon UK] *

Rustlan Oil. Yellow to amber colored liquid used as rust preventive and stamping oil; used in wire mills and steel mills to protect rust while shipping and inside storage; for metal stamping, used as a shallow stamping oil which need not be removed prior to spot welding like other oils. [Rustlan Chemical Co] ‡

rustlessiron. A rustless steel containing about 0.1% carbon. It is made in the electric furnace by means of practically carbon-free ferro-chrome.

Rust-Tap. A light yellow nonflammable liquid used as tapping oil; residual coating acts as rust preventive; used in machine shops for tapping operation of all metals except aluminum; used as drilling lubricant. [Rustlan Chemical Co] ‡

Rutaform. Phenolic, melamine and granular polyester molding powders. [Bakelite Polymers (UK) Ltd] *

Rutamod. Black entenders for epoxy and polyurethane resins. [Collinda Ltd]

Rutasolv DI. Diisopropyl naphthalene high boiling solvent. [Collinda Ltd]

Rutenol. CAS 96-69-5; 4,4-Thiobis-(6-t-butyl-3-methyl-phenol). [Collinda Ltd]

ruthenium red. Ammoniated ruthenium oxychloride, $Ru_2(OH)_2Cl_4 \cdot 7(NH_3) \cdot 3H_2O$; used as a microscopic stain, and as a reagent for pectin, plant mucin, and gum.

Ruthmol. A proprietary preparation of potassium chloride, lactose and gluten-free starch; a sodium-free table salt. [Larkhall Laboratories plc] *

Rutiox. CAS 13463-67-7; EINECS 236-675-5; A rutile titanium dioxide pigment. [Laporte Industries Ltd] *

Ruvea®. Fibers. [DuPont UK]

RVPaba Lipstick. Aminobenzoic acid; ultraviolet screen. [ICN Pharmaceuticals Inc]

RX® 1-501N. CAS 131-17-9; Diallyl phthalate, mineral-reinforced; thermoset molding material [Rogers] *

RX® 3-1-530. CAS 131-17-9; Diallyl phthalate, long glass-reinforced; thermoset molding material [Rogers] *

RX-56. Porofocon A; contact lens material. [Rynco Scientific Corp] ‡

RX® 1906. CAS 25928-94-3; Epoxy, mineral/glass-reinforced; thermoset molding material [Rogers] *

RXXL. High boiling tar acids. [Coalite Fuels & Chemicals Ltd]*

Ryax C. Center-line suspended process pump; for process pumping applications. [Sihi Pumps (UK) Ltd]

Ryax F, O. Back pull out process pump; for process pumping application. [Sihi Pumps (UK) Ltd]

Rybaferrin. A proprietary preparation of ferrous sulfate, manganese, copper, strychnine hydrochloride, thiamine, hydrochloride and nicotinic acid; a tonic. [Rybar Laboratories Ltd] *

Rybarex. A proprietary preparation of chloroxylenol, papaverine hydrochloride, atropine methonitrate, methyl sali-

---

* Trade name not verified as available

† Trade name verified as obsolete

cylate, menthol, benzocaine, pituitary extract, adrenaline, tri-iodophenol, and a saline base. [Rybar Laboratories Ltd] *

Rybarvin. A proprietary preparation of atropine methonitrate, adrenalin, papaverine hydrochloride, benzocaine, and posterior pituitary extract; used in the treatment of asthma. [Rybar Laboratories Ltd] *

Rybronsol. A proprietary preparation of phenazone iodoantipyrine, caffeine, and butethamate citrate. [Rybar Laboratories Ltd] *

Ryflex. Belt driven single-stage centrifugal pump capable of being stacked one unit on top of another to a maximum of two; for heating and cooling water applications where floor space is small and a stacked pump unit is required. [Sihi Pumps (UK) Ltd]

Rylex. Uv light absorber. [DuPont UK]

Rymel. A proprietary preparation containing ipecacuanha liquid extract, acetic acid, squill liquid extract, sodium citrate and glycerin; a cough linctus. [Rybar Laboratories Ltd] *

Rynabond. A proprietary preparation containing phenylephrine tannate, pheniramine tannate and mepyramine tannate. [Fisons plc, Pharmaceuticals Div] *

Rynacrom. A proprietary preparation of sodium cromoglycate; used in the treatment of allergic rhinitis. [Fisons plc, Pharmaceuticals Div] *

Rynite®. Polyester resin. [DuPont UK]

Ryoto Ester KA. Sugar ester, calcium carbonate, potassium carbonate, carbohydrate; quality improver for fish paste products. [Mitsubishi Kasei] *

Ryoto Ester SP. Sugar ester, monoglyceride, sorbitol, propylene glycol; batter aerating agent for sponge and pound cake. [Mitsubishi Kasei] *

Ryotol. A proprietary preparation of phenoxyethanol, phenylmercuric nitrate and phenazone. [Rybar Laboratories Ltd] *

Ryoto Sugar Ester LWA-1570. CAS 25339-99-5; EINECS 246-873-3; Sucrose laurate; o/w and w/o emulsifier, softener, conditioner, and aerating agent in foods. [Mitsubishi Kasei] *

Ryoto Sugar Ester OWA-1570. Sucrose oleate; emulsifier, conditioner, softener, detergent for foods, drugs, cosmetics; tablet lubricant. [Mitsubishi Kasei] *

Ryoto Sugar Ester P-1570, P-1670. CAS 26446-38-8;

Sucrose palmitate; emulsifier, conditioner, softener, detergent for foods, drugs, cosmetics; tablet lubricant. [Mitsubishi Kasei] *

Ryoto Sugar Ester S-170. Sucrose di, tristearate; emulsifier, conditioner, softener, detergent for foods, drugs, cosmetics; tablet lubricant. [Mitsubishi Kasei] *

Ryoto Sugar Ester S-570, S-770. CAS 27195-16-0; EINECS 248-317-5; Sucrose distearate; emulsifier, conditioner, softener, detergent for foods, drugs, cosmetics; tablet lubricant. [Mitsubishi Kasei] *

Ryoto Sugar Ester S-1170. CAS 25168-73-4; EINECS 246-705-9; Sucrose stearate; emulsifier, conditioner, softener, detergent for foods, drugs, cosmetics; tablet lubricant. [Mitsubishi Kasei] *

Ryspray. A proprietary preparation of atropine methonitrate and isoprenaline hydrochloride in an aerosol; a bronchial antispasmodic. [Rybar Laboratories Ltd] *

Rytherm. Condensate extraction pump set with low mounted tank; several sizes of set suit various sizes of industrial central heating systems, pumps extract condensate at or near 100 C, thus reducing heat loss. [Sihi Pumps (UK) Ltd]

Rythmatine. CAS 58503-79-0; Meobentine sulfate; cardiac depressant. [Burroughs Wellcome Co] *

Ryton® A-200. CAS 9016-75-5; PPS; high performance engineering thermoplastic with inherent chemical and flame resist.; for injection molding. [Phillips]

Ryton® R-4. CAS 9016-75-5; Glass fiber-reinforced PPS resin; injection molding engineering thermoplastic possessing chemical resistance and mechanical properties even at elevated temps.; dimensional stability, good moldability, and elec. properties unaffected by moisture; for injection molding; used in automotive under-the-hood components, elec./electronic parts, appliance components, pump housings and impellers, and valves, cams, etc. [Phillips]

Ryton® V-1. CAS 9016-75-5; PPS; high melt flow grade for thin coatings on metal substrates or to improve flow of highly filled PPS; for cookware, industrial release applications, pumps, valves where chemical resist. is required. [Phillips]

Ryvin. Single-stage centrifugal pump vertical in-line; for in-line circulating duties. [Sihi Pumps (UK) Ltd]

S

S.21. Nonflammable solvent. [Stowlin Ltd]

S-60 RVM. CAS 1337-76-4; Attapulgite clay; absorbent, adsorbent. [Floridin] *

S160 Beads and Powd. CAS 1333-86-4; Carbon black; for med. quality PVC and calendered systems. [Degussa]

S-201. CAS 1344-28-1; Alumina; Claus catalyst for use where H_2S is the primary compound to be removed from feed gas; offers lower pressure drop. [La Roche Chem.]

S945. Monoclinic zirconium dioxide with a zirconia content (including hafnia) of 94.5%; for manufacture of ceramic pigments, welding fluxes and insulating material. [Ferro]*

S975. Monoclinic zirconium dioxide with a zirconia content (including hafnia) of 97.5%; for manufacture of pigments for ceramics and enamels and welding fluxes. [Ferro] *

S987. Monoclinic zirconium dioxide with a zirconia content (including hafnia) of 98.7%; for manufacture of ceramic pigments. [Ferro] *

S992. Monoclinic zirconium dioxide with a zirconia content (including hafnia) of 99.2%; for manufacture of lead-zirconate-titanate piezo electric ceramics, zirconates, zirconia technical ceramics, oxygen sensors, milling media and ceramic pigments. [Ferro] *

S994. Monoclinic zirconium dioxide with a zirconia content (including hafnia) of 99.4%; for manufacture of lead-zirconate-titanate piezo electric ceramics, zirconates, zirconia technical ceramics and oxygen sensors. [Ferro]*

Sabalith®. Thermal salt mixture of inorganic nitrates ($LiNO_3$ + KNO_3); nitrite-free thermal salt for the salt bath curing in open or closed circulations, e.g., in LCM installations for the vulcanization of rubber. [Chemetall GmbH]

Sabeco Metal. A proprietary trade name for copper with 21% lead and 9% tin. §

sablon. See limo.

Sabre. CAS 34123-59-6; Suspension concentrate containing 553 g isoproturon per liter; used for annual weed control in cereals. [Schering Agrochemicals Ltd]

Sabre 1628. PC/polyester resin; engineering resin offering extra margin of heat resist. for improved processing and paintability as well as excellent chemical resist., low-temp. toughness, easy processing; applications including automotive wheelcovers, lawn and garden equip., telecommunications devices. [Dow Plastics]

Sabulite. An explosive containing ammonium nitrate, charcoal, and calcium silicide.

Sabutol. Mixture of normal butyl alcohol, isobutyl alcohol, and secondary pentyl alcohol; solvent in paints, printing inks, dyes, foundries, manufacture of butyl acetate and xanthate. [Sasolchem]

saccharase. See sucrase.

saccharin. (1,1-dioxide-1,2-benzisothiazol-3(2H)-one; 1,2-benzothiazol-3(2H)-one 1,1-dioxide; glycophenol; glycosin; neosaccharin; saccharin, insoluble; saccharinol; saccharinose; saccharol; saccharose; sycorin; sykose) Organic compound; $C_7H_5NO_3S$; CAS 81-07-2; EINECS 220-120-9; noncaloric, non-nutritive sweetener (500 times sweeter than cane sugar), pharmaceutic aid; in formulations for electroplating bath brighteners. [Aisan Chem Co Ltd; Dinoval Chem. Ltd; R.W. Greeff; Maruzen Fine Chem.; PMC Specialties; Rit-Chem; Spice King]

saccharum lactis. See lactose.

Sachsse's solution. A solution containing 18 g mercuric iodide, 25 g potassium iodide, and 80 g potassium hydroxide in a liter; used for the determination of reducing sugars.

Sachtocup. Cupreous pyrites; 25-26% Cu, 32-34% S, 28-30% Fe; filler for resin-bonded grinding wheels, brake linings, etc. [Metallgesellschaft AG]

Sachtoklar® . Primary flocculant agent. [Sachtleben Chemie GmbH]

Sachtolen®. Masterbatches and compounds. [Sachtleben Chemie GmbH]

Sachtolith® . CAS 1314-98-3; Zinc sulfide, ZnS; white inorganic pigment for paints and coatings. [Sachtleben Chemie GmbH]

Sachtoperse® HU. CAS 7727-43-7; Barium sulfate, $BaSO_4$; ultra fine solid state additive. [Sachtleben Chemie GmbH]

Sachtopyr®. Polycrystalline iron sulfide (pyrite); 47-48% S, 40-41% Fe; filler for resin-bonded grinding wheels, brake linings, etc. [Metallgesellschaft AG]

Sachtosil® . Chemical products for industrial purposes, especially pigments for paints and colors, fibers, cosmetics and catalysts. [Sachtleben Chemie GmbH]

Saci. Rust preventative. [Witco]

Sacon® . $BaSO_4$; transparent conductive pigment. [Sachtleben Chemie GmbH]

sacred bark. The bark of Rhamnus purshianus.

Saduren®. Synthetic resin solutions; precondensates based on melamine-formaldehyde; binders for bonding fiber webs, especially glass fibers. [BASF AG]

Safapryn. CAS 50-78-2, 103-90-2; A proprietary preparation of aspirin and paracetamol; an analgesic. [Pfizer International]

Safapryn-Co. A proprietary preparation of aspirin, paracetamol, and codeine phosphate; an analgesic. [Pfizer International]

Safari® . For the automotive industry. [DuPont UK]

Safe-FR. Nonhalogenated flame retardant polyolefins; high

---

loadings of magnesium hyroxide provide flame retardancy; low smoke, low toxicity, low acid gas emitted on exposure to fire. [UVtec]

Safebond 3. Modified bitumen emulsion; flooring adhesive. [Marley Floors Ltd]

Safe-Break. Glass bottles with plastic safety coating. [BDH Chemicals Ltd]

Safepak. Plastic safety containers. [BDH Chemicals Ltd]

Safety-Cool. Heavy-duty cutting and grinding fluids specifically designed to cool, lubricate and protect metal surfaces in a variety of machining applications; metalworking fluid. [Chem-Trend] †

safety dynamite. An explosive consisting of 24% nitroglycerin, 1% guncotton, and 75% ammonium nitrate.

Safety-Lube®. Water-based lubricant formulations for high-pressure aluminum and zinc die casting. [Chem-Trend]

safety nitro-powder. An explosive similar in composition to Giant powder.

Safex. Stated to be dinitro-phenyl-dimethyl-dithio-carbamate; used with zinc oxide; a proprietary super-accelerator for rubber vulcanization. §

Safezone® Cleaning Solvent & Flux Remover. Solvent cleaner. [Miller-Stephenson] *

Saffil. High-temperature inorganic fibers. [ICI Chem. & Polymers Ltd]

Saffil®. CAS 1344-28-1; High purity alumina fibers; refractory bulk fibers for high performance at elevated temps. [Thermal Ceramics]

safflower oil. (Carthamus tinctorious oil; oils, safflower) Oily liquid obtained from seeds of *Carthamus tinctorius* consisting principally of triglycerides of linoleic acid; CAS 8001-23-8; alkyd resins, paints, varnishes, medicine; salad oil blends. [Arista Industries; Clagene; Croda; Lipo]

saffron. A coloring matter obtained from the dried and powdered flowers of the saffron plant, *Crocus sativus*; used for coloring confectionery.

saffron bronze. (gold bronze). Tungsten-sodium bronze, $Na_2W_3O_9$; used as a pigment. The corresponding potassium salt is known as violet bronze or magenta bronze.

saffron oil. CAS 8001-23-8; Safflower oil from the seeds of *Carthamus tinctorius*. It has an acid value of 9.8 and a saponification value of 197.3; used in alkyd resins, paints, varnishes, medicine, dietetic foods, margarine, hydrogenated shortening.

saffron sugar. Crocase, $C_6H_{12}O_6$.

Saflex. Polyvinyl butyral film; interlayer for safety glass. [Monsanto plc]

Safoam. Endothermic chemical nucleating and blowing agent for thermoplastics which process above 340 F. [Reedy Int'l.]

safrol. CAS 94-59-7; The methylene ether of allyl-pyrocatechol, $C_6H_3 \cdot C_3H_5 \cdot (O \cdot OCH_2)$. It is found in oil of sassafras, and is obtained from red oil of camphor; used in the place of oil of sassafras, in perfumes, soaps, and medicine.

Saf-T-Side. A petroleum oil insecticide that kills by contact; used as both a dormant spray and as a summer spray; controls whiteflies, aphids, mites, mealybugs, etc.; used on vegetables, ornamentals, citrus and fruit trees. [Lawn & Garden Products Inc]

Saf-T-Sol. Aromatic solvent for paints and varnishes. [Crowley Chem.]

Sag. Silicone antifoam. [Union Carbide]

Sagatal. CAS 57-33-0; Pentobarbitone sodium solution.

[RMB Animal Health Ltd]

Sahli's reagent. A mixture of equal parts of a 48% solution of potassium iodide and an 8% solution of potassium iodate; used to test for free hydrochloric acid in stomach contents.

Sahli's stain. A solution of borax and methylene blue in water; used to stain nervous tissues and cell nuclei.

Saisan. CAS 5707-69-7; Liquid drazoxolan seed dressing. [ICI Chem. & Polymers Ltd] †

Sajji. An alkaline deposit found on the land in India; used as a soap substitute.

sakaloid. A synthetic resin. It is a polymerized sugar product obtained from sugar, dextrose, levulose, etc. It can be used for varnishes and lacquers and, when extruded, as an artificial silk.

Sakarat. CAS 81-81-2; Warfarin; a rodenticide. [Killgerm Chemicals Ltd]

Sakarat Special. CAS 3691-35-8; An oil formulation containing 2.5% of chlorophacinone; an anticoagulant rodenticide; a bait to control black rats, brown rats, house mice and voles. [Killgerm Chemicals Ltd]

sakoa oil. An oil obtained from the seeds of *Sclerocarpa caffra*. It is a nondrying oil, and has a saponification value of 193.5.

Sakresote. Creosote [Sasolchem]

sal absinthii. Salt of wormwood, potassium carbonate, K_2CO_3.

Salacetin. *See* aspirin. §

Salactol. Colorless evaporative paint containing 16.7% w/w salicylic acid BP, 16.7% lactic acid BP, 66.6% w/w flexible collodion BP; used for the topical treatment of warts, especially plantar warts. [Dermal Laboratories Ltd]

Salamac. CAS 12125-02-9; EINECS 235-186-4; Compressed blocks of ammonium chloride. [ICI Chem. & Polymers Ltd]

sal amarum. (sal Catharticum, sal Anglicum, sal Seidlitense). Magnesium sulfate, $MgSO_4$.

sal ammoniac. CAS 12125-02-9; (salmiak, muriate of ammonia). Ammonium chloride, NH_4Cl.

Salantin. *See* aspirin. §

Salargyl. A protein-silver preparation.

salarmoniac. An ammonium chloride obtained from volcanoes of Central Asia.

Salaspin. *See* aspirin. §

Salatac. Clear, viscous gel containing 12% w/w salicylic acid BP, 4% w/w lactic acid BP; used for the treatment of warts, verrucas, corns, and calluses. [Dermal Laboratories Ltd]

Salbulin Inhaler. Pressurized aerosol containing salbutamol BP, delivers 200 measured doses (100 mcg per dose). [3M Pharmaceuticals] †

Salcare®. Body-friendly polymers for hair conditioning and skin care. [Allied Colloids Ltd]

sal chalybis. *See* iron sulfate (ous).

sal commune. Sodium chloride, NaCl.

sal Cornu Cervi. Ammonium carbonate, $(NH_4)_2CO_3$.

sal Culinaris. Sodium chloride, NaCl.

Salde Uvas Picot. Effervescent antacid, for relief of digestive distress. [Richardson-Vicks Inc] *

sal digestnum Sylvii. Potassium chloride, KCl.

sal diureticum. Potassium acetate, $(C_2H_3O_2)K$.

salenixon. Crude potassium sulfate, obtained in the manufacture of nitric acid.

---

‡ Trade name and manufacturer not verified

§ Trade name without identified manufacturer

saleratus. Sodium hydrogen carbonate, $NaHCO_3$.

salesthin. Methylene chloride, CH_2Cl_2.

Sal-ethyl. CAS 118-61-6; Ethyl salicylate.

Saletin. See aspirin. §

Salfax 77. Vegetable and animal fats, aliphatic alcohols, fatty acids and cationics; hair conditioner; base for cream rinses. [Chem-Y GmbH]

Salfuride. A proprietary preparation of nifursol; a veterinary anti-protozoan. §

Salge metal. An alloy of 4% copper, 9.9% tin, 1.1% lead, and 85% zinc.

Salhar gum. A gum-resin obtained from *Boswellia serrata*.

Salicolen. Salene, a mixture of ethyl and methyl glycolic acid esters of salicylic acid.

salicylal. See salicylaldehyde.

salicylaldehyde. (salicylal; salicylic aldehyde; o-hydroxybenzaldehyde) C_6H_4OHCHO; CAS 90-02-8; EINECS 201-961-0; Analytical chemistry, perfumery, synthesis of coumarin, auxiliary fumigant, flavoring. [Janssen Chimica; Penta Mfg.; Seimi Chem.]

salicylamide. (o-hydroxybenzamide) $C_6H_4(OH)CONH_2$; CAS 65-45-2; EINECS 200-609-3; Medicine (analgesic). [Aldrich; EM Industries; R.W. Greeff; Penta Mfg.]

salicylic acid. (2-hydroxybenzoic acid; o-hydroxybenzoic acid; benzoic acid, 2-hydroxy-) Aromatic acid; HOC_6H_4COOH; CAS 69-72-7; EINECS 200-712-3; preservative for foods; manufacture of methyl salicylate, acetylsalicylic acid, etc., dyes; reagent in analytical chemistry; topical keratolytic agent. [Allchem Industries; EM Industries; Hilton Davis; Jansse Chimica; Mitsui Toatsu Chem.; PMC Specialties; Rhone-Poulenc Santé SA]

salicylic aldehyde. See salicylaldehyde.

Sali®-Prent. Cardioselective beta-receptor blocker, antihypertensive agent. [Bayer AG]

Sali®-Presinol. Pharmaceutical preparation; antihypertensive. [Bayer AG]

saliretin resins. See saliretins.

Saliretins. (saliretin resins). Resins obtained from saligenin by either heating or treating it with formaldehyde. They are similar to phenol-formaldehyde resins.

Saliter. Sodium nitrate, $NaNO_3$.

Salkowski's solution. A solution of phospho-tungstic acid; used for the detection in urine.

Sallit's speculum metal. An alloy of 64.6% copper, 31.3% tin, and 4.1% nickel.

Sally Nixon. Fused niter cake (acid sodium sulfate).

sal Martis. Ferrous sulfate, $FeSO_4$.

salmiak. See sal ammoniac.

sal mineral. Ferric oxide, Fe_2O_3.

sal Mirabil. Sodium sulfate, Na_2SO_4.

Salmocid. A proprietary preparation of polynoxylin. [Geistlich Sohne AG] *

sal niter. Potassium nitrate, KNO_3.

Salodine. A proprietary preparation; it is an iodized salt. §

Salol. (phenyl salicylate) CAS 118-55-8; $C_6H_4OH\text{-}COOC_6H_5$; used in medicine, as a uv absorber in plastics, as a laboratory reagent.

sal Prunella. Potassium nitrate, KNO_3, in balls.

sal Rupellensis. Sodium-potassium tartrate.

sal Saturni. Lead acetate, $Pb(C_2H_3O_2)_2$.

sal sedativus. Boric acid, H_3BO_3.

sal soda. Sodium carbonate, Na_2CO_3.

Salsorb®. Superabsorbent polymers for body fluids. [Allied Colloids Ltd]

sal succini. Succinic acid, $HOOC\cdot CH_2\cdot CH_2\cdot COOH$.

salt. See sodium chloride.

Salt, Amido-G. See amido-G-Acid. §

Salt, Amido-R. See amido-R-Acid. §

sal tartar. See sodium tartrate

salt cake. CAS 7757-82-6; Crude sodium sulfate, Na_2SO_4, produced in the Leblanc soda process; used in paper, pulp, detergents, sodium salts, ceramic glazes, plate and window glass.

Saltex. CAS 7647-14-5; Sodium chloride solution. [Omex Agriculture Ltd]

Saltialgine H8. CAS 9005-32-7; EINECS 232-680-1; Alginic acid; tablet disintegrant for pharmaceutical compressed tablets. [Mendell]

salt of Alembroth. (salt of wisdom, sal-Alembroth). A compound of mercuric chloride and ammonium chloride, $2NH_4Cl \cdot HgCl_2 \cdot H_2O$.

salt of amber. succinic acid, $C_2H_4(COOH)_2$.

salt of Hartshorn. Ammonium carbonate, $(NH_4)_2CO_3$.

salt of Lemery. Potassium sulfate, K_2SO_4.

salt of Norton. Platinum tetrachloride, $PtCl_4\cdot 5H_2O$.

salt of Saturn. CAS 301-04-2; (sugar of Saturn). Normal lead acetate, $Pb(C_2H_3O_2)_2$, was formerly known under these names.

salt of soda. Sodium carbonate, Na_2CO_3.

salt of Sorrel. (salts of Sonel, salts of Lemon). The two acid salts of potassium oxalate, C_2O_4HK and $C_2O_4KH \cdot C_2H_4O_2 \cdot 2H_2O$, are both sold under these names.

salt of steel. Ferrous sulfate, $FeSO_4$.

salt of Tartar. See potash.

salt of tin. Stannous chloride, $SnCl_4$.

salt of wisdom. See salt of Alembroth.

salt of wormwood. (sal Absinthii). Impure potassium carbonate, K_2CO_3, made from plant ash.

salt of saturn. See lead acetate.

salt perlate. Sodium phosphate, HNa_2PO_4.

saltpeter. (niter) CAS 7757-79-1; Potassium nitrate, KNO_3; used in explosives, matches, pyrotechnics, fertilizers, reagent for tobacco, tempering steel, curing foods.

saltpeter flour. Minute crystals of refined saltpeter, KNO_3; used in the manufacture of gunpowder.

saltpeter rot. Calcium nitrate, $Ca(NO_3)_2 \cdot 4H_2O$; causes the rapid disintegration of mortar.

saltpeter superphosphate. A fertilizer made by mixing niter with calcium superphosphate.

salufer. CAS 16893-85-9; The sodium salt of hydrofluosilicic acid, Na_2SiF_6; used as an antiseptic.

salumin. (salumen). Aluminum salicylate; $Al\text{-}(C_6H_4OHCOO)_3$; used in medicine.

salunol. An aqueous solution of sodium hypochlorite; used as a disinfectant.

Salupres. Hydrochlorothiazide, reserpine and potassium chloride; for the treatment of mild to severe hypertension. [Merck & Co Inc] †

Saluric®. CAS 58-94-6; Chlorothiazide; for the treatment of edema and hypertension. [Merck & Co Inc]

Saluron. CAS 135-09-1; Hydroflumethiazide; antihypertensive; diuretic. [Bristol Laboratories] *

Salut®. CAS 2921-88-2, CAS 60-51-5; Chlorpyrifos, dimethoate; insecticide with systemic, contact, stomach, and vapor action; used against hidden insects such as leaf miners, leaf rollers, wooly aphids. [BASF AG]

Saluthion®. Salut®

---

* Trade name not verified as available   † Trade name verified as obsolete

Salvarom. The diethyl ester of phthalic acid; used as a solvent.

Salvex. Compounded plastic material. [Mitsubishi Petrochem.] *

sal volatile. Commercial ammonium carbonate, $(NH_4)_2$-CO_3.

Salysal. CAS 552-94-3; Salicyl-salicylate.

samarium. A rare-earth metallic element; Sm; CAS 7440-19-9; EINECS 231-128-7; neutron absorber, dopant for laser crystals, metallurgical research, permanent magnets. [Aldrich; Atomergic Chemetals; Cerac; Rhone-Poulenc Basic]

samarium oxide. O_3Sm_2; CAS 12060-58-1; EINECS 235-043-6; Catalyst in the dehydrogenation of ethanol, infra-red-absorbing glass, neutron absorber, preparation of samarium salts. [Noah Chem.; Rhone-Poulenc Basic]

Samaron. Disperse dyestuffs for synthetic surfaces. [Hoechst UK] *

Sambarin. Suspension concentrate containing 500 g chlorothalonil and 250 g propiconazole per liter; a systemic fungicide for winter wheat. [Ciba-Geigy Agrochemicals]

Samite. A trade name for a carborundum product; an abrasive. §

samli. A clarified butter from East Africa.

Samorin. Isometamidium salts. [May & Baker Ltd]

Samsonite. An explosive containing nitroglycerin, collodion cotton, potassium nitrate, wood meal, and ammonium oxalate; used in coal mines.

Samson Steel. A proprietary trade name for nickel-chromium steel containing 1.25% nickel and 0.6% chromium.§

SAN. *See* styrene-acrylonitrile copolymer.

Sanachlor. Hypochlorite for sterilizing dairy equipment. [Ciba plc]

Sanaklenz. Agricultural disinfectant. [The Wellcome Foundation Ltd] *

Sanatank. Bulk milk tank sanitizer. [The Wellcome Foundation Ltd] *

Sanatogen. Casein sodium glycerophosphate. [Fisons plc, Pharmaceuticals Div] *

SAN copolymer. *See* styrene-acrylonitrile copolymer.

Sancos. A proprietary preparation of pholcodine, menthol and glycerin; a cough linctus. [Sandoz] *

Sancos Compound. A proprietary preparation of pholcodine, pseudoephedrine hydrochloride and chlorpheniramine maleate; a cough linctus. [Sandoz] *

Sanction®. For the agriculture industry. [DuPont UK]

Sancure® 776. Waterborne aliphatic urethane polymer; ink grinding vehicle forming tough, flexible films; excellent adhesion to various substrates including nylon and polyester; high modulus, med. hardness and elong. [Sanncor Industries]

Sandacid. Textile dying buffers. [Sandoz Products Ltd]

sand acid. CAS 16961-83-1; Hydrofluosilicic acid, H_2SiF_6.

sandalwood. *See* redwoods.

sandaracha. Synonym for realgar.

sandarac resin. A resin obtained from the North West African tree *Callitris quadrivalis*; used in the manufacture of spirit varnishes. Pine gum or Australian sandarac is obtained from *Callitris* species in Australia, and resembles the African variety.

sandelwood. *See* redwoods.

sanderswood. *See* redwoods.

sandiver. (glass gall). The scum formed on the surface of molten glass. It consists of calcium and sodium sulfates, with about one-tenth of its weight of glass.

Sandobet SC. CAS 68139-30-0; Cocoamidopropylhydroxy sultaine; high foaming, mild surfactant for cosmetic and toiletry applications; acid and alkali stable. [Sandoz]

Sandocal. A proprietary preparation of calcium lactate gluconate and sodium and potassium bicarbonate; used in the treatment of osteoporosis. [Sandoz] *

sandoce. Methyl-saccharin, $C_6H_3(CH_3) \cdot CO \cdot SO_2 \cdot NH$; a sweetening substance.

Sandocryl®. Specialty dye for aq. mediums. [Sandoz]

Sandofix. Fixing agents; for direct and reactive dyes. [Sandoz Products Ltd] *

Sandofluor. General water and oil repellents. [Sandoz Products Ltd]

Sandogen. Aromatic sulfonate, leveling agent. [Sandoz Products Ltd] *

Sando-K. CAS 7447-40-7; EINECS 231-211-8; A proprietary preparation of potassium chloride. [Sandoz] *

Sandol. Synthetic organic dyestuffs used on leather. [Sandoz Products Ltd]

Sandolan®. Synthetic organic acid dyestuffs; for aq. mediums, textiles, fertilizers. [Sandoz; Sandoz Products Ltd]

Sandolube. Softening/lubricating agent; improves handling of natural and synthetic fibers. [Sandoz Products Ltd]

Sandomigran. CAS 15574-96-6; Pizotyline; anabolic; antidepressant; serotonin inhibitor. [Sandoz Pharmaceuticals] *

Sandopan®. Very low foam detergent; used for removal of winding preparations from polyester fibers. [Sandoz Products Ltd]

Sandopan® DTC-100. CAS 68891-17-8; Sodium trideceth-7 carboxylate; detergent, emulsifier, wetting agent for liquid detergents, solv. cleaners, all-purpose cleaners, germicidal cleaners, shampoos, bubble baths; oil solubilizer for aq. systems, acid bowl cleaners. [Sandoz]

Sandopan® DTC-Acid. Trideceth-7 carboxylic acid; detergent, emulsifier, wetting agent for industrial and personal care, conditioning shampoos, liquid soaps, household and industrial cleaners. [Sandoz]

Sandopan® DTC Linear P. CAS 70632-06-3; Sodium C12-15 pareth-6 carboxylate; detergent, emulsifier, wetting agent, solubilizer, visc. booster for industrial, personal care, and household products; stable in alkali high temps. [Sandoz]

Sandopan® DTC Linear P Acid. CAS 88497-58-9; C12-15 pareth-7 carboxylic acid; detergent, emulsifier, wetting agent for industrial, personal care, and household use; oil solubilizer. [Sandoz]

Sandopan® JA-36. Trideceth-19 carboxylic acid; moderate foaming mild surfactant, oil solubilizer, wetting agent cosmetics/toiletries, laundry products, cleaners, industrial specialties;. [Sandoz]

Sandopan® KST. CAS 33939-65-0; Sodium ceteth-13 carboxylate; mild emulsifier, detergent, lime soap dispersant for use in sticks, bar soaps, antiperspirants, other personal care products, industrial specialties, laundry products; inhibits sodium stearate crystal formation. [Sandoz]

Sandopan® LA-8. Laureth-5 carboxylic acid; surfactant for cosmetics/toiletries, laundry products, cleaners, industrial specialties. [Sandoz]

Sandopan® LS-24. CAS 33939-64-9; Sodium laureth-13

---

‡ Trade name and manufacturer not verified   § Trade name without identified manufacturer

carboxylate; mild detergent, emulsifier, solubilizer for baby shampoos and personal care products. [Sandoz]

Sandopan® MA-18. CAS 28212-44-4; Nonoxynol-10 carboxylic acid; detergent, wetting agent, solubilizer for cosmetics/toiletries, laundry products, cleaners, industrial specialties. [Sandoz]

Sandopan® TA-10. Isosteareth-6 carboxylic acid; surfactant for cosmetics/toiletries. [Sandoz]

Sandopan® TS-10. Sodium isosteareth-6 carboxylate; surfactant for cosmetics/toiletries. [Sandoz]

Sandopec. Oil drilling auxiliaries. [Sandoz Products Ltd]

Sandoptal. CAS 77-26-9; Butalbital; sedative. [Sandoz Pharmaceuticals] *

Sandopur. Washing off assistant; dyestuff complexing and fixing agent; for printed polyamide fabrics; improvement of wet fastness of metal complex and acid milling dyes on wool. [Sandoz Products Ltd] *

Sandorin. Synthetic organic pigment colors used in printing inks and paints. [Sandoz Products Ltd]

Sandosperse. Synthetic organic pigment colors used in dispersions. [Sandoz Products Ltd]

Sandostab P-EPQ. CAS 38613-77-3; [tetrakis (2,4-di-tert-butylphenyl) 4,4-biphenylenediphosphonite]; effective processing stabilizer, secondary antioxidant for polymers including polyolefins, ABS, PS, polybutylene, PBT, and nitrile barrier resins; peroxide decomposer for plastics manufacturing.; prevents polymer yellowing; synergizes uv light stability; reduces equipment corrosion in flame-retardant applications. [Sandoz]

Sandoteric CFL. CAS 68604-73-9; EINECS 271-705-0; Cocamphohydroxypropyl sulfonate; extremely mild surfactant producing synergistic visc. increase with alkyl sulfates; weak ampholyte. [Sandoz]

Sandoteric TFL Conc. CAS 68610-38-8; Sodium oleoamphohydroxypropyl sulfonate; extremely mild surfactant producing synergistic visc. increase with alkyl sulfates; weak ampholyte. [Sandoz]

Sandotex. Low soiling antistatic agent; for finishing synthetic fiber carpets. [Sandoz Products Ltd] *

Sandoxylate® AC-46, AD-4. Alcohol ethoxylate; nonfoaming wetting agent, dispersant for cleaning products and processes; stable in alkaline and acid media. [Sandoz]

Sandoxylate® C-32. Castor oil ethoxylate; nonfoaming wetting agent, dispersant for cleaning products and processes; stable in alkaline and acid media. [Sandoz]

Sandoxylate® NT-15. Amine ethoxylate; nonfoaming wetting agent, dispersant for cleaning products and processes; stable in alkaline and acid media. [Sandoz]

Sandoxylate® SX-408. PPG-2-isodeceth-4; wetting agent for household and industrial applications, textile wet processing; intermediate for anionic surfactants. [Sandoz]

Sandoz Effervescent Compound. A proprietary preparation of caffeine, pseudoephedrine hydrochloride and paracetamol; an analgesic. [Sandoz] *

Sandozin. Powerful and economic wetting agent; for neutral and weakly acid media; many uses in textile processing. [Sandoz Products Ltd] *

Sandril. CAS 50-55-5; Reserpine; antihypertensive. [Eli Lilly & Co] †

sandscale. Impurities formed in the pan during the concentration of brine; consists of calcium carbonate, $CaCO_3$.

sandstone. A stone consisting of grains of sand cemented together by a cementing material, e.g., silica, carbonate

of iron, and iron oxide.

sanfoin. *See* orchidee.

Sangajol. A trademark for a fraction of Bormeo petroleum distillate boiling at 160-170 C; it contains cyclic hydrocarbons, and is used as a turpentine substitute and resin solvent. §

Sanguial. A preparation of blood and iron, containing 10 parts hemoglobin, 44 parts muscle albumin, and 46 parts blood salts. Prescribed for chlorosis.

Saniblanket. Urea resin-based foams; cover for refuse and hazardous waste. [Sanifoam Inc] ‡

SaniFoam. *See* Saniblanket. [Sanifoam Inc] ‡

Sani-Soil-Set. Dust laying composition. [Chevron] *

Sanitant. Dairy hygiene detergent sterilizer. [The Wellcome Foundation Ltd] *

Sanitary 1700. Silicone sealant; mildew-resist. sealant for ceramic tile, tubs, spas, showers, and plumbing fixtures. [GE Silicones]

sanitas. An aqueous liquid prepared by blowing air through warm oil of turpentine, in contact with water. It contains hydrogen peroxide and thymol, and is used as a disinfectant.

sanna. (san nai, kapur kachri, sitruti, sheduri). A Chinese drug. It is the dried roots and stems of *Hedychium spicatum*.

san nai. *See* sanna.

Sanoform. A disinfectant consisting of a mixture of the disinfecting constituents of various tar oils, with calcium chloride and magnesium chloride in a saponified form.

Sanogran® . Pigment for coloring gels, powds., solids. [Sandoz]

Sanoleum. A mixture of crude cresols with hydrocarbons; used in disinfecting urinals.

Sanoma. CAS 78-44-4; A proprietary preparation of carisoprodol; a muscle relaxant. [Pfizer International] †

Sanorex. CAS 22232-71-9; Mazindol; anorexic. [Sandoz Pharmaceuticals]

Sanoscent. (camphortar). Preparations containing camphor as the main ingredient; used as disinfectants.

Sansalid. Uredofos; anthelmintic (veterinary). [SmithKline Beecham] *

sanse. (sanse oil) The residual cakes obtained from pressed Italian olives. When dried, and the oil extracted with carbon disulfide, it gives the so-called sulfocarbon oil, or sulfur olive oils; used in the manufacture of green soap for use in the textile industries

sanse oil. *See* sanse.

Sansert. CAS 129-49-7; Methysergide maleate; vasoconstrictor. [Sandoz Pharmaceuticals] *

Sansilic 11. Silicone-free; antifoaming agent for textile processing, all fibers. [Ceca SA]

Sansorbin® . Cells. [Calbiochem Corp]

Sanspor. CAS 2939-80-2; Captafol; fungicide for use on potatoes. [ICI Chem. & Polymers Ltd]

Santac 52. CAS 9003-28-5; High molecular weight polybutene tackifier in LLDPE; cling additive for stretch cling film suitable as pallet wrap, silage wrap, food wrap, etc. [Santech]

santalwood. It contains 16% of santalin, $C_{15}H_{14}O_5$. The extract is used to color confectionery and liqueurs.

Santechem 21-21. CAS 123-77-3; Azodicarbonamide; blowing agent for polyolefins, extruded and injection molded structural foam; elminates die sink marks in injection molding. [Santech]

---

* Trade name not verified as available   † Trade name verified as obsolete

Santechem Grey F.R. P.E. Conc. Halogen compound/antimony oxide in LLDPE; flame retardant concentrate for polyethylene products. [Santech]

Santel. Water-based acrylic resins and coatings; for surface coatings for paper, board, foil and films, gloss enhancers, surface protection, barrier coatings, grease and oil resistance for paper. [ADM Tronics Unlimited Inc]

santheose. CAS 83-67-0; Theobromine; $C_5H_2(CH_3)_2 \cdot N_4O_2$.

Santiciser. A proprietary trade name for vinyl plasticizers coded as follows: M-17: Methyl phthalyl ethyl glycollate. 1-H: N-cyclohexyl paratoluenesulfonamide. 107: Di-2-ethyl hexyl phthalate. 140: Cresyl diphenyl phosphate. §

Santiciser SC. The triglycol ester of a vegetable oil fatty acid; a vinyl plasticizer. [Harwick Standard Chemical Co] ‡

Santicizer 10. A proprietary trade name for a plasticizer; o-Cresyl-p-toluene sulfonate. §

Santicizer 97. Dialkyl phthalate; plasticizer for PVC film, sheet and coatings; gives low temperature flexibility. [Monsanto plc] *

Santicizer 141. Alkylaryl phosphate; used in PVC films, sheets, extrusions, moldings, organosols and plastisols, flame retardant plasticizer. [Monsanto plc]

Santicizer 143. Modified triaryl phosphate; plasticizer compatible with PVC, vinyl nitrile elastomers, latex emulsions and cellulosic materials. [Monsanto plc]

Santicizer 148. Alkyl diaryl phosphate; flame-retardant plasticizer for PVC resins. [Monsanto plc]

Santicizer 154. Triaryl phosphate; flame-retardant plasticizer, compatible with PVC resins and vinyl nitrile rubber, PVA emulsions and cellulosics. [Monsanto plc]

Santicizer 160. CAS 85-68-7; Butyl benzyl phthalate; general purpose plasticizer for PVC, used in flooring industry. [Monsanto plc]

Santicizer 711. Dialkyl adipate; general purpose plasticizer for PVC resins, outperforms DOP. [Monsanto plc]

Santicizer DUP. Plasticizer. [Monsanto plc] *

Santion. CAS 85009-19-9; Emulsifiable concentrate of 400 g flusilazole per liter; systemic insecticide for cereals. [DuPont UK]

Santobrite. A proprietary trade name for sodium pentachlorphenate; a preservative used in paints, adhesives, etc. §

Santocel. A proprietary trade name for silica gel (qv), a porous form of silica; used as a heat insulator, drying agent, etc. §

Santochlor. A proprietary trade name for p-dichlorobenzene; used as a deodorizer, moth preventative, etc. §

Santocure. CAS 95-33-0; N-Cyclohexyl-2-benzothiazolesulfenamide; vulcanization accelerator. [Monsanto Co]*

Santocure MOR/MOR90. 4-Benzothiazole-2-sulfenyl morpholine; rubber accelerators. [Monsanto plc]

Santocure NS. CAS 95-31-8; N-t-Butylbenzothiazole-2-sulfenamide rubber. [Monsanto plc]

Santoflex. Antidegradant, antioxidant, antiozonant; rubber processing chemicals. [Monsanto Co]

Santoflex 1P. Antiozonant/antioxidant. [Monsanto plc] *

Santoflex 13, 77. Antiozonant. [Monsanto plc] *

Santoflex A. A proprietary trade name for a mixed ketone-amine and diphenyl-p-phenylene-diamine; a rubber antioxidant. §

Santoflex AW. Rubber antiozonant. [Monsanto plc] *

Santoflex B. A proprietary trade name for the condensation product of acetone and p-amino-diphenyl; a rubber antioxidant. §

Santoflex BX. A proprietary trade name for a constant composition blend of Santoflex B and diphenyl-parapheny lenediamine; a rubber antioxidant. §

Santoflex DD. DPA. Rubber antioxidant. [Monsanto plc] *

Santogard PVI. CAS 17796-82-6; N-(Cyclohexyl-thio)phthalimide; prevulcanization inhibitor for natural and synthetic rubber. [Monsanto Co] *

Santolite. A proprietary trade name for synthetic resins of the sulfonamide aldehyde type, e.g., toluene sulfonamide-formaldehyde, for use in lacquers, etc. §

Santomerse. A proprietary trade name for an alkylated aryl sulfonate; used as a wetting agent. §

Santone® 3-1-S. CAS 37349-34-1; Polyglyceryl-3 stearate; emulsifier; aeration of lipid systems; solubilizer; color/flavor dispersant in water. [Van Den Bergh Foods]

Santone® 3-1-SH. CAS 9007-48-1; Polyglyceryl-3 oleate; emulsifier and aerating agent used in food industry, textile and plastic lubricant; color dispersant. [Van Den Bergh Foods]

Santonox. 4,4´-Thiobis(6-t-butyl-m-cresol); antioxidants for polyethylene and other plastic resins. [Monsanto Co] *

Santophen 20. CAS 87-86-5; A proprietary trade name for pentachlorphenol; a preservative for paints, wood, adhesives, etc. §

Santoprene® 181-55, 181-64, 281-55, 281-64. Thermoplastic rubbers; medical grade for use in syringe sals and caps, medical tubing, sterilizable applications. [Advanced Elastomer Systems]

Santoprene® 281-87, 283-40. Thermoplastic rubber; medical grade for use in syringe sals and caps, medical tubing, sterilizable applications. [Advanced Elastomer Systems]

Santoquin. CAS 91-53-2; Ethoxyquin; feed preservative. [Monsanto Co]

Santo-Res. Wet strength paper resins; cationic retention aid for paper. [Monsanto Co] *

Santoresin. A proprietary trade name for a synthetic resin.§

santorin earth. A volcanic ash found in the island of Santorin; used to convert lime into hydraulic lime.

Santosite. CAS 7757-83-7; A proprietary trade name for anhydrous sodium sulfite; a reducing agent. §

Santotan KR. A proprietary trade name for a basic chromium sulfate, $Cr_2(SO_4)_3 \cdot (OH)_2$; a tanning agent. §

Santotrac. Synthetic hydrocarbons; high temperature lubricant. [Monsanto Co] *

Santovac. Polyphenyl ether; vacuum diffusion pump fluid. [Monsanto Co] *

Santovar. CAS 79-74-3; 2,5-Di(t-amyl) hydroquinone; antioxidant for unvulcanized rubber. [Monsanto Co]

Santowax. Mixed isomeric terphenyls; high melting hydrocarbons. [Monsanto Co] *

Santoweb. Treated cellulosic short fiber; rubber processing material. [Monsanto Co]

Santowhite. CAS 85-60-9; 4,4´-Butylidenebis(6-t-butyl-m-cresol); antioxidant for polypropylene, polyethylene, nylon molding powders, and other polymer resins. [Monsanto Co]

sanyan. A silk from a wild silkworm of Nigeria.

Sanylene. Synthetic organic pigment colors used in plastics. [Sandoz Products Ltd]

Sapamine. Finishing agent. [Ciba plc] *

Sapamine. A proprietary trade name for diethylaminoethyloleylamino acetate and similar compounds; used in

---

‡ Trade name and manufacturer not verified   § Trade name without identified manufacturer

conjunction with dyes. §

sapan wood. *See* redwoods.

Sapecron. CAS 470-90-6; Emulsifiable concentrate containing 240 g chlorfenvinphos per liter; a soil-applied organophosphorus insecticide. [Ciba-Geigy Agrochemicals]

Saphire. Lube oil. [Chevron] *

sapin. A mixture of Japan wax with heavy mineral oil (soft or liquid paraffin); a superfatting agent for soaps.

sapoform. A product containing oleic acid, alcohol, potassium hydroxide, formalin, and distilled water.

Sapogenat. Alkylphenol polyglycol detergent base. [Hoechst UK]

Sapogenat T. Range of nonionic surfactants of the tributyl phenol ethoxylate type in liquid, paste or wax form; auxiliaries in textile and paper manufacturing, domestic and industrial cleaning agents, emulsifiers and plant protection agents. [Hoechst UK]

sapogenin glycosides. *See* saponins.

sapogenin. A decomposition product ($C_{14}H_{22}O_2$) of saponin.

saponine. A name usually applied to the active constituent of Panama bark, which is used instead of soap, for washing and producing a lather. The term is also used for a boring and cutting oil.

saponins. (sapogenin glycosides) Water soluble high molecular weight glycosidal substances occurring naturally in plants; forms colloidal solutions. on shaking with water; CAS 11006-75-0; Foam producer in fire extinguishers, detergent in textile industries, sizing, substitute for soap, emulsification agent for fats and oils. [Penco of Lyndhurst]

Saponite. *See* Steatite. §

Saporin®. Feed industry additive. [BASF AG]

Sapp #4. CAS 7758-16-9; Sodium acid pyrophosphate; leavening agent for baking, cereals. [Rhone-Poulenc Food Ingreds.]

Saprol. CAS 26644-46-2; Emulsifiable concentrate containing 190 g/l triforine; a systemic insecticide. [Hoechst UK]

saprol. (disinfection oil) CAS 26644-46-2; A mixture of crude cresols, hydrocarbons, and pyridine bases; used for disinfecting lavatories.

Saquadil. Sulfaquinoxaline/diaveridine. [May & Baker Ltd]*

Saran. A range of polyvinylidene chloride plastics. [Dow UK]

Saran 313. Vinylidene chloride polymer; melt-processable extrusion resins with excellent gas and moisture barrier properties, broad chemical and ignition resist.; used for rigid multilayer coextruded containers and in extrusion/coextrusion of film and sheet. [Dow Plastics]

Saran 510. Vinylidene chloride polymer; monofilament extrusion resin with excellent gas and moisture barrier properties, broad chemical and ignition resist.; used for rigid and flexible barrier packaging of medicines, foods, and cosmetics. [Dow Plastics]

Saran F-239, F-278, F-310. Vinylidene chloride polymer; excellent gas and moisture barrier properties, broad chemical and ignition resist.; used for rigid and flexible barrier packaging of medicines, foods, and cosmetics; solv.-sol. polymer for barrier coatings. [Dow Plastics]

Saranex. Coextruded multilayered films. [Dow UK]

Saran® Filament. Polyvinylidene chloride fiber. [Asahi Chem. Industry]

Saran Wrap. Vinylidene chloride-vinyl chloride film. [Dow]

Sarapron. An emulsifiable concentrate containing 98.8 g

aminotriazole, 197.5 g atrazine and 20 g dicamba per liter; used for total weed control in non crop areas. [BP Oil Ltd]

Saratoga Steel. A proprietary trade name for a non-shrinking steel containing small quantities of manganese, chromium, tungsten, and carbon. §

Sarbox® SB 400. Carboxylated acid terminated resin; features acrylate, anhydride and carboxyl functionality; for photoimaging, inks, metal coatings, plastic coatings. [Sartomer]

sarcine. (sarkine) CAS 68-94-0; Hypoxanthine, $C_5H_4N_4O$.

sarco. A material made from elaterite (a mineral rubber); used in rubber mixings.

sarcocoll. A gum resin from *Penoea sarcocolla*, of Africa.

sarcosine. (N-methylglycine; methyl glycocoll; methyl-aminoacetic acid) CH_3NHCH_2COOH; CAS 107-97-1; EINECS 203-538-6; Synthesis of foaming antienzyme compounds for toothpaste, cosmetics, pharmaceuticals. [BASF; W.R. Grace/Hampshire; Schweizerhall; SWS Oilchemicals BV]

Sarenin. CAS 39698-78-7; Saralasin acetate; antihypertensive. [Norwich Eaton Pharmaceuticals Inc] *

Saret® 500, 515. Trifunctional crosslinking agents which minimize scorch and offer processing flexibility when used with the peroxide cure of elastomers; suitable for injection, transfer, and compression molding. [Sartomer]

Sar Gel®. Water/alcohol indicating paste showing presence of water bottoms in storage tanks containing solv., gasoline, oil, etc. [Sartomer]

Saridone. A proprietary preparation of phenacetin with caffeine and phenyldimethyl isopropyl pyrazolone; also, a proprietary preparation of isopropylantipyrine, phenacetin and caffeine; an analgesic. [Roche Products Ltd] *

sarkine. *See* sarcine.

Sarkosyl NL30. CAS 137-16-6; EINECS 205-281-5; Sodium lauroyl sarcosinate; anionic surfactant for corrosion inhibition, enzyme inhibition, bacteristatic activity; used for cosmetics, toilet goods, pharmaceuticals, particularly useful in dentifrices, hair, carpet and upholstery shampoos; specialty and alkaline detergents; window cleaners, hand dishwash; fine fabric detergents; synthetic toilet soap; emulsion polymerization; metal processing; food products. [Ciba plc] *

Sarkosyl O. Oleoyl sarcosine in acid form; anionic surfactant for corrosion inhibition, enzyme inibition, bacteristatic activity; used for cosmetics, toilet goods, pharmaceuticals, particularly useful in dentifrices, hair, carpet and upholstery shampoos; specialty and alkaline detergents; window cleaners, hand dishwash; fine fabric detergents; synthetic toilet soap; emulsion polymerization; metal processing; food products. [Rhone-Poulenc UK]

Sarlink 1000. Thermoplastic elastomer; excellent oil and chemical resist., bondability and abrasion resist.; for hose and tubing, coated fabrics, weatherstripping, diaphragms, gaskets, seals, extruded sheet and linings, grips, ducting, belts and trays. [DSM Thermoplastic Elastomers]

Sarlink 2000. Thermoplastic elastomer; low permeability to moisture and gases, excellent damping and compression creep properties; for use in tank linings, waterproofing membranes, sports grips, ball bladders, seals, gaskets, belts, hoses, tubing, rollers, medical stoppers. [DSM Thermoplastic Elastomers]

---

* Trade name not verified as available

† Trade name verified as obsolete

Sarlink 3000. Thermoplastic elastomer; high resiliency, low tensile and compression creep, excellent flexibility, weatherability; for weatherstripping, seals, o-rings, boots, bellows, switch covers, furniture parts, flashlight housings, belts, plugs, connectors, hoses, tubing. [DSM Thermoplastic Elastomers]

Saroten®. CAS 549-18-8; Amitriptyline hydrochloride; an antidepressant. [Bayer AG; Tropon Werke]

Saroul. Hand cleaners. [Brent Chemicals International plc]

Sarpol. A preparation of crude phenol (carbolic acid); a disinfectant. §

sarsaparilla. The dried root of *Smilax officinalis* and other species; used as a tonic.

Sartomer. Functional acrylics. [Sartomer]

Sascol. Collectors for mineral recovery. [Sasolchem]

Sasetone. CAS 67-64-1; Acetone; solvent in paints, varnishes, lacquer thinners, printing inks, nail polish removers, acetylene in filling of cylinders, bituminous paints, polyester resins, PVC cloth manufacture, explosives, adhesives; raw material for manufacture of methyl isobutyl ketone, diacetone alcohol, hexylene glycol and fine chemicals. [Sasolchem] *

Sasfroth. Frothers for mineral flotation. [Sasolchem]

Sasolwaks. CAS 8002-74-2; Hard, high melting point, crystalline paraffin wax; average molecular formula $C_{50}H_{102x}$ constituent in polishes, plastics to enhance gloss, color dispersion etc., insulating components in electric cables, paper conversion, chewing gums, carbon paper backing, printing inks, paints, hot-melt adhesives; lubricant in rubber, laundry machines and plastic molding. [Sasolchem]

Sasolwaks M3. CAS 8002-74-2; Semi-refined paraffin wax [Sasolchem]

sassafras. (ague tree; cinnamon wood; saloopsaxifrax) The dried bark of the root of *Sassafras variifolium*. The extract is used as a carminative.

sassy bark. (saucy bark). The bark of *Erythrophleum guineanse*.

Satco Metal. A proprietary trade name for a lead-base bearing alloy modified by the addition of tin, calcium, magnesium, mercury, aluminum, potassium lithium, all in very small quantities except tin which may rise to 1%. §

Satessa. A highly dispersed pyrogenic silica preparation; used in the textile industry as a lubricating additive; to increase strength of yarn; to increase fiber friction; to increase nonslip characteristics. [Degussa AG] *

Satexlan 20. CAS 68648-27-1; PEG-20 hydrogenated lanolin; o/w emulsifier, emollient, thickener; perfume solubilizer; imparts superfatting properties. [Croda Inc.]*

Satialgine H8. CAS 9005-32-7; EINECS 232-680-1; Alginic acid NF; tablet disintegrant for compressed tablets. [Mendell]

Satina®. Nonirritant skin and body care products for sensitive and problem skin. [Bayer AG]

Satina 44. Partially hydrogenated palm kernel oil, lecithin; kosher; coating fat for butterscotch and chocolate flavored confectionery coatings. [Van Den Bergh Foods]

satin-gloss black. See gas black.

satinite. See gypsum.

satin rouge. A variety of lamp-black used for polishing.

Satintone. CAS 1332-58-7; Very fine to coarse particle size calcined anhydrous kaolin (aluminum silicate); used as a reinforcer and extender for plastics, PVC, color concentrates, rubber, coatings. [Engelhard]

satin white. CAS 12004-14-7; A pigment consisting of gypsum mixed with alumina. A mixture of calcium sulfate with aluminum sulfate is also known under this name.

Satinwood. Satin finish alkyd paint for wood. [ICI Chem. & Polymers Ltd]

Satisfar. CAS 38260-54-7; Etrimfos; used to control pests in stored grain. [Nickerson Seeds Ltd]

sativic acid. Trihydroxy-stearic acid, $C_{17}H_{31}(OH)_3 \cdot COOH$.

Satrapol. A photographic developer containing monomethyl p-amino-phenolsulfate.

Satric. CAS 443-48-1, Metronidazole; an antiprotozoal. [Savage Laboratories] †

Satulan. CAS 8031-44-5; EINECS 232-452-1; Hydrogenated lanolin; w/o emulsifier, emollient used in personal care products. [Croda Inc.]

Saturn Glace. Polymers, fluorocarbons composition; applied to new and used vehicles. [Adasco Inc]

Saturseal. Joint sealants. [Crowley Chem.]

sauconite. A clay containing zinc.

saucy bark. See sassy bark.

Sauflon PW. Lidofilcon B; contains 79% of water; contact lens material. [Visiontech Inc] ‡

saurol. A distillation product of Meride shale. It resembles ichthyol in its therapeutic properties.

Savacort. CAS 52-21-1; Prednisolone acetate; glucocorticoid. [Savage Laboratories] †

Savall. CAS 13593-03-8; Insecticide containing quinalphos. [ICI Chem. & Polymers Ltd]

Savall. CAS 13593-03-8; Emulsifiable concentrate containing 25% w/w quinalphos per liter; an organophosphorus insecticide. [Farm Protection Ltd]

Savan. Sodium ammonium metavanadate. [Kerr-McGee Chemical Corp]

Saventrine® I.V. Isoprenaline hydrochloride; IV injection/ infusion for cardiogenic or endotoxic shock states, acute stoke Adams attack, severe bradycardia, and evaluation of congenital heart defects. [Pharmax Ltd]

Saventrine® Tablets. A proprietary preparation of isoprenaline hydrochloride; used as a second line treatment in chronic atrio-ventricular dissociation as a temporary measure where pacing is not available. [Pharmax Ltd]

Savinase. A preparation containing a proteolytic enzyme; used in the detergent industry as an additive to powder detergents to improve the detergency towards protein containing stains and as an additive to nonbuilt liquid detergents. [Novo Nordisk]

Savinyl®. Specialty dye for coloring adhesives, oils, waxes, solvs., wood stains. [Sandoz]

Savloclens. CAS 55-56-1; Chlorhexidine; broad spectrum antiseptic with added detergent properties [ICI Chem. & Polymers Ltd]

Savlodil. See Savloclens.

Savlon. A proprietary preparation of chlorhexidine and cetrimide; used for a range of antiseptics. [ICI Chem. & Polymers Ltd] †

Savlon Babycare. CAS 55-56-1; A range of products containing chlorhexidine; used principally for antiseptics for use by babies and infants. [ICI Chem. & Polymers Ltd]†

savol. A medicated soap containing salol (phenyl salicylate) with perfumes.

Savona. A soap concentrate; used to control insects in

---

‡ Trade name and manufacturer not verified   § Trade name without identified manufacturer

greenhouses. [Koppert (UK) Ltd]

savonette oil. A mixture of vegetable fatty acids and resin acids, a by-product of paper manufacture. It is recommended as a substitute for oleic acid in soap manufacture.

Savosellig REAC 4. Soaping agent for fiber reactive dyes on cotton. [Ceca SA]

Saxifragin. An explosive mixture containing 76% barium nitrate, 2% potassium nitrate, and 22% charcoal.

Saxin. Artificial sweetening agent. [The Wellcome Foundation Ltd] *

saxol. See paraffin, liquid.

Saxon. Lotion for men (woodspice and musk), aftershave skin conditioner. [Richardson-Vicks Inc] *

Saxonite. An explosive similar to Samsonite in composition. The term is also used for a mineral which is a mixture of olivine and enstatite.

Saytex®. Brominated flame retardants. [Ethyl Corp; Ethyl SA]

Saytex® 102E. CAS 1163-19-5; EINECS 214-604-9; Decabromodiphenyl oxide; flame retardant; high-purity, elec. grade for wire and cable insulation application. [Ethyl Corp]

Saytex® 111. CAS 32536-52-0; EINECS 251-087-9; Octabromodiphenyl oxide; flame retardant for ABS, HIPS, polyamides, elastomers, adhesives, and coatings; semiplasticizing additive for styrenic polymers and copolymers such as ABS. [Ethyl Corp]

Saytex® 120. CAS 58965-66-5; EINECS 261-526-6; Tetradecabromodiphenoxy benzene; flame retardant for nylon, alloys of styrenics and engineering plastics, engineering thermoplastic polyesters, ABS, crosslinked polyethylene, elastomers, and high-impact PS. [Ethyl Corp]

Saytex® 8010. Flame retardant for HIPS and other styrenics, thermoplastic polyolefins, electronic applications [Ethyl Corp]

Saytex® BCL-462. CAS 3322-93-8; EINECS 222-036-8; Dibromoethyldibromocyclohexane; flame retardant for expandable, crystalline and high-impact PS, SAN resins, adhesives, coatings, textile treatment, PU. [Ethyl Corp]

Saytex® BN-451. CAS 52907-07-0; EINECS 258-250-3; Ethylenebis dibromonorbornane dicarboximide; thermally stable flame retardant for PP, polyamides, PU elastomers and coatings; off-white powd. [Ethyl Corp]

Saytex® BT-93®. CAS 32588-76-4; EINECS 251-118-6; Ethylene bis-tetrabromophthalimide; flame retardant for high-impact PS, polyethylene, PP, thermoplastic polyesters, nylon, EPDM, rubbers, PC, ethylene copolymers, ionomer resins, textile treatment. [Ethyl Corp]

Saytex® FR-1138. Monobromopentaerythritol, dibromoneopentyl glycol, tribromoneopentyl alcohol; flame retardant for unsaturated polyesters, rigid and flexible PU foam, PU elastomers. [Ethyl Corp]

Saytex® HBCD-LM. CAS 3194-55-6; EINECS 221-695-9; Hexabromocyclododecane; low melting flame retardant; see Saytex HBCD-HM. [Ethyl Corp]

Saytex® RB-49. CAS 632-79-1; EINECS 211-185-4; Tetrabromophthalic anhydride; flame retardant; monomer for unsaturated polyester; reactive intermediate for preparation of polyols, esters, and imides. [Ethyl Corp]

Saytex® RB-100. CAS 79-94-7; EINECS 201-236-9; Tetrabromobisphenol A; reactive or additive source of bromine for flame retardancy; reactive intermediate for preparation of brominated epoxy resins, polycarbonates, and unsaturated polyesters; additive for ABS, PS, and phenolic resins; intermediate for other flame retardants. [Ethyl Corp]

Saytex® VBR. CAS 593-60-2; Monomeric vinyl bromide; flame retardant; intermediate in organic synthesis and in the manufacturing. of flame retardants, polymers, copolymers, pharmaceuticals, fumigants, and other chemicals; also used in textiles, adhesives, coatings, photographic plates and films. [Ethyl Corp]

SB 70/52P6. Mixture of ethanol and isopropanol; solvent. [Sasolchem]

SB-136. Alumina trihydrate; filler featuring flame retarding and smoke suppressing properties; resin extender in polyester, vinyls, PU, latex, neoprene foam systems, wire and cable insulation, vinyl wall and floor coverings, epoxies. [J.M. Huber/Solem]

SB-336. Alumina trihydrate; filler with flame and smoke suppression; low visc., high filler loading and glass wet-out; for spray-up or land lay-up FRP application, filament winding, panel production, resin injection, SMC/BMC/acrylic sheet rigidizing, and cast polyester parts. [J.M. Huber/Solem]

SB-632. Alumina trihydrate; filler featuring flame retarding and smoke suppressing properties; also used for plastic and rubber application; suitable only for low visc. systems or low levels of loading; particle suspension; poor processing characteristics; used in urethane foams, vinyls, wire and cable insulation, SBR belting, EPDM, EPR, adhesives, coatings, ABS, polyethylene, and XPLE. [J.M. Huber/Solem]

SBP. Special boiling point solvents. [Carless Refining & Marketing Ltd]

SB-VAC. SB-1 strain, frozen chicken herpes virus marek's vaccine; for immunization of poultry. [Intervet Inc]

SB-VAC Plus Marexine-CA. For combination use; for immunization of poultry. [Intervet Inc]

SC-10. Filled silicone-based resins; thermally conductive resin, cures to resilient solid; high thermal conductivity, thermal stability, and elec. properties; used in electronic applications requiring heat conductive encapsulant; mix ratio: 2.38 phc with Activator BA-52. [Bacon]

SC-17. Filled silicone-based compounds; thermally conductive polymers which cure to resilient solids; used in electronic applications as heat conductive encapsulant. [Bacon]

SC-53. CAS 471-34-1; Calcium carbonate; coarse ground filler for polyolefins, carpet backing, caulks, sealants, putties, as mild abrasives in cleaners. [ECC]

Scabene Lotion. CAS 58-89-9; Lindane; pediculicide; scabicide. [Stiefel Laboratories Inc] *

Scadoplast RA3L, RA350. A proprietary trade name for an adipic acid polyester vinyl plasticizer. [DSM Kunstharze GmbH] *

Scadoplast RS 20, RS 150. A proprietary trade name for a sebacic acid polyester vinyl plasticizer. [DSM Kunstharze GmbH] *

Scagliola. A stone manufactured from Keene's cement mixed with coloring matter, to which is added water containing dissolved glue or isinglass.

S-CAL. Stabilized and treated human blood cells in an artificial plasma medium; calibrator for automatic blood cell analyzers. [Coulter Electronics Ltd]

scale. Crude Scotch paraffin wax is known in commerce by

---

this name.

Scale Cleen. Dry acid descaler. [Dearborn Chemicals Ltd]*

Scalol. A photographic developer containing methyl-p-amino-phenol, $C_6H_4(OH)(NH \cdot CH_3)$, as the active constituent.

scandia. Scandium oxide, Sc_2O_3; used in the preparation of scandium fluoride.

scandium. Metallic element; Sc; CAS 7440-20-2; EINECS 231-129-2; no major industrial use; some application in semiconductor field; an artificial radioactive isotope has been used in tracer studies and leak detection. [Atomergic Chemetals; Cerac; Rhone-Poulenc]

scandium oxide. Sc_2O_3; CAS 12060-08-1; EINECS 235-042-0; Preparation of scandium fluoride. [Aldrich; Atomergic Chemetals; Cerac; Noah Chem.]

Scarab. Urea formaldehyde molding compounds. [BIP Chemicals Ltd]

Scarat. A proprietary trade name for a synthetic resin of the urea type. §

S-Carb. CAS 533-96-0; Feed grade sodium sesquicarbonate. [FMC]

scarlet vermilion. (extract of vermilion, Chinese vermilion, orange vermilion, Field's orange vermilion). Varieties of vermilion, *see* cinnabar.

Scav-Ex® 235. Synthetic pulp; fibrous processing aid for paper industry (settling of stickies in tissue and toweling, gypsum paper, boxboard, fine paper manufacturing). [Hercules]

Scav-Ox® 35%. CAS 302-01-2; Hydrazine aq. solution; corrosion protector in low-, medium- and high-pressure boilers; oxygen scavenger in feedwater. [Olin]

Scav-Ox® II. CAS 302-01-2; Catalyzed hydrazine; corrosion protector in industrial boilers. [Olin]

ScentCap. Microencapsulated fragrance oils for use in soap, clothing, personal care products, cosmetics, emollients in skin creams/powds. [M-CAP Tech. Int'l.]

Scentinel. Gas odorants. [Phillips]

Scent Off. Wax pellet containing naphthalene and essential oils; animal deterrent. [Vitax Ltd]

Scent Sticks. Flammable, scented, oil saturated, compressed wood pulp on a sandalwood stick (incense); used as a scented air freshener (incense), novelty item. [Ambrosia Scents] *

Schaeffer's acid. CAS 93-01-6; (Baum's acid, Armstrong acid). α-Naphthol-2-sulfonic acid, $C_{10}H_6(OH)(SO_3H)$, also β-naphthol-6-sulfonic acid; used as an azo dye intermediate.

Schaeffer's salt. The sodium salt of β-naphthol-6-sulfonic acid (Schaeffer's acid).

schallerite. An arseno-silicate found in the Franklin furnace. It approximates to the formula, $9MnSiO_3 \cdot Mn_3As_2O_8 \cdot 7H_2O$.

Scheele's acid. A 4% solution of hydrocyanic acid, HCN.

Scheele's green. CAS 10290-12-7; (mineral green, Swedish green). A pigment consisting of copper arsenite, $CuHAsO_3$.

scheeletine. (scheelinite). A tungstate of lead, $PbWO_4$.

scheelinite. *See* scheeletine.

scheerite. A mineral wax resembling ozokerite.

Scheiber oil. The glyceride of dehydrated ricinoleic acid; suitable for varnishes.

Scheibler's reagent. Sodium-phospho-tungstate, obtained by dissolving 100 g sodium tungstate and 70 g sodium phosphate in 500 cc water, and acidifying with nitric acid;

used as a testing reagent for alkaloids.

Scheiderite. A mixture of trinitro naphthalene and ammonium nitrate; an explosive.

Schellan solution. A colloidal solution of a synthetic resin made from urea and formaldehyde, and kept from gelatinizing by means of sodium acetate; used as a dressing material for textiles.

Schensand. Resin coated foundry sand. [Schenectady-Midland Ltd]

Schenvar. Insulating varnish. [Schenectady-Midland Ltd]

Schercamox C-AA. CAS 68155-09-9; EINECS 268-938-5; Cocamidopropylamine oxide; conditioner, detergent, wetting agent, antistat used in personal care products and light dishwashing detergents; biodeg. [Scher]

Schercamox CMA. CAS 61791-47-7; EINECS 263-180-1; Dihydroxyethyl cocamine oxide; softener, and wetting agent for cosmetics; builds visc. and stabilizes foam in personal care products; soft emollient feel on the skin; emolliency, lubricity, and slip to shave creams; conditions and prevents fly-away in hair shampoos. [Scher]

Schercamox DMC. CAS 61788-90-7; EINECS 263-016-9; Cocamine oxide; wetting agent, foam stabilizer, visc. enhancer. [Scher]

Schercamox DML. CAS 1643-20-5; EINECS 216-700-6; Lauramine oxide; antistat; emulsifier, emulsion stabilizer for used in cosmetics industry; foam booster and visc. builder for shampoos. [Scher]

Schercamox DMM. CAS 3332-27-2; EINECS 222-059-3; Myristamine oxide; wetting and foaming agent, surfactant for light duty dishwashing compounds, shampoos; emulsifier and emulsion stabilizer for mineral oils. [Scher]

Schercamox DMS. CAS 2571-88-2; EINECS 219-919-5; Stearamine oxide; skin emollient, softener, visc. controller, foam stabilizer, and hair conditioner in personal care products. [Scher]

Schercassist AC. Quaternary; dye-leveling agent for acrylics. [Scher] *

Schercemol 65. CAS 5434-57-1; Isohexyl neopentanoate; penetrating emollient with good sol. in hydro-alcoholic systems; constituent in aroma chemicals. [Scher]

Schercemol 145. Myristyl neopentanoate; light emollient; reduces tackiness in skin and hair preparations. [Scher]

Schercemol 185. CAS 58958-60-4; EINECS 261-521-9; Isostearyl neopentanoate; substantive emollient with low cloud point; aids as cloud and freeze point depressant, emulsion and freeze-thaw stabilizer; cosmetic preparations for skin care especially near eyes; low level of skin and eye irritation; binder for pigmented makeup. [Scher]

Schercemol 318. CAS 68171-33-5; EINECS 250-651-1; Isopropyl isostearate; emollient for bath oils, creams, lotions, and lipsticks; lubricity without oiliness. [Scher]

Schercemol 1688. CAS 59130-69-7, 59130-70-7; EINECS 261-619-1; Cetearyl octanoate; emollient; spreads evenly on skin imparting velvety softness; functions as waterproofing agent due to adhesion properties. [Scher]

Schercemol 1818. CAS 41669-30-1; EINECS 255-485-3; Isostearyl isostearate; substantive emollient imparting luxurious softness to skin; used in cosmetics imparting slip and lubricity, luster, and sheen; cosolv. and solubilizer in perfumes. [Scher]

Schercemol BE. CAS 18312-32-8; EINECS 242-201-8; Behenyl erucate; emollient base for lip care cosmetics,

---

‡ Trade name and manufacturer not verified   § Trade name without identified manufacturer

skin creams and lotions; chemically comparable to one of main constituents of jojoba oil; melts close to body temp.; nontoxic. [Scher]

Schercemol CM. CAS 2599-01-1; Cetyl myristate; solid emollient, lubricant and body builder. [Scher] *

Schercemol CO. CAS 59130-69-7; EINECS 261-619-1; Cetyl octanoate; solvency properties for use in make-up removers. [Scher]

Schercemol CP. CAS 540-10-3; EINECS 208-736-6; Cetyl palmitate; synthetic spermaceti wax. [Scher] *

Schercemol CS. CAS 1190-63-2; Cetyl stearate; waxy emollient for creams and lotions, thickener and body builder. [Scher] *

Schercemol DEGMS. CAS 9004-99-3; PEG-2 stearate; primary emulsifier in creams and lotions. [Scher] *

Schercemol DEIS. Decyl isostearate; emollient, lubricant and penetrant with unusual pigment-dispersing properties. [Scher] *

Schercemol DIA. CAS 6938-94-9; EINECS 248-299-9; Diisopropyl adipate; nonoily penetrating emollient, lubricant, and solv. with mild drying effects used in hydro-alcoholic cosmetic formulations. [Scher]

Schercemol DICA. Diisocetyl adipate; low viscosity emollient, useful in skin and hair preparations. [Scher] *

Schercemol DID. CAS 103213-20-3; Diisopropyl dimer dilinoleate; nonoily, glossy emollient producing a "cushiony" feel and "body" to skin and makeup preparations; improves disp. and spreading of pigments; binder for pressed powd.; offers sheen, emolliency in lip preparations; highly substantive; suitable for sun tan preparations requiring some water repellency. [Scher]

Schercemol DIS. CAS 7491-02-3; EINECS 231-306-4; Diisopropyl sebacate; nonoily emollient, lubricant, solubilizer with mild drying effects used in hydro-alcoholic personal care products; solv. and coupling properties; fast spreading action. [Scher]

Schercemol DISD. CAS 103213-19-0; Diisostearyl dimer dilinoleate; emollient offering lingering effect retained on skin after washing; used in personal care products. [Scher]

Schercemol DISF. CAS 113431-53-1; Diisostearyl fumarate; lubricant, conditioner. [Scher]

Schercemol DO. CAS 3687-46-5; EINECS 222-981-6; Decyl oleate; high molecular weight, low freeze point, nonoily lubricant, emollient, penetrant, and moisturizer for cosmetic and personal care products. [Scher]

Schercemol EE. CAS 27640-89-7; Erucyl erucate; emollient ester for use in skin, hair and suntanning preparations. [Scher] *

Schercemol EGMS. CAS 111-60-4; EINECS 203-886-9; Glycol stearate; emulsifier, opacifier, and pearlescent for cosmetic and personal care products.; thickener and visc. controller for cosmetic preparations. [Scher]

Schercemol GMIS. CAS 66085-00-5; EINECS 266-124-4; Glyceryl isostearate; emulsifier and emollient for creams and lotions. [Scher]

Schercemol GMS. Glycerol stearate; primary emulsifier for creams and lotions. [Scher] *

Schercemol ICS. CAS 25339-09-7; EINECS 246-868-6; Isocetyl stearate; nongreasy emollient used in creams; imparts elegant feel to makeup, lotions, bath preparations; remains liquid even @ low temps. [Scher]

Schercemol IDO. CAS 59231-34-4; EINECS 261-673-6; Isodecyl oleate; emollient, lubricant, penetrant with pig-

ment dispersing properties; for makeup and makeup removers. [Scher]

Schercemol ISE. CAS 977079-10-9; Isostearyl erucate; lubricating emollient for skin and bath preparations. [Scher] *

Schercemol MEL-3. CAS 84605-13-0; EINECS 283-390-7; Myreth-3 laurate; nonoily, rich, penetrating emollient for cosmetic and personal care products.; dispersibility and spreadability in bath oils; coupler in hydro-alcoholic systems; emulsifier and solubilizer in lotions. [Scher]

Schercemol MEM-3. CAS 59686-68-9; Myreth-3 myristate; nonoily, rich, penetrating emollient for cosmetic and personal care products; dispersibility and spreadability in bath oils; coupler in hydro-alcoholic systems; emulsifier and solubilizer in lotions. [Scher]

Schercemol MEP-3. CAS 84605-14-1; EINECS 293-391-2; Myreth-3 palmitate; nonoily, rich, penetrating emollient for cosmetic and personal care products; dispersibility and spreadability in bath oils; coupler in hydro-alcoholic systems; emulsifier and solubilizer in lotions. [Scher]

Schercemol MM. CAS 3234-85-3; EINECS 221-787-9; Myristyl myristate; soft, waxy emollient that melts near body temp.; visc. builder; imparts substantivity to personal care products; ease of combing of hair preparations; velvety feel on skin. [Scher]

Schercemol MP. CAS 6221-95-0; EINECS 226-300-9; Myristyl propionate; emollient for antiperspirants, body oils, creams, and lotions. [Scher]

Schercemol MS. CAS 17661-50-6; Myristyl stearate; waxy emollient for creams and lotions. [Scher]*

Schercemol NGDC. CAS 27841-06-1; Neopentyl glycol dicaprate; solvency properties for use in make-up removers. [Scher]

Schercemol NGDL. Neopentyl glycol dilaurate; emollient and skin conditioner for creams and lotions. [Scher]

Schercemol NGDO. CAS 28510-23-8; Neopentyl glycol dioctanoate; low freeze point emollient; solv. for makeup remover. [Scher]

Schercemol OHS. CAS 29383-26-4; Octyl hydroxystearate; emollient producing slip, lubricity, and tackiness reduction in skin preparations. [Scher]

Schercemol OLO. CAS 3687-45-4; EINECS 222-980-4; Oleyl oleate; nonoily emollient for cosmetic formulations contributing luster, softness, and high degree of lubricity in skin and hair preparations; cosolv. and solubilizer in perfumes; lubricant for metal working and wire drawing. [Scher]

Schercemol OP. CAS 29806-73-3; EINECS 249-862-1; Octyl palmitate; nonoily emollient ester for cosmetic and personal care products giving sheen without greasiness; anticlogging and suspending agent in antiperspirants; soft velvety feel in skin creams, lotions, and aftershaves. [Scher]

Schercemol OPG. CAS 59587-44-9; EINECS 261-819-9; Octyl pelargonate; dry, nonoily rich penetrating emollient for cosmetic and personal care products; anticlogging agent in antiperspirants; soft, luxurious feel in skin creams and aftershaves. [Scher]

Schercemol PGDP. CAS 41395-83-9; EINECS 255-350-9; Propylene glycol dipelargonate; emollient offering low f.p.; cosolv. for perfumed bath oils, creams, and lotions. [Scher]

Schercemol PGML. CAS 27194-74-7; EINECS 205-542-3;

---

* Trade name not verified as available
† Trade name verified as obsolete

Propylene glycol laurate; emollient and solv.; stable base for cosmetics; emulsion stabilizer; solubilizes and couples ingredients such as perfumes, coloring and flavoring agents, sunscreen compounds into natural fatty vegetable or mineral oils; solv. for organic pesticides for spray application; produces sprayable oils which spread well, have good adherence, and are not readily removed by rainfall; plasticizer and stabilizer in vinyl copolymers made from PVAc and PVC; defoaming agent in PVAc emulsions. [Scher]

Schercemol PGMS. CAS 1323-39-3; EINECS 215-354-3; Propylene glycol stearate; emulsifier for creams and lotions. [Scher]

Schercemol SE. Stearyl erucate; emollient wax with the look and feel of real cocoa butter. [Scher] *

Schercemol TISC. CAS 113431-54-2; Triisostearyl citrate; high visc. ester imparting gloss to lipsticks and lip gloss preparations. [Scher]

Schercemol TIST. CAS 103213-22-5; Triisostearyl trilinoleate; emollient; superior gloss and moisturizing characteristics; emolliency, shine, visc., and good binding characteristics. [Scher]

Schercemol TT. Triisopropyl trimerate; binder for pigmented products; imparts gloss and sheen in makeup and hair preparations. [Scher] *

Schercoat OE-44. Fatty oil, edible grade; coating for liquor and food glass containers. [Scher] *

Schercoat OE-44K. High purity fatty oil; protective coating for glass food containers. [Scher] *

Schercoat P-110. Modified polyethylene emulsion; protective coatings for glass bottles. [Scher] *

Schercoat PC-550. Substantive poly emulsion; lubricant for glass containers. [Scher]

Schercoat S-220. Modified vinyl resin ionomer emulsion; protective coatings for glass containers. [Scher] *

Schercoat S-330. Stabilized vinyl resin ionomer emulsion; roller coating of glass bottles to impart scuff resistance. [Scher] *

Schercodine B. CAS 60270-33-9; EINECS 262-134-8; Behenamidopropyl dimethylamine; emulsifier with conditioning properties for hair and skin preparations. [Scher]

Schercodine C. CAS 68140-01-2; EINECS 268-771-8; Cocamidopropyl dimethylamine; good foaming surfactant for hair and bath preparations; emulsifier, intermediate for betaine amphoterics. [Scher]

Schercodine I. CAS 67799-04-6; EINECS 267-101-1; Isostearamidopropyl dimethylamine; versatile liquid o/w emulsifier for creams and lotions; lubricant for hair rinses and conditioners. [Scher]

Schercodine L. CAS 3179-80-4; EINECS 221-661-3; Lauramidopropyl dimethylamine; emulsifier; surfactant, intermediate for betaine amphoterics. [Scher]

Schercodine M. CAS 45267-19-4; EINECS 256-214-1; Myristamidopropyl dimethylamine; o/w emulsifier, conditioner, visc. builder. [Scher]

Schercodine O. CAS 109-28-4; EINECS 203-661-5; Oleamidopropyl dimethylamine; o/w emulsifier; emollient conditioner for hair and skin preparations. [Scher]

Schercodine S. CAS 7651-02-7; EINECS 231-609-1; Stearamidopropyl dimethylamine; softener, emulsifier, and conditioner in hair and skin preparations. [Scher]

Schercodine T. CAS 68650-79-3; Tallamidopropyl dimethylamine; conditioner for cationic emulsions; substantivity

and thickening properties. [Scher]

Scherco Finish AL. Resin dispersion; lubricant for sewing, cutting, napping, and softening all textiles. [Scher] *

Schercolene SB. Detergent; heavy duty textile scouring agent. [Scher] *

Schercolube 707. Fatty ester ethoxylate; softener for knit goods and lubricant for nylon separator threads in sweater bodies. [Scher] *

Schercomid 304. Modified coco amide; dry-cleaning detergent. [Scher] *

Schercomid 1214. CAS 120-40-1, 7545-23-5; Lauramide DEA and diethanolamine; foam booster/stabilizer for detergent compositions; good detergency by itself and works synergistically with other surfactants; thickening agent and visc. builder; used in personal care items and cleaners for hard surfaces. [Scher]

Schercomid AME. CAS 142-26-7; EINECS 205-530-8; Acetamide MEA; solubilizer, humectant, skin and hair conditioner, intermediate, coupling agent, pigment dispersant, solubilizer. [Scher]

Schercomid CDA. CAS 68603-42-9; Cocamide DEA and diethanolamine; foam stabilizer, soil suspender, lime soap dispersant, and detergency booster for industrial and household cleaners. [Scher]

Schercomid EAC. Modified coco amide; wool scouring and fulling agent effective at low temperature. [Scher] *

Schercomid EACS-100. Mixed fatty amide; cold water textile detergent and wool fulling agent. [Scher] *

Schercomid HT-60. CAS 68783-22-2; PEG-50 hydrogenated tallow amide; thickener, detergent, emulsifier, dispersant with foam characteristic. [Scher]

Schercomid LME. CAS 5422-34-4; EINECS 226-546-1; Lactamide MEA; humectant, skin and hair conditioner, coupling agent, emollient. [Scher]

Schercomid ODA. CAS 93-83-4; EINECS 202-281-7; Oleamide DEA and diethanolamine; w/o emulsifier, pigment dispersant, conditioner, corrosion inhibitor, and visc. builder; emulsifier for aromatic and aliphatic hydrocarbon solv.; used in gel-type pine cleaners, shampoo formulations, hair conditioning agent. [Scher]

Schercomid OME. CAS 111-58-0; EINECS 203-884-8; Oleamide MEA; w/o emulsifier, conditioner, and thickener. [Scher]

Schercomid OMI. CAS 111-05-7; EINECS 203-828-2; Oleamide MIPA; thickener, foam stabilizer for shampoos; hair conditioning agent; emulsifier for mineral oil, IPP, IPM, butyl stearate, creams and lotions; imparts slip, lubrication, some emolliency, and softening effects upon the skin. [Scher]

Schercomid SAP. Apricotamide DEA; thickener, foam stabilizer for natural and herbal shampoos. [Scher]

Schercomid SCE. CAS 68603-42-9; EINECS 263-163-9; Cocamide DEA (1:1); detergent, visc. builder and foam stabilizer for cosmetic formulations, bubble baths, liquid dish wash detergents, and rug shampoos. [Scher]

Schercomid SLE. CAS 56863-02-6; EINECS 260-410-2; Linoleamide DEA (1:1); solubilizer, thickener, w/o emulsifier, conditioner, and emollient for personal care products; emulsion stabilizer for o/w emulsions. [Scher]

Schercomid SL-Extra. CAS 120-40-1; EINECS 204-393-1; Lauramide DEA (1:1); thickener, foam booster/stabilizer for hair shampoos, soaps, synthetic detergent formulations; bubble bath applications; industrial applications including manual dishwashing formulations, liquid

---

‡ Trade name and manufacturer not verified   § Trade name without identified manufacturer

heavy-duty laundry detergents, all-purpose cleaning products; emulsifier for aliphatic, aromatic hydrocarbons and oils for o/w emulsions. [Scher]

Schercomid SLS. CAS 68425-47-8; EINECS 270-355-6; Soyamide DEA (1:1); conditioner and emollient for personal care products; emulsifier for w/o systems and hydrocarbons; dispersant for pigments and mineral clays; visc. builder; emulsion stabilizer. [Scher]

Schercomid SO-A. CAS 93-83-4; EINECS 202-281-7; Oleamide DEA (1:1); w/o emulsifier, lubricant, conditioner. [Scher]

Schercomid SO-T. CAS 68155-20-4; EINECS 268-949-5; Tallamide DEA (1:1); w/o emulsifier. [Scher]

Schercomid SWG. Wheatgermamide DEA; thickener, foam stabilizer for natural and herbal shampoos. [Scher]

Schercomid TO-2. CAS 68155-20-4; EINECS 268-949-5; Tallamide DEA and diethanolamine; w/o emulsifier, visc. builder, pigment and mineral clay dispersant, corrosion inhibitor; emulsifier for aromatic and aliphatic hydrocarbon solv.; used in shampoos where it generates a creamy, luxurious foam; stabilizes foam when used with surfactants and detergents; hair conditioning agent. [Scher]

Schercopol CMS-Na. CAS 68784-08-7; EINECS 272-219-1; Disodium cocamido MEA-sulfosuccinate; nonirritating surfactant, foam-stabilizer, solubilizer, softener for personal care products; home and industrial detergent cleaning formulations; anti-irritant for other surfactants; biodeg. [Scher]

Schercopol DOS-70. CAS 577-11-7; EINECS 209-406-4; Dioctyl sodium sulfosuccinate; wetting agent, surface tension depressant. [Scher]

Schercopol DS-120. Modified ethoxylated alkylamine; dispersant for glass fibers in aq. media; for textile industry. [Scher]

Schercopol LPS. CAS 39354-45-5; EINECS 255-062-3; Disodium laureth sulfosuccinate; mild high foaming surfactant; visc. enhancer. [Scher]

Schercopol OMS-Na. CAS 68479-64-1; EINECS 270-864-3; Disodium oleamido MEA-sulfosuccinate; solubilizer; nonirritating surfactant imparting soft, emollient feel on skin and conditioning effect on hair; foamer in toiletries, hand dishwashing detergents, and personal care products. [Scher]

Schercopon 2WD. Ethoxylated sulfosuccinate; detergent and wetting agent for high electrolyte cleansers, dry-cleaning detergent. [Scher]

Schercoquat ALA. CAS 90283-04-8; Dilauryl acetyl dimonium chloride; w/o and o/w emulsifier; conditioner for hair and skin products. [Scher]

Schercoquat APAS. CAS 115340-78-8; Apricot-amidopropyl ethyldimonium ethosulfate; natural, mild conditioner; imparts good slip and shine. [Scher]

Schercoquat BAS. CAS 68797-65-9; EINECS 258-377-8; Behenamidopropyl ethyldimonium ethosulfate; conditioner for dry and over-processed hair. [Scher]

Schercoquat CAS. CAS 113492-03-8; Cocamidopropyl ethyl dimonium ethosulfate; w/o and o/w emulsifier, conditioner, antistat for hair care products. [Scher]

Schercoquat COAS. Quaternary based on canola oil and ethyl sulfate; natural, mild hair and skin conditioner; imparts good slip and velvety feel. [Scher]

Schercoquat DAS. CAS 111905-55-6; Quaternium-61; conditioner for personal care products. [Scher]

Schercoquat FOAS. CAS 113492-04-9; Safflower-amidopropyl ethyldimonium ethosulfate; quaternary effective in hair conditioners; good slip, shine, and compatibility. [Scher]

Schercoquat IALA. CAS 134112-42-8; Isostearamidopropyl laurylacetodimonium chloride; w/o and o/w emulsifier; conditioner for hair and skin care products. [Scher]

Schercoquat IAS. CAS 67633-63-0; EINECS 266-778-0; Isostearamidopropyl ethyldimonium ethosulfate; conditioner for personal care products. [Scher]

Schercoquat IB. Isostearamidopropyl alkonium chloride; liquid quaternary, possessing some bactericidal activity; used in conditioners, hair rinses and skin lotions. [Scher]*

Schercoquat IEP. CAS 84605-15-2; Isostearamidopropyl epoxypropyl dimonium chloride; conditioning quaternary offering good water-sol. and good compatible with many anionic surfactants. [Scher]

Schercoquat IIB. Isostearyl benzyl imidonium chloride; quaternary possessing bactericidal activity. [Scher] *

Schercoquat IIS. CAS 67633-57-2; EINECS 266-778-0; Isostearyl ethylimidonium ethosulfate; conditioner for personal care products. [Scher]

Schercoquat ROAS. CAS 94552-41-7; Rapeseed-amidopropyl ethyldimonium ethosulfate; conditioner for personal care products. [Scher]

Schercoquat ROEP. CAS 112324-11-5; Rapeseed-amidopropyl epoxypropyl dimonium chloride; conditioner for conditioning shampoos and hair sprays. [Scher]

Schercoquat SAS. CAS 67846-16-6; EINECS 267-360-0; Stearamidopropyl ethyl dimonium ethosulfate; conditioner for hair rinses. [Scher]

Schercoquat SOAB. Soyamidopropyl benzyldimonium chloride; conditioning agent used in hair preparations. [Scher] *

Schercoquat SOAS. CAS 90529-57-0; EINECS 291-990-5; Soyamidopropyl ethyldimonium ethosulfate; conditioner for personal care products. [Scher]

Schercoquat WOAS. CAS 115340-80-2; Wheat-germamidopropyl ethyldimonium ethosulfate; surfactant, mild conditioner imparting body, bounce, antistatic properties, shine to hair. [Scher]

Scherco Softener #1. Quaternary; softener and finishing agent for orlon and acrilan. [Scher] *

Scherco Softener #2. Quaternary; softening agent for acrylics and synthetics. [Scher] *

Schercosol DS. Chlorinated solvent; dry side rapid stain remover. [Scher] *

Schercosol NL. Modified coco amide; wet side spotter and fiber lubricant. [Scher] *

Schercosol P. Sulfated amide; protein stain remover. [Scher] *

Schercosol T. Acid-stable detergent; tannin stain remover. [Scher] *

Schercotaine APAB. CAS 133934-08-4; Apricot-amidopropyl betaine; mild detergent, conditioner, emollient, visc. enhancer. [Scher]

Schercotaine CAB. CAS 61789-40-0; EINECS 263-058-8; Cocamidopropyl betaine; detergent, wetting agent, foamer, cloud point depressant, antistat and softener in personal care products. [Scher]

Schercotaine CAB-A. Cocamidopropyl betaine, ammonium chloride; mild surfactant with higher foam than the

---

sodium counterpart, decreased defatting properties. [Scher]

Schercotaine IAB. CAS 6179-44-8; EINECS 228-227-2; Isostearamidopropyl betaine; conditioner and detergent for shampoos and emollient body treatments; visc. control agent; textile softener. [Scher]

Schercotaine MAB. CAS 59272-84-3; EINECS 261-684-6; Myristamidopropyl betaine; detergent, thickener, wetting agent with antistatic properties for cosmetic and toiletry preparations. [Scher]

Schercotaine PAB. CAS 32954-43-1; EINECS 251-306-8; Palmitamidopropyl betaine; thickening agent, good hair and skin conditioner for lotions and cream rinses. [Scher]

Schercotaine SCAB. CAS 68139-30-0; EINECS 268-761-3; Cocamidopropyl hydroxysultaine; detergent, wetting agent and foamer, cloud point depressant used in personal care products. [Scher]

Schercotaine UAB. CAS 133798-12-6; Undecylenamidopropyl betaine; surfactant with germicidal/bactericidal activity; for shampoos. [Scher]

Schercotaine WOAB. CAS 133934-09-5; Wheat germamidopropyl betaine; detergent with emulsification properties, conditioner, surfactant with vitamin E; imparts good body to hair. [Scher]

Schercotarder. Fatty amide; low temperature wool scouring and fulling agent; post-scouring agent for dyed or printed goods. [Scher] *

Schercoterge 140. Ethoxylated amide; detergent, wetting and textile scouring agent, emulsifier, wool fulling; dyeing assistant; post scouring agent; dye bath stabilizer. [Scher]

Schercoteric CY-2. CAS 7702-01-4; EINECS 231-721-0; Disodium capryloamphodiacetate; low foaming surfactant for household and industrial cleaning products. [Scher]

Schercoteric I-AA. CAS 68630-96-6; EINECS 271-929-9; Sodium isostearoamphopropionate; surfactant for cosmetic and industrial cleaners. [Scher]

Schercoteric MS. CAS 68334-21-4; EINECS 269-819-0; Sodium cocoamphoacetate; foamer, mild detergent, conditioner used in personal care products and industrial cleaners. [Scher]

Schercoteric MS-2. CAS 68650-39-5; EINECS 272-043-5; Disodium cocoamphodiacetate; mild detergent used in personal care products and industrial cleaners. [Scher]

Schercoteric MS-EP. CAS 68604-73-9; EINECS 271-705-0; Sodium cocoamphohydroxypropylsulfonate; surfactant for personal care products; low skin irritation, low cloud point. [Scher]

Schercoteric O-AA. CAS 67892-37-9; EINECS 267-569-7; Sodium oleoamphopropionate; surfactant for drycleaning industry, other industrial cleaners. [Scher]

Schercowet DOS-70. CAS 577-11-7; EINECS 209-406-4; Sodium dioctyl sulfosuccinate; emulsifier for emulsion polymerization; wetting and rewetting agent for textile wet processing. [Scher]

Schercozoline C. CAS 61791-38-6; EINECS 263-170-7; Cocoyl hydroxyethyl imidazoline; antistat, dispersant, wetting agent, emulsifier, microbicide, detergent, intermediate for quaternary ammonium compounds, primer paints, emulsion cleaning, cleaners, polishes, surface treatment, textile and leather processing, agriculture, and cosmetic. [Scher]

Schercozoline I. CAS 68966-38-1; EINECS 273-429-6;

Isostearyl hydroxyethyl imidazoline; surfactant, softener, antistat, dye assistant for textiles, paper, cutting oils, metal lubricants, polishes, cosmetics, agriculture, corrosion inhibitors, building materials. [Scher]

Schercozoline L. CAS 136-99-2; EINECS 205-271-0; Lauryl hydroxyethyl imidazoline; surfactant, softener, dye assistant, antistat for textiles, paper, cutting oils, metal lubricants, polishes, cosmetics, corrosion inhibitors, building materials; intermediate for quaternaries. [Scher]

Schercozoline O. CAS 95-38-5; EINECS 248-248-0; Oleyl hydroxyethyl imidazoline; surfactant, softener, dye assistant, antistat, w/o emulsifier, corrosion inhibitor, intermediate for quaternary ammonium compounds, textiles, paper, cutting oils, metal lubricants, polishes, cosmetics, agriculture, building materials. [Scher]

Schercozoline S. CAS 95-19-2; EINECS 202-397-8; Stearyl imidazoline; surfactant, softener, and antistatic agent. [Scher] *

Schericur. A proprietary preparation of hydrocortisone and clemizole- hexa-chlorophane for dermatological use. [Schering Chemicals Ltd] *

Schering PC 4. 250 mcg levonorgestrel (in 500 mcg norgestrel) and 50 mcg ethinylestradiol. [Schering Health Care Ltd]

Scheriproct Ointment. A proprietary preparation of prednisolone hexanoate and dibucaine hydrochloride; used in the treatment of hemorrhoids. [Schering Health Care Ltd]

Scheroba Oil. Isostearyl-erucyl erucate; similar properties to jojoba oil, but has advantages of low price, product consistency and availability. [Scher] *

Scherpol LSB. Ethoxylated alcohol; nonfoaming, jet-dyeing assistant for polyester; minimizes subsequent smoke formation. [Scher] *

Schersoftoil P. Ester of natural oils; winding lubricant applied in a package dye machine. [Scher] *

Schiff's reagents. a) Consists of a solution of rosaniline hydrochloride, decolorized by sulfur dioxide, and is used to test for aldehydes. b) Furfuraldehyde and hydrochloric acid, employed for testing for urea. c) Concentrated sulfuric acid, followed by ammonia, used as a test for cholesterol.

Schimose. See Lyddite.

schlempe. Beet sugar waste. It is the thick brown liquor remaining after the extraction of all possible sugar. It is also called Vinasse.

Schlichte®. Sizes for cellulose fibers and synthetic fibers, fiber blends and filaments. [BASF AG]

Schlippe's salt. Sodium-thio-antimonate, $Na_3SbS_4 \cdot 9H_2O$.

Schneiderite. An explosive. It contains 88% ammonium nitrate, 11% dinitronaphthalene, and 1% resin.

schoenite. (Schonite). Potassium-magnesium sulfate, $K_2SO_4 \cdot MgSO_4 \cdot 6H_2O$.

Scholine. A proprietary preparation of suxamethonium chloride; short-acting muscle relaxant. [Allen & Hanburys Ltd] *

Schollkopf's acids. CAS 82-75-7; 1-Naphthol-4, 8-disulfonic acid, and 1-naphthylamine-8-sulfonic acid.

Schonberg's alloy. A die-casting alloy containing 87% zinc, 10% tin, and 3% copper.

schonite. See schoenite.

schorl. (shorle). A black tourmaline.

schorl rock. An aggregate of black tourmaline and quartz.

---

‡ Trade name and manufacturer not verified   § Trade name without identified manufacturer

Schou oil. (Paalsguardoil). An emulsifier made from soyabean oil.

schraufite. A fossil resin found in Carpathian sandstone.

schreibersite. (dyslytite). An iron nickel phosphide found in meteorites. A chromium sulfide has also been called Schreibersite.

Schultze's reagents. a) Phospho-antimonic acid, made from sodium phosphate and antimony pentachloride, an alkaloidal reagent. b) Consists of 25 parts dry zinc chloride, 8 parts potassium iodide, 8 1/2 parts water, and iodine. It gives a blue color with cellulose.

Schultze's stain. A microscopic stain. It consists of equal parts of a 2% solution of β-naphthol sodium and a 2% solution of dimethyl-p-phenylenediamine hydrochloride. The solutions are mixed and filtered.

schungite. A mineral that is carbon in an amorphous form.

Schutzenberger's salt. Sodium hydro-sulfite, $NaHSO_2$.

Schweitzer's reagent. A solution of copper hydroxide, $Cu(OH)_2$, in strong ammonia; a solvent for cellulose.

schwelkohle. A brown coal of Germany. It is light brown in color.

Scian turpentine. See Chian turpentine.

Scillin. See Scillipicrin. §

Scillipicrin. Commercial name for pharmaceutical preparations of squill, the fleshy bulb of *Urginea scilla*. §

Scillotoxin. See Scillipicrin. §

Scintillase. CAS 9001-73-4; Papain. [Rhone-Poulenc UK]

Scintran. Reagents for scintillation counting. [BDH Chemicals Ltd]

Sclair®. Polyethylene resins. [DuPont UK]

sclerolac. A suggested name for hard lac resin (*qv*).

Sclomo. Sulfur chlorinated base; EP agent for use in threading and tapping operations, cutting and grinding oils. [Ferro/Keil]

Scolaban. CAS 1055-55-6; Bunamidine hydrochloride. [The Wellcome Foundation Ltd] *

Scopacron. A proprietary trade name for thermosetting acrylic resins modifiable by means of epoxy resins. [Stirene Co-Polymers Ltd] ‡

Scopacron 50, 75 and 80. A proprietary trade name for a thermosetting acrylic resin capable of crosslinking with amino and epoxy compounds; primarily intended for use with melamine formaldehyde resin for motor car top coats. [Stirene Co-Polymers Ltd] ‡

Scopacryl. A proprietary trade name for thermoplastic acrylic resin solutions used for wall paints and road marking applications. [Stirene Co-Polymers Ltd] ‡

Scopasol 550. A proprietary trade name for a water-dilutable thermosetting acrylic resin; used for high performance white gloss coatings to be applied by electrophoresis techniques. [Stirene Co-Polymers Ltd] ‡

Scopol 58M, 58SP. A proprietary trade name for a vinyl toluene modified alkyd resin. [Stirene Co-Polymers Ltd]‡

Scopol 85X. A proprietary trade name for a styrene modified alkyd resin; used for quick drying coatings with exceptional adhesion properties. [Stirene Co-Polymers Ltd] ‡

scopolamine. CAS 51-34-3; Hyoscine, $C_{17}H_{21}NO_4$, an alkaloid; used in medicine to inhibit the effects of acetylcholine.

Scopolux 221SP. A proprietary trade name for a medium oil alkyd based on linseed oil. [Stirene Co-Polymers Ltd] ‡

Scorbital. Tablets of phenobarbitone with ascorbic acid. [British Drug Houses] *

Scorchex. CAS 1309-48-4; Magnesium oxide products; used in rubber goods. [Croxton & Garry Ltd] *

Scorchguard O. CAS 1309-48-4; Magnesium oxide. [Bayer AG]

Scorchguard-bound. CAS 1309-48-4; Magnesium oxide. [Bayer plc]

Scotch cement. A cement prepared from feebly hydraulic limes, by the addition of 5% plaster of Paris, and grinding.

Scotch foundry pig. A pig iron made for foundry purposes from Scotch clay-band or black-band ores. It usually contains from 0.7-1% phosphorus, and 2.5% silicon.

scotch gin. See spirit of sweet niter.

Scotchkote® 213, 214. CAS 25928-94-3; Fusion-bonded one-part epoxy coatings; spray grade thermosetting coatings providing maximum corrosion protection for wire fabric and reinforcing steel. [3M]

Scotchlite. Glass bubbles; engineered fillers for industry. [3M]

scotch soda. Impure sodium carbonate, Na_2CO_3.

Scotch topaz. Golden topaz, a yellow variety of quartz.

Scotphos. Fertilizers containing phosphate. [Scottish Agricultural Industries plc]

Scour 1161. Low foaming multipurpose surfactant. [Catawba-Charlab]

Scour KSV, KSV Special. Emulsified solv. containing stain remover and wetting and scouring agent. [Catawba-Charlab]

scouring slag. A slag produced in making spiegel. It is black in color and contains up to 8% of oxide of iron.

Scram. Dog and cat repellent. [Chevron] *

scrap rubber. Formed by the drying of the latex on the bark at the tapping cut. It is variable in quality and color.

Scratch-Guard. Abrasion resistant coating; protective coating for various films i.e., polyester, polycarbonate, etc. [Custom Coating & Laminating Corp] *

Screen. Seed protectant. [Monsanto Co] *

Screen Star Photo Emulsion. Direct screen making emulsion for professional screen printing; for textiles, electronic circuits and paper stock. [Bond Adhesives Co] *

Screte. Sulfur concrete. [Chevron] *

screw bronze. An alloy of 93.5% copper, 5% zinc, 1% tin, and 0.5% lead.

Scripset. Styrene-maleic anhydride copolymers; paper coating and specialty coating resins. [Akzo]

Scripset 520. CAS 9011-13-6; Styrene/maleic anhydride copolymer; emulsifier, binder, sizing agent, visc. modifier, stabilizer; starch modifier; pigment dispersant, protective colloid; sizing, coating, water-paint calsomines, adhesives, printing, preparation of emulsifier paints. [Monsanto Co]

Scripset 720. CAS 9011-13-6; Styrene/maleic anhydride copolymer; emulsifier, dispersant, thickener for high solids systems, latex manufacturing. [Monsanto Co]

Scuranate. CAS 584-84-9; Toluene diisocyanate. [Rhone-Poulenc UK]

Scurane V. Polyurethane varnishes. [Rhone-Poulenc NV/ CdF Chimie AZF] *

Scuttle. Sulfonated cod liver oil; an animal repellent. [Fine Agrochemicals Ltd]

Scythe. CAS 1910-42-5; Soluble concentrate containing 200 g/l paraquat; a pre-emergence bipyridilium herbicide to control weeds in field crops and ornamentals. [Cyanamid of Great Britain Ltd]

SD-1, -2. Super dispersible rheological additive; for paints.

[Rheox Inc]

SD-376. CAS 9003-07-0; Med. impact polypropylene co-polymer; for end-use applications requiring excellent stiffness/impact and improved injection moldability; for consumer products, toys, appliances, crates and totes. [Himont]

SDDC. See sodium dimethyldithiocarbamate.

SDS. See sodium lauryl sulfate.

SE-458. A proprietary silicone rubber compound used for bonding to unprimed surfaces during the curing process. [GE Silicones] *

SeaBuffer. A dry, granular mixture of salts for maintenance of normal pH in recirculating seawater systems; marine aquarium water supplement. [Aquarium Systems Inc]

SeaCure. A solution of copper sulfate and citric acid for treatment of protozoan parasites of marine fishes; used for treatment of marine aquariums. [Aquarium Systems Inc]

SeaGarden. Soluble nutrients for algae, particularly in marine aquariums; aquarium water supplement. [Aquarium Systems Inc]

Sea-Gard® Formula FP-91. CAS 7758-29-4; Sodium tri-polyphosphate; food additive for meat, poultry, and sea-food industries. [Rhone-Poulenc Food Ingreds.]

Seair. Nontoxic, concentrated solution of neutralized resin; an admixture for concrete to increase workability, reduce bleeding of the mixing water, provide a more uniform concrete mix and reduce frost damage and scaling. [Secure Inc] †

Sealac. Sealers. [The Scottish Adhesives Co Ltd]

Seal and Heal. Fungicide and pruning paint. [May & Baker Ltd] *

Sea Legs. A proprietary preparation containing meclozine hydrochloride; a treatment for travel sickness. [The Boots Co plc]

sealite. A liquid containing glucose, corn starch, glycerol, calcium chloride, and glue; used to prevent evaporation from oil storage tanks.

Sealum. Mastic rubber tape; tape sealant for metal build-ings. [Chemseco] *

SeaTest. A series of colorimetric tests for analyzing seawa-ter; used for aquarium water testing and natural seawater testing. [Aquarium Systems Inc]

sea-water bronze. (sheathing bronze). An alloy of 32.5% nickel, 45% copper, 5.5% zinc, 16% tin, and 1% bismuth. It resists sea water.

sebacic acid. (decanedioic acid; 1,8-octanedicarboxylic acid) Organic dicarboxylic acid; $HOOC(CH_2)_8COOH$; CAS 111-20-6; EINECS 203-845-5; stabilizer; raw mate-rial in manufacture of alkyd resins, maleic and other polyesters, plasticizers, polyester rubbers, synthetic polyamide fibers. [Janssen Chimica; Penta Mfg.; Union Camp]

Sebacil®. For control of all ectoparasites, especially mange mites of domestic animals; veterinary medicine. [Bayer AG]

Sebase. Ethoxylated lanolin plus fatty alcohols and hydro-carbons; base, emollient, lubricant for o/w emulsions; visc. stabilizer for cosmetics. [Westbrook Lanolin]

Sebastlne. A dynamite explosive.

Sebizon. Sulfacetamide sodium; antibacterial. [Schering Corp]

sebkanite. A crude potassium chloride obtained by the evaporation and crystallization of the water of the salt lake in Tunis.

Sebond. Modified acrylic emulsion; used to bond new concrete to either new or old concrete. For patching and resurfacing precast architectural panels, industrial con-crete floors, highway and bridge deck repair. [Secure Inc] †

Sebrite. A clear, transparent, penetrating liquid sealer; for protecting and beautifying mechanically textured con-crete, exposed aggregate and stone surfaces. [Secure Inc] †

Secadrex. A proprietary preparation of acebutolol plus hydrochlorothiazide; for treatment of hypertension. [May & Baker Ltd] *

Seclomycin. A proprietary preparation containing strepto-mycin, benzylpenicillin sodium, and procaine penicillin; an antibiotic. [Glaxo Laboratories] *

Secol. Fibre lubricant. [Stephenson Thompson Textile Chemicals]

Secolan S-1, BA-1, BA-1G. CAS 9007-34-5; Soluble animal collagen. [RITA] *

Secolat. Disodium alkyl sulfosuccinamate as a clear yellow liquid; anionic surfactant used in latex foams. [KWR Chemicals Ltd] *

Secomine TA 02. Tallow amine ethoxylate; lubricating, wetting, antistatic agent, emulsifier, anticorrosive agent for detergents, surface treatments. [Stepan Europe]

Secomix® E40. Surfactant blend; detergent for industrial/household cleaners. [Stepan Europe]

Seconal. Secobarbital; hypnotic; sedative. [Eli Lilly & Co]

Seconal Sodium. CAS 309-43-3; Secobarbital sodium; hypnotic; sedative. [Eli Lilly & Co]

Seconesin. A proprietary preparation of quinalbarbitone and mephenesin; a hypnotic. [Crookes Laboratories] *

Secosol® AL 959. Sodium lauryl sulfosuccinate; mild foamer for shampoos, bubble baths, liquid soaps, shower gels, bath salts. [Stepan Europe]

Secosol AL 959. Disodium monolauryl sulfosuccinate as a white paste; anionic surfactant for shampoos; foam baths; creams. [KWR Chemicals Ltd] *

Secosol AL/MG 50. Anionic surfactant in which the anion is monolauryl sulfosuccinate and the cations are sodium and magnesium; supplied as a white paste; used for shampoos; foam baths; creams. [KWR Chemicals Ltd]*

Secosol ALL/40. Disodium monolauryl ether sulfosuccinate as a water-white liquid; anionic surfactant; used in foam baths, shampoos, and liquid soaps. [KWR Chemicals Ltd] *

Secosol® ALL40. Sodium laureth sulfosuccinate; mild foamer for shampoos, bubble baths, liquid soaps, shower gels, bath salts; emulsifier for emulsion polymer-ization. [Stepan Europe]

Secosol® DOS 70. CAS 577-11-7; EINECS 209-406-4; Sodium dioctyl sulfosuccinate; emulsifier, dispersant and wetting/rewetting agent for household/industrial cleaners, emulsion polymerization, paints, inks, oilfield production; textile additive. [Stepan Europe]

Secosol DOS/70. CAS 577-11-7; EINECS 209-406-4; Sodium dioctyl sulfosuccinate in liquid form; wetting and emulsifying agent. [KWR Chemicals Ltd] *

Secosol EA/40. Disodium monoalkyl ethanolamide sulfo-succinate as a clear yellow liquid; anionic surfactant for special shampoos; foam baths; liquid soaps. [KWR Chemicals Ltd] *

Secosov. Emulsifiers; for mineral oils; solvents; dispersant

---

‡ Trade name and manufacturer not verified   § Trade name without identified manufacturer

for phytosanitaires; cosmetic creams and milks; insecticides. [KWR Chemicals Ltd] *

Secoster® A. Fatty acid ethoxylate; dispersant, solv. and oil emulsifier for household/industrial cleaners, oilfield production. [Stepan Europe]

Secoster® DMS. CAS 627-83-8; EINECS 211-014-3; Glycol distearate; emollient, pearlescent, emulsifier, opacifier for creams, cleansing milks, shampoos. [Stepan Europe]

Secoster® DO 600. CAS 9005-07-6; PEG 600 dioleate; additive for cutting oils; solv., emulsifier for solvs. and oils, creams, cleansing milks, pesticides, textile lubricants, oilfield production; dispersant; for household/industrial cleaners. [Stepan Europe]

Secoster® EMS. Glycol stearate; emulsifier for creams and cleansing milks; emollient, pearlescent, and opacifier. [Stepan Europe]

Secoster® MA 300. PEG 300 abietate; additive for cutting oils; solv., emulsifier for solvs. and oils, creams, cleansing milks and pesticides; dispersant; for household/industrial cleaners. [Stepan Europe]

Secoster® MO 400. CAS 9004-96-0; PEG 400 oleate; lubricant, antistat, solvent and oil emulsifier; for household/industrial cleaners, oilfield production. [Stepan Europe]

Secoster® SDG. Glyceryl stearate; antistat and lubricant for polyolefins. [Stepan Europe]

Secosyl. CAS 137-16-6; EINECS 205-281-5; Sodium N-lauroyl sarcosinate; detergent, foaming agent, base, anticorrosion additive for rug shampoos, mild dishwash, household cleaners, personal care products; stable in hard water. [Stepan Europe]

Secosyl. CAS 137-16-6; EINECS 205-281-5; Sodium lauroyl sarcosinate as a clear yellow liquid; anionic surfactant used in shampoos and foam baths. [KWR Chemicals Ltd] *

Secretan. An alloy of from 91-95% copper, 5-9% aluminum, 1.5% magnesium, and 0.5% phosphorus.

Secretol. A fat-splitting material similar and equal to Twitchell's reagent in its action.

Secrodyl. Tablets containing dimethisterone with ethinylestradiol; used in gynecological disorders. [British Drug Houses] *

Sectral. CAS 37517-30-9; Acebutolol; anti-adrenergic. [Ives Laboratories Inc] *

Sectral. A proprietary preparation of acebutolol hydrochloride; used in the treatment of cardiac arrhythmias. [May & Baker Ltd] *

Securamide®. For the electrical industry. [DuPont UK]

Secure. Sprayable liquid resins, concrete curing compound; for application to freshly placed concrete following the finishing. For use in preventing cracking and crazing caused by rapid moisture loss due to hot, windy weather. [Secure Inc] †

Securite. It is a mixture of 26% m-dinitro-benzene and 74% ammonium nitrate. It sometimes contains dinitronaphthalene and potassium nitrate; a safety explosive for mines.

Securitol. A sodium silicate used to hasten the setting of cements.

Securon 540. Multifunctional bleaching aid for silicate bleaching; complexes metal ions and prevents formation of silicate scale buildup on equip. [Henkel/Textiles]

Securopen®. CAS 37091-65-9; Azlocillin sodium; a broad-spectrum penicillin especially active against *Pseudomonas aeruginosa*. [Bayer AG]

Sedacol. A proprietary preparation of clioquinol and phanquone; an antidiarrheal. [Zyma (UK) Ltd] ‡

Sedanox. CAS 470-90-6; Chlorfenvinphos; a soil-applied organophosphorus insecticide. [Bayer plc]

Sedaplant Richter. Polyvalent herbal extract plus anti-irritants (fennel, hops, camomile, balm mint, mistletoe, yarrow, urea, and urea deriv.) in water-alcohol medium; emollient for aq. and hydroalcoholic herbal cosmetics, skin and hair protection products; emulsified preparations. [Dr. Kurt Richter; Henkel/Cospha]

Sedatin. *See* Antipyrine. §

sedative salt. Boric acid, H_3BO_3.

Sedatussin. A proprietary preparation of cephaeline hydrochloride, sodium benzoate, syrup of squill, and syrup of tolu; a cough linctus. [Eli Lilly & Co] †

Sedeff. An effervescent preparation containing opium, bismuth, and digestive ferments.

Sedefos 75®. Glycol stearate, PEG-2 stearate, and trilaneth-4 phosphate; self-emulsifying base for cosmetics and pharmaceuticals. [Gattefosse; Gattefosse SA]

Sedestran. A proprietary preparation of stilboestrol and phenobarbitone; used in the treatment of menopausal disorders. [Ciba plc] *

Sedex. Ceramic foam filters to prevent nonmetallic inclusions in iron castings. [Foseco (F.S.) Ltd]

Sedifloc Flocculant Aids. Organic polymer, polyacrylamide, water soluble, polymers, cationic, anionic, nonionic; used for dewatering, settling and floating municipal and industrial solids found in their waste water treatment plant. [Benzsay & Harrison Inc]

Sedipol®. Higher aliphatic alcohols; antifoams for industrial and communal sewage works. [BASF AG]

Sedipur®. High molecular weight water-sol. anionic, cationic and nonionic polymers; flocculants for industrial and communal water treatment, sewage sludge treatment, and in the mining industry. [BASF AG; BASF plc]

Sednine. A proprietary preparation of pholcodine and pseudoephedrine hydrochloride; a cough linctus. [Allen & Hanburys Ltd] *

Seed Base. Powdered compost additive containing NPK 2.1:2.2:2 plus 3% Mg, 0.2% Fe and trace elements; compost additive. [Vitax Ltd]

seed-lac. Stick-lac, after washing free from the coloring matter soluble in water.

Seedox SC. CAS 22781-23-3; Bendiocarb; a contact, systemic insecticide. [Schering Agrochemicals Ltd]

Seedtect. Mixture of imazalil and thiabendazole; fungicide treatment of potatoes at planting time. [MSD Agvet]

Seekay Pitch®. A registered trade name for chlorinated naphthalene products available in various grades. §

Seenox 412S. CAS 29598-76-3; Pentaerythrityl tetrakis (β-laurylthiopropionate); antioxidant for polyolefins, thermoplastic elastomers, engineering thermoplastics; synergistic with primary antioxidants; outstanding long-term, heat aging performance. [Witco/Argus]

SEF. Modacrylic fibers. [Monsanto Co] *

Segetan. A silver cyanide with a copper complex; used as a seed preservative.

Seidlitz powder. (effervescent tartrated soda powder). Consists of 3 parts rochelle salt with 1 part sodium bicarbonate in the blue paper, and 1 part tartaric acid in the white paper.

---

* Trade name not verified as available   † Trade name verified as obsolete

Seidlitz salt. A name applied to magnesium sulfate, $MgSO_4$ · $7H_2O$ (Epsom salts), found in the mineral waters of Seidlitz.

Seidschütz salt. Native magnesium sulfate, $MgSO_4$ · $7H_2O$.

Seifert solder. An alloy of 73% tin, 21% zinc, 5% lead, 0.5% phosphorus and 0.5% tin.

Seignette salt. See potassium sodium tartrate.

Seignette's salt. See Rochelle salt.

Sekawrap. Polypropylene with various stabilizers; shrink film for presentation applications. [S Kempner Ltd]

Sekicel. Cellulose starch dispersion; for food industry. [Asahi Chem. Industry]

Selacryn. CAS 40180-04-9; Ticrynafen; diuretic; uricosuric; antihypertensive. [SmithKline Beecham] *

seladon green. See Bohemian earth.

Selar®. Barrier resin. [DuPont UK]

Selastin EL-10, EL-30, SE EM 95. Hydrolyzed animal elastin. [RITA] *

Selazate. Selenium diethyl dithiocarbamate; a proprietary accelerator. [Naugatuck (US Rubber)] *

Selbana 2001. Synthetic lubricant for wool; provides anti-static and lubricant properties. [Henkel/Textiles]

Selbax. Synthetic lubricant for textiles. [Crosfield Chemicals Ltd] †

sel d'Angleterre. Magnesium sulfate, $MgSO_4$.

Select-A-Sorb. CAS 14807-96-6; Hydrous magnesium silicate (industrial talc); filler, extender, and reinforcing agent for rubber, paper (pitch control), plastics. [R T Vanderbilt Co Inc]

Selectrol. Soluble concentrate of 18 g dicamba, 252 g MCPA and 84 g mecoprop per liter; used for weed control in cereals and grassland. [R P Adams Ltd]

Select-Trol. Soluble concentrate containing 6.6% w/w 2,4-D and 250 g mecoprop; used to control weeds in grassland. [Chemsearch (UK) Ltd]

Seleen. Selenium sulfide; Antifungal; antiseborrheic. [Abbott Laboratories] *

Selek. Sealants to prevent metal penetration between ingot molds and bottom plates. [Foseco (F.S.) Ltd] †

Selektan. A proprietary preparation of 2-hydroxy-5-iodopyridine. §

Selenac®. Selenium dialkyldithiocarbamate; accelerators and vulcanizing agents for rubber. [R T Vanderbilt Co Inc]

selenium. (colloidal selenium) A nonmetallic element; Se; CAS 7782-49-2; EINECS 231-957-4; electronics, colorant for glass (ceramics), rectifiers, relays, solar batteries. [Appleby Group Ltd; Asarco; Atomergic Chemetals; Cerac; Shinko Chem.; R.T. Vanderbilt]

selenium dioxide. (selenous acid anhydride) SeO_2; CAS 7446-08-4; EINECS 231-194-7; Analysis (testing for alkaloids), oxidizing agent, antioxidant in lubricating oils, catalyst. [Aldrich; Atomergic Chemetals; Cerac; Shinko Chem.]

selenous acid anhydride. See selenium dioxide.

Seleron. (Aeron). A group of aluminum alloys containing 85% aluminum, with copper, nickel, zinc, manganese, silicon, and lithium, as the other ingredients. They are claimed to be useful for electrical apparatus.

Selexsorb® COS. CAS 1344-28-1; Activated alumina; selective adsorbent for catalytic reforming, polyethylene and polypropylene production, isomerization processes. [Alcoa]

self-hardening steel. See Mushet steel.

Selin® O. Trioleate; basic oil for release agents, bread cutting oil. [Grünau]

Seliwanoff's reagent. A solution of 0.05 gram resorcinol in 100 cc dilute (1 : 2) hydrochloric acid. It gives a red color with fructose.

Seljut. Special processing emulsifying agent. [Crosfield Chemicals Ltd] †

Sella Acid. Acid dyes for leather. [Ciba plc] *

Sellacron. Dyes for leather. [Ciba plc] *

Sella Fast. Dyes for leather. [Ciba plc] *

Sellaflor. Dyes for leather. [Ciba plc] *

Sellasol. Synthetic tanning agents for leather. [Ciba plc] *

Sellifix Helios. Formaldehyde-free fixative for fiber-reactive dyes. [Ceca SA]

Sellig. Specialty blends of anionic surfactants; used in textile industry. [Atochem UK/Ceca]

Sellig Antimousse S. Silicone-based; antifoaming agent for textiles. [Ceca SA]

Sellig AO 6 100. CAS 9004-96-0; PEG-6 oleate; surfactant. [Ceca SA]

Sellig AO 15 100. CAS 9004-96-0; PEG-15 oleate; surfactant for emulsions, detergents. [Ceca SA]

Sellig AO 25 100. CAS 9004-96-0; PEG-25 oleate; surfactant. [Ceca SA]

Sellig DN 10 100. CAS 9014-93-1; Nonyl nonoxynol-10; surfactant. [Ceca SA]

Sellig DN 22 100. CAS 9014-93-1; Nonyl nonoxynol-22; surfactant. [Ceca SA]

Sellig HR 18 100. CAS 61791-12-6; PEG-21 castor oil; surfactant. [Ceca SA]

Sellig LA 1150. Laureth-20; surfactant for shampoos, degreaser, textile applications. [Ceca SA]

Sellig N 4 100. CAS 9016-45-9; Nonoxynol-4; surfactant. [Ceca SA]

Sellig N 5 100. CAS 9016-45-9; Nonoxynol-5; surfactant. [Ceca SA]

Sellig N 6 100. CAS 9016-45-9; Nonoxynol-6; surfactant for emulsions, petrol. products, chlorinated and aromatic solvs., greases, silicones. [Ceca SA]

Sellig N 8 100. CAS 9016-45-9; Nonoxynol-8; surfactant for emulsions, petrol. products, mineral oils, greases, silicones; base for low foam household detergents. [Ceca SA]

Sellig N 9 100. CAS 9016-45-9; Nonoxynol-9; surfactant for emulsions, mineral oils, greases, silicones; base for household, industrial and textile detergents. [Ceca SA]

Sellig N 1050. CAS 9016-45-9; Nonoxynol-40; surfactant. [Ceca SA]

Sellig N 10 100. CAS 9016-45-9; Nonoxynol-10; surfactant. [Ceca SA]

Sellig N 11 100. CAS 9016-45-9; Nonoxynol-11; surfactant. [Ceca SA]

Sellig N 12 100. CAS 9016-45-9; Nonoxynol-12; surfactant. [Ceca SA]

Sellig N 15 100. CAS 9016-45-9; Nonoxynol-16; surfactant for emulsions, oils, greases. [Ceca SA]

Sellig N 1780. CAS 9016-45-9; Nonoxynol-17; surfactant. [Ceca SA]

Sellig N 20 80. CAS 9016-45-9; Nonoxynol-19; surfactant. [Ceca SA]

Sellig N 30 70. CAS 9016-45-9; Nonoxynol-25; surfactant. [Ceca SA]

Sellig N 50 100. CAS 9016-45-9; Nonoxynol-52; surfactant. [Ceca SA]

---

‡ Trade name and manufacturer not verified   § Trade name without identified manufacturer

Sellig O 4 100. CAS 9002-93-1; Octoxynol-4; surfactant. [Ceca SA]

Sellig O 5 100. CAS 9002-93-1; Octoxynol-5; surfactant. [Ceca SA]

Sellig O 6 100. CAS 9002-93-1; Octoxynol-6; surfactant. [Ceca SA]

Sellig O 8 100. CAS 9002-93-1; Octoxynol-8; surfactant. [Ceca SA]

Sellig O 9 100. CAS 9002-93-1; Octoxynol-9; surfactant. [Ceca SA]

Sellig O 11 100. CAS 9002-93-1; Octoxynol-11; surfactant. [Ceca SA]

Sellig O 12 100. CAS 9002-93-1; Octoxynol-12; surfactant. [Ceca SA]

Sellig O 20 100. CAS 9002-93-1; Octoxynol-20; surfactant. [Ceca SA]

Sellig R 20 100. CAS 61791-12-6; PEG-20 castor oil; surfactant. [Ceca SA]

Sellig R 3395. CAS 61791-12-6; PEG-33 castor oil; surfactant. [Ceca SA]

Sellig R 3395-C435. CAS 61791-12-6; PEG-32 castor oil; surfactant. [Ceca SA]

Sellig R 3395 SP. CAS 61791-12-6; PEG-30 castor oil; surfactant. [Ceca SA]

Sellig R 4095. CAS 61791-12-6; PEG-40 castor oil; surfactant. [Ceca SA]

Sellig R 4495. CAS 61791-12-6; PEG-44 castor oil; surfactant. [Ceca SA]

Sellig S 30 100. CAS 9004-99-3; PEG-30 stearate; surfactant. [Ceca SA]

Sellig SP 25 50. Cetoleth-27; surfactant. [Ceca SA]

Sellig SP 3020. Cetoleth-30; surfactant. [Ceca SA]

Sellig SP 8 100. Cetoleth-8; surfactant for o/w emulsions, oils, organic solvs., mineral oil, paraffin; stable to electrolytes. [Ceca SA]

Sellig SP 16 100. Cetoleth-16; surfactant for emulsions. [Ceca SA]

Sellig SP 20 100. Cetoleth-18; surfactant. [Ceca SA]

Sellig SP 30 100. Cetoleth-30; surfactant. [Ceca SA]

Sellig Stearo 6. CAS 9004-99-3; PEG-6 stearate; surfactant. [Ceca SA]

Sellig SU 4 100. CAS 68439-49-6; Ceteareth-4; surfactant for emulsions, oils. [Ceca SA]

Sellig SU 18 100. CAS 68439-49-6; Ceteareth-18; surfactant. [Ceca SA]

Sellig SU 25 100. CAS 68439-49-6; Ceteareth-20; surfactant. [Ceca SA]

Sellig SU 30 100. CAS 68439-49-6; Ceteareth-32; surfactant, dispersant, household and industrial detergent base, emulsions. [Ceca SA]

Sellig SU 50 100. CAS 68439-49-6; Ceteareth-46; surfactant. [Ceca SA]

Sellig T 1790. CAS 61791-00-2; PEG-17 tallate; surfactant. [Ceca SA]

Sellig T 3 100. CAS 61791-00-2; PEG-3 tallate; surfactant. [Ceca SA]

Sellig T 14 100. CAS 61791-00-2; PEG-14 tallate; detergent base with controlled foam; solubilizer for essential oils. [Ceca SA]

Selligon SP. Softening detergent for wool and cotton. [Ceca SA]

Selligor 860 SP. For washing out of acidic dyes. [Ceca SA]

Sellogen DFL. CAS 9084-06-4; Sodium alkyl naphthalene sulfonate; wetting and dispersing agent for pesticide formulations. [Henkel/Functional Prods.]

Sellogen HR-90. CAS 9084-06-4; Sodium alkyl naphthalene sulfonate; wetting and dispersing agent for insecticides, acid and alkaline media. [Henkel/Functional Prods.]

Selora. CAS 7447-40-7; EINECS 231-211-8; A proprietary preparation of potassium chloride; used as a substitute for table salt. [Winthrop Laboratories] *

Seloxone. Soluble concentrate containing 15 g clopyralid and 510 g mecoprop per liter; a translocated herbicide for cereals and grassland. [ICI Chem. & Polymers Ltd]

Selsun. A proprietary preparation of selenium sulfide and a detergent; used as a treatment for dandruff. [Abbott Laboratories] *

Selsun Blue. Selenium sulfide; antifungal; antiseborrheic. [Abbott Laboratories] *

seltzers. Usually consist of 25 parts sodium carbonate, 5 parts sodium chloride, 6 parts sodium sulfate, and 1,000 parts water.

Selvigon. A proprietary preparation of pipazethate hydrochloride; a cough linctus. [SmithKline Beecham] *

Semap. CAS 26864-56-2; Penfluridol; antipsychotic. [McNeil Pharmaceuticals] *

Sembonit/Erostabil. Several compositions of natural caoutchouc, synthetic caoutchouc, filling material, synthetic resins, depending on resistance demands; for surface protection against corrosion and erosion of vessels, tanks, tubes, and industrial equipment. [Schaumstoff und Kunststoff GmbH] *

Semeron. CAS 1014-69-3; Wettable powder containing 25% w/w desmetryn; a triazine herbicide. [Ciba-Geigy Agrochemicals]

Semfreeze. Reactive or curable chemical materials for use in industry. [Courtaulds Aerospace Ltd]

semicoke. A fuel made from coal by low temperature carbonization. It is a smokeless fuel with a low ash.

Semilente Iletin. Insulin zinc, prompt; antidiabetic. [Eli Lilly & Co]

Semilente Insulin. Insulin zinc, prompt; antidiabetic. [Bristol-Myers Squibb Co Inc] *

seminose. Mannose, $C_6H_{12}O_6$.

Semirit. Electrically fused corundum, semifriable grade; used in production of abrasives, abrasive paper, discs and cloth. [Hüls AG]

semi-steel. A metal having properties between cast iron and cast steel; used for filter-press plates. The term is applied to grey cast irons of low carbon content.

Sempatap. Insulating material for walls, ceilings and floors; glass fiber nonwoven with SBR-latex backing; for interior acoustic and thermal insulation. [Ebnother AG] *

Sempollan. Cast polyurethane; for components required to display high strength, lasting resilience, high wear and oil resistance and a maximum useful life. [Schaumstoff und Kunststoff GmbH] *

Semtol. White mineral oils, technical grade. [Witco]

Senate. Lube oil. [Chevron] *

Senate. CAS 886-50-0, 1912-26-1; Suspension concentrate containing 250 g terbutryn and 250 g trietazine per liter; herbicide for weed control in potatoes, peas and field beans. [Schering Agrochemicals Ltd]

Sencor®. CAS 21087-64-9; Metribuzin; herbicide for control of many important grasses and broad-leaved weeds in soybeans, potatoes, tomatoes, sugarcane, alfalfa and asparagus; suitable for pre- and in some cases post-

---

emergence application. [Bayer AG]

Sencoral®. See Sencor®. [Bayer AG]

Sencorex® WG. CAS 21087-64-9; Water dispersible granular formulation containing 70% w/w metribuzin; used to control annual weeds in early and maincrop potatoes. [Bayer plc]

Sendoxan. See Endoxan. [Degussa AG] *

Sendust. A proprietary trade name for an iron-silicon-aluminum alloy. §

Seneca oil. A name given to American petroleum; used in medicine.

Senegal gum. (West African gum) A gum arabic ranking second to Khordofan gum. It is derived from *Acacia senegal* and other species of *Acacia*. It gives a good adhesive mucilage.

senegin. See struthiin.

Seneprolin. See paraffin, liquid.

Senesco. See Zoloft. [Pfizer International]

Sengite. An American explosive. It has a guncotton base and is similar to Tonite (*qv*), except that sodium nitrate replaces barium nitrate.

senna. The dried leaflets of *Cassia acutifolia*.

Sennaar gum. See Suakin gum.

Sensitizer. Diazo photosensitizer. [ABM Chemicals Ltd] *

Sensolve BEA. Butoxy ethyl acetate; solvent. [Sasolchem]

Sensolve EEA. CAS 111-15-9; Ethoxy ethyl acetate; solvent. [Sasolchem]

Sensolve EPA. Ethoxy propyl acetate; solvent. [Sasolchem]

Sensolve MPA. Methoxy propyl acetate; solvent. [Sasolchem]

Sensorcaine. CAS 14252-80-3; Bupivacaine hydrochloride; anesthetic. [Astra Pharmaceutical Products Inc]

Sentinel. Water treatment used in hot water boiler systems. [Grace Dearborn Ltd]

Sentry Cyclomethicone. CAS 69430-24-6; Cyclomethicone; pharmaceutic aid. [Union Carbide] *

Sentry Dimethicone. Dimethicone; prosthetic aid. [Union Carbide] *

Sentry Simethicone. CAS 8050-81-5; Simethicone; antiflatulent. [Union Carbide] *

Seominal. A proprietary preparation of reserpine, phenobarbitone, and theobromine. [Bayer AG] †

Sep 6. Isolating solution for waxes; dental preparation. [Bayer AG] †

SEP 55. A general purpose farm disinfectant which can be diluted with water or oil. [Coventry Chemicals Ltd]

Sepabase. Mineral demulsifier concentrates. [BASF plc]

Sepabase A Grades. High molecular weight EO/PO adduct; for dehydration of crude oil emulsions and removal of residual salts. [BASF AG]

Sepabeads® FP Series. High porous type hydrophilic polymer; for industrial purification of protein and enzyme by chromatography. [Mitsubishi Kasei] *

Sepacid® CE 5209. Quaternary ammonium compound; biocide for oilfield applications. [BASF AG]

Sepacid® CE 5265. Glutaraldehyde deriv.; biocide for oilfield applications. [BASF AG] *

Sepaclear®. For removal of residual oil from water in refineries and in crude oil production. [BASF AG; BASF plc]

Sepacorr®. Mineral oil corrosion inhibitors. [BASF plc]

Sepacorr® HT. Nitrogen-containing condensation production; corrosion inhibitor. [BASF AG] *

Sepaflood®. Water-sol. polymers; for tert. crude oil production. [BASF AG]

Sepaflux®. Pour point and/or visc. depressants for crude oils and residual oils. [BASF AG; BASF plc]

Sepakoll®. Protective colloids for drilling fluids and borehole cementing. [BASF AG; BASF plc]

Separpar® P. Paraffin inhibitor. [BASF AG]

Separan. Polyacrylamides used as flocculants. [Dow UK]

Separit. Parting powder. [Foseco (F.S.) Ltd]

Separol. Liquid parting medium. [Foseco (F.S.) Ltd]

Separol. Demulsifiers for crude oils. [BASF plc]

Separol AF 27. High molecular weight EO/PO adduct; demulsifier used in crude oil emulsions and dehydrating equipment. [BASF AG]

Sepascale®. Scale inhibitors for petrol. production and processing. [BASF AG; BASF plc]

Sepasolv® MPE. For purification of natural gas. [BASF AG]

Sepawet®. Surfactants for petrol. production and pipeline transport. [BASF AG; BASF plc]

sepia. A brownish-black pigment derived from the ink-bag of the cuttle-fish; used as water color.

Sepicide HB. Phenoxyethanol, methylparaben, ethylparaben, propylparaben, butylparaben; cosmetic preservative. [Seppic]

Sepigel 305. Polyacrylamide, C13-14 isoparaffin, laureth-7; cosmetic ingred. [Seppic]

Sepiogel A. CAS 1319-41-1; Sepiolite mineral; gelling and suspending agent; drilling muds, foundation drilling, slurry trench cut-off walls, soil admixture for impermeable barriers. [Floridin] *

Sepramar. Reagents for amino acids analyzis. [BDH Chemicals Ltd]

Sepratek. Fluid emulsion parting agent. [Foseco (F.S.) Ltd]

Septal. CAS 83601-81-4, 12427-38-2; Carbendazim + maneb; systemic fungicide for cereals. [Schering Agrochemicals Ltd]

Septex No. 1. A proprietary skin cream containing boric acid, zinc oleate and zinc oxide. [H N Norton & Co Ltd] ‡

Septex No. 2. A proprietary skin cream containing boric acid, zinc oxide, zinc oleate and sulfathiazole. [H N Norton & Co Ltd] ‡

Septra. Proprietary formulations of trimethoprim and sulfamethoxazole; tablets, suspension and IV infusion. for treatment of urinary tract infections, acute otitis media, acute exacerbations of chronic bronchitis, shigellosis and *Pneumocystis carinii* pneumonitis. [The Wellcome Foundation Ltd]

Septrin. A proprietary preparation of trimethoprim and sulfamethoxazole; an antibiotic for treatment of urinary tract infections, acute otitis media, acute exacerbations of chronic bronchitis, shigellosis, and *Pneumocystis carinii* pneumonitis. [The Wellcome Foundation Ltd]

Seqlene® 270. Reaction mixture forming sodium-α-δ-glucoheptonate, sodium-β-δ-glucoheptonate, aldobionates, and other complex carbohydrates; sequestrant forming nonionic chelates; used in metal applications; scavenger for antioxidants, bactericides; bottle washing, alkaline cleaning, textile applications. [Pfanstiehl Labs]

Sequalog. Reagents for peptide sequence and synthesis. [Schweizerhall]

Sequenase. Range of modified or genetically engineered DNA polymerases. [U.S. Biochemical]

Sequens. A proprietary preparation of mestranol (15 white

---

‡ Trade name and manufacturer not verified   § Trade name without identified manufacturer

tablets) and mestranol and chlormadinone acetate (5 peach tablets); oral contraceptive. [Eli Lilly & Co] †

Sequest-All. A dry nontoxic potable water treatment used to control minerals in water, to prevent red water, scale, build-up and corrosion in the distribution system; used for municipal water systems, irrigation systems, cooling towers, boilers, apartments, hotels. [SPER Chemical Corporation] *

Sequestrene®. Iron chelates-food additives foliar feeds. [Ciba plc] *

Sequestrene® 30A. CAS 64-02-8; EINECS 200-573-9; Tetrasodium EDTA; chelating agent used in water softening, liquid soaps, detergents, chemical cleaning, scale removal, beerstone removal, processing of textile, paper, and leather, in metal treatment, and for synthetic rubber. [Ciba-Geigy/Dyestuffs]

Sequestrene® 220. CAS 64-02-8; EINECS 200-573-9; Tetrasodium EDTA dihydrate; chelating agent used in powd. cleaning compounds; scale removal, hair rinses; processing of synthetic fibers and textiles, and industrial cleaning preparations. [Ciba-Geigy/Dyestuffs]

Sequestrene® AA. CAS 60-00-4; EDTA; chelating agent for photographic developer baths, shampoos, cosmetics, electroplating, rare earth separations, metal determinations, liquid soaps, germicides, herbicide sprays. [Ciba-Geigy/Dyestuffs]

Sequestrene® NA2. CAS 139-33-3; Disodium EDTA dihydrate; chelating agent for control of trace metal contamination in pharmaceutical and cosmetic products. [Ciba-Geigy/Dyestuffs]

Sequestrene® NA2Ca. Disodium-calcium EDTA dihydrate; chelating agent for control of trace metal contamination in pharmaceutical and cosmetic products. [Ciba-Geigy/Dyestuffs]

Sequestrene® NA2 Edetate USP. CAS 139-33-3; Disodium edetate USP; chelating agent for control of trace metal contamination in pharmaceutical manufacturing, ophthalmic solutions, high purity cosmetics; analytical reagent for metals. [Ciba-Geigy/Dyestuffs]

Sequestrene® NA3. CAS 150-38-9; Trisodium EDTA trihydrate; chelating agent used in personal care products; processing of synthetic fibers and textiles; stabilizer for resin systems; photographic baths, electrolytic and electroless plating; foam stabilizer, water treatment. [Ciba-Geigy/Dyestuffs]

Sequestrene® NAFe 13% Fe. Na(FeEDTA); chelating agent; micronutrient; animal feeds, photographic uses, polymerization catalyst for synthetic rubber. [Ciba-Geigy/Dyestuffs]

Sequestrene® NH4Fe. Ferric ammonium EDTA; chelating agent in photographic baths, blix baths; in fertilizer formulations. [Ciba-Geigy/Dyestuffs]

Sequestrene® Tetraammonium. Tetraammonium EDTA; chelating agent used in photographic developer baths, replenishment of blix baths, water treatment, and for boiler cleaning. [Ciba-Geigy/Dyestuffs]

Seracelle. A proprietary cellulose acetate packing material.§

Seractide. A corticotrophic peptide. Ala^{26}-Gly^{27}-Ser^{31}-α^{1-39}-corticotrophin.

Seradix. CAS 133-32-4; 4-Indol-3-ylbutyric acid; a root growth promoter. [Embetec Crop Protection Ltd]

Sera-Pak®. Test combinations for determining uric acid, cholesterol, HDL cholesterol, and urinary proteins in clinical chemistry. [Bayer AG]

Serax. CAS 604-75-1; Oxazepam; tranquilizer. [Wyeth Laboratories] *

Serc. CAS 5638-76-6; Betahistine; counteracts disorders of the inner ear, Menire's disease. [Duphar BV] *

Serdet DCK. Sodium alkyl ether sulfate based on a natural alcohol (C12-C14) in paste or liquid form; anionic surfactant used in shampoos and bubble bath formulations and dishwashing. [Chemische Fabriek Servo BV] *

Serdet DFK. Sodium alkyl sulfate, based on a natural alcohol (C12-C14), in liquid or paste form; detergent and emulsifier for shampoos and bubble baths; toothpaste; dishwashing; emulsion polymerization. [Chemische Fabriek Servo BV] *

Serdet DFL, DFM, and DFN. Anionic surfactants in liquid form; foaming agents for shampoos and bubble baths. [Chemische Fabriek Servo BV] *

Serdet DM and DMK. Dodecylbenzene sulfonate in acid form or as sodium salt; biodegradable primary emulsifiers used in scouring powders, liquid detergents, and emulsion polymerization. [Chemische Fabriek Servo BV] *

Serdet DML. Triethanolamine dodecylbenzene sulfonate in liquid form; biodegradable anionic surfactant used in shampoos and bubble baths. [Chemische Fabriek Servo BV] *

Serdet DNK. CAS 9014-90-8; Sodium nonylphenol 4EO-sulfate in liquid form; detergent base for liquid detergent formulations; emulsifier in emulsion polymerization. [Chemische Fabriek Servo BV] *

Serdet DPK. Sodium alkyl ether sulfate based on a synthetic alcohol (C12-C15) in liquid or paste form; anionic surfactant used in shampoo and bubble bath formulations and dishwashing. [Chemische Fabriek Servo BV] *

Serdet DSK. Sodium 2-ethylhexyl sulfate in liquid form; alkali-stable wetting agent used for latex stabilization. [Chemische Fabriek Servo BV] *

Serdolamide. Foam stabilizer; refatting agent; viscosity modifier and improver; for detergent, shampoo and bubble bath formulations. [Chem-Y, Fabriek van Chemische Producten BV] *

Serdox. Emulsifier; antistatic agent; for crude and vegetable oils; plastics; textiles processing; cosmetic emulsions. [Chem-Y, Fabriek van Chemische Producten BV] *

Serdox NNP4. Nonylphenol ethoxylate nonionic surfactant in liquid form; oil-soluble detergents; emulsifier for insecticides and herbicides; dispersing agent. [Chemische Fabriek Servo BV] *

Serdox NNP5 and NNP6. Nonylphenol ethoxylate nonionic surfactant in liquid form; emulsifier for insecticides and herbicides; oil soluble detergents. [Chemische Fabriek Servo BV] *

Serdox NNP7, NNP8.5 and NNP9. Nonylphenol ethoxylate nonionic surfactant in liquid form; for scouring of textiles; soaking assistant for leather; household and industrial detergents; emulsifier for insecticides and herbicides; plasticizer for mortar and concrete. [Chemische Fabriek Servo BV] *

Serdox NNP10 and NNP12. Nonylphenol ethoxylate nonionic surfactant in liquid form; for scouring of textiles; soaking assistant for leather; household and industrial detergents; emulsifier for insecticides and herbicides; plasticizer for mortar and concrete; paper manufacture. [Chemische Fabriek Servo BV] *

---

* Trade name not verified as available   † Trade name verified as obsolete

Serdox NNP15, NNP20 and NNP25. Nonylphenol ethoxylate nonionic surfactant in liquid or solid form; detergents and wetting agents for use at high temperatures and electrolyte concentrations; emulsifier for fatty acids and waxes; NNP20 is used as a stabilizer for synthetic latexes; NNP25 is an emulsifier for emulsion polymerization. [Chemische Fabriek Servo BV] *

Serdox NNP30 and NNP30/70. Nonylphenol ethoxylate nonionic surfactant in solid or liquid form; dyeing assistant; lime soap dispersing agent; emulsifier and stabilizer for emulsion polymerization. [Chemische Fabriek Servo BV] *

Serdox NNPQ 7/11. Nonionic surfactant of the alkylphenol ethoxylate type in liquid form; low foaming wetting agent. [Chemische Fabriek Servo BV] *

Serdox NOP9. Octylphenol ethoxylate nonionic surfactant in liquid form; for scouring of textiles; soaking assistant for leather; household and industrial detergents; emulsifier for insecticides and herbicides; plasticizer for mortar and concrete; paper manufacture. [Chemische Fabriek Servo BV] *

Serdox NOP 30/70. Octylphenol ethoxylate nonionic surfactant in liquid form; emulsifier and stabilizer used in emulsion polymerization. [Chemische Fabriek Servo BV] *

Sereen. Travel sickness tablets. [The Boots Co plc] *

Serenace. CAS 52-86-8; A proprietary preparation of haloperidol; a sedative. [G D Searle & Co Ltd] *

Serenid-D. CAS 604-75-1; A proprietary preparation of oxazepam; a sedative. [Wyeth Laboratories] *

Serenid Forte. CAS 604-75-1; A proprietary preparation of oxazepam; a tranquilizer. [Wyeth Laboratories] *

seretin. (carbon tetrachloride) CAS 56-23-5; A chlorinated hydrocarbon; CCl_4; used as a solvent, in refrigerants, as an agricultural fumigant, metal degreasing.

Serfene. Polyvinylidene chloride dispersion coatings. [Morton Int'l. Ltd]

sericine. (silk size, silk rubber). The gum surrounding the silk from the silk spinner.

Sericite FSE. CAS 12001-26-2; Mica; filler imparting softness, smooth feel, skin adhesion, and spreadability. [Presperse]

Sericite SL-012. CAS 12001-26-2, 9004-73-3; Mica, methicone. surface-treated powd. providing creamier, more lubricious feel, improved spreadability and skin adhesion; ideal for wet and dry applications. [Presperse]

Sericose. Cellulose acetate; used for making artificial silk and dope.

Serilan. A range of disperse and acid dye mixtures; for dyeing of polyester/wool blended fibers. [Yorkshire Chemicals plc] *

Serilene Dyes. Disperse dyes for atmospheric and high temp. exhaust, pad-thermosol and print applications on polyester and polyester blend fabrics. [Yorkshire Pat-Chem; Yorkshire Chemicals plc]

Serilube Series. Lubricants for jet dyeing various blended and 100% cotton fabrics. [Yorkshire Pat-Chem]

serine. (α-Amino-β-hydroxy-propionic acid) $CH_3(OH)CH(NH_2)COOH$; CAS 56-45-1; Used in biochemical research, as a dietary supplementfeed additive.

DL-serine. (+-2-amino-3-hydroxypropionic acid) $HOCH_2CH(NH_2)COOH$; CAS 302-84-1; EINECS 206-130-6; biochemical research, dietary supplement, culture media, microbiological tests, feed additive. [Degussa; R.W.

Greeff; Mitsui Toatsu Chem.; Tanabe USA]

L-serine. $HOCH_2CH(NH_2)COOH$; CAS 56-45-1; EINECS 200-274-3; Biochemical research, dietary supplement, culture media, microbiological tests, feed additive. [Degussa; Janssen Chimica; Mitsui Toatsu Chem.; Nippon Rikagakuyakuhin; U.S. Biochemical]

Serinyl. A range of disperse dyes; for dyeing of polyamide fibers. [Yorkshire Chemicals plc] *

Seriplast Dyes. Selected disperse dyes formulated for heat-transfer printing. [Yorkshire Pat-Chem]

Seripol Series. Dispersants, levelers, and retardants for disperse dyeing. [Yorkshire Pat-Chem]

Seriprint Dyes. Disperse dyes formulated for use in printing with synthetic printing thickeners. [Yorkshire Pat-Chem]

Serisol Dyes. Disperse dyes for acetate; also suitable for polyester, triacetate, nylon, and acrylic fibers. [Yorkshire Pat-Chem; Yorkshire Chemicals plc]

Seritox. Selective weedkiller. [May & Baker Ltd] *

Seritox. Mixture of dichlorprop and MCPA; used for weed control in cereals and turf. [Rhone-Poulenc Crop Protection Ltd]

Serizyme. A proprietary trade name for an enzyme used in desizing acetate fabrics and similar materials which contain protein. §

Sermag®. CAS 1309-48-4; Liquid magnesium oxide oil fuel additive. [Steetley Magnesia Products Ltd]

Sermix. Alkaloid animal feed additive. [Ciba plc] *

Sermul. Emulsifier, biodegradable; for mineral oils; pesticide formulations; white spirit and turpentine; vegetable and animal oils. [Chem-Y, Fabriek van Chemische Producten BV] *

Sermul EA54, EA151 and EA146. Anionic surfactants of the ether sulfate type in liquid form; emulsifiers for emulsion polymerization. [Chemische Fabriek Servo BV] *

Sermul EA 88. Calcium dodecylbenzene sulfonate in liquid form; biodegradable emulsifier for pesticide formulations. [Chemische Fabriek Servo BV] *

Sermul EA129. CAS 2235-54-3; EINECS 218-793-9; Ammonium lauryl sulfate in liquid form; emulsifier in emulsion polymerization. [Chemische Fabriek Servo BV] *

Sermul EA150. Sodium lauryl sulfate in paste form; emulsifier in emulsion polymerization. [Chemische Fabriek Servo BV] *

Sermul EA176. Sodium mono nonylphenol 10-EO sulfosuccinate in liquid from; emulsifier for emulsion polymerization. [Chemische Fabriek Servo BV] *

Sermul EA188, EA136 and EA205. Anionic surfactants of the phosphate ester type; supplied as liquids in acid form; emulsifiers for emulsion polymerization. [Chemische Fabriek Servo BV] *

Sernylan. CAS 956-90-1; Phencyclidine hydrochloride; anesthetic. [Parke-Davis] *

SeroClear®. Reagent. [Calbiochem Corp]

Seromycin. CAS 68-41-7; Cycloserine; antibacterial. [Eli Lilly & Co]

Serosteron. CAS 79-64-1; A proprietary trade name for dimethisterone. [British Drug Houses] *

Serotulle®. White soft paraffin chlorhexidine acetate 0.5% w/w; gauze wound dressing to prevent infection in minor wounds, injuries with minor skin loss and ulcerative lesions. [Seton Healthcare Group plc] *

Serpasil. CAS 50-55-5; Reserpine; antihypertensive. [Ciba-Geigy Corp] *

‡ Trade name and manufacturer not verified   § Trade name without identified manufacturer

Serpasil-esidrex. Antihypertensive/diuretic. [Ciba plc] *
Serpatonil. A proprietary preparation of reserpine and methyl phenidate hydrochloride. [Ciba plc] *
Serpiloid. CAS 50-55-5; Reserpine; antihypertensive. [3M Pharmaceuticals] †
Serramix CDA. An emulsifiable concentrate containing 100 g aminotriazole, 200 g atrazine and 100 g MCPA per liter; used for total weed control in non crop areas. [Denoon CDS; Powaspray (CDA) Ltd]
Serseal. Heat conserving compound for use in metal pre-treatment. [ICI Chem. & Polymers Ltd] †
serum-casein. See globulin.
Serumpro EN-10. Serum protein; cosmetics ingred.; contains all essential amino acids. [Brooks Industries]
Servamine KAC 422. N-Coco N-N-dimethyl-N-benzalkonium chloride in liquid form; cationic surfactant used as a bactericide, fungicide, sanitizer and germicide. [Chemische Fabriek Servo BV] *
Servamine KEP 4527. N(Palmityl amido propyl)N-N-N-trimethyl ammonium chloride in liquid form; cationic surfactant emulsifier with bactericide properties. [Chemische Fabriek Servo BV] *
Servamine KET 350. N-(Tall oil amido propyl)N-N-dimethyl amine based on tall oil; supplied as a liquid; cationic surfactant used as an adhesion agent and corrosion inhibitor in bitumen. [Chemische Fabriek Servo BV] *
Servamine KET 4542. N-(Alkylamido propyl)N-ethyl N-N-dimethyl ammonium ethosulfate, based on tall oil, in liquid form; cationic emulsifier. [Chemische Fabriek Servo BV] *
Servamine KOO 330. Amino ethyl oleyl imidazoline in liquid form; cationic adhesion agent and corrosion inhibitor for bitumen. [Chemische Fabriek Servo BV] *
Servamine KOO 330B. Oleylamido ethyl oleyl imidazoline in liquid form; basic material in the manufacture of quaternary imidazolines. [Chemische Fabriek Servo BV]*
Servamine KOO 360. Hydroxy ethyl oleyl imidazoline in liquid form; cationic adhesion agent and corrosion inhibitor for bitumen. [Chemische Fabriek Servo BV] *
Servil® Range. Wax esters; release agents for confectionery and baked goods. [Grünau]
Servit® Range. Citric acid esters of mono and diglycerides of fatty acids with other emulsifiers; food emulsifier for manufacturing of instant dry yeast. [Grünau]
Servo Ampholyt (B) JA110. Modified imidazoline in liquid form; non eye-irritating for shampoo formulations. [Chemische Fabriek Servo BV] *
Servo Ampholyt (B) JA140. Modified imidazoline in liquid form; non eye-irritating and non skin-irritating for hair shampoos. [Chemische Fabriek Servo BV] *
Servo Ampholyt (B) JB130. Betaine structure, liquid form; mild shampoo raw material with hair stipulating properties, for baby shampoos. [Chemische Fabriek Servo BV]*
Servo Brilliant Oil B AZ 75. Sodium castor oil sulfonate in liquid form; anionic surfactant used in softener and finishing oils, and pasting oil for dyestuffs. [Chemische Fabriek Servo BV] *
Servoxyl VLA 2170. CAS 577-11-7; EINECS 209-406-4; Sodium di-2-ethylhexyl sulfosuccinate in liquid form; wetting and rewetting agent. [Chemische Fabriek Servo BV] *
Servoxyl VLB 1123. Sodium monoalkyl polyglycol ether sulfosuccinate in liquid form; raw material for high quality baby shampoos and mild hair shampoos; cleaning agent. [Chemische Fabriek Servo BV] *
Servoxyl VLE 1159. Sodium mono nonylphenol 10-EO sulfosuccinate in liquid form; emulsifier for emulsion polymerization. [Chemische Fabriek Servo BV] *
Servoxyl VP. Range of anionic surfactants of the phosphate ester type, supplied mainly as liquids in acid form; those based on alcohol and alkyl polyglycol ethers are biodegradable; used in dry-cleaning; formulation of metal cleaners, emulsion polymerization, pesticide formulation, cosmetic preparations. [Chemische Fabriek Servo BV] *
Seseal. A thermoplastic acrylic sealer; seals, hardens and dustproofs concrete. [Secure Inc] †
Seseal 8. Cure, seal, hardener and dustproofer for concrete; prevents mortar and concrete droppings from bonding to floors, reducing clean-up cost, base for mastic adhesives. [Secure Inc] †
Sesolvan® L. Dye solv. for textile dyeing. [BASF AG]
Sestrip. Form release agent; for application to all types of concrete forms prior to concrete placement to ensure release of the forms and to minimize form clean up. [Secure Inc] †
Setac. Waxes for dental laboratory technology in bead form and block form; casting and modeling waxes blue, green, sticky wax red. [Bayer AG] †
Setacin 103 Spezial. Disodium laureth sulfosuccinate; detergent for personal care products; cleaning agent. [Zschimmer & Schwarz]
Setacin F Spezial Paste. Disodium lauryl sulfosuccinate; detergent, cosmetics, washing and cleaning agent. [Zschimmer & Schwarz]
Setacure. Multifunctional acrylic monomers and prepolymers; for uv, ebc, radiation curing applications. [Synthese BV] *
Setafix. Acrylic and polyester resins; photocopy toners. [Synthese BV] *
Setair. A clear or colored solution of PVC polymer; coating zinc sprayed, shot or sand blasted steel, asbestos, concrete, brick, plywood, softboard, strawboard, chipboard and hardboard. [Llewellyn Ryland Ltd]
Setal. Saturated polyester and alkyd resins; for decorative and industrial paints. [Synthese BV] *
Setalana. A proprietary name for a natural nest silk produced by worms of the genus *Anaphe*, introduced into Germany from Africa; the fiber resembles tussah silk, but is stated to be not so strong. §
Setalin. Modified phenolic resins, modified hydrocarbon resins, water thinnable acrylic resins, acrylic dispersions, alkyd resins, modified phenolic resins, modified hydrocarbon resins; for varnishes, rotogravure inks, packaging inks and offset inks. [Synthese BV] *
Setalux. Acrylic resins; air-drying, thermosetting, isocyanate curing resins for industrial paint (automotive, refinishing and general industry). [Synthese BV] *
Setamine US. Melamine-formaldehyde resins; for oven-drying industrial paints (automotive and general industry). [Synthese BV] *
Setamol®. Dispersants for dye dispersions used in dyeing synthetic and cellulose fibers with vat dyes. [BASF AG; BASF plc]
Setarol. Unsaturated polyester resins; for fiber-reinforced polyesters. [Synthese BV] *

---

* Trade name not verified as available   † Trade name verified as obsolete

Setatack A. Acrylic polyols; for pressure-sensitive adhesives. [Synthese BV] *

Setatack AF. Thermoplastic/thermosetting acrylics; for hot-melt adhesives. [Synthese BV] *

Setatack LP. Linear polyesterpolyols; PUR elastomers and prepolymers. [Synthese BV] *

Setatack P. Polyesterpolyols; for two-pack PUR adhesives. [Synthese BV] *

Setatack T. Modified rosin esters; for lamination/paint lamination adhesives. [Synthese BV] *

Sethotope. Selenomethionine Se 75; diagnostic aid. [Bristol-Myers Squibb Co Inc] *

Setic. Waxes for dental laboratory technology in bead and block form. [Bayer AG]

Setilon. Fatty alcohol compound with emulsifiers, nonionic; softener with scrooping effect especially developed for absorbent cotton, silky finish of cotton fabrics. Softening agent for the raising of synthetic goods. [Henkel Chemicals Ltd] *

Setilose. A French cellulose acetate artificial silk.

Setilthe. Cellulose acetate molding granules. [Rhone-Poulenc UK]

Setirene. Styrenated acrylated alkyds; for quick-drying industrial paints. [Synthese BV] *

Setreat. High-solids linseed oil-based penetrating concrete sealer; for treating rough finished, porous concrete to penetrate and seal the surfaces to prevent the absorption of water, salts and other contaminants harmful to concrete. [Secure Inc] †

Setsit®. Activated dithiocarbamate blend; primary and ultra accelerators for latex compounds. [R T Vanderbilt Co Inc]

Setter 33. A solution concentrate containing 50 g benazolin, 237 g 2,4-DB and 43 g MCPA per liter; a post-emergence herbicide. [DowElanco Ltd]

Sett® . Hydrogenated triglycerides; visc. enhancer, hardstock for margarine, coating agent. [Grünau]

Sevacarb. CAS 1333-86-4; Carbon black. [Sevalco Ltd]

Sevamine KOV 4342B. Cationic surfactant composed of quaternary imidazoline in liquid form; raw material in the preparation of laundry softeners. [Chemische Fabriek Servo BV] *

Sevefilm 20. Surfactant blend; film-forming, anticorrosive, antiredeposition agent for water cooling circuit treatment products. [Stepan Europe]

Sevelyte K. Potassium lauryl amino propionate, in orange yellow liquid form; pigment dispersant for paints and inks. [KWR Chemicals Ltd] *

Seven Seas®. Cod liver oil, various nutritional supplements, herbal remedies; for relief of joint aches and pains; dietary supplement. [Seven Seas Ltd]

Sevestat ML 300. Surfactant blend; lubricant, antistat, emulsifier for textile lubricants for mineral and synthetic fibers. [Stepan Europe]

Sevin. 1-naphthyl N-methylcarbamate. A proprietary preparation of carbaryl; used as a veterinary insecticide. §

Sevin® Brand SL. Insecticide for control of insects such as grasshoppers, gypsy moths, mosquitoes, ticks, ants. [Rhone-Poulenc/Ag; W.A. Cleary]

Sewarin. CAS 81-81-2; Warfarin; a rodenticide. [Killgerm Chemicals Ltd]

SE Wax. A montan wax ester containing an emulsifier; used in the preparation of nonionic, self-polishing emulsions for floors. [Bush Beach Ltd] *

Sextone. CAS 108-94-1; Cyclohexanone. [Laporte UK Trading] *

Seymourite. A proprietary trade name for an alloy of 64% copper, 18% nickel, and 18% zinc. §

SF18. Dimethicone fluid; lubricant, antifoam, mold release; rubber and plastic lubricant; base fluid for grease; mold release for rubber, plastic, and food application; antifoam in food application and aq. defoaming formulations. [GE Silicones]

SF-20CF. CAS 9016-75-5; PPS, 20% carbon fiber-reinforced; offer strength, stiffness, lubricity, conductivity, thermal and chemical resist.; used for precision electromech. components, corrosive chemical and fluid handling equip. [Compounding Tech.]

SF69. Dimethyl silicone fluid; film modifier, antifloat, flow control, anticrating, and pigment control agent in varnishes, paints, enamels, surface coatings. [GE Silicones]

SF81. Dimethyl silicone fluid; used as low temp. damping media, heat transfer media, dielec. coolants; base fluid for silicone compounds. [GE Silicones]

SF96® . Polydimethylsiloxane; emollient, lubricant for polishes, antifoams, textiles, chemical specialties; plastic and rubber lubrication; dampening or heat transfer fluids; oil defoamer, paint additives; mold release for tires, rubber, plastics; textile softener/modifier. [GE Silicones]

SF99. Reactive polydimethylsiloxane fluid; forms water-repellent fluids with heat or heat/catalyst; used in textiles, particle treatment, magnesium oxide, and Calrod® units. [GE Silicones]

SF1023. Phenyl containing silicone fluid; flow control agent in polyester coatings; pigment dispersant, gloss aid in organic finish systems. [GE Silicones]

SF1154. Polydimethyldiphenyl siloxane; lubricant, dielectric coolant, coupler; high temp. heat transfer application; base fluid in high temp. greases; high temp. ultrasonic coupler; high temp. bath and oxide protector for solder baths; outstanding heat resistance. [GE Silicones]

SF1173. CAS 69430-24-6; Cyclomethicone; emollient, lubricant used in antiperspirants, skin care products, sunscreen products, hair conditioners, facial makeup, particle treatment. [GE Silicones]

SF1188. Dimethicone copolyol; emollient, lubricant, and release agent for textiles, cosmetics and toiletries, paint, plastic mold release, and rubber lubricants. [GE Silicones]

SF1250. Chlorophenyl methyl siloxane; lubricant, mold release agent, hydraulic systems; fluid transmission; servomotors and mechanisms; instruments, clocks, and timers; machine tool components; antifriction, rolling, and sliding mechanisms; shock absorbers and damping devices. [GE Silicones]

S-Flakes. CAS 8016-70-4; EINECS 232-410-2; Hydrogenated soybean oil. [Karlshamns]

SH 420. A proprietary preparation of norethisterone acetate; used in the treatment of breast cancer. [Schering Health Care Ltd]

Shadeacrete. Dry powdered colorants, iron oxides, ochres, umbers and composite pigments; colors for mortars, concrete roofing tiles, floor tiles, sand-lime bricks, concrete blocks, reconstructed stone, split blocks, paving slabs and cement sheets. [W Hawley & Son Ltd]

Shadow. Clay; sun reflector for protection against sunburn on agronomic and ornamental crops. [Draxel Chemical

---

‡ Trade name and manufacturer not verified   § Trade name without identified manufacturer

Company] ‡

Shaku-do. A Japanese alloy. It usually contains 94-96% copper, 3.76-4.16% gold, and 0.08-1.55% silver.

shale. A dark-grey or black mineral containing 73-80% mineral matter and 20-27% organic matter. It is a source of oil for lubricating purposes.

shale oil. CAS 68308-34-9; The tarry oil obtained by the distillation of certain bituminous shales. It contains unsaturated hydrocarbons.

Shatach. Agrochemical surfactant. [Makhteshim Chemical Works Ltd]

Shawinigan Black. Acetylene black. [Chevron] *

Shawinigan's black. See acetylene black.

Shawplas. Abrasive compounds and polishes for plastic articles; for fabricated components, turned parts, buttons, buckles, spectacle frames in acrylic, cellulose acetate, casein, polyester. [Shawplas Ltd] *

shea butter. (Bambuk butter). The fat obtained from the seeds of *Butyrospermum parkii* or *Bassio parkii.*

Shebu WS. PEG-50 shea butter; emollient, humectant, pigment, conditioner, lubricant for personal care formulations. [RITA]

Shebu, Refined. CAS 68424-60-2; Shea butter; emollient, humectant, pigment, conditioner, lubricant for personal care formulations. [RITA]

shé-chuang-tzu. It is the fruit of *Selinum monnieri,* and contains an essential oil which contains l-pinene, camphene, and bormyl-iso-valerate; a Chinese drug.

Sheduri. See sanna.

sheet-lac. See lac.

Shell 5A18Z. CAS 9003-07-0; PP homopolymer, nucleated, antistat; for injection molding; provides fast cycle times, good opticals; especially suited for cutlery. [Shell]

Shell 5A95. CAS 9003-07-0; PP homopolymer; for cast unoriented film. [Shell]

Shell 5C64. CAS 9003-07-0; PP homopolymer; for stretched tape processes which produce woven fabrics for carpet backing, industrial and geotextile fabrics. [Shell]

Shell 6A01K. CAS 9003-07-0; PP random copolymer, antistat; for injection molding of high-quality housewares where high clarity is required. [Shell]

Shell 6C20S. CAS 9003-07-0; PP random copolymer, antistat; for extrusion blow molding; especially suited for multi-layered barrier bottles. [Shell]

Shell 7C55H. CAS 9003-07-0; PP extra high impact copolymer; heat resist. product for extrusion and injection molding, especially applications requiring high impact strength at ambient and sub-freezing temps.; excellent base stock for compounding filled/reinforced grades. [Shell]

Shell D 50. CAS 1702-17-6; 2,4-D; translocated herbicide for cereals and grass. [Shell UK]

Shell DS 6C46L. CAS 9003-07-0; PP high-ethylene random copolymer; for cast unoriented film. [Shell]

Shell DS 7C04N. CAS 9003-07-0; PP impact copolymer; for cast unoriented film. [Shell]

Shell JF 6100. CAS 9003-07-0; PP homopolymer; for heat-set oriented film. [Shell]

Shell PDC 1120. CAS 9003-07-0; PP low-ethylene random copolymer; for heat-set oriented film. [Shell]

Shell WRS 6-198. CAS 9003-07-0; PP high-ethylene random copolymer; for shrink oriented film. [Shell]

Shell WRS 6-205. CAS 9003-07-0; PP low-ethylene random copolymer; for heat-set oriented film. [Shell]

shellac. See lac.

shellackose. An alcohol-soluble phenol-formaldehyde resin; used in the preparation of lacquers.

Shell Elexar. Thermoplastic rubber; used for wire and cable insulations. [Shell]

Shellflex Process and Extender Oils. Hydrocarbon solvents which generally have lower aromatic content than the Dutrex grades, covering a wide viscosity range; used as extender and process oils for natural and synthetic rubber. [Shell]

Shellite. It is a mixture of ammonium perchlorate and paraffin wax; an explosive.

shell limestone. A variety of calcium carbonate in massive form.

Shellsol D40, D60, D70. Highly refined solvents with a very low aromatic content and a very slight sweet odor; principal applications similar to white spirits, in low odor paints and metal cleaning products. [Shell] *

Shellsol E, A, AB, R. High-boiling aromatic hydrocarbons; versatile solvents used in paints, varnishes and in the preparation of agricultural chemical formulations. [Shell]*

Shellsol T. A high-boiling, isoparaffinic, aliphatic solvent with high flash point; virtually without odor; useful in odorless paints, household aerosols, fragrant polishes, and cosmetic creams; also as a catalyst carrier in polymerization reactions. [Shell] *

Shellswim 5X. A polymeric additive dissolved in xylene to facilitate its homogeneous distribution in fuel and crude oils; can be added to waxy residual fuels and waxy crude oils to permit storage, handling and pumping below their natural pour point. [Shell] *

Shellswim 11T. A polymeric additive dissolved in toluene to facilitate its effective distribution in crude oil; enables waxy crude oils to be transported at temperatures below their pour point; also facilitates restarting of a pipeline following a shut-down. [Shell] *

Shellvis 50 (SAP 150). A styrene-based hydrocarbon viscosity index improver for engine lubricants; normally supplied in the form of bales which require shredding before dissolving in lubricating oil, although also available as a concentrate or in a crumb form. [Shell] *

Shelspra. See Mexphalte. §

Sherbelizer®. CAS 9005-37-2; Propylene glycol alginate blend; used for sherberts, soft-serve mixes, dispensed milk shakes, ice cream, sour cream and dips. [Kelco]

Sherpa. Insecticide. [May & Baker Ltd] *

Sherwood oil. See petroleum ether.

Shibu-ichi. A Japanese alloy containing 51-67% copper, 32-49% silver, and traces of gold and iron.

Shield. Selective herbicide. [Murphy Chemical Ltd] †

shikimole. CAS 94-59-7; Safrole, $C_{10}H_{10}O_2$, the chief constituent of oil of sassafras.

shikon. The dried roots of *Lithospermum erythrorhizon.*

Shilajatu. An Indian mineral gum.

Shimose. See Lyddite.

Shimosite. A Japanese explosive, the chief constituent of which is picric acid.

shinnamu. A vegetable dye obtained from a species of maple found in Korea.

shio liao. A Chinese cement for marble, porcelain, etc., made from 54%, slaked lime, 6% alum, and 40% blood.

Shipley's solutions. Solutions of pyro-gallol and caustic soda in water, usually 10 cc of 1 : 1 caustic soda solution,

---

1 and 4 cc water, and 2 and 10 g pyrogallol; used for the absorption of oxygen.

Ship Shape® Resin Cleaner. N-Methyl-2-pyrrolidone, butyrolactone, deceth-6, fragrance; resin cleaner for fiberglass fabrication, furniture refinishing, rubber industry. [ISP]

Shiro Bishi®. Coke; for general grade and malleable cast iron. [Mitsubishi Kasei] *

SHL Lawn Sand Plus. Dichlorophen + ferrous sulfate; moss killer/fertilizer mixture for turf. [Sinclair Horticulture & Leisure Ltd]

SHMP. See sodium hexametaphosphate.

Shoemaker's black. Ferrous sulfate, FeSO₄.

Shoemaker's paste. A paste made by allowing the gluten from flour to putrefy, rolling it out thin, and making it into a paste; used for securing leather to leather, paper, or other material.

Shokusen SE. Sugar ester, propylene glycol, ethanol, sodium citrate; detergent for foods. [Mitsubishi Kasei] *

short oil varnishes. See long oil varnishes.

Shostakovsky Balsam. Polyvinox; a proprietary preparation of synthetic vinyl butyl ether. [Leopold Charles & Co]‡

shot lead. See shot metal.

shot metal. (shot lead, bullet metal). An alloy of lead with not more than 3% arsenic. One alloy contains 99.8% lead and 0.2% arsenic.

Shur-Coal® FCA. CAS 56-81-5; Glycerin; freeze conditioner for coal industry; pumpable temp. -25 C. [Sherex/Div. of Witco]

Si69. CAS 40372-72-3; EINECS 254-896-5; Bis(3-triethoxysilylpropyl) tetrasulfane; crosslinking/coupling agent for rubber industry. [Degussa]

Si 264. 3-Thiocyanatopropyltriethoxy silane; reinforcing agent improving properties of fillers (silicas, silicates, clays and whitings) in unsaturated polymers with double bonds (NR, IR, SBR, BR, NBR, EPDM). [Degussa]

sialonite. A mineral, Be₈Se₃.

Siapton. Liquid organic foliar feed. [ICI Chem. & Polymers Ltd]

Sibercizer C6. N-Ethyl o-/p-toluene sulfonamide. [Siba Hegner Ltd]

Siberez. Toluene sulfonamide resins. [Siba Hegner Ltd]

Sibley alloy. An alloy of 67% aluminum and 33% zinc.

Sibor. A proprietary safety glass. §

Sibutol®. CAS 55179-31-2, 3878-19-1; Bitertanol, fuberidazole; seed dressing for cereals for control of loose smuts, stinking smuts, flag smuts; especially effective against dwarf bunt of wheat and seed-borne snow mold. [Bayer AG]

Sical. An alloy of from 22-29% aluminum, 50-51% silicon, 2-4% titanium, 1% calcium, 0.2-0.3% carbon, and the remainder iron.

sicalite. See Gallatite.

siccative. Manganese borate, MnB₄O₇; used as a drying agent mixed with linseed oil and resin, for impregnating leather.

Siccatol. A range of metal-based drying agents. [Akzo Chemie UK Ltd]

Siccolam. CAS 13463-67-7; EINECS 236-675-5; Compound titanium dioxide paste; desiccant for exudatory dermatoses. [British Drug Houses] *

Sickle. Bromoxynil + fluroypyr; post-emergence contact herbicide for cereals. [DowElanco Ltd]

Siclor. CAS 1897-45-6; Tetrachloroisophthalonitrile; fungicide for the preparation of antifouling marine paints, aqueous paints, wood primers, adhesives. [Caffaro SpA] *

Sico®. Predominantly azo pigments; for baking and air-drying finishes, for letterpress, offset, flexographic and special gravure inks, for coloring plastics. [BASF AG]

Sicocab®. Organic and inorganic pigments in a cellulose acetobutyrate vehicle; for wood stains with excellent fastness. [BASF AG]

Sicodop®. Organic and inorganic pigments in DOP; for coloring plasticized PVC. [BASF AG]

Sicoflex 80. A proprietary ABS terpolymer possessing very high flow properties. [Mazzucchelli Celluloide Spa] *

Sicoflex 85. A proprietary ABS terpolymer possessing high flow properties. [Mazzucchelli Celluloide Spa] *

Sicoflex 90. A proprietary ABS terpolymer having high impact strength. [Mazzucchelli Celluloide Spa] *

Sicoflex 93. A proprietary general purpose ABS terpolymer. [Mazzucchelli Celluloide Spa] *

Sicoflex 95. A proprietary ABS terpolymer possessing high tensile strength. [Mazzucchelli Celluloide Spa] *

Sicoflex 99. A proprietary ABS terpolymer offering high resistance to heat. [Mazzucchelli Celluloide Spa] *

Sicoflex MBS. A thermoplastic material based on methyl methacrylate, butadiene and styrene. [Mazzucchelli Celluloide Spa] *

Sicoflush® A. Organic and inorganic pigments in medium-oil soy alkyd resin; for air-drying and stove-drying finishes which are compatible with alkyd resins. [BASF AG]

Sicolen®. Organic and inorganic pigments in polyethylene; for coloring polyethylene hollow articles, injection molding and extrusion articles. [BASF AG]

Sicolub®. Waxes. [BASF plc]

Sicolube DSP. Distearyl phthalate; lubricant for PVC. [BASF AG]

Sicolube E. CAS 8002-53-7; Montan wax deriv.; lubricant for PVC. [BASF AG] *

Sicolube EDS. Ethylene diamine distearyl amide; lubricant for PVC and PS. [BASF AG] *

Sicolube OA 2, OA 4. Oxidized polyethylene wax; lubricant for PVC. [BASF AG] *

Sicolube TDS. Isotridecyl stearate; lubricant for PVC. [BASF AG] *

Sicomet®. Anionic dyes, cationic dyes, fat-sol. dyes, organic pigments, inorganic pigments, pigment concentrates; colorants for cosmetics. [BASF AG]

Sicomin®. Chrome yellow and molybdate orange pigments; for paints and surface coatings, for coloring plastics, for flexographic and gravure inks for packaging materials, laminated paper coloring. [BASF AG]

Sicomix®. Pigment concentrates for surface coatings. [BASF AG]

Sicopal®. Inorganic pigments having a spinel structure based on various metal oxides; for industrial finishes with excellent fastness and for coloring plastics. [BASF AG]

Sicopharm®. Sol. dyes and pigments for coloring pharmaceuticals. [BASF AG]

Sicoplast®. Predispersed pigment mixtures for mass coloring of thermoplastics. [BASF AG]

Sicopos®. Organic and inorganic pigment concentrates for coloring PET. [BASF AG]

Sicopur®. High purity iron oxides for paper refiners. [BASF AG]

---

‡ Trade name and manufacturer not verified   § Trade name without identified manufacturer

Sicopurol®. Organic and inorganic pigment concentrates in ester polyol; for coloring polyurethane foams based on polyester or polyether polyols. [BASF AG]

Sicorin®. Anticorrosion pigments. [BASF AG]

Sicostab®. Heat stabilizers for rigid and plasticized PVC, for internal and external applications, PVC foams. [BASF AG]

Sicostyren®. Organic and inorganic pigment concentrates in polystyrene; for coloring polystyrene and styrene copolymers. [BASF AG]

Sicotan®. Inorganic mixed-phase pigments with structure of rutile titanium dioxide and other metal oxides; in combination with organic pigments for luminous hues for surface coatings and plastics with high processing temps. [BASF AG]

Sicotherm®. Cadmium sulfide/zinc sulfide or cadmium sulfide/selenide mixed crystals; for coloring plastics and paints. [BASF AG]

Sicotrans®. Transparent inorganic pigments of extremely fine particle size; for high-quality paint systems, especially metallics, and for coloring plastics. [BASF AG]

Sicoversal®. Organic and inorganic pigment concentrates; for coloring thermoplastics. [BASF AG]

Sicovinyl®. Organic and inorganic pigments in plasticized PVC; for mass coloring of plasticized PVC. [BASF AG]

Sicovit®. Sol. colorants and pigments for coloring foodstuffs. [BASF AG]

Sicromo Steel. A proprietary trade name for a chromium-silicon-molybdenum steel containing from 2.25-2.75% chromium, 0.5-1.0% silicon, 0.4-0.6% molybdenum and up to 0.15% carbon. §

Sidanyl. Polyamide film; for incorporation into laminates, well suited for thermoforming. [UCB nv Film Sector] *

Sident 15. Silica; abrasiveness and thickening agent for toothpastes. [Degussa]

Sident 22LS, 22S. Silica; nonabrasive thickening agent for toothpastes. [Degussa]

Sideraphthite. An alloy resembling silver. It contains 64.5% iron, 22.5% nickel, 4.5% each aluminum and copper, and 4% tungsten; nonoxidizable.

sidero cement. A cement in which iron ores are wholly or partly substituted for the clay.

Sidot's blende. A phosphorescent zinc sulfide.

Sidros. A proprietary preparation containing ferrous gluconate and ascorbic acid; a hematinic. [Horlicks] *

Siemensite. A refractory material produced by fusing a mixture of chromite, bauxite, magnesite, and a reducing agent in the arc furnace to obtain a slag containing from 20-40% Cr_2O_3, 25-45% Al_2O_3, 18-30% MgO, and 8-14% other constituents. §

sienna. CAS 1309-37-1; It consists of hydrated oxide of iron, mixed with a little manganese, and clay. It contains from 50-70% Fe_2O_3, 8-12% SiO_2, 2-8% Al_2O_3, 2-5% $CaSO_4$ or CaO, and water; a pigment in oil paints, stains, pastels.

Sierra Leone butter. See Lamy butter.

Sifbronze. A proprietary trade name for a brass containing some ferromanganese and tin. §

Siflox. Highly dispersed precipitated silica; applied in shoe soling materials, hoses, cable sheeting, profiles, etc. [Chemische Fabriken Oker und Braunschweig AG] *

Siflural. A trade name for a solution of aluminum fluosilicate; a disinfectant. §

Sigal. A proprietary alloy of 10% Si and 90% Al; a pigment. §

Sigma. Lube oil. [Chevron] *

Sigmalium. A proprietary trade name for an alloy of aluminum containing 1% silicon, 4% copper, and 0.7% magnesium. §

Sigmamycin. A proprietary preparation of tetracycline and troleandomycin or oleandomycin phosphate; an antibiotic. [Pfizer International]

Sigmathane. A proprietary single-pack moisture-cured polyurethane coating. [Sigma Coatings] *

Signal Red. A dyestuff; the British equivalent of Lithol red B. §

Silacros. Silicates for use in industry. [Crosfield Chemicals Ltd]

Silacto®. Activator. [Kenrich Petrochemicals]

Silain. CAS 8050-81-5; Simethicone; an antiflatulent. [Wyeth Laboratories] *

Silajit. A preparation containing benzoic and hippuric acids, gums, albuminoids, resin, and fatty acids.

Silal. A proprietary trade name for a grey iron with 5% silicon and 2.5% total carbon; stated to resist oxidation, growth, and scaling up to 750 C. §

Silamide DCA-100. Conditioner for personal care products. [Siltech]

Silamine 65. Gloss and conditioning agent for personal care products. [Siltech]

Silanca. A stainless silver with a high silver content.

silane. (silicane; silicon tetrahydride) CAS 7803-62-5; SiH_4; used as a doping agent , production of amorphous silicon.

Silantox. CAS 409-21-2; A proprietary colloidal silicon dioxide. §

silaonite. A mineral; a mixture of bismuth and bismuth trisulfide.

Silargel. A German product; it is a silver chloride-silica gel preparation, a white odorless powder containing 5% silver; an adsorbent and disinfectant for the external treatment of burns. §

Silasorb. CAS 10101-39-0; Synthetic calcium silicate; adsorbent; controls free fatty acids. [Celite]

Silastic® 21145. CAS 63394-02-5; Silicone rubber compound; flame-resist. elastomer for compression, transfer, or injection molding applications. [Dow Corning STI]

Silastic® GP-30. CAS 63394-02-5; Silicone rubber; general purpose elastomer for molding, calendering, and blending; FDA compliance. [Dow Corning STI]

Silastic® GP-950+. CAS 63394-02-5; Silicone rubber with extending filler; economical, extendable, general-purpose elastomer for extrusion, molding, calendering; FDA compliance. [Dow Corning STI]

Silastic® HE-26. CAS 63394-02-5; Silicone rubber; specialty elastomer for extrusion, molding, calendering; low modulus; FDA compliance. [Dow Corning STI]

Silastic® HS-30. CAS 63394-02-5; Silicone rubber; high strength elastomer for molding, calendering; FDA compliance. [Dow Corning STI] *

Silastic® LCS-740. CAS 63394-02-5; Silicone rubber; low compression set elastomer for molding, calendering; reversion resist.; no post cure; FDA compliance. [Dow Corning STI]

Silastic® LT-40. CAS 63394-02-5; Silicone rubber; low temp. elastomer for molding, calendering; accepts filler. [Dow Corning STI]

Silastic® NPC-40. CAS 63394-02-5; Silicone rubber; no post cure elastomer for extrusion, molding, calendering; accepts filler; FDA compliance. [Dow Corning STI]

Silastic® SPG-30. CAS 63394-02-5; Silicone rubber; specialty elastomer for molding; low-temp. sponge base; FDA compliance. [Dow Corning STI]

Silastic® TR-55. CAS 63394-02-5; Silicone rubber; tough elastomer for extrusion, molding, calendering; excellent tear and flex life; no post cure; FDA compliance. [Dow Corning STI]

Silastic® WC-50. CAS 63394-02-5; Silicone rubber; wire and cable elastomer for extrusion, molding, calendering; excellent long-term heat stability; accepts filler; no post cure. [Dow Corning STI]

Silastomer. Silicone rubbers. [Midland Silicones] ‡

Silastoseal. Room temperature curing silicone rubber sealants. [Midland Silicones] ‡

Silbamine. Silver fluoride, AgF; used in medicine as an antiseptic.

Silberit. A jewelry alloy; contains aluminum, nickel, and silver.

Silbione. Silicones for cosmetics. [Rhone-Poulenc UK]

Silbond® 40. Ethyl polysilicate; intermediate for binders used for inorganic zinc coatings, investment casting molds, cores, ceramic shapes and coatings. [Akzo]

Silbond® Condensed. CAS 78-10-4; Tetra ethyl ortho silicate 90%; intermediate for binders used for investment castings, ceramic shapes, cores, and coatings; chemical intermediate. [Akzo]

Silcar. A proprietary trade name for a pigment comprising a mixture of silicon dioxide and silicon carbide. §

Silcasil S. An inorganic filler; uses include carrier material for insecticides, auxiliary for improving the free-flowing and grinding properties of powders, thickener for liquids. [Bayer AG]

Silcat. Crosslinker. [Union Carbide]

Sil-Cell® 32. Glass microcellular filler (73% silicon dioxide, 17% aluminum oxide, 5% potassium oxide); inert, inorganic, nontoxic hollow irregular shape filler/reinforcement for manufacturing of adhesives, auto body putty, cultured marble, coatings, wall patching compounds, stucco. [Silbrico]

Silchrome 46M. A proprietary trade name for a chromium steel containing 4-6% chromium, 0.5% molybdenum, and 0.2% carbon. §

Silchrome R.A. A proprietary trade name for a steel containing 16% chromium, 1% silicon, 1% copper, and 0.12% carbon. §

Silchrome Wire. An alloy of iron with 18% chromium, 3% silicon, 3% tungsten, and 0.3% carbon. §

Silchrome. A heat-resisting alloy containing 86% iron, 9.5% chromium, 4% silicon, and 0.5% carbon.

Silcolapse. Textile auxiliary chemicals. [ICI Chem. & Polymers Ltd] †

Silcolease. Silicone coatings. [ICI Chem. & Polymers Ltd]†

Silcon. Silafilcon; contact lens material. [Dow Corning Ophthalmics Inc] *

Silcoset. Rubber curing agents. [ICI Chem. & Polymers Ltd]†

Silcron®. A fine-particle silica. [SCM] *

Silderm. A proprietary preparation of triamcinolone acetonide, neomycin sulfate and undecanoic acid; a steroid skin cream. [Lederle Laboratories] *

Sildura®. A range of silicone-rubber compositions, curable by the application of heat. [GE Silicones] *

Silene. CAS 10101-39-0; A proprietary trade name for a precipitated calcium silicate; used in rubber mixes to give

wear-resistance. §

Silesia powder. It is a mixture of 75% potassium chlorate, with pure or nitrated resin, and a little castor oil; an explosive. .

Silesite. A tin silicate with 55% tin, found in the Bolivian tin deposits.

Silester. CAS 78-10-4; Ethyl silicate. [Monsanto plc]

silex. A name applied to silica (SiO_2); used also for tripoli employed as a filler in paints. A ground flint is also known as silex.

Silflex. Cosmetic specialties. [Rhone-Poulenc UK]

Sil-fos. A proprietary trade name for a phosphor-silver brazing solder containing 80% copper, 15% silver, and 5% phosphorus. §

Silfrax. A product obtained by the action of silicon on carbon and consisting of carbon with a coating of silicon carbide and carbon; stated to be tougher and stronger than carborundum; used as a refractory material in the manufacture of pyrometer tubes for electrical heating elements. §

Sil-Free. Chemical additive to prevent silicate deposit in peroxide/silicate bleach system. [Sybron]

Silglaze® II 2800. Silicone sealant; sealant offering extended tooling time and shorter cure time; for general sealing and glazing, skylights, window installation, and in-shop window fabrication. [GE Silicones]

Silgrip® PSA590. Polydimethylsiloxane gum/polysiloxane resin reactive copolymer in toluene solution; pressure-sensitive adhesive for splicing and repair tape and platers tape for the printed circuit market. [GE Silicones]

Silgrip® PSA595. Polydimethylsiloxane gum/polysiloxane resin reactive copolymer in xylene solution; pressure-sensitive adhesive for platers tape for the printed circuit market. [GE Silicones]

Silhydrate C. Copper PCA methylsilanol; provides skin moisturization, skin restructuring for anti-aging formulations, skin treatments, cosmetic and toiletry emulsions, creams, lotions. [Exsymol]

silica. (silicon dioxide, fumed; silicon dioxide; silicic anhydride) Inorganic oxide; occurs in nature as agate, amethyst, chalcedony, cristobalite, flint, quartz, sand, tridymite; SiO_2; CAS 7631-86-9; 112945-52-5; manufacture of glass, water glass, refractories, abrasives, ceramics, enamels, petrol. products; filler in cosmetics; rubber reinforcing agent; as anticaking and defoaming agent; abrasive; thickener. [Akzo; BYK-Chemie; Cabot Carbon Ltd; Catalysts & Chemicals Industries; Chisso Am.; Degussa; Du Pont; Geltech; J.M. Huber; Nippon Silica Ind.; Nissan Chem. Ind.; PPG Industries; PQ; Unimin; U.S. Silica]

Silica FK 160. Silica; active filler for hot-vulcanizing silicone rubber. [Degussa AG]

Silica FK 320 DS. Silica; active filler for hot-vulcanizing silicone rubber. [Degussa AG]

silica gel. The name applied to a colloidal form of silica, prepared by treating sodium silicate with acetic or hydrochloric acid, washing the gelatinous silica, and drying. It is highly absorbent, and is used to absorb water vapors.

silica glass. See vitreosil.

silicam. Silicon imido-nitride, Si_2N_3H, formed when silicon diimide is heated to 900 C in an atmosphere of dry nitrogen.

Silicane. CAS 7803-62-5; Silicon hydride.

Silicargol. A colloidal silver preparation for wound treat-

---

‡ Trade name and manufacturer not verified   § Trade name without identified manufacturer

ment.

silicate cotton. *See* slag wool.

silicated soap. A soap to which water glass (sodium silicate) has been added; a detergent.

silicate of carbon. *See* gas black.

Silicex. Silicone products; antifoams, release agents, molding rubbers and sealants. [Siliconas Hispania SA]

silicic acid, calcium salt. *See* calcium silicate.

silicic acid, disodium salt. *See* sodium metasilicate.

silicic acid, sodium salt. *See* sodium silicate.

silicic anhydride. *See* silica.

Silicium. A proprietary trade name for silicon used as a pigment in the Atephen system (*qv*); a chemically resistant coating. §

silicochloroform. *See* trichlorosilane.

Silicoderm® F. A preparation for the protection and the care of the skin. [Bayer AG]

silicol. Ferro-silicon, usually containing 84% silicon, for use in the preparation of hydrogen by the action of caustic soda.

silicolloid. A natural siliceous material free from iron. Suitable for use in paper manufacture, cleansers, and tooth pastes.

silico-manganese. An alloy of silicon and manganese made in the electric arc type of furnace. It contains 60-75% manganese, 20-25% silicon, and the rest iron.

silicon. Nonmetallic element; Si; CAS 7440-21-3; EINECS 231-130-8; semiconductor in solid-state devices; organosilicon compounds; silicon carbide; alloying agent in steels, aluminum, copper, bronze, iron; cermets, refractories; halogenated silanes; deoxidizer in steel manufacture [Atomergic Chemetals; Cerac; Dow Corning; Eagle-Picher; Pechiney Electrométallurgie; Shin-Etsu Chem.]

silicon brass. An alloy of 81% copper, 14% zinc, and 3% silicon.

silicon bronze. (silicon bronze). An alloy of 97.37% copper, 1.32% tin, 1.24% zinc, and 0.7% silicon.

silicon carbide. SiC; CAS 409-21-2; EINECS 206-991-8; Abrasive for cutting and grinding metals, refractory in nonferrous metallurgy, ceramic industry, boiler furnaces, composite tubes for steam reforming operations; fibrous form in filament-wound structures and heat-resistant, high-strength composites. [Atomergic Chemetals; Carborundum; Lonza; Mitsui Toatsu; New Metals & Chems. Ltd; Showa Denko]

silicon chloride. *See* silicon tetrachloride.

silicon copper. Alloys of copper with small amounts of silicon; used for the manufacture of telephone and telegraph wires. An alloy with 10% silicon is also called silicon-copper.

silicon dioxide. *See* silica.

silicon dioxide, fumed. *See* silica.

silicone. (organosiloxane) Siloxane polymers; thermosetting siloxane polymers used as mold release for plastics and rubber, defoamers for mining, latex, ink, soaps, agriculture., food processing, as lubricants, conditioners, and emollients in personal care products; molding compounds, encapsulants. [Dow Corning; GE Silicones; Genesee Polymers; Goldschmidt; Hüls Am.; Miles; Sandoz; Shin-Etsu Chem.; Union Carbide; Wacker Silicones]

Silicone Antifoam Emulsion SE 9. Silicone emulsion; antifoam with good storage stability; recommended for food sector, as processing aid in pharmaceutical and cosmetic industries and in fermentation processes; also for manufacturing of plastics in contact with food; for degassing of monomers. [Wacker-Chemie]

silicone elastomer. (polydimethylsiloxane rubber) Room-temp. vulcanizing silicone rubber; CAS 63394-02-5; encapsulation of electronic parts, elec. insulation, gaskets, surgical membranes and implants; automotive engine components; miscellaneous mechanical parts. [Ambersil Ltd; Dow Corning Europe; GE Silicones Europe]

Silicone Emulsion E-130. Silicone emulsion; used for mold release, furniture polish emulsions, cosmetics, textile lubricants and softeners. [Wacker-Chemie]

silicon-eisen. (silicon pig). A pig iron containing from 5-15% silicon.

Silicone Systems 820. One-component silicone systems; optically clear silicone. [R.H. Carlson]

Silicone Systems 976. One-component silicone system; self-leveling high temp. silicone for elec./electronic potting applications. [R.H. Carlson]

silicon fluoride. *See* silicon tetrafluoride.

silicon nickel brass. An alloy of 81% copper, 14% zinc, 3% silicon, and 2% nickel.

silicon steel. A steel made by melting steel and ferro-silicon in crucibles; used for making sheets, springs, and acid-resisting plants.

silicon tetrachloride. (tetrachlorosilane; silicon chloride) $SiCl_4$ CAS 10026-04-7; EINECS 233-054-0; Smoke screens; manufacture of ethyl silicate, silicones, high-purity silica, fused silica glass; source of silicon, silica, and hydrogen chloride; laboratory reagent. [Air Products; Atomergic Chemetals; Chisso Am.; Degussa; Dow Corning; Hüls; PCR; Union Carbide]

silicon tetrafluoride. (tetrafluorosilane; silicon fluoride) SiF_4; CAS 7783-61-1; manufacture of fluosilicic acid, intermediate in manufacture of pure silicon, to seal water out of oil wells during drilling. [Air Prods.; Atomergic Chemetals; Hüls Am.; PCR]

Silicosehl. Room temperature vulcanizing two-part silicone rubber systems; for casting and potting requiring exact surface detail, flexibility, stability and excellent electrical characteristics over wide temperature range; chemical and weather resistant; nontoxic. [Solochart Ltd] *

silico-spiegel. An alloy of 20% manganese, 12% silicon, and the rest iron.

silico-superphosphate. A preparation made by mixing superphosphate with kieselguhr or precipitated silicic acid. It is stated to give better results on medium and light soils.

silico-titanium. A titanium-silicon alloy used in the steel industry.

Siligaz. CAS 8050-81-5; Simethicone; antiflatulent. [Menley & James Laboratories] *

Siligen®. Textile finishing auxiliaries. [BASF plc]

Siligen® FA. Antimigrant and softening agent for pigment padding, textile finishing. [BASF]

Siligen® GL. Nonfoaming softening agent for resin finishing cellulosic fibers and their blends with synthetics. [BASF]

Siligen® MA. Wetting agent for mineral application techniques. [BASF]

Siligen® TX-510. Modified silicone elastomer; textile auxiliary. [BASF]

Siligran. A range of inorganic smelted products and their

---

mixtures; used as antiscale and fluxing compounds in steel production. [Bayer AG]

Silikoftal® . Resin binders for surface coatings. [Thomas Goldschmidt Ltd]

Silikoftal® CC 3. Silicone-modified polyester resin; binder for weather-resist. lacquers applied by coil-coating. [Tego GmbH] *

Silikoftal® HTL 2. Silicone-modified polyester resin; binder for decorative high-quality coating of cookware and elec. domestic appliances. [Tego GmbH] *

Silikophen® Nonstick 50. Reactive release agent and temp.-resist. silicone resin. [Tego GmbH] *

Silikophen® P 40/W. CAS 68083-14-7; Phenylmethyl polysiloxane resin emulsion; binder for water-based high temp. anticorrosion paints resist. to 650 C, e.g., coatings applied to industrial equip., chimneys, exhaust tubes. [Tego GmbH] *

Silipact. Elastic sealants; for the construction industry and formed-in-place gaskets. [Lonza AG]

Siliporite. Molecular sieves; for drying of liquids and gases, desulfuration, separation of gases and isomers, in the chemical and petrochemical industries; drying agent for the double glazing industry and for polyurethane formulations. [Atochem UK/Ceca]

Siliset. Foundry chemical hardeners. [Foseco (F.S.) Ltd]

Silit. A material made by exposing mixed silicon, silicon carbide, and carbon, to the action of carbon monoxide at 1500 C. It is made in three qualities.   a) A material for resistance subjected to permanent losses, b) for electric heating work up to 1400 C and c) a fireproof material capable of withstanding violent changes in temperatures.

Silitonite. See Frankonite.

Silk, Anaphe. The silk obtained from a caterpillar in German East Africa; it has a specific gravity of 1.282. §

silk grass. A term applied to pineapple fiber, obtained from the pineapple plant; used for making cloth in the Phihppine Islands.

Silkin. A proprietary cellulose nitrate silk. §

Silkiol. Additive; adhesive for carded worsted and cotton spinning. [Henkel Chemicals Ltd] *

Silk Pro-Tein. CAS 96690-41-4; Hydrolyzed silk; moisturizer, substantive protein, protective barrier for elegant skin and hair preparations (lotions, creams, bath gels, shampoos, conditioners, treatment products, shave preparations, soap bars). [Maybrook]

Silksoft® Supreme. Modified amine silicone; softener giving excellent soft and slick hand. [Sybron]

silk wadding. The waste from the spinning of silk.

Sillikolloid P 87. CAS 14808-60-7, 1318-74-7; Quartz-kaolinite; filler. [Hoffmann Min.]

Sillitin N 82. See Sillikolloid P 87. [Hoffmann Min.]

Sillitin-Aktisil. Quartz-kaolinite; filler for rubber and paint, soft abrasive for polishing agent. [Hoffmann Mineral] *

Sillman bronze. An alloy of 86% copper, 10% aluminum, and 4% iron.

Silm. Photographic fixer stain remover. [May & Baker Ltd]*

Silman Steel. A proprietary trade name for a silicon steel containing 2.1% silicon, 0.85% manganese, 0.3% vanadium, 0.25% chromium, and 0.55% carbon. §

Silmar® 901R. Thermoset polyester; synthetic marble two-way resin [BP Chem. Inc.]

Silmar® S249. Thermoset polyester; ortho laminating resin. [BP Chem. Inc.]

Silmar® S585. Thermoset polyester; synthetic marble resin. [BP Chem. Inc.]

Silmar® S957. Thermoset polyester; densified matrix resin. [BP Chem. Inc.]

Silmar® S958. Thermoset polyester; ortho laminating resin. [BP Chem. Inc.]

Silmod 20A. Silicon-modified polyether; for formulating one- and two-part sealants; excellent weatherability, good long-term storage stability, good paintability, good adhesive strength. [Union Carbide]

Silmod SAT-30. Silicon-modified polyether; for the formulation of adhesives for bonding large stone or metal panels to concrete, formation of expansion joints, the construction of interior walls, joining of materials with different coefficient of thermal expansion. [Union Carbide]

Sil-o-cel. A brand of kieselguhr, also a heat insulator made from kieselguhr. §

Siloid®. A registered trademark for micron-sized silica gels.§

Silopren® . Hot air and room temperature vulcanizing silicone rubber; used for seals, gaskets, dampening components, hoses, profiled strip, electrical insulating material, production of sealants, casting and coating compounds, fabric proofings and impression coatings. [Bayer AG]

Silopren® HV. A basic compound of silicone rubber enriched with crosslinking agents, pigments etc., ready for processing; used for press-cured and extruded articles, e.g., seals, damping components, articles for food contact, hoses, profiled strip, electrically conductive compounds, electrical insulating materials. [Bayer AG; Rhein-Chemie Rheinau]

Sil-O-Wet™. Inhibited aluminum pigments; used for aqueous paints and coatings used for decorative metallic effects as well as a wide range iof protective coatings applications. [Silberline Mfg Co Inc]

siloxicon. A fireproof material resistant to the action of acids and alkalis; made by heating powdered silica with a small quantity of carbon in an electric furnace, the composition is Si_2C_2O. It is produced with silicon carbide in a carborundum furnace; used alone or with binding materials for making crucibles or muffles.

siloxide. A mixture of silica with a little titanium, or zirconium oxide.

Silpruf® 2000. Silicone sealant; for high-performance weatherproofing and glazing. [GE Silicones]

Silquat Q-100. Nonrewetting conditioner for personal care products. [Siltech]

Silres® KX. Methyl silicone resin; binder which dries to tack-free finish at room temperature, for highly pigmented paints (preferably with aluminum powd.). [Wacker Silicones]

Silres® MP 42 E. Methyl phenyl silicone resin emulsion; dries to tack-free finish at room temperature; binder for highly pigmented and gloss paints, clear varnishes; recommended for blending with water-based organic resins; high shear resist.; nonionic emulsifier. [Wacker Silicones]

Silres® MSE 100. Polymethyl siloxane, methoxy-functional; solv.-free room temperature curing binder requiring addition of catalyst; for highly pigmented paints. [Wacker Silicones]

Silres® REN 50. Methyl phenyl silicone resin; binder which dries to tack-free finish at room temperature, for highly pigmented and gloss paints, clear varnishes; good adhe-

---

‡ Trade name and manufacturer not verified   § Trade name without identified manufacturer

sion to difficult substrates. [Wacker Silicones]

Silso®. Inorganic semiconductor materials as base material for solar cells. [Wacker-Chemitronic Gesellschaft fur Elektronik-Grundstoff]

Silsoft. Elastofilcon A; contact lens material. [Dow Corning Ophthalmics Inc] *

Silsoft. Silicone textile softener. [Union Carbide]

Silteg. A range of highly and medium-active aluminum silicates; used for shoe sole materials and technical rubber articles. [Degussa]

Siltek. CO_2 process binder with exceptional breakdown and fast gassing properties. [Foseco (F.S.) Ltd]

Siltek® L Polymer. CAS 9010-79-1; Ethylene/propylene copolymer; for use in cosmetic stick formulations; offers compatibility, improved oil retention and better stick structure. [Petrolite]

Siltem® STM1300. Silicone/polyetherimide block copolymer; excellent flame resist., low smoke properties for applications where combustion and corrosion must be minimized; fast line and extrusion speeds for wire and cable manufacturing. [GE Plastics]

Siltex. Fused silica; high performance extender pigment and filler industrial coatings, polymer systems, abrasives, electrical molding and potting compounds, translucent/transparent sealants; outstanding electrical properties, low refractive index and thermal expansion. [Kaopolite]

SILTherm; Thermoset polyester; bulk and sheet molding compounds; excellent corrosion resist., superior moldability. [Industrial Dielectrics] *

SilTouch Cotton Plus. Amino functional silicone; durable softener with excellent shear stabilty. [Yorkshire Pat-Chem]

Silumin. Proprietary alloys of aluminum and silicon containing 12% silicon; they have a specific gravity of 2.63-2.65.§

Silumin-Y. A proprietary aluminum-silicon alloy with small additions of manganese and magnesium; it has high corrosion resistance. §

Siluminite. An electric insulator consisting of 75% of mineral matter (asbestos, calcium silicate, and aluminum silicate), with pitch as the binding material.

silundum. CAS 409-21-2; A product similar to carborundum; articles, such as crucibles and tubes, are made by shaping pieces of graphite, embedding them in carborundum, and subjecting them to the action of silicon vapor at high temperatures in the electric furnace. It has a high electrical resistance and is used for making electrodes.

Silva. A name used mainly in Germany for a type of artificial silk.

Silvacur. CAS 107534-96-3; Tebuconazole; fungicide with systemic properties and broad spectrum activity against rusts, leaf spot diseases, e.g., *Septoria spp.*, powdery mildew and several *Fusarium* species on cereals; whitemold, *Phoma* and various leaf spot diseases on oilseed rape. [Bayer AG]

Silvapron. CAS 1702-17-6; 2,4-D; Translocated herbicide for cereals and grass. [BP Oil Ltd]

Silvatol. Pretreatment agent. [Ciba plc] *

Silvaz. A proprietary trade name for an alloy used for the manufacture of steel; contains iron with 40-45% silicon, 6.0-6.5% vanadium, 6.0-6.5% aluminum, and 6.0-6.5% zirconium. §

Silvel. A proprietary trade name for an alloy containing 67.9% copper, 16% zinc, 6.5% nickel, 0.5% lead, 2.2% iron, and 6.8% manganese. §

silver. (silver, colloidal) Metallic element; Ag; CAS 7440-22-4; EINECS 231-131-3; manufacture of silver nitrate, etc.; sterilant; water purification; for coinage; manufacture of tableware, jewelry, ornaments; for electroplating; as catalyst; in dental alloys. [Aldrich; Asarco; Cerac; Degussa; Handy & Harman; Mariovilla SpA]

silver acetate. (acetic acid silver salt) CH_3COOAg; CAS 563-63-3; EINECS 209-254-9; laboratory reagent, oxidizing agent. [Atomergic Chemetals; Spectrum Chem. Mfg.]

silver alum. An aluminum-silver sulfate, $Al_2(SO_4)_3 \cdot Ag_2SO_4 \cdot 24H_2O$.

silver amalgam. An alloy of mercury and silver. It occurs as a mineral, but is also prepared artificially.

silver bell metal. An alloy of 40-42% copper and 58-60% tin.

Silver Bond 30. Silica; filler, flatting agent for traffic paint, exterior block fillers, mastics and adhesives, buffing compounds. [Unimin]

silver bromide. AgBr; CAS 7785-23-1; EINECS 232-076-8; Photographic film and plates, photochromic glass, laboratory reagent. [Atomergic Chemetals; Cerac; Spectrum Chem. Mfg.]

silver bronze. An alloy of 64% copper, 17% manganese, 13% zinc, 5% silicon, and 1% aluminum. An electrical resistance alloy.

silver carbonate. Ag_2CO_3; CAS 534-16-7; EINECS 208-590-3; laboratory reagent. [Aldrich; Atomergic Chemetals; Spectrum Chem. Mfg.]

Silver, China. *See* Argyrolith. §

silver chloride. Ag_2CrO_4; CAS 7783-90-5; EINECS 232-033-3; photography, photometry and optics, batteries, photochromic glass, silver plating, production of pure silver, antiseptic. [Atomergic Chemetals; Degussa; Noah Chem.]

silver foil. An alloy of from 90-97% tin, 0-2.5% copper, and 0.10% zinc, is known by this name.

silver grain. The cochineal insect killed in an oven at three months old is called silver grain.

silver grey. A Bohme dyestuff. It contains extracts of logwood and redwood, together with a chrome and iron mordant.

silverine. An alloy of 77% copper, 17% nickel, 2% iron, 2% zinc, and 2% cobalt.

silvering solutions. Usually consist of solutions of silver cyanide and ammonium cyanide in water; used for the electro-deposition of silver.

silver ink. A mixture of gum arabic and ground white mica; used for inlaying buttons.

silver iodide. AgI; CAS 7783-96-2; Photography, cloud seeding for artificial rainmaking, laboratory reagent, antiseptic. [Aldrich; Atomergic Chemetals; Cerac; Noah Chem.]

silverite. *See* nickel silvers.

silver leaf. An alloy of 91% tin, 8% zinc, 0.35% lead, and 0.2% iron. Another alloy contains 91% tin, 8.25% zinc, and 0.4% antimony.

Silverline 200. CAS 14807-96-6; Platy talc; general purpose, economical filler suited for antiblocking and industrial coatings applications. [Montana Talc]

Silverline 665. CAS 14807-96-6; Platy talc; ultrafine, high purity talc with excellent reinforcing and dielec. proper-

---

* Trade name not verified as available   † Trade name verified as obsolete

ties; for extrusion of wire and cable; additive for gloss control in paints and coatings. [Montana Talc]

silver metal. An alloy of 66.5% zinc and 33.5% silver.

silver nitrate. AgNO$_3$; CAS 7761-88-8; EINECS 231-853-9; Photographic film, catalyst for ethylene oxide, indelible inks, silver plating, silver salts, silvering mirrors, germicide, hair dyeing, antiseptic, laboratory reagent. [Accurate Chem. & Scientific; Aldrich; Degussa; Johnson Matthey plc; Spectrum Chem. Mfg.]

Silveroid. An alloy of 45% nickel, 54% copper, and 1% manganese.

silver oxide (ous). (argentous oxide) Ag$_2$O; CAS 20667-12-3; EINECS 243-957-1; Polishing glass, coloring glass yellow, catalyst, purifying drinking water, laboratory reagent. [Atochem N. Am.; Atomergic Chemetals; Degussa; Spectrum Chem. Mfg.]

silver-salt. Sodium anthraquinone monosulfonate, obtained in alizarin manufacture.

silver saltpeter. A name which has been applied to silver nitrate, AgNO$_3$.

silver solder. Variable alloys. Usually they contain silver, copper, and zinc. A soft silver solder contains 67% silver and 33% brass, and is suitable for sheet. A hard silver solder consists of 80% silver and 20% copper. Some alloys contain smaller amounts of silver, and an ordinary one of this type contains 47% copper, 47% zinc, and 6% silver; sometimes a small amount of tin is present.

Silverstone®. Nonstick finishes. [DuPont UK]

Silverstone Supra®. Polymer. [DuPont UK]

silver sulfate. (silver sulfate normal) Ag$_2$SO$_4$; CAS 10294-26-5; EINECS 233-653-7; Laboratory reagent. [Atomergic Chemetals; Noah Chem.]

silver sulfate normal. See silver sulfate.

silver sulfide. Ag$_2$S; CAS 21548-73-2; EINECS 244-438-2; Inlaying in niello metal work, ceramics. [Atomergic Chemetals; Cerac]

silver white. See Flake White.

silvestrite. A mineral; siderazote.

Silvet. Aluminum pigments for plastics; used for all plastics, automobiles, toys, bottles, etc. [Silberline Mfg Co Inc]

Silvex. See Silvet. [Silberline Mfg Co Inc]

Silvital. Fertilizer for vitalizing damaged forest. [Süd-Chemie AG] *

Silvoline. Aluminum paint; for marine and general purposes. [Llewellyn Ryland Ltd]

Silwax® S. Silicone wax; patented, highly lubricious wax for personal care applications, polishes, textile lubrication and softening, laundry products, dryer sheet softeners, synthetic lubricants, and plastics lubricants. [Siltech]

Silwet®. Silicone surfactant. [Union Carbide]

Silwet® L-77. CAS 27306-78-1; Polyalkylene oxide-modified polymethylsiloxane; surfactant, flow/leveling agent, antistat, antifog, dispersant, wetting agent, flotation agent, spreading agent for coatings, printing inks, adhesives, agriculture, automotive, cleaners, antifogging agent, mining, paper, pharmaceutical applications. [Union Carbide]

Silwet® L-720. CAS 68554-65-4; Dimethicone copolyol; surfactant; anticaking agent; slip additive for paper; also for pharmaceutical use, printing inks. [Union Carbide]

Silwet® L-7200. CAS 68937-55-3; Silicone glycol copolymer; flow, wetting, slip, dispersion, gloss agent, emulsifier for industrial coatings, household and institutional products, textiles, inks. [Union Carbide]

Silwet® L-7500. CAS 68440-66-4; Dimethicone copolyol; surfactant, antifoam, dispersant, emulsifier, leveling and flow control agent, lubricant, slip additive for adhesives, automotive, chemical processing, coatings, petrol. extraction, paper, personal care products, plastics and rubber, pharmaceutical, textile applications. [Union Carbide]

Silwet® L-7600. CAS 68938-54-5; Dimethicone copolyol; surfactant, wetting agent for adhesives, window cleaners, textiles, personal care products; internal lubricant for plastics and rubber. [Union Carbide]

Silwet® L-7605. CAS 68938-54-5; Dimethicone copolyol; defoamer, slip additive for chemical processing, coatings. [Union Carbide]

Silwet® L-7607. CAS 117272-76-1; Polyalkylene oxide-modified polymethylsiloxane; surfactant, wetting agent, leveling and flow control agent, grease cleaner, flotation and spreading agent for adhesives, agriculture, automotive specialties, chemical processing, carpet antistat, mining, metal processing, petrol. extraction, printing inks, textiles. [Union Carbide]

Silwet® L-7622. CAS 68938-54-5; Silicone glycol copolymer; flow, wetting, slip, dispersion, gloss agent, emulsifier for industrial coatings, household and institutional compounds, textiles, and inks. [Union Carbide]

silzin bronze. An alloy of copper with 10-20% zinc and 4.5-5.5% silicon.

Simadex. Selective and total herbicide. [Schering Agrochemicals Ltd]

Simanex. CAS 122-34-9; Active ingredient: simazine; 2-chloro-4,6-bis(ethylamino)-1,3,5-triazine; pre-emergence herbicide for control of weeds in a variety of crops as well as a soil sterilant. [Agan Chemical Manufacturers Ltd]

Simapron. CAS 122-34-9; Emulsifiable concentrate containing 150 g/l simazine; a triazine herbicide to control weeds and grasses in cane fruit, roses and some vegetables. [BP Oil Ltd]

Simask®. For the electrical industry. [DuPont UK]

Simax. CAS 409-21-2; Silicon carbide; used in metallurgical applications. [Lonza AG]

Simazol. Active ingredients: azolan plus simanex; multipurpose herbicidal mixture which eradicates a wide spectrum of established weeds, while preventing further weed germination for extended periods. [Agan Chemical Manufacturers Ltd]

Simchin WS. PEG-40 jojoba oil; emollient, humectant, pigment, conditioner, lubricant for personal care formulations. [RITA]

Simchin, Natural. CAS 61789-91-1; Jojoba oil; moisturizer, emollient, conditioner for skin and hair care products. [RITA]

Simeco. A proprietary preparation of aluminum hydroxide, sucrose and siroethicone; used as a gastro-intestinal sedative. [Wyeth Laboratories] *

Simetite. Sicilian amber of wine-red to garnet-red color.

Simfix. Granular mixture of diuron and simazine; used for control of weeds in woody crops and noncrop areas. [Rhone-Poulenc Environmental Prods. Ltd]

Simflex. Flexible circuitry. [Carl Freudenberg] *

Simflow. CAS 122-34-9; Simazine formulated as a flowable liquid; a soil residual herbicide suitable for selective weed control in shrubs; also used as a total weedkiller to keep pathways, bare ground and industrial installations free of

---

‡ Trade name and manufacturer not verified   § Trade name without identified manufacturer

weeds and grasses. [Rhone-Poulenc Environmental Prods. Ltd]

Simflow Plus. CAS 61-82-5, 122-34-9; A suspension concentrate containing 100 g aminotriazole and 300 g simazine per liter; used for total weed control in non crop areas and fruit orchards. [Rhone-Poulenc Environmental Prods. Ltd]

Similor. A rich-colored brass. It usually contains from 80-89% copper, 9-20% zinc, and 0-7% tin.

Simmering. Rotating shaft seal. [Carl Freudenberg] *

Simoniz. Polishes and valeting products for cars. [Spectra Brands plc]

Simplex. Sanitizers. [Thomas Goldschmidt Ltd]

Simplex Steel. A proprietary trade name for a nickel-chromium steel containing 1.25% nickel and 0.6% chromium. §

Simplotan. CAS 19387-91-8; A proprietary preparation of tinidazole; an antiprotozoal. [Pfizer International] †

Simply White. Net curtain whitener. [Dylon International Ltd]

Simrax. Face seals. [Carl Freudenberg] *

Simrit. Seals and packing rings. [Carl Freudenberg] *

Simulsol 58. CAS 9004-95-9; Ceteth-20; cosmetic emulsifier. [Seppic]

Simulsol 98. CAS 9004-98-2; Oleth-20; cosmetic emulsifier. [Seppic]

Simulsol 165. PEG-100 stearate, glyceryl stearate; cosmetic emulsifier. [Seppic]

Simulsol 989. CAS 61788-85-0; PEG-7 hydrogenated castor oil; cosmetic emulsifier. [Seppic]

Simulsol 1292. CAS 61788-85-0; PEG-25 hydrogenated castor oil; solubilizer, emulsifier for cosmetics. [Seppic]

Simulsol 5719. Ethoxydiglycol, ceteareth-16, nonoxynol-8; cosmetic emulsifier. [Seppic]

Simulsol CS. CAS 68439-49-6; Ceteareth-33; cosmetic emulsifier. [Seppic]

Simulsol M 45. CAS 9004-99-3; PEG-8 stearate; emulsifier. [Seppic]

Simulsol P4. Laureth-4; emulsifier. [Seppic]

Sinapoline. Diallyl-urea, $(C_3H_5NH)_2CO$.

Sinatron. A range of polyester resins which include orthophthalic, isophthalic, bisphenolic, neopentilic, and self-extinguishing resins. [Lonza AG]

Sinaxar. CAS 94-35-9; A proprietary preparation of styramate; used in the treatment of muscle spasm. [Armour Pharmaceutical Co] *

Sinazine. Herbicide. [Murphy Chemical Ltd] †

Sinbar®. CAS 5902-51-2; Wettable powder containing 80% w/w terbacil; a weedkiller [DuPont UK]

Sin-Chu. (Japanese brass). An alloy of 66.5% copper, 33.4% zinc, and 0.1% iron.

Sindanyo. A trade name for proprietary asbestos products.§

Sinecain. Quinine hydrochloride and antipyrine for hypodermic use.

Sinemet. Levodopa and carbidopa; for the treatment of Parkinsonism. [Merck & Co Inc]

Sinemet-Plus. Levodopa and carbidopa; for the treatment of Parkinsonism. [Merck & Co Inc] †

Sinequan. CAS 1668-19-5; A proprietary preparation of doxepin; an antidepressant. [Pfizer International]

single muriate of tin. An acid solution of stannous chloride; used as a mordant.

single nickel salt. Nickel sulfate, $NiSO_4 \cdot 7H_2O$; used in the plating trade.

Single Purpose. CAS 62-38-4; Phenylmercury acetate; organomercury fungicide seed dressing for cereals and fodder beet. [DowElanco Ltd]

Singlet. Used primarily in the treatment of coughs, colds, and upper respiratory conditions. [Dow UK] *

Singlex CIP. Heavy-duty cleaner for metallic surface. [Adeka Fine Chem.]

Singoserp. A proprietary trade name for Syrosingopine. §

sinigrin. (sinigroside; potassium myronate) A constituent of black mustard seed; $KC_{10}H_{16}NS_2O_9$.

Sinodor. A basic magnesium acetate, containing an excess of magnesium hydrate; used for disinfecting purposes.

Sinpro®. Effervescent granules for pain relief. [Bayer AG]

sin red. Potassium permanganate, $KMnO_4$.

Sinter-corundum. A proprietary preparation; it is a ceramic material produced from pure alumina at a temperature of about 1800 C; the thermal conductivity at 16 C is about twenty times as high as that of porcelain, and it is stated to be not attacked by hydrofluoric acid or hot alkali. §

Sinterit. A proprietary trade name for a form of sponge iron (qv) used for coupling packings. §

Sinterloy. A proprietary trade name for a steel powder. §

Sinthrome. A proprietary preparation of nicoumalone; an anticoagulant. [Ciba plc] *

Sintrex®. [Airex AG]

Sinvabond. Flexographic printing inks; for printing laminated film structures. [Allied-Signal Inc/Sinclair and Valentine Division] *

Sinvaset. Web heatset offset printing inks; for publication printing. [Allied-Signal Inc/Sinclair and Valentine Division] *

Siogel. CAS 409-21-2; Approximately 99.7% silicon dioxide; for dehydration of air and other gases. [Chemische Fabriken Oker und Braunschweig AG] *

sionon®. A range of foods suitable for diabetics. [Bayer AG]

Siopel. A preparation of dimethicone and cetrimide; a barrier skin cream. [ICI Chem. & Polymers Ltd]

Sioplas. Polyethylene cross linking technology. [Dow Corning Ltd]

Sipalin AOC. Dicyclohexyl adipate; a proprietary solvent for cellulose nitrate. §

Sipalin AOM. Dimethylcyclohexyl adipate; it boils at from 225-232 C, has a specific gravity of 1.011, and a flash-point of 189 C; a proprietary solvent for cellulose nitrate and plasticizer for rubber. §

Sipalin MOM. Dimethylcyclo-hexyl β-methyladipate; it boils from 216-224 C, has a specific gravity of 1.009, and flashes at 195 C; a proprietary solvent and plasticizer.§

Sipcam UK Carbosip 5G. CAS 1563-66-2; Carbofuran; a granular systemic carbamate insecticide and nematode for soil treatment. [Sipcam UK Ltd]

Sipcam UK Rover 500. CAS 1897-45-6; Chlorothalonil; a fungicide for a wide range of agricultural crops. [Sipcam UK Ltd]

Sipenol IT-50-46; CAS 61791-26-2; PEG-50 hydrogenated tallow amine. [Rhone-Poulenc Surf. Canada]

Sipernat®. Range of spray dried hydrophillic precipitated silicas; free flow/anticaking aids; as carrier substances for production of highly concentrated pulverulent formulations of liquid or paste-like active substances. [Degussa]

Sipernat® 22. Hydrated silica; adsorbent, anticaking and free-flow agents; used as aid to convert liqs. into powds.; processing aid; hydrophilic; for adhesives, sealants,

---

* Trade name not verified as available   † Trade name verified as obsolete

detergents, food, cosmetics industries. [Degussa]

Sipernat® 22S. Hydrated silica; free-flow/anticaking agent for powd. detergents, sealants, foodstuffs, pharmaceuticals, fire extinguishers. [Degussa]

Sipernat® 44. CAS 12141-46-7; Aluminum silicate; antiblocking agent in PE and PP blown films. [Degussa]

Sipernat® 50. Hydrated silica; carrier, free-flow agent, anticaking agent for cosmetics, detergents, food industries. [Degussa]

Sipernat® 283LS. Precipitated silica; for polishing of silicon wafers in elec. industry. [Degussa]

Sipernat® D17. Precipitated silica; anticaking and free-flow agent; hydrophobic; for fire extinguishers, pesticides, plastics. [Degussa]

Sipex 30. Anionic surfactant as a liquid paste; general detergent; emulsifier for emulsion polymerization. [Henkel Chemicals Ltd] *

Sipex DS. Anionic surfactant in acid form, supplied as a viscous liquid; basic material for general detergents, e.g., in dishwashers, and in other formulations requiring a cheap source of anionic active matter. [Henkel Chemicals Ltd] *

Sipilite. See Bakelite. §

Sipomer® 2M1M. CAS 2867-47-2; Dimethylaminoethyl methacrylate. [Rhone-Poulenc Surf.]

Sipomer® 2ME. CAS 60-24-2; 2-Mercaptoethanol. [Rhone-Poulenc Surf.]

Sipomer® AAE. Allyl alcohol ethoxylate; reactive intermediate for silylation; bound protective colloid; copolymerizable stabilizer. [Rhone-Poulenc Surf.]

Sipomer® AGE. CAS 106-92-3; 1-Allyloxy-2,3-epoxy propane. [Rhone-Poulenc Surf.]

Sipomer® AM. CAS 96-05-9; Allyl methacrylate; monomer for coatings, elastomers, adhesives, intermediates; contributes hardness and scratch resist.; crosslinker/hardener. [Rhone-Poulenc Surf.]

Sipomer® β-CEA. CAS 79-10-7; β-Carboxyethyl acrylate; monomer for emulsion polymerization, adhesives and coatings; improves latex properties, adhesion. [Rhone-Poulenc Surf.]

Sipomer® DCPA. CAS 12542-30-2; Dicyclopentenyl acrylate; monomer for air-drying adhesives, coatings, caulks, sealants, elastomers; crosslinks acrylics, unsaturated polyesters and alkyds. [Rhone-Poulenc Surf.]

Sipomer® DCPM. CAS 51178-59-7; Dicyclopentenyl methacrylate; monomer for concrete resurfacing, air-drying adhesives and coatings, elastomers; improves cohesive strength through crosslinking. [Rhone-Poulenc Surf.]

Sipomer® HEM-5. CAS 25736-86-1; PEG mono-methacrylate. [Rhone-Poulenc Surf.]

Sipomer® HEM-D. CAS 868-77-9; 2-Hydroxyethyl methacrylate. [Rhone-Poulenc Surf.]

Sipomer® IBOA. CAS 588-33-5; Isobornyl acrylate. [Rhone-Poulenc Surf.]

Sipomer® IBOMA. CAS 7534-94-3; Isobornyl methacrylate. [Rhone-Poulenc Surf.]

Sipomer® IDA. CAS 1330-61-6; Isodecyl acrylate. [Rhone-Poulenc Surf.]

Sipomer® MEM. CAS 3990-03-2; Monoethyl maleate. [Rhone-Poulenc Surf.]

Sipomer® TATM. CAS 2694-54-4; Triallyl trimellitate [Rhone-Poulenc Surf.]

Sipomer® TMPEO. CAS 29860-47-7; Ethoxylated trimethylol propane. [Rhone-Poulenc Surf.]

Sipomer® TMPTA. CAS 15625-89-5; Trimethylolpropane triacrylate. [Rhone-Poulenc Surf.]

Sipon®. Redesignated Rhodapon® or Rhodapex®. [Rhone-Poulenc Surf.]

Sirdate®. For the agriculture industry. [DuPont UK]

Sirius®, Sirius Supra, Sirius Supra LL. A range of direct dyestuffs; used for dyeing cellulosics. [Bayer AG; Bayer plc]

Sirlene. Feed grade propylene glycol, an emulsifying agent and general purpose food additive; as a conditioner in animal feed, a preservative, humectant, energy source, lubricant, extender, palatability improver and fines reducer; in dairy cattle, used for the prevention and treatment of acetonemia (ketosis). [Dow UK]

Sirtan. Floor polishes. [Evans Vanodine International Ltd]

Sisacan. Dronabinol; anti-emetic. [PARS Pharmaceutical Laboratories Inc] ‡

Sisellig. Antislipping agent for all fibers; antiagglomerant for short fibers. [Ceca SA]

Siseptin. CAS 53179-09-2; Sisomicin sulfate; antibacterial. [Schering Corp] *

Sistan®. CAS 137-42-8; Metam-sodium 38% w/v aqueous solution; for use as a soil sterilant in horticulture and nurseries. [Universal Crop Protection Ltd]

Sitilan. The methylcyclohexyl ester of adipic acid; a solvent for cellulose and rubber.

Sitol. A proprietary trade name for the sodium salt of m-nitrobenzenesulfonic acid; used in dyeing. §

Sitren. Hydrophobic silicone agents. [Thomas Goldschmidt Ltd]

Sit-ruti. See sanna.

Siverslice. Proteinaceous substances for use as food or food ingredients. [Courtaulds Aerospace Ltd]

Sivex. Ceramic foam filters for removal of nonmetallic inclusions from aluminum alloys. [Foseco (F.S.) Ltd]

Sivex F. Ceramic foam filters used for the production of aluminum and copper based alloys. [Foseco (F.S.) Ltd]

Size. Textile sizing agents. [BASF plc]

size. Usually consists of a starch solution containing small amounts of tallow or oil, and China clay or French chalk; to strengthen and smooth the fibers for weaving; used on yarn or cloth to bind the threads together and to weight it; in papermaking,it is used to give the properties of nonabsorption of moisture, smoothness, and sheen; a treatment for plaster surfaces to prevent subsequent absorption of paint; in powder metallurgy, final pressing on a sintered compact to get desired size or physical properties.

Size CB. CAS 9003-03-6; Polyacrylate; sizing agent for cellulosic fibers and their blends. [BASF]

Sizing Wax PA, PT, SM. PEGs; for production of water-sol. sizing auxiliaries; antistats, dyeing auxiliaries. [Hüls AG]

SK-65. A proprietary preparation of dextropropoxyphene; an analgesic. [SmithKline Beecham] *

SKA. A butadiene polymer of Soviet origin, derived from petroleum.

SK-Amitriptyline. CAS 549-18-8; Amitriptyline hydrochloride; antidepressant. [SmithKline Beecham] *

SK-Ampicillin. CAS 69-53-4; Ampicillin; antibacterial. [SmithKline Beecham] *

Skane. Mildewcide; used for coatings. [Rohm & Haas]

Skane® M-8. CAS 26530-20-1; 2-n-Octyl-4-isothiazolin-3-one; mildewcide for paints. [Rohm & Haas]

Skaterpax. CAS 3691-35-8; An oil formulation containing

---

‡ Trade name and manufacturer not verified   § Trade name without identified manufacturer

2.5% of chlorophacinone; an anticoagulant rodenticide; a bait to control black rats, brown rats, house mice and voles. [Lever Industrial Ltd]

SKB. A butadiene polymer of Soviet origin, derived from alcohol.

SK-Bamate. CAS 57-53-4; Meprobamate; sedative. [SmithKline Beecham] *

SK-Doxycycline. Doxycycline hyclate; antibacterial. [SmithKline Beecham] *

Skefron. A proprietary preparation of dichlorodifluoromethane and trichlorofluoromethane; an analgesic spray. [SmithKline Beecham] *

Skeladin. CAS 1665-48-1; A proprietary trade name for metaxalone. §

Skelleftea. A variety of Stockholm tar.

Skellite®. Solv. used as stove and lamp fuel. [Texaco]

Skellysolve. A proprietary trade name for a series of petroleum solvents. §

SK-Erythromycin. CAS 643-22-1; Erythromycin stearate; antibacterial. [SmithKline Beecham] *

SK-Estrogens. Oestrogens, esterified; Oestrogen. [SmithKline Beecham] *

SK-Furosemide. CAS 54-31-9; Furosemide; diuretic. [SmithKline Beecham] *

SK-Hydrchlorothiazide. CAS 58-93-5; Hydrochlorothiazide; diuretic. [SmithKline Beecham] *

Skin Wiz. Proprietary, polymeric film forming coating to protect molds and molded parts which releases easily when required; used for fiberglass tools, injection molds and molded or fabricated plastic or metal parts. [Axel Plastics Research Laboratories Inc]

Skiodan Sodium. CAS 126-31-8; Methiodal sodium; diagnostic aid. [Sterling Drug Inc] *

Skleron. A proprietary trade name for an aluminum alloy containing 12% zinc, 3% copper, 0.6% manganese, 0.25% silicon, and a small amount of nickel. §

Skliro Distilled. CAS 68424-43-1; EINECS 270-302-7; Lanolin acid; emollient, superfatting agent for aerosol shave foams, hand soaps; water repellent films; w/o emulsifier; stable emulsions in preparation of waterproof makeup. [Croda Inc.]

SK-Lygen. CAS 438-41-5; Chlordiazepoxide hydrochloride; sedative. [SmithKline Beecham] *

SK-Metronidazole. CAS 443-48-1; Metronidazole; antiprotozoal. [SmithKline Beecham] *

SK-Penicillin G. CAS 113-98-4; Penicillin G potassium; antibacterial. [SmithKline Beecham] *

SK-Penicillin VK. CAS 132-98-9; Penicillin V potassium; antibacterial. [SmithKline Beecham] *

SK-Phenobarbital. CAS 50-06-6; Phenobarbital; anticonvulsant; hypnotic; sedative. [SmithKline Beecham] *

SK-Potassium Chloride. CAS 7447-40-7; EINECS 231-211-8; Potassium chloride; replenisher. [SmithKline Beecham] *

SK-Pramine. CAS 113-52-0; Imipramine hydrochloride; antidepressant. [SmithKline Beecham] *

SK-Prednisone. CAS 53-03-2; Prednisone; glucocorticoid. [SmithKline Beecham] *

SK-Probenecid. CAS 57-66-9; Probenecid; uricosuric. [SmithKline Beecham] *

SK-Propantheline Bromide. CAS 50-34-0; Propantheline bromide; anticholinergic. [SmithKline Beecham] *

SK-Quinidein Sulfate. CAS 50-54-4; Quinidine sulfate;

cardiac depressant. [SmithKline Beecham] *

SK-Reserpine. CAS 50-55-5; Reserpine; antihypertensive. [SmithKline Beecham] *

SK-Soxazole. CAS 127-69-5; Sulfisoxazole; antibacterial. [SmithKline Beecham] *

SK-Tetracycline. CAS 64-75-5; Tetracycline hydrochloride; anti-amebic; antibacterial; antirickettsial. [SmithKline Beecham] *

SK-Thioridazine HC1. CAS 130-61-0; Thioridazine hydrochloride; antipsychotic; sedative. [SmithKline Beecham]*

SK-Tolbutamide. CAS 64-77-7; Tolbutamide; antidiabetic. [SmithKline Beecham] *

SK-Triamcinolone. CAS 124-94-7; Triamcinolone; glucocorticoid. [SmithKline Beecham] *

Skybond. Polyamide resin varnish; designed for structural, electrical and specialty applications where extended exposure to high temperature is required. [Monsanto]*

Skydrol. Fire resistant hydraulic fluid; distributed to commercial airlines. [Monsanto]

Skyllex. Dialkyl dimethyl ammonium chloride; supplied as a mixture of water and isopropanol suspension forming a stiff paste; cationic surfactant used in fabric softeners and as a bactericide. [Efkay Chemicals Ltd] *

SLA 1208. CAS 1317-33-5; Colloidal MoS_2 in 500 solv. refined paraffinic petrol. oil; lubricant additive for gear oils. [Acheson]

SLA 1261. CAS 1317-33-5; Colloidal MoS_2 in 150 solv. refined paraffinic petrol. oil; lubricant additive for gear oils, machine oils. [Acheson]

SLA 1262. CAS 7782-42-5; Colloidal graphite in 150 solv. refined paraffinic petrol. oil; solid lubricant additive for gear oils. [Acheson]

SLA 1275. CAS 7782-42-5; Colloidal graphite in 150 solv. refined paraffinic petrol. oil; solid lubricant additive for engine oils, chain lubes, aerosols, machine oils, etc. [Acheson]

SLA 1286. CAS 1317-33-5; Colloidal MoS_2 in 150 solv. refined paraffinic petrol. oil; lubricant additive for engine oils, chain lubes, aerosols, penetrating lubes, machine oils, assembly lubes. [Acheson]

SLA 1611. CAS 9002-84-0; Colloidal PTFE in 150 solv. refined paraffinic petrol. oil; lubricant additive for engine oils, gear oils, chain lubes, aerosols, machine oils, penetrating oils, assembly and thread lubes. [Acheson]

SLA 1612. CAS 9002-84-0; Colloidal Teflon in 150 solv. refined paraffinic petrol. oil; lubricant additive for engine oils, gear oils, chain lubes, aerosols, machine oils, penetrating oils, assembly and thread lubes. [Acheson]

SLA 2208. CAS 1317-33-5; Colloidal MoS_2 in 2500 solv. refined bright stock petrol. oil; lubricant additive for greases. [Acheson]

SLA 2239. CAS 7782-42-5; Colloidal graphite in 500 solv. refined paraffinic petrol. oil; solid lubricant additive for greases. [Acheson]

Slab Dip AC699. CAS 471-34-1; Calcium carbonate pigmented powd. with dispersing agents; rubber slab dipping system for rubber industry [Ayers Cliff] *

slack wax. A soft paraffin wax from the pressing of paraffin distillate.

Slag A. A British chemical standard. It is a basic slag containing 44.5% CaO, 16.15% SiO_2, 12.93% P_2O_5, 8.97% Fe, and 6.9% MgO.

slagbestos. A similar product to slag wool (blast furnace

---

* Trade name not verified as available   † Trade name verified as obsolete

slag).

slag sand. Blast furnace slag is run out of the furnace to fall into a running stream of water, when it is broken up into a fine sand.

slag wool. (mineral cotton, mineral wool). Blast furnace slag (essentially a glass composed of silicates of aluminum and calcium), which has had air blown through it. It resembles spun glass, and is used for packing steam pipes.

slaked lime. See calcium hydroxide.

slate black. See mineral black.

slate fust. (slate filler). A ground slate used as a filler in rubber mixings. It usually has a specific gravity of 2.7-2.8.

slate grey. (stone grey, silver grey, mineral grey). Grey pigments obtained by grinding and levigating special kinds of grey slate, which occur in Germany; used as priming paint, and for the preparation of putty. They are imitated by mixtures of white clay, blacks, ochers, and ultramarine.

slate lime. A mixture of 60% lime with 40% of calcined slate powder used in the manufacture of porous concrete.

Slax. Slag coagulants. [Foseco (F.S.) Ltd]

Slaymor. CAS 28772-56-7; Bromadiolone; a ready-to-use bait for the control of rats and house mice. [Ciba-Geigy Agrochemicals]

SLCC-D. Ammonium hydroxide solution; corrosion inhibitor for preboiler and afterboiler sections of steam generating systems; used in dairies and other facilities where use of conventional neutralizing amines is prohibited. [Drew Ind. Div.]

S-lec. Polyvinyl butyral resin; interlayer film for safety glass. [Sekisu (UK) Ltd] *

Slick. A sulfurized processing aid for glass manufacturing. [Specialty Products Co] *

slicker solder. See plumber's solder.

Slick Slide®. Lubricant. [Graphite Products Corp]

Slip-Ayd® Surface Conditioners. Polyethylene and polymeric waxes in predispersed or micronized forms; for increasing slip and film hardness, reducing blocking, marring, metal marking and improving other related surface properties. [Daniel Prods.]

Slip-Ayd. Dispersed slipping agents; for paints, inks etc. [Cornelius Chemical Co Ltd] *

Slipicone. Silicone lubricants. [Dow Corning]

Slix. Heat resisting refractory cement. [Foseco (F.S.) Ltd] †

slop wax. The wax present in the heavier wax distillates obtained in the refining of petroleum waxes. It is commonly considered unpressable, and therefore different from the paraffin wax pressed from lighter wax distillates.

Slow-Fe. A proprietary preparation of ferrous sulfate in a slow-release base; iron supplement. [Ciba plc]

Slow-FE Folic. Iron replacement. [Ciba plc] *

Slow-K. CAS 7447-40-7; EINECS 231-211-8; Potassium chloride; replenisher. [Ciba-Geigy Corp] *

Slow-Sodium. CAS 7647-14-5; A proprietary preparation of sodium chloride in sustained-release form. [Ciba plc]*

Slow Trasicor. Beta-blocker. [Ciba plc] *

SLPE. Polyethylene wire and cable coating. [BP Chemicals Ltd]

sludge acid. Sulfuric acid which has been used in the refining of petroleum.

Slue. Compressor lubricants. [Dow]

Slug. Slug killers. [Murphy Chemical Ltd] †

Slug Destroyer. CAS 108-62-3; Pellets containing 6% w/w metaldehyde; snail and slug bait. [Schering Agrochemicals Ltd]

Slug-Geta. Slug and snail bait. [Chevron] *

Slugit Liquid. CAS 108-62-3; Metaldehyde slug killer. [Murphy Chemical Ltd] †

Slugoids. CAS 108-62-3; 3% Metaldehyde slug killer pellets (containing animal repellant); used for slug/snail control. [Doff Portland Ltd]

Slug Pellets. CAS 108-62-3; Metaldehyde slug killer. [ICI Garden Products]

Slug Snail Killer. Molluscidide pellets. [Fisons plc, Horticulture Div] *

SM945. Lightly milled monoclinic zirconium dioxide with a zirconia content (including hafnia) of 94.5%; for manufacture of ceramic pigments and welding fluxes. [Ferro]*

SM975. Lightly milled monoclinic zirconium dioxide with a zirconia content (including hafnia) of 97.5%; for manufacture of ceramic pigments and welding fluxes. [Ferro]*

SM987. Lightly milled monoclinic zirconium dioxide with a zirconia content (including hafnia) of 98.7%; for manufacture of ceramic pigments. [Ferro] *

SM992. Lightly milled monoclinic zirconium dioxide with a zirconia content (including hafnia) of 99.2%; for manufacture of lead-zirconate-titanate piezo electric ceramics, zirconates, zirconia technical ceramics and oxygen sensors. [Ferro] *

SM994. Lightly milled monoclinic zirconium dioxide with a zirconia content (including hafnia) of 99.4%; for manufacture of lead-zirconate-titanate piezo electric ceramics, zirconates, zirconia technical ceramics and oxygen sensors. [Ferro] *

SM 2059. Amodimethicone; cationic emulsion of an amine-functional silicone polymer in water; silicone emulsion cures to a durable, detergent-resistant film for mold release, particle treatment, textile finishes, and polish applications. [GE Silicones]

SM 2061. Silicone dimethyl o/w emulsion; aerosol spray starch, mold release. [GE Silicones]

SM 2133, 2135. Dimethyl silicone o/w emulsion; for furniture, vinyl, auto polishes. [GE Silicones]

SM 2140. Dimethyl polysiloxane resin; mold release agent used in rubber and plastic production operations, in foundry release, aerosol spray starch; good lubricity. [GE Silicones]

S-M 5731 Process-Type Silicone Sealant. One-component silicone sealant; glazing sealant for fenestration fabrication. [Schnee-Morehead]

SMA. A proprietary milk feed for babies. [Wyeth Laboratories]

SMA® 1000. CAS 9011-13-6; 1:1 Styrene/maleic anhydride copolymer; soil release agent; used in ammoniacal water solution in carpet shampoos, paints, inks, paper coatings, commercial laundries; in emulsion polymerization, temporary coatings, oven cleaners; as leveling resin for floor polishes; dispersant. [Atochem]

SMA® 1440. CAS 9011-13-6; 1:1 Styrene/maleic anhydride copolymer partially esterified with organic alcohol; dispersant for titanium dioxide, other pigments in aq. media; used in paints, and inks. [Atochem]

SMA® 2625. CAS 9011-13-6; 2:1 Styrene/maleic anhydride copolymer partially esterified with organic alcohol; dispersant; anti-resoil agent for carpet shampoo and carpet manufacturing. [Atochem]

---

‡ Trade name and manufacturer not verified   § Trade name without identified manufacturer

SMA 3840. A proprietary styrene-maleic anhydride copolymer of low molecular weight used for coating cans and drums. [Arco Chemical Europe Inc] *

SMA 5500. A proprietary copolymer of styrene and maleic anhydride, partially esterified and of low molecular weight; used as a vehicle for thermosetting electrodeposited coatings. [Arco Chemical Europe] *

SMA 17352 A. A proprietary copolymer of styrene and maleic anhydride, of low molecular weight; used as a leveling agent in polishes. [Arco Chemical Europe] *

smalt. (saxony blue, saxon blue, king's blue, royal blue, zaffer, zaffre, bleu d'azure, bleu de saxe, azure blue). A potash glass containing oxide of cobalt; used in paint pigments, ceramic industries (pigment), coloring glass, bluing paper, starch and textiles, coloring rubber.

Smaragdine. A trade name for a solidified alcohol consisting of alcohol and gun-cotton, colored with malachite green.§

S-Maz® 20. CAS 1338-39-2; EINECS 215-663-3; Sorbitan laurate; lubricant, antistat, textile softener, process defoamer, opacifier, coemulsifier, solubilizer, dispersant, suspending agent; coupler; prepares excellent w/o emulsions; with T-Maz Series used as o/w emulsifiers in cosmetics, food formulations, industrial oils, and household products; lipophilic. [PPG/Specialty Chem.]

S-Maz® 40. CAS 26266-57-9; EINECS 247-568-8; Sorbitan palmitate; lubricant, antistat, textile softener, process defoamer, opacifier, coemulsifier, solubilizer, dispersant, suspending agent, coupler; prepares excellent w/o emulsions. [PPG/Specialty Chem.]

S-Maz® 60K. CAS 1338-41-6; EINECS 215-664-9; Sorbitan stearate; food emulsifier for vegetable and dairy products, cakes; gloss aid in chocolate, nondairy creamers. [PPG/Specialty Chem.]

S-Maz® 65K. CAS 26658-19-5; EINECS 247-891-4; Sorbitan tristearate; lubricant, antistat, textile softener, process defoamer, opacifier, coemulsifier, solubilizer, dispersant, suspending agent, coupler; prepares excellent w/o emulsions. [PPG/Specialty Chem.]

S-Maz® 80. CAS 1338-43-8; EINECS 215-665-4; Sorbitan oleate; lubricant, antistat, textile softener, process defoamer, opacifier, coemulsifier, solubilizer, dispersant, suspending agent, coupler; prepares excellent w/o emulsions; also as lubricant, rust inhibitor, and penetrant in metalworking formulations. [PPG/Specialty Chem.]

S-Maz® 83R. CAS 8007-43-0; EINECS 232-360-1; Sorbitan sesquioleate; solubilizer, emulsifier and dispersant. [PPG/Specialty Chem.]

S-Maz® 85. CAS 26266-58-0; EINECS 247-569-3; Sorbitan trioleate; lubricant, antistat, textile softener, process defoamer, opacifier, coemulsifier, solubilizer, dispersant, suspending agent, coupler; prepares excellent w/o emulsions; also as lubricant, rust inhibitor, penetrant in metalworking formulations. [PPG/Specialty Chem.]

S-Maz® 90. Sorbitan tallate; lubricant, antistat, textile softener, process defoamer, opacifier, coemulsifier, solubilizer, dispersant, suspending agent, coupler; prepares excellent w/o emulsions; also as lubricant, rust inhibitor, penetrant in metalworking formulations. [PPG/Specialty Chem.]

S-Maz® 95. Sorbitan tritallate; antistat, textile softener, lubricant, process defoamer, opacifier, coemulsifier; prepares w/o emulsions; together with T-Maz series, as o/w emulsifier in metalworking fluids and coolants, semi-

synthetic and oil-based metalworking lubricants; and coolants, cosmetics, food formulations, industrial oils, and household products. [PPG/Specialty Chem.]

SMFP. See sodium monofluorophosphate.

smithite. A mineral; silver sulfarsenite.

Smithol 22LD. Ester; metal and fiber wetting agent; demulsifier; forms continuous monomolecular film with rust prevention properties; stabilizer for chlorinated systems; for drawing, stamping, rolling lubricants; waste water stripping oils; leather treatment; textile spin finishes. [Werner G. Smith]

Smithol PEG Adipate. CAS 68647-16-5; Lubricant for water-based lubricant systems. [Werner G. Smith]

smithsonite. A mineral. See calamine.

Smithsonite. See zinc carbonate.

Smitter-Lénian. An alloy containing 72% copper, 12.75% nickel, 9.75% zinc, 2.3% iron, 2.25% tin, and 1% bismuth.

Smoke. Uncut disperse and solvent dyestuffs; for coloration of smoke grenades mainly for use by army. [Holliday Dyes & Chemicals Ltd]

smoke black. A carbon black used as a pigment; contains 99.75% carbon.

smokeless diamond powder. A 33 grain powder consisting of insoluble nitrocellulose, with 15% metallic nitrates, 6% charcoal, and 3% petroleum jelly.

Smoking Deterrent. A proprietary preparation of magnesium carbonate, lobelin sulfate and tribasic calcium phosphate. [Campana Corp] ‡

smoking salts. Impure hydrochloric acid.

smoky quartz. A quartz containing organic matter or hydrocarbons. It is usually brown in color.

S Monel. A Monel metal with 3.75% silicon used in valves, etc., which are subject to corrosion.

Smoothex. A range of glycerol esters; used as emulsifiers, stabilizers and antifoam agents in the food and cosmetic industries. [Harcros Australia] *

Smooth-On. Iron and foundry cements, epoxy adhesives and cements. Other formulations based on epoxy, polysulfide and polyurethane polymers; for maintenance and repair, structural bonding and plastic tooling applications (flexible molds, cast and laminated plastic tools). [Smooth-On Inc] *

SN-30GF/000 FR. CAS 9003-54-7; SAN, 30% glass fiber-reinforced, flame-retardant. [Compounding Tech.] *

Snac-Kote. Partially hydrogenated vegetable oil (cottonseed, soybean), sorbitan stearate, polysorbate 60; kosher; coating fat for bakery coatings. [Van Den Bergh Foods]

Snapper CDA. An emulsifiable concentrate containing 95 g aminotriazole, 190 g atrazine and 99 g 2,4-D per liter; used for total weed control in non crop areas. [ICI Agrochemicals Professional Products]

Sniafil. An Italian synthetic wool substitute. It is a product obtained in a similar way to viscose silk, but differs from it in the treatment of the viscose solution.

Sniafoam. Glass fiber-reinforced polyester foam. [Lonza AG]

Sniamid® . Nylon 6 and 66 granules; engineering plastic. [SNIA (UK) Ltd]

Sniamid® ADS 40 I. CAS 25038-54-4; Nylon 6, flame retardant; high visc. grade with high flexibility and toughness; for injection molding, film extrusion, and extrusion of sheet, rod and tube. [Nylon Corp. of Am.]

Sniamid® ASN 27 T. CAS 25038-54-4; Nylon 6; transparent grade for injection molding and extrusion including carburetor filters, brake fluid reservoirs, monofilament. [Snia Technopolimeri; Nylon Corp. of Am.]

Sniamid® SSD 300 EP 021. CAS 32131-17-2; Nylon 6/6, 30% mineral-filled; ; toughened, improved impact grade. [Snia Technopolimeri; Nylon Corp. of Am.]

Sniamid® SSD AF. CAS 32131-17-2; Nylon 6/6; reduced coefficient of friction; for applications where friction and wear are likely to occur. [Snia Technopolimeri; Nylon Corp. of Am.]

Sniasan®. ABS, SAN, or PS. [SNIA (UK) Ltd]

Sniatal®. Acetal copolymers for injection molding or extrusion applications. [SNIA (UK) Ltd]

Sniater®. PBT. [SNIA (UK) Ltd]

Snomelt. CAS 10043-52-4; Calcium chloride pellets; used for snow and ice melting, dust control and tire weighting. [Standard Tar Products Co Inc]

Snowflake P.E. CAS 471-34-1, 1317-65-3; Calcium carbonate; med. ground pigment for maximum loading in highly filled systems, e.g., polyester resins, SMC, BMC, TMC, XMC. [ECC]

Snowflake White. CAS 471-34-1, 1317-65-3; Calcium carbonate; general-purpose easy dispersing pigment for protective coatings, rubber, plastics, caulks, glazing compounds, mastics. [ECC]

Snowtack. Adhesive tackifier emulsion. [Tenneco Malros Ltd] *

Snowtack 342 A. Gum rosin acid; general-purpose tackifier for SBR and acrylic polymers. [Eka Nobel]

Snowtack SE 325 A. Gum rosin ester; modifying tackifying resin for low temp. and improved low energy substrate. [Eka Nobel]

snow white. See zinc white, and oil white.

Snow White 200 Mica. CAS 12001-26-2; 200 mesh mica; chemically inert filler with excellent whiteness; highly refractory; imparts crack resist. and improves moisture impermeability to coatings; improves flexural modulus in plastics; filler for microwave cookware. [Unimin]

S-nyl. Polyvinyl acetate homopolymer; chewing gum base. [Sekisu (UK) Ltd] *

Soa. CAS 126-14-7; A proprietary trade name for sucrose octa-acetate; used as a plasticizer. §

soap bark. The bark of Quillaia saponaria of Chile. The active principle is saponin; used to clean clothes.

soap clay. See bentonite.

Soapearl®. Pearlescent pigments for incorporation into extruded soap bars where they exhibit luxurious color effects. [Mearl]

Soarblen. Ethylene-vinyl acetate copolymer. [British Traders & Shippers Ltd]

Soarnol. Ethylene-vinyl alcohol copolymers. [British Traders & Shippers Ltd]

Sobalg FD 100 Series. Sodium alginate; food stabilizer, gellant, film-former for dairy products, desserts, beverages, fruits and vegetables. [Grindsted Prods.]

Sobalg FD 200 Series. CAS 9005-36-1; Potassium alginate; for food industry applications. [Grindsted Prods.]

Sobalg FD 300 Series. CAS 9005-34-9; Ammonium alginate; for food industry applications. [Grindsted Prods.]

Sobalg FD 460. CAS 9005-35-0; Calcium alginate; for food industry applications. [Grindsted Prods.]

Sobee. A proprietary baby feed based on soya; used in cases when an infant is intolerant of milk. [Bristol-Myers Squibb Co Inc]

Sobral®. Modified alkyds, acrylic resins, polyurethane elastomers, or vinyl-modified epoxy esters; used for surface coatings, paints, primers, stains, varnishes, and inks. [Scott Bader]

Sobral 12-101. Alkyd, chain stopped, linoleic rich oil type, xylene solv.; for quick air dry industrial finishes. [Scott Bader]

Sobral 72-625D. PU resin, linoleic rich oil type, white spirit solv.; for high build decorative/marine finishes. [Scott Bader]

Sobral 1321. Alkyd, rosin/phenolic modified, linseed oil type, naphtha solv.; for industrial air drying/stoving primers/finishes, sanding sealers. [Scott Bader]

Sobral 2911. Alkyd, D.C.O. oil-modified, xylene solv.; pure drying alkyd for general purpose air drying/stoving finishes/primers; good color retention. [Scott Bader]

Sobral 9257. Alkyd, styrene-modified, D.C.O. oil type, xylene solv.; for economical quick air dry paints. [Scott Bader]

Sobral AD-002. Alkyd, lineoleic oil-modified; pure drying alkyd for litho inks, high build decorative paints. [Scott Bader]

Sobral AN-001. Alkyd, synthetic fatty acid oil-modified, xylene solv.; nondrying alkyd for tin printing inks, nonsetting mastics. [Scott Bader]

Sobral EE-632. CAS 25928-94-3; Epoxy ester, vinyl-modified, semidrying D.C.O. oil type, xylene solv.; for automobile primers/surfacers, chemical plant protection, road line paints. [Scott Bader]

Sobral L90-20A. PU resin, elastomeric type, IPA solv.; for color-retentive lacquer finishes for flexible substrates and plastics; based on aliphatic isocyanate. [Scott Bader]

Sobrom. Methyl bromide with chloropicrin; fumigant for soil-borne diseases and stored products. [Brian Jones & Associates Ltd]

Socal. CAS 471-34-1; Precipitated calcium carbonate. [Laporte UK Trading] *

Socci 3500, 3500-WP. Antimicrobial for textile applications. [Morton Int'l./Plastics Additives]

Sochamine A 271. Coconut and lauric carboxy/sulfate; for non-eye-stinging shampoos. [Witco Chemical Ltd] *

Sochamine A 7525. Coconut dicarboxylate in liquid form; for hair and baby shampoo. [Witco Chemical Ltd] *

Sochamine A 7527. Coconut dicarboxylate in liquid form; for hard surface cleaners, hair shampoos, textile treatment, strongly acid or alkaline cleaners. [Witco Chemical Ltd]*

Sochamine A 8955. Alkyl imidazoline dicarboxylate in liquid form; for low foam detergents. [Witco Chemical Ltd] *

soda. Sodium carbonate and bicarbonate are both known by this term.

soda ash. (sodium carbonate anhydrous; soda, calcined) Na_2CO_3; CAS 497-19-8; EINECS 207-838-8; Manufacture of glass, pulp and paper, chemicals, sodium compounds, soaps and detergents, water treatment, aluminum production, textile processing, cleaning compounds, petroleum refining, sealing ponds from leakage, catalyst in coal liquefaction; a commercial variety of soda ash used for softening boiler feed water is known as 58% soda ash and contains 58% Na_2O. [FMC; General Chem.; Mallinckrodt; Norsk Hydro A/S; Solvay SA; Texasgulf]

Soda Ash Blocks. Supplementary fluxes used in the melting

---

‡ Trade name and manufacturer not verified   § Trade name without identified manufacturer

of cast iron. [Foseco (F.S.) Ltd] †

soda blue. (gas blue). Impure Prussian blues prepared by using sodium ferro-cyanide instead of the potassium salt.

soda Bordeaux mixture. Made with 6 lb copper sulfate, 2 lb caustic soda, and 50 gallons water.

soda, calcined. See soda ash.

soda glass. See soda-lime glass.

soda glass, soluble. See soluble glass.

Sodagrain. Caustic soda for use in making animal feeds. [Imperial Chemical Industries plc; ICI Chem. & Polymers Ltd]

soda-lime glass. A glass usually containing from 71-78% SiO_2, 12-17% Na_2O, 5-15% CaO, 1-4% Al_2O_3 and Fe_2O_3, and 0-2% K_2O.

sodalumite. A soda alum in cubic form.

soda lye. CAS 1310-73-2; Obtained by boiling a solution of sodium carbonate with slaked lime.

soda niter. See sodium nitrate.

soda-olein. A sulfonated castor oil.

Sodaphos. CAS 10124-56-8; Sodium hexametaphosphate. [FMC]

soda pulp. Wood pulp obtained by means of caustic soda.

Sodasorb. Soda lime; carbon dioxide absorbent. [W R Grace & Co] *

Sodastraw. Caustic soda for use in making animal feeds. [Imperial Chemical Industries plc; ICI Chem. & Polymers Ltd]

soda tar. The name applied to an alkaline solution which has been used to purify petroleum oils after they have been treated with sulfuric acid.

Sodatol. It is a mixture of sodium nitrate and trinitrotoluene; an agricultural explosive.

soda water glass. A mixture of sodium silicates.

Sodiformasal. See Formasal. §

Sodital. CAS 57-33-0; Pentobarbital sodium; hypnotic; sedative. [Baxter Health Care] *

sodium. (natrium) Metallic element; Na; CAS 7440-23-5; EINECS 231-132-9; polymerization catalyst for synthetic rubber, laboratory reagent, coolant in nuclear reactors, heat transfer agent, manufacture of tetraethyl and tetramethyl lead, sodium peroxide, sodium hydride; radioactive forms in tracer studies and medicine. [Associated Octel Co Ltd; Foseco (FS) Ltd]

sodium acetate. (sodium acetate anhydrous; acetic acid sodium salt anhydrous) CH_3COONa; CAS 127-09-3; EINECS 204-823-8; Dye and color intermediate, pharmaceuticals, cinnamic acid, soaps, photography, purification of glucose, meat preservation, medicine, electroplating, tanning, dehydrating agent, buffer, laboratory reagent, food additive. [Aldrich; EM Industries; General Chem.; Heico; Honeywill & Stein Ltd; Lonza; Niacet; Verdugt BV]

sodium acetate anhydrous. See sodium acetate.

sodium acid sulfite. See sodium bisulfite.

Sodium Aerofloat Promoter. Sodium diethyl dithiophosphate; used in the mining industry. [Cyanamid BV]

sodium alum. (aluminum-sodium sulfate; SAS; alum) $Al_2(SO_4)_3 \cdot Na_2SO_4 \cdot 24H_2O$; CAS 10102-71-3; Used in textiles as a mordant and for waterproofing, dry colors, ceramics, tanning, paper size precipitant, sugar refining, water purification.

sodium aluminate. $Na_2Al_2O_3$ or $NaAlO_2$; CAS 1302-42-7; Mordant, zeolites, water purification, sizing paper, manu-

facture of milk glass, soap and cleaning compounds. [Alcoa; Asada Chem Industry Co Ltd; Hüls AG; Laporte Absorbents; Nalco]

sodium aluminum chlorhydroxy lactate. Sodium salt of a complex of lactic acid and aluminum chlorohydrate; CAS 8038-93-5. [Reheis]

sodium aluminum phosphate. (sodium aluminum phosphate acidic) $NaAl_3H_{14}(PO_4)_8 \cdot 4H_2O$ or $Na_3Al_2H_{15}$-$(PO_4)_n$; food additive for baked products. [Monsanto; Rhone-Poulenc Basic; Whiting, Peter Ltd]

Sodium Amytal. Sodium isoamylethyl barbiturate; a proprietary preparation of araylobarbitone sodium; a hypnotic. [Eli Lilly & Co] †

sodium ascorbate. (L(+)-ascorbic acid sodium salt; vitamin C sodium salt) $C_6H_7NaO_6$; CAS 134-03-2; EINECS 205-126-1; Antioxidant in food products. [BASF; EM Industries; Hoffmann-La Roche; Pfizer Spec. Chem.; Spice King; Takeda USA]

sodium benzoate. (benzoic acid, sodium salt) sodium salt of benzoic acid; C_6H_5COONa; CAS 532-32-1; EINECS 208-534-8; fungicide; preservative in pharmaceuticals and foods, especially in slightly acidic media; clinical reagent (bilirubin assay). [Aceto; Dinoval Chem. Ltd; DSM BV; Haarmann & Reimer; Mallinckrodt]

sodium bicarbonate. (baking soda; sodium hydrogen carbonate; bicarbonate of soda; carbonic acid, monosodium salt) Inorganic salt; $NaHCO_3$; CAS 144-55-8; EINECS 205-633-8; manufacture of effervescent salts and beverages, artificial mineral water; prevention of timber mold, cleaning preparations, lab reagent, antacid, mouthwash. [EM Industries; Farleyway Chem. Ltd; FMC; ICI Spec.; Rhone-Poulenc Basic]

sodium bichromate. See sodium dichromate.

sodium biphosphate. See sodium phosphate.

sodium bisulfide. (sodium hydrosulfide; sodium sulfhydrate; sodium hydrogen sulfide) $NaSH \cdot 2H_2O$; CAS 16721-80-5; paper pulping, dyestuffs processing, rayon and cellophane desulfurizing, dehairing hides, bleaching reagent. [Nissan Chem. Ind.]

sodium bisulfite. (sodium acid sulfite; sulfurous acid, monosodium salt; sodium hydrogen sulfite) Inorganic salt; $NaHSO_3$; CAS 7631-90-5; EINECS 231-548-0; fusion of minerals to make solutions for analysis; pickling metals; carbonizing wool. [Aldrich; BASF; Du Pont; Hoechst Celanese; Hüls AG; Penreco]

sodium borate. (sodium tetraborate; sodium pyroborate; borax) $Na_2B_4O_7 \cdot 10H_2O$; CAS 1303-96-4, 1330-43-4; Heat-resistant glass, procelain enamel, detergents, herbicides, fertilizers, rust inhibitors, pharmaceuticals, leather, photography, bleaches, paint, boron compounds, flame-retardant fungicide for wood, soldering flux, cleaners, lab reagent. [Spectrum Chem. Mfg.; U.S. Borax & Chem.]

sodium borohydride. (sodium tetrahydroborate; sodium tetrahydridoborate) $NaBH_4$; CAS 16940-66-2; EINECS 241-004-4; Source of H_2 and other borohydrides; bleaching wood pulp; blowing agent for plastics; decolorizer for plasticizers. [Atomergic Chemetals; Morton Int'l.; Tennant Trading Ltd]

sodium bromate. (bromic acid, sodium salt) Inorganic salt; $NaBrO_3$; CAS 7789-38-0; EINECS 232-160-4; oxidizer, photography, medicine (sedative), preparation of bromides; with sodium bromide for dissolving gold from its ores. [Tri-K Industries]

---

* Trade name not verified as available   † Trade name verified as obsolete

sodium bromide. NaBr; CAS 7647-15-6; EINECS 231-599-9; Photography, medicine (sedative), preparation of bromides. [Ethyl; Great Lakes; Hawks Chem. Co Ltd; Morton Int'l.]

sodium butyrate. (butyric acid sodium salt) $CH_3CH_2CH_2COONa$; CAS 156-54-7; EINECS 205-857-6; Used in chemical laboratory as agent for alteration of chromatin structure; enhances transfection efficiency and improved cell survival. [Penta Mfg.]

sodium cacodylate. (sodium dimethyl arsenate; cacodylic acid sodium slat trihydrate; dimethylarsinic acid sodium salt) $(CH_3)_2AsO_2Na \cdot 3H_2O$; CAS 124-65-2; EINECS 204-708-2; Herbicide. [OxyChem; U.S. Biochemical]

sodium caprylate. (caprylic acid sodium salt) $CH_3(CH_2)_6$-COONa; CAS 1984-06-1; EINECS 217-850-5. [Aldrich; Hart Prod. Corp.; Penta Mfg.]

sodium carbonate. (soda ash; carbonic acid, disodium salt) Inorganic salt; Na_2CO_3; CAS 497-19-8; EINECS 207-838-8; detergent and cleaning products; pH control of water; intermediate in thermochemical reactions; food additive; manufacture of glass; analytical reagent. [Albright & Wilson; EM Industries; Rhone-Poulenc Basic; Solvay SA]

sodium carbonate anhydrous. See soda ash.

sodium carboxymethylcellulose. See carboxymethylcellulose sodium.

sodium caseinate. (casein, sodium salt) CAS 9005-46-3; 9004-36-3; Food additive; binder, and extender in sausage, soups, etc.; emulsifier, stabilizer. [Am. Casein; Spectrum Chem. Mfg.; Spice King]

sodium chlorate. (chlorate of soda) $NaClO_3$; CAS 7775-09-9; EINECS 231-887-4; Oxidizing agent, pulp bleaching, defoliant and herbicide; leather tanning and finishing; textile mordant. [Albright & Wilson; Atochem; Eka Nobel AB; Georgia Gulf; Kerr-McGee; OxyChem; PPG Industries]

sodium chloride. (rock salt; salt; common salt) Inorganic salt; occurs in nature as the mineral halite; NaCl; CAS 7647-14-5; EINECS 231-598-3; source of chlorine and sodium; preservative, seasoning for foods; manufacture of soaps, dyes; in freezing mixtures; dyeing and printing fabrics, glazing pottery, curing hides; metallurgy; herbicide; electrolyte replenisher, emetic, topical anti-inflammatory. [Akzo Salt; EM Industries; Heico; Morton Salt; Stan Chem Int'l. Ltd]

sodium chlorite. (chlorous acid, sodium salt) $NaClO_3$; CAS 7758-19-2; EINECS 231-836-6; Improves taste and odor of potable water; oxidizing agent, bleaching agent for textiles, paper pulp, disinfecting. [Albright & Wilson; Atochem N. Am.; Int'l. Dioxcide; Olin]

sodium chromate. (sodium chromate (VI) (nonradioactive)) $NaCrO_4 \cdot 10H_2O$; CAS 7775-11-3; EINECS 231-889-5; Inks, dyeing, paint pigment, leather tanning, other chromates, protection of iron against corrosion, wood preservative. [Atochem N. Am.; OxyChem]

sodium chromate (VI) (nonradioactive). See sodium chromate.

sodium citrate. (sodium citrate dihydrate; trisodium citrate; citric acid trisodium salt dihydrate) HOC(COONa)-$(CH_2COONa)_2 \cdot 2H_2O$; CAS 68-04-2; Soft drinks, photography, frozen desserts, meat products, detergents, cheeses, electroplating, sequestrant, buffer, nutrient for cultured buttermilk, medicine (diuretic, expectorant), anticoagulant for blood. [Cargill; Haarmann & Reimer;

Hoffmann-La Roche; Mallinckrodt; Tanabe USA; Whiting, Peter Ltd]

sodium citrate dihydrate. See sodium citrate.

sodium citrotartrate. A mixture of sodium citrate and tartrate.

sodium CMC. See carboxymethylcellulose sodium.

sodium cumenesulfonate. (1-methylethyl)benzene, monosulfo derivative, sodium salt) Substituted aromatic compound; $C_9H_{11}O_3S \cdot Na$; CAS 32073-22-6; EINECS 250-913-5; 248-938-7; hydrotrope, solubilizer, coupling agent, solution aid for liquid detergents and slurries. [Hüls AG; Manro Prods. Ltd; Ruetgers-Nease]

sodium cyanide. (cyanogran) NaCN; CAS 143-33-9; EINECS 205-599-4; Extraction of gold and silver from ores, electroplating, heat treatment of metals, making hydrogen cyanide, insecticide, metal cleaning, fumigation, dyes and pigments, nylon intermediates, chelating compounds, ore flotation. [Atochem N. Am.; Degussa; DSM UK; Du Pont; FMC; W.R. Grace Ltd; ICI Spec.; Mitsui Toatsu Chem.]

sodium dichloroisocyanurate. (sodium dichloro-s-triazine-2,4,6-trione; dichloroisocyanuric acid sodium salt) NaNC(O)NCl(O)NClCO; CAS 2244-21-5; Active ingredient in dry bleaches, water and sewage treatment, etc. [Biachem Ltd; ICI Spec.; Monsanto; Nissan Chem. Ind.; 3-V]

sodium dichloro-s-triazine-2,4,6-trione. See sodium dichloroisocyanurate.

sodium dichromate. (sodium bichromate) $Na_2Cr_2O_7 \cdot 2H_2O$; CAS 7789-12-0; EINECS 234-190-3; Colorimetry (copper determination), complexing agent, oxidation inhibitor in ethyl ether. [British Chrome & Chems.; Chemisphere Ltd; OxyChem; Rit-Chem]

sodium di(2-ethylhexyl) sulfosuccinate. See dioctyl sodium sulfosuccinate.

sodium dimethyl arsenate. See sodium cacodylate.

sodium dimethyldithiocarbamate. (SDDC; sodium N,N-dimethyl dithiocarbamate; dimethyldithiocarbamic acid, sodium salt; methyl namate) $(CH_3)_2NCS_2Na$; CAS 148-18-5; CAS 128-04-1; EINECS 204-876-7; Pesticide, fungicide, corrosion inhibitor, rubber accelerator. [Novachem; Uniroyal; R.T. Vanderbilt]

sodium dioctyl sulfosuccinate. See dioctyl sodium sulfosuccinate.

Sodium Diuril. CAS 7085-44-1; Chlorothiazide sodium; diuretic; antihypertensive. [Merck & Co Inc] *

sodium dodecylbenzenesulfonate. (sodium lauryl benzene sulfonate; dodecylbenzenesulfonic acid, sodium salt; dodecylbenzene sodium sulfonate) Substituted aromatic compound; $C_{18}H_{30}O_3S\cdot Na$; CAS 25155-30-0; 68081-81-2; 85117-50-6; EINECS 246-680-4; anionic detergent; cosmetic applications. [Du Pont; Emkay; Norman, Fox; Pilot; Stepan; Unger Fabrikker AS; Witco]

sodium dodecyl sulfate. See sodium lauryl sulfate.

sodium erythorbate. (D-erythro-hex-2-enonic acid, γ-lactone, monosodium salt; sodium isoascorbate) Sodium salt of erythorbic acid; $C_6H_8O_4 \cdot Na$; CAS 6381-77-7; antioxidant, preservative. [PMP Fermentation Prods.; Schweizerhall]

sodium ferrocyanide. (sodium ferrocyanide decahydrate; sodium hexacyanoferrate; yellow prussiate of soda) $Na_4Fe(CN)_6 \cdot 10H_2O$; CAS 13601-19-9; EINECS 237-081-9; Manufacture of sodium ferricyanide, blue pigments, blueprint paper, anticaking agent for salt, ore

---

‡ Trade name and manufacturer not verified   § Trade name without identified manufacturer

flotation, pickling metals, polymerization catalyst, photographic fixative. [Atomergic Chemetals; Degussa Ltd; Rit-Chem]

sodium ferrocyanide decahydrate. *See* sodium ferrocyanide.

sodium fluoride. NaF; CAS 7681-49-4; EINECS 231-667-8; fluoridation of municipal water, degassing steel, wood preservative, insecticide, fungicide, rodenticide, chemical cleaning, electroplating, glass manufacture, vitreous enamels, preservative for adhesives, toothpastes, disinfectant, dental prophylaxis. [Cerac; EM Industries; General Chem.; Hoechst Celanese; Solvay GmbH; Whiting, Peter Ltd]

sodium fluorosilicate. *See* sodium silicofluoride.

sodium fluosilicate. *See* sodium silicofluoride.

sodium formate. (formic acid sodium salt) HCOONa; CAS 141-53-7; EINECS 205-488-0; Reducing agent, manufacture of formic acid, oxalic acid and sodium dithionite, organic chemicals, mordant, complexing agent, analytical reagent (noble metal precipitant), buffering agent. [Aqualon; Heico; Hoechst Celanese; Perstorp Polyols; Spectrum Chem. Mfg.]

sodium glucoheptonate. $HOCH_2(CHOH)_5COONa$; Metal cleaning, bottle washing, kier boiling, caustic boil-off, paint stripping. [Belzak; Hickson Danchem; Pfanstiehl Labs]

sodium gluconate. (D-gluconic acid, monosodium salt; gluconic acid sodium salt) Sodium salt of gluconic acid; $HOCH_2CHOHCHOHCOHHCHOHCOONa$; CAS 527-07-1; EINECS 208-407-7; As sequestering agent; in metal plating, tanning of hides, mordants for fabrics, paints; rust remover; mineral source in foods and pharmaceuticals. [Akzo; Albright & Wilson; Pfizer Spec. Chem.; PMP Fermentation Prods.; Rit-Chem.]

sodium glutamate. *See* MSG.

sodium hexacyanoferrate. *See* sodium ferrocyanide.

sodium hexafluorosilicate. *See* sodium silicofluoride.

sodium hexametaphosphate. (SHMP; metaphosphoric acid, hexasodium salt; Grahams salt; sodium phosphate glass) Inorganic salt; $(NaPO_3)_6$; CAS 10124-56-8; EINECS 233-343-1; food products, water treatment, cleaning formulations. [Albright & Wilson; Calgon; Farleyway Chem. Ltd; FMC; Monsanto; Rhone-Poulenc Basic]

sodium hyaluronate. *See* hyaluronic acid.

sodium hydrate. *See* sodium hydroxide.

sodium hydrogen carbonate. *See* sodium bicarbonate.

sodium hydrogen sulfide. *See* sodium bisulfide.

sodium hydrogen sulfite. *See* sodium bisulfite.

sodium hydrosulfide. *See* Sodium bisulfide.

sodium hydrosulfite. (disodium dithionite; sodium hyposulfite) Inorganic salt; NaO_2SSO_2Na; CAS 7775-14-6; 16721-80-5; EINECS 231-890-0; chemical reagent for the reduction of aldehydes and ketones to alcohols; vat dying of fibers and textiles; bleaching sugar, soap, oils; oxygen scavenger for synthetic rubbers. [Farleyway Chem. Ltd; Henkel/Organic Prods.; Hoechst Celanese; Mitsubishi Gas; Morton Int'l.; Nissan Chem. Ind.; Olin]

sodium hydroxide. (caustic soda; sodium hydrate; white caustic) Inorganic base; NaOH; CAS 1310-73-2; EINECS 215-185-5; chemical manufacture, rayon, cellophane, neutralizing agent in petrol. refining; textile processing; pulp, paper, soaps; vegetable oil refining; reclaiming rubber; food additive; etching and electroplat-

ing. [Akzo; Asahi Chem Industry Co Ltd; Asahi Denka Kogyo; Atochem N. Am.; Georgia Gulf; Georgia-Pacific; ICI Spec.; Nissan Chem. Ind.; Norsk Hydro A/S; Olin; OxyChem; PPG Industries; Rasa Ind.]

sodium hypochlorite. NaOCl · 5HOH; CAS 7681-52-9; Bleaching paper, pulp and textiles; intermediate; organic chemicals; water purification; medicine; fungicide; swimming pool disinfectant; laundering; reagent; germicide. [George Mann; Norsk Hydro A/S; Olin; OxyChem; Showa Denko]

sodium hyposulfite. *See* sodium hydrosulfite.

sodium inosinate. *See* disodium inosinate.

sodium iodate. $NaIO_3$; CAS 7681-55-2; EINECS 231-672-5; Antispetic, disinfectant, feed additive reagent. [Atomergic Chemetals; Mallinckrodt]

sodium iodide. NaI; CAS 7681-82-5; EINECS 231-679-3; Photography, solvent for iodine, organic chemicals, reagent, feed additive, cloud seeding, scintillation, expectorant. [Aldrich; EM Industries; Mallinckrodt]

sodium isoascorbate. *See* sodium erythorbate.

sodium lactate. (2-hydroxypropanoic acid, monosodium salt; lacolin) Sodium salt of lactic acid; $CH_3CHOH-COONa$; CAS 72-17-3; EINECS 200-772-0; hygroscopic agent; glycerol substitute; plasticizer for casein; corrosion inhibitor in alcoholic antifreeze; electrolyte replenisher. [Am. Biorganics; EM Industries; R.W. Greeff; Patco; RITA; Verdugt BV]

sodium laureth sulfate. (sodium lauryl ether sulfate (n=1-4); PEG (1-4) lauryl ether sulfate, sodium salt) Sodium salt of sulfated ethoxylated lauryl alcohol; $CH_3-(CH_2)_{10}CH_2(OCH_2CH_2)_nOSO_3Na$, avg. n = 1-4; CAS 1335-72-4; 3088-31-1; 9004-82-4 (generic); 13150-00-0; 15826-16-1; 68891-38-3; EINECS 221-416-0; foam stabilizer, detergent, flash foamer, wetter for detergent systems, personal care products; emulsion polymerization; shampoo base. [Chemron; Lonza; Norman, Fox; Pilot; Sandoz; Stepan; Unger Fabrikker AS; Vista]

sodium lauroyl sarcosinate. (N-methyl-N-(1-oxododecyl)glycine, sodium salt; N-lauroylsarcosine sodium salt; sodium-N-dodecanoyl-N-methylglycinate) Sodium salt of lauroyl sarcosine; $CH_3(CH_2)_{10}CONCH_3-CH_2COONa$; CAS 137-16-6; EINECS 205-281-5; foaming agent, wetting agent, detergent, lubricant, antistat, corrosion inhibitor, bacteriostat, penetrant used in dental, pharmaceutical, shampoos, depilatories, and shaving preparations, food packaging, household and industrial uses. [W.R. Grace/Hampshire; R.T. Vanderbilt]

sodium lauryl benzene sulfonate. *See* sodium dodecylbenzenesulfonate.

sodium lauryl ether sulfate. *See* sodium laureth sulfate.

sodium lauryl sulfate. (SDS; sulfuric acid, monododecyl ester, sodium salt; sodium dodecyl sulfate) Sodium salt of lauryl sulfate; $CH_3(CH_2)_{10}CH_2OSO_3Na$; CAS 151-21-3; 68585-47-7; 68955-19-1; EINECS 205-788-1; anionic detergent; surface tension depressant; emulsifier for fats; wetting agent; in textile industry, in toothpastes. [Albright & Wilson Ltd; Chemron; Dajac Labs; Du Pont; Lonza; Norman, Fox; Pilot; Sandoz; Stepan; Unger Fabrikker AS; Witco]

sodium metabisulfite. (disulfurous acid, disodium salt; sodium pyrosulfite; sodium bisulfite) Inorganic salt; $Na_2S_2O_5$; CAS 7681-57-4; EINECS 231-673-0; in foods, as preservative, laboratory reagent. [BASF; Blythe, William Ltd; EM Industries; General Chem.; Mallinckrodt]

---

* Trade name not verified as available   † Trade name verified as obsolete

sodium metaborate. NaBO$_2$; CAS 7775-19-1; Herbicide. [Ashland; U.S. Borax & Chem.]

sodium metaperiodate. See sodium m-periodate.

sodium metaphosphate. (NaPO$_3$)$_n$, n= 3-10 (cyclic) or larger (polymers); CAS 10361-03-2; dental polishing agents, detergent builders, water softening, sequestrants, emulsifiers, food additives, textile processing, laundering. [Atomergic Chemetals]

sodium metasilicate. (silicic acid, disodium salt; sodium metasilicate, anhydrous) Inorganic salt; Na$_2$SiO$_3$; CAS 6834-92-0; EINECS 229-912-9; cosmetics, laundry, dairy, and metal cleaning as soap builder and detergent, bleaching agent; as flocculant and dispersant in metallurgy and mining. [Eka Nobel AB; PQ; Rhone-Poulenc Basic]

sodium metasilicate, anhydrous. See sodium metasilicate

sodium metasilicate pentahydrate. Na$_2$OSiO$_2$ · 5H$_2$O; CAS 102-13-79-3; EINECS 229-912-9; Ingredient in detergents for food processing equipment and general cleaning in dairies, bakeries, packing houses, in laundry, textile, paper, oil, metal industries. [Oxychem; PQ]

sodium molybdate. (sodium molybdate(VI); sodium molybdate dihydrate) Na$_2$MoO$_4$·2H$_2$O; CAS 7631-95-0; 10102-40-6; EINECS 231-551-7; Reagent in analytical chemistry, paint pigment, corrosion inhibitor, catalyst in dye and pigment production, additive for fertilizers and feeds, micronutrient. [AAA Molybdenum Prods.; Cerac; Climax Molybdenum; Mallincrkodt; PMC Specialties]

sodium molybdate(VI). See sodium molybdate.

sodium molybdate dihydrate. See sodium molybdate.

sodium monofluorophosphate. (SMFP) Na$_2$PO$_3$F; CAS 7631-97-2; manufacture of toothpaste. [Albright & Wilson; Atochem N. Am.]

sodium morrhuate. The sodium salt of the fatty acids of cod-liver oil.

sodium muriate. See muriate of soda.

sodium nitrate. (soda niter; cubic niter; Chile saltpeter) NaNO$_3$; CAS 7631-99-4; EINECS 231-554-3; Oxidizing agent; solid rocket propellants; fertilizer, flux, glass manufacture, pyrotechnics, dynamites; color fixative and preservative in cured meats, fish, enamel for pottery; modifying burning properties of tobacco. [BASF; Faesy & Besthoff; Farleyway Chem. Ltd; Mallinckrodt; Nissan Chem. Ind.; Spice King]

sodium nitrite. NaNO$_2$; CAS 7632-00-0; EINECS 231-555-9; Diazotization, rubber accelerators, color fixative and preservative in cured meats, meat products, fish; pharmaceuticals, photographic and analytical reagent, dye manufacture, antidote for cyanide poisoning. [BASF; Du Pont; EM Industries; Farleyway Chem. Ltd; General Chem.; ICI Spec.; PMC Spec.]

sodium octadecanoate. See sodium stearate.

sodium 9-octadecenoate. See sodium oleate.

sodium oleate. (sodium 9-octadecenoate; 9-octadecenoic acid, sodium salt) Sodium salt of oleic acid; CH$_3$(CH$_2$)$_7$CH=CH(CH$_2$)$_7$COONa; CAS 143-19-1; EINECS 205-591-0; ore flotation, waterproofing textiles, emulsifier of oil/water systems. [Hart Prod. Corp.; Norman, Fox; Schweizerhall; Witco]

Sodium Omadine®, 40% Aq. Sol'n. CAS 3811-73-2; Sodium 2-pyridinethione in water; industrial microbiostat; chelating agent; used in aq. metal coolant and cutting fluids, latex emulsion, inks, fiber lubricants. [Olin]

Sodium para-aminohippurate. For intravenous use to measure effective renal plasma flow and tubular secretory capacity. [Merck & Co Inc]

sodium perborate. (sodium perborate anhydrous) NaBO$_3$; CAS 7632-04-4; Denture cleaner, oxygen source. [Degussa; Du Pont; Eka Nobel AB; ICI Spec.; Mitsubishi Gas]

sodium perborate anhydrous. See sodium perborate.

sodium percarbonate. (hydrogen peroxide sodium carbonate adduct) Na$_2$CO$_3$ · 1.5H$_2$O$_2$; CAS 15630-89-4; EINECS 239-707-6; Bleaching agent for domestic and industrial use, denture cleaner, mild antiseptic. [Chemoxal SA; Degussa; Interox Am.]

sodium m-periodate. (sodium metaperiodate) NaIO$_4$; CAS 7790-28-5; EINECS 232-197-6; Source of periodic acid, analytical reagent, oxidizing agent. [Atomergic Chemetals; Noah Chem.]

sodium permutite. A sodium zeolite, made artificially.

sodium peroxydisulfate. See sodium persulfate.

sodium persulfate. (sodium peroxydisulfate) Na$_2$S$_2$O$_8$; CAS 7775-27-1; EINECS 231-892-1; Bleaching agent (fats, oils, fabrics, soaps), battery depolarizers, emulsion polymerization. [Aldrich; Chemoxal SA; Degussa; FMC]

sodium phosphate. (MSP; sodium phosphate, monobasic; monosodium dihydrogen phosphate; sodium bi-phosphate) NaH$_2$PO$_4$; CAS 7558-80-7; EINECS 231-449-2; Controls pH in mildly acidic solutions; food products, water treatment, metal treatment. [Albright & Wilson; Aldrich; BritAg Industries Ltd]

sodium phosphate, dibasic. See disodium phosphate.

sodium phosphate, dibasic dihydrate. See disodium phosphate, dihydrate.

sodium phosphate glass. See sodium hexametaphosphate.

sodium phosphate, monobasic. See sodium phosphate.

sodium phosphate, tribasic. See trisodium phosphate, anhydrous.

sodium phosphate, tribasic dodecahydrate. See trisodium phosphate, dodecahydrate.

sodium polyacrylate. (polyacrylic acid, sodium salt) CAS 9003-04-7; Thickener, stabilizer, protective colloid for natural and synthetic latexes for paints, films, coatings, and adhesives; dispersant; antiredeposition agents; antiscalant. [Alco; Allchem Industries; Arakawa; Rhone-Poulenc; 3-V]

sodium propionate. (propanoic acid, sodium salt) Sodium salt of propionic acid; CH$_3$CH$_2$COONa; CAS 137-40-6; EINECS 205-290-4; food additive (preservative); fungicide. [Gist-Brocades Food Ingreds.; Niacet; Spectrum Chem. Mfg.; Verdugt BV]

sodium pyroborate. See sodium borate.

sodium pyrophosphate. (TSPP; tetrasodium pyrophosphate; sodium pyrophosphate, normal) Na$_4$P$_2$O$_7$; CAS 7722-88-5; EINECS 231-767-1; Water softener, synthetic detergent builder, dispersant, emulsifier, metal cleaner, boiler water treatment, viscosifier for drilling muds, deinking newsprint, synthetic rubber, textile dyeing, wool scouring, buffer, sequestrant, nutrient, food additive. [Farleyway Chem. Ltd; FMC; Mitsui Toatsu Chem.; Monsanto; Spectrum Chem. Mfg.]

sodium pyrophosphate, normal. See sodium pyrophosphate.

sodium pyrosulfite. See sodium metabisulfite.

sodium selenite. (sodium selenite pentahydrate) Na$_2$SeO$_3$ · 5H$_2$O; CAS 26970-82-1; EINECS 233-267-9; glass

---

‡ Trade name and manufacturer not verified   § Trade name without identified manufacturer

manufacture (color control), reagent in bacteriology, testing germination of seeds, decorating porcelain. [Atomergic Chemetals; Degussa; Noah Chem.]

sodium selenite pentahydrate. See sodium selenite.

sodium sesquicarbonate. (carbonic acid, sodium salt (2:3)) Inorganic salt; $Na_2CO_3 \cdot NaHCO_3 \cdot 2HOH$; CAS 533-96-0; detergent and soap builder; mild alkaline agent for general cleaning and water softening; bath crystal; alkaline agent in leather tanning; food additive. [FMC]

sodium silicate. (silicic acid, sodium salt; water glass; soluble glass) Sodium salt of silicic acid; Na_2SiO_3 or $Na_6Si_2O_7$ or $Na_2Si_3O_7$; CAS 1344-09-8; EINECS 215-687-4; lining Bessemer converters, acid concentrators; manufacture of grindstones, abrasive wheels; as solution: preserving eggs; fireproofing fabrics; detergent in soaps; as adhesive; waterproofing walls; in cements, and paints. [Aichi Silicate Chem Co Ltd; Asahi Denka Kogyo; Crosfield; OxyChem; PQ; Spectrum Chem. Mfg.]

sodium silicofluoride. (sodium fluorosilicate; sodium hexafluorosilicate; sodium fluosilicate) Na_2SiF_6; Fluoridation, laundry soaps, opalescent glass, vitreous enamel frits, metallurgy (aluminum and beryllium), insecticides, rodenticides, chemical intermediate, glue, leather and wood preservative, moth repellent, manufacture of pure silicon. [Faesy & Besthoff; La Roche Chem.; Mitsui Toatsu Chem.; Whiting, Peter Ltd]

sodium stannate. $Na_2SnO_3 \cdot 3H_2O$ or $Na_2Sn(OH)_6$; CAS 12058-66-1; Dyeing mordant, ceramics, glass, source of tin for electroplating, textile fireproofing, stabilizer for hydrogen peroxide, blueprint paper, laboratory reagent. [Atochem N. Am.; Blythe, William Ltd; M&T Harshaw; Spectrum Chem. Mfg.]

sodium stearate. (sodium octadecanoate; octadecanoic acid, sodium salt; stearic acid, sodium salt) CH_3-$(CH_2)_{16}COONa$; CAS 822-16-2; Waterproofing and gelling agent; toothpaste, cosmetics; stabilizer in plastics; emulsifier and stiffener in pharmaceuticals; in glycerol suppositories. [Atochem N. Am.; Magnesia GmbH; Norman, Fox; Original Bradford Soap Works; Witco]

sodium sulfate. (sodium sulfate, anhydrous) Na_2SO_4; CAS 7757-82-6; EINECS 231-820-9; Manufacture of Kraft paper, etc.; filler in synthetic detergents; processing textile fibers, dyes, tanning, pharmaceuticals, lab reagent, food additive. [Akzo Salt; Atochem N. Am.; Kemira Kemi AB; Lenzing AG; Occidental]

sodium sulfate, anhydrous. See sodium sulfate.

sodium sulfhydrate. See sodium bisulfide.

sodium sulfide. Na_2S or $Na_2S \cdot 9H_2O$; CAS 16721-80-5, 1313-82-2 (anhyd.); EINECS 215-211-5; Organic chemicals, sulfur dyes, intermediates, viscose rayon, leather depilatory, paper pulp, hydrometallurgy of gold ores, sulfiding oxidized lead and copper ores, sheep dips, photographic reagent, engraving and lithography, analytical reagent. [Aldrich; Cerac; Farleyway Chem. Ltd; PPG Industries]

sodium sulfite. (sulfurous acid, sodium salt (1:2); sulfurous acid, disodium salt) Inorganic salt; Na_2SO_3; CAS 7757-83-7; EINECS 231-821-4; paper industry; reducing agent (dyes); food preservative and antioxidant; textile bleaching; in photographic developers. [BASF; Blythe, William Ltd; EM Industries; Ferro/Grant; Indspec; Nissan Chem. Ind.; Rhone-Poulenc Basic]

sodium tartrate. (sal tartar; disodium tartrate) $Na_2C_2H_4O_6$

$\cdot 2H_2O$; CAS 868-18-8; EINECS 212-773-3; Reagent, food additive, sequestrant, stabilizer. [Novarina Srl; Schweizerhall]

sodium tetraborate. See sodium borate.

sodium tetrahydridoborate. See sodium borohydride.

sodium tetrahydroborate. See sodium borohydride.

sodium thiosulfate. (sodium thiosulfate pentahydrate; sodium subsulfite) $Na_2S_2O_3 \cdot 5H_2O$; CAS 7772-98-7; EINECS 231-867-5; Photographic fixative, chrome tanning, chlorine removal in bleaching and papermaking, extraction of silver, dechlorination of water, mordant, reagent, bleaching, reducing agent in chrome dyeing, sequestrant in salt, antidote for cyanide poisoning. [Aldrich; Blythe, William Ltd; Ferro/Grant; General Chem.; Nissan Chem. Ind.]

sodium toluenesulfonate. (methylbenzenesulfonic acid, sodium salt) Substituted aromatic compound; $CH_3CH \cdot 6H_4SO_3Na$; CAS 657-84-1; 12068-03-0; EINECS 235-088-1; hydrotropic solvent. [Hüls AG; Ruetgers-Nease]

sodium triphosphate, tripoly. See sodium tripolyphosphate.

sodium tripolyphosphate. (STPP; sodium triphosphate, tripoly; pentasodium triphosphate) $Na_5P_3O_{10}$; CAS 7758-29-4; 13573-18-7; EINECS 231-694-5; Water softener, sequestrant, deflocculating agent, food additive, texturizer. [Albright & Wilson; FMC; Kemira Kemi AB; Mitsui Toatsu Chem.; Monsanto; Rhone-Poulenc Basic]

sodium tungstate. (sodium tungstate dihydrate; sodium wolframate) $NaWO_4 \cdot 2H_2O$; CAS 10213-10-2, 13472-45-2; EINECS 236-743-4; Intermediate for tungsten compounds (e.g., phosphotungstate), reagent, fireproofing textiles, alkaloid precipitant. [Cerac; Noah Chem.; Spectrum Chem. Mfg.]

sodium tungstate dihydrate. See sodium tungstate.

sodium undecylenate. (10-undecenoic acid, sodium salt) Sodium salt of undecylenic acid; $CH_2=CH(CH_2)_8$-$COONa$; CAS 3398-33-2; bacteriostat and fungistat in cosmetics and pharmaceuticals. [Atochem N. Am.]

Sodium Versenate. CAS 139-33-3; Edetate disodium; chelating agent; pharmaceutic aid. [3M Pharmaceuticals] †

sodium wolframate. See sodium tungstate.

sodium xylenesulfonate. (dimethylbenzene sulfonic acid, sodium salt) Sodium salt of ring sulfonated mixed xylene isomers; $(CH_3)_2C_6H_3SO_3Na$; CAS 1300-72-7; EINECS 215-090-9; hydrotropic solvent, used in detergents. [Mitsubishi Gas; Pilot; Ruetgers-Nease; Stepan; Witco]

Sodos. A mixture of sodium dihydrogen phosphate and sodium bicarbonate; used in medicine.

Sodusec. Caustic soda. [ICI Chem. & Polymers Ltd]

Sofanate. A fungicide for fruit storage. [Plant Protection] *

Sofibex. CAS 7681-53-0; Sodium hypophosphite; used for surface treatment, electroless nickel plating. [Atochem UK/Ceca]

Soflens. Polymacon; contact lens material. [Bausch & Lomb, Professional Products Div] *

Sofnolite. See Sofnol Sodalime G. §

Sofnol Soda-lime G. A proprietary form of soda-lime containing a little manganic acid; it is stated to absorb much more carbon dioxide than ordinary soda-lime, and to change color as the degree of saturation is approached.§

Sofnon 105G. Imidazoline deriv.; softener for cotton, wool, and synthetic fibers. [Toho Chem. Industry]

soft amber. See Gedanite.

---

* Trade name not verified as available   † Trade name verified as obsolete

Soft-Clad®. Polyester polyol/urethane color bead particles; for use in the manufacture of synthetic leather-like coatings. [Reichhold]

Softcon. Vifilcon; contact lens material (hydrophilic). [Parke-Davis] *

soft copal. A name applied to varieties of Australian sandarac resin.

Softenol®. Fatty acid esters; for technical applications. [Dynamit Nobel Wien GmbH] *

Softenol®. Antistatics and lubricants. [Hüls UK Ltd]

Softenol® 3100. CAS 67701-26-2; C12-18 fatty acid triglyceride; lubricant for machinery used in manufacturing of foodstuffs; additive to textile and glass fiber finishing, sizing agent. [Hüls Am.]

Softenol® 3114. C14 fatty acid triglyceride; compressing aid in tableting technical substances; hydrophobing agent against moisture; lubricant for machinery, lacquer and varnish formulations; additive to liquid lubricating systems. [Hüls Am.]

Softenol® 3408. C8-10 fatty acid-1,2-propanediol ester; additive in manufacturing of textile and glass fibers; lubricant for cutting devices; mold release agent in processing of plastics. [Hüls Am.]

Softenol® 3829. C8-10 fatty acids triglyceride, modified; oil in manufacturing of aluminum foils; lubricant for machinery, medical/technical equipment, instruments; additive to cutting oils and lubricants. [Hüls Am.]

Softenol® 3991. Glyceryl stearate; lubricant, emulsifier, antistat, pigment dispersant, plasticizer, antiblocking agent for plastics, textile and leather processing aids, drilling and cutting oils, polishes, rubber; biodeg. [Hüls Am.]

Softex. Stearoyl lactylates; used as dough conditioners and antistaling additives in bread and baked goods. [Harcros Australia] *

Softex. Proprietary trade name for pure red oxide (qv). §

Softigen® 701. CAS 141-08-2; EINECS 205-455-0; Glyceryl ricinoleate; emollient; refatting and skin protecting agent; emulsifier; personal care products and pharmaceuticals. [Hüls Am.; Hüls AG]

Softigen® 767. CAS 52504-24-2; PEG-6 caprylic/capric glycerides; refatting and wetting agent, solubilizer, emollient used in cosmetics and pharmaceuticals, liquid soaps. [Hüls Am.; Hüls AG]

Softisan®. Ointment and cream bases; for creams, ointments, emulsions and lipsticks. [Hüls AG]

Softisan® 100. CAS 977056-87-3; Hydrogenated coco-glycerides; consistency regulator, emollient, ointment base for personal care products and pharmaceutical industry. [Hüls Am.]

Softisan® 378. Caprylic/capric/stearic triglyceride; ointment base, emollient, moisturizer, stabilizer; neutral ointment base with good skin compatibility and resorption characteristics, for the preparation of nonaq. ointments and creams. [Hüls Am.]

Softisan® 601. Glyceryl cocoate, hydrogenated coconut oil, and ceteareth-25; emollient; cosmetic and pharmaceutical self-emulsifying base for o/w products [Hüls Am.]

Softisan® 649. Bis diglyceryl caprylate/caprate/isostearate/stearate/hydroxystearate adipate; emollient, ointment base for creams, lipsticks, other decorative cosmetics; lanolin substitute. [Hüls Am.]

Softisan® Gel. Bis diglyceryl caprylate/caprate/isostearate/hydroxystearate adipate, propylene glycol dicaprylate/

dicaprate, stearalkonium hectorite, propylene carbonate; consistency regulator for creams; high temp. stabilizer for w/o emulsions and anhyd. skin care and decorative formulations. [Hüls Am.]

SoftMate DW. Hefilcon A; 2-hydroxyethyl methacrylate polymer with 1-vinyl-2-pyrrolidinone and ethylene dimethacrylate; contact lens material. [Barnes-Hind Inc]*

soft platinum. Commercially pure platinum, containing about 1% iridium.

Soft Resin P 65. Nonreactive polyester resin for the formulation of pigment pastes. [Bayer AG]

Softrite. A proprietary rubber softener; it has a zinc laurate base. §

Soft Touch 1052. Silicone; softener for prefinishing printed fabrics; excellent napping lubricant and shear stability. [Yorkshire Pat-Chem]

Softyne H. Amido-fatty quaternary; textile softening agent. [Hart Prods. Corp.]

soilime. A lime residue from cyanamide manufacture. It contains 50% of lime.

Soil Pests Killer. Granular insecticide. [Fisons plc, Horticulture Div] *

S-oils. Sulfur-containing oils obtained by the distillation of crude petroleum oil in the presence of sulfur. They have a strong antiseptic action against wood-destroying fungi.

Soiltex®. Nonhazardous soil conditioners in liquid and granular form for maintaining soil structure and preventing capping. [Allied Colloids Ltd]

Soil TRIGGRR. A liquid containing cytokinin, a plant growth regulator; used to increase crop yields and quality; applied to the soil; used for a wide variety of crops including corn, peanuts, sorghum, soybeans, fruits, and vegetables. [Westbridge Research Group] ‡

soja bean oil. See soya bean oil.

Sokalan® CP 2. Sodium salt of a maleic anhydride/methyl vinylether copolymer; dispersant, anti-incrustation agent. [BASF; BASF AG]

Sokalan® CP 5. Sodium salt of maleic acid/acrylic acid copolymer; dispersant, anti-incrustation agent. [BASF; BASF AG]

Sokalan® CP 7. CAS 9003-04-7; Sodium salt of a modified polyacrylic acid; auxiliary in phosphate-reduced or phosphate-free detergents to improve primary and secondary (antiredeposition and anti-incrustation) washing effects; dispersant. [BASF; BASF AG]

Sokalan® HP 22. CAS 109464-53-1; Acetic acid ethenyl ester, polymers, polymer with oxirane, graft; antiredeposition agent and soil shield polymer for laundry products. [BASF]

Sokalan® HP 50. CAS 9003-39-8; Polyvinylpyrrolidone; antiredeposition inhibitor for detergents in laundry applications; dispersant. [BASF; BASF AG]

Sokalan® HP 53. CAS 9003-39-8; Polyvinylpyrrolidone; antiredeposition inhibitor for detergents; carbon dispersant. [BASF; BASF AG]

Sokalan® PA 13 PN. Polyacrylic; dispersant for water treatment, laundry detergents, agriculture, paints, coatings, industrial and institutional cleaners. [BASF]

Sokalan® PA 15. CAS 9003-04-7; Sodium salt of polyacrylic acid; dispersant for water treatment, laundry detergents, agriculture, paints, coatings, industrial and institutional cleaners. [BASF; BASF AG]

Sokalan® PA 110 S. CAS 9003-01-4; Polyacrylic acid;

---

‡ Trade name and manufacturer not verified   § Trade name without identified manufacturer

dispersant for water treatment, laundry detergents, agriculture, paints, coatings, industrial and institutional cleaners. [BASF; BASF AG]

Sokalan® PM 10. Copolymeric carboxylate; dispersant. [BASF]

Sokoff. Industrial grease solvent. [The Wellcome Foundation Ltd] *

Sokolan. Polymeric dispersants. [BASF plc]

Solacen. CAS 4268-36-4; Tybamate; tranquilizer (minor). [Wallace Laboratories] *

Solactol. CAS 97-64-3; A proprietary trade name for ethyl lactate. §

Soladox. Chlorosulfonated polyethylene in a solvent base, with dense pigmentation; a corrosion preventative for offshore oil, all marine, roofing, any corrosive environment where complete protection is needed. [Liquid Plastics Ltd] †

Soladox 112. Chlorosulfonated polyethylene in a solvent base, with dense pigmentation; weatherproofing compound; reflects infrared radiation; for use in defense related environments. [Liquid Plastics Ltd] †

Solan. CAS 61790-81-6; PEG-75 lanolin; surfactant, emollient, conditioner, superfatting agent, emulsifier, solubilizer, foam stabilizer, plasticizer, humectant for soaps, detergent bars, shampoos, skin cleansers, hair sprays, deodorants, chemical specialties. [Croda Inc.]

Solan. CAS 2307-68-8; Emulsifiable concentrate containing 400 g/l pentanochlor; used to control weeds in horticultural crops. [Atlas Interlates Ltd]

Solan 50. CAS 61790-81-6; PEG-60 lanolin; hydrophilic surfactant, emollient, conditioner, thickener, superfatting agent, foam stabilizer, plasticizer, humectant for personal care products, soaps, pharmaceuticals, chemical specialties; fragrance solubilizer. [Croda Inc.]

Solan E. Polyethoxylated lanolin. [Croda Chem. Ltd]

Solane. Sulfur dyes. [ICI Chem. & Polymers Ltd]

Solangel 401. CAS 61790-81-6; PEG-75 lanolin; emulsifier, humectant for soap. [Croda Inc.]

Solanthrene. Vat dyes. [ICI Chem. & Polymers Ltd]

Solar. Synthetic organic direct dyestuffs. [Sandoz Products Ltd]

Solar. Nonionic spreader containing 75% polypropoxypropanol and 15% alkyl polyglycol ether; adjuvant for plant growth regulator sprays. [Ideal Manufacturing Ltd]

Solarchem® O. CAS 21245-02-3; Octyl dimethyl PABA; selective uv absorber for suncare and cosmetic products. [CasChem]

solar oil. The name given to various hydrocarbons obtained as by-products in the treatment of brown coal tar in paraffin works.

solar salt. Salt (sodium chloride) obtained by the evaporation of sea-water.

Solar Steel. A proprietary trade name for a silicon steel containing 1% silicon, 0.5% molybdenum, 0.4% manganese, and 0.5% carbon. §

Solasol. Vat dyes. [ICI Chem. & Polymers Ltd] †

Solatene. β-Carotene; uv screen. [Hoffmann-LaRoche Inc]*

Solatol. A preparation of crude phenol (carbolic acid); a disinfectant.

Solbrol®. p-Hydroxybenzoic acid ester; used for the preservation of pharmaceuticals, cosmetics, foodstuffs, and technical products. [Bayer AG]

Solcod. A proprietary trade name for sulfonated cod oil. §

Solcornol. A proprietary trade name for a sulfonated corn oil. §

Soldaflux. Soft solder flux. [Degussa Ltd]

Soldamoll. Special soft solder. [Degussa Ltd]

solder. The various alloys or mixtures which constitute solder are usually classified as hard or soft, according to their melting point. Hard solder includes brazing solder, silver solder, and gold solder, while the soft solders usually consist of tin and lead and melt below 300 C; used to join metals.

Solderel®. Solder paste for the electrical industry. [DuPont UK]

soldering acid. Hydrochloric acid, HCl.

soldering salt. Ammonium and zinc chlorides.

soldering solution. A solution of zinc chloride.

Soldis. A disinfectant containing phenolic and cresylic bodies; miscible with water.

soldo. A flux used for tinning metals. It is mixed with powdered tin.

Soledexin®. Pharmaceutical preparations for the treatment of diseases of the respiratory tract. [Cassella AG]

Soledon. Solubilized vat dyes. [ICI Chem. & Polymers Ltd]

Soledum®. Antitussive and rhinologic pharmaceuticals. [Cassella AG]

Solef®. Polyvinylidene fluoride (PVDF) homopolymers and copolymers, $(CH_2CF_2)_n$; used for high purity and corrosion resistant applications in the chemical processing and semiconductor manufacturing industries, including pipe and fittings, valves, pumps and vessels; protective coatings; wire and cable jacketing, fiber optics buffer tubing and optical fiber raceway applications. [Solvay Polymers]

Solef® 1008. CAS 24937-79-9; PVDF homopolymer; for injection molding for general use and complicated shapes or thin walls; for extrusion of tubes; transfer and centrifugal molding, film extrusion. [Solvay Polymers]

Solef® 1010. CAS 24937-79-9; PVDF homopolymer; for general extrusion of tubes, films, sheets, thin panels, injection molding, compression and transfer molding, blow molding of films and hollow objects. [Solvay Polymers]

Solef® 5008. CAS 24937-79-9; PVDF homopolymer; top coat for electrostatic powd. spraying. [Solvay Polymers]

Solef® 6010. CAS 24937-79-9; PVDF homopolymer; improved thermal stability resin for thick semi-finished items. [Solvay Polymers]

Solef® 8808. CAS 24937-79-9; PVDF homopolymer, carbon fiber-reinforced; reinforced grade for applications requiring extremely high rigidity. [Solvay Polymers]

Solef® 11008/0003. CAS 24937-79-9; PVDF copolymer; for rotational molding. [Solvay Polymers]

Solef® 11010. CAS 24937-79-9; PVDF copolymer; for use where more flexibility and very high elongation at break are required, e.g., electric and telephone cable sheathing, extrusion of sheets. [Solvay Polymers]

Sole-Mulse B. Modified ethoxylate; emulsifier for emulsion degreasers, industrial cleaners. [Calgene]

Solene. See petroleum ether.

Solenhofen stone. A porous limestone containing clay.

Solenite. An explosive. It is an Italian smokeless powder, and contains 30% nitroglycerin, 40% insoluble, and 30% soluble nitrocellulose.

Solester. Vat dyes. [ICI Chem. & Polymers Ltd] †

---

* Trade name not verified as available   † Trade name verified as obsolete

Sole Terge 8. CAS 43154-85-4; EINECS 256-120-0; Disodium oleamido MIPA-sulfosuccinate; high foaming base for bubble baths, shampoos, etc.; frother in specialized ore flotation; emulsifier and stabilizer in emulsion polymerization. [Calgene]

Solfa. Fungicide for mildew control in a wide range of crops. [ICI Chem. & Polymers Ltd]

Solfa. CAS 7704-34-9; Sulfur; used for mildew control in a wide range of crops. [Farm Protection Ltd]

Solfac. Insecticide for the control of stored product pests, vector and nuisance insects. [Bayer AG]

Solflex® 1216. Solution S/B vinyl, nonstaining stabilizer; developed for tire formulations; also suitable for molded goods. [Goodyear] *

Solfoton. CAS 50-06-6; Phenobarbital; anticonvulsant; a hypnotic and sedative. [Poythress Laboratories Inc] ‡

Solganal. Aurothioglucose; antirheumatic. [Schering Corp]

Solgen. Industrial detergent. [Crosfield Chemicals Ltd] †

Solicum. A material made from waste rubber and oil.

solidago. The dried herb of *Solidago odora*; used medicinally as a stimulant, carminative, and diuretic.

solid alcohol. A soapy mass containing about 20% water, 20% sodium stearate, and 60% alcohol.

Solidarol. Reactive dyestuffs. [Hoechst AG]

Solidegal®. Leveling agents for textile dyeings and printings. [Cassella AG]

Solidermin®. Sulfur dyestuffs; used for leather dyeing. [Cassella AG]

Solidex. Photographic developer. [May & Baker Ltd] *

solidified alcohol. A name applied to a solution of nitrocellulose in ethyl alcohol for use in heaters, etc., as a fuel.

solidified linseed oil. (oxidized linseed oil, linoxyn). A flexible solid mass obtained when linseed oil is exposed to oxidation.

Solidite. A proprietary trade name for a range of molded products made from shellac, bitumen, and fillers; used for electrical insulation. §

Solidogen®. Dyeing auxiliaries and improving agents for textile dyeing and printing. [Cassella AG]

Solidogen LT-13. Cationic resin; improves wet fastness of direct and developed dyeings on cellulosic fibers; on suede leathers for garments or gloves to increase fastness to washing and drycleaning and promote level drying. [ISP] *

Solidokoll®. Padding auxiliaries with migration resistant properties. [Cassella AG]

Soligen. A proprietary trade name for certain metallic naphtenates used as paint driers. §

Solimide® Foam. Flexible polyimide foam; fire-resistant, lightweight, low smoke, flexible insulation foams for thermal and acoustical insulation applications; flexible and resilient from -184 to 260 C. [Ethyl Corp; Ethyl SA]

Solintor. Inorganic and organic pigments. [Tennant-KVK Ltd]

Solinure. Range of soluble fertilizers. [Fisons plc, Horticulture Div] *

Solka-Floc® BW-40, BW-100, BW-200, BW-2030, UF-900-FCC & NF. CAS 9004-34-6; Cellulose; binder, diluent, disintegrant, stabilizer, absorption aid, stabilizer, tablet filler for pharmaceutical formulations. [Mendell]

Solklean™ 101. Phosphoric acid, solvents and surfactants; liquid formulated for cleaning of metallic solar absorber plates prior to painting. [Solec]

Solkote™ Hi/Sorb™-II. Selective optical coating specifically formulated for solar applications; high and low temperature air and liquid absorbers; trombe walls; photographic applications; high temperature applications. [Solec]

Sollacaro's aluminum solder. An alloy of 64% zinc, 30% tin, and 6% lead.

Solmed 100. Gloss finish film for thermoforming, heat contact and impulse welding processes. [Solvay Ind. Films]

Solo. CAS 330-55-2, 1582-09-8; Emulsifiable concentrate containing 1250 g linuron and 240 g trifluralin per liter; herbicide for winter cereals. [MTM Agrochemicals Ltd]

Solochrome. After-chrome dyes. [ICI Chem. & Polymers Ltd]

Soloform. A proprietary triiodophenol preparation. §

Solok®. For the electrical industry. [DuPont UK]

Solon Conc. CAS 64-02-8; EINECS 200-573-9; Tetrasodium EDTA; chelating of all metals, iron specialized. [Eastern Color & Chem.]

Solon Fe Special. Sodium dihydroxy-ethyl glycine type material; chelating agent for high concentrates of ion under alkaline conditions. [Eastern Color & Chem.]

Solophenyl. Direct dyes. [Ciba plc] *

Solosil. Foundry binder for the CO_2 process. [Foseco (F.S.) Ltd]

Solosin®. CAS 58-55-9; Theophylline; pharmaceutical preparations for the treatment of bronchial asthma, pulmonary and related diseases. [Cassella AG]

Solox. A proprietary trade name for an alcohol-type solvent, a fuel for alcohol lamps, blow torches, portable stoves, etc.; it contains mostly ethyl alcohol, with small quantities of ethyl acetate and petrol. §

Solozone. CAS 1313-60-6; A proprietary trade name for a sodium peroxide containing 20.5% available oxygen. §

Solpadeine. A proprietary preparation of paracetamol, codeine phosphate, and caffeine; an analgesic. [Sterling Research Laboratories] *

Solpolac. Chloropolyethylene; resin for the preparation of paints and varnishes. [Caffaro SpA] *

Solpolac. Chlorinated polyethylene. [SNIA (UK) Ltd]

Solprin. A proprietary preparation of aspirin, calcium carbonate, citric acid, and saccharin; an analgesic. [Reckitts Colours Ltd] *

Solricin® 135. CAS 7492-30-0; EINECS 231-314-8; Potassium ricinoleate; detergent, emulsifier, mild germicide, glycerized rubber lubricant, foam stabilizer in foamed rubber; making of cutting and sol. oils, household and cosmetic products. [CasChem] *

Solricin® 235. CAS 8013-05-6; EINECS 232-388-4; Potassium castor soap solution in water; emulsifier, dispersant, mild germicide; glycerized rubber lubricant; emulsifier, foam stabilizer for foamed rubber. [CasChem]

Solricin® 285. Ammonium ricinoleate solution in water; emulsifier, dispersant, rustproofing agent (leaves corrosion resistant film on exposure to air); lubricant; for both oil and water systems. [CasChem]

Solricin® 435. CAS 5323-95-5; EINECS 226-191-2; Sodium ricinoleate aq. solution; emulsifier, stabilizer, defoamer for emulsion polymerization of resins (PVC, PVAc). [CasChem]

Solricin® 535. CAS 5323-95-5; EINECS 226-191-2; Sodium ricinoleate solution in water; mild germicide; glycerized rubber lubricant; emulsifier and foam stabilizer for foamed rubber. [CasChem]

---

‡ Trade name and manufacturer not verified

§ Trade name without identified manufacturer

sol rubber. The portion of rubber which enters solution when unmilled raw rubber is treated with a solvent.

Solsaf-T-Solv™ 403. For cleaning or precleaning of aluminum. [Solec]

Solsolv™ 301. For cleaning spray equipment. [Solec]

Solsperse. Dispersants. [Imperial Chemical Industries plc]

Solstar. Computerized design programs to stimulate casting solidification prior to making patterns and designing feeding aids. [Foseco (F.S.) Ltd]

Soltair. Mixture of diquat, paraquat, and simazine; total herbicide. [ICI Chem. & Polymers Ltd]

Soltercin. A proprietary preparation of soluble aspirin, phenacetin and buto barbitone. [Cox Pharmaceuticals]†

Soltrol. Isoparaffinic solvents. [Phillips]

Solubilisant Gamma 2420. Octoxynol-11, polysorbate 20; surfactant blend. [Gattefosse]

Solubilisant Gamma 2428. PEG-40 hydrogenated castor oil, polysorbate 20, octoxynol-11; surfactant blend. [Gattefosse]

Solu-Biloptin. 3 g Calcium ipodate in powder form; used as an x-ray contrast media. [Schering Health Care Ltd]

soluble algin. Alginate of soda, obtained from seaweed.

soluble animal collagen. See collagen.

soluble castor oils. See blown oils.

soluble cream of tartar. Cream of tartar (potassium bitartrate, $C_4H_5O_6K$) dissolved in a solution of boric acid or borax.

soluble glass. (soluble soda-glass, water glass, glass liquor). A syrupy solution containing 50% of sodium silicate, Na_4SiO_4, and Na_3SiO_3; used to impregnate articles to render them fire-resistant, as an adhesive for glass and porcelain, as an adulterant in soap, dyeing, and egg preserving.

soluble glass. See sodium silicate.

soluble oil. See Turkey red oils.

soluble phenyle. A fluid containing coal-tar creosote, rosin oil, potassium oleate, and caustic soda. It gives an emulsion with water, and is used as a sheep dip.

Soluble Plant Feed. Soluble powder containing NPK 19:19:19 plus 0.2% Mg and trace elements; a fertilizer. [Vitax Ltd]

soluble potash glass. CAS 1312-76-1; Potassium silicate, K_2SiO_3.

soluble primrose. See erythrosin.

Soluble Rose Feed. Soluble powder containing NPK 16:8:32 plus 0.15% Mg and trace elements; fertilizer. [Vitax Ltd]

soluble salumin. Aluminum ammonium salicylate, $Al_2(C_6H_4(ONH_4)CO_2)_6 \cdot 2H_2O$; used as an astringent.

soluble soda glass. See soluble glass.

solublestarch. (amylodextrin). Obtained by heating starch with glycerin and adding alcohol; a soluble starch is also made by boiling starch in water and adding a little caustic soda to clear it; an emulsifying agent.

soluble tartar. Potassium tartrate, $(CHOH)_2(COOK)_2$.

Soluble Tomato Feed. Soluble powder containing NK 18:36 and trace elements; a fertilizer. [Vitax Ltd]

Solublon. Packaging materials and plastic film of polyvinyl alcohol which dissolves either in cold or hot water; for packaging for toxic, skin-irritating or strongly colored materials used in aqueous solution, to protect the handling personnel; water-soluble laundry bags for packing the contaminated lines in the hospital before laundering in hot water. [Aicello Chemical Co Ltd] *

Solubor. CAS 1330-43-4; A highly soluble form of sodium borate $Na_2B_8O_{13} \cdot 4H_2O$; used to correct boron deficiency in plants by applying either as a foliar spray, in nutrient feeds or with herbicides. [Borax Consolidated Ltd]

Solu-Coll. CAS 9007-34-5; Soluble collagen; cosmetic ingred. for moisturizing applications. [Brooks Industries]

Solu-Coll P. Procollagen; cosmetic ingred. for moisturizing applications. [Brooks Industries]

Solucort. CAS 125-02-0; Prednisolone sodium phosphate; parenteral corticosteroid for certain endocrine and nonendocrine disorders. [Merck & Co Inc]

Solu-Cortef. CAS 125-04-2; Hydrocortisone sodium succinate; glucocorticoid. [Upjohn]

Solu-Cortril. A proprietary preparation of hydrocortisone sodium hemisuccinate; a corticosteroid. [Pfizer International]

Solucryl. Acrylic solutions. [UCB (Chem) Ltd]

Solu-Delta-Cortef. CAS 1715-33-9; Prednisolone sodium succinate; glucocorticoid. [Upjohn] *

Soluene 100 and 350. 0.5 N quaternary ammonium hydroxides in toluene; cationic surfactants with the ability to solubilize a wide variety of biological samples. [Packard Instrument BV] *

Solufeed. Soluble fertilizer. [ICI Chem. & Polymers Ltd]

Soluhoba. Jojoba oil derivatives. [A & E Connock (Perfumery & Cosmetics) Ltd]

Solu Kera-Tein M. Sol. keratin; cosmetics ingred. [Maybrook]

Solulan. CAS 8027-33-6; EINECS 232-430-1; Lanolin alcohols; used in cosmetics, toiletries, and pharmaceuticals. [D F Anstead Ltd] *

Solulan® 5. Laneth-5, ceteth-5, oleth-5, and steareth-5; emulsifier, wetting agent, dispersant, lubricant, solv., conditioner, plasticizer, emollient used in personal care and dermatology products; foam stabilizer for detergent systems. [Amerchol; Amerchol Europe]

Solulan 5. Ethoxylated (5 mol) complex of lanolin alcohols and related fatty alcohols; nonionic w/o emulsifier, stabilizer for o/w systems; nontacky lubricant, emollient and moisturizer; wetting and dispersing aid for cosmetic pigments. [Amerchol Corp]

Solulan® 16. Ethoxylated (16 mol) complex of lanolin alcohols and related fatty alcohols; nonionic solubilizer, wetting agent and o/w emulsifier; excellent foam stabilizer and conditioning agent in shampoos. [Amerchol Corp]

Solulan® 75. CAS 61790-81-6; PEG-75 lanolin; emulsifier, wetting agent, dispersant, lubricant, solv., conditioner, plasticizer, emollient used in personal care and dermatology products; foam stabilizer for detergent systems. [Amerchol; Amerchol Europe]

Solulan® 97. Polysorbate 80 acetate, cetyl acetate, and acetylated lanolin alcohol; dispersant, lubricant, emollient, conditioner for cosmetics and pharmaceuticals. [Amerchol; Amerchol Europe]

Solulan® 98. Partially acetylated complex of Polysorbate 80 and Acetulan; nonionic o/w emulsifier, solubilizer and pigment wetting agent; conditioner for shampoos. [Amerchol Corp]

Solulan® C-24. Choleth-24 and ceteth-24; emulsifier, wetting agent, dispersant, lubricant, solv., conditioner, plasticizer, emollient used in personal care and dermatology products; foam stabilizer for detergent systems. [Amer-

---

* Trade name not verified as available   † Trade name verified as obsolete

chol; Amerchol Europe]

Solulan® PB-2. CAS 68439-53-2; PPG-2 lanolin alcohol ether; spreading agent, dispersant, plasticizer; emollient and conditioner for personal care products, detergents, pharmaceuticals, waxes, polishes, leather treatment. [Amerchol; Amerchol Europe]

Solulan® PB-5. Propoxylated (5 mol) lanolin alcohols; water-resistant conditioner for skin and hair; pigment wetter and glosser. [Amerchol Corp]

Solulan® PB-10. Propoxylated (10 mol) lanolin alcohols; spreading agent, pigment wetting agent, glosser and hydrophobic emollient. [Amerchol Corp]

Solulan® PB-20. Propoxylated (20 mol) lanolin alcohol; plasticizer, spreading agent, and conditioner. [Amerchol Corp]

Solu-Lastin 10. Hydrolyzed animal elastin. structural protein giving flexibility and elasticity to tissues; for cosmetics use. [Brooks Industries]

Solu-Lastin 30. Hydrolyzed elastin; structural protein giving flexibility and elasticity to the tissues; for cosmetics use. [Brooks Industries]

Solulys®. Corn steep liquor or powder; growth medium for fermentation industry. [Roquette (UK) Ltd]

Solu-Mar EN-30. Hydrolyzed marine protein; film-former and moisturizer for skin and hair cosmetics; substantive protein. [Brooks Industries]

Solu-Mar Native. Soluble marine protein; film-former and moisturizer for skin cosmetics. [Brooks Industries]

Solu-Medrol. CAS 2375-03-3; Methylprednisolone sodium succinate; glucocorticoid. [Upjohn]

Solumedrone. CAS 83-43-2; A proprietary preparation of methylprednisolone; used in the treatment of shock. [Upjohn Ltd] *

Solumin. Anionic surfactant. [Rhone-Poulenc UK]

Solumin F. A range of sodium sulfated alkylphenol ethoxylates supplied as aqueous solutions; wetting, foaming, and detergency, dispersion and emulsification, e.g., in emulsion polymerization; pigment dispersion; and cleaning formulations. [Rhone-Poulenc UK]

Solumin PFN. A range of anionic surfactants consisting of phosphate esters of ethoxylated alkylphenols, as viscous liquids; emulsion hydrotroping, corrosion inhibition and conductivity improvement agents used in emulsion polymerization and as a conductivity additive. [Rhone-Poulenc UK]

Solumin PV27. Phosphate ester of ethoxylated alcohol in yellow liquid form; anionic surfactant with good stability, detergency and corrosion resistance; used in industrial cleaners, hydrotroping and lubricants. [Rhone-Poulenc UK]

Solumin T45S. Sodium sulfated synthetic alcohol ethoxylate in aqueous solution; foaming,wetting and dispersing agent used in shampoos, bubble baths, and surgical scrubs. [Rhone-Poulenc UK]

Solumin V27SD. Sodium sulfated synthetic alcohol ethoxylate in aqueous solution; foaming, wetting, dispersing and emulsification agent. [Rhone-Poulenc UK]

so-luminum. An aluminum solder consisting of 55% tin, 33% zinc, 11% aluminum, and 1% copper.

Solumix. Mixture of toluene, xylene, and aliphatics; an aromatic solvent. [Sasolchem]

Soluphor® P. CAS 616-45-5; Pyrrolidone-2; solv. for veterinary medicine [BASF AG]

Solu-Predalone. CAS 125-02-0; Prednisolone sodium phosphate; glucocorticoid. [O'Neal, Jones & Feldman Pharmaceuticals] *

Solu-Silk 25. CAS 977077-71-6; Silk amino acids; cosmetics ingred. [Brooks Industries]

Solu-Silk Protein. CAS 96690-41-4; Hydrolyzed silk; protein. [Brooks Industries]

Solu-Soy EN-25. Hydrolyzed soy protein; cosmetic ingred. for skin and hair care products; low salt. [Brooks Industries]

Sol-U-Tein 6861. Hydrolyzed soy protein; skin/hair conditioner for permanent waves, rinses, shampoos, tonics, dressings, cleansers, face, body and hand creams and lotions, moisturizing creams. [Fanning]

Sol-U-Tein EA. CAS 977000-98-8; Albumen; binder, coagulant; used in pharmaceuticals and personal care products; dye mordant in textiles, adhesives, veneers, sizing and making papers; gilding leather; book binding; and food application. [Fanning]

Sol-U-Tein FS-1000. Hydrolyzed soy protein; skin/hair conditioner for permanent waves, rinses, shampoos, tonics, dressings, cleansers, face, body and hand creams/lotions, moisturizing creams. [Fanning]

Sol-U-Tein PS-1000. Hydrolyzed soy protein; skin/hair conditioner for permanent waves, rinses, shampoos, tonics, dressings, cleansers, face, body, and hand creams/lotions, moisturizing creams. [Fanning]

Sol-U-Tein VG. Hydrolyzed soy protein; skin/hair conditioner for permanent waves, rinses, shampoos, tonics, dressings, cleansers, face, body and hand creams/lotions, moisturizing creams. [Fanning]

Solutene. Textile auxiliary chemicals. [ICI Chem. & Polymers Ltd] †

Solutol. An alkaline solution of sodium cresol in an excess of cresol, obtained by treating cresol with caustic soda; used as a disinfectant.

Solutol® HS 15. PEG-660 hydroxystearate; solv. for injection solutions. [BASF AG]

Solu-Veg EN-35. CAS 977059-33-8; Hydrolyzed vegetable protein; cosmetic ingred. for skin and hair care products; low salt. [Brooks Industries]

Soluvit Richter. Multivitamin herbal complex (vitamins A, B, E, H, essential fatty acids, horse-chestnut extract), hydro-alcohol solubilized; broad spectrum vitamin treatment for skin and hair protection. [Dr. Kurt Richter; Henkel/Cospha]

Solux. p-Hydroxy phenylmorpholine; a proprietary trade name for a rubber antioxidant. §

Solva. Emulsifying agents used for cheese. [Giulini Corp]

Solvaperm. Dyestuffs for plastics. [Hoechst UK]

Solvatone. A mixture of approximately 80% acetone, 10% isopropyl alcohol, and 10% toluene; used as a solvent for lacquers.

Solvay® Soda. CAS 497-19-8; EINECS 207-838-8; A registered trade name for a sodium carbonate for water softening. §

Solvene. A heavy grade coal-tar naphtha; a proprietary solvent for ester gum, pitches, etc. §

Solvenol 1, 2, 226. Terpene liquids; used as thinners and antiskinning agents in paints; reclaiming agents for natural and synthetic rubbers. [Hercules] *

Solvenon® BB. Volatile solv. for surface coatings and paints; additive in cleaning and pickling agents. [BASF AG]

Solvenon® DIP. Mixture of isomers of dipropylene glycol

---

‡ Trade name and manufacturer not verified   § Trade name without identified manufacturer

monoisopropyl ethers; solv. for resins and dyes, surface coatings industry; leveling agent for water-dilutable baking finishes; film-binding assistant for dispersions; component in cleaning agents and printing inks. [BASF AG]

Solvenon® DPM. Mixture of isomeric dipropylene glycol monomethyl ethers; solv. for domestic cleaners, acid rust remover, production of printing inks; film-binding assistant for aq. polymer dispersions in paint and varnish industry. [BASF AG]

Solvenon® I. 85% Isobutyl formate, 15% isobutanol; solv. and diluent for adhesives, paints, varnishes [BASF AG]

Solvenon® IPP. 1-Isopropoxy-2-propanol, 2-isopropoxy-1-propanol; solv. for resins and dyes, surface coatings, cleaning agent for printing plates [BASF AG]

Solvenon® PC. CAS 108-32-7; Propylene carbonate; solv. for pigments and dyes, in screen printing dyes; extracting agent; washing liquid for natural and synthetic gases; intermediate for organic syntheses [BASF AG]

Solvenon® PM. CAS 107-98-2; 1-Methoxypropanol-2; solv. for cleaning agents, printing inks, surface coatings. [BASF AG]

Solvenon® PP. 1-Phenoxypropanol-2 and 2-phenoxypropanol-1; solv. in surface coatings, e.g., electrodipping varnishes by cathodic deposition; component of cleaning agents [BASF AG]

Solvent 78. Mixture of esters, alcohols, ketones, hydrocarbons; paint thinners, wash solvent. [Solrec Ltd] *

Solvent 401. Mixture of esters, alcohols, ketones, hydrocarbons; paint thinner for car refinishing. [Solrec Ltd] *

Solvent GC. Mixture of ethylene glycol acetates; solv. for printing inks and rubber stamp inks. [BASF AG]

Solvent PM. CAS 107-98-2; 1- Methoxypropan-2-ol; solv. for cosmetic industry. [BASF]

solvent naphtha. A fraction of coal tar distillation of specific gravity 0.875; also the name for the wood naphtha recovered from grey acetate of lime, prepared from the distillation of wood.

Solvent Scour 25/27. Alkanolamides and alkyl phenol ethoxylate; scouring agent and prescour for greasy and oil soiled fabrics. [Hart Chem. Ltd.]

Solveol. Cresols made soluble in water by the addition of sodium cresotinate; a disinfectant and substitute for guaiacol and creosote.

Solvesso. Aromatic solvents of high purity. [Exxon Int'l.]

Solvetek. Pigment dispersions; for coloration of solvent-based coatings. [Pacific Dispersions Inc] *

Solvethane. CAS 71-55-6; Trichloroethane. [Laporte Industries Ltd]

Solvic. PVC resins. [Laporte Industries Ltd]

Solvifog (N.R.I.). Diluent and carrier for thermal fogging of pesticides. [Makhteshim Chemical Works Ltd]

Solvigran. CAS 298-04-4; Insecticide containing disulfoton. [ICI Chem. & Polymers Ltd]

Solvochin. CAS 130-95-0; A proprietary preparation of basic quinine, 25% solution. §

Solvoclarin. Combination of special detergents, finishing agents, antistats, and deodorants; detergent for chlorinated and fluorinated hydrocarbons. [Henkel Chemicals Ltd] *

Solvol. Tetrahydro-naphthol acetate; a proprietary solvent. §

Solvtext. Solvent publications. [Exxon UK] *

Soma. CAS 78-44-4; Carisoprodol; relaxant. [Wallace Laboratories] *

Somacount. Stabilized and treated blood cells in a milk-like medium; reference controls for automated milk cell analyzers. [Coulter Electronics Ltd]

Somafix. Formaldehyde-based cell fixative; allows the automated counting of somatic cells in milk. [Coulter Electronics Ltd]

Somali gum. An acacia gum from *Acacia glaucophylla* and *Acacia abyssinica.*

Somaton. Alcoholic saline diluent; allows the counting of somatic cells in milk by automated analyzers. [Coulter Electronics Ltd]

Somatonorm. Sterile lyophilized powder of genetically produced human somatotropin for the treatment of short stature due to deficiency of growth hormone. [KabiVitrum AB] *

Sombulex. CAS 56-29-1; Hexobarbital; sedative. [3M Pharmaceuticals] †

Somepon T25. Sodium methyl cocoyl taurate; surfactant for cosmetics. [Seppic]

Somnos. CAS 302-17-0; Chloral hydrate; an effective nonbarbiturate sedative and hypnotic. [Merck & Co Inc]†

Somon. Powder containing 96% w/w sodium monochloroacetate; annual dicotyledons control in various horticultural crops. [Hortichem Ltd]

Somophyllin. CAS 317-34-0; Aminophylline; relaxant. [Fisons plc, Pharmaceuticals Div] *

Sonacide. CAS 111-30-8; Glutaral; disinfectant. [Wyeth Laboratories]

Sonalgin. Butobarbitone/codeine/paracetamol. [May & Baker Ltd] *

Sonergan. Butobarbitone/promethazine hydrochloride tablets. [May & Baker Ltd] *

Soneryl. Butobarbitone. [Rhone-Poulenc Rorer Ltd]

Sonilyl. A proprietary preparation of sulfchlorpyridazine; an antibiotic. [Mallinckrodt Inc] *

Sonnenschein's reagent. An alkaloidal reagent prepared by adding phosphoric acid to a warm solution of ammonium molybdate in nitric acid, boiling the precipitate produced in *aqua regia*, evaporating to dryness, and dissolving in 10% nitric acid.

Sonojell. Petroleum jellies. [Witco]

Sonora gum. (Arizona shellac). A variety of shellac obtained from *Larrea Mexicana.*

Sonostat 1111. Warp size lubricant and plasticizer for synthetic and gelatin type sizes designed to reduce static and buildup during weaving. [Henkel/Textiles]

Sonostat NTL. Finish for nylon carpet yarn requiring resist. to discoloration. [Henkel/Textiles]

Sontara®. Fiber. [DuPont UK]

Sontex. White mineral oils. [Pennzoil Products Co, Penreco Div]

Sontique®. Fibers. [DuPont UK]

Soothe. CAS 522-48-5; Tetrahydrozoline hydrochloride; adrenergic. [Alcon Laboratories Inc] *

Soothease. Throat drops. [Richardson-Vicks Inc] *

Sopanox. Soap antioxidant. [Monsanto plc] *

Soprodac. Food preservative. [BP Chemicals Ltd]

Soprophor® 3D33. Ethoxylated tristyrylphenol phosphate, acid form; emulsifier, dispersant for agricultural formulations. [Rhone-Poulenc Surf.; Rhone-Poulenc Geronazzo]

Soprophor® 37. Ethoxylated polystyrylphenol; emulsifier for styrene copolymer; latex stabilizer; dispersant and antioxidant for phenolic slurries. [Rhone-Poulenc

---

* Trade name not verified as available   † Trade name verified as obsolete

Geronazzo]

Soprophor® BSU. PEG-16 tristyrylphenol; emulsifier, dispersant for agricultural formulations. [Rhone-Poulenc Surf.; Rhone-Poulenc Geronazzo]

Sorane. Two-component polyurethane systems. [Avalon Chemical Co Ltd]

Sorban. A strong solution of sorbitol, HO·CH $_2$(CH·OH) $_4$· CH$_2$·OH.

Sorbax HO-40. POE sorbitol ester; emulsifier for agricultural pesticide/herbicide, emulsion polymerization, metalworking lubricants, die-cast lubricants. [Chemax]

Sorbax HO-50. POE sorbitol ester; o/w emulsifier for solvs., vegetable and petrol. oils used in the textile, agriculture, emulsion polymerization, and metal lubricant industries. [Chemax]

Sorbax PML-20. CAS 9005-64-5; Polysorbate 20; o/w emulsifier, solubilizer for perfumes, flavors; for agriculture, cosmetic, leather, metalworking, and textile industries. [Chemax]

Sorbax PMO-5. CAS 9005-65-6; Polysorbate 81; o/w emulsifier, solubilizer for perfumes and flavors; emulsifier for petrol. oils, solvs., vegetable oils, waxes, silicones in agriculture, cosmetic, leather, metalworking, and textile industries. [Chemax]

Sorbax PMO-20. CAS 9005-65-6; Polysorbate 80; o/w emulsifier, solubilizer for perfumes and flavors; emulsifier for petrol. oils, solvs., vegetable oils, waxes, silicones in agriculture, cosmetic, leather, metalworking, and textile industries. [Chemax]

Sorbax PMP-20. CAS 9005-66-7; Polysorbate 40; o/w emulsifier, solubilizer for perfumes and flavors; emulsifier for petrol. oils, solvs., vegetable oils, waxes, silicones in agriculture, cosmetic, leather, metalworking, and textile industries. [Chemax]

Sorbax PMS-20. CAS 9005-67-8; Polysorbate 60; o/w emulsifier, solubilizer for perfumes and flavors; emulsifier for petrol. oils, solvs., vegetable oils, waxes, silicones in agriculture, cosmetic, leather, metalworking, and textile industries. [Chemax]

Sorbax PTO-20. CAS 9005-70-3; Polysorbate 85; o/w emulsifier, solubilizer for perfumes and flavors; emulsifier for petrol. oils, solvs., vegetable oils, waxes, silicones in agriculture, cosmetic, leather, metalworking, and textile industries. [Chemax]

Sorbax PTS-20. CAS 9005-71-4; Polysorbate 65; o/w emulsifier, solubilizer for perfumes and flavors; emulsifier for petrol. oils, solvs., vegetable oils, waxes, silicones in agriculture, cosmetic, leather, metalworking, and textile industries. [Chemax]

Sorbax SML. CAS 1338-39-2; EINECS 215-663-3; Sorbitan laurate; emulsifier for petrol. oils, solvs., vegetable oils, waxes, silicones in agriculture, cosmetic, leather, metalworking, and textile industries. [Chemax]

Sorbax SMO. CAS 1338-43-8; EINECS 215-665-4; Sorbitan oleate; emulsifier, surfactant for o/w emulsion stabilizers and thickeners used in cosmetic, agriculture, metalworking, leather, and textile industries. [Chemax]

Sorbax SMP. CAS 26266-57-9; EINECS 247-568-8; Sorbitan palmitate; emulsifier, surfactant for o/w emulsion stabilizers and thickeners used in cosmetic, agriculture, leather, metalworking, and textile industries. [Chemax]

Sorbax SMS. CAS 1338-41-6; EINECS 215-664-9; Sorbitan stearate; emulsifier, surfactant for o/w emulsion stabilizers and thickeners used in cosmetic, agriculture,

leather, metalworking, and textile industries. [Chemax]

Sorbax STO. CAS 26266-58-0; EINECS 247-569-3; Sorbitan trioleate; emulsifier, surfactant for o/w emulsion stabilizers and thickeners used in cosmetic, agriculture, metalworking, leather, and textile industries. [Chemax]

Sorbax STS. CAS 26658-19-5; EINECS 247-891-4; Sorbitan tristearate; emulsifier, surfactant for o/w emulsion stabilizers and thickeners used in cosmetic, agriculture, leather, metalworking, and textile industries. [Chemax]

Sorbelite C. CAS 50-70-4; Crystalline sorbitol NF; provides sweetness and cooling mouthfeel to direct compression tablet manufacturing; improved flowability, excellent cohesion. [Mendell]

sorbic acid. (2,4-hexadienoic acid) Organic acid; CH$_3$CH=CHCH=CHCOOH; CAS 110-44-1; 22500-92-1; EINECS 203-768-7; animal feed preservative. [Allchem Industries; Chisso; Hoechst Celanese; Honeywill & Stein Ltd; Penta Mfg.; Spice King]

sorbic acid, potassium salt. See potassium sorbate.

Sorbichew. CAS 87-33-2; Chewable tablets containing isosorbide dinitrate; used to prevent or abort acute attacks of angina pectoris. (Sold in UK by Stuart Pharmaceuticals). [ICI Chem. & Polymers Ltd]

Sorbid. CAS 87-33-2; Isosorbide dinitrate tablets for protection against angina pectoris. (Sold in UK by Stuart Pharmaceuticals) [ICI Chem. & Polymers Ltd]

Sorbidel. CAS 50-70-4; A proprietary preparation containing sorbitol; a laxative. [Rona Laboratories] ‡

sorbimacrogol laurate 300. See polysorbate 20.

sorbimacrogol oleate 300. See polysorbate 80.

sorbimacrogol stearate 300. See polysorbate 60.

sorbimacrogol trioleate 300. See polysorbate 85.

sorbimacrogol tristearate 300. See polysorbate 65.

sorbinose. (sorbin; sorbose) CAS 87-79-6; a sugar; C$_6$H$_{12}$O$_6$.

Sorbismal. A German preparation of finely divided bismuth in oil.

Sorbistat. CAS 110-44-1; Sorbic acid. [Pfizer Ltd]

Sorbistat K. CAS 590-00-1; Potassium sorbate. [Pfizer Ltd]

sorbitan trioctadecanoate. See sorbitan tristearate.

sorbitan tristearate. (STS; anhydrosorbitol tristearate; sorbitan trioctadecanoate) Triester of stearic acid and hexitol anhydrides derived from sorbitol; C$_{60}$H$_{114}$O$_8$; CAS 26658-19-5; EINECS 247-891-4; emulsifier for foods, cosmetics, household products, industrial applications. [Henkel/Emery; ICI Spec.; Lonza]

sorbite. A constituent of iron, which is formed in the transformation of austenite; the stage following trootsite and osmondite, and preceding pearlite.

D-sorbite. See sorbitol.

sorbitol. (D-glucitol; D-sorbitol; D-sorbite) Hexahydric alcohol; CH$_2$OHHCOHHOCHHCOHHCOHCH$_2$OH; CAS 50-70-4; EINECS 200-061-5; nutrient and dietary supplement, food additive; bodying agent for paper, textile, liquid pharmaceuticals; in manufacture of sorbose, ascorbic acid, propylene glycol, synthetic plasticizers, resins; as humectant, sequestrant. [Aldrich; Cerestar UK; EM Industries; Fanning; ICI Spec.; Lipo; Lonza]

Sorbitol (EGIC). CAS 50-70-4; A proprietary preparation of sorbitol used in intravenous nutrition. [Servier Laboratories Ltd] *

Sorbitrate. CAS 87-33-2; Isosorbide dinitrate tablets for protection against angina pectoris. (Sold in UK by Stuart

---

‡ Trade name and manufacturer not verified   § Trade name without identified manufacturer

Pharmaceuticals) [ICI Pharma]

Sorbo®. CAS 50-70-4; Sorbitol; cosmetics ingred. [ICI Am.]

Sorbolene®. The trademark for a fat liquor for the leather trade; used in the tanning and dyeing process to give greater elasticity to the leather, and enables fuller shades to be obtained in dyeing. §

Sorbon S-20. CAS 1338-39-2; EINECS 215-663-3; Sorbitan laurate; emulsifier, dispersant for w/o emulsion. [Toho Chem. Industry]

Sorbon S-40. CAS 26266-57-9; EINECS 247-568-8; Sorbitan palmitate; used with Sorbon T series. [Toho Chem. Industry]

Sorbon S-60. CAS 1338-41-6; EINECS 215-664-9; Sorbitan stearate; used with Sorbon T series. [Toho Chem. Industry]

Sorbon S-66. CAS 36521-89-8; Sorbitan distearate. [Toho Chem. Industry] *

Sorbon S-80. CAS 1338-43-8; EINECS 215-665-4; Sorbitan oleate; used with Sorbon T series. [Toho Chem. Industry]

Sorbon T-20. CAS 9005-64-5; PEG-20 sorbitan laurate; emulsifier, dispersant, solubilizer for o/w emulsion. [Toho Chem. Industry]

Sorbon T-40. POE sorbitan palmitate; used with Sorbon S series. [Toho Chem. Industry]

Sorbon T-60. POE sorbitan stearate; used with Sorbon S series. [Toho Chem. Industry]

Sorbon T-80. POE sorbitan oleate; used with Sorbon S series. [Toho Chem. Industry]

Sorbon TR 814. POE sorbitol oleate; detergent, emulsifier for cosmetics. [Toho Chem. Industry]

Sorbon TR 843. POE sorbitol oleate; detergent, emulsifier for cosmetics. [Toho Chem. Industry]

Sorbonorit. CAS 7440-44-0; Highly activated granular carbon; used for solvent recovery. [Norit NV] *

Sorbosil. Thickening and polishing agents for toothpastes. [Crosfield Chemicals Ltd]

Sorbothane®. Visco-elastic polymer; features quasi-liquid properties enabling high mech. damping and energy absorp., return to original shape, stable properties over broad temp. range; for applications in shock/impact absorp., vibration isolation, acoustical damping. [Sorbothane]

Sorbsil. Dessicant silica gels. [Crosfield Chemicals Ltd]

Soreflon. A proprietary range of PTFE polymers. [Montedison UK Ltd] *

Sorel Cement. A magnesium oxychloride cement, made from magnesite (magnesium carbonate, $MgCO_3$) and magnesium chloride, $MgCl_2$. §

Sorel's gutta-percha substitutes. Substitutes containing rosin, pitch, rosin oil, slaked lime, and gutta-percha. Some are filled with china clay, and in others coal tar is used.

Sorensen's salt. Sodium phosphate, $Na_2HPO_4 \cdot 2H_2O$.

Sorexa CD. Caciferol + difenacoum; used for the control of mice in farm buildings. [Sorex Ltd]

Sorexa Plus. CAS 81-81-2; Warfarin; a rodenticide. [Sorex Ltd]

Sorex Golden Fly Bait. CAS 16752-77-5; Granules containing 1% w/w methomyl; used for fly control in livestock houses. [Sorex Ltd]

Sorex Super Fly Spray. Mixture containing pehenothrin and tetramethrin; for control of flying insects in agricultural premises. [Sorex Ltd]

Sorex Wasp Nest Destroyer. Mixture of resmethrin and tetramethrin; wasp destroyer. [Sorex Ltd]

Sorgan. Propachlor + propazine; herbicide. [Agan Chemical Manufacturers Ltd]

Soricinol 40. CAS 5323-95-5; EINECS 226-191-2; Sodium ricinoleate; mold release. [Climax Fluids Additives]

Sorlate. CAS 9005-65-6; Polysorbate 80; pharmaceutic aid. [Abbott Laboratories] *

Soromin®. Sizing agents for synthetic fiber industry. [BASF AG; BASF plc]

Soromine AT. Complex fatty amido compound; imparts an excellent hand, good body and draping qualities, and lubricity to synthetic, cellulosic and animal fibers and leathers. [ISP] *

Sorpol 320. POE alkylaryl ether, POE sorbitan alkylate, and sulfonate; emulsifier for Malathion emulsifiable concentrates and other pesticides. [Toho Chem. Industry]

Sorpur®. CAS 709-98-8; Propanil. post-emergence applied herbicide with no residual effect for control of numerous grasses and broad-leaved weeds in rice crops. [Bayer AG]

Sorrel's alloy. An alloy of 98% zinc, 1% iron, and 1% copper. Another alloy contains 80% zinc, 10% iron, and 10% copper.

Sorvall®. For medical applications. [DuPont UK]

S.O.S.®; Range of household cleaning products. [Bayer AG]

So/San 30M. Specialty product containing BTC 2125M as active biocide; fabric softener-sanitizer. [Stepan; Stepan Canada]

Sostenil. CAS 1617-90-9; A proprietary preparation of vincamine; a cerebral vasodilator. [Pfizer International]†

Sotacor. CAS 959-24-0; A proprietary preparation of sotalol hydrochloride; used in the treatment of angina pectoris. [Bristol-Myers Squibb Co Inc]

Soubieran's ammonical salt. Mercury ammonium-nitrate, $(NH_2 \cdot Hg_2O)NO_3$.

Soucol. Industrial chemical. [Crosfield Chemicals Ltd] †

Souesite. A nickel-iron alloy which occurs naturally.

Soulan's cement. Consists of 7 parts resin, 10 parts ether, 15 parts collodion, and aniline red. A semi-transparent varnish; used for sealing corks into bottles.

Southalite. A phenol-formaldehyde condensation product, with a filler of paper; used for insulating purposes.

Sovatex C1. Alkylaryl sulfonate in liquid form; anionic scouring and milling agent. [Standard Chemical Company]

Sovatex EP 5288. CAS 61791-38-6; EINECS 263-170-7; Coconut imidazoline amphoteric in liquid form; antistats; lubricants; corrosion inhibitors, detergents. [Standard Chemical Company]

Sovatex IM12H. CAS 61791-38-6; EINECS 263-170-7; Hydroxyethyl imidazoline of coconut fatty acid in the form of a semi-liquid; corrosion inhibitor; lubricant; antistatic agent; used as a base for cationic surfactants. [Standard Chemical Company]

Sovatex IM12N. Aminoethyl imidazoline of coconut fatty acid in the form of a semi-liquid; corrosion inhibitor; lubricant; antistatic agent; used as a base for cationic surfactants. [Standard Chemical Company]

Sovatex IM17H. CAS 95-38-5; Hydroxyethyl imidazoline of oleic acid in liquid form; corrosion inhibitor; lubricant; antistatic agent; used as a base for cationic surface active agents. [Standard Chemical Company]

Sovatex IM17N. Aminoethyl imidazoline of oleic acid in

---

* Trade name not verified as available   † Trade name verified as obsolete

liquid form. corrosion inhibitor; lubricant; antistatic agent; used as a base for cationic surface active agents. [Standard Chemical Company]

Sovatex MP/1. Oleyl imidazoline amphoteric in liquid form; antistats; lubricants; corrosion inhibitors; detergents. [Standard Chemical Company]

Sovatex WA. Sulfosuccinate surfactant in liquid form; concentrated anionic wetting agent. [Standard Chemical Company]

Sovermol POL 1008. Polyester polyol; for coatings, elastomers, rigid foam formulations, PU dispersions. [Henkel/Functional Prods.]

Sovermol POL 1012. Branched polyalcohol; solv.-free polyol for potting and casting, adhesives, rigid PU foams. [Henkel/Functional Prods.]

Sovprene. CAS 9010-98-4; A Russian chloroprene synthetic rubber obtained by the polymerization of acetylene to form divinyl-acetylene and then the formation of chloroprene by treating with hydrogen chloride, followed by polymerization.

Soxhlet's solution. A modified Fehling's solution. It consists of a) a solution of 34.639 g copper sulfate in 500 cc water and b) 50 g caustic soda and 173 g potassium sodium tartrate in 500 cc water; used for the determination of sugars.

Soxomide. CAS 127-69-5; Sulfisoxazole; antibacterial. [Upjohn] *

soya bean oil. (soybean oil; Chinese bean oil; soy oil) CAS 800122-7; An oil obtained from the seeds of *Soja hispida* by expression or extraction with a solvent; used as an edible oil, and is also used in soap-making, paints and varnishes, and in the linoleum industry.

Soyafluff® 200 W. CAS 68513-95-1; Soy flour; functional soy flour for food industry. [Central Soya]

Soy-Amino Quat L/O. Lauryloleylmethylamine soy amino acids; cosmetic ingredient for skin and hair care products; conditioning agent; gives excellent gloss in hairsprays; highly compatible with popular resins. [Brooks Industries]

Soyarich® 115 W. CAS 68513-95-1; Soy flour; functional soy flour for food industry. [Central Soya]

Soy-che. Zinc, iron, copper, manganese, sulfur; chelated micronutrient for soybeans. [Draxel Chemical Company]‡

Soy-Quat C. CAS 977039-11-4; Cocodimonium hydroxypropyl hydrolyzed soy protein; substantive conditioner, moisturizer for hair and skin care products (shampoos, conditioners, creams, lotions, bath products, face and body cleansers). [Maybrook]

Soy-Tein NL. CAS 70084-94-5; Hydrolyzed soy protein; protein providing moisture retentive, protective and sealing films, substantivity to hair and skin care products (shampoos, conditioners, treatment products, lotions, creams, bath products); mitigates damage due to bleaching, perming, etc. [Maybrook]

SP-33. CAS 1332-58-7; Coarse particle sized metakaolin (aluminum silicate) specially processed for polyvinyl chloride electrical insulation compounds. [Engelhard]

SP-731. Acetates/amine blend; line flushing and clean-up solv. for adhesives, resins, and urethane foam; contains no methylene chloride. [Specialty Prods.]

SP 2205. Ethylene-methyl acrylate copolymer resin; film resin for specialty coextrusions, blends. [Chevron]

SP 2207. Ethylene-methyl acrylate copolymer resin; coating and laminating resin with excellent adhesion to OPP, PVDC coated surfaces and SPT, low temp. heat seal. [Chevron]

SP-6700. Oil-modified phenol formaldehyde resin; resin in NBR, SBR, NR, and CR as plasticizer during processing; acts as hardener with 8% hexa after cure; use where scorching problems preclude use of hexa-containing materials such as SP-6600. [Schenectady]

Spac. CAS 137-40-6; Sodium propionate; feed additive (growth promoter) for swine and poultry. [Mitsubishi Kasei] *

SpaceRite S-11. Hydrated alumina; optimum spacer material for titanium dioxide pigment used in surface coatings such as trade and commercial paints, powd. coatings, inks. [Alcoa]

Spalerite. Zinc blende, ZnS.

Spallshield®. For automotive applications. [DuPont UK]

Span®. Sorbitan fatty acid esters. [ICI Am.]

Span® 20. CAS 1338-39-2; EINECS 215-663-3; Sorbitan laurate NF; emulsifier, stabilizer, thickener, lubricant, softener, antistatic agent; foods, pharmaceuticals, cosmetics, cleaning compounds, textiles. [ICI Spec. Chem.; ICI Surf. Belgium]

Span® 40. CAS 26266-57-9; EINECS 247-568-8; Sorbitan palmitate NF; emulsifier, stabilizer, thickener, lubricant, softener, antistatic agent; foods, pharmaceuticals, cosmetics, cleaning compounds, textiles. [ICI Spec. Chem.; ICI Surf. Belgium]

Span® 60, 60K. CAS 1338-41-6; EINECS 215-664-9; Sorbitan stearate NF; emulsifier, stabilizer, thickener, lubricant, softener, antistatic agent; foods, pharmaceuticals, cosmetics, cleaning compounds, textiles; also dispersant for inorganic pigments in thermoplastics. [ICI Spec. Chem.; ICI Surf. Belgium]

Span® 65. CAS 26658-19-5; EINECS 247-891-4; Sorbitan tristearate; emulsifier, stabilizer, thickener, lubricant, softener, antistatic agent; foods, pharmaceuticals, cosmetics, cleaning compounds, textiles. [ICI Spec. Chem.; ICI Surf. Belgium]

Span® 80. CAS 1338-43-8; EINECS 215-665-4; Sorbitan oleate NF; w/o emulsifier, oil additive for corrosion inhibition; fiber lubricant and softener. [ICI Spec. Chem.; ICI Surf. Belgium]

Span® 85. CAS 26266-58-0; EINECS 247-569-3; Sorbitan trioleate; emulsifier, stabilizer, thickener, lubricant, softener, antistatic agent; foods, pharmaceuticals, cosmetics, cleaning compounds, textiles. [ICI Spec. Chem.; ICI Surf. Belgium]

Spandofoam. Rigid polyurethane block foam. [Baxenden Chemicals Ltd]

Spandra Transparent Dressing. Fabric supported tecoflex polyurethane film; moisture vapor permeable, hypoallergenic dressing; a transparent intravenous dressing. [Thermedics Inc] *

Spaneph Spansules. CAS 134-72-5; A proprietary preparation of ephedrine sulfate in a sustained release form; used in the treatment of bronchospasm. [SmithKline Beecham] *

Spangite. This is phillipsite, a zeolite containing potassium.

Spanish oxide. A natural red pigment; a red oxide of iron that contains over 80% Fe_2O_3.

Spanish soap. An olive oil soap.

Spanish turpentine. See turpentine.

Spannit. CAS 2921-88-2; Chlorpyrifos; an organophos-

---

‡ Trade name and manufacturer not verified   § Trade name without identified manufacturer

phate insecticide. [Pan Britannica Industries Ltd]

Spanscour EFS. Surfactant; detergent for Spandex scouring. [CNC Int'l. L.P.]

Spanscour GR. Scour and inhibitor combination for nylon/ Spandex fabrics with a tendency to yellow due to atmospheric conditions. [CNC Int'l. L.P.]

Spanscour N20. Detergent for textile scouring. [Reilly-Whiteman]

Sparine Injection. CAS 53-60-1; 50 mg Promazine hydrochloride per ml; tranquilizer for the control of agitation and restlessness in the eldery, for the relief of nausea and vomiting particularly in labor, for use before various investigations in children and for the treatment of intractable hiccup. [Wyeth Laboratories]

Sparine Suspension. CAS 53-60-1; Suspension containing promazine embonate equivalent to 50 mg promazine hydrochloride BP per 5 ml; adjunct in the short-term management of moderate and severe psychomotor agitation, particularly agitation and restlessness in the elderly. [Wyeth Laboratories]

Sparkaloy. A proprietary trade name for silicon-manganese-nickel alloy used for spark-plug wire. §

Spark-L® HPG. CAS 9032-75-1; Fungal pectinase; enzyme for depectinization and pulp washing; for fruit juices, citrus juice, citrus oil. [Solvay Enzymes]

Sparkle Silver®. Nonleafing aluminum pigments; used for metallic colors (aesthetics), automobiles, trucks, bicycles, furniture etc. [Silberline Mfg Co Inc]

Sparkle Silver® Premier (SSP). Nonleafing aluminum pigments; aluminum pigments for paints and coatings used to achieve metallic effects most often in automotive and other decorative finishes. [Silberline Mfg Co Inc]

Sparkle Silvet. Aluminum pigments for plastics; used for all plastics, automobiles, toys, bottles etc. [Silberline Mfg Co Inc]

Sparkle Silvex. See Sparkle Silvet. [Silberline Mfg Co Inc]

Sparkolac. High gloss nitrocellulose lacquers. [The Scottish Adhesives Co Ltd]

Sparmite. CAS 7727-43-7; A bleached barium sulfate pigment. [Pfizer International] †

Spartakon. A proprietary preparation containing levamisole hydrochloride; veterinary antihelmintic (pigeons). [Janssen Pharmaceutical Ltd] *

Spartase. Potassium aspartate and magnesium aspartate; nutrient. [Wyeth Laboratories] *

Spartcide®. 2,3-Dichloro-N-4-fluorophenylmaleimide; fungicide for apple fruit spot, melanose, and scab of citrus, coffee berry disease, and pink disease on rubber. [Mitsubishi Kasei] *

Spartrix. CAS 42116-76-7; A proprietary preparation containing carnidazole; veterinary antihelmintic (pigeons). [Janssen Pharmaceutical Ltd] *

Spasmo-Dolviran. An antispasmodic analgesic. [Bayer AG]

Spasmonal. CAS 5560-59-8; A proprietary preparation of alverine citrate; used in the treatment of colonic disorders. [Norgine Ltd]

Spasor. CAS 1071-83-6; A nonresidual herbicide containing 360 g/liter glyphosate for the control of annual and perennial broad-leaved weeds and grasses; used for clearing ground prior to planting, weed control in all hard surfaces; control of floating and emergent aquatic weeds. [Rhone-Poulenc Environmental Prods. Ltd]

Spastipax. A proprietary preparation of hyoscyamine sulfate, atropine sulfate, hyoscine hydrobromide and amylobarbitone. [Nicholas Laboratories Ltd] *

spathic zinc ore. See calamine.

Spauldite. A proprietary trade name for a phenol-formaldehyde synthetic resin laminated product. §

spear pyrites. See marcasite.

Specflex. Urethane system; microcellular foam system for use in dynamic elastomers for shoe soles, industrial tires, industrial rollers, mechanical goods, wheels, power transmission belts, and sports equipment. [Dow] *

Special Black 4, 4A Beads and Powd. CAS 1333-86-4; Carbon black with oxidized surface; for water-reducible coating systems, packaging inks. [Degussa]

Special Black 100 Powd. CAS 1333-86-4; Carbon black; for paints and coatings industries. [Degussa]

Special Extender. CAS 14807-96-6; Talc, lamellar structure; extender for paints and as a general purpose filler. [Bromhead & Denison Ltd] †

Special Fat 42/44. Hydrogenated coconut oil. [Hüls Am.]

Special Oil 107. Enanthic acid triglyceride; tracer oil for butter; release agent for conveyor belts. [Hüls AG]

Special Oil 619. Triisostearin; lubricant in emulsions, lipsticks, hair care products; adds fine sheen to skin; pigment dispersant in sticks and liners. [Hüls Am.]

Specpure. Spectrographically standardized metals and chemicals. [Johnson Matthey plc]

Spectam. Spectinomycin hydrochloride; antibacterial. [Abbott Laboratories] *

Spectazole. CAS 68797-31-9; Econazole nitrate; antifungal. [Ortho Pharmaceutical Corp] *

Spectinomycin. Trobicin; an antibiotic produced by *Streptomyces spectabilis*.

Spectra®. Fibers for diverse applications including elec. transformers, medical implants. [Allied-Signal]

Spectra. Paints and maintenance products for cars. [Spectra Brands plc]

Spectraban. A proprietary preparation of isoamyl-p-N, N-dimethylaminobenzoate in ethanol; a lotion used to protect skin from uv light. [Stiefel Laboratories (UK) Ltd]*

Spectracote. Solv.-free high-performance coatings formulated for barrier, corrosion or chemical resist., concrete coating, textiles/wovens, encapsulation and adhesives uses. [Flexible Prods.]

Spectradyne® G. Chlorhexidine gluconate BP; antimicrobial for pharmaceuticals, hospital disinfectants, veterinary products, anti-plaque dental products. [Lonza Inc]

Spectraguard. Plastics materials for packaging and wrapping. [Courtaulds plc]

Spectra-Pearl®. CAS 13463-67-7, 12001-26-2; Colored titanium dioxide/mica; pearlescent giving sparkle, luster, and color to cosmetic eye, face, lip, and body makeup. [Van Dyk]

Spectra-Sorb UV 9. CAS 131-57-7; Oxybenzone; ultraviolet screen. [Am. Cyanamid]

Spectra-Sorb UV 24. Dioxybenzone; uv screen. [Am. Cyanamid]

Spectra-Sorb UV 284. CAS 4065-45-6; Sulisobenzone; uv screen. [Am. Cyanamid]

Spectra-Sorb UV 531. CAS 1843-05-6; Octabenzone; uv screen. [Am. Cyanamid]

Spectra-Sorb UV 5411. CAS 3147-75-9; Octrizole; uv screen. [Am. Cyanamid]

Spectratech® CM 10540. Dual uv inhibitor for extrusion molding. [Quantum/USI] *

---

* Trade name not verified as available   † Trade name verified as obsolete

Spectratech® CM 10608, 10634, 10778, 10779, 11013, 11056, 11513. Antiblock concentrates for film applications; FDA approved; CM 11513 also slip agent. [Quantum/USI] *

Spectratech® CM 10777, 11045, 11638, 77242. Antistat concentrates for film and sheet; CM 11638 and 77242 for elec. packaging; CM 11045 also for injection and blow molding. [Quantum/USI] *

Spectratech® CM 11014, 11126, 11172, 11174, 11194. Slip concentrates for film applications. [Quantum/USI] *

Spectratech® CM 11053, 11489, 11591. Flame retardant concentrates; CM 11053 for LDPE film, 11489 for HDPE film, 11591 for LDPE, HDPE, PP film and wire and cable. [Quantum/USI] *

Spectratech® CM 11246. Uv absorber for packaging, agricultural films, molded items. [Quantum/USI] *

Spectratech® CM 11340, KM 11264. CAS 128-37-0; BHT; antioxidant concentrate for HDPE cereal liners. [Quantum/USI] *

Spectratech® CM 11357, 11367. Uv inhibitor for agricultural film, pool coverings, greenhouse film. [Quantum/USI] *

Spectratech® CM 11698. Optical brightener for film applications. [Quantum/USI] *

Spectratech® FM 1035H. EVA foam concentrate, 20% azodicarbonamide blowing agent; for structural foam, extrusion, injection moldings. [Quantum/USI] *

Spectratech® FM 1150H. CAS 9002-88-4; LDPE foam concentrate, 20% azodicarbonamide blowing agent; structural foam, extrusion, PP sheet. [Quantum/USI] *

Spectratech® FM 1776H. PS foam concentrate, 10% sodium borohydride blowing agent; for structural foam, business machine and computer housings. [Quantum/USI] *

Spectrathene. Color and additive concentrates for polyethylene and other plastics; colorants, additives: antistat, slip, antiblock, uv inhibitors, processing aids and flame retardants. [Quantum Chemical Corp]

Spectraveil. Chemical compositions for absorbing uv light. [Tioxide Group plc]

Spectrim. Multicomponent systems used in the manufacture of polyurethane products via reaction injection molding. [Dow UK] *

Spectrim 5. Urethane system; RIM polymer for automotive applications, industrial/consumer RIM applications, and dynamic elastomers. [Dow]

Spectrim MM 310. PU composite; for structural RIM applications; excellent for load-bearing applications requiring excellent impact strength and high-temp. stability, e.g., bumper beams. [Dow]

Spectrim Polyurea HF85. Urethane system; RIM polymer for automotive applications, industrial/consumer RIM applications, and dynamic elastomers. [Dow]

Spectrobid. CAS 37661-08-8; A proprietary preparation of bacampicillin hydrochloride; an antibiotic. [Pfizer International]

Spectroflux. Buffer mixtures for spectrographic analysis. [Johnson Matthey plc]

Spectromel. Powder mixtures for spectographic analysis of relatively pure materials. [Johnson Matthey plc] †

Spectron. CAS 1698-60-8, 26225-79-6; Suspension concentrate containing 211 g chloridazon and 200 g ethofumesate per liter; a residual herbicide for beet crops. [Schering Agrochemicals Ltd]

Spectrosol. Materials for spectroscopy. [BDH Chemicals Ltd]

specularite. (specular hematite). Iron oxide, Fe_2O_3, with a bright metallic luster.

speculum metal. An alloy of 66% copper and 34% tin, with a little arsenic. An alloy of 64% copper, 32% tin, and 4% nickel. It has a specific gravity of 8.6 and a melting-point of 750 C; used for making mirrors of reflection

Speedcure BEDB. 2N-Butoxyethyl 4-(dimethylamino) benzoate; liquid photo activator. [Lambson Ltd]

Speedcure BMDS. 4-Benzoyl-4-methyl diphenyl sulfide. [Lambson Ltd]

Speedcure FDB. Ethyl-p-dimethyl aminobenzoate; uv photoinitiator. [Aceto]

Speedcure EDB. Ethyl 4-(dimethylamino)benzoate; photo activator. [Lambson Ltd]

Speedcure I.TX. Isopropylthioanthone; photo activator. [Lambson Ltd]

Speedcure ITX. Isopropyl thioxanthone; uv photoinitiator. [Aceto]

Speedway. CAS 1910-42-5; Granules containing 8% w/w paraquat; a pre-emergence bipyridilium herbicide to control weeds in field crops and ornamentals. [ICI Chem. & Polymers Ltd]

Speed X Accelerator. A proprietary rubber vulcanization accelerator containing 60% diphenylguanidine and 40% zinc oxide. §

SpeeDye. Aq. pigment dispersions; for pigment garment dyeing. [Yorkshire Pat-Chem]

Speetan SB60. Special 58/60 basic chrome tan powder. [Lancashire Chemical Works Ltd]

spelter. Zinc used in galvanizing. The term is also used for hard solder.

Spenbond®. Waterborne adhesives; for film, foil and paper lamination. [Reichhold]

Spence metal. A material obtained by melting ferrous sulfide with sulfur; used as a jointing material.

Spenkel®. Modified vegetable oil; used for manufacture of resins, copolymer resins, urethane resins, and water-thinnable paint vehicles. [Reichhold]

Spenkel® F18-M-60. Oil-modified urethane. [Reichhold]

Spenlite®. Aliphatic urethane resins; used for paints, adhesives and textiles. [Reichhold]

Spenlite® M22-X-40. Aliphatic moisture curing urethane. [Reichhold]

Spensol®. Water-reducible urethane resins; used for paints, adhesives and textiles. [Reichhold]

Spensol® F74-70. Water-disp. oil-modified urethane. [Reichhold]

spermaceti. CAS 8002-23-1; A wax obtained from the head of the sperm whale. The crude product is obtained by chilling the head and blubber oils. It consists principally of cetyl palmitate, $C_{16}H_{33}OCOC_{15}H_{31}$.

spermaceti, vegetable. See Chinese wax.

spermolin. A linseed oil product. It is a proprietary binder for sands used as cores in metal casting.

spermoline oil. A compound spindle oil used for lubrication.

Spermwax. Synthetic spermaceti cetyl ester waxes. [Robeco Chemicals Inc]

Spersol. Sodium and ammonium polyacrylates; used as antiredeposition agents in laundry detergents, as detergents, as water reducing agents in slurries and as dispersants in boiler water. [Harcros Australia] *

SPG Gelatine. CAS 9000-70-8; Gelatin powder; used for granulation of fine powders to improve free flowing and

---

‡ Trade name and manufacturer not verified   § Trade name without identified manufacturer

dispersion characteristics. [Sanofi Bio-Industries Ltd] ‡

Sphagni. An insulating material prepared from the white moss found on the Swedish peat moors.

Sphericel 110P8. Hollow glass spheres; spheres which withstand molding pressures and reduce weight in engineering grade plastic compounds and molded parts. [Potters Industries]

Spheriglass® A-Glass. Solid glass spheres; additive for thermoplastic and thermoset resins; provides improved flow properties, high resin displacement, low shrinkage and warpage, better molded parts, dimensional stability. [Potters Industries]

Spheron P-1500-030. CAS 7631-86-9, 9004-73-7; Silica, methicone; spherical hollow microbeads as carriers for sunscreens, fragrances, emollients; imparts lubricity. [Presperse]

Spidax. Specialty polyolefin resin; molding compounds. [Sumitomo Bakelite Co Ltd] *

Spiegeleisen. A ferro-manganese alloy containing from 10-35% manganese, 60-85% iron, 1% silicon, and 4-5% carbon.

Spiegler Jolle's reagent. A solution of 2 g mercuric chloride, 4 g succinic acid, and 4 g sodium chloride in 100 cc water; a reagent for albumin in urine.

Spiegler's reagent. This consists of 40 g mercuric chloride and 20 g tartaric acid dissolved in 500 cc water. To this solution is added 100 g glycerol and 50 g, sodium chloride, and the whole made up to 1,000 cc. It is used for proteins.

Spiller's resin. The oxidation products of rubber are sometimes called by this name.

spinacane. See squalane.

Spinflam. A proprietary range of flame retardants for polymers. [Montedison UK Ltd] *

Spinnaker. CAS 55219-65-3; Emulsifiable concentrate containing 250 g/l triadimenol; used to control powdery mildew, rusts and rhychosporium in winter and spring crops of cereals, beet and brassicas. [Shell UK]

Spinuvex. A proprietary range of hindered amine light stabilizers for polymers. [Montedison UK Ltd] *

Spiragas. Dimethyl silicone membrane; used for gas separations, primarily oxygen enrichment. [Allied-Signal Fluid Systems] *

Spiramycin. Rovamycin; an antibiotic produced by *Streptomyces ambofaciens*.

spirit of alum. Sulfuric acid, H_2SO_4.

spirit of Hartshorn. CAS 7664-41-7; A solution of ammonia.

spirit of niter. Spirit of nitrous ether; term is also applied to nitric acid.

spirit of red lavender. Compound tincture of lavender.

spirit of salt. CAS 7647-01-0; Strong impure hydrochloric acid.

spirit of sulfur. Sulfurous acid, H_2SO_3.

spirit of sweet niter. Spirit of nitrous ether, consisting of a solution of 1.52-2.66% of ethyl nitrite in alcohol.

spirit of sweet wine. Ethyl chloride, C_2H_5Cl; used as a local anesthetic.

spirit of tar. See oil of tar.

spirit of turpentine. See oil of turpentine.

spirit of vitriol. CAS 7664-93-9; Sulfuric acid, H_2SO_4; used in fertilizers, dyes, and pigments, chemicals, electroplating baths, nonferrous metallurgy.

spirit of wood. CAS 67-56-1; Methyl alcohol, CH_3OH; used in the manufacture of formaldehyde, acetic acid, , as a

solvent, in chemical synthesis of methyl amines, methyl chloridemethyl methacrylate.

spirit oil. The first fraction from the distillation of Yorkshire Grease (*qv*); used for making black varnish.

spirits of wine. (ethanol, silent spirit; ethyl alcohol) C_2H_5OH; CAS 64-17-5; Commercial spirits of wine contain 84% by weight of alcohol.

Spirittine. Soft wood tar creosote; a wood preservative.

spirit varnishes. Prepared by mixing resins with such solvents as methylated spirit or turpentine.

spirit vinegar. Made from potato or grain spirit. It contains up to 12% of acetic acid.

Spiro-32. Spirogermanium hydrochloride; antineoplastic. [Unimed Inc] *

Spiroflor. 3-Ethyl-2,4-dioxaspiro (5.5) undec-8-ene; complex, natural odor fragrance raw material. [Henkel/ Cospha]

Spirolite. Plastic pipe. [Chevron] *

Spirolone. CAS 52-01-7; Spironolactone; for the treatment of congestive cardiac failure, essential hypertension, hepatic cirrhosis, malignant ascites, idiopathic edema and nephrotic syndrome. [Berk Pharmaceuticals Ltd] *

Spiropitan. CAS 749-02-0; Spiperone; antipsychotic. [Janssen Pharmaceutica Inc] *

Spitfire. Mixture of cyanazine and fluroxypyr; a post-emergence herbicide. [DowElanco Ltd]

sponge iron. A finely porous form of iron obtained by reducing iron oxide at a temperature where no sintering or fusion takes place; a reagent for the precipitation of copper, lead, and other metals from solution.

Spongolit® Range. Esterified glycerides; aerating emulsifier for Madeira and sponge cakes, flavors. [Grünau; Henkel/Cospha; Henkel Canada]

Sponto® 101. Sulfonate/POE ether blend; emulsifier for phosphate toxicant mixtures. [Witco/Organics]

Sponto® 150T. Sulfonate/POE ether blend; agricultural emulsifier for highly concentrate organophosphate pesticides. [Witco/Organics]

Sponto® 168-D. Phosphoric acid mono and diesters/ alkylphenoxy polyethoxyethanol blend; emulsifier and compatibility agent for pesticides and fertilizers, high electrolyte systems. [Witco/Organics]

Sponto® AG3-55T. Sulfonate and POE ether blend; specialty emulsifier for phenoxy-ester herbicides. [Witco/ Organics]

Sponto® N-140B. Sulfonate and POE ether; emulsifier for organic phosphate insecticides. [Witco/Organics]

spoon metal. See nickel silvers.

Sporak. CAS 67747-09-5; Wettable powder containing 50% w/w prochloraz; a broad-spectrum fungicide for cereal crops. [Darmycel UK]

Sporgon. CAS 67747-09-5; Wettable powder containing 50% w/w prochloraz; a broad-spectrum fungicide for cereal crops. §

Sporocide. A wood preservative mainly consisting of potassium-o-dinitro-cresylate.

Sportak. CAS 67747-09-5; Emulsifiable concentrate containing prochloraz; a broad-spectrum fungicide for cereal crops. [Schering Agrochemicals Ltd]

Sportak Alpha. CAS 67747-09-5, 83601-81-4; A suspension concentrate containing 266 g prochloraz and 100 g carbendazim per liter; systemic fungicide for cereals. [Schering Agrochemicals Ltd]

Sportak Delta. Prochloraz, cyproconazole; fungicide.

---

* Trade name not verified as available   † Trade name verified as obsolete

[Schering Agrochemicals Ltd]

sporting ballistite. A smokeless powder consisting of 37.6% nitroglycerin and 62.3% nitrocotton.

Spotleak® 1001. 80% t-Butyl mercaptan, 20% dimethyl sulfide; odorant for natural gas to permit detection of leaks. [Atochem] *

Spotleak® 1003. 76.5% t-Butyl mercaptan, 23.5% isopropyl mercaptan; see Spotleak 1001. [Atochem] *

Spotleak® 1007. 80% t-Butyl mercaptan, 20% methyl ethyl sulfide; see Spotleak 1001. [Atochem] *

Spotleak® 1009. 79% t-Butyl mercaptan, 15% isopropyl mercaptan, 6% normal propyl mercaptan; see Spotleak 1001. [Atochem] *

Spotleak® 1044. 75% Tetrahydrothiophene, 25% t-butyl mercaptan; see Spotleak 1001. [Atochem] *

Spotleak® 2323. 50% t-Butyl mercaptan, 50% methyl ethyl sulfide; see Spotleak 1001. [Atochem] *

Spotrete 75 WDG. Thiram fungicide for control of dollar spot, brown patch, snow mold and animal repellent. [W A Cleary]

Spramex. See Mexphalte. §

Spray-Add 77. Agricultural spreader-sticker. [Chevron] *

Sprayfast. Surfactant adjuvant containing di-l-p-menthene and nonylphenol ethylene oxide condensate; coating agent for herbicides, pesticides, growth regulators and foliar feed sprays. [Mandops (UK) Ltd]

Spray Guard. Splash and spray suppressing rain flaps. [Monsanto Co] *

Spraymate Activator 90. Nonionic wetting agent containing alkylphenyl hydroxypolyoxyethylene; spreader for use in agricultural sprays. [Newman Agrochemicals Ltd]

Spraymate Bond. Extender containing 450 g/l synthetic latex; sticking agent for use with contact herbicides, fungicides, and insecticide sprays. [Newman Agrochemicals Ltd]

Spraymate LI-700. Acidifying surfactant containing 750 g/l soyal phospholids; wetting agent for systemic pesticides and foliar feeds. [Newman Agrochemicals Ltd]

SprayN Save Christmas Tree Spray. Aerosol containing di-1-p-menthene; antitranspirant spray. [Vitax Ltd]

Spraypover. Adjuvant containing 800 g/l refined mineral oil; wetting and spreading agent for use with residual herbicide sprays. [Fine Agrochemicals Ltd]

Sprayset® MEKP. CAS 1338-23-4; MEK peroxide in anhyd. ethyl acetate; catalyst for spray gun applications in the curing of polyesters. [Witco/Argus]

Spreading Agent. Complex alkyl ether. [Croda Chem. Ltd]

Spreitan. Blend of fatty acid esters and emulsifiers, nonionic; universal coning oil for mono and multifilament and for textures synthetic yarns. [Henkel Chemicals Ltd] *

Sprengel's explosives. Cakes of potassium chlorate which have absorbed combustible liquids.

Sprengsalpeter. An explosive consisting of 75% sodium nitrate, 15% brown coal, and 10% sulfur.

Sprills. Pelleted pesticide. [Chevron] *

Springbok. Bakery phosphates. [Albright & Wilson Ltd, Phosphates & Speciality Business] *

Springclene 2. Selective herbicide. [Schering Agrochemicals Ltd] *

Springcorn Extra. Soluble concentrate of 18 g dicamba, 360 g MCPA and 160 g mecoprop per liter; used for weed control in cereals and grassland. [Farmers Crop Chemicals Ltd]

Sprint. CAS 67306-03-0, 67747-09-5; Emulsifiable concentrate of 375 g fenpropimorph and 225 g prochloraz per liter; used for mildew control in cereals. [Schering Agrochemicals Ltd]

Sprodco. Textile yarn lubricant processing aid. [Specialty Products Co] *

SPS. CAS 1332-58-7; Fine kaolin clay sold in powder, bulk or aqueous slurry form; paper coating pigment. [ECC International Ltd]

Spud-Nic®. Chloropropham; herbicide. [Aceto]

Spudweed. Suspension concentrate containing 152 g prometryn and 304 g terbutryn per liter; used for weed control in peas, beans and potatoes. [Pan Britannica Industries Ltd]

Spuncote. Coatings for centrifugal dies used in the spinning of cast iron pipes. [Foseco (F.S.) Ltd]

Spurso. Dispersing agent. [Mooney Chemicals Inc]

Sputamin. A German preparation; a powder containing 80% chloramine; used as an antiseptic.

Sputolysin®. Reagent. [Calbiochem Corp]

Squadron. CAS 83601-81-4, 12427-38-2; A suspension concentrate containing 100 g carbendazim and 275 g maneb per liter; systemic fungicide for cereals. [Quadrangle Agrochemicals]

squalane. (dodecahydrosqualene; spinacane; 2,6,10,15, 19,23-hexamethyltetracosane; perhydrosqualene) Saturated branched chain hydrocarbon obtained by hydrogenation of shark liver oil or other natural oils; $[(CH_3)_2CH(CH_2)_3CH(CH_3)(CH_2)_3CH(CH_3)CH_2CH_2-]_2$; CAS 111-01-3; EINECS 203-825-6; high-grade lubricating oil, perfume fixative, gas chromatographic analysis, transformer oil; in cosmetics and pharmaceuticals as skin lubricant, in suppositories, carrier of lipid-sol. drugs, moisturizer. [Arista Industries; Robeco]

squalene. (2,6,10,15,19,23-hexamethyl-2,6,10,14,18,22-tetracosahexaene) $[(CH_3)_2C[=CHCH_2CH_2C-(CH_3)_2=CHCH_2-]_2$ CAS 111-02-4; EINECS 203-826-1; Biochemical and pharmaceutical research; a precursor of cholesterol in biosynthesis; chemical intermediate for manufacture of pharmaceuticals, organic colorants, rubber chemicals, aromatics, surfactants; bactericide. [Arista Industries; Robeco]

squill. (scilla; sea onion) Bulb of *Urginea maritima*.

SR-201. CAS 96-05-9, 123-31-9; Allyl methacrylate; with 50-185 ppm hydroquinone inhibitor; monomeric acrylic ester capable of polymerizing to hard, infusible resin that is water white, clear, and glass-like; as comonomer with vinyl-type monomers to produce crosslinked polymers; unsymmetrical crosslinking agent where a two-stage polymerizing or drying action is desired. [Sartomer]

SR-203. Tetrahydrofurfuryl methacrylate; with 100±25 ppm hydroquinone inhibitor; high boiling, low visc. monomeric ester; more reactive than methacrylates having equivalent m.w.; polymerization initiated by conventional methods (peroxide catalysts, thermally, ionizing radiation, and uv radiation); room temperature cures by promoting peroxide system with aromatic amines. [Sartomer]

SR-205. Triethylene glycol dimethacrylate; with 80±20 ppm hydroquinone inhibitor; noncorrosive, low visc., high boiling crosslinking monomeric ester; in vinyl plastisols reduces initial visc. and oil extractability, and improves ultimate hardness, heat distort., hot tear strength, and stain resistance; in cast acrylic sheet and rod, in the button, watch crystal, and contact lens industries; in synthetic rubber and ion exchange resins; in dental

---

‡ Trade name and manufacturer not verified   § Trade name without identified manufacturer

compositions. [Sartomer]

SR-206. Ethylene glycol dimethacrylate; with 40-150 ppm hydroquinone inhibitor; crosslinking agent used in emulsion polymerization, cast acrylic sheet, fiberglass-reinforced polyesters, ion exchange resins, rubber compound [Sartomer]

SR-208. Cyclohexyl methacrylate; with 80 ± 20 ppm hydroquinone inhibitor; monomer; polymerizes into hard polymers of superior optical qualities and high ref. index; one-half the shrinkage on polymerization of methyl methacrylate; can undergo vinyl polymerization and copolymerization; homopolymers are thermosetting; using azo or peroxide catalyst, can be free radically polymerized in bulk to give hard, rigid, colorless polymer; pure polymer used in optical lens applications; high boiling point, low shrinkage, and stability in castings. [Sartomer]

SR-209. Tetraethylene glycol dimethacrylate; with 75 ± 25 ppm hydroquinone inhibitor; crosslinking agent used in castings, plastisols, coatings, fibers, papers, and other fabrications. [Sartomer]

SR-211. n-Hexyl methacrylate; with 80 ± 20 ppm of hydroquinone inhibitor; polymerizes to soft polymer sol. in aliphatic hydrocarbons, copolymerized with other monomers for hard or soft compositions; used in preparation of homopolymers and copolymers by emulsion, suspension, or bulk polymerization; good aging characteristics, internal plasticization, and high sol. in aliphatic hydrocarbons; polymers and copolymers suitable for adhesives, binders, finishes and sizes for textiles, leather or paper, and waxes, emulsions, or disps. [Sartomer]

SR-212. 1,3-Butylene glycol diacrylate; with 500 ± 20 ppm hydroquinone inhibitor; low visc. monomer; curing agent; polymerizes to hard, insol., infusible, thermoset resin; exothermic polymerization reaction, initiated thermally or by common free radical initiators, i.e., high energy and uv radiation, peroxide compounds [Sartomer]

SR-220. Cyclohexyl acrylate (monomer); curing agent. [Sartomer]

SR-231. Diethylene glycol dimethacrylate; with 80 ± 20 ppm hydroquinone inhibitor; high boiling monomeric ester; polymerized by common free radical initiators, i.e., peroxidic compounds, heat, uv, and ionizing radiation. [Sartomer]

SR-239. 1,6-Hexanediol methacrylate; with 100 ± 25 ppm hydroquinone inhibitor; cross-linking agent used in casting compounds, glass fiber-reinforced plastics, adhesives, coatings, ion-exchange resins, textile products, plastisols, dental polymers, rubber compounding. [Sartomer]

SR-247. Neopentyl glycol diacrylate (monomer); with 225 ± 25 ppm p-methoxyphenol inhibitor; curing agent; diluent for uv irradiation-cured systems (coatings, printing inks, etc.). [Sartomer]

SR-256. 2-(2-Ethoxyethoxy)-ethyl acrylate; with 100 ppm MEHQ inhibitor; monofunctional monomer which can be polymerized by the use of heat, catalysts, and/or irradiation; especially sensitive to uv light, and is a good solv., making it very suitable for radiation-curable ink and coating formulations. [Sartomer]

SR-259. PEG 200 diacrylate; with 100-150 ppm hydroquinone inhibitor; cross-linking agent used in radiation-cured coatings, inks, adhesives, textile products, photoresists. [Sartomer]

SR-272. CAS 1680-21-3; Triethylene glycol diacrylate (monomer); curing agent. [Sartomer]

SR-285. Tetrahydrofurfuryl acrylate; with 500 ppm hydroquinone inhibitor; monofunctional monomeric ester of acrylic acid, polymerizes to hard, infusible, insol. thermoset resin; used in uv irradiated coatings, because of low visc. and sensitivity to uv; high boiling with low potential for crosslinking helpful in providing more flexible coating; curing agent. [Sartomer]

SR-295. Pentaerythritol tetraacrylate; with 300-400 ppm MEHQ inhibitor; cross-linking agent in adhesives, coatings, inks, textile products, photoresists, castings, or as modifiers for polyester or polymers. [Sartomer]

SR-306. Tripropylene glycol diacrylate; with 125 ppm hydroquinone inhibitor; monomer acrylic ester, polymerizes to hard, infusible resin that is clear and glass-like; used as comonomer with vinyl-type monomers to produce crosslinked polymers; polymerization reaction is mildly exothermic and; initiated by common free radical initiators, e.g., high energy and uv radiation, heat peroxides, etc.; curing agent. [Sartomer]

SR-335. CAS 2156-97-0; N-lauryl acrylate (monomer); curing agent. [Sartomer]

SR-339. 2-Phenoxyethyl acrylate; with 100 ppm HQ inhibitor; high boiling monomeric ester of acrylic acid; polymerization initiated by heat, catalysis, and/or radiation; copolymerization with other acrylic-type monomers is easily achieved; curing agent. [Sartomer]

SR-350. Trimethylolpropane trimethacrylate; with 80 ± 20 ppm hydroquinone inhibitor; cross-linking agent. [Sartomer]

SR-351. Trimethylolpropane triacrylate; with 100-150 ppm hydroquinone inhibitor; high boiling monomeric ester polymerized by common free radical initiators; curing agent. [Sartomer]

SR-379. CAS 106-91-2; Glycidyl methacrylate; with 50 ppm inhibitor; polyfunctional monomer polymerized by applications of heat, heat and peroxidic catalysts, and irradiation by uv, beta, gamma, or x-ray; in hydrogels for contact lenses and membranes, molding and casting compounds; impregnating paper, concrete, and wood, coatings and printing inks, adhesives and sealants, elastomers, etc. [Sartomer]

SR-440. CAS 29590-42-9; Isooctyl acrylate (monomer); curing agent. [Sartomer]

SR-444. Pentaerythritol triacrylate; with 300-400 ppm MEHQ inhibitor; cross-linking agent used in adhesives, coatings, inks, textile products, photoresists, castings, modifiers for polyester, fiberglass, or polymers. [Sartomer]

SR-7475. Stereospecific butadiene/styrene copolymer; blendable modifier for thermoplastic resins and asphalt. [Firestone Syn. Rubber]

SS-30. Copper flake; anti-seize compound protecting stainless steel and high-strength alloys; for elec. connections. [Jet-Lube]

SS 24049, 24519. Nickel pigment, urethane binder in water; EMC shielding coating for interior walls; protects sensitive electronic equip. [Acheson]

SS 24656. Silver/nickel pigment, polyester binder in SB-1; EMC shielding coating for plastics; protects sensitive electronic equip. [Acheson]

SSF. Sodium silico-fluoride, Na_2SiF_6.

SS Nitrocellulose. CAS 9004-70-0; Nitrocellulose; used in

---

* Trade name not verified as available   † Trade name verified as obsolete

flexographic inks where an alcohol rich solvent is desirable and is used in heat sealing coatings. [Hercules] *

S.S.T.® Sump Saver Tablets. CAS 126-11-4; Tris (hydroxymethyl) nitromethane; antibacterial agent, preservative for metalworking fluids. [Angus]

ST 52. See Honvan. [Degussa AG] *

Sta. Methanol-based antifreeze. [Chemcentral Corp]

Stabaxol. Polycarbodiimide. [Bayer plc]

Stabgel. Soil consolidation agents. [ICI Chem, & Polymers Ltd] †

Stabicol®. Blend of stabilizers and sequestrants for use in hydrogen peroxide bleaching of textiles and as a pitch control aid in papermaking. [Allied Colloids Ltd]

Stabifix. Dyeing auxiliary; fixing agent for direct dyes to improve water and wash fastness. [Henkel Chemicals Ltd] *

Stabil-9. Sodium aluminum phosphate; leavening agent for self-raising flour, corn meal, prepared mixes. [Monsanto Co] *

Stabilator A.R. Phenyl-β-naphthylamine; melts at 108 C; recommended for white mixings; a proprietary antioxidant for rubber. §

Stabileze 06. Crosslinked Gantrez with 1,9-decadiene. [ISP]

Stabilisal®. Stabilizing agents used in the dyeing and printing of textiles with sulfur dyestuffs. [Cassella AG]

Stabilite Alba. Di-o-tolyl-ethylene-diamine; a proprietary rubber vulcanization accelerator. §

Stabilizer. Potassium hydroquinone monosulfate. [Rhone-Poulenc UK]

Stabilizer 1097. Organic acid chloride in butyl acetate; stabilizer used to extend the pot life of the bonding agent system/plastisol mixture [Bayer AG; Miles; Polysar]

Stabilizer 2013-P®. CAS 622-16-2; Carbodiimide; activator; improves resistance to hydrolysis by acids, bases, and hot water in vulcanizates based on millable PU. [TSE Industries]

Stabilizer C. CAS 102-08-9; Diphenylthiourea; for economical heat stabilization of PVC containing alkalis and emulsifiers. [Bayer AG]

Stabilizer NS. Organic/inorganic; peroxide stabilizer. [Marlowe-Van Loan]

Stabilizer No. 1. A proprietary trade name for 1:3:5-isopropyl-cresol. §

Stabillin. Penicillin preparations. [The Boots Co plc] *

Stabillin V-K. A proprietary preparation containing phenoxymethylpenicillin potassium; an antibiotic. [The Boots Co plc]

Stabiloid. Colorant dispersions; for coloring of paper coating and saturation compositions, textile inks and latex paints. [Hüls Am.]

Stabilor. A range of precious metal alloys; for dentistry and dental engineering. [Degussa Ltd]

Stabilosol®. A range of Hydron dyestuffs with particular solubility properties; used for textile dyeing and printing. [Cassella AG]

Stabinol. A proprietary trade name for isobuzole. §

Stabiram. Specialty blends of quaternary ammonium compounds; bitumen emulsifiers. [Atochem UK/Ceca]

Stabismol. A proprietary solution of α-carbonyl-cyclohexanyl acetate in olive oil. §

Stablex. A stabilized bitumen used for protective coatings to resist acids.

Stabochlor. A proprietary chloride of lime specially prepared. §

Sta-Clad®. Synthetic resins. [Reichhold]

Stacol. A complex sodium borophosphate, an inorganic water-soluble resin stable to acids and alkalis.

Stadis®. Fuel oil antistatic additive. [DuPont UK]

Stadol. CAS 58786-99-5; Butorphanol tartrate; analgesic; antitussive. [Bristol Laboratories]

staff. A mixture of plaster of Paris and tow; used for moldings.

staffelite. A mineral containing calcium phosphate with calcium chloride or fluoride.

Staffordshire all mine pig. A pig iron made in Staffordshire from ore. It contains about from 0.5-0.75% of phosphorus.

Staflex CP. A mixed alkyl phthalate; vinyl plasticizer [Miles]

Stafoxil. A proprietary preparation of flucloxacillin; an antibiotic. [Brocades Pharma]

stagnine. An astyptic obtained from the spleen.

Stahl's sulfur salt. CAS 10117-38-1; Potassium sulfite, $K_2SO_3 \cdot 2H_2O$; used as a photographic developer, medicine, in food and wine as a preservative.

Stainaway L2B. Phenol sulfonate; stain-blocker for nylon 6/6 and nylon 6 carpet fibers; used in beck or continuous methods. [Am. Emulsions]

Stain-Free. Aromatic condensate; stain resist auxiliary for nylon carpet. [Sybron]

Stainguard GYS. Aromatic condensate; dye fixative for acid dyes on nylon; reserving agent for nylon and nylon/polyester blends. [Am. Emulsions]

stainless invar. A Japanese alloy of 54% cobalt, 36.5% iron, and 9.5% chromium. It has a low coefficient of expansion and with stands corrosion well.

stainless iron. This is really stainless steel, and usually contains from 0.1-0.2% carbon, 12-27% chromium, and up to 0.5% silicon.

stainless silver. This is usually an alloy of 92.5% silver with copper and antimony, and is used for tableware.

stainless steel. CAS 12597-68-1; (rustless steel). A chromium-steel alloy containing 12-15% chromium, and not more than 0.45% carbon; used for cutlery, acid pumps, turbine blades, and exhaust valves for engines. Some alloys contain 12-18% chromium, 8-12% nickel, 74-75% iron.

Stain Resist. Stain resistant treatment for carpets. [Rohm & Haas UK]

stalactites. Deposits of calcium carbonate in the form of icicles, formed when water containing calcium carbonate drips from the roofs of caves.

stalagmites. Similar deposits to stalactites, except that they are formed on the floors of caves.

Stalloy. A proprietary trade name for an alloy containing 3.5-4.0% silicon and 0.1-0.2% aluminum; has a specific resistance of about 55 michroms; its magnetic hysteresis is much lower than that of pure iron; used in the construction of cores for field and armature magnets. §

Stamere®. CAS 9000-07-1; Carrageenan; suspending agent, stabilizer, thickener, gelling agent for food industry; lubricant, emollient for pharmaceutical jellies, laxatives; tablet binder; emulsion stabilizer; pigment suspender in ceramic glazes; also for cosmetics, toothpaste, wire drawing lubricants, electroplating baths. [Meer]

Stamford powder. An explosive containing from 68-72% ammonium nitrate, 21-23% sodium nitrate, 3-4% trinitrotoluene, and 3 1/2-4 1/2% ammonium chloride.

Stamglan®. Low density polyethylene. [AKU Holland] ‡

---

‡ Trade name and manufacturer not verified   § Trade name without identified manufacturer

Stamid HT 3901. Cocamide DEA; foam stabilizer, emulsifier and thickener used in a variety of household, industrial and cosmetic formulations. [Clough]

Stamid LS 5487. CAS 68425-47-8; EINECS 270-355-6; Soyamide DEA; stabilizer, emulsifier, thickener for shampoos and bubble bath formulations. [Clough]

Stamylan HD. High density polyethylene; used in the plastics processing industry for production of crates, household articles, bottles, containers, tubes and pipes, cables, nets, packaging film, toys, etc. [DSM NV] *

Stamylan LD. Low density polyethylene; used in the plastics processing industry for production of packaging film, heavy-duty bags, extrusion-coated cardboard and paper, tubes and pipes, cable sheathing, household articles, toys, agricultural film, foamed board, etc. [DSM NV] *

Stamylan P. Polypropylene; used in the plastic processing industry for a wide variety of applications: injection molding of car components (bumpers, accumulator cases, boot linings, housings), electronic equipment, furniture and thin-walled containers (dairy products); for extrusion of film, fibers and nonwoven fabrics, tape and belts; for blow molding of bottles and containers. [DSM NV] *

Stamylex PE. Special linear low density and medium density polyethylene; used for very special high-performance applications, e.g., packaging articles such as tanks, containers, covers, special food packaging films, leisuretime products, cables, monofilaments, fasteners, caps and stoppers, and special technical applications [DSM NV] *

Stanaprin. CAS 2016-36-6; A proprietary preparation of choline salicylate; an analgesic. [Lloyd's Pharmaceuticals Ltd] *

Stancard 5219. Mineral oil-based lubricant for use as yarn lubricant for cotton dust control. [Henkel/Textiles]

Stanclere®. Organotin heat stabilizers for PVC. [Akzo Chemie UK Ltd]

Stanclere® T-883. Sulfur-containing octyltin stabilizer. [Akzo]

Stanclere® T-4817. Sulfur-containing butyltin stabilizer. [Akzo]

Standacol. United Kingdom foodstuffs colors. [Morton Int'l. Ltd] †

Standafin FCX. Extender for fluorocarbon finishes for nylon and polyproylene carpet yarn. [Henkel/Textiles]

Standalloy. Dental silver alloys and mercury for mixing amalgams; for dentistry and dental engineering. [Degussa AG] *

Standamid® CD. CAS 136-26-5; EINECS 205-234-9; Capramide DEA (2:1) and diethanolamine; detergent, foam enhancer for anionic systems; solubilizes fragrances into hydroalcoholic systems; secondary emulsifier in o/w systems, perfume stabilizer; personal care products, industrial cleaners. [Henkel/Cospha; Henkel Canada]

Standamid® KD. Cocamide DEA (1:1); foaming and thickening agent for liquid or gel shampoos and bubble baths, industrial cleaners, etc.; superfatting agent, foam stabilizer, emulsifier, intermediate; detergent and solubilizer for oily components; hair conditioner. [Henkel/Cospha; Henkel Canada]

Standamid® KDS. CAS 120-40-1; EINECS 204-393-1; Lauramide DEA (1:1); detergent, solubilizer for oils; enhances foam dens., lubricity, stability for shampoos,

skin cleansers, bath and shower products. [Henkel/Cospha; Henkel Canada]

Standamid® PD. Cocamide DEA (2:1) and diethanolamine; more efficient foam builder, less efficient visc. builder than 1:1 superamides; stabilizer; used in shampoos, bubble baths with anionics, nonionics, and amphoterics; industrial cleaning; solubilizer for oily additives. [Henkel/Cospha; Henkel Canada]

Standamid® SOMD. Linoleamide DEA (1:1); visc. builder with foam enhancement characteristics; produces especially high visc. when used with ethoxylated anionics; used in gel shampoos, bubble baths, liquid hand soaps, industrial cleaners, and formulation; where the amount of electrolyte must be kept to a min.; low-cost shampoo concentrates. [Henkel/Cospha]

Standamox CAW. CAS 68155-09-9; Cocamidopropylamine oxide; wetting agent, foam builder, stabilizer, thickener, lubricant, emollient for low irritation baby shampoos, bubble baths, skin care preparations. [Henkel/Cospha; Henkel Canada]

Standamox O1. CAS 14351-50-9; EINECS 238-311-0; Oleamine oxide; biodeg. thickener, bacteriostat, dye assistant, lubricant, softener used in industrial applications, hair products, plating compounds, lube oils. [Henkel/Cospha; Henkel Canada]

Standamox PCAW. CAS 68155-09-9; Cocamidopropyl dimethylamine oxide; detergent, emulsifier, wetting agent, foam stabilizer; stable over wide pH range. [Pulcra SA]

Standamox PL. Lauramine oxide and myristamine oxide; detergent, emulsifier, wetting agent, foam stabilizer; stable over wide pH range. [Pulcra SA]

Standamul® Conc. 1002. Cetearyl alcohol, PEG-40 hydrogenated castor oil, stearalkonium chloride; conditioner and softener used in hair care preparations. [Henkel/Emery]

Standapol® 1610. CAS 73138-79-1; Sulfated peanut oil; emulsifier, wetting agent, dispersant, antistat for nylon finishes. [Henkel/Textiles]

Standapol® 7088. Ammonium myreth sulfate, cocamide MEA; foaming surfactant base for personal care cleansing and bath products [Henkel/Cospha; Henkel Canada]

Standapol® 7092. Sodium laureth sulfate, glycol stearate; pearlizing shampoo base; excellent stability over wide temp. ranges. [Henkel/Cospha; Henkel Canada]

Standapol® A. CAS 2235-54-3; EINECS 218-793-9; Ammonium lauryl sulfate; detergent, foamer, suspending agent, base for shampoos, cleaning compounds with near neutral pH. [Henkel/Cospha; Henkel Canada]

Standapol® AP Blend. Sodium laureth sulfate, cocamide DEA, cocamidopropyl betaine; concentrate for shampoo, bath and cleansing products, liquid soaps; excellent foaming and visc. response. [Henkel/Cospha; Henkel Canada]

Standapol® CAT. Glycol stearate, lauramine oxide, propylene glycol, and ceteareth-20; pearlescent surfactant for personal care products. [Henkel/Cospha; Henkel Canada]

Standapol® CS Paste. Sodium lauryl sulfate, sodium cetyl sulfate, and laureth-3; detergent, foamer; formulated concentrate for pearlescent cream shampoos, bubble baths. [Henkel/Cospha; Henkel Canada]

Standapol® DEA. DEA-lauryl sulfate; base, detergent, foamer used in personal care products [Henkel/Cospha;

Henkel Canada]

Standapol® EA-1. CAS 67762-19-0; Ammonium laureth sulfate; surfactant for clear liquid shampoos, bubble baths. [Henkel/Cospha; Henkel Canada]

Standapol® ES-1. Sodium laureth sulfate; detergent, foamer, base for liquid shampoos and bubble baths. [Henkel/Cospha; Henkel Canada]

Standapol® LF. Sodium octyl sulfate; wetting agent for metal degreasers, hard surface cleaners, food equipment cleaners, dust control; solubilizer, hydrotrope; resistant to hard water. [Henkel/Cospha; Henkel Canada]

Standapol® MG. Magnesium lauryl sulfate; foamer with good dermatological properties for personal care products. [Henkel/Cospha; Henkel Canada]

Standapol® Pearl Conc. 7130. Glycol distearate, sodium laureth sulfate, propylene glycol, cocamide MEA, laureth-9; pearlescent surfactant used in personal care products. [Henkel/Cospha; Henkel Canada]

Standapol® S. Sodium lauryl sulfate, sodium laureth sulfate, lauramide MIPA, cocamide MEA, glycol stearate, and coceth-8; pearlescent shampoo and liquid soap base. [Henkel/Cospha; Henkel Canada]

Standapol® SCO. CAS 8002-33-3; EINECS 232-306-7; Sulfated castor oil; emulsifier and dispersant, wetting agent, foam depressant, emollient surfactant and solubilizer used in personal care products. [Henkel/Cospha; Henkel Canada]

Standapol® SH-100. CAS 56388-43-3; EINECS 260-143-1; Disodium oleamido PEG-2 sulfosuccinate; detergent used in personal care products; nonirritating shampoo base; anti-irritant for other surfactants. [Henkel/Cospha; Henkel Canada]

Standapol® SHC-101. Disodium oleamido PEG-2 sulfosuccinate and sodium lauryl sulfate; base for low-irritation shampoos for babies and adults; anti-irritant for other surfactants. [Henkel/Cospha; Henkel Canada]

Standapol® T. CAS 139-96-8; EINECS 205-388-7; TEA-lauryl sulfate; detergent, foamer, base for mild shampoos, aerosols. [Henkel/Cospha; Henkel Canada]

Standapol® WAQ-LC. CAS 151-21-3; Sodium lauryl sulfate; foaming agent, detergent, suspending agent for personal care products, liquid cleaners; low salt content for improved corrosion resistance. [Henkel/Cospha; Henkel Canada]

Standapon 4149 Conc. Concentrated scour for soaping-off dyed or printed goods. [Henkel/Textiles]

standard benzine. Light petroleum spirit of specific gravity 0.695-0.705 at 15 C., of which 95% boils between 65 and 95 C; used for the determination of asphalt in oils.

standard gold. (sterling gold). It is 22-carat gold containing 91.6% gold with 8.4% other metals, usually copper, to render it harder. American standard gold contains 90% gold and 10% copper.

standard silver. (sterling silver). Silver, 92.5% with another metal, usually copper, to harden it. American standard silver contains 90% silver and 10% copper.

stand oil. (Standöl varnish, Dicköl varnish, lithographer's varnish). Linseed oil boiled strongly, and allowed to burn until it has the desired thickness.

Standup. CAS 999-81-5; Preparation containing 40% chlormaquat; growth regulator for cereals. [L W Vass (Agricultural) Ltd] †

Stan-Fast. Chemical additive to a coloring bath for metals such as anodized aluminum; used in anodizing alumi-

num. [Reynolds Metal Co]

Staniform. A proprietary preparation of methyl stannic iodide; a remedy for boils, carbuncles, small wounds, and injuries. §

Stanlev R-276. Leveling agent for use with disperse or neutral acid dyes on nylon. [Henkel/Textiles]

Stanleys Crow Repellant. Active ingredients: refined coal tar and creosote oil; seed protectant to prevent sprout pulling by birds in newly planted corn. [Borderland Products Inc] †

Stannal. Electrolytic coloring for aluminum. [Albright & Wilson Ltd]

stannekite. A resinous hydrocarbon, $C_{20}H_{22}O_3$, found in coal deposits in Bohemia.

Stannex. Covering and cleansing fluxes for tin and tin-lead alloys. [Foseco (F.S.) Ltd] †

stannic chloride. (tin tetrachloride; tin perchloride; tin (IV) chloride anhydrous); $SnCl_4$; CAS 7646-78-8; EINECS 231-588-9; Electroconductive/electroluminescent coatings, textile dye mordant, perfume stabilizer, manufacture of fuchsin, blueprint paper, color lakes, ceramic coatings, bleaching agent for sugar, stabilizer for resins, tin salts, soap bactericide/fungicide. [Aldrich; Atochem N. Am.; Chemisphere Ltd; M&T Harshaw; Nihon Kagaku Sangyo; Witco/Argus]

Stannicide. Fungicides, bactericides and algicides. [Akzo Chemie] *

Stannicide. Formulated organotin compounds; fungicides, biocides and algicidal application in water based coatings and adhesives and in the water treatment industry. [Thomas Swan & Co Ltd] †

stannic oxide. See tin oxide (ic).

Stannine. Acid inhibitor. [Rhone-Poulenc UK]

Stanniol. An alloy of 96.2% tin, 2.4% lead, 1% copper, 0.3% nickel, and 0.1% iron.

stannite. A mineral; tin pyrites.

stannous chloride. (Tin (II) chloride anhydrous; tin crystals; tin salt; tin dichloride; tin protochloride) $SnCl_2$; CAS 7772-99-8; EINECS 231-868-0; reducing agent for intermediates, dyes, polymers, phosphors; manufacture of lakes; textile dyeing and printing; tin galvanizing; analytical reagent; silvering mirrors; antisludge for lubricants; food preservative; perfume stabilizer; soldering flux. [Aldrich; Atochem N. Am.; Blythe, William Ltd; Cerac; M&T Harshaw; Noah Chem.]

stannous oxalate. See tin (II) oxalate.

stannoxyl liquid. A solution of tin chloride, $SnCl_2$, in glycerin; used as a lotion in the treatment of boils.

stannum. See tin.

Stanolax. See paraffin, liquid.

Stanolind. See paraffin, liquid.

Stansoft 626-B. Nonyellowing towel softener imparting very soft hand and excellent absorbency. [Henkel/Textiles]

Stansperse 506. Dyebath stabilizer improving stability of softeners in textile dyebaths. [Henkel/Textiles]

Stantex Antistat F. Low melting point antistat and emulsifier for nylon finishes. [Henkel/Textiles]

Stantex PENE 20. Wetting agent for latex carpet backing. [Henkel/Textiles]

stantienite. A brown resin found with Prussian amber.

Sta-Nut EE. CAS 68334-00-9; Partially hydrogenated cottonseed oil, fatty acid mono and diglycerides; kosher; peanut butter stabilizer. [Van Den Bergh Foods]

Stanvis. Pyroxylin embedding solution for microscopy.

---

‡ Trade name and manufacturer not verified   § Trade name without identified manufacturer

[BDH Chemicals Ltd]

Stanyl®. Engineering plastics with excellent impact strength and high temperature resistance; used in the electrical and automotive industries and as material for technical yarns. [DSM NV] *

Stanyl® TE200F6. CAS 50327-22-5; Nylon 4/6, 30% chopped glass fiber-reinforced; special heat stabilizer for elec. applications; high heat engineering plastic offering outstanding wear, friction, creep, fatigue, chemical resist., toughness, and stiffness properties; used in automotive field, elec./electronic and auto under-the-hood applications, high performance mech. components. [DSM]

Stanyl® TE300. CAS 50327-22-5; Nylon 4/6; special heat stabilizer for elec. applications; see Stanyl TE200F6. [DSM] *

Stanyl® TE350. CAS 50327-22-5; Nylon 4/6, flame-retarded; special heat stabilizer for elec. applications; see Stanyl TE200F6. [DSM]

Stanyl® TQ200F6. CAS 50327-22-5; Nylon 4/6, 30% chopped glass fiber-reinforced; oil-resist. heat stabilizer; see Stanyl TE200F6. [DSM] *

Stanyl® TW300. CAS 50327-22-5; Nylon 4/6; standard heat stabilizer; see Stanyl TE200F6. [DSM]

Stanza. Fungicide. [Schering Agrochemicals Ltd]

stanzaite. A mineral. It is a variety of Andalusite.

Stapenor®, Stapenor Retard. CAS 7240-38-2; Oxacillin sodium; penicillinase-resistant antistaphylococcal penicillin; for prophylaxis and treatment of mastitis. [Bayer AG]

Staphcillin. A proprietary preparation of sodium methicillin; an antibiotic. [Bristol-Myers Squibb Co Inc]

staple fiber. (staple artificial silk, artificial wool, artificial chappe). This fiber consists of artificial threads of cellulose or cellulose compounds possessing a definite medium length. It is worked up by ordinary spinning machinery and is suitable for mixing with cotton or wool.

Stapron S SG340. CAS 9011-13-6; Styrene/maleic anhydride copolymer, rubber modified, 20% glass fiber-reinforced; amorphous engineering plastic [DSM]

Stapron S SM300. CAS 9011-13-6; Styrene/maleic anhydride copolymer, rubber modified; amorphous engineering plastic [DSM]

Star. CAS 56-81-5; Glycerin USP; humectant; in pharmaceuticals, toiletries, tobacco, alkyds, food products, explosives, cellophane, urethane foam, other industries. [Procter & Gamble]

Star DSM. CAS 919-86-8; Emulsifiable concentrate containing 580 g demeton-S-methyl per liter; a systemic organophosphorus insecticide. [Star Agrochem Ltd]

Star MCPA. CAS 94-74-6; MCPA; herbicide for cereals and grassland. [Star Agrochem Ltd]

Staralox. A trademark for abrasive goods made essentially of alumina. §

Starane. CAS 81406-37-3; Emulsifiable concentrate of 200 g fluroxypyr per liter; used for weed control in cereals and grassland. [DowElanco Ltd]

Starane 2. Selective post-emergence herbicide. [Murphy Chemical Ltd] †

star bowls. Antimony metal obtained by refining with iron. The metal containing about 91% antimony with about 7% iron is mixed with crude antimony and salt and heated. The product is known as star bowls. It contains about 99.5% antimony.

starch, corn. See corn starch.

starch glazes. Made by adding borax, powdered stearic acid, or paraffin to potato starch.

starch glue. Prepared by adding 3 pints water and 1/2 lb nitric acid to 2 1/2 lb starch, warming, then heating.

Starch Gum. See British gum.

starch gum. See dextrin.

Star Chlormequat CAS 999-81-5; Soluble concentrate containing 700 g/l chlormequat; plant growth regulator. [Star Agrochem Ltd]

starch syrups. Glucose mixed with dextrine used in the place of sugar for various purposes.

Stardrops. Household cleaner. [Thornton & Ross Ltd]

Starfol® BB. CAS 17671-27-1; Behenyl behenate; high-temp. ester lubricant with high smoke point for fiber and yarn lubrication applications [Sherex/Div. of Witco]

Starfol® CP. CAS 540-10-3; EINECS 208-736-6; Cetyl palmitate; emollient for cosmetic creams and lotions. [Sherex/Div. of Witco]

Starfol® IS. CAS 41669-30-1; EINECS 255-485-3; Isostearyl isostearate; high-temp. ester lubricant with high smoke point, low visc.; replacement for butyl stearate and mineral oil for fiber and yarn lubrication applications [Sherex/Div. of Witco]

Starfol® OO. CAS 3687-45-4; EINECS 222-980-4; Oleyl oleate; high-temp. ester lubricant with high smoke point, low visc.; replacement for butyl stearate and mineral oil in fiber and yarn lubrication applications. [Sherex/Div. of Witco]

Starfol® OS. CAS 22766-82-1; Octyldodecyl stearate; emollient and moisturizer for creams and lotions; imparts luxurious, conditioned feel to the skin without greasiness; high-temp. lubricant for textiles. [Sherex/Div. of Witco]

Starfol® Wax CG. CAS 8002-23-1; Cetyl esters; synthetic spermaceti wax; emollient for cosmetics, creams, lotions; opacifier and feel modifier. [Sherex/Div. of Witco]

Starglo. Blend of alcohols, aldehydes and nonionic wetters; tin electroplating additive. [Taskem Inc] *

Staril Tablets. CAS 88889-14-9; Fosinopril sodium. [Bristol-Myers Squibb Pharmaceuticals Ltd]

Starim. Reaction injection molding nylon, prepared by in-mold polymerization of caprolactam (feedstock for nylon) to a nylon-6 block polymer with utilization of a catalyst 2.0 agents; used for general industrial, agriculture, and automotive applications (e.g., body parts). [DSM NV] *

Starlite. A proprietary synthetic resin. §

Starpass. A proprietary urea-formaldehyde synthetic resin.§

Starplex® 90. High-purity, molecularly distilled monoglyceride prepared from edible fats or oils and glycerin with TBHQ and citric acid; provides increased hydration and improved functionality of the monoglyceride; improves softness and extends shelf life in baked goods; emulsifies and stabilizes fat in sauces and gravies; starch complexing agent. [Am. Ingredients/Patco]

Star Stran 748. Continuous filament glass strand; reinforcement for PP, PPS, PEI. [Schuller]

Starter Flowable. CAS 1698-60-8; Suspension concentrate containing 430 g chloridazon per liter; a pyridazinone herbicide for beet crops. [Truchem Ltd]

Starwax® 100. CAS 63231-60-7; Hard microcrystalline wax consisting of n-paraffinic, branched paraffinic, and naphthenic hydrocarbons; wax used in hot-melt coatings

---

and adhesives, paper coatings, printing inks, plastic modification (as lubricant and processing aid), lacquers, paints, and varnishes, as binder in ceramics, for potting/impregnant in elec./electronic components, rubber and elastomers (plasticizer, antisunchecking, antiozonant), as emulsion wax size in papermaking, as fabric softener ingredient, in emulsion and latex coatings, and in cosmetic hand creams and lipsticks. [Petrolite]

Starycide. Insect growth regulator for the control of household pest larvae (especially flea and cockroach larvae). [Bayer AG]

stasite. A mineral. Hydrated phosphate of uranium and lead, $8UO_3 \cdot 4PbO \cdot 3P_2O_5 \cdot 12H_2O$, found in Katanga.

Stasoft J. CAS 68585-05-7; Sulfated neatsfoot oil; fatliquor for chrome, chrome-alum, and chrome-bleached side leathers, production of full-grain soft upper leather and baseball glove leather [Reilly-Whiteman]

Stassfurt salts. See Abraum salts.

staszicite. A mineral from Meidzianka, containing 39% As_2O_5, 26.5% CuO, 20.8% CaO, and 7.3% ZnO.

Sta-Tac® B, T. Mixed olefin hydrocarbon resin; for pressure-sensitive adhesive, hot-melt, and laminating adhesives. [Arizona]

Statex. CAS 1333-86-4; Carbon black. [Sevalco Ltd]

Statexan®. Antistatic agents; used for the textile industry. [Bayer AG; Bayer plc]

Statexan® K1. Sulfonated aliphatic hydrocarbon; internal antistat additive or external coating for PVC, PS. [Bayer AG; Miles; Polysar]

Staticide®. Water-based topical antistat for spray, dip, or wipe-on application to any material; for industrial, commercial, and institutional facilities, clean room environments; biodeg. [ACL]

Statik-Blok® FDA-3. Polyether type; surface-active antistatic solution for food contact surfaces. [Amstat Industries]

Statil. Aldose reductase inhibitor. [ICI Chem. & Polymers Ltd] †

Stat-Kon®. Electrically conducting thermoplastic compounds. [LNP; ICI Chemicals & Polymers Ltd]

Stat-Kon® AC-1003. CAS 9003-56-9; ABS, 15% carbon fiber-reinforced; statically dissipative thermoplastic composite for protection against electrostatic discharge damage; used in electronic packaging systems and functional components. [LNP]

Stat-Kon® AS. CAS 9003-56-9; ABS, SS fiber-reinforced; statically dissipative composite for electronic packaging systems and functional components where protection from ESD damage is required. [LNP]

Stat-Kon® C. Carbon powd. grade; statically dissipative composite for electronic packaging systems and functional components where protection from ESD damage is required. [LNP]

Stat-Kon® DC-1002 FR. PC, 10% carbon fiber-reinforced; statically dissipative composite for electronic packaging systems and functional components where protection from ESD damage is required. [LNP]

Stat-Kon® FE. CAS 9002-88-4; HDPE; extrusion grade, statically dissipative thermoplastic composite for protection against electrostatic discharge damage; used in electronic packaging systems and functional components. [LNP]

Stat-Kon® M-1 HI. CAS 9003-07-0; High impact PP; statically dissipative thermoplastic composite for protec-

tion against electrostatic discharge damage; used in electronic packaging systems and functional components. [LNP]

Stat-Kon® OC-1006. CAS 9016-75-5; PPS with 30% carbon fiber reinforcement; statically dissipative thermoplastic composite for protection against electrostatic discharge damage; used in electronic packaging systems and functional components. [LNP]

Stat-Kon® P. CAS 25038-54-4; Nylon 6, carbon powd. grade; statically dissipative composite for electronic packaging systems and functional components where protection from ESD damage is required. [LNP]

Stat-Kon® PDX-84440. CAS 32131-17-2; Nylon 6/6; statically dissipative thermoplastic composite for protection against electrostatic discharge damage; used in electronic packaging systems and functional components installed close to sensitive elec./electronic devices. [LNP]

Stat-Kon® R. CAS 32131-17-2; Carbon powder grade nylon 6/6; statically dissipative thermoplastic composite for protection against electrostatic discharge damage; used in electronic packaging systems and functional components. [LNP]

Stat-Kon® RC-1002. CAS 32131-17-2; Nylon 6/6, 10% carbon fiber-reinforced; statically dissipative thermoplastic composite for protection against electrostatic discharge damage; used in electronic packaging systems and functional components. [LNP]

Stat-Kon® RF-15. CAS 32131-17-2; Carbon powder grade nylon 6/6 with 15% glass fiber reinforcement; statically dissipative thermoplastic composite for protection against electrostatic discharge damage; used in electronic packaging systems and functional components. [LNP]

Stat-Kon® W. CAS 26062-94-2; PBT polyester, carbon powd. grade; statically dissipative composite for electronic packaging systems and functional components where protection from ESD damage is required. [LNP]

Stat-Kon® ZC-1003. CAS 25134-01-4; Modified PPO, 15% carbon fiber-reinforced; statically dissipative thermoplastic composite for protection against electrostatic discharge damage; used in electronic packaging systems and functional components. [LNP]

Stat-Rite® C-2300. Chain extended low molecular weight polyoxirane; static dissipative polymer for alloying wide variety of thermoplastics. [BFGoodrich/Spec. Polymers]

statuary bronze. A variable alloy. It usually contains from 75-95% copper, 1-10% tin, 0-5% zinc, 0-6% lead, 0.12-0.34% phosphorus, and 0.19-0.7% nickel.

statuary marble. Marble, $CaCO_3$, with a crystalline or saccharoid structure.

Stature. Static control additive. [Dow]

Status®. For agricultural applications. [DuPont UK]

Statyl. A proprietary preparation of methyl benzoquate; a veterinary antiprotozoan. §

Staufen. PVC film. [ICI Chem. & Polymers Ltd] †

Stauffer N-1386®. CAS 3064-70-8; Bis(trichloromethyl) sulfone; industrial biocide for control of algae, bacteria and fungi; slimicide for paper/paperboard production; preservative for adhesives, latexes, secondary oil well recovery. [Akzo]

staurolite. A mineral that is a basic aluminum ferrous iron silicate, $HFeAl_5Si_2O_{13}$.

staurotide. See staurolite.

---

‡ Trade name and manufacturer not verified   § Trade name without identified manufacturer

Staybelite®. Hydrogenated rosin; used as a modifier for wax-elastomer ethylene adhesive compositions; used in electrical cable paper saturants, in ceramic ink vehicles, in metal resinates and soldering fluxes. [Hercules; Hercules Ltd]

Staybrite. A proprietary trade name for stainless steels containing chromium and nickel; they usually contain 18% chromium, 8% nickel, 74% iron, sometimes with molybdenum and occasionally with titanium and tungsten; they possess extreme malleability and are very resistant to corrosion. §

Stay Off. CAS 7784-25-0; Aluminum ammonium sulfate; bird and animal repellent. [Synchemicals Ltd; Vitax Ltd]

Stcherbokov's solder. An aluminum solder. It contains 49% zinc, 46% tin, and 1.5% aluminum.

steadite. Iron-phosphorus eutectic, consisting of about 61% iron-phosphide, Fe_3P, with iron, a constituent of cast iron. The same name has been applied to a basic calcium-silico phosphate, $3(CaO \cdot P_2O_5) \cdot 2CaO$ $(2CaO \cdot SiO_2)$, found in the basic slag of the Thomas-Gilchrist process for the dephosphorization of iron.

Stead's reagent. A reagent consisting of 100 cc methyl alcohol, 18 cc water, 2 cc concentrated hydrochloric acid, 1 g copper chloride ($CuCl_2 \cdot 2H_2O$), 4 g magnesium chloride ($MgCl_2 \cdot 6H_2O$); an etching reagent used in the examination of steels.

Steamate. Boiler water treatment. [Grace Dearborn Ltd]

steam black. See logwood.

steamed bone meal. A fertilizer consisting of crushed bones, which have been treated with superheated steam and benzene, to remove fat and glue and contains about 1% nitrogen.

Steamfilm FG. CAS 124-30-1; EINECS 204-695-3; Octadecylamine aq. emulsion; corrosion inhibitor for control of corrosion caused by carbon dioxide and oxygen in the afterboiler section of steam-generating systems by forming a nonwettable, monomolecular film on metal surfaces. [Drew Ind. Div.]

steam glue. (Russian steam glue). A preparation of glue made by treating glue with nitric acid.

steapsin. (pancreatic lipase) Fat-digesting enzyme contained in the pancreatic juice.

Stearal. CAS 112-92-5; EINECS 204-017-6; Stearyl alcohol; emollient, auxiliary emulsifier, texturizer; nonoily, velvety feel; higher visc. in emulsions. [Amerchol Corp]

stearalkonium chloride. (stearyl dimethyl benzyl ammonium chloride; octadecyl dimethyl benzyl ammonium chloride; N,N-dimethyl-N-octadecylbenzenemethanaminium chloride) Quaternary ammonium salt; $C_{27}H_{50}N$ · Cl; CAS 122-19-0; EINECS 204-527-9; cosmetics, surfactants, antimicrobials. [Ferrosan Fine Chem. A/S; Lonza; Mason; McIntyre; Sherex]

stearamide. (octadecanamide; stearic acid amide; Amide C_{18}) Aliphatic amide; $CH_3(CH_2)_{16}CONH_2$; CAS 124-26-5; EINECS 204-693-2; slip/antiblock agent for LDPE, HDPE, PP. [Akzo; Astor Wax; Chemax; Croda Universal; Henkel/Emery; Syn. Prods.; Witco/Humko]

Stearex. CAS 57-11-4; EINECS 200-313-4; A trademark for a standardized stearic acid; it is a commercially pure, free fatty acid prepared for rubber manufacture; two grades: double pressed stearic acid, and single pressed stearic acid; the single pressed grade contains more oleic acid.§

steargillite. A variety of the mineral montmorillonite.

stearic acid. (n-octadecanoic acid; carboxylic acid C_{18}) Fatty acid; $CH_3(CH_2)_{16}COOH$; CAS 57-11-4; EINECS 200-313-4; cosmetics, chemicals, dispersant, softener in rubber compounds, food packaging, suppositories, ointments. [Akrochem; Akzo; Henkel/Emery; Lonza; Syn. Prods.; Unichema; Witco/Humko]

stearic acid amide. See stearamide.

stearic acid, calcium salt. See calcium stearate.

stearic acid, lead salt. See lead stearate.

stearic acid, potassium salt. See potassium stearate.

stearic acid, sodium salt. See sodium stearate.

stearin pitch. (candle pitch, candle tar). A pitch obtained in the sulfuric acid treatment of fats. After distillation in steam of the washed acids (stearic, palmitic, and oleic), stearin pitch remains to the extent of 2%.

Stearite. CAS 57-11-4; EINECS 200-313-4; A proprietary trade name for synthetic stearic acid produced by hydrogenation of certain oils. §

stearopodis. Magnesium stearate; used in the preparation of soap and face creams.

stearosan. A compound of santalol and stearic acid.

stearyl alcohol. (n-octadecanol; 1-octadecanol; C18 linear alcohol) Fatty alcohol; $CH_3(CH_2)_{16}CH_2OH$; CAS 112-92-5; EINECS 204-017-6; perfumery, cosmetics, intermediate, surfactants, lubricants, resins, antifoam agent. [Aarhus Oliefabrik A/S; Amerchol; Chemron; Croda; Ethyl; Lipo; Lonza; M. Michel; Procter & Gamble; Sherex; Vista]

stearyl erucamide. (N-octadecyl-13-docosenamide) Substituted aliphatic amide; $CH_3(CH_2)_7CHCH(CH_2)_{11}$-$CONHCH_2(CH_2)_{16}CH_3$; CAS 10094-45-8. [Croda Universal; Witco/Humko; Zeeland]

stearyl/lauryl thiodipropionate. $S(CH_2CH_2COOC_{12}H_{25})$-$(CH_2CH_2COOC_{18}H_{37})$; CAS 131-0352-1. [Evans Chemetics]

stearyl methacrylate. (octadecyl methacrylate) CH_2C-$(CH_3)COOC_{18}H_{37}$; CAS 32360-05-7; Lube oil additive, pour point depressant, paper coatings, textile finishes, paints, varnishes, pressure-sensitive adhesives. [CPS; Rohm & Haas; Sartomer]

Steclin. CAS 64-75-5; Tetracycline hydrochloride; antiamebic; antibacterial; antirickettsial. [Bristol-Myers Squibb Co Inc] *

Stedbac®. CAS 122-19-0; EINECS 204-527-9; Stearalkonium chloride; hair conditioner, emulsifier imparting softness, antistatic properties to cream rinse formulations. [Zeeland]

steel. Iron containing combined carbon up to 1.5%. High carbon steels contain from 0.5-1.5% carbon, and mild steels from a trace to 0.5% carbon.

Steel 0. A British chemical standard; a nickel steel containing 0.325% carbon, 0.590% manganese, and 3.985% nickel.

Steel 01. A carbon steel containing 0.333% carbon, 0.162% silicon, 0.032% sulfur, 0.031% phosphorus, 0.617% manganese, 0.024% arsenic, 0.162% nickel, 0.017% chromium, and 0.037% copper.

Steel A2. A British chemical standard carbon steel containing 0.037% carbon, 0.034% silicon, 0.020% sulfur, 0.008% phosphorus, 0.043% manganese, 0.031% arsenic, 0.059% nickel, 0.013% chromium , 0.067% copper, 0.04% oxygen, 99.72% iron.

Steel B4. A British chemical standard carbon steel contain-

---

ing 0.400% carbon, 0.026% silicon, 0.046% sulfur, 0.103% phosphorus, 0.735% manganese, and 0.140% arsenic.

Steel C. A British chemical standard carbon steel containing 0.093% carbon.

Steel E. A standard steel containing 0.115% carbon and 0.491% manganese; used as the colorimetric standard for the determination of carbon in steels containing more than 0.100% carbon.

Steel F. A German steel containing 0.67-1.1% silicon and 0.1-0.14% carbon.

Steel H. A British chemical standard carbon steel containing 0.428% carbon, 0.047% sulfur, and 0.035% phosphorus.

Steel I. A British chemical standard carbon steel containing 0.521% carbon and 0.726% manganese.

Steel M. A British chemical standard. It is a carbon steel containing 0.228% carbon and 0.057% silicon.

Steel N. A British chemical standard carbon steel containing 0.17% carbon, 0.117% silicon, 0.034% sulfur, 0.037% phosphorus, 0.432% manganese, and 0.029% arsenic.

Steel N1. A carbon steel containing 0.153% carbon, 0.176% silicon, 0.050% sulfur, 0.036% phosphorus, 0.527% manganese, 0.030% arsenic, 0.260% nickel, and 0.04% copper. It is a British chemical standard.

Steel P. A high silicon and phosphorus steel. It is a British chemical standard.

Steel R. A British chemical standard. It is a carbon steel containing 0.786% carbon, 0.053% sulfur, and 0.914% manganese.

Steel S1. A British chemical standard. It is a carbon steel containing 0.921% carbon and 0.051% phosphorus.

Steel T. A British chemical standard nickel steel containing 3.367% nickel.

Steel U. A British chemical standard carbon steel containing 1.203% carbon, 0.472% manganese, and 0.608% nickel.

Steel V. A British chemical standard alloy steel containing 0.548% carbon, 0.161% silicon, 0.063% sulfur, 0.024% phosphorus, 0.542% manganese, 0.861% chromium, and 0.273% vanadium.

Steel V2A. A rustless steel containing iron with 20% chromium, 7% nickel, and 0.2% carbon.

Steel W. A British chemical standard. It is an alloy steel containing 0.695% carbon, 0.187% silicon, 0.075% sulfur, 0.028% phosphorus, 0.101% manganese, 0.44% nickel, 3.01% chromium, 0.791% vanadium, 4.76% cobalt, and 16.21% tungsten.

Steel W2. A British chemical standard high-speed alloy steel containing 0.17% carbon, 0.14% silicon, 0.051% sulfur, 0.220% manganese, 3.29% chromium, 0.82% vanadium, 16.12% tungsten, 4.35% cobalt, 0.43% nickel, and 0.55% molybdenum.

Steel Guard. Petroleum base; rust inhibitors for steel, industrial, commercial and vehicle applications. [Adasco Inc]

steelite. An explosive consisting of potassium chlorate, mixed with oxidized resin, and a little castor oil.

steel ore. A variety of cinnabar containing 75% mercury.

Steinazid SBU 185. Undecylenic acid alkylolamide sulfosuccinate in liquid form; antidandruff agent, fungicidal, and bacteriostatic additive for shampoos, hair lotions, foam baths, etc. [Rewo Chemicals Ltd] *

Stelabid Tablets. Combination of isopropamide iodide and trifluoperazine hydrochloride; for treatment of gastro-

intestinal disorders in which hypersecretion and/or painful spasms are a problem with condition complicated by emotional factors. [SmithKline Beecham] *

Steladex. A proprietary preparation of trifluoperazine dihydrochloride and dexamphetamine sulfate. [SmithKline Beecham] *

Stelazine. Trifluoperazine hydrochloride; antipsychotic; sedative. [SmithKline Beecham] *

Stelazine Tablets, Spansule Capsules, Syrup, Injection and Concentrate. Trifluoperazine hydrochloride (phenothiazine tranquilizer); for treatment of schizophrenia and other psychotic states; in low dosage for treatment of anxiety states. [SmithKline Beecham] *

Stelex. Ceramic foam filter for steel and reactive alloy castings. [Foseco (F.S.) Ltd]

Stellak. Chemicals for boiler water treatment. [Steetley Chemicals Ltd] *

Stellar 500. CAS 14087-96-6; Talc; for plastics applications where good color and brightness are required; antiblocking agent in polyolefin films. [Cyprus Industrial Min.]

Stellite®. Cobalt-base alloys; excellent wear resist. with limited corrosion resist. [Haynes Int'l.]

Stellited metal. Metals treated with an alloy consisting chiefly of chromium, tungsten, and cobalt (stellite). The metals treated are usually steel, cast iron, malleable iron, and semi-steel. It is an economical method for these treated metals are rendered suitable for wear-resisting parts of machinery.

Stellos. CAS 7778-54-3; Calcium hypochlorite; used for chlorination of water. [Stella-Meta Filters]

Stellox 380EC. An emulsifiable concentrate containing 190 g bromoxynil and 190 g ioxynil per liter; a post-emergence contact herbicide for cereal crops. [Ciba-Geigy Agrochemicals]

Stelogen. Flux for degassing steel. [Foseco (F.S.) Ltd]

Stelopack. Uphill teeming flux for killed steel ingots. [Foseco (F.S.) Ltd]

Stelorit. Covering and cleansing fluxes for steels. [Foseco (F.S.) Ltd] †

Stelotol. Powder flux for uphill teeming. [Foseco (F.S.) Ltd]

Stelpur. Insulating feeding sleeve with Stelex ceramic foam filter incorporated within it. [Foseco (F.S.) Ltd]

Stemetil. Prochlorperazine dimaleate or mesylate; an antiemetic and sedative. [Rhone-Poulenc Rorer Ltd]

Stemex. CAS 1597-82-6; Paramethasone acetate; a glucocorticoid. [Syntex Laboratories Inc] *

Stempor DG. CAS 83601-81-4; Carbendazim; systemic fungicide. [ICI Agrochemicals]

Stenol 1618. Saturated fatty alcohols (45-55% C16, 45-55% C18); intermediate for surfactant manufacturing. [Henkel/Emery]

Stenorol. CAS 64924-67-0; Halofuginone hydrobromide; antiprotozoal. [Roussel UCLAF, Fine Chemicals] *

Stental. CAS 50-06-6; Phenobarbital; anticonvulsant; hypnotic; sedative. [Wyeth Laboratories] *

Stentor Steel. A proprietary trade name for a non-shrinking steel containing 1.6% manganese, 0.25% silicon, and 0.9% carbon. §

Steol®. Alkyl ether sulfates; detergents, emulsifiers, foaming agents. [Stepan] *

Steol 3 OS. Sodium lauryl ether sulfate as a clear yellow viscous liquid; anionic surfactant used in shampoos, foam baths, and liquid detergents. [KWR Chemicals Ltd]*

---

‡ Trade name and manufacturer not verified   § Trade name without identified manufacturer

Steol® 4N. CAS 9004-82-4; Sodium laureth sulfate; detergent, emulsifier, foamer, wetting agent used in personal care products; car wash, dishwash; textile mill applications; emulsion polymerization. [Stepan; Stepan Canada]

Steol 4N. Sodium fatty ether sulfate as a nearly water white liquid; for shampooo, bubble bath, liquid cleaner. [KWR Chemicals Ltd] *

Steol 7T. Triethanolamine fatty ether sulfate as a pale yellow liquid; very mild base for shampoos and bubble baths. [KWR Chemicals Ltd] *

Steol® CA-460. CAS 67762-19-0; Ammonium laureth sulfate; detergent, emulsifier, foamer, dispersant, and wetting agent used in shampoos, dishwashers, car washers, textile mill applications, emulsion polymerization. [Stepan; Stepan Canada]

Steol CA-460 and KA-460. Ammonium fatty ether sulfate in liquid form; for shampoo, bubble baths, dish detergents, and degreasers. [KWR Chemicals Ltd] *

Steol® COS 433. CAS 9014-90-8; Sodium nonoxynol-4 sulfate; emulsifier for acrylics, SBR, vinyl chloride, butyl rubber. [Stepan Canada]

Steol® CS-130. Sodium laureth sulfate; surfactant for personal care applications. [Stepan; Stepan Canada]

Steol CS-460 and KS-460. Sodium fatty ether sulfate in liquid form; for shampoo, bubble baths, dish detergents and degreasers. [KWR Chemicals Ltd] *

Steol CS-760 and 7N. Sodium fatty ether sulfate as a pale yellow liquid; very mild base for shampoos and bubble baths. [KWR Chemicals Ltd] *

Steol FA. Ammonium fatty ether sulfate as a pale yellow liquid; for general detergent uses and manufacture of gypsum board. [KWR Chemicals Ltd] *

Steol® OS 28. CAS 9004-82-4; Sodium laureth sulfate; detergent, foaming agent for all-purpose and specialty household/industrial cleaners. [Stepan Europe]

Stepan. Emollient; for bath oils; antiperspirants etc. [KWR Chemicals Ltd] *

Stepan C-40. Methyl laurate; intermediate for manufacturing of detergents, emulsifiers, wetting agents, stabilizers, lubricants, plasticizers, resins, and textile specialties; lubricant for metalworking formulations. [Stepan; Stepan Canada]

Stepan C-65. Methyl palmitate-oleate; intermediate for manufacturing of detergents, emulsifiers, wetting agents, stabilizers, lubricants, plasticizers, resins, and textile specialties; lubricant for metalworking formulations. [Stepan; Stepan Canada]

Stepan C-68. Methyl oleate/stearate; intermediate for manufacturing of detergents, emulsifiers, wetting agents, stabilizers, lubricants, plasticizers, resins, and textile specialties; lubricant for metalworking compounds. [Stepan; Stepan Canada]

Stepan TAB®-2. Dihydrogenated tallow phthalic acid amide; emulsifier, suspending agent for silicones, zinc pyrithione, sulfur, selenium sulfide, coal tar, oil extracts; opacifier; especially for use in conditioning and antidandruff shampoos. [Stepan; Stepan Europe]

Stepanate®. Hydrotropes; coupling agent, cloud point depressant. [Stepan] *

Stepanate® AXS. CAS 26447-10-9; EINECS 247-710-9; Ammonium xylene sulfonate; hydrotrope, solubilizer, coupler in detergent field; heavy duty cleaners, wax strippers, dishwashing detergents; solv. or fluidizer;

freeze-thaw stabilizer; additive for high electrolyte or brine systems. [Stepan; Stepan Canada]

Stepanate® SCS. CAS 32073-22-6; Sodium cumene sulfonate; coupler or solubilizer for liquid cleaners; detergent slurries. [Stepan; Stepan Canada]

Stepanate® SXS. CAS 1300-72-7; EINECS 215-090-9; Sodium xylene sulfonate; solubilizer, coupling agent, cloud point depressant, visc. reducer for industrial and household detergents, textile applications. [Stepan; Stepan Canada]

Stepanflo. Surfactant; for enhanced oil recovery. [Stepan]*

Stepanflote® 85L. Sodium alkyl ether sulfate; flotation reagent for molybdenum ore. [Stepan; Stepan Canada]

Stepanform® 1440. Blend; foaming agent, air entrainer for cellular concrete. [Stepan; Stepan Canada]

Stepanform® 1750. Blend; foamer for drilling applications. [Stepan; Stepan Canada]

Stepanhold® Extra. CAS 26589-26-4; PVP/ethyl methacrylate/methacrylic acid terpolymer; hair care fixative for super-hold formulations. [Stepan; Stepan Canada; Stepan Europe]

Stepanhold® R-1. CAS 26589-26-4; Acrylates/PVP copolymer; hair fixative for hair sprays, setting lotions, conditioners. [Stepan; Stepan Canada]

Stepan-Mild® LSB. Sodium lauryl sulfoacetate, disodium laurethsulfosuccinate; surfactant for shampoos, hand soaps, bubble baths, facial cleansers, baby products, sensitive skin products. [Stepan; Stepan Canada; Stepan Europe]

Stepan-Mild® SL3. Disodium laurethsulfosuccinate; surfactant for low-irritation shampoos, bubble baths, dishwashing detergents. [Stepan; Stepan Canada]

Stepanol®. Alkyl sulfates; mild detergent, foaming agent. [Stepan] *

Stepanol® AEG. Ammonium lauryl sulfate, ammonium laureth sulfate, cocamidopropyl betaine, cocamide DEA; base for liquid hand and body soaps, shampoos. [Stepan; Stepan Canada]

Stepanol® AEM. Ammonium laureth sulfate, cocamide MEA; concentrate for shampoo and bath products [Stepan; Stepan Canada]

Stepanol® AM. CAS 2235-54-3; EINECS 218-793-9; Ammonium lauryl sulfate; detergent, foamer used in personal care products; rug and upholstery shampoos; household, metal, and industrial cleaners; fruit washing; insecticides; textile and leather processing; pharmaceuticals. [Stepan; Stepan Canada]

Stepanol AM. Ammonium fatty alcohol sulfate as a pale yellow liquid; for shampoo; bubble bath; liquid detergents. [KWR Chemicals Ltd] *

Stepanol DEA. Diethanolamine fatty alcohol sulfate as a pale yellow liquid; for shampoo. [KWR Chemicals Ltd] *

Stepanol® LX. DEA lauryl sulfate, DEA lauraminopropionate, sodium lauraminopropionate; mild blended concentrate for shampoos, baby soaps, bubble baths; foaming power; substantive to hair and skin. [Stepan; Stepan Canada]

Stepanol ME. Sodium fatty alcohol sulfate as a white powder; for powdered detergents. [KWR Chemicals Ltd]*

Stepanol® ME Dry. CAS 151-21-3; Sodium lauryl sulfate; detergent, foamer used in personal care products; rug and upholstery shampoos; household, metal, and industrial cleaners; fruit washing; insecticides; textile and

---

leather processing; pharmaceuticals. [Stepan; Stepan Canada]

Stepanol® MG. Magnesium lauryl sulfate; detergent, foamer used in personal care products; rug and upholstery shampoos; household, metal, and industrial cleaners; fruit washing; insecticides; textile and leather processing; pharmaceuticals. [Stepan; Stepan Canada]

Stepanol Mg. Magnesium fatty alcohol sulfate as a pale yellow liquid; for rug and upholstery shampoos. [KWR Chemicals Ltd] *

Stepanol® SPT. CAS 139-96-8; EINECS 205-388-7; TEA lauryl sulfate; mild detergent, foaming agent for household cleaners, liquid soaps. [Stepan Europe]

Stepanol WA, WAC, WAQ. Ionic surfactants from the Stepanol WA range; for shampoos, bubble baths, liquid and paste detergents. [KWR Chemicals Ltd] *

Stepanol WA-100. Sodium fatty alcohol sulfate as a white powder; anionic surfactant used as a dentifrice, and in the pharmaceutical industry. [KWR Chemicals Ltd] *

Stepanol® WA Extra. CAS 151-21-3; Sodium lauryl sulfate; detergent, foamer used in personal care products; rug and upholstery shampoos; household, metal, and industrial cleaners; fruit washing; insecticides; textile and leather processing; pharmaceuticals. [Stepan; Stepan Canada]

Stepanol WAT. Triethanolamine fatty alcohol sulfate as a nearly water white liquid; for shampoos. [KWR Chemicals Ltd] *

Stepanon CG. Surfactant; brine tolerant foamer for acid or alkaline media. [Stepan; Stepan Canada]

Stepan Pearl Series. Surfactant blend; pearlescent, satining or opacifying agent, conditioning agent for shampoos, liquid soaps, bubble baths, shower gels. [Stepan Europe]

Stepanquat® 6585. Dipalitmoylethyl hydroxyethylmonium methosulfate; mild surfactant for cream rinses and conditioners. [Stepan; Stepan Canada]

Stepantan®. Alkyl sulfonate; dispersant. [Stepan] *

Stepantan A. Anionic surfactant in powder form; dispersant, tanning agent. [KWR Chemicals Ltd] *

Stepantan® AS-12. Sodium alpha olefin sulfonate; foamer for soft and hard waters, fresh water, and moderate brine conditions. [Stepan; Stepan Canada]

Stepantan® DS-40. Sodium dodecylbenzene sulfonate; detergent, wetting agent, foaming agent; used specifically in dust control applications. [Stepan; Stepan Canada]

Stepantan® DT-60. TEA dodecylbenzenesulfonate; detergent, wetting agent, foaming agent; used specifically as air entraining agent in concrete applications. [Stepan; Stepan Canada]

Stepantan® H-100. Dodecylbenzene sulfonic acid; detergent intermediate; emulsifier for oils, solvs., waxes and oil field applications.; air entraining agent and foamer for cellular concrete; wetting agent. [Stepan; Stepan Canada]

Stepantan NP 80. Anionic surfactant in powder form; dispersant for phyto-sanitary products and wettable powders. [KWR Chemicals Ltd] *

Stepantex® B-29. Sodium octyl sulfate; wetting and mercerizing agent for textiles. [Stepan; Stepan Canada]

Stepantex® CO-30. CAS 61791-12-6; PEG-30 castor oil; lubricant, emulsifier for fiber finish and wet processing. [Stepan; Stepan Canada]

Stepantex Q90B. Dialkyl methoxysulfate as an amber viscous liquid; cationic surfactant used in textile softeners. [KWR Chemicals Ltd] *

Stepantex® TD14. Tall oil ester; biodeg. textile auxiliary [Stepan; Stepan Canada]

Stepantex® VS 90. Diester quaternary ammonium methyl sulfate; good hand, excellent rewet and antistatic properties for household and commercial rinse-added fabric softeners; textile processing aid for natural and synthetic fibers. [Stepan; Stepan Canada]

Stepfac® 8170. Ethoxylated nonylphenol phosphate; hydrotrope for nonionics; emulsifier for agriculture, emulsion polymerization, oils, metalworking lubricants, corrosion inhibitors, pigment dispersants; heavy-duty industrial/ household alkali cleaners; compatibility agent for liquid fertilizers. [Stepan; Stepan Canada]

Stepfac® 8171. Acid form; compatibility agent for agricultural formulations. [Stepan; Stepan Canada]

Stepfac® PN 10. Phosphate ester; emulsifier, wetting and dispersing agent for agricultural wettable powds. and flowables. [Stepan Europe]

Step-Flow 21. Nonionic dispersant; surfactant for aq. agricultural flowables. [Stepan; Stepan Europe]

Steposol®. Alkyl ether sulfate; foaming agent. [Stepan] *

Steposol® CA-60H. Ammonium ether sulfate; high flash point foamer for heavy brine conditions. [Stepan; Stepan Canada]

Steposol® CA-207. Ammonium ether sulfate; foaming agent used for oilfield applications; also for gypsum board, cellular concrete, air drilling, foam cleaners; excellent for heavy brine conditions; stable in soft or hard water. [Stepan; Stepan Canada; Stepan Europe]

Stepsperse® DF-100. Surfactant blend; dispersant for agricultural flowables and dry flowables. [Stepan; Stepan Europe]

Stepwet® DF-60. Surfactant blend; wetting agent for agricultural flowables and dry flowables. [Stepan; Stepan Europe]

Sterane. CAS 50-24-8; Prednisolone; glucocorticoid. [Pfizer Inc] †

Sterane IA, IM. CAS 52-21-1; Prednisolone acetate; glucocorticoid. [Pfizer Inc] †

Sterbon. See Carbora. §

sterculia gum. See karaya gum.

Sterculia Kernals. See olives of Java.

Stereon® 840A. Stereospecific SBR block copolymer; rubber modifier for thermoplastic resins, especially for HIPS, flame-retardant HIPS, and PP. [Firestone Syn. Rubber]

Stereon® 881. Butadiene-styrene copolymer; offers low gel, high gloss, toughness, clarity, and processing ease for injection molded items including medical devices, containers; toys, food containers; FDA compliance. [Firestone Syn. Rubber]

Stereon® 900. Multiblock styrene-butadiene copolymer; designed to blend easily with general purpose PS to yield extruded and thermoformed products with excellent clarity, sparkle, impact resist., and flexibility; FDA compliance. [Firestone Syn. Rubber]

stereotype plate. An alloy of 85% lead and 14% antimony, sometimes with the addition of a little tin.

steresol. An antiseptic varnish made by dissolving 270 parts purified shellac, 10 parts benzoin, 10 parts balsam of tolu, 100 parts phenol, 6 parts oil of cinnamon, and 6

---

‡ Trade name and manufacturer not verified   § Trade name without identified manufacturer

parts saccharin in alcohol, to make 1000 parts.

Sterethox. CAS 75-71-8; Sterilizer containing dichlorodifluoromethane. [ICI Chem. & Polymers Ltd]

Sterets Pre-Injection Swabs®. CAS 67-63-0; EINECS 200-661-7; Isopropyl alcohol 70%; used for injection site cleansing. [Seton Healthcare Group plc]

Steribath. An antiseptic solution containing an iodophore. [ICI Chem. & Polymers Ltd] †

Steridex. Fungicidal water-based elastomeric protective coating, applied by brush or spray; for totally eradicating mold growth and bacteria in all hygiene sensitive environments - hospitals, food factories, breweries. [Liquid Plastics Ltd] †

Steriflux. A range of sterile nonpyrogenic intravenous infusions. [The Boots Co plc] *

SteriLine 200. CAS 14807-96-6; Platy talc USP; general purpose, coarse talc for use in dusting and baby powds., soaps, antiperspirant sticks, color extensions. [Montana Talc]

SteriLine 665. CAS 14807-96-6; Platy talc USP; ultrafine talc for cosmetic applications requiring high oil absorp., gloss control, and smoothness; visc. control additive in pharmaceutical excipient applications. [Montana Talc]

Sterilite®. Emulsifiers, tar oils and high boiling tar acids; fungicidal winter wash for fruit; disinfectant for animal health field. [Coventry Chemicals Ltd]

Sterilite Hop Defoliant. CAS 120-12-7; Anthracene oil; used for chemical stripping in hop vines. [Coventry Chemicals Ltd]

Sterillium. Synthetic phenolic germicides in a detergent base; a disinfectant for laundry use. [William Pearson Ltd] *

Sterisafe. Food sterilizer treatments. [Grace Dearborn Ltd]

Sterisheen. Modified acrylic semi-gloss, tough, flexible, waterbased coating; applied by brush or spray; for totally eradicating mold, micro-organisms, fungi, and bacteria in all hygiene sensitive environmentshospitals, food factories, breweries, kitchens. [Liquid Plastics Ltd]

Sterisil. A proprietary trade name for hexetidine. §

Sterisol. CAS 5980-31-4; Hexedine; antibacterial. [Parke-Davis] *

Steritile® Plus. A two-component, water-based, acrylic epoxy coating; forms a tough, durable surface; easy to clean and resists gowth of mold and bacteria; applied by brush, roller airless spray; ideal for hospitals, food factories, breweries and kitchens. [Liquid Plastics Ltd]

Steriwipe®. Cetrimide 0.15% and chlorhexidine acetate 0.015%; used for topical disinfection, cleansing minor wounds and abrasions. [Seton Healthcare Group plc]

Sterline. An alloy of 68% copper, 17-18% nickel, 13-14% zinc, 0.75-0.8% iron, and 0-0.8% lead. It is a nickel silver (German silver).

Sterling. CAS 1333-86-4; Carbon black. [Cabot Carbon Ltd]

Sterling Gold. See Standard gold.

sterling solder. An alloy of 61.6% tin, 15.2% zinc, 11.2% aluminum, 8.3% lead, 2.5% copper, and 1.2% antimony.

Sterlite. A proprietary trade name for a nickel brass containing 25% nickel, 20% zinc, and small amounts of iron, manganese, silicon, and carbon. §

Sterlith. A trademark for materials of the refractory and abrasive type; they consist essentially of crystalline alumina. §

Sternite. Phenol formaldehyde resin; molding powders. [Manchem Ltd] *

Sternite. Phenolic and polystyrene molding materials. [Sterling Moulding Materials] †

Stero WW. Wool grease with low free fatty acids; lubricant, emulsifier, rust inhibitor; base for wire drawing and metalworking compounds, component for metal polishes, cationic fat liquors, leather stuffing compounds; fingerprint inhibitor in rust preventatives. [Climax Fluids Additives]

Sterocoll. Alkali-soluble dispersion. [BASF plc]

Sterotabs. Water treatment tablets. [The Boots Co plc] *

Sterotex®. CAS 68334-28-1; Hydrogenated vegetable oil; binder and internal lubricant for pressed powds. [Karlshamns]

Sterotex® HM. CAS 68334-28-1; Hydrogenated vegetable oils; lubricant in pharmaceutical tableting, powd. compression applications. [Karlshamns]

Sterovum. Ethynerone; antifertility agent. [Merck & Co Inc]†

Sterox. Dodecylphenol-ethylene oxide condensate (alkylaryl polyoxyethylene ether); nonionic surface active agent. [Monsanto Co]

Sterox DF, DJ. Anionic surface active agent. [Monsanto plc]

Steroxin-Hydrocortisone. A proprietary preparation of chlorquinaldol and hydrocortisone; used in dermatology as an antibacterial agent. [Ciba plc] *

Steroxol. Chlorinated detergent. [Rhone-Poulenc UK]

Sterpon. A proprietary polyester laminating resin. [Convert (Ets G)] *

Ster-Zac. A proprietary preparation of hexachlorophene; a topical antiseptic. [Hough, Hoseason & Co Ltd] *

STFF. See sodium tripolyphosphate.

St. Helen's powder. An explosive consisting of 92-95% ammonium nitrate, 2-3% aluminum powder, and 3-5% trinitrotoluene.

stibium. CAS 7440-36-0; Latin name for antimony, Sb.

Stick-lac. See lac.

Stickstoffoxydbaryt. Barium nitrite, $Ba(NO_2).H_2O$.

Stiedex. An oily cream available in three grades: 0.25%, LP 0.05% and LPN 0.05%; all grades contain desoxymethasone in an oily cream base; LPN 0.05% also contains neomycin; for the treatment of a wide range of acute inflammatory and allergic conditions and for chronic skin disorders. [Stiefel Laboratories (UK) Ltd] *

St. Ignatius bean. The seed of *Strychnos ignatii*.

Stik-It. Nonionic wetting agent for use in a wide range of fungicides and insecticides. [Quadrangle Agrochemicals]

Stillingia oil. An oil obtained by crushing the kernel of *Stillingia sebifera*.

Stilphostrol. Diethylstilbestrol diphosphate; estrogen. [Miles Pharmaceuticals]

Stimate Injection. Desmopressin acetate; antidiuretic. [Armour Pharmaceutical Co] *

Stimplete. A proprietary preparation of phenobarbitone dexamphetamine sulfate, thiamine hydrochloride, riboflavine, pyridoxine hydrochloride, nicotinamide, and alcohol. [Wyeth Laboratories] *

Stimufol. Soluble fertilizer. [ICI Chem. & Polymers Ltd]

Sting. Herbicide. [Monsanto plc]

stink quartz. (fetid quartz). A quartz which has a bad odor, due to organic matter.

stink-stone. (oil-stone). A bituminous schist found in the Tyrol. A source of ichthammol.

Stirene. Polystyrene. [Dow]

Stiresol®. Styrenated glycerol phthalate resin solution.

[Reichhold]

Stirling's gentian violet. A microscopic stain. It contains 5 g gentian violet, 10 cc 95% alcohol, 2 cc aniline, and 88 cc water.

St. Johns Wort Oil CLR. Fatty oil extract of St. Johns wort blossoms; general skin protection, especially for sensitive skin. [Dr. Kurt Richter; Henkel/Cospha]

St Joseph. CAS 50-78-2; Aspirin; analgesic; antipyretic; antirheumatic. [Schering Corp]

St Joseph Cough Syrup. Dextromethorphan hydrobromide; an antitussive. [Schering Corp]

Stockalite. A proprietary product; it is a very highly refined china clay used as a filler for tires, cables, and high grade mixes. §

Stockholm. (pine tar). A tar obtained principally from pinewood distillation. It is obtained from *Pinus sylvestris* and other species of *Pinus*; used as a preservative paint for ships and roofing and as a rubber softener. It has antioxidant properties.

Stockholm pitch. Pine-wood tar pitch. It is soluble in alkalies, and is used in the preparation of varnishes, in the rubber and gutta-percha trades, and in the preparation of impervious cements.

Stoco. A proprietary bituminous plastic. §

Stoddard Solvent. A proprietary trade name for a refined petroleum product for dry cleaning. §

Stoffertite. A calcium phosphate, $CaHPO_4 \cdot 5H_2O$. It occurs in guano.

stoic metal. An alloy similar in composition to Invar.

Stoke's reagent. A reducing agent prepared by dissolving 30 g ferrous sulfate and 20 g tartaric acid in 1 liter of water. When required for use, strong ammonia is added until the precipitate first formed is redissolved.

Stoller Flowable Sulphur. CAS 7704-34-9; Suspension concentrate containing 720 g/l sulfur; a protectant fungicide. [Stoller Chemicals Ltd]

Stomahesive. A proprietary preparation containing gelatin pectin, carboxy-methyl cellulose and polyisobutylene on a protective film; used for the protection of skin around surgical stomata. [Bristol-Myers Squibb Pharmaceuticals Ltd]

Stomosan. Ethylamine phosphate; used in medicine.

Stomp. CAS 40487-42-1; Emulsifiable or suspension concentrate containing pendimethalin; a dinitroaniline herbicide for cereals and bush fruit. [Cyanamid of Great Britain Ltd]

Stomp H. CAS 40487-42-1; Emulsifiable or suspension concentrate containing pendimethalin; a dinitroaniline herbicide for cereals and bush fruit. [Hortichem Ltd]

stone, green. See Bohemian earth.

Stoner A500. Citrus distillates; biodeg. cleaner/degreaser for cleaning built-up, baked-on residue from molds, tools, machinery, and other equip. [Stoner]

Stoner E800. Proprietary nonsilicone release blend; mold release lubricant for molds for natural rubber, nitrile, silicone, neoprene, SBR, and other compounds. [Stoner]

Stoner K206. Proprietary silicone release blend; mold release for release of plastics, rubber, and waxes. [Stoner]

stone's bronze. An alloy of 87% copper, 11% tin, and 2% phosphor-copper.

stone wax. A name applied to carnauba wax.

Stonite. An explosive consisting of 68% nitroglycerin, 20%

kieselguhr, 8% potassium nitrate, and 4% wood meal.

Stoodite. A proprietary trade name for a high manganese steel. §

Stop. CAS 7783-47-3; Stannous fluoride; dental caries prophylactic. [Oral-B Laboratories Inc] *

Stora. A proprietary trade name for a Swedish charcoal iron used for making malleable iron. §

Storaid Dust. CAS 117-18-0, 148-79-8; Dustable powder containing 6% w/w tecnazene and 1.8% w/w thiabendazole; a protectant fungicide and potato sprout suppressant. [MSD Agvet]

Storalon. See Carbura. §

storax. (styrax). An oleoresin, the product of the tree *Liquidambar orientalis*. The crude material contains 20-30% water and fragments of bark, etc. Prepared storax is used as a drug.

storax calamita. The powdered bark of *Liquidambar styraciflua*, most of the resin being first extracted. The product has no connection with storax.

Storite. CAS 148-79-8; Thiabendazole; a systemic insecticide. [MSD Agvet]

Storite SS. CAS 117-18-0, 148-79-8; Water soluble powder containing 300 g tecnazene and 100 g thiabendazole per liter; a protectant fungicide and potato sprout suppressant. [MSD Agvet]

Storm®. Bentazon, acifluorfen; for post-emergence control of broadleaf weeds in soybeans and peanuts [BASF AG]

Stortex. Malt extract. [ABM Chemicals Ltd] *

Stovarsol. Acetarsol. [May & Baker Ltd] *

Stowite. An explosive containing 58-61% nitroglycerin, 4.5-5% nitrocotton, 18-20% potassium nitrate, 6-7% wood meal, and 11-15% ammonium oxalate.

Stoxil. CAS 54-42-2; Idoxuridine; antiviral. [SmithKline Beecham] *

Strandex. Chemical compositions for use in industry as additives in polymer and plastics processing, all containing metal compounds. [Associated Lead Manufacturers Ltd] *

Strandol. Mould lubricant for continuous casting of steel billets. [Foseco (F.S.) Ltd] †

strass. (paste). A kind of glass used to imitate precious stones. It is made from 100 parts sand, 40 parts minium, 24 parts potassium carbonate, 20 parts borax, and 12 parts potassium nitrate. This gives a colorless product, and various oxides are added to color it.

Strassburg turpentine. The oleoresin from the silver fir, *Pinus picea*.

Strasser solder. An alloy of 62% tin, 12% zinc, 4% aluminum, 8% lead, 5% copper, 5% bismuth, and 4% cadmium.

Strata-Fire. A fuel additive used to reduce engine wear, improve engine performance, increase fuel economy and reduce emission of air pollutants; used for all types of internal combustion engines, both diesel and gasoline. [SN Corp/Appropriate Technology Ltd]

Stratos®. Cycloxydim; post-emergence graminicide against annual and perennial grasses; selective in broadleaf crops, e.g., sugar beet, cotton, soybean, vegetables, onions. [BASF AG]

Stratton. Solvent stripper. [Dow]

Stratyl. A proprietary polyester laminating compound. [Pechiney-St Gobain] *

Strawlink. Mineral/vitamin animal feed supplement for straw. [ICI Chem. & Polymers Ltd]

---

‡ Trade name and manufacturer not verified   § Trade name without identified manufacturer

Strelax. Road nosing compounds. [ICI Chem. & Polymers Ltd]

Strenes Metal. A proprietary trade name for a nickel-chromium-molybdenum cast iron. §

Strepolin. A proprietary preparation containing streptoroycin sulfate; an antibiotic. [Glaxo Laboratories]*

Strepsils. Proprietary preparations containing 2,4-dichlorobenzyl alcohol and amylmetacresol; antibacterial. [The Boots Co plc]

Streptets. A proprietary preparation of zinc bacitracin, polymixin B sulfate, and neomycin sulfate; antibiotic lozenges. [Wyeth Laboratories] *

Streptohydrazid. CAS 5667-71-0; Streptonicozid; antibacterial. [Pfizer Inc] †

Streptokinase. Kabikinase; an enzyme obtained from cultures of various strains of *Streptococcus haemolyticus* and capable of changing plasminogen into plasmin. §

Streptonivin. CAS 303-81-1; A proprietary trade name for novobiocin. §

Streptorex. CAS 3810-74-0; A proprietary preparation of streptomycin sulfate; an antibiotic. [Pfizer International]†

Streptotriad. A proprietary preparation of streptomycin, sulfadiazine, sulfathiazole and sulfadimidine; used in the treatment of dysentery. [May & Baker Ltd] *

Stresnil. CAS 1649-18-9; A proprietary preparation containing azaperone; veterinary sedative (pigs). [Janssen Pharmaceutical Ltd] *

Stretonex. CAS 3810-74-0; A proprietary preparation of streptomycin sulfate; an antibiotic. [Pfizer International]†

strewing smalt. The coarsest powdered smalt (*qv*).

Striate. Ion control resins. [Rohm & Haas]

Strim. Rimming agent for steel ingots. [Foseco (F.S.) Ltd]

Stripcote. Liquid parting agents and release agents. [Foseco (F.S.) Ltd]

stripping salt. See Abraum salts.

Strobane. A trademark for an insecticide and acaricide; it is based on polychlorinated terpine and contains 66% chlorine. §

Strodex® MO-100. 2-Ethylhexyl polyphosphoric ester acid anhydride; emulsifier; pigment dispersant used in oil-based paints and leather coating specialties. [Dexter]

Strodex® MOK-70. Phosphated alcohol; dispersant/stabilizer for carbon black pigments, zinc pigments, lead silicate in exterior latex paints; wetting and rewetting agent; rust inhibitor. [Dexter]

Strodex® P-100. Phosphated coester of alcohol and aliphatic ethoxylate acid anhydride; emulsifier; dispersant for pigments in polar and nonpolar solvs.; coupler for emulsifiers in aq. and nonaq. systems; used in metal and tile cleaners; pigment grinding aid. [Dexter]

Strodex® PK-90. Phosphated coester of alcohol and aliphatic ethoxylate; emulsifier, dispersant, wetting agent for extender pigments in latex paints, barium sulfate, and iron oxides; oxidation-corrosion inhibitor; used in heavy-duty alkaline cleaners. [Dexter]

Strodex® SE-100. Phosphated aliphatic ethoxylate acid anhydride; dispersant, emulsifier and coemulsifier in aq. and nonaq. systems. [Dexter]

Strodex® Super V-8. Phosphate ester, potassium salt; patented emulsifier, detergent, dispersant, wetting agent used in textile processing applications. [Dexter]

Stromba. CAS 10418-03-8; A proprietary preparation of stanozolol; an anabolic agent. [Bayer AG]

Stronscan-85. CAS 10476-85-4; Strontium chloride Sr 85; radioactive agent. [Abbott Laboratories] *

strontia. CAS 1314-11-0; Strontium oxide, SrO; used in the manufacture of strontium salts, pyrotechnics, greases, and soaps.

strontium. Metallic element; Sr; CAS 7440-24-6; alloys of strontium used in electron tubes as a 'getter' to combine chemically with active gases and to hold inactive gases by adsorption. [Atomergic Chemetals; Degussa; Noah Chem.]

strontium carbonate. $SrCO_3$; CAS 1633-05-2; EINECS 216-643-7; Catalyst, in radiation-resistant glass for color TV tubes, ceramic ferrites, pyrotechnics. [Atomergic Chemetals; Cerac; Mallinckrodt; Solvay GmbH]

strontium chloride. (strontium chloride hexahydrate) $SrCl_2$ or $SrCl_2 \cdot 6H_2O$; CAS 101476-85-4; strontium salts, pyrotechnics, electron tubes. [Aldrich; Atomergic Chemetals; Hoechst Celanese]

strontium chloride hexahydrate. See strontium chloride.

strontium nitrate. $Sr(NO_3)_2$; CAS 10042-76-9; EINECS 233-131-9; pyrotechnics, marine signals, railroad flares, matches. [Bernardy Chimie SA; Hoechst Celanese; Noah Chem.; Solvay GmbH]

strontium titanate. $SrTiO_3$; Electronics, electrical insulation. [Atomergic Chemetals; Ferro/Transelco; TAM Ceramics]

Strotope. CAS 10042-76-9; Strontium nitrate Sr 85; radioactive agent. [Bristol-Myers Squibb Co Inc] *

Struktol® 40 MS Flakes. Modified mixture of rubber-compatible nonhardening synthetic resins; resin plasticizer; homogenizing agent for elastomer blend compounds. [Struktol]

Struktol® Activator 73. Mixture of zinc salts of aliphatic and aromatic carboxylic acids; vulcanization activator for natural rubber. [Struktol]

Struktol® HP 55. Process aid for the tire industry; improves mixing and milling characteristics without adversely affecting cured properties, abrasion resist., or tear strength [Struktol]

Struktol® PE H-100. CAS 9002-88-4; Low molecular weight polyethylene homopolymer wax; improves flow and processability in natural and synthetic elastomers; provides release from equip. with improved pigment dispersion and finish to molded articles. [Struktol]

Struktol® SU 109. CAS 7704-34-9; Coated insoluble sulfur; used for rubber compounding. [Struktol]

Struktol® T.M.Q. Polymerized 2,2,4-trimethyl-1,2-dihydroquinoline; antioxidant in rubber tires, belts, mech. goods, sponge and retreading; minimally discoloring. [Struktol]

Struktol® WB 300. Blend of high molecular aliphatic and aromatic polyesters; plasticizer for nitrile rubber and chloroprene; end uses including petrol hoses, printing rollers, oil seals, etc.; processing aid. [Struktol]

struthiin. (githagin, polygalin, polygallic acid, senegin). Saponin, $C_{19}H_{30}O_{10}$, a glucoside found mainly in the common soapwort.

Strycin. CAS 3810-74-0; Streptomycin sulfate; antibacterial. [Bristol-Myers Squibb Co Inc] *

Stryden Forte Rapid. Streptomycin-penicillin bovine intramammary. [RMB Animal Health Ltd]

STS. See sorbitan tristearate.

ST-Size. Modified rosin emulsion size. [Hercules] *

stucco. A specially hard plaster which can be polished. There are two kinds: a) made from plaster-of-Paris; and

---

* Trade name not verified as available   † Trade name verified as obsolete

b) made from lime. They are usually mixed with size.

Stugeron. CAS 298-57-7; A proprietary preparation of cinnarizine; an antinauseant. [Janssen Pharmaceutical Ltd] *

Stugeron Forte. CAS 298-57-7; A proprietary preparation containing cinnarizine; for peripheral arterial disease. [Janssen Pharmaceutical Ltd] *

Stuk. Adhesives for footwear industry. [Avalon Chemical Co Ltd]

stupp. A mercurial soot condensed in the chambers during the treatment of mercury ores. It contains about 20% mercury as metal, and sulfate.

Sturcal®. CAS 471-34-1; Precipitated calcium carbonate $CaCO_3$; pharmaceutical, dentifrice. [Rhone-Poulenc Sturge Lifford]

Sturcarb. Whiting. [Rhone-Poulenc Sturge Lifford] †

Stycast® 1090. CAS 25928-94-3; Epoxy resin filled with glass microballoons; low-dens. syntactic foams which withstands high hydrostatic pressure; for room temperature cure; converts to thermoset solid by addition of a catalyst. [Emerson & Cuming] *

Stycast® 1210. CAS 25928-94-3; Filled epoxy resin; casting resin adapts to single-cure impregnation and potting, and exceeds MIL-Spec thermal shock requirements; not for room temperature cure; converts to thermoset solid by addition of second compound. [Emerson & Cuming]*

Stycast® 1266. CAS 25928-94-3; Unfilled epoxy resin; room temperature curing, high clarity casting epoxy; converts to thermoset solid by addition of second compound. [Emerson & Cuming] *

Stycast® 1467. CAS 25928-94-3; Filled epoxy resin; fire retardant casting resin offers machinability with high strength; self-extinguishing; for room temperature cure; converts to thermoset solid by addition of catalyst. [Emerson & Cuming] *

Stycast® 2850-FT. CAS 25928-94-3; Filled epoxy resin; low expansion, high thermal conductivity, good thermal shock, and high temp. resistance; castings can be machined by grinding; for room temperature cure; converts to solid thermoset by addition of catalyst. [Emerson & Cuming]

Stycond-109. Conductive/static dissipative thermoplastic composite. [United Composites]

Stygene Series. Polynuclear aromatic polymers derived from specially prepared petrol. stream; resins in rubber compounding, joint cements, plastic compounds, protective coatings, fiber board, inks, epoxy potting compounds, adhesives, insecticides, briquettes, floor tile, etc.; soft grades as rubber plasticizers; hard grades as rubber extenders; carrier for insecticidal toxicants; in calendered and extruded goods, mech. goods. [Chemfax]

Stylac® ABS. CAS 9003-56-9; Styrene-acrylonitrile-butadiene copolymer [Asahi Chem. Industry]

Stylac® AS. CAS 9003-54-7; Styrene-acrylonitrile copolymer [Asahi Chem. Industry]

Styphen I. A mixture of stirenated phenols; a proprietary antioxidant. [Corning Glass Works, Zircoa Products] *

styphnic acid. (trinitroresorcinol ; 2 : 4-dihydroxy-1 : 3 : 5-trinitro-benzene) $C_6H_3O_8N_3$; CAS 82-71-3; Used in explosives as a priming agent.

stypticine. Cotarnine hydrochloride; used medicinally as a styptic.

Styquin. CAS 54400-62-3; Butamisole hydrochloride;

anthelmintic [Am. Cyanamid] *

styracin. Cinnamyl cinnamate, $C_6H_5 \cdot CH:CH \cdot CH_2 \cdot O \cdot CO \cdot CH:CH \cdot C_6H_5$.

Styrafil®. A trade name for flame-retardant polystyrene. §

Styraloy 22, 22A. A proprietary trade name for an elastomeric styrene derivative. §

Styramic H.T. and M.T. Proprietary trade names for polystyrene thermo-plastics possessing a higher softening point than usual; they are stated to be polydichlorstyrenes. §

styrene-acrylonitrile copolymer. (SAN; SAN copolymer; ACS) CAS 9003-54-7; Injection molding and extrusion resin for cosmetic packaging, fan blades, toys and games, business machines, interior refrigerator parts, medical parts, beverage tumblers, food containers, tableware, dinnerware, containers, automotive parts, cassettes. [BASF; LNP; Monsanto; Reichhold/Emulsion Polymers]

styrene polymer. See polystyrene.

Styresol 13-031. Styrene monomer-modified alkyd. [Reichhold]

Styrid. Styrene suppressant for fiberglass molding; reduces emissions and odors without affecting interlaminar bonding. [Specialty Prods.]

Styrocell. Expanded and expandable polystyrene. [Shell UK]

Styrochrom®. Polystyrene masterbatches; antistatic finishing for consumer goods; improves surface appearance and scratch resist. of housings and tech. parts. [BASF AG]

Styrodur®. CAS 9003-53-6; Extruded rigid PS foam; contains flame retardant; used for thermal insulation of roofs, floors, walls and ceilings in domestic, industrial and farm buildings, as frost protection of subsoils below roads, aircraft runways, etc., for thermal and low temp. insulation in refrigerated vehicles, containers. [BASF AG; BASF plc]

Styrofan®. Polymer dispersions based on styrene; gloss binders for paper and board coating; raw material for production of laminating adhesives; binders for bonding fiber webs. [BASF AG; BASF plc]

Styrofill®. Loose padding and filling material for cushioning and packaging uses. [BASF AG; BASF plc]

Styroflex. A proprietary trade name that is a flexible polymer of styrene. §

Styrofoam Brand Insulation. CAS 9003-53-6; Extruded rigid PS foam insulation board; avail. in residential sheathing, tongue and groove, score board, and square edge material grades in various thicknesses and sizes; provides high strength, high resist. to moisture and water vapor, long-term R-value to the building industry. [Dow]*

styrogallol. Dihydroxyanthracoumarin, $C_{16}H_8O_5$, a yellow dyestuff.

styrol. (styrene monomer, styrolene; cinnamene, cinnamol; vinyl benzene; phenylethylene). $C_6H_5 \cdot CH:CH_2$; CAS 100-42-5; EINECS 202-851-5;used in the manufacture of polystyrene, SBR, ABS, and SAN resins.

Styrolit®. For producing formed foam plastic parts, for full-mold-casting and for thin-walled formed parts. [BASF AG]

Styrolux®. CAS 9003-55-8; S/B block copolymer; used for injection molding, extrusion, thermoforming, and blow molding, primarily for transparent, impact-resist. packaging material, domestic goods, toys, tech. parts, medical applications. [BASF AG; BASF plc]

---

‡ Trade name and manufacturer not verified   § Trade name without identified manufacturer

styrolyl alcohol. (styryl alcohol). $C_6H_5 \cdot CHOH \cdot CH_2OH$; CAS 93-56-1; Phenyl-glycol; used in perfumery.

Styromol. Refractory coatings for expanded polystyrene patterns used in Replicast FM process. [Foseco (F.S.) Ltd]

Styron. General purpose and high impact polystyrene resins used in packaging, housewares, toys, medical and electronics. [Dow UK] *

Styron 421. CAS 9003-53-6; HIPS; offers excellent extrusion performance, excellent rigidity, good impact, good gloss, excellent heat resist. for extrusion, thermoforming, injection molding, and blow molding; end uses including dinnerware, cups, packaging, etc.; FDA compliance. [Dow Plastics]

Styron 478. CAS 9003-53-6; HIPS; offers enhanced gloss with a good balance of flow and toughness for injection and blow molding of toys, housewares, and small and large appliances; FDA compliance. [Dow Plastics]

Styron 479. CAS 9003-53-6; HIPS; offers good gloss for injection and blow molding of toys and small battery cases; FDA compliance. [Dow Plastics]

Styron 697. CAS 9003-53-6; PS; general purpose resin with high heat resist., excellent clarity; designed for high-speed, in-line extrusion thermoforming of clear, strong, lightweight food containers; FDA compliance. [Dow Plastics]

Styron 6075. CAS 9003-53-6; PS; ignition-resist. grade; offers flame retardance, high heat resist., high modulus, good processability for injection molding applications. [Dow Plastics]

Styron 6087 SF. CAS 9003-53-6; PS structural foam resin; ignition-resist. grade; offers flame retardance, excellent moldability for structural foam applications. [Dow Plastics]

Styronal®. Polymer dispersions based on butadiene and styrene; binders for paper and board coating. [BASF AG]

styrone. CAS 104-54-1; Cinnamyl alcohol; $C_6H_5 \cdot CH:CH \cdot CH_2OH$.

Styroplus®. CAS 9003-55-8; S/B copolymer; for sealable packaging film and high impact strength lids for packing containers; highly resist. to thermoforming. [BASF AG]

Styropor. Expandable PS. [Mitsubishi Petrochem.] *

Styropor® F. CAS 9003-53-6; Expandable PS; difficultly flamm. insulating material for thermal insulation, for structural insulating units and system solutions in building trade; for noise absorption; for blocks and boards as road foundations; for insulation of deep-freeze rooms, refrigerators; profiles, strips, boards for furnishing and decorative purposes. [BASF AG; BASF plc]

Styropor® FH. CAS 9003-53-6; Expandable PS; difficulty flamm. foam for the production of packaging, pallets, boards, maritime articles; resist. to oils, fats, and petrol. fractions free from aromatic hydrocarbons. [BASF AG; BASF plc]

Styropor® P. CAS 9003-53-6; Expandable PS; load bearing, shock absorbent, and thermal insulation packaging for machinery, appliances, glass, porcelain, chemical, pharmaceuticals, food, etc.; toys, seating; moldings for full mold casting; aggregates for lightweight concrete; drainage boards for agricultural and building trades. [BASF AG; BASF plc]

Styrothane 5329/5330. Aromatic urethane coating; rapid curing, solventless, sprayable coating meeting EPA emission levels for VOC; used for protection of EPS,

Styrofoam, plywood, urethane and phenolic foam boardstocks. [Futura Coatings] *

Styvex 22000 NA. CAS 9003-53-6; PS, 20% glass-reinforced. [Ferro/Engineering Thermoplastics]

Styvex 32000 BK. CAS 9003-54-7; SAN, 20% glass-reinforced. [Ferro/Engineering Thermoplastics] *

Styvex 40007 BKL2. CAS 9003-56-9; ABS, self-lubricated; high-impact grade. [Ferro/Engineering Thermoplastics]*

Styvex 42023 NAFR. CAS 9003-56-9; ABS, 10% glass fiber-reinforced; flame retardant thermoplastic [Ferro/Engineering Thermoplastics] *

Styvex 72001 NA. CAS 25134-01-4; Modified PPO, 20% glass-reinforced. [Ferro/Engineering Thermoplastics] *

styxol. See Uba.

Suakin gum. (talca gum, talka gum, Sennaar gum). A brittle variety of gum acacia from Acacia fistula. It gives a ropy mucilage.

subacetate of lead. CAS 1335-32-6; Monobasic lead acetate; $(C_2H_3O_2)_2Pb + PbO + H_2O$.

suberic acid. (dicarboxylic acid C_8; octanedioic acid) $HOOC(CH_2)_6COOH$; CAS 505-48-6; EINECS 208-010-9; Intermediate for synthesis of drugs, dyes, and high polymers. [BASF; Hüls AG; Penta Mfg.]

Sublaprint. Uncut disperse dyestuffs; for the production of inks for printing heat transfer papers. [Holliday Dyes & Chemicals Ltd]

Sublimaze. CAS 437-38-7; A proprietary preparation of fentanyl; an analgesic. [Janssen Pharmaceutica Inc; Janssen Pharmaceutical Ltd]

sublimed blue lead. A pigment produced by heating mixed ores of zinc and lead in a furnace with an air blast. It usually contains 50% lead sulfate, 20% lead oxide, PbO, 11% lead sulfide, PbS, 8% lead sulfite, and 3% zinc oxide.

Sublimed Blue Lead. Basic blue lead sulfate; lubricating aid and friction aid for the manufacturing of brake linings and clutch facings; lubricant additive in oil and high-pressure greases; rust inhibitive pigment for structural steel. [Eagle-Picher] *

sublimed white lead. A white lead manufactured from mixed ores of galena and zinc blende. They are roasted in the presence of an air blast, and the lead sulfate, lead oxide, and zinc oxide formed is collected in large chambers. The average composition is 75% lead sulfate, 20% lead oxide, and 5% zinc oxide; used as a pigment.

sublimoform. A mercury-formaldehyde preparation; used as a seed preservative.

sublimo-phenol. Chloro-phenolate of mercury; used in antiseptic surgery.

subox. A protective coating paint consisting of a suspension of colloidal lead in linseed oil.

substitute of tartar. See superargol.

Suburban Propane (and Design). Liquefied petroleum gas. [Quantum Chemical Corp]

Sub-Vitralen. High molecular weight high density polyethylene sheet; used for orthopedic splints. [Stanley Smith & Co Plastics Ltd]

Sucaryl. Saccharin sodium; sweetener. [Abbott Laboratories] *

Succinellite. Succinic acid obtained from amber.

succinic acid. (butanedioic acid; amber acid; ethylene succinic acid) Dicarboxylic acid; $HOOCCH_2CH_2COOH$; CAS 110-15-6; EINECS 203-740-4; organic synthesis;

---

* Trade name not verified as available   † Trade name verified as obsolete

manufacture of lacquers, dyes, esters for perfumes, photography, in foods as a sequestrant, buffer, neutralizing agent. [Am. Biorganics; Du Pont; General Chem.; Hüls AG; Mallinckdrodt]

succinic anhydride. $C_4H_4O_3$; CAS 108-30-5; EINECS 203-570-0; Manufacture of chemicals, pharmaceuticals, esters; hardner for resins, starch modifier in foods. [Hüls AG; Humphrey; Penta Mfg.; Schweizerhall]

succinimide. (2,5-diketopyrrolidine) $C_4H_5NO_2$; CAS 123-56-8; EINECS 204-635-6; Growth stimulants for plants, organic synthesis. [Chemie Linz UK; Penta Mfg.; Schweizerhall]

succinite. Baltic amber which contains succinic acid. Also the name for a mineral, a lime-aluminum garnet.

succinol. An oil obtained by the distillation of amber.

Suchar. A decolorizing carbon used for sugar juices; prepared from waste sulfite-cellulose liquors.

Sucker Plucker Concentrate. Fatty alcohol mixture; contact tobacco sucker control. [Draxel Chemical Company]‡

Sucker Stuff. Potassium salt of maleic hydrazide; systemic control of tobacco suckers. [Draxel Chemical Company]‡

Suconox-4®. N-Butyryl-p-aminophenol; processing aid for thermoplastics. [Zeeland]

Suconox-9®. N-Pelargonoyl-p-aminophenol; processing aid for thermoplastics. [Zeeland]

Suconox-12®. N-Lauroyl-p-aminophenol; processing aid for thermoplastics. [Zeeland]

Suconox-18®. N-Stearoyl-p-amino phenol; processing aid and antioxidant in thermoplastics. [Zeeland]

Sucostrin. CAS 71-27-2; Succinylcholine chloride; blocking agent. [Bristol-Myers Squibb Co Inc] *

Sucramine. (Lyons sugar). The ammonium salt of saccharin; used in France as a sweetening substance.

sucrase. (saccharase, invertin). Invertase, an enzyme which decomposes saccharose into glucose and fructose; used in the production of invert sugar for candy and syrups, and as an analytical reagent for sucrose.

sucrate of hydrocarbonate of lime. See sucro-carbonate of lime.

Sucrene. See dulcine.

sucro-carbonate of lime. (sucrate of hydrocarbonate of lime). A complex compound of lime, calcium sucrate, and calcium carbonate formed in the production of sugar from the beet, when carbon dioxide gas is passed into a solution of sucrate of lime.

Sucro Ester 7. CAS 25168-73-4; Saccharose distearate; food emulsifier; o/w emulsifier wetting agent, crystallization inhibitor for vegetable oils and fats; inhibitor of thermal denaturation of proteins. [Gattefosse SA]

Sucro Ester 15. CAS 25168-73-4; Saccharose palmitate; food emulsifier; o/w emulsifier wetting agent, crystallization inhibitor for vegetable oils and fats; inhibitor of thermal denaturation of proteins. [Gattefosse SA]

sucrol. See dulcine.

sucro-levulose. See levulose.

sucrose. (β-D-fructofuranosyl-α-D-glucopyranoside; saccharose; sugar; cane sugar; beet sugar; table sugar) Disaccharide; $C_{12}H_{22}O_{11}$; CAS 57-50-1; EINECS 200-334-9; sweetening agent in food, pharmaceuticals; in fermentation; as flavor, preservative, antioxidant (in form of invert sugar); granulation agent and excipient for tablets; in plastics and cellulose industry, rigid polyure-

thane foams, manufacture of ink. [Am. Biorganics; Mallinckrodt; Mendell; Pfanstiehl Labs; Vista]

suction gum. (suction powder). Powdered gum tragacanth.

Sucuaryl. A proprietary trade name for sodium cyclamate.§

Sudafed. Pseudoephedrine hydrochloride; adrenergic. [Burroughs Wellcome Co]

Sudafed Co (Sudafed Sinus). A proprietary formulation of pseudoephedrine hydrochloride and paracetamol; for temporary relief of upper respiratory congestion, pyrexia and pain, e.g., in the common cold, influenza, sinusitis and nasopharyngitis. [The Wellcome Foundation Ltd]

Sudafed DM. A proprietary formulation of pseudoephedrine hydrochloride and dextromethorphan hydrobromide; for temporary relief of nonproductive cough associated with congestion of the upper respiratory tract. [The Wellcome Foundation Ltd]

Sudafed Expectorant. A proprietary formulation of pseudoephedrine hydrochloride and guaifenesin; for temporary relief of productive cough accompanied by congestion of the upper respiratory tract. [The Wellcome Foundation Ltd]

Sudafed Nasal Spray. CAS 2315-02-8; A proprietary preparation of oxymetazoline hydrochloride; for the relief of nasal congestion. [The Wellcome Foundation Ltd]

Sudafed Tablets/Syrup/12 Hour Capsules. Proprietary formulations of pseudoephedrine hydrochloride; for temporary relief of nasal congestion associated with allergic rhinitis, vasomotor rhinitis and the common cold. [The Wellcome Foundation Ltd]

Sudan®. Oil- and fat-sol. dyes; for marking mineral oil products, e.g., engine fuels, heating oil, lubricating greases; also for coloring shoe polishes, floor polishes and waxes, for producing smoke dyes. [BASF AG]

Sudan I. (carminaph, oil orange). CAS 842-07-9; A dyestuff. It is benzeneazo-β-naphthol, $C_{16}H_{12}N_2O$; used for coloring oils and varnishes.

Sudan II. CAS 3118—97-6; Xyleneazo-β-naphthol, $C_{18}H_{16}N_2O$; used for coloring oils and varnishes.

Sudan III. CAS 85-86-9; Benzeneazo-benzeneazo-β-naphthol, $C_{22}H_{16}N_4O$; used for coloring oils and varnishes.

Sudan IV. CAS 85-83-6; o-Tolueneazo-o-toluene-azo-β-naphthol; used as a dyestuff.

Südflock. Inorganic precipitants, flocculants and adsorbents for the purification of industrial and domestic effluents. [Süd-Chemie AG] *

Sudol®. Vegetable soaps and xylenols; general purpose clear soluble disinfectant for hospital and animal health fields. [Coventry Chemicals Ltd]

Sudranol. Polyethylene wax emulsions. [Suddeutsche Emulsions GmbH]

Sufatone SCS/B. Concentrated cationic surfactant in paste form; softening and antistatic agent used for all fabrics, particularly synthetics including acrylics. [Standard Chemical Company] †

Sufatone SCS/CL. Cationic surfactant in liquid form; general purpose mild softening and antistatic agent used for all fibers. [Standard Chemical Company] †

Sufatone SMC/L. Cationic surfactant in liquid form; softening agent for most fibers particularly wool and acrylics. [Standard Chemical Company]

Sufatone SMC/W. Concentrated cationic surfactant in the form of a semi-liquid; mild softening agent for most fibers, particularly wool and chlorinated wool. [Standard Chemical Company] †

---

‡ Trade name and manufacturer not verified   § Trade name without identified manufacturer

Sufenta. CAS 60561-17-3; A proprietary preparation containing sufentanil citrate; analgesic supplement to anesthesia. [Janssen Pharmaceutica Inc; Janssen Pharmaceutical Ltd]

Suffa. CAS 7704-34-9; Sulfur; fungicide for fruit and vegetable crops. [Draxel Chemical Company] ‡

Sufrexal. A proprietary preparation containing ketanserin tartrate; antihypertensive, serotonin antagonist. [Janssen Pharmaceutical Ltd] *

sugamo. A Japanese seaweed suggested for use in papermaking.

sugar. See sucrose.

sugar cane wax. A wax obtained from the dried filter press cake from sugar mills by benzine extraction; the African cake contains 14-17% of wax, the Java cake 4%. The wax obtained is not a pure product, but is a mixture of wax and fatty material with 7% glycerin.

sugar charcoal. (lampblack). Amorphous carbon.

sugar house black. A bone black pigment. It is a by-product of the sugar mills.

sugar of lead. CAS 301-04-2; Normal lead acetate, $(CH_3COO)_2Pb+3H_2O$; used as a mordant in dyeing and printing, and for the preparation of lead salts and paints.

sugar of lead. See lead acetate.

Sugartab®. CAS 57-50-1; Sucrose (90-93%), invert sugar; inert base for directly compressible pharmaceutical tablets. [Mendell]

Sugracillin. CAS 113-98-4; Penicillin G potassium; antibacterial. [Upjohn] *

Suhler white copper. A nickel silver alloy of 40% copper, 32% nickel, 25% zinc, 2.6% tin, and 0.6% cobalt.

Suicalm. CAS 1649-18-9; Azaperone; antipsycotic. [Janssen Pharmaceutica Inc] *

Suladrin. CAS 4299-60-9; Sulfisoxazole diolamine; antibacterial. [Alcon Laboratories Inc] *

Sulamyd Sodium. Sulfacetamide sodium; antibacterial. [Schering Corp] *

Sulf-10® Ophthalmic Solution 10%. Sulfacetamide sodium; for the treatment of conjunctivitis, corneal ulcer, and other superficial ocular infections due to susceptible microorganisms, and as adjunctive treatment in systemic sulfonamide therapy of trachoma. [Iolab Corp]

Sulfacet-R. Sulfacetamide sodium; antibacterial. [Dermik Laboratories Inc] *

Sulfacide. Acid dyes. [ICI Chem. & Polymers Ltd] †

Sulfactol. CAS 7772-98-7; Sodium thiosulfate; antidote to cyanide poisoning. [Sterling Drug Inc] *

Sulfads®. CAS 120-54-7; Essentially dipentamethylene thiuram hexasulfide; ultra accelerator for NR and synthetic rubbers; vulcanizing agent. [R.T. Vanderbilt]

Sulfa-Hitech® 0382. DMDS-based; solv. used to dissolve sulfur in the production of sour gas wells, in sour-gas pipelines, and in refinery and chemical plant flowlines. [Atochem] *

Sulfalar. CAS 127-69-5; Sulfisoxazole; an antibacterial. [Parke-Davis] *

sulfamic acid. (amidosulfonic acid) NH_2SO_3H; CAS 5329-14-6; EINECS 226-218-8; Metal/ceramic cleaning, nitrite removal in azo-dyeing, gas-liberating compositions, organic synthesis, analytical standard, chlorine stabilizer, bleaching paper pulp and textiles, catalyst for urea-formaldehyde resins, sulfonating agent, pH control. [General Chem.; PMC Specialties; Schweizerhall; Transol Chem. UK Ltd]

Sulfamin. Anionic surfactant in liquid form; for liquid cleaners. [Berol Kemi (UK) Ltd] *

sulfammonium. A solution of sulfur in liquid ammonia to form a purple solution.

Sulfamylon. A proprietary preparation of mafenine acetate in a cream; used in the treatment of infected burns. [Winthrop Laboratories] *

sulfanilamide. (4-aminobenzenesulfonamide) $C_6H_8N_2O_2S$; CAS 63-74-1; EINECS 200-563-4; used in vetinary medicine. [R.W. Greeff; Schweizerhall; Tanabe USA]

sulfanilic acid. (p-aminobenzenesulfonic acid; p-anilinesulfonic acid) $C_6H_7NO_3S$; CAS 121-57-3; EINECS 204-482-5; dyestuffs, organic synthesis, medicine, reagent. [Am. Cyanamid; Penta Mfg.; 3-V; U.S. Biochemical]

sulfanilic acid. CAS 121-57-3; EINECS 204-482-5; Aniline-p-sulfonic acid, $C_6H_4NH_2SO_3H$; used on medicine, organic synthesis, and dyestuffs.

m-sulfanilic acid. See metanilic acid.

Sulfanol. Sulfur dyes. [ICI Chem. & Polymers Ltd]

Sulfarine. A mixture of magnesium sulfate with 15% sulfuric acid; used against potato scab.

Sulfasan. CAS 103-34-4; 4,4´-Dithiodimorpholine; vulcanizing agent for natural and synthetic rubbers. [Monsanto Co]

Sulfasuxidine. CAS 116-43-8; Succinylsulfathiazole; antibacterial. [Merck & Co Inc] †

sulfated oils. See Turkey red oils.

sulfate pulp. Wood pulp obtained by the treatment of wood with alkali liquors containing sodium sulfate.

Sulfathalidine. CAS 85-73-4; Phthalylsulfathiazole; antibacterial. [Merck & Co Inc] †

Sulfatine. A mixture of 73% sulfur, 20% lime, and 7% copper sulfate; a fungicide used against black rot.

Sulfato de Cobre Valles. Crystallized copper sulfate; for manufacture of agricultural fungicides and many industrial products. [Industrias Quimicas Del Valles SA] *

Sulfatol CL. Sodium sulfated fatty alcohol in paste form; anionic surfactant which is stable to hard water and disperses lime soaps; used for various industrial applications including textiles and leather. [Standard Chemical Company]

Sulfatol E3. Sodium lauryl ether sulfate (coconut/palm kernel C12-C14 alcohol) as a clear almost colorless low viscosity liquid; for production of all types of liquid and lotion shampoos and bubble baths; also a raw material for light duty liquid detergents, dishwashing detergents and auto shampoos. [Efkay Chemicals Ltd] *

Sulfatol LS3. Combined sulfated fatty alcohol and nonionic surfactant in liquid form; detergent and washing off liquid for textiles etc. [Standard Chemical Company]

Sulfatol LX/B. Sodium sulfated fatty alcohol in paste form; detergent, wetting, leveling and softening agent for industrial applications including textiles and leather. [Standard Chemical Company]

Sulfatol PD/B. Potassium sulfated fatty alcohol in paste form; detergent with good water solubility, for industrial applications including textiles and leather. [Standard Chemical Company]

Sulfatol TL/B. Anionic surfactant with a blend of cations and a sulfated fatty alcohol as anion; detergent for industrial applications including textiles and leather. [Standard Chemical Company]

Sulfatryl. Trisulfapyrimidines (oral suspension); mixture of sulfadiazine, sulfamerazine, and sulfamethazine; antibacterial. [Wallace Laboratories] *

Sulfetal 4105. Sodium isooctyl sulfate; low-foaming detergent, wetting agent in acid and alkaline media; for industrial and metal cleaners. [Zschimmer & Schwarz]

Sulfetal C 38. CAS 151-21-3; Sodium lauryl sulfate; light duty detergent, dishwashing agent, cold wash agent. [Zschimmer & Schwarz]

Sulfetal CJOT 38. CAS 21142-28-9; MIPA-lauryl sulfate; basic material for hair shampoos, bath additives, cosmetics, liquid detergents. [Zschimmer & Schwarz]

Sulfetal FA 40. Sodium isooctyl sulfate, modified; low foaming detergent and wetting agent; used in industrial cleaners, metal and tank cleaning; highly resistant to alkalies. [Zschimmer & Schwarz]

Sulfetal KT 400. CAS 139-96-8; EINECS 205-388-7; TEA-lauryl sulfate; for detergents, cosmetics, hair shampoos. [Zschimmer & Schwarz]

Sulfetal MG 30. Magnesium lauryl sulfate; for detergents, cosmetics. [Zschimmer & Schwarz]

Sulfetal TC 50. Sodium tallow sulfate and sodium lauryl sulfate; detergent; washing and cleaning agents; hand cleaning pastes. [Zschimmer & Schwarz]

Sulfidal. CAS 7704-34-9; Colloidal sulfur powder; used in powders, lotions, and ointments for treatment of skin diseases. [Hüls Am.] †

sulfide dyestuffs. A class of dyestuffs prepared by the fusion of organic amines and other substances with sulfur and sodium sulfide. They are used for cotton dyeing, and are usually fixed by oxidizing agents.

sulfide white. *See* lithopone.

Sulfiformin. Formaldehyde-sulfurous acid, $H \cdot CH_2 \cdot SO_3H$; an antiseptic; a 1% solution has been used for spraying vines.

Sulfil®. A registered trade name for flame retardant polysulfone. §

sulfite carbon. A decolorizing carbon used for sugar juices. It is prepared from sulfite-cellulose liquors.

sulfite pulp. Wood pulp obtained by means of calcium bisulfide. It is made by digesting the disintegrated wood under pressure with the calcium bisulfite, which gives a mass of cellulose fibers free from lignocellulose amounting to about 45% of the wood.

sulfite turpentine. A by-product obtained from the pulping of spruce by the sulfite process. The main constituent is p-cymene.

sulfite turpentine oil. (cellulose turpentine oil). A by-product obtained in the manufacture of cellulose. When decolorized it resembles turpentine oil.

Sulfochem 436. CAS 9051-57-4; Ammonium nonoxynol-4 sulfate; high-foaming detergent for dishwashing, carwash, and carpet shampoo formulations; emulsion polymerization surfactant for SBR, acrylic, vinyl acrylic systems. [Chemron]

Sulfochem ALS. CAS 2235-54-3; EINECS 218-793-9; Ammonium lauryl sulfate; surfactant, mild detergent for use in low pH systems; foaming and suspending agent; for personal care and industrial applications. [Chemron]

Sulfochem DLS. DEA-lauryl sulfate; surfactant foamer with soap-like characteristics; for shampoos, bubble baths, skin cleaners, syndet bars, shower gels. [Chemron]

Sulfochem EA-1. CAS 67762-19-0; Ammonium laureth sulfate; surfactant for low pH shampoo systems; mild-

ness, high flash foam, low cloud point, visc. response; for shampoos, cleansers, bath products, gels. [Chemron]

Sulfochem EA-2, EA-3. CAS 67762-19-0; Ammonium laureth sulfate; surfactant for low pH shampoo systems; mildness, high flash foam, low cloud point, visc. response. [Chemron]

Sulfochem EA-60. CAS 67762-19-0; Ammonium laureth sulfate; surfactant for shampoo systems, bubble baths, cleansers. [Chemron]

Sulfochem EA-70. CAS 67762-19-0; Ammonium laureth sulfate (2 mole); surfactant for toiletries and cosmetics. [Chemron]

Sulfochem ES-1. Sodium laureth-1 sulfate; surfactant for personal care cleansers. [Chemron]

Sulfochem ES-2. CAS 9004-82-4; Sodium laureth-2 sulfate; flash foamer and detergent for personal cleansing products, specialty cleaning products [Chemron]

Sulfochem ES-3. Sodium laureth-3 sulfate; surfactant for personal cleansing products [Chemron]

Sulfochem ES-60. Sodium laureth-3 sulfate; surfactant for shampoo systems, bubble baths, cleansers. [Chemron]

Sulfochem ES-70. CAS 9004-82-4; Sodium laureth-2 sulfate; surfactant for toiletries, cosmetics, specialty industrial compounds. [Chemron]

Sulfochem K. CAS 4706-78-9; EINECS 225-190-4; Potassium lauryl sulfate; detergent, foamer. [Chemron]

Sulfochem MG. Magnesium lauryl sulfate; mild detergent, foamer for bubblebath, shampoos, cleansing preparations, carpet shampoos. [Chemron]

Sulfochem MLS. MEA-lauryl sulfate; surfactant foamer with soap-like characteristics; for shampoos, bubble baths, skin cleaners, shower gels, syndet bars. [Chemron]

Sulfochem SAC. Sodium lauryl sulfate; foamer for lotion and paste shampoos; detergent base for pearlescent shampoos, bubble baths, shower gels, cleansers. [Chemron]

Sulfochem SLC. Sodium lauryl sulfate; surfactant for clear shampoos, bath products, cleaners. [Chemron]

Sulfochem SLN. CAS 151-21-3; Sodium lauryl sulfate; foamer, dispersant, wetting agent, detergent for dry blends used in cleaning compounds, carpet shampoos, shampoo concentrates, bubble baths, cosmetic cleansers. [Chemron]

Sulfochem SLP-95. CAS 151-21-3; Sodium lauryl sulfate; foamer, dispersant, wetting agent, detergent for high-act., cleaning concentrates, dentifrices, high purity cleansers. [Chemron]

Sulfochem SLS. CAS 151-21-3; Sodium lauryl sulfate; detergent, foamer, suspending agent for hard surface cleaners, carpet shampoos, upholstery cleaners, spot removers, personal care products. [Chemron]

Sulfochem SLX. CAS 151-21-3; Sodium lauryl sulfate; detergent, foaming and suspending agent for rug and upholstery shampoos, spot removers. [Chemron]

Sulfochem TLS. CAS 139-96-8; EINECS 205-388-7; TEA-lauryl sulfate; surfactant foamer, wetting agent with soap-like characteristics; for liquid soaps and shampoos, industrial applications; good tolerance to hard water. [Chemron]

sulfoform. Triphenylstibine sulfide, $(C_6H_5)_3SbS$.

sulfogenol. A crude mineral oil obtained from bituminous shale, is saturated with sulfur, and sulfonated. It is the ammonium salt of the sulfonated product and has similar properties to ichthyol.

---

‡ Trade name and manufacturer not verified   § Trade name without identified manufacturer

Sulfokyl DAS. Alkylbenzene sulfonate; anionic; yellowish liquid; washing agent for all fibers, particularly for wool; gives excellent washing effects at low temperatures. [Thor Chemicals (UK) Ltd]

Sulfomyl. A proprietary preparation of mafenide propionate; used in the form of anti-infective eyedrops. [Winthrop Laboratories] *

Sul-fon-ate AA-10. CAS 25155-30-0; Sodium dodecylbenzene sulfonate; wetting agent; air pollution control; cement, food, commercial laundry and industrial industries; cosmetics; fertilizers, insecticides; leather, paper, petrol., and rubber processing; metal cleaning, electroplating, etching, pickling; mining. [Boliden Intertrade]

Sul-fon-ate OA-5R. CAS 67998-94-1; Sulfonated oleic acid, sodium salt; surfactant. [Boliden Intertrade]

sulfonated oils. See Turkey red oils.

sulfonic acid. Organic compound containing one or more sulfo radicals; dispersant, wetting agent, leveling agent, protective colloid for dyestuffs. [Yorkshire Chem. plc]

Sulfonsol. Trisulfapyrimidines (oral suspension); mixture of sulfadiazine, sulfamerazine, and sulfamethazine; an antibacterial. [Merrell Dow Pharmaceuticals Inc] *

3,3'-sulfonyldianiline. See 3,3'-diaminodiphenyl sulfone.

4,4'-sulfonyldianiline. See 4,4'-diaminodiphenyl sulfone.

Sulfopon 101, 101 Special. CAS 151-21-3; Sodium lauryl sulfate; surfactant for personal care products and light-duty detergents. [Henkel KGaA; Pulcra SA]

Sulfopon 101/POL. CAS 151-21-3; Sodium lauryl sulfate; surfactant for emulsion polymerization. [Pulcra SA]

Sulfopon LS. Sodium lauryl sulfate (C12-C18) in liquid/paste form; for cream shampoos and bubble baths. [Henkel Chemicals Ltd] *

Sulfopon O 680. CAS 1847-55-8; EINECS 217-430-1; Sodium oleyl sulfate; low foaming surfactant. [Henkel/Functional Prods.]

Sulfopon P-40. CAS 151-21-3; Sodium lauryl sulfate; dispersant and emulsifier for acrylates, styrene acrylic, vinyl chloride, vinyl acetate copolymers; also for cream shampoos, specialty cleaners, rug shampoos. [Pulcra SA]

Sulfose. Trisulfapyrimidines (oral suspension); mixture of sulfadiazine, sulfamerazine and sulfamethazine; antibacterial. [Wyeth Laboratories] *

Sulfosept oil. The next higher fraction to thiosept oil (*qv*); used as an insecticide.

Sulfosoft. Anionic surfactant in acid form; for detergent production. [Berol Kemi (UK) Ltd] *

Sulfotex 110. Sodium n-decyl sulfate; wetting agent and emulsifier used in cleaning formulations, rug shampoos; used in sealing food containers; effective in cold and hard water; biodeg. [Henkel/Cospha; Henkel Canada]

Sulfotex LAS-90. Sodium dodecylbenzene sulfonate; surfactant for all-purpose cleaners, laundry products, hard surface cleaners; penetrates and removes grease. [Henkel/Cospha; Henkel Canada]

Sulfotex LCX. CAS 151-21-3; Sodium lauryl sulfate; detergent, foaming agent, wetting agent, and emulsifier; primary surfactant for rug and upholstery shampoos, hard surface cleaners; biodeg. [Henkel/Cospha; Henkel Canada]

Sulfotex LMS-E. CAS 9004-82-4; Sodium lauryl ether sulfate; detergent, wetting agent, foamer for industrial applications, liquid dishwashing detergents; biodeg. [Henkel/Cospha; Henkel Canada]

Sulfotex OA. Sodium octyl sulfate; foaming agent, wetting agent for animal glue on paper and paperboard, high electrolyte concentrates, and food processing; detergent and dispersant in industrial cleaners; mercerizing agent in dyeing of fibers; biodeg. [Henkel/Cospha; Henkel Canada]

Sulfotex OT. CAS 67762-19-0; Ammonium lauryl ether sulfate; detergent, emulsifier, foaming agent used in household and industrial detergents, car shampoos; biodeg. [Henkel/Cospha; Henkel Canada]

Sulfotex SXS-40. CAS 1300-72-7; EINECS 215-090-9; Sodium xylene sulfonate; hydrotrope, solubilizer for liquid detergents, pine oil in water, inks to prevent gumming. [Henkel/Cospha; Henkel Canada]

Sulfotex T-65. TEA-dodecylbenzene sulfonate. [Henkel/Cospha]

sulfourea. Thiourea, CH_4N_2S.

Sulframin. Surfactant for cosmetics, toiletries, pharmaceutical, processing, agriculture, and other industries. [Baxenden Chemicals Ltd]

Sulframin 14-16 AOS. Sodium olefin sulfonate (C14-C16) as a clear amber liquid; foaming agent and detergent for cosmetic and household uses. [Witco Chemical Ltd] *

Sulframin 33. Sodium olefin sulfonate and sodium ether sulfate as a clear amber liquid; foaming agent and detergent for industrial and household liquid detergents. [Witco Chemical Ltd] *

Sulframin 1250. Sodium alkylaryl sulfonate in liquid form; anionic surfactant. [Witco Chemical Ltd] *

sulfur. (brimstone; sulfur) Nonmetallic element; S; CAS 7704-34-9; EINECS 231-722-6; sulfuric acid manufacture, petroleum refining, dyes and chemicals, fungicide, insecticides, explosives, detergents, rubber vulcanization. [Norsk Hydro A/S; Shell; Texaco BV]

sulfurated antimony. (antimony crocus, saffron of antimony). A mixture of antimony pentasulfide, Sb_2S_5, with a little oxide, Sb_4O_6, and some free sulfur; formerly used in making tartar emetic.

sulfurated potash. (liver of sulfur). A mixture of sulfides, mainly $K_2S_2O_3$ and K_4S_3 obtained by heating potassium carbonate with 1/2 its weight of sulfur. When fresh and carefully prepared it is the color of liver, and was called liver of sulfur; is used in the leather industry as a depilating agent.

sulfur dioxide. SiO_2; CAS 7446-09-5; EINECS 231-195-2; Chemicals, sulfite paper pulp, ore and metal refining, soybean protein, intermediates, solvent extraction, bleaching oils and starch, sulfonation of oils, disinfectant, fumigant, food additive, reducing agent, antioxidant. [Air Prods.; Boliden Intertrade; Hoechst Celanese; Outokumpu Oy; Rhone-Poulenc Basic]

sulfuretted hydrogen. Hydrogen monosulfide, H_2S.

Sulfur-F. CAS 7704-34-9; Sulfur; micronized flowable disp. of elemental sulfur. [W A Cleary]

sulfur gold. Antimony pentasulfide, Sb_2S_5; used for vulcanizing and imparting a red color to rubber.

sulfur hypochlorite. A mixture of sulfur and sulfur chloride; used in rubber vulcanizing.

sulfuric ether. (phosphoric ether). Diethyl ether.

sulfuric acid. (hydrogen sulfate; battery acid; electrolyte acid) Inorganic acid; H_2SO_4; CAS 7664-93-9; EINECS 231-639-5; fertilizers, chemicals, dyes and pigments, laboratory reagents, electroplating baths. [Akzo; Amax; Am. Cyanamid; Boliden Intertrade; Du Pont; Metallgesellschaft AG; Nissan Chem. Ind.; OxyChem;

---

* Trade name not verified as available   † Trade name verified as obsolete

Pasminco Europe; Rasa Ind.; Rhone-Poulenc Basic]

sulfuric acid, magnesium salt (1:1). *See* magnesium sulfate.

sulfuric acid, monododecyl ester, sodium salt. *See* sodium lauryl sulfate.

sulfurite. A name applied to a sulfur from Java, which contained 29% arsenic.

sulfur olive oils. A name for the oil dissolved out from residual olive oil cake by means of carbon disulfide. It is also called sulfocarbon oil. It is rich in stearin.

sulfurous acid, disodium salt. *See* sodium sulfite.

sulfurous acid, monosodium salt. *See* sodium bisulfite.

sulfurous acid, sodium salt (1:2). *See* sodium sulfite.

sulfur soap. Usually a yellow medicated soap to which has been added about 10% powdered sulfur.

sulfur waste. The residue from the distillation of iron pyrites.

SulfuSorb8. Activated carbon; for vapor phase applications. [Calgon Carbon] *

Sulindal. CAS 38194-50-2; Sulindac; antirheumatic. (Spain) [Merck & Co Inc]

sulisobenzone. *See* benzophenone-4.

Sulla. CAS 651-06-9; Sulfameter; antibacterial. [Wyeth Laboratories] *

Sulmet. Industrial detergent. [Crosfield Chemicals Ltd] †

Sulperazone®. Sultamicillin; antibiotic. [Pfizer International]

Sul-Perm® 10. Sulfur bases; replacement for sulfurized sperm oil products; lubricant for gear oils, slideway oils, cutting oils [Ferro/Keil]

Sul-Perm® 110. Sulfur bases; replacement for sulfurized sperm oil products; used in industrial oils including slideway oils, industrial gear lubricants; compatible with heavy metal salts used as antioxidants and corrosion inhibitors. [Ferro/Keil]

Sul-Perm® C. Sulfur base; replacement for sulfurized sperm oil products; used in grease applications; imparts lubricity and film strength to EP greases. [Ferro/Keil]

Sulphamagna. A proprietary preparation of streptomycin sulfate, phthalyl sulfathiazole, sulfadiazine and activated attapulgite. [Wyeth Laboratories] *

Sulphamezathine. A proprietary preparation containing sulfadimidine sodium. [ICI Chem. & Polymers Ltd]

Sulphatol 33. Sodium lauryl sulfate, derived from the C12-C14 fraction of coconut/palm kernel fatty alcohol; clear liquid or thin paste; raw material for liquid cream and egg shampoos; emulsifier for cosmetic products. [Efkay Chemicals Ltd] *

Sulphatol 33 MO. Monoethanolamine lauryl sulfate, derived from the C12-C14 fraction of coconut/palm kernel fatty alcohol; clear pale yellow liquid; raw material for clear, oil and other liquid shampoos. [Efkay Chemicals Ltd] *

Sulphatol B6. Ammonium/triethanolamine lauryl sulfate, derived from the C12-C14 fraction of coconut/palm kernel fatty alcohol; clear, golden yellow viscous liquid; raw material for all types of liquid shampoos and bubble baths. [Efkay Chemicals Ltd] *

Sulphatriad. A proprietary preparation of sulfathiazole, sulfadiazine and sulfamerazine; an antibiotic. [May & Baker Ltd] *

Sulphex. A proprietary preparation of sulfathiazole and hydroxyamphetamine hydrobromide. [SmithKline Beecham] *

Sulphol. Sulfur dyestuffs. [James Robinson & Co Ltd]

Sulphol High Fast. Sulfur and sulfurized vat dyes; liquid dispersions; used to dye cellulosic fibers alone and in blends with polyester; compatible with disperse and vat dyes. [James Robinson & Co Ltd]

Sulphonic Acid LS. Alkylbenzene sulfonic acid; base for wide variety of detergents; can be neutralized with metallic bases or amines. [Hart Chem. Ltd.]

Sulphonol. A range of acid dyes; for dyeing of wool and similar fibers. [Yorkshire Chemicals plc] *

Sulphophone. A trademark for a mixture of zinc sulfide and calcium sulfate; it is an analogous product to lithopone.§

Sulphosol. Solubilized sulfur dyestuffs. [James Robinson & Co Ltd]

Sulphramin B and TPB. Alkylaryl sulfonic acid in liquid form; emulsifier for powder and liquid detergents. [Witco Chemical Ltd] *

sulphur. *See* sulfur.

Sulsol. A proprietary trade name for a colloidal sulfur preparation for horticultural purposes. §

Sulveol DC. Alkyl polyglycol ether; nonionic; yellowish pourable paste; rapid wetting agent, fully effective with excellent emulsifying and dispersing properties in cold and hot liquors. [Thor Chemicals (UK) Ltd]

sumacel. A diatomaceous earth containing 80% SiO_2, 5-3% Fe_2O_3 and Al_2O_3, 2.02% CaO, and 8.16% H_2O; used as a filtering medium for sugars.

sumach. (sumac). The dried and finely powdered leaves and shoots of species of *Rhus*; used for tanning leather, also for dyeing and printing.

sumach wax. *See* Japan tallow.

sumalban. Alban obtained from Sumatra gutta-percha.

sumaphos. A mixture of diatomaceous earth and acid phosphate, containing 36.22% P_2O_5.

Sumatra wax. *See* Java wax.

Sumet processed lead. An alloy of 70-80% copper and 15-30% lead.

Sumibond PA. Phenolic resin; for adhesives. [Sumitomo Bakelite Co Ltd] *

Sumicidin. CAS 51630-58-1; Emulsifiable concentrate of 100 g fenvalerate per liter; a pyrethroid insecticide. [Shell UK]

Sumicool. EVA resin; for mats. [Sumitomo Bakelite Co Ltd]*

Sumiflex. Polyvinyl chloride family; molding compounds. [Sumitomo Bakelite Co Ltd] *

Sumikon. A proprietary range of phenolic molding materials. [Sumitomo Bakelite Co Ltd] *

Sumikon AM. Diallyl phthalate resin; molding compounds. [Sumitomo Bakelite Co Ltd] *

Sumikon EM, EME. Epoxy resin; molding compounds. [Sumitomo Bakelite Co Ltd] *

Sumikon IM. Polyimide resin; molding compounds. [Sumitomo Bakelite Co Ltd] *

Sumikon PM. Phenolic resin; molding compounds. [Sumitomo Bakelite Co Ltd] *

Sumikon TM. Polyester resin; molding compounds. [Sumitomo Bakelite Co Ltd] *

Sumikon VM. Polyvinyl chloride resin; molding compounds. [Sumitomo Bakelite Co Ltd] *

Sumilac PC. Phenolic resin; industrial resins. [Sumitomo Bakelite Co Ltd] *

Sumilite CEL. Composite resin; for sheets. [Sumitomo Bakelite Co Ltd] *

Sumilite EI. Epoxy resin; for laminate materials. [Sumitomo Bakelite Co Ltd] *

Sumilite EL. Epoxy resin; for laminated sheets. [Sumitomo Bakelite Co Ltd] *

---

‡ Trade name and manufacturer not verified

§ Trade name without identified manufacturer

Sumilite ELC. Epoxy resin; copper clad laminates. [Sumitomo Bakelite Co Ltd] *

Sumilite FS. PES, PEI resins; for sheets. [Sumitomo Bakelite Co Ltd] *

Sumilite IL. Polyimide resin; for laminated sheets. [Sumitomo Bakelite Co Ltd] *

Sumilite ILC. Polyimide resin; for copper clad laminates. [Sumitomo Bakelite Co Ltd] *

Sumilite ILI. Polyimide resin; for laminate materials. [Sumitomo Bakelite Co Ltd] *

Sumilite NS. Polypropylene resin; for sheets. [Sumitomo Bakelite Co Ltd] *

Sumilite PL. Phenolic resin; for laminated sheets. [Sumitomo Bakelite Co Ltd] *

Sumilite PLC. Phenolic resin; for copper clad laminates. [Sumitomo Bakelite Co Ltd] *

Sumilite Resin PR. Resorcinol resin; for adhesives. [Sumitomo Bakelite Co Ltd] *

Sumilite Resin PR. Phenolic resin; industrial resins. [Sumitomo Bakelite Co Ltd] *

Sumilite Resin PR. Epoxy resin; industrial resins. [Sumitomo Bakelite Co Ltd] *

Sumilite STS. Polystyrene resin; for sheets. [Sumitomo Bakelite Co Ltd] *

Sumilite TFC. Polyimide and other resins; for copper clad laminates. [Sumitomo Bakelite Co Ltd] *

Sumilite TFP. Polyimide and other resins; for flexible printed circuit boards. [Sumitomo Bakelite Co Ltd] *

Sumilite VSL. Polyvinyl chloride and metal foil; for sheets. [Sumitomo Bakelite Co Ltd] *

Sumilite VSS. Polyvinyl chloride resin; for sheets. [Sumitomo Bakelite Co Ltd] *

Sumine® 2005. CAS 100-46-9; N-Benzylamine. [Zeeland]

Sumine® 2015. Tertiary catalyst; epoxy curative. [Zeeland]

Suminet. Polyethylene resin; materials for land improvements. [Sumitomo Bakelite Co Ltd] *

Sumipipe. Polyethylene resin; materials for land improvement. [Sumitomo Bakelite Co Ltd] *

Sumitac EA. Epoxy resin; for adhesives. [Sumitomo Bakelite Co Ltd] *

Sumitac GA. Polyurethane resin; for adhesives. [Sumitomo Bakelite Co Ltd] *

Sumitac VA. Polyvinyl chloride resin; for adhesives. [Sumitomo Bakelite Co Ltd] *

Sumithrin. Emulsifiable concentrate containing 103 g/l pehenothrin; a pyrethroid insecticide. [Sumito Chemical (UK) plc]

Summit. CAS 55219-65-3; Triadimenol; fungicide to control powdery mildew, rusts, and rhychosporium in winter and spring crops of wheat, barley, oats, and rye. [Bayer AG]

Summit. Suspension concentrate containing 150 g terbutryn and 200 g trifluralin per liter; for weed control in winter cereals. [Ashlade Formulations Ltd]

Sumquat® 2355. CAS 56-37-1; Benzyl triethyl ammonium chloride. [Hexcel]

Sumquat® 6020. CAS 124-03-8; Cetethyldimonium bromide. [Hexcel]

Sumquat® 6030. CAS 57-09-0; EINECS 200-311-3; Cetrimonium bromide. [Hexcel]

Sumquat® 6045. CAS 107-64-2; EINECS 203-508-2; Distearyldimonium chloride. [Hexcel]

Sumquat® 6050. CAS 122-18-9; Cetalkonium chloride. [Hexcel]

Sumquat® 6110. CAS 1119-97-7; EINECS 214-291-9; Myrtrimonium bromide. [Hexcel]

Sumquat® 6210. CAS 122-19-0; EINECS 204-527-9; Stearalkonium chloride. [Hexcel]

Sumycin. A proprietary preparation of tetracycline base buffered with potassium metaphosphate; an antibiotic. [Bristol-Myers Squibb Pharmaceuticals Ltd]

Sunaptic Acids. High molecular weight naphthenic acids; corrosion inhibitor, oil well drilling mud formulations, emulsifiers, foundry binders. [Sun Refining & Marketing Co] ‡

Sunaptol. Textile auxiliary chemicals. [ICI Chem. & Polymers Ltd]

Sunaptol NP55. Nonionic surfactant of the nonylphenol ethoxylate type in liquid form; emulsifier and intermediate used in natural waxes, mineral oils; ethoxysulfate manufacture; emulsifiable solvents for metal degreasing; general cleaning applications. [Pechiney Ugine Kuhlmann Ltd] ‡

Sunaptol NP65 and NP70. Nonionic surfactants of the nonylphenol ethoxylate type in liquid form; low temperature wool scouring; emulsifier for silicone and mineral oils; emulsifier and stabilizer for kerosine based hand cleaning gels. [Pechiney Ugine Kuhlmann Ltd] ‡

Sunaptol NP80 and NP95. Nonylphenol ethoxylate nonionic surfactant in liquid form; detergency, wetting and emulsifying agent used in wool scouring; metal cleansing; mineral oils, usually in combination with a hydrophobic surfactant. [Pechiney Ugine Kuhlmann Ltd] ‡

Sunaptol NP100. Nonylphenol ethoxylate nonionic surfactant in liquid form; for solubilization of essential oils and perfumes. [Pechiney Ugine Kuhlmann Ltd] ‡

Sunaptol NP140. Nonylphenol ethoxylate nonionic surfactant in the form of a white waxy solid; iodine complexing for iodofor sterilizers; production of self emulsifiable oleines. [Pechiney Ugine Kuhlmann Ltd] ‡

Sunaptol NP350. Nonylphenol ethoxylate nonionic surfactant in the form of a white solid; ready molding into solid block for detergent blocks and tablets; emulsifier and stabilizer for vinyl acetate and acrylic polymer emulsions. [Pechiney Ugine Kuhlmann Ltd] ‡

Sunbrite. Coke. [ICI Chem. & Polymers Ltd]

sun bronze. An alloy of from 40-60% copper, 30-40% tin, and 10% aluminum; used in jeweler's work. The name is also used for an alloy of from 50-60% cobalt, 30-40% copper, and 10% aluminum.

Suncide®. CAS 114-26-1; Propoxur; insecticide used for treatment of sucking and biting insects. [Bayer AG]

Sundora. A proprietary trade name for cellulose acetate. §

Sunett. CAS 33665-90-6; Acesulfame; sweetening agent. [Hoechst AG]

sunflower seed oil. (oils, sunflower seed) CAS 8001-21-6; Oil expressed from seeds of the sunflower, *Helianthus annuus*; emollient for skin and hair care products and makeups; adds conditioning, spread, and sheen to personal care products; used in preparation of margarine. [Arista Industries; Lipo; Penta Mfg.]

Sunfort®. Photosensitive dry film resist. [Asahi Chem. Industry]

Sunfural®. Anticancer agent. [Asahi Chem. Industry]

Sungard. CAS 4065-45-6; Sulisobenzone; uv screen. [Miles Pharmaceuticals] *

Sunimac ECR. Epoxy resin; industrial resins. [Sumitomo Bakelite Co Ltd] *

Sunimac GCR. Polyurethane resin; industrial resins.

---

* Trade name not verified as available   † Trade name verified as obsolete

[Sumitomo Bakelite Co Ltd] *

Sunnis. Refrigeration oil, premium quality. [Witco]

Sunnol. A range of alkyl aryl sulfonates; used as scouring agents, dyeing assistants, and in emulsion polymerization. [Harcros Australia] *

Sunolith. A proprietary trade name for a pigment containing 71% barium sulfate and 29% zinc sulfide. §

Sunproof. Blend of waxes for all types of stock to inhibit static atmospheric cracking and frosting. [Uniroyal] *

Sunproofing Wax 1343. Complex hydrocarbon mixture; antiozonant, anticracking, and sunchecking wax for the rubber industry. [Frank B. Ross]

Sunrabin®. Anticancer agent. [Asahi Chem. Industry]

SunShade 18-89. uv stabilizer for polyethylene sections greater than 22 mils in thickness. [Santech]

Sunshine. Resistant glaze decorating colors for porcelain, bone china and earthenware. [Degussa Ltd]

Suntec®-HD. CAS 9002-88-4; High-density polyethylene [Asahi Chem. Industry]

Suntec®-LD. CAS 9002-88-4; Low-density polyethylene [Asahi Chem. Industry]

suntei tallow. A white sweetish fat expressed from the seeds of *Palaquium oleosum*.

Sunveil®. High-shrink polystyrene film. [Asahi Chem. Industry]

Sunvex®. Water gel explosive. [Asahi Chem. Industry]

Sun Wrap®. Polyvinylidene chloride wrapping film. [Asahi Chem. Industry]

Supaclean. Universal power cleaner. [Kalon Chemicals Ltd]

Suparamin 30, 120. Blend; wet strength resin for paper industry. [Toho Chem. Industry]

Suparen. Fermentation-derived rennet. [Pfizer Ltd] *

Suparex. Superplasticizers and auxiliaries; used for the construction industry. [GE Plastics Ltd]

Supec® G401. CAS 9016-75-5; PPS crystalline polymer; high-strength, high-performance resin with outstanding heat resist. and chemical resist.; ideal metal replacement material; inherently flame retardant; used for industrial, elec./electronic, and aircraft applications. [GE Plastics]

Super A. CAS 68-26-8; Vitamin A; antixerophthalmic. [Upjohn] *

Super-A. Chemical products for general industrial use; organometallic chemicals, organic chemicals, catalysts, styrene butadiene polymers, urethane and epoxy polymers, urethane and epoxy curatives and activated carbons. [Anderson Development Company] *

Super D. CAS 8001-69-2; Cod liver oil; source of vitamins A and D. [Upjohn] *

Super AD-IT. Di(phenylmercuric) dodecenyl succinate; preservative and fungicide for aqueous coating compositions. [Hüls Am.] †

Superadoplast. Textile softener used in padding machines. [Ceca SA]

Super Alkyd® 574-75TK. Alkyd resin, air dry; very fast set to touch air dry short oil used in drum enamels and other fast air dry applications; low temp. bakes. [Thibaut & Walker] *

Superam. A fertilizer obtained by neutralizing the acids of ordinary super phosphate with ammonia gas.

superargol. (tartar cake, substitute of tartar). Preparations containing simply acid sodium sulfate. Some contain oxalates, and others contain tartaric acid and sulfuric acid.

Super-ascoloy. A ferrous alloy containing 8% nickel and 18% chromium.

Superba. CAS 1333-86-4; A proprietary trade name for a carbon black. §

superbasique metal. A modification of cast iron. It is resistant to alkalis.

Super Beckacite® 2000. Terpene phenolic; tackifier resin for adhesives. [Arizona]

Super Beckamine®. Synthetic resins. [Reichhold]

Super Beckosol®. Synthetic resins. [Reichhold]

Superbond®. High-performance carrier for bonding high ring crush or recycled medium and liner. [Crain Processing]

Super Bowl. Toilet cleaner. [Momar Industrial Services Ltd]

Superbrillantoline®. Hydromethylabietate; raw material for cosmetics. [Laserson & Sabetay]

super bronze. An alloy of from 57-69% copper, 1.2-5.1% aluminum, 1.3-2% iron, 21-37% zinc, and 3-3.2% manganese.

Supercadoplast. Textile softener to facilitate raising; for use in winch and autoclave. [Ceca SA]

Super Cat. Metal salts of organic acids; fuel oil additives. [Hüls Am.]

super cement. An ordinary Portland cement to which has been added a waterproofing material.

Super-Ceram. Three-part casting system consisting of resin, curing agent, and ceramic grain; for injection and compression dies having low shrinkage and good dimensional stability; low thermal expansion enables embedment of reinforcement rods and cooling or heating coils. [Magnolia Plastics]

Superchlon. Chlorinated polymers. [British Traders & Shippers Ltd]

Superclear 80-N. Aq. colloidal disp. of a branched polysaccharide macromolecule containing active carbonyl and hydroxyl groups; natural gum for textile dyeing and printing; water modifier controlling mobility of discrete dye particles in aq. systems; hinders migration of dyestuffs during drying of fabrics. [Henkel/Textiles]

Super-cliffite. Explosives. No. 1 contains 10% nitroglycerin, 1% collodion cotton, 60% ammonium nitrate, 16% sodium chloride, 11% ammonium oxalate, and 6% wood meal. No. 2 has the sodium chloride increased to 20% and the sodium oxalate reduced to 6%.

Supercoat®. CAS 1317-65-3; Calcium carbonate; coated ultrafine ground pigment offering easy incorporation and dispersion in plastics with improved physical properties [ECC]

super cobalt. See cobalt.

Supercol® Guar Gum. CAS 9000-30-0; Guar gum. [Hercules]

Supercore® S13F. CAS 9005-25-8; Modified corn starch; used in stucco slurry to protect the critical core-to-paper bond from breakdown during drying. [Grain Processing]

Super Corona. Refined anhyd. lanolin USP; superfatting emollient and emulsifier for cosmetics and pharmaceuticals; improves spreading, penetration, and aesthetic properties of these products; plasticizer in aerosol hairsprays; stable w/o emulsions. [Croda Inc.]

Supercut. Engineering cutting fluids. [Lambson Ltd]

Super Die. A proprietary trade name for a tool steel containing 10.5% chromium, 1% tungsten, and 1% silicon. §

Super-excellite. An explosive containing 73.5-77% ammonium nitrate, 6.5-8% potassium nitrate, 2-4% wood meal, 3.5-5% nitro-glycerin, and 9-11% ammonium oxalate.

---

‡ Trade name and manufacturer not verified   § Trade name without identified manufacturer

Superfast Power Pack. Liquid epoxy resin and hardener; fast setting system; general purpose adhesive. [Wessex Resins & Adhesives Ltd]

Superfiltchar. A proprietary product; it is an active decolorizing carbon made from sawdust. §

Superfine Lanolin Anhydrous USP. Lanolin; superfatting emollient with some emulsifying properties. [Croda Inc.]

Superfloc. High molecular weight polymers based on acrylamide in powder, solution or emulsion form; nonionic, anionic, cationic. [Cyanamid BV]

Superfloc C507. Melamine formaldehyde resin. [Cyanamid BV]

Superfloc C521. Monomethylamine-epichlorohydrin condensation products. [Cyanamid BV]

Superfloc C567, C573, C577, C581. Demethylamine-epichlorohydrine condensation products. [Cyanamid BV]

Superforcite. A Belgian gelatin dynamite containing 64% nitroglycerin.

Super Glue. Cyanoacrylate adhesive; for bonding two similar or dissimilar materials in seconds. [Novest Inc] ‡

Super Green. Various granular fertilizer blends; fertilizers for lawns, gardens, and flowers. [Horn's Crop Service Center] *

Supergreen. Lawn fertilizer. [May & Baker Ltd] *

Super Hartolan. CAS 8027-33-6; EINECS 232-430-1; Lanolin alcohol; spreading agent, dispersant, stabilizer, plasticizer, o/w emulsifier, emulsion stabilizer, and emollient for cosmetics and pharmaceuticals. [Croda Inc.; Croda Chem. Ltd.]

Superinone. CAS 25301-02-4; Tyloxapol; detergent. [Sterling Drug Inc] *

Superior alloy. A heat-resisting alloy containing 78% nickel, 19.5% chromium, 2% manganese, and 0.5% iron.

Superior® Granulated. CAS 7647-14-5; Sodium chloride; evaporated granulated salt refined by closed pan (vacuum) evaporation of purified brine. [Akzo Salt]

Superite. An explosive consisting of 80-84% ammonium nitrate, 9-11% potassium nitrate, 2-5% starch, and 3.5-4.5% nitroglycerin.

Superjet. CAS 1333-86-4; A carbon black pigment. [Pfizer International] †

Super-karma. An alloy wire containing 80% nickel and 20% chromium.

Superkleen C. Polymer disp.; imparts durable hydrophilic and soil release properties to 100% polyester. [CNC Int'l. L.P.]

Super-kolax No. 2. An explosive containing nitroglycerin, collodion cotton, potassium nitrate, barium nitrate, wood meal, starch, and ammonium oxalate.

Super Lacolene. Aliphatic solv. [Ashland]

Super-ligdynite. A coal mine explosive containing from 15-17% nitroglycerin, 15-17% ammonium nitrate, 23-25% sodium nitrate, 10-12% flour, 19-21% wood pulp, and 9-11% sodium chloride.

Superlit. A proprietary synthetic resin. §

Superlite. Tin (IV) oxide. [Keeling & Walker Ltd] *

Superlock. Anaerobic; use to seal and lock parts in place. [ITW Devcon]

Superloid®. CAS 9005-34-9; Refined ammonium alginate; gum used as gelling agent, thickener, emulsifier, film-forming agent, suspending agent, and stabilizer in food, pharmaceutical, and industrial applications; stabilizer in paper and textile industry. [Kelco]

Super Lubestine. CAS 14807-96-6; Talc; paint extender. [Bromhead & Denison Ltd]

Super Lubracon. A water-based food industry lubricant, with cleaning and antimicrobial properties; bottle conveyor lubricant, can conveyor lubricant, keg conveyor lubricant, crate conveyor lubricant. [Harshaw Chemicals Ltd] *

Supermagnesia 66. Mineral fiber-free thermal insulator. [The Chemical & Insulating Co Ltd]

Supermite®. CAS 1317-65-3; Calcium carbonate; ultrafine pigment developing superior physical properties and surface gloss in plastics, elastomers, and coatings. [ECC]

Super Moss Killer & Lawn Fungicide. Dichloropren fungicide/moss killer. [Murphy Chemical Ltd] †

Super Mosstox. CAS 97-23-4; A liquid formulation containing 34% dichlorophen; controls moss in fine turf, footpaths, hard tennis courts, playgrounds, roof and other affected hard surfaces. [Burts & Harvey]

Super Mosstox. CAS 97-23-4; Dichlorophen; fungicide, bactericide, and algicide used as a moss-killer. [Rhone-Poulenc Environmental Prods. Ltd]

Superneutral metal. A silicon-iron alloy; used for nitric acid plants.

Super Nevtac® 99. Synthetic polyterpene resin; synthetic tackifier used in adhesives, coatings, rubber products, concrete-curing compounds, and caulking compounds. [Neville]

Super Nickel. A proprietary trade name for alloys of 20-30% nickel with 70-80% copper; they are corrosion resisting. §

Superol. CAS 56-81-5; Glycerin USP; humectant; in pharmaceuticals, toiletries, tobacco, alkyds, food products, explosives, cellophane, urethane foam, other industries. [Procter & Gamble]

Superox®. Oxidation catalysts for synthetic resins. [Reichhold]

Superpalite. (diphosgene, green cross gas) CAS 76-02-8; Trichloromethyl chloroformate, $ClCOOCCl_3$; a military poison gas.

Super-Pflex® 100. CAS 471-34-1; Surface-modified precipitated calcium carbonate; filler; provides excellent extrusion processing, impact strength retention, and enhanced color stability on outdoor exposure in rigid PVC siding and profiles. [Pfizer]

Superphosphate. CAS 1314-56-3; Soluble powder containing 18% phosphorus pentoxide; phosphate fertilizer for use throughout the garden. [Vitax Ltd]

superphosphate. (mineral superphosphate, superphosphate of lime). It consists of mono-calcium phosphate, $CaH_4(PO_4)_2$, mixed with calcium sulfate, and contains 25-28% soluble phosphate; used as a fertilizer.

superphosphate of lime. See superphosphate.

superphosphoric acid. See polyphosphoric acid.

Superpolystate. CAS 9004-99-3; PEG-6 stearate SE; self-emulsifying gelling base for pharmaceutical and cosmetic lotions. [Gattefosse SA]

Superprill. CAS 57-13-6; EINECS 200-315-5; Urea (prilled); fertilizer. [Columbia Nitrogen Corporation] *

Super-Pro 5A. TEA coco-hydrolyzed collagen, sorbitol; detergent, conditioner, emulsifier, moisturizer, foamer used in personal care products. [Inolex]

Super-Quench. Metal working oil. [Chevron] *

Super Refined Almond Oil NF. Almond oil; emollient; provides elegant skin feel and promotes spreading in

---

* Trade name not verified as available

† Trade name verified as obsolete

creams, lotions, bath oils. [Croda Inc.]

Super Refined Apricot Kernel Oil NF. Apricot kernel oil; emollient; lubricant and softener in nail oils; conditioner for hair care products. [Croda Inc.]

Super Refined Avocado Oil. CAS 8024-32-6; Avocado oil; emollient; provides elegant skin feel and promotes spreading in creams, lotions, bath oils; emollient for hair care products. [Croda Inc.]

Super Refined Babassu Oil. Babassu oil; emollient for sunscreen products. [Croda Inc.]

Super Refined Coconut Oil. CAS 8001-31-8; EINECS 232-282-8; Coconut oil; emollient; improved color and odor for sunscreen products. [Croda Inc.]

Super Refined Crossential EPO. Evening primrose oil; emollient; contains essential fatty acids vital to health of skin. [Croda Inc.]

Super Refined Grapeseed Oil. CAS 8024-22-4; Grapeseed oil; emollient. [Croda Inc.]

Super Refined Menhaden Oil. CAS 8002-50-4; Menhaden oil; emollient; contains essential fatty acids vital to health of skin. [Croda Inc.]

Super Refined Mink Oil. Mink oil; emollient oil for makeup remover systems. [Croda Inc.]

Super Refined Olive Oil. CAS 8001-25-0; Olive oil; cosmetic emollient; lubricant for hair care products. [Croda Inc.]

Super Refined Orange Roughy Oil. Orange roughy oil; emollient, softener, spreading agent in skin care products. [Croda Inc.]

Super Refined Peanut Oil. CAS 8002-03-7; Peanut oil; emollient in skin care products; also for nutritional supplements, pharmaceutical delivery systems. [Croda Inc.]

Super Refined Safflower Oil USP. CAS 8001-23-8; Safflower oil; emollient. [Croda Inc.]

Super Refined Sesame Oil. CAS 8008-74-0; Sesame oil; emollient oil used in skin care preparations, nutritional supplements, pharmaceutical delivery systems. [Croda Inc.]

Super Refined Shark Oil. CAS 68990-63-6; Shark liver oil; emollient, moisture repellent for skin protection. [Croda Inc.]

Super Refined Soybean Oil. CAS 8001-22-7; Soybean oil; emollient oil used in skin care preparations, nutritional supplements, pharmaceutical delivery systems. [Croda Inc.]

Super Refined Sunflower Oil. CAS 8001-21-6; Sunflower seed oil. [Croda Inc.]

Super Refined Wheat Germ Oil. CAS 8006-95-9; Wheat germ oil; emollient; contains essential fatty acids vital to health of skin; used in facial creams; moisturizer in hair care products. [Croda Inc.]

Super-rippite. A smokeless powder containing from 51-53 parts nitroglycerin, 2-4 parts nitrocotton, 13.5-15.5 parts potassium nitrate, 15.5-17.5 parts dried borax, and 7-9 parts potassium chloride.

Super-rippite No. 2. An explosive for coal mines containing 51% nitroglycerin, 3% nitrocotton, 11% potassium perchlorate, 24% borax, and 10% potassium chloride.

Super-Sat. CAS 8031-44-5; EINECS 232-452-1; Hydrogenated lanolin; plasticizer, emollient; cosmetics and pharmaceuticals; makeups, night creams, shaving creams. [RITA]

Super-Sat AWS-4. CAS 68648-27-1; PEG-20 hydroge-

nated lanolin; emollient, emulsifier, plasticizer for emulsion systems. [RITA]

Superseal. A high performance EPDM rubber-based single layer roofing membrane. [Feb Ltd]

Supersevtox. Selective herbicide. [Schering Agrochemicals Ltd]

Supersoft NI. Silicone emulsion; synthetic softener for elasticized fabrics; excellent resist. to yellowing. [CNC Int'l. L.P.]

Supersol ADM. Blend; wire drawing lubricant. [Rhone-Poulenc Surf.]

Supersol ICS. Textile solv. scour to remove wax and oil; caustic stable. [Sybron]

Super Solan Flaked. CAS 61790-81-6; PEG-75 lanolin; emollient, conditioner, superfatting agent, solubilizer. [Croda Inc.]

Supersols. Wire drawing lubricants. [Rhone-Poulenc UK]

Super Solvitax®. Cod liver oil B. vet C; dietary supplement for animals; conditioning oil; for relief of stiffness in animals. [Seven Seas Ltd]

Super-Sorb C Water Absorbant. Copolymer acrylamide sodium acrylate; increases water holding capacity of soils and horticultural media; used in greenhouses, nurseries, and landscaping. [Aquatrols Corp of Am.] *

Super-Sorb F Water Absorbant. Copolymer acrylamide sodium acrylate; holds water as a gel around plant roots; used for transplanting and tranporting of bare root plant material, in reforestation, landscaping, and crop production. [Aquatrols Corp of Am.] *

Supersorbon. Molded activated carbon for the recovery of solvents. [Degussa AG] *

supersoy. A high grade soya bean flour.

Superspray. CAS 55-56-1; Chlorhexidine teat spray. [Ciba plc] *

Super Sta-Tac® 80. Specialty hydrocarbon based on mixed olefins; for pressure-sensitive adhesive, hot-melt, and sealants; tackifier for S-I-S and S-B-S polymers. [Arizona]

Super Sta-Tac® 100. Specialty hydrocarbon based on mixed olefins; for pressure-sensitive adhesive, hot-melt, and sealants. [Arizona]

Super Sterol Ester. C10-30 cholesterol/lanosterol esters; emollient, lubricant, and moisturizer for dry skin, cosmetics, pharmaceuticals. [Croda Inc.]

Superstyrex. See polystyrene. §

Super Sulfur No. 1. The oxidized zinc salt of dimethyl dithiocarbamic acid; a proprietary rubber vulcanization accelerator. §

Super Sulfur No. 2. CAS 19010-66-3; Lead dimethyldithiocarbamate; a proprietary rubber vulcanization accelerator. §

Supersurf AFX. Blend; wetting agent for fourth generation nylon. [Am. Emulsions]

Supersurf FL/4. Penetrant, wetting/rewetting agent, emulsifier for bleaching, dyeing, finishing. [Am. Emulsions]

Supertac. Sealant adhesive. [Crowley Chem.]

Super Thane® 975-70. Oil-modified urethane; water-reducible urethane for general-purpose varnish applications. [Thibaut & Walker]

Supertherm 2003. CAS 25928-94-3; One-part epoxy adhesive; ultra-high thermally conductive diamond-filled epoxy adhesive for semiconductor die-attach and other micro-electronic applics where very high thermal conductivity and excellent elec. insulation over wide temp.

---

‡ Trade name and manufacturer not verified   § Trade name without identified manufacturer

range are required. [Tra-Con]

Superthin. Tetrafilcon A; contact lens material. [Am. Optical] *

Super Tin® 4L. CAS 76-87-9; Triphenyltin hydroxide solution; flowable fungicide for pecans, potatoes, sugar beets; restricted use. [Griffin]

Supertox. Selective weedkiller. [May & Baker Ltd] *

Supertox. Mixture of 2,4-D and mecoprop; used to control weeds in grassland. [Rhone-Poulenc/Agri]

superturpentine. Spirits of turpentine specially rectified *in vacuo*. It boils at 155C, and distills completely below 160C.

Super Verdone. Soluble concentrate containing 72 g 2,4-D, 12 g dicamba and 48 g ioxynil per liter; used to control weeds in turf. [ICI Chem. & Polymers Ltd]

Super Vilex. Denatonium saccaride; bittering aversive agent. [Atomergic Chemetals Corp]

Super Weedex. Total weedkiller. [Murphy Chemical Ltd] †

Super Wet. Wetting agent and emulsifier for pesticides; soil penetrant; enhances pesticide and fertilizer efficiency. [W A Cleary]

Super White Fonoline®. White petrolatum USP; low melting point grade with superior snow white color, exhibiting elegant feel and texture; for premium cosmetic formulations and food related applications. [Witco/Sonneborn]

Superwipes. Chlorhexidine/cetrimide udder wipes. [Ciba plc] *

Suplex. CAS 13684-63-4; Emulsifiable concentrate containing 114 g/l phenmedipham; used for weed control for beet crops. [Universal Crop Protection Ltd]

Supplex®. Fibers. [DuPont UK]

Suppocire AIP. Saturated polyglycolized glycerides. [Gattefosse SA]

Supra EF. CAS 14807-96-6; Platy talc; for cosmetic applications including dusting and pressed powds., creams and lotions, antiperspirants, bath and loose powds. [Luzenac Am.]

Suprac. Decolorizing and absorptive activated carbons; used for dry cleaning and chemical purification. [Lancashire Chemical Works Ltd]

Supracen®. A range of acid wool dyes with outstanding leveling power and very good lightfastness. [Bayer AG; Bayer plc]

Supracet. Disperse dyestuffs; for dyeing of acetate, triacetate and nylon fibers and blends. [Holliday Dyes & Chemicals Ltd]

Supracide. Organophosphorus insecticide. [Ciba plc] *

Supradyn. Proprietary multivitamin preparation containing the B complex vitamins, vitamins A, C, D and E, biotin, nicotinamide, folic acid, (in addition to minerals and trace elements Ca, Fe, Mg, Mn, Cu, Zn, Mo and P). [Roche Products Ltd] *

Supraene®. CAS 111-02-4; Purified squalene; natural emollient. [Robeco]

Supraene. CAS 111-02-4; Squalene. [A & E Connock (Perfumery & Cosmetics) Ltd]

Suprafino A. CAS 14807-96-6; Talc; for cosmetic applications including antiperspirants, eye shadows, soaps, creams and lotions, nail polish, aerosols. [Cyprus Industrial Min.]

Suprafix Paste. Vat dyes. [Bayer plc]

Suprafrax. A clay with a high percentage of alumina; used as a furnace lining.

Supragil® GN. Sodium phenyl sulfonate; dispersant, sus-

pending agent for pesticide wettable powds. and water-disp. gran.; nonfoaming. [Rhone-Poulenc Geronazzo]

Supragil® MNS/90. CAS 26264-58-4; EINECS 247-561-6; Sodium methyl naphthalene sulfonate condensate; surfactant, dispersing/suspending agent for pesticide wettable powds. and water-disp. granules. [Rhone-Poulenc Surf.; Rhone-Poulenc France; Rhone-Poulenc Geronazzo]

Supragil® NK. CAS 25417-20-3; Sodium dibutyl naphthalene sulfonate; wetting agent for pesticides; emulsifier, wetting and dispersing agent, foamer for dyes, pigments, emulsion polymerization, industrial use. [Rhone-Poulenc Surf.; Rhone-Poulenc France; Rhone-Poulenc Geronazzo]

Supragil® WP. CAS 1322-93-6; Sodium diisopropyl naphthalene sulfonate; wetting agent, dispersant, stabilizer without detergent properties, foaming agent used in textiles, paints, pesticides, latex emulsions. [Rhone-Poulenc Surf.; Rhone-Poulenc Geronazzo]

Supramin®. A range of acid wool dyestuffs with superior fastness to water, washing and perspiration. [Bayer AG; Bayer plc]

Supranol®. Acid wool dyes; used in dyeings which are fast to washing, water and seaweed. [Bayer AG; Bayer plc]

Suprapal®. Styrene copolymer; for roadmarking paints, gravure and flexographic printing inks, paper coatings, zinc dust primers. [BASF AG]

Suprarenin. CAS 51-42-3; Epinephrine bitartrate; adrenergic. [Sterling Drug Inc] *

supraresen. The residue obtained when dammar is prepared for use in lacquers and is soluble in hydrocarbons; used in the varnish industry.

Suprasec. Isocyanates for general application. [ICI Chem. & Polymers Ltd]

Suprathion. CAS 950-37-8; Active ingredient: methidathion; organophosphorous insecticide with high degree of insecticidal activity; also efficient in controlling scales. [Makhteshim Chemical Works Ltd]

Suprefact. CAS 68630-75-1; Buserelin acetate; luteinizing hormone-releasing factor; gonad-stimulating principle. [Hoechst-Roussel Pharmaceuticals Inc] *

Supressin. See Cardura. [Pfizer International]

Suprex. See Catalpo. §

Suprexcel. Fast-to-light direct cotton dyestuffs. [Holliday Dyes & Chemicals Ltd] †

Suprex white. A highly purified precipitated calcium carbonate for use as a rubber filler in the place of blanc fixe. It has a specific gravity of 2.7.

Suprilent. CAS 395-28-8; Isoxsuprine; promotes blood flow rate. [Duphar BV] *

Suprofix. High speed photographic fixer. [May & Baker Ltd]*

Suprol. Photographic developer. [May & Baker Ltd] *

Supronal®. Sulfonamide mixture; chemotherapeutic drug for the treatment of bacterial infections in animals. [Bayer AG]

Supronic. Nonionic surfactant. [Rhone-Poulenc UK]

Supronic B10, B25, B5O, B75 and B1OO. Polyoxy alkylated polyalkylene glycols; low-foam surfactants. [Glover (Chemicals) Ltd] ‡

Supro-Tein R. Sodium cocoyl hydrolyzed rice protein, sorbitol; biodeg., mild, foaming substantive protein for conditioning, cleansing, thickening in hair and skin care products (baby shampoos, ethnic shampoos, hairstying products, lotions, hand/face cleansers, bath products, depilatories). [Maybrook]

Supro-Tein V. TEA-coco-hydrolyzed animal protein, sorbitol; mild surfactant with foam stability; emulsifier, solubilizer, moisturizer; for ethnic and mild shampoos, conditioners, lotions, hair and skin mousses, liquid soap, shave creams. [Maybrook]

Surbex T. A proprietary multi-vitamin preparation. [Abbott Laboratories]

Surco. Alkanolamides. [Millmaster-Onyx UK]

Surcol®. Acrylic and vinyl polymers in bead form for use in paints, adhesives, polishes, printing inks, lacquers, concrete sealers. [Allied Colloids Ltd]

Surcopur®. CAS 709-98-8; Propanil; post-emergence applied herbicide with no residual effect for control of numerous grasses and broad-leaved weeds in rice crops. [Bayer AG]

Sure-Curd. A standardized solution of fermentation derived milk clotting enzyme elaborated by *Endothia parasitica*; used in the manufacture of cheese, especially Swiss and Italian varieties. [Pfizer International]

Suresperse® 911. Low-foaming antifoulant for dispersing oil, lint, biological matter, silt, mud, textile fibers, tobacco dust and other debris in industrial air washers; also removes deposits. [Drew Ind. Div.]

Sure-Step. CAS 10043-52-4; Chip calcium chloride. [Schaefer Salt & Chemical Co]

Suretex® 920. Static control agent for air washers for use in textile mills. [Drew Ind. Div.]

Surexin. CAS 1740-22-3; Pyrinoline; cardiac depressant. [McNeil Pharmaceuticals] *

Surf Ac 820. Alkylaryl polyethoxyethanol and n-butanol; nonionic biodegradable surfactant. [Draxel Chemical Company] ‡

Surfacaine. Cyclomethycaine sulfate; anesthetic. [Eli Lilly & Co] *

Surfactant N-42. Nonylphenol ethoxylate nonionic surfactant in the form of a clear oil; general surfactant. [Rohm & Haas UK] *

Surfactant XQS20. Phophate ester in free acid form; aqueous solution; wetting agent and detergent; textile lubricant; antistatic agent; industrial emulsifiers. [Rohm & Haas UK] *

Surfactol® 13. CAS 1323-38-2; Castor oil, modified; wetting agent, emulsifier, solubilizer, dispersant, wax plasticizer, mold release agent, antifoamer for textiles, leather, paints, household, cosmetics, dyeing, tanning, finishing, sizing, making of cutting and sol. oils, oilfield chems.; biodeg. [CasChem]

Surfactol® 318. CAS 61791-12-6; PEG-5 castor oil; emulsifier for oils, waxes; solubilizer for fragrances; emollient; used in textiles, paints, household, cosmetics, dyeing, tanning, finishing, sizing, insecticides, herbicides, fungicides, kier boiling, making of cutting and sol. oils, dispersing waxes, pigments, resins, rewetting dried skins. [CasChem]

Surfactol® Q1. Ricinoleamidopropyltrimonium chloride, propylene glycol; surfactant, emollient, refatting agent, antistat, emulsifier for skin and hair care products, clear shampoos. [CasChem]

Surfactol® Q2. Hydroxystearamidopropyl trimonium chloride, propylene glycol; surfactant, emollient, emulsifier, refatting agent, antistat for conditioners, liquid hand soaps, shampoos, skin lotions. [CasChem]

Surfactol® Q3. Hydroxystearamidopropyl trimonium metho-

sulfate, propylene glycol; surfactant, emulsifier, emollient, antistat for aerosols and mousses, liquid hand soaps, shampoos, skin lotions; compatible in aerosol systems without corroding cans. [CasChem]

Surfactol® Q4. CAS 112324-16-0; Ricinoleamidopropyl ethyldimonium ethosulfate; emollient, emulsifier for personal care products (shampoos, conditioners, aerosols). [CasChem]

Surfadone LP-100. Caprylyl pyrrolidone. [ISP]

Surfadone LP-300. Lauryl pyrrolidone. [ISP]

Surfadone QSP. Dimethyl stearamidopropyl [(2-pyrrolidonyl) methyl] ammonium chloride. [ISP]

Surfadone WSP. Stearamidopropyl pyrrolidonylmethyl dimonium chloride. [ISP]

Surfageen. Anionic surfactant of the alkyl ether phosphate type in solid form; emulsifier for emulsion polymerization; deinking of waste paper; rust inhibitor. [Chem-Y, Fabriek van Chemische Producten BV] *

Surfageen S30. Fatty alcohol ether sulfosuccinate as a clear liquid; detergent, foaming, wetting and emulsifying agent used as a mild raw material for shampoos and foam baths. [Chem-Y, Fabriek van Chemische Producten BV]*

Surfagene FAD 106. Sodium nonoxynol-6 phosphate; wetting agent for weakly acidic and alkaline cleaners, metalworking cooling lubricants. [Chem-Y GmbH]

Surfagene FAZ 109. Nonoxynol-9 phosphate; emulsifier for emulsion polymerization. [Chem-Y GmbH]

Surfagene FDD 402. Laureth-2 phosphate; emulsifier for cosmetic formulations, oil baths; additive for mineral collectors. [Chem-Y GmbH]

Surfagene FGD 600. CAS 69331-39-1; DEA-cetyl phosphate. [Chem-Y GmbH] *

Surfagene FGZ 608. Ceteth-8 phosphate. [Chem-Y GmbH] *

Surfagene FHD 704 NV. Sodium steareth-4 phosphate. [Chem-Y GmbH] *

Surfagene FMD-12-03 KAM. Trideceth-3 phosphate. [Chem-Y GmbH] *

Surfagene FPG 50. Sodium glycereth-1 polyphosphate. [Chem-Y GmbH] *

Surfagene FPT. TEA phosphate ester; scale inhibitor for cooling water systems. [Chem-Y GmbH]

Surfagene MB 1705. Disodium undecylenamido MEA-sulfosuccinate. [Chem-Y GmbH] *

Surfagene S 30. CAS 39354-45-5; Disodium laureth-3 sulfosuccinate; mild raw material for personal care products, baby products, shampoo, foam baths; mild to skin and eyes; biodeg. [Chem-Y GmbH]

Surfak. CAS 128-49-4; Docusate calcium; stool softener. [Hoechst-Roussel Pharmaceuticals Inc] *

Surf-A-Seis. A suface explosive; used for surface energy sources in portable seismic operations. [Hercules] *

Surfonic®. A series of p-nonylphenol ethoxylates; surfactants in agricultural chemicals, industrial cleaners, heavy-duty detergents, paper industry. [Texaco]

Surfonic® HDL. CAS 9016-45-9; Nonoxynol-8 (86%) and triethanolamine (14%); biodeg. emulsifier, wetting agent, dry cleaning detergent, penetrant, solubilizer, lime soap dispersant, antifoamer used in agriculture, cosmetics, industrial cleaners, ceramics, concrete, dust control, wallpaper removal, photographic film developing; fire fighting, emulsion polymerization, indirect food additives, cutting oil emulsifiers; stabilizer for rubber

---

‡ Trade name and manufacturer not verified   § Trade name without identified manufacturer

latex and drilling-mud additives; degreaser for leather industry. [Texaco]

Surfonic® JL-80X. Alkoxypolyalkoxyethanol; biodeg. surfactant used as emulsifier, wetting agent, detergent, penetrant, solubilizer, dispersant for household detergents, industrial products, agricultural sprays, dry cleaning, metal cleaners, ceramics, concrete, textile processing, paper manufacturing. [Texaco]

Surfonic® L12-3. C10-12 pareth-3; biodeg. surfactant for detergent, laundry prespotters, hard surface cleaners, emulsifiers, personal care products, agricultural pesticides. [Texaco]

Surfonic® L24-2. CAS 68439-50-9; C12-14 pareth-2; biodeg. surfactant for detergents, laundry prespotters, hard surface cleaners, personal care products, agricultural pesticides. [Texaco]

Surfonic® L46-7. C14-16 pareth-7; biodeg. surfactant, detergent, emulsifier, foamer, penetrant, intermediate for detergent, laundry prespotters, hard surface cleaners, personal care products, agricultural pesticides. [Texaco]*

Surfonic® LF-17. CAS 69013-18-9; Primary alcohol-EO adduct, modified; biodeg. low foaming wetting agent, detergent for aq. systems, metal cleaners, latex paints, textiles, paper, rinse aids, industrial and home mech. dishwashing compounds; defoamer in some systems. [Texaco]

Surfonic® N-10. CAS 9016-45-9; Nonoxynol-1; biodeg. emulsifier, wetting agent, detergent, penetrant, solubilizer, dispersant for household cleaners, textile, agriculture, metal cleaning, petrol, cosmetic, latex paint, cutting oil, janitorial supply industries. [Texaco]

Surfynol®. A range of acetylenic glycols; used as wetting agents with very low foam for paints. [Air Prods. & Chems. Inc]

Surfynol® 61. CAS 107-54-0; 3,5-Dimethyl 1-hexyn-3-ol; surfactant, wetting agent used for paper coatings, inks, floor polishes, and glass cleaning formulations; cleaner in silicon wafer industry. [Air Prods. & Chems. Inc]

Surfynol® 82. CAS 78-66-0; 3,6-Dimethyl-4-octyne-3,6-diol; surfactant, defoamer, wetting agent, visc. reducer used in aq. systems, pesticide concentrates, shampoos, vinyl plastisols, agriculture, aq. starch solutions, flexographic inks, electroplating baths; detergent for radiator cleaners, descaling compounds. [Air Prods. & Chems. Inc]

Surfynol® 82S. 3,6-Dimethyl-4-octyne-3,6-diol on amorphous silica carrier; defoamer/wetting agent in pesticide wettable powds., electroplating baths, cement, plastics, coatings; solubilizer and clarifier in shampoos. [Air Prods. & Chems. Inc]

Surfynol® 104. CAS 126-86-3; Tetramethyl decynediol; defoamer and dye dispersant in paints, inks, dyestuffs, pesticides; surfactant in rinse aids; substrate pigment wetting agent for industrial coatings and adhesives; wetting agent for industrial cleaners; visc. reducer for vinyl dispersions. [Air Prods. & Chems. Inc]

Surfynol® 104A. CAS 126-86-3; Tetramethyl decynediol and 2-ethyl hexanol; defoamer, wetting agent for pesticides, coatings, dyestuffs, aq. systems; lubricity additive for metalworking formulations. [Air Prods. & Chems. Inc]

Surfynol® 104BC. CAS 126-86-3; Tetramethyl decynediol in 2-butoxy ethanol; wetting and defoaming agent in water-based systems, e.g., coatings, adhesives, inks, cements, metalworking fluids, latex dipping and paper coatings. [Air Prods. & Chems. Inc]

Surfynol® 104E. CAS 126-86-3; Tetramethyl decynediol and ethylene glycol; wetting agent, defoamer, dispersant, visc. stabilizer. [Air Prods. & Chems. Inc]

Surfynol® 104H. CAS 126-86-3; Tetramethyl decynediol in ethylene glycol; wetting and defoaming agent in water-based systems, e.g., coatings, adhesives, inks, cements, metalworking fluids, latex dipping and paper coatings. [Air Prods. & Chems. Inc]

Surfynol® 104PA. CAS 126-86-3; Tetramethyl decynediol in IPA; wetting and defoaming agent in water-based systems, e.g., coatings, adhesives, inks, cements, metalworking fluids, latex dipping and paper coatings. [Air Prods. & Chems. Inc] *

Surfynol® 104PG. CAS 126-86-3; Tetramethyl decynediol in propylene glycol; wetting and defoaming agent in water-based systems, e.g., coatings, adhesives, inks, cements, metalworking fluids, latex dipping and paper coatings. [Air Prods. & Chems. Inc] *

Surfynol® 104S. CAS 126-86-3; Tetramethyl decynediol on amorphous silica; wetting agent, defoamer, dispersant, visc. stabilizer for agricultural formulations. [Air Prods. & Chems. Inc]

Surfynol® 420. Ethoxylated tetramethyl decynediol (1.3 EO); wetting agent, defoamer, dispersant for aq. coatings, inks, adhesives, agriculture, electroplating, oilfield chems., paper coatings. [Air Prods. & Chems. Inc]

Surfynol® 440. CAS 9014-85-1; PEG-3.5 tetramethyl decynediol; water-based industrial finishes; defoamer, rewetting, and leveling agent for paperboard coatings, agricultural formulations; metal cleaning and plating bath additive. [Air Prods. & Chems. Inc]

Surfynol® 465. CAS 9014-85-1; PEG-10 tetramethyl decynediol; wetting agent, defoamer for aq. coatings, inks, adhesives; surfactant for emulsion polymerization; electroplating additive. [Air Prods. & Chems. Inc]

Surfynol® 485. CAS 9014-85-1; PEG-30 tetramethyl decynediol; wetting agent, defoamer for aq. coatings, inks, adhesives, agriculture, electroplating, oilfield chems., paper coatings. [Air Prods. & Chems. Inc]

Surfynol® CT-136. Proprietary blend; wetting agent, defoamer, grind aid and dispersant for water and glycol-based inks and pigments. [Air Prods. & Chems. Inc]

Surfynol® D-101, D-201. Nonsilicone; defoamer/antifoamer for aq. systems, coatings, adhesives; effective for PVAc, ethylene-vinyl acetate systems. [Air Prods.; Air Prods. Nederland BV]

Surfynol® DF-08. Surfactant; wetting agent, defoamer for waterborne coatings, inks, adhesives, and latex dipping. [Air Prods.; Air Prods. Nederland BV]

Surfynol® DF-34. Proprietary blend; wetting agent, defoamer for water-based coatings, inks, and adhesives; dewebbing agent and defoamer for latex gloves and other dipped goods. [Air Prods. & Chems. Inc]

Surfynol® DF-37. Silicone-free organic defoamer; defoamer minimizing web formation during latex glove, waterborne coating dipping processes, and other aq. systems (inks, adhesives, coatings, agriculture, cement, metalworking fluids); wetting agent for low energy substrates. [Air Prods.; Air Prods. Nederland BV]

Surfynol® DF-58. Organo-modified silicone-based defoamer; self-emulsifying defoamer used in water-based applics, e.g., inks, coatings, adhesives, cements, latex-

cements, metalworking fluids, latex dipping and paper coatings. [Air Prods. & Chems. Inc]

---

containing formulations. [Air Prods. & Chems. Inc]

Surfynol® DF-60. Silicone-based defoamer; for aq. coating and ink systems, especially for grinding and dispersion stage. [Air Prods. & Chems. Inc]

Surfynol® DF-70. Organic defoamer; defoamer for aq. systems, printing inks, coatings, adhesives, especially acrylic and styrene-acrylic formulations. [Air Prods. & Chems. Inc]

Surfynol® DF-75. Silicone-free; defoamer for water-based systems, e.g., inks, adhesives, coatings, overprint varnishes, paper coatings; especially effective in acrylic systems. [Air Prods.; Air Prods. Nederland BV]

Surfynol® DF-110, DF-110S. Silicone-free acetylenic derivs.; defoamer, de-air entrainment agent for inks, metalworking fluids, coatings, cement, ceramics, adhesives. [Air Prods.; Air Prods. Nederland BV]

Surfynol® DF-110D, DF-110L. CAS 126-86-3; Higher molecular weight acetylenic glycol in dipropylene glycol; defoamer, de-air entrainment aid for water-based coatings, inks, adhesives, and highly pigmented systems (concrete, paper coatings, grouts, ceramics). [Air Prods. & Chems. Inc]

Surfynol® DF-210. Silicone-free organic defoamer; defoamer for aq. formulations, printing inks, adhesives, and coatings. [Air Prods. & Chems. Inc]

Surfynol® DF-574. Organo and organo-modified silicone compound; rapid knockdown defoamer for aq. coatings and inks. [Air Prods. & Chems. Inc]

Surfynol® DF-695. Silicone emulsion; defoamer for aq. ink systems. [Air Prods. & Chems. Inc]

Surfynol® GA. Acetylenic diol; pigment wetting agent, grinding aid in coatings and other pigmented systems. [Air Prods. & Chems. Inc]

Surfynol® PC. CAS 126-86-3; Acetylenic glycol; defoamer used in paper coating and adhesive latexes; antishock agent for paper coatings. [Air Prods. & Chems. Inc]

Surfynol® PG-50. CAS 126-86-3; Tetramethyl decynediol in propylene glycol; wetting and defoaming agent in water-based systems, e.g., coatings, adhesives, inks, cements, metalworking fluids, latex dipping and paper coatings. [Air Prods. & Chems. Inc] *

Surfynol® PSA-204. Surfactant; low or nonfoaming wetting agent for pressure-sensitive adhesives. [Air Prods.; Air Prods. Nederland BV]

Surfynol® SE. Acetylenic diol; wetting agent, foam control agent in pressure-sensitive adhesives, aq. lubricants, water-based paints, inks, dye processing, agricultural formulations, and paper coatings. [Air Prods. & Chems. Inc]

Surfynol® TG. CAS 126-86-3; Tetramethyl decynediol and ethylene glycol; pigment and substrate wetting agent used in latex and water-reducible paints, adhesives, paper coatings, and pigmented aq. systems. [Air Prods. & Chems. Inc]

Surfynol® TG-E. CAS 126-86-3; Acetylenic glycol and propylene glycol; low foaming and wetting agent used in pesticide formulations. [Air Prods. & Chems. Inc]

Surgam. CAS 33005-95-7; Tiaprofenic acid; used for rheumatoid arthritis, osteo-arthritis. [Roussel Laboratories Ltd]

Surgicel. Oxidized regenerated cellulose; hemostatic. [Johnson & Johnson Products Inc] *

Surlyn®. A trademark for a range of ionomer resins. [DuPont UK]

Surmabond Lining. Pigmented solvent-free epoxy systems; for seamless flooring and tank lining composition. [Surmak Products Ltd]

Surmabond Roadway. Tar modified solvent-free epoxy system with special aggregate; light-weight screed for footbridges, concrete floors. [Surmak Products Ltd]

Surmabond Screeding. Pigmented solvent-free epoxy system with special aggregates; chemical resistant heavy duty, light-weight screed and concrete repair material. [Surmak Products Ltd]

Surmafil. Bituminous mixture with liquid resins and aggregates; for instant repair of roads and parking areas, cold applied. [Surmak Products Ltd]

Surmaglaze U12. Moisture cured polyurethane; clear or colored dust-proofing sealer for concrete, stone, and timber floors. [Surmak Products Ltd]

Surmaplast. Solvent-free epoxy knifing systems; gap filling of blowholes in castings, repair of petrol, oil, and water tanks; quick setting. [Surmak Products Ltd]

Surmaseal 101. Chlorinated rubber paint; acid resistant paint for steelwork, concrete floors, and swimming pools. [Surmak Products Ltd]

Surmaseal 102. Solvent-based pigmented epoxy paint, polyamide cured; chemical and wear-resistant floor coating, also for corrosion protection of steel. [Surmak Products Ltd]

Surmatar. Solvent-based epoxy pitch coating; heavy duty anticorrosive coating. [Surmak Products Ltd]

Surmax. Avilamycin. [Dista Products Ltd]

Surmax® CS-504. Surfactant; alkaline-stable surfactant for formulating detergent concentrates; moderate foamer, wetting agent, detergent; for metalworking formulas, paint strippers, tire cleaners, transportation cleaners, dairy and food plant cleaners, paper felt washing, sanitizers, wax strippers; biodeg. [Chemax]

Surmontil. CAS 739-71-9; Trimipramine; an antidepressive drug. [May & Baker Ltd]

Surolan. A proprietary preparation containing miconazole nitrate, prednisolone; used for veterinary ear and skin infections (cats and dogs). [Janssen Pharmaceutical Ltd] *

Surophosphate. (Dasag). A German fertilizer made from sewage, other waste material, and peat.

Surpassol 53. Secondary emulsifier for oil muds. [Climax Fluids Additives]

Surpassol NT-57. Secondary emulsifier for oil muds; low toxicity. [Climax Fluids Additives]

Survac®. Polymer. [DuPont UK]

Susadrin. CAS 55-63-0; Nitroglycerin; vasodilator. [Merrell Dow Pharmaceuticals Inc] *

Suscard®. Glyceryl trinitrate; a proprietary preparation as a slow-release buccal tablet; for treatment of angina pectoris, LVF, congestive cardiac failure. [Pharmax Ltd]

Suscardia. A proprietary preparation of isoprenaline hydrochloride; used as a cardiac stimulant. [Pharmax Ltd] *

suSCon® Blue. CAS 2921-88-2; 140 g/kg Chlorpyrifos; for control of white grub in sugarcane. [Incitec Ltd]

suSCon® Green. CAS 2921-88-2; 100 g/kg Chlorpyrifos; for control of black vine weevil in hardy ornamental stock; pasture grub in pasture. [Incitec Ltd]

Suscovax. Pig samonellosis vaccine. [The Wellcome Foundation Ltd] *

Susini. An alloy of aluminum containing from 1.5-4.5% copper, 0.5-1.5% zinc, and 1-8% manganese.

Suspend-Ayd® Gels. Dispersions of modified Bentones; for pigment suspension and flow control of solv.-thinned coatings. [Daniel Prods.]

Suspendex. Expanded polystyrene foam loose-fill used for cushioning and packaging applications. [Dow UK] *

Sus-Phrine. CAS 51-43-4; Epinephrine; adrenergic. [Berlex Laboratories Inc] *

Sustac®. A proprietary preparation of glyceryl trinitrate; a vasodilator used in treatment of angina pectoris. [Pharmax Ltd]

Sustain. Floor polishes. [Evans Vanodine International Ltd]

Sustamycin. CAS 64-75-5; A proprietary preparation of tetracycline hydrochloride; an antibiotic. [MCP Pharmaceuticals] *

Sustane® 1-F. CAS 25013-16-5; BHA; preservative and antioxidant for foods, flavors, cosmetics, vitamins, oils, waxes, essential oils, tallow, sausage, chewing gum base, shortening, lard, food packaging materials, potatoes, and cereals; inhibits oxidation reaction of oils and fats in the presence of air and retards rancidity and off-flavors caused by oxidation. [UOP]

Sustane® 3. Propylene glycol, BHA, propyl gallate, citric acid, ratio 70:20:6:4; preservative and antioxidant used in snack foods, cosmetics, and spices. [UOP]

Sustane® 20. TBHQ, citric acid, propylene glycol, ratio 20:10:70; preservative and antioxidant used for edible fats and oils. [UOP]

Sustane® BHA. CAS 25013-16-5; BHA; antioxidant, stabilizer for fats, oils, and other foods. [UOP]

Sustane® PG. CAS 121-79-9; Propyl gallate; preservative and antioxidant for fats and oils. [UOP]

Sustane® TBHQ. CAS 1948-33-0; Mono-tert.-butyl hydroquinone; preservative and antioxidant for foodstuffs and meat products; color stable and useful as substitute for reactive antioxidants that tend to form purple complexes with iron or copper. [UOP]

Sustilan® N. Fiber preserving agent; dyeing and printing auxilliary. [Bayer AG]

Sutermeister's stain. For paper a) Contains 1.3 g iodine and 1.8 g potassium iodide in 100 cc water, and b) consists of a clear saturated solution of calcium chloride.

Suttocide® A. CAS 70161-44-3; Sodium hydroxymethylglycinate; antimicrobial preservative for cosmetics. [Sutton Labs]

Suva®. For the chemical industry. [DuPont UK]

suXon. See suSCon Blue. [Incitec Ltd]

SVC. Acetarsol vaginal compound. [May & Baker Ltd] *

SW 5063. Racephenicol; an antibacterial. [Sterwin Chemicals Inc] *

Swale. Coatings and adhesives; used for flexible packaging. [Swale Coatings & Inks Ltd]

Swale powder. An explosive containing potassium perchlorate, nitro-glycerin, collodion cotton, ammonium oxalate, wood meal, and a little nitrotoluene.

Swalite. An explosive for coal mines, similar to Swale powder (qv).

Swan. A paper-based grade of Tufnol industrial laminates. [Tufnol Ltd]

Swardsman. 25-5-5, compound fertilizer. [Kemira Ince Ltd]

Swedelec. A Swedish charcoal iron with a high magnetic permeability.

Swedex AR58P-15AC. Diacetylated tartaric cid ester of mono and diglycerides; emulsifier. [Australian Bakels]

Swedex SSL-5AC. CAS 25383-99-7; EINECS 246-929-7;

Sodium stearoyl-2-lactylate with 5% anticaking agent; emulsifier. [Australian Bakels]

Swedish factory tar. A tar obtained from waste wood in charcoal kilns, as a by-product in charcoal burning.

Swedish liquid resin. See talloel.

Swedish turpentine. See turpentine.

Sweeta. CAS 81-07-2; Saccharin; pharmaceutical aid. [Bristol-Myers Squibb Co Inc] *

sweet bark. (sweet wood bark, Eleuthera bark). Cascarilla; used for extracting cascarilla oil and as an ingredient in insecticides, etc.

sweet birch oil. See methyl salicylate.

Sweetex. Low calorie sweetener. [The Boots Co plc] *

Sweetex Plus. Low calorie sweetener. [The Boots Co plc]*

Sweetrex®. Dextrose, fructose, maltose, isomaltose, other polysaccharides; directly compressible chewable tablet base with high sweetness, coolness, and mouthfeel. [Mendell]

sweet-water. Consists of glycerin and water, obtained in the distillation of crude glycerol.

Sweetzyme. An immobilized glucose isomerase produced from a selected strain of *Bacillus coagulans*; Type Q developed specially for long-term use in a continuous fixed-bed column process for production of fructose syrup; Type A used for batch operation characterized by much higher residence times and enzyme cost compared to the fixed-bed operation. [Novo Nordisk]

Swim clear. CAS 7778-54-3; Calcium hypochlorite; used for swimming pools. [Schaefer Salt & Chemical Co]

Swipe 560 EC. An emulsifiable concentrate containing 56 g bromoxynil, 56 g ioxynil and 448 g mecoprop per liter; a post-emergence contact herbicide for cereal crops. [Ciba-Geigy Agrochemicals]

Swirl. Adjuvant containing 590 g/l refined mineral oil; for use as a wetting agent with arylalanine herbicides. [Shell UK]

Swiss Polyamid Grilon. Copolyamide; used for tape weaving, filter fabric, and filter manufacturing process, string, embroidery etc. [EMS-Chemie AG]

Swiss Polyamid Grilon. Polyamide 6; used for clothes, nonwovens, technical fabrics, and paper felts. [EMS-Chemie AG]

Swiss Polyester Grilene. Polyester; used for woven and knitted fabrics, home textiles and sewing threads, technical applications, various nonwovens and fiber-fill. [EMS-Chemie AG] †

SWS-101. Dimethylpolysiloxane terminated with non-reactive trimethylsiloxy groups; features inertness, heat and oxidative stability, excellent elec. properties for dielec. coolants, brake fluids, lubricants, auto care products, release, heat transfer, aerosols, damping media, household products, antifoam agents, cosmetics, shock absorbers. [Wacker Silicones]

SWS-290. Silica-filled dimethylpolysiloxane compound; features inertness, resist. to most organic salts, diluted acids and alkalies, water, oxidation; used in water repellents, sealants, coatings for elec. equip., lubricants for rubber and plastic, release agent for rubber/plastic molding, damping media. [Wacker Silicones]

SWS-725. Silicone rubber base; all-purpose material designed to accept nonreinforcing fillers; features low compression set, good heat resist., excellent extrusion qualities. [Wacker Silicones]

SWS-03314. CAS 69430-24-6; Cyclomethicone; slip agent, lubricant, release agent for personal care industry; emol-

---

lient for skin care products; plasticizer for hair spray resins; replacement for IPM in aerosol antiperspirants. [Wacker Silicones]

SWS-06545u. CAS 63394-02-5; Silicone elastomer; wire and cable compound. [Wacker Silicones]

SWS-7532u. CAS 63394-02-5; Silicone rubber compound; nonmilling specialty compound used as wire and cable insulation; normally catalyzed with Cadox TS-50; high strength, good elec. resistance, heat age retention of properties, high modulus, good shelf life, and good extrusion rates; meets requirments of UL Code 62 Class 22, UL Code 44, Class SA, and MIL-W-8777. [Wacker Silicones]

SWS-7655u. CAS 63394-02-5; Silicone rubber compound; high tear strength material; high resilience, high green strength, good calenderability, and extrusion; when properly cured, meets specifications of ZZR-765, Class 3b, and AMS-3347; recommended catalyst levels including 1.2 phr of TS-50, 0.8 phr of Varox, or 0.4 phr of DiCup R. [Wacker Silicones]

SWS-7675u. CAS 63394-02-5; Silicone rubber compound; uncatalyzed high strength and high tear compound; high resilience and modulus, high grn. strength, low surface tack, and good mold flow and calenderability; when properly cured, meets specifications of ZZR-765, Class 3b, Grade 70, and AMS-3349. [Wacker Silicones]

SWS-7865u. CAS 63394-02-5; Silicone rubber compound; easy processing conductive compound for gaskets for EMI/RFI shielding. [Wacker Silicones]

Sybol. CAS 29232-93-7; Contains pirimiphos-methyl; garden insecticide. [ICI Garden Products]

sycorin. *See* saccharin.

sycose. *See* saccharin.

Sydex. Soluble concentrate containing 125 g 2,4-D and 250 g mecoprop per liter; used to control weeds in grassland. [Synchemicals Ltd]

Syford. CAS 1702-17-6; 2,4-D; translocated phenoxy herbicide applied to cereals and grass. [Synchemicals Ltd]

sykose. *See* saccharin.

Syl. A proprietary preparation of dimethicone 350, benzalkonium solution, and nitrocellulose; used in dermatology as an antibacterial agent. [Lloyd, Hamol] *

Sylade. Silage preservative. [ICI Chem. & Polymers Ltd]

Sylfam 2082. Imidazoline; corrosion inhibitor for oilfield servicing, refinery operations and oil transport pipes, surfactant in metalworking formulations, pigment dispersant, chemical intermediate in the synthesis of amines. [Arizona]

Sylfan 20. Corrosion inhibitor for oilfield servicing, refinery operations and oil transport pipes, surfactant in metalworking formulations, pigment dispersant, chemical intermediate in the synthesis of amines. [Arizona]

Sylfat® D-1. Monomeric fatty acid distillate from dimerization of tall oil fatty acid; dispersant for drilling muds, component of cutting oils, grinding compounds, textile drawing lubricants, defoamers, plasticizers, thickeners in greases and lubricant oil formulations, oilfield applications. [Arizona]

Sylfat® DX, MM. Tall oil fatty acid ester; for plasticizers, extenders, surfactants in grinding and cutting oils, specialty lubricant additives, corrosion inhibitors, specialty solvs. for printing inks, metalworking, and oil well servicing. [Arizona]

Sylfat® RD-1. Monomeric fatty acid distillate from dimeriza-

tion of tall oil fatty acid; dispersant for drilling muds, component of cutting oils, grinding compounds, textile drawing lubricants, defoamers, plasticizers, thickeners in greases and lubricant oil formulations, oilfield applications. [Arizona]

Sylgard® 170. Two-part silicone system; low visc., flame retardant elec./electronic insulating resin with temp. rating to 200 C; cure can be heat accelerated; mix ratio: 1:1 by wt. [Dow Corning]

Sylgard® 182, 184. Two-part silicone system; clear, repairable elec./electronic insulating resin with temp. rating to 200 C, Sylgard 184 for quicker cure; mix ratio: 100:10 by wt. [Dow Corning]

Syllact. Psyllium husk; laxative. [G D Searle & Co] *

Syl-off®. Silicone release coatings for paper and films. [Dow Corning]

Syloid 72. A proprietary trade name for a silica gel for addition (2%) to plasticized vinyls to prevent plate out. §

Sylopal. A proprietary preparation of dimethicone, magnesium oxide, and aluminum hydroxide; a gastro-intestinal sedative. [H N Norton & Co Ltd] ‡

Sylphane S. PVC shrink film; for sales display and bundling. [UCB nv Film Sector] *

Sylphrap. A proprietary trade name for a regenerated cellulose transparent sheet. §

Syltherm®. Silicone heat transfer fluid. [Dow Corning]

Syltherm® 444. Silicone-based; heat transfer liquid for flat plate and concentrating collectors. [Dow Corning]

Syltherm® XLT. Dimethylsiloxane polymer; for use as low-temp., liquid-phase heat transfer medium for -73 to 260 C service; ideal for single-fluid processing, heating, and cooling systems in the pharmaceutical and fine chemical industries. [Dow Corning]

Sylvacote® K. Corrosion inhibitor for oilfield servicing, refinery operations and oil transport pipes, surfactant in metalworking formulations, pigment dispersant, chemical intermediate in the synthesis of amines. [Arizona]

Sylvadym® M-35. Tall oil fatty acid-based dimer acid; for amidation and esterification reactions, manufacturing of imidazolines and amidoamine derivs. for use as corrosion inhibitors in oilfield applications, chemical intermediate in production of polyamide resins, alkyds, epoxies, lubricant additives, and quaternaries. [Arizona]

Sylvamid®. Nonreactive polyamide resins. [Arizona]

Sylvan. CAS 534-22-5; α-Methylfuran, C_5H_6O; a constituent of wood tar.

Sylvania Cellophane. A proprietary trade name for regenerated cellulose. §

Sylvaros® 20. Tall oil rosin; printing ink binder as resin or salt, paper sizing agent, emulsifier for SBR polymerization as soap, tackifier resin in adhesives, imidazoline modifier in corrosion inhibitors, elastomer modifier in emulsion polymerization, dust control additive; film former/plasticizer in lacquers and varnishes. [Arizona]

Sylvaros® 315. Lower acid number tall oil rosins; for manufacturing of paper size, intermediate in rosin deriv. production, printing ink binders as resins, tackifier resin in sealants and mastics, as starting point rosin for resin esters. [Arizona]

Sylvaros® R, RX. Polymerized rosin; tackifier for pressure-sensitive adhesives, construction adhesives, pick-up gums for labeling, adhesion promoter for difficult-to-bond substrates. [Arizona]

Sylvatac® AC. Modified tall oil rosin; for manufacturing of paper size, intermediate in rosin deriv. production, printing ink binders as resins, tackifier resin in sealants and mastics, as starting point rosin for resin esters. [Arizona]

sylvic acid. CAS 514-10-3; Impure abietic acid; a major active ingredient of rosin; $C_{19}H_{29}COOH$; used as varnish driers, in soaps, and in the fermentation industry.

Sylvid®. A range of silica fillers for plastics processing. [W R Grace & Co] *

Symalit GM 20 PP. CAS 9003-07-0; 20% Glass mat-reinforced polypropylene; composite material featuring high stiffness, good impact and toughness, high energy absorp.; for automotive parts. [Symalit]

Symax. Laminates of nomex, presspaper, leatheroid, paper, melinex, mylar and kapton; slot liner and closure material used in the manufacture and repair of electrical motors, transformers, and other electrical equipment. [Fothergill Tygaflor Ltd] *

Symax. Electrical insulation material. [Courtaulds Advanced Materials (Holdings) Ltd]

Symcor. CAS 51274-83-0; Tiamenidine hydrochloride; antihypertensive. [Hoechst-Roussel Pharmaceuticals Inc] *

Symel. Extruded silicone elastomer sleeving, fusible silicone rubber tapes and fabrics; for high temperature electrical applications such as lead out wires, marker sleeves, and peristaltic pumps. [Fothergill Tygaflor Ltd]*

Symel. Electrical insulation material. [Courtaulds Advanced Materials (Holdings) Ltd]

Symmetrel. CAS 665-66-7; A proprietary preparation of amantadine hydrochloride; used in the treatment of Parkinsonism. [Ciba plc] *

Symmetrel®. CAS 665-66-7; Amantadine hydrochloride; an antiviral. [DuPont Merck Pharmaceuticals]

Sympathy®. For the agricultural industry. [DuPont UK]

Sympatol. A proprietary preparation of oxedrine tartrate used in the treatment of cardiac disorders. [Lewis Laboratories] ‡

Symphony®. For the agricultural industry. [DuPont UK]

Synacril. Acrylic dyes. [ICI Chem. & Polymers Ltd]

Synacthen. A proprietary preparation of tetracosactrin. [Ciba plc] *

Synacthen Depot. Adrenocortical. [Ciba plc] *

Synacto. Emulsifiers and corrosion preventives. [Exxon UK]*

Synadryn. A proprietary preparation of prenylamine lactate; for treatment of angina pectoris. [Hoechst UK] *

Synalar. CAS 67-73-2; A proprietary preparation of fluocinolone acetonide for dermatological use. [ICI Chem. & Polymers Ltd]

Synalar-HP. CAS 67-73-2; Fluocinolone acetonide; glucocorticoid. [Syntex Laboratories Inc]

Synandone. CAS 67-73-2; A proprietary trade name for preparations of fluocinolone acetonide [ICI Chem. & Polymers Ltd]

Synandrets. CAS 58-18-4; Methyltestosterone; androgen. [Pfizer Inc] †

Synandrol. CAS 57-85-2; Testosterone propionate; androgen. [Pfizer Inc] †

Synandrol F. CAS 58-22-0; Testosterone; androgen. [Pfizer Inc] †

Synandrotabs. CAS 58-18-4; Methyltestosterone; androgen. [Pfizer Inc] †

Synanthic. CAS 53716-50-0; Oxfendazole; anthelmintic. [Syntex Laboratories Inc] *

Synaqua. Water soluble resins. [Cray Valley Ltd]

Synasol. CAS 64-17-5; EINECS 200-578-6; Denatured ethyl alcohol. [Union Carbide]

Syn-Chek 1203. Chlorinated lubricity and EP additive for metalworking lubricants, drawing and cutting oils. [Ferro/Keil]

Synchemicals Dalapon. CAS 75-99-0; Water soluble powder containing 85% dalapon; a translocated herbicide. [Synchemicals Ltd]

Synchemicals Total Weed Killer. CAS 61-82-5, 122-34-9; A suspension concentrate containing 53 g aminotriazole and 110 g simazine per liter; used for total weed control in non crop areas and fruit orchards. [Synchemicals Ltd]

Synchrocept. CAS 54120-61-5; Prostalene; prostaglandin. [Syntex Laboratories Inc] *

Syncillin. A proprietary preparation of potassium phenethicillin; an antibiotic. [Bristol-Myers Squibb Co Inc]

Syncillin®. Antibiotic. [Bayer AG]

Synclyst. Refining catalysts. [Crosfield Chemicals Ltd]

Syncol®. A range of sizes based on polyacrylic acid with specific properties and application to all types of nylon. [Allied Colloids Ltd]

Syncrolube. Fatty acid esters; for use as plastics lubricants. [Croda Surf. Ltd]

Syncrowax. Synthetic waxes. [Croda Surf. Ltd]

Syncrowax AW1-C. C18-36 acid; emulsifier, emollient, opacifier. [Croda Inc.]

Syncrowax BB4. Synthetic beeswax; emulsifier, emollient, opacifier; also suspending agent for anhyd. systems, auxiliary w/o emulsifier, thickener for oils and waxes; used in creams and sticks. [Croda Inc.]

Syncrowax ERL-C. C18-36 acid glycol ester; emulsifier, emollient, opacifier; also lubricant, stabilizer, suspending agent for anhyd. systems, thickener, reducer of bleeding and sweating; gloss improver; sticks, creams. [Croda Inc.]

Syncrowax HGL-C. C18-36 acid triglyceride; emulsifier, emollient, opacifier; also lubricant, suspending agent, strength improver, stabilizer, gloss improver; used in cosmetic makeup. [Croda Inc.]

Syncrowax HR-C. CAS 18641-57-1; Glyceryl tribehenate; emulsifier, emollient, opacifier; also suspending agent, thickener, gloss improver used in personal care products. [Croda Inc.]

Syncrowax HRS-C. Glyceryl tribehenate, calcium behenate; emulsifier, emollient, opacifier; also suspending agent for anhyd. systems, auxiliary w/o emulsifier, gellant, thickener for oils and waxes. [Croda Inc.]

Syndane. CAS 57-74-9; Chlordane; used to control earthworms in turf. [Synchemicals Ltd]

Syndite. An explosive consisting of 10-22% nitroglycerin, 0.1-0.3% collodion cotton, 45-49% ammonium nitrate, 7-9% sodium nitrate, 2-5% glycerin, 2-5% starch, and 26-28% sodium chloride.

Syndraw. Water reducible lubricants containing no mineral oil; additives include surfactants, lubricity additives, fatty acid soaps, synthetic corrosion inhibitors, biocides and polymers. [Franklin Oil Corporation (Ohio)]

Synektan. Leather dyes. [ICI Chem. & Polymers Ltd]

Synemol. CAS 67-73-2; Fluocinolone acetonide; glucocorticoid. [Syntex Laboratories Inc] *

Synergol. Mixture of dichlorophen, 4-indol-3-ylbutyric acid, and 1-naphthylacetic acid; used to promote rooting of

---

cuttings. [Hortichem Ltd]

Synerone. CAS 57-85-2; Testosterone propionate; androgen. [Merrell Dow Pharmaceuticals Inc] *

Syn-Fab DC-1. Emulsifiable hydrocarbon used as a dye carrier. [Crowley Chem.]

Syn Fac®. Bisphenol-A polyols; polyester resin intermediates, reactive diluents. [Milliken]

Syn Fac® 222. Aryl polyoxyether; low foaming, high surface tension; dispersant for latex paint, textile printing systems; fiber finish component, dispersant for degreasing solvs. used in cleaning textiles and metals; scouring aid for cellulosics; intermediate for anionics. [Milliken]

Syn Fac® 334. Aryl polyoxyether; dispersant/emulsifier for pigments, insecticides, solvs., and cleaning compounds. [Milliken]

Syn Fac® 8009, 8017. Aromatic polyether polyols; reactive diluent, dispersant for waterborne coatings and other applications. [Milliken]

Syn Fac® TEA-97. Ethoxylated amine; solv. and dispersant in the dye industry. [Milliken]

Synfluid. Synthetic base stocks for lubes and hydraulic fluids. [Chevron] *

Syngesterone. CAS 57-83-0; Progesterone; progestin. [Pfizer Inc] †

Syngestrets. CAS 57-83-0; Progesterone; progestin. [Pfizer Inc] †

Syngran. CAS 122-34-9; Granules containing 2% w/w simazine; a triazine herbicide to control weeds and grasses in cane fruit, roses, and some vegetables. [Synchemicals Ltd]

Synkad® 100. Borate salt; corrosion inhibitor for metalworking coolants. [Ferro/Keil]

Synkad® 200. Boramide; corrosion inhibitor for metalworking coolants. [Ferro/Keil]

Synkad® 303. Carboxylic acid salt; corrosion inhibitor for synthetic grinding fluids. [Ferro/Keil]

Synkad® 6000. Complex deriv. of a carboxylate salt; rust and corrosion inhibitor for synthetic drawing and stamping fluids, synthetic and semi-synthetic cutting and grinding fluids, hydraulic fluids, and for protection of parts during short-term indoor storage. [Ferro/Keil]

Synkavit. Vitamin K analog preparations. [Roche Laboratories; Roche Products Ltd]

Synkayvite. CAs 131-13-5; Menadiol sodium diphosphate; vitamin. [Hoffmann-LaRoche Inc] *

Synkrolith. Sodium hexafluoroaluminate (synthetic cryolite), Na_3AlF_6; 55% F, 13-14% Al, 32-33% Na; powder; filler for synthetic resin-bonded grinding wheels, brake linings, etc. [Metallgesellschaft AG]

Syn Lube. Water soluble, biodegradable lubricants; fiber lubricants, textile softeners, yarn processing aids. [Milliken] *

Synmold. Phenol formaldehyde resin; molding powders. [Manchem Ltd] *

Synmold. Phenolic molding powders. [Sterling Moulding Materials] †

Syn-O-Ad® 8412. CAS 126-73-8; Tributyl (n) phosphate; antiwear and EP agents in non-crankcase lubricants; ferrous metal passivator; fluid base stock where inhibition of flame propagation is desired. [Akzo] *

Syn-O-Ad® 8475M. CAS 25155-23-1; Trixylenyl (mixed) phosphate; antiwear and EP agents in non-crankcase lubricants; ferrous metal passivator; fluid base stock where inhibition of flame propagation is desired. [Akzo]*

Syn-O-Ad® 8478. CAS 56803-37-3; Butylated triphenyl phosphate; antiwear and EP agents in non-crankcase lubricants; ferrous metal passivator; fluid base stock where inhibition of flame propagation is desired. [Akzo]*

Syn-O-Ad® 8479. CAS 29761-21-5; Isodecyl diphenyl phosphate; antiwear and EP agents in non-crankcase lubricants; ferrous metal passivator; fluid base stock where inhibition of flame propagation is desired. [Akzo] *

Syn-O-Ad® 8480. CAS 28108-99-8; Propylated triphenyl phosphate; antiwear and EP agents in non-crankcase lubricants; ferrous metal passivator; fluid base stock where inhibition of flame propagation is desired. [Akzo]*

Syn-O-Ad® 8484. CAS 1330-78-5; Tricresyl phosphate; antiwear and EP agents in non-crankcase lubricants; ferrous metal passivator; fluid base stock where inhibition of flame propagation is desired. [Akzo] *

Syn-O-Ad® 8485. CAS 56803-37-3; Butylated triphenyl phosphate; antiwear and EP agents in non-crankcase lubricants; ferrous metal passivator; fluid base stock where inhibition of flame propagation is desired. [Akzo]*

Syn-O-Ad® P-310. CAS 3658-48-8; Dioctyl (2-ethylhexyl) phosphite; used as amine salts in mineral and synthetic base stocks; as load-carrying additives with secondary activity as low-temp. stabilizers and metal deactivators [Akzo] *

Syn-O-Ad® P-312. CAS 102-85-2; Tributyl phosphite; antioxidant and antiwear agent in gear and transmission oils [Akzo] *

Syn-O-Ad® P-316. CAS 1809-19-4; Dibutyl phosphite; used as amine salts in mineral and synthetic base stocks; as load-carrying additives with secondary activity as low-temp. stabilizers and metal deactivators [Akzo] *

Syn-O-Ad® P-374. CAS 301-13-3; Trioctyl (2-ethylhexyl) phosphite; antioxidant and antiwear agent in gear and transmission oils [Akzo] *

Syn-O-Ad® P-399. CAS 101-02-0; Triphenyl phosphite; antioxidant and antiwear agent in gear and transmission oils [Akzo] *

Syn-O-Ad® P-408. Tridecyl acid phosphate; corrosion inhibitor for iron alloys in lubricants; as EP additives when neutralized. [Akzo] *

Syn-O-Ad® P-412. Octyl (2-ethylhexyl) acid phosphate; corrosion inhibitor for iron alloys in lubricants; as EP additives when neutralized. [Akzo] *

Syn-O-Ad® P-415. Amyl (iso) acid phosphate; corrosion inhibitor for iron alloys in lubricants; as EP additives when neutralized. [Akzo] *

Syn-O-Ad® P-417. Butyl (n) acid phospahte; corrosion inhibitor for iron alloys in lubricants; as EP additives when neutralized. [Akzo] *

Synocryl. Thermoplastic and thermosetting acrylic resins. [Cray Valley Ltd]

Synocure. Crosslinking acrylic resins. [Cray Valley Ltd]

Synogist. A proprietary preparation of sodium sulfosuccinate and undecylenic monoalkyl amide; used as a shampoo for dandruff. [Maltown] ‡

Synolac. Alkyds, unsaturated polyester resins, epoxy esters. [Cray Valley Ltd]

Synolide. Polyamide resins. [Cray Valley Ltd]

Synotex 800. Cyclized rubber resins for coatings. [Daniel Prods.]

Synouryn. Dehydrated castor oil products. [Akzo Chemie UK Ltd]

Synova. Mixed coal tar products with different viscosity with

---

‡ Trade name and manufacturer not verified   § Trade name without identified manufacturer

and without fillers; used for protective paint, roof paint, injection paint, etc. [Caramba Chemie GmbH] *

Synox. Soluble concentrate containing 75 g ioxynil and 225 g mecoprop per liter; used for weed control in turf. [Synchemicals Ltd]

Synperonic. Range of nonionic surfactants. [ICI Chem. & Polymers Ltd]

Synperonic 3S27 and 3S60S. Sodium 3EO sulfate of Synprol in liquid form; emulsifier, detergent, wetting and foaming agent, stable in high electrolyte concentrations and hard water, but tends to hydrolyze in acid solution; used in liquid household detergents; industrial and domestic cleaning formulations; emulsifying systems; shampoos and bubble baths. [ICI Chem. & Polymers Ltd] *

Synperonic 3S60A. Ammonium 3EO sulfate of Synprol in liquid form; emulsifier, detergent, wetting, and foaming agent, stable in high electrolyte concentrations and hard water, but tends to hydrolyze in acid solution; used in liquid household detergents; industrial and domestic cleaning formulations; emulsifying systems; shampoos and bubble baths. [ICI Chem. & Polymers Ltd] *

Synperonic N, NX, NXP and NDB. Nonylphenol ethoxylate nonionic surfactants in liquid form; general purpose detergents and wetting agents for textile processing, metal treatment, dust suppression and general cleaning applications; emulsification of medium polarity oils and solvents. [ICI Chem. & Polymers Ltd] *

Synperonic NP4, NP5 and NP6. Nonylphenol ethoxylate nonionic surfactants in liquid form; oil-soluble detergents and emulsifiers; used as intermediates for sulfation and phosphorylation to give anionic detergents, lubricants and antistatic agents; emulsifying agents for wide range of oils, waxes and solids; compatible with all other surfactants. [ICI Chem. & Polymers Ltd] *

Synperonic NP8 and NP9. Nonylphenol ethoxylate nonionic surfactants in liquid form; water-soluble, high performance detergents and wetting agents; used in textile scouring; emulsifiers for medium polarity oils and solvents. [ICI Chem. & Polymers Ltd] *

Synperonic NP10 and NP12. Nonylphenol eyhoxylate nonionic surfactants in liquid form; water-soluble detergents, detergent additives, solubilizers, dispersants, and stabilizers. [ICI Chem. & Polymers Ltd] *

Synperonic NP13 and NP15. Nonylphenol ethoxylate nonionic surfactants in liquid or paste form; used in conjunction with an oil-soluble anionic surfactant, they are good emulsifiers for a range of solvents, agrochemical pesticides and herbicides. [ICI Chem. & Polymers Ltd] *

Synperonic NP20 and NP30. Nonylphenol ethoxylate nonionic surfactants in liquid or solid form; solubilizing agents and emulsifiers or coemulsifiers for highly polar substrates. [ICI Chem. & Polymers Ltd] *

Synperonic OP. Range of nonionic surfactants of the octylphenol ethoxylate type in liquid form; water-soluble general purpose detergents, wetting agents and emulsifiers with good solution properties in the presence of alkalis and at higher temperatures. [ICI Chem. & Polymers Ltd] *

Synpro®. Metallic soaps of naturally occurring fatty acid; includes metals: calcium, zinc, magnesium, aluminum, barium and cadmium and fatty acids: stearic, palmitic and lauric; plastic lubricants and stabilizers for thermoplastics and thermosets; used for PVC, ABS, polysty-

rene, polyolefins, and phenolics; also as lubricants in powdered metals and cosmetics; as thixotropic agents and bodying materials in greases, oils and oil well drilling muds, as well as paints. [Synthetic Products Company]*

Synpro® 8. CAS 557-05-1; Zinc stearate; lubricant for PVC, polyolefins, PS, ABS. [Syn. Prods.]

Synpro® 15F. CAS 1592-23-0; EINECS 216-472-8; Calcium stearate; process aid, lubricant for PVC, polyolefins, PS, ABS. [Syn. Prods.]

Synpro® 90. CAS 557-04-0; Magnesium stearate; process aid, lubricant for PVC, polyolefins, PS, ABS. [Syn. Prods.]

Synpro® 303. CAS 300-92-5; Aluminum distearate; process aid, lubricant for PVC, polyolefins, PS, ABS. [Syn. Prods.]

Synpro® 404. CAS 637-12-7; Aluminum tristearate; process aid, lubricant for PVC, polyolefins, PS, ABS. [Syn. Prods.]

Synpro® 505 USP. CAS 637-12-7; Aluminum monostearate; process aid, lubricant for PVC, polyolefins, PS, ABS. [Syn. Prods.]

Synpro® Aluminum Octoate. Aluminum octoate; process aid, lubricant for PVC, polyolefins, PS, ABS. [Syn. Prods.]

Synpro® Barium Stearate. CAS 6865-35-6; Barium stearate; process aid, lubricant for PVC, polyolefins, PS, ABS. [Syn. Prods.]

Synpro® Cadmium Stearate. CAS 2223-93-0; Cadmium stearate; process aid, lubricant for PVC, polyolefins, PS, and ABS. [Syn. Prods.]

Synpro® Calcium Pelargonate. Calcium pelargonate; process aid, lubricant for PVC, polyolefins, PS, and ABS. [Syn. Prods.]

Synpro® Stannous Stearate. Stannous stearate; process aid, lubricant for PVC, polyolefins, PS, and ABS. [Syn. Prods.]

Synprol. Detergent alcohols C9-11, C13-15. [ICI Chem. & Polymers Ltd]

Synprolam. Range of synthetic fatty amines and derivative. [ICI Chem. & Polymers Ltd]

Synprolam 35. Synthetic (C13/C15) alkyl primary amine in liquid form; surfactant intermediate; corrosion inhibitor; flotation agent; fertilizer anticaking agent. [ICI Chem. & Polymers Ltd] *

Synprolam 35A. Acetic acid salt of a synthetic (C13/C15) alkyl primary amine in solid form; emulsifier; fertilizer anticaking agent; mineral flotation. [ICI Chem. & Polymers Ltd] *

Synprolam 35 BQC. Benzyl quaternary ammonium chloride of a synthetic (C13/C15) dimethyl tertiary amine, in liquid form; emulsifier; general sanitizer; biocide; corrosion inhibitor; textile dyeing auxilliary; timber preservative. [ICI Chem. & Polymers Ltd] *

Synprolam 35 DM. Synthetic (C13/C15) dimethyl tertiary amine in liquid form; cationic surfactant intermediate. [ICI Chem. & Polymers Ltd] *

Synprolam 35 DMA. Acetic acid salt of a synthetic (C13/C15) dimethyl tertiary amine; emulsifier; biocide; timber preservative. [ICI Chem. & Polymers Ltd] *

Synprolam 35 N3. N-(C13/C15) alkyl-1,3-propane diamine in liquid form; corrosion inhibitor; bitumen adhesion agent/emulsifier. [ICI Chem. & Polymers Ltd] *

Synprol Sulphate. Anionic surfactant as a cream viscous liquid; for shampoos, bubble baths, liquid detergents and

---

* Trade name not verified as available   † Trade name verified as obsolete

emulsifying systems. [ICI Chem. & Polymers Ltd] †

Synpron. Proprietary mixtures of various metallic soaps and salts of organic acid, antioxidants, organophosphites and lubricants supplied as solids or liquids; heat and light stabilizers for PVC flexible and rigid compounds. [Synthetic Products Company] *

Synpron 241. Phosphite chelator; auxiliary PVC heat stabilizer providing crisp initial color and long-term stability to rigid and plasticized nontoxic formulations. [Syn. Prods.]

Synpron 1009. CAS 77-58-7; Dibutyltin dilaurate; stabilizer providing long term stability in bottle compounds, PVC pipe, injection molded fittings. [Syn. Prods.]

Synpron 1027. Antimony mercaptide; PVC heat stabilizer especially for pipe and conduit applications; recommended for use with calcium stearate for optimum stabilization; may be used in NSF potable water pipe at 1.0 phr maximum. [Syn. Prods.]

Synpron 1032 and 1033. A proprietary range of liquid-antimony mercaptides used as heat stabilizers. [Dart Industries Inc] *

Synpron 1538. Zinc soap; PVC stabilizer for low odor and taste applications, e.g., refrigerator gasketing. [Syn. Prods.]

Synpron 1800. Barium/cadmium; extrusion stabilizer for clear applications. [Syn. Prods.]

Synpro-Ware. Dispersions of rubber or plastic chemicals in elastomers, silicones, pastes, wetted powder or pellet form; chemical dispersion for ease of handling, incorporation and safety for use in rubber, wire and cable and plastic compounding. [Synthetic Products Company] *

Synresin RD 461. A proprietary blocked, one-component polyurethane resin; it is thermosetting and is used as a rubber flock adhesive. [Synres International NV, Holland] *

Synsoft; Polymacon; contact lens material. [Syntex Ophthalmics Inc] *

Synsolve. General name for range of proprietary cleaning and maintenance products; used for many areas of routine and maintenance cleaning. [Synthite Ltd]

Synstryp. General name for a range of proprietary paint removers and surface coating products; for removal of paint-surface coatings. [Synthite Ltd]

Syntase® 62. CAS 131-57-7; EINECS 205-031-5; Benzophenone-3; uv absorber for protection of PS, PVC, methacrylate polymers, and polyesters, against uv degradation over prolonged exposure; useful in protecting clear varnishes and lacquers, linseed oil-based alkyds, and phenolic coatings intended for use on uv-sensitive surfaces. [Rhone-Poulenc Surf.]

Syntase® 230. CAS 4065-45-6; EINECS 223-772-2; Benzophenone-4; uv absorber used in water-based cosmetics, including sun tan lotions, body creams, shampoos, hair sprays, and hair dyes, and in wool fabrics. [Rhone-Poulenc Surf.]

Syntergent 55-A. Fulling and scouring agent for wool and wool/synthetic blends; provides excellent alkaline stability, effective felting action and detergency. [Henkel/Textiles]

Syntetrin. CAS 751-97-3; A proprietary preparation of rolitetracycline; an antibiotic. [Bristol-Myers Squibb Co Inc]

Syntex. A proprietary trade name for an oil modified alkyd resin (qv). §

Synteze. Proprietary mixture; penetrating oil. [Synthite Ltd]†

Synthacalk. Polysulfide sealant; used for general caulking and sealing in the construction industry, vertical joints, and perimeters of doors and windows, etc. [Pecora Corporation]

Synthacryl. Thermoplastic and thermosetting acrylic resins; used for automotive stoving finishes, industrial stoving finishes, and pressure sensitive adhesives. [Resinous Chemicals Ltd]

Synthamel. Air drying and stoving finishes. [ICI Chem. & Polymers Ltd] †

Synthamica. SiO_2· Al_2O_3·MgO·K ₂O F + 3% impurities; additive to oil, paint, grease, plastics, glass and other inorganic materials which require a high dielectric or high heat resistant material; a coating for welding rods requiring special atmospheric controls; a thermal barrier and high electrical resistivity. [Mykroy/Mycalex] *

Synthane. A proprietary trade name for phenol-formaldehyde synthetic resin laminated products and other plastics. §

Synthappret®. Special products for the nonfelting finishing of wool; used for the textile industry. [Bayer AG]

Synthaprufe. Pitch rubber emulsion for damp proofing. [Thomas Ness Ltd] *

Syntharesin®. Products for obtaining nonslip effects; used by the textile industry. [Bayer AG]

Synthasil. Colorless silicone water repellant for masonry brick, etc. [Thomas Ness Ltd] *

Synthawax. Hydrogenated castor oil. [Unilever] *

synthecite. It is a distillate from vulcanized rubber containing vegetable oils and waxes; used as a rubber softener.

Synthemul®. Emulsifiable synthetic resins and synthetic resin emulsions. [Reichhold]

Synthemul® 40-422. Styrene-acrylic emulsion polymer; industrial air-dry maintenance coatings; outstanding adhesion, chemical resist., corrosion resist. [Reichhold/Emulsion Polymers] *

Synthemul® 40-425. Solution acrylic; thermosetting; general industrial grade with high gloss, hardness/flexibility balance and solv. system versatility. [Reichhold/Emulsion Polymers] *

Synthemul® 40850-00. Vinyl-acrylic emulsion polymer; for nonwoven industry; excellent recovery, good filler acceptance; for filtration media, high loft automotive carpet. [Reichhold/Emulsion Polymers] *

synthe-plastic. A reaction product of a terpene base; a rubber plastic containing no pitches or waxes.

Synthetic Bone Ash. Calcium hydroxyapatite, tricalcium phosphate; used to coat molds when casting molten purified copper and copper aloys. [Murlin Chemical Inc]

synthetic oil of bitter almond. See benzaldehyde.

synthetic Peru balsam. See Perugen.

Synthetic Rutile. CAS 13463-67-7; EINECS 236-675-5; 94% Titanium dioxide. [Kerr-McGee Chemical Corp] *

synthin. A product obtained by heating synthol (qv), at 400 C in an autoclave. A liquid results which contains saturated hydrocarbons and sulfuric acid; used as a liquid fuel.

Synthite. A proprietary trade name for formaldehydes. [Synthite Ltd]

synthocarbone. A specially prepared charcoal for use as a fuel.

synthol. A liquid fuel containing hydrocarbons, acids, alcohols, aldehydes, and esters. It is obtained by reducing carbon monoxide in water gas at high temperatures and

---

‡ Trade name and manufacturer not verified   § Trade name without identified manufacturer

under pressure, using iron borings coated with potassium carbonate as a contact material.

Syntholvar. A proprietary trade name for extruded polyvinyl chloride. §

Synthroid. CAS 55-03-8; Levothyroxine sodium; thyroid hormone. [The Boots Co plc]

Syntocinon. CAS 50-56-6; A proprietary preparation containing oxytocin; used for promoting uterine contraction during and following labor. [Sandoz]

Syntol K77 and N77. Sodium lauryl ether sulfate (synthetic C12- C15 alcohol) as a pourable gelled paste or a low viscosity liquid; for production of all types of liquid and lotion shampoos and bubble baths; also a raw material for light duty liquid detergents, dishwashing detergents and auto shampoos. [Efkay Chemicals Ltd] *

Syntometrin. A proprietary preparation of ergometrine maleate and oxytocin; used in the treatment of postpartum hemorrhage. [Sandoz] *

Synton PAO-100. CAS 37309-58-3; Polydecene; synthetic lubricant. [Uniroyal]

Syntopon 8. Series of nonionic surfactants of the octylphenol ethoxylate type in liquid form; surfactants with range of properties and uses. [Witco Chemical Ltd]*

Syntopon A,B,C and D. Nonylphenol ethoxylate in liquid form; nonionic surfactant. [Witco Chemical Ltd] *

Syntopon F,G and N. Nonylphenol ethoxylate in solid form; nonionic surfactant. [Witco Chemical Ltd] *

Syntopressin. CAS 50-57-7; A proprietary preparation of lypressin. [Sandoz] *

Syntox Total Weed Killer. CAS 61-82-5, 122-34-9; A suspension concentrate containing 53 g aminotriazole and 100 g simazine per liter; used for total weed control in non crop areas and fruit orchards. [Syntex Manufacturing Ltd]

Syntroil. A series of polyol ester base fluid oils with various organic compounds; for various OEM applications, primarily small electric motors. [Ultrachem Inc] *

Syntron. Phenol formaldehyde resin; molding powders. [Manchem Ltd] *

Synvaren. A proprietary trade name for a phenol formaldehyde resin adhesive. §

Synvarol. A proprietary trade name for an urea formaldehyde resin adhesive. §

Synwax. CAS 8002-74-2; Synthetic wax from long-chain saturated fatty acids. [Reilly-Whiteman]

Syrian asphalt. A natural asphalt containing about 100% bituminous matter. It has a specific gravity of about 1.06, a meltingpoint of about 100 C, and practically no mineral matter.

Syrtussar. A proprietary preparation of dextromethorphan hydrobromide, phenylpropanolamine hydrochloride, sodium citrate, citric acid, and chloroform; a cough linctus. [Armour Pharmaceutical Co] *

Syscor. CAS 63675-72-9; Pharmaceutical preparation containing nisolidipine; also known as Baymycard; coronary selective calcium antagonist, especially for the treatment of angina pectoris. [Bayer AG]

Systamex. Oxfendazole-based veterinary anthelmintic. [The Wellcome Foundation Ltd] *

SysTec® 1998. CAS 23564-05-8; Thiophanate-methyl; a broad-spectrum systemic fungicide which controls a variety of diseases for use on turf and ornamentals. [Regal Chemical Company]

Systemic Fungicide. Garden fungicide. [Murphy Chemical Ltd] †

Systemic Insecticide. CAS 60-51-5; Dimethoate insecticide. [Murphy Chemical Ltd] †

Systhane. CAS 88671-89-0; Myclobutanil; light yellow solid; broad spectrum systemic fungicide. [Hoechst UK; Pan Britannica Industries Ltd; Rohm & Haas UK]

Systol M. Mixture of cymoxanil and mancozeb; used to control potato blight. [Quadrangle Agrochemicals]

Systral. CAS 77-38-3; Chlorphenoxamine; tablets, cream and jelly; antihistaminic and antiallergic. [Degussa AG]*

Sytam. Systemic organo phosphorous insecticide. [Murphy Chemical Co Ltd] †

Sytobex. CAS 68-19-9; Cyanocobalamin; vitamin. [Parke-Davis] *

Syton. Silica sol. [Monsanto plc]

Sytron. A proprietary preparation of sodium iron edetate; a hematinic. [Parke-Davis] *

T

T-64. CAS 1344-28-1; Tabular alumina; used for high-temp. refractory bricks, shapes, monolithic liners; as catalyst support where high purity, low porosity and high-temp. strength are desired. [Alcoa]

T-1061. CAS 1344-28-1; Tabular alumina; converted by heat to corundum state; very hard and dense; used in high temp. refractories in steel and aluminum industries, manufacturing of tech. ceramics, investment castings, as bed support in catalytic converters. [La Roche Chem.]

T1750. CAS 4766-57-8; Tetra-n-butoxysilane; for heat exchange applications; as dielec. fluids; lubricant in airborne radar. [Hüls Am.]

T1920. Tetrakis (2-ethylbutoxy) silane; for heat exchange applications; as dielec. fluids; lubricant in airborne radar. [Hüls Am.]

T4250. Phenyltris(trimethylsiloxy) silane; low visc. silicone fluid with high control over volatility, visc., dens. and ref. index, and unique solv. properties; used as a lubricant. [Hüls Am.]

Tabasan. A proprietary preparation of ephedrine, theobromine, and salicylamide; an antiasthmatic. [Ayrton Saunders plc] ‡

Tabbyite. See wurtzillite.

Tabcin. An effervescent drug for colds associated with headaches, articular pains, fever. [Bayer AG]

Tabloid. CAS 154-42-7; See Lanvis. [The Wellcome Foundation Ltd]

Tab Rybar Co. A proprietary preparation of isoprenaline sulfate, methylephedrine hydrochloride, butethamate citrate, and theophylline; a bronchial antispasmodic. [Rybar Laboratories Ltd] *

TABS. p-Menthadiene biodegradable solvent. [British Traders & Shippers Ltd]

TAC-3™. CAS 76-25-5; Sterile triamcinolone acetonide suspension; corticosteroid. [Allergan, Inc]

Tacamahac oil. See laurel nut oil.

Tacamahac resin. (West Indian anime resin). A resin obtained from various plants, usually from *Calophyllum* species.

Tacaryl. CAS 1982-37-2; Methdilazine; antipruritic. [Westwood Pharmaceuticals Inc] *

TACC 104. Silicone; thermally conductive, nondripping heat sink grease to reduce impedance of air gap between semiconductor and heat sink. [TACC Int'l.]

TACC 524. Two-component acrylic adhesive; nonflamm., fast-cure, structural adhesive for bonding wide variety of dissimilar substrates; excellent peel and tensile strength with steel, aluminum, ABS, FRP, PVC, PC, glass, wood, and ceramic. [TACC Int'l.]

TACC 700-82. One-component epoxy; self-leveling, heat-cured sealing compound; features rapid cure at moderate temps. to give high heat and chem. resist.; for filter media sealing or bonding ceramics or metals. [TACC Int'l.]

TACC AR-1001. Epoxy adhesive; high strength adhesive; variable curing agent level for rigid to flexible systems. [TACC Int'l.]

TACC CR-3200. One-component polyurethane; conformal coating varnish which forms a resilient high gloss finish; excellent salt spray resist., moisture barrier for ICs and PCBs. [TACC Int'l.]

TACE. CAS 569-57-3; Chlorotrianisene; estrogen. [Merrell Dow Pharmaceuticals Inc] *

Tachigaren 70. Hymexazol; fungicide for pelleting sugar beet seed. [Sumito Chemical (UK) plc]

tachiol. (Tachyol). Silver fluoride, AgF.

Tachostyptan. CAS 9035-58-9; A proprietary preparation of thromboplastin; used to control bleeding. [Consolidated Chemicals Ltd] *

Tachryrate. Methyl-1 methyl-3-cyclohexene carboxylate. [Quest Int'l. UK Ltd]

tachyiite. A dark volcanic glass.

tachyol. See tachiol.

Tacitin. CAS 10085-81-1; A proprietary preparation of benzoctamine hydrochloride; a tranquilizer. [Ciba plc] *

Tack. CAS 1333-86-4; Carbon black pastes; used for simple and dust-free dyeing of paints, lacquers, paper, cardboard, plastics, synthetic fibers, printing inks, and mineral binders. [Degussa AG] *

Tackidex. Dextrinified starch; used for a wide range of applications in the adhesives and food industries according to viscosity requirements. [Roquette (UK) Ltd]

tackol. A mixture of oils and resins; used as a rubber plasticizer.

Tacolyn. Tackifying resin dispersions. [Hercules Ltd]

Tactel. Polyamide textile fibers. [ICI Chem. & Polymers Ltd]

Tactesse. Polyamide carpet fibers. [Imperial Chemical Industries plc]

Tactix. Epoxy resins used for matrix resins for the fabrication of aerospace components. [Dow UK] *

Tactix 123. CAS 25928-94-3; High-purity liq. bisphenol A epoxy; thermoset performance polymer for aerospace applications. [Dow Plastics]

Tactix 556. CAS 25928-94-3; Hydrocarbon epoxy; thermoset performance polymer for aerospace applications. [Dow Plastics]

Tactix 742. CAS 25928-94-3; Tris (hydroxyphenyl) methane-based epoxy; high thermal oxidative stability; performance polymer for aerospace applications. [Dow Plastics]

---

‡ Trade name and manufacturer not verified   § Trade name without identified manufacturer

Tactix H31. Thermoset performance polymer; epoxy curing agent, hardener. [Dow Plastics]

taffy. A residue from the neutralization of the mixed organic acids produced by the fermentation of kelp-seaweed in the production of acetone. It consists chiefly of calcium propionate.

Tafigel PUR 40. Polyurethane thickener for gloss emulsion paints, wood preservative stains, and adhesives. [Münzing Chemie GmbH]

Taflite 900. A proprietary range of weather-resistant high-impact polystyrenes made from an EPDM graft polymerized with styrenes and dispersed as spherical microgels in polystyrene phases. [Mitsui Toatsu] *

Tag. Aqueous emulsion of natural and synthetic waxes; fruit coating wax. [Makhteshim Chemical Works Ltd] †

Tagamet Tablets, Syrup, Injection and Infusion. CAS 51481-61-9; Cimetidine; H_2-receptor antagonist for treatment of peptic ulcer, oesophageal reflux, dispeptic symptons, Mendelson's syndrome and Zollinger-Ellison syndrome. [SmithKline Beecham] *

Tagat®. Polyoxyethylene glycerol fatty acid esters; solubilizers for water insoluble substances such as flavors, perfumes, vitamin oils; dispersing and antistatic agents for technical purposes. [Goldschmidt Ltd]

Tagat® I. CAS 69468-44-6; PEG-30 glyceryl isostearate; preparation of o/w emulsions; solubilizer for flavors, perfumes, vitamin oils; dispersant and antistat. [Goldschmidt; Goldschmidt AG]

Tagat® I2. CAS 69468-44-6; PEG-20 glyceryl isostearate; preparation of o/w emulsions; solubilizer for flavors, perfumes, vitamin oils; dispersant and antistat. [Goldschmidt; Goldschmidt AG]

Tagat® L. CAS 51248-32-9; PEG-30 glyceryl laurate; preparation of o/w emulsions; solubilizer for flavors, perfumes, vitamin oils; dispersant and antistat. [Goldschmidt; Goldschmidt AG]

Tagat® L2. CAS 51248-32-9; PEG-20 glyceryl laurate; preparation of o/w emulsions for pharmaceuticals; solubilizer for flavors, perfumes, vitamin oils; dispersant and antistat. [Goldschmidt; Goldschmidt AG]

Tagat® O. CAS 51192-09-7; PEG-30 glyceryl oleate; preparation of o/w emulsions; solubilizer for flavors, perfumes, vitamin oils; dispersant and antistat. [Goldschmidt; Goldschmidt AG]

Tagat® O2. CAS 51192-09-7; PEG-20 glyceryl oleate; preparation of o/w emulsions for pharmaceuticals; solubilizer for flavors, perfumes, vitamin oils; dispersant and antistat. [Goldschmidt; Goldschmidt AG]

Tagat® R40. CAS 61788-85-0; PEG-40 hydrogenated castor oil; solubilizer for water-insol. substances, e.g., essential oils, perfumes, vitamins, cosmetic/pharmaceutical active ingredients; coemulsifier for o/w emulsions. [Goldschmidt; Goldschmidt AG]

Tagat® R60. CAS 61788-85-0; PEG-60 hydrogenated castor oil; solubilizer for water-insol. substances, e.g., essential oils, perfumes, vitamins, cosmetic/pharmaceutical active ingredients; coemulsifier for o/w emulsions. [Goldschmidt; Goldschmidt AG]

Tagat® R63. CAS 61788-85-0; PEG-60 hydrogenated castor oil and propylene glycol; solubilizer for water-insol. substances, e.g., essential oils, perfumes, vitamins, cosmetic/pharmaceutical active ingredients; coemulsifier for o/w emulsions. [Goldschmidt; Goldschmidt AG]

Tagat® S. CAS 51158-08-8; PEG-30 glyceryl stearate; preparation of o/w emulsions; solubilizer for flavors, perfumes, vitamin oils; dispersant and antistat. [Goldschmidt; Goldschmidt AG]

Tagat® S2. CAS 51158-08-8; PEG-20 glyceryl stearate; preparation of o/w emulsions for pharmaceuticals; solubilizer for flavors, perfumes, vitamin oils; dispersant, and antistat. [Goldschmidt; Goldschmidt AG]

Tagat® TO. CAS 68958-64-5; PEG-25 glyceryl trioleate; preparation of o/w emulsions; solubilizer for flavors, perfumes, vitamin oils; dispersant and antistat; refatting agent for hair/bath preparations. [Goldschmidt; Goldschmidt AG]

Tagit®. Reagent. [Calbiochem Corp]

ta-hong. A lead glass containing ferric oxide; used by the Chinese as a red enamel on porcelain.

TAHP-80. CAS 3425-61-4; 80% t-Amyl hydroperoxide solution in t-amyl alcohol/water; initiator. [Witco/Argus]

Taifun. Alkaline floor cleaner. [Gansow UK Ltd]

tailor's chalk. This material consists of French chalk (magnesium silicate) mixed with a little China clay.

Tak. Mold sealing compound. [Foseco (F.S.) Ltd]

Taka-diastase. An enzyme from the fungus Eurotium orzoe, grown on rice.; a proprietary preparation containing Aspergillus oryzae enzymes; an antacid. [Parke-Davis]*

Taka-Sweet®. Glucose isomerase; enzyme for production of fructose syrups from glucose; immobilized. [Solvay Enzymes]

Taka-Therm® L-340. CAS 9000-92-4; Bacterial α-amylase; enzyme for starch liquefaction and textile desizing; to produce high fructose corn syrup; in liquefaction of starch fermentation media. [Solvay Enzymes]

Takatol. p-Aminophenol, $C_6H_4OH \cdot NH_2$.

Takazyma. A proprietary preparation containing takadiastase (qv) magnesium carbonate, bismuth carbonate, ginger and calcium carbonate; an antacid. [Parke-Davis] *

takizolit. A red micro-crystalline kaolin found in Japan, and having the composition, $2Al_2O_3 \cdot 7SiO_2 \cdot 7H_2O$. It also contains appreciable amounts of rare earth oxides.

ta-kong. A lead glass containing ferric oxide and used by the Chinese as a red enamel.

Taktene 220. CAS 9003-17-2; Sol'n. polybutadiene, nonstaining; used in passenger car and truck tires for improved treadwear in blends with SBR/NR; excellent low temp. properties; also for golf centers, footwear, belting, hose, floor tile, molded and extruded goods; mill processable; FDA compliance; vulcanized with sulfur-accelerator systems or with peroxide for maximum resilience. [Miles; Polysar]

Taktic. Animal health insecticide. [Schering Agrochemicals Ltd]

Talbor's powder. Cinchona bark in powder form.

talc. (hydrous magnesium silicate; industrial, cosmetic, or platy talc; talcum) Native, hydrous magnesium silicate sometimes containing small portion of aluminum silicate; $Mg_3Si_4O_{10}(OH)_2$ or $3MgO \cdot 4SiO_2 \cdot HOH$; CAS 14807-96-6; EINECS 238-877-9; ceramics, cosmetics, pharmaceuticals; as filler and pigment in rubber, paints, soaps, etc.; dusting agent, lubricant, electrical insulation. [Cyprus Industrial Min.; Pfizer; L.A. Salomon; R.T. Vanderbilt; Whittaker, Clark & Daniels]

Talc MS. CAS 14807-96-6; Talc; cosmetic ingredient contributing soft, smooth feel, skin adhesion, and spreadability. [Presperse]

---

* Trade name not verified as available   † Trade name verified as obsolete

talca gum. See Suakin gum.

Talcid®. Hydrotalcite; an antacid preparation. [Bayer AG]

Talc Micro-Ace P-2. CAS 14807-96-6; Talc; cosmetic ingredient contributing soft, smooth feel, skin adhesion, and spreadability. [Presperse]

talcum. See talc.

Talent. Herbicide. [May & Baker Ltd] *

talide. Tungsten carbide material.

talipot. (raw palmira root flour). A starch obtained from a palm, Corypha umbraculifera.

Talisman. CAS 15545-58-9; Suspension concentrate containing 500 g chlorotoluron per liter; a cereal herbicide. [Farmers Crop Chemicals Ltd] †

talite. A siliceous earth containing 84% silica with small quantities of oxides of iron and aluminum; used as a rubber filler.

talka gum. See Suakin gum.

talloel. A Swedish liquid resin obtained as a by-product in the production of cellulose from Swedish fir by the soda process; consists mainly of resin acids, and is closely related to rosin.

tall oil. CAS 8002-26-4; By-product of sulfate pulp manufacture; contains 2.2% material soluble in petroleum ether, 12.4% unsaponifiable matter, 30.4% resin acid, and 54.9% fatty acids. The resin acid consists of abietic acid, and the fatty acids contain oleic, linoleic, and linolenic acids; used as paint vehicles, in oil well drilling muds, lubricants, and greases, asphalt derivatives; in rubber reclaiming, and as chemical intermediates.

tallow. (beef tallow; mutton tallow) Fat derived from fatty tissue of sheep or cattle; consists primarily of fatty acid glycerides; CAS 61789-97-7; EINECS 263-099-1; soap stock, leather dressing, candles, greases, manufacture of stearic and oleic acids, animal feeds; adherent in tire molds. [Atlas Refinery; Norman Fox; Geo. Pfaus Sons; Reilly-Whiteman; Witco/Humko]

Tallow Amine. CAS 61790-33-8; EINECS 263-125-1; Tallow primary amine, tech.; surfactant, wetting agent, ore flotation agent, asphalt emulsifier, corrosion inhibitor, anticaking agent, pigment modifier; oil production; petrol. production additives. [Exxon/Tomah]

tallow clays. Clays containing varying proportions of zinc silicate.

Tallow Diamine. CAS 61791-55-7; Tallow diamine; surfactant, wetting agent, ore flotation agent, asphalt emulsifier, corrosion inhibitor, anticaking agent, pigment modifier; oil production; petrol. product additives. [Exxon/Tomah]

tallow seed oil. Stillingia oil, obtained from the seeds of Stillingia sebifera.

Tallow Tetramine. CAS 68911-79-5; Tallow tetramine; surfactant, wetting agent, ore flotation agent, asphalt emulsifier, corrosion inhibitor, anticaking agent, pigment modifier; oil production; petrol. product additives. [Exxon/Tomah]

Tallow Triamine. CAS 61791-57-9; Tallow triamine; surfactant, wetting agent, ore flotation agent, asphalt emulsifier, corrosion inhibitor, anticaking agent, pigment modifier; oil production; petrol. product additives. [Exxon/Tomah]

Tally® 100 Plus. Glyceryl stearate, PEG-20 glyceryl stearate, hydrogenated soybean oil; emulsifier, dough strengthener and crumb softener used in breads; also as textile lubricant and fabric softener. [Van Den Bergh Foods]

Talon. CAS 56073-10-0; Rodenticide containing brodifacoum. [ICI Agrochemicals; ICI Chem. & Polymers Ltd]

Talon. CAS 2921-88-2; Emulsifiable concentrate containing 228g chlorpyrifos per liter; broad-spectrum insecticide with many crop uses. [Farmers Crop Chemicals Ltd]

Talotalo gum. (kau drega). A gum somewhat resembling gutta-percha, from Fiji.

ta-lou. The Chinese term for a glass flux used for enameling on porcelain. It is mainly a silicate of lead with a little copper.

Talpen. CAS 47747-56-8; A proprietary preparation of talampicillin; an antibiotic. [SmithKline Beecham] *

Talpex. Titanium-aluminum organic complex; thixotropic agent for latex paints. [Manchem Ltd] *

Talpheno. CAS 50-06-6; Phenobarbital; anticonvulsant; hypnotic, sedative. [Merrell Dow Pharmaceuticals Inc] *

Talstar. CAS 82657-04-3; An emulsifiable concentrate containing 100 g bifenthrin per liter; a residual herbicide for the control of weeds in winter cereals, oilseed rape and peas. [DowElanco Ltd]

Talunex. CAS 20859-73-8; Aluminum phosphide; used for gasing of rabbits and moles. [Kommer-Brookwick Ltd]

Talusin. CAS 466-06-8; A proprietary trade name for proscillaridin. §

talwaan. A tanning material. It is the root of Elephantorrhiza burchelli.

Talwin. CAS 64024-15-3; Pentazocine hydrochloride; analgesic. [Sterling Drug Inc] *

T.A.M. Thenoyl methionine; cosmetic ingredient rich in organic sulfur; contributes to formation of disulfur bridges in skin tissues; for scalp treatments, prevention of hair fallout. [Exsymol]

tamarac. The dried bark of Larix larcina, an American larch. The extract is used as an astringent and stimulant.

Tamaron®. CAS 10265-92-6; Methamidophos; broad spectrum systemic insecticide with stomach and contact action used for treatment of sucking and biting insects and spider mites on a wide range of crops including cotton, tobacco, vegetables, potatoes, sugar beets. [Bayer AG]

Tambocor™ Injection. CAS 54143-56-5; Each ampoule contains 15 ml of a solution of flecainide acetate. [3M Pharmaceuticals] †

Tambocor™ Tablets. CAS 54143-56-5; Each tablet contains flecainide acetate 100 mg; anti-arrhythmic. [3M Pharmaceuticals]

tambookie grass. The product of Hyperrhenice glauca; used for paper-making.

Tamclad 7200. A proprietary PVC organosol. [Tamite Industries Inc] ‡

Tamguard 840, 840H and 840S. A proprietary range of PVC plastisols used for coating electroplating racks; Shore hardnesses are A90, D35 and A70 respectively. [Tamite Industries Inc] ‡

Tamol®. Dispersing agents; used as dispersing agents. [Rohm & Haas]

Tamol® 731-25%. CAS 26426-80-2; Isobutylene/maleic anhydride copolymer; dispersant for dyes and pigments. [Rohm & Haas]

Tamol® 819. CAS 9084-06-4; Sodium naphthalene sulfonate; dispersant for cement, pigments, carbon black. [Rohm & Haas] *

---

‡ Trade name and manufacturer not verified   § Trade name without identified manufacturer

Tamol® 850. Sodium polymethacrylate; low foaming dispersant for inorganic pigments; for caulks, sealants, paper industry, paints. [Rohm & Haas]

Tamol® L Conc. CAS 9084-06-4; Sodium naphthalene sulfonate; dispersant, tanning agent for leather; secondary dispersant for polymerization of rubber, in paper/pulp slurries. [Rohm & Haas]

Tamol® SN. CAS 9084-06-4; Sodium naphthalene sulfonate; dispersant, tanning agent for leather. [Rohm & Haas]

Tamolan®. Synthetic laking agent for alkaline flexographic inks. [BASF AG]

tampicin. A resin, obtained from *Ipomoea simulans*.

Tamsil 8. Silica; filler for finishes, enamels, maintenance paints, plastic film antiblocks, urethane rubber, and polishes. [Unimin]

Tamsil Gold Bond. Silica; filler for traffic paint, interior/exterior coatings, maintenance and marine coatings, mastics and adhesives, elec. epoxy compounds, buffing compounds. [Unimin]

Tamtam. An alloy of 78% copper and 22% tin.

Tanabond STA. Durable, pressure sensitive, aq. screen printing table adhesive. [Sybron]

Tanabron W. Wetter, detergent with good oil and wax emulsification; textile auxiliary; caustic stable. [Sybron]

tanacetone. (6-ketosabinane; thujone) CAS 546-80-5; A terpene type ketone; $C_{10}H_{16}O$; used as a solvent.

Tanafresh HFO. Odor masking agent. [Sybron]

Tanal. Chromium/aluminum complexes; used for tanning. [Lancashire Chemical Works Ltd]

Tanalev® 221. Anionic polymer; direct dye leveler with retarding properties. [Sybron]

Tanalith. Copper/chrome/arsenate waterborne wood preservative to prevent fungal decay and insect attack; for pressure treated timber for construction, fencing, agriculture and any application where timber requires protection. [Hickson & Welch Ltd]

Tanalon® EFA. Orthochlorotoluene biphenyl; carrier giving good dye yield; can be used atmospherically and under pressure. [Sybron]

Tanalube® RF. Lubricant for rayon and polyester; prevents crease and rope marks. [Sybron]

Tanapal® LD-3, LD-3T. Leveling agent for disperse dyes. [Sybron]

Tanapel® 54. High molecular weight thermosetting resin; produces very durable water repellent textile finishes; works well with fluorocarbons. [Sybron]

Tanapon NF-200. Phosphate ester; low foaming wetting agent for alkaline cleaning compounds, pigments, adhesive. [Sybron]

Tanapure® AC. Leveling agent for acid dyes on nylon fibers. [Sybron]

Tanasoft® PNL. Quaternary fatty amine; durable softener producing slick full hand. [Sybron]

Tanassist® JCR. Low foaming dye migrator for pressure equip. [Sybron]

Tanastat® PH. Antistat for all fibers at low use levels. [Sybron]

Tanaterge® SCP. Nonenzymatic desizing agent for cotton and poly/cotton woven goods; stable to caustic and hard water. [Sybron]

Tanatex® Nostick. Specialty additive preventing polymer buildup on dry cans or equip. in textile industry. [Sybron]

tanatol. *See* uba.

Tanavol® URC. Trichlorobenzene/biphenyl; carrier for textile industry repair work; effective for atmospheric and pressure equip. [Sybron]

Tanawet® AR. Nonrewetting wetting agent for water repellent and fluorocarbon textile finishes. [Sybron]

Tanawet® RCN. Wetter and scouring agent for textile processing, peroxide bleaching; removes oils and waxes. [Sybron]

Tanbase®. CAS 1309-48-4; Magnesium oxide for leather tanning. [Steetley Magnesia Products Ltd]

Tancolin. A prorietary preparation of dextromethorpan hydrobromide, theophylline, sodium citrate, citric acid, ascorbic acid and glycerin; a cough linctus. [Maws Pharmacy Suppliers] ‡

Tandacote. CAS 129-20-4; A proprietary preparation of oxyphenbutazone; an anti-inflammatory drug. [Ciba plc]*

Tandaigesic. A proprietary preparation of oxyphenbutazone and paracetamol; an anti-inflammatory and analgesic drug. [Ciba plc] *

Tandearil. CAS 129-20-4; Oxyphenbutazone; anti-inflammatory; antirheumatic. [Ciba-Geigy Corp] *

Tandem. CAS 58138-08-2; A line of herbicides based primarily on tridiphane. [Dow UK]

Tandem 5K, 8. CAS 9005-67-8; Mono and diglycerides, polysorbate 60 with BHA, citric acid; food emulsifier, conditioner/softener for yeast-raised baked goods. [Witco/Humko]

Tandem 11H K. Hydrated mono- and diglycerides, polysorbate 60 with sodium propionate and lactic acid; food emulsifier, conditioner/softener for yeast-raised baked goods. [Witco/Humko]

Tanderil. CAS 129-20-4; A proprietary preparation of oxyphenbutazone. [Ciba plc] *

tanekaha. The bark of *Phyllocladus trichomanoides*; used in tanning leather.

Tangantangan oil. *See* castor oil.

Tanigan®. Range of tanning materials comprising advanced syntans which make the tanning process safe and economical and improve the quality of the leather. [Bayer AG]

tanked oil. Linseed oil from which the moisture and other matter has settled out. It has a higher value than the ordinary oil.

tannacetin. *See* acetannin.

tannal. (tannalum). A basic aluminum tannate, $Al(OH)_2 \cdot (C_{14}H_9O_9)+5H_2O$; an astringent used as a dusting powder.

tannaline films. Gelatin films hardened by formaldehyde; used for photographic purposes.

tannalum. *See* tannal.

tanner's wool. (Glover's wool). A wool pulled from the carcasses of slaughtered sheep with the assistance of lime. It does not dye well.

Tannesco. Tanning agents for leather. [Ciba plc] *

Tannex® MGP. Organic stabilizer/chelate for peroxide bleaching (textiles). [Sybron]

tannic acid. (glycerite; tannin; gallotannic acid; digallic acid; gallotannin) Organic acids, mixture; occurs in the bark and fruit of many plants, e.g., oak species, sumac; $C_{76}H_{52}O_{46}$; CAS 1401-55-4; EINECS 276-638-0; mordant in dyeing; manufacture of ink, imitation tortoise shell; sizing paper; printing fabrics; tanning; clarifying beer; in photography; as coagulant in rubber; analytical

---

* Trade name not verified as available   † Trade name verified as obsolete

chemistry reagent; astringent. [Burlington Bio-Medical; Crompton & Knowles; Fuji Chem. Ind.; Mallinckrodt; Thiem; Ulrich GmbH]

tannin. See tannic acid.

Tanolin. A proprietary trade name for a basic chromium chloride for use in chrome tanning baths. §

Tanret's reagent. To a solution of 1.35 g mercuric chloride in 25 cc water is added a solution of 3.32 g potassium iodide in 25 cc water. This is made up to 60 cc with water and 20 cc glacial acetic acid.

Tansel. A specially prepared salt for curing hides. §

Tansul. Clarifying agent; for beer. [Rheox Inc]

Tansul-7. A substance for the protection and stabilization against freezing of malt drinks like beer. [Rheox Inc]

tantalic acid anhydride. See tantalum oxide.

tantalum. Element; Ta; CAS 7440-25-7; EINECS 231-135-5; capacitors, chemical equipment, dental and surgical instruments, rectifiers, vacuum tubes, furnace components, high-speed tools, catalyst, 'getter' alloys in electron tubes, sutures, body implants, electronic circuitry, thin-film components. [Aldrich; Atomergic Chemetals; Cabot; Cerac; NRC]

tantalum carbide. TaC; cutting tools and dies, cemented carbide tools. [Atomergic Chemetals; Cerac; Noach Chem.]

tantalum oxide. (tantalum pentoxide; tantalic acid anhydride) Ta_2O_5; CAS 1314-61-0; EINECS 215-238-2; Production of tantalum, tantalum carbide; optical glass; piezoelectric and laser applications; dielectric layers in electronic circuits. [Aldrich; Atomergic Chemetals; Cabot; Cerac; NRC]

tantalum pentoxide. See tantalum oxide.

tantcopper. A copper alloy analogous to tantiron.

tantiron. An alloy of 84% iron, 15% silicon, and 1% carbon. It has a specific gravity of 6.8 and is acid-resisting.

Tao. CAS 2751-09-9; A proprietary preparation of troleandomycin; an antibiotic. [Pfizer International]

Tap Aid. ASTM S-215 Oil, 1.1.1. trichlorethane, mask odor No 3; for small hole, drilling, reaming and tapping; excellent for wire drawing and grinding. [Doyle Specialties] *

Tapar. CAS 103-90-2; Acetaminophen; analgesic; antipyretic. [Parke-Davis] *

Tapazole. CAS 60-56-0; Methimazole; thyroid inhibitor. [Eli Lilly & Co]

tap cinder. The basic silicate of iron constituting the slag, and flowing through the tap-hole of the puddling furnace.

Taquence. Taq DNA sequencing kit. [U.S. Biochemical]

tara. The tannin from the pods of Coesalpinia tinctoria.

Taractan. CAS 113-59-7; Proprietary preparation of chlorprothixene; major tranquilizer of thioxanthene group. [Roche Laboratories; Roche Products Ltd]

Tarband. A proprietary bandage impregnated with zinc oxide and coal tar paste; used in the treatment of eczema. [Seton Healthcare Group plc]

tar camphor. CAS 91-20-3; Naphthalene, $C_{10}H_8$; used as an intermediate, moth repellant, fungicide, smokeless powder, cutting fluid, lubricant, synthetic resins, textile chemicals, antiseptic.

tar camphor. See naphthalene.

Tarcortin. A proprietary preparation of coal tar and hydrocortisone; for dermatological use (ezcema). [Stafford-Miller] ‡

Tardex. A range of brominated organics; flame retardant

additives. [ISC Chemicals Ltd] *

Tardocillin® 1200. Benzathine penicillin; a long acting penicillin. [Bayer AG]

Tardomyocel®, Tardomyocel L. Long-acting broad-spectrum antibiotic for animals. [Bayer AG]

Tardrox. A proprietary preparation of chlorhydroxyquinolone and tar for dermatological use. [Carlton Laboratories (UK) Ltd] *

tari. See white tan.

Tarmac. A proprietary preparation of blast furnace slag, refined tar, and other ingredients; used for road dressing.§

Tarmex. A proprietary name for a combination of prepared tar and mexphalte; used for road dressing. §

tarnovicite. See tarnowitzite.

tarnowitzite. (plumbocalcite). A mineral, (Ca·Pb) CO $_3$.

tar oil. See oil of tar.

tarola. A coal-tar product used as a sheep dip.

Taroma. Hydrated vegetable protein flavors. [Giulini Corp]

Tarot®. For the agricultural industry. [DuPont UK]

Tarslag. A proprietary preparation of cold blast slag which has been treated with a bituminous compound; used as a road dressing. §

tartar. See argol.

tartar cake. See superargol.

tartar emetic. (tartrated antimony).Potassium-antimonyltartrate, $C_4H_4O_6(SbO)K+1/2H_2O$; used in medicine as an emetic, and in dyeing as a mordant.

tartar emetic powder. (tartar emetic substitute, antimony mordant). Mixtures of tartar emetic and zinc sulfate.

tartar emetic substitute. See tartar emetic powder.

tartaric acid. (DL-tartaric acid anhydrous; dihydroxysuccinic acid) $HOOC(CHOH)_2COOH$; CAS 133-37-9; EINECS 205-105-7; Manufacture of cream of tartar, tartar emetic, acetaldehyde; sequestrant, tanning, effervescent beverages, baking powder, fruit esters, ceramics, galvanoplastics, photography, textiles, silvering mirrors, coloring metals, acidulant in foods. [Bromhead & Denison Ltd; R.W. Greeff; Mallinckrodt; Penta Mfg.; Rit-Chem.]

DL-tartaric acid anhydrous. See tartaric acid.

tartarline. (acid potassium sulfate; potassium hydrogen sulfate potassium bisulfate) $KHSO_4$; CAS 7646-93-7; used as a substitute for tartaric acid for industrial purposes; conversion of wine lees and tartrates into potassium bitartrate; in the manufacture of fertilizers.

tartars. Raw materials which contain more than 40% tartaric acid are termed tartars.

tartar substitute. See superargol.

tartrated iron. Iron and potassium tartrate; used medicinally.

tartrated soda. See Rochelle salt.

tartrate lime. See limo.

tartrazine. CAS 1934-21-0; A dyestuff, $C_{16}H_{10}O_{10}N_4S_2Na_4$. used to dye silk and wool.

Tarvia. A proprietary trade name for a specially refined coaltar. §

tarwar. The bark of Cassia auriculate. A tanning material.

tasar silk. See Tussar silk.

Taski TR101. Highly concentrated dry foam carpet and upholstery shampoo. [Lever Industrial]

Taslan®. Fibers. [DuPont UK]

Tasmaderm. CAS 56281-36-8; Motretinide; keratolytic. [Hoffmann-LaRoche Inc] *

---

‡ Trade name and manufacturer not verified   § Trade name without identified manufacturer

Tasnon®. Anthelmintic. [Bayer AG]

Tasprin. CAS 50-78-2; Proprietary preparation of soluble aspirin. [Unichem] *

tasteless salts. Sodium phosphate, HNa_2PO_4.

Taterfos®. CAS 7758-16-9; Sodium acid pyrophosphate; additive for processed foods. [Rhone-Poulenc Food Ingreds.]

taurine. (2-aminoethanesulfonic acid) $NH_2CH_2CH_2SO_3H$; CAS 107-35-7; EINECS 203-483-8; Biochemical research, pharmaceuticals, wetting agents. [Chemisphere Ltd; Mitsui Toatsu Chem.; Penta Mfg.; Schweizerhall; Tanabe USA]

Tavist. CAS 14976-57-9; Clemastine fumarate; antihistaminic. [Dorsey Pharmaceuticals]

Taxol. A proprietary preparation of pancreatin, bile salts, aloes, and agar used in the treatment of constipation. [Cox Pharmaceuticals] †

Taylor. A proprietary trade name for a phenol-formaldehyde synthetic resin laminated product. §

Taylorite. See bentonite.

Taylor Oil. A patented binding material obtained by boiling raw linseed oil with driers (litharge), then forcing air through the oil when heated to 300 F, and finally heating it for some time at 500-600 F. §

Taylors Lawn Sand. CAS 10028-22-5; Ferrous sulfate; used for moss control in turf. [Rigby Taylor Ltd]

Taylor solder. An alloy of 60% tin, 12% lead, 12% silver, 8% zinc, 4% aluminum, and 4% copper.

Tazidime. CAS 72558-82-8; Ceftazidime; antibiotic. [Eli Lilly & Co]

Tazoline. A proprietary preparation of antazoline hydrochloride, octaphonium chloride, titanium dioxide and calaroine. [Rybar Laboratories Ltd] *

TBAB. CAS 1643-19-2; Tetrabutyl ammonium bromide. [Hexcel]

TBAB. See tetrabutyl ammonium bromide.

TBEP. CAS 78-51-3; Tributoxyethylphosphate; nonfoaming emulsifier, wetting agent, dispersant, leveling agent; acid and electrolyte stable. [Rhone-Poulenc Surf.; Rhone-Poulenc France]

TBEP. See tributoxyethyl phosphate.

TBHP. See t-butyl hydroperoxide.

TBHP-70. CAS 75-91-2; t-Butyl hydroperoxide solution with di-t-butyl peroxide and small amounts of t-butyl alcohol and water as diluents; an initiator. [Witco/Argus]

TBHQ. See t-butyl hydroquinone.

TBTO. See tributyltin oxide.

Tc 99m Lungaggregate. Technetium Tc 99 m aggregated albumin; diagnostic aid; radioactive agent. [Medi-Physics Inc] *

T-Carb. CAS 471-34-1; Calcium carbonate. [Harcros]

TCC. 3,4,4´-Trichlorocarbanilide; bacteriostatic agents for bar soaps. [Monsanto Co] *

tcha-lau. A blue powder containing copper; used by the Chinese for obtaining a blue color on porcelain.

TCP. Trade name for tricresyl phosphate, a plasticizer for cellulose lacquers, and polyvinyl chloride; it has a specific gravity of 1.185-1.189, a boiling range of 430-440 C and a flash point of 215 C; the term also applies to an aqueous solution of trichlorophenyl-iodo-methyl-salicyl; an antiseptic and germicide. §

TCP. See calcium phosphate, tribasic.

TCP. See tricresyl phosphate.

TDA-1. Phase transfer catalyst. [Rhone-Poulenc UK]

T-Det® 25-3A. Ammonium C12-15 alkyl ether sulfate (3 EO); high foaming detergent, general industrial cleaner, dishwashes, laundry products, car wash. [Harcros]

T-Det® 25-3S. Sodium C12-15 alkyl ether sulfate (3 EO); high foaming detergent, general industrial cleaner, dishwashes, laundry products, car wash. [Harcros]

T-Det® C-40. CAS 61791-12-6; PEG-40 castor oil; emulsifier, solubilizer, degreaser, lubricant, dispersant, penetrant used in leather, paper, textile and metal processing, rubber, paint; dispersant for pigment slurries; leveling agent, defoamer, stabilizer; wax and polish preparations. [Harcros]

T-Det® D-150. CAS 9014-93-1; PEG-150 dinonyl phenyl ether; surfactant used in hard surface cleaners. [Harcros]

T-Det® DD-5. CAS 9014-92-0; Dodoxynol-5; emulsifier, surfactant for aromatic and aliphatic hydrocarbon solvs., solvent and emulsion cleaners, agriculture, degreasers; dry-cleaning soap additive. [Harcros]

T-Det® EPO-61. EO/PO block copolymer; wetting agent, defoamer for paper manufacturing; water treating compounds, metal cleaning, rinse aids, mech. dishwash, textile; crude oil demulsifier; emulsifier and wetting agent in agricultural toxicant formulations. [Harcros]

T-Det® N-4. CAS 9016-45-9; Nonoxynol-4; detergent, wetting agent; agricultural toxicant formulations; intermediate for household detergents; dry cleaning soap additive, sludge dispersant additive in fuel oils; gasoline additive; solv. cleaner for metal processing; color development aid, stabilizer and emulsifier for latex paints. [Harcros]

T-Det® N-40. CAS 9016-45-9; Nonoxynol-40; emulsifier for emulsion polymerization, asphalt, elevated temp. applications. [Harcros]

T-Det® N-1007. CAS 9016-45-9; Nonoxynol-100; emulsifier for asphalt, emulsion polymerization, textiles, high temp. and high electrolyte applications. [Harcros]

T-Det® O-4. CAS 9002-93-1; Octoxynol-4; surfactant for agriculture, leather and metal processing, paint, dry-cleaning detergents, degreasing applications. [Harcros]

T-Det® O-407. CAS 9002-93-1; Octoxynol-40; detergent, emulsifier, stabilizer for various applications. [Harcros]

T-Det® RQ1. Block copolymer; low foam surfactant for mech. dishwash, metal cleaners, rinse aids, hard surface cleaners, textile, paper; defoamer. [Harcros UK]

TDI. See toluene diisocyanate.

TEA. See triethanolamine.

Teaberry Oil. CAS 119-36-8; EINECS 204-317-7; Methyl salicylate, $C_6H_4OH \cdot COO \cdot CH_3$; used as a flavor in foods and beverages, pharmaceuticals, and in perfumery.

TEA-lauryl sulfate. (sulfuric acid, monododecyl ester, compound with 2,2´,2´´-nitrilotris[ethanol] (1:1); triethanolammonium lauryl sulfate; triethanolamine lauryl sulfate) Triethanolamine salt of lauryl sulfuric acid; $CH_3(CH_2)_{10}$-$CH_2OSO_3H \cdot N(CH_2CH_2OH)_3$; CAS 139-96-8; 68908-44-1; EINECS 205-388-7; emulsifier, detergent, wetting agent, dispersant, foaming agent used for household cleaning products, cosmetics, emulsion polymerization. [Chemron; Lonza; Norman, Fox; Sandoz; Stepan]

tea-lead. An alloy of from 97-99% lead and 1-3% zinc. Also an alloy of lead with 2% tin used for wrapping tea.

teal oil. See gingelly oil.

Team. CAS 61791-44-4; Cationic surfactant containing 800 g/l ethoxylated tallow amine; for use as a wetting agent for phosphonoglycine herbicides. [Monsanto plc]

---

* Trade name not verified as available   † Trade name verified as obsolete

Tear-Efrin. CAS 61-76-7; Phenylephrine hydrochloride; adrenergic. [CooperVision Inc] ‡

Tearisol® Sterile Lubricating Eye Drops. CAS 9004-65-3; Hydroxypropyl methylcellulose 0.5%; for the temporary relief of burning and irritation due to dryness of the eye or to exposure to wind or sun. [Iolab Corp]

Tears Plus®. Polyvinyl alcohol, povidone; ocular lubricant. [Allergan, Inc]

Teatcote Plus. A proprietary preparation of polyhexanide; a veterinary antibacterial. §

Tebol 88. CAS 71-36-3; EINECS 200-751-6; t-Butyl alcohol, 88%. [Arco Chemical Europe]

Tebol 99. CAS 71-36-3; EINECS 200-751-6; High purity tertiary butyl alcohol; solv., cosolvent, compatibilizer, coupling agent, processing aid for pharmaceuticals, personal care products, aq. coatings and adhesives, agricultural formulations, polymer processing, cleaners/disinfectants; chlorinated hydrocarbon stabilizer. [Arco Chemical Co; Arco Chemical Europe]

Tec. A proprietary trade name for cellulose acetate varnish resins. §

Tecagg. Very low density mineral aggregates comprising foamed waste products or clays; density 0.3 - 0.8; used for insulating building products, refractory insulation. [Filtec Ltd]

Tecali onyx. See onyx of Tecali.

Tecane. Selective herbicide. [Schering Agrochemicals Ltd]*

Teccel. Hollow glass microspheres, expanded minerals; white, mono or multicellular lightweight fillers of density 0.15 - 0.6 gm/cc; used for explosives, deep submergence buoyancy, paints, and cultured marble. [Filtec Ltd]

Tec-Char. A proprietary trade name for a granular charcoal.§

Tecfil. Name of a range of lightweight mineral fillers in particular cenospheres; hollow ceramic microspheres derived from fly-ash; composed primarily of silica and alumina, size 5-300 μm, density 0.5 to 0.8 gm/cc; frequently shortened to 'T' with a suffix, e.g., T.300; for low density cementing, auto noise attenuation, auto underbody coating, low density plasters, refractory insulation, thermoset resin extender. [Filtec Ltd]

Tecgran. CAS 117-18-0; Granules or dispersible powder containing tecnazene; protectant fungicide and potato sprout suppressant. [Atlas Interlates Ltd]

Techmate. Resilient molded foam used for cushion packaging applications. [Dow UK] *

TechneColl. Technetium Tc 99m sulfur colloid; radioactive agent. [Mallinckrodt Inc] *

TechneScan MAA. Albumin, aggregated; diagnostic aid. [Mallinckrodt Inc] *

TechneScan PYP. CAS 15578-26-4; Stannous pyrophosphate; diagnostic aid. [Mallinckrodt Inc] *

TechneScan SSC. Stannous sulfur colloid; diagnostic aid. [Mallinckrodt Inc] *

Technyl. Nylon injection molding compounds. [Rhone-Poulenc UK]

Techroline. Gasoline additive. [Chevron] *

Techron. Gasoline additive. [Chevron] *

Techster. Polybutylonterephthalate injection molding compounds. [Rhone-Poulenc UK]

Techtron™ PPS. CAS 9016-75-5; PPS; stock shapes for demanding applications requiring resist. to high heat and hostile chem. environments; for high pressure liq. chromatography, chem. processing, automotive, elec./electronic, industrial, consumer goods, medical devices. [Polymer Corp.]

Teclam®. For the electrical industry. [DuPont UK]

Tecnocin. A proprietary range of curing agents for fluorocarbon rubbers. [Montedison UK Ltd] *

Tecnoflon®. A proprietary range of fluorocarbon based rubbers. [Montedison UK Ltd] *

Tecnoflon® FOR 45C2/R. Fluoroelastomer copolymer; with curative; low cross-link dens. polymer providing good hot tear resist. for complicated shapes, for low modulus and high elongation items; used for valve stem seals, bellows, custom molded goods. [Ausimont] *

Tecnoflon® FOR-65BI/R. Fluoroelastomer copolymer; with curative; for compression or transfer molding of general purpose items. [Ausimont] *

Tecnoflon® FOR-LHF. Fluoroelastomer polymer; with curative; for low-hardness articles (Shore A45-50) with good low-temp. resist.; used as processing aid when blended with standard copolymers or terpolymers; excellent processing and extrudability. [Ausimont] *

Tecnoflon® NH, NM, NMB, NML, NMLB. Fluoroelastomer copolymer; without curative; polymers for use alone or together to achieve required visc., mold flow, tear strength for custom formulations; curable with M1 (2-5 phr) and M2 (1-2 phr) for general-purpose compounds; also curable with diamines. [Ausimont] *

Tecnoflon® P-1, P-2, P-2HV, P-40. Fluoroelastomer polymer; without curative; peroxide-curable grades with improved chem. and acid resist., improved resist. to steam and bases, and amine-containing fluids; P-40 developed with excellent flow characteristics for extrusion. [Ausimont] *

Tecnoflon® TN-LATEX. Fluoroelastomer polymer; with curative; 70% FKM solids emulsion used to make high-temp. chem.-resist. coatings; applied by spraying, dipping, or silk screening; excellent substitute for solv.-based fluoroelastomer fabric coatings; resistant to oils and corrosive chem. [Ausimont] *

Tecnoprene. A proprietary range of filled polymer molding granules. [Montedison UK Ltd] *

Tecoflex Polyurethane. Linear segmented aliphatic polyether polyurethane, medical grade elastomer; for medical products, tubing. [Thermedics Inc] *

Tecpril. High quality foamed clay aggregates, in form of white, regular pellets; density 0.1 - 0.6 gm/cc; for high tech insulation products, fireproof composites, aerospace/hydrospace composites. [Filtec Ltd]

Tecquinol® Tech. Grade. CAS 123-31-9; Hydroquinone, tech.; antioxidant for synthetic latexes, fats, oils, monomers, polyester resins. [Eastman] *

Tecsol®. Denatured alcohol. [Eastman]

Tectilon. Acid dyes. [Ciba plc] *

Tecto. CAS 148-79-8; Thiabendazole; a systemic fungicide. [MSD Agvet]

Tecto 60%. CAS 148-79-8; Wettable powder containing 60% w/w thiabendazole; a systemic fungicide. [MSD Agvet]

Tectrode. Cathodic protection. [Imperial Chemical Industries plc]

TEDA-D007. Stannous octoate; standard catalyst for PU flexible foam; also for slabstock and elastomer shoe soles. [Tosoh]

TEDA-L33. CAS 280-57-9; Triethylenediamine in dipropylene glycol; gelling catalyst for PU slabstock and

---

‡ Trade name and manufacturer not verified   § Trade name without identified manufacturer

molded flexible, rigid, and elastomer shoe shoe applications. [Tosoh]

TEDA-T411. CAS 77-58-7; Dibutyltin dillaurate; standard catalyst for PU elastomer applications; also for slabstock, semirigid, and rigid applications. [Tosoh]

Tedimon. Isocyanates. [Montedison UK Ltd] *

Tedion V-18. CAS 116-29-0; Tetradifon; emulsifiable concentrate containing 125 g propiconazole and 350 g tridemorph per liter; selective acaricide for use against mite infestation in orchards, citrus fruit plantations, hop fields, groundnut plantations, vegetable plots, cotton fields and on ornamental plants; red spider mite control in horticultural crops. [Duphar BV]; Hortichem Ltd

Tedlar®. Clear or pigmented polyvinyl fluoride film with high resistance to weathering; generally chemically inert. [DuPont UK]

Tedral. A proprietary preparation of theophylline, ephedrine hydrochloride, and phenobarbitone; a bronchial antispasmodic. [Warner] *

Tedur®. Polyphenylene sulfide resins. [Bayer AG; Miles]

Tedur® KU1-9510-1. CAS 9016-75-5; PPS, 40% glass-reinforced; offers high strength, heat distort. temp., excellent rigidity, good elec. properties, and outstanding chem. resist.; suggested for parts exposed to high service temps. and mech. loads in chemically aggressive environments, e.g., under-the-hood automotive, elec./ electronic (connectors, halogen lamp sockets, switch housings, fuse boxes, capacitor components), industrial/ mech. (internal pump components, heavy equip. parts); as replacement for metals and thermosets. [Miles]

Tedur® KU1-9511. CAS 9016-75-5; PPS, 45% glass fiber-reinforced; injection molding grade. [Bayer AG] *

Tedur® KU1-9530. CAS 9016-75-5; PPS, 60% mineral/ glass fiber-reinforced; embedding material. [Miles]

Tedur® KU1-9552. CAS 9016-75-5; PPS, 45% glass fiber/ mineral-reinforced; conductive injection molding grade. [Bayer; Miles]

Tedur® KU1-9561. CAS 9016-75-5; PPS, 60% mineral/ glass fiber-reinforced; reflector grade. [Miles]

Teebrix. Preformed inserts for extending bottom plate lives. [Foseco (F.S.) Ltd]

Teefroth. Polyglycol ethers; froth flotation agent. [ICI Chem. & Polymers Ltd]

Teejel. A proprietary preparation of choline salicylate and cetalkonium chloride; an analgesic gel for teething pain. [Napp Laboratories Ltd] *

teel oil. See gingelly oil.

Teepol CM44. An aqueous solution of sodium salt of C_9-C_{13} alkyl benzene sulfonate; mainly used as a constituent for other detergent blends. [Shell] *

Teepol FC5. A high foam detergent for repackers, being a solution of a primary alcohol ethoxysulfate. [Shell] *

Teepol GD53. A highly biodegradable, clear, pale amber liquid containing formalin as a preservative; specially formulated for dishwashing and general cleaning. [Shell]*

Teepol HB6. A highly biodegradable aqueous solution of the sodium salts of C_9-C_{13} primary alcohol sulfate; contains formalin as a preservative; used as a solubilizer and emulsifier component of hard surface cleaners and germicidal cleaners. [Shell] *

Teepol PB. Sodium primary alcohol ethoxy sulfate as a clear aqueous solution; foaming agent used in gypsum wallboard manufacture. [Shell UK] *

teerlack. Coal tar pitch.

Teevax. A proprietary preparation of crotamiton and halopyramine; used in the treatment of pruritis. [Ciba plc]*

Tefaire®. Fibers. [DuPont UK]

Teflon®. Polytetrafluoroethylene (PTFE) plastic material having good resistance to high temperatures. [DuPont UK]

Teflon® F.E.P. Hexafluoropropylene copolymers. [DuPont UK]

Tefose® 63. PEG-6-32 stearate and glycol stearate; self-emulsifying base for cosmetics, o/w pharmaceutical ointments; excellent skin and mucosal tolerance; especially for anti-mycotic preparations. [Gattefosse; Gattefosse SA]

Tefose® 1500. CAS 9004-99-3; PEG-6-32 stearate; self-emulsifying base for o/w cosmetic or pharmaceutical emulsions. [Gattefosse; Gattefosse SA]

Tefose® 2000. PEG-6 stearate, ceteth-20, steareth-20; self-emulsifying base for o/w cosmetic or pharmaceutical emulsions. [Gattefosse; Gattefosse SA]

Tefose® 2561. PEG-6 stearate, glyceryl stearate, and ceteth-20; base for o/w cosmetic or pharmaceutical emulsions. [Gattefosse; Gattefosse SA]

Tefzel® 200. A proprietary ETFE fluoro-polymer resin extruded for use as wire insulation. [DuPont UK]

Tegamine® 18. CAS 7651-02-7; EINECS 231-609-1; Stearamidopropyl dimethylamine; surfactant, conditioner for hair care and bath products; auxiliary emulsifier for creams and lotions; provides skin feel. [Goldschmidt]

Tegamine® Oxide WS-35. CAS 68155-09-9; Cocamidopropylamine oxide; foam builder/stabilizer for shampoos, bath preparations, liq. detergents; visc. builder. [Goldschmidt]

Tegiloxan®. Methylsilicone oils; antifoams for mineral oils, in rubber and plastics industry; lubricant for tire production; additive for polishes. [Goldschmidt] *

Tegin®. Glycerol fatty acid partial esters. [Thomas Goldschmidt Ltd]

Tegin® 4011. CAS 31566-31-1; Glyceryl stearate; emulsifier for pharmaceuticals. [Goldschmidt; Goldschmidt AG]

Tegin® O. CAS 25496-72-4; Glyceryl mono/dioleate; emulsifier for w/o emulsions for pharmaceuticals. [Goldschmidt; Goldschmidt AG]

Tegin® V. Glyceryl stearate SE; self-emulsifying grade forming o/w emulsiosn for creams and lotions. [Goldschmidt]

Teginacid. Glycerol mono-distearates with other nonionics. [Thomas Goldschmidt Ltd]

Tegison. CAS 54350-48-0; Etretinate; antpsoriatic. [Hoffmann-LaRoche Inc] *

Teglac. A proprietary trade name for an alkyd synthetic varnish and lacquer resin. §

TegMeR® 703. PEG-3 diheptanoate; lubricant for aluminum can industry. [C.P. Hall]

TegMeR® 803. Triethylene glycol di-2-ethylhexoate; plasticizer for adhesives. [C.P. Hall]

TegMeR® 804 Special. PEG-4 di-2-ethylhexoate; lubricant for aluminum can industry. [C.P. Hall]

TegMeR® 903. PEG-3 dipelargonate; lubricant for aluminum can industry. [C.P. Hall]

tegmin. An emulsion made from 1 part yellow wax, 2 parts acacia, and 3 parts water. It also contains 5% zinc oxide, and a little wool fat ; used as a surgical dressing.

---

* Trade name not verified as available   † Trade name verified as obsolete

Tego® IMR 918. Polysiloxane polyoxyalkylene block copolymer; additive in production of RIM parts and integral skin foams of high density. [Goldschmidt AG]

Tego® Airex 900. Polysiloxane modified with organic groups; deaerator for solv. containing and solv.-free systems based on epoxide and unsaturated polyester. [Tego GmbH] *

Tego® Airex 960. Silicone modified organic polymer; deaerator for medium and high polar coating systems, especially PU paints; also for high solids alkyds, ES spray systems, and urethane-polyesters. [Tego GmbH]*

Tego®-Amid S 18. CAS 7651-02-7; EINECS 231-609-1; Stearamidopropyl dimethylamine; emulsifier for creams and lotions; conditioner for hair care products. [Goldschmidt; Goldschmidt AG]

Tegoamin®. Amine catalysts. [Thomas Goldschmidt Ltd]

Tegoamin® 33. Triethylene diamine; catalyst for the manufacturing of PU foams and elastomers; solution.; sol. in water; sp.gr. 1.033 ± 0.005. [Goldschmidt] *

Tegoamin® DMEA. CAS 108-01-0; Dimethylethanolamine; catalyst for the production of flexible PU foams. [Goldschmidt] *

Tegoamin® PMD. Tertiary amine; catalyst for the manufacturing of flexible PU foams; for slabstock and molded foams. [Goldschmidt] *

Tegoamin® SMP. Tertiary amine; catalyst for the manufacturing of conventional flexible PU slabstock and molded foams (hot-cured foam). [Goldschmidt] *

Tego® Antiflamm® N. Flame retardant for PU foams. [Goldschmidt] *

Tego® Antifoam. Organic antifoam for waste water treatment. [Goldschmidt]

Tego®-Betaine C. CAS 61789-40-0; Cocamidopropyl betaine; surfactant used as foam stabilizer and visc. builder in personal care products, dishwash, liq. soap. [Goldschmidt]

Tego®-Betaine HS. Cocamidopropyl betaine and glyceryl laurate; for nonirritating shampoos, baby care products; good refatting properties. [Goldschmidt; Goldschmidt AG]

Tego®-Betaine L-7. CAS 61789-40-0; Cocamidopropyl betaine; surfactant, detergent, emulsifier, foam stabilizer, visc. builder in low-irritation personal care products, pharmaceuticals. [Goldschmidt; Goldschmidt AG]

Tego®-Betaine L-90. Lauramidopropyl betaine; surfactant, foam stabilizer, visc. builder for shampoos, bath, dishwash, liq. soap, and cream and lotion products. [Goldschmidt]

Tego®-Betaine L-5351. CAS 61789-40-0; Cocamidopropyl betaine; low-salt surfactant for use in salt-critical formulations, e.g., hair dyes, fixatives, reactive products, aerosols, electrolyte-sensitive formulations. [Goldschmidt; Goldschmidt AG]

Tego®-Betaine S. Cocamidopropyl betaine; surfactant, foam stabilizer, visc. builder for shampoos, bath, dishwash, liq. soap, cream and lotion products. [Goldschmidt]

Tego® Care 150. Glyceryl stearate, steareth-25, ceteth-20, stearyl alcohol; wax-like o/w balanced emulsifier systems for creams and lotions; optimized for ease of formulation. [Goldschmidt; Goldschmidt AG]

Tego® Care 300. Glyceryl stearate, steareth-25, ceteth-20; wax-like o/w balanced emulsifier systems for creams and lotions; optimized for ease of formulation. [Goldschmidt]

Tego® Care 450. Stearyl glucoside; o/w emulsifier producing stable emulsions with cosmetics oils and waxes; emulsifier for creams, lotions, hair conditioners, sun products; for systems with pH of 6-9. [Goldschmidt]

Tegochrome® 22. 2,2-Ethyl dithiodiethanol; organic intermediate for producing color developers in the photographic industry. [Goldschmidt] *

Tegocoll®. One and two-component PU adhesives; for bonding of glass, metal, and thermoplastics to other substrates, for use with PU foam, in wood processing industry. [Goldschmidt] *

Tegocolor®. Pigment disps. in a polyether polyol; heatstable color pastes for polyether PU foams. [Goldschmidt] *

Tego® Dispers 610. Higher molecular weight unsaturated polycarboxylic acid; wetting and dispersion additive to counter sedimentation and flooding of pigments; produces selective flocculation of pigments and extenders; stabilizer for pigment dispersions; used in binder systems such as alkyd, acrylate, polyester/melamine, acrylate, polyisocyanate, nitrocellulose paints, chlorinated polymers. [Tego GmbH] *

Tego® Dispers 630. Salt of a higher molecular weight polycarboxylic acid and amine deriv.; wetting and dispersion additive against sediment, sagging, and flooding; produces selective flocculation of pigments and extenders, stabilizes pigment dispersion; for binder systems such as alkyd, acrylate, polyester/melamine, epoxy; chlorinated polymers. [Tego GmbH] *

Tegodont®. Chlorinated agent for water treatment. [Goldschmidt] *

Tegodor. Disinfectants. [Thomas Goldschmidt Ltd]

Tego® Effect L 104. Polyurethane-based solution.; thickener for aq. polymer disps. [Tego GmbH] *

Tego® Emulsion 3454. Silicone emulsion; release agent for production of food pkg. [Goldschmidt AG]

Tego® Emulsion ASL. Silicone emulsion; release agent for natural and synthetic rubbers and the foundry industry. [Goldschmidt AG]

Tego® Emulsion PK. Silicone emulsion; release agent for coating of building seals; avoids stress cracking in acrylic glass. [Goldschmidt AG]

Tego® Flow 425. CAS 68937-54-2; Polysiloxane-polyether copolymer; flow and leveling additive for clear water and solv.-based paint systems (alkyd, acrylate, PU, alkydmelamine, PU-acrylic), building protection paints, wood/furniture varnishes. [Tego GmbH] *

Tego® Foamex 800. CAS 68937-54-2; Polysiloxanepolyether copolymer o/w emulsion; defoamer for waterbased emulsion paints and water-thinnable systems, PU paints, PU-acrylics, furniture paint, wood varnish, adhesives, dispersion paints, polymer dispersions. [Tego GmbH] *

Tego® Foamex 3062. CAS 67762-96-3; Dimethicone copolyol; defoamer for water-based emulsion paints and water-thinnable systems, building protection coatings (acrylic, styrene-acrylic, PVA), wood/furniture varnishes, printing inks. [Tego GmbH] *

Tego® Foamex KS 10. Paraffin-based mineral oil, polysiloxane polyether copolymer; defoamer for emulsion paints, building protection coatings (styrene-acrylate, PVAc, vinyl propionate, S/B), wood/furniture varnishes. [Tego GmbH] *

‡ Trade name and manufacturer not verified   § Trade name without identified manufacturer

Tego® Foamex N. CAS 8050-81-5; Simethicone; defoamer concentrate for med., high solid and solv.-free paint systems, building protection coatings, anticorrosive coatings, high visc. paints. [Tego GmbH] *

Tegoglas. Glass coating agents. [Thomas Goldschmidt Ltd]

Tegoglätte. A litharge having smaller particles than the ordinary type; used in rubber mixings.

Tego® Glide 100. Dimethicone copolyol; surfactant, mar resistant and flow additive for water and solv.-based paints (alkyd, saturated polyester, polyacrylate), building protection coatings, anticorrosive paints, car finishes, printing inks. [Tego GmbH] *

Tego® Glide 410. Dimethicone copolyol; mar resistant additive for solv.-based paints, primers, extenders, automotive paints, building protection coatings, anticorrosion paints, wood/furniture varnishes, industrial paints, lacquers, aq. and solv.-based printing inks; reduces surface tension, improves leveling. [Tego GmbH] *

Tego® Hammer 300000. Methyl silicone oil; hammer finish additive for solv.-based paints; low surface tens. [Tego GmbH] *

Tego® Heat Conductive Paste Z. Conductive paste for fitting diodes, transistors and embedding thermocouples. [Goldschmidt AG]

Tegold. Disinfectants. [Thomas Goldschmidt Ltd]

Tegomag. Permanent magnet alloys. [Thomas Goldschmidt Ltd]

Tegoman. Sanitizers. [Thomas Goldschmidt Ltd]

Tegomuls®. Mono-diglycerides of edible fatty acids. [Thomas Goldschmidt Ltd]

Tegomuls® 19. CAS 61789-10-4; Lard mono/diglyceride; food emulsifier; for shortenings. [Goldschmidt AG]

Tegomuls® B. Glyceryl monodistearate; food emulsifier; for margarine, ice cream, pasta; self-emulsifying. [Goldschmidt AG]

Tegomuls® P 411. 1,2-Propylene glycol monodistearate; food emulsifier; for whipped sponge mixtures, toppings, shortenings. [Goldschmidt AG]

Tego®-Pearl B-48. Cocamidopropyl betaine, glycol distearate, cocamide DEA, cocamide MEA; pearlescent and opacifier for hair care products. [Goldschmidt; Goldschmidt AG]

Tego®-Pearl S-33. Sodium C14-16 olefin sulfonate, glycol distearate, cocamidopropyl betaine, sorbitan laurate; pearlescent for shampoos, bath and shower preparations with excellent superfatting properties; cold processable. [Goldschmidt AG]

Tegopen. CAS 61-72-3; A proprietary preparation of cloxacillin; an antibiotic. [Bristol-Myers Squibb Co Inc]

Tego® Phobe 1030. Solv.-free silicone resin; additive to regulate the uptake of foundation solution. during offset printing. [Tego GmbH] *

Tego® Phobe L 1004. Fluorocarbon resin aq. emulsion; permanent water and soil repellent for leather finishing. [Tego GmbH] *

Tegopren®. Organo modified siloxanes; surfactants used as antistats, wetting and leveling agents, emulsifiers, dispersants, and for the improvement of lubricity; additive for polishes. [Goldschmidt]

Tego® Release Agent M 379. Organic modified polysiloxane; internal release agent for thermosetting resins. [Goldschmidt AG]

Tegosil®. Silicone-based; aerosol for release of injection molded plastic and rubber moldings. [Goldschmidt] *

Tego® Silicone Acrylate RC. Silicone acrylate; for release coatings to be cured by electron beam or uv-radiation in the paper and film industries. [Goldschmidt]

Tego® Silicone Paste A. Lubricant with low load bearing properties for screw threads, rubber and plastic seals and gaskets. [Goldschmidt AG]

Tegosipon®. Silicone; antifoam for manufacturing of stomach and intestinal preparations. [Goldschmidt]

Tegosivin® HL 250. Solv.-free low molecular weight silane-modified siloxane; for prep. of impregnation solutions. for treatment of fresh concrete. [Goldschmidt AG] *

Tegosoft® CI. Cetearyl isononanoate; cosmetic esters for emolliency, skin softening, and moisture retention in hair and skin care products, creams and lotions, hair conditioners/glossing systems, sunscreen preparations, soaps. [Goldschmidt]

Tegosoft® CO. CAS 59130-69-7; EINECS 261-619-1; Cetyl octanoate; cosmetic esters for emolliency, skin softening, and moisture retention in hair and skin care products, creams and lotions, hair conditioners/glossing systems, sunscreen preparations, soaps. [Goldschmidt]

Tegosoft® CT. CAS 65381-09-1; Caprylic/capric triglycerides; cosmetic esters for emolliency, skin softening, and moisture retention in hair and skin care products, creams and lotions, hair conditioners/glossing systems, sunscreen preparations, soaps. [Goldschmidt]

Tegosoft® DO. CAS 3687-46-5; EINECS 222-981-6; Decyl oleate; cosmetic esters for emolliency, skin softening, and moisture retention in hair and skin care products, creams and lotions, hair conditioners/glossing systems, sunscreen preparations, soaps. [Goldschmidt]

Tegosoft® EE. Octyl octanoate; cosmetic esters for emolliency, skin softening, and moisture retention in hair and skin care products, creams and lotions, hair conditioners/glossing systems, sunscreen preparations, soaps. [Goldschmidt]

Tegosoft® GC. CAS 68201-46-7; PEG-7 glyceryl cocoate; mild surfactant for shampoos, bath products and personal cleansers; refatting and conditioning agent; foam booster/stabilizer; rinses clean. [Goldschmidt]

Tegosoft® Liquid. EINECS 261-619-1; Cetearyl octanoate; cosmetic esters for emolliency, skin softening, and moisture retention in hair and skin care products, creams and lotions, hair conditioners/glossing systems, sunscreen preparations, soaps. [Goldschmidt]

Tegosoft® Liquid M. Cetearyl octanoate and isopropyl myristate; cosmetic esters for emolliency, skin softening, and moisture retention in hair and skin care products, creams and lotions, hair conditioners/glossing systems, sunscreen preparations, soaps. [Goldschmidt]

Tegosoft® M. CAS 110-27-0; EINECS 203-751-4; Isopropyl myristate; cosmetic esters for emolliency, skin softening, and moisture retention in hair and skin care products, creams and lotions, hair conditioners/glossing systems, sunscreen preparations, soaps. [Goldschmidt]

Tegosoft® OP. CAS 29806-73-3; EINECS 249-862-1; Octyl palmitate; cosmetic esters for emolliency, skin softening, and moisture retention in hair and skin care products, creams and lotions, hair conditioners/glossing systems, sunscreen preparations, soaps. [Goldschmidt]

Tegosoft® P. CAS 142-91-6; EINECS 205-571-1; Isopropyl palmitate; cosmetic esters for emolliency, skin softening, and moisture retention in hair and skin care products, creams and lotions, hair conditioners/glossing systems,

---

* Trade name not verified as available

† Trade name verified as obsolete

sunscreen preparations, soaps. [Goldschmidt]

Tegosoft® S. CAS 112-10-7; Isopropyl stearate; cosmetic esters for emolliency, skin softening, and moisture retention in hair and skin care products, creams and lotions, hair conditioners/glossing systems, sunscreen preparations, soaps. [Goldschmidt]

Tegosoft® SH. CAS 66009-41-4; Stearyl heptanoate; cosmetic esters for emolliency, skin softening, and moisture retention in hair and skin care products, creams and lotions, hair conditioners/glossing systems, sunscreen preparations, soaps. [Goldschmidt]

Tegostab®. Foam stabilizers for polyurethanes. [Thomas Goldschmidt Ltd]

Tegostab® B 1048. Silicone foam stabilizer; stabilizer for continuously laminated PU boardstock, refrigeration and pipe insulation, rigid block foams. [Goldschmidt AG] *

Tegostab® B 2219. Polysiloxane-polyoxyalkylene copolymer; stabilizer for continuously laminated PU boardstock, rigid block foams. [Goldschmidt AG] *

Tegostab® B 4900. Polysiloxane-polyoxyalkylene block copolymer; foam stabilizer with wide processing latitude for hot-cured PU foam. [Goldschmidt AG] *

Tegostab® B 8406. Polysiloxane-polyether copolymer surfactant; stabilizer for PU foams with integral skins, especially shoe sole systems. [Goldschmidt AG] *

Tegostab® BF 2270. Polysiloxane-polyoxyalkylene block copolymer; foam stabilizer for manufacturing of flexible polyether polyurethane foams (hot cured foams). [Goldschmidt AG] *

Tegotain D. Cocamidopropyl betaine; mild surfactant. [Goldschmidt AG]

Tegotens 4100. Palmitic/stearic acid mono/diglycerides; surface coating for expanded polystyrene beads containing a propellant; antistat for PE/PP; fabric softener; coemulsifier. [Goldschmidt AG]

Tegotens I. POE glyceryl isostearate; solubilizer, coemulsifier, wetting agent. [Goldschmidt AG]

Tegotrenn® LH 157 A, 525. Release agent for demolding of flexible hot-cured PU foam. [Goldschmidt] *

Tegovakon. Sandstone strengtheners. [Thomas Goldschmidt Ltd]

Tego® Wet KL 245. Dimethicone copolyol; substrate wetting and spreading additive for solv. and water-based systems, wood/furniture varnishes, industrial/household appliance paints, heat-resist. coatings. [Tego GmbH] *

Tegretol. CAS 298-46-4; A proprietary preparation of carbamezepine; an analgesic. [Ciba plc] *

Tegul. A proprietary sulfur jointing compound for bell and spigot pipes; it contains sulfur and sand. §

Tegula. Bitumen sheet; for underneath stretcher strip. [Vedag GmbH] *

Teka Oil. A proprietary trade name for an extract from stand oil (qv) from which bases and acids have been removed.§

Tekblend. One-shot high yield grout. [Foseco (F.S.) Ltd]

Tekemail. Classified ceramic glaze granules. [BASF AG]

Tekfoam. Lightweight high yield grout for cavity filling. [Foseco (F.S.) Ltd]

Teknol. Photographic developer. [May & Baker Ltd] *

Tekpak-Tekbent. Reactor part of anhydrite/bentonite mixture. [Foseco (F.S.) Ltd]

Tekpak-Tekcem. Alumina cement for pump packing. [Foseco (F.S.) Ltd]

Tekstim 8504. Corrosion inhibitor for use in acid cleaning

applications where corrosion inhibition on copper is critical. [Exxon]

Tekstim 8741. Surfactant, self-demulsifying detergent for truckwash applications. [Exxon/Tomah]

Telconax. A patented insulating compound made from selected bitumen, waxes, and rubber. §

Telconite. A proprietary insulating material made in various colors. §

Telconstan. A non-magnetic nickel-copper alloy prepared in induction furnaces and having exceptional purity and very low temperature coefficient of resistance; used for resistances where standard of resistance with temperature is important. §

Telcothene®. Polythene powder, tube, and sheet. [Telcon Plastics Ltd] *

Telcovin®. Polyvinyl chloride tube and sheet. [Telcon Plastics Ltd] *

Teldrin. CAS 2438-32-6; Chlorpheniramine maleate; antihistaminic. [Menley & James Laboratories] *

Teleblock. A proprietary range of thermoplastic rubbers. [Phillips] *

telegraph bronze. (telegraph metal, electric metal). An alloy of 80% copper, 7.5% lead, 7.5% zinc, and 5% tin.

Telepaque. CAS 96-83-3; Iopanoic acid; diagnostic aid. [Sterling Drug Inc] *

Telloy®. CAS 13494-80-9; Tellurium powder; vulcanizing agent for rubber. [R T Vanderbilt Co Inc]

Tellurac®. Tellurium diethyldithiocarbamate; accelerator for elastomers. [R T Vanderbilt Co Inc]

telluretted hydrogen. Hydrogen telluride, H_2Te.

Tellurit. Chill producing mold dressing containing tellurium. [Foseco (F.S.) Ltd]

tellurium. Nonmetallic element; Te; CAS 13494-80-9; EINECS 236-813-4; alloys, secondary rubber vulcanizing agent, manufacture of iron and steel casting; coloring agent for glass and ceramics; thermoelectric devices. [Asarco; Atomergic Chemetals; Cabot; Cerac; R.T. Vanderbilt]

tellurium dioxide. See tellurium oxide.

tellurium lead. An alloy containing 0.05% tellurium with lead; resists sulfuric acid.

tellurium oxide. (tellurium dioxide; tellurous acid anhydride) TeO_2; CAS 7446-07-3; EINECS 231-193-1. [Asarco; Atomergic Chemetals; Cerac; Noah Chem.]

Tellurium Tubes. Copper tubes containing pre-determined quantities of tellurium. [Foseco (F.S.) Ltd] †

tellurous acid anhydride. See tellurium oxide.

Telmin. CAS 31431-39-7; A proprietary preparation containing mebendazole; veterinary antihelmintic (horses). [Janssen Pharmaceutical Ltd] *

Telogen. 1:2 Metal complex disperse dyes for fast dyeing of polyamide fibers. [Bayer plc] †

Telon®. Selected acid dyestuffs; for the dyeing of polyamide fibers and wool/polyamide blends. [Bayer AG; Bayer plc]

Telone. CAS 542-75-6; Agricultural products containing 1,3-dichloropropene as the active ingredient; soil fumigants added to soil prior to planting to control soil pests such as nematodes which feed on the roots of plants and reduce yields. [DowElanco Ltd]

Telopar. CAS 68813-55-8; Oxantel pamoate; anthelmintic. [Pfizer Inc] †

Telsit. A gelatin explosive containing from 10-15% dinitrotoluene or liquid trinitrotoluene.

Teluran. Acrylonitrile-butadiene-styrene polymers and re-

---

‡ Trade name and manufacturer not verified   § Trade name without identified manufacturer

lated products. [BASF plc] *

Telvar®. For the agriculture industry. [DuPont UK]

Temadex. Veterinary skin dressing. [The Wellcome Foundation Ltd] *

Temaril®. CAS 4330-99-8; Trimeprazine tartrate; antipruritic. [Allergan, Inc]

Temasept I. Dibromo/tribromo salicylanilide; antimicrobial for resins, latex emulsions, plastics. [Hexcel]

Temasept IV. CAS 87-10-5; Tribromosalicylanilide; antimicrobial for resins, latex emulsions, plastics. [Hexcel]

Tembind A 002. CAS 8061-53-8; Ammonium lignosulfonate; emulsifier and stabilizer for asphalt; road dust suppressant; coal dust suppressant; binder and pelletizing agent. [Temfibre]

Temetex. CAS 2607-06-9; Proprietary topical corticosteroid containing diflucortolone as active ingredient. [Roche Products Ltd] *

Temlock. A proprietary trade name for a board made from wood fibers impregnated with resin and subjected to pressure. §

Tempaloy. A patented alloy of approximate composition, 95% copper, 4% nickel, and 1% silicon. §

temper. Alloys of arsenic and lead or copper, and tin; used as hardening materials for shot or pewter.

Temperite Alloys. A proprietary trade name for alloys of lead, tin, and cadmium. §

Tempo. Soft antacid, for relief of acid indigestion. [Richardson-Vicks Inc] *

Tempo. Proprietary cellulose esters. §

Tempo. Herbicide containing linuron and terbutryn. [ICI Chem. & Polymers Ltd]

Tempo. Suspension concentrate containing 150 g linuron and 150 g terbutryn per liter; used for control of annual dicotyledons in potatoes. [Farm Protection Ltd]

Tempra. CAS 103-90-2; Acetaminophen; an analgesic and antipyretic. [Mead Johnson & Co] *

Tempro. A water-based acrylic copolymer temporary coating and remover. [ICI Chem. & Polymers Ltd]

Temsperse S 001. CAS 8061-51-6; Sodium lignosulfonate; dyestuff dispersant, water reducer, slurry thinner; binder; concrete admixtures; emulsion stabilizer. [Temfibre]

Tem-Tuf. Aluminious metal sheet. [Reynolds Metal Co]

Tenac®. Polyacetal homopolymer and copolymer resin [Asahi Chem. Industry]

Tenacite. See Bakelite. §

Tenacity. Fluxes for silver alloy brazing. [Johnson Matthey plc]

Tenamine 1. N-Butylated-p-aminophenol; a proprietary antioxidant. [Eastman] *

Tenamine 2. CAS 101-96-2; N,N´-Di-sec-butyl-p-phenylenediamine; a proprietary antioxidant. [Eastman] *

Tenamine 3. 2,6-Di-t-butyl-p-cresol; a proprietary antioxidant [Eastman] *

Tenasco. A proprietary trade name for synthetic fiber resembling nylon. §

Tenase® 1200, L-340, L-1200. CAS 9000-92-4; Bacterial α-amylase; enzyme for starch liquefaction; for syrups, textile desizing. [Solvay Enzymes]

Tenatine. Thermoplastic polyester molding compounds. [Ciba plc] *

Tenavoid. A proprietary preparation of bendrofluazide and meprobamate; prophylaxis for premenstrual syndrome. [Leo Laboratories] *

Tenaxatex VA 632. A proprietary trade name for a high

molecular weight vinyl acetate homopolar water emulsion containing 55% solids; used as an adhesive base. [H A Smith] *

Tenaxatex VA 956. A proprietary trade name for a vinyl acetate/acrylate copolymer emulsion containing 55% solids; a medium viscosity adhesive base. [H A Smith] *

Tenax metal. A zinc alloy containing from 0.35-2.56% copper, 0.2-4.42% aluminum, 0-0.35% iron, and up to 1.2% lead; used for the manufacture of guide rings.

Tenax Wax. Mineral grease, wax and resin formulated as a grafting wax. [Vitax Ltd]

Tenazit. A proprietary trade name for laminated bakelite or similar synthetic resin. §

Tendrelle. Nylon. [ICI Chem. & Polymers Ltd]

Tendril®. For the agriculture industry. [DuPont UK]

Tenephrol. A proprietary preparation of lithium iodide (31% solution). §

Tenif. Proprietary preparation of atenolol and nifedipine; for treating hypertension. [Imperial Chemical Industries plc]

Tenite® 105-MS. CAS 9004-35-7; Cellulose acetate; good processibility and chem. resist.; for tool handles, optical applications, toothbrushes, personal care items, tubing, pipe, medical devices, automotive parts, appliance parts, toys, sporting goods, sheeting, furniture profiles. [Eastman]

Tenite® 154DF. CAS 9002-88-4; Polyethylene; for film extrusion. [Eastman]

Tenite® 264-MH. CAS 9004-36-8; Cellulose acetate butyrate; good processibility and chem. resist.; for tool handles, optical applications, toothbrushes, personal care items, tubing, pipe, medical devices, automotive parts, appliance parts, toys, sporting goods, sheeting, furniture profiles. [Eastman]

Tenite® 360-H2. CAS 9004-39-1; Cellulose acetate propionate; good processibility and chem. resist.; for tool handles, optical applications, toothbrushes, personal care items, tubing, pipe, medical devices, automotive parts, appliance parts, toys, sporting goods, sheeting, furniture profiles. [Eastman]

Tenite® Cellulosic Acetate. CAS 9004-35-7; Cellulose acetate; thermoplastic ester offering toughness and clarity in virtually unlimited color range at reasonable price; easy moldability. [Eastman] *

Tenite® Cellulosic Butyrate. Cellulose butyrate; thermoplastic ester offering toughness and clarity in virtually unlimited color range at reasonable price; easy moldability. [Eastman] *

Tenite® Cellulosic Propionate. CAS 9004-39-1; Cellulose propionate; thermoplastic ester offering toughness and clarity in virtually unlimited color range at reasonable price; easy moldability. [Eastman] *

Tenite® PET 9902. CAS 25038-59-9; PET polyester; for extrusion into crystallizable sheet that can be thermoformed into food trays for microwave or conventional ovens. [Eastman]

Tenite® Polyethylene. CAS 9002-88-4; Polyethylene; for injection molding, film extrusion, extrusion coating, profile and other specialty extrusions. [Eastman]

Tennafast. Inorganic and organic pigments. [Tennant-KVK Ltd]

Tennal. A proprietary trade name for certain aluminum alloys for casting purposes. §

Tennalaks. Inorganic and organic pigments. [Tennant-KVK Ltd]

---

* Trade name not verified as available   † Trade name verified as obsolete

Tenncol. Saponified rosin size. [Tenneco Malros Ltd] *

Tennessee phosphates. Mineral phosphates containing from 60-70% calcium phosphate; used as a fertilizers.

Tenoban. Arecoline-acetarsol. [The Wellcome Foundation Ltd] *

Tenoran. CAS 1982-47-4; Chloroxuron; urea herbicide for on strawberries and ornamentals. [Ciba-Geigy Agrochemicals]

Tenoret. Atenolol and chlorthiadone; antihypertensive and diuretic. (Sold in UK by Stuart Pharmaceuticals) [ICI Chem. & Polymers Ltd]

Tenoretic. Atenolol and chlorthalidone; combined antihypertensive and diuretic. (Sold in UK by Stuart Pharmaceuticals). [ICI Pharma]

Tenormin. CAS 29122-68-7; Atenolol; antihypertensive. (Sold in UK by Stuart Pharmaceuticals) [ICI Pharma]

Tenox® 2. Propylene glycol, BHA, propyl gallate, citric acid; food-grade antioxidant. [Eastman]

Tenrez. Rosin esters. [Tenneco Malros Ltd] *

Tensabit. Emulsifier; alkaline emulsion; for bitumen. [Tensia SA] *

Tensactol. Self-emulsifying wax; for cosmetics and pharmaceuticals. [Tensia SA] *

Tensadal. Reviving and softening agent; for cotton and cellulose fibers. [Tensia SA] *

Tensagex. Liquid blend; for baby shampoos. [Hickson & Welch Ltd]

Tensagex BV. Anionic surfactant in the form of a low viscosity liquid; for liquid detergents; liquid shampoos; bubble baths. [Hickson & Welch Ltd] *

Tensagex DMY. Sodium alkyl ether sulfate as a low viscosity liquid; low irritant anionic surfactant used in baby shampoos and foam baths. [Hickson & Welch Ltd] *

Tensagex DP24. Sodium alkylphenol ether sulfate in liquid form; wetting and dispersing agent for pigments and metal degreasing. [Hickson & Welch Ltd] *

Tensagex EOC. Sodium lauryl alcohol/2.5EO sulfate in liquid or gel form; high foaming anionic surfactant used in shampoos, bubble baths, liquid detergents and emulsion polymerization. [Hickson & Welch Ltd] *

Tensagex SPDL. Sodium alcohol ether sulfate (C12-C15) in liquid form; high foaming anionic surfactant used in shampoos, bubble baths, liquid detergents and emulsion polymerization. [Hickson & Welch Ltd] *

Tensami 1/05. Lecithin, xanthan gum; natural emulsifier. [Alban Muller]

Tensami 3/06. Casein, xanthan gum; natural emulsifier. [Alban Muller]

Tensami 4/07. Soy protein, xanthan gum; natural emulsifier. [Alban Muller]

Tensami 8/09. Corn oil, egg yolk extract; natural emulsifier. [Alban Muller]

Tensami 10/06. Saponins, xanthan gum; natural emulsifier. [Alban Muller]

Tensamina. Adhesion improver; for bitumen on wet stones. [Tensia SA] *

Tensamine C, O, S, and SH. Cationic surfactants in the form of primary amines, with the alkyl portion being coconut, oleic, tallow, and hydrogenated tallow respectively; liquid or solid form; for emulsification, pigment dispersion, synthesis intermediate. [Tensia SA] *

Tensaminox. Foaming agents; dispersing agents; thickening agents; for shampoos; liquid detergents; pigments, textiles. [Tensia SA] *

Tensanyl. A proprietary preparation of bendrofluazide, reserpine, and potassium chloride; an antihypertensive. [Leo Laboratories] *

Tensarane SBTE. TEA-dodecylbenzene sulfonate, supplied as a liquid; anionic surfactant, wetting agent and dyeing auxiliary for cottons. [Tensia SA] *

Tensaryl. Low foaming detergent base; for light duty detergents. [Tensia SA] *

Tensaryl 40CC, 50B, 80B and 82F. Sodium tetrapropylene benzene sulfonate as powder, beads or flakes; detergent, anticaking agent, and wetting agent, for emulsion polymerization, wettable powders, and various industrial uses. [Tensia SA] *

Tensaryl DF90. Sodium dodecylbenzene sulfonate; foaming agent for fine fabrics washing. [Tensia SA] *

Tensaryl DX54Sp. and DX62. Sodium dodecylbenzene sulfonate; DX54Sp also contains perborate; anionic surfactant for heavy duty products. [Tensia SA] *

Tensaryl KD. Tetrapropylene benzene sulfonate as a viscous liquid; intermediate for liquid and powder detergents when biodegradability is not requested. [Tensia SA] *

Tensaryl L48. Sodium dodecylbenzene sulfonate; low foaming, anionic surfactant for fine fabrics washing. [Tensia SA] *

Tensaryl S30P and S70P. Sodium tetrapropylene benzene sulfonate in the form of a liquid or paste with low salt content; anionic surfactant used for emulsion polymerization. [Tensia SA] *

Tensaryl SB Ca. Calcium dodecylbenzene sulfonate in powder form; water-insoluble anionic surfactant used as a co- emulsifier in organic systems. [Tensia SA] *

Tensaryl SB. Straight chain dodecylbenzene sulfonate in acid form; supplied as a viscous liquid; raw material for the manufacture of liquid, pasty or solid surfactants. [Tensia SA] *

Tensaryl SB85P. Sodium dodecylbenzene sulfonate in paste form; wetting and dispersing agent with low salt content; used in emulsion polymerization. [Tensia SA] *

Tensaryl SBD. Triethanolamine dodecylbenzene sulfonate in liquid form; liquid detergent for dishwashing, textiles and car shampoos. [Tensia SA] *

Tensatil D100. Sodium octylsulfate in liquid form; anionic surfactant used as a wetting agent in polymerization. [Tensia SA] *

Tensatil DA120. Ammonium octyl/decyl-sulfate in liquid form; anionic surfactant used as a wetting and foaming agent in froth flotation. [Tensia SA] *

Tensatil DB120. Short chain alcohol sulfate in liquid form; anionic surfactant used as a wetting agent in electrolyte solutions. [Tensia SA] *

Tensatil DEH120. Sodium 2-ethylhexyl sulfate in liquid form; anionic surfactant used as a wetting agent in metal cleaning preparations and styrene polymerization. [Tensia SA] *

Tensiamix. Range of low foaming detergent base; for heavy duty detergents. [Tensia SA] *

Tensianol. Sensitive skin special washing composition; for cosmetics; synthetic toilet bars. [ICI plc] *

Tensibet 50. N-Alkylbetaine in liquid form; foam booster and detergent with antistatic effect; used in cosmetics and baby shampoos. [Tensia SA] *

Tensibet 55. Alkylamidobetaine in liquid form; foam stabilizer and thickening agent which is mild to the skin; used

---

‡ Trade name and manufacturer not verified   § Trade name without identified manufacturer

in lauryl ether sulfate formulations. [Tensia SA] *

Tensid. An antihypertensive preparation.

Tensidef. Antifoaming agents; for paper industry. [Tensia SA] *

Tensidye. Nonionic liquid; wetting agent for cosmetics. [Tensia SA] *

Tensilac 39. A proprietary rubber vulcanization accelerator; it is a soft form of ethylidene-aniline. §

Tensilac 40. A proprietary rubber vulcanization accelerator; it is a resinous condensation product. §

Tensilac 41. A proprietary rubber vulcanization accelerator; it is a hard form of ethylidene-aniline. §

Tensilite. An aluminum bronze. It contains from 64-67% copper, 3.1-4.4% aluminum, 0-1.2% iron, 2.5-3.8% manganese, and 24-29% zinc.

Tensilon. CAS 116-38-1; A proprietary preparation of edrophonium chloride used in the diagnosis of myasthenia gravis. [Roche Products Ltd] *

Tensimul. Emulsifier; for oils, solvents, natural and synthetic waxes. [Tensia SA] *

Tensiofix. Range of emulsifiers; for pesticide formulations. [Tensia SA] *

Tensioquat C50. Benzalkonium chloride in aqueous solution; cationic surfactant used in disinfection and emulsification. [Tensia SA] *

Tensioquat C75. Benzalkonium chloride in liquid IPA form; cationic surfactant used in the paint industry. [Tensia SA]*

Tensiorex. Conditioning agent; concentrated pearling agent; for after shampoo; shampoos. [Tensia SA] *

Tensiostat. Antistatic agent; for cellulosic fibers and polymers. [Tensia SA] *

Tensipar. Antifoaming agent; for industrial uses including food. [Tensia SA] *

Tensitex. Wetting agent; for caustic lye and mercerizing. [Tensia SA] *

Tensloy. A proprietary trade name for an alloy of iron with approximately 1.5% nickel and 0.5% chromium. §

Tensocide. Antiseptic; for paper mills. [Tensia SA] *

Tensol. Cements for vinyl and acrylic sheets. [ICI Chem. & Polymers Ltd]

Tensol. Photographic activator/stabilizer chemicals. [May & Baker Ltd] *

Tensol. A proprietary trade name for a dispersing and emulsifying agent containing a sulfonated ether. §

Tensoleate. Pigment grinding aid; dispersing agent; for paint industry. [Tensia SA] *

Tensoline. Emulsifier; leveling and dispersing agent; for oiling and fulling of wool; dyeing of woollen and acrylic fibers. [Tensia SA] *

Tensomel. Superamide; foam stabilizer; additive; anticorrosion; for shampoos; liquid detergents; bubble baths; cutting oils. [Tensia SA] *

Tensomin. Leveling agent; for dyeing woolen, acrylic, polyamide, and polyester fibers; dyeing with acid and metallizing dyes. [Tensia SA] *

Tensopac. Wetting and dispersing agents; for paper mill. [Tensia SA] *

Tensopane D. Series of nonionic surfactants of the octylphenol ethoxylate type; emulsification; detergent compounding; industrial cleaning; metal pickling; stabilizer in emulsion polymerization. [Tensia SA] *

Tensophene 2D30. Dinonylphenol ethoxylate nonionic surfactant; emulsifier; detergent; metal cleaning; emulsion polymerization. [Tensia SA] *

Tensophene D12, D15 and D18. Nonionic surfactants of the nonylphenol ethoxylate type; for foam control; dispersion; oil emulsifier; coupling agent. [Tensia SA] *

Tensophene H10, I10, DT, D36, D42EC, D45, D60 and D90. Nonionic surfactants of the nonylphenol ethoxylate type; used for all types of detergent; metal cleaning; emulsion polymerization. [Tensia SA] *

Tensopol 12A and 12P. Sodium dodecyl sulfate in needle or powder form; anionic surfactant used in pharmaceuticals, toothpastes and shampoos. [Hickson & Welch Ltd]*

Tensopol 30E and LDS. Sodium dodecylbenzene sulfonate; LDS also contains optical dye; foaming detergents. [Hickson & Welch Ltd] *

Tensopol A.7 and USP. Sodium lauryl sulfate in the form of needles or powder; detergency and foaming agent for toothpastes, shampoos, pharmaceuticals, emulsion polymerization and pigment dispersion. [Hickson & Welch Ltd] *

Tensopol ACL and PCL. Sodium lauryl sulfate in the form of needles or powder; detergency, foaming, and wetting agent for shampoos, emulsion polymerization, and pigment dispersion. [Hickson & Welch Ltd] *

Tensopol AG and MG. Magnesium lauryl sulfate in the form of needles or powder; anionic surfactant used in shampoos and toothpastes. [Hickson & Welch Ltd] *

Tensopol DX85 and FL. Sodium alcohol sulfate in liquid or paste form; foaming agent for liquid shampoos, including carpet types; latex rug backing. [Hickson & Welch Ltd] *

Tensopol LT. CAS 139-96-8; EINECS 205-388-7; Triethanolamine lauryl sulfate in liquid form; anionic surfactant used in liquid shampoos, bubble baths and hair lotions. [Hickson & Welch Ltd] *

Tensopol N. CAS 2235-54-3; EINECS 218-793-9; Ammonium lauryl sulfate in liquid form; anionic surfactant used in liquid shampoos and bubble baths. [Hickson & Welch Ltd] *

Tensopol SPK. CAS 4706-78-9; EINECS 225-190-4; Potassium lauryl sulfate in powder form; anionic surfactant used as a base in synthetic toilet bars. [Hickson & Welch Ltd] *

Tensopol VAL. Sodium lauryl sulfate in liquid form; liquid detergents, shampoos, bubble baths, and carpet shampoos. [Hickson & Welch Ltd] *

Tensoprene. Low foaming surfactant; rinse aid, detergent for dishwashing formulations; industrial cleaners. [Tensia SA] *

Tensostat. Bactericide and fungicides; for paper mills. [Tensia SA] *

Tensovax. Emulsifier, solubilizer; wetting agent; for oils, perfumes, vitamins; metal working; textile specialities. [Tensia SA] *

Tensovyl. Dispersing agent; for post-tinctorial washing of polyester fibers. [Tensia SA] *

Tensuccin D8. CAS 577-11-7; EINECS 209-406-4; Sodium dioctyl sulfosuccinate in liquid form; anionic surfactant used in textile desizing; dry cleaning; cosmetics. [Tensia SA] *

Tensuccin H724 and H925. Disodium alkyl ether hemiester sulfosuccinate in water solution; anionic surfactant used as a polymerization stabilizer. [Tensia SA] *

Tensuccin HS40. Disodium stearic hemiester sulfosuccinate in the form of a paste; anionic surfactant used as a dispersing agent. [Tensia SA] *

---

* Trade name not verified as available   † Trade name verified as obsolete

Tensuccin ML, MO, and MS. Anionic surfactants of the sulfosuccinamate type, in liquid or soft paste form; foaming agent used in latex for carpet backing and foam insulation. [Tensia SA] *

Tensyl 30. CAS 137-16-6; EINECS 205-281-5; Sodium lauroyl sarcosinate in liquid form; anionic surfactant used in cosmetics, toothpastes, and baby shampoos. [Tensia Ltd] ‡

Tensynvac. 1133 Strain viral arthritis; for immunization of poultry. [Intervet Inc]

Tentor. CAS 50-33-9; A proprietary preparation of phenylbutazone. [RP Drugs] *

Tenuate. CAS 134-80-5; A proprietary preparation of diethylpropion hydrochloride; an anti-obesity agent. [Richardson-Vicks Inc] *

Tenuate Dopan. CAS 134-80-5; A proprietary preparation of diethylpropion hydrochloride in a slow release base. [Richardson-Vicks Inc] *

Tenvate. CAS 134-80-5; Diethylpropion hydrochloride; anorexic. [Merrell Dow Pharmaceuticals Inc] *

TEP. See triethyl phosphate.

TEPA. See tetraethylenepentamine.

Tepanil™ Tablets. CAS 134-80-5; Diethylpropion hydrochloride; appetite suppressant. [3M Pharmaceuticals]

Tepanil™ Ten-tab Tablets. CAS 134-80-5; Diethylpropion hydrochloride; appetite suppressant. [3M Pharmaceuticals]

Tephal. A wetting agent and detergent; used in the textile industry for pretreatment, desizing, and dyeing. [Degussa AG] *

Tepperite. A proprietary polystyrene. §

Teralan. Dyes for transfer polyester/wool. [Ciba plc] *

Teraprint. Dyes for transfer printing. [Ciba plc] *

Terasil. Disperse dyes. [Ciba plc] *

Terate 101, 131. Balsamic resins derived from petroleum aromatic hydrocarbons; used in adhesives and sealants, in alkyd coatings, in molded goods and in rubber compounding. [Hercules] *

Terate 202. Thermoplastic resins; aromatic polyester polyols derived from polycarbomethoxy-substituted diphenyls, polyphenyls, and benzyl esters of the toluate family; used to extend reactive polyurethane and polyisocyanurate urethane polyols in the manufacture of rigid urethane foam. [Hercules] *

Terate 203. Thermoplastic resins; aromatic polyester polyols derived from polycarbomethoxy-substituted diphenyls, polyphenyls, and benzyl esters of the toluate family; used to extend reactive polyurethane and polyisocyanurate urethane polyols in the manufacture of rigid urethane foam. [Hercules] *

Terate 204. Thermoplastic resins; aromatic polyester polyols derived from polycarbomethoxy-substituted diphenyls, polyphenyls, and benzyl esters of the toluate family; used to extend reactive polyurethane and polyisocyanurate urethane polyols in the manufacture of rigid urethane foam. [Hercules] *

Terathane. Polyether glycol. [DuPont UK]

Terathane® 650. Polytetramethylene ether glycol; functions as a soft segment in PU resins. [DuPont] *

Terbalin. Active ingredients: terbutrex plus triflurex; selective pre-emergence herbicidal mixture. [Agan Chemical Manufacturers Ltd]

terbia. See terbium oxide.

terbium. Element; Tb; CAS 7440-27-9; EINECS 231-137-6;

phosphor activator, dope for solid-state devices. [Aldrich; Atomergic Chemetals; Cerac; Rhone-Poulenc Basic]

terbium oxide. (terbia) Tb_2O_3; CAS 12037-01-3; EINECS 234-856-3. [Atomergic Chemetals; Cerac; Rhone-Poulenc Basic]

Terblend®. Styrene copolymer blends. [BASF plc]

Terblend® S. ASA/PC blend; thermoplastic polymer for injection molding of car instrument covers, tail light assemblies, etc., housings for small appliances, transformer housings, switchgear for house wiring, meter housings, etc. [BASF AG]

Terbufos. S-t-Butylthio-methyl-o,o-diethyl phosphorodithioate; a pesticide. §

Terbutol. CAS 98-54-4; p-t-Butylphenol. [ICI Chem. & Polymers Ltd] †

Terbutrex. Active ingredient: terbutryne; 2-tert-butylamino-4-ethylamino-6-methylthio-1,3,5-triazine; pre-emergence and post-emergence weed control. [Agan Chemical Manufacturers Ltd]

Terbytex. Softening and antistatic agent; for natural and synthetic fibers. [Tensia SA] *

Tercin. A proprietary preparation of aspirin, phenacetin and butobarbitone; an analgesic. [Cox Pharmaceuticals] †

Tercoton. Dyes for polyester/cotton. [Ciba plc] *

terebene. Acid-isomerized turpentine consisting of a mixture of dipentene and other hydrocarbons.

terebine. (liquid drier, Japan drier). Made by heating oxides of lead and manganese with linseed oil or rosin, or mixtures of the oil and rosin, and thinning with turpentine or turpentine substitute; a drier for paints.

Terephane. A proprietary name for polyethylene terephthalate film. French Origin §

Terethane®. For the chemical industry. [DuPont UK]

Terfenol. Rosin derivatives; for paints and inks. [Resinas Sinteticas SA] *

Terfluzin. CAS 117-89-5; Trifluoperazine. [Rhone-Poulenc Rorer Ltd]

Terfonyl. Trisulfapyrimidines (oral suspension); mixture of sulfadiazine, sulfamerazine, and sulfamethazine; an antibacterial. [Bristol-Myers Squibb Co Inc] *

Terg-A-Zyme®. Alkylaryl sulfonate, lauryl alcohol sulfate, phosphate, carbonate, and protease enzyme; biodeg. detergent, wetting agent, sequestering and synergistic agents; used in hospitals, laboratories, dairies; cleaning agent in dairy and pollution processing. [Alconox]

Tergenol 1122. Blend; heavy-duty detergent for heavily soiled woven and knitted fabrics; stable to acids, alkalies, and oxidizing agents. [Hart Chem. Ltd.]

Tergitex KW. Polyether solvent blend; for oil removal from garnetted wool and knitted goods. [Scher]*

Tergitol®. A series of biodegradable nonionic intermediates comprising ethoxylates and ethoxysulfates of linear secondary alcohols; used in the production of biodegradable detergents. [Union Carbide (UK) Ltd] *

Tergitol® 15-S-3. CAS 68131-40-8; C11-15 pareth-3; biodeg. detergent, emulsifier, wetter, defoamer for aq. systems, intermediate used in textiles, solv. cleaners, drycleaning, metalworking fluids, water treatment, oilfield chems., pulp/paper deinking, latex emulsions, plastics antistat, agriculture. [Union Carbide]

Tergitol® 24-L-45. CAS 68439-50-9; C12-14 pareth-6 (6.3 EO); biodeg. surfactant, detergent, wetting/spreading agent, emulsifier, foaming agent, intermediate, dispers-

---

‡ Trade name and manufacturer not verified   § Trade name without identified manufacturer

ant for household and industrial cleaners, textile wet processing, paper processing, and agricultural formulations. [Union Carbide]

Tergitol® 26-L-3. C12-16 pareth-3; surfactant, emulsifier, intermediate for sulfation; used for prewash spotters, coning oil, hydrocarbon-based cleaners, agriculture; as sulfated product in cosmetics, hand dishwash, light duty detergents. [Union Carbide]

Tergitol® D-683. CAS 37251-69-7; Alkoxylated alkylphenol; emulsifier for fiber finishing operations; dispersant for pigments in resins, plastics, and for abrasives in hard surface cleaners; improves wetting of oil-based materials in coatings and adhesives. [Union Carbide]

Tergitol® Min-Foam 1X. CAS 68551-14-4; C11-15 alcohols reacted with EO and PO; biodeg. surfactant, foam depressant, wetting agent, detergent for household/industrial cleaners, drycleaning, textile processing, metal cleaning, circuit board cleaners, leather, paper deinking. [Union Carbide]

Tergitol® TMN-3. CAS 60828-78-6; Isolaureth-3; emulsifier, wetting agent, coupler, penetrant, leveling agent for textile processing, lubricants, water treatment, solv. cleaners/degreasers, metalworking fluids, drycleaning, oilfield chems., pulp/paper deinking; defoamer for aq. systems; intermediate for anionic surfactants used in household, industrial, and personal care products. [Union Carbide]

Tergitol® XD. CAS 9038-95-3; PPG-24-buteth-27; emulsifier, dispersant, stabilizer for agricultural insecticides/herbicides, latex polymerization, iodophor manufacturing for germicidal cleaning, latex paints, dye pigments, leather; emulsifier for silicone oils, diacyl peroxides. [Union Carbide]

Tergitol® XH. CAS 9038-95-3; EO/PO copolymer; emulsifier, dispersant for agriculture, latex polymerization, iodophor manufacturing for germicidal cleaning, latex paints, dye pigments, leather finishes, toilet bowl cleaners; emulsifier for silicone oils, diacyl peroxides. [Union Carbide]

Tergolix. Leather fat liquoring agents. [Sandoz Products Ltd]

Tergraf. Rosin derivatives; for paints and inks. [Resinas Sinteticas SA] *

Tergum. Rosin derivatives; for paints, inks, adhesives, chewing gum. [Resinas Sinteticas SA] *

Teric 9A2. CAS 68439-46-3; C9-11 ethoxylate (2 EO); surfactant for sulfation feedstock, polishes and waxes, drycleaning, solv. cleaners/degreasers. [ICI Australia]

Teric 12A2. CAS 68131-39-5; C12-15 pareth-2; intermediate for phosphate ester production, sulfation; surfactant for solv. cleaners/degreasers; fiber lubricant/antistat for textile spinning. [ICI Australia]

Teric 12M2. CAS 61791-14-8; PEG-2 cocamine; dispersant, emulsifier, stabilizer; wetting agent of hydrophobic surfaces; used in metal, stone, paper and textile processing; formulation of lubricants and dye bath auxiliaries; emulsion stabilization; fat liquoring compounds. [ICI Australia]

Teric 13A5. CAS 24938-91-8; Trideceth-5; emulsifier for agriculture; filter cake dewatering; polishes/waxes; solv. cleaners/degreasers. [ICI Australia]

Teric 15A11. C14-15 pareth-11; in biodeg. detergent formulation, low foam laundry products, bottle washing, metal soaking. [ICI Australia]

Teric 16A16. CAS 68439-49-6; Ceteareth-16; wetting agent, emulsifier, detergent, solubilizer, dispersant in hydrophobic conditions; dye and pigment carriers; textile applications; manufacturing of wax emulsions and polishes for household and industrial use; antistat. [ICI Australia]

Teric 16M2. CAS 61791-24-0; PEG-2 soya amine; wetting agent, dispersant, emulsifier for waxes and fats; formulation of leather dressing and metal cleaning compounds, fiber lubricant applications in the bldg. industry. [ICI Australia]

Teric 17A2. Cetoleth-2; emulsifier, manufacturing of textile lubricants, solv. and waterless hand cleaners; coemulsifier; detergent additive in petrol. oils; intermediate for anionic surfactants. [ICI Australia]

Teric 17M2. CAS 61791-44-4; PEG-2 tallow amine; wetting agent and dispersant; dewatering agent in elec. components and road aggregates; emulsifier for fats, waxes, mineral oils, and metal working lubricants; corrosion inhibitor. [ICI Australia]

Teric 18M5. CAS 26635-92-7; PEG-5 stearamine; wetting agent, dispersant; emulsion stabilizer; emulsifier used in agricultural toxicants and processing of textiles, paper, leather, and bldg. board; corrosion inhibitor in lubricants and greases; softener and antistat in solv. cleaning of textiles and compound of plastics. [ICI Australia]

Teric BL8. Synthetic alcohol polyalkylene oxide deriv.; detergent; wetting agent and dispersant in printing inks, textiles, wool scouring, dust suppression, household and industrial surface cleaning; emulsifier of solvs., greases and oils; biodeg. [ICI Australia]

Teric DD5. CAS 9014-92-0; Dodoxynol-5; emulsifier, coemulsifier in solv. emulsion cleaners; formulation of concs. for metal lubricants, cutting, milling and grinding aids, textile processing aid and mineral drilling lubricants. [ICI Australia]

Teric G9A5. CAS 68439-46-3; C9-11 pareth-5; emulsifier for solvs.; formulation of hard surface cleaners, degreasers and dispersants in industrial and domestic applications [ICI Australia]

Teric G12A4. CAS 68131-39-5; C12-15 pareth-4 (4.5 EO); wetting agent and dispersant, detergent; dust suppression; aq. pigment dispersion; coemulsifier for w/o-type emulsion concs.; liq. and powd. detergents for industrial and domestic laundry, dishwashing, metal cleaning, sanitizing, textile processing aids, abrasive cleaners; intermediate in manufacturing of sulfate and phosphate surfactants. [ICI Australia]

Teric LA4. CAS 68131-39-5; C12-15 pareth-4 (4.5 EO); wetting agent and dispersant, detergent; dust suppression; aq. pigment dispersion; coemulsifier for w/o-type emulsion concs.; liq. and powd. detergents for industrial and domestic laundry, dishwashing, metal cleaning, sanitizing, textile processing aids, abrasive cleaners; intermediate in manufacturing of sulfate and phosphate surfactants. [ICI Australia]

Teric N2. CAS 9016-45-9; Nonoxynol-2; defoamer, wetting agent and dispersant in solv. and oil-based systems; coemulsifier for o/w emulsions; emulsifier for w/o emulsions; used in agricultural toxicant products, industrial solv.-cleaning systems; surface coating preparations; intermediate for production of anionics. [ICI Australia]

Teric OF4. CAS 9004-96-0; PEG-4 oleate; emulsifier used in o/w emulsions of vegetable and mineral oils, fats,

---

waxes and solvs.; formulation of cutting oils; metal lubricant in metal cleaning formulations and in solv. degreasers; textile and paper finishing applications; antistat in textile processing and manufacturing of synthetic fibers. [ICI Australia]

Teric PE61. POP + 5.7 EO; demulsifier, intermediate, emulsifier, detergent, wetting agent, and dispersant; pigments in latex paints; detergent sanitizer and alkaline cleaner; dewatering aid for treatment of crude oil emulsions; defoamer in paper pulp liquors, starch sizing, glues, and boiler water systems. [ICI Australia]

Teric PEG 200. CAS 25322-68-3; PEG 200; biodeg.; binder for glazing; intermediate for PEG esters, methacrylate resins, PU foams; plasticizer/solv. for cork; toiletries; metalworking lubricants; paints/resins; paper/film; printing inks; textile emulsifier. [ICI Australia]

Teric PEG 300. CAS 25322-68-3; PEG 300; pesticide solubilizer/carrier; visc. modifier for brake fluids; intermediate for PEG esters, PU foams; plasticizer/solv.; cosmetics; metalworking lubricants; paints/resins; paper/film; pharmaceuticals; printing inks; rubber; and textile auxiliary [ICI Australia]

Teric PEG 400. CAS 25322-68-3; PEG 400; biodeg.; emulsifier, antistat for textiles; rubber; inks; cosmetics; pharmaceuticals; paper/film; pesticide solubilizer/carrier; intermediate for PEG esters, PU foams; plasticizer/solv. for cork; metalworking lubricants; paints/resins. [ICI Australia]

Teric PEG 600. CAS 25322-68-3; PEG 600; biodeg.; emulsifier for textiles; pesticide solubilizer/carrier; binder for ceramics; intermediate for PEG esters, PU foams; cosmetics; pharmaceuticals; metalworking lubricants; resins; paper/film; inks; rubber. [ICI Australia]

Teric PEG 800. CAS 25322-68-3; PEG 800; biodeg.; textile auxiliary; pharmaceutical tableting; pesticide solubilizer/carrier; intermediate for PEG esters; toiletries; metalworking lubricants. [ICI Australia]

Teric PEG 1000. CAS 25322-68-3; PEG 1000; biodeg.; intermediate for PEG esters; cosmetics/toiletries/soaps; metalworking lubricants; wood processing. [ICI Australia]

Teric PEG 1500. PEG 1500; biodeg.; textile finishing and sizing; wood processing; latex production; printing inks; pharmaceuticals; paper/film. [ICI Australia]

Teric PEG 3350. PEG 3350; biodeg.; intermediates for PEG esters; spinning aid for textiles; pharmaceuticals. [ICI Australia]

Teric PEG 4000. CAS 25322-68-3; PEG 4000; biodeg.; binder/plasticizer for ceramics; intermediate for copolymers, PEG esters; cosmetics; pharmaceuticals; metalworking lubricants and electropolishes; resins; paper/film; printing inks; rubber antistat, release, compounding aid; textile auxiliary [ICI Australia]

Teric PEG 6000. CAS 25322-68-3; PEG 6000; biodeg.; cosmetics; pharmaceuticals; thickener for inks; wood processing; binder/plasticizer for ceramics; intermediate for copolymers, PEG esters; metalworking lubricants; resins. [ICI Australia]

Teric PEG 8000. CAS 25322-68-3; PEG 8000; biodeg.; intermediate for copolymers, PEG esters; cosmetics; pharmaceuticals; thickener for inks; release for rubber molding. [ICI Australia]

Teric PEG 12000. CAS 25322-68-3; PEG 12000; biodeg. [ICI Australia]

Teric PPG 400. PPG 400; biodeg.; intermediate for surfactants, ethers, esters; metalworking lubricants; solv. for paints/varnishes, printing inks, vegetable oils; textile lubricants. [ICI Australia]

Teric PPG 1000. PPG 1000; biodeg.; intermediate for surfactants, ethers, esters; antifoam for ceramics, rubber; hydraulic brake fluids; dyeing; metalworking lubricants; plasticizer for plastics, latex coagulant; textile lubricant. [ICI Australia]

Teric PPG 1650. PPG 1650; biodeg.; intermediate for surfactants, ethers, esters; antifoam for ceramics, rubber; lubricant/softener for leather; metalworking lubricant; demulsifier for petrol. industry; plasticizer for plastics. [ICI Australia]

Teric PPG 2250. PPG 2250; biodeg.; intermediate for surfactants, ethers, esters; hydraulic brake fluids; cosmetics; dyeing; lubricant/softener for leather; metalworking lubricant; demulsifier for petrol.; latex coagulant; rubber release agent; solv. for vegetable oils. [ICI Australia]

Teric PPG 4000. PPG 4000; biodeg.; intermediate for surfactants, ethers, esters; cosmetics; lubricant/softener for leather; metalworking lubricants; demulsifier for petrol. industry; mold release for rubber. [ICI Australia]

Teric X5. CAS 9002-93-1; Octoxynol-5; wetting agent and dispersant, detergent used in alkaline and metal cleaners, agricultural powds., emulsions, pigment and wax dispersions; textile auxiliaries; paints. [ICI Australia]

Terinda. Polyester yarns. [ICI Chem. & Polymers Ltd]

Terlac. Rosin derivatives; for paints and inks. [Resinas Sinteticas SA] *

Terlan. Polyamide resins; for adhesive/molding applications. [The Terrell Corporation] *

Terluran®. CAS 9003-56-9; ABS polymers; injection molding, easy-flow injection molding, heat-resist., high impact, extrusion, transparent, flame-retardant, reduced gloss, glass-reinforced, and electroplating grades avail. [BASF AG; BASF plc]

Terluran. ABS color compounds. [Norsk Hydro Polymers Ltd] ‡

Terluran 846 L. A proprietary ABS of medium rigidity and toughness used for injection molding, extrusion and thermoforming. [BASF plc] *

Terluran 886. A tough grade of ABS. [BASF plc] *

Terluran 8760 Galvano. A special grade of ABS used for electroplating. [BASF plc] *

Terlux®. CAS 9003-56-9; clear ABS [BASF]

Term-X (Various Herbicides/Insecticides). Do-it-yourself products for household and commercial use and application. [Puma Chemical Co Inc] *

Termamyl®. CAS 9000-92-4; A liquid enzyme preparation containing an outstandingly heat-stable α-amylase expressed in and produced by a selected strain of *Bacillus licheniformis*; used in the starch, alcohol, brewing, sugar, and textile industries. [Novo Nordisk]

Termamyl® 120L. CAS 9000-92-4; Bacterial α-amylase; enzyme for liquefaction of gelatinized starch in production of dextrose, high fructose, and other syrups; used in alcohol, brewing, and textile industries. [Novo Nordisk]*

Termex. Laminates of nomex, presspaper, leatheroid, paper, melinex, mylar and kapton; slot liner and closure material used in the manufacture and repair of electrical motors, transformers and other electrical equipment. [Fothergill Tygaflor Ltd] *

---

‡ Trade name and manufacturer not verified   § Trade name without identified manufacturer

Termex. Plastics used in soles and heels of footwear, in manufacture of heel and toe stiffening components of footwear. [Courtaulds Advanced Materials (Holdings) Ltd]

Terminate. *Bacillus thuringiensis* wettable powder; applied by spray to control larvae of *Lepidopteran* insects. [Westbridge Research Group] ‡

ternary steels. Alloy steels containing one special element in addition to the iron and carbon.

Terne metal. An alloy of 80% lead, 18% tin, and 2% antimony.

Terne plate. An alloy of lead and tin, coated on iron plate, and intended for use in roofing.

Terohane. Polyester film. [Foseco (F.S.) Ltd]

Terolut. CAS 152-62-5; Dydrogesterone; progestative, counteracts complaints caused by hormonal disorders in women. [Duphar BV] *

Terpal®. Soluble concentrate containing 155g 2-chlorethylphosphonic acid and 305g mepiquat chloride per liter; plant growth regulator for use in cereals. [BASF plc; BASF AG; Clifton Chemicals Ltd]

Terpal® CC, M. Chlormequat chloride, cholinechloride, ethephon; bioregulator for improved resistance against lodging in barley, rye, wheat, and flax. [BASF AG]

Terpalin. A proprietary preparation of eucalyptol, menthol terpine hydrate, and codeine phosphate; a cough syrup. [H N Norton & Co Ltd] ‡

Terpanol. Dipentene substitute. [Crowley Chem.]

Terpenato. Rosin derivatives; for paints and inks. [Resinas Sinteticas SA] *

terpene resin. *See also* dipentene, pinene. Thermoplastic resin for use in hot-melt adhesives and coatings, as masticatory agents in chewing gum. [Arizona; Cardolite; Hercules; Langely Smith Ltd]

terpestrol. A powder containing lactose with 5% oil of turpentine.

Terpex D,K-3,S. A proprietary trade name for terpene vinyl plasticizers. [Glidden Co] *

Terphane®. Polyester film; high quality film for demanding applications incl. food pkg., magnetic tape, elec. cable insulation, graphics arts. [Rhone-Poulenc/Film Div.]

Terpigol. A proprietary trade name for terpinyl monoethylene glycol ether. [Nipa Laboratories Ltd] *

terpine. A turpentine substitute that is a product of the distillation of petroleum.

α-terpineol. $C_{10}H_{17}OH$; CAS 98-55-5; Solvent for hydrocarbon materials, mutual solvent for resins, cellulose esters and ethers, perfumes, soaps, disinfectant, antioxidant, and flavoring agent. [Aldrich; Hercules; Quest Int'l.; SCM Glidco Organics]

terpinol. CAS 2451-01-6; A mixture of terpenes containing terpinene, dipentene, terpineol, and cineol.

Terpinoxo. Oxygen forming compound for rubber. [Crowley Chem.]

Terpoin. A proprietary preparation of eucalyptol, terpin hydrate, codeine phosphate, menthol and guaifenesin; a cough linctus. [Hough, Hoseason & Co Ltd] *

Terposol No. 3. A proprietary trade name for a solvent consisting of terpene methyl ethers. §

Terposol No. 8. A proprietary trade name for a solvent consisting of terpene glycol ethers. §

Terpurile. Wetting-out agents consisting of soaps with organic solvents. §

terra alba. *See* gypsum.

Terra-Bron. A proprietary preparation of oxytetracycline, ipecacuanha and ephedrine hydrochloride. [Pfizer International] †

terra catechu. *See* cutch.

Terra-Cortril. A proprietary preparation of oxytetracycline hydrochloride, hydrocortisone; an anti-infective and anti-inflammatory. [Pfizer International]

Terracote. Coatings for dies, molds, chills etc. [Foseco (F.S.) Ltd]

terra-cotta. A building material made from clay.

Terracur® P. Granular insecticide; used for treatment of biting insects and nematodes. [Bayer AG]

Terradust. Brass foundry 'caster's flour'. [Foseco (F.S.) Ltd]†

Terrafen®. Special fertilizer incorporated into a gel which operates on the ion exchange principle; for long-term fertilization of all types of houseplants. [Bayer AG]

terra fullonica. Fuller's earth.

Terragloss. Ultraviolet curing coatings for printed paper and board. [Swale Coatings & Inks Ltd]

terra Japonica. *See* cutch.

Terraklene. Herbicide. [ICI Chem. & Polymers Ltd]

Terralacke. Water-based coatings for printed packaging. [Swale Coatings & Inks Ltd]

Terram. Nonwoven civil engineering fabric. [ICI Chem. & Polymers Ltd] †

terra Merita. *See* turmeric.

Terramix CDA. An emulsifiable concentrate containing 99 g aminotriazole, 190 g atrazine and 95 g 2,4-D per liter; used for total weed control in non crop areas. [Denoon CDS; Powaspray (CDA) Ltd]

Terramycin. CAS 79-57-2; A proprietary preparation of oxytetracycline; an antibiotic. [Pfizer International]

Terramycin Hydrochloride. CAS 2058-46-0; Oxytetracycline hydrochloride; antibacterial; antirickettsial. [Pfizer Inc] *

Terramycin Pediatric Drops. Oxytetracycline calcium; antibacterial. [Pfizer Inc]

Terraneb SP Turf Fungicide. CAS 2675-77-6; 65% Chloroneb wettable powder; for control of snow mold (*Typhula*) and pythium blight. [Kincaid Enterprises Inc]*

Terra Nova. Lead/cadmium free on glaze colors. [Degussa Ltd]

Terranox. An emulsifiable concentrate containing 105g aminotriazole, 207g atrazine and 100g 2,4-D per liter; used for total weed control in non crop areas. [Agri-Technics Ltd]

Terrapaint. Coatings for foundry molds and cores. [Foseco (F.S.) Ltd]

terra ponderosa. CAS 7727-43-7; Barium sulfate, $BaSO_4$.

Terrapowder. Ferrous mold and core dressings. [Foseco (F.S.) Ltd] †

terrar. A preparation from earthy zirconia, in Brazil; used as an opacifying agent in enamels and glazes.

Terra-Systam. Systemic organo-phosphorus insecticides. [Murphy Chemical Co Ltd] †

Terrathion. CAS 298-02-2; Granules containing 10% w/w phorate; an organophosphorus insecticide. [Farmers Crop Chemicals Ltd]

terra verte. A pigment. It is a green earthy material found in the Mendip Hills. It consists of a species of ocher, and is essentially silica with oxide of iron and small quantities of other oxides.

Terravest 801. A stereospecific, low molecular weight

---

* Trade name not verified as available   † Trade name verified as obsolete

polybutadiene; consolidates soil against erosion and binds dust, e.g., on heaps of all sorts. [Hüls AG]

Terr-O-Gas. Methylbromide with chloropicrin. [Great Lakes Europe]

terroline. See paraffin, liquid.

Terry's stain. A microscopic stain. It contains 20 cc of a 1% aqueous solution of methylene blue, 20 cc of a 1% solution of potassium carbonate, and 60 cc of water. This is boiled, cooled, and 10 cc of a 10% solution of acetic acid added, and the whole made up with water to 100 cc.

Terset. Bromoxynil + ioxynil + isoproturon + mecoprop; a contact herbicide for use in cereal crops. [Rhone-Poulenc Crop Protection Ltd]

Tertroxin. CAS 55-06-1; A proprietary preparation of liothyronine sodium; thyroid hormone preparation. [Glaxo Laboratories]

Terylene. Polyethylene terephthalate produced from dimethyl terephthalate and ethylene glycol; synthetic polyester textile fiber, resistant to most dry-cleaning solvents, possessing good wear resistance. [ICI Chem. & Polymers Ltd]

Tesal. Propylene glycol stearate SE; self-emulsifying base for cosmetic/pharmaceutical ointments, lotions, creams. [Gattefosse]

Tescol®. Vinyl or acrylic copolymers; continuous filament polyester sizes for textiles [Allied Colloids Ltd]

Teslac. CAS 968-93-4; Testolactone; antineoplastic. [Bristol-Myers Squibb Co Inc] *

Tessalon. CAS 104-31-4; Benzonatate; antitussive. [Endo Laboratories Inc] *

testalin. An aluminum soap, made by treating ordinary soap with aluminum sulfate; used for the cementing together of sandstone to form a solid block.

Testate. CAS 315-37-7; Testosterone enanthate; an androgen. [Savage Laboratories] †

testifas oil. A fraction of petroleum distillation; used as a burning oil.

Testred. CAS 58-18-4; Methyltestosterone; an androgen. [ICN Pharmaceuticals Inc] *

Tesuloid. Technetium Tc 99m sulfur colloid; a radioactive agent. [Bristol-Myers Squibb Co Inc] *

TETA. See triethylenetetramine.

Tet-Aktiv. Tetanus vaccine. [Bayer AG]

Tetanol. A proprietary preparation of calcium hevulinate. §

tetjamer. An aluminum bronze. It contains from 86-93% copper, 5-10% aluminum, 1-3% silicon, and 0.72-0.98% iron.

Tetmosol. CAS 95-05-6; A proprietary preparation of monosulfiram; used as an insecticide in soap or solution form, for either human or veterinary use. [ICI Chem. & Polymers Ltd]

Tetrabor®. CAS 12069-32-8; Boron carbide; abrasives, lapping media, nozzles, building materials, drills, saws, bearings, shafts, files, rasps, and whetstones, all made wholly or principally of boron carbide. [Elektroschmelzwerk Kempten GmbH]

tetrabutyl ammonium bromide. (TBAB; tetra-N-butylammonium bromide) Quaternary ammonium salt; $[CH_3(CH_2)_3N[CH_3(CH_2)_3]_3]^+Br^-$; CAS 1643-19-2; EINECS 216-699-2. [Aldrich; Hawks Chem Co Ltd; Schweizerhall; Zeeland]

Tetracarnit. A mixture of pyridine and its homologues with Turkey red oil or similar substances; used as a wetting-out agent to assist the penetration of textiles by liquids.

Tetrachel. CAS 64-75-5; A proprietary preparation of tetracycline hydrochloride; an antibiotic. [Berk Pharmaceuticals Ltd] *

tetrachloroethene. See perchloroethylene.

tetrachloroethylene. See perchloroethylene.

tetrachloromethane. See carbon tetrachloride.

tetrachlorosilane. See silicon tetrachloride.

Tetracyn P. A proprietary preparation containing tetracycline hydrochloride, sodium hexametaphosphate; an antibiotic. [Pfizer International]

Tetracyn S.F. A proprietary preparation of tetracycline with vitamin supplements; an antibiotic. [Pfizer International]

tetradecanoic acid. See myristic acid.

tetradecanoic acid, methyl ester. See methyl myristate.

tetradecanoic acid, 1-methylethyl ester. See isopropyl myristate.

1-tetradecanol. See myristyl alcohol.

tetradecyl alcohol. See myristyl alcohol.

tetradecyl 2-hydroxypropanoate. See myristyl lactate.

tetradine. See lycetol.

tetraethylenepentamine. (TEPA; 1,4,7,10,13-pentaazatridecane) $NH_2(CH_2CH_2NH)_3CH_2CH_2NH_2$; CAS 112-57-2; EINECS 203-986-2; Solvent for sulfur, acid gases, various resins and dyes; saponifying agent for acidic materials; manufacture of synthetic rubber; dispersant in motor oils; intermediate for oil additives. [Tosoh; Union Carbide]

tetraethyl orthosilicate. See ethyl silicate.

Tetraflon. A proprietary polytetrafluoroethylene (PTFE). [Nitto Chemical Industry Co Ltd] *

tetrafluoroethene homopolymer. See polytetrafluoroethylene.

tetrafluoroethylene polymer. See polytetrafluoroethylene.

tetrafluorosilane. See silicon tetrafluoride.

tetraform. A specially pure carbon tetrachloride.

tetrahydrofuran. (THF) $CH_2CH_2CH_2CH_2O$; CAS 109-99-9; EINECS 203-726-8; Solvent, Grignard reactions, reductions, and polymerizations; chemical intermediate and monomer. [Arco Europe; Ashland; BASF; Du Pont; Great Lakes; Hüls UK; Janssen Chimica; QO]

tetrahydrofurfuryl alcohol. (THFA; tetrahydrofuryl carbinol; tetrahydro-2-furanmethanol; tetrahydro-2-furancarbinol) Cyclic alcohol; $C_5H_{10}O_2$; $C_4H_7OCH_2OH$; CAS 97-99-4; EINECS 202-625-6; solvent for vinyl resins, dyes for leather, chlorinated rubber, cellulose esters, coupling agent, solvent-softener for nylon. [Penta Mfg.; QO; Schweizerhall]

tetrahydrofurfuryl methacrylate. CAS 2455-24-5; Anaerobic adhesives and sealants, printed circuit boards, artificial finger nails, modifier for hard rubber rolls, wire and cable coatings, screen printing inks, emulsion polymerization, plastic modifier, EB-curable coatings. [CPS; Sartomer]

tetrahydrofuryl carbinol. See tetrahydrofurfuryl alcohol.

tetrahydronaphthalene. (1,2,3,4-tetrahydronaphthalene; tetralin) $C_{10}H_{12}$; CAS 119-64-2; EINECS 204-340-2; Chemical intermediate, solvent for greases, fats, oils, waxes; substitute for turpentine. [Du Pont; Hüls AG]

1,2,3,4-tetrahydronaphthalene. See tetrahydronaphthalene.

tetrahydro-1,4-oxazine. See morpholine.

tetrahydro-2H-1,4-oxazine. See morpholine.

tetrahydrothiophene. (thiophane; tetramethylene sulfide; thiolane) C_4H_8S; CAS 110-01-0; EINECS 203-728-9;

---

‡ Trade name and manufacturer not verified   § Trade name without identified manufacturer

Solvent, intermediate, fuel gas odorant. [Atochem N. Am.]

2,2´,4,4´-tetrahydroxy benzophenone. *See* benzophenone-2.

Tetrakal. CAS 7320-34-5; EINECS 230-785-7; Tetrapotassium pyrophosphate. [Albright & Wilson Ltd]

Tetralex Plus. Soluble concentrate of 18 g dicamba, 252 g MCPA and 84 g mecoprop per liter; used for weed control in cereals and grassland. [Shell UK]

Tetralide® . 7-Acetyl 1,1,3,4,4,6-hexamethyltetraline; musky odor used in fragrances. [Bush Boake Allen Ltd]

tetralin. *See* tetrahydronaphthalene.

tetraline. An old name for tetrachloroethane.

Tetralin Extra. A mixture of tetralin (*qv*) and dekalin (*qv*).

tetralitbenzol. A mixed fuel for internal combustion engines. It contains 50% benzol, 25% tetralin, and 25% of 95% alcohol.

Tetralite. *See* Tetryl. §

Tetralol. CAS 530-91-6; Tetrahydro-β-naphthol; used as an antiseptic.

Tetralon®. A range of sequestering agents based on NTA, EDTA, and DTPA; mainly used in soaps and detergents. [Allied Colloids Ltd]

Tetramet-125. Thyroxine I 125; radioactive agent. [Abbott Laboratories] *

tetramethylammonium hydroxide. $(CH_3)_4N(OH)$; CAS 75-59-2; EINECS 200882-9. [Aldrich; Fluka; Janssen Chimica]

tetramethylene dibromide. *See* 1,4-dibromobutane.

tetramethylene glycol. *See* 1,4-butanediol.

tetramethylene sulfide. *See* tetrahydrothiophene.

tetramethylolmethane. *See* pentaerythritol.

Tetranol. Sodium butyl oleate; wetting and leveling agent for dyeing processes; assistant for vat dyeing; emulsifier and detergent in scouring processes. [Sandoz]

Tetranyl. 2, 3, 4, 6-tetranitro-aniline; an explosive.

tetra-paper. (tetra-base-paper) Paper which has been treated with dimethyl or tetramethyl-p-phenylene-di-amine; used in testing for ozone.

tetraphosphate. (tetra). Produced by mixing natural phosphate rock powder with 6% of a powder containing equal parts of the carbonates of calcium, sodium, and magnesium, with a little sulfate of soda.

tetrapotassium pyrophosphate. *See* potassium pyrophosphate.

tetrapropylene. *See* dodecene-1.

tetrasodium EDTA. (EDTA Na₄; edetate sodium; tetrasodium edetate; ethylene diamine tetraacetic acid, sodium salt) Substituted amine; $(NaOOCCH_2)_2N-CH_2CH_2N(CH_2COONa)_2$; CAS 64-02-8; EINECS 200-573-9; general-purpose chelating agent. [W.R. Grace/ Hampshire; Rhone-Poulenc Basic]

tetrasodium edetate. *See* tetrasodium EDTA.

tetrasodium pyrophosphate. *See* sodium pyrophosphate.

Tetraterge D-101, NFF. Detergent for textile scouring. [Reilly-Whiteman]

Tetrathal. Tetrachlorophthalic anhydride; flame retardant for polyester resins and polyols. [Monsanto Co]

Tetrawet DWN. Wetting agent for textile processing of cellulosics and blends. [Reilly-Whiteman]

Tetrazets. Bacitracin, tyrotricin, neomycin, and benzocaine; for the relief of minor mouth and throat infections. [Merck & Co Inc] †

tetrazolium chloride. (TTC; tetrazolium salt; 2,4,5-triphenyltetrazolium chloride) $CN_4Cl(C_6H_5)_3$; In germination and viability testing of seeds. [Dajac Labs; U.S. Biochemical]

tetrazolium salt. *See* tetrazolium chloride.

Tetrex. CAS 1336-20-5; Tetracycline phosphate complex; an antibacterial. [Bristol Laboratories] *

Tetrex Bidcaps. CAS 1336-20-5; A proprietary preparation containing tetracycline phosphate complex; an antibiotic. [Bristol-Myers Squibb Co Inc]

Tetron. CAS 7722-88-5; Tetrasodium pyrophosphate. [Albright & Wilson Ltd, Phosphates & Speciality Business]

Tetrone. Rubber accelerator. [DuPont UK]

Tetronic® 50R1. EO/PO ethylene diamine block copolymer; surfactant series functioning as emulsion stabilizers, solubilizers, dispersants, wetting agents, antistats, penetrants, plasticizers, defoaming agents, demulsifiers in the petrol., paint, paper, cement, ink, cosmetic, drug, plastic, detergent, and metalworking industries; rubber activator; R series for low foaming applications. [BASF]

Tetronic® 150R1. CAS 107397-59-1; EO/PO ethylene diamine block copolymer; *see* Tetronic 50R1. [BASF]

Tetronic® 304. CAS 11111-34-5; Poloxamine 304; emulsifier, thickener, wetting agent, dispersant, solubilizer, stabilizer for cosmetics and pharmaceuticals; demulsifier in petrol. industry; detergent ingredient; antistat for polyethylene and resin molding powds.; metal treatment; emulsion polymerization; used in latex-based paints, aq.-based synthetic cutting fluids and vulcanization of rubber. [BASF]

Tetronic® 504. CAS 11111-34-5; Poloxamine 504; *see* Tetronic 304. [BASF]

Tetronic® 701. CAS 11111-34-5; Poloxamine 701; *see* Tetronic 304. [BASF]

Tetronic® 702. CAS 11111-34-5; Poloxamine 702; *see* Tetronic 304. [BASF]

Tetronic® 704. CAS 11111-34-5; Poloxamine 704; *see* Tetronic 304. [BASF]

Tetronic® 707. CAS 11111-34-5; Poloxamine 707; *see* Tetronic 304. [BASF]

Tetronic® 901. CAS 11111-34-5; Poloxamine 901; *see* Tetronic 304. [BASF]

Tetronic® 904. CAS 11111-34-5; Poloxamine 904; *see* Tetronic 304. [BASF]

Tetronic® 908. CAS 11111-34-5; Poloxamine 908; *see* Tetronic 304. [BASF]

Tetronic® 1101. CAS 11111-34-5; Poloxamine 1101; *see* Tetronic 304. [BASF]

Tetronic® 1102. CAS 11111-34-5; Poloxamine 1102; *see* Tetronic 304. [BASF]

Tetronic® 1104. CAS 11111-34-5; Poloxamine 1104; *see* Tetronic 304. [BASF]

Tetronic® 1107. CAS 11111-34-5; Poloxamine 1107; *see* Tetronic 304. [BASF]

Tetronic® 1301. CAS 11111-34-5; Poloxamine 1301; *see* Tetronic 304. [BASF]

Tetronic® 1302. CAS 11111-34-5; Poloxamine 1302; *see* Tetronic 304. [BASF]

Tetronic® 1304. CAS 11111-34-5; Poloxamine 1304; *see* Tetronic 304. [BASF]

Tetronic® 1307. CAS 11111-34-5; Poloxamine 1307; *see* Tetronic 304. [BASF]

Tetronic® 1501. CAS 11111-34-5; Poloxamine 1501; *see* Tetronic 304. [BASF]

Tetronic® 1502. CAS 11111-34-5; Poloxamine 1502; *see* Tetronic 304. [BASF]

Tetronic® 1504. CAS 11111-34-5; Poloxamine 1504; *see* Tetronic 304. [BASF]

Tetronic® 1508. CAS 11111-34-5; Poloxamine 1508; *see* Tetronic 304. [BASF]

Tetrosan 3,4 D. Alkyl dimethyl 3:4 dichlorobenzyl ammonium chloride in liquid form; disinfectant, deodorant, and germicide with high biocidal activity; veterinary, pharmaceutical and agricultural uses. [Millmaster-Onyx UK] *

Tetroxone. Selective weedkiller containing bromoxynil, dichlorprop, ioxynil and MCPA as potassium salts. [ICI Chem. & Polymers Ltd]

Tetryl. A trade name for trinitrophenylmethylnitramine, $C_6H_2(NO_2)_3 \cdot NCH_3.NO_2$; used in explosives; adetonator known as tetryl contains 0.4 gram tetranitrophenylmethylnitramine, and 0.3 gram of a mixture of 87.5% mercury fulminate and 12.5% potassium chlorate. §

Texacar® EC. CAS 96-49-1; Ethylene carbonate; solv. for organic and inorganic materials; Rule 66 exempt; also used as reactant and plasticizer in fibers and textiles, plastics and resins, aromatic hydrocarbon extraction, electrolytes, hydraulic brake fluids. [Texaco]

Texacar® PC. CAS 108-32-7; Propylene carbonate; solv. for organic and inorganic materials; Rule 66 exempt; also used as reactant and plasticizer in fibers and textiles, hydraulic fluids, plastics and resins, gas treating, aromatic hydrocarbon extraction, metal extraction, surface coatings, foundry sand binders, lubricants, electrolytes, personal care products; gellant for clays in greases and cosmetics. [Texaco]

Texacat® DD. CAS 34745-96-5; Tertiary amine; catalyst for semiflexible foam urethanes. [Texaco]

Texacat® DMDEE. 2,2′-Dimorpholinodiethylether; catalyst for polyurethane flexible and rigid foam, one-component foam, slabstock foam applications. [Texaco]

Texacat® DME. CAS 108-01-0; N,N-Dimethylethanolamine; catalyst for polyurethane rigid foam, flexible ether slabstock. [Texaco]

Texacat® DMP. N,N′-Dimethylpiperazine; catalyst for polyurethane molded high resilience flexible foam, ester slabstock. [Texaco]

Texacat® DPA. N,N-(Dimethyl)-N′,N′-diisopropanol-1,3-propanediamine; catalyst for rigid PU foam, pkg. foam. [Texaco]

Texacat® NEM. CAS 100-74-3; N-Ethylmorpholine; catalyst for flexible and rigid PU foam, polyester slabstock. [Texaco]

Texacat® NMM. CAS 109-02-4; N-Methylmorpholine; catalyst for polyester urethane flexible foam, high rise rigid foam panels. [Texaco]

Texacat® TD-33. CAS 280-57-9; Triethylenediamine in propylene glycol; general purpose catalyst for producing polyurethane foam systems (flexible and rigid). [Texaco]

Texacat® ZF-10. N,N,N′-Trimethyl-N′-hydroxyethyl-bisaminoethylether; catalyst for polyurethane molded high resilience flexible foam, ether slabstock, pkg. foam. [Texaco]

Texacat® ZF-22. Bis-(2-dimethylaminoethyl)ether, dipropylene glycol; general purpose catalyst for flexible and rigid PU foam, polyether slabstock. [Texaco]

Texacat® ZR-50. N,N-Bis-(3-dimethylaminopropyl)-N-isopropanolamine; catalyst for flexible and rigid PU foam, pkg. foam. [Texaco]

Texacat® ZR-70. CAS 1704-62-7; 2-(2-Dimethyl-aminoethoxy)ethanol; catalyst for polyurethane molded high resilience flexible foam, ether slabstock, pkg. foam. [Texaco]

Texaco BQ. An insecticide having a petroleum base; used for killing the boll weevil. §

Texacure EA-24. CAS 112-24-3; High boiling mixture of triethylenetetramine components; epoxy curing agent featuring rapid room temperature cure, good adhesive and mech. properties, good elec. and insulating properties, good chem. resist. [Texaco]

Texadril 2010. EO/PO block copolymer; low foaming wetting agent and coemulsifier. [Henkel/Cospha; Henkel/Functional Prods.]

Texalon 600A NU. CAS 25038-54-4; Nylon 6 resin, nucleated; used for very small parts, rigid applications. [Texapol]

Texalon 1000A. CAS 25038-54-4; Nylon 6 resin; high impact, tough grade. [Texapol]

Texalon 1200A BK-11. CAS 32131-17-2; Nylon 6/6 resin with 2% carbon black; weather-resistant resin for use in wire tires and outdoor parts; uv resistant. [Texapol]

Texalon 1200A HR-2 BK-16. CAS 32131-17-2; Nylon 6/6 resin; hydrolysis-resist. grade. [Texapol]

Texalon 1308 A. CAS 32131-17-2; Nylon 6/6 resin, impact-modified; used for automotive applications, mechanical housings. [Texapol]

Texalon 1600A Nat. CAS 9008-66-6; Nylon 6/10 resin; low moisture. [Texapol]

Texalon GF 600A (6-33). CAS 25038-54-4; Nylon 6 resin, glass-filled; general purpose glass-filled resin used for automotive grade mechanical parts. [Texapol]

Texalon GF 1200A (13-40). CAS 32131-17-2; Nylon 6/6 resin, glass-filled; general purpose resin used for metal replacement in automotive applications. [Texapol]

Texalys. Modified starch; used in textile industry in precoating compounds and also in laminating double backs. [Roquette (UK) Ltd]

Texamine 84(L). Alkanolamide and ethanolamine alkylbenzene sulfonate; raw material for manufacturing of liq. detergents. [Zohar Detergent Factory]

Texanol® Ester-Alcohol. CAS 25265-77-4; 2,2,4-Trimethyl-1,3-pentanediol monoisobutyrate; slow-evaporating solv. used as coalescing agent in latex finishes and water-base inks, PVAc latexes, PVAc-acrylic latices, and acrylic, EVA, and B/S latexes; used as solv. in electrodeposition coatings, lacquer coatings; defoamer in waterborne systems and drilling muds for oil industry; chemical intermediate for production of esters; used as stain-resistance plasticizers, lubricant base stocks, solvs., synthetic detergents, and herbicides. [Eastman]

Texaphor 277. Quaternary ammonium salt; suspending agent for paints with high chemical and water resistance requirements. [Henkel Canada]

Texapon ALS. CAS 2235-54-3; EINECS 218-793-9; Ammonium lauryl sulfate; base for shampoos, shower baths, and bubble baths. [Henkel KGaA]

Texapon ASV. Sodium laureth sulfate, magnesium laureth sulfate, sodium laureth-8 sulfate, sodium oleth sulfate, magnesium oleth sulfate; extremely mild surfactant base for baby shampoos, facial cleansers, and foam baths. [Henkel/Cospha; Henkel Canada; Henkel KGaA]

Texapon CS Paste. Sodium lauryl sulfate, sodium myristyl sulfate, sodium cetyl sulfate, sodium stearyl sulfate, and

---

‡ Trade name and manufacturer not verified   § Trade name without identified manufacturer

laureth-10; concentrate for creamy shampoos, foam baths; has pearlescent effects. [Henkel/Cospha; Henkel Canada; Henkel KGaA]

Texapon DEA. DEA-lauryl sulfate; detergent, foamer for personal care products. [Henkel/Cospha; Henkel Canada]

Texapon EA-1. CAS 67762-19-0; Ammonium laureth sulfate; surfactant, visc. builder, solubilizer in personal care products. [Henkel/Cospha; Henkel Canada]

Texapon ES-1. Sodium laureth sulfate; surfactant and solubilizer for personal care products. [Henkel/Cospha; Henkel Canada]

Texapon EVR. Sodium lauryl sulfate, sodium laureth sulfate, lauramide MIPA, cocamide MEA, glycol stearate, and laureth-10; detergent, foamer; base for pearlescent shampoos. [Henkel Canada; Henkel KGaA]

Texapon IES. MIPA-laureth sulfate and cocamide DEA; concentrate for liq. foam bath prep. with high essential oil content. [Henkel KGaA]

Texapon K-12. CAS 151-21-3; Sodium lauryl sulfate; emulsifier for emulsion polymerization; detergent for high-solids cleansers; also for dispersants and wettable powds. [Henkel/Cospha; Henkel/Functional Prods.]

Texapon K-1296. CAS 151-21-3; Sodium lauryl sulfate; wetting agent and detergent for cleaning formulations; additive for mech. latex foaming. [Henkel/Cospha; Henkel/Functional Prods.]

Texapon L20C. Ammonium amine lauryl sulfate in liquid form; basic material for liquid shampoos. [Henkel Chemicals Ltd] *

Texapon L-100. CAS 151-21-3; Sodium lauryl sulfate; wetting and cleansing agent, dispersant used in dentifrice manufacturing, high-solids cleansers, wettable powds. [Henkel/Cospha]

Texapon LS Highly Conc. CAS 151-21-3; Sodium lauryl sulfate (C12-14); foaming agent for acrylate dispersions, carpet and upholstery cleaners. [Henkel/Functional Prods.]

Texapon MG. CAS 67702-21-4; Magnesium laureth sulfate; mild surfactant for liq. shampoo and bath preparation base, baby shampoos. [Henkel/Cospha; Henkel KGaA]

Texapon MG 3. Magnesium lauryl sulfate and disodium laureth sulfosuccinate; for shampoos, shower baths, bubble baths. [Henkel KGaA]

Texapon N 25. CAS 9004-82-4; Sodium laureth (2) sulfate; base for personal care products; dishwashing agent; fire-fighting foam concs.; solubilizer for perfumes. [Henkel KGaA; Pulcra SA]

Texapon NA. CAS 67762-19-0; Ammonium lauryl ether sulfate; detergent for liq. shampoos and bubble baths. [Henkel KGaA]

Texapon NSE. CAS 9004-82-4; Sodium laureth sulfate; surfactant for emulsion polymerization. [Pulcra SA]

Texapon NSF. Sodium lauryl ether sulfate in liquid form, with extremely low salt contents; basic material for cosmetic preparations such as shampoos, bubble bath preparations, etc. [Henkel Chemicals Ltd] *

Texapon NSO. CAS 9004-82-4; Sodium lauryl ether (2) sulfate; detergent for liq. shampoos, bubble baths, shower gels, manufacturing of household detergents; wetting agent and detergent for textile fibers, especially wool and blends. [Henkel KGaA; Pulcra SA]

Texapon OT Highly Conc. Needles. CAS 151-21-3; Sodium lauryl sulfate C12-C18; detergent, shampoo base;

for bubble bath, soaps; emulsifier for emulsion polymerization, additive for mech. latex foaming, carpet and upholstery cleaners. [Henkel/Functional Prods.; Henkel KGaA]

Texapon PLT-227. CAS 9004-82-4; 1335-72-4; Sodium laureth-2 sulfate; for shampoos and liq. detergents. [Pulcra SA]

Texapon PNA-127. CAS 32612-48-9; Ammonium lauryl ether (1) sulfate; wetting agent for acid or gel-like shampoos, light duty detergents, window cleaners; stable foam; low eye and skin irritation. [Pulcra SA]

Texapon QLV. Sodium laureth sulfate; low salt surfactant for shampoos, cleansing preparations, light duty detergents. [Henkel Canada]

Texapon SB-3. Disodium laurethsulfosuccinate; mild surfactant, foamer for bubble baths, baby shampoos, cleansing preparations. [Henkel Canada; Henkel KGaA]

Texapon SBN. Sodium laureth sulfate and disodium laureth sulfosuccinate; detergent base for personal care products. [Henkel KGaA]

Texapon SG. Sodium laureth sulfate, PEG-8, cocamide MEA, glycol distearate, and glycerin; detergent for emulsion and pearl shampoos. [Henkel KGaA]

Texapon SH 100. CAS 56388-43-3; EINECS 260-143-1; Disodium oleamide PEG-2 sulfosuccinate; nonirritating detergent base for shampoos and skin cleansers. [Henkel/Cospha; Henkel Canada]

Texapon T 42. CAS 139-96-8; EINECS 205-388-7; TEA lauryl sulfate; detergent, emulsifier for hair and rug shampoos, bubble baths, shower gels, fire fighting foams. [Henkel KGaA; Pulcra SA]

Texapon VHC Needles. CAS 151-21-3; Sodium lauryl sulfate; wetting agent, foamer, emulsifier, detergent and cosmetic base for personal care products, scouring agents, pigment dispersions, and emulsion polymerization. [Henkel/Cospha; Henkel/Functional Prods.; Henkel Canada; Henkel KGaA]

Texapon ZHC Needles. CAS 151-21-3; Sodium lauryl sulfate; foaming agent, dispersant, wetting agent for foaming bubble baths, cosmetic cleansing creams and emulsions; air entraining agent. [Henkel/Cospha; Henkel KGaA]

Texapret®. Fillers and stiffeners for textile finishing. [BASF AG; BASF plc]

Texgas. Distributorship services in the field of equipment using liquefied petroleum gas. [Quantum Chemical Corp]

Texi TD. Replacement for sodium hydrosulfite for reduction clearing, stripping dyes, and equip. cleaning in textile industry. [CNC Int'l. L.P.]

Texicote®. Polyesters, vinyl acetate copolymers, or polystyrene emulsions; used for powder coatings, paints, adhesives, fabric stiffeners, and binders. [Scott Bader]

Texicote® 03-001. PVAc homopolymer emulsion stabilized with PVAL; base emulsion for wet bond adhesives; high visc. and wet tack; base for ceramic tile adhesives, for wood and general-purpose adhesives in the building industry, and as a cement additive; can be compounded with PVAL, resin ester emulsions, and plasticizers, e.g., DIBP, to increase wet-tack visc., flexibility, and adhesion. [Scott Bader]

Texicote® 03-019. VA/acrylic copolymer emulsion; general purpose flat and silk finish paints. [Scott Bader]

Texicote® 1000, 1050. Polyester powd. coating; med. and

---

* Trade name not verified as available   † Trade name verified as obsolete

fast reactivity resp., excellent gloss, high flow. [Scott Bader]

Texicryl® . Emulsion polymers and copolymers based on acrylic acid esters; used for binders, coatings, membranes, thickeners, and dispersion agents; paints, textiles, adhesives, and paper coating. [Scott Bader]

Texicryl® 13-002. Acrylic/methacrylic copolymer emulsion; emulsion for surface coating applications, e.g., in self-textured finishes providing good color, stain resistance, and good flexibility, in high-build one-coat finishes, in acrylic wood primers, in paints with exterior durability; added to cement/sand screeds to improve intercoat adhesion, chemical and abrasion resistance; as binder for pigmented leatherboard coatings and fabric backings having wet abrasion resistance and flexibility. [Scott Bader]

Texicryl® 13-011. Carboxylated acrylic copolymer emulsion; aq. emulsion of alkali-sol. acrylic copolymer used in water-based printing inks, in the manufacturing of high gloss and other types of emulsion paint, and as a binder for nonwoven fabrics. [Scott Bader]

Texicryl® 13-030. Styrene acrylic copolymer emulsion; used as quality binder for water-based paints (varying from gloss finish to high PVC low-cost paints), crack fillers, textured coatings, and ceramic tile adhesives. [Scott Bader]

Texicryl® Additive 87-1280. Acrylic polymer, water-borne; transfer characteristics controller for water-based flexographic inks. [Scott Bader]

Texicryl® Ecobinder. Styrene-acrylic copolymer emulsion; media for nonpolluting water-based paints for interior and exterior use. [Scott Bader]

Texicryl® Hyperbinder. Styrene-acrylic copolymer emulsion; for cost-effective high pigment loaded interior and exterior matte paints. [Scott Bader]

Texigel® . Polyacrylate; gel thickeners. [Scott Bader]

Texileather. A proprietary trade name for pyroxylin-coated leather cloth. §

Texin 480-A. CAS 9009-54-5; Thermoplastic polyester PU; for injection molding, extrusion (hose, tubing, profiles, wire and cable, film and sheet), blow molding. [Miles]

Texin 985-A. CAS 9009-54-5; Thermoplastic polyether PU; for injection molding, extrusion (hose, tubing, profiles, wire and cable, film and sheet), blow molding. [Miles]

Texin 3203, 3215, 4203, 4206, 4210, 4215. Thermoplastic PU/PC blends; for injection molding; some are paintable formulations. [Miles]

Texin 5286. CAS 9009-54-5; Thermoplastic polyether polyurethane; medical grade for flexible tubing and film applications. [Miles]

Texipol® . Inverse polymer emulsions; used for paper, textiles, adhesives, paints, thickening agents and dispersants. [Scott Bader]

Texipol® 63-002. Acrylamide copolymer emulsion; thickener for acrylic flocking adhesives and carpet backing compositions; for use at pH 5.5-9.0. [Scott Bader]

Texlin® 300. CAS 112-24-3; Triethylenetetramine; improves yields and reduces processing time for many applications; intermediate in asphalt additives, corrosion inhibitors, epoxy curing agents, surfactants in fabric softener and textile additives; in paper industry, petrol. products. [Texaco]

Texlin® 400. CAS 112-57-2; Tetraethylenepentamine; lube oil additive; intermediate in asphalt additives, corrosion inhibitors, epoxy curing agents, surfactants, and in the paper industry. [Texaco]

Texlin® 500. CAS 68173-73-7; Polyethylenepolyamine; used in fuel and lube oil industry, in asphalt chemistry, corrosion inhibitors, and ore flotation applications. [Texaco]

Texoderm. A cellulose product; an imitation leather.

Texofor. Nonionic surfactant. [ABM Chemicals Ltd] *

Texofor A and B. A proprietary range of higher fatty alcohol-based polyoxy-alkylene condensates used as nonionic surfactants. [Glover (Chemicals) Ltd] ‡

Texofor C. A proprietary range of unsaturated fatty acid-based polyoxy- alkylene condensates used as nonionic surfactants. [Glover (Chemicals) Ltd] ‡

Texofor D. A proprietary range of glyceride oil-based polyoxyalkylene condensates; used as nonionic surfactants. [Glover (Chemicals) Ltd] ‡

Texofor E and ED. A proprietary range of saturated fatty acid-based polyoxy-alkylene condensates; used as nonionic surfactants. [Glover (Chemicals) Ltd] ‡

Texofor FN, FP and FX. Range of nonionic surfactants of the alkylphenol ethoxylate type; used for industrial and household cleaners; emulsion polymerization; agriculture, etc. [Rhone-Poulenc UK]

Texofor G. A proprietary unsaturated fatty acid-based polyoxyalkylene condensate; used as a nonionic surfactant. [Glover (Chemicals) Ltd] ‡

Texofor J4. A linear fatty alcohol ethoxylate; proprietary biodegradable nonionic emulsifier. [Glover (Chemicals) Ltd] ‡

Texofor M. A proprietary range of unsaturated fatty acid-based polyoxy- alkylene condensates; used as nonionic surfactants. [Glover (Chemicals) Ltd] ‡

Texofor N. A proprietary range of fatty alcohol-based polyoxyalkylene condensates; used as nonionic surfactants. [Glover (Chemicals) Ltd] ‡

Texofor P. A proprietary range of complex amide-based polyoxyalkylene condensates; used as nonionic surfactants. [Glover (Chemicals) Ltd] ‡

Texofor T. A proprietary range of higher fatty alcohol-based polyoxy- alkylene condensates; used as nonionic surfactants. [Glover (Chemicals) Ltd] ‡

Texogent. Blend of surfactants and solvents; used for degreasing, cleaning, antispotting in textile and engineering industries. [ABM Chemicals Ltd] *

Texowax. Water soluble waxy coatings; release lubricant; softening; used for paper coating; plastics; textiles; and adhesives. [ABM Chemicals Ltd] *

Texox® PPG-400. PPG-400; intermediate yielding esters; useful as lubricants, defoaming agents in rubber and pharmaceuticals, solvs. and humectant modifiers for inks, plasticizers, and functional fluids. [Texaco]

Texox® WL-440. EO/PO derivs.; functional fluids used as lubricants, plasticizers, solvs., coupling agents, frothing agents, heat transfer fluids, intermediates, defoaming agents; used in solder reflow applications, boiler defoaming, ore flotation, inks, dyes. [Texaco]

Texsolve. Aliphatic hydrocarbon solvents. [Texaco] *

Texsolve B. Highly refined commercial hexane; solv. used for vegetable oil and pharmaceutical extraction, compounding rubber cements and sealants, and polyolefin production. [Texaco]

Texsolve C. Mixed heptane fraction; solv. used for compounding rubber cements and sealants, extraction of oils

---

‡ Trade name and manufacturer not verified   § Trade name without identified manufacturer

and fats, and recrystallization of organic chemicals. [Texaco]

Texsolve E. Hexane-heptane fraction; solv. used in rubber tire manufacturing, rubber cements, sealants, inks, lacquers, and adhesives. [Texaco]

Texsolve S-66. Exempt mineral spirits with maximum aromatic content; solv. used in paint and protective coatings, dry- cleaning, degreasing, wood treating, charcoal lighter fluid; meets or exceeds standards for mineral spirits and Stod. [Texaco]

Texsolve V. CAS 8030-30-6; VM&P naphtha; solv. used in paints, coatings, rubber compounding, sealants, and chemical absorption. [Texaco]

Textamine Carbon Detergent K. Fatty amino complex; emulsifier for room temperature solv. degreasing formulations; aids in carbon removal from aircraft, diesels, and metal surfaces. [Henkel/Cospha; Henkel Canada]

Textamine T-1. CAS 61791-39-7; EINECS 263-171-2; 1-Hydroxyethyl-2-tall oil imidazoline; corrosion inhibitor, emulsifier, wetting agent for metalworking fluids; dispersant. [Henkel/Cospha; Henkel Canada]

Textase. A diastase preparation.

Textile Resin 2309 Conc. Highly reactive crosslinking agent for finishing of cellulosics or blends with synthetics or wool. [BASF]

Textile Resin NF-U. Crosslinking agent for resin finishing cellulosic fibers and their blends with synthetics. [BASF]

Textile Wax. Textile preparation and finishing agent. [BASF plc]

Textile Wax W. Wax-like substance for addition to textile sizing and finishing liquors. [BASF]

Textol 80 (L). Ethanolamine alkylbenzene sulfonate; raw material for manufacturing of liq. detergents. [Zohar Detergent Factory]

Textone®-50. Sodium chlorite/sodium nitrate blend; for bleaching textiles and stripping dyestuffs for natural and synthetic fibers. [Olin]

Textulite. A proprietary trade name for phenol formaldehyde laminated synthetic resin and molded compounds. §

Texzyme. Protein digesting enzyme liquid. [PMP Fermentation Products Inc]

TFC. Thin film composite spiral wound membrane; for reverse osmosis water treatment. [Allied-Signal Fluid Systems] *

tfol. An argillaceous earth containing free gelatinous silica; used as a soap.

TF Solvent. Mild cleaning agent for removal of flux residue, oil, grease and contaminants from magnetic tape heads, thermal print heads, optical equip., contacts, relays, and PCB assemblies. [Chemtronics]

TG-8. Triethylene glycol dicaprylate. [Ruco Div] *

T-gas. The commercial mixture of ethylene oxide and carbon dioxide; used as an insecticide.

TG Buffer. Trisglycine solution or powder. [Am. Research Prods.] ‡

TG-SDS Buffer. Trisglycine-SDS solution or powder. [Am. Research Prods.] ‡

Thaio Green No. 1. A halogenated copper phthalocyanine green; extremely light and heat-fast. [Reckitts Colours Ltd] *

Thalamonal. A proprietary preparation of fentanyl and droperidol; used as a premedication in anesthesia. [Janssen Pharmaceutical Ltd] *

Thalazole. CAS 85-73-4; A proprietary preparation contain-

ing phthalylsulfathiazole. [May & Baker Ltd] *

thallium. Metallic element; Tl; CAS 7440-28-0; EINECS 231-138-1; thallium salts, mercury alloys, low-melting glasses, rodenticides, photoelectric applications, electrodes in dissolved oxygen analyzers. [Aldrich; Atomergic Chemetals; Cerac; Noah Chem.]

Thallium. Thallous chloride Tl 201; diagnostic aid; radioactive agent. [Amersham Corp] *

thallium alum. A double sulfate of thallium and aluminum, $Tl_2SO_4·Al_2(SO_4)_3·24H_2O$.

Thalo Blue No. 1. An alpha, solvent sensitive, red shade phthalocyanine blue pigment; used for solventless printing inks. [Reckitts Colours Ltd] *

Thalo Blue No. 2. A beta, solvent stable, green shade phthalocyanine blue pigment. [Reckitts Colours Ltd] *

THAM. CAS 77-86-1; Tromethamine; alkalizer. [Abbott Laboratories] *

THAM. See tris (hydroxymethyl) aminomethane.

Thancat. A series of amine-type urethane catalysts; catalysts for production of flexible and rigid urethane foams, elastomers and sealants. [Texaco]

Thanecure®. CAS 14239-75-9; Zinc chloride with benzothiazyl disulfide; vulcanization activator for sulfur-curable millable urethane elastomer. [TSE Industries]

Thanol. A series of urethane polyols; intermediates for production of flexible and rigid urethane foams, elastomers and sealants. [Texaco]

Thanol E-2103. Urethane polyol; adhesives and sealants raw materials. [Eastman]

thao. A gelatinous preparation made in Cochin China from seaweed. It has frequently appeared in England under the names of Japanese or Chinese isinglass, and is used for the same purposes as isinglass.

Thapsia resin. The resin of *Thapsia garganica* root. It contains caprylic acid and thapsic acid.

Thawpit. CAS 56-23-5; A preparation of carbon tetrachloride used for cleaning materials. §

The Chemistry to Compete. Polyolefins specialty resins, colorants, and compounds for use by the plastics industries as additives to polyolefins and other petrochemicals for use in further manufacturing in a wide variety of industries. [Quantum Chemical Corp]

Theelin. CAS 53-16-7; Estrone; estrogen. [Parke-Davis]*

Theic. Tris-(2-hydroxyethyl)isocyanurate; used in the manufacture of heat-resistant wire lacquers. [BASF plc]

theine. (guaranine; caffeine; methyltheobromine) CAS 58-08-2; EINECS 200-362-1; $C_8H_{10}N_4O_2·HOH$; used as a stimulant in medicine and beverages.

theine. See caffeine.

The Little Chemical Giant. Synthetic olefin polymers or copolymers in particulate form, denatured alcohol and ethyl alcohol in industry or agriculture. [Quantum Chemical Corp]

theobroma oil. See cocoa butter.

theobromine. (3,7-dimethylxanthine) $C_7H_8O_2N_4$; CAS 83-67-0; EINECS 201-494-2; The active principle of the cocoa bean.

theobromose. The lithium compound of theobromine, $C_7H_7N_4O_2Li$.

theocal. A double salt of calcium theobromine and calcium lactate; used as a diuretic.

Theocalcin. A proprietary preparation of theobromine and calcium salicylate. §

Theodrex. CAS 317-34-0, CAS 21645-51-2; Each tablet

---

* Trade name not verified as available   † Trade name verified as obsolete

contains aminophylline BP (195 mg) and dried aluminum hydroxide gel BP (260 mg). [3M Pharmaceuticals] *

Theogardenal. Theobromine and phenobarbitone. [May & Baker Ltd] *

Theograd. CAS 58-55-9; A proprietary preparation of theophylline; a bronchodilator. [Abbott Laboratories] *

Theolair™ Liquid. CAS 58-55-9; Theophylline oral solution; asthma treatment. [3M Pharmaceuticals.]

Theolair™-SR Tablets. CAS 58-55-9; Anhydrous theophylline; sustained-release tablets for asthma treatment. [3M Pharmaceuticals.]

Theolair™ Tablets. CAS 58-55-9; Theophylline; asthma treatment. [3M Pharmaceuticals.]

Theominal. A proprietary preparation of phenobarbitone and theobromine; an antihypertensive. [Bayer AG] †

Theonar. A proprietary preparation of theophylline and noscapine; a bronchodilator. [MCP Pharmaceuticals] *

Theophen. A proprietary preparation of butethamate citrate, amylobarbitone, ephedrine hydrochloride, and theophylline; a bronchial antispasmodic. [Rybar Laboratories Ltd] *

Theophyllisilane C. Methylsilanol theophyllinacetate alginate; ingredient for slimming and anti-aging formulations, cosmetic and health products. [Exsymol]

Thephorin. A proprietary preparation of phenindamine hydrogen tartrate. [Roche Products Ltd] *

Therabloat. Poloxalene; liquid nonionic surfactant polymer of the polyethylene-polypropylene glycol type; pharmaceutic aid. [Norden Laboratories Inc] *

Theralax. CAS 603-50-9; Bisacodyl; laxative. [SmithKline Beecham] *

Therban®. Hydrogenated NBR; high abrasion resistance, good to excellent hot air resistance and resistance to ozone, technical oils, brake and hydraulic fluids, and crude oil containing H_2S and amines; used for technical rubber goods employed in mineral oil exploration and production, automotive components, cables. [Bayer AG; Bayer plc]

Therban® 1706. Saturated hydrocarbon ACN copolymer; specialty elastomer offering resist. to heat, oils, ozone, abrasion, hydrogen sulfide, amines; improved processing during mixing, extrusion, and injection molding; applics incl. automotive and industrial areas, e.g., seals, gaskets, power transmission belting, membranes, cable jackets, extruded profiles, rubber linings, hoses, bellows, sleeves, oil field parts. [Bayer; Miles]

Theriodide-131. CAS 7790-26-3; Sodium iodide I 131; used as an antineoplastic; diagnostic aid; and radioactive agent. [Abbott Laboratories] *

Therlo. An electrical resistance alloy containing 85% copper, 13% manganese, and 2% aluminum.

Thermaflo. PVC compounds; a wide range of moldings and extrusions including applications in footwear, building, cable, automotive, etc. [Evode Plastics Ltd] *

Thermalate H320. Fiberglass-reinforced thermoset polyester composite; for high temp. mold and platen thermal applications. [Haysite Reinforced Plastics]

thermal black. See carbon black.

Thermalene. An mixture of acetylene and vaporized oils; used for production at high temperatures, in cutting and welding metals.

Thermalloy. A patented form of thermit containing 50% iron oxide, 27% aluminum, and 23% sulfur; the name appears to be applied also to an alloy containing 66.5%

nickel, 30% copper, and 2% iron; it has a magnetic permeability which decreases at higher temperature; an alloy containing 75-85% iron, 10-20% chromium, 2-6% silicon, 0.5-1% manganese, 0.5-1% tungsten, and 0.2-2% carbon is also known as Thermalloy. §

Thermalloy A. A proprietary trade name for an alloy containing 67.5% nickel, 0.15% carbon, 0.15% silicon and 30% copper. §

Thermalloy B. A proprietary trade name for an alloy containing 57.8% nickel, 0.15% carbon, 0.15% silicon, and 40% copper. §

thermatomic carbon. A fine carbon produced by cracking natural gas into carbon and hydrogen by passing the gas over heated brickwork; used as a rubber filler.

Thermax® Floform N-990. CAS 1333-86-4; Medium thermal carbon black; used in rubber products with high loadings of carbon black, in metallurgy, specialty, and refractory applications. [Cancarb; R.T. Vanderbilt] *

Thermax® Stainless. CAS 1333-86-4; Medium thermal carbon black; reinforcer for all elastomers, and black stocks. [CanCarb; R.T. Vanderbilt]

Thermax® Stainless Floform N-907. CAS 1333-86-4; Med. thermal carbon black; nonstaining product used in seals and gaskets. [Cancarb; R.T. Vanderbilt] *

Thermazote. A trademark for an expanded thermosetting plastic, manufactured in densities between 7 and 30 lb/ft^3; it is nonflammable and odorless and withstands temperatures as high as 300C; has a low thermal conductivity; used in construction, etc. §

Therm-Chek. For heat and light stabilizers for vinyl halide compositions, in Int Class 1. [Ferro] *

Thermex. Heat transfer media. [ICI Chem. & Polymers Ltd]

Thermexo. Highly exothermic metal producing compounds. [Foseco (F.S.) Ltd]

Thermica. $SO_2AL_2O_3MgOK_2OF$ + 3% impurities; additive to oil, paint, grease, plastics, glass and other inorganic materials which require a high dielectric or high heat resistant material; acoating for welding rods requiring special atmospheric controls; a thermal barrier and high electrical resistivity. [Mykroy/Mycalex] *

Thermid® EL-5010, EL-5512. Polyimide; interlayer dielec. coating with exceptional adhesion and low moisture absorp. for the fabrication of multichip modules. [Ablestik]

Thermilin. Heat transfer fluid. [Monsanto Co]

Therminol®. Heat transfer fluids. [Monsanto Co] *

Therminol® 44. Modified ester-based fluid; liq. phase heat transfer fluid featuring low temp. fluidity, high flash point; for use in nonpressurized systems. [Monsanto Co]

Therminol® 66. Modified polyphenyl; liq. phase heat transfer fluid for use in nonpressurized systems; high temp., low pressure fluid. [Monsanto Co]

Therminol® VP-1. 73.5% diphenyl oxide, 26.5% biphenyl; liq./vapor phase heat transfer fluid for use in nonpressurized systems; ultra high temp. [Monsanto Co]

Thermisilio. A proprietary trade name for a chemical-resisting iron-silicon alloy in which the brittleness has been diminished. §

Thermisilizid. A Swedish iron-silicon alloy of the acid-resisting type.

Thermit®. A world-wide registered trademark for aluminothermic mixtures consisting essentially of nearly equal parts of powdered aluminum and metal oxides, usually

---

‡ Trade name and manufacturer not verified   § Trade name without identified manufacturer

iron or manganese oxides; these mixtures burn with a high temperature; used for welding metals; also used as an ingredient in incendiary bombs. §

Thermit. A bearing metal containing lead, antimony (20%) and small quantities of tin, nickel, and copper. §

Thermit manganese. Manganese metal made by the Thermit reduction method. It contains approximately 98% manganese.

Thermit metal. A German bearing metal containing: 14-16% antimony, 5-7% tin, 0.8-1.2% copper, 0.7-1.5% nickel, 0.3-0.8% arsenic, 0.7-1.5% cadmium, and 72-78.5% lead.

Thermlo F. An organic polysulfide; a proprietary rubber vulcanization accelerator. §

Thermocal. Heat transfer medium. [ICI Chem. & Polymers Ltd]

Thermocal B. Monoethylene glycol-based; corrosion inhibiting heat transfer fluid for industrial cooling and refrigeration installations and moder. high temp. heat exchange systems. [ICI Australia]

Thermocast. A proprietary trade name for an ethyl cellulose composition. §

Thermocomp®. Filled and reinforced thermoplastic compounds. [LNP; ICI Chemicals & Polymers Ltd]

Thermocomp® AF-1004. CAS 9003-56-9; ABS, 20% glass fiber-reinforced; superior toughness, strength, hardness, and high temp. resistance when compared to glass-fortified PS or SAN; used in business machine components, photographic equipment, and automotive and appliance applications where combined toughness and dimensional stability are needed. [LNP]

Thermocomp® BF-1006. CAS 9003-54-7; SAN, 30% glass fiber-reinforced; thermoplastic offering low mold shrinkage. [LNP]

Thermocomp® BF-1006FR. CAS 9003-54-7; SAN, 30% glass fiber-reinforced, flame-retardant; thermoplastic characterized by unequalled stiffness and low temp. creep resistance, processability, and lowest mold shrinkage of any thermoplastic; produces extremely high quality, tight tolerance molded part; with excellent surface finish and "metal-like" end-use performance; used in the appliance, business machine, and camera industries as materials for housings, frames, and internal components. [LNP]

Thermocomp® CF-1004. CAS 9003-53-6; PS, 20% glass fiber-reinforced; improved strength, stiffness, dimensional stability; natural and pigmented. [LNP]

Thermocomp® DF-1004. PC, 20% glass fiber-reinforced; used as replacement for metal in mechanical, structural, and elec. applications, (elec. connector block, computer insulator mount, instrument face plate, desk calculator housing, electronic terminal block); glass fibers increase strength and dimensional stability and retain inherent flame retardancy of the base resin; they lower water absorp., dramatically increase deflection temp., and substantially increase stiffness. [LNP]

Thermocomp® EC-1006. CAS 61128-46-9; Polyetherimide, 30% PAN carbon fiber. [LNP]

Thermocomp® GF-1004. CAS 25135-51-3; Polysulfone, 20% glass fiber-reinforced; high temp. engineering thermoplastic for structural/insulating applications; mechanical and thermal properties, dimensional stability, increased flamm. resistance, resistance to environmental stress-cracking, dip solderability, increased fatigue endurance, moldability, etc. [LNP]

Thermocomp® HF-1006. CAS 25035-04-5; Nylon 11, 30% glass fiber-reinforced; low moisture absorp. and good dimensional stability, but with slightly reduced mechanical properties compared to the other fortified nylon series. [LNP]

Thermocomp® IF-1002. Nylon 6/12, 10% glass fiber-reinforced; reinforced nylon similar to fortified nylon 6/10 versions (QF Series), exhibits the same moisture, dimensional stability, and impact characteristics. [LNP]

Thermocomp® JF-1004. PES, 20% glass fiber-reinforced; outstanding high temp. creep and hydrolysis resistance, self-extinguishing properties, good impact strength and low mold shrinkage. [LNP]

Thermocomp® KB-1008. CAS 105-57-7; Acetal, 40% glass bead-fortified; outstanding solv. resistance of base polymer with improved dimensional control. [LNP]

Thermocomp® MF-1002. CAS 9003-07-0; PP, 10% glass fiber-reinforced; resistance to heat, water absorp., and chemical attack; used in pipe fittings, pump housings, and heater components; certain formulations in MF series meet the requirements of MIL-P-46109. [LNP]

Thermocomp® NF-1004. CAS 9003-53-6; Styrenic copolymer, 20% glass fiber-reinforced; improved elevated temp. performance, deflection temp. 30-50 F higher than other conventional glass-reinforced styrenic systems, retains dimensional stability, stiffness, and low moisture absorp.; used in camera housings, business machine parts, swimming pool pumps, water softener parts, and interior automotive parts; NF Series compounds also avail with internal lubricants for use in bearings and wear applications. [LNP]

Thermocomp® OC-1006. CAS 9016-75-5; PPS, 30% PAN carbon fiber-reinforced; offers flex. mod. (stiffness) in addition to generally improved properties; offers reduced mold shrinkage and thermal coeff. of expansion, and increased thermal conductivity; also offers high temp. stability and extreme chemical and solv. resistance inherent in the base resin. [LNP]

Thermocomp® PC-1006. CAS 25038-54-4; Nylon 6, 30% PAN carbon fiber-fortified; reinforced thermoplastic. [LNP]

Thermocomp® QF-1006FR. CAS 9008-66-6; Nylon 6/10, 30% glass fiber-reinforced, flame-retardant; for applications requiring increased chem. resist. and lower water absorp. [LNP]

Thermocomp® RC-1002. CAS 32131-17-2; Nylon 6/6, 10% PAN carbon fiber-fortified; offers flexural mod. (stiffness), and improved physical properties; compared to glass fiber-reinforced nylon 6/6, RC Series offers reduced mold shrinkage and thermal coefficient of expansion, and greatly increased thermal conductivity; these systems capitalize on the high HDT, toughness, solv. resistance, and moldability inherent in the base resin. [LNP]

Thermocomp® SF-1006. Nylon 12, 30% glass fiber-reinforced; good dimensional stability and lowest moisture absorp. of nylons, slightly reduced mechanical properties compared to other fortified nylon series. [LNP]

Thermocomp® TF-1004. CAS 9009-54-5; Thermoplastic PU, 20% glass fiber-reinforced; tough thermoplastic; resists wear, abrasion, creep, and exposure to petrol. products; for gears, gaskets, bushings, bearings, and other applications where outstanding toughness and wear resistance are needed. [LNP]

Thermocomp® VF-1002. Super tough nylon (Zytel), 10% glass fiber-reinforced; offers strength, stiffness, toughness, chemical and thermal resistance, and notched and unnotched Izod impact values not formerly achievable in a resin system; used in tool housings and other industrial components; retains moldability and surface finish of glass-reinforced nylons. [LNP]

Thermocomp® WC-1006. CAS 26062-94-2; PBT polyester, 30% PAN carbon fiber-reinforced; offers flex. mod. (stiffness), and generally improved properties, reduced mold shrinkage and thermal coefficient of expansion, and increased thermal conductivity in comparison to glass-fortified polyester; capitalizes on low moisture absorpt., moldability, and balanced frictional and general mechanical properties inherent in base resin. [LNP]

Thermocomp® XF-1004. Amorphous nylon, 20% glass fiber-reinforced; combines the features of crystalline and amorphous materials the solv. resistance of nylon with the dimensional control possible from PC. [LNP]

Thermocomp® YF-1002. Polyester elastomer based on Hytrel, 10% glass fiber-reinforced; very tough, reinforced thermoplastic elastomer with resistance to abrasion and superior resistance to deformation at elevated temps.; resistance to common fuels and lubricating oils; used in hostile environments at high and low temps.; lubricated versions are suitable gear materials, especially when noise reduction may be required; for moderate and low temp. applications (body and drive train parts for snowmobiles, chain saw parts, shaft seals, and cross country skis). [LNP]

Thermocomp® ZF-1004. CAS 25134-01-4; Modified PPO, 20% glass fiber-reinforced; tough, rigid thermoplastic, high moisture resistance and dimensional stability; for appliance and business machine components, water softener and industrial filters, pump housings and impellers, radio, TV, and elec./electronic insulator units; meets requirements of MIL-P-46131. [LNP]

Thermodek. Dry roof screed system. [Vencel Resil Ltd]

Thermoflex. A proprietary trade name for di-p-methoxydiphenylamine; an antioxidant. [Imperial Chemical Industries plc; Du Pont (UK) Ltd] *

Thermoflex A. Contains 50 parts of phenyl-β-naphthylamine, 25 parts of methoxydiphenylamine and 25 parts of diphenyl-p-phenylene diamine; an antioxidant. [Imperial Chemical Industries plc; Du Pont (UK) Ltd] *

Thermoguard® 505. CAS 1163-19-5; EINECS 214-604-9; Decabromodiphenyloxide; flame retardant [Atochem]

Thermoguard® 8218. Decabromodiphenyloxide concentrate in polyethylene; flame retardant [Atochem]

Thermoguard® CPA. CAS 7440-36-0; Antimony; flame retardant for use as replacement for antimony oxide in many formulated plastics especially in flexible PVC applications such as wire and cable insulation and jacketing. [Atochem N. Am./Plastics Additives]

Thermoguard® FR. Antimony-containing compound; flame retardant for vinyls and other plastics. [Atochem N. Am./Plastics Additives]

Thermoguard® L. CAS 1327-33-9; Antimony oxide; flame retardant for use with a halogen-containing compound; flame retardant pigment for PVC; also for use with chlorinated organics for producing flame-retardant polyesters and polyethylene compounds. [Atochem N. Am./Plastics Additives]

Thermoguard® UF. CAS 1327-33-9; Antimony oxide; flame retardant; superfine grade [Atochem]

Thermolastic®. Extrudable thermoplastic rubber-like materials based upon styrene butadiene copolymers not requiring vulcanization. [Shell] *

Thermolin. Chlorinated phosphate ester additive for flexible foams. [Olin]

Thermoloft®. Fiber. [DuPont UK]

Thermonit. A refractory cement made in the electric furnace; it is stated to be used as a paint or mortar, and to resist high temperatures; Keramonit is the cement reinforced with metal mesh. §

Thermoplast®. Special dyes sol. in plastics; for mass dyeing of thermoplastic and thermosetting plastics, e.g., styrene polymers (PS, S/B, SAN, ABS), rigid PVC, polymethacrylate, cellulose derivs., polycarbonates, unsaturated polyester, etc. [BASF AG]

Thermoplaste. Plastics dyes. [ICI Chem. & Polymers Ltd]

Thermoprene. Products obtained by heating rubber with either an organic sulfonyl chloride or an organic sulfonic acid at 125-135 for several hours. p-Toluene-sulfonyl chloride and p-toluene- sulfonic acid are suitable reagents; used in making protective paints resistant to acids and alkalis.

Thermorun. Thermoplastic elastomer. [Mitsubishi Petrochem.] *

Thermoseal. Heat seal lacquers. [The Scottish Adhesives Co Ltd]

Thermoset 100. CAS 25928-94-3; Two-part epoxy; adhesive system for tooling applications. [Thermoset Plastics]

Thermoset 300/No. 65 Hardener. CAS 25928-94-3; Two-part epoxy; room temperature-curing system with low visc., high gloss; moisture insensitive during cure; for elec./electronic insulation applications; black [Thermoset Plastics]

Thermoset 310. CAS 25928-94-3; Two-part epoxy; oven-curing, semirigid, tough system with very good shock resist.; for elec./electronic insulation applications; black. [Thermoset Plastics]

Thermoset DC-232. CAS 25928-94-3; One-part epoxy; oven-curing impregnating system with long shelf life for a single-component system.; for elec./electronic insulation applications. [Thermoset Plastics]

Thermoset ME-101. Electrically conductive resin; general purpose, low temp. curing. [Thermoset Plastics]

Thermoset ME-177. One-part dielec. resin; chip-on-board encapsulant, long room temperature life. [Thermoset Plastics]

Thermoset SC-102. Two-part silicone system; clear, dielectric gel elec./electronic insulating resin; temp. rating to 200 C; mix ratio: 1:1 by wt. [Thermoset Plastics]

Thermoset SC-113. One-part silicone system; protective sealer for brush, dip, or spray application for elec./electronic insulation; gray. [Thermoset Plastics]

Thermoset UR-101. Two-part PU; room temperature-curing, tough potting system with good abrasion resist., excellent low temp. flexibility; for elec./electronic insulation applications; black. [Thermoset Plastics]

Thermoset UR-105. Two-part PU; room temperature-curing potting system with excellent low temp. flexibility; repairable; softer version of UR-101; for elec./electronic insulation applications; black. [Thermoset Plastics]

Thermotex. Thermit bottom plate patching compound.

---

‡ Trade name and manufacturer not verified   § Trade name without identified manufacturer

[Foseco (F.S.) Ltd]

Thermount®. Fiber. [DuPont UK]

Theromolite 139. Sulfur-containing organotin; heat stabilizer for weatherable coextruded siding, window profiles, and other rigid PVC applications. [Atochem N. Am.]

Theromolite 380. Sulfur-containing organotin; heat stabilizer for weatherable coextruded siding, window profiles, and other rigid PVC applications. [Atochem N. Am.]

THF. See tetrahydrofuran.

THFA. CAS 97-99-4; Tetrahydrofurfuryl alcohol. [Great Lakes]

THFA. See tetrahydrofurfuryl alcohol.

Thial. Hexamethylenetetramine hydroxymethylsulfonate; used as an antiseptic.

thiamine. (vitamin B$_1$) C$_{12}$H$_{17}$ClN$_4$OS; EINECS 200-425-3; Medicine, nutrient, enriched flours; available as thiamine hydrochloride and thiamine mononitrate. [Hoffmann-La Roche SA; Honeywill & Stein Ltd]

thiamine dichloride. See thiamine hydrochloride.

thiamine hydrochloride. (thiamine dichloride; vitamin B$_1$; aneurine hydrochloride) C$_{12}$H$_{17}$N$_4$OSCl · HCl; CAS 67-03-8; EINECS 200-641-8; Medicine, nutrient, enriched flours. [Aldrich; BASF; EM Industries; Hoffmann-La Roche; Takeda USA]

Thiate®. Trimethylthiourea, 1,3-diethylthiourea or 1,3-dibutylthiourea; curing agents for elastomers. [R T Vanderbilt Co Inc]

Thiate E. CAS 2489-77-2; Trimethyl thiourea; a proprietary accelerator. [K & K Greeff Chemicals Ltd] *

Thiazamide. Sulfathiazole. [May & Baker Ltd] *

Thiazina. A proprietary preparation containing thiacetazone, isoniazid; an antituberculous agent. [Smith & Nephew Pharmaceuticals Ltd] *

thiazoline. (thiazylamine). CAS 96-50-4; Aminothiazole, C$_3$H$_4$N$_2$S.

thiazylamine. See thiazoline.

Thibenzole. CAS 148-79-8; A proprietary trade name for thiabendazole. §

thickened mineral oils. Mineral oils which have been thickened by dissolving soap, usually aluminum soap.

thickened oils. See blown oils.

Thi-Di-Mer. Trisulfapyrimidines (oral suspension); mixture of sulfadiazine, sulfamerazine, and sulfamethazine; antibacterial. [Merrell Dow Pharmaceuticals Inc] *

Thiel-Stoll solution. A saturated solution of lead chlorate, Pb(ClO$_4$)$_2$. It has a density of 2.6 and is used for the determination of the specific gravity of minerals.

2-thienyl chloride. See 2-chlorothiophene.

Thiersch's antiseptic solution. A solution containing salicylic acid and boric acid.

Thiery's solution. A solution of picric acid for the treatment of burns.

thiet-sie. A resinous substance used as a varnish by the Burmese.

Thimerosal. A proprietary trade name for thiomersal. §

Thinners. Formulated thinners for use with spirit based coatings. [Foseco (F.S.) Ltd]

Thinoline. (vulcanized oils). Vulcanized linseed oil; used in rubber mixings.

Thinsec. CAS 63-25-2; A suspension concentrate containing 450 g carbaryl per liter. contact insecticide and fruit thinner for apples. [ICI Agrochemicals]

Thinsol. Lacquer thinners of various glosses. [Sasolchem]

thioacetamide. CH$_3$CSNH$_2$; CAS 62-55-5; EINECS 200-

541-4; Replacement for gaseous hydrogen sulfide in qualitative analysis. [Aceto; Burlington Bio-Medical; Penta Mfg.]

thiocamf. A liquid formed by exposing camphor to the action of sulfur dioxide; used as a disinfectant as it evolves sulfur dioxide on exposure to air.

thiocarbamide. See thiourea.

thiocarbonyl chloride. See thiophosgene.

Thioctacid. α-Liponacid; tablets and ampoules; for liver protection. [Degussa AG] *

Thiodan. CAS 115-29-7; Endosulfan; insecticide and acaricide. [Hoechst UK]

Thiodet. Residual thiosulfate test kit. [May & Baker Ltd] *

2,2'-thiodiacetic acid. See thiodiglycolic acid.

2,2'-thiodiethanol. See thiodiglycol.

thiodiethylene glycol. See thiodiglycol.

thiodiglycol. (thiodiethylene glycol; 2,2'-thiodiethanol; bis(2-hydroxyethyl) sulfide; dihydroxyethyl sulfide) HOCH$_2$CH$_2$·S·CH $_2$CH$_2$OH; CAS 111-48-8; EINECS 203-874-3; Intermediate for elastomers and antioxidants, solvent for dyes in textile printing. [Morton Int'l.]

thiodiglycolic acid. (2,2'-thiodiacetic acid) S(CH$_2$COOH)$_2$; CAS 123-93-3; EINECS 204-663-9; Analytical reagent. [Witco/Argus]

3,3'-thiodipropionic acid. (dicarboxylic acid; propanoic acid, 3,3'-thiobis) HOOCCH$_2$CH$_2$SCH$_2$CH$_2$COOH; CAS 111-17-1; Antioxidant in food packaging, soaps, plasticizers, lubricants, fats, and oils. [Evans Chemetics; Janssen Chimica; Witco/Argus]

thiodipropionic acid, dilauryl ester. See dilauryl thiodipropionate.

thioethylene glycol. See 2-mercaptoethanol.

Thiofide. Benzthiazyl disulfide; rubber accelerator. [Monsanto plc]

Thiofluor™. N,N-Dimethyl-2-mercaptoethylamine hydrochloride, C$_4$H$_{11}$NS·HCl; nucleophile used in preparation of o-phthaldehyde reagent for fluorescence detection in HPLC. [Pickering Laboratories Inc]

thiofuran. See thiophene.

thiofurfuran. CAS 110-02-1; EINECS 203-729-4; (thiophene; thiofuran) C$_4$H$_4$S; used in organic synthesis, pharmaceutical manufacture, as a dye, solvent

thioglycolic acid. (2-mercaptoacetic acid) Organic acid; HS · CH$_2$COOH; CAS 68-11-1; EINECS 200-677-4; in chemical analysis for the spectrophotometric determination of palladium; cosmetics (intermediates for hair-waving, depilatories), vinyl stabilizer intermediate, reaction intermediate for radiation-cured plastics; reagent for iron; manufacture of thioglycolates. [Atochem N. Am.; Chemische Fabrik GmbH; EM Industries; Evans Chemetics; Witco/Argus]

Thiokol® 2135. Polysulfide; one-part, fast-cure joint sealant for sealing, caulking, and glazing applications on buildings. [Morton Int'l./Polymer Systems]

Thiokol® 2153, 2157. Polysulfide latex; caulk usable alone or with MC 2027, a chem. resist. masonry sealer or R-2100 radon barrier coating; excellent adhesion to masonry surfs.; good water resist.; 2153 is self-leveling version; 2157 is gun grade, nonsag version. [Morton Int'l./Polymer Systems]

Thiokol® FEC-2232. Blend of liq. polysulfide and epoxy resin; flexibilized coating for use in secondary containment, drum storage, and truck unloading areas, and in fuel storage tanks; abrasion and chem. resist. [Morton

---

* Trade name not verified as available   † Trade name verified as obsolete

Int'l./Polymer Systems]

Thiokol® FES-2258. CAS 25928-94-3; Two-component flexible epoxy sealant; crack injection system for making concrete walls and floors water-tight. [Morton Int'l./ Polymer Systems]

Thiokol® LP. Liquid polysulfide polymers with SH-terminals; used as a basis polymer for the production of sealants for insulating glass, building joints, caulking ship decks, etc. [Thiokol GmbH]

Thiokol® MC-2027. Single-component polysulfide water disp.; excellent barrier coating and moisture barrier for concrete/masonry surfs.; provides a film with excellent resist. to common solvs., chem., water and salt spray; applications incl. highway, coastal, construction applications. [Morton Int'l./Polymer Systems]

Thiokol® R-2100. Polysulfide water disp.; radon barrier coating for application to porous masonry walls. [Morton Int'l./Polymer Systems]

Thiokol® RLP-2078. Two-component reinforced polysulfide system; elastomeric coating curing by chem. reaction to a tough, flexible, chem. resist. lining that acts as a leakproof barrier; for concrete surfaces; bridges nonstructural cracks. [Morton Int'l./Polymer Systems]

thiolane. See tetrahydrothiophene.

Thiolim. Hypo eliminator. [May & Baker Ltd] *

Thiolite. CAS 7778-18-9; An insulator prepared from formaldehyde, cresol, and sulfur chloride.

Thiolyte®. Reagent. [Calbiochem Corp]

Thionalide. A commercial name for thioglycollic acid, β-amino- naphthalide, HS·CH $_2$CO·NH·C $_{10}$H$_7$; used as an analytical reagent.

Thionex. CAS 115-29-7; Active ingredient: endosulfan; a chlorinated cyclic sulfurous acid ester having broad-spectrum insecticidal activity of long-lasting effect. [Makhteshim Chemical Works Ltd]

Thionex. (tetramethyl-thiuram-monosulfide) CAS 97-74-5; An ultra-accelerator for rubber vulcanization.

Thionol®. A registered trade name for certain dyestuffs. §

Thionoline. (hydroxyaminoiminodiphenyl-sulfide) $C_{12}H_8$-N_2OS.

thiophane. See tetrahydrothiophene.

thiophene. (thiofuran) C_4H_4S; CAS 110-02-1; EINECS 203-729-4; Organic synthesis (condenses with phenol and formaldehyde, copolymerizes with maleic anhydride), solvent, dye, pharmaceutical manufacture. [Atochem N. Am.; Penta Mfg.]

thiophenol. (phenyl mercaptan) C_6H_5SH; CAS 108-98-5; EINECS 203-635-3; pharmaceutical synthesis. [Aldrich; ICI Am.; Janssen Chimica; Schweizerhall]

Thiophor Bronze 5G. A dyestuff obtained by the fusion of p-phenylene-diamine and p-aminoacetanilide with sulfur. §

Thiophor Indigo. A dyestuff obtained by heating the indophenol derivative from α-naphthol and p-amino-dimethyl aniline with sodium sulfide and sulfur. §

thiophosgene. (thiocarbonyl chloride) $CSCl_2$; CAS 463-71-8; EINECS 207-341-6; Organic synthesis. [Aldrich; Fine Organics Ltd; Fluka; Pfaltz & Bauer]

Thioprene®-48. An elastomeric mercaptan-terminated polymer; used for sealing glass. [Polymeric Systems Inc] *

thiosept. A product containing sulfur obtained from oil shale; used in salves.

thiosept oil. A distillation product of shale oil containing

sulfur.

Thioset® M. CAS 15535-29-2; Ethanolamine sulfite. [Evans Chemetics]

Thiostab. A proprietary pure sodium thiosulfate in ampoules. §

Thiostop E. Sodium diethyl dithiocarbamate (25-30% aq. solution); an ultra-accelerator for NR and SBR latexes; an activator for guanidine type accelerators. [Uniroyal] *

Thiostop N. Sodium dimethyldithiocarbamate (40% aq. solution); a nonstaining and nondiscoloring polymerization short-stop for SBR and similar rubbers. [Uniroyal] *

Thiostop N. Sodium dimethyl dithiocarbamate (40% aq. solution); an ultra-accelerator for NR and SBR latexes. [Uniroyal] *

Thiosulfil. CAS 144-82-1; Sulfamethizole; antibacterial. [Wyeth Laboratories]

Thiotan. Reserving agent; used for polyamide/elastomer fibers. [Sandoz Products Ltd]

Thiotax. CAS 149-30-4; 2-Mercaptobenzothiazole; vulcanization accelerator. [Monsanto Co]

Thiotepa. CAS 52-24-4; Injection containing triethylene-thiophosphoramide; antineoplastic agent. [Lederle Laboratories]

thiourea. (sulfourea; thiocarbamide) NH_2CSNH_2; CAS 62-56-6; EINECS 200-543-5; Photography, photocopy papers, organic synthesis (intermediate, dyes, drugs, hair preparations), rubber accelerator, analytical reagent, amino resins, mold inhibitor. [Allchem Industries; Bechem Chemie BV; Dajac Labs; Fairmount; R.W. Greeff]

thiourea dioxide. (formammidinesulfinic acid; aminoimino-methanesulfinic acid) $NH_2C(:NH)SO_2H$; CAS 1758-73-2; EINECS 217-157-8. [Allchem Industries; Arol Chem Prods.; Degussa]

Thiovanic® Acid. CAS 68-11-1; Thioglycolic acid. [Evans Chemetics]

Thiovanol®. CAS 96-27-5; Thioglycerin; stabilizer for acrylonitrile polymers; crosslinking agent for hard high-gloss coatings; accelerator for epoxy-amine condensation reactions; reducing agent; used in hair waving and straightening, hair dyes, depilatories, textiles, furs, pharmaceuticals, surfactants, foam stabilizing additives for detergents, shampoos, insecticides, pesticides, fungicides, and dessicants. [Evans Chemetics]

Thiovit. CAS 7704-34-9; Sulfur; a protectant fungicide. [Pan Britannica Industries Ltd]

thioxydant lumire. Ammonium persulfate, $(NH_4)_2S_2O_8$.

Thisol. Bitumen emulsions; for industrial applications. [Vedag GmbH] *

Thissirol. An aqueous solution of about 57% castor oil soap and 29% chloroxylenol mixture; used as a bactericide.

thitsi. (Burma black varnish). A natural lacquer that is the sap of the black varnish tree, *Melanorrhoea visitata*.

Thiurad. Bis(dimethylthiocarbamyl)disulfide; tetramethyl-thiuram disulfide; vulcanization accelerator. [Monsanto Co]

Thiuretic. CAS 58-93-5; Hydrochlorothiazide; diuretic. [Parke-Davis] *

Thixatrol. An organic derivative of castor oil; a rheological additive designed to impart thixotropy, viscosity, and antisettling properties; used in solvent systems (paints, inks, caulks, mastics, plastisols). [Rheox Inc]

Thixcin. An organic derivative of castor oil; a rheological additive designed to impart thixotropy, viscosity, and

---

‡ Trade name and manufacturer not verified   § Trade name without identified manufacturer

antisettling properties; used in solvent systems, paints, solvent-free epoxies, mastics, inks, cosmetics. [Rheox Inc]

Thixolan. Viscosity modifiers. [Harcros UK]

Thixomen. Thixotropic additive. [ICI Chem. & Polymers Ltd]

Thixon® 300. Vulcanizable bonding system; one-coat for bonding fluoroelastomers to metals when equal parts by vol are mixed. [Morton Int'l./Industrial Adhesives]

Thixon® 511-T. Vulcanizable bonding system; cover coat for bonding NR, SBR, CR, BR, IR, CSM, NBR, EPDM, and IIR rubbers. [Morton Int'l./Industrial Adhesives]

Thixon® 753. Vulcanizable bonding system; aq. primer and one-coat for bonding NBR; especially designed for phosphatized metals. [Morton Int'l./Industrial Adhesives]

Thixon® 957. Water-based overcoat adhesive; for bonding a variety of elastomers during vulcanization. [Morton Int'l./Industrial Adhesives]

Thixon® 2000. Solv.-based adhesive; used as one-coat adhesive or two-coat system applied over Thixon P-15 adhesive primer; bonds various elastomers incl. natural rubber, SBR, neoprene, EPDM, butyl, and nitrile. [Morton Int'l./Industrial Adhesives]

Thixon® OSN-2. Vulcanizable bonding system; one-coat for bonding NR, SBR, CR, BR, IR, CO/ECO, CSM, and ACM to most substrates; excellent for post-vulcanization bonding. [Morton Int'l./Industrial Adhesives]

Thixon® P-15. Vulcanizable bonding system; primer for all Thixon cover coat adhesives and excellent one-coat for bonding NBR; resists severe environmental conditions; excellent sprayability. [Morton Int'l./Industrial Adhesives]

Thixseal. Rheological additives for imparting special properties to coating compositions; used for solvent-based sealants, caulks and thick film coatings. [Rheox Inc]

Thomasite. A compound, $6CaO \cdot P_2O_5Fe_2SiO_4$. It is a constituent of the basic slag of the Thomas-Gilchrist process for the dephosphorization of iron.

Thomas meal. Ground slag obtained from the Thomas process for iron; used as a fertilizer.

Thomas phosphate. See Thomas slag.

Thomas slag. (Belgian slag) Basic slag containing phosphorus; by-product of steel manufacture.

Thompsons. Water seal. [Sterling Roncraft]

Thonzide. CAS 553-08-2; Thonzonium bromide; detergent. [Parke-Davis] *

Thoracin. A proprietary preparation of phenylethyl nicotinate, guaiacol furoate, tetrahydrofurfuryl salicylate, camphor, and eucalyptol. [Lloyd, Hamol] *

Thoragol. CAS 15585-70-3; A proprietary preparation of bibenzonium bromide; a cough linctus. [Lloyd, Hamol] *

Thoran. An alloy of 96% tungsten with 4% carbon.

Thorazine. CAS 50-53-3; Chlorpromazine; an anti-emetic and antipsycotic. [SmithKline Beecham]

Thoren. A technical diamond substitute. It is an alloy made from tungsten and tungsten carbide.

thoria. (thorium anhydride; thorium oxide; thorium dioxide) ThO_2; CAS 1314-20-1; Used in ceramics, gas mantles, nuclear fuel, medicine, nonsilica optical glass.

Thornel® Carbon Fiber T600/50C 12K. Carbon fiber; continuous length, high strength, high modulus fiber consisting of 12,000 filaments in a one-ply construction; its treated surface increases interlaminar shear strength in a resin matrix composite. [Amoco Chemical Co]

Thoroclear. A silicone-based water repellent coating for limestone. [Standard Dry Wall Products Inc] ‡

thoron. CAS 10034-92-2; Radon; used in cancer treatment and medical research.

Thorosheen®. An acrylic paint for masonry. [Standard Dry Wall Products Inc] ‡

Thorotrast. A proprietary colloidal thorium dioxide preparation. §

Thorowet G-40 3230. CAS 577-11-7; EINECS 209-406-4; Sodium dioctyl sulfosuccinate; wetting and rewetting agent; stable to dilute acids and alkali; dewatering agent for mining ores. [Clough]

Thorquest 39. CAS 64-02-8; EINECS 200-573-9; Tetrasodium EDTA; anionic liquid; complexing agent, particularly for bivalent metal ions. [Thor Chemicals (UK) Ltd]

Thor-stabilizator BF. Organic/inorganic complex; anionic; yellowish clear liquid; silicate-free stabilizer for peroxide bleaching. [Thor Chemicals (UK) Ltd]

Thorstat ASA. Phosphoric acid ester; anionic; clear liquid; antistatic agent for textile and fiber industry. [Thor Chemicals (UK) Ltd]

Thoulet's solution. A concentrated solution of potassium and mercury iodides in water; used to determine the density of minerals.

Thovaline. A proprietary preparation of talc, kaolin, zinc oxide, and cod-liver oil; used in dermatology. [Ilon Laboratories] ‡

Thowless solder. An alloy of tin and zinc with small amounts of aluminum and silver.

Three Elephant Boric Acid. 99.8% minimum H_3BO_3; two technical granular grades (granular and fine granular), and powdered grade. [Kerr-McGee Chemical Corp]

Three Elephant Pyrobor Dehydrated Borax. 99% Minimum $Na_2B_4O_7$; standard (-20 to +200 US mesh) and fine (-100 to +325 US mesh) technical grades. [Kerr-McGee Chemical Corp]

Three Elephant V-Bor Refined Pentahydrate Borax. 99.8% Minimum $Na_2B_4O_7 \cdot 5H_2O$; standard grade. [Kerr-McGee Chemical Corp]

L-threonine. (α-amino-β-hydroxybutyric acid) $C_4H_9NO_3$; CAS 72-19-5; EINECS 200-774-1; Nutrition and biochemical research, dietary supplement. [Am. Biorganics; Dajac Labs; Degussa; Janssen Chimica; U.S. Biochemical]

Thresh's reagent. Potassium bismuth iodide; used for testing alkaloids.

Thripstick®. CAS 52820-00-5; Deltamethrin; a pyrethroid insecticide. [Aquaspersions Ltd; Hoechst UK]

thrombase. A clotting enzyme.

Thrombo-Enelbin. Pharmaceutical preparations for the treatment of varices, thrombosis, accident traumata such as hematoma, bruises, contusions, strains; scar aftertreatment. [Cassella AG]

Thronothane Hydrochloride. CAS 637-58-1; Pramoxine hydrochloride; anesthetic. [Abbott Laboratories] *

Throsil. A proprietary preparation of cetalkonium chloride and amethocaine hydrochloride; an antiseptic mouth lozenge. [Cox Pharmaceuticals] †

thsing-hoa-liao. The Chinese name for a cobaltiferous aluminic silicate; used in the manufacture of porcelain.

thulia. See thulium oxide.

thulium oxide. (thulia) Tm_2O_3; CAS 12036-44-1; EINECS 234-851-6; Source of thulium metal. [Atomergic Chemetals; Cerac; Rhone-Poulenc Basic]

Thuricide HP. Bacillus thuringiensis; a bacterial insecticide

---

* Trade name not verified as available   † Trade name verified as obsolete

for control of caterpillars. [Atlas Interlates Ltd]

Thurmalox. A line of silicon-based, heat and corrosion resistant coatings for protection of metal structures or vessels subjected to high temperatures up to 1600F (870C); for stacks, breechings, furnaces, heat exchangers, exhaust manifolds, kilns, chemical process equipment, prevention of stress-corrosion crackng of stainless steel, wood stoves, barbecue grills, solar collector panels. [Dampney Company Inc] *

Thurston's alloy. An alloy of 80% zinc, 14% tin, and 6% copper.

Thwaites' solution. A mixture of alcohol, creosote, and chalk in water; used to preserve animal tissues.

T-Hydro. CAS 75-91-2; t-Butyl hydroperoxide solution. [Arco Chemical Co]

Thylogen Maleate. CAS 59-33-6; Pyrilamine maleate; antihistaminic. [William H Rorer Inc] *

thyme camphor. (isopropyl-m-cresol; thymic acid; thymol) $(CH_3)_2CHC_6H_3(CH_3)OH$; CAS 89-83-8; Used in the prevention of mold and mildew, in flavoring and perfumery, as a preservative and antioxidant.

thymene. The residual oils obtained from the preparation of thymol; used as a inexpensive perfume for soaps.

thymic acid. See thyme camphor.

thymine. (5-methyluracil; 5-methyl-2,4-dioxypyrimidine) $C_5H_6N_2O_2$, CAS 65-71-4; EINECS 200-616-1; Obtained by the hydrolysis of nucleic acids; used in biochemical research.

thymol. (5-methyl-2-(1-methylethyl) phenol; 6-isopropyl-m-cresol; 2-isopropyl-5-methylphenol) Substituted phenol; $(CH_3)_2CHC_6H_3(CH_3)OH$; CAS 89-83-8; EINECS 201-944-8; antibacterial and antifungal agent, perfumery, microscopy, preservative, antioxidant, flavoring, lab reagent, synthetic menthol. [Haarmann & Reimer; Janssen Chimica; Quest Int'l.]

Thymoxane. 3,3-Dimethyl-1,5-dioxaspiro[5,5]undecane; fresh thyme, leathery odor; used in fragrances. [Bush Boake Allen Ltd]

Thyodene. Analytical reagent HS 382200-00-0 for iodine and iodometry; as a white-water soluble powder it is superior to starch solution, it is stable and used direct from bottle to solutions to be titrated. [Campbell Williams & Co]

thyol. A substitute for ichthyol (qv) obtained by treating tar oils with sulfur.

Thyractin. Thyroglobulin; thyroid hormone. [Sterling Drug Inc] *

Thyrar. Thyroid hormone. [USV Pharmaceutical Corp] ‡

thyroidectin. Dried antithyroid serum.

Thyrolar. Liotrix; a mixture of liothyronine sodium and levothyroxine sodium in a ratio of 1:1 in terms of biological activity, or in a ratio of 1:4 in terms of weight; thyroid hormone. [USV Pharmaceutical Corp] ‡

Thyroprotein. Thyroglobulin; thyroid hormone. [Parke-Davis] *

Thytropar. A proprietary preparation of thyrotropin. [Armour Pharmaceutical Co] *

Tiazyme. Mixtures of the enzymes trypsin and chymotrypsin; pharmaceutical. [Hans Rahn & Co] *

Tiberal Roche. CAS 16773-42-5; Proprietary preparation of ornidazole; anti-infective agent. [Roche Products Ltd] *

Tibirox. Proprietary combination antibacterial product containing tetroxoprim and sulfadiazine. [Roche Products Ltd] *

Tibond. Anodes. [ICI Chem. & Polymers Ltd]

Tibricol. CAS 21829-25-4; A proprietary preparation of nifedipine; for the treatment of angina and hypertension. [Pfizer International]

Ticar. CAS 34787-01-4; A proprietary preparation of ticarcillin; an antibiotic. [SmithKline Beecham]

Ticelgesic. CAS 103-90-2; A proprietary preparation of paracetamol; an analgesic. [Unichem] *

Ticevite. A proprietary preparation of vitamins A, D, E, and B complex. [Unichem] *

Ticillin V.K. CAS 87-08-1; A proprietary preparation of Penicillin V; an antibiotic. [Unichem] *

Ticipect. A proprietary preparation of diphenhydramine hydrochloride, ammonium chloride, sodium citrate, menthol and chloroform; a cough syrup. [Unichem] *

Tico. An electrical resistance alloy containing 67.5% iron, 30.5% nickel, and small quantities of manganese and copper.

Tidolith. See Cryptone. §

Tienam. Imipenem and cilastatin sodium; broad-spectrum β-lactam antibiotic. [Merck & Co Inc]

Tiers argent. An alloy containing 66.6% aluminum and 33.3% silver.

Tifolic. A proprietary preparation of ferrous fumarate and folic acid; used in the treatment of anemia in pregnancy. [Unichem] *

Tiform. Unit anode systems. [ICI Chem. & Polymers Ltd]

T-I-Gammagee. Tetanus immune globulin; immunizing agent. [Merck & Co Inc] *

Tigan. CAS 554-92-7; Trimethobenzamide hydrochloride; anti-emetic. [SmithKline Beecham]

Tigason. CAS 54350-48-0; Proprietary retinoid preparation containing etretinate as active ingredient; antipsoriatic. [Roche Products Ltd] *

Tiglyssin. A proprietary trade name for tigloidine hydrobromide; used in the treatment of muscular spasm. [Duncan Flockhard Ltd] *

Tiguvon®. Veterinary preparation; for use on domestic animals against warble infestation and lice. [Bayer AG]

tihydroxysuccinic acid. See tartaric acid.

Til. Compounds of titanium. [Tioxide Group plc]

Tilcom. Organic compounds of titanium and zirconium; catalysts, cross-linkers, thixatropes for emulsion paints and adhesion promoters; especially for printing inks onto plastic film. [Tioxide Group plc]

tile ore. An earthy variety of native cuprous oxide.

Tilite. Self-sinking aluminum grain refiner. [Foseco (F.S.) Ltd]

Tillantina. Seed dressing for control of fungal diseases on cereals, rice, cotton, and vegetables. [Bayer AG]

til oil. See gingelly oil.

Tilt Turbo. Emulsifiable concentrate containing 125 g propiconazole and 350 g tridemorph per liter; for control of mildew and rust in barley and wheat. [Ciba-Geigy Agrochemicals]

Timacar. CAS 26921-17-5; Timolol maleate; ophthalmic solution effective in lowering intraocular pressure; may be used for patients with chronic open-angle glaucoma, aphakic glaucoma, secondary glaucoma, or elevated intraocular pressure. (Denmark) [Merck & Co Inc]

Timail®. A range of transparent and colored enamels; for use in the surface coating of metals, particularly steel and aluminum. [Bayer AG]

Timang Steel. A proprietary trade name for a high manga-

---

‡ Trade name and manufacturer not verified   § Trade name without identified manufacturer

nese steel. §

Timbo. The root rind of a variety of *Conchocarpus.*; a narcotic. §

Tim-Bor® . CAS 12280-03-4; Disodium octaborate tetrahydrate; insecticide for use by professional pest control operators; effective against decay fungi and wood infesting pests (termites, carpenter ants); for all interior and exterior wood and wood-foam composite structural components, lumber. [U.S. Borax & Chem.; Borax Consolidated Ltd]

Timborised. Preserved timber. [Borax Consolidated Ltd]

Timbrel. CAS 69633-04-1; Emulsifiable concentrate containing 480 g/l triclopyr; herbicide to control perennial and woody weeds. [DowElanco Ltd]

Timbrelle. Polyamide carpet yarns. [ICI Chem. & Polymers Ltd]

Time Bomb. Trade name for total release fogger insecticide; household insecticide controlling flying insects, roaches, fleas etc. [Colonial Products Inc]

Timica. Titanium dioxide coated mica to meet cosmetic standards. [Cornelius Chemical Co Ltd] *

Timica®. CAS 13463-67-7, 12001-26-2; Titanium dioxide/mica blends; for frosted and iridescent effects in lipsticks, cream makeups, nail enamels, pressed powds. [Mearl]

Timodine. A proprietary preparation of nystatin, hydrocortisone, benzalkonium chloride and dimethicone; used in the treatment of infected eczema. [Lloyd, Hamol] *

Timolate. CAS 26921-17-5; Timolol maleate; anti-adrenergic. [Merck & Co Inc] *

Timonox. CAS 1327-33-9; Antimony oxide. [Anzon Ltd] *

Timonox Blue Star. CAS 1327-33-9; A proprietary preparation of pure antimony oxide; the arsenic amounts to 0.0018%. §

Timoptic. CAS 26921-17-5; Timolol maleate; anti-adrenergic; ophthalmic solution effective in lowering intraocular pressure; may be used for patients with chronic open-angle glaucoma, aphakic glaucoma, secondary glaucoma, or elevated intraocular pressure. (U.S.) [Merck & Co Inc]

Timoptol®. See Timoptic.

Timpilo®. Timolol maleate and pilocarpine; indicated for the reduction of elevated intraocular pressure in patients whose intraocular pressure is not adequately controlled on monotherapy with a beta blocker or pilocarpine or when concomitant therapy is appropriate. [Merck & Co Inc]

tin. (stannum) Element; Sn; CAS 7440-31-5]; EINECS 231-141-8; Tin plate, anodes, corrosion-resistant coatings, manufacture of chemicals. [Aldrich; Atomergic Chemetals; Cerac; M&T Harshaw; Noah Chem.]

Tinactin. CAS 2398-96-1; Tolnaftate; antifungal. [Schering Corp]

Tinaderm. CAS 2398-96-1; A proprietary preparation of tolnaftate; a skin fungicide. [Glaxo Laboratories] *

tin amalgam. A tin-mercury alloy containing 44-51% tin. It is prepared by electrolysis.

Tinamul® . Partial glycerides; for prep. of bakery goods, margarine, ice cream, release agents, sausages, mayonnaise, toffees, snacks. [Hüls AG]

tin ash. (stannic acid; stannic oxide; stannic anhydride; tin peroxide; tin dioxide) SnO_2; Used as a polishing powder.

tin bronze. An alloy of 89% copper and 11% tin.

tincal. (tinkal). An impure borax.

tin (II) chloride anhydrous. See stannous chloride.

tin crystals. See tin salts.

tin crystals. See stannous chloride.

Tindal. Acetophenazine maleate; antipsychotic. [Schering Corp] *

tin dichloride. See stannous chloride.

tin dioxide. See tin oxide (ic).

Tineafax. Preparations containing zinc undecenoate or tolnaftate; for the prophylaxis and treatment of athletes foot and other ringworm infections. [The Wellcome Foundation Ltd]

Tinegal. Dyeing and printing assistant. [Ciba plc] *

tinkal. See tincal.

Tinman. Mixture of fentin hydroxide, maneb, and zinc; used for control of potato blight. [Chiltern Farm Chemicals Ltd]

Tinoclarite. Bleaching stabilizers. [Ciba plc] *

Tinofil. Dispersed pigments. [Ciba plc] *

Tinol. A proprietary preparation of paracetamol and diphenhydramine hydrochloride; an analgesic. [Unichem] *

Tinopal® . Fluorescent whitening agents for paper and detergents. [Ciba plc] *

Tinopal® 5BM-GX. Stilbene type; whitening agent for powd. and liq. anionic and nonionic detergent. [Ciba-Geigy]

Tinopal® AMS-GX. Stilbene type; whitening agent for powd. anionic and nonionic detergents or laundry soaps. [Ciba-Geigy]

Tinopal® CBS-X. Disodium distyrylbiphenyl disulfonate; highly soluble, low dusting, lightfast, chlorine stable whitener for cotton and other cellulosics; used in anionic and nonionic laundry detergents, dry bleaches, fabric softeners, commercial laundry products, toilet bar soaps, liq. products, and products for use at low wash temps. [Ciba-Geigy]

tin ore. A mineral. It is tinstone, SnO_2.

Tinorex. Proofing agents. [Ciba plc] *

Tinosol. Dyeing and printing assistants. [Ciba plc] *

Tinovetin. Biodegradable detergent; highly effective wetting and scouring agent; used for all textile processing; scouring, wetting and emulsifying greases. [Ciba plc] *

tin (II) oxalate. (stannous oxalate; oxalic acid tin (II) salt) SnC_2O_4; CAS 814- 94-8; EINECS 212-414-0; Catalyst for stannous esterification; dying and printing textiles. [Atochem N. Am.; Atomergic Chemetals]

tin oxide (ic). (tin dioxide; stannic oxide; cassiterite) Inorganic oxide; SnO_2; CAS 1317-45-9; 18282-10-5; EINECS 242-159-0; polishing glass and metals; tin salts, catalyst, ceramic glazes and colors, putty, perfume, cosmetic preparations (fingernail polish), manufacture of special glasses. [Atomergic Chemetals; Cerac; Goldschmidt]

tin prepare liquor. Sodium stannate, $Na_2SnO_3.3H_2O$; used as a mordant for dyes in calico and printing.

tin protochloride. See stannous chloride.

tin salt. See stannous chloride.

tin salts. (tin crystals; stannous chloride) $SnCl_2$; used as a wool mordant for dyeing cochineal scarlet, for dyeing blacks on silk, for weighting silk, and for calico printing.

Tinset. CAS 60607-34-3; A proprietary preparation containing oxatomide; for allergic conjunctivitis and rhinitis, other antihistamine indications, food allergy. [Janssen Pharmaceutical Ltd] *

Tintacrete®. Dry powdered colorants, iron oxides, ochers, umbers and composite pigments sold in small packages for DIY trade; colors for mortars, concrete roofing tiles,

---

* Trade name not verified as available   † Trade name verified as obsolete

floor tiles, sand-lime bricks, concrete blocks, reconstructed stone, split blocks, paving slabs, and cement sheets. [W Hawley & Son Ltd]

Tint-Ayd. Pigment dispersions; used for paints, inks, etc. [Cornelius Chemical Co Ltd] *

Tinuvin®. A trademark for uv light absorbers for incorporation in plastics materials. [Ciba plc] *

Tinuvin® 622LD. CAS 65447-77-0; Polymeric hindered amine; dimethyl succinate polymer with tetramethyl hydroxy-1-hydroxyethyl piperidine; light and heat stabilizer for polyolefins, PP injection molded bars, HDPE, linear LDPE plaques, PP tape, LDPE film, PP multifilament, ABS polymer systems, thermoplastic polyolefins, flexible PVC, food pkg. [Ciba-Geigy/Additives]

Tinuvin® 770. CAS 52829-07-9; EINECS 258-207-9; Bis (2,2,6,6-Tetramethyl-4-piperidinyl) sebacate; light stabilizer for polyolefins, incl. natural and pigmented PP multifilament slit film, polyethylene, ethlene-propylene copolymer and terpolymer, PU, ABS, impact PS, and other styrenic polymers and copolymers; synergistic with Tinuvin P in ABS, impact PS, and PU; useful with costabilizers such as high-performance phenolic antioxidants. [Ciba-Geigy/Additives]

Tinuvin® P. A substituted benzotriazole derivative having a peak absorption at 340 mμ; recommended for PVC, polystyrene, and acrylics. [Ciba plc] *

tin white. (stannic hydroxide) $Sn(OH)_4$; a pigment used in enamel and glass-making.

tiodipropionic acid, distearyl ester. See distearyl thiodipropionate.

Tioga Adhesion Promoter 30-6-600. Conductive water-based adhesion promoter for polypropylene, TPO, and TPR. [Tioga Coatings]

Tiona®. CAS 13463-67-7; EINECS 236-675-5; Range of titanium dioxide pigments, including both anatase and rutile crystal forms, with surface treated grades for enhanced performance; opacifying and whitening of all paint systems, plastics and floorcoverings, paper, textiles, inks, ceramics, rubber and vitreous enamels. [SCM Chemicals Europe]

Tiona® HSS. CAS 13463-67-7; EINECS 236-675-5; Anatase titanium dioxide, high solids aq. disp.; provides opacity and brightness for highest quality fine papers. [SCM]

Tiona® RCL-4. CAS 13463-67-7; EINECS 236-675-5; Rutile titanium dioxide, alumina surface treatment; plastics grade additive for faster processing, superior dispersion, higher tint strength, outstanding resist. to PE yellowing. [SCM]

Tiona® RCL-535. CAS 13463-67-7; EINECS 236-675-5; Titanium dioxide; coatings grade providing high durability, haze-free gloss, opacity, and tinting strength with low oil absorp. [SCM]

Tiona® RCS-P. CAS 13463-67-7; EINECS 236-675-5; Rutile titanium dioxide aq. disp.; for fine paper, paperboard, and specialty paper applications. [SCM]

Tioveil. Uv protection product. [Tioxide Group plc]

Tiox. CAS 61570-90-9; Tioxidazole; anthelmintic. [Schering Corp] *

Tioxide. CAS 13463-67-7; EINECS 236-675-5; Titanium dioxide pigment; used for decorative and industrial paints, plastics, paper, printing inks, ceramics, and manmade fibers. [Tioxide Group plc]

TIP. A proprietary preparation; it is tetraiodophenolphthalein

for use in cholecystography. §

TIPA. See triisopropanolamine.

Tipoff. CAS 86-87-3; 1-Naphthylacetic acid; a plant growth regulator used to control suckering in fruit trees. [ICI Agrochemicals]

Ti-pure®. Titanium pigments. [DuPont UK]

Ti-Pure® R-103. CAS 13463-67-7; EINECS 236-675-5; Titanium dioxide; adhesive grade with improved rheology. [DuPont]

Tirucalli gum. A product of an Indian plant of the Euphorbia species. It somewhat resembles gutta-percha.

Tisco Steel. A proprietary trade name for a high manganese steel containing up to 15% manganese. §

Tised. CAS 57-53-4; A proprietary preparation of meprobamate; a sedative. [Unichem] *

Tisept® Solution. Cetrimide 0.15% and chlorhexidine gluconate 0.015%; antiseptic with detergent properties for swabbing in obstetrics, disinfecting and cleansing traumatic and surgical wounds and burns. [Seton Healthcare Group plc]

Ti-Sphere AB-15155A. CAS 13463-67-7, 7631-86-9; Titanium dioxide, silica; spherical powd. imparting spreadability and smooth creamy feel to cosmetic powds.; may be used as nonchemical sunscreen. [Presperse]

Tissalys. Modified starch; used in textile industry for sizing natural, artificial, and synthetic fibers. [Roquette (UK) Ltd]

Tissier's metal. An alloy of 97% copper, 2% zinc, and 1% arsenic.

TIS-U-SOL. Pentalyte; a combination of sodium chloride, potassium chloride, magnesium sulfate, sodium phosphate dibasic, potassium phosphate monobasic; electrolyte combination used to prepare a physiologic irrigation solution intended for use on wounds and open tissue surfaces. [Travenol Laboratories Inc] *

Tisyn®. CAS 12141-46-7; Aluminum silicate, thermo-optic; high hiding pigment for paint formulations, paper coatings, plastics, rubber, water and solv. systems. [Burgess Pigment]

Titan. CAS 13463-67-7; EINECS 236-675-5; Design for titanium dioxide pigments; used for paint, paper, ink, plastics, ceramics and glass. [Rheox Inc] †

Titan. CAS 999-81-5; Soluble concentrate containing 667 g/l chlormequat chloride; plant growth regulator. [Schering Agrochemicals Ltd]

Titan cements. Cements obtained by fusing a mixture of titaniferous iron ore, limestone, and coke. It consists essentially of calcium titanate ($CaTiO_3$) with small amounts of ferrites, aluminates, and calcium silicate, together with from 2-10% ferric oxide.

Titan Design. CAS 13463-67-7; EINECS 236-675-5; Design for titanium dioxide pigments; used for paints, plastics, inks, paper, ceramics and glass. [Rheox Inc] †

titanic acid anhydride. See titanium dioxide.

titanic anhydride. See titanium dioxide.

titaniferous iron. See ilmenite.

titaniferous iron ore. See ilmenite.

Titanital. The trade name for a proprietary titanium white; the golden seal brand contains from 95-98% titanium oxide. TiO_2 and the silver seal grade is a mixture of 80% titanium dioxide with 20% zinc oxide. §

Titanite. A proprietary aluminum-manganese alloy containing titanium. §

---

‡ Trade name and manufacturer not verified   § Trade name without identified manufacturer

Titanite No. 1. An explosive consisting of 85-88% ammonium nitrate, 6-8% trinitro-toluene, and 4.5-6.5% charcoal.

titanium. Metallic element; Ti; CAS 7440-32-6; EINECS 231-142-3; alloys; as structural material in aircraft, jet engines, marine equipment, chemical equipment, surgical instruments, orthopedic appliances, food-handling equipment; x-ray tube targets; abrasives; cermets; electrodeposited and dipped coatings. [Cerac; Inco Alloys Int'l.; New Metals & Chems.]

titanium alloy. A ferro-titanium is called by this name.

titanium dioxide. (titanic anhydride; titanic acid anhydride; titanium oxide) Inorganic oxide; TiO_2; CAS 13463-67-7; EINECS 236-675-5; white pigments in paints, paper, rubber, etc.; opacifying agent, cosmetics; radioactive decontamination of skin. [Bayer NV; British Traders & Shippers; Degussa; Du Pont; Ferro/Transelco; Kerr-McGee; Miles; SCM]

Titanium Dioxide 110. CAS 13463-67-7, 7787-59-9; Titanium dioxide, bismuthoxychloride; inorganic colorant. [Presperse]

Titanium Dioxide P25. CAS 13463-67-7; EINECS 236-675-5; Titanium dioxide; catalyst carrier for fixed bed catalyst; heat stabilizer for HCR silicone rubber and flame retardant; uv absorber for sunscreen lotions. [Degussa]

titanium oxide. See titanium dioxide.

Titanium Putty. Titanium reinforced epoxy resin; for repairing worn or gouged parts, rebuilding wear surfaces and reseating worn or oversized bearings. [ITW Devcon]

titanoferrite. A variety of the mineral ilmenite.

Titanox. CAS 13463-67-7; EINECS 236-675-5; Titanium dioxide pigments; used for paint, paper, plastics, ink, ceramics and glass. [Rheox Inc]

Titanox Design. CAS 13463-67-7; EINECS 236-675-5; Design for titanium dioxide pigments; used for paint, plastics, paper, glass and ceramics. [Rheox Inc]

Titanox RA-39. CAS 13463-67-7; EINECS 236-675-5; A stearate coated titanium dioxide pigment easily dispersible in polystyrene and polyolefins. [Laporte Industries Ltd]

Titanweiss C, Extra T, Standard T, Standard A. CAS 13463-67-7; EINECS 236-675-5; Trade names for titanium dioxide pigments extended with calcium or barium sulfates. §

Titite. A proprietary rubber cement, partly made from rubber; it is waterproof, and is used for mending cloth, paper, rubber, leather, and wood. §

Title®. For agriculture. [DuPont UK]

Ti-tone. A titanium lithopone containing 15% titanium dioxide, 25% zinc oxide, and 60% barium sulfate. Its specific gravity is 4.25, and it has a covering power 60% greater than ordinary lithopone.

Titralac. Each tablet contains calcium carbonate BP 420 mg and glycine 180 mg [3M Pharmaceuticals] *

Titus®. For agriculture. [DuPont UK]

Tixogel. Antisettling and thickening agents for paints, varnishes, lubricants, adhesives, coatings, putties and cosmetics. [Süd-Chemie AG]

Tixogel LAN. Stearalkonium bentonite (10%), lanolin oil (65%), isopropyl palmitate (22%), propylene carbonate (3%); for lipsticks, suntan products, creams and lotions. [United Catalysts]

Tixogel OMS. Quaternium-18 bentonite (10%), mineral spirits (87%), propylene carbonate (3%); for eye makeup, mascara, eyeshadow. [United Catalysts]

Tixogel VP. CAS 68953-58-2; Quaternium-18 bentonite; for low to medium polarity systems, e.g., antiperspirants; as suspension aid for active ingredients. [United Catalysts]

Tixogel VSP. Quaternium-18 bentonite (17.5%), cyclomethicone (79.5%), propylene carbonate (2.5%), water (0.5%); for suntan products, creams, lotions. [United Catalysts]

Tixogel VZ. Stearalkonium bentonite; for high polarity and oxygenated solv. systems such as nail lacquers. [United Catalysts]

Tixoton. Bentonites with high swelling properties for the drilling and building industry. [Süd-Chemie AG]

Tixylix. Pediatric cough linctus. [Rhone-Poulenc Rorer Ltd]

Tizit. An alloy of 40-80% tungsten, 4-15% titanium, 4% chromium, 2-4% carbon, 1-5% cerium, and 3-40% iron.

TKP. See potassium phosphate, tribasic.

TKPP. See potassium pyrophosphate.

TL 4190. Water-based polyurethane laminating adhesive; fast wet tack, high adhesion to various metal foils and films incl. polyester, PU, and vinyl, low foam coatability; for paper, film, fabric, or foil constructions. [Mace Adhesives & Coatings]

TLA-111B. Alkyl zinc dithiophosphate; antioxidant, antiwear agent for petrol. products. [Texaco]

TLA-227. Methacrylate; economical visc. index improver and pour point depressant with moderate shear stability and high thickening power; for petrol. products. [Texaco]

TLA-256. CAS 61789-86-4; Slightly basic calcium sulfonate; detergent for petrol. products. [Texaco]

T-lim. Modified rosins in aqueous emulsions; sizing agent for paper and paperboard. [Hercules] *

T-Maz® 20. CAS 9005-64-5; Polysorbate 20; emulsifier, solubilizer, wetting agent, antistat, stabilizer, dispersant, visc. modifier, suspending agent used in the food, cosmetic, drug, textile, and metalworking industries. [PPG/Specialty Chem.]

T-Maz® 28. CAS 9005-64-5; PEG-80 sorbitan laurate; emulsifier and solubilizer of essential oils, wetting agent, visc. modifier, antistat, stabilizer and dispersant used in food, cosmetic, drug, textile, and metalworking industries. [PPG/Specialty Chem.]

T-Maz® 40. CAS 9005-66-7; Polysorbate 40; emulsifier, solubilizer, wetting agent, antistat, stabilizer, dispersant, visc. modifier, suspending agent used in the food, cosmetic, drug, textile, and metalworking industries. [PPG/Specialty Chem.]

T-Maz® 60. CAS 9005-67-8; Polysorbate 60; emulsifier, solubilizer, wetting agent, antistat, stabilizer, dispersant, visc. modifier, suspending agent used in the food, cosmetic, drug, textile, and metalworking industries. [PPG/Specialty Chem.]

T-Maz® 61. CAS 9005-67-8; Polysorbate 61; emulsifier, solubilizer, wetting agent, antistat, stabilizer, dispersant, visc. modifier, suspending agent used in the food, cosmetic, drug, textile, and metalworking industries. [PPG/Specialty Chem.]

T-Maz® 65. CAS 9005-71-4; Polysorbate 65; emulsifier, solubilizer, wetting agent, antistat, stabilizer, dispersant, visc. modifier, suspending agent used in the food, cosmetic, drug, textile, and metalworking industries. [PPG/Specialty Chem.]

T-Maz® 80. CAS 9005-65-6; Polysorbate 80; emulsifier,

---

* Trade name not verified as available

† Trade name verified as obsolete

solubilizer, wetting agent, antistat, stabilizer, dispersant, visc. modifier, suspending agent used in the food, cosmetic, drug, textile, and metalworking industries. [PPG/Specialty Chem.]

T-Maz® 81. CAS 9005-65-6; Polysorbate 81; emulsifier, solubilizer, wetting agent, antistat, stabilizer, dispersant, visc. modifier, suspending agent used in the food, cosmetic, drug, textile, and metalworking industries. [PPG/Specialty Chem.]

T-Maz® 85. CAS 9005-70-3; Polysorbate 85; emulsifier, solubilizer, wetting agent, antistat, stabilizer, dispersant, visc. modifier, suspending agent used in the food, cosmetic, drug, textile, and metalworking industries. [PPG/Specialty Chem.]

T-Maz® 90. PEG-20 sorbitan tallate; emulsifier, solubilizer, wetting agent, antistat, stabilizer, dispersant, visc. modifier, suspending agent used in the food, cosmetic, drug, textile, and metalworking industries. [PPG/Specialty Chem.]

T-Maz® 95. PEG-20 sorbitan tritallate; emulsifier, solubilizer, wetting agent, visc. modifier, antistat, stabilizer, dispersant for food, cosmetic, drug, textile, and metalworking industries. [PPG/Specialty Chem.]

TMBAC. See benzyl trimethyl ammonium chloride.

T metal. An alloy of 95% aluminum, 4% magnesium, 0.5% silicon, 0.5% iron, and 0.1% copper.

TMP. Amine salts of organic acids, aromatic acid, aromatic and aliphatic petroleum distillate; for use with propanil herbicide in spraying rice to control evaporation, prevent crystallization of propanil and control drift. [Stull Chemical Company] *

TMPD® Glycol. CAS 144-19-4; 2,2,4-Trimethyl-1,3-pentanediol; resin intermediate. [Eastman]

TMTM. CAS 97-74-5; Tetramethyl thiuram monosulfide; nonstaining, nondiscoloring, fast-curing accelerator for use alone or in combination in NR, SBR, NBR, butyl rubber, neoprene, and reclaim rubber. [Akrochem]

T-Mulz® 66H. Phosphate ester, potassium salt; hydrotrope for heavy-duty liq. alkaline cleaners. [Harcros]

T-Mulz® 596. Phosphate ester, free acid; emulsion polymerization surfactant. [Harcros]

T-Mulz® 1158. Phosphate ester, free acid; emulsifier for mineral oil, metal processing, solv. cleaners, textile processing, agricultural formulations, corrosion inhibition; antistat; oilfield applications. [Harcros]

T-Mulz® AO2. Blend; emulsifier for spray oils used in conj. with pesticides. [Harcros]

T-Mulz® Mal 5. Calcium alkylaryl sulfonate/POE ether blend; high flash emulsifier for 5 lb/gal Malathion. [Harcros]

TNT. See Trotyl.

TNX. Abbreviation for tetranitroxylene.

Tobias acid. (2-naphthylamine-1-sulfonic acid; 2-amino-1-naphthalene-sulfonic acid) $C_{10}H_9NO_3S$; CAS 81-16-3; EINECS 201-331-5; Used as an azo dye intermediate and as an optical brightener.

Tobin bronze. Alloys of 59-83% copper, 3-48% zinc, 0.9-12.4% tin, 0.31-2.14% lead, and 0.1-0.8% iron. One alloy contains 58.79% copper, 40.43% zinc, and 0.88% tin.

Tobrex. CAS 32986-56-4; Tobramycin; antibacterial. [Alcon Laboratories Inc] *

Tochlorine. CAS 127-65-1; A proprietary preparation; it is chloramine-T. §

Tocopherex. CAS 59-02-9; Vitamin E supplement. [Bristol-Myers Squibb Co Inc] *

tocopherol. See vitamin E.

Tocopherol Oil CLR. Vitamin E carrier with natural tocopherols in soya oil medium; general skin care products. [Dr. Kurt Richter; Henkel/Cospha]

Today Sponge. CAS 9016-45-9; Nonoxynol-9 (1 g) as a spermicide; contraceptive; may also help to prevent the transmission of sexual disease during intercourse. [Whitehall Laboratories; Whitehall Laboratories Ltd]

Toffix. Special hard fat; used for manufacture of caramels and chewing sweets. [Dynamit Nobel Wien GmbH]

Toffix®. Fats containing an emulsifier; for prep. of toffees and chewing sweets. [Hüls AG]

Tofranil. CAS 113-52-0; A proprietary preparation of imipramine hydrochloride; an antidepressant. [Ciba plc]*

Togocoll. One and two-component polyurethanes, sealers, glazing, and structural adhesives, primers, underbody coatings, windshield-adhesives and corrosion preventative coatings (waxes) for automotive industries. [EMS-Chemie AG]

Togoplast. PVC plastisols; used for automotive underbody coatings, sealants and adhesive systems. [EMS-Chemie AG]

Togotec. Corrosion protective wax coatings for automotive box section and underbody protection. [EMS-Chemie AG]

Togotherm. Polyurethane foam systems; for filling car body box section. [EMS-Chemie AG]

Toho Me-PEG Series. Methoxy polyethylene glycol (m.w. 225, 350, 550, 705, 1000); base material for surfactant, synthetic resin, plasticizer, lubricating industries; wetting, softening, penetrating, lubricating and cleaning agent for textile, paper, ink, pigments, etc. [Toho Chem. Industry]

Toho PEG Series. Polyethylene glycol (m.w. 200, 300, 400, 600, 1000, 1500, 1540, 2000, 4000); base material for surfactant, synthetic resin, plasticizer, lubricating industries; wetting, softening, penetrating, lubricating, and cleaning agent for textile, paper, ink, pigments, etc. [Toho Chem. Industry]

Toho Salt A-5. Anionic complex; dyeing assistant, dispersant for disperse dyestuffs. [Toho Chem. Industry]

Tohol N-220. Cocamide DEA; foam stabilizer and thickener for shampoo, detergent, toothpaste. [Toho Chem. Industry]

Toisin's solution. A microscopic stain used for staining white blood corpuscles; based on methyl violet.

Toku Bishi®. Coke; for ductile and high quality cast iron. [Mitsubishi Kasei] *

Tokuthion®. Prothiofos; insecticide especially effective against leaf-eating caterpillars. [Bayer AG]

Tolan. (diphenyl-acetylene) $C(C_6H_5)$ $C(C_6H_5)$; CAS 501-65-5; EINECS 207-926-6; Used in organic synthesis.

Tolanase. CAS 1156-19-0; A proprietary preparation of tolazamide; an oral hypoglycemic agent. [Upjohn; Upjohn Ltd]

Tolcide MBT. CAS 6317-18-6; Methylenebis thiocyanate; biocide for use in water treatment, paper, antifoulant paint, leather, timber preservation. [Albright & Wilson UK]

Tolectin. CAS 64490-92-2; Tolmetin sodium; anti-inflammatory. [McNeil Pharmaceuticals] *

Toleron. CAS 141-01-5; Ferrous fumarate; hematinic. [Mallinckrodt Inc] *

Tolgard. Flame retardants. [Albright & Wilson Ltd]

---

‡ Trade name and manufacturer not verified   § Trade name without identified manufacturer

Tolinase. CAS 1156-19-0; Tolazamide; antidiabetic. [Upjohn]

Tolite. See Trotyl.

Tolkan. Herbicide. [May & Baker Ltd] *

Tolkan. CAS 34123-59-6; Suspension concentrate containing 500 g isoproturon per liter; used for annual weed control in cereals. [Rhone-Poulenc Crop Protection Ltd]

Tolkan 500. CAS 34123-59-6; Suspension concentrate containing 500 g isoproturon per liter; used for annual weed control in cereals. [Farmers Crop Chemicals Ltd]

Tollen's reagent. A solution of ammoniacal silver nitrate containing free caustic soda; used to test for aldehydes.

Tolnate. CAS 1225-65-6; A proprietary preparation of prothipendyl hydrochloride; a sedative. [SmithKline Beecham] *

Tolochrome. Photographic color developer. [May & Baker Ltd] *

Tolonate. Aliphatic diisocyanates. [Rhone-Poulenc UK]

Toloy 45. An alloy of 45% nickel and 20% chromium with other materials; used where stress corrosion resistance is required. The material conforms to BS.1648 Grade H.

Tolplaz. Specialty plasticizers. [Albright & Wilson Ltd]

tolu balsam. The oleoresin of *Myroxylon toluifera*, of South America.

toluene. (methylbenzene; phenylmethane; toluol) Aromatic compound; $C_5H_5CH_3$; CAS 108-88-3; EINECS 203-625-9; aviation gasoline additive; solvent for paint; diluent and thinner in nitrocellulose lacquers; adhesive solvent in plastic toys; manufacture of benzoic acid, benzaldehyde, explosives, dyes; in extraction of various principles from plants. [Ashland; Chevron; Exxon; Fina; Mitsubishi Petrochem.; Mitsui Petrochem. Ind.; Mobil; Phillips 66; Shell; Texaco; Unocal]

toluene diisocyanate. (TDI; toluene 2,4-diisocyanate; 2,4-tolylene diisocyanate; 2,4-diisocyanatotoluene) $CH_3C_6H_3(NCO)_2$; CAS 584-84-9; EINECS 209-544-5; In manufacture of polyurethane foams, elastomers, and coatings. [Bayer Hispania Industrial SA; ICI Polyurethanes; Nippon Polyurethane Ind.; Olin]

toluene 2,4-diisocyanate. See toluene diisocyanate.

p-toluenesulfonamide. (PTSA; toluene-4-sulfonamide) $CH_3C_6H_4SO_2NH_2$; CAS 70-55-3; EINECS 200-741-1; Organic synthesis; plasticizers, resins; fungicide and mildewcide in paints and coatings. [Allchem Industries; Honeywill & Stein Ltd; ICI Spec.; Rit-Chem.; Unitex]

toluene-4-sulfonamide. See p-toluenesulfonamide.

toluene sulfonic acid. (4-methylbenzenesulfonic acid) Substituted aromatic acid; $C_6H_4(SO_3H)(CH_3)$; CAS 104-15-4; 70788-37-3; EINECS 203-180-0; dyes, organic synthesis, acid catalyst. [Boliden Intertrade; BYK-Chemie; Eastman; Ferro/Grant; Manro Prods. Ltd; Nissan Chem. Ind.; PMC Spec.; Ruetgers-Nease; Witco]

toluhydroquinone. (methylhydroquinone) $CH_3C_6H_3(OH)_2$; CAS 95-71-6; EINECS 202-443-7; Antioxidant, polymerization inhibitor. [Eastman]

m-toluic acid. (3-methylbenzoic acid; m-toluylic acid) $C_6H_4CH_3COOH$; CAS 99-04-7; EINECS 202-723-9; Organic synthesis, to form N,N-diethyl-m-toluamide, a broad-spectrum insect repellent. [Mitsubishi Gas; Witco/Argus]

o-toluic acid. (2-methylbenzoic acid; o-toluylic acid) $C_6H_4CH_3COOH$; CAS 118-90-1; EINECS 204-284-9; Bacteriostat. [Mitsubishi Gas]

p-toluic acid. (4-methylbenzoic acid; p-toluylic acid) $C_6H_4CH_3COOH$; CAS 99-94-5; EINECS 202-803-3; Agricultural chemicals, animal feed supplement. [Hüls Am.; Nat'l. Starch & Chem.; Penta Mfg.]

α-toluic acid. See phenylacetic acid.

m-toluic acid diethylamide. See diethyl toluamide.

toluol. Obsolete name for toluene (*q.v.*).

p-toluoyl chloride. (4-methylbenzoyl chloride) C_8H_7ClO; CAS 874-60-2; EINECS 212-864-8. [Aldrich; James River]

Tolurex. CAS 15545-58-9; Active ingredient: chlortoluron; 3-(3-chloro-p-tolyl)-1,1-dimethylurea; selective pre- and post-emergence herbicide in winter cereals for control of annual grasses and broad leaved weeds. [Agan Chemical Manufacturers Ltd]

Tolurgan. CAS 15545-58-9; Chlortoluron; herbicide. [Agan Chemical Manufacturers Ltd]

toluylene. CAS 103-30-0; EINECS 203-098-5; (stilbene) $C_6H_5 \cdot CH : CH \cdot C_6H_5$; used in the manufacture of optical bleaches and dyes.

m-toluylic acid. See m-toluic acid.

o-toluylic acid. See o-toluic acid.

p-toluylic acid. See p-toluic acid.

p-tolyl aldehyde. $CH_3C_6H_4CHO$; CAS 104-87-0, 1334-78-7; EINECS 203-246-9; Perfumes, pharmaceutical, and dyestuff intermediate, flavoring agent. [BASF; Mallinckrodt; Mitsubishi Gas]

2,4-tolylene diisocyanate. See toluene diisocyanate.

tolyltriazole. CAS 29385-43-1; Corrosion inhibitor for copper, brass, bronze, ferrous metals; in metalworking fluids. [Dinoval Chem. Ltd; PMC Spec.; Sandoz]

Tomah AO-14-2. Bishydroxyethylisodecyloxypropylamine oxide; foam stabilizers/boosters in liq. detergents, shampoos, hard surface cleaners, laundry detergents; grease emulsifier, soil suspension aid; forms synergistic surfactant base for built household, institutional and industrial cleaners with quaternaries and nonionics. [Exxon/Tomah]

Tomah AO-728 Special. Amine oxide; detergent, foam booster/stabilizer for industrial and household detergents, dishwash, personal care products. [Exxon/Tomah]

Tomah BExM-1. Modifier for clay stabilized coal tar and asphalt emulsions. [Exxon]

Tomah DA-14. CAS 72162-46-0; N-Isodecyloxypropyl-1,3-diaminopropane; intermediate for textile foaming agents, surfactants, ethoxylates, agricultural chemicals; corrosion inhibitor for metalworking fluids; additive for fuels, lubricants, petrol. refining; crosslinking agent for epoxy resins; bactericidal properties. [Exxon/Tomah]

Tomah DA-16. N-Isododecyloxypropyl-1,3-diaminopropane; intermediate for textile foaming agents, surfactants, ethoxylates, agricultural chemicals; corrosion inhibitor for metalworking fluids; additive for fuels, lubricants, petrol. refining; crosslinking agent for epoxy resins. [Exxon/Tomah]

Tomah DA-17. N-Isotridecyloxypropyl-1,3-diaminopropane; intermediate for textile foaming agents, surfactants, ethoxylates, agricultural chemicals; corrosion inhibitor for metalworking fluids; additive for fuels, lubricants, petrol. refining; crosslinking agent for epoxy resins. [Exxon/Tomah]

Tomah E-14-2. CAS 34360-00-4; Bis-(2-hydroxyethyl) isodecyloxypropylamine; emulsifier, corrosion inhibitor, lubricant used in mineral acid inhibition, textile process-

---

* Trade name not verified as available   † Trade name verified as obsolete

ing. [Exxon/Tomah]

Tomah E-14-5. PEG-5 isodecyloxypropylamine; emulsifier, corrosion inhibitor, lubricant used in mineral acid inhibition, textile processing. [Exxon/Tomah]

Tomah E-17-2. Bis-(2-hydroxyethyl) isotridecyloxypropylamine; emulsifier, corrosion inhibitor, lubricant used in mineral acid inhibition, textile processing. [Exxon/Tomah]

Tomah E-18-2. Bis (2-hydroxyethyl) octadecyloxypropylamine; emulsifier, corrosion inhibitor, lubricant used in mineral acid inhibition, textile processing. [Exxon/Tomah]

Tomah E-18-5. PEG-5 stearyloxypropylamine; emulsifier, corrosion inhibitor, lubricant used in mineral acid inhibition, textile processing. [Exxon/Tomah]

Tomah E-19-2. Bis (2-hydroxyethyl) linear alkyloxypropylamine; surfactant to modify emulsification, surface tension, solubility; for acid thickeners, antistats, petrol. production and refining, agricultural adjuvants, textile processing aids, corrosion inhibition, detergent boosters; chem. intermediate. [Exxon/Tomah]

Tomah E-24-2. PEG-2 Guerbet C20 alcohol amine; surfactant, corrosion inhibitor. [Exxon/Tomah]

Tomah E-DT-3. PEG-3 1,3-diaminopropane; surfactant for acid thickeners, antistat, cationic emulsification, petrol. production and refining, agricultural adjuvants, textile processing aids, corrosion inhibition, detergent boosters; chem. intermediate. [Exxon/Tomah]

Tomah E-S-2. CAS 61791-24-0; PEG-2 soyamine; emulsifier, corrosion inhibitor, lubricant used in mineral acid inhibition, textile processing. [Exxon/Tomah]

Tomah E-T-2. CAS 61791-44-4; PEG-2 tallowamine; surfactant for acid thickeners, antistat, cationic emulsification, petrol. production and refining, agricultural adjuvants, textile processing aids, corrosion inhibition, detergent boosters; chem. intermediate. [Exxon/Tomah]

Tomah PA-10. Hexyloxypropylamine; corrosion inhibitor for metalworking fluids; antistat; flotation collector; additive for fuel, lubricant, petrol. refining; intermediate for surfactants, textile foaming agents, ethoxylates, and agricultural chem.; crosslinking agent for epoxy resins. [Exxon/Tomah]

Tomah PA-12EH. CAS 5397-31-9; 2-Ethylhexyloxypropylamine; detergent intermediate. [Exxon/Tomah]

Tomah PA-13i. CAS 29317-52-0; Isononyloxypropylamine; detergent intermediate. [Exxon/Tomah]

Tomah PA-14. CAS 7617-78-9; Isodecyloxypropylamine; emulsifier; corrosion inhibitor for metalworking fluids; antistat; flotation collector; additive for fuel, lubricant, petrol. refining; intermediate for surfactants, textile foamers, ethoxylates, and agricultural chem.; crosslinking agent for epoxies. [Exxon/Tomah]

Tomah PA-14 Acetate. Isodecyl oxypropyl amine acetate; patented emulsifier, gellation/wetting agent for clays, fillers, and fibers in organic coatings and sealants, e.g., roof coatings, tile adhesives, caulks, pipe coatings, automotive undercoatings, alkyd paints, foundry coatings, polymers/elastomers; bactericidal properties. [Exxon/Tomah]

Tomah PA-16. Isododecyloxypropylamine; corrosion inhibitor for metalworking fluids; antistat; flotation collector; additive for fuel, lubricant, petrol. refining; intermediate for surfactants, textile foamers, ethoxylates and agricultural chem.; crosslinking agent for epoxies. [Exxon/Tomah]

Tomah PA-17. Isotridecyloxypropylamine; corrosion inhibitor, cationic emulsification, and replacement for oleyl and soya amines; chemical intermediate. [Exxon/Tomah]

Tomah PA-19. CAS 68610-26-4; Linear C12-15 alkyloxypropylamine; corrosion inhibitor for metalworking fluids; antistat; flotation collector; additive for fuel, lubricant, petrol. refining; intermediate for surfactants, textile foamers, ethoxylates, and agricultural chemicals; crosslinking agent for epoxies. [Exxon/Tomah]

Tomah PA-24. Isoarachidyloxypropylamine; Guerbet C20 alcohol primary amine; detergent intermediate; experimental product for research and development. [Exxon/Tomah]

Tomah PA-1214. Octyl/decyloxypropylamine; corrosion inhibitor for metalworking fluids; antistat; flotation collector; additive for fuel, lubricant, petrol. refining; intermediate for surfactants, textile foaming agents, ethoxylates and agricultural chem.; crosslinking agent for epoxy resins. [Exxon/Tomah]

Tomah Q-14-2. CAS 125740-36-5; Isodecyloxypropyl dihydroxyethyl methyl ammonium chloride; quaternary used as acid corrosion inhibitor, plastics and textile antistat, and emulsifier; bactericidal properties. [Exxon/Tomah]

Tomah Q-17-2. Isotridecyloxypropyl dihydroxyethyl methyl ammonium chloride, IPA; emulsifier; boosts efficiency of nonionic surfactants; used in hard surface cleaners, laundry, transportation cleaners; bactericidal properties. [Exxon/Tomah]

Tomah Q-18-2. Octadecyl dihydroxyethyl methyl ammonium chloride, IPA; quaternary surfactant for use as emulsifiers, antistats, corrosion inhibitors, nonionic detergency booster in laundry products, in nonbutyl cleaning systems. [Exxon/Tomah]

Tomah Q-24-2. Methyl dihydroxyethyl isoarachidaloxypropyl ammonium chloride, IPA; Guerbet C20 alcohol dihydroxyethyl ammonium chloride; detergency booster with biocidal activity; experimental material. [Exxon/Tomah]

Tomah Q-2C. CAS 61789-77-3; EINECS 263-087-6; Dicoco dimonium chloride, IPA; detergency booster with biocidal activity. [Exxon/Tomah]

Tomah Q-311. Monosoya amidoamine quaternary; detergency booster with biocidal activity. [Exxon/Tomah]

Tomah Q-511. Monococo amidoamine quaternary; detergency booster with biocidal activity. [Exxon/Tomah]

Tomah Q-C-15. CAS 61791-10-4; PEG-15 cocomonium chloride; quaternary surfactant for use as emulsifiers, antistats, corrosion inhibitors, nonionic detergency booster in laundry products, in nonbutyl cleaning systems. [Exxon/Tomah]

Tomah Q-D-T. Tallow dimethyl trimethyl propylene diammonium chloride, IPA; quaternary used as an acid corrosion inhibitor, plastics and textile antistat, and emulsifier. [Exxon/Tomah]

Tomah Q-DT-HG. Tallow diamine quaternary in hexylene glycol; quaternary surfactant for use as emulsifiers, antistats, corrosion inhibitors, nonionic detergency booster in laundry products, in nonbutyl cleaning systems. [Exxon/Tomah]

Tomah Q-S. CAS 61790-41-8; EINECS 263-134-0; Soya trimethyl ammonium chloride, IPA; quaternary surfactant for use as emulsifiers, antistats, corrosion inhibitors, nonionic detergency booster in laundry products and in

---

‡ Trade name and manufacturer not verified   § Trade name without identified manufacturer

nonbutyl cleaning systems. [Exxon/Tomah]

Tomah Q-ST-50. CAS 112-03-8; EINECS 203-929-1; Steartrimonium chloride; quaternary surfactant for use as emulsifiers, antistats, corrosion inhibitors, nonionic detergency booster in laundry products, in nonbutyl cleaning systems. [Exxon/Tomah]

Tomahawk. CAS 70124-77-5; An emulsifiable concentrate containing flucythrinate; a pyrethroid insecticide for the control of aphids, whitefly, caterpillars, and red spider mite. [Fisons plc, Horticulture Div] *

Tomaset. Active ingredient: N-m-tolylphthalamic acid; a flower and fruit plant growth regulator. [Agan Chemical Manufacturers Ltd]

Tomato Setting Spray. Aerosol spray containing 2-naphthyloxyacetic acid; setting spray for tomatoes. [Vitax Ltd]

Tombac. An alloy usually containing 89% copper, 5.5% zinc, and 5.5% tin.

Tombasil. A proprietary trade name for an alloy consisting mainly of tombac metal with silicon. §

Tombel. Mixture of quinalphos and thiometon; for control of caterpillers and aphids. [Hortichem Ltd]

Tomophan. A proprietary viscose packing material. §

Tomorite. Liquid fertilizer. [Fisons plc, Horticulture Div] *

Toncan. A corrosion-resisting alloy containing pure iron, copper, and molybdenum.

Toncas metal. An alloy of 29% nickel, 36% copper, 7.1% iron, 7.1% zinc, 7.1% lead, 7.1% tin, and 7.1% antimony; used for ornamental work.

Tone. Caprolactone derivatives. [Union Carbide]

Tonite. (Potentite). An explosive consisting of mixtures of granulated gun cotton and barium nitrate.

tonka-bean camphor. CAS 91-64-5; Coumarin, found in species of *Dipteryx*; used as a deodorizing agent and in pharmaceutical preparations.

Tonocard. CAS 41708-72-9; Tocainide; cardiac depressant (anti-arrhythmic). [Merck & Co Inc]

Tonophosphan. CAS 575-75-7; A proprietary preparation of toldimfos sodium; a source of phosphorus used for veterinary purposes. §

Tonox® 22. CAS 101-77-9; Crude methylene dianiline; curing agent for epoxy resins. [Uniroyal]

Tonox® R. CAS 101-77-9; p,p´-Diaminodiphenylmethane; epoxy curative. [Uniroyal]

tonsil. See Frankonite.

Tonsil. For the adsorptive decolorization and purification of oils and fats, hydrocarbons, waxes, and other liquid intermediate products. [Süd-Chemie AG]

tonsil L80. See bentonite.

Toolife. A broad line of industrial fluids, both oil and synthetic in nature, water soluble and straight. [Specialty Products Co] *

Tool Life. Water-based synthetic cutting fluid (extreme pressure); for all metal removing machining operations where physical and chemical extreme pressure assistance is required to assist functional cooling properties on all metals, except magnesium. [Sumner Oil Industries] ‡

Top 7 Mosaic and Pebble. Expanded polystyrene veneer 7mm thick; decorative veneers for domestic ceilings. [Vencel Resil Ltd] *

Topal. A reflux suppressant and antacid. [Imperial Chemical Industries plc]

Topanex 100BT. CAS 2440-22-4; EINECS 219-470-5; 2-

(2´-Hydroxy-5´-methylphenyl) benzotriazole; uv absorber, light stabilizer, antioxidant protecting plastics (PVC, styrenics, acrylics, unsaturated polyesters), lacquers; effective at 290-380 nm. [ICI Am.]

Topanex 500H. 1,5-Dioxaspiro[5.5]undecane 3,3-dicarboxylic acid, bis(2,2,6,6-tetramethyl-4-piperidinyl)ester; uv stabilizer. [ICI Polymer Additives]

Topanol®. A range of antioxidants. [ICI Chem. & Polymers Ltd]

Topanol® 205. 2,2-Bis[4-(2-(3,5-di-t-butyl-4-hydroxyhydrocinnamoyloxy)) ethoxyphenyl] propane; high molecular weight hindered phenolic antioxidant for retardation of thermal and oxidative degradation in polyolefins, styrenics, and engineering polymers. [ICI Polymer Additives]

Topanol® CA. CAS 1843-03-4; 1,1,3-Tris (2-methyl-4-hydroxy-5-t-butyl phenyl) butane; antioxidant for polyolefin and styrenic polymers, plasticizers, hot melt adhesives, SBR latexes, polyamides, polyesters; FDA approved. [ICI Polymer Additives]

Topanol® LVT 11. CAS 1843-03-4; 1,1,3-Tris (2-methyl-4-hydroxy-5-t-butyl phenyl) butane in phthalate plasticizer; antioxidant, stabilizer for PVC wire and cable compounds. [ICI Polymer Additives]

Topanol® LVT 600; Antioxidant dispersion. [ICI Polymer Additives]

Topanol® M. CAS 101-96-2; N,N-Di-sec-butyl-p-phenylenediamine, commercial grade. [ICI Chem. & Polymers Ltd]

Topas 100. Emulsifiable concentrate containing 100 g/l penconazole; for control of powdery mildew. [Ciba-Geigy Agrochemicals]

Topas C 50WP. Captan + penconazole; protectant fungicide for apple and pear trees. [Ciba-Geigy Agrochemicals]

Topaz. Cooling water treatment. [Grace Dearborn Ltd]

Topbraun®. Suntan product without a sun protection factor. [Bayer AG]

Topclip Dridress. Animal health organophosphorus wound dressing. [Ciba plc] *

Topclip Fly and Scab Dip. Organophosphorus/phenols sheep dip. [Ciba plc] *

Topclip Foot Rot Aerosol. Antiseptic sheep treatment. [Ciba plc] *

Topclip Formalin. Formaldehyde for sheep foot rot control. [Ciba plc]

Topclip Gold Shield. Scab approved organophosphorus sheep dip. [Ciba plc]

Topclip Marker Aerosols. Sheep marker dyes. [Ciba plc] *

Topclip Marker Fluid. Sheep marker fluid. [Ciba plc] *

Topclip Parasol. CAS 66841-24-5; Cypermethrin pour-on; lice control for sheep and cattle. [Ciba plc]

Topclip Scab Dip. Organophosphorus/phenols sheel dips. [Ciba plc] *

Topclip Sheep Dip. Organophosphorus sheep dip. [Ciba plc] *

Topclip Vaccines. Various sheep vaccines. [Ciba plc] *

Topclip Wormer. Organophosphorus anthelmintic for sheep and pigs. [Ciba plc] *

Top-Cop. Mixture of copper sulfate and sulfur; a contact fungicide to control mildew. [Stoller Chemicals Ltd]

Topex. CAS 94-36-0; Buffered acne medication, 10% benzoyl peroxide; to help clear pimples without overdrying the skin. [Richardson-Vicks Inc] *

---

* Trade name not verified as available   † Trade name verified as obsolete

Topexane. Antibacterial face wash to help clear acne blemishes. [Richardson-Vicks Inc] *

Topfix. A range of epoxy, polyurethane and cyanoacrylate structural adhesives. [Atochem UK/Ceca]

Tophet. An electrical resistance alloy containing 61% nickel, 10% chromium, 26% iron, and 3% manganese.

Tophet A. A proprietary trade name for 80% nickel, 20% chrome resistance wire. §

Tophet C. A proprietary trade name for nickel chrome iron resistance wire. §

Topicort. CAS 382-67-2; Desoximetasone; anti-inflammatory. [Hoechst-Roussel Pharmaceuticals Inc]

Topicycline. CAS 64-75-5; Tetracycline hydrochloride; anti-amebic; antibacterial; and antirickettsial. [Norwich Eaton Pharmaceuticals Inc] *

Topostasin. CAS 9002-04-4; Proprietary preparation of thrombin; local hemostatic. [Roche Products Ltd] *

Toppel. CAS 66841-24-5; Insecticide containing cypermethrin. [ICI Chem. & Polymers Ltd]

Toppel. CAS 66841-24-5; Emulsifiable concentrate containing 100 g cypermethrin per liter; a pyrethroid insecticide. [Farm Protection Ltd]

Topshot. Bentazone + cyanazine + 2,4-DB; an herbicide. [Shell UK]

Topsol. Textile combing lubricant. [Crosfield Chemicals Ltd]

Topsyn. CAS 356-12-7; Fluocinonide; glucocorticoid. [Syntex Laboratories Inc] *

Topup. CAS 61791-44-4; Cationic surfactant containing 800 g/l ethoxylated tallow amine; for use as a wetting agent for phosphonoglycine herbicides. [Farmers Crop Chemicals Ltd]

Torapron. An emulsifiable concentrate containing 95 g aminotriazole, 190 g atrazine and 99 g 2,4-D per liter; used for total weed control in non crop areas. [BP Oil Ltd]

torbanite. A variety of cannel coal.

Torbutrol. CAS 58786-99-5; Butorphanol tartrate; analgesic; antitussive. [Bristol Laboratories] *

Tordon. CAS 1918-02-1; Herbicides based primarily on picloram; broadleaf and brush killers for forestry, grain and corn. [Dow UK]

Tordon 22K. CAS 1918-02-1; Soluble concentrate containing 240 g/l picloram; a picolinic herbicide to control woody weeds in noncrop areas. [Chipman Ltd]

Torecan. CAS 1420-55-9; A proprietary trade name for thiethylperazine or its salts; an anti-emetic. [Sandoz] *

Torelle. CAS 5598-52-7; Fospirate; anthelmintic. [Dow UK]*

Torgyl. CAS 443-48-1; Metronidazole injection. [RMB Animal Health Ltd]

Toric Contact Lens. Hefilcon B; 2-hydroxyethyl methacrylate with 1-vinyl-2-pyrrolidinone and ethylene dimethacrylate; contact lens material. [Bausch & Lomb, Professional Products Div] *

Torlon® 4203L. Poly(amide-imide) resin; high-strength resin for connectors, switches, relays, valve seats, mechanical linkages, bushings, wear rings, insulators, cams, ball bearings, rollers, and thermal insulators. [Amoco; Polymer Corp.]

Torlon® 4275. Polyamide-imide engineering resins filed with 20% graphite powder; for bearings, thrust washers, wear pads, strips, piston rings, seals, vanes and valve seats. [Amoco Chemical Co] *

Torlon® 4301. Polyamide-imide engineering resins filled with 12% graphite powder, 3% fluorocarbon; for bearings, thrust washers, wear pads, strips, piston rings, seals, vanes, valves, seats. [Amoco Chemical Co] *

Torlon® 4347. Poly(amide-imide) resin, 12% graphite powd.; wear-resistant resin for reciprocating motion or bearings subject to high loads at low speeds, e.g., for bearings, thrust washers, wear pads, strips, piston rings, seals. [Amoco Chemical Co]

Torlon® 5030. Poly(amide-imide) resin, 30% glass fiber-reinforced; high-strength resin used for burn-in sockets, gears, valve plates, fairings, tube clamps, impellors, rotors, housings, back-up rings, terminal strips, insulators, brackets. [Amoco; Polymer Corp.]

Torlon® 7130. Polyamide-imide engineering resins filled with 30% graphite fiber, 1% fluorocarbon; for metal replacements, housings, mechanical linkages, gears, fasteners, spline linears, cargo rollers, brackets, valves, labyrinth seals, fairings, tube clamps, standoffs, impellers shrouds, potential use for EMI shielding. [Amoco Chemical Co] *

Torlon® 7330. Poly(amide-imide) resin proprietary blend with carbon fibers and fluorocarbons; high-strength resin used for sliding vanes; potential use for EMI shielding. [Amoco Chemical Co]

tormentil. The dried rhizome of *Potentilla tormentilla*; used for tanning and as an astringent.

Tormol. A nickel-chromium-molybdenum steel highly resistant to shock and fatigue. §

Tormosyl. CAS 40507-23-1; Fluproquazone; analgesic. [Sandoz Pharmaceuticals] *

Tornac. Hydrogenated NBR. [Bayer plc]

Tornac B 3850. Hydrogenated (98%) nitrile rubber; non-staining stabilizer; offers superior resist. to heat, ozone, and nonpolar HC fluids incl. those containing aggressive additives; vulcanizable by sulfur and peroxides. [Miles; Polysar]

Tornado. CAS 63-25-2; A suspension concentrate containing 450 g carbaryl per liter; contact insecticide and worm killer. [ICI Agrochemicals Professional Products]

Tornalate. CAS 30392-41-7; Bitolterol mesylate; bronchodilator. [Sterling Drug Inc] *

Tornesit. A trade name for a protective coating base prepared by the chlorination of rubber. §

Tornusil. CAS 13597-65-4; A proprietary two-pack moisture cured inorganic zinc silicate primer. [Sigma Coatings] *

Toro. CAS 15545-48-9; Suspension concentrate containing 500 g chlorotoluron per liter; a contact urea herbicide for cereal crops. [Sipcam UK Ltd]

toron. A sulfur-terpene compound prepared by heating turpentine with sulfur. It is a black viscid liquid or semi-solid; used for waterproofing cloth, preparing rubberized cloth, and for attaching or coating metal surfaces with rubber.

Torqseal. Anaerobic thread locking fluid; suitable for thread locking and securing studs and bearings; easily undone with normal hand tools. [Hermetite Products Ltd] *

Torque. Fenbutatin; an acaricide. [ICI Agrochemicals]

Torrax. Malt flour. [Rhone-Poulenc UK]

Tosmilen. CAS 56-94-0; A proprietary preparation of demecarium bromide; eye drops used in the treatment of glaucoma. [Astra Chemicals Ltd] ‡

Totablan. Hydrogen peroxide bleaching stabilizer and wetting agent. [Ceca SA]

Totacillin. CAS 69-53-4; Ampicillin; antibacterial. [SmithKline Beecham]

Totacol. Herbicide based on diuron and paraquaternary [ICI

---

‡ Trade name and manufacturer not verified   § Trade name without identified manufacturer

Chem. & Polymers Ltd]

Totocillin®. CAS 69-53-4; Ampicillin; penicillinase-resistant broad-spectrum preparation for the treatment of mastitis in cattle, sheep and goats; also known as Cervantal. [Bayer AG]

Totolin. A proprietary preparation of phenylpropanolamine hydrochloride and guaifenesin; a cough linctus. [Galen Ltd] *

Totomycin. CAS 64-75-5; A proprietary preparation containing tetracycline hydrochloride; an antibiotic. [The Boots Co plc]

Totril. Herbicide. [May & Baker Ltd] *

Totril. CAS 1689-83-4; Ioxynil; contact herbicide for use in onion crops. [Embetec Crop Protection Ltd]

Toucas metal. An alloy of 35.75% copper, 28.56% nickel, 7.1% zinc, 7.2% tin, 7.1% lead, 7.2% antimony, and 7.1% iron.

Touch & Go. A proprietary preparation of camphor, ether, cajuput oil, and tolu balsam; a toothache solution. [Ayrton Saunders plc] ‡

Touchpong gum. See Pouckpong gum.

touchstone. See Lydian stone.

tough copper. Commercial copper, containing impurities such as arsenic.

toughened caustic. Consists of 95% silver nitrate and 5% potassium nitrate, fused together.

Tough Gel. Ion exchange resins. [Dow]

Tournant Oil. A commercial brand of olive oil obtained from fermented marc of expressed olives; contains free fatty acids, and is used as a Turkey- red oil. §

Tournay's metal. An alloy of 82.5% copper and 17.5% zinc.

tournesol. See litmus.

tous-les-mois starch. (Queensland arrowroot). The starch from the rhizomes of *Canna edulis*.

Tova. A proprietary preparation of cyclical hormones consisting of 16 white tablets of ethinyl estradiol and 5 pink tablets of ethinyl estradiol and dimethisterone; used for control of menstruation. [British Drug Houses] *

Toval®. Fibers. [DuPont UK]

Tower Brick. Sodium phosphates, wetting agents, and corrosion inhibitors; used in open recirculating cooling water systems to prevent lime scale and corrosion from fouling up the system. [Delaware Chemical Corp] *

Tower Treat. Liquid algicide/biocide; used for cooling water treatment. [Schaefer Technologies Inc]

toxilic anhydride. See maleic anhydride.

Toximul®. Sulfonate/nonionic blend; agricultural blender. [Stepan]

Toximul® 600. Sulfonate/nonionic blend; general purpose emulsifier for pesticides. [Stepan; Stepan Canada]

Toximul® 8240. CAS 61791-12-6; PEG-36 castor oil; emulsifier for agricultural formulations. [Stepan; Stepan Canada; Stepan Europe]

Toximul® 8320. Butyl EO/PO block copolymer; emulsifier component, flowable surfactant, intermediate for pesticides. [Stepan; Stepan Canada]

Toximul® D. Sulfonate/nonionic blend; matched pair emulsifier with Toximul H-HF for pesticide formulations; dispersant, stabilizer, hydrophobic agent. [Stepan; Stepan Canada; Stepan Europe]

Toximul® H-HF. Sulfonate/nonionic blend; matched pair emulsifier with Toximul D for pesticides; dispersant, stabilizer, hydrophilic agent. [Stepan; Stepan Canada; Stepan Europe]

Toximul® SEE-340. PEG-20 sorbitan tritallate; emulsifier for agricultural formulations. [Stepan; Stepan Canada; Stepan Europe]

Toximul® TA-2. CAS 61791-44-4; PEG-2 tallowamine; emulsifier for agricultural formulations. [Stepan; Stepan Canada; Stepan Europe]

Toyocat®-DMA. CAS 108-01-0; N,N'-Dimethylethanolamine; catalyst for PU slabstock and molded flexible, semirigid, and rigid applications. [Tosoh]

Toyocat®-DMCH. N,N'-Dimethylcyclohexylamine; catalyst for PU rigid applications. [Tosoh]

Toyocat®-DT. N,N,N',N'', N''-Pentamethyldiethylenetriamine; blowing catalyst for PU slabstock and molded flexible and rigid applications. [Tosoh]

Toyocat®-ET. Bis(dimethylaminoethyl) ether, dipropyleneglycol; blowing catalyst for PU slabstock and molded flexible and rigid applications. [Tosoh]

Toyocat®-HPW. N-Methyl-N'-hydroxyethylpiperazine, 10% water; reactive catalyst for PU slabstock and molded flexible, semirigid applications. [Tosoh]

Toyocat®-MR. N,N,N',N'-Tetramethylhexanediamine; nonthermosensitive catalyst for PU slabstock and molded flexible, semirigid and rigid applications. [Tosoh]

Toyocat®-NEM. CAS 100-74-3; N-Ethylmorpholine; catalyst for PU molded flexible, semirigid, and rigid applications. [Tosoh]

Toyocat®-NP. N,N',N'-Trimethylaminoethyl piperazine; thermosensitive catalyst for PU slabstock and molded flexbile, rigid, and elastomer shoe sole applications. [Tosoh]

Toyocat®-TE. N,N,N',N'-Tetramethylethylenediamine; balanced blowing/gelling catalyst for PU molded flexible, semirigid, and rigid applications. [Tosoh]

Toyocerin®. Probiotic feed additive. [Asahi Chem. Industry]

TP-35 Solvent. Precision cleaner/defluxer; removes residual moisture from electronic equip.; for critical cleaning of PCBs, components, magnetic tape heads, optical lenses, precision instruments; safe for most plastics. [Chemtronics]

TPG. Abbreviation for triphenylguanidine.

TPL. CAS 1314-60-9; Colloidal antimony pentoxide. [Laurel Industries Inc]

TPP. See triphenyl phosphate.

TPP. See triphenyl phosphite.

TPS 20. Diterdodecyl trisulfide; EP additive for gear oils, metalworking fluids compat. with copper and silver alloys, and rolling greases. [Atochem SA; Atochem UK Ltd]

TPS 27. Diternonyl trisulfide; EP additive for gear oils, metalworking fluids compat. with copper and silver alloys, and rolling greases. [Atochem SA; Atochem UK Ltd]

TPS 32, 37. Diterdodecyl pentasulfide; EP additive for metalworking fluids; recommended for clear premium EP lubricants; not compat. with copper and silver alloys. [Atochem SA; Atochem UK Ltd]

TPS 327. Ditertio nonyl pentasulfide. [Atochem UK Ltd]

TPX-80CNI. Polymethylolpentane, 80% carbonyl iron powd. [Compounding Tech.] *

Tra-Bond 2151. CAS 25928-94-3; Two-part epoxy; thermal conductive elec. insulating compound; used for staking transistors, diodes, resistors, integrated circuits, other heat-sensitive components to printed circuit boards; bonds readily to itself and to metals, silica, alumina, other

---

ceramics, glass, plastics, etc.; provides excellent resist. to salt solutions., mild acids and alkalis, petrol. solvs., lubricating oils, alcohol. [Tra-Con]

Tra-Bond F113. CAS 25928-94-3; Epoxy adhesive; high-impact fiber optic adhesive for bonding opto-electronic lens displays, SMA connectors; excellent glass-glass bonds, superior wicking. [Tra-Con]

Trabuk. An alloy containing 87.5% tin, 5.5% nickel, 5% antimony, and 2% bismuth. It resists vegetable acids.

Tra-Cast 3103. CAS 25928-94-3; Epoxy casting compound; general purpose. [Tra-Con]

Tracervial-131. CAS 7790-26-3; Sodium iodide I 131; antineoplastic; diagnostic aid and radioactive agent. [Abbott Laboratories] *

Tracey B. Trace element complex containing boron. [Mandops (UK) Ltd]

Tracey C. Trace element chelate containing copper. [Mandops (UK) Ltd]

Tracey M Plus. Trace element chelate containing copper and manganese. [Mandops (UK) Ltd]

Tracey MG. Trace element chelate containing magnesium. [Mandops (UK) Ltd]

Tracey SSS. Foliar feed containing sulfur and nitrogen. [Mandops (UK) Ltd]

Trachine. Fowl laryngotracheitis; for immunization of poultry. [Intervet Inc]

trachyte. A volcanic rock composed of felspar with some hornblende and mica.

Tracilon. Triamcinolone diacetate; glucocorticoid. [Savage Laboratories] †

Track. Suspension concentrate containing 500 g/l metazachlor; for weed control in brassicas and ornamental crops. [Kommer-Brookwick Ltd]

Tracker. CAS 1918-00-9; Dicamba; benzoic herbicide to control bracken. [Shell UK]

Tracrium Injection. CAS 64228-81-5; A proprietary formulation of atracurium besylate; an adjunct to general anesthesia, to facilitate endo-tracheal intubation and to provide skeletal muscle relaxation during surgery or mechanical ventilation. [The Wellcome Foundation Ltd]

Tractium. CAS 64228-81-5; Atracurium besylate; relaxant. [Burroughs Wellcome Co] *

Tradenal. CAS 466-06-8; Proscillaridin; cardiotonic. [Knoll Pharmaceutical Co] *

Tra-Duct 2902. CAS 25928-94-3; Silver paste epoxy; conductive silver paste epoxy adhesive; maximum use temp. to 110 C. [Tra-Con]

Traffaid 30 B. Barium/calcium lignosulfonate; antiscumming agent in brick and tile manufacturing. [Borregaard LignoTech]

tragasol. A gum obtained by steeping locust-bean kernels in water; used as a binding material.

Tramacin. CAS 76-25-5; Triamcinolone acetonide; glucocorticoid. [Johnson & Johnson Products Inc] *

Tramisol® . Levamisole hydrochloride; veterinary anthelmintic. [Am. Cyanamid]

Trancopal. CAS 80-77-3; A proprietary preparation of chlormezanone; a sedative. §

Trancoprin. Tablets containing chlormezanone and aspirin; an analgesic with a mild tranquilizer/muscle relaxant. [Winthrop Laboratories] *

Trandate. CAS 32780-64-6; Labetalol hydrochloride; anti-adrenergic. [Duncan Flockhard Ltd]

Tranquo-Adamon®. The trade name for an antispasmodic and tranquillizer. §

Transclene. Transformer dielectric solution. [Occidental Chemical Corp]

Transcutol. CAS 111-90-0; EINECS 203-919-7; Ethyldiglycol; solv. for active ingredients in pharmaceutical preparations; cosurfactant for microemulsions. [Gattefosse SA]

Transderm-Nitro. CAS 55-63-0; Nitroglycerin; vasodilator. [Ciba-Geigy Corp] *

Trans Gard. Blend of petroleum additives and petroleum distillates; automatic transmission stop leak and fluid conditioner. [Gard Corporation] *

Transjojoba. Jojoba oil derivatives. [A & E Connock (Perfumery & Cosmetics) Ltd]

Translink® . CAS 1332-58-7; Fine particle size surface treated calcined kaolin (aluminum silicate); used as a reinforcer and extender for rubber, wire and cable, plastics, gel coats. [Engelhard]

Translink® 37. Calcined aluminum silicate with vinyl functional surface treatment; reinforcing extender; used in crosslinked polyethylene and the ethylene-propylene rubbers used in elec. compounds. [Engelhard]

Translink® 555. Calcined aluminum silicate with aminosilane surface treatment; reinforcing extender. [Engelhard]

Transol. Transformer oils. [Carless Refining & Marketing Ltd]

Transoxide. CAS 1309-37-1; Transparent iron oxides. [PMC Specialities International Ltd]

Transpafill. Precipitated aluminum silicate; nonreactive transparent filler for printing inks. [Degussa; Degussa AG]

Transpalene® . CAS 9003-07-0; PP; for high clarity and gloss pkg.; for thermoforming, injection and blow molding. [Neste]

transpar. Lactic acid and buffered lactic acid mixtures.

Transpex. Acrylic sheet used for double glazing. [Imperial Chemical Industries plc]

Transpulmin Balsam. Quninine base, eucalyptus oil, menthol and camphor; expectorant, antitussive, secretolytic, and secretomotoric; balsam. [Degussa AG] *

Transpulmin Syrup. Thiophenylpyridylamine-10-carboxylic acid piperidino-ethoxyethylester HCl (pipazetat), peppermint oil, anise oil, eucalyptus oil, liquorice extract, guaiacol glycerin ether, polyoxyethylene hexadecyl ether N-dimethylamino-isopropyl-thiophenyl-pyridylamine HCl (Isothypendyl); expectorant, antitussive, secretolytic, and secretomotoric syrup. [Degussa AG] *

Transvasin. A proprietary preparation of tetrahydrofurfuryl salicylate, ethyl nicotinate, N-hexyl nicotinate, and ethyl p-aminobenzoate. [Lloyd, Hamol] *

Transvasin®. Tetrahydrofurfuryl salicylate, hexyl nicotinate, ethyl nicotinate; for relief of aches, pains, strains, rheumatic pain. [Seton Healthcare Group plc]

Tranxene. A proprietary preparation of potassium clorazepate; an anxiolytic. [Boehringer Ingelheim Ltd]

Traseolide. 5-Acetyl-3-isopropyl-1,12,6-tetramethyl-indane. [Quest Int'l. UK Ltd]

Tra-Shield 2867. CAS 25928-94-3; Silver-filled solvent-borne epoxy coating; conductive coating for EMI shielding, EMI gasket flange coating, surface ground, antistatic charge dissipation, and corona shielding applications. [Tra-Con]

Trasicor. CAS 6452-73-9; A proprietary preparation of

---

‡ Trade name and manufacturer not verified   § Trade name without identified manufacturer

oxprenolol hydrochloride; used in the treatment of angina pectoris and hypertension. [Ciba plc] *

Trasidrex. Beta-blocker/diuretic. [Ciba plc] *

trass. A volcanic material found on the bank of the Rhine; used in Holland as an addition to lime, to convert it into hydraulic lime.

Trastan LS. CAS 8061-52-7; Low sugar calcium lignosulfonate. [Climax Fluids Additives]

Trasylol®. CAS 9087-70-1; Aprotinin; a proteinase inhibitor. [Bayer AG]

Traton®. For agriculture. [DuPont UK]

Traumon®. Topical anti-inflammatory and antirheumatic agent. [Bayer AG]

Travamin. CAS 9015-54-7; Protein hydrolysate; replenisher. [Travenol Laboratories Inc] *

Travase. Sutilains; enzyme. [The Boots Co plc]

Travel Calm. Tablet containing hyoscine hydrobromide; used for travel sickness. [The Boots Co plc]

travertine. (calc sinter, calcareous tufa). A limestone deposited by calcareous springs.

Travogyn Pessaries. Isoconazole nitrate(300 mg); vaginal suppository. [Schering Health Care Ltd]

Traytuf Ultra-Clear. CAS 25038-59-9; PET polyester; for high purity, high strength packaging with optimum transparency. [Goodyear]

Tread-Brite. Embossed aluminum plate. [Reynolds Metal Co]

Treadfast®. Various PVA, SBR, acrylic, epoxy, polyurethane formulations; flooring adhesives for PVC, rubber, carpets, etc. [Tremco Ltd]

treble superphosphate. This consists of mono-calcium phosphate, containing 48-49% P_2O_5 (41-42% water soluble P_2O_5); used as a fertilizer.

Trecalmo®. Tranquilizer. [Bayer AG; Tropon Werke]

Trecator-SC. CAS 536-33-4; Ethionamide; antibacterial. [Ives Laboratories Inc] *

Tredalat®. Pharmaceutical preparation for the treatment of hypertension and angina pectoris. [Bayer AG]

Tree Bug-Lok Adhesive. Polyisobutylene; for the control of Gypsy moth caterpillars; nontoxic, ecologically safe, holds forever, traps any crawling insect including ants and cankerworms. [TACC Int'l.] *

tree copal. A name applied to white Zanzibar copal.

tree gum. (wood gum; xylan) $C_6H_{10}O_5$.

tree wax. See Chinese wax.

Treflan. CAS 1582-09-8; Emulsifiable concentrate containing 480 g/l trifluralin; dinitroaniline herbicide to control annual weeds and grasses. [DowElanco Ltd]

trefol. A name applied to amyl salicylate.

Trefsin®. Thermoplastic elastomers. [Exxon UK]

Trefsin® 3201-50, 3201-60. Thermoplastic elastomer; low-gas permeability rubber for medical, industrial and consumer goods; general purpose grade. [Advanced Elastomer Systems] *

Tre-Hold. Growth regulator. [A H Marks & Co Ltd] *

Trelacon. Tylosin tartrate. [Dista Products Ltd]

Trelit. A proprietary pyroxylin plastic. §

Trem-LF-40. Sodium alkylallyl sulfosuccinate; polymerizable surfactant, solubilizer; provides low foaming emulsions with improved water resistance. [Henkel]

Tremin. CAS 52-49-3; Trihexyphenidyl hydrochloride; anticholinergic; antiParkinsonian. [Schering Corp] *

Tremvac. Calnek strain avian encephalomyelitis; for immunization of poultry. [Intervet Inc]

Tremvac-FP. Calnek strain avian encephalomyelitis-fowl pox; for immunization of poultry. [Intervet Inc]

Trenbest 500. Trimethyl hydroxyethyl ammonium ester of a carboxyglycerol phosphoric acid; mold release agent for the plastic and caoutchouc industry; used for molding and extrusion powds., laminated plastics, injection molding powds., ABS. [Lucas Meyer]

Trenchmaster; Expanded polystyrene foundation formwork system for building sites. [Vencel Resil Ltd]

Trench's flameless explosive. An explosive containing ammonium nitrate.

Trend®. For agriculture. [DuPont UK]

Trenimon. CAS 68-76-8; A proprietary preparation of triaziquone; an antimitotic. [FBA Pharmaceuticals] ‡

Trentadil. CAS 20684-06-4; A proprietary preparation of bamifylline hydrochloride. [Armour Pharmaceutical Co]*

Trental. A proprietary preparation of oxypentifylline; used in the treatment of peripheral vascular disease. [Hoechst UK] *

Trenyline. Triphenylguanidine; a proprietary trade name for rubber vulcanization accelerator. §

Trescatyl. CAS 536-33-4; Ethionamide. [May & Baker Ltd]*

Tretobond®. Various building, laminating and structural adhesives based on PVA, neoprene, SBR, epoxy, polyurethane, etc; used for building, laminating, and structural bonding. [Tremco Ltd]

Tret-o-Lite. A patented preparation for the destruction of petroleum emulsions; it consists of 83% sodium oleate, 5.5% sodium resinate, 5.0% sodium silicate, 4.0% phenol, and 1.5% paraffin wax. §

Trevintex. CAS 14222-60-7; Prothionamide. [May & Baker Ltd] *

Trevira. Polyester fiber. [Hoechst UK]

TRH-Roche. Proprietary preparation of protirelin; thyroid function test. [Roche Products Ltd] *

Tri. An abbreviation for trichloroethylene, C_2HCl_3; used as a solvent.

Triabon® 16-8-12-4. Complex fertilizer for substrates regardless of their pH, and greenhouse crops. [BASF AG]

triacetin. (glyceryl triacetate; 1,2,3-propanetriol triacetate; triacetyl glycerol) Triester of glycerin and acetic acid; $C_3H_5(OCOCH_3)_3$; CAS 102-76-1; EINECS 203-051-9; plasticizer; fixative in perfumery; manufacture of cosmetics; specialty solv.; manufacture of celluloid, photographic films; topical antifungal. [Eastman; MTM Spec. Chem. Ltd; Penta Mfg.; Spectrum Chem. Mfg.; Unichema]

Triacetin. Glycerol triacetate; used in the manufacture of cigarette filter tips, plasticizer for paint and varnish systems, fixative for use in perfumes, auxiliary for use in foundries. [Bayer AG]

triacetyl glycerol. See triacetin.

triacylglycerol lipase. See lipase.

Tri-Ad. Gasoline additive. [Chevron] *

Triadcortyl Cream. Triamcinolone acetonide, nystatin, gramicidin, and neomycin sulfate in cream base; used in dermatology as an antibacterial agent. [Bristol-Myers Squibb Pharmaceuticals]

Triadcortyl Ointment. Triamcinolone acetonide, nystatin, gramicidin, and neomycin sulfate in ointment base. [Bristol-Myers Squibb Pharmaceuticals Ltd]

Triadimefon. CAS 43121-43-3; 1-(4-Chloro-phenoxy)-3,3-dimethyl-1-(1, 2, 4-triazol-1-yl) butan-2-one; used as a pesticide.

---

* Trade name not verified as available   † Trade name verified as obsolete

Triadine® 3. Hexahydro-1,3,5-tris (2-hydroxyethyl)-5-triazine; bactericide for metalworking fluids. [Olin]

Triadine® 10. Sodium pyrithione (6.4%) and hexahydro-1,3,5-tris (2-hydroxyethyl)-s-triazine (63.6%); broad spectrum microbiostat effective against gram-positive and gram-negative bacteria and fungi; used in aq.-based metalworking fluids. [Olin]

Triafol. CAS 9012-09-3; Cellulose triacetate film; particularly suitable for coil insulation. [Bayer AG]

Triagran®. Bentazon, dichorprop, MCPA; post-emergence herbicide for control of broad-leaved weeds in winter and spring cereals. [BASF AG]

triallylcyanurate. $(CH_2:CHCH_2OC)_3N_3$; CAS 101-37-1; EINECS 202-936-7; Polymers as monomers and modifier; organic intermediate. [Akzo; Am. Cyanamid; Degussa; Nat'l. Starch & Chem.]

triallyl isocyanurate. Co-agent improving peroxide-induced crosslinking of rubber. [Akzo]

Triameen T. CAS 61791-57-9; N-Tallowalkyl dipropylene triamine; industrial surfactant. [Akzo]

Triamolone 40. Triamcinolone diacetate; glucocorticoid. [O'Neal, Jones & Feldman Pharmaceuticals] *

Triamonide 40. CAS 76-25-5; Triamcinolone acetonide; glucocorticoid. [O'Neal, Jones & Feldman Pharmaceuticals] *

triaryl phosphate. $(CH_2:CHCH_2O)_3PO$; CAS 68937-41-7; Flame retardant plasticizer for PVC; compatibilizing agent; intermediate; catalyst carrier and pigment vehicle for polyurethane; processing aid in rubber belting and mech. goods. [Akzo; Ashland; FMC; Monsanto]

Triatox. See Amitraz.

Triatrix. See Amitraz.

s-triazine-2,4,6-triol. See isocyanuric acid.

Tribase. Intermediate for production of dyes. [BASF AG]

Tribiotic Spray. Topical antibacterial in aerosol form containing neomycin sulfate BP 500,000 units, bacitracin zinc BP 10,000 units, polymyxin B sulfate BP 150,000 units. [3M Pharmaceuticals] †

Tribonol. Powder coating for direct application to sand molds and cores. [Foseco (F.S.) Ltd]

Tri-Borne. Paint dip. [ICI Chem. & Polymers Ltd] †

Tribovax. Combined cattle vaccine. [The Wellcome Foundation Ltd] *

Tribrissen. Trimethoprim/sulfadiazine. [The Wellcome Foundation Ltd] †

Tribunil®. CAS 18691-97-9; Methabenzthiazuron; broad-spectrum herbicide for pre- and post-emergence application. [Bayer AG; Bayer plc]

Tribute. Mixture of dicamba, MCPA and mecoprop; used for weed control in cereals and grassland. [Chipman Ltd]

tributoxyethyl phosphate. (TBEP; ethanol, 2-butoxy-phosphate (3:1)) $[CH_3(CH_2)_3O(CH_2)_2O]_3PO$; CAS 78-51-3; primary plasticizer for most resins and elastomers; floor finishes and waxes; flame-retarding agent; latex paints; as defoamer. [Akzo; Albright & Wilson; FMC]

tributyl phosphate. (tri-n-butyl phosphate) $(C_4H_9)_3PO_4$; CAS 126-73-8; EINECS 204-800-2; Heat-exchange medium; solv. extraction of metal ions; plasticizer for cellulose esters, lacquers, plastics, vinyl resins; antifoam agent; dielectric. [Akzo; Chemron; FMC]

tributyl phosphite. (tributyl (n) phosphite) Phosphorus derivative.; $(CH_3CH_2CH_2CH_2O)_3P$; CAS 102-85-2; EINECS 203-061-3; additive for greases and extreme-pressure lubricants; stabilizer for fuel oils and polya-

mides; gasoline additive. [Albright & Wilson; Janssen Chimica]

tributyltin oxide. (HBD; TBTO; bis(tributyltin) oxide; hexabutyldistannoxane) $(CH_3CH_2CH_2CH_2)_3Sn-OSn(CH_2CH_2CH_2CH_3)_3$; CAS 56-35-9; EINECS 200-268-0; Bactericide, fungicide, intermediate. [Akzo; Atochem; KMZ Chem. Ltd]

Tricaderm. A proprietary preparation of triamcinolone acetonide, salicylic acid, and benzalkonium chloride; used in the treatment of eczema. [Bristol-Myers Squibb Pharmaceuticals Ltd] *

tricalcium citrate. See calcium citrate.

tricalcium orthophosphate. See calcium phosphate, tribasic.

tricalcium phosphate. See calcium phosphate, tribasic.

Tricap. Triethyleneglycol dicaprylate/caprate. [Croda Chem. Ltd] *

tricarbimide. See cyanuric acid.

1,2,4-trichlorobenzene. $C_6H_3Cl_3$; CAS 120-82-1; EINECS 204-428-0; Solvent in chemical manufacture, dyes, intermediates, dielectric fluid, synthetic transformer oils, lubricants, heat-transfer media, insecticides. [Ashland; Stan Chem Int'l. Ltd]

trichloro-t-butyl alcohol. See chlorobutanol.

trichlorocyanuric acid. See trichloroisocyanuric acid.

trichloroethane. (1,1,1-trichloroethane; ethane, 1,1,1-trichloro-; methylchloroform) Halogenated aliphatic hydrocarbon; CH_3CCl_3; CAS 71-55-6; EINECS 201-166-9; solvent for cleaning precision instruments, metal degreasing, pesticide, textile processing. [Asahi Chem Industry Co Ltd; Asahi-Penn; Ashland; Atochem N. Am.; Chemoxy Int'l. plc; PPG Industries; Vulcan]

1,1,1-trichloroethane. See trichloroethane.

trichloroethene. See trichloroethylene.

trichloroethylene. (trichloroethene; ethinyl trichloride) $CHCl:CCl_2$; CAS 79-01-6; EINECS 201-167-4; Metal degreasing; extraction solvent for oils, fats, waxes; solvent for cellulose esters and ethers; drycleaning; solvent dyeing; manufacture of organic chemicals, pharmaceuticals. [Ashland; Asahi-Penn; Atochem N. Am.; General Chem.; Nickerson Chem. Ltd; PPG Industries]

trichloroethylsilane. See ethyltrichlorosilane.

trichlorofluoromethane. (Fluorocarbon-11; fluorotrichloromethane) Cl_3CF; CAS 75-69-4; EINECS 200-892-3; Refrigerant; solvent; fire extinguishers; chemical intermediate; blowing agent. [Aldrich; Atochem]

trichloroisocyanuric acid. (isocyanuric chloride; trichlorocyanuric acid; N,N′,N′′-trichloroisocyanuric acid; 1,3,5-trichloro-s-triazine-2,4,6-trione) OCNCl-CONCICONCI; CAS 87-90-1; EINECS 201-782-8; Bleaching agent, dishwashing compounds, disinfectant in swimming pools, bactericide, algicide, deodorant; active ingredient in household cleaners (dry). [Allchem Industries; Chlor Chem Ltd; Monsanto; Nissan Chem. Ind.; 3-V]

N,N′,N′′-trichloroisocyanuric acid. See trichloroisocyanuric acid.

trichloromethane. See chloroform.

1,1-trichloro-2-methyl-2-propanol. See chlorobutanol.

N-trichloromethylthio-4-cyclohexene-1,2-dicarboximide. See captan.

N-trichloromethylthiotetrahydrophthalimide. See captan.

trichloromonosilane. See trichlorosilane.

---

‡ Trade name and manufacturer not verified   § Trade name without identified manufacturer

trichlorosilane. (trichloromonosilane; silicochloroform) Cl₃HSi; CAS 10025-78-2; EINECS 233-042-5; Organic synthesis; purification of silicon; intermediate. [Chisso Am.; Hüls Am.; PCR]

trichlorosilyl)ethylene. See vinyltrichlorosilane.

trichloro-s-triazinetrione. See trichloroisocyanuric acid.

trichlorotrifluoroethane. (1,1,2-trichloro-1,2,2-trifluoroethane) ClCF₂CCl₂F; CAS 76-13-1; EINECS 200-936-1; refrigerant; dry-cleaning solvent; fire extinguishers; manufacture of chlorotrifluoroethylene; blowing agent; polymer intermediate; solvent drying; drying electronic parts and precision equipment. [Air Prods.; Allied-Signal; Atochem N. Am.; PCR]

1,1,2-trichloro-1,2,2-trifluoroethane. See trichlorotrifluoroethane.

trichlorovinylsilane. See vinyltrichlorosilane.

Trichrome. Trivalent chromium plating process. [Engelhard Technologies Ltd]

Tricicatrin. A proprietary formulation containing neomycin sulfate, bacitracin zinc, nystatin, and hydrocortisone; for treatment of skin conditions in which bacterial or candidial infections are present or are likely to occur. [The Wellcome Foundation Ltd]

Triclene. CAS 79-01-6; A proprietary trade name for trichlorethylene used in dry- cleaning. [Occidental Chemical Corp]

1,3,5-tricloro-s-triazine-2,4,6-trione. See trichloroisocyanuric acid.

Tricloryl. CAS 7246-20-0; A proprietary preparation of triclofos sodium (the monosodium salt of trichlorethyl phosphate.); a sedative. [Glaxo Laboratories] *

Triclos. CAS 7246-20-0; Triclofos sodium; hypnotic; sedative. [Merrell Dow Pharmaceuticals Inc] *

Tricoid. Cine film cement. [May & Baker Ltd] *

tricresol. A purified mixture of the three cresols. It contains about 35% ortho, 20% meta, and 25% para-cresol.

tricresyl phosphate. (TCP; phosphoric acid, tris(methylphenyl) ester; tritolyl phosphate) (CH₃C₆H₄O)₃PO; CAS 1330-78-5; 78-30-8; 68952-35-2; EINECS 215-548-8; plasticizer, fire retardant for plastics, air filter medium, waterproofing, additive to extreme pressure lubricants. [Akzo; Chemron; Daihachi Chem. Ind.; FMC]

tricyanide. See cyanuric acid.

tridecanol. See tridecyl alcohol.

tridecyl alcohol. (tridecanol; C13 linear primary alcohol) Synthetic alcohol; commercial mixture of isomers; C₁₂H₂₅CH₂OH; CAS 68526-86-3; 112-70-9; EINECS 203-998-8; esters for synthetic lubricants, detergents, antifoam agents, other tridecyl compounds; perfumery. [Allchem Industries; Penta Mfg.]

tridecyl phosphite. (C₁₀H₂₁O)₃P; Chemical intermediate, stabilizer for polyvinyl and polyolefin resins. [GE Specialty; Stave]

Tridesilon. CAS 638-94-8; Desonide; anti-inflammatory. [Miles Pharmaceuticals]

Tridia. A proprietary preparation of neomycin sulfate, clioquinol, and kaolin; an antidiarrheal. [Crookes Laboratories] *

Tridil®. CAS 55-63-0; Nitroglycerin; vasodilator. [DuPont Merck Pharmaceuticals]

Tridione. CAS 127-48-0; Trimethadione; anticonvulsant. [Abbott Laboratories]

trieline. CAS 79-01-6; EINECS 201-167-4; A term for trichloroethylene, C₂HCl₃; used as an extraction solvent for oils, fats, waxes, drycleaning, diluent in paints and adhesives, chemical intermediate.

trien. See triethylenetetramine.

trientine. See triethylenetetramine.

Tri-Ethane. CAS 71-55-6; Stabilized 1,1,1-trichloroethane. [PPG Industries]

triethanolamine. (TEA; 2,2´,2´´-nitrilotris(ethanol); trolamine; tris(-2-hydroxyethyl)amine; trihydroxytriethylamine) Alkanolamine; N(CH₂CH₂OH)₃; CAS 102-71-6; EINECS 203-049-8; intermediate in manufacture of surfactants, textile specialties, waxes, polishes, toiletries, cutting oils, fatty acid soaps (drycleaning), cosmetics, household detergents, emulsions; solvent for casein, shellac, dyes; dispersion agent, water repellent. [Aldrich; Hüls AG; OxyChem; Schweizerhall; Texaco; Union Carbide]

triethanolamine hydrochloride. (OHCH₂CH₂)₃N·HCl; CAS 637-39-8; EINECS 211-284-2. [Arol Chem. Prods.; Janssen Chimica; Spectrum Chem. Mfg.; U.S. Biochemical]

triethanolamine lauryl sulfate. See TEA-lauryl sulfate.

triethoxysilyl)ethylene. See vinyltriethoxysilane.

triethoxyvinylsilane. See vinyltriethoxysilane.

triethylenetetramine. (TETA; N,N´bis(2-aminoethyl)-1,2-ethanediamine; trien; trientine) NH₂(C₂H₄NH)₂-C₂H₄NH₂; CAS 112-24-3; EINECS 203-950-6; Detergents and softening agents, synthesis of dyestuffs, pharmaceuticals, rubber accelerator; as thermosetting resin; epoxy curing agent; lubricating oil additive; analytical reagent for Cu, Ni. [Rit-Chem; Texaco; Tosoh; Union Carbide]

triethyl phosphate. (TEP; o-phosphoric acid, triethyl ester) (C₂H₅O)₃PO; CAS 78-40-0; EINECS 201-114-5; Intermediate for agricultural. insecticides; in floor polishes, unsaturated polyesters, lubricants; as plasticizer, solvent, and catalyst. [Eastman; Miles]

triethyl phosphite. (C₂H₅O)₃P; CAS 122-52-1; EINECS 204-552-5; synthesis, plasticizer, stabilizers, lube and grease additives. [Akzo; Albright & Wilson; ICI Am.; Janssen Chimica]

Tri-Farmon. CAS 330-55-2, 1582-09-8; Herbicide containing linuron and trifluralin. [ICI Chem. & Polymers Ltd]

Trifarmon. CAS 330-55-2, 1582-09-8; Suspension concentrate containing 160 g linuron and 320 g trifluralin per liter; herbicide for winter cereals. [Farm Protection Ltd]

triflic acid. See trifluoromethanesulfonic acid.

Triflucan. See Diflucan. [Pfizer International]

Trifluco. See Diflucan. [Pfizer International]

trifluoroacetic acid. CF₃COOH; CAS 76-05-1; EINECS 200-929-3; Strong nonoxidizing acid, laboratory reagent, solvent, catalyst. [Aldrich Ltd; Allied-Signal; Atomergic Chemetals; Janssen Chimica; PCR; Solvay GmbH]

2,2,2-trifluoroethanol. (trifluoroethyl alcohol) CF₃CH₂OH; CAS 75-89-8; EINECS 200-913-6. [Aldrich; Schweizerhall; Solvay GmbH]

trifluoroethyl alcohol. See 2,2,2-trifluoroethanol.

trifluoromethanesulfonic acid. (triflic acid) CF₃SO₃H; CAS 1493-13-6; Catalyst or reactant; polymerization of epoxies, styrenes, and THF, alkylation and some acylation reactions; pharmaceuticals, explosives, dyes, and intermediates; electrolytes; formation of biaryls; polymerization reactions. [Schweizerhall]

Triflurex. CAS 1582-09-8; Active ingredient: trifluralin; 2,6-

---

* Trade name not verified as available   † Trade name verified as obsolete

dinitro-N,N-dipropyl-4-trifluromethylaniline; selective pre-emergence herbicide. [Agan Chemical Manufacturers Ltd]

Trifolex-Tra. Soluble concentrate containing 34 g MCPA and 216 g MCPB per liter; for control of weeds in undersown cereals and grassland. [Shell UK]

Trigard. Copper corrosion inhibitors. [Ciba plc] *

Trigard. CAS 1582-09-8; Emulsifiable concentrate containing 480 g/liter trifluralin EC; residual herbicide for soil incorporation with many crop uses. [Farmers Crop Chemicals Ltd]

Trigger. Herbicide. [May & Baker Ltd] *

triglycine. See nitrilotriacetic acid.

triglycollamic acid. See nitrilotriacetic acid.

Trigonal®. Uv catalyst for polyester resins. [Akzo Chemie UK Ltd]

Trigonal® 12. Aromatic ketone; uv initiator; for printing ink and paper coating applications. [Akzo] *

Trigonal® 14. Benzoin ether; initiator used in uv light curing of polyester resins; long shelf life in polyester formulation; uv sensitizer for lacquers and FRP laminates. [Akzo] *

Trigonal® 15. Benzoin ether; general uv sensitizer; mainly used for curing of putties; improved stability in presence of fillers. [Akzo] *

Trigonal® 121. Ketone mixture; uv initiator used in combination with an amine accelerator for uv curing of FRP laminates. [Akzo] *

Trigonox®. Organic peroxides. [Akzo Chemie UK Ltd]

Trigonox® 17. n-Butyl 4,4-di(t-butylperoxy)valerate; initiator for cure of unsaturated polyester resins [Akzo] *

Trigonox® 21. CAS 3006-82-4; t-Butyl peroctoate; initiator for acrylates, styrenics, and LDPE polymerization; used where presence of water is objectionable. [Akzo] *

Trigonox® 22-BB80. CAS 3006-86-8; 1,1-Di-t-butylperoxy cyclohexane in butyl benzyl phthalate; initiator for SMC and BMC formulations. [Akzo] *

Trigonox® 23. CAS 26748-41-4; t-Butyl peroxyneodecanoate [Akzo] *

Trigonox® 25-C75. CAS 927-07-1; t-Butyl peroxypivalate in odorless mineral spirits; initiator for LDPE polymerization, styrenics, and PVC. [Akzo] *

Trigonox® 29. CAS 6731-36-8; 1,1-Di(t-butylperoxy)-3,5,5-trimethylcyclohexane. [Akzo] *

Trigonox® 36-C75. CAS 3851-87-4; Bis (3,5,5-trimethylhexanoyl) peroxide; initiator for LDPE polymerization, styrenics, acrylates, PVC polymerization. [Akzo]*

Trigonox® 41-C75. CAS 109-13-7; t-Butyl peroxyisobutyrate. [Akzo] *

Trigonox® 42. CAS 13122-18-4; t-Butyl peroxy-3,5,5-trimethyl hexanoate; initiator for cure of unsaturated polyester resins [Akzo] *

Trigonox® 42 PR. CAS 13122-18-4; t-Butyl peroxy-3,5,5-trimethyl hexanoate; initiator for cure of unsaturated polyester resins [Akzo] *

Trigonox® 42S. CAS 13122-18-4; t-Butyl peroxy-3,5,5-trimethylhexanoate. [Akzo] *

Trigonox® 61, 63, 67. Ketone peroxide mixture; initiator for cure of unsaturated polyester resins [Akzo] *

Trigonox® 93. CAS 614-45-9; t-Butyl peroxybenzoate; initiator for cure of unsaturated polyester resins [Akzo] *

Trigonox® 97-C75. CAS 22313-62-8; t-Butyl peroxy-2-methylbenzoate, odorless mineral spirits; initiator for

styrenics. [Akzo] *

Trigonox® 99-B75. CAS 26748-47-0; Cumyl peroxyneodecanoate [Akzo] *

Trigonox® 101. CAS 78-63-7; 2,5-Dimethyl-2,5-di (t-butyl peroxy hexane); initiator for polyester when molding at elevated temps.; also for crosslinking of olefin copolymers, chlorinated polyethylene, EPDM, SBR, and vinyls; initiator for free radical polymerizations of styrene; crosslinking agent for olefin copolymers, chlorinated polyethylene, EPDM, SBR, and vinyls. [Akzo] *

Trigonox® 111-B40. CAS 2155-71-7; Di-t-butyl peroxyphthalate. [Akzo] *

Trigonox® 121. CAS 686-31-7; t-Amyl peroxy-2-ethylhexanoate. [Akzo] *

Trigonox® 123-C75. CAS 68299-16-1; t-Amyl peroxyneodecanoate, odorless mineral spirits; initiator for PVC. [Akzo] *

Trigonox® 125-C75. CAS 29240-17-3; t-Amyl peroxypivalate, odorless mineral spirits; initiator used to provide faster curing of BMC and SMC; initiator for LDPE and PVC polymerization. [Akzo] *

Trigonox® 127. CAS 4511-39-1; t-Amyl peroxybenzoate. [Akzo] *

Trigonox® 141. CAS 13052-09-0; 2,5-Dimethyl-2,5-(2-ethylhexanoylperoxy)hexane. [Akzo] *

Trigonox® 145. CAS 1068-27-5; 2,5-Dimethyl-2,5-di(t-butylperoxy)-hexyne-3. [Akzo] *

Trigonox® 151-C70. 2,4,4-Trimethylpentyl-2-peroxyneodecanoate. [Akzo] *

Trigonox® 161. Mixture; initiator for cure of unsaturated polyester resins. [Akzo] *

Trigonox® 169-OP50. CAS 16580-06-6; Di-t-butyl peroxyazelate. [Akzo] *

Trigonox® 239A. CAS 80-15-9; Cumene hydroperoxide; initiator for cure of unsaturated polyester resins. [Akzo]*

Trigonox® A-80. CAS 75-91-2; t-Butyl hydroperoxide. [Akzo] *

Trigonox® ACS-M28. CAS 3179-56-4; Acetyl cyclohexane sulfonyl peroxide in DMP; reactive initiator avail. for use in producing PVC. [Akzo] *

Trigonox® ADC. CAS 19910-65-7; Peroxydicarbonate; initiator for low- and medium-density polyethylene; initiator for PVC polymerization. [Akzo] *

Trigonox® ADC-NS60. CAS 78350-78-4; Peroxydicarbonate mixture. [Akzo] *

Trigonox® B. CAS 110-05-4; Di-t-butyl peroxide; initiator for LDPE polymerization; lower molecular weight at higher temps.; high-temp. peroxide used as finishing initiator for styrenics; crosslinking agent for olefin copolymers; volatile liq. best suited when used in injected extruder application. [Akzo] *

Trigonox® BPIC. CAS 2372-21-6; t-Butyl peroxy isopropyl carbonate, odorless mineral spirits; initiator for BMC and SMC molding; initiator for free radical polymerizations of styrene and acrylates; cross-linking agent for elastomers incl. SBR, urethanes, EPR, EPDM, and nitrile rubbers. [Akzo] *

Trigonox® C. CAS 614-45-9; t-Butyl perbenzoate; initiator for elevated-temp. polyester cures, for LDPE polymerization, and high-temp. polymerization of acrylic emulsion polymers; used in BMC and SMC molding in temp. range 275-325 F. [Akzo] *

Trigonox® D-E50. CAS 2167-23-9; 2,2-Di-(t-butylperoxy) butane. [Akzo] *

---

‡ Trade name and manufacturer not verified   § Trade name without identified manufacturer

Trigonox® EHP. CAS 16111-62-9; Bis(2-ethylhexyl) peroxydicarbonate; versatile, highly reactive initiator for PVC production; high molecular weight reduces reactor fouling. [Akzo] *

Trigonox® F-C50. CAS 107-71-1; t-Butyl peracetate, odorless mineral spirits; initiator for low- and medium-dens. polyethylene and for styrenics. [Akzo] *

Trigonox® GS. A range of liquid organic peroxides for the high temperature curing of unsaturated polyester resins. [Akzo Chemie UK Ltd]

Trigonox® HM. MIBK peroxide; initiator for cure of unsaturated polyester resins. [Akzo] *

Trigonox® K-95. CAS 80-15-9; Cumene hydroperoxide. [Akzo] *

Trigonox® KSM. CAS 3006-82-4, 6731-36-8; t-Butyl peroctoate and 1,1-bis (t-butylperoxy)-3,3,5-trimethylcyclohexane; SMC/BMC initiator offering good shelf life in compound and molding temp. range of 230-290 F; polyester cures. [Akzo] *

Trigonox® M-50. CAS 3736-26-3; Diisopropylbenzene monohydroperoxide. [Akzo] *

Trigonox® RM. A range of liquid organic peroxides for free radical polymerization. [Akzo Chemie UK Ltd]

Trigonox® SBP. CAS 19910-65-7; Di(s-butyl) peroxydicarbonate; initiator for LDPE polymerization; fastest conversion to polymer; used for production of PVC. [Akzo] *

Trigonox® T. CAS 3457-61-2; t-Butyl cumyl peroxide; crosslinking agent for EPDM, SBR, Neoprene, Hypalon; low reactivity peroxide useful as a finishing initiator at high temps. for styrenics; a synergist for some halogen-containing flame retardants. [Akzo] *

Trigonox® TAHP-W85. CAS 3425-61-4; t-Amyl hydroperoxide [Akzo] *

Trigonox® TMPH. CAS 5809-08-5; 2,4,4-Trimethylpentyl-2-hydroperoxide. [Akzo] *

Trihyde. Alumina trihydrate; fire retardant for PVC, rubber, polyester, epoxy resins. [Croxton & Garry Ltd] *

trihydrate. See magnesium sulfate.

1,2,3-trihydroxybenzene. See pyrogallol.

1,3,5-trihydroxybenzene. See phloroglucinol.

3,4,5-trihydroxybenzoic acid. See gallic acid.

3,4,5-trihydroxybenzoic acid, propyl ester. See propyl gallate.

2,6,8-trihydroxypurine. See uric acid.

2,4,6-trihydroxy-1,3,5-triazine. See cyanuric acid.

2,4,6-trihydroxy-1,3,5-triazine. See isocyanuric acid.

trihydroxytriethylamine. See triethanolamine.

triisooctyl phosphite. $(C_8H_{17}O)_3P$; CAS 25103-12-2; Intermediate; insecticides; lubricant additive; specialty solvents; stabilizer for acrylics, nylon, unsaturated polyester, PVC; improves antiwear and antifriction properties. [Albright & Wilson; GE Specialty; Stave]

triisopropanolamine. (TIPA; 1,1',1''-nitrilotris-2-propanol; tris(2-hydroxypropyl)amine) Aliphatic amine; $N(CH_2CHOHCH_3)_3$; CAS 122-20-3; EINECS 204-528-4; emulsifying agent. [Aldrich; Ashland]

Trik. Aminotriazole + 2,4-D + diuron; used for total weed control in non crop areas. [Smyth-Morris Chemicals Ltd]

Triklone. CAS 79-01-6; Trichloroethylene solvents. [ICI Chem. & Polymers Ltd]

trikresol. See tricresol.

trilactine. A preparation of lactic acid bacilli.

Trilafon. CAS 58-39-9; Perphenazine; antipsychotic.

[Schering Corp]

Trilaurin. CAS 538-24-9; Glyceryl tri-laurate, $C_3H_5(OOC \cdot C_{11}H_{23})_3$.

Trilene. CAS 79-01-6; Trichlorethyllene, a general anesthetic. [ICI Chem. & Polymers Ltd]

Trilene® 65. CAS 25034-71-3; Ethylene-propylene-diene terpolymer; produces liq. elastomers; used alone or with solid elastomers in thermosetting and thermoplastic applications.; features low visc., crosslinking, oxidation resist., good elec. properties, ozone and uv resist. [Uniroyal]

Trilin®, 10G. CAS 1582-09-8; Trifluralin; herbicide. [Griffin]

Trilisate. Choline magnesium trisalicylate; for treatment of arthritis. [Napp Laboratories Ltd] *

Trilite. See Trotyl.

Trillat's reagent. An acetic acid solution of tetramethyl-diaminodiphenylmethane, $[(CH_3)_2N \cdot C_6H_4] \cdot CH_2$.

Trilombrin. CAS 22204-24-6; A proprietary preparation of pyrantel pamoate; an antihelmintic. [Pfizer International]

Trilon®. Complexing agents, water softeners. [BASF plc]

Trilon® A. CAS 5064-31-3; Trisodium NTA; low molecular weight general purpose chelate; complexes iron in acid pH range; detergent builder for nonphosphate liqs.; in soaps, detergents, water treatment, metal finishing and plating, pulp and paper manufacturing, synthesis of polymers, photography, textiles, chemical cleaning. [BASF]

Trilon® A-92. CAS 5064-31-3; Trisodium NTA monohydrate; chelating agents used in soaps, detergents, water treatment, metal finishing and plating, pulp and paper manufacturing, synthesis of polymers, photographic products, textiles, chemical cleaning for scale removal. [BASF]

Trilon® BD. Disodium EDTA; chelating agent. [BASF] *

Trilon® B-FA. Ferric ammonium EDTA; chelating agent for photographic applications. [BASF] *

Trilon® B Powd. CAS 64-02-8; EINECS 200-573-9; Tetrasodium EDTA; general purpose chelate; complexes most common metals over pH range, iron at acidic pH; chelating agents used in soaps, detergents, water treatment, metal finishing and plating, pulp and paper manufacturing, synthesis of polymers, photographic products, textile, chemical cleaning for scale removal. [BASF]

Trilon® BS. CAS 60-00-4; EDTA; chelating agent; used where sodium ion is undesirable; for soaps, detergents, water treatment, metal finishing and plating, pulp and paper manufacturing, synthesis of polymers, photographic products, textiles, chemical cleaning for scale removal. [BASF]

Trilon® C Liq. CAS 140-01-2; Pentasodium pentetate; chelating agent used when higher metal chelate stability is required; used in peroxide bleach systems. [BASF]

Trilon® D Liq. CAS 139-89-9; Trisodium HEDTA; general purpose chelate for control of iron in pH range of 6.5-9.5, Ca, Mg; used in soaps, detergents, water treatment, metal finishing and plating, pulp and paper manufacturing, synthesis of polymers, photographic products, textiles, chemical cleaning. [BASF]

Trim. CAS 82-68-8; Quintozene in dry granular form; used for controlling soil borne diseases in tulips and flower crops. [Wheatley Chemical Co Ltd] *

Trimangol. CAS 12427-38-2; Wettable powder containing maneb; a dithiocarbamate fungicide to control blight, rusts, and mildew. [Pennwalt Chemicals Ltd]

---

* Trade name not verified as available   † Trade name verified as obsolete

Trimanzone. Mixture of ferbam, maneb and zineb; a fungicide. [Pennwalt Chemicals Ltd]

Trimaran. CAS 1582-09-8; Emulsifiable concentrate containing 480 g/l trifluralin; dinitroaniline herbicide to control annual weeds and grasses. [Ashlade Formulations Ltd]

Trimastan. Mixture of fentin acetate and maneb; used for control of potato blight. [Pennwalt Chemicals Ltd]

Trimax. Magnesium trisilicate; antacid. [Sterling Drug Inc]*

trimellitic acid. (1,2,4-benzene-tricarboxylic acid) $C_6H_3(COOH)_3$; CAS 528-44-9; Used in organic synthesis.

Trimene Base. Ethyl chloride, ammonia and formaldehyde reaction product; high temperature accelerator with a medium to long curing range; prevents sagging of stock in early stages of air curing; it is a latex foam stabilizer which prevents foam collapse by causing gelling to take place at a higher pH; used in natural and SBR rubbers and latexes. [Uniroyal] *

trimesitinic acid. Trimesic acid, (1, 3, 5-benzene-tricarboxylic acid), $C_6H_3(COOH)_3$.

trimethoxysilyl)benzene. See phenyltrimethoxysilane.

trimethylbenzylammonium chloride. See benzyl trimethyl ammonium chloride.

3,3,5-trimethyl-2-cyclohexen-1-one. See isophorone.

trimethylolpropane. (1,1,1-tris(hydroxy)propane; 2-ethyl-2-hydroxymethyl-1,3-propanediol; hexaglycerol) $CH_3CH_2C(CH_2OH)_3$; CAS 77-99-6; EINECS 201-074-9; Conditioner, manufacture of varnishes, alkyd resins, synthetic drying oils, urethane foams and coatings, silicone lube oils, lactone plasticizers, textile finishes, surfactants, epoxidation products. [Hoechst Celanese; Mitsubishi Gas; Perstorp Polyols]

2,2,4-trimethyl-1,3-pentanediol monoisobutyrate. Ester alcohol; $(CH_3)_2CHCH(OH)C(CH_3)_2CH_2OOCCH(CH_3)_2$; CAS 25265-77-4; intermediate; manufacture of plasticizers, surfactants, pesticides, and resins. [Chisso Am.; Eastman]

trimethylsilylmethyl chloride. See chloromethyltrimethylsilane.

α,α,α´-trimethyltrimethyleneglycol. See hexylene glycol.

1,3,7-trimethylxanthine. See caffeine.

Trimetso. Industrial detergent. [Crosfield Chemicals Ltd] †

Triminol. CAS 63284-71-9; Emulsifiable concentrate containing 90 g/l nuarimol; pyrimidine fungicide to control mildew. [DowElanco Ltd]

Tri-Minulet. 21 Tablet pack containing gestodene and ethinylestradiol in varying dosages; oral contraceptive. [Wyeth Laboratories]

Trimmit. Grass growth regulator. [ICI Chem. & Polymers Ltd]

Trimogal®. White uncoated tablets containing 100 mg or 200 mg trimethoprim BP; for the treatment of acute urinary tract infections and long term prophylaxis of recurrent urinary tract infections; respiratory tract infections, in particular acute and chronic bronchopneumonia and pneumonia caused by organisms sensitive to trimethoprim; particularly useful for patients sensitive to sulfonamides. [Lagap Pharmaceuticals Ltd]

Trimopan. CAS 738-70-5; Trimethoprim; for urinary tract infections, respiratory tract infections. [Berk Pharmaceuticals Ltd] *

Trimox. CAS 26787-78-0; Amoxicillin; antibacterial. [Bristol-Myers Squibb Co Inc] *

Trimpex. CAS 738-70-5; Trimethoprim; antibacterial.

[Hoffmann-LaRoche Inc] *

Trinamide. Triple sulfonamide association. [May & Baker Ltd] *

Trinidad asphalt. A natural asphalt obtained from the Trinidad pitch lake. The crude pitch contains from 40-46% bitumen, 24-30% mineral matter (clay), and 21-29% water.

trinitroglycerin. See nitroglycerin.

Trinitrol. CAS 55-63-0; Erythrol tetranitrate, C_4H_6 $(NO_3)_4$.

Trinol. A proprietary preparation of benzhexol; used for Parkinsonism. §

Trinoram. Alkyl propylene triamines based on coco and tallow alkyl chains; intermediate for chemical synthesis. [Atochem UK/Ceca]

Trinordiol Tablets. 21 Tablet pack containing varying amounts of levonorgestrel and ethinylestradiol BP; oral contraceptive. [Wyeth Laboratories]

trioctyl phosphate. (tris(2-ethylhexyl) phosphate) $[CH_3(CH_2)_3CH(C_2H_5)CH_2O]_3PO$; CAS 78-42-2; EINECS 201-116-6; Solvent, antifoaming agent, plasticizer. [Miles]

Triodine. Iodophor low foaming sanitizer. [Ciba plc] *

Triolam. CAS 19387-91-8; A proprietary preparation of tinidazole; an antiprotozoal. [Pfizer International] †

triolein. (glyceryl trioleate; olein; 9-octadecenoic acid, 1,2,3-propanetriyl ester) Triester of glycerin and oleic acid; $C_{57}H_{104}O_6$; CAS 122-32-7; EINECS 204-534-7; lubricant, emollient, emulsifier for cosmetics, metals, leather, textiles; carbon source in antibiotic culture broths.

Trioleotope. Triolein I 131; radioactive agent. [Bristol-Myers Squibb Co Inc] *

Trione. Ready-to-use ninhydrin reagent solution containing organic modifier and buffer; used for amino acid analysis, amine analysis. [Pickering Laboratories Inc]

Triostam. Trivalent sodium stibogluconate injection. [The Wellcome Foundation Ltd] †

Triox. Vegetation killer. [Chevron] *

Trioxitol. Ethyltrigol. [Shell UK]

Trioxone. Herbicide. [ICI Chem. & Polymers Ltd]

2,6,8-trioxypurine. See uric acid.

trip. Ferric oxide, Fe_2O_3.

Tripart® Acer. Soyal phospholipids 750 g/liter; an adjuvant for use with a wide range of crop protection products. [Tripart Farm Chemicals Ltd]

Tripart® Arena 6. CAS 117-18-0; Tecnazene (6% w/w); used to control dry rot in both ware and seed potatoes, and sprouting in ware potatoes. [Tripart Farm Chemicals Ltd]

Tripart® Arena 6 + TBZ. CAS 117-18-0, 148-79-8; Dustable powder containing (6% w/w) tecnazene and thiabendazole (1.8%); a protectant fungicide and potato sprout suppressant. [Tripart Farm Chemicals Ltd]

Tripart® Arena Granules. CAS 117-18-0; Tecnazene (5 or 10% w/w); used to control dry rot in both ware and seed potatoes, and sprouting in ware potatoes. [Tripart Farm Chemicals Ltd]

Tripart® Arena Plus. CAS 117-18-0, 83601-81-4; Tecnazene (6% w/w) and carbendazim (2% w/w); protectant fungicide and sprout suppressant for stored potatoes. [Tripart Farm Chemicals Ltd]

Tripart® Beta. CAS 13684-63-4; Phenmedipham 118 g/liter (12.1% w/w) in isophorone; a post-emergence herbicide for the control of a wide range of broad-leaved weeds in sugar beet, red beet, fodder beet and mangels. [Tripart

‡ Trade name and manufacturer not verified   § Trade name without identified manufacturer

Farm Chemicals Ltd]

Tripart® Beta 2. CAS 13684-63-4; Emulsifiable concentrate containing 114 g/l phenmedipham; for weed control for beet crops. [Tripart Farm Chemicals Ltd]

Tripart® Brevis. CAS 999-81-5; Chlormequat chloride 700 g/liter (61.7% w/w) aqeuous solution; plant growth regulator for use in cereals to reduce lodging and in wheat and barley. [Tripart Farm Chemicals Ltd]

Tripart® Chlormequat 5C. CAS 999-81-5, 67-48-1; Mixture of chlormequat chloride 460 g/liter (40.8% w/w) and choline chloride 320 g/liter (28.3% w/w); plant growth regulator for cereals and ornamentals. [Tripart Farm Chemicals Ltd]

Tripart® Chlormequat 460. CAS 999-81-5; Soluble concentrate containing 460 g/l chlormequat; plant growth regulator. [Tripart Farm Chemicals Ltd]

Tripart® Cropspray 11E. Adjuvant containing 99% highly refined mineral oil; wetting agent for translocated herbicide sprays. [Tripart Farm Chemicals Ltd]

Tripart® Defensor FL. CAS 83601-81-4; A liquid formulation containing 500 g carbendazim per liter as a suspension concentrate; a broad-spectrum systemic fungicide for control of several diseases in a variety of crops. [Tripart Farm Chemicals Ltd]

Tripart® Faber. CAS 1897-45-6; Chlorothalonil 500 g/liter (40.4% w/w); a nonsystemic broad-spectrum fungicide for use in a wide variety of crops. [Tripart Farm Chemicals Ltd]

Tripart® Gladiator. CAS 1698-60-8; Chloridazon 430 g/liter (36.1% w/w); used for the control of annual broad-leaved weeds in sugar beet, fodder beet and mangels. [Tripart Farm Chemicals Ltd]

Tripart® Granular. CAS 1918-16-7; Propachlor; a preemergence granular herbicide for various horticultural crops. [Tripart Farm Chemicals Ltd]

Tripart® Imber. CAS 7704-34-9; Suspension concentrate containing 720 g/l (52.4% w/w) sulfur; used for disease prevention and growth enhancement in cereals, oil seed rape, sugar beet, and others. [Tripart Farm Chemicals Ltd]

Tripart® Legion. CAS 83601-81-4, 12427-38-2; Carbendazim 50 g and maneb 320 g/liter mixture; for use as a fungicide in cereals. [Tripart Farm Chemicals Ltd]

Tripart® Lentus. A liquid containing 450 g/liter synthetic latex; for use as an agricultural sticking and extending agent. [Tripart Farm Chemicals Ltd]

Tripart® Liquid Manganese. CAS 7439-96-5; Manganese 140 g/liter (10.5% w/w) minimum; for use on crops that are suffering from manganese deficiency or grown on manganese-deficient soils. [Tripart Farm Chemicals Ltd]

Tripart® Ludorum 700. CAS 15545-48-9; Chlortoluron 700 g/liter (58.5% w/w); used for the control of black grass, wild oats, and other annual grasses and a range of broadleaved weeds in a range of named wheats, winter barleys, durum wheats and triticale. [Tripart Farm Chemicals Ltd]

Tripart® Mensa. A concentrated complex of manganese salts; for use on crops that are suffering from manganese deficiency or grown on manganese-deficient soils. [Tripart Farm Chemicals Ltd]

Tripart® Minax. Alkyl phenol ethylene oxide 900 g/liter (87.3% w/w); for use with fungicides, insecticides, and herbicides to improve spreading and penetration; also for use in cleaning and washing out spraying machinery.

[Tripart Farm Chemicals Ltd]

Tripart® Mini Slug Pellets. CAS 108-62-3; Metaldehyde (6% w/w); for the control of slugs and snails in all agricultural and horticultural crops. [Tripart Farm Chemicals Ltd]

Tripart® Nex. CAS 1563-66-2; Carbofuran (5% w/w); a specially formulated microgranule insecticide with systemic action for the control of a range of pests in sugar beet, fodder beet, mangels, brassicas, etc. [Tripart Farm Chemicals Ltd]

Tripart® Ratio. CAS 82558-50-7; Isoxaben 125 g/liter; a residual herbicide for the control of broad-leaved weeds in winter and spring cereals, grass leys and herbage seed crops. [Tripart Farm Chemicals Ltd]

Tripart® Senator Flowable. Copper oxychloride + maneb + sulfur; a fungicide for wheat and barley which also stimulates yield. [Tripart Farm Chemicals Ltd]

Tripart® Sentinel. CAS 1918-16-7; Propachlor 480 g/liter (43.2% w/w); used for the control of germinating weeds in onion, leek, brassica crops and strawberry. [Tripart Farm Chemicals Ltd]

Tripart® Systemic Insecticide. CAS 919-86-8; Emulsifiable concentrate containing 500 g demeton-S-methyl per liter; a systemic organophosphorus insecticide. [Tripart Farm Chemicals Ltd]

Tripart® Trifluralin 48 EC. CAS 1582-09-8; Trifluralin 480 g/liter (45.5% w/w); used for the control of certain germinating broad-leaved weeds and annual grasses in beans, brassicae, cabbage, and a wide variety of other crops. [Tripart Farm Chemicals Ltd]

Tripart® Ultrafaber. CAS 1897-45-6; Chlorothalonil 720 g/liter (54% w/w); a nonsystemic broad-spectrum fungicide for use in a wide variety of crops. [Tripart Farm Chemicals Ltd]

Tripart® Victor. A liquid formulation containing 80 g carbendazim, 150 g chlorothalonil and 200 g maneb per liter as a suspension concentrate; eradicant fungicide for use on cereals. [Tripart Farm Chemicals Ltd]

tripentaerythritol. $C_{15}H_{35}O_8$; hard resins, varnishes, fast-drying tall oil vehicles. [Allchem Industries]

Triperidol. CAS 749-13-3; A proprietary preparation of trifluperidol; a sedative. [Janssen Pharmaceutical Ltd] *

triphan. See iriphan.

triphenyl phosphate. (TPP; phosphoric acid, triphenyl ester) $PO(OC_6H_5)_3$; CAS 115-86-6; EINECS 204-112-2; Fire-retarding agent; noncombustible substitute for camphor in celluloid; plasticizer for cellulose acetate and nitrocellulose; in lacquers and varnishes; impregnating roofing paper. [Monsanto]

triphenyl phosphine. (triphenyl phosphorus; phosphorus triphenyl) $(C_6H_5)_3P$; CAS 603-35-0; EINECS 210-036-0; Organic synthesis, phosphonium salts, other phosphorus compounds, polymerization initiator. [Atochem; BASF; Janssen Chimica; Morton Int'l.]

triphenyl phosphite. (TPP) $(C_6H_5O)_3P$; CAS 101-02-0; EINECS 202-908-4; Chemical intermediate, stabilizer systems for resins, metal scavenger, diluent for epoxy resins, antioxidant and antiwear agent in gear and transmission oils. [Dover; GE Specialty; Stave; Witco/Argus]

triphenyl phosphorus. See triphenyl phosphine.

2,4,5-triphenyltetrazolium chloride. See tetrazolium chloride.

Triplastic. Pepared by mixing trinitro-toluene, together with some lead nitrate and chlorate, with a gelatin made from

---

* Trade name not verified as available   † Trade name verified as obsolete

dinitrotoluene and nitrocellulose; an explosive.

Triplastite. An explosive consisting of a mixture of di- and trinitrotoluene 70 parts, guncotton 1.2 parts, and lead nitrate 28.8 parts.

Triplematic. Metering, proportioning, mixing and dispensing machines for multicomponent reactive resin systems. [Hardman] *

triple salts. (trisalytes) Used in the electro-deposition of metals. They consist of the cyanide of the metal to be deposited, potassium cyanide, and potassium sulfite.

Triple Sulfas. Trisulfapyrimidines (oral suspension); mixture of sulfadiazine, sulfamerazine, and sulfamethazine; antibacterial. [Lederle Laboratories USA] *

Triplevac. B₁ type, B₁ strain, Newcastle with Massachuttes and Connecticut types; for immunization of poultry. [Intervet Inc]

Triplopen. A proprietary preparation containing benethamine penicillin, procaine penicillin, and benzylpenicillin sodium; an antibiotic. [Glaxo Laboratories]

Tripoli. CAS 1317-95-9; (Tripoli powder, rotten stone). A mineral. It consists mainly of silica associated with small quantities of alumina and iron oxide, but the composition varies. The variety of tripoli powder found in Derbyshire, is called rotten stone; used as an abrasive.

tripolite. See infusorial earth.

tripotassium phosphate. See potassium phosphate, tribasic.

tripropylene glycol. See also PPG-3; $HO(C_3H_6O)_2$-C_3H_6OH; CAS 24800-44-0, 13987-01-4; Intermediate in resins, plasticizers, pharmaceuticals, insecticides, dyestuffs, and mold lubricants. [Union Carbide]

tripsa. Tribasic phosphate of soda; used for the prevention of incrustation on boilers.

Triptafen. CAS 549-18-8, CAS 58-39-9; A proprietary preparation of amitriptyline hydrochloride and perphenazine; an antidepressant. [Allen & Hanburys Ltd]

Triptil. Portiptyline hydrochloride; antidepressant. [Merck & Co Inc] *

Tripwite. CAS 7722-84-1; Hydrogen peroxide for tripe dressing. [Interox Chemicals Ltd]

trisalytes. See triple salts.

Tris Amino® . CAS 77-86-1; Tris (hydroxymethyl) aminomethane; pigment dispersant, neutralizing amine, corrosion inhibitor, acid-salt catalyst, pH buffer, chemical and pharmaceutical intermediate, solubilizer. [Angus]

Tris Amino® Conc. Tris (hydroxymethyl) aminomethane; chemical intermediate, resin synthesis, neutralizing amine in cosmetics, buffer for enzyme and diagnostic testing, pharmaceutical buffer and solubilizer. [Angus]

Tris Amino® Molecular Biology Grade. Tris (hydroxymethyl) aminomethane; chemical intermediate, resin synthesis, neutralizing amine in cosmetics, buffer for enzyme and diagnostic testing, pharmaceutical buffer and solubilizer. [Angus]

Tris Amino® Ultra Pure Standard. Tris (hydroxymethyl) aminomethane; chemical intermediate, resin synthesis, neutralizing amine in cosmetics, buffer for enzyme and diagnostic testing, pharmaceutical buffer and solubilizer. [Angus]

tris buffer. See tris (hydroxymethyl) aminomethane.

Triscal. A proprietary preparation of calcium and magnesium carbonates; an antacid. [Nicholas Laboratories Ltd] *

Trisec. Drying additives. [ICI Chem. & Polymers Ltd]

Trisem. Trisulfapyrimidines (oral suspension); mixture of sulfadiazine, sulfamerazine, and sulfamethazine; an antibacterial. [SmithKline Beecham] *

tris(2-ethylhexyl) phosphate. See trioctyl phosphate.

tris hydrochloride. See tris (hydroxymethyl) aminomethane hydrochloride.

tris (hydroxymethyl) aminomethane. (THAM; tromethamine (CTFA); 2-amino-2-(hydroxymethyl)-1,3-propanediol; tris buffer) $(CH_2OH)_3CNH_3$; CAS 77-86-1; EINECS 201-064-4; Emulsifying agent for oils, fats, waxes; absorbent for acidic gases, medicine; chemical intermediate. [Aldrich; Angus; Dajac Labs; W.R. Grace/ Hampshire; Heico; Janssen Chimica; Sigma]

tris (hydroxymethyl) aminomethane hydrochloride. (tris hydrochloride) $NH_2C(CH_2OH)_3 \cdot HCl$; CAS 1185-53-1; EINECS 214-684-5. [Am. Biorganics; Janssen Chimica; U.S. biochemical]

1,1,1-tris(hydroxy)propane. See trimethylolpropane.

tris(2-hydroxypropyl)amine. See triisopropanolamine.

Tris Nitro® . CAS 126-11-4; Tris (hydroxymethyl) nitromethane; antibacterial agent, preservative for water treatment, metalworking fluids, oil production, deodorizing; formaldehyde releaser. [Angus]

Tris Nitro®, 50% Aq. CAS 126-11-4; Tris (hydroxymethyl) nitromethane; chemical intermediate, registered pesticide for use as an antimicrobial in metalworking fluids, cooling water, oil production, drilling muds; formaldehyde donor; deodorant for chemical toilets. [Angus]

trisodium citrate. See sodium citrate.

trisodium orthophosphate. See trisodium phosphate, anhydrous.

trisodium phosphate, anhydrous. (TSP-O; sodium phosphate, tribasic; trisodium orthophosphate; phosphoric acid, trisodium salt) Inorganic salt; $H_3O_4P \cdot 3Na$; CAS 7601-54-9; EINECS 231-509-8; For pH adjustment in food systems; cleaning compounds, water treatment, textiles. [Albright & Wilson; Kemira Kemi UK; Monsanto]

trisodium phosphate, dodecahydrate. (TSP-12; sodium phosphate, tribasic dodecahydrate) Sodium phosphate; $Na_3PO_4 \cdot 12H_2O \cdot \approx \frac{1}{5}NaOH$; CAS 10101-89-0; EINECS 231-509-8; food products, cleaning compounds, water treatment. [Albright & Wilson; Rhone-Poulenc Basic]

Trisophone. Nonionic surfactant. [Rhone-Poulenc UK]

Trisoralen. Trixsalen; pigmentation agent. [ICN Pharmaceuticals Inc]

Tristar. CAS 1582-09-8; Emulsifiable concentrate containing 480 g/l trifluralin; dinitroaniline herbicide to control annual weeds and grasses. [Pan Britannica Industries Ltd]

Tri-Star Antifoam #27. Synergistic blend of organic chemicals, 100% active, water dispersible; for rapid knockout of existing foam and the prevention of foam formation in systems containing detergents, in latex emulsions, in industrial processes, paints, glues, paper coating formulations, water disposal systems, sewage plants, dye baths, textile printing, paper manufacturing, oil well drilling muds, refineries and petrochemical plants. [Tri-Star Chemical Co Inc] *

Tri-Star Padding Compounds. Modified polyvinyl acetate emulsions; fast setting padding compounds. [Tri-Star Chemical Co Inc] *

Tri-Star White Glues. Modified polyvinyl acetate emulsions; used for packaging, bookbinding, labeling, and wood-

---

‡ Trade name and manufacturer not verified   § Trade name without identified manufacturer

working, etc. [Tri-Star Chemical Co Inc] *

Trisul. Leather oils. [Hodgson Chemicals Ltd]

Tri-Sulfameth. Trisulfapyrimidines (oral suspension); mixture of sulfadiazine, sulfamerazine, and sulfamethazine; antibacterial. [USV Pharmaceutical Corp] ‡

tritane. CAS 519-73-3; Triphenyl-methane; $(C_6H_5)_3CH$; used as dyes.

Trithac. Fungicide. [Murphy Chemical Ltd] †

Trithion. Insecticide seed treatment. [Murphy Chemical Ltd]†

triticum aestivum germ oil. See wheat germ oil.

Tritiotope. Tritiated water; radioactive agent. [Bristol-Myers Squibb Co Inc] *

Tritisol. Soluble wheat protein; film-forming conditioning, moisturizing protein improving skin firmness; for hair and skin care preparations, permanent waves, activated conditioners; binder in mascara formulations. [Croda Inc.]

Tritole. See Trotyl.

Tritolo. See Trotyl.

tritolyl phosphate. See tricresyl phosphate.

Triton. See Trotyl.

Triton® CF-10. Alkylaryl polyether; low-foaming detergent for mechanical dishwashing, rinse aids, laundering, metal and dairy cleaning, textile wetting; nylon dyeing assistant; defoamer for food soils. [Union Carbide; Union Carbide Europe]

Triton® DF-12. Polyethoxylated alcohol, modified; biodeg. low-foaming detergent for mech. dishwashing, rinse additive, automatic laundering, metal cleaning, floor scrubbing, dairy equip. cleaners, textile wetting; stable in acid, caustic solutions. [Union Carbide; Union Carbide Europe]

Triton® GR-5M. CAS 577-11-7; EINECS 209-406-4; Dioctyl sodium sulfosuccinate; high speed wetting and rewetting agent, emulsifier, dispersant for paints, textiles. [Union Carbide; Union Carbide Europe]

Triton® H-55. Phosphate ester, potassium salt; hydrotrope/ solubilizer for built liq. concs.; as surfactant for alkaline builder solutions. [Union Carbide]

Triton® N-57. CAS 9016-45-9; Nonoxynol-5; detergent, emulsifier for solv. cleaners; intermediate. [Union Carbide; Union Carbide Europe]

Triton® QS-44. Phosphate surfactant, free acid; hydrotrope, detergent, wetting agent; solubilizer for nonionic surfactants in alkaline cleaning baths, metal cleaning and for nonionic and anionic surfactants in built concs. [Union Carbide; Union Carbide Europe]

Triton® RW-20. Alkylamine ethoxylate (2 EO); emulsifier, detergent, degreaser for secondary oil recovery, waste treatment, transport cleaners, pipeline/refinery equip./ chem. plant cleaning, metal cleaning, metalworking fluids, textiles. [Union Carbide; Union Carbide Europe]

Triton® X-15. CAS 9002-93-1; Octoxynol-1; surfactant, coupling agent, emulsifier for industrial/household cleaners, emulsion polymerization, agriculture, latex stabilizer. [Union Carbide; Union Carbide Europe]

Triton. Range of alkylaryl polyether alcohol surfactants; used for cosmetics, household products, and industrial cleaners. [Union Carbide; Union Carbide Europe]

Tritox. An aqueous concentrate containing MCPA, mecoprop, and dicamba; a selective herbicide for the control of broad-leaved weeds in turf. [Fisons plc, Horticulture Div] *

Triumphnetzer ZSG. Diisooctyl sulfosuccinate; wetting agent for textile and chem-tech products, household/ industrial cleaners, drycleaning, textile, ceramic, and varnish industries. [Zschimmer & Schwarz]

Trivax. A proprietary vaccine used to give protection against diphtheria, tetanus, and pertussis. [The Wellcome Foundation Ltd]

Trivexin. Braxy-blackleg-pulpy kidney vaccine. [The Wellcome Foundation Ltd] *

Trixene. Moisture-curing urethane prepolymers for surface coatings, adhesives, mastics and sealants; fully reacted urethane polymers in solution; blocked isocyanates for heat-activated systems; moisture scavengers; solvent and water-based acrylic polymers. [Baxenden Chemicals Ltd]

Trixidin. An emulsion of antimony trioxide, containing 30% Sb_2O_3.

Trizol. See Diflucan. [Pfizer International]

Trobicin. Spectinomycin hydrochloride; antibacterial. [Upjohn]

Trocinate. CAS 548-68-5; Thiphenamil hydrochloride; relaxant. [Poythress Laboratories Inc] ‡

Trocor. Mechanically resistant material (corundum); aluminum oxide; admixtures for the building industry. [Hüls AG]

Trodax. CAS 1689-89-0; Nitroxynil. [RMB Animal Health Ltd]

Trogamid T. Special polyamide, thermoplastic molding compound; injection, blow and extrusion molding compounds used in various branches of industry. [Dynamit Nobel Wien GmbH] *

Trogamid T G35. Special polyamide, glass fiber reinforced; injection molding compounds for electrical engineering, electronics, telecommunication, mechanical and apparatus engineering, precision mechanics. [Dynamit Nobel Wien GmbH] *

troilite. Haidinger's name for the ferrous sulfide which occurs in meteorites.

Trojan SC. CAS 1698-60-8; Suspension concentrate containing 430 g chloridazon per liter; a pyridazinone herbicide for beet crops. [Schering Agrochemicals Ltd]

trolamine. See triethanolamine.

Trolene. CAS 299-84-3; Ronnel; insecticide. [Dow UK] *

Trolit F. A proprietary pyroxylin product. §

Trolit S and Special. Proprietary phenol-formaldehyde resin molding compounds. §

Trolit W. A proprietary cellulose acetate product. §

Trolite. A synthetic resin of the phenol formaldehyde type. It is a term also applied to trinitro-toluene.

Trolon. Phenolic resins; main fields of application are molded laminated materials, wood working, metal casting, abrasives, friction linings, molded plastics. [Hüls AG]

Trolone. CAS 61-56-3; Sulthiame; anticonvulsant. [3M Pharmaceuticals]

Trolovol. CAS 52-67-5; D-Penicillamin; tablets; antirheumatic. [Degussa AG] *

Trolovol®. CAS 52-67-5; Penicillamine; for the treatment of rheumatoid arthritis. [Bayer AG]

Troluoil. A proprietary trade name for a petroleum solvent.§

tromethamine (CTFA). See tris (hydroxymethyl) aminomethane.

Tromexan. CAS 548-00-5; A proprietary preparation of ethyl bicoumacetate; an anticoagulant. [Ciba plc] *

Trona Anhydrous Sodium Sulphate. CAS 7757-82-6;

Minimum purity 99% Na_2SO_4 in fine, standard, coarse, and special coarse granulations. [Kerr-McGee Chemical Corp]

Trona Boron Tribromide. 99.8% minimum BBr_3. [Kerr-McGee Chemical Corp]

Trona Boron Trichloride. CAS 10294-34-5; 99.9% minimum BCl_3. [Kerr-McGee Chemical Corp]

Tronacarb Sodium Bicarbonate. CAS 144-55-8; White granular solid, industrial and animal feed grades. [Kerr-McGee Chemical Corp]

Trona Elemental Boron. Amorphous type, standard grade, fine, dark brown powder meeting specifications PA-PD-451, OS11608 and MIL-B-51092 (ORD); boron content is 90-92%. [Kerr-McGee Chemical Corp]

Tronalight Light Soda Ash. Na_2CO_3, extracted from the brines of Searles Lake, CA. [Kerr-McGee Chemical Corp]

Tronalight Light Soda Ash. Grades: Dense-99.7% mineral Na_2CO_3; Granular-97.7% mineral Na_2CO_3; Light-98.3% min. Na_2CO_3 (dry basis). [Kerr-McGee Chemical Corp]*

Tronamang Electrolytic Manganese Metal. Chip form; grades: Low-Hy (0.005% H_2), Extra Low-Hy (0.001% H_2), and Nitor-6 nitrided (6% N_2). [Kerr-McGee Chemical Corp]

Tronamang-75 Manganese Aluminum Briquettes. 75% manganese and 25% aluminum in briquette form for aluminum alloying. [Kerr-McGee Chemical Corp]

Trona Muriate of Potash. White agricultural grade. 60.5% minimum K_2O in coarse, standard, and fine grades. [Kerr-McGee Chemical Corp]

Trona Potassium Chloride. CAS 7447-40-7; EINECS 231-211-8; High purity white industrial grade; 96.8% KCl (61.8% K_2O equivalent). [Kerr-McGee Chemical Corp]

Trona Potassium Sulphate. CAS 7778-80-5; High purity white industrial grade. 50% K_2O minimum in granular and standard grades. [Kerr-McGee Chemical Corp]

Trona Salt Cake. Minimum purity 98% Na_2SO_4. [Kerr-McGee Chemical Corp]

Trona Soda Ash. Grades: dense-99.7% min. Na_2CO_3; Granular-97.7% min. Na_2CO_3; light-98.3% min. Na_2CO_3 (dry basis). [Kerr-McGee Chemical Corp]

Trona Sulphate of Potash. Standard white agricultural grade with 50% minimum K_2O. [Kerr-McGee Chemical Corp]

Tronolane. CAS 637-58-1; Pramoxine hydrochloride; anesthetic. [Abbott Laboratories] *

Tronox Titanium Dioxide Pigments, Chloride Process. CAS 13463-67-7; EINECS 236-675-5; Nine grades for paint, plastics, printing inks and paper applications. [Kerr-McGee Chemical Corp] *

troostite. A mineral, $2(Zn \cdot Mn)O \cdot SiO_2$. It is also the name for a constituent of steel tempered at a high temperature. It occurs in the transformation of austenite, the stage following marten site, and preceding sorbite.

Tropax. Analgesic. [Bayer AG; Tropon Werke]

Trophysan. A proprietary preparation of aminoacids, minerals, and vitamins in sorbitol used for intravenous feeding. [Servier Laboratories Ltd] *

Tropium. CAS 58-25-3; A proprietary preparation of chlordiazepoxide; a tranquilizer. [DDSA Pharmaceuticals Ltd] *

Tropoderm. Topical anti-inflammatory. [Bayer AG; Tropon Werke]

Tropotox. CAS 94-81-5; Soluble concentrate containing

400 g/l MCPB; for control of weeds in undersown cereals and grassland. [Rhone-Poulenc Crop Protection Ltd]

Tropotox Plus. Soluble concentrate containing 37.5 g MCPA and 262.5 g MCPB per liter; for control of weeds in undersown cereals and grassland. [Rhone-Poulenc Crop Protection Ltd]

Trosiplast. Rigid and plasticized PVC compounds; injection, blow and extrusion molding compounds used in various branches of industry. [Dynamit Nobel Wien GmbH] *

Trosiplast M. Compound PVC; for hollow body blow molding, injection molding, profile extruding, calendering. [Dynamit Nobel Wien GmbH] *

Trosiplast S. Suspension PVC; further processing into injection molding and extrusion molding compounds. [Dynamit Nobel Wien GmbH] *

Trosyd. A proprietary preparation of ticonazole; an antifungal. [Pfizer International]

Trotis. CAS 66063-05-6; Pencycuron; a phenylurea fungicide to control black scurf in potatoes. [Bayer AG]

trotter oil. See Neatsfoot oil.

Trotyl. (Trolite; Trilite; Tritolo; Trinol; Tolite; Triton; Tritole; T.N.T.). Trinitrotoluene, $CH_3 \cdot C_6H_2(NO_2)_3$; an explosive constituent.

Troykyd® D44. Stearate-modified hydrocarbon; defoamer for aq. systems, latex paints, pigment disps., adhesives, caulks, sealants, cement compounds. [Troy]

Troykyd® Perma-Dry. Cobalt compound complexed with rare earth and alkaline earth metals; lead-free; feeder drier for paints. [Troy]

Troysan® 174. 2-[(Hydroxymethyl) amino] ethanol; preservative for protection against bacterial spoilage in aq. systems, paints/coatings, resin emulsions, pigment slurries, adhesives, joint cements, and metalworking fluids. [Troy]

Troysan® 192. 2-[(Hydroxymethyl) amino]-2-methylpropanol; preservative used in aq. systems. [Troy]

Troysan® Polyphase® EC17. 3-iodo-2-propynyl butyl carbamate; fungicide, mildewcide for paints, leather protection, inks, metalworking fluids, paper and textile coatings, adhesives, caulks, sealants, and plastic coatings. [Troy]

Troysol AFL. Polymeric ester blend; surfactant, defoamer, antifloat agent, air release agent for nonaq. coatings (alkyd, acrylic, epoxy, polyester, lacquer), printing inks. [Troy]

Troysol LAC. Alkyl surfactant; substrate wetting agent, flow control additive, leveling agent for aq. systems, paints, printing inks, adhesives, and coatings for polyethylene and wax coated film and pkg. [Troy]

Troysol S366. Modified siloxane copolymer; surface active flow/leveling agent, substrate wetting agent for aq. and nonaq. systems, paints/coatings, alkyd, acrylic, epoxy, urethane, polyester, lacquer, printing inks. [Troy]

Troythix A. Polymerized, chemically inert organic ester; bodying agent for trade sales and industrial coatings thinned with hydrocarbons, ketone, and ester solv.; imparts high thixotrophy resistance to sagging, pigment settling, and penetration. [Troy]

TRS Rubber. A proprietary air dried fast-curing rubber. [Mitsui Co Ltd] *

Trubin. Macrolide antibiotic; especially useful against CRD in poultry and enzootic pneumonia in pigs; growth promoter. [Bayer AG]

---

‡ Trade name and manufacturer not verified   § Trade name without identified manufacturer

Trucal. CAS 471-34-1; Calcium carbonate; foodstuffs for animals, additives, included in Class 31 for use in animal foodstuffs but not including cereals or cereal products. [Tilcon Ltd] *

Trucarb. CAS 471-34-1; Calcium carbonate; industrial limestone powders and granules for use in the manufacture of carpets, paints, glues, glass, PVC products, floor covering, mastics, agrochemicals, ceramics, roofing felt, rubber, resins, pigments, and pharmaceuticals. [Tilcon Ltd] *

Truchem Quintex. Emulsifiable concentrate containing 30 g chlorpropham, 30 g fenuron and 130 g propham per liter; an herbicide for beet crops. [MTM Agrochemicals Ltd]

Tru-Color. Color anodized aluminum. [Reynolds Metal Co]†

Tru-Flake® Salt. CAS 7647-14-5; Sodium chloride; flat rectangular crystalline salt prepared by compacting vacuum pan salt. [Akzo Salt]

Trufree. A range of special flours for dietary use; gluten-free, wheat-free flour for low salt diets. [Larkhall Laboratories plc] *

Trugreen 0-0-2. Micronutrient formulation which promotes and stimulates chlorophyll production; chelating agent with seven elements. [W A Cleary]

Trulime. Hydrated lime $Ca(OH)_2$ used for horticulture/agriculture, alkalinity, building industry, civil engineering (soil stabilization and soil modification), leather processing, organic and inorganic chemicals, petrochemicals, plasterwork, sewage and water treatment. [Tilcon Ltd] *

Trump. Suspension concentrate containing 236 g isoproturon and 236 g pendimethalin per liter; used for annual weed control in winter wheat, rye, and barley. [Cyanamid of Great Britain Ltd]

Truozine. Trisulfapyrimidines (oral suspension); mixture of sulfadiazine, sulfamerazine, and sulfamethazine; antibacterial. [Abbott Laboratories] *

Trustan®. Mixture of cymoxanil and mancozeb; systemic fungicide to control potato blight. [DuPont UK]

Truxal. Neuroleptic. [Bayer AG; Tropon Werke]

Truzone. CAS 7722-84-1; Hydrogen peroxide for the hairdressing trade. [Interox Chemicals Ltd] *

Trycite. Polystyrene film used for packaging. [Dow UK]

Trycol® 5874. CAS 24938-91-8; Trideceth-14; hydrophilic general purpose emulsifier; for agricultural applications. [Henkel/Emery]

Trycol® 5882. Laureth-4; coemulsifier for silicone in polishes, mold release agents; emulsifier in industrial lubricants, agriculture, textile applications; intermediate for shampoo bases; biodeg. [Henkel/Emery; Henkel/Textile]

Trycol® 5888. CAS 9005-00-9; Steareth-20; emulsifier, solubilizer, solv. emulsifier in textile dye carriers, agricultural formulations; stabilizer in latexes; used in dyeing assistants, fruit coatings. [Henkel/Emery; Henkel/Textile]

Trycol® 5940. CAS 24938-91-8; Trideceth-6; detergent, wetting agent, emulsifier, dispersant, foam builder, solubilizer, coupling agent, rewetting agent for institutional, industrial, and household cleaners, degreasers, cutting oils, wool scouring, agricultural applications; intermediate. [Henkel/Emery; Henkel/Textile]

Trycol® 5950. CAS 26183-52-8; Deceth-4; wetting agent and penetrant for textile and agricultural applications; intermediate. [Henkel/Emery; Henkel/Textile]

Trycol® 5971. CAS 9004-98-2; Oleth-20; emulsifier, dispersant, solubilizer; textile lubricant; intermediate for shampoo base. [Henkel/Emery; Henkel/Textile]

Trycol® 6940. CAS 9016-45-9; Nonoxynol-5; emulsifier, detergent, wetting agent; moderate to low foam; for agricultural and textile formulations. [Henkel/Emery; Henkel/Textile]

Trycol® 6956. CAS 9002-93-1; Octoxynol-40; surfactant for textile use; salt-free. [Henkel/Textiles]

Trycol® 6985. CAS 9014-93-1; Nonyl nonoxynol-8; emulsifier in textile dye carrier applications, insecticides, wax emulsions; foam control agent; spreading agent in pigment printing; post-stabilizer in emulsion polymerization; intermediate. [Henkel/Emery; Henkel/Textile]

Trydet 2610. CAS 9004-99-3; PEG-23 stearate; emulsifier, thickener for agricultural formulations. [Henkel/Emery]

Trydet 2644. CAS 56002-14-3; PEG 400 isostearate; surfactant, lubricant for fiber lubricants, textile processing aids, fabric softeners. [Henkel/Emery; Henkel/Textile]

Trydet 2676. CAS 9004-96-0; PEG-10 oleate; emulsifier, lubricant for pesticides, metal cleaners, textile detergents and dyeing assistants, leather; rewetting agent for paper. [Henkel/Emery; Henkel/Textile]

Trydet 2682. CAS 61791-00-2; PEG-16 tallate; detergent, emulsifier, textile leveling agent; coemulsifier. [Henkel/Emery; Henkel/Textile]

Trydil. A safe gel heavy duty hand cleanser [Borax Consolidated Ltd] *

Tryfac® 5552. Phosphate ester, free acid form; surfactant intermediate; salts used as emulsifiers for aliphatic and aromatic solvs., agricultural formulations, textile processing. [Henkel/Emery; Henkel/Textile]

Tryfac® 5553. Phosphate ester (potassium salt of Tryfac 5552); emulsifier, wetting agent, detergent, antistat; used in heavy duty cleaners, metalworking compounds, agriculture, textile applications; corrosion inhibitor, dispersant. [Henkel/Emery; Henkel/Textile]

Tryfac® 5556. CAS 51811-79-1; Complex phosphate ester, free acid form; wetting agent, dispersant, antistat in textile processing; solv. emulsifier for textile scours, detergents, pesticides; drycleaning detergent; also used in emulsion polymerization. [Henkel/Emery; Henkel/Textile]

Tryfac® 5573. CAS 12751-23-4; Phosphate ester, free acid form; mold release agent, antistat, dispersant, emulsifier. [Henkel/Emery]

Trylon® 6735. Alkyl polyether; low-foaming emulsifier, wetting agent for industrial, institutional, and consumer detergents, textile scouring, bleaching and jet dyeing systems; coemulsifier for solvs. [Henkel/Emery; Henkel/Textile]

Trylox® 5900. CAS 61791-12-6; PEG-5 castor oil; emulsifier, dispersant, carrier, foam control agent, lubricant for paints, paper coatings, dye carriers, agricultural formulations. [Henkel/Emery]

Trylox® 5921. CAS 61788-85-0; PEG-16 hydrogenated castor oil; emulsifier, lubricant, softener; used in fabric softeners and aerosol fabric sprays. [Henkel/Emery; Henkel/Textile]

Trylox® 6746. CAS 57171-56-9; PEG-40 sorbitol hexaoleate; o/w emulsifier, dispersant, wetting agent, lubricant, plasticizer, solubilizer for household/industrial/institutional products, metal lubricants, textile, and cosmetic use. [Henkel/Emery; Henkel/Textile]

---

* Trade name not verified as available

† Trade name verified as obsolete

Trylox® 6753. PEG-20 sorbitol; humectant, plasticizer; intermediate; used in surfactant solutions.; emulsifier for textile and cosmetic use. [Henkel/Emery]

Trylube 7602. Finish for carpet staple providing effective antistatic properties and low fuming. [Henkel/Textiles]

Trymeen® 6601. CAS 61791-14-8; PEG-10 cocamine; coemulsifier, dispersant, antistat, emulsifier, softener, lubricant for textile applications, industrial lubricants, pesticides; substantive to metals, fibers, clays, etc. [Henkel/Emery; Henkel/Textile]

Trymeen® 6606. CAS 61791-44-4; PEG-15 tallow amine; emulsifier, antiprecipitant for dye baths, agricultural formulations; leveling agent; intermediate for quaternaries; antistat for synthetic fiber processing. [Henkel/Emery; Henkel/Textile]

Trymeen® 6617. CAS 26635-92-7; PEG-50 stearyl amine; emulsifier, antistat for metal buffing compounds, latex rubber compounding, agricultural formulations; anticoagulant; lubricant, leveling agent for textile applications. [Henkel/Emery; Henkel/Textile]

Trymeen® 6620. PEG-30 oleamine; surfactant for textile use. [Henkel/Textiles]

Trymeen® 6657. Stearic acid diamide; lubricant/softener; hydrophobic component with mineral oils and fatty esters in formulating fabric finishes; maintains good whiteness retention during fabric or yarn processing. [Henkel/Emery; Henkel/Textile]

Trymeen® OAM 30/60. PEG-30 oleamine; surfactant for textile use. [Henkel/Textiles]

Trymeen® SAM-50. CAS 26635-92-7; PEG-50 stearyl amine; antistat, emulsifier in metal buffing compounds; lubricant for fiberglass; leveling agent for dye applications. [Henkel/Emery; Henkel/Textile]

Trymeen® TAM-15. CAS 61791-44-4; PEG-15 tallow amine; emulsifier; antiprecipitant for mixed dye baths; leveling agent for acid dyes; migrating agent for dispersed dyes; intermediate for quaternary ammonium compounds; antistat for processing synthetic fibers. [Henkel/Emery; Henkel/Textile]

Trymer. Rigid polyisocyanurate bunstock used in the manufacture of insulation. [Dow UK]

Trymex Cream and Ointment. CAS 76-25-5; Triamcinolone acetonide USP; indicated for the relief of the inflammatory and pruritic manifestations of corticosteroid responsive dermatoses. [Altana Inc] ‡

trypsin. Proteolytic enzyme; formed in the small intestine by the action of a peptidase, enterokinase, on the pancreatic cell production, trypsinogen; CAS 9002-07-7; EINECS 232-650-8; proteolytic enzyme; dehairing of hides. [Am. Biorganics; Unibios SpA; U.S. Biochemical; Worthington Biochemical]

Tryptanol. CAS 549-18-8; Amitriptyline hydrochloride; antidepressant with sedative properties. [Merck & Co Inc]

Tryptizol®. CAS 549-18-8; Amitriptyline hydrochloride; antidepressant with sedative properties. [Merck & Co Inc]

DL-tryptophan. (indole-α-aminopropionic acid; 1-α-amino-3-indolepropionic acid) $C_{11}H_{12}N_2O_2$; CAS 54-12-6; EINECS 200-194-9; nutrition and research, medicine, dietary supplement, cereal enrichment. [Dajac Labs; Degussa; Penta Mfg.; Tanabe USA]

Trypure Novo. CAS 9002-07-7; A proprietary preparation of trypsin; used to promote wound healing and as an inhalation to reduce sputulo viscosity. [British Drug Houses] *

Trysul. Triple sulfa vaginal cream; indicated for the treatment of hemophilus vaginalis vaginitis. [Altana Inc] ‡

TSE-2000. Diphenylmethane diisocyanate in methylene chloride; one-coat adhesive to bond millathane to a variety of substrates during vulcanization. [TSE Industries]

tse-leou. (ting-yu). An oil expressed from Chinese tallow seeds (seeds of Sapium sebiferum).

tse-hong. A mixture of white lead, aluminia, ferric oxide, and silica; used by the Chinese for painting on porcelain.

TSE Mold Release®. Glycol surfactant; mold release especially for Viton parts. [TSE Industries]

T-siloxide. Product of silica fused with 0.1-2% titania; a silica glass.

tsing-lieu. A red pigment used in porcelain painting consisting of a mixture of stannic and plumbic silicates, with copper oxide or cobalt and gold.

T-Size. Rosin emulsion size. [Hercules]

TSP-12. See trisodium phosphate, dodecahydrate.

TSP-O. See trisodium phosphate, anhydrous.

TSPP. See sodium pyrophosphate.

TTC. See tetrazolium chloride.

Tuads®. Tetrabutylthiuram disulfide, tetramethylthiuram disulfide or 60:40 blend methyl and ethyl; rubber accelerators, sulfur donors, accelerators, vulcanizing agents. [R T Vanderbilt Co Inc]

Tuasol 100. CAS 87-10-5; Tribromsalan; disinfectant. [Merrell Dow Pharmaceuticals Inc] *

Tubania. Jeweler's alloys of varying composition, usually containing copper or brass, antimony, tin, and bismuth. The English alloy contains 12 parts brass, 12 parts tin, 12 parts antimony and 12 parts bismuth. German Tubania consists of 4 parts copper, 3 1/4 parts tin, and 42 parts antimony.

Tubarine. CAS 57-94-3; A proprietary formulation of tubocurarine chloride pentahydrate; a neuromuscular blocking agent used during general anesthesia to relax skeletal muscles to facilitate a wide range of surgical procedures. [Burroughs Wellcome Co; The Wellcome Foundation Ltd]

Tubazole. TBZ and iodophor as a foggable solution; used for controlling various diseases in stored potatoes. [Wheatley Chemical Co Ltd] *

Tubazole. Mixture containing nonylphenoxypoly (ethyleneoxy)ethanol-iodine complex and thiabendazole; used for controlling various diseases in stored potatoes. [Dean Agrochemicals Ltd]

Tubazole M. TBZ and iodophor as a sprayable solution; for controlling various diseases in stored potatoes. [Wheatley Chemical Co Ltd] *

Tuberculin Old Tine Test. Stainless steel disc with four prongs that have been dipped in old tuberculin; intradermal test for tuberculosis. [Lederle Laboratories]

Tubergran. CAS 82-68-8; Quintozene in dry granular form; for controlling common scab and rhizoctonia in growing potatoes. [Wheatley Chemical Co Ltd] *

Tubodust. CAS 117-18-0; Dustable powder containing 6% w/w tecnazene; protectant fungicide and potato sprout suppressant. [Farmers Crop Chemicals Ltd]

Tubostore. CAS 117-18-0; Granules containing 5% w/w tecnazene; protectant fungicide and potato sprout suppressant. [Farmers Crop Chemicals Ltd]

Tubotin. Fungicide. [May & Baker Ltd] *

---

‡ Trade name and manufacturer not verified   § Trade name without identified manufacturer

Tubotox. CAS 88-85-7; Dinoseb. [May & Baker Ltd] *
Tuc-tur metal. A nickel silver. It contains from 59-61% copper, 21-28% zinc, 12-18% nickel, and 0.3% iron.
Tuex. CAS 137-26-8; Tetramethylthiuram disulfide; short curing range in natural with normal to high sulfur. fast curing SBR and is flat curing; accelerator with sulfur for nitrile, butyl and EPDM rubbers. [Uniroyal] *
Tufcote. Acrylic emulsion finish. [DuPont UK]
Tufcote E-50SM. Polyether urethane foam, 1-mil aluminized surface; high acoustical performance for absorbers, barriers, and composite products. [E-A-R]
Tufdene®. BR, SBR thermoplastic elastomers. [Asahi Chem. Industry]
Tuf-Draw. Lubricant containing mineral oil; may contain additives such as emulsifiers, corrosion inhibitors, biocides, surfactants, and lubricating additives. [Franklin Oil Corporation (Ohio)]
Tufflake™. Degradation resistant aluminum pigments; specifically formulated for automotive OEM coatings systems. [Silberline Mfg Co Inc]
Tuff Stuff. Two-component flexible adhesive. [Hardman]
Tufftride. Metal surface treatment for wear and fatigue. [Degussa Ltd]
Tuflin HS-7066 NT7. CAS 9002-88-4; LLDPE; blown film resin for grocery sacks, shipping sacks, heavy-duty film, trash bags, liners. [Union Carbide]
Tuf-Lube. Fluorocarbon dry surface lubricant, release agents. [Specialty Products Co] *
Tufnol. Laminated plastics materials bonded with synthetic resins incorporating fillers such as cotton fabric, paper and asbestos fabric; the materials are used for electrical and mechanical components in most manufacturing industries. [Tufnol Ltd]
Tufprene®. BR, SBR thermoplastic elastomers. [Asahi Chem. Industry]
Tufseal. A trademark for a range of polymerisable mixtures of asphalt, polyols, and isocyanates; used as adhesives. [Robertson Co] ‡
Tufset. A rigid polyurethane used for engineering purposes. [Tufnol Ltd]
Tuf Stuf. Two-part epoxy putty; for filling and bonding most materials; when set (rock hard) can be drilled, tapped, filed, sanded, contoured, painted and polished; resistant to oil, water, fuel, bleach and dilute acids. [Hermetite Products Ltd] *
Tuftane. High performance polyester and polyether based urethane films; for high performance applications such as fabric lamination, belting, protective covers and similar uses where the durability of urethane is required. [Lord Corporation (UK) Ltd]
Tuftec®. Hydrogenated thermoplastic elastomer. [Asahi Chem. Industry]
Tugon. CAS 52-68-6; Trichlorfon; insecticide for control of adult flies and fly larvae. [Bayer AG]
Tugon OKO. Insecticide for indoor control of crawling and flying insects in public health. [Bayer AG]
Tugon sp 80. Controls flies and their larvae. [Bayer AG]
Tuinal. A proprietary preparation of quinalbarbitone sodium and amylobarbitone sodium; a hypnotic. [Eli Lilly & Co]
Tula metal. An alloy of silver, copper, and lead.
Tullanox HM-250. Hydrophobic precipitated silica, modified by organic silazane compound; high surface area, high water repellency; provides reinforcement, rheology control, corrosion resist., anticaking, thickening to silicone

sealants, coatings, powds., polyester resins, liq. systems, elastomers, elec. insulation, defoamers; carrier for catalysts; filler/additive for plastics, paints, coatings, inks, pharmaceuticals, cosmetics, fertilizers, metals, adhesives, toners. [Tulco]
Tumbleblite. Systemic fungicide. [Murphy Chemical Ltd] †
Tumblebug. Garden insecticide. [Murphy Chemical Ltd] †
Tumblemoss. Moss killer preventer. [Murphy Chemical Ltd]†
Tumbleslug. Slug and snail killer. [Murphy Chemical Ltd] †
Tumbleweed Gel. Weedkiller for spot application. [Murphy Chemical Ltd] †
Tumenol. See Thiol. §
Tumeson. A proprietary preparation of prednisolone, sulfonated distillate of shale oils, and titanium dioxide for dermatological use. [Hoechst UK] *
Tuncast. Tunning refractory; used for lining tundishes. [Foseco (F.S.) Ltd]
Tundak. Preformed insulating refractory cones for lining nozzle wells in tundishes. [Morton Int'l. Ltd]
tunga resin. A neutral glycerol-rosin ester, made with the aid of tung oil as esterifying catalyst.
tung oil. (Chinese wood oil). The oil obtained by pressure from the seeds of Aleurites cordata and Aleurites fordii of China and Japan. The seeds contain from 40-53% of oil.
Tungophen® B. Condensation product of substituted phenols and formaldehyde resins; for improving the hardness and through-drying of alkyd paints and varnishes. [Bayer AG]
tungsten. (wolfram) Metallic element; W; CAS 7440-33-7; EINECS 231-143-9; high-speed tool steel, alloys, filaments for electric light bulbs, contact points, x-ray and electron tubes, welding electrodes, heating elements, rocket nozzles, sheet steel, chemical apparatus, high-speed rotors, solar energy devices. [Aldrich; Atomergic Chemetals; Cerac; Noah Chem.]
tungsten blue. A colloidal solution of the blue oxide of tungsten (ditungsten pentoxide). It may be used for dyeing silk.
tungsten brass. (wolfram brass). An alloy of 60% copper, 22% zinc, 14% nickel, and 4% tungsten. An alloy containing 60% copper, 34% zinc, 2.8% aluminum, 2% tungsten, 0.7% manganese and 0.15% tin is also known by this name.
tungsten bronze. (wolfram bronze). An alloy made by fusing potassium tungstate with pure tin; used for decorative purposes. It is also the name for an alloy of 95% copper, 3% tin, and 2% tungsten. An alloy containing 90% copper and 10% tungsten is also known by this name.
tungsten (VI) fluoride. See tungsten hexafluoride.
tungsten hexafluoride. (tungsten (VI) fluoride) WF_6; CAS 7783-82-6; EINECS 232-029-1; Vapor-phase deposition of tungsten, fluorinating agent. [Air Prods.; Akzo; Atochem N. Am.; Atomergic Chemetals]
tungsten oxide (ic). (tungstic oxide; tungsten trioxide; tungstic acid anhydride; wolframic acid anhydrous) WO_3; CAS 1314-35-8; EINECS 215-231-4; Forms metals by reduction, alloys, preparation of tungstates for x-ray screens, fireproofing fabrics, yellow pigment in ceramics. [Atomergic Chemetals; Cerac; Climax Molybdenum]
tungsten steel. A very hard alloy of steel and tungsten. It usually contains from 5-8% tungsten, often 4% chromium, and 1.25% carbon; used for armor plates, projec-

---

* Trade name not verified as available   † Trade name verified as obsolete

tiles, firearms, and high speed tools. Tool steels contain 1-4% tungsten and a rifle barrel steel contains 3-6% tungsten.

tungsten steel, high. These steels usually contain from 80-85% iron with more than 14% tungsten. Some alloys contain from 77- 81% iron, 15-18% tungsten, 3-4% chromium, and 0.15- 0.35% silicon.

tungsten steel, low. These alloys usually contain about 96% iron with 1.5-2% tungsten, 0.5-1% chromium, and 0.15-0.35% silicon.

tungsten trioxide. *See* tungsten oxide (ic).

tungstic acid. (wolframic acid; orthotungstic acid) H_2WO_4; CAS 7783-03-1; EINECS 231-975-2; Textiles (mordant, color resist), plastics, tungsten metal, wire, etc. [Am. Int'l. Chem.; Atomergic Chemetals; Noah Chem.]

tungstic acid anhydride. *See* tungsten oxide (ic).

tungstic oxide. *See* tungsten oxide (ic).

tungstophosporic acid. *See* phosphotungstic acid.

turacine. A red coloring matter contained in the feathers of the turaco birds of Africa. The coloring matter contains 8% copper.

turbadium bronze. An alloy of 46% copper, 44% zinc, 5% lead, 2% nickel, 1.5% manganese, and small quantities of tin and aluminum; used for propeller castings.

Turbair. CAS 60-51-5; Dimethoate; an organophosphorus insecticide and acaricide. [Pan Britannica Industries Ltd]

Turbair Grain Store Insecticide. Mixture of fenitrothion, permethrin, and resmethrin; used to control insects in grain stores. [Pan Britannica Industries Ltd]

Turbair Permethrin. CAS 52645-53-1; Permethrin; a pyrethroid insecticide. [Pan Britannica Industries Ltd]

Turbair Roval. CAS 36734-19-7; Iprodione; a fungicide with protectant activity for glass house crops. [Pan Britannica Industries Ltd]

Turbex. Nonionic surfactant. [ABM Chemicals Ltd] *

turbiston bronze. An alloy containing 55% copper, 41% zinc, 2% nickel, 1% aluminum, 0.84% iron, and 0.16% manganese. It resists sea water.

Turboclean. A blend of detergents and surfactants with inhibitors; a cleaning fluid for compressors of gas turbine engines. [The Kent Chemical Company Ltd]

Turbo-Grass . Mineral and plant extracts in a water base containing cytokinin, B-vitamin, morphogenic and porphyrin activity to aid in increased plant metabolism and yield; for all agricultural, horticultural, and forestry products. [SN Corp/Appropriate Technology Ltd] †

Turbonit. A German synthetic varnish paper product; used for electrical insulation. §

Turfclear. CAS 83601-81-4; A liquid formulation containing 500 g carbendazim per liter as a suspension concentrate; systemic fungicide. [Fisons plc]

turgoids. A name applied to substances such as textile fibers, hide, tissue, leather, and wood fibers, which swell in water but do not dissolve.

Turgum® S. Turpene-resin acid blend; plasticizer and conditioner for SBR; retarder for high, intermediate, and super abrasion furnace black/natural rubber stocks. [Whitney & Oettler] *

Turkey red oils. (sulfonated castor oils, monopol oil, soluble castor oil, sulfated oil, red oil, oleine). CAS 72-48-0; Viscid, transparent liquids; used in the preparation of cotton fiber for printing, alizarin dye assistant, textiles, leather, manufacture of soaps.

turkey umber. *See* umber.

Türkischrotöl 100%. Sodium sulforicinoleate; solubilizer for cosmetics. [Zschimmer & Schwarz]

turmeric. CAS 8024-37-1; (Indian saffron, terra Merita, curcuma). A natural dyestuff obtained from the underground stems of rhizome of *Curcuma longa* and *C. rotunda*. The dyeing principle is curcumin, $C_{21}H_{20}O_6$. It dyes cotton greenish-yellow, and is also a coloring matter for wool, silks, oil, butter, cheese, curry powder, wood, and wax.

Turnbull's blue. (Gmelin's blue). Ferrous ferricyanide, $Fe_3[Fe(CN)_6]_2$.

turpenteen. *See* turpentyne.

turpentinc. The exudation (gum) from from incisions made in certain varieties of pine, fir, and larch. The terms American, French, German, Mexican, Portuguese, and Spanish turpentine are usually used for the balsam turpentine; German, Finnish, Polish, Russian, and Swedish turpentine, often refer to wood turpentine, but German, Finnish, Swedish oil; cab refer to refined sulfite turpentine oil.

turpentine. CAS 8006-64-2; Volatile essential oil whose chief constituents are pinene and diterpene ; used as a solvent for paints, varnishes, and lacquers.

turpentyne. (turpenteen). A turpentine substitute composed of rosin spirit, shale spirit, petroleum spirit, and coal tar naphtha.

turpeth mineral. (turbith mineral, Queen's yellow). A yellow basic sulfate of mercury, $HgSO_4 \cdot 2HgO$.

Turpex. Finishing agents. [Ciba plc] *

turps. *See* oil of turpentine.

Turrisin. Milk product stabilizer. [Giulini Corp]

Tursione. Biocidal preparation. [BDH Chemicals Ltd]

Tusadin. A proprietary agent for protection against frost in motor engines. §

Tusana. A proprietary preparation of liquid extracts of cocillana, ipecacuanha, squill, senega and senna and glycerin and dextromethorphan hydrobromide; a cough linctus. [The Boots Co plc]

tussar silk. (tasar silk). The product of the caterpillar of *Antheraca paphia* of India.

Tussend. Used primarily in the treatment of coughs, colds and upper respiratory conditions. [Dow UK] *

Tussiex. A proprietary preparation of ammonium chloride, sodium citrate, ephedrine hydrochloride, pholcodine, and menthol; a cough linctus. [Crookes Laboratories] *

Tussifan S. A proprietary preparation of belladonna extract, potassium citrate, ipecacuanha, squill, anise oil and chloroform spirit; a cough linctus. [H N Norton & Co Ltd]‡

Tussiplegyl. An antitussive preparation. [Bayer AG] †

Tutania. Alloys. An English type contains 91% tin, 8% lead, 0.7% copper, and 0.3% zinc; and another 80%tin, 165 antimony, 2.7% copper, and 1.3% zinc; the German alloy contains 92% antimony, 7% tin, and 1% copper; another consists of 62% antimony,

Tutenag. (Tutenague, Tutenay). A nickel silver. It consists of from 44- 46% copper, 16-40% zinc, and 15-40% nickel.

Tutia. *See* tutty powder.

Tutol®. A registered trade name for certain explosives. §

Tutol No. 2. An explosive similar to Rexite (*qv*); it contains sodium nitrate instead of potassium nitrate, and 12% of the explosive base is replaced by sodium chloride. §

Tutor. Glass fabric based products having fire resistance. [Rentokil Ltd]

tutty powder. (tutia). An impure oxide of zinc, formed during

---

‡ Trade name and manufacturer not verified   § Trade name without identified manufacturer

the smelting of lead ores containing zinc. Sometimes a mixture of blue clay and copper filings is sold under this name.

Tween®. Polyexyethylene sorbitan fatty acid esters; surfactants. [ICI Am.]

Tween® 20. CAS 9005-64-5; Polysorbate 20 NF; solubilizer; o/w emulsifier; detergent for shampoos; antistat and fiber lubricant used in textile industry; flavor emulsifier. [ICI Spec. Chem.; ICI Surf. Belgium]

Tween® 21. CAS 9005-64-5; Polysorbate 21; emulsifier, solubilizer; antistat, fiber lubricant for textiles. [ICI Spec. Chem.; ICI Surf. Belgium]

Tween® 40. CAS 9005-66-7; Polysorbate 40 NF; emulsifier, solubilizer; textile antistat, fiber lubricant. [ICI Spec. Chem.; ICI Surf. Belgium]

Tween® 60, 60K. CAS 9005-67-8; Polysorbate 60 NF; food emulsifier, dough strengthener for food industry; flavor emulsifier and dispersant; foaming agent for beverages. [ICI Spec. Chem.; ICI Surf. Belgium]

Tween® 61. CAS 9005-67-8; Polysorbate 61; emulsifier, solubilizer for perfume, flavor, vitamin oils. [ICI Spec. Chem.; ICI Surf. Belgium]

Tween® 65, 65K. CAS 9005-71-4; Polysorbate 65; emulsifier for cake mixes, icings, fillings, coffee whiteners, frozen desserts, whipped toppings; antifoam; 65K is kosher grade. [ICI Spec. Chem.; ICI Surf. Belgium]

Tween® 80, 80K. CAS 9005-65-6; Polysorbate 80 NF; emulsifier, solubilizer for food, vitamins, oils; antifoam; wetting agent, detergent for cleaning contact lenses, skin care products; deflocculant. [ICI Spec. Chem.; ICI Surf. Belgium]

Tween® 81. CAS 9005-65-6; Polysorbate 81; emulsifier, solubilizer for perfume, flavor, vitamin oils. [ICI Spec. Chem.; ICI Surf. Belgium]

Tween® 85. CAS 9005-70-3; Polysorbate 85; emulsifier, solubilizer for perfume, flavor, vitamin oils; floating bath oils. [ICI Spec. Chem.; ICI Surf. Belgium]

Twent®. For skin impurities. [Bayer AG]

Twinspan. Mixture of chlorpyrifos and disulfoton; a systemic and fumigant insecticide for brassica crops. [Pan Britannica Industries Ltd]

Twin-Tak. Herbicide. [May & Baker Ltd] *

Twitchells reagent. Benzenestearo-sulfonic acid; used in the decomposition of fats.

Two Cubed Eight. CAS 1309-37-1; A gamma ferric oxide for magnetic media use. [Pfizer International] †

Two-stage phenolic resin. *See* novolac resin.

Twosward. Fertilizers. [ICI Chem. & Polymers Ltd] †

Tybrite. Coextruded packaging films. [Dow]

Tycel®. Urethane laminating adhesives; for lamination of many polymeric materials to rigid and flexible substrates. [Lord Corporation (UK) Ltd]

Tycel® 7000 Series. Laminating adhesives for bonding applications in decorative and protective building, automotive, and general purpose laminations. [Lord]

Tychem®. Synthetic polymers; chemicals and curing agents for polymers and elastomers. [Reichhold]

Tygacell. Composite molding with carbon, glass, aramid and ceramic reinforcement; structural components for racing cars and power boats, components for civil aircraft, space defence, medical x-ray equipment and high performance applications. [Fothergill Tygaflor Ltd] *

Tygadure. High performance insulated wires and cables and optical fibers; used for avionics and electronics,

industrial applications, short haul data communications, secure and hazardous environments. [Courtaulds Advanced Materials (Holdings) Ltd]

Tygaflon. Fluorocarbons with fiber, metal and other fillers; molded and machined custom components for all industries; used for high performance electrical applications, expansion joints and shaft covers. [Courtaulds Advanced Materials (Holdings) Ltd]

Tygaflor. Fluoronated coated glass and aramid fabrics; for process conveying belt systems used in the baking and packaging industries and other industrial uses, lightweight membrane roofs and radomes. [Courtaulds Advanced Materials (Holdings) Ltd]

Tygafluor. A proprietary trade name for an aqueous dispersion of PTFE (*qv*) with a curing temperature of 90-140 C.§

Tygalam. Composite molding with carbon, glass, aramid, and ceramic reinforcement; structural components for racing cars and power boats, components for civil aircraft, space defense, medical x-ray equipment, and high performance applications. [Fothergill & Harvey plc]

Tygan. Polyvinylidene chloride coated fabric; filtration, insect screening, glare-reducing blinds. [Courtaulds Advanced Materials (Holdings) Ltd]

Tygatape. Engineered, high performance PTFE and silicone tapes; masking applications in light/heavy engineering, electrical and electronics, aerospace. [Courtaulds Advanced Materials (Holdings) Ltd]

Tygavac. Materials for vacuum bag molding of composite components, TFE aerosol release sprays; Aerospace, automotive, medical engineering, industrial applications. [Courtaulds Advanced Materials (Holdings) Ltd]

Tyglas. Glass fiber woven fabrics; for electrical insulation, filtration, reinforced plastics, thermal insulations, industrial plant applications. [Courtaulds Advanced Materials (Holdings) Ltd]

Tygon F. A proprietary trade name for a furan resin. §

Tylac®. Synthetic rubber, latex or synthetic resin emulsions. [Reichhold]

Tylac® 68151-00. CAS 9003-18-3; Butadiene acrylonitrile emulsion polymer; for nonwoven industry; excellent oil and other hydrocarbon resist.; excellent pigment acceptance; suggested for synthetic leather, saturant for needle punched fabrics. [Reichhold/Emulsion Polymers] *

Tylac® 68202-00. CAS 9003-55-8; Styrene butadiene latex; multipurpose adhesive base for laminating cloth to cloth; firm film. [Reichhold/Emulsion Polymers] *

Tylenol. CAS 103-90-2; Acetaminophen; and analgesic and antipyretic. [McNeil Consumer Products Co]

Tyllanex. Terbuthylazine; 2-t-butylamino-4-chloro-6-ethylamino-1,3,5-triazine; herbicide. [Agan Chemical Manufacturers Ltd]

Tylnatrin. Sodium acetylsalicylate, $C_2H_3O \cdot OC_6H_4 \cdot COONa$.

Tylose® C, CB Series. CAS 9004-32-4; Sodium CMC; binder in pencil leads; thickener in batteries, rubber industry, cosmetics, foodstuffs, pharmaceuticals, tobacco and textile industry; dispersant, emulsifier for insecticidal, fungicidal and herbicidal products; plasticizer in ceramics; surface sizing in paper industry; press aid and lubricant in welding electrodes. [Hoechst Celanese/Colorants & Surf.; Hoescht AG]

Tylose® C-p, CB-p. CAS 9004-32-4; Cellulose gum; cos-

---

metic thickener. [Hoechst Celanese/Colorants & Surf.]

Tylose® H Series. CAS 9004-62-0; Hydroxyethylcellulose; binder, thickener, plasticizer, visc. control agent, protective colloid in ceramics, emulsion polymerization, tobacco, and textile industry, agriculture, cosmetics, soaps, and hand cleaning pastes. [Hoechst Celanese/Colorants & Surf.; Hoechst AG]

Tylose® MHB. CAS 9032-42-2; Methyl hydroxyethylcellulose; binder, thickener, pigment, foam, and filler stabilizer, dispersant, emulsifier, plasticizer, visc. control and sedimenting aid, and protective colloid used in coatings, paints, resins, mining, batteries, insecticides, fungicides, herbicides, rubber, textile, and leather industry, ceramics, suspension polymerization, and pharmaceuticals. [Hoechst Celanese/Colorants & Surf.]

Tylose® P, P-x, PS-x, P-z Series. Methyl hydroxycellulose; binder in plasters, adhesive, and troweling compounds. [Hoechst Celanese/Colorants & Surf.]

tylosin. CAS 1401-69-0; An antibiotic derived from an actinomycete resembling *Streptomyces fradioe*.

Tymahist. A proprietary preparation of chlorprophenpyridamine . [Mason Pharmaceuticals Inc] *

Tymtran. CAS 71247-25-1; Ceruletide diethylamine; stimulant. [Adria Laboratories Inc] *

Tynex®. Polymer. [DuPont UK]

Tyox®. For the electrical industry. [DuPont UK]

Typar®. Fibers. [DuPont UK]

Type 798 Roving. Highly filamentized glass bulked roving sized with silane binder; for use in polyester and vinyl ester resin systems for filament winding and pultrusion applications. [PPG/Fiber Glass]

type metal. A variable alloy of lead and antimony, frequently with the addition of tin, and sometimes copper or bismuth. The lead is present to the extent of from 50-93% the antimony 4-30% the tin 2-40% copper 0-5% and bismuth 0-29%.

typewriter metal. An alloy of 57% copper, 20% nickel, 20% zinc, and 3% aluminum.

Typly. Rubber and metal bonding agents. [Anchor Chemical Group plc]

Ty-Ply BN. One-coat adhesive for NBR. [Lord]

Typophor®. Dye base preparations in olein; for brightening of printing inks. [BASF AG]

Typro®. Fibers. [DuPont UK]

Tyramine. (phenomime; p-hydroxy-phenylethylamine) $HOC_6H_4CH_2CH_2NH_2$; CAS 51-67-2; EINECS 200-115-8; The base is a constituent of ergot.

Tyrenka. A proprietary trade name for a synthetic fiber resembling nylon. §

Tyril 125. CAS 9003-54-7; SAN resin; blending resin. [Dow Plastics]

Tyril 880. CAS 9003-54-7; SAN resin; strong, transparent thermoplastic with good chem. and heat resist., toughness, load-bearing strength, easy processing; injection molding resin for cosmetic containers, industrial battery cases and caps, high-pressure filter housings, pkg., appliances, blood aspirators, other medical parts, connectors, tumblers, dinnerware, utensils. [Dow Plastics]

Tyril 880B. CAS 9003-54-7; SAN random copolymer resin; high heat, high strength resin; superior chemical resistance to other grades and high visc.; offers the lowest melt flow making it superior for injection blow molding; avail. in a wide range of transparent, translucent, and opaque colored cylindrical pellets. [Dow Plastics]

Tyril 1011. CAS 9003-54-7; SAN resin; automotive resin; *see also* Tyril 880. [Dow Plastics]

Tyrimide. A proprietary preparation containing isopropamide iodine; an antispasmodic. [SmithKline Beecham] *

Tyrin Brand. Chlorinated polyethylene resins and elastomers used in a wide variety of applications including roofing membranes, wire and cable, automotive tubing, ignition wire. [Dow UK] *

Tyrin CM 0136. Chlorinated polyethylene elastomer; used in extruded and calendered goods, e.g., hose, cable jacketing, linings, gasketing, o-rings; excellent heat aging, ozone resist., good weathering, good oil, flame, and grease resist.; outstanding chem. resist.; vulcanizable with organic peroxides, aminothiadiazole systems, or radiation. [Dow Plastics]

Tyrin CM 566. Chlorinated polyethylene elastomer; used in cable jacketing, extruded goods requiring low moisture, excellent elec. properties; vulcanizable with organic peroxides, aminothiadiazole systems. [Dow Plastics]

Tyrin CM 0836. Chlorinated polyethylene elastomer; vulcanizable with organic peroxides, aminothiadiazole systems. [Dow Plastics]

Tyrin CM 3615. Chlorinated polyethylene elastomer; thermoplastic resin which can be fabricated as the primary polymer, as a major alloy component, or as a modifying agent imparting impact resist. to other resins (e.g., SAN, PP, PE, PVC); excellent chem. resist., easy processing; used for siding, molded goods, extruded profiles, hoses, sheet, lining, tubing; an impact modifier grade. [Dow Plastics]

Tyrite® 7412. One-component urethane adhesive; structural moisture-cure adhesive; high performance hose adhesive for Hytrel®, nylon, polyester, Kevlar®, thermoplastics and TPU. [Lord]

Tyrite® 7660. One-component urethane adhesive; structural moisture-cure adhesive with high solids, grab and flexibility; initial high grab, crosslinks over time to form a structural bond to foam, plastic, fabric, rubber, and prepared metals. [Lord]

Tyrol-2, 32B, 6, CEP. Flame retardant materials for plastics. [Stauffer Chemical] *

Tyrolean earth. *See* Bohemian earth.

Tyrolite. (cupriferous calamine, copper froth). A basic copper arsenate of green color, found in the Tyrol; $Cu_3As_2O_8$ · $2Cu(OH)_2$·$7H_2O$.

Tyrosilane C. Copper acetyl tyrosinate methylsilanol; tanning activator and anti-aging action for cosmetic and health products. [Exsymol]

tyrosinase. CAS 9002-10-2; EINECS 232-653-4; An oxidizing enzyme containing copper which occurs in plant and animal tissue; it turns peeled potatoes black when exposed to air; used in medicine as an antihypertensive.

tyrosine. (β-p-hydroxyphenylalanine; α-amino-β-p-hydroxyphenylpropionic acid) Nonessential amino acid; $C_6H_4OHCH_2CHNH_2COOH$; CAS 60-18-4 (L-); 556-03-6 (DL); 556-02-5 (D-); EINECS 200-460-4(L);209-113-1(DL);209-112-6(D); growth factor in nutrition; biochemical research; dietary supplement. [Am. Biorganics; Degussa; Nippon Rikagakuyakuhin; Penta Mfg.; U.S. Biochemical]

tyrothricin. An antibiotic produced by a strain of *Bacillus brevis*.

Tyrozets. Tyroethrocin and benzocaine; for minor mouth

---

‡ Trade name and manufacturer not verified   § Trade name without identified manufacturer

and throat infections. [Merck & Co Inc]

tysonite. a) A mineral that is a fluoride of cerium metallic elements with thorium. b) A blend of Gibsonite and vulcanized vegetable oils.

Tytanpol. CAS 13463-67-7; EINECS 236-675-5; Titanium dioxide. [British Traders & Shippers Ltd]

Tyvek®. Spunbonded polyolefin. [DuPont UK]

Tyvek Practik®. Fibers. [DuPont UK]

Tyvek Protech®. Fibers. [DuPont UK]

Tyzine. CAS 522-48-5; A proprietary preparation of tetrahydrozoline hydrochloride; a nasal decongestant. [Pfizer International]

Tyzor®. Organic titanate. [DuPont UK]

U

U35. Polyester-based aliphatic urethane aq. disp.; lightfast, med. soft polymer producing a dry surface feel; used as skincoat for transfer coating applications; excellent adhesion to plasticized PVC. [BFGoodrich/Spec. Polymers] *

U 46® . 2,4-D, MCPA, dichlorprop, mecoprop; for weed control in cereals, maize, sugar cane, rice, grassland, forestry, perennial crops, tree crops. [BASF AG]

Uba. (styxol, tanatol). Preparations containing sodium fluosilicate, Na_2SiF_6, as the main ingredient.

U Blue 104. CAS 1317-97-1, 7787-59-9; Ultramarine blue, bismuthoxychloride; inorg. colorant. [Presperse]

UBOB®. p-Aminodiphenylamine; intermediate. [Uniroyal]

Ubretid. CAS 15876-67-2; A proprietary trade name for distigmine bromide; used in the treatment of urinary retention. [Berk Pharmaceuticals Ltd] *

UBS. Dry cleaning soap and paint remover. [S & D Chemicals Ltd] *

UC-5500 Series. Two-component polyurethane coatings; for high gloss maintenance coatings, e.g., exterior coatings for process equip., transport vessels, tanks, structural steel in corrosive environments; excellent weathering, flexibility, good abrasion resist. [Heresite Protective Coatings]

Ucar® Acrylic 503. Modified acrylic emulsion; surfactant for high-quality exterior, interior, and semigloss house paints. [Union Carbide] *

Ucarcide® 225. CAS 111-30-8; Glutaral; preservative, antimicrobial for cosmetic, toiletry, and chemical specialty products. [Union Carbide]

Ucare® Polymer JR-30M, JR-125, JR-400, LR-30M, LR-400, SR-10. Polyquaternium-10; hair fixative; for cosmetics, toiletries, hair and skin care prods. [Amerchol Corp]

Ucar® Latex 100. Acrylic latex; for caulks, sealants. [Union Carbide] *

Ucar® Latex 130. CAS 9003-20-7; PVAc homopolymer emulsion; multipurpose, large particle size emulsion with high pigment-binding capacity; for ready-to-use tape joint compounds, wall patching compounds, low-cost mastic compounds, and tub and tile caulks. [Union Carbide] *

Ucar® Latex 148. Styrene-acrylic latex; for mastics. [Union Carbide] *

Ucar® Latex 351. Vinyl/acrylic emulsion; high-solids latex for interior and exterior paints. [Union Carbide] *

Ucar® Latex 405. PVA homopolymer latex; for building industry applications. [Union Carbide] *

Ucar® Phenoxy Resin PKHM-301. Phenoxy resin; used for flexible and rigid pkg.; provides films with high chemical resist. without loss of toughness and durability. [Union Carbide] *

Ucarsil® FR-1A, FR-1B. Organosilicon chemical; additives allowing the processing of up to 70% alumina trihydrate into polyolefins. [Union Carbide]

Ucar® Vehicle 435. Acrylic; for trade paints, industrial finishes. [Union Carbide] *

Ucar® Vehicle 451. Styrene-acrylic; for industrial finishes. [Union Carbide] *

Ucecoat. Polyurethanes for textile coatings. [UCB (Chem) Ltd]

Ucecryl. Acrylic emulsions. [UCB (Chem) Ltd]

Ucefix. Modified acrylic copolymers or polyurethanes for adhesives. [UCB (Chem) Ltd]

Uceflex. Thermoplastic polyurethanes. [UCB (Chem) Ltd]

Ucelone. Dispersions of silicone resins; acrylic emulsions. [UCB (Chem) Ltd]

Uchatius bronze. (steel bronze). An alloy containing 92% copper and 8% tin.

Ucicline. Boiler and cooling water treatment. [Laporte Industries Ltd]

Ucinite. A proprietary trade name for a phenol-formaldehyde resin laminated product. §

Ucipol. Boiler and cooling water treatment. [Laporte Industries Ltd]

Ucon®. Synthetic lubricants and fluids. [Union Carbide (UK) Ltd]

Ucon® 50-HB-400. PPG-9-buteth-12; emollient. [Amerchol Corp]

Ucon® Fluid AP. CAS 9003-13-8; PPG-14 butyl ether. [Amerchol Corp]

Ucon® Hydrolube HP-5046. Water-glycol; hydraulic fluid designed to operate at pressures to 7000 psi; superior fire retardancy. [Union Carbide]

Ucrete. A proprietary cement-modified polyurethane resin used for flooring. [ICI Chem. & Polymers Ltd]

Ucuhuba fat. A fat obtained from the seeds of *Myristica bicuhyba* and contains 92% fatty acids.

Udel. A proprietary polysulfone a high-performance, high-temperature thermoplastic resin used for injection molding and extrusion. [Union Carbide (UK) Ltd] *

Udel® GF-110. CAS 25135-51-3; Polysulfone, glass-reinforced; resin with high tensile strength, excellent electricals, good chemical resist., water and steam resist.; useful as alternative to metals in engineering plastics; used for printed circuit boards, chip carriers, connectors, elec. housings, process equip. for severe environments, sterilizable medical devices, food service equip., automotive, plumbing fixtures, microwave cookware. [Amoco Chemical Co]

---

‡ Trade name and manufacturer not verified   § Trade name without identified manufacturer

Udel® P-1700. CAS 25135-51-3; Polysulfone resin; tough, rigid, high-strength thermoplastic polymer for elec./electronic, automotive/aerospace applications, consumer products, medical, process and sanitary pipe, plumbing components, milking machine parts, pollution control equipment; P-1700 is a general purpose grade for injection molding or extrusion; FDA compliance. [Amoco Chemical Co]

Udet 950. Sodium alkylaryl sulfonate; fast dissolving, high active sulfonate for detergent powders. [Degussa]

Udikral. ABS polymers; for injection molding, sheet extrusions. [GE Plastics ABS Ltd] *

Udilo oil. See laurel nut oil.

Ufablend DC. Surfactant blend; biodeg. high foaming detergent conc. for manufacturing of liq. detergents, dishwash, hard surface cleaners; effective in hard and soft water. [Unger Fabrikker AS]

Ufacem. Surfactant blends; used for cement and concrete admixtures. [Unger Fabrikker AS]

Ufacid K. Linear dodecylbenzene sulfonic acid; intermediate for manufacturing of sulfonates used in detergent powds., liqs., emulsifiers; biodeg. [Unger Fabrikker AS]

Ufacid KA. 2-Phenyl alkylbenzene sulfonic acid; intermediate for manufacturing of sulfonates used in liq. detergents; biodeg. [Unger Fabrikker AS]

Ufanon K-80. Cocamide DEA (2:1); biodeg. detergent, foaming agent, wetting agent, thickener, foam stabilizer for liq. detergents, shampoos, cosmetics, leather industry; confers some corrosion protection. [Unger Fabrikker AS]

Ufapol. Sulfonated and sulfated surfactants; used for emulsion polymerization. [Unger Fabrikker AS]

Ufapore GP. Anionic surfactant blends; foaming agent for gypsum board. [Unger Fabrikker AS]

Ufarol. Fatty alcohol sulfate, neutralized with different amines; active ingredient in shampoos, bath products and liquid detergents.

Ufarol Am 30. CAS 2235-54-3; EINECS 218-793-9; Ammonium lauryl sulfate; biodeg. detergent, wetting and foaming agent for shampoos, bath products, general-purpose detergents, laundry cleaners, carpet shampoos, furniture cleaning, textiles, leather, paints; stable in hard water and alkali, moderately stable in acids. [Unger Fabrikker AS]

Ufarol Na-30. CAS 151-21-3; Sodium lauryl sulfate; biodeg. detergent, wetting, and foaming agent for shampoos, bath products, general-purpose detergents, laundry cleaners, carpet shampoos, furniture cleaning, textiles, leather, paints; stable in hard water and alkali, moderately stable in acids. [Unger Fabrikker AS]

Ufarol TA-40. CAS 139-96-8; EINECS 205-388-7; TEA lauryl sulfate; biodeg. surfactant for shampoos, bath products, carpet shampoos, furniture cleaning, laundry, textiles; mild to hair and scalp. [Unger Fabrikker AS]

Ufaryl. Sodium alkylbenzene sulfonate powder; drum dried, active ingredient in detergent powders; used as emulsifier in herbicide and pesticide systems; air entrainment agent in cement. [Unger Fabrikker AS]

Ufaryl DB80. Sodium dodecylbenzene sulfonate, branched; detergent, wetting agent, foaming agent, emulsifier for light and heavy-duty detergents, dairy, metal, floor, vehicle and bottle cleaners; wetting agent in insecticides, metal pickling, printing inks, paper processing, textiles; partially biodeg. [Unger Fabrikker AS]

Ufasan 35. Linear sodium dodecylbenzene sulfonate; biodeg. detergent, wetting agent, foaming agent for liq. detergents, dishwash, hair and car shampoos and in plastics, metal, agriculture, polish, textiles, mining, oil and cement industries; stable in hard water and acids. [Unger Fabrikker AS]

Ufasan IPA. Isopropylamine alkylbenzene sulfonate; biodeg. raw material for detergents, dishwash, cleaning liqs., car shampoos, industrial cleaners and in plastics, metal, agriculture, polish, textiles, mining, oil and cement industries; emulsifier in solv.-based waterless hand and industrial cleaners; stable in hard water, diluted acids and alkalis. [Unger Fabrikker AS]

Ufasan TEA. TEA dodecylbenzene sulfonate, linear; biodeg. surfactant for liq. detergents, dishwash, hair and car shampoos and in plastics, metal, agriculture, polish, textiles, mining, oil and cement industries; stable in hard water, diluted acids and alkalis. [Unger Fabrikker AS]

Ufasoft 75. Quat. ammonium imidazoline methosulfate; surfactant, softener, fabric conditioner, antistat, bactericide for laundry products. [Unger Fabrikker AS]

Uffelmann's reagent. A lactic acid reagent, prepared by adding a ferric chloride solution to a 2% phenol solution until it is of a violet color. The color of the reagent is changed to deep yellow by the addition of lactic acid.

Uformite®. Synthetic resins; used for bonding agents. [Reichhold]

Uformite® 21-806. Urea formaldehyde resin; similar to Uformite 21-805 but supplied in high-flash naphtha to give less odor and extremely high visc.; produces baking enamels of low solids content, in automotive enamels and primers, roller-coating and metal-decorating finishes; metal bond-coats and flexible tube enamels; and for any industrial baking enamel in which butanol solv. has proved troublesome; solv.-reducible type. [Reichhold]

Uformite® 27-802. CAS 9003-08-1; Melamine formaldehyde resin; high-quality, general-purpose thermosetting resin offering fast curing speed coupled with intercoat adhesion; used as a crosslinker with hydroxyl-functional thermosetting acrylics to achieve exterior durability; in enamels for automobiles, appliances, and metal furniture; solv.-reducible type. [Reichhold]

Ufoxane 2. CAS 8061-51-6; Purified, desulfonated and fractionated high molecular weight sodium lignosulfonate; dispersant for textile dyestuffs. [Borregaard LignoTech]

U-Gencin. CAS 1405-41-0; Gentamicin sulfate; a complex antibiotic substance, produced by *Micromonospora purpurea nsp*; three components: sulfates of gentamicin C_1, gentamicin C_2, and gentamicin C_{1A}; an antibacterial. [Upjohn] *

Ugikral RA, RB and SN. Proprietary trade names for ABS terpolymers. §

Ugurol®. An antifibrinolytic. [Bayer AG]

uhligite. A mineral that is a titanate of zirconium, calcium, and aluminum; found in East Africa.

Ulcatite. A proprietary trade name for a rubber vulcanization accelerator containing hexamethylenetetramine, benzthiazyl disulfide, and diphenylguanidine. §

Ulcedal. A proprietary preparation of deglycyrrhizinized liquorice extract used in the treatment of peptic ulcers. [Boehringer Ingelheim GmbH] *

Ulcerban. CAS 54182-58-0; Sucralfate; anti-ulcerative.

---

* Trade name not verified as available   † Trade name verified as obsolete

[Marion Merrell Dow Inc] *

Ulco. A metal used as a substitute for Babbitt metal. It contains 98- 99% lead, the rest being barium and calcium.

Ulcogant®. CAS 54182-58-0; Sucralfate. [E Merck]

Ulcony. An alloy of 65% copper and 35% lead.

ulexine. CAS 485-35-8; Cytisine, $C_{10}H_{14}N_2O$, an alkaloid found in laburnum and furze.

Ulmal. A proprietary trade name for an alloy containing aluminum with 10% magnesium, 1% silicon and 0.5% manganese. §

ulmite. A name for the dark-colored fiber covering the grains of sandstone found on the coast of New South Wales. In its properties it resembles those of humus, obtained from brown peat.

Ulon. A proprietary polyurethane elastomer. [Unitex Ltd]

Ultandren. CAS 76-43-7; A proprietary preparation of fluoxymesterone. [Ciba plc] *

Ultara®. Fibers. [DuPont UK]

Ultem®. Polyetherimide; used for plastic components for automotive, electrical, electronics, lighting, medical, packaging, audio, etc. [GE Plastics Ltd]

Ultem® 1000. CAS 61128-46-9; Amorphous polyetherimide resin; thermoplastic resin; superior properties with outstanding processability, machinability, and finishing characteristics; for automotive and aircraft applications exposed to gasoline and oils, for laboratory ware and automotive heat transfer systems, elec./electronic applications, computer circuitry, and microwave components, etc.; it is an unreinforced grade. [GE Plastics; Polymer Corp.]

Ultem® 2100. CAS 61128-46-9; Polyetherimide resin, 10% glass-reinforced; the glass reinforced resins provide greater rigidity and improved dimensional stability while retaining processability. [GE Plastics; Polymer Corp.]

Ultem® 2212. CAS 61128-46-9; Polyetherimide resin, 10% short glass-reinforced; low warpage grade for dimensional stability. [GE Plastics]

Ultem® 4000. CAS 61128-46-9; Amorphous thermoplastic polyetherimide resin, reinforced, internal lubricant; bearing-grade resin for applications requiring enhanced wear resistance. [GE Plastics]

Ultem® 6000. CAS 61128-46-9; Thermoplastic polyetherimide copolymer resin; high temp. grade; processable on std. injection molding equipment. [GE Plastics]

Ultem® CRS5001. CAS 61128-46-9; Polyetherimide, thermoplastic; highest chemical resist. in series, excellent performance in hot water; also for extrusion. [GE Plastics]

Ultem® CRS5011. CAS 61128-46-9; Polyetherimide, thermoplastic; improved chemical resist., processability and performance in hot water. [GE Plastics]

Ultem® FXU100. CAS 61128-46-9; Polyetherimide structural foam resin, unreinforced; high-performance engineering structural foam; used in the aircraft and automotive industries and applications requiring resist. to chemical attack. [GE Plastics]

Ultem® LTX100A. Polyetherimide/PC blend; offers heat/chemical resist. and impact strength [GE Plastics]

Ultimem. Plastic (or other polymeric material) membranes used for separation and related liquid treatments. [ICI Chem. & Polymers Ltd]

Ulto Accelerator. Zinc salt of a complex dithiocarbamate; a proprietary rubber vulcanization accelerator. §

Ultra®. Bismuth-based; highly lustrous pearl pigments avail. in nitrocellulose paste dispersions for use in frosted nail enamels. [Mearl]

Ultra 206. Organic brightener system; used for bright nickel electroplating (high leveling, fast brightening). [Harshaw Chemicals Ltd]

Ultra-DMC. A proprietary trade name for a vulcanization accelerator. It is dimethylamine dimethyldithiocarbamate. §

Ultra Anhydrous Lanolin HP-2060. CAS 8006-54-0; Lanolin; cosmetic grade for lip and other products where odor, taste and color are critical; moisturizer; self-emulsifying. [Henkel/Cospha; Henkel Canada] *

Ultrabase. White soft paraffin (10%), liquid paraffin (10%), and stearyl alcohol (8%). [Schering Health Care Ltd]

Ultrablend®. PBT/PC blends; used for tech. parts in the automotive and elec. industries. [BASF AG]

Ultrablend® S. PBT/ASA blends; used for injection molding for automotive and elec./electronics industries; high resist. to heat distort., high gloss, good outdoor performance, no tendency to environmental stress cracking. [BASF AG]

Ultracast PE-35, PE-60. CAS 89339-41-3; PDI-PTMEG; polyurethane prepolymer producing PU elastomer with high hydrolytic stability and excellent dynamic performance, e.g., for oil field service, rollers, gaskets, pumps, wheels, rollers, bearings, shock absorbers. [Air Prods./Polyurethanes] *

Ultracef. CAS 66592-87-8; Cefadroxil; an antibacterial. [Bristol Laboratories] *

Ultracene. A proprietary rubber vulcanization accelerator; it is a guanidine derivative. §

Ultrachem Assembly Fluid 1. A tacky polymer; used for assembly of o-rings used in helicopter transmissions, jet turbines, pumps, etc. [Ultrachem Inc] *

Ultracortenol. A proprietary preparation of prednisolone pivalate. [Ciba plc] *

Ultracut. Water-reducible lubricants containing mineral oil, but when diluted in water forms a micro-emulsion which is translucent to transparent; additives include emulsifiers, corrosion inhibitors, surfactants, biocides, and lubricating additives. [Franklin Oil Corporation (Ohio)]

Ultradil. A proprietary preparation of fluocortolone pivalate BP and fluocortolone hexanoate; used in the treatment of eczema. [Schering Health Care Ltd]

Ultradur®. Polybutylene terephthalate granules; used for injection molding (electrical engineering components, key buttons). [BASF plc]

Ultradur® B 2550. CAS 26062-94-2; Thermoplastic polyester resin (PBT); low-visc. resin used for coating paper and board for pkg. oven-ready deep-freeze and convenience foods; primarily for extrusion. [BASF]

Ultradur® B 4300 G10. CAS 26062-94-2; Thermoplastic polyester resin (PBT), 50% glass fiber-reinforced; used for engineering parts with particularly great rigidity; primarily for injection molding. [BASF] *

Ultradur® B 4500. CAS 26062-94-2; Thermoplastic polyester resin (PBT); med.-visc. resin offering high precision for engineering parts, e.g., cams, number barrel mechanisms and poppet valves; primarily for injection molding. [BASF] *

Ultradur® KR 4001. CAS 26062-94-2; Thermoplastic polyester resin (PBT), 25% mineral-reinforced; used for housings and front panels for household appliances;

‡ Trade name and manufacturer not verified   § Trade name without identified manufacturer

primarily for injection molding. [BASF] *

Ultrafast® 830 Liq. Uv absorber for improvement of lightfast-ness and photo degradation properties of polyester fibers; especially suitable in combination with cationic dyes. [BASF]

ultraferran. A colloidal iron.

UltraFine® II. CAS 1309-64-4; Antimony trioxide; flame retardant for liq. systems, e.g., unsaturated polyesters, epoxies, PU, phenolics, textile treatments; produces minimum loss of physical properties in ABS, etc. [Laurel Industries]

Ultra-Flat. Aluminum alloy sheet. [Reynolds Metal Co] †

Ultraflex®. CAS 63231-60-7; Microcryst. wax; plastic wax offering high ductility, flexibility at very low temps., pro-vides protective barrier properties against moisture va-por and gases; uses incl. hot-melt laminating adhesives for papers, films, and foils; hot-melt coatings; in antisun-checking agents in rubber goods, elec. insulating agents, leather treating agents, water repellents for textiles, rustproof coatings, cosmetic ingredients, and as plasti-cizer for waxes; used in crayons, dental compounds, chewing gum base, and candles. [Petrolite]

Ultraflo. A heat-stable multi-active betaglucanase prepara-tion prodiuced by a selected strain of *Humicola insolens*; used in the mashing process of beer to degrade β-glucan for better filtration. [Novo Nordisk]

Ultrafloc. CAS 9000-07-1; Tabletted carrageenan. [Rhone-Poulenc UK]

Ultraform®. Polyoxymethylene/polyacetal granules; used for injection molding, mechanical engineering, cassette hubs. [BASF plc]

Ultraform. Polyacetal. [Degussa Ltd]

Ultraform® H 2320. Acetal copolymer resin; high molecular weight resin for extrusion of thin tubes and panels and thick stock for machining; also suitable for thick injection moldings. [BASF]

Ultraform® 2200 C4X. Acetal copolymer (POM), carbon fiber-reinforced; higher strength and stiffness than glass-reinforced grades, good elec. conductivity, excellent resist. to fuels and solvs., good frictional properties; used for gas pumps, sliding bearings. [BASF] *

Ultraform® N 2200 G5. Acetal copolymer resin, glass-reinforced; high strength and rigidity injection molding compound. [BASF] *

Ultraform® N 2211 PVX. Acetal copolymer resin, lubricated; injection molding compound with lubrication; permits low coeff. of friction and reduced wear with no adverse effect on mech. properties. [BASF] *

Ultraform® N 2320. Acetal copolymer resin; rapidly solidi-fying, general purpose injection molding production. [BASF] *

Ultraform® N 2320 BK 11005 MO. Acetal copolymer resin, MoS₂ lubricated; injection molding compound for the production of friction bearings. [BASF] *

Ultraform® W 2320. Acetal copolymer resin; very easy flow, rapidly solidifying product for injection molding to meet extreme demands on processing. [BASF] *

Ultraglaze® 4400. Two-part silicone sealant; high-strength, high-modulus structural glazing sealant. [GE Silicones]

Ultrahold. Resins for hydrocarbon propellant hairsprays. [BASF plc]

Ultrahold 8. Acrylates/acrylamide copolymer. hair spray resin with outstanding hold and humidity resistance. [BASF; BASF AG]

Ultralanum. Fluocortolone pivalate and fluocortolone hexanoate [Schering Health Care Ltd]

Ultralen® SP 3700 S. CAS 25038-59-9; PET, light-stabi-lized; spinning polymer for production of bright textile yarns [BASF AG]

Ultralen® SP 3705. CAS 25038-59-9; PET, light-stabilized; spinning polymer for production of semi-dull textile yarns [BASF AG]

Ultralente Iletin. Insulin zinc, extended; antidiabetic. [Eli Lilly & Co]

Ultralente Insulin. Insulin zinc, extended; antidiabetic. [Bristol-Myers Squibb Co Inc] *

ultra-light alloys. Alloys having a specific gravity below 2 are known by this name. Magnesium-aluminum-zinc and magnesium-copper are alloys of this type.

Ultralin. A rosin-fatty acid reaction product.

Ultralog. Grade of chemical reagents with 99.8% to 100% purity. [Schweizerhall]

Ultralumin. A jeweler's alloy containing more than 90% aluminum, with nickel, copper, and some rare earth metals of the thorium group. It is specially resistant to sea water.

ultramarine. (lapis-lazuli blue, oriental blue, brilliant ultra-marine, French blue, new blue, permanent blue, French ultramarine, soda ultramarine). A blue coloring matter formerly prepared from the rare mineral lapis-lazuli, by powdering and washing. It is prepared artificially by fusing together kaolin and sulfur with soda or with a mixture of sodium sulfate and charcoal.

ultramarine ash. In obtaining ultramarine from lapis-lazuli, a blue product is first yielded, then a pale blue, and finally a pale bluish-grey material, which is called ultramarine ash.

ultramarine, soda. *See* ultramarine.

ultramarine yellow. *See* barium chromate.

Ultramid®. Polyamides. [BASF; BASF plc]

Ultramid® A3. CAS 32131-17-2; Nylon 6/6 resin; low-visc. resin for extrusion of monofilaments and bristles. [BASF]

Ultramid® A3G5. CAS 32131-17-2; Nylon 6/6 resin, 25% glass fiber-reinforced; for machine parts and housings of great rigidity and dimensional stability. [BASF] *

Ultramid® A3K. CAS 32131-17-2; Nylon 6/6 resin, stabi-lized; for injection molding of all kinds of engineering parts subjected to high loads, e.g., bearings and gears, and dielectric parts, e.g., terminal strips and cable bind-ers. [BASF] *

Ultramid® A3X1G10. CAS 32131-17-2; Nylon 6/6 resin, 50% glass fiber-reinforced, flame-retarded; for parts with utmost rigidity and enhanced fire resistance. [BASF] *

Ultramid® A4. CAS 32131-17-2; Nylon 6/6 resin; med.-visc. resin for extrusion; stock for machining, monofilaments for spiral zip fasteners, bristles, and injection-molded engineering parts. [BASF] *

Ultramid® B3. CAS 25038-54-4; Nylon 6 resin; low-visc. resin for extrusion of monofilaments, bristles, fishing lines, nets, film, stretched tapes and coatings. [BASF] *

Ultramid® B3EG10, B3WG10. CAS 25038-54-4; Nylon 6 resin, 50% glass fiber-reinforced, stabilized; for machine parts, housings, and sheathed magnets of very great rigidity, dimensional stability, and resistance to high-temp aging; B3EG10 preferred for dielectrics. [BASF] *

Ultramid® B3G10. CAS 25038-54-4; Nylon 6 resin, 50% glass fiber-reinforced; for machine parts, housings of very great rigidity, dimensional stability. [BASF] *

---

* Trade name not verified as available　　　　　† Trade name verified as obsolete

Ultramid® B3K. CAS 25038-54-4; Nylon 6 resin, stabilized; for injection molding of machine parts, mainly with wall thicknesses > 2 mm. [BASF] *

Ultramid® B5. CAS 25038-54-4; Nylon 6 resin; high-visc. resin for extrusion of tubes, sections, extruded stock of very high impact strength for machining, tubular film, panels, tapes, thick monofilaments, and blow moldings, e.g., tanks for hydraulic fluids. [BASF] *

Ultramid® B35G3. CAS 25038-54-4; Nylon 6 resin, 15% glass fiber-reinforced; for housings of enhanced impact strength, e.g., for car outer rear-view mirrors and wheels for cross-country bicycles. [BASF] *

Ultramid® BS 400 S. CAS 25038-54-4; Polyamide 6 (polycaprolactam); fiber grade polymer for production of textile yarn. [BASF AG]

Ultramid® BS 416. CAS 25038-54-4; Polyamide 6 (polycaprolactam), light-stabilized; fiber grade polymer for production of dull textile yarn. [BASF AG]

Ultramid® BS 3300. CAS 25038-54-4; Polyamide 6 (polycaprolactam); fiber grade polymer for production of industrial yarns. [BASF AG]

Ultramid® C35. Copolyamide 6/66 coextruded with polyethylene; composite film for packing foodstuffs, multi-ply blow moldings, extremely tough and flexible fishing lines. [BASF] *

Ultramid® KR 4205. Nylon resin, stabilized, flame-retarded; for injection molding of dielectric parts with improved fire resistance, e.g., contactor bases and plug boards. [BASF] *

Ultramid® S3. CAS 9008-66-6; Nylon 6/10 resin; med. visc. resin for extruded stock for machining, tubular, and flat film, monofilaments and bristles. [BASF] *

Ultramite. CAS 471-34-1; Calcium carbonate; ultrafine wet-ground product avail. in slurry form for water-based systems, primarily coatings. [ECC]

Ultramoll® I, II, III. A range of polyadipates; polymeric plasticizers. [Bayer AG; Bayer plc]

Ultramoll® M. Polyadipate; a polymeric plasticizer for use in the formulation of flame retardant pastes for polyester based Moltopren. [Bayer AG]

Ultramoll® PP. Polyphthalate; polymeric plasticizer for PVC; resistant to oil and bitumen, largely resistant to migration; also for PVC plastisols. [BASF AG]

Ultramoll® TGN. Polyphthalate; polymeric plasticizer for PVC; relatively low viscosity; particularly suitable for use in pigment mill bases and as a desensitizer for peroxides. [BASF AG]

Ultranox® 236. CAS 96-69-5; 4,4′-Thio-bis (2-t-butyl-5-methylphenol); antioxidant for use in adhesives, rubber articles for repeated use, polymers incl. polyolefins, PVC, acrylic ethyl cellulose; antioxidant for lubricants, cutting oils, water-sol. oils, hydraulic oils. [GE Specialty]

Ultranox® 246. CAS 119-47-1; 2,2′-Methylene-bis-(4-methyl-6-t-butylphenol); stabilizer, antioxidant for PP, polyethylene, polyoxymethylene copolymer, polyoxymethylene homopolymer, styrenics, and rubber. [GE Specialty]

Ultranox® 257. CAS 68610-51-5; Polymeric sterically hindered phenol; butylated reaction product of p-cresol and dicyclopentadiene; antioxidant for wh., transparent, and colored natural and synthetic rubber goods; especially for latex applications such as carpet backing, dipping articles; also for EVA-based hot melts, thermoplastic rubber, and styrenics. [GE Specialty]

Ultranox® 276. CAS 2082-79-3; Octadecyl 3,5-di-t-butyl-4-hydroxyhydrocinnamate; high molecular weight antioxidant/stabilizer for styrenics, polyolefins, PVC, urethane and acrylic coatings, adhesives, and elastomers; effective replacement for BHT in polyolefins. [GE Specialty]

Ultranox® 626. CAS 26741-53-7; Bis (2,4-di-t-butylphenyl) pentaerythritol diphosphite/0.5-1.2% triisopropanolamine; antioxidant for polyolefin, PVC, PET, styrenics, ABS, and PC polymers; stabilizer. [GE Specialty]

Ultranox® 627A. CAS 26741-53-7; Bis (2,4-di-t-butylphenyl) pentaerythritol diphosphite; antioxidant providing color stability, hydrolytic stability, reduction in polymer degradation during processing of ABS, PVC, PC. [GE Specialty]

Ultranyl®. Polyamide and polyphenylene ether alloy; used in injection molding for tech. parts, e.g., in automotive engineering and for office machines. [BASF AG; BASF plc]

Ultrapas. Melamine resin molding compounds; for manufacture of tableware, bathroom and publicity items, screw tops. [Dynamit Nobel Wien GmbH] *

Ultrapek® KR 4176. Polyaryletherketone; low visc., easy flow for injection molding, blow molding, film extrusion, and spinning fibers. [BASF AG]

Ultrapek® KR 4177 G4. Polyaryletherketone, 20% glass fiber-reinforced; general-purpose injection molding grade; also suitable for extrusion. [BASF AG] *

Ultrapen. CAS 1245-44-9; A proprietary preparation containing propicillin (as potassium salt); an antibiotic. [Pfizer International] †

Ultra-Pflex®. CAS 471-34-1; Ultrafine surface-treated precipitated calcium carbonate; impact modifier in rigid PVC applications. [Pfizer]

Ultraphan. Acetate film. [Lonza AG]

Ultraphor® CW. Optical brighteners for cotton. [BASF]

Ultraphor® SFG Liq., SFR Liq. Optical brighteners for polyester fibers. [BASF]

Ultraprene®. CAS 9002-86-2; PVC elastomer; features conformability, resist. to compression set, abrasion, high tensile strength, outstanding weatherability; for architectural glazing seals, weatherstripping, wire/cable applications, tool handle grips, automotive interior and exterior parts, hoses, toys. [Teknor Apex]

Ultraproct Suppositories. Fluocortolone pivalate, fluocortolone hexanoate, dibucaine hydrochloride; suppository used in the treatment of hemorrhoids. [Schering Health Care Ltd]

Ultra Pure. CAS 302-01-2; Hydrazine propellant; propellant for mono and bi-propellant systems. [Olin]

Ultrasil® VN3SP. High purity silica; catalyst carrier; reinforcement for rubber compounding. [Degussa]

Ultrasol. Solvents doubly distilled in glass. [British Drug Houses] *

Ultrason® E 1010. Polyethersulfone; very easy flow grade for moldings with adverse ratios of runner length to wall thickness. [BASF AG]

Ultrason® E 1010 G2. Polyethersulfone, 10% glass fiber-reinforced; very easy flow for moldings with adverse ratios of runner length to wall thickness. [BASF AG] *

Ultrason® E 2010. Polyethersulfone; easy flow, general-purpose grade for injection molding; also suitable for extrusion, film extrusion, blow molding. [BASF AG] *

Ultrason® E 3010. Polyethersulfone; high visc. grade for injection molding, extrusion, film extrusion, blow mold-

---

‡ Trade name and manufacturer not verified   § Trade name without identified manufacturer

ing. [BASF AG] *

Ultrason® E 6010. Polyethersulfone; very high visc. grade for extrusion, film extrusion, blow molding. [BASF AG] *

Ultrason® KR 4101. Polyethersulfone, 30% mineral-reinforced; easy-flow, general-purpose injection molding grade; also suitable for extrusion. [BASF AG] *

Ultrason® S 1010. CAS 25135-51-3; Polysulfone; very easy flow grade for moldings with adverse ratios of runner length to wall thickness. [BASF AG]

Ultrastyr. A proprietary range of high impact polystyrene resins. [Montedison UK Ltd] *

Ultrasul. CAS 144-82-1; Sulfamethizole; antibacterial. [Alcon Laboratories Inc] *

Ultratard. Insulin zinc, extended; antidiabetic. [Bristol-Myers Squibb Co Inc] *

Ultratex. Finishing agent. [Ciba plc] *

Ultrathane. Hot, cast, thermosetting polyurethane from various isocyanates combined with polyesters or polyethers and crosslinked with polyols or diamines to form elastomeric products; used for solid tires, roller coverings, squeegee blades, abrasion resistant linings, noise reduction parts, consumables for printing, box making, engineering, mining, quarrying, leisure and materials handling industries etc. [Watts Urethane Products Ltd]

Ultrathene®. Polyolefin copolymers. [Quantum Chemical Corp]

Ultrathene® UE 630-000. EVA copolymer; resins with good flow and flexibility, toughness, processability, and low temp. properties; used in flexible and semiflexible parts. [Quantum/USI]

Ultrathene® UE 632-000. EVA copolymer; extrusion coating resin; film extrusion [Quantum/USI] *

Ultrathene® UE 659-04. EVA copolymer; for adhesives and coatings; FDA compliance. [Quantum/USI] *

Ultra Touch. Water, glycerin, stearic-TEA, butyl stearate, parasepts, fragrance; moisturizing hand and body lotion. [Virkler Chemical Co] *

Ultravist. CAS 73334-07-3; Aqueous solutions of iopromide; x-ray contrast media. [Schering Health Care Ltd]

Ultravon. Pretreatment agent. [Ciba plc] *

Ultravon AN. A proprietary trade name for a fatty acid amide derivative; it emulsifies oils and fats more effectively and maintains the handle and color of wool better than conventional detergents; a special detergent primarily for wool scouring. §

Ultrawet. High foaming versatile nonionic surfactant; for liquid detergent formulations. [Cornelius Chemical Co Ltd] *

Ultrawet DS. Anionic surfactant supplied as cream colored flakes; detergency, wetting, sudsing, dispersing and emulsifying agent for speciality cleaning and industrial processing. [Cornelius Chemical Co Ltd] *

Ultrawet K and AOK. Anionic surfactant supplied as cream colored flakes; various industrial and heavy-duty household detergents. [Cornelius Chemical Co Ltd] *

Ultrax®. Self-reinforcing liq. cryst. polymers; for extrusion and injection molding; very high dimensional stability of moldings, low water absorp., low thermal expansion coefficient. [BASF AG]

Ultra Zinc DMC. CAS 137-30-4; Zinc dimethyldithiocarbamate; a proprietary trade name for a rubber vulcanization accelerator. §

Ultrazine CA. CAS 8061-52-7; Calcium lignosulfonate; used in gypsum board manufacturing [Borregaard LignoTech]

Ultrazine NA. CAS 8061-51-6; Purified high molecular weight sodium lignosulfonate; high dispersing efficiency for textile dyestuffs, pesticides, gypsum board, and concrete additives. [Borregaard LignoTech]

Ultrazym. A purified pectolytic enzyme preparation produced from a selected strain of *Aspergillus niger*, can be used in any case where the aim is breaking down of soluble and insoluble pectins with varying degrees of esterification for reduction of viscosity, clarification, maceration of plant tissue and depectinization. [Novo Nordisk]

Ultrmoll NR. Polybutadiene-acrylonitrile; improves abrasion resistance and dimensional stability of PVC profiles used for shoe soles, automotive profiles, cables, window frames and jointing profiles. [Bayer AG]

Ultrmoll PP. Polyphthalate; low viscosity polymeric plasticizer for PVC with high resistance to oil and bitumen. [Bayer AG]

Ultrmoll PU. Linear polyurethane; compatible with PVC giving high abrasion resistance and used for shoe soles, cables, film, injection moldings and profiles. [Bayer AG]

Ultrmoll TGN. Polyphthalate; low viscosity polymeric plasticizer for PVC for use in pigment mill bases and as a desensitizer for peroxides. [Bayer AG]

Ultroil. A sulfonated vegetable oil; a proprietary trade name for a wetting agent for textiles. §

Ultrol®. [Calbiochem Corp]

Ultryl 6010. A proprietary nonstabilized PVC resin of the suspension type. [Phillips Petroleum Int'l.] *

Ultryl 6500. A proprietary PVC polymer of the suspension type; used in the production of glass-clear film, tube, etc. [Phillips Petroleum Int'l.] *

Ultryl 6800. A proprietary plasticizer-free PVC resin used in the manufacture of pipe and profiles. [Phillips Petroleum Int'l.] *

Ultryl 7100. A proprietary PVC resin of the suspension type with easy processing properties. [Phillips Petroleum Int'l.] *

Ultryl 7150. A proprietary PVC resin of the suspension type containing additives to give high clarity. [Phillips Petroleum Int'l.] *

Ulvio Cocoa. A German proprietary food material prepared by exposing cocoa to uv radiation. §

umber. (umber brown, mineral brown, velvet brown, chestnut brown, manganese velvet brown, umber umber). Mineral varieties of umber. They are ochers colored brown by oxides of manganese, and containing varying amounts of clay. Burnt umber is umber calcined at low heat; used in paint pigments, wallpaper, wall coverings.

umbrathor. A solution of thorium dioxide.

umbrenal. A 25% solution of lithium iodide in ampoules.

Umbrite A. An explosive containing 49% nitroguanidine, 38% ammonium nitrate, and 13% silicon.

Umbrite B. An explosive containing 37.5% nitroguanidine, 49.5% ammonium nitrate, and 13% silicon.

umburana seed. The product of *Amburana Claudii*, used in Brazil for perfuming tobacco.

umea tar. A pale Swedish pine-wood tar. It is a good variety of Stockholm tar produced in the Umea district.

UN-28 and UN-32. Fertilizer solutions containing 28% and 32% nitrogen respectively; designed for direct agricultural use as a three-way source of nitrogen. [Hercules]*

---

* Trade name not verified as available   † Trade name verified as obsolete

Unacid. See Unasyn. [Pfizer International]

Unads® . CAS 97-74-5; Tetramethylthiuram monosulfide. [R.T. Vanderbilt]

Unakalm. CAS 27223-35-4; Ketazolam; tranquilizer. [Upjohn] *

Unal. CAS 123-30-8; A photographic developer. It is Rodinal in a solid form, containing, besides p-aminophenol, the ingredients necessary for solidification.

Unamide® C-5. CAS 61791-08-0; PEG-6 cocamide; foam stabilizer, visc. builder, emulsifier for shampoos, light and heavy-duty detergents; stable over broad pH range. [Lonza Inc]

Unamide® C-72-3. Cocamide DEA (2:1); visc. builder, foam stabilizer; used for light duty liqs., industrial, household hard surface cleaners, surfactant, emulsifier, corrosion inhibitor, lubricant and personal care products. [Lonza Inc]

Unamine® C. CAS 61791-38-6; EINECS 263-170-7; Coco hydroxyethyl imidazoline; emulsifier, surfactant, fungicide; textile antistat; leather treating; base for quats.; acid detergent and wetting agent; corrosion inhibitor; pigment flushing agent. [Lonza Inc]

Unamine® O. Oleyl hydroxyethyl imidazoline; emulsifier, demulsifier; antistat for drycleaning fluids; water displacing agent; corrosion inhibitor; base for quats.; pigment flushing agent. [Lonza Inc]

Unasyn® IM/IV. CAS 69-53-4; Ampicillin; antibiotic. Also known as Bactesyn (Indonesia), Becalin (Greece), Duocid (Turkey), Unacid (Germany, Switzerland), Unasyna (L. America), Unasyne (France). [Pfizer International]

Unasyn® Oral. CAS 68373-14-8; Sulbactam; antibiotic. [Pfizer International]

Unburn. A proprietary antiburn cream containing benzocaine, hexachlorophane, orthophenyl phenol, menthol and lanolin. [Leeming] *

10-undecenoic acid. See undecylenic acid.

11-undecenoic acid. See undecylenic acid.

undecenoic acid. See undecylenic acid.

10-undecenoic acid, sodium salt. See sodium undecylenate.

undecylenic acid. (undecenoic acid; 10-undecenoic acid; 11-undecenoic acid) Aliphatic acid; $CH_2=CH-(CH_2)_8COOH$; CAS 112-38-9; EINECS 203-965-8; perfumery, flavoring, plastics, modifying agent, medicine (antifungal agent); intermediate in chemical synthesis, polyamide plastics and fibers, synthetic floors. [Atochem; CasChem; Schweizerhall]

Undeen. Insecticide; used for treatment of sucking and biting insects. [Bayer AG]

Unden® . CAS 114-26-1; Propoxur; insecticide used for treatment of sucking and biting insects, e.g., aphids, mealybugs, scales, leafhoppers, caterpillars on vegetables, pome and stone fruit, cocoa, rice, oil palms and other crops. [Bayer AG]

Undene® . Insecticide with fast killing activity and good penetrating action particularly for the control of sucking insects and bugs, cocoa flat bugs, leafhoppers and other pests; for use on cereals, pome and stone fruit, citrus fruit, rice, and sugar cane. [Bayer AG]

Unger A-50. softener, antistat for rayon and cotton fibers and yarns; stable in hard water; med. stability in acids and alkalis. [Unger Fabrikker AS]

Ungerol. Sodium fatty alcohol ether sulfate; active ingredient in shampoos, bath products and liquid detergents, emulsifying agent in polymerization. [Unger Fabrikker AS]

Ungerol AM3-75. CAS 67762-19-0; Ammonium lauryl ether (3 EO) sulfate; biodeg. detergent, wetting and foaming agent, emulsifier used in liq. detergents, car shampoos, bath foams; stable in hard water, moderately stable in acids. [Unger Fabrikker AS]

Ungerol N2-28. CAS 9004-82-4; Sodium laureth sulfate (2 EO); biodeg. detergent, wetting and foaming agent, emulsifier for liq. detergents, shampoos, bath products, wallboard manufacturing, textiles, drilling auxiliary; excellent stability in hard water, alkalis; moderately stable in acids. [Unger Fabrikker AS]

Unguentum. A proprietary preparation of silicilic acid, liquid paraffin, soft paraffin, cetostearyl alcohol, polysorbate-40, glycerol, oil, sorbic acid and propylene glycol; a protective skin cream. [E Merck] †

Uni G. Motor oil. [Chevron] *

Unibind Series. Polymer; binding agent for coal and mineral transportation or storage to prevent wind blown losses. [Hart Chem. Ltd.]

Uni-Cal 66. Colorant dispersions; for coloring of nonaqueous industrial and maintenance coating compositions. [Hüls Am.]

Unicast. One-component epoxy casting resins incl. general purpose, thermally conductive and fire retardant. [Emerson & Cuming]

Unicell. Lightweight waterproof mortars for high build and overhead applications; used for repairs to concrete and stonework. [Ronacrete Ltd]

Unicoat. One-component conformal coatings for electronics, dip coats for resistors and capacitors, and spray or brush coatings for general purpose applications. [Emerson & Cuming]

Unicor. Corrosion inhibitors. [Universal-Matthey Products Ltd]

Unicrop 6% Mini Slug Pellets. CAS 108-62-3; Pellets containing 6% w/w metaldehyde; snail and slug bait. [Universal Crop Protection Ltd]

Unicrop Leatherjacket Pellets. γ-HCH; an organochlorine insecticide. [Universal Crop Protection Ltd]

Unicrop Thianosan. CAS 137-26-8; Thiram; fungicide with animal repellent properties. [Universal Crop Protection Ltd]

Unicrop Zineb. CAS 12122-67-7; Zineb; a dithiocarbamate fungicide. [Universal Crop Protection Ltd]

Unicrylic. Solvent and acrylic sealant; used for glazing of windows, panels and general caulking. [Pecora Corporation]

Unidem. Demulsifiers. [Universal-Matthey Products Ltd] *

Unidene. Solution butadiene styrene rubbers. [EniChem Elastomers Ltd]

Unidiarea. A proprietary preparation of neomycin sulfate, clioquinol, and attapulgite; an antidiarrheal. [Unigreg Ltd] †

Unidri A-74. Dewatering agent for vacuum filtration of minerals. [Hart Chem. Ltd.]

Unidri M-75. Low foaming surfactant for dewatering metallic oxides, coal and other minerals. [Hart Chem. Ltd.]

Uniflex® 192. Octyl tallate; plasticizer for synthetic rubbers; metalworking. [Union Camp] *

Uniflex® 300. Polymeric plasticizer; plasticizer for PVC; high temp. elec. applications; high grade vinyl upholstery;

---

‡ Trade name and manufacturer not verified   § Trade name without identified manufacturer

adhesives for elec. tapes or wall coverings; gasketing and tubing; low migration. [Union Camp] *

Uniflex® BYO. CAS 142-77-8; EINECS 205-559-6; Butyl oleate; monomeric plasticizer and processing aid for plastics, rubber; textile lubricant, mold lubricant, in waterproofing agents, polishes, and metalworking lubricants. [Union Camp] *

Uniflex® BYS-Tech. CAS 123-95-5; EINECS 204-666-5; n-Butyl stearate; ester with low visc., good color, low odor for plasticizers, textile fiber lubricants, metalworking oils. [Union Camp] *

Uniflex® DBS. CAS 109-43-3; Dibutyl sebacate; monomeric plasticizer for plastics and rubber (cellulosics, PVC food wraps, nitrile and neoprene rubbers). [Union Camp] *

Uniflex® DCA. CAS 105-97-5; Dicapryl adipate; monomeric plasticizer for plastics and rubber. [Union Camp] *

Uniflex® DCP. Dicapryl phthalate; monomeric plasticizer for plastics and rubber. [Union Camp] *

Uniflex® DOA. CAS 103-23-1; Dioctyl adipate; monomeric plasticizer for plastics and rubber. [Union Camp] *

Uniflex® DOS. Dioctyl sebacate; monomeric plasticizer for plastics, rubber, and lubricants. [Union Camp] *

Uniflex® EHT. Octyl tallate; plasticizer for synthetic rubbers; in metalworking. [Union Camp] *

Uniflex® IBYS. CAS 646-13-9; Isobutyl stearate; for textile fiber lubricants, metalworking oils. [Union Camp] *

Uniflex® TEG-810. Triethylene glycol capric/caprylate; plasticizer for nitrile rubber. [Union Camp] *

Uniflood. Refinery and oilfield chemicals. [Universal-Matthey Products Ltd] *

Uniflot SP 100 Series. Sulfhydril; copper sulfide collectors for copper/zinc flotation operations. [Hart Chem. Ltd.]

Uniflu. A proprietary preparation of paracetamol, codeine phosphate, diphenhydramine hydrochloride, phenylephrine hydrochloride, and caffeine; a remedy for colds and flus. [Unigreg Ltd]

Unifog. Insecticidal formulation. [Mitchell Cotts Chemicals Ltd]

Unifroth G. Glycol blend; frother for flotation operations where selectivity is required. [Hart Chem. Ltd.]

Unifume. CAS 137-42-8; Metam-sodium 40% w/v aqueous solution; for use as a soil sterilant in agriculture and horticulture. [Universal Crop Protection Ltd]

Unigel. CAS 55-63-0; A nitroglycerin dynamite; for construction and building industry, explosives, mining, petroleum, and related industries. [Hercules] *

Unigest. A proprietary preparation of aluminum hydroxide and dimethicone; an antacid/antiflatulent. [Unigreg Ltd]

Unihepa. A proprietary preparation of di-methionine, choline, inositol, thiamine, pyridoxine, biotin, vitamin E, cyanocobalamin, panthenol and folic acid; used to counteract senile muscular degeneration. [Unigreg Ltd] †

Unihib®. Organic phosphates; for use as corrosion inhibitors in cooling towers and boilers. [Lonza AG]

Unihib® 106. CAS 2809-21-4; 1-Hydroxyethylidene-1,1-diphsophonic acid; dispersant, scale inhibitor for cooling tower, boiler treatment, oilfield applications. [Lonza Inc]

Unihib® 305-LC. CAS 6419-19-8; EINECS 229-146-5; Aminotrimethylene phosphonic acid; dispersant, corrosion and scale inhibitor for cooling tower, boiler treatment, oilfield applications. [Lonza Inc]

Unihib® 314. CAS 2235-43-0; EINECS 218-791-8; Pentasodium amino tri(methylene phosphonate); dispersant, corrosion and scale inhibitor for cooling tower,

boiler treatment, oilfield applications. [Lonza Inc]

Unihib® 905. Diethylene triamine pentamethylene phosphonic acid; dispersant, corrosion and scale inhibitor for cooling tower, boiler treatment, oilfield applications. [Lonza Inc]

Unihib® 1324. Diammonium salt of cocamine-di-(methylene phosphonic acid); dispersant, corrosion and scale inhibitor for cooling tower, boiler treatment, oilfield applications. [Lonza Inc]

Unihib® 1704. Bis-hexamethylene triamine phosphonic acid; dispersant, corrosion and scale inhibitor for cooling tower, boiler treatment, oilfield applications. [Lonza Inc]

Unilab Surgibone. Surgibone; prosthetic aid. [Unilab Inc] *

Unilin® 425 Alcohol. C20-40 alcohols; functional polymer for modification of PP, PVC, polyethylene, PS, and high-performance engineering resins; acts as antioxidant, heat stabilizer, uv stabilizer, or visc. depressant; promotes emollient protective films onto the skin; in cosmetic creams and lotions; used in hot-melt and solv.-based coatings; textile/leather lubricants and finishes; chemical intermediate; defoamer for pulp/paper processing. [Petrolite]

Unilink® 450. 4,4´-Bis (sec-butylamino) diphenyl methane; used as a chain extender in the production of specialty polyurethane plastics. [UOP Inc]

Unilink® 4100, 4200. Aromatic diamine; chain extender for PU elastomers. [UOP]

Uniloy Chrome Steels. A proprietary trade name for alloys containing 4-6% chromium, 0.1-0.25% carbon, up to 0.6% manganese and 0.4-0.6% molybdenum and 1.0-1.25% tungsten. §

Unimate® BYS. CAS 123-95-5; EINECS 204-666-5; n-Butyl stearate; skin emollient and solv. used in lipsticks, creams, lotions. [Union Camp] *

Unimate® DCA. CAS 105-97-5; Dicapryl adipate; cosmetic ester. [Union Camp] *

Unimate® DIPA. CAS 6938-94-9; EINECS 248-299-9; Diisopropyl adipate; cosmetic ester. [Union Camp] *

Unimate® DIPS. CAS 7491-02-3; EINECS 231-306-4; Diisopropyl sebacate; skin emollient, solubilizer, coupler, spreading agent for creams, lotions, bath oils, aerosol toilet preparations. [Union Camp] *

Unimate® EHP. CAS 29806-73-3; EINECS 249-862-1; 2-Ethylhexyl palmitate; skin emollient for lotions and creams. [Union Camp] *

Unimate® IPM. CAS 110-27-0; EINECS 203-751-4; IPM; skin emollient, lubricant, carrier; used in lotions, creams, antiperspirants, bath oils. [Union Camp] *

Unimate® IPP. CAS 142-91-6; EINECS 205-571-1; IPP; skin emollient, lubricant, carrier; used in lotions, creams, antiperspirants, bath oils. [Union Camp] *

Unimin Series. Blend; selective iron sulfide depressant for use in base metal sulfide separation. [Hart Chem. Ltd.]

Unimite. Ammonia dynamite; used for wet hole blasting in hard rock for operations such as quarrying and coal stripping. [Hercules] *

Unimoll®. Phthalates plasticizers. [Bayer plc]

Unimoll® 66 M. CAS 84-61-7; Dicyclohexyl phthalate; plasticizer for use in formulation of delayed tack heat sealable coatings. [Bayer; Miles; Polysar]

Unimoll® BB. CAS 85-68-7; Benzyl butyl phthalate; monomeric plasticizer. [Bayer AG]

Unimoll® DB. CAS 84-74-2; Dibutyl phthalate; monomeric plasticizer. [Bayer AG] †

---

* Trade name not verified as available   † Trade name verified as obsolete

Unimoll® DM. CAS 131-11-3; Dimethyl phthalate; fixative for perfumes, plasticizer for cellulose acetate and acetate butyrate. [Bayer AG] †

Unimycin. CAS 2058-46-0; A proprietary preparation of oxytetracycline hydrochloride; an antibiotic. [Unigreg Ltd] †

Union Carbide® A-137. CAS 2943-75-1; EINECS 220-941-2; Octyltriethoxysilane; coupling agent, crosslinking agent providing durability, gloss, hiding power to coatings. [Union Carbide]

Union Carbide® A-151. CAS 78-08-0; EINECS 201-081-7; Vinyltriethoxysilane; coupler promoting bonding between org. and inorg. components of a system; crosslinking agent; provides durability, gloss, hiding power to coatings. [Union Carbide]

Union Carbide® A-162. CAS 2031-67-6; EINECS 217-983-9; Methyltriethoxysilane; coupling agent, crosslinking agent providing durability, gloss, hiding power to coatings. [Union Carbide]

Union Carbide® A-163. CAS 1185-55-3; EINECS 214-685-0; Methyltrimethoxysilane; coupling agent, crosslinking agent providing durability, gloss, hiding power to coatings. [Union Carbide]

Union Carbide® A-171. CAS 2768-02-7; EINECS 220-449-8; Vinyltrimethoxysilane; coupling agent, crosslinking agent providing gloss, hiding power to coatings. [Union Carbide]

Union Carbide® A-172. CAS 1067-53-4; EINECS 213-934-0; Vinyl tris-(2-methoxyethoxy) silane; coupling agent, crosslinking agent providing gloss, hiding power to coatings. [Union Carbide]

Union Carbide® A-174. CAS 2530-85-0; EINECS 219-785-8; γ-Methacryloxypropyltrimethoxysilane; coupling agent, crosslinking agent providing gloss, durability, hiding power, adhesion promotion to coatings. [Union Carbide]

Union Carbide® A-186. CAS 3388-04-3; EINECS 222-217-1; β-(3,4-Epoxycyclohexyl) ethyltrimethoxy silane; coupling agent, crosslinking agent, adhesion promoter for coatings; provides durability. [Union Carbide]

Union Carbide® A-187. CAS 2530-83-8; EINECS 219-784-2; γ-Glycidoxypropyltrimethoxy silane; coupling agent, crosslinking agent, adhesion promoter for coatings; provides durability. [Union Carbide]

Union Carbide® A-189. CAS 4420-74-0; EINECS 224-588-5; γ-Mercaptopropyltrimethoxy silane; coupling agent, crosslinking agent, adhesion promoter for coatings; provides durability; features active hydrogen reaction, chain transfer, end blocking. [Union Carbide]

Union Carbide® A-1100. CAS 919-30-2; EINECS 213-048-4; γ-Aminopropyltriethoxy silane; coupling agent, crosslinking agent, adhesion promoter for coatings; provides durability; features active hydrogen reaction, end blocking. [Union Carbide]

Union Carbide® A-1106. CAS 58160-99-9; EINECS 261-145-5; Aminoalkyl silicone solution; finish for woven fiberglass; coupling agent for glass fiber sizes; filler treatments; additive to water-sol./disp. resins incl. vinyl and acrylic latex, epoxies, and phenolic binder disps. for coatings, adhesives, foundry, and insulation binders. [Union Carbide]

Union Carbide® A-1110. CAS 13822-56-5; EINECS 237-511-5; γ-Aminopropyltrimethoxysilane; coupling agent, crosslinking agent, adhesion promoter for coatings; fea-

tures active hydrogen reaction. [Union Carbide]

Union Carbide® A-1120. CAS 1760-24-3; EINECS 212-164-2; N-β-(Aminoethyl) γ-aminopropyltrimethoxy silane; coupling agent, crosslinking agent, adhesion promoter for coatings; features active hydrogen reaction. [Union Carbide]

Union Carbide® A-1160. CAS 116912-64-2; EINECS 245 876-7; γ-Ureidopropyltriethoxysilane; coupling agent, crosslinking agent, adhesion promoter for coatings; features active hydrogen reaction. [Union Carbide]

Union Carbide® XLP-57D11. Polyester modifier; low profile additive giving improved deep color pigmentation when used in conjunction with orthophthalic-modified rigid polyester resins. [Union Carbide]

Union Carbide® Y-11343. Low molecular weight silicone/ aminofunctional silane coupling agent; crosslinker, adhesion promoter for one- and two-pack RTV silicone sealant systems. [Union Carbide]

Unipel. Pelleted fertilizer. [Chevron] *

Unipen. A proprietary preparation of carindacillin sodium; an antibiotic used in the treatment of genito-urinary tract infections. [Pfizer International]

Uniperol® EL. Oxyethylated vegetable oil; surfactant, dispersant, leveling agent, emulsifier for fatty acids, fatty oils, waxes in textile dyeing and printing; stable to hard water. [BASF/Fibers; BASF AG]

Uniperol® O. Fatty alcohol-polyglycol ether; emulsifier, dyeing auxiliary, leveling and wetting agent, dispersant, detergent used in textile processing. [BASF/Fibers; BASF AG]

Uniperol® W, W Flakes. Aliphatic ethoxylate; surfactant, dispersant, leveling agent, protective colloid; for wool and synthetic fibers; sl. wetting action, no detergency; stable to acids, alkalies, water-hardening salts, heavy metal ions, electrolytes. [BASF/Fibers; BASF AG]

Unipertan P-24. Hydrolyzed animal collagen, tyrosine, riboflavin; raw material for cosmetics, toiletries, and pharmaceuticals. [Lipo]

Unipertan P-242. Hydrolyzed animal collagen, tyrosine, adenosine triphosphate; raw material for cosmetics, toiletries, and pharmaceuticals. [Lipo]

Uniphyllin. CAS 58-55-9; Theophylline in a controlled release tablet; for treatment of asthma. [Napp Laboratories Ltd] *

Uniplast. A proprietary trade name for phenol-formaldehyde molding compound. §

Uniplex 80. CAS 77-93-0; Triethyl citrate; plasticizer for food pkg. [Unitex]

Uniplex 82. CAS 77-89-4; Acetyl triethyl citrate; plasticizer for food pkg. [Unitex]

Uniplex 84. CAS 77-90-7; Acetyl tributyl citrate; plasticizer for indirect and direct food contact applications; milling lubricant for aluminum foil or sheet steel for use in cans for beverage and food products; in PVC toys, cellulose nitrate films, aerosol hair sprays, and dairy product cartons. [Unitex]

Uniplex 108. CAS 1077-56-1, 80-39-7; N-Ethyl o/p-toluene sulfonamide; plasticizer for nylon, shellac, cellulose acetate, protein materials, PVAc adhesives, and nitrocellulose lacquers. [Unitex]

Uniplex 110. CAS 131-11-3; Dimethyl phthalate; solv. and plasticizer for cellulose acetate butyrate compositions; solv. for org. catalysts; plasticizer and solv. for aerosol hair sprays. [Unitex]

---

‡ Trade name and manufacturer not verified   § Trade name without identified manufacturer

Uniplex 150. CAS 84-74-2; Dibutyl phthalate; solv./plasticizer in fingernail polish, nail polish remover, hair sprays, org. peroxide catalysts, adhesives, coatings; compatible with cellulosics, methacrylate, PS, PVB, vinyl chloride, urea-formaldehyde, melamine-formaldehye, phenolics. [Unitex]

Uniplex 155. CAS 84-69-5; Diisobutyl phthalate; plasticizer for thermoplastic and thermoset resins; solv./plasticizer in cellophane, resin coated sand for foundry casting, org. peroxides. [Unitex]

Uniplex 171. CAS 70-55-3, 88-19-7; o,p-Toluene sulfonamide; plasticizer for thermoplastic and thermoset resins; imparts gloss and wetting to melamine, urea and phenolic resins. [Unitex]

Uniplex 173. CAS 70-55-3; p-Toluene sulfonamide; plasticizer for thermoplastic and thermoset resins; imparts gloss and wetting to melamine, urea and phenolic resins. [Unitex]

Uniplex 214. CAS 3622-84-2; N-Butylbenzene sulfonamide; plasticizer for polyamide resins, enhancing low temp. flexibility; for fishing line, parts, adhesives for nonwoven interlinings, auto fuel line, coil air hoses, other high performance applications. [Unitex]

Uniplex 225. CAS 35325-02-1; N-(2-Hydroxypropyl) benzenesulfonamide; plasticizer for polyamide, polyurethane, polyacrylic, cellulose ester; excellent antistatic properties and resist. to extraction by water and drycleaning solvs.; for paints, lacquers; stabilizer for pigmented unsaturated polyester resins. [Unitex]

Uniplex 250. CAS 84-61-7; Dicyclohexyl phthalate; heat activated plasticizer for heat seal applications such as food wrappers/labels, pharmaceutical labels and other applications where delayed heat activated adhesive is required; for printing ink formulations for paper, vinyl, textiles, and other substrates. [Unitex]

Uniplex 260. CAS 614-33-5; Glyceryl tribenzoate; polymer modifier; plasticizer; for heat seal applications, lacquers, films, in PVAc-based adhesives, cellophane coatings, nitrocellulose coatings, nail lacquer formulations, printing inks, polishes; extrusion and injection molding processing aid. [Unitex]

Uniplex 270. CAS 1459-93-4; Dimethylisophthalate; modifies clarity and melting pt. of PET resins used in films, blow-molded bottles, etc. [Unitex]

Uniplex 310. CAS 71486-48-1; Cyclohexyl isooctyl phthalate. [Unitex]

Uniplex 504. CAS 591-71-7; Pentaerythritol tetraacetate. [Unitex]

Uniplex 552. CAS 4196-86-5; Pentaerythritol tetrabenzoate; plasticizer for adhesives intended for heat seal applications [Unitex]

Uniplex 600. CAS 1338-51-8; Toluenesulfonamide-formaldehyde resin; modifier and adhesion promoter for synthetic and natural resins used in adhesives and coatings applications; extender in polyamide resins. [Unitex]

Uniplex 680. CAS 1338-51-8; o,p-Toluenesulfonamide-formaldehyde resin in butyl acetate; used in nail polish formulations in conjunction with nitrocellulose to improve durability and gloss; also in PVAc adhesives to impart quick tack and green strength, in formulations to bond cellophane to itself or paper, or aluminum foil to itself. [Unitex]

Uniplex 809. CAS 9004-93-7; PEG di-2-ethylhexoate; plasticizer with low volatility and excellent heat resist.; for

polyester and polyamide engineering plastics; improves mold release. [Unitex]

Unipor. Pour point depressants. [Universal-Matthey Products Ltd] *

Uniquat. Alkyl imidazoline benzyl quaternary ammonium compounds; germicidal applications in hospitals, institutions, and industrial water treatment. [Lonza AG]

Unique. A smokeless powder; designed for use in light through heavy shotshell loads; can also be used in handgun loads. [Hercules] *

Uni-Rez® 221. Glyceryl rosinate; used in clear and pigmented lacquers, sealers, cold-cut additives, aluminum paints, wh. and light-tint interior paints and enamels. [Union Camp] *

Uni-Rez® 709. Maleic resin; used in aq. flexographic inks, emulsion floor polishes, varnishes with good color and gloss retention. [Union Camp] *

Uni-Rez® 1039. Limed rosin solution with 5% calcium oxide; additive to low cost paints, varnishes, gloss enamels; as letdown vehicle. [Union Camp] *

Uni-Rez® 1502. Polyamide resin in xylene solution; curing agent; provides good adhesion to metals, concrete, many plastics, good film resiliency, toughness, and chemical resist.; used in epoxy coatings formulated to meet air pollution regulations. [Union Camp] *

Uni-Rez® 1548. Polyamide resin; used in flexographic printing inks for plastic substrates. [Union Camp] *

Uni-Rez® 1552. Polyamide adhesive resin; designed for bonding porous substrates. [Union Camp] *

Uni-Rez® 2100. Reactive polyamide (condensation product of polymerized fatty acids with polyalkyl polyamines); visc. resin used with epoxy resins to formulate coatings and adhesives, and impart outstanding adhesion, flexibility, impact resistance, and chemical resistance to coatings; reacts with epoxy resins at room temperature imparting resiliency and toughness to the cured resin; can be used in 1:1 mixing ratio with epoxy resins. [Union Camp] *

Uni-Rez® 2355. Reactive polyamide; epoxy curing agent for adhesives, coatings, and floor toppings. [Union Camp]*

Uni-Rez® 2620. Polyamide adhesive resin; used for hot melt bonding of leather and other porous substrates. [Union Camp] *

Uni-Rez® 2800. Low visc. amidoamine; used in highly filled systems incl. adhesives for concrete, grouting compounds, concrete floor toppings, high solids coatings; gives med. cure rates with liq. epoxy resin. [Union Camp] *

Uni-Rez® 7003. Maleic resin; used in nitrocellulose lacquers, printing inks; promotes hardness, high gloss, and adhesion. [Union Camp] *

Uni-Rez® 7705. Aq. resin solution; used in aq. gravure inks for use on paper and board substrates. [Union Camp] *

Uni-Rez® 9002. Phenolic resin, rosin-modified; used in sheet-fed lithographic inks, hot melt adhesives, and surface coatings. [Union Camp] *

Uni-Rez® A-800 Light. Maleic glyceryl rosinate; used in lacquers and soft oil varnishes; gives excellent hardness, gloss, and fast dry. [Union Camp] *

Uniroid. A proprietary preparation of hydrocortisone and cinchocaine hydrochloride; used in the treatment of hemorrhoids. [Unigreg Ltd]

Uniscrub®. CAS 18472-51-0; 4% Chlorhexidine gluconate solution; antiseptic cleansing solution. [Seton

---

* Trade name not verified as available

† Trade name verified as obsolete

Healthcare Group plc]

Unisept® Solution. CAS 18472-51-0; 0.5% Chlorhexidine gluconate solution; antibacterial agent for general antiseptic purposes. [Seton Healthcare Group plc]

Uniset® D-124F. SMD adhesive for dielec. applications; for pin transfer or syringe dispense; high strength, IR cure. [Emerson & Cuming]

Unisiv® 3A. Molecular sieve in carrier oil; water scavenger in PU coatings, adhesives, elastomers, and other moisture-sensitive systems. [UOP]

Unisize HA-70. Polyvinyl alcohol, tackified; derived from super hydrolyzed grades; tackified grade for specialty paper applications. [Air Prods./Polymers]

Unislip 1753. CAS 112-84-5; EINECS 204-009-2; Erucamide; slip and antiblock agent; mold release for PP. [Unichema]

Unislip 1759. CAS 301-02-0; EINECS 206-103-9; Oleamide; lubricant, slip and antiblock agent. [Unichema]

Unisol. Textile auxiliary chemicals. [ICI Chem. & Polymers Ltd] †

Unisom. CAS 562-10-7; Doxylamine succinate; antihistaminic. [Pfizer Inc]

Unispar 40. Feldspar; filler, flatting agent with low oil absorp.; for traffic paint, interior/exterior architectural coatings, protective, maintenance and marine coatings, mastics and adhesives. [Unimin]

Unisperse-E. Pigment pastes for emulsion paints. [Ciba plc]

Unisperse-P. Aqueous pigment dispersions for wallpaper printing mass coloration of paper and paper coating. [Ciba plc]

Unistab D-33. 1,3,5-Trimethyl-2,4,6-tris(3,5-di-t-butyl-4-hydroxybenzyl) benzene (40%) in polymeric organosilicon; antioxidant for polyolefin processing. [Union Carbide] *

Uni-Tac® 70. Noncrystallizing rosin; tackifier for SBR, natural rubber, butyl rubber, ethylene-vinyl acetate, and other polymers; for water and solv.-based construction adhesives, pressure-sensitive, sealant, hot melt and rubber compounding. [Union Camp] *

Uni-Tac® R85. Glyceryl rosinate; tackifier for use in EVA, SBS, SIS and other hot melts, pressure-sensitive adhesives, in rubber compounding, sealants, coatings; stabilized to provide good heat and aging stability. [Union Camp] *

Uni-Tac® R99. CAS 8050-26-8; Pentaerythrityl rosinate; tackifier for EVA, SBS, SIS, other hot melts, construction adhesives, in rubber compounding, sealants, coatings; stabilized to provide good heat and aging stability. [Union Camp] *

Unite. Chemically modified polyolefins containing maleic anhydride functionality; compatibilizers in polymer blends and alloys (incl. recycled plastics), as polymeric coupling agent in reinforced and filled polymers (especially PP), as adhesive agent for bonding polyolefins to various substrates. [Aristech]

United. CAS 1333-86-4; Carbon black. [Cabot Carbon Ltd]

Unitex. CAS 9006-04-6; Centrifuged ammoniated NR latex; used for latex foam, dipped goods, surface coatings, adhesives, extruded thread, molded and cast products; produces very light colored latex films. [Guthrie Latex]

Unithane. Urethane polymers. [Cray Valley Ltd]

Unithox 450. C20-40 pareth-10; component for water-based PU mold release agents; metalworking additive.

[Petrolite]

Unitop. Cuprimyxin; antibacterial; antifungal. [Hoffmann-LaRoche Inc] *

Unitreat. Refinery and oilfield chemicals. [Universal-Matthey Products Ltd] *

Univadine. Dyeing and printing assistant. [Ciba plc] *

Unival DMDA-6200. Blow molding resin for household and industrial chemicals containers. [Union Carbide]

Univan. A nickel-vanadium steel.

universal balsam. Consists of 1 part camphor, 6 parts lead acetate, 16 parts beeswax, and 48 parts rape oil.

Univest. A stereospecific, low-molecular weight polybutadiene; a universal binder for almost all types of sand materials; neither cement nor water needed. [Hüls AG]

Univol U304. A proprietasy trade name for a mixture of distilled C20/22 acids. [UOP Inc] *

Univol U308. CAS 124-07-2; EINECS 204-677-5; A proprietary trade name for 90% caprylic acid. [UOP Inc] *

Univol U310. CAS 334-48-5; A proprietary trade name for 90% capric acid. [UOP Inc] *

Univol U312. A proprietary trade name for a mixture of caprylic and capric acids. [UOP Inc] *

Univol U314/a/b. CAS 143-07-7; A proprietary trade name for 90/98% lauric acids. [UOP Inc] *

Univol U320. CAS 544-63-8; EINECS 208-875-2; A proprietary trade name for 90% myristic acid. [UOP Inc] *

Univol U332. CAS 57-10-3; EINECS 200-312-9; A proprietary trade name for 90% palmitic acid. [UOP Inc] *

Univol U334. CAS 57-11-4; EINECS 200-313-4; A proprietary trade narne for 90% stearic acid. [UOP Inc] *

Univol U342. CAS 112-86-7; A proprietary trade name for 85/90% erucic acid. [UOP Inc] *

Univol U344. CAS 112-85-6; A proprietary trade name for 85/90% behenic acid. [UOP Inc] *

Uniwax 1747. CAS 3061-75-4; EINECS 221-304-1; Behenamide; lubricant, mold release agent for EPDM and PVC processing. [Unichema]

Uniwax 1750. CAS 124-26-5; EINECS 204-693-2; Stearamide; antiblock additive for polyolefin films; lubricant, release agent for rubber compounding, and PVC processing. [Unichema]

Uniwax 1760. CAS 110-30-5; EINECS 203-755-6; Ethylene bis-stearamide; lubricant, slip and antiblock agent, release agent for polyolefins, chloroprene rubber, and PVC. [Unichema]

Unna's stain. A microscopic stain. It contains 0.15 g methyl green, 0.5 g pyronin, 5 cc 95% alcohol, 20 cc glycerin, and the whole made up to 100 cc with 2% carbolic acid solution.

Unna's zinc paste. A paste made from gelatin, zinc oxide, glycerin, and water.

Unocal 76 Res 701. PVDC emulsion polymer; laminating adhesive, excellent adhesion, solv. urethane replacement, flame retardant textile backcoating. [Unocal/Polymers]

Unocal 76 Res 777. PVDC emulsion polymer; latex vehicle for interior flat and semigloss flame retardant coatings. [Unocal/Polymers]

Unocal 76 Res 1026. Styrene acrylic emulsion polymer; for elastomeric roof coatings with excellent cold temp. flexibility. [Unocal/Polymers]

Unocal 76 Res 1300. Styrene acrylic emulsion polymer; extremely tough and firm-flexible high performance polymer with outstanding water resist. for textile applications.

---

‡ Trade name and manufacturer not verified   § Trade name without identified manufacturer

[Unocal/Polymers]

Unocal 76 Res 3114. Acrylic emulsion polymer; for textile finishing, nonwoven binder, flocking, and laminating adhesive. [Unocal/Polymers]

Unocal 76 Res 3512. Acrylic emulsion polymer; self-cross-linking acrylic exhibiting good adhesion to nylon, PP, and polyester films; nonblocking primer for films. [Unocal/Polymers]

Unocal 76 Res 4305. Carboxylated SBR emulsion polymer; binder for textiles and nonwovens. [Unocal/Polymers]

Unocal 76 Res 6004. Acrylic emulsion polymer; excellent flexibility and wet adhesion in exterior and interior coatings. [Unocal/Polymers]

Unocal 76 Res 6206. CAS 108-05-4; Vinyl acetate homopolymer emulsion; for adhesives, wood glue. [Unocal/Polymers]

Unocal 76 Res 6255. Vinyl acrylic emulsion polymer; for compounding high peel, high tack pressure-sensitive adhesives. [Unocal/Polymers]

Unocal 76 Res 6272. Acrylic emulsion polymer; compounding base for repositionable high peel and tack pressure-sensitive adhesives. [Unocal/Polymers]

Unocal 76 Res 6931. Vinyl acrylic emulsion polymer; soft binder with good tensile properties for textiles and nonwovens. [Unocal/Polymers]

Unocal 76 Res 7800. Vinyl acrylic emulsion polymer; for papercoating applications. [Unocal/Polymers]

Unocal 76 Res 9410. Carboxylated SBR emulsion polymer; for compounding low temp. pressure-sensitive tapes and label systems. [Unocal/Polymers]

Unocal 76 Res S-55. CAS 108-05-4; Vinyl acetate homopolymer emulsion; for textile finishing and adhesives. [Unocal/Polymers]

U/o. CAS 511-13-7; Chlophedianol hydrochloride; antitussive. [3M Pharmaceuticals] †

Uografin. Aqueous solutions of sodium and meglumine salts of diatriazoate; x-ray contrast media. [Schering Health Care Ltd]

UOP. Antioxidants. [Universal-Matthey Products Ltd] *

UOP. Catalysts. [Universal-Matthey Products Ltd] *

UOP. Antiozonants. [Siba Hegner Ltd]

UOP 88. N,N′-Dioctyl-p-phenylene diamine; a proprietary antioxidant [UOP Inc] *

UOP 288. N,N′-Bis-(1-methyl/peptyl)-p-phenylene diamine; a proprietary antioxidant. [UOP Inc] *

UP 13600. CAS 1317-38-0; Precipitated cupric oxide. [Am. Chemet]

Upgrade. Soluble concentrate containing 360 g/l chlormequat and 180 g 2-chloroethylphosphoric acid per liter; plant growth regulator for winter wheat. [Rhone-Poulenc Crop Protection Ltd]

Upilex. Film. [Imperial Chemical Industries plc]

U Pink 113. CAS 977058-55-1, 7787-59-9; Ultramarine pink, bismuthoxychloride; inorg. colorant. [Presperse]

Upixon. An anthelmintic drug. [Bayer AG]

Uplees powder. An explosive containing 62-65% ammonium nitrate, 12 1/2-14 1/2% sodium nitrate, 4-6% trinitro-toluene, 13 1/2% ammonium chloride, and 2-4% starch.

Up-Start. Plant starter. [Chevron] *

UR-20CF/000. PU, 20% carbon fiber-reinforced. [Compounding Tech.] *

UR-20GF/000. PU, 20% glass fiber-reinforced. [Compounding Tech.] *

UR-30CF/000 Foamed. PU foam, 30% carbon fiber-reinforced. [Compounding Tech.] *

Urac. A proprietary trade name for urea-formaldehyde adhesives. §

uracil. (2,4-dihydroxypyrimidine; 2,4-dioxypyrimidine) $C_4H_4N_2O_2$; CAS 66-22-8; EINECS 200-621-9; Biochemical research. [Aldrich Ltd; PCR; Penta Mfg.; Schweizerhall]

Uracil Mustard. A proprietary preparation of uramustine; used in the treatment of leukemia. [Upjohn Ltd] *

uracil-1-β-D-ribofuranoside. See uridine.

Uracryl. A trademark for a range of acrylic synthetic resins in emulsion form. [Unilever] *

Uradal. See Adalin. §

Uradex. Active ingredients: diurex plus uragan; selective pre-emergence herbicide mixture. [Agan Chemical Manufacturers Ltd]

Uradil. A proprietary range of resins dispersible in water. Uradil 580/585 used for air-drying and storing. Uradil 587/588 are acrylic resins cross-linked by water-thinnable amino resins. Uradil 503 and 415 are nonoxidizing oil-free polyesters. §

Uragan. CAS 314-40-9; Active ingredient: bromacil; 5-bromo-3-sec-butyl-6-methyluracil; versatile herbicide for control of established annual and perennial broadleaf weeds and grasses and brush. [Agan Chemical Manufacturers Ltd]

Uralane® 5774-A/B. Two-component urethane adhesive; high strength, self-extinguishing adhesive curing at room or elevated temps. to form tough, impact resist. bonds to thermoplastics or metal substrates; for aircraft industry. [Ciba-Geigy/Furane]

uraline. Chloral-urethane, $CCl_3 \cdot CH(OH) \cdot NH \cdot CO_2 \cdot C_2H_5$; a hypnotic.

Uralite 3103. Ester PU; non-MDA; room temperature/ elevated temp. curing compound. [Hexcel]

Uralite 3109/3741. Ether PU; non-MDA; room temperature/ elevated temp. curing compound. [Hexcel]

Uralite 3111. Two-component urethane elastomer; room temperature-curing urethane elastomer; low visc., long pot life for easy vacuuming, soft, flexible, yet tough; reproduces the finest detail; will flex to release from moderate undercuts without tearing; for making molds for use with water-emulsified polyesters; used in molds, washers and gaskets, shock-mount pads, flexible production parts, and roller faces; mix ratio: 100A/46B by wt. [Hexcel]

Uralite 3150, 3152. Two-component urethane elastomer; low moisture sensitivity casting elastomer; high abrasion and impact resistance, fast cure and quick demold, optimum hardness for cold set boxes, good adhesion to many substrates; contains no TDI or MOCA; used in cold set core box liners, military training aids, industrial parts, roller facings and casters, metal forming pads, potting and encapsulating, gaskets and engine mountings, abrasion-resistant pads and bumpers; mix ratio: 100A/ 26B by wt. [Hexcel]

Uralite 3530. PU; adhesive; mix ratio 100:47. [Hexcel]

Uralite. A proprietary trade name for urea-formaldehyde. §

Uralloy® Hybrid Polymer LP-2035. PU polymer in styrene monomer; low profile additive designed for use in fiber-reinforced unsaturated polyester composites; controls shrinkage, provides automotive Class A surface appearance, offers easier processing. [Olin]

---

uramil. Amidomalonylurea, $C_4H_5N_3O_3$.

Uramon. A proprietary trade name for a fertilizer containing 43% nitrogen in the form of urea or similar compounds.§

uranine. (fluoresceine) CAS 518-47-8; EINECS 208-253-0; The sodium or potassium salt of fluoresceine, $C_{20}H_{10}O_5Na_2$; dyes silk and wool yellow.

Uranox. A high nickel stainless steel.

Urantoin. CAS 67-20-9; A proprietary preparation of nitrofurantoin; an antibiotic. [DDSA Pharmaceuticals Ltd] *

urea. (carbamide; carbonyldiamide) Organic compound.; $CO(NH_2)_2$; CAS 57-13-6; EINECS 200-315-5; fertilizer; animal feed; resins and plastics; stabilizer explosives; in paper industry to soften cellulose; in ammoniated dentifrices; diuretic; antiseptic; deodorizer; penetrant. [Air Prods.; Atochem SA; Bio-Rad Labs; Chisso Am.; EM Industries; Heico; Mallinckrodt; Mitsui Toatsu Chem.; Nissan Chem. Ind.; Norsk Hydro AS; OxyChem; Showa Denko]

urea amidohydrolase. See urease.

urea and thiourea resins. (carbamide; carbonyl diamide) CAS 57-13-6; Resins obtained by the reaction between urea or thiourea and formaldehyde; used in fertilizers, animal feed, plastics, as a stabilizer in explosives, medicine, adhesives, pharmaceuticals, cosmetics, dentifrices, sulfamic acid production; viscosity modifier for starch or casein-based paper coatings.

urea-bromine. A combination of urea and calcium bromide, $4CO(NH_2)_2 \cdot CaBr_2$, and containing 36% bromine.

urea-formaldehyde resin. Amino resin; thermosetting resin; pigment-grinding medium; aids adhesion and toughness of coatings; wet strength resin in paper treatment; in automotive enamels and primers, metal decorating finishes; modifier for water-sol. polymers. [Akzo; Am. Cyanamid; Bakelite GmbH; Cargill; DSM UK; Georgia-Pacific; Hercules; Sybron]

urea glue. A glue formed from the condensate of urea and formaldehyde; used in conjunction with a hardener.

Ureaphil. CAS 57-13-6; EINECS 200-315-5; A proprietary preparation of urea used as an osmotic diuretic and as an abortifacient. [Abbott Laboratories] *

ureaphos. A fertilizer containing phosphate of ammonia and urea.

urease. (urea amidohydrolase) Enzyme which hydrolyzes urea to ammonium carbonate; CAS 9002-13-5; EINECS 232-656-0; in determination of urea in urine, blood, and other body fluid. [Accurate Chem. & Scientific; Schweizerhall; Worthington Biochemical]

Urecholine. CAS 590-63-6; Bethanechol chloride; cholinergic. [Merck & Co Inc]

Urecoll®. Urea-formaldehyde resins; used without hardener to produce adhesives for paper processing; with hardener to produce binders for gran. materials, bonding fiber webs. [BASF AG; BASF plc]

Ureflex 6005. Two-component polyurethane elastomer systems; elastomer formulated for high abrasion resist., tear strength and resist. to oxygen and common industrial chemicals; used in gaskets, o-rings, potted components, custom mold work. [Flexible Prods.]

Ureflex 6011. Two-component polyurethane elastomer system; re-enterable potting compound for gaskets, o-rings, potted components, custom mold work. [Flexible Prods.]

Ureit. A German name for urea-formaldehyde resins.

Ureka. A mixture of diphenylguanidine and mercaptobenzo-thiazole. used as a rubber vulcanization accelerator.

Ureka B. A blend similar to Ureka with a portion of D.P.G. replaced by Guantal. [General Electric] *

Ureka C. Benzothiazylthiobenzoate; rubber vulcanization accelerator. [General Electric] *

Ureol® 6414A/5117B. Polyurethane resin/hardener system; for pattern making, core boxes, hammer form tools, explosion molding dies. [Ciba-Geigy Plastics UK] *

Urepan®. PU elastomer; for building, furniture, shoes, automotive and mech. engineering, sports surfacing, sporting goods, textiles, elec. industry, domestic appliances. [Bayer AG]

Ureresolve. Glycol-based solv.; solv. used hot (175-205 C) for removing heavily encrusted urethanes. [Dynaloy]

Uresamin. Etherified urea/formaldehyde resins; used for stoving finishes and acid curing lacquers. [Resinous Chemicals Ltd]

Uresolve 411. Flamm. solv. for urethane coatings, RTV and silicone oils; for electronic use (conformal coating removal, cable assembly repair, depotting). [Dynaloy]

urethane. (ethyl urethane; ethyl carbamate) $NH_2 \cdot COOC_2H_5$; CAS 51-79-6; EINECS 200-123-1; used as an intermediate for pharmaceuticals and pesticides; in biochemical research and medicine.

UrethHALL® 2050. Polyester polyol; for urethanes [C.P. Hall]

Urethon. Plastic film for sunblinds. [May & Baker Ltd] *

Uretix. A proprietary urea formaldehyde molding material. [Nisshin Boseki] *

Urexpan. Polyurethane sealant; self-leveling sealant for caulking dead-level horizontal joints subject to heavy foot and vehicular traffic. [Pecora Corporation]

Urex® Tablets. CAS 5714-73-8; Methenamine hippurate; an antibacterial and urinary antiseptic. [3M Pharmaceuticals]

uric acid. (2,6,8-trihydroxypurine; 2,6,8-trioxypurine; lithic acid; uric oxide) $C_5H_4N_4O_3$; CAS 69-93-2; EINECS 200-720-7; Organic synthesis. [Biosynth AG; Burlington Bio-Medical; ICI Spec.; Spectrum Chem. Mfg.]

uric oxide. See uric acid.

uridine. (uracil-1-β-D-ribofuranoside; D-ribosyl uracil) $C_9H_{12}N_2O_6$; CAS 58-96-8; EINECS 200-407-5; Biochemical research. [Am. Biorganics; R.W. Greeff; Penta Mfg.]

Urine-Pak. Test combinations for determining uric acid, cholesterol, HDL cholesterol and urinary proteins in clinical chemistry. [Bayer AG]

urisol. See hexamine.

Urispas. Favoxate hydrochloride; a relaxant. [SmithKline Beecham]

Uristix. A proprietary test strip impregnated with a citrate buffer, tetra bromophenol blue, glucose oxidase, peroxidase and o-toluidine; used for the detection of protein and glucose in urine. [B C Ames] *

uritone. See hexamine.

Urobac. A proprietary preparation of carindacillin sodium; an antibiotic used in the treatment of genito-urinary tract infections. [Pfizer International] †

Urobilistix. A proprietary test strip impregnated with p-dimethylaminobenzaldehyde in an acid buffer; used to detect urobilinogen in urine. [B C Ames] *

Uro-Binotal®. CAS 69-53-4; Ampicillin; for treatment of acute urinary tract infections. [Bayer AG]

Urobiotic. A proprietary preparation of oxytetracycline,

---

‡ Trade name and manufacturer not verified   § Trade name without identified manufacturer

sulfamethiazole, and phenazopyridine for therapy of genito-urinary infections. [Pfizer International]

uroformine. See hexamine.

Uro-Hexoids. A proprietary preparation of hexamine and lithium benzoate tablets; product now discontinued. [British Drug Houses] †

urokinase. A plasminogen activator isolated from human urine.

Urolucosil. CAS 144-82-1; A proprietary preparation containing sulfamethizole; a urinary antiseptic. [Warner] *

Uromat PE. A proprietary trade name for an aqueous dispersion based on titanium dioxide used for delustering synthetic fiber fabrics. [Ciba plc] *

urometin. See hexamine.

Uromide. A proprietary preparation of sulfacarbamide and phenazopyridine; a urinary antispasmodic. [Consolidated Chemicals Ltd] *

Uromiro(n). CAS 440-58-4; Iodamide; diagnostic aid. [Bracco Industria Chimica SpA] *

Uromitexan. Mesna; ampoule form; uroprotector. [Degussa AG] *

Uropen. CAS 5321-32-4; Hetacillin potassium; antibacterial. [Bristol Laboratories] *

Uroplas®. A range of urea molding compounds; used for electrical field, sanitary field, electrotechnical industry and pottery. [AMC SPREA S.p.A.]

Uropol. A proprietary preparation of tetracycline phosphate complex, sulfamethizole and phenazopyridine hydrochloride; a urinary antibiotic. [Bristol-Myers Squibb Co Inc]

urotropine. See hexamine.

Urotuf®. Synthetic resins and resin solutions. [Reichhold]

Urovist. Diatrizoate meglumine; diagnostic aid. [Berlex Laboratories Inc] *

Urovist Sodium 300. CAS 737-31-5; Diatrizoate sodium; diagnostic aid. [Berlex Laboratories Inc] *

Ursol®. Oxidation bases; for dyeing furs. [BASF AG] *

Urtal. An acrylonitrile-butadiene-styrene resin.

Urtenol. A combination of oils for the textile trades possessing penetration and detergent properties.

Usacert. High strength water soluble powder food colors certified by FDA (USA); used for coloring of foodstuffs and pharmaceuticals. [Morton Int'l. Ltd] †

Usagran. Granular food colors, FDA certified. [Morton Int'l. Ltd]

Usalake. Water insoluble, aluminum lake food colors; used for coloring of foodstuffs and pharmaceuticals. [Morton Int'l. Ltd] †

USI in Oval. CAS 64-17-5; EINECS 200-578-6; Ethyl alcohol. [Quantum Chemical Corp]

USI in Oval. Ethyl ether. [Quantum Chemical Corp]

USI in Oval. Polyolefin resins. [Quantum Chemical Corp]

Usol Copper Green. Wood preservative formulated from copper naphthenate for use in solvent or water reducible bases; used for brushing, dipping, soaking or mopping of liquid to various wood species. [Standard Tar Products Co Inc]

usoline. See paraffin, liquid.

Usol Organiclear. Proprietary wood preservatives to prevent wood rot, decay, mold and termite attack; used for brushing, dipping, soaking or mopping of liquid to various wood species. [Standard Tar Products Co Inc]

Usol Zinclear. Zinc naphthenate wood preservative formulated in solvent or water reducible bases; used for brushing, dipping, soaking or mopping of liquid to various wood species. [Standard Tar Products Co Inc]

USP®. Organic peroxide catalyst. [Witco/Argus]

USP®-90MD. CAS 15667-10-4; 80%solution of 1,1-di (t-amyl peroxy) cyclohexane in odorless mineral spirits; initiator for heat-cured polyester resin cures and acrylates. [Witco/Argus]

USP®-240. CAS 37187-22-7; Ketone peroxidesolution from acetylacetone; used for rapidly curing polyester resins with low levels of cobalt accelerators; DOT org. peroxide label is not required. [Witco/Argus]

USP®-245. CAS 13052-09-0; 2,5-Dimethyl-2,5-di(2-ethyl hexanoyl peroxy) hexane; catalyst for heated curing of polyester resin systems; features rapid cures and outstanding surface finishes. [Witco/Argus]

USP®-355M. CAS 3851-87-4; 75%solution of 3,5,5-trimethyl hexanoyl peroxide in odorless mineral spirits; initiator. [Witco/Argus]

USP®-400P. CAS 3006-86-8; 80% solution of 1,1-di (t-butyl peroxy) cyclohexane in butyl benzyl phthalate; initiator useful in heat-cured polyester resin systems where improved flow and pot life are critical. [Witco/Argus]

USP®-690. CAS 3006-82-4, 3006-86-8; t-Butyl peroxy 2-ethyl hexanoate and 1,1-di (t-butyl peroxy) cyclohexane in a phthalate plasticizer; single initiator for elevated temp. curing of unsaturated polyester resins and compounds over a broad temp. range. [Witco/Argus]

USP®-800. CAS 75-91-2; 70% t-butyl hydroperoxide with water as major diluent; initiator. [Witco/Argus]

Uspulun. Seed dressing for control of fungal diseases on cereals, rice, cotton, and vegetables. [Bayer AG]

Uspulun. A material containing sodium sulfate, sodium hydroxide, aniline, and mercury-chloro-phenol; used as a fungicide.

Ustilan®. CAS 30043-49-3; Ethidimuron; herbicide for total weed control on noncrop land. [Bayer AG]

Ustinex®. Products with combinations of different herbicidal compounds (aminotriazole, diuron, methabenzthiazuron, phenoxies); herbicides used for control of weeds on paths, open spaces, parks and sports grounds. [Bayer AG]

U-T-C. Colorant dispersions; for coloring of nonaqueous coating compositions. [Hüls Am.]

Uteplex. CAS 63-39-8; A proprietary preparation of uridine-5-triphosphoric acid; used to relieve muscle spasm. [Rona Laboratories] ‡

Utica Steel. A proprietary trade name for a die steel; it contains 1.4% tungsten, 1.25% carbon, 0.4% chromium and 0.2% vanadium. §

Uticillin. A proprietary preparation of carfecillin; an antibiotic. [SmithKline Beecham] *

Uticillin VK. CAS 132-98-9; Penicillin V potassium; antibacterial. [Upjohn] *

Uticort. CAS 22298-29-9; Betamethasone benzoate; glucocorticoid. [Parke-Davis] *

Utimox. CAS 26787-78-0; Amoxicillin; antibacterial. [Parke-Davis] *

Utinor. CAS 70458-96-7; Norfloxacin; urinary antiseptic. [Merck & Co Inc]

Utocyl. Antibiotic sulfonamide veterinary ethical. [Ciba plc]*

Utopar. CAS 26652-09-5; Ritodrine; uterospasmolytic, prevents premature birth. [Duphar BV] *

Uval. CAS 4065-45-6; Sulisobenzone; ultraviolet screen. [Dorsey Pharmaceuticals] *

---

* Trade name not verified as available   † Trade name verified as obsolete

Uvaseb 770. CAS 52829-07-9; EINECS 258-207-9; Bis (2,2,6,6-tetramethyl-4-piperidyl) sebacate; light stabilizer for polyolefins, ABS, PUR acrylics, PS and styrene copolymers, thermoplastic elastomers, ethylene-propylene copolymers, polyamides. [Enichem Synthesis SpA]

Uvasil 125. Polymethyl propyl 3-oxy-[4(2,2,6,6-tetramethyl) piperidinyl] siloxane absorbed on 30% fumed silica; uv-stabilizer for PP, PE, PS, ABS, polyamides, TPU, etc. [Enichem Synthesis SpA]

Uvasil 299. CAS 102089-33-8 (monomer); Polymethyl propyl 3-oxy-[4(2,2,6,6-tetramethyl) piperidinyl] siloxane; uv-stabilizer for PP, PE, PS, ABS, polyamides, TPU, etc. [Enichem Synthesis SpA]

Uvazol 236. CAS 3896-11-5; 2(2'-Hydroxy-3'-t-butyl-5'-methylphenyl)-5-chlorobenzotriazole; strong uv absorber protecting polyolefins, styrenics, acrylics, PVC, unsaturated polyesters, elastomers, PC, and PU. [Enichem Synthesis SpA]

Uvazol 237. CAS 3864-99-1; EINECS 223-383-8; 2-(3',5'-Di-t-butyl-2'-hydroxyphenyl)-5-chlorobenzotriazole; strong uv absorber protecting polyolefins, PVC, unsaturated polyester, acrylics, ABS. [Enichem Synthesis SpA]

Uvazol 311. CAS 3147-75-9; 2(2'-Hydroxy-5'-t-octylphenyl) benzotriazole; strong uv absorber protecting PNMA, PC, PS, ABS, unsaturated polyesters. [Enichem Synthesis SpA]

Uvazol P. CAS 2440-22-4; EINECS 219-470-5; 2(2'-Hydroxy-5'-methylphenyl) benzotriazole; strong uv absorber protecting styrenics, acrylics, polyesters, PC, PVC, unsaturated elastomers. [Enichem Synthesis SpA]

UV-Chek. For uv stabilizers and absorbers for polyolefins and related polymers, in Int'l. Class 1. [Ferro] *

Uvecryl®. Uv and electron beam curable materials. [UCB (Chem) Ltd]

Uvecryl® P 36. Acrylated deriv. of benzophenone; reactive photoinitiator for low odor, uv-cured coatings. [UCB Radcure]

Uvekol. Ultraviolet curable resins for glass laminating. [UCB (Chem) Ltd]

Uvilon. An anthelmintic also known as Upixon. [Bayer AG]†

Uvi-Nox 1494. CAS 4306-88-1; Diisobutyl nonyl phenol. [Rhone-Poulenc Surf.]

Uvinul®. UV protecting compounds. [BASF plc]

Uvinul® 400. CAS 131-56-6; EINECS 205-029-4; Benzophenone-1; uv absorber; used for polyester, acrylics, PS, in outdoor paints and coatings, varnishes, colored liq. toiletries and cleaning agents, filters for photographic color films and prints, and rubber-based adhesives. [BASF] *

Uvinul® 408. CAS 1843-05-6; EINECS 217-421-2; 2-Hydroxy-4-n-octoxy-benzophenone; uv absorber and stabilizer for polyethylene, PP, plasticized and rigid PVC, and other polymers; offers good compatibility, max. protection, and min. color and as a low order of toxicity. [BASF] *

Uvinul® 490. CAS 1341-54-4; Benzophenone-11; uv absorber; used in NC lacquer, fluorescent paint, inks, and for protecting furniture woods, colored liq. toiletries and cleaning agents, isocyanate systems, and butyrate metal lacquers. [BASF] *

Uvinul® D-49. CAS 131-54-4; EINECS 205-027-3; Benzophenone-6; most economical of the near-uv absorbers; greater heat stability and more sol. (in chlorinated and aromatic solvs.); gives broad protection to plastics, coatings, textiles such as PVC, chlorinated polyesters, epoxies, acrylics, urethanes, cellulosics, and oil-based paints and varnishes. [BASF] *

Uvinul® D-50. CAS 131-55-5; EINECS 205-028-9; Benzophenone-2; commercial uv absorber with the broadest uv absorp. spectrum; retards fading of pigments and dyestuffs; prolongs the life of polymeric materials; photostabilizes cosmetic formulations; and minimizes discoloration of synthetic rubber of plastic latices. [BASF] *

Uvinul® DS-49. CAS 3121-60-6; EINECS 221-498-8; Benzophenone-9; sulfonated deriv. of Uvinul D-49; uv absorber in cosmetic formulations to prevent fading of colors and visc. changes caused by uv light; in textiles and water-based paints. [BASF] *

Uvinul® M-40. CAS 131-57-7; EINECS 205-031-5; Benzophenone-3; uv absorber; similar to Uvinul 400 except for higher sol. in aromatic solvs.; good weather resistance in resins and plastics; stabilizes PVC and polyesters against uv-light degradation; used in NC lacquers, varnishes, and oil-based paints. [BASF] *

Uvinul® M-493. CAS 1341-54-4; Benzophenone-11. [BASF] *

Uvinul® MS-40. CAS 4065-45-6; EINECS 223-772-2; Benzophenone-4; sulfonated, water-sol. deriv. of Uvinul M-40; uv absorber in sunscreen products and in hair sprays and shampoos for dyed and tinted hair; for leather and textile fibers. [BASF] *

Uvinul® N-35. Etocrylene; noncolor contributing uv absorber; does not contain aromatic hydroxyl groups; effective under varying pH conditions; for NC lacquers and PVC; used in alkaline systems such as urea-formaldehyde and epoxyamine formulations, and in cosmetics. [BASF] *

Uvinul® N-539. CAS 6197-30-4; Octocrylene; Synthetic Ethylhexyl-2-cyano-3,3-diphenylacrylate; uv absorber; in flexible and rigid PVC; used in NC lacquers, varnishes, vinyl flooring, and oil-based paints; in aerosol and oil-based suntan lotions; nonreactive with metallic driers. [BASF] *

Uvinul® O-18. CAS 118-60-5; Octyl salicylate. [BASF] *

Uvinul® P 25. CAS 15716-30-0; PEG-25 PABA; uv absorber for sunscreen products. [BASF] *

Uvinul® T 150. Octyl triazone; uv filter. [BASF] *

U Violet 109. CAS 12769-96-9, 7787-59-9; Ultramarine violet, bismuthoxychloride; inorg. colorant. [Presperse]

Uvistat 12, 24, 247, 2211. A proprietary trade name for a series of additives for protecting plastics against uv light; they have the general formula R_1, C_6H_5-CO LTD-C_6H_4OH-OR; Uvistat 247 is particularly effective for the stabilization of polyolefins. [Octel Chemicals Ltd] †

Uvistat Aftersun. A moisturizer. [Windsor Healthcare Ltd]

Uvistat Babysun Aftersun. A proprietary preparation containing calamine; a moisturizer. [Windsor Healthcare Ltd]

Uvistat Sun Cream, Sun Block, Ultrablock, Lotion, Lipscreen, Babysun. Proprietary preparations containing combinations of 2-ethyl hexyl p-methoxycinnamate (uvb sunscreen), 4-5-butyl-1-4'-methoxydibenzoylmethane (uva sunscreen), 2-hydroxy-4-methoxy-4'-methylbenzophenone (uva sunscreen); and titanium dioxide (a physical barrier blocking both uva and uvb); sunscreen products. [Windsor Healthcare Ltd]

---

‡ Trade name and manufacturer not verified   § Trade name without identified manufacturer

Uvitex. A proprietary trade name for a series of fluorescent brighteners for incorporation in soap-based and synthetic detergents; Uvitex SFC is a stilbenic derivative giving high intensity whites on cellulosic fibers; Uvitex SK is a benzoxazole derivative effective on a wide variety of fibers, stable to hypochlorite and chlorisocyanate; Uvitex SOF is also a benzoxazole derivative; Uvitex ERN Conc. P is a benzoxazole derivative, a fluorescent brightener for polyester fibers. [Ciba plc] *

Uvitex MA. A proprietary trade name for an imidazole derivative fluorescent brightener for acrylic fibers; for application in the dope before spinning. [Ciba plc] *

Uvitex MP. A proprietary trade name for a heterocyclic-stilbene type fluorescent brightener for polyamid fibers. [Ciba plc] *

V

V-19. CAS 35634-74-3; 2-Phenylazo-4-methoxy-2,4-dimethylvaleronitrile; polymerization initiator. [Wako Pure Chem. Ind.; Wako Chem. USA]

V-30. CAS 10288-28-5; 1-[(1-Cyano-1-methylethyl)azo] formamide; polymerization initiator. [Wako Pure Chem. Ind.; Wako Chem. USA]

V-40. CAS 2094-98-6; 1,1′-Azobis(cyclohexane-1-carbonitrile); polymerization initiator. [Wako Pure Chem. Ind.; Wako Chem. USA]

V-59. CAS 13472-08-7; 2,2′-Azobis(2-methylbutyronitrile); polymerization initiator. [Wako Pure Chem. Ind.; Wako Chem. USA]

V-60. CAS 78-67-1; 2,2′-Azobis(2-methylpropionitrile); polymerization initiator. [Wako Pure Chem. Ind.; Wako Chem. USA]

V-65. CAS 4419-11-8; 2,2′-Azobis(2,4-dimethylvaleronitrile); polymerization initiator. [Wako Pure Chem. Ind.; Wako Chem. USA]

V-70. CAS 15545-97-8; 2,2′-Azobis(4-methoxy-2,4-dimethylvaleronitrile); polymerization initiator. [Wako Pure Chem. Ind.; Wako Chem. USA]

V-90®. CAS 7758-23-8; Monocalcium phosphate, anhydrous; leavening agent for baking, cereal. [Rhone-Poulenc Food Ingreds.]

V-601. CAS 2589-57-3; Dimethyl 2,2′-azobis (2-methylpropionate); polymerization initiator. [Wako Pure Chem. Ind.; Wako Chem. USA]

V-1065. RTV silicone; silicone rubber recommended when using polyester, castable PU resins, waxes and low-melt alloys (pewter); excellent for print pads. [Perma-Flex Mold]

V-9415. CAS 8007-18-9; NiSbTi; pigment for thermoplastic and thermoset resins, especially high temp. engineering resins, PVC siding and profile, and industrial finishes [Ferro/Color] *

VA-044. CAS 27776-21-2; 2,2′-Azobis[2-(2-imidazolin-2-yl)propane]dihydrochloride; polymerization initiator. [Wako Pure Chem. Ind.; Wako Chem. USA]

VA-058. CAS 102843-39-0; 2,2′-Azobis[2-(3,4,5,6-tetrahydropyrimidin-2-yl) propane] dihydrochloride; polymerization initiator. [Wako Pure Chem. Ind.; Wako Chem. USA]

VA-060. CAS 118585-13-0; 2,2′-Azobis[2-[1-(2-hydroxyethyl)-2-imidazolin-2-yl] propane] dihydrochloride; polymerization initiator. [Wako Pure Chem. Ind.; Wako Chem. USA]

VA-080. CAS 104222-32-4; 2,2′-Azobis[2-methyl-N-[1,1-bis(hydroxymethyl)-2-hydroxyethyl] propionamide]; polymerization initiator. [Wako Pure Chem. Ind.; Wako Chem. USA]

VA-086. CAS 61551-69-7; 2,2′-Azobis[2-methyl-N-(2-hydroxyethyl)-propionamide]; polymerization initiator. [Wako Pure Chem. Ind.; Wako Chem. USA]

VA-088. CAS 3682-94-8; 2,2′-Azobis(2-methylpropionamide) dihydrate; polymerization initiator. [Wako Pure Chem. Ind.; Wako Chem. USA]

VA-545. CAS 88684-42-8; 2,2′-Azobis(2-methyl-N-phenylpropionamidine) dihydrochloride; polymerization initiator. [Wako Pure Chem. Ind.; Wako Chem. USA]

VA-545. CAS 124960-38-9; 2,2′-Azobis[N-(4-chlorophenyl)-2-methylpropionamidine]dihydrochloride; polymerization initiator. [Wako Pure Chem. Ind.; Wako Chem. USA]

Vacrel®. Soldermask; for the electrical industry. [DuPont UK]

Vacsol. Organic solvent wood preservative; for pressure treatment for joinery and carcassing timber used above ground contact. [Hickson & Welch Ltd]

Vactran. High grade chemicals for vacuum evaporation. [British Drug Houses] *

vacuum salt. A pure salt, NaCl, obtained by boiling brine under a vacuum.

vacuum silicon iron. Alloys containing about 0.15 or 3.4% silicon, made by melting *in vacuo*; contain about 0.01% carbon and have remarkable magnetic properties.

Vaderm. CAS 66734-13-2; Alclometasone dipropionate; anti-inflammatory. [Schering Corp] *

Valadol. CAS 103-90-2; Acetaminophen; analgesic; antipyretic. [Bristol-Myers Squibb Co Inc] *

Valanone B. 3-β-m-4-Oxo-2,2,7,7-tetramethyltricyclo [6,2,1,0,3,8] undecane. [Bush Boake Allen Ltd]

Valclene®. Drycleaning fluid. [DuPont UK]

valerian. CAS 8057-49-6; The dried rhizome of *Valeriana officinalis* containing starch, a resinous substance, and essential oil of valerian.

valerianic acid. *See* n-valeric acid.

n-valeric acid. (carboxylic acid C_5; valerianic acid; n-pentanoic acid) $CH_3(CH_2)_3COOH$; CAS 109-52-4; EINECS 203-677-2; intermediate for flavors and perfumes, ester-type lubricants, plasticizers, pharmaceuticals, and vinyl stabilizers. [Aldrich; Hoechst Celanse; Union Carbide]

valerone. CAS 108-83-8; (diisobutyl ketone) $C_4H_9\cdot CO\cdot C_4H_9$.

Valfor® 100. CAS 1344-00-9; Sodium silicoaluminate; ion exchange and selective absorp./adsorp.; detergent builder; anticaking agent. [PQ Corp.]

Valinate®. For the agriculture industry. [DuPont UK]

valine. CAS 72-18-4; EINECS 200-773-6; (α-amino-isovaleric acid) $(CH_3)_2CH\cdot CH(NH_2)COOH$. A protein

---

aminoacid; used as a dietary suplement, in biochemical research.

L-valine. (α-aminoisovaleric acid) $C_5H_{11}NO_2$; CAS 72-18-4; EINECS 200-773-6; dietary supplement, culture media, biochemical and nutritional research. [Am. Biorganics; Degussa; R.W. Greeff; Mitsui Toatsu Chem.]

Valium. CAS 439-14-5; A proprietary preparation of diazepam; a sedative. [Roche Laboratories; Roche Products Ltd]

Valium Roche. Anti-anxiety agent. [Roche Products Ltd] *

Valledrine. CAS 4330-99-8; Trimeprazine tartrate cough linctus. [May & Baker Ltd] *

Vallergan. CAS 4330-99-8; Trimeprazine tartrate. [Rhone-Poulenc Rorer Ltd]

Vallex. Cough linctus. [May & Baker Ltd] *

Valmid. CAS 126-52-3; Ethinamate; sedative. [Eli Lilly & Co]†

Valmidate. CAS 126-52-3; A proprietary preparation of ethinamate; a sedative. [Eli Lilly & Co] †

Valoid. CAS 303-25-3; Cyclizine hydrochloride. for treatment of nausea and vomiting. [The Wellcome Foundation Ltd]

valonia. The acorn cups of *Quercus oegilops* contain about 35% of tannin and are used in the leather industry.

Valox®. Polybutylene terephthalate; used for plastic components for automotive, electrical, electronics, lighting, medical, packaging, audio etc. [GE Plastics Ltd]

Valox® 210HP, 220HP, 230HP, 260HP, 280HP. CAS 26062-94-2; PBT polyester resin; engineering thermoplastics for healthcare products; FDA compliance; provides dimensional stabilty, high surface gloss, chemical resist. to most oils and greases. [GE Plastics]

Valox® 325. CAS 26062-94-2; PBT polyester resin; general purpose injection molding resin with dimensional stability, high surface gloss, chemical resist., low water absorp., low coeffecient of friction and mold release. [GE Plastics]

Valox® 508. CAS 26062-94-2; PBT polyester alloy, 30% glass-reinforced; offers improved flatness, appearance, good flow. [GE Plastics]

Valox® 701. CAS 26062-94-2; PBT polyester resin, 35% glass/mineral-filled; reinforced resin combining superior elec. properties with improved stiffness and flatness; used in business machine components, keyboards and panels, circuit breakers, industrial motor controls, high voltage TV componentry, card guides, and automotive ignition parts (rotors and adaptor rings). [GE Plastics]

Valox® 780. CAS 26062-94-2; PBT polyester resin, 40% glass/mineral reinforced; superior electric performance, improved flatness, UL94 V-0 recognition, reduced shrinkage, good flow. [GE Plastics]

Valox® 815. CAS 26062-94-2; PBT polyester alloy, 15% glass-reinforced; offers improved surface appearance while maintaining elec. performance, superior heat resistance, and dimensional stability of standard glass-reinforced PBT grades; used in decorator lamp fixtures, business machine bezels, mechanical and elec. copier components where chemical resistance to lubricants and toners is required, ribbon cartridges, lighting fixtures for corrosive environments, and some automotive aftermarket products. [GE Plastics]

Valox® 9215. CAS 25038-59-9; PET polyester resin, 15% glass-reinforced; for applications requiring higher heat. [GE Plastics]

Valox® 9715M. Polycyclohexylene terephthalate, 15% glass-reinforced, flame-retardant; high heat grade. [GE Plastics]

Valox® DR48. CAS 26062-94-2; PBT polyester resin, 15% glass-reinforced, flame-retardant; provides low moisture absorp., dimensional stability in harsh environments, excellent mech. strength and stiffness; good mold-fill charactersitics and improved flow for molding thin sections. [GE Plastics]

Valox® FV608. Thermoplastic polyester foam resin; engineering structural foam resin features high dist. temp., chemical and solv. resistance, good fatigue endurance, high flex. strength and modulus, and good moldability; used for electronic enclosures, processing control instruments, and exterior transportation applications. [GE Plastics]

Valox® HV7075. PBT/PET/IM blend, 68% mineral-reinforced; injection moldable ceramic/ivory/metal replacement with excellent surface gloss, improved impact. [GE Plastics]

Valpin®. For medical applications. [DuPont UK]

Valpin 50. CAS 80-50-2; Anisotropine methylbromide; an anticholinergic. [Endo Laboratories Inc] *

Valray Alloy 1. A trademark for an alloy of 20% chromium and the balance nickel with controlled manganese, carbon and silicon. [Wiggin Alloys Ltd] ‡

Valtox. Granular systemic carbamate insecticide and nematode containing 5% w/w carbofuran; used to control a wide range of soil and seedling pests including cabbot root fly, cabbage stem weevil, flea beetle, early aphids in brassicas, turnip root fly, fruit fly, millipedes, symphilids, beet leaf miner, springtails, wireworms, free-living nematodes, potato cyst eelworm, carrot fly and carrot willow aphid. [Bayer plc]

valve bronze. An alloy of from 83-89% copper, 4-5% tin, 3-7% zinc, and 3-6% lead.

Valvoline®. A trade name for lubricating oils; they are American petroleum products. §

Valyl®. A registered trademark currently awaiting re-allocation by its proprietors. [Cassella AG] *

Vam. A proprietary preparation of vinyl and ethyl ethers; general anesthetic. [May & Baker Ltd] *

Vamac®. Ethylene acrylic elastomer. [DuPont UK]

Vamin Series. Parenteral solutions; crystalline amino acids for intravenous nutrition. [KabiVitrum AB] *

Vamitox. Herbicide. [May & Baker Ltd] *

vanadic acid anhydride. *See* vanadium pentoxide.

vanadium. Metallic element; V; CAS 7440-62-2; EINECS 231-171-1; target material for x-rays, manufacture of alloy steels. [Atomergic Chemetals; Cerac; Noah Chem.]

vanadium alum. An ammonium-vanadium sulfate, $(NH_4)_2SO_4 \cdot V_2(SO_4)_3 \cdot 24H_2O$.

vanadium brass. An alloy of 70% copper, 29.5% zinc, and 0.5% vanadium.

vanadium bronze. Metavanadic acid, HVO_3; used as a pigment in the place of gold bronze; also the name for an alloy of 61% copper, 38.5% zinc, and 0.5% vanadium.

vanadium chloride (ous). (vanadous chloride; vanadium dichloride) VCl_2; CAS 10213-09-9; Strong reducing agent, purification of hydrogen chloride from arsenic. [Atomergic Chemetals; Noah Chem.]

vanadium dichloride. *See* vanadium chloride (ous).

vanadium-manganese brass. An alloy containing 58.56% copper, 38.54% zinc, 1.48% aluminum, 1% iron, 0.48%

---

* Trade name not verified as available   † Trade name verified as obsolete

manganese, and 0.03% vanadium.

vanadium-molybdenum steels. Alloys containing from 0.1-1.0% carbon, 0.52-6.0% molybdenum and 0.1-1.0% vanadium.

vanadium (V) oxide. See vanadium pentoxide.

vanadium pentoxide. (vanadium (V) oxide; vanadic acid anhydride) V_2O_5; CAS 1314-62-1; EINECS 215-239-8; Catalyst for oxidation of sulfur dioxide in sulfuric acid manufacture, ferrovanadium, catalyst for organic reactions, ceramic coloring material, vanadium salts, inhibiting uv transmission in glass, photographic developer, textile dyeing. [Aldrich; Atomergic Chemetals; Cerac; Kerr-McGee; Shinko Chem.]

vanadium steel. An alloy of steel with vanadium.

vanadous chloride. See vanadium chloride (ous).

Vanair. A proprietary preparation of benzoyl peroxide and sulfur used in the treatment of acne. [Carter-Wallace Ltd]*

Vanall (K). Sorbitan stearate, mono and diglycerides, polysorbate 60 hydrated blend with propylene glycol, lactic acid, sodium proprionate preservatives; emulsifier used in the food industry for cake formulations. [Am. Ingredients/Patco]

Van-Amid. Epoxy curing agents; used as curing agents for Vanoxy resins. [R T Vanderbilt] †

Vanate®. A series of ethylene diamine tetraacetic acid compounds; chelating agents. [R T Vanderbilt]

Vanax®. Broad range of sulfenamide, dithiocarbamate, thiourea, thiadiazine, isophthalate, guanidine, and aldehyde-amine accelerators; used in natural, synthetic, and latex rubbers as both primary and secondary accelerators. [R T Vanderbilt]

Vanax® 552. CAS 98-77-1; Piperidinium pentamethylene dithiocarbamate; accelerator for NR, SR, cements, and latexes; peptizer for sulfur-modified G-type neoprenes. [R.T. Vanderbilt]

Vanax® 808. CAS 34562-31-7; Butyraldehyde-aniline condensation product; accelerator for NR, SBR, CR, IIR, and latexes; activator for acidic accelerators; also for reclaims, hard rubber stocks, CR cements containing. litharge. [R.T. Vanderbilt]

Vanax® 833. CAS 68411-19-8; Butyraldehyde-monobutylamine condensation product; accelerator for NR, SR, latexes, and reclaim; also in self-curing CR cements. [R.T. Vanderbilt]

Vanax® A. CAS 103-34-4; 4,4´-Dithiodimorpholine in paraffin wax; vulcanizing agent; sulfur donor for NR and synthetic rubbers; functions as primary accelerator for NR, IR, SBR, NBR, IIR elastomers, and as primary and secondary accelerator in EPDM; powd. [R.T. Vanderbilt]

Vanax® CPA. CAS 71172-17-3; Dimethylammonium hydrogen isophthalate; accelerator for W and T-type neoprenes; used in press cured, injection molded, and LCM stocks. [R.T. Vanderbilt]

Vanax® DOTG. CAS 97-39-2; N,N´-Di-ortho-tolylguanidine; accelerator for NR and SR; secondary accelerator. [R.T. Vanderbilt]

Vanax® DPG. CAS 102-06-7; N,N´-diphenyl guanidine; accelerator for NR and SR; secondary accelerator. [R.T. Vanderbilt]

Vanax® MBM. CAS 3006-93-7; m-Phenylenedimaleimide; accelerator; coagent in peroxide-cured polymers. [R.T. Vanderbilt]

Vanax® NS. CAS 95-31-8; N-t-Butyl-2-benzothiazole-

sulfenamide; accelerator; delayed action accelerator for natural and synthetic rubbers; lower scorch and faster cure than Durax; used in tires, mechanical and extruded goods. [R.T. Vanderbilt]

Vanax® PML. CAS 16971-82-7; Di-ortho-tolylguanidine salt of dicatechol borate; accelerator for NR and SR stocks, CR, neoprenes for wire and cable and mechanical goods; activator and mild antioxidant in NR and SBR. [R.T. Vanderbilt]

Van Bac®. Blend of mutated microorganisms; used for waste treatment in paper industry. [R T Vanderbilt]

Vanbeenol. CAS 121-32-4; Ethyl vanillin. [Bush Boake Allen Ltd]

Vancenase. CAS 5534-09-8; Beclomethasone dipropionate; glucocorticoid. [Schering Corp]

Vanceril. CAS 5534-09-8; Beclomethasone dipropionate; glucocorticoid. [Schering Corp]

Vanchem®. A series of organic compounds; corrosion inhibitors, chemical intermediates, adhesion promoters, and crosslinking agents for petroleum and other industries. [R T Vanderbilt]

Vanchem® HM-50. Aromatic polyisocyanate in monochlorobenzene; adhesion promotion or primer or crosslinking agent. [R T Vanderbilt]

Vanchem® HM-4346. Aromatic polyisocyanate in toluene; adhesion promotion or primer or crosslinking agent. [R T Vanderbilt]

Vancide®. A series of antimicrobial agents and fungicides; industrial preservatives for latex, paint, paper, cutting fluids, coolants, ceramics, plastics, household products, agriculture etc. [R T Vanderbilt]

Vancide® 51. Sodium dimethyldithiocarbamate (27.6%) and sodium 2-mercaptobenzothiazole (2.4%); fungicide for use as a preservative in latex and starch paste; bactericide for sol. cutting fluids and coolants; paper mill slimicide; used in petrol. storage tanks, recirculating cooling towers, paper and paperboard, and cotton fabric. [R.T. Vanderbilt]

Vancide® 51Z. Zinc dimethyldithiocarbamate (87%) and zinc 2-mercaptobenzothiazole (7.5%); fungicide for use in neoprene compositions; also for preservation of adhesives, for industrial cooling water slime control, in sanitizing cleansing compounds, for textile mildew and bacterial growth inhibition, and as mold inhibitor in caulking compounds. [R.T. Vanderbilt]

Vancide® 89. CAS 133-06-2; Captan; fungicide for natural and synthetic rubber compounds containing susceptible plasticizers; industrial preservative for vinyl, polyethylene, paint, lacquer, soap, wallpaper flour paste. [R.T. Vanderbilt]

Vancide® MZ-96. CAS 137-30-4; Zinc dimethyldithiocarbamate; antimicrobial, preservative for starch and synthetic latex adhesives and food packaging adhesives. [R.T. Vanderbilt]

Vanclay. CAS 1332-58-7; Kaolin; adsorbent. [R T Vanderbilt]

Vancocin Hydrochloride. CAS 1404-93-9; Vancomycin hydrochloride; antibacterial. [Eli Lilly & Co]

Vancorr®. A series of metal salts of alkylaryl sulfonate in solvent; corrosion inhibitor for paint. [R T Vanderbilt]

Vancote®. Antimicrobial agent; industrial preservative for coatings. [R T Vanderbilt]

Vancure D.A.A. CAS 95-33-0; N-Cyclohexyl-2-benzthiazyl sulfonamide; a proprietary accelerator. [K & K Greeff

---

‡ Trade name and manufacturer not verified   § Trade name without identified manufacturer

Chemicals Ltd] *

Vandar®. Elastomer modified polybutylene terephthalate. [Hoechst UK]

Vandar® 2100. CAS 26062-94-2; Thermoplastic alloy (elastomer-modified PBT); outstanding ductility, high impact, excellent paintability, chemical resist., very low water absorp.; designed to withstand strong abuse in service; used for automotive body components and housings, furniture, ski boots, appliances, clips, fasteners; 2100 is unfilled grade with max. impact strength at room/low temps. [Hoechst Celanese]

Vandex®. CAS 7782-49-2; Selenium powder; vulcanizing agent for rubber. [R T Vanderbilt]

Vandike 7085. A proprietary trade name for a vinyl acetate-butyl acrylate copolymer emulsion used in emulsion paints. [BOC Group plc] *

Vandike 7086. A proprietary trade name for a vinyl acetate-butyl acrylate copolymer emulsion; used in the manufacture of "nondrip" emulsion paints. [BOC Group plc] *

Vandike P.360. A proprietary trade name for a vinyl acetate-dioctyl maleate copolymer emulsion used in water-based adhesives. [BOC Group plc] *

Vandrox®. Carboxymethyl starch; viscosity increasing agent and water retention agent for coatings. [R T Vanderbilt]

vandura silk. (Gelatin Silk). An artificial silk prepared from gelatin and formaldehyde. It is also the name for a silk made from casein and formaldehyde.

Vanease (K). Hydrogenated polysorbate 80, glyceryl lactyl palmitate, sodium carboxymethyl cellulose, sodium propionate, glyceryl tristearate, sodium benzoate, acetic acid; hydrated emulsifier used in food industry to increase aeration, smoothness and stability of icings and fillings. [Am. Ingredients/Patco]

Van Ermengem's stain. Solution a) contains 1 g of osmic acid, 20 g tannin, 150 cc distilled water, and 8 drops glacial acetic acid. Solution b) is a 0.25-0.5% silver nitrate solution. Solution c) contains 6 g tannin, 1 g gallic acid, 20 g sodium acetate, and 700 cc water; a microscopic stain.

Vanesta®. Polyoxyethylene fatty acid ester; starch dispersion stabilizer for use in coatings and laminating adhesives. [R T Vanderbilt]

Vanex. Blended organic and inorganic acids in combination with surfactants, wetting agents, and inhibitors; cleaner for brick in new construction that may contain vanadium or metals. [Nova Chemical Inc] *

Vanfre®. A series of proprietary formulations; processing aids for elastomers. [R T Vanderbilt]

Vanfre® DFL. Processing aid for mold lubrication; corrosion inhibitor. [R.T. Vanderbilt]

Vanfre® IL-1. Sodium alkyl sulfates; internal lubricant for CR, NBR, CSM, and NR; improves flow charactersitics of highly loaded compounds; improves release from mill rolls; provides smoother extrusions; disperses easily. [R.T. Vanderbilt]

Vangard. CAS 13684-63-4; 114 g/liter phenmedipham EC; selective herbicide for annual weeds in cultivated beet species and strawberries. [Farmers Crop Chemicals Ltd]

Van Gel®. CAS 1327-43-1; Processed magnesium aluminum silicate (smectite); used as thickener, viscosity stabilizer, dispersion adjuster and mineral filler for rubber, paint, household products, ceramics and plastics.

[R T Vanderbilt]

Van Gel® B. Magnesium aluminum silicate (smectite); thickener and visc. stabilizer for dispersions; for industrial and agricultural uses. [R.T. Vanderbilt]

Vanguard. CAS 13684-63-4; Emulsifiable concentrate containing 114 g/l phenmedipham; used for weed control for beet crops. [Farmers Crop Chemicals Ltd]

vanilla. CAS 121-33-5; The cured unripe fruit of *Vanilla planifolia*; used as an aromatic and flavoring material.

vanillin. (benzaldehyde, 4-hydroxy-3-methoxy-; 4-hydroxy-3-methoxybenzaldehyde) Substituted aromatic aldehyde; methyl ether of protocatechuic aldehyde; $(CH_3O)(OH)C_6H_3CHO$; CAS 121-33-5; EINECS 204-465-2; perfumes, flavorings, pharmaceuticals, lab reagent, source of L-dopa. [Penta Mfg.; Schweizerhall; Trafford Chem. Ltd]

Vanisperse CB. Fractionated sodium salt of oxylignin; high dispersing power; broad range of applications. [Borregaard LignoTech]

Vanitox. Selective weed killer. [May & Baker Ltd] *

Vankalite. A proprietary trade name for a beryllium-copper alloy used for setting diamonds in drills. §

Vanlube®. A full line of antioxidants, antiwear, extreme pressure additives, metal deactivators, and friction reducers for petroleum lubricants; multifunctional additives for lubricants, functional fluids, and fuels. [R T Vanderbilt]

Vanobid. CAS 1403-17-4; Candicidin; antifungal. [Merrell Dow Pharmaceuticals Inc] *

Vanodine. Iodophors. [Evans Vanodine International Ltd]

Vanox®. A full line of amine and phenol rubber antioxidants; antioxidants for rubber, elastomers and polymers. [R T Vanderbilt]

Vanox® 3C. CAS 101-72-4; N-Isopropyl-N´-phenyl-p-phenylenediamine; antioxidant, antiozonant for dk. colored NR and SR. [R.T. Vanderbilt]

Vanox® 6H. CAS 101-87-1; N-Cyclohexyl-N´-phenyl-p-phenylenediamine; antioxidant, antiozonant for dk. colored NR and SR. [R.T. Vanderbilt]

Vanox® 12. CAS 101-67-7; p,p´-Dioctyldiphenylamine; antioxidant for elastomers used in adhesives, hot melts. [R.T. Vanderbilt]

Vanox® 1290. CAS 35958-30-6; 2,2´-Ethylidene bis(4,6-di-t-butylphenol); antioxidant; oxidative inhibitor for polymers; process stabilizer for polyolefins; stabilizer for PU and PS. [R.T. Vanderbilt]

Vanox® 1320. CAS 17540-75-9; 2,6-Di-t-butyl-4-sec-butyl phenol; antioxidant for PU foam; oxidation inhibitor and scorch preventer for mfg. and storage of bun stock; stabilizer for PU foam. [R.T. Vanderbilt]

Vanox® AM. CAS 9003-79-6; Diphenylamine and acetone low-temp. reaction product; antioxidant for NR and SR. [R.T. Vanderbilt]

Vanox® AT. CAS 68411-20-1; Butyraldehyde-aniline condensation product; antioxidant for CR, NR, SBR, EPDM stocks, and latexes; activator for thiuram and thiazoles. [R.T. Vanderbilt]

Vanox® GT. CAS 27676-62-6; Tris(3,5-di-t-butyl-4-hydroxy benzyl) isocyanurate; antioxidant for PP, polyethylene, PU, and polymers. [R.T. Vanderbilt]

Vanox® MTI. CAS 53988-10-6; 2-Mercaptotoluimidazole; antioxidant for NR, SR; synergist with Agerite antioxidants. [R.T. Vanderbilt]

Vanox® NBC. CAS 13927-77-0; Nickel di-n-butyldithio-

---

* Trade name not verified as available † Trade name verified as obsolete

carbamate; antioxidant-antiozonant for SBR, NBR, CR, CSM, and ECO. [R.T. Vanderbilt]

Vanox® ODP. CAS 101-67-7; Dioctylated diphenylamine; general purpose antioxidant for all elastomers. [R.T. Vanderbilt]

Vanox® PCX. CAS 128-37-0; 2,6-Di-t-butyl-4-methylphenol; oxidation inhibitor in soaps and cosmetics. [R.T. Vanderbilt]

Vanox® SKT. CAS 34137-09-2; 3,5-Di-t-butyl-4-hydroxyhydrocinnamic acid triester of 1,3,5-tris (2-hydroxyethyl)-s-triazine-2,4,6-(1H,3H,5H)-trione; antioxidant; stabilizer for polyolefins; hot melt and food packaging applications. [R.T. Vanderbilt]

Vanox® SWP. CAS 85-60-9; 4,4´-Butylidene-bis-(6-t-butyl-m-cresol); antioxidant for natural and synthetic latexes; nonstaining. [R.T. Vanderbilt]

Vanox® ZMTI. CAS 61617-00-3; Zinc 2-mercaptotoluimidazole; antioxidant for NR and SR, EPDM, nitrile stock; synergist with Agerite antioxidants. [R.T. Vanderbilt]

Vanoxy. Epoxy resins; uses include protective coatings, laminates, adhesives, castings, tooling, flooring, surfacing, potting and encapsulating. [R T Vanderbilt]

Vanplast® . Plasticizers and peptizers for rubber. [R T Vanderbilt]

Vanplast® 201. Barium dinonylnaphthalene sulfonate; corrosion inhibitor used in automotive rubber protective coatings. [R T Vanderbilt] †

Vanquin. A proprietary preparation containing viprynium carbonate; an anthelmintic. [Parke-Davis] *

Vanquish. Bactericidal detergent. [ICI Chem. & Polymers Ltd]

Vanseal® . A series of sarcosinate type surfactants; surfactants for cosmetic and toiletry applications. [R T Vanderbilt]

Vanseal® 35. CAS 61791-59-1; EINECS 263-193-2; Sodium cocoyl sarcosinate; biodeg. industrial grade surfactant with outstanding mildness, lather building, and conditioning properties; compatible with cationics; used for soaps, bath gels, shampoos, shaving creams, dentifrices, textile and leather processing. [R.T. Vanderbilt]

Vanseal® CS. CAS 68411-97-2; EINECS 270-156-4; Cocoyl sarcosine; biodeg. surfactant, foaming and wetting agent, detergent, foam booster for soaps, bath gels, shampoos, shaving creams, dentifrices, rug shampoos, oven cleaners, dishwash, textile/leather processing; offers tolerance to hard water, mildness. [R.T. Vanderbilt]

Vanseal® LS. CAS 97-78-9; EINECS 202-608-3; Lauroyl sarcosine; biodeg. surfactant, foaming and wetting agent, detergent, foam booster for soaps, bath gels, shampoos, shaving creams, dentifrices, rug shampoos, oven cleaners, dishwash, textile/leather processing; offers tolerance to hard water, mildness. [R.T. Vanderbilt]

Vanseal® NACS-30. CAS 61791-59-1; EINECS 263-193-2; Sodium cocoyl sarcosinate; biodeg. surfactant, foaming and wetting agent, detergent, foam booster for soaps, bath gels, shampoos, shaving creams, dentifrices, rug shampoos, oven cleaners, dishwash, textile/leather processing; offers tolerance to hard water, mildness. [R.T. Vanderbilt]

Vanseal® NALS-30. CAS 137-16-6; EINECS 205-281-5; Sodium lauroyl sarcosinate; biodeg. surfactant, foaming and wetting agent, detergent, foam booster for soaps, bath gels, shampoos, shaving creams, dentifrices, rug shampoos, oven cleaners, dishwash, textile/leather processing; offers tolerance to hard water, mildness. [R.T. Vanderbilt]

Vanseal® OS. CAS 110-25-8; EINECS 203-749-3; Oleoyl sarcosine; biodeg. surfactant, foaming and wetting agent, detergent, foam booster for soaps, bath gels, shampoos, shaving creams, dentifrices, rug shampoos, oven cleaners, dishwash, textile/leather processing; offers tolerance to hard water, mildness. [R.T. Vanderbilt]

Vanseb® . Sulfur, salicylic acid; dandruff shampoo. [Allergan, Inc]

Vanseb-T® . Coal tar solution, sulfur, salicylic acid; tar dandruff shampoo. [Allergan, Inc]

Vansil. CAS 21738-42-1; A proprietary preparation of oxamniquine; an antihelmintic. [Pfizer International]

Vansil® . CAS 13983-17-0; Calcium silicate (wollastonite); functional filler, reinforcing agent, bright color used in adhesives, ceramics, elastomers, insulating materials, cosmetics, plastics, paint, resins, sealants, and wallboards. [R T Vanderbilt]

Vansil® W-10. CAS 13983-17-0; Wollastonite; extender pigment for solvent-thinned and latex paints. [R.T. Vanderbilt]

Vanstay®. Stabilizers for polyvinyl chloride. [R T Vanderbilt]

Vantac. Acrylic emulsion pressure sensitive adhesives. [Rhone-Poulenc UK]

Vantalc® . CAS 14807-96-6; Hydrous magnesium silicate (industrial talc); extender and filler used primarily in paints; also used in paper, rubber and plastics. [R T Vanderbilt]

Vantard® . A series of retarder compounds; retarder to control scorch in processing of elastomers. [R T Vanderbilt]

Vantoc. Industrial disinfectants and bactericides. [ICI Chem. & Polymers Ltd] *

Vantoc AL. Aqueous blend of higher alkyl trimethyl ammonium bromide; pale, straw colored liquid; bactericide in the brewing and food processing industries. [ICI Chem. & Polymers Ltd]

Vantoc CL. CAS 139-07-1; EINECS 203-351-5; Lauryl dimethyl benzyl ammonium chloride in aqueous solution; used for general disinfection of plants and equipment in brewing, soft drinks and foodstuffs industries. [ICI Chem. & Polymers Ltd]

Vantocil. Biocides and bactericides. [ICI Chem. & Polymers Ltd]

Vantropol. Detergent/sanitizer. [ICI Chem. & Polymers Ltd]

Vanwax® . Protective wax blends; used for sunlight and ozone protection of rubber. [R T Vanderbilt]

Vanzak® . Nonionic surfactants; used in pulp and paper manufacturing for pitch control. [R T Vanderbilt]

Vanzyme®. CAS 9000-92-4; α-Amylase; starch converting enzymes used in paper manufacturing. [R T Vanderbilt]

Vapo-Iso. Isoproterenol hydrochloride; adrenergic. [Fisons Corp] *

Vapona. CAS 62-73-7; A proprietary preparation of dichlorvos; an insecticide. §

Vaporole. A proprietary formulation of amyl nitrite; indicated for rapid relief of angina pectoris due to coronary artery disease. [The Wellcome Foundation Ltd]

varac. See kelp.

Varamide® A-7. CAS 93-83-4; EINECS 202-281-7; Oleamide DEA (2:1); rust inhibitor and base for o/w emulsifiers, detergents, anticorrosive cleaners, and

---

‡ Trade name and manufacturer not verified   § Trade name without identified manufacturer

thickener for waterless hand cleaners; degreasers; emulsifier for oils in fiber and yarn lubricants; foam suppressor in solv. and dye carrier emulsions. [Sherex/Div. of Witco]

Varamide® C-212. CAS 68140-00-1; Cocamide MEA (1:1); foam booster and stabilizer in aq. systems; degreaser; hair conditioning agent; visc. modifier; for household and industrial applications. [Sherex/Div. of Witco]

Varamide® LO-1. Linoleamide DEA (1:1); thickener and foam stabilizer for shampoos, baby bath, and hand soap; conditioning agent. [Sherex/Div. of Witco]

Varamide® MA-1. Refined cocamide DEA (1:1); foam stabilizer and booster, thickener; basic liq. superamide for shampoos, bubble bath, and dishwasher; low cost equivalent to lauric superamide; does not require melting; gives higher visc. and foam stability than conventional 2:1 alkanolamides in liq. detergent systems. [Sherex/Div. of Witco]

Varamide® ML-1. CAS 120-40-1; EINECS 204-393-1; Lauramide DEA (1:1); thickener and foam stabilizer for shampoo, bubble bath, and hand laundry detergent; gives the highest visc., foam level, and stability of the superamides in series. [Sherex/Div. of Witco]

Varamide® T-55. Tallow MEA ethoxylate; detergent base for floor and hard-surface cleaners; tolerant to high builder levels and hard water. [Sherex/Div. of Witco]

Varbian. Cardiac stimulant. [Ciba plc] *

Varcum® 1198. Alkyl phenolic resin; heat-reactive resin; curing agent for butyl rubber; modifier for pressure-sensitive adhesives based on SBR and natural rubber. [OxyChem/Durez]

varech. See vraic.

Variamine. Azoic dyestuffs. [Hoechst AG]

Variclene. Green aqueous gel containing 0.5% w/w brilliant green BP, 0.5% w/w lactic acid BP; an aid in the topical treatment of venous and other types of skin ulcers. [Dermal Laboratories Ltd]

Vari-Cut. Fiberod product approximately 1/8 in. diameter cut to lengths between 1/4 in. and 4 in.; thermoplastic molding compounds for compression, transfer and injection molding. Application areas include automotive, appliances, equipment, sporting, etc. [Polymer Composites] *

Varidase. A proprietary preparation streptokinase and streptodornase; a fibrinolytic drug. [Lederle Laboratories] *

Varifoam® SXC. TEA-lauryl sulfate, cocamidopropyl hydroxysultaine, lauramide DEA, methylparaben; high-foaming, cost-effective shampoo conc. [Sherex/Div. of Witco]

Varine C. CAS 61791-38-6; EINECS 263-170-7; Coco hydroxyethyl imidazoline; emulsifier, anticorrosive, raw material for surfactant product; shampoo base, penetrating oils, antistats, corrosion inhibitors, paints, printing inks, textiles, adhesives. [Sherex/Div. of Witco]

Varine O®. Oleyl hydroxyethyl imidazoline; emulsifier, anticorrosive for automotive body panels, raw material; shampoo base, penetrating oils, antistats, corrosion inhibitors, paints, printing inks, textiles, adhesives. [Sherex/Div. of Witco]

Varine O Acetate. Oleic imidazoline acetate; anticorrosive. [Sherex/Div. of Witco]

Varine T. CAS 61791-39-7; EINECS 263-171-2; Tall oil hydroxyethyl imidazoline; anticorrosive for automobile industry. [Sherex/Div. of Witco]

Variotin. CAS 19504-77-9; A proprietary trade name for pecilocin; used in the treatment of fungal skin infections. [Leo Laboratories] *

Variquat® 50MC. Benzalkonium chloride; germicide, algicide, disinfectant, sanitizer, deodorant; used in pesticides and mfg. of sanitizers; food processing, dairy, restaurant, industrial and household products. [Sherex/Div. of Witco]

Variquat® 50ME. Dimethyl alkyl (C12-C16) benzyl ammonium chloride (50%), ethyl alcohol (7.5%) in water; specialty quaternary, germicide used for disinfection and sanitizing for hospitals, beautician instruments, food processing plants. [Sherex/Div. of Witco]

Variquat® 638. CAS 70750-47-9; PEG-2 cocomonium chloride, IPA; detergent booster, antistat, emulsifier for hard surface cleaners, other liq. detergents, textiles; plating bath foam blanket; base for hair conditioners, creme rinses, antistats; coemulsifier. [Sherex/Div. of Witco]

Variquat® B200. CAS 56-93-9; Benzyltrimethyl ammonium chloride; dispersant, dye leveler and retarder, emulsifier used in textile industry. [Sherex/Div. of Witco]

Variquat® K1215. CAS 61791-10-4; PEG-15 cocomonium chloride; emulsifier, antistat for personal care products; textile antistat. [Sherex/Div. of Witco]

Varisoft® 2TD. CAS 68910-56-5; Ditridecyl dimethyl ammonium chloride, aq. ethanol; softener, conditioner, base for hair conditioners and cream rinses; especially for Afro-Amer. hair. [Sherex/Div. of Witco]

Varisoft® 5 TD. Ethoxylated di C12-18 ammonium chloride in propylene glycol; hair conditioner, antistat, emulsifier, softener for textured hair; good rinseability, excellent manageability and shine. [Sherex/Div. of Witco]

Varisoft® 110. Dihydrogenated tallowamidoethyl hydroxyethylmonium methosulfate, IPA/water; non-yellowing fabric softener conc. for home and commercial laundries, textile processing. [Sherex/Div. of Witco]

Varisoft® 110-PG. Dihydrogenated tallowamidoethyl hydroxyethylmonium methosulfate in aq. propylene glycol; base for hair conditioners and creme rinses; antistat. [Sherex/Div. of Witco]

Varisoft® 136-100P. Proprietary; quaternary for fabric softeners. [Sherex/Div. of Witco]

Varisoft® 250. CAS 112-02-7; EINECS 203-928-6; Cetrimonium chloride; base for hair conditioners/creme rinses; antistat, surfactant, emulsifier; hair grooming aids. [Sherex/Div. of Witco]

Varisoft® 300. CAS 112-02-7; EINECS 203-928-6; Cetrimonium chloride; base for hair conditioners and creme rinses; imparts softness and manageability without greasiness. [Sherex/Div. of Witco]

Varisoft® 355. CAS 112-02-7; EINECS 203-928-6; Cetrimonium chloride; base for hair conditioners and creme rinses; imparts softness and manageability without greasiness. [Sherex/Div. of Witco]

Varisoft® 432-100. CAS 1812-53-9; Dicetyl dimonium chloride; coemulsifier; antistat, conditioner for hair care preps., creme rinses. [Sherex/Div. of Witco]

Varisoft® 445. Methyl (1) hydrogenated tallow amidoethyl (2) hydrogenated tallow imidazolinium methosulfate; fabric softener conc. for home and commercial laundries, textile processing. [Sherex/Div. of Witco]

Varisoft® 461. CAS 61789-18-2; EINECS 263-038-9; Coco-trimonium chloride, IPA; base for hair conditioners and

---

cream rinses; used in hot oil treatments. [Sherex/Div. of Witco]

Varisoft® 462. CAS 61789-77-3; EINECS 263-087-6; Dicocodimonium chloride, aq. IPA; base for hair conditioners and creme rinses. [Sherex/Div. of Witco]

Varisoft® 470. Ditallowdimonium chloride, aq. IPA; base for hair conditioners and cream rinses; softens hair. [Sherex/Div. of Witco]

Varisoft® 471. Tallowtrimonium chloride, IPA; base for hair conditioners and cream rinses. [Sherex/Div. of Witco]

Varisoft® 475. CAS 86088-85-9; Quaternium-27, IPA; fabric softener conc. for home and commercial laundries, textile processing; hair conditioner. [Sherex/Div. of Witco]

Varisoft® 3690. Methyl (1) oleylamidoethyl (2) oleyl imidazolinium methosulfate, IPA; quaternary for laundry detergent-softeners. [Sherex/Div. of Witco]

Varisoft® BT-85. CAS 17301-53-0; Behentrimonium chloride; antistat, suspending agent for body and hand creams and lotions, hair conditioner. [Sherex/Div. of Witco]

Varisoft® BTMS. Cetearyl alcohol, behenyltrimonium methosulfate; self-emulsifying wax for hair and skin formulations. [Sherex/Div. of Witco]

Varisoft® CRC. Cetearyl alcohol, dicetyl dimonium chloride, stearamidopropyl dimethylamine; cream rinse conc. for formulating cream rinses and conditioners. [Sherex/Div. of Witco]

Varisoft® C SAC. Cetearyl alcohol, stearalkonium chloride, PEG-40 castor oil; conc. for formulating cream rinses and conditioners; based on Varisoft SDAC. [Sherex/Div. of Witco]

Varisoft® CTB-40. CAS 57-09-0; EINECS 200-311-3; Cetrimonium bromide; base for hair conditioners and creme rinses; antistat; conditioner for skin creams and lotions. [Sherex/Div. of Witco]

Varisoft® DHT. CAS 68002-59-5; Quaternium-18, IPA; hair conditioner and antistat in cream rinse concs. [Sherex/Div. of Witco]

Varisoft® LAC. CAS 112-00-5; EINECS 203-927-0; Lauryltrimonium chloride; base for hair conditioners and cream rinses. [Sherex/Div. of Witco]

Varisoft® OIMS. Quaternium-81, aq. IPA; base for hair conditioners, creme rinses, antistats; coemulsifier; curl retention properties. [Sherex/Div. of Witco]

Varisoft® PIMS. Quaternium 27 in propylene glycol; base for hair conditioners and creme rinses. [Sherex/Div. of Witco]

Varisoft® SDAC-W. CAS 122-19-0; EINECS 204-527-9; Stearalkonium chloride; base for hair conditioners; imparts softness, manageability, antistatic properties. to hair. [Sherex/Div. of Witco]

Varisoft® ST-50. CAS 112-03-8; EINECS 203-929-1; Steartrimonium chloride aq. alcohol solution; substantive quaternary imprating softness and manageability to hair. [Sherex/Div. of Witco]

Varisoft® TA-100. CAS 107-64-2; EINECS 203-508-2; Distearyldimonium chloride; coemulsifier, antistat, conditioner for hair and skin care products, pigmented cosmetics. [Sherex/Div. of Witco]

Varisoft® TC-90. CAS 52467-63-7; Tricetylmonium chloride; base for hair conditioners, creme rinses, 2-in-1 shampoos; antistat. [Sherex/Div. of Witco]

Varisoft® TIMS. CAS 68122-86-1; Quaternium-27; base for creme rinses and conditioners; antistat, substantivity agent, conditioner. [Sherex/Div. of Witco]

Varisoft® TSC. CAS 112-03-8; EINECS 203-929-1; Steartrimonium chloride; base for hair conditioners. [Sherex/Div. of Witco]

Variton. CAS 62-97-5; Diphemanil methylsulfate; anticholinergic. [Schering Corp] *

Varitox. Sodium trichloroacetate. [May & Baker Ltd] *

varnish. A mixture of boiled oil with various gum resins, and oil of turpentine; used as protective coatings.

Varnodag. A trademark for a varnish made from phenolformaldehyde synthetic resin with colloidal graphite. §

varnoline. A petroleum distillate used as a lubricant. §

var oil. See aix oil. §

Varonic® 32-E20. CAS 9004-98-2; Oleth-20; emulsifier, stabilizer. [Sherex/Div. of Witco]

Varonic® 63 E20. CAS 68439-49-6; Ceteareth-20; emulsion stabilizer, emulsifier, solubilizer, moisturizer, and emollient for creams and lotions; surfactant for stick formulations and hair conditioners. [Sherex/Div. of Witco]

Varonic® BD. Cetearyl alcohol, ceteareth-20; self-emulsifying wax, visc. modifier for hair conditioners, creams and lotions; emulsifier, solubilizer. [Sherex/Div. of Witco]

Varonic® BG. Stearyl alcohol, ceteareth-20; self-emulsifying waxe, visc. modifier for hair conditioners, creams and lotions; emulsifier, solubilizer. [Sherex/Div. of Witco]

Varonic® DM55. Methyl capped glycol ether; solv. with low toxicity and excellent grease cutting properties for surface cleaners; textile scouring. [Sherex/Div. of Witco]

Varonic® K202. CAS 61791-14-8; PEG-2 cocamine; emulsifier, antistat; corrosion inhibitor in metal finishing (e.g., as cutting oil additives); detergent, antifouling, antistalling, and deicing agent in gasoline; also in textile lubricants, oil field emulsification. [Sherex/Div. of Witco]

Varonic® LI-42. CAS 51158-08-8; PEG-20 glyceryl stearate; low-irritation detergent, emulsifier, lubricant, solubilizer for household and industrial applications, personal care products. [Sherex/Div. of Witco]

Varonic® LI-48. PEG-80 glyceryl tallowate; emulsifier, solubilizer, thickener, dispersant, antiirritant surfactant used in household and industrial applications, personal care products. [Sherex/Div. of Witco]

Varonic® LI-63. CAS 68201-46-7; PEG-30 glyceryl cocoate; low-irritation detergent, emulsifier, solubilizer for soaps, specialized lubricants, personal care products. [Sherex/Div. of Witco]

Varonic® Q202. PEG-2 oleamine; anticorrosive emulsifier for metalworking, grinding oils. [Sherex/Div. of Witco]

Varonic® Q-230. PEG-30 oleamine; compatiblizer or antiprecipitant in dye bath of acid and cationic dyes; mild dye leveler and/or stripping agent for acid dyes; hydrophilic emulsifier. [Sherex/Div. of Witco]

Varonic® S202. CAS 10213-78-2; PEG-2 stearamine; polymer additive and emulsifier. [Sherex/Div. of Witco]

Varonic® T202. CAS 61791-44-4; PEG-2 tallowamine; lubricant, softener, scouring aid, dye leveler and antistat for textiles; in synthetic latex paints; emulsifier for latex, dyes, and oils; dispersant; acid cleaners; process modifier in polymer industry; raw material for quaternary and amphoteric surfactants. [Sherex/Div. of Witco]

Varonic® T-220. CAS 61791-44-4; PEG-20 tallowamine; acid dye leveler for nylon; antiprecipitant for mixed dye baths; migrating agent for dispersed dyes; wool lubricant and antistat; antistat for spin finishes. [Sherex/Div. of Witco]

Varonic® U-215. CAS 61791-26-2; PEG-15 hydrogenated tallowamine; nylon leveling agent; antiprecipitant for mixed dye baths; migrating agent for dispersed dyes. [Sherex/Div. of Witco]

Varox®. Organic peroxides; curing agents and crosslinking agents for elastomers and polymers. [R T Vanderbilt]

Varox® 365. Lauramine oxide; detergent, foam booster/ stabilizer for anionic surfactants for shampoo and detergent systems, textiles; hypochlorite-stable. [Sherex/Div. of Witco]

Varox® 1770. CAS 68155-09-9; Cocamidopropylamine oxide; mild detergent, foam booster for liq. soaps, shampoos, textile scouring. [Sherex/Div. of Witco]

Varsol®. General purpose hydrocarbon solvents. [Exxon Int'l.; Exxon UK] *

Varsol® 1. CAS 74742-48-9; Aliphatic hydrocarbon solv. [Exxon] *

Varsol® 18. CAS 64741-92-0; 64742-48-9; Aliphatic hydrocarbon solv. [Exxon] *

Varsulf® S-1333. Disodium ricinoleamido MEA-sulfosuccinate aq. disp.; detergent, refatting agent used in dishwash, liq. soaps, and personal care products; anti-irritant for other surfactants. [Sherex/Div. of Witco]

Varsulf® SBF-12. Disodium lauryl sulfosuccinate; low-irritation foaming agent for shampoos, bubble bath, body cleansers; some conditioning and moisturizing effects; detergent for fine fabric wash systems. [Sherex/Div. of Witco]

Varsulf® SBFA-30. Disodium laureth sulfosuccinate aq. disp.; detergent, refatting agent used in dishwash, fine fabric wash, liq. soaps, and personal care products. [Sherex/Div. of Witco]

Varsulf® SBL-203. CAS 25882-44-4; EINECS 247-310-4; Disodium lauramido MEA-sulfosuccinate aq. disp.; detergent, refatting agent used in dishwash, light duty detergents, personal care products, rug and upholstery shampoos. [Sherex/Div. of Witco]

Vasagel. Thickening agent; used to thicken automatic dishwashing detergents. [Production Chemicals Ltd]

Vascardin. A proprietary preparation of sorbide nitrate; a vasodilator used for angina pectoris. [Nicholas Laboratories Ltd] *

Vascoloy-Ramet D. A proprietary trade name for a corrosion-resisting alloy of 80% tantalum carbide, 20% tungsten and nickel. §

Vasconite BT. A suspension of magnesium compounds and combustion catalysts; a proprietary anticorrosion agent added to poor-quality boiler fuels. [Gamlen Chemical Industries SA] *

Vascuals. CAS 59-02-9; Vitamin E supplement. [USV Pharmaceutical Corp] ‡

Vasculit. CAS 5716-20-1; The sulfate of 2-n-butylamino-1-(4-hydroxyphenyl)ethanol (bamethan sulfate). [Boehringer Ingelheim GmbH] *

Vasocidin® Ophthalmic Solution and Ointment. Sulfacetamide sodium and prednisolone sodium phosphate; indicated for steroid-responsive inflammatory ocular conditions for which a corticosteroid is indicated and where superficial bacterial ocular infection or a risk of bacterial ocular infection exists. [Iolab Corp]

Vasoclear. CAS 550-29-2; Naphazoline hydrochloride; adrenergic. [CooperVision Inc] ‡

Vasocon. Naphazoline hydrochloride; adrenergic. [CooperVision Inc] ‡

Vasocort. A proprietary preparation of hydrocortisone, hydroxyamphetamine hydrobromide and phenylephrine hydrochloride; used as a nasal spray in cases of allergic rhinitis. [SmithKline Beecham] *

Vasodilan. CAS 395-28-8; Isoxsuprine; promotes blood flow rate. [Duphar BV] *

Vasogen®. Zinc oxide, calamine, dimethicone; used for treatment of diaper rash. [Pharmax Ltd]

Vasogen. A trademark for an ointment vehicle; an oxygenated petroleum. Parogen. §

Vasolastine. A proprietary preparation of lipoxydase citrogenase, amino acid oxydase, and tyrosinase complex. [FAIR Laboratories] ‡

vasoliment. See Parogen.

Vasomotal. CAS 5638-76-6; Betahistine; counteracts disorders of the inner ear, Menire's disease. [Duphar BV] *

Vasosulf® Ophthalmic Solution. Sulfacetamide sodium and phenylephrine hydrochloride; indicated for the treatment of conjunctivitis, corneal ulcer, and other superficial ocular infections due to susceptible microorganisms, and an adjunctive in systemic sulfonamide therapy of trachoma. [Iolab Corp]

Vasotec. CAS 76095-16-4; Enalapril maleate; for the treatment of hypertension and congestive heart failure. (U.S.) [Merck & Co Inc]

Vasotran. CAS 579-56-6; A proprietary preparation of isoxsuprine hydrochloride; used in the treatment of peripheral vascular disease. [Bristol-Myers Squibb Co Inc]

Vasoxine Injection. CAS 61-16-5; A proprietary formulation of methoxamine hydrochloride; counteracts systemic hypotension. [The Wellcome Foundation Ltd]

Vasoxyl. CAS 61-16-5; Methoxamine hydrochloride; adrenergic. [Burroughs Wellcome Co]

Vasoxyl Injection. CAS 61-16-5; A proprietary formulation of methoxamine hydrochloride; for supporting, restoring or maintaining blood pressure during anesthesia. [The Wellcome Foundation Ltd] †

Vassgro DSM. CAS 919-86-8; Emulsifiable concentrate containing 500 g demeton-S-methyl per liter; a systemic organophosphorus insecticide. [L W Vass (Agricultural) Ltd] †

Vassgro Flowable Sulphur. CAS 7704-34-9; Suspension concentrate containing 720 g/l sulfur; a protectant fungicide. [L W Vass (Agricultural) Ltd]

Vassgro Mini Slug Pellets. CAS 108-62-3; Pellets containing 6% w/w metaldehyde; snail and slug bait. [L W Vass (Agricultural) Ltd]

Vassgro Spreader. Nonionic wetting agent containing nonylphenol ethylene oxide condensates; for use with herbicide and insecticide sprays. [L W Vass (Agricultural) Ltd]

Vasylox. CAS 61-16-5; Methoxamine hydrochloride. [The Wellcome Foundation Ltd] †

Vatensol. CAS 551-48-4; A proprietary preparation of guanoclor sulfate; an antihypertensive drug. [Pfizer International] †

Vaucher's alloy. An alloy of 75% zinc, 18% tin, 4.5% lead, and 2.5% antimony.

Vauquelin's salt. A compound obtained by treating palladium chloride with ammonia, $[Pd(NH_3)_4]Cl_2 \cdot PdCl_2$.

Vaycron. Elastomer. [Hydro Polymers Ltd]

Vazo®. Vinyl polymerization catalyst. [DuPont UK]

VC20. Iodophor concentrate. [Evans Vanodine International Ltd]

---

* Trade name not verified as available   † Trade name verified as obsolete

V-Cil-K. A proprietary preparation containing phenoxy-methylpenicillin potassium; an antibiotic. [Eli Lilly & Co]†

V-Cil-K Sulpha. A proprietary preparation of phenoxy-methylpenicillin potassium and sulfadimidine; an antibiotic. [Eli Lilly & Co] †

V-Cillin. CAS 87-08-1; Penicillin V; antibacterial. [Eli Lilly & Co]

V-Cillin K. CAS 132-98-9; Penicillin V potassium; antibacterial. [Eli Lilly & Co]

Vebonol, Anabolic steroid veterinary ethical. [Ciba plc] *

Vec. A proprietary trade name for a vinylidene chloride synthetic resin. §

Vecortenol. Steroid veterinary ethical. [Ciba plc] *

Vecortenol-Vioform. Steroid quinoline topical veterinary ethicals. [Ciba plc] *

Vectal. Selective and total herbicide. [Schering Agrochemicals Ltd]

Vector®. Styrenic block copolymers (SIS, SBS); thermoplastic elastomers for adhesives, asphalt modification, polymer modification, molding, extrusion, health care products. [Dexco Polymers]

Vectra® A115. Liq. crystal polymer, low glass reinforced; features high tensile, chemical resist., virtually zero mold shrinkage, excellent elec. properties, easy processing for molding and extrusion; used in electronics, fiber optics, automotive, aircraft/aerospace, chemical processing, industrial, mfg. fields, encapsulation of electronic components; A115 is general purpose grade with very easy flow. [Hoechst Celanese/Engineering Plastics]

Vedacoll. Bitumen mastic. [Vedag GmbH] *

Vedacolor. Lacquer with an aqueous solvent and artificial resin base; for roof surface. [Vedag GmbH] *

Vedafix. Skirting rails. [Vedag GmbH] *

Vedaflex. Modified bitumen membrane; for sealing flat roofs and sloped roofs without additional surface protection. [Vedag GmbH] *

Vedaform. Bitumen shingles in different shapes and colors; for steep roofing. [Vedag GmbH] *

Vedag BM. Emulsion for the improvement of cement mortar. [Vedag GmbH] *

Vedagit. Bitumen plus filler; pebble-bedding compound for cold application. [Vedag GmbH] *

Vedagolan. Bitumen emulsions. [Vedag GmbH] *

Vedagully System. PU-integrated hard foam; for roof inlets. [Vedag GmbH] *

Vedagum. Bituminous crevice filler, approved for underlays; for hot application. [Vedag GmbH] *

Vedalith Facade System. Glass fiber-reinforced cement; bracket-mounted, respiratory, heat insulating system facade, shock-resistant, noncombustible. [Vedag GmbH] *

Vedaphalt. Bitumen emulsions; for road surfaces treatment. [Vedag GmbH] *

Vedaphon. Antidrone materials; for motor cars. [Vedag GmbH] *

Vedapor. Hard foam together with insulation strips; for roll form insulation. [Vedag GmbH] *

Vedapurit. PS or PU hard foam, sandwiched or non-sandwiched; for hard foam insulating boards. [Vedag GmbH] *

Vedasin. Universal sealing compound; for cellars, subgrade purposes. [Vedag GmbH] *

Vedastar. Domelight. [Vedag GmbH] *

Vedatect. Bitumen roof sheets with various inserts. [Vedag GmbH] *

Vedatex. Bitumen solvent glue; for steam barriers and heat insulating material. [Vedag GmbH] *

Vedathene. Self-adhesive sheet; for waterproofing systems. [Vedag GmbH] *

Vedatherm. Shingle thermal insulating board; for insulating sloped roofs. EPS with chipboard (V 100 G). [Vedag GmbH] *

Vedril. Polymethacrylates. [Montedison UK Ltd] *

Veecote®. CAS 12269-78-2; Hydrated aluminum silicate mineral (pyrophyllite); extender pigment for paints and coatings. [R T Vanderbilt]

Veegum®. Complex colloidal magnesium aluminum silicate derived from natural smectite clays; thickener, visc. modifier, stabilizer for emulsions, suspensions, sol'ns., liqs., creams, and pastes, cosmetics, toiletries, toothpaste, pharmaceuticals, paints, textile finishes, chemical specialties, industrial applications; suspending agent for powd. and pigments; binder for inorg. powd. and pigments; disintegrating agent in tablets; pigment dispersant; improves spreadability of lotions, creams, and ointments. [R.T. Vanderbilt]

Veegum® PRO. Tromethamine magnesium aluminum silicate; emulsion stabilizer and suspending agent for cosmetics, pharmaceuticals, veterinary products, chemical specialties, household products; superior soap and surfactant compatibility. [R.T. Vanderbilt]

veepa oil. (veppam oil, neem oil). Margosa oil, obtained from the seeds of *Melia Azadirachta.*

Veetids. CAS 132-98-9; Penicillin V potassium; an antibacterial. [Bristol-Myers Squibb Co Inc]

Vegamino 30-SF. Vegetable amino acids; moisture binding agent substantive to skin and hair; used in perms and for chemically treated hair. [Brooks Industries]

vegetable alkali. Potassium hydroxide, KOH.

vegetable black. A very light lamp-black containing 99% carbon.

vegetable butter. (lactine, vegetaline, cocoaline, laureol, nucoline, albene, palmine, cocose). Names for an edible fat prepared from coconut oil and palm nut oil; used in chocolate manufacture as a substitute for cocoa butter.

vegetable calomel. The resin of *Podo-phyllum.*

vegetable casein. Legumin, found in leguminous seeds.

vegetable ethiops. A form of charcoal obtained by the incineration of *Fuci.*

vegetable gelatin. Agar-agar.

vegetable glue. (aparatine). A glue obtained by treating starch with alkali; an adhesive.

vegetable gum. *See* British gum.

vegetable ivory. (Corajo). Tagua nut, the fruit of *Phytelephas macrocarpa* of South America.

vegetable jelly. Pectin (*qv*), found in vegetable juices.

vegetable oil. (oils, vegetable) CAS 68956-68-3; Expressed oil of vegetable origin consisting primarily of triglycerides of fatty acids; paints as drying oils, shortening, salad dressings; rubber softeners; dietary supplements. [Arista Industries; Jojoba Growers & Processors; Karlshamns; Lipo; Mendell; A.E. Staley Manufacture]

vegetable oil, hydrogenated. *See* hydrogenated vegetable oil.

vegetable pepsin. *See* papain.

vegetable rouge. Carthamin, the coloring matter of *Carthamus tinctorius* mixed with French chalk; used as a

---

‡ Trade name and manufacturer not verified   § Trade name without identified manufacturer

cosmetic.

vegetable soda. The general name for the ash of soda plants (land plants).

vegetable spermaceti. *See* Chinese wax.

vegetable tallow. The name applied to vegetable fats similar to tallow, such as Chinese tallow and Malabar tallow.

Vegetable Turf and Ornamental Weeder. CAS 1861-32-1; Dacthal; a preemergence sprayable herbicide that can be used in vegetable gardens on ornamentals and in turf areas; controls spurge. [Lawn & Garden Products Inc]

vegetable wax. *See* Japan tallow.

vegetable wool. A product obtained from green pine and fir cones by processes of fermentation, washing, and disintegration. It is mixed with cotton for the production of yarns.

vegetaline. The name given to a preparation of lactic acid; used in tanning processes for the removal of lime. It is obtained from the drainage water of preserve manufacture by evaporation and contains from 8.6-9.6% lactic acid.

Vegetoil. Vegetable oil plus emulsifiers; maximizes performance of pesticides. [Draxel Chemical Company] ‡

Vegolysen. Hexamethonium bromide. [May & Baker Ltd] *

Vegolysin T. A proprietary preparation of hexamethoniura tartrate. [May & Baker Ltd] *

velampishin. *See* wood-apple gum.

Velan. A proprietary product; it is a complex organic compound soluble in water which renders fabric fibers water repellant. §

Velbacil. CAS 37661-08-8; A proprietary preparation of bacampicillin hydrochloride; an antibiotic. [Pfizer International]

Velban. CAS 143-67-9; Vinblastine sulfate; antineoplastic. [Eli Lilly & Co]

Velbe. CAS 143-67-9; A proprietary trade name for the sulfate of vinblastine; used in the treatment of malignant diseases. [Eli Lilly & Co] †

Velcorin®. Dimethyl dicarbonate; an organic preservative for the cold sterilization of soft drinks based on fruit juices. [Bayer AG]

Velicren. Acrylic fiber; for sewing threads, carpets, and fur fabrics. [SNIA (UK) Ltd] †

Velon. A proprietary trade name for a vinylidene chloride synthetic resin. §

Velosan. A proprietary rubber vulcanization accelerator; it is aldehyde ammonia. §

Velosef. A proprietary preparation of cephraorine; an antibiotic. [Bristol-Myers Squibb Pharmaceuticals Ltd] *

Veloset. Catalysts for sodium silicate bonded sands. [Foseco (F.S.) Ltd]

Velpak. Seaweed based crop stimulant. [Bayer AG]

Velpar. CAS 51235-04-2; Soluble concentrate of 240 g hexazinone per liter; used for control of weeds in forestry plantations. [Selectokil Ltd]

Velpar®. CAS 51235-04-2; Soluble concentrate of 240 g hexazinone per liter; used for control of weeds in forestry plantations. [DuPont UK]

Velpeau's caustic powder. A caustic consisting of burnt alum and powdered savin tops.

Velsan® D8P-3. Isopropyl PPG-2-isodeceth-7-carboxylate; noncomedogenic emollient that helps to solubilize cosmetic actives; provides emulsion visc., wetting, and softer feel for skin care, sun care, hair care, bath, and pigmented products. [Sandoz]

Velsan® D8P-16. Cetyl PPG-2 isodeceth-7 carboxylate. emollient for skin care, sun care, hair care, pigmented products, bath products. [Sandoz]

Velsan® P8-3. Isopropyl C12-15 pareth-9-carboxylate; emollient for cosmetics providing emulsion visc., wetting, and softer feel to skin care, sun care, bath, and pigmented products. [Hercules; Sandoz]

Velsan® P8-16. Cetyl C12-15-pareth-9-carboxylate; emollient for cosmetics providing emulsion visc., wetting, and softer feel to skin care, sun care, bath, and pigmented products. [Sandoz]

Veltol. CAS 118-71-8; Maltol. [Pfizer Ltd] *

Veltol Plus. Ethyl maltol. [Pfizer Ltd]

Velva Coat. Foundry core or mold coating utilized to improve the surface finish of cast metals. [Ashland Chemical Company] *

Velva Dri. Foundry core and mold coatings implying refractories and chlorinated solvents; foundry core or mold coating utilized to improve the surface finish of cast metals. [Ashland Chemical Company] *

Velvalite. Foundry core and mold coatings implying refractories and alcohol; foundry core or mold coating utilized to improve the surface finish of cast metals. [Ashland Chemical Company] *

Velvaplast. Foundry core and mold coatings implying refractories and water solvent; foundry core or mold coating utilized to improve the surface finish of cast metals. [Ashland Chemical Company] *

Velva Wash. See Velvaplast. [Ashland Chemical Company]*

velvet black. A variety of gas carbon black.

velvet brown. *See* umber.

Velvetex. CAS 1333-86-4; A proprietary carbon black (thermatomic carbon) in a soft form used in rubber mixings. §

Velvetex® AB-45. CAS 68424-94-2; EINECS 270-329-4; Coco-betaine; surfactant, conditioner, emulsifier, solubilizer used in industrial use, liq. detergents, cleansing emulsions, personal care products; visc. builder, gelling agent; lime soap dispersant; frothing agent. [Henkel/ Cospha; Henkel Canada]

Velvetex® BA-35. Cocamidopropyl betaine; dispersant, foaming and wetting agent, antistat for household and industrial detergents; lime soap dispersant; biodeg.; stable in strong acid and alkaline solution. [Henkel/ Cospha; Henkel Canada]

Velvetex® CDC. Disodium cocoamphodiacetate; mild, high foaming surfactant for low-irritation shampoos, conditioners, body cleansers. [Henkel/Cospha; Henkel Canada]

Velvetol 77-19. Blend; for textile applications. [Rhone-Poulenc Surf.]

Velvet Veil 310. CAS 12001-26-2, 7631-86-9; Mica, silica; provides lubricious feel to skin for pressed and loose powds.; light stable; produces soft focus optical blurring of wrinkles and blemishes. [Presperse]

Venelbin®. Pharmaceuticals for the treatment of vein diseases. [Cassella AG]

Venice soap. An olive oil soap.

Venice turpentine. The oleo-resin of the larch, *Pinus larix.* It contains 20-22% essential oil and 74-80% rosin. A substance is often sold under this name consisting of a mixture of rosin oil, rosin, and turpentine; used in the

---

* Trade name not verified as available   † Trade name verified as obsolete

varnish industry.

Venit. Proteinaceous substances used as food or as a food ingredient. [Courtaulds plc]

Venite. A proprietary pyroxylin product. §

Ventilago Madraspanta. (Ouralpatti, Pitti, Lokandi). An Indian dyestuff obtained from a climbing shrub.

Ventolin. CAS 51022-70-9; Albuterol sulfate; bronchodilator. [Allen & Hanburys Ltd]

Venzar®. CAS 2164-08-1; Wettable powder containing 80% lenacil; used for control of annual dicoyledons and meadow grass in beet, fruit herbaceous perennials. [DuPont UK]

VeoVa. Resin intermediate. [Shell UK]

VeoVa 10 Monomer. A vinyl monomer containing the tertiary versatic structure can be copolymerized with vinyl acetate and used in all types of emulsion paints. [Shell] *

Vepesid. CAS 33419-42-0; Etoposide; antineoplastic. [Bristol Laboratories] *

veppam oil. See veepa oil.

Veracolate. A proprietary preparation of cascara, bile salts, and phenolphthalein; a laxative. [Warner] *

Veractil. A proprietary preparation of methotrimeprazine maleate; a tranquilizer. [May & Baker Ltd] *

Veractil. CAS 60-99-1; Methotrimeprazine. [Rhone-Poulenc Rorer Ltd]

Veracur. A proprietary preparation of formaldehyde used to treat warts. [Typharm] *

Verafil. Thermoplastic molding compounds. [Ciba plc] *

Veragel® 200. Aloe vera gel. [Dr. Madis Labs]

veratrine. CAS 62-59-9; 8051-02-3; A mixture of various alkaloids obtained from the seeds of *Veratrum sabadilla*. It causes sneezing and irritation when inhaled.

veratrole. (1,3-dimethoxybenzene; pyrocatechol dimethyl ether) CAS 91-16-7; EINECS 202-045-3; $C_8H_{10}O_2$; Used in medicine as an antiseptic.

Verazinc. CAS 7733-02-0; Zinc sulfate; astringent (ophthalmic). [O'Neal, Jones & Feldman Pharmaceuticals] *

Vercyte. CAS 54-91-1; Pipobroman; antineoplastic. [Abbott Laboratories] *

Verdalia A. Tricyclodecenyl methyl ether. [Quest Int'l. UK Ltd]

Verdict. A line of herbicides based primarily on haloxyfop. [Dow UK] *

verdigris. Basic copper accetate; it is usually a mixture of mono-, di- and tri-acetates of copper. Green verdigris consists chiefly of the basic acetate, $2(C_2H_3O_2)_2Cu_2O$. Blue verdigris consists mainly of the basic acetate, $(C_2H_3O_2)_2Cu_2O$. The various forms of verdigris are used in dyeing and calico printing and for the preparation of oil and water colors.

verdigris green. See verdigris.

Verdilyn. 1-Ethoxy-1-phenylethoxyethane. [Quest Int'l. UK Ltd]

Verdinal. 3,5,5-Trimethyl hexanal. [Quest Int'l. UK Ltd]

verditer blue. (verditer green, bremen green, mineral blue, bremen blue). An anhydrous basic copper carbonate, produced by the addition of sodium carbonate to a hot solution of copper sulfate or nitrate. Verditer green is an intermediate product; used for paper-staining; copper hydrate and copper carbonate are both sold under these names.

verditers. Highly basic copper carbonates.

Verdiviton. A proprietary preparation of sodium, calcium, potassium, and manganese glycerophosphates, and vitamin B complex. [Bristol-Myers Squibb Pharmaceuticals Ltd] *

Verdiviton Elixir. Multi vitamin and mineral mix in aqueous solution. [Bristol-Myers Squibb Pharmaceuticals Ltd] †

Verdley. Range of composts and soil conditioners based on peat or bark. [ICI Chem. & Polymers Ltd]

Verdone 2. Contains mecoprop and 2,4-D; selective lawn weedkiller. [ICI Garden Products]

Verdone CDA. Emulsifiable concentrate containing 6.7% 2,4-D, and 13.3% mecoprop; used to control weeds in grassland. [ICI Agrochemicals Professional Products]

Verdoracine. 4-Isopropyl-1-methyl-2-propenyl benzene. [Quest Int'l. UK Ltd]

Verdoxan. 2,2,5,5-Tetramethyl-4-isopropyl-1,3-dioxane; fragrance raw material for green or fruity notes. [Henkel/Cospha]

Verilite. An alloy of 96% aluminum, 2.5% copper, 0.7% nickel, 0.4% silicon, and 0.3% manganese.

Verinor. Precious metal dental casting alloy. [Degussa Ltd]

Veritas. Lube oil. [Chevron] *

Veriviton Elixir. Multi vitamin and mineral mixture in an aqueous solution. [Bristol-Myers Squibb Pharmaceuticals Ltd]

verjuice. The old name for the very sour juice of unripe green grapes, and of crabapples. It contains tartaric, racemic, and malic acids.

Vermadax. Combined ovine anthelmintic and fasciolacide. [RMB Animal Health Ltd]

vermiculite. CAS 1318-00-9; Hydrated magnesium-iron-aluminum silicate; lightweight concrete aggregate, insulation, sound conditioning, fireproofing, plaster, soil conditioner, fertilizer additive, seed bed, refractory, lubricant, oil-well drilling mud, filler in plastics, rubber and paint; absorbent; packing; carrier; animal feed additive [Filter-Media]

vermilion. See cinnabar.

vermilionettes. Red pigments. They are combinations of white lead or zinc white, and eosin.

Vermox. CAS 31431-39-7; A proprietary preparation containing mebendazole; antihelmintic. [Janssen Pharmaceutical Ltd] *

Vernafine. Organic pigment pastes; for paints. [Colour-Chem Ltd] *

Vernalin. Speciality dyestuffs for dyeing all types of leathers. [Colour-Chem Ltd] *

Vernamine Binders. Binders based on synthetic resin dispersions based on acrylic and other monomers; for leather. [Colour-Chem Ltd] *

Vernaminol Liquors. Synthetic oil derived from wax and hydrocarbons and their derivatives, free from fatty acids; for leather. [Colour-Chem Ltd] *

Vernasein. Extremely fine dispersions of organic pigments in suitable aqueous medium; for leather. [Colour-Chem Ltd] *

Vernasol. Disperse dyes; for polyester and polyester component in blended fibers and fabrics. [Colour-Chem Ltd]*

Vernatan. Synthetic tanning agents, comprising replacement syntans, resin tanning agents and acrylic syntans; for leather. [Colour-Chem Ltd] *

Vernol®. Specially pure nerol (cis-3,7-dimethylocta-2,6-dien-1-ol); sweet, fresh, citrus-rose odor used in fragrances. [Bush Boake Allen Ltd]

Vernol Liquors. Fat liquors derived from natural vegetable

---

‡ Trade name and manufacturer not verified   § Trade name without identified manufacturer

oils, animal fats, and synthetic esters; for leather. [Colour-Chem Ltd] *

Vernonite. Proprietary trade name for acrylic synthetic resins for denture bases, etc. §

Verofix®. A range of reactive dyestuffs; for the dyeing of wool and polyamide. [Bayer AG]

Versa-TL 3. Dispersant, crystal modifier, antiscalant, sludge preventive in boilers and cooling towers; calcium phosphate inhibitor; iron control aid; on-line cleaner; thermally stable. [Hart Chem. Ltd.]

Versa-TL® 4, TL 7. Sulfonated styrene/maleic anhydride. [Alco Chemical Corp]

Versa-TL 125. Polyelectrolyte polymer; dispersant; provides electroconductive properties for electrophotography and electrography; antistat for films and fibers. [Hart Chem. Ltd.]

Versabacs. A polycellular carpet-backing system. [Dow UK]*

Versaflex® 1. Versatate acrylic terpolymer emulsion; emulsion designed to prevent paint blistering and peeling soon after application and before the paint film is leached of its water extractables; used in latex paints for interior flats, primer sealers, exterior masonry, exterior wood, stains, and in caulks, sealants, and mastics. [W R Grace/Organics]

Versal 150. CAS 1344-28-1; Alumina; high-purity, high-performance catalytic grades; as catalyst support in washcoat slurries, extrudates, spheres, or tablets; also in abrasives, antislip paper coatings, polishes, ceramic binding agents and raw materials, scavengers. [La Roche Chem.]

Versalon® 1140. A trade name for a polyamide resin used as an adhesive between plasticized vinyl resins and metal. §

Versamag® DC. CAS 1309-42-8; Magnesium hydroxide; inorg. filler providing flame retardancy and smoke suppression to elastomers, plastics, and thermosets incl. EPDM, PP, PE, PVC; used in wire and cable compounds, conduit/tubing, film and sheet. [Morton Int'l./Plastics Additives]

Versamid®. Polyamide resins, solid thermoplastic and liquid reactable. [Cray Valley Ltd]

Versapen. CAS 3511-16-8; Hetacillin; antibacterial. [Bristol Laboratories] *

Versapen K. CAS 5321-32-4; Hetacillin potassium; antibacterial. [Bristol Laboratories] *

Versatic. Saturated tertiary monocarboxylic acid. [Shell UK]*

Versatint Fugitive tints. Water soluble, biodegradable colorants; tinting for fiber and yarn identification. [Milliken] *

Versatyl-42. Octyl acrylamide/acrylates copolymer; hairspray polymer for systems containing large proportion of hydrocarbon propellant; for aerosol and pump hairsprays, setting lotions, spritzes. [Nat'l. Starch]

Versed. Misonidazole; antiprotozoal. [Hoffmann-LaRoche Inc] *

Versene AG. Chelated micronutrients used by growers to provide their crops with these necessary nutrients: zinc, mananaganese, iron, copper and magnesium; these elements stimulate growth hormones and contribute to plant-crop health and yields. [Dow UK] *

Versene Acid. CAS 60-00-4; Edetic acid; pharmaceutical aid. [Dow UK] *

Versene CA. Edetate calcium disodium; chelating agent. [Dow UK] *

Versenol AG. Chelated micronutrients used by growers to provide their crops with these necessary nutrients - zinc, mananaganese, iron, copper and magnesium; these elements stimulate growth hormones and contribute to plant-crop health and yields. [Dow UK] *

Versicaine. Lignocaine hydrochloride. [May & Baker Ltd] *

Versicane. Lignocaine hydrochloride. [Rhone-Poulenc Rorer Ltd]

Versicon Conductive Polymer. Inherently conductive polymer for electromagnetic shielding, cable shielding, electrostatic control, conductive coatings, resistive heating, sensors, and other commercial applications [Allied-Signal]

Versiflex. A proprietary trade name for a transparent vinyl chloride acetate. §

Versilan. Strong soil penetrating action, detergency, and emulsification; hard surface cleaners; metal degreasing; laundry powders; raw wool scouring. [Henkel Chemicals Ltd] *

Versilan. Formulated surfactants. [Harcros UK]

Versilan MX134. EO/PO copolymer, modified; biodeg. moderate foaming, good wetting surfactant for highly built liqs., acid cleaners, steam cleaning, vehicle cleaning, general purpose floor and wall cleaners, descalers, concrete and aluminum cleaners, derusting agents. [Harcros UK]

Versilok® 201. Acrylic adhesive; structural adhesive bonding unprepared metals, plastics, and ceramics; excellent environmental and chemical resist.; semiflexible, medium cure, medium visc. [Lord]

Versilube® F50. Chlorophenyl methyl siloxane; lubricating fluid with excellent oxidative and thermal stability; for lubricating sliding, rolling, and antifriction mechanisms in hydraulic systems, vacuum pumps, servo motors, instruments, controls, and bearings. [GE Silicones]

Versilube® G-321. Dimethyl-diphenyl silicone grease; provides lubrication at temps. from -73 C to 204 C; for ball bearings used in maintenance-neglected locations. [GE Silicones]

Versilyt. Formulated surfactants. [Harcros UK]

Verstarktes chromammonit. An explosive containing 70 parts ammonium nitrate, 10 parts potassium nitrate, 12.5 parts trinitro-toluene, 7 parts chromium-ammonium alum, and 6 parts petroleum jelly.

Verstran. CAS 2955-38-6; Prazepam; sedative. [Parke-Davis] *

Vertal 92. CAS 14807-96-6; Platy talc; reinforcement in black PP compounds; dusting/parting agent for rubbers; filler in autobody compounds, caulks, putties, and sealants. [Luzenac Am.]

Vertal 200. CAS 14807-96-6; Talc; for general industrial, high impact applications. [Luzenac Am.]

Vertalec. *Verticillium lecanii;* a fungal parasite to control aphids and whitefly. [Koppert (UK) Ltd]

Vertan. Chelating agents; Vertan 600 controls the hardness of boiler feedwater; other Vertan chelating agents are used for removing water hardness deposits and metal oxide scale from industrial process equipment. [Dow UK]*

Vert d'eau. A jeweler's alloy containing 60% gold and 40% silver.

Verteion. 4-Methyl-2-(1-phenyl ethyl)-1-3-dioxolane.

---

* Trade name not verified as available   † Trade name verified as obsolete

[Quest Int'l. UK Ltd]

Vertifume. Fumigant containing carbon tetrachloride and carbon disulfide plus a fire inhibitor; used for controlling insects infesting stored grains; may be applied as a liquid to the grain mass or via a closed recirculating system; treated grain must be aired thoroughly before use. [Dow UK] *

Vertigon. CAS 84-02-6; A proprietary preparation of prochlorperazine maleate; an anti-emetic. [SmithKline Beecham] *

Vertigon Spansule Capsule. CAS 84-02-6; Pro-chlorperazine maleate; sustained-release preparation for use as tranquilizer, anti-emetic and vestibular sedative. [SmithKline Beecham] *

Vertocinth. 1-Ethoxy-1-(2-phenyl ethoxy) ethane; leafy green, floral odor used in fragrances. [Bush Boake Allen Ltd]

Verton®. Long fiber-reinforced thermoplastic compounds. [LNP; ICI Chemicals & Polymers Ltd]

Verton® OF-700-10. CAS 9016-75-5; PPS, 50% long glass fiber-reinforced; long fiber composite offering optimum mech. and thermal properties for demanding structural components in automotive, ordnance, chemical processing, and office equip. markets. [LNP]

Verton® RF-700-10 EM HS. CAS 32131-17-2; Nylon 6/6, 50% long glass fiber-reinforced. [LNP]

Verton® RF-7007 HS. CAS 32131-17-2; Nylon 6/6, 35% long glass fiber-reinforced; long fiber composite offering optimum mech. and thermal properties for demanding structural components in automotive, ordnance, chemical processing, and office equip. markets. [LNP]

Vertrel®. For electrical applications. [DuPont UK]

Verv®. CAS 5793-94-2; Calcium stearoyl-2-lactylate; starch and protein complexing agent, softener for use in yeast-leavened bakery products; conditioning agent in dehydrated potatoes. [Am. Ingredients/Patco]

Vesadin. Sulfadimidine dodium 33.3% injection. [RMB Animal Health Ltd]

vesaloin. See hexamine.

vesalvine. See hexamine.

Vespel®. A proprietary trade name for polyimide in the form of prefabricated parts. [DuPont UK]

Vesprin. CAS 1098-60-8; Triflupromazine hydrochloride; antipsychotic. [Bristol-Myers Squibb Co Inc]

Vestagon® B 31. Cyclic amidine; high gloss epoxy and hybrid powd. coatings. [Hüls Am.]

Vestamelt. A range of thermoplastic copolyesters; hot-melt adhesives for bonding textiles. [Hüls AG]

Vestamid®. A large range of polyamides and copolyamides; suitable for injection molding, extrusion, thermoforming, blow molding, rotational molding and fluidized bed coating. [Hüls AG]

Vestamid® D 14. Polyamide 6/12; molding compound. [Hüls AG] *

Vestamid® E40M-S3. Polyamide 12 elastomer; stabilized against hot air and sunlight; more flexible and more impact resist. than corresponding plasticized grades. [Hüls AG] *

Vestamid® L 1600. Polyamide 12; basic grade homopolymer suitable for injection moldings requiring easy melt flow. [Hüls AG] *

Vestamid® L 1833. Polyamide 12, long glass fiber-filled; molding compound. [Hüls AG] *

Vestamid® X 3500. Polyamide 12; antistatic and elec.

conductive molding compound. [Hüls AG] *

Vestamid® X 4178. Polyamide 12; self-extinguishing molding compound. [Hüls AG] *

Vestamin® IPD. Isophorone diamine; epoxy curative for coatings, adhesives, castings and composites. [Hüls Am.]

Vestamin® TMD. Trimethylhexamethylenediamine; epoxy curative for flexible coatings, adhesives, castings and composites. [Hüls Am.]

Vestanat® IPDI. CAS 4098-71-9; Isophorone diisocyanate; raw material for mfg. of light stable polyurethanes. [Hüls Am.]

Vestanat® T 1890 L. Polyisocyanate prepolymer; prepolymer for crosslinking of two-component polyurethane paints. [Hüls Am.]

Vestanat® TMDI. Trimethylhexamethylene diisocyanate; raw material for mfg. of flexible, light stable polyurethanes. [Hüls Am.]

Vestar®. Chemical for treating wearing apparel, threads, yarns and textile fabrics. [Dow Corning]

Vestenamer®. Polyoctylene; suitable for the manufacture of rubber blends for injection moldings, extruded products (profiles, hoses), calendered articles and tires. [Hüls AG]

Vestenamer® 6213. Polyoctenamer; nonstaining stabilizer; used as a blend component for other rubbers to improve plasticity, enhance filler incorporation and dispersion, improve flowability in extrusion, injection and compression molding, increase surface smoothness in calendering operations; also for hose production, tires, covulcanization. [Hüls Am.; Hüls AG]

Vesticoat® UT 647. Polyester-based isocyanate-terminated prepolymer; prepolymer for flexible moisture curing one-component PU coatings. [Hüls Am.]

Vestiform. High molecular polymethacrylate for processing polyvinyl chloride; particularly for improving the deep-draw properties of films and sheets. [Hüls AG] *

Vestinol. A range of alkyl phthalates; primary plasticizers for polyvinyl chloride and paints etc. [Hüls AG]

Vestinol AH. CAS 117-81-7; Dioctyl phthalate; a proprietary trade name for a vinyl plasticizer. [Hüls AG] *

Vestodur. A range of polybutylene terephthalate with and without various additives; Vestodur without fillers is suitable for injection molding and extrusion and Vestodur containing fillers is suitable for injection molding. [Hüls AG]

Vestogrip. Polyisoprene rubber. [Hüls AG]

Vestolen. High density polyethylene and polypropylene. [Hüls AG]

Vestolen® A3512. CAS 9002-88-4; HDPE; narrow molecular weight distribution polyethylene offering lowest dens. and highest stress cracking resistance in series, and high notched impact strength; used for tubular film, injection molded parts, and hollow moldings with high stress-cracking resistance; R 3512 F film grade avail. for high-strength shock-resistant film having a low permeability and high transparency, and suitable for the production of laminated as well as sterilized film. [Hüls AG] *

Vestolen® AS. A trademark for a flame resistant HDPE building material. [Hüls AG] *

Vestolen® AX4304. CAS 9002-88-4; HDPE; narrow molecular weight distribution grade. [Hüls AG] *

Vestolen® EM. A range of elastomer-modified polypropylene; wide range of possible uses in the automobile industry: bumper coverings, bumper corners and instru-

---

‡ Trade name and manufacturer not verified   § Trade name without identified manufacturer

ment panels. [Hüls AG] *

Vestolen® P2000 CR. CAS 9003-07-0; PP homopolymer; used for extrusion of fibers. [Hüls AG] *

Vestolen® P2000. CAS 9003-07-0; PP homopolymer; used for injection molding of packaging. [Hüls AG] *

Vestolen® P6000. CAS 9003-07-0; PP homopolymer; used for injection molding of automotive, elec., other tech. articles, packaging, medical goods. [Hüls AG] *

Vestolen® P6700. CAS 9003-07-0; PP block copolymer; used for injection molding of battery cases. [Hüls AG] *

Vestolen® P7032 G. CAS 9003-07-0; PP, reinforced; used for extrusion of sheets, compression molding of sheets, injection molding of automotive, elec., other tech. goods, and household appliances. [Hüls AG] *

Vestolit®. Polyvinyl chloride. [Hüls AG]

Vestolit® B 7021. CAS 9002-86-2; Microsuspension PVC; for paste processing; more suitable for dip coating and casting than pastes based on the Vestolit E series; suitable for mixing with grades of the E series for spread-coating. [Hüls AG] *

Vestolit® B 7090. CAS 9003-22-9; Vinyl chloride/vinyl acetate copolymer; for paste processing of floor coverings, carpet backings, adhesives/laminating coatings, and spray coating. [Hüls AG] *

Vestolit® E 6003, E 6503, E 7003. CAS 9002-86-2; Emulsion PVC; resin emulsion preferable used for processing with plasticizer; E 6003 used for calendering of floor coverings; E 6503 used for calendering of film/sheet and floor coverings, and extrusion of sections and hosepipes; E 7003 used for calendering of floor coverings and extrusion of sections and hosepipes. [Hüls AG] *

Vestolit® HI. High impact polyvinyl chloride; used for window sections, films, profiles and sheets. [Hüls AG] *

Vestolit® LF HI-SP 5735. CAS 9002-86-2; Rigid PVC; easy-flow for injection molding. [Hüls AG] *

Vestolit® M 5867. CAS 9002-86-2; Mass PVC; resin for processing without plasticizer and containing no residues of suspension aids; M 5867 is easy-flowing grade used for the production of injection-molded parts and hollow moldings by extrusion of structural foam furniture; and construction sections and thick sheet. [Hüls AG] *

Vestolit® P 1330 K. CAS 9002-86-2; PVC homopolymer; for paste processing for leathercloth, floor coverings, canvas materails, unsupported layers/films, dipping, casting, spray coating. [Hüls AG] *

Vestolit® S 6058. CAS 9002-86-2; Suspension PVC; resin suspension for processing without plasticizer; used for processing by injection molding (e.g., fittings), for the extrusion of thin-walled pipes and sections (e.g., elec. conduits, drinking straws), and for the production of rigid film and sheet. [Hüls AG] *

Vestopal. A range of unsaturated polyester resins; used for hand lay-up and compression molding, industrial moldings, for polymer-concrete and in fiber spraying processes. [Hüls AG]

Vestoplast. A range of predominantly amorphous olefin copolymers; used in hot melts, adhesives, anticorrosion strips, putties, sealing compounds, and road marking compounds. [Hüls AG]

Vestopren. Thermoplastic rubber; a rubber concentrate for improving the impact stength of polyolefins, especially polypropylene. [Hüls AG]

Vestoran® 1100. Polyphenylene ether; molding compound offering reduced shrinkage and warpage; produces di-

mensionally stable moldings which resist hydrolysis; used for instrument engineering, elec./electronic engineering, parts with large surface areas in automotive engineering and office equip. industry. [Hüls AG] *

Vestosint. Nylon 12 fluidized bed coating powders. [Hüls AG]

Vestowax. A range of Fischer-Tropsch waxes; suitable for use in hot melts, printing inks, lacquers and as aids for processing rubber and plastics. [Hüls AG]

Vesturit. A range of saturated polyesters for lacquers; used as binder components for stoving finishes containing solvents, for 'medium' and 'high' solids paints, for water soluble stoving finishes and as lacquer resins for highly flexible coatings. [Hüls AG] *

Vestypor. A range of expandable polystyrene; suitable for a wide range of molding applications. [Hüls AG]

Vestyron. A range of polystyrenes; suitable for a wide range of molding applications. [Hüls AG]

Vestyron 550. Polystyrene containing low residual monomer; a packaging material. [Hüls AG] *

Vestyron 551. Polystyrene packaging materials with low residual monomer and exceptional stress cracking resistance in contact with oils. [Hüls AG] *

Vestyron X984 and X1260AK. Polystyrenes with high impact resistance. [Hüls AG] *

Vesulong. Sulfonamide veterinary ethicals. [Ciba plc] *

Vetalar. CAS 1867-66-9; Ketamine hydrochloride; anesthetic. [Parke-Davis] *

Vetalog. CAS 76-25-5; Triamcinolone acetonide; a glucocorticoid. [Bristol-Myers Squibb Co Inc] *

Vetalog Injection. CAS 76-25-5; Triamcinolone acetonide in aqueous vehicle. [Ciba plc] *

Vetalog Plus Cream. CAS 76-25-5, 8067-69-4; Triamcinolone acetonide and halquinol in cream base. [Ciba plc]*

Vetanabol. Anabolic steroid veterinary ethical. [Ciba plc] *

Vetibenzamine. Antihistamines. [Ciba plc] *

Vetidrex. Diuretic veterinary ethicals. [Ciba plc] *

Vetol. A proprietary pure vegetable oil palm product. §

VF-077. CAS 19706-80-0; 2,2´-Azobis[2-(hydroxymethyl) propionitrile]; polymerization initiator. [Wako Pure Chem. Ind.; Wako Chem. USA]

Viacutan. CAS 53370-43-7; A proprietary preparation of methargen in a cream base; antiseptic skin cream. [Octel Chemicals Ltd] †

Vialon Fast Dyes. 1:2 Metal complex dyes for dyeing and printing polyamide fibers. [BASF; BASF plc]

Vi-Alpha. CAS 68-26-8; Vitamin A; antixerophthalmic. [Lederle Laboratories USA] *

Via Rasa. It is the calcium salt of p-toluenechlorosulfonamide; an insecticide.

Vibalt. CAS 68-19-9; Cyanocobalamin; vitamin. [J B Roerig]*

Vibatex. Finishing agent. [Ciba plc] *

Vibazine. Buclizine hydrochloride; antinauseant. [Pfizer Inc]

Vibrabond. Acrylic polymer cement flooring for heavy duty and chemical resistance. [Prodorite Ltd]

Vibrac. A nickel chrome steel.

Vibrac steel. A nickel - chromium - molybdenurn steel.

Vibramycin. CAS 17086-28-1; A proprietary preparation of doxycycline; an antibiotic. [Pfizer International]

Vibra-Tabs. Doxycycline hyclate; antibacterial. [Pfizer Inc]

Vibrathane®. A range of polyether-based and polyester-based prepolymers; offer high abrasion resistance, chemical resistance, and electrical properties.

---

* Trade name not verified as available　　　　　　　　　† Trade name verified as obsolete

[Uniroyal]*
Vibrathane® 5004. Polyester PU; millable urethane with excellent resist. to abrasion, fuel, oil, and ozone, low temp. flexibility, high load bearing, high tensile and tear strength, low compression set; used for o-rings, seals, gaskets, for the automotive, aerospace, aircraft, and industrial molded goods market; peroxide curable. [Uniroyal]
Vibrathane® 8007; Ester-MDI PU prepolymer; non-MBCA, castable; FDA wet food approval; for food contact articles. [Uniroyal]
Vibrathane® B602. PU prepolymer (polyether-TDI); castable, high resilience; for mining, slurry parts, linings. [Uniroyal]
Vibratussal. A proprietary preparation of doxycycline hyclate and codeine; an antibiotic and antitussive. [Pfizer International]
Vibratussan. A proprietary preparation of doxycycline hyclate and codeine; an antibiotic and antitussive. [Pfizer International] †
Vibrin® E-010-01. Bisphenol-A epoxy-based vinyl ester resin in styrene; corrosion-resist. resin for hand lay-up, spray-up, filament winding and pultrusion processes; for chemical tanks, pipe, fume handling equip., equip. for desulfurization processes. [Owens-Corning Fiberglas]
Vibrin® E-085. CAS 68002-44-8; Epoxy novolac-based vinyl ester resin in styrene; corrosion-resist. resin for hand lay-up, spray-up, filament winding and pultrusion processes; resist. to chems., oxidation and heat; for chemical tanks, pipe, fume handling equip., equip. for desulfurization processes. [Owens-Corning Fiberglas]
Vibrin® E-701. Two-stage isophthalic unsaturated polyester resin; corrosion-resist. resin for tanks, vessels, sm. diameter pipe, grating and air handling equip. [Owens-Corning Fiberglas]
Vibrin® E-750. Bisphenol-A fumarate-based unsaturated polyester; corrosion-resist. resin for chemical tanks, small diameter pipe, fume handling equip. subject to high temps. [Owens-Corning Fiberglas]
Vibriomune. A proprietary cholera vaccine. [Duncan Flockhard Ltd] *
Vibrocil. A proprietary preparation of dimethindine hydrogen maleate, phenylephrine hydrochloride, and neomycin sulfate. [Zyma (UK) Ltd] ‡
VIC®. Polyester and acrylic urethane; high performance, low temp. cure, corrosion-resist. coatings for three-dimensional objects, plastics, metal, heat-sensitive substrates. [Ashland] *
Vicalloy. A proprietary trade name for a high permeability alloy containing iron with 36-62% cobalt and 6-16% vanadium. §
Vichy salt. Sodium hydrogen carbonate, $NaHCO_3$.
Vicks Medinite. Night time colds medicine, for relief of symptons. [Richardson-Vicks Inc] *
Viclan. Polyvinylidene chloride copolymer resins and latexes. [ICI Chem. & Polymers Ltd]
Vicmos powder. A smokeless 33-grain powder.
Vicol®. Acrylic polymers designed particularly for application to cotton and spun cellulosic yarns and for sizing continuous filament viscose. [Allied Colloids Ltd]
Vicron. Highly refined calcite. [Pfizer International] †
Vicryl. Polyglactin; surgical suture material, absorbable. [Ethicon Inc] *
Victor bronze. An alloy of 58.5% copper, 38.5% zinc, 1.5%

aluminum, 1% iron, and 0.03% vanadium.
Victor Cream®. CAS 7758-16-9; Sodium acid pyrophosphate; leavening agent for baking, cereals. [Rhone-Poulenc Food Ingreds.]
Victoria Silver. *See* nickel silvers.
Victorium. A proprietary trade name for lignin thermoplastic materials. §
Victor metal. An alloy of 50% copper, 34.3% zinc, 15.4% nickel, 0.28% iron, and 0.11% aluminum; used for sand castings and marine work.
Victor Powder. A trademark for a smokeless powder containing nitroammonium nitrate, wood meal, and potassium chloride. §
Victors. Cough drops. [Richardson-Vicks Inc] *
Victory®. CAS 63231-60-7; Microcrystalline wax; plastic wax offering high ductility, flexibility at very low temps.; provides protective barrier properties against moisture vapor and gases; used in hot-melt adhesives; hot-melt coatings; in antisunchecking agents in rubber goods, elec. insulating agents, leather treating agents, water repellents for textiles, rustproof coatings, cosmetic ingredients, and as plasticizer for other petrol. waxes used in crayons, dental compounds, chewing gum base, and candles. [Petrolite]
Victrex. High temperature polymers such as polyetheretherketone and polyethersulfone. [ICI Chem. & Polymers Ltd]
Victron. A proprietary trade name for polystyrene and vinylite resins. §
Vicure® 10. CAS 22499-12-3; Isobutyl benzoin ether; photosensitizer for uv curable systems, e.g., coatings, inks, graphic arts. [Akzo]
Vicure® 55. CAS 15206-55-0; Methyl phenylglyoxalate; high purity photoinitiator for uv curable systems, especially acrylate-based formulations. [Akzo]
Viczsal. An ammoniacal solution of copper and zinc phenolates. A wood preservative.
Vidal's caustic powder. A caustic consisting of burnt alum and powdered savin tops.
Vi-Daylin. A proprietary preparation containing vitamins A, D, C, thiamine, riboflavine, nicotinamide and pyridoxine. [Abbott Laboratories] *
Videne Disinfectant Solution. CAS 25655-41-8; Preoperative skin antiseptic containing povidone-iodine USP. [3M Pharmaceuticals] †
Videne Disinfectant Tincture. CAS 25655-41-8; Preoperative skin antiseptic containing povidone-iodine USP. [3M Pharmaceuticals] †
Videne Powder. CAS 25655-41-8; Topical antiseptic containing povidone-iodine USP. [3M Pharmaceuticals] †
Videne Surgical Scrub. CAS 25655-41-8; Surgical scrub containing povidone-iodine USP. [3M Pharmaceuticals]†
Videobil. CAS 41473-08-9; Iopronic acid; diagnostic aid. [Bracco Industria Chimica SpA] *
Videocolangio. Iodoxamic acid; diagnostic aid. [Bracco Industria Chimica SpA] *
Vi-Dom-A. CAS 68-26-8; Vitamin A; antixerophthalmic. [Miles Pharmaceuticals] *
Vidopen. CAS 69-53-4; A proprietary preparation of ampicillin; an antibiotic. [Berk Pharmaceuticals Ltd] *
Vidox. CAS 1314-13-2; Zinc oxide; for rubber, paint, ceramics. [Manchem Ltd] *
Vienna caustic. Potassium hydroxide with lime.

---

‡ Trade name and manufacturer not verified   § Trade name without identified manufacturer

Vienna cement. A metallic cement made from 86% copper and 14% mercury; an imitation gold.

Vienna paste. A mixture of lime and potash.

Viennese tombac. An alloy of 97% copper and 2.8% zinc.

Vigantol®. CAS 67-97-0; Vitamin D$_3$ in oily solution; veterinary preparation to prevent rickets and osteomalacia. [Bayer AG]

Vigantol® E Comp. Vitamins A, D$_3$,and H; used against vitamin deficiency diseases in veterinary medicine. [Bayer AG]

Vigazoo. Sublenox; growth stimulant. [Am. Cyanamid]

Vigil. CAS 75736-33-3; Broad spectrum fungicide containing diclobutrazol. [ICI Chem. & Polymers Ltd]

Vigilan. CAS 8038-43-5; Lanolin oil; emulsifier; skin and hair substantive emollient and moisturizer; solubilizer for immiscible fluids; o/w and/or w/o emulsions; adds elegance, lubricity, and hydration to makeup and facial preparations; spreads rapidly. [Fanning]

Vigilan AWS. CAS 68458-58-8; PPG-12-PEG-65 lanolin oil; o/w emulsifier; plasticizer, emollient, solubilizer, and wetting agent in personal care products; disperses to form milky emulsion in water used in bath oils. [Fanning]

Vigorite. See Bakelite. §

Vigorite. A safety explosive for mines. It consists of 30% nitroglycerin, 49% potassium chlorate, 7% potassium nitrate, 9% wood pulp and 5% magnesium carbonate.

Vi-Grow. Chemical products used for physical conditioning of soil. [Courtaulds plc]

Vikane. Fumigant based primarily on sulfuryl fluoride is used specifically for the control of drywood termites and wood boring beetles infesting wood in structures, furniture and lumber; odorless, colorless, noncorrosive, and does not react to produce malodors. [Dow UK] *

Viking. Tires. [Chevron] *

Vikro. Proprietary alloys containing from 63-65% nickel, 13-23% chromium, 0.5-1% silicon, up to 1% manganese and carbon, and the balance iron. §

Vileda. Household cloths. [Carl Freudenberg] *

Viledon Compact. Nonwoven table cover. [Carl Freudenberg] *

Viledon Filter. Nonwoven air filter. [Carl Freudenberg] *

Vilene. Freudenberg nonwovens. [Carl Freudenberg] *

Vilit. A range of soluble vinyl chloride copolymers; used as binders for paints, heat sealable lacquers for aluminum foils, for coatings on metal, concrete and cardboard. [Hüls AG]

Villescon. A proprietary preparation of prolintane with vitamins. [Boehringer Ingelheim Ltd]

Viluite. Synonym for Grossular. §

Vimlite. A proprietary trade name for a cellulose acetate plastic. §

Vinac® 1000 DEV. CAS 9003-20-7; Polyvinyl acetate homopolymer emulsion; base for wood adhesives. [Air Prods./Polymers]

Vinac® XX-210, XX-220, XX-230, XX-240 . CAS 9003-20-7; Polvyinyl acetate homopolymer emulsion; base for adhesive compounding for carton sealing, white glues, wood veneering. [Air Prods./Polymers]

Vinaccia tartar. Tartar obtained from the manufacture of wines.

vinaconic acid. (propene-3,3-dicarboxylic acid; vinyl-malonic acid) $CH_2{:}CH{\cdot}CH{\cdot}(COOH)_2$.

Vinacron. Polyvinyl chloride plastisol dispersion; used for protective and decorative coatings, rotomolding, spread coating and dip molding. [Loes Enterprises Inc]

Vinacryl. Acrylic copolymer emulsions. [Vinamul Ltd]

Vinacryl 4001/B. A proprietary trade name for a 50% concentrated acrylic copolymer emulsion used as a cement additive. [Vinyl Products] *

Vinacryl 4005. A proprietary acrylic copolymer emulsion soluble in alkali. [Vinyl Products] *

Vinacryl 4152, 4500/X, 4501/X. Proprietary trade names for vinyl acrylic copolymer emulsions used as adhesives. [Vinyl Products] *

Vinacryl 4160. A proprietary trade name for an acrylic copolymer emulsion used as a 46.5% concentrate for a binder in paper board manufacture. [Vinyl Products] *

Vinacryl 4260. A proprietary emulsion of poly-2-ethoxyethyl methacrylate. [Vinyl Products] *

Vinacryl 4290. A proprietary polybutyl methacrylate emulsion. [Vinyl Products] *

Vinacryl 4320. A proprietary self-reactive vinyl acrylic copolymer emulsion used in the finishing of textiles. [Vinyl Products] *

Vinacryl 4322. A proprietary self-reactive vinyl acrylic copolymer emulsion. [Vinyl Products] *

Vinacryl 4450. A proprietary trade name for a vinyl-acrylic copolymer; used in crack filling compounds. [Vinyl Products] *

Vinacryl 4512. A proprietary acrylic polymer emulsion. [Cray Valley Ltd] †

Vinacryl 7170, 7172, and 7175. A proprietary range of styrene acrylic copolymer emulsions. [Vinyl Products] *

Vinacryl R3929, R3940. Proprietary trade names for a 55% concentrated vinyl-acrylic copolymer emulsions used for nonfray carpet backings. [Vinyl Products] *

Vinal. A proprietary trade name for a synthetic vinyl resin. §

Vinalak 5150. A proprietary self-reactive acrylic polymer solution in isopropyl acetate. [Vinyl Products] *

Vinamul. Vinyl acetate homopolymer and copolymer emulsions. [Vinamul Ltd]

Vinamul 3240. A proprietary vinyl acetate-ethylene copolymer emulsion. [Vinyl Products] *

Vinamul 3250. A proprietary vinyl acetate-ethylene copolymer emulsion containing a nonionic emulsifying system. [Vinyl Products] *

Vinamul 6000. A proprietary vinyl acetate emulsion of the unsaturated acid copolymer type, soluble in alkali. [Vinyl Products] *

Vinamul 6050. A proprietary vinyl acetate-vinyl caprate-unsaturated acid terpolymer emulsion, internally plasticized. [Vinyl Products] *

Vinamul 6208. A proprietary trade name for a 50% concentrated vinyl-acrylic copolymer emulsion used for wallpaper grounding, printing and overcoating. [Vinyl Products]*

Vinamul 6275. A proprietary trade name for a 55% concentrated vinyl-acrylic copolymer emulsion used in the production of emulsion paints. [Vinyl Products] *

Vinamul 6705. A proprietary ethylene grafted vinyl acetate copolymer emulsion. [Vinyl Products] *

Vinamul 6888. A proprietary modified vinyl acetate-acrylate copolymer emulsion. [Vinyl Products] *

Vinamul 6930. A proprietary trade name for 52% concentrated vinyl acetate - Veo Va 911 copolymer emulsion used for emulsion paints. [Vinyl Products] *

Vinamul 7700. A proprietary polystyrene emulsion. [Vinyl Products] *

---

* Trade name not verified as available

† Trade name verified as obsolete

Vinamul 7715. A proprietary polystyrene emulsion containing 15% dibutyl phthalate in a nonvolatile plasticizer. [Vinyl Products] *

Vinamul 8400. A proprietary trade name for a 50% emulsion of polyvinyl acetate; used for adhesives. [Vinyl Products]*

Vinamul 8430. A proprietary polyvinyl acetate emulsion. [Vinyl Products] *

Vinamul 8460 and 9000. Proprietary polyvinyl acetate emulsions. [Vinyl Products] *

Vinapol. Vinyl acetate homopolymer and copolymer powders. [Vinamul Ltd]

Vinapol 1000. A proprietary polyvinyl acetate powder dispersible in water. [Vinyl Products] *

Vinapol 1030. A proprietary polyvinyl acetate powder plasticized with 10% dibutyl phthalate and dispersible in water. [Vinyl Products] *

Vinapol 1070. A proprietary finely-divided polyvinyl acetate powder. [Vinyl Products] *

Vinapol 1088. A proprietary acrylic processing and used in the processing of rigid PVC compounds. [Vinyl Products]*

Vinapol R.3800, R.3863. A proprietary trade name for a water dispersible polyvinyl acetate powder. [Vinyl Products] *

Vinapol R3626. A proprietary trade name for a water-dispersible alkali soluble vinyl acetate powder. [Vinyl Products] *

Vinatex. Polyvinyl chloride plastisols. [Hydro Chemicals Ltd]

Vinavil. Vinyl acetate, homopolymers and copolymers. [Montedison UK Ltd] *

Vincapront. CAS 1617-90-9; A proprietary preparation of vincamine; a cerebral vasodilator. [Pfizer International]

Vincaron. See Vincapront. [Pfizer International] †

Vinchel 11. A proprietary metal-free liquid organic complex with a stabilizer system of barium and cadmium soaps; used as a chelating agent in PVC compounds. [Vinyl Products] *

Vinchel 20. A proprietary metal-free liquid organic complex with a stabilizer system of barium and cadmium soaps and tribasic barium sulfate; used as a chelating agent in PVC compounds. [Vinyl Products] *

Vinchel 22. A proprietary zinc-based chelating agent for PVC compounds; a basic lead carbonate stabilizing system is used. [Vinyl Products] *

Vinchel 35. A proprietary zinc-based chelating agent for PVC compounds; a stabilizer system of barium and cadmium soaps. [Vinyl Products] *

Vinco 99A. A proprietary stabilizer for PVC and PVA of the liquid barium/cadmium/zinc complex type. [NL Victor Wolf Ltd] *

Vinco 99G. A stabilizer similar to Vinco 99A used with paste-grade resins in rotational molding. [NL Victor Wolf Ltd] *

Vinco 248. A proprietary stabilizer of the liquid barium and cadmium complex type; used with PVC polymers sensitive to zinc. [NL Victor Wolf Ltd] *

Vinco 249C. A proprietary stabilizer for PVC with a liquid barium, cadmium, and zinc base. [NL Victor Wolf Ltd] *

Vinco 265. A proprietary stabilizer for PVC pastes with a liquid barium, cadmium, and zinc base. [NL Victor Wolf Ltd] *

Vinco A33. A proprietary liquid potassium/zinc complex used as a stabilizer and initiator in the production of expanded PVC. [NL Victor Wolf Ltd] *

Vinco A183. A proprietary liquid complex used as a stabliser and initiator in the production of expanded PVC. [NL Victor Wolf Ltd] *

Vindex. Bromoxynil + clopyralid; herbicide mixture for weed control in cereals. [Quadrangle Agrochemicals]

vinegar naphtha. (ethyl acetate) $CH_3COOC_2H_5$; CAS 141-78-6; EINECS 205-500-4; Used as a solvent in coatings and plastics, in organic synthesis, pharmaceuticals, and synthetic fruit essences.

vinegar salts. (calcium acetate; grey aceate; lime acetate) $Ca(C_2H_3O_2)_2 \cdot H_2O$; CAS 64-19-7; EINECS 200-540-9; Used as an antipyretic in medicine.

Vinex 1003. CAS 9002-89-5; Polyvinyl alcohol copolymer resin; thermoplastic resin for extrusion into cast or tubular blown film, injection or blow molding into bottles, melt spinning into fiber for industrial or personal care applications [Air Prods. & Chems. Inc]

Vinex 2019. CAS 9002-89-5; Polyvinyl alcohol copolymer resins; thermoplastic for melt blown fiber, injection molding, extrusion, lamination. [Air Prods. & Chems. Inc]

Vinex 2025. CAS 9002-89-5; Polvyinyl alcohol copolymer resin; thermoplastic for extrusion into cast or tubular blown film, injection or blow molding into bottles, melt spinning into fiber for industrial or personal care articles. [Air Prods. & Chems. Inc]

Vinex 2034. CAS 9002-89-5; Polyvinyl alcohol copolymer resins; thermoplastic for cast film, sheet, blown film, melt spun fiber, blow molding. [Air Prods. & Chems. Inc]

Vinex 2144. CAS 9002-89-5; Polvyinyl alcohol copolymer resin; thermoplastic resin for extrusion into cast or tubular blown film, injection or blow molding into bottles, melt spinning into fiber for industrial or personal care applications. [Air Prods. & Chems. Inc]

Vinex 5030; CAS 9002-89-5; Polyvinyl alcohol copolymer resin; thermoplastic resin for extrusion into cast or tubular blown film, injection or blow molding into bottles, melt spinning into fiber for industrial or personal care applications. [Air Prods. & Chems. Inc]

vinic ether. (ethyl ether; diethyl ether) $C_2H_5O \cdot C_2H_5$; CAS 60-29-7; EINECS 200-467-2; Used in smokeless powders; organic synthesis; a solvent for nitrocellulose, alkaloids, fats, and waxes; analytical chemistry.

Vinidur®. Vinyl chloride/acrylate graft copolymers; produces very high to high impact, weather resist. parts preferably by extrusion, but also by calendering, injection molding, e.g., profiles for outdoor applications (window frames), pipes, panels, films. [BASF AG; BASF aplc]

Vinisil. CAS 9003-39-8; Povidone; pharmaceutic aid (dispersing and suspending agent). [Abbott Laboratories] *

Vinnapas®. Homopolymer or copolymer, polyvinyl acetate solutions, dispersions, solid resins; used for lacquers and adhesives, paints. [Wacker-Chemie GmbH]

Vinnapas® A 50. CAS 108-05-4; VA homopolymer disp.; binder for fabrics, glass fibers; textile auxiliary and coating; paper coatings. [Wacker-Chemie] *

Vinnapas® EN 426. VA/ethylene copolymer disp. with reactive groups; binder for fabrics, glass fiber; textile auxiliary and coating. [Wacker-Chemie] *

Vinnapas® EP 177. Vinyl acetate/ethylene copolymer disp.; for textile finishing/softening. [Wacker-Chemie] *

Vinnapas® LL 364. Styrene/acrylic disp.; for textile finishing. [Wacker-Chemie] *

Vinnathen. A proprietary trade name for a vinyl acetate/ethylene copolymer which can be crosslinked with per-

---

‡ Trade name and manufacturer not verified

§ Trade name without identified manufacturer

oxides; can be used for cable jackets. §

Vinnol®. A proprietary grade of polyvinyl chloride. [Wacker-Chemie GmbH; Wacker Chemicals Ltd]

Vinnol® 50. CAS 9003-22-9; Vinyl chloride/vinyl acetate copolymer disp.; binder for fabrics, glass fiber; textile auxiliary and coating; and paper coatings. [Wacker-Chemie] *

Vinnol® LL 352. Vinyl chloride/ethylene copolymer disp.; for paper, carpet, and textile industries. [Wacker-Chemie] *

Vinofan®. Polymer dispersions based on vinyl compounds; for building adhesives and packaging adhesives; binders for paper and board coatings, for bonding fiber webs, and for textile coating. [BASF AG]

Vinoflex®. CAS 9002-86-2; PVC; after stabilization can be processed into rigid parts for extrusion, calendering, and injection molding (film, pipes, profiles, hollow articles, and panels); with addition of plasticizers produces flexible products such as film, cable insulation, cable sheaths, hoses, and floor coverings. [BASF AG]

Vinoflex. PVC; used for extrusion (pipes, profiles and cables), calendering (plasticized and UPVC films), injection molding (fittings), blowmolding (bottles). [BASF plc]

Vinoflex® 377. A proprietary PVC-emulsion homopolymer used in the production of PVC film. [BASF plc] *

Vinoflex® 516. A proprietary PVC suspension-type homopolymer. It is an easily-flowing powder used in the extrusion of rigid PVC. [BASF plc] *

Vinoflex® 526. A proprietary PVC homopolymer similar to Vinoflex 516. [BASF plc] *

Vinoflex® 534. A proprietary PVC suspension-type homopolymer used in the extrusion of high-quality cables. [Anic Agricoltura Spa] ‡

Vinoflex® 535. A proprietary PVC suspension-type homopolymer in the form of an easy-flowing powder with porous particles; used in the making of soft, calendered products. [BASF plc] *

Vinoflex® 719. A proprietary PVC suspension-type polymer used in the making of rigid and tough, weather-resistant products. [BASF plc] *

Vinsalyn. A range of thermoplastic resins; used as binders and/or binder extenders in industrial adhesives and mastics. [Hercules] *

Vinsol. Dark pine resin. [Hercules Ltd]

Vinsol. A proprietary trade name for the black residue from the extraction of rosin with solvents; used as an insulating varnish. §

Vinsol Emulsion. An emulsion of aliphatic hydrocarbon insoluble resin; used as a modifier of water-based adhesives and coatings. [Hercules] *

Vinsol Resin. Wood rosin derivative. [Langley Smith & Co Ltd]

Vinuran®. Polymers for modifying PVC; improves PVC properties incl. impact strength, resist. to heat deformation and decomposition, processing, and deep-drawing behavior; used for hot water pipes, impact-resist. profiles, films, hollow articles, panels. [BASF AG; BASF plc]

Vinychlon®. A series of Japanese vinylchloride polymers. [Mitsui Co Ltd] *

vinyl acetal polymers, butyrals. See polyvinyl butyral.

vinyl acetate. (acetic acid, ethenyl ester; ethenyl acetate; acetic acid, vinyl ester) Unsaturated ester; $CH_3COO-CH=CH_2$; CAS 108-05-4; EINECS 203-545-4; in polymerized form for plastic masses, films, and lacquers. [Exxon Belgium; Hoechst Celanese; Quantum/USI;

Union Carbide]

Vinyl Acetate Monomer. CAS 108-05-4; Vinyl acetate; intermediate for industrial and consumer products incl. PVAc to produce paints, adhesives and coatings; PVAL (adhesives, coatings, packaging films); polyvinyl acetals (insulation, safety glass interlayer); EVA copolymers (films, coatings, adhesives). [Quantum/USI] *

vinyl acetyl polymers, butyrates. See polyvinyl butyral.

vinyl compounds and polymers. See also polyvinyl chloride, vinyl chloride, vinyl acetate. A compound having the vinyl grouping ($CH_2=CH$); basis for varieties of plastics including vinyl chloride, vinyl acetate, etc. [

vinyl ester resin. Thermoset used in chemical processing industry, pulp and paper mills, corrosive-resist. applications. [Ashland; Cook Composites & Polymers; Dajac Labs; Union Carbide]

Vinylite. A proprietary trade name for polyvinyl acetate, polyvinyl chloride-acetate and polyvinyl chloride synthetic resins. §

Vinylite V. A proprietary name for an interpolymer of PVC and PVA (qv). §

Vinylite X. A proprietary trade name for polyvinyl chloride-acetate. §

Vinyloid. A proprietary trade name for a polyvinyl acetate resin. §

vinyl pyridine. $C_5H_4NCH=CH_2$; CAS 1337-81-1; Used for adhesives, tire cord, and industrial goods dips. [Penta Mfg.; Raschig; Reilly Industries]

Vinylseal. A proprietary trade name for vinyl acetate resin adhesives. §

vinyltrichlorosilane. (trichlorosilyl)ethylene; trichlorovinylsilane) $H_2C=CHSiCl_3$; CAS 75-94-5; EINECS 200-917-8; Intermediate for silicones; coupling agent in adhesives and bonds. [Hüls Am.; PCR; Schweizerhall; Union Carbide]

vinyltriethoxysilane. (triethoxysilyl)ethylene; triethoxyvinylsilane) $C_8H_{18}O_3Si$; CAS 78-08-0; EINECS 201-081-7; Intermediate, especially when acidic by-products are undesirable; filler. [Hüls Am.; PCR; Union Carbide]

Vinyon. A proprietary trade name for vinyl resins for textile fibers. §

Vinyon Fiber. A proprietary trade name for a material manufactured from polyvinyl chloride-acetate. §

vinyzene. CAS 58-36-6; Condensation product of epoxidized soy bean oil and 10, 10'-oxybisphenoxyarsine; used as a fungicide and bactericide.

Vinyzene® BP-505 DIDP. 5% 10,10'-Oxybisphenoxarsine in diisodecyl phthalate; antimicrobial, bacteriostat, fungistat for PVC, PU, other plastics, synthetic rubber; recommended for film and sheet, extruded profiles, plastisols, molded goods, organosols, fabric coatings. [Morton Int'l./Plastics Additives]

Vinyzene® IT-3000 DIDP. 2-n-Octyl-4-isothiazolin-3-one in diisodecyl phthalate carrier; fungicide for plastics (PVC, PU); recommended for vinyl film, sheeting, extruded profiles, plastisols, molded goods, organosols, fabric coatings, urethane shoe soles, and foams. [Morton Int'l./Plastics Additives]

Vinyzene® SB-1 EAA. 10,10'-Oxybisphenoxarsine in ethylene-acrylic acid copolymer carrier; fungicide, bactericide for polyolefins; for interior automotive parts, exterior drainage hose, underground cable jacketing, buoyant floats. [Morton Int'l./Plastics Additives]

Viocin. CAS 37883-00-4; A proprietary preparation contain-

---

* Trade name not verified as available   † Trade name verified as obsolete

ing viomycin sulfate; an antituberculous agent. [Pfizer International] †

Vioflor. A proprietary preparation of volatile hydrocarbons; used to deodorize turpentine substitutes. §

Vioform. Clioquinol; anti-amebic; anti-infective, topical. [Ciba-Geigy Corp]

Vioform-Hydrocortisone. A proprietary preparation of clioquinol and hydrocortisone used in dermatology as an antibacterial agent. [Ciba plc]

Vioform Powder. Treatment of skin condition. [Ciba plc] *

Vioglaze. Uv curing coatings; gloss varnish for roller coating paper and board and roller coating for flooring. [Coates Coatings Ltd]

violet copper. (red copper). Reduced copper, prepared by reducing copper oxide.

violet phosphorus. The coarse-grained red variety is metallic or violet phosphorus.

violet powder. Perfumed starch powder.

violet root. Orris root.

violettol. A mixture of 10% ionone and 90% salicyl aldehyde; used to strengthen natural violet perfume.

Violetton. Trade names for the commercial ionones; used as perfumes. §

violet tungsten. Potassium tritungstate, $K_2W_3O_9.W_2O_5$; used as a pigment.

violuric acid. (isonitrosobarbituric acid; alloxan-5-oxime) $C_4H_3N_3O_4$; CAS 26351-19-9; EINECS 201-741-4.

Viomycin. An antibiotic produced by certain strains of *Streptomyces griseus var. purpureus.* §

Viomycin Sulphate. CAS 37883-00-4; A proprietary preparation containing viomycin sulfate; an antituberculous agent. [Parke-Davis] *

Vionactane. A proprietary preparation of viomycin pantothenate and sulfate; an antituberculous agent. [Ciba plc] *

Vionate Powder. Multivitamin and mineral mix for veterinary use. [Ciba plc] *

Vipla. Emulsion polyvinyl chloride homopolymers. §

Viplan. Polyvinyl chloride paste polymers. [European Vinyls Corporation Ltd]

Viplex 680-P. CAS 64742-04-7; Heavy paraffinic distillate solvent extract; secondary vinyl plasticizer. [Crowley Chem.]

Viplex 885. CAS 68477-29-2; Highly aromatic (99%) petrol. oil; plasticizer/extender for epoxy systems, coatings, potting/encapsulation; promotes low system visc., better wetting of fillers and pigments, higher loading levels; modifier for polyesters. [Crowley Chem.]

Viplex 895-BL. CAS 64741-81-7; Polynuclear aromatic hydrocarbon; secondary vinyl plasticizer. [Crowley Chem.]

Vipophan PVC. Shrink film. [Lonza AG]

Vira-A. CAS 5536-17-4; Vidarabine; antiviral. [Parke-Davis]

Virazole. CAS 36791-04-5; Ribavirin; antiviral. [ICN Nutritional Biochemicals Corp] *

Virco. Textile dyeing auxiliaries. [Virkler Chemical Co] *

Virexen. Idoxuridine in DMSO solution; pharmaceutical speciality; antiviral for the treatment of herpes simplex and herpes zoster. [Laboratorios Vinas SA] *

virginiamycin. An antibiotic produced by *Streptomyces virginioe.*

Virginia silver. *See* nickel silvers.

Vironex. Folpet 40%, cymoxanil 4%; wettable powder used as protective fungicide for foliage application to ornamental and crop plants. [Industrias Quimicas Del Valles SA] *

Viroptic Ophthalmic Solution. CAS 70-00-8; A proprietary formulation of trifluridine; for treatment of primary keratoconjunctivitis and recurrent epithelia keratitis due to Herpes simplex virus types 1 and 2. [The Wellcome Foundation Ltd]

Virormone. CAS 57-85-2; Injectable testosterone propionate; used as replacement therapy in castrated adults and in those who are hypogonadal due to either pituitary or testicular disease; also for the control of carcinoma in post menopausal women. [Paines & Byrne Ltd] *

Visacor. Cardiovascular preparations. [ICI Chem. & Polymers Ltd] †

Viscalex®. Acrylic-based thickening agents for use in aq. adhesives, paints, latex compounds, cosmetics, and ceramics. [Allied Colloids Ltd]

Viscasil®. Dimethicone; defoamer, release agent, emollient in cosmetics, polishes, paint additives, and mechanical devices; lubricant in rubber or plastic-to-metal applications; suggested for auto polish, skin care, hair care, textile softeners, antifoams for petrol. refining, rubber and plastic mold release, film modifier in coatings, damping in mechanical/elec. applications. [GE Silicones]

Visc-Ayd® 812. With Visc-Ayd Activator 814 as bodying agent for solv.-thinned coatings incl. trade sales paints, house paints, enamels, and flat alkyds. [Daniel Prods.]

Vischem A series of diester and polyolester synthetic based greases with various thickeners; high temperature greases for all kinds of bearings, worm gears, slides, etc. [Ultrachem Inc] *

viscoid. A mixture of viscose and clay with powdered horn or zinc oxide.

Viscolas®. Viscoelastic urethane material; formulated to absorb and dissipate shock, provide good resiliency, rapid recovery, good damping; for industrial work gloves, tool grips, shoe inserts, modular orthotics, bike seat cushions, exercise equip. [E-A-R]

viscolith. The hard mass obtained when a viscose solution coagulates.

Viscoloid. A proprietary trade name for pyroxylin plastics. §

Viscoplex. Solutions in mineral oil of long chain fatty alcohol/ acrylic/methacrylic acid esters; pour point depressants and viscosity index improvers for mineral oils. [Cornelius Chemical Co Ltd] *

viscose. The sodium salt of cellulose xanthate, obtained by the action of carbon disulfide and alkali upon cellulose. In thin sheets; it is used as a substitute for glass and celluloid; as a thickening and dressing substance, as a partial substitute for resin glue in paper manufacture.

viscose silk. An artificial silk produced when viscose (*qv*) is forced through narrow orifices into ammonium chloride solution.

Viscosin. A proprietary refined oil tar. §

Viscosine. *See* Valvoline. §

Visco-Stab. Visc. stabilizer for cuprous oxide-containing paints based on bioMeT 300 Series antifoulant polymers. [Atochem N. Am.] *

Vi-Siblin. A proprietary preparation of ispaghula husks and thiamine; a laxative. [Parke-Davis] *

Visken. CAS 13523-86-9; A proprietary preparation of pindolol; used in the treatment angina pectoris. [Sandoz]

Visor. Deodorant for industrial processes and products. [CPL Group Ltd]

Visqueen. Polyethylene film. [ICI Chem. & Polymers Ltd] †

---

‡ Trade name and manufacturer not verified   § Trade name without identified manufacturer

Vista®. For medical applications. [DuPont UK]

Vista LPA, LPA-140, LPA-210. Mixtures of hydrotreated isoparaffins and naphthenes with very low levels of aromatics; solv. for food applications, pesticides, coatings, household products, water, paper, and mining chemicals; used as textile lubricants, chemical process solvs., freeze pt. depressants, printing inks solvs. [Vista]

VistaFlex®. Thermoplastic elastomer; for parts requiring good surface appearance, replacing thermoplastic olefin blends, styrene-block TPEs, and plasticized PVC. [Advanced Elastomer Systems]

Vistaflex. Thermoplastic elastomer. [Exxon Int'l.; Exxon UK]

Vistal. Polyethylene films in widths up to 12 m; films based on polyethylene LD, MD, LLD in thicknesses of 10 to 800 microns; shrink and stretch films for packaging, bundling and palletizing, coextruded films, embossed and repellent films for industrial, horticulture and agriculture applications; peelable films for easy opening packaging, special films for incorporating into laminates, for balloons for space research and for building and road construction. [UCB nv Film Sector] *

Vistalon. Polyolefin; ethylene propylene copolymers. [Exxon Int'l.; Exxon UK] *

Vistalon 719. CAS 9010-79-1; Ethylene/propylene copolymer; EPM rubber for modifying polyolefins. [Exxon] *

Vistalon 2200. EPDM terpolymer; used alone or in blends (especially in butyl rubber) where controlled cure rate and easy processing are needed; imparts excellent ozone and heat resist.; excellent processing during extrusion, molding, or calendering. [Exxon] *

Vistalon 3666. EPDM terpolymer, 75 phr paraffinic oil; accepts high loadings of carbon black and oil to yield low cost, fast extruding compounds with low tension set and good snappiness. [Exxon] *

Vistalon 3708. EPDM terpolymer; high green strength, high extrusion rate, extra high tensile strength; maintains good physical properties when highly extended; used for medium and high hardness extrusion compounds, hose, weatherstrip, molded goods; sulfur-curable. [Exxon] *

Vistalon 7000. EPDM terpolymer; fast extruding; used in hose and tubing, elec. insulation, thermoplastic polyolefins, weatherstrips, molded goods; sulfur-curable; fast curing. [Exxon] *

Vista-Marc. Etafilcon A; contact lens material. [Vistakon Inc]*

Vistanex®. Semi-solid polyisobutylene. [Exxon Int'l.; Exxon UK]

Vistanex® MML-80, MML-100, MML-120, MML-140. CAS 9003-27-4; Polyisobutylene used as polymeric additive for rubbers to impart stability, inertness, ozone, and chemical resist., elec. inertness; also useful in molded and extruded products and coated materials. [Exxon] *

Vistaril. CAS 68-88-2; A proprietary preparation of hydroxyzine; an ataractic. [Pfizer International]

Vistaril Pamoate. CAS 10246-75-0; Hydroxyzine pamoate; tranquilizer. [Pfizer Inc]

Vistone. Mild extreme pressure oiliness agent. [Exxon UK]*

Vistron. See polystyrene. §

Vita-E. A proprietary preparation of D-α-tocopherol used in the treatment of vascular disorders. [Bioglan Laboratories] *

Vita-E Gelucaps. A proprietary preparation of D-α-tocopherol acetate; used in the treatment of vascular disorders. [Bioglan Laboratories] *

Vita-Cos. Wheat germ glycerides. [CasChem]

Vitafeed 101. Soluble powder containing NK 26:26 and trace elements; fertilizer. [Vitax Ltd]

Vitafeed 102. Soluble powder containing NK 18:36 and trace elements; fertilizer. [Vitax Ltd]

Vitafeed 103. Soluble powder containing NK 13:43 and trace elements; fertilizer. [Vitax Ltd]

Vitafeed 111. Soluble powder containing NPK 19:19:19 plus 0.2% Mg and trace elements; fertilizer. [Vitax Ltd]

Vitafeed 301. Soluble powder containing NK 36:12 plus 0.15% Mg and trace elements; fertilizer. [Vitax Ltd]

Vitalum. A trademark for materials of the abrasive class and consisting essentially of alumina. §

Vitamalt. A proprietary preparation containing vitamins A, B, C, and D. §

vitamin A. (retinol; vitamin A alcohol; axerophthol) $C_{20}H_{30}O$; CAS 68-26-8; EINECS 200-683-7; Dietary supplement. [Am. Biorganics; BASF; Hoffmann-La Roche; Penta Mfg.]

vitamin A alcohol. See vitamin A.

vitamin B. See nicotinic acid.

vitamin B_1. See thiamine.

vitamin B_2. See riboflavin.

vitamin B_5 calcium salt. See calcium pantothenate.

vitamin B_6. (3-hydroxy-4,5-dimethylol-2-methylpyridine; pyridoxine) $CH_3C_5HN(OH)(CH_2OH)_2$; CAS 65-23-6; Used in medicine and nutrition.

vitamin B_{12}. (cyanocobalamin; cyanocon(III)alamin; α-(5,6-dimethylbenzimidazolyl)cyanocobamide; cobalamin) $C_{63}H_{88}CoN_{14}O_{14}P$; CAS 68-19-9; EINECS 200-680-0; Dietary supplement; deficiency in man causes pernicious anemia and neural degeneration. [EM Industries; Hoffmann-La Roche; Roussel Uclaf; Schweizerhall]

Vitamin B Complex CLR. Yeast extract with B vitamins in water-alcohol medium; products for treatment of greasy hair, dandruff and oily skin. [Dr. Kurt Richter; Henkel/Cospha]

vitamin C. (ascorbic acid.) CAS 50-81-7; EINECS 200-066-2; $C_6H_8O_6$; Deficiency in man causes scurvy. It probably plays a part in the production of collagen in the tissues; used in nutrition, color fixing, flavoring, dietary supplement.

vitamin C sodium salt. See sodium ascorbate.

vitamin D_2. (calciferol; ercalciol; ergocalciferol) CAS 50-14-6; EINECS 200-673-2; A fat soluble vitamin, deficiency of which causes rickets, in children and osteomalacia in adults; $C_{28}H_{44}O$; used as a dietary supplement.

vitamin D_3. (calciol; cholecalciferol) CAS 67-97-0; EINECS 200-673-2; A fat soluble vitamin, deficiency of which causes rickets, in children and osteomalacia in adults; used as a dietary supplement.

vitamin D_4. A synthetic vitamin derived from 22-dihydroergosterol by irridation with uv light. It has less activity than D_2 and D_3.

vitamin E. (tocopherol) $C_{29}H_{49}OOH$; CAS 59-02-9; EINECS 200-412-2; A fat soluble vitamin; used as an antioxidant, in meat curing, nutrient.

Vitamin F Alcohol-Soluble CLR. Complex of essential free fatty acids, hydro-alcohol solubilized; products for treatment of dry skin and hair. [Dr. Kurt Richter; Henkel/Cospha]

Vitamin F Ethyl Ester CLR. Complex of essential esterified fatty acids in lipophilic medium; products for treatment of dry skin and hair. [Dr. Kurt Richter; Henkel/Cospha]

---

* Trade name not verified as available   † Trade name verified as obsolete

Vitamin F Forte CLR. Complex of essential free fatty acids in lipophilic medium; products for treatment of dry skin and hair. [Dr. Kurt Richter; Henkel/Cospha]

Vitamin F Glyceryl Ester CLR. Complex of essential esterified fatty acids in lipophilic medium; products for treatment of dry skin and hair. [Dr. Kurt Richter; Henkel/Cospha]

Vitamin F Water-Soluble CLR. Complex of essential free fatty acids, hydro-alcohol solubilized; products for treatment of dry skin and hair. [Dr. Kurt Richter; Henkel/Cospha]

vitamin G. See vitamin B_2.

vitamin K_1. CAS 84-80-0; EINECS 201-564-2; (phytomenadione; 2-methyl-3-phytyl-1,4-naphthoquinone) A fat soluble vitamin; deficiency gives rise to hemorrhage; $C_{31}H_{46}O_2$;

vitamin K_2. A fat soluble vitamin. 2-methyl-3-difarnesyl-1,4-napthaquinone; synthesized in the gut by bacteria.

vitamin M. (folic acid; folacin; PGA; pteroylglutamic acid) CAS 59-30-3; EINECS 200-419-0; Considered a member of the vtamin B complex; $C_{19}H_{19}N_7O_6$; used in medicine, nutrition, and as a food additive.

vitamin P. (rutin trihydrate; quercetin 3-rutinoside) $C_{27}H_{30}O_{26}.3H_2O$; CAS 153-18-4; EINECS 205-814-1.

vitamin U. (DL-methionine methylsulfonium chloride) $C_{16}H_{14}ClNO_2S$; CAS 3493-12-7; EINECS 222-484-4; A vitamin extracted from cabbage.

Vitaminets. Proprietary multivitamin preparation containing the B complex vitamins, vitamins A, C, D, and E, biotin, nicotinamide, and various minerals or trace elements (Ca, Fe, Mg, Mn and P). [Roche Products Ltd] *

Vitapet®. Fish oils, fish liver poils, vegetable oils; dietary supplement for domestic animals and birds conditioning oil. [Seven Seas Ltd]

Vitaplant CLR Oil-Soluble N. Calendula extract and lipoid extract of pig skins in oil medium; products for aging, damaged and sun-burned skin. [Dr. Kurt Richter; Henkel/Cospha]

Vitaplant CLR Water-Soluble. Echinacea extract and aloe juice in water-alcohol medium; products for aging, damaged and sun-burned skin. [Dr. Kurt Richter; Henkel/Cospha]

Vitapup®. Vitamins and minerals, skimmed milk powder; fortified milk feed for puppies. [Seven Seas Ltd]

Vitapyrena. Medicated hot lemon drink powder, for symptomatic relief of colds. [Richardson-Vicks Inc] *

Vitavax RS Flowable. Carboxin + γ-HCH + thiram; a fungicide and insecticide dressing for oilseed rape. [Uniroyal Chemical Ltd]

Vitax Q4. Powder fertilizer containing NPK 5.3:7.5:10 plus 1.8% Mg and trace elements; all-purpose base and top dressing fertilizer. [Vitax Ltd]

Vitax Micro Gran. CAS 10028-22-5; Ferrous sulfate; used for moss control in turf. [Vitax Ltd]

Vitax Turf Tonic. CAS 10028-22-5; Ferrous sulfate; used for moss control in turf. [Vitax Ltd]

Vita Zinc. CAS 1314-13-2; Low purity zinc oxide; animal feed supplement. [Manchem Ltd] *

Vitazyme® A-Plus. Retinyl palmitate polypeptide; natural protein complexed vitamins for cosmetics use. [Brooks Industries]

Viten. Vital wheat gluten; used for flour fortification (bread making). [Roquette (UK) Ltd]

Vitesse®. For agricultural applications. [DuPont UK]

Vitexol® K. Emulsified mixture of aliphatic hydroxyl compounds and a neutral phosphoric acid ester; defoamer for dyeing of cotton/polyester knitgoods in jet dyeing machines. [BASF]

Viton®. Fluoroelastomer. [DuPont UK]

Viton® A. Fluoroelastomer; general purpose, high temp. and fluid resistant polymer for rubber goods which must withstand extremes of heat and corrosive fluids and chemicals; used in roll covers, sol'n. coatings, gaskets, seals, tubing; curable with Diak curing agents. [DuPont; Du Pont (UK) Ltd] *

Viton® A-35. Fluoroelastomer; low visc. analog of Viton A which is easier and safer to process; offers vulcanizate properties similar to those of Viton A except for tensile strength and modulus which are slightly lower; used in high-solids sol'n. coatings and for blending with Viton A or Viton A-HV for visc. control. [DuPont; Du Pont (UK) Ltd] *

Viton® AHV. A copolymer similar to Viton A but possessing greater strength at high temperatures; Mooney viscosity 180 (100 s). [DuPont UK] *

Viton® B. Fluoroelastomer; general purpose, high temp. and fluid resistant elastomer with slightly improved thermal stability and fluid resistance over Viton A; used for o-rings, shaft seals, sol'n. coatings, roll covers, tubing, hose lining; may be cured with Diak curing agents. [DuPont] *

Viton® B-50. Fluoroelastomer; low visc. analog of Viton B which is easier to process; vulcanizate properties are similar to those of Viton B except for tensile strength and elong. which are slightly lower; used in high solids sol'n. coatings and for blending with Viton B for visc. control. [DuPont] *

Viton® C-10. Fluoroelastomer; very low visc. form of Viton A, especially useful in sol'n. applications such as sealants, high solids coatings, dipped goods, and adhesives; used in dry polymer applications where its low visc. permits easy mixing and molding of highly loaded, high durometer stocks; vulcanizate properties similar to those of Viton A except for tensile strength and modulus which are slightly lower. [DuPont] *

Viton® Curative No. 20. 33% Organophosphonium salt and 67% fluoroelastomer; used in combination with Viton Curative No. 30 or No. 40 as a curing system for Viton fluoroelastomer. [DuPont] *

Viton® Curative No. 30. 50% Dihydroxy aromatic compound and 50% fluoroelastomer; used in combination with Viton Curative No. 20 as a curing system for Viton fluoroelastomers. [DuPont] *

Viton® Curative No. 40. 33% Benzophenone compound and 67% fluoroelastomer; used in combination with Viton Curative No. 20 as a curing system for Viton fluoroelastomers. [DuPont] *

Viton® E-430. A proprietary fluoroelastomer with good processing and storage properties. [DuPont UK] *

Viton® Free Flow HD. 50% Flucropolymer/50% HDPE blend; processing additive for improved performance in HDPE resins. [DuPont]

Viton® Free Flow TA. 50% Fluoropolymer/50% talc partitioning agent blend; processing additive designed for reduced interactions with talc antiblocks. [DuPont]

Viton® GFLT. Fluoroelastomer; premium grade offering hydrocarbon fuel and methanol resist.; excellent resist. to hot water, steam, aq. acids, and solvs.; recommended

‡ Trade name and manufacturer not verified   § Trade name without identified manufacturer

for o-rings, lip seals, molded shapes; peroxide-curable. [DuPont] *

Viton® LN. A waxy semi-solid fluoroelastomer of the Viton type used as a plasticizer to improve molding and extrusion characteristics. [DuPont UK] *

Vitradur. Ultra high molecular weight high density polyethylene; used for paper and textile accessories, bunker linings, and skating rinks. [Stanley Smith & Co Plastics Ltd]

Vitrafix. Coordination compound for glass laminates. [ICI Chem. & Polymers Ltd] †

Vitrafos®. CAS 10124-56-8; Sodium hexametaphosphate; food additive for baking, cereals, meat, poultry, dairy/ cheese, processed foods. [Rhone-Poulenc Food Ingreds.]

Vitrahose. Polyvinyl chloride; tubes and hoses for many applications. [Stanley Smith & Co Plastics Ltd] †

Vitralen. Ultra high molecular weight high density polyethylene sheet; used for orthopedic splints. [Stanley Smith & Co Plastics Ltd]

Vitralene. Polypropylene sheet and block; used for anti-acid fabrications, cutting boards. [Stanley Smith & Co Plastics Ltd]

Vitralex. Acrylonitrile/butadiene/styrene (ABS), polyvinylidene fluoride (PVDF) sheet and block; used for anti-acid fabrications. [Stanley Smith & Co Plastics Ltd]

Vitrapad. Polyvinylchloride or polyolefine based sheet and block; used for cutting boards for all industries. [Stanley Smith & Co Plastics Ltd]

Vitraplas. Polyolefins; used for general purpose plastics applications. [Stanley Smith & Co Plastics Ltd]

Vitrathene. Polyethylene sheet, rod, block and massive castings; used for radiation protection, chemically resistant plant, electronic and radar insulation, orthopedics, textile and paper trade accessories, bunker linings, cutting boards and tabletops. [Stanley Smith & Co Plastics Ltd]

Vitre-colloid. A proprietary cellulose acetate. §

Vitreo-colloid. A proprietary trade name for cellulose acetate plastic. §

vitreon. A porcelain used in the condensing plant for nitric acid.

vitreosil. (silica glass). CAS 60676-86-0; Fused Silica, SiO_2; used as an ablative material in rocket engines, as fibers in reinforced plastics; special camera lenses.

Vitride®. Sodium bis(2-methoxyethoxy) aluminum hydride; reducing agent. [Zeeland]

vitriolated soda. Obsolete term for sodium sulfate, Na_2SO_4.

vitriolated tartar. Obsolete term for potassium sulfate, K_2SO_4.

vitriol stone. Impure ferric sulfate obtained by the oxidation of pyrites; used in the manufacture of fuming sulfuric acid.

Vitrite. Multivitamin syrup; dietary supplement. [Marfleet Refining Co] *

Vitrite. Polyolefin-based sheet and block; used for nuclear shielding and radiation protection. [Stanley Smith & Co Plastics Ltd]

Vitromail®. A range of inorganic decoration colors; available in the form of silk screen pastes, thermoplastics, and powder for use in the enameling and glass industries. [Bayer AG]

Vitrone. Polyvinyl chloride sheet; used for thermoformed packaging, chemically resistant fabrications. [Stanley

Smith & Co Plastics Ltd]

Vivactil®. CAS 1225-55-4; Protriptyline hydrochloride; an antidepressant. (U.S.) [Merck & Co Inc]

Vivalan. CAS 35604-67-2; A proprietary preparation of viloxazine hydrochloride; an antidepressant. [ICI Chem. & Polymers Ltd]

Vivatec. CAS 83915-83-7; Lisinopril; for the treatment of hypertension. [Merck & Co Inc]

Vivazid. Lisinopril, hydrochlorothiazide; for the treatment of hypertension in patients for whom combination therapy is appropriate. (Denmark) [Merck & Co Inc]

Vivol. Cloth and textile oils. [Crosfield Chemicals Ltd] †

Vizor. CAS 2164-08-1; Herbicide containing lenacil. [ICI Chem. & Polymers Ltd]

Vizor. CAS 2164-08-1; Suspension concentrate containing 440 g lenacil per liter; used for control of annual dicoyledons and meadow grass in beet, fruit herbaceous perennials. [Farm Protection Ltd]

Vlemasque. Sulfurated lime; a solution of lime, sublimed sulfur, and water; scabicide. [Dermik Laboratories Inc] *

Vlem-Dome. See Vlemasque. [Miles Pharmaceuticals] *

Vlieseline. Nonwovens for the apparel industry. [Carl Freudenberg] *

V.M. and P. Naphtha. Varnish-maker's and painter's naphtha, a deodorized petroleum product.

Vocol. Zinc-O,O-di-n-butylphosphorodithioate; vulcanization accelerator. [Monsanto Co; Monsanto plc]

Vogan. A vitamin A concentrate. §

Vogel's alloy. An alloy containing 8 parts copper, 1 part zinc, 2 parts tin, and 1 part lead; used for polishing steel.

Voidform. Cut polystyrene shapes and molded cylinders; for formation of voids within reinforced concrete structures and for providing profiles with cast concrete. [Vencel Resil Ltd]

Voidmaster. Molded expanded polystyrene panels; used between prestressed concrete beams to provide floor insulation. [Vencel Resil Ltd]

Voidox® 100 %; A modified fatty acid derivative of a substituted phenol; a food grade antioxidant. [Guardian Chemical] *

Vol. Ammonium carbonate, $(NH_4)_2CO_3$.

Volan®. For the chemical industry. [DuPont UK]

Volathion®. CAS 14816-18-3; Phoxim; foliage- and soil-applied insecticide for control of lepidopterous larvae, beetles and their larvae and locusts on a wide range of crops including cotton, maize and vegetables. [Bayer AG]

volatile alkali. (alkaline air, spirit of Hartshorn). Ammonia, NH_3.

volatile liniment. Liniment of ammonia.

volatile oil of bitter almonds. Benzaldehyde, $C_6H_5 \cdot CHO$.

volatile salt. Ammonium carbonate, $(NH_4)_2CO_3$.

Volaton®. CAS 14816-18-3; Phoxim; foliage- and soil-applied insecticide for control of lepidopterous larvae, beetles, and their larvae and locusts on a wide range of crops including cotton, maize, and vegetables. [Bayer AG]

volcanite. A mixture of selenium and sulfur found in the Lipari Islands.

Volck. Oil insecticide. [Chevron] *

Volckmann's solution. A solution containing thymol, alcohol, glycerin, and water.

Volclay HPM-20. Sodium bentonite, tech.; suspending agent, viscosifier, gellant, and binder for household

---

* Trade name not verified as available   † Trade name verified as obsolete

products, automotive products, aerosols, paints, and enamels. [Am. Colloid]

Vole. A cotton fabric grade of Tufnol industrial laminates. [Tufnol Ltd]

Volenite. A rubber substitute made from rosin, oil, and some fibrous material. §

Volidan and Volidan 21. Megestrol acetate with ethinylestradiol tablets; oral contraceptives. [British Drug Houses] *

Volital. CAS 2152-34-3; A proprietary preparation of pemoline; central nervous stimulant. [LAB Ltd] ‡

Volkite. A molded rubber product used for electrical insulation. §

Volomite. An abrasive consisting of tungsten carbide. §

Volpar. Spermicidal contraceptive containing phenylmercuric acetate; gels, paste and foaming tablets. [British Drug Houses] *

Volpo. Polyethoxylated fatty alcohols. [Croda Chem. Ltd]

Volpo 3. CAS 9004-98-2; 52581-71-2; Oleth-3; emollient, lubricant, emulsifier, solubilizer for cosmetic application; emulsifier in astringent creams and lotions; clear gel formation; superfatting in shampoos, foaming bath preparations, and Carbopol gels; used in cold waves, depilatories, and hair straighteners; solubilizer for bromo acids in lipsticks and liq. rouge; spreading agent for bath oils. [Croda Inc.]

Volpo 5. CAS 9004-98-2; Oleth-5; emollient, lubricant, emulsifier, solubilizer for cosmetic applications. [Croda Inc.]

Volpo 10. CAS 9004-98-2; Oleth-10; emollient, lubricant, emulsifier, solubilizer for cosmetic applications. [Croda Inc.]

Volpo 20. CAS 9004-98-2; Oleth-20; emollient, lubricant, emulsifier, fragrance solubilizer for cosmetic applications. [Croda Inc.]

Volpo 25 D 3. CAS 68131-39-5; PEG-3 C12-15 ether; emulsifier, dispersant, wetting agent, gelling agent, scouring and solubilizing agent for industrial and cosmetic applications. [Croda Chem. Ltd.]

Volpo 25 D 10. CAS 68131-39-5; PEG-10 C12-15 ether; emulsifier, dispersant, wetting agent, gelling agent, scouring and solubilizing agent for industrial and cosmetic applications. [Croda Chem. Ltd.]

Volpo CS-3. CAS 68439-49-6; Ceteareth-3; emulsifier, dispersant, wetting agent, and gellant, solubilizer for industrial and cosmetic applications. [Croda Chem. Ltd.]

Volpo L4. CAS 9002-92-0; Laureth-4; general purpose emulsifier and dispersant. [Croda Chem. Ltd.]

Volpo N3. CAS 9004-98-2; Oleth-3, distilled; emulsifier, dispersant, wetting agent, gelling agent, scouring and solubilizing agent for industrial and cosmetic applications. [Croda Chem. Ltd.]

Volpo O3. CAS 9004-98-2; Oleth-3; emulsifier, dispersant, wetting agent, gelling agent, scouring and solubilizing agent for industrial and cosmetic applications. [Croda Chem. Ltd.]

Volpo S-2. CAS 9005-00-9; Steareth-2; emulsifier for o/w systems, cosmetics; stable over wide pH range. [Croda Inc.]

Volpo T-3. CAS 24938-91-8; Trideceth-3; emulsifier, dispersant, wetting agent, gelling agent, scouring and solubilizing agent for industrial and cosmetic applications. [Croda Chem. Ltd.]

Voltalef. PCTFE; for molded parts. [Atochem Inc] *

Voltaren. CAS 15307-79-6; Diclofenac sodium; anti-inflammatory. [Ciba-Geigy Corp] *

Voltarol. Anti-inflammatory/analgesic. [Ciba plc] *

Voltarol Retard. Anti-inflammatory /analgesic. [Ciba plc] *

Voltoids®. A trade name for compressed tablets of ammonium chloride used in the preparation of voltaic cells. §

Volucon. Standard volumetric concentrates. [May & Baker Ltd] *

Volucon. Volumetric concentrates for rapid preparation of laboratory solutions. [Rhone-Poulenc Laboratory Prods. Ltd]

Volunteered. Mixture of dalapon and di-1-p-menthene; a grass weed herbicide. [Mandops (UK) Ltd]

Volusol. Standard volumetric solutions. [May & Baker Ltd]*

vomiting salt. CAS 7446-20-0; EINECS 231-793-3; Zinc sulfate, $ZnSO_4$; used in medicine.

Von Forster powder. A gelatinized nitrocellulose flake powder, with a little calcium carbonate.

Vonges dynamite. An explosive containing 75% nitroglycerin, 20.8% randanite (decomposed felspar), 3.8% quartz, and 0.4% magnesium carbonate.

Vontrol. CAS 3254-89-5; Diphenidol hydrochloride; antiemetic. [SmithKline Beecham]

Von Vetter's solution. An aqueous solution of glycerin, sugar, and potassium nitrate; used to preserve anatomical specimens.

Voran. A proprietary polyether-based polyurethane elastomer crosslinked with diamine. [Dow UK] *

Voranate 3071. Specialty isocyanate; for PU industry for appliances and specialty foams. [Dow]

Voranate T-80, Type I, Type II. CAS 584-84-9; TDI; for PU industry for flexible slabstock foam, molded flexible foam. [Dow]

Voranol. Polyether polyols; used with isocyanates in the formation of urethanes. [Dow UK] *

Voranol 202, 225. Polyether polyol; for PU industry for appliances and rigid foams. [Dow]

Voranol 234-630. Polyether polyol; triol functionality, high reactivity for PU industry for industrial/ consumer RIM and structural polymers, dynamic elastomers, adhesives, binders, coatings, and sealants. [Dow]

Voranol 3741. Polyether polyol; for flexible slab foam. [Dow]*

Voranol 4148. Polyether polyol; triol functionality for PU industry for molded flexible foam. [Dow]

Vorite 63. Urethane prepolymer (TDI-based). [CasChem] *

Vorite 105. Polymerized castor oil; pigment wetting/dispersing agent; plasticizer for resins, gums, polymers; lubricant, penetrant; coupling solv.; adhesion promoter; for cellulose lacquers, inks, adhesives, industrial lubricants, polishes, caulks, leather dressing, hydraulic fluids, rubber compounding, gasket cement. [CasChem]

Vortel. A proprietary preparation of chlorprenaline, ethomoxane and methapyrilene. [Eli Lilly & Co] †

Vossenblue 705-81. Iron blue pigment; for publication gravure inks. [Degussa]

Vostec. Boiler water treatment. [Grace Dearborn Ltd]

VPA No. 3. Processing aid providing improved mold release for Viton® fluoroelastomers cured with bisphenol or peroxide systems; for use for o-rings, seals, molded shapes, diaphragms, and tubing. [DuPont]

VPC®. Polyester urethane; high performance coatings for high speed continuous sheet stock, vinyl, paper, metal, plastic. [Ashland] *

---

‡ Trade name and manufacturer not verified § Trade name without identified manufacturer

VPI. Vapor phase corrosion inhibitors. [Contract Chemicals (Knowsley) Ltd]

VR-110. CAS 39198-34-0; 2,2´-Azobis(2,4,4-trimethyl-pentane); polymerization initiator. [Wako Pure Chem. Ind.; Wako Chem. USA]

VR-160. CAS 927-83-3; 2,2´-Azobis(2-methylpropane); polymerization initiator. [Wako Pure Chem. Ind.; Wako Chem. USA]

VR-500. Air dry phenolic coatings; thin film coating for structural steel, construction equip., sewage plants, marine finishes, heating systems. [Heresite Protective Coatings]

vraic. (varech). French names for kelp (qv).

VR Coving. Expanded polystyrene; coving for domestic properties. [Vencel Resil Ltd]

V-thane. Solid polyurethane elastomers [Hallam Polymer Engineering Ltd] *

Vuelite. A proprietary trade name for a cellulose acetate plastic; Vuelite-reinforced is a transparent cellulose acetate sheet reinforced with wire mesh. §

Vuepak. A proprietary trade name for an acetate wrapping material. §

Vulcabest. Incombustible cladding panels. [Vulcan Plastics Ltd]

Vulcaboard. Vandal resistant panel. [Vulcan Plastics Ltd]

Vulcabond®. Bonding agent for PVC. [Akzo Chemie UK Ltd]

Vulcabond® C 10. rubber to metal adhesion promoter; specially designed to improve adhesion between rubber and steel cord in radial tires [Akzo] *

Vulcabond® E. Polymeric phenolic compound in 3N aq. ammonia; textile bonding agent for rubber to polyester [Akzo] *

Vulcabond® N 15. Rubber to metal adhesion promoter. [Akzo] *

Vulcabond® TX. Isocyanate compound in solv. rubber to textile bonding agent. [Akzo] *

Vulcabond® VP. 25% Solution of an isocyanurate-type polymer of TDI in dibutyl phthalate; bonding agent added to PVC plastisols to improve adhesion to a reinforcing substrate, e.g., nylon and polyester textiles. [Akzo] *

Vulcabrite. Stainless steel faced cladding panels. [Vulcan Plastics Ltd]

Vulcaflex. A complex substituted secondary amine of the dimethoxy-diphenylamine type; an antioxidant which offers protection against flex-cracking in rubber materials. [Vulnax International Ltd] *

Vulcaid. A proprietary litharge of small particle size. §

Vulcaid 27. Zinc butyl xanthate; a proprietary trade name for a rubber vulcanization accelerator. §

Vulcaid 28. CAS 103-49-1; Dibenzylamine; a proprietary trade name for a rubber vulcanization accelerator. §

Vulcaid 33. A liquid amine condensation product; a proprietary trade name for an antioxidant. §

Vulcaid 44. A proprietary trade name for an antioxidant; a naphthol-amine reaction product. §

Vulcaid 55. An acetaldehyde-aniline condensation product; a proprietary trade name for an antioxidant. §

Vulcaid 111. Butyraldehyde-aniline; a proprietary trade name for a rubber vulcanization accelerator. §

Vulcaid 222. CAS 97-74-5; Tetramethyl-thiuram monosulfide; a proprietary trade name for a rubber vulcanization accelerator. §

Vulcaid 444B. Heptaldehyde-aniline; a proprietary trade name for a rubber vulcanization accelerator. §

Vulcaid DPG. CAS 102-06-7; Diphenyl-guanidine; proprietary trade name for a rubber vulcanization accelerator.§

Vulcaid LP. Lead pentamethylene dithiocarbamate; proprietary rubber vulcanization accelerator. §

Vulcaid P. Piperidine pentamethylene dithiocarbamate; melts at 172 C and is soluble in water, alcohol, and benzene; a rubber vulcanization accelerator. §

Vulcaid ZP. Zinc pentamethylenedithiocarbamate; a proprietary rubber vulcanization accelerator. §

Vulcalap. GRP shiplap weatherboard. [Vulcan Plastics Ltd]

Vulcalon. GRP faced cladding panels. [Vulcan Plastics Ltd]

Vulcalucent. GRP translucent sheets and domelights. [Vulcan Plastics Ltd]

Vulcamel. Butyraldehyde-ammonia; a proprietary a rubber vulcanization accelerator. §

Vulcamin. Aluminum faced cladding panels. [Vulcan Plastics Ltd]

Vulcan. Herbicide containing clopyralid and bromoxynil. [ICI Chem. & Polymers Ltd] †

Vulcan® XC72R. CAS 1333-86-4; Carbon black; for coloring conductive plastics. [Cabot; Cabot Carbon Ltd]

Vulcan Bronze. A proprietary trade name for a bearing bronze containing 1.0% silicon with iron and nickel. §

Vulcanex. A proprietary rubber vulcanization accelerator; it is a Schiff's base. §

Vulcaniline. CAS 138-89-6; p-Nitroso-dimethylaniline; a proprietary trade name for a rubber vulcanization accelerator. §

Vulcanine. A patented material made from rubber, asbestos, litharge, lime, sulfur, and zinc oxide. §

Vulcanite. See Ebonite; it is also the name for a nitroglycerin explosive. §

vulcanized fiber. (hard fiber; red fiber; grey fiber; vegetable fiber; whalebone fiber; Egyptian fiber; fiberoid; horn fiber) A material made by treating sheets of paper with zinc chloride solution and subjecting the Gelatinized sheets to pressure. The paper is sometimes mixed with glycerin and vulcanized oils; used for making valve discs, brake blocks, and tubes.

vulcanized oils. See Thinoline.

vulcanized paper. See vulcanized fiber.

Vulcanol. A proprietary rubber vulcanization accelerator. §

Vulcanox Crack and Joint Sealant. Polyurethane; poured or extruded into cracks and joints as a moisture-proof sealant; cures to a solid with rubber band elasticity. [Metalcrete Mfg Co]

Vulcan powder. An explosive containing 30% nitro-glycerin, 52.5% sodium nitrate, 7% sulfur, and 10.5% charcoal.

Vulcaplas. A proprietary trade name for an organic polysulfide synthetic elastic material. §

Vulcaplast. GRP faced, aluminum framed cladding panels. [Vulcan Plastics Ltd]

Vulcapont. A proprietary rubber vulcanization accelerator; it consists of equal parts of Thionex and Vulcanol. §

vulcasbeston. A mixture of rubber and asbestos; a heat and electrical insulator.

Vulcase. A proprietary preparation of colloidal sulfur. §

Vulcastab® EFA. Formaldehyde, ammonia, ethyl chloride condensate; secondary gelling agent and stabilizer for latex foams [Akzo] *

Vulcastab® LW. Ethylene oxide and fatty alcohol condensate; stabilizer for latex rubber processing [Akzo] *

Vulcastab® T. Ammonium polymethacrylate solution stabilizer and thickener for latex rubber processing [Akzo] *

Vulcasteel. Steel faced cladding panels. [Vulcan Plastics Ltd]

Vulcastop. Short stoppers used in water emulsion polymerization processes; for synthetic elastomers production. [Vulnax International Ltd] *

Vulcatuf. Bandit resistant panel. [Vulcan Plastics Ltd]

Vulcawall; Thermally broken curtain wall cladding system. [Vulcan Plastics Ltd]

Vulcazol. Furfuramide, a vulcanization accelerator.

Vulcoferran. A trademark for linings principally of rubber or ebonite for chemical apparatus. §

Vulcogene. Thiocarbanilide; a proprietary trade name for a rubber vulcanization accelerator. §

Vulcogene ND. CAS 102-06-7; Diphenylguanidine; a proprietary trade name for a rubber vulcanization accelerator. §

Vulcoid. A proprietary trade name for a phenolic resin impregnated vulcanized fiber. §

Vulconex. Ethylidineaniline; a proprietary trade name for a rubber vulcanization accelerator. §

Vul-Cup. CAS 25155-25-3; α,α'-Bis(t-butylperoxy) diisopropylbenzene; crosslinking agent for rubber and plastics. [Hercules; Hercules Ltd]

Vul-Cup 40KE and R. CAS 25155-25-3; Di-(t-butylperoxyisopropyl)benzene; used as vulcanizing and polymerization agents. [Hercules] *

Vulkacit®. Range of accelerators. [Bayer AG; Bayer plc]

Vulkacit® 470. A rubber vulcanization accelerator; it is a condensation product of homologous acroleins with aromatic bases. [Bayer AG] *

Vulkacit® 576. A condensation product of homologous acroleins aromatic bases; a rubber vulcanization accelerator most suitable for regenerated rubber. [Bayer AG]*

Vulkacit® 774. The dithiocarbamate of cyclohexyl ethylamine; a proprietary trade name for an ultra-accelerator for rubber. [Bayer AG] *

Vulkacit® 1000. CAS 93-69-6; o-Tolylbiguanide; a rubber vulcanization accelerator. [Bayer AG] *

Vulkacit® A. Aldehyde-ammonia, a vulcanization accelerator. [Bayer AG] *

Vulkacit® BP. A rubber vulcanization accelerator; it is a paste, and consists of a mixture of bases. [Bayer AG] *

Vulkacit® CA. A rubber vulcanization accelerator; it is thiocarbanilide. [Bayer AG] *

Vulkacit® CRV/LG. 3-Methyl-thiazolidinethione-2; accelerator for polychloroprene elastomers, halobutyl rubbers especially chlorobutyl; used for mech. goods, cables, hoses, membranes, fabric proofings, vulcanizing sol'ns.; beige-brown pellets. [Miles; Polysar]

Vulkacit® CT. A proprietary trade name for a rubber vulcanization accelerator. [Bayer AG] *

Vulkacit® CZ/EGC. CAS 95-33-0; Benzothiazyl-2-cyclohexyl sulfenamide; fast accelerator giving delayed onset of cure; used for tires, dynamically stressed technical goods, technical moldings and extrudates. [Miles; Polysar]

Vulkacit® D/C. CAS 102-06-7; Diphenyl guanidine; accelerator for NR, IR, BR, SBR, NBR, CR, with very slow onset of cure; used alone for bulky goods, e.g., tires, buffers, roll covers; secondary accelerator with mercaptos for mech. goods, extrudates and moldings, footwear, fabric proofings, cable sheathings and insulation. [Miles; Polysar]

Vulkacit® DM/C. CAS 120-78-5; Dibenzothiazyl disulfide,

mineral oil coated; delayed action, semi-ultra accelerator for NR, IR, BR, SBR, NBR, CR, IIR, and chlorosulfonated polyethylene; used for mech. goods, tires, conveyor belts, cables, hoses, rubber footwear, expanded rubber goods, proofed fabrics, intricate molded goods; yellowish. coated powd. [Miles; Polysar]

Vulkacit® DZ/EGC. CAS 95-33-0; Benzothiazyl 2-dicyclohexyl sulfenamide; accelerator with delayed onset of cure for dynamically stressed rubber, e.g., NR, IR, BR, SBR, NBR; used for large tires, conveyor belts, shock absorbers, intricate molded goods with long flow periods. [Miles; Polysar]

Vulkacit® FP. Methylene-p-toluidine; a proprietary trade name for a rubber vulcanization accelerator. [Bayer AG]*

Vulkacit® H. CAS 100-97-0; Hexamethylenetetramine, a vulcanization accelerator. [Bayer AG] *

Vulkacit® M. CAS 149-30-4; Mercaptobenzothiazole; a proprietary rubber vulcanization accelerator. [Bayer AG]*

Vulkacit® Merkapto/C. CAS 149-30-4; 2-Mercaptobenzothiazole, mineral oil coated; semi-ultra accelerator for NR, IR, BR, SBR, NBR, IIR, and EPDM; used alone in bulky goods or in combination for molded and extruded goods, hoses, conveyor belts, tires, footwear, cables, expanded rubber goods, proofed fabrics, and latex articles. [Miles; Polysar]

Vulkacit® Merkapto/MGC. CAS 155-04-4; Zinc salt of 2-mercaptobenzothiazole, mineral oil coated; see Vulkacit Merkapto/C. [Miles; Polysar]

Vulkacit® P. Piperidinepiperidyldithio-formate, a vulcanization accelerator. [Bayer AG] *

Vulkacit® P Extra. The zinc salt of ethylphenyl-dithiocarbaminic acid, a rubber vulcanization accelerator. [Bayer AG] *

Vulkacit® TR. A rubber vulcanization accelerator; it is a mixture of the free bases of polyamines of ethylene. [Bayer AG] *

Vulkacit® Thiuram. CAS 137-26-8; Tetramethylthiuram disulfide; a rubber vulcanization accelerator. [Bayer AG]*

Vulkadur®. Reinforcing resin. [Bayer AG; Bayer plc]

Vulkalent®. Retarders. [Bayer AG; Bayer plc]

Vulkalent® TM. CAS 136-85-6; Blend of 4- and 5-methylbenzotriazole; prevulcanization retarder for sulfur-modified CR and halobutyl rubbers; also effective in NBR when Vulkacit DM/sulfur curing systems are used; applications incl. conveyor belting, hose, and molded goods. [Miles; Polysar]

Vulkanol®. Synthetic plasticizers. [Bayer AG; Bayer plc]

Vulkanol® 88. Dibutyl methylene bisthioglycolate; plasticizer to improve low-temp. flexibility and elastic behavior of vulcanizates; used for nitrile and chloroprene rubber. [Bayer; Miles; Polysar]

Vulkanol® 90. Di-2-ethylhexyl ester thiodiglycolate; plasticizer for natural and synthetic rubber incl. NBR, NBR/PVC blends, CR, SBR, BR, chlorosulfonated polyethylene, and chlorinated polyethylene; used for hoses, seals, shock absorbers, roll covers, and conveyor belting. [Miles; Polysar]

Vulkanol® OT. Ether thioether; plasticizer to improve low-temp. flexibility and elastic behavior of vulcanizates; used for nitrile and chloroprene rubber; applications incl. hoses, seals, mech. goods. [Bayer; Miles; Polysar]

Vulkanox®. Antioxidants. [Bayer AG; Bayer plc]

---

‡ Trade name and manufacturer not verified   § Trade name without identified manufacturer

Vulkanox® 4010 NA. CAS 101-72-4; N-isopropyl-N´-phenyl-p-phenylene diamine; staining and discoloring antioxidant/antiozonant for protection of rubber from ozone attack, oxidation, heat aging, flexcracking, and rubber poisons; suitable for natural and synthetic rubbers; used for dynamically stressed goods, tech. goods, spring components, conveyor and transmission belting, seals, insulation. [Miles; Polysar]

Vulkanox® 4020. N-(1,3-dimethylbutyl)-N´-phenyl-p-phenylene diamine; staining stabilizer for cold- and hot-polymerized emulsion and solution SBR and BR; protects vulcanizates against heat, oxidation, and flexcracking. [Miles; Polysar]

Vulkanox® BKF. CAS 119-47-1; 2,2´-Methylene-bis(4-methyl-6-t-butylphenol); nonstaining antioxidant for natural and synthetic rubbers; used in transparent, bathing, surgical tech., and latex goods, fabric proofings; stabilizer for hot- and cold-polymerized emulsion-SBR, BR, and NBR, and solution-BR, IR, and ABS. [Miles; Polysar]

Vulkanox® DDA. Styrenated diphenyl amine; staining and discoloring antioxidant for natural and synthetic rubbers, exp. CR, giving protection against oxidation, heat, flexcracking, and rubber poisons; used for inner tubes, sponge rubber, seals/gaskets, soles, latex goods, heat-resistant articles; stabilizer for cold and hot-polymerized NBR and NBR or SBR latexes. [Miles; Polysar]

Vulkanox® HS/LG. 2,2,4-Trimethyl-1,2-dihydroquinoline, polymerized; antioxidant for tech. goods and rubber products; used in latex. [Miles; Polysar]

Vulkanox® KB. CAS 128-37-0; 2,6-Di-t-butyl-p-cresol; nondiscoloring/nonstaining antioxidant for light colored and transparent natural and synthetic rubber goods; used in fabric proofings, toys, bathing, latex, and dipped goods; stabilizer in emulsion and solution-polymerized elastomers (SBR, NBR, BR, IR); costabilizer for ABS. [Miles; Polysar]

Vulkanox® MB-2/MGC. 4- and 5-Methylmercaptobenzimidazole, mineral oil coated; nondiscoloring/nonstaining antioxidant for natural and synthetic rubbers; synergist; protects against rubber poisons; sensitizing agent for latex. [Miles; Polysar]

Vulkanox® OCD/SG. Octylated diphenyl amine; weakly staining and discoloring antioxidant for CR, NR, IR, SBR, BR, NBR, protecting against rubber poisons; used for dynamically stressed goods, tires, air hoses, belts, sponge rubber, footwear, and profiles. [Miles; Polysar]

Vulkanox® ZMB2/C5. Zinc salt of 4- and 5-methylmercaptobenzimidazole; nonstaining and nondiscoloring antioxidant for natural and synthetic rubber; synergist; used mainly for heat-resistant thiuram-cured goods, white, colored, and transparent goods and latex foam; sensitizer for latexes. [Miles; Polysar]

Vulkaresen. Phenolic resins, resole types. [Hoechst AG]

Vulkasil®. Silica fillers. [Bayer AG; Bayer plc]

Vulkazon®. Antioxidants. [BASF AG]

Vulkazon® AFS/LG. Bis (1,2,3,6-tetrahydrobenzaldehyde) pentaerythritol acetal; nonstaining antiozonant for CR, IIR, CIIR, BIIR; used for hoses, extruded profiles, cables, sheeting, proofed goods; beige to gray powd. [Miles; Polysar]

Vulklor. Tetrachloro-p-benzoquinone; combines with R6 to form an effective bonding system for compounds featuring steel cord reinforcement; used to activate GMF; also functions as a vulcanizing agent without sulfur; used in natural, SBR, nitrile, butyl and chlorobutyl rubbers. [Uniroyal] *

Vulkollan®. High quality polyurethane elastomer. [Bayer AG; Bayer plc]

Vulkollan® 18W. Water-crosslinked PU elastomer; exhibits superior balance of properties incl. ultimate tensile strength, tear strength, compression set, and high wear resist. under wet abrasion conditions; used in pump diaphragms, coupling elements, solid tires, rollers, pump housings, seals, wipers, bushes, and split sections for antifriction bearings and flexible couplings. [Bayer; Miles]

vulpinite. A variety of anhydrite mixed with silica.

Vultac® 2. CAS 68555-98-6; Alkyl phenol disulfide; NR, NBR, SBR vulcanizer; plasticizer; NBR, SBR tackifier; accelerates SBR. [Atochem] *

Vultex. A trademark for vulcanized rubber latex preserved with ammonia. §

Vulvan. CAS 57-85-2; Testosterone propionate; androgen. [Baxter Health Care] *

Vybar® 103. CAS 8002-74-2; Synthetic wax; ethylene-derived hydrocarbon polymer; lubricant, anticaking agent, modifier; used in paraffin; used in candles to replace stearic acid; opacifies the candle and imparts resistance to thermal shock; pigment disp. to wet inorg. pigments and fillers at high loading levels; suitable for hot melt inks, mold release compounds, plastic lubricants, protective coatings, polishes, slip and antimar additives. [Petrolite]

Vybex 22008 BKFR. CAS 26062-94-2; PBT, 10% glass fiber-reinforced; flame retardant thermoplastic [Ferro/Engineering Thermoplastics] *

Vybex 22028 BKFR. CAS 26062-94-2; PBT, 15% glass fiber-reinforced; flame retardant thermoplastic [Ferro/Engineering Thermoplastics] *

Vybex 40001 NA. PET/PC alloy; high-impact grade. [Ferro/Engineering Thermoplastics] *

Vybex 40004 NAFC. PET/PC alloy; food contact grade. [Ferro/Engineering Thermoplastics] *

Vycel. Petrol.-derived surfactant; patented surfactant. [Crowley Chem.]

Vydate®. CAS 2135-22-0; Granules containing 10% w/w oxamyl; systemic insecticide and nematicide. [DuPont UK]

Vydax®. Fluorocarbons. [DuPont UK]

Vydon. A proprietary vinyl resin. §

Vydyne. Nylon and nylon copolymer resins including 66, 69 and 66/6 copolymers; designed for molding and extrusion applications. [Monsanto Co] *

Vyflex E. Environmentally friendly outdoor thermoplastic coating power; used for traffic engineering, fencing posts and panels. [Plascoat Systems Ltd]

Vyflex NT80S. A proprietary PVC powder coating material, containing no toxic metals and supporting no microbiological growth; can be safely brought into contact with drinking water. [Plastic Coatings Ltd] *

Vygen. A trademark for PVC resins. §

Vykamol 83G. Sorbitan ester/polysorbate blend. [Croda Surf. Ltd]

Vyloc. Oven wall catalysts. [DuPont UK]

Vylor®. Polymer. [DuPont UK]

Vylox®. Chemicals used in paper/paperboard products. [Reichhold]

---

* Trade name not verified as available   † Trade name verified as obsolete

Vynamon. Pigments for polyvinyl chloride. [ICI Chem. & Polymers Ltd]

Vynathene® . Vinyl acetate copolymers in liquid form; industrial chemicals for use as pour point depressant additives in fuels and lubricants. [Quantum Chemical Corp]

Vynel N. Polymer emulsion; hand builder; imparts full, resilient hand to cotton and blended fabrics. [Yorkshire Pat-Chem]

Vyram® . Thermoplastic elastomer; general-performance rubber replacement with temp. and oil resistance equivalent to EPDM, SBR, and natural rubber. [Advanced Elastomer Systems]

W

W2 Beta. A proprietary name for a special Angelo shellac. [Zinsser NV] *

W180. Ester/quaternary amine blend; softener, rewetting agent for cotton, nylon, and polyester fibers. [Hart Chem. Ltd.]

Wachsemulsion 1864. Carnauba/nonionic wax emulsion; mfg. of cleaning agents with glossing effect. [Zschimmer & Schwarz]

Wackenroder's solution. A solution obtained by passing sulfuretted hydrogen through an aqueous solution of sulfurous acid. The solution contains $H_2S_4O_6$, $H_2S_3O_6$, $H_2S_5O_6$, and colloidal sulfur.

Wackerschellak. An artificial shellac obtained by the condensation and polymerization of acetaldehyde. §

wad. (bog manganese). An earthy variety of hydrated manganese oxide, MnO_2.

Wafex. CAS 8061-52-7; Unfermented calcium lignosulfonate; industrial binder, dispersant, and emulsifier. [Borregaard LignoTech]

Wafolin. Calcium/magnesium lignosulfonate; pellet binder in animal feeds; contributes some nutritive value. [Borregaard LignoTech]

Wagner's reagent. A solution of 2 g potassium iodide in 100 cc water.

Wahnerit. A German synthetic varnish paper product used for electrical insulation. §

wahoo bark. The root bark of *Euonymus atropurpureus*.

Wakal® A Range. Alginates; gelling agents for desserts and filling creams. [Grünau]

Wakal® J Range. CAS 9000-40-2; Locust bean gum; thickener and stabilizer for food industry. [Grünau]

Wakal® K Range. CAS 9000-07-1; Carrageenan; gelling agents for desserts and filling creams. [Grünau]

Wakefield grease. See Yorkshire grease.

Waksol A. Waxy oil. [Sasolchem]

walchowite. See retinite.

walkerite. A clay of the fuller's earth type.

Walker's earth. Fuller's earth.

Walkover Moss Killer. CAS 10028-22-5; Ferrous sulfate; used for moss control in turf. [Walkover Sprayers Ltd]

Wallkyd®. Synthetic resins, solutions. [Reichhold]

Wallkyd® 11-024. Flat alkyd. [Reichhold]

Wallpol®. Synthetic resins. [Reichhold]

Wallpol® 40-136. Vinyl acrylic emulsion copolymer; general purpose coatings polymer. [Reichhold/Emulsion Polymers] *

Wallpol® 40-165. Vinyl acetate acrylic emulsion copolymer; high performance emulsion for interior semigloss enamels; excellent scrub resist. and wet adhesion. [Reichhold/Emulsion Polymers] *

wall saltpeter. See nitrocalcite.

walnut oil. (oils, walnut) CAS 8024-09-7; 84604-00-2; Oil derived from the nut meats of walnuts, *Juglans spp.*; emollient for makeup, skin, and hair care products [Arista Industries; Pacific Anchor; Penta Mfg.]

Walsrode Powder. A proprietary smokeless powder containing 98.6% nitrocotton and 1.4% volatile matter. §

Wando Steel. A proprietary non-shrinking steel containing 1.05% manganese, 0.5% chromium, 0.5% tungsten, and 0.95% carbon. §

Wanin AM. CAS 8061-53-8; Unfermented ammonium lignosulfonate; binder, dispersant, and auxiliary tanning agent. [Borregaard LignoTech]

W.A. powder. An American smokeless powder. It is a guncotton-nitroglycerin powder, with barium and potassium nitrates.

waras. (wars, warrus). A resinous powder which covers the seed pods of *Flemingia congesta*, of India; used in Arabia as a dye.

Warburg's tincture. A quinine preparation containing aloes, rhubarb, gentian, camphor, and oils; used in India for the treatment of malaria.

Warcodet D. Alkylphenol ethoxylate nonionic surfactant as a colorless viscous liquid; detergent and emulsifier with wetting, penetrating and soil suspending properties used for textiles. [Warwick International Ltd] *

Warcodet K54. Sodium alkylaryl sulfonate; anionic surfactant in the form of a clear golden brown liquid; detergent with wetting, penetrating and soil suspension properties; used in general purpose detergents and in the textile industry eg for scouring. [Warwick Chemical Ltd] *

Warcodet V. Phosphate ester in the form of sodium/potassium salts; anionic surfactant for wetting, scouring and washing off. [Warwick International Ltd] *

Warcodye CLP. Low foam cationic surfactant; leveling wool and polyamide with acid reactive dyes. [Warwick International Ltd] *

Warcodye RWL. Ethoxylated amine based amphoteric; leveler used in reactive dyes on wool. [Warwick International Ltd] *

Warcosoft WSC. Cationic surfactant in liquid form; permanent softener for use with acrylics. [Warwick International Ltd] *

Warcowet O. CAS 577-11-7; EINECS 209-406-4; Sodium dioctyl sulfosuccinate in liquid form; wetting agent with good penetration, stable to boiling, most effective at 30-60 C; used in textiles e.g., in scouring, bleaching, package dyeing, piece dyeing and printing. [Warwick International Ltd] *

Wardol. Nonionic emulsifiers. [Courtaulds Chemicals,

---

* Trade name not verified as available   † Trade name verified as obsolete

Leek]

Ward's stone. A concrete composed of limestone and Portland cement.

Wareflex® 650. Dibutoxyethoxyethyl adipate; plasticizer. [Sartomer]

Warefog. CAS 101-21-3; Chlorpropham as a foggable solution; for controlling sprouting in ware potatoes. [Wheatley Chemical Co Ltd] *

Warefog. CAS 101-21-3; Chlorpropham as a foggable solution; potato sprout control. [Dean Agrochemicals Ltd]

warfarin. CAS 129-06-6; A proprietary preparation of warfarin sodium; an oral anticoagulant. [Boehringer Ingelheim Ltd]

warfarin. CAS 81-81-2; Warfarin; a rodenticide. [Battle, Hayward & Bower Ltd]

Wargonin Compact. Chemically desugared sodium/calcium lignosulfonate, < 1% sugar content; water-reducing, strength-increasing, air-excluding additive for concrete. [Borregaard LignoTech]

Wargotan. CAS 8061-52-7; Unfermented calcium lignosulfonate with low calcium and iron content; auxiliary tanning agent and dispersant. [Borregaard LignoTech]

Warmaline. Expanded polystyrene veneer 2mm thick; lining for domestic walls for use beneath wallpapers. [Vencel Resil Ltd]

Warne's metal. An alloy of 26% nickel, 37% tin, 26% bismuth, and 11% cobalt.

Warrior. Residual herbicide. [ICI Chem. & Polymers Ltd]

warrus. See waras.

Wars. See waras.

wasa tar. A wood tar of a similar type to Stockholm tar, and sometimes sold under this name.

Wasc. Nonionic surfactant consisting of nonylphenol ethoxylate in liquid form; liquid cleaners for hard water. [Berol Kemi (UK) Ltd] *

wash. The fermented wort of the distilleries.

Wash-saver. Laundry bleach. [Interox Chemicals Ltd] *

Wasp Destroyer. CAS 63-25-2; Carbaryl insecticde. [Murphy Chemical Ltd] †

Waspend. Contains pirimiphos-methyl and synergized pyrethrins; pestkiller for flying or crawling pests. [ICI Garden Products]

Watchmaker's alloy. An alloy of 59% copper, 40% zinc, and 1.2% lead.

Water Brite. CAS 10124-56-8; Sodium hexametaphosphate; used for water treatment. [Flexibulk Ltd]

Waterez®. Synthetic resins. [Reichhold]

water gas. A general name for a mixture of gases obtained by the decomposition of steam by incandescent carbon. It usually contains from 43-44% carbon monoxide, 48-49% of hydrogen, 3-4% of carbon dioxide, and 3-4% of nitrogen.

water-glass. See soluble glass.

water glass. See sodium silicate.

Watergy. Water energy management. [Nalfloc Ltd]

Waterlily. MDI-based water-blown polyurethane film. [Imperial Chemical Industries plc]

Water Lock® A-100. Starch/acrylates/acrylamide copolymer; superabsorbent polymer; able to absorb or immobilize large quantities of aq. fluids, such as alkalies, dilute acids, and body fluids. [Grain Processing]

water mica. Clear transparent Muscovite (potash mica).

water of Saturn. A dilute solution of lead subacetate.

Watsonite. A proprietary material used as a mica substitute. It is made from scrap mica with a binding agent. §

wattle bark. (mimosa). A tanning material obtained from species of *Acacia*. The amount of tannin varies from 12-49%.

Watt's and Li's solution. A solution used for the electrodeposition of iron. It contains 150 g ferrous sulfate, 75 g ferrous chloride, 120 g ammonium sulfate, and 1,000 cc water.

Wave. Vinyl acrylic emulsion. [Air Prods. & Chems. Inc]

Wax C. N,N´-Distearyl ethylene diamine; a high melting point amide wax; recommended as an Internal lubricant in processing ABC, PVC, and PS. [Hoechst UK] *

wax butter. (wax oil). A thick oil obtained by the distillation of beeswax. It consists mainly of cerotene $C_{27}H_{54}$, melissine $C_{30}H_{60}$, and palmitic acid; formerly used medicinally externally and internally.

Waxemul. Wax emulsion. [Tenneco Malros Ltd] *

Waxene. A proprietary rubber vulcanization accelerator. §

Waxenol® 801. CAS 65591-14-2; Arachidyl propionate; binder; emollient, solubilizer, and ingredient in cosmetic and toiletries; lubricant for pressed powds.; metal working lubricant and corrosion inhibitor for metal surface [CasChem]

Waxenol® 810. CAS 3234-85-3; EINECS 221-787-9; Myristyl myristate; emollient with unusual afterfeel; especially for stick preps. [CasChem]

Waxenol® 815. CAS 540-10-3; EINECS 208-736-6; Cetyl palmitate; emollient additive; internal lubricant and binder in pressed powds.; in metalworking lubricant coatings. [CasChem]

Waxenol® 821. Synthetic beeswax; lipophilic emulsifier, emollient. [CasChem]

Waxenol® 822. Arachidyl behenate; emollient for sticks, creams, lotions; barrier properties for enhanced moisturization; gloss and film-forming properties. [CasChem]

Waxigel. Modified maize starch (pregelatinized). [Roquette (UK) Ltd] †

Waxilys. Maize starch; used for foods, textiles, board and paper. [Roquette (UK) Ltd]

Waxit. A wetting agent used by gold- and silversmiths in the jewelry and dentistry industries. [Degussa Ltd]

Waxolan P-5. Propoxylated (5 mol) lanolin wax; enhances lubricity and rigidity of lipsticks. [Amerchol Corp]

Waxoline. Oil and wax soluble dyes. [ICI Chem. & Polymers Ltd]

Waxsol. CAS 577-11-7; EINECS 209-406-4; A proprietary preparation of dioctyl sodium sulfosuccinate; ear drops for wax removal. [Norgine Ltd]

wax spirit. A colorless watery distillate from beeswax containing acetic and propionic acids.

wax tailings. The remaining petroleum distillate after the paraffin wax has been removed.

Wayfarer. CAS 61791-44-4; Cationic wetting agent containing 80% tallow amine ethoxylate; spreader for phosphonoglycine herbicide sprays. [Service Chemicals Ltd]

Wayfos A. Phosphate acid, acid form; antistat, emulsifier, wetting agent, detergent, coupling agent, solubilizer, lubricant, corrosion inhibitor; for alkaline built cleaners, textiles, plastics, metal, emulsion polymerization, and agricultural applications. [Olin]

Wayfos M-60. Aromatic phosphate ester, free acid; corro-

---

sion inhibitor, hydrotrope; drycleaning detergent; emulsifier for pesticides, emulsion polymerization. [Olin]

Wayhib® S. TEA phosphate ester, sodium salt; detergent builder, sequestrant, corrosion inhibitor, pipeline scale inhibitor, water circulating systems; for air conditioning and boiler treatment compounds. [Olin]

WB 200SL. A single layer microcracked chromium process. [Hanshaw Chemicals] ‡

Wear-Dated. Textile products. [Monsanto Co] *

Weathershield. Decorative range of exterior paint products. [ICI Chem. & Polymers Ltd]

Webas. Bitument emulsion for road surface treatment. [Vedag GmbH] *

Webert Alloy. A proprietary copper silicon alloy containing small amounts of manganese. §

Webnerite. A mineral that is a variety of andorite.

Wecobee® M. CAS 68334-28-1; Hydrogenated vegetable oil; used in personal care products, pharmaceuticals, food industry. [Stepan/PVO; Stepan Europe]

Wedel's oil. Consists of 1 part bergamot, 4 parts camphor, and 32 parts oil of almonds.

Wedl's stain. A microscopic stain containing 1 g orseille, 20 cc absolute alcohol, 5 cc 60% acetic acid, and 40 cc water.

Weed and Brushkiller. Emulsifiable concentrate containing 144 g 2,4-D, 32 g dicamba and 144 g mecoprop per liter; an herbicide to control perennial and woody weeds. [Synchemicals Ltd]

Weedazin. Total herbicide. [Schering Agrochemicals Ltd]

Weedazol TL. CAS 61-82-5; Amitrole; translocated herbicide. [Bayer plc]

Weed Be Gone 45. Emulsifiable concentrate of ethyl ester of 2,4-D acid; broadleaf herbicide for corn, sugar cane and wheat crops and pasture land. [Invequimica & CIA SCA]

Weed Be Gone 50. Amine salt of 2,4-D acid; broadleaf herbicide for corn, sugar cane, wheat, and barley crops. [Invequimica & CIA SCA]

Weed-B-Gon. Weed killer. [Chevron] *

Weed Ender. CAS 75-60-5; Cacodylic acid; a post-emergence contact weed killer for use on nonplanted areas and the elimination of unwanted vegetation around plants, trees, patios, fences, driveways, etc. [Lawn & Garden Products Inc]

Weedex. Herbicide. [Murphy Chemical Ltd] †

Weedex S2. CAS 122-34-9; Granules containing 2% w/w simazine; a soil-acting herbicide. [Hortichem Ltd]

Weed Gun!; Ready-for-use herbicide spray. [ICI Chem. & Polymers Ltd]

Weed Hoe. CAS 2163-80-6; MSMA; post-emergence herbicide on turf to control established crabgrass, dollis grass, and nutselse. [Lawn & Garden Products Inc]

Weedkill. Aminotriazole + 2,4-D + diuron + simazine; used for total weed control in non crop areas. [Dermaglen Ltd]

Weedmaster. CAS 1698-60-8; Suspension concentrate containing 430 g chloridazon per liter; a pyridazinone herbicide for beet crops. [Portman Agrochemicals Ltd]

Weedol. CAS 1910-42-5, 85-00-7; Weedkiller containing paraquat and diquat for gardeners. [ICI Garden Products]

Weedone® DPC. Post-emergent herbicide (2,4-DP and 2,4-D) for control of annual and perennial broadleaf weeds on golf courses and other ornamental turf area. [Rhone-Poulenc/Ag; W.A. Cleary]

Weed Stopper. CAS 19044-88-3; Oryzalin; a pre-emergence herbicide for use on ornamentals, trees, roses, flower beds, bulbs, and warm season turf; controls annual grasses any many broadleaf weeds; may be tank-mixed with Roundup. [Lawn & Garden Products Inc]

Wegin. Ketone resins and formaldehyde scavengers. [British Traders & Shippers Ltd]

Weichharz 398A. A proprietary trade name for a nondrying alkyd made from adipic acid and trimethylene glycol. [Chemische Werke, Albert] *

Weichmacher 90. A polymer from acetylene, glycerin, and ethylene oxide. §

Weichmacher 238S. A proprietary trade name for an ester formed from adipic acid and C_4-C_9 synthetic fatty acids with pentaerythritol. [Chemische Werke, Albert] *

Weichmacher 333A. A proprietary trade name for an ester formed from adipic acid and C_4-C_9 synthetic fatty acids with pentaerythritol; lower fatty acid content than 238S. [Chemische Werke, Albert] *

Weigert's stain. A microscopic stain. It is made by dissolving fuchsine and resorcinol in ferric chloride solution.

weighted silk. Silk impregnated with various inorganic and organic substances in order to increase the weight. Tannin and metallic salts are often used.

Weighter Finish 585. Blend of starch with org. weighting agents and humectants; weighter finisher for cellulosic fibers. [Arol Chem. Prods.]

Weisalloy. A proprietary sheet aluminum alloy. §

Weiss-Kupfer. See nickel silvers.

Weldox. Adhesives compositions. [Hardman]

Welgum. Alginates for ceramics, electrodes, and water treatment. [Alginate Industries Ltd] *

Welladyne. Iodophor detergent germicide for cleaning. [Ciba plc] *

Wellbrom. Oil well fluid components. [Ethyl Corp]

Wellbutrin. CAS 31677-93-7; Bupropion hydrochloride; an antidepressant. [Burroughs Wellcome Co] *

Wellcome Brand Athropine Sulphate Injection. A proprietary formulation of atropine sulfate; for preanesthetic medications to inhibit secretory glands and to suppress both vagal activity associated with the use of halogenated hydrocarbons during inhalation anesthesia and reflex excitation arising from mechanical stimulation during surgery. [The Wellcome Foundation Ltd]

Wellcome Brand Scopolamine Hydrobromide Injection. A proprietary formulation of scopolamine hydrobromide; for preanesthetic sedation and in conjunction with analgesic agents for obstetric amnesia or for calming delirium. [The Wellcome Foundation Ltd]

Wellcovorin. Proprietary formulations of leucovorin calcium; tablets and injection; for the prophylaxis and treatment of undesired hematopoietic effects of folic acid antagonists. [The Wellcome Foundation Ltd]

Welldorm. CAS 480-30-8; A proprietary preparation of dichloralphenazone; a hypnotic. [Smith & Nephew Pharmaceuticals Ltd] *

Wellferon. Interferon; protein formed by the interaction of animal cells with viruses capable of conferring on animal cells resistance to virus infection; antineoplastic; antiviral. [Burroughs Wellcome Co; The Wellcome Foundation Ltd]

Welltex 300 F. CAS 8061-52-7; Fermented calcium lignosulfonate; paper sizing agent. [Borregaard LignoTech]

Welmet. A proprietary chromium nickel-molybdenum steel. §

---

* Trade name not verified as available

† Trade name verified as obsolete

Welvic. A trademark for a range of plasticized and unplasticized polyvinyl chlorides used in the manufacture of cables, flooring, pipes, etc. [ICI Chem. & Polymers Ltd] †

WEP® 662P. Unsaturated polyester resin; water-extendable resin for casting plaques, figurines, other decorative items; designed to be extended with water and cured at room temperature. [Ashland]

Wepco. Caulking, waterproofing, and cementitious materials. [Weatherguard/Marbleloid Products Inc] *

Werderol. A formic acid preservative used for fruit preparations.

WcsBio. A liquid, biodegradable microbiocide containing sodium salts of dimethyl and ethylene dithiocarbamates; used to prevent fermentation in drilling fluid and corrosion in producing wells. [Westbridge Research Group] ‡

Wescodyne. Iodophor disinfectants. [Ciba plc] *

WesLoTemp. A liquid, sodium polyacrylate polymer-based drilling fluid additive used as a dispersant/deflocculant; used in clay-based fresh water drilling fluid systems subject to low temperatures. [Westbridge Research Group] ‡

Wessalith. Sodium aluminum silicate; phosphate replacement in laundry detergents. [Degussa AG] *

Wessalon, 50, 50S, S. Amorphous precipitated silica; for pesticides industry. [Degussa; Degussa AG]

WesScaleStop. A liquid, acrylic homopolymer scale inhibitor; used in fresh water-based drilling fluid systems. [Westbridge Research Group] ‡

Wessell's silver. A nickel silver alloy of 51-65% copper, 19-32% nickel, 12-17% zinc, and 2% silver.

WesSperse. A dry, blended sodium polyacrylate/chrome free lignosulfonate drilling fluid additive used as a dispersant/deflocculant; used in clay-based drilling fluid systems subject to calcium, magnesium, and chloride contamination and high temperature. [Westbridge Research Group] ‡

West African copaiba. Illurin balsam, an oleo-resin, is known by this name; used as a substitute for balsam of copaiba.

West African gum. A gum arabic resembling Senegal gum, obtained from *Acacia nilotica*.

WesTemp. A liquid, sodium polyacrylate polymer-based drilling fluid additive used as a dispersant/deflocculant; used in clay-based fresh water drilling fluid systems subject to high temperatures. [Westbridge Research Group] ‡

WesTemp K+. A liquid, potassium polyacrylate polymer-based drilling fluid additive used as a dispersant/deflocculant; used in clay-based fresh water drilling fluid systems subject to high temperatures. [Westbridge Research Group] ‡

Westfalite No. 3. An explosive consisting of 58-61% ammonium nitrate, 13-15% potassium nitrate, 4-6% trinitrotoluene, and 20-22% ammonium chloride.

Westhin. A liquid, sodium polyacrylate polymer-based drilling fluid additive used as a dispersant/deflocculant; used in clay-based fresh water drilling fluid systems. [Westbridge Research Group] ‡

WesThin K+. A liquid, potassium polyacrylate polymer-based drilling fluid additive used as a dispersant/deflocculant; used in clay-based fresh water drilling fluid systems. [Westbridge Research Group] ‡

Westo-Flocs. Polymer flocculants; for wastewater. [Garvey Chemical Corp] *

Weston®. Processing aids for thermoplastics. [GE Specialty] *

Weston® 399. CAS 26523-78-4; Trisnonylphenyl phosphite (contains 0.75% wt. triisopropanolamine); stabilizer used in epoxies, hot-melt adhesives, PU, polyester, SBR, PP; in molding and extrusion of PP, HDPE, LDPE, HIPS, PC, ABS, PVC, polyesters, in calendering of ABS, PVC; in film applications of PP, PE, PVC; fiber applications of PP, polyesters. [GE Specialty]

Weston® 430. CAS 36788-39-3; Tris(dipropylcneglycol) phosphite; stabilizer for hot-melt adhesives, PU, polyesters; used in molding, extrusion, and film applications in PP, HDPE, LDPE, PVC, and polyesters; also useful for PP fiber applications and calendering of PVC. [GE Specialty]

Weston® 439. CAS 93356-94-6; Poly 4,4´ isopropylidene-diphenol neodol 25 alcohol phosphite; chelating agent in conjunction with metallic soaps to enhance the clarity and color of PVC formulations; improves color and light stability and clarity of PVC formulations; FDA approved. [GE Specialty]

Weston® 474. CAS 68610-62-8; Tris Neodol-25 phosphite; stabilizer for hot-melt adhesives, PU, polyesters; used in molding, extrusion, and film applications in PP, HDPE, LDPE, PVC, and polyesters; also useful for PP fiber applications and calendering of PVC. [GE Specialty]

Weston® 491. CAS 80584-87-8; Diphenyl didecyl (2,2,4-trimethyl-1,3-pentanediol) diphosphite; stabilizer for PU, hot melt adhesives; used in molding and extrusion of PP, HDPE, LDPE, PC, ABS, PVC; used in film applications for HDPE, LDPE, and PP; used in calendering of PVC, molding of PU, and fiber PP applications. [GE Specialty]

Weston® 494. CAS 68133-13-1; Diisooctyl octylphenyl phosphite; stabilizer for PU, hot melt adhesives; used in molding and extrusion of PP, HDPE, LDPE, PC, ABS, PVC; used in film applications for HDPE, LDPE, and PP; used in calendering of PVC, molding of PU, and fiber PP applications. [GE Specialty]

Weston® 600. CAS 26544-27-4; Diisodecyl pentaerythritol diphosphite; stabilizer for hot-melt adhesives, PU, polyesters; used in molding, extrusion, and film applications in PP, HDPE, LDPE, PVC, and polyesters; also useful for PP fiber applications and calendering of PVC. [GE Specialty]

Weston® 618. CAS 3806-34-6; Distearyl pentaerythritol diphosphite; color and heat stabilizer; improves the processing and thermal stability of PP; synergistic when used with light stabilizers such as benzophenones, benzotriazoles, and hindered amines; used in olefin polymers, PS, and rubber-modified PS for food contact and packaging applications. [GE Specialty]

Weston® 619. CAS 3806-34-6; Distearyl pentaerythritol diphosphite (contains 1.0% wt. triisopropanolamine); color and heat stabilizer; improves the processing and thermal stability of PP; synergistic when used with light stabilizers such as benzophenones, benzotriazoles, and hindered amines; used in olefin polymers and PS. [GE Specialty]

Weston® DHOP. CAS 80584-86-7; Poly (dipropyleneglycol) phenyl phosphite; stabilizer for PU, hot melt adhesives; used in molding and extrusion of PP, HDPE, LDPE, PC, ABS, PVC; used in film applications for HDPE, LDPE, and PP; used in calendering of PVC, molding of PU, and

---

‡ Trade name and manufacturer not verified   § Trade name without identified manufacturer

fiber PP applications. [GE Specialty]

Weston® DLP. CAS 21302-09-0; Dilauryl phosphite; stabilizer. [GE Specialty]

Weston® DOPI. CAS 36116-84-4; Diisooctyl phosphite; stabilizer for hot-melt adhesives, PU, polyesters; used in molding, extrusion, and film applications in PP, HDPE, LDPE, PVC, and polyesters; also useful for PP fiber applications and calendering of PVC. [GE Specialty]

Weston® DPDP. CAS 26544-23-0; Diphenyl isodecyl phosphite; stabilizer for flexible and rigid PVC; reacts principally by chelation of metallic chlorides during PVC compounding; for use in conjunction with primary heat stabilizers. [GE Specialty]

Weston® DPP. CAS 4712-55-4; Diphenyl phosphite; stabilizer used in epoxies, hot-melt adhesives, PU, polyester, SBR, PP; in molding and extrusion of PP, HDPE, LDPE, HIPS, PC, ABS, PVC, polyesters, in calendering of ABS, PVC; in film applications of PP, PE, PVC; fiber applications of PP, polyesters. [GE Specialty]

Weston® DSP. CAS 19047-85-9; Distearyl phosphite; stabilizer for hot-melt adhesives, PU, polyesters; used in molding, extrusion, and film applications in PP, HDPE, LDPE, PVC, and polyesters; also useful for PP fiber applications and calendering of PVC. [GE Specialty]

Weston® DTDP. CAS 36432-46-9; Di-tridecyl phosphite; stabilizer. [GE Specialty]

Weston® EGTPP. CAS 101-02-0; Triphenyl phosphite (contains 0.5% wt. triisopropanolamine); reactive diluent for epoxy applications incl. adhesives, coatings, laminates, potting and soldering compounds, tooling; visc. reducer. [GE Specialty]

Weston® EHDPP. CAS 15647-08-2; Ethylhexyl diphenyl phosphite; color and processing stabilizer in ABS, PC, PU, coatings, PET fiber; secondary stabilizer improving color and heat stability of PVC. [GE Specialty]

Weston® ODPP. CAS 26401-27-4; Diphenyl isooctyl phosphite; stabilizer for PU, hot melt adhesives; used in molding and extrusion of PP, HDPE, LDPE, PC, ABS, PVC; used in film applications for HDPE, LDPE, and PP; used in calendering of PVC, molding of PU, and fiber PP applications. [GE Specialty]

Weston® PDDP. CAS 25550-98-5; Phenyl diisodecyl phosphite; stabilizer for PU, hot melt adhesives; used in molding and extrusion of PP, HDPE, LDPE, PC, ABS, PVC; used in film applications for HDPE, LDPE, and PP; used in calendering of PVC, molding of PU, and fiber PP applications. [GE Specialty]

Weston® PNPG. CAS 3057-08-7; Phenyl neopentylene glycol phosphite; stabilizer for PU, hot melt adhesives; used in molding and extrusion of PP, HDPE, LDPE, PC, ABS, PVC; used in film applications for HDPE, LDPE, and PP; used in calendering of PVC, molding of PU, and fiber PP applications. [GE Specialty]

Weston® PTP. CAS 13474-96-9; Heptakis (dipropyleneglycol) triphosphite; stabilizer for hot-melt adhesives, PU, polyesters; used in molding, extrusion, and film applications in PP, HDPE, LDPE, PVC, and polyesters; also useful for PP fiber applications and calendering of PVC. [GE Specialty]

Weston® TDP. CAS 25448-25-3; Triisodecyl phosphite; stabilizer for hot-melt adhesives, PU, polyesters; used in molding, extrusion, and film applications in PP, HDPE, LDPE, PVC, and polyesters; also useful for PP fiber applications and calendering of PVC. [GE Specialty]

Weston® THOP. CAS 80584-85-6; Tetraphenyl dipropyleneglycol diphosphite; stabilizer for PU, hot melt adhesives; used in molding and extrusion of PP, HDPE, LDPE, PC, ABS, PVC; used in film applications for HDPE, LDPE, and PP; used in calendering of PVC, molding of PU, and fiber PP applications [GE Specialty]

Weston® TIOP. CAS 25103-12-2; Triisooctyl phosphite; stabilizer for hot-melt adhesives, PU, polyesters; used in molding, extrusion, and film applications in PP, HDPE, LDPE, PVC, and polyesters; also useful for PP fiber applications and calendering of PVC. [GE Specialty]

Weston® TLP. CAS 3076-63-9; Trilauryl phosphite; stabilizer for hot-melt adhesives, PU, polyesters; used in molding, extrusion, and film applications in PP, HDPE, LDPE, PVC, and polyesters; also useful for PP fiber applications and calendering of PVC. [GE Specialty]

Weston® TLTTP. CAS 1656-63-9; Trilauryl trithio phosphite; stabilizer for hot-melt adhesives, PU, polyesters; used in molding, extrusion, and film applications in PP, HDPE, LDPE, PVC, and polyesters; also useful for PP fiber applications and calendering of PVC. [GE Specialty]

Weston® TNPP. CAS 26523-78-4; Trisnonylphenyl phosphite; stabilizer used in epoxies, hot-melt adhesives, PU, polyester, SBR, PP; in molding and extrusion of PP, HDPE, LDPE, HIPS, PC, ABS, PVC, polyesters, in calendering of ABS, PVC; in film applications of PP, PE, PVC; fiber applications of PP, polyesters. [GE Specialty]

Weston® TPP. CAS 101-02-0; Triphenyl phosphite; stabilizer used in epoxies, hot-melt adhesives, PU, polyester, SBR, PP; in molding and extrusion of PP, HDPE, LDPE, HIPS, PC, ABS, PVC, polyesters, in calendering of ABS, PVC; in film applications of PP, PE, PVC; fiber applications of PP, polyesters. [GE Specialty]

Weston® TSP. CAS 2082-80-6; Tristearyl phosphite; stabilizer for hot-melt adhesives, PU, polyesters; used in molding, extrusion, and film applications in PP, HDPE, LDPE, PVC, and polyesters; also useful for PP fiber applications and calendering of PVC. [GE Specialty]

Westoran®. A trade name for a cleaning agent for cotton; it contains emulsified hydrocarbons, and is also used as an insecticide. §

Westphalite I. A safety explosive for mines, consisting of 95% ammonium nitrate and 5% resin.

Westphalite II. (Westphalite, Improved). A safety explosive for mines, containing 92% ammonium nitrate, 3% potassium nitrate, and 5% resin.

Westrol®. A trade name for a cleaning liquid for cotton; it contains oils with a solvent; soaps containing trichlorethylene; a degreasing agent. §

Westropol®. A trade name for a cleaning and degreasing agent. §

Westrosol®. CAS 79-01-6; A registered trade name for a preparation of trichlorethylene, $CHCl : CCl_2$. §

West System Brand Products. Liquid epoxy resin, hardeners and accessories; used for wood encapsulation, bonding, filling and fairing. [Wessex Resins & Adhesives Ltd]

Westvaco® Resin 90. CAS 8050-09-7; Stabilized tall oil rosin; emulsifier for synthetic rubber emulsion polymerization; tackifier, stabilizer, plasticizer. [Westvaco]

WesVis. A liquid, ammonium polyacrylate invert emulsion drilling fluid additive used as a viscosifier, bentonite extender, selective flocculant and hole sweep; used in water-based drilling fluid systems. [Westbridge Re-

---

* Trade name not verified as available   † Trade name verified as obsolete

search Group] ‡

Wet 6. Dental preparation; wetting agent for waxes. [Bayer AG] †

weta material. A porcelain substitute consisting of fine, uniformly distributed carborundum particles with silicates and metals of the iron series, cobalt and nickel, and sinters after firing at 1400 C.

Wetanol. A proprietary trade name for a wetting agent for textiles, etc.; it is a modified sulfated fatty acid ester. §

Wetfix. Range of cationic surfactants comprised of amino groups and selected aliphatic hydrocarbon chains; liquid form, heat stable; promotes and retains adhesion between asphalt and aggregate; incorporated into asphalt which will eventually be used for surface dressing, asphaltic macadams and asphaltic concrete. [Thomas Swan & Co Ltd] †

Wetstrez®. Synthetic resins. [Reichhold]

Wetter-dynamite. A safety explosive for mines, consisting of 53% nitroglycerin, 14% kieselguhr, and 33% magnesium sulfate.

Wetter-dynammon. An Austrian explosive containing 94% ammonium nitrate, 2% potassium nitrate, and 4% charcoal.

Wetteren powder. A guncotton powder, containing a little calcium carbonate, gelatinized with amyl acetate.

Wetter-Fulminite. An explosive containing ammonium nitrate.

Wetter Nobelit B. A gelatinous permitted explosive (group P1). [Dynamit Nobel Wien GmbH] *

Wetting Agent FCGB. Sodium laureth phosphate; wetting agent for metal surfaces; useful in cyanide copper plating baths to produce clean finishes. [Henkel/Functional Prods.]

Wettol® D 1. Sodium salt of phenolsulfonic acid condensation product; emulsifier, wetting agent, and dispersant for formulation of wettable powds. for crop protection. [BASF AG]

Wettol® EM 1. Calcium alkylaryl sulfonate; emulsifier, wetting and dispersing agent for formulation of pesticides. [BASF AG]

Wetz. Surfactant with antifoam. [Draxel Chemical Company]‡

Weyl and Zeitler's solution. A solution used to absorb oxygen. It consists of pyrogallol in sodium hydroxide solution.

W.G.S. Hydrogenated Fish Glyceride 117. Triester of long chain fatty acids and glycerin; used in wax compounds, textile softeners and sizes, yarn lubricants, grease sticks, polishes, crayons, candles, leather stuffing, wire drawing compounds, paper coatings, and plastics. [Werner G. Smith]

W.G.S. R-60 Z-5. Fish oil deriv. containing 40% C-20 and C-22 acids; wetter for hot metal surfaces; high water absorptive capacity; easy emulsification; for marine lubricants, Degras replacement in metalworking oils and rust preventatives; scavenger in chlorinated paraffin mixtures. [Werner G. Smith]

W.G.S. Synaceti 116 NF/USP. CAS 8002-23-1; Cetyl esters; emollient for cosmetics; slip aid for inks; gloss/slip aid for varnish; processing aid and lubricant for plastics; in binder formulations for pencils. [Werner G. Smith]

Whale. A cotton fabric grade of Tufnol industrial laminates. [Tufnol Ltd]

wheat germ oil. (oil of wheat germ; triticum aestivum germ oil) CAS 8006-95-9; Oil obtained by expression or extraction of wheat germ; a vitamin E concentrate; emollient for skin and hair care products; gloss agent in makeups; moisturizer. [Arista Industries; Croda; Penta Mfg.; Provital]

Wheat Germ Oil CLR. Fatty oil of wheat germ, natural vitamin E carrier; products for general skin protection. [Dr. Kurt Richter; Henkel/Cospha]

Wheat-Pro EN-20. Hydrolyzed wheat protein; cosmetic ingredient for skin and hair care products [Brooks Industries]

Wheat-Tein NL. CAS 100864-25-1; Hydrolyzed wheat protein; substantive protein, film-former, anti-irritant, protectant, moisturizer for skin and hair care products (shampoos, conditioners, styling products, creams, lotions, hand soaps). [Maybrook]

Wheeler's solution. A mixture of pyrogallol in potassium hydroxide for the absorption of oxygen.

whetstone. (oilstone, honestone). Hard rocks, usually siliceous in character; used for sharpening tools. Suitable rocks include hornstone, sandstone, slate, lydian stone, schist, etc.

Whipcide. Phthalovyne (mono[1-ethyl-1-methyl-2-propynyl] phthalate). [Pitman-Moore] *

white acid. A mixture of hydrofluoric acid and ammonium fluoride; used for etching glass

white alloy. An alloy of 10% cast iron, 10% copper, and 80% zinc. This name is also applied to alloys containing 49-53% copper, 23-24% zinc, 22-24% nickel, and 2% iron. They are nickel silvers.

white argol. See argol.

white bole. See China clay.

white button alloy. A nickel-silver containing from 49-53% copper, 23-24.5% zinc, 22-24% nickel, and 2-2.5% iron.

white cast iron. A good variety of cast iron. It usually contains 97% iron and 3% carbon, mainly in the uncombined state.

white caustic. CAS 1310-73-2; Colorless sodium hydroxide.

white caustic. See sodium hydroxide.

white clay. See China clay.

white copper. A nickel silver usually containing 70% copper, 18% zinc, and 12% nickel. See nickel silvers.

white copperas. (cinquinolite) A mineral. It is ferric sulfate, $Fe_2(SO_4)_3 \cdot 9H_2O$. The name white copperas is also used for zinc sulfate.

White Cosmetic. A trade name for basic nitrate or mixture of basic nitrates obtained by adding water to bismuth nitrate. §

white dammar. Manila copal resin, obtained from *Vateria indica*, is known by this name.

white drying oil. Linseed oil which has been bleached.

White Gold. Water soluble polymer for oilfield use. [BP Chemicals Ltd] †

white gold. Various alloys are known by this term. A jeweler's alloy of gold whitened by means of silver, is called white gold. An alloy of 90% gold, and 10% palladium, and a platinum substitute, containing 59% nickel, and 41% gold are both known by this name.

white gunpowder. A mixture of 2 parts potassium chlorate and 1 part each potassium ferrocyanide and sugar. An ingredient of explosives.

White House Cement Paint. Colored cement; decorative finish for stonework, masonry, etc. [Calder Colours

---

‡ Trade name and manufacturer not verified   § Trade name without identified manufacturer

(Ashby) Ltd]

white insect wax. (white lac). Arjun wax of India produced by the insect *Ceroplastes ceriferus.*

white iron pyrites. *See* marcasite.

white lac. *See* white insect wax.

white lead. (Ceruse) CAS 1319-46-6; A basic carbonate of lead, the composition of which is variable; used as a pigment; *also see* lead carbonate (basic)

White Lead, Freeman's. *See* Freeman's nonpoisonous white lead. §

white metal. An alloy of 54% copper, 24% nickel, and 22% zinc. It is a nickel silver. It is also the name applied to bearing metals.

white ocher. Ordinary clay is known by this name.

white oils. (egg oils). A liniment usually containing turpentine, acetic acid, and eggs. Sometimes ammonia and camphor are added.

white ozokerite wax. *See* ceresin.

white paste. Copper sulfocyanide, $Cu_2(CNS)_2$.

white precipitate. (ammoniated mercury, Lemery's white precipitate). CAS 10124-48-8; Mercury ammonium chloride, NH_2HgCl; used for the preparation of cinnabar and in medicine.

white pyrites. (efflorescent pyrites). A variety of iron pyrites, FeS.

white sennaar gum. *See* picked Turkey gum.

white solder. An alloy of 10% nickel, 45% copper, and 45% zinc. A soldering alloy.

white spirit. A turpentine substitute. It is usually a petroleum product, having flash point and degree of evaporation similar to turpentine.

White Swan. Lanolin BP. [Croda Chem. Ltd]

white tan. (tari, teri) *Coesalpinia digyna,* containing 30-50% tannin.

white tar. (naphthalene; tar camphor) $C_{10}H_8$; CAS 91-20-3; EINECS 202-049-5; Used as an intermediate, mothrepellant, fungicide, and tanning perservative.

white tellurium. (Krennerite, bunsenine). A mineral. It contains from 25-29% gold, 2.7-14.6% silver, and 2.5-19.5% lead, as tellurides.

Whitetex. CAS 1332-58-7; Coarse particle size calcined kaolin (aluminum silicate); used as a reinforcer and extender in plastics, PVC, rubber, coatings. [Engelhard]

white tung oil. Tung oil obtained by the cold pressing method.

white vitriol. *See* zinc sulfate.

white wax. White beeswax.

Whitex. A filler for plastics; calcined clay.

white zinc. CAS 3486-35-9; Zinc carbonate; $ZnCO_3$; used in ceramics, cosmetics, pharmaceuticals, and in topical antiseptics.

Whiting. (chalk). Finely ground calcium carbonate; used as filler in rubber, plastics, and paper coatings.

Whitworth's steel. Steel which has been subjected to high pressures to eliminate blow-holes.

Wiborg phosphate. A German fertilizer made by heating mineral phosphate with soda. It consists mainly of a tetraphosphate.

Wichmann's substitute. A material made from casein and albumen.

Wickenol® 101. CAS 110-27-0; EINECS 203-751-4; Isopropyl myristate; emollient, solubilizer, and lubricant for use in cosmetic and toilet preparations. [CasChem]

Wickenol® 105. Isopropyl myristate, isopropyl palmitate;

see Wickenol 101. [CasChem]

Wickenol® 111. CAS 142-91-6; EINECS 205-571-1; Isopropyl palmitate; *see* Wickenol 101. [CasChem]

Wickenol® 127. CAS 112-10-7; Isopropyl stearate; emollient, cosolv., and lubricant. [CasChem]

Wickenol® 131. Isopropyl isostearate; lubricant, emollient, solubilizer. [CasChem]

Wickenol® 136. IPP, IPM, and isopropyl stearate; emollient, conditioner, moisturizer, solubilizer, vehicle, and solv. for cosmetic and toilet preparations. [CasChem]

Wickenol® 139. CAS 61789-91-1; Synthetic jojoba oil; emollient, plasticizer, lubricant, coupler for hair and skin care products; replacement for natural jojoba oil. [CasChem]

Wickenol® 141. CAS 110-36-1; Butyl myristate; emollient, solubilizer, and lubricant for use in cosmetic and toilet preparations; plasticizer. [CasChem]

Wickenol® 142. Octyldodecyl myristate; emollient ingredient, plasticizer for cosmetic and pharmaceutical preparations. [CasChem]

Wickenol® 143. CAS 3687-45-4; EINECS 222-980-4; Oleyl oleate; lubricant used in cosmetic and toilet preparations; replacement for sperm oil in addition to functioning as cosmetic additive. [CasChem]

Wickenol® 144. CAS 59231-34-4; EINECS 261-673-6; Isodecyl oleate; emollient, wetting agent and pigment binder for cosmetics. [CasChem]

Wickenol® 151. CAS 42131-25-9; Isononyl isononanoate; emollient, moisturizer, pigment wetter; silky emolliency and solv. charactersitics; for hair care products. [CasChem]

Wickenol® 152. CAS 41395-89-5; Isodecyl isononanoate; silky emollience and solv. char. for skin and hair care products; pigment wetter. [CasChem]

Wickenol® 153. Isotridecyl isononanoate; silky emollience and solv. char. for skin and hair care products; pigment wetter, moisturizer. [CasChem]

Wickenol® 155. CAS 29806-73-3; EINECS 249-862-1; Octyl palmitate; emollient, moisturizer, pigment wetter/dispersant; increases water vapor porosity of fatty components used in cosmetic and topical pharmaceutical preparations. [CasChem]

Wickenol® 156. CAS 22047-49-0; EINECS 244-754-0; Octyl stearate; emollient, moisturizer, pigment wetter/dispersant; increases water vapor porosity of fatty components used in cosmetic and topical pharmaceutical preparations. [CasChem]

Wickenol® 158. CAS 103-23-1; Dioctyl adipate; emollient, moisturizer, pigment wetter/dispersant, cosolv.; increases water vapor porosity of fatty components used in cosmetic and topical pharmaceutical preparations. [CasChem]

Wickenol® 159. CAS 2915-57-3; Dioctyl succinate; emollient, moisturizer, pigment wetter/dispersant; increases water vapor porosity of fatty components used in cosmetic and topical pharmaceutical preparations. [CasChem]

Wickenol® 160. CAS 59587-44-9; EINECS 261-819-9; Octyl pelargonate; emollient, moisturizer, pigment wetter/dispersant; increases water vapor porosity of fatty components used in cosmetic and topical pharmaceutical preparations; improves stick formulations. [CasChem]

Wickenol® 161. Dioctyl adipate, octyl stearate, octyl palmitate; emollient, moisturizer, pigment wetter/dispersant; increases water vapor porosity of fatty components used in cosmetic and topical pharmaceutical preparations. [CasChem]

Wickenol® 163. Dioctyl adipate, octyl stearate, octyl palmitate; emollient, solubilizer; fragrance enhancer. [CasChem]

Wickenol® 171. Octyl hydroxystearate; emollient, moisturizer, pigment wetter/dispersant; increases water vapor porosity of fatty components used in cosmetic and topical pharmaceutical preparations; refatting agent, counterirritant, cosolv., solubilizer. [CasChem]

Wickenol® 174. Myristyl octanoate; emollient; provides soft satiny skin afterfeel. [CasChem]

Wickenol® 506. CAS 1323-03-1; Myristyl lactate; emollient with smooth, satiny afterfeel. [CasChem]

Wickenol® 535. Wheat germ glycerides; hydrophilic/hydrophobic emollient, emulsifier, skin lubricant; anti-irritant. [CasChem]

Wickenol® 545. Glucose glutamate; substantive humectant, skin conditioner, moisturizer, emulsifier, surfactant, thickener used in personal care products; enhances lather in surfactant systems. [CasChem]

Wickenol® 550. CAS 9050-36-6; Maltodextrin; absorbent for lipophilic materials for powd. bath applications; food-grade carrier for flavors. [CasChem]

Wickenol® 707. CAS 9035-85-2; PPG-30 cetyl ether; all-purpose fluid, nongreasy emollient with hydroalcoholic compatibility; provides coupling and emulsion stability; foam modifier; enhances sheen and manageability. [CasChem]

Wickenol® 727. CAS 68439-53-2; PPG-30 lanolin alcohol ether; nongreasy emollient, solubilizer, bodying agent; also low odor derivative. [CasChem]

Wiegold alloy. It is a brass containing aluminum, and resembles gold in appearance. It is said to consist of 67.73% copper, 32% zinc, and 0.27% aluminum, but some analysts state that it contains 0.25-0.5% lead; a dental alloy

Wij's solution. CAS 7790-99-0; Iodine trichloride (9.4 g) and iodine (7.2 g) are dissolved separately in glacial acetic acid, and the solutions added together; used for the determination of the iodine value of fats and oils.

Wilcoloy. A proprietary tungsten carbide material. §

wild ginger oil. The oil of *Asarum canadense*.

Wiles. Fertilizers. [BritAg Industries Ltd] *

Wilhelmit. A German explosive containing potassium chlorate and mineral oil.

Wilkinite. (jelly rock). CAS 1302-78-9; A colloidal clay; used as a substitute for China clay as a paper filter.

wilkinite. *See* bentonite.

Wills Metallic 'O' Rings. Metallic sealing systems; extreme pressure and temperature sealing for fluid and vacuum service. [Fothergill Tygaflor Ltd] *

Wilmil. Alloys similar in composition to silumin. §

Wilmot's aluminum solder. An alloy of 86% tin and 14% bismuth.

Wilpo. CAS 1197-21-3; Phentermine hydrochloride; appetite suppressant. [Dorsey Pharmaceuticals] *

Wingstay® 100. Mixed diaryl p-phenylenediamine; antioxidant/antiozonant for tires, camelback, mech. goods; stabilizer for SBR, polybutadiene, and neoprene. [Goodyear; R.T. Vanderbilt]

Wingstay® C. Butylated d:(dimethylbenzyl) phenol; stabi-

lizer for nonstaining SBR and nitrile polymers. [Goodyear; R.T. Vanderbilt]

Wingstay® S. Styrenated phenol; nonstaining, nondiscoloring antioxidant for foam rubber, wh. sidewalls, kitchen and drug sundries, garden hose, shoe soles; stabilizer for SBR and BR. [Goodyear; R.T. Vanderbilt]

Wingstay® T. A blend of substituted phenols; a proprietary antioxidant. [Goodyear] *

Wingtack® 10. C5 hydrocarbon resin; plasticizer/tackifier for pressure sensitives, hot melts and sealants. [Goodyear]*

Wingtack® Extra. C5 hydrocarbon resin; tackifier resin for adhesives; superior resist. to uv light and heat aging. [Goodyear] *

Win-Kinase. CAS 9039-53-6; Urokinase; plasminogen activator. [Sterling Drug Inc] *

Winnofil. Precipitated calcium carbonate surface treated with calcium stearate; used for giving tires high tensile strength. [ICI Chem. & Polymers Ltd]

Winstrol. CAS 10418-03-8; Stanozolol; androgen. [Sterling Drug Inc] *

winter oils. Lubricating oils which remain liquid at low temperatures.

Wintomylon. CAS 389-08-2; Nalidixic acid; antibacterial. [Sterling Drug Inc] *

Witafrol®. Antifoam agents; used in mining industry, food industry, water engineering, and the paper industry. [Hüls AG]

Witafrol® 7420. CAS 26402-26-6; Caprylic/capric glycerides; surfactant for pharmaceutical, cosmetic, nutritional fields; as emulsifier, solubilizer, dispersant, plasticizer, lubricant, consistency regulator, skin/mucous membrane protectant, refatting agent, penetrant, carrier, adsorp. promoter, antifoam. [Hüls Am.]

Witafrol® 7440. Hydrophilic fatty acid ester; foam preventer for food industry. [Hüls AG]

Witafrol® 7456. Mixt. of hydrophilic fatty acid esters; food antifoam; pronounced/spontaneous retardation effect on foam formation caused by proteins. [Hüls AG]

Witafrol® 7480 N. Hydrophilic fatty acid ester; defoamer and foam preventive for sugar, dairy industries, jams, fruit flavors/juices, seasonings, and other food products. [Hüls AG]

Witafrol® 7490. Hydrophilic fatty acid ester; biodeg. antifoam agent for sugar industry. [Hüls AG]

Witafrol® 7497 N. Hydrophilic fatty acid ester; foam preventer for juice extraction/purification, external water circulation systems of sugar factories; biodeg. [Hüls AG]

Witamol. Plasticisers; used in PVC processing, imitation leather, heat resistant cables and sheets, special products with high resistance to cold and weather, application to lacquer. [Hüls AG]

Witarix®. Hard fats for chocolate. [Hüls AG]

Witarix® 212. CAS 68990-82-9; Hydrogenated palm kernel oil; fats for chocolate and confectionery products with soft consistency; also for personal care products. [Hüls Am.]

Witarix® 250. Cocoglycerides; fats for chocolate and confectionery products with soft consistency; also for personal care applications [Hüls Am.]

Witarix® 440. CAS 8016-70-4; EINECS 232-410-2; Hydrogenated soybean oil; fats for chocolate and confectionery products with soft consistency; also for personal care applications. [Hüls Am.]

---

‡ Trade name and manufacturer not verified   § Trade name without identified manufacturer

Witarix® 450. Hydrogenated peanut oil; fats for chocolate and confectionery products with soft consistency; also for personal care applications. [Hüls Am.]

Witbreak DGE-128A. Glycol ester; base for crude oil demulsifier. [Witco/Organics; Witco SA]

Witbreak DPG-15. POE glycol; oilfield surfactant, demulsifier. [Witco/Organics]

Witbreak DRA-21. Oxyalkylated phenolic resin; base for crude oil demulsifier. [Witco/Organics]

Witbreak DTG-62. Polyoxyalkylene glycol; oilfield surfactant; demulsifier. [Witco/Organics]

Witbreak RTC-326. Polymeric amine; oilfield surfactant, reverse demulsifier. [Witco/Organics]

Witcamide. Surfactant for cosmetics, toiletries, pharmaceutical, processing, agricultural and other industries. [Baxenden Chemicals Ltd]

Witcamide® 61. CAS 111-05-7; EINECS 203-828-2; Oleamide MIPA and isopropanolamine; hair conditioner, emulsifier, lubricant; cosmetics and toiletries. [Witco/Organics; Witco SA]

Witcamide® 70. CAS 111-57-9; EINECS 203-883-2; Stearamide MEA; opacifier, conditioner, lubricant, thickener, gelling agent, mold release agent, binder; cosmetics and toiletries; base for antiperspirant and makeup sticks. [Witco/Organics; Witco SA]

Witcamide® 128T. Cocamide DEA; detergent, foam booster/stabilizer, visc. modifier, substantive conditioner. [Witco/Organics]

Witcamide® 5195. CAS 120-40-1; EINECS 204-393-1; Lauramide DEA; visc. modifier, foam stabilizer, lubricant, conditioner, lubricant, emulsifier, wetting agent, thickener, penetrant, dye dispersant, scouring aid, antistat; cosmetics and toiletries; industrial foamer and stabilizer; metal processing; textile surfactant; aerosol formulations. [Witco/Organics]

Witcamide® Coco Condensate. Cocamide DEA, diethanolamine; detergent, foam stabilizer, lubricant, conditioner. [Witco/Organics]

Witcamide® MAS. CAS 14351-40-7; EINECS 238-310-5; Stearamide MEA-stearate; opacifier, conditioner, lubricant, gelling agent for personal care, household, and institutional liq. soaps; used as partial or total replacement for vegetable waxes in polishes; coating agent for paper and textiles; mold release agent for industrial processing; additive for raising melting points of petrol. waxes or glyceride waxes and fats; ingredient of insulating coatings or barriers, and of water-repellent compounds. [Witco/Organics; Witco SA]

Witcamine® 209. 1-(2-Aminoethyl)-2-n-alkyl-2-imidazoline; emulsifier, lubricity and wetting agent; penetrant, dye dispersant, scouring aid, antistat, corrosion inhibitor intermediate in petrol. industry, metal processing, textiles. [Witco/Organics; Witco SA]

Witcamine® 211. Mixed 1-(2-aminoethyl)-2-n-alkyl-2-imidazoline; intermediate for oil or water sol. salts used as corrosion inhibitors, bactericides, softeners; for petrol. industry. [Witco/Organics]

Witcamine® 6606. CAS 61791-44-4; PEG-15 tallow amine; antistat, dispersant, o/w emulsifier, lubricant, substantivity and wetting agent. [Witco/Organics]

Witcamine® 6622. PEG-30 oleamine; antistat, dispersant, o/w emulsifier, lubricant, substantivity and wetting agent. [Witco/Organics]

Witcamine® AL42-12. Cationic surfactant in the form of tall oil imidazoline; antistatic and corrosion inhibitor for use in carwash and wax formulations. [Witco Chemical Ltd]

Witcamine® E-607. N(Lauroyl colamino formyl methyl) pyridinium chloride; cationic surfactant used in deodorants, after shave and hair rinses. [Witco Chemical Ltd]*

Witcamine® E-607S. N(Stearoyl colamine formyl methyl) pyridinium chloride; cationic surfactant used in hair conditioners and as a nonirritant emollient. [Witco Chemical Ltd] *

Witcamine® RAD 0500. POE rosin amine; oilfield surfactant; corrosion inhibitor intermediate, scouring agent, softener, dye assistant, emulsifier for acid-stable emulsions; inhibitor for HCl in acid pickling. [Witco/Organics; Witco SA]

Witcizer 100. CAS 142-77-8; EINECS 205-559-6; Proprietary trade name for butyl oleate. §

Witcizer 312. CAS 117-81-7; Proprietary trade name for dioctyl phthalate. §

Witcizer 313. CAS 27554-26-3; Proprietary trade name for diisooctyl phthalate. §

Witcizer 412. CAS 103-23-1; Proprietary trade name for dioctyl adipate. §

Witco® 1298. Dodecylbenzene sulfonic acid; detergent intermediate, o/w emulsifier, solubilizer, wetting agent, and detergent for household products, metal cleaning; emulsion polymerization surfactant for latex stabilization and pigment dispersion. [Witco/Organics]

Witco® 1298 HA. Surfactant blend; surfactant for insecticides. [Witco/Organics]

Witco TX. Modified toluene sulfonic acid in liquid form; anionic surfactant used as a catalyst. [Witco Chemical Ltd] *

Witcobond. Solvent based polyurethane adhesives for laminating fabrics and foam; solvent based polyurethanes for laminating films and boards; water-based polyurethane coatings for textiles, leather, glass fiber sizing, paints, lacquers, and others. [Baxenden Chemicals Ltd]

Witcobond® W-160, W-170. PU aq. disp.; used as high-performance adhesives or coatings; crosslinkable; solventless. [Witco/Organics] *

Witcobond® W-232, W-234. Aliphatic PU aq. colloidal disp.; highly functional, protective top finishes for PVC, other plastics, metal, and wood; used in automotive, paint, textile, and other industrial markets; yields coatings with excellent resist. to abrasion, hydrolysis, oxidative discoloration, impact, solvs., and staining. [Witco/Organics] *

Witcobond® W-240. Self-crosslinking PU aq. colloidal disp.; produces nondiscoloring, high-performance, protective top finishes for metal, rigid plastics, and wood; such coatings exhibit excellent resist. to abrasion, hydrolysis, oxidative discoloration, impact, solvs., and staining; also useful for formulating primers for these substrates. [Witco/Organics] *

Witcobond® XW. CAS 25928-94-3; Epoxy aq. disp.; used in waterborne coatings; exhibits good noncryst. properties, freeze-thaw stability, low foaming tendencies, good chemical and water resist. [Witco/Organics] *

Witcodet 100. Formulated product; detergent conc. base for dishwash, carwash, laundery, all-purpose, and shampoo formulations. [Witco/Organics]

Witcodet 100 and P280. Alkylaryl sulfonate in liquid form; anionic surfactant used for liquid detergents, e.g., windscreen washer. [Witco Chemical Ltd] *

---

* Trade name not verified as available   † Trade name verified as obsolete

Witcodet AE. Pearlized blend; detergent conc. for bubble bath, hand and body soaps, shampoo. [Witco/Organics]

Witcoflex. Solvent based polyurethane textile coatings. [Baxenden Chemicals Ltd]

Witcolate 1050. Sodium C12-15 pareth sulfate; detergent, detergent base, emulsifier, foamer, and wetting agent for the detergent industry. [Witco/Organics]

Witcolate 6400. CAS 151-21-3; Sodium lauryl sulfate; detergent, foaming agent, and wetting agent. [Witco/Organics]

Witcolate 6430. CAS 2235-54-3; EINECS 218-793-9; Ammonium lauryl sulfate; detergent, foaming agent, and wetting agent. [Witco/Organics]

Witcolate 6434. CAS 139-96-8; EINECS 205 380-7; TEA-lauryl sulfate; detergent, wetting agent, foaming agent. [Witco/Organics]

Witcolate 6450. CAS 9004-82-4; Sodium laureth (1) sulfate; detergent, foaming agent, wetting agent. [Witco/Organics]

Witcolate 6465. Sodium 2-ethylhexyl sulfate; detergent, dispersant, wetting agent. [Witco/Organics]

Witcolate 7093. Sodium deceth sulfate; industrial detergent, foamer and wetting agent for pressure-spray applications; electrolyte tolerance. [Witco/Organics]

Witcolate AE-3. Ammonium C12-15 pareth sulfate; detergent, wetting agent, emulsifier, foamer for the detergent industry. [Witco/Organics]

Witcolate AE. CAS 67762-19-0; Ammonium laureth sulfate; detergent, foaming agent, wetting agent. [Witco/Organics]

Witcolate AM. CAS 2235-54-3; EINECS 218-793-9; Ammonium lauryl sulfate; detergent, foaming agent, wetting agent. [Witco/Organics]

Witcolate C. CAS 151-21-3; Sodium lauryl sulfate; detergent, foaming agent, wetting agent. [Witco/Organics]

Witcolate D51-51. Nonoxynol-4 sulfate; antistat for synthetic fibers and polymer products; surfactant, wetting agent and dispersant; emulsifier for polymerization of acrylics, vinyl acetate, vinyl acrylics, styrene, SAN, styrene acrylic, and vinyl chloride. [Witco/Organics]

Witcolate D51-53. Nonoxynol-10 sulfate; emulsifier for acrylic, vinyl acetate, vinyl acrylic, styrene, SAN, styrene acrylic, and vinyl chloride polymerization. [Witco/Organics]

Witcolate D-510. Sodium 2-ethylhexyl sulfate; detergent, wetting agent and penetrant for industrial use and polymerization reactions; dispersant for bleaching powds.; lime soap and grease dispersant. [Witco/Organics]

Witcolate ES-2. CAS 9004-82-4; Sodium laureth (2) sulfate; detergent, foaming agent, and wetting agent. [Witco/Organics]

Witcolate LCP. CAS 151-21-3; Sodium lauryl sulfate; detergent, foaming agent, and wetting agent. [Witco/Organics]

Witcolate LES-60A. CAS 67762-19-0; Ammonium laureth sulfate; deterent, foaming agent, and wetting agent. [Witco/Organics]

Witcolate LES-60C. CAS 9004-82-4; Sodium laureth sulfate; detergent, foaming agent, and wetting agent. [Witco/Organics]

Witcolate NH. CAS 2235-54-3; EINECS 218-793-9; Ammonium lauryl sulfate; detergent, foaming agent, and wetting agent. [Witco/Organics]

Witcolate SE-5. Sodium C12-15 pareth sulfate; detergent, detergent base, foamer, o/w emulsifier, wetting agent for industrial and household detergents, textiles. [Witco/Organics]

Witcolate TLS-500. CAS 139-96-8; EINECS 205-388-7; TEA-lauryl sulfate; detergent, foaming agent, and wetting agent. [Witco/Organics]

Witcolate WAC-GL. CAS 151-21-3; Sodium lauryl sulfate; detergent, foaming agent,and wetting agent. [Witco/Organics]

Witconate 30DS. Sodium dodecylbenzene sulfonate; detergent, foaming agent, and wetting agent. [Witco/Organics]

Witconate 60T. TEA-dodecylbenzenesulfonate; detergent, wetter, emulsifier, foaming agent; and base for household, industrial, and cosmetic/toiletry specialty compounds. [Witco/Organics]

Witconate 93S. Amine dodecylbenzene sulfonate; detergent, detergent base, emulsifier, foamer, wetting agent, dispersant, and solubilizer for the detergent industry; biodeg. [Witco/Organics]

Witconate 605A. Calcium dodecylbenzene sulfonate; industrial detergent, dispersant, o/w and w/o emulsifier, lubricant, wetting agent, and demulsifier; used in agriculture and oilfield applications. [Witco/Organics]

Witconate AOK. CAS 68439-57-6; EINECS 270-407-8; Sodium C14-16 olefin sulfonate; detergent, foaming agent, wetting agent. [Witco/Organics]

Witconate D24-25. Calcium alkylaryl sulfonate; oil soluble emulsifier. [Witco Chemical Ltd] *

Witconate DS. CAS 1322-98-1; EINECS 215-347-5; Sodium decylbenzenesulfonate; detergent, foaming agent, and wetting agent. [Witco/Organics]

Witconate LXH. Branched TEA-dodecylbenzene sulfonate; detergent, foaming agent, wetting agent, and dispersant. [Witco/Organics]

Witconate NIS. CAS 1562-00-1; EINECS 216-343-6; Sodium isethionate; detergent, foaming agent, and wetting agent. [Witco/Organics]

Witconate P10-59. Isopropylamine dodecylbenzene sulfonate; emulsifier, solubilizer, detergent, and wetting agent for oil-based systems; dispersant in oil and water-based systems; used in dry-cleaning surfactants; hydrotrope for liq. detergents; oilfield demulsifier. [Witco/Organics; Witco SA; Witco Israel]

Witconate PTSA. Toluene sulfonic acid in crystal form; anionic surfactant with hydrotrope properties used as a catalyst. [Witco] *

Witconate SCS 45%. CAS 32073-22-6; Sodium cumene sulfonate; hydrotrope, solubilizer, coupler and processing aid in detergent mfg. and industrial processes; antiblocking and anticaking agent in powd. products; formulates shampoos, aerosols, cutting oils, glue; textile finishing. [Witco/Organics]

Witconate STS. Sodium toluene sulfonate in liquid form; anionic surfactant. [Witco Chemical Ltd] *

Witconate SXS 40%. CAS 1300-72-7; EINECS 215-090-9; Sodium xylene sulfonate; hydrotrope, solubilizer, coupler and processing aid in detergent mfg. and industrial processes; antiblocking and anticaking agent in powd. products; formulates shampoos, aerosols, cutting oils, glue; textile finishing. [Witco/Organics]

Witconate TX Acid. Modified toluene sulfonic acid; wetting agent, hydrotrope, coupler and solubilizer for liq. detergents; anticaking aid in dry neutralization; catalyst in

---

‡ Trade name and manufacturer not verified   § Trade name without identified manufacturer

organic reactions. [Witco/Organics; Witco SA]

Witconol. Emollient; for bath oils, hair preparations; cosmetic preparations. [Witco Chemical Ltd] *

Witconol 14. CAS 9007-48-1; Polyglyceryl-4 oleate; w/o and o/w emulsifier, lubricant, emollient, spreader, sticker, and antifoamer for industrial use, aerosols. [Witco/Organics; Witco SA]

Witconol 2301. CAS 112-62-9; EINECS 203-992-5; Methyl oleate; defoamer, lubricant, moisture barrier. [Witco/Organics]

Witconol 2326. CAS 123-95-5; EINECS 204-666-5; Butyl stearate; lubricant. [Witco/Organics]

Witconol 2380. CAS 1323-39-3; EINECS 215-354-3; Propylene glycol stearate; o/w emulsifier, lubricant, and opacifier. [Witco/Organics]

Witconol 2400. Glyceryl stearate; o/w emulsifier, lubricant. [Witco/Organics]

Witconol 2421. CAS 111-03-5; Glyceryl oleate; defoamer, o/w emulsifier, lubricant, moisture barrier. [Witco/Organics]

Witconol 2500. CAS 1338-43-8; EINECS 215-665-4; Sorbitan oleate; w/o emulsifier, lubricant, coupling agent. [Witco/Organics]

Witconol 2503. CAS 26266-58-0; EINECS 247-569-3; Sorbitan trioleate; w/o emulsifier, lubricant, coupling agent. [Witco/Organics]

Witconol 2620. CAS 9004-81-3; PEG-4 laurate; o/w emulsifier, lubricant. [Witco/Organics]

Witconol 2622. CAS 9005-02-1; PEG-4 dilaurate; o/w emulsifer, lubricant. [Witco/Organics]

Witconol 2640. CAS 9004-99-3; PEG-8 stearate; o/w emulsifier, lubricant. [Witco/Organics]

Witconol 2642. CAS 9005-08-7; PEG-8 distearate; o/w emulsifier, lubricant. [Witco/Organics]

Witconol 2648. CAS 9005-07-6; PEG-8 dioleate; o/w emulsifier, lubricant, defoamer. [Witco/Organics]

Witconol 2720. CAS 9005-64-5; Polysorbate 20; o/w emulsifier, dispersant, visc. modifier, coupling agent. [Witco/Organics]

Witconol 2722. CAS 9005-65-6; Polysorbate 80; o/w emulsifier, dispersant, visc. modifier, coupling agent. [Witco/Organics]

Witconol 5906. CAS 61791-12-6; PEG-30 castor oil; dispersant, o/w emulsifier, lubricant. [Witco/Organics]

Witconol 6903. CAS 9005-70-3; Polysorbate 85; o/w emulsifier, dispersant, lubricant, and coupling agent. [Witco/Organics]

Witconol APM. CAS 63793-60-2; PPG-3 myristyl ether; lubricant, emulsifier, wetting agent, penetrant, dye dispersant, scouring aid, antistat, solv. coupler; metal processing; textile surfactant; emollient oil for cosmetics and toiletries, solubilizer. [Witco/Organics]

Witconol APS. CAS 25231-21-4; PPG-11 stearyl ether; lubricant, emulsifier, wetting agent, penetrant, dye dispersant, scouring aid, antistat, solv. coupler; synthetic oils for personal care products; metal processing; textile surfactant; emollient oil for cosmetics and toiletries, solubilizer. [Witco/Organics]

Witconol EGMS. Glycol stearate; opacifier, conditioner. [Witco/Organics]

Witconol H31A. CAS 9004-96-0; PEG-8 oleate; lubricant and plasticizer in oils and polymers; o/w emulsifer for mineral and vegetable oils, and solvs., cosmetic and industrial applications; improves flow and leveling of coatings, increases spreadability of personal care products; defoamer. [Witco/Organics; Witco SA; Witco Israel]

Witconol H33. CAS 9005-07-6; PEG-8 dioleate; industrial detergent; defoamer, o/w emulsifier, lubricant. [Witco/Organics]

Witconol MST. Glyceryl stearate; emulsifier for cosmetic, pharmaceutical, aerosol formulations; internal lubricant, plasticizer, and emulsifier in industrial applications; flow control agent for polymerization reactions; dispersant. [Witco/Organics; Witco SA]

Witconol NP-100. CAS 9016-45-9; Nonoxynol-10; pigment dispersant, emulsifier, latex stabilizer, and leveling agent for polymerization reactions. [Witco/Organics]

Witconol NS-108LQ. Polyalkoxylated alkyphenol; wetting agent, dispersant for agriculture aq. suspension concs. [Witco/Organics; Witco SA]

Witconol RHT. Glyceryl stearate SE; lubricant, plasticizer, o/w emulsifier used in industrial applications. [Witco/Organics]

Witcor 3630. Imidazoline; oilfield surfactant, corrosion inhibitor. [Witco/Organics]

Witcor CI-1. Complex surfactant; oilfield surfactant, corrosion inhibitor. [Witco/Organics]

Witcor PC200. Amine alkylaryl sulfonate; oilfield surfactant for paraffin inhibition. [Witco/Organics]

Witepsol®. Neutral hard fats based on mixtures of triglycerides; used for preparation of suppositories. [Hüls AG]

Witepsol® E75, E76, E85, H5, H12, H15, H19, H32, H35, H37, H39, H42, H175, H185. CAS 977056-87-3; Hydrogenated cocoglycerides; suppository bases for hydrophilic and lipophilic drugs. [Hüls Am.]

Witflow 60. Alkylaryl polyether alcohol; pigment wetting agent, spreading agent, flow modifier for paints/coatings. [Witco/Organics]

Witflow 901. Sodium diester sulfosuccinate; pigment wetting agent, dispersant, flow modifier for paints/coatings. [Witco/Organics]

Witflow 916. CAS 9004-96-0; PEG-400 oleate; leveling agent, flow modifier, and defoamer for paints/coatings. [Witco/Organics]

Witflow 950. Polypropoxy quaternary ammonium chloride; pigment wetting, grinding and suspension aid for oil-based paints/coatings. [Witco/Organics]

Witflow 990. CAS 9005-64-5; Polysorbate 20; o/w emulsifier, antistatic finishing agent for paints/coatings. [Witco/Organics]

Witflow 991. CAS 9005-65-6; Polysorbate 80; pigment dispersant for paints/coatings. [Witco/Organics]

Withnell powder. An explosive containing 88-92% ammonium nitrate 4-5% trinitro-toluene, and 4-6% flour.

Witocan®. Fats based on coconut and palm kernel oils; hard fat with neutral flavor; melts very rapidly in the mouth without leaving any unpleasant or greasy taste; prepares products which are easy to cut and bite and remain firm. [Hüls AG]

Witsol. Aliphatic ink solvents. [Witco]

Wittenburg weather dynamite. An explosive, consisting of 25% nitro-glycerin, 34% potassium nitrate, 38.5% rye meal, 1% wood meal, 1% barium nitrate, and 0.5% sodium bicarbonate.

Wittol wax. A wax possessing similar properties to beeswax. It is a proprietary material, and is suitable for acid-proof linings. §

---

* Trade name not verified as available   † Trade name verified as obsolete

woad. A dark, clay-like preparation made from the leaves of the woad plant, *Isatis tinctoria*; used for the purpose of exciting fermentation in the indigo vat.

Wolfaid. Saturated and unsaturated polyesters (liquid and solid); processing aids for PVC flooring, pigment dispersing aid for unsaturated polyester. [NL Victor Wolf Ltd] *

Wolfamid. Nonreactive polyamide resins; for packaging inks, cold seal release lacquers, thermographic systems, thixotropic alkyd resins. [NL Victor Wolf Ltd] *

Wolfert. A rubber substitute consisting of felt which has been impregnated with a vulcanized oil.

Wolfin 18. Thermoplastic, bitumen proof insulating foil for structures; for flat roofs, foundation insulation, protection against leakage oil, protection against seepage and pressure water. [Degussa AG] *

Wolfkur. Reactive polyamide resins, polyamino amides, and amine adducts; epoxy curing agents for coatings and adhesives. [NL Victor Wolf Ltd] *

Wolflex. Saturated polyesters; polymeric plasticizers for PVC plastics. [NL Victor Wolf Ltd] *

Wolf N Lamid IG. A proprietary polyamide resin soluble in alcohol; used as a base for varnishes. [NL Victor Wolf Ltd] *

Wolfol. Saturated polyesters with hydroxyl groups; polyester polyols for polyurethanes used in elastomers and adhesives. [NL Victor Wolf Ltd] *

wolfram. *See* tungsten.

wolframic acid. *See* tungstic acid.

wolframic acid anhydrous. *See* tungsten oxide (ic).

Wolframium. An alloy of 98% aluminum, 1.4% antimony, 0.4% copper, 0.1% tin, and 0.04% tungsten.

Wollastocoat®. CAS 13983-17-0; Family of surface-modified wollastonite products: high aspect ratio grades, milled grades and fine particle size grades; $CaSiO_3$; used in polymer composites, coatings, adhesives, elastomers and friction products. [NYCO® Minerals Inc]

Wollastokup®. CAS 13983-17-0; Chemically coupled wollastonite, high aspect ratio and fine particle sizes: CaO, SiO_2, Fe_2O_3, Al_2O_3, MnO, MgO, and TiO_2; used for polymer composites, high performance coatings, adhesives, elastomers, and friction products. [Malvern Minerals]

wollastonite. Calcium metasilicate; CAS 13983-17-0; inert filler, extender pigment for paints. [Am. Colloid; Cornelius Chemical Co Ltd; Nyco Min.; R.T. Vanderbilt]

Wollaston's cement. Consists of 1 part beeswax, 4 parts resin, and 5 parts plaster of Paris; used for fossils.

Wolle. An abbreviation for collodium-wolle nitrocellulose in various viscosities.

wood alcohol. *See* methyl alcohol.

wood-apple gum. (katbel-ki-gond, velampishin, kapithamia piscum). The gum of *Feronia elephantum*.

wood-cloth. Strips of wood treated with sulfurous acid or alkaline bisulfite, making the fiber stronger.

wood ether. (methyl ether; dimethyl ether) CH_3OCH_3; CAS 115-10-6; EINECS 204-065-8; Used in refrigerants, as a solvent, extraction agent, a catalyst and stabilizer in polymerization.

wood flour. (wood meal; wood flock) Finely powdered wood, usually white pine; used as a rubber, linoleum, or soap filler; as a filler in dynamite; in fur cleaning; in polishing agents

wood oil. The final fractions obtained in the distillation of wood spirit, containing high boiling ketones.

wood potash. Potash salts obtained from the ash of certain woods.

wood pulp, bleached. *See* cellulose.

wood rosin. Rosin obtained from the stumps and top wood of felled trees which are useless as timber. It is usually of yellow pine.

Wood's alloys. Low-melting alloys of bismuth, tin and lead, usually containing cadmium. One alloy contains 50% bismuth, 27% lead, 13% tin and 10% cadmium; another contains 50% bismuth, 25% lead, and 25% tin; a third, contains 50% bismuth, 25% lead, 12.5% tin, and 12.5% cadmium.

wood sugar. (xylose) $C_6H_{10}O_5$; CAS 58-86-6; EINECS 200-400-7; Used in dyeing, tanning, and as a source of ethanol.

wood sugar. *See* D(+)-xylose.

Woodthane WTP 311-6. Rigid polyurethane systems; for dispensing froth foams through low or high pressure pour machines or hand mix; for furniture, mirror and picture frames, decorative molding, and carvings. [Flexible Prods.]

Woodtones. Woodstain for interior use. [ICI Chem. & Polymers Ltd]

Woodwelder. Radio frequency generator; curing most commonly used glues in all companies connected with the woodworking industry. [Tregarne Ltd]

wool fat. *See* lanolin.

wool milk. An emulsion, obtained from the treatment of wool fat with caustic soda, and dilution. Lanolin is obtained from this wool milk.

wool oil. An impure oleic acid used for oiling wool, and for making lubricants and soaps.

wool pitch. A pitchy material obtained as a residue after the distillation of wool grease (Yorkshire grease).

wool wax. *See* lanolin.

wool wax alcohol. *See* lanolin alcohol.

Worm Ender. A natural occurring biological insecticide used for the control of worms and caterpillars on vegetables, ornamentals, fruit trees, shade trees, etc.; may be used up to the day of harvest. [Lawn & Garden Products Inc]

wormseed. The flower-heads of *Artemisia maritima*; used in medicine.

wormwood. (absinthium) CAS 11015-0; The dried leaves and flowering-tops of *Artemisia absinthium*.

wort. Malt is crushed and heated with water until the starch is converted into sugar by the diastase in the malt. The resulting liquid is known as wort.

Wovco SP. Trademark for a range of polyethylene plastics reinforced with carbon fiber. [Worcester Valve Co, Haywards Heath] *

WRA 1000 Varnish. Liquid polyurethane resin and hardener; used for varnishing directly onto wood or on top of epoxy coatings; has a uv inhibitor. [Wessex Resins & Adhesives Ltd] †

WRA Epoxy Resin Underwater Series. Liquid and putty epoxies; specifically formulated for underwater use; for general bonding, filling and fairing. [Wessex Resins & Adhesives Ltd]

WRA System 100 Laminating Composition. Liquid epoxy resin and hardener; laminating resin specifically for use with glass cloth, carbon fiber, aramid and hybrids. [Wessex Resins & Adhesives Ltd] †

WRA System 17. Thixotropic resin and hardener; general purpose adhesive (bonding, laminating), bonding wood,

---

‡ Trade name and manufacturer not verified   § Trade name without identified manufacturer

concrete, most metals, stone, china, GRP, unglazed ceramics, rubber. [Wessex Resins & Adhesives Ltd]

WRA System 80. PVA crosslinking liquid; general purpose wood glue; waterproof; complies with DIN68602 Section B and BS4071. [Wessex Resins & Adhesives Ltd] †

WR Base. Fatty melamine condensate; used to make wax-melamine textile water repellents. [Clark]

Wresacryl. Hydroxy acrylic resins for use with polyisocyanates; used for automotive paints and industrial finishes. [Resinous Chemicals Ltd]

Wresilac. Etherified melamine/formaldehyde resins; used for stoving finishes and acid curing lacquers. [Resinous Chemicals Ltd]

Wresinate. Metal resinates; for marine antifoulings; improves gloss and drying in alkyd finishes. [Resinous Chemicals Ltd]

Wresinite. Maleinized rosin esters; used for nitrocellulose lacquers, printing inks and to improve gloss in air drying paints. [Resinous Chemicals Ltd]

Wresinol. Alkyd resins (oil modified polyesters); used for air drying decorative paints, air drying, and stoving industrial finishes. [Resinous Chemicals Ltd]

Wresinyl. Alkyl phenol/formaldehyde resins, terpene phenolic resins; used for oil varnishes and adhesives. [Resinous Chemicals Ltd]

Wresipol. Unsaturated polyester resin in styrene; used for glass fiber reinforced laminate, casting and potting. [Resinous Chemicals Ltd]

Wright's stain. A microscopic stain for white blood corpuscles consisting of 1 g methylene blue eosin mixture in 600 cc methyl alcohol.

Wukonil. Paraffin wax emulsions. [Suddeutsche Emulsions GmbH]

Wurster's blue. CAS 100-22-1; An oxidation product of tetramethyl-p-phenylenediamine; used as an indicator.

wurtzillite. (Tabbyite, egenite, eonite). An asphaltic mineral, soluble hot in water.

Wyamine Sulfate. CAS 6190-60-9; Mephentermine sulfate; adrenergic. [Wyeth Laboratories] *

Wyamycin E. CAS 41342-53-4; Erythromycin ethylsuccinate; an antibacterial. [Wyeth Laboratories] *

Wyamycin S. CAS 643-22-1; Erythromycin stearate; an antibacterial. [Wyeth Laboratories]

Wycillin. CAS 6130-64-9; Penicillin G procaine; an antibacterial. [Wyeth Laboratories]

Wydase. CAS 9001-54-1; Hyaluronidase; a spreading agent. [Wyeth Laboratories] *

Wymox. CAS 26787-78-0; Amoxicillin; an antibacterial. [Wyeth Laboratories]

Wyovin. CAS 67-92-5; A proprietary preparation containing dicyclomine hydrochloride; an antispasmodic. [Wyeth Laboratories] *

Wytensin. CAS 23256-50-0; Guanabenz acetate; an antihypertensive. [Wyeth Laboratories]

Wytox® 312. CAS 26523-78-4; Tris (nonylphenyl) phosphite; antioxidant for PE, PP, PS, PVC, ABS, nylon, food packaging; processing, and color stabilizer. [Uniroyal]

Wytox® 312. CAS 26523-78-4; Trisnonylphenyl phosphite; a non staining, nondiscoloring, low volatility antioxidant for polyolefins, vinyl chloride polymers, high impact polystyrenes, etc. [National Polychemicals] ‡

Wytox® 335. A modified polymeric phosphite stabilizer for emulsion type styrene-butadiene polymers; outstanding suppression of gel build-up; exceptionally resistant to hydrolysis. [National Polychemicals] ‡

Wytox® ADP. Alkylated diphenylamine; an antioxidant for rubber for protection against heat aging and flex cracking. [National Polychemicals] ‡

Wytox® BHT. Alkylated p-cresol; a nonstaining antioxidant for plastics. [National Polychemicals] ‡

Wytox® LT. CAS 123-28-4; Dilauryl thiodipropionate; an antioxidant suitable for use in plastics of the polyolefine and ABS type in contact with food. [National Polychemicals] ‡

X

X2B. A proprietary hard rubber. §

X50-S. Si69 and N330 carbon black (1:1); reinforcing filler for rubber industry. [Degussa]

X-743. Aromatic plasticizer; aromatic, chemically inert, nonsaponifiable plasticizer exhibiting low reactivity; used in adhesives (mastic, pressure sensitive), rubber (cements, mechanical and molded goods, tires), and caulking compounds. [Neville]

Xact®. For agricultural applications. [DuPont UK]

Xala. A name for borax.

xametrin. See hexamine.

Xanax. CAS 28981-97-7; Alprazolam; sedative. [Upjohn]

Xanco-Frac®. Heteropolysaccharide product; gum for use as a viscosifier in oil field hydraulic fracturing fluids; suspending agent. [Kelco/Oil Field Prods.]

Xanflood®. CAS 11138-66-2; Industrial-grade xanthan gum; foam stabilizer, flocculant, suspending, gelling agent, rheology modifier, lubricant for industrial applications especially for secondary and tert. oil recovery. [Kelco]

Xantalgin®. Alginate impression material; dental preparation. [Bayer AG]

xanthan. See xanthan gum.

xanthan gum. (corn sugar gum; xanthan) CAS 11138-66-2; EINECS 234-394-2; High molecular weight hetero polysaccharide gum produced by a pure-culture fermentation of a carbohydrate with xanthomonas campestris; thickening and suspending agent; in drilling fluids, ore flotation, foods, and pharmaceuticals. [Kelco Int'l; Meer]

Xanthano®. Impression plaster; for dentistry. [Bayer AG]

xanthene. (dibenzopyran, tricyclic) $C_{13}H_{10}O$; CAS 92-83-1; EINECS 202-194-4; Organic synthesis, fungicide. [Aldrich; Schweizerhall]

xanthine. (2,6-dihydroxypurine; dioxopurine) $C_5H_4N_4O_2$; CAS 69-89-6; EINECS 200-718-6; organic synthesis, medicine. [Am. Biorganics; Dajac Labs; U.S. Biochemical]

xanthophyll. (lutein; carotenoid) $C_{40}H_{56}O_2$; CAS 127-40-2; EINECS 204-840-0; The yellow pigment occurring in green vegetation and some animal products.

Xanthopicrin. A yellow coloring matter from the bark of Xanthoxylum caribgum.

xanthopicrite. A yellow resin from Xanthoxylum species.

Xantopren®. Precision Impression material on an elastomer basis in several different viscosities; used in dentistry. [Bayer AG]

Xantopren® Function. Special impression material for functional impressions; used in dentistry. [Bayer AG]

Xantopren® Plus. Organic silicone polymer for precision impressions; used in dentistry. [Bayer AG]

Xantygen®. Thermoplastic impression material in rods and plates; for dentistry. [Bayer AG]

XAS 10961.01. Polybutylene oxide polyol; high hydrophobicity polyol for formultion of improved moisture barrier polyurethane adhesives, sealants and elastomers in construction, marine and electronics markets. [Dow Plastics]

XEA 9361. Two-component adhesive; room-temperature curing adhesive with high shear and peel strength and flexibility; for general purpose bonding and applications requiring high elong., e.g., sealing and cryogenic applications [Hysol Aerospace Prods.]

Xenacryl®. Acrylic resin; used for surface coatings, adhesives and sealants. [Baxenden Chemicals Ltd]

Xenalak®. Acrylic polymer in solution; used for surface coatings. [Baxenden Chemicals Ltd]

Xeneisol 133. A proprietary trade name for the 133 isotope of xenon [Mallinckrodt Inc] *

xenene. See biphenyl.

Xenith. Blend of polymers, amine, and solvents; zinc electroplating additive. [Taskem Inc] *

o-xenol. See o-phenylphenol.

Xenomatic. Xenon Xe 133; radioactive agent. [Mallinckrodt Inc] *

xenon. A heavy inert gaseous element present in minute quantities in the atmosphere; Xe; CAS 7440-63-3; luminescent tubes, flash lamps in photography, fluorimetry, lasers, tracer studies, and anesthesia. [Air Prods.; Electrochem Ltd; Liquid Air Corp.]

Xenon Xe 133-VSS. Xenon Xe 133; radioactive agent. [Medi-Physics Inc] *

Xenoy®. Polymer alloys; used for plastic components for automotive, electrical, electronics, lighting, medical, packaging, audio etc. [GE Plastics Ltd]

Xenoy® 2230. Thermoplastic alloy, PC-based, impact-modified; alloy offering dimensional stability, mechanical performance. [GE Plastics]

Xenoy® 5220. Thermoplastic alloy, PC-based, impact-modified; thermoplastic resin offering excellent impact strength, chemical resistance. [GE Plastics]

Xenoy® 5770. PC/polyester alloy, 20% glass/mineral-reinforced; impact-modified grade with chemical resist., good surface and low warpage; for lawn and garden applications. [GE Plastics]

Xenoy® 6120. Thermoplastic alloy, impact-modified; alloy providing superior chemical resistance and excellent mechanical and elec. properties. [GE Plastics]

Xenoy® DX2735. Thermoplastic alloy; unreinforced blend offering good chemical and heat resist., excellent low-temp. impact performance. [GE Plastics]

Xenoy® DX6125. Thermoplastic engineering structural foam resin; used for material handling, industrial, and recreational applications requiring toughness, chemical resist., heat resist., and load carrying capabilities. [GE Plastics]

Xeroderm® S100, L67. Silicone-based water repellent agent, especially for chrome leathers and chrome suede; used for leather industry. [Bayer AG; Bayer plc]

Xerol. A proprietary trade name for glyceryl monostearate.§

X.F.L.X. A substituted ethylene mixed with secondary aromatic amines. [C.P. Hall] *

X-lte. A proprietary alloy containing 37-39% nickel and 17-19% chromium with iron. §

X.L. High boiling tar acids. [Coalite Fuels & Chemicals Ltd]*

XL-All Insecticide. CAS 54-11-5; Nicotine; alkaloid insecticide. [Synchemicals Ltd]

XP-4. Phosphate animal feed supplement. [FMC]

XP Pastes. Plasticizer dispersions of organic pigments. [Reckitts Colours Ltd]

X-Prep. Sennocides A and B; laxative/bowel evacuant. [Napp Laboratories Ltd] *

XR 7. Semipermanent mold release agent; forms a tough, durable film with excellent adhesion to mold surface; gives easy and multiple releases with most thermosetting rubber and plastic materials. [Releasomers]

XSA 80. CAS 25321-41-9; EINECS 246-839-8; Xylene sulfonic acid; catalyst for resins. [Hart Chem. Ltd.]

XT® Polymer 250. Acrylic multipolymer; polymer sheet for thermoformed packaging applications requiring transparency and toughness at an economical price; suited for the food market where grease-proofness, good gas barrier charactersitics, and freedom from odor or taste transfer are mandatory; can be hot filled; contains no plasticizers to leach out; used in tubs for margarine and dairy products, lids, portion packs, candy trays, meat trays, etc.; FDA approved for food contact use; 250 grade has lower impact strength, but is more rigid than 375 [Cyro Industries]

XT® Polymer 375. Acrylic multipolymer; polymer sheet for thermoformed packaging applications requiring transparency and toughness at an economical price; suited for the food market where grease-proofness, good gas barrier charactersitics, and freedom from odor or taste transfer are mandatory. [Cyro Industries]

X-Tan® Special C. Textile auxiliary which suppresses chlorine release in chlorite bleaching; prevents metal corrosion. [Sybron]

Xtol. CAS 61790-12-3; EINECS 263-107-3; Tall oil fatty acid. [Georgia-Pacific]

Xtrusorb 600, 700. Activated carbon; for vapor phase applications. [Calgon Carbon] *

Xuprin. CAS 395-28-8; Isoxsuprine; promotes blood flow rate. [Duphar BV] *

XX 601. A proprietary brand of zinc oxide used in rubber compounding. §

XXX-1. Lubricant for food processing conveyors. [CasChem]

Xydar® G-540 LCP. Liquid crystal polymer, glass-reinforced; injection molding resin for elec./electronic applications; high strength and stiffness, high heat deflection temp., inherent flame resist., low warpage, excellent weld line strength. [Amoco Chemical Co]

Xylan 330. A proprietary aerosol form of PTFE used as a

mold-releasing agent in plastics processing. [Whitfield Plastics] *

Xylan 1052. A proprietary lubricant based on PTFE, for use under extreme pressures and in extremes of temperature. [Whitfield Plastics] *

xylene. (methyl toluene; dimethylbenzene; xylol) Aromatic compound.; commercial mixture of 3 isomers: o-, m-, and p-xylene (1,3-dimethylbenzene, 1,2-dimethylbenzene, 1,4-dimethylbenzene); $C_6H_4(CH_3)_2$; CAS 1330-20-7; solvent; raw material for production of benzoic acid, phthalic anhydride, dyes, other organics; aviation gasoline; protective coating; solvent for alkyd resins, lacquers, enamels, and rubber cements. [Ashland; Crowley; Exxon; Fina; Mallinckrodt; Mitsubishi Petrochem.; Mitsui Petrochem. Ind.; Mobil; Shell; Texaco; Unocal]

Xylene®. Specialty dye for aq. mediums. [Sandoz]

xylene musk. An artificial musk perfume. It is trinitro-t-butyl-m-xylene.

xylenol. (dimethylphenol; hydroxydimethylbenzene; dimethylhydroxybenzene) $(CH_3)_2C_6H_3OH$; CAS 1300-71-6; Disinfectants, solvents, pharmaceuticals, insecticides, fungicides, plasticizers, rubber chemicals, additives to lubricants and gasoline, manufacture of polyphenylene oxide, wetting agents, and dyestuffs. [Coalite Chem. Div.; Crowley; Crowley Tar Prods.]

Xyligen®. Active ingredients for wood preservative formulations; protective agents against wood-destructive fungi; additives to glues to protect wood-based materials. [BASF AG]

Xylite. A proprietary rubber vulcanization accelerator; it is a tarry diphenylguanidine. §

Xylocaine. CAS 137-58-6; Lidocaine; anesthetic. [Astra Pharmaceutical Products Inc]

Xylocard. A proprietary preparation of lionocaine hydrochloride; used in the treatment of cardiac arrhythroias. [Astra Chemicals Ltd] ‡

Xylock 225. A proprietary condensation product of phenols with an aryl alkyl ether; used as a high-performance, heat-stable molding resin. [Albright & Wilson Ltd] *

Xylok. High performance resins. [Albright & Wilson Ltd, Phosphates & Speciality Business] *

xylol. Commercial xylene. It consists of a mixture of about 60% m-xylene, 10-25% o- and p-xylene, ethylbenzene, and small quantities of trimethyl-benzene, paraffin, and thioxene.

xylol. See xylene.

Xylon FR. A proprietary nylon containing a flame-retarding additive. [Dart Industries Inc] *

Xylonite. See Celluloid. §

Xyloproct. A proprietary preparation of lignocaine, aluminum acetate, zinc oxide and hydrocortisone acetate; used in the treatment of hemorrhoids. [Astra Chemicals Ltd] ‡

xyloquinone. Dimethylbenzoquinone, $C_6H_2(CH_3)_2O_2$.

xylorcinol. CAS 527-55-9; Dimethylorcinol, $C_6H_2(CH_3)_2 \cdot (OH)_2$.

D(+)-xylose. (wood sugar) $C_5H_{10}O_5$; CAS 58-86-6; EINECS 200-400-7; Dyeing, tanning, diabetic food, source of ethanol. [Aldrich; Am. Biorganics; Penta Mfg.]

Xylose. A proprietary benzyl cellulose. §

Xyron®. Modified polyphenylene ether resin. [Asahi Chem. Industry]

---

* Trade name not verified as available   † Trade name verified as obsolete

Y

yaba bark. The bark of *Andira excelsa*.

Y alloy. An aluminum alloy containing 4% copper, 2% nickel, and 1.5% magnesium with small amounts of iron and silicon. It has a specific gravity of 2.8; used for die-cast pistons, etc.

Yaltox®. CAS 1563-66-2; A free flowing granule containing 5% w/w carbofuran; soil-applied insecticide for control of soil and seedling pests including cabbage root fly, cabbage stem weevil, flea beetle, cabbage stem flea beetle, early aphids in brassicas, turnip root fly, frit fly, millipedes, symphilids, beet leaf miner, springtails, wierworms, free living nematodes, potato cyst eelworm, carrot fly and carrot willow aphid. [Bayer AG; Bayer plc]

yama-mai silk. A silk produced by the Japanese oak caterpillar of Japan, China, and India.

yara-yara. *See* Nerolin.

Yarmor. Pine oil terpene liquids; antiskinning agents in protective coatings, for pigment grinding aids and as an antifoam agent. [Hercules]

yarmor. *See* pine oil.

yarrow. The dried leaves of *Achillea millefolium;* a tonic.

yeast. (barm) CAS 68876-77-7; Fungus with unicellular growth form; fermentation of sugar, molasses, cereals for alcohol; brewing; baking; food supplement; protein biosynthesis; source of vitamins, enzymes, nucleic acids; biochemical research. [Champlain Industries; Gist-Brocades Food Ingreds.]

yeast extract. (extract of yeast) CAS 8013-01-2; .Saccharomycetacae; fermentation of sugars, etc., brewing, baking, food supplement. [Atomergic Chemetals; Champlain Industries; Gist-Brocades Food Ingreds.]

Yeast Lactase L-50,000. Yeast lactase; enzyme for hydrolyzing lactose in dairy products (milk, whey, cheese, yogurt), pharmaceuticals (digestive aids). [Solvay Enzymes]

Yellow 201. CAS 1309-37-1, 7787-59-9; Iron oxides, bismuthoxychloride; inorg. colorant. [Presperse]

yellow bark. The bark of *Cinchona calisaya*.

yellow carmine. (Italian pink, yellow lake). Pigments prepared by precipitating the glucoside quercitrin with alumina.

yellow catechu. *See* cutch.

yellow cross gas. *See* mustard gas.

yellow earth. *Sec* ocher.

yellow gold. An alloy of 53% gold, 25% silver, and 22% copper.

yellow liquors. The drainage from alkali waste heaps.

yellow precipitate. (yellow mercury oxide) CAS 21908-53-2; HgO. Ammonium phosphomolybdate is also given this name.

yellow precipitate. *See* mercury oxide (ic), red and yellow.

yellow prussiate of potash. (ferro prussiate of potassium; potassium ferrocyanide) $K_4Fe(CN)_6$; CAS 13943-58-3; Used for tempering steel, engraving, and as a lab reagent

yellow prussiate of soda. *See* sodium ferrocyanide.

yellow soda ash. A soda ash (sodium carbonate) containing traces of iron oxide.

Yellowstone. Wyoming sodium bentonite; used for steel production and foundries. [Bromhead & Denison Ltd]

Yellow Sulphur. CAS 7704-34-9; Sulfur powder; garden fungicide. [Vitax Ltd]

yellow wax. A viscous, semi-solid, difficultly volatile substance obtained by the distillation of the still residues of petroleum. It contains anthracene and other hydrocarbons.

yenshee. The dregs and carbonized opium which remains after smoking containing from 1-10% morphine.

Yeoman. Lanolin BP. [Croda Chem. Ltd]

yerba mate. (Paraguay tea; Jesuit's tea; St. Batholomew's tea) Leaves of a tree found in Paraguay, containing caffeine and tannin.

Yieldmaster®. Fibers. [DuPont UK]

ylang-ylang oil. Essential oil distilled from the flowers of the *Conanga odorata*; used in perfumery.

Yodoxin. CAS 83-73-8; Iodoquinol; anti-amebic. [Glenwood Inc]

yolk powder. *See* lecithin.

Yoloy. A proprietary alloy; it is a steel containing 1% copper, 2% nickel, and up to 0.2% carbon. §

Yomesan®. CAS 50-65-7; Niclosamide; anthelmintic preparation; used against tapeworm infestation in humans, dogs, and cats. [Bayer AG]

Yonckite. A Belgian explosive consisting of ammonium perchlorate, ammonium nitrate, sodium nitrate, and trinitrotoluene or nitronaphthalene.

Yoracryl Dyes. Cationic dyes for acrylic, modified acrylic and other cationic dyeable fibers such as cationic dyeable polyester and nylon. [Yorkshire Pat-Chem; Yorkshire Chemicals plc]

York Krystal Kleer Castor Oil. Castor oil; used in lipsticks as dye solv. and to impart gloss and emollience. [United Catalysts]

Yorkshire grease. (Wakefield grease). The recovered fatty acids from wool grease.

York USP Castor Oil. Castor oil USP; used in lipsticks as dye solv. and to impart gloss and emollience. [United Catalysts]

yperite. *See* mustard gas.

Y-Pof. Solvent based graffiti remover. [Kalon Chemicals Ltd]

YSE-Cure B-001. Amine adduct; epoxy curing agent for adhesives and coatings. [Ajinomoto]

YSE-Cure B-002. Amine adduct; epoxy curing agent for coating, flooring, casting. [Ajinomoto]

YSE-Cure B-002W. Amine adduct; epoxy curing agent for coating, flooring, casting; anti-crystallization version of B-002. [Ajinomoto]

YSE-Cure B-003. Amine adduct; epoxy curing agent for coating, flooring, casting; clear colorless version of B-002. [Ajinomoto]

YSE-Cure C-002. Amine adduct; epoxy curing agent for lining and adhesives; acid resist. [Ajinomoto]

YSE-Cure F-100. Amine; raw material for epoxy curing agents and imide compounds. [Ajinomoto]

YSE-Cure LX-1N. Modified amine; epoxy curing agent for lining, potting; heat and chemical resist. [Ajinomoto]

YSE-Cure LX-1N. Modified amine; epoxy curing agent for adhesives, grout and coating; flexible. [Ajinomoto]

YSE-Cure N-001, N-002. Amine adduct; epoxy curing agent elec. potting applications. [Ajinomoto]

YSE-Cure PX-3. Modified amine; epoxy curing agent for grout and lining (tar-epoxy); inexpensive. [Ajinomoto]

YSE-Cure QX-2, QX-3. Thiourea condensate; epoxy curing agent for lining, adhesives, and putty; fast curing. [Ajinomoto]

YSE-Cure RX-2. Accelerated modified amine; epoxy curing agent for lining and adhesives; fast curing. [Ajinomoto]

YSE-Cure RX-3. Accelerated modified amine; epoxy curing agent for lining and flooring; fast curing. [Ajinomoto]

YSE-Cure S-002. Amine adduct solution; epoxy curing agent; primer for metal, concrete, mortar, and wood; excellent adhesion. [Ajinomoto]

Y-Tack. Polyglycol ether and polyacrylic copolymerides in aqueous solution; warp and weft treatment chemicals which provide a combination of film forming yarn cover, intra fiber bonding, lubrication and static control. [Thomas Swan & Co Ltd] †

ytterbia. See ytterbium oxide.

ytterbium. Metallic element; Yb; CAS 7440-64-4; EINECS 231-173-2; lasers, dopant for garnets, portable x-ray source, chemical research. [Aldrich; Atomergic Chemetals; Cerac; Rhone-Poulenc Basic]

Ytterbium Yb-169 DTPA. CAS 12111-24-9; Pentetate calcium trisodium Yb 169; radioactive agent. [3M] †

ytterbium oxide. (ytterbia) Yb_2O_3; CAS 1314-37-0; EINECS 215-234-0; Special alloys, dielectric ceramics, carbon rods for industrial lighting, catalyst, and special glasses. [Atomergic Chemetals; Noah Chem.; Rhone-Poulenc Basic]

yttria. See yttrium oxide.

yttrium. Metallic element; Y; CAS 7440-65-5; EINECS 231-174-8; nuclear technology, iron and other alloys, deoxidizer for vanadium and other nonferrous metals, microwave ferrites, coating on high-temperature alloys, and special semiconductors. [Atomergic Chemetals; Cerac; Noah Chem.; Rhone-Poulenc Basic]

yttrium oxide. (yttria) Y_2O_3; CAS 1314-36-9; EINECS 215-233-5; Phosphors for color TV tubes, yttrium-iron garnets for microwave filters, stabilizer for high-temperature service materials. [Atomergic Chemetals; New Metals & Chems. Ltd; Noah Chem.; Rhone-Poulenc Basic; Shin-Etsu Chem.]

Yukalon. Trademark for a proprietary grade of polyethylene. [Mitsubishi Petrochem.] *

Yutopar. CAS 26652-09-5; Ritodrine; uterospasmolytic, prevents premature birth. [Duphar BV] *

Z

zaccatila. See cochineal.

Zaditen. CAS 34580-14-8; Ketotifen fumarate; anti-asthmatic. [Sandoz Pharmaceuticals] *

Zadstat. CAS 443-48-1; Metronidazole; antibiotic. [Lederle Laboratories] *

zaffer. See smalt.

zakin rubber. A rubber-like substance prepared from glue or similar material.

Zamak Alloys. Proprietary alloys of zinc with aluminium and sometimes small amounts of copper and magnesium; copper-aluminium-zinc alloys, suitable for die castings contain: 3.9-4.3 % aluminium, 0.9-2.9 % copper, 0.003-0.06 % magnesium, and the remainder zinc; they have a specific gravity of 6.64-6.7. §

Zam-Buk. Antiseptic ointment. [Fisons plc, Pharmaceuticals Div] *

Zam Metal. A proprietary alloy of zinc with aluminium and magnesium. §

Zanchol. CAS 519-95-9; A proprietary trade name for florantyrone. §

Zanil. CAS 2277-92-1; A proprietary preparation of oxyclozanide; a veterinary anthelmintic. §

Zanosar. CAS 18883-66-4; Streptozocin; antineoplastic. [Upjohn]

Zap-Oglobin. . Reagent which destroys red blood cells to leave white blood cells in suspension for analyzis; used for white blood cell and hemoglobin determination. [Coulter Electronics Ltd]

Zapon®. Metal complex dyes; for air-drying, acid-hardening, amine-hardening transparent lacquers, baking finishes, polyurethane lacquers, peroxide-hardening polyester lacquers, and wood stains. [BASF AG]

Zaponin. Reagent which destroys red blood cells to leave white blood cells in suspension for analysis; used for white blood cell determination using semi-automated cell counters. [Coulter Electronics Ltd] †

zapoto gum. See chicle.

Zarontin. CAS 77-67-8; Ethosuximide; anticonvulsant. [Parke-Davis]

Zaroxolyn. CAS 17560-51-9; A proprietary preparation of metolazone; an antihypertensive. [Pennwalt Corp] *

Zauberin. A material containing Chloramine T; a detergent and bleaching agent. §

ZB2335. CAS 12513-27-8; Zinc borate for flame retardancy. [Borax Consolidated Ltd]

zea mays oil. See corn oil.

Zeasorb. A proprietary preparation of microporous cellulose hexachlorophane, chloroxylenol, aluminum dihydroxyallantoinate and purified talc; used in dermatology. [Stiefel Laboratories Inc; Stiefel Laboratories (UK) Ltd]

Zedox. CAS 1314-23-4; Zirconium oxide. [Anzon Ltd]*

Zeeospheres® 200. Silica-alumina ceramic; strong, hard, inert, thick-walled hollow spheres for use as filler for a variety of plastic resins in injection molding, extrusion, SMC, BMC, RTM, compression molding, potting/encapsulating, adhesives, tooling, casting, flooring, grouting, sealants, mastics, coatings, films, and other applications. [Zeelan]

Zeese. Honey substitute for diabetics containing sorbitol, fructose, water, citric acid, permitted .color E150 and flavoring. [LAB Ltd] ‡

Zeiodelite. A mixture obtained by stirring 24 parts powdered glass into 20 parts melted sulfur; used as a cement, and for taking casts.

Zeise's salt. A salt, [Pt(C₂H₄)Cl₃]K, formed when potassium chloride is added to a solution of platinous chloride saturated with ethylene.

Zeklan. Chemical products for use in industry. [Courtaulds plc]

Zelco metal. An alloy of 83% zinc, 15% aluminum, and 2% copper.

Zelcon. Fabric conditioner. [DuPont UK]

Zelec. Antistatic agent. [DuPont UK]

Zeller's ointment. Ammoniated mercury ointment.

Zellwonet. A proprietary viscose packing material. §

Zelulone. A proprietary artificial yarn made from wood pulp.§

Zemdrain®. Fibers. [DuPont UK]

Zemid® 610. CAS 9002-88-4; Toughened polyethylene resin, mineral-reinforced; balanced impact resist. and stiffness even at low temps.; for industrial and consumer products, e.g., equip. shields, shrouds, lawn mower decks, snow shovels, instrument cases, swimming pool equip., snowboards. [DuPont]

Zemid® 650. CAS 9002-88-4; Toughened HDPE resin, min.-reinforced; good processability in blow molding, high stiffness and toughness. [DuPont]

Zenadrid. CAS 53-03-2; Prednisone; glucocorticoid. [Syntex Laboratories Inc] *

Zendium. CAS 7681-49-4; Sodium fluoride; dental caries prophylactic. [Oral-B Laboratories Inc] *

Zenith. Organic brightener system; used for bright nickel electroplating (high leveling, fast brightening). [Engelhard Technologies Ltd]

Zenker's fluid. A solution containing 2.5g potassium chromate, 1g sodium sulfate, 5g mercuric chloride, 5cc glacial acetic acid, and 100cc water.

Zennapron. Mixture of 2,4-D and mecoprop; used to control weeds in grassland. [BP Oil Ltd]

Zentralin. Dimethyldiphenylurea, $(CH_3)_2(C_6H_5)_2CON_2$; used for explosives.

---

‡ Trade name and manufacturer not verified   § Trade name without identified manufacturer

Zentralit I. Diethyldiphenylurea.

Zentralit II. Dimethyldiphenylurea.

Zeolex. Synthetic sodium aluminum silicate; flow promoter, pigment extender. [Cornelius Chemical Co Ltd] *

zeolites. CAS 1318-02-1; Hydrated alumimium silicates containing alkali or alkaline earth metals occurring naturally and used in ion-exchangers, absorbents, dessicants, and solar collectors.

zeolite synthetic. CAS 68989-22-0; Water softener, detergent builder, cracking catalyst, adsorbents, desiccants, in solar collectors (as heating and cooling agents). [Crosfield Chemie BV; Ethyl; Miles; PQ; Sybron]

Zeolum. Crystalline hydrous alumino-silicate of alkali metals or alkaline earth metals; molecular sieve showing strong selective adsorption capacity; for petrochemical industry, drying of gases and liqs., separation and purification processes. [Tosoh] *

Zeonet A. CAS 108-80-15; Isocyanuric acid; curing agent for Nipol AR and HyTemp 4050 series polyacrylate elastomers. [Zeon]

Zeonet B. Trimethyl octadecyl ammonium bromide; accelerator for Nipol AR and HyTemp 4050 series polyacrylate elastomers. [Zeon]

Zeonet U. CAS 102-07-8; N,N´-Diphenylurea; retarder to slow down the cure of Nipol AR and HyTemp 4050 series elastomers. [Zeon]

Zeonex 280. Amorphous polyolefin; offers low water absorp., transmittance, high heat resist. for optical applics. (optical disks, plastic lenses). [Nippon Zeon/Zeonex]

Zeospan 303, 306. Polyether elastomer; oil-resist., high-elasticity rubber with excellent balance of ozone resist. and cold resist. [Zeon]

Zeothix. A ground silica; thickening agent. [Cornelius Chemical Co Ltd] *

Zeotokol. A coarse dolerite (an igneous rock composed essentially of labradorite and anorthite, with augite and sometimes olivine) ground up; and used as a fertilizer.

Zepel®. Fabric fluoridizer. [DuPont UK]

Zephiran Chloride. Benzalkonium chloride; pharmaceutic aid. [Sterling Drug Inc] *

Zephirol®. Benzalkonium chloride; a preparation for the disinfection of hands and instruments. [Bayer AG]

Zephron®. For the chemical industry. [DuPont UK]

Zephyr. Inks used in the screen printing process; for billboards, signs and displays. [Allied-Signal Inc/Sinclair and Valentine Division] *

Zerex. A proprietary trade name for a polyvinyl alcohol antifreeze compound. §

Zerofil. A proprietary rock wool (qv) which has been treated with asphalt. §

Zerogen® 10. CAS 1309-42-8; Halogen-free magnesium hydroxide composition; flame retardant/smoke suppressant for thermoplastics and elastomeric formulations. [J.M. Huber/Solem]

Zerogen® 60. CAS 1309-42-8; Halogen-free magnesium hydroxide composition; medium particle sized flame retardant/smoke suppressant for thermoplastics and elastomeric formulations. [J.M. Huber/Solem]

Zerol. Refrigeration fluid. [Chevron] *

Zerone. A proprietary trade name for a methanol and polyvinyl alcohol antifreeze product. §

Zeronox. Catalyst for the reduction of nitrogen oxides. [Hüls AG]

Zerotherm. Self drying coatings for sand molds and cores. [Foseco (F.S.) Ltd]

Zerox. CAS 302-01-2; Hydrazine solutions; used for water treatment. [Schering Industrial Products Ltd]

Zestoretic. Proprietary preparation of lisinopril and hydrochlorthiazide; for treating hypertension. [Imperial Chemical Industries plc]

Zestril. CAS 83915-83-7; Proprietary preparation of lisinopril; for treating cardiac failure and hypertension. [Imperial Chemical Industries plc]

Zetabon. Plastic coated steel and aluminum used in wire and cable for armor, corrosion and lightning protection. [Dow UK]

Zetag®. High molecular weight cationic polyelectrolytes for sewage and industrial effluent treatment supplied in microbead, powder or liquid form; typical applications include primary sedimentation processes, sludge dewatering, sludge thickening, phosphate removal processes. [Allied Colloids Ltd]

Zetar. Coal tar; anti-eczematic. [Dermik Laboratories Inc] *

Zetax®. CAS 155-04-4; Zinc 2-mercaptobenzothiazole; secondary accelerator in latex foam curing systems. [R.T. Vanderbilt]

Zetesap. Base for soap-free "soap" tablets. [Surfachem Ltd]

Zetesol 100. MIPA-laureth sulfate, laureth-4, cocamide DEA; detergent and emulsifier for personal care products; for foam bath preps. with high oil concs. [Zschimmer & Schwarz]

Zetesol 856 D. MIPA C12-15 pareth sulfate; detergent for cosmetics, shampoos, bath preps., and liq. synthetic soap. [Zschimmer & Schwarz]

Zetesol AP. Ammonium C12-15 pareth sulfate, propylene glycol; detergent for cosmetics, shampoos, bath preps., liq. hand cleaners, dishwash, household cleaners. [Zschimmer & Schwarz]

Zetesol NL. CAS 9004-82-4; 1335-72-4; Sodium laureth sulfate; detergent for personal care, household and industrial cleaners. [Zschimmer & Schwarz]

Zetpol® 1010. Hydrogenated nitrile rubber; elastomer for fuel hose, fuel diaphragm, Freon packing, in-tank insulator applics. [Zeon]

Zetpol® 2000L. Acrylonitrile-ethylene-butadiene terpolymer; elastomer for o-rings, packings, gaskets, oil seals, oil hose, well head seals, blow out preventor, and water pump seal applications. [Zeon]

Zetpol® 2020. Hydrogenated nitrile rubber; elastomer for oil seals, synchronous belt, rolls, oil hose, and drilling pipe protector applications. [Zeon]

Zetpol® ZSC-2295. Nitrile rubber, zinc oxide-methacrylic acid modified; used for very high durometer, high abrasion resist. compounds. [Zeon]

Zettnow's stain. A microscopic stain. Solution a) contains 10 g tannic acid to which has been added 30 cc of a 5% solution of tartar emetic. Solution b) contains 1g silver sulfate in 250 cc water. Take 50 cc and add ethylamine until precipitate redisssolves.

Zettyn. CAS 122-18-9; Cetalkonium chloride; anti-infective, topical. [Sterling Drug Inc] *

Zeus. An alloy of 20% silver and 80% copper; used for fuse wire.

Zewa SL 2. CAS 8061-51-6; Virtually desugared sodium lignosulfonate (< 1% sugar content); water-reducing concrete additive; dispersant in pesticides. [Borregaard LignoTech]

---

* Trade name not verified as available

† Trade name verified as obsolete

Zewakol. Modified lignosulfonates; extenders for U/F resins in particle board mfg. [Borregaard LignoTech]

Zewalon FN. CAS 8061-51-6; Sodium lignosulfonate; auxiliary tanning agent. [Borregaard LignoTech]

Zewaphosphate. A phosphate fertilizer.

Zewa powder. Sodium lignin sulfonate obtained by evaporation of waste sulfite lyes. It has detergent and water softening properties.

Zidan. CAS 12122-67-7; Zincb; fungicide, Insecticide. [Makhteshim Chemical Works Ltd]

Zidanit. CAS 12122-67-7; Zineb; fungicide, insecticide. [Makhteshim Chemical Works Ltd]

Zienam. Imipenem and cilastatin sodium, broad-spectrum β-lactam antibiotic. [Merck & Co Inc]

zigueline. A red oxide of copper.

Zilloy. A proprietary zinc alloy containing zinc with 1% copper, 0.01% magnesium, and lead and cadmium in addition; the rolled sheets are suitable for building purposes. §

Zimalium. Alloys containing 74-93.5% aluminum, 2.8-14.8% zinc, and 3.7-11.2% magnesium.

Zimate®. Zinc diamyl, dibutyl, diethyl and dimethyl dithiocarbamates; rubber accelerators for NR and latex. [R T Vanderbilt Co Inc]

Zimco. CAS 121-33-5; Vanillin; pharmaceutic aid (flavor). [Sterwin Chemicals Inc] *

Zinamide. CAS 98-96-4; Pyrazinamide; antituberculous agent [Merck & Co Inc] *

Zinar®. Rosin-based resinate; tackifier resin for adhesives; [Arizona]

zinc. Metallic element; Zn; CAS 7440-66-6; EINECS 231-175-3; alloys, galvanizing iron and other metals, and fungicides. [Aldrich; Cerac; Cuproquim; Ferro/Bedford; Pasminco Europe; Zinc Corp. of Am.]

Zincaband. A proprietary zinc paste bandage; used for subacute and chronic eczema and lichenification. [Seton Healthcare Group plc]

Zincalium. See Zimalium.

zinc ammonium chloride. $ZnCl_2 \cdot 2NH_4Cl$; EINECS 238-687-6; welding, soldering flux, dry batteries, and galvanizing. [Blythe, William Ltd; Du Pont]

zinc anhydride. A variety of lithopone, also known as zinc barytes and consisting of a mixture of calcium and barium sulfates and zinc oxide.

Zincazol. Zinc-α-phenylbiguanide; a proprietary rubber vulcanization accelerator. §

zinc-baryta white. See lithopone.

zinc barytes. See zinc anhydride.

zinc bende. (black Jack; blende; rosinjack; pseudo-galena). A mineral that is zinc sulfide, ZnS; used as a pigment.

zinc borate. (boric acid, zinc salt) B_2O_3; CAS 1332-07-6; 120007-67-9; EINECS 233-471-8; Medicine, fireproofing textiles, fungistat, mildew inhibitor. [BA Chem. Ltd; Climax Performance; U.S. Borax & Chem.]

Zinc Borate 2335. $2ZnO \cdot 3B_2O_3 \cdot 3.5H_2O$; A specialty flame retardant additive to plasticized PVC and other polymers to reduce afterglow and smoke. [Borax Consolidated Ltd] *

zinc bromide. (zinc bromide anhydrous) $ZnCl_2$; CAS 7699-45-8; EINECS 231-718-4; Photographic emulsions, manufacture of rayon, radiation viewing shield. [Atomergic Chemetals; Cerac; Ethyl; Great Lakes; Hoechst Celanese; Ryvan Ltd]

zinc bromide anhydrous. See zinc bromide.

zinc carbonate. (carbonic acid, zinc salt (1:1); Smithsonite) Inorganic salt; $ZnCO_3$; CAS 3486-35-9; EINECS 222-477-6; accelerator-activator for transparent natural and synthetic rubber goods, adhesives; as pigment; fireproofing filler for rubber and plastics; topical antiseptics, cosmetics, and lotions. [Allchem Industries; Harcros Durham; Nihon Kagaku Sangyo; Spectrum Chem. Mfg.]

zinc chloride. $ZnCl_2$; CAS 7646-85-7; EINECS 231-592-0; Catalyst, dehydrating and condensing agent in organic synthesis, fireproofing, preserving food, electroplating, antiseptic denaturant for alcohol. [Allied-Signal; Atochem N. Am.; Blythe, William Ltd; EM Industries; Malalinckrodt]

zinc chromate. $ZnCrO_4 \cdot 7H_2O$; EINECS 236-878-9; Yellow pigments. [Colores Hispania SA; Landers-Segal Color]

zinc chrome. (citron yellow). A pigment that is zinc chromate, $ZnCrO_4$.

zinc chrome yellow. See zinc yellow.

zinc dust. See blue powder.

zinc flowers. See philosopher's wool.

zinc fluoride. ZnF_2; CAS 7783-49-5; EINECS 232-001-9; Phosphors, ceramic glazes, wood preservative, electroplating, and organic fluorination. [Atochem N. Am; Atomergic Chemetals; Cerac; Hoechst Celanese; Noah Chem.]

zinc formosul. CAS 24887-06-7; A basic zinc-formaldehyde-sulfoxylate; used in fat-splitting.

zinc fume. See blue powder.

zinc gluconate. Zinc salt of gluconic acid; $C_{12}H_{22}O_{14}Zn$; CAS 4468-02-4; mineral source for pharmaceutical and food products. [Akzo; Atomergic Chemetals]

zinc grey. The name originally used for zinc dust employed for painting on iron. The term is now used for finely ground zinc blende. A mixture of zinc oxide with finely divided charcoal is sold under this name. It is produced in the manufacture of zinc.

Zincidone®. CAS 15454-75-8; Zinc PCA; cicatrizing agent, tissue hardening agent; for dermatological soap, shampoo, shower gel, spray or stick deodorant, nutritive cream. [UCIB]

zinc lactate. $Zn(C_3H_5O_3)_2 \cdot 3H_2O$; CAS 16039-53-5. [Patco]

zinc naphthenate. $Zn(C_6H_5COO)_2$; EINECS 234-409-2; Drier and wetting agent in paints, varnishes, resins; insecticide, fungicide, mildew preventive; wood preservative; waterproofing textiles; and insulating materials. [King Industries]

zinc nitrate. (zinc nitrate hexahydrate) $Zn(NO_3)_2.6H_2O$; CAS 10196-18-6; EINECS 231-943-8; Acidic catalyst, latex coagulant, reagent, intermediate, mordant. [Blythe, William Ltd; Mallinckrodt; Nihon Kagaku Sangyo]

zinc nitrate hexahydrate. See zinc nitrate.

zincocalcite. A calcite containing zinc carbonate.

zinc octadecanoate. See zinc stearate.

Zincofol. Fungicide. [Chevron] *

zincolith. See lithopone.

Zinc Omadine. CAS 13463-41-7; Pyrithione zinc; an antibacterial; antifungal; and antiseborrheic. [(E R Squibb & Sons Inc).]

Zinc Omadine®, 48% Fine Particle Disp. CAS 13463-41-7; Zinc pyrithione; antimicrobial inhibiting growth of Gram-negative and Gram-positive bacterial, fungi, mold and yeast; used in aq. metal coolant and cutting fluids, PVC plastics. [Olin]

Zinc Omadine®, Powd. CAS 13463-41-7; Zinc pyrithione;

---

‡ Trade name and manufacturer not verified   § Trade name without identified manufacturer

antidandruff agent for shampoos; inhibits growth of Gram-positive and Gram-negative bacteria and fungi; cosmetic preservative. [Olin]

zinconal. See Ektogan.

Zincon Dandruff Shampoo. CAS 13463-41-7; Pyrithione zinc; antibacterial, antifungal, and antiseborrheic. [Lederle Laboratories USA]

Zincov™. Inhibitor. [Calbiochem Corp]

zinc oxide. (Chinese white; zinc white; flowers of zinc) Inorganic oxide; ZnO; CAS 1314-13-2; EINECS 215-222-5; uv absorber; accelerator activator; pigment in white paints, cosmetics, driers, dental cements; mold inhibitor in paints; in manufacture of opaque glass, enamels, tires, printing inks, porcelains; reagent in analytical chemistry; as flame retardant. [Am. Chemet; Asarco; Eagle Zinc; General Chem.; Harcros Durham; Mallinckrodt; Zinc Corp. of Am.]

Zinc Oxide No. 185. CAS 1314-13-2; Tertiary zinc oxide; accelerator activator, pigment, reinforcing agent for rubber and applics. not requiring a high degree of purity; also for prod. of zinc compounds. [Eagle Zinc]

Zinc Oxide No. 318. CAS 1314-13-2; American process zinc oxide, lead-free; slow curing fine particle size pigment and reinforcing agent for rubber industry; also for mfg. of various zinc compounds. [Eagle Zinc]

Zinc Oxide Transparent. CAS 1314-13-2; Highly disperse precipitated zinc oxide; vulcanization accelerator activator for transparent rubber goods, in natural and syn. rubbers; acid acceptor in polychloroprene adhesives; light-colored reinforcing filler; used for vulcanizates needing high elasticity or transparency, and vulcanizates cured in hot air or with sulfenamides, and vulcanizates cured with metal oxides and without sulfur, [Bayer AG; Miles; Polysar]

zinc perhydrol. See Ektogan.

zinc phosphate. $Zn_3(PO_4)_2$; CAS 7779-90-0; In dental cements, phosphors. [Calgon; Colores Hispania SA; Hammond Lead Prods.; Pasminco Europe; G. Whitfield Richards; Witco/Allied-Kelite]

zinc powder. CAS 1314-13-2; Zinc oxide, ZnO.

zinc pyrithione. (bis[1-hydroxy-2(1H)-pyridinethinato-O,S]-(T-4) zinc; zinc 2-pyridinethiol-1-oxide; pyrithione zinc) Aromatic salt; $C_{10}H_8N_2O_2S_2Zn$; CAS 13463-41-7; antidandruff agent for shampoos; cosmetic preservative; antimicrobial for plastics, metalworking fluids. [Allchem Industries; Olin]

Zincrex. Flux for zinc and alloys. [Foseco (F.S.) Ltd]

zinc spar. See calamine.

zinc stannate. CAS 12036-37-2; Fire retardant and smoke suppressant for plastics. [Atomergic Chemetals; Blythe, William Ltd]

zinc stearate. (zinc octadecanoate; octadecanoic acid, zinc salt) Zinc salt of stearic acid; $Zn(C_{18}H_{35}O_2)_2$; CAS 557-05-1; EINECS 209-151-9; in cosmetics, pharmaceuticals, lacquers, ointments, tablet manufacture; mold release agent for plastic; filler, antifoamer; flatting agent in lacquers; as a drying lubricant and dusting agent for rubber; waterproofing agent for concrete, paper, and textiles. [Allchem Industries; Ferro/Grant; Magnesia GmbH; Mallinckrodt; Norac; Stave; Syn. Prods.]

zinc sulfate. (zinc sulfate heptahydrate; white vitriol; white copperas; zinc vitriol) $ZnSO_4 \cdot 7H_2O$; CAS 7446-20-0; EINECS 231-793-3; Rayon, manufacture, dietary supplement, animal feeds, mordant, wood preservative, analytical reagent. [Aldrich; Mallainckrodt]

zinc sulfate heptahydrate. See zinc sulfate.

zinc sulfate, monohdyrate. $ZnSO_4 \cdot H_2O$; rayon manufacture, agricultural sprays, chemical intermediate, dyestuffs, electroplating. [Allchem Industries; Blythe, William Ltd; EM Industries]

zinc sulfide. Inorganic salt; ZnS; CAS 1314-98-3; EINECS 215-251-3; Pigment; in white and opaque glass, plastics, dyeing, paints, linoleum, leather, dental rubber; fungicide; anhydrous in x-ray screens, and TV screens. [Cerac; Chemson GmbH; Noah Chem.; Ore & Chem. Corp.; Sachtleben GmbH]

zinc sulfide grey. (calamine white). A pigment that is a dense zinc oxide used for painting iron, and is artificially made by tinting lithopone with ochers and charcoal.

zinc sulfide white. See lithopone.

Zinctrace. CAS 7646-85-7; Zinc chloride; an astringent and dentin desensitizer. [Armour Pharmaceutical Co] *

zinc vitriol. CAS 7733-02-0; 7446-20-0; EINECS 231-793-3; Zinc sulfate, $ZnSO_4$; used in rayon manufacture, in animal feeds, wood preservative, analytical reagent.

zinc white. (Chinese white, Permanent white, snow white). CAS 1314-13-2; A pigment. It is zinc oxide, ZnO. Chinese white is a very dense oxide, and snow white a very pure one; used as an activator, accelerator, and reinforcing agent in rubber manufacture; dietary supplement; animal feed; cosmetics.

zinc yellow. (zinc chrome yellow, buttercup yellow, lemon yellow). A pigment consisting of zinc chromate, $ZnCrO_4$.

zineb. (zinc dithiocarbamate) $Zn(CS_2NHCH_2)_2$; CAS 12122-67-7; Used as an insecticide and fungicide.

Zinkan. A proprietary combination of aluminium coated with zinc; obtained by rolling at elevated temperatures. §

Zinkgrau. Inexpensive, off-color zinc oxide pigment.

Zinkoxyd Activ®. CAS 1314-13-2; Highly disperse precipitated zinc oxide; vulcanization accelerator activator for rubber goods based on natural and synthetic elastomers and latex applications; suitable at high levels for dynamically stressed articles, e.g., buffers and rollers, and in low concs. in transparent and translucent goods. [Bayer AG; Miles; Polysar]

Zinkoyd Aktiv. Technical grade oxides and zeolites; for use as an additive in the paint and rubber industries and as a catalyst in the chemical industry. [Bayer AG]

Zink Pyrion. CAS 13463-41-7; Zinc pyrithione; used as an antidandruff agent, preservative, antibacterial, and antimicrobial. [Ruetgers-Nease]

Zinnal. A proprietary dual metal consisting of aluminium sheet coated on both sides with tin. §

Zinol. Powder flux for treating dross on hot-dip Galvanizing baths. [Foseco (F.S.) Ltd] †

Zinox. CAS 1314-13-2; Precipitated zinc oxide; used as a paint pigment. [Am. Chemet]

Zinsser's insulating wax. a) Consists of beeswax ; b) consists of shellac, rosin, and oxide of iron.

Zintox. A proprietary trade name for an agricultural spray containing basic zinc arsenate. §

Zipan. CAS 58-33-3; Promethazine hydrochloride; an antiemetic and antihistaminic. [Savage Laboratories] †

Zipcillin. Veterinary intramammary procaine penicillin. [The Wellcome Foundation Ltd] *

Zip Grip. Cyanoacrylate; for bonding closely matted surfaces and maintenance, production, and prototype bonding. [ITW Devcon]

---

* Trade name not verified as available   † Trade name verified as obsolete

Ziploc. Brand plastic storage bags. [Dow UK]
Zippo. A trade name for an aluminium solder for joining aluminium to itself, to copper, zinc, tin, or brass. §
Zircomplex. Ammoniacal zirconium complex; thixotrope in emulsion paints. [Rhone-Poulenc UK]
zirconia. See zirconium oxide.
zirconic anhydride. See zirconium oxide.
zirconium boride. (zirconium diboride) ZrB$_2$; Refractory for aircraft and rocket applications, thermocouple protection tubes, high-temp. electrical conductor, cutting-tool component, coating tantalum, cathode in high-temp. electrochemical systems; oxidation-resistant composites. [Atomergic Chemetals; Boride Ceramics & Composites Ltd; Cerac; Noah Chem.]
zirconium (IV) chloride. See zirconium tetrachloride.
zirconium diboride. See zirconium boride.
zirconium dioxide. See zirconium oxide.
zirconium fluoride. See zirconium tetrafluoride.
zirconium hydride. ZrH$_2$; CAS 7704-99-6; EINECS 231-727-3; Vacuum tube getter, powder metallurgy, source of hydrogen, metal-foaming agent, nuclear moderator, reducing agent, and hydrogenation catalyst. [Atomergic Chemetals; Cerac; Degussa; Morton Int'l.]
zirconium oxide. (zirconia; zirconium dioxide; zirconic anhydride) ZrO$_2$; CAS 1314-23-4; EINECS 215-227-2; Unstabilized: prod. of piezoelectric crystals, high-frequency induction coils, ceramic glazes, glasses, heat-resistant fibers; hydrous: odor absorbent, and poison ivy treatment. [Atomergic Chemetals; Ferro/Transelco; Hüls Am.; Magnesium Elektron Ltd; TAM Ceramics; Zircar Prods.]
zirconium silicate. (silicic acid, zirconium salt (1:1); zircon) ZrSiO$_4$; CAS 14940-68-2; Glaze opacifier; stabilizes color shades; used in white and colored glazes for sanitary ware, wall tile, glazed brick, structural tile, stoneware, dinnerware, special porcelains, refractory compositions, epoxy formulations, and encapsulating resins. [Atochem N. Am.; Atomergic Chemetals; Du Pont; TAM Ceramics]
zirconium tetrachloride. (zirconium (IV) chloride) CAS 10026-11-6; EINECS 233-058-2; Source of the pure metal, analytical chemistry, water repellents for textiles, tanning agent, zirconium compounds, special catalysts (Friedel-Crafts, Ziegler). [Atomergic Chemetals; Noah Chem.]
zirconium tetrafluoride. (zirconium fluoride) ZrF$_4$; CAS 7783-64-4; Component of molten salts for nuclear reactors. [Atochem N. Am.; Atomergic Chemetals; Cerac]
Zircosil. Zirconium silicates. [Anzon Ltd] *
Zircosol P. Recycled paper additive. [Magnesium Elektron Ltd]
Zircotan. Synthetic tanning agent. [Rohm & Haas]
Zirex®. Rosin-based resinate; tackifier resin for adhesives; [Arizona]
Zirgel®. Thixotropic gelling agent. [Magnesium Elektron Ltd]
Zirkonal. Zirconium aluminum compounds. [Giulini Corp]
Zirmax®. Zirconium hardener. [Magnesium Elektron Ltd]
Zirmel®. Zirconium, magnesium, rare-earth and other chemicals, powders and alloys; for use in industry, including metalworking. [Magnesium Elektron Ltd]
Ziscon. (Ziskon). An alloy of 60% aluminum and 40% zinc.
Zisium. An alloy of from 82-83% aluminum, 1-3% copper, 15% zinc, and 0-1% tin.
Ziskon. See Ziscon.

Zisnet F-PT. CAS 638-16-4; 2,4,6-Trimercapto-s-triazine; curing agent for epichlorohydrin rubber; used in place of ethylene thiourea and red lead; gives improved heat resist., less mold fouling, reduced toxicity; oil treated to reduce dusting. [Zeon]
Zithate®. CAS 1000-90-4; Zinc isopropyl xanthate; accelerator for vulcanization of rubber. [R T Vanderbilt Co Inc]
Zithromax®. CAS 83905-01-5; Azithromycin; antibiotic. Also known as Acilef (Greece), Acithrocin (Germany), Azitrocin (Guatemala), Azitrocina (Guatemala), Azitromax (Sweden), Azromax (Indonesia), Zitromax (Benelux, Finland). [Pfizer International]
Zitrilon® 10%. Zinc chelate with 10% Zn; for foliar application; prevents and cures zinc deficiency. [BASF AG]
Zitro®. Rosin-based resinate; tackifier resin for adhesives; [Arizona]
Zitromax. See Zithromax. [Pfizer International]
ZMBT. CAS 155-04-4; Zinc salt of 2-mercaptobenzothiazole; semi-ultra accelerator. [Akrochem]
Z-M-L. Zinc salt of mercaptobenzenethiazole with laurex; a proprietary rubber vulcanizing accelerator. §
Zoamix. Coccidiostat; often used on an exchange program with Coyden; the main purpose in rotating Zoamix coccidiostat with Coyden coccidiostat is to prevent poultry from developing a resistance to the latter product. [Dow UK] *
Zoaquin. Diiodohydroxyquinoline. [May & Baker Ltd] *
Zodiac. Organic brightener system; used for bright nickel electroplating. [Harshaw Chemicals Ltd]
Zoflora. Concentrated floral disinfectant. [Thornton & Ross Ltd]
Zohar 60 SD(L). Sodium alkylbenzene sulfonate and builders; spray-dried conc. for detergent powd. manufacturing. [Zohar Detergent Factory]
Zohar Automat SD. Sodium alkylbenzene sulfonate and builders; spray-dried built powd. for machine washing. [Zohar Detergent Factory]
Zohar EGMS. Glycol stearate; coemulsifier for o/w emulsions; opacifier, pearlescent, emollient, superfatting agent for mfg. of shampoos, liq. toilet soaps, bath products. [Zohar Detergent Factory]
Zohar GLST. Glyceryl stearate; emulsifier, thickener, superfatting agent. [Zohar Detergent Factory]
Zohar KAL SD. Sodium alkylbenzene sulfonate and builders; spray-dried built powd. for machine washing. [Zohar Detergent Factory]
Zoharconc A.D. Aircraft detergent conc. [Zohar Detergent Factory]
Zoharconc DIS. Disinfectant liq. conc. [Zohar Detergent Factory]
Zoharconc Dead Sea. Foaming bath conc. (with Dead Sea minerals). [Zohar Detergent Factory]
Zoharconc FC. Floor cleaner conc. without wax. [Zohar Detergent Factory]
Zoharconc FS. Domestic fabric softener conc. [Zohar Detergent Factory]
Zoharconc J-Super. Textile softener conc. for denim. [Zohar Detergent Factory]
Zoharconc RA. Rinse aid conc. [Zohar Detergent Factory]
Zoharex A-10. Fatty acid polyglycol ester; detergent for laundry liqs. [Zohar Detergent Factory]
Zoharfoam. Foaming agent for drilling use. [Zohar Detergent Factory]
Zoharlab. Linear sodium alkylbenzene sulfonate; raw mate-

---

‡ Trade name and manufacturer not verified   § Trade name without identified manufacturer

rial for manufacturing of detergents. [Zohar Detergent Factory]

Zoharpon DT-80. Alkanolamine lauryl sulfate; raw material for shampoo and foam baths. [Zohar Detergent Factory]

Zoharpon ETA 27. CAS 9004-82-4; Sodium lauryl ether sulfate (2 EO); raw material for personal care products, lt. duty detergents. [Zohar Detergent Factory]

Zoharpon LAA. CAS 2235-54-3; EINECS 218-793-9; Ammonium lauryl sulfate; raw material for clear body and hair shampoos. [Zohar Detergent Factory]

Zoharpon LAD. DEA lauryl sulfate; raw material for shampoos, cosmetics, and light duty detergents. [Zohar Detergent Factory]

Zoharpon LAEA 253. CAS 67762-19-0; Ammonium lauryl ether sulfate (3 EO); raw material for shampoos, cosmetics, and light duty detergents. [Zohar Detergent Factory]

Zoharpon LAM. MEA-lauryl sulfate; raw material for shampoos and foam baths. [Zohar Detergent Factory]

Zoharpon LAS Special. Sodium lauryl sulfate; raw material for shampoos. [Zohar Detergent Factory]

Zoharpon LAT. CAS 139-96-8; EINECS 205-388-7; TEA lauryl sulfate; raw material for mild shampoos and skin care preps. [Zohar Detergent Factory]

Zoharpon LMT42. Sodium lauryl methyl taurate; mild detergent, foamer, dispersant used in soap, syndet toilet bars, shampoos, and bubble baths. [Zohar Detergent Factory]

Zoharpon MgES. CAS 67702-21-4; Magnesium lauryl ether sulfate (2 EO); raw material for mild shampoos and cosmetic preps. [Zohar Detergent Factory]

Zoharpon MgS. Magnesium lauryl sulfate; raw material for mild shampoos and cosmetic preps. [Zohar Detergent Factory]

Zoharquat 50. Benzalkonium chloride; disinfectant, fungicide, bacteriocide and algicide. [Zohar Detergent Factory]

Zoharsoft 90. Quat. imidazoline deriv.; fabric softener base for domestic and commercial laundry use. [Zohar Detergent Factory]

Zoharsoft DAS. Fatty acid-amine deriv.; softener base for textile industry for all fibers (cotton, rayon, acetates, wool, nylon). [Zohar Detergent Factory]

Zoharsyl L-30. CAS 137-16-6; EINECS 205-281-5; Sodium lauroyl sarcosinate; raw material for manufacturing of hair shampoos, conditioners, toothpastes, carpet and upholstery shampoos; anticorrosive properties. [Zohar Detergent Factory]

Zohartaine AB. CAS 683-10-3; Lauryl betaine; foam booster, mild ingredient for shampoos and detergents; industrial foamer. [Zohar Detergent Factory]

Zohartaine TM. Dihydroxyethyltallow glycinate; thickener and anticorrosive agent for tech. acid formulations, industrial cleaners; component of mild shampoos. [Zohar Detergent Factory]

Zoharteric D-2. Disodium cocoamphodiacetate, sodium trideceth sulfate; component of mild, nonirritating conditioning shampoos, bubble baths, baby products. [Zohar Detergent Factory]

Zoharteric D-SF. Disodiumcocoamphopropionate; component of extra mild, nonirritating shampoos, bubble baths; detergent for heavy-duty household and industrial cleaners with tolerance for alkalies and electrolytes. [Zohar Detergent Factory]

Zoharteric DJ. Disodium lauroamphodiacetate; component of mild, nonirritating conditioning shampoos, bubble baths; detergent for specialty cleaners. [Zohar Detergent Factory]

Zoharteric LF. CAS 13039-35-5; EINECS 235-907-2; Sodium capryloamphoacetate; surfactant and wetting agent for low-foam heavy-duty household and industrial cleaners; tolerance for alkalies and electrolytes. [Zohar Detergent Factory]

Zoharteric LF-SF. CAS 68815-55-4; Disodium capryloamphodipropionate; surfactant and wetting agent for low-foam heavy-duty household and industrial cleaners; high tolerance for alkalies and electrolytes. [Zohar Detergent Factory]

Zoharteric M. Sodium cocoamphoacetate; component for personal care products; detergent for specialty cleaners. [Zohar Detergent Factory]

Zoharteric M-2. Sodium cocoamphoacetate, sodium trideceth sulfate; component of mild, nonirritating conditioning shampoos, bubble baths, and baby products. [Zohar Detergent Factory]

Zoladex. Anticancer preparations. [ICI Pharma]

Zoldine® MS-52. CAS 137796-06-6; 4-Ethyl-2-methyl-2-(3-methylbutyl)-1,3-oxazolidine; moisture scavenger; urethane crosslinker. [Angus]

Zoldine® ZT-55. CAS 6542-37-6; Oxazolidine; cross-linking agent for resorcinol phenol-formaldehyde or protein-based resin systems; raw material for synthesis; used in hair care products. [Angus]

Zoloft®. Sertaline; for treatment of depression. Also known as Lustral (Great Britan, Ireland), Lustran (Great Britain, Ireland), Senesco (Germany), Serlain (Benelux), Serlin (Belelux), Serline (France), Sertral (Germany). [Pfizer International]

Zolone. CAS 2310-17-0; Emulsifiable concentrate containing 350 g/l phosalone; an organophosphorus insecticide and acaricide. [Hortichem Ltd]

Zolone. CAS 2310-17-0; Emulsifiable concentrate containing 350 g/l phosalone; an organophosphorus insecticide and acaricide. [Rhone-Poulenc Crop Protection Ltd]

Zolyse. Chymotrypsin; enzyme [Alcon Laboratories Inc] *

Zomax. CAS 64092-49-5; Zomepirac sodium; an analgesic and anti-inflammatory. [McNeil Pharmaceuticals] *

Zonarez® 7115, 7115 LITE. Polyterpene resin produced from dipinene; bright, clear, pale-colored, low m.w. polymers imparting high levels of tack and adhesion to elastomeric and polymeric materials; in mfg. of pressure-sensitive adhesives, rubber cements, hot-melt adhesives/coatings, can sealants, caulking; inks, paints, concrete waterproofing agents, varnishes, chewing and bubble gum bases, moisture-resist. soft gelatin capsules and powds. of ascorbic acid and its salts; LITE resins feature water-white color; FDA compliance. [Arizona]

Zonarez® Alpha 25. Polyterpene resin based on α-pinene; clear, light-colored thermoplastic polymers imparting tack and str. to adhesives; increases hard resin loading; co-tackifier for the mfg. of pressure-sensitive adhesives, solv.-based adhesives, emulsion adhesives, and hot-melt adhesives/coatings; especially designed as co-tackifier for manufacturing of adhesive systems based on thermoplastic block copolymer such as Kraton®; plasticizing and softening properties; FDA compliance. [Arizona]

Zonatac®. Modified polyterpene resins; tackifying resins for adhesives and sealants. [Arizona] *

Zonatac® 105, 501. Modified terpene hydrocarbon resin;

---

* Trade name not verified as available   † Trade name verified as obsolete

light-colored, low odor thermoplastic resins for use in hot-melt adhesives and coatings, hot-melt pressure-sensitive adhesives, and solv.-based thermoplastic elastomer pressure-sensitive adhesive systems; FDA compliance. [Arizona]

Zonatac® 105 LITE, 501 LITE. Modified terpene hydrocarbon resin; light-colored, low odor thermoplastic resins for use in hot-melt adhesives and coatings, hot-melt pressure-sensitive adhesives, and solv.-based thermoplastic elastomer pressure-sensitive adhesive systems; FDA compliance. [Arizona]

Zonester® 55. Glycerol ester of Arizona DR-24 disproportionated tall oil rosin; thermoplastic resin ester used as tackifier for difficult to-tackify elastomers such as SBR as well as natural and syn. latexes; used in pressure-sensitive adhesives, contact cements, latex adhesives; where good flexibility at low temps. is desirable. [Arizona]

Zonester® 65. CAS 8050-26-8; Pentaerythritol ester of disproportionated tall oil rosin; thermoplastic resin ester used in SBR latex, natural rubber, and neoprene adhesives. [Arizona]

Zonester® 85. Glyceryl ester of tall oil rosin; thermoplastic resin ester used in chewing gums, hot-melt adhesives, hot-melt coatings, mastic adhesives, contact cements, and pressure-sensitive adhesives. [Arizona]

Zonite. CAS 7681-52-9; A proprietary trade name for sodium hypochlorite solution for disinfecting and antiseptic uses.§

Zonolite®. A proprietary trademark for a mineral product made by the heat treatment of vermiculite, a mineral which is similar to a crude mica; used in the manufacture of building materials and for use as high temperature insulation. §

Zonyl®. Fluorochemical surfactant. [DuPont UK]

Zonyl® FSA. Fluorochemical surfactant; wetting agent, emulsifier, dispersant, corrosion inhibitor, leveling agent for adhesives, agric., polishes, polymerization, pigment grinding, cleaners, coatings and paints, fire fighting, paper, ink, oil, plastics, and textile industries. [DuPont]

Zonyl® FSE. Fluorochemical surfactant; leveling agent for emulsion, pigment dispersant, wetting and foaming agent for corrosive media. [DuPont]

Zonyl® UR. Fluorochemical surfactant; wetting agent, antifoam, corrosion inhibitor; leveling agent for floor polishes. [DuPont]

zootic acid. (hydrocyanic acid; prussic acid; hydrogen cyanide; formonitrile) HCN; CAS 74-90-8; Used in the manufacture of acrylonitrile, acrylates, dyes, and pesticides.

Zopaque. CAS 13463-67-7; EINECS 236-675-5; A proprietary form of titanium oxide for rubber mixing. §

Zoramide CM. CAS 68140-00-1; Cocamide MEA; foam booster, thickener, superfatting agent. [Zohar Detergent Factory]

Zoramox. Coconut amido alkyl amine oxide; wetting agent, visc. booster/stabilizer for shampoos, bubble baths; visc. builder for low pH shampoos, other liq. detergents. [Zohar Detergent Factory]

Zorbax®. For the medical industry. [DuPont UK]

Zorite. A proprietary alloy containing 35% nickel, 15% chromium, 1.75% manganese, and 0.5% manganese, and 0.5% carbon with iron. §

Zoroxin. CAS 70458-96-7; Norfloxacin; urinary antiseptic.

[Merck & Co Inc]

Zovirax. Proprietary formulations of acyclovir or acyclovir sodium; for the management of herpes simplex virus infections and treatment of herpes zoster virus infections. [The Wellcome Foundation Ltd]

Z.P.D. The zinc salt of pentamethylene-dithiocarbamic acid; an ultra-rubber vulcanization accelerator.

Z-Siloxide. A trade name for a compound obtained by fusing silica with 0.1-2.0% zirconia; it is a silica glass. §

Z Span Spansule Capsule. CAS 7733-02-0; Zinc sulfate monohydrate; sustained-release preparation for use when inadequate diet calls for supplementary zinc and treatment of zinc deficiency where indicated. [SmithKline Beecham] *

Zumisite. Yeast food. [ABM Chemicals Ltd] *

Zusolat 1004. Fatty alcohol polyglycol ether (4 EO); washing and cleansing agent. [Zschimmer & Schwarz]

Zusolat 1005/85. Fatty alcohol polyglycol ether (5 EO); dispersant, emulsifier, wetting agent for mfg. of washing and cleaning agents, dishwash, household/industrial and metal cleaners, textile, leather, and paper auxiliary. [Zschimmer & Schwarz]

Zusomin C 108. CAS 61791-14-8; PEG-8 cocamine; basic material and component for dyeing and textile auxiliaries; intermediate for quaternization; metal cleaners. [Zschimmer & Schwarz]

Zusomin O 105. PEG-5 oleamine; basic material for textile and dyeing auxiliaries, intermediate for quaternization, metal cleaners. [Zschimmer & Schwarz]

Zusomin S 110. CAS 26635-92-7; PEG-10 stearamine; basic material for dyeing and textile auxiliary; intermediate for quaternization, metal cleaners. [Zschimmer & Schwarz]

Zusomin TG 102. CAS 61791-44-4; PEG-2 tallowamine; basic material and component for dyeing and textile auxiliary; intermediate for quaternization and metal cleaners. [Zohar Detergent Factory]

Zs-04-xxyyAU. Anisotropically conductive flip chip adhesives; microelectronic adhesives curable at moderate temps. [Zymet]

Zyklon. CAS 74-90-8; A proprietary hydrogen cyanide fumigant for use against insect and vermin pests. §

Zylar® 90-495. Methyl methacrylate butadiene styrene terpolymer; features excellent gamma ray recovery, high clarity, abuse/scratch resist.; for medical devices, office accessories, small appliances, and toys. [Novacor]

Zylar® 91. Acrylic; clear impact acrylic for extrusion with exc. gamma ray recovery, high clarity, abuse/scratch resist.; for medical devices, thermoformed packaging. [Novacor]

Zylar® 93-541. Methyl methacrylate butadiene styrene terpolymer; clear impact acrylic characterized by exc. moldability, abuse resist.; used for displays, medical devices, office accessories, small appliances, toys. [Novacor]

Zylar® ST 94-561. Alloy; clear impact resist. alloy with high clarity, strong and toughness and superior molding charactersitics; for medical devices, safety devices, and toys. [Novacor]

Zyloprim Tablets. CAS 315-30-0; A proprietary formulation of allopurinol; for the management of primary and secondary gout, recurrent calcium oxalate calculi and leukemia and in patients with lymphoma and other malignancies who are receiving therapy which causes elevations

---

in serum and urinary uric acid levels. [The Wellcome Foundation Ltd] *

Zyloric. CAS 315-30-0; A proprietary preparation of allopurinol; used in treatment of gout. [The Wellcome Foundation Ltd] *

zymase. An alcohol-producing enzyme secreted by yeast cells that decomposes grape sugar.

zymin. (permanent yeast). A product obtained by partially drying ordinary yeast, immersing it in acetone for 15 minutes, which kills yeast, drying on filter paper, and washing with ether. It produces alcohol from grape sugar.

zymocasein. A phospho-protein obtained from yeast.

Zymogen. A commercial product consisting of a nitrogenous substance that provides food for yeast in fermentation. §

Part II

Manufacturers Directory

Manufacturers Directory

International telephone country codes for represented countries are:

Argentina	54	Hungary	36	Russia	7
Australia	61	India	91	Saudi Arabia	966
Austria	43	Ireland	353	Singapore	65
Belgium	32	Israel	972	South Africa	27
Brazil	55	Italy	39	Spain	34
Canada	1	Japan	81	Sweden	46
China	86	Korea (South)	82	Switzerland	41
Denmark	45	Mexico	52	Taiwan	886
Finland	358	Netherlands	31	United Kingdom	44
France	33	Norway	47	United States	1
Germany	49	Pakistan	92	Venezuela	58
Greece	30	Poland	48	Yugoslavia	38
Hong Kong	852	Portugal	351		

AAA Molybdenum Products, Inc.
7233 W. 116th Place, Broomfield, CO 80020 USA (Tel.: 303-460-0844; 800-443-6812; FAX 303-460-0851)

Aarhus Oliefabrik A/S
Postboks 50, Bruunsgade 27, DK-8100 Aarhus C, Denmark (Tel.: 86-12 60 00; FAX 86-18 38 39; Telex: 64341)

Abatron, Inc.
33 Center Dr., Gilberts, IL 60136 USA (Tel.: 708-426-2200; 800-445-1754; FAX 708-426-5966)

Abbott Laboratories
Abbott Laboratories, 1400 Sheridan Rd., N. Chicago, IL 60064 USA (Tel.: 708-937-8800; 800-323-9597; FAX 708-937-6676)

Abbott Laboratories/Chemical & Agricultural Products Division, 14th Sheridan Road, North Chicago, IL 60064 USA (Tel.: 312-937-5841; 800-323-9065; FAX 312-937-3679)

Abbott Laboratories Ltd., Div. of Abbott Laboratories, Queenborough, Kent, ME11 5EL, UK (Tel.: 0795 580099; FAX 0795 580404; Telex: 96347)

Abbott SA, Div. of Abbot Laboratories, 2 rue du Bosquet, B-1348 Louvain, Belgium (Tel.: 10-47 53 11; FAX 10-47 55 75; Telex: 59334 ABBOTT B)

ABCD Petrochimica. *Address unavailable*

Abco Industries Ltd.
200 Railroad St., PO Box 335, Roebuck, SC 29376 USA (Tel.: 803-576-6821; 800-476-4476; FAX 803-576-9378; Telex: 628 17731)

Ablestik, Div. of National Starch
10 Finderne Ave., Bridgewater, NJ 08807 USA (Tel.: 908-685-5126; FAX 908-707-3778)

ABM Chemicals Ltd.
Poleacre Lane, Woodley, Stockport, Cheshire, SK6 1PQ, UK (Tel.: 061-430-4391; FAX 061-430-4364)

Abril Industrial Waxes
Sturmi Way, Village Farm Industrial Estate, Pyle, Mid Glamorgan, CF33 6NU, UK (Tel.: 0656 744362; FAX 0656 742471)

Accro-Seal, A Kalplas Co.
316 W. Briggs St., PO Box 210, Vicksburg, MI 49097-0210 USA (Tel.: 616-649-1014; FAX 616-649-1067)

Accurate Chemical & Scientific Corp.
300 Shames Dr., Westbury, NY 11590 USA (Tel.: 516-333-2221; 800-645-6264; FAX 516-997-4948; Telex: 4972582)

Ace Chemicals Ltd.
Loanwath Road, Gretna, Dumfriesshire, UK (Tel.: 0461-37572)

Aceto Chemical Co., Inc.
1 Hollow Lane, Suite 201, Lake Success, NY 11042-1215 USA (Tel.: 516-627-6000; FAX 516-627-6093; Telex: 62662)

Acheson
Acheson Colloids Co., 1600 Washington Ave., PO Box 611747, Port Huron, MI 48061-1747 USA (Tel.: 313-984-5581; 800-255-1908)

Acheson Colloids (Canada) Ltd., PO Box 665, Brantford, Ontario, N3T 5P9, Canada (Tel.: 519-752-5461)

Acheson Industries (Europe) Ltd., Div. of Acheson Industries Inc, Sun Life House, 85 Queens Rd., Reading, Berkshire, RG1 4PT, UK (Tel.: 0734-588844; FAX 0734-574897)

Acheson (Japan) Ltd., No. 119, Mitsui Seimei Bldg., Ito-machi, Chuo-ku, Kobe 650, Kobe Port, 538, Japan (Tel.: (078) 332-3601; FAX (078) 391-5903; Telex: 5623230)

ACL Inc.
1960 E. Devon Ave., Elk Grove Village, IL 60007 USA (Tel.: 708-981-9212; 800-782-8420; FAX 708-981-9278; Telex: 4330251)

Acorga Ltd., Div. of ICI
PO Box 42, Hexagon House, Blackley, Greater Manchester, M9 3DA, UK (Tel.: 061 740 1460; FAX 061 721 4794)

Active Organics, Inc.
6849 Hayvenhurst Ave., Van Nuys, CA 91406 USA (Tel.: 818-786-3310; FAX 818-786-3313)

R P Adams Ltd.
Arpal Works, Riverside Road, Selkirk, TD7 5DU, Scotland (Tel.: 0750-21586; FAX 0705-21506)

Adasco Inc.
2029 South Broadway, Building A, Geneva, OH 44041 USA (Tel.: 216-466-2114)

ADC Resins
2410 Peninsula Road, Oxnard, CA 93030 USA

Addagrip Surface Treatments UK Ltd.
Bird-in-Eye Hill, Uckfield, East Sussex, TN22 5HA, UK (Tel.: 0825-761333; FAX 0825-768566)

Adhesives Research, Inc.
PO Box 100, Glen Rock, PA 17327 USA (Tel.: 717-235-7979; FAX 717-235-8320; Telex: 84-0454)

ADM Tronics Unlimited Inc.
224-S Pegasus Ave, Northvale, NJ 07647 USA (Tel.: 201-767-6040; FAX 201-784-0620)

Adria Laboratories Inc.
PO Box 16529, 582 W Goodale Blvd, Columbus, OH 43216-6529 USA (Tel.: 614-764-8100)

Adshead Ratcliffe & Co. Ltd.
Derby Road, Belper, Derbyshire, DE5 1WJ, UK (Tel.: 0773-826661; FAX 0773-821215; Telex: 377184 Arbo G)

Advanced Elastomer Systems
Advanced Elastomer Systems, L.P., 260 Springside Dr., PO Box 5584, Akron, OH 44334-0584 USA (Tel.: 216-668-3600, 216-668-8242)

Advanced Elastomer Systems Canada Inc., Box 787, Streetsville Postal Station, Mississauga, Ontario, L5M 2G4, Canada (Tel.: 416-826-9575)

Advanced Elastomer Systems NV/SA, Ave. de Tervuren 270-272, PO Box 1, B-1150 Brussels, Belgium (Tel.: 2-761-41-11)

Monsanto Japan Ltd., Room 520, Kokusai Bldg., 1-Marunouchi 3-Chome, Chiyoda-Ku, Tokyo, 100, Japan (Tel.: (03) 287-1251; FAX (03) 287-1250; Telex: J 22614)

Advanced Polymer Coatings, Inc.
PO Box 127, West Point, PA 19486 USA (Tel.: 215-794-5466; FAX 215-794-5468)

Advanced Web Products, Inc.
PO Box 117, Ingomar, PA 15127 USA (Tel.: 412-367-8820; FAX 412-367-8848)

AEA Technology (UK AEA)
B465, Harwell Laboratories, Didcot, Oxfordshire, OX11 0RA, UK (Tel.: 0235 435555; FAX 0235 432859; Telex: 83135 ATOMHA G)

AEI Compounds, AEI Cables Ltd., Power Cable Div.

Gravesend, Kent, DA11 9AF, UK (Tel.: 474-564466; FAX 474-564386)

Agan Chemical Manufacturers Ltd.

PO Box 262, Ashdod, Israel

AG Chemische Fabrik. *Address unavailable*

Agfa-Gevaert

Agfa-Gevaert AG, Div. of Bayer AG, Septestraat 27, W-5090 Leverkusen, Germany (Tel.: 214-7030; FAX 214-7146; Telex: 31223)

Agfa-Gevaert Ltd., Div. of Bayer AG, 27 Great West Rd, Brentford, Middlesex, TW8 9AX, UK (Tel.: 081-560 2131; FAX 081-847 5803; Telex: 28154 AGUK G)

Agfa-Gevaert NV, Septestraat 27, B-2510 Mortsel, Belgium (Tel.: 03-4442111; FAX 03-4447094)

Agrichem (International) Ltd.

Industrial Estate, Station Road, Whittlesey, Cambridgeshire, PE7 2EY, UK (Tel.: 0733-204019; FAX 0733-204162)

Agri-Technics Ltd.

Muston Gorse, Redmile, Nottingham, NG13 0GN, UK (Tel.: 0949-42255; FAX 0949-43407)

Aicello Chemical Co Ltd.

45 Koshikawa, Ishimaki-honmachi, Toyohashi City, Aichi Pref, 441 11, Japan (Tel.: (0532) 88-0611; FAX (0532) 88-5102; Telex: 4322129 AICELL J)

Aichi Silicate Chemical Co., Ltd.

92-2, Bogane-cho, Seto-shi, Aichi, 489, Japan (Tel.: (0561) 83-8711; FAX (0561) 83-0561)

Airex AG

5634 Sins, Switzerland (Tel.: 042 660056; FAX 042 661707; Telex: 868975)

Air Liquide Hellas SA

26-28 Asklipiou St, GR-10679 Athens, Greece (Tel.: 1-360 84 11; FAX 1-362-54 31; Telex: 216444)

Air Products and Chemicals

Air Products and Chemicals, Inc., 7201 Hamilton Blvd., Allentown, PA 18195 USA (Tel.: 215-481-4911; 800-345-3148; FAX 215-481-5900; Telex: 275425)

Air Products and Chemicals, Inc., Polymer Chemicals Div., Allentown, PA 18195 USA (Tel.: 215-481-6799; 800-345-3148; FAX 215-481-5900)

Air Products and Chemicals, Inc., Polyurethane Chemicals Div, 7201 Hamilton Blvd., Allentown, PA 18195-1501 USA (Tel.: 800-345-3148; FAX 215-481-5900)

Air Products Nederland B.V., Herculesplein 359, PO Box 85075, NL 3508 AB Utrecht, The Netherlands (Tel.: 30-511828)

Air Products Pacific, Inc., Sakurabashi Yachiyo Bldg., 5-6, Umeda 2-Chome, Kita-Ku, Osaka, 530, Japan

Aisan Chemical Co., Ltd.
5th Fl., No. 2 Horiuchi Bldg., 5-20, Meieki 4-chome, Nakamura-ku, Nagoya, 450, Japan (Tel.: (052) 571-9156; FAX (052) 571-9159; Telex: 04427746 AISAN J)

Aiscondel SA
Aragon 182, 08011 Barcelona, Spain   (Tel : 93-323 1020; FAX 93-323-7921; Telex: 97887 ctin e)

R. Aitken
123 Harmony Road, Govan, Glasgow, GS1 3NB, UK (Tel.: 041-440-0033; FAX 041-440 2744)

Ajinomoto
Ajinomoto Co., Inc., 15-1, Kyobashi 1-chome, Chuo-ku, Tokyo, 104, Japan (Tel.: (03) 5250-8111; Telex: J22690)

Ajinomoto USA, Inc., Glenpointe Centre West, 500 Frank W. Burr Blvd., Teaneck, NJ 07666-6894 USA (Tel.: 201-488-1212; FAX 201-488-6472; Telex: 275425 (AJNJ))

Akrochem Chemical Co.
255 Fountain St., Akron, OH 44304 USA (Tel.: 216-535-2108)

AKU Holland. *Address unavailable*

Akzo
Akzo België NV, Div. of Akzo NV, Marnixlaan 13, B-1050 Bruxelles, Belgium (Tel.: 2-518 04 07; Telex: 22071 ORGABE B)

Akzo Chemie BV, Div. of Akzo NV, Postbus 247, NL-3800 AE Amerfoort, The Netherlands (Tel.: 33-676767; FAX 33-676150; Telex: 79276)

Akzo Chemie UK Ltd., 1-5, Queens Road, Hersham, Walton-on-Thames, Surrey, KT12 5NL, UK (Tel.: 0932-247891; FAX 0932-231204; Telex: 21997)

Akzo Chemie GmbH, Div. of Akzo NV, Postfach 100132, Phillippstrasse 27, W-5160 Düren, Germany (Tel.: 2421-492261; FAX 2421-492487; Telex: 833911)

Akzo Chemie Italia SpA, Div. of Akzo NV, Via E Vismara 80, I-20020 Arese, Italy (Tel.: 2-938 08 71; FAX 2-938 08 16; Telex: 332526)

Akzo Chemicals Inc./Chemicals Div., 300 S. Riverside Plaza, Chicago, IL 60606 USA (Tel.: 312-906-7500; 800-257-8292; FAX 312-906-7680; Telex: 25-3233)

Akzo Chemicals Ltd., 100 University Ave., Suite 906, Toronto, Ontario, M5J IV6, Canada

Tosoh Akzo Corp., Uruma-Kowa Bldg., 11-37, Akasaka 8-chome, Minato-ku, Tokyo, 107, Japan (Tel.: (03) 3479-8911; FAX (03) 3479-8915; Telex: 2222985 TSCTOK)

Akzo Chemicals Ltd., PO Box 80, Parramatta N.S.W., 2150, Australia

Akzo Engineering BV, Div. of Akzo NV, Postbus 9300, Velperweg 76, NL-6800 SB Arnhem, The Netherlands (Tel.: 85-662714; FAX 85-665140; Telex: 45204 ENKA NL)

Akzo Engineering Plastics Sweden AB, Box 10, S-424 21 Angered, Sweden (Tel.: 031-94 36 90; FAX 031 94 37 84; Telex: 27756)

Akzo Engineering Plastics, Inc., PO Box 3333, 2267 West Mill Rd., Evansville, IN 47732 USA (Tel.: 812-424-3831; 800-457-3764; FAX 812-424-0892)

Akzo Resins BV, Div. of Akzo NV, Ringersweg 5, NL-4612 PR Bergen op Zoom, The Netherlands (Tel.: 1640-76911; FAX 1640-76258; Telex: 78284)

Akzo Salt Co., Abington Executive Park, Clarks Summit, PA 18411 USA (Tel.: 717-587-5131; FAX 717-586-6278; Telex: 756470)

Alba International Inc.

508 Clearwater Dr., N. Aurora, IL 60542 USA (Tel.: 708-897-4200; 800-669-9333; FAX 708-377-5330)

Alban Muller International

212, rue de Rosny, 93100 Montreuil, France (Tel.: 1-48-58-30-25; FAX 1-48-58-03-71; Telex: 236030 F)

Chemische Werke, Albert

Albertstrasse 10-14, 6202 Wiesbaden-Biebrich, Germany (Tel.: 06121 6741; Telex: 04186655 ACQA D)

Albion Group Ltd. *Address unknown*

Albright & Wilson, Div. of Tenneco Inc.

Albright & Wilson Ltd., European Hdqtrs., PO Box 3, 210-222 Hagley Rd. West, Oldbury, Warley, West Midlands, B68 0NN, UK (Tel.: 21-429-4942; FAX 21-420-5151; Telex: 336291)

Albright & Wilson Ltd., Detergents Div, P O Box 3, A.S.W. House, 210-222 Hagley Road West, Oldbury, Warley, West Midlands, B68 ONN, UK (Tel.: 021-420 4156; FAX 021-420 5111)

Albright & Wilson Ltd., Phosphates & Speciality Business, P O Box 3, A.S.W.House, 210-222 Hagley Road West, Oldbury, Warley, West Midlands, B68 ONN, UK (Tel.: 021-429-4942)

Marchon France SA, BP 19, Han sur Meuse, F-55300, St Mihiel, France (Tel.: 29-91 7300; FAX 29-91 7399)

Marchon Espanola SA, Carretera Montblack Km 2, 4, Alcover (Tarragona), Spain

Marchon Italiana SpA, Casella Postale No. 30, 1-46043, Castiglione delle Stivier, Italy

Albright & Wilson Americas, PO Box 26229, Richmond, VA 23260-6229 USA (Tel.: 804-550-4300; 800-446-3700; FAX 804-550-4385)

Albright & Wilson Am. (Canada), 2 Gibbs Rd., Islington, Ontario, M9B 1R1, Canada (Tel.: 416-234-7000; 800-268-2520; FAX 416-237-1064)

Albright & Wilson Ltd. Japan, No. 2 Okamotoya Bldg. 6 Fl., 1-24, Toranomon 1-chome, Minato-ku, Tokyo, 105, Japan (Tel.: (03) 3508-9461; FAX (03) 3591-0733)

Albright & Wilson (Australia) Ltd., PO Box 20, Yarraville, Victoria, Australia (Tel.: 3-688-7777; FAX 3-688-7788)

Tenneco Malros Ltd., Albright & Wilson Resins & Organics Division, Rockingham Works, Avonmouth, Bristol, BS11 0YT, UK

Tenneco Organics Ltd., Albright & Wilson Resins & Organics Division, Avonmouth Works, Avonmouth, Bristol, Avon, BS11 0YT, UK (Tel.: 0272-823611; FAX 0272-824443)

Alcan Chemicals

Alcan Chemicals, Div. of Alcan Aluminum Corp., 3690 Orange Place, Suite 400, Cleveland, OH 44122-4438 USA (Tel.: 216-765-2550; 800-321-3864; FAX 216-765-2570)

Alcan Chemicals Ltd., British Alcan Aluminum PLC, Chalfont Park, Gerrards Cross, Buckinghamshire, SL9 0QB, UK (Tel.: 0753-887373; FAX 0753 889667; Telex: 847343)

Alchemie Ltd.

Brookhampton Lane, Kineton, Warwickshire, CV35 0JA, UK (Tel.: 0926-641600; FAX 0926-641698; Telex: 311890)

Alco Chemical Corp., Div. of National Starch & Chem.

909 Mueller Dr., PO Box 5401, Chattanooga, TN 37406 USA (Tel.: 615-629-1405; 800-251-1080; FAX 615-698-8723; Telex: 755002)

Alcoa Industrial Chemicals Div.

PO Box 300, Bauxite, AR 72011 USA (Tel.: 501-776-4981; 800-643-8771; FAX 501-776-4685; Telex: 536447)

Alcon Laboratories Inc.

6201 South Freeway, Ft Worth, TX 76134 USA (Tel.: 817-293-0450; Telex: 758320)

Alconox Inc.

215 Park Ave. So., New York, NY 10003 USA (Tel.: 212-473-1300; FAX 212-353-1342)

Alcon Universal. *See* Alcon Laboratories Inc.

Aldo Products Co. Inc.

18005 Lisa Lane, Brookfield, WI 53005 USA

Aldrich

Aldrich Chemical Co Inc., PO Box 355, Milwaukee, WI 43201 USA (Tel.: 414-273-3850; 800-558-9160; Telex: 26843 ALDRICH MI)

Aldrich Chemical Co Ltd., The Old Brickyard, New Rd., Gillingham, Dorset, SP8 4JL, UK (Tel.: 0747 822211; FAX 0747 823779; Telex: 417238)

Alfa Chemicals Ltd., Distributor

Alfa House, 15 Moor Park Avenue, Preston, Lancashire, PR1 6AS, UK (Tel.: 0772-58969; FAX 0772-53207)

Algemene Industriele. *Address unavailable*

Alginate Industries Ltd. *See* Kelco International Ltd

Allchem Industries Inc.

4001 Newberry Rd, Suite E-3, Gainesville, FL 32607 USA (Tel.: 904-378-9696; FAX 904-338-0400; Telex: 509540 ALLCHEM UD)

Allen & Hanburys Ltd.

Glaxo House, Berkeley Avenue, Greenford, Middlesex, UB6 0NN, UK

Allergan

Allergan, Inc., 2525 Dupont Drive, P.O. Box 19534, Irvine, CA 92715-9534 USA (Tel.: 714-752-4500)

Allergan Pharmaceuticals Inc., 2525 Dupont Drive, PO Box 19534, Irvine, CA 92713-9534 USA (Tel.: 714-752-4500)

Allied Colloids

Allied Colloids Ltd., Div. of Allied Colloids Group plc, PO Box 38, Low Moor, Bradford, West Yorkshire, BD12 0JZ, UK (Tel.: 0274-671267; FAX 0274-606499; Telex: 51646)

Allied Colloids Inc., 2301 Wilroy Rd., PO Box 820, Suffolk, VA 23434 USA (Tel.: 804-934-3700; FAX 804-934-3989)

Allied Products Corporation. *Address unknown*

Allied-Signal

Allied-Signal Inc., PO Box 2332R, Columbia Rd. & Park Ave., Morristown, NJ 07960 USA (Tel.: 201-455-2000; 800-446-1800; FAX 201-445-4807)

Allied-Signal Inc./A-C® Performance Additives, PO Box 2332R, Morristown, NJ 07962-2332 USA (Tel.: 201-455-2145; 800-222-0094; FAX 201-455-6154; Telex: 990433)

Allied-Signal Inc./Engineered Plastics, PO Box 2332, Morristown, NJ 07962-2332 USA (Tel.: 201-455-5010)

Allied-Signal Inc/.Sinclair and Valentine Division, 2520 Pilot Knob Road, St Paul, MN 55120 USA

Allied-Signal Inc/.Planarization & Diffusion Products Div, 1090 S Milpitas Blvd., Milpitas, CA 95035 USA

Allied-Signal Fluid Systems, 10124 Old Grove Road, San Diego, CA 92131 USA

Allied-Signal Inc/Water Treatment Div.

Allied Corp. Int'l. NV-SA, International House, Bickenhill Lane, Birmingham, B37 7HQ, UK

NV Allied Corp. Int'l. SA, Haasrode Research Park, B-3001 Heverlee, Belgium (Tel.: 016-211-211; FAX 016-203-269; Telex: 26283)

Allied Chemical International NV, Div. of Allied-Signal, Haasrode Research Park, Grauwmeer 1, B-3030 Leuven, Belgium (Tel.: 16-20 21 11; Telex: 26283 ALCHEM B)

Allied Chemical International Corp., PO Box 99067, Tsimshatsui Post Office, Hong Kong

Alma Paint & Varnish Co. *Address unverified*

Ontario, Canada

Alox Corp.

3943 Buffalo Ave., PO Box 517, Niagara Falls, NY 14302 USA (Tel.: 716-282-1295; FAX 716-282-2289)

Altana Inc. *Address unknown*

Altex Chemical Co. Ltd.

Clayfield Works, Slaithwaite, Huddersfield, West Yorkshire, BD12 0JZ, UK

Alzo Inc.
6 Gulfstream Blvd., Matawan, NJ 07747 USA (Tel.: 908-446-3270; FAX 908-446-5225)

Amax Inc.
200 Park Ave 33rd Fl, New York, NY 10066 USA (Tel.: 212-856-4200; FAX 212-856-5986)

Amax Asia Ltd. *See under* Climax

Ambersil Ltd.
Wylds Road, Castlefield Industrial Estate, Bridgewater, Somerset, TA6 4DD, UK (Tel.: 0278-424200; FAX 0278-425644; Telex: 46796)

Ambrosia Scents
524 E Ohio, Princeton, IN 47670 USA

AMC SPREA S.p.A.
Sede Amministrativa e Comercicale, Via Pasubio 37, 21040 Venegono Superiore (VA), Milan, Italy (Tel.: 0331-827500; FAX 0331-827492)

Amerchol
Amerchol Corp., PO Box 4051, 136 Talmadge Rd., Edison, NJ 08818 USA (Tel.: 908-248-6000; 800-367-3534; FAX 908-287-4186; Telex: 833472)

Amerchol Europe, Havenstraat 86, B-1800 Vilvoorde, Belgium (Tel.: 2-252-4012; FAX 2-252-4909; Telex: 846-69105)

Ameribrom, Inc., Div. of Dead Sea Bromine
52 Vanderbilt Ave., New York, NY 10017 USA (Tel.: 212-286-4000; FAX 212-286-4475; Telex: RCA 220531)

American Biorganics, Inc.
2236 Liberty Dr, Niagara Falls, NY 14304 USA (Tel.: 716-283-1434; 800-648-6689; FAX 716-283-1570; Telex: 926074)

American Bio-Synthetics Corp.
710 W. National Ave., Milwaukee, WI 53204 USA (Tel.: 414-384-7017)

American Casein Co.
109 Elbow Lane, Burlington, NJ 08016 USA (Tel.: 609-387-3130; FAX 609-387-7204)

American Chemet Corp.
PO Box 437, 400 County Line Rd., Deerfield, IL 60015 USA (Tel.: 708-948-0800; FAX 708-948-0811; Telex: 72-4301)

American Chemical Services
PO Box 190, Griffith, IN 46319 USA (Tel.: 219-924-4370)

American Colloid Co.
1500 W. Shure Dr., Arlington Hts., IL 60004-1434 USA (Tel.: 708-392-4600; FAX 708-506-6199; Telex: 4330321)

American Critical Care. See Baxter Healthcare Corp

American Cyanamid
AmericanCyanamid/Corporate Headquarters, One Cyanamid Plaza, Wayne, NJ 07470 USA (Tel.: 201-831-4111; 800-922-0187; FAX 201-839-8847)

American Cyanamid/Agricultural Div., One Cyanamid Plaza, Wayne, NJ 07470 USA (Tel.: 201-831-2000; 800-443-0443)

American Cyanamid/Engineered Materials Dept., 1300 Revolution St., Havre de Grace, MD 21078 USA (Tel.: 301-939-1910; FAX 301-939-0930)

American Cyanamid/Polymer Additives Dept., One Cyanamid Plaza, Wayne, NJ 07470 USA (Tel.: 201-831-2000)

American Cyanamid/Process Chems., Fine Chems., Textile Chem, One Cyanamid Plaza, Wayne, NJ 07470 USA (Tel.: 800-438-5615)

American Cyanamid/Textiles, One Cyanamid Plaza, Wayne, NJ 07470 USA (Tel.: 201-831-4111; 800-922-0187; FAX 201-839-8847)

Cyanamid Canada Inc./Carbide Products Div., 88 McNabb St., Markham, Ontario, L3R 6E6, Canada (Tel.: 416-470-3600; FAX 416-470-3852; Telex: 06-966602)

Cyanamid of Great Britain Ltd/Crop Protection Div, Div. of American Cyanamid, Cyanamid House, Fareham Road, PO Box 7, Gosport, Hants, PO13 0AS, UK (Tel.: 0329-224000; FAX 0329-220213)

Cyanamid BV, Postbus 1523, NL 3000 BM Rotterdam, The Netherlands (Tel.: 010-4116340; FAX 010-4136788; Telex: 23554)

Cyanamid India Ltd., Nyloc House, 254-D2 Dr. Annie Besant Rd., Bombay, 400 025, India

Cyanamid Taiwan Corp., 8/F Union Commercial Bldg., 137, Nanking E. Rd., Sec. 2, Taipei, Taiwan, R.O.C.

Cyanamid Quimica do Brasil Ltda., Av. Imperatriz Leopoldina, 86, Sao Paulo, Brazil

American Emulsions Co.
PO Box 3787, Dalton, GA 30721 USA (Tel.: 404-226-7028; FAX 404-278-5183)

American Ingredients Co.
14622 S. Lakeside Ave., Dolton, IL 60419 USA (Tel.: 708-849-8590; 800-821-2250; FAX 816-561-0422)

American International Chemical, Inc.
27 Strathmore Rd, Natick, MA 01760 USA (Tel.: 508-655-5805; 800-238-0001; FAX 508-655-0927; Telex: 948342)

American Laboratories Inc.
4410 South 102 St., Omaha, NE 68127 USA (Tel.: 402-339-2494; 800-445-5989; FAX 402-339-0801; Telex: 3735593 CACOMA)

American Lecithin Co., Div. of Rhône-Poulenc Rorer/Nattermann
33 Turner Rd., Danbury, CT 06813-1908 USA (Tel.: 203-790-2700; FAX 203-790-2705)

American Maize Products Co.
1100 Indianapolis Blvd., Hammond, IN 46320-1094 USA (Tel.: 219-659-2000; 800-348-9896)

American McGaw, American Hospital Supply Corp.
2525 McGaw Ave, Irvine, CA 92714-5895 USA

American Norit Co., Inc.
1050 Crown Pointe Pkwy., Suite 1500, Atlanta, GA 30338 USA (Tel.: 404-512-4610; 800-641-9245; FAX 404-512-4622)

American Optical Corp.
Soft Contact Lens Business, 14 Mechanic St, Southbridge, MA 01550 USA (Tel.: 508-765-9711)

American Research Products Co. *Address unknown*

American Synthetic Rubber Corp.
PO Box 32960, 4500 Camp Ground Rd., Louisville, KY 40232 USA (Tel.: 502-449-8300; 800-262-9253; FAX 502-449-8468)

Americhem, Inc.
225 Broadway East, PO Box 375, Cuyahoga Falls, OH 44222-0375 USA (Tel.: 216-929-4213; FAX 216-929-4144)

AmeriHaas, Joint venture of AmeriBrom and Rohm & Haas. *See* Ameribrom and Rohm & Haas

Ameripol Synpol Co.
146 S. High St., 7th floor, Akron, OH 44308-1493 USA (Tel.: 800-962-TECH)

Amersham Corp.
Amersham Corp., 2636 S Clearbrook Dr., Arlington Heights, IL 60005 USA (Tel.: 708-593-6300)

Amersham International plc, Amersham Place, Little Chalfont, Amersham, Buckinghamshire, HP7 9NA, UK (Tel.: 0494 544000; FAX 0494 542266; Telex: 83818 ACTIVL G)

B. C. Ames
131 Loxington Street, Box 70, Waltham, MA USA

Amoco Corp.
Amoco Chemicals Corp., 200 East Randolph Dr., Chicago, IL 60601 USA (Tel.: 312-856-3806; 800-621-8888)

Amoco Chemical Co./Engineering Resins Department, Subsidiary of Amoco Corp., 200 E Randolph Drive, Chicago, IL 60601 USA (Tel.: 312-856-3200; FAX 312-856-2460)

Amoco Chemical (Europe) SA, Div. of Amoco Corp., 15, Rue Rothschild, CH-1211 Geneva 21, Switzerland (Tel.: 22-31-02-81; Telex: 422787)

Amoco Performance Products, Japan Ltd., 10th Floor, Tonichi Bldg., 2-31 Roppongi 6-Chome, Mianto Ku, Tokyo, 106, Japan

Ampacet Corp.
660 White Plains Rd., Tarrytown, NY 10591 USA (Tel.: 914-631-6600; FAX 914-631-7197)

Amspec Chemical Corp.
Foot of Water St., Gloucester City, NJ 08030 USA (Tel.: 609-456-3930; 800-5AMSPEC; FAX 609-456-6704; Telex: 136714 GLCY)

Amstat Industries, Inc.
3012 N. Lake Terrace, Glenview, IL 60025-5794 USA (Tel.: 708-998-6210; FAX 708-998-6218)

Anaquest, Div. of BOC Group Inc.
2005 W Beltline Highway, Madison, WI 53713 USA (Tel.: 608-273-0019)

Anar Chemical Co.
1765F Cortland Court, Addison, IL 60101 USA (Tel.: 708-953-1660; 800-344-1660; FAX 708-953-1698)

Anchor Chemical Group plc, Div. of Air Products & Chemicals Inc.
Clayton House, Clayton, Manchester, M11 4SR, UK (Tel.: 061-223-2451; FAX 061-223-5488)

Anchor Chemical (UK) Ltd. *See* Anchor Chemical Group plc

Anderson Development Co.
1415 E. Michigan St., Adrian, MI 49221 USA (Tel.: 517-263-2121; FAX 517-263-1000)

Andrea Aromatics
PO Box 3091, Princeton, NJ 08543-3091 USA (Tel.: 609-695-7710; FAX 609-392-8914)

Andrulex Trading Ltd.
Unit 34, Saffron Court, Southfields Industrial Estate, Laindon, Basildon, Essex, SS15 6SS, UK (Tel.: 0268 416441; FAX 0268 541639; Telex: 99339 RULEX G)

Anedco, Inc.
10429 Koenig Rd., Houston, TX 77034 USA (Tel.: 713-484-3900; FAX 713-484-3931)

Anglo Speciality Adhesives. *See* Evode Speciality Adhesives Ltd

Angus
Angus Chemical Co., 1500 E. Lake Cook Rd., Buffalo Grove, IL 60089 USA (Tel.: 708-215-8600; 800-362-2580; FAX 708-215-8626; Telex: 275422 ANGUS UR)

Angus Chemie GmbH, 19, Moorgate St., Rotherham, Yorkshire, S60 2DA, UK (Tel.: 0709-377743; FAX 0709-370596; Telex: 547159 ANGUK G)

Angus Chemie GmbH, Le Bonaparte, Centre d'Affaires Paris-Nord, F-93153 Le Blanc Mesnil, Paris, France (Tel.: 1-48-65-73-40; FAX 1-48-65-73-20; Telex: ANGUS 232089 F)

Angus Chemie GmbH, Huyssenallee 5, W-4300 Essen 1, Germany (Tel.: 201-233531; FAX 201-238661; Telex: 8571563 ANGE D)

Angus Chemical (Singapore) Pte. Ltd., Blk 265, Serangoon Central Dr., #04-263, 1955, Singapore (Tel.: 382-4468; FAX 286-0739; Telex: RS 28521 ANGUS)

Anic Agricoltura SpA. *Address unknown*

D F Anstead Ltd.

Radford Way, Billericay, Essex, CM12 0DE, UK (Tel.: 0277-63 00 631; FAX 0277 631356)

Ansul Co.

1 Stanton St, Marinette, WI 54143 USA (Tel.: 715-735-7411; Telex: 263433)

Antec International Ltd.

Windham Road, Chilton Industrial Estate, Sudbury, Suffolk, CO10 6XD, UK (Tel.: 0787-77305; FAX 0787-310846; Telex: 987495)

Antigen International Ltd. *Address unknown*

Anzon Ltd.

Cookson House, Willington Quay, Wallsend, Tyne & Wear, NE28 6UQ, UK

Apollo Chemicals Ltd.

Sandy Way, Amington Ind. Estate, Tamworth, Staffs, B77 4DS, UK (Tel.: 0827-54281; FAX 0827-53030)

Appleby Group Ltd.

Brigg Rd., Scunthorpe, South Humberside, DN16 1AQ, UK (Tel.: 0724 282211; FAX 0724 280338; Telex: 527328)

Aqualon

Aqualon Co., A Hercules Inc. Co., PO Box 15417, 2711 Centreville Rd., Wilmington, DE 19850-5417 USA (Tel.: 302-996-2000; 800-345-8104; FAX 302-996-2049; Telex: 4761123)

Aqualon Canada Inc., 5407 Eglinton Ave. West, Suite 103, Etobicoke, Ontario, M9C 5K6, Canada (Tel.: 416-620-5400)

Aqualon (UK) Ltd., Genesis Centre, Garrett Field, Birchwood, Warrington, Cheshire, WA3 7BH, UK (Tel.: 925-830077)

Aqualon France, 44, Ave. de Chatou, F-92508 Rueil Malmaison Cedex, France (Tel.: 1-4751-2919)

Aqualon GmbH, PO Box 130125, Paul Thomas Strasse 58, D-4000 Düsseldorf 13, Germany (Tel.: 0211-7491-0)

Aquarium Systems Inc.

8141 Tyler Boulevard, Mentor, OH 44060-4889 USA (Tel.: 216-255-1997; 800-822-1100; FAX 216-255-8994)

Aquaspersions Ltd.

Charlestown Works, Charlestown, Hebden Bridge, West Yorkshire, HX7 6PL, UK (Tel.: 0422-843715; FAX 0422-845067; Telex: 517723 AKWA G)

Aquatrols Corp. of America

1432 Union Avenue, Pennsauken, NJ 08110 USA

Aquitaine-Organico. *Address unknown*

Arakawa Chemical Industries Ltd.

3-7, Hiranomachi 1-chome, Chuo-ku, Osaka, 541, Japan (Tel.: (06) 209-8580; FAX (06) 209-8542; Telex: 5222296 ARKOSA J)

Arakawa Chemical (USA) Inc., 625 N. Michigan Ave., Suite 1700, Chicago, IL 60611 USA (Tel.: 312-642-1750; FAX 312-642-0089; Telex: 26-5514)

Arcmann-Denmark A/S

Strandparken 15, DK 8000 Aarhus C, Denmark (Tel.: 06-12 65 44)

Arco

Arco Chemical/Headquarters, Research & Engineering Center, 3801 West Chester Pike, Newtown Sq., PA 19073 USA (Tel.: 215-359-2000; 800-321-7000)

Arco Chemical Canada Inc., 100 Consilium Pl., Suite 306, Scarborough, Ontario, M1H 3E3, Canada

Arco Chemical Pan American, Inc., Paseo de la Reforma, 390 Decimo Piso, 06600 Mexico City, Mexico

Arco Chemical Europe Inc., Bridge Ave., Maidenhead, Berkshire, SL6 1YP, UK (Tel.: 628-775000)

Arco Chemical Europe, Weenahuis, Weena 141, NL-30 13 CK Rotterdam, The Netherlands (Tel.: 010-401 0400; FAX 010-411 4849)

Arco Chemical Asia/Pacific Ltd., Toranomon 37 Mori Bldg., 5th Floor, 5-1 Toranomon 3-Chome, Minato-Ku, Tokyo, 105, Japan

Argus Division. *See under* Witco Corp.

Arista Industries, Inc.

1082 Post Rd., Darien, CT 06820 USA (Tel.: 203-655-0881; 800-255-6457; FAX 203-656-0328; Telex: 996493)

Aristech Chemical Corp.

600 Grant St., Room 1028, Pittsburgh, PA 15219-2704 USA (Tel.: 412-433-2747; 800-526-4032; FAX 412-433-1816)

Arizona Chemical Co, Div. of International Paper

1001 E. Business Hwy. 98, Panama City, FL 32401 USA (Tel.: 904-785-6700; 800-526-5294; FAX 904-785-2203; Telex: 514411)

Arlington Mills Inc.
1430 East Davis Street, Arlington Heights, IL 60005 USA (Tel.: 708-392-5700; FAX 708-392-0698)

Armillatox Ltd.
The Old Malt House, Well Street, Brassington, Derbyshire, DE4 4HJ, UK (Tel.: 0629-85-205; FAX 0773-590681)

Armour Hess Chemicals. *Address unknown*

Armour Pharmaceutical Co, Subsidiary of Rhône-Poulenc Rorer Inc.
500 Arcola Road, P.O. Box 1200, Collegeville, PA 19422 USA (Tel.: 215-454-8000; 800-72-RORER; FAX 215-454-8940)

Arol Chemical Products Co.
649 Ferry St., Newark, NJ 07105 USA (Tel.: 201-344-1510; FAX 201-344-7127)

Arveta SA. *Address unverified*
Basel, Switzerland

Asada Chemical Industry Co., Ltd.
180, Miya, Shikama-ku, Himeji-shi, Hyogo, 672, Japan (Tel.: (0792) 35-1911; FAX (0792) 35-1915)

Asahi Chemical Industry Co., Ltd.
Hibiya Mitsui Bldg., 1-2, Yuraku-cho 1-chome, Chiyoda-ku, Tokyo, 100, Japan (Tel.: (03) 3507-2730; FAX (03) 3507-2495; Telex: 222-3518 BEMBRGJ)

Asahi Denka Kogyo K.K.
Furukawa Bldg. 3-14, Nihonbashi Muro-machi 2-chome, Chuo-ku, Tokyo, 103, Japan (Tel.: (03) 5255-9017; FAX (03) 3270-2463; Telex: 222-2407 TOKADK)

Adeka Fine Chemical Co., Ltd.
Yoko Bldg., 4-5, Hongo 1-chome, Bunkyo-ku, Tokyo, 113, Japan (Tel.: (03) 5689-8681; FAX (03) 5689-8680)

Asahi Glass Co. Ltd.
Chiyoda Bldg, 1-2, Marunouchi 2-chome, Chiyoda-ku, Tokyo, 100, Japan (Tel.: (03) 3218-5257; Telex: J24616 ASAGLAS)

Asahi Glass America, Inc., 1185 Ave. of the Americas, 30th floor, New York, NY 10036 USA (Tel.: 212-764-3155; FAX 212-764-3384; Telex: 275830 ASAGL UR)

Asahi-Penn Chemical Co. Ltd., Joint venture of Asahi Glass Co Ltd/PPG Industries Inc.
Shuwa Kioicho TBR Bldg., 5-7, Kojimachi, Chiyoda-ku, Tokyo, 102, Japan (Tel.: (03) 3234-0561; FAX (03) 3234-0304)

Asahi Yukizai Kogyo Co., Ltd.
2-5955, Nakanose-cho, Nobeoka-shi, Miyazaki, 882, Japan (Tel.: (0982) 33-3311; FAX (0982) 21-8606)

Asarco Inc.
180 Maiden Lane, New York, NY 10038 USA (Tel.: 212-510-2000; Telex: ITT 420585)

A/S Cheminova. *See* Cheminova Agro A/S

B. F. Ascher & Co. Inc.
15501 W. 109th Street, Lenexa, KS 66219 USA (Tel.: 913-888-1880; FAX 913-888-2250)

Ashe Chemicals
Ashetree Works, Kingston Road, Leatherhead, Surrey, UK

Ashlade Formulations Ltd., Div. of Willmot Industries Ltd.
Ness Road, Slade Green, Erith, Kent, DA8 2LD, UK (Tel.: 0322-331671; FAX 0322-332279; Telex: 896727 AGFORM G)

Ashland
Ashland Chemical Inc., Subsidiary of Ashland Oil, Inc., 5200 Paul G. Blazer Memorial Pkwy., Dublin, OH 43017 USA (Tel.: 614-889-3333)

Drew Industrial Div., One Drew Plaza, Boonton, NJ 07005 USA (Tel.: 201-263-7800; 800-526-1015 x7800; FAX 201-263-4483; Telex: DREWCHEMS BOON)

Ashland Chemical Co/Resin & Chemicals Division, Subsidiary of Ashland Oil Inc., 2620 Royal Windsor Drive, Mississauga, Ontario, L5J 4ET, Canada

Ashland-Südchemie-Kernfest GmbH, Reisholzstrasse 16, 4010 Hilden, Germany (Tel.: 021-711030; FAX 0211-7110335; Telex: 8582889)

Ashley Polymers
5114 Fort Hamilton Parkway, Brooklyn, NY 11219 USA (Tel.: 718-851-8111; FAX 718-972-3256)

Associated Adhesives. *Address unknown*

Associated Electrical Industries (GEC). *Address unknown*

Associated Lead Manufacturers Ltd.
Crescent House, P O Box 19E, Newcastle upon Tyne, NE99 1GE, UK (Tel.: 0632-610161)

Associated Octel Co Ltd.
20 Berkeley Sq., London, W1X 6DT, UK (Tel.: 071-499 6030; FAX 071-491 2332; Telex: 25107)

H. A. Astlett & Co.
8 King St. East, Toronto, Ontario, M5C 1B5, Canada (Tel.: 416-366-2747; 800-387-1152 (U.S.); FAX 416-366-6389; Telex: 06-218801)

Astor Chemical Ltd., Div. of Associated British Industries plc

Tavistock Road, West Drayton, Middlesex, UB7 7RA, UK (Tel.: 0895-445511; FAX 0895-449199; Telex: 28559 ASTORW G)

Astor Wax Corp.

200 Piedmont Ct., Doraville, GA 30340 USA (Tel.: 404-448-8083; FAX 404-840-0954)

Astra Chemicals

Astra Chemicals GmbH, Div. of Astra AB, Postfach 249, Tinsdaler Weg 183, DW-2000 Wedel, Germany (Tel.: 4103-70 80; FAX 4103-70 82 93)

Astra Chemicals Ltd. *Address unknown*

Astra Pharmaceutical Products Inc.

50 Otis St, Westboro, MA 01581-4500 USA (Tel.: 508-366-1100; 800-225-6333; FAX 508-366-7406)

Astro Industries Inc., Subsidiary of Borden Inc.

PO Box 2559, 114 Industrial Boulevard, Morganton, NC 28655 USA (Tel.: 704-584-3800; 800-872-7876; FAX 704-584-3885)

Atlas Chemical Industries (UK) Ltd., Subsidiary of ICI

Cleeve Road, Leatherhead, Surrey, KT22 7SW, UK

Atlas Interlates Ltd., Div. of Allied Colloids Group plc

Gladden Place, Skelmersdale, Lancashire, WN8 9SX, UK (Tel.: 0695-33535; FAX 0695-50268; Telex: 51646 ALCOLL G)

Atlas Refinery, Inc.

142 Lockwood St., Newark, NJ 07105 USA (Tel.: 201-589-2002; FAX 201-589-7377; Telex: 138-425 ATLASOIL)

Atochem

Atochem, Groupe elf aquitaine, 4, cours Michelet, La Défense 10, F-92091 Paris Cedex 42, France (Tel.: 49-00-8080; FAX 49-00-7447; Telex: 611922 ATO F)

Ceca SA, Div. of Atochem, 22, place de l'Iris, La Défense 2, Cedex 54, 92062 Paris-La Défense, France (Tel.: 147-96-9090; FAX 147-96-9234; Telex: 611444 ckd)

Elf Atochem UK Ltd., Colthrop Lane, Thatcham, Newbury, Berkshire, RG13 4LW, UK (Tel.: 0635-70000; FAX 0635-61212; Telex: 847689 ATOKEM G)

Elf Atochem UK, Ceca Specialities Div, Rowan Court, 56 High St., Wimbledon Village, London, SW19 5EE, UK (Tel.: 081-9 46 77 74; FAX 081-94738 73; Telex: 928041)

Pennwalt Plastics Ltd., Div. of Atochem, Cherwell House, St. Clements, Oxford, OX4 1BD, UK (Tel.: 0865-726961)

Atochem Deutschland GmbH, Niederlassung Düsseldorf, Uerdiger Strasse 5, Postfach 30 01 52, D-4000 Düsseldorf 30, Germany (Tel.: 0211-4552-0; FAX 0211-14552-112; Telex: 8584682)

Pennwalt Italia SpA, Via del Porto, I-28040 Marano-Ticino, Italy (Tel.: 321-9791; FAX 321-979246; Telex: 200136 PENNTI I)

Atochem North America Inc./Plastics Dept., Three Parkway, Philadelphia, PA 19102 USA (Tel.: 215-587-7000; 800-328-2811; FAX 215-587-7497)

Atochem North America Inc., Three Parkway, Philadelphia, PA 19102 USA (Tel.: 215-587-7000; 800-225-7788; FAX 215-587-7591)

Atochem North America Inc./Fine Chemicals Group, Org. Chems., PO Box 59209, Philadelpha, PA 19102 USA (Tel.: 215-587-7711; 800-628-4453; FAX 215-587-7875)

Atochem North America Inc./Organic Peroxides Div., 1740 Military Rd., PO Box 1048, Buffalo, NY 14240 USA (Tel.: 716-877-1740; 800-558-5575; FAX 716-877-1541)

Atochem North America Inc./Plastics Additives Div., 4 King James South, 24500 Center Ridge Rd., Room 180, Cleveland, OH 44145 USA (Tel.: 216-835-5030; FAX 216-835-9185)

Atochem North America Inc./Wire Mill Products Dept., 43 James St., Homer, NY 13077 USA (Tel.: 607-749-2652)

Pennwalt Corp/Prescription Div, PO Box 1710, Rochester, NY 14603 USA

Elf Atochem Canada, PO Box 278, Oakville, Ontario, L6J 5A3, Canada (Tel.: 416-827-9841; FAX 416-827-7913)

Atochem Yoshitomo, Ltd., 6-9 Hiranomachi, 2-Chome, Chuo-Ku, Osaka, 541, Japan (Tel.: 201-1161)

Pennwalt Japan Ltd., Div. of Atochem, Sumitomo Jimbo-Cho Bldg., 3-25, Kanda Jimbo-Cho, Chiyoda-Ku, Tokyo, 101, Japan (Tel.: (03) 262-8481)

Atochimie. *Address unknown*

Atohaas North America, Inc., Joint venture between Elf Atochem SA and Rohm & Haas Co. *See under* Atochem and Rohm & Haas

Atomergic Chemetals Corp.

222 Sherwood Ave., Farmingdale, NY 11735-1718 USA (Tel.: 516-694-9000; FAX 516-694-9177; Telex: 6852289)

Atomic Energy Establishment. *See* AEA Technology

AT Plastics

AT Plastics Inc., 1419 East Blvd., Suite K, Charlotte, NC 28203 USA (Tel.: 704-331-9161; 800-342-2990; FAX 704-343-0019)

AT Plastics Inc., 142 Kennedy Rd. S., Brampton, Ontario, L6W 3G4, Canada (Tel.: 416-452-6756; FAX 416-451-1677; Telex: 063690056)

Ausimont USA Inc.

44 Whippany Rd., Morristown, NJ 07962-1838 USA (Tel.: 201-292-6250; 800-323-AUSI; FAX 201-292-0886)

Australian Bakels (Pty) Ltd.

PO Box 147, Lidcombe N.S.W. 2141, Australia (Tel.: 648-3833; FAX (02) 647-2663; Telex: 23064)

Australian Synthetic Rubber Co. Ltd.

Maidstone Street, PO Box 33, Altona, 3018, Australia

Avalon Chemical Co. Ltd., Div. of ICI Polyurethanes
Hitchen Lane, Shepton Mallet, Somerset, BA4 5TZ, UK (Tel.: 0749-343061; FAX 0749-344221)

Avebe
Avebe BV, Avebeweg 1, NL-9607 PT Foxhol, The Netherlands (Tel.: 5980-42234; FAX 5980-97892; Telex: 53018)

Avebe UK Ltd., Thornton Hall, Thornton Curtis, South Humberside, DN39 6XD, UK (Tel.: 0469-32222; FAX 0469-31488)

Avebe America Inc., 4 Independence Way, Princeton, NJ 08540 USA (Tel.: 609-520-1400; FAX 609-520-1473; Telex: 0820713)

Avicon Inc.
6201 S Freeway, PO Box 85, Ft Worth, TX 76101 USA

Avitrol Corp.
7644 E 46th Street, Tulsa, OK 74145 USA (Tel.: 918-622-7763; 800-633-5069; FAX 918-622-2527)

Avon Packers Ltd.
Salisbury Road, Downton, Wilts, SP5 3JJ, UK (Tel.: 0725-22822; FAX 0725 22840)

Axel Plastics Research Laboratories Inc.
PO Box 855, 58-20 Broadway, Woodside, NY 11377 USA (Tel.: 718-672-8300; FAX 718-565-7447; Telex: 429033 AXELLAB)

Ayer's Cliff Chemical Products Ltd.
PO Box 322, Ayer's Cliff, Quebec, J0B 1C0, Canada (Tel.: 514-649-0503; FAX 514-922-5166)

Ayerst Laboratories. *See* Wyeth Laboratories

Ayrton Saunders plc. *Address unknown*

BA Chemicals Ltd., Div. of Alcan
Chalfont Park Gerrards Cross, Buckinghamshire, SL9 0QB, UK (Tel.: 0753-887373; FAX 0753-889602; Telex: 847343)

Bacon Industries Inc.
192 Pleasant St., Watertown, MA 02172 USA (Tel.: 617-926-2550)

Bakelite
Bakelite GmbH, Div. of Rütgerswerke AG, Postfach 7154, Gennaer Strasse 2-4, W-5860 Iserlohn 7, Germany (Tel.: 2374-510; FAX 2374-51409; Telex: 827255 BGLH)

Bakelite Polymers (UK) Ltd., Syer House, Stafford Court, Stafford Park Road, Stafford Park, Telford, Shropshire, TF3 3BD, UK

Baker Laboratories. *Address unverified*
Miami, FL USA

Baker Performance Chemical

Baker Performance Chemical, Inc., A Baker Hughes Co., 3920 Essex Lane, Houston, TX 77027 USA (Tel.: 713-599-7400; 800-231-3606; FAX 713-599-7460; Telex: ITT-4620058)

Baker Performance Chemicals Ltd., Div. of Baker Hughes Inc., Erimus House, Queens Square, Middlesbrough, Cleveland, TS2 1AA, UK (Tel.: 0642 221441; FAX 0642 225108; Telex: 587267 BPC G)

Baker Rubber Inc.

700 West Chippewa, PO Box 2438, South Bend, IN 46680 USA (Tel.: 219-291-5101; FAX 219-291-9152)

Bamberger Polymers

Bamberger Polymers Inc., 690 Jersey Ave., Bldg. #4, New Brunswick, NJ 08901 USA (Tel.: 908-214-8867; 800-888-8959; FAX 908-214-8868)

Bamberger Polymers (Canada) Inc., 62 Selby Rd., Brampton, Ontario, L6W 3L4, Canada (Tel.: 416-456-0851)

Bamberger Polymers (Europe) BV, Steurweg 2, 4941 VR Raamsdonksveer, The Netherlands (Tel.: 1621-20240)

Bamberger Polymers De Mexico S.A., Rio Lerma 143-103, Col. Cuauhtemoc, Mexico, DF

Barclay Chemicals UK

28 Howard Street, Glossop, Derbyshire, SK13 9DD, UK (Tel.: 0457-853386; FAX 0457-853557)

Otto Bärlocher GmbH, Chemische Werke München

Postfach 500108, Riesstrasse 16, DW 8000 München, Germany (Tel.: 0891-4880; FAX 0891-488312)

Barnes-Hind Inc.

895 Kifer Road, Sunnyvale, CA 94086 USA

Barron Chemicals Ltd. *Address unknown*

BASF

BASF AG, ESA/WA-H 201, D-6700 Ludwigshafen, Germany (Tel.: 0621-60-99603; FAX 0621-60-41787; Telex: 469499-0 BAS D)

BASF plc, PO Box 4, Earl Road, Cheadle Hulme, Cheadle, Cheshire, SK8 6QG, UK (Tel.: 061-485-6222; FAX 061-486-0891; Telex: 669211 BASFCH G)

BASF plc/Agricultural Division, Lady Lane, Hadleigh, Ipswich, Suffolk, IP7 6BQ, UK (Tel.: 0473-822531; FAX 0473-827450)

BASF Belgium S.A., Ave. Hamoir-laan 14, B-1180 Brussels, Belgium

BASF S.A. Compagnie Francaise, MC-NT, 140, Rue Jules Guesde, 92303 Levallois-Perret, France

BASF Espanola S.A., Apartado 762, Barcelona, 8, Spain

BASF Corp., 100 Cherry Hill Rd., Parsippany, NJ 07054 USA (Tel.: 201-316-3000; 800-669-BASF; FAX 201-402-1832)

BASF Corp./Fibers Div., Textile Colors & Chemicals, 9401 Arrowpoint Blvd., Suite 200, Charlotte, NC 28273 USA (Tel.: 800-247-0557; FAX 704-527-3503)

BASF Canada Ltd., PO Box 430, Montreal, Quebec, H4L 4V8, Canada

BASF India, Ltd., Maybaker House, S.K. Ahire Marg., PO Box 19108, Bombay, 400 025, India

BASF Japan Ltd., 3-3, Kioicho, Chiyoda-ku, Tokyo, 102, Japan (Tel.: (03) 3238-2300; Telex: 222-2130 BASFTK)

Battle, Hayward & Bower Ltd.

Victoria Chemical Works, Crofton Drive, Allenby Road Industrial Estate, Lincoln, LN3 4NP, UK (Tel.: 0522-29206/41241; FAX 0522-38960)

Bausch & Lomb, Professional Products Div.

1 Lincoln First Sq, Rochester, NY 14604 USA (Tel.: 716-338-6000)

Baxenden Chemicals Ltd., Div. of Witco Chemicals Inc.

Paragon Works, Accrington, Lancashire, BB5 2SL, UK (Tel.: 0254-872278; FAX 0254-871247; Telex: 63141)

Baxter Health Care

Baxter Health Care Corp., One Baxter Pkwy, Deerfield, IL 60015 USA (Tel.: 708-948-2000)

Baxter Healthcare Ltd., Wallingford Rd, Compton, Newbury, Berkshire, RG16 0QW, UK (Tel.: 0635 200000; FAX 0635 578010)

Bayer

Bayer AG, Bayerwerk, D-5000 Leverkusen, Germany (Tel.: (0214)30-1; FAX (0214)30 6 63 41)

Bayer plc, Bayer House, Strawberry Hill, Newbury, Berkshire, RG13 1JA, UK (Tel.: 0635-39000; FAX 0635-563393; Telex: 847205 BAYNEW G)

Bayer UK Ltd., Agrochem Business Group, Eastern Way, Bury St Edmunds, Suffolk, IP32 7AH, UK (Tel.: 0284-763200; FAX 0284-702810)

Bayer Antwerpen NV, Div. of Bayer AG, Kanaaldok 1, B-2040 Antwerp 4, Belgium (Tel.: 3-540 30 11; FAX 3-541 69 36; Telex: 71175 BAYANT B)

Bayer Hispania Industrial SA, Pablo Claris 196, E-80037 Barcelona, Spain (Tel.: 3-218 45 50; FAX 3-217 41 49; Telex: 54482 BAYIN E)

Bay Resins, Inc.

PO Box 630, Route 313, Millington, MD 21651 USA (Tel.: 410-928-3083; FAX 410-928-5412)

BDH Chemicals Ltd.

Broom Road, Parkstone, Poole, Dorset, BH12 4NN, UK (Tel.: 0202-745520; FAX 0202-738299)

Beacon Chemical Co. Inc.

PO Box 2500, 125 Macquiesten Pkwy., Mt. Vernon, NY 10550 USA (Tel.: 914-699-3400; FAX 914-699-2783)

Beardow & Adams (Adhesives) Ltd.

32 Blundell's Road, Bradville, Milton Keynes, Bucks, MK13 7HF, UK (Tel.: 0908-315474; FAX 0908-222435)

Bechem Chemie BV
Postbus 62, Kanaalweg 2, NL-2100 AB Heemstede, The Netherlands (Tel.: 43-36 74 74; FAX 43-36 83 64; Telex: 41436)

Beecham Pharmaceuticals. See SmithKline Beecham

Beecham Research Laboratories. See SmithKline Beecham

Bell Flavors & Fragrances, Inc.
500 Academy Dr., Northbrook, IL USA (Tel.: 312-291-8300; 800-323-4387; Telex: 910-686-0653)

Belzak Corp.
850 Bloomfield Ave., Clifton, NJ 07012 USA (Tel.: 201-773-0602)

Bengue & Co. Ltd.
Syntex House, St Ives Road, Maidenhead, Berks, SL6 1RD, UK (Tel.: 0628-33191)

Benzsay & Harrison Inc.
Railroad Ave, PO Box 459, Delanson, NY 12053 USA (Tel.: 518-895-2311; FAX 518-895-8475)

Berger Elastomers
Portland Road, Newcastle-upon-Tyne, NE2 1BL, UK

Berger Jenson & Nicholson Ltd.
Berger Decorative Paints Division, Petherton Road, Hengrove, Bristol, BS99 7JA, UK

Bergvik Sales Ltd.
Manor House, 1, The Crescent, Leatherhead, Surrey, KT22 8DH, UK (Tel.: 0372-363433; FAX 0372-360025)

Berk Chemicals Ltd., Div. of Berk Ltd.
PO Box 56, Priestley Rd, Basingstoke, Hampshire, RG24 9QB, UK (Tel.: 0256 29292; FAX 0256 64711; Telex: 858371)

Berk Pharmaceuticals Ltd.
St Leonards House, St Leonards Road, Eastbourne, Sussex, BN21 3YG, UK (Tel.: 0323 501111)

Berk Spencer Acids Ltd., Div. of Rhône-Poulenc SA
Abbey Mills Chemical Works, Canning Road, Stratford, London, E15 3NX, UK (Tel.: 081-534 5162; FAX 081-519 5455; Telex: 897563)

Berlex Laboratories Inc.
300 Fairfield Road, Wayne, NJ 07470 USA (Tel.: 201-694-4100; 800-221-1756; FAX 201-305-5365; Telex: 132789 BERLICHM WYNE)

Bernardy Chimie SA, Div. of Société des Produits Chimiques d'Harbonnieres
Thenioux, F-18100 Vierzon, France (Tel.: 48 52 00 80; FAX 48 852 04 61; Telex: 760328)

Bernel Chemical Co., Inc.
174 Grand Ave., Englewood, NJ 07631 USA (Tel.: 201-569-8934; FAX 201-569-1741)

Berol Nobel
Berol Nobel AB, Box 11536, S-10061 Stockholm, Sweden (Tel.: 8-743-4000; FAX 8-644-3955; Telex: 10513 benobl s)

Berol Nobel Ltd., Div. of Berol Nobel AB, 23 Grosvenor Road, St Albans, Hertfordshire, AL1 3AW, UK (Tel.: 0727-41421; FAX 0727-41529; Telex: 23242)

Berol Kemi (UK) Ltd., 55/57 Clarendon Road, Watford, Hertsfordshire, WD1 1SP, UK

Berol Nobel SA, Div. of Berol Nobel AB, Rue Gachard 88, Bte 9, B-1050 Bruxelles, Belgium (Tel.: 2-640-5065; FAX 2-640-6997; Telex: 62812)

Berol Nobel Inc., Meritt 8 Corporate Park, 99 Hawley Lane, Stratford, CT 06497 USA

Bevaloid Ltd., Subsidiary of May & Baker
PO Box 3, Flemingate, Beverley, North Humberside, HU17 0NW, UK

BFGoodrich. *See under* Goodrich

Biachem Ltd.
Boundary House, 91-93, Charterhouse St., London, EC1M 6HR, UK (Tel.: 071-250 1905; FAX 071-250-1913; Telex: 267136)

BioContacts Inc. *Address unknown*

Bioglan Laboratories
1 The Cam Centre, Wilbury Way, Hitchin, Hertfordshire, S94 OTW, UK (Tel.: 0462-438444; FAX 0462-421242)

Bio-Rad Laboratories
Bio-Rad Laboratories Inc., 3300 Regatta Blvd., Richmond, CA 94804 USA (Tel.: 415-232-7000; 800-227-5589; FAX 415-232-4257; Telex: 71-3720184)

Bio-Rad Laboratories NV, Div. of Bio-Rad Laboratories Inc., Begoniastraat 5, B-9810 Eke, Belgium (Tel.: 91-85 55 11; FAX 91-85 65 54)

Biosynth
Biosynth International, PO Box 541, Skokie, IL 60076 USA (Tel.: 708-674-5160; FAX 708-674-8885)

Biosynth AG, Postfach 125, CH-9422 Staad, Switzerland (Tel.: 71-430190; FAX 71-425859; Telex: 882929)

BIP Chemicals Ltd., Div. of T&N plc
PO Box 6, Popes Lane, Oldbury, Warley, West Midlands, B69 4PD, UK (Tel.: 021-552-1551; FAX 021-552-4267; Telex: 337261)

Bisco Prods., A Dow Corning Affiliate
2300 East Devon Ave., Elk Grove Village, IL 60007-6120 USA (Tel.: 708-640-0800; 800-237-2068; FAX 708-640-3838)

Blagden Chemicals Ltd.
AMP House, Dingwall Road, Croydon, Surrey, CR9 3QU, UK (Tel.: 081-681 2341; FAX 081-689 5851; Telex: 24285)

Blew Chemical Co.
PO Box 501, Palos Heights, IL 60463 USA (Tel.: 708-448-5780; FAX 708-448-5781)

Blue Line Chemical Co. *Address unverified*
St Louis, MO USA

Blueminster Ltd.
Unit 17, Chaucer Business Park, Kemsing, Sevenoaks, Kent, TN15 6PJ, UK (Tel.: 0732-61858; FAX 0732-63800)

Blythe, William & Co Ltd., Div. of Holliday Chemical Holdings plc
Holland Bank Works, Bridge St., Church, Accrington, Lancashire, BB5 4PD, UK (Tel.: 0254 872872; FAX 0254 872000; Telex: 63142 BLYCO G)

BMC, Inc./Bulk Molding Compounds Inc.
3N497 N. 17th St., St. Charles, IL 60174 USA (Tel.: 708-377-1065; FAX 312-377-7395)

BOC
BOC Group plc, Chertsey Road, Windlesham, Surrey, GU20 6HJ, UK (Tel.: 0276-77222; FAX 0276-71333; Telex: 859363)

BOC Ltd., Div. of BOC Group plc, The Priestley Centre, 10 Priestley Road, Surrey Research Centre, Guildford, Surrey, GU2 5XY, UK (Tel.: 0483-579857; FAX 0483-505211)

Bock, Bruno Chemische Fabrik KG
Eichholzer Strasse 23, W-2095 Marschacht, Germany (Tel.: 4176-1397/8; FAX 4176-1396; Telex: 2189219)

Boehringer Ingelheim
Boehringer Ingelheim KG, Binger Strasse 172, Postfach 200, D-6507 Ingelheim, Germany (Tel.: 6132 773666; FAX 6132-773755; Telex: 79122 BI W)

Boehringer Ingelheim Ltd., Affiliate of Boehringer Ingelheim KG, Ellesfield Avenue, Bracknell, Berkshire, RG12 8YS, UK (Tel.: 0344 424600; FAX 0344 424600; Telex: 847634 BILWBT G)

Boehringer Mannheim GmbH, Postfach 310120, Sandhofer Strasse 116, W-6800 Mannheim 31, Germany (Tel.: 621-7591; FAX 621-7592; Telex: 463193 BM D)

Boehringer Ingelheim Pharmaceuticals Inc., 900 Ridgebury Road, P.O. Box 368, Ridgefield, CT 06877-0368 USA

Boliden Intertrade Inc.

3400 Peachtree Rd. NE, Suite 401, Atlanta, GA 30326 USA (Tel.: 404-239-6700; 800-241-1912; FAX 404-239-6701; Telex: 981036)

Bonar Polymers Ltd.

Horndale Avenue, Newton Aycliffe, County Durham, DL5 6YE, UK (Tel.: 0325-300 990; FAX 0325 314925)

Bond Adhesives Co.

301 Frelinghuysen Avenue, Newark, NJ 07114 USA

The Boots Co.

The Boots Co plc, D110 Main Office, Beeston, Nottingham, Nottinghamshire, NG2 3AA, UK (Tel.: 0602 591 7642; FAX 0602 591 680)

Boots Pharmaceuticals, Subsidiary of Boots Company plc, Suite 200, 300 Tri-State International Center, Lincolnshire, IL 60069-4415 USA (Tel.: 708-405-7505; FAX 708-405-7505)

Borax Consolidated Ltd., Div. of The RTZ Corp plc

Borax House, Carlisle Place, London, SW1P 1HT, UK (Tel.: 071-834 9070; FAX 071-973 9079; Telex: 920453 BORAXL G)

Borden Inc.

Borden (UK) Ltd., Div. of Borden Inc., North Baddesley, Southampton, Hamsphire, SO52 9ZB, UK (Tel.: 0703 732131; FAX 0703 738656; Telex: 47212)

Borderland Products Inc. *Defunct company*

Borg-Warner Chemicals. *See* GE Plastics

Boride Ceramics & Composites Ltd., Div. of Rhône-Poulenc SA

Lake Rd., Leeway Industrial Estate, Newport, Gwent, NP9 0SR, UK (Tel.: 0633-276014)

Borregaard

Borregaard Industries Ltd., Div. of Orkla-Borregaard A/S, PO Box 162, N-1701 Sarpsborg, Norway (Tel.: 9-11 80 00; FAX 9-11 87 70; Telex: 71400)

LignoTech (U.K.) Ltd., Clayton Rd., Birchwood, Warrington, Cheshire, WA3 6QQ, UK (Tel.: 0925 824511; FAX 0925 812186)

LignoTech USA, Inc., 100 Highway 51 South, Rothschild, WI 54474-1198 USA (Tel.: 715-359-6544; FAX 715-355-3648)

Borregaard LignoTech, 81 Holly Hill Lane, Greenwich, CT 06830 USA

LignoTech Canada Inc., 1950, Rue Léon Harmel, Quebec, P.Q., GIN HK3, Canada (Tel.: 418-684-3000; FAX 418-684-3005)

Bos Chemicals Ltd.

Paget Hall, Tydd St Giles, Wisbech, Cambs, PE13 5FL, UK (Tel.: 0945-870118; FAX 0945-870264)

Bostik Inc.
Boston St., Middleton, MA 01949 USA (Tel.: 508-777-0100; 800-726-7845; FAX 508-774-7376)

Bostik Ltd., Div. of Total Cie Francaise des Pétroles SA
Ulverscroft Road, Leicester, Leicestershire, LE4 6BW, UK (Tel.: 0533-510015; FAX 0533-531943; Telex: 34625)

J. C. Bottomley, Div. of Croda Colours Ltd.
Brookfoot, Brighouse, West Yorkshire, WDG 2QZ, UK (Tel.: 0484 714574; FAX 0484 717718; Telex: 517296)

BP Chemicals
BP Chemicals Ltd., Belgrave House, 76 Buckingham Palace Rd., London, SW1W 0SU, UK (Tel.: 071-581-1388; FAX 071-581-6411; Telex: 266883)

BP Chemicals Ltd., Subsidiary of The British Petroleum Co plc, Britannic House, 1 Finsbury Circus, London, EC2M 7BA, UK (Tel.: 071-496-4000; FAX 071-496-5656; Telex: 888811 BPLDN X G)

BP Oil UK Ltd., Div. of The British Petroleum Co plc, BP House, Breakspear Way, Hemel Hempstead, Hertfordshire, HP2 4UL, UK (Tel.: 0442-232323; FAX 0442-225225; Telex: 827711)

BP Chemicals Inc., 200 Public Square 7-A, Cleveland, OH 44114-2375 USA (Tel.: 216-586-4141; 800-447-2724; FAX 216-586 8510; Telex: 6873199)

BP Performance Polymers, Newburg Rd., PO Box 400, Hackettstown, NJ 07840 USA (Tel.: 908-852-1100; 800-526-4636; FAX 908-850-7282; Telex: 510-235-2736)

Bracco Industria Chimica SpA
Casella Postale 12064, Via E Folli 50, I-20134 Milan, Italy (Tel.: 2-2 1771; FAX 2-21773; Telex: 311185)

Braunschweig AG. *See* Chemische Fabriken Oker und Braunschweig AG

Bray Health & Leisure Ltd.
25 Park Road, Faringdon, Oxfordshire, SN7 7BP, UK (Tel.: 0367-240736; FAX 0367-242625; Telex: 445784 KEMSUN)

Breamhurst Ltd. *Address unknown*

Brent Chemicals International plc
Ridgeway, Ivor, Buckinghamshire, SL0 9JJ, UK (Tel.: 0753 651812; FAX 0753 652460; Telex: 847693 BRENT G)

Bretagne Chimie Fine SA, Div. of Guyomarc'h, Diana Div.
Boisel, F-56140 Pleucadeuc, France (Tel.: 97 26 91 21; FAX 97 26 90 46; Telex: 951084 BCFF)

Brian Jones & Associates Ltd. *See* Jones, Brian & Associates Ltd.

Bridge Pharmaceuticals, Div. of SmithKline-Beecham plc
Welwyn Garden City, Herts, AL7 IEY, UK

Bridgestone. *See* Firestone Synthetic Rubber

James Briggs Ltd.
Salmon Fields, Royton, Oldham, Lancashire, OL2 6HZ, UK (Tel.: 061-627 0101; FAX 061-627 0971)

R. F. Bright Enterprises Ltd.
Enterprise House, London Road, West Kingsdown, Sevenoaks, Kent, TN15 6AP, UK (Tel.: 0474-852852; FAX 0474 853944)

Bristol Laboratories
PO Box 4500755, Princeton, NJ 08543-4500 USA (Tel.: 609-897-2000; FAX 315-432-4804)

Bristol-Myers Squibb
Bristol-Myers Squibb Co. Inc., 345 Park Avenue, New York, NY 10154-0037 USA (Tel.: 212-546-4000; FAX 212-546-5664)

Bristol-Myers Co. Ltd., Swakeleys House, Milton Rd, Ickenham, Uxbridge, UB10 8NS, UK (Tel.: 0895 639911; FAX 0895 636975; Telex: 925374)

Bristol-Myers Squibb Pharmaceuticals Ltd., Div. of Bristol-Myers Squibb Co, BMS House, 141-149 Staines Road, Hounslow, Middlesex, TW3 3JA, UK (Tel.: 081-572-7422; FAX 081-577-1756; Telex: 261537)

BritAg Industries Ltd., Div. of ICI plc
Waterfront House, Skeldergate Bridge, York, North Yorkshire, YO1 1DR, UK (Tel.: 0904-611800; FAX 0904-627473; Telex: 57827 BRITAG G)

British Celanese. *See* Hoechst UK Ltd.

British Cellophane. *Address unknown*

British Chrome & Chemicals Ltd., A Div. of Harcros Chemicals UK Ltd.
Urlay Nook, Eaglescliffe, Stockton-on-Tees, Cleveland, TS16 0QG, UK (Tel.: 0642-787755/781937; FAX 0642-781935; Telex: 58363 Dicrom G)

British Drug Houses. *See* Merck Sharp & Dohme Ltd.

British Geon. *See* BP Chemicals Ltd.

British Nova Works Ltd.
57-61 Lea Road, Southall, Middlesex, UB2 5QB, UK (Tel.: 081 574 6531; FAX 081 571 7572)

British Solvent Oils, Subsidiary of Carless Refining & Marketing Co
Metcalf Drive, Altham Ind. Est., Altham, Accrington, Lancashire, BB5 5TU, UK (Tel.: 0282-79112; FAX 0282-78660; Telex: 63300)

British Traders & Shippers Ltd., Div. of Linton Park plc
6-7 Merrielands Crescent, Dagenham, Essex, RM9 6SL, UK (Tel.: 081-595 4211; FAX 081-593 0933; Telex: 897438)

British Wax Refining Co. Ltd.
29 St John's Rd, Redhill, Surrey, RH1 6DT, UK (Tel.: 0737 761242; FAX 0737 761472)

Brocades Pharma
Brocades House, Pyrford Road, West Byfleet, Surrey, KT14 6RA, UK (Tel.: 0932-355535/ 342291; FAX 0932-353458)

Brockhues AG. *See* Chemische Werke Brockhues AG

Broemmel Pharmaceuticals, Div. of Riker Laboratories Inc.
19901 Nordhoff St, Northridge, CA 91324 USA

Bromhead & Denison Ltd., Div. of M. W. Hardy & Co (Holdings) Ltd.
7 Stonebank, Welwyn Garden City, Hertfordshire, AL8 6NQ, UK (Tel.: 0707-331031; FAX 0707- 325012; Telex: 261886 BROMDN G)

Bromine & Chemicals Ltd., Div. of Dead Sea Bromine
6 Arlington Street, St James's, London, SW1A 1RE, UK (Tel.: 071-493-9711; FAX 071-493- 9714; Telex: 23845)

Brooks Industries Inc.
70 Tyler Place, South Plainfield, NJ 07080 USA (Tel.: 908-561-5200; FAX 908-561-9174)

Broomchemie BV. *See* Eurobrom

Brotherton Specialty Products Ltd.
Calder Vale Rd, Wakefield, West Yorkshire, WF1 5PH, UK (Tel.: 0924 371919; FAX 0924 290408; Telex: 556320 BROKEM G)

Brown Butlin Ltd.
Brook House, Ruskington, Sleaford, Lincs, NG34 9EP, UK (Tel.: 0526-832771; FAX 0526- 832967)

James M. Brown Ltd., Div. of Tennants Consolidated Ltd.
Boving Works, Napier Street, Fenton, Stoke on Trent, Staffordshire, ST4 4NX, UK (Tel.: 0782 744171; FAX 0782 744473; Telex: 367381 JMBRUN G)

Brunner Mond & Co. Ltd.
PO Box 4, Mond House, Northwich, Cheshire, CW8 4DT, UK (Tel.: 0606 724365; FAX 0606 781353)

Buckeye Cellulose Corp. *See* Procter & Gamble Co.

Buckton Scott Ltd.
Black Horse House, Bentalls, Pipps Hill Estate, Basildon, Essex, SS14 3BX, UK (Tel.: 0268 531308; FAX 0268 531316; Telex: 995923)

Building Adhesives Ltd.
Longston Road, Trentham, Stoke-on-Trent, ST4 8JB, UK (Tel.: 0782-659921; FAX 0782-643909)

Burgess Pigment Co.
PO Box 349, Sandersville, GA 31082 USA (Tel.: 912-552-2544; 800-841-8999; FAX 912-552-1772; Telex: 804523)

Burlington Bio-Medical Corp.
222 Sherwood Ave., Farmingdale, NY 11735-1718 USA (Tel.: 516-694-9000; FAX 516-694-9177; Telex: 6852289)

Burlington Chemical Co., Inc.
PO Box 111, Burlington, NC 27216 USA (Tel.: 919-584-0111; 800-334-8550; FAX 919-584-3548; Telex: 9102502503)

Burmah-Castrol Chemicals Ltd., Div. of Burmah-Castrol plc
Burmah Castrol House, Pipers Way, Swindon, Wiltshire, SN3 1RE, UK (Tel.: 0793-511521; FAX 0793-612524; Telex: 449221)

Burroughs Wellcome Co.
3030 Cornwallis Rd, Research Triangle Park, NC 27709 USA (Tel.: 919-248-3000; 800-722-9292; FAX 919-248-8375)

Burts & Harvey
Crabtree Manorway North, Belvedere, Kent, DA17 6BQ, UK

Bush Beach Ltd., Div. of Degussa AG
Paul Ungerer House, Earl Rd, Stanley Green, Handforth, Wilmslow, Cheshire, SK9 3RL, UK (Tel.: 061-485-8231; FAX 061-485-6445; Telex: 669583 BBLMCH G)

Bush Boake Allen Ltd.
Blackhorse Lane, Walthamstow, London, E17 5QP, UK (Tel.: 081-531-4211; FAX 081-527-2360)

Butese. *Address unknown*
Italy

BV Nekami. *See* Nekami Maasgroeven BV

BXL Plastics Ltd.
Huddersfield Rd., Darton, Barnsley, South Yorkshire, S75 5NA, UK (Tel.: 0226-390039; FAX 0226-390125; Telex: 54157)

Byk-Chemie

Byk Chemie GmbH, Div. of Altana Industrie Aktien und Anlagen AG, Abelstrasse 14, W-4230 Wesel, Germany (Tel.: 281-6700; FAX 281-65735; Telex: 812772)

Byk Gulden Lomberg Chem Fabrik GmbH, Div. of Altana Industrie Aktien und Anlagen AG, Postfach 6500, Byk Guldenstrasse 2, W 7750 Konstanz, Germany (Tel.: 7531-840; FAX 7531-84474; Telex: 733354 BYK D)

Byk-Chemie USA, 524 S. Cherry St., PO Box 5670, Wallingford, CT 06492-7656 USA (Tel.: 203-265-2086; FAX 203-284-9158; Telex: 643378)

Cabot Corp.

Cabot Corp./Cab-O-Sil Div., PO Box 188, Tuscola, IL 61953 USA (Tel.: 217-253-3370; 800-222-6745; FAX 217-253-4334; Telex: 910-663-2542)

Cabot GmbH/Cab-O-Sil Div., PO Box 1766, D-7888 Rheinfelden, Germany (Tel.: 7623-9090; FAX 7623-90932; Telex: 773451)

Cabot Carbon Ltd., Div. of Cabot Corp., Lees Lane, Stanlow, South Wirral, Merseyside, L65 4HT, UK (Tel.: 051-355 36 77; FAX 051-356-07 12; Telex: 629261 CABLAK G)

Cabot Plastics Ltd., Gate St., Dunkinfield, SK16 4RU, UK (Tel.: 061-330-5051; FAX 061-308-2641)

Cabot Plastics International, Les Pléiades A3, Avenue des Pléiades 11, B-1200 Brussels, Belgium (Tel.: 2-774 1411; FAX 02 771 2362; Telex: 20278)

Cabot Plastics Italiana SpA, Z.I. 38055 Grigno (TN), Italy

Cabot Plastics Hong Kong Ltd., 914 Sun Plaza, 28 Canton Rd., Tsim Sha Tsui, Kowloon, Hong Kong

Cadillac Plastic & Chem. Co.

Cadillac Plastic & Chem. Co., 143 Indusco Court, PO Box 7035, Troy, MI 48007-7035 USA (Tel.: 313-583-1200; 800-488-1200; Telex: 810-232-1886)

Cadillac Plastic (Canada) Ltd., 91 Kelfield St., Rexdale (Toronto), Ontario, M9W 5A4, Canada (Tel.: 416-249-8311; FAX 416-249-0148; Telex: 06-989-322)

Cadillac Plastic Ltd., Rivermead Drive, Westlea, Swindon, Wiltshire, SN5 7YT, UK (Tel.: 0793 514 949; FAX 0793 511 762; Telex: 449537)

Cadillac Plastic GmbH, Alfred-Nobel-Strasse 17, D-6806 Viernheim, Germany (Tel.: 011-49-6204-70960; FAX 011-49-6204-1050; Telex: 841-4-67402)

Cadillac Plastics Pacific Group Pty. Ltd., PO Box 145, Ermington, Sydney, NSW 2115, Australia (Tel.: 2-748-3722; Telex: 790-126-743)

Caffaro SpA

Via Privata Vasto No 1, I-20121 Milan, Italy (Tel.: 2 62011; FAX 2-6201230; Telex: 312065 CFFARO I)

Cairn Chemicals Ltd.

Cairn House, Elgiva Lane, Chesham, Buckinghamshire, HP5 2JD, UK (Tel.: 0494 786066; FAX 0494 791816; Telex: 837075)

Calaire Chimie SNC
23-25 Ave. Morane Saulnier, F-92366 Meudon La Foret, France (Tel.: 46 31 48 55; FAX 46 31 51 90)

Calbiochem-Behring Corp.
10933 N Torrey Pines Rd, La Jolla, CA 92307 USA

Calbiochem Corp.
PO Box 12087, San Diego, CA 92119 USA

Calder Colours (Ashby) Ltd.
Nottingham Road, Ashby-de-la-Zouch, Leics., LE6 5DR, UK (Tel.: 0530-412885; FAX 0530-417315; Telex: Calder 341313)

Calgene Chemical Inc.
602 E. Algonquin Rd., Des Plaines, IL 60016 USA (Tel.: 708-298-4000; 800-432-7187; FAX 708-298-1519; Telex: 206168)

Calgon Corp.
PO Box 1346, Pittsburgh, PA 15230 USA (Tel.: 412-777-8000; FAX 412-777-8927)

Calgon Carbon Corp.
PO Box 717, Pittsburgh, PA 15230-0717 USA (Tel.: 412-787-6700; 800-422-7266; FAX 412-787-6713; Telex: 6711837 CCC)

Callaway Chemical Co.
6003 Hamilton Rd., PO Box 2335, Columbus, GA 31993-3599 USA (Tel.: 706-576-2000; FAX 706-576-6455)

Callery Chemical Co., Div. of Mine Safety Appliances Co.
PO Box 429, Pittsburgh, PA 15230 USA (Tel.: 412-967-4141; FAX 412-967-4140; Telex: ITT 4900010457)

Cal Polymers, Inc.
2115 Gaylord St., Long Beach, CA 90813 USA (Tel.: 213-436-7372)

Calumet Lubricants Co.
2780 Waterfront Pkwy. E. Drive, Suite 200, Indianapolis, IN 46214 USA (Tel.: 317-328-5660; FAX 317-328-5668)

Cambridge Animal & Public Health Ltd., Subsidiary of Schering Agrochemicals Ltd.
Hauxton, Cambridge, Cambridgeshire, CB2 5HV, UK (Tel.: 0223 870312; FAX 0223 872 142; Telex: 81654)

Camie-Campbell, Inc.
9225 Watson Industrial Park, St. Louis, MO 63126-1581 USA (Tel.: 314-968-3222; 800-325-9572; FAX 314-968-0741)

Campana Corp. *Address unverified*
Batavia, IL USA

Campbell Williams & Co.
14 St Neots Road, Abbotsley, Huntingdon, Cambridgeshire, PE19 4UU, UK (Tel.: 0767-7500)

Camtex Fabrics Ltd., Subsidiary of ICI Chemicals & Polymers
Blackwood Road, Lilyhall, Workington, Cumbria, CA14 4JJ, UK (Tel.: 0900 602646; FAX 0900 66827; Telex: 64109)

Cancarb Ltd.
1702 Brier Park Cresc. N.W., PO Box 310, Medicine Hat, Alberta, T1A 7G1, Canada (Tel.: 403-527-1121; FAX 403-529-6093; Telex: 03-824866)

Cape Insulation Cape Asbestos Co. *See* Charter Consolidated plc

Caramba Chemie GmbH
Wanheimer Strasse 334/6, D-4100 Duisburg 1, Germany

The Carborundum Co., Subsidiary of of BP America Inc.
PO Box 156, 345 3rd St, Niagara Falls, NY 14302 USA (Tel.: 716-278-2000; FAX 716-278-2900)

Carbo-Tech GmbH, Div. of Bergwerksverband GmbH
Postfach 130140, Franz-Fischer-Weg 61, W-4300 Essen 13, Germany (Tel.: 201-17204; FAX 201-1721260)

Carboxyl Chemicals Ltd.
Coburg Dock, Queens Gate, Liverpool, L3 4AP, UK (Tel.: 051-708 6929; FAX 051-708 6929)

Cardolite Corp.
500 Doremus Ave., Newark, NJ 07105-4805 USA (Tel.: 201-344-5015; 800-322-7365; FAX 201-344-1197; Telex: 325446)

Cargill, Inc.
Box 5630, Minneapolis, MN 55440 USA (Tel.: 612-475-6478; Telex: CGL MPS 290625)

Cargo Fleet Chemical Co Ltd.
Eaglescliffe Industrial Estate, Eaglescliffe, Stockton, Cleveland, UK

Carless Refining & Marketing Ltd., Div. of Repsol SA
St. James House, Eastern Road, Romford, Essex, RM1 3NL, UK (Tel.: 0708-755557; FAX 0708-753890; Telex: 261071 CRAM G)

Carless Solvents Ltd. *See* Carless Refining & Marketing Ltd.

R H Carlson Co./Northern Labs
41 Chestnut St., Greenwich, CT 06830 USA (Tel.: 203-531-5500; 800-243-5404; FAX 203-531-0032)

Carlton Laboratories (UK) Ltd.
4 Manor Parade, Salvington Road, Durrington, Worthing, W.Sussex, BN13 2JP, UK

G. W. Carnrick Co Ltd. *Address unknown*

Carpenter Technology
PO Box 14662, 101 W. Bern St., Reading, PA 19612-4662 USA (Tel.: 215-371-2000)

R. E. Carroll, Inc.
1570 N. Olden Ave., Trenton, NJ 08638 USA (Tel.: 609-695-6211; 800-257-9365; FAX 609-695-0102)

Carter-Wallace Ltd.
Wear Bay Road, Folkestone, Kent, CT19 6PG, UK (Tel.: 0303-850661)

CasChem Inc.
40 Ave. A, Bayonne, NJ 07002 USA (Tel.: 201-858-7900; 800-CASCHEM; FAX 201-437-2728; Telex: 710-729-4466)

Cassella AG, Div. of Hoechst AG
Hanauer Landstrasse 526, W-6000 Frankfurt am Main 61, Germany (Tel.: 069-410901; FAX 069-41092100; Telex: 411208 cass d)

Caswell & Co Ltd.
Chelsea Works, St Michaels Road, Kettering, Northamtonshire, NN15 6AU, UK (Tel.: 0536-518340; FAX 0536-310059; Telex: 312305 HONDEL G)

Catalyst Resources, Inc.
2190 North Loop West, Suite 400, Houston, TX 77018 USA (Tel.: 713-682-5300; FAX 713-957-6839; Telex: 910-881-7119)

Catalysts & Chemicals Industries Co., Ltd.
Nippon Bldg., 6-2, Ohte-machi 2-chome, Chiyoda-ku, Tokyo, 100, Japan (Tel.: (03) 3270-6086; FAX (03) 3246-0617; Telex: 2223480 PETCATJ)

Catawba-Charlab, Inc.
5046 Pineville Rd., PO Box 240497, Charlotte, NC 240497 USA (Tel.: 704-523-4242; FAX 704-522-8142)

Catomance Ltd.
96 Bridge Rd East, Welsyn Garden City, Hertfordshire, AL7 1JW, UK (Tel.: 0707 324373; FAX 0707 372191; Telex: 267418 CATAC G)

CDA Chemicals Ltd.
The Granary, Billingborough, Sleaford, Lincs, NG34 0QS, UK (Tel.: 0529-240456; FAX 0529-240053)

CDF Chimie Nederland BV
Locatellikade 1, NL-1076 AZ Amsterdam, The Netherlands (Tel.: 20-664 74 11; Telex: 16099)

CdF Chimie. *See* Noroxo

Ceca. *See under* Atochem

Celanese Corp. *See* Hoechst Celanese Corp.

Celite Corp.

Celite Corp., PO Box 519, Lompoc, CA 93438-0519 USA (Tel.: 805-735-7791; FAX 805-735-5699; Telex: 62776493 ESL UD)

Celite (UK) Ltd., Livingston Rd., Hessle, North Humberside, HU13 OEG, UK (Tel.: 482 64 52 65; FAX 482 64 11 76; Telex: 592160)

Manville Great Britain Ltd., Regal House, 1st Floor, London Rd., Twickenham, Middlesex, TW13QE, UK (Tel.: 81-891-0813; FAX 81 892-9325; Telex: 928635 MANVIL G)

Celite Corp., 9 rue du Colonel de Rochebrune, BP 240, F-92500 Rueil Malmaison, France (Tel.: 1-47 49 05 60; FAX 1-47 08 30 25; Telex: 631969 CELITE F)

Celite Italiana srl, Viale Pasubio No. 6, I-20154 Milano, Italy (Tel.: 654531; FAX 39 2 29005439; Telex: 311136 MANVII)

Manville Japan Ltd., Toranomon Asahi Bldg., 10F, 1-11-3, Nishi-shimbashi, Minato-ku, Tokyo, 105, Japan (Tel.: (03)580-3791; FAX (03) 580-3793; Telex: 2222487 MVL J)

Celite Pacific, 2nd Floor, Shui On Centre, 8 Harbor Road, Hong Kong (Tel.: 582 5609; FAX 802 4275)

Celtic Chemicals Ltd.

Gas Works Industrial Estate, Victoria Rd, Port Talbot, West Glamorgan, SA12 6DB, UK (Tel.: 0639 886236; FAX 0639 893147; Telex: 94013871 CCPT G)

Centerchem Inc.

225 High Ridge Rd, Stamford, CT 06905-3036 USA (Tel.: 203-975-9800; FAX 203-975-8777; Telex: 233322 UW)

Central Soya

Central Soya Co., Inc./Chemurgy Div., PO Box 2507, Fort Wayne, IN 46801-2507 USA (Tel.: 219-425-5100; 800-348-0960; FAX 219-425-5301; Telex: 49609682)

Central Soya, PO Box 5063, 3008 AB Rotterdam, Netherlands (Tel.: 10-42-39-600; FAX 10-42-30-897; Telex: 20041 CNSOY NL)

Stern Lecithin & Soja GmbH & Co. KG, Div. of Central Soya, An der Alster 81, D-2000 Hamburg 1, Germany (Tel.: 40/2 48 30-01; FAX 40/280 34 27; Telex: 2 162 052 star d)

Century Pharmaceuticals Inc.

10377 Hague Road, Indianapolis, IN 46256 USA

Cerac, Inc.

PO Box 1178, 407 N. 13th St., Milwaukee, WI 53201 USA (Tel.: 414-289-9800; FAX 414-289-9805; Telex: RCA 286122)

Cerestar UK Ltd.

Trafford Park Rd, Trafford Park, Manchester, Lancashire, M17 1PA, UK (Tel.: 061-872 5959; FAX 061-848 9034; Telex: 667022)

Certified Laboratories Ltd.
PO Box 70, Oldbury, Warley, W. Midlands, B69 4AD, UK (Tel.: 021-525-6678; FAX 021-500-5386)

CFPI, Compagnie Francaise de Produits Industriels
BP No 75, 28 Boulevard Camelinet, F-92233 Gennevilliers, France (Tel.: (1) 40 85 5050; FAX (1) 47 92 2545; Telex: 610626 F)

Champlain Industries Inc.
25 Styertowne Rd., Clifton, NJ 07012 USA (Tel.: 201-778-4900; 800-222-4904; Telex: 3725769)

Charter Consolidated plc
40 Holborn Viaduct, London, EC1P 1AJ, UK (Tel.: 071-353-1545; FAX 071-583 2950; Telex: 929582)

Chemax, Inc.
PO Box 6067, Highway 25 South, Greenville, SC 29606 USA (Tel.: 803-277-7000; 800-334-6234; FAX 803-277-7807; Telex: 570412 IPM15SC)

Chemcentral Corp.
7050 W 71 St, PO Box 730, Bedford Park, IL 60499-0730 USA (Tel.: 708-594-7000; 800-331-6174; FAX 708-594-7022)

Chemetall GmbH
Reuterweg 14, Postfach 10 15 01, D-6000 Frankfurt am Main 1, Germany (Tel.: 069-159-0; FAX 069-159-3018; Telex: 41225-0 mgf)

Chemetals Inc.
711 Pittman Rd., Baltimore, MD 21226 USA (Tel.: 410-636-7100; 800-876-3464; FAX 410-636-7113)

Chemfax Inc.
3 Rivers Rd., PO Box 1390, Gulfport, MS, 39502 USA (Tel.: 601-863-6511; FAX 601-863-6547)

Chemical Combine
6005 Stara Zagora, SHK, Bulgaria

The Chemical Co.
PO Box 436, 19 Narragansett Ave, Jamestown, RI 02835 USA (Tel.: 401-423-3100; FAX 401-423-3102)

Chemical Dynamics Corp.
3001 Hadley Road, South Plainfield, NJ 07080 USA

The Chemical & Insulating Co. Ltd.
West Auckland Road, Darlington, Co Durham, DL3 0UR, UK

Chemicals International plc. *Address unknown*

Chemical Spraying Co. Ltd.

Unit 5, Henson Rd, Yarm Road Industrial Estate, Darlington, Co Durham, DL1 4QD, UK (Tel.: 0-325 380637)

Chemie Linz

Chemie Linz GmbH, Div. of Chemie Holding AG, ST Peter-Strasse 25, A-4021 Linz, Austria (Tel.: 732-59160; FAX 732-5916155; TElex: 21324 LINZ A)

Chemie Linz UK Ltd., Div. of Chemie Linz GmbH, 12 The Green, Richmond, Surrey, TW9 1PX, UK (Tel.: 081-948 6966; FAX 081-332 2516; Telex: 924941)

Chemie Linz North America, Inc., 65 Challenger Rd, Ridgefield Park, NJ 07660 USA (Tel.: 201-641-6410; FAX 201-641-2323; Telex: 853211 Chemie Linz)

Cheminova Agro A/S

PO Box 9, DK-7620 Lemvig, Denmark (Tel.: 9783 41 00; FAX 9783 45 55)

Chemische Fabriek Servo BV. *See* Servo Delden BV

Chemische Fabrik GmbH & Co.

Postfach 1260, Hauptstrasse 4, W-6719 Kleinkarlbach, Germany (Tel.: 6359-8010; FAX 6359-801209; Telex: 451244 SPIES D)

Chemische Fabriken Oker und Braunschweig AG

Postfach 1328, IM Schleeke 77, DW-3380 Goslar, Germany (Tel.: 05321-7510; FAX 05321-751 192)

Chemische Fabrik Weyl GmbH. *See* Weyl GmbH

Chemische Werke Brockhues AG

Postfach 60, Muhlstrasse 118, Walluf, Germany (Tel.: 06123 71071; FAX 06123 72336)

Chemische Werke Hüls. *See* Hüls AG

Chemisphere Ltd.

38 King St., Chester, Cheshire, CH1 2AH, UK (Tel.: 0244 320878; FAX 0244 320858; Telex: 61398 CHEMSPR G)

Chemlease Inc.

PO Box 386, Groveport, OH 43125 USA (Tel.: 614-836-2331; FAX 614-836-2868)

Chemoxal SA

75 quai d'Orsay, F-75321 Paris Cedex 7, France (Tel.: 40 62 55 55; FAX 45 55 03 89; Telex: 270779)

Chemoxy International

Chemoxy International plc, Div. of Suter plc, All Saints Refinery, Cargo Fleet Rd, Middlesbrough, Cleveland, TS3 6AF, UK (Tel.: 0642 248555; FAX 0642 244340; Telex: 587185)

Chemoxy International Ltd., Div. of Chemoxy International plc, All Saints Refinery, Cargo Fleet Road, Middlesbrough, Cleveland, TS3 6AF, UK (Tel.: 0642-248555; FAX 0642-244340; Telex: 587185 Cemint)

Chemplex Chemicals, Inc.

201 Route 17 #300, Rutherford, NJ 07070 USA (Tel.: 201-935-8903; FAX 201-935-9051)

Chemron Corp.

PO Box 2299, Paso Robles, CA 93447 USA (Tel.: 805-239-1550; FAX 805-239-8551)

Chemsal Chemicals & Co. KG, Joint venture of Chem-Y GmbH and Salim

Kupferstrasse 1, PO Box 100262, D-4240 Emmerich 1, Germany (Tel.: 02822 711-0; FAX 02822 18294; Telex: 8125124)

Chemsearch (UK) Ltd.

Landchard House, Victoria Street, West Bromwich, W. Midlands, B70 8ER, UK (Tel.: 021-525-1666; FAX 021-500-5386)

Chemseco

4800 Blue Parkway, Kansas City, MO 64130 USA

Chemson

Chemson GmbH, Post Box 12, A-9601 Arnoldstein, Austria (Tel.: 04255 2226; FAX 04255 2435; Telex: 45596)

Chemson Ltd., Hayhole Works, Willington Quay, Wallsend, Tyne and Wear, NE28 0PB, UK (Tel.: 091 258 5892; FAX 091 258 1549; Telex: 537726)

Chemson Polymer Additiv GmbH, Postfach 43, A 9601 Arnoldstein, Austria (Tel.: 04255 222 6324; FAX 04255 222 6350)

Chemtech (Crop Protection) Ltd.

The Arable Centre, Winterbourne Monkton, Swindon, Wilts, SN4 9NW, UK (Tel.: 06723-591)

Chem-Trend A/S

Smedeland 14, PO Box 13 84, DK-2600 Glostrup, Denmark (Tel.: 42 45 6711; FAX 43 63 03 50; Telex: 33187 CMTREND DK)

Chem-Trend Inc.

1445 W. McPherson Park Dr., PO Box 860, Howell, MI 48844-0860 USA (Tel.: 517-546-4520; 800-248-4056)

Chemtronics Inc.

8125 Cobb Center Dr., Kennesaw, GA 30144 USA (Tel.: 800-645-5244; FAX 404-424-4267; Telex: 968567)

Chem-Y

Chem-Y Fabriek Van Chemische Producten B.V., PO Box 50, 2410 AB, Bodegraven, Netherlands (Tel.: 028-22/711-0; FAX 49282218294; Telex: 8125 124)

Chem-Y GmbH, Kupferstrasse 1, D4240 Emmerich, Germany (Tel.: 2822/7110; FAX 2822/ 18294; Telex: 8125124)

Chevron Chemical Co.

PO Box 3766, Houston, TX 77253 USA (Tel.: 713-754-2000; 800-231-3260; Telex: 762799)

Chiltern Farm Chemicals Ltd.

11 High Street, Thornborough, Buckingham, Buckinghamshire, MK18 2DF, UK (Tel.: 0280-822400; FAX 0280-822082)

Chipman Ltd.

Portland Building, Portland Street, Staple Hill, Bristol, BS16 4PS, UK (Tel.: 0272-574574; FAX 0272-563461)

Chisso

Chisso Corp., Tokyo Bldg., 7-3, Marunouchi 2-chome, Chiyoda-ku, Tokyo, 100, Japan (Tel.: (03) 3284-8411; Telex: 02225212 CHISSO J)

Chisso America Inc., 1185 Ave of the Americas, New York, NY 10036 USA (Tel.: 212-302-0500; FAX 212-302-0643; Telex: WU 147029 CHISSO NYK)

Chlor Chem Ltd.

Site 7, Kidderminster Rd., Cutnall, Gren, Droitwich, Worcestershire, WR9 0NS, UK (Tel.: 0299 23561)

Church & Dwight Co. Inc./Specialty Prods. Div.

Box CN5297, 469 N. Harrison St., Princeton, NJ 08543-5297 USA (Tel.: 609-497-7116; 800-221-0453; FAX 609-497-7176; Telex: 752226)

Ciba-Geigy

Ciba-Geigy AG, CH-4002, Basel, Switzerland (Tel.: 061 223 1341; FAX 061 231 7422; Telex: 668083)

Ciba-Geigy Marienberg GmbH, Div. of Ciba-Geigy AG, Postfach 1253, D-6140 Bensheim 1, Germany (Tel.: 6254 79 0; FAX 6254 79505; Telex: 468141)

Ciba plc, Div. of Ciba-Geigy AG, Hulley Rd, Macclesfield, Cheshire, SK10 2NX, UK (Tel.: 0625-421933; FAX 0625-619637; Telex: 667336)

Ciba Additives, Div. of Ciba-Geigy AG, Tenax Road, Trafford Park, Manchester, Lancashire, M17 1WT, UK (Tel.: 061-872-2323; FAX 061-873 7271; Telex: 666177)

Ciba Agrochemicals, Div. of Ciba-Geigy AG, Whittlesford, Cambridge, Cambridgeshire, CB2 4QT, UK (Tel.: 0223-833621; FAX 0223-835211; Telex: 81325)

Ciba Dyestuffs & Chemicals (UK), Ashton New Road, Clayton, Manchester, M11 4AR, UK (Tel.: 061-223-1341; FAX 061-223 0944)

Ciba Pharmaceuticals, Wimblehurst Road, Horsham, W Sussex, RH12 4AB, UK (Tel.: 0403-5 01 01; FAX 0403- 5 66 43)

Ciba Plastics (UK), Duxford, Cambridge, CB2 4QA, UK (Tel.: 0223-832121; FAX 0223-838404; Telex: 0223-81101)

Ciba-Geigy Corp., 444 Saw Mill River Rd., Ardsley, NY 10502 USA (Tel.: 914-478-3131; 800-431-1874; FAX 914-478-3480)

Ciba-Geigy Corp./Additives Div., Seven Skyline Dr., Hawthorne, NY 10532-2188 USA (Tel.: 914-785-4461; 800-431-1900)

Ciba-Geigy Corp./Dyestuffs & Chemicals Div., PO Box 18300, Greensboro, NC 27419 USA (Tel.: 800-334-9481; FAX 919-632-7098)

Ciba-Geigy Corp./Furane Aerospace Prods., 5121 San Fernando Road West, Los Angeles, CA 90039-1071 USA (Tel.: 818-247-6210; FAX 818-507-0167)

Ciba-Geigy Pharmaceutical Co., 556 Morris Ave., Summit, NJ 07901 USA (Tel.: 908-277-5000)

Ciba-Geigy Corp./Plastics Div., Seven Skyline Dr., Hawthorne, NY 10532 USA (Tel.: 914-347-6600; 800-222-1906)

Ciba-Geigy (Japan) Ltd., 10-66, Miyuki-cho, Takarazuka-shi, Hyogo, 665, Japan (Tel.: (0797) 74-2472; FAX (0797) 74-2472; Telex: 5645684 CIGYTZ J)

Cilag AG

Hochstrasse 201-209, CH-8201 Schaffhausen, Switzerland (Tel.: 53-82 91 11; FAX 53-82 94 43)

Cincinnati Milacron Chemicals Inc. *See* Morton International

Cindu Chemicals BV, Div. of Cindu BV

Hoogovens Groep BV, Postbus 9, NL-1420 AA Uithoorn, The Netherlands (Tel.: 02975-4 55 33; FAX 02975-4 52 77; Telex: 12215)

Clark Chemical Inc.

25 Trammel St., Marietta, GA 30064 USA (Tel.: 404-514-8909; FAX 404-514-8906)

Clayton Aniline Co Ltd., Div. of Ciba-Geigy AG

PO Box 2, Ashton New Road, Clayton, Manchester, Lancashire, M11 4AP, UK (Tel.: 061-223-1391; FAX 061-223-4315; Telex: 668808)

Cleanacres Ltd.

Adoversford, Cheltenham, Glos., GL54 4LZ, UK (Tel.: 0242-820481; FAX 0242-820807)

W. A. Cleary Chemical Corp.

Southview Industrial Park, 178 Route #522 Suite A, Dayton, NJ 08810 USA (Tel.: 908-329-8399; 800-524-1662; FAX 908-274-0894)

Clevite Corp. *Address unverified*

Cleveland, OH USA

Clifton Chemicals Ltd.

119 Grenville Street, Edgeley, Stockport, Cheshire, SK3 9EU, UK (Tel.: 061-476-1128; FAX 061-476-1280)

Climax Performance

Climax Performance Materials Corp./Corporate Headquarters, 101 Merritt 7 Corporate Park, PO Box 5113, Norwalk, CT 06856-5113 USA (Tel.: 203-845-2951; 800-323-3231; FAX 203-845-2953)

Climax Molybdenum Co., Div. of Climax Performance, 1370 Washington Pike, Bridgeville, PA 15017 USA (Tel.: 412-257-1560; FAX 412-257-0540)

Climax Fluids Additives, Div. of Climax Performance, 7666 West 63rd St., Summit, IL 60501 USA (Tel.: 708-458-8450; FAX 708-458-0286)

Climax Molybdenum UK Ltd., Div. of Amax Inc., 50-52 Great Eastern St., London, EC2A 3EP, UK (Tel.: 071 739 6422; FAX 071 729 1964; Telex: 27316)

Climax Molybdenum BV, Div. of Amax Holdings BV, Postbus 1130, NL-3180 AC Rozenburg, The Netherlands (Tel.: 15933; Telex: 23673 MOLY NL)

Amax Asia Ltd., Fuji Bldg., Room 311, 2-3 Marunouchi 3-Chome, Chiyoda-Ku, Tokyo, 100, Japan (Tel.: (03) 3-201-2911; FAX (03) 3-201-2912)

Clough Chemical Co., Ltd.

178 St. Pierre, PO Box 1017, St-Jean-sur-Richelieu, Quebec, J3B 7B5, Canada (Tel.: 514-346-6848; 800-363-9284; FAX 514-346-7263)

CNC International, Limited Partnership

PO Box 3000, Woonsocket, RI 02895 USA (Tel.: 401-769-6100; FAX 401-769-4509)

Coalite Chemicals Div.

PO Box 152, Buttermilk Lane, Bolsover, Chesterfield, Derbyshire, S44 6AZ, UK (Tel.: 0246 826816; FAX 0246 240309; Telex: 547624)

Coalite Fuels & Chemicals Ltd. See Coalite Chemicals Div.

Coal Products Ltd., Div. of British Coal

PO Box 16, Mill Lane, Wingerworth, Chesterfield, Derbyshire, S42 6JJ, UK (Tel.: 0246 277001; FAX 0246 212212; Telex: 547061)

Coates Coatings International, Div. of Total CFP

Station Lane, Witney, Oxfordshire, OX8 6XZ, UK (Tel.: 0993-702969; FAX 0993 775579; Telex: 497133 COTINS G)

Colebrand Ltd.

20 Warwick St, Regent St, London, WI4 6BE, UK (Tel.: 071 439 1000; FAX 071 734 3358; Telex: 261495)

Collinda Ltd.

Collinda House, 25 Ottways Lane, Ashstead, Surrey, KT21 2PZ, UK (Tel.: 0372-278416; FAX 0372-278559; Telex: 946817)

Colloids Ltd.

Dennis Road, Widnes, Cheshire, WA8 0SL, UK (Tel.: 051-424 7424-7; FAX 051-423-3553; Telex: 629164 COLOID)

Colonial Products Inc.
1830 10th Avenue North, Lake Worth, FL 33461 USA

Colorcon
Moyer Blvd., West Point, PA 19486 USA (Tel.: 215-699-7733)

Colorcon Ltd.
Murray Road, St Paul's Cray, Orpington, Kent, BR5 3QY, UK (Tel.: 0689-838301; FAX 0689-878342; Telex: 896271)

Colores Hispania SA
Josep Pla 149, 08019 Barcelona, Spain (Tel.: 3 307 13 50; FAX 3 303 2505; Telex: 54116 COHI)

Colour-Chem Ltd.
Ravindra Annexe, Dinshaw Vachha Road, 194 Churchgate Reclamation, Bombay, 400 020, India

Colourex Ltd.
Plot 7 Wimbledon Avenue, Brandon, Suffolk, IP27 0NZ, UK (Tel.: 0842-811693)

Colours Div. of Yorkshire. *See* Yorkshire Chemicals plc

Columbia Nitrogen Corp., Subsidiary of Arcadian Corp.
PO Box 1483 (13), Augusta, GA 30913 USA

Compounding Ingredients Ltd.
Unit 217, Walton Summit Centre, Bamber Bridge, Preston, Lancashire, PR5 8AL, UK (Tel.: 0772-322888; FAX 0772 315853; Telex: 677621 CIL G)

Compounding Technology Inc.
13435 Estelle St., Corona, CA 91720 USA (Tel.: 714-371-7701; 800-325-1564; FAX 714-371-7724)

Conap, Inc.
1405 Buffalo St., Olean, NY 14760 USA (Tel.: 716-372-9650; FAX 716-372-1594)

Concept Pharmaceuticals Ltd.
The Old Coach House, Amersham Hill, High Wycombe, Bucks, HP13 6NQ, UK

A & E Connock (Perfumery & Cosmetics) Ltd.
232 W. Cummings Park, Woburn, MA 01801-6330 USA

Consolidated Chemicals Ltd.
Abbey Road, The Industrial Estate, Wrexham, Clwyd, LL13 9PW, UK (Tel.: 0978 661 351; FAX 0978 661 673; Telex: 61497 CONCEM G)

Continental Pharma Inc., Div. of Monsanto Co.
Industrie Park, Roosveld 2, B-3400 Landen, Belgium (Tel.: 11-83 1313; FAX 11-83 2253; Telex: 39824 CONFAR B)

Continental Polymers Inc.

2225 East Del Amo Blvd., Compton, CA 90220 USA (Tel.: 310-637-2103; 800-441-3943; FAX 310-637-2415; Telex: 4720686 CPIINC)

Contract Chemicals (Knowsley) Ltd.

Penrhyn Road, Knowsley Industrial Park South, Prescot, Merseyside, L34 9HY, UK (Tel.: 051 548 8840; FAX 051 548 6548; Telex: 628659)

Convert, G SA

Usine du Grand Moulin, F-01101, Oyonnax, France (Tel.: 74 7746 22; FAX 74 77 83 08; Telex: 340948)

Cook Composites & Polymers Co.

PO Box 419389, Kansas City, MO 64141-6389 USA (Tel.: 816-391-6000; FAX 816-391-6122)

Cook-Waite Laboratories Inc.

90 Park Ave, New York, NY 10016 USA

Cool-Amp Conducto-Lube Co.

15834 Upper Boones Ferry Rd, Lake Oswego, OR 97035 USA (Tel.: 503-624-6426; FAX 503-624-6436)

Coopers Animal Health Ltd. *See* Coopers Pitman-Moore Europe

Coopers Pitman-Moore Europe

Crewe Hall, Western Road, Crewe, Cheshire, CW1 IYR, UK (Tel.: 0270 580131; Telex: 22134)

CooperVision Pharmaceuticals Inc. *See* Iolab Corporation

Coperbo/Companhia Permambucana de Boracha Sintética

Av. Brig. Faria Lima, 1541-17° andar-Cjs. 17L e 17M, CEP 01451-907 Sao Paulo, Brazil (Tel.: 011-815-0568; FAX 011-211-4333; Telex: (11)81-713)

Copolymer Rubber & Chemical Corp., A DSM Co.

Scenic Highway, PO Box 2591, Baton Rouge, LA 70821 USA (Tel.: 504-355-5655; 800-535-9960; FAX 504-355-8056; Telex: 586419)

Cornelius Chemical Co. Ltd., Cornelius Chemical Group Ltd.

St. James's House, 27-43 Eastern Road, Romford, Essex, RM1 3NN, UK (Tel.: 0708 722300; FAX 0708 768204; Telex: 885589 CORNEL G)

Corning

Corning Inc., 31501 Solon Road, Solon, OH 44139 USA (Tel.: 216 248 0500; Telex: 985408)

Corning Ltd., Wear Glass Works, Div. of Corning Inc., Sunderland, Tyne & Wear, SR4 6EJ, UK (Tel.: 091-567-6222; FAX 091-567-0116)

Corn Products/Unit of CPC Int'l.

6500 Archer Rd., Summit-Argo, IL 60501 USA (Tel.: 708-563-2400; FAX 708-563-6852; Telex: 708-563-6763)

Coulter Electronics Ltd.

Northwell Drive, Luton, Beds, LU3 3RH, UK (Tel.: 0582-491414; FAX 0582-490390; Telex: 825074)

Courtaulds

Courtaulds plc, Patents Dept., PO Box 111, 72 Lockhurst Lane, Coventry, CV6 5RS, UK (Tel.: 0203-688771; FAX 0203-583837)

Courtaulds Acetate, Nelson Acetate Works, Caton Rd., Lancaster, Lancashire, LA1 3PF, UK (Tel.: 0524 66111; FAX 0524 846384)

Courtaulds Advanced Materials Ltd., Div. of Courtaulds Fibres Ltd., 72 Lockhurst Lane, Coventry, West Midlands, CV6 5RS, UK (Tel.: 0203 582000; FAX 0203 687328; Telex: 312171)

Courtaulds Aerospace Ltd., Div. of Courtaulds Fibres, 72 Lockhurst Lane, Coventry, West Midlands, CV6 5RS, UK (Tel.: 0203 583067; FAX 0203 583826)

Courtaulds Chemicals, Leek, Macclesfield Road, Leek, Staffs, ST13 8UZ, UK

Courtaulds Engineering Ltd., Div. of Courtaulds Fibres, PO Box 11, Foleshill Rd, Coventry, West Midlands, CV6 5AB, UK (Tel.: 0203 688771; FAX 0203 687925; Telex: 312171)

Courtaulds Fibres Ltd., Div. of Courtaulds plc, Ash Rd, Wrexham Ind. Est., Wrexham, Clwyd, L13 9UH, UK (Tel.: 0978-661945; FAX 0978-661243; Telex: 61114)

Courtaulds Water Soluble Polymers, Div. of Courtaulds plc, P O Box 5, Spondon, Derby, Derbyshire, DE2 7BP, UK (Tel.: 0332-661422; FAX 0332-661078; Telex: 37391 CHMPLS G)

Coventry Chemicals Ltd.

Woodhams Road, Siskin Drive, Coventry, West Midlands, CV3 4FX, UK (Tel.: 0203-639739; FAX 0203-639717; Telex: 312435)

Cow Proofings Ltd. *Address unknown*

Cox Chemicals Ltd.

Overley Hill, Wellington, Telford, Shropshire, TF6 5HD, UK (Tel.: 0952 86333; FAX 0952 86207)

Cox Pharmaceuticals Ltd., Subsidiary of Hoechst UK Ltd.

Whiddon Valley, Barnstaple, Devon, EX32 8NS, UK (Tel.: 0271-75001; FAX 0271 46106)

CP International Chemicals Ltd.

Northgate House, 21 Northgate End, Bishops's Stortford, Hertfordshire, CM23 2ET, UK (Tel.: 0279 506330; FAX 0279 755873; Telex: 818423 CPCHEM G)

CPL Group Ltd.

Barrington Hall, Hatfield Broad Oak, Bishops Stortford, Hertfordshire, CM22 7LE, UK (Tel.: 027971 8573; FAX 027971 8527; Telex: 817427 CPL G)

CPS

CPS Kemi Aps, Hejreskovv, 22, 3490 Kvistagaard, Denmark (Tel.: 2 890533; FAX 42 23 80 77)

CPS Chemical Co Inc., PO Box 162, Old Bridge, NJ 08857 USA (Tel.: 908-727-3100; FAX 908-727-2260; Telex: 844532-CPSOLDB)

Cray Valley

Cray Valley Products Inc., Box 247A, Stuyvesant, NY 12173 USA (Tel.: 518-828-4383; FAX 518-828-4382)

Cray Valley Ltd., Farnborough, Kent, BR6 7EA, UK (Tel.: 0689-853311; FAX 0689-862843)

Cri-Tech, Inc.

85 Winter St., Hanover, MA 02339 USA (Tel.: 617-826-5600; FAX 617-826-5770)

Croda

Croda Adhesives Ltd., Div. of Croda International plc, Winthorpe Rd, Newark, Nottinghamshire, NG24 2AL, UK (Tel.: 0636-646711; FAX 0636-605187; Telex: 37579)

Croda Application Chemicals Ltd., Div. of Croda International plc, Cowick Hall, Snaith, Goole, North Humberside, DN14 9AA, UK (Tel.: 0405-860551; FAX 0405-860205; Telex: 57601)

Croda Chemicals Ltd., Div. of Croda International plc, Cowick Hall, Snaith Goole, North Humberside, DN14 9AA, UK (Tel.: 0405-8605551; FAX 0405-860205; Telex: 57601)

Croda Food Products Ltd., Div. of Croda International plc, Cowick Hall, Snaith, Goole, North Humberside, DN14 9AA, UK (Tel.: 0405 860551; FAX 0405 860205; Telex: 57601)

Croda Resins Ltd., Div. of Croda International plc, Crabtree Manorway South, Belvedere, Kent, DA17 6BA, UK (Tel.: 081-311 9109; FAX 081-310 9878; Telex: 896384)

Croda Surfactants Ltd., Cowick Hall, Snaith, Goole, North Humberside, DN14 9AA, UK (Tel.: 0405 860551; FAX 0405 860205; Telex: 57601)

Croda Universal Ltd., Div. of Croda International plc, Cowick Hall, Snaith, Goole, North Humberside, DN14 9AA, UK (Tel.: 0405 860551; FAX 0405 860205; Telex: 57601)

Croda Italiana Srl, Via Grocco, N917 27036, Mortara (PV), Italy

Croda Inc., 7 Century Dr., Parsippany, NJ 07054-4698 USA (Tel.: 201-644-4900; FAX 201-644-9222)

Croda Canada Ltd., 78 Tisdale Ave., Toronto, Ontario, M4A 1Y7, Canada (Tel.: 416-751-3571; FAX 416-751-9611)

Croda Japan KK, Aceman Bldg., 1-10, Tokuicho 1-chome, Chuo-ku, Osaka, 540, Japan (Tel.: (06) 942-1791; FAX (06) 942-1790; Telex: 5233117)

Croda Chemicals Group Pty. Ltd., PO Box 1012, Richmond, North Victoria, 3121, Australia

Croda do Brazil Ltda, Rua Croda 230 Distrito Industrial, CEP 13.053, Campinas/SP-C.P. 1098, Brazil

Crompton & Knowles

Crompton & Knowles Corp./Dyes & Chems. Div., PO Box 33188, Charlotte, NC 28233 USA (Tel.: 704-372-5890; FAX 704-372-1522)

Crompton & Knowles Tertre SA, Div. of Crompton & Knowles Corp., 141 ave de la Reine, B-1210 Brussels, Belgium (Tel.: 2-216 2045; FAX 2-242 84 83; Telex: 24595 CKBR B)

Crookes Laboratories, Subsidiary of Boots plc

1 Thane Road, Nottingham, Nottinghamshire, NG2 3AA, UK (Tel.: 0602 592 584; FAX 0602 595 508)

Crosfield

Crosfield Chemicals Ltd., Div. of Unilever plc, PO Box 26, Warrington, Cheshire, WA5 1AB, UK (Tel.: 0925 416100; FAX 0925 59828; Telex: 627067)

Crosfield Chemie BV, Postbus 1, NL-6245 ZG Eijsden, The Netherlands (Tel.: 4409-9333; FAX 4409-3995; Telex: 56884)

Crosfield Chemicals Inc., 101 Ingalls Ave., Joliet, IL 60435 USA (Tel.: 815-727-3651; 800-727-3651; FAX 815-727-5312)

Crowley

Crowley Chemical Co., 261 Madison Ave., New York, NY 10016 USA (Tel.: 212-682-1200; 800-424-9300; FAX 212-953-3487; Telex: 12-7662)

Crowley Tar Products Co., Inc., 261 Madison Ave., New York, NY 10016 USA (Tel.: 212-682-1200)

Crown Metro Inc.

PO Box 5857, Greenville, SC 29606 USA (Tel.: 803-299-1331; 800-368-1331; FAX 803-299-1678; Telex: 805113)

Crown Technology, Inc./Chemical Div.

7513 E. 96 St., PO Box 50426, Indianapolis, IN 46250 USA (Tel.: 317-845-0045; FAX 317-845-9086)

Croxton & Garry Ltd.

Curtis Rd. Industrial Estate, Dorking, Surrey, RH4 1XA, UK (Tel.: 0306-886688; FAX 0306-887780; Telex: 859567/8 cand g)

Crucible Chemical Co Inc.

PO Box 6786, Donaldson Center, Greenville, SC 29606 USA (Tel.: 803-277-1284; 800-845-8873; FAX 803-299-1192)

Cuproquim Corp.

9601 Katy Freeway, Suite 350, Houston, TX 77024-1333 USA (Tel.: 713-464-1103; 800-488-2224; FAX 713-464-1421; Telex: 910 240 6712 PDC)

Cusi. Address unknown

Custom Coating & Laminating Corp.

715 Plantation Street, Worcester, MA 01605 USA

Custom Fibers

Customer Fibers International, 28130 Ave. Crocker, #311, Valencia, CA 91355 USA (Tel.: 805-295-0990; 800-321-5324; FAX 805-295-5148)

Custom Fibers Europe, 13 Rassau Indust. Esatate, Ebbw Vale, Gwent, Wales, UK (Tel.: 495-350-655)

Cutter Laboratories, Miles Laboratories Inc.

4th & Parker, PO Box 1986, Berkeley, CA 94701 USA

Cuyahoga Plastics
1265 Babbitt Rd., Cleveland, OH 44132-2798 USA (Tel.: 216-261-2744; FAX 216-261-3537)

Cyanamid. *See* American Cyanamid

Cyanamid Fothergill Ltd.
Abenbury Way, Wrexham Industrial Estate, Wrexham, LL13 9UF, UK

Cyclo Corp. *See* Rhône-Poulenc Surfactants & Specialties

Cyprus Industrial Minerals
9100 East Mineral Circle, PO Box 3299, Englewood, CO 80155 USA (Tel.: 303-643-5484; FAX 303-643-5168)

Cyro
Cyro Industries, 100 Valley Rd., PO Box 950, Mt. Arlington, NJ 07856 USA (Tel.: 201-770-3000; 800-631-5384; FAX 201-770-6117)

Cyro Canada Inc., 360 Carlingview Dr., Etobicoke, Ontario, M9W 5X9, Canada (Tel.: 416-675-9433)

Daicel-Hüls Ltd.
Shin-Onarimon Bldg., 5th Fl., 17-19, Shimbashi 6-chome, Minato-ku, Tokyo, 105, Japan (Tel.: (03) 5470-1803; FAX (03) 5470-1811; Telex: 222-7385 DAHU J)

Daihachi Chemical Industry Co., Ltd.
Sanyo Nissei Kawaramachi Bldg., 2-7, Kawaramachi 2-chome, Chuo-ku, Osaka, 541, Japan (Tel.: (06) 201-1455; FAX (06) 201-1458)

Daiki Chemical Co. Ltd.
1-378 Handa Nishi-Machi, Sakai City, Osaka Pref, 590, Japan (Tel.: 072273-3331)

Daikin Kogyo Co. *Address unknown*

Daikin Industries Co. Ltd.
Umeda Center Bldg, 4-12 Nishi 2-chome, Nakazaki, Kita-ku, Osaka, 530, Japan (Tel.: (06) 373 1201; FAX (06) 373 4390; Telex: J 63930)

Dainippon
Dainippon Ink & Chemicals Inc., 7-20, Nihonbashi 3-Chome, Chuo-ku, Tokyo, 103, Japan (Tel.: (03) 3272-4511; FAX (03) 3278-8558; Telex: 02222977 DIC J)

Dainippon Pharmaceutical Co, 3-25 Dosho-machi, Higashi-ku, Osaka, 541, Japan (Tel.: (06) 203-5321)

Dajac Laboratories Inc.
1111 Street Rd, Suite 212, Southampton, PA 18966 USA (Tel.: 215-364-6111; FAX 215-364-6867)

Dales Pharmaceuticals Ltd.
Snaygill Industrial Estate, Keighly Road, Skipton., North Yorkshire, BD23 2RW, UK

Dampney Co. Inc.
85 Paris Street, Everett, MA 02149 USA (Tel.: 617-389-2805)

Danbert Chemical Co. *Address unverified*
Oak Brook, IL USA

Daniel Products Co.
400 Claremont Ave., Jersey City, NJ 07304 USA (Tel.: 201-432-0800; FAX 201 432-0266; Telex: 126-304)

Darmex Corp.
71 Jane Street, Roslyn, NY 11577 USA

Darmycel UK
Station Road, Rustington, Littlehampton, West Sussex, BN16 3RF, UK (Tel.: 0903-775111; FAX 0903-787022)

Dart Industries Belgium NV
Pierre Corneliskaai 35, B-9300 Aalst, Belgium (Tel.: 53-70 14 11; FAX 53-78 27 26; Telex: 12301 BELTUP B)

Datac Adhesives Ltd.
Globe Lane Industrial Estate, Dukinfield, Cheshire, SK16 4XE, UK (Tel.: 061-339 8400; FAX 061-343 2713; Telex: 669975 DATAC G)

Daubert Chemical Co., Inc.
4700 S. Central Ave., Chicago, IL 60638 USA (Tel.: 708-496-7350)

Davis & Geck
RD2 Km 47 Hm 4, Manati 00701, PR, USA

Dawood Hercules Chemicals Ltd.
Corporate Office, 35A Shahrahe Abdul Hameed Bin Baadees, PO Box 1, Lahore, Pakistan

Dax Products Ltd.
PO Box 119, Nottingham, NG3 5ED, UK (Tel.: 0602-609996; FAX 0602-604620)

DCS Color & Supply Co., Inc.
2011 S. Allis St., Milwaukee, WI 53207 USA (Tel.: 414-769-2580; FAX 414-769-2598)

DDSA Pharmaceuticals Ltd.
310 Old Brompton Road, London, SW5 9JQ, UK (Tel.: 071-373 7884; FAX 071-370 4321)

Dead Sea Bromine Co. Ltd., Bromine Compds. Ltd.
Makleff House, PO Box 180, Beer-Sheva, 84101, Israel (Tel.: 057-667222; FAX 057-32444; Telex: 5335, 5343)

Dean Agrochemicals Ltd.
19 Whitehouse Gardens, York, YO2 2DZ, UK (Tel.: 0904-629657)

Deanshanger Oxides Ltd. *See* Harcros Pigments Europe

Dearborn Chemicals Ltd.
Foundry Lane, Widnes, Cheshire, WA8 8UD, UK

de Beers Laboratories Inc. *Address unverified*
Broadview, IL USA

Degussa
Degussa AG, Postfach 11 05 33, Weissfrauenstrasse 9, W-6000 Franfurt am Main 11, Germany (Tel.: 069-21801; FAX 069-218 32 18; Telex: 412220 DG)

Degussa Ltd., Div. of Degussa AG, Winterton House, Winterton Way, Macclesfield, Cheshire, SK11 0LP, UK (Tel.: 061-4866211; FAX 0625 502096)

Degussa Corp., 65 Challenger Rd., Ridgefield Park, NJ 07660 USA (Tel.: 201-641-6100; FAX 201-807-3200; Telex: 221420 degus ur)

Jan Dekker BV
Postbus 10, NL-1520 AA Wormerveer, The Netherlands (Tel.: 75-2782 78; FAX 75-21 38 83; Telex: 19273)

Delabole Slate, Div. of R.T.Z. Mining & Exploration Ltd.
Pengelly Road, Delabole, Cornwall, PL33 9AZ, UK (Tel.: 0840-212242; FAX 0840 212948)

Delandale Laboratories Ltd.
37 Old Dover Road, Canterbury, Kent, CT1 3JF, UK (Tel.: 0227-766353; FAX 0227-470626; Telex: 96275 Dellab G)

J.W.S. Delavau Co. Inc.
2140 Germantown Ave., Philadelphia, PA 19122 USA (Tel.: 215-235-1100)

Delaware Chemical Corp.
PO Box 126, Daleville, IN 47334 USA

Denoon CDS
The Campbell House, Bilbrough, York, YO2 3PN, UK (Tel.: 0937-835515)

Deosan Ltd.
Weston Favell Centre, Northampton, NN3 4PD, UK (Tel.: 0604-414000; FAX 0604-406809)

De Pree Co. *Address unverified*
Holland, MI USA

Dermaglen Ltd., Div. of Ellis + Everard plc
PO Box 60, Northampton, Northamptonshire, NN4 9JH, UK (Tel.: 0604-766361; FAX 0604-701238)

The Dermalex Co. Ltd., Div. of Labaz Sanofi Ltd.
Floats Road, Wythenshawe, Manchester, M23 6NF, UK

Dermal Laboratories Ltd.
Tatmore Place, Gosmore, Hitchin, Herts, SG4 7QR, UK (Tel.: 0462-458866; Telex: 826802 DERMAL G)

Dermlk Laboratories Inc.
500 Arcola Road, P.O. Box 1200, Collegeville, PA 19426-0107 USA (Tel.: 312-687-7440)

Desarrollo Quimico Industrial SA
(Dequisa), Serrano 16, E 28001 Madrid, Spain (Tel.: 1-431-1343; FAX 1-431 3714; Telex: 45955 DQMo E)

Dexco Polymers, A Dow/Exxon Partnership
12012 Wickchester, Houston, TX 77079 USA (Tel.: 713-870-7750)

Dexine Rubber Co. Ltd.
Spotland Bridge Works, Rochdale, Lancs, OL12 6AU, UK (Tel.: 0706-40011; FAX 0706-527714)

Dexter
Dexter Chemical Corp., 845 Edgewater Rd., Bronx, NY 10474 USA (Tel.: 212-542-7700; FAX 212-991-7684; Telex: 127061)

Dexter Corp./Frekote Products, One Dexter Dr., Seabrook, NH 03874 USA (Tel.: 603-474-5541; FAX 603-474-5545; Telex: 6817306 HYSEA)

Dexter Corp./Hysol Engineering Adhesives, One Dexter Dr., Seabrook, NH 03874 USA (Tel.: 603-474-5545; FAX 603-474-5545)

Diaflex Ltd.
Kirkhaw Lane, Pontefract, West Yorkshire, WF11 8RD, UK (Tel.: 0977 675270)

Diamalt GmbH
Georg-Reismuller-Strasse 32, W-8000 Munich 50, Germany (Tel.: 89-81060; FAX 89-8106513; Telex: 525316)

Dimex Ltd., Div. of Ellis + Everard plc
46 Peckover St, Bradford, West Yorkshire, BD1 5BD, UK (Tel.: 0274 308052; FAX 0274 737058; Telex: 51305)

Dinoval Chemicals Ltd., Div. of Pointing Ltd.
Stamford House, Stamford New Rd, Altrincham, Cheshire, WA14 1BL, UK (Tel.: 061-941 4254; FAX 061-926 8173; Telex: 668839)

Dista Products
Dista Products Ltd., Kingsclare Road, Basingstoke, Hampshire, RG21 2XA, UK (Tel.: 0256-52011; FAX 0256-485170)

Dista Products Ltd., Fleming Road, Speke, Liverpool, Merseyside, L24 9LN, UK (Tel.: 051-486 3939; FAX 051 486 8750; Telex: 627178 DISTA G)

Dista Products Co., Lilly Research Laboratoies, Indianapolis, IN 46285 USA (Tel.: 317-276-040003714)

Distillers M G Ltd., Div. of Messer Griesheim GmbH

Cedar House, 39 London Road, Reigate, Surrey, RH2 9QE, UK (Tel.: 0737-241133; FAX 0737-241842; Telex: 917280 DICOOL G)

Distillex Ltd.

Unit 117 & 120, Clydesdale Pl, Moss Side Industrial Estate, Leyland, Preston, Lancashire, PR5 3QS, UK (Tel.: 0772 454129; FAX 0772 622258; Telex: 67200 DISTIL G)

D.J. Enterprises

PO Box 31366, Cleveland, OH 44131 USA (Tel.: 216-524-3879)

D.L. Forster Ltd. *See under* Forster Ltd., D.L.

Doak Pharmacal Co. *Address unknown*

Doff Portland Ltd.

Bolsover Street, Hucknall, Nottingham, NG15 7TY, UK (Tel.: 0602-632842; FAX 0602-638657; Telex: 37605 COMIND G)

Dome/Hollister-Stier, Subsidiary of Bayer plc

Strawberry Hill, Newbury, Berkshire, UK

Dorsey

Dorsey Laboratories, 59 Route 10, East Hanover, NJ 07936 USA (Tel.: 201-503-7500; FAX 201-503-8265)

Dorsey Pharmaceuticals, Div. of Sandoz Inc., 4000 Monroe Rd, Charlotte, NC 28205 USA (Tel.: 704-331-7000; FAX 704-377-1064)

Doulton. *Address unknown*

Dover Chemical Corp.

Davis at West 15th St., PO Box 40, Dover, OH 44622 USA (Tel.: 216-343-7711; 800-321-8805/6; FAX 216-364-1579; Telex: 983466)

Dow

Dow Chemical U.S.A., 2020 W.H. Dow Center, Midland, MI 48674 USA (Tel.: 517-636-1000; 800-441-4DOW

Dow Plastics, 2040 W.H. Dow Center, Midland, MI 48674 USA (Tel.: 800-441-4DOW; FAX 517-638-9942)

Dow Chemical Canada Inc., 1086 Modeland Rd., PO Box 1012, Sarnia, Ontario, N7T 7K7, Canada (Tel.: 519-339-3131; 800-363-6250)

Dow Chemical Co Ltd., Div. of The Dow Chemical Co, Lakeside House, Stockley Park, Uxbridge, Middlesex, UB10 1BE, UK (Tel.: 081-848-8688; FAX 081-848-5400; Telex: 934626)

DowElanco Ltd., Div. of Dow Chemical Co Ltd., Latchmore Court, Brand Street, Hitchen, Herts, SG5 1HZ, UK (Tel.: 0462-457272; FAX 0462-453906; Telex: 82414)

Dow Chemical Europe S.A., Bachtobelstrasse 3, CH-8810 Horgen, Switzerland (Tel.: 1-728-2111; FAX 1-728-2935; Telex: 826940)

Dow Chemical Pacific Ltd., 39th Floor, Sun Hung Kai Centre, 30 Harbour Rd., Wanchai, PO Box 711, Hong Kong

Dow Quimica S.A., Sao Paulo, Brazil

Dow Corning Corp.

Dow Corning Corp., PO Box 0994, Midland, MI 48686-0994 USA (Tel.: 517-496-4000; 800-248-2481; FAX 517-496-5849; Telex: 227450)

Dow Corning Ophthalmics Inc., Midland, MI 48640 USA

Dow Corning STI, 47799 Halyard Dr., Suite 99, Plymouth, MI 48170 USA

Dow Corning STI, Eastern Region, 594 Pepper St., Monroe, CT 06468 USA

Dow Corning Ltd., Div. of Dow Corning Corp., Kings Court, 185 Kings Rd, Reading, Berkshire, RG1 4EX, UK (Tel.: 0736-507251; FAX 0736-575051)

Dow Corning France SA, Div. of Dow Corning Corp., Le Britannia A10, 20 Bid E Deruelle, 69432 Lyon Cedex 03, France (Tel.: 78 60 51 48; FAX 78 62 78 98; Telex: 300537)

Dow Corning Kabushiki Kaisha, 507-1, Kishi Yamakita-cho, Ashigarakami-gun, Kanagawa, 258-01, Japan (Tel.: (0465) 76-3108; FAX (0465) 75-1064)

Doyle Specialties

9800 Cozycroft Avenue, Chatsworth, CA 91311 USA

Draxel Chemical Company. *Address unknown*

Drew Industrial Div. *See under* Ashland

Drexel Chemical Company

PO Box 9306, Memphis, TN 38109 USA (Tel.: 901-774-4370; FAX 901-774-4666)

Dry Branch Kaolin

Rt. 1, Box 468D, Dry Branch, GA 31020 USA (Tel.: 201-851-2820)

DSM

DSM NV, Postbus 6500, NL 6401, JH Heerlen, The Netherlands (Tel.: 45-78 81 11; FAX 45-74 06 80; Telex: 56018)

Andeno BV, Div. of DSM Chemicals BV, Posbtus 81, NL-5900 AB Venlo, The Netherlands (Tel.: 77-89 92 81; FAX 77-89 93 00; Telex: 58310)

DSM Engineering Plastics, Div. of DSM NV, Poststraat 1, 6135 KR Sittard, PO Box 43, NL-6130 AA Sittard, Netherlands (Tel.: 46 77 01 23; FAX 46 77 04 00)

DSM Resins UK Ltd., PO Box 8, Ellesmere Port, South Wirral, L65 0HB, UK (Tel.: 051 355 6170; FAX 051 357 1282; Telex: 628213)

DSM United Kingdom Ltd., Div. of DSM NV, Kingfisher House, Kingfisher Walk, Redditch, Worcestershire, B97 4EZ, UK (Tel.: 0527-68-254; FAX 0527-68-949; Telex: 339861)

DSM Deutschland GmbH & Co., Tersteegenstrasse 77, 4 Düsseldorf 30, Germany (Tel.: 0211 454940; FAX 0211 4370917; Telex: 2114365)

DSM Kunstharze GmbH, Div. of DSM NV, Postfach 1351, Am Kreisforst 1, DW 4470 Meppen, Germany (Tel.: 5931-15160; FAX 5931-156101)

DSM France SA, Immeuble Périsud, 5, Rue Legeune, 92128 Montrouge (Cédex), France

DSM Résines France, Div. of DSM NV, BP No 21, 119 Rue Salvator Allendé, F-95872 Bezons Cedex, France (Tel.: 39-47-09 25; FAX 3947-50 76; Telex: 606723 DSMRES F

DSM Resins Espana SA, C'an Jané-Coll de la Manya, 08400 Granollers, Barcelona, Spain (Tel.: 93 849 55 22; FAX 93 849 85)

DSM Chemicals North America, Inc., 4751 Best Road, Ste 140, Atlanta, GA 30337 USA (Tel.: 404-766-3179; 800-825-4376; FAX 404-766-3540)

DSM Thermoplastic Elastomers Inc., 690 Mechanic St., Leominster, MA 01453 USA (Tel.: 508-534-1010; 800-524-0120; FAX 508-534-1005)

DSM Engineering Plastics, N. Am. Headquarters, PO Box 3333, 2267 West Mill Road, Evansville, IN 47732-3333 USA (Tel.: 812-424-3831; 800-333-4237; FAX 812-424-0892)

DSM Engineering Plastics, 233 Arvin Ave., Stoney Creek, Ontario, L8E 2L9, Canada (Tel.: 416-662-1866; FAX 416-662-3493)

DSM Japan K.K., 7F., Shin Kokusai Bldg., 4-1, Marunouchi 3-chome, Chiyoda-ku, Tokyo, 100, Japan (Tel.: (03) 3217-8941; FAX (03) 3201-5074)

Duncan Flockhart

Duncan Flockhart & Co. Ltd., Owned by Glaxo plc, Stockley Park West, Uxbridge, Middlesex, UB11 1B7, UK (Tel.: 081-990-9333; FAX 081-990-4325; Telex: 946442)

Duncan Flockhart Ltd., Associate of Glaxo plc, Cobden Street, Montrose, Angus, DD10 8FB, Scotland

Dunlop

Dunlop Adhesives, Div. of BTR plc, Chester Road, Birmingham, West Midlands, B35 7AL, UK (Tel.: 021-373 8101; FAX 021-384 2826; Telex: 334316)

Dunlop Rubber, Subsidiary of BTR plc, Silvertown House, Vincent Square, London, SW1P 2PL, UK (Tel.: 071-834 3848; FAX 071-834 3879)

Duphar BV. See Solvay Duphar BV

Du Pont

E. I. DuPont de Nemours & Co., Inc., 1007 Market St., Wilmington, DE 19898 USA (Tel.: 302-774-7573; 800-441-9442; FAX 302-774-7573; Telex: 6717325)

DuPont Merck Pharmaceuticals Co, Barley Mill Plaza, Wilmington, DE 19880-0025 USA (Tel.: 302-992-5000; 800-441-9861; FAX 302-892-8530; Telex: 4419861)

DuPont-NEN Medical Products, Div. of E I Du Pont de Nemours & Co Inc., 1007 Market St, Wilmington, DE 19898 USA (Tel.: 302-774-1000; FAX 302-774-7321)

DuPont-NEN Medical Products, 331 Treble Code Rd, North Billerica, MA 01862 USA

DuPont Canada Inc., Box 2200, Streetsville, Mississauga, Ontario, L5M 2H3, Canada (Tel.: 416-821-5612)

DuPont S.A. de C.V., Homero 206, Col. Polanco, Mexico, 5, D.F. Mexico

DuPont (UK) Ltd., Div. of E I Du Pont de Nemours & Co, Wedgewood Way, Stevenage, Hertsfordshire, SG1 4QN, UK (Tel.: 0438 734000; FAX 0438 734154; Telex: 825591 DUPONT G)

DuPont (UK) Ltd./Agricultural Products Dept, Wedgewood Way, Stevenage, Hertsfordshire, SG1 4QN, UK (Tel.: 0438-734000; FAX 0438-734154)

DuPont de Nemours (France) S.A., 137 rue de L'Université, F-75334 Paris, France (Tel.: 45 50 65 50; FAX 47 53 09 65; Telex: 206772)

DuPont de Nemours International S.A., Polymer Products Dept., 2, Chemin du Pavillon, PO Box 50, Le Grand Saconnex, CH-1218 Geneva, Switzerland (Tel.: 22-717-51-11; FAX 22-717 60 77; Telex: 415777 DUP CH)

DuPont Far East Inc., Kowa Bldg. No. 2, 11-39 Akasaka 1-Chome, Minato-Ku, Tokyo, 107, Japan (Tel.: 585-5511)

Durez Div. *See under* Oxychem

Durham Chemicals Ltd.

Birtley, Chester Le Street, Durham, DH3 1QX, UK (Tel.: 091-410-2361; FAX 091-410 6005)

Durham Raw Materials. *See* Durham Chemicals Ltd.

Dwight Products, Inc.

10 Stuyvesant Ave., Lyndhurst, NJ 07071 USA (Tel.: 201-438-3377; FAX 201-438-0594)

Dylon Industries, Inc.

7700 Clinton Rd., Cleveland, OH 44144 USA (Tel.: 216-651-1300; 800-237-8246; FAX 216-651-1777)

Dylon International Ltd., Div. of Mayborn Group

Worsley Bridge Rd, Sydenham, London, SE26 5HD, UK (Tel.: 081 650 4801; FAX 081 658 8735; Telex: 21516)

Dymax Corporation, Div. of Dymax Engineering Adhesives

51 Greenwoods Road, Torrington, CT 06791 USA (Tel.: 203 482 1010; FAX 0203 496 0608; Telex: 179060 ACTC UT)

DynaGel Inc.

Wentworth Ave. & Plummer St., Calumet City, IL 60409 USA (Tel.: 708-891-8400; FAX 708-891-8432; Telex: 211666)

Dynaloy, Inc.

7 Great Meadow Lane, Hanover, NJ 07936 USA (Tel.: 201-887-9270; FAX 201-887-3678; Telex: 642033)

Dynamit Nobel Wien GmbH

Postfach 74, A 1015 Wien, Austria

Dynamold Inc.

2905 Shamrock Avenue, PO Box 9617, Fort Worth, TX 76107 USA

Dynochem UK Ltd., Div. of Dynoindustrier AS

Duxford, Cambridge, Cambridgeshire, CB2 4QB, UK (Tel.: 0223-837370; FAX 0223-832386; Telex: 81411 DYNOCH G)

Eagle-Picher Industries, Inc./Chemicals Dept

PO Box 550, C & Porter Sts., Joplin, MO 64801 USA (Tel.: 417-623-8000; FAX 417-782-1923)

Eagle Zinc Co.

30 Rockefeller Plaza, New York, NY 10112 USA (Tel.: 212-582-0420; FAX 212-582-3412)

E-A-R Specialty Composites

7911 Zionsville Rd., Indianapolis, IN 46268 USA (Tel.: 317-872-1111; FAX 317-692-3111)

Eastern Color & Chemical Co.

35 Livingston St., PO Box 6161, Providence, RI 02904 USA (Tel.: 401-331-9000; FAX 401-331-2155)

Eastman Kodak

Eastman Chemical Co., PO Box 431, Kingsport, TN 37662 USA (Tel.: 615-229-2000; 800-EASTMAN; FAX 615-229-1064; Telex: 6715569)

Eastmanchem, Inc., 155 Gordon Baker Rd., Suite 213, Willowdale, Ontario, M2H 3N7, Canada (Tel.: 416-497-7222)

Kodak Mexicana, Calzada de Tlalpan No. 2980, Mexico D.F., 04870, Mexico

Eastman Chemical (UK) Ltd., Subsidiary of Eastman Kodak, Hemel Hempstead, PO Box 66, Kodak House, Station Road, Herts, HP1 1JU, UK (Tel.: 0442-241171; FAX 0442 241177; Telex: 826502)

Kodak Ltd., Div. of Eastman Kodak Co., PO Box 66, Kodak House, Station Rd, Hemel Hempstead, Hertfordshire, HP1 1TU, UK (Tel.: 0442 61122; FAX 0442 40609; Telex: 825101)

Kodak Ltd/Manufacturing Div., Div. of Eastman Kodak Co, Acornfield Road, Kirkby, Liverpool, Merseyside, L33 7UF, UK (Tel.: 051-546-2101; Telex: 629640)

Eastman Chemical International AG, Hertizentrum 6, CH-6300 Zug 6, Switzerland (Tel.: 042 23 25 25; FAX 042 21 12 52; Telex: 868 824)

Eastman Chemical International, 24 Rue Villiot, F 75012, Paris, France (Tel.: 140-014126; FAX 140-014294)

Eastman Japan Ltd., Nishi-Shinbashi Mitsui Bldg., 1-24-14 Nishi-Shinbashi, Minato-Ku, Tokyo, 105, Japan

Eastman Chemical Brasileira Ltda., Rua George Eastman, 213, Caixa Postal 225, Sao Paulo, Brazil (Tel.: 011-543-5122)

Ebnöther AG, Div. of Ebnöther Holding AG

CH-6203 Sempach-Station, Switzerland (Tel.: 98 91 11; FAX 98 22 46; Telex: 868378 EBAG CH)

ECC International

ECC International Ltd., Div. of ECC Group plc, John Keay House, St Austell, Cornwall, PL25 4DJ, UK (Tel.: 0726 74482; FAX 0726 623019; Telex: 45526 ECCSAU G)

ECC International SA, Div. of ECC Group plc, 2 rue du Canal, B-4551 Lixhe, Belgium (Tel.: 79 82 79; FAX 79 98 11; Telex: 41700)

ECC International, 5775 Peachtree-Dunwoody Rd. NE, Suite 200G, Atlanta, GA 30342 USA (Tel.: 404-843-1551; 800-843-3222; FAX 404-843-8872; Telex: 6827225)

ECC International/Calcium Prods., 5775 Peachtree-Dunwoody Rd 200G, Atlanta, GA 30342 USA (Tel.: 205-249-4905; 800-251-6327; FAX 205-245-0383)

ECC Japan Ltd., Div. of ECC International, 10th Central Bldg., 10-3, Ginza 4-chome, Chuo-ku, Tokyo, 104, Japan (Tel.: (03) 3546-8250; FAX (03) 3546-8255; Telex: J28915)

Efkay Chemicals Ltd.

Banderway House, 156-162 Kilburn High Rd, London, NW6 4JD, UK (Tel.: 071-625 4445; FAX 071-328 9101; Telex: 298912)

Eka Nobel

Eka Nobel AB, Div. of Nobel Industrier Sverige AB, S-445 01 Bohus, Sweden (Tel.: 31-58 70 00; FAX 31-98 17 74; Telex: 2435)

Eka Nobel Ltd., Div. of Eka Nobel AB, Unit 304 Worle Pkwy., Summer Lane, Worle, Weston-Super-Mare, Avon, BS22 OWA, UK (Tel.: 0934-522244; FAX 0934 522577; Telex: 444351)

Eka Nobel Inc./Int'l. Resins Div., Div. of Nobel Industries, 7 Beeches Lane, Woodstock, CT 06281 USA (Tel.: 203-963-2061)

Eka Nobel (Australia) Pty Ltd., Div. of Nobel Industries, 22 Commercial Dr., Dandenong, Victoria, 3175, Australia (Tel.: 3-706-4488)

Elastogran Kunststoff-Technik GmbH, Div. of BASF

Worms Plant, Flosshafenstasse 40, D-6520 Worms, Germany (Tel.: 06241-844-0; FAX 06241-844-113; Telex: 467748 ektwo d)

Electrochem Ltd.

Unit 11, Newfield Industrial Estate, Tunstall, Stoke On Trent, Staffordshire, ST6 5PD, UK (Tel.: 0782 822 058; FAX 0782 822 350; Telex: 848668 GASKEM G)

Electronic Materials Inc.

PO Box 1014, New Milford, CT 06776 USA (Tel.: 203-355-3749)

Elektroschmelzwerk Kempten GmbH

Postfach 609, Herzog-Wilhelm-Strasse 16, W-8000 Munich 33, Germany (Tel.: 89-51201; FAX 89-591671; Telex: 7522749 ESWM D)

Elf Atochem. *See under* Atochem

Elkem Chemicals Inc., Div. of Elkem-American Carbide Co

Park West Office Center, PO Box 37, Pittsburgh, PA 15230 USA (Tel.: 412-778-3600; FAX 412-787-1097)

Eli Lilly & Co. *See under* Lilly

Elliott Ltd., Thomas

Hast Hill, Hayes, Bromley, Kent, UK (Tel.: 081-462-2271)

Embetec Crop Protection Ltd.

Springfield House, Kings Road, Harrogate, N. Yorkshire, HG1 5JJ, UK (Tel.: 0423-509731; FAX 0423-509736; Telex: 57486)

E/M Corporation, Subsidiary of Great Lakes Chemical Corporation

PO Box 2400, 2801 Kent Ave., W. Lafayette, IN 47906 USA (Tel.: 317-497-6346; 800-428-7802; FAX 317-497-6348)

Emco Services Inc.

PO Box 2191, Taunton, MA 02780 USA (Tel.: 508-823-8852; FAX 508-822-1931)

Emerson & Cuming

Emerson & Cuming Inc., Div. of W.R. Grace, 25 Hartwell Ave., Lexington, MA 02173 USA (Tel.: 800-DIE BOND

Emerson & Cuming Inc., 77 Dragon Court, Woburn, MA 01888 USA (Tel.: 617-938-8630; 800-933-4318; FAX 617-933-4318)

Emerson & Cuming (UK) Ltd., Legion House, 838 Uxbridge Road, Hayes, Middlesex, UB4 0RP, UK (Tel.: 081-561 1796; FAX 081-561 4055)

EM Industries, Inc./Fine Chems. Div.

5 Skyline Drive, Hawthorne, NY 10532 USA (Tel.: 914-592-4350; FAX 914-592-9469)

Emkay Chemical Co.

319-325 Second St., PO Box 42, Elizabeth, NJ 07206 USA (Tel.: 908-352-7053; FAX 908-352-6398)

EMS-Chemie

EMS-Chemie AG, Selnaustrasse 16, CH-8039 Zurich, Switzerland (Tel.: 1-201 5411; FAX 1-201 0155; Telex: 815312)

EMS-Grilon (UK) Ltd., Subsid. of EMS-Chemie AG, Switzerland, Drummond Road, Astonfields Industrial Estate, Stafford, ST16 3EL, UK (Tel.: 0785-59121; FAX 0785-213068; Telex: 36254)

EMS-American Grilon Inc., Subsid. of EMS-Chemie AG, Switzerland, PO Box 1717, Corporate Way Road, Sumter, SC 29151-1717 USA (Tel.: 803-481-3172; 800-845-8501; FAX 803-481-3820; Telex: 805077)

EMS-Japan Corp., Kikuchi Bldg., 4-6 Nihonbashi Hongoku-cho, 3-Chome, Chuo-ku, Tokyo, 103, Japan

Emulan Inc.

3726 Roosevelt Rd., PO Box 582, Kenosha, WI 53141 USA (Tel.: 414-654-0734; FAX 414-654-3410)

Emulsion Systems Inc.
70 East Sunrise Hwy., Valley Stream, NY 11581-1233 USA (Tel.: 516-825-3232; 800-ESI-CRYL; FAX 516-825-3233)

Endo Laboratories Inc., Subsidiary of E. I. du Pont de Nemours
1000 Stewart Ave, Garden City, NY 11530 USA

Energen Corp.
2101 6th Avenue N, Birmingham, AL, 35203 USA (Tel.: 205-326 2700)

Engelhard Corp.
Performance Minerals Group, 101 Wood Ave., Iselin, NJ 08830-0770 USA (Tel.: 908-205-5000; FAX 908-205-6711; Telex: 219984 ENGL UR)

Engelhard Technologies Ltd. *See* Engelhard Corp.

English Woodlands Ltd.
Hoyle Depot, Graffham, Petworth, W Sussex, GU28 0LR, UK (Tel.: 07986-574; FAX 042879-4711)

EniChem
EniChem Elastomeri Srl, Milanofiori Strada 3, Palazzo B1, 20090 Assago, Milan, Italy (Tel.: 02 5201; FAX 0039 2 52026088; Telex: 310246)

EniChem Elastomers Ltd., Div. of EniChem Elastomeri Srl, Charleston Rd, Hardley, Hythe, Southampton, Hampshire, SO4 6YY, UK (Tel.: 0703 894919; FAX 0703 894334; Telex: 47519)

EniChem America, Inc., 1211 Ave. of the Americas, New York, NY 10036 USA (Tel.: 212-382-6531; FAX 212-382-6520; Telex: 6801159 ENICHEM)

EniChem Synthesis SpA
Via Medici del Vascello 40, I-20138 Milano, Italy (Tel.: 02 5203 9218; FAX 02 5203 9450; Telex: 310246)

Enka Glanzstoff AG
Postfach 1001 49, Enka-Haus, Kasinostrasse, DW 5600 Wuppertal 1, Germany (Tel.: 0202-3210)

Enso-Gutzeit OY
PO Box 309, SF-00101 Helsinki, Finland (Tel.: 0-16291; FAX 0-1629471; Telex: 121438 ENSO SF)

Envhy Ltd. *Address unknown*

Enzypharm Biochemicals Ltd. *Address unknown*

Epoleon Corp.
Epoleon Corp., Div. of Aikoh Co., Ltd., D.S. Bldg. 1-39 Ikenohata 2-Chome, Taito-Ku, Tokyo, 110, Japan (Tel.: (03) 823-1111)

Epoleon Corp. of Am., 2858 Carson St., Suite 121, Torrance, CA 90503 USA (Tel.: 213-316-4242; 800-448-6367; FAX 213-316-4463)

Epoxy Technology, Inc.
14 Fortune Dr., Billerica, MA 01821 USA (Tel.: 508-667-3805; 800-227-2201; FAX 508-663-9782)

Essex Specialty Products Inc./Subsidiary of Dow Chemical Co.
1135 Broad St., Clifton, NJ 07015 USA (Tel.: 201-773-6300; FAX 201-778-3280; Telex: 62953879)

Estron Chemical, Inc.
P.O. Box 127, Highway 95, Calvert City, KY 42029 USA (Tel.: 502-395-4195; FAX 502-395-5070)

Ethicon Inc.
Rt 22, PO Box 151, Somerville, NJ 08876 USA (Tel.: 908-218-0707)

Ethox Chemicals, Inc.
PO Box 5094, Sta. B, Greenville, SC 29606 USA (Tel.: 803-277-1620; FAX 803-277-8981)

Ethyl
Ethyl Corp., 451 Florida Blvd., Baton Rouge, LA 70801 USA (Tel.: 504-388-7040; 800-535-3030; FAX 504-388-7686; Telex: 586441, 586431)

Ethyl Canada, Inc., 350 Burnhamthorpe Rd. West, Suite 600, Mississauga, Ontario, L5B 3JI, Canada (Tel.: 416-566-9222; FAX 416-566-99962)

Ethyl Corp. UK, Goldlay House, 114 Parkway, Chelmsford, Essex, CM2 7PP, UK (Tel.: 245-287-577)

Ethyl SA, Div. of Ethyl Corp., London Rd, Bracknell, Berkshire, RG12 2UW, UK (Tel.: 0344 780378; FAX 0344 778360; Telex: 848291)

Ethyl SA, Div. of Ethyl Corp., 523 Ave. Louise, Box 19, B-1050 Brussels, Belgium (Tel.: 2-642-4411; FAX 2-648-4336; Telex: 22549)

Ethyl Japan, Christy Bldg. 2/F, 1-22 Moto Azabu 3-Chome, Minato-Ku, Tokyo, 106, Japan

Ethyl Asia Pacific Co., #13-06 PUB Bldg., Devonshire Wing, 111 Somerset Rd., Singapore 0923, Singapore

Eurobrom
Eurobrom BV, PO Box 158, NL-2280 AD Rijswijk, Netherlands (Tel.: 070 340 84 08; FAX 070 399 90 35; Telex: 32137)

Broomchemie BV, Div. of Eurobrom BV, Postbus 318, Frankrijkweg 5, NL-4530 AH Terneuzen, The Netherlands (Tel.: 1150-89000; FAX 1150-17291; Telex: 55450 BCHEM NL)

European Colour (Pigments) Ltd.
Bankfield Street, Stockport, SK5 7PB, UK (Tel.: 061 480 3891; FAX 061-480-9852; Telex: 669657 COLOUR)

European Vinyls Corp.

European Vinyls Corp. (Deutschland) GmbH, Emil-von-Behring-Strasse 2, D-6000 Frankfurt M.50, Germany (Tel.: 0 69 58 01 01; FAX 069 5801 640; Telex: 4189387)

European Vinyls Corp. UK Ltd., Norton House, Crown Gate, Runcorn, Cheshire, WA7 2UX, UK (Tel.: 0928 714482; FAX 0928 715101; Telex: 269415 EVC UK)

Eval Co. of America

1001 Warrenville Rd., Suite 201, Lisle, IL 60532 USA (Tel.: 708-719-4610; 800-423-9762; FAX 708-719-4622)

Evans Chemetics. See Grace, W.R.

Evans Medical Ltd.

Langhurst, Horsham, W Sussex, RH12 4QD, UK (Tel.: 0403 41400; FAX 0403 61101)

Evans Vanodine International plc

Brierley Rd, Walton Summit Centre, Bamber Bridge, Preston, Lancashire, PR5 8AH, UK (Tel.: 0772 322200; FAX 0772 626000; Telex: 677688)

Evode Group plc

Evode Plastics Ltd., Div. of Evode Group plc, Wanlip Road, Syston, Leicester, Leicstershire, LE7 8PD, UK (Tel.: 0533-696 752; FAX 0533-693209; Telex: 342527)

Evode Speciality Adhesives Ltd., Div. of Evode Group plc, Wanlip Road, Syston, Leicester, Leicstershire, LE7 8PD, UK (Tel.: 0533-606001; FAX 0533-692411; Telex: 34485)

Exchem Industries Ltd., Div. of Exchem plc

Great Oakley Works, Great Oakley, Harwich, Essex, CO12 5JW, UK (Tel.: 0255 880824; FAX 0255 880429; Telex: 98569)

Expancel

Expancel, Div. of Nobel Industries, 1519 Johnson Ferry Rd., Suite 200, Marietta, GA 30062 USA (Tel.: 404-971-8005; FAX 404-578-1359)

Expancel, Div. of Nobel Industries, PO Box 13000, S-85013 Sundsvall, Sweden (Tel.: 60-134000; FAX 60-569518; Telex: 71399 expancel s)

Expansia SA

BP 6, F-30390 Aramon, France (Tel.: 66 57 01 01; FAX 66 57 01 48; Telex: 480191)

Exsymol

4 Ave. Prince Hereditaire Albert, Zone F-Bloc C, MC 98000, Monaco (Tel.: 93-30-13-08; FAX 93-50-43-47)

Exxon

Exxon Chemical Co., PO Box 3272, Houston, TX 77253-3272 USA (Tel.: 713-870-6000; 800-526-0749; FAX 713-870-6661; Telex: 794588)

Exxon Chemical Co./Tomah Products, 1012 Terra Dr., PO Box 388, Milton, WI 53563 USA (Tel.: 608-868-6811; 800-441-0708; FAX 608-868-6810; Telex: 910-280-1401)

Exxon Chemical Ltd., Div. of Exxon Corp., 4600 Parkway, Solent Business Park, Whiteley, Fareham, Hampshire, PO15 7AZ, UK (Tel.: 0489-884400; Telex: 47437)

Deutsche Exxon Chemical GmbH, Dompropst Ketzer-Str. 1-9, 5000 Köln 1, Germany (Tel.: 221 16150; FAX 221 160 5320)

Exxon Chemical Belgium, Div. of Exxon Corp., Vorstlaan 280, B-1160 Brussels, Belgium (Tel.: 2-674 41 11; FAX 2-674 41 29; Telex: 22364)

Exxon Chemical International Inc., Div. of Exxon Corp., Mechelsesteenweg 363, B-1950 Kraainem, Belgium (Tel.: 2-769-3111; FAX 2-769 32 25; Telex: 24733)

Exxon Chemical Holland BV, Div. of Exxon Corp., 's Gravelandse 298, NL-3125 BK Schiedam, The Netherlands (Tel.: 10-488 19 11; FAX 10-488 13 88; Telex: 22402 ECSCH NL)

Exxon Chemical Mediterranea SpA, Div. of Exxon Corp., Via Paleocapa 7, I-20121 Milano, Italy (Tel.: 2 88031; FAX 2 8803231; Telex: 311561 ESSOCH I)

Exxon Chemical Japan Ltd., TBS Kaikan Bldg., 3-3, Akasaka 5-Chome, Minato-Ku, Tokyo, 107, Japan (Tel.: (03) 582-9243; Telex: 22846)

Fabriquimica S.R.L.
Calle 32 No. 3313, San Martin 1650, Argentina (Tel.: 1-755-7290)

Faesy & Besthoff, Inc.
143 River Rd., Edgewater, NJ 07020 USA (Tel.: 201-945-6200; FAX 201-945-6145)

Fairfax Biological Laboratory Inc.
PO Box 300, Electronic Road, Clinton Corners, NY 12514 USA (Tel.: 914-266-3705; FAX 914-266-4892)

FAIR Laboratories. *Address unknown*

Fairmount Chemical Co., Inc.
117 Blanchard St., Newark, NJ 07105 USA (Tel.: 201-344-5790; 800-872-9999; FAX 201-690-5298; Telex: 138905)

The Fanning Corp.
1775 W. Diversity Pkwy., Chicago, IL 60614-1009 USA (Tel.: 312-248-5700; FAX 312-248-6810; Telex: 910-221-1335)

Farbwerke Hoechst, Div. of Hoechst AG
Brueningstrasse 50, Hoechst, Frankfurt-am-Main 80, Germany (Tel.: 069 3050; FAX 069 303666; Telex: 412340 HOEAG D)

Fargro Ltd.
Toddington Lane, Littlehampton, West Sussex, BN17 7PP, UK (Tel.: 0903-721591; FAX 0903-730737; Telex: 877575)

Farley Health Products. *Address unknown*

Farleyway Chemicals Ltd.
Ham Lane, Kingswinford, West Midlands, DY6 7JU, UK (Tel.: 0384 400 222; FAX 0384 400 020; Telex: 339528 FAR G)

Farmers Crop Chemicals Ltd.
County Mills, Worcester, Worcestershire, WR1 3NU, UK (Tel.: 0905 27733; FAX 0905 27574)

Farmitalia Carlo Erba SpA, Div. of Erbamont NV
Casella Postale 10519, I-20100 Milan, Italy (Tel.: 2-69-951; FAX 2-69 959938; Telex: 330314 ERBA I)

FarmItalia (Farmaceutici Italia)
Viale E Bezzi 24, I-20146 Milano, Italy

Farm Protection Ltd., Subsidiary of ICI
Woolmead House West, Bear Lane, Farnham, Surrey, GU9 7UB, UK (Tel.: 0252-734201; FAX 0252-736333)

Fatsco. *Address unknown*

FBA Pharmaceuticals. *Address unknown*

FBC Ltd. *See* Schering Agrochemicals Ltd.

Feb Ltd.
Albany House, Swinton Hall Road, Swinton, Manchester, M27 1DT, UK (Tel.: 061-794 7411; FAX 061-793 4529; Telex: 668369 FEB UK G)

The Feldspar Corp.
One West Pack Square, Suite 700, Asheville, NC 28801 USA (Tel.: 704-254-7400; FAX 704-255-4909)

Ferguson & Menzies Ltd.
312 Broomloan Road, Glasgow, G51 2JW, UK (Tel.: 041-445 3555; FAX 041-425 1079)

Fermenta ASC Europe Ltd.
Bayheath House, 4 Fairway, Petts Wood, Orpington, Kent, BR5 1EG, UK (Tel.: 0689-74011; FAX 0689-74085)

Fernox Manufacturing Co. Ltd.
Britannia Works, Clavering, Saffron Walden, Essex, CB11 4QZ, UK (Tel.: 0799 550811; FAX 0799 550853; Telex: 818292 FERNOX G)

Ferro Corp.
Ferro Corp., World Headquarters, 1000 Lakeside Ave., Cleveland, OH 44114 USA (Tel.: 216-641-8580)

Ferro Corp./Bedford Chemical Div., 7050 Krick Rd., Bedford, OH 44146 USA (Tel.: 216-641-8580; 800-321-9946; Telex: 98-165)

Ferro Corp./Color Div., 4150 E. 56th St., PO Box 6550, Cleveland, OH 44101 USA (Tel.: 216-641-8580; FAX 216-641-8831; Telex: 98-0165)

Ferro Corp./Engineering Thermoplastics Div., 7500 East Pleasant Valley Rd., Independence, OH 44131 USA (Tel.: 216-641-8580; FAX 216-524-0493)

Ferro Corp./Grant Chemical Div., 7050 Krick Rd., Bedford, OH 44146 USA

Ferro Corp./Keil Chemical Div., 3000 Sheffield Ave., Hammond, IN 46320 USA (Tel.: 219-931-2630; FAX 219-931-6318; Telex: 725484)

Ferro Corp./Transelco Div., Box 217, Penn Yan, NY 14527 USA (Tel.: 315-536-3357; FAX 315-536-8091; Telex: 97 8373)

Ferro Corp./International Office, Alexander House, Crown Gate, Runcorn, Cheshire, WA7 2UP, UK (Tel.: 0928-71-9737)

Ferrosan Fine Chemicals A/S, Div. of Novo Nordisk A/S
Posboks 230, Københavnsvej 216, DK-4600 Køge, Denmark (Tel.: 53-65 40 55; FAX 53-65 72 25; Telex: 43516 FERRO DK)

Fiberite West Coast Corp. *Address unverified*
Orange, CA USA

Fieldspray, Div. of Nilco Chemical Co Ltd.
Stewart Road, Kingsland Industrial Park, Basingstoke, Hants, RG24 0GX, UK (Tel.: 0256-474600; FAX 0256-450603)

Fillite USA Inc.
State Rt 2 & Industrial Lane, P O Box 3074, Huntington, WV, 25702 USA

Fillite (Runcorn) Ltd.
Goddaed Road, Astmoor Industrial Estate, Runcorn, Cheshire, WA7 IQF, UK (Tel.: 0928-566661; FAX 0928-572380)

Filtec Ltd.
Suite A, Constance House, Constance Ind. Est., Waterloo Road, Widnes, Cheshire, WA8 0QR, UK (Tel.: 051-495 1988; FAX 051-420 1407; Telex: 628151 Filtec G)

Filter-Media Inc.
3603 Westcenter Dr., Houston, TX 77224-9156 USA (Tel.: 713-780-9000; FAX 713-781-4320; Telex: 775144 FEMCO HOU)

Fina
Fina Chemicals, Div. of Petrofina SA, Nijverheldsstraat, 52 Rue de l'Industrie, B-1040 Brussels, Belgium (Tel.: 2-288-9111; FAX 32-2-288-3388; Telex: 21 556 PFINA B)

Oleofina, 4, Rue Jacques de Lalaing, B-1040 Brussels, Belgium (Tel.: 02-288-91-11; FAX 288-32-99)

Fina plc, Fina House, 1 Ashley Ave, Epsom, Surrey, KT18 5AD, UK (Tel.: 0372-726226; FAX 0372-744520; Telex: 894317)

Fina Oil and Chemical Co., 8350 North Central Expressway, PO Box 2159, Dallas, TX 75221 USA (Tel.: 214-750-2400; 800-344-FINA

Oleofina Far East, 138 Cecil St., #17-01, Cecil Court, Singapore 0106)

Fine Agrochemicals Ltd.

3 The Bull Ring, Worcester, Worcestershire, WR2 5AA, UK (Tel.: 0905-748444; FAX 0905-748440; Telex: 335603)

Fine Dyestuffs & Chemicals Ltd.

Oak Road off Clough Road, Stoneferry, Hull, North Humberside, UK (Tel.: 0482 41377; Telex: 52468)

Fine Organics Ltd., Div. of Laporte plc

Seal Sands, Middlesbrough, Cleveland, TS2 1UB, UK (Tel.: 0642 546666; FAX 0642 546046; Telex: 58639 FINORG G)

Finetex Inc.

418 Falmouth Ave., PO Box 216, Elmwood Park, NJ 07407 USA (Tel.: 201-797-4686; FAX 201-797-6558; Telex: 710-988-2239)

Represented by: Pennine Chemical Ltd., Kent Works, Thomas St., Conglton, Cheshire, CW12 1QZ, UK

Firestone Synthetic Rubber & Latex Co., Div. of Bridgestone/Firestone Inc.

PO Box 26611, 381 W. Wilbeth Rd., Akron, OH 44319 USA (Tel.: 216-379-7759; 800-282-0222; FAX 216-379-7483)

Firmenich

Fermenich Inc., PO Box 5880, Princeton, NJ 08543 USA (Tel.: 609-452-1000; 800-257-9591; FAX 609-951-8716; Telex: 21-99-15)

Firmenich SA, PO Box 239, CH-1211 Geneva 8, Switzerland (Tel.: 22-42 42 00; FAX 22-43 73 22; Telex: 423181 FICO CH)

Fischer Instrumentation (GB) Ltd.

Arnhem Road, Bone Lane Industrial Estate, Newbury, Berks, RG14 5RU, UK

G. Fiske & Co. Ltd.

64 Sheen Rd, Richmond, Surrey, TW9 1UF, UK (Tel.: 081-948 5811; FAX 081-948 7059; Telex: 925878)

Fisons

Fisons plc, Fison House, Princes Street, Ipswich, Suffolk, IP1 1QH, UK (Tel.: 0473-232525; FAX 0473-231540; Telex: 98240 FISONS G)

Fisons plc, Horticulture Div., Paper Mill Lane, Bramford, Ipswich, Suffolk, IP8 LBZ, UK (Tel.: 0473-830492; FAX 0473-830386)

Fisons plc, Pharmaceuticals Div., 12 Derby Road, Loughborough, Leicestershire, LE11 088, UK (Tel.: 0509-611001; FAX 0509-611116)

Fisons Corp., Subsidiary of Fisons plc, 755 Jefferson Rd, Rochester, NY 14623 USA (Tel.: 716-475-9000; 800-334-6433)

C.B. Fleet Co. Inc.

4615 Murray Road, Lynchburg, VA 24502-2235 USA (Tel.: 804-528-4000)

Sam Fletcher Agricultural Specialists, Div. of Banks Odam Dennick Ltd.

Fleet Road Industrial Estate, Holbeach, Spalding, Lincs, PE12 7EG, UK (Tel.: 0406-22207; FAX 0406-26525)

Flexchemie BV

Westmolendijk 27, NL-2985 XJ Ridderkerk, The Netherlands (Tel.: 26388; FAX 28211; Telex: 28255)

Flexible Products

Flexible Products Co./Polyurethane Div., 1007 Industrial Park Dr., Box 3190, Marietta, GA 30061 USA (Tel.: 404-428-2684; 800-662-3270; FAX 404-421-6495)

Flexible Products Int'l. Polyurethane, Excelsiorlaan 40, B-1930 Zaventem, Belgium (Tel.: 011-322-725-7879; FAX 011-322-725-7887)

Flexibulk Ltd.

Davidson House, Upper St. John Street, Lichfield, Staffordshire, WS14 9DU, UK (Tel.: 0543-253461; FAX 0543-253466; Telex: 335476)

Flexshield Chemical Mfg.

128 E Campbell, Chandler, AZ, 85225 USA

Floridin Co.

Floridin Co., PO Box 510, 1101 N. Madison St., Quincy, FL 32351-0510 USA (Tel.: 904-627-7688; 800-228-1131; FAX 904-875-4408; Telex: 4931835 FLOQYUI)

Floridin Co./Administrative Office, 5380 Capital Circle NW, Tallahassee, FL 32303 USA (Tel.: 904-562-5005; FAX 904-562-5935)

Flowering Plants Ltd.

55 Well Street, Buckingham, Bucks, MK18 1EP, UK (Tel.: 0280-813764; FAX 0280-822238)

Fluka

Fluka Chemical Corp., 980 South Second St., Ronkonkoma, NY 11779 USA (Tel.: 516-467-0980; 800-FLUKA-US; FAX 800-441-8841; Telex: 96-7807)

Fluka Chemicals Ltd., Peakdale Rd, Glossop, Derbyshire, SK13 9XE, UK (Tel.: 0457 862518; Telex: 669960)

Fluka Sarl, BP 1114, 68052 Mulhouse Cedex, France (Tel.: 89 61 87 47; Telex: 881236 F

Fluorocarbon Co. Ltd.

Caxton Hill, Hertford, Herts, SG13 7NH, UK (Tel.: 0992 550731; FAX 0992 584697; Telex: 81435)

FMC Corp.

FMC Corp./Chemical Products Group, 1735 Market St., Philadelphia, PA 19103 USA (Tel.: 215-299-6000; FAX 215-299-5999; Telex: 685-1326)

FMC Corp. Canada, 11475 Cote de Liesse Rd., Dorval, Quebec, H9P 1B3, Canada

FMC Corp (UK) Ltd., Commercial Road, Bromborough, Merseyside, Lancs, L62 3NL, UK (Tel.: 051-334-8085; FAX 051-334-8501)

FMC Corp Belgium, Ave. Louise 523, Box 1, B-1050 Brussels, Belgium

FMC Far East Ltd., Onarimon Yusen Bldg., 23-5, Nishi-Shinbashi 3-chome, Minato-ku, Tokyo, 105, Japan (Tel.: (03) 3431-9117; FAX (03) 3433-7014; Telex: J22973)

Fordath Engineering Co. *Address unverified*
West Bromwich, UK

Ford Smith & Co. Ltd.
Lyndean Industrial Estate, Felixstowe Road, Abbey Wood, London, SE2 9SG, UK (Tel.: 081-310-8127; FAX 081-310-9563)

D. L. Forster Ltd.
12 The Ongar Road Trading Estate, Great Dunmow, Essex, CM6 1EU, UK (Tel.: 0371-865201; FAX 0371-876541)

Fortafix Ltd.
First Drove, Fengate, Peterborough, PE1 5BJ, UK (Tel.: 0733-61136; FAX 0733-315393)

Foseco (FS) Ltd.
Foseco (FS) Ltd., Div. of Foseco plc, Tamworth, Staffordshire, B78 3TL, UK (Tel.: 0827-289999; FAX 0827-250806)

Foseco Ltd./Metallurgical Div, Tamworth, Staffordshire, B78 3TL, UK

Fosroc Ltd.
Timber Treatments Div, Fieldhouse Lane, Marlow, Buckinghamshire, SL7 1LS, UK (Tel.: 0628 46644; FAX 0628 476757; Telex: 847057)

Fothergill & Harvey plc
PO Box 3, Church Street, Littleborough, Lancashire, OL15 8HG, UK (Tel.: 0706-7015; FAX 0706-70576)

Fothergill Tygaflor Ltd.
PO Box 2, Summit, Littleborough, Lancashire, OL5 0LT, UK (Tel.: 0706-78837; FAX 0706-70203)

Samuel Fox & Co. Ltd. *Address unknown*

FPT Industries Ltd.
The Airport, Portsmouth, Hampshire, PO3 5PE, UK (Tel.: 0705 662391; FAX 0705 670899; Telex: 86106)

Franklin Industrial Minerals
Franklin Industrial Minerals, 612 Tenth Ave. North, Nashville, TN 37203 USA (Tel.: 615-259-4222; FAX 615-726-2693)

Franklin Industrial Minerals, 821 Tilton Bridge Rd., S.E., Dalton, GA 30721 USA (Tel.: 404-277-3740; FAX 404-277-9827)

Franklin Mineral Products Co., Div. of Mearl Corp.

PO Drawer 390, Hartwell, GA 30643 USA (Tel.: 706-376-3174; FAX 706-376-3044)

Franklin Oil Corp. (Ohio)

Box 46030, Cleveland, OH 44146-0030 USA (Tel.: 216-232-3000)

Freeman Chemical Corp. See Cook Composites & Polymers Co

Carl Freudenberg

Postfach 1369, D-6940 Weinheim (Bergstr), Germany

Fries & Fries, Div. of Mallinckrodt

1199 Edison Dr., Cincinnati, OH 45216 USA (Tel.: 513-948-8000; 800-543-4643; Telex: 21-4348)

Frodingham Cement Co. Ltd.

Brigg Road, Scunthorpe, Humberside, DN16 1AW, UK

Fry's Metals Ltd., Div. of Cookson Group plc

Tandem House, Beddington Farm Rd, Croydon, Surrey, CR9 4BT, UK (Tel.: 081-665 6666; FAX 081-665 6196; Telex: 265732)

Fudow Company, Ltd.

11-26, Nishi Rokugo 4-chome, Ohta-ku, Tokyo 144, Japan Tel.: (03) 3737-0611; FAX (03) 3738-0554)

Fuji Chemical Industrial Co., Ltd.

1570 Nakanoshima, Wakayami-shi, Wakayama, 640, Japan (Tel.: (0734) 23-1247; FAX (0734) 31-3005)

H. B. Fuller

H. B. Fuller Co., 3530 Lexington Ave. North, St. Paul, MN 55126 USA (Tel.: 612-481-1588; 800-468-6358; FAX 612-481-1863)

H. B. Fuller UK Ltd., Amber Business Centre, Greenhill Lane, Leabrooks, Derbyshire, DE55 4BR, UK (Tel.: 0773-608877; FAX 0773-602673)

H. B. Fuller Japan Co., Ltd., 700, Matsushima-cho, Hamamatsu, Shizuoka, 430, Japan (Tel.: (053) 425-0751; FAX (053) 426-1672; Telex: 4225155 HBF HJ)

Furon

Furon Co., 386 Metacom Ave., Bristol, RI 02809 USA (Tel.: 401-253-2000; 800-336-3534; FAX 401-253-8211)

Furon/CHR Div., 407 East St., New Haven, CT 06509-9988 USA (Tel.: 203-777-3631; 800-525-2523)

Futura Coatings, Inc.

9200 Latty Ave., Hazelwood, MO 63042 USA (Tel.: 314-521-4100; FAX 314-521-7255)

GAF Corp. See ISP Technologies

Galderma Laboratories, Inc.
3000 Altamesa Blvd., Suite 300, Fort Worth, TX 76133 USA (Tel.: 817-551-8664)

Galen Ltd.
19 Lower Seagoe Industrial Estate, Portadown, Craigavon, Armagh, BT63 5QA, N. Ireland

Gamlen Industries SA, Div. of W Canning plc
62-70 Rue Yvan Tourgueneff, F-78380 Bougival, France (Tel.: 39 18 92 34; FAX 39 18 22 05; Telex: 695355)

Gansow (UK) Ltd., Div. of Gansow GmbH
Yelvertoft Rd, Kingthorpe, Northampton, Northamptonshire, NN2 7TQ, UK (Tel.: 0604 719031)

Gard Corp.
2727 Roe Lane, Kansas City, KS 66103 USA (Tel.: 913-236-5000; FAX 913-432-8309; Telex: 510-100-6629)

Garvey Chemical Corp.
PO Box 127, 1300 N 7th, St Joseph, MO 64502 USA

Gattefosse
Gattefosse SA, 36 Chemin de Genas, BP 603, 69800 Saint Priest, France (Tel.: 78-90-63-11; FAX 78-90-4567; Telex: 340 240)

Gattefosse Corp., 189 Kinderkamack Rd., Westwood, NJ 07675 USA (Tel.: 201-573-1700; FAX 201-573-9671)

Represented by: Alfa Chemicals Ltd., Broadway House, 7-9 Shute End, Workingham, Berkshire, RG11 1BH, UK

Gaylord Chemical Co.
PO Box 1209, Slidell, LA 70459-1209 USA (Tel.: 504-649-5464; 800-426-6620; FAX 504-649-0068; Telex: 629-20-353)

G. D. Searle & Co. *See* Searle Chemicals Inc

Geeco, Div. of McKechnie Consumer Products Ltd.
Gore Road Industrial Estate, New Milton, Hants, BH25 6SE, UK (Tel.: 0425-614600; FAX 0425-619463)

Geistlich Sohne AG
CH-6110 Wolhusen, Switzerland (Tel.: 71 03 33; Telex: 866133)

Geltech, Inc.
Two Innovation Dr., Alachua, FL 32615 USA (Tel.: 904-462-2358; FAX 904-462-2993)

Genentech Inc., Subsidiary of Roche Holdings Inc.
460 Point San Bruno Blvd, South San Francisco, CA 94080 USA (Tel.: 415-266-1000)

General Chemical

General Chemical Corp., 90 East Halsey Rd., PO Box 393, Parsippany, NJ 07054-0393 USA (Tel.: 201-515-0900; 800-631-8050; FAX 201-515-4461; Telex: 362262)

General Chemical Canada Ltd., 201 City Center Dr., Mississauga, Ontario, L5B 3A3, Canada (Tel.: 416-896-9595; 800-668-0433; FAX 416-276-6594)

General Electric (GE)

GE Company, 3135 Easton Tpke., Fairfield, CT 06431 USA (Tel.: 800-626-2004)

General Electric Co, 21800 Tungsten Rd, Cleveland, OH 44117 USA (Tel.: 216-266-2451; FAX 216-266-3372)

GE Specialty Chemicals, 5th and Avery Sts., Parkersburg, WV, 26102 USA (Tel.: 304-424-5411)

GE Europe, Cyprusweg 2, 1044 AA Amsterdam, The Netherlands (Tel.: 20-5806911)

GE Hong Kong, 15/F Convention Center, No. 1 Harbor Rd., Wanchai, Hong Kong (Tel.: 5-8105616)

General Electric Co./Plastics Div., One Plastics Ave., Pittsfield, MA 01201 USA (Tel.: 413-448-7110; 800-845-0600; Telex: 926430)

GE Plastics-Canada, 2300 Meadowvale Blvd., Mississauga, Ontario, L5N 5P9, Canada (Tel.: 416-858-5790; FAX 416-858-5798)

General Electric Plastics BV, Plasticslaan 1, PO Box 117, 4600 AC Bergen op Zoom, The Netherlands (Tel.: 1640 32911; FAX 1640 43949)

GE Plastics Ltd., Div. of General Electric Co, Birchwood Science Park, Warrington, Cheshire, WA3 6DA, UK (Tel.: 0925-811522)

GE Plastics Japan Ltd., Shionogi Honcho Kyodo Bldg., 3-7-2, Nihonbashi-Honcho, Chuo-Ku, Tokyo, 103, Japan (Tel.: (03) 5695-4877; FAX (03) 5695-4860)

General Electric Co./Silicone Products Div., 260 Hudson River Rd., Waterford, NY 12188 USA (Tel.: 518-237-3330; 800-255-8886)

GE Silicones, Old Hall Rd, Sale, Cheshire, M33 2HG, UK (Tel.: 061-905 5000; FAX 061-905 5022)

GE Silicones Europe, Postbus 117, Plasticslaan 1, NL-4600 AC Bergen op Zoom, The Netherlands (Tel.: 1640-32291; FAX 1640-32708; Telex: 78421)

General Electric-Huntsman Corp. (GEH), Div. of Huntman, One Noryl Ave., Selkirk, NY 12158 USA (Tel.: 518-475-5734; FAX 518-475-5583)

Genesee Polymers Corp.

G-5251 Fenton Rd., PO Box 7047, Flint, MI 48507-0047 USA (Tel.: 313-238-4966; FAX 313-767-3016)

Genesis Polymers, Affiliate of Nova Corp. of Alberta (Novacor)

2550 Busha Hwy., Marysville, MI 48040 USA (Tel.: 313-364-5555; 800-627-1221; FAX 313-364-4670)

Genstar Stone Products Co.

Executive Plaza IV, 11350 McCormick Rd., Hunt Valley, MD 21031 USA (Tel.: 301-527-4000; FAX 301-527-4535)

Geoliquids, Inc.

1618 Barclay Blvd., Buffalo Grove, IL 60089 USA (Tel.: 708-215-0938; FAX 708-215-9821)

George Mann & Co. *See under* Mann & Co., Inc., George

Georgia Gulf Corp./PVC Div.

PO Box 629, Plaquemine, LA 70765-0629 USA (Tel.: 504-685-1200; 800-PVC-VYCM

Georgia Marble Co.

1201 Roberts Blvd., Bldg. 100, Kennesaw, GA 30144-3619 USA (Tel.: 404-521-4711)

Georgia-Pacific

Georgia-Pacific Corp., 133 Peachtree St. N.E., PO Box 105605, Atlanta, GA 30348 USA (Tel.: 404-220-6185)

Georgia-Pacific Corp., 300 Laurel St., PO Box 1236, Bellingham, WA 98225 USA (Tel.: 206-733-4410; FAX 206-676-7217)

Georgia-Pacific Chemical Div., 1754 Thorne Rd., Tacoma, WA 98421 USA (Tel.: 206-572-8181)

Georgia-Pacific Resins Inc., 2883 Miller Rd, Decatur, GA 30035 USA (Tel.:404-593-6895; FAX 404-593-6801; Telex: 804600)

Gerhardt Pharmaceuticals

P O Box 777, London, SW19 5DY, UK (Tel.: 081-944 0505)

Geronazzo S.p.A. *See* Rhône-Poulenc Geronazzo SpA

Gist-Brocades

Gist-Brocades Food Ingredients, Inc., 2200 Renaissance Blvd., Suite 150, King of Prussia, PA 19406 USA (Tel.: 215-272-4040; 800-662-4478; Telex: 216902)

Gist-Brocades SpA, Via Milano 42, I-27045 Casteggio, Italy (Tel.: 383-8931; FAX 383-805397; Telex: 321197 VINAL I)

Giulini Corp.

105 E. Union Ave., Boundbrook, NJ 08805 USA (Tel.: 201-469-6504; FAX 201-469-8418; Telex: 700179)

Givaudan

Givaudan SA, 5 Chemin de la Perfumiere, CH-1214 Vernier, Switzerland (Tel.: 22-780 91 11; FAX 22-780 91 50)

Givaudan Iberica SA, Plá d'en Batlle, Sant Celoni, E-08470 Barcelona, Spain (Tel.: 3-867 06 00; FAX 3-867 03 19; Telex: 94000)

Givaudan K.K., 11-11, Shoutou 2-chome, Shibuya-ku, Tokyo, 150, Japan (Tel.: (03) 3481-5120; FAX (03) 5478-7047)

Glaxo Inc.

Five Moore Dr., Research Triangle Park, NC 27709 USA (Tel.: 919-248-2100)

Glaxo Laboratories
Glaxo House, Berkeley Avenue, Greenford, Middlesex, UB6 0NN, UK (Tel.: 071-493-4060; FAX 081-966-8330)

Glenwood Inc.
83 N. Summit St, PO Box 518, Tenafly, NJ 07670 USA (Tel.: 201-569-0050)

Glessner Corp. Inc. (GGI Products) DBA
1301 Sansome Street, San Francisco, CA 94111 USA

Glidden Co. See ICI America Holdings Inc./Paints Group

Glo-Mold, Inc.
3261 Copley Rd., Copley, OH 44321 USA (Tel.: 216-668-6513)

Glover (Chemicals) Ltd. Address unknown

GLS Plastics
740 B Industrial Dr., Cary, IL 60013-1962 USA (Tel.: 708-516-8300; 800-457-TPRS; FAX 708-516-8361)

Glucona
Posbtus 975, NL-3800 AZ Amersfoort, The Netherlands (Tel.: 643911; FAX 637448; Telex: 79322 AKZO NV)

Goldschmidt
Th. Goldschmidt AG, Goldschmidtstrasse 100, Postfach 101461, D-4300 Essen 1, Germany (Tel.: 0201-173-2947; FAX 201-173-2160; Telex: 857170)

Goldschmidt Ltd., Subsidiary of Goldschmidt AG, Tego House, Victoria Road, Ruislip, Middlesex, HA4 0YL, UK (Tel.: 081-422 7788; FAX 081-864 8159; Telex: 923146)

Goldschmidt Chemical Corp., 914 E. Randolph Rd., PO Box 1299, Hopewell, VA 23860 USA (Tel.: 804-541-8658; 800-446-1809; FAX 804-541-2783; Telex: 710-958-1350)

Tego Chemie Service GmbH, Goldschmidstr. 100, Postfach 101461, D-4300 Essen 1, Germany (Tel.: 0201-1732571; FAX 0201-1732639; Telex: 85717-20tgd)

Tego Chemie Service USA, PO Box 1299, 914 E. Randolph Rd., Hopewell, VA 23860 USA (Tel.: 804-541-8658; 800-446-1809; FAX 804-541-2783)

Goldschmidt Japan KK, Th., Rm. 1113, Shuwa Kioi-cho TBR Bldg. No. 7, 5-Chome, Koji-machi, Chiyoda-ku, Tokyo, 102, Japan

Goldsmith & Eggleton
Goldsmith & Eggleton, Inc., 2550 Gilchrist Rd., PO Box 1784, Akron, OH 44309 USA (Tel.: 216-733-7565; FAX 216-733-7560)

Goldsmith & Eggleton Spain, Plaza Urquinaona, 6, 10D, EDIF Torre Urquinaona, 08010 Barcelona, Spain

BFGoodrich Co.

BFGoodrich Co., Specialty Polymers & Chem. Div., 9921 Brecksville Rd., Brecksville, OH 44141 USA (Tel.: 216-447-5000; 800-331-1144; FAX 216-447-5720; Telex: 423313)

BFGoodrich/Geon Vinyl Div., 6100 Oak Tree Blvd., Cleveland, OH 44131 USA (Tel.: 216-447-6000; 800-438-4366)

BFGoodrich Canada, 195 Columbia St. West, Waterloo, Ontario, N2J 4N9, Canada

BFGoodrich (UK) Ltd., The Lawn, 100 Lampton Road, Hounslow, Middlesex, TW3 4EB, UK (Tel.: 081-570 4700; FAX 081-570 0850)

BFGoodrich Chemical (Deutschland) GmbH, Goerlitzer Str. 1, 4040 Neuss 1, Germany

Goodrich-Gulf Chemicals. *See* Goodrich Co., BF

Goodyear

Goodyear Tire & Rubber Co., 1485 E. Archwood Ave., Akron, OH 44316 USA (Tel.: 216-796-8755; FAX 216-796-2617; Telex: 640550 Gdyr)

Goodyear Canada Inc., 45 Raynes Ave., Bowmanville, Ontario, Canada

Goodyear Chemicals Europe, Ave. des Tropiques, Z.A. de Courtaboeuf, 91952 Les Ullis Cedex, France (Tel.: 1 64463660; FAX 1 64465551; Telex: 602895F)

Goodyear Int'l. Corp., Sankaido Bldg., 1-9-13 Akasaka, Minato-Ku, Tokyo, 107, Japan (Tel.: (03) 582-0926; FAX (03) 582-1877)

Gordon International Corp.

Bossington Farms Ltd., Houghton, Stockbridge, Hants, SO20 6LY, UK (Tel.: 0794-388265)

GP66 Chemical Corp.

PO Box 8832, Baltimore, MD 21224 USA

W.R. Grace & Co

W.R. Grace/Dearborn Div., 300 Genesse St., Lake Zurich, IL 60047 USA (Tel.: 708-438-1800; FAX 708-540-1588)

W.R. Grace/Evans Chemetics, 55 Hayden Ave., Lexington, MA 02173 USA (Tel.: 617-861-6600 x2331; FAX 617-862-3869; Telex: 200076 GRLX UR)

W.R. Grace/Evans Chemetics, Griffin Lane, Aylesbury, Bucks, HP19 3BP, UK (Tel.: 0296 84877; FAX 0296 393122)

W.R. Grace/Hampshire & Polymer Prods., 55 Hayden Ave., Lexington, MA 02173 USA (Tel.: 617-861-6600 x2314)

W.R. Grace/Nitroparaffins Unit, 55 Hayden Ave, Lexington, MA 02173 USA (Tel.: 617-861-6600 x 2827; FAX 617-863-8070; Telex: 200076 GRLX UR)

W.R. Grace/Organic Chemicals Div., 55 Hayden Ave., Lexington, MA 02173 USA (Tel.: 617-861-6600; 800-232-6100; FAX 617-862-3869; Telex: 200076)

Grace & Co. of Canada Ltd., W.R., 3455 Harvester Rd., Unit #7, Burlington, Ontario, L7N 3P2, Canada (Tel.: 416-681-0285)

W.R. Grace Ltd., Northdale House, North Circular Road, London, NW10 7UH, UK (Tel.: 081-965-0611; Telex: 25139)

Grace Dearborn Ltd., Div. of W.R. Grace, Waterside Lane, Ditton, Widnes, Cheshire, WA8 8UD, UK (Tel.: 051 424 5351; FAX 051 423 2722; Telex: 627341)

Grace NV, Div. of W.R. Grace, Nijverheidsstraat 7, B-2260 Westerlo, Belgium (Tel.: 014 57 56 11; FAX 014 58 55 30; Telex: 31500)

Grace Rexolin Chem. AB, Box 622, S-25106 Helsingborg, Sweden (Tel.: 42261460; FAX 42260051; Telex: 72353)

Graesser Laboratories Ltd. *See* Nipa Laboratories Ltd.

Grain Processing Corp.

1600 Oregon St., Muscatine, IA 52761 USA (Tel.: 319-264-4265; FAX 319-264-4289; Telex: 46-8497)

Graphite Products Corp.

5756 Warren-Sharon Road, PO Box 29, Brookfield, Trumbull, OH 44403 USA (Tel.: 216-394-1617; 800-321-7521; FAX 216-394-2389; Telex: 271512 GPIUR)

Great Lakes

Great Lakes Chemical Corp., PO Box 2200, W. Lafayette, IN 47906 USA (Tel.: 317-497-6100; 800-428-7947)

Great Lakes Chemical (Europe) Ltd., P O Box 44, Oil Sites Road, Ellesmere Port, South Wirral, L65 4GD, UK (Tel.: 051-356 8489; FAX 051-356 8490)

Great Lakes Chem. (Europe) Ltd./QO Chems. Div., Div. of QO, Units 8 and 9, Twelve O'Clock Court, 21 Attercliffe Road, Sheffield, S4 7 WW, UK (Tel.: 0742-767842; FAX 44-742-758472; Telex: 851-54462 GLCEQO G)

Great Lakes-QO Chemicals Europe Sarl, 238 route de l'Empereur, F-92500 Rueil Malmaison, France (Tel.: 47 32 06 22; FAX 47 32 08 88; Telex: 204555)

K & K Greeff Chemicals Ltd., Div. of Beijer Industries AB,

Suffolk House, George Streeet, Croydon, Surrey, CR9 3QL, UK (Tel.: 081-686 0544; FAX 081 686 4792; Telex: 28386)

R. W. Greeff & Co. Inc.

1445 E Putnam Ave, Old Greenwich, CT 06870 USA (Tel.: 203-637-4371; FAX 203-637-5618)

Greene, Tweed & Co.

Box 305, Kulpsville, PA 19443 USA (Tel.: 215-256-9521; FAX 215-256-0189)

Griffin Corp.

PO Box 1847, Rock Ford Rd., Valdosta, GA 31601 USA (Tel.: 912-242-8635; 800-237-1854)

Grindsted

Grindsted Products A/S, Edwin Rahrs Vej 38, DK-8220 Brabrand, Denmark (Tel.: 06-25-3366; FAX 06-25-1077; Telex: 64177)

Grindsted Products Ltd., Northern Way, Bury St. Edmunds, Suffolk, IP32 6NP, UK (Tel.: 284769631)

Grinsted France S.A.R.L., Parc D'Activités de Tissaloup, Ave. Jean D'Alembert, F-78190 Trappes, France

Grindstedvaerket GmbH, Roberts-Bosch Strabe, D-2085 Quickborn, Deutschland, Germany

Grindsted Products Inc., 201 Industrial Pkwy., PO Box 26, Industrial Airport, KS 66031 USA (Tel.: 913-764-8100; 800-255-6837; FAX 913-764-5407; Telex: 4-37295)

Grindsted do Brazil, Ind. Ecom Ltda., Rodovia Regisé Bitten Court, KM 275, 5 Cx. Postal 172, 06800 Embú S.P., Brazil

Grünau, Chemische Fabrick Grünau GmbH, A Henkel Group Co.

Robert-Hansen Str. 1, Postfach 1063, W-7918 Illertissen, Bavaria, Germany (Tel.: (07303)13-0; FAX (07303)13206; Telex: 719114 gruea-d)

Guardian Chemical, A Div. of United Guardian Inc.

230 Marcus Blvd, P O Box 2500, Smithtown, NY 11787 USA (Tel.: 516-273-0900; 800-645-5566; FAX 516-273-0858)

Guest Industrials. *Address unknown*

Guthrie Latex, Inc.

7400 N. Oracle, Suite 330, Tucson, AZ, 85704 USA (Tel.: 602-742-3087; FAX 602-575-0511; Telex: 187150 guth ut)

Haagen Chemie BV. *See* Harcros Chemicals BV

Haarmann & Reimer Corp.

Haarmann & Reimer Ltd., Div. of Bayer UK Ltd., Denison Rd, Selby, North Yorkshire, YO8 8EF, UK (Tel.: 0757 703691; FAX 0757 701468)

Haarmann & Reimer Ltd., Fieldhouse Lane, Marlow, Buckinghamshire, SL7 1NA, UK (Tel.: 0628 472051; FAX 06288 472238; Telex: 848 859)

Haarmann & Reimer GmbH, Postfach 1253, D-3450 Holzminden, Germany (Tel.: 0 55 31/7011; FAX 055 31/7016 49; Telex: 965 330)

Haarmann & Reimer Corp., PO Box 175, 70 Diamond Road, Springfield, NJ 07081 USA (Tel.: 201-467-5600; 800-422-1559; FAX 201-912-0499; Telex: 219134)

Haarmann & Reimer Corp./Food Ingreds. Div., 1127 Myrtle St., PO Box 932, Elkhart, IN 46515 USA (Tel.: 219-264-8716; FAX 219-262-7056)

Hadfield, George & Co. *Address unknown*

Haifa Chemicals Ltd.

Marketing Department, PO Box 1809, Haifa, Israel

Hallam Polymer Engineering Ltd.

Callywhite Lane, Dronfield, Sheffield, South Yorkshire, S18 6XR, UK (Tel.: 0246-415511; FAX 0246-414818; Telex: 54378)

C. P. Hall Co.

7300 South Central Ave., Chicago, IL 60638-0428 USA (Tel.: 708-594-6000; 800-321-8242; FAX 708-458-0428)

Halstab, Div. of Hammond Lead Products

3100 Michigan St., Hammond, IN 46323 USA (Tel.: 219-844-3980; FAX 219-844-7287; Telex: 72-5481)

Haltermann

Haltermann AG, Ferdinand Strasse 55-57, W-2000 Hamburg 1, Germany (Tel.: 040-33380; FAX 040 3338214; Telex: 2161898)

Haltermann NV, Kentenislaan 3, B-9130 Kallo, Belgium (Tel.: 03 750 0211; FAX 03 775 0261; Telex: 33705)

Hammond Lead Products Inc.

PO Box 6408, 5231 Hohman Ave., Hammond, IN 46325-6408 USA (Tel.: 219-931-9360; FAX 219-931-2140)

Hampshire & Polymer Products. *See under* Grace, W.R.

Handy & Harman

850 Third Ave., New York, NY 10022 USA (Tel.: 212-752-3400; FAX 212-207-2614; Telex: 126288)

Hanshaw Chemicals. *Address unknown*

Harbison-Carborundum Corp. *Address unknown*

Harcros

Harcros Chemicals UK Ltd., Specialty Chemicals Div., Lankro House, PO Box 1, Eccles, Manchester, M30 0BH, UK (Tel.: 061-789-7300; FAX 061-788-7886; Telex: 667725)

Harcros Durham Chemicals, Div. of Harcros Chemicals UK Ltd., Birtley, Chester Le Street, Co Durham, DH3 1QX, UK (Tel.: 091-410 2361; FAX 091-410 6005; Telex: 53618 DURHAM G)

Harcros Chemicals BV, Haagen House, PO Box 44, 6040 AA Roermond, The Netherlands (Tel.: 04750-9-1777; FAX 04750-1-7489)

Harcros Chemicals Scandia ApS, Vesterbrogade 14A, 1620 Copenhagen V, Denmark (Tel.: 31 21 42 00; FAX 31 21 42 27; Telex: 16 152 LANKRO DK)

Harcros Chemicals France Sarl, BP 40, 441220 St. Laurent, Nouan, France

Harcros Pigments Europe, Deanshanger, Milton Keynes, MK19 6HA, UK (Tel.: 0908-562051; FAX 0908-566431)

Harcros Chemicals Inc., 5200 Speaker Rd., PO Box 2930, Kansas City, KS 66106-1095 USA (Tel.: 913-321-3131; FAX 913-621-7718; Telex: 477266)

Harcros Industrial Chemicals, 3-5 Alan Street, Rydalmere, NSW 2116, Australia

Lankro Chemicals, 140 Chia Hsin Bldg., 96 Chung Shan N. Rd., Sec. 2, Taipei 10449, Taiwan, R.O.C.

Hardman Inc., A Harcros Chemical Group Co.

600 Cortlandt St., Belleville, NJ 07109 USA (Tel.: 201-751-3000; FAX 201-751-8407; Telex: TWX: 710-995-4940)

Harker Stagg. *Address unknown*

Harlow Chemical Co. Ltd.
  Central Road, Templefields, Harlow, Essex, CM20 2AH, UK (Tel.: 0279-436211; FAX 0279-444025; Telex: 817528 HARCO G)

Harshaw Chemicals Ltd.
  Industrial Finishing Division, P.O. Box 4, Daventry, Northamptonshire, NN11 4HF, UK (Tel.: 032 722161; Telex: 31354)

Hart Chemicals Ltd.
  256 Victoria Rd. South, Guelph, Ontario, N1H 6K8, Canada (Tel.: 519-824-3280; FAX 519-824-0755; Telex: 06956537)

Hart Products Corp.
  173 Sussex St., Jersey City, NJ 07302 USA (Tel.: 201-433-6632)

Harwick Chemical Corp.
  60 S. Seiberling St., PO Box 9360, Akron, OH 44305-0360 USA (Tel.: 216-798-9300; FAX 216-798-0214; Telex: TWX: 810-431-2126)

Harwick Standard Chemical Co. *Address unknown*

A. C. Hatrick Chemicals Pty Ltd.
  49-61 Stephen Road, Botany Bay, Botany, NSW 2019, Australia (Tel.: 666-9331)

Hawks Chemical Co. Ltd.
  2 Tower St, Hyde, Cheshire, SK14 1JW, UK (Tel.: 061-367 9441; FAX 061-367 9443; Telex: 665408)

W. Hawley & Son Ltd.
  Colour Works, Duffield, Derbyshire, DE56 4FG, UK (Tel.: 0332-840294; FAX 0332-842570; Telex: 37286 HAWLEY G)

Haworth (ARC) Ltd. *Address unknown*

Haynes International
  Haynes Int'l. Ltd., PO Box 10, Parkhouse St., Openshaw, Manchester, M11 2ER, UK (Tel.: 061-230-7777; FAX 061-223-2412; Telex: 667611)
  Haynes Int'l. S.A.R.L., 43 Rue de Bellevue, Boite Postale No. 47, 92101 Boulogne Cedex, France (Tel.: (01)48-25-22-76; FAX (01)48250251)
  Haynes Int'l. Inc., 1020 West Park Ave., PO Box 9013, Kokomo, IN 46904-9013 USA (Tel.: 800-342-9637; 800-354-0806; FAX 317-456-6905; Telex: 272280)

Haysite Reinforced Plastics, An Alco Industries Co.
  Sales & Marketing Division, 5599 New Perry Highway, Erie, PA 16509 USA (Tel.: 814-868-3691; FAX 814-864-7803)

H.B. Fuller. *See under* Fuller, H.B.

Heico Chemicals, Inc., A Cambrex Co.

Route 611, PO Box 160, Delaware Water Gap, PA 18327-0160 USA (Tel.: 717-420-3900; 800-34-HEICO; FAX 717-421-9012)

Henkel

Henkel KG, Postfach 1100, Henkelstrasse 67, W-4000 Düsseldorf, Germany (Tel.: 211-7970; FAX 211-7974008; Telex: 858170)

Henkel KGaA/Cospha, Postfach 1100, D-4000, Dusseldorf 1, Germany (Tel.: 0211-797-2289; FAX 0211-798-7696; Telex: 085817-0)

Henkel KGaA/Dehydag, Postfach 1100, D-4000 Düsseldorf 1, Germany (Tel.: 0211 797-4221; FAX 0211 798-8558; Telex: 085817-122)

Henkel Ltd., Div. of Henkel KG, Henkel House, 292-308 Southbury Road, Enfield, Middlesex, EN1 1TS, UK (Tel.: 081-804 3343; FAX 081-805 0398; Telex: 922708 HENKEL G)

Henkel Ltd/Adhesives, 292-308 Southbury Road, Enfield, Middlesex, EN1 1TS, UK (Tel.: 081-804 3343; FAX 081-443 2777)

Henkel Ltd/Organic Products Div., Merit House, The Hyde, Edgeware Rd., London, NW9 5AB, UK

Henkel Belgium NV, Div. of Henkel KG, 66 ave du Port, B-1210 Brussels, Belgium (Tel.: 2-423 17 11; FAX 2-428 34 67; Telex: 21294 HENKEL B)

Henkel France SA, Div. of Henkel KG, BP 309, 150 rue Gallieni, F-92102 Boulogne Billancourt, France (Tel.: 46 84 90 00; FAX 46 84 90 90; Telex: 633177 HENKEL F)

Henkel-Nopco SA/Process Chem. Div., 185 Ave. de Fontainebleau, 77310 St. Fargeau, Ponthierry, France (Tel.: 60-65-9090; FAX 60-65-7880; Telex: 692027)

Henkel Corp./Coating Chemicals, 300 Brookside Ave., Ambler, PA 19002 USA (Tel.: 800-445-2207)

Henkel Corp./Cospha, 300 Brookside Ave., Ambler, PA 19002 USA (Tel.: 215-628-1476; 800-531-0815 (sales); FAX 215-628-1450)

Henkel Corp./Functional Products, 300 Brookside Ave., Ambler, PA 19002 USA (Tel.: 215-628-1466; 800-654-7588; FAX 215-628-1155)

Henkel Corp./Organic Products Div., 300 Brookside Ave., Ambler, PA 19022 USA (Tel.: 215-628-1000; 800-922-0605; FAX 215-628-1200)

Henkel Corp./Polymers Div., Fine Chemicals Div., 5325 South 9th Ave., La Grange, IL 60525-3602 USA (Tel.: 708-579-6150; 800-237-4037; FAX 312-579-6152)

Henkel Corp./Textile Chemicals, 11709 Fruehauf Dr., Charlotte, NC 28273-6507 USA (Tel.: 800-634-2436; FAX 704-587-3804)

Henkel Canada Ltd., 2290 Argentia Rd., Mississauga, Ontario, L5N 6H9, Canada (Tel.: 416-542-7550; FAX 416-542-7588)

Henkel Argentina S.A., Avda. E. Madero Piso 14, 1106 Capital Federal, Argentina

Henkel Corp./Emery Group

Henkel Corp./Emery Group, 11501 Northlake Dr., Cincinnati, OH 45249 USA (Tel.: 513-530-7300; 800-543-7370; FAX 513-530-7581)

Henkel Corp./Emery Group/OPG, 3300 Westinghouse Blvd., Charlotte, NC 28217 USA (Tel.: 800-634-2436)

Henkel Corp./Emery Chemicals Ltd., 365 Evans Ave., Toronto, Ontario, M8Z 1K2, Canada (Tel.: 416-259-3751)

Henkel Corp./Emery Group Japan, PO Box 191, World Trade Center, 2-4-1 Hamamatsu-cho, Minato-Ku, Tokyo, 105, Japan (Tel.: 4355611-2)

Henley Chemicals, Inc., Div. of Boehringer Ingelheim
50 Chestnut Ridge Rd., Montvale, NJ 07645 USA (Tel.: 201-307-0422; 800-635-3558; FAX 201-307-0424)

Henry Doubleday Research Association
Ryton-on-Dunsmore, Coventry, CV8 3LG, UK (Tel.: 0203-303517; FAX 0203-639229)

Hental KGaA
Postfach 101100, D-4000 Düsseldorf, Germany (Tel.: 211 7970; FAX 211 798 7696)

Herbert Laboratories, Dermatology Div. of Allergan Pharmaceuticals
2525 Dupont Drive, Irvine, CA 92713 USA (Tel.: 714-252-4500)

Hercules
Hercules Inc., Hercules Plaza-6205SW, Wilmington, DE 19894 USA (Tel.: 302-594-6500; 800-247-4372; FAX 302-594-5400; Telex: 835-479)

Hercules Ltd., Div. of Hercules Inc., 31 London Road, Reigate, Surrey, RH2 9YA, UK (Tel.: 0737-242434; FAX 0737-224288; Telex: 25803)

Hercules BV, 8 Veraartlaan, PO Box 5822, 2280 HV Rijswijk, The Netherlands (Tel.: 070-150-000; Telex: 31172)

Heresite Protective Coatings Inc.
PO Box 249, 822 S. 14th St., Manitowoc, WI 54220 USA (Tel.: 414-684-6646)

Hermetite Products Ltd., Retail Div.
Tavistock Road, West Drayton, UB7 7RA, UK

Heterene Chemical Co., Inc.
PO Box 247, 792 21st Ave., Paterson, NJ 07543 USA (Tel.: 201-278-2000; FAX 201-278-7512; Telex: 883358)

Hexcel
Hexcel Corp., 11711 Dublin Blvd., Dublin, CA 94566 USA (Tel.: 415-828-4200)

Hexcel Chemical Products, Hexcel Corp, Chemical Products Div, 215 North Centennial St, Zeeland, MI 49464 USA (Tel.: 616-772-2193; FAX 616-772-7344)

Hexcel Chemical Products Ltd., Div. of Hexcel Corp., Seal Sands Road, Seal Sands, Middlesborough, TS2 1UB, UK (Tel.: 0642-546546; FAX 0642-546068)

Hickson Danchem Corp.

PO Box 4000, Danville, VA 24543 USA (Tel.: 804-797-8100; FAX 804-799-2814; Telex: 940103 WU PUBTLXBSN)

Hickson & Welch Ltd., Div. of Hickson International plc

Wheldon Road, Castleford, West Yorkshire, WF10 2JT, UK (Tel.: 0977-556565; FAX 0977-518058; Telex: 55378 HIKSON G)

High Duty Alloys (Properties) Ltd.

154 Edinburgh Avenue, Slough, Berks, SL1 4UD, UK (Tel.: 0753-72778)

High Point Chemical Corp.

PO Box 2316, 243 Woodbine St., High Point, NC 27261 USA (Tel.: 919-884-2214; 800-727-2214; FAX 919-884-5039)

Hill Brothers Chem. Co.

1675 N. Main St., Orange, CA 92667 USA (Tel.: 714-998-8800; 800-821-7234; FAX 714-998-6310)

Hilton Davis Chemical Co.

2235 Langdon Farm Rd., Cincinnati, OH 45237 USA (Tel.: 513-841-4000; 800-477-1022; FAX 800-477-4565)

Himont

Himont U.S.A. Inc., Three Little Falls Centre, 2801 Centerville Rd., PO Box 15439, Wilmington, DE 19850-5439 USA (Tel.: 302-996-6000; 800-545-7719)

Himont Advanced Materials, 2727 Alliance Dr., PO Box 26248, Lansing, MI 48909 USA (Tel.: 517-336-9600; 800-874-4668)

Himont Canada Inc., 3 Robert Speck Pkwy., Mississauga, Ontario, L4Z 2G5, Canada (Tel.: 416-848-8800)

Hi Temp Lubricants Inc.

7019 Corporate Way, Dayton, OH 45459 USA

Hitox Corp. of America

PO Box 2544, Corpus Christi, TX 78403 USA (Tel.: 512-882-5175; FAX 512-882-6948)

HMC Wheels. *Address unknown*

Hodag Corp. *See* Calgene Chem. Inc.

Hodgson Chemicals Ltd., Div. of BTP plc

Chantry Lane, PO Box 7, Beverley, North Humberside, HU17 0NN, UK (Tel.: 0482 881133; FAX 0482 871888; Telex: 592406)

Hoechst AG

Hoechst AG, Postfach 80 03 20, D-6230 Frankfurt am Main 90, Germany (Tel.: 069-305-03113/7043; FAX 069-303665/66; Telex: 6990936)

Celanese GmbH, Hordenbachstrasse #40, D-5600 Wuppertal 21, Germany

Celanese France S.a.r.l., Z.I. Le Broteau 69540, Irigny, France

Hoechst Chemicals (UK) Ltd., Div. of Hoechst AG, Hoechst House, Salisbury Road, Hounslow, Middlesex, TW4 6JH, UK (Tel.: 081-570 7712; FAX 081-577 1854; Telex: 22284)

Hoechst (UK) Ltd/Agriculture Div, Promark Div, Div. of Hoechst AG, East Winch Hall, East Winch, King's Lynn, Norfolk, PE32 1HN, UK (Tel.: 0553-841581; FAX 0553-841090)

Hoechst (UK) Ltd./Films Div., Div. of Hoechst AG, Hoechst House, Salisbury Rd., Hounslow, Middlesex, TW4 6JH, UK (Tel.: 081 570 7712; FAX 081 572 4854; Telex: 23284)

Hoechst (UK) Ltd./Polymers Div., Div. of Hoechst AG, Walton Manor, Wlaton, Milton Keynes, Bucks, MK7 7A3, UK (Tel.: 0908 665050; FAX 0908 680516; Telex: 826300)

Hoechst Celanese Plastics Ltd., 78-80 St. Albans Rd., Watford, Herts, WD2 4AP, UK (Tel.: 923-33616)

Hoechst Celanese/Int'l. Headqtrs., 26 Main St., Chatham, NJ 07928 USA (Tel.: 201-635-2600; 800-235-2637; FAX 201-635-4330; Telex: 136346)

Hoechst Celanese/Colorants & Surfactants Div., 5200 77 Center Dr., Charlotte, NC 28217 USA (Tel.: 704-527-6000; 800-255-6189; FAX 704-559-6323)

Hoechst Celanese/Engineering Plastics Div., 26 Main St., Chatham, NJ 07928 USA (Tel.: 201-635-2600; FAX 201-635-4300)

Hoechst Celanese/Fine Chemicals Div., Portsmouth Tech. Ctr., 3340 West Norfolk Rd., Portsmouth, VA 23703 USA (Tel.: 804-483-7320; FAX 804-483-7460)

Hoechst Celanese/Spec. Polymers/Waxes Div., PO Box 58160, Houston, TX 77258 USA (Tel.: 713-474-6026; 800-242-UHMW; FAX 713-474-6025)

Hoechst Celanese/Specialty Chem. /Polymer Additives Group, 5200 77 Center Dr., PO Box 1026, Charlotte, NC 28217 USA (Tel.: 704-559-6027; FAX 704-559-6305)

Hoechst Canada Inc., Div. of Hoechst AG, 800 Blvd Rene Levesque O, Montreal, Quebec, PQH38121, Canada (Tel.: 514-871-5511)

Hoechst Canada Inc./Plastic Prods. Dept., 4045 Cote Vertu Blvd., Montreal, Quebec, H4R 1R6, Canada (Tel.: 514-333-3500; FAX 514-331-1526)

Celanese Mexicana S.A., Av. Revolucion No. 1425, Mexico 20, D.F., Mexico

Hoechst Japan, 10-33, 4-Chome-Akasaka, Minato-Ku, Tokyo, Japan

Hoechst Celanese Far East Ltd., 801 Hong Kong Club Bldg., 3A Chater Rd., Central, Hong Kong

Hoechst-Roussel Pharmaceuticals Inc.

Route 202-206 North, P.O. Box 2500, Somerville, NJ 08876-1258 USA (Tel.: 201-231-2000; 800-451-4455)

Hoffmann-La Roche

Hoffmann-La Roche Inc., 340 Kingsland St., Nutley, NJ 07110 USA (Tel.: 201-235-8080; 800-526-0189; FAX 201-235-8023)

Hoffmann-La Roche SA, Grenzacherstrasse 124, CH-4002 Basle, Switzerland (Tel.: 61-688 11 11; FAX 61-691 93 91; Telex: 962292 HLR CH)

Hoffmann Mineral, Div. of Franz Hoffmann & Söhne KG

Postfach 1460, Münchener Str 75, W-8858 Neuburg/Donau, Germany (Tel.: 8431/53-0; FAX 8431/53-330; Telex: 55223 hond-d)

Holliday Dyes & Chemicals Ltd.

PO Box B22, Leeds Road, Huddersfield, West Yorkshire, HD2 1UH, UK (Tel.: 0484-421841; FAX 0484-515328)

Holt Lloyd

Holt Lloyd Corp., 4647 Hugh Howell Road, Tucker, GA 30084 USA

Holt Lloyd Ltd., Lloyds House, Alderley Rd, Wilmslow, Cheshire, SK9 1QT, UK (Tel.: 0625 526838; FAX 0625 526962; Telex: 669737)

Hommel

Chemische Werke Hommel GmbH, Postfach 1263, Klosterrunstrasse 17, W-7840 Mülheim 1, Germany (Tel.: 7631-5547; FAX 7631-15028; Telex: 772942 CHWH D)

Hommel SA, CP 269, CH-1260 Nyon, Switzerland (Tel.: 631111; FAX 621744; Telex: 419988 ZYMA CH)

Hommel Pharmaceuticals, Industriering 34, CH-8134, Aduswil, Zurich, Switzerland

Honeywell Atlas. *Address unknown*

Honeywill & Stein Ltd., Div. of BP Chemicals Ltd.

Times House, Throwley Way, Sutton, Surrey, SM1 4AF, UK (Tel.: 081-770 7090; FAX 081-770 7295; Telex: 946560 BPCLGH G)

Horace Cory PLC. *See* European Colour plc

Horlicks, Div. of SmithKline-Beecham

New Horizon Court, Brentford, Middlesex, TW8 9EP, UK (Tel.: 081 975 2000)

Hormel & Co., Geo. A.

501 16th Ave. NE, Box 800, Austin, MN 55912 USA (Tel.: 507-437-5676; FAX 507-437-5120)

Horn's Crop Service Center

P O Box 326, Bellevue, OH 44811 USA

Hortichem Ltd.

1 Edison Road, Churchfields Industrial Estate, Salisbury, Wilts, SP2 7NU, UK (Tel.: 0722-20133; FAX 0722-26799)

Hough, Hoseason & Co. Ltd.

20-22 Chapel Street, Levenshulme, Manchester, Lancashire, M19 3PT, UK (Tel.: 061-224 3271; FAX 061-257 2076; Telex: 666976)

Howlett Adhesives Ltd.

Horsley Road, Off Kingsthorpe Road, Northampton, NN2 6LL, UK (Tel.: 0604-712977; FAX 0604-791471)

J. M. Huber Corp.

J. M. Huber Corp./Chemicals Div., PO Box 310, Havre de Grace, MD 21078 USA (Tel.: 301-939-3500; FAX 301-939-0394)

J. M. Huber Corp./Clay Div., One Huber Rd., Macon, GA 31298 USA (Tel.: 912-745-4751; 800-TRY-HUBER; FAX 912-745-1116; Telex: 544438)

J. M. Huber Corp./Solem Div., 4940 Peachtree Industrial Blvd., Norcross, GA 30071 USA (Tel.: 404-441-1301; FAX 404-368-9908)

Hubron

Hubron Ltd., Albion St. Works, Failsworth, Manchester, M35 0FP, UK (Tel.: 061 681 2691; FAX 061 683 4658; Telex: 666852)

Hubron Rubber Chemicals, Albion Works, Albion Street, Failsworth, Manchester, M35 0FP, UK (Tel.: 061-681 2691; FAX 061-683 4658)

Huels AG. See Hüls AG

Hüls AG

Hüls AG, Postfach 1320, D-4370 Marl 1, Germany (Tel.: 02365-49-1; FAX 02365-49-2000; Telex: 829211-0)

Hüls UK Ltd., Edinburgh House, 43-51 Windsor Rd., Slough, Berks, SL1 2HL, UK (Tel.: 0753-71851; FAX 0753-820480)

Hüls UK Ltd., Featherstone Rd, Wolverton Mill South, Milton Keynes, Buckinghamshire, MK12 5TB, UK (Tel.: 0908 226 444; FAX 0908 224 950; Telex: 826500)

Hüls France SA, Div. of Hüls AG, 49-51 Quai de Dion Bouton, 92815 Puteaux Cedex, France (Tel.: 1 49 06 55 00; FAX 1 47 73 97 65; Telex: 611868)

Hüls America Inc., Turner Place, PO Box 365, Piscataway, NJ 08855-0365 USA (Tel.: 908-981-5377; FAX 908-981-5497)

Hüls Canada, Inc., 235 Orenda Rd., Brampton, Ontario, L6T 1E6, Canada (Tel.: 416-451-3810)

Humber Fertilisers plc

P O Box 27, Stoneferry, Hull, HU8 8DQ, UK (Tel.: 0482-20458; FAX 0482-212825)

Humphrey Chemical Co Inc., A Cambrex Co.

Devine St., North Haven, CT 06473-0325 USA (Tel.: 203-230-4945; 800-652-3456; FAX 203-287-9197; Telex: 994487)

Huntington Laboratories, Inc.

970 East Tipton St., Huntington, IN 46750 USA (Tel.: 219-356-7073; 800-537-5724; FAX 219-356-6485)

Octavius Hunt Ltd., Member of Chemring Group plc

5 Dove Lane, Redfield, Bristol, BS5 9NQ, UK (Tel.: 0272-555304; FAX 0272-557875; Telex: 449134 OCHUNT G)

Huntsman

Huntsman Chemical Corp., 2000 Eagle Gate Tower, Salt Lake City, UT, 84111 USA (Tel.: 801-532-5200; FAX 801-355-6629)

Huntsman Chemical Co. of Canada, Inc., Bellevue St., C.P./PO Box 231, Mansonville, Quebec, J0E 1X0, Canada

Huntsman Chemical Co. Ltd., Carrington, Urmston, Manchester, M31 4AJ, UK (Tel.: 061-775-5321; FAX 44-61-775-0105)

Huntsman Pacific Chem. Corp., 42-1 Hsu Chang St., Taipei, 100, Taiwan, R.O.C.

Huntsman Expandable Polystyrene

5100 Bainbridge Blvd., Chesapeake, VA 23320 USA (Tel.: 804-494-2500; FAX 804-494-2770; Telex: 9103334117)

Hybri-Chem Inc.

212 West Taft Ave., Orange, CA USA (Tel.: 714-921-2300; FAX 714-921-9643)

Hydro Chemicals Ltd., Div. of Norsk Hydro AS

Immingham, Grimsby, South Humberside, DN40 2NS, UK (Tel.: 0469 571602; FAX 0469 571603; Telex: 52292)

Hydro Plast AB

S-444, 83 Stenungsund, Sweden (Tel.: 303 87798; FAX 303 771884)

Hydro Polymers, Div. of Norsk Hydro A/S

Hydro Polymers Ltd./Performance Prods. Div., Div. of Norsk Hydro A/S, New Lane, Havant, Hampshire, PO9 2NQ, UK (Tel.: 0705 486350; FAX 0705 452430; Telex: 86720)

Hydro Polymers Ltd./Vinyls Div., Aycliffe Ind. Est., Newton Aycliffe, Co. Durham, DL5 6EA, UK (Tel.: 0325 300555; FAX 0325 300215; Telex: 58322)

Hyland Therapeutics, Div. of Travenol Laboratories Inc.

444 W Gelnoaks Blvd, Glendale, CA 91202 USA

Hynson Westcott & Dunning, Div. of Becton Dickinson & Co

Charles & Chase Sts, Baltimore, MD 21201 USA

Hysol Aerospace Prods., Div. of Dexter

2850 Willow Pass Rd., PO Box 312, Pittsburg, CA 94565 USA (Tel.: 415-687-4201; FAX 415-687-4205; Telex: TWX: 910-387-0363)

Hytak Ltd.

Greenhill Industrial Estate, Leabrooks, Derby, DE55 4BR, UK

ICC Industries Inc.

720 Fifth Ave., New York, NY 10019 USA (Tel.: 212-903-1730; Telex: 212-903-1794)

ICI plc

ICI plc, 9 Millbank, London, SW1 3JF, UK (Tel.: 071 834 4444; FAX 071 834 2040)

ICI plc, Chemicals & Polymers Group, PO Box 90, Wilton, Middlesborough, Cleveland, TS6 8JE, UK (Tel.: 0642 454144; FAX 0642 432444; Telex: 587 461)

ICI Chemicals & Polymers Ltd., Div. of ICI plc, PO Box 13, The Heath, Runcorn, Cheshire, WA7 3JF, UK (Tel.: 0928-514444; FAX 0928-576675; Telex: 629655 ICIMOH G)

ICI Deutsche GmbH, Postfach 500728, Emil-von-Behring-Strasse 2, W-6000 Frankfurt am Main, Germany (Tel.: 69-5801-00; FAX 69-6802345; Telex: 416974 ICI D)

ICI Belgium SA, Div. of ICI plc, Everslaan 45, B-3078 Everberg, Belgium (Tel.: 2-758 92 11; FAX 2-759 77 22; Telex: 21332 ICIEVB B)

ICI Moscow, Krasnopresnenskaya Naberezhnaya 12, Moscow, Russia (Tel.: 95-253 2056; FAX 95-230 2044; Telex: 413241 ICIMO SU)

ICI Americas, Inc., Subsidiary of ICI plc, New Murphy Rd. & Concord Pike, Wilmington, DE 19897 USA (Tel.: 302-886-3000; 800-456-3669; FAX 302-886-2972; Telex: 4945649)

ICI Japan Ltd., Osaka Green Bldg., 1,3-Chome Kitahama, Higashi-Ku, Osaka, 541, Japan

ICI Australia Operations Pty. Ltd./ICI Surfactants Australia, ICI House, 1 Nicholson St., Melbourne, 300, Australia (Tel.: 3-665-7111; FAX 61-03-665-7009; Telex: 30192)

ICI Acrylics-KSH Inc., 10091 Manchester Rd., St. Louis, MO 63010 USA (Tel.: 314-996-3111; 800-325-9577; FAX 800-44-0803)

ICI Advanced Materials UK, PO Box 6, Shire Park, Bessemer Rd., Welwyn Garden City, Herts, AL7 1HD, UK (Tel.: 0707-323400; FAX 0642-432444)

ICI Advanced Materials, 475 Creamery Way, Exton, PA 19341 USA (Tel.: 800-854-8774)

ICI Advanced Materials, Rt. 202 & Murphy Rd., Wilmington, DE 19897 USA (Tel.: 302-886-3290; FAX 302-886-5333)

ICI Agrochemicals Professional Products, Woolmead House East, Woolmead Walk, Farnham, Surrey, GU9 7UB, UK (Tel.: 0252-733919; FAX 0252-736222)

ICI Agrochemicals, Subsidiary of ICI plc, Fernhurst, Haslemere, Surrey, GU27 3JE, UK (Tel.: 0428-644061; FAX 0428-652922; Telex: 858270 ICIPPF G)

ICI Fiberite Molding Materials, 501 East 3rd St., Winona, MN 55987 USA (Tel.: 507-454-3611)

Fiberite Europe, Div. of ICI, Industriestrasse 1, D-7524 Oestringen, Germany (Tel.: 7253-81870)

Kasei Fiberite Ltd., Mitsubishi Bldg., 11th Floor, 5-2, Marunouchi 2-Chome, Chiyoda-Ku, Tokyo, Japan

ICI Films/Polyester Polymer Group, Wilmington, DE 19897 USA

ICI Fluoropolymers, 475 Creamery Way, Exton, PA 19341 USA (Tel.: 215-363-4746; 800-842-8739)

ICI Garden Products, A Business Unit of ICI Agrochemicals, Fernhurst, Haslemere, Surrey, GU27 3JE, UK (Tel.: 0428-645454; FAX 0428-657222)

ICI Petrochemicals & Plastics Div, PO Box 90, Wilton, Middlesbrough, Cleveland, North Humberside, TS6 8JE, UK (Tel.: 0642 454144; FAX 0642 432398)

ICI Pharma, Wilmington, DE 19897 USA (Tel.: 302-886-2231)

ICI Polymer Additives, PO Box 751, Wilmington, DE 19897 USA (Tel.: 302-886-3564; 800-822-8215 x 3564; FAX 302-886-5267)

ICI Polyurethanes Group/Formulated Products Div., 6555 Fifteen Mile Rd., Sterling Heights, MI 48077 USA (Tel.: 313-826-7660; 800-553-8624)

ICI Resins BV, Div. of ICI plc, Postbus 123, Sluisweg 12, NL-5140 AC Waalwijk, The Netherlands (Tel.: 04160-8 99 11; FAX 04160-8 99 22; Telex: 35079 ICI RW NL)

ICI Resins US, 730 Main St., Wilmington, MA 01887-0677 USA (Tel.: 508-658-6600; 800-225-0947; FAX 508-657-7978)

ICI Speciality Chemicals Ltd., Div. of ICI plc, Cleeve Rd, Leatherhead, Surrey, KT22 7SW, UK (Tel.: 0372 376122; FAX 0372 386245; Telex: 8954510 ICILHD G)

ICI Specialty Chemicals, Concord Pike & New Murphy Rd., Wilmington, DE 19897 USA (Tel.: 302-886-3000; 800-822-8215; FAX 302-886-2972)

ICI Surfactants (Belgium), Everslaan 45, B-3078 Everberg, Belgium (Tel.: 2-758-9361; FAX 2-758-9686)

ICI Surfactants Ltd. (UK), Smith's Rd., Bolton, Lancashire, BL3 2QJ, UK (Tel.: 204-21971/4; FAX 204-363676; Telex: 667841)

LNP Engineering Plastics/An ICI Company, 475 Creamery Way, Exton, PA 19341 USA (Tel.: 215-363-4500; 800-854-8774)

LNP Plastics UK, Unit 25 Monkspath Business Park, Solihull, West Midlands, B90 4NX, UK (Tel.: 021-744-9922)

LNP Plastics Nederland B.V., Ottergeerde 24, Raamsdonksveer, The Netherlands

ICN Biomedicals Ltd., Div. of ICN Biomedicals Inc.
PO Box 17, Second Ave, Industrial Estate, Irving, KA12 8NB, Scotland (Tel.: 0294 74242; FAX 0294 311415; Telex: 77231 FLOW G)

ICN Nutritional Biochemicals Corp.
26201 Miles Road, Cleveland, OH 44128 USA

ICN Pharmaceuticals Inc.
ICN Plaza, 3300 Hyland Ave, Costa Mesa, CA 92626 USA (Tel.: 714-545-0100; 800-556-1937)

Ideal Manufacturing Ltd.
Atlas House, Burton Road, Finedon, Wellingborough, Northants, NN9 5HX, UK (Tel.: 0933-681616; FAX 0933-681042)

IGI Boler, Inc.
85 Old Eagle School Rd., Wayne, PA 19087 USA (Tel.: 215-687-9030; 800-852-6537)

IGI Petroleum Specialities
164 Sheridan St., Perth Amboy, NJ 08861 USA (Tel.: 201-826-0140; 800-323-6549; FAX 201-826-0641)

Ilon Laboratories. *Address unknown*

IMC Fertilizer, Inc.
501 E. Lange St., Mundelein, IL 60060 USA (Tel.: 708-949-3700; 800-323-5523)

I M I (Kynoch) Ltd. *See* Imperial Chemical Industries

Imperial Chemical Industries plc

PO Box 6, Shire Park, Bessemer Road, Welwyn Garden City, Hertfordshire, AL7 1HD, UK (Tel.: 0707-323400; FAX 0707-337332; Telex: 94028500 ICIC G)

Incitec Ltd.

PO Box 140, Morningside, Queensland, 4170, Australia (Tel.: 07-867-9300; FAX 07-867 9310; Telex: 40747)

Inco Alloys

Inco Alloys Int'l. Inc., 3200 Riverside Dr., Huntington, WV, 25720 USA (Tel.: 304-526-5100; FAX 304-526-5441; Telex: 886413)

Inco Europe Ltd., 1-3 Grosvenor Place, London, SW1X 7EA, UK (Tel.: 071-235 2040; FAX 071-235 4358; Telex: 22621 INCOLN G)

Indspec Chemical Corp.

Indspec Chemical Corp., 411 Seventh Ave., Suite 300, Pittsburgh, PA 15219 USA (Tel.: 412-765-1200; FAX 412-765-0439; Telex: 199187 Indspec)

Indspec Chemical Corp./European Sales, Gebouw de Goudsesingel, Kipstraat 8-10, 3011 RT Rotterdam, The Netherlands (Tel.: 10-4-120-122; Telex: 21253 Indsp NL)

Industrial Adhesives Company

Bond-Plus Adhesives & Coatings, 2632 W Washington Blvd, Chicago, IL 60612 USA

Industrial Adhesives Ltd.

Moor Road, Chesham, Bucks, HP5 1SB, UK (Tel.: 0494-784444; FAX 0494-791903; Telex: 837356)

Industrial Dielectrics, Inc.

PO Box 357, Noblesville, IN 46060 USA (Tel.: 317-773-1766)

Industrial Fibers, Inc.

2889 North Nagel Court, Lake Bluff, IL 60044 USA (Tel.: 708-295-0046; FAX 708-295-0520)

Industrial Waxes Ltd.

Tubs Hill House, London Rd, Sevenoaks, Kent, TN13 1BL, UK (Tel.: 0732 741888; FAX 0732 464466; Telex: 957054)

Industrias Quimicas del Valles SA

Avenida Rafael de Casanova 81, Mollet del Vallés, E-08100 Barcelona, Spain (Tel.: 3-570 56 96; FAX 3-593 80 11; Telex: 52170)

Ingasetter Ltd.

391 Union St, Aberdeen, AB1 2BX, UK (Tel.: 0224 580641; FAX 0224 574264)

Innovative Engineering of Michigan, Inc.

941 West Round Lake Rd., DeWitt, MI 48220 USA (Tel.: 517-669-1591; FAX 516-669-9120)

Innoxa (England) Ltd.
Beauty House, Hawthorne Road, Eastbourne, East Sussex, BN23 6QX, UK (Tel.: 0323-641244; FAX 0323-639375)

Inolex Chemical Co.
Jackson & Swanson Sts., Philadelphia, PA 19148-3497 USA (Tel.: 215-271-0800; 800-521-9891; FAX 215-271-2621; Telex: 834617)

Insulgard Ltd.
Southwater Business Park, Worthing Road, Southwater, Nr Horsham, West Sussex, RH13 7HE, UK (Tel.: 0403 730032; FAX 0403 731783)

Interferon Sciences Inc.
783 Jersey Avenue, New Brunswick, NJ 08901 USA

International Bio-Synthetics
International Bio-Synthetics Inc., PO Box 241068, Charlotte, NC 28224 USA (Tel.: 704-527-9000; 800-438-1361; FAX 704-527-8184)

International Biosynthetics Ltd., Hale Bank, Widnes, Cheshire, WA8 8NS, UK (Tel.: 051-424 3671; FAX 051-420 1301; Telex: 629491 WBCHEM)

International Chemical Co. Ltd.
11 Chenies Street, London, WCI 7ET, UK (Tel.: 081-6260544)

International Coatings Co. Inc.
13929 166th Street, Cerritos, CA 90701 USA (Tel.: 310-926-0747; FAX 310-926-9486)

International Dioxcide Inc.
136 Central Ave., Clark, NJ 07066 USA (Tel.: 908-499-9660; FAX 908-388-3648; Telex: 642582 PROLINE)

International Flavors & Fragrances
1515 Hwy 36, Union Beach, NJ 07735 USA (Tel.: 908-264-4500; FAX 908-888-2595; Telex: 275284)

International Gallium GmbH
Postfach 1860, Alustrasse 50, W-8460 Schwandorf, Germany (Tel.: 61586; FAX 50807)

International Sourcing Inc.
121 Pleasant Ave., Upper Saddle River, NJ 07458 USA (Tel.: 201-934-8900; 800-772-7672; FAX 201-934-8291; Telex: 697-2957 INSOURC)

International Specialty Products, Inc. See ISP

International Synthetics Ltd. See International Bio-Synthetics Ltd.

Interox

Interox Chemicals Ltd., PO Box 7, Warrington, Cheshire, WA4 6HB, UK (Tel.: 0925 51277; FAX 0925 232207; Telex: 627834 LAPORT G)

Interox America, 3333 Richmond Ave., Houston, TX 77098 USA (Tel.: 713-522-4155; 800-468-3769; FAX 713-524-9032; Telex: 166307)

Intervet Inc.

405 State Street, P O Box 318, Millsboro, DE 19966 USA (Tel.: 302-934-8051; FAX 302-934-6087)

Interwood Ltd.

Stafford Avenue, Hornchurch, Essex, RM11 2ER, UK (Tel.: 0708-452591; FAX 0708-457813; Telex: 896801)

Intracrop Ltd.

The Crop Centre, Waterstock, Oxford, OX9 1LJ, UK (Tel.: 08447-377; FAX 08447-267)

Invequimica & CIA SCA

PO Box 3227, Medellin, Columbia

Iolab Corp., A Johnson & Johnson Co.

500 Iolab Drive, Claremont, CA 91711 USA (Tel.: 714-624-2020; 800-423-1871; FAX 714-399-1646; Telex: 188668)

Ionac Chemical Co, Div. of Sybron Chemicals Inc.

Birmingham Rd, Birmingham, NJ 08011 USA (Tel.: 609-893-1100)

Irathane International Ltd. *Address unknown*

ISC Alloys Ltd. *See* Pasminco Europe

ISC Chemicals Ltd.

St Andrews Road, Avonmouth, Bristol, BS11 9HP, UK (Tel.: 0272-823631; FAX 0272-822688)

Ishihara Sangyo Kaisha

Ishihara Sangyo Kaisha, Ltd., 3-22, Edobori 1-chome, Nishi-ku, Osaka, 550, Japan (Tel.: (06) 444-1451; FAX (06) 445-7798; Telex: 5622-774)

ISK Europe SA, Div. of Ishihara Sangyo Kaisha Ltd., Tour ITT, 480 avenue Louise, B-1050 Brussels, Belgium (Tel.: 2-646 34 90; FAX 2-648 34 72; Telex: 23362 ISK B)

ISP

International Specialty Products, Inc., 1361 Alps Rd, Wayne, NJ 07470 USA (Tel.: 201-628-4000; FAX 201-628-4117; Telex: 130374)

ISP Technologies Inc., PO Box 1006, Bound Brook, NJ 08805 USA (Tel.: 908-271-0111; 800-622-4423)

GAF (Great Britain) Ltd., Chemical & Industrial Products, Tilson Road, Roland Thorn Ind. Estate, Wythenshawk, Manchester, M23 9PH, UK (Tel.: 061-998 1122; FAX 061-998 6218)

GAF Europe, 40 Alan Turing Rd., Surrey Research Park, Buildford, Surrey, UK

ITW Devcon, An Illinois Tool Works Co.
30 Endicott St., Danvers, MA 01923 USA (Tel.: 508-777-1100; 800-933-8266; FAX 800-765-4329)

Ivax Industries Inc.
Textile Products Div., PO Box 10027, Rock Hill, SC 29731 USA (Tel.: 803-366-9411; 800-343-7872; FAX 803-366-7256)

Ives Laboratories Inc.
685 Third Ave, New York, NY 10017 USA

Ernst Jager GmbH, Jager Fabrik Chemischer Rohnstoffe GmbH
Postfach 130380, Oerschbachstrasse 35-39, DW-4000 Düsseldorf 13, Germany (Tel.: 0211 792271; FAX 0211-793144; Telex: 8582127 JAEG D)

James River Corp./Berlin-Gorham Group
650 Main St., Berlin, NH 03570 USA (Tel.: 603-752-4600)

Janssen Chimica
Janssen Chimica, Div. of Janssen Pharmaceutica, Janssen Pharmaceuticalaan 3, B-2440 Geel, Belgium (Tel.: 14-60 42 00; FAX 14-60 42 20; Telex: 34103)

Janssen Pharmaceutical Ltd., Grove, Wantage, Oxon, OX12 0DQ, UK (Tel.: 0235-772121; FAX 0235-772901)

Janssen Pharmaceutica Inc., Subsidiary of Johnson & Johnson, 1125 Trenton-Harbourton Road, P.O. Box 200, Titusville, NJ 08560-0200 USA (Tel.: 609-730-2000; 800-253-3682; FAX 609-730-3044)

Spectrum Chemical Mfg Corp, Distributor, 14422 S San Pedro St, Gardena, CA 90248 USA (Tel.: 310-516-8000; 800-772-8786; FAX 310-516-9843; Telex: 182395 BIOSPECT)

Janssens NV
PO Box 129, Europark-Oost II 15, B-9100 Sint-Niklass, Belgium (Tel.: 3-776 43 11; FAX 3-776 80 02)

Japan Synthetic Rubber Co., Ltd.
11-24, Tsukiji 2-chome, Chuo-Ku, Tokyo, 104, Japan (Tel.: (03) 5565-6500; FAX (03) 5565-6636; Telex: 0252-2638 JSRCTKJ)

Jarchem Industries Inc.
414 Wilson Ave., Newark, NJ 07105 USA (Tel.: 201-344-0600; FAX 201-344-5743; Telex: 362-660)

Jenkins Laboratories Inc. *Address unverified*
Auburn, NY USA

Jetco Chemicals, Inc.

PO Box 1898, Corsicana, TX 75110 USA (Tel.: 903-872-3011; 800-477-5353; FAX 903-872-4216; Telex: 75110)

Jet-Lube

Jet-Lube, Inc., 4849 Homestead Rd. 77028, PO Box 21258, Houston, TX 77226 1258 USA (Tel.: 713-674-7617; 800-JET-LUBE; FAX 713-678-4604; Telex: 775393)

Jet-Lube (UK), Inc., Unit 5E & D, Broomfield Industrial Estate, Montrose, DD10 8SY, Scotland (Tel.: 674-77804; FAX 674-73481; Telex: 76547)

J.M. Huber. *See under* Huber, J.M.

John & E Sturge Ltd. *See* Haarmann & Reimer Ltd. (Selby)

Johnson & Johnson Products Inc.

One Johnson/Johnson Plaza, New Brunswick, NJ 08933 USA (Tel.: 908-524-0400)

Johnson Matthey

Johnson Matthey plc, Johnson Matthey Technology Centre, Blount's Court, Sonning Common, Reading, Berks, RG4 9NH, UK (Tel.: 0734 722811; FAX 0734 723236)

Johnson Matthey plc, Orchard Rd, Royston, Hertfordshire, SG8 5HE, UK (Tel.: 0763 244161; FAX 0763 245014; Telex: 817351)

Johnson Matthey SA, Div. of Johnson Matthey plc, BP 50240, 13 rue de la Perdrix, F-95956 Roissy CDG, France (Tel.: 48 63 22 99; Telex: 230195)

Johnsons of Hendon Ltd.

Hempstalls Lane, Newcastle, Staffordshire, ST5 0SW, UK (Tel.: 0782 717100; FAX 0782 717707)

Jojoba Growers & Processors Inc.

2267 S. Coconino Dr., Apache Junction, AZ, 85220 USA (Tel.: 602-982-1125; FAX 602-982-4183; Telex: 754858)

Brian Jones & Associates Ltd.

Fluorocarbon Building, Caxton Hill, Hertford, Hertfordshire, SG13 7NH, UK (Tel.: 0992-553065; FAX 0992-551873)

Jonk BV, Div. of Witco

Postbus 5, Wezelstraat 12, NL-1540 AA Koog a/d Zaan, The Netherlands (Tel.: 75-283 854; FAX 75-210 811; Telex: 19270)

Jotun

Jotun Polymer A/S, Div. of Jotun A/S, Postboks 2061, N-3200 Sandefjord, Norway (Tel.: 34 61000; FAX 34 64614; Telex: 21526 JOTUN N)

Jotun Polymer UK Ltd., Div. of Jotun Polymer A/S, 54 Willow Lane, Mitcham, Surrey, CR4 4NA, UK (Tel.: 081 648 4684; FAX 081 640 6432; Telex: 944038 JOTUN G)

Jotun (Deutschland) GmbH, Winsbergring 25, D-2000 Hamburg 54, Germany (Tel.: 040 853136 0; FAX 040 85 62 34)

J R Technology Ltd.
81 North End, Meldreth, Royston, Herts, SG8 6NU, UK (Tel.: 0763-260721; FAX 0763-262002)

Jungbunzlauer AG, Div. of Montana AG
Postfach 546, Schwarzenbergplatz 16, A-1010 Vienna, Austria (Tel.: 222 50200; FAX 222 502008; Telex: 133396 JUBA A)

Kabi Pharmacia
800 Centennial Ave, P.O. Box 1327, Piscataway, NJ 08855-1327 USA (Tel.: 908-457-8000; 800-526-3619; FAX 908-457-8283)

KabiVitrum
KabiVitrum AB, Lindhagensgaten 133, S-112 87 Stockholm, Sweden (Tel.: 8-13 80 00; FAX 8-5480 20; Telex: 17065 KABI S)

KabiVitrum NV, 1001 chaussée d'Alsemberg, B-1180 Brussels, Belgium (Tel.: 2-370 4747; FAX 2-377 29 16; Telex: 23450 CCP B)

KabiVitrum Ltd. *Address unknown*

Kalon Chemicals Ltd., Div. of Kalon Group plc
Bassington Industrial Estate, Cramlington, Northumberland, NE23 8AD, UK (Tel.: 0670 713411; FAX 0670 590235; Telex: 537565 KALON G)

Kao Corp. S.A
Puig dels Tudons, 10, 08210 Barbera Del Valles, Barcelona, Spain (Tel.: 3-729-0000; FAX 3-718-9829; Telex: 59749)

Kaopolite, Inc.
2444 Morris Ave., Union, NJ 07083 USA (Tel.: 908-789-0609; FAX 908-851-2974)

Karlshamns
Karlshamns Lipids for Care, S-374 82, Karlshamn, Sweden (Tel.: 454-823-00; FAX 454-129-11; Telex: 4500 fopart s)

Karlshamns Lipids for Care, PO Box 569, Columbus, OH 43216 USA (Tel.: 614-299-3131; 800-526-4547; FAX 614-299-8279; Telex: 245494 capctyprdcol)

K&B Antipollution Ltd.
Thame, Oxfordshire, UK

Keeling & Walker Ltd.
Whielden Road, Stoke-on-Trent, Staffordshire, ST4 4JA, UK (Tel.: 0782-744136; Telex: 36225)

Kelco
Kelco International Ltd., Westminster Tower, 3 Albert Embankment, London, SE1 7RZ, UK (Tel.: 071-735-0333; FAX 071-735-1363; Telex: 23815 KAILIL G)

Kelco, Div. of Merck & Co Inc., 8355 Aero Drive, PO Box 23576, San Diego, CA 92123-1718 USA (Tel.: 619-467 6547; 800-535-2656; FAX 619-467 6520; Telex: WUD 695228)

Kelco/Oil Field Prods. Div. *Address unknown*

Kemira Ince Ltd.

Ince, Chester, CH2 4LB, UK (Tel.: 051 357 2777; FAX 051 357 2144; Telex: 627407 KEMCH G)

Kemira Oy

Kemira Kemi AB, Div. of Kemira Oy, Box 902, Industrigatan 83, S-251 09 Helsingborg, Sweden (Tel.: 42-17 10 00; FAX 42-14 06 35; Telex: 72185 KEMWATS S)

Kemira Kemi (UK) Ltd., Div. of Kemira Oy, Orm House, 2 Hookstone Park, Harrogate, North Yorkshire, HG2 8QT, UK (Tel.: 0423 885005; FAX 0423 885939; Telex: 57215 KEMFOS)

Kemira Polymers, Station Road, Birch Vale, Stockport, Cheshire, SK12 5BR, UK (Tel.: 0663-746518; FAX 0663-746605)

S. Kempner Ltd.

498-500 Honeypot Lane, Stanmore, Middlesex, HA7 1JZ, UK (Tel.: 081 952 5262; FAX 081 952 8061)

Kemtron International Inc.

P O Box 2508, 289 Sherman Avenue, Newark, NJ 07114 USA (Tel.: 201-623-7787)

Keno Gard (UK) Ltd. *Address unknown*

Kenrich Petrochemicals, Inc.

140 E. 22nd St., PO Box 32, Bayonne, NJ 07002-0032 USA (Tel.: 201-823-9000; 800-LICA KPI; FAX 201-823-0691; Telex: 12-5023)

Kensol Corporation

PO Box 3179, Allentown, PA 18106 USA (Tel.: 800-776-0612)

The Kent Chemical Company Ltd.

George House, Bridewell Lane, Tenterden, Kent, TN30 6HS, UK (Tel.: 058-06 4244; FAX 058-06 5652)

Kerr-McGee Chemical Corp.

Kerr-McGee Ctr., PO Box 25861, Oklahoma City, OK 73125 USA (Tel.: 405-270-1313; 800-654-3911; FAX 405-270-3123; Telex: 747-128)

Key Polymer Corp.

Jacob's Way, Lawrence Ind. Park, Lawrence, MA 01842 USA (Tel.: 508-683-9411; FAX 508-686-7729; Telex: 940 103)

Killgerm Chemicals Ltd.

Denholme Drive, Ossett, West Yorkshire, WF5 9BW, UK (Tel.: 0924-277631; FAX 0924-265033)

Kincaid Enterprises, Inc.

PO Box 549, Plant Rd., Nitro, WV, 25143 USA (Tel.: 304-755-3377; FAX 304-755-4547)

King Industries, Inc.
Science Rd., Norwalk, CT 06852 USA (Tel.: 203-866-5551; 800-431-7900; FAX 203-866-1268; Telex: 710-468-0247)

Kinglsey & Keith Chemical Corp. See K & K Greeff Chemical Group Ltd.

Kingston Technologies, Inc.
2235-B Route 130, Dayton, NJ 08810 USA (Tel.: 908-274-2288)

Kinnis & Brown. See K&B Antipollution Ltd.

KMZ Chemicals Ltd.
48 Station Rd, Stoke D'Abernon, Cobham, Surrey, KT11 3BN, UK (Tel.: 0932 866426; FAX 0932 867099; Telex: 929624)

F. E. Knight Inc.
8 Perry Drive, Foxboro, MA 02035 USA

Knoll AG
Knoll AG, Div. of BASF AG, Postfach 210805, W-6700 Ludwigshafen, Germany (Tel.: 621-5890; FAX 621-5892950; Telex: 464823)

Knoll AG, Div. of BASF AG, Oristalstrasse 65, CH-4410 Liestal, Switzerland (Tel.: 061-9125 05; FAX 061-91 2563; Telex: 966000 KNL CH)

Knoll Pharmaceutical Co.
30 N Jefferson Rd, Whippany, NJ 07981 USA (Tel.: 201-887-8300; 800-526-0710)

Koch Chemical Co., Div. of Koch Refining Co.
4111 E 47 St N, Wichita, KS 67220 USA (Tel.: 800-835-1121)

Koch Chemicals Ltd.
2 Marshgate Drive, Hertford, Hertfordshire, SG13 7JY, UK (Tel.: 0992 553781; FAX 0992 586961; Telex: 817136 KOCHEM G)

Kodak Ltd. See under Eastman Kodak

Kommer-Brookwick Ltd.
88 Westlaw Place, Whitehill Estate, Glenrothes, Fife, KY6 2RZ, UK (Tel.: 0592-630052; FAX 0592-630109)

Koninklijke Maastrichtsche
Zinkwit Maastrichtsche, 1 Rocourstraat 28, NL-6245 AD Eijsden, Netherlands (Tel.: 04409-9222; FAX 04409-3995)

Koninklijke Nederlandsche Gist-En Spiritusfabriek. Address unknown

Koppers Co Inc. Defunct company

Koppers Industries, Inc.

436 Seventh Ave, Room 1600, Koppers Bldg, Pittsburgh, PA 15219 USA (Tel.: 412-227-2001; 800-321-9876; FAX 412-227-2202)

Koppert (UK) Ltd.

1 Wadhurst Business Park, Faircriouch Lane, Wadhurst, East Sussex, TN5 6PT, UK (Tel.: 0892-884411; FAX 0892-882469)

Koster Keunen

Koster Keunen, Inc., 90 Bourne Blvd., PO Box 447, Sayville, NY 11782 USA (Tel.: 516-589-0456; FAX 516-589-0120)

Koster Keunen Europe, Bladel, The Netherlands (Tel.: 4977-87902)

Kraeber GmbH & Co.

Hochallee 80, W-2000 Hamburg 13, Germany (Tel.: 40-445061; FAX 40-455163; Telex: 17403443)

Kronos, Inc.

Kronos, Inc., PO Box 60087, 3000 N. Sam Houston Pkwy. East, Houston, TX 77205 USA (Tel.: 713-987-6300; 800-866-5600; FAX 713-987-6358)

Kronos Canada, Inc., 4 Place Ville-Marie, Suite 500, Montreal, Quebec, H3B 4M5, Canada (Tel.: 514-397-3501; FAX 514-393-1186; Telex: 5268667)

Kronos Ltd., Div. of Kronos Inc., St Ann's House, Wilmslow, Cheshire, SK9 1HG, UK (Tel.: 0625 529511; FAX 0625 533123; Telex: 669055)

Kureha Chemical Industry Co., Ltd.

9-11, Nihonbashi Horidome-cho 1-chome, Chuo-ku, Tokyo, 103, Japan (Tel.: (03) 3249-4666; FAX (03) 3661-1277; Telex: 252-2737 KUREHA J)

KWR Chemicals Ltd., Div. of Berk Ltd.

Priestly Road, Basingstoke, Hampshire, RG24 9QH, UK (Tel.: 0256-810155; FAX 0992-760103; Telex: 261585)

Kyowa Chemical Industry Co., Ltd.

305, Yashima-Nishimachi, Takamatsu-shi, Kagawa, 761-01, Japan (Tel.: (0878) 41-9179; FAX (0878) 41-6147; Telex: 5822220)

Kyowa Hakko Kogyo Co. Ltd.

Ohtemachi Bldg., 6-1, Ohtemachi 1-chome, Chiyoda-ku, Tokyo, 100, Japan (Tel.: (03) 3201-7211; FAX (03) 3284-1968; Telex: J24543 HKKYOWA)

LAB Ltd. *Address unknown*

Laboratorios Vinas SA

Planta 5A, Provenza 386, E-08025 Barcelona, Spain (Tel.: 3-207 05 12; FAX 3-207 1932)

Laevosan GmbH

Postfach 316, Estermannstrasse 17, A-4021 Linz, Austria (Tel.: 732-278231; FAX 732-278231400; Telex: 21803 LAEV A)

Lagap Pharmaceuticals Ltd.

Woolmer Way, Bordon, Hants, GU35 9QE, UK (Tel.: 0420-478301; FAX 0420-474427)

Lakeland Laboratories Ltd.

Peel Lane, Astley Green, Tyldesley, Manchester, M29 7FE, UK (Tel.: 0942-873555; FAX 0942-884409; Telex: 67413)

Lambson Ltd.

Aire & Calder Works, Cinder Lane, Castleford, West Yorkshire, WF10 1LU, UK (Tel.: 0977 510511; FAX 0977 603049; Telex: 557661 AGRAM G)

Lanaetex Products, Inc.

151-157 Third Ave., Elizabeth, NJ 07206 USA (Tel.: 201-351-9700; FAX 201-351-8753; Telex: 3792268 TLAP1)

Lancashire Chemical Works Ltd.

High Street West, Glossop, Derbyshire, SK13 8ES, UK (Tel.: 0457-860006; FAX 0457-868394)

Landers-Segal Color Co. Inc.

84 Dayton Ave., Passaic, NJ 07055 USA (Tel.: 201-779-5001; FAX 201-779-8948; Telex: 6971185)

Langley Alloys Ltd. *Address unknown*

Langley Smith & Co. Ltd.

36 Spital Square, London, E1 6DY, UK (Tel.: 071-247 7473; FAX 071-375 1470; Telex: 883013)

Lankro Chemicals Ltd. *See* Harcros Chemicals UK Ltd

Lanstar Ltd.

Liverpool Road, Cadishead, Manchester, M30 5DT, UK (Tel.: 061-775 2644; FAX 061-776 1077)

Laporte

Laporte plc, Laporte House, Kings Way, Luton, LU4 8EW, UK (Tel.: 0582-21212; FAX 0582-31818)

Laporte Absorbants, Subsidiary of Laporte plc, PO Box 2, Moorfield Rd, Widnes, Cheshire, WA8 0JU, UK (Tel.: 051-495-2222; FAX 051-420-4088)

Laporte Industries Ltd., Subsidiary of Laporte plc, Laporte Inorganics Division, PO Box 2, Moorfield Road, Widnes, Cheshire, WA8 0JU, UK

Laporte Inc., 3701 Algonquin Rd., Suite 390, Rolling Meadows, IL 60008 USA (Tel.: 708-670-8870; FAX 708-670-8874)

Laporte Inc., Southern Clay Prods. Inc., PO Box 44, Gonzales, TX 78629 USA (Tel.: 800-324-2891; FAX 512-672-3930)

Laporte UK Trading. *See* Laporte plc

Larkhall Laboratories plc
225 Putney Bridge Road, London, SW15 2PY, UK (Tel.: O81-874 1130; FAX 081-871 0066; Telex: 928345)

La Roche Chemicals Inc.
PO Box 1031 Airline Hwy., Baton Rouge, LA 70821 USA (Tel.: 504-356-8463; 800-524-2586; FAX 504-356-8551; Telex: 196137)

La Roche Industries Inc.
1100 Johnson Ferry Rd. NE, Atlanta, GA 30342 USA (Tel.: 404-851-0300; FAX 404-851-0476)

Laserson & Sabetay Ets
BP57, Avenue des Grenots, F-91151 Etampes Cedex, France (Tel.: 64 943124; FAX 64 949897; Telex: 601532 LASAROM)

Laurel Industries
30000 Chagrin Blvd., Cleveland, OH 44124-5794 USA (Tel.: 216-831-5747; 800-221-1304; FAX 216-831-8479)

Laur Silicone Rubber Compounding Inc.
4930 S. M-18, Box 509, Beaverton, MI 48612 USA (Tel.: 517-435-7706; FAX 517-535-7707)

Lawn & Garden Products Inc.
P O Box 5317, Fresno, CA 93755 USA (Tel.: 209 225 4770; FAX 209 225 1319)

Lederle Laboratories
Lederle Laboratories USA, Div. of American Cyanamid Co, One Cyanamid Plaza, Wayne, NJ 07470 USA (Tel.: 914-735-2815)

Lederle Laboratories USA, Professional Services Dept., Pearl River, NY 10965 USA (Tel.: 914-735-2815)

Lederle Laboratories, Subsidiary of Cyanamid of Great Britain Ltd., Cyanamid House, Fareham Road, Gosport, Hants, PO13 0AS, UK (Tel.: 0329-224000; FAX 0329-220213)

Leeming, Div. of Pfizer Inc.
100 Jefferson Rd, Parsippany, NY 07054 USA

Leepoxy Plastics, Inc.
3324 Ferguson Rd., Fort Wayne, IN 46809 USA

Lenzing AG
Abt. P84, A-4860 Lenzing, Austria (Tel.: 7672-701-3602; FAX 7672-74823; Telex: 26606 lenfa a)

Leo Fay Ltd. (trading as Layson Ltd)
35 Tatton Court, Kingsland Grange, Warrington, Cheshire, WA1 4RR, UK (Tel.: 0925-814025; FAX 0925-837814)

Leo Laboratories
Longwick Road, Princes Risborough, Aylesbury, Buckinghamshire, HP17 9RR, UK (Tel.: 084-44 7333; FAX 084-44 2278)

Leopold Charles & Co. *Address unknown*

Leuchstoffwerk GmbH
IM Klingenbuhl 8, W-6900 Heidelberg, Germany (Tel.: 6221-73965; FAX 6221-776482; Telex: 461721)

Lever Industrial Co.
7450 Industry Dr., North Charleston, SC 29418 USA (Tel.: 803-767-0540; FAX 803-552-6337)

Lever Industrial Ltd., Subsidiary of Unilever plc
P O Box 20, Cressex Industrial Estate, High Wycombe, Buckinghamshire, HP12 3TL, UK (Tel.: 0494-461234; FAX 0494-462565; Telex: 83116)

Leverton-Clarke Ltd.
Unit 16, Sherrington Way, Lister Rd Industrial Estate, Basingstoke, Hampshire, RG22 4DQ, UK (Tel.: 0256 810393; FAX 0256 479324; Telex: 858558 LEVCOS G)

Lewis Laboratories. *Address unknown*

LignoTech. *See under* Borregaard

Eli Lilly & Co.
Lilly Corporate Center, Indianapolis, IN 46285 USA (Tel.: 317-276-2000; FAX 317-276-6876)

Lin-Pac International
Morr Lane Trading Estate, Sherburn-in-Elmet, Leeds, West Yorkshire, LS25 6ES, UK (Tel.: 0977-685 114)

Lipha SA
126 Avenue Carton de Wiart, B-1090 Brussels, Belgium (Tel.: 2-424 3240; FAX 2-424 2219; Telex: 65354 LIPHAB)

Lipo Chemicals
Lipo Chemicals, Inc., 207 19th Ave., Paterson, NJ 07504 USA (Tel.: 201-345-8600; FAX 201-345-8365; Telex: 130117)

Represented by: Blagden Chemicals Ltd., Div. of Blagden Industries, AMP House, Dingwall Rd., Croydon, Surrey, CR9 3QU, UK (Tel.: 081 681 2341; FAX 081 688 5851; Telex: 24285)

Liquid Air Corp.
2121 North California Blvd., Suite 350, Walnut Creek, CA 94596 USA (Tel.: 415-977-6500; Telex: ITT 470020)

Liquid Crystal Lens Co.
29 Bolinas Rd, Fairfax, CA 94930 USA

Liquid Nitrogen Processing Corp. *See* LNP *under* ICI

Liquid Plastics Ltd.
> PO Box 7, London Road, Preston, Lancashire, PR1 4AJ, UK (Tel.: 0772-59781; FAX 0772 202627; Telex: 67169 LPL-G)

J. H. Little. *Address unknown*

L & K Fertilisers Ltd.
> Saxilby, Lincoln, LN1 2LS, UK

Llewellyn Ryland Ltd.
> Balsall Heath Works, Haden Street, Birmingham, W Midlands, B12 9DB, UK (Tel.: 021-440 2284; FAX 021-440 0281)

Lloyd, Hamol Ltd., Reckitt & Colman Pharmaceutical Div.
> Clerk Green, Batley, West Yorkshire, WF17 5RU, UK

Lloyd's Pharmaceuticals Ltd., Reckitt & Colman Pharmaceutical Div.
> Clerk Green, Batley, West Yorkshire, WF17 5RU, UK

LNP. *See under* ICI

Loes Enterprises Inc.
> 1457 Iglehart Avenue, St Paul, MN 55104 USA (Tel.: 612-646-1385; FAX 612-646-3057)

Longfield Chemicals
> Tarvin Road, Frodsham, Warrington, Cheshire, WA6 6UZ, UK

Lonza
> Lonza AG, A member of the A-L Alusuisse-Lonza Group, Münchensteinerstrasse 38, CH-4002 Basle, Switzerland (Tel.: 61-316 8111; FAX 61-316 8733; Telex: 965960 lon ch)
>
> Lonza (UK) Ltd., Imperial House, Lypiatt Road, Cheltenham, Gloucestershire, GL50 2QJ, UK (Tel.: 242 513211; FAX 242 222294; Telex: 43152)
>
> Lonza France SARL, Div. of Alusuisse, 10-12 rue des Trois Fontanot, F-92000 Nanterre, France (Tel.: 47 75 87 08; FAX 47 78 06 27; Telex: 613647)
>
> Lonza G+T Ltd./Graphites & Technologies, CH-5643 Sins, Switzerland (Tel.: 042-66 01 11; FAX 042 66 23 16; Telex: 862-652)
>
> Lonza Inc., 17-17 Route 208, Fair Lawn, NJ 07410 USA (Tel.: 201-794-2400; 800-777-1875 (tech.); FAX 201-703-2028)

Lord Corp.
> Lord Corp., Elastomer Prods. Div., Industrial Adhesives, 2000 West Grandview Blvd., PO Box 10038, Erie, PA 16514-0038 USA (Tel.: 814-868-3611; 800-CHEMLOK; FAX 814-864-3452; Telex: 291935)
>
> Lord Corp (UK) Ltd., Stretford Motorway Industrial Estate, Barton Dock Road, Stretford, Manchester, M32 0ZH, UK (Tel.: 061 865 8048; FAX 061 865 0096; Telex: 665701 HUGMAN)

Represented by: Qyadra Chemical Ltd., 2121 Argentia Rd., Suite 303, Mississauga, Ontario, L5N 2X4, Canada

Lowi Chem. Corp.

7 West Bowery St., Suite 811, Akron, OH 44308 USA (Tel.: 216-762-8614; FAX 216-762-1135)

LRC Products Ltd.

North CIrcular Road, London, E4 8QA, UK

Lubrizol

Lubrizol Corp., 29400 Lakeland Blvd., Wickliffe, OH 44092 USA (Tel.: 216-943-4200; FAX 216-943-5337; Telex: 4332033)

Lubrizol Ltd., Waldron House, 57-63 Old Church St., London, SW3 5BS, UK (Tel.: 351-3311-20; FAX 351-3310)

Lubrizol France SA, 25 quai de France, F-76100 Rouen, France (Tel.: 35 72 04 09; Telex: 180641)

Lubrizol GmbH, Bogenallee 10, 2000 Hamburg 13, Germany

Lubrizol Japan Ltd., No. 23 Mori Bldg., 5th Floor, 23-7, Toranomon 1-chome, Minato-ku, Tokyo, 105, Japan (Tel.: (03) 3504-1345; FAX (03) 3504-1340; Telex: 26814)

Lucas Meyer GmbH & Co. *See under* Meyer GmbH & Co., Lucas

Lucta SA

Km 12, Carretera Masnou-Granollers, Montornés del Vallés, E-08170 Barcelona, Spain (Tel.: 3-845 93 00; Telex: 52485 LUCTA E)

Luitpold-Werk

Hayes Gate House, 27 Uxbridge Road, Hayes, Middlesex, UB4 0JN, UK

Lukens International Corp. *Address unknown*

Luzenac

Luzenac, Inc., 1075 North Service Rd. W., Suite 14, Oakville, Ontario, L6M 2G2, Canada (Tel.: 416-825-3930; FAX 416-825-3932)

Luzenac America, Inc., 9000 E. Nichols Ave., Englewood, CO 80112 USA (Tel.: 800-325-0299)

Mace Adhesives & Coatings Co., Inc.

Roberts Rd., PO Box 37, West Main Branch, Dudley, MA 01571 USA (Tel.: 508-943-9052; FAX 508-943-6527)

Macfarlan Smith Ltd.

Wheatfield Road, Midlothian, Edinburgh, EH11 2QA, Scotland (Tel.: 031-337 2434; FAX 031-337 9813; Telex: 727271)

Mach-1 Compounding

775 E. Highland Rd., Macedonia, OH 44056 USA (Tel.: 216-467-8108)

MacPherson Polymers

Station Road, Birch Vale, Nr Stockport, Cheshire, SK12 5BR, UK (Tel.: 0663 746518; FAX 0663 746605; Telex: 669258)

Macsil Inc.

PO Box 29276, 1326 Frankford Avenue, Philadelphia, PA 19125 USA (Tel.: 215-739-7300)

Dr. Madis Labs Inc.

375 Huyler St., South Hackensack, NJ 07606 USA (Tel.: 201-440-5000; FAX 201-342-8000; Telex: 134-200)

Magie Bros. Oil/Div. Pennzoil Products Co.

9101 Fullerton Ave., Franklin Park, IL 60131 USA (Tel.: 708-455-4500; 800-MAGIE 47; FAX 708-455-0383)

Magnesia GmbH

Postfach 2168, Kurt-Höbold-Strasse 6, W-2120 Lüneburg, Germany (Tel.: 4131-52011; FAX 4131-53050; Telex: 2182159)

Magnesium Elektron Ltd., Subsidiary of British Alcan Aluminum plc

Regal House, London Road, Twickenham, Middlesex, TW1 3QA, UK (Tel.: 081-892 4488; FAX 081-891 5744; Telex: 8811765 MELLDN G)

Magnolia Plastics, Inc.

5547 Peachtree Industrial Blvd., Chamblee, GA 30341 USA (Tel.: 404-451-2777; FAX 404-451-5376)

Makhteshim Chemical Works Ltd.

PO Box 60, Beer-Sheva, 84100, Israel (Tel.: 057-666611; FAX 057-33304; Telex: 5276)

Mallinckrodt

Mallinckrodt Inc., Imcera Group Inc., Mallinkdrodt & 2nd Street, PO Box 5439, St Louis, MO 63147 USA (Tel.: 314-539-1214; FAX 314-539-1251)

Mallinckrodt Specialty Chemicals, 16305 Swingley Ridge Dr., Chesterfield, MO 63017 USA (Tel.: 314-895-2000; 800-325-7155; FAX 314-530-2562)

Mallinckrodt Canada Inc., 7500 Trans Canada Hwy., Pointe Claire, Quebec, H9R 5H8, Canada (Tel.: 514-695-1220)

Mallinckrodt GmbH, Postfach 1268, Industriestrasse 19-21, W-6110 Dieburg, Germany (Tel.: 6071-20040; FAX 6071-200444; Telex: 4191823 MCD D)

Maltown. *Address unknown*

Malvern Minerals Co.

PO Box 1238, Hot Springs, AR 71901-1238 USA (Tel.: 501-623-8893; FAX 501-623-5113; Telex: 536-120)

Manchem Ltd. *See* Rhône-Poulenc Chemicals Ltd.

Mandops (UK) Ltd.
48 Leigh Road, Eastleigh, Hampshire, SO5 4DT, UK (Tel.: 0703-641826; FAX 0703-629106)

Mandoval Ltd.
Mark House, The Square, Lightwater, Surrey, GU18 5SS, UK (Tel.: 0276 71617; FAX 0276 76910; Telex: 858094)

George Mann & Co., Inc.
123 Georgia Ave., PO Box 9066, Providence, RI 02940 USA (Tel.: 401-781-5600; 800-556-2426; FAX 401-941-0830)

Manox Ltd.
Manox House, Coleshill Street, Miles Platting, Manchester, M10 7AA, UK

Manro Products Ltd.
Bridge St., Stalybridge, Cheshire, SK15 1PH, UK (Tel.: 061-338-5511; FAX 061-303-2991; Telex: 668442)

Manville. *See* Celite Corp.

H. Marcel Guest Ltd.
Riverside Works, Collyhurst Road, Manchester, M10 7RU, UK (Tel.: 061-205 7631; FAX 061-205 4829)

Marchon. *See* Albright & Wilson

Marfleet Refining Co. Ltd.
Hedon Road, Marfleet, Hull, North Humberside, HU9 5NJ, UK

Marine Magnesium Co.
995 Beaver Grade Rd., Coraopolis, PA 15108 USA (Tel.: 412-264-0200; FAX 412-264-9020)

Marion Merrell Dow Inc.
9300 Ward Parkway, P.O. Box 8480, Kansas City, MO 64114-0480 USA (Tel.: 800-552-3656)

Mariovilla SpA
Casella Postale 70, I-28013 Gattico, Italy (Tel.: 322-88001; FAX 322-880022; Telex: 223338 MAVILL I)

A. H. Marks & Co Ltd.
Wyke Lane, Wyke, Bradford, West Yorkshire, BD12 9EJ, UK (Tel.: 0274-675231; FAX 0274-691176; Telex: 51653 MARKS G)

Marley Adhesives Ltd.
Bath Road, Beenham, Reading, Berks, RG7 5PU, UK

Marley Floors Ltd.
Lenham, Maidstone, Kent, ME17 2DE, UK (Tel.: 0622 858877; FAX 0622 858944; Telex: 966491 MFLOOR G)

Marlowe-Van Loan Corp.

PO Box 1851, High Point, NC 27261 USA (Tel.: 919-886-7126; 800-422-4MVL; FAX 919-889-6663; Telex: TWX: 510-926-1589)

Marlow Foods Ltd., Subsidiary of Zeneca Group plc

9 Station Road, Marlow, Buckinghamshire, SL7 1NG, UK (Tel.: 0628 890850; FAX 0628 890549)

Martin Marietta Magnesia Specialties

Executive Plaza II, Hunt Valley, MD 21030 USA (Tel.: 301-527-3700; 800-648-7400; FAX 301-527-3861; Telex: 710-862-2630)

Martinswerk GmbH, Member of Lonza Group

PO Box 1209, D-5010 Bergheim, Germany (Tel.: (0 22 71) 90 2-0; FAX (02271) 90 27 10; Telex: 888 712)

Maruzen Kasei

Maruzen Kasei Co., Ltd., 14703-10, Mukaihigashi-cho, Onomichi-shi, Hiroshima, 722, Japan (Tel.: (0848) 44-2200; FAX (0848) 44-2217)

Maruzen Fine Chemicals Inc., Div. of Maruzen Kasei Co Ltd., 525 Yale Ave, Pitman, NJ 08071 USA (Tel.: 609-589-4042; FAX 609-582-8894; Telex: 333812 MARFINE)

Marval Industries Inc.

315 Hoyt Ave., Mamaroneck, NY 10543 USA (Tel.: 914-381-2400)

Mason Chemical Co. Ltd.

Carolyn House, Dingwall Rd., Croydon, CR0 9XF, UK (Tel.: 081 686 5625; FAX 081-686 1408; Telex: 929278 MASTON G)

Mason Chemical Co.

5253 West Belmont Ave., Chicago, IL 60641 USA (Tel.: 312-282-0200; 800-362-1855; FAX 312-282-0821)

Mason Pharmaceuticals Inc.

4425 Jamboree, Newport Beach, CA 92660 USA (Tel.: 714-851-6860)

Master Bond Inc.

154 Hobart St., Hackensack, NJ 07601 USA (Tel.: 201-343-8983)

Maws Pharmacy Suppliers. *Address unknown*

Maxicrop International Ltd., Div. of Stimero Ltd.

Weldon Road, Corby, Northamptonshire, NN17 IUS, UK (Tel.: 0536-402182; FAX 0536-204254; Telex: 34565 MAXINT G)

Maxwell Hart Ltd.

17 Adlington Court, Birchwood, Warrington, WA3 6PL, UK (Tel.: 0925-825501; FAX 0925-812712)

May & Baker Ltd.
Rainham Road South, Dagenham, Essex, RM10 7XS, UK (Tel.: 018-592-3060)

Maybrook Inc.
570 Broadway, PO Box 68, Lawrence, MA 01842 USA (Tel.: 508-682-1853; FAX 508-682-2544)

Mayco Oil & Chemical Co.
775 Louis Dr., PO Box 2809, Warminster, PA 18974-0357 USA (Tel.: 215-672-6600; 800-523-3903; FAX 215-443-7094)

Mayfair Chemicals Ltd.
81 Morshead Mansions, Morshead Road, London, SW9, UK (Tel.: 081 286 0274)

Mazzucchelli Celluloide SpA
Mazzucchelli 1949 SpA, Via S & P Mazzucchelli 7, Castiglione Olona, I-21043 Varese, Italy (Tel.: 826211; FAX 826213; Telex: 330609 SIC I)

M-CAP Technologies Int'l.
PO Box 7136, Wilmington, DE 19803-0136 USA (Tel.: 302-695-5616; FAX 302-695-5681)

McGhan NuSil Corp.
1150 Mark Ave., Carpinteria, CA 93013 USA (Tel.: 805-684-8780; FAX 805-684-2365)

McIntyre Chemical Co., Ltd.
1000 Governors Hwy., University Park, IL 60466 USA (Tel.: 708-534-6200; FAX 708-534-6216)

McKechnie Chemicals Ltd.
PO Box 4, Tanhouse Lane, Widnes, Cheshire, WA8 0PG, UK (Tel.: 051 424 2611; FAX 051 424 4221; Telex: 629435)

McLaughlin Gormley King Co.
8810 10th Ave. N., Minneapolis, MN 55427 USA (Tel.: 612-544-0341; Telex: 290 544 MACK GOVY)

McLube, Div. of McGee Industries Inc.
9 Crozerville Rd., Aston, PA 19014 USA (Tel.: 215-459-1890; 800-2-MCLUBE; FAX 215-459-9538; Telex: 910- 3807571 MCLUBE)

McNeil Consumer Products Co, Div. of McNeil-PPC Inc.
Camp Hill Rd, Fort Washington, PA 19304 USA (Tel.: 215-233-7000)

McNeil Pharmaceutical, Div. of McNEILAB Inc.
Spring House, PA 19477-0776 USA (Tel.: 215-628-5000)

MCP Pharmaceuticals
Parkway, Kirkton Campus, Livingstone, West Lothian, Scotland

Mead Johnson & Co.
P.O. Box 4500, Princeton, NJ 08543-4500 USA (Tel.: 609-897-2000; 800-321-1335)

Mearl
Mearl Corp., 41 E. 42nd St., New York, NY 10017 USA (Tel.: 212-573-8500; FAX 212-557-0495; Telex: 421841)

Mearl International BV, Postbus 31, Emrikweg 18, NL-2040 AA Zandvoort, The Netherlands (Tel.: 23-318 058; FAX 23-315 365; Telex: 41492)

Represented by: Cornelius Chemical Co Ltd., St. James House, 27-43 Eastern Rd., Romford, Essex, London, RM1 3NN, UK

Mechema Chemicals Ltd.
Talbot Wharf Chemical Works, Port Talbot, West Glamorgan, SA13 1RL, UK (Tel.: 0639-88 10 81; FAX 0639-88 07 05)

Meclec Ltd.
6, Towerfield Close, Shoeburyness, Essex, SS3 9QP, UK (Tel.: 0702-295047; FAX 0702-297784)

Medi-Physics Inc., Subsidiary of Amersham International plc
2636 Clearbrook Drive, Arlington Heights, IL 60005 USA (Tel.: 708-593-6300)

Medway Packaging Ltd
Medway House, New Hythe Lane, Larkfield, Aylesford, Kent ME20 6SH, UK (Tel.: 0622 717855; FAX 0622 716360)

Meer Corp.
PO Box 9006, 9500 Railroad Ave., N. Bergen, NJ 07047-1206 USA (Tel.: 201-861-9500; FAX 201-861-9267; Telex: 219130)

Meggle Marketing GmbH
Postfach 40, Megglestrasse 6-12, W-8090 Wasserburg 2, Germany (Tel.: 8071-730; FAX 8071-73449; Telex: 525137)

Edward A Mendell Co., Inc., A Penwest Company
2981 Rt. 22, Patterson, NY 12563-9970 USA (Tel.: 914-878-3414; 800-431-2457; FAX 914-878-3484; Telex: 4971034)

Menley & James Laboratories
1500 Spring Garden St, Philadelphia, PA 19101 USA

Mercian Minerals & Colours Ltd. *Address unknown*

Merck AG
E Merck, Postfach 4119, Frankfurter Strasse 250, W-6100 Darmstadt 1, Germany (Tel.: 06151-72-0; FAX 06151-72-3684; Telex: 419328-0 em d)

Merck Ltd., Div. of Merck AG, Merck House, Poole, Dorset, BH15 1TD, UK (Tel.: 0202 669700; FAX 0202 665599; Telex: 41186 TETRA G)

Merck & Co Inc., PO Box 2000, Rahway, NJ 07065-0900 USA (Tel.: 908-594-4000; FAX 908-594-5431; Telex: 138825)

Merck & Co Inc., Professional Information, West Point, PA 19486 USA (Tel.: 215-652-7300)

Merck Japan Ltd., Arco Tower, 8-1, Shimomeguro 1-chome, Meguro-ku, Tokyo, 153, Japan (Tel.: (03) 5434-4700; FAX (03) 5434-4705; Telex: MJTKO J 2226868)

Merck Sharp & Dohme Ltd., Div. of Merck & Co Inc., Hertford Road, Hoddesdon, Hertfordshire, EN11 9BU, UK (Tel.: 0992-467272; FAX 0992-468175; Telex: 261915 MSDHOD G)

Merck Sharp & Dohme, Div. of Merck & Co Inc., West Point, PA 19486 USA

Merix Chemical Co.
2234 E. 75th St., Chicago, IL 60649 USA (Tel.: 312-221-8242; FAX 312-221-3047)

Merlec Co. *Address unknown*

Merquinsa
Gran Vial, 17, E-08160 Montmeló (Barcelona), Spain (Tel.: 572 11 00; FAX 572 09 34; Telex: 94236)

Merrell Dow Pharmaceuticals Inc., Subsidiary of Dow Chemical
2110 E Galbraith Road, Cincinnati, OH 45237 USA (Tel.: 513-948-9111)

Metablen Co. BV, Joint Venture of Atochem and Mitsubishi Rayon
22, Place des Vosges, La Défense 5, Cédex 54, 92062 Paris la Défense, France (Tel.: 14904 1515; FAX 14904 1020; Telex: 611 352 F)

Metalcrete Mfg. Co.
10330 Brecksville Road, Cleveland, OH 44141 USA (Tel.: 216-526-5600; FAX 216-526-5601)

Metallgesellschaft AG
Postfach 10 15 01, Reuterweg 14, D-6000 Frankfurt am Main 1, Germany (Tel.: 069-159-2675; FAX 069-159-3287; Telex: 41225-0 mgf)

Métaux Précieux SA
2 avenue du Vignoble, CH-2009 Neuchatel, Switzerland (Tel.: 38-21 21 51; Telex: 952502)

Metco N. America Inc., Div. of Metablen
Three Parkway, Philadelphia, PA 19102 USA (Tel.: 215-587-7000)

Lucas Meyer GmbH & Co.
PO Box 261665, D-2000 Hamburg 26, Germany (Tel.: 40-789-550; FAX 40-789-8329; Telex: 2163220 myer d)

M. Michel & Co., Inc.
90 Broad St., New York, NY 10004 USA (Tel.: 212-344-3878; FAX 212-344-3880; Telex: 421468)

Micro-Biologicals Ltd.
Salisbury Street, Fordinbridge, Hants, SP6 1AE, UK (Tel.: 0425-52205; FAX 0425-54325)

Microcel Technology, Inc.
1 Ethel Road, Suite 108, Edison, NJ 08817 USA (Tel.: 908-287-1122; FAX 908-287-1030)

Microcide Ltd.
Shepherds Grove, Stanton, Bury St. Edmunds, Suffolk, IP31 2AR, UK (Tel.: 0359-51077; FAX 0284-706410)

Microfluidics Corp., Subsidiary of Biotechnology Development Corp.
90 Oak St., PO Box 9101, Newton, MA 02164-9101 USA (Tel.: 617-969-5452; 800-370-5452; FAX 617-965-1213)

Microporous Materials Ltd., Div. of Unilever
24 The Green, Braunston, Daventry, Northhamptonshire, NN11 7HW, UK (Tel.: 0788 890900; FAX 0788 890503; Telex: 31370)

Midkem Agrochemicals/Midkem Group plc, Div. of Ellis + Everard plc
20 Rothersthorpe Avenue, Northampton, Northamptonshire, NN4 9JH, UK (Tel.: 0604-764027; FAX 0604-701238; Telex: 311732)

Midland Silicones. *Address unknown*

Midland-Yorkshire Tar Distillers. *Address unknown*

Midwest Elastomers Inc.
700 Industrial Dr., PO Box 412, Wapakoneta, OH 45895 USA (Tel.: 419-738-9634; FAX 419-738-4504)

Midwest Rubber Reclaiming Div.
PO Box 2349, E. St. Louis, IL 62202-2349 USA (Tel.: 618-337-6400)

Miles Inc.
Miles Inc., Div. of Bayer AG, One Mellon Center, 500 Grant Street, Pittsburgh, PA 1, 5219-2502 USA (Tel.: 412-394 5500; FAX 412-394 5578)

Miles Inc./Organic Products Div., Bldg. 14, Mobay Rd., Pittsburgh, PA 15205-9741 USA (Tel.: 412-777-2000; 800-662-2927; FAX 412-777-7840; Telex: 1561261)

Miles Inc./Polysar Rubber Div., Div. of Bayer AG, 2603 W. Market St., Akron, OH 44313 USA (Tel.: 216-836-0451; FAX 216-836-4614)

Miles Inc./Polymers Div., Div. of Bayer AG, Mobay Rd., Pittsburgh, PA 15205-9741 USA (Tel.: 412-777-2000; 800-526-4550)

Miles Pharmaceuticals, 400 Morgan Lane, Westhaven, CT 06516 USA (Tel.: 203-937-2000; 800-468-0894; FAX 412-394-5578)

Miller-Stephenson Chem. Co., Inc.
George Washington Hwy., Danbury, CT 06810 USA (Tel.: 203-743-4447; 800-992-2424)

Milliken Chemicals
PO Box 1927, Spartanburg, SC 29304 USA (Tel.: 803-573-2200; FAX 803-573-2430; Telex: 810-282-2580)

Millmaster-Onyx UK. *Address unknown*

Milton Roy Co., Subsidiary of Sunstrand Corp.
4949 Harrison Avenue, Rockford, IL 61108 USA (Tel.: 815-226 2500)

Minas de Gádor SA
General Zabala 24, E-28002 Madrid, Spain (Tel.: 1411 03 55; FAX 1562 28 30; Telex: 41204 GADOR E)

Mineral Research & Development Corp.
One Woodlawn Green, Charlotte, NC 28217 USA (Tel.: 704-525-2771; 800-334-0417; FAX 704-527-8232)

Minnesota Mining & Manufacturing Co. *See* 3M

Mirachem Srl
Via Guido Rossa 12, I-40133 Bologna, Italy (Tel.: 51-73 71 11; FAX 51-73 64 40; Telex: 510011)

Mirfield Sales Services Ltd.
Moorend House, Moorend Lane, Dewsbury, West Yorks, WF13 4QQ, UK (Tel.: 0484-842851; FAX 0484-847066)

Misemer Pharmaceuticals Inc.
4553 South Campbell Street, Springfield, MO 65804 USA

Mission Pharmacal Co.
1325 East Durango, San Antonio, TX 78210 USA (Tel.: 512-533-7118)

Mitchell Cotts Chemicals Ltd., Div. of Suter plc
PO Box No 6, Steanard Lane, Mirfield, West Yorkshire, WF14 8QB, UK (Tel.: 0924-493861/6; FAX 0924-490972; Telex: 577016 KEMCOTTS-G)

Mitsubishi Gas
Mitsubishi Gas Chemical Co., Inc., Mitsubishi Bldg., 5-2, Marunouchi 2-chome, Chiyoda-ku, Tokyo, 100, Japan (Tel.: (03) 3283-5000; FAX (03) 214-0938; Telex: 222-2624 MGCHO J)

Mitsubishi Gas Chemical Co., Inc., 520 Madison Ave., 9th Floor, New York, NY 10022 USA (Tel.: 212-752-4620; FAX 212-758-4012; Telex: 649545 MGC UR NYK)

Represented by: Franklin Polymers, Inc., 2 Bala Plaza, Suite 300, Bala Cynwyd, PA 19004 USA (Tel.: 215-668-4353; FAX 215-667-8174)

Mitsubishi Kasei Corp.
Mitsubishi Kasei Corp., Mitsubishi Bldg., 5-2, Marunouchi 2-chome, Chiyoda-ku, Tokyo, 100, Japan (Tel.: (03) 3283-6254; Telex: BISICH J 24901)

Mitsubishi Kasei America, Inc., 81 Main St., Suite 401, White Plains, NY 10601 USA (Tel.: 914-761-9450; FAX 914-681-0760)

Mitsubishi Kasei, Europe Office, Am Seestern, Niederkasseler LohWeg 8, 4000 Duesseldorf 11, Germany

Mitsubishi Petrochemical Co., Ltd.

Mitsubishi Bldg., 5-2, Marunouchi 2-chome, Chiyoda-ku, Tokyo, 100, Japan (Tel.: (03) 3283-5700; FAX (03) 3283-5472; Telex: 222-3172)

Mitsui Co Ltd., Div. of North American Brine Resources

Enterprise Plaza, 5600 North May Ave, Oklahoma City, OK 73112 USA (Tel.: 405-842-2233; FAX 405-842-9901)

Mitsui Petrochemicals

Mitsui Petrochemical Industries, Ltd., Kasumigaseki Bldg., 2-5, Kasumigaseki 3-chome, Chiyoda-ku, Tokyo, 100, Japan (Tel.: (03) 3580-2012; FAX (03) 3593-0027; Telex: J22984 MIPECA)

Mitsui Petrochemicals (America) Ltd., 250 Park Ave., Suite 950, New York, NY 10017 USA (Tel.: 212-682-2366; FAX 212-490-6694; Telex: 7105814089)

Mitsui Petrochemical Industries, Ltd. (UK), Temple Court, 11 Queen Victoria St., London, EC4N 4SB, UK (Tel.: 236-2463; FAX 236-2220)

Mitsui Toatsu

Mitsui Toatsu Chemicals, Inc., 2-5, Kasumigaseki 3-chome, Chiyoda-ku, Tokyo, 100, Japan (Tel.: (03) 3592-4594; FAX (03) 3592-4282; Telex: 2223622 MTCHEM J)

Mitsui Toatsu Chemicals, Inc., NY Office, Two Grand Central Tower Bldg., 34th Fl., 140 E. 45 St., New York, NY 10017 USA (Tel.: 212-867-6330; FAX 212-867-6315; Telex: 127057 MTC NYK)

Mitsui Toatsu Chemicals, Inc., London Office, 13 Charles II St., London, SW1Y 4QU, UK (Tel.: 71-976-1180; FAX 71-976-1185; Telex: 8953938 MITOL G)

Mobay Chemical Co. *See* Miles Inc.

Mobil Chemical Co.

Mobil Chemical Co., PO Box 250, Edison, NJ 08817-0250 USA (Tel.: 908-805 3800)

Mobil Chemical Co./Films Div., 1150 Pittsford-Victor Rd., Pittsford, NY 14534 USA (Tel.: 800-654-3436 (NY); 800-828-6381)

Mobil Chemical Co./Polystyrene Business Group, Rt. 27 & Vinyard Rd., Box 3029, Edison, NJ 08818-3029 USA (Tel.: 908-321-3500; 800-922-0380; FAX 908-321-3501)

Mobil Plastics Europe, Zoning Industriel de Latour, B-6761 Virton, Belgium (Tel.: 63-21 32 11; FAX 63-21 34 24)

Momar Industrial Services Ltd.

Barracks Road, Sand Lane Industrial Estate, Stourport on Severn, Worcestershire, DY13 9QB, UK (Tel.: 0299 827232; FAX 0299 827608)

Mona Industries Inc.

PO Box 425, 76 E. 24th St., Paterson, NJ 07544 USA (Tel.: 201-345-8220; 800-553-6662; FAX 201-345-3527; Telex: 130308)

Monmouth Plastics, Inc.

Box 921, 814 Asbury Ave., Asbury Park, NJ 07712 USA (Tel.: 201-775-5100; 800-526-2820; FAX 201-775-9068)

Monomer-Polymer & Dajac Laboratories, Inc., Div. of MTM Research
PO Box 28, Trevose, PA 19053 USA (Tel.: 215-938-1750; FAX 215-938-7410)

Monsanto
Monsanto Chemical Co., 800 N. Lindbergh Blvd., St. Louis, MO 63167 USA (Tel.: 314-694-1000; 800-325-4330; FAX 314-694-7625; Telex: 44-7282)

Monsanto/Detergents & Phosphates Div., 800 N. Lindbergh Blvd., St. Louis, MO 63167 USA (Tel.: 314-694-1000; 800-325-4330; FAX 314-694-7625; Telex: 650 397 7820)

Monsanto Plastics, 800 N. Lindbergh Blvd., St. Louis, MO 63167 USA (Tel.: 800-325-4330)

Monsanto Canada Inc., 2330 Argentia Rd., Box 787, Streetsville Postal Station, Mississauga, Ontario, L5M 2G4, Canada (Tel.: 416-826-9222)

Monsanto plc, Monsanto House, Chineham Court, Great Binfield Rd., Basingstoke, Hampshire, RG24 0UL, UK (Tel.: 0256-572-88; FAX 0256-54995; Telex: 858837)

Monsanto plc, Agriculture, Thames Tower, Burleys Way, Leicster, LE1 3TP, UK (Tel.: 0533-620864; FAX 0533-530320)

Monsanto Europe SA, Ave. de Tervuren 270-272, B-1150 Brussels, Belgium (Tel.: 2-761-41-11; FAX 2-761-40-40; Telex: 62927 Mesab)

Monsanto, Detergents Division, Rue Laid Burniat, 1348 Louvain la Neuve, France

Monsanto Japan Ltd., Room 520, Kokusai Bldg., 1-1, Marunouchi 3-chome, Chiyoda-Ku, Tokyo, 100, Japan (Tel.: (03) 3287-1251; FAX (03) 3287-1250; Telex: J22614)

Montana Talc Co.
28769 Sappington Rd., Three Forks, MT, 59752 USA (Tel.: 406-285-3286; FAX 406-285-3530)

Montedipe SpA
Via Rosellini 15-17, I-20124 Milan, Italy (Tel.: 2-63331; FAX 2-63338943; Telex: 310679)

Montedison
Montedison SpA, Casella Postale 10528, Foro Buonoparte 31, I-20121 Milano, Italy (Tel.: 02-633 31/627 01; FAX 02-805 0969)

Montedison UK Ltd., 7/8 Lygon Place, Ebury Street, London, SW1W 0JR, UK (Tel.: 071-730 3405)

Montedison Belgio SA, Div. of Montedison SpA, 8 rue de l'Industrie, B-1400 Nivelles, Belgium (Tel.: 67-21 86 82; FAX 67-21 86 82; Telex: 57536 MONTED B)

Montefluos SpA
Via Principe Eugenio, 1/5, I-20155 Milano, Italy (Tel.: 02-6270-3427; FAX 02-6270-3948; Telex: 310679 Monted I)

Mooney Chemicals Inc.
2301 Scranton Rd., Cleveland, OH 44113 USA (Tel.: 216-781-8383; 800-321-9696)

Morflex, Inc.
2110 High Point Rd., Greensboro, NC 27403 USA (Tel.: 919-292-1781; Telex: 910240 7846)

Morgan Refractories Ltd.

PO Box 110, 515 S 9th, Canon City, CO 81212 USA

Morton International

Morton International Inc., 100 N Riverside Plaza, Randolph Street, Chigago, IL 60606-1598 USA (Tel.: 312-807 2000; FAX 312-807 3150)

Morton International, Inc./Industrial Adhesives, 10 S. Electric St., West Alexandria, OH 45381 USA (Tel.: 513-839-4612; 800-348-8846; FAX 513-839-5615)

Morton International, Inc./Plastics Additives, 150 Andover St., Danvers, MA 01923 USA (Tel.: 508-774-3100; 800-621-2847; FAX 508-777-5672)

Morton International, Inc./Polymer Systems, 100 North Riverside Plaza, Chicago, IL 60606 USA (Tel.: 312-807-3127)

Morton International, Inc./Specialty Chemicals Group, 150 Andover St, Danvers, MA 01923 USA (Tel.: 508-774-3100; FAX 508-777-7606)

Morton Salt, 100 North Riverside Plaza, Chicago, IL 60606 USA (Tel.: 312-807-2562; FAX 312-807-2228; Telex: 25-4433)

Morton International Ltd./Specialty Chem, Ind. Chem. & Addit., 7900-A Taschereau Blvd., Suite 106, Brossard, Quebec, J4X 1C2, Canada (Tel.: 514-466-7764; FAX 514-466-7771)

Morton International NV SA, Chaussee de la Hulpe 130, Boite 5, B-1050 Brussels, Belgium (Tel.: 2-6602909; FAX 2-6604702; Telex: 23708)

Morton International Ltd., Greville House, Hibernia Road, Hounslow, Greater London, TW3 3RX, UK (Tel.: 081-570 7766; FAX 081-570 6943; Telex: 262002)

Morton International Inc., Room 2501 Dominion Centre, 37-59A Queen's Road East, Wanchai, Hong Kong

Mosselman NV

80 boulevard Industriel, B-1070 Brussels, Belgium (Tel.: 5243947; FAX 5200158; Telex: 23533 MOSS B)

MSD Agvet

Hertford Road, Hoddesdon, Herts, EN11 9BU, UK (Tel.: 0992-467272; FAX 0992-467270)

M&T Harshaw

M&T Harshaw, Div. of Atochem, Two Riverview Dr., PO Box 6768, Somerset, NJ 08875-6768 USA (Tel.: 908-302-3500; FAX 908-271-8960)

M&T Harshaw Ltd., William St, West Bromwich, West Midlands, B70 0BG, UK (Tel.: 021-557 3949; FAX 021-557 5607; Telex: 338790)

MTM plc

MTM plc, Rudby Hall, Hutton Rudby, Yarm, Cleveland, TS15 0JN, UK (Tel.: 0642 677595; FAX 0642 700667; Telex: 787737)

MTM Speciality Chemicals Ltd., Div. of MTM plc, Station Rd, Cheddleton, Leek, Staffordshire, ST13 7EF, UK (Tel.: 0538 361302; FAX 0538 361330; Telex: 367443)

MTM Agrochemicals Ltd., 18 Liverpool Road, Great Sankey, Warrington, Cheshire, WA5 1QR, UK (Tel.: 0925-33232; FAX 0925-52679)

Multibase Inc.

3835 Copley Rd., Copley, OH 44321 USA (Tel.: 216-867-5124; 800-343-5626; FAX 216-666-7419)

Multicrom SA

Almirante Brown 778, 1704 Ramos Mejia, Argentina (Tel.: 658 8091-94; FAX 541 656 3905; Telex: 18102 MULTI AR)

Multitherm Corp.

125 S. Front St., Colwyn, PA 19023 USA (Tel.: 215-461-6442; 800-225-7440; FAX 215-532-1289; Telex: 510-669-0032)

Münzing Chemie GmbH

Salzstrasse 174, D-7100 Heilbronn, Germany (Tel.: 07131/1586-0; FAX 07131/1586-25; Telex: 728614)

Murlin Chemical Inc.

Balligomingo Road, West Conshohocken, PA 19428 USA (Tel.: 215-825-1165; FAX 215-825-8659)

Murphy Chemical Co. Ltd.

Paper Mill Lane, Bramford, Ipswich, Suffolk, IP8 4BZ, UK (Tel.: 0473 830492; FAX 0473 830046)

Mydrin Ltd.

Albion Road, Carlton Ind. Estate, Barnsley, South Yorkshire, S71 3PL, UK (Tel.: 0226 723661; FAX 0226-728298; Telex: 547445)

Mykroy/Mycalex

Mykroy/Mycalex, Spaulding Composites Co. Inc., 1560 Sherman Ave, Suite 1100, Evanston, IL 60201 USA

Mykroy/Mycalex Ceramics, 125 Clifton Blvd., Clifton, NJ 07011 USA (Tel.: 201-779-8866; FAX 201-779-2013)

Naarden International Chemicals Div. *See* Quest International

Nalco

Nalco Chemical Co., One Nalco Center, Naperville, IL 60563-1198 USA (Tel.: 708-305-1000; 800-527-7753)

Nalco Chemical BV, Div. of Nalco Chemical Co, Postbus 5131, NL-5004 EC Tilburg, The Netherlands (Tel.: 13-63 55 55; FAX 13-67 45 45; Telex: 52532)

Nalco Ltd., Div. of Nalco Chemical Co, PO Box 11, Winnington Ave, Northwich, Cheshire, CW8 4DX, UK (Tel.: 0606 74488; FAX 0606 79557; Telex: 668663)

Nalfloc Norge A/S

Postboks 13, Drammensveien 126, N-0277 Oslo 2, Norway (Tel.: 2-44 20 27; FAX 2-55 33 82; Telex: 71633 ICINO N)

Naphtachimie SA
BP 2, F-13117 Martigues-Lavera, France (Tel.: 42 07 71 23; FAX 42 07 70 07; Telex: 410084)

Napp Laboratories Ltd.
Cambridge Science Park, Milton Road, Cambridge, CB4 4BH, UK

National Casein Co.
601 W. 80 St., Chicago, IL 60620 USA (Tel.: 312-846-7300; FAX 312-487-5709)

National Lead Co. *See* NL Industries

National Polychemicals. *Address unverified*
Wilmington, MA USA

National Starch & Chemical
National Starch & Chemical Corp., Box 6500, 10 Finderne Ave., Bridgewater, NJ 08807 USA (Tel.: 908-685-5000; 800-726-0450; FAX 908-685-5005)

Represented by: National Adhesives & Resins Ltd., Braunston Daventry, Northante, NN11 7JL, UK

National Starch & Chemical (Holdings) Ltd., Canon House, 27 London End, Beaconsfield, Buckinghamshire, HP9 2HN, UK (Tel.: 0494 677966; FAX 0494 673960; Telex: 848386 NATADH G)

National Starch & Chemical Ltd./Adhesives Div., Div. of National Starch & Chemical (Holding) Ltd., Galvin Rd, Slough Trading Estate, Slough, Berkshire, SL1 4DF, UK (Tel.: 0753-533494; FAX 0753-539338; Telex: 848386)

Natrochem, Inc.
PO Box 1205, Exley Ave., Savannah, GA 31498 USA (Tel.: 912-236-4464; FAX 912-236-1919)

Naugatuck (U S Rubber). *See* Uniroyal Inc.

Nayler Chemicals Ltd.
Old Rolling Mill, Edge Green Rd, Ashton in Makerfield, Wigan, Lancashire, WN4 8SL, UK (Tel.: 0942 720988; FAX 0942 271251; Telex: 669367 BREEZE G)

Nekami Maasgroeven BV, Div. of Carmeuse SA
Postbus 436, NL-2800 AK Gouda, The Netherlands (Tel.: 01820-27255; FAX 01820-26264)

Nelson Research
1001 Health Sciences Road West, Irvine, CA 92715 USA

Nepera
Nepera, Inc., Route 17, Harriman, NY 10926 USA (Tel.: 914-782-1200; FAX 914-783-9713; Telex: 510-249-4847)

Nepera, Ltd., 4 Heath Sq., Boltro Rd., Haywards Heath, W. Sussex, RH16 1BL, UK (Tel.: 0444-441270; FAX 0444 441127)

Neste Oy

Neste OY Chemicals, PO Box 20, Keilaniemi, SF-02150, Espoo 15, Finland (Tel.: 0-4501; FAX 0-4504985; Telex: 124641 NESTE)

Neste Composite Materials, Läkkisepänkuja 5, 00620 Helsinki, Finland (Tel.: 0-450-5293; FAX 0-450-5260)

Neste Polyeten AB, Div. of Neste Oy, Box 44, S-44486 Stenungsund, Sweden (Tel.: 303-86000; FAX 303-81034; Telex: 2402)

Neste Chemicals UK Ltd., Div. of Neste Oy, Neste House, Water Lane, Wilmslow, Cheshire, SK9 5AR, UK (Tel.: 0625 537390; FAX 0625 535218; Telex: 668398)

Neste Chemicals International NV, Div. of Neste Oy, Bazellaan 1, B-1140 Brussels, Belgium (Tel.: 2-729 42 11; FAX 2-216 05 40; Telex: 62233)

Neville Chemical Co.

2800 Neville Rd., Pittsburgh, PA 15225-1496 USA (Tel.: 412-331-4200; FAX 412-777-4234)

Newman Agrochemicals Ltd.

Cavendish House, Barton, Cambs, CB3 7AR, UK (Tel.: 0223-263777; FAX 0223-264038)

New Metals & Chemicals Ltd.

Abbey Chambers, Highbridge St, Waltham Abbey, Essex, EN9 1DF, UK (Tel.: 0992 711111; FAX 0992 768393; Telex: 28816)

Niacet Corp.

PO Box 258, 400 47th St., Niagara Falls, NY 14304 USA (Tel.: 716-285-1474; 800-828-1207; FAX 716-285-1497; Telex: 6730170)

Nichia Kagaku Kogyo K.K.

PO Box 6 Anan, Tokushima, 774, Japan (Tel.: (0884) 22-2311; FAX (0884) 23-1802; Telex: 5867790 NICHIA J)

Nicholas Laboratories Ltd.

225 Bath Road, Slough, Berkshire, SL1 4AU, UK (Tel.: 0753 570340; FAX 0753-523971)

Nickerson Chemicals Ltd., Div. of Nickerson Investments Ltd.

Mill St East, Dewsbury, West Yorkshire, WF12 9BQ, UK (Tel.: 0924 453886; FAX 0924 458995)

Nickerson Seeds Ltd.

JNRC, Rothwell, Lincs, LN7 6DT, UK (Tel.: 0472-89471; FAX 0472-89602)

Nihon Emulsion Co., Ltd.

Minami 5-32-7, Koenji, Suginami-ku, Tokyo, Japan (Tel.: (03) 314-3211; FAX (03) 312-7207; Telex: 2322358 EMALEX J)

Nihon Kagaku Sangyo Co., Ltd.

20-5, Shitaya 2-chome, Taito-ku, Tokyo, 110, Japan (Tel.: (03) 3876-3131; FAX (03) 3876-3278; Telex: J 28318 NIKKASAN)

Nipa Laboratories

Nipa Laboratories, Inc., 3411 Silverside Rd., 104 Hagley Bldg., Wilmington, DE 19810 USA (Tel.: 302-478-1522; FAX 302-478-4097; Telex: 905030)

Nipa Laboratories Ltd., Div. of BTP plc, Llantwit Fadre, Pontypridd, Mid Glamorgan, CF38 2SN, Wales UK (Tel.: 0443-205311; FAX 0443-207746; Telex: 497111)

Nippon Nyukazai Co., Ltd.

Yoshizawa Bldg., 9-19, Ginza 3-chome, Chuo-ku, Tokyo, 104, Japan (Tel.: (03) 3543-8571; FAX (03) 3546-3174)

Nippon Polyurethane Industry Co., Ltd.

Kotohira Kaikan Bldg., 2-8, Toranomon 1-chome, Minato-ku, Tokyo, 105, Japan (Tel.: (03) 3508-0611; FAX (03) 3580-9304; Telex: 222-3066 NIPOLY J)

Nippon Rikagakuyakuhin Co., Ltd.

2-12, Nihonbashi-Honcho 4-chome, Chuo-ku, Tokyo, 103, Japan (Tel.: (03) 3241-3557; FAX (03) 3242-3345; Telex: 2223651 NITIRJ J)

Nippon Silica Industrial Co., Ltd.

Toso-Kyobashi Bldg., 2-4, Kyobashi 3-chome, Chuo-ku, Tokyo, 104, Japan (Tel.: (03) 3273-1641; FAX (03) 3272-3879)

Nippon Zeon

Nippon Zeon Co., Ltd., Zeonex Div., 43 Mori Bldg., 3-13-16 Mita, Minato-ku, Tokyo, 108, Japan (Tel.: (03) 3769-8630; FAX (03) 3769-2840)

Nippon Zeon of America, Inc., Div. of Nippon Zeon Co Ltd., 50 Main St., White Plains, NY 10606 USA (Tel.: 914-683-8500; FAX 914-683-8546)

Nippon Zeon Europe GmbH, Div. of Nippon Zeon Co Ltd., Königs Alle 94, 4000 Düsseldorf 1, Germany

Nissan Chemical Industries, Ltd.

Kowa-Hitotsubashi Bldg., 7-1, Kanda-Nishiki-cho 3-chome, Chiyoda-ku, Tokyo, 101, Japan (Tel.: (03) 3296-8111; FAX (03) 3296-8360; Telex: 222-3071)

Nisshinbo Industries Inc.

3-10 Nihonbashi, Yokoyama-cho, Chuo-ku, Tokyo, 103, Japan (Tel.: (03) 3665 8833; Telex: 28405)

Nitta Gelatin Inc.

8-12, Honmachi 1-chome, Chuo-ku, Osaka, 541, Japan (Tel.: (06) 266-1911; Telex: 522-2882 NITTAO J)

Nitto Chemical Industry Co., Ltd.

New Marunouchi Bldg., 5-1, Marunouchi 1-chome, Chiyoda-ku, Tokyo, 100, Japan (Tel.: (03) 3271-0251; FAX (03) 3287-2725; Telex: 222-2338)

Nitto Electric Industrial Co. Ltd.

Corporate Planning Department, 1-2 Shimohozumi 1-chome, Ibaraki, Osaka, 567, Japan

NL Chemicals (UK) Ltd.
St Ann's House, Wilmslow, Cheshire, SK9 1HG, UK (Tel.: 0625-529511)

NL Industries Inc. *See* Rheox Inc.

NL Victor Wolf Ltd.
St Ann's House, Wilmslow, Cheshire, SK9 1HG, UK (Tel.: 0625-529511; FAX 0625-533123)

Noah Chem. Div., Div. of Noah Technologies Corp.
7001 Fairgrounds Pkwy, San Antonio, TX 78238 USA (Tel.: 512-680-9000; FAX 512-521-3323)

Nobel Chemicals
Nobel Chemicals AB, S-691 85 Karlskoga, Sweden (Tel.: 586-83000; FAX 586-32147; Telex: 73346 NOBELC S)

Nobel Chemicals Ltd., Div. of Nobel Chemicals AB, 39 High St, Billericay, Essex, CM12 9BA, UK (Tel.: 0277 631407; FAX 0277 631463; Telex: 996743)

The Norac Co. Inc.
405 S. Motor Ave., Azusa, CA 91702 USA (Tel.: 818-334-2908; FAX 818-334-3512; Telex: 882552 NORAC CO AZSA)

Norda Inc. *Address unknown*

Norden Laboratories Inc.
601 W Cornhusker Highway, Lincoln, NE 68501 USA

Nordox Industrier AS
Østensjøveien 13, N-0661 Oslo 6, Norway (Tel.: 2-659090; FAX 2-641208; Telex: 76614)

Nordson (UK) Ltd.
Thames Industrial Estate, Wenman Road, Thame, Oxon, OX9 3XB, UK (Tel.: 0844-213171; FAX 0844-215358)

Norgine Ltd.
116-120 London Road, Headington, Oxford, Oxfordshire, OX3 9BA, UK (Tel.: 0865-750717; FAX 0865-68826; Telex: 837547 NOROXF G)

Norit
Norit NV, Postbus 105, NL-3800 AC Amersfoort, The Netherlands (Tel.: 648911; FAX 617429; Telex: 79040)

Norit UK Ltd., Div. of Norit NV, Clydesmill Place, Cambuslang Industrial Estate, Glasgow, Lanarkshire, G32 8RF, UK (Tel.: 041-641-8841; FAX 041-641- 0742; Telex: 777880 CACTUS G)

Norman, Fox & Co.
5511 S. Boyle Ave., PO Box 58727, Vernon, CA 90058 USA (Tel.: 213-583-0016; 800-632-1777; FAX 213-583-9769)

Norma Products Ltd.
Arnham Road, Newbury, Berkshire, RG14 5RU, UK (Tel.: 0635 521 880; FAX 0635 523 418)

Norold Composites Inc.
4255 Sherwoodtown Blvd., 3rd Floor, Mississauga, Ontario, L4Z 1Y5, Canada (Tel.: 416-279-2740; FAX 416-848-0455)

Noroxo
Chemin De Vermelles, BP 19, F-62440 Harnes, France (Tel.: 121-78 62 41; FAX 121-78-6785)

Norplex
PO Box 1448, La Crosse, WI 54601 USA

Norsk Hydro AS
Bygdoyalle 2, N 0240 Oslo 2, Norway (Tel.: 243 2100; FAX 243 2308; Telex: 78350 HYDRO N)

Norsk Hydro Polymers Ltd. *Address unknown*

North American Brine Resources
Enterprise Plaza, 5600 North May Avenue, Oklahoma City, OK 73112 USA (Tel.: 405-842-2233; FAX 405-842 9901)

H. N. Norton & Co. Ltd. *Address unknown*

Norton Chemical Process Products
PO Box 350, Akron, OH 44309 USA (Tel.: 216-673-5860; FAX 216-673-5868; Telex: 433-8012)

Norwegian Talc (UK) Ltd., Div. of Ernström Mineral AB
205 Cotton Exchange Bldg, Old Hall St, Liverpool, Merseyside, L3 9LA, UK (Tel.: 051-236 6435; FAX 051-227 5903; Telex: 627012 NORDOL G)

Norwich Eaton Pharmaceuticals Inc., Subsidiary of Proctor & Gamble Co.
PO Box 191, 27 Eaton Ave, Norwich, NY 13815-1799 USA (Tel.: 607-335-2565; 800-448-4878; FAX 607-335-2098)

Novachem Corp.
PO Box 6379, High Point, NC 27262 USA (Tel.: 919-885-0041; FAX 919-885-4964)

Nova Chemical Inc.
PO Box 12599, 1548 Erie, Kansas City, MO 64116 USA

Novacor Chemicals
Novacor Chemicals Ltd., PO Box 2535, Station M, Calgary, Alberta, T2P 2N6, Canada (Tel.: 403-290-8977; 800-661-1548; FAX 403-264-6012; Telex: 038-25775)

Novacor Chemicals Inc., One Gatehall Dr, Parsippany, NJ 07054 USA (Tel.: 201-267-1400; FAX 201-455-0041)

Novamont North America, Inc., Co. of Montedison Group
1114 Ave. of the Americas, Suite 3300, New York, NY 10036 USA (Tel.: 212-997-7035)

Nova Polymers, Inc.
4004 East Morgan Ave., Evansville, IN 47715 USA (Tel.: 812-476-0339)

Novarina Srl
Via Pinerolo 35, I-10060 Bibiana, Italy (Tel.: 121-55724; Telex: 211591)

Novatec Plastics & Chemicals Co., Inc.
PO Box 597, 275 Industrial Way West, Eatontown, NJ 07724 USA (Tel.: 908-542-6600; 800-PVC-NOVA; FAX 908-389-0431)

Novest Inc/Pro Seal Products Division. *Address unknown*

Novon Prods., Div. of Warner-Lambert Co.
5500 Forest Hills Rd., Box 1205, Rockford, IL 61105-1205 USA (Tel.: 815-282-5696; 800-35-NOVON; FAX 815-282-5941)

Novo Nordisk A/S
Novo Nordisk A/S, Bioindustrial Group, Novo Allé, DK-2880 Bagsvaerd, Denmark (Tel.: 4444-8888; FAX 4444-6088; Telex: 37173)

Novo Industri A/S, Subsidiary of Novo Nordisk A/S, Novo Alle, DK-2880 Bagsvaerd, Denmark (Tel.: 2-98 23 33; FAX 2-98 27 33; Telex: 37173 NOVO DK)

Novo Nordisk Bioindustries UK Ltd., 4 St. Georges Yard, Castle St., Farnham, Surrey, GU9 7LW, UK (Tel.: 0252 711212; FAX 0252 711187)

Novo Nordisk Bioindustrials Inc., 33 Turner Rd., Danbury, CT 06813-1907 USA (Tel.: 800-251-6686; FAX 203-790-2748)

Novo Nordisk Bioindustry Ltd., Div. of Novo Nordisk A/S, Makuhari Techno Garden CB-6, 3, Nakase 1-chome, Chiba, 261-01, Japan (Tel.: (0472) 96-6767; FAX (0472) 96-6760)

NRC Inc.
45 Industrial Pl., Newton, MA 02164 USA (Tel.: 617-969-7690; FAX 617-332-3947)

Nuodex Espanola SA
Gran Via 55, E-28013 Madrid, Spain (Tel.: 1-241 35 03; FAX 1-247 53 79; Telex: 44818 NOUD E)

NYCO® Minerals Inc.
34 Mountain View Drive, Willsboro, NY 12996-0368 USA (Tel.: 518-963-4262; FAX 518-963-4187; Telex: 957014)

William F. Nye, Inc.
PO box 8927, New Bedford, MA 02742-8927 USA (Tel.: 508-996-6721; FAX 508-997-5285)

Nylon Corp. of America (Nycoa), Div. of SNIA
333 Sundial Ave., Manchester, NH 03103 USA (Tel.: 603-627-5150; 800-851-2001; FAX 603-627-5154)

Oak International Inc.
1160 White Street, P O Box 837, Sturgis, MI 49091 USA (Tel.: 616 651 979-0; FAX 616 651 7849)

Oakite Products, Inc.
50 Valley Rd., Berkeley Hts., NJ 07922 USA (Tel.: 908-464-6900; 800-526-4473; FAX 908-464-6031)

Occidental
Occidental Chemical Corp., 5005 LBJ Freeway, Dallas, TX 75244 USA (Tel.: 214-404-3925; 800-752-5151)

Occidental Chemical Corp., 360 Rainbow Blvd. South, PO Box 728, Niagara Falls, NY 14302 USA (Tel.: 716-286-3000; FAX 716-286-3441)

Occidental Chem. Europe, Holidaystraat 5, B-1920 Diegem, Belgium (Tel.: 2 721 24 20; FAX 2 721 45 66; Telex: 23046)

Occidental Chemical Asia Ltd., Toranomon 34 Mori Bldg., 9th Floor, 25-5 Toranomon 1-Chome, Minato-ku, Tokyo, 105, Japan

Occidental Chemical Corp. (Australia Pty. Ltd.), Suite 15, Gateway Court, 81-91 Military Rd., Neury Bay, NSW 2089, Australia

Octel Chemicals Ltd.
Halebank, Widnes, Cheshire, WA8 8NS, UK (Tel.: 051 424 3671; FAX 051 420 1301; Telex: 629491)

Oils & Soaps Ltd.
Rutland St, Bradford, West Yorkshire, BD4 7EA, UK (Tel.: 0274 393286; FAX 0274 309143)

Oleofina. *See* Fina

Olin Chemicals
Olin Chemicals, 120 Long Ridge Rd., PO Box 1355, Stamford, CT 06904 USA (Tel.: 203-356-2000; 800-243-9171)

Olin Chemicals Ltd., Suite 7, Kidderminster Rd., Cutnall Green, Worchestershire, WR9 0NS, UK

Olin Europe SA, 108-110 Blvd. Haussmann, 75008 Paris, France (Tel.: 1-293-3210)

Olin Japan Inc., Shiozaki Bldg., 7-1 Hirakawa-Cho 2-Chome, Chiyoda-ku, Tokyo, 102, Japan (Tel.: (03) 263-4615)

Olin Australia Ltd., 1-3 Atchison St., PO Box 141, St. Leonards 2065, N.S.W., Australia

Olin Brasil Limitada, Rua Galeno de Castro, 165, Jurubatuba, Santo Amaro, 04696 Sao Paulo, SP, Brazil

Olin Hunt
Olin Hunt Specialty Products Inc., 5 Garret Mountain Plaza, West Paterson, NJ 07424 USA (Tel.: 201-977-6195; 800-367-4868; FAX 201-977-6110)

Olin Hunt Specialty Products NV, Europark-Noord 17-18, B-2700 Sint-Niklaas, Belgium (Tel.: 7801511; FAX 780153; Telex: 32328 NVHUNT NL)

Omex Agriculture Ltd.
Bardney Airfield, Tupholme, Lincoln, Lincolnshire, LN3 5TP, UK (Tel.: 0526 398661; FAX 0526 398434)

O'Neal, Jones & Feldman Pharmaceuticals
2510 Metro Blvd, Maryland Heights, MO 63043 USA

Onyx Chemical Co., Millmaster Onyx Group. *Address unverified*
Jersey City, NJ 07302 USA

Optech Inc.
7310 S Alton Way, Englewood, CO 80112 USA

Optrex. *See* Boots plc

Oral-B Laboratories
Oral-B Laboratories Inc., 170 S Whisman Rd, Mountain View, CA 94041 USA

Oral-B Laboratories Ltd., Gatehouse Road, Aylesbury, Buckinghamshire, HP19 3ED, UK (Tel.: 0296-432601; FAX 0296-434283)

Orange Chemicals Ltd.
34 St Thomas Street, Winchester, Hampshire, SO23 9HJ, UK (Tel.: 0962-842525; FAX 0962-841101; Telex: 477361 ORANGE G)

Ore & Chemical Corp., Div. of Chemetall
520 Madison Ave., New York, NY 10022 USA (Tel.: 212-715-5236; FAX 212-715-5291; Telex: ITT 422681)

Organon Inc.
375 Mt Pleasant Ave, West Orange, NJ 07052 USA (Tel.: 201-325 4500)

Original Bradford Soap Works Inc.
200 Providence St., West Warwick, RI 02893-0907 USA (Tel.: 401-821-2141; FAX 401-821-5960; Telex: 952 240)

Orkem UK Ltd.
Colthrop Way, Thatcham, Newbury, Berkshire, RG13 4LW, UK (Tel.: 0635 70000; FAX 0635 61212; Telex: 847689)

Orkla Exolon A/S & Co.
PO Box 25, N-7301 Orkanger, Norway (Tel.: 74-805000; FAX 74-82040)

Ortho Pharmaceutical Corp., Subsidiary of Johnson & Johnson
PO Box 300, Raritan, NJ 08869 USA (Tel.: 908-218-6000)

Otsuka Chemical Co., Ltd.
Otsuka Chemical Co., Ltd., 2-27, Ote-dori 3-chome, Chuo-ku, Osaka, 540, Japan (Tel.: (06) 943-7711; FAX (06) 946-0860; Telex: J63586 JPOTSUKA)

Otsuka Chemical Co., Ltd., 747 Third Ave., 26th Floor, New York, NY 10168 USA (Tel.: 212-826-4374; FAX 212-826-5094)

Outokumpu Oy
PO Box 280, SF-02101 Espoo, Finland (Tel.: 0-4211; FAX 0-4213888; Telex: 124441 OKHI SF)

Owens-Corning Fiberglas Corp.
Fiberglas Tower, Toledo, OH 43659 USA (Tel.: 419-248-8000; FAX 419-248-6712)

OxyChem®
OxyChem®, 5005 LBJ Freeway, Dallas, TX 75244 USA (Tel.: 214-404-3800; 800-752-5151; FAX 214-404-3669; Telex: 229835)

OxyChem/Alathon Polymers Div., PO Box 27702, Houston, TX 77227-7702 USA (Tel.: 800-521-2905)

OxyChem/Durez Div., 673 Walck Rd., North Tonawanda, NY 14120 USA (Tel.: 716-696-6000; 800-733-3339)

OxyChem/Polymers & Plastics, PO Box 1772, Berwyn, PA USA (Tel.: 215-251-1000)

OxyChem/Vinyls Div., 300 Berwyn Park, Suite 300, Berwyn, PA 19312 USA (Tel.: 215-251-1000; FAX 215-251-5873)

Pacific Anchor
Pacific Anchor Chemicao Corp., Subsidiary of Air Products & Chemicals Inc., 5701 S. Eastern Ave., Suite 530, Los Angeles, CA 90040 USA (Tel.: 213-725-1800; 800-423-4391; FAX 213-725-1915)

Pacific Anchor Chemical Corp., Subsidiary of Air Products & Chemicals Inc., 1224 Mindon Road, Cumberland, RI 02864 USA (Tel.: 401-333 4100; FAX 401-333 4630)

Pacific Chemical Industries Pty Ltd.
6 Grand Avenue, Camellia, N.S.W., 2142, Australia

Pacific Dispersions Inc.
4615 Ardine Street, Cudahy, CA 90201 USA

Pacific Petroleums (Quebec). *Address unknown*

Packard Instrument BV, Div. of Canberra Industries Inc.
Postbus 9403, NL-9703 LP Groningen, The Netherlands (Tel.: 50-41 33 60; FAX 50-42 22 92; Telex: 53799 PCHEM NL)

Pafra Ltd.
Adhesive Division, Bentalls, Basildon, Essex, SS14 3BU, UK (Tel.: 0268-280606)

Paines & Byrne Ltd.
Pabryn Laboratories, Perivale, Greenford, Middlesex, UB6 7HG, UK (Tel.: 081-997 1143; FAX 081-997 0327; Telex: 267763)

M. S. Paisner Inc.
53 Beaumont St., PO Box 358, Canton, MA 02021 USA (Tel.: 617-828-2040; FAX 617-828-2202)

Pan Britannica Industries Ltd.
Britannica House, Waltham Cross, Herts, EN8 7DY, UK (Tel.: 0992-23691; FAX 0992-26452)

Pantasote Polymers Inc.
26 Jefferson St., Passaic, NJ 07055 USA (Tel.: 201-777-8500; FAX 201-777-3370)

Paratherm Corp.
1050 Colwell Rd., Conshohocken, PA 19428 USA (Tel.: 215-941-4900; 800-222-3611; FAX 215-941-9191)

Parke-Davis
Parke-Davis, Div. of Warner-Lambert Co., 201 Tabor Rd, Morris Plains, NJ 07950 USA (Tel.: 201-540-2000; 800-223-0423; FAX 201-540-2248)

Parke-Davis, Specialty Chemical Div. of Warner-Lambert Co, 188 Howard Ave, Holland, MI 49424 USA (Tel.: 616-392-2375; FAX 616-392-8914)

Parke-Davis, Div. of Warner-Lambert Co, Mitchell House, Southampton Road., Eastleigh, Hants, SO5 5RY, UK

PARS Pharmaceutical Laboratories Inc. *Address unknown*

Pasminco Europe Ltd./ISC Alloys Div., Div. of Pasminco Ltd.
Alloys House, PO Box 36, Willenhall Lane, Bloxwich, Walsall, West Midlands, WS3 2XW, UK (Tel.: 0922-408444; FAX 0922-710043; Telex: 338270)

Patco Products Div. *See* Am. Ingred. Co.

PBI-Gordon Corp. *Address unknown*

P B & S Chemical Co Inc. *Address unknown*

PCR Inc.
PO Box 1466, Gainesville, FL 32602 USA (Tel.: 904-376 8246; 800-331-6313)

William Pearson Ltd., Subsidiary of BTP plc
Clough Road, Hull, North Humberside, HU6 7QA, UK (Tel.: 0482-443151; FAX 0482-440444; Telex: 597028 BTPWPL G)

Peboc Ltd., Div. of Duphar BV
Industrial Estate, Llangefni, Gwynedd, LL77 7YQ, UK (Tel.: 0248 750724; FAX 0248 723890; Telex: 61379)

Pechiney
Pechiney SA, Tour Manhattan Cedex 21, F-92087 Paris La Défense, France (Tel.: 47 62 88 00; FAX 47 74 83 16; Telex: 614986 PELEC F)

Pechiney Electrométallurgie, Div. of Pechiney SA, Tour Manhattan Cedex 21, F-92087 Paris La Défense, France (Tel.: 47 62 88 00; FAX 46 74 83 16; Telex: 614986 PELEC F)

Pecora Corp.
165 Wambold Road, Harleysville, PA 19438 USA (Tel.: 215-723 6051; FAX 215-721 0286)

PEI Precision Elastomers Inc. *Address unknown*

Penco of Lyndhurst Inc.
540 New York Ave., Lyndhurst, NJ 07071 USA (Tel.: 201-935-6600; FAX 201-896-0539; Telex: 132641)

Penick Corp.
158 Mt Olivet Ave, Newark, NJ 07114 USA (Tel.: 201-621 2802; FAX 201-621 2816)

Pennwalt Corp. *See* Atochem North America Inc.

Pennwalt Chemicals Ltd. *See* Bos Chemicals Ltd.,Distributors in UK

Pennzoil Products Co. *See* Penreco

Penreco, Div. of Pennzoil Products Co.
RD 2, Box 1, Karns City, PA 16041 USA (Tel.: 412-756-0110; 800-245-3952; FAX 412-756-1050; Telex: 866-321)

Penta Manufacturing Co.
PO Box 1448, Fairfield, NJ 07007 USA (Tel.: 201-740-2300; FAX 201-740-1839; Telex: RCA 219472 PENT UR)

Pentagon Chemicals Ltd., Div. of Suter plc
Northside, Workington, Cumbria, CA14 1JJ, UK (Tel.: 0900 604371; FAX 0900 66943; Telex: 64353 PENTA G)

Pentagon Urethanes Ltd.
Northside, Workington, Cumbria, CA14 1JJ, UK

Peridot Chemicals (NJ) Inc.
1680 Rt. 23 North, Wayne, NJ 07470 USA (Tel.: 201-696-9000; 800-222-0121; FAX 201-696-2501)

Perkin-Elmer Corp.
761 Main Ave., Norwalk, CT 06859-0012 USA (Tel.: 203-762-1000; FAX 203-762-6000; Telex: 965-954)

Perma-Flex Mold Co.
1919 East Livingston Ave., Columbus, OH 43209 USA (Tel.: 614-242-8034; 800-736-6653 (orders; FAX 614-252-8572)

The Permafuse Corp.
PO Box 884, 675 Main Street, Westbury, NY 11590 USA

The Permutit Co Ltd., Div. of Thames Water Group plc
632-652 London Road, Isleworth, Middlesex, TW7 4EZ, UK (Tel.: 081 560 6431; FAX 081 568 9772; Telex: 924640 PERM G)

Person & Covey Inc.
PO Box 25018, 616 Allen Ave, Glendale, CA 91221-5018 USA (Tel.: 818-240-1030; 800-423-2341)

Perstorp Ferguson Ltd.
Aycliffe Industrial Estate, Newton Aycliffe, Co Durham, DL5 6UE, UK (Tel.: 0325-300666; FAX 0325-300385)

Perstorp Polyols, Inc.
600 Matzinger Rd, Toledo, OH 43612 USA (Tel.: 419-729-5448; 800-537-0280; FAX 419-729-3291)

Pestco Inc.
PO Box 1000, 215-225 Eighth Ave., Braddock (PGH), PA 15104 USA (Tel.: 412-271-6200; FAX 412-351-7701)

Petrolite
Petrolite Corp., Polymers Div., 6910 E. 14th St., Tulsa, OK 74112 USA (Tel.: 918-836-1601; 800-331-5516; FAX 918-834-9718)

Petrolite Ltd., Div. of Petrolite Speciality Polymers Group, Kirkby Bank Rd, Liverpool, Merseyside, L33 7SY, UK (Tel.: 051-546 2855; FAX 051 549 1858; Telex: 627293 PETLTD G)

Pfaltz & Bauer Inc.
172 E. Aurora St., Waterbury, CT 06708 USA (Tel.: 203-574-0075; 800-225-5172; FAX 203-574-3181; Telex: 996471)

Pfanstiehl Laboratories, Inc.
1219 Glen Rock Ave., PO Box 439, Waukegan, IL 60085 USA (Tel.: 708-623-0370; 800-383-0126; FAX 708-623-9173; Telex: 25 3672 PFANLAB)

Geo. Pfau's Sons., Co., Inc.
PO Box 7, Jeffersonville, IN 47131 USA (Tel.: 800-PFAU-OIL; FAX 812-283-0765; Telex: 20-4135)

Pfister Chem, Inc.
Linden Ave, Ridgefield, NJ 07657 USA (Tel.: 201-945-5400 x 208; FAX 201-945-0159; Telex: 130295)

Pfizer
Pfizer International, 235 East 42nd Street, New York, NY 10017 USA

Pfizer Inc./Chem. Div., Minerals, Pigments, Metals, 235 E. 42nd St., New York, NY 10017 USA (Tel.: 212-573-2323; 800-336-9008; FAX 212-573-2273; Telex: 420440)

Pfizer Chemicals Europe & Africa, 10 Dover Rd., Sandwich, Kent, CT13 0BN, UK (Tel.: 0304 615518; FAX 0304 615529; Telex: 966555)

Pfizer Ltd/Chemical Div, Specialty Chemicals, 10 Dover Rd, Sandwich, Kent, CT13 0BN, UK (Tel.: 0304 615502; FAX 0304 615529; Telex: 966555)

Pfizer SA, Principe de Vergara 109, E-28002 Madrid, Spain (Tel.: 1-262 11 00; Telex: 42732 PRIZ E)

Pharmaceutical Basics Inc.
8755 West Higgins Road/810, Chicago, IL 60631 USA (Tel.: 312-380 0080)

Pharmacia AB
Rapsagatan 7, S-751 82 Uppsala, Sweden (Tel.: 18-16 30 00; Telex: 76027 PHARMUP S)

Pharmacia Laboratories, Div. of Pharmacia Inc.
800 Centennial Ave, Piscataway, NJ 08854 USA

Pharmax Ltd., An International Subsidiary of Forest Laboratories Inc.
Bourne Road, Bexley, Kent, DA5 1NX, UK (Tel.: 0322-550550; FAX 0322-558776; Telex: 896314)

Phillips
Phillips 66 Co., Subsidiary of Phillips Petroleum Co, 376 Phillips Bldg. Annex, Bartlesville, OK 74004 USA (Tel.: 806-274-5236; 800-858-4327; FAX 806-274-5230)

Phillips 66 Co/Plastics Resins Div., Subsidiary of Phillips Petroleum Co., Phillips Building, Bartlesville, OK 74004 USA (Tel.: 918-661-6600; FAX 918-661-7636)

Phillips Petroleum Chemical Co., Philips Quadrant, 35 Guildford Road, Woking, Surrey, GU22 7QT, UK (Tel.: 0483-756666; FAX 0483-752371)

Phillips Petroleum Chemicals NV, Steenweg op Brussels 355, B-1900 Overijse, Belgium (Tel.: 2-689 1293; FAX 2-689 1472; Telex: 23866/22197)

Phillips Petroleum International, Ltd., Shin-Tokyo Bldg., 3-1, Marunouchi 3-chome, Chioyda-ku, Tokyo, 100, Japan (Tel.: (03) 3216-6951; FAX (03) 3216-6960; Telex: J24641 PHILPET)

Phillips Yeast Products Ltd.
Park Royal Road, London, NW10 7JX, UK

Phoenix Chemical, Inc.
322 Courtyard Dr., Somerville, NJ 08876 USA (Tel.: 201-707-0232; FAX 201-707-0186)

Piccadilly Products Ltd.
199 Piccadilly, London, W1V 9LE, UK

Pickering Laboratories Inc.
1951 Colony Street, Suite S, Mountain View, CA 94043 USA (Tel.: 415-968-9502; FAX 415-968-0749)

Pierce & Stevens Corp.
PO Box 1092, Buffalo, NY 14240-9990 USA (Tel.: 716-856-4910; 800-888-4910; FAX 716-856-7530)

Pigment & Chem. Inc.

604 Main St. E., Milton, Ontario, L9T 1R2, Canada (Tel.: 416-878-8858; Telex: 055-66339)

Pilot Chemical Co.

11756 Burke St., Santa Fe Springs, CA 90670 USA (Tel.: 310-723-0036; FAX 310-945-1877; Telex: 4991200 PILOT)

Pitman-Moore

Pitman-Moore Ltd., Brakesplan Road South, Harefield, Uxbridge, UB9 7LS, UK

Pitman-Moore, Subsidiary of Imcera Group plc, One Conway Park, 100 Field Drive, Lake Forest, IL 60045 USA (Tel.: 708-615-3770)

PJ1 Corp. *Address unknown*

Plant Protection, Subsidiary of Zeneca Group plc

Woodmead House West, Bear Lane, Farnham, Surrey, GU9 7UB, UK (Tel.: 0252 733888; FAX 0252 736333)

Plascoat Systems Ltd., A member of the BTR Nylex group of companies

Trading Estate, Farnham, Surrey, GU9 9NY, UK (Tel.: 0252 7333777; FAX 0252 725719; Telex: 858433)

Plastics Europe

Rue Jeanne D'Arc, BP23, 52101 Saint-Dizier, France (Tel.: 25 05 91 00; Telex: 840637)

Plastic Coatings Ltd.

Woodbridge Industrial Estate, Guildford, Surrey, GU1 1BG, UK (Tel.: 0483 31155; FAX 0483 33534; Telex: 859237)

Plasticolors, Inc.

2600 Michigan Ave., Ashtabula, OH 44004 USA (Tel.: 216-997-5137; FAX 216-992-3613)

Plastics & Chemicals, Inc.

PO Box 306, Cedar Grove, NJ 07009 USA (Tel.: 908-221-0002; FAX 908-221-1095; Telex: 219744)

Plough Inc. *See* Schering-Plough HealthCare Products

Plysolene Ltd.

Southwater Business Park, Worthing Road, Southwater nr. Horsham, West Sussex, RH13 7HE, UK (Tel.: 0403-730032; FAX 0403-731783; Telex: 877283 PLYSOL)

PMC

PMC, Inc./Specialties Group, 501 Murray Rd., Cincinnati, OH 45217 USA (Tel.: 513-242-3300; 800-543-2466)

PMC Specialties Group, Div. of PMC Inc., 20525 Center Ridge Road, Rocky River, OH 44116 USA (Tel.: 216-356-0700; FAX 216-356-2787; Telex: 4332035)

PMC Specialities International Ltd., Div. of PMC Specialities Group, 65B Wigmore Street, London, W1H 9LG, UK (Tel.: 071-935-4058; FAX 071-935 9895; Telex: 24358 PMCS G)

PMP Fermentation Products, Inc.
9525 W. Bryn Mawr Ave., Suite 725, Rosemont, IL 60018 USA (Tel.: 708-928-0050; 800-558-1031; FAX 708-928-0065)

Pointing Ltd.
Princess Way, Prudhoe, Northumberland, NE42 6NJ, UK (Tel.: 0661-832621; FAX 0661-835650; Telex: 537036 POINTX G)

Polimex, Inc.
204 North Dooley St., Grapevine, TX 76051 USA (Tel.: 817-481-3547; FAX 817-488-4816)

Polomyx Industries Inc.
14 Jewel Drive, Wilmington, MA 01887 USA

Polycolors, Inc.
PO Box 291, Bull Run Rd., Greene, ME, 04236 USA (Tel.: 207-946-7339)

Polymer Composites
Polymer Composites, Inc., Subsidiary of Hoechst Celanese Corp., PO Box 30010, 4610 Theurer Blvd., Winona, MN 55987 USA (Tel.: 507-454-4150; 800-526-4960; FAX 507-457-4040)

Represented by: Hoechst Celanese/Engineering Plastics, 26 Main St., Chatham, NJ 07928 USA (Tel.: 201-635-2600; 800-526-4960)

The Polymer Corp.
2120 Fairmount Ave, PO Box 14235, Reading, PA 19612-4235 USA (Tel.: 215-320 6600; 800-366-0300; FAX 800-366-0301)

Polymeric Systems Inc.
Wheatland & Mason Streets, Phoenixville, PA 19460 USA

Polysar
Polysar Inc., 17 Woodland Rd., Madison, CT 06643-2399 USA (Tel.: 203-245-0441; FAX 203-245-4004)

Polysar Inc., 690 Mechanical St., Leominster, MA 01453 USA (Tel.: 508-537-1111; FAX 508-537-2272)

Polysar Nederland BV, Westervoortsedijk 71, NL-6827 AV Arnheim, The Netherlands (Tel.: 85-65 39 11; Telex: 45593)

Polysciences Inc.
400 Valley Road, Warrington, PA 18976-2590 USA (Tel.: 215-343-6484; 800-523-2575; FAX 215-343-0214; Telex: 510-665-8542)

Polyurethane Corp. of America
PO Box 8, Everett, MA 02149 USA (Tel.: 617-389-7889)

Polyvel, Inc.
120 No. White Horse Pike, Hammonton, NJ 08037 USA (Tel.: 609-567-0080)

Poly-Version Inc. *Address unverified*
Tulsa, OK USA

Polyvinyl Chimie Holland BV. *See* ICI Resins BV

Pomosin GmbH, Div. of Copenhagen Pectin A/S
Postfach 7, Von-Herwarth-Strasse, W-2443 Grobenbrode, Germany (Tel.: 4367-8051; FAX 4367-692; Telex: 436710)

Portman Agrochemicals Ltd.
Apex House, Grand Arcade, Tally-Ho Corner, North Finchley, London, N12 0EH, UK (Tel.: 081-446-8383; FAX 081-445-6045)

Potters Industries Inc., Affiliate of The PQ Corp.
Waterview Corp. Center, 20 Waterview Blvd., Parsippany, NJ 07054 USA (Tel.: 201-299-2900; FAX 201-335-9350; Telex: 219054)

Powaspray (CDA) Ltd.
Middlefield Farm, Great Shelford, Cambridge, CB2 5AN, UK (Tel.: 0223-841995; FAX 0223-845576)

Powell Duffryn plc
Powell Duffryn House, London Road, Bracknell, Berkshire, RG12 2AQ, UK (Tel.: 0344 53101; FAX 0344 50599; Telex: 858906)

Poythress Laboratories Inc. *Address unknown*

PPF International Ltd. *See* Quest International UK Ltd

PPG
PPG Industries Inc., One PPG Place, Pittsburgh, PA 15272 USA (Tel.: 412-434-3131; 800-CHEM-PPG; Telex: 86 6570)

PPG Industries, Inc./Fiber Glass Products, One PPG Place, Pittsburgh, PA 15272 USA (Tel.: 412-434-3250; FAX 412-434-2197)

PPG Industries, Inc./Polymer Products, PO Box 98, 100 Station Ave., Stockertown, PA 18083-0098 USA (Tel.: 215-759-3690; FAX 215-759-3692)

PPG Industries, Inc./Specialty Chemicals, 3938 Porett Dr., Gurnee, IL 60031 USA (Tel.: 708-244-3410; 800-323-0856; FAX 708-244-9633; Telex: 25-3310)

PPG Canada Inc./Spec. Chem., 2 Robert Speck Pkwy., Suite 750, Mississauga, Ontario, L4Z 1H8, Canada (Tel.: 416-848-2500; FAX 416-848-2501; Telex: 06960351 canbiz miss)

PPG Industries (UK) Ltd./Specialty Chem., Carrington Business Park, Carrington, Urmston, Manchester, M31 4DD, UK (Tel.: 061-777-9203; FAX 061-777-9064; Telex: 851-94014896 mazu g)

PPG Industries (France) SA, BP 377, Écluse Folien, F-59307 Valenciennes, France (Tel.: 27 14 46 00; FAX 27 29 36 34)

PPG Industries/Asia/Pacific Ltd., Takanawa Court, 5th floor, 12-1 Takanawa 3-Chome, Minato-Ku, Tokyo, 108, Japan (Tel.: (03) 3280-2911; FAX (03) 3280-2920; Telex: 02-42719 PPGPACJ)

PPG Industries Int'l. Inc./Taiwan, Suite 601, Worldwide House, No. 131, Min Sheng East Rd., Sec. 3, Taipei, 105, Taiwan, R.O.C. (Tel.: 886-2-514-8052; FAX 886-2-514-7957)

PPG Industrial do Brazil Ltda., Edificio Grande Avenida, Paulista Ave. 1754, Suite 153, Sao Paulo, Brazil 01310 (Tel.: 011-2840433; FAX 011-2892105; Telex: 391-1139104ppgbrazil)

PQ Corp.

PO Box 840, Valley Forge, PA 19482 USA (Tel.: 215-293-7200; FAX 215-293-7456)

Presperse Inc.

PO Box 735, South Plainfield, NJ 07080 USA (Tel.: 908-756-2033; FAX 908-756-8754; Telex: 2409388)

Procter & Gamble

Procter & Gamble Co., Chemicals Div, PO Box 599, Cincinnati, OH 45201 USA (Tel.: 513-983-3928; 800-543-1580; FAX 513-983-1436; Telex: 21-4185)

Procter & Gamble Inc. Canada, 4711 Yonge St., PO Box 355, Station A, Toronto, Ontario, M5W 1C5, Canada (Tel.: 416-730-4059; FAX 416-730-4122)

Prodorite Ltd.

Unit G, Stafford Park 18, Telford, TF3 3AX, UK (Tel.: 0952 299333; FAX 0952 293933)

Production Chemicals Ltd.

Dalton Way, Middlewich Motorway Estate, Middlewich, Cheshire, CW10 0HS, UK (Tel.: 060 684 5557; FAX 060 684 6408; Telex: 667330 PCL G)

Protameen Chemicals, Inc.

375 Minnisink Rd., PO Box 166, Totowa, NJ 07511 USA (Tel.: 201-256-4374; FAX 201-256-6764; Telex: 130125)

Protan

Protan, Inc., Suite 201, 135 Commerce Way, Portsmouth, NH 03801 USA (Tel.: 603-433-1231; 800-223-9030; FAX 603-433-1348)

Protan Ltd., Westbrook House, High St, Alton, Hampshire, GU34 1EZ, UK (Tel.: 0420 82503; FAX 0420 83360; Telex: 859078)

Protan, Postboks 420, N-3002 Drammen, Norway (Tel.: 3837660)

Protective Rubber Coatings (Limpetite) Ltd.

Paynes Shipyard, Coronation Road, Bristol, BS3 1RP, UK (Tel.: 0272-662206; FAX 0272-661158)

Protex SA

B.P. 177, 6, rue Barbès, 92305 Levallois-Paris, France (Tel.: 47-57-74-00; FAX 47-57-69-28; Telex: 620987)

Provital

Centro Industrial Santiaga, Talleres 6, no. 15, Apartado Correos 78, Barcelona, Spain (Tel.: 93-718-80-12; FAX 93-718-38-30; Telex: 98476 DITT E)

Pulcra SA

Sector E C/42, Barcelona, 08040, Spain (Tel.: 3-323-5914; FAX 3-323-6760; Telex: 98301)

Puma Chemical Co. Inc.

1601 109th St, Grand Prairie, TX 75050 USA

Punati Chemical Corp.

615 S Eton, Birmingham, MI 48009 USA

Purac America, Inc.

111 Barclay Blvd., Suite 280, Lincolnshire, IL 60069 USA (Tel.: 708-634-6330; FAX 708-634-1992; Telex: 280231 PURACINC ARHT)

Purac Biochem BV, Div. of CSM NV

Postbus 21, Arkelsedijk 46, NL-4200 AA Gorinchem, The Netherlands (Tel.: 1830-41799; FAX 1830-22741; Telex: 23615 CCA NL)

The Purdue Frederick Co.

100 Connecticut Ave, Norwalk, CT 06850-3590 USA (Tel.: 203-853-0123)

PVO International Inc. *See* Stepan/PVO Dept

QO Chemicals

QO Chemicals, Inc., 2801 Kent Ave., Box 2500, West Lafayette, IN 47906 USA (Tel.: 317-497-6100; 800-621-9521; FAX 317-497-6287; Telex: 446968 QOC UD)

QO Chemicals Inc. Belgium, Industriedok 391, B-2030 Antwerp, Belgium

QO Chemicals GmbH, Hermannstrasse 54, D-6078 Neu Isenburg, Germany (Tel.: 6102-25098; FAX 6102-1620; Telex: 841-417678 qoche d)

QO Chemicals (Australia) Inc., Suite #4, 21 Drummond Place, Carlton, Victoria, 3053, Australia

Quadrangle Agrochemicals

Langthorpe, Boroughbridge, N Yorks, YO5 9BZ, UK (Tel.: 0423-322429; FAX 0845-537668)

Quantum Chemical Corp.

Quantum Chemical Corp./USI Div., 11500 Northlake Dr., PO Box 429550, Cincinnati, OH 45249 USA (Tel.: 513-530-6556; 800-543-7900; FAX 513-530-6562; Telex: 155116)

Quantum Chemical Europe BV, Lange Bunder 7, 4854 MB Bavel, The Netherlands (Tel.: 001613-6600)

Quantum Composites, Inc., Subsidiary of Premix Inc.

4702 James Savage Rd., Midland, MI 48642-8642 USA (Tel.: 517-496-2884; FAX 517-496-2333)

Ques Industries Inc.
5420 W.140th Street, Cleveland, OH 44142 USA (Tel.: 216-267-8989; FAX 216-267-8998)

Quest International
Quest International, Lindtsedijk 8, 3336 le Zwijndrecht, The Netherlands (Tel.: 78-128511; FAX 78-195279; Telex: 29477)

Quest International, Postbus 2, NL-1400 CA Bussum, The Netherlands (Tel.: 2159-99111; FAX 2159-46067; Telex: 43050 QSTN NL)

Quest International UK Ltd., Kennington Road, Ashford, Kent, TN24 0LT, UK (Tel.: 0233-644444; FAX 0233-644146; Telex: 96369)

Quest International Deutschland GmbH, Postfach 650170, Poppenbütteler Chaussee 36, W-2000 Hamburg 65, Germany (Tel.: 40-607970; FAX 40-6079710; Telex: 215196)

Quest International Fragrances Co, 400 International Dr., Mt. Olive, NJ 07828 USA (Tel.: 201-691-7100; FAX 201-691-7479; Telex: 6714933)

Quinoderm Ltd.
Manchester Road, Hollinwood, Oldham, Lancashire, OL8 4PB, UK (Tel.: 061 624 9307; FAX 061 627 0928)

Radici Novacips
Radici Novacips SpA, Via Provinciale 11, I-24020 Villa D'Ogna (BG), Italy (Tel.: 0346 224 53; FAX 0346 23730; Telex: 303344)

Radilon Inc., Div. of Radici Novacips, PO Box 18367, Spartanburg, SC 29318 USA (Tel.: 803-579-2729)

Radiol Chemicals Ltd.
Stepfield, Witham, Essex, CM8 3AG, UK

Ralco Industries, Inc.
1112 River St., PO Box 509, Woonsocket, RI 02895 USA (Tel.: 401-767-2700; 800-343-4166; FAX 401-767-2823)

RapidPurge Corp.
2285 Reservoir Ave., Trumbull, CT 06611 USA (Tel.: 203-372-5677; 800-243-4203)

Rasa Industries, Ltd.
Yaesu Dai Bldg., 1-1, Kyobashi 1-chome, Chuo-ku, Tokyo, 104, Japan (Tel.: (03) 3278-3801; Telex: 2225818 RASAKOJ)

Raschig
Raschig AG, Mundenheimer Strasse 100, D-6700 Ludwigshafen/Rhine, Germany (Tel.: (0621)56180; FAX 0621-532885; Telex: 464 877 ralu d)

Raschig Corp., 5000 Old Osborne Tpke., Box 7656, Richmond, VA 23231 USA (Tel.: 804-222-9516; FAX 804-226-1569)

Reade Metals & Minerals Corp. *Address unknown*

Reading, Green & Marvell Ltd.
117 Loverock Rd, Reding, Berkshire, RG3 1DZ, UK (Tel.: 0734 574133; FAX 0734 594339)

Reagens SpA
Via Codronchi 4, I-40016 San Giogio di Piano, Italy (Tel.: 897157; FAX 897561; Telex: 510374 REAG I)

Reagent Chemical & Research Inc.
124 River Road, Middlesex, NJ 08846 USA (Tel.: 908-469-0100)

R.E. Carroll. *See under* Carroll, R.E.

Reckitt & Colman
Reckitt & Colman plc, PO Box 26, 1-17 Burlington Lane, London, W4 2RW, UK (Tel.: 081-994 6464; FAX 081-994 8920)

Reckitt's Colours Ltd., Div. of Reckitt & Colman plc, Morley Street, Hull, North Humberside, HU8 8DN, UK (Tel.: 0482 29875; FAX 0482 223114; Telex: 597065)

Reed Plastics
Holden Industrial Park, Holden, MA 01520 USA (Tel.: 508-829-6321; FAX 508-829-2118)

Reedy International Corp.
42 First St., Keyport, NJ 07735 USA (Tel.: 908-264-1777; FAX 908-264-1189)

Regal Chemical Co.
PO Box 900, Alpharetta, GA 30201 USA (Tel.: 404-475-4837; FAX 404-475-1254)

Reheis
Reheis Inc., PO Box 609, 235 Snyder Ave., Berkeley Heights, NJ 07922 USA (Tel.: 908-464-1500; FAX 908-464-7726; Telex: 219463 RCCA UR)

Reheis Ireland, Div. of Reheis Inc., Kilbarrack Rd., Dublin, 5, Irish Republic (Tel.: 1-322621; FAX 1-392205; Telex: 32532 REHI EI)

Reichhold Chemicals
Reichhold Chemicals, Inc., Corporate Headquarters, PO Box 13582, Research Triangle Park, NC 27709 USA (Tel.: 919-544-9225; 800-448-3482; FAX 919-990-7707)

Reichhold Chemicals/Emulsion Polymers Div., PO Drawer K, Dover, DE 19903 USA (Tel.: 302-736-9100; 800-441-6461)

Reichhold Chemicals/Reactive Polymers Div., 8540 Baycenter Rd., PO Box 19129, Jacksonville, FL 32245 USA (Tel.: 904-739-2170)

Reichhold Chemie AG, CH 5212 Hausen b. Brugg, Switzerland (Tel.: 056 482222; FAX 056 482412; Telex: 825109 RCAG CH)

Reilly Industries Inc.
1510 Market Sq. Ct., 151 N. Delaware St., Indianapolis, IN 46204 USA (Tel.: 317-248-6411; FAX 317-248-6413; Telex: 27 404)

Reilly-Whiteman Inc.
801 Washington St., Conshohocken, PA 19428 USA (Tel.: 215-828-3800; 800-533-4514; FAX 215-834-7855; Telex: 5106608845)

Releasomers, Inc.
PO Box 82, Bradfordwoods, PA 15015 USA (Tel.: 412-452-4474; FAX 412-452 1965)

Remy & Co.
Postfach 110608, Baumwall 5, W-2000 Hamburg 11, Germany (Tel.: 40-3690010; FAX 40-363842; Telex: 2162378)

W. J. Rendell Ltd.
Ickleford Manor, Hitchin, Herts, SG5 3XE, UK (Tel.: 0462-432596)

Rentokil Ltd.
Felcourt, East Grinstead, West Sussex, RH19 2JY, UK (Tel.: 0342-833022; FAX 0342-326229; Telex: 95456 RNTKIL G)

Resadhesion Ltd.
53 Royce Close, West Portway Industrial Estate, Andover, Hampshire, SP10 3TS, UK (Tel.: 0264-334633; FAX 0263-332639)

Resart GmbH
Gassnerallee 40, W-6500 Mainz 1, Germany (Tel.: 6131-6310; FAX 6131-631142; Telex: 4187781)

Resinall Corp.
3065 High Rdige Rd., PO Box 8149, Stamford, CT 06903 USA (Tel.: 203-329-7100)

Resinas Sintéticas SA
Aribau, 185-6a planta, E-08021 Barcelona, Spain (Tel.: 3-200 69 11; FAX 3-209 39 00)

Resinoid Engineering Corp./Materials Div.
7557 St. Louis Ave., Skokie, IL 60076 USA (Tel.: 708-673-1050; FAX 708-673-2160)

Resinous Chemicals Ltd.
Cross Lane, Dunston, Tyne & Wear, NE11 9HQ, UK (Tel.: 091-493 2525; FAX 091-460 6270)

Restech Industries Inc.
590 S Seneca (PO Box 2747), Eugene, OR 97402 USA

Revertex Ltd.
Templefields, Harlow, Essex, CM20 2BH, UK (Tel.: 0279-429555; FAX 0279-412984; Telex: 81318)

Rewo
Rewo Chemische Werke GmbH, Postfach 1160, Max Wolf Strasse 7, W-6497 Steinau, Germany (Tel.: 06663-540; FAX 06663-54-129; Telex: 493589)

Rewo Chemicals Ltd., Div. of Schering AG, Gorsey Lane, Widnes, Cheshire, WA8 0HE, UK (Tel.: 051-495 1989; FAX 051-495 2003; Telex: 627 434)

Rexene Products Co.

5005 LBJ Freeway, Occidental Tower, Dallas, TX 75244 USA (Tel.: 214-450-9000; 800-233-1159; FAX 214-450-9028)

Reynolds Metal Co.

Reynolds Metal Bldg, PO Box 27003, Richmond, VA 23261-7003 USA (Tel.: 804-281-2000)

Hans Rhan & Co.

Dörflistrasse 120, CH-8050 Zurich, Switzerland (Tel.: 1-312 1512; FAX 1-312 2150; Telex: 822030)

Rhein-Chemie Rheinau GmbH

Postfach 810409, Müulheimer Strasse 24-28, DW-6800 Mannheim 81, Germany (Tel.: 0621-8 90 70)

Rheox

Rheox Inc., PO Box 700, Wyckoffs Mill Road, Hightstown, NJ 08520 USA (Tel.: 609-443-2500; FAX 609-443-2306)

Rheox Inc., 31 rue de l'Hôpital, B-1000 Brussels, Belgium (Tel.: 2-512-0048; FAX 2-513 24 25; Telex: 24662 EUR B)

Rhône-Poulenc SA

Rhône-Poulenc Chimie, Div. of Rhône-Poulenc SA, 25 Quai Paul Doumer, F-92408 Courbevoie Cedex, France (Tel.: 47 68 1234; FAX 47 68 23 00; Telex: 610500)

Rhône-Poulenc Agrochimie SA, Div. of Rhône-Poulenc SA, BP 9163, 14-20 rue Pierre Baizet, F-69263 Lyon 09, France (Tel.: 72 29 25 25; FAX 72 29 27 99; Telex: 310098 RHONE F)

Rhône-Poulenc Industries, 22 Avenue Montaigne, 75360 Paris Cedex 08, France

Rhône-Poulenc Santé SA, Div. of Rhône-Poulenc SA, 18 ave d'Alsace, F-92400 Courbevoie Cedex 29, France (Tel.: 47 68 12 34; Telex: 610500 RHONE F)

Rhône-Poulenc Chemicals Ltd., Div. of Rhône-Poulenc SA, Staveley, Chesterfield, Derbyshire, S43 2PB, UK (Tel.: 0246-277251; FAX 0246-280090; Telex: 577425 STAVEX G)

Rhône-Poulenc Chemicals Ltd., Div. of Rhône-Poulenc SA, Oak House, Reeds Crescent, Watford, Herts, WD1 1QH, UK (Tel.: 0923 211700; FAX 0923 211580)

Rhône-Poulenc Chemicals Ltd., Perf. Prods. Group, Div. of Rhône-Poulenc SA, Poleacre Lane, Woodley, Stockport, Cheshire, SK6 1PQ, UK (Tel.: 061-430-4391; FAX 061-430-4364; Telex: 667835)

Rhône-Poulenc Crop Protection Ltd., Div. of Rhône-Poulenc SA, Regent House, Hubert Road, Brentwood, Essex, CM14 4TZ, UK (Tel.: 0277-261414; FAX 0277-260621; Telex: 97152 RPDAGN G)

Rhône-Poulenc Environmental Products Ltd., Div. of Rhône-Poulenc SA, Regent House, Hubert Road, Brentwood, Essex, CM14 4TZ, UK (Tel.: 0277-261414; FAX 0277-260621; Telex: 97152 RPDAGN G)

Rhône-Poulenc Rorer Ltd., Div. of Rhône-Poulenc SA, Rainham Rd South, Dagenham, Essex, RM10 7XS, UK (Tel.: 081 592 3060; FAX 081-593 2140; Telex: 28691 MBDAGN G)

Rhône-Poulenc Sturge Lifford, Lifford Lane, Kings Norton, Birmingham, B30 3JW, UK (Tel.: 021 458 6671; FAX 021 451 1759)

Rhône-Poulenc NV, Kuhlmannkaai, B-9020 Ghent, Belgium (Tel.: 091 44 8891; Telex: 11275 RPCHEM B)

Rhône-Poulenc Chimie NV, Div. of Rhône-Poulenc SA, Ketenislaan 1, B-9130 Kallo, Belgium (Tel.: 37-551759; FAX 37-756995; Telex: 33706 RPKALO B)

Rhône-Poulenc Geronazzo SpA, Rhône-Poulenc SA, Via Milano 78, Ospiate Di Bollate, I-20021 Milano, Italy (Tel.: 2-350-3212; FAX 2-350-1770; Telex: 331547 GERO I)

Rhône-Poulenc Ag Co., PO Box 12014, 2 TW Alexander Dr., Research Triangle Park, NC 27709 USA (Tel.: 919-549-2101; 800-334-8577; FAX 919-549-9639; Telex: 4999378 APC RTP)

Rhône-Poulenc Basic Chemical Co., One Corporate Dr., Box 881, Shelton, CT 06484 USA (Tel.: 203-925-3300; 800-642-4200; FAX 203-925-3627)

Rhône-Poulenc, Inc./Chemicals Div., Box 125, Monmouth Junction, NJ 08852 USA (Tel.: 201-297-0100; FAX 201-297-1597)

Rhône-Poulenc, Inc./Film Div., 2754 West Park Dr., Holcomb, NY 14469 USA (Tel.: 716-657-5800; FAX 716-657-5838)

Rhône-Poulenc, Inc./Food Ingredients, CN 7500, Prospect Plains Rd., Cranbury, NJ 08512 USA (Tel.: 609-860-4600; 800-253-5052)

Rhône-Poulenc, Inc./Latex & Specialty Polymer. *Address unknown*

Rhône-Poulenc, Inc./Surfactants & Specialties, CN 7500, Prospect Plains Rd., Cranberry, NJ 08512-7500 USA (Tel.: 609-860-8300; 800-922-2189; FAX 609-860-7626)

Rhône-Poulenc, Inc./Water Soluble Polymers. *Address unknown*

Rhône-Poulenc Oil Field Chems. *Address unknown*

Rhône-Poulenc Rorer Pharmaceuticals Inc., 500 Arcola Road, Collegeville, PA 19426-2911 USA (Tel.: 215-454-8000)

Rhône-Poulenc Silicones. *Address unknown*

Rhône-Poulenc Surfactants & Specialties Canada, 2000 Argentia Rd., Plaza 3, Suite 400, Mississauga, Ontario, L5N 1V9, Canada (Tel.: 416-821-4450; FAX 416-821-9339)

Richards Co., G. Whitfield
4202-10 Main St., Philadelphia, PA 19127 USA (Tel.: 215-487-1202; FAX 215-487-3090; Telex: 845-322 GWRCO)

Richardson-Vicks
Richardson-Vicks, Inc., P.O. Box 5516, Cincinnati, OH 45201 USA (Tel.: 800-358-8707)

Richardson-Vicks Ltd., Div. of Richardson-Vicks Inc., Rusham Park, Whitehall Lane, Egham, Surrey, TW20 9NW, UK

Richter GmbH, Chemische Laboratorium Dr. Kurt
PO Box 410480, Bennigsenstrasse 25, D-1000 Berlin 41, Germany (Tel.: 30-852-70-75; FAX 30-851-18-22; Telex: 184626 clr d)

Ricon Resins, Inc.
569 24 1/4 Road, Grand Junction, CO 81505 USA (Tel.: 303-245-8148; FAX 303-245-4348)

Ridge Technologies, Inc.

PO Box 352, Ridgewood, NJ 07451-0352 USA (Tel.: 201-251-9350; 800-252-0120; FAX 201-251-9351)

Rigby Taylor Ltd.

The Riverway Estate, Portsmouth Road, Peasmarsh, Guildford, Surrey, GU3 1LZ, UK (Tel.: 0483-35657; FAX 0483-34058)

Riker Laboratories Inc. *See* 3M Pharmaceuticals

Rilsan. *See* Atochem

RITA

RITA Corp., 1725 Kilkenny Court, PO Box 585, Woodstock, IL 60098 USA (Tel.: 815-337-2500; 800-426-7759; FAX 815-337-2522; Telex: 72-2438)

Represented by: Maprecos, 4, Rue des Passe-Loups, 7770 Fontaine, Le Port, France

Rit-Chem Co. Inc.

109 Wheeler Ave., PO Box 435, Pleasantville, NY 10570 USA (Tel.: 914-769-9110; FAX 914-769-1408; Telex: 229 639 RTCH)

RJ Manufacturing Inc.

PO Box 120274, 322 Fredericksburg Rd, San Antonio, TX 78212 USA

RMB Animal Health Ltd.

Rainham Road South, Dagenham, Essex, RM1D 7XS, UK

RMc Minerals, Inc.

111 E. Drake Rd., Suite 7104, Fort Collins, CO 80525 USA (Tel.: 303-223-7790; FAX 303-226-5617)

Robeco Chemicals Inc.

99 Park Ave., New York, NY 10016 USA (Tel.: 212-986-6410; FAX 212-986-6419; Telex: 23-3053)

Robertson Co. *Address unverified*

Pittsburgh, PA USA

A. H. Robins Co.

A. H. Robins Co. Ltd., Sussex Manor Business Park, Gatwick Road, Crawley, W Sussex, RH10 2NH, UK

A. H. Robins Co. Inc., 1405 Cummings Drive, Richmond, VA 23220 USA

Robinson Brothers Ltd.

Phoenix Street, West Bromwich, West Midlands, B70 0AH, UK (Tel.: 021-553 2451; FAX 021-500 5183; Telex: 337894)

James Robinson Ltd., Div. of Holliday Chemical Holdings plc

PO Box B3, Hillhouse Lane, Huddersfield, West Yorkshire, HD1 6BU, UK (Tel.: 0484-435577; FAX 0484-435580; Telex: 51191 JROBCO G)

Robnorganic Systems Ltd.

Highworth Road, South Marston, Swindon, Wiltshire, SN3 4TE, UK (Tel.: 0793-823741; FAX 0793-827033; Telex: 449 905)

Roche Laboratories, Div. of Hoffmann-La Roche Inc.

340 Kingsland St, Nutley, NJ 07110 USA (Tel.: 201-235-3381; FAX 201-235-8023)

Roche Products Ltd./Vitamins & Fine Chemicals Div., Div. of Roche AG

40 Broadwater Road, Welwyn Garden City, Hertfordshire, AL7 3AY, UK (Tel.: 0707-328128; FAX 0707-329587; Telex: 262098 ROCHEW G)

Rocol Ltd., Div. of Morgan Crucible Co. plc

Rocol House, Swillington, Leeds, West Yorkshire, LS26 8BS, UK (Tel.: 0532-866511; FAX 0532-872159; Telex: 55249)

Roehm Ltd./Hexoran Div.

Derwent Street, Belper, Derby, Derbyshire, DE5 1WQ, UK (Tel.: 0773-822471; FAX 0773-820176; Telex: 377716)

J. B. Roerig, Div. of Pfizer Pharmaceuticals

235 East 42nd Street, New York, NY 10017 USA (Tel.: 212-573-2323; 800-533-4535)

Rogers Corp.

One Technology Drive, Rogers, CT 06263 USA (Tel.: 203-774 9605; FAX 203-774 9630)

Rohm & Haas

Rohm & Haas Co., Independence Mall West, Philadelphia, PA 19105 USA (Tel.: 215-592-3000; 800-323-4165; FAX 215-592-2285; Telex: 845-247)

Rohm & Haas Canada Inc., 2 Manse Rd., West Hill, Ontario, M1E 3T9, Canada

Rohm & Haas Co. European Operations, Chesterfield House, Bloomsbury Way 15-19, London, WC1A 2TP, UK (Tel.: 071 242 4455; FAX 071 404 4126; Telex: 24139)

Rohm & Haas (UK) Ltd., Lennig House, 2 Mason's Avenue, Croydon, CR9 3NB, UK (Tel.: 081-686-8844; FAX 081-686 8329; Telex: 917266)

Röhm GmbH Chemische Fabrik, Kirschenallee, Postfach 4242, D-6100 Darmstadt, Germany (Tel.: 6151 18 01; FAX 6151 184007)

Rohm & Haas Asia Ltd., Kaisei Bldg., 8-10 Azabudai 1-Chome, Minato-ku, Tokyo, 106, Japan

Rohm & Haas (Australia) Pty. Ltd., 969 Burke Rd., PO Box 11, Camberwell, Victoria, 3124, Australia

Rohm Tech Inc.

83 Authority Dr., Fitchburg, MA 01420 USA (Tel.: 508-342-5831; FAX 508-345-1971)

Ronacrete Ltd.
Ronac House, Selinas Lane, Dagenham, Essex, RM8 1QL, UK (Tel.: 081-593 7621; FAX 081-595 6969)

Rona Laboratories. *Address unknown*

Ronsheim & Moore Ltd., Div. of Hickson & Welch Ltd.
Wheldon Road, Castleford, West Yorkshire, WF10 2JT, UK (Tel.: 0977-556565; FAX 0977-518058; Telex: 55378)

Roquette
Roquette Corp., 1550 Northwestern Ave., Gurnee, IL 60031 USA (Tel.: 708-249-5950; 800-223-5305; FAX 708-578-1027; Telex: 687 1679 ROQ ILL)

Roquette (UK) Ltd., Pantiles House, 2 Nevill Street, Tunbridge Wells, Kent, TN2 5TT, UK (Tel.: 0892-540188; FAX 0892-510872; Telex: 957558 G)

Rorer. *See* Rhône-Poulenc Rorer

William H. Rorer Inc.
500 Virginia Drive, Fort Washington, PA 19034 USA

Frank B. Ross
Frank B. Ross Co., Inc., 22 Halladay St., PO Box 4085, Jersey City, NJ 07304-0085 USA (Tel.: 201-433-4512; FAX 201-332-3555)

Represented by: Harrisons & Crosfield (Canada) Ltd., St. Laurent, Quebec, H4M 2N6, Canada (Tel.: 514-748-7911)

Ross Laboratories, Div. of Abbott Laboratories
625 Cleveland Ave, Columbus, OH 43216 USA (Tel.: 614-624-7677)

Rostine Manufacturing & Supply Co.
PO Box 8192, 4227C W Church, Springfield, MO 65801 USA

Rostone Corp.
PO Box 7497, 2450 Sagamore Pkwy. South, Lafayette, IN 47903 USA (Tel.: 317-474-2421; 800-637-4851; FAX 317-474-8785)

Rottapharm SpA
Via Valosa di Sopra 9, I-20052 Monza, Italy (Tel.: 39-73901; FAX 39-747806; Telex: 331661 ROTTA I)

Rotuba
Rotuba Extruders, Inc., 1401 Park Ave. S., Linden, NJ 07036 USA (Tel.: 201-486-1000)

Rotuba Plastics, 1401 Park Ave. South, Linden, NJ 07036-1698 USA (Tel.: 201-486-1000; FAX 201-486-0874; Telex: 910-240-3809)

Roussel Laboratories Ltd., Div. of Roussel Uclaf SA
Broadwater Park, North Orbital Rd, Uxbridge, Middlesex, UB9 5HP, UK (Tel.: 0895 834343; FAX 0895 834578; Telex: 889180)

Roussel Uclaf, Fine Chemicals, Div. of Hoechst AG
Roussel Uclaf, Fine Chemicals, Div. of Hoechst AG, Tour Roussel Hoechst, F-92080 Paris, France (Tel.: 40 81 40 81; FAX 47 78 92 54; Telex: 610884 UCLAF F)

Roussel-Uclaf, Mc Intryre House, High Street, Berkhampsted, Herts, HP4 2DY, UK

Rowell Laboratories Inc.
Baudette, MN 56623 USA

Royale Polymers Ltd.
Poucher Street, Hilltop, Kimberworth, Rotherham, S Yorks, S61 2ET, UK

RP Drugs Ltd.
RPD House, Yorkdale Industrial Park, Braithwaite Street, Leeds, West Yorkshire, LS11 9XE, UK (Tel.: 0532-441400; FAX 0532-460738)

RRI Malaya. *Address unknown*

RTP Co.
580 E. Front St., PO Box 439, Winona, MN 55987-5439 USA (Tel.: 507-454-6900; 800-433-4787; FAX 507-454-8130)

R. T. Vanderbilt Chemical Corp. *See under* Vanderbilt, R.T.

Rubber Regenerating Co. *Address unknown*

Ruco Polymer Corp.
New South Rd., Hicksville, NY 11802 USA (Tel.: 516-931-8100; FAX 516-931-8179; Telex: 5102220856)

Ruetgers-Nease Chemical Co., Inc., Subsidiary of Rütgerswerke AG
201 Struble Rd., State College, PA 16801 USA (Tel.: 814-238-2424; FAX 814-238-1567)

Runnymede Dispersions Ltd.
Ruspidge Works, Cinderford, Glostershire, GL14 3AS, UK (Tel.: 0594 822375; FAX 0594 826251)

W. J. Ruscoe Co
219 E Miller Avenue, Akron, OH 44301 USA (Tel.: 216-253-8148; FAX 216-253-2933)

Rustlan Chemical Co. *Address unknown*

Rütgerswerke AG
Postfach 111541, Mainzer Landstrasse 217, W-6000 Frankfrut am Main 1, Germany (Tel.: 69-75921; FAX 69-7592488; Telex: 411226)

RV Chemicals Ltd.
Tanhouse Lane, Widnes, Cheshire, WA8 0TE, UK (Tel.: 051-424 6101; FAX 051-420 4330; Telex: 627097 RVC G)

RWE-DEA Chemicals UK Ltd., Div. of RWE-DEA AG

Suite 316-317, Surrey House, 34 Eden St, Kingston Upon Thames, Surrey, KT1 1ZR, UK (Tel.: 051 424 6101; FAX 051 420 4330; Telex: 627097 RVC G)

Rynco Scientific Corp. *Address unknown*

Rystan Co Inc.

PO Box 214, 47 Center Avenue, Little Falls, NJ 07424 USA (Tel.: 201-256-3737)

Ryvan Chemical Co. Ltd.

Botley Rd, Hedge End, Southampton, Hampshire, SO3 3HE, UK (Tel.: 0489 784346; FAX 0489 787810; Telex: 477663)

Sachtleben Chemie GmbH

Dr Rudolph-Sachtleben-Str. 4, D-4100 Duisburg 17, Germany (Tel.: 2066-220; FAX 2066-22 20 00)

L A Salomon Inc.

150 River Rd., Suite L-3B, Montville, NJ 07045 USA (Tel.: 201-335-8300; FAX 201-335-1236)

Salt Service and Chemical Co.

601 Chester Pike, Crom Lynne, PA 19022 USA (Tel.: 215-833-5200; FAX 215-833-2414)

Sandoz

Sandoz AG, Lichtstrasse 35, CH-4002 Basel, Switzerland (Tel.: 61-24-1111; FAX 61-24-8081; Telex: 965050)

Sandoz Chemicals (UK) Ltd., Div. of Sandoz AG, Calverley Lane, Horsforth, Leeds, West Yorkshire, LS18 4RP, UK (Tel.: 0532 584646; FAX 0532 390063; Telex: 557114)

Sandoz Products Ltd/Crop Protection Div, Div. of Sandoz AG, Norwich Union House, 16-18 Princes St, Ipswich, Suffolk, IP1 1QT, UK (Tel.: 0473 255972; FAX 0473 258252)

Sandoz Products Ltd/Pharmaceuticals Div, Div. of Sandoz AG, Frimley Business Park, Frimley, Camberley, Surrey, GU16 5SG, UK (Tel.: 0276 692255; FAX 0276 692508)

Sandoz Chemicals Corp., 4000 Monroe Rd., Charlotte, NC 28205 USA (Tel.: 704-331-7000; 800-631-8077; FAX 704-372-5787; Telex: 704-216-922)

Sandoz Pharmaceuticals, Div. of Sandoz Inc., Route 10, East Hanover, NJ 07936 USA (Tel.: 201-503-7500)

Sandoz Chemicals Corp. Canada, Dorva, Quebec, H9R 4PR, Canada

Sandoz K.K., Kobe Chamber of Commerce & Ind. Bldg., 6-1, Minatojima Nakamachi, Chuo-ku, Kobe, 650, Japan (Tel.: (078) 303-5850; Telex: 05624139 SANDOZ J)

Sanifoam Inc. *Address unknown*

Sanncor Industries, Inc.

300 Whitney St., PO Box 703, Leominster, MA 01453 USA (Tel.: 508-537-4748; FAX 508-537-8245; Telex: 701662)

Sanofi Bio-Industries Ltd. *Address unknown*

Santech Inc.
35 Jutland Rd., Toronto, Ontario, M8Z 2G6, Canada (Tel.: 416-252-5997; FAX 416-251-9315)

Sanyo Chemical Industries, Ltd.
11-1, Ikkyo Nomoto-cho, Higashiyama-ku, Kyoto, 605, Japan (Tel.: (075) 541-4311; FAX (075) 551-2557; Telex: 05422110)

San Yuan Chemical Co., Ltd.
3F, No 3, Alley 6, Lane 575, Tun Hwa South Rd., Taipei, Taiwan, Taiwan, R.O.C.

Sariaf SpA
Via San Silvestro 1, I-48018 Faenza, Italy (Tel.: 546-67 21 11; Telex: 663950)

Sartomer
Sartomer Co., Oaklands Corp. Center, 468 Thomas Jones Way, Exton, PA 19341 USA (Tel.: 215-363-4100; 800-345-8247; FAX 215-363-4140)

Sartomer International, Inc., Kingswick House, Sunninghill, Berkshire, SL5 7BH, UK (Tel.: 99023491; FAX 99026176)

Sartomer International Deutschland GmbH, Konigsallee 60F, 4000 Dusseldorf 1, Germany

Sartomer International, Inc., Bankastraat 131d, 2585 El Den Haag, The Netherlands

Sartomer International, Inc., 7500A Beach Rd., Unit No. 14-313, The Plaza, Singapore

Sasolchem, Div. of Sasol Chemical Industries (Pty) Ltd.
2 Sturdee Avenue, Rosebank, Johannesburg, 2196, South Africa (Tel.: 011-880-1322; FAX 011-880-1362; Telex: 4-25413 SA)

SAS Pharmaceuticals Ltd. *Address unknown*

Saurefabrik Schweizerhall
CH-4133 Schweizerhalle, Switzerland (Tel.: 61-8253111; FAX 61-8218027; Telex: 968016)

Savage Laboratories, Div. of Byk-Gulden Inc.
60 Baylis Road, PO Box 2006, Melville, NY 11747 USA

SBC Technology Ltd. *Address unknown*

SBD Construction Products Ltd.
Dickens House, Maulden Road, Flitwick, Bedford, MK45 5BY, UK (Tel.: 0525 718877; FAX 0525 718988; Telex: 826343 SBDG)

Schaefer Salt & Chemical Co.
1255 Magie Ave., PO Box 236, Elizabeth, NJ 07207 USA (Tel.: 908-352-7010; 800-631-7300; FAX 908-352-7329; Telex: 139423)

Schaefer Technologies Inc.
3000 Carrollton Road, Saginaw, MI 48604 USA (Tel.: 517-753-1877; 800-444-9034; FAX 517-753-3346)

Schaumstoff und Kunststoff GmbH
Eduard-Suesse 19, A-4021 Linz-Wegscheid, Obersterreich, Austria

Schenectady
Schenectady Chemicals, Inc., PO Box 1046, Schenectady, NY 12301 USA (Tel.: 518-370-4200; FAX 518-382-8129; Telex: 145457)

Schenectady Chemicals Canada Ltd., 319 Comstock Rd., Scarborough, Ontario, M1L 2H3, Canada

Schenectady-Midland Ltd., Div. of Schenectady Chemicals Inc., Four Ashes, Wolverhampton, West Midlands, WV10 7BT, UK (Tel.: 0902 790555; FAX 0902 791640; Telex: 339075 SCHMID G)

Scher
Scher Chemicals, Inc., Industrial West & Styertowne Rd., PO Box 4317, Clifton, NJ 07012 USA (Tel.: 201-471-1300; FAX 201-471-3783; Telex: 642643 Scherclif)

Represented by: Chesham Chemicals Ltd., Cunningham House, Bessborough Rd., Harrow, HA1 3DU, UK

Schering AG
Schering AG, Postfach 650311, Müllerstrasse 170-178, W-1000, Berlin 65, Germany (Tel.: 30-4680; FAX 30-4685305; Telex: 182030)

Schering Agrochemicals Ltd., Div. of Schering AG, Hauxton, Cambridge, Cambridgeshire, CB2 5HU, UK (Tel.: 0223 870312; FAX 0223 872142; Telex: 81654 SCHCAM G)

Schering Agrochemicals Ltd., Div. of Schering AG, Chesterford Park Industrial Research Station, Saffron Walden, Essex, CB10 1XL, UK (Tel.: 0799-30123; FAX 0799-30991; Telex: 817300)

Schering Industrial Chemicals Ltd., Div. of Schering AG, Gorsey Lane, Widnes, Cheshire, WA8 0HE, UK (Tel.: 051 495 1989; FAX 051 495 2003; Telex: 627434 SIPWID G)

Schering Corp., Galloping Hill Rd, Kenilworth, NJ 07033 USA (Tel.: 908-298-4000; 800-526-4099)

Schering Berlin Polymers Inc., Div. of Schering AG, 4868 Blazer Memorial Pkwy., PO Box 1227, Dublin, OH 43017 USA (Tel.: 614-793-7700; FAX 614-793-7711)

Schering Health Care Ltd., Div. of Schering AG, The Brow, Burgess Hills, West Sussex, RH15 9NE, UK (Tel.: 0444-232323; FAX 0444 246613; Telex: 87577)

Schering-Plough Ltd., Div. of Schering AG, Mildenhall, Bury St Edmunds, Suffolk, IP28 7AX, UK (Tel.: 0638 716321; FAX 0638 717858)

Schering-Plough HealthCare Products Inc., Div. of Schering-Plough Corp., 110 Allen Rd, Liberty Corner, NJ 07938 USA (Tel.: 908-709-2860; FAX 908-709-2855; Telex: 6719534)

Schering-Plough K.K., Subsid. of Schering-Plough Corp., 2-6, Awajimachi 1-chome, Chuo-ku, Osaka, 541, Japan (Tel.: (06) 201-1701; FAX (06) 201-1791)

Schnee-Morehead, Inc.
111 N. Nursery Rd., Irving, TX 75060 USA (Tel.: 214-438-9111; 800-255-9427; FAX 214-554-3939)

Schuller Mats & Reinforcements, Div. of Schuller Int'l.

PO Box 517, Toledo, OH 43697-0517 USA (Tel.: 419-878-8111; Telex: 5225243)

A. Schulman

A. Schulman Inc., 3550 West Market St., PO Box 1710, Akron, OH 44309-1710 USA (Tel.: 216-666-3751; FAX 216-668-7204)

A. Schulman Canada Ltd., 170 Attwell Dr., Suite 503, Ontario, M9W 5Z5, Canada (Tel.: 416-675-7878)

A. Schulman Inc. Ltd., Croespenmaen Industrial Estate, Crumlin, Newport, Gwent NP 14AG, South Wales, UK

Schweizerhall Inc.

10 Corporate Place South, Piscataway, NJ 08854 USA (Tel.: 908-981-8200; 800-243-6564; FAX 908-981-8282; Telex: 4754581 SUSA)

Scientific Hospital Supplies

100 Wavertree Blvd, Wavertree Technology Park, Liverpool, Merseyside, L7 9PQ, UK (Tel.: 051-708 8008)

SCM Chemicals

SCM Chemicals, 7 St. Paul St., Suite 1010, Baltimore, MD 21202 USA (Tel.: 301-783-1120; 800-638-3234; FAX 301-783-1087)

SCM Glidco Organics, PO Box 389, Jacksonville, FL 32201 USA (Tel.: 904-768-5800; 800-231-6728; FAX 904-768-2200; Telex: 441763)

SCM Chemicals Europe, PO Box 26, Grimsby, S Humberside, DN37 8DP, UK (Tel.: 0469 571000; FAX 0469 571234; Telex: 52595)

SCM Chemicals Ltd., Hop Hing Centre, 22nd Floor, 8-12 Hennessy Rd., Wanchai, Hong Kong

SCM Chemicals Ltd., PO Box 465, Auburn, N.S.W., 2144, Australia (Tel.: 2-647-2566)

ScotFoam Corp.

1500 E Second St, Eddystone, PA 19013 USA

Scott Bader

Scott Bader Co. Ltd., Wollaston, Wellingborough, Northamptonshire, NN9 7RL, UK (Tel.: 0933-663100; FAX 0933-663474; Telex: 31387)

Scott Bader SA, Div. of Scott Bader Co Ltd., 65 rue Sully, F-80044 Amiens, France (Tel.: 22 44 42 00; FAX 22 44 06 24; Telex: 150425 SCOTBAD F)

The Scottish Adhesives Co. Ltd.

9-23 Farnell Street, Glasgow, KA30 8PR, UK (Tel.: 041-332 1736; FAX 041-332 3736)

Scottish Agricultural Industries plc, Div. of ICI plc

West Mains of Ingleton, Ingliston, Newbridge, Midlothian, EH28 8ND, UK (Tel.: 031-335 3280; FAX 031-335 2280; Telex: 72268 SCOTAG G)

S & D Chemicals Ltd.

Cunningham House, Westfield Lane, Harrow, Middlesex, HA3 9ED, UK (Tel.: 081-907-8822; FAX 081-907-1798)

Seal Ltd.
46 Chesterfieid Road, Leicester, LE5 5LP, UK (Tel.: 0533-739501)

Sealocrete Ltd.
Binns Close, Coventry, West Midlands, CV4 9WE, UK (Tel.: 0203-694567; FAX 0203-466440)

G. D. Searle & Co Ltd.
Whalton Rd, Morpeth, Northumberland, NE61 3YA, UK (Tel.: 0670 51431; FAX 0670 57112; Telex: 53638 SEARLE G)

Searle Chemicals Inc., Div. of Monsanto Co
PO Box 8526, Chicago, IL 60680 USA (Tel.: 708-470-6360; FAX 708-470-6736)

John L. Seaton & Co Ltd.
Bankside, Hull, North Humberside, HU5 1RR, UK (Tel.: 0482 41345; FAX 0482 447157; Telex: 592207)

Secure Inc. *Obsolete products*

Seedcote Systems Ltd.
Telford Way, Thetford, Nofolk, IP24 1HU, UK (Tel.: 0842-66261; FAX 0842-66263)

Seimi Chemical Co., Ltd.
2-10, Chigasaki 3-chome, Chigasaki-shi, Kanagawa, 253, Japan (Tel.: (0467) 82-4131; FAX (0467) 86-2767)

Sekisu (UK) Ltd.
Ward Royal Parade, Alma Road, Windsor, Berks, SL4 3HR, UK (Tel.: 0735-856641; FAX 0735-859941)

Selectokil Ltd.
Abbey Gate Place, Tovil, Maidstone, Kent, ME15 0PP, UK (Tel.: 0622-755471; FAX 0622-692113)

Seppic
Seppic, Div. of L'Air Liquide, 75 Quai d'Orsay, F-75321 Paris Cedex 07, France (Tel.: 40 62 55 55; FAX 40 62 52 53; Telex: 202901 SEPPI F)

Seppic Inc., Subsidiary of Seppic France, 30 Two Bridges Rd., Suite 370, Fairfield, NJ 07004 USA (Tel.: 201-882-5597; FAX 201-882-5178)

Service Chemicals Ltd.
17 Lanchester Way, Royal Oak Industrial Estate, Daventry, Northamptonshire, NN11 4PH, UK (Tel.: 0327-704444; FAX 0327-711154)

Servier Laboratories Ltd.
Fulmer Hall, Windmill Road, Fulmer, Slough, Bucks, SL3 6HH, UK

Servo Delden B.V.

Postbus 1, NL-7490 AA Delden, The Netherlands (Tel.: 5407-63535; FAX 5407-64125; Telex: 44347)

Seton Healthcare Group plc

Turbiton House, Oldham, OL1 3HS, UK (Tel.: 061 652 2222; FAX 061-626-9090; Telex: 669956 SETON G)

Sevalco Ltd., Div. of Phelps Dodge Corp.

Severn Rd, Avonmouth, Bristol, Avon, BS11 0YL, UK (Tel.: 0272 822611; FAX 0272 35444; Telex: 44239 SVLCO G)

Seven Seas Ltd.

Marfleet, Hull, HU9 5NJ, UK (Tel.: 0482-275234; FAX 0482-74345; Telex: 592535)

S F C

3 Rue des Carrires, 93800 Epinay s/seine, France

Shawplas Ltd.

Alms Close, Stukeley Meadows Industrial Estate, Huntingdon, Cambridgeshire, PE18 6DY, UK (Tel.: 0480-459338; FAX 0480 45918; Telex: 329173)

Sheffield Plastics, Inc., Div. of DSM

Salisbury Rd., Sheffield, MA 01257 USA (Tel.: 800-628-5084)

Shell

Shell Chemical Co., PO Box 2463, One Shell Plaza, Houston, TX 77252-2463 USA (Tel.: 713-241-6161; FAX 713-241-4043; Telex: 762248)

Represented by for International sales: Pecten Chemicals, Inc., One Shell Plaza, Houston, TX 77252-9932 USA (Tel.: 713-241-6161)

Shell Chemicals UK Ltd., Shell-Mex House, Strand, London, WC2R0DX, UK (Tel.: 071-257 40000; FAX 071-257 1336; Telex: 21795)

Shell Chemicals UK Ltd., Heronbridge House, Chester Business Park, Wrexham Rd, Chester, Cheshire, CH4 9QA, UK (Tel.: 0244-685678; FAX 0244-685010; Telex: 21795)

Shell Chemicals UK Ltd., Agriculture Div., Heronshaw House, Ermine Business Park, Huntingdon, Cambs, PE18 6YA, UK (Tel.: 0480-414140; FAX 0480-444444)

Shell Chemicals Ireland Ltd., Gratton House, 68-72 Lower Mount Street, Dublin, Irish Republic (Tel.: 1-785177; FAX 1-767489; Telex: 93450)

Shell Chimie SA, BP 319, 23-25 ave de Republique, 7539 Paris Cedex 08, F-92506 Rueil-Malmaison, France (Tel.: 47 52 27 00; FAX 47 52 28 02; Telex: 632051 SHELL F)

Shell Nederland Chemie B.V., PO Box 187, 2501 CD The Hague, The Netherlands

Deutsche Shell Chemie GmbH, Postfach 5220, 6236 Eschborn, Taunus, Germany (Tel.: 61964740; FAX 61474502; Telex: 410458 DSCE D)

Sherex Chemical Co., Inc., Div. of Witco

PO Box 646, 5777 Frantz Rd., Dublin, OH 43017 USA (Tel.: 614-764-6500; 800-848-7370; FAX 614-764-6650; Telex: 245356)

Sherman Chemicals Ltd.
Sunderland Rd, Sandy, Bedfordshire, SG19 1QY, UK (Tel.: 0767 681131; FAX 0767 691578; Telex: 825743)

Shin-A Chemical Mfg. Co. Ltd.
Distributed by Synthetic Rubber Technologies

Shin-Etsu Chemical Co., Ltd.
6-1, Otemachi 2-chome, Chiyoda-ku, Tokyo, 100, Japan (Tel.: (03) 3246-5011; FAX (03) 3246-5350; Telex: SHINCHEM J-24790)

Shinko Chemical Co., Ltd.
No. 5 Fuji Bldg., 3-10, Nishi-Honmachi 1-chome, Nishi-ku, Osaka, 550, Japan (Tel.: (06) 533-6655; FAX (06) 533-6620)

Showa Denko K.K.
13-9, Shiba-Daimon 1-chome, Minato-ku, Tokyo, 105, Japan (Tel.: (03) 5470-3533; FAX (03) 3436-2625; Telex: J26232)

Shuman Plastics
35 Neoga St., Depew, NY 14043 USA (Tel.: 716-685-2121; FAX 716-685-3236; Telex: 91 9121)

Siba Hegner Ltd., Div. of Siba Hegner & Co Ltd.
County House, 221-241 Beckenham Road, Beckenham, Kent, BR3 4UF, UK (Tel.: 081-659 2345; FAX 081-659 1292; Telex: 946651)

3V-Sigma
1500 Harbor Blvd., Weehawken, NJ 07087 USA (Tel.: 201-865-3600; FAX 201-865-1892)

Sigma Coatings, Subsidiary of Fina plc
Fina House, Ashley Avenue, Epsom, Surrey, KT18 5AD, UK (Tel.: 03727-26226; FAX 0372-744520)

Sigri GmbH
Postfach 1160, Werner von Siemens Strasse 18, W-8901 Meitingen, Germany (Tel.: 8271-830; FAX 8271-833127; Telex: 53823 SIGRI D)

Sihi Pumps (UK) Ltd.
Broadheath, Altrincham, Cheshire, WA14 INB, UK (Tel.: 061 928 6371; FAX 061 928 3022; Telex: 667013)

Silberline Ltd.
Banbeath Rd., Leven, Fife, KY8 5HD, UK (Tel.: 0333 24734; FAX 0333 21369; Telex: 727373 SILBER G)

Silberline Mfg. Co. Inc.
Lincoln Drive, P O Box B, Tamaqua, PA 18252-0420 USA (Tel.: 717-668-6050; FAX 717-668-0197)

Silbrico Corp.

6300 River Rd., Hodgkins, IL 60525-4257 USA (Tel.: 708-354-3350; FAX 708-354-6698)

Siliconas Hispania SA

Balmes 357, E-08006 Barcelona, Spain (Tel.: 93-212 39 54; FAX 93-211 54 61)

Siltech Inc.

4437 Park Dr., Suite E, Norcross, GA 30093 USA (Tel.: 404-279-8601; FAX 404-279-8535)

Simonis BV

Postbus 33, Conradstraat 38, NL-3000 AA Rotterdam, The Netherlands (Tel.: 10-411 32 00; FAX 10-413 56 27; Telex: 22418)

Sinclair Horticulture & Leisure Ltd.

Firth Road, Lincoln, LN6 7AH, UK (Tel.: 0522-537561; FAX 0522-513609)

Sipcam UK Ltd.

100 Chalk Farm Road, London, NW1 8EH, UK

H. A. Smith & Sons Ltd.

Torrington Ave, Till Hill, Coventry, West Midlands, CV4 9GX, UK (Tel.: 0203 461111; FAX 0203 465325)

SmithKline Beecham

SmithKline Beecham plc, Clarendon Road, Worthing, West Sussex, BN14 8QH, UK

SmithKline Beecham Pharmaceuticals Ltd., Div. of SmithKline Beecham plc, Mundells, Welwyn Garden City, Hertfordshire, AL7 1EY, UK (Tel.: 0707 325111; FAX 0707 325600)

SmithKline & French SA, Div. of SmithKline Beecham plc, 89 rue de l'Institut, B-1330 Rixensart, Belgium (Tel.: 2-653-88 44; Telex: 63251 SBBIO B)

Smith Kline & French Laboratories. See SmithKline Beecham

Smith & Nephew

Smith & Nephew plc, 2 Temple Place, Victoria Embankment, London, WC2R 3BP, UK (Tel.: 071-836 7922; FAX 071-240 7088; Telex: 299494 SANACO G)

Smith & Nephew Pharmaceuticals Ltd., Div. of Smith & Newphew plc, Bampton Road, Harold Hill, Romford, Essex, RM3 8SL, UK (Tel.: 0402-349333; FAX 0402-371316)

Werner G. Smith, Inc.

1730 Train Ave., Cleveland, OH 44113 USA (Tel.: 216-861-3676; 800-535-8343; FAX 216-861-3680)

Smooth-On, Inc.

1000 Valley Rd., Gillette, NJ 07933 USA (Tel.: 201-647-5800; FAX 201-604-2224; Telex: 882833)

Smyth-Morris Chemicals Ltd.

Bassington Industrial Estate, Cramlington, Northumberland, ME23 8AD, UK (Tel.: 0670-713411; FAX 0670-590235)

SN Corp./Appropriate Technology Ltd.
3601 Garden Brook, Dallas, TX 75234 USA (Tel.: 214-243-8930; FAX 214-406-1125)

SNIA
SNIA Tecnopolimeri SpA, Via Stabilimenti 11, I-20020 Ceriano Laghetto (MI), Italy (Tel.: 02-96161; FAX 9616547; Telex: 314594 TECPOL I)

SNIA (UK) Ltd., 36 Broadway, St. James's, London, SW1H 0BH, UK (Tel.: 071-222-8696; FAX 071-222-8705; Telex: 23377)

SNPE
SNPE Chimie, 12, quai Henri-IV, F-75004 Paris Cedex 4, France (Tel.: 48 04 66 66; FAX 48 04 69 89; Telex: 220 380 SNPE F)

SNPE Ltd., Div. of SNPE, Suffolk House, George St, Corydon, Surrey, CR0 1PE, UK (Tel.: 081 649 7474; FAX 081 760 0420; Telex: 263697 SNPE G)

Société Chimique Roche
BP 170, Boulevard d'Alsace, F-68305 Saint Louis, France (Tel.: 89 69 00 20; FAX 89 69 85 05; Telex: 881528 ROCHE F)

Societ Indochine de Plantations d'Hvas (SIPH). *Address unknown*
Ivory Coast

Solaver SA
Rue de l'Invasion, B-4800 Verviers, Belgium

Solec/Solar Energy Corp.
Box 3065, Princeton, NJ 08543-3065 USA (Tel.: 609-883-7700; FAX 609-497-0182)

Solem. *See* Huber, J.M.

Solochart Ltd.
Brookhampton Lane, Kineton, Warks, CV35 0JA, UK

Solrec Ltd.
Middleton Road, Heysham, Morecambe, Lancashire, LA3 3JW, UK (Tel.: 0524-853053; FAX 0524-851284; Telex: 65107 SOLREC G)

Soltex Polymer Corp. *See* Solvay Polymers

Soluol Chemical Co.
Green Hill & Market Sts., Box 112, W. Warwick, RI 02893 USA (Tel.: 401-821-8100; FAX 401-823-6673)

Solvay
Solvay SA, 33 rue du Prince Albert, B-1050 Brussels, Belgium (Tel.: 2-509-6111; FAX 2-509-6617; Telex: 21337)

Solvay Chemicals Ltd., Unit 1, Grovelands Business Centre, Boundary Way, Hemel Hempstead, Herts, HP2 7TE, UK (Tel.: 0442-236555)

Solvay Deutschland GmbH/Solvay Catalysts GmbH, Postfach 220, W-3000 Hannover-1, Germany (Tel.: 511-857-0; FAX 511-282126; Telex: 922-755)

Solvay Enzymes GmbH & Co KG, Div. of Solvay & Cie SA, Postfach 690307, Hans Böckler Allee 20, W-3000 Hannover 1, Germany (Tel.: 511-8570; FAX 511-8572371; Telex: 922755)

Solvay Duphar BV, Postbus 900, 1380 DA Weesp, The Netherlands (Tel.: 2940-77711; FAX 2940-80253; Telex: 14232)

Solvay Enzymes, Inc., PO Box 4859, Elkhart, IN 46514-0859 USA (Tel.: 219-523-3700; 800-342-2097)

Solvay Industrial Films, 4305 Laplata Ave., Baltimore, MD 21211 USA (Tel.: 410-467-2061; FAX 410-889-0378)

Solvay Polymers Inc., Subsidiary of Solvay America Inc., 3333 Richmond Ave., PO Box 27328, Houston, TX 77227-7328 USA (Tel.: 713-522-1781; 800-231-6313; FAX 713-522-2435; Telex: 023-166307)

Sorbothane, Inc.
2144 State Route 59, PO Box 178, Kent, OH 44240 USA (Tel.: 216-678-9444; FAX 216-678-1303)

Sorex Ltd.
St Michael's Industrial Estate, Hale Road, Widnes, Cheshire, WA8 8TJ, UK (Tel.: 051-420-7151; FAX 051-495-1163)

Southeastern Clay Co.
PO Box 1055, Aiken, SC 29802 USA (Tel.: 803-648-3248; FAX 803-649-5701)

Southern Clay Prods. Div. of Laporte Inc.
1212 Church St., PO Box 44, Gonzales, TX 78629 USA (Tel.: 512-672-2891; 800-324-2891; FAX 512-672-3930)

Spartan Flame Retardants, Inc.
345 E. Terra Cotta Ave., PO Box 395, Crystal Lake, IL 60014 USA (Tel.: 815-459-8500; 800-435-5700; FAX 815-459-8560; Telex: RCA 297135)

Spaulding Composites
Spaulding Composites Co. , 310 Wheeler St., Tonawanda, NY 14150 USA (Tel.: 716-692-2000; FAX 716-692-4410)

Spaulding Composites Co. Inc., 1560 Sherman Avenue, Suite 11000, Evanston, IL 60201 USA (Tel.: 312-475-1600; FAX 312-475-1839)

Spaulding Fibre Co. See Spaulding Composites Co. Inc.

Specialty Products Co., Div. of Chemical Solvents Inc.
75 Montgomery St., PO Box 306, Jersey City, NJ 07303-0306 USA (Tel.: 201-434-4700; 800-321-8506; FAX 201-434-6052)

Spectar Ltd.
Highfield Lane, Handsworth, Sheffield, South Yorkshire, S13 9NA, UK (Tel.: 0742 691857; FAX 0742 540580)

Spectra Brands plc, Div. of Burma Castrol
Factory 475, Treloggan Industrial Estate, Newquay, Cornwall, TR7 2SX, UK (Tel.: 0637 871171; FAX 0637 878627; Telex: 45440 SPCTRA G)

Spectrum Chemical Mfg. Corp.
14422 S. San Pedro St., Gardena, CA 90248 USA (Tel.: 213-516-8000; 800-772-8786; FAX 213-516-9843; Telex: 182395)

Spencer Kellogg & Sons
120 Delaware Ave, PO Box 807, Buffalo, NY 14240 USA (Tel.: 716-852-5850)

SPER Chemical Corp.
P O Box 5566, 14770 66th North Street, Clearwater, FL 33518 USA

Sphere Laboratories (London) Ltd.
The Yews, Main Street, Chilton, Oxon, OX11 0RZ, UK (Tel.: 0235-833896)

Spice King Corp.
6009 Washington Blvd., Culver City, CA 90232-7488 USA (Tel.: 213-836-7770; FAX 213-836-6454; Telex: 664350)

Spraydex Ltd.
Moreton Avenue, Wallingford, Oxon, OX10 9DE, UK (Tel.: 0491-25251; FAX 0491-25014)

E. R. Squibb & Sons Inc. See Bristol-Myers Squibb Co. Inc.

E. R. Squibb & Sons Ltd. See Bristol-Myers Squibb Pharmaceuticals Ltd.

Stafford-Miller Continental NV
Nijverheidsstraat 9, B-2431 Oevel, Belgium (Tel.: 14-58 94 81; FAX 14-58 94 85; Telex: 33959 STAMIL B)

Stag Polymers & Sealants Ltd.
Tavistock Road, West Drayton, Middlesex, UV7 7RA, UK (Tel.: 0895 445511; FAX 0895 449199; Telex: 28559)

A. E. Staley Manufacturing Co.
2200 E. Eldorado St., Decatur, IL 62525 USA (Tel.: 217-423-4411)

Stan Chem International Ltd.
4 Kings Rd, Reading, Berkshire, RG1 3AA, UK (Tel.: 0734 580247; FAX 0734 589580; Telex: 847746)

Standard Chemical Co.
Mill Lane, Cheadle, Cheshire, SK8 2NX, UK (Tel.: 061-428-5225; FAX 061-428-0890; Telex: 666413)

Standard Dry Wall Products Inc. Address unverified
Miami, FL USA

Standard Laboratories
4815 MacCorkie Avenue SE, Charleston, WV, 25304 USA (Tel.: 304-925-2145)

Standard Oil Co. *See* BP America Inc.

Standard Tar Products Co. Inc.
2456 W Cornell Street, Milwaukee, WI 53209-6294 USA (Tel.: 414-873-7650; 800-825-7650; FAX 414-873-7737)

Standard Telecommunications Laboratories. *Address unknown*

Standex Laboratories. *Address unverified*
Columbus, OH USA

Stanley Smith & Co. Plastics Ltd.
Worple Road, Isleworth, Middlesex, TW7 7AU, UK (Tel.: 081-568 6831; FAX 081-847 5322)

Star Agrochem Ltd.
Odder Farm, Saxilby Road, Burton, Lincoln, LN1 2BB, UK (Tel.: 0522-703777)

Stauffer Chemical Co. *See* Rhône-Poulenc Basic

Stave Chemical Co.
132 Village Dr., Basking Ridge, NJ 07920 USA (Tel.: 201-766-0059)

Staveley Chemicals Ltd. *See* Rhône-Poulenc Chemicals Ltd.

Steetley
Steetley Chemicals Ltd., Berk House, PO Box 56, Basing View, Basingstoke, Hants, RG21 2EG, UK

Steetley Magnesia Products Ltd., Div. of Steetley plc, PO Box 8, Hartlepool Works, Hartlepool, Cleveland, TS24 0BY, UK (Tel.: 0429 267071; FAX 0429 266600; Telex: 58649 PCLASE G)

Steetley Minerals Ltd., Div. of Steetley plc, PO Box 2, Retford Park, Worksop, Nottinghamshire, S81 8AF, UK (Tel.: 0909-475511; FAX 0909-486532; Telex: 547901)

M. A. Steinhard Ltd./Pharmaceutical Div.
32-36 Minerva Road, London, NW10 6HJ, UK

Stella-Meta Filters
Laverstoke Mill, Whitchurch, Hampshire, RG28 7NR, UK (Tel.: 0256 895959; FAX 0256 892074; Telex: 859605 STELLA G)

Stepan Co.
Stepan Co., 22 West Frontage Rd., Northfield, IL 60093 USA (Tel.: 708-446-7500; 800-228-8312; FAX 708-501-2443)

Stepan Co./PVO Dept., 100 West Hunter Ave., Maywood, NJ 07607 USA (Tel.: 201-845-3030; FAX 201-845-6754; Telex: 710-990-5170)

Stepan Canada, 90 Matheson Blvd. W., Suite 201, Mississauga, Ontario, L5R 3P3, Canada (Tel.: 416-507-1631; FAX 416-507-1633)

Stepan Europe, BP127, 38340 Voreppe, France (Tel.: 7650-8133; FAX 7656-7165; Telex: 320511 F)

Stephenson Thompson Textile Chemicals, A Div. of Stephenson Group Ltd.
P O Box 305, Listerhills Road, Bradford, West Yorkshire, BD7 IHY, UK (Tel.: 0274 723811; FAX 0274 370108; Telex: 517100 STEBRO G)

Sterling Drug Inc. *See* Sterling Winthrop

Sterling Moulding Materials. *Defunct company*

Sterling Research Laboratories, Div. of Sterling-Winthrop Group Ltd.
Sterling-Winthrop House, Onslow Street, Guildford, Surrey, GU1 4YS, UK (Tel.: 0483-505515; FAX 0483-35432)

Sterling Roncraft, Div. of LXF Products
Thorncliffe Park, Chapeltown, Sheffield, South Yorkshire, S30 4YP, UK (Tel.: 0742 467171; FAX 0742 455629; Telex: 547096)

Sterling Ronseal Ltd.
Chapeltown, Sheffield, S30 4YP, UK (Tel.: 0742 467171; FAX 0742 455629; Telex: 547096)

Sterling Winthrop
Sterling Winthrop, Div. of Eastman Fine Chemicals, Eastman Kodak Co, 90 Park Avenue, New York, NY 10016 USA (Tel.: 212-972-4141)

Sterling-Winthrop Group Ltd., Sterling-Winthrop House, Onslow Street, Guildford, Surrey, GU1 4YS, UK (Tel.: 0483-505515; FAX 0483-35432)

Sterling Winthrop BV, Div. of Sterling-Winthrop Group Ltd., Postbus 615, Prins Berhardlaan 2, NL-2003 RP Haarlem, The Netherlands (Tel.: 23-35 26 34; FAX 23-36 17 49; Telex: 41131)

Sterwin Chemicals Inc., Div. of Imcera Group Inc.
PO Box 537, Rt 113-Dupont Highway, Millsborough, DE 19966 USA

Stevenson Brothers & Co. Inc.
PO Box 38349, 1039 West Venango St., Philadelphia, PA 19140 USA (Tel.: 215-223-2600; FAX 215-223-3597)

Stiefel Laboratories
Stiefel Laboratories Inc., 2801 Ponce de Leon Blvd, Coral Gables, FL 33134 USA (Tel.: 305-443-3807)

Stiefel Laboratories (UK) Ltd., Holtspur Lane, Woodburn Green, High Wycombe, Bucks, HP10 0AU, UK

Stirene Co-Polymers Ltd. *Address unknown*

Stockhausen, Inc.

2408 Doyle St., Greensboro, NC 27406 USA (Tel.: 919-333-3500; FAX 919-333-3545; Telex: 574405)

Stoller Chemicals Ltd.

Unit 23, Marathon Place, Moss Side Industrial Estate, Leyland, Preston, Lancashire, PR5 3QN, UK (Tel.: 0772-454443; FAX 0772-622320; Telex: 677515 STOLLR G)

Stoner Inc.

1070 Robert Fulton Hwy., PO Box 65, Quarryville, PA 17566 USA (Tel.: 717-786-7355; 800-227-5538; FAX 717-786-9088)

Stowlin Ltd.

Radnor Rd, South Wigston, Leicester, Leicestershire, LE18 4XY, UK (Tel.: 0533 785373; FAX 0533 772616; Telex: 34694)

Strahl & Pitsch, Inc.

230 Great E. Neck Rd., W. Babylon, NY 11704 USA (Tel.: 516-587-9000; FAX 516-587-9120; Telex: 221636 STRALUR)

Struktol

Struktol Co., 201 E. Steels Corner Rd., PO Box 1649, Stow, OH 44224-0649 USA (Tel.: 216-928-5188; 800-327-2709; FAX 216-928-8726)

Struktol Co., Ltd., 60 Venture Dr., Unit 23, Scarborough, Ontario, M1B 3S4, Canada (Tel.: 416-286-4040; FAX 416-286-4043)

Stuart Pharmaceuticals, Div. of ICI Americas Inc.

New Murphy Rd & Concord Pike, Wilmington, DE 19897 USA (Tel.: 302-886-3000)

Stull Chemical Company

PO Box 47907, San Antonio, TX 78265 USA

Sturge Lifford

Lifford Chemical Works, Lifford Lane, Birmingham, B30 3JW, UK (Tel.: 021-458 6671; FAX 021-451 1759)

Suchema AG

Haupstrasse 15, CH-8251 Kaltenbach, Switzerland (Tel.: 54-41 12 65; Telex: 897170 SUC CH)

Süd-Chemie AG

Postfach 20 22 40, W-8000 Munich 2, Germany (Tel.: 89-5110 0; FAX 89 5110375; Telex: 523872 SCMU D)

Süddeutsche Emulsions Chemie GmbH

Postfach 240634, Rhenaniastrasse 46, W-6800 Mannheim, Germany (Tel.: 621-851053; FAX 621-851055; Telex: 462040)

Sumito Chemical (UK) plc,
107 Cheapside, London, EC2V 6DQ, UK (Tel.: 071-796-3533; FAX 071-796-3533)

Sumitomo Bakelite Co., Ltd.
2-2, Uchisaiwaicho 1-chome, Chiyoda-ku, Tokyo, 100, Japan (Tel.: (03) 3506-7080; FAX (03) 3506-7331; Telex: 222-4394 SUMIBK J)

Sumner Oil Industries. *Address unknown*

Sun Refining & Marketing Co., Subsidiary of Sun Company Inc. *Address unknown*

Superfos Biosector A/S
Frydenlundsvej 30, DK-2950 Vedbaek, Denmark (Tel.: 42-89 31 11; FAX 42-89 14 86; Telex: 37191)

Superol BV
Postbus 38, NL-6900 AA Zevenaar, The Netherlands (Tel.: 8360-23728)

Surex International Ltd.
Unit 5, Airport Trading Estate, Biggin Hill, Westerham, Kent, TN16 3BW, UK (Tel.: 0959 76000; FAX 0959 71000; Telex: 894280 UNIREX G)

Surfachem Ltd., Div. of Surfachem Group plc
Wellington Park House, Thirsk Row, Leeds, West Yorkshire, LS1 4DP, UK (Tel.: 0532 342636; FAX 0532 445910; Telex: 556172)

Surmak Products Ltd.
99 Mabgate, Leeds, LS9 7DR, UK (Tel.: 0532 450371; FAX 0532 428701)

A. F. Suter & Co Ltd.
Swan Wharf, 60 Dace Road, London, E3 2NQ, UK (Tel.: 081-986 8218; FAX 081-985 0747)

Sutton Laboratories
Sutton Laboratories, Inc., Member of the ISP Inc. Group, 116 Summit Ave., PO Box 837, Chatham, NJ 07928-0837 USA (Tel.: 201-635-1551; FAX 201-635-4964; Telex: 710-999-5607)

Represented by: Blagden Chemicals Ltd., Div. of Blagden Industries, AMP House, Dingwall Rd, Croydon, Surrey, CR9 3QU, UK (Tel.: 081 681 2341; FAX 081 688 5851; Telex: 24285)

Swale Coatings & Inks Ltd., Div. of Brent Chemicals International plc
Taylor Rd, Trafford Park, Urmston, Manchester, M31 2TE, UK (Tel.: 061 748 7340; FAX 061 748 7685)

Swan & Co. Ltd., Thomas. *See* Thomas Swan & Co. Ltd.

Swift & Co. *Address unknown*

SWS Oilchemicals BV
Raadhuisstraat 84, NL-2101 HJ Heemstede, The Netherlands (Tel.: 23-28 39 52; FAX 23-29 46 81)

Sybron
Sybron Chemicals Inc., PO Box 125, Wellford, SC 29385 USA (Tel.: 803-439-6333; 800-677-3500; FAX 803-439-1612)

Sybron/Biochemical, Birmingham Rd., Birmingham, NJ 08011 USA (Tel.: 609-893-1100; 800-678-0020; FAX 609-894-8641)

Sybron Chemicals Canada Ltd., 120 Norfinch Dr., Unit 1, Downsview, Ontario, M3N 1X2, Canada (Tel.: 416-663-7166)

Sybron Chemie Nederland BV, Postbus 46, NL-6710 BA Ede, The Netherlands (Tel.: 31828; FAX 30236; Telex: 37078)

Symalit AG, A Co. of the Royal Dutch/Shell Group
CH-5600 Lenzburg, Switzerland (Tel.: (0)64 50 81 50; FAX (0)64 50 83 83; Telex: 981 352 syma ch)

Synchemicals Ltd.
Owen Street, Oalville, Leics, LE6 2DE, UK (Tel.: 0530-510060; FAX 0530-510299)

Synres-Almoco BV, Div. of DSM Resins BV
Postbus 18, Slachthuisweg 50, NL-3150 AA Hoek van Holland, The Netherlands (Tel.: 1747-04342; FAX 1747-02645; Telex: 31741)

Syntex Corp.
3401 Hillview Ave, Palo Alto, CA 94304 USA (Tel.: 415-855-5050)

Syntex Laboratories Inc. *See* Syntex Corp.

Syntex Manufacturing Ltd.
Mid Road, Blailinn Industrial Estate, Cumbernauld, GR7 2TL, UK (Tel.: 0236-739696)

Syntex Ophthalmics Inc. *See* Syntex Corp.

Synthese BV. *See* Akzo Resins BV

Synthetic Chemicals Ltd., Div. of Shell Chemicals UK Ltd.
Four Ashes, Wolverhampton, West Midlands, WV10 7BP, UK (Tel.: 0902 794000; FAX 0902 794300; Telex: 337306)

Synthetic Products Co., Subsidiary of Cookson America Inc.
1000 Wayside Rd., Cleveland, OH 44110 USA (Tel.: 216-531-6010; 800-321-4236; FAX 216-486-6638)

Synthetic Resins Ltd. *See* Scott Bader Co Ltd.

Synthetic Rubber Technologies

3898 Shawnee St. N.W., PO Box 639, Uniontown, OH 44685 USA (Tel.: 216-699-1256; FAX 216-699-1404)

Synthetic Surfaces Inc.

PO Box 241, Scotch Plains, NJ 07076 USA (Tel.: 908-233-6803; FAX 908-233-6844; Telex: 833231 Att: 663)

Synthite Ltd.

Synthite Ltd., Div. of Tennants Consolidated Ltd., Alyn Works, Denbigh Rd, Mold, Clwyd, CH7 1B7, UK (Tel.: 0352-752521; FAX 0352-700182; Telex: 61303)

Synthite Ltd., Ryders Green Road, West Bromwich, W Midlands, B70 0AX, UK

Synthron Inc.

PO Box 1111, Morganton, NC 28655 USA (Tel.: 704-437-8611; FAX 704-437-4126)

TACC International Corp.

Air Station Industrial Park, PO Box 535, Rockland, MA 02370 USA (Tel.: 617-878-7015; FAX 617-871-6727; Telex: 5106015650)

Taiwan Surfactant Corp.

No. 106, 8-1 Floor, Sec. 2, Chung An E. Rd., Taipei, Taiwan, R.O.C. (Tel.: 886-2-507-9155; FAX 886-2-507-7011; Telex: 27568 surfact)

Takeda USA, Inc.

8 Corporate Dr., Orangeburg, NY 10962-2614 USA (Tel.: 914-365-2080; 800-825-3328; FAX 914-365-2786; Telex: 421149)

TAM Ceramics Inc.

4511 Hyde Park Blvd., Niagara Falls, NY 14305 USA (Tel.: 716-278-9400; FAX 716-285-3026; Telex: 710-524-1659)

Tamite Industries Inc. *Address unverified*

Miami, FL USA

Tanabe USA, Inc.

PO Box 85132, San Diego, CA 92186 USA (Tel.: 619-571-8410; 800-7-TANABE; FAX 619-571-3476)

TAP Pharmaceuticals Inc.

2355 Waukegan Road, Deerfield, IL 60015 USA (Tel.: 708-317-5700; 800-622-2011)

Taskem Inc.

4542 Spring Road, Cleveland, OH 44131 USA

Tego Chemie. *See* Goldschmidt

Teknor Apex Co.

505 Central Ave., Pawtucket, RI 02861 USA (Tel.: 401-725-8000; 800-556-3864; FAX 401-725-8095; Telex: 927530)

Telcon Plastics Ltd. *See* Medway Packaging Ltd

Temfibre Inc.

Mill Rd., Temiscaming, Quebec, J0Z 3R0, Canada (Tel.: 819-627-9505; FAX 819-627-3622; Telex: 067-76281)

Tennants Consolidated Ltd.

Tennant Trading (Chemicals) Ltd., Div. of Charles Tennant (London) Ltd., Denny Avenue, Waltham Abbey, Essex, EN9 1NS, UK (Tel.: 0992 715777; FAX 0992 700449; Telex: 24329)

Charles Tennant & Co Ltd., Div. of Tennants Consolidated Ltd., 69 Grosvenor St, London, W1X 0BP, UK (Tel.: 071 493 5451; FAX 071 495 1269; Telex: 23335 TENLON G)

Tennant-KVK Ltd., Div. of Tennants Consolidated Ltd., 69 Grosvenor St, London, W1X 0BP, UK (Tel.: 071 493 5451; FAX 071 491 4922; Telex: 23335 TENLON G)

Tenneco. *See* Albright & Wilson

Tennessee Chemical Co. *See* Boliden-Intertrade Inc.

Tensia Ltd.

Postfach 1378, Kalkbreitestrasse 51, CH-8036 Zurich, Switzerland (Tel.: 1-462 28 00; FAX 1-461 20 67; Telex: 814507)

Tensia SA

28 Rue de Rénory, B-4200 Ougrée, Seraing, Belgium (Tel.: 41-37 89 15)

Tercol

Shifnal, Shropshire, UK

The Terrell Corp.

820 Woburn Street, Wilmington, MA 01887 USA

Testworth Laboratories Inc.

401 S Main Street, PO Box 91, Columbia City, IN 46725 USA (Tel.: 219 244 5137; FAX 219 244 5138)

Tetrahedron Association Inc.

5060 A Convoy St, San Diego, CA 92111 USA

Texaco

Texaco Chemical Co., PO Box 15730, Austin, TX 78761 USA (Tel.: 512-483-0053; 800-231-3107; FAX 512-483-0925; Telex: 776-408)

Texaco Chemical Co/Oxides & Specialties Division, 4800 Fournace Place, PO Box 430, Bellaire, TX 77401 USA

Texaco Ltd., 1 Knightsbrige Green, London, SW1X 7QJ, UK (Tel.: 071-584 5000; FAX 071-584 6999; Telex: 8956681 TEXACO G)

SA Texaco Belgium N.V., Int'l. Congress Center, Citadel Park, B-900 Ghent, Belgium (Tel.: 011-32-91-41-5920)

Texaco France S.A., 5, rue Bellini, Tour Arago, F-92806 Puteaux Cedex, France (Tel.: 011-33-1-47-78-1655)

Texaco Chemical Deutschland GmbH, Baumwall 5, 2000 Hamburg 11, Germany (Tel.: 011-49-40-36-3737)

Texaco Olie Matschappij BV, Weena 170, NL-3012 CR Rotterdam, The Netherlands (Tel.: 614471; Telex: 31542)

Texapol Corp.
177 Mikron Rd., Bethlehem, PA 18017 USA (Tel.: 215-759-8222; 800-523-9242; FAX 215-759-9433)

Texasgulf Inc.
3101 Glenwood Ave., Raleigh, NC 27622 USA (Tel.: 919-881-2700; Telex: 6844904)

Thermal Ceramics
2102 Old Savannah Rd., Augusta, GA 30906 USA (Tel.: 404-796-4200; 800-KAOWOOL; FAX 404-796-4398; Telex: 545423 THERMAL)

Thermedics Inc.
470 Wildwood Street, PO Box 2999, Woburn, MA 01888-1799 USA

Thermofil, Inc.
PO Box 489, 6150 Whitmore Lake Rd., Brighton, MI 48116-0489 USA (Tel.: 800-444-4408; FAX 313-227-3824)

Thermoset Plastics Inc.
5101 East 65th St., PO Box 20902, Indianapolis, IN 46220 USA (Tel.: 317-259-4161; FAX 317-252-8402)

Thibaut & Walker Co.
PO Box 296, 49 Rutehrford St., Newark, NJ 07101 USA (Tel.: 201-589-3331)

Thiokol GmbH
Sandhofer Strasse 96, 6800 Mannheim 31, Germany

Thomas Ness Ltd.
Eastwood Hall, Eastwood, Nottingham, NG16 3EB, UK (Tel.: 0773-530404; FAX 0773-530404)

Thomas Ness, North Thames Gas Board. See Thomas Ness Ltd

Thomas Swan & Co. Ltd.
Crookhall, Consett, Co Durham, DH8 7ND, UK (Tel.: 0207-505131; FAX 0207-590467; Telex: 53565)

Thomas Triantaphyllou SA. *See under* Triantaphyllou SA, Thomas

Thompson, Weinman & Co.
PO Box 130, Cartersville, GA 30120 USA

Thor Chemicals
Thor Chemicals, Inc., Brook House, 37 North Ave., Norwalk, CT 06851 USA (Tel.: 203-846-8613; FAX 203-846-4810; Telex: 888630)

Thor Chemicals (UK) Ltd., Cowley House, Earl Road, Cheadle Hulme, Cheshire, SK8 6QP, UK (Tel.: 061-486-1051; FAX 061-488-4125; Telex: 666679 THORUK G)

Thornton & Ross Ltd.
Linthwaite Laboratories, Huddersfield, West Yorkshire, HD7 5QH, UK (Tel.: 0484 842217; FAX 0484 847301; Telex: 517677 ZOFLOR G)

3M
3M Co./Engineered Materials, Industrial Specialty Div., 3M Center Bldg., St. Paul, MN 55144-1000 USA (Tel.: 612-736-9700)

3M Co./Industrial Chem. Prods. Div., 3M Center Bldg. 223-6S-04, St. Paul, MN 55144-1000 USA (Tel.: 612-736-1394; 800-541-6752)

3M Co./Performance Polymers and Additives, 3M Center Bldg., St. Paul, MN 55144-1000 USA (Tel.: 612-736-9700)

3M Canada Inc., PO Box 5757 Terminal A, 1840 Oxford St. East, London, Ontario, N6A 4T1, Canada

3M United Kingdom plc, Commercial Chemicals Div., 3M House, PO Box 1, Bracknell, Berkshire, RG12 1JU, UK

3M Belgium N.V./S.A., Canadastraat 11, 2730 Zwijndrecht, Belgium

3M Deutschland GmbH, PO Box 100422, D-4040 Neuss 1, Germany

Sumitomo/3M Ltd., Central PO Box 490, 33-1, Tamagawa-dai 2-chome, Setagaya-ku, Tokyo, 158, Japan (Tel.: (03) 3709-8111; Telex: J2468534)

3M Australia Pty. Ltd., 950 Pacific Hwy., PO Box 99, Pymble N.S.W. 2073, Australia

3M Pharmaceuticals, Bldg 275-3W-01, 3M Center, St Paul, MN 55144-1000 USA (Tel.: 612-736-4930; 800-328-0255)

3M Health Care, Morley Street, Loughborough, Leicestershire, LE11 1EP, UK (Tel.: 0509-611611; FAX: 0509-237288)

Tiarco Chemical Div./Textile Rubber & Chemical Co.
1300 Tiarco Dr., Dalton, GA 30720 USA (Tel.: 706-277-1300; FAX 706-277-3738)

TIC Gums, Inc.
4609 Richlynn Dr., Belcamp, MD 21017 USA (Tel.: 301-273-7300; FAX 301-273-6469; Telex: 221049)

Tilcon Ltd., Div. of BTR plc
Conyngham Hall, Knaresborough, North Yorkshire, HG5 9AY, UK (Tel.: 0423-862841; FAX 0423-864555; Telex: 57997)

Tillots Laboratories
Unit 24, Henlow Trading Estate, Henlow, Beds, SG16 6DS, UK

Tioga Coatings Corp.
208 Quaker Rd., Rockford, IL 61104-7088 USA (Tel.: 815-962-4200; FAX 815-962-1712)

Tioxide
Tioxide Group plc, Div. of ICI, Tioxide House, 137-143 Hammersmith Road, London, W14 0QL, UK (Tel.: 071-602 7121; FAX 081-784 0019; Telex: 920900)

Tioxide UK Ltd., Div. of ICI, Haverton Hill Road, Billingham, Cleveland, TS23 1PS, UK (Tel.: 0642-370300; FAX 0642-370290)

Toagosie Chemical Industry Co. Ltd.
14-1 Nishi Shinbashi 1-Chome, Minato-ku, Tokyo, 105, Japan (Tel.: (03) 3597-7215; FAX (03) 3597-7217)

Toho Chemical Industry Co., Ltd.
No. 2-5, Ningyo-cho 1-chome, Nihonbashi, Chuo-ku, Tokyo, 103, Japan (Tel.: (03) 3668-2271; FAX (03) 3668-2278; Telex: 252-2332 TOHO K J)

Tosoh
Tosoh Corp. 7-7 Akasaka 1-chome, Minato-ku, Tokyo, 107, Japan (Tel.: (03) 3585-6707; FAX (03) 3582-7846; Telex: J24475tosoh)

Tosoh USA Inc., 1700 Water Place, Suite 204, Atlanta, GA 30339 USA (Tel.: 404-956-1100; FAX 404-956-7368; Telex: 542272 tosoh atl)

Tosoh Canada Ltd., 1200 Sheppard Ave. East, Suite 511, Willowdale, Ontario, M2K 2S5, Canada (Tel.: 416-756-2226; FAX 416-756-2750)

Tosoh Europe BV, World Trade Centre Amsterdam, Tower C, Floor 13, Strawinskylaan 1351, 1077 XX Amsterdam, The Netherlands (Tel.: 020-644026, 020-623412; Telex: 18573tosoh nl)

Tra-Con, Inc.
55 North St., Medford, MA 02155 USA (Tel.: 617-391-5550; 800-872-2661; FAX 617-391-7380)

Trafford Chemicals Ltd., Div. of Borregaard Industries Ltd.
Clayton Rd, Risley Industrial Estate, Birchwood, Warrington, Cheshire, WA3 6QQ, UK (Tel.: 0925 824511; FAX 0925 812186; Telex: 627696)

Transene Co. Inc.
Route 1, Rowley, MA 01969 USA (Tel.: 508-948-2501; FAX 508-948-2206)

Transol Chemicals (UK) Ltd., Div. of Transol Chemicals Int'l. BV
Caledonian House, Tatton St, Knutsford, Cheshire, WA16 6AG, UK (Tel.: 0565 650386; FAX 0565 653255)

Travenol Laboratories Inc., Div. of Baxters Healthcare Corporation
One Baxter Parkway, Deerfield, IL 60015 USA

Tregarne Ltd.

9B Beehive Workshops, Cardrew Industrial Estate, Redruth, Cornwall, TR15 1SS, UK (Tel.: 0209-213103; FAX 0209-211118)

Tremco Ltd/Adhesive Systems Div

86-88 Bestobell Road, Slough, Berks, SL1 4SZ, UK (Tel.: 0753-691696; FAX 0753-822640)

T R Fastenings Ltd.

Trifix Division, Bellbrook Park, Uckfield, East Sussex, TN22 1QW, UK (Tel.: 0825-764711; FAX 0825-767526)

Triantaphyllou SA, Thomas

405 Tatoiou Ave., TK 136 71, Acharnes, Athens, Greece (Tel.: 1-807-6413; Telex: 216370)

Tri-K Industries, Inc.

466 Old Hook Rd., PO Box 312, Emerson, NJ 07630 USA (Tel.: 201-261-2800; 800-526-0372; FAX 201-261-1432)

Tripart Farm Chemicals Ltd.

Swan House, Beulah Street, Gaywood, King's Lynn, Norfolk, PE30 4DN, UK (Tel.: 0553-674303; FAX 0553-674422)

Tri-Star Chem. Co.

PO Box 38627, Dallas, TX 75238 USA (Tel.: 214-341-0054)

Troy

Troy Chemical Corp., PO Box 366, 72 Eagle Rock Ave., East Hanover, NJ 07936 USA (Tel.: 201-884-4300; FAX 201-884-4317; Telex: 138930 Troychem Nwk)

Troy Chem. Co. Ltd., 157 Overture Rd., Scarborough, Ontario, M1E 2W5, Canada (Tel.: 416-287-9116; FAX 416-287-9779)

Troy Chem. Co. UK, The Firs, Church Lane, Utterby, Louth, Lincs, LN11 0TH, UK (Tel.: 472-840451; FAX 472-840096)

Troy Chem. Co. BV, Haringsuisweg 35, 3133 KP Vlaardingen, The Netherlands (Tel.: 10-460-1777; FAX 10-460-1323; Telex: 26473 Troy NL)

Troy Chemie GmbH, Uerdingerstrasse 541, 4150 Krefeld Bockum, Germany (Tel.: 2151-59-03-38; FAX 2151-59-81-45)

Truchem Ltd.

Brook House, 30 Larwood Grove, Sherwood, Nottingham, HG5 3JD, UK (Tel.: 0602-260762; FAX 0602-671153)

TRW Space & Technology Group

1 Space Park, Redondo Beach, CA 90278 USA (Tel.: 213-535-4321)

TSE Industries, Inc.

5260 113th Ave. North, PO Box 17225, Clearwater, FL 33520-7225 USA (Tel.: 813-573-7676; 800-237-7634; FAX 813-572-0415)

Tufnol Ltd.
PO Box 376, Perry Barr, Birmingham, West Midlands, B42 2TB, UK (Tel.: 021-356 9351; FAX 021-331 4235; Telex: 339730)

Tulco, Inc.
9 Bishop Rd., Ayer, MA 01432 USA (Tel.: 508-772-4412; FAX 508-772-1751)

Typharm Ltd.
16 Parkstone Road, Poole, Dorset, BH15 2PG, UK (Tel.: 0202-666626; FAX 0202-666309)

Ucar Carbon Co., Inc.
39 Old Ridgebury Rd.-J4, Danbury, CT 06817 USA (Tel.: 203-794-3684; 800-342-3698)

UCB
UCB SA, Chemical Sector, Speciality Chems. Div., 33 rue d'Anderlecht, B-1620 Drogenbos, Belgium (Tel.: 2-371 45 11; FAX 2-378 39 44; Telex: 22342 UCBOS B)

UCB SA/Chemical Sector, 326 avenue Louise, B-1050 Brussels, Belgium (Tel.: 2-641-1411; FAX 2-640 9860; Telex: 63769)

UCB NV Filmsektor, Ottergemsesteenweg 801, PO Box 369, B-9000 Ghent, Belgium (Tel.: 091 40 32 11; FAX 091 40 88 00; Telex: 11280 SIDAC B)

UCB (Chem) Ltd., Div. of UCB SA, Star House, 69 Clarendon Road, Watford, Hertfordshire, WD1 1DJ, UK (Tel.: 0923-248011; FAX 0923-244472; Telex: 23958)

UCB Radcure Inc./U.S. Headquarters, 200 Lake Park Dr., Smyrna, GA 30080 USA (Tel.: 800-433-2873; FAX 404-319-8228)

UCIB, Distributed by SST Corp.
635 Brighton Rd., PO Box 1649, Clifton, NJ 07015 USA (Tel.: 201-473-4300; FAX 201-473-4326; Telex: RCA 219149)

UCO Optics Inc.
3000 Winton Rd, Rochester, NY 14623 USA

Ulfcar International A/S
P O Box 1020, Lindenborg, 4000 Roskilde, Denmark

Ulmer Pharmacal Co
2440 Fernbrook Lane, Minneapolis, MN 55441 USA

W. Ulrich GmbH
Postfach 54, Salzstrasse 20, W-8084 Inning, Germany (Tel.: 8143-8535; FAX 8143-7300; Telex: 527657 ULKAG D)

Ultrachem Inc.
1400 North Walnut Street, P O Box 2053, Wilmington, DE 19899 USA

Unger Fabrikker AS
PO Boks 254, N-1601 Fredrikstad, Norway (Tel.: 9-32-0020; FAX 9-32-3775; Telex: 76382 unger n)

Ungerer & Co.

4 Bridgewater Lane, PO Box U, Lincoln Park, NJ 07035 USA (Tel.: 201-628-0600; FAX 201-628-0251; Telex: 4754267)

Unibios SpA

Via S Pellico 3, I-28069 Trecate, Italy (Tel.: 321-73261; FAX 321-76816; Telex: 200329 UBS I)

Unichem plc

Unichem House, Cox Lane, Chessington, Surrey, KT9 1SH, UK (Tel.: 081 391 2323; FAX 081 974 1707)

Unichema

Unichema International, Postbus 309, NL-2800 AH Gouda, The Netherlands (Tel.: 1820-42933; FAX 1820-26877)

Unichema Chemicals Ltd., Div. of Unichema International, Bebington, Wirral, Merseycide, L62 4UF, UK (Tel.: 051-645-2020; FAX 051-645-9197; Telex: 629408)

Unichema France SA, 148 Boulevard Haussemann, 75008 Paris, France (Tel.: 1 45630863; FAX 1 42563188; Telex: 643217)

Unichema Chemie GmbH, Postfach 1280, Steintor 9, W-4240 Emmerich, Germany (Tel.: 2822-721; FAX 2822-72276; Telex: 8125113)

Unichema Chemie BV, Postbus 2, NL-2800 AA Gouda, The Netherlands (Tel.: 1820-42911; FAX 1820-42250)

Unichema North America, 4650 S. Racine Ave., Chicago, IL 60609 USA (Tel.: 312-376-9000; 800-833-2864; FAX 312-376-0095; Telex: 176068)

Unichema Japan, Sankei Bldg. 7F 708, 4-9, Umeda 2-chome, Kita-ku, Osaka, 530, Japan (Tel.: 6341-7221; FAX 6341-7725)

Unigreg Ltd.

Enterprise House, 181-189 Garth Road, Morden, Surrey, SM4 4LL, UK (Tel.: 081-330 1421; FAX 081-330 6812; Telex: 927468)

Unilab Inc.

764 Ramsey Ave, Hillside, NJ 07205 USA

Unilever plc

PO Box 68, Unilever House, Blackfriars, London, EC4P 4BQ, UK (Tel.: 071-822 5252; FAX 071-822 5951; Telex: 28395 UNIL G)

Unimed Inc.

35 Columbia Rd, Somerville, NJ 08876 USA (Tel.: 908-526-6894)

Unimin Specialty Minerals

Unimin Specialty Minerals Inc., 258 Elm St., New Canaan, CT 06840 USA (Tel.: 203-966-8880; 800-243-9004)

Unimin Specialty Minerals Inc., PO Box 33, Rt. 127, Elco, IL 62929 USA (Tel.: 618-747-2311; FAX 618-747-9318)

Union Camp

Union Camp Corp., 1600 Valley Rd., Wayne, NJ 07470 USA (Tel.: 201-628-2680; 800-628-9220; Telex: 130735)

Union Camp Corp./Chem. Prods. Div., PO Box 60369, Jacksonville, FL 32236 USA

Union Carbide

Union Carbide Corp., Old Ridgeway Road, Danbury, CT 06817 USA (Tel.: 203-794-2000; FAX 203-794-6193)

Union Carbide Chem. & Plastics Co. Inc./Specialty Chem. Div., 39 Old Ridgebury Rd., Danbury, CT 06817-0001 USA (Tel.: 203-794-2000; 800-568-4000)

Union Carbide Canada Ltd., 10455 Metropolitan E., Montreal East, Quebec, H1B 1A1, Canada (Tel.: 514-493-2610)

Union Carbide (UK) Ltd./Chemicals & Plastics, 93/95 High Street, Rickmansworth, Herts, WD3 1RB, UK (Tel.: 0923-720 366; FAX 0923-896721)

Union Carbide Europe S.A., 15 Chemin Louis-Dunant, CH-1211 Geneve 20, Switzerland (Tel.: 22-739-6111; FAX 22-739-6545; Telex: 419207)

Union Carbide Japan KK, Toranomon 45 Mori Bldg., 1-5 Toranomon, 5-Chome Minato-Ku, Tokyo, 105, Japan (Tel.: 3431-7281)

Union Carbide Brazil, Rua Dr. Eduardo De Souza Aranha, 153, Sao Paulo, 04530, Brazil

Union Derivan SA

Av. Meridiana 133, Barcelona 08026, Spain (Tel.: 2322113; FAX 2323951; Telex: 98204)

Uniroyal

Uniroyal Chemical Co., Inc., World Headquarters, Benson Road, Middlebury, CT 06749 USA (Tel.: 203-573-3880; 800-243-3024; FAX 203-573-3393; Telex: 6710383 uniroyal)

Uniroyal Chemical Ltd., Division of Uniroyal Limited, Brooklands Farm, Cheltenham Road, Evesham, Worcs, WR11 6LW, UK

Uniroyal Chemical Ltd., Kennet House, 4 Langley Quay, Waterside Drive, Slough, Berkshire, SL3 6EH, UK (Tel.: 0753 580888; FAX 0753 591352; Telex: 84 9934 UCHEM G)

Uniroyal Chimica, SpA, Via delle Industrie 40, I-104013 Latina-Scalo, Italy (Tel.: 773-43605; Telex: 680056)

Uniroyal Chemical (Singapore), Cathay Bldg., 11, Dhoby Ghaut, Suite #14-05, Singapore 0922)

Uniroyal Quimica S.A., Avenida Morumbi, 7.029 (CEP 05650) Caixa Postal 30.380, 0100 Sao Paulo, SP, Brazil

United Catalysts Inc.

PO Box 32370, Louisville, KY 40232 USA (Tel.: 502-634-7500; 800-468-7210; FAX 502-634-7727; Telex: 204190)

United Coconut Chemicals, Inc./Cocochem

UCPB Bldg., 17th Fl., Makat Ave., Makati, Metro Manila, Philippines (Tel.: 818-8361; FAX (00632) 817-2251; Telex: 66928 COCOCHEM PN)

United Composites, Inc.

Lyn Creek Facility, 2203 Webb Lynn Rd., Arlington, TX 76018 USA (Tel.: 817-468-2929; FAX 817-468-3122)

U. S. Biochemical

U. S. Biochemical Corp., PO Box 22400, Cleveland, OH 44122 USA (Tel.: 216-765-5000; 800-321-9322; FAX 216-464-5075; Telex: 980718)

U. S. Biochemical (UK), Div. of U. S. Biochemical Corp., 25 Signet Court, Newmarket Rd, Cambridge, Cambridgeshire, CB5 8LA, UK (Tel.: 0223 467064; FAX 0223 60732)

U. S. Biochemical, Div. of U. S. Biochemical Corp., Niederstedter Weg 11, D-6380 Bad Homburg, Germany (Tel.: 6172 32048; FAX 6172 304755)

U. S. Borax & Chemical Corp.

3075 Wilshire Blvd., Los Angeles, CA 90010 USA (Tel.: 213-251-5400; 800-USB-ORAX; FAX 213-251-5455)

U. S. Cosmetics

313 Lake Rd., PO Box 859, Dayville, CT 06241 USA (Tel.: 203-779-3990; 800-752-0490; FAX 203-779-3994)

U. S. Gypsum Co., DAP Inc., Subsidiary of USG Corp.

101 S. Wacker Dr., Chicago, IL 60606-4385 USA (Tel.: 312-606-4000; FAX 312-606-4093)

U. S. Industrial Chemical Corp. *See* Quantum Chemical Corp./USI Div

U. S. Silica Co.

PO Box 187, Berkeley Springs, WV, 25411 USA (Tel.: 304-258-2500; 800-243-7500; FAX 304-258-3500; Telex: 4942414)

U. S. Stoneware Co.

40 Whitney Road, Mahwah, NJ 07430 USA

Unitex Chemical Corp.

PO Box 16344, 520 Broome Rd., Greenboro, NC 27406 USA (Tel.: 919-378-0965; FAX 919-272-4312)

Unitex Ltd.

Halfpenny Lane, Knaresborough, North Yorkshire, HG5 0PP, UK (Tel.: 0423-862677; FAX 0423-868340; Telex: 57884)

Universal Crop Protection Ltd.

Park House, Maidenhead Rd, Cookham, Maidenhead, Berkshire, SL6 9DS, UK (Tel.: 0628-526083; FAX 0628-810457; Telex: 848426)

Universal-Matthey Products Ltd.

Jeffreys Road, Brimsdown, Enfield, Middlesex, EN3 7PN, UK

Universal Preserv-A-Chem Inc. *Address unknown*

Unocal

Unocal Chemicals/Chemicals Distribution Div., 1700 East Golf Rd., Schaumburg, IL 60173 USA (Tel.: 708-619-2539; 800-CHE-MS76; FAX 708-619-2515)

Unocal Chemicals/Unocal Polymers, 1700 East Golf Rd., Schaumburg, IL 60173-5862 USA (Tel.: 800-548-0162)

UOP

UOP Inc., Universal Oil Products, 25 East Algonquin Rd., Box 5017, Des Plaines, IL 60017-5017 USA (Tel.: 312-391-3300; 800-348-0832; FAX 312-391-2758)

UOP Speciality Products, Div. of UOP Inc., 10 UOP Plaza, Des Plaines, IL 60016 USA (Tel.: 312-391-2000; Telex: 253285)

UOP Ltd., Liongate, Ladymeade, Guildford, Surrey, GU1 1AT, UK (Tel.: 0483-304848; FAX 0483-304863; Telex: 858051)

Upjohn

Upjohn Co., Fine Chemical Marketing, 7000 Portage Rd., Kalamazoo, MI 49001 USA (Tel.: 616-323-5844; 800-253-8600; FAX 616-329-3604)

Upjohn Ltd., Div. of The Upjohn Co, PO Box 8, Fleming Way, Crawley, West Sussex, RH10 2NJ, UK (Tel.: 0293 531133; FAX 0293 548850; Telex: 87367)

Upjohn NV, Div. of The Upjohn Co, Rijksweg 12, B-2670 Puurs, Belgium (Tel.: 890 92 11; FAX 889 65 32; Telex: 32268 UPJOHN B)

USV Pharmaceutical Corp. *Address unknown*

UVtec, Inc.

1121 108th St., Arlington, TX 76011 USA (Tel.: 817-640-5600, FX 817-649-2630)

Vale Chemical Co

PO Box 299, 1827 W Woodlawn Street, Allentown, PA 18105 USA

VAMP Srl

Viale Teodorico 19/2, I-20148 Milan, Italy (Tel.: 2 3493231; FAX 2 3492281; Telex: 322546 vamp i)

Van Den Bergh Foods Co.

2200 Cabot Dr., Lisle, IL 60532 USA (Tel.: 708-955-5276; 800-325-7286; FAX 708-955-5497)

R. T. Vanderbilt Co Inc.

30 Winfield St, PO Box 5150, Norwalk, CT 06855-5150 USA (Tel.: 203-853-1400; FAX 203-853-1452; Telex: 6813581 RTVAN)

Van Dyk & Co., Inc., An ISP Co

11 William St., Belleville, NJ 07109 USA (Tel.: 201-450-3264; FAX 201-759-5279; Telex: 710-995-4928)

L. W. Vass (Agricultural) Ltd.

Springfield Farm, Silsoe Road, Maulden, Bedford, MK45 2AX, UK (Tel.: 0525-403041; FAX 0525-402282)

Vedag GmbH

Flinschstrasse 10-16, 6000 Frankfurt am Main 60, Germany

Veitsiluoto Oy/Forest Chemicals Industry
PO Box 196, SF-90101 Oulu 10, Finland (Tel.: 81-316 3111; FAX 81-378 575; Telex: 32125 oulpk sf)

Velsicol
Velsicol Chemical Corp., 10400 W Higgins Road, Rosemont, IL 60068 USA (Tel.: 800-843-7759; FAX 312-298-9014; Telex: 3730755)

Velsicol Chemical Ltd., Worting House, Bsingstoke, Hampshire, RG23 8PY, UK (Tel.: 0256 817640; FAX 0256 811876; Telex: 9312131051 VC G)

Vencel Resil Ltd/Head Office & Technical Dept., Div. of Shell Chemicals UK Ltd.
Arndale House, 18-20 Spital Street, Dartford, Kent, DA1 2HT, UK (Tel.: 0322-227299; FAX 0322-294250; Telex: 896027 VR DART)

Venture Chemical Products Ltd.
Boxgrove House, Little Heath Road, Tilehurst, Reading, Berks, RG3 5TX, UK

Verdugt BV
Postbus 60, Papesteeg 91, NL-4000 AB Tiel, The Netherlands (Tel.: 3440-15224; FAX 3440-11475; Telex: 47200)

Vikwood Botanicals Inc.
1817 N 5th Street, Sheboygan, WI 53081 USA

Vinamul
Vinamul Ltd., Div. of Unilever plc, MIll Lane, Carshalton, Surrey, SM5 2JU, UK (Tel.: 081 669 4422; FAX 081 669 3189; Telex: 8955487)

Vinamul BV, De Asselen Kuil 20, NL-6161 RD Geleen, The Netherlands (Tel.: 4490-89898; FAX 4490-89710; Telex: 56435)

Vinyl Products. *See* BP Chemicals

Virkler Chemical Co.
12345 Steele Creek Road, Charlotte, NC 28273 USA

Visiontech Inc. *Address unknown*

Vista
Vista Chemical Co., PO Box 19029, 900 Threadneedle, Houston, TX 77224 USA (Tel.: 713-588-3000; 800-231-8216; FAX 713-588-3236; Telex: 794557)

Vista Chemical Europe, Hilton Tower, Blvd. de Waterloo #39, B-81000 Brussels, Belgium (Tel.: 2-513-7490)

Vista Chemical Far East Inc., Kasumigaseki Bldg., 25th Floor, PO Box 110, Tokyo, 100, Japan

Vistakon Inc., Subsidiary of Johnson & Johnson
4500 S Salisbury Road, Jacksonville, FL 32216 USA (Tel.: 904-443-1000)

Vitabiotics Ltd.
3 Bashly Road, London, NW10 6SU, UK (Tel.: 081-963 0999; FAX 081-963 1880; Telex: 923811 BIOTIC G)

Vitax Ltd.
Owen Street, Coalville, Leicestershire, LE67 3DE, UK (Tel.: 0530-510060; FAX 0530-510299)

Volclay Ltd., Div. of American Colloid Co
Birkenhead Rd, Wallasey, Merseyside, L44 7BU, UK (Tel.: 051-638 0967; FAX 051-630 2764; Telex: 627029)

Vulcan Materials Co./Chemicals Div.
PO Box 530390, Birmingham, AL, 35253-0390 USA (Tel.: 205-877-3000; FAX 205-877-3448)

Vulcan Plastics Ltd.
Hosey Hill, Westerham, Kent, TN16 1TB, UK (Tel.: 0959-562304)

Vulnax International Ltd.
Vulnax International Ltd., PO Box 60, Chesford Grange, Woolston, Warrington, WA1 4SE, UK

Vulnax International Ltd., 321 Bureaux de la Colline, 92213 Saint Cloud, Cedex, Paris, France

Wacker
Wacker-Chemie GmbH, Div. L, Prinzregentenstrasse 22, W-8000 Munich 22, Germany (Tel.: 89-2109-0; FAX 89-2109-1770; Telex: 5291210 WK D)

Wacker-Chemitronic Gesellschaft fur Elektronik-Grundstoff, Postfach 1140, D-8263 Burghausen, Germany (Tel.: 089-6279 2797; FAX 089-6279 2795)

Wacker Chemicals Ltd., Div. of Wacker-Chemie GmbH, The Clock Tower, Mount Felix, Bridge Street, Walton-on-Thames, Surrey, KT12 1AS, UK (Tel.: 0932-246111; FAX 0932-240141)

Wacker Chemie Danmark A/S, Park Alle 380 A, Postboks 170, DK-2625 VAllensbaek, Denmark

Wacker Quimica Ibérica SA, Div. of Wacker-Chemie GmbH, Córcega, 303-2 3a, E-08008 Barcelona, Spain (Tel.: 3-217 59 00; FAX 3-217 57 66; Telex: 97801)

Wacker Chemicals (USA) Inc., 50 Locust Ave., New Canaan, CT 06840 USA (Tel.: 203-966-9999; Telex: 643 444)

Wacker Silicones Corp., 3301 Sutton Rd., Adrian, MI 49221-9397 USA (Tel.: 517-264-8500; 800-248-0063; FAX 517-264-8246; Telex: 510-450-2700)

Wako Pure Chemical Industries Ltd.
Wako Pure Chemical Industries Ltd., 1,2-Doshomachi 3-Chome, Chuo-ku, Osaka, 541, Japan (Tel.: (06) 203-3741; FAX (06) 222-1203; Telex: 65188 wakoos j)

Wako Chemicals GmbH, Div. of Wako Pure Chemical Industries Ltd., Nissanstrasse 2, W-4040 Neuss 1, Germany (Tel.: 2131-3110; FAX 2131-311100; Telex: 8517001 wako d)

Wako Chemicals USA, Inc., 1600 Bellwood Rd., Richmond, VA 23237 USA (Tel.: 804-271-7677; FAX 804-271-7791; Telex: 293208 wako ur(rca))

Walkover Sprayers Ltd.
21 London Road, Great Shelford, Cambridge, CB2 5DF, UK (Tel.: 0223-844024)

Wallace Laboratories, Div. of Carter-Wallace Inc.
Cranbury, NJ 08512 USA (Tel.: 609-665-6000)

Wallace & Tiernan. *Address unverified*
Bellville, NJ USA

Walton Pharmaceuticals Ltd.
Bowes House, Bowes Rd, Walton On Thames, Surrey, KT12 3HS, UK (Tel.: 0932 241032; FAX 0932 253461; Telex: 928306 WPHARM G)

Warner Co Ltd/Lime Div.
919 Conestoga Rd, #100, Rosemont, PA 19010 USA (Tel.: 215-527-1350)

Warner-Chilcott, Div. of Warner-Lambert
201 Tabor Road, Morris Plains, NJ 07950 USA (Tel.: 201-540-2000; FAX 201-540-3283)

Warner-Jenkinson Co.
2526 Baldwin St., St. Louis, MO 63106 USA (Tel.: 314-658-7469; 800-325-8110; FAX 314-658-7431; Telex: 44 7184)

Warner-Lambert
Warner-Lambert Co., 201 Tabor Rd, Morris Plains, NJ 07950 USA (Tel.: 201-540-2000)

Warner-Lambert, Div. of Parke-Davis & Co Ltd., Usk Rd, Pontypool, Gwent, NP4 1YH, UK (Tel.: 0495 762468; FAX 0495 762628; Telex: 498647)

Warwick Chemical Ltd. *See* Jotun Polymer UK Ltd.

Warwick International Ltd., Div. of Warwick International Group plc
Wortley Moor Road, Leeds, West Yorkshire, LS12 4JE, UK (Tel.: 0532-637331; FAX 0532-794795; Telex: 556325)

Watts Urethane Products Ltd.
Church Road, Lydney, Gloucester, GL15 5EN, UK (Tel.: 0594-844090; FAX 0594-843586)

WBC Technology Ltd.
Norfolk House, Gt. Chesterfield Court, Gt. Chesterfield, Saffron Walden, Essex, CB10 1PF, UK (Tel.: 0799-30146; FAX 0799-30229)

W D Service Co. Inc.
780 Creek Road, PO Box 147, Bellmawr, NJ 08031 USA (Tel.: 609-931-6100; FAX 609-931-4505)

Weatherguard/Marbleloid Products Inc.
2515 Newbold Avenue, Bronx, NY 10462 USA

Weiders Farmasøytiske A/S
Postboks 9113, Vaterland, N-0134 Oslo 1, Norway (Tel.: 2-20 54 15; FAX 2-36 40 52; Telex: 78151 WEIFA N)

The Wellcome Foundation Ltd.
The Wellcome Research Laboratories, Langley Court, South Eden Park Road, Beckenham, Kent, BR3 3BS, UK (Tel.: 081-658-2211; FAX 081-650-9862; Telex: WELLAB 23937 G)

Werner G. Smith, Inc. *See under* Smith

Wesco Technologies Ltd.
PO Box 3880, San Clemente, CA 92674-3880 USA (Tel.: 714-661-1142; 800-223-3878 (CA); FAX 714-492-6025; Telex: GRT 3718658)

Wesley-Jessen, Div. of Schering Corp.
Galloping Hill Rd, Kenilworth, NJ 07033 USA (Tel.: 908-298-4000)

Wessex Resins & Adhesives Ltd.
189-193 Spring Road, Sholing, Southampton, Hants., SO2 7NY, UK (Tel.: 0703-444744; FAX 0703-431792)

Westbridge Research Group. *Address unknown*

Westbrook Lanolin Co.
Argonaut Works, Laisterdyke, Bradford, West Yorkshire, BD4 8AU, UK (Tel.: 0274-663331; FAX 0274-667665; Telex: 51502)

Westlake Plastics Co.
PO Box 127, W. Lenni Rd., Lenni, PA 19052 USA (Tel.: 215-459-1000; FAX 215-459-1084; Telex: 83-5406)

Westo Industrial Products Ltd.
31 Pembridge Rd, London, W11 3HG, UK (Tel.: 071-727 8700; FAX 071-792 0329; Telex: 888941)

Westvaco Chemical Div.
PO Box 70848, Charleston Hts., SC 29415-0848 USA (Tel.: 803-740-2300; 800-336-2211; FAX 803-747-2270)

Westwood Pharmaceuticals Inc.
100 Forest Ave, Buffalo, NY 14213 USA

Weyl GmbH
Postfach 310160, Sanhofer Str.96, D-6800 Mannheim 31, Germany (Tel.: 0621-750 10; FAX 0621-750 1446)

Whaley Pharmaceuticals
7 Sheep Street, Rugby, Warwickshire, CV21 3BY, UK

Wheatley Chemical Co Ltd.
Langthwaite Grange Industrial Estate, South Kirkby, Pontefract, W Yorks, WF9 3AP, UK

Whitecourt Ltd.

Wincham House, 832 High Rd., London, N12 9RA, UK (Tel.: 081-446 4158; FAX 081-446 2280; Telex: 262634)

Whitehall Laboratories

Whitehall Laboratories, Div. of American Home Products Corp., 685 Third Ave, New York, NY 10017 USA (Tel.: 212-878-6000; 800-322-3129)

Whitehall Laboratories Ltd., Div. of American Home Products Corp., 22-24 Torrington Place, London, WC1E 7ET, UK (Tel.: 071-636 8080; FAX 071-580 6037)

Whitfield Chemicals Ltd.

23 Albert St, Newcastle, Staffordshire, ST5 1JP, UK (Tel.: 0782 711777; FAX 0782 717290; Telex: 367165)

Whitfield Plastics

Unit Ten, Chrystleton Court, Manor Park, Runcorn, Cheshire, WA7 1SU, UK (Tel.: 0928 571000; FAX 0928 571010)

G. Whitfield Richards Co. *See* Richards Co., G. Whitfield

Whitford Corp.

PO Box 2347, West Chester, PA 19381 USA (Tel.: 215-296-3200; FAX 215-647-4849; Telex: 83-5305)

Whiting, Peter (Chemicals) Ltd.

5 Lord Napier Place, Upper Mall, London, W6 9UB, UK (Tel.: 081-741 4025; FAX 081-741 1737; Telex: 8814670 WHICHEM G)

Whitmoyer Laboratories Inc. *Address unknown*

Whitney & Oettler

PO Box 8024, Savannah, GA 31402 USA (Tel.: 912-232-7166; Telex: 546 443)

Whittaker, Clark & Daniels

1000 Coolidge St., South Plainfield, NJ 07080 USA (Tel.: 201-561-6100; 800-732-0562; FAX 800-833-8139)

Wiggin Alloys Ltd. *Address unknown*

L. Wilcox & Co Ltd. *Address unknown*

Williams Division of Morton Thiokol Ltd. *See* Morton International Ltd.

Williams (Hounslow) Ltd. *See* Morton International Ltd.

Wilmington Chemical Co. *See* Rhône-Poulenc Inc.

Winchem Ltd.

7-11 Claremont Rd, West Byfleet, Weybridge, Surrey, KT14 6DY, UK (Tel.: 0932 340597; FAX 0932 340459; Telex: 928499)

Windsor Healthcare Ltd., Affiliate of Boehringer Ingelheim GmbH
Ellesfield Avenue, Bracknell, Berkshire, RG12 4YS, UK

Winthrop Laboratories. *See* Sterling Winthrop
Witco
Witco Corp., 520 Madison Ave., New York, NY 10022 USA (Tel.: 212-605-3680; FAX 212-486-4198)

Witco Corp/Allied-Kelite, 2701 Lake St., Melrose Park, IL 60160-3041 USA (Tel.: 800-323-9784)

Witco Corp/Argus Chem. Div., Bussey Rd., PO Box 1439, Marshall, TX 75671-1439 USA (Tel.: 903-938-5141; 800-431-1413; FAX 903-938-2647)

Witco Corp/Argus Div./Int'l. Sales, 520 Madison Ave., New York, NY 10022-4236 USA (Tel.: 212-605-3999; FAX 212-759-5739; Telex: 422186)

Witco Corp/Concarb, 10500 Richmond, PO Box 42817, Houston, TX 77242-2817 USA (Tel.: 713-978-5700; FAX 713-266-9982)

Witco Corp/Golden Bear Div., 10100 Santa Monica Blvd., Los Angeles, CA 90067-4183 USA (Tel.: 213-277-4511; FAX 213-201-0383)

Witco Corp/Humko Chem. Div., PO Box 125, Memphis, TN 38101-0125 USA (Tel.: 901-684-7000; 800-238-9150; FAX 901-682-6531; Telex: 53-928)

Witco Corp/Kendall/Amalie Div, 77 N Kendall Avenue, Bradford, PA 16701 USA

Witco Corp/Organics Div., 1000 Convery Blvd., Perth Amboy, NJ 08862-1932 USA (Tel.: 201-826-7777; 800-231-1542)

Witco Corp/Sonneborn Div., 520 Madison Ave., New York, NY 10022-4236 USA (Tel.: 212-605-3981; FAX 212-754-5676; Telex: 62470)

Witco Canada Ltd., 2 Lansing Sq., Suite 1200, Willowdale, Ontario, M2J 4Z4, Canada (Tel.: 416-497-9991)

Witco BV, 1 Canalside, Lowesmoor Wharf, Worcester, Worcestershire, WR1 2RS, UK (Tel.: 0905-21521; FAX 0905-611593)

Witco Chemical Ltd. (UK), Union Lane, Droitwich, Worcester, WR9 9BB, UK

Witco BV, PO Box 5, NL-1540 LZ Koog aan de Zaan, The Netherlands (Tel.: 75-283854; FAX 75-210811; Telex: 19270)

Witco SA, 10 Rue Cambaceres, 75008 Paris, France (Tel.: 42-65-99-03; FAX 42-65-67-61; Telex: 290233)

Witco Ltd., PO Box 10245, 26112 Haifa Bay, Israel (Tel.: 04-469-111; FAX 04-469-137; Telex: 45198)

Worcester Controls UK Ltd., Div. of BTR plc
Burell Rd, Haywards Heath, West Sussex, PH16 1TL, UK (Tel.: 0444 414133; FAX 044 459468; Telex: 87189 WOVCO G)

Worcester Valve Co. *See* Worcester Controls UK Ltd

Worthington Biochemical Corp.
Halls Mill Rd, Freehold, NJ 07728 USA (Tel.: 908-462-3838; 800-445-9603; FAX 908-308-4453; Telex: 3715614)

W. R. Grace. *See under* Grace, W.R.

Wyckoff Chemical Co. Inc.
>1421 Kalamazoo St., South Haven, MI 49090 USA (Tel.: 616-637-8474; FAX 616-637-8410)

Wyeth Laboratories
>Wyeth Laboratories Inc., Subsidiary of American Home Products Corp., P.O. Box 8299, Philadelphia, PA 19101 USA (Tel.: 215-688-4400)

>Wyeth Laboratories, Huntercombe Lane South, Taplow, Maidenhead, Berkshire, SL6 0PH, UK (Tel.: 0628-604377; FAX 0628-666368; Telex: 847640)

Yokkaichi Chemical Co., Ltd.
>1, Miyahigashi-cho 2-chome, Yokkaichi-shi, Mie, 510, Japan (Tel.: (0593) 45-1161; FAX (0593) 45-1168)

Yorkshire Chemicals plc
>Kirkstall Rd, Leeds, West Yorkshire, LS3 1LL, UK (Tel.: 0532 443111; FAX 0532 421670; Telex: 55366)

Yorkshire Pat-Chem Inc.
>11 Worley Rd., PO Box 1926, Greenville, SC 29602 USA (Tel.: 803-233-3941; 800-443-9358; FAX 803-232-3542)

Zeeland Chemicals, Inc., A Cambrex Co.
>215 N. Centennial St., Zeeland, MI 49464 USA (Tel.: 616-772-2193; 800-223-0453; FAX 616-772-7344; Telex: 226375)

Zeelan Industries Inc.
>141 East Fourth St., #220, St. Paul, MN 55101-1620 USA (Tel.: 612-292-9271; FAX 612-297-6138)

Zemmer Co Inc. *Address unverified*
>Oakmont, PA USA

Zeon Chemicals, Inc.
>Three Continental Towers, Suite 1211, 1701 Golf Rd., Rolling Meadows, IL 60008 USA (Tel.: 312-437-9770; 800-735-3388; FAX 312-437-9773)

Zinc Corp. of America
>Fourth & Delaware, Palmerton, PA 18071 USA (Tel.: 412-773-2295; 800-962-7500; FAX 412-773-2217)

Zinder SpA
>Viale Lombardia 58, I-20056 Milan, Italy (Tel.: 2-90 96 14; Telex: 312265)

Zinsser
>Zinsser Analytic (UK) Ltd., Howarth Rd., Maidenhead, Berkshire, SL6 1AP, UK (Tel.: 0628 773202; FAX 0628 72199)

>Zinsser NV. *Address unknown*

Zipperling Kessler & Co.

Postfach 1464, D-2070 Ahrensburg, Germany (Tel.: 04102-5151-0; FAX 04102-5151-69; Telex: 21 89 842)

Zircar Products Inc.

110 N. Main St., Florida, NY 10921 USA (Tel.: 914-651-4481; FAX 914-651-0441; Telex: 996608)

Zohar Detergent Factory

PO Box 11 300, Tel-Aviv, 61 112, Israel (Tel.: 03-528-7236; FAX 03-5287239; Telex: 33557 zohar il)

Zschimmer & Schwarz

Zschimmer & Schwarz GmbH & Co., Postfach 2179, 4-5 Max-Schwarz-Strasse, W-5420 Lahnstein, Germany (Tel.: 2621 121; FAX 2621-12407; Telex: 869816 ZSO D)

Zschimmer & Schwarz SARL, 10 rue Saint-Marc, F-75002 Paris, France (Tel.: 42 33 10 33; FAX 40 26 23 81; Telex: 670465 ZS F)

Zschimmer & Schwarz Italiana SpA, Casella Postale 1, I-13038 Tricerro (Vc), Italy (Tel.: 161-82 14 21; Telex: 200313 ZSI I)

Zschimmer & Schwarz Argentina S.A., Bdo. de Irigoyen 556-5 B, Buenos Aires, Argentina

Zyma (UK) Ltd. *Address unknown*

Zyma SA

Route de l'Etraz, CH-1260 Nyon, Switzerland (Tel.: 22-63 31 11; FAX 22-62 17 44)

Zymet Inc.

7 Great Meadow Lane, East Hanover, NJ 07936 USA (Tel.: 201-428-5245)

Appendices

Apendices

CAS Number-to-Tradename
Cross Reference

CAS 50-00-0	Hercules® 37M6-8
CAS 50-02-2	Aeroseb-Dex®
CAS 50-02-2	Decaderm
CAS 50-02-2	Decadron
CAS 50-02-2	Decadron Duofase
CAS 50-02-2	Decaspray®
CAS 50-02-2	Dexone
CAS 50-02-2	Hexadrol
CAS 50-03-3	Barseb
CAS 50-03-3	Colifoam
CAS 50-03-3	Cortaid
CAS 50-03-3	Cortef Acetate
CAS 50-03-3	Cortoderm
CAS 50-03-3	Cortril
CAS 50-03-3	Cortril Acetate-AS
CAS 50-03-3	Ficortril
CAS 50-03-3	HC45
CAS 50-03-3	Komed HC
CAS 50-03-3	Pabracort
CAS 50-04-4	Cortelan
CAS 50-04-4	Cortistab
CAS 50-04-4	Cortone Acetate
CAS 50-06-6	Barbivis
CAS 50-06-6	Eskabarb
CAS 50-06-6	Gardenal
CAS 50-06-6	Phenobarbitone Spansule Capsules
CAS 50-06-6	SK-Phenobarbital
CAS 50-06-6	Solfoton
CAS 50-06-6	Stental
CAS 50-06-6	Talpheno
CAS 50-10-2	Antrenyl Duplex
CAS 50-11-3	Gemonil
CAS 50-12-4	Mesantoin
CAS 50-13-5	Demerol
CAS 50-18-0	Endoxan
CAS 50-18-0	Neosar
CAS 50-21-5	Patlac® LA
CAS 50-23-7	Aeroseb-HC®
CAS 50-23-7	Barseb HC
CAS 50-23-7	Cetacourt
CAS 50-23-7	Cobadex Ointment
CAS 50-23-7	Cort-Dome
CAS 50-23-7	Cortef
CAS 50-23-7	Cortenema
CAS 50-23-7	Cortifoam
CAS 50-23-7	Dermacort
CAS 50-23-7	Dioderm
CAS 50-23-7	Efcortelan
CAS 50-23-7	Efcortelan Soluble
CAS 50-23-7	Efcortesol
CAS 50-23-7	Eldecort
CAS 50-23-7	Genacort
CAS 50-23-7	Hydrea Capsules
CAS 50-23-7	Hydrocortone
CAS 50-23-7	Hytone
CAS 50-23-7	Isopto Hydrocortisone
CAS 50-23-7	Mildison
CAS 50-23-7	Nutracort
CAS 50-23-7	Penecort®
CAS 50-23-7	Prepcort
CAS 50-23-7	Pro-cort
CAS 50-23-7	Proctocort
CAS 50-24-8	Co-Hydeltra
CAS 50-24-8	Codelcortone
CAS 50-24-8	Delta-Genacort
CAS 50-24-8	Deltacortril
CAS 50-24-8	Hydelta
CAS 50-24-8	Meti-Derm
CAS 50-24-8	Paracortol
CAS 50-24-8	Predne-Dome
CAS 50-24-8	Prednelan
CAS 50-24-8	Prednis
CAS 50-24-8	Sterane
CAS 50-28-2	Diogyn
CAS 50-28-2	Diogynets
CAS 50-28-2	Estrace
CAS 50-29-3	DDT
CAS 50-29-3	De De Tane
CAS 50-29-3	Neocid
CAS 50-33-9	Azolid
CAS 50-33-9	Butacote
CAS 50-33-9	Butazolidin
CAS 50-33-9	Tentor
CAS 50-34-0	Pro-Banthine
CAS 50-34-0	SK-Propantheline Bromide
CAS 50-35-1	Kevadon
CAS 50-41-9	Clomid
CAS 50-44-2	Purinethol
CAS 50-47-5	Pertofran
CAS 50-48-6	Limbitrol
CAS 50-49-7	Dimipressin
CAS 50-52-2	Mellaril-S
CAS 50-52-2	Melleril
CAS 50-53-3	Thorazine
CAS 50-54-4	Cin-Quin
CAS 50-54-4	Quinidex
CAS 50-54-4	SK-Quinidein Sulfate

CAS 50-55-5	Rau-Sed
CAS 50-55-5	Reserpoid
CAS 50-55-5	Sandril
CAS 50-55-5	Serpasil
CAS 50-55-5	Serpiloid
CAS 50-55-5	SK-Reserpine
CAS 50-56-6	Pitocin
CAS 50-56-6	Syntocinon
CAS 50-57-7	Diapid
CAS 50-57-7	Syntopressin
CAS 50-58-8	Metra
CAS 50-58-8	Plegine
CAS 50-59-9	Ceporin
CAS 50-63-5	Aralen Phosphate
CAS 50-63-5	Avloclor
CAS 50-65-7	Niclocide
CAS 50-65-7	Yomesan®
CAS 50-70-4	A-625/641
CAS 50-70-4	Hydex® 100 Gran. 206
CAS 50-70-4	Liponic 70-NC
CAS 50-70-4	Sorbelite C
CAS 50-70-4	Sorbidel
CAS 50-70-4	Sorbitol (EGIC)
CAS 50-70-4	Sorbo®
CAS 50-78-2	ASA
CAS 50-78-2	Cirin
CAS 50-78-2	Claradin
CAS 50-78-2	Codiphen
CAS 50-78-2	Codis
CAS 50-78-2	Easprin
CAS 50-78-2	Ecotrin
CAS 50-78-2	Entericin
CAS 50-78-2	Laboprin
CAS 50-78-2	Measurin
CAS 50-78-2	Nu-Seals Aspirin
CAS 50-78-2	Paleva
CAS 50-78-2	Percodan®
CAS 50-78-2	Safapryn
CAS 50-78-2	St Joseph
CAS 50-78-2	Tasprin
CAS 50-81-7	Cirin
CAS 50-99-7	Candex®
CAS 50-99-7	Emdex®
CAS 50-99-7	Flolys®
CAS 50-99-7	Roferose
CAS 51-12-7	Niamid
CAS 51-12-7	Niamide
CAS 51-15-0	Protopam Chloride
CAS 51-21-8	Adrucil
CAS 51-21-8	Efudex
CAS 51-21-8	Efudix
CAS 51-21-8	Fluoroplex®
CAS 51-21-8	Fluorouracil Roche
CAS 51-42-3	Asmatane Mist
CAS 51-42-3	Epitrate
CAS 51-42-3	Medihaler-Epi™
CAS 51-42-3	Suprarenin
CAS 51-43-4	Adrenalin
CAS 51-43-4	Bronkaid Mist
CAS 51-43-4	Epifrin®
CAS 51-43-4	Primatene Mist
CAS 51-43-4	Sus-Phrine
CAS 51-48-9	Riamat
CAS 51-56-9	Isopto Homatropine
CAS 51-77-4	Gefarnil
CAS 51-83-2	Isopto Carbachol
CAS 51-83-2	Miostat
CAS 51-98-9	Aygestin
CAS 51-98-9	Norlutate
CAS 52-01-7	Aldactone
CAS 52-01-7	Laractone®
CAS 52-01-7	Spirolone
CAS 52-21-1	Deltastab
CAS 52-21-1	depPredalone
CAS 52-21-1	Econopred
CAS 52-21-1	Meticortelone Acetate
CAS 52-21-1	Pred Forte®
CAS 52-21-1	Pred Mild®
CAS 52-21-1	Savacort
CAS 52-21-1	Sterane IA, IM
CAS 52-24-4	Thiotepa
CAS 52-28-8	Codiphen
CAS 52-28-8	Codis
CAS 52-28-8	Colrex Compound
CAS 52-39-1	Aldocorten
CAS 52-49-3	Antitrem
CAS 52-49-3	Artane
CAS 52-49-3	Tremin
CAS 52-51-7	Bioban® BNPD-40
CAS 52-51-7	Bronopol
CAS 52-51-7	Canguard® 409
CAS 52-51-7	Myacide® AS Plus
CAS 52-51-7	Myacide® S-1
CAS 52-51-7	Myacide® S-2
CAS 52-53-9	Berkatens
CAS 52-62-0	Ansolysen
CAS 52-67-5	Cuprimine
CAS 52-67-5	Trolovol®
CAS 52-68-6	Danex
CAS 52-68-6	Dipterex®
CAS 52-68-6	Dipterex® 80
CAS 52-68-6	Dylox®
CAS 52-68-6	Tugon
CAS 52-78-8	Nilevar
CAS 52-86-8	Brotopon
CAS 52-86-8	Fortunan
CAS 52-86-8	Haldol
CAS 52-86-8	Serenace
CAS 53-03-2	Co-Deltra
CAS 53-03-2	Delta Cortef
CAS 53-03-2	Deltacortone
CAS 53-03-2	Delta-Dome
CAS 53-03-2	Deltasone
CAS 53-03-2	Meticorten
CAS 53-03-2	Orasone
CAS 53-03-2	Paracort
CAS 53-03-2	SK-Prednisone
CAS 53-03-2	Zenadrid
CAS 53-06-5	Cortistan
CAS 53-16-7	Menformom A
CAS 53-16-7	Theelin
CAS 53-19-0	Lysodren
CAS 53-34-9	Alphadrol
CAS 53-36-1	Depmedalone 40, 80

CAS 53-36-1	Depo-Medrol
CAS 53-36-1	Depomedrone
CAS 53-36-1	Medrol Enpak
CAS 53-36-1	Medrone Veriderm
CAS 53-36-1	Mepred
CAS 53-39-4	Anavar
CAS 53-46-3	Banthine
CAS 53-60-1	Sparine Injection
CAS 53-60-1	Sparine Suspension
CAS 53-73-6	Hypertcnsin
CAS 53-86-1	Amuno
CAS 53-86-1	Imbrilon
CAS 53-86-1	Inacid
CAS 53-86-1	Indaciri
CAS 53-86-1	Indocid®
CAS 53-86-1	Indocid®-R
CAS 53-86-1	Indocin
CAS 53-86-1	Indomed
CAS 53-86-1	Indomee
CAS 53-86-1	Indoptol
CAS 54-05-7	Pfizerquine
CAS 54-11-5	Campbell's Nico-Soap
CAS 54-11-5	XL-All Insecticide
CAS 54-21-7	Alysine
CAS 54-21-7	Entrosalyl
CAS 54-21-7	Entrosalyl Standard
CAS 54-21-7	Nu-Seals Sodium Salicylate
CAS 54-31-9	Lasix
CAS 54-31-9	SK-Furosemide
CAS 54-36-4	Metopirone
CAS 54-42-2	Dendrid
CAS 54-42-2	Herplex Liquifilm®
CAS 54-42-2	Kerecid
CAS 54-42-2	Stoxil
CAS 54-49-9	Aramine
CAS 54-71-7	Adsorbocarpine
CAS 54-71-7	Almocarpine
CAS 54-71-7	Isopto Carpine
CAS 54-71-7	Pilocar® Ophthalmic Solution
CAS 54-71-7	Pilomiotin
CAS 54-85-3	Cotinazin
CAS 54-85-3	Dinacrin
CAS 54-85-3	INH
CAS 54-85-3	Isonex
CAS 54-85-3	Isonex Forte
CAS 54-85-3	Mybasan
CAS 54-85-3	Neoteben®
CAS 54-85-3	Nisapas
CAS 54-85-3	Nydrazin
CAS 54-85-3	Pycazide
CAS 54-85-3	Rimifon
CAS 54-91-1	Vercyte
CAS 54-92-2	Marsilid
CAS 55-03-8	Levothroid
CAS 55-03-8	Synthroid
CAS 55-06-1	Cytomcl
CAS 55-06-1	Tertroxin
CAS 55-38-9	Baycid®
CAS 55-38-9	Baytex®
CAS 55-38-9	Lebaycid®
CAS 55-56-1	Hibidil
CAS 55-56-1	Hibiscrub

CAS 55-56-1	Hibisol
CAS 55-56-1	Hibispray
CAS 55-56-1	Savloclens
CAS 55-56-1	Savlodil
CAS 55-56-1	Savlon Babycare
CAS 55-56-1	Superspray
CAS 55-63-0	Cascade
CAS 55-63-0	Hercol 2
CAS 55-63-0	Hercol 2X
CAS 55-63-0	Hercon® 2
CAS 55-63-0	Hercon® 2X
CAS 55-63-0	Hercosplit WR
CAS 55-63-0	Klavikordal
CAS 55-63-0	Minitran™
CAS 55-63-0	Niong
CAS 55-63-0	Nitrodisc
CAS 55-63-0	Nitronet
CAS 55-63-0	Nitrospan
CAS 55-63-0	Nitrostat
CAS 55-63-0	Susadrin
CAS 55-63-0	Transderm-Nitro
CAS 55-63-0	Tridil®
CAS 55-63-0	Unigel
CAS 55-65-2	Ismelin
CAS 55-86-7	Mustargen®
CAS 55-91-4	Floropryl
CAS 55-98-1	Myleran Tablets
CAS 56-03-1	Baquacil
CAS 56-23-5	Katharin
CAS 56-23-5	Thawpit
CAS 56-29-1	Evipal
CAS 56-29-1	Sombulex
CAS 56-35-9	bioMeT TBTO
CAS 56-35-9	Keycide® X-10
CAS 56-37-1	Sumquat® 2355
CAS 56-38-2	Bladan®
CAS 56-38-2	Folidol®-E605
CAS 56-38-2	Fosferno
CAS 56-38-2	Murphos
CAS 56-41-7	Ritalanine
CAS 56-75-7	Chloromycetin
CAS 56-75-7	Chloromycetin Intramuscular
CAS 56-75-7	Chloromycetin Kapseals
CAS 56-75-7	Chloromycetin Palmitate Suspension
CAS 56-75-7	Chloromycetin Pure
CAS 56-75-7	Chloromycetin Succinate
CAS 56-75-7	Chloromycetin Suppositories
CAS 56-75-7	Chloroptic®
CAS 56-75-7	Chloroptic S.O.P.®
CAS 56-75-7	Econochlor
CAS 56-75-7	Leukomycin®
CAS 56-81-5	Croderol G7000
CAS 56-81-5	Emery® 912
CAS 56-81-5	Glycon® G 100, G 300
CAS 56-81-5	Glyrol
CAS 56-81-5	Kemstrene® 96.0%
CAS 56-81-5	Osmoglyn
CAS 56-81-5	Pricerine 9071
CAS 56-81-5	Shur-Coal® FCA
CAS 56-81-5	Star
CAS 56-81-5	Superol

CAS 56-87-1	Laboprin
CAS 56-93-9	Hipochem Migrator J
CAS 56-93-9	Variquat® B200
CAS 56-94-0	Humorsol
CAS 56-94-0	Tosmilen
CAS 57-09-0	Acetoquat CTAB
CAS 57-09-0	Bromat®
CAS 57-09-0	Catinal HTB-70
CAS 57-09-0	Cetrimide BP
CAS 57-09-0	Rhodaquat® M242B/99
CAS 57-09-0	Sumquat® 6030
CAS 57-09-0	Varisoft® CTB-40
CAS 57-10-3	Emersol® 143
CAS 57-10-3	Glycon® P-45
CAS 57-10-3	Hystrene® 8016
CAS 57-10-3	Industrene® 4516
CAS 57-10-3	Prifrac 2960
CAS 57-10-3	Univol U332
CAS 57-11-4	Emersol® 110
CAS 57-11-4	Emersol® 6349
CAS 57-11-4	Emery® 400
CAS 57-11-4	Glycon® S-90
CAS 57-11-4	Glycon® TP
CAS 57-11-4	Hydrofol Acid 1655
CAS 57-11-4	Hystrene® 4516
CAS 57-11-4	Hystrene® 5016 NF
CAS 57-11-4	Hystrene® 7018 FG
CAS 57-11-4	Hystrene® 9718 NF FG
CAS 57-11-4	Industrene® 4518
CAS 57-11-4	Industrene® 5016
CAS 57-11-4	Industrene® 7018 FG
CAS 57-11-4	Industrene® R
CAS 57-11-4	Petrac® 270
CAS 57-11-4	Prifrac 2980
CAS 57-11-4	Pristerene 4904
CAS 57-11-4	Stearex
CAS 57-11-4	Stearite
CAS 57-11-4	Univol U334
CAS 57-13-6	Bubber Shet
CAS 57-13-6	Superprill
CAS 57-13-6	Ureaphil
CAS 57-15-8	Chloretone
CAS 57-22-7	Oncovin
CAS 57-27-2	Duromorph
CAS 57-30-7	Gardenal Sodium
CAS 57-30-7	Luminal Sodium
CAS 57-33-0	Euthatal
CAS 57-33-0	Napental
CAS 57-33-0	Nembutal
CAS 57-33-0	Nembutal Sodium
CAS 57-33-0	Palapent
CAS 57-33-0	Sagatal
CAS 57-33-0	Sodital
CAS 57-41-0	Dilantin
CAS 57-41-0	Eptoin
CAS 57-50-1	Sugartab®
CAS 57-53-4	Bamate
CAS 57-53-4	Bamo
CAS 57-53-4	Equanil Tablets
CAS 57-53-4	Meprospan
CAS 57-53-4	Miltown
CAS 57-53-4	SK-Bamate
CAS 57-53-4	Tised
CAS 57-55-6	Chillsa FE®
CAS 57-55-6	Dowfrost
CAS 57-55-6	Ilexan P
CAS 57-55-6	Monawet MO-84R2W.
CAS 57-55-6	Prolugen
CAS 57-57-8	Betaprone
CAS 57-63-6	Diogyn E
CAS 57-63-6	Estigyn
CAS 57-63-6	Estinyl
CAS 57-63-6	Feminone
CAS 57-64-7	Antilirium
CAS 57-64-7	Isopto Eserine
CAS 57-66-9	Benemid
CAS 57-66-9	SK-Probenecid
CAS 57-74-9	Syndane
CAS 57-83-0	Gesterol 50
CAS 57-83-0	Gestone
CAS 57-83-0	Lipo-Lutin
CAS 57-83-0	Membrettes
CAS 57-83-0	Nalutron
CAS 57-83-0	Protormone
CAS 57-83-0	Syngesterone
CAS 57-83-0	Syngestrets
CAS 57-85-2	Neo-Hombreol
CAS 57-85-2	Oreton Propionate
CAS 57-85-2	Synandrol
CAS 57-85-2	Synerone
CAS 57-85-2	Virormone
CAS 57-85-2	Vulvan
CAS 57-88-5	Cholesterol
CAS 57-88-5	Dastar
CAS 57-88-5	Fancol CH
CAS 57-88-5	Kathro
CAS 57-88-5	Liquid Crystal CN/9
CAS 57-94-3	Tubarine
CAS 57-96-5	Anturan
CAS 57-96-5	Anturane
CAS 58-14-0	Daraprim Tablets
CAS 58-18-4	Metandren
CAS 58-18-4	Oreton Methyl
CAS 58-18-4	Synandrets
CAS 58-18-4	Synandrotabs
CAS 58-18-4	Testred
CAS 58-20-8	depAndro 100
CAS 58-20-8	depAndro 200
CAS 58-20-8	Depovirin
CAS 58-20-8	Malogen CYP 200
CAS 58-22-0	Andro 100
CAS 58-22-0	Meretestate
CAS 58-22-0	Oreton
CAS 58-22-0	Synandrol F
CAS 58-25-3	Limbitrol
CAS 58-25-3	Tropium
CAS 58-28-6	Norpramin
CAS 58-28-6	Pertofrane
CAS 58-32-2	Persantin
CAS 58-32-2	Persantine
CAS 58-33-3	Anergan 25
CAS 58-33-3	Atosil®
CAS 58-33-3	Fellozine
CAS 58-33-3	Phenergan

CAS 58-33-3	Phensedyl
CAS 58-33-3	Remsed
CAS 58-33-3	Zipan
CAS 58-38-8	Compazine
CAS 58-39-9	Fentazin
CAS 58-39-9	Trilafon
CAS 58-39-9	Triptafen.
CAS 58-46-8	Nitoman
CAS 58-54-8	Edecril
CAS 58-54-8	Edecrina
CAS 58-54-8	Edecrin®
CAS 58-54-8	Hydromedin
CAS 58-55-9	Lasma®
CAS 58-55-9	Nuelin SA
CAS 58-55-9	Nuelin SA-250
CAS 58-55-9	Nuelin Tablets
CAS 58-55-9	Provent
CAS 58-55-9	Solosin®
CAS 58-55-9	Theograd
CAS 58-55-9	Theolair™ Liquid
CAS 58-55-9	Theolair™ Tablets
CAS 58-55-9	Theolair™-SR Tablets
CAS 58-55-9	Uniphyllin
CAS 58-56-0	Beesix
CAS 58-56-0	Gravidox
CAS 58-56-0	Hexa-Betalin
CAS 58-56-0	Hexavibex
CAS 58-56-0	Paxadon
CAS 58-85-5	Ritatin
CAS 58-89-9	Esoderm
CAS 58-89-9	Linfos
CAS 58-89-9	Scabene Lotion
CAS 58-93-5	Dichlosuric
CAS 58-93-5	Dichlotride
CAS 58-93-5	Esidrex
CAS 58-93-5	Hydro-Saluric®
CAS 58-93-5	Hydrodiuril
CAS 58-93-5	Moduretic.
CAS 58-93-5	Oretic
CAS 58-93-5	SK-Hydrchlorothiazide
CAS 58-93-5	Thiuretic
CAS 58-94-6	Chlotride
CAS 58-94-6	Diuril
CAS 58-94-6	Saluric®
CAS 59-02-9	E-Toplex
CAS 59-02-9	Ephynal
CAS 59-02-9	Eprolin
CAS 59-02-9	Epsilan-M
CAS 59-02-9	Esorb
CAS 59-02-9	Natopherol
CAS 59-02-9	Phytoforol
CAS 59-02-9	Tocopherex
CAS 59-02-9	Vascuals
CAS 59-05-2	Methotrexate
CAS 59-05-2	Mexate
CAS 59-30-3	Folicet
CAS 59-30-3	Folvite
CAS 59-30-3	Mission Prenatal
CAS 59-33-6	Dorantamin
CAS 59-33-6	Minihist
CAS 59-33-6	Pymafed
CAS 59-33-6	Thylogen Maleate
CAS 59-52-9	B.A.L
CAS 59-52-9	BAL in Oil
CAS 59-63-2	Marplan
CAS 59-66-5	Diamox
CAS 59-67-6	Niac
CAS 59-67-6	Nico-400
CAS 59-67-6	Nicobid
CAS 59-67-6	Nicocap
CAS 59-67-6	Nicolar
CAS 59-87-0	Furacin
CAS 59-92-7	Bendopa
CAS 59-92-7	Brocadopa
CAS 59-92-7	Dopar
CAS 59-92-7	Larodopa
CAS 59-97-2	Priscol
CAS 59-97-2	Priscoline Hydrochloride
CAS 60-00-4	Aroquest 75
CAS 60-00-4	Hamp-Ene® Acid
CAS 60-00-4	Kalex Acids
CAS 60-00-4	Questric Acid 5286
CAS 60-00-4	Sequestrene® AA
CAS 60-00-4	Trilon® BS
CAS 60-00-4	Versene Acid
CAS 60-24-2	Sipomer® 2ME
CAS 60-29-7	Ethyl Ether Anhydrous A.C.S
CAS 60-29-7	Ethyl Ether USP/ACS
CAS 60-31-1	Miochol® 1:100 Intraocular
CAS 60-32-2	Epsikapron
CAS 60-33-3	Emersol® 315
CAS 60-33-3	Pamolyn 200
CAS 60-33-3	Pamolyn 240
CAS 60-33-3	Pamolyn 300
CAS 60-33-3	Pamolyn 380
CAS 60-51-5	Atlas Sheriff
CAS 60-51-5	Dimethoate Bayer
CAS 60-51-5	Maktion.
CAS 60-51-5	Perfekthion®
CAS 60-51-5	Saluthion®
CAS 60-51-5	Salut®
CAS 60-51-5	Systemic Insecticide
CAS 60-51-5	Turbair
CAS 60-54-8	Bristocycline
CAS 60-54-8	Dema
CAS 60-54-8	Liquamycin
CAS 60-54-8	Panmycin
CAS 60-54-8	Panmycin Syrup
CAS 60-54-8	Pfizercycline
CAS 60-54-8	Robitet
CAS 60-56-0	Tapazole
CAS 60-57-1	Murdiel
CAS 60-99-1	Levoprome
CAS 60-99-1	Nozinan
CAS 60-99-1	Veractil
CAS 61-12-1	Nupercaine Hydrochloride
CAS 61-16-5	Vasoxine Injection
CAS 61-16-5	Vasoxyl
CAS 61-16-5	Vasoxyl Injection
CAS 61-16-5	Vasylox
CAS 61-56-3	Conadil
CAS 61-56-3	Ospolot
CAS 61-56-3	Trolone
CAS 61-57-4	Ambilhar

CAS 61-68-7	Ponstan
CAS 61-68-7	Ponstel
CAS 61-72-3	Ampiclox
CAS 61-72-3	Cloxicap
CAS 61-72-3	Cloxisyrup
CAS 61-72-3	Orbenin
CAS 61-72-3	Tegopen
CAS 61-75-6	Bretylate
CAS 61-75-6	Bretylol
CAS 61-76-7	Alcon-Efrin
CAS 61-76-7	Biomydrin
CAS 61-76-7	Famel
CAS 61-76-7	Fenox
CAS 61-76-7	Isopto Frin
CAS 61-76-7	Mydfrin
CAS 61-76-7	Neo-Synephrine Hydrochloride
CAS 61-76-7	Neophryn
CAS 61-76-7	Prefin Liquifilm
CAS 61-76-7	Tear-Efrin
CAS 61-82-5	Aminotriazole Bayer
CAS 61-82-5	Atlazin
CAS 61-82-5	Atraflow Plus
CAS 61-82-5	Azolan
CAS 61-82-5	Boroflow A/ATA
CAS 61-82-5	Boroflow S/ATA
CAS 61-82-5	CDA Simflow Plus
CAS 61-82-5	Chipman Path Weedkiller
CAS 61-82-5	Clearway
CAS 61-82-5	Herbazin Plus SC
CAS 61-82-5	Mascot Highway
CAS 61-82-5	MSS Aminotriazole
CAS 61-82-5	MSS Simazine/
	Aminotriazole 43FL
CAS 61-82-5	Orchard Herbide
CAS 61-82-5	Primatol SE 500FW
CAS 61-82-5	Simflow Plus
CAS 61-82-5	Synchemicals Total Weed Killer
CAS 61-82-5	Syntox Total Weed Killer
CAS 61-82-5	Weedazol TL
CAS 62-31-7	Dopastat
CAS 62-31-7	Dopram Injection
CAS 62-31-7	Intropin®
CAS 62-33-9	Calcium Disodium Versenate
CAS 62-38-4	Acticide PMA 100
CAS 62-38-4	Agrosan D
CAS 62-38-4	Ceresol
CAS 62-38-4	Merpectogel
CAS 62-38-4	Phe-Mer-Nite
CAS 62-38-4	Phenitol
CAS 62-38-4	Phenmerzyl Nitrate
CAS 62-38-4	PMA 18
CAS 62-38-4	PMA 60
CAS 62-38-4	Single Purpose
CAS 62-44-2	Phenin
CAS 62-51-1	Provocholine
CAS 62-54-4	Calac
CAS 62-54-4	Niacet Calcium Acetate Tech
CAS 62-73-7	Atgard
CAS 62-73-7	Dedevap®
CAS 62-73-7	Dethlac
CAS 62-73-7	Divipan
CAS 62-73-7	Mafu®

CAS 62-73-7	Nuvan
CAS 62-73-7	Vapona
CAS 62-90-8	Durabolin
CAS 62-90-8	Nandrobolic
CAS 62-97-5	Prantal
CAS 62-97-5	Variton
CAS 63-12-7	Quantril
CAS 63-25-2	Carylderm
CAS 63-25-2	Microcarb
CAS 63-25-2	Murvin 85
CAS 63-25-2	Thinsec
CAS 63-25-2	Tornado
CAS 63-25-2	Wasp Destroyer
CAS 63-39-8	Uteplex
CAS 63-68-3	Mepron
CAS 63-92-3	Dibenyline
CAS 63-92-3	Dibenyline Capsules
CAS 63-92-3	Dibenzyline
CAS 63-98-9	Phenurone
CAS 64-02-8	Aroquest 100
CAS 64-02-8	Cheelox® 100
CAS 64-02-8	Hamp-Ene® 100
CAS 64-02-8	Hamp-Ene® Na4
CAS 64-02-8	Intraquest® TA Sol'n
CAS 64-02-8	Kalex 220 Crystal
CAS 64-02-8	Kalex Liq. 50%
CAS 64-02-8	Questal Extra Powd. Conc. 0780
CAS 64-02-8	Questal Special 0860
CAS 64-02-8	Sequestrene® 30A
CAS 64-02-8	Sequestrene® 220
CAS 64-02-8	Solon Conc
CAS 64-02-8	Thorquest 39
CAS 64-02-8	Trilon® B Powd
CAS 64-17-5	Ethylol
CAS 64-17-5	Punctilious® Ethyl Alcohol
CAS 64-17-5	Pyro
CAS 64-17-5	Synasol
CAS 64-17-5	USI in Oval
CAS 64-18-6	Add F
CAS 64-18-6	Amasil®
CAS 64-31-3	MS Contin
CAS 64-31-3	MST
CAS 64-31-3	Oramorph
CAS 64-43-7	Amytal Sodium
CAS 64-55-1	Capla
CAS 64-55-1	Dormate
CAS 64-72-2	Aureomycin
CAS 64-75-5	Clinitetrin
CAS 64-75-5	Cyclopar
CAS 64-75-5	Panmycin Hydrochloride
CAS 64-75-5	Polycycline Hydrochloride
CAS 64-75-5	SK-Tetracycline
CAS 64-75-5	Steclin
CAS 64-75-5	Sustamycin
CAS 64-75-5	Tetrachel
CAS 64-75-5	Topicycline
CAS 64-75-5	Totomycin
CAS 64-77-7	Orinase
CAS 64-77-7	Pramidex
CAS 64-77-7	Rastinon
CAS 64-77-7	SK-Tolbutamide
CAS 65-29-2	Flaxedil

CAS 65-49-6	Nisapas
CAS 65-85-0	Retarder BA, BAX
CAS 66-71-7	Activ-8®, Activ-8 in Hexylene Glycol
CAS 67-03-8	Benerva
CAS 67-20-9	Dantafur
CAS 67-20-9	Furadantin
CAS 67-20-9	Macrodantin
CAS 67-20-9	Nitrex
CAS 67-20-9	Urantoin
CAS 67-43-6	Aroquest M Special
CAS 67-43-6	Chel DTPA
CAS 67-43-6	Hamp-Ex® Acid
CAS 67-43-6	Pentaquest OPAC 0201
CAS 67-45-8	Furoxone
CAS 67-48-1	Tripart® Chlormequat 5C.
CAS 67-63-0	Alcowipe®
CAS 67-63-0	Avantine
CAS 67-63-0	I.P.S
CAS 67-63-0	Sterets Pre-Injection Swabs®
CAS 67-64-1	Sasetone
CAS 67-68-5	Decap
CAS 67-68-5	Demavet
CAS 67-68-5	DMSO
CAS 67-68-5	Domoso
CAS 67-68-5	Herpid
CAS 67-73-2	Fluonid®
CAS 67-73-2	Psoranide
CAS 67-73-2	Synalar
CAS 67-73-2	Synalar-HP
CAS 67-73-2	Synandone
CAS 67-73-2	Synemol
CAS 67-92-5	Bentyl
CAS 67-92-5	Merbentyl
CAS 67-92-5	Wyovin
CAS 67-96-9	Hytakerol
CAS 67-97-0	Vigantol®
CAS 68-04-2	Cystemme
CAS 68-11-1	Thiovanic® Acid
CAS 68-12-2	Dynasolve 100
CAS 68-19-9	Anacobin
CAS 68-19-9	Berubigen
CAS 68-19-9	Betalin 12 Crystalline
CAS 68-19-9	Cobastab
CAS 68-19-9	Cyomin
CAS 68-19-9	Cytacon
CAS 68-19-9	Cytamen
CAS 68-19-9	Fermin
CAS 68-19-9	Redisol
CAS 68-19-9	Rubramin PC
CAS 68-19-9	Sytobex
CAS 68-19-9	Vibalt
CAS 68-22-4	Micronor
CAS 68-22-4	Nor-Q D
CAS 68-22-4	Norlutin
CAS 68-26-8	A-Sol
CAS 68-26-8	Homagenets Aoral
CAS 68-26-8	Ro-A-Vit
CAS 68-26-8	Super A
CAS 68-26-8	Vi-Alpha
CAS 68-26-8	Vi-Dom-A
CAS 68-35-9	Coco-Diazine
CAS 68-35-9	Codiazine
CAS 68-35-9	Eskadiazine
CAS 68-41-7	Seromycin
CAS 68-76-8	Trenimon
CAS 68-88-2	Atarax
CAS 68-88-2	Masmoran
CAS 68-88-2	Parixistil
CAS 68-88-2	Vistaril
CAS 69-09-0	Chloractil
CAS 69-09-0	Largactil
CAS 69-09-0	Megaphen®
CAS 69-09-0	Promapar
CAS 69-52-3	Pen A/N
CAS 69-52-3	Polycillin-N
CAS 69-52-3	Princillin
CAS 69-52-3	Principen/N
CAS 69-53-4	Acillin
CAS 69-53-4	Amcill
CAS 69-53-4	Amfipen
CAS 69-53-4	Ampen
CAS 69-53-4	Ampiclox
CAS 69-53-4	Ampilar®
CAS 69-53-4	Omnipen
CAS 69-53-4	Omnipen-N
CAS 69-53-4	Pen A
CAS 69-53-4	Penbritin
CAS 69-53-4	Pentrexyl
CAS 69-53-4	Pfizerpen A
CAS 69-53-4	Polycillin
CAS 69-53-4	Principen
CAS 69-53-4	SK-Ampicillin
CAS 69-53-4	Totacillin
CAS 69-53-4	Totocillin®
CAS 69-53-4	Unasyn® IM/IV
CAS 69-53-4	Uro-Binotal®
CAS 69-53-4	Vidopen
CAS 69-57-8	Crystapen
CAS 69-65-8	Osmitrol
CAS 69-65-8	Resectisol
CAS 69-72-7	Ionil
CAS 69-72-7	Retarder SAX
CAS 70-00-8	Viroptic Ophthalmic Solution
CAS 70-55-3	Uniplex 171
CAS 70-55-3	Uniplex 173
CAS 71-23-8	Pro-Gas (Gas Disclaimed)
CAS 71-27-2	Quelicin
CAS 71-27-2	Sucostrin
CAS 71-36-3	Isanol®
CAS 71-36-3	Tebol 88
CAS 71-36-3	Tebol 99
CAS 71-55-6	Baltane CF
CAS 71-55-6	Baltane D
CAS 71-55-6	Chlorothene
CAS 71-55-6	Chlorothene (VG)
CAS 71-55-6	Dabco® CS90
CAS 71-55-6	Delf® Fabric Protector
CAS 71-55-6	Distillex DS1
CAS 71-55-6	Ethana®
CAS 71-55-6	MS-170 1,1,1 Trichloroethane Solv
CAS 71-55-6	Solvethane
CAS 71-55-6	Tri-Ethane

CAS 71-58-9	Depo-Provera
CAS 71-58-9	Provera
CAS 71-58-9	Provera Dosepak
CAS 71-63-6	Crystodigin
CAS 71-67-0	Bromsulphalein
CAS 71-68-1	Dilaudid
CAS 71-73-8	Intraval
CAS 71-73-8	Intraval Sodium
CAS 71-73-8	Pentothal
CAS 71-73-8	Pentothal Sodium
CAS 72-14-0	Cerazole
CAS 72-17-3	Patlac® NAL
CAS 72-43-5	Marlate 2-MR Emulsifiable Insecticide
CAS 72-43-5	Marlate 50 WP
CAS 72-43-5	Marlate 300 Flowable
CAS 72-43-5	Marlate 400 Flowable Concentrate
CAS 72-43-5	Marlate Methoxychlor Insecticide
CAS 72-44-6	Melsed
CAS 73-22-3	Pacitron
CAS 73-48-3	Naturetin
CAS 73-49-4	Hydromox
CAS 74-31-7	Agerite® DPPD
CAS 74-31-7	JZF
CAS 74-31-7	Naugard® J
CAS 74-31-7	Permanax DPPD
CAS 74-55-5	Myambutol
CAS 74-73-3	Declomycin
CAS 74-83-9	Embafume
CAS 74-83-9	Meth-O-Gas
CAS 74-90-8	Zyklon
CAS 74-98-6	A-108
CAS 75-00-3	Ethyl Chloride BP
CAS 75-09-2	Aerothene
CAS 75-09-2	Driverit
CAS 75-09-2	M-Clean D
CAS 75-09-2	Methoklone
CAS 75-09-2	Nevolin®
CAS 75-11-6	Mi-Gee Brand
CAS 75-18-3	Exact-S®
CAS 75-25-2	Bromoform
CAS 75-28-5	A-31
CAS 75-52-5	Nitrofuel®
CAS 75-52-5	NM
CAS 75-52-5	NM-55®
CAS 75-54-7	CM8750
CAS 75-60-5	Weed Ender
CAS 75-69-4	Distillex DS6
CAS 75-69-4	Genesolv A Solvent
CAS 75-69-4	Genetron® 11
CAS 75-71-8	Genetron® 12
CAS 75-71-8	Sterethox
CAS 75-72-9	Genetron® 13
CAS 75-76-3	CT2050
CAS 75-77-4	CT2950
CAS 75-79-6	CM9000
CAS 75-91-2	Aztec® t-Butyl Hydroperoxide-70, Aq.
CAS 75-91-2	TBHP-70
CAS 75-91-2	T-Hydro
CAS 75-91-2	Trigonox® A-80
CAS 75-91-2	USP®-800
CAS 75-94-5	CV4900
CAS 75-94-5	Dynasylan® VTC
CAS 75-99-0	BH Dalapon
CAS 75-99-0	Couch and Grass Killer
CAS 75-99-0	Synchemicals Dalapon
CAS 76-03-9	Farmon TCA
CAS 76-03-9	NaTa
CAS 76-06-2	Larvacide
CAS 76-13-1	Arklone
CAS 76-13-1	Distillex DS5
CAS 76-13-1	Genesolv D Solvent
CAS 76-13-1	Genetron® 113
CAS 76-13-1	MS-180 Freon® TF Solv
CAS 76-14-2	Genetron® 114
CAS 76-25-5	Adcortyl Cream
CAS 76-25-5	Adcortyl in Orabase
CAS 76-25-5	Adcortyl Injection
CAS 76-25-5	Adcortyl Ointment
CAS 76-25-5	Adcortyl Spray
CAS 76-25-5	Adcortyl-A
CAS 76-25-5	Aristocort Acetonide
CAS 76-25-5	Kenalog Injection
CAS 76-25-5	Remiderm
CAS 76-25-5	Remiderm Cream
CAS 76-25-5	Remiderm Ointment
CAS 76-25-5	Remiderm Spray
CAS 76-25-5	Remotic Capsules
CAS 76-25-5	TAC-3™
CAS 76-25-5	Tramacin
CAS 76-25-5	Triamonide 40
CAS 76-25-5	Trymex Cream and Ointment
CAS 76-25-5	Vetalog
CAS 76-25-5	Vetalog Injection
CAS 76-25-5	Vetalog Plus Cream
CAS 76-38-0	Penthrane
CAS 76-39-1	NMP
CAS 76-39-1	NMP Conc
CAS 76-42-6	Percodan®
CAS 76-43-7	Halotestin
CAS 76-43-7	Ora-Testryl
CAS 76-43-7	Ultandren
CAS 76-74-4	Dorsital
CAS 76-87-9	Ashlade Flotin
CAS 76-87-9	Du-Ter®
CAS 76-87-9	Farmatin
CAS 76-87-9	Quadrangle Super-Tin 4L
CAS 76-87-9	Super Tin® 4L
CAS 76-90-4	Cantil
CAS 77-06-5	Activol
CAS 77-06-5	Berelex
CAS 77-06-5	Regulex
CAS 77-09-8	Evac-Q-Tabs
CAS 77-09-8	Feen-A-Mint®
CAS 77-09-8	Phenolax
CAS 77-09-8	Probilin
CAS 77-09-8	Prulet
CAS 77-09-8	Purganol
CAS 77-09-8	Purgen
CAS 77-21-4	Doriden
CAS 77-26-9	Sandoptal
CAS 77-36-1	Hygroton

CAS 77-38-3	Systral
CAS 77-41-8	Celontin
CAS 77-41-8	Celontin Kapseals
CAS 77-58-7	ADK STAB BT-11
CAS 77-58-7	Dabco® T-12
CAS 77-58-7	Metacure® T-12
CAS 77-58-7	Synpron 1009
CAS 77-58-7	TEDA-T411
CAS 77-67-8	Emeside
CAS 77-67-8	Zarontin
CAS 77-86-1	THAM
CAS 77-86-1	Tris Amino®
CAS 77-89-4	Citroflex A-2
CAS 77-89-4	Uniplex 82
CAS 77-90-7	Citroflex A-4
CAS 77-90-7	Estaflex
CAS 77-90-7	Uniplex 84
CAS 77-92-9	Citraclean
CAS 77-93-0	Citroflex 2
CAS 77-93-0	Hydagen® C.A.T
CAS 77-93-0	Uniplex 80
CAS 77-94-1	Citroflex 4
CAS 78-08-0	CV4910
CAS 78-08-0	Dynasylan® VTEO
CAS 78-08-0	Union Carbide® A-151
CAS 78-10-4	Dynasil® A
CAS 78-10-4	Silbond® Condensed
CAS 78-10-4	Silester
CAS 78-11-5	Cardiacap
CAS 78-11-5	Mycardol
CAS 78-11-5	Myocardol
CAS 78-11-5	Pentritol
CAS 78-11-5	Pentryate 80
CAS 78-11-5	Peritrate
CAS 78-21-7	Barquat® CME-35
CAS 78-44-4	Carisoma®
CAS 78-44-4	Rela
CAS 78-44-4	Sanoma
CAS 78-44-4	Soma
CAS 78-48-8	Def®
CAS 78-51-3	Amgard TBEP
CAS 78-51-3	KP-140®
CAS 78-51-3	TBEP
CAS 78-62-6	CD5600
CAS 78-62-6	EXP-49
CAS 78-63-7	Aztec® 2,5-Di
CAS 78-63-7	Esperal® 120
CAS 78-63-7	Luperco 101-P20
CAS 78-63-7	Lupersol 101
CAS 78-63-7	Polyvel CR-5F
CAS 78-63-7	Trigonox® 101
CAS 78-66-0	Surfynol® 82
CAS 78-67-1	V-60
CAS 78-78-4	Exxsol® Isopentane
CAS 79-01-6	Altene DG
CAS 79-01-6	Disparit B
CAS 79-01-6	Distillex DS2
CAS 79-01-6	Triclene
CAS 79-01-6	Triklone
CAS 79-01-6	Trilene
CAS 79-01-6	Westrosol®
CAS 79-09-4	Luprosil®

CAS 79-10-7	Sipomer® β-CEA
CAS 79-21-0	Oxymaster
CAS 79-21-0	Proxitane
CAS 79-24-3	NE
CAS 79-34-5	Acetosol
CAS 79-46-9	NiPar S-20
CAS 79-57-2	Berkmycen
CAS 79-57-2	Biostat A.1
CAS 79-57-2	Clinimycin
CAS 79-57-2	Imperacin
CAS 79-57-2	Macocyn
CAS 79-57-2	Oxydon
CAS 79-57-2	Oxymycin
CAS 79-57-2	Terramycin
CAS 79-64-1	Serosteron
CAS 79-74-3	Lowinox® AH25
CAS 79-74-3	Santovar
CAS 79-94-7	FR-1524
CAS 79-94-7	Great Lakes BA-59P
CAS 79-94-7	Saytex® RB-100
CAS 80-05-7	Parabis
CAS 80-10-4	CD5950
CAS 80-15-9	Aztec® CHP-80
CAS 80-15-9	CHP-5
CAS 80-15-9	CHP-158
CAS 80-15-9	HPC-9
CAS 80-15-9	Trigonox® 239A
CAS 80-15-9	Trigonox® K-95
CAS 80-35-3	Midicel
CAS 80-39-7	Uniplex 108
CAS 80-43-3	Aztec® DCP-R
CAS 80-43-3	Di-Cup
CAS 80-43-3	Esperal® 115RG
CAS 80-43-3	Luperco 500-40KE
CAS 80-43-3	Luperox 500R
CAS 80-43-3	Percumyl D
CAS 80-43-3	Perkadox® BC
CAS 80-43-3	Peroximon® DC-40
CAS 80-43-3	Peroximon® DC 40 MG
CAS 80-43-3	Polyvel PCL-20
CAS 80-46-6	Nipacide® PTAP
CAS 80-46-6	Orthophen® 278
CAS 80-46-6	Pentaphen® 67
CAS 80-49-9	Homapin
CAS 80-50-2	Valpin 50
CAS 80-51-3	Celogen® OT
CAS 80-77-3	Trancopal
CAS 81-07-2	Sweeta
CAS 81-13-0	Ilopan
CAS 81-13-0	Panthoderm
CAS 81-23-2	Cholan DH
CAS 81-23-2	Decholin
CAS 81-23-2	Dehydrocholin
CAS 81-23-2	Dilabil
CAS 81-81-2	Dethmor
CAS 81-81-2	Killgerm® Sewarin P
CAS 81-81-2	Kypfarin
CAS 81-81-2	Ratox
CAS 81-81-2	RCR Grey Squirrel Killer Concentrate
CAS 81-81-2	Sakarat
CAS 81-81-2	Sewarin

CAS 81-81-2	Sorexa Plus	CAS 85-70-1	Reomol 4PG
CAS 81-81-2	Warfarin	CAS 85-73-4	Sulfathalidine
CAS 82-68-8	Botrilex	CAS 85-73-4	Thalazole
CAS 82-68-8	Brabant PCNB	CAS 86-34-0	Milontin
CAS 82-68-8	Bras-sicol	CAS 86-50-0	Cotnion-Ethyl-Methyl
CAS 82-68-8	Quintozene	CAS 86-50-0	Cotnion-Methyl
CAS 82-68-8	Trim	CAS 86-50-0	Gusathion®
CAS 82-68-8	Tubergran	CAS 86-50-0	Guthion®
CAS 83-12-5	Dindevan	CAS 86-87-3	Rhizopon B
CAS 83-12-5	Hedulin	CAS 86-87-3	Rooting Powder
CAS 83-12-5	Indon	CAS 86-87-3	Tipoff
CAS 83-43-2	Medrol ADT Pak	CAS 87-08-1	G.P.V
CAS 83-43-2	Medrol Dosepak	CAS 87-08-1	Pen-Vee
CAS 83-43-2	Medrone	CAS 87-08-1	Ticillin V.K
CAS 83-43-2	Medrone Medules	CAS 87-08-1	V-Cillin
CAS 83-43-2	Solumedrone	CAS 87-10-5	Temasept IV
CAS 83-73-8	Yodoxin	CAS 87-10-5	Tuasol 100
CAS 83-79-4	Derris Dust	CAS 87-33-2	Isordil
CAS 83-79-4	FS Derris	CAS 87-33-2	Isordil Tablets
CAS 83-88-5	Beflavine Roche	CAS 87-33-2	Isordil Tembids Capsules
CAS 83-88-5	Beflavit	CAS 87-33-2	Sorbichew
CAS 83-88-5	Flavaxin	CAS 87-33-2	Sorbid
CAS 84-02-6	Vertigon	CAS 87-33-2	Sorbitrate
CAS 84-02-6	Vertigon Spansule Capsule	CAS 87-51-4	Rhizopon A, AA
CAS 84-04-8	Mornidine	CAS 87-66-1	Pyrogallol
CAS 84-17-3	DV	CAS 87-83-2	FR-705
CAS 84-36-6	Isotense	CAS 87-86-5	Santophen 20
CAS 84-61-7	KP 201	CAS 87-90-1	ACL 85
CAS 84-61-7	Morflex 150	CAS 87-90-1	ACL 90 Plus
CAS 84-61-7	Unimoll® 66 M	CAS 87-90-1	CDB 90
CAS 84-61-7	Uniplex 250	CAS 87-90-1	Queschlor
CAS 84-65-1	Morkit®	CAS 88-04-0	Ayrtol
CAS 84-66-2	Kodaflex® DEP	CAS 88-04-0	Nipacide® MX
CAS 84-66-2	Palatinol® A	CAS 88-19-7	Uniplex 171
CAS 84-69-5	Kodaflex® DIBP	CAS 88-24-4	Anti-oxidant 425
CAS 84-69-5	Palatinol 1C	CAS 88-24-4	Cyanox® 425
CAS 84-69-5	Uniplex 155	CAS 88-58-4	Eastman® DTBHQ
CAS 84-74-2	Bufa	CAS 88-85-7	Dynamyte
CAS 84-74-2	Kodaflex® DBP	CAS 88-85-7	Premerge
CAS 84-74-2	Morflex 240	CAS 88-85-7	Tubotox
CAS 84-74-2	NLA-10	CAS 89-78-1	Fancol Menthol
CAS 84-74-2	Palatinol® C	CAS 90-03-9	Myringacaine Drops
CAS 84-74-2	Palatinol® DBP	CAS 90-30-3	Additin 30
CAS 84-74-2	PX-104	CAS 90-30-3	Akrochem® Antioxidant PANA
CAS 84-74-2	Unimoll® DB	CAS 90-30-3	Naugard® PANA
CAS 84-74-2	Uniplex 150	CAS 90-33-5	Cantabiline
CAS 84-96-8	Alimemazine	CAS 90-43-7	Nipacide® OPP
CAS 85-00-7	Dukatalon	CAS 90-43-7	Preventol® O Extra
CAS 85-00-7	Katalon	CAS 91-33-8	Fovane
CAS 85-00-7	Midstream	CAS 91-53-2	Santoquin
CAS 85-00-7	Power Diquat	CAS 92-13-7	Ocusert
CAS 85-00-7	Reglone	CAS 92-88-6	Isotex® 100
CAS 85-00-7	Reglox	CAS 93-14-1	Colrex Expectorant
CAS 85-00-7	Weedol	CAS 93-14-1	Hill's Adult Expectorant
CAS 85-44-9	Retarder AK	CAS 93-14-1	Respenyl
CAS 85-44-9	Retarder ESEN	CAS 93-14-1	Robitussin
CAS 85-44-9	Retarder PX	CAS 93-30-1	Orthoxine
CAS 85-60-9	Santowhite	CAS 93-46-9	Agerite® White
CAS 85-60-9	Vanox® SWP	CAS 93-65-2	Astix
CAS 85-68-7	Santicizer 160	CAS 93-65-2	Duplosan® CMPP
CAS 85-68-7	Unimoll® BB	CAS 93-65-2	Duplosan New System CMPP
CAS 85-70-1	Morflex 190	CAS 93-69-6	Vulkacit® 1000

CAS 93-72-1	Kuron
CAS 93-82-3	Ablumide SDE
CAS 93-82-3	Alkamide® DS-280/S
CAS 93-82-3	Alkamide® HTDE
CAS 93-82-3	Amidex SE
CAS 93-82-3	Karamide ST-DEA
CAS 93-82-3	Monamid® 718
CAS 93-82-3	Nopcogen 14-S
CAS 93-83-4	Active #18
CAS 93-83-4	Alkamide® DO-280
CAS 93-83-4	Alkamide® WRS 1-66
CAS 93-83-4	Amidex O
CAS 93-83-4	Crillon ODE
CAS 93-83-4	Emid® 6545
CAS 93-83-4	Hartamide 9137
CAS 93-83-4	Incromide OD
CAS 93-83-4	Laurel SD-400
CAS 93-83-4	Mackamide MO
CAS 93-83-4	Marlamid® D 1885
CAS 93-83-4	Mazamide® O 20
CAS 93-83-4	Ninol® 201
CAS 93-83-4	Norfox® F-221
CAS 93-83-4	Rewomid® DO 280
CAS 93-83-4	Rewomid® DO 280 SE
CAS 93-83-4	Schercomid ODA
CAS 93-83-4	Schercomid SO-A
CAS 93-83-4	Varamide® A-7
CAS 94-09-7	Americaine
CAS 94-13-3	Lexgard P
CAS 94-13-3	Nipasol M
CAS 94-17-7	Cadox® TDP
CAS 94-17-7	Cadox® TS-50S
CAS 94-18-8	Nipabenzyl
CAS 94-20-2	Dabinese
CAS 94-20-2	Diabinese
CAS 94-20-2	Mellinese
CAS 94-26-8	Lexgard B
CAS 94-28-0	Kodaflex® TEG-EH
CAS 94-35-9	Sinaxar
CAS 94-36-0	Abcure S-40-25
CAS 94-36-0	Aztec® Benzoyl Peroxide-70
CAS 94-36-0	Aztec® Benzoyl Peroxide-77
CAS 94-36-0	Aztec® Benzoyl Peroxide-Dry
CAS 94-36-0	Benoxyl
CAS 94-36-0	Benox® L-40LV
CAS 94-36-0	Benzac
CAS 94-36-0	Cadet® BPO-70W
CAS 94-36-0	Cadox® 40E
CAS 94-36-0	Cadox® BPO-W40
CAS 94-36-0	Cadox® BS
CAS 94-36-0	Cadox® BTW-50
CAS 94-36-0	Clear by Design
CAS 94-36-0	Clearasil Super Strength
CAS 94-36-0	Dermoxyl®
CAS 94-36-0	Dry and Clear
CAS 94-36-0	Eloxyl
CAS 94-36-0	Florox
CAS 94-36-0	Lucidol
CAS 94-36-0	Lucidol 75FP
CAS 94-36-0	Lucidol-78
CAS 94-36-0	Lucidol GS
CAS 94-36-0	Lucipal

CAS 94-36-0	Luperco A
CAS 94-36-0	Luperco AC
CAS 94-36-0	Luperco AFR
CAS 94-36-0	Luperco AFR-250
CAS 94-36-0	Nericur Gel 5
CAS 94-36-0	Novadelox
CAS 94-36-0	PanOxyl Aquagel
CAS 94-36-0	PanOxyl Wash
CAS 94-36-0	Perlygel®
CAS 94-36-0	Persadox
CAS 94-36-0	Topex
CAS 94-74-6	Agrichem MCPA-25, 50
CAS 94-74-6	Agricorn 500
CAS 94-74-6	Agritox 50
CAS 94-74-6	Agroxone 50
CAS 94-74-6	Albar-M
CAS 94-74-6	Atlas MCPA
CAS 94-74-6	BASF MCPA Amine 50
CAS 94-74-6	BH MCPA 75
CAS 94-74-6	Campbell's MCPA 25, 50
CAS 94-74-6	Campbell's Redlegor
CAS 94-74-6	Chafer MCPA 675
CAS 94-74-6	Farmon MCPA 50
CAS 94-74-6	MSS MCPA 50
CAS 94-74-6	Phenoxylene 50
CAS 94-74-6	Power MCPA
CAS 94-74-6	Quad MCPA 50%
CAS 94-74-6	Star MCPA
CAS 94-81-5	Belmac Straight
CAS 94-81-5	Tropotox
CAS 94-82-6	Campbell's DB Straight
CAS 94-82-6	Campbell's Redlegor
CAS 94-91-7	Keromet MD 60, MD 80
CAS 94-91-7	Metal Deactivator S
CAS 95-05-6	Tetmosol
CAS 95-14-7	Preventol® CI 8
CAS 95-19-2	Atlasol KAD
CAS 95-19-2	Crodazoline S
CAS 95-19-2	Hodag C-100-S
CAS 95-19-2	Miramine® GS
CAS 95-19-2	Schercozoline S
CAS 95-25-0	Paraflex
CAS 95-31-8	BBTS
CAS 95-31-8	Delac NS
CAS 95-31-8	Perkacit® TBBS
CAS 95-31-8	Santocure NS
CAS 95-31-8	Vanax® NS
CAS 95-32-9	Morfax
CAS 95-33-0	CBTS
CAS 95-33-0	Delac S
CAS 95-33-0	Durax®
CAS 95-33-0	Furbac
CAS 95-33-0	Perkacit® CBS
CAS 95-33-0	Santocure
CAS 95-33-0	Vancure D.A.A
CAS 95-33-0	Vulkacit® CZ/EGC
CAS 95-33-0	Vulkacit® DZ/EGC
CAS 95-38-5	Crodazoline O
CAS 95-38-5	Marlowet® 5440
CAS 95-38-5	Schercozoline O
CAS 95-38-5	Sovatex IM17H
CAS 96-05-9	Ageflex AMA

CAS 96-05-9	Sipomer® AM
CAS 96-05-9	SR-201
CAS 96-20-8	AB®
CAS 96-27-5	Thiovanol®
CAS 96-48-0	Agrisynth BLO
CAS 96-48-0	Agsol Ex BLO
CAS 96-48-0	BLO®
CAS 96-48-0	Dynasolve 699
CAS 96-48-0	GBL
CAS 96-49-1	Texacar® EC
CAS 96-69-5	Lowinox® 44S36
CAS 96-69-5	Rutenol
CAS 96-69-5	Ultranox® 236
CAS 96-83-3	Telepaque
CAS 97-23-4	Acticide DDM
CAS 97-23-4	Algafen
CAS 97-23-4	Algofen
CAS 97-23-4	Anthiphen
CAS 97-23-4	Dicestal
CAS 97-23-4	Ecco MP®-2004
CAS 97-23-4	Fungo®
CAS 97-23-4	Mascot Moss Killer
CAS 97-23-4	Nuophene
CAS 97-23-4	Panacide
CAS 97-23-4	Super Mosstox
CAS 97-39-2	Accelerator DT
CAS 97-39-2	Akrochem® DOTG
CAS 97-39-2	Perkacit® DOTG
CAS 97-39-2	Vanax® DOTG
CAS 97-53-0	Phyllol
CAS 97-59-6	Fancol TOIN
CAS 97-64-3	Solactol
CAS 97-74-5	Ancazide IS
CAS 97-74-5	Monex
CAS 97-74-5	Mono Thiurad
CAS 97-74-5	Perkacit® TMTM
CAS 97-74-5	TMTM
CAS 97-74-5	Unads®
CAS 97-74-5	Vulcaid 222
CAS 97-77-8	Akrochem® TETD
CAS 97-77-8	Ancazide ET
CAS 97-77-8	Ethyl Tuads®
CAS 97-77-8	Ethyl Tuex
CAS 97-77-8	Ethylthiurad
CAS 97-77-8	Perkacit® TETD
CAS 97-78-9	Crodasinic L
CAS 97-78-9	Hamposyl® L
CAS 97-78-9	Vanseal® LS
CAS 97-90-5	Ageflex EGDMA
CAS 97-90-5	Perkalink® 401
CAS 97-99-4	QO® Tetrahydrofurfuryl Alcohol (THFA®)
CAS 97-99-4	THFA
CAS 98-00-0	QO® Furfuryl Alcohol (FA®)
CAS 98-13-5	CP0280
CAS 98-19-1	Lowinox® TBMX
CAS 98-51-1	Lowinox® PTBT
CAS 98-54-4	Lowinox® 070
CAS 98-54-4	Terbutol
CAS 98-77-1	Accelerator 2P
CAS 98-77-1	Akrochem® P.P.D
CAS 98-77-1	Vanax® 552
CAS 98-79-3	Ajidew A-100
CAS 98-79-3	Pidolidone®
CAS 98-96-4	Zinamide
CAS 99-30-9	Fumite Dicloran
CAS 99-66-1	Depakene
CAS 99-68-3	Evanacide® 3CS
CAS 99-76-3	Lexgard M
CAS 99-76-3	Nipagin M
CAS 99-77-8	Antabuse
CAS 99-79-6	Ethiodan
CAS 99-79-6	Pantopaque
CAS 100-37-8	Pennad 150
CAS 100-46-9	Sumine® 2005
CAS 100-51-6	Benatol
CAS 100-51-6	Bentalol
CAS 100-55-0	Roniacol
CAS 100-74-3	Texacat® NEM
CAS 100-74-3	Toyocat®-NEM
CAS 100-88-9	Hexamic Acid
CAS 100-97-0	Grasselerator 102
CAS 100-97-0	Vulkacit® H
CAS 101-02-0	Doverphos® 10
CAS 101-02-0	Doverphos® 10-HR
CAS 101-02-0	Lankromark® LE65
CAS 101-02-0	Syn-O-Ad® P-399
CAS 101-02-0	Weston® EGTPP
CAS 101-02-0	Weston® TPP
CAS 101-05-3	Dairene®
CAS 101-05-3	Dairin®
CAS 101-05-3	Dyrene®
CAS 101-21-3	Atlas CIPC 40
CAS 101-21-3	Campbell's CIPC 40%
CAS 101-21-3	Mirvale
CAS 101-21-3	MSS CIPC
CAS 101-21-3	Residuren
CAS 101-21-3	Residuren Extra
CAS 101-21-3	Warefog
CAS 101-26-8	Mestinon
CAS 101-26-8	Regonol
CAS 101-31-5	Cystospaz
CAS 101-34-8	Flexricin® P-8
CAS 101-34-8	Naturechem® GTR
CAS 101-37-1	Perkalink® 300
CAS 101-43-9	Ageflex CHMA
CAS 101-67-7	Naugalube® 438
CAS 101-67-7	Vanox® 12
CAS 101-67-7	Vanox® ODP
CAS 101-68-8	Desmodur® VKS-2, VKS-4, VKS-18
CAS 101-68-8	Isonate® 125M
CAS 101-68-8	Isonate® 2125M
CAS 101-68-8	Novor 950
CAS 101-68-8	Rubinate® LF-168
CAS 101-72-4	Flexzone 3C
CAS 101-72-4	Permanax IPPD
CAS 101-72-4	Vanox® 3C
CAS 101-72-4	Vulkanox® 4010 NA
CAS 101-77-9	Tonox® 22
CAS 101-77-9	Tonox® R
CAS 101-87-1	Vanox® 6H
CAS 101-96-2	Kerobit® BPD
CAS 101-96-2	Naugalube® 403

CAS 101-96-2	Tenamine 2	CAS 103-90-2	Tylenol
CAS 101-96-2	Topanol® M	CAS 103-90-2	Valadol
CAS 102-06-7	Akrochem® DPG	CAS 104-29-0	Gechophen
CAS 102-06-7	Perkacit® DPG	CAS 104-29-0	Mycil
CAS 102-06-7	Phenaldine	CAS 104-31-4	Tessalon
CAS 102-06-7	Vanax® DPG	CAS 104-46-1	Arizole® Anethole Extra
CAS 102-06-7	Vulcaid DPG	CAS 104-74-5	Dehyquart C Crystals
CAS 102-06-7	Vulcogene ND	CAS 104-74-5	Ledmin LPC
CAS 102-06-7	Vulkacit® D/C	CAS 104-75-6	Armeen® L8D
CAS 102-07-8	Zeonet U	CAS 104-76-7	Aerofroth 88
CAS 102-08-9	A-1	CAS 105-16-8	Ageflex FM-2
CAS 102-08-9	Akrochem® Thio No. 1	CAS 105-55-5	Pennzone E 0686
CAS 102-08-9	Stabilizer C	CAS 105-57-7	Acetron® GP
CAS 102-60-3	Mazeen® 173	CAS 105-57-7	Acetron® NS
CAS 102-60-3	Neutrol® TE	CAS 105-57-7	AT-20GF
CAS 102-60-3	Quadrol®	CAS 105-57-7	Cadco® Acetal
CAS 102-76-1	Enzactin	CAS 105-57-7	Delrin® 100, 500
CAS 102-76-1	Kodaflex® Triacetin	CAS 105-57-7	Delrin® 100ST, 500T
CAS 102-77-2	Accelerator MF	CAS 105-57-7	Delrin® 107, 507
CAS 102-77-2	Amax®, Amax No 1	CAS 105-57-7	Delrin® 150SA, 550SA
CAS 102-77-2	Delac MOR	CAS 105-57-7	Delrin® 570
CAS 102-77-2	OBTS	CAS 105-57-7	Delrin® 900
CAS 102-77-2	OMTS	CAS 105-57-7	Delrin® AF Blend
CAS 102-85-2	Syn-O-Ad® P-312	CAS 105-57-7	Electrafil® J-80/CF/10/TF/10
CAS 102-87-4	Armeen® 3-12	CAS 105-57-7	Thermocomp® KB-1008
CAS 103-23-1	Adimoll® DO	CAS 105-60-2	Nipro (i)
CAS 103-23-1	Good-rite® GP-223	CAS 105-74-8	Alperox-F
CAS 103-23-1	Jayflex® DOA	CAS 105-74-8	Laurox®
CAS 103-23-1	Kodaflex® DOA	CAS 105-74-8	Laurydol
CAS 103-23-1	Monoplex® DOA	CAS 105-76-0	Bisomer DBM
CAS 103-23-1	Palatinol® DOA	CAS 105-76-0	Octomer DBM
CAS 103-23-1	Plasthall® DOA	CAS 105-76-0	PX-504
CAS 103-23-1	Plastomoll® DOA	CAS 105-76-9	Bisomer DBF
CAS 103-23-1	Polycizer 332	CAS 105-95-3	Emeressence® 1150
CAS 103-23-1	PX-238	CAS 105-97-5	Uniflex® DCA
CAS 103-23-1	Uniflex® DOA	CAS 105-97-5	Unimate® DCA
CAS 103-23-1	Wickenol® 158	CAS 105-99-7	Adimoll® DB
CAS 103-23-1	Witcizer 412	CAS 105-99-7	Cetiol® B
CAS 103-24-2	Kodaflex® DOZ	CAS 106-01-4	Dermol 489.
CAS 103-24-2	Plasthall® DOZ	CAS 106-11-6	Alkamuls® SDG
CAS 103-24-2	Priplast 3018	CAS 106-11-6	Hydrine
CAS 103-34-4	Akrochem® Accelerator R	CAS 106-14-9	Ceroxin GL
CAS 103-34-4	Naugex SD-1	CAS 106-16-1	Alkamide® R-280
CAS 103-34-4	Sulfasan	CAS 106-16-1	Rewomid® R 280
CAS 103-34-4	Vanax® A	CAS 106-17-2	Flexricin® 15
CAS 103-44-6	Rapi-Cure EHVE	CAS 106-24-1	Meranol
CAS 103-49-1	Accelerator DBA	CAS 106-46-7	Paradow
CAS 103-49-1	Vulcaid 28	CAS 106-91-2	SR-379
CAS 103-90-2	Calpol	CAS 106-92-3	Ageflex AGE
CAS 103-90-2	Datril	CAS 106-92-3	Sipomer® AGE
CAS 103-90-2	Eneril	CAS 106-97-8	A-17
CAS 103-90-2	Febrilix	CAS 106-97-8	Bu-Gas
CAS 103-90-2	Nebs	CAS 107-21-1	Ilexan E
CAS 103-90-2	Paleva	CAS 107-22-2	Glyoxal 40%
CAS 103-90-2	Panadol	CAS 107-22-2	Protectol® GL 40
CAS 103-90-2	Panets	CAS 107-46-0	CH7310
CAS 103-90-2	Panok	CAS 107-46-0	H7310
CAS 103-90-2	Phenaphen	CAS 107-51-7	CO9816
CAS 103-90-2	Safapryn	CAS 107-51-7	O9816
CAS 103-90-2	Tapar	CAS 107-54-0	Surfynol® 61
CAS 103-90-2	Tempra	CAS 107-64-2	Adogen® TA-100
CAS 103-90-2	Ticelgesic	CAS 107-64-2	Adogen® TA-101

CAS 107-64-2	Arosurf® TA-100
CAS 107-64-2	Arquad® 218-75
CAS 107-64-2	Arquad® 218-100
CAS 107-64-2	Blandofen CT
CAS 107-64-2	Dehyquart DAM
CAS 107-64-2	Genamin DSAC
CAS 107-64-2	Prepagen WK
CAS 107-64-2	Sumquat® 6045
CAS 107-64-2	Varisoft® TA-100
CAS 107-71-1	Aztec® t-Butyl Peracetate-50 OMS
CAS 107-71-1	Aztec® t-Butyl Peracetate-60 OMS
CAS 107-71-1	Aztec® t-Butyl Peracetate-75 OMS
CAS 107-71-1	Esperox® 12MD
CAS 107-71-1	Lupersol 70
CAS 107-71-1	Trigonox® F-C50
CAS 107-72-2	CA0900
CAS 107-98-2	Arcosolv® PM
CAS 107-98-2	Dowanol® PM
CAS 107-98-2	Icinol PM
CAS 107-98-2	Poly-Solv® MPM
CAS 107-98-2	Solvenon® PM
CAS 107-98-2	Solvent PM
CAS 108-01-0	Tegoamin® DMEA
CAS 108-01-0	Texacat® DME
CAS 108-01-0	Toyocat®-DMA
CAS 108-03-2	NiPar S-10
CAS 108-05-4	Everflex® 81L
CAS 108-05-4	Plyamul® 40305-00
CAS 108-05-4	Unocal 76 Res 6206
CAS 108-05-4	Unocal 76 Res S-55
CAS 108-05-4	Vinnapas® A 50
CAS 108-05-4	Vinyl Acetate Monomer
CAS 108-32-7	Arconate® Propylene Carbonate
CAS 108-32-7	Solvenon® PC
CAS 108-32-7	Texacar® PC
CAS 108-46-3	Cohedur® RK
CAS 108-62-3	Farmon Mini Slug Pellets
CAS 108-62-3	Gastratox 6G Slug Pellets
CAS 108-62-3	Helarion
CAS 108-62-3	Mifaslug
CAS 108-62-3	Mini Slugit Pellets
CAS 108-62-3	PBI Slug Pellets
CAS 108-62-3	Quad Mini Slug Pellets
CAS 108-62-3	Slug Destroyer
CAS 108-62-3	Slug Pellets
CAS 108-62-3	Slugit Liquid
CAS 108-62-3	Slugoids
CAS 108-62-3	Tripart® Mini Slug Pellets
CAS 108-62-3	Unicrop 6% Mini Slug Pellets
CAS 108-62-3	Vassgro Mini Slug Pellets
CAS 108-65-6	Arcosolv® PMA
CAS 108-65-6	Dowanol® PMA
CAS 108-65-6	Ektasolve® PM Acetate
CAS 108-73-6	Phloroglucinol
CAS 108-80-15	Zeonet A
CAS 108-88-3	Black Out® Black
CAS 108-90-7	Abluton T30
CAS 108-94-1	Sextone
CAS 109-02-4	Texacat® NMM
CAS 109-13-7	Aztec® t-Butyl Peroxyisobutyrate-75 OMS
CAS 109-13-7	Lupersol 80
CAS 109-13-7	Trigonox® 41-C75
CAS 109-28-4	Chemidex O
CAS 109-28-4	Incromine OPB
CAS 109-28-4	Lexamine O-13
CAS 109-28-4	Mackine 501
CAS 109-28-4	Schercodine O
CAS 109-31-9	Priplast 3013
CAS 109-38-6	Dermol BES
CAS 109-38-6	KP-23
CAS 109-43-3	Kodaflex® DBS
CAS 109-43-3	Plasthall® DBS
CAS 109-43-3	Reomol DBS
CAS 109-43-3	Uniflex® DBS
CAS 109-46-6	Pennzone B 0685
CAS 109-86-4	Methyl Icinol
CAS 109-86-4	Poly-Solv® EM
CAS 109-99-9	Dynasolve 150
CAS 109-99-9	QO® Tetrahydrofuran (THF)
CAS 110-01-0	Pennodorant® 1013
CAS 110-05-4	Aztec® Di-t-Butyl Peroxide
CAS 110-05-4	Trigonox® B
CAS 110-25-8	Crodasinic O
CAS 110-25-8	Hamposyl® O
CAS 110-25-8	Vanseal® OS
CAS 110-27-0	Crodamol IPM
CAS 110-27-0	Emerest® 2314
CAS 110-27-0	Estergel
CAS 110-27-0	Kessco® IPM
CAS 110-27-0	Lexol IPM-NF
CAS 110-27-0	Liponate IPM
CAS 110-27-0	Promyr
CAS 110-27-0	Radia® 7190
CAS 110-27-0	Tegosoft® M
CAS 110-27-0	Unimate® IPM
CAS 110-27-0	Wickenol® 101
CAS 110-30-5	Abluwax EBS
CAS 110-30-5	Acrawax® C
CAS 110-30-5	Advawax® 290
CAS 110-30-5	Alkamide® STEDA
CAS 110-30-5	Armowax EBS
CAS 110-30-5	Glycowax® 765
CAS 110-30-5	Kemamide® W-39
CAS 110-30-5	Kemwax
CAS 110-30-5	Nopcowax 22-DS
CAS 110-30-5	Uniwax 1760
CAS 110-31-6	Advawax® 240
CAS 110-31-6	Kemamide® W-20
CAS 110-36-1	Bumyr
CAS 110-36-1	Wickenol® 141
CAS 110-44-1	Sorbistat
CAS 110-54-3	Exxsol® Hexane
CAS 110-63-4	BDO
CAS 110-63-4	Dabco® BDO
CAS 110-80-5	EE Solvent
CAS 110-80-5	Ethyl Icinol
CAS 110-80-5	Poly-Solv® EE
CAS 110-85-0	Antepar
CAS 110-90-0	Ethyl Di-Icinol
CAS 111-01-3	Cosbiol®

CAS 111-01-3	Dermane	CAS 112-00-5	Chemquat 12-50
CAS 111-01-3	Polysphere 3000 SP	CAS 112-00-5	Dehyquart LT
CAS 111-01-3	Robane®	CAS 112-00-5	Empigen® 5089
CAS 111-01-3	Romane	CAS 112-00-5	Octosol 562
CAS 111-02-4	Supraene®	CAS 112-00-5	Varisoft® LAC
CAS 111-03-5	Ablunol GMO	CAS 112-02-7	Ammonyx® CETAC, CETAC-30
CAS 111-03-5	Aldo® MO	CAS 112-02-7	Arquad® 16-29
CAS 111-03-5	Capmul® GMO	CAS 112-02-7	Arquad® 16-50
CAS 111-03-5	Kemester® 2000	CAS 112-02-7	Barquat® CT-29
CAS 111-03-5	Kessco® GMO	CAS 112-02-7	Carsoquat® CT-429
CAS 111-03-5	Mazol® 300 K	CAS 112-02-7	Chemquat 16-50
CAS 111-03-5	Mazol® GMO	CAS 112-02-7	Dehyquart A
CAS 111-03-5	Monomuls® 90-O18	CAS 112-02-7	Genamin CTAC
CAS 111-03-5	Witconol 2421	CAS 112-02-7	Incroquat CTC-30
CAS 111-05-7	Alkamide® OIP	CAS 112-02-7	Quatrex CTAC
CAS 111-05-7	Mackamide OP	CAS 112-02-7	Querton 16Cl-29
CAS 111-05-7	Rewomid® IPE 280	CAS 112-02-7	Radiaquat 6444
CAS 111-05-7	Schercomid OMI	CAS 112-02-7	Rhodaquat® M242C/29
CAS 111-05-7	Witcamide® 61	CAS 112-02-7	Varisoft® 250
CAS 111-15-9	EE Acetate	CAS 112-02-7	Varisoft® 300
CAS 111-15-9	Sensolve EEA	CAS 112-02-7	Varisoft® 355
CAS 111-27-3	Epal® 6	CAS 112-03-8	Arquad® 18-50
CAS 111-30-8	Glutarol	CAS 112-03-8	Octosol 474
CAS 111-30-8	Protectol® GDA, GT 50	CAS 112-03-8	Tomah Q-ST-50
CAS 111-30-8	Sonacide	CAS 112-03-8	Varisoft® ST-50
CAS 111-30-8	Ucarcide® 225	CAS 112-03-8	Varisoft® TSC
CAS 111-40-0	D.E.H. 20	CAS 112-04-9	CO9750
CAS 111-40-0	D.E.H. 52	CAS 112-05-0	Emery® 1202
CAS 111-57-9	Ablumide SME	CAS 112-07-2	Ektasolve® EB Acetate
CAS 111-57-9	Alkamide® S-280	CAS 112-10-7	Tegosoft® S
CAS 111-57-9	Amidex SME	CAS 112-10-7	Wickenol® 127
CAS 111-57-9	Mackamide SMA	CAS 112-18-5	Adma® 12
CAS 111-57-9	Mazamide® SMEA	CAS 112-18-5	Adma® 1214
CAS 111-57-9	Monamid® S	CAS 112-18-5	Armeen® DM12D
CAS 111-57-9	Rewomid® S 280	CAS 112-23-3	Texacure EA-24
CAS 111-57-9	Witcamide® 70	CAS 112-24-3	Texlin® 300
CAS 111-58-0	Schercomid OME	CAS 112-30-1	Epal® 10
CAS 111-60-4	Alkamuls® EGMS/C	CAS 112-34-5	Butyl Di-Icinol
CAS 111-60-4	Alkamuls® SEG	CAS 112-34-5	Dowanol® DB
CAS 111-60-4	Monthyle	CAS 112-34-5	Ektasolve® DB
CAS 111-60-4	Schercemol EGMS	CAS 112-39-0	Emery® 2216
CAS 111-66-0	Neodene® 8	CAS 112-39-0	Radia® 7120
CAS 111-66-0	Neodene® 810	CAS 112-41-4	Neodene® 6/12.
CAS 111-76-2	Butyl Icinol	CAS 112-41-4	Neodene® 12
CAS 111-76-2	Dowanol® EB	CAS 112-41-4	Neodene® 1012.
CAS 111-76-2	Ektasolve® EB	CAS 112-42-5	Neodol® 1
CAS 111-77-3	Dowanol® DM	CAS 112-42-5	Neoflex® 11
CAS 111-77-3	Ektasolve® DM	CAS 112-53-8	Emery® 3326
CAS 111-77-3	Methyl Di-Icinol	CAS 112-53-8	Emery® 3332
CAS 111-77-3	Poly-Solv® DM	CAS 112-53-8	Emery® 3357
CAS 111-86-4	Amine 8 D	CAS 112-53-8	Epal® 12
CAS 111-87-5	Emery® 3322	CAS 112-53-8	Lorol C12
CAS 111-87-5	Emery® 3324	CAS 112-53-8	Lorol C12-C14
CAS 111-87-5	Epal® 8	CAS 112-53-8	Lorol C8-C10 Special
CAS 111-87-5	Lorol C8	CAS 112-53-8	Lorol Special
CAS 111-90-0	Diethoxol	CAS 112-53-8	Philcohol 1200
CAS 111-90-0	Ektasolve® DE	CAS 112-57-2	Texlin® 400
CAS 111-90-0	Transcutol	CAS 112-61-8	Emery® 2218
CAS 111-96-6	Diglyme	CAS 112-61-8	Kemester® 4516
CAS 112-00-5	Adogen® 412	CAS 112-62-9	Emerest® 2301
CAS 112-00-5	Arquad® 12-37W	CAS 112-62-9	Emery® 2219
CAS 112-00-5	Arquad® 12-50	CAS 112-62-9	Emery® 2301

CAS 112-62-9	Kemester® 104	CAS 113-18-8	Placidyl
CAS 112-62-9	Priolube 1400	CAS 113-52-0	Efuranol
CAS 112-62-9	Witconol 2301	CAS 113-52-0	Imavate
CAS 112-69-6	Adma® 16	CAS 113-52-0	Janimine
CAS 112-69-6	Armeen® DM16D	CAS 113-52-0	Norpramine
CAS 112-69-6	Crodamine 3.A16D	CAS 113-52-0	Presamine
CAS 112-72-1	Cachalot® M-43	CAS 113-52-0	SK-Pramine
CAS 112-72-1	Dehydag Wax 14	CAS 113-52-0	Tofranil
CAS 112-72-1	Emery® 3334	CAS 113-59-7	Taractan
CAS 112-72-1	Epal® 14	CAS 113-98-4	Cilloral
CAS 112-72-1	Lanette 14	CAS 113-98-4	Liquapen
CAS 112-72-1	Lorol C14	CAS 113-98-4	Penisem
CAS 112-72-1	Philcohol 1400	CAS 113-98-4	Pentids
CAS 112-75-4	Adma® 14	CAS 113-98-4	Pfizerpen
CAS 112-75-4	Adma® 1214	CAS 113-98-4	SK-Penicillin G
CAS 112-75-4	Empigen® AH	CAS 113-98-4	Sugracillin
CAS 112-80-1	Emersol® 210	CAS 114-07-8	E-Mycin
CAS 112-80-1	Emersol® 6333 NF	CAS 114-07-8	Erthro
CAS 112-80-1	Emersol® 7021	CAS 114-07-8	Ery-Tab
CAS 112-80-1	Industrene® 104	CAS 114-07-8	ERYC
CAS 112-80-1	Pamolyn 100	CAS 114-07-8	EryDerm
CAS 112-80-1	Pamolyn 100 FG	CAS 114-07-8	Erygel®
CAS 112-80-1	Pamolyn 100 FGK	CAS 114-07-8	Erymax®
CAS 112-80-1	Pamolyn 125	CAS 114-07-8	Erythromid
CAS 112-80-1	Pamolyn® 125	CAS 114-07-8	Retcin
CAS 112-80-1	Priolene 6900	CAS 114-07-8	Robimycin
CAS 112-84-5	Armid® E	CAS 114-26-1	Baygon®
CAS 112-84-5	Crodamide E, ER	CAS 114-26-1	Blattanex®
CAS 112-84-5	Kemamide® E	CAS 114-26-1	Blattanex® 20
CAS 112-84-5	Petrac® Eramide®	CAS 114-26-1	Fumite Propoxur
CAS 112-84-5	Polydis® TR 131	CAS 114-26-1	Suncide®
CAS 112-84-5	Unislip 1753	CAS 114-26-1	Unden®
CAS 112-85-6	Hystrene® 5522	CAS 114-80-7	Prostigmin
CAS 112-85-6	Hystrene® 9022	CAS 115-10-6	Dymel®
CAS 112-85-6	Prifrac 2989	CAS 115-21-9	CE6350
CAS 112-85-6	Univol U344	CAS 115-29-7	Thiodan
CAS 112-86-7	Prifrac 2990	CAS 115-29-7	Thionex
CAS 112-86-7	Univol U342	CAS 115-32-2	Acarin
CAS 112-88-9	Neodene® 18	CAS 115-32-2	Fumite Dicofol
CAS 112-90-3	Amine OL	CAS 115-32-2	Kelthane
CAS 112-90-3	Armeen® OL	CAS 115-32-2	Mitigan
CAS 112-90-3	Armeen® OLD	CAS 115-33-3	Lavema
CAS 112-90-3	Crodamine 1.O, 1.OD	CAS 115-38-8	Mebaral
CAS 112-90-3	Jet Amine PO	CAS 115-44-6	Lotusate
CAS 112-90-3	Kemamine® P-989D	CAS 115-67-3	Paradione
CAS 112-90-3	Radiamine 6172	CAS 115-69-5	AMPD
CAS 112-92-5	Adol® 62 NF	CAS 115-70-8	AEPD®
CAS 112-92-5	Cachalot® S-56	CAS 115-70-8	AEPD®-85
CAS 112-92-5	Crodacol S70	CAS 115-77-5	Hercules® Improved Tech. PE
CAS 112-92-5	Crodacol S95NF	CAS 115-77-5	Hercules® Mono-PE
CAS 112-92-5	Dehydag Wax 18	CAS 115-83-3	Liponate PS-4
CAS 112-92-5	Emery® 3343	CAS 115-86-6	Disflamoll® TP
CAS 112-92-5	Epal® 18NF	CAS 115-86-6	Fyrol® PBR.
CAS 112-92-5	Fancol SA	CAS 115-86-6	Kronitex® TPP
CAS 112-92-5	Lanette 18 DEO	CAS 115-95-7	Phantene
CAS 112-92-5	Lipocol S	CAS 116-14-3	MS-122
CAS 112-92-5	Lorol C18	CAS 116-25-6	Dantoin® MDMH
CAS 112-92-5	Loxiol VPG 1354	CAS 116-29-0	Tedion V-18
CAS 112-92-5	Philcohol 1800	CAS 116-38-1	Tensilon
CAS 112-92-5	Rita SA	CAS 116-43-8	Sulfasuxidine
CAS 112-92-5	Stearal	CAS 117-18-0	Ashlade TCNB
CAS 112-99-2	Armeen® 2-18	CAS 117-18-0	Bygran S

CAS Number	Trade Name
CAS 117-18-0	Fumite TCNB
CAS 117-18-0	Fumite TCNB Smoke
CAS 117-18-0	Fusarex
CAS 117-18-0	Hickstor
CAS 117-18-0	Hortag Tecnazene Plus
CAS 117-18-0	Hystor 10
CAS 117-18-0	Hytec
CAS 117-18-0	Hytec Super
CAS 117-18-0	Nebulin
CAS 117-18-0	New Hickstor 6
CAS 117-18-0	New Hystor
CAS 117-18-0	Quad-Keep
CAS 117-18-0	Quad Store
CAS 117-18-0	Storaid Dust
CAS 117-18-0	Storite SS
CAS 117-18-0	Tecgran
CAS 117-18-0	Tripart® Arena 6
CAS 117-18-0	Tripart® Arena 6 + TBZ
CAS 117-18-0	Tripart® Arena Granules
CAS 117-18-0	Tripart® Arena Plus
CAS 117-18-0	Tubodust
CAS 117-18-0	Tubostore
CAS 117-39-5	Ritacetin
CAS 117-81-7	Bisoflex 81
CAS 117-81-7	Bisoflex 82
CAS 117-81-7	Kodaflex® DOP
CAS 117-81-7	Octoil
CAS 117-81-7	Palatinol® DOP
CAS 117-81-7	Plasthall® DOP
CAS 117-81-7	Plasticizer 28P
CAS 117-81-7	Polycizer 162
CAS 117-81-7	PX-138
CAS 117-81-7	Vestinol AH
CAS 117-81-7	Witcizer 312
CAS 117-82-8	Kodaflex® DMEP
CAS 117-82-8	Reomol P
CAS 117-83-9	Plasthall® 200
CAS 117-89-5	Terfluzin
CAS 118-52-5	Dantoin® DCDMH
CAS 118-52-5	Hydan
CAS 118-56-9	Filtrosol A
CAS 118-56-9	Heliophan
CAS 118-56-9	Kemester® HMS
CAS 118-60-5	Dermoblock OS
CAS 118-60-5	Escalol® 587
CAS 118-60-5	Neo Heliopan, Type OS
CAS 118-60-5	Uvinul® O-18
CAS 118-71-8	Veltol
CAS 118-79-6	FR-613
CAS 118-79-6	Great Lakes PH-73
CAS 118-92-3	AA
CAS 119-07-3	Good-rite® GP-265
CAS 119-07-3	Morflex 125
CAS 119-07-3	Morflex 175
CAS 119-47-1	Anti-oxidant 2246
CAS 119-47-1	Cyanox® 2246
CAS 119-47-1	Lowinox® 22M46
CAS 119-47-1	Ralox® 46
CAS 119-47-1	Ultranox® 246
CAS 119-47-1	Vulkanox® BKF
CAS 119-64-2	Tetralin
CAS 120-12-7	Sterilite Hop Defoliant
CAS 120-18-3	Dehscofix 918
CAS 120-32-1	Nipacide® BCP
CAS 120-36-5	Campbell's Redipon
CAS 120-36-5	MSS 2,4-DP
CAS 120-36-5	Polymone
CAS 120-40-1	Ablumide LDE
CAS 120-40-1	Alkamide® 327
CAS 120-40-1	Alkamide® LE
CAS 120-40-1	Amidex L-9
CAS 120-40-1	Amidex LD
CAS 120-40-1	Carsamide® SAL-7
CAS 120-40-1	Comperlan LD
CAS 120-40-1	Crillon LDE
CAS 120-40-1	Emalex NN-7
CAS 120-40-1	Emid® 6519
CAS 120-40-1	Empilan® LDE
CAS 120-40-1	Hartamide LDA
CAS 120-40-1	Incromide L-90
CAS 120-40-1	Incromide LLT
CAS 120-40-1	Incromide LM-70
CAS 120-40-1	Mackamide L10
CAS 120-40-1	Mazamide® 1214
CAS 120-40-1	Mazamide® L-298
CAS 120-40-1	Monamid® 150-LMWC
CAS 120-40-1	Ninol® 30-LL
CAS 120-40-1	Ninol® AX
CAS 120-40-1	Schercomid 1214
CAS 120-40-1	Schercomid SL-Extra
CAS 120-40-1	Standamid® KDS
CAS 120-40-1	Varamide® ML-1
CAS 120-40-1	Witcamide® 5195
CAS 120-47-8	Nipagin A
CAS 120-51-4	Ascabiol
CAS 120-51-4	Benylate
CAS 120-54-7	DPTT
CAS 120-54-7	Perkacit® DPTT
CAS 120-54-7	Sulfads®
CAS 120-55-8	Benzoflex 2-45
CAS 120-78-5	Altax®
CAS 120-78-5	MBTS
CAS 120-78-5	Perkacit® MBTS
CAS 120-78-5	Vulkacit® DM/C
CAS 120-97-8	Daranide
CAS 120-97-8	Oratrol
CAS 121-25-5	Amprol
CAS 121-32-4	Ethavan
CAS 121-32-4	Rhodiarome
CAS 121-32-4	Vanbeenol
CAS 121-33-5	Zimco
CAS 121-54-0	Hyamine® 1622 50%
CAS 121-54-0	Hyamine® 1622 Crystals
CAS 121-54-0	Phemerol Chloride
CAS 121-59-5	Carb-O-Sep
CAS 121-66-4	Entramin
CAS 121-75-5	Malathion 60
CAS 121-75-5	Prioderm
CAS 121-79-9	Progallin P
CAS 121-79-9	Sustane® PG
CAS 122-03-2	Cumal
CAS 122-09-8	Duromine
CAS 122-09-8	Ionamin
CAS 122-11-2	Madribon

CAS 122-14-5	Dicofen
CAS 122-14-5	Fenitex
CAS 122-14-5	Folithion®
CAS 122-14-5	Micromite
CAS 122-18-9	Sumquat® 6050
CAS 122-18-9	Zettyn
CAS 122-19-0	Ablumine 280
CAS 122-19-0	Ammonyx® 4, 4B, 485, 4002
CAS 122-19-0	Amyx A-25-S 0040
CAS 122-19-0	Carsoquat® SDQ-25
CAS 122-19-0	Catinal OB-80E
CAS 122-19-0	Cycloton® SCS
CAS 122-19-0	Emcol® 4
CAS 122-19-0	Hetquat S-20
CAS 122-19-0	Incroquat SDQ-25
CAS 122-19-0	M-Quat® B-25
CAS 122-19-0	Mackernium SDC-25
CAS 122-19-0	Miracare® SCS
CAS 122-19-0	Quatrex STC-25
CAS 122-19-0	Rhodaquat® M270C/18
CAS 122-19-0	Stedbac®
CAS 122-19-0	Sumquat® 6210
CAS 122-19-0	Varisoft® SDAC-W
CAS 122-32-7	Emerest® 2423
CAS 122-32-7	Kemester® 1000
CAS 122-32-7	Priolube 1435
CAS 122-34-9	Boroflow
CAS 122-34-9	Boroflow S/ATA
CAS 122-34-9	CDA Simflow Plus
CAS 122-34-9	Clearway
CAS 122-34-9	Gesatop
CAS 122-34-9	Herbazin 50
CAS 122-34-9	Herbazin Plus SC
CAS 122-34-9	Mascot Highway
CAS 122-34-9	MSS Simazine/ Aminotriazole 43FL
CAS 122-34-9	Primatol SE 500FW
CAS 122-34-9	Remtal
CAS 122-34-9	Simanex
CAS 122-34-9	Simapron
CAS 122-34-9	Simflow
CAS 122-34-9	Simflow Plus
CAS 122-34-9	Synchemicals Total Weed Killer
CAS 122-34-9	Syngran
CAS 122-34-9	Syntox Total Weed Killer
CAS 122-34-9	Weedex S2
CAS 122-42-9	MSS IPC 50
CAS 122-42-9	Propham
CAS 122-99-6	Emeressence® 1160 Rose Ether
CAS 122-99-6	Igepal® Cephene Distilled
CAS 122-99-6	Igepal® OD-410
CAS 122-99-6	Phenoxetol
CAS 122-99-6	Prox-onic PH-01
CAS 122-99-6	Rewopal® MPG 10
CAS 123-03-5	Acetoquat CPC
CAS 123-03-5	Merocets
CAS 123-28-4	Argus DLTDP
CAS 123-28-4	Carstab® DLTDP
CAS 123-28-4	Cyanox® LTDP
CAS 123-28-4	Evanstab® 12
CAS 123-28-4	Lankromark® DLTDP
CAS 123-28-4	Lowinox® DLTDP

CAS 123-28-4	Wytox® LT
CAS 123-31-9	Black and White Bleaching Cream
CAS 123-31-9	Eldopaque
CAS 123-31-9	Eldoquin
CAS 123-31-9	SR-201
CAS 123-31-9	Tecquinol® Tech. Grade
CAS 123-33-1	Bos MH
CAS 123-33-1	Burtolin
CAS 123-33-1	Chiltern Fazor
CAS 123-33-1	Mazide 25
CAS 123-33-1	MSS MH18
CAS 123-47-7	Endojodin®
CAS 123-63-7	Paral
CAS 123-77-3	Celogen® AZ 120, 130, 150, 180, 199
CAS 123-77-3	Ficel® AC2
CAS 123-77-3	Genitron
CAS 123-77-3	Kempore® 60/14FF
CAS 123-77-3	Porofor® ADC/E
CAS 123-77-3	Santechem 21-21
CAS 123-95-5	ADK STAB LS-8
CAS 123-95-5	Butyl Stearate C-895
CAS 123-95-5	Emerest® 2325
CAS 123-95-5	Kemester® 5510
CAS 123-95-5	Kessco® BS
CAS 123-95-5	Lexolube® BS-Tech
CAS 123-95-5	Priolube 1451
CAS 123-95-5	Uniflex® BYS-Tech
CAS 123-95-5	Unimate® BYS
CAS 123-95-5	Witconol 2326
CAS 123-99-9	Emerox® 1110
CAS 124-03-8	Bretol®
CAS 124-03-8	Sumquat® 6020
CAS 124-07-2	Emery® 657
CAS 124-07-2	Prifrac 2901
CAS 124-07-2	Univol U308
CAS 124-10-7	Emery® 2214
CAS 124-22-1	Armeen® 12
CAS 124-22-1	Armeen® 12D
CAS 124-22-1	Radiamine 6164
CAS 124-26-5	Crodamide S, SR
CAS 124-26-5	Kemamide® S
CAS 124-26-5	Petrac® Vyn-Eze®
CAS 124-26-5	Uniwax 1750
CAS 124-28-7	Adma® 18
CAS 124-28-7	Adogen® MA-108 SF
CAS 124-28-7	Amine 2M18D
CAS 124-28-7	Armeen® DM18D
CAS 124-28-7	Barlene® 18S
CAS 124-28-7	Crodamine 3.A18D
CAS 124-28-7	Kemamine® T-9902
CAS 124-30-1	Amine 18-90
CAS 124-30-1	Armeen® 18
CAS 124-30-1	Armeen® 18D
CAS 124-30-1	Armid® HTD
CAS 124-30-1	Crodamine 1.18D
CAS 124-30-1	Kemamine® P-990, P-990D
CAS 124-30-1	Steamfilm FG
CAS 124-38-9	Cardice
CAS 124-38-9	Dricold
CAS 124-38-9	Drikold

CAS 124-43-6	Exterol Ear Drops
CAS 124-68-5	AMP
CAS 124-68-5	AMP-95
CAS 124-68-5	AVT-75
CAS 124-94-7	Aristocort
CAS 124-94-7	Kenacort
CAS 124-94-7	Ledercort
CAS 124-94-7	SK-Triamcinolone
CAS 125-02-0	Codelsol®
CAS 125-02-0	Colicort
CAS 125-02-0	Hydeltrasol
CAS 125-02-0	Inflamase® Forte
	1% Ophthalmic Solution
CAS 125-02-0	Inflrnase® Mild
	1/8% Ophthalmic Solution
CAS 125-02-0	Metreton
CAS 125-02-0	Solu-Predalone
CAS 125-02-0	Solucort
CAS 125-04-2	A-hydroCort
CAS 125-04-2	Solu-Cortef
CAS 125-33-7	Mysoline
CAS 125-33-7	Mysoline
CAS 125-53-1	Daricol
CAS 125-53-1	Daricon
CAS 125-84-8	Cytadren
CAS 126-06-7	Dantoin® BCDMH
CAS 126-06-7	Dantoin® GSD-550
CAS 126-06-7	Halobrom
CAS 126-06-7	Quesbrom
CAS 126-07-8	Fulcin
CAS 126-07-8	Fulvicin-P/G
CAS 126-07-8	Fulvicin-U/F
CAS 126-07-8	Grifulvin V
CAS 126-07-8	Gris-PEG®
CAS 126-07-8	Grisactin
CAS 126-07-8	Grisovin
CAS 126-11-4	S.S.T.® Sump Saver Tablets
CAS 126-11-4	Tris Nitro®
CAS 126-11-4	Tris Nitro® , 50% Aq
CAS 126-12-5	Alidine
CAS 126-14-7	Soa
CAS 126-30-7	NPG® Glycol
CAS 126-31-8	Skiodan Sodium
CAS 126-52-3	Valmid
CAS 126-52-3	Valmidate
CAS 126-73-8	Antifoam T
CAS 126-73-8	Kronitex® TBP
CAS 126-73-8	Phos-Ad 100
CAS 126-73-8	Syn-O-Ad® 8412
CAS 126-80-7	CB2405
CAS 126-86-3	Surfynol® 104
CAS 126-86-3	Surfynol® 104A
CAS 126-86-3	Surfynol® 104BC
CAS 126-86-3	Surfynol® 104E
CAS 126-86-3	Surfynol® 104H
CAS 126-86-3	Surfynol® 104PA
CAS 126-86-3	Surfynol® 104PG
CAS 126-86-3	Surfynol® 104S
CAS 126-86-3	Surfynol® DF-110D
CAS 126-86-3	Surfynol® DF-110L
CAS 126-86-3	Surfynol® PC
CAS 126-86-3	Surfynol® PG-50

CAS 126-86-3	Surfynol® TG
CAS 126-86-3	Surfynol® TG-E
CAS 126-92-1	Niaproof® Anionic Surfactant 08
CAS 126-92-1	Rhodapon® BOS
CAS 126-99-8	Baypren® 110
CAS 126-99-8	Baypren® 110 VSC
CAS 126-99-8	Baypren® 216
CAS 126-99-8	Baypren® 310
CAS 126 99-8	Baypren® AT-H, AT-M, AT-S
CAS 126-99-8	Baypren® EM1
CAS 126-99-8	Baypren® Latex KA 8348
CAS 126-99-8	Baypren® Latex L 200Λ
CAS 126-99-8	Baypren® M1
CAS 126-99-8	Daubond DC-9300
CAS 126-99-8	Neoprene Latex 115
CAS 126-99-8	Neoprene NPG 6856
CAS 127-09-3	Niacet Sodium Acetate Anhyd. Tech
CAS 127-18-4	Distillex DS4
CAS 127-18-4	Dowper
CAS 127-18-4	Nema
CAS 127-18-4	Perclene
CAS 127-18-4	Perclene TG
CAS 127-18-4	Perklone
CAS 127-25-3	Hercolyn®
CAS 127-39-9	Aerosol® IB-45
CAS 127-39-9	Bevaloid 6423
CAS 127-39-9	Geropon® CYA/DEP
CAS 127-39-9	Monawet MB-45
CAS 127-39-9	Octosol IB-45
CAS 127-39-9	Rewopol® SBDB 45
CAS 127-48-0	Tridione
CAS 127-65-1	Tochlorine
CAS 127-68-4	Ludigol F
CAS 127-69-5	SK-Soxazole
CAS 127-69-5	Soxomide
CAS 127-69-5	Sulfalar
CAS 128-03-0	Aquatreat KM
CAS 128-12-1	Promacetin
CAS 128-37-0	Anti-Oxydant Bayer
CAS 128-37-0	Deenax
CAS 128-37-0	Lowinox® BHT
CAS 128-37-0	Nipanox® BHT
CAS 128-37-0	Ralox® BHT food grade
CAS 128-37-0	Spectratech® CM 11340, KM 11264
CAS 128-37-0	Vanox® PCX
CAS 128-37-0	Vulkanox® KB
CAS 128-39-2	Isonox® 103
CAS 128-39-2	Lowinox® 001
CAS 128-49-4	Surfak
CAS 128-62-1	Coscopin
CAS 128-62-1	Coscopin Paediatric
CAS 128-62-1	Narcotine
CAS 129-03-3	Heptuss
CAS 129-06-6	Coumadin
CAS 129-06-6	Marevan
CAS 129-06-6	Panwarfin
CAS 129-06-6	Warfarin
CAS 129-20-4	Oxalid
CAS 129-20-4	Tandacote
CAS 129-20-4	Tandearil

CAS 129-20-4	Tanderil
CAS 129-49-7	Sansert
CAS 129-51-1	Ergotrate Maleate
CAS 129-77-1	Dactil
CAS 130-61-0	Mellaril
CAS 130-61-0	SK-Thioridazine HC1
CAS 130-95-0	Solvochin
CAS 131-11-3	Kemester® DMP
CAS 131-11-3	Kodaflex® DMP
CAS 131-11-3	Palatinol® M
CAS 131-11-3	Unimoll® DM
CAS 131-11-3	Uniplex 110
CAS 131-13-5	Synkayvite
CAS 131-17-9	Dapon 35
CAS 131-17-9	Daponite Sheet
CAS 131-17-9	Nonflammable Decobest DA
CAS 131-17-9	RX® 1-501N
CAS 131-17-9	RX® 3-1-530
CAS 131-53-3	Cyasorb® UV 24
CAS 131-54-4	Uvinul® D-49
CAS 131-55-5	Uvinul® D-50
CAS 131-56-6	Uvinul® 400
CAS 131-57-7	Cyasorb® UV 9
CAS 131-57-7	Escalol® 567
CAS 131-57-7	Lankromark® LE296
CAS 131-57-7	Neo Heliopan, Type BB
CAS 131-57-7	Rhodialux A
CAS 131-57-7	Ritaphenone 3
CAS 131-57-7	Spectra-Sorb UV 9
CAS 131-57-7	Syntase® 62
CAS 131-57-7	Uvinul® M-40
CAS 132-17-2	Cogentin
CAS 132-18-3	Diafen
CAS 132-18-3	Hispril
CAS 132-18-3	Histryl Spansule Capsule
CAS 132-18-3	Lergoban
CAS 132-20-7	Daneral-SA
CAS 132-49-0	Magisal
CAS 132-69-4	Difflam Cream
CAS 132-69-4	Difflam Oral Rinse
CAS 132-92-3	Azapen
CAS 132-92-3	Celbenin
CAS 132-93-4	Broxli
CAS 132-98-9	Betapen-VK
CAS 132-98-9	Fenocin
CAS 132-98-9	Fenocin Forte
CAS 132-98-9	Ledercillin VK
CAS 132-98-9	Pen-Vee K
CAS 132-98-9	Penapar VK
CAS 132-98-9	Pfizerpen Vk
CAS 132-98-9	Robicillin Vk
CAS 132-98-9	SK-Penicillin VK
CAS 132-98-9	Uticillin VK
CAS 132-98-9	V-Cillin K
CAS 132-98-9	Veetids
CAS 133-06-2	Captan-Col
CAS 133-06-2	Captan Granular
CAS 133-06-2	Merpan
CAS 133-06-2	Orthocide
CAS 133-06-2	PP Captan 83
CAS 133-06-2	Vancide® 89
CAS 133-07-3	Folpan
CAS 133-11-9	Pheny-PAS-Tebamin
CAS 133-17-5	Hippuran I 131
CAS 133-17-5	Hipputope
CAS 133-17-5	Nephroflow
CAS 133-32-4	Chryzoplus
CAS 133-32-4	Chryzopon
CAS 133-32-4	Chryzosan
CAS 133-32-4	Chryzotek
CAS 133-32-4	Seradix
CAS 133-49-3	Akrochem® Peptizer PTP
CAS 133-49-3	Renacit® 7
CAS 133-53-9	Nipacide® DX
CAS 133-53-9	Nipacide® PX
CAS 133-67-5	Metahydrin
CAS 133-67-5	Naqua
CAS 134-03-2	Ascorbin
CAS 134-03-2	Cevalin
CAS 134-03-2	Liqui-Cee
CAS 134-09-8	Dermoblock MA
CAS 134-09-8	Neo Heliopan, Type MA
CAS 134-31-6	Cryptonol
CAS 134-50-9	Monacrin
CAS 134-72-5	Spaneph Spansules
CAS 134-80-5	Tenuate
CAS 134-80-5	Tenuate Dopan
CAS 134-80-5	Tenvate
CAS 134-80-5	Tepanil™ Tablets
CAS 134-80-5	Tepanil™ Ten-tab Tablets
CAS 135-07-9	Enduron
CAS 135-09-1	Diucardin
CAS 135-09-1	Hydrenox-M
CAS 135-09-1	Naclex
CAS 135-09-1	Saluron
CAS 135-31-9	Pyronil
CAS 136-23-2	Accelerator BZ Powder
CAS 136-23-2	Butasan Vulcanization Accelerator
CAS 136-23-2	Butazate
CAS 136-23-2	Butazate 50D
CAS 136-23-2	Butyl Zimate®
CAS 136-23-2	Octocure ZDB-50
CAS 136-23-2	Perkacit® ZDBC
CAS 136-26-5	Amidex CP
CAS 136-26-5	Monamid® 150-CW
CAS 136-26-5	Standamid® CD
CAS 136-30-1	Butyl Namate®
CAS 136-30-1	Octopol NB-47
CAS 136-36-7	Eastman® Inhibitor RMB
CAS 136-40-3	Azo-Standard
CAS 136-47-0	Anestan
CAS 136-47-0	Anethaine
CAS 136-47-0	Pontocaine Hydrochloride.
CAS 136-53-8	Octoate Z
CAS 136-60-7	Hipochem B-3-M
CAS 136-60-7	Marvanol® Carrier BB
CAS 136-77-6	Crystoids
CAS 136-85-6	Retrocure® G
CAS 136-85-6	Vulkalent® TM
CAS 136-99-2	Hodag C-100-L
CAS 136-99-2	Mackazoline L
CAS 136-99-2	Schercozoline L
CAS 137-08-6	Pantholin

CAS 137-08-6	Ritapan CAP
CAS 137-16-6	Closyl LA 3584
CAS 137-16-6	Crodasinic LS30
CAS 137-16-6	Crodasinic LS35
CAS 137-16-6	Hamposyl L-30
CAS 137-16-6	Maprosyl® 30
CAS 137-16-6	Medialan LD
CAS 137-16-6	Sarkosyl NL30
CAS 137-16-6	Secosyl
CAS 137-16-6	Tensyl 30
CAS 137-16-6	Vanseal® NALS-30
CAS 137-16-6	Zoharsyl L-30
CAS 137-20-2	Adinol OT
CAS 137-20-2	Arkopon T
CAS 137-20-2	Fenopon T-33 and T-43
CAS 137-20-2	Fenopon T-51
CAS 137-20-2	Fenopon T-77
CAS 137-20-2	Geropon® T-22/A
CAS 137-20-2	Geropon® T-33
CAS 137-20-2	Hostapon SO
CAS 137-20-2	Hostapon T Powd
CAS 137-26-8	Agrichem Flowable Thiram
CAS 137-26-8	Akrochem® TMTD
CAS 137-26-8	Ancazide ME
CAS 137-26-8	Ascot.
CAS 137-26-8	FS Thiram 15% Dust
CAS 137-26-8	Hortag Thiram
CAS 137-26-8	Hy-TL.
CAS 137-26-8	Hy-Vic.
CAS 137-26-8	Methyl Tuads®
CAS 137-26-8	Perkacit® TMTD
CAS 137-26-8	Pomarsol®
CAS 137-26-8	Rezifilm
CAS 137-26-8	Tuex
CAS 137-26-8	Unicrop Thianosan
CAS 137-26-8	Vulkacit® Thiuram
CAS 137-29-1	Akrochem® Cu.D.D
CAS 137-29-1	Cumate®
CAS 137-29-1	Methyl Cumate®
CAS 137-29-1	Perkacit® CDMC
CAS 137-30-4	AAprotect
CAS 137-30-4	Accelerator MZ Powder
CAS 137-30-4	Ancazate ME
CAS 137-30-4	Methasan
CAS 137-30-4	Methazate
CAS 137-30-4	Methyl Zimate®
CAS 137-30-4	Octocure ZDM-50
CAS 137-30-4	Perkacit® ZDMC
CAS 137-30-4	Ultra Zinc DMC
CAS 137-30-4	Vancide® MZ-96
CAS 137-40-6	Luprosil® Sodium Salt
CAS 137-40-6	Spac
CAS 137-42-8	Sistan®
CAS 137-42-8	Unifume
CAS 137-58-6	Anestacon
CAS 137-58-6	Lida-Mantle
CAS 137-58-6	Xylocaine
CAS 138-84-1	Potaba
CAS 138-86-3	Achilles Dipentene
CAS 138-89-6	Vulcaniline
CAS 139-07-1	Catinal CB-50
CAS 139-07-1	Dehyquart LDB
CAS 139-07-1	Retarder N
CAS 139-07-1	Vantoc CL
CAS 139-08-2	BTC® 824
CAS 139-08-2	BTC® 2565
CAS 139-08-2	Catigene® DC 100
CAS 139-08-2	Exameen 824 3724
CAS 139-08-2	FMB 65-15 Quat, 65-28 Quat
CAS 139-08-2	JAQ Powdered Quat
CAS 139-13-9	Hampshire® NTA Acid
CAS 139-33-3	Endrate
CAS 139-33-3	Sequestrene® NA2
CAS 139-33-3	Sequestrene® NA2 Edetate USP
CAS 139-33-3	Sodium Versenate
CAS 139-40-2	Prozinex
CAS 139-41-3	Hampshire® DEG
CAS 139-44-6	Rheocin
CAS 139-88-8	Niaproof® Anionic Surfactant 4
CAS 139-89-9	Cheelox® 120
CAS 139-89-9	Cheelox® HE-24
CAS 139-89-9	Chel DM-41
CAS 139-89-9	Emkasene 800
CAS 139-89-9	Hamp-Ol® 120
CAS 139-89-9	Hamp-Ol® Crystals
CAS 139-89-9	Kalex OH
CAS 139-89-9	Questal FEC 0800
CAS 139-89-9	Trilon® D Liq
CAS 139-96-8	Akyposal TLS 42
CAS 139-96-8	Avirol® T 40
CAS 139-96-8	Berol 480
CAS 139-96-8	Carsonol® TLS
CAS 139-96-8	DeSonol T
CAS 139-96-8	Elfan® 240T and 240T/S
CAS 139-96-8	Empicol® TL40
CAS 139-96-8	Lorol TA and TAR
CAS 139-96-8	Maprofix TLS
CAS 139-96-8	Marlinat® DFL 40
CAS 139-96-8	Neopon LT
CAS 139-96-8	Norfox® TLS
CAS 139-96-8	Nutrapon TLS-500
CAS 139-96-8	Perlankrol® ATL40
CAS 139-96-8	Rewopol® TLS 40
CAS 139-96-8	Rhodapon® LT-6
CAS 139-96-8	Standapol® T
CAS 139-96-8	Stepanol® SPT
CAS 139-96-8	Sulfetal KT 400
CAS 139-96-8	Sulfochem TLS
CAS 139-96-8	Tensopol LT
CAS 139-96-8	Texapon T 42
CAS 139-96-8	Ufarol TA-40
CAS 139-96-8	Witcolate 6434
CAS 139-96-8	Witcolate TLS-500
CAS 139-96-8	Zoharpon LAT
CAS 140-01-2	Cheelox® 80
CAS 140-01-2	Chel DTPA-41
CAS 140-01-2	Hamp-Ex® 80
CAS 140-01-2	Kalex Penta
CAS 140-01-2	Pentaquest Extra 0685
CAS 140-01-2	Pentaquest OPNA 0256
CAS 140-01-2	Polyquest 80
CAS 140-01-2	Trilon® C Liq
CAS 140-03-4	Flexricin® P-4
CAS 140-03-4	Naturechem® MAR

CAS 140-11-4	Benteine
CAS 140-56-7	Bayer 5072
CAS 140-63-6	Brolene
CAS 140-72-7	Acetoquat CPB
CAS 140-87-0	PTAL
CAS 140-93-2	Aero 343 Xanthate
CAS 141-01-5	C-Ron
CAS 141-01-5	C-Ron Forte
CAS 141-01-5	Feostat
CAS 141-01-5	Fersaday
CAS 141-01-5	Fersamal
CAS 141-01-5	Norfer
CAS 141-01-5	Toleron
CAS 141-04-8	Plasthall® DIBA
CAS 141-08-2	Aldo® MR
CAS 141-08-2	Flexricin® 13
CAS 141-08-2	Hodag GMR
CAS 141-08-2	Mazol® GMR
CAS 141-08-2	Softigen® 701
CAS 141-21-9	Avistin® PN
CAS 141-21-9	Chemical 39 Base
CAS 141-21-9	Marlamid® A 18
CAS 141-22-0	Flexricin® 100
CAS 141-22-0	P®-10 Acid
CAS 141-23-1	Cenwax® ME
CAS 141-23-1	Naturechem® MHS
CAS 141-23-1	Paricin® 1
CAS 141-24-2	Flexricin® P-1
CAS 141-53-7	Formax
CAS 141-57-1	CP0800
CAS 141-62-8	CD3780
CAS 141-62-8	D3780
CAS 141-94-6	Hexetidine
CAS 141-94-6	Oraldene
CAS 142-18-7	Ablunol GML
CAS 142-18-7	Aldo® MLD
CAS 142-18-7	Grindtek ML 90
CAS 142-18-7	Hodag GML
CAS 142-18-7	Imwitor® 312
CAS 142-18-7	Kessco® GML
CAS 142-18-7	Monomuls® 90-L12
CAS 142-26-7	Acetamide MEA
CAS 142-26-7	Amidex AME
CAS 142-26-7	Carsamide® AMEA
CAS 142-26-7	Incromectant AMEA-100
CAS 142-26-7	Lipamide MEAA
CAS 142-26-7	Mackamide AME-75, AME-100
CAS 142-26-7	Schercomid AME
CAS 142-31-4	Rhodapon® OLS
CAS 142-48-3	Hamposyl® S
CAS 142-54-1	Alkamide® LIPA/C
CAS 142-54-1	Amidex LIPA
CAS 142-54-1	Empilan® LIS
CAS 142-54-1	Monamid® LIPA
CAS 142-54-1	Rewomid® IPL 203
CAS 142-59-6	Campbell's Nabam Soil Fungicide
CAS 142-77-8	Butyl Oleate C-914
CAS 142-77-8	Emerest® 2328
CAS 142-77-8	Kemester® 4000
CAS 142-77-8	Plasthall® 503
CAS 142-77-8	Priolube 1405
CAS 142-77-8	Uniflex® BYO
CAS 142-77-8	Witcizer 100
CAS 142-78-9	Ablumide LME
CAS 142-78-9	Alkamide® L-203
CAS 142-78-9	Amidex LMMEA
CAS 142-78-9	Crillon LME
CAS 142-78-9	Empilan® LME
CAS 142-78-9	Hartamide LMEA
CAS 142-78-9	Incromide LCL
CAS 142-78-9	Mackamide LMM
CAS 142-78-9	Monamid® LMA
CAS 142-78-9	Ninol® LMP
CAS 142-78-9	Rewomid® L 203
CAS 142-82-5	Exxsol® Heptane
CAS 142-87-0	Akyporox SAL SAS
CAS 142-88-1	Coopane
CAS 142-88-1	Entacyl
CAS 142-90-5	Ageflex FM-12
CAS 142-91-6	Emerest® 2316
CAS 142-91-6	Kessco® IPP
CAS 142-91-6	Lexol IPP
CAS 142-91-6	Liponate IPP
CAS 142-91-6	Propal
CAS 142-91-6	Radia® 7200
CAS 142-91-6	Tegosoft® P
CAS 142-91-6	Unimate® IPP
CAS 142-91-6	Wickenol® 111
CAS 143-07-7	Emery® 650
CAS 143-07-7	Hystrene® 9512
CAS 143-07-7	Philacid 1200
CAS 143-07-7	Prifrac 2920
CAS 143-07-7	Univol U314/a/b
CAS 143-18-0	Emkapol PO-18
CAS 143-18-0	Marley Cement Waterproofer
CAS 143-18-0	Octosol 449
CAS 143-27-1	Amine 16D
CAS 143-27-1	Armeen® 16
CAS 143-27-1	Armeen® 16D
CAS 143-27-1	Crodamine 1.16D
CAS 143-27-1	Kemamine® P-880, P-880D
CAS 143-28-2	Adol® 85
CAS 143-28-2	Cachalot® O-15
CAS 143-28-2	Dermaffine®
CAS 143-28-2	Emery® 3312
CAS 143-28-2	Emery® 3317
CAS 143-28-2	Fancol OA-95
CAS 143-28-2	HD-Eutanol
CAS 143-28-2	HD-Ocenol 90/95
CAS 143-28-2	Lipocol O
CAS 143-28-2	Novol
CAS 143-28-2	Novol
CAS 143-33-9	Cymag
CAS 143-67-9	Velban
CAS 143-67-9	Velbe
CAS 144-19-4	TMPD® Glycol
CAS 144-29-6	Bryrel
CAS 144-29-6	Oxucide
CAS 144-34-4	Methyl Selenac
CAS 144-55-8	Bufferight
CAS 144-55-8	Col-Evac
CAS 144-55-8	Tronacarb Sodium Bicarbonate
CAS 144-75-2	Diasone Sodium Enterab
CAS 144-82-1	Methisul

CAS 144-82-1	Proklar	CAS 151-21-3	Hartenol LAS-30
CAS 144-82-1	Thiosulfil	CAS 151-21-3	Marlinat® DFK 30
CAS 144-82-1	Ultrasul	CAS 151-21-3	Naxolate WA-97
CAS 144-82-1	Urolucosil	CAS 151-21-3	Norfox® SLS
CAS 144-83-2	M & B 693	CAS 151-21-3	Nutrapon DL 3891
CAS 145-12-0	Oranabol	CAS 151-21-3	Nutrapon W 1367
CAS 146-22-5	Mogadon	CAS 151-21-3	Nutrapon WAQE 2364
CAS 146-22-5	Nitrados	CAS 151-21-3	Octosol SLS
CAS 146-22 5	Remnos	CAS 151-21-3	Perlankrol® DSA
CAS 147-24-0	Benadryl	CAS 151-21-3	Polystep® B-3
CAS 147-94-4	Alexan	CAS 151-21-3	Rewopol® NLS 15 L
CAS 147-94-4	Cytosar	CAS 151-21-3	Rhodapon® 101-10
CAS 147-94-4	Cytosar-U	CAS 151-21-3	Rhodapon® LCP
CAS 148-72-1	P V Carpine Liquifilm	CAS 151-21-3	Standapol® WAQ-LC
CAS 148-72-1	Pilagan Liquifilm®	CAS 151-21-3	Stepanol® ME Dry
CAS 148-72-1	Pilofrin	CAS 151-21-3	Stepanol® WA Extra
CAS 148-79-8	Ascot	CAS 151-21-3	Sulfetal C 38
CAS 148-79-8	Ceratotect	CAS 151-21-3	Sulfochem SLN
CAS 148-79-8	Hortag Tecnazene Plus.	CAS 151-21-3	Sulfochem SLP-95
CAS 148-79-8	Hy-TL	CAS 151-21-3	Sulfochem SLS
CAS 148-79-8	Hy-Vic	CAS 151-21-3	Sulfochem SLX
CAS 148-79-8	Hymush	CAS 151-21-3	Sulfopon 101, 101 Special
CAS 148-79-8	Hytec Super.	CAS 151-21-3	Sulfopon 101/POL
CAS 148-79-8	Mintezol	CAS 151-21-3	Sulfopon P-40
CAS 148-79-8	Storaid Dust.	CAS 151-21-3	Sulfotex LCX
CAS 148-79-8	Storite	CAS 151-21-3	Texapon K-12
CAS 148-79-8	Storite SS.	CAS 151-21-3	Texapon K-1296
CAS 148-79-8	Tecto	CAS 151-21-3	Texapon L-100
CAS 148-79-8	Tecto 60%	CAS 151-21-3	Texapon LS Highly Conc
CAS 148-79-8	Thibenzole	CAS 151-21-3	Texapon OT Highly Conc. Needles
CAS 148-79-8	Tripart® Arena 6 + TBZ.		
CAS 148-82-3	Alkeran Injection	CAS 151-21-3	Texapon VHC Needles
CAS 148-82-3	Alkeran Tablets	CAS 151-21-3	Texapon ZHC Needles
CAS 149-30-4	Accelerator Mercapto	CAS 151-21-3	Ufarol Na-30
CAS 149-30-4	Captax	CAS 151-21-3	Witcolate 6400
CAS 149-30-4	MBT®	CAS 151-21-3	Witcolate C
CAS 149-30-4	Perkacit® MBT	CAS 151-21-3	Witcolate LCP
CAS 149-30-4	Rokon	CAS 151-21-3	Witcolate WAC-GL
CAS 149-30-4	Rotax®	CAS 151-38-2	Panogen M
CAS 149-30-4	Thiotax	CAS 151-67-7	Fluothane
CAS 149-30-4	Vulkacit® M	CAS 151-67-7	Halothane M & B
CAS 149-30-4	Vulkacit® Merkapto/C	CAS 152-11-4	Cordilox
CAS 149-39-3	Hostapon STT Paste	CAS 152-11-4	Isoptin
CAS 149-44-0	Formusol®	CAS 152-43-2	Estrovis
CAS 149-44-0	Hydro AWC	CAS 152-47-6	Kelfizina
CAS 149-74-6	CM8930	CAS 152-62-5	Duphaston
CAS 150-38-9	Hamp-Ene® Na3 Liq	CAS 152-62-5	Terolut
CAS 150-38-9	Sequestrene® NA3	CAS 152-97-6	Ficoid
CAS 150-76-5	Eastman® HQMME	CAS 153-87-7	Forit
CAS 151-06-4	Lucofen S A	CAS 153-87-7	Integrin
CAS 151-21-3	Akyposal NLS	CAS 154-41-6	Nobese
CAS 151-21-3	Alscoap LN-40, LN-90	CAS 154-41-6	Propadrine
CAS 151-21-3	Calfoam SLS-30	CAS 154-42-7	Lanvis
CAS 151-21-3	Carsonol® SLS Paste B	CAS 154-42-7	Tabloid
CAS 151-21-3	Chemsalan NLS 30	CAS 154-69-8	Dibistin
CAS 151-21-3	Empicol® 0303	CAS 154-69-8	Pyribenzamine Hydrochloride
CAS 151-21-3	Empicol® LS30	CAS 154-69-8	ReCovr
CAS 151-21-3	Empicol® LXV	CAS 154-93-8	BiCNU
CAS 151-21-3	Empicol® LY28/S	CAS 155-04-4	Octocure ZMBT-50
CAS 151-21-3	Empicol® LZG 30	CAS 155-04-4	Oxaf
CAS 151-21-3	Empicol® LZP	CAS 155-04-4	Perkacit® ZMBT
CAS 151-21-3	Empimin® LR28	CAS 155-04-4	Vulkacit® Merkapto/MGC

CAS 155-04-4	Zetax®	CAS 306-21-8	Paredrine
CAS 155-04-4	ZMBT	CAS 306-83-2	Genetron® 123
CAS 155-41-9	Pamine	CAS 307-35-7	Fluorad® FX-8
CAS 275-51-4	Azilex	CAS 309-43-3	Seconal Sodium
CAS 280-57-9	Amicure® 33-LV	CAS 314-40-9	Hyvar® X
CAS 280-57-9	Dabco® 33-LV	CAS 314-40-9	Hyvar X.7
CAS 280-57-9	Dabco® Crystalline	CAS 314-40-9	Uragan
CAS 280-57-9	TEDA-L33	CAS 315-30-0	Aloral
CAS 280-57-9	Texacat® TD-33	CAS 315-30-0	Aluline
CAS 297-76-7	Femulen	CAS 315-30-0	Caplenal
CAS 298-00-0	Folidol® M	CAS 315-30-0	Zyloprim Tablets
CAS 298-00-0	Metacide	CAS 315-30-0	Zyloric
CAS 298-02-2	Terrathion	CAS 315-37-7	Andro LA 200
CAS 298-04-4	Disyston® FE-10	CAS 315-37-7	Androgyn LA
CAS 298-04-4	Solvigran	CAS 315-37-7	Delatestryl
CAS 298-14-6	K-Lyte	CAS 315-37-7	Malogen LA 200
CAS 298-46-4	Tegretol	CAS 315-37-7	Primoteston Depot
CAS 298-57-7	Cinnar	CAS 315-37-7	Testate
CAS 298-57-7	Marzine RF Tablets	CAS 316-81-4	Majeptil
CAS 298-57-7	Stugeron	CAS 317-34-0	Aminophyllin
CAS 298-57-7	Stugeron Forte	CAS 317-34-0	Phyllocontin
CAS 298-59-9	Ritalin Hydrochloride	CAS 317-34-0	Somophyllin
CAS 298-81-7	Oxsoralen-Ultra	CAS 317-34-0	Theodrex
CAS 299-27-4	Gluconal® K	CAS 318-98-9	Bedranol®
CAS 299-27-4	Jaon	CAS 318-98-9	Inderal
CAS 299-27-4	Katorin	CAS 320-67-2	Mylosar
CAS 299-28-5	Gluconal® CA A	CAS 321-55-1	Loxon
CAS 299-28-5	Gluconal® CA M	CAS 330-54-1	Direx® 4L
CAS 299-29-6	Fergon	CAS 330-54-1	Diurex
CAS 299-29-6	Gluconal® FE	CAS 330-54-1	Diuron Bayer
CAS 299-84-3	Korlan	CAS 330-54-1	Gramixel
CAS 299-84-3	Trolene	CAS 330-54-1	Gramuron
CAS 300-76-5	Bromex	CAS 330-54-1	Karmex®
CAS 300-76-5	Dibrom	CAS 330-54-1	Orchard Herbide
CAS 300-92-5	ACuflow AF-1	CAS 330-54-1	Residuren Extra.
CAS 300-92-5	Synpro® 303	CAS 330-55-2	Campbell's Trifluron
CAS 301-02-0	Armid® O	CAS 330-55-2	Chandor
CAS 301-02-0	Crodamide O, OR	CAS 330-55-2	Flint
CAS 301-02-0	Kemamide® O	CAS 330-55-2	Gramonol
CAS 301-02-0	Petrac® Slip-Eze	CAS 330-55-2	Gramonol 5
CAS 301-02-0	Polydis® TR 121	CAS 330-55-2	Janus
CAS 301-02-0	Unislip 1759	CAS 330-55-2	Linnet
CAS 301-04-2	Ledac	CAS 330-55-2	Marksman
CAS 301-12-2	Metasystox® R	CAS 330-55-2	Onslaught
CAS 301-13-3	Syn-O-Ad® P-374	CAS 330-55-2	Rotalin
CAS 302-01-2	Amerzine	CAS 330-55-2	Solo
CAS 302-01-2	Scav-Ox® 35%	CAS 330-55-2	Tri-Farmon
CAS 302-01-2	Scav-Ox® II	CAS 330-55-2	Trifarmon
CAS 302-01-2	Ultra Pure	CAS 330-55-3	Afalon
CAS 302-01-2	Zerox	CAS 330-55-3	Arresin
CAS 302-17-0	Noctec	CAS 330-55-3	Ashlade Linuron
CAS 302-17-0	Noctec Capsules	CAS 330-55-3	Atlas Linuron
CAS 302-17-0	Somnos	CAS 330-55-3	Campbell's Linuron 45%
CAS 302-41-0	Dipidolor	CAS 330-55-3	Du Pont Linuron 50, 4L
CAS 303-25-3	Valoid	CAS 330-55-3	Linex® 4L
CAS 303-42-4	Primobolan Depot	CAS 330-55-3	Linurex
CAS 303-81-1	Streptonivin	CAS 330-55-3	Linuron
CAS 304-20-1	Apresoline	CAS 330-55-3	Linuron 15
CAS 304-20-1	Apresoline Hydrochloride	CAS 330-55-3	Linuron 450 FL
CAS 304-55-2	MPI DMSA Kidney Reagent	CAS 330-55-3	Norunil
CAS 305-03-3	Leukeran Tablets	CAS 330-55-3	Rotalin
CAS 306-07-0	Eutonyl	CAS 333-36-8	Idoklon

CAS 333-41-5	Antlak
CAS 333-41-5	Basudin
CAS 333-41-5	Diazol
CAS 333-41-5	Diziktol
CAS 333-41-5	Root Guard
CAS 334-48-5	Prifrac 2906
CAS 334-48-5	Univol U310
CAS 338-98-7	Predef
CAS 339-44-6	Lycanol
CAS 340-56-7	Melsedin
CAS 346-18-9	Nephril
CAS 346-18-9	Renese
CAS 356-12-7	Fluonex™
CAS 356-12-7	Lidex
CAS 356-12-7	Metosyn
CAS 356-12-7	Topsyn
CAS 357-07-3	Numorphan
CAS 357-08-4	Narcan®
CAS 357-56-2	Palfium
CAS 358-52-1	Evipan Sodium®
CAS 360-70-3	Deca-Durabolin
CAS 360-70-3	Nandrobolic LA
CAS 361-37-5	Deseril
CAS 362-29-8	Indorm
CAS 364-62-5	Metox
CAS 364-98-7	Eudemine
CAS 364-98-7	Hyperstat
CAS 364-98-7	Proglycem
CAS 366-70-1	Matulane
CAS 378-44-9	Betacortril
CAS 378-44-9	Betacortril Forte
CAS 378-44-9	Betnelan
CAS 378-44-9	Celestone
CAS 379-79-3	Ercal
CAS 379-79-3	Ergomar
CAS 379-79-3	Ergostat
CAS 379-79-3	Femergin
CAS 379-79-3	Lingraine
CAS 379-79-3	Medihaler Ergotamine
CAS 379-79-3	Migraine Dolviran®
CAS 382-67-2	Topicort
CAS 389-08-2	Cybis
CAS 389-08-2	NegGram
CAS 389-08-2	Negram
CAS 389-08-2	Wintomylon
CAS 395-28-8	Cardilan
CAS 395-28-8	Duvadilan
CAS 395-28-8	Duvadilan Retard
CAS 395-28-8	Suprilent
CAS 395-28-8	Vasodilan
CAS 395-28-8	Xuprin
CAS 396-01-0	Dyrenium
CAS 396-01-0	Dytac
CAS 404-82-0	Ponderax PA
CAS 404-82-0	Pondimin
CAS 409-21-2	Carbogran
CAS 409-21-2	Carbogran E
CAS 409-21-2	Carbomant
CAS 409-21-2	Carborex
CAS 409-21-2	Carsilon
CAS 409-21-2	Lonsicar
CAS 409-21-2	Resilon

CAS 409-21-2	Silantox
CAS 409-21-2	Simax
CAS 409-21-2	Siogel
CAS 426-13-1	Fluor-Op® Ophthalmic Suspension USP 0.1%
CAS 426-13-1	FML Forte Liquifilm®
CAS 426-13-1	FML Liquifilm®
CAS 426-13-1	FML S.O.P.®
CAS 426-13-1	Oxylone
CAS 427-51-0	Androcur
CAS 427-51-0	Cyprostat OP
CAS 434-05-9	Primobolan
CAS 434-07-1	Adroyd
CAS 434-07-1	Anadrol
CAS 434-07-1	Anapolon
CAS 435-97-2	Liquamar
CAS 435-97-2	Marcoumar
CAS 437-38-7	Sublimaze
CAS 437-74-1	Complamin
CAS 438-41-5	Librium
CAS 438-41-5	SK-Lygen
CAS 439-14-5	Alupram
CAS 439-14-5	Atensine
CAS 439-14-5	Valium
CAS 440-58-4	Jodomiron
CAS 440-58-4	Uromiro(n)
CAS 443-48-1	Flagyl
CAS 443-48-1	Metro IV
CAS 443-48-1	Metrolyl®
CAS 443-48-1	Nidazol
CAS 443-48-1	Satric
CAS 443-48-1	SK-Metronidazole
CAS 443-48-1	Torgyl
CAS 443-48-1	Zadstat
CAS 446-86-6	Imuran
CAS 456-59-7	Cyclospasmol
CAS 463-40-1	Industrene® 120
CAS 466-06-8	Talusin
CAS 466-06-8	Tradenal
CAS 470-90-6	Birlane
CAS 470-90-6	Sapecron
CAS 470-90-6	Sedanox
CAS 471-34-1	Aeromatt
CAS 471-34-1	Amical® 85
CAS 471-34-1	Amical® 101
CAS 471-34-1	Amical® SC
CAS 471-34-1	Atomite
CAS 471-34-1	Calofil
CAS 471-34-1	Calofort®
CAS 471-34-1	Calopake®
CAS 471-34-1	Camel-CAL®
CAS 471-34-1	Camel-CARB®
CAS 471-34-1	Camel-FIL
CAS 471-34-1	Camel-FINE
CAS 471-34-1	Camel-TEX®
CAS 471-34-1	Camel-WITE®
CAS 471-34-1	Carbital®
CAS 471-34-1	CC-103
CAS 471-34-1	CP Filler
CAS 471-34-1	Drikalite®
CAS 471-34-1	Hallcote® 573
CAS 471-34-1	Hi-Pflex® 100

CAS 471-34-1	Hubercarb® Q 6-20	CAS 523-87-5	Dramamine
CAS 471-34-1	Hubercarb® W 2	CAS 525-66-6	Berkolol
CAS 471-34-1	Kotamite®	CAS 525-66-6	Paxalol
CAS 471-34-1	Macromite	CAS 526-08-9	Orisulf
CAS 471-34-1	Marble Dust	CAS 527-07-1	Asahi Aji®
CAS 471-34-1	No. 1 White	CAS 527-07-1	Gluconal® NA
CAS 471-34-1	No. 3 White	CAS 527-07-1	Naglusol
CAS 471-34-1	Opacicoat	CAS 527-09-3	Gluconal® CU
CAS 471-34-1	Opacimite	CAS 530-78-9	Arlef-100
CAS 471-34-1	SC-53	CAS 532-03-6	Robaxin
CAS 471-34-1	Slab Dip AC699	CAS 532-76-3	Cyclaine
CAS 471-34-1	Snowflake P.E	CAS 533-06-2	Myanesin
CAS 471-34-1	Snowflake White	CAS 533-74-4	Amerstat® 233
CAS 471-34-1	Socal	CAS 533-96-0	S-Carb
CAS 471-34-1	Sturcal®	CAS 536-33-4	Trecator-SC
CAS 471-34-1	Super-Pflex® 100	CAS 536-33-4	Trescatyl
CAS 471-34-1	T-Carb	CAS 536-43-6	Dyclone
CAS 471-34-1	Trucal	CAS 537-65-5	Oxynone
CAS 471-34-1	Trucarb	CAS 538-23-8	Captex® 8000
CAS 471-34-1	Ultra-Pflex®	CAS 538-23-8	Emalex O.T.G
CAS 471-34-1	Ultramite	CAS 538-23-8	Miglyol® 808
CAS 471-53-4	Biosone	CAS 538-24-9	Dynasan® 112
CAS 473-41-6	Orinase Diagnostic	CAS 538-71-6	Bradosol
CAS 479-18-5	Dilor Elixir	CAS 539-21-9	Inversal
CAS 480-30-8	Welldorm	CAS 539-68-4	Rolaids
CAS 490-55-1	Daptazole	CAS 540-10-3	Crodamol CP
CAS 497-19-8	Consal	CAS 540-10-3	Cutina® CP
CAS 497-19-8	Oxyper	CAS 540-10-3	Emalex CC-16
CAS 497-19-8	Solvay® Soda	CAS 540-10-3	Kemester® CP
CAS 501-24-6	Cardolite® NC-507	CAS 540-10-3	Kessco® 653
CAS 501-30-4	Rita KA	CAS 540-10-3	Precifac ATO
CAS 501-68-8	Nydrane	CAS 540-10-3	Rewowax CG
CAS 504-24-5	Avitrol	CAS 540-10-3	Schercemol CP
CAS 509-67-1	Ethnine	CAS 540-10-3	Starfol® CP
CAS 509-67-1	Ethrine	CAS 540-10-3	Waxenol® 815
CAS 509-67-1	Folcovin	CAS 541-02-6	CD3770
CAS 509-67-1	Hill's Adult Balsam	CAS 541-02-6	D3770
CAS 510-15-6	Benzilan	CAS 541-05-9	CH7260
CAS 511-13-7	Detigon	CAS 542-75-6	Telone
CAS 511-13-7	U/o	CAS 544-17-2	Latibon®
CAS 514-36-3	Alflorone	CAS 544-63-8	Emery® 654
CAS 514-36-3	Florinef	CAS 544-63-8	Hystrene® 9014
CAS 514-36-3	Florinef Acetate	CAS 544-63-8	Hystrene® 9514
CAS 514-65-8	Akineton	CAS 544-63-8	Philacid 1400
CAS 518-28-5	Condyline	CAS 544-63-8	Prifrac 2940
CAS 518-47-8	Fluor-Amps	CAS 544-63-8	Univol U320
CAS 518-47-8	Fluor-I-Strip	CAS 546-56-5	CO9817
CAS 518-47-8	Fluorescite	CAS 546-88-3	Lithostat
CAS 518-47-8	Ful-Glo	CAS 546-93-0	Almacarb.
CAS 518-47-8	Funduscein®-10 or -25 Injection	CAS 546-93-0	Elastocarb Tech Light,
CAS 519-95-9	Zanchol		Tech Heavy
CAS 521-12-0	Drolban	CAS 546-93-0	Elastocarb UF
CAS 521-12-0	Masterone	CAS 546-93-0	F-MA 11®.
CAS 521-18-6	Anabolex	CAS 546-93-0	Gastrils.
CAS 521-18-6	Anobolex	CAS 546-93-0	Magocarb-33
CAS 522-40-7	Honvan	CAS 546-93-0	Nidrin.
CAS 522-48-5	Anthelvet	CAS 546-93-0	R-MA 11®.
CAS 522-48-5	Murine Plus	CAS 548-00-5	Pelentan
CAS 522-48-5	Soothe	CAS 548-00-5	Tromexan
CAS 522-48-5	Tyzine	CAS 548-68-5	Trocinate
CAS 522-51-0	Dequadin	CAS 548-73-2	Droleptan
CAS 522-51-0	Labosept	CAS 548-73-2	Inapsine

CAS 549-18-8	Amitril
CAS 549-18-8	Domical
CAS 549-18-8	Elavil
CAS 549-18-8	Laroxyl
CAS 549-18-8	Lentizol
CAS 549-18-8	Saroten®
CAS 549-18-8	SK-Amitriptyline
CAS 549-18-8	Triptafen
CAS 549-18-8	Tryptanol
CAS 549-18-8	Tryptizol®
CAS 550-29-2	Albalon Liquifilm®
CAS 550-29-2	Clear Eyes
CAS 550-29-2	Comfort Eye Drops
CAS 550-29-2	Degest-2
CAS 550-29-2	Naphcon
CAS 550-29-2	Privine Hydrochloride
CAS 550-29-2	Vasoclear
CAS 550-83-4	Blockain Hydrochloride
CAS 550-83-4	Darvon
CAS 550-83-4	Dolene
CAS 550-83-4	Ravocaine Hydrochloride
CAS 551-48-4	Vatensol
CAS 551-92-8	Emtryl
CAS 552-94-3	Disalcid™
CAS 553-08-2	Thonzide
CAS 554-13-2	Eskalith
CAS 554-13-2	Liskonum Tablets
CAS 554-13-2	Lithane
CAS 554-13-2	Lithobid
CAS 554-13-2	Lithonate
CAS 554-13-2	Lithotabs
CAS 554-13-2	Phasal®
CAS 554-13-2	Priadel
CAS 554-57-4	Neptazane
CAS 554-92-7	Tigan
CAS 555-30-6	Aldomet
CAS 555-30-6	Dopamet
CAS 555-30-6	Medomet
CAS 555-30-6	Presinol®, Presinol Mite
CAS 555-31-7	Aliso
CAS 555-37-3	Neburex
CAS 555-37-3	Noruben
CAS 555-43-1	Dynasan® 118
CAS 555-43-1	Neobee® 62
CAS 555-44-2	Dynasan® 116
CAS 555-45-3	Dynasan® 114
CAS 557-04-0	Afco-Chem MGS
CAS 557-04-0	Petrac® MG-20 NF
CAS 557-04-0	Synpro® 90
CAS 557-05-1	Afco-Chem ZNS
CAS 557-05-1	Altax®.
CAS 557-05-1	Antidust 2
CAS 557-05-1	DLG-10
CAS 557-05-1	DLG-20
CAS 557-05-1	Hallcote® ZS 5050
CAS 557-05-1	Petrac® ZN-41
CAS 557-05-1	Synpro® 8
CAS 561-78-4	Nisentil
CAS 562-09-4	Phenoxine
CAS 562-10-7	Decapryn Succinate
CAS 562-10-7	Unisom
CAS 569-57-3	TACE

CAS 573-41-1	Monotheamin Pulvules
CAS 575-75-7	Tonophosphan
CAS 577-11-7	Aerosol® GPG
CAS 577-11-7	Aerosol® OT-70 PG
CAS 577-11-7	Aerosol® OT-75%
CAS 577-11-7	Aerosol® OT-MSO
CAS 577-11-7	Agrilan® AEC266
CAS 577-11-7	Anonaid TH
CAS 577-11-7	Arowet SC-75
CAS 577-11-7	Arylene M40
CAS 577-11-7	Astrowet O-70-PG
CAS 577-11-7	Astrowet O-75
CAS 577-11-7	Bevaloid 1299
CAS 577-11-7	Chemax DOSS/70
CAS 577-11-7	Complemix
CAS 577-11-7	Coprol
CAS 577-11-7	Correctol® Extra Gentle
CAS 577-11-7	Cropol 60
CAS 577-11-7	D-S-S
CAS 577-11-7	Dioctyl
CAS 577-11-7	Discol DFW
CAS 577-11-7	Disponil SUS IC 8
CAS 577-11-7	Doxinate
CAS 577-11-7	Drewfax® 0007
CAS 577-11-7	Drewfax® S-700
CAS 577-11-7	Elfanol® 883
CAS 577-11-7	Emcol® 4500
CAS 577-11-7	Emcol® DOSS
CAS 577-11-7	Empimin® OT
CAS 577-11-7	Geropon® CYA/60
CAS 577-11-7	Geropon® DOS
CAS 577-11-7	Hipochem EK-18
CAS 577-11-7	Hodag DOSS-70
CAS 577-11-7	Lankropol® KO2
CAS 577-11-7	Lankropol® KPH70
CAS 577-11-7	Leonil OS
CAS 577-11-7	Mackanate DOS-40
CAS 577-11-7	Manoxol OT, OT/P and OT/B
CAS 577-11-7	Manoxolot
CAS 577-11-7	Marlinat® DF 8
CAS 577-11-7	Mazawet® DOSS 70
CAS 577-11-7	Modane Soft
CAS 577-11-7	Monawet MO
CAS 577-11-7	Monawet MO-65-150
CAS 577-11-7	Monawet MO-70
CAS 577-11-7	Monawet MO-84R2W
CAS 577-11-7	Ninate® DS 70
CAS 577-11-7	Norval
CAS 577-11-7	Octowet 40
CAS 577-11-7	Pentasol
CAS 577-11-7	Regutol
CAS 577-11-7	Rewopol® SBDO 75
CAS 577-11-7	Rexowet ASG-81
CAS 577-11-7	Schercopol DOS-70
CAS 577-11-7	Schercowet DOS-70
CAS 577-11-7	Secosol® DOS 70
CAS 577-11-7	Servoxyl VLA 2170
CAS 577-11-7	Tensuccin D8
CAS 577-11-7	Thorowet G-40 3230
CAS 577-11-7	Triton® GR-5M
CAS 577-11-7	Warcowet O
CAS 577-11-7	Waxsol

CAS 577-48-0	Butesin Picrate		CAS 627-83-8	Kemester® EGDS
CAS 579-56-6	Vasotran		CAS 627-83-8	Kessco® EGDS
CAS 584-08-7	Pearl Dust®		CAS 627-83-8	Lexemul® EGDS
CAS 584-84-9	Scuranate		CAS 627-83-8	Lipo EGDS
CAS 584-84-9	Voranate T-80, Type I, Type II		CAS 627-83-8	Mackester EGDS
CAS 587-23-5	Mandelamine		CAS 627-83-8	Mapeg® EGDS
CAS 588-33-5	Sipomer® IBOA		CAS 627-83-8	Pegosperse® 50 DS
CAS 588-42-1	Praenitrona		CAS 627-83-8	Rewopal® PG 280
CAS 588-59-0	Eccobrite RB		CAS 627-83-8	Rita EDGS
CAS 590-00-1	Sorbistat K		CAS 627-83-8	Secoster® DMS
CAS 590-63-6	Duvoid		CAS 629-73-2	Neodene® 16
CAS 590-63-6	Urecholine		CAS 629-76-5	Neodol® 5
CAS 591-71-7	Uniplex 504		CAS 630-56-8	Delalutin
CAS 592-01-8	Cyanolime		CAS 630-56-8	Lutate
CAS 592-41-6	Neodene® 6		CAS 630-60-4	Ouabaine Arnaud
CAS 592-41-6	Neodene® 6/12		CAS 630-93-3	Diphentoin
CAS 593-60-2	Saytex® VBR		CAS 630-93-3	Epanutin
CAS 595-33-5	Megace		CAS 632-79-1	Great Lakes PHT4
CAS 595-33-5	Ovaban		CAS 632-79-1	Saytex® RB-49
CAS 596-51-0	Robinul Injection		CAS 637-07-0	Antromid-S
CAS 596-51-0	Robinul Powder		CAS 637-07-0	Atromid-S
CAS 597-09-1	NEPD		CAS 637-12-7	Duroseal
CAS 598-63-0	Halcarb 20		CAS 637-12-7	HiGel
CAS 598-63-0	Ledca		CAS 637-12-7	LoGel
CAS 603-00-9	Brontyl		CAS 637-12-7	MedGel
CAS 603-50-9	Dulco-Lax		CAS 637-12-7	Metasap® 537
CAS 603-50-9	Paxolax		CAS 637-12-7	Synpro® 404
CAS 603-50-9	Theralax		CAS 637-12-7	Synpro® 505 USP
CAS 604-75-1	Oxanid		CAS 637-58-1	Thronothane Hydrochloride
CAS 604-75-1	Serax		CAS 637-58-1	Tronolane
CAS 604-75-1	Serenid-D		CAS 638-16-4	Zisnet F-PT
CAS 604-75-1	Serenid Forte		CAS 638-23-3	Mucopront
CAS 606-17-7	Cholografin		CAS 638-38-0	Manal
CAS 611-75-6	Bisolvon		CAS 638-94-8	Tridesilon
CAS 614-33-5	Mollit B		CAS 643-22-1	Bristamycin
CAS 614-33-5	Plastic A		CAS 643-22-1	Erypar
CAS 614-33-5	Uniplex 260		CAS 643-22-1	Erythrolar
CAS 614-39-1	Procan		CAS 643-22-1	Ethril
CAS 614-39-1	Pronestyl		CAS 643-22-1	Pfizer-E
CAS 614-39-1	Pronestyl Solution		CAS 643-22-1	Pfizer-EM
CAS 614-39-1	Pronestyl Tablets		CAS 643-22-1	SK-Erythromycin
CAS 614-45-9	Aztec® t-Butyl Perbenzoate		CAS 643-22-1	Wyamycin S
CAS 614-45-9	Esperox® 10		CAS 644-62-2	Arquel
CAS 614-45-9	Polyvel CR-5T		CAS 646-13-9	Emerest® 2324
CAS 614-45-9	Trigonox® 93		CAS 646-13-9	Estol 1476
CAS 614-45-9	Trigonox® C		CAS 646-13-9	Kemester® 5415
CAS 614-87-9	Brulidine		CAS 646-13-9	Kessco® IBS
CAS 615-58-7	FR-612		CAS 646-13-9	Uniflex® IBYS
CAS 616-45-5	Soluphor® P		CAS 651-06-9	Sulla
CAS 616-91-1	Airbron		CAS 652-67-5	Ismotic
CAS 616-91-1	Mucomyst		CAS 657-27-2	Enisyl
CAS 617-86-7	CT2523		CAS 657-84-1	Eltesol® ST 40
CAS 620-99-5	Holocaine Hydrochloride		CAS 661-19-8	Dehydag Wax 22 (Lanette)
CAS 621-71-6	Dynasan® 110		CAS 661-19-8	Emery® 3304
CAS 622-16-2	Stabilizer 2013-P®		CAS 661-19-8	Loxiol VPG 1451
CAS 624-04-4	Emalex EG-di-L		CAS 661-61-6	Akyposal TIPA 45
CAS 624-04-4	Kemester® EGDL		CAS 661-61-6	Rewopal® TLS 90 L
CAS 627-83-8	Alkamuls® EGDS		CAS 665-66-7	Symmetrel®
CAS 627-83-8	Emalex EG-di-S		CAS 667-83-4	Humectant SD-35
CAS 627-83-8	Emerest® 2355		CAS 671-16-9	Natulan
CAS 627-83-8	Ethylene Glycol Distearate VA		CAS 671-88-5	Disamide
CAS 627-83-8	Genapol® PMS		CAS 681-84-5	Dynasil® M

CAS 682-01-9	CT2090
CAS 683-10-3	Dehyton® PAB-30
CAS 683-10-3	Zohartaine AB
CAS 686-31-7	Esperox® 570
CAS 686-31-7	Lupersol 575
CAS 686-31-7	Trigonox® 121
CAS 686-8-66-2	Miranol® BM Conc
CAS 693-33-4	Alkateric® PB
CAS 693-33-4	Mackam CET
CAS 693-33-4	Mirataine® BET-P-30
CAS 693-36-7	Argus DSTDP
CAS 693-36-7	Carstab® DSTDP
CAS 693-36-7	Cyanox® STDP
CAS 693-36-7	Evanstab® 18
CAS 693-36-7	Lankromark® DSTDP
CAS 693-36-7	Lowinox® DSTDP
CAS 696-59-3	Protectol® DMT
CAS 709-98-8	Sorpur®
CAS 709-98-8	Surcopur®
CAS 723-46-6	Gantanol
CAS 729-99-7	Nuprin
CAS 737-31-5	MD 50
CAS 737-31-5	Urovist Sodium 300
CAS 738-70-5	Ipral
CAS 738-70-5	Proloprim
CAS 738-70-5	Proloprim Tablets
CAS 738-70-5	Trimopan
CAS 738-70-5	Trimpex
CAS 739-71-9	Surmontil
CAS 747-36-4	Plaquenil
CAS 749-02-0	Spiropitan
CAS 749-13-3	Triperidol
CAS 751-97-3	Syntetrin
CAS 756-79-6	Fyrol® DMMP
CAS 759-94-4	Eptam 6E
CAS 762-16-3	Perkadox® SE-8
CAS 762-72-1	CA0570
CAS 763-69-9	Ektapro® EEP Solvent
CAS 768-33-2	CP0160
CAS 780-69-8	CP0320
CAS 797-63-7	Microlut
CAS 801-52-5	Regamycin
CAS 804-63-7	Coco-Quinine
CAS 804-63-7	Quinamin
CAS 804-63-7	Quine
CAS 808-26-4	Bonomycin
CAS 811-54-1	Ledfo
CAS 816-94-4	Phospholipon® SC
CAS 822-16-2	Afco-Chem B
CAS 826-39-1	Inversine
CAS 826-39-1	Mevasine
CAS 834-12-8	Ametrex
CAS 834-12-8	Amigan
CAS 834-28-6	Dibotin
CAS 834-28-6	Dipar
CAS 846-49-1	Almazine
CAS 846-49-1	Ativan Injection
CAS 846-49-1	Ativan Tablets
CAS 846-50-4	Normison Capsules
CAS 849-55-8	Arlidin
CAS 850-52-2	Regu-mate
CAS 868-77-9	Bisomer 2HEMA
CAS 868-77-9	Sipomer® HEM-D
CAS 870-72-4	Rongalit® C
CAS 871-37-4	Chembetaine OL-30
CAS 871-37-4	Incronam OD-50
CAS 871-37-4	Mackam OB-30
CAS 871-37-4	Mafo® OB
CAS 872-05-9	Neodene® 10
CAS 872-05-9	Neodene® 810.
CAS 872-05-9	Neodene® 1012
CAS 872-50-4	NMP
CAS 872-50-4	PartsPrep™ Degreaser
CAS 877-66-7	Celogen® TSH
CAS 886-50-0	Amigan.
CAS 886-50-0	Clarosan 1FG
CAS 886-50-0	Prebane 500
CAS 886-50-0	Senate
CAS 886-74-8	Maolate
CAS 894-71-3	Allegron
CAS 901-44-0	Dianol® 220.
CAS 909-39-7	Insidon
CAS 919-16-4	Cibalith-S
CAS 919-30-2	Aktisil AM
CAS 919-30-2	CA0750
CAS 919-30-2	Dynasylan® AMEO
CAS 919-30-2	Dynasylan® AMEO-P
CAS 919-30-2	Prosil® 220
CAS 919-30-2	Union Carbide® A-1100
CAS 919-86-8	Azotoz 580
CAS 919-86-8	D S M
CAS 919-86-8	Demetox
CAS 919-86-8	Metasystox® I
CAS 919-86-8	Mifatox
CAS 919-86-8	Persyst
CAS 919-86-8	Power DSM
CAS 919-86-8	Quad DSM
CAS 919-86-8	Star DSM
CAS 919-86-8	Tripart® Systemic Insecticide
CAS 919-86-8	Vassgro DSM
CAS 922-80-5	Aerosol® AY-65
CAS 927-07-1	Aztec® t-Butyl Peroxypivalate-75 OMS
CAS 927-07-1	Esperox® 31M
CAS 927-07-1	Lupersol 11
CAS 927-07-1	Trigonox® 25-C75
CAS 927-83-3	VR-160
CAS 928-24-5	Emalex EG-di-O
CAS 928-65-4	CH7332
CAS 929-77-1	Kemester® 9022
CAS 940-41-0	CP0110
CAS 944-22-9	Cudgel
CAS 944-22-9	Dyfonate
CAS 947-42-2	CD6150
CAS 950-10-7	Cytro-Lane
CAS 950-37-8	Maktion
CAS 950-37-8	Suprathion
CAS 956-90-1	Sernylan
CAS 957-51-7	Enide
CAS 959-24-0	Beta-Cardone
CAS 959-24-0	Sotacor
CAS 965-90-2	Maxibolin
CAS 968-81-0	Dymelor
CAS 968-93-4	Teslac

CAS 973-21-7	Acrex
CAS 979-32-8	Delestrogen
CAS 979-32-8	Gynogen LA 10
CAS 979-32-8	Primogyn Depot
CAS 979-32-8	Progynova
CAS 980-71-2	Dimetane
CAS 980-71-2	Dimotane Tablets
CAS 983-85-7	Havapen
CAS 985-16-0	Nafcil
CAS 992-21-2	Mucomycin
CAS 994-30-9	CT2520
CAS 995-33-5	Luperco 230-XL
CAS 995-33-5	Lupersol 230
CAS 996-50-9	CD4450
CAS 998-30-1	CT2500
CAS 999-81-5	ABM Chlormequat 40
CAS 999-81-5	ABM Chlormequat 72.5
CAS 999-81-5	Ashlade 4-60 CCC
CAS 999-81-5	Ashlade 700 CCC
CAS 999-81-5	Atlas Chlormequat 46, 700
CAS 999-81-5	CCC 700
CAS 999-81-5	Cleanacres PDR 675
CAS 999-81-5	Clifton Chlormequat 46
CAS 999-81-5	Cycogan
CAS 999-81-5	Fargro Chlormequat
CAS 999-81-5	Farmacel
CAS 999-81-5	Farmacel 645
CAS 999-81-5	Headland Swift
CAS 999-81-5	Hyquat 70, 75
CAS 999-81-5	Mandops Barleyquat B
CAS 999-81-5	Mandops Bettaquat B
CAS 999-81-5	Mandops Spring Poquat
CAS 999-81-5	MSS Chlormequat 40, 46, 60, 70
CAS 999-81-5	Portman Chlormequat 400, 460, 600,700
CAS 999-81-5	Power 64, 640, 700
CAS 999-81-5	Quadrangle Chlormequat 700
CAS 999-81-5	Standup
CAS 999-81-5	Star Chlormequat
CAS 999-81-5	Titan
CAS 999-81-5	Tripart® Brevis
CAS 999-81-5	Tripart® Chlormequat 5C
CAS 999-81-5	Tripart® Chlormequat 460
CAS 1000-50-6	CB2785
CAS 1000-90-4	Propyl Zithate®
CAS 1000-90-4	Zithate®
CAS 1002-89-7	Ammonium Stearate 33% Liq
CAS 1009-93-4	CH7250
CAS 1009-93-4	H7250
CAS 1014-69-3	Semeron
CAS 1020-84-4	CO9800
CAS 1055-55-6	Scolaban
CAS 1066-17-7	Colomycin
CAS 1066-33-7	ABC-Trieb®
CAS 1066-35-9	CD5470
CAS 1067-25-0	CP0810
CAS 1067-33-0	Metacure® T-1
CAS 1067-53-4	Aktisil VM
CAS 1067-53-4	CV5000
CAS 1067-53-4	Dynasylan® VTMOEO
CAS 1067-53-4	Union Carbide® A-172
CAS 1068-27-5	Aztec® 2,5-Tri
CAS 1068-27-5	Esperal® 230
CAS 1068-27-5	Lupersol 130
CAS 1068-27-5	Polyvel CR-L10
CAS 1068-27-5	Trigonox® 145
CAS 1071-27-8	CC3555
CAS 1071-83-6	Glyphogan
CAS 1071-83-6	Roundup
CAS 1071-83-6	Spasor
CAS 1072-35-1	Hal-Lub-N
CAS 1072-35-1	Haro® Chem P28G
CAS 1072-35-1	P-289
CAS 1072-35-1	P-51
CAS 1077-56-1	Uniplex 108
CAS 1085-98-9	Elvaron®
CAS 1085-98-9	Euparen®
CAS 1094-08-2	Lysivane
CAS 1094-08-2	Parsidol
CAS 1095-90-5	Dolophine Hydrochloride
CAS 1095-90-5	Methadose
CAS 1095-90-5	Physeptone Linctus
CAS 1095-90-5	Physeptone Tablets and Injection
CAS 1098-60-8	Vesprin
CAS 1112-39-6	CD5605
CAS 1113-02-66	Folimat®
CAS 1119-97-7	Mytab®
CAS 1119-97-7	Querton 14Br-40
CAS 1119-97-7	Rhodaquat® M214B/99
CAS 1119-97-7	Sumquat® 6110
CAS 1120-01-0	Berol 472
CAS 1120-24-7	Adma® 10
CAS 1120-28-1	Kemester® 2050
CAS 1120-36-1	Neodene® 14
CAS 1120-49-6	Armeen® 2-10
CAS 1143-38-0	Anthra-Derm
CAS 1143-38-0	Lasan
CAS 1151-11-7	Oragrafin Calcium
CAS 1156-19-0	Ronase
CAS 1156-19-0	Tolanase
CAS 1156-19-0	Tolinase
CAS 1163-19-5	FR-1210
CAS 1163-19-5	Great Lakes DE-83R
CAS 1163-19-5	Octoguard FR-01
CAS 1163-19-5	Octoguard FR-15
CAS 1163-19-5	Saytex® 102E
CAS 1163-19-5	Thermoguard® 505
CAS 1166-52-5	Progallin LA
CAS 1169-79-5	Pentovis
CAS 1177-87-3	Decadron-LA
CAS 1177-87-3	Decadronal
CAS 1181-54-0	Megaclor
CAS 1185-55-3	CM9100
CAS 1185-55-3	Dynasylan® MTMS
CAS 1185-55-3	Union Carbide® A-163
CAS 1189-098-8	Ageflex 1,3 BGDMA
CAS 1190-63-2	Schercemol CS
CAS 1194-65-6	BH Prefix D
CAS 1194-65-6	Casoron
CAS 1194-65-6	Casoron G
CAS 1194-65-6	Casoron G4
CAS 1194-65-6	Fydulan
CAS 1194-65-6	Fydumas
CAS 1194-65-6	Fydusit

CAS 1194-65-6	Prefix D
CAS 1197-21-3	Fastin
CAS 1197-21-3	Obermine Black & Yellow
CAS 1197-21-3	Wilpo
CAS 1218-35-5	Otrivin Hydrochloride
CAS 1221-56-3	Bilivist
CAS 1221-56-3	Oragrafin Sodium
CAS 1225-20-3	Angio-Conray
CAS 1225-20-3	Conray 325, 400
CAS 1225-55-4	Concordin®
CAS 1225-55-4	Vivactil®
CAS 1225-60-1	Nilergex
CAS 1225-65-6	Tolnate
CAS 1229-29-4	Adapin
CAS 1229-29-4	Curatin
CAS 1229-35-2	Dilosyn
CAS 1239-04-9	Narphen
CAS 1240-15-9	Largon
CAS 1245-44-9	Ultrapen
CAS 1249-84-9	Ornitrol®
CAS 1263-89-4	Humatin
CAS 1264-72-8	Coly-Mycin S
CAS 1300-72-7	Eltesol® SX 30
CAS 1300-72-7	Esi-Terge SXS
CAS 1300-72-7	Hartotrope SXS 40, Powd
CAS 1300-72-7	Manro SXS
CAS 1300-72-7	Naxonate® 4L
CAS 1300-72-7	Naxonate® SX
CAS 1300-72-7	Nutrol SXS 5418
CAS 1300-72-7	Pilot SXS-40
CAS 1300-72-7	Reworyl® NXS 40
CAS 1300-72-7	Stepanate® SXS
CAS 1300-72-7	Sulfotex SXS-40
CAS 1300-72-7	Witconate SXS 40%
CAS 1302-42-7	Amerfloc® 2
CAS 1302-42-7	Dynaflock
CAS 1302-42-7	Dynagrout
CAS 1302-42-7	Manfloc
CAS 1302-78-9	Ben-A-Gel®
CAS 1302-78-9	BentoPharm
CAS 1302-78-9	Brebent
CAS 1302-78-9	Bregel
CAS 1302-78-9	Fulbent
CAS 1302-78-9	Fulbond
CAS 1302-78-9	Gadorgel
CAS 1302-78-9	K 129-H
CAS 1302-78-9	Korthix
CAS 1302-78-9	Korthix H-NF
CAS 1302-78-9	MicroForm B
CAS 1302-78-9	MicroForm BCS
CAS 1302-78-9	Montigel
CAS 1302-78-9	Polargel® HV
CAS 1302-78-9	Polargel® NF
CAS 1303-96-4	Borax
CAS 1304-85-4	Mammol
CAS 1305-62-0	Edelwit
CAS 1305-62-0	Red Hot Pellets
CAS 1305-78-8	Caloxol CP2
CAS 1305-78-8	Dynacal
CAS 1305-78-8	KM Pebble Lime
CAS 1305-78-8	Rhenosorb
CAS 1305-79-9	Fertilox
CAS 1305-79-9	Oxy-Gro
CAS 1306-06-5	Alveograf
CAS 1306-06-5	Periograf
CAS 1308-38-9	Accrox
CAS 1308-38-9	Chrome Green 106
CAS 1308-38-9	M100
CAS 1309-37-1	Black 103
CAS 1309-37-1	Blackox
CAS 1309-37-1	Brown 208
CAS 1309-37-1	Deanox
CAS 1309-37-1	Disperfin
CAS 1309-37-1	Ferroxide
CAS 1309-37-1	Foundrox
CAS 1309-37-1	Hydroferrox®
CAS 1309-37-1	Kroma Red
CAS 1309-37-1	Magnox
CAS 1309-37-1	Mapico
CAS 1309-37-1	No Vein Compound
CAS 1309-37-1	Octotint 103
CAS 1309-37-1	OSO® 440
CAS 1309-37-1	OSO® 1905
CAS 1309-37-1	Pferrico
CAS 1309-37-1	Pferrisperse
CAS 1309-37-1	Pferrox
CAS 1309-37-1	Rainbow Custom Coloured Mortars
CAS 1309-37-1	Red 139
CAS 1309-37-1	Transoxide
CAS 1309-37-1	Two Cubed Eight
CAS 1309-37-1	Yellow 201
CAS 1309-42-8	Almagel.
CAS 1309-42-8	FR-20
CAS 1309-42-8	Gilumag
CAS 1309-42-8	Liquigel®.
CAS 1309-42-8	Magnaspheres®
CAS 1309-42-8	Magnifin® H10
CAS 1309-42-8	Magnifin® H10C
CAS 1309-42-8	Magnifin® H7A
CAS 1309-42-8	Magoh-S
CAS 1309-42-8	MGH-93
CAS 1309-42-8	Mucogel®.
CAS 1309-42-8	Neutramag®
CAS 1309-42-8	Versamag® DC
CAS 1309-42-8	Zerogen® 10
CAS 1309-42-8	Zerogen® 60
CAS 1309-48-4	Bislumina.
CAS 1309-48-4	Corox
CAS 1309-48-4	Dynamag
CAS 1309-48-4	Elastomag® 100
CAS 1309-48-4	Encapsulated MgO
CAS 1309-48-4	Flamarret®
CAS 1309-48-4	Insulmag®
CAS 1309-48-4	Ken-Mag®
CAS 1309-48-4	MagChem® 10B
CAS 1309-48-4	MagChem® 10—60
CAS 1309-48-4	MagChem® 20
CAS 1309-48-4	Maglite Y
CAS 1309-48-4	Maglite® D
CAS 1309-48-4	Magotex
CAS 1309-48-4	Magox® 98 HR
CAS 1309-48-4	Magox® Super Premium
CAS 1309-48-4	Magrods

CAS 1309-48-4	Plastomag® 170
CAS 1309-48-4	Rhenomag
CAS 1309-48-4	Scorchex
CAS 1309-48-4	Scorchguard O
CAS 1309-48-4	Scorchguard-bound
CAS 1309-48-4	Sermag®
CAS 1309-48-4	Tanbase®
CAS 1309-64-4	Crystic® Prefil F
CAS 1309-64-4	FireShield® H
CAS 1309-64-4	FireShield® HPM
CAS 1309-64-4	FireShield® L
CAS 1309-64-4	Octoguard FR-10
CAS 1309-64-4	Octoguard FR-15
CAS 1309-64-4	Petcat R-9
CAS 1309-64-4	UltraFine® II
CAS 1310-73-2	Pels
CAS 1312-76-1	Double White
CAS 1312-76-1	Kasil
CAS 1312-76-1	Potkem
CAS 1313-13-9	KM Manganese Dioxide
CAS 1313-13-9	Mangoxe
CAS 1313-60-6	Solozone
CAS 1313-99-1	Nico
CAS 1314-13-2	Activox
CAS 1314-13-2	Activox B
CAS 1314-13-2	Canfelzo
CAS 1314-13-2	Decelox
CAS 1314-13-2	Denzox
CAS 1314-13-2	Electrox
CAS 1314-13-2	Entrox
CAS 1314-13-2	Extrox
CAS 1314-13-2	Felzodox
CAS 1314-13-2	Finex-25
CAS 1314-13-2	Finex-25-020
CAS 1314-13-2	Fotofax
CAS 1314-13-2	Garozinc
CAS 1314-13-2	K-Zinc
CAS 1314-13-2	Ken-Zinc®
CAS 1314-13-2	Mazin®.
CAS 1314-13-2	Microx
CAS 1314-13-2	Octocure 553
CAS 1314-13-2	Pharmakon
CAS 1314-13-2	Photozinc
CAS 1314-13-2	Rubox
CAS 1314-13-2	Vidox
CAS 1314-13-2	Vita Zinc
CAS 1314-13-2	Zinc Oxide No. 185
CAS 1314-13-2	Zinc Oxide No. 318
CAS 1314-13-2	Zinc Oxide Transparent
CAS 1314-13-2	Zinkoxyd Activ®
CAS 1314-13-2	Zinox
CAS 1314-23-4	Dynazirkon
CAS 1314-23-4	Zedox
CAS 1314-56-3	Superphosphate
CAS 1314-60-9	Nyacol® A-1530
CAS 1314-60-9	TPL
CAS 1314-62-1	KM Vanadium Pentoxide
CAS 1314-98-3	Sachtolith®
CAS 1317-26-6	Riopan
CAS 1317-33-5	Moldag 200
CAS 1317-33-5	Molydag 204
CAS 1317-33-5	Molydag 206
CAS 1317-33-5	Molydag 208
CAS 1317-33-5	Molydag 210
CAS 1317-33-5	Molydag 211
CAS 1317-33-5	Molydag 214
CAS 1317-33-5	SLA 1208
CAS 1317-33-5	SLA 1261
CAS 1317-33-5	SLA 1286
CAS 1317-33-5	SLA 2208
CAS 1317-38-0	UP 13600
CAS 1317-39-1	Brown Copp
CAS 1317-39-1	Cupridan
CAS 1317-39-1	Cuprox
CAS 1317-39-1	Lolotint 97
CAS 1317-39-1	Nordox
CAS 1317-39-1	Purple Copp
CAS 1317-39-1	Purplecopp 97N Premium
CAS 1317-39-1	Red Copp
CAS 1317-39-1	Redcopp 97N Premium
CAS 1317-65-3	Atomite®
CAS 1317-65-3	Calbux
CAS 1317-65-3	Carbital® 35
CAS 1317-65-3	Carbital® 50
CAS 1317-65-3	CC-103.
CAS 1317-65-3	CP Filler.
CAS 1317-65-3	Drikalite®.
CAS 1317-65-3	Duramite®
CAS 1317-65-3	Garden Lime
CAS 1317-65-3	Kotamite®.
CAS 1317-65-3	Marble Dust.
CAS 1317-65-3	Marblemite
CAS 1317-65-3	Micro-White® 07 Slurry
CAS 1317-65-3	Micro-White® 10 Codex
CAS 1317-65-3	Micro-White® 15
CAS 1317-65-3	Micro-White® 25
CAS 1317-65-3	Micro-White® 40
CAS 1317-65-3	Micro-White® 100
CAS 1317-65-3	No. 1 White.
CAS 1317-65-3	Opacicoat.
CAS 1317-65-3	Opacimite.
CAS 1317-65-3	Snowflake P.E.
CAS 1317-65-3	Snowflake White.
CAS 1317-65-3	Supercoat®
CAS 1317-65-3	Supermite®
CAS 1317-97-1	U Blue 104
CAS 1318-74-7	Sillikolloid P 87.
CAS 1318-74-7	Sillitin N 82.
CAS 1318-93-0	Fulcat Catalysts
CAS 1318-93-0	Fulmont Activated Bleaching Earths
CAS 1318-93-0	K 129
CAS 1318-93-0	Lubrigel
CAS 1319-41-1	Sepiogel A
CAS 1321-94-4	Arosolve MN-LF
CAS 1322-93-6	Aerosol® OS
CAS 1322-93-6	Dehscofix 916
CAS 1322-93-6	Rhodacal® IN
CAS 1322-93-6	Supragil® WP
CAS 1322-98-1	Witconate DS
CAS 1323-03-1	Ceraphyl 50
CAS 1323-03-1	Wickenol® 506
CAS 1323-38-2	Surfactol® 13
CAS 1323-39-3	Aldo® PGHMS KFG

CAS 1323-39-3	Cerasynt PA
CAS 1323-39-3	CPH-52-SE
CAS 1323-39-3	Emalex PGS
CAS 1323-39-3	Emerest® 2380
CAS 1323-39-3	Grindtek PGMS 90
CAS 1323-39-3	Hodag PGMS
CAS 1323-39-3	Hodag PGS
CAS 1323-39-3	Kessco® PGMS
CAS 1323-39-3	Lipo PGMS
CAS 1323-39-3	Mazol® PGMS
CAS 1323-39-3	Pegosperse® PMS CG
CAS 1323-39-3	Promodan SP
CAS 1323-39-3	Radiamuls® PG 2201
CAS 1323-39-3	Schercemol PGMS
CAS 1323-39-3	Witconol 2380
CAS 1323-42-8	Naturechem® GMHS
CAS 1323-42-8	Paricin® 13
CAS 1323-83-7	Kessco® GDS 386F
CAS 1327-33-9	Cooksons
CAS 1327-33-9	Dechlorane A-O
CAS 1327-33-9	Thermoguard® L
CAS 1327-33-9	Thermoguard® UF
CAS 1327-33-9	Timonox
CAS 1327-33-9	Timonox Blue Star
CAS 1327-43-1	Van Gel®
CAS 1330-43-4	FB 48
CAS 1330-43-4	Solubor
CAS 1330-61-6	Ageflex FA-10
CAS 1330-61-6	Sipomer® IDA
CAS 1330-76-3	Bisomer D10M
CAS 1330-76-3	Octomer DIOM
CAS 1330-78-5	Syn-O-Ad® 8484
CAS 1330-80-9	Emalex PGO
CAS 1330-80-9	Prolein
CAS 1330-80-9	Radiamuls® PG 2206
CAS 1330-85-4	Country Fresh Disinfectant
CAS 1331-61-9	Ablusol DBM
CAS 1331-61-9	Hetsulf 50A
CAS 1331-61-9	Nansa® AS 40
CAS 1332-40-7	Cupravit®
CAS 1332-40-7	Cuprokylt
CAS 1332-40-7	Cuprosana
CAS 1332-40-7	Curenox-50
CAS 1332-40-7	FS Dricol 50
CAS 1332-40-7	Headland Inorganic Liquid Copper
CAS 1332-58-7	Anhydrol
CAS 1332-58-7	ASP®
CAS 1332-58-7	Bilt-Cote®
CAS 1332-58-7	Bilt-Plates®
CAS 1332-58-7	Buca
CAS 1332-58-7	Catalpo
CAS 1332-58-7	Continental® Clay
CAS 1332-58-7	Dixie Clay®
CAS 1332-58-7	Fiberfrax® 6000 RPS
CAS 1332-58-7	Fiberkal
CAS 1332-58-7	Huber 40C
CAS 1332-58-7	Huber 65A
CAS 1332-58-7	Huber 95
CAS 1332-58-7	Kaowool®
CAS 1332-58-7	Kayphobe-ABO
CAS 1332-58-7	Langford Clay

CAS 1332-58-7	McNamee Clay®
CAS 1332-58-7	Par Clay®
CAS 1332-58-7	Peerless®
CAS 1332-58-7	Peerless® No. 1
CAS 1332-58-7	Pharmolin
CAS 1332-58-7	Polyplate 90
CAS 1332-58-7	Satintone
CAS 1332-58-7	SP-33
CAS 1332-58-7	SPS
CAS 1332-58-7	Translink®
CAS 1332-58-7	Vanclay
CAS 1332-58-7	Whitetex
CAS 1333-39-7	Eltesol® PSA 65
CAS 1333-86-4	Addipast
CAS 1333-86-4	Black Pearls® 1100
CAS 1333-86-4	Continex® LH-10
CAS 1333-86-4	Continex® N351
CAS 1333-86-4	Derussole
CAS 1333-86-4	Diablack® A
CAS 1333-86-4	Efweko
CAS 1333-86-4	Elftex® 675
CAS 1333-86-4	Flex Carbon
CAS 1333-86-4	Furnex
CAS 1333-86-4	FW 18
CAS 1333-86-4	FW 200 Beads and Powd
CAS 1333-86-4	Ketjenblack® EC-310 NW
CAS 1333-86-4	Microsperse
CAS 1333-86-4	Modulex
CAS 1333-86-4	Mogul® L
CAS 1333-86-4	Monarch® 1100
CAS 1333-86-4	Neospectra
CAS 1333-86-4	Neotex
CAS 1333-86-4	Novacarb
CAS 1333-86-4	Octojet 104
CAS 1333-86-4	Printex
CAS 1333-86-4	Printex 25 Beads and Powd
CAS 1333-86-4	Printex P
CAS 1333-86-4	Printex U Beads and Powd
CAS 1333-86-4	Raven
CAS 1333-86-4	Regal® 400R
CAS 1333-86-4	S160 Beads and Powd
CAS 1333-86-4	Sevacarb
CAS 1333-86-4	Special Black 4, 4A Beads and Powd
CAS 1333-86-4	Special Black 100 Powd
CAS 1333-86-4	Statex
CAS 1333-86-4	Sterling
CAS 1333-86-4	Superba
CAS 1333-86-4	Superjet
CAS 1333-86-4	Tack
CAS 1333-86-4	Thermax® Floform N-990
CAS 1333-86-4	Thermax® Stainless
CAS 1333-86-4	Thermax® Stainless Floform N-907
CAS 1333-86-4	United
CAS 1333-86-4	Velvetex
CAS 1333-86-4	Vulcan® XC72R
CAS 1335-72-4	Avirol® SE 3002
CAS 1335-72-4	Texapon PLT-227
CAS 1335-72-4	Zetesol NL
CAS 1336-20-5	Tetrex
CAS 1336-20-5	Tetrex Bidcaps

CAS 1336-29-4	Clysodrast
CAS 1337-33-3	Morflex MSC
CAS 1337-76-4	Atasorb
CAS 1337-76-4	Attaclay
CAS 1337-76-4	Attacote
CAS 1337-76-4	Attaflow
CAS 1337-76-4	Attagel
CAS 1337-76-4	Attapulgus
CAS 1337-76-4	Attasorb
CAS 1337-76-4	Carrisorb
CAS 1337-76-4	Diluex®, FG
CAS 1337-76-4	Economy Flor-Dri
CAS 1337-76-4	Emcor
CAS 1337-76-4	Flor-Kleen
CAS 1337-76-4	Florco®
CAS 1337-76-4	Florco®-X
CAS 1337-76-4	Florex®
CAS 1337-76-4	Florex® Ag-Dri 6/30
CAS 1337-76-4	Florex® LVM 8/16
CAS 1337-76-4	Florex® RVM 8/16
CAS 1337-76-4	Florigel® H-Y
CAS 1337-76-4	Gelsorb B
CAS 1337-76-4	Minugel
CAS 1337-76-4	Min-U-Gel® 100
CAS 1337-76-4	Min-U-Gel® 200
CAS 1337-76-4	Min-U-Gel® AR
CAS 1337-76-4	Min-U-Gel® CW
CAS 1337-76-4	Min-U-Gel® LF
CAS 1337-76-4	Pharmasorb
CAS 1337-76-4	QA-555
CAS 1337-76-4	Refinex
CAS 1337-76-4	S-60 RVM
CAS 1337-81-1	Good-rite® 2528X10
CAS 1338-23-4	Butanox
CAS 1338-23-4	Cadox® HBO-50
CAS 1338-23-4	Cadox® L-30
CAS 1338-23-4	Esperfoam® FR
CAS 1338-23-4	Hi-Point® 90
CAS 1338-23-4	HPC-9
CAS 1338-23-4	Lupersol DDM-9
CAS 1338-23-4	Lupersol DSW
CAS 1338-23-4	Quickset® Extra
CAS 1338-23-4	Sprayset® MEKP
CAS 1338-24-5	Agenap HMW-H
CAS 1338-39-2	Ablunol S-20
CAS 1338-39-2	Alkamuls® S-20
CAS 1338-39-2	Alkamuls® SML
CAS 1338-39-2	Arlacel® 20
CAS 1338-39-2	Atmer® 100
CAS 1338-39-2	Crill 1
CAS 1338-39-2	Dehymuls SML
CAS 1338-39-2	Disponil SML 100 F1
CAS 1338-39-2	Emsorb® 2515
CAS 1338-39-2	Ethylan® GL20
CAS 1338-39-2	Glycomul® L
CAS 1338-39-2	Hetan SL
CAS 1338-39-2	Hodag SML
CAS 1338-39-2	Kemester® S20
CAS 1338-39-2	Liposorb L
CAS 1338-39-2	Montane 20
CAS 1338-39-2	Prote-sorb SML
CAS 1338-39-2	S-Maz® 20
CAS 1338-39-2	Sorbax SML
CAS 1338-39-2	Sorbon S-20
CAS 1338-39-2	Span® 20
CAS 1338-41-6	Ablunol S-60
CAS 1338-41-6	Alkamuls® S-60
CAS 1338-41-6	Alkamuls® SMS
CAS 1338-41-6	Arlacel® 60
CAS 1338-41-6	Capmul® S
CAS 1338-41-6	Crill 3
CAS 1338-41-6	Dehymuls SMS
CAS 1338-41-6	Disponil SMS 100 F1
CAS 1338-41-6	Drewsorb® 60K
CAS 1338-41-6	Durtan® 60
CAS 1338-41-6	Emalex SPE-100S
CAS 1338-41-6	Emsorb® 2505
CAS 1338-41-6	Ethylan® GS60
CAS 1338-41-6	Famodan MS Kosher
CAS 1338-41-6	Glycomul® S FG
CAS 1338-41-6	Grindtek SMS
CAS 1338-41-6	Hetan SS
CAS 1338-41-6	Hodag SMS
CAS 1338-41-6	Kemester® S60
CAS 1338-41-6	Liposorb S
CAS 1338-41-6	Montane 60
CAS 1338-41-6	Polycon S60 K
CAS 1338-41-6	Prote-sorb SMS
CAS 1338-41-6	Radiamuls® Sorb 2145
CAS 1338-41-6	Radiamuls® Sorb 2161
CAS 1338-41-6	Radiamuls® Sorb 2166
CAS 1338-41-6	S-Maz® 60K
CAS 1338-41-6	Sorbax SMS
CAS 1338-41-6	Sorbon S-60
CAS 1338-41-6	Span® 60, 60K
CAS 1338-43-8	Ablunol S-80
CAS 1338-43-8	Alkamuls® S-80
CAS 1338-43-8	Alkamuls® SMO
CAS 1338-43-8	Arlacel® 80
CAS 1338-43-8	Atmer® 105
CAS 1338-43-8	Capmul® O
CAS 1338-43-8	Crill 4
CAS 1338-43-8	Dehymuls SMO
CAS 1338-43-8	DeSotan® SMO
CAS 1338-43-8	Disponil SMO 100 F1
CAS 1338-43-8	Emalex SPO-100
CAS 1338-43-8	Emsorb® 2500
CAS 1338-43-8	Ethylan® GO80
CAS 1338-43-8	Glycomul® O
CAS 1338-43-8	Hetan SO
CAS 1338-43-8	Hodag SMO
CAS 1338-43-8	Kemester® S80
CAS 1338-43-8	Liposorb O
CAS 1338-43-8	Montane 80
CAS 1338-43-8	Polycon S80 K
CAS 1338-43-8	Prote-sorb SMO
CAS 1338-43-8	S-Maz® 80
CAS 1338-43-8	Sorbax SMO
CAS 1338-43-8	Sorbon S-80
CAS 1338-43-8	Span® 80
CAS 1338-43-8	Witconol 2500
CAS 1338-51-8	Uniplex 600
CAS 1338-51-8	Uniplex 680
CAS 1341-54-4	Uvinul® 490

CAS 1341-54-4	Uvinul® M-493	CAS 1405-10-3	Myciguent
CAS 1343-88-0	Celkate T-21	CAS 1405-10-3	Neobiotic
CAS 1343-98-2	Flexin	CAS 1405-10-3	Neomin
CAS 1344-00-9	Ketjensil® SM 405	CAS 1405-10-3	Neomix
CAS 1344-00-9	Valfor® 100	CAS 1405-10-3	Nivemycin
CAS 1344-09-8	Crystal	CAS 1405-20-5	Aerosporin Sterile Powder
CAS 1344-09-8	Gasbinda	CAS 1405-37-4	Capastat
CAS 1344-09-8	Marley Cement Dustproofer	CAS 1405-37-4	Capastat Sulfate
CAS 1344-09-8	N	CAS 1405-41-0	Brislagen
CAS 1344-28-1	A-2	CAS 1405-41-0	Garamycin
CAS 1344-28-1	Alcan AA-100	CAS 1405-41-0	Geminimycin
CAS 1344-28-1	Alcan C-70, C-71, C-72,	CAS 1405-41-0	Genoptic Liquifilm®
	C-73, C-75	CAS 1405-41-0	Genoptic S.O.P.®
CAS 1344-28-1	Alexite	CAS 1405-41-0	Gentacidin® Ophthalmic Solution
CAS 1344-28-1	Alfrax B301		and Ointment
CAS 1344-28-1	Alufrit	CAS 1405-41-0	U-Gencin
CAS 1344-28-1	Aluminum Oxide C	CAS 1405-87-4	Baciguent
CAS 1344-28-1	Alumite	CAS 1406-65-1	Amplex
CAS 1344-28-1	Brasivol	CAS 1420-03-7	Delinal
CAS 1344-28-1	C-1	CAS 1420-04-8	Bayluscid
CAS 1344-28-1	Compalox	CAS 1420-55-9	Torecan
CAS 1344-28-1	D-201	CAS 1421-68-7	Dricol
CAS 1344-28-1	Dirubin	CAS 1424-00-6	Pro-Viron
CAS 1344-28-1	Dural	CAS 1446-61-3	Amine D
CAS 1344-28-1	Dycron	CAS 1450-14-2	CH7280
CAS 1344-28-1	Flame guard	CAS 1459-93-4	Morflex 1129
CAS 1344-28-1	Italcor	CAS 1459-93-4	Uniplex 270
CAS 1344-28-1	Maftec®	CAS 1462-54-0	Mirataine® HC-Acid
CAS 1344-28-1	Martipol	CAS 1462-55-1	DV-1936
CAS 1344-28-1	Martisorb	CAS 1476-53-5	Albamycin Capsules
CAS 1344-28-1	Martoxin	CAS 1491-41-4	Maretin
CAS 1344-28-1	Purdox	CAS 1508-65-2	Ditropan
CAS 1344-28-1	Realox®	CAS 1508-75-4	Mydriacyl
CAS 1344-28-1	Rewagit	CAS 1508-76-5	Kemadrin Tablets/Injection
CAS 1344-28-1	S-201	CAS 1524-88-5	Cordran
CAS 1344-28-1	Saffil®	CAS 1528-49-0	Morflex 560
CAS 1344-28-1	Selexsorb® COS	CAS 1533-45-5	Eastobrite® OB-1
CAS 1344-28-1	T-1061	CAS 1538-09-6	Bicillin L-A
CAS 1344-28-1	T-64	CAS 1538-09-6	Diapen
CAS 1344-28-1	Versal 150	CAS 1538-09-6	Permapen
CAS 1344-40-7	EPlthal 120	CAS 1553-60-2	Dytransin
CAS 1369-66-3	Empimin® OP70	CAS 1558-33-4	CC3275
CAS 1390-65-4	Carmine 40	CAS 1561-92-8	Geropon® MLS/A
CAS 1390-65-4	Carmine 224	CAS 1562-00-1	Witconate NIS
CAS 1397-89-3	Fungilin Cream	CAS 1563-66-2	Carbodan
CAS 1397-89-3	Fungilin Lozenges	CAS 1563-66-2	Curaterr®
CAS 1397-89-3	Fungilin Ointment	CAS 1563-66-2	Nex
CAS 1397-89-3	Fungizone for Infusion	CAS 1563-66-2	Rampart
CAS 1400-61-9	Diastatin	CAS 1563-66-2	Sipcam UK Carbosip 5G
CAS 1400-61-9	Mycostatin	CAS 1563-66-2	Tripart® Nex
CAS 1400-61-9	Nystan Cream	CAS 1563-66-2	Yaltox®
CAS 1400-61-9	Nystan Pastille	CAS 1582-09-8	Autumn Kite.
CAS 1400-61-9	Nystavescent	CAS 1582-09-8	Campbell's Trifluron.
CAS 1400-61-9	Nystex	CAS 1582-09-8	Chandor.
CAS 1400-61-9	O-V Statin	CAS 1582-09-8	Devrinol T.
CAS 1402-82-0	Ecomytrin®	CAS 1582-09-8	Flint.
CAS 1403-17-4	Vanobid	CAS 1582-09-8	Janus.
CAS 1404-04-2	Ecomytrin®	CAS 1582-09-8	Linnet.
CAS 1404-88-2	Bactratycin	CAS 1582-09-8	Marksman.
CAS 1404-93-9	Vancocin Hydrochloride	CAS 1582-09-8	Onslaught.
CAS 1405-10-3	Biosol	CAS 1582-09-8	Ornamental Weeder
CAS 1405-10-3	Mycifradin	CAS 1582-09-8	Solo.

CAS 1582-09-8	Treflan		CAS 1702-17-6	Agricorn D
CAS 1582-09-8	Tri-Farmon.		CAS 1702-17-6	Albar-40
CAS 1582-09-8	Trifarmon.		CAS 1702-17-6	Albar-Super
CAS 1582-09-8	Triflurex		CAS 1702-17-6	Atlas D
CAS 1582-09-8	Trigard		CAS 1702-17-6	BH 2,4-D Ester 40
CAS 1582-09-8	Trilin®		CAS 1702-17-6	Campbell's Destox
CAS 1582-09-8	Trilin® 10G		CAS 1702-17-6	Campbell's Dioweed 50
CAS 1582-09-8	Trimaran		CAS 1702-17-6	CDA Dicotox Extra
CAS 1582-09-8	Tripart® Trifluralin 48 EC		CAS 1702-17-6	Dicotox Extra
CAS 1582-09-8	Tristar		CAS 1702-17-6	Farmon 2,4-D
CAS 1592-23-0	Afco-Chem CS		CAS 1702-17-6	For-Ester
CAS 1592-23-0	Hallcote® CSD		CAS 1702-17-6	Green Up Lawn Spot Weedkiller
CAS 1592-23-0	Nopcote C-104		CAS 1702-17-6	Green Up Weedfree Lawn
CAS 1592-23-0	Petrac® CP-11			Weedkiller
CAS 1592-23-0	Synpro® 15F		CAS 1702-17-6	Green Up Weedfree Spot
CAS 1596-84-5	B-Nine			Weedkiller for Lawns
CAS 1596-84-5	Dazide		CAS 1702-17-6	Hadranol
CAS 1597-82-6	Haldrate		CAS 1702-17-6	Lontrel
CAS 1597-82-6	Haldrone		CAS 1702-17-6	MSS 2,4-D Amine
CAS 1597-82-6	Monocortin		CAS 1702-17-6	MSS 2,4-D Ester
CAS 1597-82-6	Stemex		CAS 1702-17-6	Polymone
CAS 1607-17-6	Petrin		CAS 1702-17-6	Shell D 50
CAS 1617-90-9	Sostenil		CAS 1702-17-6	Silvapron
CAS 1617-90-9	Vincapront		CAS 1702-17-6	Syford
CAS 1617-90-9	Vincaron		CAS 1704-62-7	Texacat® ZR-70
CAS 1620-21-9	Histantin		CAS 1707-14-8	Preludin
CAS 1622-61-3	Clonopin		CAS 1715-33-9	Solu-Delta-Cortef
CAS 1622-61-3	Rivotril		CAS 1717-00-6	Genetron® 141b
CAS 1622-62-4	Rohypnol		CAS 1719-57-9	CC3270
CAS 1634-02-2	Akrochem® TBUT		CAS 1719-58-0	CV4720
CAS 1634-04-4	Driveron		CAS 1722-62-9	Carbocaine
CAS 1642-54-2	Banocide		CAS 1722-62-9	Polocaine
CAS 1642-54-2	Caritrol		CAS 1732-10-1	Emery® 2914
CAS 1642-54-2	Dirocide		CAS 1740-22-3	Surexin
CAS 1642-54-2	Franocide		CAS 1760-24-3	CA0700
CAS 1643-19-2	TBAB		CAS 1760-24-3	Dow Corning® Z-6020
CAS 1643-20-5	Empigen® OB		CAS 1760-24-3	Dynasylan® DAMO
CAS 1643-20-5	Rhodamox® LO		CAS 1760-24-3	Dynasylan® DAMO-P
CAS 1643-20-5	Schercamox DML		CAS 1760-24-3	Dynasylan® DAMO-T
CAS 1649-18-9	Stresnil		CAS 1760-24-3	Prosil® 3128
CAS 1649-18-9	Suicalm		CAS 1760-24-3	Union Carbide® A-1120
CAS 1656-63-9	Weston® TLTTP		CAS 1777-82-8	Myacide® SP
CAS 1665-48-1	Skeladin		CAS 1786-81-8	Citanest
CAS 1668-19-5	Sinequan		CAS 1808-12-4	Ambodryl
CAS 1680-21-3	SR-272		CAS 1809-19-4	Syn-O-Ad® P-316
CAS 1689-83-4	Actrilawn 10		CAS 1812-30-2	Lexotan
CAS 1689-83-4	Hobane		CAS 1812-53-9	Varisoft® 432-100
CAS 1689-83-4	Totril		CAS 1825-62-3	CT2970
CAS 1689-84-5	Hobane		CAS 1825-62-3	EXP-51
CAS 1689-89-0	Trodax		CAS 1841-19-6	Redeptin
CAS 1698-60-8	Atlas Silver		CAS 1841-19-6	Redeptin Injection
CAS 1698-60-8	Better Flowable		CAS 1843-03-4	Lowinox® CA 22
CAS 1698-60-8	Chiltern Pyrazol		CAS 1843-03-4	Topanol® CA
CAS 1698-60-8	Gladiator		CAS 1843-03-4	Topanol® LVT 11
CAS 1698-60-8	Paramin DF		CAS 1843-05-6	Cyasorb® UV 531
CAS 1698-60-8	Pyramin®		CAS 1843-05-6	Hostavin® ARO 8
CAS 1698-60-8	Spectron		CAS 1843-05-6	Lankromark® LE285
CAS 1698-60-8	Starter Flowable		CAS 1843-05-6	Lowilite® 22
CAS 1698-60-8	Tripart® Gladiator		CAS 1843-05-6	Spectra-Sorb UV 531
CAS 1698-60-8	Trojan SC		CAS 1843-05-6	Uvinul® 408
CAS 1698-60-8	Weedmaster		CAS 1847-55-8	Duponol LS
CAS 1702-17-6	Agrichem		CAS 1847-55-8	Rhodapon® OS

CAS 1847-55-8	Sulfopon O 680
CAS 1854-23-5	Bioban® P-1487®
CAS 1854-23-5	Fuelsaver®
CAS 1861-32-1	Dacthal
CAS 1861-32-1	Vegetable Turf and Ornamental Weeder
CAS 1861-40-1	Benefex
CAS 1867-66-9	Ketaject
CAS 1867-66-9	Ketalar
CAS 1867-66-9	Ketaset
CAS 1867-66-9	Ketavet
CAS 1867-66-9	Vetalar
CAS 1889-67-4	Perkadox® 30
CAS 1897-45-6	RB Chlorothalonil
CAS 1897-45-6	Bombardier
CAS 1897-45-6	Bravo 500
CAS 1897-45-6	Chiltern Ole
CAS 1897-45-6	Chlorothalonil
CAS 1897-45-6	Contact 75
CAS 1897-45-6	Daconil Turf
CAS 1897-45-6	Groutcide 75
CAS 1897-45-6	Impact Excel
CAS 1897-45-6	Jupital
CAS 1897-45-6	Nopcocide® N-40-D
CAS 1897-45-6	Nopcocide® N-96
CAS 1897-45-6	Nopcocide® N-96-S
CAS 1897-45-6	NuoCide®
CAS 1897-45-6	Power Chlorothalonil 50
CAS 1897-45-6	Repulse
CAS 1897-45-6	Siclor
CAS 1897-45-6	Sipcam UK Rover 500
CAS 1897-45-6	Tripart® Faber
CAS 1897-45-6	Tripart® Ultrafaber
CAS 1910-42-5	Dextrone X
CAS 1910-42-5	Dukatalon
CAS 1910-42-5	Gramixel
CAS 1910-42-5	Gramonol
CAS 1910-42-5	Gramonol 5.
CAS 1910-42-5	Gramoxone
CAS 1910-42-5	Gramoxone X
CAS 1910-42-5	Gramuron
CAS 1910-42-5	Scythe
CAS 1910-42-5	Speedway
CAS 1910-42-5	Weedol
CAS 1912-24-9	Alazine.
CAS 1912-24-9	Ashlade 4% At Gran
CAS 1912-24-9	Ashlade Atrazine 50 FL
CAS 1912-24-9	Atlazin
CAS 1912-24-9	Atraflow
CAS 1912-24-9	Atraflow Plus
CAS 1912-24-9	Atranex
CAS 1912-24-9	Borocil A
CAS 1912-24-9	Boroflow A
CAS 1912-24-9	Boroflow A/ATA
CAS 1912-24-9	Chipman Path Weedkiller
CAS 1912-24-9	Chlorea
CAS 1912-24-9	Gesaprim 500FW
CAS 1912-24-9	Herbazin Total
CAS 1912-24-9	Mascot Gauntlet
CAS 1912-26-1	Remtal
CAS 1912-26-1	Senate.
CAS 1918-00-9	Endox
CAS 1918-00-9	Green Up Weedfree Lawn Weedkiller
CAS 1918-00-9	Green Up Weedfree Spot Weedkiller for Lawns
CAS 1918-00-9	Tracker
CAS 1918-02-1	Tordon
CAS 1918-02-1	Tordon 22K
CAS 1918-16-7	Albrass
CAS 1918-16-7	Atlas Orange
CAS 1918-16-7	Bexton
CAS 1918-16-7	Croptex Amber
CAS 1918-16-7	Portman Propachlor 50FL
CAS 1918-16 7	Prolex
CAS 1918-16-7	Ramrod
CAS 1918-16-7	Tripart® Granular
CAS 1918-16-7	Tripart® Sentinel
CAS 1931-62-0	Esperox® 41-25A
CAS 1948-33-0	Eastman® MTBHQ
CAS 1948-33-0	Sustane® TBHQ
CAS 1954-81-4	Naptol
CAS 1975-78-6	Nitrile 10 D
CAS 1977-10-2	Loxitane
CAS 1982-37-2	Tacaryl
CAS 1982-47-4	Tenoran
CAS 2002-29-1	Locorten
CAS 2016-36-6	Stanaprin
CAS 2016-42-4	Amine 14D
CAS 2016-56-0	Acetamin 24
CAS 2016-56-0	Nopcogen 16-L
CAS 2016-57-1	Amine 12
CAS 2016-57-1	Amine 12-98D
CAS 2022-85-7	Alcobon
CAS 2022-85-7	Ancobon
CAS 2030-63-9	Lamprene
CAS 2031-67-6	CM9050
CAS 2031-67-6	Dynasylan® MTES
CAS 2031-67-6	Union Carbide® A-162
CAS 2032-65-7	Club
CAS 2032-65-7	Draza®
CAS 2032-65-7	Mesurol®
CAS 2044-56-6	Empicol® HL25
CAS 2058-46-0	Terramycin Hydrochloride
CAS 2058-46-0	Unimycin
CAS 2062-78-4	Orap
CAS 2062-84-2	Anquil
CAS 2068-78-2	Oncovin
CAS 2082-79-3	Anox® PP 18
CAS 2082-79-3	Lowinox® PO35
CAS 2082-79-3	Naugard® 76
CAS 2082-79-3	Ralox® 530
CAS 2082-79-3	Ultranox® 276
CAS 2082-80-6	Weston® TSP
CAS 2083-91-2	CD5400
CAS 2094-98-6	V-40
CAS 2116-84-9	Abil® AV 20-1000
CAS 2116-84-9	Abil® AV 8853
CAS 2116-84-9	Emalex MTS-30E
CAS 2119-75-7	Methral
CAS 2135-22-0	Blade
CAS 2135-22-0	Vydate®
CAS 2152-34-3	Cylert
CAS 2152-34-3	Volital

CAS 2152-44-5	Betatrex
CAS 2152-44-5	Betnovate
CAS 2152-44-5	Bextasol
CAS 2152-44-5	Retenema
CAS 2155-71-7	Trigonox® 111-B40
CAS 2156-97-0	Ageflex FA-12
CAS 2156-97-0	SR-335
CAS 2163-42-0	MPDiol Glycol
CAS 2163-80-6	Drexar 530
CAS 2163-80-6	Weed Hoe
CAS 2164-08-1	Venzar®
CAS 2164-08-1	Vizor
CAS 2164-17-2	Cottonex
CAS 2164-17-2	Meturon® 4L
CAS 2167-23-9	Trigonox® D-E50
CAS 2190-04-7	Acetamin 86
CAS 2190-04-7	Acetamin T
CAS 2210-79-9	DY 023
CAS 2210-79-9	Heloxy® 62
CAS 2223-93-0	Synpro® Cadmium Stearate
CAS 2224-33-1	CV5050
CAS 2224-44-4	Bioban® P-1487®
CAS 2224-44-4	Fuelsaver®
CAS 2227-17-0	Pentac Aquaflow
CAS 2235-43-0	Unihib® 314
CAS 2235-54-3	Akyposal ALS 33
CAS 2235-54-3	Avirol® A
CAS 2235-54-3	Carsonol® ALS-R
CAS 2235-54-3	DeSonol A
CAS 2235-54-3	Lorol NH
CAS 2235-54-3	Maprofix NH and NHL
CAS 2235-54-3	Marlinat® DFN 30
CAS 2235-54-3	Neopon LAM
CAS 2235-54-3	Nutrapon HA 3841
CAS 2235-54-3	Nutrapon PP 3563
CAS 2235-54-3	Octosol ALS-28
CAS 2235-54-3	Perlankrol® DAF25
CAS 2235-54-3	Polystep® B-7
CAS 2235-54-3	Rhodapon® L-22, L-22/C
CAS 2235-54-3	Sermul EA129
CAS 2235-54-3	Standapol® A
CAS 2235-54-3	Stepanol® AM
CAS 2235-54-3	Sulfochem ALS
CAS 2235-54-3	Tensopol N
CAS 2235-54-3	Texapon ALS
CAS 2235-54-3	Ufarol Am 30
CAS 2235-54-3	Witcolate 6430
CAS 2235-54-3	Witcolate AM
CAS 2235-54-3	Witcolate NH
CAS 2235-54-3	Zoharpon LAA
CAS 2240-14-4	Euvitol
CAS 2244-21-5	ACL 56
CAS 2244-21-5	ACL 59
CAS 2244-21-5	ACL 60
CAS 2244-21-5	CDB Clearon
CAS 2244-21-5	Clearon
CAS 2259-96-3	Anhydron
CAS 2259-96-3	Fluidil
CAS 2277-92-1	Zanil
CAS 2307-68-8	Croptex Bronze
CAS 2307-68-8	Solan
CAS 2310-17-0	Zolone
CAS 2311-27-5	Dehydrophen PNP 4
CAS 2315-02-8	Afrin® 12 Hour Nasal Spray
CAS 2315-02-8	Dristan Long Lasting Nasal Mist
CAS 2315-02-8	Duration® 12 Hour Nasal Spray
CAS 2315-02-8	Hazol
CAS 2315-02-8	Sudafed Nasal Spray
CAS 2344-80-1	CC3285
CAS 2358-84-1	Ageflex DEGDMA
CAS 2372-21-6	Trigonox® BPIC
CAS 2375-03-3	A-methaPred
CAS 2375-03-3	Solu-Medrol
CAS 2392-39-4	Decadron Shock-Pak®
CAS 2392-39-4	Maxidex Ointment
CAS 2398-96-1	Aftate
CAS 2398-96-1	Mycil
CAS 2398-96-1	Tinactin
CAS 2398-96-1	Tinaderm
CAS 2409-55-4	Cao
CAS 2409-55-4	Lowinox® MBPC
CAS 2425-79-8	Heloxy® 67
CAS 2426-08-6	Ageflex BGE
CAS 2426-08-6	Epirez 501
CAS 2426-08-6	Heloxy® 61
CAS 2426-54-2	Ageflex FA-2
CAS 2437-25-4	Nitrile 12
CAS 2438-32-6	Alunex
CAS 2438-32-6	Chlor-Trimeton
CAS 2438-32-6	Piriton
CAS 2438-32-6	Teldrin
CAS 2438-72-4	Parfenac
CAS 2439-01-2	Morestan®
CAS 2439-10-3	Dodine FL
CAS 2439-10-3	Dodine WP
CAS 2439-10-3	Radspor FT, 65 WP
CAS 2439-35-2	Adame
CAS 2439-35-2	Ageflex FA-1
CAS 2440-22-4	Lowilite® 55
CAS 2440-22-4	Topanex 100BT
CAS 2440-22-4	Uvazol P
CAS 2447-57-6	Fanasil
CAS 2447-57-6	Fanzil
CAS 2455-24-5	Ageflex THFMA
CAS 2489-77-2	Thiate E
CAS 2492-26-4	Nacap®
CAS 2492-26-4	Nuodex 84
CAS 2499-95-8	Ageflex FA-6
CAS 2499-95-8	Ageflex n-HA
CAS 2508-72-7	Antistin
CAS 2508-72-7	Dibistin
CAS 2530-83-8	Aktisil EM
CAS 2530-83-8	CG6720
CAS 2530-83-8	Dow Corning® Z-6040
CAS 2530-83-8	Dynasylan® GLYMO
CAS 2530-83-8	Prosil® 5136
CAS 2530-83-8	Union Carbide® A-187
CAS 2530-85-0	CM8550
CAS 2530-85-0	Dow Corning® Z-6030
CAS 2530-85-0	Dynasylan® MEMO
CAS 2530-85-0	Prosil® 248
CAS 2530-85-0	Union Carbide® A-174
CAS 2530-87-2	CC3300
CAS 2530-87-2	Dow Corning® Z-6076

CAS 2530-87-2	Dynasylan® CPTMO
CAS 2550-06-3	CC3291
CAS 2551-83-9	CA0567
CAS 2553-19-7	CD6000
CAS 2555-05-7	Agsol Ex 1
CAS 2555-05-7	Micropure® Ultra
CAS 2571-88-2	Admox® 18-85
CAS 2571-88-2	Amyx SO 3734
CAS 2571-88-2	Barlox® 18S
CAS 2571-88-2	Chemoxide ST
CAS 2571-88-2	Incromine Oxide S
CAS 2571-88-2	Mackamine SO
CAS 2571-88-2	Mazox® SDA
CAS 2571-88-2	Ninox® SO
CAS 2571-88-2	Rewominox S 300
CAS 2571-88-2	Rewominoxid S 300
CAS 2571-88-2	Schercamox DMS
CAS 2576-92-3	Bronkosol
CAS 2576-92-3	Numotac
CAS 2589-57-3	V-601
CAS 2593-15-9	Aaterra WP
CAS 2599-01-1	Schercemol CM
CAS 2607-06-9	Temetex
CAS 2608-24-4	Ancyte
CAS 2622-26-6	Neulactil
CAS 2624-44-4	Dicynene
CAS 2627-95-4	CD6210
CAS 2631-40-5	Etrofolan®
CAS 2631-40-5	Mipcin®
CAS 2642-71-9	Cotnion-Ethyl
CAS 2642-71-9	Cotnion-Ethyl-Methyl
CAS 2642-71-9	Gusathion® A
CAS 2644-64-6	Phospholipon® PC
CAS 2668-66-8	HMS Liquifilm®
CAS 2673-22-5	Aerosol® TR-70
CAS 2673-22-5	Monawet MT
CAS 2675-77-6	Chloroneb 65W Fungicide
CAS 2675-77-6	Chloroneb Systemic Flowable Fungicide
CAS 2675-77-6	Terraneb SP Turf Fungicide
CAS 2694-54-4	Sipomer® TATM
CAS 2724-58-5	Emersol® 871
CAS 2724-58-5	Prisorine 3508
CAS 2724-58-5	Proto-Lan IP
CAS 2745-8-931	Exxal® 18
CAS 2746-81-8	Moditen Enanthate
CAS 2746-81-8	Prolixin Enanthate
CAS 2751-09-9	Matromycin
CAS 2751-09-9	Matromycin-T
CAS 2751-09-9	Tao
CAS 2754-27-0	CT3254
CAS 2768-02-7	CV4917
CAS 2768-02-7	Dynasylan® VTMO
CAS 2768-02-7	Union Carbide® A-171
CAS 2778-96-3	Alkamuls® SS
CAS 2778-96-3	Emalex CC-18
CAS 2778-96-3	Hetester 412
CAS 2778-96-3	Lexol SS
CAS 2778-96-3	Liponate SS
CAS 2778-96-3	Ritachol® SS
CAS 2807-30-9	Ektasolve® EP
CAS 2809-21-4	Briquest® ADPA-60A
CAS 2809-21-4	Fostex P
CAS 2809-21-4	Unihib® 106
CAS 2825-60-7	Deflamene
CAS 2825-60-7	Fluderma
CAS 2829-19-8	Cypromin
CAS 2857-97-8	CT2928
CAS 2867-47-2	Ageflex FM-1
CAS 2867-47-2	Sipomer® 2M1M
CAS 2897-60-1	CG6710
CAS 2897-60-1	GP-137
CAS 2898-12-6	Nobrium
CAS 2915-57-3	Wickenol® 159
CAS 2919-66-6	MGA
CAS 2921-88-2	Atlas Sheriff
CAS 2921-88-2	Crossfire
CAS 2921-88-2	Dowco 179
CAS 2921-88-2	Dursban®
CAS 2921-88-2	Dursban® 2E
CAS 2921-88-2	Linfos
CAS 2921-88-2	Lorsban
CAS 2921-88-2	Pyrinex
CAS 2921-88-2	Salut®
CAS 2921-88-2	Saluthion®
CAS 2921-88-2	Spannit
CAS 2921-88-2	suSCon® Blue
CAS 2921-88-2	suSCon® Green
CAS 2921-88-2	Talon
CAS 2921-92-8	Etrynit
CAS 2939-80-2	Difolatan
CAS 2939-80-2	Merpafol
CAS 2939-80-2	Milcap.
CAS 2939-80-2	Sanspor
CAS 2943-75-1	CO9835
CAS 2943-75-1	Dynasylan® OCTEO
CAS 2943-75-1	Prosil® 9202
CAS 2943-75-1	Union Carbide® A-137
CAS 2955-38-6	Centrax
CAS 2955-38-6	Verstran
CAS 2998-57-4	Estracyt
CAS 3006-15-3	Empimin® MA
CAS 3006-82-4	Aztec® t-Butyl Peroctoate
CAS 3006-82-4	Aztec® t-Butyl Peroctoate-50 OMS
CAS 3006-82-4	Esperox® 28
CAS 3006-82-4	Lupersol P-31, P-33
CAS 3006-82-4	Trigonox® 21
CAS 3006-82-4	Trigonox® KSM
CAS 3006-82-4	USP®-690
CAS 3006-86-8	Aztec® 1,1-Bis(t-Butylperoxy) Cyclohexane-80 BBP
CAS 3006-86-8	Luperco 331-XL
CAS 3006-86-8	Lupersol 331-80B
CAS 3006-86-8	Lupersol P-31, P-33
CAS 3006-86-8	Trigonox® 22-BB80
CAS 3006-86-8	USP®-400P
CAS 3006-86-8	USP®-690
CAS 3006-93-7	HVA-2
CAS 3006-93-7	Vanax® MBM
CAS 3010-24-0	Ethoquad® 18/12
CAS 3026-63-9	Avirol® SA 4113
CAS 3026-63-9	Rhodapon® TDS
CAS 3033-77-0	Ogtac 85 V

CAS 3034-79-5	Perkadox® 20
CAS 3055-93-4	Akyporox RLM 22
CAS 3055-95-6	Mulsifan RT 23
CAS 3057-08-7	Weston® PNPG
CAS 3060-89-7	Patoran® FL
CAS 3060-89-7	Pattonex
CAS 3061-75-4	Kemamide® B
CAS 3061-75-4	Uniwax 1747
CAS 3064-70-8	Amerstat® 294
CAS 3064-70-8	Stauffer N-1386®
CAS 3069-25-8	CM8620
CAS 3069-29-2	CA0699
CAS 3076-63-9	Doverphos® 53
CAS 3076-63-9	Doverphos® TLP
CAS 3076-63-9	Weston® TLP
CAS 3081-14-9	Naugard® I-2
CAS 3088-31-1	Akyposal RLM 56 S
CAS 3093-35-4	Halciderm
CAS 3093-35-4	Halog
CAS 3097-08-3	Akyposal MGLS
CAS 3097-08-3	Rhodapon® LM
CAS 3101-60-8	Heloxy® 65
CAS 3115-05-7	Osbil
CAS 3115-49-9	Akypo NP 70
CAS 3121-60-6	Uvinul® DS-49
CAS 3121-61-7	Ageflex MEA
CAS 3129-91-7	Dichan 100
CAS 3147-75-9	Cyasorb® UV 5411
CAS 3147-75-9	Spectra-Sorb UV 5411
CAS 3147-75-9	Uvazol 311
CAS 3158-56-3	Cardolite® NC-507.
CAS 3179-56-4	Trigonox® ACS-M28
CAS 3179-76-8	CA0742
CAS 3179-80-4	Chemidex L
CAS 3179-80-4	Lexamine L-13
CAS 3179-80-4	Mackine 801
CAS 3179-80-4	Schercodine L
CAS 3179-81-5	Cerasynt 303
CAS 3194-55-6	Saytex® HBCD-LM
CAS 3198-07-0	Bronkephrine
CAS 3200-06-4	Praxilene
CAS 3234-85-3	Alkamuls® MM/M
CAS 3234-85-3	Ceraphyl 424
CAS 3234-85-3	Cetiol® MM
CAS 3234-85-3	Crodamol MM
CAS 3234-85-3	Kemester® MM
CAS 3234-85-3	Liponate MM
CAS 3234-85-3	Schercemol MM
CAS 3234-85-3	Waxenol® 810
CAS 3254-89-5	Vontrol
CAS 3270-71-1	Furamazone
CAS 3277-26-7	CT2030
CAS 3278-89-5	FR-913
CAS 3278-89-5	Great Lakes PHE-65
CAS 3290-92-4	Ageflex TM 402, 403, 404, 410, 421, 423, 451, 461, 462
CAS 3290-92-4	Ageflex TMPTMA
CAS 3290-92-4	Perkalink® 400
CAS 3296-90-0	FR-522
CAS 3319-31-1	Jayflex® TOTM
CAS 3319-31-1	Kodaflex® TOTM
CAS 3319-31-1	Palatinol® TOTM
CAS 3319-31-1	Plasthall® TOTM
CAS 3319-31-1	PX-338
CAS 3322-93-8	Saytex® BCL-462
CAS 3327-22-8	Catiomaster-C
CAS 3327-22-8	CHPTA 65%
CAS 3327-22-8	Quab 188
CAS 3327-22-8	Reagens-CF2
CAS 3332-27-2	Admox® 14-85
CAS 3332-27-2	Barlox® 14
CAS 3332-27-2	Empigen® OH25
CAS 3332-27-2	Incromine Oxide M
CAS 3332-27-2	Lilaminox M4
CAS 3332-27-2	Mazox® MDA
CAS 3332-27-2	Ninox® M
CAS 3332-27-2	Schercamox DMM
CAS 3337-71-1	Asulox
CAS 3344-18-1	Evac-Q-Mag
CAS 3347-22-6	Delan-Col
CAS 3349-06-2	Nicfo
CAS 3380-34-5	Irgasan DP 300
CAS 3383-96-8	Abate® 1-SG, 2-CG, 4-E, 5CG
CAS 3388-04-3	CE6250
CAS 3388-04-3	Union Carbide® A-186
CAS 3416-26-0	Clinium
CAS 3425-61-4	TAHP-80
CAS 3425-61-4	Trigonox® TAHP-W85
CAS 3436-44-0	Phospholipon® CC
CAS 3452-07-1	Neodene® 20
CAS 3457-61-2	Luperco 801-XL
CAS 3457-61-2	Trigonox® T
CAS 3483-12-3	Clelands Reagent
CAS 3485-14-1	Cyclapen-W
CAS 3486-35-9	Akrochem® 9930 Zinc Oxide Transparent
CAS 3505-38-2	Clistin
CAS 3511-16-8	Versapen
CAS 3521-62-8	Illosone
CAS 3521-62-8	Ilosone
CAS 3521-84-4	Cholografin Meglumine
CAS 3595-11-7	Benzedrex
CAS 3595-11-7	Dristan Inhaler
CAS 3614-30-0	Ceteprin
CAS 3614-30-0	Cetiprin
CAS 3614-69-5	Fenostil Retard
CAS 3614-69-5	Forhistal Maleate
CAS 3622-84-2	Dellatol® BBS
CAS 3622-84-2	Plasthall® BSA
CAS 3622-84-2	Plastomoll® BMB
CAS 3622-84-2	Uniplex 214
CAS 3625-06-7	Colofac
CAS 3625-06-7	Duspatal
CAS 3625-06-7	Duspatalin
CAS 3648-20-2	Jayflex® DUP
CAS 3648-20-2	PX-111
CAS 3655-00-3	Deriphat 160
CAS 3655-00-3	Miranol® H2C-HA
CAS 3655-00-3	Monateric 1188M
CAS 3658-48-8	Syn-O-Ad® P-310
CAS 3682-94-8	VA-088
CAS 3687-45-4	Cetiol®
CAS 3687-45-4	Schercemol OLO
CAS 3687-45-4	Starfol® OO

CAS 3687-45-4	Wickenol® 143	CAS 4098-71-9	Baymidur® KL 3-5001
CAS 3687-46-5	Ceraphyl 140	CAS 4098-71-9	Vestanat® IPDI
CAS 3687-46-5	Cetiol® V	CAS 4130-08-9	CV4800
CAS 3687-46-5	Schercemol DO	CAS 4130-08-9	Dow Corning® Z-6075
CAS 3687-46-5	Tegosoft® DO	CAS 4171-13-5	Axiquel
CAS 3689-24-5	Bladafum®	CAS 4196-86-5	Uniplex 552
CAS 3691-35-8	Drat	CAS 4205-90-7	Catapres-TTS
CAS 3691-35-8	Karate	CAS 4205-91-8	Catapres
CAS 3691-35-8	Ridene	CAS 4205-91-8	Dixarit
CAS 3691-35-8	Sakarat Special	CAS 4219-49-2	Lanol P
CAS 3691-35-8	Skaterpax	CAS 4253-34-3	CM8980
CAS 3704-09-4	Cheque	CAS 4268-36-4	Solacen
CAS 3710-84-7	Pennstop® 1866	CAS 4292-10-8	Lexaine® LM
CAS 3734-16-5	Cogesic	CAS 4299-60-9	Suladrin
CAS 3734-33-6	Bitrexene®	CAS 4304-40-9	Canopar
CAS 3734-33-6	Bitrex®	CAS 4306-88-1	Uvi-Nox 1494
CAS 3735-90-8	Escorpal	CAS 4330-99-8	Temaril®
CAS 3736-26-3	Trigonox® M-50	CAS 4330-99-8	Valledrine
CAS 3768-58-9	CB2100	CAS 4330-99-8	Vallergan
CAS 3775-90-4	Ageflex FM-4	CAS 4369-14-6	CA0397
CAS 3778-73-2	Holoxan	CAS 4390-04-9	Permethyl 101A
CAS 3778-73-2	Ifex	CAS 4419-11-8	V-65
CAS 3784-99-4	Monopar	CAS 4420-74-0	Aktisil MM
CAS 3806-34-6	Doverphos® S-680	CAS 4420-74-0	CM8500
CAS 3806-34-6	Doverphos® S-686, S-687	CAS 4420-74-0	Dynasylan® MTMO
CAS 3806-34-6	Weston® 618	CAS 4420-74-0	Prosil® 196
CAS 3806-34-6	Weston® 619	CAS 4420-74-0	Union Carbide® A-189
CAS 3810-74-0	Isoject-Streptomycin Injection	CAS 4468-02-4	Gluconal® ZN
CAS 3810-74-0	Streptorex	CAS 4484-72-4	CD6220
CAS 3810-74-0	Stretonex	CAS 4485-12-5	Afco-Chem LIS
CAS 3810-74-0	Strycin	CAS 4511-39-1	Esperox® 5100
CAS 3811-73-2	Sodium Omadine®, 40% Aq. Sol'n	CAS 4511-39-1	Trigonox® 127
		CAS 4618-18-2	Bifiteral
CAS 3813-05-6	Benazalox	CAS 4618-18-2	Cephulac
CAS 3818-50-6	Alcopar	CAS 4618-18-2	Chronulac
CAS 3819-00-9	Quide	CAS 4618-18-2	Duphalac
CAS 3847-29-8	Erythrocin	CAS 4618-18-2	Portalac
CAS 3847-29-8	Erythrocin Lactobionate-IV	CAS 4658-28-0	Brasoran 50 WP
CAS 3851-87-4	Trigonox® 36-C75	CAS 4682-36-4	Banflex
CAS 3851-87-4	USP®-355M	CAS 4682-36-4	Disipal (as HCl)
CAS 3858-89-7	Nesacaine	CAS 4682-36-4	Norflex™ Injectable
CAS 3864-99-1	Uvazol 237	CAS 4682-36-4	Norflex™ Tablets
CAS 3878-19-1	Sibutol®.	CAS 4706-78-9	Sulfochem K
CAS 3896-11-5	Lowilite® 26	CAS 4706-78-9	Tensopol SPK
CAS 3896-11-5	Uvazol 236	CAS 4712-55-4	Doverphos® 213
CAS 3913-02-8	Michel XO-150-12	CAS 4712-55-4	Doverphos® DPP
CAS 3930-19-6	Nigrin	CAS 4712-55-4	Weston® DPP
CAS 3963-95-9	Rondomycin	CAS 4719-04-4	Nipacide® BK
CAS 3978-86-7	Optimine	CAS 4722-98-9	Akyposal MLS 30
CAS 3990-03-2	Sipomer® MEM	CAS 4722-98-9	Empicol® LQ33/T
CAS 3999-10-8	Expandex® 5PT	CAS 4744-10-9	DMP
CAS 4008-48-4	Nibiol	CAS 4748-78-1	Ebal
CAS 4065-45-6	Rhodialux S	CAS 4766-57-8	CT1750
CAS 4065-45-6	Spectra-Sorb UV 284	CAS 4766-57-8	T1750
CAS 4065-45-6	Sungard	CAS 4810-50-8	Kalidone®
CAS 4065-45-6	Syntase® 230	CAS 4862-18-4	Hampamide B
CAS 4065-45-6	Uval	CAS 4936-47-4	Magnilor
CAS 4065-45-6	Uvinul® MS-40	CAS 5003-48-5	Benoral
CAS 4070-80-8	Pruv	CAS 5015-36-1	Medrol Stabisol
CAS 4075-81-4	Luprosil® Salt	CAS 5026-62-0	Nipagin M Sodium
CAS 4088-22-6	Amine M218	CAS 5039-78-1	Ageflex FM-1Q80MC
CAS 4088-22-6	Radiamine 6346	CAS 5039-78-1	Madquat Q-6

CAS 5064-31-3	Cheelox® NTA-Na3	CAS 5598-13-0	Reldan 50
CAS 5064-31-3	Hampshire® NTA 150	CAS 5598-52-7	Torelle
CAS 5064-31-3	Hampshire® NTA Na3	CAS 5611-51-8	Aristospan
CAS 5064-31-3	NTA	CAS 5611-51-8	Lederspan
CAS 5064-31-3	Trilon® A	CAS 5633-14-7	Dioxatrine
CAS 5064-31-3	Trilon® A-92	CAS 5634-38-8	Eclabron
CAS 5089-70-3	CC3292	CAS 5638-76-6	Betaserc
CAS 5089-70-3	Dynasylan® CPTEO	CAS 5638-76-6	Serc
CAS 5104-49-4	Ansaid	CAS 5638-76-6	Vasomotal
CAS 5131-24-8	Plondrel	CAS 5667-71-0	Streptohydrazid
CAS 5221-53-4	Milcurb	CAS 5707-69-7	Ganocide
CAS 5250-39-5	Floxapen	CAS 5707-69-7	Mil-Col
CAS 5259-88-1	Plantvax	CAS 5707-69-7	Saisan
CAS 5259-88-1	Plantvax 20	CAS 5714-04-5	Envacar
CAS 5259-88-1	Plantvax 75	CAS 5714-73-8	Hiprex
CAS 5259-88-1	Ringmaster	CAS 5714-73-8	Urex® Tablets
CAS 5274-68-0	Akyporox RLM 40	CAS 5714-82-9	Ranestol
CAS 5274-68-0	Hetoxol L-4	CAS 5716-20-1	Vasculit
CAS 5274-68-0	Mulsifan CPA	CAS 5743-34-0	Gluconal® CA M B
CAS 5283-66-9	CO9830	CAS 5786-21-0	Clozaril
CAS 5306-85-4	Arlasolve® DMI	CAS 5793-94-2	Crolactil CSL
CAS 5321-32-4	H-K Mastitis	CAS 5793-94-2	Pationic CSL
CAS 5321-32-4	Hetacin-K	CAS 5793-94-2	Radiamuls® CSL 2980
CAS 5321-32-4	Uropen	CAS 5793-94-2	Verv®
CAS 5321-32-4	Versapen K	CAS 5809-08-5	Trigonox® TMPH
CAS 5323-95-5	Emulgeen S	CAS 5836-29-3	Racumin®
CAS 5323-95-5	Solricin® 435	CAS 5870-29-1	Cyclogyl
CAS 5323-95-5	Solricin® 535	CAS 5870-29-1	Mydrilate
CAS 5323-95-5	Soricinol 40	CAS 5874-97-5	Metaprel
CAS 5333-42-6	Eutanol G	CAS 5875-06-9	Alcaine
CAS 5333-42-6	Exxal® 20	CAS 5875-06-9	Ophthaine
CAS 5333-42-6	Michel XO-150-20	CAS 5888-33-5	Ageflex IBOA
CAS 5356-84-3	CV5100	CAS 5902-51-2	Sinbar®
CAS 5370-01-4	Mexitil	CAS 5907-38-0	Diprofarn
CAS 5397-31-9	Tomah PA-12EH	CAS 5907-38-0	Novaldin
CAS 5411-22-3	Didrex	CAS 5928-84-7	Pen-Vee Drops
CAS 5422-34-4	Incromectant LMEA	CAS 5928-84-7	Pen-Vee Suspension
CAS 5422-34-4	Mackamide LME	CAS 5980-31-4	Sterisol
CAS 5422-34-4	Parapel® LAM-100	CAS 5987-82-6	Dorsacaine
CAS 5422-34-4	Schercomid LME	CAS 6001-97-4	Lankropol® KMA
CAS 5434-57-1	Schercemol 65	CAS 6108-05-0	Dalcaine
CAS 5466-77-3	Escalol® 557	CAS 6130-64-9	Aquacillin
CAS 5466-77-3	Neo Heliopan, Type AV	CAS 6130-64-9	Crysticillin
CAS 5466-77-3	Parsol® MCX	CAS 6130-64-9	Depo-Penicillin
CAS 5534-09-8	Beclovent Inhaler	CAS 6130-64-9	Diurnal-Penicillin
CAS 5534-09-8	Beconase	CAS 6130-64-9	Eskacillin
CAS 5534-09-8	Beconase Nasal Inhaler	CAS 6130-64-9	Flo-Cillin
CAS 5534-09-8	Becotide	CAS 6130-64-9	Lentopen
CAS 5534-09-8	Propaderm	CAS 6130-64-9	Parencillin
CAS 5534-09-8	Vancenase	CAS 6130-64-9	Pfizerpen-AS
CAS 5534-09-8	Vanceril	CAS 6130-64-9	Wycillin
CAS 5534-95-2	Peptavlon	CAS 6138-56-3	Pyribenzamine Citrate
CAS 5536-17-4	Vira-A	CAS 6138-79-0	Actidil
CAS 5560-59-8	Spasmonal	CAS 6138-79-0	Pro-Actidil Tablets
CAS 5560-72-5	Prondol Tablets	CAS 6179-44-8	Schercotaine IAB
CAS 5560-78-1	Falmonox	CAS 6182-11-2	Emalex PG-di-S
CAS 5578-42-7	CM8645	CAS 6190-60-9	Mephine
CAS 5585-73-9	Evadyne Tablets	CAS 6190-60-9	Wyamine Sulfate
CAS 5593-20-4	Alphatrex	CAS 6197-30-4	Neo Heliopan, Type 303
CAS 5593-20-4	Diprosone	CAS 6197-30-4	Uvinul® N-539
CAS 5593-20-4	Psorion®	CAS 6202-23-9	Flexeril
CAS 5598-13-0	Cooper Graincote	CAS 6202-23-9	Flexiban®

CAS 6221-95-0	Lonzest® 143-S
CAS 6221-95-0	Schercemol MP
CAS 6272-74-8	Emcol® E-607L
CAS 6281-42-1	DV-2301
CAS 6283-92-7	Ceraphyl 31
CAS 6317-18-6	Amerstat® 282
CAS 6317-18-6	N-948® Biocide
CAS 6317-18-6	Tolcide MBT
CAS 6381-77-7	Eribate
CAS 6381-77-7	Neo-Cebitate®
CAS 6385-02-0	Meclomen
CAS 6419-19-8	Chemphonate AMP
CAS 6419-19-8	Fostex AMP
CAS 6419-19-8	Unihib® 305-LC
CAS 6440-58-0	Glydant®
CAS 6440-58-0	Mackstat® DM
CAS 6440-58-0	Nipaguard® DMDMH
CAS 6452-73-9	Trasicor
CAS 6484-52-2	Ansax
CAS 6484-52-2	Herco-Prills
CAS 6484-52-2	Nitram
CAS 6484-52-2	Old Plantation
CAS 6500-81-8	Edecril
CAS 6500-81-8	Edecrin Sodium
CAS 6500-81-8	Edecrina
CAS 6500-81-8	Edecrin®
CAS 6500-81-8	Hydromedin
CAS 6500-81-8	Lyovac Sodium Edecrin
CAS 6506-37-2	Nulogyl
CAS 6542-37-6	Zoldine® ZT-55
CAS 6556-11-2	Palohex
CAS 6683-19-8	Anox® 20
CAS 6683-19-8	Lowinox® PP35
CAS 6683-19-8	Naugard® 10
CAS 6683-19-8	Ralox® 630
CAS 6700-39-6	Medihaler-Iso™
CAS 6700-39-6	Norisodrine Sulfate
CAS 6724-53-4	Pexid
CAS 6731-36-8	Aztec® 1,1-Bis(t-Butylperoxy)-3,3,5 Trimethyl Cyclohexane-75 DBP
CAS 6731-36-8	Aztec® 1,1-Bis(t-Butylperoxy)-3,3,5-Trimethyl Cyclohexane
CAS 6731-36-8	Luperco 231-XL
CAS 6731-36-8	Lupersol 231
CAS 6731-36-8	Peroximon® S-164/40P
CAS 6731-36-8	Trigonox® 29
CAS 6731-36-8	Trigonox® KSM
CAS 6804-07-5	Mecadox
CAS 6834-92-0	Crystamet 1020
CAS 6834-92-0	Drymet® 59
CAS 6834-92-0	Lasilso
CAS 6834-92-0	Metso
CAS 6835-16-1	Peptard
CAS 6843-66-9	CD6010
CAS 6846-50-0	Kodaflex® TXIB
CAS 6865-35-6	Synpro® Barium Stearate
CAS 6881-94-3	Ektasolve® DP
CAS 6891-44-7	Ageflex FM-1Q80DMS
CAS 6923-22-4	Monocron
CAS 6938-94-9	Ceraphyl 230
CAS 6938-94-9	Schercemol DIA
CAS 6938-94-9	Unimate® DIPA
CAS 7005-47-2	DMAMP-80
CAS 7054-25-3	Duraquin
CAS 7054-25-3	Quinaglute
CAS 7081-44-9	Cloxapen
CAS 7085-19-0	Chafer CMPP Super
CAS 7085-19-0	Cleanacres CMPP
CAS 7085-19-0	Clenecorn
CAS 7085-19-0	Clifton CMPP Amine 60
CAS 7085-19-0	Clovotox
CAS 7085-19-0	Compitox Extra
CAS 7085-19-0	Endox
CAS 7085-19-0	Green Up Lawn Spot Weedkiller
CAS 7085-19-0	Headland Charge
CAS 7085-19-0	Hymec
CAS 7085-19-0	Iso-Cornox
CAS 7085-19-0	Mascot Cloverkiller
CAS 7085-44-1	Sodium Diuril
CAS 7123-62-8	Diamine OL
CAS 7128-91-8	Ammonyx® CO
CAS 7128-91-8	Amyx CO 3764
CAS 7128-91-8	Barlox® 16S
CAS 7128-91-81	Mazox® CDA
CAS 7173-51-5	Arquad® 210-50
CAS 7173-51-5	BTC® 99
CAS 7173-51-5	Catigene® 1011
CAS 7173-51-5	FMB 210-8 Quat, 210-15 Quat
CAS 7173-51-5	Querton 210Cl-50
CAS 7173-51-5	Radiaquat 6410
CAS 7173-51-5	Radiaquat 6412
CAS 7173-51-5	Rewoquat B 10
CAS 7173-62-8	Duomeen® OL
CAS 7173-62-8	Jet Amine DO
CAS 7173-62-8	Kemamine® D-989
CAS 7179-49-9	Lincocin
CAS 7179-49-9	Linocin
CAS 7179-49-9	Mycivin
CAS 7235-40-7	Lucarotin® 10%
CAS 7240-38-2	Bactocill
CAS 7240-38-2	Prostaphlin
CAS 7240-38-2	Stapenor®, Stapenor Retard
CAS 7246-20-0	Tricloryl
CAS 7246-20-0	Triclos
CAS 7246-21-1	Bilopaque
CAS 7281-04-1	Catigene BR 80 B
CAS 7287-19-6	Cotton-Pro®
CAS 7287-19-6	Gesagard
CAS 7287-19-6	Prometrex
CAS 7297-25-8	Cardilate
CAS 7320-34-5	Tetrakal
CAS 7378-99-6	Adma® 8
CAS 7396-58-9	Amine M210D
CAS 7396-58-9	Armeen® M2-10D
CAS 7396-58-9	Dama® 1010
CAS 7396-58-9	Radiamine 6310
CAS 7398-69-8	Ageflex mDMDAC
CAS 7398-69-8	Ageflex NB-50
CAS 7398-69-8	Agefloc PC20HV
CAS 7398-69-8	Agequat C505
CAS 7398-69-8	Agestat 41
CAS 7414-83-7	Didronel
CAS 7421-40-1	Biogastrone

CAS 7421-40-1	Bioral	CAS 7491-09-0	Kasof
CAS 7421-40-1	Duogastrone	CAS 7491-14-7	Dipsal
CAS 7439-92-1	Haro® Mix CE-701	CAS 7492-30-0	Emulgeen P
CAS 7439-92-1	Haro® Mix CK-711	CAS 7492-30-0	Kricinol 35
CAS 7439-92-1	Haro® Mix MH-204	CAS 7492-30-0	Solricin® 135
CAS 7439-96-5	Tripart® Liquid Manganese	CAS 7527-91-5	Akrinol
CAS 7440-02-0	Nickel 200	CAS 7534-94-3	Ageflex IBOMA
CAS 7440-02-0	Nickel 201	CAS 7534-94-3	Sipomer® IBOMA
CAS 7440-02-0	Nickel 204	CAS 7545-23-5	Emalex NN-15
CAS 7440-02-0	Nickel 205	CAS 7545-23-5	Monamid® 150-MW
CAS 7440-02-0	Nickel 211	CAS 7545-23-5	Schercomid 1214.
CAS 7440-02-0	Nickel 212	CAS 7558-80-7	Monosorb
CAS 7440-02-0	Nickel 213	CAS 7558-80-7	Oilfos
CAS 7440-02-0	Nickel 222	CAS 7558-80-7	Phosphotope
CAS 7440-02-0	Nickel 223	CAS 7601-54-9	Avgard
CAS 7440-02-0	Nickel 229	CAS 7601-55-0	Metubine Iodide
CAS 7440-02-0	Nickel 270	CAS 7601-89-0	KM Sodium Perchlorate
CAS 7440-22-4	Collosol Argentum	CAS 7617-78-9	Tomah PA-14
CAS 7440-36-0	Thermoguard® CPA	CAS 7631-86-9	Celatom
CAS 7440-44-0	Calgon® Type SGL	CAS 7631-86-9	Celite® 110
CAS 7440-44-0	Carboraffin	CAS 7631-86-9	Celite® 209
CAS 7440-44-0	Cecarbon	CAS 7631-86-9	Celite® 270
CAS 7440-44-0	Darco	CAS 7631-86-9	Celite® HSC
CAS 7440-44-0	Degusorb	CAS 7631-86-9	Celite® R-625
CAS 7440-44-0	Hydraffin	CAS 7631-86-9	Celite® Snow Floss
CAS 7440-44-0	Hydrodarco	CAS 7631-86-9	Clarcel
CAS 7440-44-0	Metasil DA	CAS 7631-86-9	Hyflo Super-Cel
CAS 7440-44-0	Metasil W/2	CAS 7631-86-9	Spheron P-1500-030
CAS 7440-44-0	Norit	CAS 7631-86-9	Ti-Sphere AB-15155A.
CAS 7440-44-0	Norit C	CAS 7631-86-9	Velvet Veil 310.
CAS 7440-44-0	Norit PK	CAS 7631-90-5	Amersite® 2
CAS 7440-44-0	Norit R	CAS 7637-07-2	Leecure B Series
CAS 7440-44-0	Norit RO	CAS 7646-78-8	Fascat® 4400
CAS 7440-44-0	Norithene	CAS 7646-79-9	Cobatope-60
CAS 7440-44-0	Purac	CAS 7646-85-7	Zinctrace
CAS 7440-44-0	Sorbonorit	CAS 7647-14-5	Alberger® Natural Flake
CAS 7440-66-6	ADK STAB 144	CAS 7647-14-5	Betrox
CAS 7446-70-0	Anhydrol Forte	CAS 7647-14-5	Natritope Chloride
CAS 7446-70-0	Driclor	CAS 7647-14-5	Normasol®
CAS 7446-70-0	Praestol® K2001	CAS 7647-14-5	Saltex
CAS 7447-39-4	Coclor	CAS 7647-14-5	Slow-Sodium
CAS 7447-39-4	Coppertrace	CAS 7647-14-5	Superior® Granulated
CAS 7447-40-7	Emplets Potassium Chloride	CAS 7647-14-5	Tru-Flake® Salt
CAS 7447-40-7	K-Contin	CAS 7651-02-7	Adogen® S-18 V
CAS 7447-40-7	K-Lor	CAS 7651-02-7	Chemidex S
CAS 7447-40-7	K-Lyte/C1	CAS 7651-02-7	Incromine SB
CAS 7447-40-7	K.Tab	CAS 7651-02-7	Lexamine S-13
CAS 7447-40-7	Kaon-Cl	CAS 7651-02-7	Lipamine SPA
CAS 7447-40-7	Kay Ciel	CAS 7651-02-7	Mackine 301
CAS 7447-40-7	KM Potassium Chloride	CAS 7651-02-7	Mazeen® S-13
CAS 7447-40-7	Leo K	CAS 7651-02-7	Miramine® SODI
CAS 7447-40-7	Micro-K	CAS 7651-02-7	Schercodine S
CAS 7447-40-7	Nu-Seals Potassium Chloride	CAS 7651-02-7	Tegamine® 18
CAS 7447-40-7	Pfiklor	CAS 7651-02-7	Tego®-Amid S 18
CAS 7447-40-7	Sando-K	CAS 7664-38-2	Marphos
CAS 7447-40-7	Selora	CAS 7664-38-2	NFB
CAS 7447-40-7	SK-Potassium Chloride	CAS 7664-38-2	Phosphoric Acid
CAS 7447-40-7	Slow-K	CAS 7665-72-7	Ageflex TBGE
CAS 7447-40-7	Trona Potassium Chloride	CAS 7681-11-0	Embamix
CAS 7491-02-3	Schercemol DIS	CAS 7681-14-3	Hydeltra-TBA
CAS 7491-02-3	Unimate® DIPS	CAS 7681-14-3	Predalone TBA
CAS 7491-09-0	Dialose	CAS 7681-49-4	Checkmate

CAS 7681-49-4	Fluoral
CAS 7681-49-4	Fluorinse
CAS 7681-49-4	Gel II
CAS 7681-49-4	Pediaflor
CAS 7681-49-4	Pro-Portion
CAS 7681-49-4	Zendium
CAS 7681-52-9	Adeka Hypote
CAS 7681-52-9	Chlorasol®
CAS 7681-52-9	Chloros
CAS 7681-52-9	Domestos
CAS 7681-52-9	HyPure N
CAS 7681-52-9	Zonite
CAS 7681-53-0	Sofibex
CAS 7681-76-7	Dugro
CAS 7681-80-3	Perium
CAS 7681-93-8	Myprozine
CAS 7681-93-8	Natacyn
CAS 7681-93-8	Pimafucin
CAS 7681-93-8	Pimaricin
CAS 7696-12-0	Killgerm® Py-Kill W
CAS 7696-12-0	Nippon Ant and Crawling Insect Killer
CAS 7696-12-0	Nippon Fly Killer Spray.
CAS 7697-37-2	Nitraline
CAS 7702-01-4	Schercoteric CY-2
CAS 7704-34-9	Biosulphur Powder
CAS 7704-34-9	Crystex®
CAS 7704-34-9	Crystex® Regular
CAS 7704-34-9	Flotox
CAS 7704-34-9	Gofrativ
CAS 7704-34-9	Gofravik
CAS 7704-34-9	Green Sulphur
CAS 7704-34-9	Hortag Aquasulf
CAS 7704-34-9	Kumulus® DF, FL
CAS 7704-34-9	Kumulus® S
CAS 7704-34-9	Manox
CAS 7704-34-9	Octocure 456
CAS 7704-34-9	Rubbermakers Sulfur
CAS 7704-34-9	Solfa
CAS 7704-34-9	Stoller Flowable Sulphur
CAS 7704-34-9	Struktol® SU 109
CAS 7704-34-9	Suffa
CAS 7704-34-9	Sulfidal
CAS 7704-34-9	Sulfur-F
CAS 7704-34-9	Thiovit
CAS 7704-34-9	Tripart® Imber
CAS 7704-34-9	Vassgro Flowable Sulphur
CAS 7704-34-9	Yellow Sulphur
CAS 7722-76-1	Firesaife
CAS 7722-76-1	Mono Ammonium Phosphate
CAS 7722-76-1	Phosai
CAS 7722-84-1	Baquatop
CAS 7722-84-1	Genoxide
CAS 7722-84-1	Hioxyl
CAS 7722-84-1	Oxzone
CAS 7722-84-1	Percarbamid
CAS 7722-84-1	Perone
CAS 7722-84-1	Peroxal
CAS 7722-84-1	Peroxol
CAS 7722-84-1	Peroxyl
CAS 7722-84-1	Tripwite
CAS 7722-84-1	Truzone

CAS 7722-88-5	Nutrifos
CAS 7722-88-5	Tetron
CAS 7727-21-1	Anthion
CAS 7727-37-9	Nitraprill
CAS 7727-43-7	Bariform
CAS 7727-43-7	Barosperse
CAS 7727-43-7	Barotrast
CAS 7727-43-7	Bartex® 80
CAS 7727-43-7	Basofor
CAS 7727-43-7	Blanc Fixe Micro
CAS 7727-43-7	Citobaryum
CAS 7727-43-7	Esophotrast
CAS 7727-43-7	Huberbrite 1
CAS 7727-43-7	Micropaque
CAS 7727-43-7	Oratrast
CAS 7727-43-7	Raybar
CAS 7727-43-7	Sachtoperse® HU
CAS 7727-43-7	Sparmite
CAS 7732-18-5	Amino-Collagen-25, -40
CAS 7733-02-0	Verazinc
CAS 7733-02-0	Z Span Spansule Capsule
CAS 7747-35-5	Amine CS-1246
CAS 7747-35-5	Oxaban®-E
CAS 7757-79-1	Petre
CAS 7757-82-6	Crimidesa
CAS 7757-82-6	Trona Anhydrous Sodium Sulphate
CAS 7757-83-7	Santosite
CAS 7757-93-9	Calbrite
CAS 7757-93-9	Caliment
CAS 7757-93-9	Calipharm
CAS 7757-93-9	Dentplus Special
CAS 7757-93-9	Lucaphos® 40, 48
CAS 7758-01-2	Bromox
CAS 7758-16-9	B.P. Pyro®
CAS 7758-16-9	Donut Pyro®
CAS 7758-16-9	Perfection®
CAS 7758-16-9	Sapp #4
CAS 7758-16-9	Taterfos®
CAS 7758-16-9	Victor Cream®
CAS 7758-19-2	Adox 3125
CAS 7758-19-2	C-2
CAS 7758-19-2	Chloritane
CAS 7758-23-8	Cefkaphos®
CAS 7758-23-8	H.T
CAS 7758-23-8	Py-Ran
CAS 7758-23-8	Regent® 12XX
CAS 7758-23-8	V-90®
CAS 7758-29-4	Curafos® STP
CAS 7758-29-4	Flav-R-Keep FP-51
CAS 7758-29-4	Hysorb
CAS 7758-29-4	Lem-O-Fos®
CAS 7758-29-4	Polygon
CAS 7758-29-4	Rhodia-Phos
CAS 7758-29-4	Sea-Gard® Formula FP-91
CAS 7758-89-6	Cuproid
CAS 7758-94-3	Ferrofloc
CAS 7758-95-4	Leclo
CAS 7758-98-7	Basicop
CAS 7758-99-8	Kocide® Copper Sulfate Pentahydrate Crystals
CAS 7761-88-8	Avoca

CAS 7772-98-7	Ametox	CAS 7787-59-9	Brown 208
CAS 7772-98-7	Sulfactol	CAS 7787-59-9	Carmine 224
CAS 7772-99-8	Fascat® 2004	CAS 7787-59-9	Chrome Green 106
CAS 7773-01-5	Mangatrace	CAS 7787-59-9	H Chrome Green 105.
CAS 7773-06-0	Amcide	CAS 7787-59-9	M Violet 112.
CAS 7773-06-0	Root-Out	CAS 7787-59-9	Mearlite® GBU
CAS 7775-09-9	Arpal Non Selex	CAS 7787-59-9	Pearl I, II, III
CAS 7775-09-9	Atlacide	CAS 7787-59-9	Pearl Supreme UVS
CAS 7775-09-9	Centex	CAS 7787-59-9	Pearl-Glo®
CAS 7775-09-9	Defol	CAS 7787-59-9	Perlex
CAS 7775-09-9	Granular Weedkiller	CAS 7787-59-9	Perlextra
CAS 7775-09-9	KM Sodium Chlorate	CAS 7787-59-9	Titanium Dioxide 110.
CAS 7775-14-6	Blankit®	CAS 7787-59-9	U Blue 104
CAS 7775-14-6	Hybaite	CAS 7787-59-9	U Pink 113.
CAS 7775-14-6	Hybrite®	CAS 7787-59-9	U Violet 109.
CAS 7775-14-6	Hydrolin	CAS 7787-59-9	Yellow 201
CAS 7775-14-6	Hydros	CAS 7789-04-0	Phosphocol P32
CAS 7775-14-6	Hydrosulfite AWC	CAS 7789-38-0	Dyetone®
CAS 7775-14-6	Hydrosulphit®	CAS 7789-77-7	Emcompress®
CAS 7775-14-6	Luredox® BP, PO	CAS 7790-26-3	Iodotope I-125
CAS 7775-14-6	Reductone	CAS 7790-26-3	Iodotope I-131
CAS 7778-18-9	Anhydrite	CAS 7790-26-3	Iodotope Therapeutic
CAS 7778-39-4	Aresenid	CAS 7790-26-3	Oriodide-131
CAS 7778-44-1	Calcars	CAS 7790-26-3	Radiocaps-131
CAS 7778-54-3	HTH	CAS 7790-26-3	Theriodide-131
CAS 7778-54-3	Hyporit	CAS 7790-26-3	Tracervial-131
CAS 7778-54-3	Induclor	CAS 7790-92-3	HyPure A
CAS 7778-54-3	Pittabs	CAS 7790-93-4	HyPure C
CAS 7778-54-3	Pittclor	CAS 7790-98-9	KM Ammonium Perchlorate
CAS 7778-54-3	Prestochlor	CAS 8000-10-1	Nuelin Liquid
CAS 7778-54-3	Pulsar	CAS 8001-17-0	EmCon E-5
CAS 7778-54-3	Stellos	CAS 8001-21-6	Lipovol SUN
CAS 7778-54-3	Swim clear	CAS 8001-21-6	Super Refined Sunflower Oil
CAS 7778-66-7	HyPure K	CAS 8001-22-7	Lipovol SOY
CAS 7778-74-7	KM Potassium Perchlorate	CAS 8001-22-7	Super Refined Soybean Oil
CAS 7778-77-0	Beycostat 148 K	CAS 8001-23-8	Lipovol SAF
CAS 7778-77-0	Beycostat 273 P	CAS 8001-23-8	Neobee® 18
CAS 7778-80-5	Trona Potassium Sulphate	CAS 8001-23-8	Super Refined Safflower Oil USP
CAS 7779-90-0	Diroval	CAS 8001-25-0	Super Refined Olive Oil
CAS 7782-42-5	Aquadag	CAS 8001-26-1	Ilinol
CAS 7782-42-5	Dag 137	CAS 8001-31-8	Coconut Oils® 76, 92, 110
CAS 7782-42-5	Dag 154	CAS 8001-31-8	Esi-Terge 40% Coconut Oil Soap
CAS 7782-42-5	Dag 197	CAS 8001-31-8	Konut
CAS 7782-42-5	Lonza KS	CAS 8001-31-8	Super Refined Coconut Oil
CAS 7782-42-5	Prodag	CAS 8001-69-2	Atlas Tanked Cod Oil
CAS 7782-42-5	SLA 1262	CAS 8001-69-2	Mainstay®
CAS 7782-42-5	SLA 1275	CAS 8001-69-2	Super D
CAS 7782-42-5	SLA 2239	CAS 8002-03-7	Super Refined Peanut Oil
CAS 7782-49-2	Vandex®	CAS 8002-13-9	Actralube 7142
CAS 7783-20-2	Nipro (ii)	CAS 8002-23-1	Crodamol SS
CAS 7783-40-6	Magtran	CAS 8002-23-1	Dermalcare® SPS
CAS 7783-47-3	Stop	CAS 8002-23-1	Liponate SPS
CAS 7784-25-0	Curb	CAS 8002-23-1	Ritaceti
CAS 7784-25-0	Guardsman	CAS 8002-23-1	Ross Spermaceti Wax
CAS 7784-25-0	Mandops Narsty		Substitute 573
CAS 7784-25-0	Stay Off	CAS 8002-23-1	Starfol® Wax CG
CAS 7784-25-0	Stay Off	CAS 8002-23-1	W.G.S. Synaceti 116 NF/USP
CAS 7785-87-7	Mansu	CAS 8002-31-1	Fancol CB
CAS 7787-59-9	Biju®	CAS 8002-33-3	Actrasol C-50, C-75, C-85
CAS 7787-59-9	Bismica 46.	CAS 8002-33-3	Chemax SCO
CAS 7787-59-9	Bisoxyl	CAS 8002-33-3	Chemsulf SCO/75
CAS 7787-59-9	Black 103	CAS 8002-33-3	Eureka 102

CAS 8002-33-3	Laurel R-50
CAS 8002-33-3	Marvanol® SCO (50%)
CAS 8002-33-3	Nopcocastor
CAS 8002-33-3	Nopcosulf CA-60, -70
CAS 8002-33-3	Nopcosurf CA
CAS 8002-33-3	Standapol® SCO
CAS 8002-43-5	Actiflo® 68, 70
CAS 8002-43-5	Alcolec® 439-C
CAS 8002-43-5	Alcolec® 440-WD
CAS 8002-43-5	Alcolec® 495
CAS 8002-43-5	Alcolec® BS
CAS 8002-43-5	Alcolec® F-100
CAS 8002-43-5	Alcolec® S
CAS 8002-43-5	Aqualipid 95
CAS 8002-43-5	Blendmax 322
CAS 8002-43-5	Centrocap® 162SS, 162US
CAS 8002-43-5	Centrolex® C
CAS 8002-43-5	Centrol® 2FSB, 2FUB, 3FSB, 3FUB
CAS 8002-43-5	Centrol® CA
CAS 8002-43-5	Centromix® CPS
CAS 8002-43-5	Centrophase® C
CAS 8002-43-5	Centrophase® HR
CAS 8002-43-5	Centrophase® HR2B, HR2U
CAS 8002-43-5	Centrophil® K
CAS 8002-43-5	Crolec 4135
CAS 8002-43-5	Dermasome® MT
CAS 8002-43-5	Kelecin®
CAS 8002-43-5	Lecithin L-Range
CAS 8002-43-5	Lecithin Water Dispersible CLR
CAS 8002-43-5	Lipotin 100, 100J, SB
CAS 8002-43-5	Phosal
CAS 8002-43-5	PhosPho E-100
CAS 8002-43-5	PhosPho F-97
CAS 8002-43-5	PhosPho LCN-TS
CAS 8002-43-5	PhosPho S-85
CAS 8002-43-5	PhosPho T-20
CAS 8002-48-0	Nutrimalt® Range
CAS 8002-50-4	Super Refined Menhaden Oil
CAS 8002-53-7	Sicolub® E
CAS 8002-74-2	Glycolube® VL
CAS 8002-74-2	Microlube A
CAS 8002-74-2	Octowax 321
CAS 8002-74-2	Paracol® 800N
CAS 8002-74-2	Paracol® 810N
CAS 8002-74-2	Paracol® 1886
CAS 8002-74-2	Paracol® M161
CAS 8002-74-2	Paratulle®
CAS 8002-74-2	Press-Aid
CAS 8002-74-2	Ross Wax #100
CAS 8002-74-2	Ross Wax #145, 165
CAS 8002-74-2	Sasolwaks
CAS 8002-74-2	Sasolwaks M3
CAS 8002-74-2	Synwax
CAS 8002-74-2	Vybar® 103
CAS 8002-75-3	Lipovol PAL
CAS 8002-76-4	Omnopon
CAS 8005-44-5	Fanwax P
CAS 8006-13-1	Buro-sol Concentrate
CAS 8006-13-1	Domeboro
CAS 8006-54-0	Anhydrous Lanolin HP-2050
CAS 8006-54-0	Anhydrous Lanolin P80
CAS 8006-54-0	Anhydrous Lanolin P9SRA
CAS 8006-54-0	Fluilan
CAS 8006-54-0	Lantrol® HP-2073
CAS 8006-54-0	Ultra Anhydrous Lanolin HP-2060
CAS 8006-95-9	EmCon W
CAS 8006-95-9	Lipovol WGO
CAS 8006-95-9	Super Refined Wheat Germ Oil
CAS 8007-18-9	V-9415
CAS 8007-24-7	Cardolite® NC-511
CAS 8007-43-0	Arlacel® 83
CAS 8007-43-0	Crill 43
CAS 8007-43-0	Dehymuls SSO
CAS 8007-43-0	Disponil SSO 100 F1
CAS 8007-43-0	Emalex SPO-150
CAS 8007-43-0	Emsorb® 2502
CAS 8007-43-0	Glycomul® SOC
CAS 8007-43-0	Hodag SSO
CAS 8007-43-0	Liposorb SQO
CAS 8007-43-0	S-Maz® 83R
CAS 8007-69-0	Lipovol ALM
CAS 8008-20-6	Pearl
CAS 8008-74-0	Lipovol SES
CAS 8008-74-0	Super Refined Sesame Oil
CAS 8012-95-1	Fancol LAO
CAS 8013-05-6	Solricin® 235
CAS 8013-07-8	ADK CIZER O-130P
CAS 8013-07-8	Drapex® 6.8
CAS 8013-07-8	Edenol D82
CAS 8013-07-8	Epoxol 7-4
CAS 8013-07-8	Hoe S 3680
CAS 8013-07-8	Lankroflex® GE
CAS 8013-07-8	Paraplex® G-60
CAS 8013-07-8	Paraplex® G.62
CAS 8013-07-8	Plasthall® ESO
CAS 8013-07-8	PX-800
CAS 8015-86-9	Emulsion C-340
CAS 8016-70-4	Akorex
CAS 8016-70-4	Code 321
CAS 8016-70-4	Diamond D 31
CAS 8016-70-4	Famous
CAS 8016-70-4	Lipovol HS
CAS 8016-70-4	S-Flakes
CAS 8016-70-4	Witarix® 440
CAS 8018-01-7	Dithane
CAS 8018-01-7	Karamate
CAS 8018-01-7	Manzate® 200 DF
CAS 8018-01-7	Penncozeb
CAS 8020-84-6	Degras Special
CAS 8022-00-2	Campbell's DSM
CAS 8022-00-2	Metasystox 55
CAS 8024-22-4	Lipovol G
CAS 8024-22-4	Super Refined Grapeseed Oil
CAS 8024-32-6	Avocado Oil CLR
CAS 8024-32-6	Lipovol A
CAS 8024-32-6	Super Refined Avocado Oil
CAS 8027-33-6	Argobase 125
CAS 8027-33-6	Argowax Standard
CAS 8027-33-6	Ceralan®
CAS 8027-33-6	Fancol LA
CAS 8027-33-6	Hartolan
CAS 8027-33-6	Hartolite
CAS 8027-33-6	Ivarlan 3310

CAS 8027-33-6	Ritawax
CAS 8027-33-6	Solulan
CAS 8027-33-6	Super Hartolan
CAS 8028-66-8	Fancol HON
CAS 8029-44-5	Myverol® 18-85K
CAS 8029-68-3	Ichthymall
CAS 8029-68-3	Ichthyol
CAS 8029-76-3	Alcolec® Z-3
CAS 8029-76-3	Centrolene® A, S
CAS 8029-76-3	Lipotin H
CAS 8029-76-3	PhosPho 642
CAS 8029-91-2	Axol® E 61
CAS 8029-92-3	Myvacet® 9-40
CAS 8030-30-6	Aromatic Solvent 150
CAS 8030-30-6	Hi-Sol® 10, 15, 70
CAS 8030-30-6	Texsolve V
CAS 8030-78-2	Arquad® T-27W
CAS 8030-78-2	Arquad® T-50
CAS 8030-78-2	Jet Quat T-50
CAS 8031-14-9	Clorpactin XCB
CAS 8031-44-5	Fancol HL
CAS 8031-44-5	Ivarlan HL
CAS 8031-44-5	Lipolan
CAS 8031-44-5	Satulan
CAS 8031-44-5	Super-Sat
CAS 8038-43-5	Vigilan
CAS 8039-09-6	Lan-Aqua-Sol 100
CAS 8039-09-6	Lantrol® PLN
CAS 8042-47-5	Paratherm NF
CAS 8048-52-0	Burnol Acriflavine Cream
CAS 8049-47-6	Panar
CAS 8049-47-6	Pancrex V
CAS 8049-47-6	Panteric
CAS 8050-09-7	Westvaco® Resin 90
CAS 8050-26-8	Oulutac 90 D
CAS 8050-26-8	Uni-Tac® R99
CAS 8050-26-8	Zonester® 65
CAS 8050-81-5	Amersil® Simethicone
CAS 8050-81-5	Dow Corning® Antifoam A
CAS 8050-81-5	Dow Corning® Antifoam C
CAS 8050-81-5	Hodag Antifoam F-1
CAS 8050-81-5	Mazu® DF 200SP
CAS 8050-81-5	Mylicon
CAS 8050-81-5	Sentry Simethicone
CAS 8050-81-5	Silain
CAS 8050-81-5	Siligaz
CAS 8050-81-5	Tego® Foamex N
CAS 8051-30-7	Empilan® 2502
CAS 8052-48-0	Emery® 2895 Foamaster Soap L
CAS 8052-48-0	Foamaster Soap L
CAS 8061-51-6	Darvan® No. 2
CAS 8061-51-6	Dynasperse A
CAS 8061-51-6	Kelig
CAS 8061-51-6	Lignosol DXD
CAS 8061-51-6	Maracarb N-1
CAS 8061-51-6	Maracell XE
CAS 8061-51-6	Marasperse 52 CP
CAS 8061-51-6	Marasperse N-22
CAS 8061-51-6	Polyfon
CAS 8061-51-6	Polyfon® F
CAS 8061-51-6	Reax® 45A
CAS 8061-51-6	Reax® 80C

CAS 8061-51-6	Temsperse S 001
CAS 8061-51-6	Ufoxane 2
CAS 8061-51-6	Ultrazine NA
CAS 8061-51-6	Zewa SL 2
CAS 8061-51-6	Zewalon FN
CAS 8061-52-7	Additive-A
CAS 8061-52-7	Ameribond
CAS 8061-52-7	Borrebond
CAS 8061-52-7	Borresperse CA/CAF
CAS 8061-52-7	Collex G
CAS 8061-52-7	Darvan® No. 404
CAS 8061-52-7	Glutrin
CAS 8061-52-7	Goulac
CAS 8061-52-7	Lignorit
CAS 8061-52-7	Lignosite®
CAS 8061-52-7	Lignosol B
CAS 8061-52-7	Lignosol SF
CAS 8061-52-7	Marasperse GFC
CAS 8061-52-7	Norlig 11 DA
CAS 8061-52-7	Norlig A
CAS 8061-52-7	Trastan LS
CAS 8061-52-7	Ultrazine CA
CAS 8061-52-7	Wafex
CAS 8061-52-7	Wargotan
CAS 8061-52-7	Welltex 300 F
CAS 8061-53-8	Lignosol TS
CAS 8061-53-8	Tembind A 002
CAS 8061-53-8	Wanin AM
CAS 8067-32-1	Precirol ATO
CAS 8067-32-1	Precirol WL 2155
CAS 8067-69-4	Quixalud
CAS 8067-69-4	Remiderm
CAS 8067-69-4	Remiderm Cream
CAS 8067-69-4	Remiderm Ointment
CAS 8067-69-4	Remiderm Spray
CAS 8067-69-4	Remotic Capsules
CAS 8067-69-4	Vetalog Plus Cream
CAS 8068-28-8	Coly-Mycin M Parenteral
CAS 9000-07-1	Clarifloc
CAS 9000-07-1	Stamere®
CAS 9000-07-1	Ultrafloc
CAS 9000-07-1	Wakal® K Range
CAS 9000-30-0	Decorpa
CAS 9000-30-0	Jaguar® C
CAS 9000-30-0	Jaguar® Guar Gum
CAS 9000-30-0	Prinza® Range
CAS 9000-30-0	Progacyl® CP-7
CAS 9000-30-0	Supercol® Guar Gum
CAS 9000-40-2	Wakal® J Range
CAS 9000-70-8	Byco A, C, O
CAS 9000-70-8	Crodyne BY19
CAS 9000-70-8	Hydrocoll G-40
CAS 9000-70-8	Pharmagel
CAS 9000-70-8	Pronel Capsules
CAS 9000-70-8	Rousselot Gelatine
CAS 9000-70-8	SPG Gelatine
CAS 9000-92-4	Arozyme TD
CAS 9000-92-4	BAN 120 L
CAS 9000-92-4	Clarase® 5,000, 40,000
CAS 9000-92-4	Taka-Therm® L-340
CAS 9000-92-4	Tenase® 1200, L-340, L-1200
CAS 9000-92-4	Termamyl®

CAS 9000-92-4	Termamyl® 120L
CAS 9000-92-4	Vanzyme®
CAS 9001-01-8	Depot Glumorin
CAS 9001-01-8	Glumorin
CAS 9001-05-2	Catalase L
CAS 9001-05-2	Microcatalase®
CAS 9001-09-6	Discase
CAS 9001-54-1	Rondase
CAS 9001-54-1	Wydase
CAS 9001 62-1	Pancreatic Lipase 250
CAS 9001-73-4	Scintillase
CAS 9002-01-1	Kabikinase
CAS 9002-04-4	Topostasin
CAS 9002-07-7	Parenzyme
CAS 9002-07-7	Trypure Novo
CAS 9002-64-6	Paroidin
CAS 9002-72-6	Asellacrin
CAS 9002-72-6	Crescormon
CAS 9002-72-6	Humatrope
CAS 9002-83-9	Fluorolube® GR-290, GR-362, GR-470, GR-544, GR-660
CAS 9002-84-0	Electrafil® TR-1900/EC
CAS 9002-84-0	Emralon 304
CAS 9002-84-0	Emralon 8301-01
CAS 9002-84-0	Fluon® AD1, AD1L, AD1H
CAS 9002-84-0	Fluon® CDI
CAS 9002-84-0	Fluon® G170
CAS 9002-84-0	Fluorocomp® FC-101
CAS 9002-84-0	Fluorocomp® FC-144
CAS 9002-84-0	Fluorocomp® FC-174
CAS 9002-84-0	Fluorocomp® FC-182
CAS 9002-84-0	Fluoromelt®
CAS 9002-84-0	Fluorosint® 500
CAS 9002-84-0	Hostaflon® TF 1101
CAS 9002-84-0	Hostaflon® TF 1620
CAS 9002-84-0	Hostaflon® TF 2071
CAS 9002-84-0	Hostaflon® TF 5032
CAS 9002-84-0	Hostaflon® TF 5537
CAS 9002-84-0	McLube 1777
CAS 9002-84-0	Polymist® F-5
CAS 9002-84-0	PTFE-19
CAS 9002-84-0	RT/Duroid® M
CAS 9002-84-0	SLA 1611
CAS 9002-84-0	SLA 1612
CAS 9002-86-2	Advex 91025
CAS 9002-86-2	Geon® 8700A
CAS 9002-86-2	Geon® 8720
CAS 9002-86-2	Geon® 8812, 8813
CAS 9002-86-2	Geon® 8896
CAS 9002-86-2	Geon® 83457
CAS 9002-86-2	Geon® 83718
CAS 9002-86-2	Geon® 86100, 86101, and 86103
CAS 9002-86-2	Geon® 87239, 87241
CAS 9002-86-2	Geon® 87396
CAS 9002-86-2	Geon® 87420
CAS 9002-86-2	Geon® HTX 92190
CAS 9002-86-2	Geon® W015
CAS 9002-86-2	Georgia Gulf 3132 Clear 02
CAS 9002-86-2	Georgia Gulf 5006
CAS 9002-86-2	Georgia Gulf 5006 General
CAS 9002-86-2	Georgia Gulf 9105
CAS 9002-86-2	Georgia Gulf 9151
CAS 9002-86-2	Georgia Gulf 9175J
CAS 9002-86-2	Georgia Gulf 9202
CAS 9002-86-2	Georgia Gulf CL-7049
CAS 9002-86-2	Georgia Gulf EH-71L
CAS 9002-86-2	Georgia Gulf EX-240
CAS 9002-86-2	Georgia Gulf HH-1900
CAS 9002-86-2	Georgia Gulf HM-7054
CAS 9002-86-2	Georgia Gulf SP-7107
CAS 9002-86-2	Georgia Gulf UV-/160
CAS 9002-86-2	Klegecell R30
CAS 9002-86-2	Lutofan®
CAS 9002-86-2	Norvinyl DX 550
CAS 9002-86-2	Novablend® 501
CAS 9002-86-2	Novablend® 5555
CAS 9002-86-2	Ultraprene®
CAS 9002-86-2	Vestolit® B 7021
CAS 9002-86-2	Vestolit® E 6003, E 6503, E 7003
CAS 9002-86-2	Vestolit® LF HI-SP 5735
CAS 9002-86-2	Vestolit® M 5867
CAS 9002-86-2	Vestolit® P 1330 K
CAS 9002-86-2	Vestolit® S 6058
CAS 9002-86-2	Vinoflex®
CAS 9002-88-4	A-C® 6
CAS 9002-88-4	A-C® Polyethylene 6, 6A, 7, 7A, 8, 8A, 9, 9A
CAS 9002-88-4	A-C® Polyethylene 617, 617A
CAS 9002-88-4	ACtone® P
CAS 9002-88-4	ACumist B-6, B-12, B-18, C-5, C-12, C-18
CAS 9002-88-4	Akrowax PE
CAS 9002-88-4	Alathon® H5234
CAS 9002-88-4	Alathon® L5440
CAS 9002-88-4	Alathon® M5560
CAS 9002-88-4	Alathon® M6062
CAS 9002-88-4	Amsoft MDH-20
CAS 9002-88-4	Arcel Moldable Polyethylene Copolymers
CAS 9002-88-4	Arpak 4322
CAS 9002-88-4	Attane 4601
CAS 9002-88-4	Attane 4802
CAS 9002-88-4	Cabelec® 1017
CAS 9002-88-4	Cabelec® 3172
CAS 9002-88-4	Cabot® PE 9007
CAS 9002-88-4	Cadco® UHMW
CAS 9002-88-4	Compound 403/401
CAS 9002-88-4	Conductomer HDC-22HLMI-M
CAS 9002-88-4	Dowlex 2032
CAS 9002-88-4	Dowlex 2035
CAS 9002-88-4	Dowlex 2042
CAS 9002-88-4	Dowlex 2500
CAS 9002-88-4	Dowlex 3010
CAS 9002-88-4	Ecothene EC 101
CAS 9002-88-4	Electrafil® PE-90/EC
CAS 9002-88-4	Empee® FR 42 LM
CAS 9002-88-4	Empee® PE-112
CAS 9002-88-4	Empce® PE-113
CAS 9002-88-4	Epolene® C-10
CAS 9002-88-4	Epolene® C-13
CAS 9002-88-4	Epolene® N-20, N-21
CAS 9002-88-4	Esi-Cryl 1E10N
CAS 9002-88-4	Esi-Cryl 325N
CAS 9002-88-4	Esi-Cryl 1540A

CAS 9002-88-4	Forbest 410
CAS 9002-88-4	Fortiflex® A60-70-99, A60-70-119
CAS 9002-88-4	Fortiflex® B45-06R-09
CAS 9002-88-4	Fortiflex® G36-24-149
CAS 9002-88-4	Fortiflex® G38-70C
CAS 9002-88-4	Fortiflex® J36-25-142
CAS 9002-88-4	Fortiflex® K36-55-122
CAS 9002-88-4	Fortiflex® T50-200
CAS 9002-88-4	Fortiflex® XF-855
CAS 9002-88-4	Hartolon 5683
CAS 9002-88-4	HDPE 04352N
CAS 9002-88-4	HDPE 25053-P
CAS 9002-88-4	HDPE 32060C
CAS 9002-88-4	HDPE 35053
CAS 9002-88-4	HDPE IP-10
CAS 9002-88-4	HiD 9301
CAS 9002-88-4	HiD 9602
CAS 9002-88-4	HiD 9632
CAS 9002-88-4	HiD 9650
CAS 9002-88-4	Hoechst Wax PE 190
CAS 9002-88-4	Hoechst Wax PE 520
CAS 9002-88-4	Hoechst Wax PED 521
CAS 9002-88-4	Hostalen® EP 4450
CAS 9002-88-4	Hostalen® GB 6950
CAS 9002-88-4	Hostalen® GM 5010 T2
CAS 9002-88-4	Hostalen® GM 7745 HP
CAS 9002-88-4	Hostalen® GUR 5121
CAS 9002-88-4	Loobwax 0651
CAS 9002-88-4	Lupolen®
CAS 9002-88-4	Luwax A
CAS 9002-88-4	Luwax AF 30
CAS 9002-88-4	Luwax EAS 1
CAS 9002-88-4	Luwax EVA 1
CAS 9002-88-4	Marlex® CL-200
CAS 9002-88-4	Marlex® HHM 4903
CAS 9002-88-4	Marlex® HHM TR-140
CAS 9002-88-4	Marlex® HHM TR-400
CAS 9002-88-4	Marlex® HMN 4550
CAS 9002-88-4	Marlex® HXM 50100
CAS 9002-88-4	Micropoly 520
CAS 9002-88-4	Microthene® MA 530-060
CAS 9002-88-4	Microthene® MN 701-00
CAS 9002-88-4	Microthene® MP 625U
CAS 9002-88-4	Neopolen® E
CAS 9002-88-4	Nortuff RA 7020-KO
CAS 9002-88-4	Novapol® GF-0218-F
CAS 9002-88-4	Novapol® GI-2024-A
CAS 9002-88-4	Novapol® HB-L455-A
CAS 9002-88-4	Novapol® LC-0517-A
CAS 9002-88-4	Novapol® LE-0220-A
CAS 9002-88-4	Novapol® LF-0223-B
CAS 9002-88-4	Novapol® LF-Y819-D
CAS 9002-88-4	Novapol® PF-0118-B
CAS 9002-88-4	Novapol® PR-0636-UG
CAS 9002-88-4	Octowax 518
CAS 9002-88-4	Orevac® 18302
CAS 9002-88-4	PE 1017
CAS 9002-88-4	PE 4517
CAS 9002-88-4	PE 5554-H
CAS 9002-88-4	PE 5861
CAS 9002-88-4	Petrothene® GA 501
CAS 9002-88-4	Petrothene® GA 564
CAS 9002-88-4	Petrothene® GA 808-090
CAS 9002-88-4	Petrothene® HD 5903B
CAS 9002-88-4	Petrothene® LB 5003-00
CAS 9002-88-4	Petrothene® LB 6001-00
CAS 9002-88-4	Petrothene® LF 6030-00
CAS 9002-88-4	Petrothene® LP 5102-00
CAS 9002-88-4	Petrothene® LS 3150-00
CAS 9002-88-4	Petrothene® LT 5704-00
CAS 9002-88-4	Petrothene® NA 155-000
CAS 9002-88-4	Petrothene® NA 204-000
CAS 9002-88-4	Petrothene® NA 341-000
CAS 9002-88-4	Petrothene® PA 436
CAS 9002-88-4	Poligen PE
CAS 9002-88-4	Polybond® 1009
CAS 9002-88-4	Polysoft B
CAS 9002-88-4	Polywax® 500
CAS 9002-88-4	Primax UH-1060
CAS 9002-88-4	Primax UH-1080
CAS 9002-88-4	Primax UH-1250
CAS 9002-88-4	Retain PE-1001
CAS 9002-88-4	Retain PE-5009
CAS 9002-88-4	Retain RP-120
CAS 9002-88-4	Rexene® PE 6010
CAS 9002-88-4	Rexene® PE 6076
CAS 9002-88-4	RTP 701
CAS 9002-88-4	Spectratech® FM 1150H
CAS 9002-88-4	Stat-Kon® FE
CAS 9002-88-4	Struktol® PE H-100
CAS 9002-88-4	Suntec®-HD
CAS 9002-88-4	Suntec®-LD
CAS 9002-88-4	Tenite® 154DF
CAS 9002-88-4	Tenite® Polyethylene
CAS 9002-88-4	Tuflin HS-7066 NT7
CAS 9002-88-4	Vestolen® A3512
CAS 9002-88-4	Vestolen® AX4304
CAS 9002-88-4	Zemid® 610
CAS 9002-88-4	Zemid® 650
CAS 9002-888-4	Fusabond® MB-110D
CAS 9002-888-4	Luwax AH 3
CAS 9002-888-4	Mitsubishi BT002
CAS 9002-888-4	Mitsubishi ET008
CAS 9002-888-4	Mitsubishi F101A
CAS 9002-888-4	Mitsubishi JS050
CAS 9002-888-4	Mitsubishi L300
CAS 9002-888-4	Mitsubishi UF421
CAS 9002-89-5	Airvol® 103, 107
CAS 9002-89-5	Airvol® 125
CAS 9002-89-5	Airvol® 165
CAS 9002-89-5	Airvol® 321, 325, 350
CAS 9002-89-5	Airvol® MH-82, MM-14, MM-51, MM-81
CAS 9002-89-5	Airvol® SH-72, SM-73
CAS 9002-89-5	Elvanol® 20-25
CAS 9002-89-5	Elvanol® 71-30
CAS 9002-89-5	Elvanol® 90-50
CAS 9002-89-5	Release Agent NL-2
CAS 9002-89-5	Vinex 1003
CAS 9002-89-5	Vinex 2019
CAS 9002-89-5	Vinex 2025
CAS 9002-89-5	Vinex 2034
CAS 9002-89-5	Vinex 2144
CAS 9002-89-5	Vinex 5030

CAS 9002-92-0	Akyporox RLM 160
CAS 9002-92-0	Marlipal® 124
CAS 9002-92-0	Marlipal® MG
CAS 9002-92-0	Marlowet® BL
CAS 9002-92-0	Volpo L4
CAS 9002-93-1	Akyporox OP 100
CAS 9002-93-1	Akyporox OP 115 SPC
CAS 9002-93-1	Akyporox OP 200
CAS 9002-93-1	Akyporox OP 250 V
CAS 9002-93-1	Akyporox OP 400V
CAS 9002-93-1	Chemax OP-3
CAS 9002-93-1	Chemax OP-30/70
CAS 9002-93-1	Chemax OP-40
CAS 9002-93-1	Chemax OP-40/70
CAS 9002-93-1	Chemax OP-7
CAS 9002-93-1	Dehydrophen POP 4
CAS 9002-93-1	Hyonic OP-7
CAS 9002-93-1	Iconol OP-10
CAS 9002-93-1	Iconol OP-30
CAS 9002-93-1	Iconol OP-40
CAS 9002-93-1	Igepal® CA-210
CAS 9002-93-1	Igepal® CA-897
CAS 9002-93-1	Macol® OP-3
CAS 9002-93-1	Makon® OP-6
CAS 9002-93-1	Marlophen® 85
CAS 9002-93-1	Marlophen® 810
CAS 9002-93-1	Nutrol 100
CAS 9002-93-1	Prox-onic OP-09
CAS 9002-93-1	Remcopal 306
CAS 9002-93-1	Remcopal O9
CAS 9002-93-1	Remcopal O11
CAS 9002-93-1	Remcopal O12
CAS 9002-93-1	Rexol 45/1
CAS 9002-93-1	Sellig O 4 100
CAS 9002-93-1	Sellig O 5 100
CAS 9002-93-1	Sellig O 6 100
CAS 9002-93-1	Sellig O 8 100
CAS 9002-93-1	Sellig O 9 100
CAS 9002-93-1	Sellig O 11 100
CAS 9002-93-1	Sellig O 12 100
CAS 9002-93-1	Sellig O 20 100
CAS 9002-93-1	T-Det® O-4
CAS 9002-93-1	T-Det® O-407
CAS 9002-93-1	Teric X5
CAS 9002-93-1	Triton® X-15
CAS 9002-93-1	Trycol® 6956
CAS 9002-97-5	Akypopress DB
CAS 9003-01-4	Acrysol® ASE-75
CAS 9003-01-4	Acumer 1000
CAS 9003-01-4	Acusol® 445
CAS 9003-01-4	Alcogum L-15, L-26, L-28, L-31, L-35, L-36
CAS 9003-01-4	Burcotreat 900-A
CAS 9003-01-4	Carbopol® 613, 614
CAS 9003-01-4	Good-rite® K-702
CAS 9003-01-4	Good-rite® K-752
CAS 9003-01-4	Sokalan® PA 110 S
CAS 9003-03-6	Acrysol® G-110
CAS 9003-03-6	Alcogum 9639
CAS 9003-03-6	BYK®-156
CAS 9003-03-6	Size CB
CAS 9003-04-7	Acrysol® GS

CAS 9003-04-7	Acrysol® HV-1
CAS 9003-04-7	Acusol® 410N
CAS 9003-04-7	Acusol® 445N
CAS 9003-04-7	Acusol® 445ND
CAS 9003-04-7	Acusol® 480N
CAS 9003-04-7	Acusol® 860N
CAS 9003-04-7	Alcosperse 107, 124, 149, 157
CAS 9003-04-7	Alkasperse® A-20
CAS 9003-04-7	Burcosperse AP Liq
CAS 9003-04-7	Colloid 202
CAS 9003-04-7	Daxad® 37LN10-35
CAS 9003-04-7	Drewsperse® 611
CAS 9003-04-7	Good-rite® K-7058D
CAS 9003-04-7	Sokalan® CP 7
CAS 9003-04-7	Sokalan® PA 15
CAS 9003-05-8	Polyhall® 21J
CAS 9003-07-0	Acctuf 3045
CAS 9003-07-0	Adpro AP 2112-GP
CAS 9003-07-0	Adpro AP 8210-HS
CAS 9003-07-0	Amoco® 1012
CAS 9003-07-0	Amoco® 1016
CAS 9003-07-0	Amoco® 1246
CAS 9003-07-0	Amoco® 4018
CAS 9003-07-0	Amoco® 5016
CAS 9003-07-0	Amoco® 6114
CAS 9003-07-0	Amoco® 6400P
CAS 9003-07-0	Amoco® 7234
CAS 9003-07-0	Amoco® 7239
CAS 9003-07-0	Amoco® 7728
CAS 9003-07-0	Amoco® 9119
CAS 9003-07-0	Arpro 3313
CAS 9003-07-0	Astryn® 63A6-2
CAS 9003-07-0	Astryn® 63F4-2
CAS 9003-07-0	Astryn® 65F4-4
CAS 9003-07-0	Astryn® 65F5-4
CAS 9003-07-0	Astryn® 73F4-2
CAS 9003-07-0	Astryn® 73F5-2
CAS 9003-07-0	Astryn® 78F4-2
CAS 9003-07-0	Astryn® BA16G
CAS 9003-07-0	Astryn® SD068-4
CAS 9003-07-0	Bicor® 240 B, 306 B, 420 B, 470 B
CAS 9003-07-0	Cabelec® 3004
CAS 9003-07-0	Cabelec® 3464
CAS 9003-07-0	Celstran® PPG30-01-4
CAS 9003-07-0	Eastman® P4C5B-030
CAS 9003-07-0	Electrafil® J-60/CF/30
CAS 9003-07-0	Electrafil® JM-61/CF/10
CAS 9003-07-0	Electrafil® PP-60/CC/20/EC
CAS 9003-07-0	Empee PP Conc. 33
CAS 9003-07-0	Empee® PP-301
CAS 9003-07-0	Empee® PP-459
CAS 9003-07-0	Empee® PP-560
CAS 9003-07-0	Epolene® N-15
CAS 9003-07-0	Fiberfil® J-60/30/E8
CAS 9003-07-0	Fiberfil® J-60/30/FR
CAS 9003-07-0	Fiberfil® M-1492
CAS 9003-07-0	Fiberfil® PP-60/TC/40
CAS 9003-07-0	Fortilene® 1001
CAS 9003-07-0	Fortilene® 1602
CAS 9003-07-0	Fortilene® 1802
CAS 9003-07-0	Fortilene® 2104

CAS 9003-07-0	Fortilene® 3151
CAS 9003-07-0	Fortilene® 4104, 4109
CAS 9003-07-0	Fortilene® 4209
CAS 9003-07-0	Fortilene® 5801
CAS 9003-07-0	Fortilene® 9000
CAS 9003-07-0	Fusabond® MZ-109D
CAS 9003-07-0	Fusabond® MZ-203D
CAS 9003-07-0	Hercoflat® Texturing Pigments and Flatting Agent
CAS 9003-07-0	HiFax AB 6023
CAS 9003-07-0	HiFax CA45A
CAS 9003-07-0	HiFax CB 17AC
CAS 9003-07-0	HiFax ETA 3011
CAS 9003-07-0	HiFax ETA 3095
CAS 9003-07-0	HiFax ETA 5012
CAS 9003-07-0	HiFax RTA 3263E
CAS 9003-07-0	HiGlass BJ44A
CAS 9003-07-0	HiGlass PF062-2
CAS 9003-07-0	HiGlass PF072-1
CAS 9003-07-0	HiGlass SB 224-2
CAS 9003-07-0	Hoechst Wax PP 230
CAS 9003-07-0	Hostalen® PP 927
CAS 9003-07-0	Marlex® HGX-030
CAS 9003-07-0	Marlex® HGZ-120-02
CAS 9003-07-0	Marlex® HNS-080
CAS 9003-07-0	Marlex® RMX-020
CAS 9003-07-0	Metallyte 70-2, 70-4, 70-U, 80-2, 80-4, 80-U, 140-2, 140-4, 140-U
CAS 9003-07-0	Mitsubishi 4300J
CAS 9003-07-0	Neopolen® P
CAS 9003-07-0	Nepol® PP40
CAS 9003-07-0	Nortuff RA 1700-MO
CAS 9003-07-0	Nortuff RC 1700-MO
CAS 9003-07-0	OPPalyte® 233 TW, 278 TW, 350 TW
CAS 9003-07-0	Orevac® PP-C
CAS 9003-07-0	Petrothene® PP 1510-HC
CAS 9003-07-0	Petrothene® PP 2004-MR
CAS 9003-07-0	Petrothene® PP 7300-KF
CAS 9003-07-0	Petrothene® PP 8000-GK
CAS 9003-07-0	Petrothene® PP 8770-HU
CAS 9003-07-0	Plastech™ EP 8126
CAS 9003-07-0	Plastech™ PP 3344
CAS 9003-07-0	Polifil® C-10
CAS 9003-07-0	Polifil® CAS-40
CAS 9003-07-0	Polifil® GFPP-10
CAS 9003-07-0	Polifil® GFPPCC-10
CAS 9003-07-0	Polifil® M-20
CAS 9003-07-0	Polifil® RMC-10
CAS 9003-07-0	Polifil® RMT-10
CAS 9003-07-0	Polifil® T-10
CAS 9003-07-0	PP-10GF/000
CAS 9003-07-0	PPX-30GF/000 HC
CAS 9003-07-0	Pro-fax® 65F4-4
CAS 9003-07-0	Pro-fax® 65F5-4
CAS 9003-07-0	Pro-fax® 6323
CAS 9003-07-0	Pro-fax® 7523
CAS 9003-07-0	Pro-fax® 8523
CAS 9003-07-0	Pro-fax® HB-301
CAS 9003-07-0	Pro-fax® PC-072PM
CAS 9003-07-0	Pro-fax® PD-064
CAS 9003-07-0	Pro-fax® PF-101
CAS 9003-07-0	Pro-fax® SA-747M
CAS 9003-07-0	Pro-fax® SB-242
CAS 9003-07-0	Pro-fax® SD-062
CAS 9003-07-0	Pro-fax® SE-191
CAS 9003-07-0	Pro-fax® SV-256M
CAS 9003-07-0	Pro-fax® Z-39S
CAS 9003-07-0	Procond-101
CAS 9003-07-0	Rexene® 11S12
CAS 9003-07-0	Rexene® 13S10A
CAS 9003-07-0	Rexene® 14S4A
CAS 9003-07-0	Rexene® 18C3A
CAS 9003-07-0	Rexene® PP 12R10A
CAS 9003-07-0	Rexene® PP 23S2
CAS 9003-07-0	Rexene® PP 41E2
CAS 9003-07-0	Rexene® PP 9234
CAS 9003-07-0	RTP 100 GB 10
CAS 9003-07-0	SD-376
CAS 9003-07-0	Shell 5A18Z
CAS 9003-07-0	Shell 5A95
CAS 9003-07-0	Shell 5C64
CAS 9003-07-0	Shell 6A01K
CAS 9003-07-0	Shell 6C20S
CAS 9003-07-0	Shell 7C55H
CAS 9003-07-0	Shell DS 6C46L
CAS 9003-07-0	Shell DS 7C04N
CAS 9003-07-0	Shell JF 6100
CAS 9003-07-0	Shell PDC 1120
CAS 9003-07-0	Shell WRS 6-198
CAS 9003-07-0	Shell WRS 6-205
CAS 9003-07-0	Stat-Kon® M-1 HI
CAS 9003-07-0	Symalit GM 20 PP
CAS 9003-07-0	Thermocomp® MF-1002
CAS 9003-07-0	Transpalene®
CAS 9003-07-0	Vestolen® P2000
CAS 9003-07-0	Vestolen® P2000 CR
CAS 9003-07-0	Vestolen® P6000
CAS 9003-07-0	Vestolen® P6700
CAS 9003-07-0	Vestolen® P7032 G
CAS 9003-08-1	Basocoll® CM
CAS 9003-08-1	Ecco Rez M-300-7
CAS 9003-08-1	Luwipal®
CAS 9003-08-1	MEL 80-P
CAS 9003-08-1	Uformite® 27-802
CAS 9003-09-2	Gantrez® M-154
CAS 9003-11-6	Antarox® 17-R-2
CAS 9003-11-6	Antarox® E-100
CAS 9003-11-6	Antarox® L-61
CAS 9003-11-6	Antarox® PGP 23-7
CAS 9003-11-6	Berol 374
CAS 9003-11-6	Hodag Nonionic 1017-R
CAS 9003-11-6	Hodag Nonionic 1035-L
CAS 9003-11-6	Hodag Nonionic 1044-L
CAS 9003-11-6	Hodag Nonionic 1064-L
CAS 9003-11-6	Hodag Nonionic 1088-F
CAS 9003-11-6	Hodag Nonionic 2017-R
CAS 9003-11-6	Industrol® N3
CAS 9003-11-6	Macol® 1
CAS 9003-11-6	Macol® 2
CAS 9003-11-6	Macol® 4
CAS 9003-11-6	Macol® 8
CAS 9003-11-6	Macol® 15
CAS 9003-11-6	Macol® 16

CAS Number	Trade Name
CAS 9003-11-6	Macol® 18
CAS 9003-11-6	Macol® 19
CAS 9003-11-6	Macol® 23
CAS 9003-11-6	Macol® 27
CAS 9003-11-6	Macol® 32
CAS 9003-11-6	Macol® 33
CAS 9003-11-6	Macol® 34
CAS 9003-11-6	Macol® 35
CAS 9003-11-6	Macol® 40
CAS 9003-11-6	Macol® 42
CAS 9003-11-6	Macol® 44
CAS 9003-11-6	Macol® 46
CAS 9003-11-6	Macol® 72
CAS 9003-11-6	Macol® 77
CAS 9003-11-6	Macol® 85
CAS 9003-11-6	Macol® 101
CAS 9003-11-6	Macol® 108
CAS 9003-11-6	Monolan® 8000 E/80
CAS 9003-11-6	Pluronic® 10R5
CAS 9003-11-6	Pluronic® 10R8
CAS 9003-11-6	Pluronic® 17R1
CAS 9003-11-6	Pluronic® 17R2
CAS 9003-11-6	Pluronic® 17R4
CAS 9003-11-6	Pluronic® 17R8
CAS 9003-11-6	Pluronic® 25R1
CAS 9003-11-6	Pluronic® 25R2
CAS 9003-11-6	Pluronic® 25R4
CAS 9003-11-6	Pluronic® 25R5
CAS 9003-11-6	Pluronic® 25R8
CAS 9003-11-6	Pluronic® 31R1
CAS 9003-11-6	Pluronic® 31R2
CAS 9003-11-6	Pluronic® 31R4
CAS 9003-11-6	Pluronic® F108
CAS 9003-11-6	Pluronic® F127
CAS 9003-11-6	Pluronic® F38
CAS 9003-11-6	Pluronic® F68
CAS 9003-11-6	Pluronic® F68LF
CAS 9003-11-6	Pluronic® F77
CAS 9003-11-6	Pluronic® F87
CAS 9003-11-6	Pluronic® F88
CAS 9003-11-6	Pluronic® F98
CAS 9003-11-6	Pluronic® L31
CAS 9003-11-6	Pluronic® L35
CAS 9003-11-6	Pluronic® L42
CAS 9003-11-6	Pluronic® L43
CAS 9003-11-6	Pluronic® L44
CAS 9003-11-6	Pluronic® L61
CAS 9003-11-6	Pluronic® L62
CAS 9003-11-6	Pluronic® L62D
CAS 9003-11-6	Pluronic® L62LF
CAS 9003-11-6	Pluronic® L63
CAS 9003-11-6	Pluronic® L64
CAS 9003-11-6	Pluronic® L72
CAS 9003-11-6	Pluronic® L81
CAS 9003-11-6	Pluronic® L92
CAS 9003-11-6	Pluronic® L101
CAS 9003-11-6	Pluronic® L121
CAS 9003-11-6	Pluronic® L122
CAS 9003-11-6	Pluronic® P65
CAS 9003-11-6	Pluronic® P75
CAS 9003-11-6	Pluronic® P84
CAS 9003-11-6	Pluronic® P85
CAS 9003-11-6	Pluronic® P94
CAS 9003-11-6	Pluronic® P103
CAS 9003-11-6	Pluronic® P104
CAS 9003-11-6	Pluronic® P105
CAS 9003-11-6	Pluronic® P123
CAS 9003-11-6	Polykol
CAS 9003-13-8	Hodag PB-285
CAS 9003-13-8	Ucon® Fluid AP
CAS 9003-17-2	Budene® 1207
CAS 9003-17-2	Budene® 1254
CAS 9003-17-2	Cisdene® 1203
CAS 9003-17-2	Diene® 35AC
CAS 9003-17-2	Diene® 70AC
CAS 9003-17-2	E-BR® 8405
CAS 9003-17-2	E-BR® 8471
CAS 9003-17-2	Poly bd® R-45HT
CAS 9003-17-2	Polyoil Hüls 110
CAS 9003-17-2	Ricon 159
CAS 9003-17-2	Taktene 220
CAS 9003-18-3	Krynac® 19.65
CAS 9003-18-3	Krynac® 20H35
CAS 9003-18-3	Krynac® 34.140
CAS 9003-18-3	Krynac® PXL 34.17
CAS 9003-18-3	Nipol® 1000X132
CAS 9003-18-3	Nipol® 1001 CG
CAS 9003-18-3	Nipol® 1022X59
CAS 9003-18-3	Nipol® 1072
CAS 9003-18-3	Nipol® 1411
CAS 9003-18-3	Tylac® 68151-00
CAS 9003-20-7	Catomer VA
CAS 9003-20-7	Daratak® RP2000
CAS 9003-20-7	Daratak® SP1011
CAS 9003-20-7	Everflex® SP-1084
CAS 9003-20-7	Liquid Latex
CAS 9003-20-7	Quabond® 210
CAS 9003-20-7	Rovace 571
CAS 9003-20-7	Rovace 2113
CAS 9003-20-7	Ucar® Latex 130
CAS 9003-20-7	Vinac® 1000 DEV
CAS 9003-20-7	Vinac® XX-210
CAS 9003-20-7	Vinac® XX-220
CAS 9003-20-7	Vinac® XX-230
CAS 9003-20-7	Vinac® XX-240
CAS 9003-22-9	Vestolit® B 7090
CAS 9003-22-9	Vinnol® 50
CAS 9003-28-5	Actipol® E6
CAS 9003-28-5	Amoco® BR-310
CAS 9003-28-5	Amoco® CI-500
CAS 9003-28-5	Amoco® H-15
CAS 9003-28-5	Amoco® L-14
CAS 9003-28-5	Duraflex® 8410
CAS 9003-28-5	Santac 52
CAS 9003-29-6	Permethyl 104A
CAS 9003-39-8	Agrimer 15L
CAS 9003-39-8	Agrimer AL-22
CAS 9003-39-8	H₂old EP-1
CAS 9003-39-8	Luviskol® K12, K17, K30, K60
CAS 9003-39-8	Luviskol® K80, K90
CAS 9003-39-8	Peregal® ST
CAS 9003-39-8	Plasdone®
CAS 9003-39-8	Plasdone® C
CAS 9003-39-8	Plasdone® K

CAS 9003-39-8	Plasmosan
CAS 9003-39-8	Sokalan® HP 50
CAS 9003-39-8	Sokalan® HP 53
CAS 9003-39-8	Vinisil
CAS 9003-53-6	Amoco® H2R
CAS 9003-53-6	Amoco® H3E
CAS 9003-53-6	Amoco® R1
CAS 9003-53-6	Amoco® R5
CAS 9003-53-6	Dylite® D195B
CAS 9003-53-6	Dylite® R2595B EPS
CAS 9003-53-6	Edistir® FA
CAS 9003-53-6	Edistir® N 1280, N 1281
CAS 9003-53-6	Edistir® RC
CAS 9003-53-6	Edistir® RK
CAS 9003-53-6	Edistir® RKV
CAS 9003-53-6	Edistir® RV 8
CAS 9003-53-6	Edistir® SR 550, SRL 550
CAS 9003-53-6	Edistir® UT/1
CAS 9003-53-6	Edistir® UT/SF
CAS 9003-53-6	Electrafil® J-30/CF/20
CAS 9003-53-6	Empee® PS-921
CAS 9003-53-6	Huntsman 201
CAS 9003-53-6	Huntsman 240
CAS 9003-53-6	Huntsman 312
CAS 9003-53-6	Huntsman 331
CAS 9003-53-6	Huntsman 351
CAS 9003-53-6	Huntsman 474
CAS 9003-53-6	Huntsman 765
CAS 9003-53-6	Mobil 1240
CAS 9003-53-6	Mobil 2120
CAS 9003-53-6	Mobil 5350
CAS 9003-53-6	Mobil 5600
CAS 9003-53-6	Mobil 8020
CAS 9003-53-6	Mobil MX 4354
CAS 9003-53-6	Polypenco® Q200.5
CAS 9003-53-6	Polysphere 3000 SP
CAS 9003-53-6	Polystyrene 101
CAS 9003-53-6	Polystyrene 220
CAS 9003-53-6	Polystyrene 410
CAS 9003-53-6	Polystyrene P 2122
CAS 9003-53-6	Replay RP 2177
CAS 9003-53-6	Replay RP 2236
CAS 9003-53-6	Retain PS-4000
CAS 9003-53-6	RTP 401
CAS 9003-53-6	Styrodur®
CAS 9003-53-6	Styrofoam Brand Insulation
CAS 9003-53-6	Styron 421
CAS 9003-53-6	Styron 478
CAS 9003-53-6	Styron 479
CAS 9003-53-6	Styron 697
CAS 9003-53-6	Styron 6075
CAS 9003-53-6	Styron 6087 SF
CAS 9003-53-6	Styropor® F
CAS 9003-53-6	Styropor® FH
CAS 9003-53-6	Styropor® P
CAS 9003-53-6	Styvex 22000 NA
CAS 9003-53-6	Thermocomp® CF-1004
CAS 9003-53-6	Thermocomp® NF-1004
CAS 9003-54-7	Baymod® A95
CAS 9003-54-7	Darex® 165L
CAS 9003-54-7	Luran® 358 N
CAS 9003-54-7	Luran® 378 P G7
CAS 9003-54-7	Luran® KR 2556
CAS 9003-54-7	RTP 501
CAS 9003-54-7	SN-30GF/000 FR
CAS 9003-54-7	Stylac® AS
CAS 9003-54-7	Styvex 32000 BK
CAS 9003-54-7	Thermocomp® BF-1006
CAS 9003-54-7	Thermocomp® BF-1006FR
CAS 9003-54-7	Tyril 125
CAS 9003-54-7	Tyril 880
CAS 9003-54-7	Tyril 880B
CAS 9003-54-7	Tyril 1011
CAS 9003-55-8	Bayer SBR Latex 200 C
CAS 9003-55-8	KR01 K-Resin Polymer
CAS 9003-55-8	KR04 K-Resin Polymer
CAS 9003-55-8	Pliolite® S-6B, S-6F
CAS 9003-55-8	Styrolux®
CAS 9003-55-8	Styroplus®
CAS 9003-55-8	Tylac® 68202-00
CAS 9003-56-9	ABS 124ESG
CAS 9003-56-9	ABS 236F
CAS 9003-56-9	ABS 236MA
CAS 9003-56-9	ABS 301K
CAS 9003-56-9	ABS 500FR-1
CAS 9003-56-9	AS-10GF
CAS 9003-56-9	AS-15CF/000
CAS 9003-56-9	Baymod® A KU3-2086
CAS 9003-56-9	Blendex® 101
CAS 9003-56-9	Blendex® 310
CAS 9003-56-9	Claradex CH-540
CAS 9003-56-9	Conductomer ABS-22
CAS 9003-56-9	Cycolac® CKM1
CAS 9003-56-9	Cycolac® DH
CAS 9003-56-9	Cycolac® GPM4700
CAS 9003-56-9	Cycolac® GPX2800
CAS 9003-56-9	Cycolac® KCS
CAS 9003-56-9	Cycolac® KJM
CAS 9003-56-9	Cycolac® X-11
CAS 9003-56-9	Electrafil® G-1204/SS/3
CAS 9003-56-9	Electrafil® J-1200/CF/10
CAS 9003-56-9	EMI-X® PDX-A-88128
CAS 9003-56-9	Magnum 240
CAS 9003-56-9	Magnum 275
CAS 9003-56-9	Magnum 445 HQ
CAS 9003-56-9	Magnum 788HP
CAS 9003-56-9	Magnum 2610
CAS 9003-56-9	Magnum 3661
CAS 9003-56-9	Magnum 4420
CAS 9003-56-9	Magnum 9450P
CAS 9003-56-9	Magnum FG960
CAS 9003-56-9	Multibase ABS 3075
CAS 9003-56-9	Multibase ABS 3525 CL
CAS 9003-56-9	Multibase ABS 3959
CAS 9003-56-9	Novodur® L3FR
CAS 9003-56-9	Novodur® P2H-AT
CAS 9003-56-9	Novodur® P2HE
CAS 9003-56-9	Novodur® P2HGV
CAS 9003-56-9	Novodur® P2T, P2T-AT
CAS 9003-56-9	Novodur® PMTM
CAS 9003-56-9	RTP 601
CAS 9003-56-9	Stat-Kon® AC-1003
CAS 9003-56-9	Stat-Kon® AS
CAS 9003-56-9	Stylac® ABS

CAS Number	Trade Name
CAS 9003-56-9	Styvex 40007 BKL2
CAS 9003-56-9	Styvex 42023 NAFR
CAS 9003-56-9	Terluran®
CAS 9003-56-9	Terlux®
CAS 9003-56-9	Thermocomp® AF-1004
CAS 9003-79-6	Agerite® Superflex, Superflex Solid G
CAS 9003-79-6	Aminox® Flake, Powd
CAS 9003-79-6	BXA Flake
CAS 9003-79-6	Vanox® AM
CAS 9004-32-4	Aqualon® Cellulose Gum
CAS 9004-32-4	Aqualon® CMC-T
CAS 9004-32-4	Aquasorb® A250
CAS 9004-32-4	Blanose
CAS 9004-32-4	Cekol
CAS 9004-32-4	Cellufresh™
CAS 9004-32-4	Celluvisc®
CAS 9004-32-4	Courlose®
CAS 9004-32-4	Tylose® C, CB Series
CAS 9004-32-4	Tylose® C-p, CB-p
CAS 9004-34-6	Celluflow C-25
CAS 9004-34-6	CF 31,000C Coarse
CAS 9004-34-6	CF 42,500T Medium
CAS 9004-34-6	CF 70,000WDK, Ex. Superfine
CAS 9004-34-6	Elcema® F150, G250, P100
CAS 9004-34-6	Emcocel® 90M
CAS 9004-34-6	Fibra-Cel®
CAS 9004-34-6	Solka-Floc® BW-40, BW-100, BW-200, BW-2030, UF-900-FCC & NF
CAS 9004-35-7	CA-394-60S
CAS 9004-35-7	Celluflow TA-25
CAS 9004-35-7	Courtaulds
CAS 9004-35-7	Rotuba H
CAS 9004-35-7	Tenite® 105-MS
CAS 9004-35-7	Tenite® Cellulosic Acetate
CAS 9004-36-8	CAB-171-15S
CAS 9004-36-8	CAB-381-0.1, 381-0.5, 381-2, 381-20
CAS 9004-36-8	CAB-500-5
CAS 9004-36-8	Cabulite
CAS 9004-36-8	Dispercab
CAS 9004-36-8	Tenite® 264-MH
CAS 9004-39-1	CAP-482-0.5
CAS 9004-39-1	CAP-482-20
CAS 9004-39-1	Dispercap
CAS 9004-39-1	Tenite® 360-H2
CAS 9004-39-1	Tenite® Cellulosic Propionate
CAS 9004-57-3	Ethocel Standard Premium
CAS 9004-57-3	Hercules® K
CAS 9004-57-3	Hercules® N
CAS 9004-57-3	Hercules® T
CAS 9004-58-4	Aqualon® EHEC
CAS 9004-58-4	Bermacoll
CAS 9004-61-9	Biomatrix®
CAS 9004-61-9	Hyladerm®
CAS 9004-62-0	Cellosize
CAS 9004-62-0	Cellosize® HEC QP Grades
CAS 9004-62-0	Cellosize® HEC WP Grades
CAS 9004-62-0	Cellosize® Polymer PCG-10
CAS 9004-62-0	Natrosol®
CAS 9004-62-0	Natrosol 250
CAS 9004-62-0	Natrosol® Hydroxyethylcellulose
CAS 9004-62-0	Tylose® H Series
CAS 9004-64-2	Klucel®
CAS 9004-64-2	Klucel® E, G, H, J, L, M
CAS 9004-64-2	Klucel® EF
CAS 9004-65-3	Benecel® Hydroxypropyl-methylcellulose
CAS 9004-65-3	Culminal® Hydroxypropyl-methylcellulose
CAS 9004-65-3	Goniosol
CAS 9004-65-3	Lacril®
CAS 9004-65-3	Methocel® 40-202
CAS 9004-65-3	Methocel® E3 Premium
CAS 9004-65-3	Methocel® E4M
CAS 9004-65-3	Methocel® E5
CAS 9004-65-3	Methocel® F4M
CAS 9004-65-3	Methocel® K100MP
CAS 9004-65-3	Methocel® K35
CAS 9004-65-3	Tearisol® Sterile Lubricating Eye Drops
CAS 9004-67-5	Culminal® Methylcellulose
CAS 9004-67-5	Methocel
CAS 9004-67-5	Methocel® A15-LV
CAS 9004-67-5	Methocel® A4C, A4M
CAS 9004-67-5	Methocel® A4MP
CAS 9004-70-0	AS
CAS 9004-70-0	Dispercel
CAS 9004-70-0	Hercules® AS
CAS 9004-70-0	Hercules® RS
CAS 9004-70-0	Hercules® SS
CAS 9004-70-0	RS Nitrocellulose
CAS 9004-70-0	SS Nitrocellulose
CAS 9004-73-3	Finex-25-020
CAS 9004-73-3	Masil® SF-MH
CAS 9004-73-3	Nylon N-012.
CAS 9004-73-3	Sericite SL-012.
CAS 9004-73-7	Spheron P-1500-030
CAS 9004-74-4	Carbowax® MPEG 350
CAS 9004-81-3	Ablunol 200ML
CAS 9004-81-3	Alkamuls® L-9
CAS 9004-81-3	Chemax E-200 ML
CAS 9004-81-3	Chemax E-400 ML
CAS 9004-81-3	Chemax E-600 ML
CAS 9004-81-3	CPH-43-N
CAS 9004-81-3	Crodet L4
CAS 9004-81-3	Emerest® 2620
CAS 9004-81-3	Emerest® 2630
CAS 9004-81-3	Ethox ML-5
CAS 9004-81-3	Hetoxamate LA-5
CAS 9004-81-3	Hodag 20-L
CAS 9004-81-3	Hodag DGL
CAS 9004-81-3	Kessco® PEG 200 ML
CAS 9004-81-3	Lexemul® PEG-400ML
CAS 9004-81-3	Lipo DGLS
CAS 9004-81-3	Lipopeg 4-L
CAS 9004-81-3	Mapeg® 200 ML
CAS 9004-81-3	Mapeg® DGLD
CAS 9004-81-3	Nopalcol 1-L
CAS 9004-81-3	Nopalcol 4-L
CAS 9004-81-3	Pegosperse® 100 L
CAS 9004-81-3	Witconol 2620
CAS 9004-82-4	Abex® 23S

CAS 9004-82-4	Akyposal 9278 R	CAS 9004-96-0	Chemax E-1000 MO
CAS 9004-82-4	Akyposal EO 20 MW	CAS 9004-96-0	Crodet O4
CAS 9004-82-4	Akyposal MS SPC	CAS 9004-96-0	Emalex 218
CAS 9004-82-4	Akyposal RLM 70	CAS 9004-96-0	Emalex OE-6
CAS 9004-82-4	Avirol® FES 996	CAS 9004-96-0	Emerest® 2617
CAS 9004-82-4	Avirol® SE 3002	CAS 9004-96-0	Emerest® 2624
CAS 9004-82-4	Calfoam ES-30	CAS 9004-96-0	Emerest® 2660
CAS 9004-82-4	Carsonol® SES-S	CAS 9004-96-0	Empilan® BQ 100
CAS 9004-82-4	Chemsalan RLM 28	CAS 9004-96-0	Emulan® A
CAS 9004-82-4	Disponil FES 32	CAS 9004-96-0	Ethofat® O/20
CAS 9004-82-4	Empicol® ESB	CAS 9004-96-0	Ethox MO-9
CAS 9004-82-4	Empimin® KSN27	CAS 9004-96-0	Ethylan® A10
CAS 9004-82-4	Genapol® ARO	CAS 9004-96-0	Ethylan® A2
CAS 9004-82-4	Genapol® LRO Liq., Paste	CAS 9004-96-0	Ethylan® FO30
CAS 9004-82-4	Genapol® ZRO Liq., Paste	CAS 9004-96-0	Eumulgin PLT 4
CAS 9004-82-4	Hartenol LES 60	CAS 9004-96-0	Genagen O-090
CAS 9004-82-4	Marlinat® 242/28	CAS 9004-96-0	Hetoxamate MO-2
CAS 9004-82-4	Nonasol N4SS	CAS 9004-96-0	Hodag 40-O
CAS 9004-82-4	Norfox® SLES-60	CAS 9004-96-0	Hodag DGO
CAS 9004-82-4	Nutrapon ES-60 3568	CAS 9004-96-0	Kessco® PEG 200 MO
CAS 9004-82-4	Nutrapon KPC 0156	CAS 9004-96-0	Laurel PEG 400 MO
CAS 9004-82-4	Polystep® B-12	CAS 9004-96-0	Mapeg® 200 MO
CAS 9004-82-4	Rewopol® NL 2-28	CAS 9004-96-0	Marlosol® OL2
CAS 9004-82-4	Rhodapex® ES	CAS 9004-96-0	Nonex O4E
CAS 9004-82-4	Steol® 4N	CAS 9004-96-0	Nopalcol 4-O
CAS 9004-82-4	Steol® OS 28	CAS 9004-96-0	Pegosperse® 100 O
CAS 9004-82-4	Sulfochem ES-2	CAS 9004-96-0	Prox-onic OL-1/05
CAS 9004-82-4	Sulfochem ES-70	CAS 9004-96-0	Remcopal 6
CAS 9004-82-4	Sulfotex LMS-E	CAS 9004-96-0	Remcopal 207
CAS 9004-82-4	Texapon N 25	CAS 9004-96-0	Secoster® MO 400
CAS 9004-82-4	Texapon NSE	CAS 9004-96-0	Sellig AO 6 100
CAS 9004-82-4	Texapon NSO	CAS 9004-96-0	Sellig AO 15 100
CAS 9004-82-4	Texapon PLT-227	CAS 9004-96-0	Sellig AO 25 100
CAS 9004-82-4	Ungerol N2-28	CAS 9004-96-0	Teric OF4
CAS 9004-82-4	Witcolate 6450	CAS 9004-96-0	Trydet 2676
CAS 9004-82-4	Witcolate ES-2	CAS 9004-96-0	Witconol H31A
CAS 9004-82-4	Witcolate LES-60C	CAS 9004-96-0	Witflow 916
CAS 9004-82-4	Zetesol NL	CAS 9004-97-1	Hodag 40-R
CAS 9004-82-4	Zoharpon ETA 27	CAS 9004-97-1	Nopalcol 6-R
CAS 9004-83-5	Alcodet® 218	CAS 9004-98-2	Ablunol OA-6
CAS 9004-92-2	Hetoxol OL-2	CAS 9004-98-2	Akyporox RO 90
CAS 9004-93-7	Uniplex 809	CAS 9004-98-2	Akyporox RTO 70
CAS 9004-94-8	Genagen P-070	CAS 9004-98-2	Ameroxol® OE-2
CAS 9004-95-9	Akyporox RC 200	CAS 9004-98-2	Ameroxol® OE-5
CAS 9004-95-9	Brij® 52	CAS 9004-98-2	Brij® 93
CAS 9004-95-9	Emalex 103	CAS 9004-98-2	Chemal OA-4
CAS 9004-95-9	Emalex 1605	CAS 9004-98-2	Chemal OA-5
CAS 9004-95-9	Ethosperse® CA-2	CAS 9004-98-2	Chemal OA-20/70CWS
CAS 9004-95-9	Fancol CH-24.	CAS 9004-98-2	Emalex 508
CAS 9004-95-9	Hetoxol CA-2	CAS 9004-98-2	Ethal OA-10
CAS 9004-95-9	Lipocol C-2	CAS 9004-98-2	Eumulgin M8
CAS 9004-95-9	Macol® CA-2	CAS 9004-98-2	Eumulgin O5
CAS 9004-95-9	Rhodasurf® C-2	CAS 9004-98-2	Eumulgin PWM2
CAS 9004-95-9	Simulsol 58	CAS 9004-98-2	Eumulgin WM5
CAS 9004-96-0	Ablunol 200MO	CAS 9004-98-2	Genapol® O-020
CAS 9004-96-0	Acconon 400-MO	CAS 9004-98-2	Hetoxol OA-3 Special
CAS 9004-96-0	Alkamuls® 400-MO	CAS 9004-98-2	Hostacerin O-20
CAS 9004-96-0	Alkamuls® A	CAS 9004-98-2	Lipocol O-2
CAS 9004-96-0	Alphoxat O 105	CAS 9004-98-2	Macol® OA-2
CAS 9004-96-0	Chemax E-200 MO	CAS 9004-98-2	Marlipal® 1850/5
CAS 9004-96-0	Chemax E-400 MO	CAS 9004-98-2	Marlowet® WOE
CAS 9004-96-0	Chemax E-600 MO	CAS 9004-98-2	Prox-onic OA-1/04

CAS 9004-98-2	Rhodasurf® ON-870
CAS 9004-98-2	Ritoleth 2
CAS 9004-98-2	Simulsol 98
CAS 9004-98-2	Trycol® 5971
CAS 9004-98-2	Varonic® 32-E20
CAS 9004-98-2	Volpo 3
CAS 9004-98-2	Volpo 5
CAS 9004-98-2	Volpo 10
CAS 9004-98-2	Volpo 20
CAS 9004-98-2	Volpo N3
CAS 9004-98-2	Volpo O3
CAS 9004-99-3	Ablunol 200MS
CAS 9004-99-3	Ablunol DEGMS
CAS 9004-99-3	Acconon 200-MS
CAS 9004-99-3	Acconon 400-MS
CAS 9004-99-3	Alkamuls® S-65-40
CAS 9004-99-3	Alkamuls® S-65-8
CAS 9004-99-3	Alphoxat S 110
CAS 9004-99-3	Cerasynt 840
CAS 9004-99-3	Chemax E-200 MS
CAS 9004-99-3	Chemax E-400 MS
CAS 9004-99-3	Chemax E-600 MS
CAS 9004-99-3	Chemax E-1000 MS
CAS 9004-99-3	Cremophor® S 9
CAS 9004-99-3	Crodet S4
CAS 9004-99-3	Emalex 400A
CAS 9004-99-3	Emalex 805
CAS 9004-99-3	Emalex 6300 M-ST
CAS 9004-99-3	Emalex DEG-m-S
CAS 9004-99-3	Emerest® 2610
CAS 9004-99-3	Emerest® 2636
CAS 9004-99-3	Ethofat® 18/14
CAS 9004-99-3	Ethox MS-8
CAS 9004-99-3	Eumulgin PST 5
CAS 9004-99-3	Genagen S-080
CAS 9004-99-3	Hetoxamate SA-5
CAS 9004-99-3	Hodag 40-S
CAS 9004-99-3	Hodag 150-S
CAS 9004-99-3	Hodag DGS
CAS 9004-99-3	Hydrine
CAS 9004-99-3	Kemester® 5221SE
CAS 9004-99-3	Kessco® PEG 200 MS
CAS 9004-99-3	Mapeg® 200 MS
CAS 9004-99-3	Mapeg® 1500 MS
CAS 9004-99-3	Mapeg® S-40
CAS 9004-99-3	Marlosol® 183
CAS 9004-99-3	Mergital ST 30/E
CAS 9004-99-3	Myrj® 45
CAS 9004-99-3	Myrj® 52
CAS 9004-99-3	Myrj® 53
CAS 9004-99-3	Myrj® 59
CAS 9004-99-3	Nonex S3E
CAS 9004-99-3	Nopalcol 4-S
CAS 9004-99-3	Pegosperse® 50 MS
CAS 9004-99-3	Pegosperse® 100 S
CAS 9004-99-3	Pegosperse® 1750 MS
CAS 9004-99-3	Polystate C
CAS 9004-99-3	Prox-onic ST-05
CAS 9004-99-3	Ritapeg 150 DS
CAS 9004-99-3	Ritox 52
CAS 9004-99-3	Schercemol DEGMS
CAS 9004-99-3	Sellig S 30 100
CAS 9004-99-3	Sellig Stearo 6
CAS 9004-99-3	Simulsol M 45
CAS 9004-99-3	Superpolystate
CAS 9004-99-3	Tefose® 1500
CAS 9004-99-3	Trydet 2610
CAS 9004-99-3	Witconol 2640
CAS 9005-00-9	Ablunol SA-7
CAS 9005-00-9	Brij® 72
CAS 9005-00-9	Brij® 700 S
CAS 9005-00-9	Brij® 721
CAS 9005-00-9	Emalex 640
CAS 9005-00-9	Hetoxol STA-2
CAS 9005-00-9	Hodag Nonionic S-2
CAS 9005-00-9	Lipocol S-2
CAS 9005-00-9	Macol® SA-2
CAS 9005-00-9	Prox-onic SA-1/02
CAS 9005-00-9	Rhodasurf® S-2
CAS 9005-00-9	Rhodasurf® TB-970
CAS 9005-00-9	Ritox 721
CAS 9005-00-9	Trycol® 5888
CAS 9005-00-9	Volpo S-2
CAS 9005-02-1	Acconon 200-DL
CAS 9005-02-1	Emalex 200 di-L
CAS 9005-02-1	Emalex DEG-di-L
CAS 9005-02-1	Emalex TEG-di-L
CAS 9005-02-1	Emerest® 2622
CAS 9005-02-1	Emerest® 2704
CAS 9005-02-1	Ethox DL-5
CAS 9005-02-1	Hetoxamate 200 DL
CAS 9005-02-1	Hodag 22-L
CAS 9005-02-1	Kessco® PEG 200 DL
CAS 9005-02-1	Lexemul® PEG-200 DL
CAS 9005-02-1	Lipopeg 2-DL
CAS 9005-02-1	Mapeg® 200 DL
CAS 9005-02-1	Nonex DL-2
CAS 9005-02-1	Nopalcol 2-DL
CAS 9005-02-1	Pegosperse® 200 DL
CAS 9005-02-1	Witconol 2622
CAS 9005-07-6	Alkamuls® 400-DO
CAS 9005-07-6	Alkamuls® 600-DO
CAS 9005-07-6	Chemax PEG 200 DO
CAS 9005-07-6	Chemax PEG 400 DO
CAS 9005-07-6	Chemax PEG 600 DO
CAS 9005-07-6	CPH-211-N
CAS 9005-07-6	Dyafac PEG 6DO
CAS 9005-07-6	Emalex 200 di-O
CAS 9005-07-6	Emalex DEG-di-O
CAS 9005-07-6	Ethox DO-9
CAS 9005-07-6	Hodag 42-O
CAS 9005-07-6	Kessco® PEG 200 DO
CAS 9005-07-6	Lipopeg 4-DO
CAS 9005-07-6	Mapeg® 200 DO
CAS 9005-07-6	Marlosol® FS
CAS 9005-07-6	Nonex DO-4
CAS 9005-07-6	Nopalcol 6-DO
CAS 9005-07-6	Secoster® DO 600
CAS 9005-07-6	Witconol 2648
CAS 9005-07-6	Witconol H33
CAS 9005-08-7	Emalex 200 di-S
CAS 9005-08-7	Emalex 6300 DI-ST
CAS 9005-08-7	Genapol® TS Powd
CAS 9005-08-7	Hetoxamate 400 DS

CAS 9005-08-7	Hodag 602-S
CAS 9005-08-7	Kessco® PEG 200 DS
CAS 9005-08-7	Lipopeg 4-DS
CAS 9005-08-7	Lipopeg 6000-DS
CAS 9005-08-7	Mapeg® 200 DS
CAS 9005-08-7	Mapeg® 1540 DS
CAS 9005-08-7	Mapeg® 6000 DS
CAS 9005-08-7	Marlosol® BS
CAS 9005-08-7	Rewopal® PEG 6000 DS
CAS 9005-08-7	Ritapeg 400 DS
CAS 9005-08-7	Witconol 2642
CAS 9005-25-8	Control I-100
CAS 9005-25-8	D12F
CAS 9005-25-8	Pure-Dent® B700
CAS 9005-25-8	Purity® 21
CAS 9005-25-8	Supercore® S13F
CAS 9005-32-7	Kelacid®
CAS 9005-32-7	Saltialgine H8
CAS 9005-32-7	Satialgine H8
CAS 9005-34-9	Amoloid HV
CAS 9005-34-9	Amoloid LV
CAS 9005-34-9	Collatex
CAS 9005-34-9	Sobalg FD 300 Series
CAS 9005-34-9	Superloid®
CAS 9005-35-0	Sobalg FD 460
CAS 9005-36-1	Improved Kelmar®
CAS 9005-36-1	Kelmar®
CAS 9005-36-1	Sobalg FD 200 Series
CAS 9005-37-2	Concentrated Dariloid® KB
CAS 9005-37-2	Dricoid® KB, KBC
CAS 9005-37-2	Kelcoloid® D
CAS 9005-37-2	Kelcoloid® DH, DSF
CAS 9005-37-2	Kelcoloid® HV
CAS 9005-37-2	Kelcoloid® LV
CAS 9005-37-2	Kelcoloid® O
CAS 9005-37-2	Kelcoloid® S
CAS 9005-37-2	Manucol Ester E/RK
CAS 9005-37-2	Manucol Ester EX/LL
CAS 9005-37-2	Sherbelizer®
CAS 9005-38-3	Kelcosol®
CAS 9005-38-3	Kelgin® F
CAS 9005-38-3	Kelvis®
CAS 9005-49-6	Liquemin
CAS 9005-64-5	Accosperse 20
CAS 9005-64-5	Alkamuls® PSML-20
CAS 9005-64-5	Alkamuls® T-20
CAS 9005-64-5	Capmul® POE-L
CAS 9005-64-5	Crillet 1
CAS 9005-64-5	Crillet 11
CAS 9005-64-5	Disponil SML 104 F1
CAS 9005-64-5	Disponil SML 120 F1
CAS 9005-64-5	Emalex ET-2020
CAS 9005-64-5	Emsorb® 2720
CAS 9005-64-5	Emsorb® 2721
CAS 9005-64-5	Emsorb® 6915
CAS 9005-64-5	Ethsorbox L-20
CAS 9005-64-5	Ethylan® GEL2
CAS 9005-64-5	Eumulgin SML 20
CAS 9005-64-5	G-4280
CAS 9005-64-5	Glycosperse® L-10
CAS 9005-64-5	Glycosperse® L-20
CAS 9005-64-5	Hetsorb L-10
CAS 9005-64-5	Hetsorb L-20
CAS 9005-64-5	Hetsorb L-4
CAS 9005-64-5	Hetsorb L-80-72%
CAS 9005-64-5	Hodag PSML-20
CAS 9005-64-5	Liposorb L-10
CAS 9005-64-5	Liposorb L-20
CAS 9005-64-5	Montanox 20 DF
CAS 9005-64-5	Mulsifan RT 141
CAS 9005-64-5	Prox-onic SML-020
CAS 9005-64-5	Radiasurf® 7137
CAS 9005-64-5	Sorbax PML-20
CAS 9005-64-5	Sorbon T-20
CAS 9005-64-5	T-Maz® 20
CAS 9005-64-5	T-Maz® 28
CAS 9005-64-5	Tween® 20
CAS 9005-64-5	Tween® 21
CAS 9005-64-5	Witconol 2720
CAS 9005-64-5	Witflow 990
CAS 9005-65-6	Accosperse 80
CAS 9005-65-6	Alkamuls® PSMO-5
CAS 9005-65-6	Alkamuls® PSMO-20
CAS 9005-65-6	Alkamuls® T-80
CAS 9005-65-6	Capmul® POE-O
CAS 9005-65-6	Crillet 4
CAS 9005-65-6	Crillet 41
CAS 9005-65-6	DeSotan® SMO-20
CAS 9005-65-6	Disponil SMO 120 F1
CAS 9005-65-6	Drewpone® 80K
CAS 9005-65-6	Durfax® 80
CAS 9005-65-6	Emalex ET-8020
CAS 9005-65-6	Emsorb® 2722
CAS 9005-65-6	Emsorb® 6900
CAS 9005-65-6	Emsorb® 6901
CAS 9005-65-6	Ethsorbox O-20
CAS 9005-65-6	Ethylan® GEO8
CAS 9005-65-6	Ethylan® GEO81
CAS 9005-65-6	Eumulgin SMO 20
CAS 9005-65-6	Glycosperse® O-20 FG, O-20 KFG
CAS 9005-65-6	Glycosperse® O-5
CAS 9005-65-6	Hetsorb O-5
CAS 9005-65-6	Hetsorb O-20
CAS 9005-65-6	Hodag PSMO-20
CAS 9005-65-6	Hodag SVO-9
CAS 9005-65-6	Liposorb O-5
CAS 9005-65-6	Liposorb O-20
CAS 9005-65-6	Montanox 80 DF
CAS 9005-65-6	Montanox 81
CAS 9005-65-6	Mulsifan RT 146
CAS 9005-65-6	Polycon T80 K
CAS 9005-65-6	Prox-onic SMO-020
CAS 9005-65-6	Prox-onic SMO-05
CAS 9005-65-6	Radiamuls® Sorb 2157
CAS 9005-65-6	Radiasurf® 7157
CAS 9005-65-6	Ritabate 80
CAS 9005-65-6	Sorbax PMO-5
CAS 9005-65-6	Sorbax PMO-20
CAS 9005-65-6	Sorlate
CAS 9005-65-6	T-Maz® 80
CAS 9005-65-6	T-Maz® 81
CAS 9005-65-6	Tween® 80, 80K
CAS 9005-65-6	Tween® 81

CAS 9005-65-6	Witconol 2722
CAS 9005-65-6	Witflow 991
CAS 9005-66-7	Crillet 2
CAS 9005-66-7	Disponil SMP 120 F1
CAS 9005-66-7	Ethylan® GEP4
CAS 9005-66-7	G-4252
CAS 9005-66-7	Glycosperse® P-20
CAS 9005-66-7	Hetsorb P-20
CAS 9005 66 7	Hodag PSMP-20
CAS 9005-66-7	Liposorb P-20
CAS 9005-66-7	Montanox 40 DF
CAS 9005-66-7	Prox-onic SMP-020
CAS 9005-66-7	Sorbax PMP-20
CAS 9005-66-7	T-Maz® 40
CAS 9005-66-7	Tween® 40
CAS 9005-67-8	Accosperse 60
CAS 9005-67-8	Alkamuls® PSMS-20
CAS 9005-67-8	Alkamuls® T-60
CAS 9005-67-8	Capmul® POE-S
CAS 9005-67-8	Crillet 3
CAS 9005-67-8	Crillet 31
CAS 9005-67-8	Disponil SMS 120 F1
CAS 9005-67-8	Drewpone® 60K
CAS 9005-67-8	Durfax® 60
CAS 9005-67-8	Emsorb® 2728
CAS 9005-67-8	Emsorb® 6906
CAS 9005-67-8	Emsorb® 6909
CAS 9005-67-8	Ethsorbox S-20
CAS 9005-67-8	Ethylan® GES6
CAS 9005-67-8	Eumulgin SMS 20
CAS 9005-67-8	Fanwax P
CAS 9005-67-8	Glycosperse® S-20 FG, S-20 KFG
CAS 9005-67-8	Hetsorb S-20
CAS 9005-67-8	Hetsorb S-4
CAS 9005-67-8	Hodag PSMS-20
CAS 9005-67-8	Hodag SVS-18
CAS 9005-67-8	Liposorb S-4
CAS 9005-67-8	Liposorb S-20
CAS 9005-67-8	Montanox 60 DF
CAS 9005-67-8	Montanox 61
CAS 9005-67-8	Norfox® Sorbo T-60
CAS 9005-67-8	Polycon T60 K
CAS 9005-67-8	Prox-onic SMS-020
CAS 9005-67-8	Radiamuls® Sorb 2147
CAS 9005-67-8	Radiasurf® 7147
CAS 9005-67-8	Ritabate 60
CAS 9005-67-8	Sorbax PMS-20
CAS 9005-67-8	T-Maz® 60
CAS 9005-67-8	T-Maz® 61
CAS 9005-67-8	Tandem 5K, 8
CAS 9005-67-8	Tween® 60, 60K
CAS 9005-67-8	Tween® 61
CAS 9005-70-3	Alkamuls® PSTO-20
CAS 9005-70-3	Alkamuls® T-85
CAS 9005-70-3	Crillet 45
CAS 9005-70-3	Disponil STO 120 F1
CAS 9005-70-3	Emsorb® 6903
CAS 9005-70-3	Emsorb® 6913
CAS 9005-70-3	Ethsorbox TO-20
CAS 9005-70-3	Ethylan® GPS85
CAS 9005-70-3	Hetsorb TO-20
CAS 9005-70-3	Liposorb TO-20
CAS 9005-70-3	Montanox 85
CAS 9005-70-3	Sorbax PTO-20
CAS 9005-70-3	T-Maz® 85
CAS 9005-70-3	Tween® 85
CAS 9005-70-3	Witconol 6903
CAS 9005-71-4	Crillet 35
CAS 9005-71-4	Disponil STS 120 F1
CAS 9005-71-4	Drewpone® 65K
CAS 9005-71-4	Durfax® 65
CAS 9005-71-4	Ethsorbox TS-20
CAS 9005-71-4	Glycosperse® TS-20 FG, TS-20 KFG
CAS 9005-71-4	Hetsorb TS-20
CAS 9005-71-4	Hodag PSTS-20
CAS 9005-71-4	Ice No. 12K
CAS 9005-71-4	Liposorb TS-20
CAS 9005-71-4	Montanox 65
CAS 9005-71-4	Sorbax PTS-20
CAS 9005-71-4	T-Maz® 65
CAS 9005-71-4	Tween® 65, 65K
CAS 9006-04-6	DPNR
CAS 9006-04-6	Dynatex GTZ
CAS 9006-04-6	ENR 25
CAS 9006-04-6	Hyflo NS, S
CAS 9006-04-6	PA-57
CAS 9006-04-6	PA-80
CAS 9006-04-6	Unitex
CAS 9006-42-2	Polyram® DF
CAS 9006-42-2	Polyram®-Combi
CAS 9006-65-9	Fancorsil A.
CAS 9007-34-5	Actigen C
CAS 9007-34-5	Cationic Collagen Polypeptides
CAS 9007-34-5	Clearcol
CAS 9007-34-5	Collagen CLR
CAS 9007-34-5	Collagen Native Extra 1%
CAS 9007-34-5	Collasol
CAS 9007-34-5	Desamidocollagen
CAS 9007-34-5	Polygeline™
CAS 9007-34-5	Polymoist Mask
CAS 9007-34-5	Ritacollagen BA-1
CAS 9007-34-5	Secolan S-1, BA-1, BA-1G
CAS 9007-34-5	Solu-Coll
CAS 9007-48-1	Caprol® 3GO
CAS 9007-48-1	Drewpol® 3-1-O
CAS 9007-48-1	Grindtek PGE 25
CAS 9007-48-1	Hodag PGO-61
CAS 9007-48-1	Hodag PGO-101
CAS 9007-48-1	Isolan® GO 33
CAS 9007-48-1	Mazol® PGO-31 K
CAS 9007-48-1	Santone® 3-1-SH
CAS 9007-48-1	Witconol 14
CAS 9007-58-3	Exsyproteines 2%, 4%
CAS 9008-66-6	Electrafil® J-2/CF/30
CAS 9008-66-6	NI-20GF
CAS 9008-66-6	Ny-Kon® Q
CAS 9008-66-6	RTP 201B
CAS 9008-66-6	Texalon 1600A Nat
CAS 9008-66-6	Thermocomp® QF-1006FR
CAS 9008-66-6	Ultramid® S3
CAS 9008-99-8	Pea Pro-Tein BK
CAS 9009-54-5	Aspect® TPPE

CAS 9009-54-5	Autofroth
CAS 9009-54-5	Autopak
CAS 9009-54-5	Autopour
CAS 9009-54-5	Desmopan® 150
CAS 9009-54-5	Desmopan® 385
CAS 9009-54-5	Desmopan® 585
CAS 9009-54-5	Desmopan® 786
CAS 9009-54-5	Desmopan® KA 8333
CAS 9009-54-5	Electrafil® J-100/CF/30
CAS 9009-54-5	Estane® 5701 F1
CAS 9009-54-5	Estane® 58092
CAS 9009-54-5	Estane® 58300
CAS 9009-54-5	Isoplast 101
CAS 9009-54-5	Isoplast 101LGF40 Nat
CAS 9009-54-5	Isoplast 101LGF60 Blk
CAS 9009-54-5	Isoplast 302
CAS 9009-54-5	Pearlstick 46-10/06
CAS 9009-54-5	Resin 164
CAS 9009-54-5	RTP 2301A
CAS 9009-54-5	RTP 2381A
CAS 9009-54-5	Texin 480-A
CAS 9009-54-5	Texin 985-A
CAS 9009-54-5	Texin 5286
CAS 9009-54-5	Thermocomp® TF-1004
CAS 9009-99-8	Crosilk Powder
CAS 9009-99-8	Fibro-Silk Powd
CAS 9009-99-8	Ritasilk Powd
CAS 9010-01-9	Depepsen
CAS 9010-66-6	Press-Aid
CAS 9010-77-9	A-C® Copolymer 540, 540A
CAS 9010-77-9	A-C® Copolymer 580
CAS 9010-77-9	A-C® Copolymer 5120
CAS 9010-77-9	A-C® Copolymer 5180
CAS 9010-77-9	AClyn® 250
CAS 9010-77-9	AClyn® 262
CAS 9010-77-9	AClyn® 296
CAS 9010-77-9	ACuflow AF-1
CAS 9010-77-9	Primacor 4990 Dispersion
CAS 9010-79-1	Amoco® 8244
CAS 9010-79-1	Amoco® 8410
CAS 9010-79-1	Siltek® L Polymer
CAS 9010-79-1	Vistalon 719
CAS 9010-85-9	Butex
CAS 9010-85-9	CryOfine® Butyl
CAS 9010-85-9	Exxon® Butyl 077
CAS 9010-85-9	Kalar® 5214
CAS 9010-85-9	Kalar® 5263
CAS 9010-85-9	Kalene® 800
CAS 9010-85-9	Polysar Butyl 100
CAS 9010-85-9	Polysar XL 30102
CAS 9011-13-6	Celstran® SMAG30-01-4
CAS 9011-13-6	Dylark® 132
CAS 9011-13-6	Scripset 520
CAS 9011-13-6	Scripset 720
CAS 9011-13-6	SMA® 1000
CAS 9011-13-6	SMA® 1440
CAS 9011-13-6	SMA® 2625
CAS 9011-13-6	Stapron S SG340
CAS 9011-13-6	Stapron S SM300
CAS 9011-14-7	Acryloid® A-30
CAS 9011-14-7	Acryloid® B-44
CAS 9011-14-7	Lucryl®

CAS 9011-14-7	Paraloid® K-120N
CAS 9011-16-9	Agrimer VEMA-H-240
CAS 9011-16-9	Gantrez® S-95
CAS 9011-16-9	Luviform® FA 119
CAS 9012-09-3	Triafol
CAS 9012-54-8	Cellulase 4000
CAS 9012-54-8	Cellulase AC
CAS 9012-54-8	Celluzyme® 2400 T
CAS 9014-85-1	Surfynol® 440
CAS 9014-85-1	Surfynol® 465
CAS 9014-85-1	Surfynol® 485
CAS 9014-90-8	Akyposal NPS 60
CAS 9014-90-8	Akyposal NPS 100
CAS 9014-90-8	Polystep® B-27
CAS 9014-90-8	Serdet DNK
CAS 9014-90-8	Steol® COS 433
CAS 9014-92-0	Igepal® RC-520
CAS 9014-92-0	Igepal® RC-620
CAS 9014-92-0	Prox-onic DDP-09
CAS 9014-92-0	Rexol 65/4
CAS 9014-92-0	T-Det® DD-5
CAS 9014-92-0	Teric DD5
CAS 9014-93-1	Chemax DNP-18
CAS 9014-93-1	Chemax DNP-8
CAS 9014-93-1	Ethal DNP-8
CAS 9014-93-1	Hetoxide DNP-4
CAS 9014-93-1	Igepal® DM-430
CAS 9014-93-1	Igepal® DM-530
CAS 9014-93-1	Igepal® DM-710
CAS 9014-93-1	Igepal® DM-730
CAS 9014-93-1	Igepal® DM-970
CAS 9014-93-1	Macol® DNP-5
CAS 9014-93-1	Marlophen® DNP 16
CAS 9014-93-1	Prox-onic DNP-08
CAS 9014-93-1	Sellig DN 10 100
CAS 9014-93-1	Sellig DN 22 100
CAS 9014-93-1	T-Det® D-150
CAS 9014-93-1	Trycol® 6985
CAS 9015-54-7	Collagen Hydrolyzate
	Cosmetic 55
CAS 9015-54-7	Collamino 25
CAS 9015-54-7	Cropepsol
CAS 9015-54-7	Cropeptone
CAS 9015-54-7	Crotein A,C, O
CAS 9015-54-7	Crotein CAA
CAS 9015-54-7	Crotein CAA/SF
CAS 9015-54-7	Crotein O
CAS 9015-54-7	Crotein SPA
CAS 9015-54-7	Extlato®
CAS 9015-54-7	Hydrocoll AL-50, AL-55, EN-40, EN-55, EN-55-X, EN-SD, EN-SD-1M, EN-SD-10M
CAS 9015-54-7	Lexein® X-250HP
CAS 9015-54-7	Nutrilan® FPK, H, M
CAS 9015-54-7	Nutrilan® I-50
CAS 9015-54-7	Nutrilan® L
CAS 9015-54-7	Parenamine
CAS 9015-54-7	Peptein® 2000®
CAS 9015-54-7	Polypeptide 10
CAS 9015-54-7	Polypeptide 37
CAS 9015-54-7	Protogest
CAS 9015-54-7	Travamin

CAS 9016-45-9	Ablunol NP4
CAS 9016-45-9	Akyporox NP 30
CAS 9016-45-9	Akyporox NP 40
CAS 9016-45-9	Akyporox NP 90
CAS 9016-45-9	Akyporox NP 95
CAS 9016-45-9	Akyporox NP 105
CAS 9016-45-9	Akyporox NP 150
CAS 9016-45-9	Akyporox NP 200
CAS 9016-45-9	Akyporox NP 300V
CAS 9016-45-9	Akyporox NP 1200V
CAS 9016-45-9	Alkamuls® AG-900
CAS 9016-45-9	Alkasurf® NP-4
CAS 9016-45-9	Carsonon® N-4
CAS 9016-45-9	Chemax NP-1.5
CAS 9016-45-9	Chemax NP-4
CAS 9016-45-9	Chemax NP-6
CAS 9016-45-9	Chemax NP-9
CAS 9016-45-9	Chemax NP-10
CAS 9016-45-9	Chemax NP-15
CAS 9016-45-9	Chemax NP-30
CAS 9016-45-9	Chemax NP-40
CAS 9016-45-9	Conceptrol
CAS 9016-45-9	Cremophor® NP 10
CAS 9016-45-9	Cremophor® NP 14
CAS 9016-45-9	Dehydrophen PNP 4
CAS 9016-45-9	Diazopon SS-837
CAS 9016-45-9	Emalex NP-2
CAS 9016-45-9	Empilan® NP9
CAS 9016-45-9	Emulgator U4
CAS 9016-45-9	Ethal NP-1.5
CAS 9016-45-9	Ethal NP-6
CAS 9016-45-9	Ethylan® 44
CAS 9016-45-9	Ethylan® BV
CAS 9016-45-9	Ethylan® NP 1
CAS 9016-45-9	Etophen 102
CAS 9016-45-9	Etophen 114
CAS 9016-45-9	Gynol
CAS 9016-45-9	Hetoxide NP-4
CAS 9016-45-9	Hodag Nonionic E-5
CAS 9016-45-9	Hostapal N-040
CAS 9016-45-9	Hyonic NP-40
CAS 9016-45-9	Hyonic PE-100
CAS 9016-45-9	Iconol NP-30
CAS 9016-45-9	Iconol NP-40
CAS 9016-45-9	Iconol NP-50
CAS 9016-45-9	Iconol NP-70
CAS 9016-45-9	Iconol NP-100
CAS 9016-45-9	Igepal® CO-210
CAS 9016-45-9	Igepal® CO-430
CAS 9016-45-9	Igepal® CO-520
CAS 9016-45-9	Igepal® CO-530
CAS 9016-45-9	Igepal® CO-610
CAS 9016-45-9	Igepal® CO-630
CAS 9016-45-9	Igepal® CO-660
CAS 9016-45-9	Igepal® CO-710
CAS 9016-45-9	Igepal® CO-720
CAS 9016-45-9	Igepal® CO-730
CAS 9016-45-9	Igepal® CO-850
CAS 9016-45-9	Igepal® CO-880
CAS 9016-45-9	Igepal® CO-887
CAS 9016-45-9	Igepal® CO-890
CAS 9016-45-9	Igepal® CO-970
CAS 9016-45-9	Igepal® CO-997
CAS 9016-45-9	Indulin® XD-70
CAS 9016-45-9	Intercept
CAS 9016-45-9	Lutensol® AP 10
CAS 9016-45-9	Lutensol® AP 20
CAS 9016-45-9	Macol® NP-4
CAS 9016-45-9	Makon® 4
CAS 9016-45-9	Marlophen® 810N
CAS 9016-45-9	Naxonic NI-40
CAS 9016-45-9	Neutronyx® 656
CAS 9016-45-9	Nonipol 20
CAS 9016-45-9	Nutrol 600
CAS 9016-45-9	Nutrol 611
CAS 9016-45-9	Nutrol 622
CAS 9016-45-9	Nutrol 640
CAS 9016-45-9	Nutrol 656
CAS 9016-45-9	Poly-Tergent® B-150
CAS 9016-45-9	Polystep® F-1
CAS 9016-45-9	Polystep® F-95B
CAS 9016-45-9	Prox-onic NP-04
CAS 9016-45-9	Remcopal 29
CAS 9016-45-9	Remcopal 334
CAS 9016-45-9	Remcopal 349
CAS 9016-45-9	Remcopal 666
CAS 9016-45-9	Remcopal 3112
CAS 9016-45-9	Remcopal 3712
CAS 9016-45-9	Remcopal 3820
CAS 9016-45-9	Remcopal 6110
CAS 9016-45-9	Remcopal 31250
CAS 9016-45-9	Remcopal 33820
CAS 9016-45-9	Remcopal L30
CAS 9016-45-9	Remcopal NP 30
CAS 9016-45-9	Remcopal PONF
CAS 9016-45-9	Rewopal® HV 4
CAS 9016-45-9	Rexol 25/10
CAS 9016-45-9	Rexol 25/4
CAS 9016-45-9	Sellig N 4 100
CAS 9016-45-9	Sellig N 5 100
CAS 9016-45-9	Sellig N 6 100
CAS 9016-45-9	Sellig N 8 100
CAS 9016-45-9	Sellig N 9 100
CAS 9016-45-9	Sellig N 1050
CAS 9016-45-9	Sellig N 10 100
CAS 9016-45-9	Sellig N 11 100
CAS 9016-45-9	Sellig N 12 100
CAS 9016-45-9	Sellig N 15 100
CAS 9016-45-9	Sellig N 1780
CAS 9016-45-9	Sellig N 20 80
CAS 9016-45-9	Sellig N 30 70
CAS 9016-45-9	Sellig N 50 100
CAS 9016-45-9	Surfonic® HDL
CAS 9016-45-9	Surfonic® N-10
CAS 9016-45-9	T-Det® N-4
CAS 9016-45-9	T-Det® N-40
CAS 9016-45-9	T-Det® N-1007
CAS 9016-45-9	Teric N2
CAS 9016-45-9	Today Sponge
CAS 9016-45-9	Triton® N-57
CAS 9016-45-9	Trycol® 6940
CAS 9016-45-9	Witconol NP-100
CAS 9016-75-5	Celstran® PPSG30-01-4
CAS 9016-75-5	Debron 711

CAS 9016-75-5	Electrafil® J-1300/CF/30/TF/15
CAS 9016-75-5	EMI-X® OC-1008
CAS 9016-75-5	EMI-X® PDX-O-91074
CAS 9016-75-5	Fortron® 0205B4
CAS 9016-75-5	Lubricomp® 189
CAS 9016-75-5	Mitsubishi Kasei PPS 704G40
CAS 9016-75-5	Plaslube® J-1300/30/TF/15
CAS 9016-75-5	Plaslube® J-1300/CF/20/MS/10/ TF/15
CAS 9016-75-5	Polycomp® 139
CAS 9016-75-5	Polycomp® 185
CAS 9016-75-5	RTP 1301
CAS 9016-75-5	RTP 1378
CAS 9016-75-5	Ryton® A-200
CAS 9016-75-5	Ryton® R-4
CAS 9016-75-5	Ryton® V-1
CAS 9016-75-5	SF-20CF
CAS 9016-75-5	Stat-Kon® OC-1006
CAS 9016-75-5	Supec® G401
CAS 9016-75-5	Techtron™ PPS
CAS 9016-75-5	Tedur® KU1-9510-1
CAS 9016-75-5	Tedur® KU1-9511
CAS 9016-75-5	Tedur® KU1-9530
CAS 9016-75-5	Tedur® KU1-9552
CAS 9016-75-5	Tedur® KU1-9561
CAS 9016-75-5	Thermocomp® OC-1006
CAS 9016-75-5	Verton® OF-700-10
CAS 9027-33-6	Fancol LAO
CAS 9032-08-0	Aldomax GA-100
CAS 9032-08-0	Diazyme® L-200
CAS 9032-08-0	Distillase® L-200
CAS 9032-08-0	Fermenzyme® L-200
CAS 9032-42-2	Tylose® MHB
CAS 9032-75-1	Clarex® L
CAS 9032-75-1	Pearex-L®
CAS 9032-75-1	Pectinase AT
CAS 9032-75-1	Spark-L® HPG
CAS 9034-50-8	Pitressin
CAS 9035-58-9	Tachostyptan
CAS 9035-85-2	Carsonon® 169-P
CAS 9035-85-2	Fancol 707
CAS 9035-85-2	Procetyl 10
CAS 9035-85-2	Wickenol® 707
CAS 9035-99-8	Crystex® Regular
CAS 9038-41-9	Calcisorb
CAS 9038-95-3	Pluracol® W3520N
CAS 9038-95-3	Tergitol® XD
CAS 9038-95-3	Tergitol® XH
CAS 9039-53-6	Win-Kinase
CAS 9040-38-4	Aerosol® A-103
CAS 9041-93-4	Blenoxane
CAS 9043-30-5	Berol 048
CAS 9046-01-9	Crodafos T2 Acid
CAS 9046-01-9	Lubrhophos® LS-500
CAS 9046-01-9	Rhodafac® RS-610
CAS 9046-56-4	Arvin
CAS 9048-46-8	Bovinal-20
CAS 9050-36-6	Lycadex®
CAS 9050-36-6	Maltrin® M040
CAS 9050-36-6	Maltrin® M510
CAS 9050-36-6	Microduct®
CAS 9050-36-6	Mor-Rex® I-920

CAS 9050-36-6	Wickenol® 550
CAS 9051-57-4	Abex® EP-110
CAS 9051-57-4	Aerosol® NPES 458
CAS 9051-57-4	Polystep® B-1
CAS 9051-57-4	Rhodapex® CO-415
CAS 9051-57-4	Rhodapex® CO-436
CAS 9051-57-4	Sulfochem 436
CAS 9062-73-1	Glycosperse® L-20
CAS 9063-38-1	Explotab®
CAS 9064-14-6	Marlox® L 6
CAS 9067-32-7	ActiMoist
CAS 9067-32-7	Amo Vitrax®
CAS 9067-32-7	Dermasome® H
CAS 9067-32-7	Pronova
CAS 9067-32-7	Rita HA C-1
CAS 9076-43-1	Emcol® CC-42
CAS 9084-06-4	Ablusol ML
CAS 9084-06-4	Aerosol® NS
CAS 9084-06-4	Chromasist 1487A
CAS 9084-06-4	Darvan® No. 1
CAS 9084-06-4	Daxad® 11
CAS 9084-06-4	Dehscofix 912
CAS 9084-06-4	Dehscofix 920
CAS 9084-06-4	Dispersogen A
CAS 9084-06-4	Emery® 5370 Sellogen W
CAS 9084-06-4	Lomar® D
CAS 9084-06-4	Lomar® LS
CAS 9084-06-4	Nopcosant
CAS 9084-06-4	Rhodacal® Liquid
CAS 9084-06-4	Rhodacal® N
CAS 9084-06-4	Sellogen DFL
CAS 9084-06-4	Sellogen HR-90
CAS 9084-06-4	Tamol® 819
CAS 9084-06-4	Tamol® L Conc
CAS 9084-06-4	Tamol® SN
CAS 9087-53-0	Procetyl AWS
CAS 9087-61-0	Dry Flo®
CAS 9087-70-1	Trasylol®
CAS 9105-54-7	Amino-Collagen-25, -40
CAS 10024-97-2	Nitral
CAS 10025-73-7	Chrometrace
CAS 10025-78-2	Dynasylan® TCS
CAS 10026-04-7	CT1800
CAS 10028-22-5	Elliott's Lawn Sand
CAS 10028-22-5	Elliott's Moss Killer
CAS 10028-22-5	Green-up Mossfree
CAS 10028-22-5	Greenmaster Autumn
CAS 10028-22-5	Greenmaster Mosskiller
CAS 10028-22-5	Hart Lawn Sand
CAS 10028-22-5	Hart Moss Killer
CAS 10028-22-5	Maxicrop Moss Killer & Conditioner
CAS 10028-22-5	Taylors Lawn Sand
CAS 10028-22-5	Vitax Micro Gran
CAS 10028-22-5	Vitax Turf Tonic
CAS 10028-22-5	Walkover Moss Killer
CAS 10030-90-7	Ferromyn
CAS 10039-53-9	Chromitope Sodium
CAS 10039-53-9	Rachromate-51
CAS 10042-76-9	Strotope
CAS 10043-01-3	Alcan Aluminum Sulphate Liquid
CAS 10043-01-3	Alferric

CAS 10043-01-3	Aluminoferric
CAS 10043-01-3	Clar+Ion® A410P
CAS 10043-01-3	Hydrangea Colourant
CAS 10043-01-3	Lapotan
CAS 10043-35-3	Ant Flip
CAS 10043-52-4	Cal Plus
CAS 10043-52-4	Dowflake
CAS 10043-52-4	Ice Melt
CAS 10043-52-4	Jarcal
CAS 10043-52-4	Liquical
CAS 10043-52-4	Liquidow
CAS 10043-52-4	Marley Cement Accelerator
CAS 10043-52-4	Peladow
CAS 10043-52-4	Snomelt
CAS 10043-52-4	Sure-Step
CAS 10045-89-3	Hydrangea Colourant
CAS 10049-04-4	Anthium Dioxcide
CAS 10081-67-1	Naugard® 445
CAS 10085-81-1	Tacitin
CAS 10090-54-7	Liquid Code XLR
CAS 10094-45-8	Chemstat® HTSA#3
CAS 10094-45-8	HTSA #3
CAS 10094-45-8	Kemamide® E-180
CAS 10099-74-8	Ledni
CAS 10099-76-0	BSWL 202
CAS 10101-39-0	Cecasil
CAS 10101-39-0	Extrusil
CAS 10101-39-0	Keical-Ace
CAS 10101-39-0	Micro-Cel® A
CAS 10101-39-0	Micro-Cel® T-38
CAS 10101-39-0	Microcal
CAS 10101-39-0	Microcal ET
CAS 10101-39-0	Paratemp
CAS 10101-39-0	Silasorb
CAS 10101-39-0	Silene
CAS 10101-41-4	Compactrol®
CAS 10112-91-1	Calomel
CAS 10118-76-0	Acerdol
CAS 10124-56-8	Glass H
CAS 10124-56-8	Hexaphos
CAS 10124-56-8	Kalex HMP
CAS 10124-56-8	Limex G
CAS 10124-56-8	Sodaphos
CAS 10124-56-8	Vitrafos®
CAS 10124-56-8	Water Brite
CAS 10161-34-9	Finaplix
CAS 10192-93-5	Perkadox® 58
CAS 10213-78-2	Chemeen 18-2
CAS 10213-78-2	Chemstat® 273-E
CAS 10213-78-2	Ethomeen® 18/12
CAS 10213-78-2	Varonic® S202
CAS 10238-21-8	Diabeta
CAS 10238-21-8	Micronase
CAS 10246-75-0	Equipose
CAS 10246-75-0	Vistaril Pamoate
CAS 10265-92-6	Monitor
CAS 10265-92-6	Nitofol®
CAS 10265-92-6	Prodex
CAS 10265-92-6	Tamaron®
CAS 10288-28-5	V-30
CAS 10294-34-5	Trona Boron Trichloride
CAS 10310-32-4	Glyvenol
CAS 10347-81-6	Ludiomil
CAS 10377-60-3	Magnisal
CAS 10401-55-5	Liponate CRM
CAS 10401-55-5	Naturechem® CR
CAS 10416-59-8	CB2500
CAS 10416-59-8	Dynasylan® BSA
CAS 10418-03-8	Stromba
CAS 10418-03-8	Winstrol
CAS 10476-81-0	Kuronlum Bromide
CAS 10476-85-4	Stronscan-85
CAS 10525-14-1	Lanamine®
CAS 10540-29-1	Nolvadex
CAS 10591-85-2	Benzyl Tuex®
CAS 10595-72-9	Argus DTDTDP
CAS 10595-72-9	Cyanox® 711
CAS 10595-72-9	Evanstab® 13
CAS 11094-60-3	Caprol® 10G10O
CAS 11094-60-3	Hodag PGO-1010
CAS 11094-60-3	Polyaldo® DGDO KFG
CAS 11096-42-7	Biopal® NR-20
CAS 11097-59-9	Alcamizer 1
CAS 11097-59-9	Alcamizer 2
CAS 11111-34-5	Tetronic® 304
CAS 11111-34-5	Tetronic® 504
CAS 11111-34-5	Tetronic® 701
CAS 11111-34-5	Tetronic® 702
CAS 11111-34-5	Tetronic® 704
CAS 11111-34-5	Tetronic® 707
CAS 11111-34-5	Tetronic® 901
CAS 11111-34-5	Tetronic® 904
CAS 11111-34-5	Tetronic® 908
CAS 11111-34-5	Tetronic® 1101
CAS 11111-34-5	Tetronic® 1102
CAS 11111-34-5	Tetronic® 1104
CAS 11111-34-5	Tetronic® 1107
CAS 11111-34-5	Tetronic® 1301
CAS 11111-34-5	Tetronic® 1302
CAS 11111-34-5	Tetronic® 1304
CAS 11111-34-5	Tetronic® 1307
CAS 11111-34-5	Tetronic® 1501
CAS 11111-34-5	Tetronic® 1502
CAS 11111-34-5	Tetronic® 1504
CAS 11111-34-5	Tetronic® 1508
CAS 11113-70-5	M50
CAS 11125-95-4	CCA Type C Wood Preservative 50-60%
CAS 11138-66-2	Dariloid® 100
CAS 11138-66-2	Dricoid® 200
CAS 11138-66-2	Kel-Lite™ CM
CAS 11138-66-2	Kelflo®
CAS 11138-66-2	Kelgum®
CAS 11138-66-2	Keltrol®
CAS 11138-66-2	Keltrol® F
CAS 11138-66-2	Kelzan®
CAS 11138-66-2	Kelzan® AR
CAS 11138-66-2	Kelzan® S
CAS 11138-66-2	Kelzan®, D, M, XC Polymer
CAS 11138-66-2	Rhodicare XC
CAS 11138-66-2	Rhodigel®
CAS 11138-66-2	Rhodigel® EZ
CAS 11138-66-2	Rhodopol® 23
CAS 11138-66-2	Rhodopol® XGD

CAS 11138-66-2	Xanflood®
CAS 11140-78-6	Rewoteric® AM DM-35L
CAS 11141-17-6	Margosan-O
CAS 11174-62-0	Crotein Q
CAS 11174-62-0	Quat-Coll QS
CAS 11174-62-0	Quat-Pro S, S-30
CAS 12001-26-2	Alsibronz
CAS 12001-26-2	Bismica 46
CAS 12001-26-2	Kemira Phlogopite Mica
CAS 12001-26-2	Mearlin®
CAS 12001-26-2	Mearlmica® SVA
CAS 12001-26-2	Micacoat®
CAS 12001-26-2	Nyflake®
CAS 12001-26-2	Polymica 200
CAS 12001-26-2	Polymica 325
CAS 12001-26-2	Polymica 400
CAS 12001-26-2	Sericite FSE
CAS 12001-26-2	Sericite SL-012
CAS 12001-26-2	Snow White 200 Mica
CAS 12001-26-2	Spectra-Pearl®
CAS 12001-26-2	Timica®
CAS 12001-26-2	Velvet Veil 310
CAS 12001-99-9	H Chrome Green 105
CAS 12007-56-6	Meyerhofferite
CAS 12027-96-2	Flamtard H
CAS 12033-89-5	Roydazide
CAS 12036-37-2	Flamtard S
CAS 12069-32-8	Tetrabor®
CAS 12071-83-9	Antracol®
CAS 12071-83-9	Fruvit®
CAS 12111-24-9	Ytterbium Yb-169 DTPA
CAS 12122-67-7	Unicrop Zineb
CAS 12122-67-7	Zidan
CAS 12122-67-7	Zidanit
CAS 12124-97-9	FR-11
CAS 12125-02-9	Salamac
CAS 12141-46-7	Aciculite
CAS 12141-46-7	Alphatex®
CAS 12141-46-7	Altowhite LL
CAS 12141-46-7	Aluminum Silicate P820
CAS 12141-46-7	Alusil
CAS 12141-46-7	Alusil ET
CAS 12141-46-7	Burgess 30-P
CAS 12141-46-7	Burgess 2211
CAS 12141-46-7	Burgess KE
CAS 12141-46-7	Dynamullit
CAS 12141-46-7	Iceberg®
CAS 12141-46-7	Icecap® K
CAS 12141-46-7	Kaopolite® 1152
CAS 12141-46-7	Kaopolite® AB
CAS 12141-46-7	Kaopolite® SF
CAS 12141-46-7	Multex
CAS 12141-46-7	Optiwhite®
CAS 12141-46-7	Pasilex
CAS 12141-46-7	Sipernat® 44
CAS 12141-46-7	Tisyn®
CAS 12173-47-6	Hectabrite® AW
CAS 12173-47-6	Hectalite® 200
CAS 12178-41-5	Ferrosil 14
CAS 12244-57-4	Myochrysine
CAS 12269-78-2	Pyrax® A
CAS 12269-78-2	Veecote®
CAS 12280-03-4	Polybor®
CAS 12280-03-4	Tim-Bor®
CAS 12284-76-3	Bisiumina Suspension
CAS 12427-38-2	Ashlade M
CAS 12427-38-2	Ashlade Mancarb FL
CAS 12427-38-2	Campbell's MC Flowable
CAS 12427-38-2	Campbell's X-Spor
CAS 12427-38-2	Delsene® M Flowable
CAS 12427-38-2	Headland Dual
CAS 12427-38-2	Headland Spirit
CAS 12427-38-2	Hispor 45WP
CAS 12427-38-2	Legion
CAS 12427-38-2	Manguard®
CAS 12427-38-2	Manzate®
CAS 12427-38-2	MaxiMate
CAS 12427-38-2	Mazin®
CAS 12427-38-2	Multi-W FL
CAS 12427-38-2	Septal
CAS 12427-38-2	Squadron
CAS 12427-38-2	Trimangol
CAS 12427-38-2	Tripart® Legion
CAS 12513-27-8	ADK STAB 2335
CAS 12513-27-8	Borogard® ZB
CAS 12513-27-8	Firebrake® ZB
CAS 12513-27-8	ZB2335
CAS 12539-23-0	Alcamizer 1
CAS 12542-30-2	Sipomer® DCPA
CAS 12607-92-0	Glumal
CAS 12645-31-7	Marlophor® IH-Acid
CAS 12645-31-7	Rhodafac® PEH
CAS 12751-23-4	Crafol AP-201
CAS 12751-23-4	Tryfac® 5573
CAS 12769-96-9	U Violet 109
CAS 13010-47-4	CeeNU
CAS 13039-35-5	Rewoteric® AM V
CAS 13039-35-5	Zoharteric LF
CAS 13048-33-4	Ageflex HDDA
CAS 13052-09-0	Trigonox® 141
CAS 13052-09-0	USP®-245
CAS 13087-53-1	Conray, 30, 43
CAS 13087-53-1	Cysto-Conray
CAS 13103-34-9	Equipoise
CAS 13106-44-0	Ageflex FA-1Q80DMS
CAS 13121-70-5	Plictran
CAS 13121-70-5	Plictran
CAS 13122-18-4	Trigonox® 42
CAS 13122-18-4	Trigonox® 42 PR
CAS 13122-18-4	Trigonox® 42S
CAS 13127-82-7	Berol 302
CAS 13127-82-7	Chemstat® 172
CAS 13170-23-5	CD4153
CAS 13170-23-5	Dynasylan® BDAC
CAS 13194-48-4	Mocap 10G
CAS 13197-76-7	Rewoteric® AM HC
CAS 13253-82-2	Diak No. 4
CAS 13292-46-1	Rifadin
CAS 13411-16-0	Furanace
CAS 13412-64-1	Dicloxin
CAS 13412-64-1	Dycill
CAS 13412-64-1	Dynapen
CAS 13412-64-1	Pathocil
CAS 13422-51-0	Alpha-Ruvite

CAS 13422-51-0	AlphaRedisol
CAS 13422-51-0	Codroxomin
CAS 13422-51-0	Ducobee-Hy
CAS 13422-51-0	Neo-Cytamen
CAS 13434-12-6	CT3250
CAS 13457-18-6	Afugan
CAS 13457-18-6	Missile
CAS 13463-41-7	Danex
CAS 13463-41-7	Head and Shoulders
CAS 13463-41-7	Zinc Omadine
CAS 13463-41-7	Zinc Omadine®, 48% Fine Particle Disp.
CAS 13463-41-7	Zinc Omadine®, Powd.
CAS 13463-41-7	Zincon Dandruff Shampoo
CAS 13463-41-7	Zink Pyrion
CAS 13463-67-7	Bayertitan
CAS 13463-67-7	Covermark
CAS 13463-67-7	Diox DR 22
CAS 13463-67-7	Finntitan
CAS 13463-67-7	Hitox®
CAS 13463-67-7	Hombitan®
CAS 13463-67-7	Kronos®
CAS 13463-67-7	Kronos® 1000
CAS 13463-67-7	Kronos® 2020
CAS 13463-67-7	Kronos® 2073
CAS 13463-67-7	Kronos® 3020
CAS 13463-67-7	Mearlin®
CAS 13463-67-7	Octotint 138
CAS 13463-67-7	PTFE-19
CAS 13463-67-7	Rayox
CAS 13463-67-7	Runa
CAS 13463-67-7	Rutiox
CAS 13463-67-7	Siccolam
CAS 13463-67-7	Spectra-Pearl®
CAS 13463-67-7	Synthetic Rutile
CAS 13463-67-7	Ti-Pure® R-103
CAS 13463-67-7	Ti-Sphere AB-15155A
CAS 13463-67-7	Timica®
CAS 13463-67-7	Tiona®
CAS 13463-67-7	Tiona® HSS
CAS 13463-67-7	Tiona® RCL-4
CAS 13463-67-7	Tiona® RCL-535
CAS 13463-67-7	Tiona® RCS-P
CAS 13463-67-7	Tioxide
CAS 13463-67-7	Titan
CAS 13463-67-7	Titan Design
CAS 13463-67-7	Titanium Dioxide 110
CAS 13463-67-7	Titanium Dioxide P25
CAS 13463-67-7	Titanox
CAS 13463-67-7	Titanox Design
CAS 13463-67-7	Titanox RA-39
CAS 13463-67-7	Titanweiss C, Extra T, Standard T, Standard A
CAS 13463-67-7	Tronox Titanium Dioxide Pigments, Chloride Process
CAS 13463-67-7	Tytanpol
CAS 13463-67-7	Zopaque
CAS 13472-08-7	V-59
CAS 13474-96-9	Weston® PTP
CAS 13475-82-6	Permethyl 99A
CAS 13492-01-8	Parnate Tablets
CAS 13494-80-9	Telloy®
CAS 13516-27-2	Rappor
CAS 13516-27-2	Rappor Plus
CAS 13523-86-9	Visken
CAS 13528-93-3	CT2015
CAS 13530-50-2	Aluphos
CAS 13530-50-2	Phosphaljel
CAS 13557-75-0	Pationic 138C
CAS 13560-89-9	Dechlorane® Plus 25
CAS 13593-03-8	Savall
CAS 13597-65-4	Tornusil
CAS 13609-67-1	Locoid
CAS 13614-98-7	Minocin
CAS 13647-35-3	Modrenal
CAS 13682-92-3	Alminate
CAS 13682-92-3	Robalate
CAS 13684-63-4	Beetomax
CAS 13684-63-4	Beetup
CAS 13684-63-4	Betalion
CAS 13684-63-4	Betanal E
CAS 13684-63-4	Betanal Tandem
CAS 13684-63-4	Goliath
CAS 13684-63-4	Gusto
CAS 13684-63-4	Headland Dephend
CAS 13684-63-4	Pistol
CAS 13684-63-4	Pistol 400
CAS 13684-63-4	Protrum K
CAS 13684-63-4	Suplex
CAS 13684-63-4	Tripart® Beta
CAS 13684-63-4	Tripart® Beta 2
CAS 13684-63-4	Vangard
CAS 13684-63-4	Vanguard
CAS 13698-49-2	Delmate
CAS 13707-88-5	Aptine
CAS 13717-04-9	Dirame
CAS 13770-89-3	Nimate
CAS 13822-56-5	CA0880
CAS 13822-56-5	Dynasylan® AMMO
CAS 13822-56-5	Union Carbide® A-1110
CAS 13838-16-9	Ethrane
CAS 13840-33-0	HyPure L
CAS 13927-77-0	Naugard® NBC
CAS 13927-77-0	NBC
CAS 13927-77-0	Perkacit® NDBC
CAS 13927-77-0	Vanox® NBC
CAS 13952-84-6	CSC 2-Aminobutane
CAS 13983-17-0	Nyad® 1250
CAS 13983-17-0	Nycor® R
CAS 13983-17-0	Vansil®
CAS 13983-17-0	Vansil® W-10
CAS 13983-17-0	Wollastocoat®
CAS 13983-17-0	Wollastokup®
CAS 14087-96-6	Cimflx 606
CAS 14087-96-6	Stellar 500
CAS 14222-60-7	Ektebin®
CAS 14222-60-7	Trevintex
CAS 14234-82-3	Octomer DIBM
CAS 14235-86-0	Penotrane
CAS 14239-68-0	Cadmate®
CAS 14239-68-0	Ethyl Cadmate
CAS 14239-75-9	Thanecure®
CAS 14252-80-3	Marcaine
CAS 14252-80-3	Sensorcaine

CAS 14286-84-1	Dilangio
CAS 14286-84-1	Irribral
CAS 14293-44-8	Diurex
CAS 14293-44-8	Diurexan
CAS 14323-55-1	Accelerator EZ Powder
CAS 14323-55-1	Ancazate ET
CAS 14323-55-1	Ethazate
CAS 14323-55-1	Ethyl Zimate
CAS 14323-55-1	Octocure ZDE-50
CAS 14323-55-1	Perkacit® ZDEC
CAS 14324-55-1	Ethyl Zimate®
CAS 14350-96-0	Dehyton® PLG
CAS 14350-97-1	Rewoteric® AM 2L-40
CAS 14351-40-7	Cerasynt D
CAS 14351-40-7	Witcamide® MAS
CAS 14351-50-9	Ammonyx® OAO
CAS 14351-50-9	Chemoxide O
CAS 14351-50-9	Incromine Oxide OD-50
CAS 14351-50-9	Mackamine O2
CAS 14351-50-9	Mazox® ODA
CAS 14351-50-9	Standamox O1
CAS 14402-89-2	Nipride
CAS 14402-89-2	Nitropress
CAS 14450-05-6	Afilan PP
CAS 14450-05-6	Emerest® 2485
CAS 14481-60-8	Aerosol® 18
CAS 14481-60-8	Empimin® MKK
CAS 14481-60-8	Octosol A-18
CAS 14643-87-9	Ageflex ZDA
CAS 14643-87-9	Akrochem® ZDA Powd
CAS 14698-29-4	Prodoxol
CAS 14727-68-5	Jet Amine DMOD
CAS 14807-96-6	ABT-2500®
CAS 14807-96-6	Altalc 200 USP
CAS 14807-96-6	Artic Mist
CAS 14807-96-6	Beaverwhite 200
CAS 14807-96-6	Ceramitalc
CAS 14807-96-6	Cimpact 699
CAS 14807-96-6	Cimpact 710
CAS 14807-96-6	I T Talc
CAS 14807-96-6	J-13
CAS 14807-96-6	Jetfil 700C
CAS 14807-96-6	Lubestine
CAS 14807-96-6	Luzenac 8170
CAS 14807-96-6	Microbloc®
CAS 14807-96-6	MicroPflex 1200
CAS 14807-96-6	Microtuff 1000
CAS 14807-96-6	Mistron CB
CAS 14807-96-6	Mistron Vapor-R
CAS 14807-96-6	Mistron ZSC
CAS 14807-96-6	Nicron 325
CAS 14807-96-6	Nicron 665
CAS 14807-96-6	Nicron JS 422
CAS 14807-96-6	Nytal® 100
CAS 14807-96-6	Olympic
CAS 14807-96-6	PolyTalc 445
CAS 14807-96-6	Select-A-Sorb
CAS 14807-96-6	Silverline 200
CAS 14807-96-6	Silverline 665
CAS 14807-96-6	Special Extender
CAS 14807-96-6	SteriLine 200
CAS 14807-96-6	SteriLine 665
CAS 14807-96-6	Super Lubestine
CAS 14807-96-6	Supra EF
CAS 14807-96-6	Suprafino A
CAS 14807-96-6	Talc Micro-Ace P-2
CAS 14807-96-6	Talc MS
CAS 14807-96-6	Vantalc®
CAS 14807-96-6	Vertal 92
CAS 14807-96-6	Vertal 200
CAS 14808-60-7	Sillikolloid P 87
CAS 14808-60-7	Sillitin N 82
CAS 14816-18-3	Baythion®
CAS 14816-18-3	Volathion®
CAS 14816-18-3	Volaton®
CAS 14857-34-2	CD5635
CAS 14885-29-1	Ipropran
CAS 14960-06-6	Mirataine® H2C-HA
CAS 14976-57-9	Tavist
CAS 15096-52-3	Monterey Kryocide
CAS 15180-03-7	Alloferin
CAS 15206-55-0	Vicure® 55
CAS 15299-99-7	Banweed
CAS 15299-99-7	Devrinol
CAS 15299-99-7	Devrinol T
CAS 15305-07-4	Q-1301
CAS 15307-79-6	Voltaren
CAS 15310-01-7	Calirus
CAS 15310-01-7	Mascot Clearing
CAS 15317-78-9	Isobutyl Niclate®
CAS 15454-75-8	Zincidone®
CAS 15489-16-4	Fouadin
CAS 15500-66-0	Pavulon
CAS 15520-10-2	Dytek® A
CAS 15520-11-3	Perkadox® 16
CAS 15520-11-3	Perkadox® 16-W40-GB5
CAS 15521-65-0	Methyl Niclate®
CAS 15535-29-2	Thioset® M
CAS 15535-69-0	ADK STAB BT-31
CAS 15535-69-0	ADK STAB LS-2
CAS 15537-76-5	Lapudrine
CAS 15545-48-9	Dicurane 500 FW
CAS 15545-48-9	Ludorum
CAS 15545-48-9	Toro
CAS 15545-48-9	Tripart® Ludorum 700
CAS 15545-58-9	Talisman
CAS 15545-58-9	Tolurex
CAS 15545-58-9	Tolurgan
CAS 15545-97-8	V-70
CAS 15574-96-6	Sandomigran
CAS 15578-26-4	TechneScan PYP
CAS 15585-70-3	Thoragol
CAS 15599-39-0	Noxyflex
CAS 15622-65-8	Moban
CAS 15625-89-5	Ageflex TMPTA
CAS 15625-89-5	Sipomer® TMPTA
CAS 15647-08-2	Weston® EHDPP
CAS 15663-27-1	Cisplatin
CAS 15663-27-1	Istin
CAS 15663-27-1	Platinol
CAS 15667-10-4	USP®-90MD
CAS 15676-16-1	Dolmatil
CAS 15686-71-2	Ceporex
CAS 15686-71-2	Keflex

CAS 15687-27-1	Advil
CAS 15687-27-1	Brufen
CAS 15687-27-1	Fenbid Spansule Capsul
CAS 15687-27-1	Ibugel
CAS 15687-27-1	Ibular®
CAS 15687-27-1	Motrin
CAS 15687-27-1	Nuprin
CAS 15687-27-1	Nurofen
CAS 15687-27-1	Paxofen
CAS 15687-41-9	Ildamen
CAS 15716-30-0	Uvinul® P 25
CAS 15826-37-6	Nasalcrom
CAS 15826-37-6	Opticrom
CAS 15876-67-2	Ubretid
CAS 15972-60-8	Alagan
CAS 15972-60-8	Alanex
CAS 15972-60-8	Alazine
CAS 16029-98-4	CT3610
CAS 16034-77-8	Cholebrine
CAS 16051-77-7	Monit
CAS 16066-38-9	Lupersol 221
CAS 16111-62-9	Espercarb® 840
CAS 16111-62-9	Lupersol 223
CAS 16111-62-9	Trigonox® EHP
CAS 16118-49-3	Carbetamex
CAS 16118-49-3	Ronstar TX
CAS 16230-35-6	Bislumina
CAS 16230-35-6	CB2409.5
CAS 16260-09-6	Chemstat® HTSA#1
CAS 16260-09-6	HTSA #1
CAS 16260-09-6	Kemamide® P-181
CAS 16485-10-2	Fancol DL
CAS 16545-54-3	Argus DMTDP
CAS 16545-54-3	Cyanox® MTDP
CAS 16545-54-3	Evanstab® 14
CAS 16580-06-6	Trigonox® 169-OP50
CAS 16590-41-3	Nalorex®
CAS 16672-87-0	Ethrel
CAS 16693-53-1	Crodasinic LT40
CAS 16752-77-5	Lannate®
CAS 16752-77-5	Methomex
CAS 16752-77-5	Sorex Golden Fly Bait
CAS 16773-42-5	Tiberal Roche
CAS 16789-79-5	Radiamine 6240
CAS 16841-14-8	Kemamine® BQ-2802C
CAS 16889-14-8	Chemical Base 6532
CAS 16889-14-8	Lexamine 22
CAS 16960-16-0	Cortrosyn
CAS 16971-82-7	Vanax® PML
CAS 17086-28-1	Vibramycin
CAS 17090-79-8	Romensin
CAS 17109-49-8	Hinosan®
CAS 17243-57-1	Pondinil Roche
CAS 17301-53-0	Genamin KDM-F
CAS 17301-53-0	Varisoft® BT-85
CAS 17342-21-1	Dehyquart D
CAS 17440-83-4	Amilorin
CAS 17440-83-4	Midamide
CAS 17440-83-4	Midamor
CAS 17440-83-4	Moduretic
CAS 17540-75-9	Isonox® 132
CAS 17540-75-9	Vanox® 1320
CAS 17560-51-9	Diulo
CAS 17560-51-9	Zaroxolyn
CAS 17598-65-1	Cedilanid Ampoules
CAS 17617-23-1	Dalmane
CAS 17617-23-1	Paxane
CAS 17661-50-6	Hest MS
CAS 17661-50-6	Hetester MS
CAS 17661-50-6	Kemester® 1418
CAS 17661-50-6	Schercemol MS
CAS 17671-27-1	Starfol® BB
CAS 17673-56-2	Cetiol® J600
CAS 17673-56-2	Dynacerin® 660
CAS 17689-77-9	CE6345
CAS 17689-77-9	Dynasylan® ETAC
CAS 17693-51-5	Avomine
CAS 17784-12-2	Renoquid
CAS 17796-82-6	Santogard PVI
CAS 17804-35-2	Benlate®
CAS 17831-71-9	Ageflex T4EGDA
CAS 17865-32-6	CM8650
CAS 17932-62-6	CC3433
CAS 18156-74-6	CT3600
CAS 18162-48-6	CB2790
CAS 18162-84-0	CO9819
CAS 18171-19-2	CC3290
CAS 18181-70-9	Elocril
CAS 18194-24-6	Phospholipon® MC
CAS 18312-32-8	Crodamol BE
CAS 18312-32-8	Kemester® BE
CAS 18312-32-8	Schercemol BE
CAS 18323-44-9	Cleocin
CAS 18323-44-9	Dalacin C
CAS 18395-30-7	CI7810
CAS 18395-30-7	Dynasylan® IBTMO
CAS 18395-30-7	Prosil® 178
CAS 18407-94-8	Dynasil® CA
CAS 18448-65-2	Ethoquad® O/12
CAS 18467-77-1	Cutlass
CAS 18472-51-0	Acriflex
CAS 18472-51-0	Hibiclens
CAS 18472-51-0	Phisomed
CAS 18472-51-0	Uniscrub®
CAS 18472-51-0	Unisept® Solution
CAS 18507-89-6	Deccox
CAS 18559-94-9	Proventil Inhaler
CAS 18641-57-1	Syncrowax HR-C
CAS 18643-08-8	CD5636
CAS 18691-97-9	Tribunil®
CAS 18769-78-3	M9030
CAS 18883-66-4	Zanosar
CAS 18917-89-0	Magan
CAS 19010-66-3	Amyl Ledate®
CAS 19010-66-3	Methyl Ledate
CAS 19010-66-3	Super Sulfur No. 2
CAS 19044-88-3	Weed Stopper
CAS 19047-85-9	Weston® DSP
CAS 19237-84-4	Hypovase
CAS 19237-84-4	Minipress
CAS 19237-84-4	Peripress
CAS 19321-40-5	Edenor PTO
CAS 19321-40-5	Liponate PO-4
CAS 19321-40-5	Mazol® PETO

CAS 19356-17-3	Calderol
CAS 19387-91-8	Fasigyn
CAS 19387-91-8	Simplotan
CAS 19387-91-8	Triolam
CAS 19504-77-9	Variotin
CAS 19666-30-9	Ronstar 2G
CAS 19666-30-9	Ronstar TX
CAS 19706-80-0	VF-077
CAS 19910-65-7	Espercarb® 438M-60
CAS 19910-65-7	Lupersol 225
CAS 19910-65-7	Trigonox® ADC
CAS 19910-65-7	Trigonox® SBP
CAS 19937-59-8	Dosaflo
CAS 20202-66-3	M Violet 112
CAS 20324-33-8	Arcosolv® TPM
CAS 20344-49-4	Colliron
CAS 20344-49-4	Neoferrum
CAS 20354-26-1	Probe
CAS 20427-59-2	Chiltern Kocide 101
CAS 20427-59-2	Comac Parasol
CAS 20566-35-2	Great Lakes PHT4-Diol
CAS 20684-06-4	Trentadil
CAS 20830-75-5	Diganox
CAS 20830-75-5	Digibind
CAS 20830-75-5	Lanoxicaps
CAS 20830-75-5	Lanoxin
CAS 20830-75-5	Lanoxine-PG
CAS 20859-73-8	Phostoxin
CAS 20859-73-8	Power Phosphine Pellets
CAS 20859-73-8	Talunex
CAS 21087-64-9	Lexone®
CAS 21087-64-9	Sencoral®
CAS 21087-64-9	Sencorex® WG
CAS 21087-64-9	Sencor®
CAS 21142-28-9	Sulfetal CJOT 38
CAS 21245-02-3	Escalol® 507
CAS 21245-02-3	Padimate O
CAS 21245-02-3	Solarchem® O
CAS 21256-18-8	Durapro
CAS 21260-46-8	Bismate®
CAS 21260-46-8	Bismet
CAS 21302-09-0	Weston® DLP
CAS 21416-87-5	Razoxin
CAS 21542-96-1	Crodamine 3.ABD
CAS 21645-51-2	Almacarb
CAS 21645-51-2	Almagel
CAS 21645-51-2	AL terna GEL
CAS 21645-51-2	Alu-Cap™
CAS 21645-51-2	Alu-Tab™
CAS 21645-51-2	Aludrox
CAS 21645-51-2	Alugel
CAS 21645-51-2	Amphojel
CAS 21645-51-2	Colugel
CAS 21645-51-2	Dialume
CAS 21645-51-2	F-500, -1000, -3600, etc.
CAS 21645-51-2	F-1000®
CAS 21645-51-2	F-1000 Dried Gel
CAS 21645-51-2	F-2000
CAS 21645-51-2	F-2000 Dried Gel
CAS 21645-51-2	F-2100 Dried Gel
CAS 21645-51-2	F-2200 Dried Gel
CAS 21645-51-2	F-MA 11®
CAS 21645-51-2	Gastrils
CAS 21645-51-2	Hydroxal
CAS 21645-51-2	Liquigel®
CAS 21645-51-2	Martifin
CAS 21645-51-2	Martinal
CAS 21645-51-2	Martinal® OL-111 LE
CAS 21645-51-2	Martinal® ON-4608
CAS 21645-51-2	Martinal® OS
CAS 21645-51-2	Mucogel®
CAS 21645-51-2	Nidrin
CAS 21645-51-2	R-MA 11®
CAS 21645-51-2	Rehydragel® Compressed Gel
CAS 21645-51-2	Theodrex
CAS 21652-27-7	Miramine® O
CAS 21725-46-2	Fortrol
CAS 21725-46-2	Match
CAS 21738-42-1	Mansil
CAS 21738-42-1	Vansil
CAS 21810-39-9	Ageflex FA-2Q50DMS
CAS 21829-25-4	Procardia
CAS 21829-25-4	Tibricol
CAS 21908-53-2	Kankerex
CAS 22023-23-0	Jet Amine DE-13
CAS 22042-96-2	Briquest® 543-45AS
CAS 22047-49-0	Afilan EHS
CAS 22047-49-0	Cetiol® 868
CAS 22047-49-0	Lexolube® T-110
CAS 22047-49-0	Tegosoft® CO
CAS 22047-49-0	Wickenol® 156
CAS 22059-60-5	Norpace
CAS 22071-15-4	Orudis
CAS 22071-15-4	Oruvail
CAS 22131-79-9	Mervan
CAS 22151-68-4	Brevital
CAS 22151-68-4	Brevital Sodium
CAS 22195-34-2	Hylorel
CAS 22204-24-6	Antiminth
CAS 22204-24-6	Combantrin
CAS 22204-24-6	Helmex
CAS 22204-24-6	Trilombrin
CAS 22204-53-1	Equiproxen
CAS 22204-53-1	Laraflex®
CAS 22204-53-1	Naprosyn
CAS 22204-88-2	Melanate
CAS 22224-92-6	Nemacur®
CAS 22232-71-9	Mazanor
CAS 22232-71-9	Sanorex
CAS 22254-24-6	Atrovent
CAS 22254-24-6	Rinatec
CAS 22260-51-1	Parlodel
CAS 22298-29-9	Bebate
CAS 22298-29-9	Uticort
CAS 22313-62-8	Esperox® 497M
CAS 22313-62-8	Trigonox® 97-C75
CAS 22389-47-0	Equal
CAS 22389-47-0	NutraSweet
CAS 22494-42-4	Dolobid
CAS 22499-12-3	Vicure® 10
CAS 22662-39-1	Ranide
CAS 22760-18-5	Biarsan
CAS 22766-82-1	Starfol® OS
CAS 22781-23-3	Ficam

CAS 22781-23-3	Garvox 3G
CAS 22781-23-3	Seedox SC
CAS 22801-45-2	O.O.D
CAS 22832-87-7	Brentan
CAS 22832-87-7	Daktarin
CAS 22832-87-7	Dermonistat
CAS 22832-87-7	Gyno-Daktarin
CAS 22832-87-7	Micatin
CAS 22832-87-7	Monistat-Derm
CAS 22882-95-7	Ceraphyl IPL
CAS 22984-54-9	CM9220
CAS 23031-32-5	Bricanyl
CAS 23047-25-8	Gamonil®
CAS 23092-17-3	Paxipam
CAS 23103-98-2	Aphox
CAS 23103-98-2	Phantom
CAS 23103-98-2	Pirimor
CAS 23103-98-2	Power Demo
CAS 23103-98-2	Rapid
CAS 23184-66-9	Butanex
CAS 23184-66-9	Machete
CAS 23214-92-8	Adriblastina
CAS 23256-50-0	Wytensin
CAS 23277-43-2	Nubain
CAS 23288-60-0	Pertscan-99m
CAS 23327-57-3	Acupan
CAS 23383-11-1	Ferrutope
CAS 23386-52-9	Aerosol® A-196-85
CAS 23386-52-9	Octosol TH-40
CAS 23386-52-9	Rewopol® SBDC 40
CAS 23474-91-1	Esperox® 13M
CAS 23505-41-1	Fernex
CAS 23505-41-1	Primicid
CAS 23505-41-1	Primotec
CAS 23560-59-0	Hostaquick
CAS 23564-05-8	CDA Mildothane
CAS 23564-05-8	Cercobin
CAS 23564-05-8	Mildothane
CAS 23564-05-8	Mildothane Turf Liquid
CAS 23564-05-8	SysTec® 1998
CAS 23564-06-9	Nemafax
CAS 23593-75-1	Canesten®
CAS 23593-75-1	Gyne-Lotrimin
CAS 23593-75-1	Lotrimin
CAS 23593-75-1	Mycelex
CAS 23779-32-0	CT2507
CAS 23947-60-6	Milcap
CAS 23947-60-6	Milgo
CAS 23947-60-6	Milstem
CAS 23950-58-5	Campbell's Rapier
CAS 23950-58-5	Kerb 50W
CAS 23950-58-5	Kerb Propyzamide 50
CAS 23950-58-5	Rapier
CAS 24017-47-8	Hostathion
CAS 24017-47-8	Methoxone
CAS 24017-47-8	MSS CMPP
CAS 24166-13-0	Enadel
CAS 24307-26-4	Pix®, ULV
CAS 24356-60-3	Cefa-Lake
CAS 24356-60-3	Cefadyl
CAS 24390-14-5	Doxylar®
CAS 24589-78-4	CM9160

CAS 24602-86-6	Calixin®
CAS 24678-13-5	Elanone
CAS 24801-88-5	CI7840
CAS 24815-24-5	Cinnaloid
CAS 24817-92-3	Citroflex A-6
CAS 24887-06-7	Decrolin®
CAS 24887-06-7	Parolite
CAS 24887-06-7	Redusol Z
CAS 24887-06-7	Reduxol Z
CAS 24937-78-8	AT 1806M
CAS 24937-78-8	AT 4030M
CAS 24937-78-8	Baymod® L450 N
CAS 24937-78-8	Fusabond® MC-19/D
CAS 24937-78-8	Hysol® 342
CAS 24937-78-8	Hysol® DoAll (1942)
CAS 24937-79-9	Dykor 204
CAS 24937-79-9	Foraflon® 1000 HD
CAS 24937-79-9	KF Polymer® C-1000
CAS 24937-79-9	KF Polymer® T-850
CAS 24937-79-9	KF Polymer® T-1300
CAS 24937-79-9	KF Polymer® U-1000
CAS 24937-79-9	KF Polymer® W-1000
CAS 24937-79-9	Kynar® 301 F
CAS 24937-79-9	Kynar® 460
CAS 24937-79-9	Kynar® 700 Series
CAS 24937-79-9	Kynar® Flex® 2800, 2801
CAS 24937-79-9	Kynar® Flex® 2900
CAS 24937-79-9	Solef® 1008
CAS 24937-79-9	Solef® 1010
CAS 24937-79-9	Solef® 5008
CAS 24937-79-9	Solef® 6010
CAS 24937-79-9	Solef® 8808
CAS 24937-79-9	Solef® 11008/0003
CAS 24937-79-9	Solef® 11010
CAS 24938-91-8	Bio-Soft® TD 400
CAS 24938-91-8	Chemal TDA-3
CAS 24938-91-8	Ethal TDA-3
CAS 24938-91-8	Ethal TDA-6
CAS 24938-91-8	Ethal TDA-18
CAS 24938-91-8	Genapol® X-040
CAS 24938-91-8	Hetoxol TD-3
CAS 24938-91-8	Hodag Nonionic TD-15
CAS 24938-91-8	Iconol TDA-3
CAS 24938-91-8	Iconol TDA-6
CAS 24938-91-8	Iconol TDA-8
CAS 24938-91-8	Iconol TDA-9
CAS 24938-91-8	Iconol TDA-10
CAS 24938-91-8	Lipocol TD-3
CAS 24938-91-8	Macol® TD-3
CAS 24938-91-8	Merpol® SH
CAS 24938-91-8	Prox-onic TD-1/03
CAS 24938-91-8	Rhodasurf® BC-420
CAS 24938-91-8	Rhodasurf® T-95
CAS 24938-91-8	Rhodasurf® TDA-6
CAS 24938-91-8	Teric 13A5
CAS 24938-91-8	Trycol® 5874
CAS 24938-91-8	Trycol® 5940
CAS 24938-91-8	Volpo T-3
CAS 24969-11-7	Penacolite® R-2170
CAS 25013-16-5	Nipantiox
CAS 25013-16-5	Nipantiox 1-F
CAS 25013-16-5	Sustane® 1-F

CAS 25013-16-5	Sustane® BHA
CAS 25034-71-3	Trilene® 65
CAS 25035-04-5	Rilsan® BESHVO, BESVO
CAS 25035-04-5	RTP 201C
CAS 25035-04-5	Thermocomp® HF-1006
CAS 25038-54-4	Ashlene® 630-33G
CAS 25038-54-4	Ashlene® 735
CAS 25038-54-4	Ashlene® 830L
CAS 25038-54-4	Ashlene® 840
CAS 25038-54-4	Ashlene® 858
CAS 25038-54-4	Cabelec® 1015
CAS 25038-54-4	Cadco® Cast Nylon
CAS 25038-54-4	Capran® 77C
CAS 25038-54-4	Capran® Unidraw®
CAS 25038-54-4	Capron® 8202
CAS 25038-54-4	Capron® 8203C HS
CAS 25038-54-4	Capron® 8232G HS FR
CAS 25038-54-4	Capron® 8233G HS
CAS 25038-54-4	Capron® 8253
CAS 25038-54-4	Capron® 8259
CAS 25038-54-4	Capron® 8266G HS
CAS 25038-54-4	Capron® 8280
CAS 25038-54-4	Celstran® N6G30-01-4
CAS 25038-54-4	CTX-312
CAS 25038-54-4	Durethan® B 30 S, B 31 SK
CAS 25038-54-4	Durethan® B 35 F, B 38 F, B 40 F
CAS 25038-54-4	Durethan® BKV 30 H
CAS 25038-54-4	Durethan® BKV 115
CAS 25038-54-4	Durethan® BM 30 X
CAS 25038-54-4	Durethan® KL1-2402/30
CAS 25038-54-4	Durethan® RM KU 2-2501/30
CAS 25038-54-4	Electrafil® J-3/CF/30
CAS 25038-54-4	EMI-X® PC-1008
CAS 25038-54-4	EMI-X® PDX-P-90305
CAS 25038-54-4	Fiberfil® J-7/33
CAS 25038-54-4	Fiberfil® J-7/33/IT
CAS 25038-54-4	Fiberfil® NY-7
CAS 25038-54-4	Fiberfil® NY-7/VO
CAS 25038-54-4	Fiberstran® G-3/50
CAS 25038-54-4	Grilon® A23GM
CAS 25038-54-4	Grilon® BT40Z
CAS 25038-54-4	Grilon® PV-15H
CAS 25038-54-4	Grilon® PVN-15H
CAS 25038-54-4	Grilon® R47HW
CAS 25038-54-4	MC®
CAS 25038-54-4	Novamid® 1010C
CAS 25038-54-4	Novamid® 1020CA2
CAS 25038-54-4	NY-10GF
CAS 25038-54-4	NY-30CF
CAS 25038-54-4	Ny-Kon® P
CAS 25038-54-4	Nybex 12034 BKFR
CAS 25038-54-4	Nybex 12056 BKFR
CAS 25038-54-4	Nybex 13001 BKC
CAS 25038-54-4	Nybex 15011 NA
CAS 25038-54-4	Nybex 17000 NAX
CAS 25038-54-4	Nybex 42002 BKHS
CAS 25038-54-4	Nycoa® 438
CAS 25038-54-4	Nycoa® 446
CAS 25038-54-4	Nycoa® 567
CAS 25038-54-4	Nycoa® 714
CAS 25038-54-4	Nycoa® 870
CAS 25038-54-4	Nycoa® 1417
CAS 25038-54-4	Nycoa® 4015
CAS 25038-54-4	Nylatron® GS-63
CAS 25038-54-4	PA-211
CAS 25038-54-4	PA-211G13
CAS 25038-54-4	PA-211N40
CAS 25038-54-4	PA-221
CAS 25038-54-4	Plaslube® G-3/40/MS/5
CAS 25038-54-4	Plaslube® J-3/30/MS/5
CAS 25038-54-4	RTP 201A
CAS 25038-54-4	Sniamid® ADS 40 I
CAS 25038-54-4	Sniamid® ASN 27 T
CAS 25038-54-4	Stat-Kon® P
CAS 25038-54-4	Texalon 600A NU
CAS 25038-54-4	Texalon 1000A
CAS 25038-54-4	Texalon GF 600A (6-33)
CAS 25038-54-4	Thermocomp® PC-1006
CAS 25038-54-4	Ultramid® B3
CAS 25038-54-4	Ultramid® B35G3
CAS 25038-54-4	Ultramid® B3EG10, B3WG10
CAS 25038-54-4	Ultramid® B3G10
CAS 25038-54-4	Ultramid® B3K
CAS 25038-54-4	Ultramid® B5
CAS 25038-54-4	Ultramid® BS 400 S
CAS 25038-54-4	Ultramid® BS 416
CAS 25038-54-4	Ultramid® BS 3300
CAS 25038-59-9	Celstran® PETG30-01-4
CAS 25038-59-9	Cleartuf Series
CAS 25038-59-9	Crastine® XMB 1068
CAS 25038-59-9	Electrafil® J-1800/CF/30
CAS 25038-59-9	Ertalyte®
CAS 25038-59-9	Grilpet® EV-30
CAS 25038-59-9	Impet® 330
CAS 25038-59-9	Kodapak® 5214A
CAS 25038-59-9	Kodapak® PET Copolyester 13339
CAS 25038-59-9	Kodar® PETG Copolyester 6763
CAS 25038-59-9	Mitsubishi Kasei GF-PET 6010G15
CAS 25038-59-9	Petra® 130
CAS 25038-59-9	Petra® 130FR
CAS 25038-59-9	Petra® 230
CAS 25038-59-9	Petra® 242
CAS 25038-59-9	RTP 1105FR
CAS 25038-59-9	Tenite® PET 9902
CAS 25038-59-9	Traytuf Ultra-Clear
CAS 25038-59-9	Ultralen® SP 3700 S
CAS 25038-59-9	Ultralen® SP 3705
CAS 25038-59-9	Valox® 9215
CAS 25046-79-1	Pro-Diaban®
CAS 25054-76-6	Incronam OP-30
CAS 25054-76-6	Lexaine® O
CAS 25054-76-6	Mackam HV
CAS 25054-76-6	Mirataine® BET-O-30
CAS 25057-89-0	Basagran®
CAS 25057-89-0	Basagran®
CAS 25066-20-0	Chemoxide SAO
CAS 25066-20-0	Mackamine SAO
CAS 25068-26-2	Crystalor DC-6
CAS 25086-62-8	Alkasperse® M-10
CAS 25086-89-9	Agrimer VA 6
CAS 25086-89-9	Luviskol® VA 28 E
CAS 25103-12-2	Doverphos® TIOP

CAS 25103-12-2	Weston® TIOP	CAS 25213-24-5	Airvol® 203, 205
CAS 25122-46-7	Dermovate	CAS 25213-24-5	Airvol® 205S, 523S, 540S
CAS 25122-57-0	Molivate	CAS 25213-24-5	Airvol® 425, WS 42
CAS 25126-32-3	Kinevac	CAS 25213-24-5	Airvol® 523, 540
CAS 25134-01-4	Electrafil® F-1700/CF/10/A	CAS 25213-24-5	Airvol® 705, 723, 740
CAS 25134-01-4	Electrafil® G-1704/SS/5	CAS 25231-21-4	Acconon E
CAS 25134-01-4	Electrafil® J-1700/CF/10	CAS 25231-21-4	Arlamol® E
CAS 25134-01-4	MP-10CF-4CC/15T	CAS 25231-21-4	Fancol SA-15
CAS 25134-01-4	Noryl® 731	CAS 25231-21-4	Hetoxol SP-15
CAS 25134-01-4	Noryl® BN25	CAS 25231-21-4	Prostearyl 15
CAS 25134-01-4	Noryl® EM5101	CAS 25231-21-4	Prox-onic SA1-015/P
CAS 25134-01-4	Noryl® EN185	CAS 25231-21-4	Witconol APS
CAS 25134-01-4	Noryl® FN150	CAS 25265-77-4	Texanol® Ester-Alcohol
CAS 25134-01-4	Noryl® FN215X	CAS 25301-02-4	Alevaire
CAS 25134-01-4	Noryl® GTX810	CAS 25301-02-4	Superinone
CAS 25134-01-4	Noryl® HM3020	CAS 25311-71-1	Amaze®
CAS 25134-01-4	Noryl® HS1000X	CAS 25311-71-1	Amidocid®
CAS 25134-01-4	Noryl® N190X	CAS 25311-71-1	Oftanol®
CAS 25134-01-4	Noryl® PC180X	CAS 25311-71-1	Pryfon
CAS 25134-01-4	Noryl® PN235	CAS 25321-41-9	Eltesol® XA
CAS 25134-01-4	Noryl® PX0722	CAS 25321-41-9	Manro XSA
CAS 25134-01-4	Noryl® SE100	CAS 25321-41-9	Reworyl® X
CAS 25134-01-4	Noryl® SE1GFN2	CAS 25321-41-9	XSA 80
CAS 25134-01-4	Stat-Kon® ZC-1003	CAS 25322-68-3	Alkapol PEG 300
CAS 25134-01-4	Styvex 72001 NA	CAS 25322-68-3	Atpet 300
CAS 25134-01-4	Thermocomp® ZF-1004	CAS 25322-68-3	Atpet 400
CAS 25135-51-3	Electrafil® J-1500/CF/20	CAS 25322-68-3	Atpet 600
CAS 25135-51-3	Mindel® B-310	CAS 25322-68-3	Carbowax® Compound 20M
CAS 25135-51-3	Mindel® B-322	CAS 25322-68-3	Carbowax® PEG 200
CAS 25135-51-3	Mindel® M-800	CAS 25322-68-3	Carbowax® PEG 540 Blend
CAS 25135-51-3	Mindel® S-1000	CAS 25322-68-3	Carbowax® PEG 8000
CAS 25135-51-3	PF-10GF/15T	CAS 25322-68-3	Carbowax® Sentry
CAS 25135-51-3	PF-20GF	CAS 25322-68-3	Chemstat® P-400
CAS 25135-51-3	Polypenco® Polysulfone	CAS 25322-68-3	Droxol 200
CAS 25135-51-3	RTP 901	CAS 25322-68-3	Emery® 6686
CAS 25135-51-3	Thermocomp® GF-1004	CAS 25322-68-3	Hodag PEG 200
CAS 25135-51-3	Udel® GF-110	CAS 25322-68-3	Hodag PEG 300
CAS 25135-51-3	Udel® P-1700	CAS 25322-68-3	Hodag PEG 400
CAS 25135-51-3	Ultrason® S 1010	CAS 25322-68-3	Hodag PEG 540
CAS 25155-18-4	Diaparene	CAS 25322-68-3	Hodag PEG 600
CAS 25155-18-4	Hyamine® 10X	CAS 25322-68-3	Hodag PEG 1000
CAS 25155-23-1	Coalite N.T.P	CAS 25322-68-3	Hodag PEG 1450
CAS 25155-23-1	Kronitex® TXP	CAS 25322-68-3	Hodag PEG 3350
CAS 25155-23-1	Syn-O-Ad® 8475M	CAS 25322-68-3	Hodag PEG 8000
CAS 25155-25-3	Luperox 802	CAS 25322-68-3	Ilexan HT
CAS 25155-25-3	Perkadox® 14	CAS 25322-68-3	Lipoxol® 12000
CAS 25155-25-3	Retilox® F 40 MG	CAS 25322-68-3	Lutrol® E 300
CAS 25155-25-3	Vul-Cup	CAS 25322-68-3	Lutrol® E 400
CAS 25155-25-3	Vul-Cup 40KE and R	CAS 25322-68-3	Lutrol® E 1500
CAS 25155-30-0	Nansa® HS80S	CAS 25322-68-3	Lutrol® E 4000
CAS 25155-30-0	Rhodacal® DDB-40	CAS 25322-68-3	Lutrol® E 6000
CAS 25155-30-0	Sul-fon-ate AA-10	CAS 25322-68-3	Macol® E-200
CAS 25159-40-4	Incromine Oxide O	CAS 25322-68-3	Nopalcol 200
CAS 25159-40-4	Mackamine OAO	CAS 25322-68-3	Nopalcol 400
CAS 25167-32-2	Aerosol® DPOS-45	CAS 25322-68-3	Nopalcol 600
CAS 25168-05-2	Halso® 99	CAS 25322-68-3	Pluracol® E200
CAS 25168-73-4	Crodesta DKS F110	CAS 25322-68-3	Pluracol® E300
CAS 25168-73-4	Crodesta F-160	CAS 25322-68-3	Pluracol® E400
CAS 25168-73-4	Grilloten PSE141G	CAS 25322-68-3	Pluracol® E400 NF
CAS 25168-73-4	Ryoto Sugar Ester S-1170	CAS 25322-68-3	Pluracol® E600 NF
CAS 25168-73-4	Sucro Ester 7	CAS 25322-68-3	Pluracol® E1000
CAS 25168-73-4	Sucro Ester 15	CAS 25322-68-3	Pluracol® E1500

CAS 25322-68-3	Pluracol® E2000
CAS 25322-68-3	Pluracol® E4000
CAS 25322-68-3	Pluracol® E6000
CAS 25322-68-3	Pluracol® E8000
CAS 25322-68-3	Poly-G® 200
CAS 25322-68-3	Polyox® WSR 3333
CAS 25322-68-3	Polyox® WSR N-10
CAS 25322-68-3	Polytrix
CAS 25322-68-3	Prox-onic PEG-2000
CAS 25322-68-3	Rhodasurf® E 400
CAS 25322-68-3	Rhodasurf® PEG 400
CAS 25322-68-3	Rhodasurf® PEG 3350
CAS 25322-68-3	Teric PEG 200
CAS 25322-68-3	Teric PEG 300
CAS 25322-68-3	Teric PEG 400
CAS 25322-68-3	Teric PEG 600
CAS 25322-68-3	Teric PEG 800
CAS 25322-68-3	Teric PEG 1000
CAS 25322-68-3	Teric PEG 4000
CAS 25322-68-3	Teric PEG 6000
CAS 25322-68-3	Teric PEG 8000
CAS 25322-68-3	Teric PEG 12000
CAS 25332-39-2	Desyrel
CAS 25332-39-2	Molipaxin
CAS 25339-09-7	Afilan ICS
CAS 25339-09-7	Ceraphyl 494
CAS 25339-09-7	Cetiol® G16S
CAS 25339-09-7	Kemester® 5822
CAS 25339-09-7	Kessco® ICS
CAS 25339-09-7	Schercemol ICS
CAS 25339-99-5	Grilloten LSE87
CAS 25339-99-5	Ryoto Sugar Ester LWA-1570
CAS 25383-99-7	Artodan SP 55 Kosher
CAS 25383-99-7	Crolactil SSL
CAS 25383-99-7	Emplex
CAS 25383-99-7	Emulsilac S
CAS 25383-99-7	Grindtek FAL 1
CAS 25383-99-7	Pationic SSL
CAS 25383-99-7	Radiamuls® SSL 2990
CAS 25383-99-7	Swedex SSL-5AC
CAS 25389-94-0	Kantrex
CAS 25389-94-0	Kantrim
CAS 25389-94-0	Klebcil
CAS 25417-20-3	Dehscofix 917
CAS 25417-20-3	Supragil® NK
CAS 25446-78-0	Rhodapex® 674/C
CAS 25446-78-0	Rhodapex® EST-30
CAS 25448-25-3	Doverphos® 6
CAS 25448-25-3	Weston® TDP
CAS 25496-01-9	LABS 100/H.V
CAS 25496-01-9	Nansa® TDB
CAS 25496-72-4	Tegin® O
CAS 25550-98-5	Doverphos® 7
CAS 25550-98-5	Weston® PDDP
CAS 25606-41-1	Filex
CAS 25609-89-6	Luviset CA 66
CAS 25609-89-6	Resyn® 28-1310
CAS 25637-84-7	Emerest® 2419
CAS 25637-84-7	Priolube 1409
CAS 25637-99-4	FR-1206
CAS 25640-78-2	Nusolv ABP-62
CAS 25655-41-8	Betadine
CAS 25655-41-8	Pevidine
CAS 25655-41-8	PVP-Iodine
CAS 25655-41-8	Videne Disinfectant Solution
CAS 25655-41-8	Videne Disinfectant Tincture
CAS 25655-41-8	Videne Powder
CAS 25655-41-8	Videne Surgical Scrub
CAS 25717-80-0	Corvaton
CAS 25717-80-0	Corvaton®
CAS 25736-86-1	Sipomer® HEM-5
CAS 25812-30-0	Lopid
CAS 25882-44-4	Geropon® SBL-203
CAS 25882-44-4	Incrosul LMS
CAS 25882-44-4	Mackanate LM-40
CAS 25882-44-4	Rewopol® SBL 203
CAS 25882-44-4	Varsulf® SBL-203
CAS 25928-94-3	Allabond Twenty/twenty Adhesive
CAS 25928-94-3	Allabond Twenty/twenty NM
CAS 25928-94-3	Amicon® C-860-4
CAS 25928-94-3	Amicon® CT-4042-5
CAS 25928-94-3	Amicon® ECT-86
CAS 25928-94-3	Amicon® ME-868
CAS 25928-94-3	Amicon® TG-86
CAS 25928-94-3	Araldite® 2001
CAS 25928-94-3	Araldite® CY 225
CAS 25928-94-3	Araldite® ECN 1235
CAS 25928-94-3	Araldite® GT 6060
CAS 25928-94-3	Araldite® GZ 540 X-90
CAS 25928-94-3	Araldite® PT 810
CAS 25928-94-3	Araldite® PY 306
CAS 25928-94-3	Araldite® XD 897
CAS 25928-94-3	Araldite® XD 4955
CAS 25928-94-3	Araldite® XU GY 358
CAS 25928-94-3	Aratronic® 5001
CAS 25928-94-3	Aratronic® 5040
CAS 25928-94-3	Aroflint® 303-X-90
CAS 25928-94-3	Basoset® 162
CAS 25928-94-3	C-84
CAS 25928-94-3	CI-2
CAS 25928-94-3	Conapoxy® FR-1270
CAS 25928-94-3	Conapoxy® TE-1257/ Conacure® EA-08
CAS 25928-94-3	D.E.R. 317
CAS 25928-94-3	D.E.R. 362
CAS 25928-94-3	D.E.R. 383
CAS 25928-94-3	D.E.R. 642U
CAS 25928-94-3	D.E.R. 732
CAS 25928-94-3	E-3810
CAS 25928-94-3	E-3824
CAS 25928-94-3	E-9405
CAS 25928-94-3	Emcast 1510, 1511
CAS 25928-94-3	Emcast 1550, 1551
CAS 25928-94-3	Epocap® 16129 A/B
CAS 25928-94-3	Epocap® 16358 A/B
CAS 25928-94-3	Epolite 1301
CAS 25928-94-3	Epolite 1302
CAS 25928-94-3	Epolite 2315
CAS 25928-94-3	Epolite 3300
CAS 25928-94-3	Epolite 5363
CAS 25928-94-3	Epon® Resin DPL-1911
CAS 25928-94-3	Epoweld® 19157
CAS 25928-94-3	FA-1
CAS 25928-94-3	FA-8

CAS 25928-94-3	FA-14	CAS 26027-38-3	Iconol NP-100
CAS 25928-94-3	FFA-5, FFA-9	CAS 26062-79-3	Agefloc WT-20
CAS 25928-94-3	Fiberite 7669	CAS 26062-79-3	Agefloc WT-40
CAS 25928-94-3	Fiberite 7701	CAS 26062-79-3	Agequat 400
CAS 25928-94-3	Hysol® 1C Epoxi-Patch Kit	CAS 26062-79-3	M-Quat® 40
CAS 25928-94-3	Hysol® 6C Epoxi-Patch Kit	CAS 26062-79-3	Mackernium 006
CAS 25928-94-3	Hysol® EO1016	CAS 26062-94-2	Celanex® 1300A
CAS 25928-94-3	Hysol® MG1 Series	CAS 26062-94-2	Celanex® 1600A
CAS 25928-94-3	LCA-1	CAS 26062-94-2	Celanex® 2000K
CAS 25928-94-3	LCA-20	CAS 26062-94-2	Celanex® 3310
CAS 25928-94-3	LCA-127	CAS 26062-94-2	Celanex® 5300
CAS 25928-94-3	Master Bond EP11HT	CAS 26062-94-2	Celanex® J600
CAS 25928-94-3	Master Bond EP21HT	CAS 26062-94-2	Celstran® PBTG30-01-4
CAS 25928-94-3	Master Bond EP30HT	CAS 26062-94-2	Crastine® S 600
CAS 25928-94-3	Master Bond EP34CA	CAS 26062-94-2	Crastine® SG 625
CAS 25928-94-3	Master Bond EP75	CAS 26062-94-2	Crastine® SG 665 FR
CAS 25928-94-3	Master Bond Supreme 3HT	CAS 26062-94-2	Crastine® SO 653
CAS 25928-94-3	Master Bond Supreme 11HT	CAS 26062-94-2	Crastine® XB 3035
CAS 25928-94-3	Neonite® EG60/6mm, EG61/12mm	CAS 26062-94-2	Electrafil® G-1854/SS/7
		CAS 26062-94-2	Electrafil® J-1850/CF/30
CAS 25928-94-3	Nor-Mer 020	CAS 26062-94-2	EMI-X® PDX-W-88341
CAS 25928-94-3	Norcast 142 Systems	CAS 26062-94-2	Grilpet® XE3060
CAS 25928-94-3	Norcast 154FR	CAS 26062-94-2	Mitsubishi Kasei PBT 5008
CAS 25928-94-3	Norcast 1460-1	CAS 26062-94-2	Mitsubishi Kasei PBT 5010F1
CAS 25928-94-3	Norcast 3220G-1	CAS 26062-94-2	PBT-1100
CAS 25928-94-3	Norcast 3258	CAS 26062-94-2	PBT-1100G15
CAS 25928-94-3	Norcast 3705	CAS 26062-94-2	PBT-1300
CAS 25928-94-3	Norcast 4914-1	CAS 26062-94-2	PBT-1700
CAS 25928-94-3	P-11	CAS 26062-94-2	PDX-84369
CAS 25928-94-3	P-51	CAS 26062-94-2	Pocan® B 1300
CAS 25928-94-3	P-56	CAS 26062-94-2	Pocan® B 1305
CAS 25928-94-3	P-80F	CAS 26062-94-2	Pocan® B 1505
CAS 25928-94-3	P-85	CAS 26062-94-2	Pocan® KU1-7033
CAS 25928-94-3	Quatrex 1010	CAS 26062-94-2	Pocan® S 1506
CAS 25928-94-3	Redux® 501	CAS 26062-94-2	PS-30GM/000
CAS 25928-94-3	RX® 1906	CAS 26062-94-2	RTP 1001
CAS 25928-94-3	Scotchkote® 213, 214	CAS 26062-94-2	Stat-Kon® W
CAS 25928-94-3	Sobral EE-632	CAS 26062-94-2	Thermocomp® WC-1006
CAS 25928-94-3	Stycast® 1090	CAS 26062-94-2	Ultradur® B 2550
CAS 25928-94-3	Stycast® 1210	CAS 26062-94-2	Ultradur® B 4300 G10
CAS 25928-94-3	Stycast® 1266	CAS 26062-94-2	Ultradur® B 4500
CAS 25928-94-3	Stycast® 1467	CAS 26062-94-2	Ultradur® KR 4001
CAS 25928-94-3	Stycast® 2850-FT	CAS 26062-94-2	Valox® 210HP, 220HP, 230HP, 260HP, 280HP
CAS 25928-94-3	Supertherm 2003		
CAS 25928-94-3	Tactix 123	CAS 26062-94-2	Valox® 325
CAS 25928-94-3	Tactix 556	CAS 26062-94-2	Valox® 508
CAS 25928-94-3	Tactix 742	CAS 26062-94-2	Valox® 701
CAS 25928-94-3	Thermoset 100	CAS 26062-94-2	Valox® 780
CAS 25928-94-3	Thermoset 300/No. 65 Hardener	CAS 26062-94-2	Valox® 815
CAS 25928-94-3	Thermoset 310	CAS 26062-94-2	Valox® DR48
CAS 25928-94-3	Thermoset DC-232	CAS 26062-94-2	Vandar® 2100
CAS 25928-94-3	Thiokol® FES-2258	CAS 26062-94-2	Vybex 22008 BKFR
CAS 25928-94-3	Tra-Bond 2151	CAS 26062-94-2	Vybex 22028 BKFR
CAS 25928-94-3	Tra-Bond F113	CAS 26097-80-3	Equiben
CAS 25928-94-3	Tra-Cast 3103	CAS 26097-80-3	Novazole
CAS 25928-94-3	Tra-Duct 2902	CAS 26112-07-2	Nipagin M Potassium
CAS 25928-94-3	Tra-Shield 2867	CAS 26155-31-7	Paratect Bolus
CAS 25928-94-3	Witcobond® XW	CAS 26159-34-2	Anaprox
CAS 25954-13-6	Krenite	CAS 26183-52-8	Bio-Soft® FF 400
CAS 26027-37-2	Dionil® OC	CAS 26183-52-8	Chemal DA-4
CAS 26027-37-2	Ethomid® O/17	CAS 26183-52-8	Desonic® DA-4
CAS 26027-38-3	Akyporox NP 150	CAS 26183-52-8	Desonic® DA-6

CAS 26183-52-8	Ethal DA-4
CAS 26183-52-8	Genapol® DA-040
CAS 26183-52-8	Iconol DA-4
CAS 26183-52-8	Iconol DA-6
CAS 26183-52-8	Iconol DA-9
CAS 26183-52-8	Marlipal® 1012/4
CAS 26183-52-8	Oxetal D 104
CAS 26183-52-8	Prox-onic DA-1/04
CAS 26183-52-8	Rhodasurf® DA-4
CAS 26183-52-8	Trycol® 5950
CAS 26225-79-6	Betanal Tandem
CAS 26225-79-6	Nortron
CAS 26225-79-6	Spectron
CAS 26256-79-1	Deriphat 160C
CAS 26264-05-1	Arylan® PWS
CAS 26264-05-1	Kowet 3300
CAS 26264-05-1	Rhodacal® 330
CAS 26264-05-1	Rhodacal® IPAM
CAS 26264-06-2	Kowet 12
CAS 26264-58-4	Rhodacal® RM/210
CAS 26264-58-4	Supragil® MNS/90
CAS 26266-57-9	Ablunol S-40
CAS 26266-57-9	Arlacel® 40
CAS 26266-57-9	Crill 2
CAS 26266-57-9	Disponil SMP 100 F1
CAS 26266-57-9	Emsorb® 2510
CAS 26266-57-9	Glycomul® P
CAS 26266-57-9	Hodag SMP
CAS 26266-57-9	Kemester® S40
CAS 26266-57-9	Liposorb P
CAS 26266-57-9	Montane 40
CAS 26266-57-9	Prote-sorb SMP
CAS 26266-57-9	Radiasurf® 7135
CAS 26266-57-9	S-Maz® 40
CAS 26266-57-9	Sorbax SMP
CAS 26266-57-9	Sorbon S-40
CAS 26266-57-9	Span® 40
CAS 26266-58-0	Ablunol S-85
CAS 26266-58-0	Alkamuls® S-85
CAS 26266-58-0	Alkamuls® STO
CAS 26266-58-0	Arlacel® 85
CAS 26266-58-0	Atmer® 106
CAS 26266-58-0	Crill 45
CAS 26266-58-0	Disponil STO 100 F1
CAS 26266-58-0	Emsorb® 2503
CAS 26266-58-0	Ethylan® GT85
CAS 26266-58-0	Glycomul® TO
CAS 26266-58-0	Hodag STO
CAS 26266-58-0	Kemester® S85
CAS 26266-58-0	Liposorb TO
CAS 26266-58-0	Prote-sorb STO
CAS 26266-58-0	S-Maz® 85
CAS 26266-58-0	Sorbax STO
CAS 26266-58-0	Span® 85
CAS 26266-58-0	Witconol 2503
CAS 26266-77-3	Abitol®
CAS 26309-95-5	Alpha-Cillin
CAS 26309-95-5	Alpivicin
CAS 26401-27-4	Doverphos® DPIOP
CAS 26401-27-4	Weston® ODPP
CAS 26402-26-6	Imwitor® 308
CAS 26402-26-6	Imwitor® 742
CAS 26402-26-6	Imwitor® 988
CAS 26402-26-6	Witafrol® 7420
CAS 26402-31-3	Flexricin® 9
CAS 26402-31-3	Naturechem® PGR
CAS 26426-80-2	Tamol® 731-25%
CAS 26444-49-5	Disflamoll® DPK
CAS 26446-38-8	Ryoto Sugar Ester P-1570, P-1670
CAS 26447-10-9	Eltesol® AX 40
CAS 26447-10-9	Hartotrope AXS
CAS 26447-10-9	Naxonate® 4AX
CAS 26447-10-9	Stepanate® AXS
CAS 26483-35-2	Incromine Oxide B-30P
CAS 26523-78-4	Doverphos® 4
CAS 26523-78-4	Doverphos® 4-HR
CAS 26523-78-4	Lankromark® LE109
CAS 26523-78-4	Lowinox® TNPP
CAS 26523-78-4	Weston® 399
CAS 26523-78-4	Weston® TNPP
CAS 26523-78-4	Wytox® 312
CAS 26530-20-1	Micro-Chek® 11 P
CAS 26530-20-1	Pancil T
CAS 26530-20-1	Skane® M-8
CAS 26544-23-0	Doverphos® 8
CAS 26544-23-0	Weston® DPDP
CAS 26544-27-4	Weston® 600
CAS 26545-53-9	Ablusol DBD
CAS 26545-53-9	Marlopon® ADS 50
CAS 26570-10-5	Darvon-N
CAS 26589-26-4	Luviflex® VBM 35
CAS 26589-26-4	Stepanhold® Extra
CAS 26589-26-4	Stepanhold® R-1
CAS 26590-05-6	Agequat 500, -5008
CAS 26590-05-6	Mackernium 007
CAS 26590-05-6	Mirapol® 550
CAS 26597-36-4	Kemamine® Q-2802C
CAS 26635-75-6	Amidox® L-2
CAS 26635-92-7	Ethomeen® 18/15
CAS 26635-92-7	Ethomeen® 18/20
CAS 26635-92-7	Ethomeen® 18/25
CAS 26635-92-7	Ethomeen® 18/60
CAS 26635-92-7	Ethox SAM-10
CAS 26635-92-7	Hetoxamine ST-5
CAS 26635-92-7	Marlazin® S 10
CAS 26635-92-7	Prox-onic MS-05
CAS 26635-92-7	Teric 18M5
CAS 26635-92-7	Trymeen® 6617
CAS 26635-92-7	Trymeen® SAM-50
CAS 26635-92-7	Zusomin S 110
CAS 26635-93-8	Berol 28
CAS 26635-93-8	Berol 303
CAS 26635-93-8	Marlazin® OL 2
CAS 26635-93-8	Marlowet® 5400
CAS 26644-46-2	Fairy Ring Destroyer
CAS 26644-46-2	Funginex
CAS 26644-46-2	Nimrod T
CAS 26644-46-2	Saprol
CAS 26652-09-5	Pre-Par
CAS 26652-09-5	Utopar
CAS 26652-09-5	Yutopar
CAS 26658-19-5	Alkamuls® S-65
CAS 26658-19-5	Alkamuls® STS

CAS 26658-19-5	Crill 35
CAS 26658-19-5	Disponil STS 100 F1
CAS 26658-19-5	Emsorb® 2507
CAS 26658-19-5	Famodan TS Kosher
CAS 26658-19-5	Glycomul® TS KFG
CAS 26658-19-5	Grindtek STS
CAS 26658-19-5	Hodag STS
CAS 26658-19-5	Kemester® S65
CAS 26658-19-5	Liposorb TS
CAS 26658-19-5	Montane 65
CAS 26658-19-5	Prote-sorb STS
CAS 26658-19-5	Radiamuls® Sorb 2344
CAS 26658-19-5	Radiamuls® Sorb 2345
CAS 26658-19-5	S-Maz® 65K
CAS 26658-19-5	Sorbax STS
CAS 26658-19-5	Span® 65
CAS 26675-46-7	Forane
CAS 26680-54-6	Milldride® OSA
CAS 26741-53-7	Alkanox® P-24
CAS 26741-53-7	Blendex® 340
CAS 26741-53-7	Ultranox® 626
CAS 26741-53-7	Ultranox® 627A
CAS 26748-38-9	Esperox® 750M
CAS 26748-41-4	Aztec® t-Butyl Peroxy-neodecanoate-50 OMS
CAS 26748-41-4	Aztec® t-Butyl Peroxy-neodecanoate-75 OMS
CAS 26748-41-4	Esperox® 33M
CAS 26748-41-4	Lupersol 10
CAS 26748-41-4	Trigonox® 23
CAS 26748-47-0	Esperox® 939M
CAS 26748-47-0	Lupersol 188-M75
CAS 26748-47-0	Lupersol 288-M75
CAS 26748-47-0	Trigonox® 99-B75
CAS 26761-45-5	Glydexx N-10
CAS 26774-90-3	Dexacillin
CAS 26780-96-1	Agerite® MA
CAS 26780-96-1	Agerite® Resin D
CAS 26780-96-1	Flectol H
CAS 26780-96-1	Flectol Pastilles
CAS 26787-78-0	Amoxicap
CAS 26787-78-0	Amoxil
CAS 26787-78-0	Amoxisyrup
CAS 26787-78-0	Larocin
CAS 26787-78-0	Larotid
CAS 26787-78-0	Polymox
CAS 26787-78-0	Qualamox
CAS 26787-78-0	Trimox
CAS 26787-78-0	Utimox
CAS 26787-78-0	Wymox
CAS 26807-65-8	Lozol
CAS 26844-12-2	Baratol Tblets
CAS 26850-24-8	Dantocol® DHE
CAS 26864-56-2	Semap
CAS 26921-17-5	Blocadren
CAS 26921-17-5	Timacar
CAS 26921-17-5	Timolate
CAS 26921-17-5	Timoptic
CAS 26921-17-5	Timoptol®
CAS 26944-48-9	Glutril
CAS 27164-46-1	Ancef
CAS 27176-87-0	Pentine Acid 5431
CAS 27176-87-0	Rhodacal® ABSA
CAS 27177-77-1	Polystep® A-15-30K
CAS 27178-16-1	Kodaflex® DIDA
CAS 27178-16-1	Monoplex® DDA
CAS 27178-16-1	Plasthall® DIDA
CAS 27194-74-7	Schercemol PGML
CAS 27195-16-0	Crodesta DKS F10
CAS 27195-16-0	Crodesta F-10
CAS 27195-16-0	Ryoto Sugar Ester S-570, S-770
CAS 27223-35-4	Unakalm
CAS 27233-00-7	Hetester HCA
CAS 27233-00-7	Naturechem® GTH
CAS 27306-78-1	Silwet® L-77
CAS 27306-79-2	Hetoxol M-3
CAS 27306-90-7	Akypo RLM 160
CAS 27321-96-6	Fancol CH-24
CAS 27321-96-6	Forlan C-24
CAS 27503-81-7	Neo Heliopan, Type Hydro
CAS 27554-26-3	Jayflex® DIOP
CAS 27554-26-3	Morflex 100
CAS 27554-26-3	Palatinol® D10
CAS 27554-26-3	Witcizer 313
CAS 27607-77-8	CT3795
CAS 27638-00-2	Capmul® GDL
CAS 27638-00-2	Emulsynt GDL
CAS 27638-00-2	Kemester® GDL
CAS 27638-00-2	Kessco® GDL
CAS 27638-00-2	Lexemul® GDL
CAS 27640-89-7	Kemester® EE
CAS 27640-89-7	Schercemol EE
CAS 27676-62-6	Anox® IC-14
CAS 27676-62-6	Vanox® GT
CAS 27776-21-2	VA-044
CAS 27813-02-1	Bisomer 2HPMA
CAS 27841-06-1	Schercemol NGDC
CAS 27877-51-6	Dalnate
CAS 27883-12-1	Alkamide® CP-1255
CAS 27986-36-3	Akyporox NP 15
CAS 28061-69-0	Armeen® DMOD
CAS 28061-69-0	Crodamine 3.AOD
CAS 28108-99-8	Syn-O-Ad® 8480
CAS 28211-18-9	Ganex® V-220
CAS 28212-44-4	Sandopan® MA-18
CAS 28395-03-1	Bumex
CAS 28434-01-7	Reslin
CAS 28510-23-8	Schercemol NGDO
CAS 28519-02-0	Calfax DB-45
CAS 28519-02-0	Chemcogen AC
CAS 28519-02-0	Rhodacal® DSB
CAS 28631-35-8	Duplosan® DP
CAS 28631-63-2	Eltesol® CA 65
CAS 28631-63-2	Reworyl® C
CAS 28657-80-9	Cinobac
CAS 28724-32-5	Ethoquad® 18/12
CAS 28724-32-5	Ethoquad® 18/25
CAS 28772-56-7	Rentokil Deadline
CAS 28772-56-7	Slaymor
CAS 28860-95-9	Lodosin
CAS 28860-95-9	Lodosyn
CAS 28874-51-3	Ajidew N-50
CAS 28874-51-3	Nalidone®
CAS 28874-51-3	Ritamectant PCA

CAS 28911-01-5	Halcion
CAS 28961-43-5	Ageflex EOTMPTA
CAS 28981-97-7	Xanax
CAS 28984-69-2	Alkaterge®-T
CAS 29051-57-8	Pationic 122A
CAS 29094-61-9	Glibenese
CAS 29110-48-3	Akfen Tablets
CAS 29122-68-7	Tenormin
CAS 29216-28-2	Primalan
CAS 29232-93-7	Actellic
CAS 29232-93-7	Actellifog
CAS 29232-93-7	Blex
CAS 29232-93-7	Fumite Pirimiphos-Methyl Smoke
CAS 29232-93-7	Sybol
CAS 29240-17-3	Aztec® t-Amyl Peroxypivalate-75 OMS
CAS 29240-17-3	Esperox® 551M
CAS 29240-17-3	Lupersol 554-M50, 554-M75
CAS 29240-17-3	Trigonox® 125-C75
CAS 29297-22-0	Luviquat® FC 370
CAS 29297-22-0	Luviquat® FC 550
CAS 29297-22-0	Luviquat® FC 905
CAS 29297-22-0	Luviquat® HM 552
CAS 29317-52-0	Tomah PA-13i
CAS 29381-93-9	Marlopon® AT
CAS 29383-26-4	Schercemol OHS
CAS 29385-43-1	Preventol® CI 7-100
CAS 29590-42-9	Ageflex FA-8
CAS 29590-42-9	SR-440
CAS 29598-76-3	Seenox 412S
CAS 29710-25-6	Dermol OO
CAS 29761-21-5	Syn-O-Ad® 8479
CAS 29806-73-3	Bernel® Ester EHP
CAS 29806-73-3	Ceraphyl 368
CAS 29806-73-3	Kessco® Octyl Palmitate
CAS 29806-73-3	Lexol EHP
CAS 29806-73-3	Schercemol OP
CAS 29806-73-3	Tegosoft® OP
CAS 29806-73-3	Unimate® EHP
CAS 29806-73-3	Wickenol® 155
CAS 29860-47-7	Sipomer® TMPEO
CAS 29911-28-2	Dowanol® DPnB
CAS 29964-84-9	Ageflex FM-10
CAS 29973-13-5	Arylmate®
CAS 29973-13-5	Croneton®
CAS 30043-49-3	Ustilan®
CAS 30286-75-0	Oxivent
CAS 30342-62-2	Crodateric S
CAS 30364-51-3	Hamposyl® M-30
CAS 30392-41-7	Tornalate
CAS 30416-77-4	Perlankrol® PA Conc
CAS 30516-87-1	Retrovir
CAS 30525-89-4	Granuform
CAS 30657-38-6	Laurydone®
CAS 30902-17-9	Alphamine
CAS 31001-77-1	CM8450
CAS 31112-62-6	Amipaque
CAS 31394-71-5	Hodag CSA-80
CAS 31394-71-5	Lutrol® OP-2000
CAS 31430-15-6	Flubenol
CAS 31430-15-6	Fluvermal
CAS 31431-39-7	Equivurm Plus
CAS 31431-39-7	Mebatreat
CAS 31431-39-7	Mebenvet
CAS 31431-39-7	Ovitelmin
CAS 31431-39-7	Telmin
CAS 31431-39-7	Vermox
CAS 31556-45-3	Afilan TDS
CAS 31556-45-3	Emerest® 2308
CAS 31556-45-3	Kemester® 5721
CAS 31556-45-3	Lexolube® B-109
CAS 31556-45-3	Liponate TDS
CAS 31566-31-1	Alkamuls® GMS/C
CAS 31566-31-1	Geleol
CAS 31566-31-1	Imwitor® 191
CAS 31566-31-1	Tegin® 4011
CAS 31570-04-4	Alkanox® 240
CAS 31570-04-4	Hostanox® PAR 24
CAS 31570-04-4	Lowinox® 242
CAS 31570-04-4	Naugard® 524
CAS 31586-77-3	Bismucyn
CAS 31621-91-7	Ethox 1122
CAS 31621-91-7	Genagen PL-090
CAS 31677-93-7	Wellbutrin
CAS 31692-79-2	Masil® SFR 70
CAS 31694-55-0	Acconon ETG
CAS 31694-55-0	Ethosperse® G-26
CAS 31694-55-0	Hetoxide G-7
CAS 31694-55-0	Liponic EG-1
CAS 31717-87-0	F238
CAS 31717-87-0	Meltatox®
CAS 31778-15-1	Evangard® 18MP
CAS 31793-07-4	Rengasil
CAS 31799-71-0	Dionil® OC
CAS 31799-71-0	Lutensol® FSA 10
CAS 31842-01-0	Endyne
CAS 31842-61-2	Pulmadil Auto
CAS 31842-61-2	Pulmadil Inhaler
CAS 31884-77-2	Ancolan
CAS 32057-14-0	Peceol Isostearique
CAS 32073-22-6	Eltesol® SC 93
CAS 32073-22-6	Naxonate® 45SC
CAS 32073-22-6	Naxonate® SC
CAS 32073-22-6	Reworyl® NCS
CAS 32073-22-6	Stepanate® SCS
CAS 32073-22-6	Witconate SCS 45%
CAS 32131-17-2	Ashlene® 61-2M
CAS 32131-17-2	Ashlene® 520
CAS 32131-17-2	Ashlene® 520-13G
CAS 32131-17-2	Ashlene® 520BU
CAS 32131-17-2	Ashlene® 520MS
CAS 32131-17-2	Ashlene® 525-13G
CAS 32131-17-2	Ashlene® 527
CAS 32131-17-2	Ashlene® 527LD-13G
CAS 32131-17-2	Ashlene® 528BR-WO
CAS 32131-17-2	Ashlene® 528L-13G
CAS 32131-17-2	Ashlene® 541
CAS 32131-17-2	Ashlene® 541S
CAS 32131-17-2	Celanese® Nylon 1000-1
CAS 32131-17-2	Celanese® Nylon 1003-1
CAS 32131-17-2	Celanese® Nylon 1500-1
CAS 32131-17-2	Celanese® Nylon 7420
CAS 32131-17-2	Celanese® Nylon 7423
CAS 32131-17-2	Celanese® Nylon N-186

CAS 32131-17-2	Celstran® N66G30-01-4	CAS 32360-05-7	Ageflex FM-68.
CAS 32131-17-2	CTC-3300	CAS 32426-11-2	BTC® 818.
CAS 32131-17-2	Durethan® A 30 S	CAS 32440-50-9	Ganex® V-216.
CAS 32131-17-2	Electrafil® J-1/30/CF/7/H	CAS 32456-28-6	Crodateric O, O.100.
CAS 32131-17-2	EMI-X® PDX-R-89496	CAS 32492-61-8	Dianol® 220
CAS 32131-17-2	EMI-X® RC-1008	CAS 32534-81-9	FR-1205
CAS 32131-17-2	Fiberfil® J-1/30	CAS 32534-81-9	Fyrol® PBR
CAS 32131-17-2	Fiberfil® J-17/30/VO	CAS 32534-81-9	Great Lakes DE-71
CAS 32131-17-2	Fiberfil® NY-16/MF/40	CAS 32536-52-0	FR-1208
CAS 32131-17-2	Fiberstran® G-1/50	CAS 32536-52-0	Great Lakes DE-79
CAS 32131-17-2	Grilon® T300GM	CAS 32536-52-0	Saytex® 111
CAS 32131-17-2	Grilon® TV-15H	CAS 32588-76-4	Saytex® BT-93®
CAS 32131-17-2	Leona®	CAS 32612-48-9	Texapon PNA-127
CAS 32131-17-2	Maranyl® A125	CAS 32647-67-9	Millithix® 925
CAS 32131-17-2	Maranyl® A175S	CAS 32780-64-6	Labrocol®
CAS 32131-17-2	Maranyl® A360	CAS 32780-64-6	Normodyne
CAS 32131-17-2	Maranyl® TA505HS	CAS 32780-64-6	Trandate
CAS 32131-17-2	NN-10GF	CAS 32795-47-4	Merital
CAS 32131-17-2	NN-20CF	CAS 32886-97-8	Coactabs
CAS 32131-17-2	NT-15GF/000	CAS 32887-01-7	Coactin
CAS 32131-17-2	Ny-Kon® R	CAS 32954-43-1	Schercotaine PAB
CAS 32131-17-2	Nybex 22008 BKUT	CAS 32986-56-4	Tobrex
CAS 32131-17-2	Nybex 52000 NA	CAS 33005-95-7	Surgam
CAS 32131-17-2	Nycoa® 500	CAS 33089-61-1	Mitaban
CAS 32131-17-2	Nycoa® 528	CAS 33089-61-1	Mitac 20
CAS 32131-17-2	Nycoa® 5015	CAS 33125-97-2	Amidate
CAS 32131-17-2	Nylatron® 1018 HS	CAS 33125-97-2	Hypnomidate
CAS 32131-17-2	Nylatron® 1024 HS	CAS 33342-05-1	Glurenorm
CAS 32131-17-2	Nylatron® NSB-90	CAS 33386-08-2	Buspar
CAS 32131-17-2	Nylon N-012	CAS 33401-94-4	Banminth
CAS 32131-17-2	PA-111	CAS 33402-03-8	Pressonex Bitartrate
CAS 32131-17-2	PA-111CF30	CAS 33419-42-0	Vepesid
CAS 32131-17-2	PA-111G13	CAS 33564-31-7	Florone
CAS 32131-17-2	PA-121	CAS 33564-31-7	Maxiflor®
CAS 32131-17-2	Plaslube® G-1/30/SI/2	CAS 33659-28-8	Calcibronat
CAS 32131-17-2	Plaslube® J-1/30/MS/5	CAS 33665-90-6	Sunett
CAS 32131-17-2	Plaslube® J-1/33/TF/13/SI/2	CAS 33671-46-4	Clozan
CAS 32131-17-2	Plaslube® J-1/CF/15/TF/20	CAS 33703-08-1	Adimoll® DN
CAS 32131-17-2	Plaslube® J-77/30/TF/15	CAS 33703-08-1	Jayflex® DINA
CAS 32131-17-2	Plaslube® NY-1/MS/5/TF/30	CAS 33703-08-1	Plastomoll® NA
CAS 32131-17-2	Plaslube® NY-1/SI/5	CAS 33817-20-8	Pivatil
CAS 32131-17-2	Plaslube® NY-1/TF/10	CAS 33907-46-9	Naturechem® EGHS
CAS 32131-17-2	Polypenco® Nylon 101	CAS 33907-46-9	Paricin® 15
CAS 32131-17-2	RTP 200FR	CAS 33907-47-0	Naturechem® PGHS
CAS 32131-17-2	Sniamid® SSD 300 EP 021	CAS 33907-47-0	Paricin® 9
CAS 32131-17-2	Sniamid® SSD AF	CAS 33939-64-9	Akypo RLM 100 NV
CAS 32131-17-2	Stat-Kon® PDX-84440	CAS 33939-64-9	Akypo®-Soft 160 NV
CAS 32131-17-2	Stat-Kon® R	CAS 33939-64-9	Rewopol® CLN 100
CAS 32131-17-2	Stat-Kon® RC-1002	CAS 33939-64-9	Sandopan® LS-24
CAS 32131-17-2	Stat-Kon® RF-15	CAS 33939-65-0	Sandopan® KST
CAS 32131-17-2	Texalon 1200A BK-11	CAS 34014-18-1	Bushwacker
CAS 32131-17-2	Texalon 1200A HR-2 BK-16	CAS 34031-32-8	Ridaura
CAS 32131-17-2	Texalon 1308 A	CAS 34097-16-0	Cloderm
CAS 32131-17-2	Texalon GF 1200A (13-40)	CAS 34123-59-6	Arelon
CAS 32131-17-2	Thermocomp® RC-1002	CAS 34123-59-6	Autumn Kite
CAS 32131-17-2	Ultramid® A3	CAS 34123-59-6	Chiltern IPU
CAS 32131-17-2	Ultramid® A3G5	CAS 34123-59-6	Hytane
CAS 32131-17-2	Ultramid® A3K	CAS 34123-59-6	Portman Isotop
CAS 32131-17-2	Ultramid® A3X1G10	CAS 34123-59-6	Power Swing
CAS 32131-17-2	Ultramid® A4	CAS 34123-59-6	Protugan
CAS 32131-17-2	Verton® RF-700-10 EM HS	CAS 34123-59-6	Sabre
CAS 32131-17-2	Verton® RF-7007 HS	CAS 34123-59-6	Tolkan

CAS 34123-59-6	Tolkan 500
CAS 34137-09-2	Vanox® SKT
CAS 34195-34-1	Didrate
CAS 34316-64-8	Cetiol® A
CAS 34360-00-4	Tomah E-14-2
CAS 34424-98-1	Caprol® 10G4O
CAS 34424-98-1	Drewpol® 10-4-O
CAS 34424-98-1	Hodag PGO-104
CAS 34424-98-1	Hodag SVO-1047
CAS 34424-98-1	Mazol® PGO-104
CAS 34443-12-4	Esperox® C-496
CAS 34552-84-6	Maxicam
CAS 34562-31-7	Vanax® 808
CAS 34580-14-8	Zaditen
CAS 34590-94-8	Arcosolv® DPM
CAS 34590-94-8	Icinol DPM
CAS 34745-96-5	Texacat® DD
CAS 34787-01-4	Ticar
CAS 34938-91-8	Dehydol PID 6
CAS 34962-91-9	Dermol 108
CAS 35121-78-9	Flolan
CAS 35141-30-1	CT2910
CAS 35141-30-1	Dynasylan® TRIAMO
CAS 35274-05-6	Ceraphyl 28
CAS 35274-05-6	Liponate CL
CAS 35285-68-8	Nipagin A Sodium
CAS 35285-69-9	Nipasol M Sodium
CAS 35325-02-1	Uniplex 225
CAS 35367-38-5	Dimilin
CAS 35400-43-2	Bolstar
CAS 35400-43-2	Helothion
CAS 35512-33-9	Lentagran
CAS 35545-57-4	Hetoxide BN-13
CAS 35554-44-0	Imaverol
CAS 35575-96-3	Alfacron 10WP
CAS 35604-67-2	Vivalan
CAS 35607-66-0	Mefoxin
CAS 35634-74-3	V-19
CAS 35958-30-6	Isonox® 129
CAS 35958-30-6	Vanox® 1290
CAS 36116-84-4	Doverphos® DIOP
CAS 36116-84-4	Weston® DOPI
CAS 36311-34-9	Ceraphyl ICA
CAS 36311-34-9	Eutanol G16
CAS 36311-34-9	Exxal® 16
CAS 36311-34-9	Michel XO-150-16
CAS 36322-90-4	Felden
CAS 36322-90-4	Feldene
CAS 36330-85-5	Cinopal
CAS 36330-85-5	Lederfen
CAS 36409-57-1	Empicol® SLL
CAS 36409-57-1	Geropon® LSS
CAS 36432-46-9	Weston® DTDP
CAS 36445-71-3	Calfax 10L-45
CAS 36445-71-3	Dowfax 3B2
CAS 36445-71-3	Poly-Tergent® 3B2
CAS 36457-19-9	Nipagin A Potassium
CAS 36457-20-2	Nipabutyl Sodium
CAS 36483-57-5	FR-513
CAS 36521-89-8	Sorbon S-66
CAS 36637-18-0	Duranest
CAS 36653-82-4	Adol® 52 NF
CAS 36653-82-4	Cachalot® C-50
CAS 36653-82-4	Cetaffine®
CAS 36653-82-4	Cetal
CAS 36653-82-4	Crodacol C70
CAS 36653-82-4	Crodacol C95NF
CAS 36653-82-4	Dehydag Wax 16
CAS 36653-82-4	Emery® 3336
CAS 36653-82-4	Epal® 16NF
CAS 36653-82-4	Fancol CA
CAS 36653-82-4	Lanette 16
CAS 36653-82-4	Lipocol C
CAS 36653-82-4	Lorol C16
CAS 36653-82-4	Loxiol VPG 1743
CAS 36653-82-4	Philcohol 1600
CAS 36653-82-4	Rita CA
CAS 36734-19-7	CDA Roval
CAS 36734-19-7	Roval Dust
CAS 36734-19-7	Roval Flo
CAS 36734-19-7	Roval Green
CAS 36734-19-7	Roval WP
CAS 36734-19-7	Turbair Roval
CAS 36788-39-3	Weston® 430
CAS 36791-04-5	Virazole
CAS 36863-52-2	Phospholan® KPE4
CAS 37091-65-9	Securopen®
CAS 37091-66-0	Azlin
CAS 37106-97-1	Chymex
CAS 37139-99-4	Ammonyx® KP
CAS 37139-99-4	Incroquat O-50
CAS 37139-99-4	M-Quat® JO-50
CAS 37139-99-4	Mackernium KP
CAS 37187-22-7	Esperfoam® FR
CAS 37187-22-7	USP®-240
CAS 37199-81-8	Colloid 111D
CAS 37205-87-1	Marlophen® 830N
CAS 37205-87-1	Marlowet® ISM
CAS 37220-82-9	Emerest® 2421
CAS 37251-69-7	Tergitol® D-683
CAS 37294-49-8	Aerosol® A-268
CAS 37296-80-3	Colestid
CAS 37309-58-3	Synton PAO-100
CAS 37318-14-2	Radiasurf® 7423
CAS 37332-99-3	Avotan
CAS 37340-60-6	Rhodafac® LO-529
CAS 37349-34-1	Caprol® 3GS
CAS 37349-34-1	Emalex MSG-2
CAS 37349-34-1	Hodag PGS-101
CAS 37349-34-1	Hodag PGS-61
CAS 37349-34-1	Polyaldo® TGMS KFG
CAS 37349-34-1	Santone® 3-1-S
CAS 37354-45-5	Empicol® SDD
CAS 37475-88-0	ACS 60
CAS 37475-88-0	Eltesol® AC60
CAS 37475-88-0	Reworyl® ACS
CAS 37478-68-5	Crodazoline Cy
CAS 37478-68-5	Mackazoline CY
CAS 37478-68-5	Monazoline CY
CAS 37517-26-3	Piportil
CAS 37517-26-3	Piportil Depot
CAS 37517-30-9	Prent®
CAS 37517-30-9	Sectral
CAS 37661-08-8	Bacacil

CAS Number	Trade Name
CAS 37661-08-8	Daxid
CAS 37661-08-8	Spectrobid
CAS 37661-08-8	Velbacil
CAS 37883-00-4	Viocin
CAS 37883-00-4	Viomycin Sulphate
CAS 37924-13-3	Lancer
CAS 37971-36-1	Bayhibit® AM
CAS 38194-50-2	Arthrobid
CAS 38194-50-2	Artribid
CAS 38194-50-2	Clinoril®
CAS 38194-50-2	Sulindal
CAS 38260-54-7	Satisfar
CAS 38304-91-5	Loniten
CAS 38411-30-3	Marlon® A 365
CAS 38517-23-6	Acylglutamate HS-11
CAS 38566-94-8	Nipabutyl Potassium
CAS 38613-77-3	Alkanox® 24-44
CAS 38613-77-3	Sandostab P-EPQ
CAS 38720-61-5	Akypostat MA 35
CAS 38821-53-3	Anspor
CAS 38821-53-3	Eskacef
CAS 38916-42-6	Aerosol® 22
CAS 39148-24-8	Aliette
CAS 39198-34-0	VR-110
CAS 39236-46-9	Abiol
CAS 39236-46-9	Germall® 115
CAS 39300-45-3	Karathane
CAS 39354-45-5	Aerosol® A-102
CAS 39354-45-5	Schercopol LPS
CAS 39354-45-5	Surfagene S 30
CAS 39354-47-5	Emcol® 4300
CAS 39407-03-9	Marlophor® HS-Acid
CAS 39421-75-5	Galactasol® Guar Derivs
CAS 39421-75-5	Jaguar® 413
CAS 39421-75-5	Jaguar® HP 8
CAS 39421-75-5	Jaguar® HP 60
CAS 39421-75-5	Jaguar® HP-11
CAS 39464-64-7	Chemphos TC-341
CAS 39464-64-7	Lubrhophos® LM-400
CAS 39464-64-7	Rhodafac® RM-410
CAS 39464-66-9	Akypomine® MW 05
CAS 39464-66-9	Chemphos TR-510
CAS 39464-66-9	Marlophor® MO 3-Acid
CAS 39464-66-9	Rewophat EAK 8190
CAS 39464-66-9	Rhodafac® RD-510
CAS 39464-67-0	Lubrhophos® LF-200
CAS 39464-69-2	Brophos OL-3
CAS 39464-69-2	Chemfac PB-184
CAS 39464-69-2	Chemphos TR-505
CAS 39464-69-2	Chemphos TR-515
CAS 39464-69-2	Chemphos TR-541
CAS 39464-69-2	Crodafos N10 Acid
CAS 39464-69-2	Crodafos N3 Acid
CAS 39464-69-2	Crodafos N5 Acid
CAS 39464-69-2	Crodafos O2 Acid
CAS 39464-69-2	Hetphos OA-3
CAS 39464-69-2	Laurelphos 400
CAS 39464-69-2	Lubrhophos® LB-400
CAS 39464-69-2	Rhodafac® RB-400
CAS 39464-70-5	Chemfac PC-006
CAS 39464-70-5	Lubrhophos® LP-700
CAS 39464-70-5	Marlowet® 5324
CAS 39464-70-5	Rhodafac® BP-769
CAS 39515-41-8	Meothrin
CAS 39529-26-5	Caprol® 10G10S
CAS 39529-26-5	Hodag PGS-1010
CAS 39562-70-4	Bayotensin®
CAS 39562-70-4	Baypress®
CAS 39669-97-1	Chemidex P
CAS 39669-97-1	Incromine PB
CAS 39698-78-7	Sarenin
CAS 39831-55-5	Amikin
CAS 40034-42-2	Roxadyl
CAS 40180-04-9	Selacryn
CAS 40372-72-3	Aktisil PF 216
CAS 40372-72-3	CB2494
CAS 40372-72-3	Si69
CAS 40487-42-1	Stomp
CAS 40487-42-1	Stomp H
CAS 40507-23-1	Tormosyl
CAS 40716-42-5	Amidex RC
CAS 40716-42-5	Mackamide R
CAS 40716-42-5	Rodea
CAS 40754-60-7	Geropon® SBR-3
CAS 41083-11-8	Clermait®
CAS 41083-11-8	Peropal®
CAS 41183-64-6	Neoscan
CAS 41340-25-4	Lodine
CAS 41342-53-4	E-Mycin E
CAS 41342-53-4	EES
CAS 41342-53-4	Ery-Ped
CAS 41342-53-4	Erythroped
CAS 41342-53-4	Pediamycin
CAS 41342-53-4	Wyamycin E
CAS 41354-29-4	Periactin
CAS 41394-05-2	Countdown
CAS 41394-05-2	Goltix®
CAS 41395-83-9	DPPG
CAS 41395-83-9	Emerest® 2388
CAS 41395-83-9	Lexol PG 900
CAS 41395-83-9	Schercemol PGDP
CAS 41395-89-5	Wickenol® 152
CAS 41473-08-9	Bilimiro
CAS 41473-08-9	Bilimiron
CAS 41473-08-9	Oravue
CAS 41473-08-9	Videobil
CAS 41483-43-6	Nimrod
CAS 41483-43-6	Nimrod T
CAS 41484-35-9	Anox® 70
CAS 41621-49-2	Loprox
CAS 41637-38-1	CD480
CAS 41669-30-1	Iso Isotearyle WL 3196
CAS 41669-30-1	Schercemol 1818
CAS 41669-30-1	Starfol® IS
CAS 41672-81-5	Lipacide DPHP
CAS 41708-72-9	Tonocard
CAS 41859-67-0	Bezalip
CAS 42116-76-7	Spartrix
CAS 42131-25-9	Wickenol® 151
CAS 42131-28-2	Patlac® IL
CAS 42200-33-9	Corgard
CAS 42461-84-7	Banamine
CAS 42540-40-9	Kefadol
CAS 42540-40-9	Mandol

CAS 42612-52-2	Chemphos TR-510S	CAS 51481-65-3	Mezlin
CAS 42612-52-2	Rhodafac® MC-470	CAS 51630-58-1	Sumicidin
CAS 42808-36-6	Laurel SBT	CAS 51773-92-3	Lariam
CAS 43121-13-3	Monterey Bayleton	CAS 51781-21-6	Cartrol
CAS 43121-43-3	Bayleton®	CAS 51811-79-1	Dextrol OC-20
CAS 43121-43-3	Bayleton® 5	CAS 51811-79-1	Lubrhophos® LE-500
CAS 43154-85-4	Emcol® 4161L	CAS 51811-79-1	Lubrhophos® LE-700
CAS 43154-85-4	Emcol® K8300	CAS 51811-79-1	Tryfac® 5556
CAS 43154-85-4	Mackanate OP	CAS 51812-80-7	Ceraphyl 60
CAS 43154-85-4	Sole Terge 8	CAS 52019-36-0	Chemfac PD-600
CAS 43210-67-9	Panacur	CAS 52152-93-9	Cefomonil
CAS 43222-48-6	Avenge® 2	CAS 52205-73-9	Emcyt
CAS 44992-01-0	Ageflex FA-1Q75MC	CAS 52229-50-2	Gantrez® AN-119
CAS 45267-19-4	Chemidex M	CAS 52292-17-8	Arosurf® 66-E2
CAS 45267-19-4	Schercodine M	CAS 52292-17-8	Emalex 1805
CAS 47747-56-8	Talpen	CAS 52292-17-8	Hetoxol IS-2
CAS 48145-04-6	Ageflex PEA	CAS 52315-75-0	Amihope LL-11
CAS 48165-04-6	Laromer® POEA	CAS 52467-63-7	Varisoft® TC-90
CAS 48165-04-6	Melcril 4087	CAS 52504-24-2	Softigen® 767
CAS 49745-95-1	Dobutrex	CAS 52508-35-7	Atrinal
CAS 50327-22-5	Stanyl® TE200F6	CAS 52556-42-0	Cops I
CAS 50327-22-5	Stanyl® TE300	CAS 52558-73-3	Hamposyl® M
CAS 50327-22-5	Stanyl® TE350	CAS 52581-71-2	Provol 50
CAS 50327-22-5	Stanyl® TQ200F6	CAS 52581-71-2	Volpo 3
CAS 50327-22-5	Stanyl® TW300	CAS 52623-95-7	Laurelphos RH-44
CAS 50471-44-8	Fumite Ronilan	CAS 52645-53-1	Ambush
CAS 50471-44-8	Mascot Contact Turf Fungicide	CAS 52645-53-1	Cooper Coopex
CAS 50471-44-8	Power Drive	CAS 52645-53-1	Darmycel Agarifume Smoke
CAS 50471-44-8	Ronilan®, DF, FL	CAS 52645-53-1	Elimite®
CAS 50471-44-8	Ronilon	CAS 52645-53-1	Fumite Permethrin
CAS 50643-20-4	Brophos 5C10	CAS 52645-53-1	Kafil
CAS 50643-20-4	Crodafos SG	CAS 52645-53-1	Nippon Ant and Crawling Insect
CAS 50643-20-4	Hetphos SG		Killer
CAS 50700-72-6	Norcuron	CAS 52645-53-1	Nippon Ant Killer Powder
CAS 50975-76-3	CT2902	CAS 52645-53-1	Nippon Fly Killer Spray
CAS 51022-69-6	Cyclocort	CAS 52645-53-1	Nippon Ready For Use Ant and
CAS 51022-70-9	Ventolin		Crawling Insect Killer
CAS 51158-08-8	Aldosperse® MS-20 FG	CAS 52645-53-1	Nix Creme Rinse
CAS 51158-08-8	Capmul® EMG	CAS 52645-53-1	Nix Dermal Cream
CAS 51158-08-8	Cutina® E24	CAS 52645-53-1	Nix Dermal Cream
CAS 51158-08-8	Durfax® EOM	CAS 52645-53-1	Perigen
CAS 51158-08-8	Emalex GM-5	CAS 52645-53-1	Permasect
CAS 51158-08-8	Mazol® 80 MG K	CAS 52645-53-1	Picket
CAS 51158-08-8	Tagat® S	CAS 52645-53-1	Quamilin
CAS 51158-08-8	Tagat® S2	CAS 52645-53-1	Turbair Permethrin
CAS 51158-08-8	Varonic® LI-42	CAS 52668-97-0	Marlosol® FS
CAS 51178-59-7	Sipomer® DCPM	CAS 52668-97-0	Marlowet® 4702
CAS 51192-09-7	Tagat® O	CAS 52756-22-6	Commando
CAS 51192-09-7	Tagat® O2	CAS 52756-22-6	Gunner
CAS 51200-87-4	Amine CS-1135®	CAS 52756-22-6	Power Flame
CAS 51200-87-4	Canguard® 327	CAS 52756-22-6	Power Flamprop
CAS 51200-87-4	Oxaban®-A	CAS 52794-79-3	Mackamide ISA
CAS 51235-04-2	Velpar®	CAS 52820-00-5	Decis
CAS 51248-32-9	Tagat® L	CAS 52820-00-5	Thripstick®
CAS 51248-32-9	Tagat® L2	CAS 52829-07-9	Lowilite® 77
CAS 51258-15-2	Adeka GH-200	CAS 52829-07-9	Tinuvin® 770
CAS 51264-14-3	Amsidyl	CAS 52829-07-9	Uvaseb 770
CAS 51274-83-0	Symcor	CAS 52906-84-0	Clorpactin WCS-90
CAS 51333-22-3	Preferid	CAS 52907-07-0	Saytex® BN-451
CAS 51338-27-3	Hoegrass	CAS 53003-10-4	Coxistac
CAS 51481-61-9	Tagamet Tablets, Syrup, Injection	CAS 53003-10-4	Posistac
	and Infusion	CAS 53152-21-9	Buprenex

CAS Number	Trade Name
CAS 53179-09-2	Siseptin
CAS 53179-11-6	Arret
CAS 53179-11-6	Imodium
CAS 53220-22-7	Perkadox® 26-fl
CAS 53320-86-8	Laponite® D
CAS 53320-86-8	Laponite® XLG
CAS 53370-43-7	Viacutan
CAS 53597-27-6	Alnovin
CAS 53610-02-9	Akypo®-Soft 45 NV
CAS 53633-54-8	Gafquat® 734
CAS 53648-55-0	Dalgan
CAS 53716-49-7	Imadyl
CAS 53716-49-7	Rimadyl
CAS 53716-50-0	Synanthic
CAS 53780-34-0	Mowchem
CAS 53879-54-2	CD492
CAS 53894-23-8	Jayflex® TINTM
CAS 53894-23-8	Plasthall® TIOTM
CAS 53894-23-8	PX-339
CAS 53988-10-6	Vanox® MTI
CAS 54045-08-8	Marlipal® SU
CAS 54120-61-5	Synchrocept
CAS 54143-56-5	Tambocor™ Injection
CAS 54143-56-5	Tambocor™ Tablets
CAS 54143-57-6	Maxolon
CAS 54143-57-6	Parmid®
CAS 54143-57-6	Primperan
CAS 54143-57-6	Reglan
CAS 54182-58-0	Antepsin
CAS 54182-58-0	Carafate
CAS 54182-58-0	Ulcerban
CAS 54182-58-0	Ulcogant®
CAS 54340-58-8	Meptid Injection
CAS 54340-58-8	Meptid Tablets
CAS 54350-48-0	Tegison
CAS 54350-48-0	Tigason
CAS 54400-62-3	Styquin
CAS 54667-43-5	Polamine® 650
CAS 54667-43-5	Polamine® 1000
CAS 54667-43-5	Polamine® 2000
CAS 54739-18-3	Fevarin 50
CAS 54739-18-3	Floxifral
CAS 54965-24-1	Noltam
CAS 55028-70-1	Arbacet
CAS 55134-13-9	Monteban
CAS 55179-31-2	Baycor®
CAS 55179-31-2	Sibutol®
CAS 55219-65-3	Bayfidan®
CAS 55219-65-3	Spinnaker
CAS 55219-65-3	Summit
CAS 55285-14-8	Marshal 10G
CAS 55285-14-8	Marshal/suSCon
CAS 55297-96-6	Dynamutilin
CAS 55353-21-4	Resyn® 28-2913
CAS 55635-13-7	Clout
CAS 55779-18-5	Arpocox
CAS 55799-16-1	Nalzin
CAS 55819-53-9	Emcol® 3780
CAS 55819-53-9	Hetamine 5L-25
CAS 55819-53-9	Incromate SDL
CAS 55819-53-9	Lexamine S-13 Lactate
CAS 55837-27-9	Arlix
CAS 55852-13-6	Mackine 321
CAS 55852-15-8	Incromate IDL
CAS 55963-33-2	28-1898
CAS 56002-14-3	Emalex PEIS-3
CAS 56002-14-3	Emerest® 2625
CAS 56002-14-3	Ethox MI-9
CAS 56002-14-3	Olepal ISO
CAS 56002-14-3	Trydet 2644
CAS 56073-07-5	Killgerm® Ratak Cut Wheat Rat Bait
CAS 56073-07-5	Neosorexa
CAS 56073-07-5	Ratak
CAS 56073-10-0	Matikus
CAS 56073-10-0	Mouser
CAS 56073-10-0	Talon
CAS 56187-47-4	Refosporin®
CAS 56235-92-8	Ceraphyl 45
CAS 56265-06-6	Argidone®
CAS 56281-36-8	Tasmaderm
CAS 56377-79-8	Primofax
CAS 56388-43-3	Anionyx® 12S
CAS 56388-43-3	Mackanate OD-35
CAS 56388-43-3	Monamate OPA-100
CAS 56388-43-3	Standapol® SH-100
CAS 56388-43-3	Texapon SH 100
CAS 56391-57-2	Netromycin
CAS 56392-17-7	Beloc
CAS 56392-17-7	Lopresor
CAS 56392-17-7	Lopressor
CAS 56519-71-2	Captex® 800
CAS 56519-71-2	Lexol PG 800
CAS 56803-37-3	Syn-O-Ad® 8478
CAS 56803-37-3	Syn-O-Ad® 8485
CAS 56863-02-6	Schercomid SLE
CAS 56995-20-1	Katadolon
CAS 57018-52-7	Arcosolv® PTB
CAS 57171-56-9	Ethox HO-50
CAS 57171-56-9	Trylox® 6746
CAS 57432-61-8	Methergine
CAS 57569-76-3	Dermol GL-7A
CAS 57635-48-0	Akypo RO 50
CAS 57635-48-0	Akypo RO 90
CAS 57646-30-7	Fongarid
CAS 57754-85-5	Benazalox
CAS 57754-85-5	Dow Shield
CAS 57754-85-5	Format
CAS 57801-81-7	Lendorm
CAS 57808-65-8	Flukiver
CAS 57808-66-9	Motilium
CAS 57834-33-0	Givsorb® UV-1
CAS 57938-82-6	Deracyn
CAS 58068-97-6	Dynasylan® IMEO
CAS 58069-11-7	Dehyquart SP
CAS 58138-08-2	Tandem
CAS 58160-99-9	Union Carbide® A-1106
CAS 58229-88-2	ADK STAB 465
CAS 58229-88-2	Okstan XO
CAS 58353-68-7	Octosol A-1
CAS 58479-61-1	CB2805
CAS 58503-79-0	Rythmatine
CAS 58551-69-2	Prostin/15M
CAS 58767-50-3	Chemstat® 106G/90

CAS 58786-99-5	Stadol
CAS 58786-99-5	Torbutrol
CAS 58855-63-3	Brophos OL-3N
CAS 58855-63-3	Chemphos TR-505D
CAS 58855-63-3	Chemphos TR-515D
CAS 58855-63-3	Crodafos N3 Neutral
CAS 58855-63-3	Crodafos N10 Neutral
CAS 58958-60-4	Ceraphyl 375
CAS 58958-60-4	Dermol 185
CAS 58958-60-4	Schercemol 185
CAS 58965-66-5	Saytex® 120
CAS 59070-56-3	Aldosperse® ML 23
CAS 59122-46-2	Cytotec
CAS 59130-69-7	Bernel® Ester CO
CAS 59130-69-7	Emalex CC-168
CAS 59130-69-7	Schercemol 1688
CAS 59130-69-7	Schercemol CO
CAS 59130-69-7	Tegosoft® CO
CAS 59130-70-7	Schercemol 1688
CAS 59186-41-3	Dehydag Wax E
CAS 59186-41-3	Lanette E
CAS 59186-41-3	Rhodapon® EC111
CAS 59227-89-3	Azone
CAS 59231-34-4	Ceraphyl 140-A
CAS 59231-34-4	Mackester IDO
CAS 59231-34-4	Schercemol IDO
CAS 59231-34-4	Wickenol® 144
CAS 59272-84-3	Schercotaine MAB
CAS 59355-61-2	Empigen® OY
CAS 59467-70-8	Hypnovel
CAS 59559-30-7	Akypo RS 60
CAS 59587-44-9	Bernel® OPG
CAS 59587-44-9	Schercemol OPG
CAS 59587-44-9	Wickenol® 160
CAS 59686-68-9	Cetiol® 1414E
CAS 59686-68-9	Lanol 14 M
CAS 59686-68-9	Schercemol MEM-3
CAS 59703-84-3	Pipracil
CAS 59789-51-4	Actimer FR-1033
CAS 59792-81-3	Aludone®
CAS 59828-07-8	Beta-Air
CAS 59917-39-4	Eldisine
CAS 60166-93-0	Isovue
CAS 60207-90-1	Bumper
CAS 60207-90-1	Power Propiconazole
CAS 60207-90-1	Powerspire
CAS 60207-90-1	Radar
CAS 60207-90-1	Radar Propiconazole
CAS 60209-82-7	Dermol 105
CAS 60270-33-9	Chemidex B
CAS 60270-33-9	Incromine BB
CAS 60270-33-9	Mackine 601
CAS 60270-33-9	Schercodine B
CAS 60561-17-3	Sufenta
CAS 60607-34-3	Tinset
CAS 60628-96-8	Mycospor®
CAS 60719-84-8	Inocor
CAS 60828-78-6	Tergitol® TMN-3
CAS 60925-61-3	Precef
CAS 61128-46-9	Electrafil® J-1106/CF/30
CAS 61128-46-9	PDX-84367
CAS 61128-46-9	PI-20GF/000
CAS 61128-46-9	Thermocomp® EC-1006
CAS 61128-46-9	Ultem® 1000
CAS 61128-46-9	Ultem® 2100
CAS 61128-46-9	Ultem® 2212
CAS 61128-46-9	Ultem® 4000
CAS 61128-46-9	Ultem® 6000
CAS 61128-46-9	Ultem® CRS5001
CAS 61128-46-9	Ultem® CRS5011
CAS 61128-46-9	Ultem® FXU100
CAS 61197-93-1	Dormonoct
CAS 61270-78-8	Monocid
CAS 61318-91-0	Exelderm
CAS 61368-34-1	Actimer FR-803
CAS 61477-96-1	Pipril
CAS 61551-69-7	VA-086
CAS 61570-90-9	Tiox
CAS 61617-00-3	Vanox® ZMTI
CAS 61632-57-3	Magala® 0.5E
CAS 61682-73-3	Liponate PB-4
CAS 61693-08-1	Luvitol HP
CAS 61693-08-1	Panalane® L-14E
CAS 61788-40-7	Hypan® SA100H
CAS 61788-45-2	Amine HBG
CAS 61788-45-2	Amine HBGD
CAS 61788-45-2	Armeen® HT
CAS 61788-45-2	Armeen® HTD
CAS 61788-45-2	Crodamine 1.HT
CAS 61788-45-2	Jet Amine PHT
CAS 61788-45-2	Kemamine® P-970
CAS 61788-45-2	Radiamine 6140
CAS 61788-45-2	Radiamine 6141
CAS 61788-46-3	Amine KK
CAS 61788-46-3	Armeen® C
CAS 61788-46-3	Armeen® CD
CAS 61788-46-3	Jet Amine PC
CAS 61788-46-3	Kemamine® P-650D
CAS 61788-46-3	Radiamine 6160
CAS 61788-46-3	Radiamine 6161
CAS 61788-48-5	Acelan L
CAS 61788-48-5	Acetadeps
CAS 61788-48-5	Acylan
CAS 61788-48-5	Fancol Acel
CAS 61788-48-5	Modulan®
CAS 61788-48-5	Ritacetyl®
CAS 61788-49-6	Hetlan AC
CAS 61788-59-8	Emery® 2253
CAS 61788-59-8	Radia® 7117
CAS 61788-62-3	Armeen® M2C
CAS 61788-62-3	Jet Amine M2C
CAS 61788-62-3	Kemamine® T-6501
CAS 61788-62-3	Radiamine 6360
CAS 61788-63-4	Amine M2HBG
CAS 61788-63-4	Armeen® M2HT
CAS 61788-63-4	Kemamine® T-9701
CAS 61788-63-4	Radiamine 6343
CAS 61788-85-0	Akyporox CO 400
CAS 61788-85-0	Alkamuls® COH-5
CAS 61788-85-0	Arlatone® G
CAS 61788-85-0	Chemax HCO-200/50
CAS 61788-85-0	Chemax HCO-5
CAS 61788-85-0	Cremophor® RH 40
CAS 61788-85-0	Cremophor® RH 60

CAS 61788-85-0	Cremophor® WO 7
CAS 61788-85-0	Croduret 10
CAS 61788-85-0	Dehymuls HRE 7
CAS 61788-85-0	Emalex HC-5
CAS 61788-85-0	Emulsogen HEL-050
CAS 61788-85-0	Ethox HCO-16
CAS 61788-85-0	Eumulgin HRE 40
CAS 61788-85-0	Fancol HCO-25
CAS 61788-85-0	Hetoxide HC-16
CAS 61788-85-0	Hodag Nonionic GRH-25
CAS 61788-85-0	Mapeg® CO-16H
CAS 61788-85-0	Nopalcol 10-COH
CAS 61788-85-0	Prox-onic HRH-05
CAS 61788-85-0	Remcopal HC 7
CAS 61788-85-0	Remcopal HC 20
CAS 61788-85-0	Remcopal HC 33
CAS 61788-85-0	Remcopal HC 40
CAS 61788-85-0	Remcopal HC 60
CAS 61788-85-0	Simulsol 989
CAS 61788-85-0	Simulsol 1292
CAS 61788-85-0	Tagat® R40
CAS 61788-85-0	Tagat® R60
CAS 61788-85-0	Tagat® R63
CAS 61788-85-0	Trylox® 5921
CAS 61788-90-7	Barlox® 12
CAS 61788-90-7	Chemoxide WC
CAS 61788-90-7	Empigen® 5083
CAS 61788-90-7	Genaminox CS
CAS 61788-90-7	Genaminox KC
CAS 61788-90-7	Karox AO-30
CAS 61788-90-7	Mackamine CO
CAS 61788-90-7	Naxide 1230
CAS 61788-90-7	Schercamox DMC
CAS 61788-91-8	Armeen® DMSD
CAS 61788-91-8	Jet Amine DMSD
CAS 61788-93-0	Amine 2M1218D
CAS 61788-93-0	Amine 2MKKD
CAS 61788-93-0	Armeen® DMCD
CAS 61788-93-0	Jet Amine DMCD
CAS 61788-93-0	Kemamine® T-6502D
CAS 61788-95-2	Amine 2MHBGD
CAS 61788-95-2	Armeen® DMHTD
CAS 61789-05-7	Imwitor® 928
CAS 61789-08-0	Myverol® 18-50K
CAS 61789-10-4	Alphadim® 90LC
CAS 61789-10-4	Myverol® 18-40
CAS 61789-10-4	Tegomuls® 19
CAS 61789-13-7	Myverol® 18-30
CAS 61789-18-2	Adogen® 461
CAS 61789-18-2	Arquad® C-33W
CAS 61789-18-2	Arquad® C-50
CAS 61789-18-2	Chemquat C/33W
CAS 61789-18-2	Jet Quat C-50
CAS 61789-18-2	Marlazin® KC 30/50
CAS 61789-18-2	Varisoft® 461
CAS 61789-30-8	Norfox® 1101
CAS 61789-40-0	Dehyton® PK
CAS 61789-40-0	Incronam 30
CAS 61789-40-0	Rewoteric® AM B-13
CAS 61789-40-0	Schercotaine CAB
CAS 61789-40-0	Tego® -Betaine C
CAS 61789-40-0	Tego® -Betaine L-7
CAS 61789-40-0	Tego® -Betaine L-5351
CAS 61789-68-2	Ethoquad® CB/12
CAS 61789-71-7	Alkaquat® DMB-451-50, DMB-451-80
CAS 61789-71-7	Arquad® DMCB-80
CAS 61789-71-7	Marlazin® KC 21/50
CAS 61789-71-7	Querton GCI-50
CAS 61789-71-7	Querton KKBCI-50
CAS 61789-72-8	Arquad® DMHTB-75
CAS 61789-72-8	Kemamine® BQ-9702C
CAS 61789-72-8	Querton 441-BC
CAS 61789-73-9	Arquad® M2HTB-80
CAS 61789-73-9	Kemamine® BQ-9701C
CAS 61789-75-1	Kemamine® BQ-9742C
CAS 61789-76-2	Armeen® 2C
CAS 61789-76-2	Radiamine 6260
CAS 61789-77-3	Accoquat 2C-75
CAS 61789-77-3	Accoquat 2C-75H
CAS 61789-77-3	Adogen® 462
CAS 61789-77-3	Arquad® 2C-75
CAS 61789-77-3	Dye Retarder #1
CAS 61789-77-3	Jet Quat 2C-75
CAS 61789-77-3	Kemamine® Q-6502C
CAS 61789-77-3	M-Quat® 2475
CAS 61789-77-3	Radiaquat 6462
CAS 61789-77-3	Tomah Q-2C
CAS 61789-77-3	Varisoft® 462
CAS 61789-79-5	Amine 2 VT
CAS 61789-79-5	Amine 2HBG
CAS 61789-79-5	Armeen® 2HT
CAS 61789-79-5	Kemamine® S-970
CAS 61789-80-8	Adogen® 442
CAS 61789-80-8	Adogen® 442-P100
CAS 61789-80-8	Arquad® 2HT-75
CAS 61789-80-8	Kemamine® Q-9702C
CAS 61789-80-8	M-Quat® 257
CAS 61789-80-8	Querton 442
CAS 61789-80-8	Radiaquat 6442
CAS 61789-80-8	Radiaquat 6475
CAS 61789-80-8	Radiaquat 6480
CAS 61789-86-4	Lubrizol® 2152
CAS 61789-86-4	TLA-256
CAS 61789-91-1	Emalex J.J. O-V
CAS 61789-91-1	Lipovol J
CAS 61789-91-1	Simchin, Natural
CAS 61789-91-1	Wickenol® 139
CAS 61790-12-3	Acintol® 736
CAS 61790-12-3	Acintol® 2122
CAS 61790-12-3	Acintol® D25LR
CAS 61790-12-3	Acintol® D30E
CAS 61790-12-3	Acintol® DFA
CAS 61790-12-3	Acintol® EPG
CAS 61790-12-3	Acintol® FA-1
CAS 61790-12-3	Acintol® FA-2
CAS 61790-12-3	Acintol® FA-3
CAS 61790-12-3	Acofor
CAS 61790-12-3	Oulu 102
CAS 61790-12-3	Pamolyn
CAS 61790-12-3	Xtol
CAS 61790-18-9	Armeen® S
CAS 61790-18-9	Jet Amine PS
CAS 61790-31-6	Armid® HT

CAS 61790-33-8	Adogen® 170
CAS 61790-33-8	Amine BG
CAS 61790-33-8	Armeen® T
CAS 61790-33-8	Armeen® TD
CAS 61790-33-8	Crodamine 1.T
CAS 61790-33-8	Genamin TA Grades
CAS 61790-33-8	Jet Amine PT
CAS 61790-33-8	Radiamine 6170
CAS 61790-33-8	Radiamine 6171
CAS 61790-33-8	Tallow Amine
CAS 61790-35-0	Actrasol SS
CAS 61790-35-0	Eureka 392
CAS 61790-37-2	Industrene® 143
CAS 61790-38-3	Glycon® S-65
CAS 61790-38-3	Petrac® PHTA
CAS 61790-38-3	Prifac 9428
CAS 61790-41-8	Adogen® 417
CAS 61790-41-8	Arquad® S-50
CAS 61790-41-8	Jet Quat S-50
CAS 61790-41-8	Tomah Q-S
CAS 61790-50-9	Arizona DRS-40
CAS 61790-50-9	Arizona DRS-50
CAS 61790-50-9	Diprosin K-80
CAS 61790-51-0	Arizona DRS-43
CAS 61790-51-0	Diprosin N-70
CAS 61790-57-6	Acetamin C
CAS 61790-57-6	Amine Acetate KK
CAS 61790-57-6	Radiamac 6169
CAS 61790-59-8	Acetamin HT
CAS 61790-59-8	Amine Acetate HBG
CAS 61790-59-8	Radiamac 6149
CAS 61790-63-4	Marlamid® DF 1218
CAS 61790-64-5	Akypogene FP 35 T
CAS 61790-81-6	Aqualose L30
CAS 61790-81-6	Ethoxylan® 1685
CAS 61790-81-6	Ivarlan 3400
CAS 61790-81-6	Ivarlan 3406
CAS 61790-81-6	Laneto 27
CAS 61790-81-6	Lanogel® 21
CAS 61790-81-6	Lanolex L-40
CAS 61790-81-6	Solan
CAS 61790-81-6	Solan 50
CAS 61790-81-6	Solangel 401
CAS 61790-81-6	Solulan® 75
CAS 61790-81-6	Super Solan Flaked
CAS 61790-85-0	Ethoduomeen® T/13
CAS 61790-85-0	Ethoduomeen® T/20
CAS 61790-85-0	Ethoduomeen® T/25
CAS 61791-00-2	Aconol X6
CAS 61791-00-2	Actrol 6M25P
CAS 61791-00-2	Chemax TO-8
CAS 61791-00-2	Chemax TO-10
CAS 61791-00-2	Chemax TO-16
CAS 61791-00-2	EM-600
CAS 61791-00-2	Ethofat® 242/25
CAS 61791-00-2	Ethox TO-8
CAS 61791-00-2	Genagen TA-080
CAS 61791-00-2	Hetoxamate FA-5
CAS 61791-00-2	Industrol® TO-16
CAS 61791-00-2	Laurel PEG 400 MT
CAS 61791-00-2	Mapeg® 200 MOT
CAS 61791-00-2	Mapeg® TAO-15
CAS 61791-00-2	Marlosol® TF3
CAS 61791-00-2	Prox-onic TA-1/08
CAS 61791-00-2	Sellig T 3 100
CAS 61791-00-2	Sellig T 14 100
CAS 61791-00-2	Sellig T 1790
CAS 61791-00-2	Trydet 2682
CAS 61791-01-3	Ethox DTO-9A
CAS 61791-01-3	Laurel PEG 400 DT
CAS 61791-01-3	Mapeg® 200 DOT
CAS 61791-01-3	Pegosperse® 400 DOT
CAS 61791-08-0	Alkamide® C-5
CAS 61791-08-0	Amide CMA-2
CAS 61791-08-0	Amidox® C-2
CAS 61791-08-0	Empilan® LP10
CAS 61791-08-0	Empilan® MAA
CAS 61791-08-0	Eumulgin C4
CAS 61791-08-0	Eumulgin PC 2
CAS 61791-08-0	Genagen CA-050
CAS 61791-08-0	Hetoxamide C-4
CAS 61791-08-0	Rewopal® C 6
CAS 61791-08-0	Unamide® C-5
CAS 61791-10-4	Ethoquad® C/25
CAS 61791-10-4	Tomah Q-C-15
CAS 61791-10-4	Variquat® K1215
CAS 61791-12-6	Ablunol CO 10
CAS 61791-12-6	Acconon CA-5
CAS 61791-12-6	Acconon CA-9
CAS 61791-12-6	Acconon CA-15
CAS 61791-12-6	Akyporox RZO 30
CAS 61791-12-6	Alkamuls® 14/R
CAS 61791-12-6	Alkamuls® B
CAS 61791-12-6	Alkamuls® EL-620
CAS 61791-12-6	Berol 108
CAS 61791-12-6	Berol 191
CAS 61791-12-6	Berol 195
CAS 61791-12-6	Berol 198
CAS 61791-12-6	Berol 199
CAS 61791-12-6	Berol 829
CAS 61791-12-6	Chemax CO-5
CAS 61791-12-6	Chemax CO-16
CAS 61791-12-6	Cremophor® EL
CAS 61791-12-6	Emalex C-20
CAS 61791-12-6	Emulsogen EL-050
CAS 61791-12-6	Ethox CO-5
CAS 61791-12-6	Eumulgin PRT 36
CAS 61791-12-6	Eumulgin RO 40
CAS 61791-12-6	Fancol CO-30
CAS 61791-12-6	Hetoxide C-2
CAS 61791-12-6	Hetoxide C-200-50%
CAS 61791-12-6	Hodag Nonionic GR-8
CAS 61791-12-6	Incrocas 30
CAS 61791-12-6	Makon® 8240
CAS 61791-12-6	Mapeg® CO-5
CAS 61791-12-6	Marlosol® R70
CAS 61791-12-6	Marlowet® R 11
CAS 61791-12-6	Marlowet® R 40
CAS 61791-12-6	Polymene AZ
CAS 61791-12-6	Prox-onic HR-05
CAS 61791-12-6	Remcopal 40
CAS 61791-12-6	Remcopal 40 S3
CAS 61791-12-6	Remcopal 4000
CAS 61791-12-6	Remcopal 4018

CAS 61791-12-6	Ricinion
CAS 61791-12-6	Sellig HR 18 100
CAS 61791-12-6	Sellig R 3395
CAS 61791-12-6	Sellig R 3395 SP
CAS 61791-12-6	Sellig R 3395-C435
CAS 61791-12-6	Sellig R 4095
CAS 61791-12-6	Sellig R 4495
CAS 61791-12-6	Sellig R 20 100
CAS 61791-12-6	Stepantex® CO-30
CAS 61791-12-6	Surfactol® 318
CAS 61791-12-6	T-Det® C-40
CAS 61791-12-6	Toximul® 8240
CAS 61791-12-6	Trylox® 5900
CAS 61791-12-6	Witconol 5906
CAS 61791-13-7	Dehydol LT 3
CAS 61791-13-7	Genapol® C-050
CAS 61791-13-7	Genapol® GC-050
CAS 61791-14-8	Ablumox C-7
CAS 61791-14-8	Accomeen C2
CAS 61791-14-8	Accomeen C5
CAS 61791-14-8	Accomeen C10
CAS 61791-14-8	Accomeen C15
CAS 61791-14-8	Alkaminox® C-2
CAS 61791-14-8	Berol 307
CAS 61791-14-8	Berol 397
CAS 61791-14-8	Chemeen C-2
CAS 61791-14-8	Ethomeen® C/12
CAS 61791-14-8	Ethomeen® C/15
CAS 61791-14-8	Ethomeen® C/20
CAS 61791-14-8	Ethomeen® C/25
CAS 61791-14-8	Ethox CAM-2
CAS 61791-14-8	Ethylan® TN-10
CAS 61791-14-8	Eumulgin PA 12
CAS 61791-14-8	Hetoxamine C-2
CAS 61791-14-8	Mazeen® C-2
CAS 61791-14-8	Prox-onic MC-02
CAS 61791-14-8	Teric 12M2
CAS 61791-14-8	Trymeen® 6601
CAS 61791-14-8	Varonic® K202
CAS 61791-14-8	Zusomin C 108
CAS 61791-20-6	Aqualose W20
CAS 61791-20-6	Fancol LA-5
CAS 61791-20-6	Polychol 5
CAS 61791-20-6	Ritawax 5
CAS 61791-24-0	Accomeen S2
CAS 61791-24-0	Accomeen S10
CAS 61791-24-0	Accomeen S15
CAS 61791-24-0	Chemeen S-2
CAS 61791-24-0	Ethomeen® S/12
CAS 61791-24-0	Ethomeen® S/15
CAS 61791-24-0	Ethomeen® S/20
CAS 61791-24-0	Ethomeen® S/25
CAS 61791-24-0	Hetoxamine S-2
CAS 61791-24-0	Mazeen® S-2
CAS 61791-24-0	Mazeen® S-15
CAS 61791-24-0	Teric 16M2
CAS 61791-24-0	Tomah E-S-2
CAS 61791-25-1	Mirataine® TM
CAS 61791-25-1	Rewoteric® AM TEG
CAS 61791-26-2	Berol 381
CAS 61791-26-2	Berol 386
CAS 61791-26-2	Berol 391
CAS 61791-26-2	Berol 457
CAS 61791-26-2	Chemeen HT-5
CAS 61791-26-2	Ethomeen® T/15
CAS 61791-26-2	Ethomeen® T/25
CAS 61791-26-2	Prox-onic MHT-015
CAS 61791-26-2	Rhodameen® PN-430
CAS 61791-26-2	Rhodameen® T-5
CAS 61791-26-2	Rhodameen® VP-532/SPB
CAS 61791-26-2	Sipenol IT 50-4G
CAS 61791-26-2	Varonic® U-215
CAS 61791-28-4	Dehydol PTA 7
CAS 61791-28-4	Dehydol TA 11
CAS 61791-28-4	Oxetal TG 111
CAS 61791-28-4	Rewopal® TA 11
CAS 61791-29-5	Crodet C10
CAS 61791-29-5	Eumulgin PK 23
CAS 61791-29-5	Genagen C-100
CAS 61791-29-5	Nonex C5E
CAS 61791-29-5	Nopalcol 4-C
CAS 61791-29-5	Prox-onic CC-05
CAS 61791-31-9	Berol 307
CAS 61791-31-9	Chemstat® 273-C
CAS 61791-31-9	Ethokem C/12
CAS 61791-31-9	Ethomeen® C/12
CAS 61791-31-9	Norfox® DC
CAS 61791-32-0	Rewoteric® AM 2C NM
CAS 61791-38-6	Chemzoline C-22
CAS 61791-38-6	Mackazoline C
CAS 61791-38-6	Monazoline C
CAS 61791-38-6	Schercozoline C
CAS 61791-38-6	Sovatex EP 5288
CAS 61791-38-6	Sovatex IM12H
CAS 61791-38-6	Unamine® C
CAS 61791-38-6	Varine C
CAS 61791-39-7	Chemzoline T-44
CAS 61791-39-7	Hodag C-100-T
CAS 61791-39-7	Miramine® HPS-B
CAS 61791-39-7	Miramine® TO
CAS 61791-39-7	Monazoline T
CAS 61791-39-7	Textamine T-1
CAS 61791-39-7	Varine T
CAS 61791-41-1	Fenopon TK32
CAS 61791-41-1	Geropon® TK-32
CAS 61791-42-2	Geropon® TC-42
CAS 61791-44-4	Ablumox T-15
CAS 61791-44-4	Accomeen T2
CAS 61791-44-4	Accomeen T5
CAS 61791-44-4	Accomeen T15
CAS 61791-44-4	Agrisorb
CAS 61791-44-4	Berol 456
CAS 61791-44-4	Chemeen T-2
CAS 61791-44-4	Clifton Glyphosate Additive
CAS 61791-44-4	Ethomeen® T/12
CAS 61791-44-4	Ethox TAM-2
CAS 61791-44-4	Ethylan® TT-15
CAS 61791-44-4	Eumulgin PA 10
CAS 61791-44-4	Exell
CAS 61791-44-4	Frigate
CAS 61791-44-4	Genamin T-020
CAS 61791-44-4	Hetoxamine T-2
CAS 61791-44-4	Hyspray
CAS 61791-44-4	Icomeen® T-2

CAS 61791-44-4	Icomeen® T-15
CAS 61791-44-4	Jogral
CAS 61791-44-4	Lo-Dose
CAS 61791-44-4	Marlazin® T 10
CAS 61791-44-4	Mazeen® T-2
CAS 61791-44-4	Power Spray Save
CAS 61791-44-4	Prox-onic MT-02
CAS 61791-44-4	Team
CAS 61791-44-4	Teric 17M2
CAS 61791-44-4	Tomah E-T-2
CAS 61791-44-4	Topup
CAS 61791-44-4	Toximul® TA-2
CAS 61791-44-4	Trymeen® 6606
CAS 61791-44-4	Trymeen® TAM-15
CAS 61791-44-4	Varonic® T202
CAS 61791-44-4	Varonic® T-220
CAS 61791-44-4	Wayfarer
CAS 61791-44-4	Witcamine® 6606
CAS 61791-44-4	Zusomin TG 102
CAS 61791-46-6	Chemoxide T
CAS 61791-47-7	Aromox® C/12-W
CAS 61791-47-7	Schercamox CMA
CAS 61791-53-5	Duomeen® TDO
CAS 61791-55-7	Diamine BG
CAS 61791-55-7	Duomeen® T
CAS 61791-55-7	Jet Amine DT
CAS 61791-55-7	Tallow Diamine
CAS 61791-56-8	Deriphat 154
CAS 61791-56-8	Mirataine® T2C-30
CAS 61791-56-8	Monateric TDB-35
CAS 61791-57-9	Amine 660
CAS 61791-57-9	Jet Amine TRT
CAS 61791-57-9	Tallow Triamine
CAS 61791-57-9	Triameen T
CAS 61791-59-1	Closyl 30 2089
CAS 61791-59-1	Hampfoam 35
CAS 61791-59-1	Hamposyl® C-30
CAS 61791-59-1	Medialan KA
CAS 61791-59-1	Vanseal® 35
CAS 61791-59-1	Vanseal® NACS-30
CAS 61791-63-7	Diamine KKP
CAS 61791-63-7	Duomeen® C
CAS 61791-63-7	Duomeen® CD
CAS 61791-63-7	Jet Amine DC
CAS 61791-63-7	Radiamine 6560
CAS 61792-31-2	Mackamine LAO
CAS 61827-42-7	Rhodasurf® DA-530
CAS 61840-27-5	Cartaretin F-4
CAS 61849-72-7	Glucam® P-10
CAS 61849-72-7	Glucam® P-20
CAS 61849-72-7	Prox-onic MG-020 P
CAS 61901-02-8	Miranol® HM-SF Conc
CAS 61970-18-9	Armeen® SD
CAS 62265-68-3	Amenide
CAS 62449-33-6	Norfox® IM-38
CAS 62476-59-9	Blazer®
CAS 62571-86-2	Capoten
CAS 62610-77-9	Damfin
CAS 62893-20-3	Cefobid
CAS 62893-20-3	Cefobine
CAS 62893-20-3	Cefobis
CAS 62893-20-3	Pathozone

CAS 63123-11-5	Eastman® Inhibitor Poly TDP 2000
CAS 63123-11-5	Eastman® Poly TDP 2000
CAS 63148-55-0	Alkasil® NE 58-50
CAS 63148-56-1	PS181
CAS 63148-57-2	CPS120
CAS 63148-62-9	CPS034
CAS 63148-62-9	Crystal 1000
CAS 63217-13-0	Geropon® WS-25, WS-25-I
CAS 63231-60-7	Be Square® 185
CAS 63231-60-7	Forbest MW 23
CAS 63231-60-7	Fortex®
CAS 63231-60-7	Mekon® White
CAS 63231-60-7	Multiwax® 180-M
CAS 63231-60-7	Multiwax® HS
CAS 63231-60-7	Paracol® 404C
CAS 63231-60-7	Polymekon®
CAS 63231-60-7	Starwax® 100
CAS 63231-60-7	Ultraflex®
CAS 63231-60-7	Victory®
CAS 63284-71-9	Triminol
CAS 63323-46-6	Cindol
CAS 63393-82-8	Neodol® 25
CAS 63393-93-1	Amerlate® P
CAS 63393-93-1	Fancor IPL
CAS 63393-93-1	Lanesta S
CAS 63393-93-1	Ritasol
CAS 63394-02-5	B-2
CAS 63394-02-5	B-182
CAS 63394-02-5	C-715u
CAS 63394-02-5	C-920u
CAS 63394-02-5	COHRlastic® 400
CAS 63394-02-5	COHRlastic® 1867
CAS 63394-02-5	COHRlastic® 3320
CAS 63394-02-5	COHRlastic® 9041
CAS 63394-02-5	COHRlastic® R10450
CAS 63394-02-5	COHRlastic® TC100
CAS 63394-02-5	D1-SEA 210 Silicone
CAS 63394-02-5	Elastosil® LR 3001
CAS 63394-02-5	Elastosil® LR 3003/20
CAS 63394-02-5	Laur 101A
CAS 63394-02-5	Laur 676U
CAS 63394-02-5	Laur Q-1330
CAS 63394-02-5	LIM6045
CAS 63394-02-5	Norsil RTV 811
CAS 63394-02-5	Pensil® 100
CAS 63394-02-5	Rhodorsil® RS 44, RS 48
CAS 63394-02-5	RTF 762
CAS 63394-02-5	RTV 511
CAS 63394-02-5	RTV 615
CAS 63394-02-5	Silastic® 21145
CAS 63394-02-5	Silastic® GP-30
CAS 63394-02-5	Silastic® GP-950+
CAS 63394-02-5	Silastic® HE-26
CAS 63394-02-5	Silastic® HS-30
CAS 63394-02-5	Silastic® LCS-740
CAS 63394-02-5	Silastic® LT-40
CAS 63394-02-5	Silastic® NPC-40
CAS 63394-02-5	Silastic® SPG-30
CAS 63394-02-5	Silastic® TR-55
CAS 63394-02-5	Silastic® WC-50
CAS 63394-02-5	SWS-06545u

CAS 63394-02-5	SWS-7532u
CAS 63394-02-5	SWS-7655u
CAS 63394-02-5	SWS-7675u
CAS 63394-02-5	SWS-7865u
CAS 63451-23-0	Crodateric Cy
CAS 63675-72-9	Baymycard
CAS 63675-72-9	Syscor
CAS 63793-60-2	Carsonon® 144-P
CAS 63793-60-2	Hetoxol MP-3
CAS 63793-60-2	Promyristyl PM-3
CAS 63793-60-2	Witconol APM
CAS 64019-93-8	Propine®
CAS 64024-15-3	Fortral
CAS 64024-15-3	Talwin
CAS 64092-49-5	Zomax
CAS 64228-81-5	Tracrium Injection
CAS 64228-81-5	Tractium
CAS 64318-79-2	Cergem
CAS 64318-79-2	Cervagem
CAS 64485-93-4	Claforan
CAS 64490-92-2	Tolectin
CAS 64628-44-0	Alsystin
CAS 64665-57-2	Maxahibit TT-50
CAS 64741-81-7	Viplex 895-BL
CAS 64741-92-0	Varsol® 18
CAS 64742-04-7	Viplex 680-P
CAS 64742-06-9	HAN® 857
CAS 64742-14-9	Jayflex® 215
CAS 64742-14-9	Penreco 2251 Oil
CAS 64742-47-8	Exxsol® D-40
CAS 64742-47-8	Exxsol® D-60
CAS 64742-47-8	Exxsol® D-80
CAS 64742-47-8	Exxsol® D-110
CAS 64742-47-8	Exxsol® D-130
CAS 64742-47-8	Isopar® M
CAS 64742-48-9	Isopar® C
CAS 64742-48-9	Isopar® E
CAS 64742-48-9	Isopar® G
CAS 64742-48-9	Isopar® H
CAS 64742-48-9	Isopar® L
CAS 64742-48-9	Varsol® 18
CAS 64742-49-0	Exxsol® Hexane
CAS 64742-53-6	Jayflex® 210
CAS 64742-89-8	Exxsol® Heptane
CAS 64743-02-8	Neodene® 1420
CAS 64743-02-8	Neodene® 1624
CAS 64743-02-8	Neodene® 2024
CAS 64771-72-8	Norpar® 12
CAS 64924-67-0	Stenorol
CAS 64953-12-4	Moxam
CAS 65009-35-0	Lidarral
CAS 65071-95-6	Ethofat® 242/25
CAS 65271-80-9	Novantrone
CAS 65277-42-1	Nizoral
CAS 65381-09-1	Captex® 300
CAS 65381-09-1	Emalex K.T.G
CAS 65381-09-1	Labrafac Lipophile WL 1349
CAS 65381-09-1	Lexol GT 855, GT 865
CAS 65381-09-1	Liponate GC
CAS 65381-09-1	Mazol® 1400
CAS 65381-09-1	Miglyol® 812
CAS 65381-09-1	Myritol 318
CAS 65381-09-1	Neobee® M-5
CAS 65381-09-1	Tegosoft® CT
CAS 65447-77-0	Tinuvin® 622LD
CAS 65473-14-5	Exoderil
CAS 65473-14-5	Naftin®
CAS 65497-29-2	Cosmedia Guar C-261 N
CAS 65497-29-2	Hi-Care® 1000
CAS 65497-29-2	Jaguar® C-13S, C-14S
CAS 65497-29-2	N-Hance® 3000
CAS 65591-14-2	Waxenol® 801
CAS 65816-20-8	Givsorb® UV-2
CAS 65876-95-1	Penacolite® B-18-S
CAS 65996-61-4	CF 1500
CAS 66009-41-4	Crodamol W
CAS 66009-41-4	Tegosoft® SH
CAS 66063-05-6	Monceren®
CAS 66063-05-6	Trotis
CAS 66085-00-5	Imwitor® 780 K
CAS 66085-00-5	Schercemol GMIS
CAS 66085-59-4	Nimotop®
CAS 66104-23-2	Permax
CAS 66108-95-0	Omnipaque
CAS 66197-78-2	Crafol AP-53
CAS 66441-23-4	Cheetah R
CAS 66455-14-9	Neodol® 23-6.5
CAS 66455-15-0	Marlipal® KF
CAS 66455-17-2	Neodol® 91
CAS 66455-29-6	Amphoteen 24
CAS 66455-29-6	Empigen® BB
CAS 66592-87-8	Duricef
CAS 66592-87-8	Ultracef
CAS 66722-44-9	Concor®
CAS 66734-13-2	Vaderm
CAS 66794-58-9	Crillet 6
CAS 66794-58-9	Montanox 70
CAS 66794-74-9	Enkade
CAS 66841-24-5	Ambush C
CAS 66841-24-5	Ashlade Halt
CAS 66841-24-5	Chemtech Cypermethrin
CAS 66841-24-5	Chiltern Cyperkill 10
CAS 66841-24-5	Cymbush
CAS 66841-24-5	Cymperator
CAS 66841-24-5	Cyperkill
CAS 66841-24-5	Cypersect
CAS 66841-24-5	Cypertox
CAS 66841-24-5	FAL Cypermethrin 10
CAS 66841-24-5	Quadrangle Cyper
CAS 66841-24-5	Topclip Parasol
CAS 66841-24-5	Toppel
CAS 66988-04-3	Crolactil SISL
CAS 66988-04-3	Pationic ISL
CAS 67306-00-7	Patrol
CAS 67306-03-0	Corbel®
CAS 67306-03-0	Mistral
CAS 67306-03-0	Power Task
CAS 67306-03-0	Sprint
CAS 67375-30-8	Fastac
CAS 67633-57-2	Monaquat ISIES
CAS 67633-57-2	Schercoquat IIS
CAS 67633-59-4	Incroquat I-85
CAS 67633-63-0	Foamquat IAES
CAS 67633-63-0	M-Quat® 522

CAS 67633-63-0	Schercoquat IAS
CAS 67700-98-5	Amine 2M12D
CAS 67700-98-5	Empigen® AB
CAS 67701-00-2	Armeen® 3-16
CAS 67701-05-7	Hystrene® 1835
CAS 67701-26-2	Softenol® 3100
CAS 67701-27-3	Neustrene® 059
CAS 67701-30-8	Radia® 7363
CAS 67701-32-0	Alphadim® 90NLK
CAS 67701-33-1	Alphadim® 90AB
CAS 67702-21-4	Empicol® EGB
CAS 67702-21-4	Empicol® EGC
CAS 67702-21-4	Texapon MG
CAS 67702-21-4	Zoharpon MgES
CAS 67747-09-5	Mirage
CAS 67747-09-5	Octave
CAS 67747-09-5	Prelude
CAS 67747-09-5	Sporak
CAS 67747-09-5	Sporgon
CAS 67747-09-5	Sportak
CAS 67747-09-5	Sportak Alpha
CAS 67747-09-5	Sprint
CAS 67762-19-0	Avirol® AE 3003
CAS 67762-19-0	Calfoam NEL-60
CAS 67762-19-0	Carsonol® SES-A
CAS 67762-19-0	DeSonol AE
CAS 67762-19-0	Empicol® EAA
CAS 67762-19-0	Empicol® EAB
CAS 67762-19-0	Empicol® EAC
CAS 67762-19-0	Nonasol N4AS
CAS 67762-19-0	Nutrapon AL 1
CAS 67762-19-0	Nutrapon AL 60
CAS 67762-19-0	Polystep® B-11
CAS 67762-19-0	Rhodapex® AB-20
CAS 67762-19-0	Rhodapex® EA
CAS 67762-19-0	Rhodapex® EAY
CAS 67762-19-0	Standapol® EA-1
CAS 67762-19-0	Steol® CA-460
CAS 67762-19-0	Sulfochem EA-1
CAS 67762-19-0	Sulfochem EA-2
CAS 67762-19-0	Sulfochem EA-3
CAS 67762-19-0	Sulfochem EA-60
CAS 67762-19-0	Sulfochem EA-70
CAS 67762-19-0	Sulfotex OT
CAS 67762-19-0	Texapon EA-1
CAS 67762-19-0	Texapon NA
CAS 67762-19-0	Ungerol AM3-75
CAS 67762-19-0	Witcolate AE
CAS 67762-19-0	Witcolate LES-60A
CAS 67762-19-0	Zoharpon LAEA 253
CAS 67762-36-1	Industrene® 365
CAS 67762-36-1	Philacid 0810
CAS 67762-38-3	Radia® 7060
CAS 67762-39-4	Emery® 2209
CAS 67762-41-8	Exxal® L1315
CAS 67762-96-3	Tego® Foamex 3062
CAS 67774-74-7	Marlican®
CAS 67784-77-4	Ethoquad® T/12
CAS 67784-87-6	Myverol® 18-04K
CAS 67784-90-1	Miramine® C
CAS 67799-04-6	Chemidex SI
CAS 67799-04-6	Mackine 401
CAS 67799-04-6	Mazeen® DAPI
CAS 67799-04-6	Schercodine I
CAS 67846-16-6	Schercoquat SAS
CAS 67892-37-9	Schercoteric O-AA
CAS 67923-14-2	CPS340
CAS 67990-17-4	Miranate® B
CAS 67998-94-1	Sul-fon-ate OA-5R
CAS 68002-44-8	Vibrin® E-085
CAS 68002-59-5	Varisoft® DHT
CAS 68002-61-9	Radiaquat 6471
CAS 68002-71-1	Neustrene® 064
CAS 68002-72-2	Neustrene® 045
CAS 68002-79-9	Radia® 7370
CAS 68002-79-9	Radia® 7371
CAS 68002-97-1	Empilan® KB 2
CAS 68002-97-1	Rhodasurf® L-4
CAS 68003-46-3	Hamposyl® AL-30
CAS 68037-49-0	Chemstat® PS-101
CAS 68037-87-6	CPS925
CAS 68037-92-3	Amine B11
CAS 68037-93-4	Amine 2M16D
CAS 68037-97-8	Radiamine 6572
CAS 68039-13-4	Lexquat® 2240
CAS 68071-35-2	Crodafos 25 D2 Acid
CAS 68081-96-9	Empicol® AL30
CAS 68081-97-0	Empicol® ML 26/F
CAS 68083-14-7	Silikophen® P 40/W
CAS 68092-28-4	Emulamid TO-21
CAS 68122-86-1	Varisoft® TIMS
CAS 68128-59-6	Octosol A-18-A
CAS 68130-24-5	Liponate DPC-6
CAS 68130-43-8	Akypomine® BC 50
CAS 68130-47-2	Rhodafac® RA-600
CAS 68131-37-3	Maltrin® M200
CAS 68131-37-3	Maltrin® QD M600
CAS 68131-39-5	Bio-Soft® E-400
CAS 68131-39-5	Mulsifan RT 203/80
CAS 68131-39-5	Neodol® 25-3
CAS 68131-39-5	Rhodasurf® 25-7
CAS 68131-39-5	Rhodasurf® LA-3
CAS 68131-39-5	Teric 12A2
CAS 68131-39-5	Teric G12A4
CAS 68131-39-5	Teric LA4
CAS 68131-39-5	Volpo 25 D 3
CAS 68131-39-5	Volpo 25 D 10
CAS 68131-40-8	Tergitol® 15-S-3
CAS 68133-13-1	Weston® 494
CAS 68139-30-0	Amonyl 675 SB
CAS 68139-30-0	Chembetaine CAS
CAS 68139-30-0	Crosultaine C-50
CAS 68139-30-0	Lexaine® CSB-50
CAS 68139-30-0	Lonzaine® CS
CAS 68139-30-0	Mafo® CSB
CAS 68139-30-0	Mirataine® BET-CS
CAS 68139-30-0	Rewoteric® AM CAS
CAS 68139-30-0	Rewoteric® AM CAS-15
CAS 68139-30-0	Sandobet SC
CAS 68139-30-0	Schercotaine SCAB
CAS 68140-00-1	Ablumide CME
CAS 68140-00-1	Adeka Sole YA
CAS 68140-00-1	Alkamide® C-212
CAS 68140-00-1	Amidex CME

CAS 68140-00-1	Amidex KME
CAS 68140-00-1	Carsamide® CMEA
CAS 68140-00-1	Comperlan P 100
CAS 68140-00-1	Emid® 6500
CAS 68140-00-1	Empilan® CME
CAS 68140-00-1	Foamole M
CAS 68140-00-1	Incromide CM
CAS 68140-00-1	Mackamide CMA
CAS 68140-00-1	Marlamid® M 1218
CAS 68140-00-1	Mazamide® CFAM
CAS 68140-00-1	Monamide
CAS 68140-00-1	Monamid® CMA
CAS 68140-00-1	Ninol® CMP
CAS 68140-00-1	Ninol® CNR
CAS 68140-00-1	Rewomid® C 212
CAS 68140-00-1	Varamide® C-212
CAS 68140-00-1	Zoramide CM
CAS 68140-01-2	Chemidex C
CAS 68140-01-2	Chemidex WC
CAS 68140-01-2	Incromine CB
CAS 68140-01-2	Lexamine C-13
CAS 68140-01-2	Mackine 101
CAS 68140-01-2	Mazeen® DAPL
CAS 68140-01-2	Mazeen® SHCFA
CAS 68140-01-2	Miramine® CODI
CAS 68140-01-2	Schercodine C
CAS 68140-08-9	Amidex TD
CAS 68140-98-7	Alkaterge®-E
CAS 68153-28-6	Promax® HV
CAS 68153-28-6	Promosoy® 20/60
CAS 68153-32-2	Radiaquat 6470
CAS 68153-63-9	Marlamid® M 1618
CAS 68153-64-0	Nopalcol 1-TW
CAS 68154-36-9	Radiasurf® 7125
CAS 68154-97-2	Marlox® FK 64
CAS 68155-09-9	Ablumox CAPO
CAS 68155-09-9	Aminoxid WS 35
CAS 68155-09-9	Amyx CDO 3599
CAS 68155-09-9	Barlox® C
CAS 68155-09-9	Chemoxide CAW
CAS 68155-09-9	Empigen® OS/A
CAS 68155-09-9	Incromine Oxide C
CAS 68155-09-9	Mackamine CAO
CAS 68155-09-9	Mazox® CAPA
CAS 68155-09-9	Monalux CAO
CAS 68155-09-9	Ninox® FCA
CAS 68155-09-9	Patogen AO-30
CAS 68155-09-9	Rewominox B 204
CAS 68155-09-9	Rhodamox® CAPO
CAS 68155-09-9	Schercamox C-AA
CAS 68155-09-9	Standamox CAW
CAS 68155-09-9	Standamox PCAW
CAS 68155-09-9	Tegamine® Oxide WS-35
CAS 68155-09-9	Varox® 1770
CAS 68155-20-4	Monamine T-100
CAS 68155-20-4	Schercomid SO-T
CAS 68155-20-4	Schercomid TO-2
CAS 68155-24-8	Ethomid® HT/23
CAS 68155-24-8	Ethomid® HT/60
CAS 68171-33-5	Schercemol 318
CAS 68171-38-0	Emerest® 2384
CAS 68171-38-0	Hydrophilol ISO

CAS 68173-73-7	Texlin® 500
CAS 68184-04-3	Akyposal MLES 35
CAS 68186-14-1	Abalyn®
CAS 68186-34-5	Rhodafac® GB-520
CAS 68187-29-1	Acylglutamate CT-12
CAS 68187-32-6	Acylglutamate CS-11
CAS 68188-30-7	Chemidex SO
CAS 68188-30-7	Mackine 901
CAS 68201-46-7	Cetiol® HE
CAS 68201-46-7	Rewoderm® ES 90
CAS 68201-46-7	Tegosoft® GC
CAS 68201-46-7	Varonic® LI-63
CAS 68201-49-0	Albalan
CAS 68201-49-0	Fancor Lanwax
CAS 68201-49-0	Lanfrax® 1776
CAS 68201-49-0	Lanocerin®
CAS 68238-35-7	Crotein HKP
CAS 68238-35-7	Kera-Tein AA
CAS 68238-35-7	Keramino 25
CAS 68238-87-9	Emsorb® 2518
CAS 68239-42-9	Glucam® E-10
CAS 68239-43-0	Glucam® E-20
CAS 68299-16-1	Esperox® 545M
CAS 68299-16-1	Lupersol 546-M75
CAS 68299-16-1	Trigonox® 123-C75
CAS 68299-17-2	Rhodapon® CAV
CAS 68304-37-6	CD4368
CAS 68308-22-5	Hostalub® VP Ca W 2
CAS 68308-54-3	Imwitor® 191
CAS 68308-67-8	Larostat® 88
CAS 68308-67-8	M-Quat® 1033
CAS 68310-73-6	Duoquad® O-50
CAS 68334-00-9	C-Flakes
CAS 68334-00-9	Duratex
CAS 68334-00-9	Duromel
CAS 68334-00-9	Duromel B108
CAS 68334-00-9	Emvelop®
CAS 68334-00-9	Lubritab®
CAS 68334-00-9	Sta-Nut EE
CAS 68334-21-4	Empigen® CDR40
CAS 68334-21-4	Schercoteric MS
CAS 68334-28-1	Aratex
CAS 68334-28-1	BBS
CAS 68334-28-1	Cirol
CAS 68334-28-1	Creamtex
CAS 68334-28-1	Durex 500
CAS 68334-28-1	Durko
CAS 68334-28-1	Durlite F
CAS 68334-28-1	Hydrokote® 95
CAS 68334-28-1	K.L.X
CAS 68334-28-1	Kaokote F
CAS 68334-28-1	Kaola
CAS 68334-28-1	Kaomel
CAS 68334-28-1	Kaorich Beads
CAS 68334-28-1	Kaorich Gold
CAS 68334-28-1	Lipo SS
CAS 68334-28-1	Lipodan SET Kosher
CAS 68334-28-1	Magna A
CAS 68334-28-1	Optima 23B
CAS 68334-28-1	Sterotex®
CAS 68334-28-1	Sterotex® HM
CAS 68334-28-1	Wecobee® M

CAS 68359-37-5	Baythroid®
CAS 68373-14-8	Betamaze® IM/IV
CAS 68373-14-8	Unasyn® Oral
CAS 68389-70-8	Glucamate® SSE-20
CAS 68391-01-5	Arquad® B-100
CAS 68391-03-7	Querton 280
CAS 68391-07-1	Kemamine® T-9742D
CAS 68401-82-1	Cefizox
CAS 68411-00-7	Neodene® 1112
CAS 68411-19-8	Vanax® 833
CAS 68411-20-1	Vanox® AT
CAS 68411-97-2	Hamposyl® C
CAS 68411-97-2	Vanseal® CS
CAS 68412-53-3	Rhodafac® PE-510
CAS 68412-53-3	Rhodafac® RE-610
CAS 68412-54-4	Berol 02
CAS 68412-54-4	Berol 09
CAS 68412-54-4	Berol WASC
CAS 68424-43-1	Amerlate® LFA
CAS 68424-43-1	Amerlate® WFA
CAS 68424-43-1	Fancor LFA
CAS 68424-43-1	Ritalafa®
CAS 68424-43-1	Skliro Distilled
CAS 68424-59-9	Fancol Karite Extract
CAS 68424-60-2	Cetiol® SB45
CAS 68424-60-2	Fancol Karite Butter
CAS 68424-60-2	Lipex 102
CAS 68424-60-2	Shebu, Refined
CAS 68424-61-3	Radiasurf® 7150
CAS 68424-66-8	Hydroxylan
CAS 68424-66-8	Ivarlan OH
CAS 68424-66-8	OHlan®
CAS 68424-66-8	Ritahydrox
CAS 68424-94-2	Accobetaine CL
CAS 68424-94-2	Amphoteen BCM-30
CAS 68424-94-2	Chembetaine BW
CAS 68424-94-2	Chembetaine CB
CAS 68424-94-2	Incronam CD-30
CAS 68424-94-2	Lonzaine® 12C
CAS 68424-94-2	Mackam CB-35
CAS 68424-94-2	Mafo® CB 40
CAS 68424-94-2	Velvetex® AB-45
CAS 68424-95-3	FMB 302-8 Quat
CAS 68425-37-6	Philcohol 1214
CAS 68425-42-3	Incromate CDL
CAS 68425-43-4	Emcol® 1655
CAS 68425-43-4	Foamid 117
CAS 68425-43-4	Incromate CDP
CAS 68425-47-8	Alkamide® DIN-295/S
CAS 68425-47-8	Alkamide® SDO
CAS 68425-47-8	Amidex S
CAS 68425-47-8	Mackamide S
CAS 68425-47-8	Marlamid® DF 1818
CAS 68425-47-8	Schercomid SLS
CAS 68425-47-8	Stamid LS 5487
CAS 68425-50-3	Chemidex T
CAS 68439-39-4	Marlophor® DS-Acid
CAS 68439-45-2	Empilan® KA10/80
CAS 68439-46-3	Berol 260
CAS 68439-46-3	Neodol® 91-2.5
CAS 68439-46-3	Rhodasurf® 91-6
CAS 68439-46-3	Teric 9A2
CAS 68439-46-3	Teric G9A5
CAS 68439-49-6	Acconon W230
CAS 68439-49-6	Cetomacrogol 1000 BP
CAS 68439-49-6	Cremophor® A 11
CAS 68439-49-6	Cremophor® A 25
CAS 68439-49-6	Dehydol PCS 6
CAS 68439-49-6	Empilan® KM 11
CAS 68439-49-6	Emthox® 5885
CAS 68439-49-6	Emulgator E 2568 SE
CAS 68439-49-6	Eumulgin B2
CAS 68439-49-6	Hetoxol 15 CSA
CAS 68439-49-6	Hetoxol CS-4
CAS 68439-49-6	Hostacerin T-3
CAS 68439-49-6	Incropol CS-20
CAS 68439-49-6	Lipocol SC-4
CAS 68439-49-6	Macol® CSA-2
CAS 68439-49-6	Marlipal® 1618/6
CAS 68439-49-6	Marlowet® 4800
CAS 68439-49-6	Marlowet® FOX
CAS 68439-49-6	Marlowet® PW
CAS 68439-49-6	Plurafac® A-38
CAS 68439-49-6	Plurafac® A-39
CAS 68439-49-6	Prox-onic CSA-1/04
CAS 68439-49-6	Remcopal 229
CAS 68439-49-6	Remcopal 238
CAS 68439-49-6	Rhodasurf® A-1P
CAS 68439-49-6	Ritacet-20
CAS 68439-49-6	Sellig SU 4 100
CAS 68439-49-6	Sellig SU 18 100
CAS 68439-49-6	Sellig SU 25 100
CAS 68439-49-6	Sellig SU 30 100
CAS 68439-49-6	Sellig SU 50 100
CAS 68439-49-6	Simulsol CS
CAS 68439-49-6	Teric 16A16
CAS 68439-49-6	Varonic® 63 E20
CAS 68439-49-6	Volpo CS-3
CAS 68439-50-9	Genapol® 24-L-3
CAS 68439-50-9	Genapol® 42-L-3
CAS 68439-50-9	Marlipal® 24/20
CAS 68439-50-9	Surfonic® L24-2
CAS 68439-50-9	Tergitol® 24-L-45
CAS 68439-51-0	Marlox® MO 124
CAS 68439-53-2	Hetoxol PLA
CAS 68439-53-2	Solulan® PB-2
CAS 68439-53-2	Wickenol® 727
CAS 68439-57-6	Bio-Terge® AS-40
CAS 68439-57-6	Carsonol® AOS
CAS 68439-57-6	Nansa® LSS38/A
CAS 68439-57-6	Norfox® ALPHA XL
CAS 68439-57-6	Polystep® A-18
CAS 68439-57-6	Rhodacal® 301-10
CAS 68439-57-6	Witconate AOK
CAS 68439-70-3	Amine 2M14D
CAS 68439-73-6	Kemamine® D-974
CAS 68439-73-6	Radiamine 6570
CAS 68440-05-1	Empilan® CIS
CAS 68440-66-4	Silwet® L-7500
CAS 68440-90-4	CPS140
CAS 68440-90-4	PS140
CAS 68458-58-8	Fluilan AWS
CAS 68458-58-8	Ivarlan AWS
CAS 68458-58-8	Lantrol® AWS 1692

CAS 68458-58-8	Ritalan® AWS
CAS 68458-58-8	Vigilan AWS
CAS 68458-88-8	Ivarlan 3420
CAS 68458-88-8	Laneto AWS
CAS 68458-88-8	Lanexol AWS
CAS 68476-03-9	Hoechst Wax S
CAS 68476-38-0	Hostalub® WE4
CAS 68477-29-2	Viplex 885
CAS 68479-64-1	Schercopol OMS-Na
CAS 68484-43-1	Fancol WGFA
CAS 68511-41-1	Jet Amine PE 1214
CAS 68513-95-1	Emcosoy®
CAS 68513-95-1	Soyafluff® 200 W
CAS 68513-95-1	Soyarich® 115 W
CAS 68515-47-9	Jayflex® DTDP
CAS 68515-47-9	Jayflex® UDP
CAS 68515-48-0	Jayflex® DINP
CAS 68515-48-0	Palatinol® DN
CAS 68515-48-0	Palatinol® N
CAS 68515-48-0	PX-109
CAS 68515-49-1	Bisoflex 100
CAS 68515-49-1	Jayflex® DIDP
CAS 68515-49-1	Kodaflex® DIDP
CAS 68515-49-1	NLA-30
CAS 68515-49-1	Palatinol® DIDP
CAS 68515-49-1	Palatinol® Z
CAS 68515-49-1	PX-120
CAS 68515-50-4	Jayflex® DHP
CAS 68515-65-1	Mackanate CP
CAS 68515-65-1	Monamate C-1142
CAS 68515-73-1	APG® 225 Glycoside
CAS 68515-73-1	Burco NPS-225
CAS 68515-73-1	Glucopon 225
CAS 68515-73-1	Glucopon 425
CAS 68516-06-3	Propomeen C/12
CAS 68526-79-4	Exxal® 6
CAS 68526-83-0	Exxal® 8
CAS 68526-84-1	Exxal® 9
CAS 68526-85-2	Exxal® 10
CAS 68526-86-3	Exxal® 12
CAS 68526-86-3	Exxal® 13
CAS 68527-05-9	Neoflex® 9
CAS 68527-24-2	Escopol® R-020
CAS 68551-12-2	Genapol® 26-L-1
CAS 68551-13-3	Empilan® KCMP 0703/F
CAS 68551-14-4	Tergitol® Min-Foam 1X
CAS 68554-09-6	Propoquad® 2HT/11
CAS 68554-53-0	Abil®-Wax 2434
CAS 68554-53-0	Belsil SDM 6021
CAS 68554-65-4	Silwet® L-720
CAS 68555-36-2	Mirapol® A-15
CAS 68555-98-6	Vultac® 2
CAS 68584-22-5	Nansa® 1042
CAS 68584-22-5	Nansa® SSA
CAS 68584-24-7	Nansa® YS94
CAS 68584-25-8	Nansa® TS 50
CAS 68585-05-7	Stasoft J
CAS 68585-34-2	Empicol® ESA
CAS 68585-34-2	Empicol® ESB
CAS 68585-34-2	Empicol® ESC/AU
CAS 68585-34-2	Empimin® SQ25
CAS 68585-44-4	Empicol® 0031/T
CAS 68585-44-4	Empicol® DA
CAS 68585-47-4	Empicol® LX
CAS 68586-07-2	MB 450
CAS 68603-25-8	Antarox® BL-214
CAS 68603-42-9	Alkamide® 101 CG
CAS 68603-42-9	Alkamide® CDE
CAS 68603-42-9	Aminol KDE
CAS 68603-42-9	Schercomid CDA
CAS 68603-42-9	Schercomid SCE
CAS 68603-64-5	Diamine HBG
CAS 68603-64-5	Kemamine® D-970
CAS 68603-64-5	Radiamine 6540
CAS 68604-44-4	Radia® 7171
CAS 68604-71-7	Miranol® C2M-SF 70%
CAS 68604-71-7	Miranol® FBS
CAS 68604-73-9	Miranol® CS Conc
CAS 68604-73-9	Sandoteric CFL
CAS 68604-73-9	Schercoteric MS-EP
CAS 68607-29-4	Duoquad® T-50
CAS 68607-29-4	Jet Quat DT-50
CAS 68607-75-0	Abil®-Wax 9809
CAS 68607-75-0	CPS130
CAS 68607-75-0	PS130
CAS 68608-61-7	Miranol® SM Conc
CAS 68608-63-9	Miranol® DM Conc. 45%
CAS 68608-64-0	Ampholak XJO
CAS 68608-64-0	Miranol® J2M Conc
CAS 68608-64-0	Miranol® JB
CAS 68608-65-1	Ampholak XCO-30
CAS 68608-65-1	Miranol® CM Conc. NP
CAS 68608-65-1	Miranol® FA-NP
CAS 68608-66-2	Empigen® CDL60
CAS 68608-66-2	Miranol® H2M Conc
CAS 68608-66-2	Miranol® HM Conc
CAS 68608-88-8	Nansa® SBA
CAS 68610-26-4	Tomah PA-19
CAS 68610-38-8	Miranol® OS-D
CAS 68610-38-8	Sandoteric TFL Conc
CAS 68610-39-9	Miranol® JS Conc
CAS 68610-43-5	Miranol® H2M-SF Conc
CAS 68610-51-5	Ultranox® 257
CAS 68610-62-8	Weston® 474
CAS 68630-75-1	Suprefact
CAS 68630-96-6	Monateric ISA-35
CAS 68630-96-6	Schercoteric I-AA
CAS 68647-16-5	Smithol PEG Adipate
CAS 68647-53-0	Dehyton® G
CAS 68647-53-0	Miranol® 2CIB
CAS 68647-73-4	EmCon TEA TREE
CAS 68647-77-8	Chemoxide TAO
CAS 68648-27-1	Fancol HL-20
CAS 68648-27-1	Ivarlan 3450
CAS 68648-27-1	Lipolan 31
CAS 68648-27-1	Satexlan 20
CAS 68648-27-1	Super-Sat AWS-4
CAS 68648-66-8	Ceraphyl GA
CAS 68648-87-3	Alkylate 215
CAS 68649-29-6	Rhodafac® PL-620
CAS 68650-39-5	Dehyton® PG
CAS 68650-39-5	Miranol® C2M Conc. NP-PG
CAS 68650-39-5	Miranol® FB-NP
CAS 68650-39-5	Schercoteric MS-2

CAS 68650-79-3	Schercodine T
CAS 68715-87-7	Duomeen® OTM
CAS 68783-22-2	Rhodameen® HT-50
CAS 68783-22-2	Schercomid HT-60
CAS 68783-24-4	Armeen® 2T
CAS 68783-24-4	Radiamine 6270
CAS 68783-25-5	Duomeen® TTM
CAS 68783-78-8	Arquad® 2T-75
CAS 68784-08-7	Schercopol CMS-Na
CAS 68797-31-9	Ecostatin Cream
CAS 68797-31-9	Ecostatin Lotion
CAS 68797-31-9	Ecostatin Pessaries
CAS 68797-31-9	Ecostatin Powder Solution
CAS 68797-31-9	Spectazole
CAS 68797-65-9	Schercoquat BAS
CAS 68813-55-8	Telopar
CAS 68814-69-7	Amine 2MBGD-M
CAS 68814-69-7	Amine 2MOLD
CAS 68814-69-7	Armeen® DMTD
CAS 68814-69-7	Jet Amine DMTD
CAS 68815-23-6	Neutral Degras
CAS 68815-45-2	Miranol® S2M-SF Conc
CAS 68815-55-4	Amphoterge® KJ-2
CAS 68815-55-4	Miranol® J2M-SF Conc
CAS 68815-55-4	Miranol® JBS
CAS 68815-55-4	Monateric 811
CAS 68815-55-4	Monateric 1000
CAS 68815-55-4	Zoharteric LF-SF
CAS 68815-56-5	Geropon® SBFA-30
CAS 68844-77-9	Hismanal
CAS 68877-55-4	Miranol® JAS-50
CAS 68890-66-4	Octopirox
CAS 68890-92-6	Incrosul LAFS
CAS 68890-92-6	Rewolan® 5
CAS 68891-17-8	Akypo ITD 30 N
CAS 68891-17-8	Sandopan® DTC-100
CAS 68891-21-4	Berol 269
CAS 68891-38-3	Berol 452
CAS 68891-39-4	Rhodapex® CO-433
CAS 68901-05-3	Ageflex TPGDA
CAS 68901-05-3	Laromer® TPGDA
CAS 68908-44-1	Empicol® LQ33/T
CAS 68908-44-1	Empicol® TA40
CAS 68909-20-6	Cab-O-Sil® TS-530
CAS 68910-56-5	Varisoft® 2TD
CAS 68911-61-5	Epal® 20+
CAS 68911-79-5	Amine 740
CAS 68911-79-5	Jet Amine TET
CAS 68911-79-5	Tallow Tetramine
CAS 68915-25-3	Croquat HH
CAS 68915-25-3	Croquat WKP
CAS 68915-25-3	Kera-Quat WKP
CAS 68915-25-3	Quat Keratin WKP
CAS 68918-77-4	Lamepon PA-TR
CAS 68918-77-4	Lexein® A520
CAS 68919-40-4	Ampholak YCO-40
CAS 68919-40-4	Miranol® C2M Anhyd. Acid
CAS 68919-41-5	Amphoterge® K
CAS 68919-41-5	Mackam CSF
CAS 68919-41-5	Miranol® CM-SF Conc
CAS 68919-41-5	Monateric CA-35
CAS 68919-41-5	Monateric CAM-40
CAS 68920-65-0	Foam-Coll 4C
CAS 68920-65-0	Lamepon S
CAS 68920-65-0	May-Tein C
CAS 68920-65-0	Maypon 4C
CAS 68920-65-0	Monteine LCK-32
CAS 68920-66-1	Marlowet® FOX
CAS 68921-83-5	Ceraphyl 70
CAS 68936-95-8	Glucate® SS
CAS 68937-41-7	Fyrquel® EHC
CAS 68937-41-7	Kronitex® 25
CAS 68937-41-7	Kronitex® 50
CAS 68937-41-7	Kronitex® 100
CAS 68937-41-7	Kronitex® 200
CAS 68937-41-7	Kronitex® 1840
CAS 68937-54-2	Tego® Flow 425
CAS 68937-54-2	Tego® Foamex 800
CAS 68937-55-3	Abil® B 8851
CAS 68937-55-3	Abil® B 8852
CAS 68937-55-3	Abil® B 88183
CAS 68937-55-3	Abil® B 88184
CAS 68937-55-3	Silwet® L-7200
CAS 68937-90-6	Hystrene® 5460
CAS 68938-15-8	Hystrene® 5012
CAS 68938-15-8	Industrene® 223
CAS 68938-54-5	Silwet® L-7600
CAS 68938-54-5	Silwet® L-7605
CAS 68938-54-5	Silwet® L-7622
CAS 68951-72-4	Propomeen T/12
CAS 68951-89-3	Crotein ASC
CAS 68951-89-3	Pro-Tein ES-20
CAS 68951-92-8	Lamepon UD
CAS 68951-92-8	Maypon UD
CAS 68952-15-8	Pro-Tein SA-20
CAS 68952-16-9	Bio-Soft® MT 40
CAS 68952-16-9	Foam-Coll 4CT
CAS 68952-16-9	Lamepon ST40
CAS 68952-16-9	May-Tein CT
CAS 68952-16-9	Maypon 4CT
CAS 68952-16-9	Monteine LCT
CAS 68952-35-2	Kronitex® TCP
CAS 68952-98-7	Akwilox 133
CAS 68953-11-7	Foamole B
CAS 68953-58-2	Tixogel VP
CAS 68953-64-0	Ceraphyl 65
CAS 68953-96-8	Nansa® EVM50
CAS 68954-89-2	Akypo RCS 60
CAS 68954-89-2	Akypo RLM 25
CAS 68954-89-2	Akypo RLM 130
CAS 68954-89-2	Akypo RLMQ 38
CAS 68954-89-2	Akypo RS 60
CAS 68954-89-2	Akypo RS 100
CAS 68954-89-2	Akypo RT 60
CAS 68955-19-1	Empicol® LM
CAS 68955-19-1	Empicol® LMV/T
CAS 68955-19-1	Empicol® LZ
CAS 68955-19-1	Empicol® LZV
CAS 68955-20-4	Empicol® TAS30
CAS 68955-45-3	Radia® 7506
CAS 68956-08-1	Kessco® 887
CAS 68957-18-6	Akyposal DS 28
CAS 68957-18-6	Akyposal DS 56
CAS 68958-64-5	Tagat® TO

CAS 68966-38-1	Monazoline IS		CAS 70356-09-1	Parsol® 1789
CAS 68966-38-1	Schercozoline I		CAS 70458-96-7	Barazan
CAS 68987-89-3	Akypo NTS		CAS 70458-96-7	Noroxin
CAS 68987-89-3	Akypo®-Soft 100 NV		CAS 70458-96-7	Utinor
CAS 68989-00-4	Empigen® BCB50		CAS 70458-96-7	Zoroxin
CAS 68989-03-7	Rewoquat CPEM		CAS 70496-39-8	Incromide BED
CAS 68990-05-6	Axol® C 62		CAS 70592-80-2	Empigen® OB
CAS 68990-06-7	Lamegin® GLP 10, 20		CAS 70609-66-4	Hostapon KTW New
CAS 68990-58-9	Lamegin® EE		CAS 70632-06-3	Miranate® LEC
CAS 68990-59-0	Acidan N 12		CAS 70632-06-3	Sandopan® DTC Linear P
CAS 68990-59-0	Lamegin® ZE 30, 60		CAS 70693-04-8	Michel XO-150-1620
CAS 68990-63-6	Super Refined Shark Oil		CAS 70693-05-9	Exxal® 26
CAS 68990-82-9	Paramount B		CAS 70693-32-2	Liponate NPGC-2
CAS 68990-82-9	Witarix® 212		CAS 70729-87-2	Parapel® LIS
CAS 68991-88-8	Miranol® L2M-SF Conc		CAS 70750-46-8	Ampholeen BTH-35
CAS 68991-88-8	Miranol® TBS		CAS 70750-47-9	Ethoquad® C/12
CAS 69009-90-1	Nusolv ABP-103		CAS 70750-47-9	Variquat® 638
CAS 69011-84-3	Akyposal BD		CAS 70788-37-3	Eltesol® TSX
CAS 69013-18-9	Surfonic® LF-17		CAS 70851-07-9	Ampholeen BCA-30
CAS 69198-10-3	Clont®		CAS 70851-07-9	Mirataine® BET-C-30
CAS 69198-10-3	Flagyl IV		CAS 70851-08-0	Mirataine® BSC
CAS 69227-21-0	Marlox® MS 48		CAS 70851-08-0	Mirataine® CBS, CBS Mod
CAS 69331-39-1	Amphisol		CAS 70879-28-6	Rapifen
CAS 69331-39-1	Surfagene FGD 600		CAS 70914-20-4	Exxal® 7
CAS 69430-24-6	Abil® B 8839		CAS 71010-52-1	Gelrite®
CAS 69430-24-6	Abil® K 4		CAS 71010-52-1	Kelco-Gel® Gellan Gum
CAS 69430-24-6	Amersil® VS-7207		CAS 71060-61-2	Propomeen 2HT-11
CAS 69430-24-6	Belsil CM 020		CAS 71060-72-5	Arquad® 316(W)
CAS 69430-24-6	Dow Corning® 344 Fluid, 345 Fluid		CAS 71172-17-3	Vanax® CPA
			CAS 71247-25-1	Tymtran
CAS 69430-24-6	Fancorsil A		CAS 71329-50-5	Jaguar® C-162
CAS 69430-24-6	Masil® SF-V		CAS 71486-48-1	Uniplex 310
CAS 69430-24-6	Sentry Cyclomethicone		CAS 71487-00-8	Ethoquad® C/12 Nitrate
CAS 69430-24-6	SF1173		CAS 71487-01-9	Arquad® 2C-70 Nitrite
CAS 69430-24-6	SWS-03314		CAS 71566-49-9	Dermol 89
CAS 69430-36-0	Crotein ASK		CAS 71566-49-9	Kessco® Octyl Isononanoate
CAS 69430-36-0	Crotein K		CAS 71864-46-5	CD5610
CAS 69430-36-0	Crotein WKP		CAS 71888-89-6	Jayflex® 77
CAS 69430-36-0	Hydrokeratin AL-30		CAS 71902-01-7	Crill 6
CAS 69430-36-0	Kera-Tein 1000		CAS 71902-01-7	Emalex SPIS-100
CAS 69430-36-0	Nutrilan® Keratin W		CAS 71902-01-7	Emsorb® 2516
CAS 69468-44-6	Tagat® I		CAS 72160-13-5	Akypo OP 80.
CAS 69468-44-6	Tagat® I2		CAS 72160-13-5	Akypo OP 190.
CAS 69537-38-8	Akypoquat 131		CAS 72162-46-0	Tomah DA-14
CAS 69633-04-1	Garlon		CAS 72269-52-4	Dermol 489
CAS 69633-04-1	Garlon 2		CAS 72319-06-3	Lexein® A200
CAS 69633-04-1	Garlon 4		CAS 72319-06-3	Pro-Tein SM-20
CAS 69633-04-1	Timbrel		CAS 72347-89-8	Caprol® 2G4S
CAS 69712-56-7	Apatef		CAS 72558-82-8	Tazidime
CAS 69806-50-4	Fusilade		CAS 72869-62-6	Radia® 7345
CAS 70024-77-0	Ampholak XOO-30P		CAS 73049-73-7	European Elastin 10
CAS 70084-87-6	Hydrotriticum 2000		CAS 73138-79-1	Standapol® 1610
CAS 70084-94-5	Soy-Tein NL		CAS 73334-07-3	Ultravist
CAS 70124-77-5	Tomahawk		CAS 73384-59-5	Rocephin
CAS 70161-44-3	Suttocide® A		CAS 73771-04-7	EsCort
CAS 70191-76-3	Poly-Tergent® 4C3		CAS 73790-28-0	Clinafarm
CAS 70225-05-7	Liponate TDTM		CAS 73790-28-0	Fungaflor
CAS 70288-86-7	Eqvalan		CAS 73790-28-0	Magnate
CAS 70288-86-7	Ivomec		CAS 73790-28-0	Rappor Plus
CAS 70321-78-7	Lubrhophos® HR-719		CAS 73816-42-9	Meclan Cream
CAS 70331-94-1	Naugard® XL-1		CAS 74051-80-2	Checkmate
CAS 70356-03-5	Distaclor		CAS 74051-80-2	Poast®

CAS 74115-24-5	Apollo 50C
CAS 74176-31-1	Alfavet
CAS 74191-85-8	Cardura®
CAS 74223-64-6	Ally®
CAS 74381-53-6	Lupron
CAS 74623-31-7	Pluracol® W170
CAS 74623-31-7	Pluracol® W2000
CAS 74623-31-7	Pluracol® W5100N
CAS 74623-31-7	Pluracol® W660
CAS 74742-48-9	Varsol® 1
CAS 74764-40-2	Bepadin
CAS 75330-75-5	Mevacor
CAS 75330-75-5	Mevinacor
CAS 75422-21-0	Empimin® LAM30/AU
CAS 75736-33-3	Vigil
CAS 75738-58-8	Cefmax
CAS 75782-86-4	Neodol® 23
CAS 75782-87-5	Neodol® 45
CAS 76095-16-4	Innovace®
CAS 76095-16-4	Renitec
CAS 76095-16-4	Reniten
CAS 76095-16-4	Vasotec
CAS 76483-21-1	Marlophor® CS-Acid
CAS 76578-14-8	Pilot
CAS 76674-21-0	Impact
CAS 76674-21-0	Impact Excel
CAS 76738-62-0	Bonzi
CAS 76738-62-0	Cultar
CAS 76824-35-6	Pepcid®
CAS 76824-35-6	Pepcidine
CAS 76824-35-6	Pepdine
CAS 76824-35-6	Pepdul
CAS 76963-41-2	Axid
CAS 78110-38-0	Azactam
CAS 78266-06-5	Choletec
CAS 78350-78-4	Trigonox® ADC-NS60
CAS 78491-02-8	Germall® II
CAS 79645-27-5	Nebcin
CAS 79770-24-4	Iovist
CAS 79770-97-1	Propoquad® T/12
CAS 80260-73-2	Poly-Tergent® 2A1 Acid
CAS 80584-85-6	Doverphos® 11
CAS 80584-85-6	Doverphos® DPGDP
CAS 80584-85-6	Weston® THOP
CAS 80584-86-7	Doverphos® 12
CAS 80584-86-7	Weston® DHOP
CAS 80584-87-8	Weston® 491
CAS 81131-70-6	Lipostat Tablets
CAS 81161-17-3	Brevibloc®
CAS 81406-37-3	Starane
CAS 82204-94-2	Amiter LGOD
CAS 82419-36-1	Oflox®
CAS 82469-79-2	Citroflex B-6
CAS 82558-50-7	Flexidor
CAS 82558-50-7	Knot Out
CAS 82558-50-7	Ratio
CAS 82558-50-7	Tripart® Ratio
CAS 82560-54-1	Oncol
CAS 82560-54-1	Oncol 10G
CAS 82657-04-3	Talstar
CAS 83601-81-4	Ashlade M
CAS 83601-81-4	Ashlade Mancarb FL

CAS 83601-81-4	Battal
CAS 83601-81-4	Battal FL
CAS 83601-81-4	Bavistin®
CAS 83601-81-4	Bavistin® DF
CAS 83601-81-4	Campbell's MC Flowable
CAS 83601-81-4	Carbate Flowable
CAS 83601-81-4	Delsene® 50 DF
CAS 83601-81-4	Delsene® M Flowable
CAS 83601-81-4	Derosal WDG
CAS 83601-81-4	Headland Dual
CAS 83601-81-4	Hinge
CAS 83601-81-4	Hispor 45WP
CAS 83601-81-4	Legion
CAS 83601-81-4	Mascot Systemic Turf Fungicide
CAS 83601-81-4	Maxim
CAS 83601-81-4	MaxiMate
CAS 83601-81-4	Multi-W FL
CAS 83601-81-4	Septal
CAS 83601-81-4	Sportak Alpha
CAS 83601-81-4	Squadron
CAS 83601-81-4	Stempor DG
CAS 83601-81-4	Tripart® Arena Plus
CAS 83601-81-4	Tripart® Defensor FL
CAS 83601-81-4	Tripart® Legion
CAS 83601-81-4	Turfclear
CAS 83682-78-4	Phospholipid PTD
CAS 83881-51-0	Reactine®
CAS 83905-01-5	Zithromax®
CAS 83915-83-7	Carace®
CAS 83915-83-7	Coric
CAS 83915-83-7	Novatec
CAS 83915-83-7	Prinil
CAS 83915-83-7	Prinivil
CAS 83915-83-7	Vivatec
CAS 83915-83-7	Zestril
CAS 83933-91-3	Glucate® DO
CAS 84082-44-0	Incronam B-40
CAS 84501-49-5	Empimin® SDS
CAS 84507-84-1	Lamictal Tablets
CAS 84539-90-2	Radia® 7514
CAS 84604-14-8	Pristene R20
CAS 84605-13-0	Schercemol MEL-3
CAS 84605-14-1	Schercemol MEP-3
CAS 84605-15-2	Schercoquat IEP
CAS 84812-94-2	Ampholyte KKDP-60
CAS 84812-94-2	Ampholyte KKE-70
CAS 84812-94-2	Mackam 151C
CAS 84930-16-5	Nipasol M Potassium
CAS 84988-74-9	Radia® 7040
CAS 84988-79-4	Priolube 1414
CAS 84988-79-4	Radia® 7230
CAS 85005-47-6	Argonol 40
CAS 85005-55-6	Marlophen® 81N
CAS 85009-19-9	Santion
CAS 85049-34-9	Emalex PG-di-O
CAS 85049-34-9	Priolube 1429
CAS 85049-34-9	Radia® 7204
CAS 85049-36-1	Radia® 7187
CAS 85049-37-2	Radia® 7331
CAS 85068-76-4	Perfusamine
CAS 85116-87-6	Radia® 7231
CAS 85116-93-4	Radiasurf® 7175

CAS 85116-97-8	Radiasurf® 7410
CAS 85186-88-5	Radia® 7355
CAS 85186-88-5	Radiasurf® 7155
CAS 85209-91-2	ADK STAB NA-11
CAS 85251-77-0	Radiasurf® 7140
CAS 85251-77-0	Radiasurf® 7600
CAS 85408-49-7	Lilaminox M24
CAS 85408-76-0	Radia® 7051
CAS 85409-09-2	Radia® 7108
CAS 85536-04-5	Radia® 7501
CAS 85536-07-8	L.A.S
CAS 85536-07-8	Labrasol
CAS 85536-08-9	Labrafil M 2125 CS
CAS 85536-14-7	Marlon® AS3
CAS 85536-23-8	Aminol N
CAS 85536-23-8	Aminol TEC N
CAS 85586-21-6	Radia® 7110
CAS 85632-63-9	Lilamin LSP 33
CAS 85711-45-1	Radiasurf® 7156
CAS 85721-33-1	Ciprobay®
CAS 85721-33-1	Ciproxin
CAS 85736-49-8	Radiasurf® 7402
CAS 85736-49-8	Radiasurf® 7403
CAS 85736-49-8	Radiasurf® 7404
CAS 85736-49-8	Radiasurf® 7443
CAS 85736-49-8	Radiasurf® 7444
CAS 85865-69-6	Radia® 7241
CAS 86088-85-9	Incrosoft S-75
CAS 86088-85-9	Varisoft® 475
CAS 86089-12-5	Mackam RA
CAS 86089-12-5	Rewoteric® AM R40
CAS 86386-73-4	Diflucan®
CAS 86438-78-0	Mirataine® BB
CAS 87616-36-2	Ceraphyl 85
CAS 88150-42-9	Norvasc®
CAS 88497-58-9	Sandopan® DTC Linear P Acid
CAS 88671-89-0	Systhane
CAS 88684-42-8	VA-545
CAS 88889-14-9	Staril Tablets
CAS 88917-22-0	Arcosolv® DPMA
CAS 89339-41-3	Ultracast PE-35
CAS 89339-41-3	Ultracast PE-60
CAS 90052-75-8	Ceraphyl 847
CAS 90193-76-3	Radia® 7505
CAS 90268-48-7	Rewopol® B 1003
CAS 90283-04-8	Schercoquat ALA
CAS 90388-14-0	Crodafos CDP
CAS 90453-59-1	Akypo®-Muls 400
CAS 90529-57-0	Schercoquat SOAS
CAS 90624-75-2	Mirapol® AD-1
CAS 90624-76-3	Mirapol® AZ-1
CAS 90640-45-2	Diamine B11
CAS 90730-68-0	Kemamine® BQ-2982B
CAS 91031-31-1	Radia® 7266
CAS 91031-43-5	Radia® 7185
CAS 91031-48-0	Priolube 1458
CAS 91031-48-0	Radia® 7131
CAS 91031-57-1	Radia® 7510
CAS 91031-74-2	Radiasurf® 7145
CAS 91050-82-7	Radia® 7176
CAS 91052-47-0	Radiasurf® 7900
CAS 91723-32-9	Dynacet®
CAS 91723-55-6	Rewoquat W 75 H
CAS 91744-38-6	Imwitor® 369
CAS 91744-38-6	Imwitor® 370
CAS 91824-88-3	Isolan® GI 34
CAS 91995-05-0	Ampholak YCA/P
CAS 93356-94-6	Weston® 439
CAS 93455-78-8	Radiasurf® 7400
CAS 93572-63-5	Rewoquat DQ 35
CAS 93685-79-1	Permethyl 102A
CAS 93820-52-1	Rewoteric® AM KSF 40
CAS 94109-05-4	Incromide BEM
CAS 94441-92-6	Ampholak YJH-40
CAS 94552-41-7	Schercoquat ROAS
CAS 95706-86-8	Alkaterge®-T-IV
CAS 95823-35-1	Mark® 2140
CAS 96328-09-1	Mark® 5089
CAS 96492-31-8	Alcamizer 2
CAS 96690-41-4	Crosilk 10,000
CAS 96690-41-4	Ritasilk
CAS 96690-41-4	Silk Pro-Tein
CAS 96690-41-4	Solu-Silk Protein
CAS 96726-23-9	Miranol® Ester PO-LM4
CAS 97281-23-7	Radiasurf® 7270
CAS 97281-23-7	Radiasurf® 7414
CAS 97281-23-7	Radiasurf® 7417
CAS 97281-23-7	Radiasurf® 7453
CAS 97281-23-7	Radiasurf® 7454
CAS 97281-23-7	Radiasurf® 7473
CAS 97338-28-8	Ceraphyl 791
CAS 97338-28-8	Hetester HSS
CAS 97404-33-6	Radia® 7500
CAS 97404-50-7	Lanesta 3
CAS 97488-62-5	Ampholak 7TY
CAS 97488-91-0	Labrafil M 1944 CS
CAS 97593-29-8	Dimodan PVP Kosher
CAS 97593-29-8	Dimodan S
CAS 97593-31-2	Acidan
CAS 97659-50-2	Ampholak YCE
CAS 97659-51-3	Ampholak XCE
CAS 97659-53-5	Ampholak 7TX
CAS 97659-53-5	Ampholak 7TX-SD 55
CAS 97659-53-5	Ampholak 7TX-T
CAS 97659-53-5	Ampholak 7TX/C
CAS 97659-53-5	Ampholak XO7
CAS 97759-33-8	Rice-Pro EN-20
CAS 97808-04-3	Amine 780
CAS 97999-44-5	Marlowet® 5311
CAS 98073-10-0	Glucam® E-20 Distearate
CAS 99330-44-6	Akypo®-Soft 100 MgV
CAS 100085-64-1	Rewoteric® QAM 50
CAS 100864-25-1	Wheat-Tein NL
CAS 102089-33-8	Uvasil 299
CAS 102523-96-6	Abil® B 9950
CAS 102843-39-0	VA-058
CAS 103213-19-0	Schercemol DISD
CAS 103213-20-3	Bernel® Ester DID
CAS 103213-20-3	Schercemol DID
CAS 103213-22-5	Schercemol TIST
CAS 103597-45-1	Mixxim® BB/100
CAS 104222-32-4	VA-080
CAS 104909-82-2	Akypo TBP 180
CAS 105391-15-9	Akypo LF 3

CAS 105391-15-9	Akypo LF 5
CAS 105391-15-9	Akypo MB 2528S
CAS 105827-78-9	Admire
CAS 105827-78-9	Confidor
CAS 105827-78-9	Gaucho
CAS 106392-12-5	Dowfax 30C05, 30C10, 50C15
CAS 106436-39-9	Ceraphyl 55
CAS 106990-43-6	Chimassorb® 119FL
CAS 107397-59-1	Tetronic® 150R1
CAS 107534-96-3	Folicur
CAS 107534-96-3	Horizon/Horizont
CAS 107534-96-3	Raxil
CAS 107534-96-3	Silvacur
CAS 107600-33-9	Akypo LF 2
CAS 107600-36-2	Akyposal 100 DAL
CAS 107628-03-5	Akypo®-Soft KA 250 BV
CAS 107628-08-0	Akypo OP 80
CAS 107628-08-0	Akypo OP 190
CAS 107987-23-5	Quatrisoft Polymer LM-200
CAS 108419-32-5	Exxate® 800
CAS 108419-33-6	Exxate® 900
CAS 108419-34-7	Exxate® 1000
CAS 108419-35-8	Exxate® 1300
CAS 108818-88-8	Rhodafac® BG-510
CAS 109464-53-1	Sokalan® HP 22
CAS 109678-33-3	FR-1034
CAS 110152-58-4	Indulin® W-1
CAS 110615-47-9	APG® 600 Glycoside
CAS 110615-47-9	Glucopon 425
CAS 110615-47-9	Glucopon 600
CAS 111019-03-5	Crodafos CAP
CAS 111174-64-2	Quat-Pro E
CAS 111905-55-6	Schercoquat DAS
CAS 112324-11-5	Schercoquat ROEP
CAS 112324-16-0	Lipoquat R
CAS 112324-16-0	M-Quat® JN
CAS 112324-16-0	Surfactol® Q4
CAS 112945-52-5	Cab-O-Sil® L-90
CAS 113431-53-1	Schercemol DISF
CAS 113431-54-2	Schercemol TISC
CAS 113492-03-8	Schercoquat CAS
CAS 113492-04-9	Schercoquat FOAS
CAS 113976-90-2	APG® 300 Glycoside
CAS 115047-92-2	DV-1801
CAS 115340-78-8	Schercoquat APAS
CAS 115340-80-2	Schercoquat WOAS
CAS 116912-64-2	Union Carbide® A-1160
CAS 117272-76-1	Silwet® L-7607
CAS 118337-09-0	Ethanox® 398

CAS 118585-13-0	VA-060
CAS 118777-77-8	Katemul IG-70
CAS 123754-28-9	Hypan® SS201
CAS 124960-38-9	VA-545
CAS 125740-36-5	Tomah Q-14-2
CAS 125804-12-8	Dermol G-76
CAS 125804-13-9	Dermol L45
CAS 127358-81-0	Dermol DISD
CAS 129541-36-2	Katemul IGU-70
CAS 130097-36-8	Esperox® 740M
CAS 130124-24-2	Carsosoft® T-90
CAS 130124-24-2	Incrosoft T-75
CAS 133798-12-6	Schercotaine UAB
CAS 133934-08-4	Schercotaine APAB
CAS 133934-09-5	Incronam WG-30
CAS 133934-09-5	Mackam WGB
CAS 133934-09-5	Schercotaine WOAB
CAS 134112-42-8	Schercoquat IALA
CAS 136097-97-7	Kester Wax® K 48
CAS 136505-00-5	Hypan® SR150H
CAS 137796-06-6	Zoldine® MS-52
CAS 138208-68-1	Dermol ICSA
CAS 138314-11-1	Dermolan GLH
CAS 270028-82-6	Empicol® ETB
CAS 594477-57-3	FR-1025
CAS 617788-68-9	Actrasol 6092
CAS 617788-68-9	Laurel SRO
CAS 689990-06-7	Axol® L 61, L62
CAS 977000-98-8	Sol-U-Tein EA
CAS 977039-11-4	Quat-Soy CDMA-25
CAS 977039-11-4	Soy-Quat C
CAS 977053-96-5	Rewomul MG SE
CAS 977056-87-3	Softisan® 100
CAS 977056-87-3	Witepsol® E75, E76, E85, H5, H12, H15, H19, H32, H35, H37, H39, H42, H175, H185
CAS 977058-55-1	U Pink 113
CAS 977059-33-8	HVP
CAS 977059-33-8	Hydrosoy 2000
CAS 977059-33-8	Solu-Veg EN-35
CAS 977066-20-8	Proto-Lan IP
CAS 977067-77-8	Empicol® EMB
CAS 977077-71-6	Amino-Silk SF
CAS 977077-71-6	Crosilk
CAS 977077-71-6	Crosilk Liq
CAS 977077-71-6	Solu-Silk 25
CAS 977079-10-9	Schercemol ISE
CAS 1150047-92-2	CDS-1801

CAS Number-to-Chemical
Cross Reference

CAS 50-00-0	Formaldehyde		CAS 51 77-4	Gefarnate	
CAS 50-01-1	Guanidine hydrochloride		CAS 51-83-2	Carbachol	
CAS 50 02-2	Dexamethasone		CAS 51-98-9	Norethindrone acetate	
CAS 50-03-3	Hydrocortisone acetate		CAS 52-01-7	Spirono lactone	
CAS 50-04-4	Cortisone acetate		CAS 52-21-1	Prednisolone acetate	
CAS 50-06-6	Phenobarbital		CAS 52-24-4	Triethylene thiophosphoramide	
CAS 50-10-2	Oxyphenonium bromide		CAS 52-28-8	Codeine phosphate	
CAS 50-11-3	Metharbital		CAS 52-39-1	Aldosterone	
CAS 50-12-4	Mephenytoin		CAS 52-49-3	Trihexyphenidyl hydrochloride	
CAS 50-13-5	Meperidine hydrochloride		CAS 52-51-7	Bromonitropropanediol	
CAS 50-18-0	Cyclophosphamide		CAS 52-53-9	Verapamil	
CAS 50-21-5	Lactic acid		CAS 52-62-0	Pentoliniumtartrate	
CAS 50-23-7	Hydrocortisone		CAS 52-67-5	Penicillamine	
CAS 50-24-8	Prednisolone		CAS 52-68-6	Trichlorfon	
CAS 50-28-2	Estradiol		CAS 52-78-8	Norethandrolone	
CAS 50-29-3	DDT		CAS 52-86-8	Haloperidol	
CAS 50-29-3	Dichlorodiphenyl trichlorethane		CAS 52-89-1	Cysteine hydrochloride	
CAS 50-33-9	Phenyl butazone		CAS 52-90-4	Cysteine	
CAS 50-34-0	Propantheline bromide		CAS 53-03-2	Prednisone	
CAS 50-35-1	Thalidomide		CAS 53-06-5	Cortisone	
CAS 50-41-9	Clomiphenecitrate		CAS 53-16-7	Estrone	
CAS 50-44-2	Mercaptopurine		CAS 53-19-0	Mitotane	
CAS 50-47-5	Desipramine		CAS 53-34-9	Fluprednisolone	
CAS 50-48-6	Amitriptyline		CAS 53-36-1	Methylprednisolone acetate	
CAS 50-49-7	Imipramine		CAS 53-39-4	Oxandrolone	
CAS 50-52-2	Thioridazine		CAS 53-46-3	Methantheline bromide	
CAS 50-53-3	Chlorpromazine		CAS 53-60-1	Promazine hydrochloride	
CAS 50-54-4	Quinidine sulfate		CAS 53-73-6	Angiotensinamide	
CAS 50-55-5	Reserpine		CAS 53-86-1	Indomethacin	
CAS 50-56-6	Oxytocin		CAS 54-05-7	Chloroquine	
CAS 50-57-7	Lypressin		CAS 54-11-5	Nicotine	
CAS 50-58-8	Phendimetrazinetartrate		CAS 54-12-6	Tryptophan	
CAS 50-59-9	Cephaloridine		CAS 54-21-7	Sodium salicylate	
CAS 50-63-5	Chloroquinephosphate		CAS 54-31-9	Furosemide	
CAS 50-65-7	Niclosamide		CAS 54-36-4	Metyrapone	
CAS 50-70-4	Sorbitol		CAS 54-42-2	Idoxuridine	
CAS 50-78-2	Aspirin		CAS 54-49-9	Metaraminol	
CAS 50-81-7	Ascorbic acid		CAS 54-71-7	Pilocarpine hydrochloride	
CAS 50-99-7	Dextrose (glucose anhydrous)		CAS 54-85-3	Isoniazid	
CAS 51-03-6	Piperonyl butoxide		CAS 54-91-1	Pipobroman	
CAS 51-05-8	Benzocaine		CAS 54-92-2	Iproniazid	
CAS 51-12-7	Nialamide		CAS 55-03-8	Levothyroxine sodium	
CAS 51-15-0	Pralidoximechloride		CAS 55-06-1	Llothyronine sodium	
CAS 51-17-2	Benzimidazole		CAS 55-22-1	Isonicotinic acid	
CAS 51-21-8	Fluorouracil		CAS 55-38-9	Fenthion	
CAS 51-42-3	Epinephrine bitartrate		CAS 55-56-1	Chlorhexidine	
CAS 51-43-4	Epinephrine		CAS 55-63-0	Nitroglycerin	
CAS 51-48-9	Thyroxine		CAS 55-65-2	Guanethidine	
CAS 51-56-9	Homatropine hydrobromide		CAS 55-86-7	Mechlorethamine hydrochloride	

CAS 55-91-4	Isoflurophate	CAS 58-56-0	Pyridoxine hydrochloride
CAS 55-98-1	Busulfan	CAS 58-85-5	Biotin
CAS 56-03-1	Biguanide	CAS 58-86-6	Xylose
CAS 56-23-5	Carbon tetrachloride	CAS 58-89-9	Lindane
CAS 56-29-1	Hexobarbital	CAS 58-93-5	Hydrochlorothiazide
CAS 56-35-9	Tributyltin oxide (bistributyltin oxide)	CAS 58-94-6	Chlorothiazide
		CAS 58-96-8	Uridine
CAS 56-37-1	Benzyltriethyl ammonium chloride	CAS 59-02-9	VitaminE
CAS 56-38-2	Parathion	CAS 59-05-2	Methotrexate
CAS 56-40-6	Glycine	CAS 59-30-3	Folic acid
CAS 56-41-7	α-Alanine (L-form)	CAS 59-33-6	Pyrilamine maleate
CAS 56-45-1	L-Serine	CAS 59-51-8	Methionine, DL
CAS 56-65-5	Adenosine triphosphate	CAS 59-52-9	Dimercaprol
CAS 56-75-7	Chloramphenicol	CAS 59-63-2	Isocarboxazid
CAS 56-81-5	Glycerin	CAS 59-66-5	Acetazolamide
CAS 56-84-8	L-Aspartic acid	CAS 59-67-6	Nicotinic acid
CAS 56-85-9	Glutamine	CAS 59-87-0	Nitrofurazone
CAS 56-86-0	Glutamic acid	CAS 59-92-7	Levodopa
CAS 56-87-1	Lysine	CAS 59-97-2	Tolazoline hydrochloride
CAS 56-89-3	Cystine	CAS 60-00-4	Edetic acid
CAS 56-93-9	Benzyl trimethyl ammonium chloride	CAS 60-18-4	L-Tyrosine
		CAS 60-24-2	Mercaptoethanol
CAS 56-94-0	Demecarium bromide	CAS 60-29-7	Ethyl ether
CAS 57-09-0	Cetrimonium bromide	CAS 60-31-1	Acetylcholine chloride
CAS 57-10-3	Palmitic acid	CAS 60-32-2	Aminocaproic acid
CAS 57-11-4	Hydrogenated stearic acid	CAS 60-33-3	Linoleic acid
CAS 57-11-4	Stearic acid	CAS 60-51-5	Dimethoate
CAS 57-13-6	Urea	CAS 60-54-8	Tetracycline
CAS 57-15-8	Chlorobutanol	CAS 60-56-0	Methimazole
CAS 57-22-7	Vincristine	CAS 60-57-1	Dieldrin
CAS 57-27-2	Morphine	CAS 60-99-1	Methotrimeprazine
CAS 57-30-7	Phenobarbital sodium	CAS 61-12-1	Dibucaine hydrochloride
CAS 57-41-0	Phenytoin	CAS 61-16-5	Methoxamine hydrochloride
CAS 57-48-7	D-Fructose	CAS 61-56-3	Sulthiame
CAS 57-50-1	Sucrose	CAS 61-57-4	Niridazole
CAS 57-53-4	Meprobamate	CAS 61-68-7	Mefenamic acid
CAS 57-55-6	Propylene glycol	CAS 61-72-3	Cloxacillin
CAS 57-57-8	Propiolactone	CAS 61-75-6	Bretylium tosylate
CAS 57-63-6	Ethinylestradiol	CAS 61-76-7	Phenylephrine hydrochloride
CAS 57-64-7	Physostigmine salicylate	CAS 61-82-5	Amitrole
CAS 57-66-9	Probenecid	CAS 61-90-5	Leucine
CAS 57-74-9	Chlordane	CAS 62-23-7	Nitrobenzoic acid
CAS 57-83-0	Progesterone	CAS 62-31-7	Dopamine hydrochloride
CAS 57-85-2	Testosterone propionate	CAS 62-31-7	Doxapram hydrochloride
CAS 57-88-5	Cholesterol	CAS 62-33-9	Edetate calcium disodium
CAS 57-94-3	Tubocurarine chloride	CAS 62-38-2	Phenylmercuric acetate
CAS 57-96-5	Sulfinpyrazone	CAS 62-44-2	Phenacetin
CAS 58-08-2	Caffeine	CAS 62-51-7	Methacholine chloride
CAS 58-14-0	Pyrimethamine	CAS 62-54-4	Calcium acetate
CAS 58-18-4	Methyltestosterone	CAS 62-55-5	Thioacetamide
CAS 58-20-8	Testosterone cypionate	CAS 62-56-6	Thiourea
CAS 58-22-0	Testosterone	CAS 62-73-7	Dichlorvos (DDVP)
CAS 58-25-3	Chlordiazepoxide	CAS 62-90-8	Nandrolone phenpropionate
CAS 58-25-3	Chlordiazepoxide	CAS 62-97-5	Diphemanil methylsulfate
CAS 58-28-6	Desipramine hydrochloride	CAS 63-12-7	Benzquinamide
CAS 58-32-2	Dipyridamole	CAS 63-25-2	Carbaryl
CAS 58-33-3	Promethazine hydrochloride	CAS 63-39-8	Uridine triphosphate
CAS 58-38-8	Prochlorperazine	CAS 63-42-3	Lactose
CAS 58-39-9	Perphenazine	CAS 63-68-3	Methionine
CAS 58-46-8	Tetrabenazine	CAS 63-74-1	Sulfanilamide
CAS 58-54-8	Ethacrynicacid	CAS 63-92-3	Phenoxybenzamine hydrochloride
CAS 58-55-9	Theophylline	CAS 63-98-9	Phenacemide

CAS Number	Chemical
CAS 64-02-8	Tetrasodium EDTA
CAS 64-17-5	Ethyl alcohol
CAS 64-18-6	Formic acid
CAS 64-19-7	Acetic acid
CAS 64-31-3	Morphine sulfate
CAS 64-43-7	Amobarbital sodium
CAS 64-55-1	Mebutamate
CAS 64-72-2	Chlortetracycline hydrochloride
CAS 64-75-5	Tetracycline hydrochloride
CAS 64-77-7	Tolbutamide
CAS 65-29-2	Gallamine triethiodide
CAS 65-45-2	Salicylamide
CAS 65-49-6	Aminosalicylic acid
CAS 65-85-0	Benzoic acid
CAS 66-22-8	Uracil
CAS 66-71-7	Phenanthroline
CAS 67-03-8	Thiamine hydrochloride
CAS 67-20-9	Nitrofurantoin
CAS 67-43-6	Pentetic acid
CAS 67-45-8	Furazolidone
CAS 67-48-1	Choline chloride
CAS 67-56-1	Methyl alcohol
CAS 67-63-0	Isopropyl alcohol
CAS 67-64-1	Acetone
CAS 67-66-3	Chloroform
CAS 67-68-5	Dimethylsulfoxide
CAS 67-73-2	Fluocinolone acetonide
CAS 67-92-5	Dicyclomine hydrochloride
CAS 67-96-9	Dihydrotachysterol
CAS 67-97-0	Vitamin D_3
CAS 68-04-2	Sodium citrate
CAS 68-11-1	Thioglycolic acid
CAS 68-12-2	Dimethyl formamide
CAS 68-19-9	Cyanocobalamin
CAS 68-19-9	Vitamin B_{12}
CAS 68-22-4	Norethindrone
CAS 68-26-8	Vitamin A
CAS 68-35-9	Sulfadiazine
CAS 68-41-7	Cycloserine
CAS 68-76-8	Triaziquone
CAS 68-88-2	Hydroxyzine
CAS 69-09-0	Chlorpromazine hydrochloride
CAS 69-52-3	Ampicillin sodium
CAS 69-53-4	Ampicillin
CAS 69-57-8	Benzylpenicillin sodium
CAS 69-65-8	Mannitol
CAS 69-72-7	Salicylic acid
CAS 69-79-4	Maltose
CAS 69-89-6	Xanthine
CAS 69-93-2	Uric acid
CAS 70-00-8	Trifluridine
CAS 70-47-3	L-Asparagine
CAS 70-53-1	Lysine monohydrochloride
CAS 70-55-3	p-Toluene sulfonamide
CAS 71-00-1	Histidine
CAS 71-23-8	Propyl alcohol
CAS 71-27-2	Succinyl choline chloride
CAS 71-36-3	Butyl alcohol
CAS 71-41-0	n-Amylalcohol
CAS 71-43-2	Benzene
CAS 71-55-6	Trichloroethane
CAS 71-58-9	Medroxyprogesterone acetate
CAS 71-63-6	Digitoxin
CAS 71-67-0	Sulfobromophthalein sodium
CAS 71-68-1	Hydromorphone hydrochloride
CAS 71-73-8	Thiopental sodium
CAS 72-14-0	Sulfathiazole
CAS 72-17-3	Sodium lactate
CAS 72-18-4	Valine
CAS 72-19-5	Threonine
CAS 72-43-5	Methoxychlor
CAS 72-44-6	Methaqualone
CAS 73-22-3	Tryptophan
CAS 73-24-5	Adenine
CAS 73-32-5	Isoleucine
CAS 73-40-5	Guanine
CAS 73-48-3	Bendroflumethiazide
CAS 73-49-4	Quinethazone
CAS 74-31-7	Diphenyl phenylene diamine
CAS 74-55-5	Ethambutol
CAS 74-73-3	Demeclocycline hydrochloride
CAS 74-79-3	L-Arginine
CAS 74-83-9	Methyl bromide
CAS 74-86-2	Acetylene
CAS 74-87-3	Methyl chloride
CAS 74-88-4	Methyl iodide
CAS 74-90-8	Hydrogen cyanide
CAS 74-98-6	Propane
CAS 75-00-3	Ethyl chloride
CAS 75-05-8	Acetonitrile
CAS 75-07-0	Acetaldehyde
CAS 75-09-2	Methylene chloride
CAS 75-11-6	Methylene iodide
CAS 75-12-7	Formamide
CAS 75-15-0	Carbon disulfide
CAS 75-18-3	Dimethyl sulfide
CAS 75-20-7	Calcium carbide
CAS 75-21-8	Ethylene oxide
CAS 75-25-2	Bromoform
CAS 75-28-5	Isobutane
CAS 75-33-2	Isopropyl mercaptan
CAS 75-36-5	Acetyl chloride
CAS 75-52-5	Nitroisophthalic acid
CAS 75-52-5	Nitromethane
CAS 75-54-7	Methyldichlorosilane
CAS 75-56-9	Propylene oxide
CAS 75-59-2	Tetramethyl ammonium hydroxide
CAS 75-60-5	Cacodylic acid
CAS 75-69-4	Trichlorofluoromethane
CAS 75-71-8	Dichlorodifluoromethane
CAS 75-72-9	Chlorotrifluoromethane
CAS 75-76-3	Tetramethylsilane
CAS 75-77-4	Trimethylchlorosilane
CAS 75-79-6	Methyltrichlorosilane
CAS 75-85-4	t-Amylalcohol
CAS 75-89-8	Trifluoroethanol
CAS 75-91-2	Butylhydroperoxide
CAS 75-94-5	Vinyltrichlorosilane
CAS 75-99-0	Dalapon
CAS 76-03-9	TCA
CAS 76-05-1	Trifluoroacetic acid
CAS 76-06-2	Chloropicrin
CAS 76-13-1	Trichlorotrifluoroethane

CAS 76-14-2	Dichlorotetrafluoroethane
CAS 76-22-2	Camphor
CAS 76-25-5	Triamcinolone acetonide
CAS 76-38-0	Methoxyflurane
CAS 76-39-1	Nitromethyl propanol
CAS 76-42-6	Oxycodone
CAS 76-43-7	Fluoxymesterone
CAS 76-74-4	Pentobarbital
CAS 76-87-9	Fentin hydroxide
CAS 76-87-9	Triphenyltin hydroxide
CAS 76-90-4	Mepenzolate bromide
CAS 77-06-5	Gibberellic acid
CAS 77-09-8	Phenolphthalein
CAS 77-21-4	Glutethimide
CAS 77-26-9	Butalbital
CAS 77-36-1	Chlorthalidone
CAS 77-38-3	Chlorphenoxamine
CAS 77-41-8	Methsuximide
CAS 77-58-7	Dibutyltin dilaurate
CAS 77-67-8	Ethosuximide
CAS 77-71-4	DM hydantoin
CAS 77-76-9	Dimethoxy propane
CAS 77-86-1	Trishydroxymethyl aminomethane
CAS 77-86-1	Tromethamine
CAS 77-89-4	Acetyltriethyl citrate
CAS 77-90-7	Acetyltributyl citrate
CAS 77-92-9	Citric acid
CAS 77-93-0	Triethyl citrate
CAS 77-94-1	Tributyl citrate
CAS 77-99-6	Trimethylolpropane
CAS 78-08-0	Vinyltriethoxysilane
CAS 78-10-4	Ethyl silicate
CAS 78-10-4	Tetraethoxysilane
CAS 78-11-5	Pentaerythritol tetranitrate
CAS 78-21-7	Cetethyl morpholinium ethosulfate
CAS 78-30-8	Tricresyl phosphate
CAS 78-40-0	Triethyl phosphate
CAS 78-42-2	Ethylhexoic acid
CAS 78-42-2	Trioctyl phosphate
CAS 78-44-4	Carisoprodol
CAS 78-48-8	Tribufos
CAS 78-51-3	Tributoxyethyl phosphate
CAS 78-59-1	Isophorone
CAS 78-62-6	Dimethyldiethoxysilane
CAS 78-63-7	Dimethyldibutyl peroxyhexane
CAS 78-66-0	Dimethyl octynediol
CAS 78-67-1	Azobisisobutyronitrile
CAS 78-78-4	Isopentane
CAS 78-83-1	Isobutyl alcohol
CAS 78-87-5	Propylene dichloride
CAS 78-93-3	Methyl ethyl ketone
CAS 78-96-6	Isopropanolamine
CAS 79-01-6	Trichloroethylene
CAS 79-09-4	Propionic acid
CAS 79-10-7	Acrylic acid
CAS 79-10-7	Carboxyethyl acrylate
CAS 79-11-8	Monochloroacetic acid
CAS 79-14-1	Glycolic acid
CAS 79-20-9	Methyl acetate
CAS 79-21-0	Peracetic acid
CAS 79-24-3	Nitroethane
CAS 79-31-2	Isobutyric acid
CAS 79-34-5	Tetrachloroethane
CAS 79-46-9	2-Nitropropane
CAS 79-57-2	Oxytetracycline
CAS 79-64-1	Dimethisterone
CAS 79-74-3	Diamyl hydroquinone
CAS 79-94-7	Tetrabromo bisphenol A
CAS 80-05-7	Bisphenol A
CAS 80-08-0	4,4-Diaminodiphenyl sulfone
CAS 80-10-4	Diphenyldichlorosilane
CAS 80-15-9	Cumene hydroperoxide
CAS 80-35-3	Sulfamethoxypyridazine
CAS 80-39-7	Ethyltoluene sulfonamide
CAS 80-43-3	Dicumyl peroxide
CAS 80-46-6	Amylphenol
CAS 80-49-9	Homatropine methylbromide
CAS 80-50-2	Anisotropine methylbromide
CAS 80-51-3	Oxybisbenzene sulfonyl hydrazide
CAS 80-56-8	a-Pinene
CAS 80-62-6	Methylmethacrylate
CAS 80-77-3	Chlormezanone
CAS 81-07-2	Saccharin
CAS 81-13-0	Dexpanthenol
CAS 81-23-2	Dehydrocholic acid
CAS 81-81-2	Warfarin
CAS 82-68-8	PCNB
CAS 82-68-8	Quintozene
CAS 83-12-5	Phenindione
CAS 83-43-2	Methylprednisolone
CAS 83-73-8	Iodoquinol
CAS 83-79-4	Rotenone
CAS 83-88-5	Riboflavin
CAS 83-88-5	Vitamin B_2
CAS 84-02-6	Prochlorperazine maleate
CAS 84-04-8	Pipamazine
CAS 84-17-3	Dienestrol
CAS 84-36-6	Syrosingopine
CAS 84-61-7	Dicyclohexyl phthalate
CAS 84-65-1	Anthraquinone
CAS 84-66-2	Diethyl phthalate
CAS 84-69-5	Diisobutyl phthalate
CAS 84-74-2	Dibutyl phthalate
CAS 84-96-8	Trimeprazine
CAS 85-00-7	Diquat
CAS 85-44-9	Phthalic anhydride
CAS 85-60-9	Butylidene bis butylcresol
CAS 85-68-7	Benzylbutyl phthalate
CAS 85-68-7	Butylbenzylphthalate
CAS 85-70-1	Butylphthalyl butyl glycolate
CAS 85-73-4	Phthalyl sulfathiazole
CAS 85-91-6	Dimethyl anthranilate
CAS 86-29-3	Diphenyl acetonitrile
CAS 86-34-0	Phensuximide
CAS 86-50-0	Cotnion-methyl
CAS 86-50-0	Azinphos-methyl
CAS 86-87-3	Naphthyl acetic acid
CAS 87-08-1	Penicillin V
CAS 87-10-5	Tribromo salicylanilide
CAS 87-10-5	Tribromsalan
CAS 87-33-2	Isosorbide dinitrate
CAS 87-51-4	Indole acetic acid

CAS 87-66-1	Pyrogallol
CAS 87-83-2	Pentabromotoluene
CAS 87-86-5	Pentachlorophenol
CAS 87-89-8	Inositol
CAS 87-90-1	Trichloroisocyanuric acid
CAS 87-90-1	Trichlorotriazinetrione
CAS 88-04-0	Chloroxylenol
CAS 88-14-2	Furoic acid
CAS 88-24-4	Methylenebis ethylbutylphenol
CAS 88-58-4	Dibutylhydroquinone
CAS 88-85-7	Dinoseb
CAS 89-65-6	Erythorbic acid
CAS 89-78-1	Menthol
CAS 89-83-8	Thymol
CAS 90-02-8	Salicylaldehyde
CAS 90-03-9	Mercufenol chloride
CAS 90-04-0	o-Anisidine
CAS 90-05-1	Guaiacol
CAS 90-12-0	Methylnaphthalene
CAS 90-30-3	a-Phenylnaphthylamine
CAS 90-33-5	Hymecromone
CAS 90-43-7	o-Phenylphenol
CAS 90-64-2	Mandelic acid
CAS 91-10-1	Dimethoxyphenol
CAS 91-20-3	Naphthalene
CAS 91-22-5	Quinoline
CAS 91-33-8	Benzthiazide
CAS 91-53-2	Ethoxyquin
CAS 91-63-4	Quinaldine
CAS 92-13-7	Pilocarpine
CAS 92-44-4	2,3-Naphthalenediol
CAS 92-52-4	Biphenyl
CAS 92-52-4	Diphenyl
CAS 92-71-7	Diphenyloxazole
CAS 92-83-1	Xanthene
CAS 92-88-6	Biphenol
CAS 93-14-1	Guaifenesin
CAS 93-15-2	Methyleugenol
CAS 93-30-1	Methoxyphenamine
CAS 93-46-9	Dinaphthyl phenylene diamine
CAS 93-58-3	Methylbenzoate
CAS 93-65-2	Mecoprop-P
CAS 93-69-6	Tolylbiguanide
CAS 93-72-1	Silvex
CAS 93-82-3	Stearamide DEA
CAS 93-83-4	Oleamide DEA
CAS 94-09-7	Ethylaminobenzoate
CAS 94-13-3	Propyl paraben
CAS 94-17-7	Dichlorobenzoyl peroxide
CAS 94-18-8	Benzyl paraben
CAS 94-20-2	Chlorpropamide
CAS 94-26-8	Butyl paraben
CAS 94-28-0	Triethylene glycol diethyl-hexanoate
CAS 94-35-9	Styramate
CAS 94-36-0	Benzoyl peroxide
CAS 94-74-6	MCPA
CAS 94-81-5	MCPB
CAS 94-82-6	2,4-DB
CAS 94-91-7	Disalicylidene propane diamine
CAS 94-96-2	Ethylhexanediol
CAS 95-05-6	Sulfiram

CAS 95-14-7	Benzotriazole
CAS 95-19-2	Stearyl hydroxyethyl imidazoline
CAS 95-25-0	Chlorzoxazone
CAS 95-31-8	Butylbenzothiazole sulfenamide
CAS 95-32-9	Morpholinyl benzothiazole disulfide
CAS 95-33-0	Cyclohexyl benzothiazole sulfenamide
CAS 95-38-5	Oleyl hydroxyethyl imidazoline
CAS 95-45-4	Dimethylglyoxime
CAS 95-48-7	o-Cresol
CAS 95-50-1	o-Dichlorobenzene
CAS 95-51-2	o-Chloroaniline
CAS 95-71-6	Toluhydroquinone
CAS 95-88-5	Chlororesorcinol
CAS 96-05-9	Allyl methacrylate
CAS 96-13-9	Dibromopropanol
CAS 96-20-8	Aminobutanol
CAS 96-22-0	Diethyl ketone
CAS 96-26-4	Dihydroxyacetone
CAS 96-27-5	Thioglycerin
CAS 96-29-7	Methylethyl ketoxime
CAS 96-33-3	Methyl acrylate
CAS 96-43-5	Chlorothiophene
CAS 96-45-7	Ethylene thiourea
CAS 96-48-0	Butyrolactone
CAS 96-49-1	Ethylene carbonate
CAS 96-69-5	Thiobisbutyl methylphenol
CAS 96-83-3	Iopanoic acid
CAS 97-23-4	Dichlorophen
CAS 97-23-4	Dihydroxydichlorodiphenyl-methane
CAS 97-23-4	Methylenebischlorophenol
CAS 97-39-2	Ditolylguanidine
CAS 97-53-0	Eugenol
CAS 97-59-6	Allantoin
CAS 97-63-2	Ethyl methacrylate
CAS 97-64-3	Ethyl lactate
CAS 97-74-5	Tetramethylthiuram monosulfide
CAS 97-77-8	Tetraethylthiuram disulfide
CAS 97-78-9	Lauroyl sarcosine
CAS 97-88-1	Butyl methacrylate
CAS 97-90-5	Ethylene glycol dimethacrylate
CAS 97-90-5	Ethylene glycol dimethacrylate
CAS 97-99-4	Tetrahydrofurfuryl alcohol
CAS 98-00-0	Furfuryl alcohol
CAS 98-01-1	Furfural
CAS 98-10-2	Benzene sulfonamide
CAS 98-13-5	Phenyltrichlorosilane
CAS 98-19-1	Butyl xylene
CAS 98-51-1	Butyl toluene
CAS 98-54-4	Butyl phenol
CAS 98-55-5	Terpineol
CAS 98-77-1	Piperidinium pentamethylene dithiocarbamate
CAS 98-79-3	PCA
CAS 98-82-8	Cumene
CAS 98-83-9	Methylstyrene
CAS 98-86-2	Acetophenone
CAS 98-96-4	Pyrazinamide
CAS 99-04-7	m-Toluic acid
CAS 99-30-9	Dicloran

CAS 99-34-3	Dinitrobenzoic acid
CAS 99-49-0	Carvone
CAS 99-50-3	Protocatechuic acid
CAS 99-62-7	Diisopropyl benzene
CAS 99-66-1	Valproic acid
CAS 99-68-3	Carboxymethyl mercaptosuccinic acid
CAS 99-76-3	Methyl paraben
CAS 99-77-8	Disulfiram
CAS 99-79-6	Iophendylate
CAS 99-93-4	p-Hydroxyacetophenone
CAS 99-94-5	p-Toluic acid
CAS 100-01-6	Nitroaniline
CAS 100-07-2	Anisoyl chloride
CAS 100-09-4	p-Anisidine
CAS 100-10-7	Dimethylaminobenzaldehyde
CAS 100-37-8	Diethylaminoethanol
CAS 100-46-9	Benzylamine
CAS 100-47-0	Benzonitrile
CAS 100-51-6	Benzyl alcohol
CAS 100-52-7	Benzaldehyde
CAS 100-55-0	Nicotinyl alcohol
CAS 100-74-3	Ethyl morpholine
CAS 100-88-9	Cyclamic acid
CAS 100-97-0	Hexamethylenetetramine
CAS 101-02-0	Triphenylphosphite
CAS 101-05-3	Anilazine
CAS 101-21-3	Chlorpropham
CAS 101-26-8	Pyridostigmine bromide
CAS 101-31-5	Hyoscyamine
CAS 101-34-8	Glyceryl triacetyl ricinoleate
CAS 101-37-1	Triallylcyanurate
CAS 101-43-9	Cyclohexyl methacrylate
CAS 101-67-7	Dioctyl diphenylamine
CAS 101-68-8	Diphenylmethane diisocyanate
CAS 101-68-8	MDI
CAS 101-68-8	Polymethylene polyphenyl isocyanate
CAS 101-72-4	Isopropyl phenylphenylene diamine
CAS 101-77-9	Diaminodiphenyl methane
CAS 101-77-9	Methylene dianiline
CAS 101-84-8	Diphenyl oxide
CAS 101-87-1	Cyclohexyl phenylphenylene diamine
CAS 101-96-2	Dibutyl phenylene diamine
CAS 102-06-7	Diphenyl guanidine
CAS 102-07-8	Diphenyl urea
CAS 102-08-9	Diphenyl thiourea
CAS 102-60-3	Tetrahydroxy propylethylene diamine
CAS 102-71-6	Triethanolamine
CAS 102-76-1	Triacetin
CAS 102-77-2	Oxydiethylene benzothiazole sulfenamide
CAS 102-85-2	Tributyl phosphite
CAS 102-87-4	Trilaurylamine
CAS 102-92-1	Cinnamoyl chloride
CAS 103-09-0	Ethylhexyl acetate
CAS 103-11-7	Octyl acrylate
CAS 103-23-1	Dioctyl adipate
CAS 103-24-2	Dioctyl azelate

CAS 103-34-4	Dithiodimorpholine
CAS 103-44-6	Ethylhexyl vinyl ether
CAS 103-49-1	Dibenzylamine
CAS 103-82-2	Phenylacetic acid
CAS 103-83-3	Benzyldimethylamine
CAS 103-90-2	Acetaminophen
CAS 104-01-8	Methoxy phenylacetic acid
CAS 104-15-4	Toluenesulfonicacid
CAS 104-29-0	Chlorphenesin
CAS 104-31-4	Benzonatate
CAS 104-46-1	Anethole
CAS 104-74-5	Laurylpyridinium chloride
CAS 104-75-6	Ethylhexylamine
CAS 104-76-7	Ethylhexanol
CAS 104-87-0	Tolylaldehyde
CAS 105-16-8	Diethylaminoethyl methacrylate
CAS 105-34-0	Methyl cyanoacetate
CAS 105-46-4	s-Butylacetate
CAS 105-55-5	Diethylthiourea
CAS 105-57-7	Acetal
CAS 105-60-2	Caprolactam
CAS 105-74-8	Dilauroyl peroxide
CAS 105-74-8	Lauroyl peroxide
CAS 105-76-0	Dibutyl maleate
CAS 105-76-9	Dibutyl fumarate
CAS 105-95-3	Ethylene brassylate
CAS 105-97-5	Dicapryl adipate
CAS 105-99-7	Dibutyl adipate
CAS 106-01-4	Diethylene glycol diisononanoate
CAS 106-11-6	PEG-2 stearate
CAS 106-12-7	PEG-2 oleate
CAS 106-14-9	Hydroxystearic acid
CAS 106-16-1	Ricinoleamide MEA
CAS 106-17-2	Glycol ricinoleate
CAS 106-22-9	Citronellol
CAS 106-23-0	Citronellal
CAS 106-24-1	Geraniol
CAS 106-44-5	p-Cresol
CAS 106-46-7	p-Dichlorobenzene
CAS 106-47-8	p-Chloroaniline
CAS 106-50-3	p-Phenylene diamine
CAS 106-65-0	Dimethyl succinate
CAS 106-90-1	Glycidyl acrylate
CAS 106-91-2	Glycidyl methacrylate
CAS 106-92-3	Allyl glycidyl ether
CAS 106-97-8	Butane
CAS 107-03-9	Propyl mercaptan
CAS 107-06-2	Ethylene dichloride
CAS 107-13-1	Acrylonitrile
CAS 107-15-3	Ethylene diamine
CAS 107-21-1	Ethylene glycol
CAS 107-21-1	Glycol
CAS 107-22-2	Glyoxal
CAS 107-35-7	Taurine
CAS 107-41-5	Hexylene glycol
CAS 107-46-0	Hexamethyldisiloxane
CAS 107-51-7	Octamethyltrisiloxane
CAS 107-54-0	Dimethylhexynol
CAS 107-64-2	Distearyl dimonium chloride
CAS 107-71-1	Butyl peracetate
CAS 107-72-2	Amyl trichlorosilane
CAS 107-87-9	Methylpropyl ketone

CAS 107-97-1	Sarcosine	CAS 110-25-8	Oleoyl sarcosine
CAS 107-98-2	Methoxypropanol	CAS 110-26-9	Methylene bisacrylamide
CAS 107-98-2	Propylene glycol methyl ether	CAS 110-27-0	Isopropyl myristate
CAS 108-01-0	Dimethylethanolamine	CAS 110-30-5	Ethylene distearamide
CAS 108-03-2	Nitropropane	CAS 110-31-6	Ethylene dioleamide
CAS 108-05-4	Vinyl acetate	CAS 110-36-1	Butyl myristate
CAS 108-11-2	Methylamyl alcohol	CAS 110-42-9	Methyl caprate
CAS 108-18-9	Diisopropylamine	CAS 110-43-0	Methylamyl ketone
CAS 108-23-6	Isopropylchloroformate	CAS 110-44-1	Sorbic acid
CAS 108-24-7	Acetic anhydride	CAS 110-52-1	Dibromobutane
CAS 108-30-5	Succinic anhydride	CAS 110-54-3	Hexane
CAS 108-31-6	Maleic anhydride	CAS 110-63-4	Butanediol
CAS 108-32-7	Propylene carbonate	CAS 110-65-6	Butyncdiol
CAS 108-39-4	m-Cresol	CAS 110-80-5	Ethoxyethanol
CAS 108-42-9	m-Chloroaniline	CAS 110-82-7	Cyclohexane
CAS 108-45-2	m-Phenylenediamine	CAS 110-85-0	Piperazine
CAS 108-46-3	Resorcinol	CAS 110-86-1	Pyridine
CAS 108-48-5	Lutidine	CAS 110-89-4	Piperidine
CAS 108-62-3	Metaldehyde	CAS 110-90-0	Ethoxydiglycol
CAS 108-65-6	Propylene glycol methyl ether acetate	CAS 110-91-8	Morpholine
		CAS 110-97-4	Diisopropanolamine
CAS 108-73-6	1,3,5-Trihydroxybenzene	CAS 110-98-5	Dipropylene glycol
CAS 108-73-6	Phloroglucinol	CAS 111-01-3	Squalane
CAS 108-80-5	Cyanuric acid	CAS 111-02-4	Squalene
CAS 108-80-5	Isocyanuric acid	CAS 111-03-5	Glyceryl oleate
CAS 108-83-8	Diisobutyl ketone	CAS 111-05-7	Oleamide MIPA
CAS 108-87-2	Methylcyclohexane	CAS 111-14-8	Heptanoic acid
CAS 108-88-3	Toluene	CAS 111-15-9	Ethoxyethanol acetate
CAS 108-89-4	g-Picoline	CAS 111-17-1	Thiodipropionic acid
CAS 108-90-7	Chlorobenzene	CAS 111-20-6	Sebacic acid
CAS 108-91-8	Cyclohexylamine	CAS 111-27-3	Hexyl alcohol
CAS 108-93-0	Cyclohexanol	CAS 111-30-8	Glutaral
CAS 108-94-1	Cyclohexanone	CAS 111-30-8	Glutaraldehyde
CAS 108-98-5	Thiophenol	CAS 111-40-0	Diethylenetriamine
CAS 108-99-6	b-Picoline	CAS 111-41-1	Aminoethylethanolamine
CAS 109-02-4	Methyl morpholine	CAS 111-42-2	Diethanolamine
CAS 109-06-8	a-Picoline	CAS 111-46-6	Diethylene glycol
CAS 109-09-1	Chloropyridine	CAS 111-48-8	Thiodiglycol
CAS 109-13-7	Butylperoxy isobutyrate	CAS 111-55-7	Ethylene glycol diacetate
CAS 109-28-4	Oleamidopropyl dimethylamine	CAS 111-57-9	Stearamide MEA
CAS 109-31-9	Dihexyl azelate	CAS 111-58-0	Oleamide MEA
CAS 109-38-6	Butoxyethyl stearate	CAS 111-60-4	Glycol stearate
CAS 109-43-3	Dibutyl sebacate	CAS 111-66-0	Octene-1
CAS 109-46-6	Dibutyl thiourea	CAS 111-70-6	Heptyl alcohol
CAS 109-52-4	Valeric acid	CAS 111-76-2	Butoxyethanol
CAS 109-60-4	Propyl acetate	CAS 111-77-3	Methoxy diglycol
CAS 109-65-9	Butyl bromide	CAS 111-82-0	Methyl laurate
CAS 109-66-0	Pentane	CAS 111-86-4	Octanamine
CAS 109-73-9	Butylamine	CAS 111-87-5	Capryl alcohol
CAS 109-74-0	Butyronitrile	CAS 111-87-5	Caprylic alcohol
CAS 109-79-5	n-Butyl mercaptan	CAS 111-90-0	Ethoxydiglycol
CAS 109-86-4	Ethylene glycol methyl ether	CAS 111-96-6	Diethylene glycol dimethyl ether
CAS 109-86-4	Methoxyethanol	CAS 112-00-5	Laurtrimonium chloride
CAS 109-89-7	Diethylamine	CAS 112-02-7	Cetrimonium chloride
CAS 109-99-9	Tetrahydrofuran	CAS 112-03-8	Steartrimonium chloride
CAS 110-01-0	Tetrahydrothiophene	CAS 112-04-9	Octadecyl trichlorosilane
CAS 110-02-1	Thiophene	CAS 112-05-0	Pelargonic acid
CAS 110-05-4	Dibutyl peroxide	CAS 112-07-2	Butoxyethanol acetate
CAS 110-15-6	Succinic acid	CAS 112-10-7	Isopropyl stearate
CAS 110-16-7	Maleic acid	CAS 112-13-0	Decanoyl chloride
CAS 110-17-8	Fumaric acid	CAS 112-16-3	Lauroyl chloride
CAS 110-19-0	Isobutyl acetate	CAS 112-18-5	Dimethyl lauramine

CAS 112-24-3	Triethylene tetramine	CAS 118-60-5	Octyl salicylate
CAS 112-29-8	Decyl bromide	CAS 118-71-8	Maltol
CAS 112-30-1	Decyl alcohol	CAS 118-79-6	Tribromophenol
CAS 112-34-5	Butoxydiglycol	CAS 118-90-1	o-Toluic acid
CAS 112-38-9	Undecylenic acid	CAS 118-92-3	Anthranilic acid
CAS 112-39-0	Methyl palmitate	CAS 118-93-4	o-Hydroxy acetophenone
CAS 112-41-4;	Dodecene-1	CAS 119-07-3	Octyldecyl phthalate
CAS 112-42-5	Undecyl alcohol	CAS 119-36-8	Methyl salicylate
CAS 112-53-8	Lauryl alcohol	CAS 119-47-1	Methylenebis butyl methylphenol
CAS 112-55-0	n-Dodecyl mercaptan	CAS 119-53-9	Benzoin
CAS 112-57-2	Tetraethylene pentamine	CAS 119-64-2	Tetrahydronaphthalene
CAS 112-60-7	PEG-4	CAS 120-12-7	Anthracene
CAS 112-61-8	Methyl stearate	CAS 120-18-3	Naphthalenesulfonicacid
CAS 112-62-9	Methyl oleate	CAS 120-32-1	Benzylchlorophenol
CAS 112-69-6	Dimethyl palmitamine	CAS 120-36-5	Dichlorprop
CAS 112-70-9	Tridecyl alcohol	CAS 120-40-1	Lauramide DEA
CAS 112-72-1	Myristyl alcohol	CAS 120-47-8	Ethyl paraben
CAS 112-75-4	Dimethyl myristamine	CAS 120-51-4	Benzyl benzoate
CAS 112-80-1	Oleic acid	CAS 120-54-7	Dipentamethylene thiuram
CAS 112-84-5	Erucamide		hexasulfide
CAS 112-85-6	Behenic acid	CAS 120-55-8	Diethylene glycol dibenzoate
CAS 112-86-7	Erucic acid	CAS 120-78-5	Benzothiazyl disulfide
CAS 112-88-9	Octadecene-1	CAS 120-78-5	Dibenzothiazyl disulfide
CAS 112-90-3	Oleamine	CAS 120-80-9	Pyrocatechol
CAS 112-92-5	Stearyl alcohol	CAS 120-82-1	1,2,4-Trichlorobenzene
CAS 112-99-2	Dioctadecylamine	CAS 120-97-8	Dichlorphenamide
CAS 113-18-8	Ethchlorvynol	CAS 121-25-5	Amprolium
CAS 113-52-0	Imipramine hydrochloride	CAS 121-32-4	Ethyl vanillin
CAS 113-59-7	Chlorprothixene	CAS 121-33-5	Vanillin
CAS 113-98-4	Penicillin G potassium	CAS 121-47-1	Metanilic acid
CAS 114-07-8	Erythromycin	CAS 121-54-0	Benzethonium chloride
CAS 114-26-1	Propoxur	CAS 121-59-5	Carbarsone
CAS 114-80-7	Neostigmine bromide	CAS 121-66-4	Aminonitrothiazole
CAS 115-07-1	Propylene	CAS 121-69-7	Dimethyl aniline
CAS 115-10-6	Dimethyl ether	CAS 121-75-5	Malathion
CAS 115-19-5	Methylbutynol	CAS 121-79-9	Propyl gallate
CAS 115-21-9	Ethyltrichlorosilane	CAS 122-03-2	Isopropyl benzaldehyde
CAS 115-29-7	Endosulfan	CAS 122-09-8	Phentermine
CAS 115-32-2	Dicofol	CAS 122-11-2	Sulfadimethoxine
CAS 115-33-3	Oxyphenisatin acetate	CAS 122-14-5	Fenitrothion
CAS 115-38-8	Mephobarbital	CAS 122-18-9	Cetalkonium chloride
CAS 115-44-6	Talbutal	CAS 122-19-0	Stearalkonium chloride
CAS 115-67-3	Paramethadione	CAS 122-20-3	Triisopropanolamine
CAS 115-69-5	Aminomethyl propanediol	CAS 122-32-7	Triolein
CAS 115-70-8	Aminoethyl propanediol	CAS 122-34-9	Simazine
CAS 115-77-5	Pentaerythritol	CAS 122-42-9	Propham
CAS 115-83-3	Pentaerythrityl tetrastearate	CAS 122-52-1	Triethyl phosphite
CAS 115-86-6	Triphenyl phosphate	CAS 122-57-6	Benzylidene acetone
CAS 115-95-7	Linalyl acetate	CAS 122-59-8	Phenoxyacetic acid
CAS 116-14-3	Tetrafluoroethylene	CAS 122-99-6	Phenoxy ethanol
CAS 116-25-6	MDM hydantoin	CAS 123-03-5	Cetylpyridinium chloride
CAS 116-29-0	Tetradifon	CAS 123-08-0	Hydroxy benzaldehyde
CAS 116-38-1	Edrophonium chloride	CAS 123-28-4	Dilauryl thiodipropionate
CAS 116-43-8	Succinyl sulfathiazole	CAS 123-31-9	Hydroquinone
CAS 117-18-0	Tecnazene	CAS 123-33-1	Maleic hydrazide
CAS 117-39-5	Quercetin	CAS 123-42-2	Diacetone alcohol
CAS 117-81-7	Dioctyl phthalate	CAS 123-47-7	Prolonium iodide
CAS 117-82-8	Dimethoxyethyl phthalate	CAS 123-54-6	Acetyl acetone
CAS 117-83-9	Dibutoxyethyl phthalate	CAS 123-56-8	Succinimide
CAS 117-89-5	Trifluoperazine	CAS 123-63-7	Paraldehyde
CAS 118-52-5	Dichlorodimethyl hydantoin	CAS 123-72-8	Butyraldehyde
CAS 118-56-9	Homosalate	CAS 123-76-2	Levulinic acid

CAS 123-77-3	Azodicarbonamide	CAS 128-39-2	Dibutylphenol
CAS 123-86-4	Butyl acetate	CAS 128-49-4	Docusate calcium
CAS 123-91-1	Dioxane	CAS 128-62-1	Noscapine
CAS 123-92-2	Isoamyl acetate	CAS 129-03-3	Cyproheptadine
CAS 123-93-3	Thiodiglycolic acid	CAS 129-06-6	Warfarin sodium
CAS 123-94-4	Glycerylstearate	CAS 129-20-4	Oxyphenbutazone
CAS 123-95-5	Butyl stearate	CAS 129-49-7	Methysergide maleate
CAS 123-99-9	Azelaic acid	CAS 129-51-1	Ergonovine maleate
CAS 124-03-8	Cetethyl dimonium bromide	CAS 129-77-1	Piperidolate hydrochloride
CAS 124-04-9	Adipic acid	CAS 129-61-0	Thioridazine hydrochloride
CAS 124-07-2	Caprylic acid	CAS 130-95-0	Quinine
CAS 124-10-7	Methyl myristate	CAS 131-11-3	Dimethyl phthalate
CAS 124-22-1	Lauramine	CAS 131-13-5	Menadiol sodium diphosphate
CAS 124-26-5	Stearamide	CAS 131-17-9	Diallyl phthalate
CAS 124-28-7	Dimethyl stearamine	CAS 131-53-3	Benzophenone-8
CAS 124-30-1	Stearamine	CAS 131-54-4	Benzophenone-6
CAS 124-38-9	Carbon dioxide	CAS 131-55-5	Benzophenone-2
CAS 124-43-6	Urea hydrogen peroxide	CAS 131-56-6	Benzophenone-1
CAS 124-65-2	Sodium cacodylate	CAS 131-57-7	Benzophenone-3
CAS 124-68-5	Aminomethyl propanol	CAS 131-57-7	Oxybenzone
CAS 124-94-7	Triamcinolone	CAS 132-17-2	Benztropine mesylate
CAS 125-02-0	Prednisolone sodium phosphate	CAS 132-18-3	Diphenyl pyraline hydrochloride
CAS 125-04-2	Hydrocortisone sodium succinate	CAS 132-20-7	Pheniramine maleate
CAS 125-33-7	Primidone	CAS 132-49-0	Magnesium acetyl salicylate
CAS 125-53-1	Oxyphencyclimine	CAS 132-69-4	Benzydamine hydrochloride
CAS 125-84-8	Aminoglutethimide	CAS 132-92-3	Methicillin sodium
CAS 126-06-7	Bromochloro dimethyl hydantoin	CAS 132-93-4	Phenethicillin potassium
CAS 126-07-8	Griseofulvin	CAS 132-98-9	Penicillin V potassium
CAS 126-11-4	Trishydroxymethyl nitromethane	CAS 133-06-2	Captan
CAS 126-12-5	Anileridine dihydrochloride	CAS 133-07-3	Folpet
CAS 126-14-7	Sucrose octaacetate	CAS 133-11-9	Phenylamino salicylate
CAS 126-30-7	Neopentyl glycol	CAS 133-17-5	Iodohippurate sodium
CAS 126-31-8	Methiodal sodium	CAS 133-32-4	Indole butyric acid
CAS 126-33-0	Sulfolane	CAS 133-37-9	Tartaric acid
CAS 126-52-3	Ethinamate	CAS 133-49-3	Pentachlorothiophenol
CAS 126-58-9	Dipentaerythritol	CAS 133-53-9	Dichloroxylenol
CAS 126-73-8	Tributyl phosphate	CAS 133-67-5	Trichlormethiazide
CAS 126-80-7	Bisglycidoxy propyl tetramethyl-	CAS 134-03-2	Sodium ascorbate
	disiloxane	CAS 134-09-8	Menthyl anthranilate
CAS 126-86-3	Acetylenic glycol	CAS 134-20-3	Methyl anthranilate
CAS 126-86-3	Tetramethyldecynediol	CAS 134-31-6	Hydroxy quinoline sulfate
CAS 126-92-1	Sodium octyl sulfate	CAS 134-50-9	Aminacrine hydrochloride
CAS 126-99-8	Polychloroprene	CAS 134-62-3	Diethyl toluamide
CAS 127-08-2	Potassium acetate	CAS 134-72-5	Ephedrine sulfate
CAS 127-09-3	Sodium acetate	CAS 134-80-5	Diethylpropion hydrochloride
CAS 127-17-3	Pyruvic acid	CAS 135-07-9	Methyclothiazide
CAS 127-18-4	Perchloroethylene	CAS 135-09-1	Hydroflumethiazide
CAS 127-18-4	Tetrachloroethylene	CAS 135-31-9	Pyrrobutamine phosphate
CAS 127-25-3	Methyl abietate	CAS 136-23-2	Zinc dibutyl dithiocarbamate
CAS 127-39-9	Diisobutyl sodium sulfosuccinate	CAS 136-26-5	Capramide DEA
CAS 127-48-0	Trimethadione	CAS 136-30-1	Sodium dibutyl dithiocarbamate
CAS 127-65-1	Chloramine-T	CAS 136-36-7	Resorcinol benzoate
CAS 127-68-4	Sodium m-nitrobenzene sulfonate	CAS 136-40-3	Phenazopyridine hydrochloride
CAS 127-69-5	Sulfisoxazole	CAS 136-47-0	Tetracaine hydrochloride
CAS 127-91-3	β-Pinene	CAS 136-53-8	Zinc diethyl hexoate
CAS 128-03-0	Potassium dimethyl dithio-	CAS 136-60-7	Butyl benzoate
	carbamate	CAS 136-77-6	Hexyl resorcinol
CAS 128-04-1	Sodium dimethyl dithiocarbamate	CAS 136-85-6	Methyl benzotriazole
CAS 128-09-6	Chlorosuccinimide	CAS 136-99-2	Lauryl hydroxyethyl imidazoline
CAS 128-12-1	Acetosulfone sodium	CAS 137-08-6	D-Calcium pantothenate
CAS 128-37-0	BHT	CAS 137-16-6	Sodium lauroyl sarcosinate
CAS 128-37-0	Dibutylmethyl phenol	CAS 137-20-2	Sodium methyl oleoyl taurate

CAS 137-26-8	Tetramethylthiuram disulfide	CAS 142-78-9	Lauramide MEA
CAS 137-26-8	Thiram	CAS 142-82-5	Heptane
CAS 137-29-1	Copper dimethyl dithiocarbamate	CAS 142-87-0	Sodium lauryl sulfate
CAS 137-30-4	Zinc dimethyl dithiocarbamate	CAS 142-88-1	Piperazine adipate
CAS 137-30-4	Ziram	CAS 142-90-5	Lauryl methacrylate
CAS 137-40-6	Sodium propionate	CAS 142-91-6	Isopropyl palmitate
CAS 137-42-8	Metam sodium	CAS 143-07-7	Lauric acid
CAS 137-58-6	Lidocaine	CAS 143-18-0	Potassium oleate
CAS 138-22-7	Butyl lactate	CAS 143-19-1	Sodium oleate
CAS 138-84-1	Potassium aminobenzoate	CAS 143-27-1	Palmitamine
CAS 138-86-3	Dipentene	CAS 143-28-2	Oleyl alcohol
CAS 138-89-6	Nitrosodimethyl aniline	CAS 143-33-9	Sodium cyanide
CAS 139-07-1	Lauralkonium chloride	CAS 143-67-9	Vinblastine sulfate
CAS 139-08-2	Myristalkonium chloride	CAS 144-19-4	Trimethyl pentanediol
CAS 139-13-9	Nitrilotriacetic acid	CAS 144-23-0	Magnesium citrate
CAS 139-33-3	Disodium EDTA	CAS 144-29-6	Piperazine citrate
CAS 139-40-2	Propazine	CAS 144-34-4	Selenium dimethyl dithio-
CAS 139-41-3	Sodium dihydroxyethyl glycinate		carbamate
CAS 139-44-6	Trihydroxystearin	CAS 144-55-8	Sodium bicarbonate
CAS 139-88-8	Sodium myristyl sulfate	CAS 144-75-2	Sulfoxone sodium
CAS 139-89-9	Trisodium HEDTA	CAS 144-82-1	Sulfamethizole
CAS 139-96-8	TEA laurylsulfate	CAS 144-83-2	Sulfapyridine
CAS 140-01-2	Pentasodium pentetate	CAS 145-12-0	Oxymesterone
CAS 140-03-4	Methylacetyl ricinoleate	CAS 146-22-5	Nitrazepam
CAS 140-10-3	Cinnamic acid	CAS 147-24-0	Diphenhydramine hydrochloride
CAS 140-11-4	Benzyl acetate	CAS 147-85-3	L-Proline
CAS 140-31-8	Aminoethyl piperazine	CAS 147-94-4	Cytarabine
CAS 140-56-7	Dimethylaminobenzene diazo	CAS 148-18-5	Sodium dimethyl dithiocarbamate
	sodium sulfonate	CAS 148-24-3	Hydroxy quinoline
CAS 140-63-6	Propamidine isethionate	CAS 148-72-1	Pilocarpine nitrate
CAS 140-72-7	Cetyl pyridinium bromide	CAS 148-79-8	Thiabendazole
CAS 140-87-0	Tolualdehyde	CAS 148-82-3	Melphalan
CAS 140-93-2	Sodium isopropyl xanthate	CAS 149-30-4	Mercaptobenzothiazole
CAS 141-01-5	Ferrous fumarate	CAS 149-39-3	Sodium methyl stearoyl taurate
CAS 141-04-8	Diisobutyl adipate	CAS 149-44-0	Sodium formaldehyde sulfoxylate
CAS 141-08-2	Glyceryl ricinoleate	CAS 149-74-6	Methylphenyl dichlorosilane
CAS 141-20-8	PEG-2 laurate	CAS 149-91-7	Gallic acid
CAS 141-21-9	Stearamidoethyl ethanolamine	CAS 150-13-0	p-Aminobenzoicacid
CAS 141-22-0	Ricinoleic acid	CAS 150-38-9	Trisodium EDTA
CAS 141-23-1	Methylhydroxy stearate	CAS 150-76-5	Hydroquinone monomethyl ether
CAS 141-24-2	Methyl ricinoleate	CAS 151-06-4	Chlorphentermine hydrochloride
CAS 141-32-2	Butyl acrylate	CAS 151-21-3	Sodium lauryl sulfate
CAS 141-43-5	Ethanolamine	CAS 151-38-2	Methoxyethyl mercury acetate
CAS 141-53-7	Sodium formate	CAS 151-50-8	Potassium cyanide
CAS 141-57-1	Propyltrichlorosilane	CAS 151-67-7	Halothane
CAS 141-62-8	Decamethyltetrasiloxane	CAS 152-11-4	Verapamil hydrochloride
CAS 141-78-6	Ethyl acetate	CAS 152-43-2	Quinestrol
CAS 141-82-2	Malonic acid	CAS 152-47-6	Sulfalene
CAS 141-86-6	Diaminopyridine	CAS 152-62-5	Dydrogesterone
CAS 141-94-6	Hexetidine	CAS 152-97-6	Fluocortolone
CAS 142-08-5	Hydroxypyridine	CAS 153-87-7	Oxypertine
CAS 142-18-7	Glyceryl laurate	CAS 154-41-6	Phenylpropanolamine
CAS 142-26-7	Acetamide MEA		hydrochloride
CAS 142-31-4	Sodium octyl sulfate	CAS 154-42-7	Thioguanine
CAS 142-47-2	MSG	CAS 154-69-8	Tripelennamine hydrochloride
CAS 142-48-3	Stearoyl sarcosine	CAS 154-93-8	Carmustine
CAS 142-54-1	Lauramide MIPA	CAS 155-04-4	Zinc mercaptobenzothiazole
CAS 142-59-6	Nabam	CAS 155-41-9	Methscopolamine bromide
CAS 142-62-1	Caproic acid	CAS 156-54-7	Sodium butyrate
CAS 142-71-2	Copper acetate	CAS 275-51-4	Azulene
CAS 142-72-3	Magnesium acetate	CAS 280-57-9	Triethylenediamine
CAS 142-77-8	Butyl oleate	CAS 288-32-4	Imidazole

CAS 297-76-7	Ethynodiol diacetate
CAS 298-00-0	Parathion methyl
CAS 298-02-2	Phorate
CAS 298-04-4	Disulfoton
CAS 298-14-6	Potassium bicarbonate
CAS 298-46-4	Carbamezepine
CAS 298-57-7	Cinnarizine
CAS 298-59-9	Methyl phenidate hydrochloride
CAS 298-81-7	Methoxsalen
CAS 299-27-4	Potassium gluconate
CAS 299-28-5	Calcium gluconate
CAS 299-29-6	Ferrous gluconate
CAS 299-84-3	Ronnel
CAS 300-76-5	Naled
CAS 300-92-5	Aluminum distearate
CAS 300-92-5	Aluminum stearate
CAS 301-02-0	Oleamide
CAS 301-04-2	Lead acetate
CAS 301-12-2	Oxydemeton methyl
CAS 301-13-3	Trioctyl phosphite
CAS 302-01-2	Hydrazine
CAS 302-17-0	Chloral hydrate
CAS 302-41-0	Piritramide
CAS 302-84-1	DL-Serine
CAS 303-25-3	Cyclizine hydrochloride
CAS 303-42-4	Methenolone enanthate
CAS 303-81-1	Novobiocin
CAS 304-20-1	Hydralazine hydrochloride
CAS 304-55-2	Succimer
CAS 305-03-3	Chlorambucil
CAS 306-07-0	Pargyline hydrochloride
CAS 306-21-8	Hydroxyamphetamine hydrobromide
CAS 306-83-2	Dichlorotrifluoroethane
CAS 307-35-7	Perfluorooctane sulfonyl fluoride
CAS 309-43-3	Secobarbital sodium
CAS 314-40-9	Bromacil
CAS 315-30-0	Allopurinol
CAS 315-37-7	Testosterone enanthate
CAS 316-81-4	Thioproperazine
CAS 317-34-0	Aminophylline
CAS 318-98-9	Propranolol hydrochloride
CAS 320-67-2	Azacitidine
CAS 321-55-1	Haloxon
CAS 328-50-7	α-Ketoglutaric acid
CAS 330-54-1	Diuron
CAS 330-55-2	Linuron
CAS 333-36-8	Flurothyl
CAS 333-41-5	Diazinon
CAS 334-48-5	Capric acid
CAS 338-98-7	Isoflupredone acetate
CAS 339-44-6	Glymidine
CAS 340-56-7	Methaqualone hydrochloride
CAS 346-18-9	Polythiazide
CAS 356-12-7	Fluocinonide
CAS 357-07-3	Oxymorphone hydrochloride
CAS 357-08-4	Naloxone hydrochloride
CAS 357-56-2	Dextromoramide
CAS 358-52-1	Hexapropymate
CAS 360-70-3	Nandrolone decanoate
CAS 361-37-5	Methysergide
CAS 362-29-8	Propiomazine
CAS 364-62-5	Metoclopramide
CAS 364-98-7	Diazoxide
CAS 366-70-1	Procarbazine hydrochloride
CAS 371-41-5	Fluorophenol
CAS 373-02-4	Nickel acetate
CAS 378-44-9	Betamethasone
CAS 379-79-3	Ergotamine tartrate
CAS 382-67-2	Desoximetasone
CAS 389-08-2	Nalidixic acid
CAS 395-28-8	Isoxsuprine
CAS 396-01-0	Triamterene
CAS 404-82-0	Fenfluramine hydrochloride
CAS 409-21-2	Silicon carbide
CAS 409-21-2	Silicon dioxide
CAS 426-13-1	Fluorometholone
CAS 427-51-0	Cyproterone acetate
CAS 431-03-8	Diacetyl
CAS 434-05-9	Methenolone acetate
CAS 434-07-1	Oxymetholone
CAS 435-97-2	Phenprocoumon
CAS 437-38-7	Fentanyl
CAS 437-74-1	Xanthinol niacinate
CAS 438-41-5	Chlordiazepoxide hydrochloride
CAS 439-14-5	Diazepam
CAS 440-58-4	Iodamide
CAS 443-48-1	Metronidazole
CAS 446-86-6	Azathioprine
CAS 456-59-7	Cyclandelate
CAS 461-58-5	Dicyandiamide
CAS 462-06-6	Fluorobenzene
CAS 463-40-1	Linolenic acid
CAS 463-71-8	Thiophosgene
CAS 466-06-8	Proscillaridin
CAS 470-82-6	Eucalyptol
CAS 470-90-6	Chlorfenvinphos
CAS 471-34-1	Calcium carbonate
CAS 471-53-4	Enoxolone
CAS 473-41-6	Tolbutamide sodium
CAS 479-18-5	Dyphylline
CAS 480-30-8	Dichloral phenazone
CAS 490-55-1	Amiphenazole
CAS 495-69-2	Hippuric acid
CAS 497-19-8	Soda ash
CAS 497-19-8	Sodium carbonate
CAS 497-25-6	Oxazolidine
CAS 501-24-6	Pentadecyl phenol
CAS 501-30-4	Kojic acid
CAS 501-52-0	Hydrocinnamic acid
CAS 501-53-1	Benzylchloroformate
CAS 501-68-8	Beclamide
CAS 502-44-3	Caprolactone monomer
CAS 504-24-5	Aminopyridine
CAS 505-48-6	Suberic acid
CAS 506-59-2	Dimethylamine hydrochloride
CAS 506-68-3	Cyanogen bromide
CAS 506-93-4	Guanidine nitrate
CAS 509-67-1	Pholcodine
CAS 510-15-6	Chlorobenzilate
CAS 511-13-7	Chlophedianol hydrochloride
CAS 513-77-9	Barium carbonate
CAS 513-86-0	Acetylmethyl carbinol
CAS 514-36-3	Fludrocortisone acetate

CAS 514-65-8	Biperiden
CAS 518-28-5	Podophyllotoxin
CAS 518-47-8	Fluorescein sodium
CAS 519-95-9	Florantyrone
CAS 520-45-6	Dehydroaceticacid
CAS 521-12-0	Dromostanolone propionate
CAS 521-18-6	Stanolone
CAS 522-40-7	Fosfestrol
CAS 522-48-5	Tetrahydrozoline hydrochloride
CAS 522-51-0	Dequalinium chloride
CAS 523-87-5	Dimenhydrinate
CAS 525-66-6	Propranolol
CAS 526-08-9	Sulfaphenazole
CAS 526-95-4	Gluconic acid
CAS 527-07-1	Sodium gluconate
CAS 527-09-3	Copper gluconate
CAS 530-78-9	Flufenamic acid
CAS 532-03-6	Methocarbamol
CAS 532-27-4	Chloroacetophenone
CAS 532-32-1	Sodium benzoate
CAS 532-76-3	Hexylcaine hydrochloride
CAS 533-06-2	Mephenesin carbamate
CAS 533-74-4	Dimethyl tetrahydrothiadiazone-thione
CAS 533-96-0	Sodium sesquicarbonate
CAS 534-16-7	Silver carbonate
CAS 534-17-8	Cesium carbonate
CAS 536-33-4	Ethionamide
CAS 536-43-6	Dyclonine hydrochloride
CAS 536-90-3	m-Anisidine
CAS 537-65-5	Diaminodiphenylamine
CAS 538-23-8	Glyceryl trioctanoate
CAS 538-23-8	Tricaprylin
CAS 538-24-9	Trilaurin
CAS 538-71-6	Domiphen bromide
CAS 538-75-0	Dicyclohexyl carbodiimide
CAS 539-21-9	Ambazone
CAS 539-68-4	Dihydroxy aluminum sodium carbonate
CAS 540-10-3	Cetyl palmitate
CAS 540-88-5	t-Butylacetate
CAS 541-02-6	Decamethyl cyclopentasiloxane
CAS 541-05-9	Hexamethylcyclotrisiloxane
CAS 542-18-7	Cyclohexyl chloride
CAS 542-75-6	Dichloropropene
CAS 543-80-6	Barium acetate
CAS 544-17-2	Calcium formate
CAS 544-63-8	Myristic acid
CAS 546-56-5	Octaphenyl cyclotetrasiloxane
CAS 546-88-3	Acetohydroxamic acid
CAS 546-93-0	Magnesium carbonate
CAS 548-00-5	Ethylbiscoumacetate
CAS 548-68-5	Thiphenamil hydrochloride
CAS 548-73-2	Droperidol
CAS 549-18-8	Amitriptyline hydrochloride
CAS 550-29-2	Naphazoline hydrochloride
CAS 550-83-4	Propoxycaine hydrochloride
CAS 550-83-4	Propoxyphene hydrochloride
CAS 551-48-4	Guanoclor sulfate
CAS 551-92-8	Dimetridazole
CAS 552-89-6	o-Nitrobenzaldehyde
CAS 552-94-3	Salsalate

CAS 553-08-2	Thonzonium bromide
CAS 554-13-2	Lithium carbonate
CAS 554-57-4	Methazolamide
CAS 554-92-7	Trimethobenzamide hydrochloride
CAS 555-16-8	p-Nitrobenzaldehyde
CAS 555-30-6	Methyldopa
CAS 555-31-7	Aluminum isopropoxide
CAS 555-37-3	Neburon
CAS 555-43-1	Tristearin
CAS 555-44-2	Tripalmitin
CAS 555-45-3	Trimyristin
CAS 556-02-5	D-Tyrosine
CAS 556-03-6	DL-Tyrosine
CAS 556-50-3	Glycyl glycine
CAS 557-04-0	Magnesium stearate
CAS 557-05-1	Zinc stearate
CAS 561-78-4	Alphaprodine hydrochloride
CAS 562-09-4	Chlorphenoxamine hydrochloride
CAS 562-10-7	Doxylamine succinate
CAS 563-63-3	Silver acetate
CAS 569-57-3	Chlorotrianisene
CAS 573-41-1	Theophylline ethanolamine
CAS 575-44-6	1,6-Naphthalenediol
CAS 575-75-7	Toldimfos sodium
CAS 577-11-7	Dioctyl sodium sulfosuccinate
CAS 577-11-7	Docusatesodium
CAS 577-48-0	Butamben picrate
CAS 579-56-6	Isoxsuprine hydrochloride
CAS 582-17-2	2,7-Naphthalenediol
CAS 582-25-2	Potassium benzoate
CAS 583-39-1	Mercapto benzimidazole
CAS 583-52-8	Potassium oxalate
CAS 584-08-7	Potassium carbonate
CAS 584-09-8	Rubidium carbonate
CAS 584-84-9	Toluene diisocyanate
CAS 587-23-5	Methenamine mandelate
CAS 588-33-5	Isobornyl acrylate
CAS 588-42-1	Trolnitrate phosphate
CAS 588-59-0	Stilbene
CAS 590-00-1	Potassium sorbate
CAS 590-01-2	Butyl propionate
CAS 590-63-6	Bethane cholchloride
CAS 590-86-3	Isovaleraldehyde
CAS 591-50-4	Iodobenzene
CAS 591-71-7	Pentaerythrityl tetraacetate
CAS 592-01-8	Calcium cyanide
CAS 592-41-6	Hexene
CAS 593-29-3	Potassium stearate
CAS 593-60-2	Vinyl bromide
CAS 593-84-0	Guanidine thiocyanate
CAS 595-33-5	Megestrol acetate
CAS 596-51-0	Glycopyrrolate
CAS 597-09-1	Nitroethyl propanediol
CAS 598-56-1	Dimethylethylamine
CAS 598-63-0	Lead carbonate
CAS 599-61-1	3,3-Diaminodiphenylsulfone
CAS 599-64-4	Cumyl phenol
CAS 603-00-9	Proxyphylline
CAS 603-35-0	Triphenyl phosphine
CAS 603-50-9	Bisacodyl
CAS 604-75-1	Oxazepam

CAS 606-17-7	Iodipamide	CAS 738-70-5	Trimethoprim
CAS 611-75-6	Bromhexine hydrochloride	CAS 739-71-9	Trimipramine
CAS 614-33-5	Glyceryl tribenzoate	CAS 747-36-4	Hydroxychloroquin sulfate
CAS 614-39-1	Procainamide hydrochloride	CAS 749-02-0	Spiperone
CAS 614-45-9	Butyl perbenzoate	CAS 749-13-3	Trifluperidol
CAS 614-87-9	Dibromopropamidine isethionate	CAS 751-97-3	Rolitetracycline
CAS 615-13-4	Indanone	CAS 756-79-6	Dimethylmethyl phosphonate
CAS 615-58-7	Dibromophenol	CAS 759-94-4	EPTC
CAS 616-45-5	Pyrrolidone	CAS 762-16-3	Dioctanoyl peroxide
CAS 616-91-1	Acetyl cysteine	CAS 762-72-1	Allyltrimethylsilane
CAS 617-86-/	Triethylsilane	CAS 763-69-9	Ethylethoxy propionate
CAS 620-99-5	Phenacaine hydrochloride	CAS 768-33-2	Phenyldimethylchlorosilane
CAS 621-71-6	Tricaprin	CAS 771-03-9	Dehydroacetic acid
CAS 622-16-2	Carbodiimide	CAS 780-69-8	Phenyltriethoxysilane
CAS 624-04-4	Glycol dilaurate	CAS 797-63-7	Norgestrel
CAS 627-83-8	Glycol distearate	CAS 801-52-5	Porfiromycin
CAS 627-93-0	Dimethyl adipate	CAS 804-63-7	Quinine sulfate
CAS 628-63-7	Amyl acetate	CAS 808-26-4	Sancycline
CAS 629-11-8	Hexamethylene glycol	CAS 811-54-1	Lead formate
CAS 629-73-2;	Hexadecene-1	CAS 814- 94-8	Tin oxalate
CAS 629-76-5;	Pentadecyl alcohol	CAS 816-94-4	Distearoyl glycero phosphatidyl
CAS 630-56-8	Hydroxyprogesterone caproate		choline
CAS 630-60-4	Ouabaine	CAS 822-16-2	Sodium stearate
CAS 630-93-3	Phenytoin sodium	CAS 826-39-1	Mecamylamine hydrochloride
CAS 631-61-8	Ammonium acetate	CAS 834-12-8	Ametryn
CAS 632-79-1	Tetrabromophthalic anhydride	CAS 834-28-6	Phenformin hydrochloride
CAS 637-07-0	Clofibrate	CAS 846-49-1	Lorazepam
CAS 637-12-7	Aluminum stearate	CAS 846-50-4	Temazepam
CAS 637-12-7	Aluminum tristearate	CAS 849-55-8	Nylidrin hydrochloride
CAS 637-39-8	Triethanolamine hydrochloride	CAS 850-52-2	Altrenogest
CAS 637-58-1	Pramoxine hydrochloride	CAS 866-84-2	Potassium citrate
CAS 638-16-4	Trimercapto triazine	CAS 868-14-4	Potassium bitartrate
CAS 638-23-3	Carbocysteine	CAS 868-18-8	Sodium tartrate
CAS 638-38-0	Manganese acetate	CAS 868-77-9	Hydroxyethyl methacrylate
CAS 638-94-8	Desonide	CAS 870-72-4	Sodium hydroxymethane
CAS 643-22-1	Erythromycin stearate		sulfonate
CAS 644-62-2	Meclofenamic acid	CAS 871-37-4	Oleyl betaine
CAS 646-13-9	Isobutyl stearate	CAS 872-05-9	Decene-1
CAS 651-06-9	Sulfameter	CAS 872-50-4	Methyl pyrrolidinone
CAS 652-67-5	Isosorbide	CAS 874-60-2	Toluoyl chloride
CAS 657-27-2	Lysine hydrochloride	CAS 877-66-7	Toluene sulfonyl hydrazide
CAS 657-84-1	Sodium toluene sulfonate	CAS 886-50-0	Terbutryn
CAS 661-19-8	Behenyl alcohol	CAS 886-74-8	Chlorphenesin carbamate
CAS 661-61-6	TIPA lauryl sulfate	CAS 894-71-3	Nortriptyline hydrochloride
CAS 665-66-7	Amantadine hydrochloride	CAS 909-39-7	Opripramol dihydrochloride
CAS 667-83-4	Panthenylethyl ether	CAS 919-16-4	Lithium citrate
CAS 670-96-2	Phenyl imidazole	CAS 919-30-2	Aminopropyl triethoxysilane
CAS 671-16-9	Procarbazine	CAS 919-86-8	Demeton-S-methyl
CAS 671-88-5	Disulfamide	CAS 922-80-5	Diamyl sodium sulfosuccinate
CAS 681-84-5	Tetramethoxysilane	CAS 927-07-1	Butyl peroxy pivalate
CAS 682-01-9	Tetrapropoxysilane	CAS 927-83-3	Azobismethyl propane
CAS 683-10-3	Lauryl betaine	CAS 928-24-5	Glycol dioleate
CAS 686-31-7	Amyl peroxyethylhexanoate	CAS 928-65-4	Hexyltrichlorosilane
CAS 693-33-4	Cetyl betaine	CAS 929-77-1	Methyl behenate
CAS 693-36-7	Distearyl thiodipropionate	CAS 931-36-2	Ethylmethyl imidazole
CAS 693-98-1	Methyl imidazole	CAS 940-41-0	Phenethyl trichlorosilane
CAS 694-83-7	Diamino cyclohexane	CAS 944-22-9	Fonofos
CAS 696-59-3	Dimethoxy tetrahydrofuran	CAS 947-42-2	Diphenyl silanediol
CAS 709-98-8	Propanil	CAS 950-10-7	Mephosfolan
CAS 723-46-6	Sulfamethoxazole	CAS 950-37-8	Methidathion
CAS 729-99-7	Sulfamoxole	CAS 956-90-1	Phencyclidine hydrochloride
CAS 737-31-5	Diatrizoate sodium	CAS 957-51-7	Diphenamid

CAS 959-24-0	Sotalol hydrochloride
CAS 965-90-2	Ethylestrenol
CAS 968-81-0	Acetohexamide
CAS 968-93-4	Testolactone
CAS 973-21-7	Dinobuton
CAS 979-32-8	Estradiol valerate
CAS 980-71-2	Brompheniramine maleate
CAS 983-85-7	Penamecillin
CAS 985-16-0	Nafcillin sodium
CAS 992-21-2	Lymecycline
CAS 994-30-9	Triethyl chlorosilane
CAS 995-33-5	Butyl bisbutyl peroxy valerate
CAS 996-50-9	Diethylamino trimethylsilane
CAS 998-30-1	Triethoxysilane
CAS 999-21-3	Diallyl maleate
CAS 999-81-5	Chlormequat chloride
CAS 999-97-3	Hexamethyl disilazane
CAS 1000-50-6	n-Butyl dimethylchlorosilane
CAS 1000-90-4	Zinc isopropyl xanthate
CAS 1002-89-7	Ammonium stearate
CAS 1009-93-4	Hexamethylcyclotrisilazane
CAS 1014-69-3	Desmetryn
CAS 1020-84-4	Octamethyl cyclotetrasilazane
CAS 1055-55-6	Bunamidine hydrochloride
CAS 1066-17-7	Colistin
CAS 1066-30-4	Chromium acetate (ic)
CAS 1066-33-7	Ammonium bicarbonate
CAS 1066-35-9	Dimethylchlorosilane
CAS 1067-25-0	Propyltrimethoxysilane
CAS 1067-33-0	Dibutyltin diacetate
CAS 1067-53-4	Vinyltrismethoxyethoxysilane
CAS 1068-27-5	Dimethyl dibutyl peroxy hexyne
CAS 1071-27-8	3-Cyanopropyltrichlorosilane
CAS 1071-83-6	Glyphosate
CAS 1072-35-1	Lead stearate
CAS 1076-38-6	Hydroxycoumarin
CAS 1077-56-1	Ethyltoluene sulfonamide
CAS 1085-98-9	Dichlofluanid
CAS 1094-08-2	Ethopropazine hydrochloride
CAS 1095-90-5	Methadone hydrochloride
CAS 1098-60-8	Triflupromazine hydrochloride
CAS 1112-39-6	Dimethyldimethoxysilane
CAS 1119-97-7	Myrtrimonium bromide
CAS 1120-01-0	Sodium cetyl sulfate
CAS 1120-24-7	Dimethyl decylamine
CAS 1120-28-1	Methyl eicosenate
CAS 1120-36-1	Tetradecene-1
CAS 1120-49-6	Didecylamine
CAS 1122-62-9	Acetyl pyridine
CAS 1143-38-0	Anthralin
CAS 1151-11-7	Ipodate calcium
CAS 1156-19-0	Tolazamide
CAS 1163-19-5	Decabromo diphenyloxide
CAS 1166-52-5	Dodecyl gallate
CAS 1169-79-5	Quinestradiol
CAS 1177-87-3	Dexamethasone acetate
CAS 1181-54-0	Clomocycline
CAS 1185-53-1	Trishydroxy methylamino methane hydrochloride
CAS 1185-55-3	Methyltrimethoxysilane
CAS 1190-63-2	Cetyl stearate
CAS 1193-65-3	Quinuclidinone hydrochloride
CAS 1194-65-6	Dichlobenil
CAS 1197-21-3	Phentermine hydrochloride
CAS 1210-35-1	Dibenzosuberone
CAS 1218-35-5	Xylometazoline hydrochloride
CAS 1221-56-3	Ipodate sodium
CAS 1225-20-3	Iothalamate sodium
CAS 1225-55-4	Protriptyline hydrochloride
CAS 1225-60-1	Isothipendyl hydrochloride
CAS 1225-65-6	Prothipendyl hydrochloride
CAS 1229-29-4	Doxepin hydrochloride
CAS 1229-35-2	Methdilazine hydrochloride
CAS 1239-04-9	Phenazocine hydrobromide
CAS 1240-15-9	Propiomazine hydrochloride
CAS 1245-44-9	Propicillin potassium
CAS 1249-84-9	Diazacholesterol dihyrochloride
CAS 1263-89-4	Paromomycin sulfate
CAS 1264-72-8	Colistin sulfate
CAS 1300-71-6	Xylenol
CAS 1300-72-7	Sodium xylene sulfonate
CAS 1302-42-7	Sodium aluminate
CAS 1302-78-9	Bentonite
CAS 1303-28-2	Arsenic pentoxide
CAS 1303-86-2	Boron oxide
CAS 1303-96-4	Sodium borate
CAS 1304-28-5	Barium oxide
CAS 1304-29-6	Barium peroxide
CAS 1304-76-3	Bismuth trioxide
CAS 1304-85-4	Bismuth subnitrate
CAS 1305-62-0	Calcium hydroxide
CAS 1305-78-8	Calcium oxide
CAS 1305-79-9	Calcium peroxide
CAS 1306-06-5	Durapatite
CAS 1306-19-0	Cadmiumoxide
CAS 1307-96-6	Cobalt oxide (ous)
CAS 1308-04-9	Cobalt oxide (ic)
CAS 1308-38-9	Chromium (III) oxide
CAS 1309-37-1	Ferric oxide
CAS 1309-37-1	Iron oxide
CAS 1309-42-8	Magnesium hydroxide
CAS 1309-48-4	Magnesium oxide
CAS 1309-64-4	Antimony oxide
CAS 1310-58-3	Potassium hydroxide
CAS 1310-73-2	Sodium hydroxide
CAS 1312-43-2	Indium oxide
CAS 1312-76-1	Potassium silicate
CAS 1312-81-8	Lanthanum oxide
CAS 1313-13-9	Manganese dioxide
CAS 1313-27-5	Molybdenum trioxide
CAS 1313-60-6	Sodium peroxide
CAS 1313-82-2	Sodium sulfide, anhydrous
CAS 1313-96-8	Niobium oxide
CAS 1313-99-1	Nickel oxide (ous)
CAS 1314-08-5	Palladium oxide
CAS 1314-13-2	Zinc oxide
CAS 1314-23-4	Zirconium oxide
CAS 1314-35-8	Tungsten oxide (ic)
CAS 1314-36-9	Yttrium oxide
CAS 1314-37-0	Ytterbium oxide
CAS 1314-56-3	Phosphorus pentoxide
CAS 1314-60-9	Antimony pentoxide
CAS 1314-61-0	Tantalum oxide
CAS 1314-62-1	Vanadium pentoxide

CAS 1314-98-3	Zinc sulfide
CAS 1315-04-4	Antimony pentasulfide
CAS 1317-26-6	Magaldrate
CAS 1317-33-5	Molybdenum disulfide
CAS 1317-38-0	Copper oxide (ic)
CAS 1317-39-1	Copper oxide (ous)
CAS 1317-45-9	Tin oxide
CAS 1317-65-3	Calcium carbonate
CAS 1317-65-3	Limestone
CAS 1317-97-1	Ultramarineblue
CAS 1318-74-7	Kaolinite
CAS 1318-93-0	Montmorillonite
CAS 1319-41-1	Sepiolite
CAS 1322-93-6	Sodium diisopropyl naphthalene sulfonate
CAS 1322-98-1	Sodium decylbenzene sulfonate
CAS 1323-03-1	Myristyl lactate
CAS 1323-38-2	Castor oil
CAS 1323-39-3	Propylene glycol stearate
CAS 1323-42-8	Glyceryl hydroxystearate
CAS 1323-65-5	Dinonylphenol
CAS 1323-83-7	Glyceryl distearate
CAS 1327-33-9	Antimony oxide
CAS 1327-41-9	Aluminum chlorohydrate
CAS 1327-43-1	Magnesium aluminum silicate
CAS 1327-53-3	Arsenic trioxide
CAS 1330-20-7	Xylene
CAS 1330-43-4	Sodium borate
CAS 1330-61-6	Isodecyl acrylate
CAS 1330-76-3	Diisooctyl maleate
CAS 1330-78-5	Tricresyl phosphate
CAS 1330-80-9	Propylene glycol oleate
CAS 1330-85-4	Dodecylbenzyl trimonium chloride
CAS 1331-61-9	Ammonium dodecylbenzene sulfonate
CAS 1332-07-6	Zinc borate
CAS 1332-40-7	Copper oxychloride
CAS 1332-58-7	Kaolin
CAS 1333-39-7	Phenolsulfonic acid
CAS 1333-82-0	Chromium (VI) oxide
CAS 1333-84-2	Alumina hydrate
CAS 1333-86-4	Carbon black
CAS 1334-78-7	Tolylaldehyde
CAS 1335-72-4	Sodium laureth sulfate
CAS 1336-20-5	Tetracycline phosphate
CAS 1336-29-4	Bisacodyltannex
CAS 1337-33-3	Stearyl citrate
CAS 1337-76-4	Attapulgite
CAS 1337-81-1	Vinyl pyridine
CAS 1338-02-9	Copper naphthenate
CAS 1338-23-4	MEK peroxide
CAS 1338-23-4	Methyl ethyl ketone peroxide
CAS 1338-24-5	Naphthenic acid
CAS 1338-39-2	Sorbitan laurate
CAS 1338-41-6	Sorbitan stearate
CAS 1338-43-8	Sorbitan oleate
CAS 1338-51-8	Toluene sulfonamide formaldehyde resin
CAS 1341-49-7	Ammonium bifluoride
CAS 1341-54-4	Benzophenone-11
CAS 1343-88-0	Magnesium silicate
CAS 1343-98-2	Silicic acid

CAS 1344-00-9	Sodium silicoaluminate
CAS 1344-09-8	Sodium silicate
CAS 1344-28-1	Alumina
CAS 1344-28-1	Aluminum oxide
CAS 1344-40-7	Lead phthalate basic
CAS 1344-48-5	Mercury sulfide (ic)
CAS 1345-05-7	Lithopone
CAS 1345-44-6	Antimony trisulfide
CAS 1369-66-3	Dioctyl sodium sulfosuccinate
CAS 1390-65-4	Carmine
CAS 1397-89-3	Amphotericin
CAS 1398-61-4	Chitin
CAS 1400-61-9	Nystatin
CAS 1401-55-4	Tannic acid
CAS 1402-82-0	Amphomycin
CAS 1403-17-4	Candicidin
CAS 1404-04-2	Neomycin
CAS 1404-88-2	Tyrothricin
CAS 1404-93-9	Vancomycin hydrochloride
CAS 1405-10-3	Neomycin sulfate
CAS 1405-20-5	Polymyxin B sulfate
CAS 1405-37-4	Capreomycin sulfate
CAS 1405-41-0	Gentamicin sulfate
CAS 1405-87-4	Bacitracin
CAS 1406-65-1	Chlorophyll
CAS 1420-03-7	Propenzolate hydrochloride
CAS 1420-55-9	Thiethyl perazine
CAS 1421-68-7	Amidephrine mesylate
CAS 1424-00-6	Mesterolone
CAS 1446-61-3	Dehydroabietylamine
CAS 1450-14-2	Hexamethyldisilane
CAS 1459-93-4	Dimethylisophthalate
CAS 1462-54-0	Lauraminopropionic acid
CAS 1462-55-1	Dodecylthioethanol
CAS 1476-53-5	Novobiocin sodium
CAS 1491-41-4	Naftalofos
CAS 1493-13-6	Trifluoromethane sulfonic acid
CAS 1508-65-2	Oxybutynin chloride
CAS 1508-75-4	Tropicamide
CAS 1508-76-5	Procyclidine hydrochloride
CAS 1524-88-5	Flurandrenolide
CAS 1528-49-0	Trihexyl trimellitate
CAS 1533-45-5	Ethenediyl diphenylene bisbenzoxazole
CAS 1538-09-6	Penicillin G benzathine
CAS 1553-60-2	Ibufenac
CAS 1558-33-4	Chloromethylmethyl dichloro-silane
CAS 1561-92-8	Sodium methallyl sulfonate
CAS 1562-00-1	Sodium isethionate
CAS 1563-66-2	Carbofuran
CAS 1582-09-8	Trifluralin
CAS 1592-23-0	Calcium stearate
CAS 1596-84-5	Daminozide
CAS 1597-82-6	Paramethasone acetate
CAS 1600-27-7	Mercury acetate (ic)
CAS 1607-17-6	Pentrinitrol
CAS 1617-90-9	Vincamine
CAS 1620-21-9	Chlorcyclizine hydrochloride
CAS 1622-61-3	Clonazepam
CAS 1622-62-4	Flunitrazepam
CAS 1633-05-2	Strontium carbonate

CAS 1634-02-2	Tetrabutyl thiuram disulfide
CAS 1634-04-4	Methyl butyl ether
CAS 1642-54-2	Diethylcarbamazine citrate
CAS 1643-19-2	Tetrabutyl ammonium bromide
CAS 1643-20-5	Lauramine oxide
CAS 1649-18-9	Azaperone
CAS 1656-63-9	Trilauryl trithiophosphite
CAS 1665-48-1	Metaxalone
CAS 1668-19-5	Doxepin
CAS 1680-21-3	PEG-3 diacrylate
CAS 1689-83-4	Ioxynil
CAS 1689-84-5	Bromoxynil
CAS 1689-89-0	Nitroxynil
CAS 1698-60-8	Chloridazon
CAS 1702-17-6	Clopyralid
CAS 1702-17-6	2,4-D
CAS 1704-62-7	Dimethylaminoethoxyethanol
CAS 1707-14-8	Phenmetrazine hydrochloride
CAS 1715-33-9	Prednisolone sodium succinate
CAS 1717-00-6	Dichlorofluoroethane
CAS 1719-57-9	Chloromethyl dimethyl-chlorosilane
CAS 1719-58-0	Vinyl dimethyl chlorosilane
CAS 1722-62-9	Mepivacaine hydrochloride
CAS 1732-10-1	Dimethyl azelate
CAS 1740-22-3	Pyrinoline
CAS 1758-73-2	Thiourea dioxide
CAS 1760-24-3	Aminoethylaminopropyl-trimethoxysilane
CAS 1762-95-4	Ammonium thiocyanate
CAS 1777-82-8	Dichlorobenzyl alcohol
CAS 1783-96-6	D-Aspartic acid
CAS 1786-81-8	Prilocaine hydrochloride
CAS 1808-12-4	Bromodiphenhydramine hydrochloride
CAS 1809-19-4	Dibutyl phosphite
CAS 1812-30-2	Bromazepam
CAS 1812-53-9	Dicetyl dimonium chloride
CAS 1825-62-3	Trimethylethoxysilane
CAS 1841-19-6	Fluspirilene
CAS 1843-03-4	Trismethylhydroxybutylphenyl butane
CAS 1843-05-6	Benzophenone-12
CAS 1843-05-6	Octabenzone
CAS 1847-55-8	Sodium oleyl sulfate
CAS 1861-32-1	Chlorthaldimethyl
CAS 1861-32-1	Dacthal
CAS 1861-40-1	Benfluralin
CAS 1867-66-9	Ketamine hydrochloride
CAS 1889-67-4	Dimethyl diphenyl butane
CAS 1897-45-6	Chlorothalonil
CAS 1897-45-6	Tetrachloroisophthalonitrile
CAS 1910-42-5	Paraquat
CAS 1912-24-9	Atrazine
CAS 1912-26-1	Trietazine
CAS 1918-00-9	Dicamba
CAS 1918-02-1	Picloram
CAS 1918-16-7	Propachlor
CAS 1931-62-0	Butylperoxy maleic acid
CAS 1948-33-0	Butylhydroquinone
CAS 1954-81-4	Chloramben
CAS 1975-78-6	Decane nitrile

CAS 1977-10-2	Loxapine
CAS 1982-37-2	Methdilazine
CAS 1982-47-4	Chloroxuron
CAS 1984-06-1	Sodium caprylate
CAS 1854-23-5	Ethyl nitrotrimethylene dimorpholine
CAS 2002-29-1	Flumethasone pivalate
CAS 2016-36-6	Choline salicylate
CAS 2016-42-4	Myristamine
CAS 2016-56-0	Cocamine acetate
CAS 2016-57-1	Lauramine
CAS 2022-85-7	Flucytosine
CAS 2030-63-9	Clofazimine
CAS 2031-67-6	Methyl triethoxysilane
CAS 2032-65-7	Methiocarb
CAS 2044-56-6	Lithium lauryl sulfate
CAS 2058-46-0	Oxytetracycline hydrochloride
CAS 2058-58-4	D-Asparagine
CAS 2062-78-4	Pimozide
CAS 2062-84-2	Benperidol
CAS 2068-78-2	Vincristine sulfate
CAS 2082-79-3	Octadecyl dibutyl hydroxy hydrocinnamate
CAS 2082-79-3	Octadecyl dibutyl hydroxyphenyl propionate
CAS 2082-80-6	Tristearyl phosphite
CAS 2083-91-2	Dimethylaminotrimethylsilane
CAS 2094-98-6	Azobiscyclohexane carbonitrile
CAS 2116-84-9	Phenyldimethicone
CAS 2116-84-9	Phenyltrimethicone
CAS 2119-75-7	Fluperolone acetate
CAS 2135-22-0	Oxamyl
CAS 2152-34-3	Pemoline
CAS 2152-44-5	Betamethasone valerate
CAS 2155-71-7	Dibutyl peroxy phthalate
CAS 2156-97-0	Lauryl acrylate
CAS 2163-42-0	Methyl propanediol
CAS 2163-80-6	MSMA
CAS 2164-08-1	Lenacil
CAS 2164-17-2	Fluometuron
CAS 2167-23-9	Dibutyl peroxy butane
CAS 2190-04-7	Stearamine acetate
CAS 2190-04-7	Tallowamine acetate
CAS 2210-79-9	Cresyl glycidyl ether
CAS 2223-93-0	Cadmium stearate
CAS 2224-33-1	Vinyl trismethylethyl ketoxime silane
CAS 2224-44-4	Nitrobutylmorpholine
CAS 2227-17-0	Dienochlor
CAS 2235-43-0	Pentasodium aminotrimethylene phosphonate
CAS 2235-54-3	Ammonium lauryl sulfate
CAS 2240-14-4	Fencamfamin hydrochloride
CAS 2244-21-5	Potassium dichloroisocyanurate
CAS 2259-96-3	Cyclothiazide
CAS 2277-92-1	Oxyclozanide
CAS 2307-68-8	Pentanochlor
CAS 2310-17-0	Phosalone
CAS 2311-27-5	Nonoxynol-006
CAS 2315-02-8	Oxymetazoline hydrochloride
CAS 2321-07-5	Fluorescein
CAS 2344-80-1	Chloromethyltrimethylsilane

CAS 2358-84-1	Diethylene glycol dimethacrylate
CAS 2372-21-6	Butyl peroxy isopropyl carbonate
CAS 2375-03-3	Methyl prednisolone sodium succinate
CAS 2392-39-4	Dexamethasone sodium phosphate
CAS 2398-96-1	Tolnaftate
CAS 2409-55-4	p-Butylcresol
CAS 2425-79-8	Butanediol diglycidyl ether
CAS 2426-08-6	Butyl glycidyl ether
CAS 2426-54-2	Diethyl aminoethyl acrylate
CAS 2437-25-4	Lauryl nitrile
CAS 2438-32-6	Chlorpheniramine maleate
CAS 2438-72-4	Bufexamac
CAS 2439-01-2	Chinomethionate
CAS 2439-10-3	Dodine
CAS 2439-35-2	Dimethylamino ethylacrylate
CAS 2440-22-4	Drometrizole
CAS 2447-57-6	Sulfadoxine
CAS 2455-24-5	Tetrahydrofurfuryl methacrylate
CAS 2489-77-2	Trimethyl thiourea
CAS 2492-26-4	Sodium mercaptobenzothiazole
CAS 2499-95-8	Hexyl acrylate
CAS 2508-72-7	Antazoline hydrochloride
CAS 2530-83-8	Glycidoxy propyl trimethoxysilane
CAS 2530-85-0	Methacryloxypropyl trimethoxy-silane
CAS 2530-87-2	Chloropropyl trimethoxysilane
CAS 2550-06-3	Chloropropyl trichlorosilane
CAS 2551-83-9	Allyl trimethoxysilane
CAS 2553-19-7	Diphenyl diethoxysilane
CAS 2555-05-7	Methyl pyrrolidone
CAS 2571-88-2	Stearamine oxide
CAS 2576-92-3	Isoetharine hydrochloride
CAS 2589-57-3	Dimethyl azobismethyl propionate
CAS 2592-95-2	Hydroxy benzotriazole
CAS 2593-15-9	Etridiazole
CAS 2599-01-1	Cetyl myristate
CAS 2607-06-9	Diflucortolone
CAS 2608-24-4	Piposulfan
CAS 2622-26-6	Pericyazine
CAS 2624-44-4	Ethamsylate
CAS 2627-95-4	Divinyl tetramethyl disiloxane
CAS 2631-40-5	Isopropyl phenylmethyl carbamate
CAS 2642-71-9	Azinphos-ethyl
CAS 2642-71-9	Cotnion-ethyl
CAS 2644-64-6	Dipalmitoyl glycerophosphatidyl choline
CAS 2668-66-8	Medrysone
CAS 2673-22-5	Ditridecyl sodium sulfosuccinate
CAS 2675-77-6	Chloroneb
CAS 2694-54-4	Triallyl trimellitate
CAS 2724-58-5	Isostearic acid
CAS 2746-81-8	Fluphenazine enanthate
CAS 2751-09-9	Troleandomycin
CAS 2754-27-0	Trimethylsilyl acetate
CAS 2768-02-7	Vinyl trimethoxysilane
CAS 2778-96-3	Stearyl stearate
CAS 2807-30-9	Ethylene glycol propyl ether
CAS 2809-21-4	Hydroxyethylidene diphosphonic acid

CAS 2809-21-4	Etidronic acid
CAS 2825-60-7	Formocortal
CAS 2829-19-8	Rolicyprine
CAS 2857-97-8	Trimethyl bromosilane
CAS 2867-47-2	Dimethylaminoethyl methacrylate
CAS 2893-78-9	Sodium dichloroisocyanurate
CAS 2897-60-1	Glycidoxypropyl methyl diethoxysilane
CAS 2898-12-6	Medazepam
CAS 2915-57-3	Dioctyl succinate
CAS 2919-66-6	Melengestrol acetate
CAS 2921-88-2	Chlorpyrifos
CAS 2921-92-8	Propatyl nitrate
CAS 2939-80-2	Captafol
CAS 2943-75-1	Octyl triethoxysilane
CAS 2955-38-6	Prazepam
CAS 2996-92-1	Phenyl trimethoxysilane
CAS 2998-57-4	Estramustine
CAS 3006-15-3	Dihexyl sodium sulfosuccinate
CAS 3006-82-4	Butyl peroctoate
CAS 3006-86-8	Dibutyl peroxy cyclohexane
CAS 3006-93-7	Phenylene dimaleimide
CAS 3010-24-0	PEG-2 stearmonium chloride
CAS 3026-63-9	Sodium tridecyl sulfate
CAS 3033-77-0	Glycidyl trimethyl ammonium chloride
CAS 3034-79-5	Dimethyl benzoyl peroxide
CAS 3055-93-4	Laureth-2
CAS 3055-95-6	Laureth-5
CAS 3057-08-7	Phenyl neopentylene glycol phosphite
CAS 3060-89-7	Metobromuron
CAS 3061-75-4	Behenamide
CAS 3064-70-8	Bistrichloromethyl sulfone
CAS 3069-25-8	Methylaminopropyl trimethoxy-silane
CAS 3069-29-2	Aminoethylaminopropylmethyl dimethoxysilane
CAS 3076-63-9	Trilauryl phosphite
CAS 3081-14-9	Bis dimethyl pentyl phenylene diamine
CAS 3088-31-1	Sodium laureth sulfate
CAS 3093-35-4	Halcinonide
CAS 3097-08-3	Magnesium lauryl sulfate
CAS 3101-60-8	Butyl phenyl glycidyl ether
CAS 3115-05-7	Iobenzamic acid
CAS 3115-49-9	Nonoxynol-8 carboxylic acid
CAS 3121-60-6	Benzophenone-9
CAS 3121-61-7	Methoxyethyl acrylate
CAS 3129-91-7	Dicyclohexylamine nitrite
CAS 3147-75-9	Hydroxy octylphenyl benzo-triazole
CAS 3147-75-9	Octrizole
CAS 3158-56-3	Pentadecyl phenol
CAS 3179-56-4	Acetyl cyclohcxane sulfonyl peroxide
CAS 3179-76-8	Aminopropylmethyl diethoxy-silane
CAS 3179-80-4	Lauramidopropyl dimethylamine
CAS 3179-81-5	Diethylaminoethyl stearate
CAS 3194-55-6	Hexabromo cyclododecane
CAS 3198-07-0	Ethyl norepinephrine HCl

CAS 3200-06-4	Nafronyl oxalate
CAS 3234-85-3	Myristyl myristate
CAS 3254-89-5	Diphenidol hydrochloride
CAS 3270-71-1	Nifuraldezone
CAS 3277-26-7	Tetramethyl disiloxane
CAS 3278-89-5	Tribromo phenyl allyl ether
CAS 3290-92-4	Trimethylolpropane trimeth-acrylate
CAS 3296-90-0	Dibromo neopentyl glycol
CAS 3319-31-1	Trioctyl trimellitate
CAS 3322-93-8	Dibromoethyl dibromo-cyclohexane
CAS 3327-22-8	Chlorohydroxypropyl trimonium chloride
CAS 3332-27-2	Myristamine oxide
CAS 3333-67-3	Nickel carbonate basic
CAS 3337-71-1	Asulam
CAS 3344-18-1	Magnesium citrate
CAS 3347-22-6	Dithianon
CAS 3349-06-2	Nickel formate
CAS 3380-34-5	Triclosan
CAS 3383-96-8	Temephos
CAS 3388-04-3	Epoxy cyclohexylethyl trimethoxysilane
CAS 3398-33-2	Sodium undecylenate
CAS 3416-26-0	Lidoflazine
CAS 3425-61-4	Amyl hydroperoxide
CAS 3436-44-0	Dicaproyl glycerophosphatidyl choline
CAS 3452-07-1	Eicosene-1
CAS 3457-61-2	Butyl cumyl peroxide
CAS 3483-12-3	Dithiothreitol
CAS 3485-14-1	Cyclacillin
CAS 3486-35-9	Zinc carbonate
CAS 3505-38-2	Carbinoxamine maleate
CAS 3511-16-8	Hetacillin
CAS 3521-62-8	Erythromycin estolate
CAS 3521-84-4	Iodipamide meglumine
CAS 3524-68-3	Pentaerythrityl triacrylate
CAS 3546-41-6	Pyrvinium pamoate
CAS 3595-11-7	Propyl hexedrine
CAS 3614-30-0	Emepronium bromide
CAS 3614-69-5	Dimethindene maleate
CAS 3615-41-6	Rhamnose
CAS 3622-84-2	Butylbenzene sulfonamide
CAS 3625-06-7	Mebeverine
CAS 3632-91-5	Magnesium gluconate
CAS 3648-20-2	Diundecyl phthalate
CAS 3655-00-3	Disodium laurimino dipropionate
CAS 3658-48-8	Dioctyl ethylhexyl phosphite
CAS 3682-94-8	Azobismethyl propionamide dihydrate
CAS 3687-45-4	Oleyl oleate
CAS 3687-46-5	Decyl oleate
CAS 3689-24-5	Sulfotep
CAS 3691-35-8	Chlorophacinone
CAS 3704-09-4	Mibolerone
CAS 3710-84-7	Diethylhydroxylamine
CAS 3724-65-0	Crotonic acid
CAS 3734-16-5	Prodilidine hydrochloride
CAS 3734-33-6	Denatonium benzoate
CAS 3735-90-8	Phencarbamide

CAS 3736-26-3	Diisopropyl benzene monohydroperoxide
CAS 3768-58-9	Bisdimethylamino dimethylsilane
CAS 3775-90-4	Butylaminoethyl methacrylate
CAS 3778-73-2	Ifosfamide
CAS 3784-99-4	Stilbaziumiodide
CAS 3806-34-6	Distearyl pentaerythrityl diphosphite
CAS 3810-74-0	Streptomycin sulfate
CAS 3811-73-2	Sodium pyrithione
CAS 3813-05-6	Benazolin
CAS 3818-50-6	Bephenium hydroxy naphthoate
CAS 3819-00-9	Piperacetazine
CAS 3847-29-8	Erythromycin lactobionate
CAS 3851-87-4	Bistrimethylhexanoyl peroxide
CAS 3851-87-4	Trimethyl hexanoyl peroxide
CAS 3858-89-7	Chloroprocaine hydrochloride
CAS 3864-99-1	Dibutyl hydroxyphenyl chlorobenzotriazole
CAS 3878-19-1	Fuberidazole
CAS 3896-11-5	Hydroxybutylmethylphenyl chlorobenzotriazole
CAS 3913-02-8	Isododecyl alcohol
CAS 3930-19-6	Streptonigrin
CAS 3963-95-9	Methacycline hydrochloride
CAS 3978-86-7	Azatadine maleate
CAS 3990-03-2	Monoethyl maleate
CAS 3999-10-8	Phenyl tetrazole
CAS 4008-48-4	Nitroxoline
CAS 4065-45-6	Benzophenone-4
CAS 4065-45-6	Sulisobenzone
CAS 4070-80-8	Sodium stearyl fumarate
CAS 4075-81-4	Calcium propionate
CAS 4088-22-6	Distearyl methylamine
CAS 4098-71-9	Isophorone diisocyanate
CAS 4130-08-9	Vinyl triacetoxysilane
CAS 4171-13-5	Valnoctamide
CAS 4196-86-5	Pentaerythrityl tetrabenzoate
CAS 4205-90-7	Clonidine
CAS 4205-91-8	Clonidine hydrochloride
CAS 4219-49-2	Glycol palmitate
CAS 4253-34-3	Methyl triacetoxysilane
CAS 4268-36-4	Tybamate
CAS 4292-10-8	Lauramidopropyl betaine
CAS 4299-60-9	Sulfisoxazole diolamine
CAS 4304-40-9	Thenium closylate
CAS 4306-88-1	Diisobutyl nonylphenol
CAS 4330-99-8	Trimeprazine tartrate
CAS 4369-14-6	Acryloxypropyl trimethoxysilane
CAS 4390-04-9	Isohexadecane
CAS 4419-11-8	Azobisdimethyl valeronitrile
CAS 4420-74-0	Mercaptopropyl trimethoxysilane
CAS 4468-02-4	Zinc gluconate
CAS 4484-72-4	Dodecyl trichlorosilane
CAS 4485-12-5	Lithium stearate
CAS 4511-39-1	Amyl peroxy benzoate
CAS 4511-39-1	Amylperoxybenzoate
CAS 4584-46-7	Dimethylaminoethyl chloride hydrochloride
CAS 4618-18-2	Lactulose
CAS 4658-28-0	Aziprotryne
CAS 4682-36-4	Orphenadrine citrate

CAS 4691-65-0	Disodium inosinate
CAS 4706-78-9	Potassium lauryl sulfate
CAS 4712-55-4	Diphenyl phosphite
CAS 4719-04-4	Hydroxyethyl triazine
CAS 4722-98-9	MEA lauryl sulfate
CAS 4744-10-9	Dimethoxy propane
CAS 4748-78-1	Ethyl benzaldehyde
CAS 4766-57-8	Tetrabutoxysilane
CAS 4810-50-8	Potassium PCA
CAS 4862-18-4	Nitrilotriacetamide
CAS 4936-47-4	Nifuratel
CAS 5003-48-5	Benorylate
CAS 5015-36-1	Methyl prednisolone sodium phosphate
CAS 5026-62-0	Sodium methyl paraben
CAS 5039-78-1	Dimethylaminoethyl methacrylate methyl chloride
CAS 5064-31-3	Sodium NTA
CAS 5064-31-3	Trisodium NTA
CAS 5089-70-3	Chloropropyl triethoxysilane
CAS 5104-49-4	Flurbiprofen
CAS 5131-24-8	Ditalimfos
CAS 5221-53-4	Dimethirimol
CAS 5250-39-5	Floxacillin
CAS 5259-88-1	Oxycarboxin
CAS 5274-68-0	Laureth-4
CAS 5283-66-9	Octyl trichlorosilane
CAS 5306-85-4	Dimethyl isosorbide
CAS 5321-32-4	Hetacillin potassium
CAS 5323-95-5	Sodium ricinoleate
CAS 5328-37-0	L-Arabinose
CAS 5329-14-6	Sulfamic acid
CAS 5333-42-6	Octyl dodecanol
CAS 5356-84-3	Vinyl tristrimethyl siloxysilane
CAS 5370-01-4	Mexiletine hydrochloride
CAS 5392-40-5	Citral
CAS 5397-31-9	Octyloxy propylamine
CAS 5411-22-3	Benzphetamine hydrochloride
CAS 5422-34-4	Lactamide MEA
CAS 5434-57-1	Isohexyl neopentanoate
CAS 5466-77-3	Octyl methoxy cinnamate
CAS 5534-09-8	Beclomethasone dipropionate
CAS 5534-95-2	Pentagastrin
CAS 5536-17-4	Vidarabine
CAS 5560-59-8	Alverine citrate
CAS 5560-72-5	Iprindole
CAS 5560-78-1	Teclozan
CAS 5578-42-7	Methyl cyclohexyl dichlorosilane
CAS 5585-73-9	Butriptyline hydrochloride
CAS 5593-20-4	Betamethasone dipropionate
CAS 5598-13-0	Chlorpyrifos-methyl
CAS 5598-52-7	Fospirate
CAS 5611-51-8	Triamcinolone hexacetonide
CAS 5633-14-7	Benzetimide hydrochloride
CAS 5634-38-8	Guaithylline
CAS 5638-76-6	Betahistine
CAS 5667-71-0	Streptonicozid
CAS 5707-69-7	Drazoxolan
CAS 5714-04-5	Guanoxan sulfate
CAS 5714-73-8	Methenamine hippurate
CAS 5714-82-9	Triclofenol piperazine
CAS 5716-20-1	Bamethan sulfate

CAS 5743-34-0	Calcium borogluconate
CAS 5786-21-0	Clozapine
CAS 5793-94-2	Calcium stearoyl lactylate
CAS 5809-08-5	Trimethylpentyl hydroperoxide
CAS 5836-29-3	Coumetralyl
CAS 5870-29-1	Cyclopentolate hydrochloride
CAS 5874-97-5	Metaproterenol sulfate
CAS 5875-06-9	Proparacaine hydrochloride
CAS 5888-33-5	Isobornyl acrylate
CAS 5902-51-2	Terbacil
CAS 5907-38-0	Dipyrone
CAS 5928-84-7	Penicillin V benzathine
CAS 5980-31-4	Hexedine
CAS 5987-82-6	Benoxinate hydrochloride
CAS 5989-27-5	Limonene
CAS 5996-10-1	Glucose (hydrous)
CAS 6001-97-4	Dihexyl sodium sulfosuccinate
CAS 6004-24-6	Cetyl pyridinium chloride
CAS 6018-89-9	Nickel acetate
CAS 6108-05-0	Lidocaine hydrochloride
CAS 6119-70-6	Quinine sulfate
CAS 6130-64-9	Penicillin G procaine
CAS 6138-56-3	Tripelennaminecitrate
CAS 6138-79-0	Triprolidine hydrochloride
CAS 6144-28-1	Dilinoleic acid
CAS 6153-56-6	Oxalic acid
CAS 6179-44-8	Isostearamidopropyl betaine
CAS 6182-11-2	Propylene glycol distearate
CAS 6190-60-9	Mephentermine sulfate
CAS 6197-30-4	Octocrylene
CAS 6202-23-9	Cyclobenzaprine hydrochloride
CAS 6221-95-0	Myristyl propionate
CAS 6272-74-8	Lapyrium chloride
CAS 6281-42-1	Aminoethyl ethylene urea
CAS 6283-92-7	Lauryl lactate
CAS 6317-18-6	Methylenebisthio cyanate
CAS 6381-59-5	Potassium sodium tartrate
CAS 6381-77-7	Sodium erythorbate
CAS 6385-02-0	Meclofenamate sodium
CAS 6419-19-8	Aminotrimethylene phosphonic acid
CAS 6440-58-0	DMDM hydantoin
CAS 6452-73-9	Oxprenolol hydrochloride
CAS 6484-52-2	Ammonium nitrate
CAS 6500-81-8	Ethacrynate sodium
CAS 6506-37-2	Nimorazole
CAS 6542-37-6	Oxazolidine
CAS 6556-11-2	Inositol niacinate
CAS 6674-22-2	Diazabicyclo undecene
CAS 6683-19-8	Pentaerythrityl tetrakis dibutyl hydroxy phenylpropionate
CAS 6700-39-6	Isoproterenol sulfate
CAS 6724-53-4	Perhexiline maleate
CAS 6731-36-8	Bisbutyl peroxy trimethyl-cyclohexane
CAS 6804-07-5	Carbadox
CAS 6834-92-0	Sodium metasilicate
CAS 6835-16-1	Hyoscyamine sulfate
CAS 6842-15-5	Dodecene-1
CAS 6843-66-9	Diphenyl dimethoxysilane
CAS 6846-50-0	Trimethyl pentanediol diisobutyrate

CAS 6865-35-6	Barium stearate
CAS 6881-94-3	Diethylene glycol propyl ether
CAS 6891-44-7	Dimethylaminoethyl methacrylate dimethylsulfate
CAS 6915-15-7	Hydroxy succinic acid
CAS 6923-22-4	Monocrotophos
CAS 6938-94-9	Diisopropyl adipate
CAS 7005-47-2	Dimethylaminomethy propanol
CAS 7047-84-9	Aluminum stearate
CAS 7054-25-3	Quinidine gluconate
CAS 7081-44-9	Cloxacillin sodium
CAS 7085-19-0	Mecoprop
CAS 7085-44-1	Chlorothiazide sodium
CAS 7123-62-8	Oleyl propylene diamine
CAS 7128-91-8	Palmitamine oxide
CAS 7173-51-5	Didecyl dimonium chloride
CAS 7173-62-8	Oleyl propylene diamine
CAS 7179-49-9	Lincomycin hydrochloride
CAS 7235-40-7	Carotene
CAS 7240-38-2	Oxacillin sodium
CAS 7246-20-0	Triclofos sodium
CAS 7246-21-1	Tyropanoate sodium
CAS 7281-04-1	Lauralkonium bromide
CAS 7287-19-6	Prometryn
CAS 7297-25-8	Erythrityl tetranitrate
CAS 7320-34-5	Potassium pyrophosphate
CAS 7320-34-5	Tetrapotassium pyrophosphate
CAS 7378-99-6	Dimethyl octylamine
CAS 7396-58-9	Didecyl methylamine
CAS 7398-69-8	Dimethyl diallyl ammonium chloride
CAS 7414-83-7	Etidronate disodium
CAS 7421-40-1	Carbenoxolone sodium
CAS 7429-90-5	Aluminum
CAS 7439-89-6	Iron
CAS 7439-91-0	Lanthanum
CAS 7439-92-1	Lead
CAS 7439-93-2	Lithium
CAS 7439-95-4	Magnesium
CAS 7439-96-5	Manganese
CAS 7439-97-6	Mercury
CAS 7439-98-7	Molybdenum
CAS 7440-02-0	Nickel
CAS 7440-03-1	Niobium
CAS 7440-04-2	Osmium
CAS 7440-05-3	Palladium
CAS 7440-06-4	Platinum
CAS 7440-09-7	Potassium
CAS 7440-16-6	Rhodium
CAS 7440-19-9	Samarium
CAS 7440-20-2	Scandium
CAS 7440-21-3	Silicon
CAS 7440-22-4	Silver
CAS 7440-23-5	Sodium
CAS 7440-24-6	Strontium
CAS 7440-25-7	Tantalum
CAS 7440-27-9	Terbium
CAS 7440-28-0	Thallium
CAS 7440-31-5]	Tin
CAS 7440-32-6	Titanium
CAS 7440-33-7	Tungsten
CAS 7440-36-0	Antimony

CAS 7440-38-2	Arsenic
CAS 7440-39-3	Barium
CAS 7440-41-7	Beryllium
CAS 7440-42-8	Boron
CAS 7440-43-9	Cadmium
CAS 7440-44-0	Carbon
CAS 7440-45-1	Cerium
CAS 7440-46-2	Cesium
CAS 7440-47-3	Chromium
CAS 7440-48-4	Cobalt
CAS 7440-50-8	Copper
CAS 7440-52-0	Erbium
CAS 7440-55-3	Gallium
CAS 7440-56-4	Germanium
CAS 7440-57-5	Gold
CAS 7440-60-0	Holmium
CAS 7440-62-2	Vanadium
CAS 7440-64-4	Ytterbium
CAS 7440-65-5	Yttrium
CAS 7440-66-6	Zinc
CAS 7440-69-9	Bismuth
CAS 7440-70-2	Calcium
CAS 7440-74-6	Indium
CAS 7446-07-3	Tellurium oxide
CAS 7446-08-4	Selenium dioxide
CAS 7446-09-5	Sulfur dioxide
CAS 7446-20-0	Zinc sulfate
CAS 7446-70-0	Aluminum chloride
CAS 7447-39-4	Copper chloride (ic)
CAS 7447-39-4	Cupric chloride
CAS 7447-40-7	Potassium chloride
CAS 7487-88-9	Magnesium sulfate
CAS 7487-94-7	Mercury chloride (ic)
CAS 7488-54-2	Rubidium sulfate
CAS 7491-02-3	Diisopropyl sebacate
CAS 7491-09-0	Docusate potassium
CAS 7491-14-7	PPG-2 salicylate
CAS 7492-30-0	Potassium ricinoleate
CAS 7527-91-5	Acrisorcin
CAS 7534-94-3	Isobornyl methacrylate
CAS 7545-23-5	Diethanolamine
CAS 7545-23-5	Myristamide DEA
CAS 7553-56-2	Iodine
CAS 7558-79-4	Disodium phosphate
CAS 7558-80-7	Sodium phosphate
CAS 7585-39-9	Cyclodextrin
CAS 7601-54-9	Trisodium phosphate
CAS 7601-55-0	Metocurine iodide
CAS 7601-89-0	Sodium perchlorate
CAS 7601-90-3	Perchloric acid
CAS 7617-78-9	Isodecyloxy propylamine
CAS 7631-86-9	Diatomaceous earth
CAS 7631-86-9	Silica
CAS 7631-90-5	Sodium bisulfite
CAS 7631-95-0	Sodium molybdate
CAS 7631-97-2	Sodium monofluorophosphate
CAS 7631-99-4	Sodium nitrate
CAS 7632-00-0	Sodium nitrite
CAS 7632-04-4	Sodium perborate
CAS 7637-07-2	Boron trifluoride
CAS 7646-78-8	Stannic chloride
CAS 7646-79-9	Cobaltous chloride

CAS 7646-85-7	Zinc chloride	CAS 7758-87-4	Calcium phosphate tribasic
CAS 7647-01-0	Hydrochloric acid	CAS 7758-89-6	Cuprous chloride
CAS 7647-10-1	Palladium chloride (ous)	CAS 7758-94-3	Ferrous chloride
CAS 7647-14-5	Sodium chloride	CAS 7758-95-4	Lead chloride
CAS 7647-15-6	Sodium bromide	CAS 7758-98-7	Copper sulfate tribasic
CAS 7647-17-8	Cesium chloride	CAS 7758-99-8	Copper sulfate (ic)
CAS 7651-02-7	Stearamidopropyl dimethylamine	CAS 7761-88-8	Silver nitrate
CAS 7664-38-2	Phosphoric acid	CAS 7772-98-7	Sodium thiosulfate
CAS 7664-39-3	Hydrofluoric acid	CAS 7772-99-8	Stannous chloride
CAS 7664-41-7	Ammonia	CAS 7773-01-5	Manganese chloride
CAS 7664-93-9	Sulfuric acid	CAS 7773-06-0	Ammonium sulfamate
CAS 7665-72-7	Butyl glycidyl ether	CAS 7775-09-9	Sodium chlorate
CAS 7681-11-0	Potassium iodide	CAS 7775-11-3	Sodium chromate
CAS 7681-14-3	Prednisolone tebutate	CAS 7775-14-6	Sodium dithionite
CAS 7681-49-4	Sodium fluoride	CAS 7775-14-6	Sodium hydrosulfite
CAS 7681-52-9	Sodium hypochlorite	CAS 7775-19-1	Sodium metaborate
CAS 7681-53-0	Sodium hypophosphite	CAS 7775-27-1	Sodium persulfate
CAS 7681-55-2	Sodium iodate	CAS 7778-18-9	Calcium sulfate
CAS 7681-57-4	Sodium metabisulfite	CAS 7778-39-4	Arsenic acid
CAS 7681-65-4	Copper iodide (ous)	CAS 7778-44-1	Calcium arsenate
CAS 7681-76-7	Ronidazole	CAS 7778-50-9	Potassium dichromate
CAS 7681-80-3	Pentapiperium methyl sulfate	CAS 7778-53-2	Potassium phosphate tribasic
CAS 7681-82-5	Sodium iodide	CAS 7778-54-3	Calcium hypochlorite
CAS 7681-93-8	Natamycin	CAS 7778-66-7	Potassium hypochlorite
CAS 7696-12-0	Tetramethrin	CAS 7778-74-7	Potassium perchlorate
CAS 7697-37-2	Nitric acid	CAS 7778-77-0	Potassium phosphate
CAS 7699-45-8	Zinc bromide	CAS 7778-80-5	Potassium sulfate
CAS 7702-01-4	Disodium capryloamphodiacetate	CAS 7779-90-0	Zinc phosphate
CAS 7704-34-9	Sulfur	CAS 7782-41-4	Fluorine
CAS 7704-99-6	Zirconium hydride	CAS 7782-42-5	Graphite
CAS 7705-08-0	Ferric chloride	CAS 7782-44-7	Oxygen
CAS 7718-54-9	Nickel chloride (ous)	CAS 7782-49-2	Selenium
CAS 7719-12-2	Phosphorus trichloride	CAS 7782-50-5	Chlorine
CAS 7720-78-7	Iron sulfate (ous)	CAS 7782-85-6	Disodium phosphate
CAS 7722-64-7	Potassium permanganate	CAS 7783-03-1	Tungstic acid
CAS 7722-76-1	Ammonium phosphate	CAS 7783-18-8	Ammonium thiosulfate
CAS 7722-84-1	Hydrogen peroxide	CAS 7783-20-2	Ammonium sulfate
CAS 7722-88-5	Sodium pyrophosphate	CAS 7783-28-0	Ammonium phosphate dibasic
CAS 7722-88-5	Tetrasodium pyrophosphate	CAS 7783-35-9	Mercury sulfate (ic)
CAS 7726-95-6	Bromine	CAS 7783-40-6	Magnesium fluoride
CAS 7727-21-1	Potassium persulfate	CAS 7783-47-3	Stannous fluoride
CAS 7727-37-9	Nitrogen	CAS 7783-49-5	Zinc fluoride
CAS 7727-43-7	Barium sulfate	CAS 7783-61-1	Silicon tetrafluoride
CAS 7732-18-5	Collagenaminoacids	CAS 7783-64-4	Zirconium tetrafluoride
CAS 7733-02-0	Zinc sulfate	CAS 7783-70-2	Antimony pentafluoride
CAS 7738-94-5	Chromic acid	CAS 7783-82-6	Tungsten hexafluoride
CAS 7747-35-5	Ethylbicyclo oxazolidine	CAS 7783-90-5	Silver chloride
CAS 7747-35-5	Oxazolidine	CAS 7783-96-2	Silver iodide
CAS 7757-79-1	Potassium nitrate	CAS 7784-25-0	Aluminum ammonium sulfate
CAS 7757-82-6	Sodium sulfate	CAS 7784-25-0	Ammonium alum
CAS 7757-83-7	Sodium sulfite	CAS 7784-34-1	Arsenic trichloride
CAS 7757-86-0	Magnesium phosphate dibasic	CAS 7784-35-2	Arsenic trifluoride
CAS 7757-93-9	Calcium phosphate dibasic	CAS 7784-42-1	Arsine
CAS 7757-93-9	Dicalcium phosphate	CAS 7785-23-1	Silver bromide
CAS 7758-01-2	Potassium bromate	CAS 7785-87-7	Manganese sulfate
CAS 7758-02-3	Potassium bromide	CAS 7786-30-3	Magnesium chloride
CAS 7758-05-6	Potassium iodate	CAS 7787-59-9	Bismuth oxychloride
CAS 7758-11-4	Potassium phosphate dibasic	CAS 7787-69-1	Cesium bromide
CAS 7758-16-9	Sodium acid pyrophosphate	CAS 7787-70-4	Copper bromide (ous)
CAS 7758-19-2	Sodium chlorite	CAS 7789-04-0	Chromic phosphate
CAS 7758-23-8	Calcium phosphate	CAS 7789-09-5	Ammonium dichromate
CAS 7758-29-4	Sodium tripolyphosphate	CAS 7789-12-0	Sodium dichromate

CAS 7789-17-5	Cesium iodide
CAS 7789-18-6	Cesium nitrate
CAS 7789-38-0	Sodium bromate
CAS 7789-41-5	Calcium bromide
CAS 7789-75-5	Calcium fluoride
CAS 7789-77-7	Calcium phosphate dibasic dihydrate
CAS 7789-82-4	Calcium molybdate
CAS 7790-21-8	Potassium periodate
CAS 7790-26-3	Sodium iodide radioactive
CAS 7790-28-5	Sodium periodate
CAS 7790-29-6	Rubidium iodide
CAS 7790-69-4	Lithium nitrate
CAS 7790-79-6	Cadmium fluoride
CAS 7790-80-9	Cadmium iodide
CAS 7790-92-3	Hypochlorous acid
CAS 7790-93-4	Chloric acid
CAS 7790-98-9	Ammonium perchlorate
CAS 7791-11-9	Rubidium chloride
CAS 8000-10-1	Theophylline sodium glycinate
CAS 8001-17-0	Egg oil
CAS 8001-21-6	Sunflower seed oil
CAS 8001-22-7	Soybean oil
CAS 8001-23-8	Safflower oil
CAS 8001-25-0	Olive oil
CAS 8001-26-1	Linseed oil
CAS 8001-30-7	Corn oil
CAS 8001-31-8	Coconut oil
CAS 8001-54-5	Benzalkonium chloride
CAS 8001-69-2	Cod liver oil
CAS 8001-75-0	Ceresin
CAS 8001-78-3	Hydrogenated castor oil
CAS 8001-79-4	Castor oil
CAS 8002-03-7	Peanut oil
CAS 8002-09-3	Pine oil
CAS 8002-13-9	Rapeseed oil
CAS 8002-23-1	Cetyl esters
CAS 8002-31-1	Cocoa butter
CAS 8002-33-3	Sulfated castor oil
CAS 8002-43-5	Lecithin
CAS 8002-48-0	Malt extract
CAS 8002-50-4	Menhaden oil
CAS 8002-53-7	Montan wax
CAS 8002-74-2	Paraffin
CAS 8002-74-2	Petroleum wax
CAS 8002-74-2	Synthetic wax
CAS 8002-75-3	Palm oil
CAS 8002-76-4	Papaveretum
CAS 8005-44-5	Cetearyl alcohol
CAS 8006-13-1	Aluminum acetate
CAS 8006-40-4	Beeswax, white
CAS 8006-44-8	Candelilla wax
CAS 8006-54-0	Lanolin
CAS 8006-54-0	Lanolin, anhydrous
CAS 8006-54-0	Lanolin oil
CAS 8006-95-9	Wheat germ oil
CAS 8007-24-7	Pentadecenyl phenol
CAS 8007-43-0	Sorbitan sesquioleate
CAS 8007-69-0	Sweet almond oil
CAS 8008-20-6	Kerosene
CAS 8008-74-0	Sesame oil
CAS 8009-03-8	Petrolatum NF

CAS 8012-89-3	Beeswax, yellow
CAS 8012-95-1	Mineral oil
CAS 8013-01-2	Yeast extract
CAS 8013-05-6	Potassium castorate
CAS 8013-07-8	Epoxidized soybean oil
CAS 8015-86-9	Carnauba
CAS 8016-70-4	Hydrogenated soybean oil
CAS 8017-16-1	Polyphosphoric acid
CAS 8018-01-7	Mancozeb
CAS 8020-84-6	Lanolin, hydrous
CAS 8021-55-4	Ozokerite
CAS 8022-00-2	Demeton methyl
CAS 8023-79-8	Palm kernel oil
CAS 8024-09-7	Walnut oil
CAS 8024-22-4	Grapeseed oil
CAS 8024-32-6	Avocado oil
CAS 8027-32-5	Petrolatum USP
CAS 8027-33-6	Lanolin alcohol
CAS 8028-66-8	Honey
CAS 8029-43-4	Corn syrup
CAS 8029-44-5	Cottonseed glyceride
CAS 8029-68-3	Ichthammol
CAS 8029-76-3	Hydroxylated lecithin
CAS 8029-91-2	Acetylated hydrogenated lard glyceride
CAS 8029-92-3	Acetylated lard glyceride
CAS 8030-30-6	Naphtha
CAS 8030-78-2	Tallow trimonium chloride
CAS 8031-14-9	Oxychlorosene
CAS 8031-44-5	Hydrogenated lanolin
CAS 8038-43-5	Lanolin oil
CAS 8038-93-5	Sodium aluminum chloro-hydroxylactate
CAS 8039-09-6	PEG-75 lanolin
CAS 8042-47-5	Mineral oil
CAS 8048-52-0	Acriflavine
CAS 8049-47-6	Pancreatin
CAS 8050-09-7	Rosin
CAS 8050-26-8	Pentaerythrityl rosinate
CAS 8050-81-5	Simethicone
CAS 8051-30-7	Cocamide DEA
CAS 8052-10-6	Rosin
CAS 8052-48-0	Sodium tallowate
CAS 8061-51-6	Sodium lignosulfonate
CAS 8061-52-7	Calcium lignosulfonate
CAS 8061-53-8	Ammonium lignosulfonate
CAS 8067-32-1	Glyceryl ditripalmitostearate
CAS 8067-32-1	Glyceryl ditristearate
CAS 8067-69-4	Halquinol
CAS 8068-28-8	Colistimethate sodium
CAS 9000-01-5	Gum arabic
CAS 9000-07-1	Carrageenan
CAS 9000-30-0	Guar gum
CAS 9000-40-2	Locust bean gum
CAS 9000-69-5	Pectin
CAS 9000-70-8	Gelatin
CAS 9000-71-9	Casein
CAS 9000-92-4	Amylase
CAS 9000-92-4	Diastase
CAS 9001-01-8	Kallikrein
CAS 9001-05-2	Catalase
CAS 9001-09-6	Chymopapain

CAS 9001-54-1	Hyaluronidase	CAS 9004-74-4	PEG-6 methylether
CAS 9001-62-1	Lipase	CAS 9004-81-3	PEG laurate series
CAS 9001-73-4	Papain	CAS 9004-82-4	Sodium laureth sulfate
CAS 9001-75-6	Pepsin	CAS 9004-83-5	PEG-10 isolauryl thioether
CAS 9002-01-1	Streptokinase	CAS 9004-92-2	Oleth-2
CAS 9002-04-4	Thrombin	CAS 9004-94-8	PEG-7 palmitate
CAS 9002-07-7	Trypsin	CAS 9004-95-9	Ceteth series
CAS 9002-13-5	Urease	CAS 9004-96-0	PEG oleate series
CAS 9002-64-6	Parathyroid	CAS 9004-97-1	PEG ricinoleate series
CAS 9002-72-6	Somatropin	CAS 9004-98-2	Oleth series
CAS 9002-83-9	Polychlorotrifluoroethylene	CAS 9004-99-3	PEG stearate series
CAS 9002-84-0	Polytetrafluoroethylene	CAS 9005-00-9	Steareth series
CAS 9002-86-2	Polyvinylchloride	CAS 9005-02-1	PEG dilaurate series
CAS 9002-88-4	Polyethylene	CAS 9005-07-6	PEG dioleate series
CAS 9002-89-5	Polyvinylalcohol	CAS 9005-08-7	PEG distearate series
CAS 9002-92-0	Laureth series	CAS 9005-25-8	Corn starch
CAS 9002-93-1	Octoxynol series	CAS 9005-32-7	Alginic acid
CAS 9003-01-4	Polyacrylic acid	CAS 9005-34-9	Ammonium alginate
CAS 9003-03-6	Ammonium polyacrylate	CAS 9005-35-0	Calcium alginate
CAS 9003-04-7	Sodium polyacrylate	CAS 9005-36-1	Potassium alginate
CAS 9003-05-8	Polyacrylamide	CAS 9005-37-2	Propylene glycol alginate
CAS 9003-07-0	Polypropylene	CAS 9005-38-3	Algin
CAS 9003-08-1	Melamine-formaldehyde resin	CAS 9005-46-3	Casein
CAS 9003-09-2	Polyvinyl methyl ether	CAS 9005-46-3	Sodium caseinate
CAS 9003-11-6	Meroxapol series	CAS 9005-49-6	Heparin
CAS 9003-11-6	Poloxamer series	CAS 9005-64-5	Polysorbate series
CAS 9003-13-8	PPG butyl ether series	CAS 9005-64-5	PEG sorbitan laurate series
CAS 9003-17-2	Polybutadiene	CAS 9005-65-5	Polysorbate series
CAS 9003-18-3	Butadiene-acrylonitrile copolymer	CAS 9005-65-6	Polysorbate series
CAS 9003-20-7	Polyvinyl acetate	CAS 9005-66-7	PEG-80 sorbitan palmitate
CAS 9003-22-9	PVC/VA copolymer	CAS 9005-66-7	Polysorbate 40
CAS 9003-22-9	Vinylchloride/vinyl acetate copolymer	CAS 9005-67-8	Polysorbate 60
		CAS 9005-67-8	Polysorbate 61
CAS 9003-27-4	Polyisobutene	CAS 9005-70-3	Polysorbate 85
CAS 9003-27-4	Polybutene	CAS 9005-71-4	Polysorbate 65
CAS 9003-31-0	Polyisoprene	CAS 9006-04-6	Natural rubber
CAS 9003-39-8	PVP	CAS 9006-04-6	Polyisoprene
CAS 9003-53-6	Polystyrene	CAS 9006-42-2	Metiram
CAS 9003-54-7	Styrene-acrylonitrile	CAS 9006-50-2	Albumen
CAS 9003-55-8	Styrene-butadiene	CAS 9007-28-7	Chondroitin sulfate
CAS 9003-56-9	Acrylonitrile-butadiene-styrene	CAS 9007-34-5	Collagen
CAS 9003-62-7	Polyvinylbutyral	CAS 9007-34-5	Soluble collagen
CAS 9003-79-6	Diphenylamine acetone	CAS 9007-48-1	Polyglyceryl oleate series
CAS 9003-99-0	Peroxidase	CAS 9007-58-3	Hydrolyzed elastin
CAS 9004-32-4	Carboxymethyl cellulose sodium	CAS 9008-66-6	Nylon 6/10
CAS 9004-34-6	Cellulose	CAS 9008-99-8	Hydrolyzed pea protein
CAS 9004-34-6	Microcrystalline cellulose	CAS 9009-54-5	Polyurethane foam
CAS 9004-35-7	Cellulose acetate	CAS 9009-54-5	Polyurethane thermoplastic
CAS 9004-36-3	Sodium caseinate	CAS 9009-99-8	Silk powder
CAS 9004-36-8	Cellulose acetate butyrate	CAS 9010-01-9	Sodium amylosulfate
CAS 9004-39-1	Cellulose acetate propionate	CAS 9010-66-6	Corn gluten protein
CAS 9004-53-9	Dextrin	CAS 9010-77-9	Ethylene acrylic acid copolymer
CAS 9004-54-0	Dextran	CAS 9010-77-9	Ethylene copolymer
CAS 9004-57-3	Ethyl cellulose	CAS 9010-79-1	EPM rubber
CAS 9004-58-4	Ethyl hydroxyethyl cellulose	CAS 9010-79-1	Ethylene propylene copolymer
CAS 9004-61-9	Hyaluronic acid	CAS 9010-79-1	Polypropylene
CAS 9004-62-0	Hydroxyethyl cellulose	CAS 9010-85-9	Isobutylene isoprene copolymer
CAS 9004-64-2	Hydroxypropyl cellulose	CAS 9011-13-6	Styrene/MA copolymer
CAS 9004-65-3	Hydroxypropyl methylcellulose	CAS 9011-14-7	Methyl methacrylate copolymer
CAS 9004-67-5	Methyl cellulose	CAS 9011-14-7	Polymethyl methacrylate
CAS 9004-70-0	Nitrocellulose	CAS 9011-16-9	PVM/MA copolymer
CAS 9004-73-3	Methicone	CAS 9012-09-3	Cellulose triacetate

CAS 9012-54-8	Cellulase	CAS 10034-93-2	Hydrazine sulfate
CAS 9014-01-1	Protease	CAS 10035-10-6	Hydrobromic acid
CAS 9014-85-1	PEG tetramethyl decynediol series	CAS 10039-53-9	Sodium chromate
		CAS 10042-76-9	Strontium nitrate
CAS 9014-90-8	Sodium nonoxynol sulfate series	CAS 10043-01-3	Aluminum sulfate
CAS 9014-92-0	Dodoxynol series	CAS 10043-11-5	Boron nitride
CAS 9014-93-1	Nonyl nonoxynol series	CAS 10043-35-3	Boric acid
CAS 9015-54-7	Collagen amino acids	CAS 10043-52-4	Calcium chloride
CAS 9015-54-7	Hydrolyzed collagen	CAS 10045-89-3	Ferrous ammonium sulfate
CAS 9016-45-9	Nonoxynol series	CAS 10049-04-4	Chlorine dioxide
CAS 9016-75-5	Polyphenylene sulfide	CAS 10049-05-5	Chromium chloride (ous)
CAS 9027-33-6	Lanolin alcohol	CAS 10049-07-7	Rhodium chloride
CAS 9031-11-2	Galactosidase	CAS 10081-67-1	Bisdimethylbenzyl diphenylamine
CAS 9032-08-0	Amyloglucosidase	CAS 10085-81-1	Benzoctamine hydrochloride
CAS 9032-08-0	Glucoamylase	CAS 10090-54-7	Potassium pyroantimonate
CAS 9032-42-2	Methyl hydroxyethyl cellulose	CAS 10094-45-8	Stearyl erucamide
CAS 9032-75-1	Pectinase	CAS 10099-58-8	Lanthanum chloride
CAS 9034-50-8	Vasopressin	CAS 10099-59-9	Lanthanum nitrate
CAS 9035-58-9	Thromboplastin	CAS 10099-74-8	Lead nitrate
CAS 9035-85-2	PPG cetyl ether series	CAS 10099-76-0	Lead silicate
CAS 9035-99-8	Sulfur polymeric	CAS 10101-39-0	Calcium silicate
CAS 9038-41-9	Sodium cellulose phosphate	CAS 10101-41-4	Calcium sulfate dihydrate
CAS 9038-95-3	PPG-24 buteth-27	CAS 10101-89-0	Trisodium phosphate dodeca-
CAS 9039-53-6	Urokinase		hydrate
CAS 9040-38-4	Disodium nonoxynol-10 sulfosuccinate	CAS 10102-05-3	Palladium nitrate (ous)
		CAS 10102-40-6	Sodium molybdate
CAS 9041-93-4	Bleomycin sulfate	CAS 10108-64-2	Cadmium chloride
CAS 9046-01-9	Trideceth phosphate series	CAS 10112-91-1	Mercurous chloride
CAS 9046-56-4	Ancrod	CAS 10118-76-0	Calcium permanganate
CAS 9048-46-8	Serum albumin	CAS 10124-43-3	Cobalt sulfate
CAS 9050-36-6	Maltodextrin	CAS 10124-56-8	Sodium hexametaphosphate
CAS 9051-57-4	Ammonium nonoxynol sulfate series	CAS 10140-65-5	Disodium phosphate
		CAS 10141-05-6	Cobalt nitrate
CAS 9062-73-1	Polysorbate 20	CAS 10161-34-9	Trenbolone acetate
CAS 9063-38-1	Sodium starch glycolate	CAS 10190-55-3	Lead molybdate
CAS 9064-14-6	PPG-7 lauryl ether	CAS 10192-93-5	Dimethyl diphenyl hexane
CAS 9067-32-7	Sodium hyaluronate	CAS 10196-18-6	Zinc nitrate
CAS 9076-43-1	PPG-40 diethylmonium chloride	CAS 10213-09-9	Vanadium chloride (ous)
CAS 9084-06-4	Sodium polynaphthalene sulfonate	CAS 10213-10-2	Sodium tungstate
		CAS 10213-78-2	PEG-2 stearamine
CAS 9087-53-0	PPG-5 ceteth-20	CAS 10213-79-3	Sodium metasilicate pentahydrate
CAS 9087-61-0	Aluminum starch octenyl succinate	CAS 10238-21-8	Glyburide
		CAS 10246-75-0	Hydroxyzine pamoate
CAS 9087-70-1	Aprotinin	CAS 10265-92-6	Methamidophos
CAS 9105-54-7	Collagen amino acids	CAS 10288-28-5	Cyanomethylethyl azoformamide
CAS 10022-31-8	Barium nitrate	CAS 10294-26-5	Silver sulfate
CAS 10024-97-2	Nitrous oxide	CAS 10294-33-4	Boron tribromide
CAS 10025-73-7	Chromic chloride	CAS 10294-34-5	Boron trichloride
CAS 10025-73-7	Chromium chloride (ic)	CAS 10294-40-3	Barium chromate
CAS 10025-78-2	Trichlorosilane	CAS 10294-42-5	Cerium sulfate
CAS 10025-87-3	Phosphorus oxychloride	CAS 10294-54-9	Cesium sulfate
CAS 10025-91-9	Antimony trichloride	CAS 10294-56-1	Phosphorous acid
CAS 10026-04-7	Silicon tetrachloride	CAS 10310-32-4	Tribenoside
CAS 10026-04-7	Tetrachlorosilane	CAS 10323-20-3	Arabinose
CAS 10026-11-6	Zirconium tetrachloride	CAS 10347-81-6	Maprotiline hydrochloride
CAS 10026-13-8	Phosphorus pentachloride	CAS 10361-03-2	Sodium metaphosphate
CAS 10026-18-3	Cobalt fluoride	CAS 10361-37-2	Barium chloride
CAS 10028-22-5	Ferrous sulfate	CAS 10361-46-3	Bismuth nitrate
CAS 10028-24-7	Disodium phosphate dihydrate	CAS 10377-60-3	Magnesium nitrate
CAS 10030-90-7	Ferrous succinate	CAS 10401-55-5	Cetyl ricinoleate
CAS 10031-43-3	Copper nitrate (ic)	CAS 10416-59-8	Bistrimethylsilyl acetamide
CAS 10034-85-2	Hydriodic acid	CAS 10418-03-8	Stanozolol

CAS 10421-48-4	Ferric nitrate
CAS 10450-60-9	Periodic acid
CAS 10476-81-0	Strontium bromide
CAS 10476-85-4	Strontium chloride
CAS 10525-14-1	Mixed isopropanolamines myristate
CAS 10540-29-1	Tamoxifen
CAS 10591-85-2	Tetrabenzyl thiuram disulfide
CAS 10595-72-9	Ditridecyl thiodipropionate
CAS 11006-75-0	Saponins
CAS 11094-60-3	Polyglyceryl-10 decaoleate
CAS 11096-42-7	Nonoxynol-12 iodine
CAS 11097-59-9	Magnesium aluminum carbonate
CAS 11099-07-3	Glyceryl stearate
CAS 11104-88-4	Phosphomolybdic acid
CAS 11111-34-5	Poloxamine series
CAS 11113-70-5	Lead silicochromate
CAS 11120-25-5	Ammonium tungstate
CAS 11125-95-4	Chromate copper arsenate
CAS 11138-66-2	Xanthan gum
CAS 11140-78-6	Lauryl betaine
CAS 11141-17-6	Azadirachtin
CAS 11174-62-0	Stearyl trimonium hydroxyethyl hydrolyzed collagen
CAS 12001-26-2	Mica
CAS 12001-99-9	Chromium hydroxide
CAS 12007-56-6	Calcium borate
CAS 12027-06-4	Ammonium iodide
CAS 12027-96-2	Zinc hydroxy stannate
CAS 12033-89-5	Silicon nitride
CAS 12036-37-2	Zinc stannate
CAS 12036-44-1	Thulium oxide
CAS 12037-01-3	Terbium oxide
CAS 12042-91-0	Aluminum chlorohydrate
CAS 12047-27-7	Barium titanate
CAS 12049-50-2	Calcium titanate
CAS 12055-62-8	Holmium oxide
CAS 12058-66-1	Sodium stannate
CAS 12060-08-1	Scandium oxide
CAS 12060-58-1	Samarium oxide
CAS 12061-16-4	Erbium oxide
CAS 12067-99-1	Phosphotungstic acid
CAS 12068-03-0	Sodium toluene sulfonate
CAS 12069-32-8	Boron carbide
CAS 12069-69-1	Copper carbonate (ic)
CAS 12071-83-9	Propineb
CAS 12111-24-9	Pentetate calcium trisodium
CAS 12122-67-7	Zineb
CAS 12124-97-9	Ammonium bromide
CAS 12125-01-8	Ammonium fluoride
CAS 12125-02-9	Ammonium chloride
CAS 12141-46-7	Aluminum silicate
CAS 12142-33-5	Potassium stannate
CAS 12173-47-6	Hectorite
CAS 12178-41-5	Ferroaluminum silicate
CAS 12199-37-0	Magnesium aluminum silicate
CAS 12244-57-4	Gold sodium thiomalate
CAS 12269-78-2	Pyrophyllite
CAS 12280-03-4	Disodium octaborate
CAS 12284-76-3	Bismuth aluminate
CAS 12427-38-2	Maneb
CAS 12513-27-8	Zinc borate
CAS 12539-23-0	Magnesium aluminum carbonate
CAS 12542-30-2	Dicyclopentenyl acrylate
CAS 12602-23-2	Cobalt carbonate
CAS 12607-92-0	Aceglutamide aluminum
CAS 12645-31-7	Ethylhexyl phosphate
CAS 12645-31-7	Trioctyl phosphate
CAS 12751-23-4	Lauryl phosphate
CAS 12769-96-9	Ultramarine violet
CAS 13010-47-4	Lomustine
CAS 13039-35-5	Sodium capryloamphoacetate
CAS 13048-33-4	Hexanediol diacrylate
CAS 13052-09-0	Dimethyl diethylhexanoyl peroxyhexane
CAS 13087-53-1	Iothalamate meglumine
CAS 13103-34-9	Boldenone undecylenate
CAS 13103-52-1	Stearyl lauryl thiodipropionate
CAS 13106-44-0	Dimethylaminoethyl acrylate dimethyl sulfate
CAS 13121-70-5	Cyhexatin
CAS 13122-18-4	Butylperoxy trimethyl hexanoate
CAS 13127-82-7	PEG-2 oleamine
CAS 13138-45-9	Nickel nitrate (ous)
CAS 13150-00-0	Sodium laureth sulfate
CAS 13170-23-5	Dibutoxy diacetoxysilane
CAS 13194-48-4	Ethoprop
CAS 13197-76-7	Lauryl hydroxysultaine
CAS 13253-82-2	Methylenebis cyclohexylamine carbamate
CAS 13292-46-1	Rifampin
CAS 13400-13-0	Cesium fluoride
CAS 13411-16-0	Nifurpirinol
CAS 13412-64-1	Dicloxacillin sodium
CAS 13422-51-0	Hydroxocobalamin
CAS 13434-12-6	Trimethylsilyl acetamide
CAS 13453-07-1	Gold chloride
CAS 13457-18-6	Pyrazophos
CAS 13463-41-7	Pyrithione zinc
CAS 13463-41-7	Zinc pyrithione
CAS 13463-67-7	Titanium dioxide
CAS 13472-08-7	Azobismethyl butyronitrile
CAS 13472-45-2	Sodium tungstate
CAS 13474-96-9	Heptakis dipropylene glycol triphosphite
CAS 13475-82-6	Isododecane
CAS 13492-01-8	Tranylcypromine sulfate
CAS 13494-80-9	Tellurium
CAS 13516-27-2	Guazatine
CAS 13523-86-9	Pindolol
CAS 13528-93-3	Tetramethyl dichlorodisilethylene
CAS 13530-50-2	Aluminum phosphate
CAS 13557-75-0	Sodium lauroyl lactylate
CAS 13560-89-9	Dodecachloro dodecahydro dimethano dibenzocyclooctene
CAS 13573-18-7	Sodium tripolyphosphate
CAS 13593-03-8	Quinalphos
CAS 13597-65-4	Zinc silicate
CAS 13598-36-2	Phosphorous acid
CAS 13601-19-9	Sodium ferrocyanide
CAS 13609-67-1	Hydrocortisone butyrate
CAS 13614-98-7	Minocycline hydrochloride
CAS 13647-35-3	Trilostane
CAS 13682-92-3	Dihydroxy aluminum aminoacetate

CAS 13684-63-4 | Phenmedipham
CAS 13698-49-2 | Delmadinone acetate
CAS 13707-88-5 | Alprenolol hydrochloride
CAS 13717-04-9 | Propiram fumarate
CAS 13770-89-3 | Nickel sulfamate
CAS 13814-96-5 | Lead fluoborate
CAS 13822-56-5 | Aminopropyl trimethoxysilane
CAS 13838-16-9 | Enflurane
CAS 13840-33-0 | Lithium hypochlorite
CAS 13845-36-8 | Potassium tripolyphosphate
CAS 13927-77-0 | Nickel dibutyl dithiocarbamate
CAS 13952-84-6 | Aminobutane
CAS 13983-17-0 | Wollastonite
CAS 13987-01-4 | Tripropylene glycol
CAS 14007-07-9 | Chlorhexidine digluconate
CAS 14087-96-6 | Talc
CAS 14222-60-7 | Protionamide
CAS 14234-82-3 | Diisobutyl maleate
CAS 14235-86-0 | Hydrargaphen
CAS 14239-68-0 | Cadmium diethyl dithiocarbamate
CAS 14252-80-3 | Bupivacaine hydrochloride
CAS 14286-84-1 | Bencyclane fumarate
CAS 14293-44-8 | Xipamide
CAS 14323-55-1 | Zinc diethyl dithiocarbamate
CAS 14350-96-0 | Sodium lauroamphoacetate
CAS 14350-97-1 | Disodiuml auroamphodiacetate
CAS 14351-40-7 | Stearamide MEA stearate
CAS 14351-50-9 | Oleamine oxide
CAS 14371-10-9 | Cinnamic aldehyde
CAS 14402-89-2 | Sodium nitroprusside
CAS 14450-05-6 | Pentaerythrityl tetrapelargonate
CAS 14481-60-8 | Disodium stearyl sulfo-succinamate
CAS 14643-87-9 | Zinc diacrylate
CAS 14698-29-4 | Oxolinic acid
CAS 14727-68-5 | Dimethyl oleamine
CAS 14807-96-6 | Talc
CAS 14808-60-7 | Quartz
CAS 14816-18-3 | Phoxim
CAS 14857-34-2 | Dimethylethoxysilane
CAS 14885-29-1 | Ipronidazole
CAS 14940-68-2 | Zirconium silicate
CAS 14960-06-6 | Sodium lauriminodipropionate
CAS 14976-57-9 | Clemastine fumarate
CAS 15096-52-3 | Cryolite
CAS 15180-03-7 | Alcuronium chloride
CAS 15206-55-0 | Methylphenyl glyoxalate
CAS 15299-99-7 | Napropamide
CAS 15305-07-4 | Nitrosophenyl hydroxylamine aluminum salt
CAS 15307-79-6 | Diclofenac sodium
CAS 15310-01-7 | Benodanil
CAS 15317-78-9 | Nickel diisobutyl dithiocarbamate
CAS 15454-75-8 | Zinc PCA
CAS 15489-16-4 | Stibophen
CAS 15500-66-0 | Pancuronium bromide
CAS 15520-10-2 | Methyl pentamethylene diamine
CAS 15520-11-3 | Bisbutyl cyclohexyl peroxy-dicarbonate
CAS 15520-11-3 | Dibutyl cyclohexyl peroxy-dicarbonate
CAS 15521-65-0 | Nickel dimethyl dithiocarbamate

CAS 15535-29-2 | Ethanolamine sulfite
CAS 15535-69-0 | Dibutyltin maleate
CAS 15537-76-5 | Chlorproguanil hydrochloride
CAS 15545-48-9 | Chlorotoluron
CAS 15545-58-9 | Chlortoluron
CAS 15545-97-8 | Azobismethoxy dimethyl valeronitrile
CAS 15574-96-6 | Pizotyline
CAS 15578-26-4 | Stannous pyrophosphate
CAS 15585-70-3 | Bibenzonium bromide
CAS 15599-39-0 | Noxytiolin
CAS 15622-65-8 | Molindone hydrochloride
CAS 15625-89-5 | Trimethylol propane triacrylate
CAS 15630-89-4 | Sodium percarbonate
CAS 15647-08-2 | Ethylhexyl diphenyl phosphite
CAS 15663-27-1 | Cisplatin
CAS 15667-10-4 | Diamyl peroxycyclohexane
CAS 15676-16-1 | Sulpiride
CAS 15686-71-2 | Cephalexin
CAS 15687-27-1 | Ibuprofen
CAS 15687-41-9 | Oxyfedrin
CAS 15716-30-0 | PEG-25 PABA
CAS 15826-16-1 | Sodium laureth sulfate
CAS 15826-37-6 | Cromolyn sodium
CAS 15876-67-2 | Distigmine bromide
CAS 15972-60-8 | Alachlor
CAS 16029-98-4 | Trimethylsilyl iodide
CAS 16034-77-8 | Iocetamic acid
CAS 16039-53-5 | Zinc lactate
CAS 16051-77-7 | Isosorbide nitrate
CAS 16066-38-9 | Dipropyl peroxy dicarbonate
CAS 16111-62-9 | Bisethylhexyl peroxy dicarbonate
CAS 16111-62-9 | Diethylhexyl peroxy dicarbonate
CAS 16118-49-3 | Carbetamide
CAS 16230-35-6 | Bismethylbenzamidoethoxy-methylsilane
CAS 16260-09-6 | Oleyl palmitamide
CAS 16409-44-2 | Geranyl acetate
CAS 16485-10-2 | Panthenol
CAS 16545-54-3 | Dimyristyl thiodipropionate
CAS 16580-06-6 | Dibutylperoxy azelate
CAS 16590-41-3 | Naltrexone
CAS 16672-87-0 | Ethephon
CAS 16693-53-1 | TEA-lauroyl sarcosinate
CAS 16721-80-5 | Sodium bisulfide
CAS 16721-80-5 | Sodium hydrosulfite
CAS 16721-80-5 | Sodium sulfide
CAS 16731-55-8 | Potassium metabisulfite
CAS 16752-77-5 | Methomyl
CAS 16773-42-5 | Ornidazole
CAS 16789-79-5 | Dihydrogenated tallowamine
CAS 16841-14-8 | Behenalkonium chloride
CAS 16889-14-8 | Stearamidoethyl diethylamine
CAS 16940-66-2 | Sodium borohydride
CAS 16960-16-0 | Cosyntropin
CAS 16971-82-7 | Dicatechol borate ditolyl guanidine salt
CAS 17086-28-1 | Doxycycline
CAS 17090-79-8 | Monensin
CAS 17109-49-8 | Edifenphos
CAS 17243-57-1 | Mefenorex
CAS 17301-53-0 | Behentrimonium chloride

CAS Number	Chemical
CAS 17342-21-1	Lauryl pyridinium bisulfate
CAS 17440-83-4	Amiloride hydrochloride
CAS 17540-75-9	Dibutylbutyl phenol
CAS 17560-51-9	Metolazone
CAS 17598-65-1	Deslanoside
CAS 17617-23-1	Flurazepam
CAS 17661-50-6	Myristyl stearate
CAS 17671-27-1	Behenyl behenate
CAS 17673-56-2	Oleyl erucate
CAS 17689-77-9	Ethyl triacetoxysilane
CAS 17693-51-5	Promethazine theoclate
CAS 17784-12-2	Sulfacytine
CAS 17796-82-6	Cyclohexyl thiophthalimide
CAS 17804-35-2	Benomyl
CAS 17831-71-9	PEG-4 diacrylate
CAS 17865-32-6	Methyl cyclohexyl dimethoxy-silane
CAS 17932-62-6	Cyanoethyl triethoxysilane
CAS 18156-74-6	Trimethylsilyl imidazole
CAS 18162-48-6	t-Butyldimethyl chlorosilane
CAS 18162-84-0	Octyldimethyl chlorosilane
CAS 18171-19-2	Chloropropyl methyl dimethoxy-silane
CAS 18181-70-9	Iodofenphos
CAS 18194-24-6	Dimyristoyl glycerophosphatidyl choline
CAS 18282-10-5	Tin oxide
CAS 18312-32-8	Behenyl erucate
CAS 18323-44-9	Clindamycin
CAS 18395-30-7	Isobutyl trimethoxysilane
CAS 18407-94-8	Tetrakis ethoxyethoxysilane
CAS 18448-65-2	PEG-2 oleammonium chloride
CAS 18467-77-1	Dikegulac
CAS 18472-51-0	Chlorhexidine gluconate
CAS 18472-51-0	Chlorhexidine digluconate
CAS 18507-89-6	Decoquinate
CAS 18559-94-9	Albuterol
CAS 18641-57-1	Glyceryl tribehenate
CAS 18643-08-8	Dimethyl octadecyl chlorosilane
CAS 18691-97-9	Methabenzthiazuron
CAS 18769-78-3	Methyl tridecylsilane
CAS 18868-43-4	Molybdenum dioxide
CAS 18883-66-4	Streptozocin
CAS 18917-89-0	Magnesium salicylate
CAS 19010-66-3	Lead dimethyl dithiocarbamate
CAS 19044-88-3	Oryzalin
CAS 19047-85-9	Distearyl phosphite
CAS 19237-84-4	Prazosin hydrochloride
CAS 19321-40-5	Pentaerythrityl tetraoleate
CAS 19356-17-3	Calcifediol
CAS 19387-91-8	Tinidazole
CAS 19504-77-9	Pecilocin
CAS 19666-30-9	Oxadiazon
CAS 19706-80-0	Azobishydroxy methyl propionitrile
CAS 19910-65-7	Dibutylperoxy dicarbonate
CAS 19937-59-8	Metoxuron
CAS 20202-66-3	Manganese violet
CAS 20324-33-8	PPG-3 methyl ether
CAS 20344-49-4	Ferric hydroxide
CAS 20354-26-1	Methazole
CAS 20427-59-2	Copper hydroxide (ic)
CAS 20566-35-2	Tetrabromo phthalatediol
CAS 20667-12-3	Silver oxide
CAS 20684-06-4	Bamifylline hydrochloride
CAS 20816-12-0	Osmium tetroxide
CAS 20830-75-5	Digoxin
CAS 20859-73-8	Aluminum phosphide
CAS 21087-64-9	Metribuzin
CAS 21142-28-9	MIPA-lauryl sulfate
CAS 21245-02-3	Octyl dimethyl PABA
CAS 21256-18-8	Oxaprozin
CAS 21260-46-8	Bismuth dimethyl dithiocarbamate
CAS 21302-09-0	Dilauryl phosphite
CAS 21351-79-1	Cesium hydroxide
CAS 21416-87-5	Razoxane
CAS 21542-96-1	Dimethyl behenamine
CAS 21548-73-2	Silver sulfide
CAS 21645-51-2	Alumina hydrate
CAS 21645-51-2	Aluminum hydroxide
CAS 21652-27-7	Oleyl hydroxyethyl imidazoline
CAS 21725-46-2	Cyanazine
CAS 21738-42-1	Oxamniquine
CAS 21810-39-9	Diethylaminoethyl acrylate dimethyl sulfate
CAS 21829-25-4	Nifedipine
CAS 21908-53-2	Mercuric oxide
CAS 22042-96-2	Diethylene triamine pentakis methylene phosphonicacid
CAS 22047-49-0	Octyl stearate
CAS 22059-60-5	Disopyramide phosphate
CAS 22071-15-4	Ketoprofen
CAS 22131-79-9	Alclofenac
CAS 22151-68-4	Methohexital sodium
CAS 22151-68-4	Methohexitone sodium
CAS 22195-34-2	Guanadrel sulfate
CAS 22204-24-6	Pyrantel pamoate
CAS 22204-53-1	Naproxen
CAS 22204-88-2	Tramadol hydrochloride
CAS 22224-92-6	Fenamiphos
CAS 22232-71-9	Mazindol
CAS 22254-24-6	Ipratropium bromide
CAS 22260-51-1	Bromocriptine mesylate
CAS 22298-29-9	Betamethasone benzoate
CAS 22313-62-8	Butylperoxy methyl benzoate
CAS 22389-47-0	Aspartame
CAS 22494-42-4	Diflunisal
CAS 22499-12-3	Isobutyl benzoin ether
CAS 22500-92-1	Sorbic acid
CAS 22662-39-1	Rafoxanide
CAS 22760-18-5	Proquazone
CAS 22766-82-1	Octyldodecyl stearate
CAS 22781-23-3	Bendiocarb
CAS 22801-45-2	Octyldodecyl oleate
CAS 22832-87-7	Miconazole nitrate
CAS 22882-95-7	Isopropyl linoleate
CAS 22984-54-9	Methyl trismethylethyl ketoximesilane
CAS 23031-32-5	Terbutaline sulfate
CAS 23047-25-8	Lofepramin
CAS 23092-17-3	Halazepam
CAS 23103-98-2	Pirimicarb
CAS 23184-66-9	Butachlor
CAS 23214-92-8	Doxorubicin

CAS 23256-50-0	Guanabenzacetate
CAS 23277-43-2	Nalbuphine hydrochloride
CAS 23288-60-0	Sodium pertechnetate
CAS 23327-57-3	Nefopam hydrochloride
CAS 23383-11-1	Ferrous citrate
CAS 23386-52-9	Dicyclohexyl sodium sulfo-succinate
CAS 23474-91-1	Butylperoxy crotonate
CAS 23505-41-1	Pirimiphos ethyl
CAS 23560-59-0	Heptenophos
CAS 23564-05-8	Thiophanate methyl
CAS 23564-06-9	Thiophanate
CAS 23593-75-1	Clotrimazole
CAS 23779-32-0	Triethoxysilyl propyl urea
CAS 23947-60-6	Ethirimol
CAS 23950-58-5	Propyzamide
CAS 24017-47-8	Triazophos
CAS 24166-13-0	Cloxazolam
CAS 24304-00-5	Aluminum nitride
CAS 24307-26-4	Mepiquat chloride
CAS 24356-60-3	Cephapirin sodium
CAS 24390-14-5	Doxycycline hydrochloride
CAS 24424-99-5	Dibutyl dicarbonate
CAS 24589-78-4	Methyl trimethylsilyl trifluoro-acetamide
CAS 24602-86-6	Tridemorph
CAS 24678-13-5	Lenperone
CAS 24800-44-0	Tripropylene glycol
CAS 24801-88-5	Isocyanato propyltriethoxysilane
CAS 24815-24-5	Rescinnamine
CAS 24817-92-3	Acetyl trihexyl citrate
CAS 24887-06-7	Zinc formaldehyde sulfoxylate
CAS 24937-16-4	Nylon 12
CAS 24937-78-8	Ethylene/VA copolymer
CAS 24937-79-9	Polyvinylidene fluoride
CAS 24938-91-8	Trideceth series
CAS 24969-11-7	Resorcinol formaldehyde resin
CAS 24979-97-3	Polytetramethylene ether glycol
CAS 25013-16-5	BHA
CAS 25034-71-3	Ethylene propylene diene terpolymer
CAS 25035-04-5	Nylon 11
CAS 25038-54-4	Nylon 6
CAS 25038-59-9	Polyethylene terephthalate
CAS 25038-74-8	Nylon 12
CAS 25046-79-1	Glisoxepide
CAS 25054-76-6	Oleamidopropyl betaine
CAS 25057-89-0	Bentazon
CAS 25066-20-0	Stearamidopropylamine oxide
CAS 25068-26-2	Polymethyl pentene
CAS 25086-62-8	Sodium polymethacrylate
CAS 25086-89-9	PVP/VA copolymer
CAS 25103-09-7	Isooctyl thioglyconate
CAS 25103-12-2	Triisooctyl phosphite
CAS 25103-58-6	t-Dodecyl mercaptan
CAS 25122-46-7	Clobetasol propionate
CAS 25122-57-0	Clobetasone butyrate
CAS 25126-32-3	Sincalide
CAS 25134-01-4	Polyphenylene oxide
CAS 25135-51-3	Polysulfone
CAS 25154-52-3	Nonylphenol
CAS 25155-18-4	Methylbenzethonium chloride
CAS 25155-23-1	Trixylenyl phosphate
CAS 25155-25-3	Bisbutyl peroxy diisopropyl benzene
CAS 25155-25-3	Dibutyl peroxy isopropyl benzene
CAS 25155-30-0	Sodium dodecylbenzene sulfonate
CAS 25159-40-4	Oleamidopropylamine oxide
CAS 25168-05-2	Monochlorotoluene
CAS 25168-73-4	Saccharose monostearate
CAS 25168-73-4	Sucrose stearate
CAS 25190-06-1	Polytetramethylene ether glycol
CAS 25213-24-5	Polyvinyl alcohol, partially hydrolyzed
CAS 25231-21-4	PPG stearyl ether series
CAS 25265-77-4	Trimethyl pentanediol monoisobutyrate
CAS 25301-02-4	Tyloxapol
CAS 25311-71-1	Isofenphos
CAS 25321-41-9	Xylene sulfonic acid
CAS 25322-68-3	PEG series
CAS 25322-68-3	Polyethylene glycol
CAS 25332-39-2	Trazodone hydrochloride
CAS 25339-09-7	Isocetyl stearate
CAS 25339-99-5	Sucrose laurate
CAS 25383-99-7	Sodium stearoyl lactylate
CAS 25389-94-0	Kanamycin sulfate
CAS 25417-20-3	Sodium dibutyl naphthalene sulfonate
CAS 25446-78-0	Sodium trideceth sulfate
CAS 25448-25-3	Triisodecyl phosphite
CAS 25496-01-9	Tridecylbenzene sulfonic acid
CAS 25496-72-4	Glyceryl oleate
CAS 25550-98-5	Phenyl diisodecyl phosphite
CAS 25606-41-1	Propamocarb hydrochloride
CAS 25608-12-2	Potassium polyacrylate
CAS 25609-89-6	VA/crotonates copolymer
CAS 25637-84-7	Glyceryl dioleate
CAS 25637-99-4	Hexabromocyclododecane
CAS 25655-41-8	Povidone-iodine
CAS 25717-80-0	Molsidomine
CAS 25812-30-0	Gemfibrozil
CAS 25882-44-4	Disodium lauramido MEA-sulfosuccinate
CAS 25928-94-3	Epoxy resin
CAS 25954-13-6	Fosamine ammonium
CAS 26027-37-2	PEG oleamide series
CAS 26027-38-3	Nonoxynol series
CAS 26062-79-3	Polydimethyl diallyl ammonium chloride
CAS 26062-79-3	Polyquaternium-6
CAS 26062-94-2	Polybutylene terephthalate
CAS 26097-80-3	Cambendazole
CAS 26112-07-2	Potassium methyl paraben
CAS 26155-31-7	Morantel tartrate
CAS 26159-34-2	Naproxen sodium
CAS 26183-52-8	Deceth series
CAS 26225-79-6	Ethofumesate
CAS 26256-79-1	Sodium laurimino dipropionate
CAS 26264-05-1	Isopropylamine dodecylbenzene sulfonate
CAS 26264-06-2	Calcium dodecylbenzene sulfonate

CAS 26264-58-4	Sodium methylnaphthalene sulfonate
CAS 26266-57-9	Sorbitan palmitate
CAS 26266-58-0	Sorbitan trioleate
CAS 26266-77-3	Dihydroabietyl alcohol
CAS 26309-95-5	Pivampicillin hydrochloride
CAS 26401-27-4	Diphenyl isooctyl phosphite
CAS 26402-26-6	Caprylic/capric glycerides
CAS 26402-26-6	Glyceryl caprylate
CAS 26402-31-3	Propylene glycol ricinoleate
CAS 26426-80-2	Isobutylene/MA copolymer
CAS 26444-49-5	Diphenyl cresyl phosphate
CAS 26446-38-8	Sucrose palmitate
CAS 26447-10-9	Ammonium xylene sulfonate
CAS 26483-35-2	Behenamine oxide
CAS 26523-78-4	Trisnonylphenyl phosphite
CAS 26530-20-1	Octhilinone
CAS 26530-20-1	Octylisothiazolinone
CAS 26544-23-0	Diphenyl isodecyl phosphite
CAS 26544-27-4	Diisodecyl pentaerythritol diphosphite
CAS 26545-53-9	DEA-dodecylbenzene sulfonate
CAS 26570-10-5	Propoxyphene napsylate
CAS 26589-26-4	Acrylates/PVP copolymer
CAS 26590-05-6	Polyquaternium-7
CAS 26597-36-4	Dibehenyl dimonium chloride
CAS 26635-75-6	PEG-3 lauramide
CAS 26635-92-7	PEG stearamine series
CAS 26635-93-8	PEG oleamine series
CAS 26644-46-2	Triforine
CAS 26652-09-5	Ritodrine
CAS 26658-19-5	Sorbitan tristearate
CAS 26675-46-7	Isoflurane
CAS 26680-54-6	Octenyl succinic anhydride
CAS 26741-53-7	Bisdibutylphenyl pentaerythritol diphosphite
CAS 26748-38-9	Butylperoxy neoheptanoate
CAS 26748-41-4	Butylperoxy neodecanoate
CAS 26748-47-0	Cumylperoxy neodecanoate
CAS 26761-45-5	Glycidyl decanoate
CAS 26774-90-3	Epicillin
CAS 26780-96-1	Dihydrotrimethylquinoline polymer
CAS 26787-78-0	Amoxicillin
CAS 26807-65-8	Indapamide
CAS 26844-12-2	Indoramin
CAS 26850-24-8	DEDM hydantoin
CAS 26864-56-2	Penfluridol
CAS 26921-17-5	Timolol maleate
CAS 26944-48-9	Glibornuride
CAS 26952-21-6	Isooctylalcohol
CAS 26970-82-1	Sodium selenite
CAS 27164-46-1	Cefazolin sodium
CAS 27176-87-0	Dodecylbenzene sulfonic acid
CAS 27177-77-1	Potassium dodecylbenzene sulfonate
CAS 27178-16-1	Diisodecyl adipate
CAS 27194-74-7	Propylene glycol laurate
CAS 27195-16-0	Sucrose distearate
CAS 27223-35-4	Ketazolam
CAS 27233-00-7	Glyceryl triacetyl hydroxystearate
CAS 27306-79-2	Myreth-3
CAS 27306-90-7	Laureth-17 carboxylic acid

CAS 27321-96-6	Choleth-24
CAS 27503-81-7	Phenylbenzimidazole sulfonic acid
CAS 27554-26-3	Diisooctyl phthalate
CAS 27607-77-8	Trimethylsilyl trifluoromethane sulfonate
CAS 27638-00-2	Glyceryl dilaurate
CAS 27640-89-7	Erucyl erucate
CAS 27676-62-6	Trisdibutyl hydroxybenzyl isocyanurate
CAS 27776-21-2	Azobisimidazolinyl propane dihydrochloride
CAS 27813-02-1	Hydroxypropyl methacrylate
CAS 27841-06-1	Neopentyl glycol dicaprate
CAS 27877-51-6	Tolindate
CAS 27883-12-1	Linoleamide DEA
CAS 27986-36-3	Nonoxynol-1
CAS 28061-69-0	Dimethyl oleamine
CAS 28108-99-8	Propylated triphenyl phosphate
CAS 28211-18-9	PVP/eicosene copolymer
CAS 28212-44-4	Nonoxynol-10 carboxylic acid
CAS 28395-03-1	Bumetanide
CAS 28434-01-7	Bioresmethrin
CAS 28510-23-8	Neopentyl glycol dioctanoate
CAS 28519-02-0	Sodium dodecyl diphenyloxide disulfonate
CAS 28631-35-8	Dichlorprop P
CAS 28631-63-2	Cumene sulfonic acid
CAS 28657-80-9	Cinoxacin
CAS 28724-32-5	PEG-15 stearmonium chloride
CAS 28772-56-7	Bromadiolone
CAS 28860-95-9	Carbidopa
CAS 28874-51-3	Sodium PCA
CAS 28911-01-5	Triazolam
CAS 28981-97-7	Alprazolam
CAS 29051-57-8	Sodium caproyl lactylate
CAS 29094-61-9	Glipizide
CAS 29110-48-3	Guanfacine hydrochloride
CAS 29122-68-7	Atenolol
CAS 29216-28-2	Mequitazine
CAS 29232-93-7	Pirimiphos methyl
CAS 29240-17-3	Amylperoxy pivalate
CAS 29297-22-0	Polyquaternium-16
CAS 29317-52-0	Isononyl oxypropylamine
CAS 29381-93-9	TEA-dodecylbenzene sulfonate
CAS 29383-26-4	Octyl hydroxystearate
CAS 29385-43-1	Tolyltriazole
CAS 29406-96-0	Polybutadiene-1,2 (syndiotactic)
CAS 29590-42-9	Isooctyl acrylate
CAS 29598-76-3	Pentaerythrityl tetrakis lauryl thiopropionate
CAS 29710-25-6	Octyl oxystearate
CAS 29761-21-5	Isodecyl diphenyl phosphate
CAS 29806-73-3	Octyl palmitate
CAS 29911-28-2	Dipropylene glycol butyl ether
CAS 29964-84-9	Isodecyl methacrylate
CAS 29973-13-5	Ethiofencarb
CAS 30043-49-3	Ethidimuron
CAS 30286-75-0	Oxitropium bromide
CAS 30364-51-3	Sodium myristoyl sarcosinate
CAS 30392-41-7	Bitolterol mesylate
CAS 30516-87-1	Zidovudine

CAS 30525-89-4	Paraformaldehyde
CAS 30657-38-6	Lauryl PCA
CAS 30902-17-9	Midodrine hydrochloride
CAS 31001-77-1	Mercaptopropyl methyl dimethoxysilane
CAS 31112-62-6	Metrizamide
CAS 31394-71-5	PPG-26 oleate
CAS 31430-15-6	Flubendazole
CAS 31431-39-7	Mebendazole
CAS 31556-45-3	Tridecyl stearate
CAS 31566-31-1	Glyceryl stearate
CAS 31570-04-4	Trisdibutylphenyl phosphite
CAS 31586-77-3	Bismuth sodium tartrate
CAS 31621-91-7	PEG-9 pelargonate
CAS 31677-93-7	Bupropion hydrochloride
CAS 31692-79-2	Dimethiconol
CAS 31694-55-0	Glycereth series
CAS 31717-87-0	Dodemorph acetate
CAS 31778-15-1	Octadecyl mercaptopropionate
CAS 31793-07-4	Pirprofen
CAS 31799-71-0	PEG-3 oleamide
CAS 31842-01-0	Indoprofen
CAS 31842-61-2	Rimiterol hydrobromide
CAS 31884-77-2	Meclizine dihydrochloride
CAS 32057-14-0	Glyceryl isostearate
CAS 32073-22-6	Sodium cumene sulfonate
CAS 32131-17-2	Nylon 6/6
CAS 32360-05-7	Stearyl methacrylate
CAS 32426-11-2	Quaternium-24
CAS 32440-50-9	PVP/hexadecene copolymer
CAS 32534-81-9	Pentabromo diphenyloxide
CAS 32536-52-0	Octabromo diphenyloxide
CAS 32588-76-4	Ethylene bistetrabromo phthalimide
CAS 32612-48-9	Ammonium laureth sulfate
CAS 32647-67-9	Dibenzylidene sorbitol
CAS 32780-64-6	Labetalol hydrochloride
CAS 32795-47-4	Nomifensine maleate
CAS 32886-97-8	Amdinocillin pivoxil
CAS 32887-01-7	Amdinocillin
CAS 32954-43-1	Palmitamidopropyl betaine
CAS 32986-56-4	Tobramycin
CAS 33005-95-7	Tiaprofenic acid
CAS 33089-61-1	Amitraz
CAS 33125-97-2	Etomidate
CAS 33342-05-1	Gliquidone
CAS 33386-08-2	Buspirone hydrochloride
CAS 33401-94-4	Pyrantel tartate
CAS 33402-03-8	Metaraminol bitartrate
CAS 33419-42-0	Etoposide
CAS 33564-31-7	Diflorasone diacetate
CAS 33659-28-8	Calcium bromidolactobionate
CAS 33665-90-6	Acesulfame
CAS 33671-46-4	Clotiazepam
CAS 33703-08-1	Diisononyl adipate
CAS 33817-20-8	Pivampicillin
CAS 33907-46-9	Glycol hydroxystearate
CAS 33907-47-0	Propylene glycol hydroxystearate
CAS 33939-64-9	Sodium laureth carboxylate series
CAS 33939-65-0	Sodium ceteth carboxylate series
CAS 34014-18-1	Tebuthiuron
CAS 34031-32-8	Auranofin
CAS 34097-16-0	Clocortolone pivalate
CAS 34123-59-6	Isoproturon
CAS 34137-09-2	Dibutyl hydroxy hydrocinnamic acid trishydroxyethyl triazinetrione triester
CAS 34195-34-1	Hydrocodone bitartrate
CAS 34316-64-8	Hexyl laurate
CAS 34360-00-4	Bishydroxyethyl isodecyloxy propylamine
CAS 34424-98-1	Polyglyceryl-10 tetraoleate
CAS 34443-12-4	Butylperoxy ethylhexyl carbonate
CAS 34552-84-6	Isoxicam
CAS 34562-31-7	Butyraldehyde aniline condensation product
CAS 34580-14-8	Ketotifen fumarate
CAS 34590-94-8	PPG-2 methyl ether
CAS 34787-01-4	Ticarcillin
CAS 34938-91-8	Laureth-6
CAS 34962-91-9	Isodecyl octanoate
CAS 35121-78-9	Prostacyclin
CAS 35141-30-1	Trimethoxy silylpropyl diethylenetriamine
CAS 35274-05-6	Cetyl lactate
CAS 35285-68-8	Sodium ethyl paraben
CAS 35285-69-9	Sodium propyl paraben
CAS 35325-02-1	Hydroxypropyl benzenesulfonamide
CAS 35367-38-5	Diflubenzuron
CAS 35400-43-2	Sulprofos
CAS 35512-33-9	Pyridate
CAS 35545-57-4	PEG-13 betanaphthol ether
CAS 35554-44-0	Enilconazole
CAS 35575-96-3	Azamethiphos
CAS 35604-67-2	Viloxazine hydrochloride
CAS 35607-66-0	Cefoxitin
CAS 35634-74-3	Phenylazomethoxy dimethy valeronitrile
CAS 35958-30-6	Ethylidenebis dibutylphenol
CAS 36116-84-4	Diisooctyl phosphite
CAS 36311-34-9	Cetyl alcohol
CAS 36311-34-9	Isocetyl alcohol
CAS 36322-90-4	Piroxicam
CAS 36330-85-5	Fenbufen
CAS 36409-57-1	Disodium lauryl sulfosuccinate
CAS 36432-46-9	Ditridecyl phosphite
CAS 36445-71-3	Sodium decyl diphenylether disulfonate
CAS 36445-71-3	Sodium decyl diphenyloxide disulfonate
CAS 36457-19-9	Potassium ethyl paraben
CAS 36457-20-2	Sodium butyl paraben
CAS 36483-57-5	Tribromo neopentyl alcohol
CAS 36521-89-8	Sorbitan distearate
CAS 36637-18-0	Etidocaine
CAS 36653-82-4	Cetyl alcohol
CAS 36734-19-7	Iprodione
CAS 36788-39-3	Trisdipropylene glycol phosphite
CAS 36791-04-5	Ribavirin
CAS 37091-65-9	Azlocillin sodium
CAS 37091-66-0	Azlocillin
CAS 37106-97-1	Bentiromide
CAS 37139-99-4	Olealkonium chloride

CAS 37187-22-7   Acetyl acetone peroxide
CAS 37200-49-0   Polysorbate 80
CAS 37205-87-1   Nonoxynol series
CAS 37220-82-9   Glyceryl oleate
CAS 37294-49-8   Disodium isodecyl sulfosuccinate
CAS 37296-80-3   Colestipol hydrochloride
CAS 37309-58-3   Polydecene
CAS 37318-14-2   PEG-8 laurate
CAS 37332-99-3   Avoparcin
CAS 37340-60-6   Sodium nonoxynol-6 phosphate
CAS 37349-34-1   Polyglyceryl stearate series
CAS 37354-45-5   Disodium laureth sulfosuccinate
CAS 37475-88-0   Ammonium cumene sulfonate
CAS 37478-68-5   Capryl hydroxyethyl imidazoline
CAS 37517-26-3   Pipothiazine palmitate
CAS 37517-30-9   Acebutolol
CAS 37661-08-8   Bacampicillin hydrochloride
CAS 37883-00-4   Viomycin sulfate
CAS 37924-13-3   Flamprop methyl
CAS 37971-36-1   2-Phosphono butane tricarboxylic
                   acid-1,2,4
CAS 38194-50-2   Sulindac
CAS 38260-54-7   Etrimfos
CAS 38304-91-5   Minoxidil
CAS 38411-30-3   Sodium dodecylbenzene
                   sulfonate
CAS 38517-23-6   Sodium hydrogenated tallow
                   glutamate
CAS 38566-94-8   Potassium butyl paraben
CAS 38613-77-3   Tetrakis dibutyl phenyl
                   biphenylene diphosphonite
CAS 38720-61-5   Myreth-5 carboxylic acid
CAS 38821-53-3   Cephradine
CAS 38916-42-6   Tetrasodium dicarboxyethyl
                   stearyl sulfosuccinamate
CAS 39148-24-8   Fosetyl aluminum
CAS 39198-34-0   Azobistrimethyl pentane
CAS 39236-46-9   Imidazolidinyl urea
CAS 39300-45-3   Dinocap
CAS 39354-45-5   Disodium deceth sulfosuccinate
                   series
CAS 39354-45-5   Disodium laureth sulfosuccinate
CAS 39354-47-5   Disodium C12-15 pareth
                   sulfosuccinate
CAS 39407-03-9   Trioctyl phosphate
CAS 39421-75-5   Hydroxypropyl guar
CAS 39464-64-7   Nonyl nonoxynol phosphate
                   series
CAS 39464-66-9   Laureth phosphate series
CAS 39464-69-2   Oleth phosphate series
CAS 39515-41-8   Fenpropathrin
CAS 39529-26-5   Polyglyceryl-10 decastearate
CAS 39562-70-4   Nitrendipine
CAS 39669-97-1   Palmitamidopropyl dimethylamine
CAS 39698-78-7   Saralasin acetate
CAS 39831-55-5   Amikacin sulfate
CAS 40034-42-2   Rosoxacin
CAS 40180-04-9   Ticrynafen
CAS 40372-72-3   Bistriethoxy silylpropyl
                   tetrasulfane
CAS 40487-42-1   Pendimethalin
CAS 40507-23-1   Fluproquazone

CAS 40716-42-5   Ricinoleamide DEA
CAS 40754-60-7   Disodium ricinoleamido MEA
                   sulfosuccinate
CAS 41083-11-8   Azocyclotin
CAS 41183-64-6   Gallium citrate
CAS 41340-25-4   Etodolac
CAS 41342-53-4   Erythromycin ethyl succinate
CAS 41354-29-4   Cyproheptadine hydrochloride
CAS 41394-05-2   Metamitron
CAS 41395-83-9   Propylene glycol dipelargonate
CAS 41395-89-5   Isodecyl isononanoate
CAS 41473-08-9   Iopronic acid
CAS 41483-43-6   Bupirimate
CAS 41484-35-9   Thiodiethylene bisdibutyl hydroxy
                   hydrocinnamate
CAS 41621-49-2   Ciclopiroxolamine
CAS 41637-38-1   PEG-10 bisphenol A dimeth-
                   acrylate
CAS 41669-30-1   Isostearyl isostearate
CAS 41672-81-5   Dipalmitoyl hydroxy proline
CAS 41708-72-9   Tocainide
CAS 41859-67-0   Bezafibrate
CAS 42116-76-7   Carnidazole
CAS 42131-25-9   Isononyl isononanoate
CAS 42131-28-2   Isostearyl lactate
CAS 42200-33-9   Nadolol
CAS 42461-84-7   Flunixin meglumine
CAS 42540-40-9   Cefamandole nafate
CAS 42612-52-2   Sodium-4 laureth phosphate
CAS 42808-36-6   Sulfated butyl tallate
CAS 43121-13-3   Triadimefon
CAS 43154-85-4   Disodium oleamido MIPA
                   sulfosuccinate
CAS 43210-67-9   Fenbendazole
CAS 43222-48-6   Difenzoquat methyl sulfate
CAS 44992-01-0   Dimethylaminoethyl acrylate
                   methyl chloride
CAS 45267-19-4   Myristamidopropyl dimethylamine
CAS 47747-56-8   Talampicillin
CAS 48145-04-6   Phenoxyethyl acrylate
CAS 49745-95-1   Dobutamine hydrochloride
CAS 50327-22-5   Nylon 4/6
CAS 50471-44-8   Vinclozolin
CAS 50643-20-4   PPG-5 ceteth-10 phosphate
CAS 50700-72-6   Vecuronium bromide
CAS 50975-76-3   Trimethoxy silyl chloromethyl
                   phenylethane
CAS 51022-69-6   Amcinonide
CAS 51022-70-9   Albuterol sulfate
CAS 51158-08-8   PEG glyceryl stearate series
CAS 51178-59-7   Dicyclopentenyl methacrylate
CAS 51192-09-7   PEG glyceryl oleate series
CAS 51200-87-4   Dimethyloxazolidine
CAS 51200-87-4   Oxazolidine
CAS 51235-04-2   Hexazinone
CAS 51248-32-9   PEG glyceryl laurate series
CAS 51258-15-2   PPG-24 glycereth-24
CAS 51264-14-3   Amsacrine
CAS 51274-83-0   Tiamenidine hydrochloride
CAS 51333-22-3   Budesonide
CAS 51338-27-3   Diclofop methyl
CAS 51429-74-4   Phosphomolybdic acid

CAS 51481-61-9	Cimetidine
CAS 51481-65-3	Mezlocillin
CAS 51630-58-1	Fenvalerate
CAS 51773-92-3	Mefloquine hydrochloride
CAS 51781-21-6	Carteolol hydrochloride
CAS 51812-80-7	Quaternium-22
CAS 52152-93-9	Cefsulodin sodium
CAS 52205-73-9	Estramustine phosphate sodium
CAS 52229-50-2	PVM/MA copolymer
CAS 52292-17-8	Isosteareth series
CAS 52315-75-0	Lauroyl lysine
CAS 52467-63-7	Tricetylmonium chloride
CAS 52504-24-2	PEG-6 caprylic/capric glycerides
CAS 52508-35-7	Dikegulac sodium
CAS 52558-73-3	Myristoyl sarcosine
CAS 52581-71-2	PPG-50 oleyl ether
CAS 52645-53-1	Permethrin
CAS 52725-64-1	Lauramide DEA
CAS 52756-22-6	Flamprop-M-isopropyl
CAS 52794-79-3	Isostearamide DEA
CAS 52820-00-5	Deltamethrin
CAS 52829-07-9	Bistetramethyl piperidinyl sebacate
CAS 52906-84-0	Oxychlorosene sodium
CAS 52907-07-0	Ethylenebis dibromo norbornane dicarboximide
CAS 53003-10-4	Salinomycin
CAS 53152-21-9	Buprenorphine hydrochloride
CAS 53179-09-2	Sisomicin sulfate
CAS 53179-11-6	Loperamide
CAS 53220-22-7	Dimyristyl peroxy dicarbonate
CAS 53320-86-8	Sodium magnesium silicate
CAS 53370-43-7	Methargen
CAS 53597-27-6	Fendosal
CAS 53610-02-9	Sodium laureth-6 carboxylate
CAS 53633-54-8	Polyquaternium-11
CAS 53648-55-8	Dezocine
CAS 53716-49-7	Carprofen
CAS 53716-50-0	Oxfendazole
CAS 53780-34-0	Mefluidide
CAS 53894-23-8	Triisononyl trimellitate
CAS 53988-10-6	Mercapto toluimidazole
CAS 54045-08-8	Cetoleth-25
CAS 54120-61-5	Prostalene
CAS 54143-56-5	Flecainide acetate
CAS 54143-57-6	Metoclopramide hydrochloride
CAS 54182-58-0	Sucralfate
CAS 54340-58-8	Meptazinol
CAS 54350-48-0	Etretinate
CAS 54400-62-3	Butamisole hydrochloride
CAS 54667-43-5	Polytetramethylene oxide diaminobenzoate
CAS 54739-18-3	Fluvoxamine
CAS 54965-24-1	Tamoxifen citrate
CAS 55028-70-1	Arbaprostil
CAS 55134-13-9	Narasin
CAS 55179-31-2	Bitertanol
CAS 55219-65-3	Triadimenol
CAS 55285-14-8	Carbosulfan
CAS 55297-96-6	Tiamulin fumarate
CAS 55353-21-4	VA/crotonates vinyl neo-decanoate copolymer

CAS 55635-13-7	Alloxydim sodium
CAS 55779-18-5	Arprinocid
CAS 55799-16-1	Zinc hydroxy phosphite
CAS 55819-53-9	Stearamidopropyl dimethylamine lactate
CAS 55837-27-9	Piretanide
CAS 55852-13-6	Stearamidopropyl morpholine
CAS 55852-15-8	Isostearamidopropyl dimethyl-amine lactate
CAS 55963-33-2	Distarch phosphate
CAS 56002-14-3	PEG isostearate series
CAS 56073-07-5	Difenacoum
CAS 56073-10-0	Brodifacoum
CAS 56187-47-4	Cefazedone
CAS 56235-92-8	Dioctyl maleate
CAS 56265-06-6	Arginine PCA
CAS 56281-36-8	Motretinide
CAS 56377-79-8	Nosiheptide
CAS 56388-43-3	Disodium oleamido PEG-2 sulfosuccinate
CAS 56391-57-2	Netilmicin sulfate
CAS 56392-17-7	Metoprolol tartrate
CAS 56519-71-2	Propylene glycol dioctanoate
CAS 56803-37-3	Butylated triphenyl phosphate
CAS 56863-02-6	Linoleamide DEA
CAS 56995-20-1	Flupirtine
CAS 57018-52-7	Propylene glycol butyl ether
CAS 57171-56-9	PEG sorbitan hexaoleate series
CAS 57432-61-8	Methyl ergonovine maleate
CAS 57569-76-3	Glycereth-7 triacetate
CAS 57635-48-0	Oleth carboxylic acid series
CAS 57646-30-7	Furalaxyl
CAS 57754-85-5	Clopyralid
CAS 57801-81-7	Brotizolam
CAS 57808-65-8	Closantel
CAS 57808-66-9	Domperidone
CAS 57834-33-0	Ethoxy carbonyl phenylmethyl phenylformamidine
CAS 57938-82-6	Adinazolam mesylate
CAS 58068-97-6	Triethoxy silylpropyl dihydro-imidazole
CAS 58069-11-7	Quaternium-52
CAS 58138-08-2	Tridiphane
CAS 58229-88-2	Dioctyltin mercaptide
CAS 58353-68-7	Disodium dodecyloxy propyl sulfosuccinate
CAS 58479-61-1	Butyldiphenyl chlorosilane
CAS 58503-79-0	Meobentine sulfate
CAS 58551-69-2	Carboprost tromethamine
CAS 58767-50-3	Bishydroxyethyl octylmethyl ammonium toluene sulfonate
CAS 58786-99-5	Butorphanol tartrate
CAS 58855-63-3	DEA oleth phosphate series
CAS 58958-60-4	Isostearyl neopentanoate
CAS 58965-66-5	Tetradecabromo diphenoxy benzene
CAS 59070-56-3	PEG-23 glyceryl laurate
CAS 59122-46-2	Misoprostol
CAS 59130-69-7	Cetyl octanoate
CAS 59130-70-7	Cetearyl octanoate
CAS 59186-41-3	Sodium cetearyl sulfate
CAS 59227-89-3	Laurocapram

CAS 59231-34-4   Isodecyl oleate
CAS 59272-84-3   Myristamidopropyl betaine
CAS 59355-61-2   PEG-3 lauramine oxide
CAS 59467-70-8   Midazolam
CAS 59559-30-7   Steareth-7 carboxylic acid
CAS 59587-44-9   Octyl pelargonate
CAS 59686-68-9   Myreth-3 myristate
CAS 59703-84-3   Piperacillin sodium
CAS 59789-51-4   Tribromophenyl maleimide
CAS 59792-81-3   Aluminum PCA
CAS 59828-07-8   Procaterol hydrochloride
CAS 59917-39-4   Vindesine sulfate
CAS 60166-93-0   Iopamidol
CAS 60207-90-1   Propiconazole
CAS 60209-82-7   Isodecyl neopentanoate
CAS 60270-33-9   Behenamidopropyl dimethylamine
CAS 60561-17-3   Sufentanil citrate
CAS 60607-34-3   Oxatomide
CAS 60628-96-8   Bifonazole
CAS 60719-84-8   Amrinone
CAS 60828-78-6   Isolaureth-3
CAS 60925-61-3   Ceforanide
CAS 61128-46-9   Polyetherimide
CAS 61197-93-1   Loprazolam
CAS 61270-78-8   Cefonic disodium
CAS 61318-91-0   Sulconazole nitrate
CAS 61368-34-1   Tribromostyrene
CAS 61477-96-1   Piperacillin
CAS 61551-69-7   Azobismethyl hydroxyethyl
                  propionamide
CAS 61570-90-9   Tioxidazole
CAS 61617-00-3   Zinc mercapto toluimidazole
CAS 61682-73-3   Pentaerythrityl tetrabehenate
CAS 61693-08-1   Hydrogenated polyisobutene
CAS 61788-40-7   Acrylic acid acrylonitrogens
                  copolymer
CAS 61788-45-2   Hydrogenated tallowamine
CAS 61788-46-3   Cocamine
CAS 61788-48-5   Acetylated lanolin
CAS 61788-49-6   Acetylated lanolin alcohol
CAS 61788-59-8   Methyl cocoate
CAS 61788-62-3   Dicoco methylamine
CAS 61788-63-4   Dihydrogenated tallow
                  methylamine
CAS 61788-85-0   PEG hydrogenated castor oil
                  series
CAS 61788-89-4   Dilinoleic acid
CAS 61788-90-7   Cocamine oxide
CAS 61788-91-8   Dimethyl soyamine
CAS 61788-93-0   Dimethyl cocamine
CAS 61788-95-2   Dimethyl hydrogenated
                  tallowamine
CAS 61789-05-7   Glyceryl cocoate
CAS 61789-08-0   Hydrogenated vegetable
                  glyceride
CAS 61789-10-4   Lard glyceride
CAS 61789-13-7   Tallow glyceride
CAS 61789-18-2   Cocotrimonium chloride
CAS 61789-30-8   Potassium cocoate
CAS 61789-40-0   Cocamidopropyl betaine
CAS 61789-68-2   PEG-2 cocobenzonium chloride
CAS 61789-71-7   Benzalkonium chloride

CAS 61789-71-7   Cocoalkonium chloride
CAS 61789-71-7   Cocoguanidinium chloride
CAS 61789-72-8   Hydrogenated tallowalkonium
                  chloride
CAS 61789-73-9   Dihydrogenated tallow
                  benzylmonium chloride
CAS 61789-75-1   Tallowalkonium chloride
CAS 61789-76-2   Dicocamine
CAS 61789-77-3   Dicocodimonium chloride
CAS 61789-79-5   Hydrogenated ditallowamine
CAS 61789-80-8   Quaternium-18
CAS 61789-86-4   Calcium sulfonate
CAS 61789-91-1   Jojoba oil
CAS 61789-97-7   Tallow
CAS 61790-12-3   Tall oil acid
CAS 61790-18-9   Soyamine
CAS 61790-31-6   Hydrogenated tallowamide
CAS 61790-33-8   Tallowamine
CAS 61790-35-0   Sulfated tall oil
CAS 61790-37-2   Tallow acid
CAS 61790-38-3   Hydrogenated tallow acid
CAS 61790-41-8   Soytrimonium chloride
CAS 61790-50-9   Potassium rosinate
CAS 61790-51-0   Sodium rosinate
CAS 61790-57-6   Cocamine acetate
CAS 61790-59-8   Hydrogenated tallowamine
                  acetate
CAS 61790-63-4   Cocamide DEA
CAS 61790-64-5   TEA cocoate
CAS 61790-81-6   PEG lanolin series
CAS 61790-85-0   PEG tallowaminopropylamine
                  series
CAS 61791-00-2   PEG tallate series
CAS 61791-01-3   PEG ditallate series
CAS 61791-08-0   PEG cocamide series
CAS 61791-10-4   PEG-15 cocomonium chloride
CAS 61791-12-6   PEG castor oil series
CAS 61791-13-7   Coceth series
CAS 61791-14-8   PEG cocamine series
CAS 61791-20-6   Laneth series
CAS 61791-24-0   PEG soyamine series
CAS 61791-25-1   Dihydroxyethyl tallow glycinate
CAS 61791-26-2   PEG hydrogenated tallowamine
                  series
CAS 61791-26-2   PEG tallowamine series
CAS 61791-28-4   Talloweth series
CAS 61791-29-5   PEG cocoate series
CAS 61791-31-9   Bishydroxyethyl cocamine
CAS 61791-31-9   PEG-2 cocamine
CAS 61791-31-9   Cocamide DEA
CAS 61791-32-0   Disodium cocoamphodiacetate
CAS 61791-38-6   Cocoyl hydroxyethyl imidazoline
CAS 61791-39-7   Tall oil hydroxyethyl imidazoline
CAS 61791-41-1   Sodium methyl tall oil acid taurate
CAS 61791-42-2   Sodium methyl cocoyl taurate
CAS 61791-44-4   PEG tallowamine series
CAS 61791-46-6   Dihydroxyethyl tallowamine oxide
CAS 61791-47-7   Dihydroxyethyl cocamine oxide
CAS 61791-53-5   Tallowpropane diaminedioleate
CAS 61791-55-7   Tallowdiamine
CAS 61791-55-7   Tallow propylenediamine
CAS 61791-56-8   Disodium tallow iminodipropionate

CAS 61791-57-9	Tallow dipropylene triamine
CAS 61791-57-9	Tallow triamine
CAS 61791-59-1	Sodium cocoyl sarcosinate
CAS 61791-63-7	Cocodiamine
CAS 61791-63-7	Coco propylenediamine
CAS 61792-31-2	Lauramidopropylamine oxide
CAS 61827-42-7	Isodeceth-4
CAS 61840-27-5	Adipic acid dimethylamino hydroxypropyl diethylenetriamine copolymer
CAS 61849-72-7	PPG methyl glucose ether series
CAS 61901-02-8	Sodium lauroamphopropionate
CAS 61970-18-9	Soyamine
CAS 62265-68-3	Quinfamide
CAS 62449-33-6	Oleyl imidazolinium hydrochloride
CAS 62476-59-9	Acifluorfen
CAS 62571-86-2	Captopril
CAS 62610-77-9	Methacrifos
CAS 62893-20-3	Cefoperazone sodium
CAS 63123-11-5	Thiodipropionate polyester
CAS 63148-55-0	Dimethicone copoolyol
CAS 63148-56-1	Polymethyltrifluoropropylsiloxane
CAS 63148-57-2	Polymethylhydrosiloxane
CAS 63148-58-3	Phenyldimethicone
CAS 63148-62-9	Dimethicone
CAS 63148-65-2	Polyvinylbutyral
CAS 63217-13-0	Sodium dinonyl sulfosuccinate
CAS 63231-60-7	Microcrystalline wax
CAS 63284-71-9	Nuarimol
CAS 63323-46-6	Ciramadol hydrochloride
CAS 63393-82-8	C12-15 alcohols
CAS 63393-93-1	Isopropyl lanolate
CAS 63394-02-5	Silicone elastomer
CAS 63428-83-1	Nylon
CAS 63675-72-9	Nisolidipine
CAS 63793-60-2	PPG-3 myristyl ether
CAS 64019-93-8	Dipivefrin hydrochloride
CAS 64024-15-3	Pentazocine hydrochloride
CAS 64092-49-5	Zomepirac sodium
CAS 64228-81-5	Atracurium besylate
CAS 64318-79-2	Gemeprost
CAS 64485-93-4	Cefotaxime sodium
CAS 64490-92-2	Tolmetin sodium
CAS 64628-44-0	Triflumuron
CAS 64665-57-2	Sodium tolyltriazole
CAS 64741-81-7	Polynuclear aromatic hydrocarbon
CAS 64742-04-7	Paraffinic distillate
CAS 64742-14-9	Petroleum distillates
CAS 64742-47-8	C isoparaffin series
CAS 64742-49-0	Hexane
CAS 64742-89-8	Heptane
CAS 64743-02-8	C alpha olefin series
CAS 64924-67-0	Halofuginone hydrobromide
CAS 64953-12-4	Moxalactam disodium
CAS 65009-35-0	Lidamidine hydrochloride
CAS 65271-80-9	Mitoxantrone
CAS 65277-42-1	Ketoconazole
CAS 65381-09-1	Caprylic/capric triglyceride
CAS 65447-77-0	Dimethyl succinate tetramethyl hydroxy hydroxyethyl piperidine polymer
CAS 65473-14-5	Naftifine hydrochloride
CAS 65497-29-2	Guar hydroxypropyl trimonium chloride
CAS 65591-14-2	Arachidyl propionate
CAS 65816-20-8	Ethoxy carbonyl phenylethyl phenylformamidine
CAS 65876-95-1	Resorcinol-formaldehyde resin
CAS 65996-61-4	Cellulose
CAS 66009-41-4	Stearyl heptanoate
CAS 66063-05-6	Pencycuron
CAS 66085-00-5	Glyceryl isostearate
CAS 66085-59-4	Nimodipine
CAS 66104-23-2	Pergolide mesylate
CAS 66108-95-0	Iohexol
CAS 66441-23-4	Fenoxaprep ethyl
CAS 66455-14-9	C12-13 pareth-7
CAS 66455-15-0	Deceth-6
CAS 66455-17-2	C9-11 alcohols
CAS 66455-29-6	Lauryl betaine
CAS 66592-87-8	Cefadroxil
CAS 66722-44-9	Bisoprolol
CAS 66734-13-2	Alclometasone dipropionate
CAS 66794-58-9	PEG-20 sorbitan isostearate
CAS 66794-74-9	Encainide hydrochloride
CAS 66841-24-5	Cypermethrin
CAS 66988-04-3	Sodium isostearoyl lactylate
CAS 67306-00-7	Fenpropidin
CAS 67306-03-0	Fenpropimorph
CAS 67375-30-8	Alpha cypermethrin
CAS 67633-57-2	Isostearylethyl imidonium ethosulfate
CAS 67633-59-4	Isostearaminopropalkonium chloride
CAS 67633-63-0	Isostearamidopropyl ethyl dimonium ethosulfate
CAS 67700-98-5	Dimethyl lauramine
CAS 67701-00-2	Trihexadecylamine
CAS 67701-26-2	C12-18 acid triglyceride
CAS 67701-27-3	Hydrogenated tallow glycerides
CAS 67701-30-8	Triolein
CAS 67702-21-4	Magnesium laureth sulfate
CAS 67747-09-5	Prochloraz
CAS 67762-19-0	Ammonium laureth sulfate
CAS 67762-36-1	Caprylic/capric acid
CAS 67762-38-3	Methyl oleate
CAS 67762-39-4	Methyl caprylate/caprate
CAS 67762-40-7	Methyl laurate
CAS 67762-41-8	C13-15 alcohols
CAS 67762-96-3	Dimethicone copolyol
CAS 67784-77-4	PEG-2 tallowmonium chloride
CAS 67784-87-6	Hydrogenated palm glyceride
CAS 67784-90-1	Cocoyl hydroxyethyl imidazoline
CAS 67799-04-6	Isostearamidopropyl dimethylamine
CAS 67846-16-6	Stearamidopropyl ethyl dimonium ethosulfate
CAS 67892-37-9	Sodium oleoamphopropionate
CAS 67923-14-2	Dimethicone
CAS 67990-17-4	Sodium butoxyethoxy acetate
CAS 67998-94-1	Sodium oleic sulfate
CAS 68002-44-8	Vinyl ester resin
CAS 68002-59-5	Quaternium-18

CAS 68002-61-9	Tallowtrimonium chloride
CAS 68002-71-1	Hydrogenated soybean oil
CAS 68002-72-2	Hydrogenated menhaden oil
CAS 68002-79-9	Trimethylpropane trioleate
CAS 68002-97-1	Laureth series
CAS 68003-46-3	Ammonium lauroyl sarcosinate
CAS 68037-49-0	Sodium C10-18 alkyl sulfonate
CAS 68037-87-6	Polyvinylmethylsiloxane
CAS 68037-92-3	Eicosyl docosylamine
CAS 68037-93-4	Dimethyl palmitamine
CAS 68037-97-8	Oleyl propane diamine
CAS 68039-13-4	Polymethacrylamidopropyl trimonium chloride
CAS 68081-81-2	Sodium dodecylbenzene sulfonate
CAS 68081-96-9	Ammonium lauryl sulfate
CAS 68081-97-0	Magnesium lauryl sulfate
CAS 68083-14-7	Phenylmethyl polysiloxane
CAS 68122-86-1	Quaternium-27
CAS 68128-59-6	Diammonium stearyl sulfo-succinamate
CAS 68130-24-5	Dipentaerythrityl hexacaprylate hexacaprate
CAS 68130-47-2	Deceth-4 phosphate
CAS 68131-37-3	Corn syrup solids
CAS 68131-39-5	C12-15 pareth series
CAS 68131-40-8	C11-15 pareth series
CAS 68133-13-1	Diisooctyl octylphenyl phosphite
CAS 68139-30-0	Cocamidopropyl hydroxysultaine
CAS 68140-00-1	Cocamide MEA
CAS 68140-01-2	Cocamidopropyl dimethylamine
CAS 68140-08-9	Tallowamide DEA
CAS 68140-98-7	Ethylhydroxymethyl oleyl oxazoline
CAS 68153-28-6	Soy protein
CAS 68153-32-2	Ditallow dimonium chloride
CAS 68153-63-9	Tallowamide MEA
CAS 68153-64-0	PEG-2 tallowate
CAS 68154-36-9	Sorbitan laurate
CAS 68154-97-2	PPG-6 deceth-4
CAS 68155-09-9	Cocamidopropylamine oxide
CAS 68155-20-4	Tallamide DEA
CAS 68155-24-8	PEG hydrogenated tallowamide series
CAS 68171-33-5	Isopropyl isostearate
CAS 68171-38-0	Propylene glycol isostearate
CAS 68173-73-7	Polyethylene polyamine
CAS 68184-04-3	MEA-laureth sulfate
CAS 68186-14-1	Methyl rosinate
CAS 68186-34-5	Sodium oleth-7 phosphate
CAS 68187-29-1	TEA-cocoyl glutamate
CAS 68187-32-6	Sodium cocoyl glutamate
CAS 68188-30-7	Soyamidopropyl dimethylamine
CAS 68201-46-7	PEG glyceryl cocoate series
CAS 68201-49-0	Lanolin wax
CAS 68238-35-7	Keratin amino acids
CAS 68238-87-9	Sorbitan diisostearate
CAS 68239-42-9	Methyl gluceth-10
CAS 68239-43-0	Methyl gluceth-20
CAS 68299-16-1	Amylperoxy neodecanoate
CAS 68299-17-2	Sodium isodecyl sulfate
CAS 68304-37-6	Dichlorotetraisopropyldisiloxane

CAS 68308-22-5	Calcium montanate
CAS 68308-54-3	Glyceryl stearate
CAS 68308-67-8	Soyethyl dimonium ethosulfate
CAS 68310-73-6	Pentamethyl octadecenyl diammonium dichloride
CAS 68334-00-9	Hydrogenated cottonseed oil
CAS 68334-21-4	Sodium cocoamphoacetate
CAS 68334-28-1	Hydrogenated vegetable oil
CAS 68359-37-5	Cyfluthrin
CAS 68373-14-8	Sulbactam
CAS 68389-70-8	PEG-20 methyl glucose sesquistearate
CAS 68391-01-5	Benzalkoniumchloride
CAS 68391-03-7	Alkyltrimethyl ammonium chloride
CAS 68391-07-1	Dimethyl tallowamine
CAS 68401-82-1	Ceftizoxime sodium
CAS 68411-19-8	Butyraldehyde monobutylamine condensation product
CAS 68411-20-1	Butyraldehyde aniline condensation product
CAS 68411-32-5	Dodecylbenzene sulfonic acid
CAS 68411-97-2	Cocoyl sarcosine
CAS 68412-53-3	Nonoxynol phosphate series
CAS 68424-43-1	Lanolin acid
CAS 68424-59-9	Shea butter extract
CAS 68424-60-2	Shea butter
CAS 68424-61-3	Glyceryl oleate
CAS 68424-66-8	Hydroxylated lanolin
CAS 68424-85-1	Benzalkonium chloride
CAS 68424-94-2	Cocobetaine
CAS 68424-95-3	Dicapryl/dicaprylyl dimonium chloride
CAS 68425-37-6	Coconut alcohol
CAS 68425-42-3	Cocamidopropyl dimethylamine lactate
CAS 68425-43-4	Cocamidopropyl dimethylamine propionate
CAS 68425-47-8	Linoleamide DEA
CAS 68425-47-8	Soyamide DEA
CAS 68425-50-3	Tallowamidopropyl dimethylamine
CAS 68439-39-4	Butyl phosphate
CAS 68439-46-3	C9-11 pareth series
CAS 68439-49-6	Ceteareth series
CAS 68439-50-9	C12-14 pareth series
CAS 68439-50-9	Laureth-2
CAS 68439-51-0	PPG-4 laureth-2
CAS 68439-53-2	PPG lanolin alcohol ether series
CAS 68439-57-6	Sodium C14-16 olefin sulfonate
CAS 68439-70-3	Dimethyl myristamine
CAS 68439-73-6	Tallow aminopropylamine
CAS 68439-73-6	Tallow propane diamine
CAS 68440-05-1	Cocamide MIPA
CAS 68440-66-4	Dimethicone copolyol
CAS 68440-90-4	Polymethyloctylsiloxane
CAS 68440-90-4	Polyoctylmethyl siloxane
CAS 68458-58-8	PPG-12 PEG-65 lanolin oil
CAS 68458-88-8	PPG-12 PEG-50 lanolin
CAS 68476-03-9	Montan acid wax
CAS 68476-38-0	Glyceryl montanate
CAS 68479-64-1	Disodium oleamido MEA sulfosuccinate
CAS 68513-95-1	Soy flour

CAS 68515-47-9	Ditridecyl phthalate
CAS 68515-47-9	Undecyl dodecyl phthalate
CAS 68515-48-0	Diisononyl phthalate
CAS 68515-49-1	Diisodecyl phthalate
CAS 68515-50-4	Dihexyl phthalate
CAS 68515-65-1	Disodium cocamido MIPA sulfosuccinate
CAS 68516-06-3	PPG-2 cocamine
CAS 68526-79-4	Hexyl alcohol
CAS 68526-83-0	Isooctyl alcohol
CAS 68526-84-1	Isononyl alcohol
CAS 68526-85-2	Decyl alcohol
CAS 68526-86-3	Dodecyl alcohol
CAS 68526-86-3	Lauryl alcohol
CAS 68526-86-3	Tridecyl alcohol
CAS 68527-05-9	Isononyl alcohol
CAS 68551-12-2	C12-16 pareth-1
CAS 68554-09-6	Dihydrogenated tallow hydroxymethyl ethyl ammonium chloride
CAS 68554-53-0	Stearoxydimethicone
CAS 68554-65-4	Dimethicone copolyol
CAS 68555-36-2	Polyquaternium-2
CAS 68555-98-6	Alkylphenol disulfide
CAS 68584-22-5	Dodecylbenzene sulfonic acid
CAS 68584-24-7	Isopropylamine dodecylbenzene sulfonate
CAS 68585-05-7	Sulfated neatsfoot oil
CAS 68585-34-2	Sodium laureth sulfate
CAS 68585-44-4	DEA-lauryl sulfate
CAS 68585-47-4	Sodium lauryl sulfate
CAS 68586-07-2	MEA-borate
CAS 68603-42-9	Cocamide DEA
CAS 68603-64-5	Hydrogenated tallow propylene diamine
CAS 68604-44-4	Pentaerythrityl tetraoleate
CAS 68604-71-7	Disodium cocoamphodipropionate
CAS 68604-73-9	Sodium cocoamphohydroxypropyl sulfonate
CAS 68607-29-4	Pentamethy tallow propane diammonium dichloride, IPA
CAS 68607-75-0	Polymethyloctadecylsiloxane
CAS 68607-75-0	Stearyl methicone
CAS 68608-61-7	Sodium caproamphoacetate
CAS 68608-63-9	Sodium stearoamphoacetate
CAS 68608-64-0	Disodium capryloamphodiacetate
CAS 68608-65-1	Sodium cocoamphoacetate
CAS 68608-66-2	Disodium lauroamphodiacetate
CAS 68608-88-8	Dodecylbenzene sulfonic acid
CAS 68610-38-8	Sodium oleoamphohydroxypropyl sulfonate
CAS 68610-39-9	Sodium capryloampho-hydroxypropyl sulfonate
CAS 68610-43-5	Disodium lauroamphodipropionate
CAS 68610-51-5	Cresol dicyclopentadiene butylated reaction product
CAS 68610-62-8	Trisneodol-25 phosphite
CAS 68630-75-1	Buserelin acetate
CAS 68630-96-6	Sodium isostearoampho-propionate
CAS 68647-53-0	Disodium cocoamphodiacetate
CAS 68647-73-4	Tea tree oil
CAS 68647-77-8	Tallowamidopropylamine oxide
CAS 68648-27-1	PEG hydrogenated lanolin series
CAS 68648-66-8	Maleated soybean oil
CAS 68650-39-5	Disodium cocoamphodiacetate
CAS 68650-79-3	Tallamidopropyldimethylamine
CAS 68715-87-7	Trimethyloctadecenyl diaminopropane
CAS 68783-22-2	PEG hydrogenated tallowamide series
CAS 68783-24-4	Ditallowamine
CAS 68783-25-5	Trimethyl tallow diaminopropane
CAS 68783-78-8	Ditallowdimonium chloride
CAS 68784-08-7	Disodium cocamido MEA sulfosuccinate
CAS 68797-31-9	Econazole nitrate
CAS 68797-65-9	Behenamidopropyl ethyl dimonium ethosulfate
CAS 68813-55-8	Oxantel pamoate
CAS 68814-69-7	Dimethyl tallowamine
CAS 68815-45-2	Disodium caproampho-dipropionate
CAS 68815-55-4	Disodium capryloampho-dipropionate
CAS 68815-56-5	Disodium laureth sulfosuccinate
CAS 68844-77-9	Astemizole
CAS 68876-77-7	Yeast
CAS 68877-55-4	Capryloamphopropionate
CAS 68890-66-4	Piroctoneolamine
CAS 68890-92-6	Disodium laneth-5 sulfosuccinate
CAS 68891-17-8	Sodium trideceth carboxylate series
CAS 68891-38-3	Sodium laureth sulfate
CAS 68891-39-4	Sodium nonoxynol sulfate series
CAS 68901-05-3	PPG-3 diacrylate
CAS 68909-20-6	Hexamethyl disilazane
CAS 68910-56-5	Ditridecyldimonium chloride
CAS 68911-79-5	Oleotripropylene tetraamine
CAS 68911-79-5	Tallow tetramine
CAS 68915-25-3	Cocodimonium hydroxypropyl hydrolyzed keratin
CAS 68918-77-4	TEA abietoyl hydrolyzed collagen
CAS 68919-40-4	Cocoamphodipropionic acid
CAS 68919-40-4	Disodium cocoamphodipropionate
CAS 68919-41-5	Sodium cocoamphopropionate
CAS 68920-65-0	Potassium cocoyl hydrolyzed collagen
CAS 68921-83-5	Quaternium-70
CAS 68936-95-8	Methyl glucose sesquistearate
CAS 68937-41-7	Triaryl phosphate
CAS 68937-54-2	Polysiloxane polyether copolymer
CAS 68937-55-3	Dimethicone copolyol
CAS 68937-90-6	Trilinoleic acid
CAS 68938-15-8	Hydrogenated coconut acid
CAS 68938-54-5	Dimethicone copolyol
CAS 68951-72-4	PPG-2 tallowamine
CAS 68951-89-3	Hydrolyzed collagen ethyl ester
CAS 68951-92-8	Potassium undecylenoyl hydrolyzed collagen
CAS 68952-15-8	Lauroyl hydrolyzed collagen

CAS 68952-16-9	TEA cocoyl hydrolyzed collagen
CAS 68952-35-2	Tricresyl phosphate
CAS 68952-98-7	Brominated soybean oil
CAS 68953-11-7	Minkamidopropyl dimethylamine
CAS 68953-58-2	Quaternium-18 bentonite
CAS 68953-64-0	Quaternium-26
CAS 68953-96-8	Calcium dodecylbenzene sulfonate
CAS 68954-89-2	Ceteareth-7 carboxylic acid
CAS 68954-89-2	Laureth carboxylic acid series
CAS 68954-89-2	Steareth carboxylic acid series
CAS 68954-89-2	Talloweth carboxylic acid series
CAS 68955-19-1	Sodium lauryl sulfate
CAS 68955-20-4	Sodium tallow sulfate
CAS 68955-45-3	Ethylene distearamide
CAS 68956-08-1	Trimethylol propane tricaprylate/ tricaprate
CAS 68956-68-3	Vegetable oil
CAS 68957-18-6	Sodium C12-13 pareth sulfate
CAS 68957-18-6	Sodium laureth sulfate
CAS 68958-64-5	PEG-25 glyceryl trioleate
CAS 68966-38-1	Isostearyl hydroxyethyl imidazoline
CAS 68987-89-3	Sodium laureth carboxylate series
CAS 68989-00-4	Benzalkonium chloride
CAS 68989-03-7	PEG-5 cocomonium methosulfate
CAS 68989-22-0	Zeolite synthetic
CAS 68990-06-7	Hydrogenated tallow glyceride lactate
CAS 68990-58-9	Acetylated hydrogenated tallow glyceride
CAS 68990-59-0	Hydrogenated tallow glyceride citrate
CAS 68990-63-6	Shark liver oil
CAS 68990-82-9	Hydrogenated palm kernel oil
CAS 68991-88-8	Disodium tallamphodipropionate
CAS 68991-88-8	Sodium tallamphodipropionate
CAS 69009-90-1	Bismethylethyl biphenyl
CAS 69011-84-3	Sodium octoxynol-6 sulfate
CAS 69198-10-3	Metronidazole hydrochloride
CAS 69331-39-1	DEA-cetyl phosphate
CAS 69430-24-6	Cyclomethicone
CAS 69430-36-0	Hydrolyzed keratin
CAS 69468-44-6	PEG glyceryl isostearate series
CAS 69537-38-8	Behenoyl PG trimonium chloride
CAS 69633-04-1	Triclopyr
CAS 69712-56-7	Cefotetan
CAS 69806-50-4	Fluazifop butyl
CAS 70024-77-0	Disodium oleoamphodiacetate
CAS 70084-87-6	Hydrolyzed wheat protein
CAS 70084-94-5	Hydrolyzed soy protein
CAS 70124-77-5	Flucythrinate
CAS 70161-44-3	Sodium hydroxy methyl glycinate
CAS 70191-76-3	Sodium diphenylether disulfonate
CAS 70225-05-7	Tridecyl trimellitate
CAS 70288-86-7	Ivermectin
CAS 70331-94-1	Oxamidobis ethyldibutyl hydroxyphenyl propionate
CAS 70356-03-5	Cefaclor
CAS 70356-09-1	Butylmethoxy dibenzoyl methane
CAS 70458-96-7	Norfloxacin
CAS 70496-39-8	Behenamide DEA
CAS 70592-80-2	Lauramine oxide
CAS 70609-66-4	Sodium lauroyl taurate
CAS 70632-06-3	Sodium C12-15 pareth carboxylate series
CAS 70632-06-3	Sodium laureth-13 carboxylate
CAS 70693-04-8	Isostearyl alcohol
CAS 70693-05-9	Undecyl pentadecanol
CAS 70693-32-2	Neopentyl glycol dicaprylate/ dicaprate
CAS 70729-87-2	Dimethyl lauramine isostearate
CAS 70750-46-8	Dihydroxyethyl tallow glycinate
CAS 70750-47-9	PEG-2 cocomonium chloride
CAS 70788-37-3	Toluene sulfonic acid
CAS 70851-07-9	Cocamidopropyl betaine
CAS 70851-08-0	Cocamidopropyl hydroxysultaine
CAS 70879-28-6	Alfentanil hydrochloride
CAS 70914-20-4	Isoheptyl alcohol
CAS 71010-52-1	Gellan gum
CAS 71060-61-2	Bishydrogenated tallow hydroxypropyl amine
CAS 71060-72-5	Trihexadecyl methyl ammonium chloride
CAS 71172-17-3	Dimethyl ammonium hydrogen isophthalate
CAS 71247-25-1	Ceruletide diethylamine
CAS 71329-50-5	Hydroxypropyl guar hydroxy- propyl trimonium chloride
CAS 71486-48-1	Cyclohexyl isooctyl phthalate
CAS 71487-00-8	PEG-2 cocomonium nitrate
CAS 71487-01-9	Dicoconitrite
CAS 71566-49-9	Octyl isononanoate
CAS 71786-47-5	Magnesium sulfonate
CAS 71864-46-5	Dimethyl propyl dimethyl chlorosilane
CAS 71888-89-6	Diisoheptyl phthalate
CAS 71902-01-7	Sorbitan isostearate
CAS 72162-46-0	Isodecyloxypropyl diaminopro- pane
CAS 72269-52-4	Diethylene glycol dioctanoate
CAS 72319-06-3	Myristoyl hydrolyzed collagen
CAS 72347-89-8	Polyglyceryl-2 tetrastearate
CAS 72558-82-8	Ceftazidime
CAS 72869-62-6	Sorbitan tristearate
CAS 73049-73-7	Hydrolyzed elastin
CAS 73138-79-1	Sulfated peanut oil
CAS 73334-07-3	Iopromide
CAS 73384-59-5	Ceftriaxone
CAS 73771-04-7	Prednicarbate
CAS 73790-28-0	Imazalil
CAS 73816-42-9	Meclocycline sulfosalicylate
CAS 74051-80-2	Sethoxydim
CAS 74115-24-5	Clofentezine
CAS 74176-31-1	Alfaprostol
CAS 74191-85-8	Doxazosin
CAS 74223-64-6	Metsulfuron methyl
CAS 74381-53-6	Leuprolide acetate
CAS 74623-31-7	PPG buteth series
CAS 74764-40-2	Bepridil hydrochloride
CAS 75330-75-5	Lovastatin
CAS 75736-33-3	Diclobutrazol
CAS 75738-58-8	Cefmonoxime hydrochloride
CAS 75782-86-4	C12-13 alcohols

CAS 75782-87-5   C14-15 alcohols
CAS 76095-16-4   Enalapril maleate
CAS 76483-21-1   Isopropyl phosphate
CAS 76578-14-8   Quizalofop ethyl
CAS 76674-21-0   Flutriafol
CAS 76738-62-0   Paclobutrazol
CAS 76824-35-6   Famotidine
CAS 76963-41-2   Nizatidine
CAS 78110-38-0   Aztreonam
CAS 78266-06-5   Mebrofenin
CAS 78491-02-8   Diazolidinyl urea
CAS 79645-27-5   Tobramycin sulfate
CAS 79770-24-4   Iotrolan
CAS 79770-97-1   Tallowalkyl methylbishydroxy methylethyl ammonium methosulfate
CAS 80260-73-2   Dodecyl diphenyl ether disulfonic acid
CAS 80584-85-6   Tetraphenyl dipropylene glycol diphosphite
CAS 80584-86-7   Polydipropylene glycol phenyl phosphite
CAS 80584-87-8   Diphenyl didecyl trimethyl pentanediol diphosphite
CAS 81131-70-6   Pravastatin sodium
CAS 81161-17-3   Esmolol hydrochloride
CAS 81406-37-3   Fluroxypyr
CAS 82204-94-2   Dioctyl dodecyl lauroyl glutamate
CAS 82419-36-1   Ofloxacin
CAS 82469-79-2   Butyroyl trihexyl citrate
CAS 82558-50-7   Isoxaben
CAS 82560-54-1   Benfuracarb
CAS 82657-04-3   Bifenthrin
CAS 83138-08-3   Cocamidopropylbetaine
CAS 83601-81-4   Carbendazim
CAS 83682-78-4   Lauramidopropyl PEG dimonium chloride phosphate
CAS 83881-51-0   Cetirizine
CAS 83905-01-5   Azithromycin
CAS 83915-83-7   Lisinopril
CAS 83933-91-3   Methyl glucose dioleate
CAS 84082-44-0   Behenyl betaine
CAS 84501-49-5   Sodium decyl sulfate
CAS 84507-84-1   Lamotrigine
CAS 84539-90-2   Pentaerythrityl tetrabehenate
CAS 84604-00-2   Walnut oil
CAS 84604-14-8   Rosemary extract
CAS 84605-13-0   Myreth-3 laurate
CAS 84605-14-1   Myreth-3 palmitate
CAS 84605-15-2   Isostearamidopropyl epoxypropyl dimonium chloride
CAS 84812-94-2   Cocaminopropionic acid
CAS 84930-16-5   Potassium propyl paraben
CAS 84988-74-9   Butyl oleate
CAS 84988-79-4   Isobutyl oleate
CAS 85005-47-6   Isobutylated lanolin oil
CAS 85005-55-6   Nonoxynol-1
CAS 85009-19-9   Flusilazole
CAS 85049-34-9   Propylene glycol dioleate
CAS 85049-36-1   Ethyl oleate
CAS 85049-37-2   Ethylhexyl oleate
CAS 85068-76-4   Iofetamine hydrochloride

CAS 85116-87-6   Isopropyl oleate
CAS 85116-93-4   Pentaerythrityl stearate
CAS 85116-97-8   PEG-2 stearate
CAS 85117-50-6   Sodium dodecylbenzene sulfonate
CAS 85186-88-5   Sorbitan oleate
CAS 85186-88-5   Sorbitan trioleate
CAS 85209-91-2   Sodium methylenebis dibutyl phenyl phosphate
CAS 85251-77-0   Glyceryl stearate
CAS 85408-49-7   C12-14 alkyl dimethylamine oxide
CAS 85408-76-0   Butyl stearate
CAS 85409-09-2   Glyceryl tri C8C10
CAS 85409-22-9   Benzalkonium chloride
CAS 85536-04-5   Stearyl stearate
CAS 85536-07-8   PEG-8 caprylic/capric glycerides
CAS 85536-08-9   Corn oil PEG-6 esters
CAS 85536-14-7   Dodecylbenzene sulfonic acid
CAS 85536-23-8   PEG-4 rapeseedamide
CAS 85586-21-6   Methyl stearate
CAS 85632-63-9   Tallow dipropylene triamine
CAS 85666-92-8   Glyceryl stearate
CAS 85711-45-1   Pentaerythrityl oleate
CAS 85721-33-1   Ciprofloxacin
CAS 85736-49-8   PEG dioleate series
CAS 85736-49-8   PEG oleate series
CAS 85865-69-6   Isobutyl stearate
CAS 86088-85-9   Quaternium-27
CAS 86089-12-5   Ricinoleamidopropyl betaine
CAS 86386-73-4   Fluconazole
CAS 86438-78-0   Lauramidopropyl betaine
CAS 86438-79-1   Cocamidopropylbetaine
CAS 87616-36-2   Stearamidopropyl cetearyl dimonium tosylate
CAS 88150-42-9   Amlodipine
CAS 88497-58-9   C12-15 pareth-7 carboxylic acid
CAS 88671-89-0   Myclobutanil
CAS 88684-42-8   Azobismethyl phenyl propionamidine dihydrochloride
CAS 88889-14-9   Fosinopril sodium
CAS 88917-22-0   PPG-2 methyl ether acetate
CAS 89339-41-3   pPDI-PTMEG
CAS 90052-75-8   Octyldodecyl stearoyl stearate
CAS 90193-76-3   Distearyl phthalate
CAS 90268-48-7   Disodium tallow sulfosuccinamate
CAS 90283-04-8   Dilauryl acetyl dimonium chloride
CAS 90388-14-0   Cetyl diethanolamine phosphate
CAS 90453-59-1   PEG-9 stearamide carboxylic acid
CAS 90529-57-0   Soyamidopropyl ethyl dimonium ethosulfate
CAS 90624-75-2   Polyquaternium-17
CAS 90624-76-3   Polyquaternium-18
CAS 90730-68-0   Erucalkonium chloride
CAS 91031-31-1   Glycol distearate
CAS 91031-43-5   Ethyl stearate
CAS 91031-48-0   Isooctyl stearate
CAS 91031-57-1   Isononyl stearate
CAS 91031-74-2   Sorbitan stearate
CAS 91050-82-7   Pentaerythrityl tetrastearate
CAS 91052-47-0   Glyceryl stearate
CAS 91723-55-6   Quaternium-83
CAS 91744-38-6   Glyceryl stearate citrate

CAS 91824-88-3	Polyglyceryl-4 isostearate
CAS 93356-94-6	Polyisopropylidene diphenol neodol 25 alcohol phosphite
CAS 93455-78-8	PEG-2 oleate
CAS 93572-63-5	PEG-3 tallow propylene dimonium dimethosulfate
CAS 93685-79-1	Isoeicosane
CAS 93820-52-1	Sodium cocoamphopropionate
CAS 94109-05-4	Behenamide MEA
CAS 94552-41-7	Rapeseedamidopropyl ethyl dimonium ethosulfate
CAS 95823-35-1	Pentaerythrityl hexyl thio-propionate
CAS 96328-09-1	Pentaerythrityl alkyl thio-dipropionate
CAS 96492-31-8	Magnesium aluminum carbonate
CAS 96690-41-4	Hydrolyzed silk
CAS 96726-23-9	Pentaerythrityl tetralaurate
CAS 97281-23-7	Glycol stearate
CAS 97281-23-7	PEG distearate series
CAS 97281-23-7	PEG stearate series
CAS 97338-28-8	Isocetyl stearoyls tearate
CAS 97404-33-6	Cetyl palmitate
CAS 97404-50-7	Glyceryl lanolate
CAS 97488-62-5	Tallow amphopolycarboxy propionic acid
CAS 97488-91-0	Apricot kernel oil PEG-6 esters
CAS 97593-29-8	Hydrogenated palm glyceride
CAS 97593-29-8	Lard glyceride
CAS 97659-50-2	Coco iminodipropionate
CAS 97659-51-3	Coco iminodiglycinate
CAS 97659-53-5	Disodium oleoamphodiacetate
CAS 97659-53-5	Sodium carboxymethyl tallow polypropylamine
CAS 97659-53-5	Stearyl amphopolycarboxy glycinate
CAS 97759-33-8	Hydrolyzed rice protein
CAS 97808-04-3	Cocotripropylene tetraamine
CAS 98073-10-0	Methyl gluceth-20 distearate
CAS 99330-44-6	Magnesium laureth-11 carboxylate
CAS 100085-64-1	Coco betainamido ampho-propionate
CAS 100864-25-1	Hydrolyzed whea tprotein
CAS 101476-85-4	Strontium chloride
CAS 102089-33-8	Polymethyl propyloxy tetramethyl piperidinyl siloxane (monomer)
CAS 102523-96-6	Dimethicone propyl PG betaine
CAS 102843-39-0	Azobis tetrahydropyrimidinyl propane dihydrochloride
CAS 103213-19-0	Diisostearyl dimer dilinoleate
CAS 103213-20-3	Diisopropyl dimer dilinoleate
CAS 103213-22-5	Triisostearyl trilinoleate
CAS 103597-45-1	Bishydroxy octylbenzotriazolyl phenylmethane
CAS 104222-32-4	Azobismethyl bishydroxymethyl hydroxyethyl propionamide
CAS 104909-82-2	Butoxynol-19 carboxylic acid
CAS 105391-15-9	Buteth-2 carboxylic acid
CAS 105391-15-9	Hexeth-4 carboxylic acid
CAS 105827-78-9	Imidacloprid
CAS 106436-39-9	Tridecyl neopentanoate
CAS 107534-96-3	Tebuconazole
CAS 107600-33-9	Capryleth-9 carboxylic acid
CAS 107600-36-2	TIPA-laureth sulfate
CAS 107628-03-5	Sodium PEG-6 cocamide carboxylate
CAS 107628-08-0	Octoxynol-9 carboxylic acid
CAS 107987-23-5	Polyquaternium-24
CAS 108419-32-5	C8 alkyl acetate
CAS 108419-33-6	C9 alkyl acetate
CAS 108419-34-7	C10 alkyl acetate
CAS 108419-35-8	C13 alkyl acetate
CAS 109678-33-3	Tetrabromo dipentaerythritol
CAS 111019-03-5	PPG-10 cetyl ether phosphate
CAS 111174-64-2;	Triethonium hydrolyzed collagen ethosulfate
CAS 111905-55-6	Quaternium-61
CAS 112324-11-5	Rapeseedamidopropyl epoxypropyl dimonium chloride
CAS 112324-16-0	Ricinoleamidopropyl ethyl-dimonium ethosulfate
CAS 112945-52-5	Silica
CAS 113431-53-1	Diisostearyl fumarate
CAS 113431-54-2	Triisostearyl citrate
CAS 113492-03-8	Cocamidopropyl ethyl dimonium ethosulfate
CAS 113492-04-9	Saffloweramidopropyl ethyl dimonium ethosulfate
CAS 115340-78-8	Apricotamidopropyl ethyl dimonium ethosulfate
CAS 115340-80-2	Wheatgermamidopropyl ethyl dimonium ethosulfate
CAS 116912-64-2	Ureidopropyl triethoxysilane
CAS 118337-09-0	Ethylidenebis dibutylphenyl fluorophosphonite
CAS 118585-13-0	Azobishydroxyethyl imidazolinyl propane dihydrochloride
CAS 118777-77-8	Isostearamidopropyl dimethyl-amine glycolate
CAS 120007-67-9	Zinc borate
CAS 123754-28-9	Ammonium acrylates acrylo-nitrogens copolymer
CAS 124960-38-9	Azobischlorophenyl methyl propionamidine dihydrochloride
CAS 125740-36-5	Isodecyloxy propyl dihydroxy ethylmethyl ammonium chloride
CAS 125804-12-8	Glycereth-7 benzoate
CAS 125804-13-9	Glycereth-5 lactate
CAS 127358-81-0	Diisostearyl dimer dilinoleate
CAS 129541-36-2	Isostearamidopropyl dimethyl-amine gluconate
CAS 130097-36-8	Cumylperoxy neoheptanoate
CAS 130124-24-2	Quaternium-53
CAS 133798-12-6	Undecylenamidopropyl betaine
CAS 133934-08-4	Apricotamidopropyl betaine
CAS 133934-09-5	Wheatgermamidopropyl betaine
CAS 134112-42-8	Isostearamidopropyl laurylaceto dimonium chloride
CAS 136505-00-5	Acrylic acid acrylonitrogens copolymer
CAS 137796-06-6	Ethylmethyl methylbutyl oxazolidine
CAS 138208-68-1	Isocetyl salicylate

CAS 138314-11-1	Glycereth-7 hydroxystearate	CAS 977056-87-3	Hydrogenated cocoglycerides
CAS 270028-82-6	TEA-laureth sulfate	CAS 977058-55-1	Ultramarine pink
CAS 594477-57-3	Polypentabromo benzylacrylate	CAS 977059-33-8	Hydrolyzed vegetable protein
CAS 617788-68-9	Sulfated rapeseed oil	CAS 977066-20-8	Isostearoyl hydrolyzed collagen
CAS 977000-98-8	Albumen	CAS 977067-77-8	MEA-laureth sulfate
CAS 977039-11-4	Coco dimonium hydroxypropyl hydrolyzed soy protein	CAS 977077-71-6	Silk amino acids
CAS 977053-96-5	Glyceryl stearate SE	CAS 977079-10-9	Isostearyl erucate

EINECS Number-to-Chemical
Cross Reference

EINECS 200-001-8	Formaldehyde
EINECS 200-001-8	Paraformaldehyde
EINECS 200-002-3	Guanidine hydrochloride
EINECS 200-018-0	Lactic acid
EINECS 200-061-5	Sorbitol
EINECS 200-076-7	Piperonyl butoxide
EINECS 200-081-4	Benzimidazole
EINECS 200-157-7	Cysteine hydrochloride
EINECS 200-158-2	L-Cysteine
EINECS 200-194-9	DL-Tryptophan
EINECS 200-228-2	Isonicotinic acid
EINECS 200-262-8	Carbon tetrachloride
EINECS 200-268-2	Tributyltin oxide
EINECS 200-270-1	Benzyltriethyl ammonium chloride
EINECS 200-272-2	Glycine
EINECS 200-273-8	α-Alanine
EINECS 200-274-3	L-Serine
EINECS 200-289-5	Glycerin
EINECS 200-291-6	L-Aspartic acid
EINECS 200-292-1	L-Glutamine
EINECS 200-293-7	Glutamic acid
EINECS 200-294-2	L-Lysine
EINECS 200-296-3	Cystine
EINECS 200-300-3	Benzyl trimethyl ammonium chloride
EINECS 200-311-3	Cetrimonium bromide
EINECS 200-312-9	Palmitic acid
EINECS 200-313-4	Stearic acid
EINECS 200-315-5	Urea
EINECS 200-333-3	Fructose
EINECS 200-334-9	Sucrose
EINECS 200-338-0	Propylene glycol
EINECS 200-353-2	Cholesterol
EINECS 200-362-1	Caffeine
EINECS 200-400-7	D(+)-Xylose
EINECS 200-407-5	Uridine
EINECS 200-425-3	Thiamine
EINECS 200-432-1	DL-Methionine
EINECS 200-441-0	Nicotinic acid
EINECS 200-449-4	Edetic acid
EINECS 200-460-4	L-Tyrosine
EINECS 200-464-6	2-Mercaptoethanol
EINECS 200-467-2	Ethyl ether
EINECS 200-470-9	Linoleic acid
EINECS 200-522-0	L-Leucine
EINECS 200-526-2	p-Nitrobenzoic acid
EINECS 200-532-5	Phenylmercuric acetate
EINECS 200-540-9	Calcium acetate
EINECS 200-541-4	Thioacetamide

EINECS 200-543-5	Thiourea
EINECS 200-563-4	Sulfanilamide
EINECS 200-573-9	Tetrasodium EDTA
EINECS 200-578-6	Ethyl alcohol
EINECS 200-579-1	Formic acid
EINECS 200-580-7	Acetic acid
EINECS 200-609-3	Salicylamide
EINECS 200-618-2	Benzoic acid
EINECS 200-621-9	Uracil
EINECS 200-641-8	Thiamine hydrochloride
EINECS 200-655-4	Choline chloride
EINECS 200-661-7	Isopropyl alcohol
EINECS 200-662-2	Acetone
EINECS 200-663-8	Chloroform
EINECS 200-664-3	Dimethyl sulfoxide
EINECS 200-677-4	Thioglycolic acid
EINECS 200-679-5	Dimethyl formamide
EINECS 200-680-0	Vitamin B12
EINECS 200-683-7	Vitamin A
EINECS 200-712-3	Salicylic acid
EINECS 200-718-6	Xanthine
EINECS 200-720-7	Uric acid
EINECS 200-735-9	L-Asparagine , anhydrous
EINECS 200-739-0	L-Lysine monohydrochloride
EINECS 200-741-1	p-Toluenesulfonamide
EINECS 200-745-3	Histidine
EINECS 200-746-9	n-Propyl alcohol
EINECS 200-751-6	Butyl alcohol
EINECS 200-753-7	Benzene
EINECS 200-772-0	Sodium lactate
EINECS 200-773-6	L-Valine
EINECS 200-774-1	L-Threonine
EINECS 200-796-1	Adenine
EINECS 200-798-2	L-Isoleucine
EINECS 200-799-8	Guanine
EINECS 200-811-1	Arginine
EINECS 200-815-3	Polyethylene
EINECS 200-819-5	Methyl iodide
EINECS 200-835-2	Acetonitrile
EINECS 200-836-8	Acetaldehyde
EINECS 200-838-9	Methylene chloride
EINECS 200-842-0	Formamide
EINECS 200-843-6	Carbon disulfide
EINECS 200-849-9	Ethylene oxide
EINECS 200-857-2	Isobutane
EINECS 200-861-4	Isopropyl mercaptan
EINECS 200-865-6	Acetyl chloride
EINECS 200-876-6	5-Nitroisophthalic acid
EINECS 200-876-6	Nitromethane

EINECS 200-879-2	Propylene oxide
EINECS 200-882-9	Tetramethylammonium hydroxide
EINECS 200-892-3	Trichlorofluoromethane
EINECS 200-913-6	2,2,2-Trifluoroethanol
EINECS 200-915-7	t-Butyl hydroperoxide
EINECS 200-917-8	Vinyltrichlorosilane
EINECS 200-929-3	Trifluoroacetic acid
EINECS 200-936-1	Trichlorotrifluoroethane
EINECS 201-001-0	Gibberellic acid
EINECS 201-039-8	Dibutyltin dilaurate
EINECS 201-056-0	Dimethoxy propane
EINECS 201-064-4	Tris (hydroxymethyl) amino-methane
EINECS 201-069-1	Citric acid
EINECS 201-074-9	Trimethylol propane
EINECS 201-081-7	Vinyltriethoxysilane
EINECS 201-094-8	Cetethyl morpholinium ethosulfate
EINECS 201-114-5	Triethyl phosphate
EINECS 201-116-6	Trioctyl phosphate
EINECS 201-126-0	Isophorone
EINECS 201-142-8	Isopentane
EINECS 201-148-0	Isobutyl alcohol
EINECS 201-152-2	Propylene dichloride
EINECS 201-159-0	Methyl ethyl ketone
EINECS 201-162-7	Isopropanolamine
EINECS 201-166-9	Trichloroethane
EINECS 201-167-4	Trichloroethylene
EINECS 201-176-3	Propionic acid
EINECS 201-177-9	Acrylic acid
EINECS 201-178-4	Monochloroacetic acid
EINECS 201-180-5	Glycolic acid
EINECS 201-185-2	Methyl acetate
EINECS 201-188-9	Nitroethane
EINECS 201-195-7	Isobutyric acid
EINECS 201-209-1	2-Nitropropane
EINECS 201-236-9	Tetrabromobisphenol A
EINECS 201-245-8	Bisphenol A
EINECS 201-248-4	4,4´-Diaminodiphenyl sulfone
EINECS 201-297-1	Methyl methacrylate (monomer)
EINECS 201-507-1	Riboflavin
EINECS 201-507-1	Vitamin B2
EINECS 201-549-0	Anthraquinone
EINECS 201-550-6	Diethyl phthalate
EINECS 201-557-4	Dibutyl phthalate
EINECS 201-607-5	Phthalic anhydride
EINECS 201-662-5	Diphenyl acetonitrile
EINECS 201-748-2	3-Indole acetic acid
EINECS 201-762-9	Pyrogallol
EINECS 201-778-6	Pentachlorophenol
EINECS 201-781-2	Inositol
EINECS 201-782-8	Trichloroisocyanuric acid
EINECS 201-800-4	PVP
EINECS 201-803-0	Furoic acid
EINECS 201-944-8	Thymol
EINECS 201-961-0	Salicylaldehyde
EINECS 201-963-1	o-Anisidine
EINECS 201-964-7	Guaiacol
EINECS 201-966-8	α-Methylnaphthalene
EINECS 201-993-5	o-Phenylphenol
EINECS 202-041-1	2,6-Dimethoxy phenol
EINECS 202-049-5	Naphthalene
EINECS 202-051-6	Quinoline
EINECS 202-085-1	Quinaldine
EINECS 202-156-7	2,3-Naphthalenediol
EINECS 202-163-5	Biphenyl
EINECS 202-181-3	Diphenyloxazole
EINECS 202-194-4	Xanthene
EINECS 202-259-7	Methyl benzoate
EINECS 202-280-1	Stearamide DEA
EINECS 202-281-7	Oleamide DEA
EINECS 202-307-7	Propyl paraben
EINECS 202-327-6	Benzoyl peroxide
EINECS 202-377-9	Ethyl hexanediol
EINECS 202-394-1	1H-Benzotriazole
EINECS 202-397-8	Stearyl hydroxyethyl imidazoline
EINECS 202-420-1	Dimethyl glyoxime
EINECS 202-423-8	o-Cresol
EINECS 202-425-9	o-Dichlorobenzene
EINECS 202-443-7	Toluhydroquinone
EINECS 202-473-0	Allyl methacrylate
EINECS 202-480-9	2,3-Dibromo-1-propanol
EINECS 202-490-3	Diethyl ketone
EINECS 202-496-6	Methyl ethyl ketoxime
EINECS 202-500-6	Methyl acrylate (monomer)
EINECS 202-505-3	2-Chlorothiophene
EINECS 202-506-9	Ethylene thiourea
EINECS 202-509-5	Butyrolactone
EINECS 202-589-1	Eugenol
EINECS 202-592-8	Allantoin
EINECS 202-597-5	Ethyl methacrylate
EINECS 202-608-3	Lauroyl sarcosine
EINECS 202-615-1	Butyl methacrylate
EINECS 202-617-2	Ethylene glycol dimethacrylate
EINECS 202-625-6	Tetrahydrofurfuryl alcohol
EINECS 202-626-1	Furfuryl alcohol
EINECS 202-627-7	Furfural
EINECS 202-637-1	Benzenesulfonamide
EINECS 202-679-0	4-t-Butyl phenol
EINECS 202-705-0	α-Methylstyrene monomer
EINECS 202-708-7	Acetophenone
EINECS 202-723-9	m-Toluic acid
EINECS 202-751-1	3,5-Dinitrobenzoic acid
EINECS 202-760-0	Protocatechuic acid
EINECS 202-773-1	1,3-Diisopropyl benzene
EINECS 202-802-8	p-Hydroxyacetophenone
EINECS 202-803-3	p-Toluic acid
EINECS 202-810-1	p-Nitroaniline
EINECS 202-816-4	p-Anisoyl chloride
EINECS 202-818-4	p-Anisidine
EINECS 202-819-0	p-Dimethylamino benzaldehyde
EINECS 202-826-9	1,4-Diisopropyl benzene
EINECS 202-855-7	Benzonitrile
EINECS 202-859-9	Benzyl alcohol
EINECS 202-860-4	Benzaldehyde
EINECS 202-905-8	Hexamethylene tetramine
EINECS 202-908-4	Triphenyl phosphite
EINECS 202-935-1	Glycery ltriacetyl ricinoleate
EINECS 202-936-7	Triallyl cyanurate
EINECS 202-981-2	Diphenyl oxide
EINECS 203-041-4	Tetrahydroxypropyl ethylene diamine
EINECS 203-049-8	Triethanolamine
EINECS 203-051-9	Triacetin

EINECS 203-061-3	Tributyl phosphite
EINECS 203-063-4	Trilaurylamine
EINECS 203-065-5	Cinnamoyl chloride
EINECS 203-090-1	Dioctyl adipate
EINECS 203-148-6	Phenylacetic acid
EINECS 203-149-1	N-Benzyl dimethylamine
EINECS 203-166-4	p-Methoxyphenyl acetic acid
EINECS 203-180-0	Toluene sulfonic acid
EINECS 203-213-9	Cinnamic aldehyde
EINECS 203-232-2	Laurylpyridinium chloride
FINECS 203-234-3	2-Ethylhexanol
EINECS 203-246-9	p-Tolyl aldehyde
EINECS 203-246-9	Tolualdehyde
EINECS 203-288-8	Methyl cyanoacetate
EINECS 203-328-4	Dibutyl maleate
EINECS 203-351-5	Lauralkonium chloride
EINECS 203-363-5	PEG-2 stearate
EINECS 203-364-0	PEG-2 oleate
EINECS 203-366-1	Hydroxystearic acid
EINECS 203-368-2	Ricinoleamide MEA
EINECS 203-369-8	Glycol ricinoleate
EINECS 203-375-0	Citronellol
EINECS 203-376-6	Citronellal
EINECS 203-377-1	Geraniol
EINECS 203-398-6	p-Cresol
EINECS 203-400-5	p-Dichlorobenzene
EINECS 203-404-7	p-Phenylene diamine
EINECS 203-419-9	Dimethyl succinate
EINECS 203-441-9	Glycidyl methacrylate
EINECS 203-448-7	Butane
EINECS 203-455-5	n-Propyl mercaptan
EINECS 203-466-5	Acrylonitrile
EINECS 203-468-6	Ethylene diamine
EINECS 203-473-3	Glycol
EINECS 203-473-3	Polyethylene glycol
EINECS 203-483-8	Taurine
EINECS 203-489-0	Hexylene glycol
EINECS 203-508-2	Distearyl dimonium chloride
EINECS 203-528-1	Methyl propyl ketone
EINECS 203-538-6	Sarcosine
EINECS 203-542-8	Dimethylethanolamine
EINECS 203-544-9	Nitropropane
EINECS 203-545-4	Vinyl acetate
EINECS 203-558-5	Diisopropylamine
EINECS 203-564-8	Acetic anhydride
EINECS 203-570-0	Succinic anhydride
EINECS 203-571-6	Maleic anhydride
EINECS 203-577-9	m-Cresol
EINECS 203-584-7	m-Phenylene diamine
EINECS 203-585-2	Resorcinol
EINECS 203-587-3	2,6-Lutidine
EINECS 203-611-2	Phloroglucinol
EINECS 203-618-0	Cyanuric acid
EINECS 203-618-0	Isocyanuric acid
EINECS 203-620-1	Diisobutyl ketone
EINECS 203-624-3	Methyl cyclohexane
EINECS 203-625-9	Toluene
EINECS 203-626-4	γ-Picoline
EINECS 203-628-5	Chlorobenzene
EINECS 203-629-0	Cyclohexylamine
EINECS 203-630-6	Cyclohexanol
EINECS 203-631-1	Cyclohexanone
EINECS 203-635-3	Thiophenol
EINECS 203-636-9	β-Picoline
EINECS 203-640-0	p-Methyl morpholine
EINECS 203-643-7	α-Picoline
EINECS 203-646-3	2-Chloropyridine
EINECS 203-661-5	Oleamidopropyl dimethylamine
EINECS 203-677-2	n-Valeric acid
EINECS 203-686-1	Propyl acetate
EINECS 203-692-4	n-Pentane
EINECS 203-699-2	n-Butylamine
EINECS 203-700-6	Butyronitrile
EINECS 203-705-3	n-Butyl mercaptan
EINECS 203-716-3	Diethylamine
EINECS 203-726-8	Tetrahydrofuran
EINECS 203-728-9	Tetrahydrothiophene
EINECS 203-729-4	Thiophene
EINECS 203-733-6	Di-t-butyl peroxide
EINECS 203-740-4	Succinic acid
EINECS 203-742-5	Maleic acid
EINECS 203-743-0	Fumaric acid
EINECS 203-745-1	Isobutyl acetate
EINECS 203-749-3	Oleoyl sarcosine
EINECS 203-750-9	Methylene bisacrylamide
EINECS 203-751-4	Isopropyl myristate
EINECS 203-755-6	Ethylene distearamide
EINECS 203-756-1	Ethylene dioleamide
EINECS 203-766-6	Methyl caprate
EINECS 203-768-7	Sorbic acid
EINECS 203-775-5	1,4-Dibromobutane
EINECS 203-777-6	Hexane
EINECS 203-786-5	1,4-Butanediol
EINECS 203-788-6	But-2-yne-1,4-diol
EINECS 203-804-1	Ethoxyethanol
EINECS 203-806-2	Cyclohexane
EINECS 203-808-3	Piperazine
EINECS 203-809-9	Pyridine
EINECS 203-813-0	Piperidine
EINECS 203-815-1	Morpholine
EINECS 203-821-4	Dipropylene glycol
EINECS 203-825-6	Squalane
EINECS 203-826-1	Squalene
EINECS 203-827-7	Glyceryl oleate
EINECS 203-828-2	Oleamide MIPA
EINECS 203-838-7	Heptanoic acid
EINECS 203-845-5	Sebacic acid
EINECS 203-852-3	Hexyl alcohol
EINECS 203-856-5	Glutaraldehyde
EINECS 203-865-4	Diethylene triamine
EINECS 203-867-5	Aminoethylethanolamine
EINECS 203-868-0	Diethanolamine
EINECS 203-872-2	Diethylene glycol
EINECS 203-874-3	Thiodiglycol
EINECS 203-881-1	Ethylene glycol diacetate
EINECS 203-883-2	Stearamide MEA
EINECS 203-884-8	Oleamide MEA
EINECS 203-886-9	Glycol stearate
EINECS 203-893-7	Octene-1
EINECS 203-897-9	Heptyl alcohol
EINECS 203-905-0	Butoxyethanol
EINECS 203-911-3	Methyl laurate
EINECS 203-917-6	Capryl alcohol
EINECS 203-917-6	Caprylic alcohol

EINECS 203-919-7	Ethoxydiglycol
EINECS 203-927-0	Laurtrimonium chloride
EINECS 203-928-6	Cetrimonium chloride
EINECS 203-929-1	Steartrimonium chloride
EINECS 203-931-2	Pelargonic acid
EINECS 203-938-0	Decanoyl chloride
EINECS 203-941-7	Lauroyl chloride
EINECS 203-943-8	Dimethyl lauramine
EINECS 203-950-6	Triethylene tetramine
EINECS 203-955-3	Decyl bromide
EINECS 203-956-9	n-Decyl alcohol
EINECS 203-961-6	Butoxydiglycol
EINECS 203-965-8	Undecylenic acid
EINECS 203-966-3	Methyl palmitate
EINECS 203-968-4	Dodecene-1
EINECS 203-982-0	Lauryl alcohol
EINECS 203-984-1	n-Dodecyl mercaptan
EINECS 203-986-2	Tetraethylene pentamine
EINECS 203-989-9	PEG-4
EINECS 203-990-4	Methyl stearate
EINECS 203-992-5	Methyl oleate
EINECS 203-998-8	Tridecyl alcohol
EINECS 204-000-3	Myristyl alcohol
EINECS 204-007-1	Oleic acid
EINECS 204-009-2	Erucamide
EINECS 204-015-5	Oleamine
EINECS 204-017-6	Stearyl alcohol
EINECS 204-062-1	Propylene
EINECS 204-065-8	Dimethyl ether
EINECS 204-070-5	Methyl butynol
EINECS 204-072-6	Ethyltrichlorosilane
EINECS 204-104-9	Pentaerythritol
EINECS 204-112-2	Triphenyl phosphate
EINECS 204-211-0	Dioctyl phthalate
EINECS 204-284-9	o-Toluic acid
EINECS 204-288-0	o-Hydroxyacetophenone
EINECS 204-317-7	Methyl salicylate
EINECS 204-331-3	Benzoin
EINECS 204-340-2	Tetrahydronaphthalene
EINECS 204-393-1	Lauramide DEA
EINECS 204-399-4	Ethyl paraben
EINECS 204-402-9	Benzyl benzoate
EINECS 204-427-5	Pyrocatechol
EINECS 204-428-0	1,2,4-Trichlorobenzene
EINECS 204-442-7	BHA
EINECS 204-464-7	Ethyl vanillin
EINECS 204-465-2	Vanillin
EINECS 204-479-9	Benzethonium chloride
EINECS 204-482-5	Sulfanilic acid
EINECS 204-493-5	n,n-Dimethylaniline
EINECS 204-498-2	Propyl gallate
EINECS 204-516-9	Isopropylbenzaldehyde
EINECS 204-527-9	Stearalkonium chloride
EINECS 204-528-4	Triisopropanolamine
EINECS 204-534-7	Triolein
EINECS 204-552-5	Triethyl phosphite
EINECS 204-555-1	Benzylidene acetone
EINECS 204-556-7	Phenoxyacetic acid
EINECS 204-589-7	Phenoxyethanol
EINECS 204-593-9	Cetylpyridinium chloride
EINECS 204-599-1	p-Hydroxybenzaldehyde
EINECS 204-617-8	Hydroquinone

EINECS 204-626-7	Diacetone alcohol
EINECS 204-634-0	Acetyl acetone
EINECS 204-635-6	Succinimide
EINECS 204-646-6	n-Butyraldehyde
EINECS 204-648-7	2-Pyrrolidone
EINECS 204-649-2	Levulinic acid
EINECS 204-650-8	Azodicarbonamide
EINECS 204-658-1	n-Butyl acetate
EINECS 204-663-9	Thiodiglycolic acid
EINECS 204-664-4	Glyceryl stearate
EINECS 204-666-5	Butyl stearate
EINECS 204-673-3	Adipic acid
EINECS 204-677-5	Caprylic acid
EINECS 204-680-1	Methyl myristate
EINECS 204-690-6	Lauramine
EINECS 204-693-2	Stearamide
EINECS 204-694-8	Dimethyl stearamine
EINECS 204-695-3	Stearamine
EINECS 204-696-9	Carbon dioxide
EINECS 204-708-2	Sodium cacodylate
EINECS 204-781-0	Neopentyl glycol
EINECS 204-783-1	Sulfolane
EINECS 204-800-2	Tributyl phosphate
EINECS 204-822-2	Potassium acetate
EINECS 204-823-8	Sodium acetate
EINECS 204-824-3	Pyruvic acid
EINECS 204-825-9	Perchloroethylene
EINECS 204-876-7	Sodium dimethyldithiocarbamate
EINECS 204-878-8	n-Chlorosuccinimide
EINECS 204-881-4	BHT
EINECS 204-884-0	2,6-Di-t-butylphenol
EINECS 205-011-6	Dimethyl phthalate
EINECS 205-016-3	Diallyl phthalate
EINECS 205-026-8	Benzophenone-8
EINECS 205-027-3	Benzophenone-6
EINECS 205-028-9	Benzophenone-2
EINECS 205-029-4	Benzophenone-1
EINECS 205-031-5	Benzophenone-3
EINECS 205-105-7	Tartaric acid
EINECS 205-126-1	Sodium ascorbate
EINECS 205-132-4	Methyl anthranilate
EINECS 205-149-7	Diethyl toluamide
EINECS 205-234-9	Capramide DEA
EINECS 205-252-7	Butyl benzoate
EINECS 205-271-0	Lauryl hydroxyethyl imidazoline
EINECS 205-281-5	Sodium lauroyl sarcosinate
EINECS 205-285-7	Sodium methyl oleoyl taurate
EINECS 205-290-4	Sodium propionate
EINECS 205-341-0	Dipentene
EINECS 205-352-0	Myristalkonium chloride
EINECS 205-355-7	Nitrilotriacetic acid
EINECS 205-388-7	TEA-lauryl sulfate
EINECS 205-398-1	Cinnamic acid
EINECS 205-399-7	Benzyl acetate
EINECS 205-411-0	Aminoethyl piperazine
EINECS 205-447-7	Ferrous fumarate
EINECS 205-455-0	Glyceryl ricinoleate
EINECS 205-468-1	PEG-2 laurate
EINECS 205-469-7	Stearamidoethyl ethanolamine
EINECS 205-471-8	Methyl hydroxy stearate
EINECS 205-472-3	Methyl ricinoleate
EINECS 205-480-7	Butyl acrylate

EINECS 205-483-3	Ethanolamine		EINECS 208-167-3	Barium carbonate
EINECS 205-488-0	Sodium formate		EINECS 208-174-1	Acetyl methyl carbinol
EINECS 205-500-4	Ethyl acetate		EINECS 208-401-4	Gluconic acid
EINECS 205-503-0	Malonic acid		EINECS 208-407-7	Sodium gluconate
EINECS 205-507-2	2,6-Diaminopyridine		EINECS 208-534-8	Sodium benzoate
EINECS 205-520-3	1-Hydroxy-2-pyridine		EINECS 208-590-3	Silver carbonate
EINECS 205-526-6	Glyceryl laurate		EINECS 208-591-9	Cesium carbonate
EINECS 205-530-8	Acetamide MEA		EINECS 208-651-4	m-Anisidine
EINECS 205-539-7	Stearoyl sarcosine		EINECS 208-704-1	Dicyclohexyl carbodiimide
EINECS 205-541-8	Lauramide MIPA		EINECS 208-736-6	Cetyl palmitate
EINECS 205-542-3	Propylene glycol laurate		EINECS 208-750-2	Polyvinyl chloride
EINECS 205-550-7	Caproic acid		EINECS 208-760-7	t-Butyl acetate
EINECS 205-553-3	Copper acetate (ic)		EINECS 208-806-6	Cyclohexyl chloride
EINECS 205-559-6	Butyl oleate		EINECS 208-849-0	Barium acetate
EINECS 205-560-1	Lauramide MEA		EINECS 208-875-2	Myristic acid
EINECS 205-563-8	Heptane		EINECS 209-025-3	o-Nitrobenzaldehyde
EINECS 205-571-1	Isopropyl palmitate		EINECS 209-084-5	p-Nitrobenzaldehyde
EINECS 205-582-1	Lauric acid		EINECS 209-097-6	Tristearin
EINECS 205-590-5	Potassium oleate		EINECS 209-112-6	D-Tyrosine
EINECS 205-591-0	Sodium oleate		EINECS 209-113-1	DL-Tyrosine
EINECS 205-596-8	Palmitamine		EINECS 209-127-8	Glycyl glycine
EINECS 205-597-3	Oleyl alcohol		EINECS 209-151-9	Zinc stearate
EINECS 205-599-4	Sodium cyanide		EINECS 209-183-3	Polyvinyl alcohol
EINECS 205-633-8	Sodium bicarbonate		EINECS 209-254-9	Silver acetate
EINECS 205-634-3	Oxalic acid		EINECS 209-386-7	1,6-Naphthalenediol
EINECS 205-702-2	Proline		EINECS 209-406-4	Dioctyl sodium sulfosuccinate
EINECS 205-711-1	8-Hydroxyquinoline		EINECS 209-478-7	2,7-Naphthalenediol
EINECS 205-736-8	2-Mercaptobenzothiazole		EINECS 209-481-3	Potassium benzoate
EINECS 205-788-1	Sodium lauryl sulfate		EINECS 209-502-6	2-Mercaptobenzimidazole
EINECS 205-792-3	Potassium cyanide		EINECS 209-529-3	Potassium carbonate
EINECS 205-857-6	Sodium butyrate		EINECS 209-530-9	Rubidium carbonate
EINECS 206-019-2	Imidazole		EINECS 209-544-5	Toluene diisocyanate
EINECS 206-074-2	Potassium D-gluconate		EINECS 209-691-5	Isovaleraldehyde
EINECS 206-075-8	Calcium gluconate		EINECS 209-719-6	Iodobenzene
EINECS 206-103-9	Oleamide		EINECS 209-812-1	Guanidine thiocyanate
EINECS 206-130-6	DL-Serine		EINECS 209-940-8	n,n-Dimethyl ethylamine
EINECS 206-156-8	Potassium sodium tartrate		EINECS 210-036-0	Triphenyl phosphine
EINECS 206-376-4	Capric acid		EINECS 210-410-3	2-Indanone
EINECS 206-736-0	p-Fluorophenol		EINECS 211-014-3	Glycol distearate
EINECS 206-761-7	Nickel acetate		EINECS 211-020-6	Dimethyl adipate
EINECS 206-864-7	Nalidixic acid		EINECS 211-074-0	Hexamethylene glycol
EINECS 206-991-8	Silicon carbide		EINECS 211-162-9	Ammonium acetate
EINECS 207-069-8	Diacetyl		EINECS 211-185-4	Tetrabromophthalic anhydride
EINECS 207-122-5	Diazepam		EINECS 211-284-2	Triethanolamine hydrochloride
EINECS 207-312-8	Dicyandiamide		EINECS 211-546-6	Behenyl alcohol
EINECS 207-321-7	Fluorobenzene		EINECS 211-748-4	Cetyl betaine
EINECS 207-334-8	Linolenic acid		EINECS 211-765-7	2-Methyl imidazole
EINECS 207-341-6	Thiophosgene		EINECS 211-776-7	1,2-Diaminocyclohexane
EINECS 207-355-2	Camphor		EINECS 211-797-1	Dimethoxy tetrahydrofuran
EINECS 207-431-5	Eucalyptol		EINECS 212-164-2	Aminoethyl aminopropyl
EINECS 207-439-9	Calcium carbonate			trimethoxysilane
EINECS 207-757-8	Glucose		EINECS 212-227-4	Dehydroacetic acid
EINECS 207-806-3	Hippuric acid		EINECS 212-308-4	Dimethyl anthranilate
EINECS 207-838-8	Soda ash		EINECS 212-359-2	Quinine sulfate
EINECS 207-838-8	Sodium carbonate		EINECS 212-414-0	Tin (II) oxalate
EINECS 207-924-5	Hydrocinnamic acid		EINECS 212-773-3	Sodium tartrate
EINECS 207-925-0	Benzyl chloroformate		EINECS 212-782-2	2-Hydroxyethyl methacrylate
EINECS 207-938-1	ε-Caprolactone monomer		EINECS 212-806-1	Oleyl betaine
EINECS 208-010-9	Suberic acid		EINECS 212-864-8	p-Toluoyl chloride
EINECS 208-046-5	Dimethylamine hydrochloride		EINECS 213-048-4	Aminopropyl triethoxysilane
EINECS 208-051-2	Cyanogen bromide		EINECS 213-085-6	Diamyl sodium sulfosuccinate
EINECS 208-060-1	Guanidine nitrate		EINECS 213-234-5	2-Ethyl, 4-methyl imidazole

EINECS 213-658-0	Diallyl maleate
EINECS 213-668-5	Hexamethyldisilazane
EINECS 213-695-2	Ammonium stearate
EINECS 213-934-0	Vinyl trismethoxyethoxysilane
EINECS 214-060-2	4-Hydroxycoumarin
EINECS 214-291-9	Myrtrimonium bromide
EINECS 214-292-4	Sodium cetyl sulfate
EINECS 214-355-6	2-Acetyl pyridine
EINECS 214-604-9	Decabromodiphenyloxide
EINECS 214-684-5	Tris (hydroxymethyl) amino-methane hydrochloride
EINECS 214-685-0	Methyl trimethoxysilane
EINECS 214-776-5	3-Quinuclidinone hydrochloride
EINECS 214-912-3	Dibenzosuberone
EINECS 215-090-9	Sodium xylene sulfonate
EINECS 215-125-8	Boron oxide
EINECS 215-127-9	Barium oxide
EINECS 215-128-4	Barium peroxide
EINECS 215-134-7	Bismuth trioxide
EINECS 215-137-3	Calcium hydroxide
EINECS 215-138-9	Calcium oxide
EINECS 215-146-2	Cadmium oxide
EINECS 215-160-9	Chromium oxide (ic)
EINECS 215-168-2	Ferric oxide
EINECS 215-170-3	Magnesium hydroxide
EINECS 215-171-9	Magnesium oxide
EINECS 215-175-0	Antimony oxide
EINECS 215-175-0	Antimony trioxide
EINECS 215-181-3	Potassium hydroxide
EINECS 215-185-5	Sodium hydroxide
EINECS 215-193-9	Indium oxide
EINECS 215-200-5	Lanthanum oxide
EINECS 215-202-6	Manganese dioxide
EINECS 215-204-7	Molybdenum trioxide
EINECS 215-211-5	Sodium sulfide
EINECS 215-213-6	Niobium oxide
EINECS 215-218-3	Palladium oxide
EINECS 215-222-5	Zinc oxide
EINECS 215-227-2	Zirconium oxide
EINECS 215-231-4	Tungsten oxide (ic)
EINECS 215-233-5	Yttrium oxide
EINECS 215-234-0	Ytterbium oxide
EINECS 215-236-1	Phosphorus pentoxide
EINECS 215-238-2	Tantalum oxide
EINECS 215-239-8	Vanadium pentoxide
EINECS 215-251-3	Zinc sulfide
EINECS 215-255-5	Antimony pentasulfide
EINECS 215-263-9	Molybdenum disulfide
EINECS 215-269-1	Copper oxide (ic)
EINECS 215-279-6	Limestone
EINECS 215-290-6	Lead carbonate (basic)
EINECS 215-293-2	Cresylic acid
EINECS 215-347-5	Sodium decylbenzene sulfonate
EINECS 215-350-1	Myristyl lactate
EINECS 215-354-3	Propylene glycol stearate
EINECS 215-355-9	Glyceryl hydroxystearate
EINECS 215-481-4	Arsenic trioxide
EINECS 215-548-8	Tricresyl phosphate
EINECS 215-549-3	Propylene glycol oleate
EINECS 215-559-8	Ammonium dodecylbenzene sulfonate
EINECS 215-607-8	Chromium oxide (ous)
EINECS 215-661-2	Methyl ethyl ketone peroxide
EINECS 215-663-3	Sorbitan laurate
EINECS 215-664-9	Sorbitan stearate
EINECS 215-665-4	Sorbitan oleate
EINECS 215-681-1	Magnesium silicate
EINECS 215-687-4	Sodium silicate
EINECS 215-691-6	Alumina
EINECS 215-691-6	Aluminum oxide
EINECS 215-696-3	Mercury sulfide (ic), red and black
EINECS 215-713-4	Antimony trisulfide
EINECS 215-724-4	Carmine
EINECS 215-744-3	Chitin
EINECS 215-911-0	Hexamethyldisilane
EINECS 216-343-6	Sodium isethionate
EINECS 216-472-8	Calcium stearate
EINECS 216-491-1	Mercury acetate (ic)
EINECS 216-643-7	Strontium carbonate
EINECS 216-699-2	Tetrabutyl ammonium bromide
EINECS 216-700-6	Lauramine oxide
EINECS 217-157-8	Thiourea dioxide
EINECS 217-175-6	Ammonium thiocyanate
EINECS 217-234-6	D-Aspartic acid
EINECS 217-421-2	Benzophenone-12
EINECS 217-430-1	Sodium oleyl sulfate
EINECS 217-752-2	t-Butyl hydroquinone
EINECS 217-830-6	Decane nitrile
EINECS 217-850-5	Sodium caprylate
EINECS 217-983-9	Methyl triethoxysilane
EINECS 218-163-3	D-Asparagine, anhydrous
EINECS 218-463-4	Lauryl acrylate
EINECS 218-791-8	Pentasodium aminotrimethylene phosphonate
EINECS 218-793-9	Ammonium lauryl sulfate
EINECS 218-827-2	Carvone
EINECS 219-031-8	Fluorescein
EINECS 219-058-5	Chloromethyl trimethylsilane
EINECS 219-440-1	Lauryl nitrile
EINECS 219-470-5	Drometrizole
EINECS 219-784-2	Glycidoxy propyl trimethoxysilane
EINECS 219-785-8	Methacryloxy propyl trimethoxy-silane
EINECS 219-787-9	3-Chloropropyl trimethoxysilane
EINECS 219-919-5	Stearamine oxide
EINECS 219-989-7	1-Hydroxybenzotriazole
EINECS 220-120-9	Saccharin
EINECS 220-219-7	Ditridecyl sodium sulfosuccinate
EINECS 220-336-3	Isostearic acid
EINECS 220-449-8	Vinyl trimethoxysilane
EINECS 220-476-5	Stearyl stearate
EINECS 220-941-2	Octyl triethoxysilane
EINECS 221-066-9	Phenyl trimethoxysilane
EINECS 221-304-1	Behenamide
EINECS 221-416-0	Sodium laureth sulfate
EINECS 221-498-8	Benzophenone-9
EINECS 221-661-3	Lauramidopropyl dimethylamine
EINECS 221-695-9	Hexabromocyclo dodecane
EINECS 221-787-9	Myristyl myristate
EINECS 221-838-5	Copper nitrate (ic)
EINECS 222-036-8	Dibromoethyl dibromo-cyclohexane
EINECS 222-059-3	Myristamine oxide
EINECS 222-068-2	Nickel carbonate, basic

EINECS 222-217-1	Epoxy cyclohexylethyl trimethoxysilane
EINECS 222-477-6	Zinc carbonate
EINECS 222-848-2	Magnesium gluconate
EINECS 222-899-0	Disodium laurimino dipropionate
EINECS 222-980-4	Oleyl oleate
EINECS 222-981-6	Decyl oleate
EINECS 223-095-2	Denatonium benzoate NF
EINECS 223-383-8	Dibutyl hydroxyphenyl chlorobenzotriazole
EINECS 223-445-4	Hydroxybutyl methylphenyl chlorobenzotriazole
EINECS 223-772-2	Benzophenone-4
EINECS 223-795-8	Calcium propionate
EINECS 223-819-7	Distearyl methylamine
EINECS 224-588-5	Mercaptopropyl trimethoxysilane
EINECS 224-970-1	Dimethylaminoethyl chloride hydrochloride
EINECS 225-190-4	Potassium lauryl sulfate
EINECS 225-268-8	Ethyl benzaldehyde
EINECS 225-805-6	3-Chloropropyl triethoxysilane
EINECS 226-159-8	Dimethyl isosorbide
EINECS 226-191-2	Sodium ricinoleate
EINECS 226-214-6	L-Arabinose
EINECS 226-218-8	Sulfamic acid
EINECS 226-300-9	Myristyl propionate
EINECS 226-394-6	Citral (cis and trans)
EINECS 226-546-1	Lactamide MEA
EINECS 227-813-5	Limonene
EINECS 228-227-2	Isostearamidopropyl betaine
EINECS 228-464-1	Lapyrium chloride
EINECS 229-146-5	Aminotrimethylene phosphonic acid
EINECS 229-347-8	Ammonium nitrate
EINECS 229-713-7	Diazabicyclo undecene
EINECS 229-912-9	Sodium metasilicate
EINECS 230-429-0	Palmitamine oxide
EINECS 230-528-9	Oleyl propylene diamine
EINECS 230-636-6	Carotene
EINECS 230-698-4	Lauralkonium bromide
EINECS 230-785-7	Tetrapotassium pyrophosphate
EINECS 230-990-1	Didecyl methylamine
EINECS 231-072-3	Aluminum
EINECS 231-096-4	Iron
EINECS 231-099-0	Lanthanum
EINECS 231-100-4	Lead
EINECS 231-102-5	Lithium
EINECS 231-104-6	Magnesium
EINECS 231-105-1	Manganese
EINECS 231-106-7	Mercury
EINECS 231-107-2	Molybdenum
EINECS 231-111-4	Nickel
EINECS 231-113-5	Niobium
EINECS 231-114-0	Osmium
EINECS 231-115-6	Palladium
EINECS 231-116-1	Platinum
EINECS 231-119-8	Potassium
EINECS 231-125-0	Rhodium
EINECS 231-128-7	Samarium
EINECS 231-129-2	Scandium
EINECS 231-130-8	Silicon
EINECS 231-131-3	Silver

EINECS 231-132-9	Sodium
EINECS 231-135-5	Tantalum
EINECS 231-137-6	Terbium
EINECS 231-138-1	Thallium
EINECS 231-141-8	Tin
EINECS 231-142-3	Titanium
EINECS 231-143-9	Tungsten
EINECS 231-146-5	Antimony
EINECS 231-148-6	Arsenic
EINECS 231-149-1	Barium
EINECS 231-150-7	Beryllium
EINECS 231-151-2	Boron
EINECS 231-152-8	Cadmium
EINECS 231-154-9	Cerium
EINECS 231-155-4	Cesium
EINECS 231-157-5	Chromium
EINECS 231-158-0	Cobalt
EINECS 231-159-6	Copper
EINECS 231-160-1	Erbium
EINECS 231-163-8	Gallium
EINECS 231-164-3	Germanium
EINECS 231-165-9	Gold
EINECS 231-169-0	Holmium
EINECS 231-171-1	Vanadium
EINECS 231-173-2	Ytterbium
EINECS 231-174-8	Yttrium
EINECS 231-175-3	Zinc
EINECS 231-177-4	Bismuth
EINECS 231-179-5	Calcium
EINECS 231-180-0	Indium
EINECS 231-193-1	Tellurium oxide
EINECS 231-194-7	Selenium dioxide
EINECS 231-195-2	Sulfur dioxide
EINECS 231-198-1	Rubidium iodide
EINECS 231-208-1	Aluminum chloride, anhydrous
EINECS 231-211-8	Potassium chloride
EINECS 231-298-2	Magnesium sulfate
EINECS 231-299-8	Mercury chloride (ic)
EINECS 231-301-7	Rubidium sulfate
EINECS 231-306-4	Diisopropyl sebacate
EINECS 231-314-8	Potassium ricinoleate
EINECS 231-426-7	Myristamide DEA
EINECS 231-442-4	Iodine
EINECS 231-448-7	Disodium phosphate
EINECS 231-449-2	Sodium phosphate
EINECS 231-493-2	Cyclodextrin
EINECS 231-509-6	Trisodium phosphate
EINECS 231-512-4	Perchloric acid
EINECS 231-548-0	Sodium bisulfite
EINECS 231-551-7	Sodium molybdate
EINECS 231-554-3	Sodium nitrate
EINECS 231-555-9	Sodium nitrite
EINECS 231-588-9	Stannic chloride
EINECS 231-589-4	Cobalt chloride (ous)
EINECS 231-592-0	Zinc chloride
EINECS 231-595-7	Hydrochloric acid
EINECS 231-596-2	Palladium chloride (ous)
EINECS 231-598-3	Sodium chloride
EINECS 231-599-9	Sodium bromide
EINECS 231-600-2	Cesium chloride
EINECS 231-601-8	Antimony pentachloride
EINECS 231-609-1	Stearamidopropyl dimethylamine

EINECS 231-633-2	Phosphoric acid	EINECS 232-076-8	Silver bromide
EINECS 231-634-8	Hydrofluoric acid	EINECS 232-077-3	α-Pinene
EINECS 231-635-3	Ammonia	EINECS 232-094-6	Magnesium chloride
EINECS 231-639-5	Sulfuric acid	EINECS 232-130-0	Cesium bromide
EINECS 231-659-4	Potassium iodide	EINECS 232-131-6	Copper bromide (ous)
EINECS 231-667-8	Sodium fluoride	EINECS 232-143-1	Ammonium dichromate
EINECS 231-672-5	Sodium iodate	EINECS 232-145-2	Cesium iodide
EINECS 231-673-0	Sodium metabisulfite	EINECS 232-146-8	Cesium nitrate
EINECS 231-674-6	Copper iodide (ous)	EINECS 232-160-4	Sodium bromate
EINECS 231-679-3	Sodium iodide	EINECS 232-164-6	Calcium bromide
EINECS 231-694-5	Sodium tripolyphosphate	EINECS 232-188-7	Calcium fluoride
EINECS 231-714-2	Nitric acid	EINECS 232-197-6	Sodium m-periodate
EINECS 231-718-4	Zinc bromide	EINECS 232-218-9	Lithium nitrate
EINECS 231-721-0	Disodium capryloamphodiacetate	EINECS 232-222-0	Cadmium fluoride
EINECS 231-722-6	Sulfur	EINECS 232-223-6	Cadmium iodide
EINECS 231-727-3	Zirconium hydride	EINECS 232-240-9	Rubidium chloride
EINECS 231-729-4	Ferric chloride	EINECS 232-277-0	Olive oil
EINECS 231-749-3	Phosphorus trichloride	EINECS 232-281-2	Corn oil
EINECS 231-760-3	Potassium permanganate	EINECS 232-282-8	Coconut oil
EINECS 231-765-0	Hydrogen peroxide	EINECS 232-293-8	Castor oil
EINECS 231-767-1	Sodium pyrophosphate	EINECS 232-306-7	Sulfated castor oil
EINECS 231-778-1	Bromine	EINECS 232-307-2	Lecithin
EINECS 231-781-8	Potassium persulfate	EINECS 232-315-6	Paraffin
EINECS 231-784-4	Barium sulfate	EINECS 232-348-6	Lanolin
EINECS 231-793-3	Zinc sulfate	EINECS 232-360-1	Sorbitan sesquioleate
EINECS 231-810-4	Ethylbicyclo oxazolidine	EINECS 232-373-2	Petrolatum
EINECS 231-818-8	Potassium nitrate	EINECS 232-384-2	Mineral oil
EINECS 231-820-9	Sodium sulfate	EINECS 232-388-4	Potassium castorate
EINECS 231-821-4	Sodium sulfite	EINECS 232-391-0	Epoxidized soybean oil
EINECS 231-829-8	Potassium bromate	EINECS 232-410-2	Hydrogenated soybean oil
EINECS 231-830-3	Potassium bromide	EINECS 232-430-1	Lanolin alcohol
EINECS 231-831-9	Potassium iodate	EINECS 232-440-6	Hydroxylated lecithin
EINECS 231-836-6	Sodium chlorite	EINECS 232-452-1	Hydrogenated lanolin
EINECS 231-837-1	Calcium phosphate (monobasic)	EINECS 232-475-7	Rosin
EINECS 231-842-9	Copper chloride (ous)	EINECS 232-519-5	Gum arabic
EINECS 231-847-6	Copper sulfate (ic)	EINECS 232-536-8	Guar gum
EINECS 231-853-9	Silver nitrate	EINECS 232-541-5	Locust bean gum
EINECS 231-867-5	Sodium thiosulfate	EINECS 232-553-0	Pectin
EINECS 231-868-0	Stannous chloride	EINECS 232-554-6	Gelatin
EINECS 231-871-7	Ammonium sulfamate	EINECS 232-555-1	Casein
EINECS 231-887-4	Sodium chlorate	EINECS 232-567-7	Diastase
EINECS 231-889-5	Sodium chromate	EINECS 232-619-9	Lipase
EINECS 231-890-0	Sodium hydrosulfite	EINECS 232-627-2	Papain
EINECS 231-892-1	Sodium persulfate	EINECS 232-629-3	Pepsin
EINECS 231-900-3	Calcium sulfate (anhydrous)	EINECS 232-650-8	Trypsin
EINECS 231-906-6	Potassium dichromate	EINECS 232-656-0	Urease
EINECS 231-907-1	Potassium phosphate, tribasic	EINECS 232-668-6	Peroxidase
EINECS 231-908-7	Calcium hypochlorite	EINECS 232-674-9	Cellulose
EINECS 231-912-9	Potassium perchlorate	EINECS 232-675-4	Dextrin
EINECS 231-913-4	Potassium phosphate	EINECS 232-677-5	Dextran
EINECS 231-915-5	Potassium sulfate	EINECS 232-678-0	Hyaluronic acid
EINECS 231-943-8	Zinc nitrate	EINECS 232-680-1	Alginic acid
EINECS 231-955-3	Graphite	EINECS 232-697-4	Collagen
EINECS 231-956-9	Oxygen	EINECS 232-734-4	Cellulase
EINECS 231-957-4	Selenium	EINECS 232-752-2	Protease
EINECS 231-975-2	Tungstic acid	EINECS 232-864-1	β-Galactosidase
EINECS 231-982-0	Ammonium thiosulfate	EINECS 232-885-6	Pectinase
EINECS 231-984-1	Ammonium sulfate	EINECS 233-038-3	Chromium chloride (ic)
EINECS 232-001-9	Zinc fluoride	EINECS 233-042-5	Trichlorosilane
EINECS 232-021-8	Antimony pentafluoride	EINECS 233-046-7	Phosphorus oxychloride
EINECS 232-029-1	Tungsten hexafluoride	EINECS 233-047-2	Antimony trichloride
EINECS 232-033-3	Silver chloride	EINECS 233-054-0	Silicon tetrachloride

EINECS 233-058-2	Zirconium tetrachloride
EINECS 233-062-4	Cobalt fluoride (ic)
EINECS 233-072-9	Iron sulfate (ic)
EINECS 233-109-9	Hydriodic acid
EINECS 233-110-4	Hydrazine sulfate
EINECS 233-113-0	Hydrobromic acid
EINECS 233-131-9	Strontium nitrate
EINECS 233-139-2	Boric acid
EINECS 233-140-8	Calcium chloride
EINECS 233-163-3	Chromium chloride (ous)
EINECS 233-165-4	Rhodium chloride
EINECS 233-237-5	Lanthanum chloride
EINECS 233-245-9	Lead nitrate
EINECS 233-265-8	Palladium nitrate (ous)
EINECS 233-267-9	Sodium selenite
EINECS 233-343-1	Sodium hexametaphosphate
EINECS 233-471-8	Zinc borate
EINECS 233-653-7	Silver sulfate
EINECS 233-657-9	Boron tribromide
EINECS 233-658-4	Boron trichloride
EINECS 233-660-5	Barium chromate
EINECS 233-662-6	Cesium sulfate
EINECS 233-708-5	D-Arabinose
EINECS 233-792-3	Bismuth nitrate
EINECS 234-190-3	Sodium dichromate
EINECS 234-319-3	Magnesium aluminum carbonate
EINECS 234-325-6	Glyceryl stearate
EINECS 234-364-9	Ammonium tungstate
EINECS 234-394-2	Xanthan gum
EINECS 234-409-2	Zinc naphthenate
EINECS 234-717-7	Ammonium iodide
EINECS 234-851-6	Thulium oxide
EINECS 234-856-3	Terbium oxide
EINECS 234-975-0	Barium titanate
EINECS 235-015-3	Holmium oxide
EINECS 235-042-0	Scandium oxide
EINECS 235-043-6	Samarium oxide
EINECS 235-045-7	Erbium oxide
EINECS 235-087-6	Phosphotungstic acid
EINECS 235-088-1	Sodium toluenesulfonate
EINECS 235-113-6	Copper carbonate (ic)
EINECS 235-183-8	Ammonium bromide
EINECS 235-185-9	Ammonium fluoride
EINECS 235-186-4	Ammonium chloride
EINECS 235-714-3	Cobalt carbonate (ous)
EINECS 235-907-2	Sodium capryloamphoacetate
EINECS 236-487-3	Cesium fluoride
EINECS 236-623-1	Gold chloride
EINECS 236-675-5	Titanium dioxide
EINECS 236-743-4	Sodium tungstate
EINECS 236-813-4	Tellurium
EINECS 236-878-9	Zinc chromate
EINECS 237-029-5	Cerium sulfate
EINECS 237-066-7	Phosphorous acid
EINECS 237-081-9	Sodium ferrocyanide
EINECS 237-396-1	Nickel sulfamate
EINECS 237-511-5	Aminopropyl trimethoxysilane
EINECS 237-558-1	Lithium hypochlorite
EINECS 238-310-5	Stearamide MEA stearate
EINECS 238-311-0	Oleamine oxide
EINECS 238-430-8	Pentaerythrityl tetrapelargonate
EINECS 238-479-5	Disodium stearyl sulfosuccinamate
EINECS 238-687-6	Zinc ammonium chloride
EINECS 238-877-9	Talc
EINECS 238-878-4	Quartz
EINECS 239-707-6	Sodium percarbonate
EINECS 240-263-0	Dithiothreitol
EINECS 240-458-0	Geranyl acetate
EINECS 240-795-3	Potassium metabisulfite
EINECS 240-865-3	Behenalkonium chloride
EINECS 240-924-3	Stearamidoethyl diethylamine
EINECS 241-004-4	Sodium borohydride
EINECS 242-159-0	Tin oxide (ic)
EINECS 242-201-8	Behenyl erucate
EINECS 243-885-0	Tetrabromo phthalatediol
EINECS 243-957-1	Silver oxide (ous)
EINECS 244-058-7	Osmium tetroxide
EINECS 244-344-1	Cesium hydroxide
EINECS 244-438-2	Silver sulfide
EINECS 244-492-7	Aluminum hydroxide
EINECS 244-654-7	Mercury oxide (ic), red and yellow
EINECS 244-754-0	Octyl stearate
EINECS 245-629-3	Dicyclohexyl sodium sulfo-succinate
EINECS 245-876-7	Ureidopropyl triethoxysilane
EINECS 246-240-1	Di-t-butyl dicarbonate
EINECS 246-376-1	Potassium sorbate
EINECS 246-584-2	Oleamidopropyl betaine
EINECS 246-613-9	Isooctyl thioglyconate
EINECS 246-619-1	t-Dodecyl mercaptan
EINECS 246-680-4	Sodium dodecylbenzene sulfonate
EINECS 246-684-6	Oleamidopropylamine oxide
EINECS 246-705-9	Sucrose stearate
EINECS 246-839-8	Xylene sulfonic acid
EINECS 246-868-6	Isocetyl stearate
EINECS 246-873-3	Sucrose laurate
EINECS 246-929-7	Sodium stearoyl lactylate
EINECS 247-036-5	Tridecylbenzene sulfonic acid
EINECS 247-144-2	Glyceryl dioleate
EINECS 247-310-4	Disodium lauramido MEA sulfosuccinate
EINECS 247-561-6	Sodium methyl naphthalene sulfonate
EINECS 247-568-8	Sorbitan palmitate
EINECS 247-569-3	Sorbitan trioleate
EINECS 247-669-7	Propylene glycol ricinoleate
EINECS 247-710-9	Ammonium xylene sulfonate
EINECS 247-784-2	DEA-dodecylbenzene sulfonate
EINECS 247-891-4	Sorbitan tristearate
EINECS 248-133-5	Isooctyl alcohol
EINECS 248-248-0	Oleyl hydroxyethyl imidazoline
EINECS 248-289-4	Dodecylbenzene sulfonic acid
EINECS 248-296-2	Potassium dodecylbenzene sulfonate
EINECS 248-299-9	Diisopropyl adipate
EINECS 248-317-5	Sucrose distearate
EINECS 248-470-8	Isostearyl alcohol
EINECS 248-586-9	Glyceryl dilaurate
EINECS 248-666-3	Hydroxypropyl methacrylate
EINECS 248-938-7	Sodium cumene sulfonate
EINECS 249-862-1	Octyl palmitate
EINECS 250-151-3	Sodium myristoyl sarcosinate
EINECS 250-651-1	Isopropyl isostearate

EINECS 250-696-7	Tridecyl stearate
EINECS 250-705-4	Glyceryl stearate
EINECS 250-913-5	Sodium cumene sulfonate
EINECS 251-087-9	Octabromo diphenyloxide
EINECS 251-118-6	Ethylenebistetrabromo phthalimide
EINECS 251-306-8	Palmitamidopropyl betaine
EINECS 251-732-4	Glycol hydroxystearate
EINECS 251-734-5	Propylene glycol hydroxystearate
EINECS 252-964-9	Isocetyl alcohol
EINECS 253-149-0	Cetyl alcohol
EINECS 253-407-2	Glyceryl oleate
EINECS 253-458-0	PEG-8 laurate
EINECS 253-519-1	Ammonium cumene sulfonate
EINECS 253-521-2	Capryl hydroxyethyl imidazoline
EINECS 254-495-5	Polyglyceryl-10 decastearate
EINECS 254-896-5	Bistriethoxysilylpropyltetrasulfane
EINECS 255-051-3	Ricinoleamide DEA
EINECS 255-062-3	Disodium laureth sulfosuccinate
EINECS 255-350-9	Propylene glycol dipelargonate
EINECS 255-485-3	Isostearyl isostearate
EINECS 256-120-0	Disodium oleamido MIPA sulfosuccinate
EINECS 256-214-1	Myristamidopropyl dimethylamine
EINECS 258-007-1	Myristoyl sarcosine
EINECS 258-193-4	Isostearamide DEA
EINECS 258-207-9	Bistetramethyl piperidinyl sebacate
EINECS 258-250-3	Ethylenebisdibromo norbornane dicarboximide
EINECS 258-377-8	Behenamidopropyl ethyl dimonium ethosulfate
EINECS 259-837-7	Stearamidopropyl dimethylamine lactate
EINECS 260-081-5	Arginine PCA
EINECS 260-143-1	Disodiumoleamido PEG-2 sulfosuccinate
EINECS 260-410-2	Linoleamide DEA
EINECS 261-521-9	Isostearyl neopentanoate
EINECS 261-526-6	Tetradecabromo diphenoxy benzene
EINECS 261-619-1	Cetearyl octanoate
EINECS 261-619-1	Cetyl octanoate
EINECS 261-673-6	Isodecyl oleate
EINECS 261-684-6	Myristamidopropyl betaine
EINECS 261-819-9	Octyl pelargonate
EINECS 262-134-8	Behenamidopropyl dimethylamine
EINECS 262-976-6	Hydrogenated tallowamine
EINECS 262-977-1	Cocamine
EINECS 262-979-2	Acetylated lanolin
EINECS 262-980-8	Acetylated lanolin alcohol
EINECS 262-988-1	Methyl cocoate
EINECS 262-991-8	Dihydrogenated tallow methylamine
EINECS 263-016-9	Cocamine oxide
EINECS 263-020-0	Dimethyl cocamine
EINECS 263-027-9	Glyceryl cocoate
EINECS 263-038-9	Cocotrimonium chloride
EINECS 263-058-8	Cocamidopropyl betaine
EINECS 263-080-8	Benzalkonium chloride
EINECS 263-080-8	Cocoalkonium chloride
EINECS 263-081-3	Hydrogenated tallowalkonium Cl
EINECS 263-085-5	Tallow alkonium chloride
EINECS 263-087-6	Dicoco dimonium chloride
EINECS 263-089-7	Hydrogenated ditallowamine
EINECS 263-090-2	Quaternium-18
EINECS 263-099-1	Tallow
EINECS 263-107-3	Tall oil acid
EINECS 263-123-0	Hydrogenated tallowamide
EINECS 263-125-1	Tallowamine
EINECS 263-130-9	Hydrogenated tallow acid
EINECS 263-134-0	Soytrimonium chloride
EINECS 263-147-1	Cocamine acetate
EINECS 263-155-5	TEA cocoate
EINECS 263-163-9	Cocamide DEA
EINECS 263-170-7	Cocoyl hydroxyethyl imidazoline
EINECS 263-171-2	Tall oil hydroxyethyl imidazoline
EINECS 263-179-6	Dihydroxyethyl tallowamine oxide
EINECS 263-180-1	Dihydroxyethyl cocamine oxide
EINECS 263-189-0	Tallow propylene diamine
EINECS 263-191-1	Tallow dipropylene triamine
EINECS 263-193-2	Sodium cocoyl sarcosinate
EINECS 263-195-3	Cocopropylene diamine
EINECS 263-218-7	Lauramidopropylamine oxide
EINECS 264-119-1	Isopropyl lanolate
EINECS 266-124-4	Glyceryl isostearate
EINECS 266-368-1	C12-14 alkyl dimethyl betaine
EINECS 266-533-8	Sodium isostearoyl lactylate
EINECS 266-778-0	Isostearamidopropyl ethyl dimonium ethosulfate
EINECS 266-778-0	Isostearylethyl imidonium ethosulfate
EINECS 266-948-4	Triolein
EINECS 267-015-4	Methyl oleate
EINECS 267-101-1	Isostearamidopropyl dimethyl-amine
EINECS 267-360-0	Stearamidopropyl ethyl dimonium ethosulfate
EINECS 267-569-7	Sodium oleoamphopropionate
EINECS 268-093-2	Trimethyl propane trioleate
EINECS 268-130-2	Ammonium lauroyl sarcosinate
EINECS 268-215-4	Eicosyl docosylamine
EINECS 268-761-3	Cocamidopropyl hydroxysultaine
EINECS 268-771-8	Cocamidopropyl dimethylamine
EINECS 268-772-3	Tallowamide DEA
EINECS 268-910-2	Sorbitan laurate
EINECS 268-938-5	Cocamidopropylamine oxide
EINECS 268-949-5	Tallamide DEA
EINECS 269-084-6	TEA cocoyl glutamate
EINECS 269-087-2	Sodium cocoyl glutamate
EINECS 269-220-4	Lanolin wax
EINECS 269-819-0	Sodium cocoamphoacetate
EINECS 269-919-4	Benzalkonium chloride
EINECS 269-922-0	Alkyl trimethyl ammonium chloride
EINECS 270-156-4	Cocoyl sarcosine
EINECS 270-302-7	Lanolin acid
EINECS 270-312-1	Glyceryl oleate
EINECS 270-315-8	Hydroxylated lanolin
EINECS 270-325-2	Benzalkonium chloride
EINECS 270-329-4	Coco-betaine
EINECS 270-331-5	Didecyl dimonium chloride
EINECS 270-351-4	Coconut alcohol
EINECS 270-355-6	Soyamide DEA

EINECS 270-407-8	Sodium C14-16 olefin sulfonate
EINECS 270-416-7	Tallowaminopropylamine
EINECS 270-664-6	Montan acid wax
EINECS 270-864-3	Disodium oleamido MEA sulfosuccinate
EINECS 271-102-2	Disodium cocamido MIPA sulfosuccinate
EINECS 271-694-2	Pentaerythrityl tetraoleate
EINECS 271-696-6	Hydrogenated tallow propylene diamine
EINECS 271-705-0	Sodium cocoamphohydroxypropyl sulfonate
EINECS 271-792-5	Disodium capryloamphodiacetate
EINECS 271-793-0	Sodium cocoamphoacetate
EINECS 271-929-9	Sodium isostearoampho-propionate
EINECS 272-043-5	Disodium cocoamphodiacetate
EINECS 272-219-1	Disodium cocamido MEA sulfosuccinate
EINECS 272-787-0	Oleo tripropylene tetraamine
EINECS 272-897-9	Disodium cocoampho-dipropionate
EINECS 273-429-6	Isostearyl hydroxyethyl imidazoline
EINECS 273-612-0	Acetylated hydrogenated tallow glyceride
EINECS 273-613-6	Hydrogenated tallow glyceride citrate
EINECS 274-022-6	Coco guanidinium chloride
EINECS 274-267-9	Disodium oleoamphodiacetate
EINECS 274-845-0	Dihydroxyethyl tallow glycinate
EINECS 274-923-4	Cocamidopropyl betaine
EINECS 276-638-0	Tannic acid
EINECS 276-951-2	Sorbitan tristearate
EINECS 283-078-0	Pentaerythrityl tetrabehenate
EINECS 283-390-7	Myreth-3 laurate
EINECS 284-219-9	Cocaminopropionic acid
EINECS 284-863-0	Butyl oleate
EINECS 284-868-8	Isobutyl oleate
EINECS 285-203-4	Propylene glycol dioleate
EINECS 285-206-0	Ethyl oleate
EINECS 285-207-6	Ethylhexyl oleate
EINECS 285-540-7	Isopropyl oleate
EINECS 285-547-5	Pentaerythrityl stearate
EINECS 285-550-1	PEG-2 stearate
EINECS 286-074-7	Sorbitan oleate
EINECS 286-074-7	Sorbitan trioleate
EINECS 286-344-4	Sodium methylenebis dibutyl-phenyl phosphate
EINECS 286-490-9	Glyceryl stearate
EINECS 287-011-6	C12-14 alkyl dimethylamine oxide
EINECS 287-039-9	Butyl stearate
EINECS 287-075-5	Glyceryl tri C8-C10
EINECS 287-089-1	Benzalkonium chloride
EINECS 287-484-9	Stearyl stearate
EINECS 287-824-6	Methyl stearate
EINECS 288-048-0	Tallow dipropylene triamine
EINECS 288-305-7	Pentacrythrityl olcatc
EINECS 288-459-5	PEG dioleate series
EINECS 288-459-5	PEG oleate series
EINECS 288-668-1	Isobutyl stearate
EINECS 290-580-3	Distearyl phthalate
EINECS 290-850-0	Disodium tallow sulfosuccinamate
EINECS 291-990-5	Soyamidopropyl ethyl dimonium ethosulfate
EINECS 292-932-1	Glycol distearate
EINECS 292-945-2	Ethyl stearate
EINECS 292-951-5	Isooctyl stearate
EINECS 292-960-4	Isononyl stearate
EINECS 292-979-8	Sorbitan stearate
EINECS 293-029-5	Pentaerythrityl tetrastearate
EINECS 293-208-8	Glyceryl stearate
EINECS 293-391-2	Myreth-3 palmitate
EINECS 296-473-8	Kaolin
EINECS 297-364-8	PEG-2 oleate
EINECS 297-495-0	PEG-3 tallow propylene dimonium dimethosulfate
EINECS 306-522-8	Glycol stearate
EINECS 306-522-8	PEG distearate series
EINECS 306-522-8	PEG stearate series
EINECS 306-797-4	Cetyl palmitate
EINECS 306-998-7	Tallow amphopolycarboxy propionic acid
EINECS 307-455-7	Cocoimino dipropionate
EINECS 307-456-2	Cocoimino diglycinate
EINECS 307-458-3	Disodium oleoamphodiacetate
EINECS 307-458-3	Sodium carboxymethyl tallow polypropylamine
EINECS 307-458-3	Stearyl amphopolycarboxy glycinate
EINECS 307-919-9	Coco tripropylene tetraamine
EINECS 309-206-8	Cocobetainamido ampho-propionate

Manufacturer Successors

Former company	Present company
ABM Chem (Leeds)	Rhône-Poulenc Surfactants & Specialties
Alcolac	Rhône-Poulenc Surfactants & Specialties
Alginate Industries Ltd.	Kelco International Ltd.
Alkaril	Rhône-Poulenc Surfactants & Specialties
Allied Corp	Allied-Signal Inc.
Dr. Angle GmbH	Rhône-Poulenc Surfactants & Specialties
Anglo Speciality Adhesives.	Evode Speciality Adhesives Ltd.
Atomic Energy Establishment.	AEA Technology
Ayerst Laboratories.	Wyeth Laboratories
Baxter Travenol	Boots Pharmaceuticals
Beecham Pharmaceuticals.	SmithKline Beecham
Beecham Research Laboratories.	SmithKline Beecham
Blagden Campbell Chemicals	Blagden Chemicals Ltd.
Borg-Warner Chemicals.	GE Plastics
Braunschweig AG.	Chemische Fabriken Oker und Braunschweig AG
British Ceca	Elf Atochem UK Ltd, Ceca Specialities
British Celanese.	Hoechst UK Ltd.
British Drug Houses.	Merck Sharp & Dohme Ltd.
British Geon.	BP Chemicals Ltd.
Buckeye Cellulose Corp.	Procter & Gamble Co.
BV Nekami.	Nekami Maasgroeven BV
Cape Insulation Cape Asbestos Co.	Charter Consolidated plc
Carless Solvents Ltd.	Carless Refining & Marketing Ltd.
CdF Chimie.	Noroxo
Celanese Corp.	Hoechst Celanese Corp.
Cincinnati Milacron Chemicals Inc.	Morton International
Clintwood Chemical Co.	Rhône-Poulenc Surfactants & Specialties
CooperVision Pharmaceuticals Inc.	Iolab Corp.
Cow & Gate	Nutricia Holdings Ltd.
Cyclo Corp.	Rhône-Poulenc Surfactants & Specialties
Daishowa	Borregaard Lignotech
Deanshanger Oxides Ltd.	Harcros Pigments Europe
Dow Agriculture	DowElanco Ltd.
Elder Pharmaceuticals	ICN Pharmaceuticals Inc.
Emser Industries	EMS-American Grilon Inc.
FBC Ltd.	Schering Agrochemicals Ltd.
Freeman Chemical Corp.	Cook Composites & Polymers Co.

GAF Corp. .... ISP
GAF Corp (surfactants) .... Rhône-Poulenc Surfactants & Specialties
Geronazzo S.p.A. .... Rhône-Poulenc Geronazzo SpA
Glidden Co. .... ICI America Holdings Inc/Paints Group
Graesser Laboratories Ltd. .... Nipa Laboratories Ltd.
Haagen Chemie BV. .... Harcros Chemicals BV
G. M. Haynes .... Rhône-Poulenc Surfactants & Specialties
Hermadex Ltd. .... Whitecourt Ltd.
Horace Cory PLC. .... European Colour plc
International Synthetics Ltd. .... International Bio-Synthetics Ltd.
ISC Alloys Ltd. .... Pasminco Europe
John & E Sturge Ltd. .... Haarmann & Reimer Ltd. (Selby)
Kinglsey & Keith Chemical Corp. .... K & K Greeff Chemical Group Ltd.
Kinnis & Brown. .... K&B Antipollution Ltd.
Lancashire Tar Distillers Ltd. .... Lanstar Ltd.
Lankro Chemicals Ltd. .... Harcros Chemicals UK Ltd.
Manchem Ltd. .... Rhône-Poulenc Chemicals Ltd.
Marion Laboratories .... Marion Merrell Dow Inc.
Mars .... Rhône-Poulenc Surfactants & Specialties
Merck Sharpe & Dohme .... Merck & Co Inc.
Miranol .... Rhône-Poulenc Surfactants & Specialties
Mobay Chemical Co. .... Miles Inc.
Murphy Chemicals Ltd .... DowElanco Ltd.
Murphy Chemicals Ltd (home & garden products) .... Fisons plc, Horticulture Div.
Naarden International Chemicals Div. .... Quest International
National Lead Co. .... Rheox Inc.
Naugatuck (U.S. Rubber). .... Uniroyal Inc.
NL Industries Inc. .... Rheox Inc.
Nuodex Inc .... Hüls America Inc.
Optrex. .... Boots plc
Owen-Galiderma Laboratories .... Galderma Laboratories
Pennwalt Corp. .... Atochem North America Inc.
Pharmacia Laboratories .... Kabi-Pharmacia
Plough Inc. .... Schering-Plough HealthCare Products
Polyvinyl Chimie Holland BV .... ICI Resins BV
PPF International Ltd. .... Quest International UK Ltd.
PVO .... Stepan/PVO Dept.
Refined Onyx .... Rhône-Poulenc Surfactants & Specialties
Riker Laboratories Inc. .... 3M Pharmaceuticals
Rilsan. .... Atochem
A H Robins Co Ltd .... Wyeth Laboratories
Rorer. .... Rhône-Poulenc Rorer
Smith Kline & French Laboratories. .... SmithKline Beecham
Soltex Polymer Corp .... Solvay Polymers Inc.
E. R. Squibb & Sons Inc. .... Bristol-Myers Squibb Co Inc.
E. R. Squibb & Sons Ltd. .... Bristol-Myers Squibb Pharmaceuticals Ltd.

Standard Oil .... BP America Inc.
Stauffer Chemical Co. .... Rhône-Poulenc Basic
Staveley Chemicals Ltd. .... Rhône-Poulenc Chemicals Ltd.
Sterling Drug Inc. .... Sterling Winthrop
Synthese BV. .... Akzo Resins BV
Synthetic Resins Ltd. .... Scott Bader Co Ltd.
Telcon Plastics Ltd. .... Medway Packaging Ltd.
J. C. Thompson & Co (Duron) Ltd .... Stephenson Thompson Textile Chemicals
U. S. Ethicals .... Rhône-Poulenc-Rorer Pharmaceuticals
U. S. Industrial Chemical Corp. .... Quantum Chemical Corp/USI Div.
Vinyl Products .... BP Chemicals
Williams Div. of Morton Thiokol Ltd. .... Morton International Ltd.
Williams (Hounslow) Ltd. .... Morton International Ltd.
Wilmington Chemical Co. .... Rhône-Poulenc Inc.